D1479567

Fire Protection Handbook

Fire Protection Handbook™

Sixteenth Edition

ARTHUR E. COTE, P.E., Editor-in-Chief
JIM L. LINVILLE, Managing Editor

National Fire Protection Association
Quincy, Massachusetts

Editor-in Chief: Arthur E. Cote
Managing Editor: Jim L. Linville
Associate Editor: John S. Petraglia
Consulting Editor: Gordon P. McKinnon
Editorial Staff: Gene A. Moulton
Marion Cole
Copyeditors: Catherine G. Cronin
Stanley T. Dingman
Parrish Dobson
Artwork: Kathy Linnehan
George Nichols
Frank Lucas
Text Processing: Elizabeth K. Carmichael
Debra A. Rose
Louise Grant
Interior Design: K. Sulier Associates
Cover Design: George Nichols
Production Coordinator: Donald C. McGonagle
Composition: William Byrd Press
Printed By: R.R. Donnelley & Sons

First Impression, March 1986
Fifty thousand Copies

Previous Editions

First 1896
Second 1901
Third 1904
Fourth 1909
Fifth 1914
Sixth 1921
Seventh 1924
Eighth 1935
Ninth 1941
Tenth 1948
Eleventh 1954
Twelfth 1962
Thirteenth 1969
Fourteenth 1976
Fifteenth 1981

Library of Congress Catalog Card Number 62-12655
NFPA Number: FPH1686
ISBN: 0-87765-315-1
The National Fire Protection Association
Batterymarch Park, Quincy, MA 02269

Dedication

To the professional men and women of the fire protection community whose untiring devotion to the task of protecting life and property from the ravages of fire has made this world a better place.

CONTENTS

INTRODUCTION

Since its first edition exactly 90 years ago the *Fire Protection Handbook* has endeavored to fulfill the needs of the fire protection community for a single-source handbook on the state of the art in fire protection and fire prevention practices.

It was originally known as the Crosby-Fiske *Handbook of the Underwriter's Bureau of New England*, and first published in 1896, the same year that the National Fire Protection Association was founded. The original authors were Everett U. Crosby and Henry A. Fiske, who were joined by H. Walter Forster in 1918. In 1935 Crosby, Fiske, and Forster donated all rights to their handbook to the NFPA, and NFPA has published all successive editions since that 8th edition.

The *Fire Protection Handbook* has changed significantly in the last nine decades. While the body of knowledge in the field of fire protection has proliferated, the HANDBOOK has kept pace, expanding from less than 200 pages in the first edition to almost 1800 pages in this 16th edition. As the most pressing concerns of fire protection have evolved, from city-wide conflagrations in the last 1800s through building fires and more recently to room fires, the number of subjects covered by the *Handbook* has greatly increased. This is evidenced by the expansion of the text from the short, running commentary that comprised the first edition to material organized into 173 chapters in this volume. Today, there are nearly as many chapters as there were pages in 1896!

Production of the *Fire Protection Handbook* through 16 editions has involved literally thousands of fire protection experts from within and outside NFPA. However, in addition to its founders, over the last half-century four individuals were especially responsible for establishing the HANDBOOK as the "bible" for fire protection practitioners.

Robert S. Moulton, late NFPA Technical Secretary who, during his 40 years of service, edited the 9th, 10th, and 11th editions;

George H. (Hitch) Tryon, late NFPA Assistant Vice President who edited the 12th and 13th editions;

Richard E. Stevens, retired NFPA Vice President and Chief Engineer who, during his 35 years with the Association, contributed to five editions of the HANDBOOK and served as chief technical consultant for the 14th and 15th editions; and

Gordon P. (Mac) McKinnon, retired NFPA Editor-in-Chief, who has been directly involved in the preparation of five editions of the HANDBOOK over a quarter century. He served as Editorial Coordinator of the 12th edition, Managing Editor of the 13th edition, Editor of the 14th and 15th editions, and Consulting Editor for this edition.

In offering this Edition of the *Fire Protection Handbook*, the editors solicit suggestions for improvements in the interest of continuing to make future editions increasingly useful to all concerned. Every effort has been made to assure that the text is consistent with the best available information on current fire protection practices. However, the National Fire Protection Association, as a body, is not responsible for the contents, as there has been no opportunity for the membership to review the *Handbook* prior to its publication. If readers discover errors or omissions, the editors would appreciate those shortcomings being called to their attention.

Arthur E. Cote, P.E.

PREFACE

Without a doubt, supervising the editing of the 16th edition of the *Fire Protection Handbook* has been the most challenging and complicated task I've had the pleasure of attempting during my 20 (plus) years in technical book publishing. When you finally have the time to stop and think about it, the facts are mind boggling: 173 chapters by almost 200 authors; hundreds of reviewers to keep track of; more than 2,000 illustrations to be rendered, captioned, cropped, and checked; tens of thousands of metric conversions to make; close to a million words to edit; a readership that ranges from fire fighters and fire protection engineers to university professors and research scientists; a breakneck schedule with a small but dauntless staff augmented by part-time help, and freelancers; an end-product with more pages than most major magazines publish in an entire year; but, most of all, the haunting knowledge that this book is different because it's *The* FIRE PROTECTION HANDBOOK, and a misplaced comma or a poorly worded sentence could result in an innocent person's injury or death. There are times when you simply have to strive for excellence and the HANDBOOK is one of those times.

This 16th edition of the *Fire Protection Handbook* is a major revision. It has a new Editor-in-Chief, a new editorial staff, a new design, and many new authors. Not only has it been thoroughly updated to reflect current fire protection and fire prevention practices, it also reflects the personalities and priorities of its staff. The changes made in this edition are too numerous to list individually, but in general:

1. There are 22 sections of which three are new: "Education for Fire Protection," "Hazardous Wastes and Materials," and "Fire Modeling and Analysis."
2. Every chapter has been thoroughly updated, and there are more than 50 new or totally revised chapters.
3. The "Life Safety" section has been revised and reorganized to parallel the *Life Safety Code* ®; it includes 16 new chapters.
4. More than 60 authors have contributed to the HANDBOOK for the first time.
5. The chapter Bibliographies have been expanded and updated.
6. Metric equivalents for all values have been added to the text, illustrations, and tables.
7. The book has been completely redesigned and reorganized to reflect its status as the foremost professional reference book on fire protection.

The result is a book with more than 600 pages of "new" material and a grand total of approximately 1,750 pages.

Acknowledgments

Each edition of any book as massive and as important as the *Fire Protection Handbook* is an enormous undertaking requiring the cooperation of hundreds of people, and this 16th edition is no exception. In one way or another virtually every NFPA staff member has been involved; many made generous personal sacrifices; all deserve accolades. One member of our team, however, deserves special mention. Louise Grant not only keyboarded every word, (sometimes three or four times) but did so with enthusiasm and skill. If this book is good it owes much to Louise. Likewise, over the five year period between this and the previous edition, many conscientious readers have called errors and inconsistencies to our attention. We wish we had time and/or space to mention them all. Nothing is perfect, and so we continue to solicit your comments. We do try, we do care, and we do promise to respond to each and every letter.

Obviously, no one person can know everything about a subject as enormous and complex as fire protection. Every word in this edition was reviewed by the Editor-in-Chief and other appropriate NFPA staff members. Selected chapters were reviewed by qualified outside experts from all over the world. Therefore, the editors wish to acknowledge and thank all of those experts who contributed their time and energy to improving the Handbook. Special thanks are due Dr. Jack Snell, Director, U.S. National Bureau of Standards, Center for Fire Research; Mr. Paul M. Fitzgerald, Chief Operating Officer, Factory Mutual Research Corporation; Mr. James A. Lambert, Vice-President Engineering, Industrial Risk Insurers: and Mr. Jack A. Bono, President, Underwriters Laboratories Inc. Their willingness to allow their respective staffs to devote time to participate as both reviewers and authors of many of the chapters in this Handbook is gratefully acknowledged. Without their help the book would have suffered. Special thanks is also due Harold E. (Bud) Nelson, of the U.S. National Bureau of Standards, Center for Fire Research and Past-President of the Society of Fire Protection Engineers, whose eagerness to share his personal knowledge of the subject and that of his colleagues in the SFPE led us to many of the new authors in this edition, and many of its technical reviewers.

The following experts from outside the NFPA staff gave generously of their time and experience by reviewing chapters from the 15th edition and offering suggestions for revisions as a prelude to preparing the new edition:

Louis G. Almgren, Norman Alvarez, Mario Antonette, Wayne E. Ault, Carl Baldasserra, Irwin Benjamin, Vytenis Babrauskas, Boris Bresseler, Edward Budnick, Jim Byers, John Campbell, Kenneth Charter, Al Clark, Leonard Cooper, Iving Deutsch, John E. Echternacht, Richard Gann, Richard Gewain, Thomas E. Goonan, Matrin Hertzburg, Clayton Huggett, David C. Jeanes, John D. Jensen, Rolf H. Jensen, Peter Johnson, John Krasney, Jane Lataille, Barry M. Lee, Bernard Levin, T.T. Lie, Richard E. Master, Harold E. Nelson, John O'Neill, William Parker, Patrick E. Phillips, Robert G. Planer, James Quintiere, John Rockett, Robert Rosenberg, Gerald J. Rosicky, John Ed Ryan, Merwin Schulkin, Jack Snell, Thomas E. Seymoure, James G. Spence, Richard Stevens, Roger K. Sweet, Richard Thornberry, R. Brady Williamson, Rexford Wilson, James Winger, and Reginald Wright.

And finally, thanks to Catherine Cronin, whose diligent work gave the *Handbook* any degree of consistency it may have; Stanley Dingman, who was fast and accurate; John Petraglia, who deserves special mention for hard work and dedication to accuracy; John Titus, of Worcester Polytechnic Institute, who prepared the expanded Bibliographies; Peter Johnson, of the Central Investigation and Research Laboratory, Melbourne, Australia, who served as metrication consultant, general technical reviewer, and chief critic; Joan Croce, who gave up Thanksgiving and Christmas to prepare the Index; and last (but not least), Gordon P. (Mac) McKinnon who stuck it out for one more year to hold my hand through this his final (and my first) edition of the *Fire Protection Handbook*. Your comments and suggestions for improving the 17th Edition will be appreciated.

Jim L. Linville
Managing Editor

SECTION 1

THE FIRE PROBLEM

ASSESSING THE MAGNITUDE OF THE FIRE PROBLEM

Arthur E. Cote, P.E.

In *America Burning*, the Report of the National Commission on Fire Prevention and Control NCFPC (1973), the problem of fire in the United States was defined and put into proper perspective: "Indifferent to fire as a national problem, Americans are similarly careless about fire as a personal threat. It takes the careless or unwise action of a human being, in most cases, to begin a destructive fire. In their home environments, Americans live their daily lives amid flammable materials, close to potential sources of ignition. Though Americans are aroused to issues of safety in consumer products, firesafety is not one of their prime concerns. And often when fire strikes, ignorance of what to do leads to panic behavior and aggravation of the hazards, rather than to successful escape."

The fire problem must be addressed as a personal threat; personal behavior before and during fire emergencies must be changed so that individuals will know how to *prevent* fires, to *protect* themselves in the event a fire does occur, and to *persuade* others to prevent and protect themselves from fire.

An important milestone of the late 1970s was introduction into the nation's schoolrooms of firesafety education based upon sound behavioral principles and good learning techniques. NFPA's Learn Not to Burn campaign, which includes both a curriculum and a public information program, is a good example of thoughtfully prepared educational campaigns with specific far-reaching goals. The *Learn Not to Burn Curriculum®*, in particular, offers the real prospect of reaching all elementary and middle school children by teaching and reinforcing 25 key firesafety behaviors that can be retained for a lifetime. Most dramatically, fire prevention education is moving away from the sporadic and negative, or threatening, methods of promoting firesafety—with mixed results at best—to a more positive and long-lasting approach. While the positive effects on the population as a whole may take years to realize, through 1985 more than 200 cases of lives being saved as a direct result of NFPA's Learn Not to Burn campaign have been recorded.

Mr. Cote is the Assistant Vice President/Standards of the National Fire Protection Association, and is Secretary to the NFPA Standards Council.

Since its beginning, NFPA has sought quantitative measures of the fire problem. Table 1-1A presents a summary of historical statistics published by NFPA during the fifty year period up to 1984. Some of the sharp changes in fire loss estimates evident in Table 1A have resulted from changes in statistical methods rather than from changes in the fire problem.

During 1977 and 1978, NFPA further modified its survey techniques to utilize modern statistical design principles, in the process greatly improving its estimates of the size of the national fire problem.

U.S. CIVILIAN FIRE DEATHS

By most measures the number of civilian (non fire service) fire deaths and the rate of such deaths per million population have declined by about a third since the publication of *America Burning* (NCFPC 1973), which set a goal of cutting the fire problem in half in a generation. America appears to be on schedule to achieve that goal. Fire deaths seem to have declined fairly steadily and significantly since 1977, as shown in Table 1-1B.

Despite sustained declines in U.S. fire deaths, however, the U.S. and Canada still have the highest fire death rates of all developed countries in the free world.

In the late 1970s, a series of reports by the Georgia Institute of Technology showed that fire fatality rates per million population in the U.S. and Canada are at least twice the rates in Japan and western Europe (the only other areas examined). No single cause has been identified to explain this pattern, but analyses have shown that, while the U.S. does well in holding down the average severity of fires, it fares very badly in holding down the fire incident rate.

From 1977 through 1984, an average of 6,633 civilians died each year from fire in the United States. Of this total, an average of 5,634 died in structure fires (85 percent of all fire deaths). Most of the structure fire deaths—96 percent, or an annual average of 5,384 fatalities—occurred in fires in residential structures.

TABLE 1-1A. Fifty-Year Record of U.S. Fire Losses

Years	No. of Fires	No. of Building Fires	Deaths[1]	Deaths per 100,000 Population[15]	Dollar Loss	Loss in 1984 Dollars[15] Total	per Capita
1926	350,000[2]	N/A	10,000[3]	8.5	$ 560,548,642[4]	3.3 Billion	28
1936	643,000[5]	N/A	10,000[6]	7.8	303,700,000[5]	2.3 Billion	14
1946	608,000[7]	528,000[8]	10,000[9]	7.1	580,000,000[7]	3.1 Billion	22
1956	1,940,150[10]	824,400[10]	12,000[11]	7.1	1,231,576,000[10]	4.7 Billion	28
1966	2,396,550[12]	970,800[12]	12,100[13]	6.2	1,860,500,000[12]	5.9 Billion	30
1977[14]	3,264,500	1,098,000	7,395	3.4	4,709,000,000	8.1 Billion	37
1984[14]	2,343,000	848,000	5,240	2.5	6,707,000,000	6.7 Billion	28

Notes:
[1] Death estimates prior to 1966 are not specific to a year, but are typical values published by NFPA during the respective periods.
[2] *Fire Protection Handbook,* 8th Edition, 1935, p. 1. Figure is typical of the period, but not specific to a year, and only includes reported fires.
[3] *Fire Protection Handbook,* 8th Edition, 1935, p. 8.
[4] *NFPA Quarterly,* Vol 21, No. 1, p. 14, July 1927. National Board of Fire Underwriters' estimate.
[5] "Fires and Losses by Occupancy and Cause," Dr. Wayne E. Nolting, *NFPA Quarterly,* Vol. 31, No. 2, October 1937, p. 151.
[6] *Fire Protection Handbook,* 9th Edition, 1941, p. 6.
[7] "Fires and Fire Losses Classified, 1946," *NFPA Quarterly,* Vol. 41, No. 2, p. 135, October 1947.
[8] "Fires and Fire Losses Classified, 1947," *NFPA Quarterly,* Vol. 42, No. 2, p. 103, October 1948.
[9] *Fire Protection Handbook,* 10th Edition, 1948, p. 1.
[10] "Fires and Fire Losses Classified, 1956," *NFPA Quarterly,* Vol. 51, No. 2, p. 143, October 1957.
[11] *Fire Protection Handbook,* 12th Edition, 1962, p. 1–2.
[12] "Fires and Fire Losses Classified, 1966," *Fire Journal,* Vol. 61, No. 5, p. 84, September 1967.
[13] "Fire Protection Developments in 1966," Percy Bugbee, *Fire Journal,* Vol. 61, No. 2, p. 43, March 1967.
[14] The estimates are based on data reported to the NFPA by fire departments that responded to the 1977 and 1984 NFPA National Fire Experience Survey. Dollar loss figures represent direct property loss only.
[15] Population and consumer price index figures taken from Statistical Abstract of the United States and Historical Statistics of the United States, Colonial Times to 1970 (published 1976). Both published by the U.S. Census Bureau.

TABLE 1-1B. U.S. Civilian Fire Deaths (1977–1983)*

	1977	1978	1979	1980	1981	1982	1983	1984
(Homes)‡	5,865	6,015	5,500	5,200	5,400	4,820	4,670	4,075
(Other Residential Structures)§	270	170	265	246	140	120	150	165
(Non-Residential Structures)‖	370†	165	205	229	220	260	270	285
(Structures Total)	6,505	6,350	5,970	5,675	5,760	5,200	5,090	4,525
(Vehicles)††#**	740	1,255	1,535	740	840	695	725	630
(All Other)††	150	105	70	90	100	125	105	85
	7,395	7,710	7,575	6,505	6,700	6,020	5,920	5,240

* Estimates are based primarily on data reported to the NFPA by Fire Departments that respond to the NFPA's Annual Survey of Fire Departments for U.S. Fire Experience.
† Includes the Beverly Hills Supper Club fire that took the lives of 165 people.
‡ Homes include one and two family dwellings, mobile homes and apartments.
§ Public residential properties including hotels and motels.
‖ Includes public assembly, educational, institutional, store and office, industry, utility, storage and special structure properties.
Vehicle fire deaths prior to 1980 were unverified as to whether a reported vehicle death was caused by impact or fire, which may partially account for the high estimates of vehicle deaths for 1978 and 1979.
** Includes highway vehicles, trains, boats, ships, aircraft, farm vehicles and construction vehicles.
†† Includes outside properties with value, as well as brush, rubbish, and other outside locations.

U.S. CIVILIAN FIRE INJURIES

Annual estimates of civilian injuries indicate little change from year to year over the last few years and show that on the average about 30,000 civilians are injured by fire annually in the U.S. Over the eight year period from 1977 to 1984, residential structure fires accounted for approximately 69 percent of all civilian injuries reported to fire departments. Table 1-1C shows civilian fire injuries in the U.S. from 1977 through 1984.

Many fire injuries are not reported to the fire service. These include minor injuries that did not require immediate medical attention, injuries that occurred at small fires the fire department did not attend, or injuries to persons who may not have been transported to medical facilities by the fire service. For these reasons, the annual estimates for civilian injuries are probably on the low side.

NUMBER OF FIRES AND PROPERTY LOSS

Total fires attended by public fire departments during the 1977 through 1984 period averaged 2,752,100 annually. Over the same period, the number of structure fires averaged 994,000 annually, accounting for 36 percent of all fires. Also for the eight year period, vehicle fires accounted for an annual average of 473,600 fires or 17 percent of all fires, while outside fires and others accounted for an annual average of 1,284,500 fires or 47 percent of all fires. An average of 704,200 fires occurred annually in residential properties, accounting for 71 per-

TABLE 1-1C. U.S. Civilian Injuries (1977–1983)*

	1977	1978	1979	1980	1981	1982	1983	1984
Residential Structures†	22,600	21,260	20,450	21,100	20,375	21,100	21,450	19,275
Non-Residential Structures‡	3,710	3,725	4,400	3,625	5,325	4,475	4,700	3,750
Structures (Total)	26,310	24,985	24,850	24,725	25,700	25,575	26,150	23,025
Vehicles§	3,515	3,720	5,175	4,075	3,400	3,425	3,800	3,600
Other#	1,365	1,120	1,300	1,400	1,350	1,525	1,325	1,500
Total	31,190	29,825	31,325	30,200	30,450	30,525	31,275	28,125

* Estimates are based primarily on data reported to the NFPA by Fire Departments that respond to the NFPA's Annual Survey of Fire Departments for U.S. Fire Experience.
† Includes one and two family dwellings, apartments, hotels and motels and other residential properties.
‡ Includes public assembly, educational, institutional, stores and offices, industry, utility, storage, and special structure properties.
§ Includes highway vehicles, trains, boats, ships, aircraft, farm vehicles and construction vehicles.
# Includes outside properties with value, as well as brush, rubbish, and other outside locations.

TABLE 1-1D. Fires Attended by Fire Departments in the U.S. (1977–1984)

	1977	1978	1979	1980	1981	1982	1983	1984
Residential Structures	750,000	730,500	721,500	757,500	733,000	676,500	641,500	623,000
Non-Residential Structures	348,000	331,500	315,000	307,500	294,500	270,000	227,000	225,000
Total Structure Fires	1,098,000	1,062,000	1,036,500	1,065,000	1,027,500	946,500	868,500	848,000
Outside Fires (with value)	130,000	110,000	80,500	86,500	81,000	54,000	49,500	45,000
Vehicles	508,000	503,000	495,500	471,500	466,500	443,000	447,000	454,500
Brush Grass, Wildland (no value)	734,000	582,500	614,500	718,500	711,000	522,500	467,500	487,500
Rubbish	406,000	296,000	342,000	397,000	341,000	309,500	288,000	303,000
All Other	388,500	264,000	276,500	249,500	266,500	262,500	206,000	205,000
TOTAL FIRES	3,264,500	2,817,500	2,845,500	2,988,000	2,893,500	2,538,000	2,326,500	2,343,000

cent of all structure fires over the eight-year period. Table 1-1D shows the number of fires attended by fire departments in the U.S. from 1977 through 1984. Over the 1980-1984 period, the number of structure fires followed a downward trend similar to that noted for all fires to which fire departments responded. The high occurred in 1980 with 1,065,000 fires, and the numbers have decreased each year since. The overall drop in structure fires over the five year period was 20 percent. Figure 1-1A shows the downward trend of all fires from 1980 through 1984.

Direct property damage from fire increased over the 1977-1984 period. In 1977, total fires resulted in an estimated property damage of $4,709,000,000 compared with $6,707,000,000 in 1984, for an increase of 42 percent over the eight year period. For structure fires, an estimated $4,125,000,000 in property damage occurred in 1977 compared with $5,891,000,000 in property damage in 1984—an increase of 43 percent over the eight year period. Structure fires accounted for an average of 88 percent of all fire property loss annually during the 1977-1984 period. Table 1-1E shows direct property loss in the U.S. from 1977 through 1984. Figure 1-1B shows the average property loss per structure in the U.S. from 1980 through 1984.

Over the 1980-1984 period, the average property loss for structures increased 36 percent, from $5,121 in 1980 to $6,947 in 1984. However, as Figure 1-1B shows, when the property loss is adjusted for inflation, there is hardly any change (just 8 percent) in the average loss per structure fire over the five year period.

It can be seen that the annual total number of fires has declined in recent years from a peak of more than three million in 1977. Total direct damage has increased, but this increase has been primarily due to inflation. Measured in dollars adjusted for inflation, average loss per fire has increased only slightly. With the decline in the number of fires, total damage in adjusted dollars has actually declined in recent years.

Rough estimates of the total cost of fire in the U.S. indicate that the cost of fire departments, the cost of built-in fire protection in new construction, and the net of fire insurance premiums over losses each involves costs on the same order of magnitude as the direct property damage due to fire.

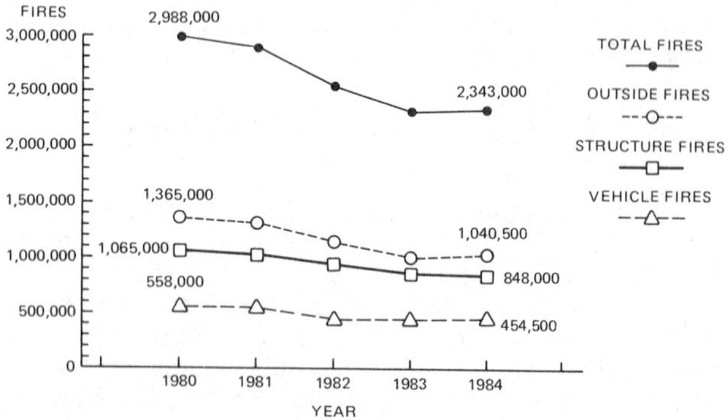

FIG. 1-1A. Estimates of fires by type in the United States from 1980 to 1984.

Leading Classes of Fire Ignitions

The four leading classes of fire ignitions are as follows:

A. *Smoking material* related ignitions, account for the largest share of civilian fire deaths, although the number

TABLE 1-1E. Direct Property Loss in the U.S. (1979–1984) × $1,000,000

	1977	1978	1979	1980	1981	1982	1983	1984
Residential Structures	2,179	2,192	2,529	3,042	3,259	3,253	3,306	3,440
Non-Residential Structures	1,946	1,830	2,435	2,412	2,717	2,478	2,520	2,451
Total Structure Loss	4,125	4,022	4,964	5,454	5,976	5,731	5,826	5,891
Outside Fires (with value)	144	32	38	61	61	52	35	29
Vehicles	332	391	682	685	594	591	694	749
All Other	108	53	66	54	45	58	43	38
Total Property Loss	4,709	4,498	5,750	6,254	6,676	6,432	6,598	6,707

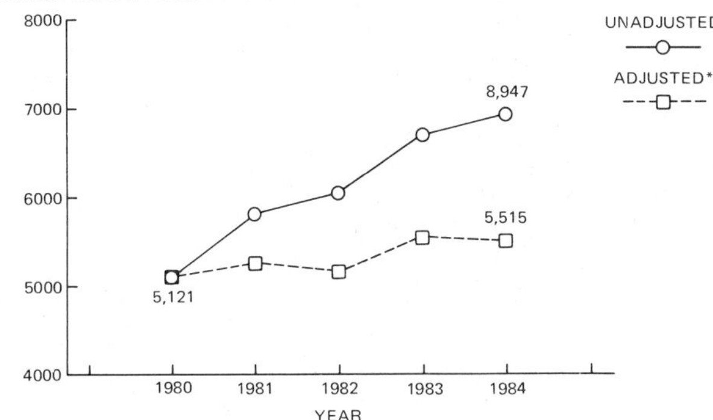

FIG. 1-1B. Average property loss per structure fire in the United States from 1980 to 1984.

and percentage of deaths due to these causes has been declining in recent years.

B. *Incendiary and suspicious* ignitions, account for the largest share of direct property damage, even in residential properties. The number of fires and dollars of loss in structures have been declining in recent years, however.

C. *Heating equipment* related ignitions, account for the largest share of residential fires. Auxiliary heating equipment related ignitions are the principal factor in the higher fire death rates in rural areas and in the South.

D. *Cooking equipment* related ignitions, account for the largest share of residential fire injuries to civilians and by far the largest share of fires that go unreported to fire departments.

MULTIPLE-DEATH FIRES

Most of the visibility of the fire problem comes from "multiple-death" incidents—fires that kill three or more persons. Deaths in such fires number approximately a thousand per year, considerably more than the combined U. S. totals for earthquakes, tornadoes, hurricanes, floods, and other natural disasters.

Table 1-1F shows the U.S. experience for both residential and non-residential multiple death fires (fires causing three or more deaths) from 1980 through 1984.

Table 1-1G presents the number of catastrophic multiple-death fires resulting in 10 or more deaths over the 1980-1984 period. A number of different types of residential properties experienced this kind of incident. Roughly

TABLE 1-1F. Multiple Death Fires 1980–1984

Year	Number of Multiple-Death Fires	Number of Deaths in Multiple-Death Fires*	Fires Causing 10 or More Deaths	Deaths in Fires Killing 10 or More
1984	244	1,007	5	84
1983	259	986	2	34
1982	266	1,111	7	116
1981	296	1,179	6	95
1980	326	1,356†	6	168†

* Includes both civilian and fire-fighter deaths.
† Includes 85 deaths in the MGM Grand Hotel Fire.
Note: The information presented in this Table for 1980 to 1984 may differ from previously published statistics because of additional information which has become available to the NFPA.

65 percent of both the incidents and deaths during this five year period occurred in residential occupancies.

DIVIDING THE FIRE PROBLEM INTO MANAGEABLE TERMS

There are numerous ways to look at the fire problem. One convenient method is that used in NFPA's Long Range Plan (NFPA 1979) which divides the American environment into eight major components and three topical fire problem areas:

Major Environmental Components

1. The Home Environment
2. The Mobile Environment
3. The American Community
4. The Industrial Environment
5. The Recreational Environment
6. Our Forest and Wildlands
7. Agricultural (Rural) America
8. Our American Heritage

Topical Fire Problem Areas

A. Public Apathy
B. Economic Impact of Protection
C. Incendiarism, Arson, and Suspicious Fires

Each component represents a portion of the overall fire problem in the United States.

THE HOME ENVIRONMENT

The American fire problem is strikingly concentrated in the Home Environment, made up of one and two family dwellings (including mobile homes) and apartments. In

TABLE 1-1G. Fires Resulting in 10 or More Deaths, 1980–1984

Property Use	Location	Date	Number of Deaths
Hotel*	Las Vegas, Nev.	Nov. 21, 1980	85
Hotel Meeting Facility*	Harrison, N.Y.	Dec. 4, 1980	26
Boarding Home Facility*	Bradley Beach, N.J.	July 26, 1980	24
Apartment Building	Salt Lake City, Utah	Dec. 18, 1980	12
Metal Manufacturing Plant	Brooklyn, N.Y.	July 24, 1980	11
Apartment Building	Chicago, Ill.	Oct. 28, 1980	10
Boarding Home Facility*	Morganville, N.J.	Jan. 9, 1981	31
Residential Hotel*	Chicago, Ill.	March 14, 1981	19
Coal Mine	Whitwell, Tenn.	Dec. 8, 1981	13
Single-Family Dwelling	East Saint Louis, Ill.	Jan. 11, 1981	11
Apartment Building	Hoboken, N.J.	Oct. 24, 1981	11
Motorhome	San Bernardino, Calif.	June 26, 1981	10
Jail*	Biloxi, Miss.	Nov. 8, 1982	29
Residential Hotel*	Los Angeles, Calif.	Sept. 4, 1982	24
Apartment Building	Waterbury, Ct.	July 5, 1982	14
Aircraft Post-Crash Fire*	Bakersfield, Calif.	Oct. 17, 1982	14
Residential Hotel	Hoboken, N.J.	April 30, 1982	13
Hotel*	Houston, Tex.	March 6, 1982	12
Single-Family Dwelling	Baltimore, Md.	May 14, 1982	10
Passenger Aircraft*	Cincinnati, Ohio	June 2, 1983	23
Fireworks Factory	Benton, Tenn.	May 27, 1983	11
Coal Mine	Orangeville, Utah	Dec. 19, 1984	27
Refinery	Romeoville, Ill.	July 23, 1984	17
Rooming House*	Beverly, Mass.	July 4, 1984	15
Hotel*	Paterson, N.J.	Oct. 18, 1984	15
Single-Family Dwelling	Philadelphia, Pa.	Feb. 8, 1984	10
Total: 26 Incidents			497 Deaths

* Incident was investigated by NFPA's Fire Investigations and Applied Research Division.

the home, fire is the second leading cause of accidental deaths, after falls. The Home Environment accounts for approximately:

80 percent of all fire deaths,
67 percent of all fire injuries,
70 percent of all structure fires, and
61 percent of all structure fire property loss reported to municipal fire departments in the U.S.

Key facts about home fires include the following:

1. Smoking materials are the leading cause of fatal home fires. Nearly one-third of the victims of home fires lose their lives in fires started by smoking materials, generally in upholstered furniture, mattresses or bedding.
2. Heating and cooking equipment is the second leading cause of fatal home fires. These fires involve the ignition of structural components, clothes, other soft goods, creosote, and cooking materials.
3. Another major cause of fatal home fires involves the ignition of bedding, upholstered furniture, clothes, or other soft goods.
4. There has been a sharp rise in home fires involving portable and area heating equipment, particularly wood-burning equipment, according to analyses of national fire incident data done by the Federal Emergency Management Agency's U.S. Fire Administration (FEMA/USFA). The number of fires involving solid-fueled heating equipment (nearly all of it wood-burning) in one and two family dwellings nearly doubled between 1977 and 1980.
5. In 60 percent of all home fires reported to fire departments, flame is confined to the area around the point of origin; but 65 percent of the deaths and direct property damage occur in fires where flame ultimately extends beyond the floor of origin.
6. Of the victims of fatal home fires, an average of 26 percent are aged 9 years or less, and 21 percent are aged 65 years or more.

The number of civilian fire deaths in the home, after reaching its high level of 6,015 deaths in 1978, showed an overall decrease of 32.2 percent from 1978–1984. Figure 1-1C shows the overall downward trend in fire deaths in the home since 1978.

This impressive decrease is probably due in part to recent home firesafety efforts, including the passage of numerous state laws and local ordinances requiring the installation of smoke detectors in residential properties.

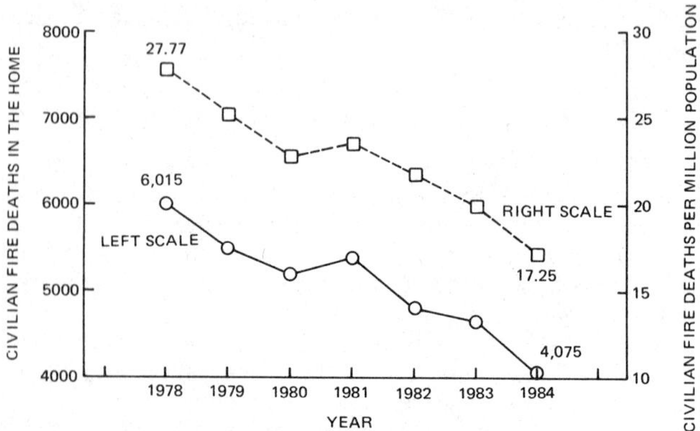

SOURCE: NFPA'S ANNUAL SURVEY OF FIRE DEPARTMENTS FOR U.S. FIRE EXPERIENCE (1978-84)

FIG. 1-1C. Civilian fire deaths and rates in the home in the United States from 1978 through 1984.

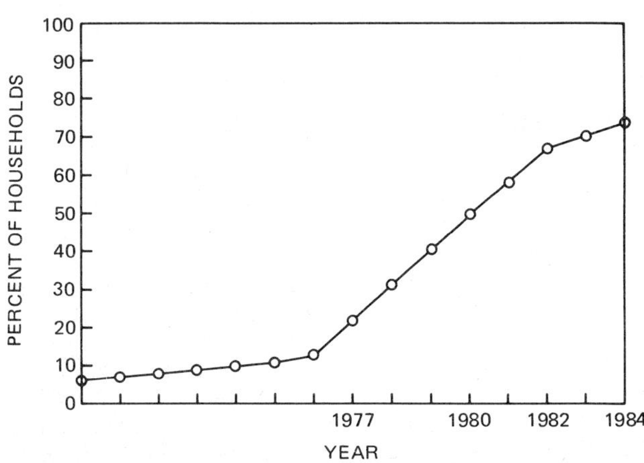

FIG. 1-1D. Detector coverage grew significantly from 1970 through 1983.

Figure 1-1D shows the remarkable growth in detector coverage during the past period between 1970 and 1984. The years 1977, 1980, and 1982 are those in which the U.S. Fire Administration (USFA) conducted national surveys. The growth in coverage before 1977 was estimated using manufacturers' sales figures included in the first survey. A 1985 Louis Harris poll provides the latest benchmark: 74 percent of U.S. households now have detectors. Clearly, the detector explosion is unchallenged as the fire protection improvement of our time.

The various detector laws are summarized in Figures 1-1E and 1-1F. No simple figure can adequately capture the pattern of detector legislation, inasmuch as state laws are modified by local laws, and there is considerable variation in requirements from one place to another at both levels. New laws are being passed nearly every month.

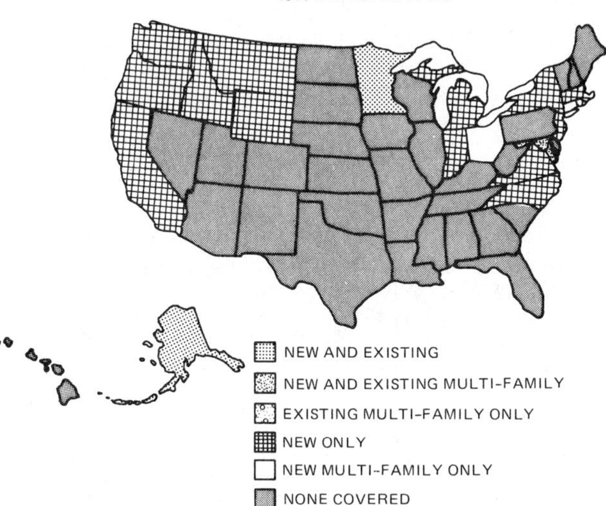

FIG. 1-1E. In 1977, in 31 states there were no detector laws.

The differences in the breadth and depth of requirements from 1977 through 1983 can be readily seen. In 1977, there were 31 states without smoke detector requirements at all for dwellings and apartments. By 1983, the number of states with no home smoke detector requirements had dropped to 13 (Hall 1985).

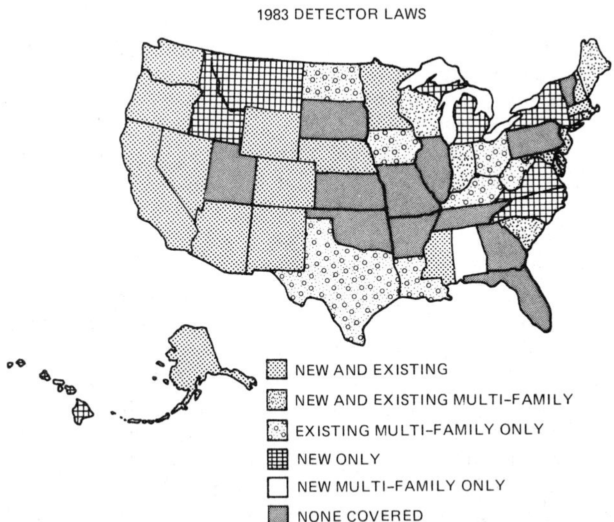

FIG. 1-1F. By 1983, only 13 states were still without any detector laws.

In spite of the decrease in civilian fire deaths in the home, these fires still accounted for 92 percent of all structure fire deaths, or an annual average of 5,193 fatalities from 1977 through 1984, meaning that home firesafety remains the key to any major improvements in the overall fire death picture in the U.S.

Home firesafety education is vital. Topics such as the major causes of fatal home fires, the use and maintenance of smoke detectors, and escape planning must continue to be brought to the attention of dwelling and apartment residents of all ages.

Recent research has shown that residential sprinkler systems using "quick-response" residential sprinklers and domestic water supplies are effective and are becoming more affordable. Several communities have begun to require residential sprinkler systems in new homes, and in some cases in existing homes as well.

THE MOBILE ENVIRONMENT

This component covers all forms of transportation: water, rail, air, and road. It includes public as well as private and commercial transportation.

America is a mobile society. The habitual use of the automobile is a way of life and is the primary means of transportation for a large part of the American public. The fire problem in automobiles is particularly difficult to analyze since many fires occur as the result of highway crashes. In these incidents, it is often a highly technical and complex task to sort out the respective contributions of impact and fire to resulting deaths, injuries, and property damage.

Based on data reported by fire departments nationwide, an average of 490,000 vehicle fires, resulting in an average of 725 deaths and 3,700 injuries, occurred annually from 1980 through 1984. Largely because of the difficulties of separating the effects of fire from those of impact, these statistics are subject to considerable year to year fluctuation. In spite of these fluctuations, however, annual fire deaths in vehicles consistently rank third, exceeded only by deaths in one and two family dwellings and in apartments. Over the 1980 to 1984 period, vehicle fires accounted for an average of 10 percent of all property loss from fire in the U.S.

A new fire protection problem is the increasing use of automobile fuels other than gasoline and diesel (e.g., LP-Gas, LNG, and CNG). These new fuels introduce hazards with which the public and authorities are not totally familiar.

Changes in transportation preference, such as the growing use of public transportation, have opened up new problems for fire protection. The fire problems of public transportation center around equipment design and materials as well as growing dependency on automation.

Transportation of raw materials and finished goods throughout the nation and in U.S. coastal waters by common carriers represents the potential for substantial economic loss if fire destroys the materials transported or the vehicles in which they are being moved. This type of transportation at the interstate level is subject to federal regulation.

Another key area is hazardous materials. This problem not only poses the potential for costly loss, but, more important, carries with it substantial risk to public safety. Hazardous materials often are transported over land, even through highly congested urban communities. The fire record shows an abundance of examples of disastrous accidents involving hazardous materials transportation.

Air transportation is also regulated by the federal government, with regard to the level of firesafety in aircraft design and fire potentials following ground impact. Other fire protection concerns associated with air transportation include fuel servicing, aircraft maintenance, design of airport facilities, and aircraft rescue and fire fighting techniques.

THE AMERICAN COMMUNITY

The American Community represents all of those activities that accompany community living, and the many fire problems inherent in those activities. In terms of property uses, the American Community includes places of assembly, educational facilities, institutions, stores and offices, hotels and motels, and board and care facilities.

Community activities often concentrate large numbers of people, creating the risk of large loss of life should a fire occur in the properties which make up the American Community. For this reason, properties in the American Community are generally subject to legally binding fire codes such as the NFPA *Life Safety Code®*.

Figure 1-1G below presents the 20th century record in the United States for building fires taking at least 25 lives. Although some fires in industrial and storage properties and a number of fires involving vehicles are included, the record is dominated by properties in the American Community, such as theatres, night clubs, schools, institutions, and hotels.

Except for the Depression and World War II periods which presented unique problems for code enforcement and compliance, the record shows considerable progress during the century against major loss-of-life building fires. This progress is particularly impressive considering that during the 80 years covered by Figure 1-1G, the U.S. population nearly tripled and great numbers of buildings addressed by the fire codes were constructed. These build-

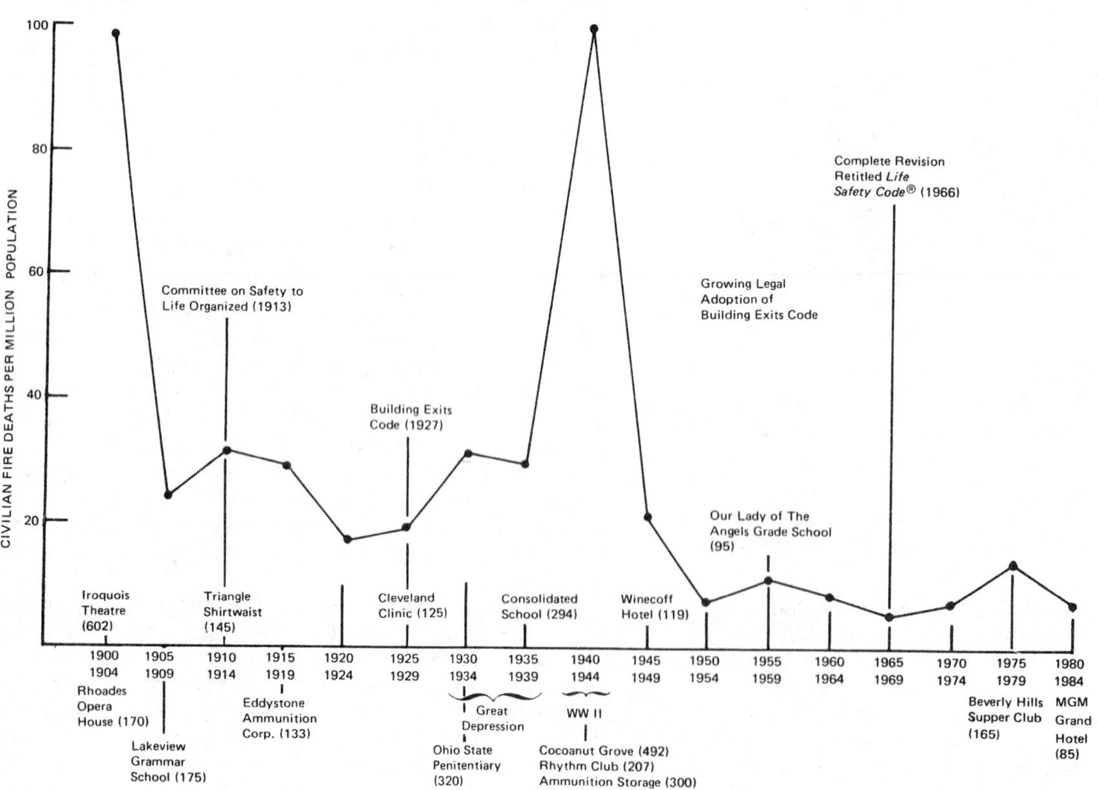

FIG. 1-1G. Major loss of life in 20th century U.S. building fires taking at least 25 lives.

ings include facilities that concentrate very large numbers of people.

Major Life Loss Fires of Significance in the American Community

Historically, fires resulting in a major loss of life have resulted in important changes in building and fire codes and in standard fire protection or prevention practices. In the early 1900s four building fires—the Rhoades Opera House in Boyertown, PA (1903), the Iroquois Theatre in Chicago (1903), the Lakeview Grammar School in Collinwood, OH (1908), and the Triangle Shirtwaist Factory in New York City (1911)—were largely responsible for the appointment in 1913 of the NFPA Committee on Safety to Life. The opening summary of the "Origin and Development of 101" in the current *Life Safety Code* (NFPA 101) states: "For the first few years of its existence, the Committee devoted its attention to a study of the notable fires involving loss of life and in analyzing the causes of this loss of life. This work led to the preparation of standards for the construction of stairways, fire escapes, etc., for fire drills in various occupancies, and for the construction and arrangement of exit facilities for factories, schools, etc., which form the basis of the present Code."

The 1937 fire at the Consolidated School in New London, TX, tragically pointed out the need for state laws to protect public buildings not subject to municipal ordinance and inspection. Then, in the 1940s, a series of multiple-death fires—including those at The Rhythm Club, The Cocoanut Grove, and the La Salle, Canfield, and Winecoff Hotels—focused national attention upon the need for adequate exits and other firesafety features in hotels and public buildings. These fires resulted in major changes to the "Building Exits Code" (as the *Life Safety Code* was then known) over a period of almost two decades. NFPA 102, *Standard for Assembly Seating, Tents, and Air-Supported Structures* was the result of still another multiple death fire of the 1940s—the 1944 Hartford, CT circus tent fire in which 168 people were killed.

Three hospital fires—St. Anthony's in Effingham, IL, in 1949 (74 killed); Mercy Hospital in Davenport, IA, in 1950 (41 killed); and Hartford, CT, Hospital in 1961 (16 killed)—moved hospital administrators and fire prevention officials across the nation to assess the quality of construction and fire protection systems in hospitals.

The Our Lady of Angels School fire in Chicago on Dec. 1, 1958 probably resulted in the swiftest action in the wake of any major fire since World War II. Within days of the fire, state and local officials throughout the nation ordered fire inspections of schools, and within one year it was reported that major improvements in life safety had been made in 16,500 schools across the country. Improvements in the frequency and quality of exit drills and inspections, in the storage of combustible supplies, and in the disposal of waste materials were also reported in almost every community where schools were surveyed.

The effects of more recent fires such as the Beverly Hills Supper Club (165 killed), the MGM Grand Hotel (85 killed), and others are still being felt in the fire protection community.

Fire Causes in the American Community

Cigarettes and open flames are the leading sources of ignition for *accidental* fatal fires in institutions. A particular fire problem for these properties is the ignition of clothing being worn. Institutions are predominantly penal and health care facilities and thus must deal with people with the reduced capability for self-preservation that results from physical restraint or from physical or mental impairment. Such people must rely heavily on the institution's supervisory personnel for their protection against fire.

The ignition of *accidental* fatal hotel and motel fires is dominated by cigarettes igniting bedding or upholstered furniture in guest rooms or lounge areas. *Incendiary and suspicious* fires and fires due to mechanical failures or malfunctions are predominant in public assembly and educational occupancies. Many fire fatalities in the American Community are also nonaccidental, as a result of *arson*.

The Boarding Home Fire Problem

Deinstitutionalization has severely worsened the problem of multiple-death fires in boarding homes. Mental hospitals began deinstitutionalizing chronic mental patients in the mid-1950s, and the effects of the move began to show up in boarding home fires in the late 1970s. During 1974-1984, roughly 300 people died in boarding home fires killing three or more persons. Residents of such homes have a risk of dying in a multiple death fire that is five times the risk for residents of other residential properties (measured as deaths in multiple death fires per million population).

Analysis of recent fatal boarding home fires reveals a significant problem with electrical systems and arson, coupled with the lack of basic fire protection provisions in these buildings. Contributory factors include inadequate means of egress, combustible interior finishes, unenclosed stairways, lack of automatic detection or sprinkler systems, and lack of emergency training for the staff and residents. Many of the facilities were either licensed for other than a boarding home (such as a hotel) or were unlicensed, "underground" boarding homes. None of the facilities were provided with automatic sprinkler protection.

There is no particular mystery about how to reduce the number of fatal boarding home fires. Enclosed stairs, providing two ways out, avoiding the use of combustible interior finishes, compartmentation, providing automatic detection and sprinkler systems, and training of staff and residents in emergency procedures would greatly reduce the number of fatalities.

Table 1-1H includes the boarding home fires with three or more deaths from 1978 through 1984.

THE INDUSTRIAL ENVIRONMENT

After the Home Environment, the Industrial Environment is the second largest contributor to property loss in American fires. For the years 1980 to 1984, an annual average of 99,000 structure fires accounted for $1.25 billion in property loss in Industrial Environment fires attended by municipal fire departments. Since industrial fires that are handled by fixed suppression systems or private fire brigades may go unreported to municipal fire departments, these statistics understate the direct fire damage. In addition, indirect loss such as business failure or interruption occurs and can be significant in particular fires, although it is difficult to measure accurately.

TABLE 1-1H. Boarding Home Fires With Three or More Deaths, 1978–1984

Date	Occupancy	Civilians Killed	Civilians Injured
2/1/78	Custodial Care Home* Nashville, TN	4	0
2/22/78	Boarding Home Prichard, AL	3	3
4/16/78	Convalescent Home* San Francisco, CA	3	1
5/13/78	Putnam Farms Halfway House* Athens, TX	4	0
5/21/78	Boarding Home* Birmingham, AL	3	1
7/5/78	Lanys Guest House* Asbury Park, NJ	4	0
1/26/79	Boarding Home* Jackson, TN	3	0
4/1/79	Marietta Foster Home* Connellsville, PA	10	2
4/2/79	Wayside Inn Boarding House* Farmington, MO	25	9
4/11/79	1715 Lamont Street* Washington, DC	10	4
11/11/79	Coats Boarding Home* Pioneer, OH	14	0
1/11/80	Sports News Boarding Home Johnson City, TN	4	1
7/26/80	Brinley Inn Boarding Home* Bradley Beach, NJ	24	4
8/16/80	Boarding Home Honolulu, HI	3	0
10/4/80	Little Friends, Inc.* Naperville, IL	3	0
10/20/80	Boarding House St. Louis, MO	3	0
11/30/80	Donahue Foster Home* Detroit, MI	5	0
12/21/80	Rooming House* Corinth, MS	4	0
1/1/81	Boarding House Decatur, GA	3	1
1/9/81	Beachview Rest Home* Keansburg, NJ	31	10
1/18/81	916-18 No. Eighth Street* Camden, NJ	3	1
2/3/81	Keifer's Quarters* Point Pleasant Beach, NJ	7	0
4/15/81	Boarding Home Pottstown, PA	3	2
8/8/81	Boarding Home Orlando, FL	4	0
12/19/81	31 Carmen Street* Patchogue, NY	4	0
1/10/82	Boarding House Ironton, OH	3	0
1/11/82	Boarding Home Newark, NJ	3	0
4/3/82	Boarding Home Elizabeth, NJ	3	1
4/8/82	Boarding Home Philadelphia, PA	4	3
10/28/82	Perrys' Domiciliary Care Home* Pittsburgh, PA	5	1
11/27/82	Jones Dwg.* Great Valley, NY	3	0
12/16/82	Rooming House Boston, MA	4	0
1/9/83	Rooming House Detroit, MI	3	0
1/17/83	Rooming House Boston, MA	4	2
2/7/83	Silver Leaves Group Home* Eau Claire, WI	6	0
3/13/83	Shannons Foster Care Home* Gladstone, MI	5	3
4/19/83	Central Community Home* Worcester, MA	7	2
5/8/83	Boarding Home Tifton, GA	3	0
8/31/83	Anandale Village* Lawrenceville, GA	8	0
10/4/83	Boarding House Austin, TX	4	0
10/21/83	Boarding House Cleveland, OH	3	1
12/5/83	Dana House* Cincinnati, OH	7	3
12/31/83	Rooming House Newark, NJ	4	0
2/4/84	Boarding House Montgomery, AL	5	0
4/21/84	Boarding House Savannah, Ga	4	0
7/4/84	Boarding House* Beverly, MA	15	9

* These are incidents in which NFPA has evidence that elderly occupants or former mental patients were tenants.

Over many years, a great deal of attention and fire protection technology have been directed toward the Industrial Environment. As a result, compared with other environments, fire deaths and injuries in industry are relatively low.

The voluntary, consensus standards making system has worked well in the Industrial Environment, as in other areas. The use of consensus firesafety standards by government agencies has further enhanced industrial firesafety.

Newly developed statistics on the ignition of fires in the Industrial Environment have focused attention on key issues and created an opportunity for further improvements in industrial fire safety. Important problems, based principally on the latest statistical compilations are:

Arson. This is the leading cause of fire loss in storage properties and in industrial and manufacturing properties.

Ignition of fires by *electrical systems* arcing, overloading, or otherwise failing.

Ignition of fires by failure of or misuse of *heating equipment* (e.g., failed bearings, hydraulic line ruptures).

Ignition of fires by careless use of *cutting and welding torches.*

Table 1-1I shows the leading ignition scenarios for large loss fires in manufacturing properties from 1974 to 1983.

Table 1-1J shows the leading ignition scenarios for large loss fires in storage properties from 1974 to 1983.

THE RECREATIONAL ENVIRONMENT

Leisure time for the average American family has increased dramatically since World War II. The continuing

TABLE 1-1I. Ignition Scenarios of Manufacturing Properties, United States, 1974–1983*

Equipment Involved		Heat of Ignition	Ignition Factor	Percent of Fires	Percent of Dollar Loss
Fixed wiring Appliances Panelboard, etc. Processing equip. Other electrical distribution equip. Unspecified elec. distribution equip. Special equipment Other equipment Unknown	(29%) (13%) (11%) (11%) (11%) (9%) (3%) (9%) (4%)	Electrical equipment arcing, overloaded	Mechanical failure, malfunction	25	25.4
		Open flame or incendiary device	Incendiary	11	10.2
Processing equip. Service equipment Separate motor Elec. distribution equipment Appliances Other equipment Unknown	(41%) (21%) (15%) (8%) (6%) (3%) (4%)	Heat from hot object (friction, molten or hot material)	Mechanical failure, malfunction	8	8.9
		Torch (cutting, welding)	Cutting, welding too close to combustibles	8	9.0
Processing equip. Heating equipment Other	(53%) (32%) (15%)	Fuel fired, fuel powered equipment	Mechanical failure, malfunction	7	9.1
Processing equip. Other Unknown	(35%) (3%) (62%)	Spontaneous ignition, chemical reaction	Operational deficiency	6	3.3
Processing equip. Other Unknown	(76%) (18%) (6%)	Heat from hot object (friction, molten or hot material)	Operational deficiency	4	2.9
Processing equip. Heating equip. Service equipment	(70%) (15%) (15%)	Fuel fired, fuel powered equipment	Operational deficiency	3	1.5
		Cigarette	Carelessly discarded	3	2.6

* Based on 476 fires with known ignition scenarios resulting in a dollar loss of $500,000 or more reported in the NFPA Fire Incident Data Organization (FIDO) data. All losses have been adjusted to 1974 dollars using the Producer Price Index; total adjusted dollar loss was $584,611,000.

† Out of 74 distinct scenarios in this data set, these nine scenarios accounted for 75 percent of the fires and 72.9 percent of the loss.

trend toward even fewer working hours holds significant implications for fire protection planning, as it is logical to assume that the hours at risk for potential life loss, injuries, and property loss from fire have increased in the same or greater relationship.

The increased use of recreational occupancies such as recreational vehicles, camping trailers, and tents means more hours at risk than in the past; and increased consumer demands and use for leisure time products creates a potential increase in fire hazards in recreational activities.

OUR FORESTS AND WILDLANDS

A part of our great American heritage is that created by nature. Our forests and wildlands represent resources which have been nationally recognized in legislation as a part of our heritage only since the days of President Theodore Roosevelt (1901-1909).

Our forests and open lands provide raw materials, recreational facilities, and scenic breaks from the urban sprawl.

Great portions of the American landscape—wildlands, forests, and deserts—are under control of the U.S. Department of Agriculture. Departments of this federal agency have their own regulations which include fire prevention and suppression programs. Fire protection through systematic application of these programs has resulted in a laudable reduction in fire losses of these lands.

The states and private enterprise have also done a commendable job in protecting forest lands through their own fire prevention programs.

The present concern of Americans for their environment has resulted in a "return to nature" as more people try to escape urban blight. Homes are being built in the most remote areas. Examples of the problems created by this trend abound in the dwellings on the hillsides and in the canyons around Los Angeles, CA: beautiful homes in

TABLE 1-1J. Ignition Scenarios of Storage Properties, United States, 1974–1983*

Equipment Involved		Heat of Ignition	Ignition Factor	Percent of Fires	Percent of Dollar Loss
		Open flame or incendiary device	Incendiary	24	16.6
Fixed wiring Other electrical distribution equip. Separate motor Heating equip. Appliances Other Unknown	(32%) (35%) (13%) (7%) (5%) (7%) (1%)	Electrical equipment arcing, overloaded	Mechanical failure, malfunction	20	18.2
		Torch (cutting, welding)	Cutting, welding too close to combustibles	6	5.9
		Spontaneous ignition, chem. reac.	Operational deficiency	5	4.0
Conveyer Service equip. Other special equipment Other	(37%) (19%) (13%) (31%)	Heat from hot object (friction, molten or hot matrl.)	Mechanical failure, malfunction	5	4.6
		Cigarette	Carelessly discarded	4	5.0
Heating equip. Other	(92%) (8%)	Fuel fired, fuel powered equipment	Mechanical failure, malfunction	3	2.8
		Hostile fire	Property too close	3	2.4
Heating equip. Unknown	(90%) (10%)	Fuel fired, fuel powered equipment	Combustibles too close	3	1.8

* Based on 357 fires with known ignition scenarios resulting in a loss of $500,000 or more reported in the NFPA Fire Incident Data Organization (FIDO) data. All losses have been adjusted to 1974 dollars using the Producer Price Index; total adjusted loss was $411,115,000.

† Out of 56 distinct scenarios in this data set, these nine scenarios accounted for 73 percent of the fires and 61.3 percent of the loss.

beautiful settings—but their proximity to one another and their vulnerability to fast spreading wildfires, compounded by the popular untreated wood shingle roof, pose a conflagration potential unmatched by most urban areas.

In NFPA's recent annual studies of fires with $500,000 or more damage "large loss fires," brush and forest fires have played a prominent role. In 1978, there were two such fires with a total loss of $43.4 million; in 1979 there were 10 fires with a total loss of $30.1 million; in 1980 there were 5 fires with a total loss of $46 million; in 1981 there were 4 fires with a total loss of $42 million, and in 1982 there were 2 fires with a total loss of $10.5 million. These loss figures include damage to exposed structures and vehicles, the value of watershed lost, and the value of crops (e.g., timber) destroyed.

Table 1-1K shows four of the worst large loss fires reported to NFPA from 1972 through 1981, all involving brush and forest fires in the State of California.

Moreover, brush and forest fires regularly take fire fighter lives. In the years 1978 through 1981, a total of 50 fire fighters lost their lives to these blazes.

It must also be pointed out that the fire service responds to many "nuisance" brush and grass fires. Although they cause negligible dollar losses, it is a serious economic burden to maintain public protection forces needed to combat such fires. In addition, their high frequency pulls costly suppression forces out of service—forces not available for those fires posing a more serious threat to life and property.

TABLE 1-1K. Major Brush and Forest Fires*

		Constant Dollars	Adjusted to 1981 Dollars
November 24, 1980	San Bernadino, California	$38.03	42.06
June 22, 1982	Saint Helena, California	$35.67	35.67
October 23, 1978	Los Angeles, California	$35.11	48.98
July 26, 1977	Santa Barbara, California	$35.00	52.57

* Loss × $1,000,000.

Environmental protection regulations which control open air burning have helped to reduce the number of brush, grass, and woods fires.

Our forest, grass, and wildlands—the great outdoors—are an important natural resource. They must be protected from the ravages of fire. The continued success of this effort requires that federal, state, local, and private agencies work together in areas of regulation, public awareness, and fire suppression techniques.

AGRICULTURAL (RURAL) AMERICA

Fire departments in Rural America face special challenges and problems, including:

Protection of widely scattered populations.

Long runs to respond to a fire.

Inadequate water supply for fire fighting.

Limited budgets for apparatus and training.

Limited opportunity to develop skills in actual fires fighting, due to the small number of fires in any one community.

In addition, there are areas of Rural America with no organized fire protection at all.

Rural areas have the highest fire death and property loss rates. Refinements in NFPA survey techniques in the mid-1970s provided the first hard data on the size of the overall rural fire problem, hitherto hidden by the large number of small communities among which the fires are scattered. Based on 1977 to 1980 data, the rate of fire deaths per million population in communities of less than 2,500 population is roughly double the rate in all other communities and is nearly 50 percent higher than the death rate in the largest cities (more than one million population). Property damage due to fire averages $10 per person per year in middle-sized cities, $15 per person per year in the largest cities, and $40 per person per year in rural communities (Klem 1984).

The residential death rate by community size for 1977 to 1980 is shown in Figure 1-1H.

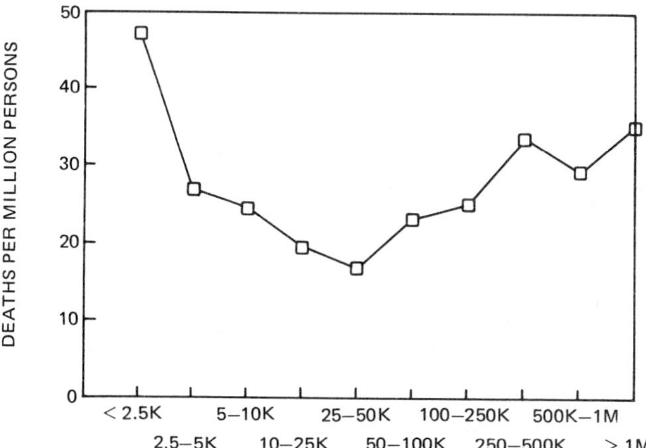

SOURCE: FEMA ANALYSIS OF NFPA SURVEY DATA

FIG. 1-1H. Residential death rate by community size, 1977 to 1980 average.

The residential property loss rate by community size for 1977 to 1980 adjusted to 1980 dollars is shown in Figure 1-1I.

OUR AMERICAN HERITAGE

Our American Heritage includes historic buildings, museums, art galleries, and memorial structures. Such structures often attract large numbers of visitors, which may introduce the potential for large loss of life.

Since many of these structures are owned and operated by nonprofit charitable organizations, funds for fire protection are often limited. In addition, many historic structures are located in remote areas where public protection is minimal at best.

Museums and art galleries incorporate the special hazards of work rooms for restoration and storage.

We tend to forget the vulnerability to fire of these vestiges of our heritage that were so important to our forefathers and will be important to future generations. To give a fire loss figure for these structures would be deceptive because the actual loss sustained when artifacts and ancient structures burn cannot be measured in dollars alone.

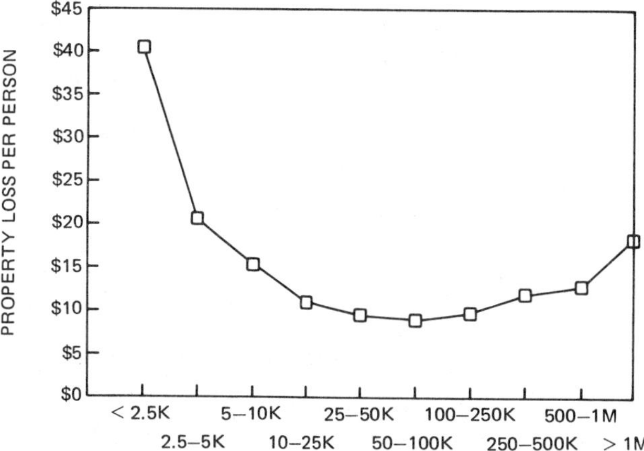

SOURCE: FEMA ANALYSIS OF NFPA SURVEY DATA

FIG. 1-1I. Residential property loss rate by community size, 1977 to 1980, average adjusted to 1980 dollars.

During the 1980s, fires have damaged or destroyed a long list of historic buildings. Included are the Paul Revere House in Boston, MA; the Wayside Inn in Sudbury, MA; Sutters Mills in Sacramento, CA; the Daniel Boone Dwelling in Defiance, MO; and the Franklin Roosevelt Home in Hyde Park, NY.

TOPICAL FIRE PROBLEM AREAS

Public Apathy

As stated in the opening paragraphs of this chapter, public apathy plays a large role in keeping U.S. fire statistics as high as they are. Many adults have the notion that "fire only happens to the other guy." Until fire strikes their home and their family, they largely ignore fire prevention information, fail to install and maintain smoke detectors, don't practice a fire escape plan for use in emergencies, and generally omit the other steps necessary to keep their homes safe from fire. For many, once Fire Prevention Week comes and goes each October, not much thought is given to firesafety until the next campaign a year later.

Fire departments, schools, civic organizations, and many others work throughout the year to keep firesafety part of the normal routine of people of all ages, but efforts must continue if public apathy on the subject is to be overcome.

Economic Impact of Protection

Fire suppression by public fire departments is an important and vitally necessary service. It is, however, last resort action. Prevention, detection, automatic extinguishment, and restraints against spread of fire are, in that order, the logical steps that should precede public service suppression.

The preventive and remedial actions taken by public authorities to mitigate the losses of life and property from fire are a major interactive component of the fire problem.

Although the majority of such actions are necessitated by failure to effectively control the root cause of fire, there exist courses of action, open to public authorities, that could eliminate much of the problem.

Principal courses of action for public authorities include:

A. Fire Prevention Education and Awareness
B. Code Adoption and Enforcement
C. Fire Suppression

The cost of operating public fire protection services in the U.S. runs several billion dollars per year, over 90 percent of which is expended on suppression activities.

In addition, fire fighting is one of the most dangerous of all occupations. In the years ahead, the challenge of fire fighting safety will become even greater as fire departments confront new technologies and hazards, while feeling the impact of reduced manning and fewer funds for the purchase and maintenance of equipment.

An increased emphasis by public services on preventive measures, together with increased effectiveness of suppression techniques, can help change the balance of the fire problem and facilitate the elimination of many root causes.

The heavy burden that reliance on public remedial action places on the taxpayer points up the desirability of shifting the burden into known, more cost effective lines of action. The balance between private protection and publicly provided service should constantly be reevaluated, and the emphasis changed where beneficial results can be achieved.

Quite apart from the cost of public protection is the increasing demand for evidence of cost effectiveness and cost benefit of firesafety codes and standards. This is brought about by the heavy expenditures that property owners sometimes must make in order to comply with the requirements of the codes and standards. More precise engineering of protection requirements recognizing the economic implications of "over designed" systems is one of the major fire protection challenges of the 1980s.

Incendiarism, Arson, and Suspicious Fires

Incendiary fires are set fires, and they can be grouped by motive as fraud fires, vandalism fires, spite fires, political fires, "pyro" fires, crime cover-up fires, vanity fires, and "psycho" fires.

Suspicious fires have many of the characteristics of incendiary fires but are not conclusively judged to be incendiary; for most purposes they are included in estimates of the size of the arson problem. Arson is the crime of setting an incendiary fire, and most incendiary fires are arson offenses. The rare exceptions would be fires set by those incapable of criminal intent, such as mentally ill persons unable to understand the nature of their actions.

Since 1977, the number of structure fires set deliberately or suspected of having been set deliberately has declined every year but one. During the period from 1980 to 1984, incendiary and suspicious structure fires decreased 28.5 percent as shown in Figure 1-1J.

When adjusted for inflation, the average loss per incendiary or suspicious fires remained steady from 1980 to 1984, except in 1980, which was considerably worse than the other years. The average loss per incendiary or suspicious fire from 1980 to 1984 is shown in Figure 1-1K.

Also, the number of multiple death fires of incendiary or suspicious origin and the number of deaths from these fires has decreased fairly steadily from 1980 through 1984.

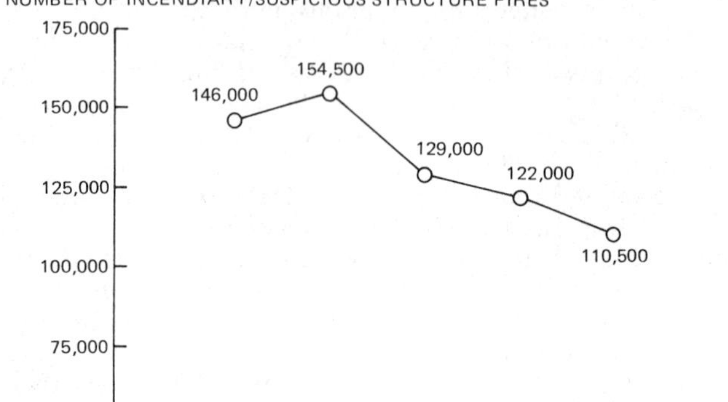

SOURCE: NFPA'S ANNUAL SURVEY FOR U.S. FIRE EXPERIENCE (1980–1984)

FIG. 1-1J. Number of incendiary or suspicious structure fires in the United States from 1980 to 1984.

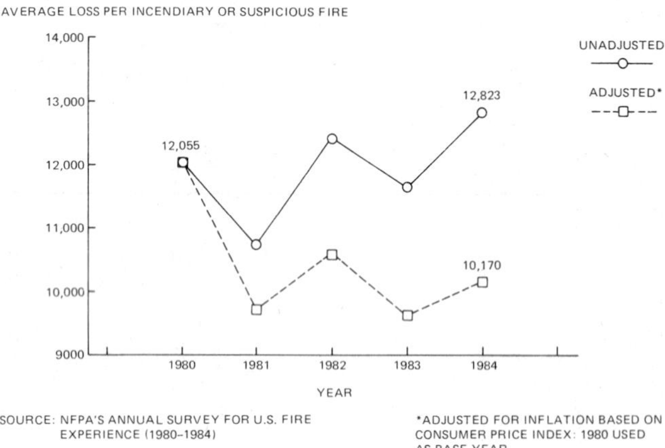

SOURCE: NFPA'S ANNUAL SURVEY FOR U.S. FIRE EXPERIENCE (1980–1984)

*ADJUSTED FOR INFLATION BASED ON CONSUMER PRICE INDEX: 1980 USED AS BASE YEAR.

FIG. 1-1K. Average property loss per incendiary or suspicious structure fire in the United States from 1980 to 1984.

The number of incendiary and suspicious multiple death fires from 1980 to 1984 is shown in Figure 1-1L.

The number of multiple death fire deaths from incendiary and suspicious fires from 1980 to 1984 is shown in Figure 1-1M.

Incendiary and suspicious fires are of particular concern in part because they account for a large share of the largest multiple death fires, and also because they constitute the largest single cause of fire loss overall.

During the 1980 to 1984 period, incendiary fires that were started in a means of egress resulted in nearly half the deaths that occurred during fires in properties where people commonly gather. In fact, 44.8 percent of the incendiary or suspicious multiple death fires for the entire period began in an egress, as shown in Figure 1-1N.

Many of these fires probably grew rapidly, blocking the exits and trapping people who were unfamiliar with usable secondary exits. This sequence of events has been specifically cited in studies of fatal lounge and hotel fires. Other properties in which these fires occurred included places of public assembly, educational facilities, institutions, dwellings, stores, and offices.

As Figure 1-1O indicates, a combustible or flammable liquid was the first material ignited in more than two-

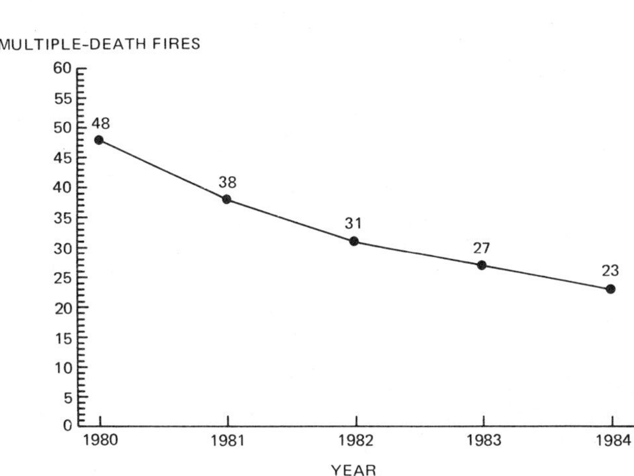

FIG. 1-1L. Incendiary and suspicious multiple death fires for 1980 to 1984.

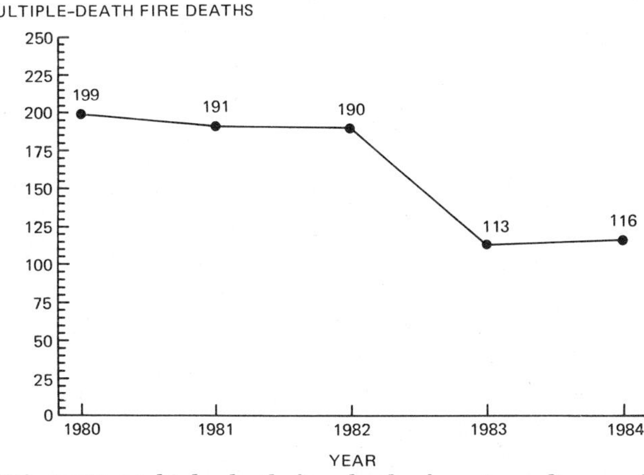

FIG. 1-1M. Multiple death fire deaths from incendiary and suspicious fires for 1980 to 1984.

thirds of the incendiary and suspicious multiple death fires and deaths from 1980 to 1984 in which the material ignited was reported. This liquid was usually an accelerant applied directly by the firesetter, although it was occasionally poured into a cloth or thrown onto a wall. This percentage is much higher than the percentage found in all incendiary and suspicious fires regardless of size of loss or number of casualties. For all incendiary and suspicious multiple-death fires, the materials ignited most frequently after flammable or combustible liquids were rubbish and papers.

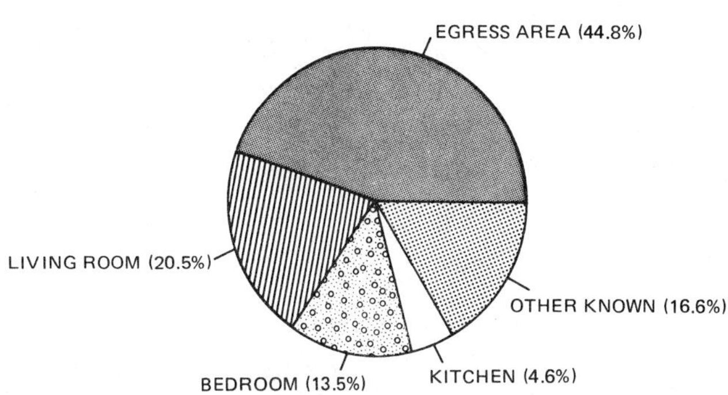

BASED ON 761 DEATHS IN 157 FIRES: EXCLUDES 39 DEATHS IN 7 FIRES FOR WHICH NO AREA OF ORIGIN WAS REPORTED.

FIG. 1-1N. Distribution of fire deaths by area of origin for incendiary and suspicious multiple death fires for 1980 to 1984.

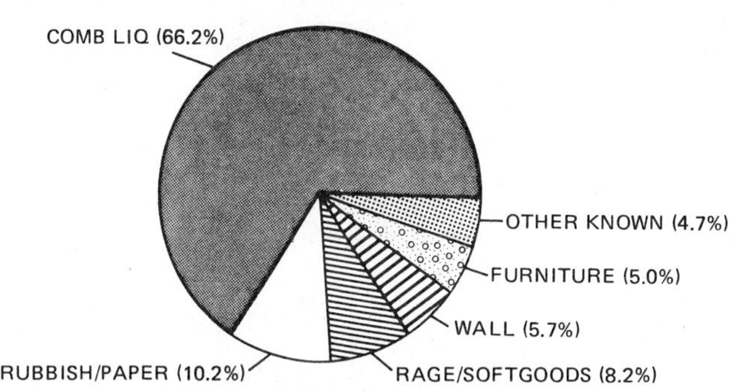

BASED ON 619 DEATHS IN 130 FIRES: EXCLUDES 181 DEATHS IN 34 FIRES FOR WHICH NO MATERIAL FIRST IGNITED WAS REPORTED.

COMB LIQ = COMBUSTIBLE AND FLAMMABLE LIQUIDS.

FIG. 1-1O. Distribution of fire deaths by material first ignited in incendiary and suspicious multiple-death fires for 1980 to 1984.

Bibliography

References Cited

Hall, John R., Jr. 1985. "A Decade of Detectors: Measuring the Effect." *Fire Journal.* Sept. 1985.

Klem, Thomas J. 1984. "The Rural Fire Picture." *Fire Journal.* Sept 1984.

NFPA. 1979. *Planning a Firesafe Future.* NFPA's Long-Range Plan. National Fire Protection Association. Quincy, MA.

NCFPC. 1973. *America Burning.* Report of the U.S. National Commission on Fire Prevention and Control. U.S. Government Printing Office. Washington, DC.

NFPA Codes, Standards, Recommended Practices and Manuals. (See the latest *NFPA Codes and Standards Catalog* for availability of current edition of the following document.)

NFPA 101, *Code for Safety to Life from Fire in Buildings and Structures.*

Additional Readings

Curtis, Martha H., Hall, John R., Jr., and LeBlanc, Paul R. "Analysis of Multiple-Death Fires in the United States During 1984." *Fire Journal,* July 1985.

Karter, Michael J., Jr. "Fire Loss in the United States During 1984." *Fire Journal,* Sept. 1985.

O'Brien, Anthony, and Redding, Donald. "Large-Loss Fires in the United States During 1983." *Fire Journal.* Nov. 1984.

The 1984 Fire Almanac. National Fire Protection Association, Quincy, MA. 1983.

HUMAN BEHAVIOR AND FIRE

Revised by Dr. John L. Bryan

How one reacts during a fire emergency is related to: (1) the role assumed, previous experience, education, and personality; (2) the perceived threat of the fire situation; (3) the physical characteristics and means of egress available within the structure; and (4) the actions of others who are sharing the experience. Post event analysis of behavior has described actions as adaptive or nonadaptive, participative or inhibited, and altruistic or individualistic. Detailed interview and questionnaire studies over 30 years have established that instances of nonadaptive (panic) type of behavior are rare events that occur under specific conditions. Most behavior in fire incidents is determined by information analysis, resulting in cooperative and altruistic actions.

INTRODUCTION

The characteristics of the behavior of people individually and within groups have been determined primarily by research studies in which individuals were interviewed at the time of the fire incident by fire department personnel (Bryan 1977; Wood 1972). It must be recognized that an individual's behavior in a fire incident is affected by the variables of the building in which the fire incident occurs and the appearance of the fire incident at the time of detection. For example, response of the occupants will vary if they perceive an odor of smoke rather than visible flames, and dark acrid smoke completely obscuring a corridor. Variables of the fire protection provided for the building may also be critical to the individual's perception of the threat involved in the fire incident. Obviously, in life threatening situations, the most important individual decisions and behavior occur prior to the arrival of the fire department, in the early stages of the fire incident. Studies of health care facilities have indicated the importance of this early behavior:

> "In the process of investigating these case studies we have come to believe that the period between detection of the fire and the arrival of the fire department is the most crucial life saving period in terms of the first compartment (the area in direct contact with the room of origin and the fire)" (Lerup et al 1978).

Thus, the behavior of the individuals intimately involved with the initiation of the fire incident is critical not only for themselves, but often for the other occupants of the building. It should be recognized that the altruistic behavior observed in most fire incidents (with the interaction of the occupants and the fire environment in a deliberate, purposeful manner) appears to be the general mode of reaction. The nonadaptive flight or panic type behavioral reaction is apparently an unusual behavior in fire incidents.

Awareness of the Fire Incident

Obviously, the way in which an individual is alerted to the presence of a fire may determine the degree of threat perceived. With vocal alerting systems in buildings, variations in voice quality, pitch, or volume, as well as the content of the message, tend to provide threat cues (Keating and Loftus 1981). Most of the participants were alerted initially to the fire incident by the odor of the smoke. However, when the two categories "notified by family" and "notified by others" are combined, personal notification becomes the most common means of initial perception of fire, as indicated in Table 1-2A. The category of noise includes noise from persons moving downstairs and through corridors, plus miscellaneous noise sources including the breaking of glass and the arrival of fire department apparatus.

Table 1-2B presents a comparison of the means of awareness of the British population (Wood 1972) and the U.S. population (Bryan 1977). The number of stimuli was reduced because the British study had fewer categories, and the U.S. population responses have been adapted to the British categories. There was only one significant difference in the means of awareness between the two populations: 15 percent of the British population became aware of the fire incident upon observing flame, contrasted with 8.1 percent of the U.S. population.

Dr. Bryan is Professor and Chairman of the Department of Fire Protection Engineering, College of Engineering, University of Maryland, College Park, MD.

TABLE 1-2A. Means of awareness of the fire incident

Means of awareness	Participants	Percent
Smelled smoke	148	26.0
Notified by others	121	21.3
Noise	106	18.6
Notified by family	076	13.4
Saw smoke	052	09.1
Saw fire	046	08.1
Explosion	006	01.1
Felt heat	004	00.7
Saw/heard fire department	004	00.7
Electricity went off	004	00.7
Pet	002	00.3
N = 11	569	100.0

TABLE 1-2B. Comparison of the British and United States populations relative to means of awareness of the fire incident

Means of awareness	British percent	U.S. percent	$P_1 - P_2$	$SE_{P_1 - P_2}$	CR
Flame	15.0	08.1	6.9	1.64	*4.21**
Smoke	34.0	35.1	1.1	2.27	0.48
Noises	09.0	11.2	2.2	1.41	1.56
Shouts & told	33.0	34.7	2.7	2.25	1.20
Alarm	07.0	07.4	0.4	1.23	0.33
Other	02.0	02.8	0.8	0.70	1.14
	2193	569			

* Critical Ratio significant at or above the 1 percent level of confidence.

A study of the NFPA recommended smoke detector noise level of 75 dBA indicates that individuals with hearing impairments, taking sleeping pills or on medication may require a detector noise level exceeding 100 dBA (Berry 1978). (See NFPA 74, *Standard for the Installation, Maintenance, and Use of Household Fire Warning Equipment.*) Flashing or activated lights are effective fire signals in occupancies populated primarily by hearing impaired persons (Cohen 1977). The 1981 edition of NFPA 101, *Life Safety Code®*, permitted the flashing of exit signs along with activation of an audible fire alarm system for the first time.

A study with 24 male subjects on their awakening from a smoke detector audible alarm signal and their identification of fire cues found the subjects slept through the alarm signals at a signal/noise ratio of 10 dBA and consistently failed to identify the awakening cue or radiant heat and smoke odor cues as fire warnings (Kahn 1984). Other researchers have indicated the alarm signal to noise ratio is attenuated by physical surroundings (Nober et al 1981). Thus, a signal passing through a ceiling or wall may be reduced by 40 dBA and by 15 dBA in passing through a door. In addition, the signal could be masked by a typical residential air conditioner noise level of 55 dBA.

The recognition of ambiguous threat cues as signaling an emergency condition may be inhibited by the presence of other people. Experimental studies of the inhibition of adaptive reaction to emergencies (Latane and Darley 1968) created an experimental situation involving college students. While the students were completing a written questionnaire, the experimenter would introduce smoke into the room through a small vent in the wall. If the student left the room and reported the smoke, the experiment was terminated. If the student had not reported the presence of the smoke within a six minute interval from the time smoke was first noticed, the experiment was considered completed. Students alone in the room reported the smoke in 75 percent of the cases. When two passive non-committal persons joined each student, only 10 percent of the groups reported the smoke. When the total experimental group consisted of three naive subjects, in only 38 percent of the groups did one individual report the smoke. Of the 24 persons involved in the eight naive subject groups, only one person reported the smoke within the first four minutes of the experiment. In the single subject situation, 55 percent of the subjects had reported the smoke within two minutes and 75 percent in four minutes.

It was reported in the study that noticing the smoke was apparently delayed by the presence of other persons, with the median being five seconds for single subjects but 20 seconds in both of the group conditions. These results undoubtedly reflect the constraints that people accept regarding their behavior in public places. The performance of naive subjects in the passive confederate situation was reported as follows:

> "The other nine stayed in the waiting room as it filled up with smoke, doggedly working on their questionnaires, and waving the fumes away from their faces. They coughed, rubbed their eyes, and opened the window but did not report the smoke."

It has been suggested that while trying to interpret the emergency potential of ambiguous threat cues, the individual is influenced by the behavioral reaction of others. Should these others remain passive and seem to interpret the situation as a nonemergency, the individual will tend to have this interpretation modified by this inhibiting social influence (Latane and Darley 1968). This behavioral experiment may help explain the reported tendency of people to (1) disregard threat cues, or (2) interpret them as being nonthreatening when the threat situation occurs where there are many other people such as in a restaurant, motion picture theater, or department store. These experimental results may be of assistance in explaining the incidents of calls received by fire departments minutes or even hours after the incident was first detected. In the report of the Arundel Park Hall fire (Bryan 1957), several of the sample population indicated that when they entered the hall after observing the fire from outside the building, they warned their friends and suggested they should leave, but were laughed at and their warning apparently disregarded.

The processes of social inhibition, diffusion of responsibility, and mimicking were indicated to be primarily responsible for the inhibition of adaptive and assistance behavior in emergency situations. The inhibition of behavior in the early stages of a fire incident, when the cues are relatively ambiguous, may lead toward nonadaptive flight behavior because the time available for evacuation has been expended. It is sometimes difficult to get occupants of a building to evacuate because of social inhibition and diffused responsibility. The tendency to adopt cues for behavior from others is a well documented occurrence in fire incidents in restaurants, other public assembly occupancies, and hotels.

DECISION PROCESSES OF THE INDIVIDUAL

Seven processes have been identified that an individual may utilize in attempting to structure and evaluate situational threat cues (Withey 1962). Six of these processes are presented in Figure 1-2A, in the following manner: recognition, validation, definition, evaluation, commitment, and reassessment.

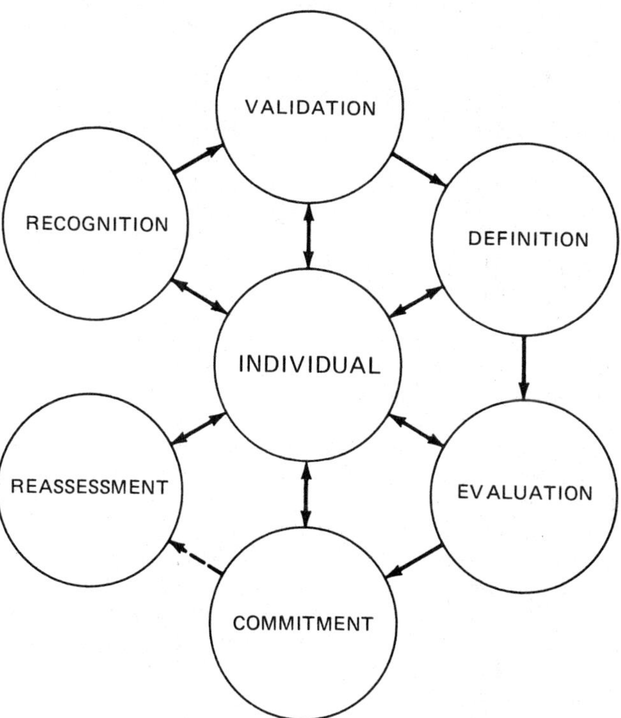

FIG. 1-2A. Decision processes of the individual in a fire incident.

Recognition

The process of recognition occurs when the individual perceives cues that indicate a threatening fire incident. The cues may be of a very ambiguous nature and not clearly indicative of a severe fire situation. The clues usually are, however, of a continuous nature with an increasing intensity due to the dynamics of flame, heat, and smoke production. It was also reported that the usual predisposition of the individual is to recognize threat cues in terms of the most probable occurrences, usually in relation to past experience and in the form of an optimistic wish. The optimistic wish aspect of the response to the warnings may be a direct result of the individual's concept of his/her personal invulnerability.

The problem of threat recognition appears to be important for fire protection. The adaptive action involved in the initiation of the fire alarm, the evacuation of building occupants, and the suppression of the fire may be delayed or postponed if individuals do not perceive the cues as indicative of an emergency fire situation. The ambiguous nature of threat cues indicates that only large amounts of smoke or sudden and threatening flames are recognized as indicative of a threatening fire situation by individuals other than those who have specialized fire prevention or fire protection education and experience.

Validation

The process of validation consists of attempts by the individual to determine the seriousness of the threat cues, usually by reassurance of the mild nature of the threat and its improbability. However, when the cues are significantly ambiguous, the individual attempts to obtain additional information. In other words, the person is aware that something is happening but is not sure exactly what it is. This process of validation may be conducted by questioning other nearby individuals. In studies of the explosion of a fireworks plant in Houston, TX, it was found that of the 139 persons interviewed, 85 individuals—or 61 percent of the population—obtained information on the source and nature of the explosion and smoke from someone they saw or from someone who telephoned and told them (Killian et al 1956). The presence of others during the threat recognition and validation process was found to possibly inhibit or influence the behavioral responses of the individual.

Definition

The definition process essentially consists of an attempt by the individual to relate the information concerning the threat to some of the variables, such as the qualitative nature, the magnitude of deprivation of the threat, and time context. The generation of stress and anxiety in the individual appears to be most severe before structure or meaning is determined for the situation, although it is apparent the situation requires interpretation. The role concept of the individual is one of the critical factors in the situation relative to the personalization of the threat and the physical environment. The most important physical aspects in relation to the definition process are the generation, intensity, and propagation of the smoke, flames, and thermal exposure.

Evaluation

The process of evaluation may be described as the cognitive and psychological activities required for the individual to respond to the threat. The individual's ability to reduce the stress and anxiety levels becomes the essential psychological factor. In the threat situation created by a fire incident, evaluation is the process involved in the decision to react by fight or flight. With evaluation, an initial decision for an overt behavioral response is completed. Because of the time context of the generation and propagation of the fire, the mental processes up to and including the process of evaluation may have to be accomplished within a time frame of several seconds. Variables of the physical environment are an important source of information for the decision process of individuals involved in the formulation of adaptation, escape, or defense plans. Additional determinants may be the location of the individual relative to the egress routes, other people, the untenable effects of the fire, and the behavior of other individuals. During this process of evaluation, the individual may decide to leave the building (flight) or to use a portable fire extinguisher (fight). During this time, the individual is particularly susceptible to the actions and communications of others. Thus, the behavioral reactions of observed individuals may be mimicked, resulting in mass adaptive or nonadaptive behavior rather than selective individualized behavior. The situation described by NFPA relative to the delayed alarm for a fire in an auto sales and service agency in 1971, indicates what may have been a situation of mimicked behavior becoming normative group behavior, as described below (NFPA 1971):

> "About 10 p.m., the fire department received an alarm from a street fire alarm box. When fire fighters arrived the 150 by 200 ft (46 by 61 m), one and two story building of wood frame and hollow block construction was well alight and nearly 300 spectators were watching the fire in 10°F (−12°) weather. An investigation revealed the fire had been burning for about 90 minutes before the fire department was notified."

In studies of nonadaptive group behavior, the concept of this mode of behavior being directly dependent upon the individual's perception of the reward structure of a situation has been developed (Mintz 1951). People in a building who are confronted with a fire threat situation would probably initially perceive a reward structure conducive to cooperative and adaptive behavior responses; in such cases, everyone should be able to reach and proceed through the available exits. However, the reward structure perceived by some of the individuals more remotely located from the egress routes could result in competitive behavior. With only cooperative behavior, it would be perceived as impossible for some of the individuals to reach an exit in time to escape the deprivation effects of the fire threat. Once the pattern of competitive behavior is initiated by one or more individuals, the behavior pattern of the group may become one of intense, individual competition for the escape routes.

In the evaluation process, an individual's cultural influences and assumption of a particular role may be very important in formulation of defense or escape plans. It is believed that the individual in a familiar role that is also suitable for the threat situation will experience less anxiety and respond with more adaptive behavior, than individuals with an unfamiliar role confronted with the occurrence of an unfamiliar threat.

Commitment

The process of commitment consists of the mechanisms utilized by the individual to initiate the behavioral activity required to fulfill the defense plans conceptualized in the evaluation process. This overt response to the threat of fire results in success or failure. If the response fails, the individual then immediately becomes involved in the next process of reassessment and commitment. If the action results in success, the anxiety and stress aspects of the situation are reduced and relieved for the individual, although the general fire situation may have increased in severity.

Reassessment

The process of reassessment and overcommitment is the most stressful of the individual's processes because of the failure of previous attempts to adjust to the threat situation. Thus, more intense effort goes into the behavioral reactions, and the individual tends to become less selective in the choice of response. As successive failures are encountered, the individual becomes more frustrated. The possibility of injury and risk increases with a greater activity level and with less probability of success, as was demonstrated in the Arundel Park Hall fire situation. There, the number of people who selected windows as a means of escape increased as people became involved in their second escape attempts.

In analyzing behavior of the individual involved in the processes of recognition, validation, definition, evaluation, commitment, and reassessment, it must be remembered that all of these processes are dynamic; they are constantly being modified in relation to their magnitude, velocity, and intensity. A person's usual psychological and physiological activities will probably be at a below normal level during the recognition process, when concentration is on perception of the threat cues. During the process of validation and definition of the threat, there will be overt communication with adjacent members of the threatened population. The period of hyperactivity appears to occur initially during the process of commitment, and to become intense during the process of reassessment and recommitment. Stress generation will increase with each successive stage, since the primary motivation of the behavioral activity, is stress reduction. The appearance, the proximity, the propagation, the time, and the toxic gases of the fire threat will also tend to predispose the individual to a higher level of behavioral activity again depending upon the individual's perception of these threat variables. During the process of reassessment and recommitment, the individual's activity level may assume the hyperactive mode of frantic activity, or may be expressed in the catastrophic state of complete physical immobility with a complete loss of ability to communicate coherently. These individuals appear to perceive the threat situation as above their level of adaptability. The stress generation is too severe and they give up completely. Thus they cease to make any attempt at an adaptive behavior, and adopt a total retreat from the situation through the mechanism of psychological withdrawal. These behavioral dynamics are presented in Figure 1-2B.

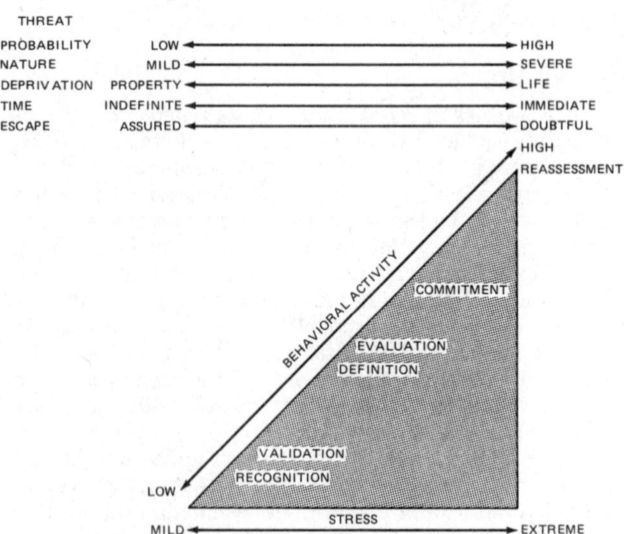

FIG. 1-2B. Dynamics of behavioral activity of the individual.

A conceptual model of the decision processes of the individual similar to some of the concepts previously discussed has been developed. Instead of the six processes referenced previously, only three processes have been utilized: (1) recognition/interpretation; (2) behavior, with either action or inaction; and (3) the outcome of the action that involves the evaluation and long term effect of the behavior (Breaux et al 1976). The behavior evaluation is similar to the process of reassessment of the conceptual model. Both the recognition/interpretation process and the behavior process involve factors critical to the decision processes. Experience and immediate circumstances all have an impact on the recognition/interpretation process. It has been emphasized that the individuals in the fire incident may not know right away that they are involved in a fire, and may not know where, in relation to their location, the fire is developing or their specific location relative to the egress routes in the building. This conceptual model is presented as Figure 1-2C.

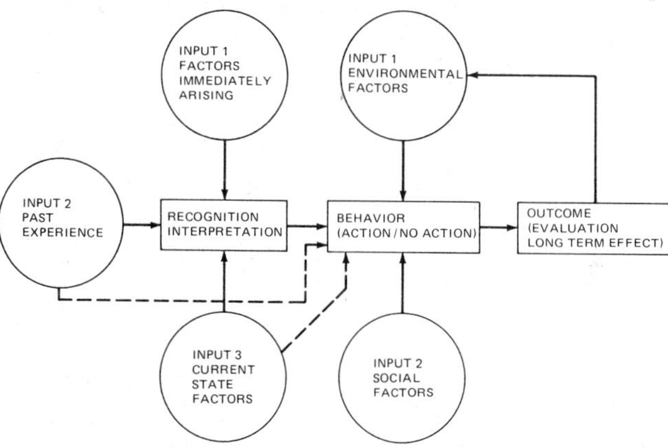

FIG. 1-2C. Preliminary heuristic systems model of behavior in fires.

The conceptual model just described has been modified into one involving three phases: (1) the detection of cues, (2) the definition of the situation, and (3) coping behavior. In addition, tentative determinants of the behavior have been developed that increases the probability of detection and of fire suppression (Bickman et al 1977). The similarity between the two models can be observed by comparing Figure 1-2C with Figure 1-D.

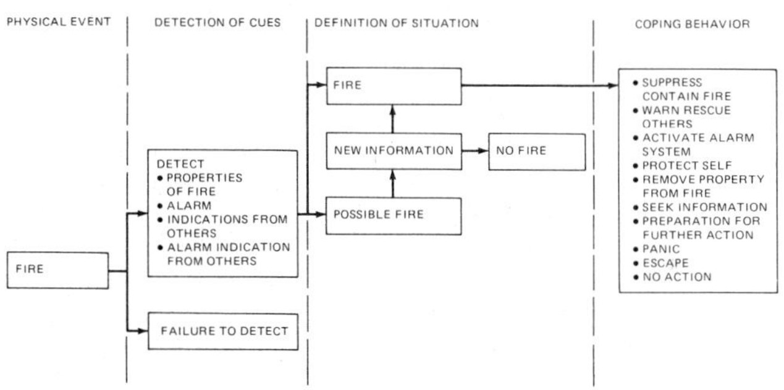

FIG. 1-2D. Flow diagram model of human behavior in a fire emergency.

BEHAVIOR ACTIONS OF OCCUPANTS

In a study (which involved 952 fire incidents and 2,193 individuals interviewed by fire department personnel at the fire scene in Great Britain) it was found that the most frequent responses to fire involved evacuation of the building, fighting or containing the fire, and alerting of other individuals or the fire brigade (Wood 1972). The identical type of broad categorization of behavior was found in a similar study, which involved interviewing 584 participants in 335 fire incidents in the U.S. Interviews were conducted by fire department personnel who used a structured questionnaire at the scene of the fire incident (Bryan 1977).

Next is an examination of initial actions to fire, as found in parallel in the studies of British and U.S. populations. (See Table 1-2C.) The behavior of the individuals

TABLE 1-2C. Comparison of the first actions of British and U.S. populations

Actions	British percent	U.S. percent	$P_1 - P_2$†	$SE_{P_1 - P_2}$‡	CR#
Notified others	8.1	15.0	6.9	1.38	*5.00***
Searched for fire	12.2	10.1	2.1	1.51	1.39
Called fire dept.	10.1	9.0	1.1	1.40	0.79
Got dressed	2.2	8.1	5.9	0.85	*6.94***
Left building	8.0	7.6	0.4	1.27	0.31
Got family	5.4	7.6	2.2	1.11	*1.98**
Fought fire	14.9	10.4	4.5	1.63	*2.76***
Left area	1.8	4.3	2.5	0.70	*3.57***
Nothing	2.1	2.7	0.6	0.69	0.87
Had others call F.D.	2.8	2.2	0.6	0.76	0.79
Got personal property	1.2	2.1	0.9	0.55	1.64
Went to fire area	5.6	2.1	3.5	1.01	*3.47***
Removed fuel	1.2	1.7	0.5	0.53	0.94
Enter building	0.1	1.6	1.5	0.30	*5.00***
Tried to exit	1.6	1.6	0	0	0
Closed door to fire area	3.1	1.0	2.1	0.76	*2.76***
Pulled fire alarm	2.7	0.9	1.8	0.70	*2.57**
Turned off appliances	4.1	0.9	3.2	0.85	*3.20***
N = 18	2193	580			

* Critical Ratios significant at or above the 5 percent level of confidence.
** Critical Ratios significant at or above the 1 percent level of confidence.
† British percent minus U.S. percent
‡ Standard Error
\# Critical Ratio

varied relative to their sex; female and male behavior divided along culturally determined primary group roles. Thus, the males were predominately active in fighting the fire, while the females were predominately concerned with alerting others and assisting them to leave the building.

It should be noted that there were ten statistically significant differences between the British and the U.S. populations. The U.S. population was predominant in five categories of first actions: "notified others," "got dressed," "got family," "left area," and "entered the building." A greater percentage of the British population engaged in the first actions of "fought fire," "went to fire area," "closed door to fire area," "pulled fire alarm," and "turned off appliances."

The general classification of the three early actions for British and the U.S. populations alike were categorized as "evacuation," "reentry," "fire fighting," "moved through smoke," and "turned back" behavior. The behavioral comparisons of the two populations are presented in Table 1-2D. There was a statistically significant difference in every category except "moved through smoke."

An interesting aspect of the U.S. study involves the modifications in the first, second, and third behavioral actions of the participants. Table 1-2E presents the three actions for the sample U.S. population, totaling 584 individuals. It should be noted how the action of "notifying others" accounted for 15 percent of the first actions, but by the time of the third actions, accounted for only 5.8 percent. A similar reduction in frequency can be observed with the action of "searching for the fire." This action shows a reduction in the activity from 10.1 percent as the first action to 0.8 percent for the third action. The behavioral actions of "getting dressed" and "got family" presented this same tendency toward a reduction in the frequency of the action with the progression from the first to the third action, with the time progression during the fire incident. In contrast, an increase in frequency from the first to the third actions may be noted with "called fire department," "left the building," and "fought the fire."

TABLE 1-2D. Comparison of the behavior of the British and U.S. populations

Behavior	British percent	U.S. percent	$P_1 - P_2$†	$SE_{P_1 - P_2}$‡	CR#
Evacuation	54.5	80.0	25.5	2.30	*11.09***
Reentry	43.0	27.9	15.1	2.30	*6.57***
Fire fighting	14.7	22.9	8.2	1.74	*4.71***
Moved through smoke	60.0	62.7	2.7	2.29	1.18
Turned back	26.0	18.3	7.7	2.01	*3.83***
	2193	584			

** Critical Ratios significant at or above the 1 percent level of confidence.
† Britishy percent minus U.S. percent
‡ Standard Error
# Critical Ratio

TABLE 1-2E. Summary of the first, second and third actions of the occupants

Actions	1st action (percent)	2nd action (percent)	3rd action (percent)
Notified others	15.0	09.6	05.8
Searched for fire	10.1	02.4	00.8
Called fire department	09.0	14.6	12.7
Got dressed	08.1	01.8	00.3
Left building	07.6	20.9	35.9
Got family	07.6	05.9	01.4
Fought fire	04.6	05.7	11.5
Got extinguisher	04.6	05.3	01.6
Left area	04.3	02.8	01.1
Woke up	03.1	00.0	00.0
Nothing	02.7	00.0	00.0
Had others call F.D.	02.2	04.0	04.1
Got personal property	02.1	03.8	00.8
Went to fire area	02.1	01.0	00.0
Removed fuel	01.7	01.0	01.1
Entered building	01.6	00.8	01.1
Tried to exit	01.6	02.4	00.5
Went to fire alarm	01.6	01.8	01.1
Telephoned others	01.2	00 6	01.1
Tried to extinguish	01.2	01.8	01.9
Closed door to fire area	01.0	00.2	00.3
Pulled fire alarm	00.9	00.6	00.5
Turned off appliances	00.9	00.6	00.3
Checked on pets	00.9	01.4	00.5
Awaited F.D. arrival	00.0	01.0	03.6
Went to balcony	00.2	00.8	02.7
Removed by F.D.	00.0	00.0	01.6
Opened doors/windows	00.2	00.4	01.1
Other	03.9	08.0	06.6
N = 29	100.0	100.0	100.0
Range	0–87	0–106	0–131
Percent of participant population	99.3	86.6	62.9

Behavior According to Sex

The differences between the first actions of the participants according to the sex of the participants has been examined. Table 1-2F presents the initial actions of the U.S. study population relative to the sex of the participants.

There are significant statistical differences between males and females in the categories of "searched for fire," "called the fire department," "got the family," and "got extinguishers." Male participants were predominant in fire fighting activities. Thus, 14.9 percent of the males "search for fire" as opposed to 6.3 percent of the females; and 6.9 percent of the males "got extinguishers" as opposed to 2.8 percent of the females. Females differed significantly from males in the warning and evacuation activities. Thus, 11.4 percent of the females "called the fire department" as their initial action, as opposed to 6.1 percent of the male participants. In relation to the evacuation behavior, 10.4 percent of the females left the building as the first action, contrasted with 4.2 percent of the male participants. The cultural role influence on female participants is probably explicitly indicated in the concern for other family members, with the indication that 11 percent of the females "got the family" as the first action, while only 3.4 percent of males engaged in this initial action. It should be noted that the male actions of "searching for" or "fighting the fire" were matched by the female actions of "alarm initiating" and "evaluation behavior" (Bryan 1977). This behavior has also been observed in health care and educational situations.

Behavior in Hotel Fire Incidents

The fire protection of high rise buildings, and their occupants, was severely tested by the MGM Grand Hotel fire in Clark County, NV, on November 21, 1980 (Best and Demers 1982) and the subsequent fire at the Las Vegas Hilton Hotel on February 10, 1981 (Demers 1982). Both of these hotel fires involved injuries and fatalities among the guests. NFPA conducted an intensive questionnaire study of the guests registered in the MGM Grand Hotel for the evening of November 20-21, 1980 (Bryan 1982).

The MGM Grand Hotel fire was discovered by a hotel employee who entered the unoccupied deli-restaurant located on the casino level of the hotel at approximately 7:10 on the morning of November 21, 1980. As instructed, the hotel telephone operator immediately notified the Clark County Fire Department, at approximately 7:18. The telephone operators were forced from their switchboard positions by the smoke immediately after they had initiated an announcement on the public address system, at approximately 7:20, for the evacuation of the casino area. The fire quickly reached a flashover condition in the deli area, immediately spread from east to west through the main casino area, and extended out the west portico doors on the casino level immediately following the arrival of the initial fire department personnel.

TABLE 1-2F. The first actions of the occupants relative to the sex of the occupant

First action	Male (percent)	Female (percent)	$P_1 - P_2$†	$SE_{P_1 - P_2}$‡	CR#
Notified others	16.3	13.8	2.5	2.98	0.83
Searched for fire	14.9	06.3	8.6	2.51	*3.43***
Called fire dept.	06.1	11.4	5.3	2.41	*2.19**
Got dressed	05.8	10.1	4.3	2.30	1.87
Left building	04.2	10.4	6.2	2.22	*2.79***
Got family	03.4	11.0	7.6	2.22	*3.42***
Fought fire	05.8	03.8	2.0	1.77	1.13
Got extinguishers	06.9	02.8	4.1	1.77	*2.31**
Left area	04.6	04.1	0.5	1.70	0.29
Woke up	03.8	02.5	1.3	1.45	0.90
Nothing	02.7	02.8	0.1	1.38	0.72
Had others call F.D.	03.4	01.3	2.1	1.23	1.71
Got personal property	01.5	02.5	1.0	1.17	0.85
Went to fire area	01.9	02.2	0.3	1.20	0.25
Removed fuel	01.1	02.2	1.1	1.08	1.02
Entered building	02.3	0.09	1.4	1.02	1.37
Tried to exit	01.5	01.6	0.1	1.05	0.09
Went to fire alarm	01.1	0.19	0.8	1.02	0.78
Telephoned others	00.8	01.6	0.8	0.91	0.87
Tried to extinguish	01.9	00.6	1.3	0.91	1.43
Closed door to fire area	00.8	01.3	0.5	0.87	0.57
Pulled fire alarm	01.1	00.6	0.5	0.75	0.66
Turned off appliances	00.8	00.9	0.1	0.79	0.12
Checked on pets	00.8	00.9	0.1	0.79	0.12
Other	06.5	02.5	4.0	1.70	2.35*
N =25	262	318			

* Critical Ratios significant at or above the 5 percent level of confidence.
** Critical Ratios significant at or above the 1 percent level of confidence.
† Male percentage minus female percentage
‡ Standard Error
\# Critical Ratio

An addition to the hotel was being constructed adjacent to the west end of the building. Construction workers there helped warn and evacuate guests and assisted in fire fighting. The heat and smoke rapidly extended from the casino area through seismic joints, elevator shafts, and stairways throughout the 21 residence floors of the hotel. The heat was intense enough on the 26th (top) floor to activate automatic sprinklers in the lobby area adjacent to the elevator shafts.

Due to the rapid early evacuation of the telephone staff, guests were not alerted by the hotel public address system or local fire alarm system. Guests alerted early in the fire incident, and those already awake and dressed, were able to escape before the smoke conditions became untenable on the upper floors. Guests alerted later remained in their rooms or moved to other rooms, usually with other occupants. The fire itself did not extend above the casino level, with the exception of its extension of a rather minor nature into two guest rooms on the fifth floor. The fire resulted in 85 fatalities and injuries to 778 guests and 7 hotel employees. Seventy-nine body locations were documented: 14 on the casino level, 29 in guest rooms, 21 in corridors and lobbies, 9 in stairways, and 5 fatalities in elevators. The victims were found on the casino level and the 16th and upper floors, with the majority between the 20th and the 25th floors.

Figure 1-2E is a diagram of the guest floor arrangement of the MGM Grand Hotel which was used in the occupant questionnaire study conducted by NFPA (Bryan 1982). Of the nine victims found in the stairways, two were in Stairway 1 at the extreme south end of the south wing at the 17th floor; six were between the 20th and 23rd floor in Stairway 2 at the central end of the south wing; and one victim was found at the ground floor level of Stairway 4 at the extreme west end of the west wing. There are various estimates of the number of guests and fire department personnel who suffered injuries at the MGM Grand Hotel fire. Morris indicated that 619 people were taken to hospitals and another 150 were treated at the Las Vegas Convention Center, where the survivors were transported from the hotel (Morris 1981).

The MGM Grand Hotel tragedy was a unique fire incident from several aspects. First, it was the second most serious hotel fire in the history of the U.S., being surpassed only by the Winecoff Hotel fire in Atlanta, GA on December 7, 1946, when 119 died. Second, it was the first high rise fire in the U.S. in which helicopters evacuated large numbers of people—about 300 were evacuated in this manner; the fire department rescued approximately 900 people by other means.

Shortly after the MGM Grand Hotel fire, NFPA prepared a four page, 28 item questionnaire, including the floor plan of the guest rooms. A total of 1,960 questionnaires were mailed and approximately 28 percent of these were returned. While 455 of the responding individuals indicated a willingness to be interviewed.

The age of the questionnaire population ranged from 20 to 84 years, with an average age of 45. The population consisted of 331 males and 222 females, with one respondent not indicating a sexual classification. One hundred three guests indicated they were alone at the time they became aware of the fire within the hotel. The presence of other people, especially if they belong to the individual's primary group, appears to be a determinant of the response of many individuals in residential fire situations.

The initial five actions of the 554 guests as elicited from the NFPA questionnaire study are presented in Table 1-2G. Notice that the five most frequent first actions were "dressed," "opened door," "notified roommates," "partially dressed" and "looked out window." Guests involved in the first actions were predominately engaged in determining the degree of threat to themselves. Only 7.9 or approximately 8 percent of the study population initiated or attempted to initiate their own evacuation with such

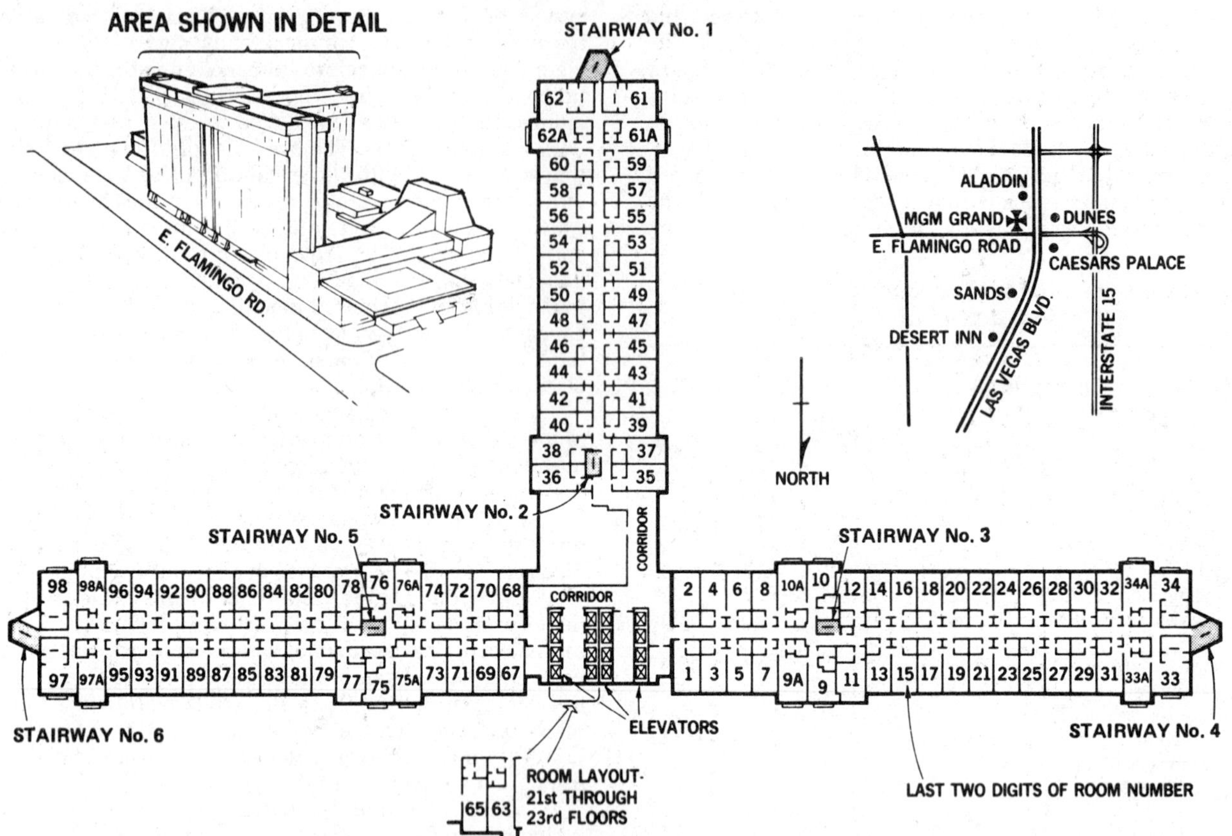

FIG. 1-2E. Residential floor diagram of the MGM Grand Hotel.

actions as "attempted exit," "went to exit," and "left room." A total of 16 individuals, 2.9 percent of the population, initiated actions to improve the room as an area of refuge: "wet towels for face" and "put towels around door." The actions of the guests responding to the NFPA questionnaire study could be classified in general as evacuation actions or refuge processes. Actions relating to evacuation behavior appeared to be initiated early if the egress passages were clear of smoke, or if the smoke was not perceived as personally threatening. However, if the smoke was heavy, the guests apparently decided to stay in their rooms or other rooms and initiate actions to prevent smoke migration into the rooms of refuge to protect themselves from the smoke.

Further examination of Table 1-2G indicates the five most popular actions reported by guests as second actions were: "opened door," "dressed," "went to exit," "partially dressed," and "secured valuables." Approximately 19 percent of the study population were still involved in the dressing actions prior to initiating evacuation or refuge procedures.

Third actions of guests in the study population generally progressed to evacuation, attempted evacuation, and notification. Approximately 25 percent of this population were involved in evacuation actions, and approximately 10 percent in attempted evacuations as identified by the third actions of "attempted to exit" and "returned to room." The alerting and notification actions are identified as third actions of "notified occupants" and "notified other room."

The fourth actions of the guests in the study population indicate a progression to evacuation, attempted evacuation, and self-protection or room refuge procedural actions. The most frequent fourth action of the guests was "went to exit" (approximately 16 percent of this population). However, when one combines the guests involved with this action with those utilizing the actions of "went down stairs," "went to another exit," "left hotel," and "left room," a total of 151 guests (approximately 30 percent of the fourth action guest population) were involved in evacuation actions. The process of guests forming convergence clusters was noted in this hotel fire. This action involved individuals clustering together in rooms they considered areas of refuge, with the cluster individuals characterized as usually not knowing each other prior to the fire incident. The fourth actions of "went to other room" and "went to other room/others" are explicit indicators of the formation of the convergence clusters.

The fifth actions of the guests were primarily for self-protection, including improvement of the room as an area of refuge, and evacuation behavior. The evacuation actions would consist of the fifth actions of: "went up stairs to roof," "left hotel," and "left room." Guests involved with these evacuation actions consisted of 175 individuals (approximately 40 percent of the study population). Those unable to evacuate, and thus vitally con-

TABLE 1-2G. First five actions of guests in the MGM Grand Hotel fire

Actions	Percent of Population				
	first	second	third	fourth	fifth
Dressed	16.8	11.6	6.5	—	—
Opened door	15.9	11.7	6.7	3.4	—
Notified roommates	11.6	3.0	—	—	—
Dressed partially	10.1	7.5	4.5	—	—
Looked out window	9.7	5.7	—	—	—
Got out of bed	4.5	—	—	—	—
Left room	4.3	5.4	8.1	2.4	2.0
Attempted to phone	3.4	3.6	—	2.8	—
Went to exit	2.5	10.3	9.5	16.1	6.7
Put towels around door	1.6	2.5	3.0	6.8	7.7
Felt door for heat	1.3	2.3	—	—	—
Wet towels for face	1.3	3.7	6.3	4.6	7.9
Got out of bath	1.1	—	—	—	—
Attempted to exit	1.1	3.0	5.8	—	—
Secured valuables	—	6.8	4.3	—	—
Notified other room	—	3.4	2.2	—	—
Attempted to exit	—	—	—	4.3	—
Returned to room	—	—	3.9	8.4	4.1
Went down stairs	—	—	3.9	5.4	21.3
Left hotel	—	—	3.4	2.6	2.0
Notified occupants	—	—	3.0	—	—
Went to another exit	—	—	—	3.6	4.8
Went to other room	—	—	—	3.6	3.6
Went to other room/others	—	—	—	3.4	8.7
Looked for exit	—	—	—	2.4	—
Broke window	—	—	—	—	4.3
Offered refuge in room	—	—	—	—	1.8
Went up stairs to roof	—	—	—	—	2.9
Went to balcony	—	—	—	—	1.8
Other	14.8	19.5	28.9	30.2	20.1
Total (percent):	100.0	99.1	96.9	90.6	79.6
No. of guests:	554	549	537	502	441

cerned with refuge procedures, utilized the fifth actions of: "went to other room/others," "wet towels for face," "put towels around door," "broke window," "returned to room," "went to other room," "offered refuge in room" and "went to balcony." Approximately 40 percent of the fifth action study population were involved in the refuge procedures and self-protection actions.

Convergence Clusters

The phenomenon of convergence cluster formation was first noticed in a study of occupant behavior in a high rise apartment building fire in 1979 (Bryan 1979a). The clusters appear to involve convergence of occupants of the building involved in the fire into specific rooms perceived as areas of refuge. In the MGM Grand Hotel fire, guests tended to select rooms on the north side of the east and west wings, and rooms on the east side of the south wing. In addition, guests reported that people had converged in the rooms with balconies and doors leading out to the balconies because of ventilation, reduced smoke, improved visibility, and communication that the balconies offered. The guests who reported their participation in this behavior in rooms with other people either estimated the number of persons in the room or connecting rooms, or indicated only that "others" or "other persons" were present.

Table 1-2H lists the rooms identified by guests as areas of refuge for numerous persons other than the original occupants. This table also presents estimates of the length of time that the cluster was maintained in the rooms—usually until evacuation assistance was obtained, or until the occupants were notified by fire or rescue personnel that evacuation was possible. The numbers in the two right hand columns indicate the total number of persons in the clusters for the total number of rooms identified on the floor. The smallest number of people identified as a cluster involved three persons, and the largest cluster involved 35 persons.

The greatest number of rooms used by convergence clusters, and thus the largest population participating in convergence clusters, was located on the 17th floor of the hotel. (No convergence clusters were identified by guests as occurring on the 6th, 21st or 26th floors.) The clusters appear to serve as an anxiety and tension reducing mechanism for individuals confronted with a threatening situation. The action of "offered refuge in room," previously identified in the discussion of the fifth actions, is a positive indication of the occurrence of a convergence cluster.

TABLE 1-2H. Summary of rooms, time duration and number of guests reported in convergence clusters

Floor	Room Number	Time (hours)	Persons no.	Persons percent
7	731	0.6	3	0.7
8	827, 840	1.5–1.75	14*	3.3
9	927	2.5	5	1.2
10	1009A, 1025, 1034,1060	1–2	53	12.7
11	1129, 1115	1.5–2	30*	7.2
12	1261, 1225, 1233A	2–3	53	12.7
14	1433A, 1461A, 1451, 1416A	1.5–2	8*	1.9
15	1501, 1533A, 1510	2–3	38*	9.1
16	1643, 1625, 1633, 1629, 1627, 1615	2–3.5	35*	8.4
17	1725, 1775, 1731, 1719, 1762, 1756, 1733A	2–2.5	84	20.1
18	1819, 1802, 1850	2–3	20	4.8
19	1929, 1919, 1962A, 1962, 1964, 1925	2–3.5	13*	3.1
20	2027, 2013, 2030	2.5–3.5	25	6.0
22	2213, 2221,2229	2–3	13	3.1
23	2329, 2314, 2342, 2331,2308, 2340	2.5–3.25	20*	4.8
24	2446	3.5	4	0.9
25	2512, 2509A	3.5	*	0
Total: 17	57		418	100.0
Range 7–25	1–7	0.6–3.5	3–84	0–20.1

* Persons indicated only as "Others."

In addition to the detailed human behavior study of the MGM Grand Hotel fire (Bryan 1983a), NFPA conducted a questionnaire study of guests' behavior in the Westchase Hilton Hotel fire in Houston, TX, March 6, 1982, in which 12 people died (Bryan 1983b).

The classic types of nonadaptive behavior in a fire incident involve the disregarding of adaptive actions, or behavior that might facilitate the evacuation of others or limit the propagation of the smoke, heat, or flame from the fire. Nonadaptive behavior ranges from the single act of leaving a room of fire origin without closing the door to the room, thus allowing the fire to spread throughout the structure and endanger the lives of all the occupants, to the more generalized concept of the individual fleeing from the fire incident without regard for others, and perhaps inflicting injuries on others in what is often termed panic.

Nonadaptive behavior may be an omission such as forgetting to close a door, or may involve an action that, although well meaning and positive in intent, results in negative consequences. When the results of behavior are extinguishment of the fire and elimination of the threat, the behavior may be said to be adaptive. However, the same behavior is sometimes ineffective because the fire was a more severe threat than was first perceived. In such cases, the time might have been more effectively utilized to warn others and to notify the fire department. Thus, some behavior that appears to be nonadaptive is really unsuccessful behavior that would have seemed most adaptive if it had been successful. Injuries suffered by people in relation to fire incidents may be cues to nonadaptive or risk behavior by the individual.

Panic Behavior

One concept always discussed following fire incidents such as the Beverly Hills Supper Club fire (Best 1978), in which multiple fatalities occur, is panic behavior. One classical definition of panic is:

> A sudden and excessive feeling of alarm or fear, usually affecting a body of persons, originating in some real or supposed danger, vaguely apprehended, and leading to extravagant and injudicious efforts to secure safety.

According to this definition, panic is a flight or fleeing type of behavior that involves extravagant and injudicious effort and is likely not to be limited to a single individual but to be transmitted and adopted by a group of people. From simulation experiments, a panic type of behavior reaction has been defined (Schultz 1968) in the following manner:

> "A fear induced flight behavior which is nonrational, nonadaptive, and nonsocial, which serves to reduce the escape possibilities of the group as a whole."

The concept of panic is often used to explain the occurrence of multiple fatalities in fires even when there is no physical, social, or psychological evidence showing that competitive, injudicious flight behavior actually took place. Media representatives and public officials often label various types of fire incident behavior as panic. The evidence accumulated from interviews with participants and questionnaires completed by occupants provided no evidence of the classical group type of panic behavior with competitive flight for the exits in the Beverly Hills Supper Club fire (Kentucky 1977).

It has been indicated that panic as a concept is primarily a description rather than an explanation of behavior. The concept is used to support the introduction of requirements in fire and building laws or ordinances to provide for the firesafety of occupants. The difference also has been shown between use of the concept to describe other persons' behavior in a fire incident, and the use by someone engaged in the behavior to indicate that individual's own high state of concern and anxiety (Sime 1980). As has been indicated, just because an individual identifies behavior as being associated with a panic reaction, this does not necessarily identify the behavior as being the classical panic type of response. The outcome of the behavior, as previously discussed, affects its labeling. It was indicated that the behavior of people in a fire is most likely to be misinterpreted when the outcome of the fire incident has been unfortunate.

The use of the concept of panic must be separated from use of the terms "anxiety" or "fear." The concept of self-destructive or animalistic panic responses to stimuli such as the presence of smoke has not been supported by the research on human behavior in fire incidents. As has been pointed out (Sime 1980; Quarantelli 1979; Bryan 1977; Wood 1972; Keating and Loftus 1981; Keating 1982), it is rare to have panic behavior in which the flight is characterized by competition among the participants and resultant personal injuries.

In an interview study of 100 participants in single family dwelling fires, no instances of panic behavior were found, but primarily altruistic helpful behavior was found instead (Keating 1982).

Reentry Behavior

The study of the 1956 Arundel Park Hall fire documented the initial examination of the phenomenon of reentry behavior (Bryan 1957). Some older codes and regulations affecting design of the means of egress appeared to be based on the assumption that pedestrian traffic only moves away from the fire area and away from the area or floor of the building involved. Conversely, the Arundel Park Hall study indicated that approximately one-third of the survivors interviewed had reentered the building.

Thus, it has become apparent that doors, stairways, and corridors often will be subjected to two way movement of occupants and others. The occupant who, after leaving the building safely, turns around and reenters is often completely aware of the fire in the building and of the specific portions of the building involved in the area of fire origin and smoke propagation. Table 1-2I presents the number of participants who reentered Arundel Park Hall during the fire, from the interviewed population of 61

TABLE 1-2I. Behavior of occupants who reentered in the Arundel Park fire, relative to sex, and reentry reasons

Sex	Reentered and left same exit	Reentered and left different exit	Stated reason for reentrance
M	1		Turn off kitchen stoves
M	1	1	Tell people to leave
M	3		To help
M	1		Assist people
M	2	3	Find wife
M	2	2	Assist fire fighting
M & 1 F		5	No stated reason
21 M & 1 F	10	12	

persons. Note the reasons for the reentry behavior and the fact that the reentry participants were predominately male.

The Arundel Park Hall fire incident occurred in an assembly occupancy being utilized for a church sponsored oyster roast—a family type of affair. The primary group cultural role of father or husband therefore apparently was a critical variable in the predominant aspect of the reentry behavior in the population interviewed, and may have resulted in the fact that the reentry participants were mostly male. The argument can reasonably be considered that reentry behavior is not nonadaptive behavior, since reentry behavior is often used to assist or rescue persons remaining or believed to be remaining in the building. This type of behavior is often used by parents whose children are missing during a fire incident. The behavior is often undertaken in a rational, deliberate, and purposeful manner, without the emotional anxiety and self-anxiety characteristics usually associated with nonadaptive behavior. However, reentry behavior has been considered nonadaptive, since reentry of people into a burning building is often nonadaptive in relation to the efficient and effective evacuation by other people through the same means of egress selected for reentry.

TABLE 1-2J. Reasons for reentry of the occupants

Reasons	Participants	Percent
Fight fire	36	22.2
Obtain personal property	28	17.2
Check on fire	18	11.0
Notify others	13	8.0
Assist F.D.	12	7.4
Retrieve pets	12	7.4
Call F.D.	9	5.5
Assist evacuation	4	2.5
Taken to hospital	3	1.8
Turn power back on	2	1.2
Rescue from balcony	1	0.6
Help injured family member	1	0.6
Turned off gas	1	0.6
Open windows	1	0.6
Close door	1	0.6
No apparent danger	1	0.6
Entered non-danger area	1	0.6
Job responsibility	1	0.6
Due to fire	1	0.6
Told to by others	1	0.6
Not reported	16	9.8
N = 21	163	100.0
Range = 1–36	Percent of participant population = 27.9	

The reasons elicited from participants in reentry behavior in the Project People study (Bryan 1977) in the U.S. are presented in Table 1-2J. It would seem that 162 people from the total study population of 584, or 27.9 percent, engaged in reentry behavior. The most popular reason given for reentry behavior was "to fight the fire," followed by "to obtain personal property," "to check on the fire," "to notify others," "to assist the fire department," and "to retrieve pets." These six reasons accounted for approximately 73 percent of this reentry behavior.

In Table 1-2K, which compares the reentry behavior of the British and U.S. study populations, that all of the reasons were significantly different with the exception of the item "save personal effects." The U.S. population was predominant with the reentry reasons of: "save personal effects," "call the fire department," "rescue pets," "notify others," "assist fire department," and "assist the evacuation." The British population was predominant in: "fight the fire," "observe the fire," "shut doors," "await fire department," and "fire not severe."

Occupant Fire Fighting Behavior

Occupants who engaged in fire fighting behavior during fire incidents were predominately male; this behavior now appears to be a culturally determined and expected aspect of the male role. However, it should be noted that in the "Project People" study of 335 U.S. fire incidents, approximately 23 percent of the study population of 584 individuals were involved in occupant fire fighting behavior. Of these, 37.3 percent were females. Of the 134 individuals who participated in fire fighting behavior, 50 were female and 84 male.

TABLE 1-2K. Comparison of reasons for reentry behavior of British and U.S. study populations

Reasons	British	U.S.	$P_1 - P_2$†	$SE_{P_1 - P_2}$‡	CR#
Fight fire	36.0	22.2	13.8	4.02	*3.43***
Observe fire	19.0	11.0	08.0	3.25	*2.46**
Save personal effects	13.0	17.2	04.2	2.91	1.44
Shut doors	10.0	00.6	09.4	2.38	*3.95***
Await fire department	09.0	00.0	09.0	2.26	*3.98***
Call fire department	02.0	05.5	03.5	1.32	*2.65***
Rescue pets	02.0	07.4	05.4	1.40	*3.86***
Fire not severe	05.0	01.2	03.8	1.74	*2.18**
Notify others	00.0	08.0	08.0	0.92	*8.69***
Assist fire department	00.0	07.4	07.4	0.88	*8.41***
Assist evacuation	00.0	02.5	02.5	0.54	*4.63***
N = 11	943	163			

* Critical Ratios significant at or above the 5 percent level of confidence.
** Critical Ratios significant at or above the 1 percent level of confidence.
† British minus U.S. populations
‡ Standard Error
# Critical Ratio

Their ages ranged from a 7 year old girl to an 80 year old man. Distribution of the participants by sex and age is presented in Table 1-2L. The majority of those involved in fire fighting behavior were between 28 and 37 years old—or approximately 30 percent of the fire fighting behavior population.

The statistically significant differences in the behavioral actions of males and females were the actions of "got extinguisher" and "fought fire." Approximately 15 percent of the female population reacted by obtaining extinguishers. Similarly, approximately 26 percent of the male population fought the fire when they first became aware of it, as contrasted with approximately 10 percent of the female participants. The women predominately notified the fire department first: approximately 33 percent of the females, compared with 25 percent of the males, reacted to the fire incident by notifying the fire department first, as indicated in Table 1-2M.

Occupant fire fighting behavior appears most prevalent in occupancies in which the individuals are emotion-

TABLE 1-2L. Age and sex of the occupants engaging in fire fighting behavior

Sex	Participants	Percent
Male	84	62.7
Female	50	37.3
Total	134	100.0
Age		
07–17	08	05.9
18–27	31	23.1
28–37	41	30.6
38–47	27	20.1
48–57	16	11.9
58–67	02	01.5
68–80	03	02.2
Unknown	06	04.7
Total	134	100.0

Percent of participant population = 22.9

TABLE 1-2M. Sexual differences of the occupants engaging in fire fighting and notifying the fire department

Action	Male (percent)	Female (percent)	$P_1 - P_2$†	$SE_{P_1 - P_2}$‡	CR#
Searched for fire	17.2	09.1	08.1	4.23	1.91
Got extinguisher	15.6	06.0	09.6	3.95	*2.43**
Fought fire	25.6	09.7	15.9	4.83	*3.29***
Removed fuel	03.4	03.1	00.3	2.17	0.14
Tried to extinguish	05.3	02.8	02.5	2.49	1.00
Went to fire area	03.1	02.8	00.3	2.07	0.14
Total	70.2	33.5	36.7	6.01	*6.11***
N =	184	101			
Called F.D.	25.6	33.0	07.4	5.83	1.27
Had others call F.D.	09.2	07.5	01.7	3.27	0.52
Went to fire alarm	03.8	03.8	0	0	0
Pulled fire alarm	01.9	01.6	00.3	1.65	0.18
Total	40.5	45.9	05.4	6.31	0.85
N =	106	146			

* Critical Ratios significant at or above the 5 percent level of confidence.
** Critical Ratios significant at or above the 1 percent level of confidence.
† Male percent minus female percent
‡ Standard Error
# Critical Ratio

ally and economically involved: in their homes, or where such behavior is an assigned role as a result of training. A total of 285 individuals at some time during the incident engaged in one of the six actions defined as fire fighting behavior, and a total of 252 individuals participated in one of the four actions relative to notification of the fire department.

In the study of residential fire incidents in Berkeley, CA, a total of 180 persons were involved in extinguishing and fire fighting behavior. This study surveyed a population different from that of Project People, since the 1,411 Berkeley households with 208 fire incidents included incidents not reported to the fire department: these accounted for approximately 80 percent of the total 208. The majority of the unreported fire incidents had been extinguished by the occupants or by the occupants assisted by neighbors (Crossman et al 1975). Table 1-2N presents the

TABLE 1-2N. Occupants extinguishing residential fires in Berkeley, CA

Fire Suppressed by	Percent
Person engaged in heat-using activity	52.8
Other member(s) of household	28.3
Friends and neighbors	8.9
Fire department	18.9
Burnout	6.1
Total	115.0
Single individual	80.7
Group effort	19.3
Total	100.0

percentage distribution from the study of the individuals responsible for extinguishing the fire. Six percent of these fire incidents self-extinguished, and 52 percent of the fires were extinguished by the individual engaged in the heat using activity that created the fire incident. Thus, it appears that only the fire incidents in the Project People study that were judged uncontrollable by the occupants resulted in notification of the fire department. Similarly, approximately 85 percent of the fire occurrences in the National Fire Prevention and Control Administration National Household Survey (NFPCA 1976) had not been reported to the fire department.

The types of occupancies in which equipment provided within the occupancy was used to fight the fire are presented in Table 1-2O. The rather high frequency of

TABLE 1-2O. Distribution of occupancies in which fire fighting equipment was utilized by occupants in fire fighting behavior

Occupancy	Incidents	Percent
Dwelling (1 family)	23	35.9
Apartment (20 units)	18	28.1
Restaurant	3	4.8
Apartment (20 units)	3	4.8
Manufacturing	2	4.8
Hotel and motel	2	3.2
School	3	3.2
Billiard center	1	1.5
City club	1	1.5
Hospital	1	1.5
Dwelling (2 family)	1	1.5
College dormitory	1	1.5
Service station	1	1.5
Office	1	1.5
Photographic laboratory	1	1.5
Other	2	3.2
N = 16	64	100.0

M = 4.00 SE_M = 6.55 SD = 6.55 SE_{SD} = 1.16

residential occupancies—64 percent being either single family dwellings or apartments—may be a variable of the population of this study (Bryan 1977), as well as representative of many urban areas where fire problems are essentially residential.

In the Project People Study, 107 of the 584 participants did not voluntarily leave the building after becoming aware of the fire incident. Their reasons for staying in the building are presented in Table 1-2P. Fifty-two of the participants—approximately 49 percent of the population staying in the building—reported they remained because they wished to engage in fire control or fire fighting activities. The other most frequent reasons were to notify others of the fire or because the occupant's way out of the building was blocked by smoke. It is apparent that approximately 15 percent of the study population voluntarily remained within the fire incident building.

TABLE 1-2P. Reasons elicited from occupants for not leaving the fire building

Reason	Participants	Percent
Fight fire	52	48.7
Notify others	7	6.5
Blocked by smoke	7	6.5
Blocked by fire	5	4.7
Overcome by smoke	5	4.7
Search for fire	3	2.8
Needed help	2	1.9
Secure property	2	1.9
Afraid of fire spread	2	1.9
No fire in area	1	0.9
Help others	1	0.9
Does not know	1	0.9
No response to F.D.	1	0.9
Home	1	0.9
Return to area	1	0.9
Not reported	16	15.0
N = 15	107	100.0
Range = 1–52	Percent of participant population = 15.6	

Occupant's Movement Through Smoke

Often related to fire fighting behavior, and a definite component of evacuation behavior in many fire incidents (Bryan 1977; Wood 1972), is the movement of occupants through smoke. The principal variables influencing the occupant's decision to move through smoke appear to be recognition of the location of the exit and thus being able to estimate the travel distance required, the appearance of the smoke, the smoke density, and the presence or absence of heat with the smoke (Bryan 1983a, Bryan 1983b). To achieve evacuation, occupants have moved through smoke, even for extended distances under conditions of extremely limited visibility at personal risk, and sometimes have been forced to turn back without completing the evacuation (Bryan 1977; Wood 1972; Bryan 1983a; 1983b).

Table 1-2Q compares the distance moved through smoke for the 1,316 persons in the British study and the 322 persons in the U.S. study. Sixty percent of the British study population and 62.7 percent of the Project People participants reported they moved through smoke. It is thus apparent that building occupants will move through smoke in an evacuation process. An important variable may be the smoke density or the visibility distance of the occupants during their evacuation process.

TABLE 1-2Q. Comparison of the distance moved through smoke for the British and U.S. populations

Distance moved (ft)§	British (percent)	U.S. (percent)	$P_1 - P_2$†	$SE_{P_1 - P_2}$‡	CR#
0–2	03.0	02.3	00.7	1.02	00.69
3–6	18.0	08.4	09.6	2.23	*04.30***
7–12	30.0	17.1	12.9	2.71	*04.76***
13–30	19.0	45.5	26.5	2.62	*10.11***
31–36	05.0	02.0	03.0	1.25	*02.40**
37–45	04.0	04.1	00.1	1.19	00.08
46–60	05.0	11.0	06.0	1.47	*04.08***
60	15.0	09.6	05.4	2.10	*02.57**
	1316				

* Critical Ratios significant at or above the 5 percent level of confidence.
** Critical Ratios significant at or above the 1 percent level of confidence.
† British percent minus U.S. percent
‡ Standard Error
# Critical Ratio
§ 1 ft = 0.3 m

Table 1-2R presents the visibility distance of the British and U.S. occupants as they moved through smoke in evacuating the fire incident buildings. They reported their movement through smoke under relatively high smoke density conditions, with visibility below 12 ft (3.7 m) for 64 percent of the British population and for 47.6 percent of the U.S. population.

Visibility distance for both the British and U.S. population at the time participants were forced to turn back is presented in Table 1-2S. Comparison of Table 1-2S with Table 1-2R reveals that very few participants turned back when the visibility distance exceeded 31 ft (9.4 m). The greater percentage of participants turned back at the reduced visibility levels. Comparing the visibility distance below 12 ft (3.7 m) in Table 1-2S, note that 91 percent of the British study population and 76.4 percent of the U.S. population turned back when visibility distances were less than 12 ft (3.7 m).

TABLE 1-2R. Comparison of the visibility distance for the British and U.S. populations when moving through smoke

Visibility distance (ft)§	British (percent)	U.S. (percent)	$P_1 - P_2$†	$SE_{P_1 - P_2}$‡	CR#
0–2	12.0	10.2	01.8	1.99	0.90
3–6	25.0	17.2	07.8	2.65	*2.94***
7–12	27.0	20.2	06.8	2.73	*2.49**
13–30	11.0	31.7	21.7	2.24	*9.69***
31–36	03.0	02.2	00.8	1.03	0.78
37–45	03.0	03.7	00.7	1.08	0.65
46–60	03.0	07.4	04.4	1.21	*3.64***
60+	17.0	07.4	09.6	2.24	*42.9***
	1316	322			

* Critical Ratios significant at or above the 5 percent level of confidence.
** Critical Ratios significant at or above the 1 percent level of confidence.
† British percent minus U.S. percent
‡ Standard Error
# Critical Ratio
§ 1 ft = 0.3 m

HANDICAPPED OR IMPAIRED OCCUPANTS

Fire problems involving occupancies, such as nursing homes and hospitals, designed for permanently or temporarily disabled persons, appear to be matched on the basis of building design, adequate staff training, and ability to protect the occupants in place until evacuation is possible. An extensive study of human behavior in health care facilities (Bryan et al 1980) indicated the nursing staff performed their professional roles related to patient responsibility even in situations with a high degree of personal risk.

The few studied fire incidents involving handicapped persons in occupancies other than health care facilities have primarily been in residential occupancies. In two of these cases, handicapped individuals were assisted by other occupants to evacuate successfully. One instance involved a wheel chair user (Bryan 1983a) and the other a blind person (Bryan et al 1979). Handicapped people may have a variety of limitations that increase their risk in a fire situation: (1) sensory problems such as deafness and blindness, (2) mobility problems such as the need for a wheelchair, and (3) intellectual problems such as mental retardation. It has also been indicated that many handicapped persons with mobility problems are concerned about their personal risk in high rise office and residential buildings where use of elevators is not allowed in a fire. In such situations, adequate areas of refuge must be provided for handicapped as well as nonhandicapped occupants (Levin and Nelson 1981).

TABLE 1-2S. Comparison of the visibility distance for the British and the U.S. populations relative to the turned back behavior

Visibility distance (f)§	British (percent)	U.S. (percent)	$P_1 - P_2$†	$SE_{P_1 - P_2}$‡	CR#
0–2	29.0	31.8	02.8	5.31	0.53
3–6	37.0	22.3	14.7	5.57	*2.64**
7–12	25.0	22.3	02.7	5.02	0.54
13–30	06.0	17.6	11.6	3.07	*3.78**
31–36	00.5	01.2	00.7	0.90	0.77
37–45	01.0	0	01.0	1.10	0.91
46–60	00.5	04.7	04.2	1.16	*3.62**
60+	01.0	0	01.0	1.10	0.91
	570	85			

* Critical Ratios significant at or above the 1 percent level of confidence.
† British percent minus U.S. percent
‡ Standard Error
# Critical Ratio
§ 1 ft = 0.3 m

A study of a number of evacuation drills in high rise office buildings in Canada has indicated that approximately three percent of the occupants will be unable to use the stairs due to permanent or temporary conditions limiting their mobility (Pauls 1977). The study population included occupants with heart conditions and individuals recovering from surgery, other illnesses, and accidents.

SUMMARY

Behavior in fires can be understood as a logical attempt to deal with a complex, rapidly changing situation in which minimal information for action is available. It is suggested that the goals of codes should be "reoriented to increase the likelihood of informed decisions being made by people in fires" (Canter et al 1978). Examination of behavior in the Beverly Hills Supper Club fire led to the recommendation that "firesafety education should consider and be based on people's erroneous conceptions about distance being related to safety, and the time needed to escape from a fire emergency" (Pauls and Jones 1980). More than a decade of detailed systematic research on human behavior in fires has resulted in the following consensus (Sime 1980) of the behavior of most persons: "Despite the highly stressful environment, people generally respond to emergencies in a "rational," often altruistic manner, in so far as is possible within the constraints imposed on their knowledge, perceptions and actions by the effects of the fire. In short, 'instinctive panic' type reactions are not the norm."

The relationship between the physical and social environment in which behavior occurs is complex. The situation is complicated by the individual's perception of ambiguous fire cues, which is primarily influenced by the person's revelant training and previous fire experience, if any. It must be recognized that fire cues are a product of a rapidly changing dynamic process which is constantly altering the decision choices of the building occupant. This decision dilema has been summarized, "What is an appropriate action at one stage may be quite inappropriate a minute later."

Bibliography

References Cited

Berry, Charles H. 1978. "Will Your Smoke Detector Wake You?" *Fire Journal*. Vol 72, No 4. pp 105-108.

Best, Richard, and Demers, David P. 1982. "Fire at the MGM Grand." *Fire Journal*. Vol 76, No 1. pp 19-37.

Best, Richard L. 1978. "Tragedy in Kentucky." *Fire Journal*. Vol 72, No 1. pp 18-35.

Bickman, L., et al. 1977. "A Model of Human Behavior in a Fire Emergency." *NBS-GCR-78-120*. Center for Fire Research. National Bureau of Standards, Washington, DC.

Breaux, John, et al. 1976. *Psychological Aspects of Behavior of People in Fire Situations*. University of Surrey, Guilford, England.

Bryan, John L. 1983a. *An Examination and Analysis of the Dynamics of the Human Behavior in the MGM Grand Hotel Fire*. Revised Edition. National Fire Protection Association, Quincy, MA.

Bryan, John L. 1983b. *An Examination and Analysis of the Dynamics of the Human Behavior in the Westchase Hilton Hotel Fire*. National Fire Protection Association, Quincy, MA.

Bryan, John L. 1982. "Human Behavior in The MGM Grand Hotel Fire." *Fire Journal*. Vol 76, No 2. pp 37-41, 44-48.

Bryan, John L., et al. 1980. "The Determination of Behavior Response Patterns in Fire Situations, Project People II, Final Report —Incident Reports, Aug 1977 to June 1980." *NBS-GCR-80-297*. Center for Fire Research, Washington, DC.

Bryan, John L. 1977. *Smoke As a Determinant of Human Behavior in Fire Situations*. Department of Fire Protection Engineering. University of Maryland, College Park, MD.

Bryan, John L., and DiNenno, Philip J. 1979a. "An Examination and Analysis of the Dynamics of the Human Behavior in the Fire Incident at the Georgian Towers on January 9, 1979." *NBS-GCR-79-187*. Center for Fire Research. National Bureau of Standards, Washington, DC.

Bryan, John L., et al. 1979b. "An Examination and Analysis of the Dynamics of the Human Behavior in the Fire Incident at the Taylor House on April 11, 1979." *NBS-GCR-80-200*. Center for Fire Research, Washington, DC.

Bryan, John L. 1971. *Human Behavior Factors and the Fire Occurrence in Buildings*. International Fire Protection Engineering Institute. Department of Fire Protection Engineering. University of Maryland, College Park, MD.

Bryan, John L. 1957. *A Study of the Survivors' Reports on the Panic in the Fire at Arundel Park Hall, Brooklyn, Maryland, on January 29, 1956*. Fire Protection Curriculum. University of Maryland, College Park, MD.

Canter, David, et al. 1978. *Human Behavior in Fires*. Department of Psychology. University of Surrey, Guilford, England.

Cohen, Hal C. 1982. "Fire Safety for the Hearing Impaired." *Fire Journal*. Vol 76, No 1. pp 70-72.

Crossman, Edward R., et al. 1975. "FIRRST: A Fire Risk and Readiness Study of Berkeley Households." *UCBFRG/WP 75-5*. University of California, Berkeley, CA.

Demers, David P. 1982. "Investigation Report on the Las Vegas Hilton Hotel Fire." *Fire Journal*. Vol 76, No 1. pp 52-57.

Kahn, Michael J. 1984. "Human Awakening and Subsequent Identification of Fire-Related Cues." *Fire Technology*. Vol 20, No 1.

Keating, John P. 1982. "The Myth of Panic." *Fire Journal*. Vol 76, No 3. pp 57-61, 147.

Keating, John P., and Loftus, Elizabeth F. 1981. "The Logic of Fire Escape." *Psychology Today*. pp 14-19.

Kentucky State Police. 1977. *Investigative Report to the Governor, Beverly Hills Supper Club Fire*. Kentucky State Police, Frankfort, KY.

Killiam, R. M., et al. 1956. *A Study of Response to the Houston, Texas, Fireworks Explosion*. Disaster Study No. 2. Publication 391. National Academy of Science, Washington, DC.

Latane, Bibb, and Darley, John M. 1968. "Group Inhibition of Bystander Intervention in Emergencies." *Journal of Personality and Social Psychology*. Vol 10, No 3. pp 215-221.

Lerup, Lars, et al. 1978. "Human Behavior in Institutional Fires and Its Design Complications." *NBS-GCR-77-93*. Center for Fire Research. National Bureau of Standards, Washington, DC.

Levin, Bernard N., and Nelson, Harold E. 1981. "Firesafety and Disabled Persons." *Fire Journal*. Vol 75, No 5. pp 35-40.

Mintz, A. 1951. "Nonadaptive Group Behavior." *Journal of Abnormal and Social Psychology*. Vol 46. pp 150-159.

Morris, Gary P. 1981. "Preplan was the Key to MGM Rescue as EMS Helped Thousands of Hotel Fire Victims." *Fire Command*. Vol 48, No 6. pp 20-21. N

FPA. 1971. "Bimonthly Fire Record." *Fire Journal*. National Fire Protection Association. Vol 65, No 3. p 51.

NFPCA. 1976. National Fire Prevention and Control Administration. *Highlights of the National Household Fire Survey*. United States Fire Administration, Washington, DC.

Nober, E. H., et al. 1981. "Waking Effectiveness of Household Smoke and Fire Detector Devices." *Fire Journal*. Vol 75, No 4. pp 86-91, 130.

Pauls, J. L., and Jones, B. K. 1980. "Research in Human Behavior." *Fire Journal*. Vol 74, No 3. pp 35-41.

Pauls, J. L. 1977. "Movement of People in Building Evacuations." in Conway, D. J. ed. *Human Response to Tall Buildings*. Dowden, Hutchinson and Ross, Stroudsburg, PA.

Paulsen, R. L. 1984. "Human Behavior and Fires: An Introduction." *Fire Technology*. Vol 20, No 2. pp 15-27.

Quanrantelli, E. L. 1979. *Panic Behavior in Fire Situations: Findings and a Model From the English Language Research Literature*. Disaster Research Center. Ohio State University, Columbus, OH.

Sime, Jonathan D. 1980. "The Concept of Panic." in Canted, David. ed. *Fire and Human Behavior*. John Wiley & Sons, NY.

Swartz, Joseph A. 1979. "Human Behavior in the Beverly Hills Fire." *Fire Journal*. Vol 73, No 3. pp 73-74, 108.

Withey, Stephen B. 1962. "Reaction to Uncertain Threat." in *Man and Society in Disaster*. Edited by G. W. Baker and Dwight W. Chapman. Basic Books, NY.

Wood, Peter G. 1972. *The Behavior of People in Fires*. Building Research Establishment. Fire Research Note 953. Fire Research Station. Boreham Wood, Hertfordshire, England.

NFPA Codes, Standards, Recommended Practices and Manuals. (See the latest *NFPA Codes and Standards Catalog* for availability of current editions of the following documents.)

NFPA 74, *Standard for the Installation, Maintenance, and Use of Household Fire Warning Equipment*.

NFPA 101, *Code for Safety to Life from Fire in Buildings and Structures*.

SECTION 2

FIRE LOSS INFORMATION

FIRE LOSS INVESTIGATION

Richard L. P. Custer

This chapter discusses fire loss investigation by detailing organizations involved in fire loss investigations; and the reconstruction and failure analysis process. The remaining two pertinent chapters within this fire loss information section are: Chapter 2, "Collecting Fire Data;" and Chapter 3, "Use of Fire Loss Information."

Additionally, three sections in this HANDBOOK—Section 4, "Characteristics and Behavior of Fire;" Section 7, "Firesafety in Building Design and Construction;" and Section 9, "Hazards to Life in Structures"—discuss allied topics. Notably, a discussion of a fault tree for managing fire risk can be found in Section 7, Chapter 2, "Systems Concepts for Building Firesafety;" while fire resistive construction information can be found in Section 7, Chapter 5, "Classification of Building Construction."

INTRODUCTION

The data gathered during a fire loss investigation has several applications, ranging from filling out the fire department incident report to a detailed engineering reconstruction and failure analysis of the incident. Understanding the fire cause and origin can lead to targeted inspection, public education programs, or perhaps to proposed code changes. A complete failure analysis can reveal factors concerning the fuel, building, and fire suppression that resulted in the ultimate extent of flame and smoke movement. The codes and design standards involved would be included in this analysis as well as construction, operation, and maintenance practices. The results of a failure analysis provide input to insurance loss adjustment and underwriting, to improved design practice, and to civil and criminal litigation.

In most fire incidents, cause and origin are expressed in terms of the heat of ignition, the materials involved, and the area of origin. In the past, only the large life or property loss fires received the attention of a complete investigation. In these cases, the reconstruction and failure analysis is often conducted months or years after the fire when only fragmentary evidence (if any) remains. In these situations the depth and accuracy of the fire department report can be critical. Recently, insurance claims, subrogation activities, and greatly increased litigation have been concerned with smaller and smaller losses. Thus more detailed examination is needed of all fires where a loss occurred, almost regardless of the extent of the loss.

ORGANIZATIONS INVOLVED IN FIRE LOSS INVESTIGATION

The organizations involved in fire investigation can be separated into two types: public agencies (those required to investigate fires by law at the local, state and federal levels); and private sector organizations.

Local and State

Of the public agencies, the local fire department generally has the primary responsibility to document the fire scenario and provide an initial determination of the cause and origin. If the department uses NFPA 901, *Uniform Coding for Fire Protection* (hereinafter referred to as NFPA 901), as the basis of a reporting system, factors that affected the ultimate course of the fire, such as fire and smoke spread factors and the performance of fire protection features, may be included.

When an investigation by the fire department determines that the fire may be of incendiary or suspicious origin, other organizations may be assigned the legal responsibility to make the final determination of fire cause and to continue the investigation pursuant to possible criminal action. In much of the United States and in all Canadian provinces, the investigation responsibility for incendiary fires is vested in the state or provincial fire marshal's office. In states where this is not the case, the local fire marshal, or the local police or fire department, may have investigation responsibility, either legally or by delegation from the state.

Federal

Federal agencies are also involved in investigation of fires. These agencies are generally charged with investigation of incidents to determine compliance with federal

Mr. Custer is Associate Director, and Associate Professor of Fire Protection Engineering at the Center for Firesafety Studies, Worcester Polytechnic Institute, Worcester, MA.

regulations, and to provide input for the regulatory process to reduce human and property losses.

The National Transportation Safety Board (NTSB), for example, conducts complete reconstruction and failure analysis of accidents (including fire related incidents) involving air, rail, and highway transportation. Investigations of pipeline incidents and maritime incidents are under the authority of the Department of Transportation (DOT). Other Federal agencies involved in fire investigations include the Occupational Safety and Health Administration (OSHA), within the Department of Labor; the Bureau of Mines (BoM), within the Department of the Interior; the Federal Bureau of Investigation (FBI); the Bureau of Alcohol, Tobacco and Firearms (ATF), within the Department of Justice; the Consumer Product Safety Commission (CPSC); the Nuclear Regulatory Commission (NRC); the Federal Emergency Management Agency (FEMA); and the National Bureau of Standards (NBS), within the Department of Commerce.

In addition to investigating fires under their jurisdiction, these federal agencies serve as resources for lab analysis and technical information to state and local fire officials.

Committees of the U.S. House of Representatives and the Senate are often very interested in the results of these investigations, particularly when the incidents under investigation deal with timely public policy topics such as standards for elderly care facilities and prison firesafety.

Private Sector Organizations

Other than for fires under government jurisdiction, most of the complete reconstruction and failure analyses are conducted by private sector organizations for education, improved design, insurance, and litigation purposes.

The National Fire Protection Association's (NFPA) fire investigations program is a long standing data collection and engineering analysis activity. The purpose of the program is to collect, analyze, and report detailed fire experience data through on site investigations. This activity assists NFPA in analyzing and documenting facts about fires for technical or educational value; distributing the information to the fire protection community to help prevent future similar fire losses; publishing of investigation reports in NFPA's *Fire Journal* and *Fire Command*; providing technical assistance to state and local officials as an integral service of the NFPA investigation process; determining important "lessons learned" for input to NFPA Technical Committees and technical programs; and contributing to an increasing fire incident data base for analysis purposes.

In recent years, NFPA fire investigations activity also has included an analysis of human behavior patterns in selected fire incidents involving large properties, where discovery of the fire and subsequent evacuation occurred in sufficient time to permit analysis. The human behavior studies have been conducted through on site interviews of fire survivors or mailed surveys, and have generally been carried out by NFPA staff in combination with leading university authorities on human behavior in fire.

Since 1972, NFPA has conducted many fire investigations in cooperation with the U.S. Fire Administration (Federal Emergency Management Agency), and the National Bureau of Standards under a continuing cost sharing agreement. In 1983, NFPA signed a Memorandum of Understanding for joint, on scene investigation of selected building fires with representatives of three model building code organizations. In this program, the Council of American Building Officials (CABO) coordinates the activities of the three model building code groups—the Building Officials and Code Administrators International (BOCA), the International Conference of Building Officials (ICBO), and the Southern Building Code Congress International (SBCCI)—with NFPA. The results of most of these special on site investigations are published as special studies or as reports in various fire protection periodicals.

The insurance industry often reconstructs a fire to evaluate the merits of claims submitted or to identify parties for subrogation action. The reconstruction may be conducted by insurance company personnel, an adjustment bureau, or a consultant.

Individual consultants or consulting firms perform reconstruction and failure analysis for a variety of clients, including the insurance industry. Consultants are also retained for litigation purposes by law firms, to evaluate fire cause and origin, code compliance, conformance with design standards, and other technical aspects of fire loss.

When major financial consequences are linked to investigators' conclusions, it is not uncommon for two or more parties to retain investigators, and for two or more theories of fire origin and development to be presented. The need to resolve such disputes is a major reason for increasing interest in application of state of the art scientific models, procedures and test results to fire investigations.

THE RECONSTRUCTION AND FAILURE ANALYSIS PROCESS

The complete reconstruction and failure analysis process involves developing a time history for the incident and establishing the factors that led to ignition and increased or lessened the severity of the loss. The reconstruction includes the prefire history as well as the transfire events, i.e., those occurring from the point of established burning to extinguishment. Failure analysis is concerned with the design, construction, and performance aspects of the incident. It is also important in the reconstruction to document the successes, i.e., whether fire protection features inhibited the spread of fire within a portion of the building.

Time is the framework for reconstruction and failure analysis, and can be viewed in two scales. These scales are: the macro scale, where time is in increments of days, weeks and years; and the micro scale, where the increments are in seconds, minutes, and hours. Macro time is pre fire, while micro time may apply to pre or transfire.

The best way to organize the reconstruction and failure analysis information is through the use of timelines or event sequences (Benner 1975; Ferry 1981). In constructing a timeline, it is important to establish reference events. A reference event can be described as an event that is well documented in time and space, and serves as a point of reference because it can be assumed to have created a set of specific conditions or initiated a sequence of actions with significant impact on the fire. The issuance of a building permit or the time of arrival of the fire department apparatus are good examples of reference events. Using time of arrival as a reference event, other events—for which the time of occurrence is unknown—

can be placed approximately in time either before or after arrival of the fire department apparatus. Other examples of typical reference events are explosions and the collapse of walls or roofs.

A single timeline can be adequate for incidents having few events. For more complex situations, separate timelines (on the same scale) can be used for the history of the building, evolution of the "model codes," installation and maintenance history of a suppression system, human activity, or any other time related aspect of the incident.

Prefire timeline information can be obtained from building permits, plans and specifications, inspection reports, fire and building codes, and fire research and engineering literature. Sources of information for use in the development of transfire timelines include the fire department incident report, recordings of fire department radio transmissions during the incident, and interviews with fire fighters and other witnesses.

It should be pointed out that in many cases, only parts of the reconstruction and failure analysis process are carried out. Multiple organizations may be involved in the investigation, often at different times. Since people may have to rely upon information gathered by others, an understanding of the whole process will facilitate obtaining the most complete and accurate data at each step.

The basic steps of reconstruction and failure analysis are examination of the scene, collection of background data, testing, reconstruction and failure analysis.

Examination of the Scene

Examination of the fire scene is critical to the entire reconstruction and analysis process. It is here that the physical and photographic evidence is gathered and the fire damage is documented.

As many photographs as possible should be taken at the scene and each should be documented on a diagram with a description of the direction of view and contents of the photograph. The techniques of fire scene photography have been well reported (Lyons 1979; Kodak 1976; 1977) and can be applied to reconstruction and failure analysis.

In addition, maps and diagrams should be prepared that include dimensions and interior layout of the building, the location and degree of fire damage, location and types of fuel materials, and the locations of victims. The location of fire protection features such as fire walls, detectors, and suppression systems should also be noted.

Often buildings will be modified considerably over their life times. An individual reviewing the original plans could receive a false impression of the building at the time of the fire. Thus it is important to record the architectural, construction, and fire protection features of a structure both as built and at the time of the fire.

When fire suppression systems are present, the position of valves, locations of discharge nozzles, and location and type of detectors used to operate the system should be noted. With special systems such as Halon, CO_2, or dry chemical, agent storage tanks should be checked to determine whether or not the system discharged as designed.

Careful study of the damage patterns, combined with on site interviews and a knowledge of the ignition sources and fuel materials present immediately prior to the fire should be used by an investigator to determine the area of origin, and if possible, the source and form of heat of ignition and the type and form of the materials first ignited (NFPA 901). Burn patterns such as a "V," low burns, and multiple "points of origin" can often be misleading and should only be used in establishing cause and origin when they are unambiguous or are supported by other physical evidence. In visualizing conditions in the area of origin, it is often very helpful to physically reconstruct the scene by replacing carpet, furniture, and other items in their original locations.

While interviews are conducted at many different times in the reconstruction and failure analysis process, those taken at or close to the time of the incident can be very useful, particularly in describing the course of the fire. Whenever possible, events described in interviews should be related to reference events.

Collection of Background Data

While some of the background data (such as activities immediately prior to the fire) can be gathered at the scene very soon after the fire, most of the information needed for the investigation will be gathered later.

Building plans, permits, and the codes in effect at the time of approval should be obtained, along with the specifications and drawings for the building fire protection features. If any major modifications were made to the building or its fire protection systems, or if there has been a change of occupancy, codes, plans, and permits should be obtained again. These materials should be reviewed as part of the reconstruction process, since these changes may have influenced the course of the fire.

Other background data include instruction or training manuals, maintenance records, data on construction and interior finish materials, previous fire history of the building or equipment involved, the amount of any previous fire damage, the location of contents, and the physical and burning characteristics of the contents. This step also involves obtaining detailed statements from the building occupants to determine the circumstances leading up to ignition and established burning.

Testing

Three types of tests are involved in reconstruction and failure analysis: (1) standard tests of flammability properties; (2) standard analysis of unknown materials; and (3) special tests designed to establish burning behavior under conditions related to a specific fire scenario.

Properties of flammability that may be determined by test include flame spread of materials (ASTM 1981a, 1981b); fabric flammability (CFR 1953; CFR 1981); flash point (ASTM 1978, 1979); and rate of heat release (ASTM 1983; Babrauskas 1982) among others. Analysis of unknowns is performed by analytical laboratories using techniques such as gas chromatography and mass spectroscopy to identify unknown materials, particularly possible accelerants. Other useful analysis techniques include metallurgical or X-ray analysis.

In collecting samples for testing, it is important to know the sizes and/or amounts of the sample needed for each test, as well as the conditions under which the samples should be transported. This information can be obtained through a testing laboratory or by reviewing a copy of the test method.

Although special tests, sometimes called "demonstrations," can often be extremely revealing in understanding the course of a particular fire, results of these demonstrations can be misleading or questionable in whether or not the test is truly representative of the actual event. Factors such as ventilation, arrangement of the fuel, or even the point of ignition can have a significant effect upon the outcome of the test. In other words, a great deal must be known about the specific conditions at the time of the fire, and these conditions must be reasonably reproduced in the demonstration. Special tests or demonstrations should be conducted under laboratory conditions (to permit the control of variables such as ambient temperature, relative humidity, and wind effects) by personnel familiar with full scale fire testing.

Reconstruction

The fire reconstruction is an organized, step by step portrayal of the most likely course of the fire under study, from the ignition sequence through established burning, to the limits of flame and smoke movement at the time of extinguishment. It is based upon the facts developed through on scene investigation, background studies, and testing. The reconstruction should provide an explanation of the speed and direction of fire growth and smoke spread within the compartment or area of origin and beyond. It also should explain the effects of barriers and automatic suppression systems on fire spread and growth, and the effects of manual extinguishment. The reconstruction should also include an evaluation of the performance of life safety features of the building relative to injury of loss of life.

In sorting through the facts to develop the reconstruction, it is often useful to apply the system safety techniques of Failure Modes and Effects Analysis (FMEA) and Fault Tree Analysis (Henley and Kumamoto 1981). These methods help identify alternative sequences of events that could lead to the observed fire situation, such as development of room flashover or fire breaching a fire wall. The alternatives can then be evaluated against the evidence to determine the most likely scenario. These techniques are particularly valuable where complex or interrelated machinery, buildings, or human activities are involved.

Analysis

Analysis of the ignition sequence is concerned with three specific items: the size, nature, and source of the ignition energy; the physical and chemical fire properties of the materials ignited; and the circumstances (human or mechanical) that brought the energy in contact with the fuel. In selecting the most likely ignition source, the energy available from a candidate source must be compared with the ignition energy requirements of the fuel in its given form. For example, given the same material and heat source, ignition is more likely if the material is present in finely divided form, than in a solid mass. Energy available at the source (temperature, thermal radiation, etc.), distance from the fuel to the energy source, and time of exposure to the source of energy are also factors that must be considered in ignition sequence analysis.

The circumstances which result in ignition generally involve human factors or equipment failure. Careless welding, poor housekeeping, and improper selection or misuse of materials are examples of human factors. Valve rupture and combustion control failure could be considered equipment failures. Ultimately, most equipment failures can be traced to design, manufacturing, installation, misuse, or maintenance problems.

If established burning is achieved after ignition, it must be determined why the fire continued to full room involvement (flashover) or, in the absence of compartmentation (such as in a warehouse or industrial building), why it reached a certain extent. In answering these questions, a number of factors need to be evaluated, including the rates of heat release of the fuels, continuity of the fuels (proximity of fuel packages to each other), and location of fuels relative to walls, compartment ceiling height, ventilation and interior finish.

Heat release rate is one of the more important factors in analysis. In many instances, a fire reported to have grown or spread "unusually fast" is characterized as suspicious or incendiary when, in fact, the growth rate may have been due to a combination of the burning properties of the materials and the nature of the compartment in which the fire burned. A number of studies have been carried out to characterize the burning rates of materials and furnishings. Data can be found in NBS reports (Babrauskas 1977; 1979; 1982; Lee 1985; Walton and Twilly; 1984) and in Appendix C of NFPA 72E, *Standard on Automatic Fire Detectors*. The roles of heat release rate, fire location, and ceiling height on fire development also has been reported (Alpert and Ward 1983). Alpert's work includes methods to estimate ceiling temperatures and ceiling gas velocities. Guidance on estimating the likelihood that a given fire size and room ventilation combination can develop flashover also is provided in NBS reports (Babrauskas 1980, 1981).

The performance of automatic suppression systems in controlling the growth and spread of the fire also must be evaluated in the reconstruction process. Here, the factors to be reviewed include the appropriateness of the agent being used; the system response time, as related to the detection devices used, relative to ceiling height and fire growth rate; location of the agent discharge nozzles relative to the source of the fire; agent availability (rate of discharge, pressure, and adequacy of supply); and conformance of the system design and installation with NFPA standards, industry design, and installation manuals. For example, a sprinkler system may have failed to control a fire due to a very high ceiling that delayed the response time, combined with low water pressure that could not produce the needed water flow rate.

Another component of the reconstruction is barrier performance. Barrier performance evaluation focuses on the nature of the barrier (rated or not rated), barrier construction, the thermal stresses to which it was exposed, and the manner in which the barrier failed. Barriers can fail either by passage of flame or hot gases through a small opening, a hot spot failure, or by movement of flame and hot gases through open doors, or by barrier collapse resulting in a massive failure. Failure of opening protection, poor construction practices, modifications to the barrier after construction (pokethroughs, etc.) and fire stresses that exceed those assumed in the design can be significant transporters of products of combustion beyond design limits.

Manual suppression activities, whether by a municipal fire department or a plant fire brigade, can have a significant effect on the outcome of a fire. In the absence of fire suppression systems, the response time of the fire department for a given fire will influence the size of the fire at the time of agent application. For ease of analysis, the response time can be broken down into the following segments: detection time, alarm transmission time, dispatch time, travel time, and time from arrival of fire suppression personnel to agent application. Problems resulting in delays in any of the above time segments will extend the limits of flame and smoke travel. Examples of factors that could cause delays are the absence of automatic detectors, an unoccupied building, or an improperly designed detection system. Equipment failure or out of service fire apparatus can also delay response time. Weather, terrain, traffic, or condition of apparatus can extend travel time to the incident. Frozen hydrants, blocked access, or shortage of personnel are also examples of factors that could delay application of fire suppression agents.

The reconstruction process also must consider the performance of the building life safety system. This includes the detection and alarm system, the building egress system, and smoke control system. In addition to the factors discussed above affecting detector response, the audibility of alarms plays a role in life safety. Guidance on this topic can be found in work by Nober (Nober et al 1984) and in NFPA 72F, *Standard for the Installation, Maintenance and Use of Emergency Voice/Alarm Communication Systems*. Building egress factors include the capabilities of the occupants and the number, capacity, and protection of the egress routes. The movement of smoke in buildings is related to the temperature of the fire gases, differences between inside and outside temperatures (stack effect), effects of the building heating and air conditioning systems, and effects of wind.

Failure Analysis

In a firesafety context, failure can be defined as a fire event that results in personal injury, death, property, or monetary losses, or in conditions that prevent buildings or mechanisms from functioning as designed. In fire failure analysis, the objective is to use on site and background data, testing results, and reconstruction to identify the primary and contributing causes of the failures and their sources. The sources of failures include basic design, material or equipment selection (design), material or equipment defects (manufacturing), construction or assembly, testing, post construction modifications, service conditions (use), and unanticipated conditions (abuse).

The analysis is made by comparing the causes of failures with the codes, standards, design practice, and the state of the art technology present at the time of the fire. In the analysis, it is important to identify noncompliance with the standard practices, but it is even more important to note where compliance with the standards failed to provide the expected or needed degree of firesafety. In either event, the failure analysis should provide recommendations for how the failures could have been prevented in the fire, or how the knowledge gained can be applied to preventing similar failures in the future.

Bibliography

References Cited

Alpert, R. L., and Ward, E. J. 1983. "Evaluation of Unsprinklered Fire Hazards." *Technology Report 83-2*. Society of Fire Protection Engineers. Boston, MA.

ASTM. 1978. *Test for Flash and Fire Points by Cleveland Opened Cup, ASTM D92-78*. American Society for Testing and Materials. Philadelphia, PA.

ASTM. 1979. *Test for Flash Point by Tag Closed Tester, ASTM D56-79*. American Society for Testing and Materials, Philadelphia, PA. 1979.

ASTM. 1981a. *Test for Surface Burning Characteristics of Building Materials. ASTM E84-81*. American Society for Testing and Materials. Philadelphia, PA.

ASTM. 1981b. *Test for Flammability of Materials Using a Radiant Heat Energy Source. ASTM E162-81*. American Society for Testing and Materials. Philadelphia, PA.

ASTM. 1983. *Standard Test Method for Heat and Visible Smoke Release Rate for Materials and Products. ASTM E906-83*. American Society for Testing and Materials. Philadelphia, PA.

Babrauskas, V. 1977. "Combustion of Mattresses Exposed to Flaming Ignition Sources Part I: Full Scale Tests and Hazard Analysis." *NBSIR 77-1290*. National Bureau of Standards. Gaithersburg, MD.

Babrauskas, V. 1979. "Full Scale Burning Behavior of Upholstered Chairs." *NBSTN 1003*. National Bureau of Standards. Gaithersburg, MD.

Babrauskas, V. 1980. "Estimating Room Flashover Potential." *Fire Technology*. Vol 16. No 2. p 94.

Babrauskas, V. 1981. "Will the Second Item Ignite?" *Fire Safety Journal*. Vol 4. 1981/1982. p 281.

Babrauskas, V. et al. 1982. "Upholstered Furniture Heat Release Rates Measured With a Furniture Calorimeter." *NBSIR 82-2604*. National Bureau of Standards. Gaithersburg, MD.

Benner, Ludwig. 1975. "Accidental Investigations: Multilinear Event Sequencing Methods." *Journal of Safety Research*. Vol 7. No 2. 1975. p 67.

CFR. 1953. *Standard For the Flammability of Clothing Textiles (CS-191-53)*. 16CFR1610. Code of Federal Regulations.

CFR. 1981. *Standard for the Flammability of Childrens Sleepwear: Sizes 0 through 6X (FF3-71)* 16CFR1615. Code of Federal Regulations.

Ferry, Ted. S. 1981. "Modern Accident Investigation and Analysis." *An Executive Guide*. John Wiley & Sons. NY.

Henley, Ernest J. and Kumamoto, Hiromitsu. 1981. *Reliability Engineering and Risk Assessment*. Prentice Hall. Englewood Cliffs, NJ.

KODAK. 1976. *Using Photography to Preserve Evidence. Pamphlet M-2*. Eastman Kodak Company. Rochester, NY.

KODAK. 1977. *Fire and Arson Photography*. Pamphlet M-67. Eastman Kodak Company. Rochester, NY.

Lee, Billy T. 1985. "Heat Release Rate Characteristics of Some Combustible Fuel Sources in Nuclear Power Plants." *NBSIR 85-3195*. National Bureau of Standards. Gaithersburg, MD.

Lyons, Paul R. 1978. *Techniques of Fire Photography*. National Fire Protection Association. Quincy, MA.

Nober, E. H. et al. 1984. "Waking Effectiveness of Household Smoke and Fire Detection Devices." *NBSGCR-80-284*. National Bureau of Standards. Gaithersburg, MD.

Walton, William D. and Twilley, William H. 1984. "Heat Release and Mass Loss Rate for Selected Materials." *NBSIR 84-2960*. National Bureau of Standards. Gaithersburg, MD.

NFPA Codes, Standard and Recommended Practices and Manuals. (See the latest *NFPA Codes and Standards Catalog* for availability of current editions of the following documents.)

NFPA 72A, *Standard for the Installation, Maintenance and Use of Local Protective Signaling Systems for Guard's Tour, Fire Alarm and Supervisory Service.*

NFPA 72B, *Standard for the Installation, Maintenance and Use of Auxiliary Protective Signaling Systems for Fire Alarm Service.*

NFPA 72C, *Standard for the Installation, Maintenance and Use of Remote Station Protective Signaling Systems.*
NFPA 72D, *Standard for the Installation, Maintenance and Use of Proprietary Protective Signaling Systems.*
NFPA 72E, *Standard on Automatic Fire Detectors.*
NFPA 72F, *Standard for the Installation, Maintenance and Use of Emergency Voice/Alarm Communication Systems.*
NFPA 72G, *Standard for the Installation, Maintenance and Use of Notification Appliances for Protective Signaling Systems.*
NFPA 72H, *Guide for Testing Procedures for Local, Auxiliary, Remote Station and Proprietary Protective Signaling Systems.*
NFPA 901, *Uniform Coding for Fire Protection.*
NFPA 902M, *Fire Reporting Field Incident Manual.*
NFPA 903M, *Fire Reporting Property Survey Manual.*
NFPA 904M, *Incident Follow-up Report Manual.*
NFPA 907M, *Manual on the Investigation of Fires of Electrical Origin.*

Additional Readings

Berrin, E. R., *Investigative Photography*, Society of Fire Protection Engineers, Boston, MA, 1977.
Building Materials Directory, Underwriters Laboratories, Inc., Northbrook, IL.
deHaan, J. D., *Kirk's Fire Investigation*, 2nd ed., John Wiley, NY, 1983.
Dennett, M. F., *Fire Investigation: A Practical Guide for Fire Students and Officers, Insurance Investigators, Loss Adjusters, and Police Officers*, Pergamon Press, Elmsford, NY, 1980.
FIFI-Fire Investigation Field Investigation, National Fire Protection Association, Quincy, MA.
Fire Investigation Handbook, National Bureau of Standards, U.S. Department of Commerce, Washington, DC, 1980.
Fire Investigator's Guide to NFPA 1031, National Fire Protection Association, Quincy, MA, 1983.
Fire Resistance Directory, Underwriters Laboratories, Inc., Northbrook, IL.
Fire Tests Performance, American Society for Testing and Materials, Philadelphia, PA.
Franklin, Frederick F., "A Survey of Electrical Systems," *Fire Journal*, Vol. 78, No. 2, Mar. 1984, pp. 41-44.
Lathrop, James K., ed., *Life Safety Code Handbook*, 3rd ed., National Fire Protection Association, Quincy, MA, 1985.
Mills, John F., "The Fire Fighter's Role in Fire Investigation," *Fire Command*, Vol. 47, No. 2, Feb. 1980, p. 25.
Phillips, Calvin, and McFadden, David, *Investigating the Fireground*, R. J. Brady, Bowie, MD, 1982.
Roblee, C., and McKechnie, A., *Investigation of Fires*, Prentice-Hall, Inc. Englewood Cliffs, NJ, 1981.

COLLECTING FIRE DATA

Revised by Carl E. Peterson

Data on individual fires provide a valuable tool for fire service manangement. Aggregated data on the fire experience of a community, state, or nation is also helpful, permitting a fire department to assess the effectiveness of fire prevention and suppression methods and to suggest areas for improvement and requirements for additional resources. Use of fire data is discussed further in Chapter 3 of this section, "Use of Fire Loss Information."

Data collected at the state and national level support meaningful comparisons among communities and regions. To support this activity, a fire department needs to produce fire reports that work both as self-standing descriptions of individual incidents and as components in the creation of broad based fire incident data bases.

The following sections discuss fire reporting and methods for collecting data from fire reports to form broad based data systems.

THE FIRE REPORT

A fire report is the written documentation that a fire occurred. It may be as brief as a basic fact statement or as lengthy as an extensive discussion of the fire, supported by photographs, witness statements, laboratory test results, and physical evidence. The length and complexity of the report will depend upon the size and nature of the fire, the local fire service manager's need for specific data, and the resources available for obtaining information and completing reports.

The fire report should include, at some level of detail, a time staged description of the circumstances related to the initiation, discovery, growth, and termination of the fire along with a description of the casualties or the damage resulting from the incident. This report should be in the words of the fire officer and must be complete, so persons who were not at the fire scene can understand what happened.

Mr. Peterson is Manager, Fire Service Management Systems at NFPA.

Purpose of the Fire Report

There are three basic purposes of a fire report at the local level: (1) it is the legal record of the fact that the fire occurred, and provides official notification to those who may be legally required to know of the incident (e.g., the state or local fire marshal). It reports facts about the particular property affected, why the fire occurred, how building components and fire protection devices performed, casualties or damage that resulted, and fire department action; (2) it provides information to senior officers and fire department managers so they are kept informed about what is happening within their area of responsibility. This allows them to evaluate the performance of their units at the incident and to talk intelligently about the incident; and (3) the report provides data on the fire problem to fire service management so they can track trends, gauge the effectiveness of fire prevention and fire suppression measures presently in practice, evaluate the impact of new methods, and indicate those areas that may require further attention.

The first two purposes can be served by any report that is an accurate description of the incident. The third purpose, however, requires that information be collected in a consistant format that will permit meaningful aggregation of the data from reports on many incidents.

It is also important that a single report serve the data needs of all potential users. The data needed at the state and national levels must be provided from what is collected locally. At the same time, it is important that the locally collected data also have a visible, significant use at the local fire service level. If the data are collected only for the benefit of those outside the local area, the motivation and commitment to quality and completeness may diminish, with a resulting reduction in the usefulness of the data.

Uniformity in Fire Reporting

To maintain uniformity in fire reporting, the NFPA Technical Committee on Fire Reporting has developed NFPA 901, *Uniform Coding for Fire Protection*, (hereinafter referred to as NFPA 901). This standard establishes basic definitions and terminology for use in fire reporting and a means of classifying data so that they can be aggregated either manually or automatically.

NFPA 901 provides the common language used by nearly all large scale (e.g., state, national) data bases in the United States and many others around the world. It is recognized that every fire department will not want to collect every data element; likewise, there may be additional data elements that a fire department wishes to collect. Therefore, a fire department that uses NFPA 901 will find that it is in a position to most efficiently contribute data to larger data bases and, in turn, to use data from these larger systems in its management of the local fire problem. It also will find that the quality, uniformity, and usefulness of information within the fire department will improve. NFPA 901 contains data elements that provide a classification of practically every type of property (whether fixed or mobile), a description of a specific structure prior to an incident, a description of the ignition sequence including the area of origin of the fire, conditions found upon arrival, what action was taken, and why the fire grew or stayed as small as it did. There are also data elements for describing injuries or fatalities to both civilians and fire fighters as a result of the fire incident, extent of damage, loss of property, and the investment in personnel provided by the community to control the incident. Useful analysis can be adapted to the particular data elements adopted by the fire department, no matter how few or how many.

FIRE REPORTING SYSTEMS

There is a major difference between a fire report and a fire reporting system. A fire report is only part of a fire reporting system. A complete fire reporting system has to address three stages of operation: fact finding, fact processing, and fact use. Each of these is discussed below.

Fact Finding

The traditional functions of a fire report can be satisfied with a minimal written narrative of the basic facts of the incident. To serve as input to a fire reporting system, however, an incident report must be clearly structured and must use uniform definitions and terminology. The collection of such information requires not only the form or forms upon which to record the information desired, but also a set of instructions and related training regarding the forms so the information is provided in a uniform manner, and a procedure is established for forwarding the forms to a central point.

Equally important, however, is ensuring that the persons responsible for reporting the data are trained and capable of investigating a fire to determine fire scenario and cause.

Fact Processing

Once data have been recorded, they must be processed into a form useful for legal, statistical, planning, and management purposes. The first step in information processing is to check the incident reports for accuracy, clarity, consistency, and completeness. A procedure of quality control screening and follow up corrections is needed for this purpose. For best results, manual screening of the reports is always needed, even when computerized edit checks are employed. The second step is to combine information about one incident from several reports into a composite record. The third step is to create a fire fact file of all records of all reported incidents. This fire fact file then becomes the basic source of information about past incidents. Use of the file will largely determine the facts that must be recorded on an individual incident report.

In the past several years, fire departments have begun using computers to assist with fact processing. Computer programs can quickly check the consistency and completeness of data about an incident. These same computers build and maintain a data base, which becomes one part of the fire fact file.

Fact Use

Once an incident file has been created, whether as a paper file or a computerized data base, it has many potential uses. At a minimum, the file should meet all the informational needs of the local fire service. Legal and statistical information required to manage the department, spot trends in fire incidence, and provide documentation for program evaluation and fire research will be readily accessible.

Small fire departments have special problems because they may have a low fire incidence level. This means they may be able to analyze their fire problems only by working with several other small departments within a geographic region. Therefore, it is especially important that small fire departments use uniform terminology and uniform coding in collecting information so data from different fire services can be merged.

Benefits of a Fire Reporting System

At the local level, a fire department can derive many benefits from a good fire incident reporting system, particularly if it is based on NFPA 901. Some of the following uses involve no more than totaling data from the system. Others require more extensive analysis. Many of these benefits also apply to users at the state and national levels.

Describing a Community's Fire Problem: It is possible to pinpoint where fires are occurring, what factors are most responsible for ignitions, and what casualties and damage are occurring as a result of fires. With the problem placed in proper perspective, the most serious and vulnerable aspects of the fire problem can be tackled first.

Supporting Budget Requests: In this era of increasing concern about taxes, municipal officials are quick to cut budgets and slow to add new programs. Frequently, fire department managers do not have the statistics to support their requests for additional funds. Good statistics will put the fire problem in perspective with other municipal concerns, and help community officials realize the consequences of budget cuts or the value of new programs for the fire department.

Supporting Code Refinements: A good data base permits fire departments to identify and describe fire incidents that would have developed differently or might not have occurred at all if certain code changes had been in place. Loss statistics from other areas with more stringent codes also can be used for comparison. Estimation of the likely impact of a code change can involve complex analysis, however, and no incident data base can address all the subtleties of code impact.

Evaluating Code Enforcement Programs: It is not sufficient to have codes on the books if they are not properly enforced. In evaluating loss experience, it is possible to see whether certain losses are occurring because existing codes are not being properly enforced. The reason for the improper enforcement then can be analyzed and corrected.

Evaluating Public Fire Education Programs: Not all problems can be solved by establishing and enforcing codes. There are certain aspects of the fire problem which can best be controlled by public education programs—educating people about the dangers of fire, how to reduce them, and how to react when hazardous situations arise. Such programs require considerable time and money. It is important to know the exact problem that needs to be addressed. Appropriate evaluation criteria must also be in place to measure whether an educational program is in fact helping to solve that aspect of the problem.

Planning Future Fire Protection Needs: Many communities and fire departments are becoming very active in planning and are developing master plans. It is essential that the fire service be involved in such planning. A good data base will allow a fire department to compute fire rates relative to population and building inventory. These, with other characteristics of the community fire problem and planning, will support better fire protection in the future based on changing demography and planned community growth. It will also provide input to decisions on the type and level of fire protection a community will provide so requirements can be established for developers who construct properties that exceed fire department capabilities.

Improving Allocation of Resources: It is not always possible for a fire department to grow at the rate necessary to protect a changing community. In fact, it may not be necessary for a department to expand. Proper analysis may show where a redeployment of existing resources can provide the same level of protection or even improve the level of protection within a community.

Scheduling Nonemergency Activities: Training sessions, in-service inspections, and other activities are important aspects of a fire department's function. A fire department which tracks times and severity of fires can schedule these activities when they are least likely to be interrupted by emergency calls, or when the normal delay from such activities will have the least impact on emergencies.

Regulating Product Safety: Particularly at the national and state levels, a fire reporting system can be useful in measuring the size and severity of problems associated with various types of consumer products. By identifying the most common ways these products become involved in fire, more accurate precautions and information can be disseminated to consumers.

Support for Major Fire Engineering Analysis: The NFPA Systems Concepts Committee and other users of sophisticated engineering models will depend upon the output of fire reporting systems on a continuing basis. Each time a method of fire defense works well and fire loss and danger are confined to a small area, success will increase confidence in that particular method of fire defense. Conversely, each time a method of fire defense fails, as indicated by injuries, deaths or an expensive loss, this failure needs to be recorded so that method of fire defense can be reevaluated.

DESIGNING A FIRE REPORTING SYSTEM

When a new fire reporting system is to be designed or an existing system redesigned, the first step should be an analysis of the need for and use of the data which will be collected. Some of the questions that should be answered include:

1. What information is this system expected to produce, who will use it, and what characteristics will they want to track?
2. What data elements are necessary to provide the information needed?
3. How quickly is the information needed and how long after the incident should the information be kept available?
4. Who will be expected to collect the data, assemble it, and screen it for errors?
5. Who will be expected to summarize and analyze the data? Will they use their own analysis? To what other uses will the analysis be put?
6. What savings or efficiencies could be realized if the data were available?
7. Is any of the needed data already being collected for another use?

The next steps are to design the form for data collection, develop written procedures for completing the form, compiling and editing the forms, and filing, storage, and disposal of the data collection forms.

Form Design

Several things should be considered when designing the form. These include:

1. The form should be designed to apply to local needs. However, existing forms should be used where appropriate, because forms designed from scratch may have problems that have already been identified and solved by others.
2. Enough space to write the answer should be included on the form. Too little space results in illegible handwriting. Never squeeze out available space for answers; another page should be added to the form or some of the data elements deleted.
3. If a typewriter is to be used to complete the form, correct spacing between lines is essential. Also consider aligning the answer spaces so tabular settings can be used.
4. Standard sizes should be used for the form. Forms with odd sizes require special file cabinets, which are more expensive; oversized forms are more difficult to use in the field.
5. If more than one copy of a report will be needed, consider using NCR (no carbon required) paper. Carbon paper is messy and inconvenient.
6. Different colors for different forms help distinguish the forms at a distance and can facilitate filing.
7. If data is to be keypunched from the form for computer storage, the form design should be coordinated with the person responsible for the keypunching so that data entry will proceed smoothly.
8. Data should be recorded in the order it will most likely be collected and information on related topics kept together.

9. If the whole form need not be completed in all cases, the form can be subdivided. Provide brief instructions on the form as to when each section should be used.
10. Consider providing space at the top or right margin for case identifiers (e.g., address, file number, date or whatever other key is used for filing) so that forms can be readily identified by users thumbing through the file.
11. The form should be kept single sided, if possible.

The written procedures for completing the form should define the intended use of the form. Procedures should cover, item by item, the kind of information sought, the format in which it should be recorded, and, if it is to be classified, the appropriate classifications. If the form includes a remarks section, the type of information and extent of detail expected in this section should be defined. Samples of properly completed forms can improve the proper understanding and use of the form.

If the data is to be routinely summarized into management reports, the written procedures should define how this is to be done, how often, and which persons should receive copies of the summaries.

Training

Actual implementation of the use of the new form must be preceded by training for those individuals who will be responsible for completing the form. The actual training should be the responsibility of the training division in the department. They should work closely with who ever developed the reporting system to develop the training materials. Written instructions should be used as the basis of the training. Questions raised during the training may point to a need for additional written instructions. Once the initial training has been accomplished, periodic retraining should be scheduled as part of the normal retraining processes. All persons promoted or assigned to positions where they will be responsible for completion of the forms should receive special report processing training.

Organizing to Compile Reports

In many fire departments, the data on various aspects of a fire may come from different sources. The communications office will be able to provide data on alarm receipt and dispatch. The officer in charge will be responsible for the incident report. Each company officer will be in the best position to record the role of his/her company and the resources used in an incident. If there were casualties, data on the injuries or fatalities may best be provided by still other personnel. As the fire is investigated, the fire marshal or fire investigation unit may develop details which the officer in charge could not determine at the time of the fire.

All of this data is often recorded on separate forms by the individuals with direct knowledge of a certain part of the fire incident. It is extremely important that all these forms be compiled into a complete fire report and that the data on the various forms be checked for completeness and consistency.

One division within the department (frequently the fire marshal) should be given the responsibility for the compilation of all fire reports. Clerical personnel should be properly trained in recording fire data. Some of the responsibilities associated with compiling reports include:

1. Ensuring that all forms are submitted in a timely fashion.
2. Ensuring that all forms are properly completed and are legible.
3. Screening the forms for obvious errors, omissions, or conflicting data.
4. Returning the forms to the author for completion or correction when necessary and ensuring they are returned again from the field.
5. Preparing the forms for keypunching and forwarding the batches (if appropriate).
6. Filing the complete report.
7. Updating the report and the files as new or additional information is received.

Detailed written procedures should be developed for the clerical personnel outlining how they should perform each step in report compilation. These procedures should address such issues as how the reports are to be edited, processed, and filed, and within what time frame.

It is important that everyone responsible for completing the forms understand the procedure which the clerical staff uses and the schedule they are expected to follow in completing, correcting, and forwarding the forms. Delays in submitting forms will result in delays in assembling the complete report and in making it available for use.

Quality Control

An important aspect of any record keeping function is ensuring quality in the data collected. This starts with proper training of the person who is responsible for completing the forms. Written instructions should explain how forms are completed. All forms should be screened as part of the process of compiling them into a master record and preparing them for keypunching. This screening can be done by a trained clerical person. At least periodically, a supervisor should spot check a sample of the forms to ensure they are completed properly. All forms with errors or omissions should be flagged and returned to the originator for correction or completion as appropriate. This reinforces training because the individual will see where the mistakes are and can avoid them in the future. If a person continues to make the same mistakes, special training or other action may be necessary to improve the quality of the reports submitted. Incomplete or poor quality reports waste time and defeat the whole purpose for collecting data.

Record Retention

An important aspect of any record keeping system is determining how long to keep the records. The answer varies with the purpose of the record and to some extent, with local and state legal requirements for record retention.

Obviously, records require storage space, and storage space costs money. Likewise, the more voluminous the files become, the more likely material will be lost or misfiled.

There should be a written policy which details the length of time a report is to be retained in the active files, the length of time it needs to be kept in inactive storage, and when it can be discarded. The city attorney's office can help determine legal requirements for retaining records; that office should review the final written policy prior to its implementation. There may be different requirements for computerized records which require less space and are useful for multiyear analysis.

Setting Standards Within the Department

The fire department should have certain standards which apply to all its record keeping operations. This will help make the information more uniform from report to report and application to application. Standardization aids all users of the report, makes analysis more accurate, and is essential if data is to be automated. Some areas where department standards should be developed include:

1. Methods to enter names of persons (first name first, or last name first) on records.
2. Recording addresses of buildings with multiple or ambiguous addresses and nonstructure locations such as on highways or street intersections.
3. Designating apparatus (e.g., company assigned, serial number, shop number).
4. Designating employees (e.g., name, badge number, social security number).
5. The common abbreviations that are acceptable to use.

NFPA 901 can be used as a basis of terminology, definition and classification if data is to be classified when it is reported.

Data Analysis

In many fire reporting systems, better than 90 percent of the reporting effort is devoted to data collection and very little effort is devoted to data analysis. Strictly adding things up is not data analysis.

Data analysis involves determining what the data is really showing relative to the community or the project supported by the data. Data analysis often involves studying the interrelationship between various data elements or various data bases. For example, studying the fire problem by tracking the census and relating it to socioeconomic data for the census tracks leads to a better understanding of the problem and the population groups who are most affected. Likewise, comparing fire losses over the years to inflation rates and increasing property values may show that fire protection is presently more effective, even though in raw numbers the total loss is greater.

Data analysis is a special skill which should be sought out and used by a fire department. With such analysis, the data is often more meaningful in defining the problem and targeting a solution.

Collecting Data for Special Studies

No one can anticipate all the data that may be needed to understand the fire problem. Once a problem has been identified through an ongoing fire reporting system a special study may be needed to gather the data required to further define and help solve the problem. The steps in conducting a special study are:

1. Identify the type and sources of data needed to further define the problem. Assess the feasibility and costs of the study and proceed in the most efficient way (e.g., sampling may be appropriate). Previous work, along with the results achieved, should be researched to solve this or similar problems.
2. Develop a mechanism or procedure to collect the needed data. For example, this can be a special form completed by the officer in charge of the incident, or a request to include certain data in the remarks section of a report whenever a certain condition is encountered. Alternatively, the procedure may stipulate merely that the officer in charge notify a certain person or agency whenever the officer encounters the particular problem or condition.
3. Collect and analyze the data. Develop a methodology to reduce or solve the problem, based on the results of data analysis and any additional documentation.
4. Terminate the special study. When the methodology is implemented, monitor the progress of the problem solving methodology using the normal reporting system. Return to steps 1, 2, or 3 as appropriate if the problem is not being reduced or solved.

Periodic Review

Over time, information needs within a fire department change. As these needs change, it is not uncommon for new reporting requirements to be added. New forms are introduced or existing forms are expanded. Much less common, however, is a reduction in forms or requested information.

At least yearly and certainly before any form is reprinted, a careful review should be made to determine if the original purpose for the form is still valid and if all the information is still needed and useful in its present format.

Do not perpetuate data gathering if the data is not serving a useful purpose. It is easy to fall into a trap of justifying data collection on the basis of "it may be needed someday." This approach generally results in poor quality data, if and when used, because lack of data use results in lack of quality control in collecting the data.

STATUS OF UNIFORM FIRE REPORTING

There has been and continues to be considerable work performed in developing fire reporting systems based on NFPA 901. The NFPA Fire Reporting Committee also has established a basic system published in NFPA 902M, *Fire Reporting Field Incident Manual* (hereinafter referred to as NFPA 902M). Figure 2-2A is an example of a form used in NFPA 902M. The system provides the necessary instructions for completion of an incident report, supplementary reports on civilian or fire fighter casualties suffered at incidents, and EMS reports which can be used by those fire departments providing Emergency Medical Services. Work sheets are provided for summarizing incident data. This system can be used manually or adapted for electronic data processing.

NFPA also has an automated incident based reporting system which is available to fire departments. This system is known as the Uniform Fire Incident Reporting System (UFIRS) and is available with the necessary computer programs to edit and process incident data onto master files and to generate summary reports. Both of the systems are receiving widespread fire department use at the local level.

The U.S. Fire Administration (USFA), within the Federal Emergency Management Agency (FEMA) has de-

BASIC INCIDENT REPORT 902F

Fill In This Report In Your Own Words ____________________ Fire Department | ☐ Revised Report

A	FD ID	Incident No.	Index No.	Mo. / Day / Year	Alarm Time	Time on Scene / Time Last Unit Clear
B	Location/Address	City/Town	Zip Code	Property No.		
C	Occupant Name (Last, First, MI)	Telephone No.	Room or Apt.			
D	Owner Name (Last, First, MI)	Address	Telephone No.			
E	Method of Alarm to Fire Department	Type of Incident				
F	Type of Action Taken	District	Shift	No. Alarms	Mutual Aid ☐ Rec'd ☐ Given ☐ N/A	
G	General Property Use	Specific Property Use	County	Census Tract		
H	No. Injuries* Fire Service	Other Emerg.	Civilian	No. Fatalities* Fire Service	Other Emerg.	Civilian
I	No. Fire Service Personnel Responsed	No. Engines Responded	No. Aerial Apparatus Responded	No. Other Vehicles Responded		
J	Condition of Fire upon Arrival of First Unit	Time from Alarm to Agent Application	Area of Fire Origin			
K	Equipment Involved in Ignition	Year	Make	Model	Serial No.	
L	Form of Heat of Ignition	Material First Ignited Form/Use	Type			
M	Ignition Factor	Method of Extinguishment				
N	Property Damage Classification	No. Buildings Damaged	Termination Stage			
O	Construction Type	No. of Stories	Level of Origin			
P	Structure Status	No. of Occupants at Time of Incident				
Q	Material Generating Most Flame Form/Use	Type	Factor Contributing to Flame Travel			
R	Material Generating Most Smoke Form/Use	Type	Avenue of Smoke Travel			
S	Detector Type	Detector Power Supply	Detector Performance			
T	Sprinkler System Performance	No. of Sprinkler Heads Operated				
U	Extent of Flame Damage	Extent of Smoke Damage	Extent of Extinguishing Agent Damage			
V	Mobile Property Type	Year	Make	Model	Serial No.	License No.
W	No. of Private Acres Burned	No. of Federal Acres Burned	No. of Other Public Acres Burned			
X	Member Making Report	Date	Officer in Charge (Name, Position, Assignment)	Date		
Y	Remarks:					

☐ Remarks continued on reverse side.

Side labels: COMPLETE ON ALL INCIDENTS (rows A–I); COMPLETE IF FIRE TYPE OF INCIDENT (TI) 10-19 (rows J–W): ON ALL FIRES TI 10-19 (rows J–N); FOR STRUCTURE FIRE TI 11-13 (rows O–U); TI 12-14 (row V); TI 15 (row W); COMPLETE ON ALL INCIDENTS (rows X–Y).

*A Form 902G must be completed for each Fire Casualty.

This form is for use with NFPA 902M, *Field Incident Manual.* Users should also refer to NFPA 901, *Uniform Coding for Fire Protection,* for information on fire reporting systems and classifications for information entered on this form.

FIG. 2-2A. Basic Incident Report Form recommended by NFPA's Technical Committee on Fire Reporting.

veloped the National Fire Incident Reporting System (NFIRS) which is an automated system based on the work of the NFPA Fire Reporting Committee as published in NFPA 901. The system now has been installed in approximately 40 states, the District of Columbia, and a number of larger fire departments.

NFIRS is also an incident based system. Data is reported on an incident by incident basis. Computer programs edit the data, maintain a data base, and generate a variety of summary reports. The data collected is similar to that collected by the basic system published in NFPA 902M.

At the state level, NFIRS provides for the collection of written reports on incidents to which local communities responded. Communities with automated data systems based on NFPA 901 can participate by providing their data on computer tape to the state. UFIRS is an example of one such system that is providing automated data to state systems. Summary reports are then generated on both individual community and statewide fire experience.

At the national level, NFIRS provides for the merging of data bases from individual states to form a national data base. FEMA/USFA analyzes this data base and publishes the analysis. The data base is also available to businesses or individuals who wish to perform their own analyses.

Bibliography

NFPA Codes Standards, and Recommended Practices and Manuals. (See the latest *NFPA Codes and Standards Catalog* for availability of current editions of the following documents.)

NFPA 901, *Uniform Coding for Fire Protection.*
NFPA 902M, *Fire Reporting Field Incident Manual.*
NFPA 903M, *Fire Reporting Property Survey Manual.*
NFPA 904M, *Incident Follow-up Report Manual.*

Additional Readings

National Fire Incident Reporting System Handbook, Version IV, National Fire Information Council, as sponsored by the U.S. Fire Administration, Federal Emergency Management Agency, 1984. (Computer programs and system documentation available from the U.S. Fire Administration).

Schaenman, Philip S., "Data Collection, Processing, and Analysis," *Managing Fire Services*, Bryan, John L. and Picard, Raymond C., eds., International City Management Association, Washington, DC, 1979, pp. 455-499.

Uniform Fire Incident Reporting System, National Fire Protection Association, Quincy, MA, 1977. (A series of documents and computer programs for reporting and analyzing fire incident information.)

Weiner, Myron E., *Computers for Fire Departments: Current Technology and Typical Systems*, U.S. Fire Administration, Washington, DC, 1979.

USE OF FIRE LOSS INFORMATION

Dr. John R. Hall, Jr.

The preceding chapter examined techniques for collecting data on fires. This chapter addresses uses of that collected data in fire protection decision making and planning.

Since the last edition of this HANDBOOK, the quantity and quality of fire data available for analysis has grown at an unprecedented rate. In the early 1970s, even the best fire departments typically had fire incident records only in narrative form. Neither the method of storage (in file cabinets, by incident number) nor the method of description (phrases selected by fire officers, without standardized coded descriptions or even a standard list of characteristics to be documented) facilitated analysis of the community's leading fire problems. At the state level, very few states had any fire experience data base. At the national level, the National Fire Protection Association (NFPA) prepared estimates of U.S. incidents, deaths, injuries, and losses from its annual survey, and provided analyses of fire causes and other detailed factors from its Fire Incident Data Organization (FIDO) data base. The Fire Reporting Committee of NFPA had developed NFPA 901, *Uniform Coding for Fire Protection* (hereinafter referred to as NFPA 901), a standard fire incident reporting format, but few fire departments were using it.

In 1977, NFPA modified its survey techniques to take account of modern statistical design principles and, in the process, greatly improved estimates of the size of the national fire problem. At about the same time, the new National Fire Prevention and Control Administration (now the U.S. Fire Administration, a branch of FEMA, the Federal Emergency Management Agency) launched the National Fire Incident Reporting System (NFIRS). Based on NFPA's 901 standard, the NFIRS system provided the framework for a large scale, truly national fire experience data base. Since 1980, NFIRS has had the participation, in whole or in part, of the majority of states and receives standardized reports on roughly one-third of all fire incidents. (As sources of the principal fire data bases, NFPA and FEMA also have helped each other over the years, with FEMA providing financial support and technical suggestions on NFPA data bases and NFPA providing developmental work and analysis services on NFIRS for FEMA.) Other countries, such as Australia, have adopted their own versions of NFIRS, making possible much more valid international comparisons.

A parallel development has enlarged the capacity for analysis of the newly generated fire data. The advent of powerful microcomputers and the increasing affordability and flexibility of computer power in general have made a significant impact upon the fire service and other fire data users. This is the first edition of this HANDBOOK that could be written with the expectation that most readers will have access to valid national and local data bases in forms suitable for analysis, and to the equipment and skills required to analyze that data.

This chapter will first address the techniques of analyzing fire data, including examples demonstrating the amount of information we already have obtained as a result of the fire community's new data collection and retrieval capabilities. The major data bases now available to analysts then will be reviewed. See Table 2-3A for a sample of the many potential data users and uses.

USING DATA TO CHARACTERIZE THE FIRE PROBLEM

Table 2-3B lists ten major findings on the nature of the fire problem that have emerged from analyses of the past decade. Many of these analyses could not have been conducted without the new and refined data bases that emerged in the late 1970s. This list is not arranged in any order of priority, nor is it necessarily a list of what everyone would consider the ten most important data based findings. The list does, however, illustrate the ability of data and analysis to identify emerging problems, describe those problems in ways that can help the design of corresponding programs, and track the impact of those programs.

Top-Down and Topic-Driven Analysis

Most characterizations of the fire problem begin with some measures of the problem's size. Why? Implicit in these measurements is the idea that fire is a big enough problem to justify concern and attention. Each year fire

Dr. Hall is Director, Fire Analysis Division, for NFPA.

kills thousands of persons, injures tens of thousands more, and destroys billions of dollars worth of property. Many problems causing far less loss of life and property, such as floods and tornadoes, receive much more attention and have much greater resources devoted to the mitigation of their effects. When the fire community wishes to pursue a program that requires the cooperation of others, first they must attract their attention. The big *numbers* on fire—incidents, deaths, injuries and property damage—accomplish this.

Top-down analysis is an approach that begins with the big numbers on fire, then subdivides the totals into their major parts. The top-down approach provides the broadest perspective on the overall fire problem. It differs from *topic-driven analysis*, which begins with interest in certain issues involving only certain types of fires. For example, a top-down analysis of fires in residences might identify smoking materials as the leading factor in civilian fatalities in residences, accounting for one-third of the cases for which the fire cause was reported. On the other hand, topic-driven analysis focused on smoking-material fires—e.g., a study of the likely impact of a self-extinguishing cigarette or of a public education program aimed at smokers—would involve more than just residences. Although both topic-driven and top-down analyses might be interested in the role of smoking-related fires in residential deaths, the topic-driven analysis also might be interested in the involvement of smoking materials in nonresidential fire fatalities, or in property damage in residential fires. Even though smoking related fires do not account for the largest shares of the latter two problems (incendiary and suspicious causes lead in both cases), they are still part of the overall smoking-fire problem; any change that affects the smoking-fire problem may affect all smoking related fires.

In a top-down analysis, the basic rule is to look for the biggest parts of the problem. Why bother subdividing the problem? Why not simply attack the whole problem? The answer is that resources usually are too limited to permit an attack on the whole problem and choices must be made. The use of data to identify the largest manageable parts of the problem is the first step in making those choices. Where totals were used to show the overall size of the problem, percentages are used to indicate the largest parts.

Use of percentages to identify the largest parts of the fire problem is easier than it sounds. Judgment is involved in selecting the categories for which percentages will be calculated. For example, suppose an analysis showed the biggest subgroup of a particular occupancy group's fire problem consisted of fires involving electrical short circuits. This finding does not fit well into most strategies for attacking the fire problem. If one is going to examine every piece of equipment that could be subject to short circuits, one might as well also check for other electrical faults in that equipment. The available strategy may dictate a broader category, covering fires involving all electrical

TABLE 2-3A. Typical Users and Uses of Fire Experience Results

User	Use
1. Fire protection planners at national and state levels	Comparing fire to other problems competing for resources. Planning the most effective deployment of resources to make the greatest impact on the fire problem. Monitoring fire protection progress.
2. Local fire service	Targeting fire prevention and suppression programs. Backing up budget requests. Enacting and enforcing fire codes. Developing community firesafety education program. Monitoring progress.
3. Technical firesafety standards committees	Identifying needs for standards. Keeping standards current. Monitoring the performance of standards. Analyzing the benefits and costs of standards.
4. Communicators/educators	Reporting of fire problems. Educating for firesafety.
5. Committees developing fire fighter safety standards, individuals, fire departments working to improve fire fighter safety	Identifying reasons for fire fighter deaths and injuries. Developing safety and training programs. Designing protective equipment.
6. Researchers	Establishing firesafety research priorities. Designing research programs.
7. Insurers	Loss prevention. Risk selection and underwriting.
8. Industry	Worker safety. Loss prevention.
9. Product developers	Identifying needed products and markets. Modifying products to improve firesafety.
10. Building designers, architects, managers	Assuring a firesafe design.
11. Legislators, regulators, enforcers, courts	Effective and equitable firesafety laws and regulations.

equipment faults. Or maybe it is easier to design strategies around particular types of electrical equipment—e.g., fixed wiring, heating systems, ranges, and ovens—and address all fires involving such equipment, whether the equipment or the user were at fault.

Take another example. Which of the following findings is more useful: (1) 20 percent of all home fires begin in the kitchen; or (2) 15 percent of home fires involve cooking equipment? The latter finding identifies a smaller part of the problem, and uses a completely different categorization (equipment involved in ignition versus area of origin), but it locates the problem much more specifically. In a kitchen, there could be dozens of fire hazards to check out; with cooking equipment, most homes have only a few.

Finally, suppose a categorization includes equipment involvement, material ignited, and the human elements that brought them together. Such a categorization will produce detailed scenarios, but the "leading" category might account for only two to four percent of all fires. It would be necessary to develop strategies for several dozen scenarios in order to make any noticeable impact upon the total fire problem. If that is so, it may make more sense to select scenarios that fit together logically and can be addressed by variations of the same strategy—i.e., use of

TABLE 2-3B. Ten Major U.S. Data Based Findings of the Last Decade, 1974–1983

1. *Rapid growth in home smoke detector usage.* Four surveys of American households tracked the growth in smoke detector usage from roughly one home in 20 at the beginning of the 1970s to roughly three of four homes in 1984. National fire incident data, however, tracked a much slower growth in detector presence in homes with fires—less than one house in four in 1982.

2. *Sharp rise in home fires involving heating equipment, particularly wood burning equipment.* According to analyses of national fire incident data done by the Federal Emergency Management Agency's U.S. Fire Administration (FEMA/USFA), the number of fires involving solid fueled heating equipment (nearly all of it wood burning) in one and two family dwellings nearly doubled between 1977 and 1980. The increase from 1979 to 1980 was so great that it more than offset a decline in the number of fires due to all other causes combined.

3. *1970s growth in incendiary/suspicious fire problem reversed in 1980s.* During the decade that ended in 1977, the number of incendiary or suspicious fires in structures in the U.S. rose 254 percent to a peak of 177,000 fires, while the property damage due to those fires rose to nine times the 1968 dollar loss. Since 1977, the number of fires has declined every year but one to roughly two-thirds the peak total. Property damage peaked in 1980 and has declined since then, as average property damage per fire has become fairly stable. This decline probably reflects lower rates of inflation and a halt in the shift from the typically less costly vandalism/revenge motivated fire setting to the more costly arson for hire scenarios.

4. *Large share of fire fighter fatalities attributed to heart attacks and activities away from the fireground.* NFPA data showed that, since 1977 at least, the more familiar fire fighter fatality scenarios of burns, smoke exposure, falls, and structural collapse have been overshadowed by stress related heart attacks and accidents away from the fireground. These findings have implications for physical fitness testing and conditioning programs, and training in such areas as operating or riding on fire vehicles. Incidents during vehicle response to and return from a call contributed nearly one-fourth of all fire service deaths in 1983, compared to just over one-half that occurred at the firegrounds.

5. *Rural areas have the highest fire death and property loss rates.* Refinements in the NFPA survey techniques in the mid 1970s provided the first hard data on the size of the overall fire problems, hitherto hidden by the large number of small communities among which the fires are scattered. Based on 1977–80 data, the rate of fire deaths per million population in communities of less than 2,500 population is roughly double the rate in all other communities and is nearly 50 percent higher than the death rate in the largest cities (more than one million population). Property damage due to fire averages $10 per person per year in middle sized cities, $15 per person per year in the largest cities, and $40 per person per year in rural communities.

6. *Smoking materials igniting upholstered furniture, mattresses or bedding produce the most deadly residential fire scenarios.* In recent years smoking related fire deaths have accounted for nearly a third of all residential fire deaths where the cause of fire was known. Seven of every eight smoking related fire deaths where the form of material ignited was known involved upholstered furniture, mattresses or bedding. No other fire scenario of comparable specificity comes close to accounting for this large a share of the residential fire fatality problem.

7. *U.S. and Canadian fire death rates are highest of all developed countries in the free world.* In the late 1970s, a series of reports by Georgia Institute of Technology for FEMA showed that fire fatality rates per million population in the U.S.A. and Canada are at least twice the rates in Japan and western Europe (the only other areas examined). No single cause has been identified to explain this pattern, but analyses have shown that the U.S. does well in holding down the average severity of fires (i.e., quick response and suppression) but very badly in holding down the fire incident rate (i.e., prevention).

8. *The states of the Old South have the highest fire death rates.* The number of civilian fire deaths per million population in the states of the South is 50–100 percent higher than the rate in any other region. Only Alaska and the District of Columbia outside the South have comparably high rates. The leading factor in these high rates is the high death rate in fires involving portable and area heating equipment.

9. *Deinstitutionalization has severely worsened the problem of multiple death fires in boarding homes.* Mental hospitals began deinstitutionalizing chronic mental patients in the mid-1950s, and the effects of the move began to show up in boarding home fires in the late 1970s. During 1978–84, roughly 300 persons died in boarding home fires killing three or more persons. Residents of such homes have a risk of dying in a multiple death fire that is five times the risk for residents of other residential properties (measured as deaths in multiple death fires per million population).

10. *Some fire problems, including children playing with matches, are not as important as had been thought.* Fire prevention messages traditionally have devoted a substantial share of attention to fires caused by children playing with matches. Yet these fires account for less than one in twenty of U.S. fires and fire deaths. Playing with matches is not even the leading cause of fire fatalities for children, accounting for less than one fourth of all such deaths, however, it is the leading cause of fire deaths for very young children, accounting for three of every eight victims aged two or less. Fire data can be used not only to pinpoint previously unsuspected problems but also to reduce the perceived urgency and seriousness of what had been considered leading problems.

less specific categories—than to select several dozen distinct scenarios and tackle each one separately. To put it another way, the scenario technique of analyzing several fire characteristics simultaneously is most useful as a second stage analysis, performed after one characteristic has been used to establish basic clustering.

Therefore, in selecting categories for calculating percentages, it is important to match the structure of the categories to the structure of the strategies available to attack the fire problem. Remember that the size of the problem being attacked is no more or less important than the anticipated leverage on the problem, e.g., a 7 percent reduction in a problem that causes 500 deaths a year is worth just as much as a 70 percent reduction in a problem that causes 50 deaths a year. (This is discussed at greater length later in this chapter under the heading "Using Data in Program and Strategy Analysis.") So be sure when you pick out the biggest parts of the problem that they collectively account for a sizable share of the total, but also remember that some parts of the problem may be much more preventable than others.

Figures 2-3A and 2-3B illustrate the effective use of percentages (some implicitly presented through the display of numbers) to describe patterns in the 1983 fire fighter fatality problem. (These figures also illustrate the effective use of graphics to bring potentially dull numbers to life.) In Figure 2-3A, which shows the percentage of fire fighter fatalities by type of duty, the fact that "Fire Fighting" is the leading type of duty is not as interesting as the fact that its lead is so small. "Response/Return," which should involve only controllable dangers and relatively little risk, accounts for nearly one-fourth of all fire fighter deaths. Figure 2-3B implicitly demonstrates that as fire fighter age increases, the leading share of deaths dramatically shifts to heart attacks. Such patterns have direct implications for targeting of physical fitness programs.

Analysis by Cause or Property Type

The dimensions most frequently chosen for top-down and topic-driven analyses are fire cause and property type, because most fire prevention strategies are structured along those lines. "Analysis by cause" actually refers to a variety of analyses addressed to different aspects of the ignition factors. The question of how to categorize, already

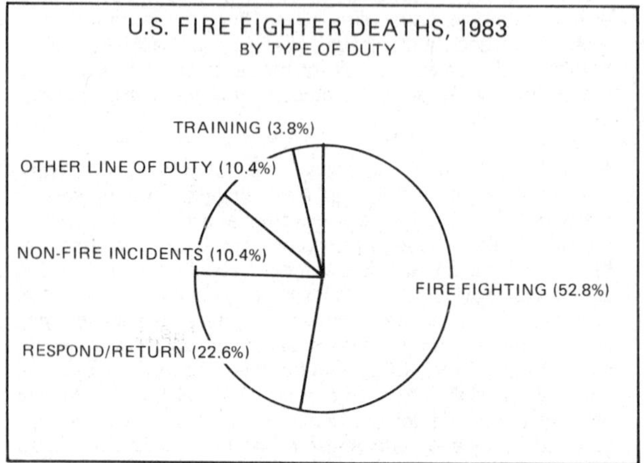

FIG. 2-3A. U.S. Fire Fighter Deaths, 1983.

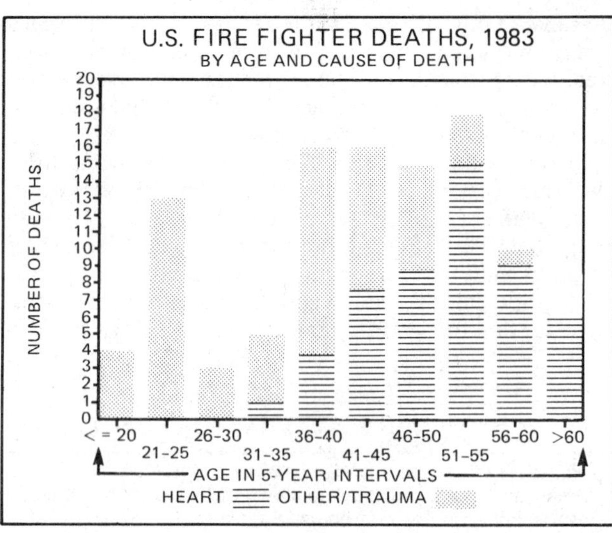

FIG. 2-3B. U.S. Fire Fighter Deaths, 1983.

discussed in general terms, becomes especially difficult in analyses of cause related information.

Even the question of how to refer to cause based descriptions can be a problem. For example, suppose one observes that roughly one-third of all civilian fire fatalities in residential properties with fire cause reported were smoking-related. Does this mean that cigarettes (the principal smoking material involved, by far) were to blame for all those deaths? Not necessarily. If the smokers in question had been careful not to smoke in bed and to always place cigarettes in sturdy ashtrays, few of the deaths would have occurred. Or if the bedding, mattresses, pillows and upholstered furniture that are the first items ignited in most smoking related fires had been more flame resistant, many or most of the deaths would have been avoided. It is also true that if the cigarettes used had been designed to self-extinguish more quickly, many or most of the deaths would not have occurred. A decision on which dimension to use in categorizing fire causes—the source of heat, items ignited, or the behaviors that brought them together—should be made in a way that supports analyses of workable prevention strategies. Assessments of blame are tasks for lawyers or philosophers; they do not help in the search for ways to reduce fire problems.

Fire incident data coded according to NFPA 901 format, which is used in most major fire data bases, contains five fire cause related elements of information: the form of heat producing ignition; equipment involved in ignition, if any; type of material that ignited (e.g., wood, plastic, fabric); the form in which that material appeared (e.g., chair, floor covering, structural beam); and ignition factors—the human or equipment failures that brought together the ignition heat and ignited material. Together, these five dimensions can define billions of distinct fire-cause categories, a clearly unworkable partitioning of the problem. In the face of all this detail, several approaches have been developed to extract major patterns. See Table 2-3C for examples of how different approaches can affect the description of "leading cause."

One approach is to divide fire causes into incendiary, suspicious, all known causes that are accidental in nature, and unknown. For many property types, such a partition sufficiently identifies the major fire causes to justify concentrating on arson problems. If not, a different approach is needed.

A second approach has been to use one of the NFPA 901 cause dimensions to sort the fires. One can use the first digit alone to create 10 categories, or both digits to create nearly 100 categories for each dimension. This approach often proves frustrating, however, because the structure of the coding elements often does not coincide with the issues to be analyzed. For example, suppose the Ignition Factor dimension is used. Incendiary and suspicious fires will be easy to identify. Fires caused by children playing, however, involve two code values (36 and 48) that do not have the same first digit and therefore do not fall into a natural common grouping. An analysis using "Ignition Factors" also will produce a large number of entries under "abandoned, discarded material." A novice analyst might not recognize that these are nearly all smoking related fires and can best be characterized in this manner.

If "Ignition Factor" has problems when used by itself, what about the other dimensions? Problems arise here because incendiary and suspicious fires are not isolated. Under "Form of Heat of Ignition," arson fires set with matches (the most common type) are indistinguishable from accidental fires involving matches.

A third approach has been to use scenarios based on two or three of the NFPA 901 cause dimensions. This avoids the tendency, in the one dimensional sort, for major problems to be mixed together (e.g., arson and accidental match fires) or disguised (e.g., smoking-related fires listed as discarded or abandoned material). The use of even two dimensions of NFPA 901, however, or the use of three dimensions with first digit only, creates thousands of cause categories, and again one is faced with the likelihood that even the largest parts of the total will not represent a very large share of the total.

The fourth approach has been to use a small number of major, easily recognizable categories based on NFPA 901 data elements, but not necessarily grouped along the lines laid out in the NFPA 901 coding structure. The most widely used scheme based on this approach is a set of approximately a dozen categories developed at the USFA. This scheme uses a hierarchical process to sort fires into categories. For example, one will start with "Ignition Factor" and remove all the incendiary, suspicious, and children-playing fires. Then the analyst may switch to "Form of Heat of Ignition" and use it to separate fires involving smoking materials.

Criticisms of this or any other hierarchical sorting approach focus on the reasonableness of the sorting priorities and the usefulness of the categories produced. If a fire is caused by a child playing with cigarettes, should that be categorized as a "children-playing" fire or a "smoking related" fire? Is it useful to have a category called "Open Flame" that combines matches, bonfires, undoused embers, and cutting and welding torches? For most of the categories, the number of fires falling into each category is not greatly dependent upon the sorting order. The most striking exception is match-related fires, which are scattered among the incendiary/suspicious, children-playing and open flame categories. As for usefulness, the categories that seem to be most mixed are also categories that account for relatively little of the fire problem in residential properties, the setting for which the categories were designed to be most helpful. For nonresidential properties, particularly manufacturing and industry, the usefulness of the categories may be more questionable.

The last approach, which can be used as a refinement of any of the others, is nesting of cause categories. Start

TABLE 2-3C. How "Leading Cause" of Fires Can Depend on How Data is Sorted*

Emphasis on Accidental vs. Intentional		*Emphasis on Heat Source*	
Incendiary	9%	Smoking materials involved	32%
Suspicious	5%	Electrically powered equipment involved	24%
Accidental	77%	Matches involved	11%
Unknown	9%	Fueled equipment involved	7%
		Other known	13%
		Unknown form of heat	11%
Emphasis on Equipment		*Emphasis on Item Ignited*	
No equipment involved	44%	Rubbish or trash	21%
Cooking equipment involved	10%	Bedding, mattress, pillow or linen	21%
Appliances involved	8%	Electrical wires	11%
Electrical system involved	8%	Cooking materials	6%
Other known equipment type	14%	Supplies or stock	6%
Unknown equipment	16%	Other known	28%
		Unknown form of item	6%
Emphasis on Behavior		*Hierarchical Sorting Approach*	
Abandoned material	26%	Smoking related	32%
Mechanical failure or malfunction	20%	Incendiary or suspicious	14%
Other misuse of heat source	15%	Cooking equipment involved	10%
Incendiary or suspicious	14%	Appliances involved	8%
Operational deficiency	8%	Electrical system involved	8%
Other known	8%	Matches, lighters, other open flame	6%
Unknown ignition factor	9%	Other equipment	13%
		Other known or unknown	10%

* Figures in some groupings may not sum to 100 percent because of rounding errors.
Source: 1980–82 NFIRS data on hospital fires.

TABLE 2-3D. Ignition Scenarios of U.S. Properties, 1974–1983

Equipment Involved	Heat of Ignition	Ignition Factor	Percent of Fires	Percent of Dollar Loss
Fixed wiring (37%) Lighting fixtures (28%) Panelboards, etc. (10%) Other electrical (24%) Unknown (1%)	Electrical equipment arcing, overloaded	Mechanical failure, malfunction	36	43.9
	Open flame or incendiary device	Incendiary	28	25.9
Heating (87%) Other equipment (13%)	Fuel fired fuel powered equipment	Mechanical failure, malfunction	5	4.6
	Open flame or incendiary device	Suspicious	5	3.7
	Cigarettes	Carelessly discarded	4	3.5

Based on 280 fires with known ignition scenarios each resulting in a loss of $500,000 or more reported in the NFPA Fire Incident Data Organization (FIDO) data. All losses have been adjusted to 1974 dollars using the Producer Price Index; total adjusted loss was $229,743,000.

Out of 41 distinct scenarios in this data set, these five scenarios accounted for 78 percent of the fires and 81.6 percent of the dollar loss.

with an initial sorting of fires based on either a single dimension of NFPA 901, a scenario structure, or a hierarchically based structure. Then, for each category of the initial sort that has a sufficient number of fires to be worthy of further analysis, use another dimension of NFPA 901 to provide more detail. Table 2-3D shows this nesting approach applied to an initial scenario sort for large loss fires in store properties during 1974-83. Nesting represents the top-down approach at its best—it provides usable levels of detail without sacrificing an overview of the largest parts of the total.

Another matter to consider when analyzing fire data by cause is how to handle incidents for which the cause was unknown or unreported. A sizable share of fires are reported without cause information, particularly severe fires involving deaths or significant loss. Note that several points in this section have been illustrated with the observation that roughly one-third of all civilian fire fatalities in residential properties with known cause involved smoking materials. Just under one-fourth of all civilian fire fatalities in residential properties were coded as involving smoking materials. A calculation of percentages based only on cases where cause was reported implicitly assumes (in the absence of contrary evidence) that the cause profile of the fires reported without known cause would look the same (if those causes were known) as the cause profile of the fires reported with known causes. (The two types of percentages are sometimes referred to as causes based on "allocating unknowns over known causes" and causes based on "unallocated unknowns.")

Why is it desirable to allocate unknowns? Suppose two neighboring states have relative cause profiles that look the same, except that fire deaths due to unknown causes appear as 10 percent in one state and 40 percent in the other. Table 2-3E shows the results. State B seems to be doing better than State A with respect to every known cause, but if unknowns are allocated, both states have the same problems to the same degree. If it is necessary or desirable to compare two groups, differences in rates of unknowns need to be taken into consideration.

Why is it difficult to allocate unknowns? Many knowledgable individuals believe that assessments of probable cause are subject to persistent biases. In fact, no credible evidence, consistent among a variety of fire departments, exists to support the notion that fire officers consistently label fires as suspicious, smoking-related, electrical in origin (all very popular suspicions in certain quarters) or anything else when they really are not sure of the cause.

TABLE 2-3E. The Problem of Comparisons with Unknown Cause Fires Unallocated

	Percentage of Civilian Fire Fatalities in Residences	
Cause Category	State A	State B
Smoking	36%	24%
Heating	18%	12%
Incendiary/Suspicious	18%	12%
Other Known	18%	12%
Unknown	10%	40%

A more serious argument goes as follows: Probable cause is most difficult to assess in the largest fires because more of the evidence of fire origin is destroyed, and the known-cause fires indicate that cause profiles are different for larger versus smaller fires. For example, incendiary/suspicious and electrical distribution system fires tend to have greater loss per fire, and incendiary/suspicious and smoking-related fires are more likely than other fires to result in deaths. Therefore, if unknowns are to be allocated, the allocation should take into account the cause profiles for fires of comparable size. This is a fire data analysis issue around which no consensus has yet formed. While the ideal solution would be to find ways to reduce the rate of unknowns, there also may be better ways to arrive at estimates in the face of the unknowns that do exist.

Analyses by property use are much more straightforward. Most analyses are concerned with distinctions covered by the "Fixed or Specific Property Use" or "Mobile Property Type" categories of NFPA 901. About the only major property class requiring use of both scales is mobile homes, and there is little difficulty with unknown or unreported property types. However, unknowns are a data problem for many other fire characteristics, not just cause and property type. Many questions analysts would like to have answered on all reported fires are not asked at all, because fire officers cannot be expected to be able to answer them.

A problem can arise in the size of the data base available for analyses aimed at particular property uses. In 1983, according to NFPA, all nonresidential structures combined accounted for 227,000 fires, 270 civilian deaths, 4,700 civilian injuries, and $2.52 billion in property damage. Typically, approximately one-third of all fires are represented in the NFIRS, which has individual incident reports and can be used to analyze patterns by specific property use. There are, however, several hundred separate categories of nonresidential properties. A topic-driven analysis focused on one of them might be able to draw only upon a few dozen incidents per year, which are not enough to generate statistical confidence in the results. An analyst who discovers that there is a very small data base available on the property use of interest should do two things. One is to keep the analysis of these fires fairly limited; if each fire represents 2 to 3 percent of the total number of fires being analyzed, there is no point in conducting fine structure analysis. The other is to reconsider the appropriateness of focusing on this particular property use. If it does not have a large data base it is not suffering many fires. Even the most successful strategy applied to so small a problem can have only minor impact. There may be more appropriate classes of property in which to attack the fire problem.

Rates and Measures of Fire Risk

Having addressed the usefulness of numbers and percentages as measures of the fire problem, we can turn to the third and last type of measure, which is rates. A rate is a ratio consisting of a measure of the size of a fire problem divided by a measure of the size of the group affected by that problem. The U.S. suffers on the order of 20 civilian fire deaths in the home per million persons. That is a rate consisting of the number of home fire deaths divided by the number of persons who live at home. The average fire related property loss in residential properties in rural communities is $40 per person per year. That is a ratio consisting of the total property damage in residential properties in rural communities divided by the number of persons living in those communities.

Rates provide measures of relative fire risk. They can be used, therefore, in any analysis where the size of the group affected by a problem may change. For example, as has been said, the U.S. suffered about 20 civilian fire deaths in the home per million persons in 1983. Because the total population growth adds about two million persons to the population each year, the number of civilian fire fatalities will be about 40 deaths higher in 1984 than in 1983 unless increased firesafety lowers the rate. Along similar lines, the National Safety Council's estimates of fire deaths show total deaths declining by 32 percent from 1971 to 1983, while the fire death rate per million persons declined 39 percent in the same period. Increased firesafety is best measured by the decline in the fire death rate; population growth produced a smaller decline in the death toll.

Risk measures do the best job of bringing the fire problem down to a personal level. Rural communities do not account for a majority of the country's fire deaths, but they have by far the highest fire death rates compared to communities of larger size. Person for person, their citizens are in the most danger from fire. Occupants of mobile homes suffer a substantially higher rate of fire fatalities per million persons than do occupants of conventional one and two family dwellings. Because there are comparatively few mobile homes, deaths there do not constitute a large share of the total fire fatality problem, but the individuals living in those mobile homes are more at risk than their counterparts elsewhere and should be concerned. If you as an individual want to know how much and what kind of danger you face from fire, you want to know a measure of risk for people like yourself.

Risk measured by rates is of particular interest if there is the potential for large scale shifts from one group to another of substantially higher or lower risk. In the early 1980s, analysts determined that solid fueled heating devices, principally wood burning stoves, accounted for an unusually high risk, i.e., there was a high ratio of fires involving those devices divided by homes using those devices. At the time, it appeared that the generation long decline in use of these devices was to be reversed, under the pressure of higher prices for oil and natural gas. This was an ideal opportunity to use data to guide action; at a time when decisions were being made, most people did not know the unusual dangers posed by their choices. As an example of risk moving in the other direction, analysts using fire incident data from the early growth years of smoke detector usage were able to demonstrate that the rate of deaths per home fire was half as great for homes with detectors as for homes without them. Here again, decisions with substantial impact on risk were being made regarding equipment, so it was possible to get people's attention and encourage the move toward detectors.

For some analyses, it may be difficult or impossible to obtain the information needed to construct the appropriate rate. One example is an analysis comparing the dangers of fire fighting with risks of other professions. The ratio should be number of on duty fire service deaths to number of exposed fire service persons. No one knows how many volunteer fire fighters there are, however. Even if the number were known, some allowance should be made for the much lower number of hours per person volunteers spend on duty; but then that might mean differences in work week should be considered for career fire service personnel.

Another example is a comparison of fire fatality risks in homes versus hotels. Numbers of persons cannot be used for both, because length of stay is obviously much shorter for hotels than for homes. Numbers of units might work better if one could obtain that data and if occupancy rates could be ignored. The problems of selecting the right denominator and obtaining data for it usually can be solved to some degree of acceptability, but the process is often complex and challenging.

Another problem in constructing rates can occur if the group of interest is defined too narrowly. A household, for example, might like to be able to calculate a risk index that reflects its smoking and drinking habits, its number of children, the ages of its members, its types of equipment and how they are fueled or powered, and many other characteristics. Today this would pose an insoluble problem. There is not enough data to calculate the numbers of households with all those characteristics (so there can be no denominator), and the list of descriptors defines a class so narrow that the number of fires experienced by households of that precise type would be very small and subject to considerable statistical uncertainty. One could separately measure the effects of some of these characteristics, and then combine the results, but such a calculation would involve modeling assumptions that are already known to be false. Considerable statistical modeling and research will be needed to make progress on questions of this type.

USING DATA TO IDENTIFY TRENDS

While the analyses described above provide a "snapshot" of the fire problem, some questions need a "moving picture." Those questions demand trend analysis. Is a part of the fire problem getting better or worse? Is the character of the fire problem changing? If changes are occurring, do they track with corresponding changes in product use, property use, fire service practices, fire prevention activities, codes and regulations, or other elements of the environment? (For this last question, one may need a full fledged strategy or program analysis, as described below under "Using Data in Program and Strategy Analysis.")

All trend analyses must address the question of whether or not the past is a consistent guide to the present, let alone the future. It would be nice to be able to say that a trend indicates whether fire risk is going up or coming down. However, a simple trend calculation may not provide this information if the base of comparison has been shifting during the years covered. For example, start with a calculation of the number of fire deaths per year in mobile homes. If the number dropped over time, that might mean that mobile homes were becoming safer or it might mean that fewer people were choosing to live in mobile homes. A trend calculation based on the number of fire deaths per million persons living in mobile homes would factor out the latter possibility and show unequivocally whether mobile homes were becoming safer.

Another problem in trend calculations might be a shift in the sampling technique used to create the data base. Suppose that while analyzing fire deaths per million population, as in the previous paragraph, the state of Alaska was added to the data base in the middle of the trend period. The increase in the number of persons would not by itself bias the results because, all other things being equal, numbers of deaths and persons would rise together. But in this case, all other things are not equal. Alaska has an unusually large percentage of households living in mobile homes, so this will exert a sizeable influence on the results, and it presents some additional problems (including a severe climate and above average rates of alcoholism) that differ from those of other states. There is no way to know how much of any change was due to changes in mobile home firesafety and how much was due to the addition of Alaska to the data base. In the discussion below of "Available Major Data Bases on Fire," more will be said as to which data bases are better designed for trend analysis.

Trends in Fire Fatalities and the Protection of Life

Figure 2-3C shows the seven year trend in civilian fire deaths in the home starting with 1977, when the NFPA survey was upgraded. The NSC, which since 1949 has been using a consistent estimating procedure based on

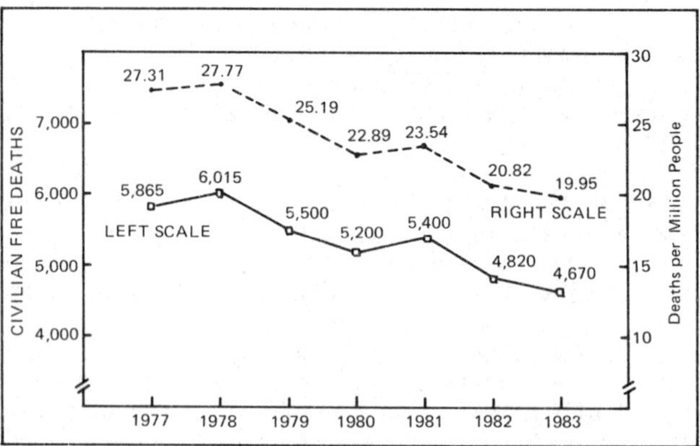

FIG. 2-3C. Civilian fire deaths and rates in the home, 1977-1983.

state health department records of death certificates, shows a steady decline in the rate of fire deaths per million population. The cumulative drop since 1949 is more than half, and other NSC figures, using slightly different estimating methods, suggest that since 1910 the rates may have dropped by three-fourths to four-fifths. (Note that the NSC method produces a different, lower estimate of fire deaths than the NFPA method.)

The most complete and consistent long term trends of life loss due to fire are those that focus on fires involving large numbers of fatalities per incident. Figure 2-3D traces the rate of deaths per 10 million persons in building fires killing at least 25 persons each, at 5 year intervals from 1900 through 1984. A cutoff of 25 fatalities is used because such fires are the ones most likely to involve factors addressed by fire codes and also because even the poor reporting conditions of the early years are likely to have produced a complete listing of incidents of this magnitude.

Milestones during the century include the formation in 1913 of NFPA's Committee on Safety to Life which produced several standards addressing exiting provisions and firesafety in factories. The advent of these standards in the 1910s was followed by a marked decline in life loss in major incidents in the 1920s.

These gains were wiped out, temporarily, during the years of the Great Depression and World War II. Fires directly related to wartime hazards did not play a major role in this increase in life loss, but there may have been a diversion of resources from fire code enforcement and compliance to wartime production priorities. (Similarly, the loss of resources during the Depression years may have had a disproportionate impact on resources available for code enforcement and compliance.)

Following World War II, the rate of deaths per ten million persons in building fires killing 25 or more persons dropped to a new low. The rate is so low now that a single fire like the 1977 Beverly Hills Supper Club fire can double or triple the rate for the entire five year period. In other words, the random effects of a single fire are now sufficient to overshadow any trend, because our progress has been so great.

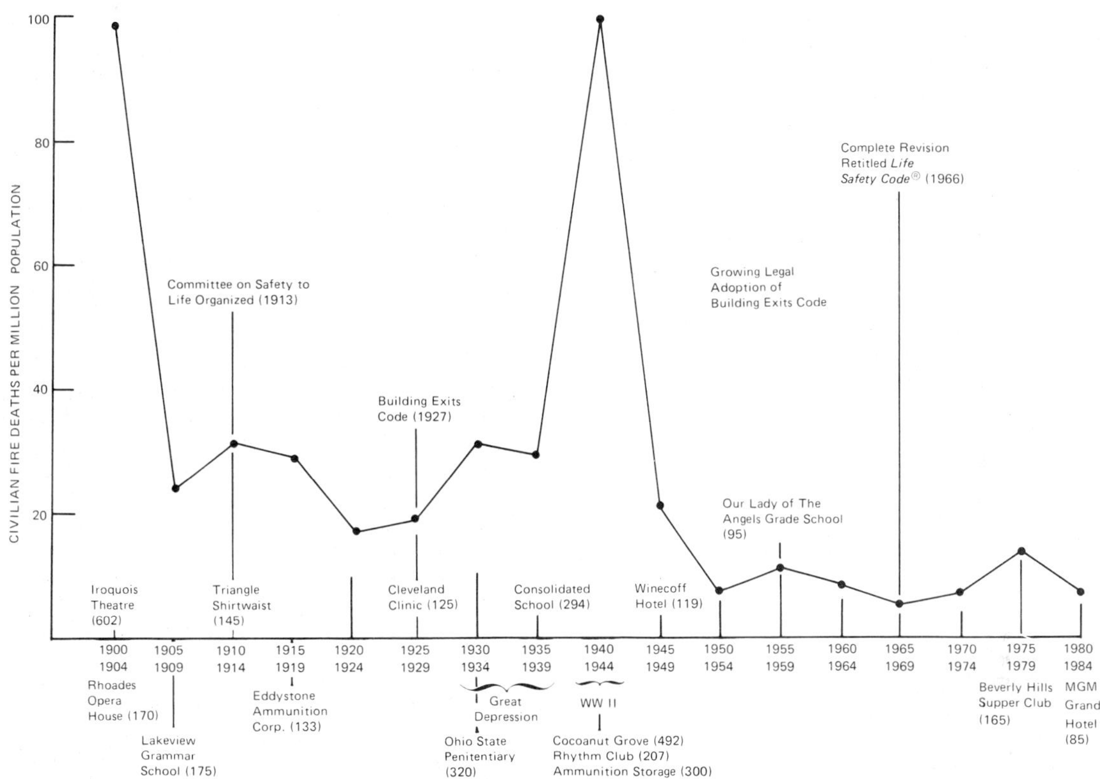

FIG. 2-3D. Major loss of life in 20th century U.S. building fires taking at least 25 lives.

Table 2-3F provides another view of the same trend. This table shows the three incidents in each of the last nine decades that involved the largest loss of life, plus any additional incidents that killed at least 100 persons. Note how this table reinforces Figure 2-3D's picture of dramatic reductions in loss of life in major multiple death fires. The last three decades have seen one fire that killed at least 100 persons—the Beverly Hills Supper Club fire in Kentucky. The previous three decades (1924-1953) had 13 fires that size. The three prior decades (1894-1923) had 9 fires that each killed at least 100 persons. Also, the last three decades have had no fires that killed at least 200 persons, while the preceding six decades averaged two such fires per decade.

Several of the fires on this list provided the impetus for code changes that reduced the likelihood of further incidents of that size. NFPA's earliest standards included a number with obvious roots in particular turn of the century fires. The 1912 standard on exit drills in factories, schools, department stores, and theaters had its origins in the four building fires listed under the two decades 1894-1913. New York City's Triangle Shirtwaist Company fire in 1911, in particular, also spurred the adoption of NFPA's 1918 standard on safeguarding factory workers from fires.

More recently, the 1944 circus tent fire in Hartford, CT, focused attention on unconventional structures that were not buildings but still involved major life safety hazards. NFPA 102, *Standard for Assembly Seating, Tents and Air-Supported Structures*, was developed to address hazards in these structures. In the same decade, further refinements were made to the NFPA code on building exits as a result of fires like the Winecoff Hotel tragedy in Atlanta, GA, in 1946 because the old code had not been widely adopted into law.

The ability to develop codes that will be translated into law and will receive vigorous enforcement is a key to fire prevention success. The list of major fires points to several instances where this kind of thorough follow through has occurred. The devastating fire in Chicago's Our Lady of the Angels Grade School in 1958 was the last of five major life loss school fires that appear on the list. The specific impact of that fire was an increase in code requirements, including the addition of requirements on interior finishes in educational occupancies, reflecting the role of combustible ceiling tiles in the Chicago fire, which involved a building that met the less restrictive codes of its day. Equally important, this fire galvanized the fire service and motivated the educational community and the public to provide the resources necessary to ensure tough standards and close, continuing safety supervision of schools. School fires today kill comparatively few persons. It now takes decades to accumulate a school death toll comparable to that of the Our Lady of the Angels fire.

Just as the 1960s was the decade for a strong attack on the school fire fatality problem, the 1970s was the decade for similar attention to fatal nursing home fires. The 1963 Golden Age Nursing Home fire in Ohio was the third largest of the decade of 1954-1963, and the Marietta, OH, nursing home fire in 1970 was the second largest life loss building fire of the decade of 1964-1973. That decade saw most large life loss incidents occurring in nonbuilding structures and in vehicles. Requirements of NFPA 101® *Life Safety Code* (hereinafter referred to as NFPA 101)

TABLE 2-3F. Deadliest U.S. Fires of the Last Nine Decades, 1894–1983

	Number Killed
1974–1983	
1. Beverly Hills Supper Club, Southgate, KY (May 28, 1977)	165
2. MGM Grand Hotel, Las Vegas, NV (Nov. 21, 1980)	85
3. County Jail, Columbia, TN (June 26, 1977)	42
1964–1973	
1. Silver mine, Kellogg, ID (May 2, 1972)	91
2. Missile silo, near Searcy, AR (Aug. 9, 1965)	53
3. DC8, Anchorage, AL (Nov. 28, 1970)	47
1954–1963	
1. Our Lady of the Angels Grade School, Chicago, IL (Dec. 1, 1958)	95
2. Indiana State Fairgrounds Coliseum, Indianapolis, IN (Oct. 10, 1963)	74
3. Golden Age Nursing Home, Fitchville Township, OH (Nov. 23, 1963)	63
1944–1953	
1. S.S. Grandcamp, Texas City, TX (Apr. 16, 1947)	468
2. Munitions depot, Port Chicago, CA (July 17, 1944)	300
3. Ringling Brothers Circus, Hartford, CT (July 6, 1944)	168
4. East Ohio Gas Co., Cleveland, OH (Oct. 20, 1944)	136
5. Winecoff Hotel, Atlanta, GA (Dec. 7, 1946)	119
6. C.W. & F. Coal Co., West Frankfort, IL (Dec. 21, 1951)	119
7. Centralia Coal Co., Centralia, IL (May 25, 1947)	111
1934–1943	
1. Cocoanut Grove Lounge, Boston, MA (Nov. 28, 1942)	492
2. Consolidated School, New London, TX (Mar. 18, 1937)	294
3. Rhythm Club, Natchez, MS (Apr. 23, 1940)	207
4. S.S. Morro Castle, off New Jersey coast (Sept. 8, 1934)	135
1924–1933	
1. Ohio State Penitentiary, Columbus, OH (Apr. 21, 1930)	320
2. Cleveland Clinic, Cleveland, OH (May 15, 1929)	125
3. Babbs Switch School, Hobart, OK (Dec. 24, 1924)	36
1914–1923	
1. Forest fire, MN (Oct. 12, 1918)	559
2. Eddystone Ammunition Corp., Eddystone, PA (Apr. 10, 1917)	133
3. Cleveland School, Beulah, SC (May 17, 1923)	77
1904–1913	
1. General Slocum excursion steamer, New York, NY (June 15, 1904)	1,030
2. San Francisco, CA, earthquake and fire (Apr. 18, 1906)	500
3. Lakeview Grammar School, Collinwood, OH (Mar. 4, 1908)	175
4. Triangle Shirtwaist Co., New York, NY (March 25, 1911)	145
1894–1903	
1. Iroquois Theater, Chicago, IL (Dec. 30, 1903)	602
2. Steamship, Hoboken, NJ (June 30, 1900)	326
3. Rhoades Opera House, Boyertown, PA (Jan. 12, 1903)	170

were expanded to cover such hazards as carpets, and the Federal government provided a major impetus to code compliance by including NFPA 101 in its requirements for Medicare and Medicaid eligibility.

The target occupancies of the 1980s probably are the board and care homes that have changed their character as a result of the widespread deinstitutionalization of chronic mental patients. Persons needing care are coming in large numbers to facilities not designed for this purpose. Already these facilities have suffered dozens of multiple-death fires, including 2 that killed at least 25 persons each—1 in Farmington, MO, in 1979 (25 deaths) and 1 in Keansburg, NJ, in 1981 (31 deaths).

A common element of the worst life loss fires of recent years has been the involvement of properties that are on the fringes of code enforcement. The 1965 Searcy, AR, missile silo fire and the 1977 Columbia, TN, county jail fire both involved government properties, which typically are outside the jurisdiction of local fire departments. The 1977 Beverly Hills Supper Club fire, the 1963 Golden Age Nursing Home fire, and the 1972 silver mine fire in Idaho all occurred in small communities where resources for fire code enforcement typically are very limited. These patterns suggest a need to focus attention as much on innovations in the code enforcement process as on modifications to the consensus codes themselves.

Event driven changes in the fire codes primarily involve the fire community. However, getting those changes adopted into law and backing the law with strong enforcement programs require the cooperation of elected officials and the public—who are concerned about many subjects in addition to fire. Recognizing this, some fire chiefs have prepared legal packages targeted at particular problems, then held them in reserve until a major fire provides the visibility and sense of urgency required for adoption.

In view of this strategy, it is worth noting that "newsworthiness" is not always synonymous with incident severity. Two examples will suffice. The third most deadly fire of the decade of 1954-1963 was the Golden Age Nursing Home fire, which killed 63 persons; it occurred, however, on the day after President John F. Kennedy was assassinated. This coincidence of timing sharply reduced the visibility of the fire. By contrast, the 1937 "Hindenburg" zeppelin fire killed "only" 36 persons, a death toll barely a third the size of even the fourth largest fire of that decade and only one-eighth the size of the New London, TX, school fire in the same year. The zeppelin fire, however, was broadcast "live" and occurred in full view of newsreel cameras, so it may be the most widely heard and seen fatal U.S. fire of all time. It meant the end to the rigid airship industry.

These examples also have a lesson for analysis of trends in fire deaths, injuries, or incidents. The more severe the type of incident being tracked, the more random year to year fluctuation can be expected. This makes it important to look at trends over an appropriate length of time. For fires as severe as those discussed in the last few paragraphs, only decade by decade comparisons suffice to separate real trends from random ups and downs. For total U.S. civilian fire deaths or even NFPA's annual study of multiple death fires (those killing three or more persons), year to year tracking is meaningful. For a single large city, however, year to year tracking of fire deaths will be subject to considerable statistical noise; and for town and rural areas, there is no good way to track fatality trends except as part of a larger aggregation, such as the state or the rural portion of the entire region.

Trends in Property Damage Due to Fire

The first fire protection standards promulgated by NFPA were directed principally at property protection, reflecting the Association's founding by men in the business of insuring property against loss due to fire. Assessments of the success of strategies for property protection are more difficult than for life protection, however, because data on fire damage are subject to several gaps and quirks that do not affect data on fire deaths. While all indicators of fire—incidents, deaths, injuries, and property damage—are clouded by the absence of information on fires not reported to local fire departments, property damage is the only indicator of severity that is left unreported in a significant minority of reported fires. Estimates of property damage involve more guesswork than estimates of persons killed or injured. Some very large loss fires are never reported to local fire departments, because the affected firms are able to handle them with on site resources. No comparable omissions occur among multiple death fires.

Trend analysis must take into account the effect of inflation. Table 2-3G shows NFPA's estimates for total

TABLE 2-3G. U.S. Trends in Property Damage Due to Fire, 1979–1983

Year	Total Estimated Damage	Adjusted to 1979 Dollars
1979	$5.750 billion	$5.750 billion
1980	$6.254 billion	$5.509 billion
1981	$6.676 billion	$5.328 billion
1982	$6.432 billion	$4.845 billion
1983	$6.598 billion	$4.823 billion

national property damage due to fire in the years 1979-1983, both as published and as recalculated in terms of 1979 dollars, using the U.S. Bureau of Labor Statistics' Consumer Price Index to remove the effects of inflation. When inflation is not removed, the loss totals, expressed in what are called "constant dollars," show a 15 percent increase in property damage in the four years from 1979-1982. When inflation is removed, the loss totals, now expressed in "real dollars," show a 16 percent decrease in property damage for those four years. Also significant is the effect of changes in the number of fires, because total damage figures involve the total number of fires and the average dollar loss per fire. From 1979 to 1983 the average loss per fire in constant dollars rose 40 percent, but the average loss per fire in real dollars rose only 2 percent.

To put this even more dramatically, suppose an energetic, creative, intelligent, far sighted fire chief had begun a massive ten year program in 1972 to reduce by half the property damage due to fire in his city. Then suppose he had succeeded. Because of the effects of inflation, his actual reported loss would have shown a nine percent increase in property damage over that decade.

This same effect makes it difficult to compare large loss experience over several years. In the early 1930s, the NFPA defined a large loss fire as one involving at least $250,000 in direct property damage. In 1978, a large loss fire was redefined as one involving at least $500,000, and in the early 1980s the amount became at least $1,000,000 in direct property damage. Each minimum represented the cutoff point for the largest several hundred fires of its time.

Table 2-3H lists the 20 largest loss fires of 1972-1981, first on the basis of their published losses in constant dollars, then on the basis of real 1981 dollars. Four fires on the second list do not appear on the first list. In those cases, four fires from the beginning of the decade (1974 and 1975) are calculated to have been more costly, in real terms, than four fires from the end of decade (1979 and 1981) with higher reported losses. The third largest fire on the first list, a 1975 fire involving a telephone exchange, is at the top of the second list. The eleventh largest fire on the first list, a 1973 fire involving an ink manufacturer, is the third largest fire on the second list. The Louisville sewage system fire that tied for eighth on the first list is last on the second list.

Since the end of the period covered by these two lists, two fires have had higher reported losses than the largest fire on the list—a 1982 retail warehouse fire, estimated at $100 million in loss, and a 1982 bank building fire with losses estimated at $91-93 million. Yet barring any later adjustments in the loss estimates for these fires, neither was as costly, in real dollar terms, as the 1975 telephone exchange fire.

Difficulties in comparison become more pronounced when points even further back in history are considered. The 1947 chemical plant fire in Texas City, TX, had a reported loss of $60 million. Using the Producer Price Index, that translates into more than $200 million today. The 1906 earthquake and fire in San Francisco had a reported loss of $350 million. That could mean on the order of $3 billion today—or nearly half the loss attributed to all reported fires in the United States. At this point, comparisons can become very sensitive to assumptions. Was the property lost in the Texas City fire sufficiently similar to the items covered by the Producer Price Index? Or should a more industry specific price index be used? Were the estimates of loss for the San Francisco fire made to the nearest $10 million or even the nearest $50 million? Then the 1980s version of that loss might be accurate to only the nearest $100 million or $500 million, which would mean its range of uncertainty would be higher than the total loss of any recent fire.

The lesson in all this is that trend analysis of loss figures can be very tricky and needs to be done with an eye toward the effects of inflation and population growth (which also affects trends in numbers of fires, deaths, and injuries). Keep in mind that loss trends have a tendency to look bad even if real progress is being made.

USING DATA IN PROGRAM AND STRATEGY ANALYSIS

Earlier in this chapter, techniques were described for use of data to analyze the present (size and characteristics of the fire problem) and the past (trends in the size and nature of the fire problem). Program and strategy analysis is the most decision relevant use of data because it tries to project the future, and, in particular, the ways in which the future will be different if a particular program or strategy is or is not adopted.

TABLE 2-3H. Largest-Loss U.S. Fires Reported to NFPA, 1972–1981

Constant Dollars		
1. Tank ship for crude oil, Deer Park, TX	(Sept. 1, 1979)	$ 77 million
2. Conflagration, Lynn, MA	(Nov. 28, 1981)	$ 70 million
3. Central telephone exchange, New York, NY	(Feb. 27, 1975)	$ 60 million
4. Hotel, Las Vegas, NV	(Nov. 21, 1980)	$ 50 million
4. Refinery, Borger, TX	(Jan. 20, 1980)	$ 50 million
6. Chemical manufacturer, New Castle, DE	(Oct. 21, 1980)	$ 45.75 million
7. Grain elevator, Galena Park, TX	(Feb. 22, 1976)	$ 42 million
8. Pumping station, Fairbanks, AK	(July 8, 1977)	$ 40 million
8. Sewage system, Louisville, KY	(Feb. 14, 1981)	$ 40 million
10. Brush and dwellings fire, San Bernardino, CA	(Nov. 24, 1980)	$ 38.03 million
11. Ink manufacturer, Chicago, IL	(May 27, 1973)	$ 37 million
12. Brush and dwellings fire, Saint Helena, CA	(June 22, 1981)	$ 35.67 million
13. Plumbing supply manufacturer, New Orleans, LA	(July 8, 1980)	$ 35.25 million
14. Brush fire, Los Angeles, CA	(Oct. 23, 1978)	$ 35.11 million
15. Brush and dwellings fire, Santa Barbara, CA	(July 26, 1977)	$ 35 million
16. Refinery, Texas City, TX	(May 30, 1978)	$ 32.43 million
17. Tank ship for crude oil, Port Neches, TX	(Apr. 19, 1979)	$ 32 million
18. Grain elevator, Westwego, LA	(Dec. 22, 1977)	$ 30 million
18. Aircraft on runway, Los Angeles, CA	(Mar. 1, 1978)	$ 30 million
18. General warehouse, Edison, NJ	(Jan. 17, 1979)	$ 30 million
18. Shopping Mall, Greenville, SC	(Dec. 13, 1981)	$ 30 million
Adjusted to 1981 Dollars ("Real" Dollars)		
1. Central telephone exchange, New York, NY	(Feb. 27, 1975)	$101.52 million
2. Tank ship for crude oil, Deer Park, TX	(Sept. 1, 1979)	$ 89.86 million
3. Ink manufacturer, Chicago, IL	(May 27, 1973)	$ 75.81 million
4. Conflagration, Lynn, MA	(Nov. 28, 1981)	$ 70 million
5. Grain elevator, Galena Park, TX	(Feb. 22, 1976)	$ 67.20 million
6. Pumping station, Fairbanks, AK	(July 8, 1977)	$ 60.08 million
7. Hotel, Las Vegas, NV	(Nov. 21, 1980)	$ 55.30 million
7. Refinery, Borger, TX	(Jan. 20, 1980)	$ 55.30 million
9. Brush and dwellings fire, Santa Barbara, CA	(July 26, 1977)	$ 52.57 million
10. Chemical manufacturer, New Castle, DE	(Oct. 21, 1980)	$ 50.60 million
11. Brush fire, Los Angeles, CA	(Oct. 23, 1978)	$ 48.98 million
12. Refinery, Texas City, TX	(May 30, 1978)	$ 45.24 million
13. Grain elevator, Westwego, LA	(Dec. 22, 1977)	$ 45.06 million
14. Railroad yard, Decatur, IL	(July 19, 1974)	$ 44.35 million
15. Transportation vehicle storage, Philadelphia, PA	(Oct. 23, 1975)	$ 42.30 million
15. Commercial aircraft, Jamaica, NY	(Nov. 12, 1975)	$ 42.30 million
17. Brush and dwelling fire, San Bernardino, CA	(Nov. 24, 1980)	$ 42.06 million
18. Aircraft on runway, Los Angeles, CA	(Mar. 1, 1978)	$ 41.85 million
19. Foam rubber manufacturing plant, Shelton, CT	(Mar. 1, 1975)	$ 40.61 million
20. Sewage system, Louisville, KY	(Feb. 14, 1981)	$ 40 million

Source: NFPA's Fire Incident Data Organization (FIDO)

The most fundamental question to be answered in using data for program or strategy analysis is whether past fire experience is a good guide to future fire problems. Sometimes it is; sometimes it is not. However, intelligent use of data with other sources of information is an essential part of the best possible projection of the future. After all, what will we use if not data? Models are tied to reality by validation, which is based on experience of the past. Expert judgment is formed, at least in part, by the experience of the past. The issue is not whether to use the past as a guide; it is how best to characterize the important changes that will occur in the future.

Often the most important aspect of the future will be a different mix of the elements already in place in the present. For example, home smoke detector usage has grown at a phenomenal rate in the last 15 years, with most of the growth occurring in or after 1976. Fire incident data from the early years of growth in detector usage showed that smoke detectors reduce by half the risk (measured by the number of deaths per 1,000 fires) that a person will die if he or she has a fire. At that point, there were uncertainties about the applicability of this finding to the future. Would the life saving impact of detectors remain the same when detectors were installed in poorer households with less educated occupants?

The answer turned out to be yes. But the form of the question shows how data can be used intelligently to try to project the future. Start with the future predicted by a simple projection of recent trends. Using judgment, creativity, and brainstorming, try to identify what aspects of the present environment might be changing in ways that would modify the simple projection. Try to obtain data on the speed of those changes. Try to analyze the fire data in hand to determine the sensitivity of your conclusions to those changes. Combine these results to produce a revised projection.

Most analyses to date have found that environmental changes relevant to fire occur slowly and so produce relatively modest changes in what are called baseline projections, i.e., projections of the size and character of the fire problem if no new strategies are adopted. For example, the average age of the U.S. population is rising. This means a drop in the percentage of the population in the high death rate years of infancy, a rise in the percentage in the high death rate elderly years, and other shifts along the age spectrum. By the end of the century the cumulative effect will be large, but on a year to year basis it produces only a small predicted change in the national fire death rate.

An analysis of the projected impact of a strategy or program involves these five steps:

1. Identify the part of the fire problem that the strategy can affect and measure the size of that problem. For example, a change in the flame resistance of upholstered furniture covering materials would affect only fires involving those materials. Analysis will show the change would affect primarily those fires that begin with ignition of such materials, in as much as the number of severe fires that begin with upholstered furniture appears to be much larger than the number that begin elsewhere and become severe primarily because of spread to upholstered furniture. A change to self-extinguishing cigarettes would affect only fires whose form of heat of ignition was cigarettes, and so on.
2. Estimate the likely percentage reduction in this target fire problem if the strategy or program were adopted, and prepare estimates for each of the measures of fire severity (deaths, injury, property loss) because the impact probably will be different for each measure. If the strategy or program is already in use, (as was true for smoke detectors), then fire incident data analysis may produce these estimates. If the strategy or program is not in use or is in very limited use, as is true for home sprinklers, then some combination of modeling, laboratory tests, and expert judgment will be needed to produce the estimates. In either case, it makes sense to develop a range of estimates, from optimistic to pessimistic.
3. Estimate how much of the target population will adopt the program or strategy, and how quickly. This typically is a marketing question, and the results can be surprising. Most analyses of the growth in smoke detector coverage were too pessimistic on both how much and how fast. At the same time, most analyses were too optimistic about the installation of detectors in homes most likely to have fires. While about three-fourths of U.S. households now have detectors, less than one-fourth of home fires are reported to occur in homes with detectors. This discrepancy is not unusual for strategies that work through the marketplace. The first purchasers tend to be more affluent and better educated; these people traditionally have lower fire rates than the rest of the population. Also, speed of adoption will reflect the normal life of a product. Changes in cigarettes can be implemented for all cigarettes in months, while changes in the upholstered furniture ignited by cigarettes may take decades to work their way through the whole population.
4. Estimate how often the strategy will be defeated in practice. For example, what proportion of detector equipped homes will have their detectors out of service because worn out batteries were not replaced? Note that if your method in step (2) involved the use of actual fire incident data, your estimate already may include the effects of defeating, or "attenuation" (the latter a term that does not have the connotation of deliberate sabotage, which "defeating" may have for some people).
5. Combine the measures of fire problem size from step (1) and the percentages from steps (2) through (4) to produce estimates of the net percentage reduction in the fire problem and of the new size of the fire problem. With these results, decide how valuable this strategy or program would be and whether to press for its adoption. In some forums, this decision will require a parallel calculation of the cost of the strategy or program, followed by some kind of comparison of the costs and loss reductions.

Analysis can be used not only to make yes or no decisions on adopting a strategy, or to select the strategy with the biggest project impact, but also to fine tune the design of the strategies themselves. Two examples will illustrate how this can be done.

One city fire department had a small scale home inspection program, in which fire companies used slack periods to offer home inspections in their first-due areas. Since no one was home at about half the houses visited, this program offered a good opportunity to compare the fire experience of the homes inspected with the fire experience of the similar homes where no one had been home when inspections were offered. Results of the analysis showed no inspection impact, but, more importantly, showed that the homes chosen for inspection were in areas with some of the lowest fire incident rates in the city. In other words, the program was somehow being targeted at the neighborhoods needing it least. Why? Once the data analysis identified a hitherto unsuspected pattern, the explanation was fairly simple: inspection effort would be greater in areas with more company slack time, which in turn would tend to be areas requiring fewer company runs, including fewer fires. The analysis pointed to a need to redesign the home inspection program so it would target fire prone neighborhoods.

Unexpected results of analysis are not the only indicators of the need for design improvements. While designing a public education campaign aimed at heating safety, the Seattle, WA, Fire Department conducted a day by day analysis of fire incidence. By comparing that analysis with historical data on Seattle weather, the analysts were able to identify a significant pattern of cold snaps followed three days later by a sharp rise in heating system fires. Vulnerable heating systems, it seemed, would develop fires after sustained heavy use. This analytical discovery had a direct bearing upon the design of the public education campaign, because it suggested that firesafety messages should be released as soon as a cold snap started. People would already be thinking about their heating systems when the messages appeared, so attention would be greater; also, the need to listen and act at once could be underlined by reference to the established pattern. What is more, the three day lag meant there was time to get out the messages after the cold snap started but before the fires would begin.

The common thread of these two examples is the use of analysis to narrow the target of a program or strategy in order to maximize impact and conserve scarce program resources. In the first case, the geographical targeting was affected; in the second case, the timing benefited from analysis.

Another consideration in analysis is the need to clearly identify which of many important objectives are to be given priority. For example, is it more important to reduce the fire loss actually being suffered, thereby targeting properties where most loss now occurs? Or is it more important to reduce the potential for catastrophic fire loss, thereby targeting properties with sizeable numbers of lives or property value at risk, even if their actual loss has been relatively small? Also, is it most important to reduce numbers of fires, numbers of deaths and injuries, or numbers of dollars in direct property loss? In apartment fires, for example, a focus on fires might mean targeting cooking related fires, a focus on deaths might mean targeting smoking related fires, and a focus on property loss might mean targeting incendiary and suspicious fires.

Finally, remember that tracking analysis continues to be useful even after a decision has been made and a strategy or program has been implemented. As noted earlier, most of the estimates used in an analysis involve uncertainty; therefore, it may be useful to see whether the future unfolds as projected. If not, some further refinements in the design of a strategy or program may be warranted. Even early termination of a program might be indicated.

AVAILABLE MAJOR DATA BASES ON FIRES

Three major data bases are available to analyze patterns in U.S. fire experience—the annual NFPA survey of fire departments; the FEMA/USFA National Fire Incident Reporting System (NFIRS); and the NFPA Fire Incident Data Organization (FIDO). Together, these three data bases can provide valid, detailed information on national and regional fire problems, overall or by specific property type and cause. The characteristics of the three data bases and the best ways of using each are the subject of this section.

Annual NFPA Survey of Fire Departments

The NFPA survey is based on a stratified random sample of roughly 3,000 U.S. fire departments (or just over one of every ten fire departments in the country). The survey collects the following information: (1) the total number of fire incidents, civilian deaths, and civilian injuries, and the total estimated property damage (in dollars), for each of the major property use classes defined by the NFPA 901 standard for fire incident reporting; (2) similar tallies, specifically for incendiary and suspicious fires separated only into structure versus vehicle; (3) the number of on duty fire fighter injuries, by type of duty and nature of injury or illness; (4) information on the type of community protected (e.g., county versus township versus city) and the size of the population protected, which is used in the statistical formula for projecting national estimates from sample results; and (5) leads on multiple death and large loss fires and fire fighter fatalities, which if cited are then captured under FIDO.

The totals in (1) and the special incendiary and suspicious fire results in (2) are analyzed and reported in NFPA's annual study, "Fire Loss in the United States," which traditionally appears in the September issue of *Fire Journal* magazine. The fire fighter injury information in (3) is analyzed and reported in NFPA's annual report "U.S. Fire Fighter Injuries," traditionally published in the November or December issue of *Fire Command* magazine.

The NFPA survey begins with the NFPA Fire Service Inventory, a computerized file of nearly 30,000 U.S. fire departments, which is the most complete and thoroughly validated such listing in existence. The survey is stratified by size of population protected to reduce the uncertainty of the final estimate. Small, rural communities protect fewer people per department and are less likely to respond to the survey, so a larger number must be surveyed to obtain an adequate sample of those departments. (NFPA also makes follow up calls to a sample of the smaller fire departments that do not respond, to confirm that those that did respond are truly representative of fire departments their size.) On the other hand, large city departments are so few in number and protect such a large proportion of the population that it makes sense to survey all of them. Most respond, resulting in excellent precision for their part of the final estimate.

These methods have been used in the NFPA survey since 1977 and represent a state of the art approach to sample surveying. Because of the attention paid to representativeness and appropriate weighting formulas for projecting national estimates, the NFPA survey provides a valid basis for measuring national trends in fire incidents, civilian deaths and injuries, and direct property loss, as well as for determining patterns and trends by community size and major region.

FEMA/USFA's National Fire Incident Reporting System (NFIRS)

FEMA/USFA's NFIRS provides annual computerized data bases of fire incidents, with data classified according to a standard format based on NFPA 901. Roughly three-fourths of all states have NFIRS coordinators, who receive fire incident data from participating fire departments and combine the data into a state data base. These data are then transmitted to FEMA/USFA. Participation by the states, and by local fire departments within participating states, is voluntary. NFIRS captures roughly one-third of all U.S. fires each year. More than one-third of all U.S. fire departments are listed as participants in NFIRS, although not all of these departments provide data every year.

One of the strengths of NFIRS is that it provides the most detailed incident information of any national data base not limited to large fires. NFIRS is the only data base capable of addressing national patterns for fires of all sizes by specific property use and specific fire cause. (The NFPA survey separates fewer than 20 of the hundreds of property use categories defined by NFPA 901 and provides no cause related information except for incendiary and suspicious fires.) NFIRS also captures information on the construction type of the involved building, avenues and extent of flame spread and smoke spread, and performance of detectors and sprinklers.

One weakness of NFIRS is that its voluntary character produces annual samples of shifting composition. Despite the fact that NFIRS draws on three times as many fire departments as the NFPA survey, the NFPA survey is more suitable as a basis of projecting national estimates because its sample is truly random and is systematically stratified to be representative.

Analyses based on NFIRS have been widely calculated only since 1982, the year of publication of the second edition of FEMA/USFA's *Fire in the United States*—the first study based primarily on NFIRS. Because consensus on how best to address the weaknesses of NFIRS does not yet exist, the next few years may see further revisions in the formulas used to calculate statistics from this system. In the meantime, most analysts use NFIRS to calculate percentages, e.g., the percentage of residential fires that occur in apartments, or the percentage of apartment fire deaths that involve discarded cigarettes. Some analysts combine NFIRS based percentages with NFPA survey based totals to produce estimates of numbers of fires, deaths, injuries, and dollar loss for subparts of the fire problem. This is the simplest approach now available to compensate in the area where NFIRS is weak.

NFPA's Fire Incident Data Organization (FIDO) System

The NFPA FIDO System is a computerized data base that provides the most detailed incident information available, short of a full scale fire investigation. The fires covered are those deemed to be of high technical interest. The system that identifies fires for inclusion in FIDO is believed to provide virtually complete coverage of incidents reported to fire departments involving three or more civilian deaths, one or more fire fighter deaths, or large dollar loss (redefined periodically to reflect the effects of inflation, and defined since 1980 as $1,000,000 or more in direct property damage).

FIDO also captures a selection of smaller incidents as technical interests dictate. These are useful primarily because of the type of property involved (e.g., high rise buildings), the presence of hazardous materials, or the performance of detectors or sprinkers.

The FIDO System covers these fires from 1971 to date, contains information on more than 53,000 fires, and adds 3,000 to 4,000 fires per year. NFPA learns of fires that may be candidates for FIDO through a newspaper clipping service, insurers' reports, state fire marshals, NFIRS, respondents to the NFPA annual survey, and other sources. Once notified of a candidate fire, NFPA seeks standardized incident information from the responsible fire department and solicits copies of other reports prepared by concerned parties, such as the fire department's own incident report and results of any investigations.

The strength of FIDO is its depth of detail on individual incidents. Information captured by FIDO, but not by NFIRS, includes types and performance of all built in systems for detection, suppression, and smoke and flame control; detailed information on factors contributing to flame and smoke spread; estimates of time between major events in fire development (e.g.; ignition to detection, detection to alarm); reasons for any unusual delay at various points; indirect loss and detailed breakdowns of direct loss; and escapes, rescues, and numbers of occupants. (Building height—necessary to analysis of high rise building fires—will be included in Version IV of NFIRS but to date has been available only in FIDO.) Additional uncoded information often is available in the hard copy FIDO files, which are indexed for use in research and analysis.

One weakness of FIDO is that it mostly covers larger incidents. Many questions can best be answered by comparing the characteristics of large and small fires that involved similar types of properties and similar causes of ignition. FIDO does not permit such comparisons.

FIDO supports three annual NFPA reports: "U.S. Fire Fighter Deaths," typically published in the May or June issue of *Fire Command*; "Multiple-Death Fires in the United States," usually published in the July issue of *Fire Journal*; and "Large-Loss Fires in the United States," typically published in the November issue of *Fire Journal*. FIDO also supports the anecdotal summaries publishedin the "Bimonthly Fire Record" of *Fire Journal* and in the annual study cited earlier on U.S. fire fighter injuries.

Comparing Estimates Using Different Data Bases or Analytic Approaches

Occasionally a situation arises for which two different numbers exist, derived from different sources or different analytical approaches. For nonanalysts who may not know the source of either number, such a situation can prove frustrating and can encourage a cynical attitude about the arbitrariness of all fire statistics. Even analysts familiar with the sources of both numbers may have difficulty pinning down the precise reasons for any discrepancies, and deciding how important the differences are and which number is best. It is very important, therefore, for all data users—and others who expect to confront arguments based on fire statistics—to understand how and why estimates can differ. For the most part, variations involve the ways different data sources and estimates deal with the inevitable gaps in coverage that affect all sources of fire incident information.

No fire data base can possibly capture all instances of unwanted fires. Few data bases—and none of the three discussed in this section—cover fires that are not reported to fire departments. (Special studies for some property types do provide one-time estimates of the size and composition of the unreported fire problem. No fire data base even captures all the fires reported to fire departments, but some data bases, notably the NFPA survey, are built on samples designed to be representative of all fire departments.

By their nature, fire data bases are biased in favor of "failures" rather than "successes." The fire that is controlled so quickly it does not need to be reported to a fire department is not captured by the data bases that cover reported fires. Analyses of the impact of devices and procedures that provide early detection or suppression also need to allow for the phenomenon of missing "success" stories. Moreover, data bases like FIDO that provide the most detail on the building's features and their performance are limited to the largest of the reported fires.

There is also the issue of quality control for a data base. For data bases with limited depth of detail, like the NFPA survey, or limited breadth of coverage, like FIDO which is confined to large fires, it is possible to invest considerable effort in assuring that each report is as complete and accurate as possible. Follow up calls can be used to fill gaps and check possible odd anwers. For a data base with the depth and breadth of NFIRS, however, the same level of quality control effort has not been possible. Consequently, NFIRS is missing more entries and has more that are dubious. The tradeoff between data quality and data quantity is never easy; an analyst needs to be aware of the strength and dependability of the sources before conducting an analysis.

Two examples indicate how differences in fire data bases and assumptions can produce different results. The U.S. Department of Justice, estimates the size of the nation's arson problem through its Uniform Crime Reports (UCR) based on reports from law enforcement agencies. NFPA, through its annual survey of fire departments, estimates the size of the nation's fire problem due to incendiary or suspicious causes. For 1983, the UCR estimate of the arson problem in structures was 297 fires per million population. For the same year, NFPA estimates were 524 incendiary or suspicious structure fires per million population and 311 structure fires per million population for incendiary fires alone. The UCR arson estimate and the NFPA incendiary fire estimate differ by about five percent, which is within the range of statistical uncertainty for an estimate based on a survey the size of NFPA's.

Several points are illustrated by this example. First, the UCR estimate is close to the NFPA incendiary-only estimate, because the UCR definition of arson approximates NFPA's definition of incendiary. NFPA and other fire organizations, however, traditionally regard the combination of incendiary and suspicious fires as the best indicator of the nation's problem with intentionally set fires. The UCR and NFPA systems produce roughly the same estimates where they are attempting to measure the same thing. However, because they differ on how to handle the more ambiguous fires, the "arson" numbers they release may appear significantly different to the casual reader or listener.

Second, it makes sense that the UCR estimate would be the lower of the two because fire officers see every reported fire, while police departments see only those fires that might involve crimes. A fire department may conclude that a fire is incendiary, while the corresponding police department either disagrees or is not notified; but the reverse is much less likely to occur. Also, definitions of arson vary from state to state.

Third, if there are differences in the ways that fire departments and police departments define structures, any such differences appear to have negligible impact on these calculations.

The second example of differing estimates involves the 1981 estimates made by FEMA and NFPA of total U.S. civilian fire fatalities. The NFPA estimate, based on the survey, was 6,700 deaths. The FEMA estimate, based on a multipart procedure that started with death certificate information reported to the National Center for Health Statistics, was 7,600 deaths. At least five significant differences in methods help to account for this discrepancy.

First, NFPA estimates of annual death totals are subject to some uncertainty because they are based upon sample surveys. Using accepted statistical principles, NFPA specified a range for the true value of 6,200 to 7,200. In other words, if someone else had run the same kind of survey and had produced an estimate that was higher or lower by up to 500 deaths, that would be within the range of precision achievable by a survey of that size. The difference would be the result of random variations in the representativeness of the fire departments selected for the sample.

Second, all fire fatality estimating procedures have difficulty with deaths that could have resulted from multiple causes. The principal examples of this problem are vehicle accidents followed by fire. It must be either determined or estimated whether traumatic injuries in the crash or subsequent fire related injuries caused death. When FEMA began making estimates, one of its principal objections to NFPA's figures were that they tended to include all crash/fire fatalities. As a result, the two organizations pursued different courses to deal with this problem. NFPA began making follow up inquiries on vehicle fire deaths, permitting the fatalities of suspect cause to be screened out prior to calculation of national projections. The initiation of this screening reduced estimated total fire deaths by nearly one-third from their previous level. FEMA chose to address this problem by use of a correction formula that assumed vehicle fire deaths would continue to be a constant percentage of structure fire deaths and/or total vehicle deaths. This approach was less satisfactory because it could not detect changes in the relative size of the problem.

Third, the NFPA procedure cannot capture fatalities suffered in fires not reported to fire departments. These deaths principally involve clothing ignitions in which the fire is smothered by a civilian, but not before it has caused fatal injuries. The death certificate approach captures these deaths, which numbered on the order of 250 to 400 per year. A recent unpublished NFPA analysis of clothing-ignition fire deaths showed that those reported now number on the order of 375. Since analysis of death certificates also indicates the total number of this type of fire death has decreased, it may be that the recent expansion of the fire service into emergency medical service has sharply reduced the number of fire deaths that go unreported. In any event, the principal strength of the death certificate approach is that it collects deaths in unreported fires, which no system based on fire department records can do.

Fourth, an approach based on death certificates experienced difficulty in 1981 concerning arson deaths, then classified as murder with no indication of the role of fire. Subsequently, the death certificate coding was expanded to increase the number of factors that can be included, so an arson incident coded as murder now can also be coded as involving fire. As of 1981, however, arson deaths were still impossible to identify on death certificates and FEMA had to use other sources to make these estimates. (Suicides by fire could have posed a similar problem, but are rare in the U.S. In Japan, however, where immolation by fire accounts for about a third of all fire deaths, the decision to include or exclude these victims makes a large difference in the totals.)

The fifth and final difference has to do with timeliness and the methods used to achieve it. As with NFIRS, the death certificate records are assembled on a state by state basis with wide variations in speed of delivery. In order to issue an estimated national total in a timely fashion, FEMA was forced to use other than death certificates sources for about one state in four, including some of the largest states. The need to work around late states was a major reason that FEMA abandoned this approach after 1981. Death certificate tracking still has advantages as an after the fact check on the accuracy of survey based estimates, but it cannot by itself provide timely estimates.

SPECIAL DATA BASES ON FIRE

Many special data bases exist. Some concern aspects of the fire problem not well covered by the existing major data bases. Others contain information that may be useful in analysis. The examples cited below are not exhaustive.

Occupational

In addition to NFPA's data bases on fire fighter fatalities and injuries, the International Association of Fire Fighters (IAFF) annually releases totals of on duty fatalities and injuries among career members of the fire service.

Also, fire injuries suffered on the job in some industries, e.g., the grain storage industry, have been among the occupational injury subjects studied by the National Institute for Occupational Safety and Health (NIOSH). Published reports on these studies are available.

Civilian Injuries

The U.S. Consumer Product Safety Commission (CPSC) maintains a computerized data base drawn from a sample of hospital emergency room cases. The National Electronic Injury Surveillance System (NEISS), dating back to 1972, focuses on product related injuries in the home. Reports on fire casualties related to consumer products (about five percent of the NEISS total) are published annually. Information collected is similar to that collected for casualties under NFPA 901 but is much more detailed regarding the type of product involved. NEISS is particularly useful for analysis of serious injuries due to electrical shock or burns not caused by fire.

Severe burns receiving specialized care are addressed to some extent in annual surveys by the American Burn Association. Admissions and some other factors are tabulated for patients passing through the nation's specialized burn care units. Some of these survey results have been reprinted in NFPA's *Fire Almanac*.

Transportation

Accidents, including fires, are tabulated by several government agencies. The National Transportation Safety Board (NTSB) publishes accident reports on aircraft and railway accidents and on highway accidents involving hazardous materials. The information collected tends to emphasize the circumstances of the accident with little discussion of the ensuing fire. Some of these reports, however, have unpublished supplements addressing issues, e.g., human factors in escape, and often contain much more fire related information. The computerized portions of the records have the least information on fires. Human factors studies became routine in the mid 1960s; computerized records date from 1964.

The National Highway Traffic Safety Administration (NHTSA) established a computerized file on fatal motor vehicle highway accidents, beginning in 1975. Statistical summary reports are issued annually.

The U.S. Coast Guard (USCF) collects reports on accidents involving recreational boats and commercial vessels. As with the other special data bases on vehicle accidents, there is little coded, standardized information on the cause and development of these fires.

International data bases also exist for accidents involving aircraft. Narrative summaries ranging from one sentence to a few paragraphs appear in the *World Airline Accident Summary*, published by the Air Registry Board in England. Standardized narrative reports, usually abstracted from original accident reports prepared by national air safety organizations, are published by the International Civil Aviation Organization, headquartered in Canada, in its *Airline Accident Digest* series.

Forests and Wildlands

The U.S. Forest Service issues annual reports titled *Wildfire Statistics* and *National Forest Fire Report*, and the U.S. Department of the Interior's (DOI) Bureau of Land Management issues an annual report of *Public Lands Statistics*. The U.S. Forest Service reports cover fires occurring on national, state and private forests, including numbers, estimates of the extent of damage (acres burned), and profiles of causes. The DOI reports address fires on the lands it owns and administers. Information addressed includes cause, extent of damage, rate of spread, and method of suppression. These data bases add records on thousands of fires each year.

Military

As with transportation and forest fires, fires on military installations tend to fall outside the jurisdiction of the local fire departments covered by the nation's major data bases (NFIRS and the NFPA survey). The various branches of the military historically have compiled their own fire incident data bases and have produced a number of analyses. Beginning in 1985, the Naval Safety Center at Norfolk, VA, is the receiving point for fire reporting for all the U.S. military services. Incident records from the Center also are submitted to NFIRS.

Electrical

The International Association of Electrical Inspectors (IAEI) and Underwriters Laboratories (UL) maintain data bases generated from clipping files, covering not only electrical fires but also the closely related subject of electrical shocks. The IAEI data is published annually in *IAEI News*. Both organizations discourage the use of this data for statistical analysis because clipping file records are not representative of incidents of all sizes. By concentrating on large incidents, however, these data bases can be valuable for certain uses.

International

The World Health Organization's *Statistical Annual* includes information on fire death rates by country. It can be difficult, however, to obtain data for many countries for the same year. The World Fire Statistics Centre in England also prepares annual studies with international fire statistic comparisons, covering not only fire fatalities but also property damage and estimates of national expenditures on the various elements of fire protection (e.g., fire departments, insurance, built in fire protection).

Special Studies

Some special studies have produced one time data bases or statistics that provide continuing value for fire analysts. In 1985 the CPSC will have completed its survey of unreported home fires and their characteristics; the resulting data base will be of use for many years. CPSC also has conducted a number of special projects, including in depth investigations of samples of home fires involving such equipment as electrical systems or alternative heating systems. Reports have been published on all these studies. A similar in depth study of a sample of mobile home fires was conducted in the late 1970s by the Department of Housing and Urban Development (HUD).

As of 1984, the NFPA fire investigation program, supported by FEMA and the National Bureau of Standards, had generated on the order of 250 in depth incident reports. No one property use accounted for enough incidents to form a statistically significant data base. However, some issues, such as those involving patterns in major fires in buildings with large life exposures, might be able to draw on this data base for statistical significance. These incident reports contain unequaled depth of detail on fire development, smoke spread, human factors in escape, and the performance of fire protection systems and features.

SECTION 3 EDUCATION FOR FIRE PREVENTION

FIRESAFETY EDUCATION

Diane C. Roche

Firesafety education is designed to develop or change the attitudes and behaviors of men, women, and children toward fire. It encompasses a wide spectrum of programs and activities presented to audiences as diverse as school children, senior citizens, homeowners, preschoolers, apartment dwellers, handicapped people, employees, hospital and nursing home staffs, and church, service and civic organizations. Firesafety education topics may include home fire escape planning, babysitting firesafety, cooking fires, clothing fires, juvenile fire setters, first aid for burns, home fire hazard inspections, scald prevention, smoke detectors, fire extinguishers, and home sprinkler systems, to name a few.

Whether a firesafety education program is large or small, it takes planning to be effective. From a short 30 second public service announcement to an extensive year long program, firesafety education efforts must be well planned—targeted, continual, and measurable.

There are too many possible programs, approaches, and subjects to consider here, so this chapter will limit itself to discussing the planning and implementation processes that can be used for any program. The information should provide anyone with the foundation for beginning or expanding successful firesafety education efforts in the community.

Effective firesafety education programs do not just happen. They are well thought out. They have measurable goals and objectives. And they are targeted at real problems.

The suggested process for firesafety education program development includes the following steps:

1. Initial Planning Phase.
   a. Establish responsibility and support.
   b. Form a planning team.
   c. Identify local fire problems.
   d. Define goals and objectives.
2. Design and Implementation Phase.
   a. Conduct audience/market research.
   b. Develop program strategies.
   c. Develop action plans for program objectives.
   d. Make a program proposal.
   e. Prepare teaching aids and train instructors.
   f. Conduct pilot tests.
3. Evaluation Phase.
   a. Provide for program documentation.
   b. Determine effectiveness.
   c. Revise action plans and objectives.

Ms. Roche is Assistant to the Director of Public Information for the city of Virgnia Beach, VA. Named Fire Educator of the Year, 1981, by the U.S. Fire Administration, she is a member and past board director of the National Fire and Burn Education Association and past chairman of the Firesafety Education Section of the International Society of Fire Service Instructors.

INITIAL PLANNING PHASE

Establish Responsibility and Support

Before beginning to design an individual program or a program for an entire community, fire departments should clarify who will be responsible for firesafety education. The fire chief normally makes the assignment, thus demonstrating strong support for such efforts. One person, a small group, or an entire division might be responsible for developing programs. While a different person or group might be appointed to implement them, the same people may do both jobs. It is important that those with the responsibilities also have the sufficient authority, time, and resources to get the job done.

Firesafety education responsibilities and tasks are assigned in various ways. In some fire departments, programs are conducted by career and volunteer fire fighters as part of their regular shift duties. Some fire chiefs assign personnel from the prevention bureau, or hire a specialist in education to develop and implement programs. Others enlist civilians, such as community volunteers, teachers, nurses, or senior citizens. Some fire departments use a combination of all of these people for their educational work. Whoever is responsible, it is important that the entire fire department, and eventually the community, supports them.

For fire personnel to actively support and encourage firesafety education, they need a working knowledge of the process involved. Fire departments should include an introduction to firesafety education in rookie fire schools and during in-service training. All fire personnel should also be encouraged to take a course in firesafety education at the National Fire Academy or at state and local fire schools and workshops offered by professional fire service organizations.

Form a Planning Team

After responsibilities are clarified, a planning team should be formed. This group should be responsible for the initial planning of community firesafety education efforts and can later serve in an advisory capacity. Planning teams have several advantages: they divide the work load, contribute creative ideas and objective opinions, supply the expertise needed from outside the fire service, and provide general support. Members of the team should include appropriate fire personnel, educators, community leaders, and others who can provide knowledge, expertise, influence, and resources. The ideal size for this group is 10 persons, but if a larger group is needed to include everyone, smaller sub-groups or committees can be formed to expedite work.

An example of a local firesafety education planning team might be: fire chief, fire fighters, fire prevention personnel, a teacher, a doctor and a nurse (burn specialists, if possible), a lawyer, a representative of civic clubs, a representative of service organizations, a media representative, a social worker, a member of the crime prevention team, an arson investigator, and a representative of the building industry.

Identify Local Fire Problems

Since effective firesafety education programs are aimed at correcting or reducing specific fire problems, one of the first tasks of the planning team is to identify the local fire problem. How lifesaving information is presented to the public depends on the types of fire problems occurring in the community. Develop a profile of the types and frequency of fires, causes of ignition, places of origin, types and behavior of victims, and fire patterns by neighborhood. Before people can be taught to stop starting fires, what they are doing to carelessly or intentionally start fires needs to be identified. In addition, the types of people causing fires need to be determined so that the program can be designed for a target audience. Where the fires are starting also needs to be identified. Is the biggest problem house fires, car fires, industrial fires, fires in nursing homes or in schools?

To identify the fire problems, the planning team will need fire department records as well as records of hospitals, insurance companies, and state agencies. During this phase of development, the value of an organized reporting system is appreciated. The National Fire Incident Reporting System (NFIRS) provides detailed information that can be used to analyze local problems.*

*NFIRS—the National Fire Incident Reporting System—is a computer based system which builds a data base at a local or state level. Copies of these local and state data bases are merged to form a national data base which is maintained by the U.S. Fire Administration, Federal Emergency Management Agency. Because the same definitions are used by fire departments of all sizes in all parts of the country, the computerized information forms a reliable base from which to draw local comparisons and determine local problem areas.

Once the statistical information is obtained, it must be analyzed to determine problem areas or trends. The National Fire Protection Association's *Firesafety Educator's Handbook* (NFPA 1983) includes, in the "Message" section, instructions and examples for determining local fire problems. It is recommended that the planning team study this information.

Define Goals and Objectives

Once the planning team has identified the specific local fire problems, it can set goals and objectives and, eventually, develop action plans based on those goals and objectives. This goal setting is similar to deciding where you want to go on a road map, then determining the fastest, most economical route with the fewest obstacles along the way. Although you always have your final goal in mind, your route (your objectives) requires most of your attention and energy. You begin to focus your attention on finishing each portion of the trip that brings you nearer your final destination.

Before goals and objectives are set, be certain the team distinguishes between them. The following explanation should help clarify each term.

	Goal	Objective	Action Plan
	Ultimate End	Interim End	Means to the End
Time frame	Future	1–2 years	Less than one year
Responsibility for completion	Entire department's organization	Program Manager Regional Leader	Individual staff
Emphasis	Ultimate benefits to community. Future direction of the department	Your program's measurable results that constitute progress toward your goal	Your resources—how you will allocate time, money, and staff to achieve your objectives
Specifics	Not specific. Does not include target completion dates and measurement of success	Specific results, including target completion dates and measurement of success	Specific results, including responsibility for carrying out the plan.

Goals are achieved through completion of specific objectives, and objectives are completed through individual steps in an action plan. Goals are what should be ultimately achieved; objectives are the programs needed to achieve your goal. Action plans are the detailed steps to be completed to accomplish the objectives. Some objectives consist of on-going programs, while others are one-time efforts. Goals, objectives, and action plans form a pyramid with development beginning at the top and completion beginning at the bottom. (See Fig. 3-1A.)

FIG. 3-1A. Development pyramid.

The need for goals and objectives appears at various levels and planning stages in your program design. First, discuss the overall goals and program objectives. Then take a look at the behavioral objectives involved in your program content.

The planning team should develop a major goal for its overall firesafety education efforts and should continually work toward it. Program objectives should be designed to bring the goal closer. In the design and implementation stage of this planning process, the team will need to work specificity and detail into the action plans. The following is an example of an overall goal for firesafety education: to reduce the number of fires and the resulting deaths, injuries, and property damage by educating the public in the subjects of hazards of fire, methods of fire and burn prevention, and reactions necessary for a successful escape from fire.

The planning team should choose realistic objectives based on the scenarios of typical fire problems in their community. Keep in mind that, even when fire departments target their programs, they still receive many requests for programs dealing with general firesafety information and skills. These programs should be presented because they help reduce future fire problems by providing prevention and survival information.

The following are sample objectives for a community which determined that its major fire problems included a high rate of cooking fires in homes, home fire deaths due to improper reactions (especially by children), and a disregard for fire prevention at local nursing homes.

Objective—Reduce Cooking Fires

To present a lesson on cooking firesafety to each of the 20 cooking classes held in the four high schools. Do this by the end of April.

To produce, with the local radio and television stations, a 15- and a 30- second public service announcement telling how to prevent cooking fires and how to deal with one if it happens. To get stations to air these messages twice a day, at breakfast and dinner times, for a two month period.

To write a newspaper article on cooking firesafety to appear in the food section of the local newspaper within one week.

Objective—Reduce House Fire Deaths, Especially of Children

To provide firesafety lessons during October/November and April/May for all children enrolled in local preschools and daycare centers.

To provide four firesafety lessons to all fifth grade students (approximately 2,000) in public schools beginning in September and ending in June.

To provide annual firesafety assembly programs for grades K–4 in the 22 public elementary schools during January, February, and March.

To include firesafety messages for children in ongoing programs conducted by the Parks and Recreation Department and the public libraries.

Objective—Reduce Nursing Home Fires

To conduct three lessons in fire prevention and fire survival for all three shifts at the four local nursing homes within the next month.

DESIGN AND IMPLEMENTATION PHASE

After the program goals and objectives have been defined, the design of the program(s) can begin. During this critical "drawing board" stage, basic decisions are made that will influence the program's effectiveness and public acceptance. Firesafety education should be designed to be functional (reducing the fire problem) as well as appealing and motivational to the public.

Firesafety educators should give their target audiences a reason to listen to them and should motivate their audiences to respond to lifesaving suggestions. Since programs should convey the message in an efficient and effective manner, both audiences and the product should be well known to firesafety educators. Firesafety educators need to "sell" firesafety information to the public. They are the fire department "sales force" and programs should be designed with that in mind.

Conduct Audience/Market Research

One of the first steps in designing firesafety education programs is to research the attitudes, characteristics, learning abilities, and preferences of the audience. Too often, firesafety education programs are designed without regard for the audience's willingness or ability to listen and learn. Know the target audience. Speak to audience members, to those who already know or work with them, to researchers who study them, and learn all you can. In the *Firesafety Educator's Handbook* (NFPA 1983) is a section titled "Market" which describes people and their ability to learn during different stages of their development. After referring to this section, determine the following:

Learning Abilities: Determine at what level and rate the audience usually accepts information. In what form is the information usually presented—written, spoken, or pictorially? Do audience members learn best in small or large groups? Maturity levels should also be considered. It is not wise, for instance, to present a lesson on the proper use of portable fire extinguishers to a class of primary school children, who are not mature enough to decide when it is safe to use one.

Attitudes and Behaviors: How does the group usually behave? Are they apathetic, friendly, hostile, hospitable, conservative, liberal? Do they usually respond well to people from outside their own group? Do they have any preferences that may affect their willingness to listen? Realize that some religious or social groups prefer a particular demeanor and/or dress code. Some teachers prefer that guest speakers maintain a particular level of discipline during firesafety lessons in their classrooms. Unfortunately, some groups may even enjoy "roasting" the guest speaker. Don't cater to the whims of any one group, but be aware of their particular attitudes and behaviors before appearing before them.

Accessibility: How easy is it to reach the audience? How does one gain access to their attention and time? Research their interests and/or occupations and find out where and when they meet as a group. Determine who will be the best contact person in setting up arrangements with this audience. The frequency with which the audience meets in a group will affect their accessibility. Generally, children are more accessible in school than in a social club which may meet only once a week or month. Find out also how accessible the audience can be through the media. Do they watch television, listen to the radio, read the newspaper?

Group Size: Before planning any programs, determine the potential size of the audience. How many people are involved in the group or in similar or associated groups? Make sure to have enough resources and personnel to handle the anticipated demand. Good news travels fast! Once a program is presented to a particular group and there are 10 more such groups in the area, there may well be 10 requests for similar programs. Conversely, if a program is planned for a particular group, involving many hours and some money for the design and preparation, and it turns out there are only a couple such groups in the area, there would not be much demand for repeat performances. Do homework. Determine potential group size in advance.

Current Knowledge of Firesafety: Determine how much information the audience may already have received on the subject. If the majority have the same level of knowledge due to past lessons or due to their occupation, adjust the program to avoid repetition. Never take for granted that the audience has attained a given level of understanding of firesafety. Check the records for previous program contact.

How can all of this audience/market research be obtained? Here are some suggestions:

Talk with members of the group before making a formal presentation.

Get in touch with caretakers or providers of the audience. For instance, teachers of preschool and school age children; nurses in nursing homes; recreation specialists working with the elderly; parents or relatives of handicapped people; adult education specialists; social workers or psychologists who may work with juvenile firesetters; and special education teachers.

Get in touch with groups providing service for the audience, e.g., for a group of senior citizens, the American Association of Retired Persons, the Mayor's Council on Aging, or the administration of a drop-in center for the elderly.

Coordinate with other governmental or community leaders who visit these groups with other messages—the crime prevention officer, substance abuse specialists, Department of Parks and Recreation. What were their experiences with the audience?

It is important to gear the program content and approach to the audience's needs, while at the same time maintaining the integrity of the firesafety message. Find out in advance where the communication opportunities or barriers exist. Anticipation and preparation are preferable to learning by mistakes or "playing it by ear."

Develop Program Strategies

Now that the audience has been sized up, it's time to develop a strategy to reach them with firesafety messages. What will the approach be? Where, when, and how often will the audience receive the messages? What format will be used? These questions can be answered by learning about the following items:

Existing Networks: There is usually a communication network of some kind already in use by any target audience. There are places where they meet or publications from which they already obtain information. Some of these meeting places, such as schools, may require attendance, while others such as social groups are totally voluntary. Utilizing existing networks will usually be easier and more effective than attempting to arrange special networks for firesafety education. Examples of some existing networks are school classrooms, civic leagues, service organizations, social clubs, safety committees in industry, fairs, and exhibits.

When a network does exist, integrating a firesafety program within that network as much as possible is important. For example, the *Learn Not to Burn Curriculum®* has been integrated into existing school subjects; this makes firesafety information attractive to the teachers because of the dual benefits. It is also true that many times the existing networks are looking for interesting and informative programs to fill their needs. Many television and radio talk shows need as many as 30 interesting guests every week for 52 weeks of the year! Utilize such existing networks when possible and when appropriate.

Point of Contact: Contact the target audience at a time and place when they will be most likely to accept the firesafety message, when they will be most "open." For example, avoid visiting a civic league on the night they are debating a heated local issue. Instead, visit them on the night of their banquet, when many people who might not otherwise attend are present and the general mood is hospitable. Avoid visiting senior citizens centers when card games, bingo, or other activities have been scheduled. Avoid scheduling school programs within two weeks of major holidays, so as not to compete with school plays and classroom parties. School programs scheduled from January through March are much more acceptable, as are those during September and October with Fire Prevention Week in the spotlight. Avoid scheduling preschool programs in the very early morning or after lunch, so as not to conflict with breakfast and naps.

A place of business or industry is another good point of contact, especially if a safety committee is functioning and management encourages such participation.

Schedule programs and firesafety messages as conveniently as possible for the target audience and determine where the best location will be—at home, work, school, or a recreation site. Where is the audience most likely to learn and retain your firesafety message?

Appropriate Format: Determine how to convey the firesafety message to the target audience. There are as many formats as there are firesafety educators, but some of the more common formats include: lectures, demonstrations, printed materials, television, radio, public service announcements, press releases, newsletters, mobile vans, puppet shows, seminars, films, slide programs, videotape programs, posters, speeches, displays, and newspaper articles. The choice of format(s) will depend on such factors as:

Size of the audience.
Available resources and cost of materials.
Preparation time.
Length of program.
Level of audience attention and involvement.
Age of the audience.
Nature of the firesafety message.

For a listing of the advantages and disadvantages of available formats, see the *Firesafety Educator's Handbook*, "Methods" section (NFPA 1983).

Decision to Adopt, Adapt or Create Your Program: It is important that firesafety educators do not isolate themselves or their programs from outside influences. It is not always necessary to create entirely new programs. Other fire departments or other communities have dealt with some of the same audiences and messages, and it often is possible to adopt or adapt an existing program. Nationally recognized programs such as NFPA's "Learn Not To Burn" have been carefully designed to allow local community adaptation to fit local problems and circumstances.

A listing of some recent programs appears in the "Materials" section of the *Firesafety Educator's Handbook* (NFPA 1983). In addition, most fire departments will share their materials at little or no cost. National and regional firesafety education conferences provide outstanding opportunities for exchange of program information and networking.

Develop Action Plans For Program Objectives

The initial planning phase was for development of networking program goals and objectives. Now that further information has been developed and implementation nears, it is time to develop action plans to meet the objectives. Action plans are the steps necessary to achieve the objectives. The plans should be very specific, and assigned to appropriate personnel. The plans can then serve as a reminder of what needs to be accomplished.

The following are samples of action plans for two of the objectives listed earlier dealing with programs for school children:

Objective: To provide four firesafety lessons to every fifth grade student (approximately 2,000) in public schools beginning in September and ending in June.

Action Plan

Steps	Target Date	Responsible Person
Revise program format.	7/01	Bob and Nancy
Develop and order handout materials.	11/15	Bill, Bob and Nancy
Schedule schools.	6/01	Bill
Provide the schedule to the school board for approval and distribution.	7/15	Bill
Contact each principal to verity the schedule four weeks prior to program.	as stated	Bob and Nancy
Present the assembly program	per scheduel	Bob and Nancy
Ask the nearest fire company to arrange for a fire fighter to assist with the program.	two weks prior to the program	Bob and Nancy
Develop an evaluation tool for this program.	11/04	Bill

Make a Program Proposal

Before proceeding with design and implementation, it is necessary to receive permission from the decision maker regarding the target audience or group. Before printing booklets, ordering audiovisual materials, and training instructors, be sure the audience is willing and ready for the program. It may also be necessary to receive permission from the fire department or planning team before going any further.

Contact the decision maker for your audience—Superintendent of Schools, teacher, club president, television or radio news or program director, newspaper editor—and present a program proposal to this key person. Program proposals will vary in length and detail according to the audience and program. School systems, for instance, may want detailed lesson plans, including a list of audiovisual materials and handouts. Clubs and service organizations, on the other hand, will probably want only a general description of the program content. Television and radio stations may want to see script ideas.

If possible, do not mail program proposals. Make an appointment to see the decision maker for the audience and make the proposal in person. Be prepared. Be confident. Never approach the decision maker with "I think we would like to do something with you." Have precise information on exactly what you want to achieve and how you plan to go about it.

In general, proposals should include:

1. Statement of the need for the program and how the audience would benefit. Explain the fire problem in local terms.
2. Behavioral objectives.
3. A format description including:
   a. proposed number and length of lessons.
   b. topics to be covered.
   c. audiovisual materials to be used.
   d. sample printed handouts, or a rough design and text for folders and booklets.
4. Cost estimate if the audience will be sharing any program expenses.
5. Description of evaluation tools used in the program.

Once there is a decision to proceed with the program, begin to prepare for the pilot test phase. If the decision is not favorable, find out what the objections were and attempt to reconcile them.

Prepare Teaching Aids and Train Instructors

After receiving permission from the intended audience to present the program(s), prepare for program implementation. It is strongly recommended that a pilot phase be included in the implementation process. Whether preparation is for a pilot phase or a full program, it will be necessary to prepare teaching aids and train program instructors/presenters.

Preparing Teaching Aids: Steps in preparing teaching aids should include the following:

Develop program outlines. Decide what the program content should be. Apply to the content what has been learned about the audience, strategy, and objectives. Develop a topical outline and a narrative outline. These can be used to train the presenters and as reference notes during the program. Narrative outlines indicate how the topics should be taught and provide a detailed description of the program content including audiovisual aids. Program outlines help assure that all presenters will be teaching from the same materials.

Produce/order the written materials and audiovisual programs. Since the first phase of implementation will include a pilot test, materials should be produced and ordered with this in mind. If 10,000 brochures are printed and the pilot test indicates that they are not effective, valuable resources have been wasted. Once they have passed the pilot test, all print and audiovisual materials should be ordered well in advance of scheduled program dates.

Obtain audiovisual equipment. Buy, borrow, or lease enough film projectors, slide projectors, screens, video players, etc., to operate all programs. Make sure that all equipment is in good working condition; keep extra projector bulbs on hand.

Prepare a teaching aid checklist for each type of program. For example, include in a checklist for a Home Escape Plan Lesson these items:

Program outline.
Film.
Film projector including take-up reel.
Extra projector bulbs.
Screen.
Extension cord with 3-prong adapter.
Brochures.
Smoke detector with fresh batteries.
Folding escape ladder.
Sample evacuation on floor plan.

Training Instructors: The program instructor/presenter may or may not be the program designer. In many fire departments, they are one and the same person. Some fire departments provide formal training in firesafety education for prevention and suppression forces alike. Conditions definitely vary from area to area, but the general training requirements include:

Instructional Techniques: Firesafety educators need not be certified teachers but they should have some training in instructional techniques and communication skills. Many fire departments and state training offices give such courses.

Knowledge of subject matter. Initial training may be obtained in several ways. The National Fire Academy at the National Emergency Training Center in Emmitsburg, MD, offers a residential course in Public Fire Education. The International Fire Service Training Association (IFSTA) publishes a manual, *Public Fire Education* (IFSTA No. 606), which is used by many fire departments for this initial training. The *Firesafety Educator's Handbook* (NFPA 1983) is another valuable resource. Several states offer certification in Public Fire Education.

Many fire departments utilize civilians in their firesafety education efforts. If these presenters do not have a knowledge of the fire service, they should be provided with information and training to familiarize them with the basics of fire suppression and operation of a fire department.

After the initial training takes place, continual professional development should occur. This will keep the firesafety educator current with the state of the art. There are numerous conferences, seminars, and workshops on local, state, and national levels that can supply this training. Membership in national organizations concerned with firesafety education will keep the firesafety educator informed about training opportunities, as well as current literature in the field. These organizations include the NFPA Education Section; the International Society of Fire Service Instructors' Fire Safety Educators Section; and the National Fire and Burn Education Association.

Sometimes scheduling a firesafety program involves no more than a telephone call and a follow-up letter of confirmation. At other times, scheduling can be a complicated and extensive task, depending on the number and nature of the programs and the accessibility of the audience. All programs should be as convenient as possible for the audience. The following information should be obtained at the time of scheduling:

Name and type of audience.
Program location and time.
Kind of firesafety program requested (ask for a copy of the printed program if available in advance).
Name and telephone numbers of the contact person.
Size of group, age range.
Kind of room in which presentation will be made.

a. Type of seating arrangement.
b. Availability of electrical outlets.
c. Availability of projector screen.

Conduct Pilot Tests

Pilot tests are necessary to validate the program elements and teaching aids before full implementation of a program takes place. This ensures the effectiveness, accuracy, and appeal of the program before large amounts of time and resources are committed to the project.

Pilot tests vary in depth and duration. They may involve presenting a program to a sample audience a few times to see how members react to the materials and to the approach; or pilot tests may last as long as a year and involve many sample audiences and a great deal of evaluation. Whatever type of pilot test is appropriate, be sure that the sample audiences are representative of the total target audience and that all conditions are as realistic as possible. The pilot audience can supply feedback regarding:

Relevance and clarity of the program message.
Readability and appeal of the printed materials.
Effectiveness and impact of the audiovisual aids.
Ability of the instructor/presenter.
Appropriateness of the format and the point of contact.
Ability of the program to achieve the objectives.

EVALUATION PHASE

Provide for Program Documentation

After pilot testing and full implementation, it will be necessary to evaluate the program's effectiveness. If the program was aimed at reducing the high rate of cooking fires, did it succeed?

Before such questions can be answered, two tools are needed. First, refer to the information gathered in the identification phase. This information becomes the baseline data with which to make comparisons. Next is needed documentation that indicates how many programs were conducted, how many people were reached, and how many personnel it took to do the job. Also needed is information from any pre- and post-testing or any surveys or questionnaires completed by the target audience. To evaluate a media program, collect information concerning the number of viewers, listeners, or readers exposed to the message. Most television and radio stations and newspapers are willing to supply these figures. Also, keep clippings of articles and copies of public service announcements as a part of the media support documentation.

A sample firesafety education report follows.

Firesafety Education Programs

Type of Program	Approx. Length of Program	Number of Programs Completed: Reg.*	OT†	Number of People in Audience
School:				
5th Grade	50–60 minutes	64		419
Assembly (K–4)	60 minutes	none		
Private Schools	60 minutes	none		
Preschools/Day Care	60 minutes	1		55
Babysitting	50 minutes	6		124
General:				
Service Clubs	1 hour		1	25
Dormitory Firesafety	1 hour	2		230
Senior Citizen	1 hour		1	40
Juvinile Firesetters	1 hour	1		3
Fire Extinguishers	1 hour		3	65
Displays	7 hours, 6 hours, 25 hours		3	(est.) 3,400
	TOTALS	74	8	4,361

Total Mileage: 1747

*Number of programs completed during regular working hours
†Number of programs completed during overtime hours

Determine Effectiveness

Ultimately, the planning team will need to report to the fire chief or others concerning the effectiveness of the firesafety education programs. If the programs are not achieving the objectives, then alternate plans and programs will be needed. Reports using baseline data and documentation should be prepared for presentation. In addition, testimonials or letters of support from the target audience can be used to illustrate the program's effectiveness. Not all programs will result in a reduction of fires; it is very difficult to count how many fires the program may have prevented. What can be shown is a gain in knowledge or awareness which will lead to fewer fires in the future. Demonstrate a change in behavior or attitude which will affect the fire incidence rate. All such information should be analyzed.

Revise Action Plans and Objectives

Action plans and program objectives provide the basis for well organized and effective programs. Once revised and approved, the plans and objectives should be used as management tools. However, they are never static and should be revised periodically to reflect changes in the local fire problem, goals, program content, or resources (financial, technological, material).

The information contained in this chapter about planning and implementing firesafety education programs can be used to good effect by fire personnel, whether they are career, volunteer, full time, or part time. The process can be very lengthy for extensive programs, or accomplished in a relatively short amount of time for more limited programs.

What is important is that the fire service plans for firesafety education programs. It is essential that this planning include members of the community beyond the fire service. Well targeted, organized, effective, measurable programs supported by the fire department and the community are those most successful at saving lives, preventing injuries, and reducing property damage. Firesafety education works when it is well planned and supported.

LEARN NOT TO BURN

Since the mid 1970s, the National Fire Protection Association has spearheaded a number of national projects in public firesafety education for people of all ages. Under the umbrella slogan "Learn Not to Burn," now also being used by many fire departments and other organizations for their educational efforts, NFPA has developed programs and activities ranging from public service anouncements on television to the *Learn Not to Burn Curriculum* for children from kindergarden through the eighth grade. For three years, NFPA also conducted the Learn Not to Burn Competition to recognize and publish information about outstanding fire prevention education programs in communities, industry, health care facilities, educational institutions, government at all levels, and the military.

More than 20 television "spots" featuring actor Dick Van Dyke have been distributed free of charge to TV stations throughout the United States and, on request, to other stations throughout the world. Each 30-second spot focuses on a single aspect of fire and burn prevention, illustrating—often with humor—what *to* do rather than what *not* to do: Van Dyke rolls back and forth on the floor to show how to put out a clothing fire; and he crawls on hands and knees to demonstrate escape from a smoky fire.

Radio, print and audiovisual materials have augmented the TV spots in NFPA's Learn Not to Burn media campaign.

Twenty-five key firesafety behaviors are the core of the *Learn Not to Burn Curriculum*, which during its first five years was introduced in an estimated 37,000 classrooms nationwide. In some jurisdictions, use of the Curriculum has been mandated at the state or county level as well as in local school systems.

The most comprehensive firesafety program available nationally for school children, the Curriculum organizes life saving material to be taught year-round, using state of the art teaching methods. Lessons in the ring binder Curriculum manual are arranged in three levels, for grades kindergarten through two; three through five; and six through eight.

The 25 specific behaviors, which can be integrated into the on-going subject areas of the classroom, are intended to teach children to prevent fires; to protect themselves, their families, and friends from fire; and to persuade others to be mindful of the need for firesafe behavior at all times. Each behavior is outlined for the teacher in terms of three objectives—attitude, knowledge, and behavior—and suggests participatory affirming activities that are within the subjects normally taught at that grade level (e.g., arithmetic, science, English, social studies, art and music). Each Curriculum manual contains firesafety background information, evaluation instruments, and an extensive reference guide to supporting materials.

Implementation of the Curriculum is facilitated by a national network of 10 Regional Representatives, each an expert in teaching firesafety. The "Regional Reps" are available to work with fire departments, fire marshals, parent-teacher organizations, and other groups toward adoption of the Curriculum for classroom teaching. The "Reps" also work with individual teachers to integrate firesafety information into existing school subjects.

For students from the ninth grade through high school, another, less structured, program is "Firesafety for the Rest of Your Life." Information in this phase of the Learn Not to Burn teaching effort is applicable to all age groups, including the elderly.

By starting now to educate youngsters in school, it is foreseen that as today's students grow up, they will carry with them throughout their lives a healthy respect for fire, along with factual information on how to prevent fire, and what to do if one occurs. As present day students pass along good firesafety behaviors to their own children, successive generations of Americans should experience far fewer preventable fires and should be in a better position to save themselves and others in case of fire.

The best measure of success of the Learn Not to Burn effort is in the number of people saved from burn injury and even from death because of what they learned from the program's fire prevention messages and classroom lessons. During the first 10 years of Learn Not to Burn, NFPA received documented reports of more than 200 lives saved because, in a fire emergency, people remembered what they had learned from the media campaign, the Curriculum, and other parts of the Learn Not to Burn campaign.

Learn Not to Burn activities are coordinated through the NFPA Public Education Programs Department.

Bibliography

References Cited

NFPA. 1983. *Firesafety Educator's Handbook: A Comprehensive Guide to Planning, Designing and Implementing Firesafety Programs.* National Fire Protection Association, Quincy, MA. 1983.

Learn Not to Burn Curriculum®. National Fire Protection Association, Quincy, MA. 1979.

Osterhout, Connie, ed. *Public Fire Education* (IFSTA No. 606). International Fire Service Training Association, Stillwater, OK. 1979.

Additional Readings

"Chuckie's Tale Is Fire Prevention." *Fire Engineering*, Feb. 1976.

"Classifying Juvenile Arsonists." *Fire Command*, Oct. 1985.

"Cooperation Adds Success to Idea for School Fire Safety Program," *Fire Engineering*, Sept. 1975.

"Fire Course for School Pupils," *Fire Engineering*, Oct. 1973.

Fire Education and the News, U.S. Fire Administration, Federal Emergency Management Agency, Washington, DC, Apr. 1982.

Fire Education Planning for Youthful Firesetters, The Children's Museum, Boston, MA, and Whitewood Stamps Inc., Newton, MA, prepared for the National Fire Prevention and Control Administration, Washington, DC, Oct. 1976.

"Fire Prevention Bureau: One Way to Overcome the Manpower Crunch," *Fire Command*, Jan. 1974.

"Fire Prevention Promoted by Volunteer Unit," *Fire Engineering*, Oct. 1973.

"Fire Safety Messages Go Home in Grocery Bags," *Fire Engineering*, Sept. 1972.

"Grocery Bags Carry Safety Messages in Oregon and Washington," *Fire News*, June 1972.

"Home Fires Come First in Lutheran Safety Drive," *Fire Engineering*, May 1974.

"How a Small Fire Department Developed an Imaginative Fire Prevention Program," *Fire Chief*, Dec. 1976.

Master Planning for Public Fire Education, The Children's Museum, Boston, MA, and Whitewood Stamps Inc., Newton, MA, prepared for the National Fire Prevention and Control Administration, Washington, DC. (undated).

"Penny-wise Ben Franklin Talks on Fire Prevention," *Fire Engineering*, Aug. 1971.

"Reaching the Public," Column appearing in most issues of Bimonthly *Fire Journal*, 1965 to date.

"Santa Ana Fire Prevention Program Results in Significant Fire Loss Reduction," *Fire News*, May 1974.

"School Activities Can Get Pupils Excited About Fire Prevention," *Fire Engineering*, Sept. 1974.

FIRE PREVENTION PRACTICES IN COMMERCE AND INDUSTRY

Philip Blye and James Yess

Large or even small fires at an industrial plant can severely interrupt production, disrupt the lives of those who depend upon the plant for a livelihood, and weaken the economy of the community in which the plant is located. Although the primary purpose of any fire prevention strategy is to protect lives, corporations have special responsibilities to their own viability and that of the community with respect to loss prevention. A fire occurring in an industrial setting is usually far more threatening to the community than residential fires, because of the materials and chemicals used in manufacturing processes. In addition to this immediate threat, ripple effects follow in the wake of a serious industrial fire. Among these consequences are:

1. Suspension of production can force customers to seek alternate suppliers with whom they remain even after the damage from the fire has been corrected and production has been resumed.
2. Business records may be destroyed, requiring a costly, time consuming process of reconstructing accounts, mailing lists, sales records, inventories, etc.
3. While repairs are in progress, employees that are laid off may find permanent employment elsewhere.
4. Insurance may be insufficient to cover property reconstruction and equipment replacement, thus forcing the corporation to draw from its assets to cover these costs.
5. If the fire spreads to neighboring properties, the owners may sue to recover damages.
6. Large claims paid by insurance companies are often the stimuli for increasing insurance premiums.
7. In some cases, the cumulative effect of these possibilities may leave no alternative other than bankruptcy and permanent closing of the plant.

The community also directly suffers from an industrial fire, through the initial cost of extinguishing the fire to other, more long lasting results such as environmental damage, lost corporate taxes, personal income removed from the local economy, and unemployment compensation payments. Sound fire prevention practices help to protect a company and the community from the many difficulties resulting from an industrial fire.

OBJECTIVES OF A FIRE PREVENTION PROGRAM

The primary goal of a fire prevention program is to reduce or eliminate fire in the workplace by heightening the firesafety awareness of all employees. While it usually is the specific responsibility of one person in the plant to institute, teach, and monitor sound fire prevention practices, the primary goal of a fire prevention program is to provide all employees with the information necessary to recognize hazardous conditions and take appropriate action before such conditions result in a fire emergency.

Several objectives that collectively help the plant manager attain the overall goal of firesafety flow from this primary goal. These objectives are discussed individually below.

Role of the Plant Firesafety Officer

Regardless of the size of a manufacturing plant, a specific individual should be responsible for the development and implementation of the fire prevention program. In large facilities, a firesafety officer administers a firesafety committee comprised of representatives from the various divisions of the company. The committee helps the firesafety officer devise and present fire prevention programs. The firesafety officer should also be in charge of coordinating code compliance with the local fire department, establishing and rehearsing evacuation plans, and ensuring that fire suppression and alarm systems are properly installed, inspected, and maintained. Sometimes the firesafety officer is also the plant fire chief, in charge of the industrial fire brigade.

The firesafety officer must be knowledgeable about the fire hazards inherent in the type of industry in which the company is engaged. Specific manufacturing processes are

Capt. Blye is chairperson of the Department of Fire Science Technology at Massasoit Community College in Brockton, MA. He is also currently employed in the fire service. Dr. Yess is an educator and administrator who has developed educational materials for NFPA.

associated with identifiable fire and explosion risks. The duty of the firesafety officer is to become aware of these risks and to protect employees and property from them. Additionally, the officer must be familiar with those behaviors of employees that may create fire hazards. Failure to act, as well as overt action, can be identified and corrected. The officer must also be cognizant of what periods of the day, month and year a factory is more vulnerable to fire, and establish procedures to protect against this vulnerability. It may be that large shipments of highly combustible materials enter the plant at predictable times; if so, these are the periods for heightened vigilance. Close communication with local, county, and state fire officials will also keep the firesafety officer informed about current fire prevention strategies. One of the best sources of firesafety training for employees can be no further away than the local fire department.

Once a fire prevention plan has been adopted at an industrial site, the firesafety officer must regularly monitor the plan's effectiveness in preventing fires. The only sure way to monitor such a plan is by frequent observations of the physical facilities and the behavior of the employees at work and during nonwork periods. The firesafety officer must conduct regular inspections, covering important basic aspects of plant operations. Features to be inspected include:

1. Smoking habits of employees, in the plant as well as in the administrative offices.
2. Electrical equipment, wiring, and controls.
3. Fire alarm systems.
4. Fire extinguishing equipment.
5. Integrity of plant construction.
6. Heating, ventilating, and air conditioning (HVAC) systems.
7. Storage of combustible material.
8. Housekeeping practices.
9. Use of flammable and combustible liquids and gases.
10. Security.
11. Industrial processes such as painting, cutting, welding, and creating flammable dust.
12. Removal of wastes from industrial processes.

Other important functions of the firesafety officer are recordkeeping and filing of incident reports. The reports can be used to document the need for improvements in operating procedures, employee behavior, and equipment. In general, the company should establish a mechanism for regular review of the inspection logs and fire incident reports by management level personnel with sufficient authority to take quick corrective action.

Essentials of a Fire Prevention Training Program

The person who should coordinate a plant fire prevention training program is the firesafety officer, who may also have the title of plant fire chief. This individual is the most familiar with the firesafety precautions and vulnerable aspects of a factory. It takes a special talent, however, to operate successful fire prevention training programs because the focus of attention is on the intangible aspects of human beings as workers. The aim is to create an attitude of safety in the workplace and recognition on the part of employees that protecting the plant from fire also protects their lives and livelihood.

The human factor, whether an individual acts deliberately or indirectly through ignorance or negligence, is the major cause of fires. Therefore, a concerted effort to educate workers in fire prevention will go a long way in protecting lives and property. Employees must realize that a fire at their workplace may not only cause them physical harm but may mean loss of income until repairs from the fire are completed and production resumed. In some cases, the option of returning to the same job never presents itself because a serious fire at an industrial site forces the company to go out of business.

Usually, training programs based on what workers might perceive as a remote threat to job security are less effective than a clear commitment on the part of the company to become more safety conscious. Concerned supervisors, short workshops on fire prevention, helpful suggestions made during inspections by the firesafety officer, and awards presented to employees who help to spot or correct a fire hazard all help create the desired fire conscious attitude in employees.

The success of a fire prevention program can be measured by factors such as the reduction of fire incidents, the number and quality of suggestions made by employees to reduce fire hazards, and the results of surveys which compare the level of employee firesafety awareness prior to and after the initiation of a training program.

The test of a successful fire prevention program is nowhere more in evidence than in the inspection logs of the firesafety officer. Through accurate and regularly kept inspection records, the firesafety officer will be able to see the long term results of fire prevention efforts and how they might be further improved.

It is important to realize that the fire prevention program of an industrial plant cannot be limited to those locations—the production and storage areas—most often identified with industrial fires. Because factories also house offices where serious fires can originate, office personnel must be included in any employee fire prevention training program. The general office staff's awareness of firesafety can be heightened by comparing office fire hazards with those at home involving ordinary combustibles. The elements of a fire prevention program for office personnel may include advice about:

1. Use and disposal of smoking materials.
2. Checking for frayed electrical cords and faulty appliances.
3. Proper use of electrical equipment and appliances.
4. Unplugging coffee pots and other heat producing equipment (when not in use), and portable heaters at the end of each work day.
5. Correct storage of flammable and combustible materials.

All employees should be instructed on the locations and proper use of fire extinguishers in their work areas. They should also all know how to activate the building's fire alarm system, and be familiar with evacuation plans and routes. Failure to give this instruction is an indication that the fire prevention program is deficient in some respect.

Operators of mechanical equipment have a special responsibility to themselves and others with respect to firesafety. A fire prevention program should alert them to:

1. Variations in the operating pressure and temperature of equipment from levels recommended by the manufacturers.
2. Potential sources of ignition from cutting, welding, faulty wiring, friction (due to such factors as belts operating at the wrong tension or poorly lubricated bearings); and mechanical sparks caused by the moving metallic parts of a faulty or poorly maintained machine.
3. Corrosion of motor parts, or build up of dust or lint around machinery.
4. Leakage of lubricants from motors or drip pans onto floors and walls.

Employees should be urged to report any of these and other hazardous conditions to their supervisor. Preventing such conditions requires adherence to a schedule of regular periodic equipment inspection, maintenance, and testing.

Plant supervisors, firesafety officers, and factory workers must also be prepared to prevent fires in the general plant setting. The fundamental point to remember in all fire prevention activities is that all sources of combustion must be separated from all sources of ignition. Sources of combustion could be an undetected gas leak, a high concentration of dust or flammable vapors, a flammable liquid spill, oil soaked rags, or improperly stored trash. Ignition sources often are faulty electrical wiring, equipment or controls.

Another common source of ignition, however, is not associated with the plant or its equipment but with the habits of the workers. Failure to observe smoking restrictions in the workplace or careless disposal of ashes, cigarette butts, and matches cause deaths, injuries, and the destruction of property worth millions of dollars each year. A sound fire prevention program limits smoking to places designated as safe for this activity and ensures that there is an abundant supply of properly designed ash trays and other noncombustible receptacles that are emptied frequently.

Many fires in the workplace can be prevented through good housekeeping practices. These practices are relatively inexpensive and uncomplicated. While a well trained custodial staff is the key to safe housekeeping methods, all employees can make significant contributions to fire prevention where housekeeping is involved. General precautions include the following:

1. Trash and other waste should be stored in covered metal containers.
2. Work areas should be kept clean and free of flammable debris.
3. Hazardous materials should only be stored in designated locations and in properly capped or ventilated containers, depending upon the material.
4. Most flammable liquids should be kept tightly covered; gasoline requires vented containers.
5. Gas cylinder valves should be closed when not in use.
6. Trash should be prevented from collecting in corners, under machinery, in stairwells, and in other out of the way locations.
7. Oily or greasy deposits and condensates should be wiped up.

An indirect, yet contributing, factor involved with industrial fires is the abuse of controlled substances by some employees. Alcoholism and other drug abuse can cause the employee to be careless about fire hazards or can prevent the employee from properly reacting in an emergency situation. Part of any fire prevention program should be an awareness of how the intoxicated employee can be a threat to the safety of other workers in the plant. Companies should refer employees with these problems to programs for appropriate treatment.

Fire Prevention Through Safe Plant Operations

Efficiency and economy as well as fire prevention are enhanced through the regular maintenance and repair of equipment used in the manufacturing process and in other facets of plant operation. As part of the normal operating procedures of a factory, a team of safety inspectors should check for properly operating valves; electrical wiring and motors; and ventilating, heating, and air conditioning equipment. Flues and vents must be kept free of dirt and combustibles. A strict maintenance schedule should be observed for such items as motors, switches, wiring, fans, bearings, chains, and conveyor belts. Loose fittings and supports should be tightened. The atmosphere of the plant should be monitored regularly with a flammable vapor indicator to detect any potentially dangerous conditions. Fire doors should be checked so they move freely and can close fully. Combustible wastes should be removed from the building every day and more frequently if necessary. Safe shutdown procedures must be followed at the end of a work day or in an emergency.

Security Measures That Enhance Fire Prevention

During the regular working day, a plant is susceptible to fire hazards associated with production and with the expenditure of high levels of electrical or mechanical energy. However, it is wrong to assume that a plant is less at risk from fire at off periods such as evenings, weekends, and vacations; this is why the plant maintenance crew and security force need to be totally integrated into the fire prevention plan. The custodians and security guards must be trained to notice and correct hazards or sound the alarm. Often it is their diligence that spots an overheated motor, faulty wiring, or a smoldering cigarette.

The security force especially must protect the plant from intruders and potential arsonists. Security personnel making their rounds throughout the facility should pay special attention to places where a person might hide or where combustibles are generally stored. Every point of possible entry by an intruder should be checked, including doors, windows, gates, and loading docks.

Bibliography

NFPA Codes, Standards, Recommended Practices and Manuals. (See the latest *NFPA Codes and Standards Catalog* for availability of current editions of the following documents.)

NFPA 1, *Fire Prevention Code.*

NFPA 10, *Standard for Portable Fire Extinguishers.*

NFPA 27, *Recommendations for Organization, Training and Equipment of Private Fire Brigades.*

Additional Readings

Conducting Fire Inspections: A Guidebook for Field Use, National Fire Protection Association, Quincy, MA, 1982.

Employee Evacuation: Action for Survival. (Film and Videotape), National Fire Protection Association, 1984.

Linville, J. L., ed., *Industrial Fire Hazards Handbook*, National Fire Protection Association, 2nd ed., 1984.

NFPA Inspection Manual: A Guide to Property Inspection for Fire Protection and Life Safety, National Fire Protection Association, 5th ed., 1982.

TRAFFIC AND EXIT DRILLS

Revised by Dr. John L. Bryan

Traffic and exit drills are important components of effective firesafety. During a fire or an emergency, fire fighters, police, fire brigade staff, industry management personnel, and other people who perform essential duties must be able to move to locations where they are needed and where they can perform their necessary functions. In addition, in fire or emergency situations, it is usually necessary to quickly and efficiently evacuate building occupants.

To achieve evacuation it is necessary to devise and implement plans and practices that move people to safe areas in or outside the building of fire origin and provide access for various emergency personnel. Implicit in such plans and practices is the need for pedestrian and vehicular traffic control. Even well drilled people are likely to be panicky and behave erratically in times of real emergency unless guided and directed. Without direction and control, evacuees may actually hamper or nullify safety procedures.

This chapter examines traffic control and exit drills with special emphasis on health care facilities. Additional helpful information on traffic and exit drills can be found in Section 7, Chapter 3, "Concepts of Egress Design." Also, Section 9 "Hazards to Life in Structures," contains specific information regarding life safety in various occupancies.

TRAFFIC CONTROL

Emergency traffic control can be divided into two categories: (1) external—on public streets and highways, and (2) internal—inside private property. Both external and internal traffic control can be made more effective with careful preparation.

External Traffic

The traffic direction on public streets and highways is usually the responsibility of the local law enforcement agency. Occasionally, however, a facility may employ private security personnel or guards to help ensure the safety of employees and property. This should be done only after consultation with and authorization by the appropriate law enforcement officials.

Regardless of who is responsible for external traffic control on public streets and highways, employees in the fire risk prevention and control management program of the facility must be able to reach the location of their assignment.

Many large companies provide members of their emergency control forces with identification approved by both the local law enforcement agency and the Civil Defense authorities. This identification allows passage through roadblocks into emergency or disaster areas and it may also be used to gain emergency admittance to restricted areas inside the facility. With external vehicle and pedestrian traffic control for safety, external traffic planning should include primary and alternate approach routes for fire equipment and other emergency vehicles.

NFPA 601, *Standard for Guard Service in Fire Loss Prevention*, and NFPA 601A, *Standard for Guard Operations in Fire Loss Prevention*, provide information on guard service and operations which may be involved in traffic firesafety management for facilities.

Internal Traffic

Most police departments will not enter private property to control traffic unless prior arrangements have been made to do so. Generally, if there is a private security force, traffic control for both general and emergency movement is usually part of their responsibility. If there is no security force, company personnel should be assigned and trained to control both pedestrian and vehicular traffic.

One person should be assigned to the entrance designated for use by fire department and other emergency control forces to direct them to the fire scene. In a large facility, it may be advisable to have an employee accompany incoming emergency forces to the specific area and building involved.

Personnel should also be assigned and trained to control access to the emergency scene, and admit only identified emergency personnel. This precaution will prevent unnecessary exposure to personnel and will minimize general confusion.

John L. Bryan is chairman, Department of Fire Protection Engineering, University of Maryland, College Park, MD.

Planning and preparing before emergencies occur should not be limited to traffic control. If the facility has underpasses, overpasses, tunnels, or overhead piping across access drives, the fire department should be invited to bring in its largest piece of fire apparatus to ensure that there is adequate clearance through the facility.

FIRE EXIT DRILLS

Well marked exits do not ensure life safety during a fire. Exit drills are necessary so that occupants will know how to make an efficient and orderly escape according to NFPA 101, *Life Safety Code®* (hereinafter referred to as NFPA 101) which contains detailed information on exit drills in individual occupancies. Exit drills are required in schools and health care facilities and are common in industries with high hazards. Some form of exit drill should be conducted wherever or whenever possible to avoid confusion and ensure the evacuation of all occupants during a fire (Keating et al 1978). Personnel should be assigned to check exits for availability, to search for stragglers, to count occupants once they are outside the fire area, and to control reentry into the building before it is safe. (See Fig. 3-3A.)

FIG. 3-3A. Orderly evacuation of school building during fire exit drills.

This reentry behavior for rescue purposes of persons involved in fire incidents should be noted. In a study of 335 fire incidents involving 584 persons, it was found that 163 persons, approximately 28 percent of the study population, reentered the fire buildings following evacuation (Bryan 1975). Approximately 10½ percent of those persons who reentered the buildings did so to alert or assist other persons.

Determining when and what area to evacuate is probably the most important decision in a fire emergency. Any area at all affected by heat, flame, or smoke should be evacuated; in case of doubt, the entire building should be evacuated.

The fire loss prevention and control management staff are responsible for planning exit drills. Plans should be discussed with both middle and line management to ensure understanding and cooperation. If there is no fire loss prevention and control manager, the plant, facility, or building manager should assume this responsibility or assign it to a staff member.

All employees should recognize the evacuation signal and know the exit route they are to follow (Keating and Loftus 1977). Upon hearing the signal, they should shut off equipment and report to a predetermined assembly point. Primary and alternate routes should be established and all employees should be trained to use either route (Keating et al 1978; Herz et al 1978).

The problem with audible evacuation signals is the conditioning of the population within the occupancy to ignore the signal due to numerous false alarms. An investigation of an apartment building fire found that the alarm system actuation to initiate evacuation behavior was ignored by many of the building occupants due to the conditioning effect of numerous prior false alarms: 44 percent of the building occupants believed the alarm signal was a false alarm (Scanlon 1978).

When employees are assembled, the manager or supervisor of each area should account for all personnel. Missing employees should be immediately reported to the fire loss prevention and control manager and responding fire department personnel so that search and rescue efforts can be initiated. Only trained fire fighting search and rescue personnel with adequate protective equipment should be permitted to reenter an evacuated area.

After each exit drill a meeting of the responsible managers should be held to evaluate the success of the drill and to solve any problems that may have arisen.

One significant improvement to the traditional concept of fire drills in educational occupancies was suggested by a study of fire drills conducted in such occupancies (Phillips 1979). The concept of smoke drills has been established, whereby the occupants are instructed to move through the simulated smoke areas in a crouched position. Students have transferred the smoke drill concept to fire incidents in residential occupancies, with effective results. Obviously, the utilization of smoke drill training may be effective in a fire incident and should be used where applicable.

The timing of drills depends upon the nature of the operation in the facility. Generally, drills conducted a few minutes before the lunch break have been found to minimize loss of time and production. The frequency of drills should be determined by the degree of hazard present and by the complexity of shutdown or evacuation procedures.

If a facility does not maintain a security organization that is responsible for daily inspection of emergency exits and designated evacuation routes, one employee in each area should be assigned this task. Maintenance of doors, panic hardware, exit lights, and emergency illumination should be given high priority, and repairs must be made without delay.

Research has found that for multistoried office buildings, a trained group of floor wardens is the most effective means of monitoring the evacuation of occupants (Keating et al 1978). Adequate training for floor wardens or other personnel is necessary and must be specifically developed to include the procedures of the emergency evacuation plan for the facility.

The lecture method has been used to convey the essential features of the emergency plan to employees in health care facilities (Herz et al 1978). However, it was also found the emergency plan in the facility studied was too general and ambiguous.

The most serious problem with using building monitors is the turnover of personnel due to employee transfers, reassignments, or resignations. Effective evacuation planning and preparation should assign specific responsibilities to staff positions (rather than individuals) within an organization. This assures a continuity of performance despite personnel changes.

A content and time evaluation of fire drill behavior by staff in six nursing homes concluded that a training program of the most modest type can produce changes in both knowledge and behavior of evacuation and fire emergency procedures (Bickman et al 1979). A total of 339 nursing home staff participated in the study that matched a group of 37 persons receiving the training with a control group of 49 persons not receiving the training. The high rate of personnel turnover which appears to be rather typical in nursing home facilities was noted. Following the presentation of the training program, staff members were evaluated on a written knowledge test and their behavior was observed during the conduct of a drill of the emergency plan.

The evacuation signal should be familiar to all employees. Vocal alarm systems (VAS) (Keating et al 1978; Keating and Loftus 1979) reduce the need for employee perception and recognition of a signal, since the system provides vocal communication to the areas designated for evacuation. The 1981 edition of NFPA 101 recommended the use of the VAS in the assembly occupancies. An evaluation of VAS systems in nine buildings found that familiarization with the system or initial activation did not significantly affect the egress behavior of populations (Keating et al 1982). In addition, the investigation determined the evacuation drills were valuable because they gave floor or area wardens an opportunity to rehearse their procedures.

The use of an alerting tone in the frequency range of 2,000 to 4,000 Hz for the VAS is recommended prior to the verbal announcement, which should be specific for the audience and the facility (Keating et al 1978; Shavit 1978).

Health Care Facilities Drills

Fire drills in health care institutions are usually conducted as a part of the orientation program for new employees. Later the drills are supplemented with in service training for the staff personnel, including the emergency procedures. Fire drills in many facilities are conducted once a month on each shift. The training for the drills typically involves instruction and practice for the staff personnel in the various means of moving nonambulatory patients, the procedures for alerting the facility staff, and the method of notifying the fire department. Many facilities have fire departments provide training one or twice a year in the operation of portable fire extinguishers. Some actually provide staff personnel with experience in the operation of extinguishers on external fires.

However, most health care facilities prefer to train their personnel. Most adopt the philosophy that it is the staff's responsibility to assure the safe evacuation of patients initially to an area of refuge, and then to the exterior if necessary. The control of the fire is limited to preventing the spread of heat and smoke by the closing of doors. Doors are closed to protect the occupants and to inhibit or restrict the propagation of the smoke and heat throughout the facility. Staff personnel have effectively accomplished evacuations of numerous patients under fire conditions, or have protected the patients in their rooms by closing doors (Bryan and DiNenno 1980).

Thus, the evacuation process may be considered in four sequential phases: (1) the manpower supply phase, (2) the patient preparation phase, (3) the patient removal phase, and (4) the rest and recovery phase (Archea 1979). This approach focuses on the occupants in the fire threatened area and the patients in or adjacent to the fire area. Removing immediately threatened patients, closing doors to the room of fire origin and to adjacent patient rooms, would be compatible with this four phase approach to evacuation.

A detailed report on the fire evacuation organization, training, and drills involved in a 502 bed acute care teaching hospital with 2,500 employees has been published (Elliot and Scheidt 1983).

The fire problem in boarding homes which has recently come to public attention, is thought to result from the deinstitution of patients from mental health facilities and an apparent lack of firesafety standards in these facilities. The 1985 edition of NFPA 101 includes a new chapter (21) devoted to the fire protection and life safety requirements for board and care facilities.

The ultimate evaluation of fire drill and emergency plans has two factors: the performance of the occupants in a fire incident, and the effectiveness of the behaviors used in accordance with the fire drills or the fire emergency plan. In a World Trade Center fire incident report (Lathrop 1976), building occupants tended to attempt to verify their beliefs about the threat of fire by physical clues, primarily smoke in the occupants' area. The report also indicated public address messages were not sufficient to alleviate spontaneous evacuations when occupants saw smoke on their floor. Further, occupant evacuation was reported (9th through 22nd floors) due to the perception and concern that a valid fire threat existed. In actuality, the fire did not require such an extensive evacuation.

Successful evacuation of personnel from the two floors above and below a fire in a 28 story high rise college dormitory has been reported (Nygren 1972). Finally, to allow free evacuation flow of the occupants down the stairways, and allow fire department personnel to move up the stairs, the stairs have been marked for occupant and fire department movement. The fire department movement stair has a red circle on the door, 6 in. (152 mm) diameter, an is also utilized for ventilation. The occupant stair is marked with a 6 in. (152 mm) green circle.

Bibliography

References Cited

Archea, John. 1979. "The Evacuation of Non-Ambulatory Patients from Hospital and Nursing Home Fires: A Frame Work for a Model." *NBSIR 79-1906*. Center for Fire Research, Gaithersburg, MD.

Bickman, L. E., et al. 1979. "An Evaluation of Planning and Training for Fire Safety in Health Care Facilities—Phase Two." *NBS-GCR-79-179*. Center for Fire Research, Gaithersburg, MD.

Bryan, John L. 1977. *Smoke as a Determinant of Human Behavior in Fire Situations (Project People)*. Department of Fire

Protection Engineering. University of Maryland, College Park, MD.
Bryan, John L., and DiNenno, Philip J. 1980. "Human Behavior in a Nursing Home Fire." *Fire Journal*. Vol 73, No 3. pp 82-87, 126-127, 141-143.
Elliott, Shirley, and Scheidt, James. 1983. "Hospital and Fire Department Unite to Design New Code Red Program." *Fire Journal*. Vol 73, No 4. pp 47-54.
Herz, E., et al. 1978. "The Impact of Fire Emergency Training on Knowledge of Appropriate Behavior in Fires." *NBS-GCR-78-137*. Center for Fire Research, Gaithersburg, MD.
Holton, David. 1981. "Boarding Homes—The New Residential Fire Problem?" *Fire Journal*. Vol 74, No 2. pp 53-56.
Keating, John P., et al. 1978. *An Evaluation of the Federal High Rise Emergency Evacuation Procedures*. Department of Psychology. University of Washington, Seattle, WA.
Keating, John P., and Loftus, Elizabeth F. 1977. "Vocal Emergency Alarms in Hospitals and Nursing Facilities: Practice and Potential." *NBS-GCR-77-102*. Center for Fire Research, Washington, DC.
Lathrop, James K. 1976. "Two Fires Demonstrate Evacuation Problems in High-Rise Buildings." *Fire Journal*. Vol 70, No 1. pp 65-70.
Nygren, Ronald G. 1972. "Alarm Signaling and Evacuation in A High-Rise University Resident Hall." *Fire Journal*. Vol 66, No 2. pp 5-6, 11.
Phillips, Anne W. 1979. *To Keep Them Safe*. National Smoke, Fire and Burn Institute, Inc., Brookline, MA.
Scanlon, Joseph. 1978. *Human Behavior in a Fatal Apartment Fire—Research Problems and Findings*. Emergency Communications Research Unit. Carleton University, Ottawa, Ohio.
Shavit, Gideon. 1978. "Evacuation: Testing the Effect of Voice-Message Formats." *ASHRAE Journal*. pp 38-41.

NFPA Codes, Standards, Recommended Practices and Manuals. (See the latest edition of *NFPA Codes and Standards Catalog* for availability of current editions of the following documents.)

NFPA 72A, *Standard for the Installation, Maintenance and Use of Local Protective Signaling Systems for Guard's Tour, Fire Alarm and Supervisory Service.*
NFPA 101, *Code for Safety to Life from Fire in Buildings and Structures.*
NFPA 601, *Standard for Guard Service in Fire Loss Prevention.*
NFPA 601A, *Standard for Guard Operations in Fire Loss Prevention.*

SECTION 4
THE CHARACTERISTICS AND BEHAVIOR OF FIRE

CHEMISTRY AND PHYSICS OF FIRE

Revised by Dr. John deRis

This chapter presents basic definitions of some of the physical properties and chemical terms applicable to the chemistry and physics of fire; it also discusses combustion, the principles of fire, heat measurement, heat transfer, and heat energy sources (sources of ignition).

The material contained in this chapter does not attempt to offer a comprehensive course of instruction on the subject, but is intended to present basic background reference material applicable to this and other sections of this HANDBOOK.

The remaining chapters in this section discuss explosions, combustion products and life safety, and fire and explosion control.

BASIC DEFINITIONS AND PROPERTIES

Atoms: The basic particles of chemical composition are called atoms. Atoms are extremely minute. Substances composed of only one type of atom are called elements. An atom has a compact core or nucleus around which electrons (negatively charged units of matter) travel in orbit. The nucleus is made up of protons (positively charged) and neutrons (no charge). Substances which have loosely bound electrons generally are good electrical and thermal conductors. Those with rigidly bound electrons (so that no easy transfer of electrons is possible) are good insulators.

Molecules: Combined groups of atoms are called molecules. Molecules composed of two or more different kinds of atoms are called compounds.

Chemical Formula: The chemical formula shows the number of atoms of the various elements in the molecule, and indicates their arrangement. The chemical formula for propane is $CH_3CH_2CH_3$ where C stands for carbon and H for hydrogen. (See Table 4-1A for symbols of other elements.)

Atomic Number of an Element: The number of electrons or protons of a particular atom determines the atomic number of that element, and its place in the periodic table.

Dr. deRis is Manager, Basic Research at Factory Mutual Research Corporation in Norwood, MA.

Atomic Weight of an Element: The atomic weight of an element is proportional to the weight of its atom. The atomic weight of carbon (isotope C-12) is arbitrarily assigned the value of 12.000. Table 4-1A gives the atomic weights of elements.

Molecular Weight of a Compound: The molecular weight of a compound is the sum of the atomic weights of all atoms in its molecule. The subscripts following an element's symbol indicate the number of atoms of that element present in the compound.

Gram Molecular Weight: The gram molecular weight of a substance (in grams) is equal to its molecular weight.

Specific Gravity: Specific gravity is the ratio of the weight of a solid or liquid substance to the weight of an equal volume of water. The scales of the most commonly used hydrometers are based on a specific gravity of 1 for water at 4°C [1 cc (0.06 cu in.) of water at 4°C (39.2°F) weighs 1 g (0.0353 oz)].

Gas Specific Gravity: Gas specific gravity is the ratio of the weight of a gas to the weight of an equal volume of dry air at the same temperature and pressure. It may be evaluated by the following: gas specific gravity equals its molecular weight divided by 29, where 29 is the composite molecular weight of dry air.

Buoyancy: The upward force exerted on a body or volume of fluid by the ambient fluid surrounding it. If a volume of a gas has positive buoyancy, then it is lighter than the surrounding gas and will tend to rise. If it has negative buoyancy, it is heavier and will tend to sink. The buoyancy of a gas depends upon both its molecular weight (e.g., gas specific gravity) and its temperature.

If a flammable gas with a gas specific gravity greater than 1 escapes from its container, it will typically sink to a low level and will travel considerable distances to potential sources of ignition. Carbon dioxide (CO_2) which has a molecular weight of 44, is heavier than air and, when discharged from an extinguisher, will accumulate near the ground. Generally, the higher the temperature of a gas, the lighter or less dense it is. Thus, hot products of combustion tend to rise.

TABLE 4-1A. Chemical Elements

Based on the assigned relative atomic mass of the Carbon-12 isotope equal to 12.00. Most elements consist of isotope mixtures. Elements with atomic weights in parentheses are unstable isotopes.

Element	Symbol	Atomic No.	Atomic Weight
Actinium	Ac	89	(227)
Aluminum	Al	13	26.9815
Americium	Am	95	(243)
Antimony, stibium	Sb	51	121.75
Argon	Ar	18	39.948
Arsenic	As	33	74.9216
Astatine	At	85	(210)
Barium	Ba	56	137.34
Berkelium	Bk	97	(247)
Beryllium	Be	4	9.0122
Bismuth	Bi	83	208.980
Boron	B	5	10.81
Bromine	Br	35	79.904
Cadmium	Cd	48	112.40
Calcium	Ca	20	40.08
Californium	Cf	98	(251)
Carbon	C	6	12.011
Cerium	Ce	58	140.13
Cesium	Cs	55	132.9055
Chlorine	Cl	17	35.453
Chromium	Cr	24	51.996
Cobalt	Co	27	58.9332
Columbium, see niobium			
Copper	Cu	29	63.546
Curium	Cm	96	(247)
Dysprosium	Dy	66	162.50
Einsteinium	E	99	(254)
Erbium	Er	68	167.26
Europium	Eu	63	151.96
Fermium	Fm	100	(257)
Fluorine	F	9	18.9984
Francium	Fr	87	(223)
Gadolinium	Gd	64	157.2
Gallium	Ga	31	69.72
Germanium	Ge	32	72.59
Gold, aurum	Au	79	196.9665

Element	Symbol	Atomic No.	Atomic Weight
Hafnium	Hf	72	178.49
Helium	He	2	4.0026
Holmium	Ho	67	164.9303
Hydrogen	H	1	1.0080
Indium	In	49	114.82
Iodine	I	53	126.9045
Iridium	Ir	77	192.2
Iron, ferrum	Fe	26	55.85
Krypton	Kr	36	83.8
Lanthanum	La	57	138.905
Lawrencium	Lr	103	(257)
Lead, plumbum	Pb	82	207.2
Lithium	Li	3	6.941
Lutetium	Lu	71	174.97
Magnesium	Mg	12	24.305
Manganese	Mn	25	54.9380
Mendelevium	Md	101	(256)
Mercury, hydrargyrum	Hg	80	200.59
Molybdenum	Mo	42	95.94
Neodymium	Nd	60	144.2
Neon	Ne	10	20.179
Neptunium	Np	93	237.0482
Nickel	Ni	28	58.71
Niobium, columbium	Nb	41	92.9064
Nitrogen	N	7	14.0067
Nobelium	No	102	(254)
Osmium	Os	76	190.2
Oxygen	O	8	15.9994
Palladium	Pd	46	106.4
Phosphorus	P	15	30.9738
Platinum	Pt	78	195.09
Plutonium	Pu	94	(244)
Polonium	Po	84	(210)
Potassium, kalium	K	19	39.10

Element	Symbol	Atomic No.	Atomic Weight
Praseodymium	Pr	59	140.9077
Promethium	Pm	61	(145)
Protoactinium	Pa	91	231.0359
Radium	Ra	88	226.0254
Radon	Rn	86	(222)
Rhenium	Re	75	186.2
Rhodium	Rh	45	102.9055
Rubidium	Rb	37	85.4678
Ruthenium	Ru	44	101.07
Samarium	Sm	62	150.4
Scandium	Sc	21	44.9559
Selenium	Se	34	78.96
Silicon	Si	14	28.086
Silver, argentum	Ag	47	107.868
Sodium, natrium	Na	11	22.9898
Strontium	Sr	38	87.62
Sulfur	S	16	32.06
Tantalum	Ta	73	180.9479
Technetium	Tc	43	98.9062
Tellurium	Te	52	127.60
Terbium	Tb	65	158.9254
Thallium	Tl	81	204.37
Thorium	Th	90	232.0381
Thulium	Tm	69	168.9342
Tin, stannum	Sn	50	118.69
Titanium	Ti	22	47.90
Tungsten	W	74	183.8
Uranium	U	92	238.029
Vanadium	V	23	50.9414
Xenon	Xe	54	131.30
Ytterbium	Yb	70	173.04
Yttrium	Y	39	88.9059
Zinc	Zn	30	65.38
Zirconium	Zr	40	91.22

In the case of a spill of liquefied natural gas (LNG), the vapors are heavier than air because of their extremely low temperature—despite the fact that the gas specific gravity of methane (CH_4), the principal component, is less than 1 (its molecular weight is 16). LNG spills can be dangerous because the vapors can spread over a wide area with little dilution.

Vapor Pressure and Boiling Point

Because molecules of a liquid are always in motion (the amount of motion depends upon the temperature of the liquid), molecules are continually escaping from the free surface of the liquid to the space above. Some molecules remain in this space while others, due to random motion, collide with the liquid and are "recaptured."

If the liquid is in an open container, molecules (collectively called vapor) escape from the surface, and the liquid evaporates. If, on the other hand, the liquid is in a closed container, the motion of the escaping molecules is confined to the vapor space. As an increasing number of molecules strike and reenter the liquid, a point of equilibrium eventually is reached where the rate of escape of molecules from the liquid equals the rate of their return to the liquid. The pressure exerted by the escaping vapor at the point of equilibrium is called vapor pressure. It is measured in pounds per square inch absolute (psia) or kilopascals (kPa).*

As the temperature of a liquid increases, its vapor pressure approaches and ultimately exceeds atmospheric

*Absolute pressure equals the total force exerted against a unit of area. It is measured in pounds per square inch (psi) or Pascals (Newtons per square meter). It often is expressed in fractions or multiples of atmospheric pressure, or in terms of the height of a column of liquid (usually mercury) which will balance the absolute pressure. In situations where pressure gages are used, absolute pressure is determined by adding gage pressure to atmospheric pressure. Atmospheric, or ambient, pressure equals 14.7 psia, 101.3 kPa, or 30 in. of mercury. A kilopascal equals 1000 Pascals.

pressure. At the temperature at which vapor pressure equals atmospheric pressure, boiling takes place and the opposition to evaporation exerted by the atmosphere is neutralized.

Although the vapor pressure of a liquid varies with its temperature, a great many variables affect the rate at which it actually evaporates when exposed to air. These variables include atmospheric temperature and pressure, air movement, incident radiation, specific heat, and latent heat of evaporation.

Vapor-air Specific Gravity (vasg): Vapor-air specific gravity is the weight of a vapor-air mixture resulting from the vaporization of a liquid at equilibrium temperature and pressure, as compared with the weight of an equal volume of air under the same conditions. The specific gravity (density) of a vapor-air mixture thus depends upon the ambient temperature, the vapor pressure of the liquid at the temperature, and the molecular weight of the liquid. At temperatures well below the boiling point of a liquid, the vapor pressure of the liquid may be so low that the vapor-air mixture, consisting mostly of air, has a density which approximates that of pure air, i. e., a vapor-air specific gravity near 1. As the temperature of the liquid increases to the boiling point, the rate of vaporization increases, the vapor displaces the surrounding air, and the vapor-air mixture specific gravity approaches that of the pure vapor specific gravity.

A vapor-air mixture with a density significantly above that of air at the ambient temperature will seek lower levels. On the other hand, diffusion and mixing by convection will limit the distance of travel of mixtures having densities near or less than 1.

The vapor-air density of a substance at ambient temperature may be calculated as:

Let P equal the ambient pressure, p the vapor pressure of the material at ambient temperature, and s its pure vapor specific gravity. Then,

$$vasg = \frac{ps}{P} + \frac{P - p}{P}$$

The first term, $\frac{ps}{P}$, is the contribution of the vapor to the specific gravity of the mixture; the second term, $\frac{P - p}{P}$, is the contribution of air.

EXAMPLE: Find the vapor-air density at 100°F (38°C) and atmospheric pressure for a flammable liquid whose vapor specific gravity is 2, and whose vapor pressure at 100°F (38°C) is 3.0 in. (10.1 kPa) of Hg (mercury), or one-tenth of atmospheric pressure.

$$vasg = \frac{(76)(2)}{760} + \frac{760 - 76}{760} = 0.2 + 0.9 = 1.1$$

Endothermic and Exothermic Chemical Reactions

Heat of reaction is the energy that is absorbed or released when a given reaction takes place. In endothermic reactions, the new substances formed contain more energy than the reacting materials, so energy must be absorbed by the reaction. Exothermic reactions produce substances with less energy than that of the reacting materials, so energy is released by the reaction. While energy may appear in different forms, it is usually in the form of heat, added to or released from a chemical reaction.

COMBUSTION

Combustion is an exothermic, self-sustaining reaction involving a solid, liquid, and/or gas-phase fuel. The process is usually (but not necessarily) associated with the oxidation of a fuel by atmospheric oxygen with the emission of light. Generally, solid and liquid fuels vaporize before burning. Sometimes a solid can burn directly by glowing combustion or smoldering. Gas-phase combustion usually occurs with a visible flame. If the process is confined so that a rapid pressure rise occurs, it is called an explosion.

Oxidation Reactions

Oxidation reactions in fires are exothermic; that is, one of the products of the reaction is heat. These reactions are often complex, and many are not completely understood. However, certain helpful statements can be made.

For an oxidation reaction to take place, a combustible material (fuel) and an oxidizing agent both must be present. Fuels include innumerable materials not already in their most highly oxidized state. Whether or not a particular material can be further oxidized depends upon the chemistry of the material. For practical purposes, material consisting primarily of carbon and hydrogen can be oxidized. Most combustible solid organic materials and flammable liquids and gases contain large percentages of carbon and hydrogen.

The most common oxidizing material is oxygen in the air. Air is approximately one-fifth oxygen and four-fifths nitrogen. Certain chemicals that readily release oxygen under favorable conditions [(i.e., sodium nitrate ($NaNO_3$) and potassium chlorate ($KClO_3$)] are among the less common but well known oxidizing agents met in fires. A few combustible materials, such as cellulose nitrate, contain oxygen combined in their molecules so that partial combustion may occur without oxygen from any outside source.

Combustion also may occur, however rarely, in special cases in an atmosphere of chlorine, carbon dioxide, nitrogen, and some other gases without oxygen being involved. For example, zirconium dust can be ignited in carbon dioxide.

Ignition (Piloted Ignition and Autoignition): Ignition is the process of initiating self-sustained combustion. If ignition is caused by the introduction of some small external flame, spark, or glowing object (ember), it is piloted ignition. If it occurs without the assistance of an external pilot source, it is autoignition.

The ignition temperature of a substance is the minimum temperature to which it must be heated for it to ignite. Usually the piloted ignition temperature of a substance is considerably lower than its autoignition temperature.

In general, both fuel and oxygen molecules must be first excited to some activated state before they can interact chemically to produce heat. This excitation can be induced by other excited molecules from a nearby flame or spark, or by raising the general temperature of the fuel. After chemically interacting, the fuel and oxygen produce other excited molecules as well as heat. If there is sufficient fuel and oxygen as well as a sufficient number of excited species, ignition will occur as a chain reaction wherein the rate of production of activated molecules exceeds their natural rate of decay. Once ignition has occurred, it will continue until all the available fuel or oxidant has been consumed, or until the flame is extinguished by cooling, or by reducing the number of excited molecules, or by some other means.

In general, a self-sustained ignition can occur only in those situations which are capable of supporting self-sustained combustion. For example, if the ambient pressure (or ambient oxidant concentration) is insufficient for sustaining combustion, it also will be insufficient for ignition.

For most combustible solids and liquids the initiation of the flame reaction occurs in the gas or vapor phase. (Exceptions are pure carbon, certain metals, and certain smoldering cases which involve direct solid surface oxidation.) Also, for most solids or liquids, sufficient thermal energy must first be supplied to convert a part of the fuel to vapor, so one has a combustible gas-phase mixture. In such situations, one can usually identify a minimum solid or liquid temperature which is capable of supplying a combustible mixture near the fuel surface. This minimum temperature is called the piloted ignition temperature, since a pilot is needed to ignite the gas mixture. (For flammable liquids, this is called the flash point.)

In practice, the piloted ignition temperatures of solids and liquids are influenced by the rate of air flow (oxidant), the rate of heating, and the size and shape of the solid or liquid. As a result, reported piloted ignition temperatures depend somewhat upon the specific test methods.

In general, gas mixture ignition temperatures depend upon the composition, ambient pressure, mixture volume, and shape of the vessel, as well as the nature and energy of the pilot. For a given fuel-air mixture there is a minimum pressure below which ignition does not occur. As the temperature increases, less and less pilot energy is required to ignite the mixture until, at a sufficiently high temperature, the mixture will ignite spontaneously. This temperature is referred to as the autoignition (or spontaneous ignition) temperature.

The autoignition temperature of a material may differ substantially with changing conditions. In addition to composition and pressure, other variables known to affect autoignition temperatures are the shape and size of the space where the ignition occurs, rate and duration of heating, type and temperature of the ignition source, and catalytic or other effect of materials which may be present. Because there are differences in test methods (size and shape of containers and methods of heating), different ignition temperatures are sometimes reported for the same substance.

The effect of percentage composition of the fuel-oxygen mixture is shown by the following autoignition temperatures for pentane-air mixtures: 548°C (1,018.4°F) for 1.5 percent pentane in air; 502°C (935.6°F) for 3.75 percent pentane, and 476°C (888.8°F) for 7.65 percent pentane. The following ignition temperatures for carbon disulfide demonstrate the effect of size of space containing the ignitible mixture: in a 200 cm^3 (12 cu in.) flask the ignition temperature was 120°C (248°F); in a 1,000 cm^3 (61 cu in.) flask 110°C (230°F), and in a 10,000 cm^3 (610 cu in.) flask 96°C (204.8°F).

Explosions

Explosions generally occur in situations where the fuel and oxidant have been allowed to mix intimately before ignition. As a result, the combustion reaction proceeds very rapidly without being delayed by the need for first mixing fuel and oxidant. If premixed gases are confined, their tendency to expand with combustion can cause a rapid pressure rise or explosion.

Fires, in contrast, generally occur in situations where the mixing of fuel and oxidant is controlled by the combustion process itself. As a result, the burning rate per unit volume is much lower for fires, and the very rapid increase in pressure characteristic of explosions is not encountered.

For ignition to be possible, an adequate fuel concentration must be available in the particular oxidizing atmosphere. Once ignition occurs, sustaining of combustion requires a continued supply of fuel and oxidant. In the case of combustible gases, vapors, mists, foams, or solid dusts, two types of mixtures—homogeneous (uniform) or heterogeneous (nonuniform)—can exist within the atmosphere. A homogeneous mixture is one in which the components are intimately and uniformly mixed so that a small-volume sample is representative of the whole mixture. The composition of a flammable homogeneous mixture lies between the limits of flammability of the combustible gas, vapor, mist, foam, or dust in the particular atmosphere at a specified temperature and pressure.

The Limits of Flammability: The limits of flammability are the extreme concentration limits of a combustible in an oxidant through which a flame, once initiated, will continue to propagate at the specified temperature and pressure. For example, hydrogen-air mixtures will propagate flames for concentrations between 4 and 74 percent by volume of hydrogen at 70°F (21°C) and atmospheric pressure. The smaller value is the lower (lean) limit, and the larger value is the upper (rich) limit of flammability. When the mixture temperature is increased, the flammability range widens; when the temperature is decreased, the range narrows. (See Fig. 4-1A). A decrease in temperature can cause a flammable mixture to become nonflammable by placing it either above or below the limits of flammability for the specific environmental conditions.

Note in Figure 4-1A that for liquid fuels in equilibrium with their vapors in air, a minimum temperature exists for each fuel above which there is sufficient vapor to form a flammable vapor-air mixture. There is also a maximum temperature above which the fuel vapor concentration is too high to propagate flame. These minimum and maximum temperatures are, respectively, the lower and upper flash points in air. For temperatures below the lower flash point there is insufficient fuel vapor in the gas-phase to sustain a homogeneous ignition. The flash point temperatures for a combustible liquid increase with environmental pressure.

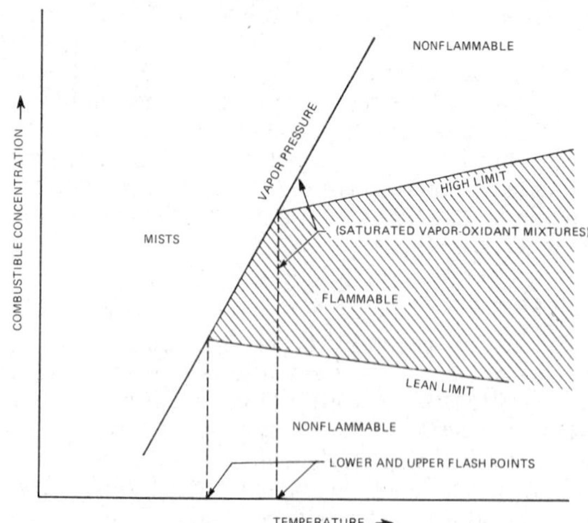

FIG. 4-1A. Saturated vapor-oxidant mixtures should be associated with the sloping vapor pressure line rather than just the upper flash point vertical line.

Fire Point: Fire point is the lowest temperature of a liquid in an open container at which vapors evolve fast enough to support continuous combustion. The fire point is usually a few degrees above the lower flash point. For typical fuels, the minimum rate of vaporization required to support combustion is approximately 4 $(g/m^2)/sec$.

It should be emphasized that, if a source of ignition has been established, fires can spread over liquids whose temperatures are considerably below their lower flash points. In such situations the ignition source or fire itself heats the liquid surface locally so that its temperature rises above the fire point.

Catalysts, Inhibitors, and Contaminants

Catalyst: A catalyst is a substance which greatly affects the rate of a chemical reaction but is not permanently changed by the reaction itself. For example, platinum in a car's catalytic converter causes residual fuel to burn without permanently consuming the platinum catalyst.

Inhibitors: Inhibitors, also called stabilizers, are chemicals which may be added in small quantities to an unstable material to prevent a vigorous reaction. For example, premature polymerization of styrene monomer is inhibited by the addition of at least 10 ppm (parts per million) of tertiary-butyl-catechol (TBC). Fire retardant chemicals usually act as inhibitors. For example, small quantities of chlorine or bromine-related compounds, when added to plastics, can reduce plastics ignitability and ability to support the spread of small flames. Fire retardant chemicals are often less effective in large fires.

Contaminants: Contaminants are foreign materials not normally found in a substance. From a fire hazard standpoint, some contaminants (such as sand in calcium chloride) may be harmless. Inert contaminants such as aluminum oxide or calcium carbonate, when present in sufficient quantities in a plastic, can significantly reduce flammability of the plastic. Dangerous contaminants function by acting as catalysts, or by themselves entering into a potentially dangerous reaction.

Stable and Unstable Materials

Stable Materials: Stable materials are materials which normally have the capacity to resist changes in their chemical composition despite exposure to air, water, heat, shock, or pressure. Stable materials however, may burn. Most solids fall into this category.

Unstable Materials: Unstable materials polymerize, decompose, condense, or become self-reactive when exposed to air, water, heat, shock, or pressure. For example, decomposition of pure gaseous acetylene, hydrazine, or ethylene oxide can result in violent explosions.

PRINCIPLES OF FIRE

Considerable technical knowledge exists concerning the ignition, burning and fire spread characteristics of combustible materials (solids, liquids, and gases). However, most of this knowledge is for very simple geometric arrangements and thus is inadequate for predicting ignition and resultant burning in realistic situations. Nevertheless, insight can be gained from an understanding of these simplified situations.

Perhaps the simplest combustion situation is the burning of premixed, fuel-air gas mixtures involved in explosions. Much is known about premixed flames because of the obvious advantages of conducting controlled experiments. The flammability limits and associated burning rates have been catalogued for most common vapor and gas mixtures (Zabetakis 1965). In addition, scientists now can accurately calculate the burning rates for simple hydrocarbon fuel-air mixtures in terms of their multiple individual chemical reactions (Westbrook and Dryer 1981).

Other simple situations involve the steady burning of fuel droplets or small samples of simple plastics. In such situations the burning takes place in gas-phase laminar diffusion flames where the rate of combustion is controlled by the rate of fuel and air supply.

In contrast to explosions, fires occur in situations where fuel and oxidant are initially unmixed. Their burning rates are restricted primarily by the supply of fuel and oxidant (air) to the flames rather than the basic chemical reaction rates within the flames. In fires, the basic gas-phase combustion process usually occurs along thin flame sheets, called "diffusion flames," which separate regions rich in fuel vapor from regions rich in oxidant. Fuel vapor and oxidant diffuse toward each other from opposite sides of the flame sheet where they combine to produce combustion products and heat, which in turn diffuse away from the flame sheet.

When diffusion flames are small—e.g., from a burning match or candle—they typically appear quite smooth and steady. They are called "laminar" diffusion flames. If the fire is allowed to grow, the flames become unstable and wiggle around in search of more fuel or oxidant. Eventually, as the fire grows, the flame motion becomes truly random, in which case they are called "turbulent" diffusion flames.

Scientists have obtained a relatively clear understanding of small fires involving laminar diffusion flames. For example, they can calculate flame spread rates (Williams 1977), and steady burning rates (Kim et al 1971) of small solid fuels in terms of the basic combustion properties for a variety of simple geometries (e.g., smooth flat surfaces, cylinders, etc.). In these situations, the burning rates are controlled by the convective heat transfer from the flame to the solid fuel which, in response, gasifies and supplies fuel vapors to the flames. The oxidant (air) is supplied to the flames by the upward flow induced by the buoyant hot combustion products. The buoyant flow also may enhance the convective heat transfer from the flames to the solid fuel. In the case of a spreading flame, the spread rate is governed by the forward heat transfer from the flames to the as yet uninvolved fuel which must be preheated before it can provide fuel vapors to the flames. (An analysis of many different fire spread situations which shows how one might predict their spread rates is contained in Williams 1977.)

Larger fires involving turbulent diffusion flames are less well understood because of difficulties in describing the turbulent gas motion, and the flame radiation which is usually the dominant form of heat transfer for larger fires. Experience and measurements have shown that this enhanced role of flame radiation in larger fires often alters the relative flammability ranking of fuels in comparison to their small scale flammability rankings.

The study of these larger (hazardous) scale fires is at the forefront of current fire research. In recent years, scientists, using diverse knowledge of fire processes, have prepared complex computer models which predict the fire development from ignition through its various stages of growth, to full room involvement (flashover), propagation to adjacent rooms, or possibly even to other buildings.

Ignition and Combustion

To illustrate the many physical and chemical processes involved in fires, it is necessary to first discuss the ignition, burning, and eventual extinction of a wood slab in a typical situation, such as a fireplace (Browne 1958).

1. Suppose the wood slab is initially heated by thermal radiation. As its surface temperature approaches the boiling point of water, gases (principally steam) slowly evolve from the wood. These initial gases have little, if any, combustible content. As the slab temperature increases above the boiling point of water, the "drying" process penetrates deeper into the wood interior.
2. With continued heating, the wood surface begins to discolor when the surface temperature approaches 300°C (572°F). This discoloring is visible evidence of pyrolysis, the chemical decomposition of matter through the action of heat. When wood pyrolyzes, it releases combustible gases while leaving behind a black carbonaceous residue called char. This pyrolysis process penetrates deeper into the wood slab interior as the heating continues.
3. Soon after active pyrolysis begins, combustible gases typically evolve rapidly enough to support gas-phase combustion. However, combustion will occur only if there is a pilot flame or some other source of chemically activated molecules sufficient to support a piloted ignition. If no such pilot is present, the wood surface often must be heated to a much higher temperature before autoignition occurs.
4. Once ignition occurs, a diffusion flame rapidly covers the pyrolyzing surface. The diffusion flame shields the pyrolyzing surface from direct contact with oxygen. Meanwhile, the flame heats the fuel surface and causes an increase in the rate of pyrolysis. If the original radiant heat source is withdrawn at ignition, the burning will continue provided the wood slab is thin enough (less than 2 cm or ¾ in.). Otherwise the flames will go out, because the slab surface loses too much heat by thermal radiation and thermal conduction into its interior. If there is an adjacent parallel wood surface (or insulating material) facing the ignited slab, much of the surface radiation loss can be recaptured and returned, so the ignited slab can continue burning even after the withdrawal of the initial heat source. This explains why one cannot burn a single large log in a fireplace, but must use several logs to recapture the radiant heat losses.
5. As the burning continues, a char layer builds up. This char layer, which is a good thermal insulator, restricts the flow of heat to the wood interior, and consequently tends to reduce the rate of pyrolysis. The pyrolysis rate also will decrease when the supply of unpyrolyzed wood runs out. When the pyrolysis rate decreases to the point of not being able to sustain gas-phase combustion, oxygen from the air will come into direct contact with the char, permitting it to undergo direct glowing combustion, provided the radiant heat losses are not too large.
6. This scenario presumes an ample (but not excessive) supply of air (oxidant) for combustion. If there were insufficient oxidant to burn the available fuel vapor, the excess vapors would travel with the flow and possibly burn where they eventually would find sufficient oxidant. For example, this happens when fuel vapors emerge and burn outside a window of a fully involved but underventilated room fire. Generally, underventilated fires produce large amounts smoke and toxic products (e.g., carbon monoxide).

If, on the other hand, one was to impose a forced draft onto the pyrolyzing surface, the oxidant supply may exceed that required for complete combustion of the fuel vapors. In this case the excess oxidant can cool the flames sufficiently to suppress their chemical reaction and extinguish them, as happens, for example, when one blows out a match. In the case of larger fires with ample supply of fuel vapors, imposing a forced draft on them simply increases their rate of burning by increasing the flame-to-fuel-surface heat transfer which in turn enhances the fuel supply rate.

7. Following the ignition of a certain portion of our wood slab, the flames are likely to spread over the entire fuel array. Flame spread can be thought of as a continuous succession of piloted ignitions where the flames themselves provide the heat source. One commonly observes that upward flame spread is much more rapid than downward or horizontal flame spread. This is because hot flames generally travel upward and contribute their heat over a greater area in an upward direction. Thus each successive "upward ignition" adds a much greater burning area to the fire than a corresponding "downward" or "horizontal" ignition.

Generally, materials which ignite easily (rapidly) also propagate flames rapidly. The ignitability of a material is controlled by its resistance to heating (thermal inertia), and the temperature rise required for it to begin to pyrolyze. Materials with low thermal inertia, such as foamed plastics or balsa wood, heat rapidly when subjected to a given heat flux. These materials are often easy to ignite and can cause very rapid flame spread. On the other hand, dense materials such as ebony wood tend to have relatively high thermal inertias and are difficult to ignite.

8. The burning rates of larger, more hazardous fires are principally governed by the radiant heat transfer from the flames to the pyrolyzing fuel surface (deRis 1979). This flame radiation comes primarily from the luminous soot particles in the flames. Combustibles which tend to produce copious amounts of soot or smoke (such as polystyrene) also tend to support more intense fires, despite the fact that their fuel vapors burn less completely, as evidenced by their higher smoke output.

Well ventilated fires generally release less smoke than poorly ventilated ones. In well ventilated fires, the surrounding air can gain speedy access to the unburned fuel vapors and soot before the fuel vapors cool down by radiation. Poorly ventilated fires can release copious amounts of smoke and incomplete products of combustion, such as carbon monoxide. In poorly ventilated fires, fuel vapors have insufficient air to burn completely before cooling off and leaving the fire area.

Fires occurring in oxygen-enriched atmospheres have higher flame temperatures, increased fractions of heat release by radiation, and increased burning rates per unit fuel area. These higher flame temperatures generally cause a much greater conversion of fuel vapors into soot, resulting in significantly increased smoke release rates. For example, a well ventilated methanol fire typically burns with a blue (i.e., soot free) flame in normal air. However, a similarly well ventilated methanol fire can burn with a brightly luminous smoky flame in an oxygen-enriched atmosphere. This sensitivity to ambient oxygen concentration significantly increases the flame radiation, burning rates, and resultant fire hazard.

Flammability Properties of Solid and High Fire Point Liquid Combustibles

The above discussion can be summarized by describing the combustible properties which contribute to fire hazards in typical fire situations:

Heat of Combustion: This is a measure of the maximum amount of heat that can be released by the complete combustion of a unit mass of combustible material.

Stoichiometric Oxidant: The mass of oxidant required for complete combustion of a unit mass of combustible is called the stoichiometric oxident requirement. Combustibles with large stoichiometric oxidant requirements often produce large flame heights which in turn present greater fire spread hazards. The stoichiometric oxidant requirements for typical (organic) combustibles is approximately proportional to their heat of combustion, so that organic combustibles all release approximately the same amount of heat per unit mass of consumed oxidant.

Heat of Gasification: The heat required to vaporize a unit mass of combustible material which is initially at ambient temperature. This quantity is very important because it determines the amount of combustible vapor supplied to a fire in response to a given supply of heat to the pyrolyzing surface. Sometimes the fire hazard of plastics is reduced by adding inert fillers which increase their effective heats of gasification.

Ignitability (Piloted): Ignitability is inversely proportional to the time it takes for a given applied heat flux to raise the surface temperature of a material to its piloted ignition temperature. This property is important both for ignition and fire spread.

Char Formation: Char is a black carbonaceous residue. Many materials such as wood develop a char layer during pyrolysis. The insulating properties of such char layers can be very effective in reducing burning rates by restricting the flow of heat to the unpyrolyzed material. Intumescent paints simulate formation of char by swelling and forming an insulating protective layer when heated.

Soot Formation: Combustibles whose flames produce significant amounts of soot are generally more hazardous because the soot increases the flame radiation which influences the burning rate. The soot itself also contributes to smoke damage.

Inhibitors: The addition of small quantities of chemical inhibitors to the fuel or oxidant can impede gas-phase reactions. Such flame inhibitors can be quite effective in retarding ignition and flame spread involved in small fires. Flames also can be inhibited by introducing additives to solid combustibles. This promotes char formation as well as higher concentrations of water vapor in the pyrolysis gases.

Melting: Combustibles which tend to melt are often more hazardous, because the molten combustible can increase the pyrolyzing surface area. The molten material itself also can be a hazard.

Toxicity: Carbon monoxide usually is the principal toxicant produced by a fire. It is typically present in pyrolysis gases and combustion products which have not been completely oxidized. Some materials, especially those containing elements other than C, H, and O (e.g., polyvinylchloride, polyurethane), can produce additional toxicants.

Geometry: Last, but not least, the geometry of a material strongly influences its flammability. Thin materials usually ignite more readily and spread flames more rapidly. Upward flame spread is more rapid than downward or horizontal flame spread. Finally geometric arrangements which provide ample air (oxidant) access while providing shielding to prevent radiant losses are usually the most hazardous.

Flammability Principles

In summary, the science of fire protection rests upon the following principles:

1. An oxidizing agent, a combustible material, and an ignition source are essential for combustion.
2. The combustible material must be heated to its piloted ignition temperature before it will ignite or support flame spread.
3. Subsequent burning of a combustible is governed by the heat feedback from the flames to the pyrolyzing or vaporizing combustible.
4. The burning will continue until:
   a. The combustible material is consumed or;

b. The oxidizing agent concentration is lowered to below the concentration necessary to support combustion or;
c. Sufficient heat is removed or prevented from reaching the combustible material to prevent further fuel pyrolysis or;
d. The flames are chemically inhibited or sufficiently cooled to prevent further reaction.

All the material presented in this HANDBOOK for the prevention, control or extinguishment of fire is based upon these principles.

HEAT MEASUREMENT

The temperature of a material is the condition which determines whether it will transfer heat to or from other materials. Heat always flows from higher to lower temperatures. Temperature is measured in degrees.

Temperature Units

Celsius: A Celsius (or Centigrade) degree (°C) is 1/100 the difference between the temperature of melting ice and boiling water at 1 atmosphere pressure. On the Celsius scale, zero is the melting point of ice; 100 is the boiling point of water. This unit is approved by the International System of Units (SI).

Kelvin: A Kelvin degree or Kelvin (K) is the same size as the Celsius degree. However, the zero on the Kelvin scale (sometimes called Celsius Absolute) is −459.67°F (−273.15°C). Zero on the Kelvin scale is the absolute lowest achievable temperature; hence the Kelvin scale provides us with so called absolute temperatures. The Kelvin also is an approved SI unit.

Fahrenheit: A Fahrenheit degree (°F) is 1/180 the difference between the temperature of melting ice and boiling water at 1 atmosphere pressure. On the Fahrenheit scale, 32 is the melting point of ice (0°C); 212 is the boiling point of water (100°C).

Rankine: A Rankine degree (°R) is the same size as the Fahrenheit degree. On the Rankine scale zero is −459.67°F (273.15°C), so that the Rankine scale also provides an absolute temperature.

Fahrenheit and Rankine degrees are not approved SI units and their use is greatly discouraged.

Heat Units

Joule: The amount of heat energy provided by one watt flowing for 1 second is called a Joule. This is an approved SI unit.

Calorie: The amount of heat required to raise the temperature of one gram of water one degree Celsius [measured at 59°F (15°C)] is called a calorie. One calorie equals 4.183 Joules.

British thermal unit (Btu): The amount of heat required to raise the temperature of 1 pound of water one degree Fahrenheit [measured at 60°F (15.5°C)] is called the British thermal unit. One Btu equals 1,054 Joules (252 calories), so that we can infer that 1.054 kilowatts would heat 1 pound of water one degree Fahrenheit in 1 second.

Btu and calories are not approved SI units.

Heat energy has quantity as well as potential (intensity). For example, consider the following analogy:

Two water tanks stand side by side. If the first tank holds twice as many gallons as the second, then the first tank can hold twice the quantity of water as the second. But if the level of water in the two tanks is equal, then their pressures or potentials are equal. If the bottoms of the two tanks are connected by a pipe, water will not flow from one to the other because both tanks have the same equilibrium pressure. In a similar manner, one body may hold twice the quantity of heat energy (measured in Joules or Btu) as a second body. But if the potentials or temperatures of the energies of the bodies are equal, no heat energy will flow from one body to the other when they are brought into contact, since the bodies are at equilibrium. If a third body at a lower temperature were brought into contact with the first body, heat would flow from the first to the third until both body temperatures became equal. The amount or quantity of heat flowing until this equilibrium is reached depends upon the heat retention capacities of each body involved.

Essentially, ignition is a matter of increasing temperature by adding heat, whereas physical fire extinguishment usually is accomplished through reduction of temperature by removal of heat. Chemical extinguishment works by another mechanism—by interrupting chemical reactions that are important in the combustion process (Westbrook and Dryer 1981).

Temperature Measurement

Devices that measure temperature depend upon physical change (expansion of a solid, liquid, or gas), change of state (solid to liquid), energy change (changes in electrical potential energy, i.e., voltage), or changes in thermal radiant emission and/or spectral distribution. The principles of operation of the more common temperature measuring devices are discussed here.

Liquid Expansion Thermometers: These thermometers consist of a tube (partially filled with a liquid), which measures expansion and contraction of the liquid with changes in temperature. The tube is calibrated to permit reading of the level of the liquid in degrees of a temperature scale.

Bimetallic Thermometers: Bimetallic thermometers contain strips of two metals (that are laminated) with different coefficients of expansion. As the temperature changes, the two metals expand or contract to different extents, causing the strip to deflect. The amount of deflection is measured on a scale which is calibrated in degrees of temperature.

Solid Fusion: Solid fusion makes use of the melting or fusion point of a solid (metal, chemical, or chemical mixture) to indicate whether or not a hot object is above or below the melting point of the solid. The eutectic metal in a sprinkler link melts in a fire environment, which activates the sprinkler.

Thermocouples: Thermocouples consist of a pair of wires of different metals connected to each other at one end (the sensing end). The other ends are connected to a voltmeter. When the sensing end is at a different temperature from the voltmeter, a voltage is set up, the magnitude depends in part upon the temperature difference between the two ends. The voltmeter can be calibrated to give readings in degrees of temperature.

Pyrometers: Pyrometers measure the intensity of radiation from a hot object. Since intensity of radiation depends upon temperature, pyrometers can be calibrated to give readings in degrees of temperature. Optical pyrometers measure the intensity of a particular wavelength of radiation.

Specific Heat

The specific heat of a substance describes the amount of heat it absorbs as its temperature increases. More precisely, it is the amount of thermal energy required to raise the unit mass of a substance one temperature degree. It is measured in J/g(°C) or cal/g(°C) or Btu/lb(°F). Water has a specific heat of 1 cal/g°C = 1 Btu/lb°F.

The specific heats of various substances vary over a considerable range; for most common substances, except water, they are less than 1 cal/g°C. Specific heat figures are significant in fire protection because they indicate the relative quantity of heat needed to raise the temperature to a point of danger, or the quantity of heat that must be removed to cool a hot substance to a safe temperature.

One reason for the effectiveness of water as an extinguishing agent is that its specific heat is higher than that of most other substances.

Latent Heat

Heat is absorbed by a substance when it is converted from a solid to a liquid, or from a liquid to a gas. This thermal energy is called latent heat. Conversely, heat is released during conversion of a gas to a liquid or a liquid to a solid.

Latent heat is the quantity of heat absorbed by a substance in passing between liquid and gaseous phases (latent heat of vaporization), or between solid and liquid phases (latent heat of fusion). It is measured in Joules per unit mass. The latent heat of fusion of water (normal atmospheric pressure) at the freezing or melting point of ice 32°F (O°C) is 333.4 J/g; the latent heat of vaporization of water at its boiling point of 212°F (100°C) is 2257 J/g (or 970.3 Btu/lb or 539.5 cal/g). The larger heat of vaporization of water is another reason for the effectiveness of water as an extinguishing agent. It requires 3 million Joules to convert 1 kilogram of ice at 32°F (O°C) to steam at 212°F (100°C). The latent heats of most other common substances are substantially less than that of water. Thus, the heat absorbed by water is subtracted from the burning system so it cannot be used to propagate the flame further by vaporizing more liquid, or by pyrolyzing more solid.

HEAT TRANSFER

Transfer of heat governs the ignition, burning, and extinguishment of most fires. Heat is transferred by one or more of three methods: (1) conduction, (2) radiation, or (3) convection.

Conduction: Heat transferred by direct contact from one body to another is a process called conduction. Thus, a steam pipe in contact with wood transfers its heat to the wood by actual contact (in this example, the pipe is the conductor).

The quantity of heat energy transferred by conduction through a body or in a given time is a function of the temperature difference and the conductance of the path involved. Conductance depends on thermal conductivities, cross-sectional areas normal to the flow path, and length of flow path. The rate of heat transfer, then, is simply the quantity of heat per unit time, while the heat "flux" is the quantity of heat per unit cross-sectional area per unit time. The thermal conductivity of a material is the heat energy flux resulting from a unit temperature gradient (falloff of one degree per unit of distance).

The conduction of heat through air or other gases is independent of pressure in the usual range of pressures. It approaches zero only at very low pressures. No heat is conducted in a perfect vacuum. Solids are better heat conductors than gases: the best commercial insulators consist of fine particles or fibers, with the spaces between the particles filled with air.

Heat conduction cannot be completely stopped by any "heat-insulating" material. Thus, the flow of heat is unlike the flow of water, which can be stopped by a solid barrier. Heat-insulating materials have a low heat conductivity. No matter how thick the insulation, solidly insulating the space between the source of heat and the combustible material may be insufficient to prevent ignition. If the rate of heat conduction through the insulating material is greater than the rate of dissipation from the combustible material, the temperature of the combustible material may increase to the point of ignition. For this reason there should always be an air space or some manner of carrying the heat away by convection, rather than relying solely upon the heat-insulating material for protection. By the transmission of heat over a long period, fires have occurred through as much as a 2 ft (0.6 m) thickness of solid concrete.

For heat conduction, the most important physical properties of a material are thermal conductivity (k), density (ρ), and specific heat (c). Unfortunately, the last two quantities are usually listed separately, although it is their product (ρc) that is of any interest in the field of heat conduction. Heat conduction is a measure of the amount of heat necessary to raise the unit volume of the material by unit temperature. A typical unit might be Joules per cubic centimeter per degree Celsius, [(J/(cm^3 · °C), and a possible name for the quantity is thermal capacity per unit volume.

The thermal conductivity of a material is a measure of the rate of flow of heat through a unit area of the material with unit temperature gradient. Unit temperature gradient means that in the direction of heat flow, the temperature is falling off one degree per unit distance. A typical unit of thermal conductivity is J/(cm · sec · °C).

Thermal conductivity and thermal capacity per unit volume are rarely of much importance individually. In fact, the solution to heat conduction problems is so complex that it cannot be adequately presented here. However, one or two interesting features can be appropriately mentioned. Probably the most useful feature is concerned with the time constant of a thickness (x) of a material. Thus, if the surface of a material is suddenly exposed to a temperature rise, then the temperature at a depth (x) within the material will begin to change substantially at a certain time.

$$t = \frac{x^2 \rho C}{K}$$

The dimensions of the time expression are given as:

$$\frac{cm^2 \cdot (J/cm^3 \cdot °C)}{J/(cm \cdot sec \cdot °C)} = sec$$

Centimeters, Joules, and degrees Celsius cancel out, leaving the result of the dimension in seconds, i.e., time. This is a time constant; thus, the higher the number, the slower the transfer. We see that the time required for a thermal wave to penetrate a body increases with the square of its thickness.

Convection: By convection, heat is transferred by a circulating medium—either a gas or a liquid. Thus, heat generated in a stove is distributed throughout a room by initially heating the air in contact with the stove by conduction; the circulation of heated air through the room to distant objects is heat transfer by convection. Finally, heat is transferred from the air to the objects by conduction. Heated air expands and rises, and for this reason heat transfer by convection often occurs in an upward direction, although air currents can be made to carry heat by convection in any direction, e.g., by use of a fan or blower.

Radiation: Radiation is a form of energy traveling through space or materials as an electromagnetic wave such as light, radio waves or X-rays. All radiant energy waves travel at the same speed in a vacuum as the speed of light. On arrival at a body, they are absorbed, reflected, or transmitted. Visible light consists of wavelengths between 0.4×10^{-6} to 0.7×10^{-6} m (violet to red). Emission from combustion processes occurs principally in the infrared region (wavelengths longer than the red wavelength). Our eyes see only the tiny fraction which is emitted within the visible region.

A candle flame is a common example of radiation. Air heated by the flame rises upward while cooler air moves in toward the candle to supply the flame with more oxygen, thus sustaining the burning process. If a hand is held to the side of the flame, it would experience a sensation of warmth. This energy is called radiant heat or radiation.

In the case of small fires, such as one involving a candle, most of the heat leaves the combustion zone by vertical convection, e.g., a hand held over the flame instead of beside it will sense more heat. However, larger, more hazardous fires release about equal amounts of radiative and convective energy. Radiated energy is more dangerous because a stationary surface near the fire will absorb essentially all the radiation incident upon it, while most of the convected energy flows past the surface as it moves away with the gas stream.

Radiation energy travels in straight lines. Obviously, the heat received from a small source would be less than that received from a large radiating surface, provided the sources emit comparable energies per unit area. (See Fig. 4-1B.)

Heat from the sun passes through the near vacuum of space until it reaches the earth and is absorbed by a tangible object or substance. Radiant heat also passes freely through symmetrical diatomic molecules such as hydrogen (H_2), oxygen (O_2), and nitrogen (N_2). Thus, there is no absorption of heat by air except by tangible air constituents (particulates, including smoke) or contaminants (generally unsymmetrical molecules) such as water vapor, carbon monoxide, carbon dioxide, sulfur dioxide, or hydrocarbons. Usually the concentrations of these absorptive components of air are low enough so that the radiation absorbed by them is of minor significance. However, at very large scales, the water vapor and carbon dioxide in the atmosphere can absorb appreciable amounts of radiation. This helps to explain why forest fires or large LNG fires are less hazardous on days when the humidity is high. Since water drops absorb almost all the incident infrared radiation, mists or water sprays are very effective attenuators of radiation.

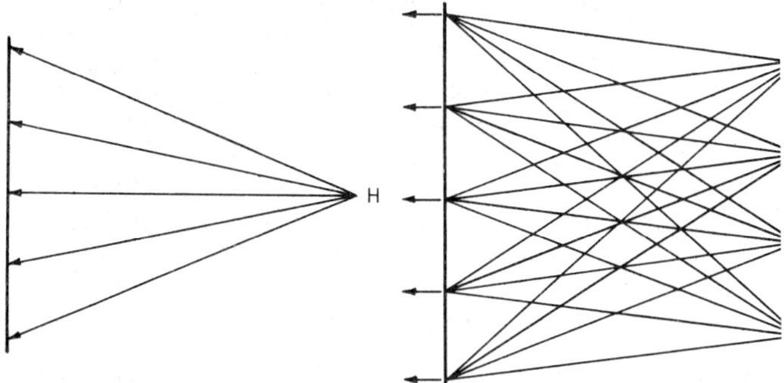

FIG. 4-1B. A comparison of heat absorption by surfaces of similar area from a pinpoint source (left), and a large radiating surface (right).

When two bodies face each other and one body is hotter than the other, radiant energy will flow from the hotter body to the cooler body until they both have the same temperature (i.e., thermal equilibrium). The ability to absorb this radiated heat is a function of the kind of surface of the cooler body and the area of the radiating surface of the hotter body. If the receiving surface is shiny or polished, it will reflect most of the radiant heat away; if it is black or dark in color, it will absorb most of the heat. Most nonmetallic materials are effectively "black" to infrared radiation, despite the fact that they may appear light or colored to visible radiation. Some substances like water or glass are transparent to visible radiation and will allow it to pass through them with minimal absorption; however, both liquid water and glass are opaque to most infrared wavelengths. Glass greenhouses and solar panels operate on the principle of being transparent to the sun's visible radiation, while at the same time being opaque to the infrared radiation attempting to escape from the greenhouse or solar panel.

Shiny metallic materials are excellent reflectors of radiant energy. For example, aluminum foil often is used together with fiberglass in building insulation. Sheet metal is often used beneath stoves or on heat-exposed walls.

The Stefan-Boltzmann law states that the radiant emission per unit area from a black surface is proportional to the fourth power of its absolute temperature. The law can be expressed by the formula:

$$q = \epsilon\sigma T^4$$

where q is the radiant emission per unit surface area, ϵ is the surface emissivity correction factor (which is typically close to 1 for most nonmetallic materials in the infrared wavelength region, and thus can usually be ignored), σ is the Stefan-Boltzmann proportionality constant [equal to 5.67×10^{-12} W/(cm$^2 \cdot$ K^4)], and T is the absolute temperature expressed in Kelvin. To appreciate the importance of this fourth power we consider the following situation.

EXAMPLE: A heater is designed to operate safely with a 500°F (260°C) outside surface temperature. How much more heat will it radiate if this outside surface temperature is allowed to increase, say by 212°F (118°C) up to 680°F (360°C), or even increase by 464°F (258°C) to a total maximum to 932°F (500°C)?

We must first convert the temperatures from degrees Celsius to their absolute values in Kelvins by adding 273 to them. Next we make the usual assumption that the surface is black, and then evaluate the radiant emission per unit area for safe operation [500°F (260°C)] as:

$$q = \sigma(260 + 273)^4 = 5.67 \times 10^{-12} \times (533)^4 = 0.46 \text{ Watts/cm}^2$$

The corresponding radiant emissions per unit areas at the higher temperatures are:

$$q = \sigma(360 + 273)^4 = 0.91 \text{ Watts/cm}^2$$

$$q = \sigma(500 + 273)^4 = 2.0 \text{ Watts/cm}^2$$

Thus we see that by increasing the stove temperature by only 212°F (100°C) one approximately doubles its emission from 0.46 to 0.91 W/cm^2. This is the dramatic effect of the fourth power. If heat radiation was merely a function of the first power of the absolute temperature, then it would increase by only about 20 percent as the absolute temperature increased from 533 to 633 K. Finally, if one were so careless to allow the stove to reach 932°F (500°C), it would emit 2.0 W/cm^2 which is sufficiently high to ignite many typical home furnishings.

Because residential coal or wood stoves can sometimes undergo a temperature "runaway" if given too much air, it is important to limit the maximum received radiant heat transfer by keeping all near by furnishings away from the stove. The radiant energy transmitted from a point like source to a receiving surface will vary inversely as the square of their separation distance. If a stove is small relative to its distance to nearby objects, then it behaves like a point source; doubling its separation distance will decrease the incident radiant heat (per unit area) by a factor of four. However, if the nearby object is close to the stove, the stove appears like a large surface and small changes in separation will have little or no effect on the severity of the radiative heat transfer (e.g., a large oven within a few centimeters from a combustible wall). In this case one must protect the wall by some other means—for example, a shiny metal coated noncombustible board.

Generally, scientists and engineers have a good understanding of radiation between solid surfaces. They also can reliably estimate radiant heat transfer absorbed or emitted by gases of a known composition and temperature. However, they have difficulty in estimating the radiation from flames because most of this radiation comes from the soot in the flames and there have been very few measurements of soot in flames. This lack of soot data often prevents the estimate of fire burning rates. There is, however, one useful rule of thumb. The total radiant output from flames (higher than about 200 mm or 8 in) is usually about 30 to 40 percent of the maximum heat output assuming complete combustion (deRis 1979). A comparable amount of energy leaves by convection, with the remaining fraction leaving in the form of incomplete combustion.

HEAT ENERGY SOURCES OR SOURCES OF IGNITION

Since fire prevention and extinguishment depend upon the control of heat, it is important to be familiar with the more common ways in which heat energy can be produced. Four sources of heat energy may be considered: (1) chemical, (2) electrical, (3) mechanical, and (4) nuclear.

Chemical Heat Energy

Oxidation reactions usually produce heat. They are the source of the heat which is of primary concern to fire protection engineers.

Heat of Combustion: The heat of combustion is the amount of heat released during a substance's complete oxidation (combustion, i.e., conversion to carbon dioxide and water). Heat of combustion, commonly referred to as calorific or fuel value, depends upon the kinds and numbers of atoms in the molecule as well as their arrangement in the molecule. Calorific values are commonly expressed in Joules per gram, but are sometimes reported in Btu per pound or calories per gram (1 Btu/lb = 2.32 J/g; and 1 cal = 4.18 J/g). In the case of fuel gases, calorific values are commonly reported in Btu/cu ft. Calorific values are used in calculating fire loading, but do not necessarily indicate relative fire hazard, since the fire hazard depends upon rate of burning as well as upon the total amount of heat produced.

Also, heat is produced in incomplete or partial oxidations which occur at some stage in almost all accidental fires and in spontaneous heating by oxidation. For almost all compounds of carbon and hydrogen, or of carbon, hydrogen and oxygen (which include substances of vegetable or petroleum origin), the heat of oxidation, whether complete or partial, depends upon the amount of oxygen consumed. For these common substances (i.e., coal, natural gas, common plastics, oils, wood, cotton, sugar, vegetable and mineral oils), the heat of oxidation is approximately 100 Btu/cu ft (3.7 kJ/m^3) of air consumed, regardless of the completeness of combustion. For this reason, the heat produced in a fire, or in a spontaneous heating oxidation, often is limited by the air (oxygen) supply.

Spontaneous Heating: Spontaneous heating is the process whereby a material increases in temperature without drawing heat from its surroundings. Spontaneous heating of a material to its ignition temperature results in spontaneous ignition or spontaneous combustion. The fundamental causes of spontaneous heating are few, but the conditions are many and varied under which these fundamental factors may operate to create a dangerous condition. Three conditions which have much to do with whether or not an oxidation reaction will cause dangerous heating are rate of heat generation, air supply, and insulation properties of the immediate surroundings.

If exposed to the atmosphere, most all organic substances capable of combination with oxygen will oxidize at some critical temperature with evolution of heat. The rate of oxidation at normal temperatures is usually so slow; however, the released heat is transferred to surroundings as rapidly as it is formed, with the result that there is no appreciable temperature increase in the combustible material being oxidized. This, however, is not true of all combustible materials since certain oxidation reactions at normal temperatures (e.g., oxidation of powdered zirconium in air) generate heat more rapidly than it can be dissipated, with spontaneous ignition being the result.

In order for spontaneous ignition to occur, sufficient air must be available to permit oxidation, yet not so much air that the heat is carried away by convection as rapidly as it is formed. An oily (vegetable oil) rag which might heat spontaneously in the bottom of a wastebasket would not be expected to do so if hung on a clothesline where air movement would remove heat; nor would it be likely to heat up if it were sealed in a tightly packed bale of rags. On the other hand, because of more air and the insulating effect of the bale, a loosely packed bale might provide ideal conditions for heating. Because of the many possible combinations of the interrelated factors of air supply and insulation, it is impossible to predict with certainty whether a material will heat spontaneously.

In the presence of air, substances subject to oxidation will first form products of partial oxidation which may act as catalysts to further oxidation. For example, olive oil which has turned rancid from exposure to the air will oxidize at a higher rate than fresh, pure oil.

Additional heat can initiate spontaneous heating in some combustible materials not subject to this phenomenon at ordinary temperatures. In these instances, preheating increases the rate of oxidation sufficiently so that more heat is produced than can be lost. Many fires have been caused by the spontaneous heating of foam rubber following preheating in a dryer.

A common cause of heating of agricultural crops appears to be oxidation by bacteria, one product of which is heat. Since most bacteria cannot live at temperatures much above the range of 160 to 175°F (70 to 80°C), continued heating of agricultural products to their ignition temperatures is thought to be due to rapid oxidation initiated after bacteriological preheating.

The moisture content of agricultural products has a definite influence on the spontaneous heating by bacteria. Wet or improperly cured hay is very likely to heat in barn lofts. Experience has indicated that such heating may result in ignition within a period of two to six weeks after storage. Alfalfa meal which has been exposed to rain and then stored in bins or piles is very susceptible to spontaneous heating. Soybeans stored in bins have been known to sustain what is called "bin burn" (i.e., the beans next to the bin walls are charred due to the moisture condensation on the inside surfaces of the wall and the self heating of the beans). Other agricultural products susceptible to spontaneous heating are those with a high content of oxidizable oils, such as cornmeal feed, linseed, rice bran, and pecan meal.

Heat of Decomposition: The heat of decomposition is the heat released by the decomposition of compounds requiring the addition of heat for their original formation from the elements. Since most chemical compounds are produced by exothermic reactions, heat of decomposition is not a common phenomenon. Compounds formed from endothermic reaction often are unstable. When decomposition is started by heating a substance above a critical temperature, decomposition continues with the liberation of heat. Cellulose nitrate is well known for its tendency to decompose with the liberation of dangerous quantities of heat. The chemical action responsible for this effect in many commercial and military explosives is the rapid decomposition of an unstable compound.

Heat of Solution: The heat of solution is the heat released when a substance is dissolved in a liquid. Most materials release heat when dissolved, although the amount of heat is usually not sufficient to have any significant effect on fire protection. In the case of some chemicals (such as concentrated sulfuric acid), the heat evolved may be sufficient to be dangerous. The chemicals that react with water in this manner are not themselves combustible, but the liberated heat may be sufficient to ignite nearby combustible materials.

In contrast to most materials, ammonium nitrate absorbs heat when dissolved in water. (It is said to have a negative heat of solution.) Some first aid products, for use where cold is recommended, consist of dry ammonium nitrate in watertight packages. These packages become cold when water is added.

Electrical Heat Energy

When current flows through a conductor, electrons are passed along from atom to atom within the conductor with frequent collisions with atomic particles on the way. The better conductors, such as copper and silver, have the most easily removed outer electrons so that the force or voltage required to establish or maintain any unit electric current (or electron flow) through the conductor is less than for substances composed of more tightly bound electrons. Thus, the electrical resistance of any substance depends upon atomic and molecular characteristics; the electrical resistance is proportional to the energy required to move a unit quantity of electrons through the substance against the forces of electron capture and collision. This energy expenditure appears in the form of heat.

Resistance Heating: Resistance heating is characterized by a rate of heat generation proportional to the resistance and the square of the current. Since the temperature of the conductor resulting from resistance heating depends upon dissipation of heat to the surroundings, bare wires can carry more current than insulated wires without heating dangerously, and single wires can carry more current than closely grouped wires, or wires bundled into a cable.

The heat generated by incandescent and infrared bulbs is due to the resistance of the filaments in the bulbs. Material of very high melting point is used for the "white-hot" filaments of incandescent lamps, and destruction of the filament by oxidation is prevented by partial evacuation of the bulb and by removal of oxygen. The filaments of infrared lamps operate at a much lower temperature (a "red" heat); the most efficient infrared lamp reflectors are gold because gold is one of the best reflectors of infrared radiation.

Dielectric Heating: Whenever atoms are subjected to electric potential gradients from external sources, the arrangement of the atom (or of a molecule of several atoms) is distorted. A tendency exists for electrons to move in the direction of the positive potential, and for protons to move in the opposite direction. This is true whether the externally applied potential is due to a battery, generator, or is the result of a magnetic field. Even though the external potential is insufficient to break away any electrons, the distortion of the normal atomic or molecular arrangement represents an energy expenditure. This is of practically no consequence if the external force is unidirectional, but can be substantial if the potential is pulsating or alternating. For example, the heating of a dielectric (a good insulator) may be considerable if the frequency of alternation of external potential becomes high.

Induction Heating: Whenever a conductor is subject to the influence of a fluctuating or alternating magnetic field, or whenever a conductor is in motion across the lines of force of a magnetic field, potential differences appear in the conductor. These potential differences result in the flow of current with attendant resistance heating in the conductor. For rapidly changing or alternating potentials, additional energy is expended and, as the polarity changes, this energy appears as heat energy due to the mechanical and electrical distortion of the molecular structure. This latter type of heating increases with the frequency of alternation. Food in a microwave oven, for example, is heated by the molecular friction induced by absorbed microwave energy.

A useful form of induction heating is created by passing a high frequency alternating current through a coil surrounding the material to be heated.

An alternating current passing through a wire can induce a current in another wire parallel to it. If the wire in which a current is induced does not have adequate current carrying capacity for the size of the induced current, resistance heating will occur. In this example, the heating is due primarily to resistance to flow, and only in a small degree to molecular friction.

Leakage Current Heating: Since all available insulating materials are imperfect insulators, there is always some current flow when the insulators are subjected to substantial voltages. This flow is commonly referred to as a leakage current and is usually not important from the standpoint of heat generation. However, if the insulating material is not suited for the service, or the material is too thin (for reasons of economy, space saving, or attempts to attain the maximum capacity in a condenser), leakage currents may exceed safe limits, resulting in heating of the insulator with consequent deterioration of the material and ultimate breakdown.

Heat from Arcing: Arcing occurs when an electric circuit which is carrying current is interrupted either intentionally (as by a knife switch), or accidentally (as when a contact or terminal becomes loosened). Arcing is especially severe when motor or other inductive circuits are involved. The temperatures of arcs are very high, and the heat released may be sufficient to ignite combustible or flammable material in the vicinity. In some instances, the arc may melt the conductor, and scatter molten metal. One requirement of an intrinsically safe electrical circuit is that arcing, due to accidental current interruption, will not release sufficient energy to ignite the hazardous atmosphere in which the circuit is located.

Static Electricity Heating: Static electricity (sometimes called frictional electricity) is an electrical charge that accumulates on the surfaces of two materials that have been brought together and then separated. One surface becomes charged positively, the other negatively. If the substances are not bonded or grounded, they will eventually accumulate sufficient electrical charge so that a spark discharge may occur. Static arcs are ordinarily of very short duration, and do not produce sufficient heat to ignite ordinary combustibles such as paper. Some, static arcs however, are capable of igniting flammable vapors and gases, and clouds of combustible dust. Fuel flowing in a pipe can generate enough static electricity of sufficient energy to ignite a flammable vapor.

Heat Generated by Lightning: Lightning is the discharge of an electrical charge from a cloud to an opposite charge on another cloud or on the ground. Lightning passing between a cloud and the ground can develop very high temperatures in any material of high resistance in its path, such as wood or masonry.

Mechanical Heat Energy

Mechanical heat energy is responsible for a significant number of fires each year. Frictional heat is responsible for most of these fires, although there are a few notable examples of ignition by the mechanical heat energy released by compression.

Frictional Heat: The mechanical energy used in overcoming the resistance to motion when two solids are rubbed together is known as frictional heat. Any friction generates heat. The danger depends upon the available mechanical energy, the rate at which the heat is generated, and its rate of dissipation. Some examples of frictional heat are the heat caused by friction of a slipping belt against a pulley, and the hot metal particles (sparks) thrown off when a piece of foreign metal enters a grinding mill.

Friction Sparks: Friction sparks include the sparks which result from the impact of two hard surfaces, at least one of which is usually metal. Some examples of friction sparks often reported as responsible for fires are sparks from dropping steel tools on a concrete floor, from falling tools striking machinery (or piping), from tramp metal in grinding mills, and from shoe nails on concrete floors.

Friction sparks are formed in the following manner: heat, generated by impact or friction, initially heats the particle. Then, depending upon the ease of oxidation and the heat of combustion of the metal particle, the freshly exposed surface of the particle may oxidize at the elevated temperature with the heat of oxidation increasing the temperature of the particle until it is incandescent.

Although the temperatures necessary for incandescence vary with different metals, in most cases they are well above the ignition temperatures of flammable materials [e.g., the temperature of a spark from a steel tool approaches 2,500°F (1400°C); sparks from copper-nickel alloys with small amounts of iron may be well above 500°F (300°C)]. However, the ignition potential of a spark depends upon its total heat content; thus, the particle size has a pronounced effect on spark ignition. The practical danger from mechanical sparks is limited by the fact that usually they are very small and have a low total heat content, even though each spark may have a temperature of 2,000°F (1100°C) or higher. Mechancial sparks cool quickly and start fires only under favorable conditions, such as when they fall into loose dry cotton, combustible dust, or explosive materials. Larger particles of metal, able to retain their heat longer, are not usually heated to dangerous temperatures. Although the hazard of ignition of flammable vapors or gases by friction sparks is often overemphasized, it is best to avoid the use of grinding wheels and other sources of mechanical sparks in areas where any flammable liquids, gases, or vapors are, or may be present. The possibility of ignition due to some unusual condition should not be overlooked.

Nickel, monel metal, and bronze have a very slight spark hazard; stainless steel has a much lower spark hazard than ordinary tool steel. Special tools of copper-beryllium and other alloys are designed to minimize the danger of sparks in hazardous locations. Such tools cannot, however, completely eliminate the danger of sparks because a spark may be produced under several conditions. Little or no benefit is gained by using nonsparking hand tools in place of steel to prevent explosions of hydrocarbons (*NFPA 1959*). Leather, plastic, and wooden tools are, however, free from the friction spark hazard.

Heat of Compression: The heat of compression is the heat released when a gas is compressed. This is also known as the diesel effect. The temperature of a gas increases when compressed and this has found practical application in diesel engines in which heat of compression eliminates the need for a spark ignition system. Air is first compressed in a diesel engine cylinder, after which an oil spray is injected into the compressed air. The heat released when the air is compressed is sufficient to ignite the oil spray.

Tests have shown that wood will ignite when a pressurized jet of air is directed into a cavity in a block of wood. Apparently, compression waves set up in the cavity are converted to heat, which raises the wood to its ignition temperature. When pipe fittings are substituted for the wood, an oil film on the inside surface of the fittings can be ignited.

Nuclear Heat Energy

Nuclear heat energy is energy released from the nucleus of an atom. The nucleus is composed of matter held together by tremendous forces which can be released when the nucleus is bombarded by energized particles. Nuclear energy is released in the form of heat, pressure, and nuclear radiation. In nuclear fission, energy is released by splitting the nucleus; in nuclear fusion, energy is released by the joining of two nuclei.

The energy released by bombardment of the nucleus is commonly a million times greater than that released by ordinary chemical reaction. Instantaneous release of large quantities of nuclear heat energy results in an atomic explosion. Controlled release of nuclear energy is a source of heat for everyday use (i.e., steam generation for electric generating stations).

Bibliography

References Cited

Browne, F. L. 1958. "Theories of the Combustion of Wood and its Control," Report No 2136. Forest Products Laboratory. U.S. Dept. of Agriculture. Madison, WI.

deRis, J. 1979. "Fire Radiation—A Review." Seventeenth Symposium (International) on Combustion. The Combustion Institute. Pittsburgh, PA. pp 1003-1016.1.

Kim, J. S., de Ris, J., and Kroesser, F. W. 1971. "Laminar Free-Convective Burning of Fuel Surfaces." Thirteenth Symposium (International) on Combustion. The Combustion Institute. Pittsburgh, PA. pp 949-961.

NFPA. 1959. "Friction Spark Ignition of Flammable Vapors." *NFPA Quarterly*. 1959. Vol 53, No 2. Oct 1959. pp 155-157.

Westbrook, C. K., and Dryer, F. L. 1981. "Chemical Kinetics and Modeling of Combustion Processes." Eighteenth Symposium (International) on Combustion. The Combustion Institute. Pittsburgh, PA.

Williams, F. A. 1977. "Mechanisms of Fire Spread." Sixteenth Symposium (International) on Combustion. The Combustion Institute. Pittsburgh, PA. pp 1281-1294.

Zabetakis, M. G. 1965. "Flammability Characteristics of Combustible Gases and Vapors." Bulletin 627. Bureau of Mines. U.S. Dept of Interior.

Additional Readings

Babrauskas, V., "Will the Second Item Ignite?" *NBSIR 81-2271*, National Bureau of Standards, Washington, DC, 1981.

Burgoyne, J. H., and Roberts, A. F., "The Spread of Flame Across a Liquid Surface. Parts 1-3," *Proc. Royal Society, A308*, 39, 55, 69, 1968.

Carslaw, H. S., and Jaeger, J. C., *Conduction of Heat in Solids*, 2nd ed., Oxford Univ. Press, London, England, 1959.

Cato, R. J., et al., "Effect of Temperature on Upper Flammability Limits of Hydrocarbon Fuel Vapors in Air," *Fire Technology*, Vol. 3, No. 1, Feb. 1967, pp. 14-19.

Caydon, A. G., and Woldhard, H. G., *Flames, Their Structure, Radiation, and Temperature*, 3rd ed., Chapman and Hall, London, England, 1970.

Downing, A. G., "Frictional Sparking of Cast Magnesium, Aluminum and Zinc," *NFPA Quarterly*, Vol. 57, No. 3, Jan. 1964, pp. 235-245.

Drysdale, D. D., "Aspects of Smoldering Combustion," *Fire Prevention Science and Technology*, No. 23, 1980, p. 18.

Drysdale, D. D., *Fire Dynamics*, John Wiley & Sons, New York, NY, 1985.

Emmons, H. W., "Fire and Fire Protection," *Scientific American*, 231, 1974, pp. 21-27.

Friedman, R., "Ignition and Burning of Solids," *ASTM STP 614* (1977), p. 91.

Fristrom, R. M., and Westenberg, A. A., *Flame Structure*, NY, McGraw-Hill Inc., 1965.

Gaydon, A. G., and Wolfhard, H. G., *Flames: Their Structure, Radiation and Temperature*, 4th ed., Chapman and Hall, London, England, 1979.

Heskestad, G., "Engineering Relations for Fire Plumes," *TR 82-8*, Society of Fire Protection Engineers, Boston, MA, 1982.

Hirano, T., Noreikis, S. E., and Waterman, T. E., "Postulations of Flame Spread Mechanisms," *Combustion and Flame*, Vol. 22, 1974, p. 253.

Huggett, C., "Rate of Heat Release—Implications for Engineering Decisions," *Proceedings*, Engineering Applications of Fire Technology Workshop, Society of Fire Protection Engineers, Boston, MA, 1980, pp. 233-145.

Johnson, R. H., and Grunwald, E., *Atoms, Molecules and Chemical Change*, 2nd ed., Prentice-Hall Inc., Englewood Cliffs, NJ, 1965.

Kanury, A. M., *Introduction to Combustion Phenomena*, Gordon and Breach, New York, NY, 1977.

Kelley, C. Stuart, and Frickel, Robert, "Graybody Radiation from Conical Flames," *Fire Technology*, Vol. 8, No. 1, Feb. 1972, pp. 24-32.

Magee, R. S., and McAlevy, R. F., "The Mechanism of Flame Spread," *Journal of Fire and Flammability*, Vol. 2, 1971, p. 271.

Rasbash, D. J., and Drysdale, D. D., "Fundamentals of Smoke Production," *Fire Safety Journal*, Vol. 5, 1982, p. 77.

Rasbash, D. J., and Drysdale, D. D., "Theory of Fire and Fire Processes," *Fire and Materials*, 7,2,79, 1983.

Rohsenow, W. M., and Choi, H. Y., *Heat, Mass and Momentum Transfer*, Prentice-Hall Inc., Englewood Cliffs, NJ, 1961.

Selwood, S. W., *General Chemistry*, 4th ed., Holt, Rinehart and Winston, NY, 1965.

Semat, H., *Fundamentals of Physics*, 4th ed., Holt, Rinehart and Winston, NY, 1966.

Sirignano, W. A., "A Critical Discussion of Theories of Flame

Spread Across Solid and Liquid Fuels," *Combustion Science and Technology*, Vol. 6, 1972, p. 95.

Spalding, D. B., *Some Fundamentals of Combustion*, Butterworths, London, England, 1955.

Tamanini, F., "Reaction Rates, Air Entrainment and Radiation in Turbulent Fire Plumes," *Combustion and Flame*, Vol. 30, 1977, pp. 85-101.

Tuve, Richard L., *Principles of Fire Protection Chemistry*, National Fire Protection Association, Boston, MA, 1976.

Van Dolah, R. W., et al., "Flame Propagation, Extinguishment and Environmental Effects on Combustion," *Fire Technology*, Vol. 1. No. 2, May 1965, pp. 138-145.

Van Dolah, R. W., et al., "Ignition or the Flame-Initiating Process," *Fire Technology*, Vol. 1 No. 1, Feb. 1965, pp. 32-41.

Van Name, F. W. Jr., *Elementary Physics*, 1st ed., Prentice-Hall Inc., Englewood Cliffs, NJ, 1966.

Walker, I. K., "The Role of Water in the Spontaneous Heating of Solids," *Fire Research Abstracts and Reviews*, Vol. 9, No. 1, 1967, pp. 5-22.

Weast, R. C., ed., *Handbook of Chemistry and Physics*, Chemical Rubber Co., OH.

Williams, F. A., *Combustion Theory*, Addison-Wesley, Reading, MA, 1965.

EXPLOSIONS

William J. Cruice

The term "explosion," in its most widely accepted sense, means a bursting associated with a loud, sharp noise and an expanding pressure front, varying from a supersonic shock wave to a relatively mild wind. Unfortunately, the term also has been extended to signify chemical or physical/chemical events that produce explosions. In addition, terms denoting specific chemical reactions are used as synonyms for "explosion." This blurring of meanings has caused confusion. Nevertheless, certain terms incorporating the word "explosion" that actually signify causative events must be used, since many such terms have relatively wide acceptance and reasonably specific meanings.

This chapter discusses Explosions by detailing Fundamental Concepts, Classifications by Sources, Blast Effects and Damage Potential, Physical Explosions, Gases and Vapors, Dusts and Mists, Thermal Explosions and Runaway Reactions, Deflagration/Detonation, Assessment of Explosion Potential, and finally Countermeasures.

Other chapters within this section "The Characteristics and Behavior of Fire" discuss the related topics of Chemistry and Physics of Fire, and Theory of Fire and Explosion Control.

For specific information see the relevant chapters in Section 5: Chapter 5, "Gases," and Chapter 9, "Dusts."

FUNDAMENTAL CONCEPTS OF AN EXPLOSION

An explosion is defined as a rapid release of high pressure gas into the environment. The primary key word is "rapid;" the release must be sufficiently fast that energy contained in the high pressure gas is dissipated in a shock wave. The second key term is "high pressure," which signifies that at the instant of release the gas pressure is above the pressure of the surroundings.

Note that the basic definition is independent of the source or mechanism by which the high pressure gas is produced. An explosion may result from the overpressurization of a containing vessel or structure by physical means (bursting of a balloon), physical/chemical means (boiler explosion), or a chemical reaction (combustion of a gas mixture). Certain explosions result from chemical reactions that proceed so quickly the high pressure gas is generated instantaneously, even though no confining vessel or structure exists (detonation of high explosives). The classification of explosions by their source of high pressure gas is extremely useful. The fundamental concept of an explosion, then, is that of a sudden release of high pressure gas and its dissipation of energy in the form of a shock wave.

The released high pressure gas comes to equilibrium with the surroundings, creating certain effects while doing so. The characteristics of the effects of the gas on the surroundings depend upon (1) the rate of release, (2) the pressure at release, (3) the quantity of gas released, (4) directional factors governing the release, (5) mechanical effects coincident with release, and (6) the temperature of the released gas. The latter two considerations are relatively straightforward in basic concepts. The striking of surrounding elements by high temperature gases can cause severe heat damage, including direct damage to surfaces, thermal distortion, and fires in combustibles. Projectiles launched in the release process can impinge and cause crushing, piercing, and/or other structural failures. However, some explosions do not involve high temperature gases and projectiles; the characteristics of the pressure wave are determined by the first four considerations.

Pressure equilibriates at the velocity of sound; in ambient air, the velocity of sound is approximately 1,100

TABLE 4-2A. Effect of Shock Waves (Physiological)

Physiological Effect	Peak Overpressure, psi	(kPa)
Knock personnel down	1	(7)
Eardrum rupture	5	(34)
Lung damage	15	(100)
Threshold for fatalities	35	(240)
50% Fatalities	50	(345)
99% Fatalities	65	(450)

William J. Cruice is vice president, Hazards Research Corporation, Rockaway, NJ.

TABLE 4-2B. Effect of Shock Waves (Structural)

Structural Element	Failure	Peak Overpressure, psi	(kPa)
Glass windows, large and small	Shattering usually, occasional frame failure	0.5–1	(3.5–7)
Corrugated asbestos siding	Shattering	1–2	(7–14)
Corrugated steel or aluminum paneling	Connection failure followed by buckling	1–2	(7–14)
Wood siding panels, standard house construction	Usually failure occurs in main connections allowing a whole panel to be blown in	1–2	(7–14)
Concrete or cinder block wall panels 8 or 12 in. (200 or 300 mm) thick (not reinforced)	Shattering of the wall	2–3	(14–21)
Self-framing steel panel building	Collapse	3–4	(21–28)
Oil storage tanks	Rupture	3–4	(21–28)
Wooden utility poles	Snapping failure	5	(34)
Loaded rail cars	Overturning	7	(48)
Brick wall panel, 8 or 12 in. (200 or 300 mm) thick (not reinforced)	Shearing and flexure failures	7–8	(48–55)

feet per second (300 m/s). For energy to be dissipated as a shock wave, the velocity of release must be sonic or supersonic. The initial shock wave spreads radially from the point of release. Its form near the point of release is of very short duration (milliseconds or microseconds) and includes a high intensity impulse. As the distance from the point of release increases, the intensity (amplitude) decreases and the duration (period) increases until, far in the field, the impulse has a form similar to a mild breeze.

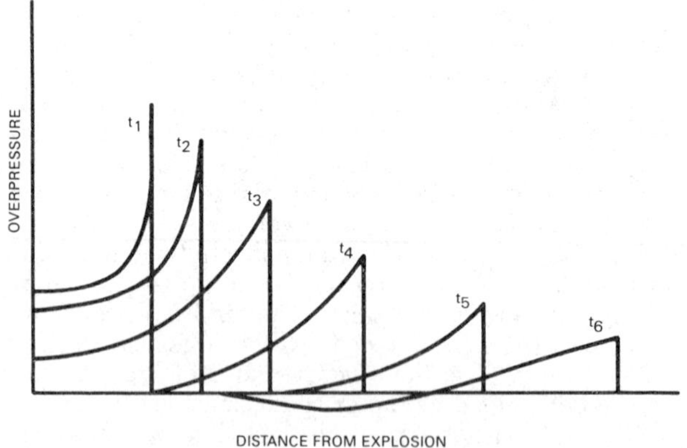

FIG. 4-2A. Decay of shock wave with distance. (Wave form at various times.)

The initial intensity (amplitude) of the shock wave is determined by the pressure of the gas at the instant of release. The overall energy of the shock wave is related to the total quantity of gas released and the pressure and temperature at the instant of release. An estimate of the overall energy available in a compressed gas is obtained by multiplying the pressure by the volume. In the event that high pressure gas is generated following the initial release, the total energy released increases, but the shock wave intensity does not; the duration (period) of the impulse is longer.

If the gas is uniformly released simultaneously in all directions, the shock wave travels in an expanding sphere. However, the velocity of release must be at least sonic, and only certain types of events can bring about a uniform release at sonic or supersonic velocities. The vast majority of explosions involve some form of mechanical confinement (vessel or other structure) in which gas accumulates to high pressure. When its pressure exceeds the strength of the confining vessel, the vessel fails at its weakest point, and the shock wave travels in the direction of the failure. Thus, vessel failure explosions create a pressure wave that is generally not uniform in all directions.

Since the form of the initial shock wave impulse is dependent upon the total energy of release and is usually highly directional, the terms "near field" and "far field" are not numerically definable. "Near field" refers to the radial distance in which shock effects (shattering) predominate, whereas "far field" refers to the area beyond that, where observed damage corresponds to wind loading effects.

CLASSIFICATION OF EXPLOSIONS BY SOURCE

Physical versus Chemical

The basic difference between explosions caused by a high pressure gas is between physical and/or physical/chemical sources and chemical reaction sources. In certain cases, all of the high pressure gas is produced by mechanical means or by events that do not involve a basic change of the chemical substance. Gas may be raised to high pressure mechanically; by external heating of gases, liquids, or solids; or a superheated liquid may be suddenly released by mechanical means and create high pressure by flash vaporization. None of these events involve changes in the basic chemical nature of the substances involved; the entire process of generating high pressure, releasing and creating the effects of explosions can be understood in terms of basic physics. These events are commonly called "physical explosions."

In other cases, generation of the high pressure gas is the result of chemical reactions where the fundamental nature of the product(s) differ substantially from the material(s) previously present (reactants). The most common chemical reaction involved in explosions is combustion, in which a fuel (such as methane) mixed with air is ignited and burns to produce carbon dioxide, water vapor, and other products. Many other chemical reactions also give rise to high pressure gases. Explosions result from decomposition of pure substances, detonation, combustion, hydration, corrosion, and various interactions of two or more chemical substances in greater or lesser degrees of mixture. Any chemical reaction can cause an explosion if gaseous products are produced by the reaction, if substances not otherwise involved are vaporized by heat released by the reaction, or if gases already present are raised to higher temperatures by heat released by the reaction.

Chemical reactions are classified thermodynamically as either exothermic (heat releasing), or endothermic (heat absorbing). Whether heat is released or absorbed depends upon the conditions under which the substance(s) react. However, since a chemical reaction proceeds faster at higher temperatures and slower at lower temperatures, a chemical reaction that absorbs heat will cool and self-extinguish if external heat is not continually supplied. Thus, endothermic reactions can result in explosions only in special circumstances, where the reactant is brought into contact with a copious supply of external heat, and the reaction itself produces gaseous products. Examples of such reactions include certain decarboxylations, pyrolyses, and other forced decompositions.

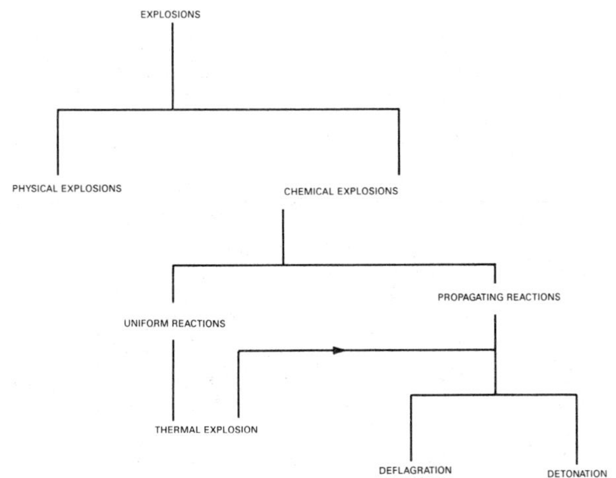

FIG. 4-2B. Classification of hazardous reactions.

Exothermic reactions, on the contrary, produce a temperature increase in the reaction mass, causing the reaction to proceed more rapidly. If heat is produced in the reaction mass faster than it can be lost to the surroundings, exothermic reactions easily become self-sustaining and self-accelerating. Such reactions can quickly become uncontrollable, even under carefully designed conditions, as well as under accidental conditions where few or no control mechanisms exist. Since heat is released, high pressure gas can be created by vaporizing reactants, products or surroundings, or by heating gases already present, even if no gases are directly produced by the reaction. Accordingly, nearly all chemical explosions result from exothermic reactions.

Uniform versus Propagating Reactions

Chemical reactions can be classified as *uniform* reactions, in which the chemical transformations occur essentially throughout the entire reaction mass, and *propagating* reactions, in which there is a clearly defined reaction zone separating the unreacted materials from the products of reaction, which moves through the reaction mass.

Uniform Reactions: The rate of a uniform reaction depends only upon the temperature and the concentration(s) of the reacting agents and is the same throughout the reaction mass.

As the temperature of the mass is raised, the energy releasing reaction proceeds more rapidly and eventually leads to observable self-heating, in which the heat generated exceeds that lost to the surroundings from the reaction mass (in most cases, the walls of a vessel). Since heat is generated throughout the reaction mass, but lost more slowly from the center than from the outside surface, the center becomes hotter and its reaction rate increases.

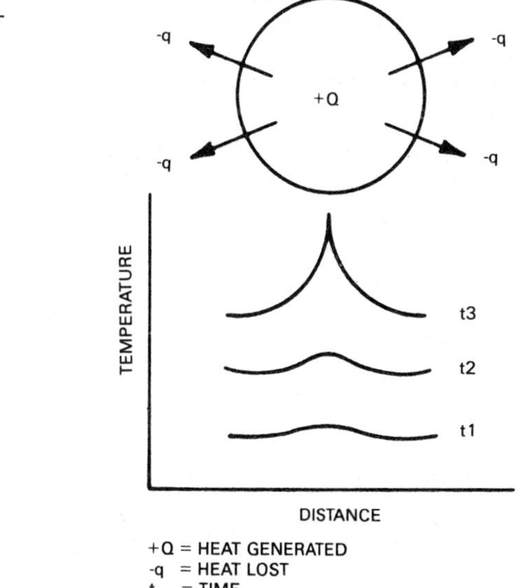

FIG. 4-2C. Growth of thermal explosion (+Q = heat generated, −q = heat lost, t = time).

This rapidly increasing reaction rate in the center continues until the reactant(s) in the hot zone are either entirely consumed or the center erupts to dissipate the high temperature.

Uniform reactions occur in solids, liquids, and gases. Gases undergo convection and diffusion quite readily; hot material can be transferred quickly from the center to the periphery, where heat can be lost more easily to the surroundings. Liquids also undergo convection and diffusion, although the effective rate of transfer is lower than with gases. Solids transfer heat exclusively by conduction and have the same higher heat production per unit volume as liquids. In general, solids present the greatest heat transfer problem. However, since mixtures of solids can react only at contact points between the reacting agents, most uniform reactions of solids occur with compounds that decompose, rather than with mixtures.

Pure liquids and true solutions can react rapidly and continuously. Liquid mixtures that form two separate phases, and liquids that react with undissolved solids, react only at the interface; the bulk rate of the reaction depends upon the surface area of globules or particles of the reacting phases where contact occurs. Stirring improves heat transfer from the center, which reduces the reaction rate of the system. Stirring also can increase the reaction rate by bringing fresh, unreacted materials together (especially in multiphase systems such as slurries and immiscible liquids). The exact effect of agitation depends upon the specific system and the conditions existing within it.

Usually, uniform reactions generate gas too slowly to produce high pressure in the absence of a confining structure. A confined reaction can generate high pressure gas by producing gaseous products, by vaporizing reactants or other materials, or by raising the temperature of gases present in the vessel. If a reaction produces a sufficiently high pressure, the confining vessel will rupture and cause an explosion.

In practice, most chemical systems capable of uniform reaction can undergo alternative reaction paths, whether of the same or different products. The rates of reaction are quite different—largely determined by temperature as shown in Figure 4-2D. Most uniform reactions, if not quenched by early release and/or dissipation of the high temperature center, undergo a transformation to propagating reactions.

Propagating Reactions: A mixture of hydrogen and oxygen can be stored at normal room temperatures for extensive periods with no sign of chemical reaction. However, most mixtures of hydrogen with oxygen react violently if an igniter is applied. The reaction begins at the igniter and travels through the mixture. Three distinct zones can be observed: the reaction zone (flame), the product zone (behind the flame), and the unreacted zone (in front of the flame). This is a classic illustration of a propagating reaction. (See Fig. 4-2E.)

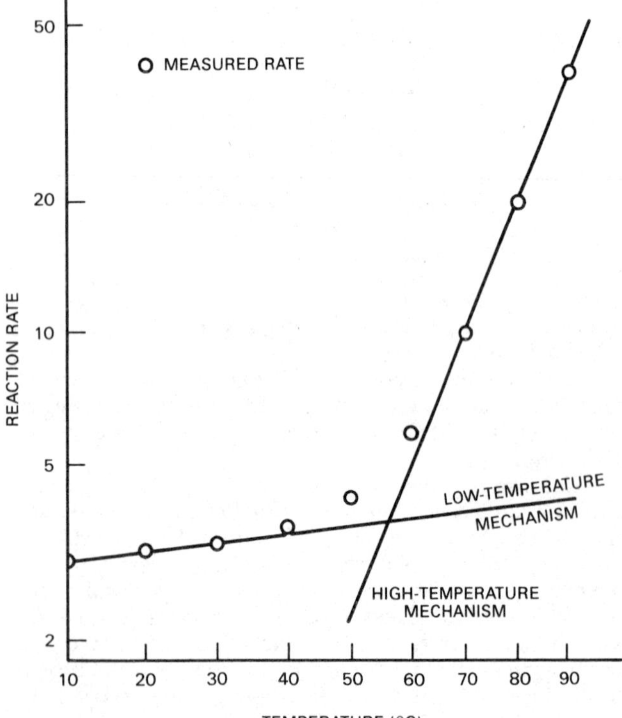

FIG. 42D. Change in dominant reaction path with temperature.

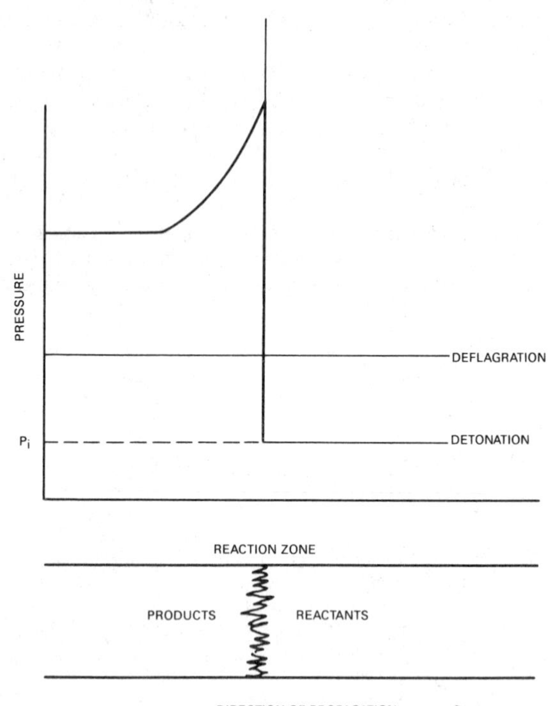

FIG. 4-2E. Instantaneous pressure profiles of propagating reactions.

A propagating reaction is always exothermic. The reaction is started by the creation of a relatively small high temperature zone, whether by an external igniter (match, spark, shock) or by heat buildup in the core of a uniform reaction system. For the reaction to propagate, the reaction core, which is activated by the igniter, must sufficiently raise the temperature of the surrounding material so that it reacts in like manner. The higher the initial temperature of the system, the easier it is to ignite, and the more likely it is to support a propagating reaction, since relatively less energy transfer is required to initiate the surrounding unreacted material.

Heat not transferred to the unreacted material is retained in the products of reaction or lost to the surroundings. The temperature of the products is determined by dividing the total heat retained by the heat capacity. Any pressure increase is determined by the vapor pressure of the system and conventional P-T-V relationships for any gases present or produced (directly or indirectly), by the reaction.

Since a propagating reaction begins at some specific point and travels through the reaction mass, the rate of energy release is related to the propagation velocity or rate of travel of the reaction zone. Propagation velocities vary from nearly zero to several times the speed of sound, depending upon composition, temperature, pressure, degree of confinement, and other considerations. Since pressure equilibrates at the speed of sound in the unreacted materials, a critical difference exists between subsonic reactions (deflagrations) and supersonic reactions (detonations).

Deflagrations and detonations propagate in gases, liquids, solids, pure compounds, and single phase and multiphase mixtures. In many cases, a propagating reaction occurs only under some form of external confinement, as the reaction otherwise dissipates energy too rapidly to be self-sustaining. Confinement required for propagation varies widely with the rate of reaction and the physical state of the materials. It is necessary only that the reactants be kept in place long enough to react, and that the reaction zone retain energy for sufficient time to ignite unreacted material. Relatively weak confinement can be adequate for these purposes when the reaction time is only a few milliseconds or microseconds. Detonations and deflagrations can propagate in paper, plastic, glass and metal containers, pipes, reactors, tanks, drums, boreholes, wells, open pits, buildings, caves, tunnels, etc. Many materials sustain violent propagating reactions with no external confinement.

BLAST EFFECTS AND DAMAGE POTENTIAL

The primary procedures for estimating the damage potential of an explosion relate to the intensity of the shock wave produced by the release of the high pressure gas. The most common means of estimating involves comparing explosions with TNT detonations, as much information exists on TNT-produced shock waves. However, the specific means of relating a potential explosion to TNT detonation differs with the type of explosion assumed.

Unconfined Materials: Unconfined materials may be analyzed by conventional thermodynamics to ascertain the anticipated heat release from a total reaction of most stable products. Once the available energy is determined, an assumption is made that the material can detonate, and the energy potential is considered to be released from an energy-equivalent mass of TNT. The intensity of the assumed shock wave then is obtained as a function of distance using standard references.

It is essential to realize that this only provides an *estimate* of the possible damage potential *if the material can* detonate. Many materials will not support a detonation when unconfined. Whether or not a material can support a detonation must be determined by experimentation, using appropriately scaled samples. If the material cannot detonate, a very conservative estimate of deflagration pressure can be obtained by assuming that the thermal energy is released, and that the gases are formed at thermodynamic pressures within the geometry of the subject charge. A comparison then can be made using an energy equivalent mass of TNT.

Confined Materials: Confined materials must be analyzed in three ways. First, the charge should be evaluated in the same manner as for unconfined materials. The effective pressure front must be sufficient to rupture the confinement, or the external damage will not occur due to an expanding shock front. Damage, however, may result from physical displacement of the confining structure due to unequal loading of internal shock waves.

Second, an analysis should be performed based upon thermodynamic pressure. The thermodynamic pressure must be sufficient to rupture the confining structure, or external damage from an expanding shock front will not be experienced. Conventional constructions (buildings, rooms, etc.) generally fail at quite low pressures; vessels generally fail at approximately four times the allowed working pressure. In all cases, the ultimate failure pressure of the confinement should be obtained from a competent structural engineer, with appropriate allowances for temperature, corrosion, and other weakening factors.

The third method of estimation is based upon equating the failure of the confinement to a shock front from a TNT charge. Vessel failure should be assumed and the amount of TNT required to produce a shock wave of intensity equal to the ultimate failure pressure of the vessel at the vessel wall should be determined, (assuming the charge to be at the geometric center of the vapor space of the vessel). The shock wave then can be projected radially, as in prior cases.

In all cases of confined materials, the pressure/shock wave estimates must be followed by an intelligent evaluation of the projectile effects, as ruptured vessels or building elements invariably produce some projectiles.

The initial estimation of the damage potential is frequently based upon the worst possible conditions of energy release, reaction velocity, etc. Using experimental data, the estimates can be refined. However, the confidence that can be placed in any estimate must be inversely proportional to the cost of being incorrect; conservative estimates of damage potentials are preferred, unless very strong evidence is cited to demonstrate that the worst case cannot occur.

Most combustions and explosions generate and release high temperature gases and/or condensed phase materials. These hot products can cause severe heat damage to personnel, structures, and adjacent equipment. Secondary fires and process upsets also are common. In assessing the damage potential associated with combustions and explosions, it is essential to consider the heat effects associated with the high temperature products of reaction.

PHYSICAL EXPLOSIONS

Physical explosions are releases of high pressure gas that do not involve chemical reactions, although most physical explosions involve vaporization.

Most physical explosions involve a confining vessel such as a boiler, gas cylinder, or other tank. High pressure in the vessel is created by a mechanical compression of the gas, heating of the contents, or introduction of high pressure gas from another vessel. When the pressure reaches the ultimate strength of the weakest portion of the vessel, failure occurs. In some cases, relatively small elements of the overall vessel assembly may be the weakest portion, and these elements are expelled as projectiles. In other cases, the actual vessel walls or seams fail and the vessel is flung open with great violence.

Damage to surroundings depends primarily upon the failure mode. If small elements fail but the vessel remains essentially intact, the expelled projectiles are as dangerous as bullets or cannonballs; gas release, however, is highly directional and is controlled by the diameter(s) of the hole(s) created when the small elements are expelled. Under these conditions, the damage to the surroundings may be limited to projectile penetration, scorching or other harmful effects of hot gases, and/or minor displacement of weak structural elements. If the expulsion of small parts and the release of gases is unbalanced, the vessel will be thrust in the direction opposite to the gas expulsion, and may be toppled, hurled, or otherwise moved. Such an event can produce damage by crushing objects in the vessel's path, and/or can lead to the collapse of buildings or other structures if load bearing elements are displaced or destroyed.

The failure of vessel walls or seams generally releases major projectiles and causes severe thrust of the vessel body in the direction opposite to the gas release. However, the gas release is extremely rapid under such circumstances, and a generally severe and broad shock wave is produced. The damage potential may be crudely estimated by multiplying the volume of gas in the vessel by the actual pressure at the moment of failure, although the accuracy of the estimate may vary considerably with the compressibility and the condensibility of the gas (which cools on expansion from the original vessel volume). The pressure wave from the rupture of the vessel is highly directional, traveling largely in the direction of release and creating equivalent pressure effects at much greater distances in that direction than in others, but substantial pressure effects are created in essentially all directions.

If the vessel contains a superheated liquid (a liquid at a temperature above its normal boiling point, or a liquefied gas such as ammonia or carbon dioxide), flash vaporization of the liquid occurs when the vessel ruptures. A sufficient amount of liquid vaporizes to cool the discharged material to its normal boiling temperature, and the vaporized material adds significantly to severe pressure effects, depending upon the total amount of vaporizing per unit of time. This phenomenon is called a BLEVE (Boiling Liquid Expanding Vapor Explosion).

An analogous phenomenon is the flash vaporization of a liquid (or, rarely, a solid) suddenly brought into contact with another substance at a temperature far above the ordinary boiling point of the liquid. Such events include spillage of cryogenic liquids or cold liquefied gases into the normal environment, or flash vaporization of liquids by hot metals or minerals. In such cases, the liquid receives heat from the hotter surface at a sufficient rate that a high pressure gas is created instantly at the hot surface. If the interface between the cold liquid and the hot surface is large, vaporization can occur quickly enough to create a shock wave, spreading in all directions. Actually, the vaporization rate is usually significantly lower, but the rapid generation of gas frequently creates pressures high enough to destroy buildings or vessels. Sudden introduction of water into high temperature boiler tubes, heat exchangers, or heat transfer fluid tanks can produce severe explosions.

In the event that the discharged material can burn in air or otherwise chemically react with the environment into which it is discharged, a severe danger of secondary explosion exists.

GASES AND VAPORS

The most common form of chemical reaction that produces high pressure gases from other gases or vapors is the combustion of gaseous fuels in air. However, other oxidizing gases (such as oxygen, chlorine, fluorine and a variety of gaseous compounds and mixtures) may be substituted for normal air, frequently producing more severe combustion processes. Certain gases (acetylene, ethylene, ethylene oxide, butadiene, nitrous oxide, and others) can propagate a decomposition flame under proper conditions of temperature and pressure in the absence of any other gas. This latter phenomenon is generally called disproportionation.

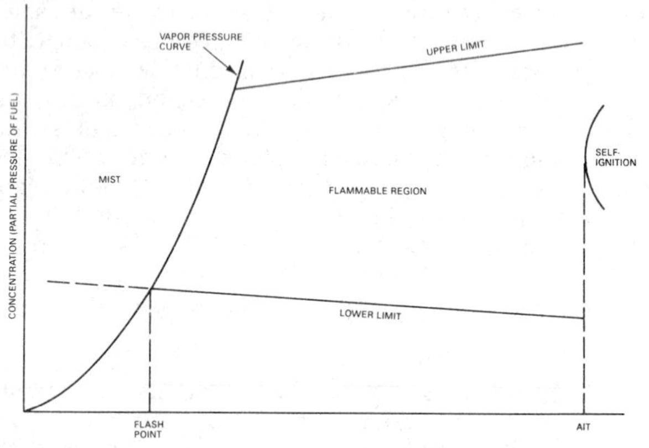

FIG. 4-2F. Flammability limits as a function of temperature.

For the general case of combustion involving a fuel gas and an oxidizing gas (such as air), mixtures are flammable only within a certain range of compositions. A minimum fuel proportion is required to sustain combustion; there is also a maximum fuel proportion above which combustion cannot be self-sustaining. Near these limiting fuel proportions, flames propagate through the mixture rather slowly, but in the middle of the flammable range, combustion velocities can be faster than the speed of sound.

In general, increasing the temperature and/or pressure of the mixture widens the flammable range and increases the combustion velocity throughout the flammable range. Generally, limits must be determined experimentally for the specific mixture, temperature, and pressure of interest.

The great majority of combustible gas mixtures are stable at ordinary temperatures and pressures. The combustion reaction must be initiated by some means. Once ignition occurs, the combustion reaction is self-sustaining by transferring heat and activated agents from the burned gas (products) to the unburned gas (reactants) at the flame front (reaction zone), and the reaction propagates from the ignition point to the physical limits of the combustible mixture.

Gas mixtures are generally more difficult to ignite near the flammable limits and most easily ignited somewhat above the precisely balanced (stoichiometric) mixture. In general, gas mixtures are very easily ignited anywhere in the flammable range. The concept of minimum energy requirement for ignition is very difficult to apply in practical situations, particularly since the minimum energy input required varies with the type of ignition source. The

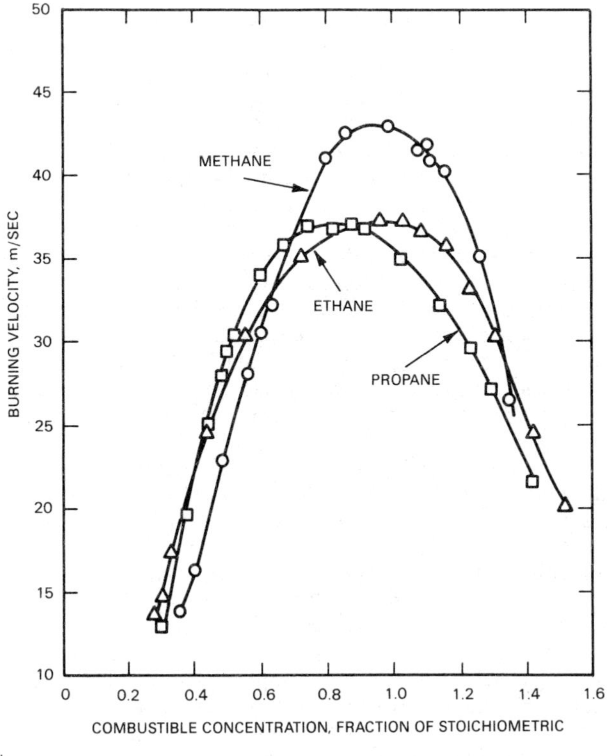

FIG. 4-2G. Combustion velocity versus concentration.

minimum electrical discharge energy required to initiate combustion is very different from the minimum shock energy, for example, and neither relate well to the energy requirements for hot surface ignition. Some mixtures (e.g., chlorine/fuel) can be ignited by radiant energy (visible light, ultraviolet light, etc.) of the correct frequency.

Most reactive gas mixtures are capable of self-ignition under specific conditions of temperature and pressure. At any given temperature, certain molecules are energetic enough to react when they meet; ordinarily such reactions and the heat liberated by them are not detectible. At some temperature, the intermolecular reactions are frequent enough so the gas mixture begins to self-heat and eventually reaches its kindling temperature. Propagating combustion then is initiated if the overall mixture is flammable at the temperature and pressure conditions attained.

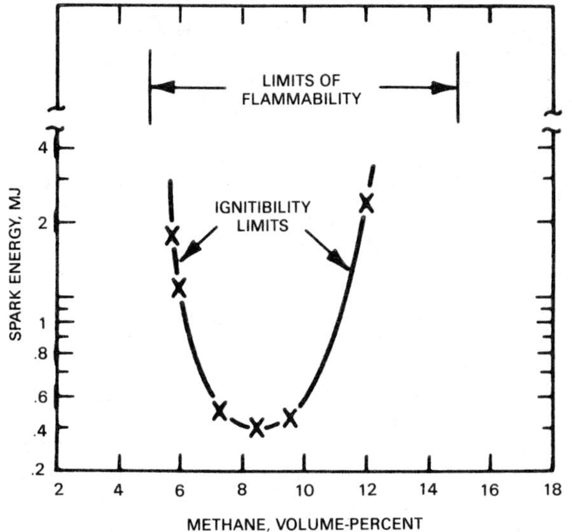

FIG. 4-2H. Ignition energy versus concentration.

Autoignition depends upon the specific mixtures of gases, the volume and geometry of the vessel, the materials of construction, and the initial temperature and pressure of the mixture and the surroundings. Published autoignition temperatures are specific to the method of determination and cannot be used indiscriminately. The existence of thermal and catalytic wall effects makes application of published autoignition temperatures to practical situations a delicate and potentially very dangerous activity.

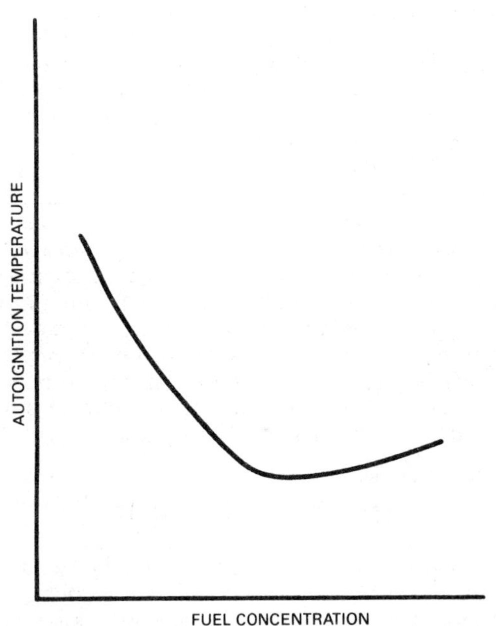

FIG. 4-2I. Autoignition temperature versus concentration.

Pressure generated by the combustion of gas mixtures results primarily from the heat liberated and the consequent high temperature of the product gases. In most gas combustions, heat is lost to the surroundings by radiation from the flame, convection currents, etc. Generally, the highest pressures are attained when combustion occurs rapidly. For most confined subsonic combustions, the maximum pressure produced is approximately ten times the initial pressure, since the flame temperature is limited by dissociation reactions. This ratio may differ if the starting mixture is composed of compressible gases or if the fuel, oxidizer, or an otherwise inert component of the mixture can decompose and/or yield a significant change in the number of moles of gas in the system.

Conversely, if the gas mixture is unconfined or if confinement is breached, the burning gas expands as a fireball at normal atmospheric pressure, and the maximum fireball volume is approximately ten times the initial volume of the mixture. As in the case of pressure generation, expansion of the mixture is least near the limiting concentrations and greatest near the middle of the range.

Gases that propagate decomposition flames generally exhibit minimum absolute concentrations (minimum reactive gas pressure and minimum ratio of gas to other gases in a mixture) below which the flame does not propagate. However, most such gases do not have upper limits, but are flammable from the lean limit to 100 percent. The characteristics of pressure generation, combustion velocity, fireball size, etc., are specific to each gas and do not relate readily to principles applicable to conventional fuel/oxidizer mixtures.

Certain gases and mixtures are inherently capable of supersonic, shockwave driven reaction (detonation) under appropriate conditions of temperature and pressure. For flammable gas/oxidizer mixtures, the detonable range is largely dependent upon the source of ignition, but always includes the stoichiometric and the fastest burning ratios for the normal (subsonic) combustion reaction. Gases capable of sustaining a detonation ordinarily do not reach actual supersonic conditions unless initiated by a high intensity shock wave. However, in certain confinement geometries, such as pipes, where the length exceeds the diameter by a factor of ten or more, a conventional combustion reaction can self-accelerate to the point where a transition from deflagration to detonation occurs if the mixture is in the detonable range of compositions. The final pressure generated in detonation is the same as the final pressure from deflagration, but the transient pressures created by the shock wave in the detonation process can be more than double the final pressure, and the effects on objects in the path of the shock wave can characteristically be more than four times the final pressure. Accordingly, vessels capable of containing deflagration maximum pressures can sustain partial or total destruction if the gas mixture detonates rather than deflagrates.

DUSTS AND MISTS

Dusts and mists (finely divided liquids) can generate high pressure gas by combustion in air or another reactive gaseous environment. However, since the necessary chemical reactions can occur only at the interface of the dispersed particles or droplets with the surrounding gas, the rates of pressure generated are limited by the available surface area of the dispersed material. For any given mass of dust or mist, the surface area increases as the particle or droplet diameter decreases, so the violence of combustion for any given substance increases as the particles or droplets become finer. These dispersed condensed phases are quite similar in combustion behavior.

Combustion can occur with virtually any particle or droplet, but in practice most explosion hazards with dispersed materials are encountered with 20 mesh (840 micron) sizes or smaller. As the particle or drop size decreases, the dispersion is more easily created and is relatively more stable and enduring. Accordingly, finer particles or droplets constitute graver hazards by making dispersions more likely, by remaining dispersed for a longer time, and by burning more rapidly than coarser particles. For most combustible agents, however, any ignition of a dispersion in air or other reactive gas can create pressures sufficient to destroy process equipment and structures.

There is a minimum uniform concentration for a propagating combustion to occur. There is no reliable upper concentration limit for dusts or mists, however, since the reaction is surface area controlled. Most upper limits actually represent that concentration of dust or mist that renders some experimental igniter ineffective, rather than a true thermodynamic limiting concentration.

As a first step, the combustion process involves vaporization of volatile combustibles from the surface of the particle. Accordingly, coal with less than approximately 20 percent by weight volatiles (excluding moisture) does not propagate a dust combustion, but clay particles coated with volatile organics can produce dust explosions. In this context, "volatiles" includes materials that can be pyrolized or otherwise vaporized at high temperatures (soaps, flowing agents, silicones, etc.) as well as low-boiling solvents, etc.

The rate of pressure rise and the maximum pressure attained in dust combustions and mist combustions increases as the concentration of dispersed particles increases, generally passing through a maximum at a concentration that depends upon particle size, then slowly decreases as the concentration increases further. A substantial amount of data has been published for a wide variety of dusts based on specific apparatus configurations. However, while useful for comparison with other dusts under the same conditions, the numerical values can be very misleading if improperly used. In particular, combustions of dispersed particulates are highly subject to turbulence, which greatly increases the rate of pressure rise actually experienced. Maximum pressures are about ten times the starting pressure but are seldom attained.

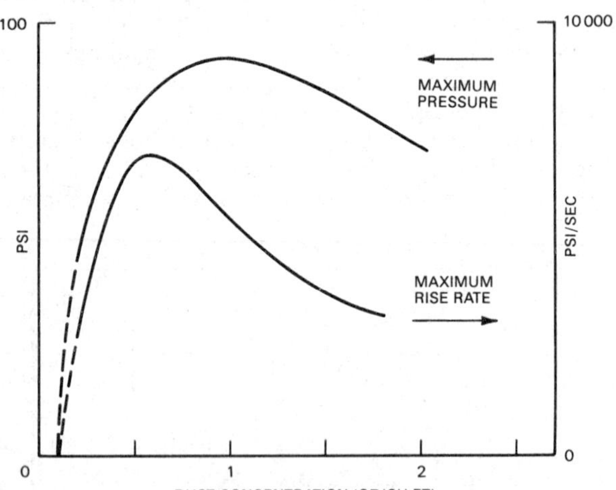

FIG. 4-2J. Severity of dust explosion versus concentration.

Since dusts and mists frequently occur as dispersions in lightly constructed equipment, and since, in practice, there is usually a large supply of quiescent material that can be hurled into the air by the initial combustion, dust combustions and many mist combustions can self-propagate for large distances into areas where no flammable mixture previously existed. Dust explosions are frequently described as rolling explosions, characterized as sounding like distant thunder. Rolling explosions generally are initiated by a primary dust explosion which creates and ignites secondary dust dispersions, which in turn create and ignite further dust dispersions.

Ignition of dust and/or mist dispersions in air commonly are achieved by an electric discharge, open flame, or hot surface. The energy required for ignition is generally higher than that needed to ignite vapor/air mixtures, but is low compared to the energy sources available in the environment. Dusts and mists frequently cause abrasion, binding, and consequent generation of high temperature surfaces or flames in mechanical equipment, which can in turn ignite the dispersion. Moving dust or mist particles frequently create electrostatic accumulations which eventually discharge and ignite the dispersion. Most insidious in dusty environments is the layering of particles on surfaces which become hot and initiate a combustion in the layer. A small puff of wind from the combustion in the layer creates adjacent airborne dispersions which are ignited by hot products from the original combustion. The resulting cloud formation can travel throughout an entire dusty area, creating widespread and massive destruction.

The presence of flammable vapors in dust dispersions, even far below the normal lower limit for the vapors, recently has been found to greatly enhance the ignitability and velocity of dust combustions. This synergistic effect emphasizes the uniqueness of many chemical systems and operating environments. Assumptions of behavior not experimentally or experientially confirmed or supported can be disastrous.

Effects of dust and mist explosions on the surroundings are analyzed in essentially the same manner as gas explosions. The same principles of vessel failure, structural failure, pressure generation, fireball generation, and secondary effects apply. However, unlike gaseous combustions, dust and mist combustions tend to create substantial amounts of condensed phase products at very high temperatures, i.e., hot solids, tars, gums, oils, etc. These hot products tend to adhere to surfaces they contact, creating severe heat damage and frequently causing additional fires.

THERMAL EXPLOSIONS AND RUNAWAY REACTIONS IN CONDENSED PHASES

Thermal explosions and runaway reactions begin as uniform reactions, whether in reaction vessels, storage vessels, pools, or piles of material. All follow essentially the same pattern of heat accumulation and reaction rate acceleration. At some environmental and system temperatures, depending upon the size of the reaction mass, nonuniformities in temperature and the reaction rate lead to a change in the favored reaction mechanism so that it becomes a propagating reaction. If high pressure gases are released, explosive effects are created in the environment, their form being determined by the rate at which the gas is released. In some cases, the change in mechanism may result in a detonation, with disastrous effects.

The general thermal runaway is frequently accelerated by either or both of two other chemical phenomena:

1. Autocatalytic reactions are those in which the reaction products chemically accelerate the initial reaction in which they were formed. The chemical acceleration often is more important than the self-heating effect.
2. The critical concentration effect can take two basic forms. In both cases, the initial reaction creates a metastable condition in the mass by slowly generating reactive products. In one form, these reactive products are capable of rapid exothermic decomposition with the reaction rate being highly dependent upon the temperature. In the other form, the reactive products are capable of further exothermic reaction with the unreacted portion of the original chemical(s) present, and this form also is highly dependent upon temperature. In either event, the reactive products may accumulate in the overall mass until a slight increase in overall temperature produces an immense surge of reaction, rapid liberation of enormous quantities of heat and/or gaseous products, and usually a disastrous explosion.

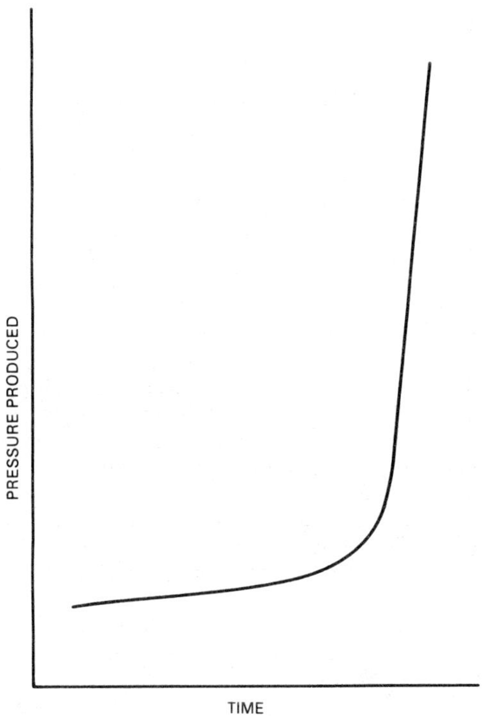

FIG. 4-2K. Auto acceleration of thermal decomoposition.

Thermal explosions and runaway reactions differ from combustions of gases, mists and dusts in several ways; these differences create significantly greater explosion potentials. First, the amount of material physically present in a unit volume is much greater in a condensed phase system, and the total heat available within the reacting mass is many times greater than that for dispersions in air. Second, the attainable pressure is generally much higher for condensed phases because the higher temperature reaction mechanisms usually produce large amount of gaseous products in a limited vapor space. Third, the reaction mass generally swells and forms gas bubbles with thermal expansion of the condensed phase. In some cases the thermal expansion alone can hydraulically fail a vessel that theoretically could contain the gases produced. This produces a BLEVE effect when the confinement is ruptured. Fourth, whether because of vessel rupture or boilover, large quantities of hot chemicals often are discharged into the environment. Under such circumstances, as the chemicals are usually combustible in air, catastrophic secondary explosions and fires occur.

Estimating the potential effects of thermal explosions and runaway reactions may be approached by the methods previously described, but such methods have limitations. It often is assumed that the ultimate pressure of a confining vessel is that at which it ruptures under any circumstances. In fact, the physical rupturing of a vessel requires a certain amount of time. If the reaction taking place in the vessel creates high pressure faster than the rate at which the vessel expands, the actual gas pressure upon release may be much higher than the pressure at which the vessel will fail under sustained load. Specific instances have occurred in which the effects of environmental damage clearly indicate that the actual gas pressure at release was two to ten times the maximum pressure the vessel would be expected to hold. Accordingly, if thermodynamic estimations of the damage potential predict significantly worse effects than would be anticipated from a rupturing vessel, it may be foolhardy to rely on the simple rupture model.

It is unwise to predicate reaction paths in the absence of complete experimental information. Various evaluation methods are available to assess the actual severity and sensitivity of chemical systems under thermal explosion conditions; all the methods have limitations. Nevertheless, experimental assessments are essential for properly evaluating potential hazards of runaways. Furthermore, it is unrealistic to assess only the theoretical or desired compositions; process upsets in composition also must be anticipated and evaluated.

Thermal explosions and runaway reactions usually are associated with nitration, polymerization and diazotization processes, and with chemical structures involving unsaturation or N-N, N-O, Cl-O, O-O, N-Cl bonds, among others. Generally, any reaction that is self-sustaining under existing environmental conditions theoretically can give rise to a thermal explosion. In chemical processing, this includes any reaction that does not require a continuous application of heat to proceed to completion. All such reactions must be assessed for thermal explosion/runaway reaction potential, with due regard for the fact that process equipment, controls, and personnel are subject to various failures and errors.

DEFLAGRATION/DETONATION IN CONDENSED PHASES

Theoretically, any exothermic reaction can be the basis for a deflagration or a detonation, provided the heat of reaction is directed in sufficient quantity into the unreacted materials. Accordingly, these features of the chemical system and its environment that promote such energy transfer favor propagation, whereas those that promote loss of the energy do not.

Propagating reactions deliver energy to the unreacted materials or to the surroundings. The amount of energy delivered is determined by the distance that the energy must travel and the ability of the energy to be transferred to the next reacting element. Consider a column of material (such as a full pipe) initiated at one end. The reaction zone may be represented as a cross sectional element, with the radius being the distance the energy must travel to reach the surroundings. When the energy reaches the outer edge of the reaction zone, its loss to the surroundings is determined by the confinement characteristics (strength, heat capacity, etc.) of the containment vessel. Thus, propagation is generally favored when charges with larger diameters and stronger confinement vessels are used. There is a minimum diameter of charge below which a reaction will not propagate. This characteristic—the critical diameter—is specific to each chemical system and the effective characteristics of confinement. The principle of critical diameter applies equally to unconfined materials (solids, gels, etc.), although the critical diameter for an unconfined material is generally greater.

Propagating reactions that initiate in condensed phases are commonly associated with moving equipment such as pumps and fast acting valves, but they also begin from thermal decomposition in localized elements. Localized high temperature conditions, whether arising from cavitation, friction, adiabatic compression, self-heating, or external sources such as welding, electrical phenomena, mechanical impingement, fire, etc., can give rise to a propagating reaction if the chemical system and surroundings favor it. The effects of the reaction depend upon the velocity of propagation.

Deflagration of condensed phases produces higher pressures than deflagration of gases, mists or dusts in air, as there is more energy released per unit of volume. The propagation proceeds by mass transfer (movement of heat and activated agents into the unreacted materials) at velocities ranging from millimeters per hour to hundreds of meters per second. The reaction zone is very hot and composed of gasified matter, much of which loses heat to confining pipes and vessels. In high energy, low velocity systems, the confinement frequently weakens due to excessive heating and causes localized failures. Deflagrations are extremely pressure sensitive, since loss of confinement permits products and reactants to vaporize and thus permits energy from the reaction zone to be lost as heat of vaporization. Accordingly, many deflagrations can be extinguished by releasing the pressure through venting mechanisms or by failure of confinement. The concept of critical diameter applied to deflagration relates more to heat capacity and thermal conductivity of the confinement vessel than to strength or rigidity. Since the reaction proceeds at mass transfer rates, heat losses to the walls are important, but the heat may travel through the walls to the unreacted material, sensitizing the system. Use of critical diameter principles in precluding deflagration is not widespread since other, more reliable methods are available.

The detonation of condensed phases invariably produces extremely high pressures in the confinement vessel. The reaction proceeds in the unreacted material by shock compression at velocities ranging from slightly supersonic to several times the velocity of sound. There is a large impulse in the direction of propagation. The classic testing method for detonability of liquids involves placing a mild steel plate 0.375 in. (9.5 mm) thick above the 50 mL (1.7 oz) sample and using a high explosive charge to initiate the sample. If the sample propagates a detonation, a hole of the same diameter as the sample charge is produced in the unrestrained plate. (The test is usually done in a bunker, as in the open the plate may stay airborne for over 10 seconds.)

Lateral pressures (at right angles to the direction of propagation) frequently are millions of pounds per square inch (1 million PSI equals 6895 MPa), enduring at such levels for millionths of a second at any given point. Such tremendous impulses cause shattering of the confinement vessel, which flies apart, propelled by the still-high pressure of the products. Usually, the shock wave in the chemical system passes the fragmenting confinement vessel faster than the parts of the vessel can physically move, and detonations cannot be quenched by conventional venting methods. Propagation can be interrupted only by reducing the diamter of the unreacted material below the critical diameter in the specific confinement vessel, or by breaking the continuity of the chemical system creating the shock wave. Since the shock wave is traveling thousands of meters per second, the latter option is rarely feasible. The use of critical diameter traps with certain chemicals is well established, however.

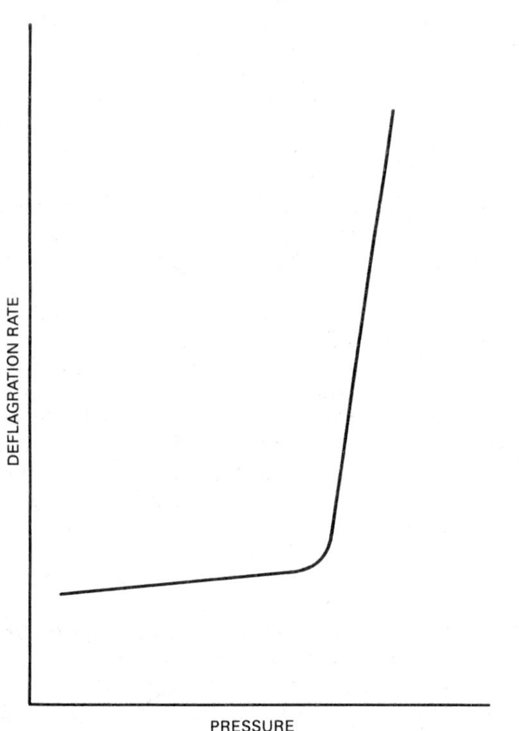

FIG. 4-2L. Effect of initial pressure on deflagration rate.

Detonations obviously are more dangerous than deflagrations. Theoretically, however, any material that is exothermic can sustain a detonation under some set of circumstances. In keeping with theory, it is frequently found that condensed-phase deflagrations can self accelerate under confinement to a point where a deflagration becomes a detonation. The transition is attributed to the preheating of unreacted materials by energy transmissions through confinement vessels and by pressure-piling and precursor wave effects in the unreacted material itself, as well as to physical changes such as bubble formation, etc.

Propagating reactions in condensed phase systems are particularly hazardous in that large masses of liquids, solids, and slurries frequently are connected to operating equipment by pipelines, conveyors, etc. Accordingly, it is possible for a minor malfunction (e.g., pump cavitation, auger friction, valve compression) to initiate a reaction that can deflagrate or detonate through the mass transfer system to storage vessels or transportation tanks with catastrophic results. The most insidious condition occurs when a deflagration reaction propagates in piping that is small enough to prevent transition to detonation. If the deflagration enters a vessel, it may easily become a detonation because the vessel diameter is no longer the controlling factor. Under this condition, the entire volume of material in the vessel will probably detonate because the shock wave travels through the material faster than the material can be moved by the pressure.

ASSESSMENT OF EXPLOSION POTENTIAL

All explosions involve materials that release energy to the surroundings. All accidental explosions result from abnormal conditions or "upsets" in a system at equilibrium. Assessing the explosion potential involves two distinct sets of information. The first involves the fundamental nature of the substances being processed, stored, handled, used, or transported. The second includes the characteristics of the specific equipment in which the substances are present. While the data sets are distinct, it is necessary to consider the effects of each on the other, so the actual assessment process involves a high degree of iteration. Consideration must be given to the environment in which the system is located, including both abnormal energy inputs and the effects of system abnormalities.

Properties of Materials

This information may best be understood in terms of two basic concepts corresponding to two basic questions: *Severity* —Type and power of reactions, i.e., "What can it do?" *Sensitivity* —Initiation modes, i.e., "What can cause it?" This distinction appears obvious, but confusion and error in explosion assessment results if the distinction is not made.

Severity: The various types of reactions have been discussed with particular characteristics highlighted. For assessment, however, it is inadequate to qualitatively describe reactions. It becomes essential to determine the maximum pressures and temperatures of reactions, rates of increase of pressure and temperature, heat of reaction, etc., and the conditions under which such reactions can occur. These determinations should be independent of the conditions under which handling, etc., is expected to occur, but maximum attention is often given to those conditions that are expected. Such practice can be risky because all such characteristics and conditions are not predictable. Preferred practice includes a determination of the material properties under extreme conditions, at least initially.

Severity is a more critical element in assessing subject materials for several reasons. First, if the materials cannot undergo an explosion-generating reaction, questions regarding sensitivity do not apply. Second, if the material can react to produce explosive forces, it is reasonable to anticipate that input energy can somehow concentrate to initiate the reaction. Third, since severity experiments employ strong rather than marginal initiators, severity experiments are more reliable than sensitivity experiments.

Usually the first step in assessing severity involves comparing the material structures with those of materials having known properties. The thermodynamic potential is evaluated by manual or computerized calculations. (The manual process is arduous; use of CHETAH* is far superior.) In the final analysis, however, the results of such efforts must be tested by experimentation.

Sensitivity: All initiation mechanisms involve an input of energy into the subject materials. If a specific form of potential initiation energy is chosen, the minimum amount of such energy required to produce a reaction can be determined. However, it is essential to realize that the minimum energy requirement varies widely with the type of initiation energy chosen. There is no reliable quantitative relationship between sensitivity to electrical sparks and sensitivity to mechanical impact, for example. Relationships between different types of energy inputs are complicated in that any source delivers energy in multiple forms, and only those forms that can be absorbed by molecular bonds are effective in initiating chemical reactions. Even under carefully controlled conditions, sensitivity determinations are apparatus-dependent. Applying minimum energy values to real situations presumes that the same general phenomenon (electric arc, for example) will exhibit the same distribution of energy forms in the laboratory and in the field. Such an assumption is never 100 percent reliable, and can be wrong.

In practice, energy inputs that create localized high temperatures are those most likely to initiate a chemical reaction. The most commonly evaluated forms of energy input are flames, electrical discharges, hot surfaces, mechanical compression, and shock wave compression. Other commonly recognized sources of initiation energy are special cases of the above, i.e., friction which creates hot surfaces and/or mechanical compression. Such initiation sources often are evaluated separately, but the measurements thus obtained are so highly apparatus-dependent that applying them to real situations is possible only by analogy.

Assessing thermal sensitivity is actually a severity procedure (i.e., responds to "What can it do?") rather than a sensitivity procedure. Although materials are ranked as more or less thermally sensitive according to the temperature at which self accelerating decomposition appears, any determined threshold temperature for self acceleration is highly apparatus-dependent. Further, thermal decomposition is dependent only upon the mass temperature and how it is attained. Thus, there is no way of measuring a discrete energy input that initiates uniform thermal decomposition. Essentially all sensitivity evaluations refer to propagating reactions of one form or another.

Thermal stability evaluations are, however, the most comprehensive and reliable assessments that can be performed, primarily since the reactivity is independent of the mode of energy input. The thermal stability of a chemical system, properly evaluated, provides more widely applicable information than any other evaluation procedure. Generally, if the subject materials can release energy, the phenomenon will appear in a proper thermal stability evaluation.

*The ASTM Chemical Thermodynamic and Energy Release Evaluation Program.

Assessing severity and sensitivity characteristics depends heavily upon selecting materials for evaluation. The potentials of the system must be probed, or the exercise can yield a false sense of security. Consideration must be given not only to those materials and ratios that are planned, but also to those that may arise from malfunctions, human errors, and normal variations in the operating system.

Properties of the System

The "system" used here refers to the operating equipment in which the materials are or can be found. Any system includes *active* or operating elements, and *passive* or nonoperating elements.

Active Elements: This term refers to units that use moving parts such as pumps, blowers, grinders, valves, stirrers, classifiers, etc. These represent paths by which external energy is supplied to the materials under normal operating conditions. Further, most such equipment has a built in capacity which is greater than required for normal use, and, in the event of malfunction, can usually input energy at levels far above normal and can input energy in form(s) quite different from normal. (A blower or fan, for example, normally supplies energy for transporting gas; the energy is moderate and supplied discretely through the motion of the blades. If a blade is distorted, however, energy is concentrated in the form of molten particles from the fan blade or housing, which are highly efficient ignition sources.)

Assessing the explosion potential requires determining the type and intensity of energy transmitted to the materials and comparing it with the sensitivity characteristics of the materials. First, the analysis is performed with respect to the normal operation of active elements over the range of materials that could be affected. All active elements should be found safe in the normal mode by substantial margins. The types and intensities of energy input possible under various failure modes are then assessed, and the results compared again with the sensitivity of the materials. This analysis must indicate further safety margins, or action must be taken to prevent any failure mode from supplying excessive energy inputs, or, ultimately, action must be taken to protect personnel from the effects of the reaction(s) that may be produced.

Assessment of active element failure must be succeeded by the opposite consideration, i.e. the results expected when the active elements fail to function. Pump or valve failure may prevent the transfer of hot materials and consequently result in excessive high temperature exposure. Filter failure may produce dangerous material concentrations downstream. Fan failure may permit accumulation of flammable vapors. Assessing this phase requires a good working knowledge of the system and all its elements, so that a failing element can be translated into terms of the effects on other parts of the system. The ultimate objective is to assure that the system is so designed that all such failures cannot lead to an explosion, or to assure that proper design features are incorporated to control explosive effects and protect personnel and surroundings.

Passive Elements: The nonoperating elements of the system do not supply energy to the materials. Instead, these elements confine materials and the energy that the materials carry or release. Passive elements thereby affect the type of reaction that can occur as well as the magnitude and form of pressure that is released. Passive elements also can create conditions under which energy in the materials can be concentrated and/or transformed, creating potential ignition conditions. Examples of such phenomena include generation of electrostatic charges in flowing materials, adiabatic compression of vapors in deadended pipes, and catalytic effects of container surfaces. Passive elements frequently can be corroded and/or dissolved by the contents, altering the composition of the materials that subsequently reach other points in the system. The geometry of passive elements can produce collection points for highly sensitive materials that tend to separate from the normal flow, thereby creating localized concentration gradients that may cause a disaster. The passive elements of the system cannot be ignored while assessing the explosion potential. Examining the passive elements of the system must be guided by a consideration of the characteristics of the materials to be contained and of the active elements. The passive elements must be appropriate in strength, geometry, and chemical properties for the contents under the conditions of energy inputs that can prevail during normal and malfunctioning conditions.

The Environment

Here "environment" refers to the immediate surroundings of the system that can interact with materials in the system. The immediate environment often is treated as an extension of the operating system because environmental features also can be active or passive. Since all interactions are by definition abnormal, the evaluation procedures are often judgmental.

The environment can contribute energy to the system by lightning, welding or tapping, external fire or explosion, impingement of moving machinery or vehicles, and many other means. Some such inputs can be precluded by design, others by protective structures, and still others by administrative controls. For all practical purposes, any environmental energy input to the system that can be credibly anticipated must be considered, as much harm can be produced.

Environmental effects such as loss of water, power, gas, etc., must be considered, and frequently present more severe problems than environmental energy sources. While utility losses frequently will be anticipated when considering the functional failure of active elements, it is desirable to separately assess the effects of losing utilities. This separate assessment requires evaluating the simultaneous failure of multiple functioning units, a condition often overlooked or considered unlikely, but an obvious result of utility losses.

Finally, the environment (such as a building) should be considered as regards the escape of materials from the system, both by leakage and by explosive breaches of containment. In assessing the effects of leaks, concern should be given to the risk of fire/explosion within the escaped materials alone or as they are mixed with air. Explosive containment breaches create severe potentials for secondary explosions of the expelled materials in air; however, they also create failures in adjacent structures and systems, including the loss of utilities in otherwise unrelated areas. Assessing the explosion potential must address such secondary effects which can change a minor system upset into a disaster.

COUNTERMEASURES

Despite the best efforts of design and operating personnel, human errors occur, equipment breaks, and instruments fail, frequently producing potentially explosive conditions. However, by properly assessing explosion potential, the character and severity of such upsets and resultant explosive reactions can be determined, and countermeasures can be designed into the operating system. In this context, countermeasures mean system elements or operations designed to cope with explosive or potentially explosive reaction, rather than preventive measures.

Countermeasures can be grouped into five general types: (1) containment, (2) quenching, (3) dumping, (4) venting, and (5) isolation.

Containment: In many cases it is feasible to design the operating system to contain the maximum credible pressures and/or other forces that can be produced by the anticipated explosive reaction. Because of the design, there is no sudden release of pressure to the environment, there is no explosion, properly speaking.

The prime benefits of containment design are that it is *passive* (i.e., no operating function is involved), and it is clean (i.e., no dissipation of materials occurs). The primary disadvantage is that, since the energy released in the shock wave is closely related to the failure pressure, containment places a very high premium on accuracy of forecasted intensity. If the reaction should be intense enough to breach containment design, the resulting explosion will be far worse than would result from a normal operating design.

Containment is most easily practiced in cases involving gas phase combustions, where maximum pressures of up to ten or twenty times the initial pressure are the rule. Containment can also be practiced in many cases involving dust and mist combustions, and certain low intensity condensed phase runaway or thermal decomposition reactions.

Containment of normally intense thermal runaway/decomposition reactions is extremely difficult, because the overall volume of the reactants and the maximum pressures that can be attained to render containment design impractical from both engineering and economic viewpoints. Practical containment for normally intense thermal reactions is limited to relatively low volume systems in bench scale, pilot scale or custom processing operations. The same considerations apply to deflagration in condensed phases.

Containment of condensed phase detonations is practical only in very limited cases, as the pressure and impulse characteristics are such that extremely heavy equipment is needed, and the lifetime of such equipment is nevertheless generally very brief.

Quenching: Quenching of potentially or actually explosive conditions involves either heat removal or chemical inhibition. Heat removal may be performed by external means (emergency additional heat sinking capacity of equipment) or by introduction of a new material to absorb heat of reaction. Chemical inhibition procedures involve introduction of new material(s) into the chemical system to abate reaction by dilution or by removing active chemical species. Quenching is attractive in many cases because no release to the environment is involved, but is practical only when the function (heat removal, dilution, etc.) requires a short time relative to the time-to-explosion.

The most common quenching application is the use of flame arrestors to prevent propagation of vapor/air combustions; wherein heat from the flame front is lost to the flame arrestor, the reaction zone is cooled and the reaction rate drops sharply. To be effective, these passive units must have a high heat-sinking capacity relative to the heat output of the flame, must have a high surface area in contact with the flame, and must be large enough to cool the gas mixture below the temperature at which reignition can occur. Flame arrestors are generally designed to cope with a specific range of flames. It is essential in selection of flame arrestors that the designer be confident that the chosen unit will in fact cope with the potential flame.

Gas, dust, and mist combustions can also frequently be combated by dilution with carbon dioxide, water, mist, steam, certain dry powders, etc., to create enough internal heat-sinking capacity to abate or extinguish the flame front. Use of halogenated hydrocarbon suppressants for extinguishment of gas, dust, and mist combustions involves chemical interference with the flame propagation mechanism as well as heat-sinking and dilution effects. In all cases, operation of an injection system is needed to introduce the quenchant. This system must operate fast enough to catch the flame front, an operation involving a detection subsystem and a discharge subsystem. Reliability of these subsystems must be considered in the assessment of explosion potential. Further, it must be considered in design whether the introduction of the quenchant will independently create sufficient pressure and/or other forces to cause damage, possibly equivalent in severity to the potential damage from the 'prevented explosion.' This latter consideration applies particularly to many applications in powder handling equipment, storage vessels, and building type structures.

Quenching by introduction of heat absorbing, diluting, or chemically interfering reagents into condensed phase systems is invariably a more time consuming process, generally applicable only to uniform reactions caught early in the self-accelerating stage. In most condensed phase systems, the mass of reagents is large, and a correspondingly large mass of quenchant must be introduced and thoroughly mixed to be effective. In certain cases, chemical inhibitors can be effective in relatively small concentrations, but the required extent of mixing is also very high. Selection of quenchants is a highly technical matter, but certain generally desirable characteristics may be mentioned. The quenchant should have a high effective heat capacity, fluidity, and miscibility with the reagent mixture, but should also have a low vapor pressure even at high temperatures. It is worth restating that, in all cases, the quenching process requires operation of subsystems for detection and injection; the reliability of these subsystems must be considered in a reassessment of explosion potential after the quenchant system has been designed.

Quenching is generally inapplicable in coping with propagating reactions in condensed phases, with the exception of critical diameter traps for deflagration in pipelines. Introduction and proper mixing of quenchants are generally too slow to be valuable for deflagration or detonation processes, especially in pipelines or other similar geometry confinement.

Dumping: Dumping essentially means the release of the reaction mixture itself (as opposed to venting, which usually means release of gas). Dumping is ordinarily a relatively slow process applicable to the early stages of a self-accelerating uniform reaction, and, as ordinarily understood, implies the transfer of the reaction mixture to an alternate containment of some kind. The term is more commonly applied to condensed phases, but in actuality the procedure is often used with gas phase systems which may be 'blown down' in the event of an upset condition.

Dumping does not stop the reaction, of course, but merely transfers the problem to a (presumably) more favorable location for other treatment. Most commonly, condensed phases are dumped to a cold quenchant filled vessel; this vessel must itself be capable of withstanding or coping with a potentially explosive condition in the event that the quench process does not proceed properly. Gas mixtures may be dumped through scrubbers, into catch tanks and/or to the environment, depending on the gases involved. Again, receivers must be examined relative to the consequences of continued reaction after dumping. Further, dumping is again an active process requiring operation of subsystems whose reliability must be factored into the overall assessment of explosion potential.

Venting: Venting refers specifically to the release of gas from a vessel or confining structure in a controlled manner. The venting may be performed by manually operated elements, automated elements, preloaded elements or fully passive elements such as burst discs, rupture panels or blowout walls. This countermeasure is useful for coping with gas, mist, and dust combustions, uniform or propagating condensed phase reactions, and most conditions leading to physical explosions. The essential requirements are that the vent system be sufficiently fast acting to become fully operational in a timely manner, and that the vent be capable of releasing gas from the confining structure at the maximum rate at which gas is being formed by the potentially explosive situation.

Venting of all-gas systems is generally fairly well understood in practical terms. The rate of gas generation is reasonably easy to establish, and the equations describing the fluid dynamics are common currency in engineering. The situation changes markedly when the confining structure contains a liquid near or above its normal boiling point or a liquid with substantial amounts of gas dissolved under pressure. The sudden release of pressure essentially invariably results in sudden boiling, swelling and foaming of the liquid—and the vented matter is commonly a mixture of gas and liquid obeying a radically different fluid dynamics.* In many cases, essentially all of the liquid is expelled from the tank in the venting process. Accordingly, even when the gas phase can safely be vented to

*Venting of process vessels involving such two phase flow has been extensively investigated by a consortium of commercial firms acting as the Design Institute for Emergency Relief Systems (DIERS), under the auspices of the American Institute of Chemical Engineers. A number of papers based on this activity have recently been published in *Chemical Engineering Progress*, Vol. 81, No. 8 (August, 1985). A DIERS Project Manual is scheduled for publication in 1986.

atmosphere, provision must be made for effluent liquid as part of the overall design of the vent system.

In addition to the engineering design of the venting process elements, there are two other critical aspects of vent system design that must be considered:

– The venting process releases material from the original location to another location. Venting frequently includes discharge of hot gases, liquids and/or solids which may burn or otherwise react with materials in the receiving location. Venting frequently includes discharge of flammables, and improper selection of a receiving location can produce a hazardous condition, possibly of greater severity than the original. Venting of 'harmless' gases into occupied areas can reduce oxygen levels and cause personnel injuries. In general, venting should discharge to the outside or to locations designed specifically to cope with the vented materials.

– The venting process produces dynamic effects on the equipment from which material is discharged and on equipment elements in the receiving area. Since the discharge is generally not uniform in all directions, the discharging element must be capable of withstanding the thrust effect which is equal and opposite in direction to the discharge. The discharging element must also be capable of withstanding other dynamic effects of venting such as vibration, internal materials flows, and sustained flowing pressure during the venting process. In the receiving location, equipment and structural elements are subjected to shock and/or pressure effects from the venting process, load-bearing requirements to sustain the mass of discharged material and (in cases of rupture or blowout elements) potential projectile effects. Accordingly, the overall design of the venting system extends beyond speed and capacity of the vent element to inherently include consideration of the effects of venting on the relieved location and on the receiving location, although different aspects of the overall design are frequently handled by different specialists.

Isolation: Isolation encompasses separation of an element from surroundings that may be adversely affected by an explosion. Separation may be achieved by remote location of the potentially hazardous element, or by blast resistant structures designed to deflect, abate, or contain blast waves, missiles and/or expelled materials.

Isolation simply by remote location is practical for facilities that routinely perform potentially hazardous work, such as explosives facilities, etc. Site selection must consider the magnitude of the potential explosion, blast range, and missile range relative to presently and foreseeably occupied areas, as well as those other considerations generally applicable to site selection activities. The same considerations apply to isolation by distance of a unit process from the balance of a plant. It is important to realize that distance is of primary benefit for the primary explosion only. Expelled materials may travel well beyond the pertinent blast range, and consideration of possible secondary effects is critical. Evaluation of secondary explosion potentials is especially important where expelled materials include, for example, flammable gas or atomized liquids but assessment of effects on the overall environment is important even when no secondary explosion potential exists.

Isolation by blast resistant structures requires a complex design process. The magnitude of the explosion must be assessed, as well as the form of the blast wave and the missiles and product materials produced. The structure must then be designed to withstand the shock or impulse load of the explosion, the static pressure created by gas released within the structure, the range and depth of penetration of missiles, and any secondary effects of expelled materials on the blast resistant structure itself (combustion of expelled flammable vapors, for example).

Blast resistant structures are seldom designed to fully contain effects of an explosion, but are employed to deflect or abate effects of the blast wave and missiles. Full containment of all explosion effects is practical only when the potentially explosive element is quite small. Certain bench scale operations can be performed using full containment by blast resistant structures, and certain flow reactors are small enough that full containment is feasible, but, generally, potentially explosive elements are too large to permit practical full containment design. Accordingly, secondary effects of deflected blast waves, missiles, and expelled materials must be anticipated in the design of facilities incorporating blast resistant structures.

Blast resistant structures may be constructed of reinforced concrete, steel, timber, earth, sand, or a wide variety of other materials. Many designs utilize multiple elements of different materials to achieve specific effects under specific circumstances. Timber, earth, and sand are frequently used in construction of simple deflecting walls or revetments, often employed for relatively short duration conditions and/or for separation of elements in remotely sited facilities. Construction of steel or reinforced concrete blast resistant structures is more common in industrial process operations. A substantial body of information on blast resistant structures has been compiled for military purposes (U. S. Army 1969) which is available to non military designers for general application. An excellent short treatment has been recently published in *Chemical Engineering Progress* (Tunkel 1983).

In certain cases, where the potentially explosive condition is large and the environmental elements requiring protection are small, blast resistant structures may be used to enclose key environmental elements rather than to confine the effects of explosion. Typical of such strategy is the use of blast resistant control rooms in refineries and other large process plants. In such cases, the overall design must also consider personnel safety as regards oxygen supply, thermal loads, and toxic effects of expelled materials.

It is essential to understand that a blast resistant structure is ordinarily rated in terms of a specific charge loading at a specific location. Use of a blast resistant structure for purposes other than that for which it was originally designed requires reassessment of blast effects to be sustained. Any alteration of the system within the blast resistant structure that involves movement of potentially explosive elements may render the structural design inadequate for continued protection. In particular, alterations that move elements closer to walls and/or corners increase resultant stresses of explosion on the structure, and failure of the structure may occur in the event of an explosion.

CONCLUSION

The foregoing is a necessarily brief summation of countermeasures technology. Generally speaking, countermeasures design incorporates more than one form of countermeasure in a single operating system, to provide intermediate levels of response. A reaction vessel may well be equipped with a pressure relief valve, a dump valve, and a rupture disc, and yet may also be placed in a blast resistant structure if the severity of the maximum credible reaction so warrants.

The most important single criterion in the selection of countermeasures is ability of the system design to reliably perform the necessary functions in the available time. In general, it is preferred to buttress use of active countermeasures with passive countermeasures so that, in the event of failure of the active countermeasure, a second line of protection exists. Obviously, any system of countermeasures must be properly inspected, exercised and maintained on a regular basis. Countermeasures can be frustrated by inadvertence. Circumstances change in the environment, and new technology continually becomes available. Finally, installation of countermeasures does not mean the system is 'safe.' Safety is maintained only by constant vigilance.

Bibliography

References Cited

Tunkel, Steven J. "Barricade Design Criteria." *Chemical Engineering Progress*. Vol. 79, No. 9, Sept. 1983. pp. 50-55.

U.S. Army. 1969. "Structures to Resist the Effects of Accidental Explosions," U.S. Army Technical Manual, No. 5-1300. Department of Defense. Washington, DC.

NFPA Codes, Standards, Recommended Practices and Manuals. (See the latest *NFPA Codes and Standards Catalog* for availability of current additions of the following documents.)

NFPA 61A, *Standard for Prevention of Fire and Dust Explosions in Facilities Manufacturing and Handling Starch.*

NFPA 61B, *Standard for the Prevention of Fires and Explosions in Grain Elevators and Facilities Handling Bulk Raw Agricultural Commodities.*

NFPA 61C, *Standard for the Prevention of Fire and Dust Explosions in Feed Mills.*

NFPA 61D, *Standard for the Prevention of Fire and Dust Explosions in the Milling of Agricultural Commodities for Human Consumption.*

NFPA 65, *Standard for the Processing and Finishing of Aluminum.*

NFPA 68, *Guide for Explosion Venting.*

NFPA 69, *Standard on Explosion Prevention Systems.*

Additional Readings

Baker, Wilfred E., *Explosions in Air*, University of Texas Press, Austin, TX, 1973.

Bartknecht, W., *Explosions —Course Prevention Protection*, Springer-Verlag, NY, 1982.

Beausoleil, Robert W., et al. "Field Investigation of Natural Gas Pipeline Accident, Canterbury Woods, Annandale, Virginia," *Fire Journal*, Vol. 68, No. 3, May 1974, pp. 77-84.

Burgoyne, J. H., "The Flammability of Mists and Sprays," *Second Symposium on Chemical Process Hazards*, American Institute of Chemical Engineers, NY, 1963.

Cole, R. H., *Underwater Explosions*, Dover Publications, Inc., New York, NY, 1965, p. 1. (Original Copyright by Princeton University Press, 1948.)

Cook, M. A., *The Science of High Explosives*, Van Nostrand, Reinhold, NY, 1958.

Cross, J., and Farrer, D., *Dust Explosions*, Plenum Press, NY, 1982.

Davenport, J. A., "A Study of Vapor Cloud Incidents" *Loss Prevention*, Vol. 11, American Institute of Chemical Engineers, New York, NY, 1977.

Decker, D. A., "An Analytical Method for Estimating Overpressure from Theoretical Atmospheric Explosions," NFPA Annual Meeting, May 1974.

DiMeo, Michael, "Explosion in Fremont," *Fire Journal*, Vol. 70, No. 4, July 1976, pp. 15-19, 22.

Eichel, F. G., "Electrostatics," *Chemical Engineering*, Mar. 13, 1967, pp. 153-167.

Field, P., *Dust Explosions*, Elsevier, NY, 1982.

Frank-Kamenetsky, D. A., "On the Mathematical Theory of Thermal Explosions," *Acta Physicochim USSR*, Vol. 16, 1942, p. 347.

Glasstone, S., ed., *The Effects of Nuclear Weapons*, U. S. Atomic Energy Commission, Washington, DC, Apr. 1962.

Gray, P., and Lee, P. R., "Thermal Explosion Theory," *Oxidation and Combustion Reviews*, Vol. 2, 1967.

Haase, H., "Electrostatic Hazards: Their Evaluation and Control," *Verlagcheme*, NY, 1977.

Kletz, T. A., "Unconfined Vapor Cloud Explosions," *Loss Prevention*, Vol. 11, 1977.

"Lessons from a L. P. Gas Utility Plant Explosion and Fire," *Fire Command*, Apr. 1972.

Lewis, B., and Von Elbe, G., *Combustion, Flame, and Explosions of Gases*, 2nd ed., Academic Press, NY, 1961.

"Mist and Spray Explosions," *Chemical Engineering Progress*, Vol. 53, 1957.

Nagy, J., and Verakis, H. C., *Development and Control of Dust Explosions*, Marcel Dekker, NY, 1983.

Palmer, K. N., *Dust Explosions and Fires*, Chapman & Hall, London, Distributed by Halsted Press—John Wiley & Sons, NY, 1973.

Sharry, J. A., and Walls, W. L., "L. P. Gas Distribution Plant Fire," *Fire Journal*, Jan. 1974.

Stull, D. R., "Fundamentals of Fire and Explosions," *AICHE Monograph Series No. 10*, Vol. 73, American Institute of Chemical Engineers, NY, 1977.

Thomas, P. H., "Some Approximations in the Theory of Self-Heating and Thermal Explosion," *Trans. Faraday Soc.*, Vol. 56, 1960, p. 833.

Urbanski, T., *Chemistry and Technology of Explosives*, Vol. 1, Pergamon Press, NY, 1964, Vol. 2, 1967, Vol. 3, 1967.

Walls, W. L., "Just What is a BLEVE?" *Fire Journal*, Nov. 1978.

Zabetakis, M. G., "Flammability Characteristics of Combustible Gases and Vapors," *Bulletin 627*, U.S. Bureau of Mines, Pittsburgh, PA, 1965.

Zabetakis, M. G., *Safety with Cryogenic Fluids*, Plenum Press, NY, 1967.

COMBUSTION PRODUCTS AND THEIR EFFECTS ON LIFE SAFETY

Dr. Gordon E. Hartzell

This chapter discusses fire gases, heat, visible smoke, and the development of toxic hazard. The remaining chapters in this section discuss other relevant components of the characteristics and behavior of fire.

For detailed discussions of allied topics consult the following chapters in this HANDBOOK: Section 1, Chapter 2, "Human Behavior and Fire;" Section 5, Chapter 1, "Fire Hazards of Materials: an Overview;" and Section 9, Chapter 1, "Assessing Life Safety in Buildings."

Exposure to the products of combustion presents numerous hazards to humans. Predominant among these hazards are effects from heat, impaired vision due to smoke density or eye irritation, narcosis from inhalation of asphyxiants, and irritation of the upper and/or lower respiratory tracts.

These effects, often occurring simultaneously in a fire, contribute to physical incapacitation, loss of motor coordination, faulty judgment, disorientation, restricted vision, and panic. The resulting delay or prevention of escape may lead to subsequent injury or death from further inhalation of toxic gases and/or the suffering of thermal burns. Survivors from a fire also may experience post exposure pulmonary (lung) complications and burn injuries which can lead to delayed death.

Assessment of the overall physiological and behavioral effects of human exposure to fire and its combustion products is an extremely difficult and complex task. For example, effects of inhaled toxicants normally depend upon the "dose" of the insult received and this, in turn, is greatly influenced by changes in both rate and depth of breathing. Some fire gas components depress respiration, whereas others stimulate it. The physical exertion and possible panic involved in escaping from a fire may increase the respiratory minute volume (RMV) of a person as much as 10 to 20 fold. All these factors may operate simultaneously to influence the "dose" of toxicants.

In general, several important factors are integral to our complete understanding of the effects of products of combustion on life safety.

Dr. Hartzell is Director, Department of Fire Technology, Southwest Research Institute, San Antonio, TX.

Acute Exposure Assessment

Most available toxicological data relevant to humans has been developed specifically for assessment of long term exposures in the workplace. These data generally are not applicable for evaluation of the acute effects from brief exposure to the relatively high concentrations of combustion products which may be present in a fire.

Development of Toxicological Data

Development of the desired toxicological data from scientific experiments under controlled laboratory conditions usually cannot be accomplished with human subjects, since the experiments themselves would be hazardous to humans. Research in this field normally must be limited to the exposure of experimental animals, usually rodents, or in some cases, nonhuman primates.

Extrapolation

Reasonably reliable extrapolation to human exposures may be made, providing that both qualitative and quantitative differences in toxicological effects between laboratory animals and humans are understood. Sometimes this is not the case. Consequently, assessment of the toxicological effects of the exposure of humans to combustion products often involves considerable speculation and a variety of opinions, even among the experts (Rawls 1984).

FIRE GASES

Smoke most often is defined as the airborne solid and liquid particulates and fire gases evolved when a material undergoes pyrolysis or combustion (ASTM 1982). The fire gases have received the most attention, while knowledge of the effects of inhalation of particulates and aerosols from smoke is still quite limited.

Although a wide variety of fire gases may be generated, the toxicant gases are usually separated into three basic classes: the asphyxiants or narcosis producing toxicants; the irritants, which may be sensory or pulmonary; and those toxicants exhibiting other or unusual specific toxicities.

In pharmacologic terminology, a "narcotic" is a drug which induces unconsciousness (narcosis) with loss of pain. In combustion toxicology, the term refers primarily to asphyxiant toxicants that are capable of resulting in central nervous system depression, with loss of consciousness and ultimate death. Effects of the asphyxiant or narcosis producing toxicants depend upon the accumulated doses, i.e., both concentration and time of exposure. The severity of the effects increases with increasing doses. Although many asphyxiants may be produced by combustion of materials, only carbon monoxide (CO) and hydrogen cyanide (HCN) have been measured in sufficient concentrations in fire gases to cause significant acute toxic effects.

Irritant effects, produced by essentially all fire atmospheres, are normally considered by combustion toxicologists as being of two types: (1) sensory irritation, including irritation of the eyes and the upper respiratory tract; and (2) pulmonary irritation, which affects the lungs. Most irritants produce signs and symptoms characteristic of both sensory and pulmonary irritation.

Eye irritation, an immediate effect which depends only upon the concentration of an irritant, may be underestimated in its ability to impair escape of a victim from a fire situation. Nerve endings in the cornea are stimulated, causing pain, reflex blinking, and tearing. Severe irritation also may lead to subsequent eye damage. Victims may shut their eyes, which can partially alleviate these effects but also may impair their escape from a fire.

An understanding of the effects of irritants on the respiratory system is complicated by significant differences in respiratory physiology between research rodents, and humans. Research using nonhuman primates has improved our ability to compare effects on rodents with those expected for human subjects.

Airborne irritants enter the upper respiratory tract, where nerve receptors are stimulated, causing burning sensations in the nose, mouth, and throat, along with the secretion of mucus. Sensory effects are primarily related to the concentration of the irritant and do not normally increase in severity as the exposure time is increased.

A decrease in respiratory rate is a significant response by rodents to sensory irritants (Kane et al 1979). The effect, which lasts throughout exposure to an irritant, is a protective mechanism for rodents and reduces the penetration of irritants into the lower respiratory tract. Although the accompanying decrease in RMV reduces oxygen available to the lungs, the rodent is remarkably tolerant of such oxygen deficiency (hypoxia). The characteristic reflex response of rodents upon exposure to sensory irritants has been used as a laboratory tool to identify and quantify irritant effects and has been useful in setting industrial hygiene standards. However, its relevance to incapacitating and lethal effects in primates, including humans, has not been demonstrated, and its value for assessing toxic hazards in fires is highly questionable (Potts & Lederer 1978).

Another protective mechanism characteristic only of rodents involves the nasal passages, which absorb water soluble irritants such as hydrogen chloride, sulfur dioxide, and ammonia. A several fold increase in sensitivity of mice toward hydrogen chloride has been observed when the irritant is administered through a tracheal cannula, thus bypassing the nasal scrubbing mechanism (Alarie 1980).

Human nasal passages are simple, with relatively small surface area. Moreover, human mouth breathing is common, particularly under conditions of stress; thus the minimal human protective scrubbing mechanism for irritant gases is largely bypassed. Furthermore, the reflex decrease in respiratory rate exhibited by rodents appears to be only a transient response with primates. Monkeys and baboons exposed to irritant smoke atmospheres appear to exhibit only a brief initial decrease in respiratory rate, followed by increases in both respiratory rate and RMV (Purser & Woolley 1983). These are characteristic responses of these species to a pulmonary irritant. Thus, following signs of initial sensory irritance in primates, significant amounts of inhaled irritants are quickly taken into the lungs with the symptoms of pulmonary or lung irritation being exhibited. Lung irritation often is characterized by coughing, bronchoconstriction, and increased pulmonary flow resistance. Tissue inflammation and damage, pulmonary edema, and subsequent death often follow exposure to high concentrations, usually after 6 to 48 hours. Furthermore, exposure to pulmonary irritants appears to increase susceptibility to bacterial infection. Unlike sensory irritation, the effects of pulmonary irritation are related both to the concentration of the irritant and to the duration of the exposure.

Carbon Monoxide

Although carbon monoxide (CO) is not the most toxic of gases, it is always one of the most abundant and, therefore, is the major threat in most fire atmospheres. Under controlled burning conditions, the carbon of most organic materials can be oxidized completely by supplying excess oxygen. In the uncontrolled burning of an accidental fire, the availability of oxygen is never ideal and some of the carbon is incompletely oxidized to carbon monoxide. In a confined smoldering fire, the ratio of carbon monoxide (CO) to carbon dioxide (CO_2) is usually greater than in a well ventilated, free burning fire.

The toxicity of CO is primarily due to its affinity for the hemoglobin in blood. The carbon monoxide content of blood can readily be measured and is expressed as percent carboxyhemoglobin (COHb) saturation. Even partial conversion of hemoglobin to (OHb) results in a decreased supply of oxygen to body tissues (hypoxia).

There exists no minimum blood COHb saturation associated with death, below which it can be certain that a victim died from other causes or toxicants. The actual blood COHb saturation levels associated with both incapacitation and death vary quite widely over the general population and depend upon many factors. With some persons having preexisting functional impairments, even quite low COHb saturations are likely to be dangerous. The very young, the elderly, the physically disabled, those under the influence of alcohol, drugs, or medication, and those with heart diseases are particularly susceptible. One study has shown that those persons under the age of 9 and over 60, collectively, comprised 66 percent of the fire deaths (Harland & Woolley 1979).

From studies involving both rats and nonhuman primates, along with a reasonable amount of human exposure data, it is felt that any COHb saturation above approximately 30 percent would be potentially hazardous to most humans. A saturation of about 50 percent is likely to be lethal to many individuals (Kaplan & Hartzell 1984a).

Analysis of data from victims of improperly operated gas heaters showed a mean COHb saturation of 49.5 percent with a standard deviation of 14.0 percent (Packham 1984). In essence, some of the victims died at COHb levels as low as about 35 percent, while others survived at saturations as high as 64 percent.

In terms of CO concentrations required to reach hazardous COHb levels, a simple rule of thumb may be used. Any exposure in which the product of concentration (ppm) x time (minutes) exceeds aproximately 35,000 ppm-min is likely to be dangerous (Kaplan & Hartzell 1984a). Thus, a 10 min exposure to 3,500 ppm of carbon monoxide would be expected to be hazardous and possibly incapacitating to many people. The concentration x time rule of thumb must be applied with caution at high concentrations, since progressively lower doses can be tolerated as the concentration is increased. It is, however, reasonably applicable for the range of CO concentrations normally generated in fires.

Extensive investigations examining human fire fatalities with respect to exposure to toxic atmospheres have shown carbon monoxide to be the primary toxicant (Halpin and Berl 1978, Harland and Woolley 1979). In approximately half of the fire fatality cases studied, blood COHb levels resulting from inhalation of carbon monoxide were found sufficiently high as to be lethal. In an additional 30 percent of the victims, combinations of carbon monoxide with preexisting heart disease and/or alcohol intoxication were considered to be the cause of death. Of the fire fatalities in which alcohol was a factor, data showed that 88 percent of the victims had sufficient alcohol levels in the blood to be classified as legally intoxicated.

Hydrogen Cyanide

Hydrogen cyanide (HCN) is produced from the burning of materials which contain nitrogen (N). Natural and synthetic materials such as wool, silk, acrylonitrile polymers, nylons, polyurethanes and urea containing resins are included.

Hydrogen cyanide is a very rapidly acting toxicant which is approximately 20 times more toxic than carbon monoxide. It does not combine appreciably with hemoglobin, but the toxicant inhibits the use of oxygen by cells (histotoxic hypoxia).

Data relating symptoms in humans to various concentrations of HCN are very limited. One widely used descriptive account of hydrogen cyanide intoxication of humans reports that 50 ppm may be tolerated for 30 to 60 min without difficulty, 100 ppm for that same period is likely to be fatal, 135 ppm may be fatal after 30 min, and 181 ppm may be fatal after 10 min (Kimmerle 1974). Since incapacitation normally occurs at one-third to one-half the lethal dose, the data suggest that doses for incapacitation by HCN range from approximately 2,500 ppm-min at 100 ppm to 750 ppm-min at 200 ppm. Using a rule of thumb analogous to that for carbon monoxide, it would appear that a product of hydrogen cyanide concentration (ppm) x time (minutes) in the range of about 1,500 ppm-min would likely be hazardous to humans (Kaplan & Hartzell 1984a). With hydrogen cyanide, in particular, progressively lower doses can be tolerated as the concentration is increased. Therefore, the concentration x times rule of thumb must be applied with caution at high concentrations. It is, however, reasonably applicable for concentrations of hydrogen cyanide generally found in fire atmospheres.

The role of hydrogen cyanide as a causative agent in human fire fatalities is considerably less clear than that of carbon monoxide. Documented cases are rare in which hydrogen cyanide alone can be shown to be the primary toxicant. Blood can be analyzed in the laboratory for cyanide, but the procedure is more complex than that for carbon monoxide. Analyses must be interpreted with caution, due both to uncertainties in the accuracy of the data and also because cyanide is normally present in blood as a result of destruction of body tissue. It is generally regarded that blood cyanide concentrations greater than 1.0 microgram per milliliter are indicative of possibly significant toxicological effects due to hydrogen cyanide (Halpin and Berl 1978). Blood cyanide levels greater than 3.0 micrograms per millileter are generally lethal. In one study, hydrogen cyanide was found at elevated levels in the blood of 70 percent of the fire victims, with possible toxic levels of blood cyanide found in 13 percent of the victims (Halpin and Berl 1978). However, significant levels of blood cyanide also were normally found associated with high levels of carboxyhemoglobin saturation (Halpin and Berl 1978; Harland and Woolley 1979). Thus, the contribution of each to death could not be determined with confidence.

Although there is no evidence for the synergistic effects between hydrogen cyanide and carbon monoxide, the question of additivity between the two toxicants remains unresolved. Lethality data obtained on rats by some investigators suggest apparent additivity. However, other combustion toxicologists report that the two toxicants act independently, with perhaps only a weakly additive effect (Purser et al in press).

Carbon Dioxide

Carbon dioxide (CO_2) is usually evolved in large quantities from fires. While not particularly toxic at observed levels, moderate concentrations of carbon dioxide increased both the rate and depth of breathing, thereby increasing the RMV. This condition contributes to the overall hazard of a fire gas environment by causing accelerated inhalation of toxicants and irritants. The rate and depth of breathing are increased about 50 percent by 2 percent carbon dioxide. If 4 percent carbon dioxide is breathed, the RMV is approximately doubled, but the effect may be scarcely noticed by an individual. Any further increase in carbon dioxide from 4 percent up to 10 percent produces a correspondingly greater RMV, and at 10 percent, the RMV may be 8 to 10 times the resting level. The subject also may have symptoms of dizziness, faintness, and headache (Brobeck 1973).

Insufficient Oxygen

Oxygen (O_2) is consumed from the atmosphere during combustion. When oxygen drops from its usual level of 21 percent in air to approximately 17 percent, a person's motor coordination is impaired. When oxygen drops into the range of 14 to 10 percent, a person is still conscious but may exercise faulty judgment and will be quickly fatigued. In the range of 10 to 6 percent oxygen, a person loses consciousness and must be revived with fresh air or oxygen within a few minutes to prevent death (Reinke and Reinhardt 1973). During periods of exertion, increased oxygen demands may result in oxygen deficiency symptoms at oxygen levels.

Acrolein

Acrolein is a particularly potent irritant, both sensory and pulmonary, which has been demonstrated to be present in many fire atmospheres (Burgess et al 1979). It is formed from the smoldering of all cellulosic materials and also from pyrolysis of polyethylene (Potts et al 1978). Acrolein is extremely irritating, with concentrations as low as a few parts per million irritating to the eyes and possibly psychologically incapacitating. Surprisingly, studies with nonhuman primates have shown that concentrations up to 2,780 ppm for 5 min were not physically incapacitating during exposure (Kaplan et al 1984). However, pulmonary complications caused by even lower concentrations did result in death within hours after the exposure.

Hydrogen Chloride

Hydrogen chloride is formed from the combustion of chlorine containing materials, the most notable of which is polyvinyl chloride (PVC). It is both a potent sensory irritant and also a strong pulmonary irritant. Concentrations as low as 75 ppm are extremely irritating to the eyes and the upper respiratory tract, and impairment from a behavioral standpoint has also been suggested. However, hydrogen chloride gas has been found not to be physically incapacitating to nonhuman primates subjected to concentrations up to 17,000 ppm for 5 min (Kaplan et al 1984b). The toxicant was reported to cause post exposure death at doses which did not appear to incapacitate. Comparable studies have not been conducted using actual PVC smoke, however, and it is claimed that other irritants also may be present from PVC in a real fire atmosphere (Barrow et al 1976). Furthermore, the question remains as to the extent of respiratory dysfunction and susceptibility to infection caused by exposure to hydrogen chloride and PVC smoke.

Other Toxicants

Other toxicants formed in a fire depend upon many variables. The principal variables are:

Chemical composition of the burning material,
Amount of oxygen available for combustion, and
Temperature.

Sulfur dioxide, ammonia, nitrogen oxides, hydrogen fluoride, hydrogen bromide, isocyanates, phosphorous compounds, and a wide variety of volatile hydrocarbons have been identified in smoke (Kaplan et al 1983). In general, effects of acute exposure to these toxicants have not been sufficiently characterized and quantified to enable assessment of their hazards when present in fire atmospheres.

Testing for Toxicity of Combustion Products

Typically, tests for the toxicity of smoke produced by a burning material involve some quantitative measurement in the laboratory of the toxic potency. A concentration/response relationship is determined by measuring the response of animals, usually rodents, exposed over a fixed time to different concentrations of a combustion atmosphere. This is accomplished by conducting a series of experiments in which the quantity of combusted material or the flow rate of diluting air is varied in order to produce the different concentrations. The number of animals showing a response such as lethality or incapacitation will increase as the exposure concentration is increased. In combustion toxicology, concentration traditionally has been expressed either as the mass of test material used per chamber unit volume (the material charge), or the material mass loss per chamber unit volume (the combustion product concentration). When the percent of animals responding within a specified time is plotted as a function of the logarithm of the concentration, a straight line is typically approximated. From this plot, the concentration which will produce a response or effect in 50 percent of the animals within the specified time is obtained by statistical methods. This concentration, commonly termed the EC_{50}, is a measure of the toxic potency of the smoke. The EC_{50} is a general term and may be used for any observed response of the animal. When lethality is the observed response, the term LC_{50} is used to denote the concentration that produces death in 50 percent of the animals. Similarly, the IC_{50} designates the concentration necessary to incapacitate 50 percent of the animals.

Some test methods measure the rapidity of action of the combustion atmosphere rather than its toxic potency in causing either lethality or incapacitation. In these methods, the times are recorded at which animals die or are incapacitated when exposed to a fixed concentration of material or combustion atmosphere. From these data, a mean time-to-death or time-to-incapacitation is determined to characterize the toxicity of the smoke. If time to the effect (either death or incapacitation), is measured for different exposure concentrations, a time/concentration relationship may be determined.

There are at least five major test methods for assessing the performance of materials with regard to toxicity of combustion products (Kaplan et al 1983). Often referred to DIN, FFA, NBS, PITT and USF, the methods all have been used with a sufficient number of materials to achieve some significant level of attention in the combustion toxicology literature.

DIN Method: The DIN 53 436 method (West Germany-Deutches Institut fur Normung) specifies only the combustion apparatus. The operating procedure is still a prestandard, and description of the animal response model is designated as a draft. The DIN method is characterized by the use of a moving annular tube furnace operated at a constant temperature, with smoke being diluted with air before rats are exposed. Concentration/response relationships are normally obtained, with concentration varied by dilution of the smoke with air.

FAA Method: The FAA method (U.S. Federal Aviation Administration) uses a tube furnace operated at a constant temperature in a static mode. Times to incapacitation and death for rats exposed in motor driven rotary cages serve as end points.

NBS Method: The NBS test (U.S. National Bureau of Standards) is a static system using a cup type furnace as the combustion device, operated just below and just above the autoignition temperature of the specimen. Concentration-response relationships are determined using rats, with concentration being controlled by varying sample weight. There is also a radiant heat furnace modification of this method which controls concentration by varying specimen area and/or irradiation time.

PITT Method: The PITT methodology (U.S. University of Pittsburgh) exposes mice to smoke produced in a dynamic system from ramped heating of a specimen in a thermobalance furnace. Bioassays include concentration/response and time-to-death for lethality and respiratory rate depression to assess irritants.

USF Method: The USF Method (U.S. University of San Francisco) is also known as the Dome Chamber Method. In a small dome shaped chamber, mice are exposed to smoke produced from a tube furnace using either constant or ramped heating. Although concentration/response relationships can be determined by varying sample weight, most data involve various degrees of time-to-incapacitation and also death.

These five test methods do not differ greatly in the chemical analyses included in the procedures. Each method normally employs measurement of CO, CO_2, and O_2. Additional selected gases, principally HCN and HCl, are optionally measured by most of the methods. Measurement of blood COHb saturation levels of exposed animals is routinely made by only the DIN and the NBS tests.

In general, laboratory smoke toxicity tests are unable to demonstrate useful and practical quantitative differences between most materials—differences which can be used confidently to choose one material over another in the interest of improving firesafety. Experience has shown that most common materials, both natural and synthetic, do not differ widely in the toxicity of smoke produced from combustion. Smoke toxicity values generally cover a two orders of magnitude range, as compared with about a nine orders of magnitude range for the broad spectrum of chemicals known to toxicologists. Differences that are observed between materials, even though of statistical significance, are of questionable practical significance from the perspective of impacting life safety hazard in a fire.

Smoke from all fires is toxic, and the development of toxic hazard depends more upon the growth of the fire than upon the materials involved.

HEAT

The burning of most materials is an exothermic chemical oxidation process. Energy from the process is evolved as heat, which possesses both convective (hot gases) and radiative components. The radiative component represents energy released in the visible and infrared portions of the spectrum and is seen as flame or luminosity of a fire.

Heat poses a significant physical danger to humans. If the total heat energy reacting with the body surpasses the capability of the physiological defense processes to compensate, a series of events will occur ranging from minor injury to death. The effects of exposure to heated air are greatly augmented by the presence of moisture in the fire atmosphere. With higher moisture content, transfer of heat energy is more efficient and the body is less able to rid itself of the heat burden. Moisture can be present in a fire environment as the result of natural humidity, from the combustion itself, and from the application of water for extinguishment.

If excessive heat is conducted rapidly to the lungs, a serious decline in blood pressure may result along with capillary blood vessel collapse leading to circulatory failure. Severe heat also may cause fluid buildup in the lungs. In fire tests conducted by the National Research Council of Canada, (NRCC), 300°F (140°C) was taken as the maximum survivable breathing air temperature (Shorter et al 1960). A temperature this high can be endured for only a short period and not at all in the presence of moisture. It has been suggested that fire fighters should not enter *any* hostile atmosphere without full protective clothing and masks.

In school fire tests conducted at Los Angeles in 1959, a temperature of 150°F (65°C) at the 5 ft (1.5 m) level was selected as the temperature beyond which teachers and children could not be expected to enter a corridor from a relatively cool room (NFPA 1959). This selection assumed exposure to dry air, and only for the brief period of time necessary to reach exits.

Skin tissue burns are commonly classified as first, second, or third degree burns. First degree burns involve only the outer layer of the skin and are characterized by abnormal redness, pain, and sometimes a small accumulation of fluid underneath the skin. Second degree burns penetrate more deeply into the skin. The burned area is moist and pink; skin blisters and there is usually a considerable amount of subcutaneous fluid accumulation. Third degree burns are usually dry, charred, or pearly white. If a large percentage of the body skin tissue has suffered third degree burns, post exposure consequences may be extremely critical.

Studies have shown that a skin surface temperature as low as 160°F (71°C) will result in second degree burns if contact is maintained for 60 seconds (Woodson 1981). If the surface temperature is increased, the time to produce second degree burns decreases. For example, burns result in 30 seconds at 180°F (82°C), and in 15 seconds at 212°F (100°C).

Before human skin can absorb sufficient heat to raise its surface temperature, the body's heat dissipating capabilities must be defeated. The body dissipates heat by evaporative cooling (perspiration) and by circulation of blood. The cooling effect of evaporation of skin moisture may compensate for the effect of heat on skin up to 140°F (60°C) or more in dry air. This limit is lower in moist air. The time for the skin temperature to increase depends upon the exposure temperature, which, in most fires, increases very rapidly. Under these conditions, the skin temperature may increase faster than the defense mechanisms can function. The minimum temperature at which this occurs has not been established. Calm dry air heated above the range of 280 to 320°F (137 to 160°C), or an equivalent of radiant heat, will cause extreme pain to unprotected skin.

Exposure to excessive heat also may cause death due to hyperthermia without the production of actual burns. Hyperthermia occurs if the body absorbs heat faster than it can be dissipated by evaporation of surface moisture and outward radiation. The entire body temperature is thereby elevated sufficiently above normal to cause damage, particularly to the central nervous system.

Shock often is observed in fire victims, and may appear after exposure to heat, irritants, or oxygen deficient atmospheres containing significant levels of carbon monoxide. These conditions also may cause an increase in heart rate such that death due to heart failure may occur in the case of a preexisting cardiac condition.

VISIBLE SMOKE

In addition to fire gases, smoke also consists of finely divided particulate matter and suspended liquid droplets known as aerosols. Such carbonaceous matter is formed from the burning of most materials under the conditions of incomplete combustion normally present in fires. Since the average size of the particulates and aerosols is about the same as the wavelength of visible light, the light scatters and vision through smoke is obscured. Petroleum derived materials, especially those from aromatic hydrocarbons, are prone to the formation of dark, sooty smoke. There is, however, no relationship between the color of visible smoke and the toxicity of fire gases that may be present.

Since smoke obscures the passage of light, visibility to exits may be blocked, thus impairing escape of victims from a fire. The development of quantities of smoke sufficient to impair egress can be very rapid and is usually the first hazard to occur in a fire. As evidenced in nearly every one of the Los Angeles school fire tests, smoke in the corridors arising from fires in the basement reached untenable levels before temperatures attained hazardous conditions (NFPA 1959). In the tests, smoke, as it pertained to visibility, was the principal hazard. Although smoke frequently provides an early warning of fire, it also contributes to creation of panic because of its blinding and irritating effects.

Smoke particulates and aerosols also can be harmful when inhaled, and long exposure may cause damage to the respiratory system. Smoke particulates are often of a sufficiently small size that enable them to be inhaled deeply into the lungs, where absorbed toxicants may produce damage to the respiratory system. These effects have not been sufficiently studied as to enable a full understanding of their consequences.

DEVELOPMENT OF TOXIC HAZARDS

Full scale fire tests are useful for placing the development of toxic hazard in perspective. Four such tests were reported by Southwest Research Institute in which fires were set in a fully furnished simulated hotel room with attached corridor and remote room (Fig. 4-3A) (Grand et al 1985). During the tests, the door from the fire room to the corridor was fully open. The door between the corridor and the remote room was open approximately 1 in. (25 mm) until about 3 minutes after flashover, when it was closed to prevent the development of excessive concentrations of toxicants.

The sequence of events that occurred in the burn room is shown as a composite illustration in Figure 4-3B. The tests were initiated with a smoldering fire in the chair closest to the sofa. During the smoldering phase of approximately 19 minutes, no conditions were attained which would be considered hazardous to life. Following flaming ignition of the chair, the fire progressed rapidly to flashover within about 8 minutes, accompanied by life threatening conditions of temperature, carbon monoxide, hydrogen cyanide, and oxygen depletion in the room of origin. After flashover in the burn room, tenability in the remote room also quickly deteriorated, as shown in Figure

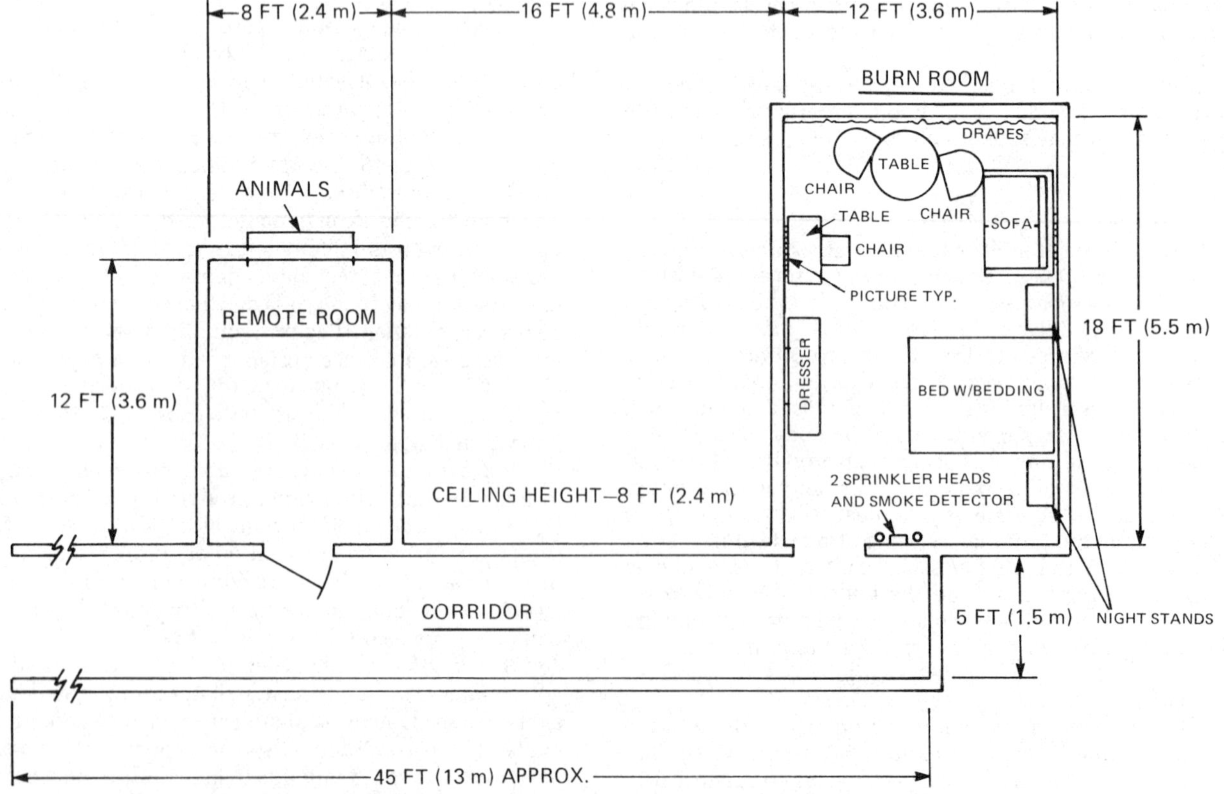

FIG. 4-3A. Burn room, corridor, and remote room full scale test facility at Southwest Research Institute.

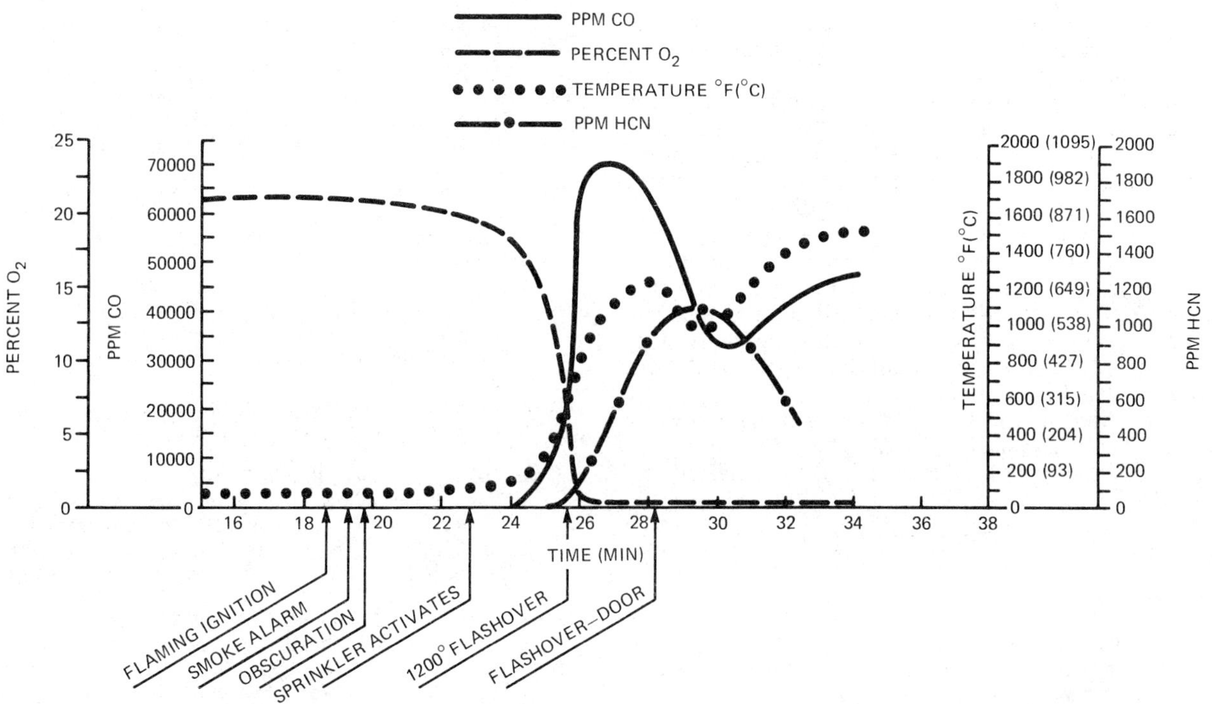

FIG. 4-3B. Composite illustration of events and development of hazardous conditions at 5½ ft (1.7 m) level in the burn room during fully furnished room fires conducted at SWRI.

4-3C, first with visual obscuration by smoke, followed by rapidly increasing concentrations of toxic gases. Test animals (rats) in the remote room became incapacitated within about 2 minutes, with death occurring from carbon monoxide asphyxiation approximately 11 minutes after flashover. From our knowledge of carbon monoxide intox-

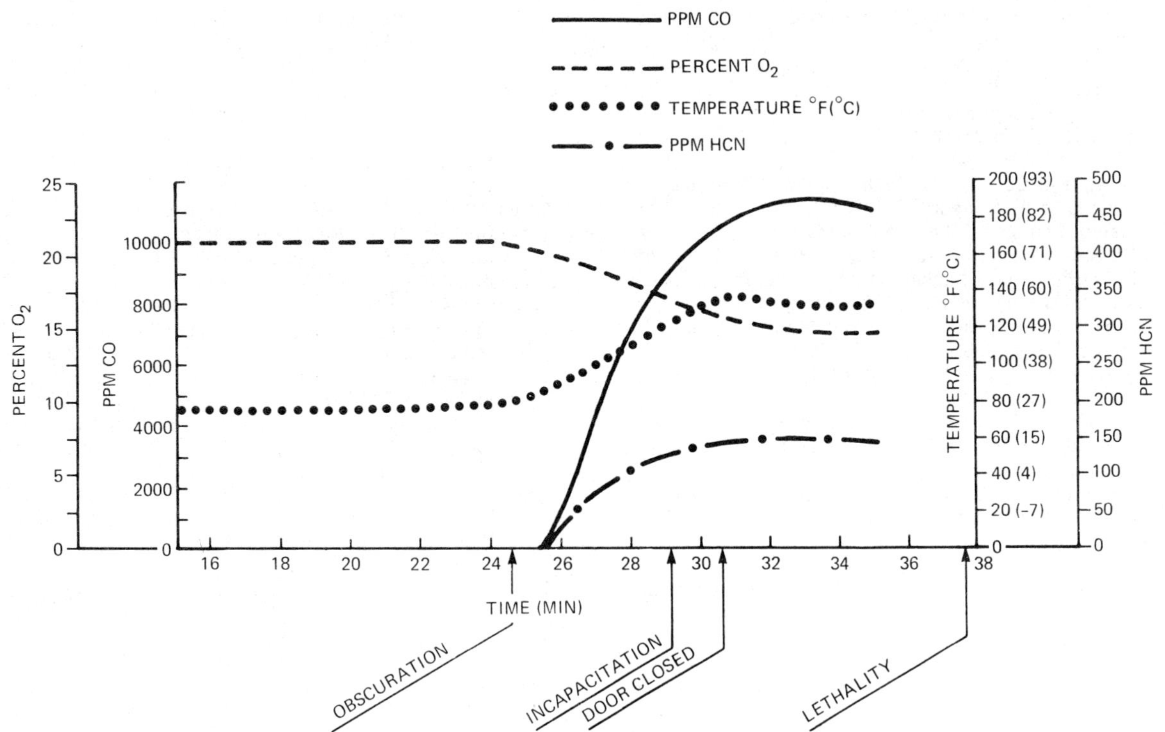

FIG. 4-3C. As in Figure 4-3B, but for remote room.

ication, it is likely that humans in the remote room would have been incapacitated and killed in approximately the same time intervals.

It was considered significant in these tests that the toxic hazard did not develop until flashover was approached in the burn room. After that point, smoke obscuration and life threatening conditions increased very rapidly in the burn room, corridor, and the remote room. The smoke detectors employed in these tests provided adequate warning (five minutes) prior to the development of acutely toxic concentrations of combustion products. The sprinkler activation times demonstrated that a properly installed and working sprinkler system, by controlling the fire at an early stage, would have prevented any significant toxic threat from developing.

Combustion products produced in a fire are always toxic and extremely hazardous to life safety. Control of the fire is most important to reduction of the toxic hazard.

Bibliography

References Cited

Alarie, Y. 1980. *Proceedings of the Inhalation Toxicology and Technology Symposium*. Oct 23-24, 1980. Leon, B. K. J., ed. Ann Arbor Science Publishers. p 207.

ASTM 1982. "Standard Terminology Relating to Fire Standards." *ASTM E176-82*. American Society for Testing and Materials, Philadelphia, PA.

Barrow, C. S., Alarie, Y., and Stock, M. F. 1976. "Sensory Irritation Evoked by the Thermal Decomposition Products of Plasticized Poly-Vinyl Chloride." *Journal of Fire and Materials*. Vol 1. pp 147-153.

Brobeck, J. R., ed. 1973. *Best and Taylor's Physiological Basis of Medical Practice*. 9th Ed. Williams and Wolkins, Baltimore, MD. Ch 6. p 52.

Burgess, W. A., Treitman, R. D., and Gold, A. 1979. *Air Contaminants in Structural Firefighting*. Harvard School of Public Health, Boston, MA.

Grand, A. F. et al., 1985. "An Evaluation of Toxic Hazards from Full-Scale Furnished Room Fire Studies." ASTM Symposium on Application of Fire Science to Fire Engineering, Denver, CO. June 26-27, 1984. *ASTM Special Technical Publication (STP)*. 1985 (in press).

Halpin, B. M., and Berl, W. G. 1978. "Human Fatalities from Unwanted Fires." *Report NBS-GCR-79-168*. U.S. National Bureau of Standards, Washington, DC.

Harland, W. A., and Woolley, W. E. 1979. "Fire Fatality Study." Building Research Establishment Information Paper. University of Glasgow. IP 18/79. Glasgow, Scotland.

Kane, L. E., Barrow, C. S., and Alarie, Y. 1979. "A Short-Term Test to Predict Acceptable Levels of Exposure to Airborne Sensory Irritants." *American Industrial Hyg. Association Journal*. Vol 40. pp 207-229.

Kaplan, H. L, Grand, A. F., and Hartzell, G. E. 1983. *Combustion Toxicology: Principles and Test Methods*. Technomic, Lancaster, PA.

Kaplan, H. L., and Hartzell, G. E. 1984a. "Modeling of Toxicological Effects of Fire Gases: I. Incapacitating Effects of Narcotic Fire Gases." *Journal of Fire Sciences*. Vol 2. pp 286-305.

Kaplan, H. L., et al. 1984b. "A Research Study of the Assessment of Escape Impairment by Irritant Combustion Gases in Postcrash Aircraft Fires." Southwest Research Institute. *Final Report, DOT/FAA/CT-84/16*. U.S. Department of Transportation. Federal Aviation Administration, Atlantic City, NJ.

Kimmerle, G. 1974. "Aspects and Methodology for the Evaluation of Toxicological Parameters During Fire Exposure." *The Journal of Fire and Flammability Combustion Toxicology Supplement*. Vol 1. Feb 1974.

NFPA. 1959. *Operation School Burning*, Quincy, MA. p 25.

Packham, S. C. 1984. "Forensic Applications of Combustion Toxicology." *Proceedings, Toxic Hazards From Fire Workshop*, Washington, DC. Nov 15-16, 1984. Technomic, Lancaster, PA.

Potts, W. J., and Lederer, T. S. 1978. "Some Limitations in the Use of the Sensory Irritation Method as an End-Point in Measurements of Smoke Toxicity." *Journal of Combustion Toxicology*. Vol 5. pp 182-195.

Potts, J. W., Lederer, T. S., and Quast, J. F. 1978. "A Study of the Inhalation Toxicity of Smoke Produced Upon Pyrolysis and Combustion of Polyethylene Foams, Part I: Laboratory Studies." *Journal Combustion Toxicology*. Vol 5. pp 408-433.

Purser, D. A., and Woolley, W. D. 1983. "Biological Studies of Combustion Atmospheres." *Journal Fire Sciences*. Vol 1, No 2. pp 140-141.

Purser, D. A., Grimshaw, P. and Berrill, K. R., "Intoxication by Cyanide in Fires: A Study in Monkeys Using Polyacrylonitrile." *Arch. Evn. Health* (in press).

Rawls, R. L. 1984. "Fighting Fire Toxicity: Experts Hold Conflicting Views." *Chemical and Engineering News*. July 9, 1984. pp 20-22.

Reinke, R. W., and Reinhardt, C. F. 1973. "Fires, Toxicity and Plastics." *Modern Plastics*. Feb 1973. pp 94-98.

Shorter, G. W., et al. 1960. "The St. Lawrence Burns." *NFPA Quarterly*. Vol 53, No 4. Apr. 1960. pp 300-316.

Woodson, W. E. 1981. *Human Factors Design Handbook*. McGraw-Hill Inc., New York, NY. p 812.

Additional Readings

Alarie, Y., and Anderson R., "Toxicologic Classification of Thermal Decomposition Products of Synthetic and Natural Polymers," *Toxicology and Applied Pharmacology*, Vol. 57, 1981, pp. 181-188.

Alarie, Y., and Barrow, S., "Toxicity of Plastic Combustion Products, Toxicological Methodologies to Assess the Relative Hazards of Thermal Decomposition Products from Polymeric Materials," *NTIS PB -267 233/5ST*, U.S. National Bureau of Standards (NBS), Washington, DC, 1977.

Anderson R. C., and Alarie, Y., "An Attempt to Translate Toxicity of Polymer Thermal Decomposition Products into a Toxicological Hazard Index and Discussion on the Approaches Selected," *Journal of Combustion Toxicology*, Vol. 5, 1978, pp. 476-484.

Anderson R. C., and Alarie, Y., "Approaches to the Evaluation of the Toxicity Decomposition Products of Polymeric Materials Under Stress," *Journal of Combustion Toxicology*, Vol. 5, 1978, pp. 214-221.

Anderson R., and Alarie, Y., "Screening Procedure to Recognize 'Supertoxic' Decomposition Products from Polymeric Materials Under Thermal Stress," *Journal of Combustion Toxicology*, Vol. 5, 1978, pp. 54-63.

Birky, M. M. et al., "Physiological and Toxicological Effects of the Products of Thermal Decomposition from Polymeric Materials," *NBS Special Publication 411*, NBS, Washington, DC, Nov. 1974, pp. 105-124.

Birky, M. M., "Philosophy of Testing for Assessment of Toxicological Aspects of Fire Exposure," *Journal of Fire and Flammability/Combustion Toxicology*, Vol. 3, 1976, pp. 5-23.

Birky, M. M., "Hazards Characteristics of Combustion Products in Fires: The State of the Art Review," National Aeronautical and Space Administration, *CR −135088, Final Report*, NASA, Washington, DC, 1977.

Birky, M. M., "Test Method for Interlaboratory Comparison of Combustion Product Toxicity, Jan. 18, 1979," *NBSIR 80 −2077*, NBS, Washington, DC, 1980.

Birky, M. M., et al., "Development of Recommended Test Method for Toxicological Assessment of Inhaled Combustion Products," *NTIS PB81 −110884*, NBS, Washington, DC, 1981.

Clarke, Frederic B., "Toxicity of Combustion Products: Current Knowledge," *Fire Journal*, Vol. 77, No. 5, Sept. 1983, pp. 84-108.

Gaume, J. Bartek, P., and Rostami, H., "Experimental Results on Time of Useful Function (TUF) After Exposure to Mixtures of Serious Contaminants," *Aerospace Medicine*, Vol. 42, No. 9, 1971.

Halpin, Byron M., et al., "A Fire Fatality Study," *Fire Journal*, Vol. 69, No. 3, May 1975.

Halpin, B., Fisher, R. S., and Caplan, Y. H., *International Symposium on Toxicity and Physiology of Combustion Products*, University of Utah, Salt Lake City, UT, 1976.

Hilado, Carlos J., and Huttlinger, Patricia A., "Toxic Hazards from Common Materials," *Fire Technology*, Vol. 17, No. 3, Aug. 1981, pp. 177-182.

Kimmerle, G., "Aspects and Methodology for the Evaluation of Toxiocological Parameters During Fire Exposure," *Journal of Fire and Flammability/Combustion Toxicology*, Vol. 1, 1974, pp. 4-51.

Levin, B. et al., "Further Development of a Test Method for the Assessment of the Acute Inhalation Toxicity of Combustion Products," *NBSIR 82 −2532*, NBS, Washington, DC, June 1982.

Paabo, M., Birky, M. M., and Womble, S. E., "Analysis of Hydrogen Cyanide in Fire Environments," *Journal of Combustion Toxicology*, Vol. 6, 1979, pp. 99-108.

Petajan, J., "Survival Response During Fire Exposure," in Physiological and Toxicological Aspects of Combustion Products, *International Symposium of National Academy of Sciences*, National Research Council, Salt Lake City, UT, 1976.

Potts, W., and Lederer, T., "Some Limitations in the Use of the Sensory Irritant Method as an End-Point in Measurement of Smoke Toxicity," *Journal of Combustion Toxicology*, Vol. 5, 1978, pp. 182-195.

"Procedures for the Evaluation of the Toxicology of Household Products," *Bulletin 1138*, National Academy of Science, Washington, DC, 1977.

Pryor, A. J., Johnson, D. E., and Jackson, N. N., *Hazards of Smoke and Toxic Gases Produced in Urban Fires*, Southwest Research Institute, San Antonio, TX, 1969.

Punderson, J., "A Closer Look at Cause and Effect in Fire Fatalitites −The Role of Toxic Fires," *Journal of Fire and Materials*, Vol. 5, No. 1, 1981.

"Report of the Committee on The Toxicity of the Products of Combustion to the Standards Council of the National Fire Protection Association," *Fire Journal*, Vol. 77, No. 2, Mar. 1983, pp. 21-27.

Robison, M. M., Wagner, P. E., and Fristrom, R. M., "The Accumulation of Gases on an Upper Floore During Fire Buildup," *Fire Technology*, Vol. 8, No. 4, Nov. 1972, pp. 278-290.

Seader, J. D., and Einhorn, I. N., "Some Physical, Chemical Toxicological and Physiological Aspects of Fire Smokes," *Sixteenth Combustion Symposium*, The Combustion Institute, Pittsburgh, PA, 1976.

Williams, S., and Clarke, F., "Combustion Product Toxicity: Dependence on the Mode of Product Generation," *Journal of Fire and Materials*, Vol. 6, Nos. 3 and 4, 1982.

Yuill, Calvin H., "Smoke: What's In it?" *Fire Journal*, Vol. 66, No. 3, May 1972, pp. 47-55.

Zikria, B. et al., "Smoke and Carbon Monoxide Poisoning in Fire Victims," *Journal of Trauma*, Vol. 12, No. 6, 1972.

THEORY OF FIRE AND EXPLOSION CONTROL

Revised by Walter Haessler

This chapter provides an overview of the theory and fundamentals of fire and explosion control, extinguishment, and explosion prevention. Four basic means are detailed: cooling, oxygen depletion, fuel removal, and chemical flame inhibition. In addition, environmental considerations are also discussed.

Related specific discussions can be found in Chapter 1 of this section, "Chemistry and Physics of Fire;" Chapter 2, "Explosions;" and Chapter 3, "Combustion Products and Life Safety." Section 11 of this HANDBOOK devotes 12 chapters to special fire protection and prevention problems. Section 18 discusses "Water-Based Extinguishing Systems" while Section 19 "Special Fire Suppression Agents and Systems" explains such agents as carbon dioxide, halogens, dry chemicals, foams, dry powders, etc.

Generally, combustion is defined as a reaction which is a continuous combination of a fuel (reducing agent) with certain elements, prominent among which is oxygen in a free or combined form (oxidizing agent). While oxygen may be the most common oxidizing agent, this combination exists with other elements, including fluorine and chlorine. The common factor in all these reactions is that they are exothermic (i.e., chemical energy trapped in the original molecules is converted into the form of thermal energy).

Certain metals have unique combustion characteristics that react without oxygen. Under certain conditions, magnesium, aluminum, and calcium can, "burn" in a pure atmosphere of nitrogen. Furthermore, many materials, when exposed to sufficiently elevated temperatures, will decompose directly by themselves, emitting light and heat. Examples of these materials are hydrazine (N_2H_4), diborane (B_2H_6), nitromethane (CH_3NO_2), hydrogen peroxide (H_2O_2), and ozone (O_3).

For purposes of this chapter, however, fire is referred to in its most usual form, which involves rapid oxidation at temperatures above 1,500°F (815.5°C), accompanied by the evolution of highly heated gaseous products of combustion and the emission of visible and invisible radiations.

Mr. Haessler is an instructor, author, lecturer, and consultant in the field of fire protection.

The combustion process occurs in two modes: flaming (including explosions), and flameless surface (including glow and deep seated glowing embers). The requirements for sustained burning are illustrated in the system diagram shown in Figure 4-4A. As shown, the flaming mode is

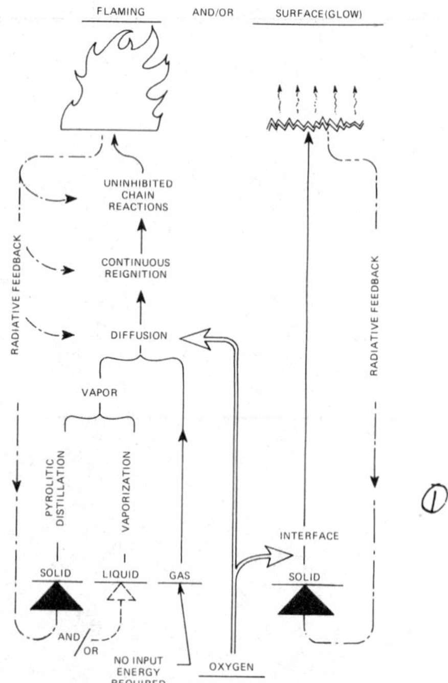

FIG. 4-4A. Basic fire system modes: flaming, surface.

associated with relatively high burning rates. These burning rates are expressed in terms of heat energy per unit time released from the originally bound chemical energy. These rates, together with the weight-time rate and specific heats of effluent gaseous combustion products, determine flame temperature.

Interestingly, theoretical flame temperatures for different gases burning in air, (with no excess air present), do not vary appreciably despite large differences in heat of combustion. For example, hydrogen with a heating value of 319 Btu/cu ft (11 885.94 kJ/m^3) at 60°F (15.5°C) and 1 atmosphere has a theoretical flame temperature of about 4,200°F (2315°C). Benzene, with a heating value more than ten times as great, (3,675 Btu per cu ft or 136 930.5 kJ/m^3) gives a temperature of 4,300°F (2371°C). This is because gases with the higher value require larger amounts of air for combustion. Most hydrocarbon flame temperatures (ideal combustion, no excess air), vary between 3,500 and 4,200°F (1926 and 2315°C). Higher temperatures for these gases and vapors would require air preheating and/or oxygen-enriched air. Usual fire conditions result in air deficiency, incomplete combustion, and somewhat lower flame temperatures.

An approximate analysis of hydrocarbon liquid fuel pan fires shows that approximately two-thirds of the heat release passes off to the surrounding environment as sensible heat of the effluent, and one-third as radiative heat flux. For equilibrium conditions, the heat energy generated and the heat energy lost to the environment, both of which are measured on a time basis, must balance. If the former is in excess, the fire will grow. Conversely, if the latter is in excess, the fire will diminish. The process is highly heat dependent. One method of fire control does, indeed, upset this heat balance, and that is the application of water streams, the mainstay of the fire service. The complexity of the flaming mode provides more fire control options which can be used individually or in combination. This is in sharp contrast to the flameless surface glowing mode, in which only three control options exist, (used singly or in combination).

While the combustion process is extraordinarily complex, sufficient information has been gained to make illustration of the process possible. (See Fig. 4-4B.) The flaming

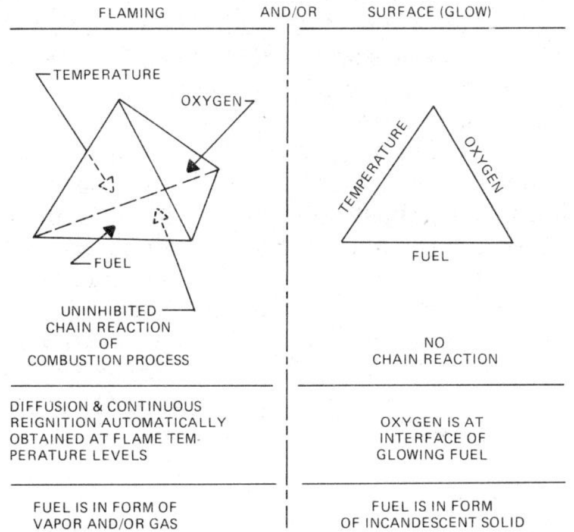

FIG. 4-4B. Basic fire system mode requirements: flaming, surface.

mode can be conceived of as a tetrahedron, in which each of the four sides is contiguous with the other three sides and each side represents one of the four basic requirements of combustion: fuel, temperature, oxygen, and uninhibited combustion chain reactions (Haessler 1974).

As shown on the right side of the illustration, the surface combustion mode can be symbolized correctly in the form of the traditional triangle, in which each of the three sides is contiguous with the other two sides and each side representing one of the three basic requirements: fuel, temperature, oxygen.

These two modes may occur singly or in combination; they are not mutually exclusive. Flammable liquids and gases burn in the flaming mode only. Most solid plastics can be construed as "frozen flammable liquids," and as such will melt with sufficient thermal feedback prior to burning. However, these fuels must vaporize and be diffused with oxygen immediately before burning.

Examples in which both modes exist are: solid carbonaceous fuels such as coal; solid carbohydrates such as sugars and starches; solid cellulosic/lignins such as wood, straw, brush, and similar vegetable materials; and thermosetting plastics, which do not melt. With these materials, the early stages of combustion start in the flaming mode due to destructive distillation, with a gradual transition toward the surface combustion mode during which both modes are in action simultaneously. Ultimately, the flaming mode is terminated with the residual surface combustion mode existing alone.

Examples in which the surface combustion mode exists alone are: pure carbon and other readily oxidizable nonmetals such as sulfur and phosphorus; as well as readily oxidizable metals such as magnesium, aluminum, zirconium, uranium, sodium, and potassium. These metals burn with characteristically higher temperatures ranging from 5,000 to 6,000°F (2760 to 3315°C), as compared to temperatures ranging from 3,000 to 3,500°F (1648 to 1926°C) obtained with the atmospheric burning of hydrocarbons.

From the foregoing, it is apparent that within the flaming mode there are four separate and distinct means of fire and explosion control, as compared with the surface combustion (glow) mode in which there are only three separate and distinct means of fire control. (Note that explosion control is not, *per se*, directly involved unless burning gases are present. Such gases could initiate a flaming mode.)

EXTINGUISHMENT BY COOLING

Under fire conditions, water—applied as a straight stream (for range and/or powerful drenching action) or in a wide angle spray pattern—is the most effective means of removing heat from ordinary combustible materials such as wood, straw, paper, cardboard, and other materials used in building construction and furnishings. This extinguishing mechanism depends upon cooling the solid fuel, thereby reducing and ultimately stopping the rate of release of combustible vapors and gases.

This cooling action also results in the formation of steam (particularly noticeable in wide angle spray patterns), which partially dilutes the ambient oxygen concentration in compartment or structural fires. Due to its low density, this effect is transitory. Therefore the rapid diffusibility and short residence time of steam within the immediate fire area are significant only in a secondary sense. In the case of outside fires, this effect is nonexistent, as will be noted later.

The efficiency of an extinguishing agent as a cooling medium depends upon its specific and latent heat, and its boiling point. The superior properties of water can be attributed to its relatively high values of specific heat, latent heat, and availability. However, water is heavy and constitutes a burden when it has to be hauled for any distance. Its cooling action is performed by means of sequentially conducting, evaporating, and convecting heat away from solid surfaces which are either burning or are hot from exposure. The capabilities of water can be summarized as:

1. One gpm (3.78 L/min) of water can be expected to absorb 10,000 Btu/min (10 560 kJ/min) when applied at 60°F (15.5°C) and fully vaporized and superheated to 500°F (260°C).
2. Water expands to approximately 2,500:1, greatly reducing the oxygen in closed spaces.
3. Water can induct air, depending upon the chosen stream. At a 30 degree spray setting and at 100 psi (689.5 kPa) nozzle pressure, about 30 cfm (9.14 m^3) of air is induced into the water stream per gpm (1 gpm equals 3.78 L/min) of flow, i.e., at 100 gpm (378.5 L/min), 3,000 cfm (914 m^3) of air is induced. This ventilation can be beneficial or harmful, depending upon its use. It is utilized in the formation of high expansion foam.
4. One gpm (3.78 L/min) of water can be expected to extinguish an interior compartment fire involving 100 cu ft (2.8 m^3) of ordinary combustibles.
5. Finally, water can be more effective by the addition of: (1) surfactants to promote soaking and penetration; (2) thickening agents to retard runoff and penetration; (3) ammonium phosphates, alkali carbonates, and alkali borates to leave a residual fire retardant coating; and (4) foam concentrates to form foam blankets on solids and most liquids.

Because heat is continuously being carried away by radiation, conduction, and convection, it is only necessary that the water absorb a small proportion of the total heat being evolved by the fire in order to extinguish it by cooling. However, water must always reach the burning fuel directly. Good visibility is needed to do this correctly unless, as with automatic sprinklers, discharge occurs in the early stages of a fire.

In high hazard areas, high piled storage areas, high rise structures, and other places of difficult accessibility for fire fighting, automatic fire protection systems become vital to safety.

EXTINGUISHMENT BY OXYGEN DILUTION

As stated, oxygen may be present as free gaseous oxygen in the atmosphere (20.9 percent O_2: 78.1 percent N_2: 1.0 percent argon, CO_2 and other gases), or combined in the form of hypochlorites, chlorates, perchlorates, nitrates, chromates, oxides, and peroxides. The term "dilution" can be applied only to the gaseous state, because in the combined state, oxygen is locked into the molecule and no dilution is possible. Hence, chemicals in this category will always present a high level of hazard, and oxygen dilution is useless in combatting fires having high concentrations of these materials. Because equal volumes of gases (and gase mixture) contain the same number of molecules, it is possible to compute gas densities from their molecular weights, as well as to rationalize that the percentage of oxygen in a space will be reduced when "foreign" gases (such as carbon dioxide or nitrogen) are artificially injected into the same space. This dilution of oxygen also is accomplished by the formation of steam, which is generated by the application of water in compartment fires. The necessary degree of oxygen dilution varies greatly with the particular fuel or combinations thereof. Furthermore, solid fuels have their own special minimum oxygen requirements. For instance, wood is known to continue burning in the surface (glow) mode, following its earlier flaming combustion mode, at oxygen levels as low as 4 to 5 percent. Acetylene, an unstable hydrocarbon, requires the oxygen concentration to be below 4 percent. On the other hand, stable hydrocarbon gases and vapors usually will not burn when the oxygen level is lowered to below 15 percent.

Fires in closed spaces will, of course, consume oxygen. However, this cannot be relied upon to achieve self-extinguishment, since combustion in oxygen deficient atmospheres results in the copious generation of flammable gases due to incomplete combustion. Inadvertent open entry or improper ventilation involving such spaces becomes an invitation to an explosion or, as fire fighters services call this dreaded phenomenon, "backdraft."

A typical example of the effective use of the oxygen dilution principle is when carbon dioxide is discharged in total flooding of closed or semiclosed spaces. In local application of carbon dioxide systems (and in the discharge of portable carbon dioxide extinguishers) another flame characteristic, namely flame velocity, is suppressed, which varies with different fuels. A carbon dioxide discharge plume entrains air, the residual velocity of which, if properly applied, with its carbon dioxide content will overcome the flame velocity dynamically and result in rapid extinguishment through the combined action of oxygen dilution and flame "blow-out."

EXTINGUISHMENT BY FUEL REMOVAL

A chemist defines a fuel as a reducing agent. A reducing agent is a substance which by losing one or more electrons, can reduce an oxidizing agent. In the process, the oxidizing agent gains the corresponding electrons and the reducing agent becomes oxidized. It is necessary to realize that oxidation and reduction always take place simultaneously. In the presence of sufficiently powerful oxidizing agents, most substances can be considered as reducing agents (i.e., substances which can be oxidized). Conversely, in the presence of sufficiently active reducing agents, most substances can be considered as oxidizing agents (i.e., substances which can be reduced). Ordinarily a substance is not classified as an oxidizing or reducing agent if it manifests its oxidizing or reducing properties only in the presence of extremely active substances of the opposite kind.

Figure 4-4A shows that for the flaming mode of combustion it is necessary for solid and liquid fuels first to be vaporized, with the solid fuels being distilled pyrolytically, and the liquid fuels being evaporated. In some instances (e.g., thermoplastics), the solids are melted or fused, after which they vaporize. Fuel gases, however, do not undergo such changes before ignition.

Figure 4-4A also shows that the surface or glowing mode of combustion does not require this gasification, because combustion occurs directly at the solid interface with the air. Thus, burning rates are small compared with the flaming mode. Some examples are wood, charcoal, coke, and combustible metals.

From the preceding, it is obvious that many materials which can be classed as fuels have different factors which influence the type of fire controls. These factors include wide ranges of ignition temperatures, lower and upper flammable limits in air, flash points (if liquid), solubility in water, and overtones of chemical activity (decomposition). For solid fuels, another important influence on fire intensity [Btu per unit of time (1 Btu = 1.056 kJ)] is the fuel array (as exemplified in dusts, splinters, shavings, logs, or timbers), and whether that array is horizontal, vertical, cribbed, or piled. In each condition, the same fuel will have entirely different burning characteristics.

From a chemical standpoint fuels can be categorized as:

1. Carbon and other readily oxidizable nonmetals such as sulfur, phosphorus, and arsenic;
2. Compounds rich in carbon and hydrogen (hydrocarbons);
3. Compounds containing carbon, hydrogen, and oxygen—such as alcohols, aldehydes, organic acids, cellulose, and lignin (wood and vegetable materials); and
4. Many metals and their alloys (including sodium, potassium, magnesium, aluminum, zinc, titanium, zirconium, and uranium).

Fuel removal can be accomplished directly by removing the fuel, indirectly by shutting off fuel vapors from combustion in the flaming mode, or (in the nonflaming mode) by covering the glowing fuel. What follows are examples of these types of fuel removal in different circumstances:

1. Large tanks of flammable liquids can be extinguished by expediently pumping out the burning tank and transferring the liquid to some other empty tank.
2. If (as above), the flammable liquid has a flash point higher than the ambient storage temperature, and it is not possible to transfer the liquid to an empty tank, then proper agitation of the liquid to raise the cooler bottom portion to the top (and correspondingly to displace the hot upper layer to the bottom of the tank) will result in starving the flames from the vapors. (See Fig. 4-4A.)
3. When gas fires result from broken lines, cracked flanges, or blown packings and gaskets, safe extinguishment can be obtained only by shutting off the gas (usually by closing the valves). Figure 4-4A shows that the fuel flow into the combustion zone is not under the control of the radiative feedback, as with flammable solids and liquids. Instead, with flammable gases, the fuel flow is controlled by such mechanical entities as the size of rupture, the pressure of the gas, its specific gravity, and other essentials.
4. Standard forest fire fighting procedure includes the tactic of bulldozing a firebreak across the path of an advancing flame front, to clear the path of all fuel and halt the fire advance.
5. The only practical method of extinguishing deep seated fires in silos and piles of solid combustibles is by removal of fuel.
6. Burning material (wood or vegetable) can be coated with metaphosphoric acid obtained via the thermal decomposition of monoammonium phosphate (multipurpose dry chemical) or diammonium phosphate originally in water solution. The metaphosphoric acid is a glassy infusible substance, is very adhesive, and imparts a fire retarding character to the originally burning fuel.
7. Liquid and solid burning fuels can be covered with blankets of thick fire fighting foams produced by the aeration of water and foam concentrate solutions. This method of attack has become standard procedure for fires in aircraft crashes, large tank farms, and oil tankers. Special foam concentrate formulations are required for polar solvents (water soluble).
8. Another new version of the method described in Example 7 uses a very thin film on hydrocarbon fuels. The film, known as aqueous film-forming foam (AFFF), is established by applying a water solution of perfluorocarboxylic acid onto the burning surface of the liquid. The surface tension of the fuel exceeds the aggregate surface tensions of the film.
9. Burning metals can be covered by various materials that are inert respective to the particular metal involved. Such materials include dry petroleum coke, various inorganic salts used singly or in eutectic proportions, sand, coal, foundry fluxes, soda ash, and similar materials.
10. The use of water into which gelling or slurry producing agents (which greatly retard water runoff) have been introduced is a technique that has been used successfully in fighting fires involving forests, brush, and other solid combustibles such as wood and vegetable materials. Variations of the mixture range from thin slimes to viscous layers. To date, the principal use of this method has been in forest fire fighting and on some lumberyard fires.
11. Nonwater-soluble flammable liquids (such as carbon disulfide), when burning, are heavier than water. They can be extinguished by applying a light spray of water (so as not to agitate the fuel) which will float on top, thereby covering the fuel. On the other hand, water soluble fuels such as alcohols, ketones, and aldehydes can be watered sufficiently to raise the flash point and lower the fuel vapor pressure to a point where fuel vapors rising from the liquid phase are not sufficient to accumulate and form a flammable vapor-air mixture.
12. A technique often used in combatting liquid grease fires involving unsaturated animal and vegetable oils and fats is the application of alkaline dry chemicals or alkaline solutions which, upon contact with the burning surface, generate a saponification reaction (as in making soap). The light froth containing steam causes carbon dioxide and glycerine bubbles to float on top of the burning oil. Because the liquid grease cannot burn, the fire is extinguished.

EXTINGUISHMENT BY CHEMICAL FLAME INHIBITION

Fire extinguishment by chemical flame inhibition applies to the flaming mode only. Extinguishment by cooling, oxygen dilution, and fuel removal are applicable extinguishing methods for all classes of fires of both the flaming and glowing modes. (See Figs. 4-4A and 4-4B.) The chemical flame inhibition method is only partially understood and is the subject of continuing research. Although there is no doubt that flames can be inhibited chemically, such information is mostly empirical. The outstanding effect of the chemical flame inhibition method is the extreme rapidity and high efficiency with which flames can be extinguished. These virtues can be more fully appreciated when it is realized that this method, when properly executed, is the only means by which an explosion can be prevented in a flammable gas/air (or even a gas/oxygen) mixture after ignition has occurred. Activating methods vary from very simple to very sophisticated applications which utilize highly responsive fire detection apparatus.

Before flame inhibition can be discussed, it is necessary to realize that the combustion reactions proceed in a complex series, characteristic of a chain reaction. For example, Figure 4-4C illustrates the branched chain com-

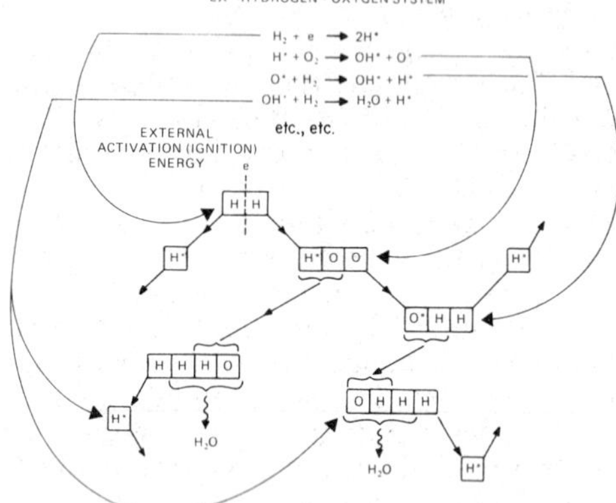

FIG. 4-4C. Basic combustion chain reactions (branched type).

bustion reaction of the hydrogen-oxygen system. This reaction is the simplest (and the most rapid) of all combustion types. Following the initial splitting of the hydrogen molecule, the individual hydrogen atoms (active H* species) interreact with oxygen molecules to produce active OH* and O* species. Note that the active species are formed as products and are consumed as reactants, and as such have a dual personality; thus, they can be called "chain carriers." Although the example was taken for hydrogen, the same phenomenon holds for fuels containing hydrogen in chemical union with carbon and other elements. While the $2H_2 + O - 2H_2O$ reaction expresses the weight and volume relationships, it does not express the chemical kinetics involved. As the result of much research, it has been found that flame velocity is dependent upon the concentration of the active OH* species and upon the pressure at which the reaction proceeds. For example, for the hydrogen-oxygen flame the highest flame velocities (16 in. per sec or/0.6 cm/sec) at atmospheric pressure have been determined. For other hydrogen bearing carbonaceous fuels, the hydroxyl concentration decreases with resulting lower flame velocities. For fuels not containing hydrogen, the active species O* becomes the determinant of flame velocity. (An example of the advantage obtained when air velocity exceeds flame velocity is illustrated by the simple act of blowing out a match or candle flame. On a larger scale, a flaming oil gusher can be extinguished with explosives.)

Extinguishment by flame inhibition is possible only when the active species OH*, H*, and O* are not allowed to fulfill their role in sustaining the flame. Extinguishing agents which accomplish this do so without the accompanying action of other flame extinguishment methods such as cooling, oxygen dilution, fuel removal, or covering. These agents extinguish flames efficiently and quickly; however, as stated previously, they do not combat glowing fires except under certain conditions (such as long "soaking" periods following flame extinguishment).

The exact manner in which the active species are interfered with in achieving flame extinguishment is uncertain at present. However, it is known that substances having this property fall into the following three categories:

1. Gaseous and liquid halogenated hydrocarbons wherein the effectiveness increases as higher order halogens are used. Some examples presently in use are:

Bromotrifluoromethane	$CBrF_3$	Halon 1301
Bromochlorodifluoromethane	$CBrClF^2$	Halon 1211
Dibromotetrafluoroethane	$CBrF_2CBrF_2$	Halon 2402

There are many more examples, but the necessary criteria of stability and acceptable levels of toxicity become limiting factors. The bromine containing agents are much more effective and much less toxic than the chlorine containing agents, such as carbon tetrachlorides.

2. Alkali metal salts wherein the cationic portion is sodium or potassium, and the anionic portion is a bicarbonate, carbamate, or halide. Some examples presently in use are:

Sodium bicarbonate	So-called "regular dry chemical"
Potassium bicarbonate	Trade name "Purple K"
Potassium carbamate	Trade name "Monnex"
Potassium chloride	Trade name "Super K"

There are many more substances, but the difficulties caused by hydrate formation (extreme hygroscopicity and toxicity) become limiting factors. The most effective salt ever found is potassium oxalate, but the preceding restrictions prohibit its use.

3. Ammonium salts (the most prominent of which is monoammonium phosphate) wherein cationic ammonium radical (NH_4^{+*}) and the anionic phosphate radica1 ($H_2PO_4^-$) are formed, with the latter absorbing an H^+ active radical becoming orthophosphoric acid, which dehydrates to metaphosphoric acid.

Upon injection of these substances into flames, the substances thermally dissociate into their anionic and cationic free radicals, and catalyze the union of the OH^- and H^+ combustion reaction chain carriers, thereby mitigating their influence upon the continuation of the flame. By this action the flame becomes inhibited, and extinguishment is accomplished when proper amounts of the agent are applied. Agents working in this manner also are referred to as negative catalysts. With time, more and more of these substances will become available.

ENVIRONMENTAL CONSIDERATIONS

The preceding parts of this chapter presented the four basic requirements of the flaming mode and the three basic requirements of the surface, or glowing, mode of burning. Interest was centered on the nature of the fuel, thermal balance, degree of ventilation, and chemical interplay occurring within the flames. The influence of radiative feedback was recognized as a factor of major importance in determining fire intensity (Btu or kJ/min). This latter factor is greatly influenced by the environmental effect of whether the fire is internal (compartment, or the so called structural type), or external (outside, open air type).

The confinement of a fire within a space interferes with the dissipation of heat, gases, and smoke, with the result that the radiative feedback to the seat of the fire is greatly enhanced and initial burning rates build up briskly. As contrasted with outside fires, the environment is underventilated in a general sense. However, the following two situations can exist:

1. Under partially ventilated conditions within a partially enclosed area (such as a room with open doors and windows), carbon dioxide will first be produced until ceiling temperatures of 1,200 to 1,500°F (648 to 815°C) are obtained. Beyond this temperature the carbon dioxide reacts with free carbon to produce carbon monoxide according to the reaction $C + CO_2 \rightarrow 2CO$. Although this reaction is endothermic, it is a prolific source of carbon monoxide and at O_2/CO ratios between 3 and 1 will form lower flammability mixtures, depending upon the prevailing CO_2 concentration. This condition causes the "flashover phenomenon."
2. Under sealed, nonventilated conditions, almost the same general chemical conditions exist except that the oxygen concentration is much lower, which together with the prevailing carbon dioxide concentration prevents ignition of the carbon monoxide at the lower flammable limits. Instead, the O_2/CO is much lower, and at levels of 0.06 to 0.07 constitutes the upper flammable limit. Under these conditions, improper ventilation at low levels can cause an explosion known as "backdraft."

In both cases, the wide flammability range of carbon monoxide/air mixtures is responsible for the results. At temperatures in excess of 1,200°F (648°C) the carbon dioxide reacts with the incandescent free carbon (present as smoke) to produce carbon monoxide at twice the rate that the carbon dioxide is consumed; at 1,500°F (815°C) this reaction is virtually complete. At temperatures in excess of 1,800°F (982°C) the water gas reaction prevails and the incandescent carbon reacts with the ever present water vapor to produce even more carbon monoxide with free hydrogen being given off as well ($C + H_2O \rightarrow CO + H_2$). Again, more gas volumes are being released than are being consumed. This accelerated pace is further heightened by ever increasing temperatures, and danger is further increased by the extremely wide flammability ranges of carbon monoxide and hydrogen.

The peculiar fire behavior within compartments, whether partially ventilated or completely unventilated, dictates that fire attack should begin before ceiling temperatures reach 1,200 to 1,400°F (648 to 760°C), because at these temperatures the fire gases (together with the incandescent smoke particles) react to form flammable or potentially explosive gas mixtures. Human survivability is no longer possible when these conditions are reached. Structural fires demand early detection and prompt response to achieve rescue and minimize fire loss. It is significant that properly sprinklered installations have never experienced "flashover" or "backdraft."

In the following chapters of this HANDBOOK, the individual means and systems of fire combat, whether manual or automatic, will be presented in detail. It is necessary to understand that attack can be accomplished by many well chosen combinations, and that no single answer will suffice, since fire presents many "aspects" and "personalities." The fire protection engineer in the planning phase, the safety director and inspector in the installation and maintenance phases, and the combat fire officer and fire fighters in their training, planning, and attack phases all have vital roles in fire protection. All three groups must do their utmost to initially prevent fire in the original planning stage.

Bibliography

References Cited

Haessler, W. M. 1974. *Extinguishment of Fire*. Revised Edition. National Fire Protection Association, Quincy, MA.

NFPA Codes, Standards, Recommended Practices and Manuals. (See the latest *NFPA Codes and Standards Catalog* for availability of current editions of the following documents.)

NFPA 68, *Guide for Explosion Venting.*

NFPA 69, *Standard on Explosion Prevention Systems.*

NFPA 204M, *Guide for Smoke and Heat Venting.*

NFPA 495, *Code for the Manufacture, Transportation, Storage and Use of Explosive Materials.*

Additional Readings

Application Guide for Explosion Suppression Systems, Fenwall Inc., Ashland, MA, Nov. 1979.

Fire Prevention and Control Master Planning. Introductory Summary, National Fire Prevention Control Administration, Fire Safety and Research Office, Washington, DC, 1978.

Groothuizen, T. M., and H. J. Pasman, "Explosions in Liquids and Solids," *Loss Prevention*, No. 9, 1975, pp. 91-96.

Hirst, R., Savage, N., and Booth, K., "Measurement of Inerting Concentrations," *Fire Safety Journal*, Vol. 4, No. 3, 1981/82, p. 147.

Klueg, Eugene P., "Liquid Nitrogen as a Powerplant Fire Extinguishant," *Fire Technology*, Vol. 5, No. 3, Aug. 1969, pp. 197-202.

McHale, Edward T., "Life Support Without Combustion Hazards," *Fire Technology*, Vol. 10, No. 1, Feb. 1974, pp. 15-24.

Smith, P. L, "Foam and Fire, Urethanes and the Environment," *Proceedings* (Conference held Sept. 21-22, 1976, at the Skyline Hotel, Hayes, Middlesex), Plastics and Rubber Institute, London, England, 1976.

Stull, D. R., *Fundamentals of Fire and Explosion*, Monograph Series No. 10, Vol. 73, American Institute of Chemical Engineers, NY, 1977.

Tucker, D. M., Drysdale, D. D., and Rasbash, D. J., "The Extinction of Diffusion Flames Burning in Various Oxygen Concentrations by Inert Gases and Bromotrifluoromethane," *Combustion and Flame*, 1981.

Wagner, H. G., "Gas Explosions, Their Origin and Effects," (Translated from *Chemie Ing. Tech.*, 1975, Vol. 47, No. 6, pp. 236-241.) United Kingdom Atomic Energy Authority, Risley Translation 2861, Risley, England, 1975.

Weldon, George E., "Damage-Limiting Construction," *Fire Technology*, Vol. 9, No. 4, Nov. 1973, pp. 263-270.

SECTION 5

FIRE HAZARDS OF MATERIALS

FIRE HAZARDS OF MATERIALS: AN OVERVIEW

Dr. Frederic B. Clarke

This section of the HANDBOOK is devoted to the specific fire problems posed by various kinds of materials. Solving such problems involves preventing ignition or, if ignition occurs, minimizing the size of the fire. In addition, increased attention must be paid to understanding and controlling the effects of smoke. Such effects include smoke's toxic and chemical characteristics as well as its ability to hamper vision.

This chapter will survey the various kinds of fire hazards, ways of characterizing those hazards, and some of the methods used to reduce them. The most commonly encountered classes of materials, and their special fire problems, are dealt with in detail in the succeeding chapters of this section.

Ordinary burning means oxidation—the chemical reaction of a material with oxygen to form relatively simple compounds called oxides. A material will burn if the reaction with oxygen results in the release of energy, which usually appears as heat and light. The first descriptor of the hazard of a material might be the amount of heat it produces when it burns, i.e., the heat of combustion. The greater the heat of combustion, the greater its potential as a fuel, and the hotter it burns.

The amount of heat produced on burning is determined almost entirely by the chemical composition of the material. The speed with which the heat is produced, on the other hand, is determined by the physical form of the material. Thus, excelsior burns more quickly than a solid block of wood, although equal weights of the two will produce the same overall amount of heat.

BURNING OF MATERIALS

Organic Materials

Any discussion of the fire hazards of materials must center upon those materials which are organically derived, i.e., those which are based on carbon. This is simply because organic materials as a class are ubiquitous. The simplest organic compounds, such as methane and propane, are common commercial fuels. They are also building blocks for more complex materials. Organic liquids are fuels, solvents, and chemical intermediates, but the richest variety of materials are organic solids. They include everything from specific compounds, such as aspirin, to common but complex everyday materials like wood, paper, textiles, and most plastics. All of these materials have carbon as their principal constituent; almost all contain hydrogen; and many contain oxygen, nitrogen, and other elements in varying amounts. Most organic materials are also excellent fuels. The common products of combustion of organic materials are water, which is the oxide of hydrogen, and carbon dioxide.

A more extensive tabulation of the heats of combustion of organic materials and more details on how to calculate them are contained in Chapter 11 of this Section. Strictly speaking, the heats of combustion outlined in that chapter are based on an idealized reaction path, such as complete oxidation of a hydrocarbon to carbon dioxide and water. In practice, these conditions often do not hold. For example, there may be insufficient oxygen available to convert all of the carbon to carbon dioxide; some carbon monoxide may be formed, or soot may remain. Thus, as a practical matter, the heats are upper limits of the actual amount of heat to be expected for the combustion of materials under normal conditions.

Examination of the heat of combustion tables in Chapter 11 of this Section will show that while the heat of combustion is quite different for different organic materials, the heat produced per equivalent of oxygen consumed is the same within about ten percent. This fact, sometimes called Thornton's Rule (Thorton 1917), allows one to use oxygen consumption as a reasonable measure of the heat produced by a burning material.

Solid organic materials fall into two broad classes: hydrocarbon based and cellulose based. The former are derivatives of the unoxidized hydrocarbon building blocks: $-CH_2-$, or -CH-. The latter are based on a partly oxidized carbon unit: -CH(OH)-. In a sense, cellulose based material is already partly burned in its natural state, so that when the two classes are combusted to carbon dioxide and water, they consume different amounts of oxygen and produce different amounts of heat.

Dr. Clarke is President of Benjamin/Clarke Associates, Inc. of Kensington, MD, a firm specializing in fire hazard analysis.

$$-CH_2- + 3/2(O_2) \longrightarrow CO_2 + H_2O$$
$$-CHOH- + O_2 \longrightarrow CO_2 + H_2O$$

For equivalent amounts of oxide formed, the hydrocarbon based materials consume 50 percent more oxygen and thus produce about 50 percent more heat. On a weight basis, the difference is greater; hydrocarbons produce over twice the heat, pound for pound, as do cellulosics.

Until the middle 1950s, this fact was of limited significance to life safety. Naturally occurring hydrocarbon materials were almost entirely industrial chemicals or commercial fuels, and the civilian population exposed to such fires was relatively small. However, the development of modern plastics changed all that. Most plastics are made from hydrocarbon, not cellulosic, materials. Their wide use in building materials and consumer goods, such as furniture, has meant increasingly that the burning properties of hydrocarbons are of consequence to the firesafety of the general population.

Polymers

The linking of small molecules into long chains is called polymerization. The products are called polymers. Most of the simplest polymers are manmade and are what are commonly referred to as "plastics." Polyethylene, a long, relatively unbranched hydrocarbon chain, is produced by the end-to-end linkage of ethylene molecules, C_2H_2. A typical "molecule" of polyethylene will consist of several thousand ethylene units joined together. The repeating units used to make up the polymer are called monomers.

$$\text{+}CH_2 - CH_2\text{++}CH_2 - CH_2\text{++}CH_2 - CH_2\text{+}$$

Original Ethylene Molecules

Naturally occurring polymers are drawn from a smaller class of monomers than are manmade materials. However, they have a wide variety of structural variations, and these, more than chemical differences, account for the differing properties of such natural materials as wood, paper, and cotton. All of these are cellulosic materials; they have a polymeric backbone based on the glucose molecule:

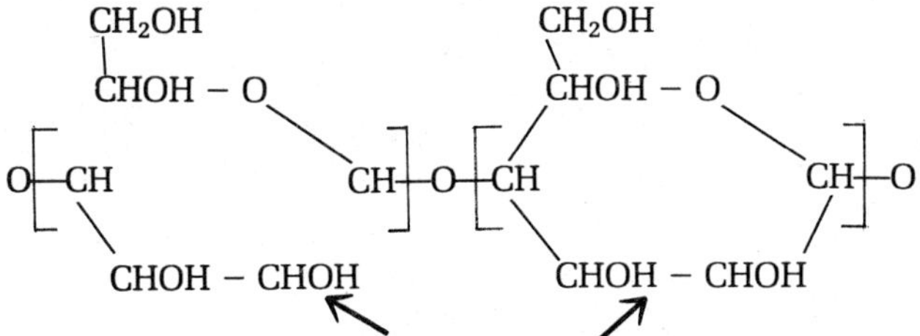

```
     CH2OH                        CH2OH
     |                             \
     CHOH - O                      CHOH - O
   [          \        ]   [      /          \        ]
 O-[ CH          CH  ]-O-[ CH                 CH ]-O
   [ \                ]   [ \                 /   ]
      CHOH - CHOH              CHOH - CHOH
```

Original Glucose Molecules

There are other natural polymer families as well, protein based materials for example, but the cellulosics are particularly widespread.

Both manmade and natural polymers have a major distinction from simple solids, liquids, or gases: in order to be vaporized, the solid material itself must be decomposed to some extent. The polymeric molecules are too large to be transported "as is" into the vapor phase. Actual chemical bonds must be broken, and this consumes energy. This fact provides an extra opportunity to measure and to control the hazard components of solid materials.

Rate of Burning

The rate at which the fire products are produced is always an important component of fire hazard whether the concern is for heat, smoke, or toxic gases. In the later sections of this chapter, there will be more to say about the measurement of this factor, but it is instructive to examine some of the general physical facts which contribute to the burning rate.

The burning material may be gaseous, liquid, or solid, but the oxygen (normally free oxygen in the air) is usually in a gaseous state. For the necessary chemical actions to occur, the fuel and oxygen must be brought into contact at a molecular level, and this in turn means that burning is generally a vapor-phase phenomenon.

The rate of burning is a function of how fast the chemical reaction of oxidation occurs, as well as the speed at which the vaporized fuel and the oxygen are delivered to the combustion zone. In premixed flames, i.e., those in which mixing has occurred before the combustion is initiated, the burning rate is controlled only by the inherent rate at which the substances combine. This is generally quite fast; flames propagate under premixed conditions at several feet per second (one ft per sec equals 0.3 m/sec). It is for this reason that the contact of air and combustible vapors is so dangerous; the process, once started, is virtually impossible to interrupt except in closed spaces specifically equipped for that purpose.

A more common mode of burning is the diffusion flame, shown schematically in Figure 5-1A. In this case, vaporized fuel mixes with oxygen in the combustion zone. The rate of burning is essentially controlled by the rate at which the two components arrive in the heated combustion zone. Once they arrive, combustion is, by comparison, instantaneous.

Since gases mix with one another readily, the burning of a gaseous fuel, such as hydrogen or methane in air, is a rapid process. However, the burning of the liquid or solid requires first that the fuel be converted to the gaseous state (volatilization). This process requires the input of an appreciable amount of heat energy, often from the fire itself, and is almost always slow when compared to the rate of oxidation. As a practical matter, therefore, the rate of volatilization of a material strongly affects its rate of burning. Liquids and solids are more concentrated fuels than are gases. Propane, for example, burns completely according to the following expression:

$$C_3H_8 + 5O_2 \longrightarrow 3CO_2 + 4H_2O$$

At any given pressure, one volume of propane gas consumes five volumes of oxygen, or about 25 volumes of air. By contrast, liquefied propane is some 300 times denser than propane gas under normal conditions. This means that one volume of the liquid will produce 300 times as much heat, but it will also require 300 times as much air. Thus, the burning of highly volatile liquids and solids may also be affected by the rate of delivery of oxygen to the combustion zone, especially under poorly ventilated conditions.

Except for hypergolic combinations of materials (those which ambient temperatures are sufficent to ignite), the initiation of the combustion reaction requires some added heat. This heat must be sufficient to vaporize enough of the fuel to initiate reaction and to accelerate the chemical combustion reaction to a rate where it can sustain itself. The heat required for ignition is greatly dependent upon the physical state of the materials and the heat transfer properties of its environment. Thus, combustible materials present in a dust, i.e., with a large surface-to-volume ratio, may burn very readily or explode, while the same material in a solid block may be very difficult to ignite at all.

EXTREMES OF BURNING BEHAVIOR

The extremes of burning behavior differ principally from normal burning in the time scale over which burning occurs. At the fast end of the spectrum are explosions, while at the slow end are the phenomena of auto-oxidation and smoldering. Explosions are a source of substantial hazard and are discussed in detail in several chapters in this volume. The slow processes, however, are less well known. Auto-oxidation is the combination of a material with oxygen at a rate too slow to produce the heat and light normally associated with fire. Metals are particularly prone to auto-oxidation, demonstrated by the rusting of iron, the anodization of aluminum, and the tarnishing of silver. Many organic materials are also prone to auto-oxidation. The deterioration of rubber and plastics over time is often the result of slow oxidative processes. Antioxidants can be added to such materials in order to increase their durability and working life. Such additives operate either by being themselves oxidized in preference to the organic material, or by chemically inhibiting the oxidation reaction.

Self-heating, and eventual spontaneous ignition, can occur if the heat produced by the auto-oxidation is not removed. This is especially true in porous solids, such as coal, where air can diffuse into the interior of the material, and yet the heat produced is effectively contained by the insulating properties of the material.

Smoldering is a distinct burning process, quite different from flaming. It has been thoroughly investigated and many aspects of smoldering are still unclear. One characteristic, however, is that only a small fraction of combustible materials will smolder; those that do appear to be able to produce a porous char during the course of combustion. This char allows air to reach the smoldering region. Compared to flaming, smoldering is a slow process. Temperatures in the smolder zone are typically 900-1600°F (483 to 871°C) (Clarke 1983); temperatures in a flame typically reach 2700°F (1482°C). Burning may in fact occur on the surface of the fuel, so the importance of a stable char structure can be appreciated. When materials smolder, the degree of combustion is usually less complete than when the same materials burn under flame.

Most smolderers are organic; all are solids, and, with the exception of carbon itself, none are pure compounds. They tend to be complex natural or synthetic polymers which can yield both volatile fuel species and a rigid, porous char structure.

MEASURING AND CONTROLLING HAZARDS

The vastly different burning characteristics of gases, liquids, and solids pose different types of hazards. The methods for measuring and controlling them are also different. Table 5-1A shows the different kinds of measurement techniques available for different components of hazard. It is most convenient to think of hazard control in terms of controlling first the likelihood of ignition, then the control or containment of the fire spread, and finally management of the fire impact if ignition and spread cannot be prevented.

Fire Hazards of Gases and Dusts

As mentioned above, gases and dusts burn rapidly once ignited, so avoidance of ignition is very desirable. The principal measurement tool for doing this is the determination of flammability limits, i.e., the concentra-

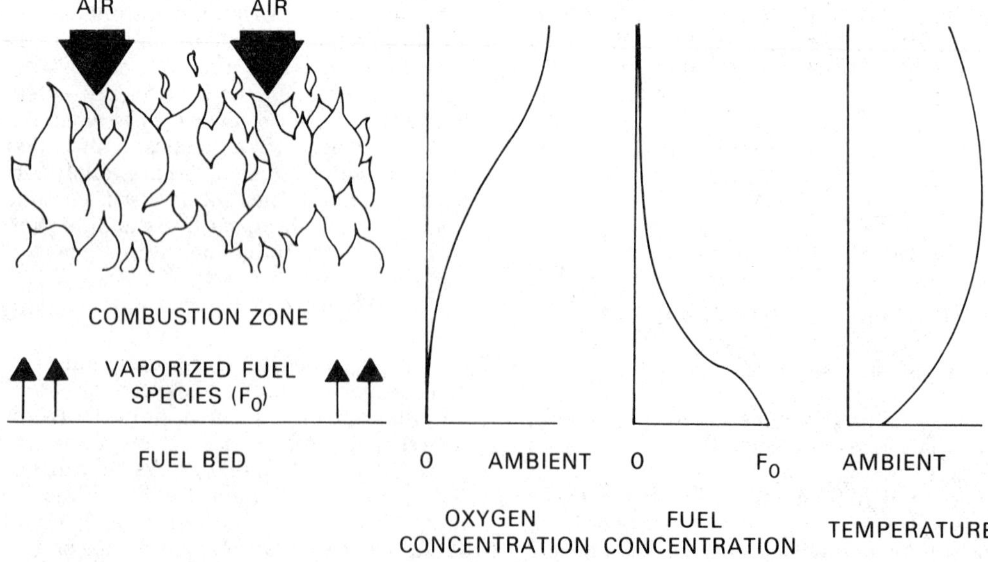

FIG. 5-1A. A schematic view of a burning surface, showing profiles of oxygen, fuel, and temperature as a function of distance above surface.

TABLE 5-1A. Means of Measuring and Controlling Fire Hazards of Materials

Hazard Component	Ignition: Measurement	Ignition: Ignition Control	Spread and Growth: Measurement	Spread and Growth: Control of Spread and Growth	Fire Impact: Measurement	Fire Impact: Control of Fire Impact
MATERIAL Gases	Ignitability Limits	1. Storage and Handling Safeguards 2. Inerting of Atmospheres	Flammability Limits, Density, Diffusion Coefficient	Tank Venting Procedures	Placarding, NFPA Hazardous Materials Identifiers	1. Emergency Response and Evacuation 2. Placarding, NFPA Hazardous Materials Identifiers
Liquids	Flash Point	1. Handling Safeguards 2. Hazard Classification Systems	Volatility	1. Ventilation and Flame Arrest 2. Storage and Tank Separation	Placarding NFPA Hazardous Materials Identifiers	1. Emergency Response and Evacuation 2. Placarding, NFPA Hazardous Materials Identifiers
Solids 1. Textiles, Cushioning Materials	1. Ease of Ignition Tests and Small-Scale Flame Spread Tests	1. Flame resistive Materials and Treatments 2. Protective Layers	1. Supported Flame Spread 2. Rate-of-Heat Release	1. Material Selection and Assembly 2. Detection and Suppression	1. Toxicity of Combustion Products (under development) 2. Smoke Generation	1. Breathing Apparatus for Fire Fighters 2. Smoke Control Systems
2. Structural Materials, Building Components and Finishes	Same	Same	Flame Spread Rate-of-Heat Release	1. Fire Resistivity 2. Low Combustibility 3. Fire-resistive Coatings 4. Detection and Suppression	1. Fire Endurance (Stability, Integrity, Insulation) 2. Smoke and Toxic Combustion Products	Same, Plus Construction Design: Compartmentation; Separation

tion range of a particular gas or dust (usually in air) within which combustion will occur. It is possible to reduce the likelihood of ignition of a gas or dust by adding chemical inhibitors to raise the lower flammability limit. In practice, however, this is rarely done, either because of expense or because it alters otherwise desirable properties.

Ignition control is generally effected by stringent storage and handling safeguards; the bulk of the discussion in the chapters on flammable gases are concerned with these procedures. It is also possible in closed environments to reduce the amount of oxygen available and thereby to raise the effective flammability limit. Conversely, atmospheres which are enriched in oxygen beyond normal atmospheric concentrations offer special explosion hazards for vapors and dusts. The intensity of a fire resulting from gases, vapors, or dust is a function of the density of gas or the concentration and particle size of the dust. Very light gases, e.g., hydrogen, and very finely divided dust are rapidly dispersed and can lead to the onset of explosive conditions in a correspondingly shorter period of time than more dense materials. Fire and explosion control efforts concentrate on containing the products and, under specialized conditions, providing for automatic suppression of an incipient explosion. This is done by rapidly introducing a suppressant, such as water vapor or a chemical inhibitor, into the exploding cloud. To be effective, automatic suppression must occur in a time scale of a few hundredths of a second.

In recent years there has been a growing concern over the possiblity of large flammable gas fires in populated areas. With vast quantities of gaseous fuels now being transported either in bulk or by pipeline, the likelihood of such occurrences increases. The principal means by which the impact of such accidents is minimized is the use of hazardous material identification systems, for example, placarding and NFPA 704, *Standard System for the Identification of the Fire Hazards of Materials*. In the absence of a major breakthrough in the prevention or suppression of such fires, the most appropriate means of minimizing fire impact is the advanced planning, preparedness, and response of trained emergency personnel.

Fire Hazards of Liquids

Since burning actually occurs in the vapor phase, the most hazardous combustible liquids are those with a high vapor pressure, or volatility. An empirical measure which combines volatility with the heat producing capabilities of the vapor is the flashpoint determination. The flashpoint is simply the temperature at which a liquid gives off vapors which can be ignited under specified laboratory conditions. Flashpoint determinations give rise to hazard classification systems, the most severe hazard being afforded by those liquids with the lowest flashpoints. As in the case of gases, however, the principal means of controlling the ignition of combustible liquids is in handling safeguards. Elaborate procedures exist to minimize the escape of flammable vapors in the handling of volatile liquids and to avoid sources of ignition.

If a fire involving such materials is initiated, means are also available to prevent the supply of additional combustible fuel to the fire. These include designs for the venting of storage tanks in order to minimize explosions, and flame arrestors to guard against the travel of the flame into the storage area itself. Storage and separation criteria for bulk storage of flammable liquids are also used to minimize the likelihood that additional supplies of flammable liquids will come in contact with the fire. The intensity of the fire once begun is a function both of the volatility and of the amount of heat released when the fuel burns. Thus, heavy oils or tars may be difficult to ignite, but can burn readily once underway. A portion of the heat produced in the flame radiates back to the fuel surface and vaporizes more fuel. For most common organic liquids, the heat required to vaporize a given amount of material is a small percentage of the heat of combustion.

Some fire fighting techniques interfere with the passage of fuel from the liquid to the vapor phase. These include cooling the liquid to slow vaporization and the use of foam to cover the liquid surface.

Solids

The part of Table 5-1A relating to solids concentrates on structural materials; it ignores the large group of chemicals which are solid but which are primarily found in industrial environments. Unless these materials are present as dusts, foams, or in other forms which present high surface areas, their fire hazards are similar to those of liquids.

In Table 5-1A, solid materials are divided into two major classes: flexible materials, such as textiles and cushioning; and structural materials, which can include everything from steel and concrete to wood and synthetic structural plastic foams. For both of these classes of materials there is a variety of tests to determine their susceptibility to ignition. However, since ignition requires the volatilization of some of the solid fuel, ignition behavior is strongly dependent upon the amount of heat applied to the surface. Therefore, different ignition tests often give different results depending on the size of the ignition source. The same principle applies to tests which attempt to measure flame spread over small samples. In many such tests, a sample receives no radiant heat load except that available from its own burning. Therefore, the larger the amount of sample burning, the greater the heat transferred to the unburned area ahead of the flame and a large sample will appear to have a higher flame spread than a small one. There are a variety of such tests principally distinguished by the geometry necessary to support the sample in a configuration which approximates its real use.

The fire hazards posed by inorganic structural materials are most likely to be passive. For example, steel can lose its strength, concrete can crack and spall, and glass can break and melt on exposure to high temperature.

Flame Resistant Treatments

Improvements in ignition control of materials have come about as a result of flame resistant treatments developed for both natural and synthetic polymerics. It is also possible to design polymeric systems in which a substantial fraction of the material is unavailable as fuel. One example is the dilution of the polymer with a large amount of alumina trihydrate, a nonburning, heat absorbing material. Another example is the inclusion of halogen atoms, such as chlorine or fluorine, in the monomers from which plastics are made. A third approach is the development of polymers based on materials other than carbon. These include silicone polymers, and phosphonitrilic whose composition is principally phosphorus and nitrogen.

It is important to recognize that claims for flame resistance are tied closely to the test method which is used to evaluate fire performance. Materials which resist ignition or behave acceptably in small scale tests may be wholly inappropriate for use when more severe or substantially different fire conditions are encountered. Figure 5-1B, for example, shows that the burning rate depends

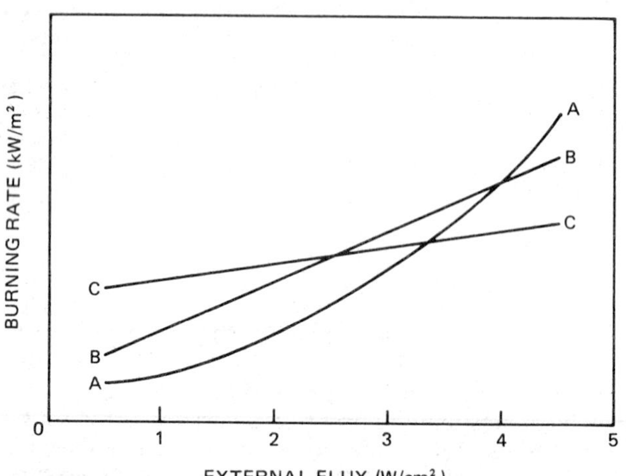

FIG. 5-1B. Burning behavior of various materials as a function of external flux (W/cm²).

heavily on imposed heat load (external flux). The relative performance of materials under a radiant heat load of 1 or 2 W/cm^2 may be different from their performance at 5 W, an external flux representative of what might be imposed on an exposed surface in a well developed room fire. This makes it important that materials be tested under conditions which duplicate those expected in real fire scenarios whenever possible.

Flame spread tests are probably the best known fire performance tests. The most widely used of these are the Steiner Tunnel Test—NFPA 255, and ASTM E84; and the radiant panel test—ASTM E162. These tests attempt to simulate spread of fire across a plane surface and may include the imposition of a known external radiant flux.

The role of heat release rate in determining fire hazards has recently been recognized. No standard tests yet exist, although several are under consideration. These tests measure the speed with which heat is produced, generally as a function of the imposed radiant energy. Because of the importance of an imposed flux in real fire hazard, these measurements promise to be of wide utility in predicting fire hazards. Ideally, fire resistant treatments should be detectable as diminished heat release rate measurements. At present, this appears to be true only for a few materials, notably wood impregnated with fire resistive compounds. Generally, decreased heat release is effected by the selection of materials which are inherently of low combustibility, such as concrete, metal, glass, and the like. Some success has been obtained in recent years with fire resistant coatings which can lower flame spread.

SMOKE

Because smoke* is usually the primary threat to life safety, it receives a great deal of attention. Traditionally, smoke has been characterized simply by its ability to obscure light. However, there has been increasing interest in the identity and effects of combustion products themselves as synthetic materials have become common components of building construction and furnishings. Because of the variability of their chemical composition, smoke from synthetic materials is more likely to contain constituents not often encountered when natural materials were the rule. The real question, however, is not whether new materials produce toxic combustion products—they do—but whether these products represent a different kind of, or more serious hazard than those of traditional materials. There is no general answer to this question—it depends on the material in question and the use to which it is put. Moreover, the tests used to measure the toxicity of smoke often give results which are more dependent on test conditions than on the material tested (Clarke 1983). The amount of smoke produced and the chemistry of the combustion products (i.e., the smoke's toxicity) are greatly influenced by the manner in which the smoke is generated. The heating rate or combustion temperature of the sample, the availability of oxygen, whether the sample smolders or flames—all of these are important factors. Therefore, using test conditions to predict full scale smoke conditions must be done with considerable care.

Although these measurement techniques are far from perfect, their development means that it is becoming possible to predict how a material, burning in a given scenario, will influence the hazard to life of that fire. This emerging field of fire protection, known as fire hazard analysis, requires information on the burning properties of a material, and its conditions of use, as well as smoke obscuration and smoke toxicity data. Fire hazard analysis is described in more detail in Chapter 2 of Section 21.

*Smoke is defined here as the *total* airborne effluent from heating or burning a material. Thus, expressions such as "smoke and toxic gases" are, by this definition, redundant.

Bibliography

References Cited

Clarke, F. 1983. "Toxicity of Combustion Products: Current Knowlege." *Fire Journal*. Vol 77, No. 5. pp 84ff.

Thorton, W. M. 1917. *Philosophical Magazine*. Vol 33. pp 196-203.

NFPA Codes, Standards, Recommended Practices and Manuals. (See the latest *NFPA Codes and Standards Catalog* for availability of current editions of the following documents.)

NFPA 49, *Hazardous Chemicals Data*.

NFPA 255, *Standard Method of Test of Surface Burning Characteristics of Building Materials*.

NFPA 258, *Standard Research Test Method for Determining the Smoke Generation of Solid Materials*.

NFPA 701, *Standard Methods of Fire Tests for Flame Resistant Textiles and Films*.

NFPA 704, *Standard System for the Identification of the Fire Hazards of Materials*.

Additional Readings

Burge, S., and Tipper, C., "The Burning of Polymers," *Combustion and Flame*, Vol. 13, 1969, pp. 495-505.

Drysdale, Dougal D., *Fire Dynamics*, Wiley, NY, 1985.

Friedman, R., "Quantification of Threat From a Rapidly Growing Fire in Terms of Relative Material Properties," *Fire and Materials*, Vol. 2, 1978, pp. 27-33.

Jiggett. C., "Estimation of Rate of Heat Release by Means of Oxygen Consumption," *Fire and Materials*, Vol. 4, 1980, pp. 61-65.

Kanury, A. M., *Introduction to Combustion Phenomena*, Gordon and Breach, NY, 1977.

Krause, R., and Gann, R., "Range of Heat Release Measurements Using Oxygen Consumption," *Jornal of Fire and Flammability*, Vol. 12, 1980, p. 117.

Kuryla, W., and Papa, A.. *Flame Retardancy of Polymeric Materials*, Vols. 1-4, Marcel Dekker, NY, 1978.

Lyons, J., *The Chemistry and Uses of Fire Retardants*, Wiley and Sons, NY, 1970.

Moussa, N., Garris, C., and Toong, T., *16th Symposium (International) on Combustion*, Combustion Institute, Pittsburgh, 1976.

Ohlemiller, T., Bellan, J., and Rogers, F., *Combustion and Flame*, Vol. 36, 1979, pp. 197-215.

Rashbash, D. J., and Drysdale, D. D., "Fundamentals of Smoke Production," *Fire Safety Journal*, Vol. 5, 1983, pp. 77-86.

Spalding, D. B., *Combustion and Mass Transfer*, Pergamon Press, London, 1979.

"Standard Test Method for Flashpoint and Firepoint by the Cleveland Open Cup," *ASTM D92*, American Society for Testing and Materials, Philadelphia, PA.

"Standard Test Method for Flashpoint by the Pensky-Martens Closed Tester," *ASTM D93*, American Society of Testing and Materials, Philadelphia.

Tewarson, A., "Heat Release in Fires," *Fire and Materials*, Vol.4, 1980, pp. 185-191.

Thomas, D., "Self-Heating and Thermal Ignition—A Guide to its Theory and Applications," *Ignition, Heat Release and Noncombustibility of Fuels*, American Society of Testing and Materials, Philadelphia, 1972.

WOOD AND WOOD BASED PRODUCTS

Revised by Harold O. Beals

Wood is one of the most useful materials of construction. It is employed for framing, sheathing, siding, shingles, underlayment, concrete formwork, and interior finish. Wood based building products include fiberboard, insulation board, ceiling tile, plywood, flakeboard, strandboard, and particleboard. Wood based materials are found in furniture, cabinets and in manufactured housing.

Wood is the major raw material source for paper and paper based products such as cardboard, box board, and corrugated board used in containers. Paper, used alone or combined with metal foils and plastics, is employed in the manufacture of many hundreds of products, including writing and wrapping papers, magazines, books, bags, wall coverings, and labels. As the source material for most cellulose based products, it is found in textiles, explosives, transparent films and coatings, plastics, and in many food products.

The products made from wood and wood based material are legion. As their grades often overlap, it is sometimes difficult to separate those made from wood, paper, or cellulose.

Wood and its cellulose derivatives are frequently involved in fire, and an understanding of the behavior of this versatile material under fire exposure is important. The discussion of wood in this section will be limited to solid wood and wood based products, including paper and paper products, and their ignition and burning characteristics. Those interested in a more detailed treatment on the subject should consult the list of references at the end of this chapter.

Wood based cellulosic fiberous materials used in textiles, including acetates and various rayons, are discussed in Chapter 3 of this Section. Other reconstituted cellulose materials used in explosives and propellents are discussed in Chapter 7. Plastics and films made from cellulose are covered in Chapter 8. Fire hazards of finely divided wood particles, such as sander dust, are discussed in Chapter 9. Cellulose based lacquers and paints are covered in Section 10, Chapter 7. Manufacturing and process information on various wood products is found in Section 10, Chapter 19. Fire hazards of wood fuels are covered in Section 11, Chapter 8. Biomass and standing timber hazards are discussed in Section 12, Chapter 9. Products of combustion, their importance on the fire scene, their physiological effects on persons trapped in the fire environment, and the degree to which they affect rescue and fire fighting efforts are discussed in Section 4, Chapter 3. Fire retardant treatments are discussed in Section 12, Chapter 6.

Harold O. Beals is Associate Professor of Wood Products, School of Forestry, Auburn University, AL.

THE NATURE OF WOOD AND WOOD BASED PRODUCTS

Wood products made from solid wood include lumber, laminated beams, large timbers, mine timbers, railroad ties, poles, and pilings. Other solid wood products are cooperage (barrels), boxes, crates, and other containers. Hardwoods are used for firewood and in the manufacture of furniture. Wood products made from thin sheets of wood (veneer) bonded together with adhesives are plywood and laminated wood.

Wood based products are those in which the wood has been mechanically reduced in size to form flakes, strands, splinters, fibers, or particles and bonded with adhesives or other binders to make flat panels or shapes. These products include flakeboard, oriented strand board (OSB), medium density fiberboard (MDF), other fiberboards and hardboards, cement excelsior boards, particleboard, and some insulation boards and ceiling tile.

Paper and paper based products differ from wood based products in that no adhesive is used to bond the paper fibers together. Most papers are made by a chemical process that removes the lignin and hemicellulose from the wood and leaves the cellulose fibers intact. An exception is newsprint, in which the fibers are separated by a mechanical or grinding process—no chemicals are used. Paper and paper based products may be made without additives (absorbent papers), but most contain binders (wet strength resins), fillers (clay), pigments (titanium dioxide), colorants (dyes), sizes (rosin, starch), or other materials to provide improved appearance or performance. Some papers are coated with wax (paraffin) or various plastic materials for use in container applications. Many paper products are printed with colored inks for use as containers or labels. Paper and paper based products are often classified into such catagories as kraft or sulfate (brown papers), used for wrapping and container applications; sulfite (white papers), used for writing, book and magazine papers; and groundwood, used for tissue and newsprint. Paper products are generally available in the form of thin sheets or rolls. Paper based products are thicker and, when graded, can overlap into wood based materials, such as the fiberboards, and it is sometimes difficult to place such products into either category.

Cellulose based products are made from purified wood cellulose obtained from certain special grades of paper pulp called dissolving pulp. Wood has been used for dissolving pulp since about 1920, replacing cotton linters which were formerly used for this purpose. Most cellulose based products are not discernible as being made from wood or paper, but they often have the odor of burnt paper when combusted. Cellulose based products include the nitrocellulose lacquers, films, and plastics; acetate films; textiles and plastics; rayon textiles; cellophane; and smokeless powder used in small arms cartridges. An expanding market for purified wood cellulose fibers is in disposible sanitary applications such as baby diapers and feminine protection products. Purified cellulose is a common ingredient of many food products where it is used to thicken liquids and to improve the texture of foods. It is commonly found in diet aids where it serves as a bulking agent.

Wood cellulose can be broken down to glucose using appropriate chemical techniques to make molasses suitable for cattle feed supplements, or fermented to make ethyl alcohol (grain alcohol) for use as a gasoline additive (super unleaded). The hemicellulose portion of wood can also be fermented to make methyl-ethyl-ketone (MEK) or butanol (butyl alcohol). New techniques may soon allow it to be fermented to make ethyl alcohol.

Destructive distillation of wood produces charcoal, methanol (wood alcohol), acetic acid, formic acid, and wood creosote, as well as a low Btu fuel that can be used to power internal combustion engines.

Turpentine and rosin, known as naval stores, are produced from standing pine trees treated with paraquat, from old stumps, and as a byproduct in the manufacture of kraft paper by the sulfate process. Both turpentine and rosin are highly flammable and may cause spontaneous ignition under certain conditions.

Wood and wood based products are combustible—they char, smolder, ignite, and burn when the thermal environment is conducive to such reactions. Rarely do they self-ignite except in certain forms and under certain conditions of storage and moisture content. Normally a spark, open flame, contact with hot surfaces, or exposure to thermal radiation is required for ignition. Even then, ignition will be a function of the intensity of the ignition source and the time of exposure as well as other factors to be considered later in this chapter.

Dry wood burns with the characteristic odor of wood smoke. The smoke is bluish or dark when the wood is green or wet and low temperature or insufficient air is present; but, with high temperature and excess air, the smoke is often almost invisible. Burning wood produces smoke containing glowing particles and large amounts of a gray ash. Wood burned under conditions of incomplete combustion in which little or no air is present produces charcoal and a number of condensible vapors in a process known as destructive distillation. Burning wood in modern airtight stoves is considered destructive distillation. When the tars and creosote are condensed in the chimney, they are easily ignited and burn at very high temperatures, often causing house fires. Charcoal burns with little smoke and uniform heat. Wood is considered to be a clean fuel because it contains no sulfur.

Fire retardant chemicals can be, and frequently are, added or applied to wood and wood based products. When properly formulated and used, fire retardant chemicals reduce the combustibility of these products. Depending on the formulations used, flame and glow can be reduced once the outside source of heat is removed. Use of fire retardant treated lumber and plywood has greatly increased in recent years due to a greater awareness of firesafety.

To resist rot and insect damage, railroad crossties, utility poles, and pilings are usually pressure treated with various toxic chemicals (wood preservatives) containing creosote (made from coal tar) and pentacholorphenol in solutions of oil similar to diesel or fuel oil, or water soluble salt solutions containing arsenic, chromium, copper, or phenol. Waterborne salts are usually used to treat lumber and plywood but are commonly used for treating utility poles. Wood products freshly treated with creosote or pentacholorphenol in oil solutions are easily ignited because of the oil; however, after the wood ages and the oil evaporates, ignition is more difficult.

Wood and wood based products are often used in combination with other materials to provide a greater degree of fire protection. An insulated, plywood faced wood stud wall, for instance, will provide a greater degree of fire resistance than one without insulation. Also, it is possible to construct fire rated interior partitions using gypsum board and wood framing members. The relatively slow burning or char rate of heavier wood members provides a degree of protection because of the insulating effects of the char buildup. Finishes or additives used to improve the properties of many wood and paper based materials such as binders, adhesives, fillers, sizes, etc., can often modify the fire performance of these materials as compared to bare wood. Coatings applied to wood or wood based products may cause them to ignite easier and to burn more rapidly than the unfinished wood.

During the 1970s there were pronounced changes in the types and use of wood and wood based building products, due primarily to a desire for a more "natural" appearance in homes by more external use of wood materials (wood shingles or siding) and enactment of new legislation designed to encourage conservation and save energy. In many parts of the country, a return to wood as a primary fuel (burned in fireplaces and stoves) and to log homes with shingle roofs has created greater fire risks and air pollution problems. All of these factors have important bearings on the performance of wood and wood based materials in a building fire today.

CHEMICAL COMPOSITION OF WOOD

Wood consists primarily of carbon, hydrogen, and oxygen, with lesser percentages of nitrogen and other elements, but it contains no sulfur which is often present in other fuels. (See Table 5-2A.)

Table 5-2A. Chemical Composition of Dry Woods

	Constituents, Percent by Weight				
Species	Carbon	Hydrogen	Oxygen	Nitrogen	Ash
Oak	50.16	6.02	43.26	.09	0.37
Ash	49.18	6.27	43.19	.07	0.57
Elm	48.99	6.20	44.25	.06	0.50
Beech	49.06	6.11	44.17	.09	0.57
Birch	48.88	6.06	44.67	.10	0.29
Pine	50.31	6.20	43.08	.04	0.37
Poplar	49.37	6.21	41.60	.96	1.86
Calif. redwood	53.50	5.90	40.30	.10	0.20
Western hemlock	50.40	5.80	41.40	.10	2.20
Douglas fir	52.30	6.30	40.50	.10	0.80

Although wood is a complex of many substances, cellulose $[(C_6H_{10}O_5)_x]$ represents its major component. Wood contains about 50 percent cellulose by weight, approximately 25 percent hemicellulose in the form of sugars, 24 percent lignin, and 1 percent extractives (in the form of minerals, gums, resins, and other materials). Some pieces of wood may contain flammable extractives in the form of resins, e.g., lighter wood southern pine, but most woods contain no extracts of any consequence. The ash formed from burning wood is high in mineral content, primarily potash. Wood also contains moisture, the amount present depending upon the degree of previous drying. Removal of moisture above the fiber saturation point has little effect on wood other than to reduce its weight, but removal of moisture below the fiber saturation point increases the strength properties of wood.

The major difference between various woods depends largely on density (specific gravity), which is related to cell wall thickness or strength properties, and on infiltrated materials or extracts that contribute to the color and other properties of wood. Some of these extracts consist of flammable resinous products that may cause spontaneous ignition in certain wood products, i.e., piled chips of southern pine used in making paper.

Bark is an important residue often used for fuel in many wood industries. Bark contains large amounts of tannin (used in tanning leather) and suberin (cork) which makes it resistant to penetration by water. Commercial cork, obtained from the bark of the cork oak, is resistant to high temperatures and ignition and burns with difficulty. Cork contains a natural adhesive that allows small particles to be bonded under heat and pressure to make composition cork. Bark and cork are generally drier than wood.

Wood based products, paper, and paper based products generally contain much less moisture than wood and are easier to ignite and burn than solid wood. Many fiberous cellulosic based materials such as bagasse (spent sugar cane stalks) jute, hemp, or sisal have burning characteristics similar to wood based materials. Often these and other similar products are impregnated, pressure treated, or coated with a variety of chemicals which may enhance or reduce the ignition and burning rate of such materials.

Wood based and paper based products are frequently bonded, coated, or combined with various plastic resins (including urea, melamine, phenolformaldehyde, nitrocellulose, vinyl chloride, urethane, isocyanates, latex, or acrylics) that may modify their burning characteristics and products of combustion.

VARIABLES INFLUENCING IGNITION AND BURNING CHARACTERISTICS OF WOOD AND WOOD BASED PRODUCTS

Physical Form

The influence of physical form on the ignition and burning characteristics of wood and wood based products is illustrated by the fact that wood kindling will flame and burn from relatively small heat sources while heavier wood logs will resist ignition to a considerable degree. Paper is easier to ignite than small pieces of wood. The reason for this is that as the size of the particle diminishes, the ratio of surface area to volume (mass) increases. Thus there is both greater exposure of fuel to air and less mass to conduct heat away from the particle; consequently, heat does not readily dissipate within the material. It follows that small, thin forms of wood (such as excelsior), or other combustible solids (such as paper), will burn more readily than larger objects of the same material.

Although paper is considered to be easy to ignite, large, compact masses of paper such as books or cut and packaged stacks of paper do not burn readily due to charring and insufficient air in the interior of the mass, as well as the influence of additives such as binders, colorants, sizes, etc., in the paper. However, contrary to the general opinion that compact paper in rolls will not burn, large rolls of paper or paper based products stacked on end may unwind during a fire, exposing long, single sheets of paper which are easily ignited and burn readily.

Very finely divided wood particles such as sander dust or wood flour (made from nut shells) are prone to explosion and flash fire as are other carbonaceous dusts. But sawdust, which has a much larger particle size, generally does not present such a hazard under most conditions. Fires in dust control, dryer, or exhaust systems are a constant hazard in most wood manufacturing operations due to ignition of finely divided wood particles.

Thermal (Heat) Conductivity

Thermal conductivity (K factor—the inverse of insulating value, R factor) of a particular material greatly influences how a material will burn. Thermal conductivity is a measure of the rate at which absorbed heat will flow through the mass material. For instance, wood is a poor heat conductor, and thus has a high insulating value. Steel and aluminum are about 350 and 1,000 times more conductive, respectively. The low thermal conductivity of wood finds common illustration in the fact that a wood match burning at one end can be held between the fingers, while a steel needle of similar form, heated at one end, cannot be held with equal comfort.

The thermal conductivity of wood is dependent on the direction of heat flow with respect to the axis of grain orientation in wood, the moisture content of the wood, and the density (specific gravity) of the wood. Thermal conductivity is two to three times greater along the grain in the longitudinal direction than across the grain. Thermal conductivity of wood in any direction is about one-third greater for moisture content above 30 percent than for drier wood.

Thermal conductivity of wood is further illustrated by the fact that the structural framing of an unsprinklered building of heavy timber construction will withstand fire exposure for a considerably longer time than the unprotected framing of a light metal framed building. This is explained, in part, by the insulating effect of the char that develops as wood burns. More importantly, unprotected steel will absorb heat from the fire much faster and will eventually reach its yield point. Investigations in the United States, Great Britain, Japan, and elsewhere indicate a char rate of approximately 37 mm (1.5 in.) per hour with an accelerated rate during the first five to ten minutes of exposure (Imaizumi 1963).

Moisture Content

Wood and cellulose are hygroscopic materials and they have an affinity for water in both the liquid and vapor form. Wood may contain moisture in the form of: (1) liquid water (free water) in the cell cavities above the fiber saturation point (FSP), which is usually considered to be 25 to 30 percent moisture content; (2) absorbed or bound water located in the cell walls below the fiber saturation point; and (3) water of constitution, which is the water contained in the various components that make up wood. This water cannot be removed without destroying the physical structure of wood. Both the free water and the absorbed water have an important bearing on the physical and mechanical properties of wood as well as on the ignition and burning of wood. As wood comes from the living tree (green wood), it contains a great deal of moisture in the form of liquid water, but, as it dries, air replaces much of the water and moisture remains in the form of vapor in the cell cavity or as absorbed water in the cell wall. Dry wood, however, always contains some moisture. Construction lumber is considered to be dry at 19 percent moisture content, but wood used for furniture is dried to 6 percent moisture content or less. Moisture Content (MC) in wood is calculated as a percentage of the oven dry weight by the formula:

$$MC = \frac{\text{Original Weight} - \text{Oven Dry Weight}}{\text{Oven Dry Weight}}$$

Wood reaches an equilibrium moisture content (EMC) with its surroundings at a given temperature and relative humidity. In a heated building, the EMC of wood used in its construction is generally slightly less than 10 percent moisture content in the winter months and about 12 percent moisture content in the summer. Thus a newly constructed building generally contains wetter wood than one that has been built for some time. Figure 5-2B illustrates conditions under which equilibrium will be established at three different temperatures.

Moisture content has an important bearing on the electrical properties of wood. Dry wood is a good insulator but wet wood is a conductor. The electrical resistance of wood increases in approximate proportion to the moisture content of wood, from 25 percent or the fiber saturation point (FSP) to about 6 percent moisture content where it becomes too high to measure with a resistance moisture meter. Within this range, the moisture content of wood can be determined with reasonably accurate instruments. The most accurate determination of moisture content in wood is generally performed by oven drying and weighing. However, the moisture in wood is not uniformly distributed, and wood is often drier or wetter at the surface than in its interior. Surface layers which control ignition and flame spread respond much faster to changes in atmospheric conditions than the interior; thus bulk moisture contents can sometimes be misleading.

The rate of moisture loss or gain depends a great deal upon the size and shape of the wood product. Some finely divided cellulosic materials absorb or lose moisture very rapidly and their combustibility can vary from low to high within just a few hours. Examples are shredded newspaper, tissues, and loose forest litter (duff). Dense and compact combustibles such as books or large timbers change moisture content much more slowly. Rate of moisture loss depends on the relative humidity, the temperature, and the air velocity. Moisture loss is quite slow in air drying where there is no control over air velocity, temperature, or relative humidity; lumber may require 30 days or more to dry in such an environment. But in a kiln, where humidity, temperature and air velocity can be adjusted, lumber can be dried in less than 24 hours. As a general rule of thumb, the amount of time required to dry lumber varies with the square of the thickness. Thus a 1 in. (25.4 mm) board can be air dried in a few weeks, but a railroad crosstie or large timber requires 24 to 30 months. If wood is dried below the EMC it will reach in service, it will regain moisture slowly until it reaches that EMC. (See Fig. 5-2A.)

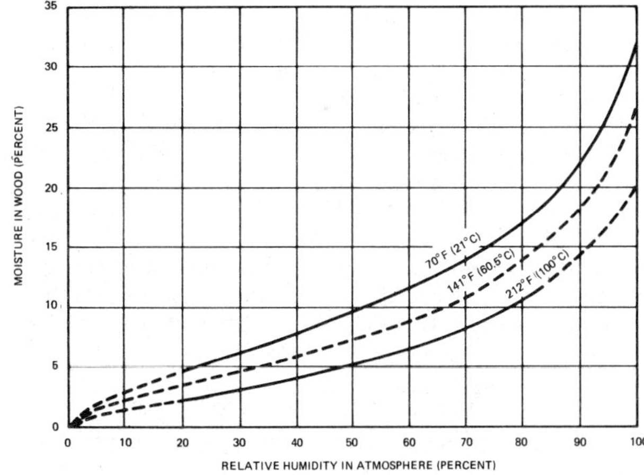

FIG. 5-2A. Equilibrium moisture content of wood.

Ignition and burning tests in the laboratory show that the behavior of combustible solids of the same size, shape, and chemical composition varies markedly depending upon the moisture content. In research conducted by the British Department of Scientific and Industrial Research (now the Ministry of Technology), the influence of the moisture content of oak and western red cedar on the time necessary for these woods to ignite was studied; the results are shown in Figure 5-2B (FOC 1952). Radiation intensity in this figure is expressed in terms of the number of calories falling on one square centimeter of the exposed wood in one sec, using a test apparatus developed by the Ministry.

Once a fire is well under way, the significance of the moisture factor decreases since the heat radiation and

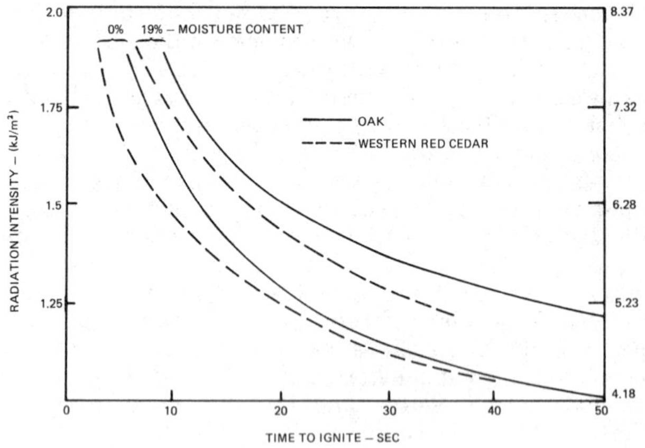

FIG. 5-2B. Effect of moisture content on the ignition of wood.

therefore the rate of pyrolysis increases. Under such conditions wood with a moisture content of 50 percent or more will burn.

Conflagration experience in the United States illustrates the effect of moisture content on structural fire susceptibility. In many conflagrations, it has been noted that prolonged dry weather preceded the outbreak of fire. This dryness had a noticeable effect on the susceptibility of shingles, as well as other combustibles present, to ignition from flying brands and thermal radiation. The tragic Peshtigo (Wisconsin) forest fire of 1871 followed a prolonged period of extremely dry weather. It occurred the same weekend as the great Chicago fire and destroyed 17 towns and claimed 1,152 lives.

There is a time lag between changes in the fuel moisture content of cellulosic materials and changes in atmospheric humidity. Most solid cellulosic combustibles are not immediately absorptive of airborne water vapor; conversely, the moisture content of most solids will be retained for a longer period of time than air retains its moisture in suspension. Thus the humidity in the ambient air is not always a reliable index of the moisture content of ordinary combustibles.

Ignition of Wood and Wood Products

The thermal degradation of wood is a complex process. Four stages of decomposition of wood on exposure to heat have been described (Beall and Eichner 1970):

Temperature	Reaction
392°F (200°C)	Production of water vapor, carbon dioxide, formic and acetic acids—all noncombustible gases.
392°-536°F (200-280°C)	Less water vapor, some carbon monoxide—still primary endothermic reaction.
536°-932°F (280°-500°C)	Exothermic reaction with flammable vapors and particulates. Some secondary reaction from charcoal formed.
over 932°F (500°C)	Residue principally charcoal with notable catalytic action.

Ignition temperatures may vary, depending on the analytical method and test equipment used. With proper test equipment some exothermic reaction can be observed in the first stage just listed. In the third stage, ignitable gases are produced and the pilot ignition temperature is reached. When the decomposition has progressed to the point where the evolved gases no longer insulate the charcoal layer from oxygen, spontaneous ignition may result.

Specific ignition temperatures of wood are difficult to determine because of the many variables involved. Leading research groups have attempted to pinpoint the ignition temperature of wood, but the results vary greatly. In one series of tests using untreated, airdried wood blocks 1¼ by 1¼ by 4 in. (32 × 32 × 102 mm), and of nine different species, the Forest Products Laboratory found that the specimens could be heated to temperatures varying from 315 to 385°F (157 to 196°C) for 40 min in a stream of heated air held at a constant temperature without igniting. In another series of tests by the same laboratory, using the same test procedures but higher temperatures, 34 untreated softwood and hardwood species ignited in 4½ to 6 min at temperatures varying from 608 to 660°F (320 to 349°C) for the softwood, and from 595 to 740°F (313 to 393°C) for the hardwood. It was noted that a fairly close relationship existed between the specific gravity of the wood under test and the ignition temperature. In general, low density species ignited at lower temperatures than high density species (Fleischer 1960). The National Bureau of Standards tested three conifers and two hardwoods and reported ignition temperatures from 378 to 428°F (192 to 220°C) in small samples varying from shavings to match size.

The ignition and charring temperatures of wood were recorded at the Forest Products Laboratory (McNaughton 1944). His findings, represented in Table 5-2B, show the length of time before wood specimens maintained at the specified constant temperatures evolved combustible gases in sufficient quantities to be ignited by a pilot flame located about ½ in. (12.7 mm) above the specimen. The specimens, 1¼ by 1¼ by 4 in. (32 × 32 × 102 mm) in size and oven-dry, were heated in a vertical quartz chamber 3 in. (76 mm) in diameter by 10 in. (254 mm) long which was maintained at constant temperature by an electric furnace.

C. R. Brown reported on ignition temperatures of some materials as shown in Table 5-2C (Brown 1934). In his investigations, specimens about 2½ in. (63.5 mm) long and weighing 3 g (0.10 oz) were suspended in a vertical glass container 2½ in. (63.5 mm) in diameter and 4¹³⁄₁₆ in. (122 mm) high. The container was heated by an electric furnace, the temperature being increased at a predetermined rate. Air preheated to the temperature of the furnace was passed through the container at measured rates of flow. The results were found to be affected by the size of the sample, the rate of heating, and the rate of air flow over the specimens.

In 1947 the National Bureau of Standards reported ignition temperatures of wood as shown in Table 5-2D. In these tests the specimens of wood in the form of shavings were heated in a glass test tube in the presence of a measured flow of heated air at optimum velocity. The ignition temperature was taken as the lowest temperature at which the exothermic oxidation reaction culminated in an ignition as indicated by flaming or glowing combustion.

Table 5-2B. Time Required to Ignite Wood Specimens

Wood	No Ignition in 40 min °F	No Ignition in 40 min °C	Exposure Before Ignition, By Pilot Flame, Minutes 356°F (180°C)	392°F (200°C)	437°F (225°C)	482°F (250°C)	572°F (300°C)	662°F 350°C)	752°F (400°C)
Long leaf pine	315	157	14.3	11.8	8.7	6.0	2.3	1.4	0.5
Red oak	315	157	20.0	13.3	8.1	4.7	1.6	1.2	0.5
Tamarack	334	167	29.9	14.5	9.0	6.0	2.3	0.8	0.5
Western larch	315	157	30.8	25.0	17.0	9.5	3.5	1.5	0.5
Noble fir	369	187	—	—	15.8	9.3	2.3	1.2	0.3
Eastern hemlock	356	180	—	13.3	7.2	4.0	2.2	1.2	0.3
Redwood	315	157	28.5	18.5	10.4	6.0	1.9	0.8	0.3
Sitka spruce	315	157	40.0	19.6	8.3	5.3	2.1	1.0	0.3
Basswood	334	167	—	14.5	9.6	6.0	1.6	1.2	0.3

Table 5-2C. Ignition Temperatures of Woods-Brown

Wood	Self-Ignition Temperature °F	°C
Western red cedar	378	192
White pine	406	207
Long leaf pine	428	220
White oak	410	210
Paper birch	399	204

Table 5-2D. Ignition Temperatures of Woods-NBS

Wood	Self-Ignition Temperature °F	°C
Short leaf pine	442	228
Long leaf pine	446	230
Douglas fir	500	260
Spruce	502	261
White pine	507	264

Wood in contact with steam pipes or a similar constant temperature source over a very long period of time may undergo a chemical change resulting in the formation of charcoal which is capable of heating spontaneously. It has been suggested that 212°F (100°C) is the highest temperature to which wood can be *continually* exposed without risk of ignition (McGuire 1969).

To summarize, ignition temperatures of wood vary widely depending on such variables as:

1. The specific gravity or density of the sample.
2. The physical characteristics of the sample; i.e., its size, its form.
3. The moisture content.
4. The rate and period of heating.
5. The nature of the heat source.
6. The air (oxygen) supply and its velocity.

Rate and Period of Heating

The rate and duration of heating influence the susceptibility of wood and wood products to ignition. Compared to flammable liquids or gases, combustible solids are usually less hazardous because they do not vaporize readily nor do they emit flammable vapors at normal ambient temperatures and atmospheric pressures. Thus, ignition of solid combustibles usually requires long enough contact between the heat source and the material to permit release of ignitable vapors. This contact can be either momentary or prolonged, with actual ignition depending on the variables just discussed. An exception is glowing combustion that may occur without vaporization of some materials, e.g., charcoal.

Ignition will occur when a material is heated above its autoignition temperature. Thus, if moderate heat is applied long enough to raise a combustible to its ignition temperature, it may cause ignition. However, high heat applied momentarily to a combustible may not cause ignition. For example, a steam pipe in contact with wood over a period of time, may result in ignition, whereas a gas fired blowtorch held momentarily on a painted wood surface may blister the paint but not ignite the wood, despite the fact that the torch flame temperature is much higher than the ignition temperature of the wood. However, because the precise reaction in both cases cannot be determined in advance, it is common practice to avoid direct contact between steam pipes and adjacent combustibles and to condemn as unsafe the practice of burning paint off wood surfaces. The illustrations do, however, vividly show the influence of the rate and period of heating on the ignition of the same type of combustible in roughly the same shape and form.

The shape and form of a sample of combustible material not only influence its susceptibility to ignition, as explained earlier in this chapter, they also affect the reaction of the combustible to any given rate and period of heating. Likewise, the moisture content of the combustible, the air space between the heat source and the material, and other variables affect the potential for ignition.

Fire experience further illustrates these factors. It is common experience for fire departments to arrive at a fire scene after a delayed alarm and find a wood structure seriously involved and threatening adjacent combustible structures. If water curtains cannot be established in time, ignition of the adjacent exposed property may occur. This is a result of the direct relationship between ignition and the rate and period of heat exposure. In actual fires, accurate measurements of such rates and periods of heating are, of course, not practical. Laboratory fire tests, however, do confirm what takes place. (See the preceeding discussion on ignition temperatures of wood.)

Spontaneous Heating and Ignition

Spontaneous heating and ignition of wood and wood based products usually result only from contamination or induced artificial heating. For instance, cellulosic materials which are clean and dried will not heat spontaneously; on the other hand, those soaked with certain drying oils (linseed oil, tung oil, turpentine) and stored in unventilated spaces can heat spontaneously. If this process is allowed to continue, ignition results. Similarly, a measure of artificial heating can cause certain combustibles to self-heat to their ignition temperature. For example, if wood fiberboards are not subjected to a direct, artificial heat source either during manufacture or when installed in structures, they will not heat spontaneously. On the other hand, tests have shown that heating such fiberboards at relatively moderate temperatures can set up an exothermic reaction leading to their ignition (Mitchell 1951). In one such test, a 22 in. (559 mm) high stack of wood fiberboards ⅛ in. (3 mm) thick subjected to a heat source of 228°F (109°C) was ignited in 96 hours.

An adequate amount of air (oxygen) is essential if combustion is to continue. Combustion engineers have suggested a formula for the minimum volume of air required for complete combustion of solid fuels with air at 62°F (17°C):

$$V_m = 147\left[C + 3\left(H - \frac{O}{8}\right)\right] \text{ cu ft}$$

where

V_m = minimum volume of air in cu ft,
C = part by weight of carbon in 1 lb of fuel,
H = parts by weight of hydrogen in 1 lb of fuel, and
O = parts by weight of oxygen in 1 lb of fuel.

While this formula may have little direct application in fire protection engineering, smoldering fires which are starved of adequate air (oxygen) are frequently encountered by firemen, so that the significance of this factor must not be overlooked.

Wood and wood based products are not susceptible to the same type of spontaneous heating from microbiological action that is prevalent in certain agricultural products.

Rate of Combustion

In most fires involving combustible solids, there is a time delay between ignition and rapid combustion. This statement might be challenged because of the exceptional speed of some fires, particularly many dwelling fires which result in fatalities. In most such cases, however, investigation shows that detection was delayed or that flammable liquids, gases, or similar materials were involved. The rate of combustion is significantly influenced by the physical form of the combustible, the air supply present, the moisture content, and allied factors; but there is the fundamental need for progressive vaporization of the solid by heat exposure for complete combustion to proceed.

Speed of Flame Propagation

Speed of flame propagation over the surface of a combustible critically affects the severity of a fire in many cases. The 1959 Los Angeles School Fire Tests (LAFD 1959) showed that under controlled test conditions, flames spread over combustible surfaces with such speed that observers had to flee the sudden release of heat and flame. In 1949 the British Department of Scientific and Industrial Research conducted full scale dwelling fire tests on the influence of combustible wallboards vs the use of plaster finish (FOC 1950). The record of these tests graphically illustrates the observed time-temperature curves in the living rooms of the two houses. Originating in an easy chair in the living room, the fires in their initial stages were identical in both dwellings. The loading of combustible materials was roughly proportional to the loadings observed in occupied dwellings in England. It should be noted that in the United States and Canada, earlier ignition and faster flame spread on the combustible surfaces could be expected since winter heating in these countries results in lower relative humidities and thus lower moisture content of the combustibles. Mobile homes, because of their construction and design, are especially prone to high speed flame propagation.

Compared with flammable vapors or gases, the speed of flame propagation across most combustible solids is slow; the presence of readily ignitable vapors with the former materials is the primary reason for this difference. Also, pyrolysis gases from combustible solids must be mixed in proper proportions with air; thus flame spread is frequently influenced by the need of these gases to find an adequate air supply to be progressively consumed.

Amount of Fuel Contributed

The amount of fuel contributed by a substance is normally measured by its heat of combustion. Table 5-2E shows some figures for the heats released by various wood products and also gives comparative figures for some other materials. It will be noted that the heat of combustion of most dry wood products is less per pound than that of the other fuels listed. This is of significance for the evaluation of a measured quantity of combustibles to the fire exposure it creates. It is common practice in fire protection engineering to weigh or estimate the weight of combustibles in a fire area to determine, for instance, the needed fire resistance of structural elements or the optimum quantity of water required per square foot or meter of floor area for sprinkler discharge. Reasoned judgments are frequently

Table 5-2E. Heat of Combustion of Various Wood and Wood Based Products and Comparative Substances

Substance	Heating Valve	
	Btu per lb	kJ/kg
Wood sawdust (oak)	8,493†	19 755
Wood sawdust (pine)	9,676†	22 506
Wood shavings	8,248†	19 185
Wood bark (fir)	9,496†	51 376
Corrugated fiber carton	5,970†	13 866
Newspaper	7,883†	18 336
Wrapping paper	7,106†	16 529
Petroleum coke	15,800	36 751
Asphalt	17,158	39 910
Oil (cottonseed)	17,100	39 775
Oil (paraffin)	17,640	41 031

† Dry

necessary in applying this principle. Libraries, for example, have a phenomenal fire load. While there have been very serious fire losses in these properties, strict application of the calculated fire load to the sprinkler requirements would not be appropriate in most of these cases because of the physical arrangement of the combustibles. On the other hand, a relatively low loading is frequently misused by those seeking to depreciate the need and value of sound fire protection.

The amount of fuel contributed by one combustible vs another, or different forms of the same combustible can also be used to evaluate the desirability of utilizing the material in a given situation. This is particularly true for measuring the fuel contributed by certain forms of building materials.

Heat Release Rate

The total heat release rate of the materials of construction and furnishing is an important factor in determining whether a room will reach flashover conditions. If the total heat release rate in megawatts exceeds $0.75\ A \sqrt{h}$ the room is likely to flashover (Babrauskas 1980).

Here A and h are the area and height of the window or door opening. A better approximation takes the wall and ceiling and their thermal properties into account (McCaffrey et al. 1980). The heat release rate of wood depends on the incident radiant flux, moisture content, thickness, orientation, the boundary conditions at the rear face, and the oxygen concentration of the ambient air. It is also a function of time characterized by an early peak followed by a slow decay until flaming ceases. A second peak will also occur if the rear face of the specimen is insulated. For an oven dry vertical pine board of nominal 1 in. thickness (one inch equals 25.4 mm) exposed to a thermal radiation level of 60 kW/m^2 in normal air with the rear face exposed to a water cooled plate, a peak heat release rate of 250 kW/m^2 in 19 min was reached when the flame went out (Parker and Long 1972). A series of wood products with moisture contents of about 8 percent exhibited peak heat release rates between 90 and 170 kW/m^2 under the same conditions. The heat release rate of wood products increases linearly with the density (Chamberlain 1983).

Bibliography

References Cited

Babrauskas, V. 1980. "Estimating Room Flashover." *Fire Technology*. Vol 16, No 2. May 1980.

Beall, F. C., and Eichner, H. W. 1970. *Thermal Degradation of Wood Components: A Review of the Literature*. Report No 130. Forest Products Laboratory, Madison, WI.

Bierberdorf, F. W., and Yuill, C. H. 1963. *An Investigation of the Hazards of Combustion Products in Building Fires*. U.S. Public Health Service Contract No PH86-62-208. Final Report.

BOCA. 1970. *The BOCA Basic Building Code*. 5th ed. Sec 922. Building Officials and Code Administrators International, Chicago.

Brenden, J. J. 1970. *Determining the Utility of a New Optical Test Procedure for Measuring Smoke from Various Wood Products*. Research Paper 137. Forest Products Laboratory, Madison, WI.

Brown, C. R. 1934. "The Ignition Temperatures of Solid Materials." *NFPA Quarterly*. Vol 28, No 2. pp 135-145.

Chamberlain, D. L. 1983. *Heat Release Rate Properties of Wood-Based Materials*. NBSIR 83-2597.

Ferguson, G. E. 1933. "Fire Gases." *NFPA Quarterly*. Vol 27.

FOC. 1950. *Fire Research Report for the Year 1950*. Department of Scientific and Industrial Research and Fire Offices' Committee. Her Majesty's Stationery Office, London.

FOC. 1952. *Fire Research Report for the Year 1952*. Department of Scientific and Industrial Research and Fire Offices' Committee. Her Majesty's Stationery Office, London.

Fleischer, H. O. 1960. *The Performance of Wood in Fire*. Report No 2202. Forest Products Laboratory, Madison, WI.

ICBO. 1970. *Uniform Building Code*. Vol 1, Sec 4202b. International Conference of Building Officials, Whittier, CA.

Imaizumi, Katsuyoshi. 1963. "Stability in Fire of Protected and Unprotected Glued Laminated Beams." *Transactions*. No 263. Chalmers University of Technology, Gotenburg, Sweden. pp 46-71.

LAFD. 1959. *Operation School Burning No 1*. Los Angeles Fire Department. National Fire Protection Association, Quincy, MA.

McCaffrey, B. J., Quintiere, J. G., and Harkelroad, M. F. 1980. "Estimating Room Temperatures and the Likelihood of Flashover Using Fire Data Correlations." *Fire Technology*. Vol 17, No 2.

McGuire, J. H. 1969. "Limited Safe Surface Temperatures for Combustible Materials." *Fire Technology*. Vol 5, No 3. pp 237-241.

McNaughton, G. C. 1944. *Ignition and Charring Temperatures of Wood*. Mimeo No R1464. Forest Products Laboratory, Madison, WI.

Mitchell, N. D. 1951. "New Light on Self-Ignition," *NFPA Quarterly*. Vol 45, No 2. pp 165-172.

NFPA. 1952. "Fire Gas Research Report." *NFPA Quarterly*, Vol 45, No 3.

NSF. 1973. "Answers to Burning Questions—Reducing the Hazards of Fire." *Mosaic*. National Science Foundation, Washington, DC.

Parker, W. J., and Long, M. E. 1972. "Development of a Heat Release Rate Calorimeter at NBS." *Ignition, Heat Release, and Noncombustibility of Materials*. ASTM STP 502. American Society for Testing Materials. Philadelphia, PA. pp 135-151.

NFPA Codes, Standards, Recommended Practices and Manuals. (See the latest *NFPA Codes and Standards Catalog* for availability of current editions of the following documents.)

NFPA 255, *Method of Test of Surface Burning Characteristics of Building Materials*.

NFPA 258, *Standard Research Test Method for Determining Smoke Generation of Solid Materials*.

Additional Readings

Amaro, A. J. et al., "Thermal Indices from Heat Release Rate Calorimetry," Speech presented at the Western States Section of the Combustion Institute, Oct. 1974.

Atreya, A., "Pyrolysis, Ignition, and Fire Spread on Horizontal Surfaces of Wood," *Ph.D. Thesis*, Harvard University, Cambridge, MA, May 1983.

Beall, F. C., *Specific Heat of Wood —Further Research Required to Obtain Meaningful Data*, Research Note FPL-0184. Forest Products Laboratory, Madison, WI.

Beyreis, J. R., Monsen, H. W., and Abbasi, A. F., "Properties of Wood Crib Flames," *Fire Technology*, Vol 7, No 2, May 1971, pp 145-155.

Browne, F. L., *Theories of the Combustion of Wood and Its Control*, Report No. 2136; Forest Products Laboratory, Madison, WI, 1958.

Chamberlain, David L, "Heat Release Rates of Lumber and Wood Products," *ASTM Special Technical Pub. 816*, American Society for Testing and Materials, Philadelphia, PA, 1982.

Chamberlain, D. L., *Heat Release Rate Properties of Wood-Based Materials*, NBSIR 82-2597, National Bureau of Standards, Washington, DC, 1983.

Fitzgerald, Warren E., "Quantification of Fires," *Journal of Fire and Flammability*, Vol. 9, 1978.

"Flammability of Cellulosic Materials," *Fire and Flammability Series*, Vol. 2, Technomic Publishing Company, Westport, CT, 1973.

Graf, S. H., *Ignition Temperatures of Various Papers, Woods and Fabrics*, Bulletin No. 26, Oregon State University, Corvallis, OR, Mar. 1949.

Gross, Daniel, and Robertson, A. F., "Self-Ignition Temperatures of Materials from Kinetic-Reaction Data," Research Paper 2909, *Journal of Research*, National Bureau of Standards, Vol. 61, No. 5, Nov. 1958.

Hadvig, S., *Charring of Wood in Building Fires*, Technical University of Denmark, Lyngby, Denmark, 1981.

Hall, Keith F., "Wood Pulp," *Scientific American*, Vol 230, No 4, April 1974, pp 52-62.

Ignition, Heat Release and Noncombustibility of Materials, STP 502, 1972, American Society for Testing and Materials, Philadelphia, PA.

Interior Finish and Fire Spread, National Fire Protection Association, Quincy, MA, 1977.

Kanury, A. M., "Ignition of Cellulosic Solids—A Review," *Fire Research Abstracts and Review*, Vol. 14, No. 1, 1972, pp 24-52.

Knudson, R. M., and Schniewind, A. P., "Performance of Structual Wood Members Exposed to Fire," Forest Products Journal, Vol. 25, No. 2, 1975, pp. 23-32.

Koch, P., *Utilization of the Southern Pines, Vol. 1*, U.S. Department of Agriculture Forest Service.

Kollmann, F. F. P., and Cote, W. A., Jr., *Principles of Wood Science and Solid Wood Technology*. Springer-Verlag, New York, 1968, pp. 240-257.

Koohyar, A. N., Welker, J. R., and Sliepcevich, C. M., "The Irradiation and Igniton of Wood by Flame," *Fire Technology*, Vol. 4, No. 4, Nov. 1968, pp. 284-291.

MacLean, J. D., "Thermal Conductivity of Wood," *Heating/Piping/Air Conditioning*, Vol 13, No. 6, 1941, pp. 380-391.

Magee, R. S., and McAlery, R. F., "The Mechanism of Flame Spread," *Journal of Fire and Flammability*, Vol. 2, Oct. 1971, p. 271.

R. R. McNeil, "Fire Impact on Wood: The Next Five Years," *Wood and Fiber*, Vol. 9. No. 1, pp. 2-12.

Moisture in Materials in Relation to Fire Tests, STP 385, 1965, American Society for Testing and Materials, Philadelphia, PA.

Parker, W. J., *Thermal Hardening Considerations Pertaining to Residential Areas*, USNRDL-TR-984, U.S. Naval Radiological Defense Laboratory, San Francisco, Feb. 1966.

Peterson, Alvin O., "Sound-Deadening Board Hazard," *Fire Journal*, Vol. 68, No. 4, July 1974, pp 100-102.

Quintiere, J., "A Simplified Theory for Generalizing Results from a Radiant Panel Rate of Flame Spread Apparatus," *Fire and Materials*, Vol. 5, No. 4, 1981, p. 52.

Schaffer, E. L., "Smoldering Initiation in Cellulosics Under Prolonged Low-level Heating," *Fire Technology*, Vol. 16, No. 1, 1980, pp. 22-28.

———, *Charring Rate of Selected Woods —Transverse to Grain*, Research Paper FPL 69, Forest Products Laboratory, Madison, WI, 1967.

———, *Review of Information Related to the Charring Rate of Wood*, Research Note FPL-0145, Forest Products Laboratory, Madison, WI, 1980.

Sensenig, D. L., *An Oxygen Consumption Technique for Determining the Contribution of Interior Wall Finishes to Room Fires*, NBS Technical Note 1128, National Bureau of Standards, Washington, DC, 1980.

Simms, D. L., "Experiments on the Ignition of Cellulosic Materials by Thermal Radiation," *Combustion and Flame*, Dec. 1961.

——— "On the Pilot Ignition of Wood by Radiation," *Combustion and Flame*, Sept. 1963.

Smith, Edwin E., "Release Rate Tests and Their Application," *Journal of Fire and Flammability*, Vol. 8, 1977.

Smith, P., and Thomas, P. H., "The Rate of Burning of Wood Cribs," *Fire Technology*, Vol. 6, No. 1, Feb. 1970, pp. 29-38.

"Smoke and Products of Combustion," *Fire and Flammability Series*, Vol. 2, Technomic Publishing Company, Westport, CT, 1973.

Stamm, A. J., *Wood and Cellulose Science*, Ronald Press Company, NY, 1964.

Steinhagen, P. A., *Thermal Conductive Properties of Wood: A Literature Review*, Research Note FPL-0184, Forest Products Laboratory, Madison, WI, 1977.

Theories of the Combustion of Wood and Its Control, Report No. 2136, Forest Products Laboratory, Madison, WI, 1958.

White, Robert H., and Schaffer, E. L., "Application of CMA Program to Wood Charring," *Fire Technology*, Vol. 14, No. 4, Nov. 1978, p. 279.

———, "Transient Moisture Gradient in Fire-Exposed Wood Slab," *Wood and Fiber*, May 1980.

———, *Thermal Characteristics of Thick Red Oak Flakeboard*, Research Paper FPL 407, Forest Products Laboratory, Madison, WI, 1981.

"Wood and Wood Products," *Fire Safety Aspects of Polymeric Materials*, National Materials Advisory Board, National Academy of Sciences, Technomic Publishing, Westport, CT, 1977.

Wood Fire Behavior and Fire Retardant Treatment, (A Review of the Literature), Canadian Wood Council, Ottawa, Nov. 1966.

Wood Handbook: Wood as an Engineering Material, U.S. Dept. of Agriculture, Forest Service, Handbook 72, 1974.

FIBERS AND TEXTILES

Revised by John F. Krasny and Stephen B. Sello

This chapter discusses fibers and textiles as commonly used materials, as well as their flammability, their end use, and the fire retardant treatments used on them. Also detailed in this chapter are special environments where textiles are used, and the testing methods employed in evaluating and creating standards for the use of fibers and textiles.

Much allied information with more specific discussion can be found elsewhere in this HANDBOOK. Section 2, "Fire Loss Information," discusses the NFPA Fire Incident Data Organization system and the USFA/FEMA-National Fire Incident Reporting System in detail. Section 7, Chapter 7 discusses the related topic of interior finish. Section 12, Chapters 3 and 4, explains oxygen enriched atmospheres and medical gases which potentially have significant bearing on fibers and textiles. Flammability of upholstered items, such as mattresses and upholstered furniture, are discussed in Section 12, Chapter 15.

Textiles are an important part of daily life. Clothing, chairs, carpets, and curtains are examples of the ways textiles are used. Almost all textiles are combustible. This fact combined with the prevalence of textile products in human activities explains the frequency of textile related fires and the many deaths and injuries that result.

Studies of fatal fires continue to show that textiles provide the primary fuel for the leading ignition scenarios that account for more fire deaths than any other combustible material. An analysis of 1980-82 fatal structure fires in the National Fire Incident Reporting System (NFIRS) operated by the U.S. Fire Administration identified that for fires where both the type of material and the form of material first ignited were known, 51 percent of the fatal incidents involved a fabric. A similar analysis of multiple death fires in the National Fire Protection Association's Fire Incident Data Organization* identified textiles as the primary fuel involved in 41 percent of the multiple death fires studies with known types and forms of material first ignited.†

Table 5-3A shows the rankings of the fabric products involved in fatal structure fires and the frequencies of fires and deaths reported in NFIRS for 1980 through 1982. Similar statistics are shown in Table 5-3B which lists fabric products by their ranking according to their involvement in multiple death fires reported in FIDO from 1971 through 1984. Table 5-3C profiles the types of fabrics first ignited. Included within the cotton category are the polyester-cotton blends.

TABLE 5-3A. Fatal Structure Fires Involving Textiles as the Primary Fuel*

Fabric Product	Percent of Fires	Percent of Deaths
Upholstered furniture	45	46
Bedding	34	32
Clothing on a person	8	7
Clothing not worn	5	5
Other fabric material	8	10

* Based on 1,285 fatal fires accounting for 1,545 civilian deaths.
Source: USFA/FEMA NFIRS Data 1980–1982

TEXTILES —GENERAL

Fibers are the basic components of all textiles. These are usually spun into yarns and then woven or knitted into a variety of fabrics, e.g., plain or patterned fabrics or pile, such as flannels, towels, artificial furs, and carpets. Fibers also can be processed directly into useful structures such as nonwoven fabrics or battings. Fabrics are usually scoured, bleached, and dyed and/or finished with chemi-

Mr. Krasny is a Textile Technologist in the Center for Fire Research, National Bureau of Standards, Gaithersburg, MD. Dr. Sello is a professor in the Department of Chemistry, Polytechnic Institute of New York.

*In operation since 1971, NFPA's computer based Fire Incident Data Organization (FIDO) system contains records of individual fires, collected with the primary objective of studying life safety. Consequently, technical data, e.g., details of fire ignition and spread, are collected on as many fatal fires as possible.

†Clothing ignition deaths are underrepresented in the FIDO system because these deaths often result from fires where the fire department was not called. NFPA estimates, from data collected in surveys of fire departments, that the correct proportion of clothing ignition deaths is 6.5 percent of all United States fire deaths.

TABLE 5-3B. Multiple Death Structure Fires Involving Textiles as the Primary Fuel*

Fabric Product	Percent of Fires	Percent of Deaths
Upholstered furniture	49	47
Bedding	27	27
Clothing not on a person	11	13
Other fabric material	12	13

* Based on 702 multiple death fires, (three or more deaths), accounting for 2,776 civilian deaths for the years 1971 through most of 1984.
Source: NFPA FIDO Data 1971–1984

TABLE 5-3C. Type of Fabric First Ignited in Fatal Structure Fires*

Fabric Type	Percent of Fires	Percent of Deaths
Cotton, rayon	25	22
Artificial fibers	16	16
Wool, wool mixture	1	1
Fabric type not reported	9	9

* Based on 2,534 fatal structure fires accounting for 3,277 civilian deaths, where the type and form of material first ignited was known. Percents in this data-set refer to only fabric related incidents.
Source: USFA/FEMA NFIRS Data 1980–1982.

cals which give them a stiffer or softer hand, durable press characteristics, luster, or flame or weather resistance.

Nonwoven fabrics are made from various fibers by gluing them together, either before they are fully hardened in the spinning process, or by means of latex adhesives. They are widely used in disposables and other areas; some are treated with flame retardant.

The ignitability, rate of flame spread and heat release, total heat produced, shrinkage, and ease of extinction of textiles depends on their fiber content, weight, and construction. The finish generally does not affect these properties, unless it is flame retardant.

Common textile fibers are listed in Table 5-3D, along with some of their burn hazard properties, major trade names, and end uses. The data have been collected from a variety of sources (MMFPA 1978; Gottlieb and Beck 1959; Einsele 1972.) The descriptions of burning behavior in this table and the discussions throughout this chapter are based on small scale tests and may be misleading when large fires or strong radiation are encountered. In such cases, the burning characteristics of fabrics may change. Even fabrics which are considered flame retardant may burn readily.

Natural fibers include cellulosic and protein fibers. Artificial fibers are produced allowing the emerging filaments to harden (coagulate). The resulting products can be used as "filament yarns," or can be cut into shorter lengths and spun into "staple yarns." The staple yarns can contain more than one type of fiber, e.g., blends of polyester and cotton. Many different textile properties can be attained by blending various fibers in this manner.

Natural Fibers

The most important cellulosic fiber is cotton. Cotton consists of more than 90 percent cellulose $(C_6H_{10}O_5)_x$. Other plant fibers, such as jute, flax (linen), hemp, and sisal also are basically cellulose in composition. Cotton and the other plant fibers are combustible; the ignition temperature of cotton fiber is 752°F (400°C). When ignited they produce heat, smoke, carbon dioxide, carbon monoxide, water, and numerous other compounds. The plant fibers char but do not melt.

Protein fibers, such as silk, wool, and other hair fibers derived from animals are chemically different from cotton because they consist of complex protein molecules containing high percentages of nitrogen as well as carbon, hydrogen, oxygen, and small amounts of sulfur. Wool supports combustion with difficulty, and, other factors being equal, is more difficult to ignite, burns more slowly, and is easier to extinguish than cotton.

Artificial Fibers

Artificial fibers can be divided into "regenerated" or "reconstituted" fibers and fully synthetic fibers. The major regenerated fiber in use today is viscose rayon. Rayon is made from dissolved cellulose and is similar to cotton in its charring and burning properties. Acetate is made by the reaction of cellulose and acetic acid; it melts before burning. It is more similar in its burning characteristics to the thermoplastic fibers (nylon, olefin, and polyester) than to cellulose.

The major fully synthetic fibers are acrylics (which char, intumesce, and burn when exposed to heat), and the thermoplastic fibers, such as nylon, olefin (including polypropylene and the less popular polyethylene fibers), and polyester. These consist of carbon, hydrogen, and oxygen atoms. Acrylics and nylon also contain nitrogen atoms and, like wool, produce hydrogen cyanide when burned. The thermoplastics do not generally char but shrink away from heat sources and melt, form holes near the heat source, and/or ablate. If such fabrics burn at all, the upward flame spread tends to be slow. Molten polymer runs downward, moving the flames sideward and downward, or forms drops which may or may not flame. Compared to fabrics containing cellulose, thermoplastics are more difficult to ignite, and the flames tend to cover smaller areas, travel more slowly, or self-extinguish.

There are, however, situations in which thermoplastics can burn readily. This occurs when they are stiffened by finishes or heavy dye application and lose their ability to evade flames and ablate. It also occurs in blends of thermoplastic and char forming fibers (even when the latter are flame retardant), e.g., in the popular polyester/cotton blends which burn much like 100 percent cotton than 100 percent polyester. A thermoplastic fiber fabric in contact with a char forming fabric, e.g., a polyester/cotton skirt in contact with a nylon slip (McCullough and Noel 1978), or a polyester curtain in contact with a foam lining, can burn readily. If a burning outer garment is not in contact with a nylon or polyester slip, the latter may not ignite but become limp, shrink, make close contact with, and possibly stick to the skin, causing injury even without burning. Finally, if an ablated thermoplastic fabric still flames when it falls, it can ignite garments or furnishings on a lower level.

However, with such exceptions, thermoplastics can be considered relatively safe with respect to small flame ignition, as compared to cellulosics and if used as single layer apparel. Thus, accidents in which children's nightwear was the first item to ignite have been essentially eliminated since the introduction of the *Children's Nightwear Standard* (CPSC 1974), and the subsequent change from cotton to primarily thermoplastic nightwear

TABLE 5-3D. Fire Hazard Properties of Common Textile Fibers

Fiber Designation	Temperatures, °C (°F) Decomp.	Melting	Ignition	Burning	Burning Behavior	Trade Names	End Uses
A. Natural Fibers							
Cellulosic: cotton, hemp, jute, linen, sisal, etc.	305–320 (580–610)	—	255–400 (490–750)	850 (1560)	char, burn, sometimes afterglow	—	apparel, furnishings, towels, cordage
Protein: wool, mohair cashmere, camel hair, etc.	230 (450)	—	570–600 (1060–1110)	940 (1720)	char, intumesce, burn less readily than cellulosics	—	apparel, blankets, carpets, furniture covers
B. Artificial Fibers							
Acetate	300 (570)	260 (500)	440–525 (820–1000)	960 (1760)	melts and burns	Ariloft, Avron, Celanese, Chromspun, Estron	apparel, lingerie, furnishings
Acrylic	285–310 (540–590)	— —	460–560 (860–1040)	850 (1560)	chars, intumesces, burns	Acrilan, Creslan, Orlon, Zefran	apparel, furnishings, carpets, blankets
Nylon	315–420 (600–790)	215–255 (420–490)	450–570 (840–1060)	875 (1600)	melts, ablates, burns	ACE, Anso, Antron, Blue "C", Caprolan, Cantrece, Celanese, Cordura, Courtaulds, Enkalon, Ultron	apparel, lingerie, furnishings, carpets, cordage, industrial uses
Olefin (polypropylene)	400 (750)	165 (330)	500–570 (930–1060)	840 (1540)	melts, ablates, burns	Herculon, Marquesa, Marvess, Patlon	knitted apparel, carpets, cordage, furniture covers, industrial uses
polyester	360–400 (680–750)	250–300 (480–570)	450–560 (840–1040)	700–725 (1290–1330)	melts, ablates, burns	A.E.C., Avlin, Blue "C", Dacron, Encron, Fortrel, Hofil, Kodel, Trevira	apparel, lingerie, furnishings, carpets, blankets, cordage, industrial uses
Rayon (viscose)	290 (550)	—	420 (790)	850 (1560)	chars, burns	Avril, Coloray, Enkron, Fibro, Xantel	apparel, lingerie, furnishings
Spandex	305–355 (581–671)	230–250 (446–482)	415 (780)	NA	melts, burns	Lycra, Vyrene	in apparel lingerie where stretch is desired
Triacetate	300 (580)	310 (590)	440–540 (820–1000)	880 1620	melts, burns	Arnel	apparel, lingerie

Furnishings: upholstery, bedding, curtains and drapes
NA—not available

(Bolieu 1977; Crikelair 1975), at no additional cost of the garments (Beckwith 1979). On the other hand, thermoplastics, even if they are hard to ignite, should not be used in garments which are intended to protect against heat, such as fire fighter or industrial uniforms, because of their tendency to shrink, melt, and stick when exposed to heat.

In addition to the burn behavior, Table 5-3D lists temperatures which have been cited for onset of decompositon, melting, ignition, and the burning temperature of the flames. The ranges are often quite wide for any one fiber because of the differences in methods for measuring such temperatures as well as the variety of polymer compositions used in each fiber group for various end uses. In the cases of nylon, there are two major types, Nylon 6 and Nylon 66, varying in the structure of the base molecule. For the ignition temperature of cotton, one finds two values in the literature, 490 and 750°F (255 and 400°C); it is possible that the lower value was obtained with aged cotton. (It is well known that the ignition temperature of wood decreases with aging, and that *may* hold for cotton as well.) The ignition temperature of blends tends to be similar to that of the component with the lower ignition temperature. Thus, measured by one method, the ignition temperature for 14 different fiber types ranged from 752°F (400°C) for cotton to 1040°F (560°C) for wool; those with 50/50 blends of each of these fibers with cotton, from 770 to 860°F (410 to 460°C) (Gottlieb and Beck 1959).

Noncombustible Textiles

Noncombustible fabrics include those made entirely from inorganic materials. Table 5-3E lists noncombustible fibers. Glass fabrics listed by testing laboratories for use as draperies are woven from uncoated glass yarns which do not burn or propagate flame. If a sufficient quantity of combustible coating or decorative material is applied to glass fabrics or to other noncombustible fabrics, however,

TABLE 5-3E. Noncombustible Fibers

Glass	Metal
Beta Fiber	Stainless steel
E-Glass	Super alloy
Quartz	Refractory-Whiskers
Carbonaceous Residue	Alumina
Carbon	Zirconia
Graphite	Boron

the fabrics will support continued flaming. The inherent brittleness of glass and the other fibers listed in Table 5-3E eliminates them from use in clothing fabrics except for heat protective garments. They are used as reinforcement for plastics, and in the case of the relatively low priced glass fibers, for curtains and draperies.

High Temperature and Flame Retardant Textiles

Several of the fibers which resist high temperatures and are used in textiles are listed in Tables 5-3F (AIRC 1984; Carborundum 1974) and 5-3G along with their ignition temperatures, Limited Oxygen Index (LOI) and some other physical properties.

continue to burn after the ignition source is removed. The higher the LOI of a fabric, the greater the probability that it will cease to burn once the ignition source is removed. Fabrics with high ignition temperatures and high LOIs can be expected to be safe for clothing and furnishings since they will not ignite readily and will not continue to burn after an accidental ignition source is removed.

As previously discussed, many synthetic fabrics shrink when exposed to temperatures approaching their melting or decomposition temperatures. When shrinkage brings the fabric into contact with the skin, the insulating layer of air is eliminated and the amount of heat transferred to the skin is increased significantly. Thermal shrinkage is a particularly important property to consider when selecting fabric for use in protective clothing. Another important property of fabrics used in protective clothing is the ability to maintain integrity during a fire. Fabrics that weaken by charring split apart and fall away or burn completely, leaving the skin directly exposed to heat.

FLAME RETARDANT TREATMENTS

The flame retardant treatment of theater scenery, curtains, and draperies in places of public assembly is commonly required by local fire authorities. Treated fabrics are

TABLE 5-3F. High Temperature, Flame and Chemical Resistant Fibers

	Ryton	*Stilan*	*Nomex*[4,6]	*Kevlar*[4,6]	PBI[6]	2080 Polymide	Kynol[5,6] Phenolic
Softening point							
°C	285	>315					
°F	545	>600		Decomposes[2]			NA
Upper Use Temp. (50% tenacity)							
°C	200	260	230	260	>315	>230	
°F	400	500	450	500	>600	>550	NA
Strength (g/denier)	2–4	4–6	3–5	18–20	3–5	3–4	1.5
Elongation (%)	20–40	20–40	20–40	3–4	25–40	30–40	35
Chemical Resistance							
hot acid	VG	G	P	P	VG	VG	P
hot base	VG	VG	P	F	G	P	P
hydrolysis	VG	VG	F	F–P	VG	VG	NA
solvents	VG	VG	G	VG	VG	VG	VG
Dry heat resistance	G	VG	G	G	VG	VG	VG
Flammability (LOI)[3]	34	26	30	29	38–49	37	29–31
Smoke density*	NA	NA	5	5	3	7	low

*1 Smoke density is a relative scale where, for example, wool = 46 and cotton = 26.
2 Decomposes rather than melts.
3 LOI—Limiting Oxygen Index, i.e., the percent oxygen in an oxygen/nitrogen volume mixture that just sustains combustion. Wool = 28; cotton = 19.
4 Generic name: aramid
5 Novoloid
6 Presently used in structural firefighter's protective gear.
VG = very good; G = good; F = fair; P = poor; NA = not available

Table with minor additions reproduced by courtesy of Albany International Research Co. (formerly FRL), Dedham, MA.

The Limiting Oxygen Index (LOI) is a way to measure the tendency of a fabric, once ignited, to continue burning after the ignition source is removed. LOI is defined as the minimum volume concentration of oxygen in a mixture of oxygen and nitrogen that will just support sustained combustion of the material when ignited in a vertical position at its uppermost edge (ASTM 1963). Cotton fabrics, for example, have a LOI of 19 percent which means that if the oxygen is reduced below 19 percent the cotton will not

used in hotels, hospitals, and other occupancies in the interest of preservation of lives and property.

Many businesses specialize in the flame retardant treatment of theater scenery, draperies, and other fabrics, using standard chemicals. Such organizations are listed in the yellow pages of a telephone directory, under "flame" or "fire resistance." However, there are some people in this field who do not treat fabrics properly or who use relatively ineffective chemicals. It is therefore advisable to

TABLE 5-3G. Flame Retardant (FR) Fibers

Fiber Type	LOI	Ignit. Temp. °C/°F	Burning Behavior	Trade Names
PVC	37	500/930	chars, shrinks at low temps.	*Rhovyl, Leavyl, Teviron*
FR Rayon	31	—	chars	*PFR*
Matrix	29–32	500-600/ 930–1100	chars at 220/430, shrinks	Kohjin, Cordelan
Modacrylic	27–30	315/600	chars, shrinks at .140–180/280–360	*SCF, Verel, Dynel*
FR Polyester	28		melts at 250/480	*Trevira* 271, *Heim*
FR wool	32–34	570–600/ 1060–1100	chars, intumesces	—
FR cotton	28–32	290/450	chars	—

deal only with businesses of known reliability, or, if dealing with an unknown concern, to have treated fabrics tested for adequacy of treatment.

The effects of chemical treatments in reducing the flammability of combustible fabrics are varied and complex, and all phases are not fully understood. There are five different ways in which chemicals or mixtures retard spread of flame and afterglow: (1) they generate noncombustible gases that tend to exclude oxygen from the burning surface, (2) radicals or molecules from degradation of the flame retardant chemical react endothermically and interfere with chain reactions in the flame, (3) the flame retardant chemical decomposes endothermically, (4) a nonvolatile char or liquid is formed by the chemical, which reduces the amount of oxygen and heat that can reach the fabric, and (5) finely divided particles are formed that change the combustion reactions. Usually a flame retardant chemical or mixture affects flammability in more than one of these ways. A recent review of flame retardant treatments for textiles can be found in F. B. Gordon's article in *Fire Safety Journal* (Gordon 1981). Flame retardant treatments can increase the LOI of cotton from 19 to 27 or 32, and of wool from 25 to 32 or 34.

Hundreds of different chemicals have been suggested as flame retardants for textiles. It must be emphasized that no one system works for all fibers, as can be seen from Table 5-3H. In addition, some chemical systems change textile properties, such as resistance to sunlight, absorption, color, and flexibility; or lose some of their effectiveness when exposed to high temperatures or repeated launderings. Much progress has been made in overcoming most of these undesirable effects. Loss of strength, for example, has become a minor problem in correctly treated flame retardant cotton fabrics. However, fear that some flame retardants may be skin irritants or carcinogens, and the often high cost of treatment, are still deterrents to wider use.

It must be emphasized that treatments which protect against ignition of textiles by small flames do not necessarily impart resistance to smoldering ignition by cigarettes. A case in point are flame retardant treated cotton fabrics which generally lower the charring temperature of cotton and often decrease, rather than increase, the cigarette ignition resistance.

TABLE 5-3H. Flame Retardant Chemicals for Textiles

Fiber	Flame Retardant Chemical
Cotton	Tetrakis (hydroxy methyl) phosphonium salt insolubilized with ammonia gas
Cotton-rayon (disposable, nondurable finish)	Diammonium phosphate Ammonium sulfamate Boron compounds
Rayon (modified fiber)	Hexapropoxy phosphazene
Polyester (modified fiber)	Oligomeric phosphonate
Polyester, acetate, nylon	decabromo-diphenyl ether (DBDPO) and antimony oxide
Nylon (nondurable finish)	Thiourea
Wool	Ti, Zr compounds Dibromo-terephthalic acid
Modacrylics (modified fibers)	Vinylchloride, vinylidene chloride, vinylbromide as co-monomer

RELATIONSHIP OF FLAMMABILITY AND END USE OF TEXTILES

So far this discussion of textile flammability has been confined primarily to the textiles themselves. The end use of the textile—ordinary or heat protective clothing, curtains, carpets, etc.—is very important when evaluating the flammability of a textile.

Clothing

Congress passed the *Flammable Fabrics Act* in 1953 (CPSC 1967), after a rash of serious fire accidents in which rayon pile fabrics made into children's cowboy chaps or sweaters were involved. All wearing apparel sold since 1953 has to pass a test described in the CPSC *Standard for the Flammability of Clothing Textiles* (CPSC 1953), which will be described below under fabric flammability tests. It excluded the rayon pile fabrics, and such other extremely flammable fabrics as very high pile cotton flannels, stiffened nylon nets, and very light fabrics. However, all normal apparel fabrics must pass this test.

Several studies indicated that additional protection of young and old populations would be desirable. For example, in a study of 463 fire fatalities which occurred during 1972 through 1977 in the State of Maryland (Berl and Halpin 1979), deaths from clothing ignitions comprised 5.5 percent of the fatalities. Of those victims, 21.6 percent were 9 years old and younger and 20.5 percent were 60 years and older. Persons aged 50 years and older were more likely to become fire casualties. The number of fire deaths in this age group was approximately twice the predicted number, based on 1970 census data for Maryland. Factors which contributed to the increased proportion of fire fatalities at both ends of the age spectrum were inability to rapidly and successfully cope with the fire situation and lowered tolerance to toxic combustion products. This applied especially to the older population, where increased cardiopulmonary weakness contributed to lower tolerance to toxic exposures.

In addition, it was shown that many victims of garment fires wore nightwear at the time of the accident. Consequently, a children's nightwear standard was promulgated for sizes 0 to 6x in 1971, followed by a standard for sizes 7 to 14 three years later (CPSC 1974). At the time of promulgation, this standard strained the state of the art, but thermoplastic and flame retardant garments were soon produced which passed this rather severe test. The results have been unexpectedly beneficial.

The style or type of garment is an important factor in evaluating its fire hazard. For example, a full length gown containing several yards of flammable fabric could be a death trap whereas a tightly fitting blouse or shirt made of the same material might be far less dangerous. Not only would the gown present much more fuel for burning, but its size and shape would make it far more vulnerable to contact with typical ignition sources. Flared skirts, large bows, and full sleeved kimonos are other examples of easily ignitable garments. The availability of oxygen on both sides of loosely fitting garments accelerates flame spread.

Regardless of flammability, clothing cannot catch fire unless the fabric comes into contact with a source of ignition through some act of carelessness, negligence, or ignorance. In a study of 4,493 one and two fatality fires which occurred from 1972 through 1978 and which resulted in 5,405 deaths, the most frequent sources of heat of ignition in clothing fires were cooking equipment, matches, lighters, and candles. These deaths most often resulted from misuse of the heat of ignition and misuse of the material ignited.

Curtains and Draperies

Curtains and draperies can ignite from such ignition sources as electrical malfunction or waste paper basket fires. Flames rapidly spread to walls and ceilings. It has been shown that large differences exist among the potential of various domestic curtain materials to ignite wall covering materials and produce heat and smoke (Moore 1978). Curtains in public occupancies are regulated by local authorities, and conforming materials do not seem to present major problems during the initial stages of a fire. The test most prescribed by regulatory authorities for curtains and draperies is NFPA 701, *Standard Methods of Fire Tests for Flame-Resistant Textiles and Films.*

Carpets

Until 1960 there was very little fire experience to indicate that carpets were a significant factor in fires. Consequently, floor coverings were customarily exempted from coverage in building codes regulating life safety from fire. Since 1960, although there was a small number of fires reported in which carpets spread fire, those that have been reported indicate that carpets can be the prime material involved in the development or spread of a fire. In 1970 a fire in a California dwelling was first seen by an occupant when it was a few feet in diameter on carpeting in front of the fireplace (Fire Journal 1970a, 1970b). As the occupant applied water with a garden hose to the burning carpet, the fire spread rapidly on the carpet and enveloped the entire room within ten minutes after discovery, including furniture and wood paneling. (The additional readings section of this chapter cites many *Fire Journal* articles which further discuss such incidents.) Because of such experiences, carpet flammability is currently being regulated at the federal, state, and municipal levels.

Among the many factors in carpet construction which may affect ease of ignition and ability to spread fire are: fiber type, area unit weight density (or, loops per square inch), pile height, backing material, padding, type of adhesive, single versus double back, and type of dye. (Carpet flammability tests for residential and special occupancies are discussed later in this chapter.)

Heat Protective Clothing

Heat protective clothing should not burn, melt, or disintegrate on exposure to heat or flame. It should be an effective thermal barrier, should not shrink excessively when heated, and should be durable and comfortable. Detailed reviews of research and development on heat protective garments and methods to measure their properties will be found in other publications (Brewster and Barker 1983; ASTM 1984). Abbott and Schulman have made an analysis of exposure conditions encountered by fire fighters, and protective materials for routine and emergency conditions (Abbott and Schulman 1976).

For structural fire fighting, turnout coats are generally worn in the U.S. Those coats consist of a water and abrasion resistant outer shell, a vapor barrier to prevent penetration of water and chemicals, and an inner liner to provide further insulation against heat and cold. The common materials for the outer shell are Nomex® and flame retardant cotton, and, more recently, blends of PBI and Kevlar®, glass and Kevlar, Nomex and Kevlar, etc. Among the vapor barriers used are neoprene coated polyester/cotton fabrics or Goretex® fabrics which are water vapor permeable but do not permit liquid water permeation. Nomex needled or batting materials, flame retardant cotton or wool fabrics, etc., are popular inner liners. Similar materials are used in bunker pants, gloves, and hoods used to protect the ears and neck. European fire departments frequently use wool outer shells in their turnout coats; wool tends to char at lower temperatures than Nomex.

TEXTILE FLAMMABILITY TESTS

The severity of the hazard of a fabric in a particular type of product, such as a dress, curtain, or carpet depends greatly upon ease of ignition of the product, the speed with which flame spreads, the amount of heat generated and the rate at which it is released, susceptibility to melting, and the quantity and composition of the smoke and gaseous products of combustion. No individual test method gives a complete evaluation of the hazard and there are no nationally recognized test methods to measure some of the hazards. Many of the more important test methods for textile products have been summarized (Hilado 1973: Krasny 1982; Troitzsch 1983). With the exception of the test methods for mattresses and upholstered furniture, those in common use in the United States are described below.

Clothing

The Federal Flammable Fabrics Act was signed into law in 1953 and amended in 1967 (CPSC 1967). The Act prohibits the sale of all wearing apparel and fabrics subject to the Act which are classified "highly flammable" when tested according to Part 1610, which was formerly designated Commercial Standard CS 191-53 (CPSC 1953). The specified test is designed to measure ease of ignition and rate of flame spread. It does this by impinging a small gas flame for 1 sec against the lower part of the upper face of a 2 by 6 in. (51 by 152 mm) specimen that is suspended at a 45 degree angle from the horizontal in a test chamber. If the fabric is ignited by the 1 sec exposure, the time required for the flame to spread 5 in. (127 mm) on the fabric is used to classify fabrics. This test has been successful in excluding dangerously flammable garments from interstate commerce. It does not discriminate among other apparel fabrics, even though large differences exist between them. For example, cotton/polyester terry fabrics and cotton flannels that are cut into gowns, robes, and pajamas gave considerably higher burn injury estimates (percentage of body surface burned weighted by depth of burn) than estimates obtained with almost 60 other common apparel fabrics when burned on an instrumented mannequin (Bercaw 1977). The same fabrics ranked high on the flammability scale when tested in twelve laboratory fabric flammability tests, but easily passed the Part 1610 requirements. One hundred percent nylon and polyester fabrics also passed the test, as well as some blends containing wool or modacrylic fibers, which produced very low injury estimates on the mannequin and low flammability results in the other laboratory tests.

The test method in NFPA 702, *Standard for the Classification of the Flammability of Wearing Apparel*, is the same as that in Part 1610 with two important exceptions. The Committee that drafted the NFPA wearing apparel standard had noted that many fabrics, particularly those without raised fiber surfaces, were not ignited when the top surface was touched by a small test flame for 1 sec. Those fabrics could not be evaluated for rate of flame spread. Furthermore, the Committee concluded that the lower edge of a fabric should be easier to ignite than the surface. Consequently, the test flame in NFPA 702 is brought in contact with the lower edge of all samples and, in the case of those without raised fiber surface, is held there until ignition occurs. Many fabrics which do not ignite in Part 1610 ignite in the NFPA 702 test (Weaver 1976).

As mentioned earlier, more stringent standards were introduced for children's sleepwear in the early 1970s; Parts 1615 and 1616 (CPSC 1974) are much more restrictive than Part 1610. Fabric samples 10 by 3½ in. (254 by 89 mm) are suspended vertically in a cabinet and subjected to a test flame along the bottom edge. A fabric passes the test if the average char length of five specimens does not exceed 7 in. (178 mm), and no specimen has a char length of 10 in. (254 mm). The barrel of the test flame burner is ¾ in. (19 mm) below the lower edge of the sample and the test flame is adjusted to extend 1½ in. (38 mm). The sample is exposed for 3 sec. Char length is the distance from the fabric's lower edge to the end of the void or tear in the charred, burned, or damaged area. The tear is made by lifting the sample by one of the corners near the lower edge after weights had been attached to the opposite corner near the lower edge. Residual flame time is calculated as the time from the removal of the burner until final extinction of flame, smoldering, and afterglow. The small scale test method in NFPA 701, *Standard Methods of Fire Tests for Flame Resistant Textiles and Films*, is quite similar to the federal regulations.

The U.S. thus has two apparel flammability standards—the children's sleepwear standard which has greatly reduced the incidence of accidents but restricted the choice of fabrics; and the general apparel standard which has eliminated inordinately flammable, limited use fabrics, but does not significantly affect fabric choice. However, it permits many highly flammable fabrics to be used without regard to garment configuration. Attempts to further categorize fabric flammability hazards have been made; some have based such categorization on flame spread rate measurements (Weaver 1976; Nielsen and Richards 1969). Flame spread rate measured on remainders of garments involved in fire accidents, however, did not correlate with the size of the victim's burn injury (Vickers et al 1973). Researchers have developed methods for measuring the rate of heat release from burning fabrics (Miller 1976; Miller and Meiser 1978; Braun et al 1976). The latter paper suggested that fabrics with a low rate of heat release could be used in all garments; those with short ignition times and higher rate of heat release could be used only in tightly fitting garments; and finally suggested the elimination of fabrics with short ignition time (½ sec or less) and high heat release rate.

Heat Protective Clothing

NFPA 1971, *Standard on Protective Clothing for Structural Fire Fighters*, employs three methods to evaluate the firesafety of turnout coats:

1. An ignition test: Vertical Test Method 5903.2, Flame Resistance of Cloth (GSA 1971) which is quite similar to the Children's Sleepwear Test except that it uses a different gas, with a flame temperature of 1550°F (843°C), and a 12 sec ignition time. Maximum char length is 4 in. (102 mm), after a flame time of 2 sec.
2. A heat stability test: The fabrics used in the ensemble must not char, separate, or melt when exposed in a forced air oven at 482°F (250°C) for 5 min.
3. A heat protection test: The Thermal Protection Performance (TPP) test, which is described below.

In recent years, a test has been developed which measures actual heat transfer through the ensemble and relates it to burn injury potential, rather than relying on the ensemble thickness (Shalev and Barker 1984). A horizontal specimen cut from the ensemble is placed over a heat source delivering 2 (cal/cm^2)/sec (84 kW/m^2 or 7.4 Btu/sq ft per sec). The heat source consists of two gas flames and nine quartz heaters adjusted to obtain a 50:50 ratio of convective and radiative heat. The time until an incipient second degree burn to skin would occur in contact with the inner layer is measured; this is based on work correlating the rate of heat and the total heat delivered to skin with burn injury severity (Stoll and Chianta 1969). This time measure, multiplied by two [because of the exposure to (2 cal/cm^2)/sec] is called the thermal protective performance (TPP). This test probably has replaced the ensemble thickness requirement in NFPA 1971. A minimum TPP rating of 35, well within the state of the art, has been proposed as the acceptance criterion.

The Health and Safety Division of the International Association of Fire Fighters has written guidelines for procurement of fire fighter protective clothing, citing, among other items of interest, OSHA (Occupational Safety and Health Administration) standards and military procurement specifications as of 1983 (Smith et al 1983).

Other methods for evaluating heat protective garments include ignition temperature measured by bringing fabric specimens into contact with Calrod heaters with adjustable temperature (Kaswell 1972), and the previously discussed LOI determination (ASTM 1963). In addition, another TPP test like the one described above, but using a 84 kW/m^2 gas flame rather than a combination of flame and radiant heat, is listed by the American Society for Testing and Materials (ASTM 1982). Fabrics exposed to other heat sources on one side and in contact with artificial skin or other thermosensors can be ranked according to protective value (ASTM 1982; Schoppe et al 1982; Braun et al 1980). Tests for the tendency of molten metals to adhere to protective clothing, for thermal protective performance of materials when in contact with hot surfaces, and for resistance to penetration by chemicals have been developed or are under development.

Carpets and Rugs

Carpets and rugs are currently regulated in the U.S. by Federal regulations (CPSC 1970). Separate parts of the regulations apply to particular sizes. In testing these carpets and rugs a 9 by 9 in. (229 by 229 mm) sample is placed on the floor of a test chamber, and a flat steel plate with a 8 in. (203 mm) diameter hole in the center is placed on the sample. A methanamine tablet is placed in the center of the hole and ignited. The sample passes the test if the charred portion does not extend beyond 1 in. (25.4 mm) of the edge of the hole at any point. Seven of eight specimens of other than small carpets and rugs must pass if carpets made of the tested material are to be sold. In the case of small carpets and rugs, those that do not pass the test may be sold if they carry a permanent label stating that they failed to pass and should not be used near an ignition source.

Hospitals and other institutions receiving federal financial aid are required to comply with flammability limits for carpets and other floor coverings used on patient occupied areas and exit ways. The flame spread rating of floor coverings must not exceed 75, as determined by the ASTM *Test for Surface Burning Characteristics of Building Materials* (ASTM 1970). This test is similar to NFPA 255, *Method of Test of Surface Burning Characteristics of Building Materials.*

The Appendix of NFPA 101, *Life Safety Code®*, suggests that where limitations are to be placed on the flammability of floor coverings, the test method in NFPA 255 is to be used. NFPA 255 is commonly referred to as the Steiner Tunnel Test. Objections have been raised regarding its use to evaluate floor coverings because positioning the test specimen on top of the tunnel bears no resemblance to its position when in place in a building. The method is also criticized because it may result in delamination, melting, and dripping of test specimens which removes combustible material from the reach of the burner. The Canadian carpet test places the specimen on the floor of a tunnel (CGSB 1972).

In an effort to more correctly assess the flame spread hazard of flooring materials, the concept of critical radiant flux was developed. In their work on model corridor fire tests, the National Bureau of Standards found that the radiant energy impacting on a floor covering had a significant influence on the propagation of flame across the floor covering. This work led to the development of the flooring radiant panel test.

The Flooring Radiant Panel Test (FRPT) as embodied in NFPA 253, *Standard Method of Test for Critical Radiant Flux of Floor Covering Systems Using a Radiant Heat Energy Source*, subjects a horizontal floor covering specimen to a radiant energy flux from a gas fired panel at a 30° angle. The radiant energy flux decreases along the length of the specimen according to a standard energy flux vs distance profile. A pilot flame ignites the specimen at the end with the highest radiant energy flux. The flame front is allowed to travel until it is no longer reinforced by the radiant energy flux and extinguishes. The heat flux at this point of the specimen is called the "critical radiant flux." For corridors and exitways in such occupancies as hospitals and nursing homes, a minimum of 0.45 W/cm^2 is recommended, except for corridors and exitways in other occupancies (except one and two family houses—where the recommendation is 0.22 W/cm^2), (Benjamin and Davis 1978). The critical heat flux of carpets which are believed to have aggrevated the initial fire conditions, e.g., the Baptist Tower fire, was found to be 0.1 W/cm^2 (Willey 1973). However, the above recommendations are well within the state of the art of carpet manufacture.

Textiles in Aircraft

The flammability of textiles in the crew and passenger compartments of aircraft is regulated by the Federal Aviation Administration. The Vertical Test Method 5903 is used (GSA 1971); average char length for three specimens must not exceed 8 in. (203 mm), afterburn time must be less than 15 sec, and the time that drippings may continue to flame must be 5 sec. In addition, the FAA has recently specified a new test for passenger seats (FAA 1983). For the purpose of this test, a unique oil burner which has the following specifications is used: (1) kerosene consumption, 2 gal per hr. (7.57 L/h); (2) nozzle diameter, 6 in. (152 mm); distance of nozzle from side of seat to be tested, 4 in. (102 mm); (4) heat delivery to seat specimen, 10 Btu per sq ft per sec (115 kW/m^2); (5) temperature at specimen, 1850°F (1010°C). At least 3 seats must be tested. For at least half of the number tested, the flame spread must not reach the side of the cushion opposite the burner, the average weight loss must not exceed 10 percent, and there must be no flaming melt drip below the cushion. Among the seat assemblies which pass this test are those which contain the widely used wool/nylon fabric; an interlayer consisting of aluminized Nomex fabric which prevents pyrolysis gases from venting towards the flames (vents are provided at the back of the seats where flames are not expected to reach in a survivable accident); and polyurethane foam. An interliner made from Vonar neoprene type foam sheet also makes it possible to pass this test, but the lighter aluminized Nomex fabric is more cost effective in this application. Other interliner materials are under development.

Textiles in Motor Vehicles

Materials used in the occupant compartments of passenger cars, multipurpose passenger vehicles, trucks, and buses in the U.S. must not burn or transmit a flame front across the surface at a rate of more than 4 in. (102 mm) per min when tested according to the Motor Vehicle Safety Standard (Department of Transportation). In that standard, a specimen is suspended in a horizontal position in a test chamber, and one end is ignited by a Bunsen burner flame. The specimen is held in a U-shaped clamp and the assembly is suspended so that the top of the barrel of the burner is ¾ in. (17 mm) below the exposed edge of the specimen. The burner flame is adjusted to 1½ in. (38 mm) high and the flame temperature to that of a natural gas flame. The specimen is exposed to the test flame for 15 sec. Timing begins when the flame from the burning specimen reaches a point 1½ in. (38 mm) from the open end of the specimen and timing ends when the flame has progressed to 10 in. (254 mm) or self extinguishes. Burn rate must not exceed 4 in. (102 mm) per min.

Tents, Tarpaulins, Air supported Structures

NFPA 102, *Standard for Assembly Seating Tents, and Air-supported Structures*, requires that the textiles used in these structures meet the appropriate flame resistance requirements when tested by NFPA 701. Covered by this standard are: (1) tents and air supported structures used for assembly; (2) tents and air supported structures in which animals are stabled; (3) tents and air supported structures located in a portion of the premises used by the public; (4) tents and air supported structures in places of assembly in or about which fuel burning devices are located; and (5) all tarpaulins used in connection with (1) through (4).

The small scale test in NFPA 701 is similar to the previously described test that is used to evaluate children's sleepwear (CPSC 1974). The large scale test in the NFPA Standard uses specimens 7 ft (2.135 m) long and 5 or 25 in. (0.127 or 0.645 m) wide, depending upon whether the specimen is to be tested as a flat sheet or is to be hung in folds. The flat specimen is suspended vertically in an open ended metal stack and its lower edge is exposed to a 11 in. (280 mm) high flame from a Bunsen burner, the top of whose barrel is 4 in. (102 mm) below the bottom of the specimen. A folded specimen is tested with its folds ½ in. (102 mm) apart in the same metal stack and with the same burner arrangement. For both single sheet and folded specimen, the test flame is applied for 2 min, then withdrawn, and the duration of flaming combustion recorded. Length of char—the distance from the top of the test flame to the top of the charred area resulting from spread of flame or afterglow—is also recorded. To pass the test, duration of flaming combustion of single sheets cannot exceed 2 sec and the char length cannot exceed 10 in. (254 mm). For specimens in folds, maximum duration of flaming permitted is 2 sec, and maximum char length is 35 in. (840 mm), but afterglow may spread in the folds.

In addition, the Canvas Products Association International has developed CPAI-84, *A Specification for Flame Resistant Materials Used in Camping Tentage* (CPAI 1984). The test for wall and top material is essentially the above described 5903.2 test for flooring, The Methanamine Tablet Carpet Test. Maximum char length for walls and tops varies from 4.5 in. (114 mm) for light fabrics to 9 in. (228 mm) for heavy fabrics. For floors, the Part 1630 criteria apply (CPSC 1970). Provisions for leaching, weathering, and simulated sunlight exposure are part of this voluntary standard. Conforming tents carry the CPAI label.

Curtains and Draperies

Curtains and draperies for public occupancies are generally regulated by local or state fire authorities, and the most widely used test is described in NFPA 701. The small scale part of this method excludes certain thermoplastic fabrics which develop unacceptable char length because in the specimen frames, they cannot shrink freely away from the flames and ablate as they can in real life situations. The American Society for Testing and Materials' *Standard Test Method of Apparel Fabrics by Semi-Restraint Methods* (ASTM 1959), has been suggested to replace the small scale NFPA 701. In this test, the 15 by 6 in. (381 by 152 mm) specimens are attached to a bar on top of the test apparatus, and are somewhat restrained from movement by chains which loosely connect the bottom corners of the specimens to fixed points. With the specimen frames removed, more thermoplastic materials pass the test. As in the present version of NFPA 701, materials which fail the small test can still be qualified if they pass the large test.

TEXTILES IN SPECIAL ENVIRONMENTS

Certain ambient atmospheric conditions influence the behavior of treated textiles in fire situations. They are: (1) oxygen enriched atmospheres; (2) compressed air atmospheres; (3) atmospheres where flammable inhalation anesthetizing agents may be present; and (4) exposure to weather.

Action of Treated Textiles in Oxygen Enriched Air

Some industrial processes require work in atmospheres enriched above the normal oxygen content. Such atmospheres present a high fire hazard, particularly to the clothing worn by the workers. The effect of flame retardant treatments on the combustibility of the fabrics has been studied (Turner and Segal 1965).

An indication of the resistance to burning of flame retardant textiles in oxygen enriched atmospheres may be inferred from the Limiting Oxygen Indices in Table 5-3G.

The Textile Fibers Department of the Du Pont Company has conducted research on their Nomex high temperature resistant nylon fiber in both high pressure air environments and oxygen enriched atmospheres at greater than normal pressures. Nomex based fabric was found to retain its flame retardant qualities under air pressure of four atmospheres (59 psig or 405 kPa). At this pressure and in higher oxygen concentrations (40 percent and greater), Nomex did support combustion, but at a substantially slower rate than did untreated, readily combustible fabrics. NASA has further improved the state of the art in this area, and uses materials such as PBI, and flame retardant cotton, etc., in the oxygen rich atmospheres.

Fabrics used for hospital operating room garments, or in other atmospheres with flammable gas contamination, should be electrically conductive to prevent accumulation of static electricity. Such accumulations, if discharged to ground, could provide a spark of sufficient energy to ignite the gas.

Weather Exposed Textiles

Flame retardant treatments for tents, awnings, tarpaulins, and other fabrics exposed to the weather must not leach out over a period of time. If water soluble chemicals are used, the treatments must be renewed at frequent intervals. Both the NFPA 701 and CPAI tests for tentage specify methods for determination of the effects of weathering.

A wide variety of chemicals may be used for treating fabrics exposed to the weather. Many of the formulas used include chlorinated paraffin, chlorinated synthetic resins, or chlorinated rubber, in combination with various water insoluble metallic salts, plasticizers, stabilizers, synthetic

mixtures are not water soluble and are used with hydrocarbon solvents, or are suspended in aqueous emulsions. The effective application of such treatments calls for techniques and equipment ordinarily available only for factory processing or to businesses specializing in flame retardant applications. Application of paint to these treated fabrics may lessen their flame resistance.

Bibliography

References Cited

Abbott, N. J., and Schulman S. 1976. "Protection From Fire: Nonflammable Fabrics and Coatings." *Journal of Coated Fabrics*. Vol 6. July 1976. pp 48-64.

AIRC. 1984. *Chemical, High Temperature and Fire Resistant Fibers*. Albany International Research Co. (formerly FRL). Newsletter 11/1. Spring 1984. Dedham, MA.

ASTM. 1984. International ASTM Symposium on the Performance of Protective Clothing. *ASTM Special Technical Publication* (in press). ASTM. Philadelphia, PA.

ASTM. 1982. *TPP of Materials for Clothing by Open Flame Method*. ASTM D4108. ASTM, Philadelphia, PA.

ASTM. 1970. *Test for Surface Burning Characteristics of Building Materials*. ASTM E84-70. ASTM, Philadelphia, PA.

ASTM. 1963. *Method for Measuring the Minimum Oxygen Concentration to Support Candle-Like Combustion of Plastics (Oxygen Index)*. ASTM D2863. ASTM. Philadelphia, PA.

ASTM. 1959. *Standard Test Method for Flammability of Apparel Fabrics by Semi-Restraint Method*. ASTM D3659. ASTM, Philadelphia, PA.

ASTM. *Committee F-23 on Protective Clothing*. ASTM, Philadelphia, PA.

Beckwith, O. P. 1979. "Current Status of Children's Sleepwear Manufacturing and Marketing." *Proceedings, Information Council on Fabric Flammability*, Atlanta, GA. pp 114-122.

Benjamin, I. A., and Davis, S. 1978. "Flammability Testing of Carpet." *NBSIR 78-1436*. National Bureau of Standards. Washington, DC.

Bercaw, J. R. 1977. "Cooperative Program on General Apparel Flammability: Status Report." *Proceedings, Information Council on Fabric Flammability*, NY. pp 175-190.

Berl, W. G., and Halpin, B. M. 1979. "Human Fatalities from Unwanted Fires." *Fire Journal*. Vol 73, No 5. Sept 1979. pp 105-123.

Bolieu, S. 1977. "Children's Sleepwear Flammability Standards: Have They Worked?" *Proceedings, Information Council on Fabric Flammability*, NY. pp 1-5.

Braun, E. et al. 1980. "Measurement of the Protective Value of Apparel Fabrics in a Fire Environment." *Journal of Consumer Product Flammability*. Vol 7. pp 15-25.

Braun, E. et al. 1976. *Back-up Report for the Proposed Standard for the Flammability of General Wearing Apparel*. NBSIR 76-1072. National Bureau of Standards, Washington, DC.

Brewster, E. P., and Barker, R. L. 1983. "A Summary of Research on Heat Resistant Fabrics for Protective Clothing." *American Industrial Hygiene Association Journal*. Vol 44. pp 123-130.

Carborundum. 1974. *Kynol Flame Resistant Fibers*. The Carborundum Co. Sanborn, NY.

CPAI. 1984. *A Specification for Flame Resistant Materials Used in Camping Tentage*. Canvas Products Association International, St. Paul, MN.

CGSB. 1972. *Standard Method of Test for Surface Burning Characteristics of Flooring and Floor Covering Materials*. B.54.9-1972. Canadian Government Specification Board, Ottawa, Ontario.

CPSC. 1974. Code of Federal Regulations Part 1615. *Standard for the Flammability of Children's Sleepwear: Size 0 through 6x (FF 3-71); Part 1616 Sizes 7 through 14 (FF5-74)*. U.S. Consumer Product Safety Commission, Washington, DC.

CPSC. 1970. Code of Federal Regulations Part 1630. *Standard for the Surface Flammability of Carpets and Rugs (FFI-70); Part 1631- Standard for the Surface Flammability of Small Carpets and Rugs (FF2-70)*. U.S. Consumer Product Safety Commission, Washington, DC.

CPSC. 1967. *Flammable Fabrics Act of 1953* (as ammended and revised 1967). U.S. Consumer Product Safety Commission, Washington, DC.

CPSC. 1953. *Standard for the Flammability of Clothing Textiles*. (formerly CS 191-53); *and* Part 1611. *Standard for the Flammability of Vinyl Plastic Film*. (formerly CS 192-53) Code of Federal Regulations. Part 1610. U.S. Consumer Product Safety Commission, Washington, DC.

Crikelair, G. F. 1975. "A Win for the Team." *Proceedings, Information Council on Fabric Flammability*, NY. pp 218-233.

DOT. Code of Federal Regulations. Part 571. *Motor Vehicle Safety Standard No. 302, Flammability of Interior Materials-Passenger Cars, Multipurpose Passenger Vehicles, Trucks, and Buses*. U.S. Department of Transportation, Washington, DC.

Einsele, U. 1972. "Uber das Brennverhalten und den Brennmechanismus von Synthesefasern." *Melliand Textilberichte*. Vol 53. pp 1395-1402.

FAA. 1983. *Flammability Requirements for Aircraft Seat Cushions*. 14 CFR Parts 25, 29, and 1231. Oct 11, 1983. Federal Aviation Administration, Washington, DC.

Fire Journal. 1970a. "Burning Incense Fell on Carpet." *Fire Journal*. Vol 64, No 4. July 1970. p 50.

Fire Journal. 1970b. "Polyester Carpet Fire, Suisun, CA." *Fire Journal*. Vol 64, No 5. pp 88, 89.

Gordon, B. F. 1981. "Flame Retardants and Textile Materials." *Fire Safety Journal*. Vol 4. pp 109-123.

Gottlieb, I. M., and Beck, L. R., Jr. 1959. "Thermal Stability, Flammability, and Melting Characteristics of Textile Substrates." *Annals NY Academy Science*. Vol 82. pp 724-743.

GSA. 1971. Flame Resistance of Cloth; Vertical Method 5903.2. *Federal Test Method Standard No. 191*. General Services Administration, Washington, DC.

Hilado, C. J. 1973. *Flammability Test Methods Handbook*. Technomic Publishing Company, Inc., Westport, CT.

Kaswell, E. R. 1972. "Some Thoughts and Information on Nonflammable Products." *Journal of the American Association of Textile Chemist and Colorists*. Vol 4, No 1. pp 33-40.

Krasny, J. F. 1982. "Flammability Evaluation Methods for Textiles." *Flame Retardant Polymeric Materials*. Vol 3. M. Lewin, S. M. Atlas, and E. M. Pearce, Eds. Plenum, NY.

Krasny, J. F., Singleton, R. W., and Pettengill, J. 1982. "Performance Evaluation of Fabrics Used in Fire Fighter's Turnout Coats," *Fire Technology*. Vol 18. pp 309-318.

Martin, J. R., and Miller, B. 1978. "The Thermal and Flammability Behavior of Polyester-Wool Blends." *Textile Research Journal*. Vol 48. 1978. pp 97-103.

McCullough, E. A., and Noel, C. J. 1978. "Flammability Characteristics of Layered Fabric Assemblies." *Proceedings, Information Council on Fabric Flammability*, NY. pp 175-184.

Miller, B. 1976. "A New Concept for Monitoring the Burning Behavior of Fabric." *Proceedings, Information Council on Fabric Flammability*, NY. pp 198-208.

Miller, B., and Meiser, C. H., Jr. 1978. "Heat Emission form Burning Fabrics, Potential Harm Ranking." *Textile Research Journal*. Vol 48. pp 238-243.

MMFPA. 1978. *Man-Made Fibers Fact Book*. Man-Made Fibers Producers Association, Inc., Washington, DC.

Moore, L. D. 1978. *Full Scale Burning Behavior of Curtains and Draperies*. NBSIR 78-1448. U.S. National Bureau of Standards, Washington, DC.

Nielsen, E. B., and Richards, H. R. 1969. "A Proposed Method for Measuring Rate-of-Burn." *Journal of the American Association of Textile Chemists and Colorists*. Vol 1. pp 270-277.

Schoppe, M. M., Welsford, J. M., and Abbott, N. J. 1982. *Resistance of Navy Shipboard Work Clothing Materials to Extreme Heat.* Tech Rep No 148. Navy Clothing and Textile Research Facility, Natick, MA.

Shalev, I., and Barker, R. L. 1984. "Protective Fabrics: A Comparison of Laboratory Methods for Evaluating Thermal Protective Performance in Convective/Radiant Exposure." *Textile Research Journal.* Vol 54. pp 648-654.

Sharman, L. J., and Tovey, H. 1972. "Current Status and National Priorities for Flammable Fabric Standards." *Proc. Information Council on Fabric Flammability,* NY. pp 64-306.

Smith, M. J., et al. 1983. *Procuring Fire Fighter Protective Clothing.* Health and Safety Division. Department of Research. International Association of Fire Fighters. AFL-CIO-CLC. Washington, DC.

Stoll, A. M., and Chianta, M. A. 1969. "Method and Rating Systems for Evaluation of Thermal Protection." *Aerospace Medicine Journal.* Vol 40. pp 1232-1238.

Troitzsch, J. 1983. *International Plastics Flammability Handbook: Principles—Regulations—Testing and Approval.* MacMillan, New York, NY.

Turner, H. L., and Segal, L. 1965. "Fire Behavior and Protection in Hyperbaric Chambers." *Fire Technology.* Vol 1. No 4. pp 269-277.

Vickers, A., Krasny, J., and Tovey, H. 1973. "Some Apparel Fire Hazard Parameters." *Proceedings, Information Council on Fabric Flammability,* NY. pp 205-226.

Weaver, J. W. 1976. "Rate of Burning of Apparel Fabrics." *Journal of the American Association of Textile Chemists Colorists.* Vol 8. pp 176-181.

Willey, A. E. 1973. "Fire, Baptist Towers Housing for the Elderly." *Fire Journal.* Vol 67, No 3. pp 15-21, 103.

NFPA Codes, Standards, Recommended Practices and Manuals. (See the latest *NFPA Codes and Standards Catalog* for availability of current editions of the following documents.)

NFPA 102, *Standard for Assembly Seating, and Air-Supported Structures.*

NFPA 253, *Standard Method of Test for Critical Radiant Flux of Floor Covering Systems Using a Radiant Heat Energy Source.*

NFPA 255, *Standard Method of Test for the Surface Burning Characteristics of Building Materials.*

NFPA 701, *Standard Methods of Fire Tests for Flame-Resistant Textiles and Films.*

NFPA 702, *Standard for the Classification of the Flammability of Wearing Apparel.*

NFPA 1971, *Standard on Protective Clothing for Structural Fire Fighters.*

Additional Readings

"Baled Fibers," *Loss Prevention Data Sheet 8-7,* Factory Mutual Engineering Corp., Norwood, MA.

Beninate, John V., et al., "Wool: Its Effect on Flame Retardant Properties of Blend Fabrics," *Journal of Fire Sciences,* Vol. 1, No. 2, Mar./Apr. 1983, pp. 145-154.

Bhatnagar, V. M., ed., "Advances in Fire Retardant Textiles," and "Flammability of Apparel," Vols 5 & 7, respectively, *Progress in Fire Retardancy Series,* Technomic, Westport, CT, 1975.

Butler, M. J., and Slater, J. A., eds., *Fire Safety Research* (Proceedings of a Symposium held at the National Bureau of Standards, Gaithersburg MD, August 22, 1973.), NBS Special Pub. 411, Nov. 1974, pp. 1-104.

Carroll-Porczynski, C. Z., *Flammability of Composite Fabrics,* First American Ed., Chemical Publishing Co., NY, 1984.

Corbman, Bernard P., *Textiles: Fiber to Fabric,* 5th ed., McGraw-Hill, NY, 1975.

"Hospitals and Flammable Fabrics: California Case History," *Fire Journal,* Vol. 66, No. 3, May 1972, pp. 18-24.

Johnson, Robert F., "Camping Tent Flammability—What the Record Shows," *Fire Journal,* Vol. 71, No. 4, July 1977, pp. 50-55.

Krasny, J. F., Singleton, R. W., and Pettengill, J., "Performance Evaluation of Fabrics Used in Fire Fighter's Turnout Coats," *Fire Technology,* Vol. 18, No. 4, 1982, pp. 309-318.

Martin, J. R., and Miller, B., "The Thermal and Flammability Behavior of Polyester-Wool Blends," *Textile Research Journal,* Vol. 48, 1978, pp. 97-103.

Peterson, Barbara A., and King, Rosalie R., "Laboratory Testing of Pile Fabrics—A Proposed Flammability Standard," *Fire Technology,* Vol. 16, No. 2, May 1980, pp. 142-149.

Sharman, L. J., and Tovey, H., "Current Status and National Priorities for Flammable Fabric Standards," *Proceedings, Information Council on Fabric Flammability,* NY, 1972, pp. 64-306.

Slater, J. A., Buchbinder, B. and Tovey, H., "Matches and Lighters in Flammable Fabric Incidents: The Magnitude of the Problem," *NBS Tech Note 750,* National Bureau of Standards, Gaithersburg, MD, Dec. 1972.

Sund, J. L., and King, R. R., "Longitudinal Wear Study of Four Work Shirts in Ferrous Metal Operations," *Fire Technology,* Vol. 19, No. 3, Aug. 1983, pp. 163-169.

Yuill, C. H., "Floor Coverings: What is the Hazard?," *Fire Journal,* Vol. 61, No. 1, Jan. 1967, pp. 11-19.

———, "The Flammability of Floor Coverings," *Journal of Fire and Flammability,* Vol. 1, 1970, pp. 64-70.

FLAMMABLE AND COMBUSTIBLE LIQUIDS

Revised by Martin F. Henry

The discussion of flammable and combustible liquids presented in this chapter focuses on the fire hazards of these materials. After these fire hazards are discussed, the NFPA classification system for flammable and combustible liquids is presented. The physical and fire characteristics of these materials is discussed next, including a special section on the burning characteristics of liquids. Methods to prevent fires while using flammable and combustible liquids completes the chapter.

Information about flammable and combustible liquids not contained in this chapter is found elsewhere in this HANDBOOK. Section 4, Chapter 1, "The Chemistry and Physics of Fire," includes information about the physical and fire characteristics of liquids and other materials. Section 12, Chapter 10, "Control of Electrostatic Ignition Sources," discusses the control of static electricity as an ignition source; Section 12, Chapter 11, "Special Systems for Explosion Damage Control," discusses explosion venting; and Section 12, Chapter 13, "Air Moving Equipment," discusses blower and exhaust systems for flammable and combustible liquids. Section 17, Chapter 1, "Water and Water Additives for Fire Fighting," includes information on wetting agents to be used in fighting flammable and combustible liquids fires.

HAZARDS OF FLAMMABLE AND COMBUSTIBLE LIQUIDS

Strictly speaking, flammable and combustible liquids do not "cause" fires; they are merely contributing factors. Rather, a spark or some other ignition source causes a fire or explosion in the presence of flammable vapors. Thus it is the vapor of a flammable or combustible liquid, rather than the liquid itself, which ignites or explodes when mixed with air in certain proportions in the presence of a source of ignition. For example, the proportions (flammable range) for carbon disulfide are from about 1 to 44 percent carbon disulfide vapors in air by volume; for ethyl alcohol, from about 4 to 19 percent by volume; and for gasoline, from about 1.4 to 7.6 percent by volume. Therefore, storing flammable and combustible liquids in the proper closed containers, and minimizing the exposure of the liquid to air while in use are fundamental safety measures.

Explosions of flammable vapor-air mixtures near the lower or upper limits of the flammable range are less intense than those occurring in intermediate concentrations of the same mixture. Flammable vapor-air explosions occur most frequently in confined spaces such as containers, tanks, rooms, or buildings. The violence of flammable vapor explosions depends upon the nature of the vapors and enclosure containing the mixture as well as on the quantity of vapor-air mixture.

Distinct from an explosion of a flammable vapor-air mixture inside a tank or container is the overpressuring of a tank which results in a rupture. As with vapor explosions, pressure ruptures vary in intensity. Any closed container may rupture violently when exposed to a severe fire.

Fire and explosion prevention measures are based on one or more of the following techniques or principles: (1) exclusion of sources of ignition, (2) exclusion of air (oxygen), (3) storage of liquids in closed containers or systems, (4) ventilation to prevent the accumulation of vapor within the flammable range, and (5) use of an atmosphere of inert gas instead of air. Extinguishing methods for flammable and combustible liquid fires involve shutting off the fuel supply, excluding air by various means, cooling the liquid to stop evaporation, or a combination thereof.

Gasoline is the most widely used flammable liquid. It is commonly known that gasoline generates flammable vapors at ambient temperatures. There are many other volatile flammable products. A comprehensive list of the more commonly used flammable or combustible liquids, including the characteristics of each, appears in NFPA 325M, *Fire Hazard Properties of Flammable Liquids, Gases, and Volatile Solids.*

Flash point, though commonly accepted as the most important criterion of the relative hazard of flammable and combustible liquids, is not the only factor used to evaluate the hazard. The ignition temperature, flammable range, rate of evaporation, reactivity when contaminated or exposed to heat, density, and rate of diffusion of the vapor

Martin Henry is Flammable Liquids Specialist on the staff of NFPA.

also effects the hazard. The flash point and other factors that determine the relative susceptibility of a flammable or combustible liquid to ignition have comparatively little influence on the liquid's burning characteristics after fire has burned for a short time.

The use of flammable and combustible liquids produced by chemical and petrochemical companies is increasing rapidly. Although many of these products can be classed as normal or stable liquids, others introduce the problem of instability or reactivity.

The storage, handling, and use of unstable (reactive) flammable or combustible liquids require special attention. It may be necessary to increase the distances to property lines and between tanks or to provide extra fire protection. For example, it would be poor practice to locate heat reactive and water reactive flammable or combustible liquid tanks adjacent to each other. In the event of a ground fire, water applied to the heat reactive tank for protection might penetrate the tank of water reactive liquid and cause a violent reaction.

CLASSIFICATION OF FLAMMABLE AND COMBUSTIBLE LIQUIDS

For fire protection purposes, an arbitrary division between liquids and gases has been established based upon the definition of a flammable liquid found in NFPA 321, *Standard on Basic Classification of Flammable and Combustible Liquids.* Liquids are defined as those fluids with a vapor pressure not exceeding 40 psi absolute (275 kPa) which is approximately 25 psi gage pressure (172 kPa) at 100°F (38°C). Another arbitrary division also has been established between liquids and solids for the purposes of this classification system. A liquid is additionally defined as one which has a fluidity greater than that of 300 penetration asphalt.

The following classification system for flammable and combustible liquids (from NFPA 321) is based on the division of flammable liquids into three categories. It is anticipated that in most areas the indoor temperature could reach 100°F (38°C) at some time during the year. Therefore, all liquids with flash points below 100°F (38°C) are called Class I liquids. In some areas, the ambient temperature could exceed 100°F (38°C), requiring only a moderate degree of heating to bring the liquid to its flash point. Based on this concept, an arbitrary division of 100 to 140°F (38 to 60°C) was established for liquids in this flash point range to be known as Class II liquids. Since liquids with flash points higher than 140°F (60°C) would require considerable heating from a source other than ambient temperatures, they have been identified as Class III liquids.

Flammable Liquids

Flammable liquids have flash points below 100°F (38°C) and vapor pressures not exceeding 40 psia at 100°F (275 kPa at 38°C). They are classified as Class I liquids, and may be subdivided as follows:

(a) Class IA liquids include those with flash points below 73°F (23°C) and with boiling points below 100°F (38°C).

(b) Class IB liquids include those with flash points below 73°F (23°C) and with boiling points at or above 100°F (38°C).

(c) Class IC liquids include those with flash points at or above 73°F (23°C) and below 100°F (38°C).

Combustible Liquids

Liquids with flash points at or above 100°F (38°C) are referred to as combustible liquids and may be subdivided as follows:

1. Class II liquids have flash points at or above 100°F (38°C) and below 140°F (60°C).
2. Class IIIA liquids have flash points at or above 140°F (60°C) and below 200°F (93°C).
3. Class IIIB liquids have flash points at or above 200°F (93°C).

Other Classification Systems

There are other classification systems for flammable and combustible liquids. In some, the break points between classes of liquids are different, or the open cup flash point tester is used to determine the flash point. In other classification systems, the solubility of the liquid with water is considered.

Solids with Flash Points

Many combustible chemicals that are solids at 100°F (38°C) or above are classified as solids. When heated, the solid becomes liquid and gives off flammable vapors; flash points can then be determined. In their liquid state, these solids should be treated as liquids with similar flash points. Some manufactured solids such as paste waxes or polishes may contain varying amounts of flammable liquids. The flash point and amount of liquid in such manufactured solid materials will indicate the degree of hazard.

PHYSICAL PROPERTIES OF LIQUIDS

In liquids, molecules move freely among themselves, but they do not tend to separate from one another in the way that molecules in gases do. Liquids, unlike gases, are only slightly compressible and are incapable of indefinite expansion. They differ from solids in the ease with which their molecules move against another, causing them to adapt to the shape of the containing vessel. Some liquids are very viscous, however, and no sharp definition can be drawn between these liquids and solids. Neither is there a sharp line of demarcation between liquids and gases. Materials can exist in either state, depending upon temperature and pressure conditions. Liquids tend to become gases as their temperatures increase or their pressures decrease. On the other hand, gases tend to become liquids as their temperatures decrease or their pressures increase. However, a material can only exist as a gas, regardless of pressure, if its temperature is above the so called critical temperature. (The *critical temperature* is an intrinsic property of a material and is the temperature above which the material can exist only in the gaseous state.) In all cases, the actual state of a liquid is determined by the combination of temperature and pressure.

PHYSICAL CHARACTERISTICS OF FLAMMABLE AND COMBUSTIBLE LIQUIDS

Specific Gravity

Since the specific gravity of water equals one, a liquid with a specific gravity of less than one will float on water (unless it is water soluble). A specific gravity greater than one indicates that water will float on the liquid. This can be an important consideration in fighting a flammable or combustible liquid fire.

Vapor Density Ratio

Vapor density is the weight per unit volume of a pure gas or vapor. For fire protection purposes, vapor density is reported in terms of the ratio of the relative weight of a volume of vapor to the weight of an equal volume of air under the same conditions of temperature and pressure. In this respect, the vapor density ratio is similar to the specific gravity of a liquid, except that air is used as the standard in place of water. Air is taken as unity and the vapor density is reported as a ratio. A vapor density of 3 (3:1) indicates that the vapor is three times as dense or heavy as air.

$$\text{Vapor density ratio} = \frac{\text{molecular weight (MW) of the material}}{\text{composite molecular weight of air}} = \frac{MW}{29}$$

Vapor density ratios are reported at equilibrium temperature and atmospheric pressure conditions. Unequal or changing conditions will appreciably change the density of any vapor.

Generally, pure vapors are seldom found except when the liquid is normally stored above its boiling point. For those liquids that are stored or handled below their boiling points, the vapor located above the liquid will be a mixture of vapor and air.

For liquids with boiling points above the prevailing ambient temperature at atmospheric pressure, the vapor density may be a misleading figure. For example, the boiling point of ethyl acetate is 171°F (77°C). At 70°F (21°C) its vapor pressure is about 2.35 psi (16.2 kPa) or 0.16 atm (atmospheres). Thus, an equilibrium mixture of ethyl acetate vapor and air at 70°F (21°C), as might be found in a closed vessel, would have a composition of 16 percent ethyl acetate vapor and 84 percent air. The "theoretical" vapor density of ethyl acetate is 3.0. The "actual" density of the "vapor-air mixture" which is produced by ethyl acetate at 70°F (21°C) and at atmospheric pressure is:

$$.16 \times 3.0 + .84 \times 1 = 1.32$$

Vapor densities are ordinarily used only as an indication of the settling or rising tendency of a vapor.

Vapor Pressure

Where a liquid is present in a closed container with an atmosphere of vapor-air mixture above the liquid surface, the percentage of vapor in the mixture may be determined from the vapor pressure. The percentage of vapor is directly proportional to the relationship between the vapor pressure of the liquid and the total pressure of the mixture. For example, acetone at a temperature of 100°F (38°C) has a vapor pressure of 7.6 psia (52 kPa). Assuming total pressure at 14.7 psia (101 kPa), the proportion of acetone vapor present will be 7.6 divided by 14.7, or 52 percent. Vapor pressures of petroleum liquids usually are determined by the Reid method as recommended by the American Society for Testing and Materials, ASTM Standard D-323. This method gives the vapor pressure in psia (kPa) at 100°F (38°C) which differs slightly from the true vapor pressure. Vapor pressure figures for many substances will be found in chemical handbooks. If the closed cup flash point of a liquid and the vapor pressure at the flash point temperature are known, the lower flammable limit for the vapor (at the flash point temperature) in percent by volume at normal atmospheric pressure may be calculated as follows:

$$LFL = \frac{V}{0.147}$$

$$\text{in SI: } LFL = \frac{V}{1.01}$$

where *LFL* is percent vapor by volume at the lower flammable limit, and *V* is vapor pressure, psia (kPa), at flash point temperature. At other pressures,

$$LFL = \frac{100V}{P}$$

where *P* is the ambient pressure in psia (kPa).

In a mixture of volatile flammable liquids, the effects of the two vapor pressures, one upon the other, generally depend upon whether the two liquids are completely miscible, partly miscible, or completely immiscible. If the two liquids are completely miscible, they lower each other's vapor pressures; if they are nearly completely immiscible, the vapor pressure of the mixture is the sum of the partial pressure of each liquid layer (Dalton's Law of Partial Pressure); if both liquids are partially miscible, the relationships are more complex.

Evaporation Rate

Evaporation rate is the rate at which a liquid is converted to the vapor state at any given temperature and pressure. The differing rates of mixture evaporation are of primary concern in fire protection. In general, as the boiling point decreases, the vapor pressure and the evaporation rate increase.

Viscosity

The viscosity of a liquid is a measure of its resistance to flow resulting from the combined effects of adhesion and cohesion, i.e., a measure of internal friction in a fluid. Although there are several recognized devices for determining viscosity, the measurement principles are generally the same. These involve measuring the time required for a predetermined quantity of liquid at a specified temperature to flow into a receiving flask or through an orifice of a prescribed size.

Latent Heat of Vaporization

Latent heat of vaporization is the heat which is absorbed when one gram of liquid is transformed into vapor at the boiling point under one atmosphere pressure, and is expressed in calories per gram (cal/g) or in Btu per pound.

Solubility in Water and Surface Tension

Solubility in water and surface tension are other characteristics of a liquid which concern the fire protection field. Fires in liquids that are water soluble can be extinguished by dilution of the liquid with water, or by use of "alcohol type" foams for extinguishment. The use of wetting agents affects the surface tension of a liquid and in some cases aids in fire extinguishment.

FIRE CHARACTERISTICS OF LIQUIDS

Flash Point

A vapor mixed with air in proportions below the lower limit of flammability may burn at the source of ignition, i.e., in the zone immediately surrounding the source of ignition without propagating (spreading) flame away from the ignition source. The flash point of a liquid corresponds roughly to the lowest temperature at which the vapor pressure of the liquid is just sufficient to produce a flammable mixture at the lower limit of flammability. (See Fig. 5-4A.)

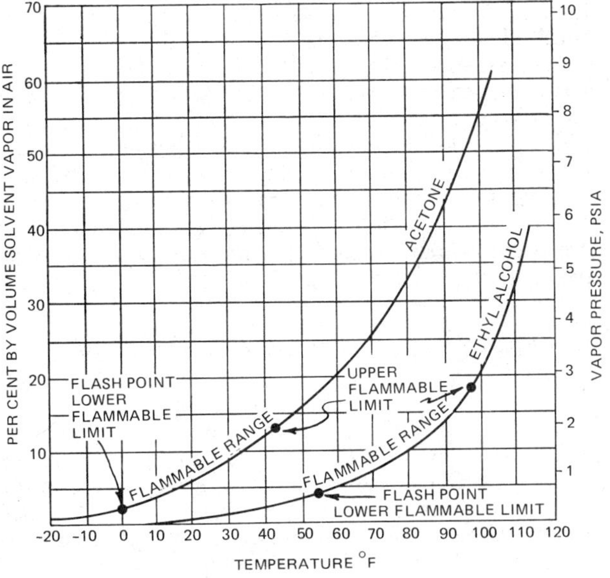

FIG. 5-4A. Relationship between flash point, flammable limits, temperature, and vapor pressure for acetone and ethyl alcohol. Liquid, vapor, and air in equilibrium in a closed container at normal atmospheric pressure. (One psia equals psi + 14.7 × 6.985; 5/9 (°F − 32) equals °C.)

There are several types of test apparatus used for determining flash point. (See Fig. 5-4B.) NFPA 321 specifies the Tagliabue (Tag) Closed Tester for testing liquids except for certain viscous or film forming materials with a flash point at or below 200°F (93°C), as described in ASTM D56. The Pensky-Martens Closed Tester, ASTM D93, is specified for testing liquids with flash points above 200°F (93°C) or for certain viscous or film forming materials.

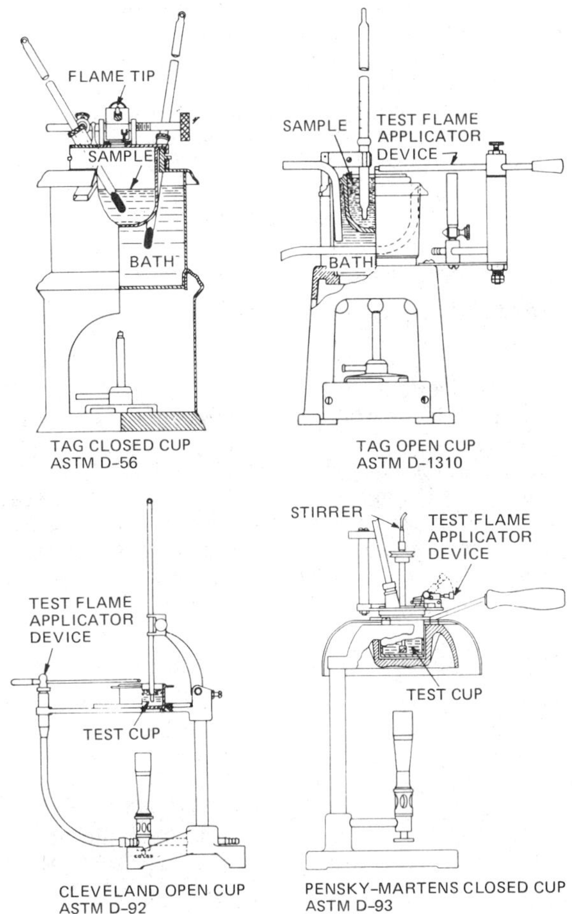

FIG. 5-4B. Four of the commonly used testers for determining flash points of flammable or combustible liquids. The material to be tested is slowly heated, and at periodic intervals a test flame is applied to the vapor space. Flash point is the temperature at which a flash of fire is seen when the test flame is applied. The full details of conducting tests for each type of testing apparatus are given in the applicable standard of the American Society for Testing and Materials or by the manufacturer.

ASTM D3278 is used for testing paints, enamels, lacquers, varnishes and related products and their components with flash points between 32 and 230°F (0 and 18°C).

Open cup flash points are sometimes used in grading flammable liquids in transportation and are determined by the Tagliabue (Tag) Open Cup Apparatus, ASTM D1310, or the Cleveland Open Cup Apparatus, ASTM D92. Open cup flash points represent conditions with the liquid in the open and are generally higher than the closed cup flash point figures for the same substances. However, closed cup flash point figures are generally higher than the actual lower temperature limit of flammability (flash point) when tested in a vertical tube utilizing the principle of upward flame propagation.

It is essential to realize that flash point varies with pressure and with the oxygen content of the atmosphere as well as with the purity of the product being tested, the method of test, and the skill of the operator conducting the test. It is quite possible to have a flammable mixture far below the stated flash point value if the pressure is considerably less than one atmosphere.

Ignition Temperature (Autoignition Temperature, Autogenous Ignition Temperature)

The autoignition temperature reported for a flammable liquid is generally the temperature to which a closed or nearly closed container must be heated in order that the liquid in question, when introduced into the container, will ignite spontaneously and burn. Time lags of a minute or more are frequently involved. The standard testing method for autoignition temperatures of petroleum products is described in ASTM D2155, *Autoignition Temperature of Liquid Petroleum Products.*

Ignition Temperatures and Molecular Weights

Within a given hydrocarbon series, such as the straight chain series running from normal methane down through normal decane, ignition temperature decreases as molecular weight or carbon chain length increases, all other factors being equal. (See Fig. 5-4C.) Thus pentane, $CH_3(CH_2)_3CH_3$ (Molecular Weight 72.1), has a higher ignition temperature than hexane, $(CH_3CH_2)_4CH_3$ (Molecular Weight 86.2).

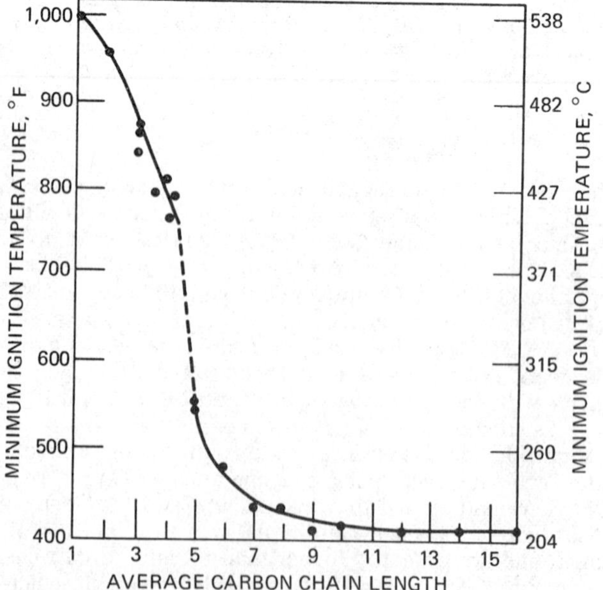

FIG. 5-4C. Minimum ignition temperatures of hydrocarbons of various carbon chain lengths.

Boiling Point

The temperature at which the equilibrium vapor pressure of a liquid equals the total pressure on the surface is known as the boiling point. The boiling point is entirely dependent on the total pressure. (Boiling point increases with an increase in pressure.) Theoretically, any liquid may be made to boil at any desired temperature by sufficiently altering the total pressure on its surface. Similarly, unless decomposition takes place, any liquid may be made to boil at any desired pressure by sufficiently changing its temperature.

The temperature at which a liquid boils when under a total pressure of one atmosphere (14.7 psia or 101 kPa) is termed the normal boiling point.

The above definitions apply to pure materials or constant boiling mixtures. Most flammable liquids and gases on the market today are mixtures and do not follow the physical laws governing pure materials. The boiling points of mixtures are reported as distillation curves. The ten percent point of a distillation performed in accordance with ASTM D86, *Standard Method of Test for Distillation of Petroleum Products*, is used as the boiling point in most NFPA standards.

Flammable (Explosive) Limits

The term "lower flammable limit" (LFL) describes the minimum concentration of vapor to air below which propagation of a flame will not occur in the presence of an ignition source. The "upper flammable limit" (UFL) is the maximum vapor to air concentration above which propagation of flame will not occur. If a vapor to air mixture is below the lower flammable limit, it is described as being "too lean" to burn, and if it is above the upper flammable limit, it is "too rich" to burn.

When the vapor to air ratio is somewhere between the lower flammable limit and the upper flammable limit, fires and explosions can occur. The mixture is then said to be within its flammable or explosive range. When the mixture happens to be in the intermediate range between the LFL and UFL (synonymous with LEL—lower explosive limit and UEL—upper explosive limit), the ignition is more intense and violent than if the mixture were closer to either the upper or lower limits.

Calculating Vapor Volume and Flammable Mixtures

It is frequently helpful to calculate the volume of air required to provide dilution sufficient to prevent the formation of an ignitible mixture as, for example, in the design of a ventilating system for a drying oven. This can be readily done when the quantity of solvent supplied is known or can be reasonably estimated.

For example, consider a process in which acetone vapor is released. The flammable range for acetone vapor in air is from 2.6 to 12.8 percent by volume. Any mixture containing more air than 100 minus 2.6, or 97.4 percent of air—equivalent to 97.4 divided by 2.6, or approximately 37 volumes of air to one volume of vapor—will be too lean to be ignitible. Hence, if the volume of vapor can be estimated, it is a simple matter to determine the required air volume.

The volume of vapor produced from 1 gal of solvent can be calculated from the specific gravity of the liquid and the vapor density as follows:

Cubic feet of vapor from 1 gal of liquid =

$$\frac{8.33 \times \text{specific gravity of liquid } (H_2O = 1)}{0.075 \times \text{vapor density of vapor (air} = 1)}$$

where 8.33 is the weight of 1 gal of water in pounds and 0.075 is the weight of 1 cu ft of air in pounds.

Stated more simply:

$$\text{Vapor equivalent of one gal} = 111 \times \frac{Sp.\ G.}{V.D.}.$$

If the vapor density is not known, it can readily be calculated from the molecular weight.

Again taking acetone as an example (Sp.G. =.792; V. D. =2) we have:

$$\text{Vapor equivalent of one gal is: } 111 \times \frac{.792}{2} = 44 \text{ cu ft.}$$

The volume of air required to dilute vapor from 1 gal of acetone to below the lower flammable limit is:

44 ×37 =1,628 cu ft

If the rate of evaporation of acetone into a given space (oven, etc.) should be 1 gpm, it would require 1,628 cfm of uncontaminated ventilating air to keep the vapor concentration below the lower flammable limit.

For S.I. units, the volume of vapor (m^3) produced by 1 liter of solvent can also be calculated from the specific gravity of the liquid and the vapor density as follows:

The cubic meter vapor equivalent of 1 liter (L) of liquid is:

$$L = 0.83 \times \frac{Sp.\ G.}{V.D.}$$

For acetone, vapor equivalent of 1 liter (L) is:

$$L = \frac{0.83 \times 0.792}{2} = 0.33 \text{ m}^3$$

In S.I. units the volume of air required to dilute 1 liter of acetone to below the lower flammability limit is:

$0.33 \times 37 = 12.2 \text{ m}^3$

For different evaporation rates, the air quantity would be proportional. In practice and as a safety factor, a substantial excess of air ventilation would be applied because of the inevitable nonuniformity of the atmosphere within the enclosure.

Where it is desirable to keep a space "too rich" to be ignitible, a similar procedure (using the upper flammable limit) may be used to determine the maximum air quantity which can be tolerated. This is not, however, a recommended procedure, since it is necessary to enter and pass through the flammable range.

Variation in Hazard with Temperature and Pressure

The temperature and pressure to which a particular liquid is subjected must often be considered in the practical application of the flammable limits of flammable liquids. It can be shown that these factors have a pronounced effect on the fire and explosion hazard and are as important in safe handling as are the flammable limits themselves. A liquid which has a flash point above room temperature and, therefore, is not expected to produce flammable vapors will do so if its temperature is raised to its flash point or higher.

The extent to which a liquid will vaporize, when the opposing vapor is that of the liquid itself, is reduced when the pressure opposing this vaporization is increased as the opposing pressure is decreased. In the same manner, at a higher temperature the liquid will have a higher vapor pressure and will tend to vaporize in greater amounts. When at a given temperature and pressure the liquid has vaporized to the point where no further vaporization will occur unless conditions are changed, a state of equilibrium is said to have been reached. It is evident that equilibrium cannot exist in other than a closed system. In the open air, a vaporizing liquid would continue to vaporize until the supply was exhausted. The pressure-temperature effect, therefore, is applicable only in tanks, pipes, and processing equipment in which the liquid and vapor-air mixture approach equilibrium.

Figures 5-4D and 5-4E show the effect of temperature and pressure on flammable limits as determined by F. C. Mitchell and H. C. Vernon. In using these charts, it should be noted that they represent equilibrium conditions, and that prior to the establishment of equilibrium, values may be different (Mitchell and Vernon 1939). For example, when a liquid is introduced into a pressure container with air at a temperature such that the resultant vapor-air mixture will be above the upper flammable limit (according to the chart), some time will elapse before sufficient evaporation has taken place to reach this condition; meanwhile, the mixture may be flammable or explosive.

It should also be noted that the pressures developed, or the intensity of an explosion, will vary with the initial pressure of the vapor-air mixture. With high initial pressures, the pressures are greater, as in the familiar example of the gasoline engine. With low initial pressures, the pressures are relatively lower.

The lower limits of flammability of solvent vapors are affected by the temperature. This is a safety factor in installations such as industrial ovens, where pressure is not a factor but where temperature needs to be taken into account in hazard evaluation. In such industrial ovens, mechanical or forced ventilation is usually employed to dilute the concentration of the vapors below their lower flammable limits in air and to remove them from the oven. Information on the lower flammable limits of the vapors at elevated temperatures is needed in order to determine the amount of ventilation required for safe operation of an oven. (See Table 5-4A.)

Energy Required for Ignition of Vapors

The principal sources of ignition of flammable liquids include flames, hot surfaces, electrical or frictional sparks, and adiabatic compression.

Flames: Except for extremely small examples produced under laboratory conditions, flames are unfailing sources of ignition for flammable vapor and air mixtures that are within the flammable range. Flames must be capable of heating the vapor to its ignition temperature in the pres-

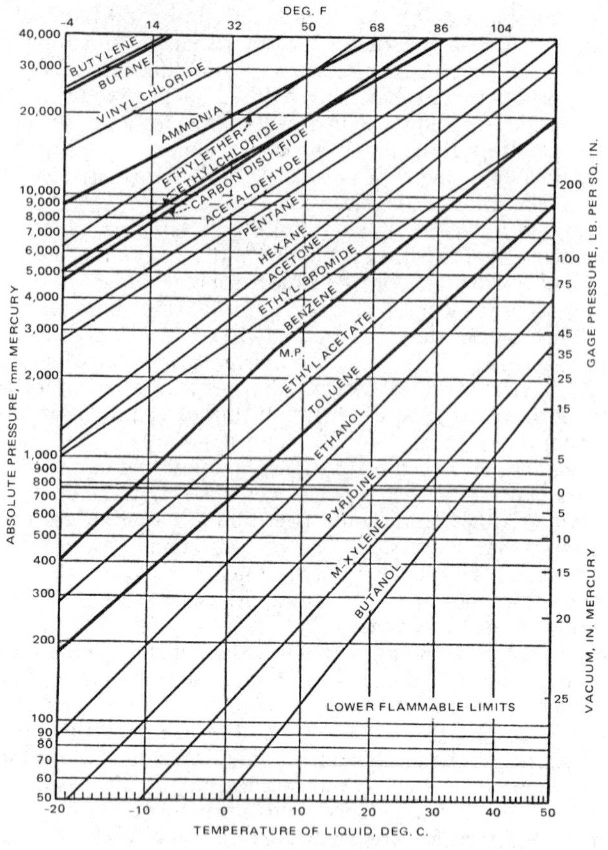

FIG. 5-4D. Variation of lower flammable limits with temperature and pressure. This chart is applicable only to flammable liquids or gases in equilibrium in a closed container. Mixtures of vapor and air will be too "lean" to burn at temperatures below and at pressures above the values shown by the line on the chart for any substance. Conditions represented by points to the left and above the respective lines are accordingly nonflammable. Points where the diagonal lines cross the zero gage pressure line (760 mm of mercury absolute pressure) indicate flash point temperatures at normal atmospheric pressure.

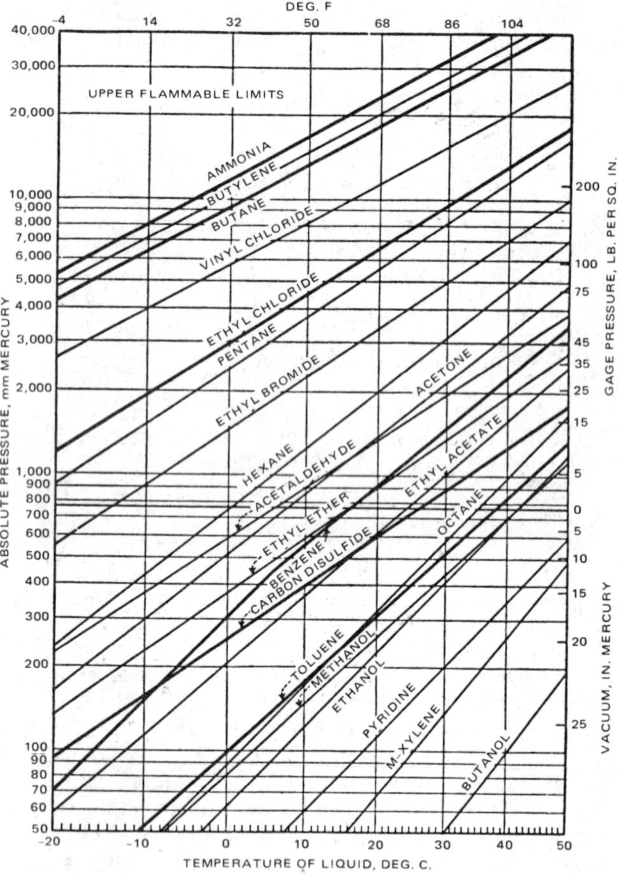

FIG. 5-4E. Variation of upper flammable limits with temperature and pressure. This chart is applicable only to flammable liquids or gases in equilibrium in a closed container. Mixtures of vapor and air will be too "rich" to be flammable at temperatures above and pressures below the values shown by the lines on the chart for any substance. Conditions represented by points to the right of and below the respective lines are accordingly nonflammable.

ence of air in order to be a source of ignition. For some liquids and solids, it will be necessary for the flame to be of sufficient duration and heat to volatilize the fuel and to ignite the released vapors. Once ignited, the radiated heat from the burning vapors perpetuates the burning process.

Electrical, Static, and Friction Sparks: These must have sufficient energy to ignite flammable vapor-air mixtures. Electrical sparks from most commercial electrical supply installations are above flame temperatures and will usually ignite flammable mixtures. However, friction sparks may fail to ignite a flammable mixture because the spark may be of such short duration as to fail to heat the vapor to its ignition point. Also, not only must an electrical spark have sufficient intensity and length, but some variable conditions must also exist within certain limits before ignition will occur. The nature of the ignition points and surfaces as well as the composition, temperature, and pressure of vapor-air mixtures are the principal variables in calculating ignition. Most of these same factors and variables also apply to static electrical sparks and frictional sparks as a source of ignition because in either case the spark must be of sufficient duration, intensity, or both, to create sufficient heat to cause ignition.

Hot Surfaces: These can be a source of ignition if they are large enough and hot enough. The smaller the heated surface, the hotter it must be to ignite a mixture. The larger the heated surface in relation to the mixture, the more rapidly ignition will take place and the lower the temperature necessary for ignition. A flammable liquid, however, must remain in contact with a hot surface for a sufficient length of time to form a vapor-air mixture within its flammable range. For example, a single drop of a low viscosity highly volatile flammable liquid that falls on the surface of an electric hot plate at 2,000°F (1093°C) may ignite. A hot exhaust pipe in the open seldom ignites a flammable mixture even though the surface temperature may be considerably above the environmental ignition temperature test method.

Adiabatic Compression: This has been the cause of several destructive explosions. When controlled, adiabatic

TABLE 5-4A. The Effect of Elevated Temperature on the Lower Flammable Limit of Combustible Solvents as Encountered in Industrial Ovens‡

Solvent	Flash Pt Closed Cup	Lower Flammable Limit Percent Vapor by Volume at Initial Temperature, °F (°C)						
		Room Temp.	212 (100)	392 (200)	437 (225)	482 (250)	572 (300)	662 (350)
Acetone	3	2.67	2.40	2.00*	—	—	—	—
Amyl Acetate, Iso	77	—	1.00	0.82	—	0.76*	—	—
Benzene	−4	1.32	1.10	0.93	—	—	0.80*	—
Butyl Alcohol, Normal	100	—	1.56	1.27	1.22*	—	—	—
Cresol, Meta-Para	202	—	1.06†	0.93	—	0.88*	—	—
Cyclohexane	−4	1.12	1.01	0.83*	—	—	—	—
Cyclohexanone	111	—	1.11	0.96	0.94	0.91*	—	—
Ethyl Alcohol	54	3.48	3.01	2.64	—	2.47	2.29*	—
Ethyl Lactate	131	—	1.55	1.29	—	1.22*	—	—
Gasoline	−45	1.07	0.94	0.77*	—	—	—	—
Hexane, Normal	−15	1.08	0.90	0.72*	—	—	—	—
High-Solvency Petroleum Naphtha	36	1.00	0.89	0.74	0.72	0.69*	—	—
Methyl Alcohol	52	6.70	5.80	4.81	—	4.62	4.44*	—
Methyl Ethyl Ketone	21	1.83	1.70	1.33*	—	—	—	—
Methyl Lactate	121	—	2.21	1.86	1.80	1.75*	—	—
Mineral Spirits, No. 10	104	—	0.77	0.63*	—	—	—	—
Toluene	48	1.17	0.99	0.82	—	—	0.72*	—
Turpentine	95	—	0.69	0.54*	—	—	—	—
V. M. & P. Naptha	28	0.92	0.76	0.67*	—	—	—	—

* Rapid and extensive thermal decomposition and oxidation reactions in vapor-air mixture at this temperature.
† Lower limit determined at 302°F (150°C).
‡ From NFPA *Quarterly*, April 1950; UL Bulletin of Research, No. 43.

compression is the basis of diesel engine operation. A rapidly compressed flammable mixture will be ignited when the heat generated by the compressing action is sufficient to raise the flammable vapor to its ignition temperature.

Behavior of Mixed Liquids

The behavior of mixed liquids varies considerably, depending upon physical characteristics of the liquid and environmental conditions. However, the vapor pressure or evaporation rate of mixed liquids is particularly important in the prevention of fires. These factors are significant when a mixture of liquids is listed as nonflammable or has a high flash point but becomes flammable under conditions of use. As a specific example, sufficient carbon tetrachloride can be added to gasoline so that the mixture has no flash point. However, while standing in an open container, the carbon tetrachloride will evaporate more rapidly than the gasoline. Over a period of time, therefore, the residual liquid will first show a high flash point, then a progressively lower one, until the flash point of the final ten percent of the original sample will approximate that of the higher boiling fractions of gasoline.

In order to evaluate the fire hazard of such liquid mixtures, fractional evaporation tests can be conducted at room temperature in open vessels. After evaporation of appropriate fractions such as 10, 20, 40, 60, and 90 percent of the original sample, flash point tests can be conducted on the residue. The results of such tests indicate the class into which the liquid should be placed if the conditions of use are such as to make it likely that appreciable evaporation will take place.

With other liquid mixtures the flash point may increase as evaporation occurs.

Noncompatible Materials

Noncompatible materials of construction or equipment are those materials with which an unstable flammable liquid or its normal reaction products will combine, or whose products of corrosion are contaminants. For example, if mercury is a harmful contaminant, then mercury thermometers should not be used. If a chlorinated compound slowly releases hydrochloric acid, aluminum should not be used. The aluminum chloride formed by this acid may act as a catalyst. There are many materials that will chemically react with each other; when there is a question, reference should be made to NFPA 491M, *Manual of Hazardous Chemical Reactions.*

BURNING CHARACTERISTICS OF LIQUIDS

Since it is the vapors from flammable liquids which burn, the ease of ignition as well as the rate of burning can be related to such properties as the vapor pressure, flash point, boiling point, and evaporation rate. Liquids with vapors in the flammable range above the liquid surface at the stored temperature will have a rapid rate of flame propagation. Flammable and combustible liquids with flash points above temperatures at which they are stored have a slower rate of flame propagation because it is necessary for the heat of the fire to sufficiently heat the liquid surface and form a flammable vapor-air mixture before the flame will spread through the vapor. There are many variables affecting the rate of flame propagation and burning, including environmental factors, wind velocity, temperature, heat of combustion, latent heat of vaporization, and barometric pressure.

Liquid hydrocarbons normally burn with an orange flame and emit dense clouds of black smoke. Alcohols normally burn with a clean blue flame and produce very little smoke. Certain terpenes and ethers burn with considerable ebullition (boiling) of the liquid surface, making it difficult to extinguish fires involving these substances.

Boilover, Slopover, or Frothover

Three specific conditions—boilover, slopover, and frothing—deserve special mention in regard to fires in open topped tanks containing various types of oils.

Boilover: Boilover is a phenomenon that may occur spontaneously during a fire in an open top tank containing most types of crude oils and some synthetically produced heavy oil mixtures. This may occur when the roof of a tank is blown off by an explosion, usually caused by lightning. After a long period of quiescent burning, there is a sudden overflow or ejection of some of the residual oil in the tank. It is caused by boiling water that forms a quickly expanding steam-oil froth. The frothing results from the existence of the following three conditions, the absence of any one of which will prevent the occurrence:

1. The tank must contain free water or water-oil emulsion at the tank bottom. This situation will normally prevail in tanks storing crude oil.
2. The oil must contain components having a wide range of boiling points, such that when the lighter components have been distilled off and burned at the surface, the residue, with a temperature of 300°F (149°C) or higher, is more dense than the oil immediately below. This residue sinks below the surface and forms a layer of gradually increasing depth which advances downward at a rate substantially faster than the rate of regression of the burning surface. This sets up the so called "heat wave" which is the result of localized settling of a part of the hot surface oil until it reaches the colder oil below. There is no heat conduction from the burning surface downward.
3. Sufficient content of heavy ends are present in the oil to produce a residue which can form a tough persistent froth of oil and steam.

What is believed to be the most serious loss of life tank fire in history occurred on December 19, 1982 in Tacoa, Venezuela. The storage tank, located a few miles northwest of Caracas in the small seaside village of Tacoa, supplied fuel to an electric generating plant which provided power to Caracas's population of three million. Two workers died upon ignition of the tank. After six hours of intense burning, the contents of the tank erupted into an extremely violent boilover. The final toll—more than 150 people dead, scores more injured, others still unaccounted for, and damage estimated at $50 million. Forty of those who died were fire fighters, which also makes the Tacoa incident the gravest of its kind in terms of loss of life of fire personnel. (See Fig. 5-4F.)

FIG. 5-4F. Emergency personnel on the scene of the fuel oil storage tank fire boilover in Tacoa, Venezuela, December 19, 1982. At this point the fire had been declared under control. The boilover occured minutes after this photo was taken, and all of the people shown here were reportedly killed.

Slopover: Slopover can result when a water stream is applied to the hot surface of a burning oil, provided the oil is viscous and its temperature exceeds the boiling point of water. Since only the surface oil is involved, a slopover is a relatively mild occurrence.

Frothover: Frothover refers to the overflowing of a container not on fire when water boils under the surface of a viscous hot oil. A typical example would be when hot asphalt is loaded into a tank car containing some water. The first asphalt is cooled by contact with the cold metal, and at first, nothing may happen. But when the water becomes heated and starts to boil, the asphalt may overflow the tank car.

A similar situation can arise when a tank containing a water bottom or wet emulsion is used for storing oil slops or residium at temperatures below 200°F (93°C), and the tank receives a substantial addition of hot residium at a temperature of 300°F (149°C) or higher. After enough time has elapsed for the effect of the hot oil to reach the water in a tank, a prolonged boiling action can take place, which can remove a tank roof and spread froth over a wide area.

Burning Rates of Liquids

The burning rate of flammable liquids will vary somewhat similar to the rate of flame propagation. Gasoline, a compound of light and heavy fractions, will burn more rapidly at first while the lighter fractions are burning, while the heavier fractions will burn at a rate approaching that of kerosene. The burning rate for gasoline is 6 to 12 in. (150 to 300 mm) of depth per hr, and for kerosene the rate is 5 to 8 in. (130 to 200 mm) of depth per hr. For example, a pool of gasoline, ½ in. (12.7mm) deep, could be expected to burn itself out in 2½ to 5 min.

In a series of tests conducted at the U.S. Bureau of Mines (Burgess, Strasser, and Grumer 1961), the burning rates of several liquids and gases were determined and found to approach "a maximum and constant value with increasing pool diameter. This constant burning rate is proportional to the ratio of the net heat of combustion to the sensible heat of vaporization."

Based upon these observed burning rates in petroleum tank fires, an estimate can be made of the extent of area that will be involved in fire from a petroleum spill. When a spill is burning, the fire area will first be small and then spread to a point of equilibrium where it will burn as fast as it is released. At a burning rate of 1 ft per hr (0.3 m/hr), 1 gal per min (3.8 L/min) of spillage will reach an equilibrium burning area of about 8 sq ft (0.74 m^2). Thus, a 10 gpm (38 L/min) spill rate will be in equilibrium with 80 sq ft (7.4 m^2) of burning area; 100 gpm at 800 sq ft (380 L/min at 74 m^2), etc. For liquids having higher burning rates, the area would be smaller; and for those liquids having lower burning rates, the area would be larger. Also, the terrain will have an influence on the shape of the burning area.

FIRE PREVENTION METHODS

Whenever flammable and combustible liquids are stored or handled, the liquid is usually exposed to the air at some stage in the operation, except where the storage is confined to sealed containers that are not filled or opened on the premises or where handling is in closed systems and vapor losses are recovered. Even when the storage or handling is in a closed system, there is always the possibility of breaks or leaks which permit the liquid to escape. Therefore, ventilation is of primary importance to prevent the accumulation of flammable vapors. It is also good practice to eliminate sources of ignition in places where low flash point flammable liquids are stored, handled, or used, even though no vapor may ordinarily be present.

Whenever possible in manufacturing processes involving flammable or combustible liquids, equipment such as compressors, stills, towers, and pumps should be located in the open. This will lessen the fire potential created by the escape and accumulation of flammable vapors. Gasoline and almost all other flammable liquids produce heavier than air vapors which tend to settle on the floor or in pits or depressions. Such vapors may flow along the floor or ground for long distances, be ignited at some remote point, and flash back. The removal of such vapors at the floor level (including pits) is usually the proper method of ventilation. Convection currents of heated air or normal vapor diffusion may carry even heavy vapors upward, and in such instances, ceiling ventilation may also be desirable. Ventilation to eliminate flammable vapors may be either natural or artificial. Although natural ventilation has the advantage of not being dependent on manual starting or on power supply, it is not so easily controlled as is mechanical ventilation, where natural ventilation depends upon temperature and wind conditions. Mechanical ventilation should be used wherever there are extensive indoor operations involving flammable and combustible liquids.

Explosion Venting

In rooms or buildings where possible explosions of flammable vapors may occur, it is recommended that relief through explosion venting be provided for at least Class 1A liquids and unstable liquids.

Substitution of Nonflammable Liquids

The hazard created by flammable liquids may be avoided or reduced by the substitution of relatively safe materials. Such materials should be stable, have a low toxicity, and be either nonflammable or have a high flash point. For example, trichloroethylene, though higher in price, is nonflammable at ordinary temperatures, and may for some uses be substituted for a more hazardous flammable solvent. Tetrachloroethylene (perchloroethylene) is another nonflammable liquid. However, even though these products are less toxic than carbon tetrachloride, they should be used only in well ventilated areas.

There are several commercial type stable solvents available which have flash points from 140 to 190°F (60 to 88°C) and which have a comparatively low degree of toxicity.

Specially refined petroleum products, first developed as "Stoddard Solvent" but now sold by different companies under a variety of trade names, have solvent properties approximating gasoline, but with fire hazard properties similar to those of kerosene. A danger in their use lies in the possibility that persons believing that they are using a safe solvent without fire hazard may neglect ordinary precautions which would be observed with a liquid such as kerosene. When heated to above their flash point (about 100°F or 38°C), these solvents produce vapors as flammable as those of gasoline at its flash point temperature.

Several other types of commercial solvents which are mixtures of liquids with differing rates of evaporation are available. Some are mixtures of gasoline or one of the various naphthas and a chlorinated solvent having a different and often higher evaporation rate than the flammable solvent which, over a period of time, would leave the original low flash point solvent. These mixtures create a toxicity hazard as well as a fire hazard, and their use in open containers should be discouraged.

Solvents and solvent vapors are toxic in varying degrees, and ventilation is almost always necessary to keep vapor concentration within safe limits.

Bibliography

References Cited

Burgess, D. S., Strasser, A., and Grumer, J. 1961. "Diffusion Burning of Liquid Fuels in Open Trays." *Fire Research Abstracts and Reviews.* Vol 3. 1961. pp 177-192.

Mitchell, F. C., and Vernon, H. C. 1938. "Effect of Pressure on Explosion Hazards." *NFPA Quarterly.* Vol 31 No 4. Apr. 1938. pp 306-313.

NFPA Codes, Standards, Recommended Practices and Manuals. (See the latest *NFPA Codes and Standards Catalog* for availability of current editions of the following documents.)

NFPA 30, *Flammable and Combustible Liquids Code.*

NFPA 30A, *Automotive and Marine Service Station Code.*

NFPA 31, *Standard for the Installation of Oil Burning Equipment.*

NFPA 321, *Standard on Basic Classification of Flammable and Combustible Liquids.*

NFPA 325M, *Fire Hazard Properties of Flammable Liquids, Gases, and Volatile Solids.*

NFPA 327, *Procedures for Cleaning or Safeguarding Small Tanks and Containers.*

NFPA 385, *Standard for Tank Vehicles for Flammable and Combustible Liquids.*

NFPA 491M, *Manual of Hazardous Chemical Reactions.*

Additional Readings

Accident Prevention Manual for Industrial Operations, 7th ed., National Fire Safety Council, Chicago, 1974.

Alger, R. S., et al., "Some Aspects of Structures of Turbulent Pool Fires," *Fire Technology,* May 1979.

Atallah, Sami, and Allan, Donald S., "Safe Separation Distances from Liquid Fuel Fires," *Fire Technology,* Vol. 7, No. 1, Feb. 1971, pp. 47-56.

Babrauska V., "Estimating Large Pool Fire Burning Rates," *Fire Technology,* Nov. 1983.

Baker, W., *Explosions in Air*, University of Texas Press, Austin, 1973.

Blinov, V. I., and Khudiakov, G. N., "Certain Laws Governing Diffusive Burning of Liquids," *Academiia Nauks, SSR Doklady*, 1957, pp. 1094-1098.

Burgess, D. S., and Zabetakis, M. G., "Fire and Explosion Hazards Associated with LNG," *Report 6099*, U.S. Bureau of Mines, Washington, DC, 1962.

Cline, D. D., and Koenig, L. N., "The Transient Growth of an Unconfined Pool Fire," *Fire Technology*, Vol. 19, No. 3, Aug. 1983, pp. 149-162.

Condensed Guide to Chemical Hazards, U.S. DOT-U.S. Coast Guard, Washington, DC. 1974.

"Controlling the Power of Flammable Liquids," *Pamphlet No. P7045*, Factory Mutual Engineering Corp., Norwood, MA.

De Ris, J., and Orloff, L., "A Dimensionless Correlation of Pool Burning Data," *Combustion and Flame*, Vol. 18, 1972, pp. 381-388.

DiNenno, Philip J., "Simplified Radiation Heat Transfer Calculations from Large Open Hydrocarbon Fires," TR 82-9, Society of Fire Protection Engineers, Boston, MA, 1982.

Donakowski, T. D., "Is Liquid Hydrogen Safer Than Liquid Methane?" *Fire Technology*, Vol. 17, No. 3, Aug. 1981, pp. 183-188.

"Evaluation of the Fire Hazard of Water-Borne Coatings," *Scientific Circular 804*, National Paint and Coatings Association, Washington, DC, 1977.

Factory Mutual Engineering Corporation, *Handbook of Industrial Loss Prevention*, 2nd ed., McGraw-Hill, NY, 1967.

"The Fearsome Fireballs of Flammable Liquids," *The Sentinel*, Vol. 35, No. 6, Nov.-Dec. 1979, pp. 3-9.

Flash Point Index of Trade Name Liquids, 9th ed., National Fire Protection Association, Quincy, MA, 1978.

Gann, Richard G., and Manka, Michael J., "Ignitability of Decomposed Transformer Fluids," *Fire Technology*, Vol. 18, No. 3, Aug. 1982, pp. 251-258.

Glassman, I., and Dryer, F. L. "Flame Spreading Across Liquid Fuels," *Fire Safety Journal*, Vol. 3, Nos. 2-4, Jan.-Mar., 1981.

Goodall, D. G., and Ingle, R., "The Ignition of Flammable Liquids by Hot Surfaces," *Fire Technology*, Vol. 3, No. 2, May 1967, pp. 115-128.

Gugan, K., *Unconfined Vapor Cloud Explosions*, Gulf Publishing, Houston, TX, 1978.

Handbook of Organic Industrial Solvents, 4th ed., National Association of Mutual Casualty Companies, Chicago, IL, 1972.

Hawley, G. G., ed., *The Condensed Chemical Dictionary*, 8th ed., Van Nostrand Reinhold Co., NY, 1971.

"Hazardous Materials Identification System," *Raw Materials Rating Manual*, National Paint and Coatings Association, Washington, DC.

Henry, Martin F., ed., *Flammable and Combustible Liquids Code Handbook*, 2nd ed., National Fire Protection Association, Quincy, MA, 1984.

Hottel, H. C., "Review: Certain Laws Governing Diffusive Burning of Liquids" (by Blinov and Khudiakov), *Fire Research Abstracts and Reviews*, Vol. 1, No. 2, 1959, p. 41.

Huffman, K. G., Walker, J. R., and Sliepcevich, C. M., "Interaction Effects of Multiple Pool Fires," *Fire Technology*, Vol. 5, No. 3, Aug. 1969, pp. 225-232.

Johnson, Donald M., "New Developments in Bulk Storage of Flammable Liquids," TR 81-1, Society of Fire Protection Engineers, Boston, MA, 1981.

Kirk, R. E., and Othmer, D. F., eds., *Encyclopedia of Chemical Technology*, 2nd Ed., 22 Vol., Interscience Encyclopedia, Inc., NY, 1963-1970.

Linville, J., ed., *Industrial Fire Hazards Handbook*, 2nd ed., NFPA, Quincy, MA, 1984.

"LNG Safety Research Program," *Report IS 3-1*, American Gas Association, 1974.

May, W. G., and McQueen, W., "Radiation from Large Liquified Natural Gas Fires," *Combustion Science and Technology*, Vol. 7, 1973, pp. 51-56.

Mellan, Ibert, *Industrial Solvents Handbook*, Noyes Data Corp., Park Ridge, NJ, 1977.

Merck Index of Chemicals and Drugs, 9th ed., Merck & Co., Rahway, NJ, 1976.

Minzer, G. A., and Eyre, J. A., "Large Scale LNG and LPG Pool Fires," *I. Chem. E. Symposium Series No. 71*, 1982, pp. 147-163.

Modak, A. T., "Ignitability of High Fire-Point Liquid Spills," *EPRI NP-1731*, Electric Power Research Institute, Palo Alto, CA, Mar. 1981.

Moorehouse, J., "Scaling Criteria for Pool Fires Derived from Large Scale Experiments," *I. Chem. E. Symposium Series No. 71*, 1982, pp. 165-179.

Mudan, Krishna S., "Thermal Radiation Hazards from Hydrocarbon Pool Fires," *Prog. Energy and Combustion Sciences*, Vol. 10, 1984, pp. 59-80.

Perry, J. H., and Chilton, C. H., eds., *Chemical Engineers' Handbook*, 5th ed., McGraw-Hill, NY, 1974.

Pingree, Daniel, "Looking at Fire Hazards: Hydraulic Fluids," *Fire Journal*, Vol. 59, No. 6, Nov. 1965, p. 23.

"Procedures for Working with Substances that Pose Hazards because of Flammability or Explosibility," In: *Prudent Practices for Handling Hazardous Chemicals in Laboratories*, National Academy Press, Washington, DC, 1981.

Rasbash, D. J., et al., "Properties of Fires of Liquids," *Fuel*, Vol. 35, 1956.

Safety and Fire Protection Committee, *C.M.A. Guide for Safety in the Chemical Laboratory*, 2nd ed., Van Nostrand Reinhold Co., NY, 1972.

Sawyer, W., "Relationship of Flash Points of Solvents, Resin Solutions, and Paints," *Journal of Coatings Technology*, Vol. 49, 1977, pp. 52-55.

Sax, N. I., *Dangerous Properties of Industrial Materials*, 5th ed., Van Nostrand Reinhold Co., NY, 1979.

Shimy, A. A., "Calculating Flammability Characteristics of Hydrocarbons and Alcohols," *Fire Technology*, Vol. 6, No. 2, May 1970, pp. 135-139.

Spalding, D. B. "The Combustion of Liquid Fuels," *Fourth Symposium (International) on Combustion*, 1953, pp. 847-864.

Stevens, A., "Flammable Liquids—Why the Hazard?" *Journal of Chemical Education*, Vol. 56, No. 3, Mar. 1979, pp. 119A-124A.

Thorne, P. F., "Flashpoints of Mixtures of Flammable and Non-Flammable Liquids," *Fire and Materials*, Vol. 1, 1976, pp. 134-140.

Threshold Limit Values, American Conference on Governmental Industrial Hygienists, Cincinnati, OH, 1974.

Title 46, "Shipping," Parts 146 to 149; Title 49, "Transportation," Parts 171 to 178, Code of Federal Regulations, U.S. Government Printing Office, Washington, DC.

Van Dolah, R. W., et al., "Flame Propagation, Extinguishment and Environmental Effects on Combustion," *Fire Technology*, Vol. 1, No. 2, May 1965, pp. 138-145.

Vervalin, C. H., ed., *Fire Protection Manual for Hydrocarbon Processing Plants*, Gulf Publishing Co., Houston, TX, 1973.

Weast, R. C., ed., *Handbook of Chemistry and Physics*, Chemical Rubber Co., Cleveland, OH, 1974-1975.

Welker, J. R., Pipkin, O. A., and Sliepcevich, C. M., "The Effect of Wind on Flames," *Fire Technology*, Vol. 1, No. 2, May 1965, pp 122-129.

Welker, J. R., and Sliepcevich, C. M., "Bending of Wind-Blown Flames from Liquid Pools," *Fire Technology*, Vol. 2, No. 2, May 1966, pp 127-135.

Welker, J. R., and Sliepcevich, C. M., "Burning Rates and Heat Transfer from Wind-Blown Flames," *Fire Technology*, Vol. 2, No. 3, Aug. 1966, pp 211-218.

Zimmerman, O. T., and Lavine, Irvin, *Handbook of Material Trade Names*, Industrial Research Service, Dover, NH, 1953 (plus supplements).

GASES

Wilbur L. Walls, PE

The term "gas" describes the physical state of a substance that has no shape or volume of its own but will take the shape and fill the entire volume of whatever container or other enclosure it occupies. This is in contrast to a liquid, which has no shape of its own but does have volume, and to a solid, which has both its own shape and volume. Gases are composed of extremely minute particles in constant motion. This motion affects the properties and behavior of gases; e.g., the higher the temperature, the more rapid the motion.

This chapter presents a definition of the substance known as gas and then outlines the various ways gases are classified. The hazards that gases present in storage differ from those presented when gases escape—both kinds of hazards are discussed along with the control measures to follow in an emergency. The chapter concludes with a summary of the properties and behavior of some common gases. For each gas, information is provided about its classification, chemical properties, physical properties, usage, hazards inside containers, hazards when released from containment, and emergency control.

Additional information about gases may be found in the following chapters of this HANDBOOK: Section 4, Chapter 2, "Explosions"; Section 11, Chapter 5, "Storage and Handling of Gases"; Section 12, Chapter 3, "Oxygen-Enriched Atmospheres"; Section 12, Chapter 11, "Special Systems for Explosion Damage Control"; and Section 19, Chapter 4, "Foam Extinguishing Agents and Application Systems."

GASES DEFINED

Because all substances can exist as gases depending upon the temperature and pressure applied to them, the term gas as used in this chapter is applied only to substances that exist in the gaseous state at so called "normal" temperature and pressure (NTP) conditions [approximately 70°F (21°C) and 14.7 psia (101 kPa)]. However, even at normal or near normal temperatures and pressures many substances can exist as either liquids or gases. The term "gas" is not precisely defined in NFPA standards. However, NFPA standards partially define a flammable liquid as a liquid having a vapor pressure not exceeding 40 psia at 100°F (275 kPa at 38°C). For comparative purposes, any substance or mixture of substances which in its liquid state exerts a vapor pressure greater than 40 psia at 100°F (275 kPa at 38°C) can be considered a gas.

CLASSIFICATION OF GASES

Effective handling of the great number and variety of gases in commerce and in our environment (we breathe a gasous mixture called air) requires that gases be classified. These classifications are based on certain common denominators which reflect the chemical and physical properties of gases and the primary uses of gases.

Classification by Chemical Properties

The chemical properties of gases are of primary fire protection concern due to their ability to react chemically with other materials (or within themselves) to produce potentially hazardous quantities of heat or reaction products, or to produce physiological effects hazardous to humans.

Flammable Gases: In NFPA usage, any gas which will burn in the normal concentrations of oxygen in the air is considered a flammable gas. Flammable gases will burn in air the same way flammable liquid vapors burn in air, i.e., each gas will burn only within a certain range of gas-air mixture compositions (the flammable or combustible range) and will ignite only at or above a certain temperature (the ignition temperature).

In a few instances, consideration of the width of the flammable range or the magnitude of the lower limit of flammability or both has resulted in classification of an ostensibly flammable gas as nonflammable because the chances of fire are low under certain conditions. Anhydrous ammonia is notable in this respect. While it is a flammable gas in the context of the previous paragraph, it is classified by the DOT (U.S. Department of Transportation) as a nonflammable gas for purposes of transportation in interstate commerce.

Wilbur Walls was the NFPA Gases Field Service Engineer from 1962 until his retirement in 1984.

Although flammable liquid vapors and flammable gases exhibit similar combustion characteristics, the term "flash point," which describes a common and useful combustion property of flammable liquids, has no practical significance for flammable gases. The flash point is basically a measure of the temperature at which a flammable liquid produces sufficient vapors for combustion and this temperature is always below the normal boiling point. A flammable gas exists normally at a temperature exceeding its normal boiling point, even when the gas is in the liquid state (as it often is in shipment and storage). Thus the gas exists at a temperature which not only exceeds its flash point, but which is usually well above it.

Nonflammable Gases: Nonflammable gases are those that will not burn in any concentration of air or oxygen. A number of these gases, however, support combustion while others suppress combustion. Those gases that support combustion are often referred to as oxidizers or oxidizing gases. They are generally either oxygen or mixtures of oxygen and other gases, such as oxygen-helium or oxygen-nitrogen mixtures, or certain gaseous oxides, such as nitrous oxide. These mixtures contain considerably more oxygen than is present in the oxygen-nitrogen mixture that comprises air.

Gases that will not support combustion are generally known as inert gases. Among the most common are nitrogen, argon, helium, and other rare gases in the atmosphere as well as carbon dioxide and sulfur dioxide. (There are some metals, however, that are combustible and can react vigorously in carbon dioxide or nitrogen atmospheres, e.g., magnesium.)

Reactive Gases: Most gases can be made to react chemically with some other substance under some conditions. Therefore the term "reactive gas" is used to distinguish gases that will either react with other materials or within themselves (producing potentially hazardous quantities of heat or reaction products) by a chemical reaction other than burning (combustion) and under reasonably anticipated initiating conditions of environmental heat, shock, etc.

Fluorine is an example of a highly reactive gas because it reacts with practically all organic and inorganic substances at normal temperatures and pressures, often fast enough to result in flaming. Another example is the reaction between chlorine (a nonflammable gas) and hydrogen (a flammable gas), which can produce flames.

Several gases can rearrange themselves chemically when subjected to reasonably anticipated conditions of environmental heat and shock, including fire exposure to their containers, with the production of potentially hazardous quantities of heat or reaction products. Examples of these gases are acetylene, methyl acetylene, propadiene, and vinyl chloride. For transport or storage these gases are usually mixed with other substances or placed in special containers to stabilize them against reasonably anticipated reaction initiators.

Toxic Gases: Certain gases can present a serious life hazard if they are released into the atmosphere. These gases, which are poisonous or irritating when inhaled or contacted, include chlorine, hydrogen sulfide, sulfur dioxide, ammonia, and carbon monoxide, among others. The presence of these gases may complicate fire fighting efforts by exposing fire fighters to toxic hazard.

Classification by Physical Properties

These properties are of primary fire protection concern because they affect the physical behavior of gases while they are inside containers and after any accidental release from containers. Until they are used, gases must be completely confined in containers during transportation, transfer, and storage. In any situation, the quantity (weight) of gas is important since gases are inherently lighter than liquids or solids. It is a matter of practical economic necessity and the ease of usage that gases be packaged in containers that contain as much gas as is practical. These requirements have resulted in transportation and storage of gases in the liquid as well as the gaseous state. Distinguishing between gases in a liquid or gaseous state is important for the application of sound fire prevention and protection practices.

Compressed Gases: For purposes of this chapter, a compressed gas is one that exists solely in the gaseous state under pressure at all normal atmospheric temperatures inside its container. The pressure basically depends on the pressure to which the container was originally charged, and on how much gas remains in the container, although the gas temperature will have some effect. There are no universally defined lower or upper limits to the container pressure. In the United States, the lower limit is customarily considered to be 25 lb per sq in. gage (273 kPa) at normal temperatures (70 to 100°F or 21 to 38°C). The upper limit is restricted only by the economics of container construction and is usually in the range of 1,800 to 3,600 psig (12 512 to 24 923 kPa).

A container of compressed gas is still rather limited in the weight of gas it can hold. For example, the largest common portable cylinder of compressed oxygen contains only about 20 lb (9 kg) of oxygen, or about 245 cu ft (6.9 m^3) of oxygen measured at normal temperature and pressure (NTP) of 70°F (21°C) and 14.7 psia (101 kPa).

Liquefied Gases: For purposes of this chapter, a liquefied gas is one that, at normal atmospheric temperatures inside its closed container, exists partly in the liquid state and partly in the gaseous state; and under pressure as long as any liquid remains in the container. The pressure basically depends on the temperature of the liquid, although the quantity of liquid can affect this under some conditions.

A liquefied gas is much more concentrated than a compressed gas. For example, the aforementioned compressed oxygen cylinder could hold about 116 lb (53 kg) of liquefied oxygen, or about 1,400 cu ft (39.6 m^3) of oxygen measured at NTP, which is about six times greater. This comparison is valid only for illustrative purposes because compressed oxygen and liquefied oxygen will not be found in the same type of container.

Cryogenic Gases: For purposes of this chapter, a cryogenic gas is a liquefied gas which exists in its container at temperatures far below normal atmospheric temperatures, but usually slightly above its boiling point at NTP, and at correspondingly low to moderate pressures. A principal reason for this distinction with respect to a liquefied gas is that a cryogenic gas cannot be retained indefinitely in a container. Heat from the atmosphere, which can be slowed but not prevented from entering the container, tends to continually raise the container pressure. If the gas is confined, the resulting pressure could greatly exceed any feasible container strength.

Some pertinent physical properties of liquefied gases, including cryogenic gases, are given in Table 5-5A. It is emphasized that these descriptions of gas classifications by physical properties pertain only to usage in this chapter. Somewhat different descriptions of the terms compressed gas, liquefied gas, and cryogenic gas may be found in codes and regulations, particularly the DOT Hazardous Materials Regulations applicable to interstate transportation. For example, many liquefied gases are classified as compressed gases by the DOT.

Classification by Usage

Standard and code making organizations and general industry often classify gases by their principal use. This classification scheme is not as precise as the preceding scheme and there is much overlap in uses with the gases.

Fuel Gases: These gases are flammable gases customarily used for burning with air to produce heat which in turn is used as a source of heat (comfort and process), power, or light. By far the principal and most widely used fuel gases are natural gas and the liquefied petroleum gases (butane and propane).

Industrial Gases: These include the entire range of gases classified by chemical properties customarily used in industrial processes, for welding and cutting, heat treating, chemical processing, refrigeration, water treatment, etc.

Medical Gases: By far the most specialized usage classification, medical gases are used for medical purposes such as anesthesia and respiratory therapy. Oxygen and nitrous oxide are common medical gases.

BASIC HAZARDS OF GASES

A systematic evaluation of gas hazards distinguishes between the hazards presented by a gas when it is confined in a container and the hazards presented when it escapes from a container—even though both of these hazards may be simultaneously present in a single incident.

Hazards Within Containers

Gases expand when heated, producing an increase in pressure on a container which can result in gas release and/or cause container failure. In addition, containers can lose their strength and fail during a fire. Compressed and liquefied gases are affected somewhat differently when heated.

A compressed gas (solely in the gaseous state) simply attempts to expand and follow the classic gas behavior laws. No actual gas follows these laws exactly, but the laws of Boyle and Charles are sufficiently accurate to predict the behavior of compressed gases under commonly encountered conditions. It is essential, however, that a consistent set of U.S. customary or SI units be used in calculations using the so called "ABSOLUTE" values for temperature and pressure. In the formulas, T =absolute temperature (deg F +459; or deg C +273), P =absolute pressure (gage pressure in lb per sq in. +14.7 psi; or kPa), and V =volume in cu ft or m^3 or any other suitable unit of volume.

Table 5-5A. Physical Properties of Cryogenic Gases

(Compiled from data in the *Handbook of Compressed Gases*, Compressed Gas Association)

	Normal Boiling Point								Normal Conditions 70°F, 14.7 psi (21°C, 101 kPa)			
	Temp		Liquid Density		Gas Density		Latent Heat		Gas Density		Cu Ft Gas from 1 cu ft Liquid	m3 Gas from 1 m3 Liquid
Name	°F	°C	lb/cu ft	kg/m3	lb/cu ft	kg/m3	Btu/lb	kJ/kg	lb/cu ft	kg/m3		
Air	−317.8	−194.3	54.6	87.5	—	—	88.2	205.1	0.075	0.120	728	726
Argon	−302.6	−158.9	86.98	139.3	0.356	0.570	70.2	163.3	0.103	0.165	842	842
Carbon monoxide	−312.7	−191.5	51.12	81.9	—	—	92.8	215.8	0.073	0.117	706	706
Ethylene	−154.8	−103.7	35.4	56.7	0.130	0.208	208.0	483.8	0.072	0.115	487	487
Fluorine	−306.6	−188.1	94.1	150.7	0.363	0.581	74.1	172.3	0.098	0.160	961	961
Helium	−452.1	−268.9	7.8	12.5	0.106	0.170	10.3	23.9	0.01	0.016	754	754
Hydrogen	−423.0	−252.8	4.43	7.1	0.084	0.134	192.7	448.2	0.005	0.008	850	850
Methane	−258.7	−161.5	26.5	42.4	0.111	0.178	219.2	509.8	0.042	0.067	636	636
Nitrogen	−320.4	−195.8	50.46	80.8	0.288	0.461	85.7	199.3	0.072	0.115	696	696
Oxygen	−297.4	−183.0	71.27	114.2	0.296	0.474	91.7	213.3	0.083	0.133	861	861

Boyle's Law

Boyle's Law states that the volume occupied by a given mass of gas varies inversely with the absolute pressure if the temperature is not allowed to change, or

$$PV = \text{constant}$$

Charles' Law

Charles' Law states that the volume of a given mass of gas is directly proportional to the absolute temperature if the pressure is kept constant. Thus:

$$\frac{V}{T} = \text{constant}$$

Therefore, for most gases, within practical working limits, the relation between temperature, pressure, and volume may be closely approximated by the following formula:

$$\frac{T_1}{T_2} = \frac{P_1 \times V_1}{P_2 \times V_2}$$

T_1, P_1, and V_1 refer to initial conditions; T_2, P_2, and V_2 refer to changed conditions to be determined.

EXAMPLE 1: Assume a 10 cu ft cylinder of compressed gas at 70°F has a gage pressure of 1000 psi. Assume that the temperature is increased to 150°F. What will be the pressure?

$$T_1 = 70 + 459 = 529$$
$$P_1 = 1000 + 14.7 = 1014.7$$
$$V_1 = 10$$
$$T_2 = 150 + 459 = 609$$
$$P_2 = \text{to be determined}$$
$$V_2 = 10$$

Substituting in the formula:

$$\frac{529}{609} = \frac{1014.7 \times 10}{P_2 \times 10}$$

$$P_2 = 1014.7 \times \frac{609}{529} = 1170$$

Therefore, gage pressure will be 1170 —14.7, or 1155.3 psi.

EXAMPLE 2: Assume a 2.0 m^3 cylinder of compressed gas at 20°C has a gage pressure of 1200 kPa. Assume that the gas is further compressed into a 1.0 m^3 cylinder and maintained at 50°C. What will be the gage pressure?

$$T_1 = 20 + 273 = 293$$
$$P_1 = 1200 + 101 = 1301$$
$$V_1 = 2.0$$
$$T_2 = 50 + 273 = 323$$
$$P_2 = \text{to be determined}$$
$$V_2 = 1.0$$

Substituting in the formula:

$$\frac{293}{323} = \frac{1301 \times 2.0}{P_2 \times 1.0}$$

$$P_2 = 2868$$

A liquefied gas, including a cryogenic gas (which is partly in the liquid state) exhibits a more complicated behavior because the net end result of heating is the combination of three effects. First, the gas phase is subject to the same effect as for a compressed gas. Second, the liquid attempts to expand, compressing the vapor. Third, the vapor pressure of the liquid increases as the temperature of the liquid increases. The combined result of these effects is an increase in pressure when the container is heated.

A most serious pressure rise can occur if the liquid expansion results in the container becoming full of liquid (the initial gas phase condensing). If this happens, a small amount of additional heating results in a large increase in pressure. For this reason, it is vital never to place more liquefied gas in the liquid phase into a container than can be accommodated—leaving a gas space—if the liquid temperature is raised to a level commensurate with the expected ambient temperatures. The proper quantity varies considerably with the liquefied gas and the factors which affect expected temperature rises, such as the temperature of the liquid when placed into the container, the size of the container, and whether the container is insulated or installed above or below ground. The quantities permitted are commonly expressed as "filling densities" or "loading densities" and are specified for specific gases (in some cases, groupings of similar gases) by codes, standards, and regulations. Filling densities expressed in terms of weight are absolute values, i.e., the designated weight of gas may always be placed into the container. Filling densities expressed in terms of volume, however, must always be qualified by specifying the liquid temperature.

Table 5-5B shows an example of how filling densities are expressed for liquefied gases—in this case, liquefied petroleum gases stored at normal temperatures in uninsulated containers. Note that larger quantities of higher specific gravity materials are permitted (reflecting the fact that these liquids tend to expand less) and that larger containers can be filled more than smaller ones can (reflecting the fact that it takes longer for them to absorb heat from atmospheric temperatures or solar radiation). Also, underground containers can be filled even more (reflecting that fact that their ambient temperatures are relatively constant and well below summer atmospheric temperatures).

Containers of compressed or liquefied gases can represent high levels of potential energy release due to the concentration of matter by compression or liquefaction. Container failure releases this energy—often extremely rapidly and violently—with a simultaneous release of gas to the surroundings and propulsion of the container or container pieces. Compressed gas container failures are distinguished more by the flying missile hazard than by the results of gas release because these containers contain lesser quantities of gas. Liquefied gas container failures can release larger quantities of gas.

Overpressure Relief Devices

Spring loaded pressure relief valves or bursting discs, or sometimes both, are provided on most compressed and liquefied gas containers to limit container pressure to a level the container can safely withstand, although fusible plugs are sometimes used on smaller containers. The start-to-discharge pressure settings of these devices are related to the strength of the container. In most cases, the relieving capacity (in terms of gas flow rate through them) is based upon consideration of heat input rates resulting from fire exposure because this is generally the largest anticipated source of heat. In some instances, such as underground or insulated containers, other sources of overpressure may be the dominant effect of fire exposure.

As noted, the relieving capacity of these devices is based upon discharge of gas. In the case of liquefied gas containers exposed to fire, it is possible to have conditions such that liquid will be discharged instead of gas, i.e., when the container is tipped over. Under such conditions, the relieving capacity will be reduced—in some cases as much as 60 to 70 percent. With the possible exception of the fire exposure condition, this reduction is of little

TABLE 5-5B. Maximum Permitted Filling Density
(LP-G Stored at Normal Temperatures in Uninsulated Containers)

Specific Gravity at 60°F (15.6°C)	Above ground Containers 0 to 1,200 gal.* % of WWC†	Above ground Containers 0 to 1,200 gal.* Vol. % at 60°F‡	Above ground Containers Over 1,200 gal.* % of WWC†	Above ground Containers Over 1,200 gal.* Vol. % at 60°F‡	All Underground Containers % of WWC†	All Underground Containers Vol. % at 60°F‡
.496—.503	41	82.0	44	88.0	45	90.0
.504—.510	42	82.6	45	88.5	46	90.5
.511—.519	43	83.5	46	89.4	47	91.3
.520—.527	44	84.3	47	90.0	48	91.8
.528—.536	45	84.7	48	90.3	49	92.2
.537—.544	46	85.2	49	90.7	50	92.5
.545—.552	47	85.6	50	91.2	51	93.0
.553—.560	48	86.4	51	91.8	52	93.6
.561—.568	49	87.0	52	92.3	53	94.1
.569—.576	50	87.5	53	92.7	54	94.4
.577—.584	51	87.9	54	93.1	55	94.8
.585—.592	52	88.3	55	93.5	56	95.1
.593—.600	53	88.9	56	94.0	57	95.6

* Total Water Capacity, U.S. Gallons (1200 U.S. Gallons equals 4.54 m^3)
† WWC—Water Weight Capacity
‡ 60°F = 15.6°C

practical significance. Even in the case of fire exposure, this situation is not as critical as it may appear because prevention of container failure under fire exposure conditions requires other safeguards in addition to the overpressure relief device.

Instances of failure of overpressure devices are rather rare even though inspection, maintenance, and replacement of these devices is essentially unregulated and they are subject to potentially harmful influences.

Containers of certain poisonous and highly toxic gases do not have overpressure relief devices because the overall hazard of a prematurely operating or leaking relief device outweighs the hazard of container failure from overpressure. It is also the practice in some countries, especially in Asia and the Mediterranean area, not to provide overpressure protection on liquefied gas containers—a practice which reflects the widespread use and installation of such containers inside buildings, including homes, and commercial and institutional buildings. In such cases, the hazard of gas release indoors due to operation of the relief device is felt to outweigh the hazard of container failure due to overpressure.

The Liquefied Gas BLEVE

Instances where containers of liquefied gases fail and break into two or more pieces are common enough to warrant treatment in some detail. These failures are described as Boiling Liquid-Expanding Vapor Explosions, or BLEVEs (pronounced "blevey"), a type of pressure release explosion.

All liquefied gases are stored in containers at temperatures above their boiling points at NTP and remain under pressure only so long as the container remains closed to the atmosphere. This pressure ranges from less than 1 psi (6.895 kPa) for some cryogenic gas containers to several hundred psi (kPa) for noncryogenic liquefied gas containers at normal storage temperatures. If the pressure is reduced to atmospheric, such as through container failure, the substantial heat which is in effect "stored" in the liquid causes very rapid vaporization of a portion of the liquid, to a degree directly proportional to the temperature difference between that of the liquid at the instant of container failure and the normal boiling point of the liquid. For many liquefied flammable gases, this temperature difference at normal atmospheric temperatures can result in vaporization of about one third of the liquid in the container.

Because overpressure relief devices are set to discharge at pressures corresponding to liquid temperatures above normal atmospheric temperatures (to prevent premature operation), the liquid temperature is higher than this if container failure occurs when a relief device is functioning. Therefore, more liquid is vaporized under these conditions—often over one half of the liquid in the container. This is the usual situation when a container fails from fire exposure. The remaining liquid unvaporized is refrigerated by the "self-extraction" of heat when the pressure is reduced to atmospheric and is cooled to near its normal boiling point.

Liquid vaporization is accompanied by a large liquid-to-vapor expansion. (See Table 5-5A.) It is this expansion process which provides the energy for propagation of cracks in the container structure, propulsion of pieces of the container, rapid mixing of the vapor and air resulting in the characteristic fireball upon ignition by the fire which caused the BLEVE, and atomization of the remaining cold liquid. Many of the atomized droplets burn as they fly through the air. However, it is not uncommon for the cold liquid to be propelled from the fire zone too quickly for ignition to occur and fall to earth still in liquid form. In one case, dissolved spots in asphalt paving were noted up to ½ mile (0.8 km) from the site of an LP-Gas BLEVE. In other BLEVEs, fire fighters have been cooled by cold liquid passing in their vicinity.

Reduction of internal pressure to atmospheric level in a container results from structural failure of the container. Failure is most often due to weakening of the container metal from flame contact; however, this will happen if the container is punctured or fails for any other reason.

As shown in Figure 5-5A, the strength of carbon steel

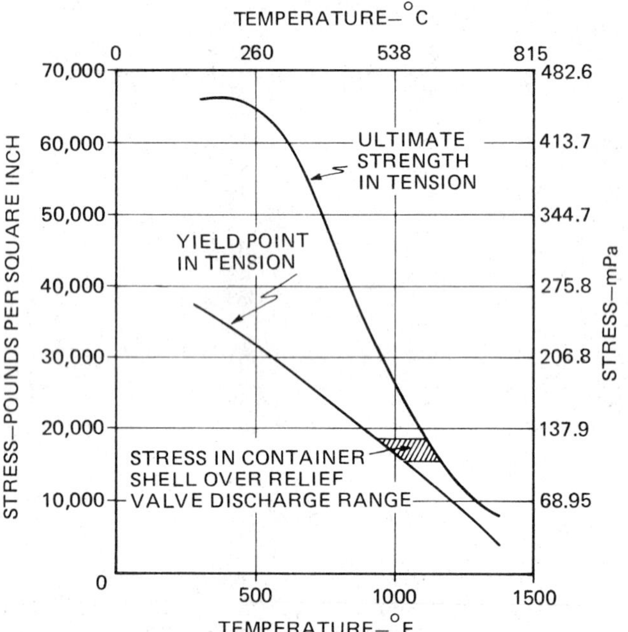

FIG. 5-5A. Behavior of liquefied gas container metal (carbon steel) when exposed to fire.

steadily decreases with temperature increases above about 400°F (204°C). Figure 5-5A is based upon a typical low carbon steel. The curves will vary quantitatively with other steels, but the loss of strength with increasing temperature is valid for all common metals and the critical temperatures are well below those attainable in a fire.

Figure 5-5A also shows why entirely satisfactory performance of a spring loaded relief valve to design parameters cannot prevent a BLEVE. By its nature, such a valve cannot reduce the pressure to atmospheric but only to a point somewhat below its start-to-discharge pressure. Therefore, the liquid will always be at a temperature above its normal boiling point, pressure will remain inside the container, and the container structure will be stressed in tension. This stressed area is shown as a shaded area in Figure 5-5A for a common type of LP-Gas container having a pressure relief valve set for 250 psi (172 kPa). Again, while this area will vary for different steels or pressure vessel relief valve design characteristics, it is evident that if the metal is heated above this range (which is quite possible in the event of direct flame contact), the metal will not withstand the stress and the container will fail.

It is extremely difficult to significantly heat the container metal where it is in contact with liquid because the liquid conducts the heat away from the metal and acts as a heat absorber. For example, when the relief valve cited in this example is discharging, the propane liquid stays in the 120 to 140°F (49 to 60°C) range. As a result, the metal temperature is well within safe limits. This situation does not exist for the metal in the vapor space of the container, as vapor is relatively nonheat conductive and has little heat absorbing capacity.

In most BLEVEs where the failure is due to metal overheating, it originates in the metal of the vapor space and is characterized by both the metal stretching and thinning out and the appearance of a longitudinal tear which progressively gets larger until a critical length is reached. At this point, the failure becomes brittle in nature and propagates at sonic velocity through the metal in both longitudinal and circumferential directions. As a result, the container comes apart in two or more pieces.

Magnitude of a BLEVE: Although most liquefied gas BLEVEs that involve container failure result from fire exposure, a few BLEVEs have occurred due to container failures from other causes, such as corrosion or impact from an outside force. Impact failures are particularly noticeable in transportation accidents involving railcars and cargo vehicles. In these cases, the BLEVE generally occurs simultaneously with impact. In one instance, however, a 30,000 gal (113.5 m^3) tank car of LP-Gas was only severely weakened by impact during derailment and did not BLEVE until more than 40 hrs later. The tank car had been lifted and moved without incident in the interim. At the time of failure, however, the internal pressure was increasing as the ambient temperature was rising.

The size of a BLEVE depends basically upon the weight of the container pieces and upon how much liquid vaporizes when the container fails. This is analogous in many respects to the performance of rockets as far as propulsion of container parts is concerned. Most liquefied gas BLEVEs occur when containers are from slightly less than ½ to about ¾ full of liquid. The liquid vaporization-expansion-energy to container-piece weight ratio is such that pieces are propelled for distances up to approximately ½ mile (0.8 km). Deaths from such missiles have occurred up to 800 ft (244 m) from larger containers. Fireballs several hundred feet in diameter are not uncommon, and deaths from burns have occurred to persons as much as 250 ft (76 m) from the larger containers.

A major LP-Gas disaster with multiple BLEVEs occurred on November 19, 1984 near Mexico City, Mexico. According to a Swedish investigator who cooperated with the NFPA, and contrary to earlier published reports, this incident occurred at about 5:00 A.M. on a Monday morning while the terminal was being filled by pipeline from a refinery in another part of the country. A failure in a component of the 12 in. (25.4 mm) fill line resulted in a line break and the development of a large LP-Gas cloud which ignited upon reaching a ground level flare about 20 minutes later. The resulting explosion and fire is believed to have been the major factor in the 500 deaths reported. The flame issuing from the broken pipeline impinged on one of the four 10,000 barrel (1590 m^3) spheres which BLEVE'd about 10 minutes later. In all, approximately 15 BLEVEs were noted on seismic charts among the four 10,000 barrel (1590 m^3) spheres and 48 horizontal tanks.

The two 15,000 barrel (2384 m^3) spheres remained on their foundations but were severely damaged. (See Fig.5-5B.) The four 10,000 (1590 m^3) barrel spheres were found in pieces both on and off the terminal site. All 48 horizontal tanks were damaged in varying degrees, with major sections as far as 3,600 ft (1.1 km) from the site. Figure 5-5C illustrates the spectacular dimensions a BLEVE can reach.

Time Intervals for Fire Caused BLEVEs: The time between initiation of flame contact and a BLEVE varies

FIG. 5-5B. Two of the 15,000 barrel (2384 m³) spheres involved in the Mexico City LP-Gas disaster. Note how far out of vertical they are. This is due to partial failure of the container supports.

FIG. 5-5C. The fireball formed in a BLEVE involving LP-Gas railroad tank cars at Crescent City, IL, on June 21, 1970. The elevated water tank at lower right is a point of reference in visualizing the dimensions of the fireball.

because it depends upon such widely varying factors as the size and nature of the fire as well as the container itself. Uninsulated containers located above ground can BLEVE in the absence of water cooling in a matter of a very few minutes in the case of small containers to a few hours for very large containers. A study of such LP-Gas storage containers, ranging in size from 1,000 to 30,000 gal (3.8 to 113 m^3), showed a time range from 8 to 30 min with 58 percent occurring in 15 min or less. Data on insulated containers is meager because only cryogenic containers and some reactive gas containers are usually insulated. However, there is no doubt that insulation designed for fire exposure conditions can delay BLEVE times significantly. In one case involving an insulated LP-Gas railroad tank car, the BLEVE did not occur until 20½ hr of fire exposure—undoubtedly an extreme example. In comparison, fire tests on LP-Gas railroad tank cars, a BLEVE occurred in 93 min in the insulated case as opposed to 25 min for the uninsulated tank.

Protection Against a BLEVE: Protection for an uninsulated liquefied gas container which can be exposed to fire is provided by the application of water so that a film of water coats portions of the container not in internal contact with liquid. The methods used can range from hose streams to installation of water spray fixed systems.

The preceding data on BLEVE magnitude, time intervals, and protection is applicable only to unreactive liquefied flammable gases. The additional chemical energy in reactive gases introduces chemical factors not present in the purely physical and combustion phenomena just described, even though the outward appearance and general effects are similar. The hazard can be increased by evolution of internal heat of chemical reaction (which adds to the heat from the exposing fire) and the inability of externally applied cooling water to cool the liquid inside the container.

Combustion Within Containers

A less frequent but significant hazard of gas stored in containers is the danger of container failure from overpressure, resulting from combustion of the gas while inside the container. Flammable gas-air or gas-oxygen mixtures are seldom intentionally provided in containers, but can be established accidentally. Most of these explosions have occurred in industrial and medical gas applications where oxygen or compressed air is often used in conjunction with flammable gases. Where such a possibility is inherent in a process, e.g., an oxygen-fuel gas metal cutting system, provisions are made to prevent the occurrence of such mixtures in containers. More generally, explosions can be prevented only by education and training in proper container filling procedures. Consumers of industrial and medical gases seldom have the expertise needed to fill containers safely.

Combustion When Released from Containers

When gases are released from their containers, the hazards vary according to the chemical and physical properties of the gas and the nature of the environment into which they are released. All gases, with the exception of oxygen and air, present a hazard to life if they displace the breathing air. Nitrogen, helium, argon, and other odorless and colorless gases, are particularly hazardous because they are not detectable by the human senses. The minimum oxygen concentration in air for survival is about 6 to 10 percent (compared with the normal 21 percent) by volume, but even at higher concentrations judgment and coordination are affected.

Toxic or Poisonous Gases: These gases present obvious life hazards. They are especially dangerous when released during a fire because they can impede fire fighting efforts by either preventing or delaying access to the site by fire fighters.

Oxygen and Other Oxidizing Gases: Although these gases are nonflammable, they can make combustibles ignite at lower temperatures, accelerate combustion, and start fires by causing flames in fuel burning appliances to extend beyond their combustion chambers.

Liquefied Gases, Including Cryogenic Gases: Because of their low temperatures, these gases present a hazard to persons and property when they escape as liquids. Contact with cold liquid can cause frostbite which can be severe if the exposure is prolonged. The properties of many structural materials, particularly carbon steel and plastics, are affected by low temperatures and embrittlement, which could lead to structural failure.

Flammable Gases: Because of their prevalence, the behavior of flammable gases when released from their containers is of major interest. Released flammable gases present two basic hazards—combustion explosions and fire. Failure to distinguish between the circumstances surrounding these two hazards can result in misapplication of protective measures.

Combustion Explosions

It is useful to consider a flammable gas combustion explosion as occurring in the following sequence:

1. A flammable gas or the liquid phase of a liquefied flammable gas is released from its container, piping, or equipment (including the normal operation of an overpressure relief device). If liquid escapes, it rapidly vaporizes and produces the potentially large quantities of vapor associated with this liquid-to-vapor transition.
2. The gas mixes with the air.
3. With certain proportions of gas and air (in the flammable or combustible range), the mixture is ignitable and will burn. (See Table 5-5C.)
4. When ignited, the flammable mixture burns rapidly and produces heat rapidly.
5. The heat is absorbed by everything in the vicinity of the flame and by the very hot gaseous combustion products.
6. Nearly all materials expand when they absorb heat. The one material in the vicinity of the flame or hot gaseous combustion products that expands most when it is heated is air. Referring to the "gas laws" cited earlier in this chapter, it will be noted that air expands to double its original volume for every 459°F (237°C) it is heated.
7. If the heated air is not free to expand because, for example, it is confined in a room, the result is a rise in pressure in the room.
8. If the room structure is not strong enough to withstand the pressure, some part of the room will suddenly and abruptly move and depart from its original position, and a bang, woosh, boom, or other noise will be heard. This activity, in part, describes an explosion. Because the source of pressure is combustion, this kind of explosion is called a combustion explosion. It is also called less accurately a room explosion or vapor-air explosion.

A combustion explosion requires that a quantity of flammable gas-air mixture accumulates in an enclosure. In addition, the quantity/strength-of-enclosure relationship

TABLE 5-5C. Combustion Properties of Common Flammable Gases

Gas	Btu per Cu Ft (Gross)	mJ/m³ (Gross)	Limits of Flammability Percent by Volume in Air		Specific Gravity (Air = 1.0)	Air Needed to Burn 1 Cu Ft of Gas Cu Ft	Air Needed to Burn 1 m³ of Gas m³	Ignition Temp	
			Lower	Upper				°F	°C
Natural gas									
High inert type *Note 1*	958–1051	35.7–39.2	4.5	14.0	.660–.708	9.2	9.2	—	—
High methane type *Note 2*	1008–1071	37.6–39.9	4.7	15.0	.590–.614	10.2	10.2	900–1170	482–632
High Btu type *Note 3*	1071–1124	39.9–41.9	4.7	14.5	.620–.719	9.4	9.4	—	—
Blast furnace gas	81–111	3.0–4.1	33.2	71.3	1.04–1.00	0.8	0.8	—	—
Coke oven gas	575	21.4	4.4	34.0	.38	4.7	4.7	—	—
Propane (commercial)	2516	93.7	2.15	9.6	1.52	24.0	24.0	920–1120	493–604
Butane (commercial)	3300	122.9	1.9	8.5	2.0	31.0	31.0	900–1000	482–538
Sewage gas	670	24.9	6.0	17.0	0.79	6.5	6.5	—	—
Acetylene	1499	208.1	2.5	81.0	0.91	11.9	11.9	581	305
Hydrogen	325	12.1	4.0	75.0	0.07	2.4	2.4	932	500
Anhydrous ammonia	386	14.4	16.0	25.0	0.60	8.3	8.3	1204	651
Carbon monoxide	314	11.7	12.5	74.0	0.97	2.4	2.4	1128	609
Ethylene	1600	59.6	2.7	36.0	0.98	14.3	14.3	914	490
Methyl acetylene, propadiene, stabilized *Note 4*	2450	91.3	3.4	10.8	1.48	—	—	850	454

Note 1: Typical composition CH_4 71.9–83.2%; N_2 6.3–16.20%
Note 2: Typical composition CH_4 87.6–95.7; N_2 0.1–2.39
Note 3: Typical composition CH_4 85.0–90.1; N_2 1.2–7.5
Note 4: MAPP® Gas

must be such that the strength of some part of the enclosure must be exceeded by the pressure building potential of the mixture. If the enclosure is strong enough to withstand the pressure, a combustion explosion could not occur, because it is the manner in which the enclosure performs that basically determines whether such an explosion occurs. However, few enclosures are strong enough to withstand such pressure.

If an enclosure was full of a flammable gas-air mixture at atmospheric pressure, the enclosure would have to withstand about 60 to 110 psi (400 to 750 kPa) in order to remain intact to prevent a combustion explosion. If a reactive flammable gas is involved or oxygen enrichment occurs, even higher pressures are possible. Conventional structures are capable of withstanding pressures only in the order of ½ to 1 psi (3.5 to 7 kPa). This wide pressure disparity shows clearly that conventional structural enclosures are vulnerable even if they are not full of a flammable gas-air mixture. Experience supports this conclusion, and it has been estimated that most combustion explosions of conventional structures occur with less than 25 percent of the enclosure occupied by the flammable mixture. This fact should not be overlooked—it is erroneous to assume that a flammable gas-air mixture needs to completely fill a room or building before an explosion can occur.

The mechanics of gas accumulation in a structure are affected by the rate of gas release, whether the gas is in the liquid or gas phase, the density of the gas, and the ventilation in the structure. Classic diffusion laws are of little significance under actual conditions because the combination of extremely slow release rates and airtight structures seldom happens. For this reason, and the fact that most flammable gas-air mixtures are about 90 percent air and, thus, have about the same density as air, the density of the gas itself (i.e., whether it is lighter or heavier than air) is seldom a significant factor in gas combustion explosions involving structures.

Due to the large and rapid flammable gas-air mixture potential of liquefied gases, codes and standards impose severe limitations on the handling of such gases indoors. Considering the fireball aspect of the BLEVE, it is evident that an indoor fireball behaves similarly to an ignited gas-air mixture accumulation; the results of a BLEVE of a container located indoors can be very similar to a combustion explosion.

Combustion Explosion Safeguards

Basic combustion explosion prevention safeguards are designed to limit flammable gas-air mixture accumulation in a structure. Fundamental safeguards are the use of rugged containers and equipment which minimize the chance of leakage, the minimal quantity of gas released by emergency flow control devices, and the use of limiting orifices to minimize the quantity released. Burners are often equipped with flame failure devices which shut off the flow of gas if the flame is extinguished for any reason.

Many gases are colorless and odorless, and it is common to odorize the more widely used fuel gases to increase chances of leak detection. This is particularly true for the common fuel gases, natural gas and LP-Gas. Codes, standards and regulations customarily require that these gases be odorized so that they are detectable by humans at gas concentrations in air not exceeding one fifth of the lower limit of flammability. The odorants are usually volatile organic liquids containing chemically combined sulfur and have a characteristic gassy odor.

While it is an effective safeguard, odorization has limitations. The functioning detector (a human's sense of smell) is not always present, such as when the premises are vacant or the occupants are asleep. In addition, all odors deaden the sense of smell if inhaled long enough, and senses of smell vary. Odorants can be scrubbed from the gas by some soils, a factor where leakage stems from underground piping. Mixture accumulations can also be limited by ventilation systems in structures. These systems, however, are usually in industrial operations, and even then only rather nominal release rates can be handled by practical ventilation systems because of the necessity to condition the make up ventilation air for comfort purposes.

Ignition source control is also fundamental to combustion explosion prevention. However, this safeguard is also limited mainly to industrial operations where the many flammable gases burned in heat producing equipment provide an inherent ignition source.

In addition to combustion explosion prevention safeguards, the severity of the explosion can be reduced by special structural design whereby some elements of the structure are designed to dislocate at lower pressures, and other elements are designed to stay in place at the lower pressures which result. This is known as "explosion-venting" (not to be confused with "ventilation") and is covered in some detail in NFPA 68, *Guide for Explosion Venting.* Such building design is not practical for most ordinary buildings, however, so this practice is essentially restricted to certain industrial structures.

Flammable Gas Fires: The flammable gas fire can be considered as an aborted combustion explosion wherein an explosive quantity of flammable gas-air mixture does not accumulate because the mixture is either ignited too quickly or a confining structure is not present. As would be expected when flammable gas escapes outdoors, fires usually occur. However, if a massive release occurs, it is possible for the air or surrounding buildings to comprise enough confinement to lead to a type of combustion explosion known as an open air explosion or space explosion. Liquefied noncryogenic gases are subject to this phenomenon as are hydrogen, ethylene, and some reactive gases that have an extremely rapid rate of flame propagation. An example of the intentional provision of prompt ignition to achieve a fire instead of a combustion explosion is the use of pilots in gas burning equipment such as range ovens, water heaters, boilers, and furnaces.

Gas fire prevention safeguards include many of the combustion explosion prevention safeguards. However, unlike the combustion explosion, the destructive effects of the gas fire can be minimized by control measures applied after the fact.

GAS EMERGENCY CONTROL

Controllable gas emergencies resulting from the escape of gas from containers present two basic forms of hazard: (1) toxic, inert, or oxidizing gases can present hazards to persons or property and unignited flammable gases can be ignited, possibly explosively ("no fire" emergencies); and (2) gas fires can present thermal hazards to persons or property ("fire" emergencies) and, if such fires expose gas containers, introduce the possibility of container failure and the BLEVE. A fire in any combustible material also presents the hazard of container failure from fire exposure.

Control of "No Fire" Emergencies

Control generally consists of directing, diluting, and dispersing the gas to prevent contact with persons, preventing it from infiltrating structures if release is outdoors, and avoiding its contact with ignition sources while, if possible, simultaneously stopping the flow of escaping gas. Gas direction, dilution, and dispersion require the use of a carrier fluid; air, steam, and water have proven to be practical. The use of air, for all practical purposes, is limited to indoor situations and is an extension of the ventilation safeguards against a combustion explosion. Steam has been distributed through systems of fixed nozzles around outdoor ethylene processing equipment and where the large quantities of needed steam are available.

Water in the form of a spray, applied from hoses or monitor nozzles or by fixed water spray systems, is the most common carrier fluid. The use of hose streams is largely a fire department operation because of the personnel requirements. Fixed water spray systems for this purpose are designed differently from the more conventional spray systems used to control fire. To date, such systems have been provided at a few outdoor ethylene and Liquefied Natural Gas (LNG) facilities.

The physical properties of the escaping gas affect control techniques. With compressed gases, the density of the gas is an important factor. When such gases are colorless and odorless, control tactics may be complicated because instruments may be needed to define the extent of the hazardous area.

Liquefied gases possess an inherent visible indicator of their location because the refrigerating effect of their vaporization condenses water vapor from the air and produces a visible fog. The fog roughly defines the gas area, but invisible ignitible gas-air mixtures often extend for several feet (1 ft = 0.3 m) beyond the extremities of the visible fog.

Because noncryogenic liquefied gases contain considerable heat for vaporization, they will often vaporize so rapidly from contact with the air or ground that they will not exist in the liquid phase once they escape—at least not to the extent that a pool will form. The lower vapor pressure noncryogenic liquefied gases, such as butane and chlorine, and those with high latent heats of vaporization, such as anhydrous ammonia, are exceptions. Even the higher vapor pressure gases, such as propane, will pool when ambient temperatures are well below freezing.

The cryogenic liquefied gases, on the other hand, must obtain nearly all the heat for vaporization from ground or air contact and, therefore, will characteristically form a pool if the leak continues for a long enough time. In such cases, application of a carrier fluid will increase the vaporization rate if applied to the liquid—an often undesirable effect.

The gas produced near the source of vaporization of a liquefied gas is always heavier than air at normal temperatures because of the low temperature of the gas at that point. This, together with the associated water fog, tends to cause even normally lighter than air gases to hug the ground for some distance.

The use of foam, particularly high expansion foam, to control the flow of gas produced by a vaporizing cryogenic gas which is normally lighter than air, has been investigated to some extent. A sufficiently thick blanket of foam can warm the gas to the point where it will rise above the blanket instead of flowing horizontally near the ground. However, this is applicable only to established pools of vaporizing liquid and not to the initial high vaporization rate escape phase.

Some gases will react chemically with the carrier fluid—particularly if the carrier is steam or water. Chlorine is a notable example—the reaction produces hydrochloric acid. However, the principal problem in this respect is enlargement of the leak as the metal reacts with the acid.

A special type of "no fire" control for indoor application to flammable gases is the explosion suppression system. Actually, such systems do permit ignition to occur, but arrest the combustion of the flammable gas-air mixture before the pressure needed for a combustion explosion is obtained.

Control of Fire Emergencies

Control of fire emergencies is generally the control of heat from the fire by the application of water while, if possible, stopping the flow of escaping gas. Many gas fires can be extinguished by conventional extinguishing agents, including carbon dioxide, dry chemical, and the halogenated agents. However, the potential conversion of a gas fire into a combustion explosion if gas continues to escape after extinguishment must be recognized. The generally accepted practice is to limit actual extinguishment by agent application to small leaks which will not present a hazard if they reignite.

Methods of water application are the same as described for the "no fire" emergencies, i.e., hose streams, monitor nozzle streams, and fixed water spray systems as well as sprinkler systems. The selection of the particular application method, or combination of methods, requires a sound fire protection analysis of the existing conditions. This is particularly true where water is used to prevent a BLEVE because of the limited time available if uninsulated containers are involved. This, together with the specialized tactics needed for safety of emergency personnel, has been shown to severely tax the capabilities of hose stream application in many instances. Manual actuation of fixed systems (water spray or sprinkler) that have nozzles and dry piping in the fire area is questionable because the immediate intensity of a gas fire can quickly damage the piping before the water is turned on.

Conventional automatic sprinkler protection is limited to indoor or roofed over areas. However, sprinklers have been effective in greatly reducing the number of overpressure relief devices on cylinders that operate during a fire. In turn, the number of cylinders that may fail from contact with the burning relief device effluents is reduced; sprinkler spacing and water density must be tailored to this hazard.

Foam can control a fire in a pool or tank of cryogenic gas, but cannot extinguish it. The degree of control depends upon the extent to which the foam can cover the liquid and the length of time the foam application can be maintained.

SPECIFIC GASES

Acetylene

Classifications: Reactive, flammable, compressed, industrial.

Chemical Properties: Acetylene is comprised of carbon and hydrogen joined by a triple chemical bond which is responsible for its reactivity. In the liquid or solid states, or in the gaseous state at moderate or high pressures, acetylene can decompose rapidly with the formation of carbon and hydrogen and evolution of heat. Decomposition can be initiated by heat. Decomposition of liquid or solid acetylene also can be initiated by mechanical impact. In a confined space, the heated decomposition gases can cause overpressure and container, piping or equipment failure.

As an indication that a hazardous, heat producing reaction can occur in the absence of air, an upper flammable limit of 100 percent is often found in the literature rather than the 81 percent given in Table 5-4C. This technical misapplication of the flammable range concept has caused confusion.

Acetylene can react with certain metals to produce metallic acetylides, extremely shock sensitive explosive compounds which, if detonated even in small quantities, can initiate acetylene decomposition. Copper and some copper alloys are notable in this respect and their use must be avoided in most acetylene piping and equipment. However, these metals can be used in some components of acetylene systems under certain conditions which reflect the reaction kinetics involved. Torch tips, for example, fall into this category.

Acetylene is not toxic and has been used as an anesthetic. Pure acetylene is odorless, but acetylene in general use has a characteristic odor due to minor impurities inherent in its generation from calcium carbide or derivation from other hydrocarbons.

Physical Properties: Although acetylene is considered a compressed gas, the usual acetylene container used in transportation and storage does not exclusively contain acetylene in the gaseous phase and is unique in this respect. To assure stability under reasonably anticipated thermal and mechanical impact conditions, acetylene cylinders are filled with a porous-mass packing material containing very small pores or cellular spaces so that any volume of gas therein is correspondingly small. This limits the decomposition energy available and restricts communication between spaces. In addition, the mass is saturated with acetone—a flammable liquid in which acetylene is very soluble. In this manner, acetylene gas can be compressed in solution in a manner similar to that of carbon dioxide in water to produce carbonated water. When the pressure is reduced, e.g., by opening a cylinder valve, the gas escapes from solution and thus leaves the container in the gaseous state. Currently, a charging pressure maximum of 250 psig at 70°F (1724 kPa at 21°C) is recognized by DOT and CTC Regulations. There will be a variation of about 2.5 psig rise or fall per °F of temperature change (9 kPa 1°C). For each atmosphere (14.7 psi or 101.3 kPa) of pressure, acetone will dissolve about 25 times its own volume of acetylene so that at 250 psig (1724 kPa) acetone will dissolve about 425 volumes of acetylene.

An acetylene cylinder is illustrated in Figure 5-5D.

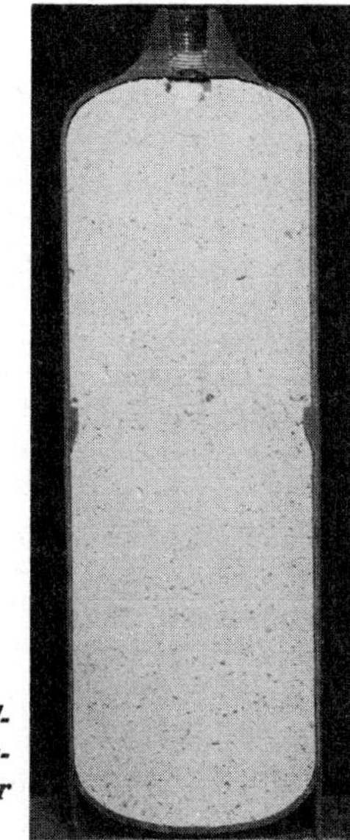

FIG. 5-5D. Cutaway of a typical acetylene cylinder showing porous filler material. Cylinder valve and packing for well at top of cylinder are not shown.

Usage: Acetylene is used primarily in chemical processing and as a fuel gas in oxygen-fuel gas cutting and welding operations. Occasionally it is manufactured at the consuming location by reacting calcium carbide (a solid) and water in acetylene generators and piped directly to gas holders or to the point of use.

Hazards Inside Containers: Acetylene is shipped in DOT and CTC cylinders as described previously and stored in these or in low pressure gas holders. It is either handled in hoses or piping systems.

Acetylene cylinders are indirectly protected against overpressure by heat actuated devices, usually fusible plugs of eutectic alloys which have low melting points similar to those alloys employed in automatic sprinklers. Unlike spring loaded pressure relief valves, operation of fusible plugs results in complete reduction to atmospheric pressure because both the compressed gas and acetone are released at the same time. Due to this mode of overpressure protection, acetylene cylinders are not subject to a BLEVE, which by definition cannot occur if the liquid contents are not above their boiling points at NTP. Under fire exposure conditions and during charging operations in acetylene cylinder charging plants, however, internal decomposition has occurred which has resulted in BLEVE effects, such as cylinder rupture and propulsion accompanied by small fireballs.

Because susceptibility to decomposition is directly related to pressure, i.e., the higher the pressure, the easier it is to initiate decomposition and the more violent the effects, an acceptable degree of stability in most piping systems containing gaseous acetylene is achieved by limiting the pressure. In general, this pressure does not exceed 15 psi (103 kPa). In applications requiring higher pressures, particularly in acetylene cylinder charging plants, special piping design features can be employed to control this hazard.

The use of acetylene in either the liquid or solid phases is especially hazardous and is prohibited by standards and codes covering conventional applications.

Hazards When Released from Containment: When acetylene is released from containment, it presents combustion explosion and fire hazards. Because of its reactivity, it is easier to ignite than most flammable gases and burns more rapidly. The latter effect increases the severity of combustion explosions and increases the difficulty of providing explosion venting. Acetylene is only slightly lighter than air. (See Table 5-5C.)

Because of acetylene's reactivity, the design of electrical equipment for use in acetylene atmospheres is unique to this material, and Group A electrical equipment (Article 500, *National Electrical Code®*) is devoted exclusively to acetylene. In practice, electrical equipment is avoided in the more likely acetylene release areas or the release potentials are reduced to a level where Group C or D equipment can be used. As a result, available Group A equipment is extremely limited in variety.

Emergency Control: Because of the limited quantities of acetylene in conventional shipping modes and storage practices, "no fire" emergencies seldom occur. Fire control measures are similar to those involved with any flammable, nontoxic gas—that is, application of water to containers and stopping the flow of escaping gas if possible.

Anhydrous Ammonia

Classifications: Flammable, liquefied (including cryogenic), industrial.

Chemical Properties: Anhydrous ammonia is comprised of nitrogen and hydrogen. While often referred to simply as "ammonia," this term is more appropriately used to describe a solution of anhydrous (which means "without water") ammonia in water. The nitrogen component is inert in the combustion reaction and accounts for the somewhat limited flammability of anhydrous ammonia as manifest by its high lower flammability limit and low heat of combustion. (See Table 5-5C.)

Moist ammonia will vigorously attack copper and zinc and many of their brass or bronze alloys. This has caused some problems in agricultural areas where, because of their similar vapor pressures, anhydrous ammonia and propane are often handled in the same storage tanks, transport containers, and associated equipment at different times of the year. LP-Gas equipment customarily utilizes brass, bronze, copper, and zinc extensively for valves, gages, piping, regulators, etc. This equipment, if converted to anhydrous ammonia service, must be changed to steel. Subsequently anhydrous ammonia must be carefully purged from a container when it is reconverted to LP-Gas service.

The characteristic pungent odor and irritant properties of anhydrous ammonia serve as warnings for this relatively toxic gas. However, the effectiveness of the warning depends upon the rate of release. Large clouds of ammonia gas have been rapidly produced from large liquid leaks, so that people have been trapped and killed before they could evacuate the area.

Physical Properties: At its normal boiling point of −28°F (−33°C), anhydrous ammonia has a liquid density of 42.6 lb per cu ft (682.4 kg/m^3), a gas density of 0.055 lb per cu ft (0.88 kg m^3), and a latent heat of vaporization of 589.3 Btu/lb (1371 kJ/kg). At NTP, the gas has a density of about 0.045 lb per cu ft (0.72 kg/m^3), and vaporization of 1 cu ft (0.3 m^3) of liquid will produce about 885 cu ft (25 m^3) of gas.

Usage: Anhydrous ammonia is used primarily as an agricultural fertilizer, as a refrigerant, and as a source of hydrogen for metal heat treating and semi-conductor manufacturing special atmospheres.

Hazards Inside Containers: Anhydrous ammonia is shipped in DOT and CTC cylinders, DOT Specification cargo trucks and railroad tank cars and barges, and is stored in cylinders, American Society of Mechanical Engineering (ASME) Code tanks, and in cryogenic form in insulated American Petroleum Institute (API) tanks.

Anhydrous ammonia containers with a capacity of less than 165 lb (79 kg) are not required to be equipped with overpressure protection devices. This reflects the toxicity of the gas and is a safety trade off between the hazards of container rupture from overpressure and gas release due to operation of an overpressure protective device, especially indoors where the smaller containers are often found.

BLEVEs of uninsulated anhydrous ammonia containers are infrequent, reflecting the limited flammability of the gas which minimizes the probability of it supplying the exposing fire. What fires have occurred, have been in other combustibles.

Hazards When Released from Containment: Anhydrous ammonia presents combustion explosion and fire hazards (as well as a toxicity hazard) when released from containment. However, its high lower limit of flammability and low heat of combustion reduce these hazards substantially.

If anhydrous ammonia is released outdoors, it is difficult for it to reach the lower flammability limit concentration except for small zones in the immediate vicinity of the leak. Even where large quantities of liquid are released, ignitible concentrations tend to be in discontinuous pockets and this, together with the low heat of combustion, reduces the possibilities of sustained burning. Experience indicates that similar circumstances apply when the release occurs indoors in conventional and reasonably well ventilated buildings. In unusually tight rooms, such as refrigerated process or storage areas, however, the release of liquid or large quantities of gas can result in the accumulation of hazardous quantities of a flammable mixture and result in a combustion explosion. In such cases, even though the low heat of combustion produces lower pressures than most flammable gases, the pressure is enough to do major structural damage. As a result of these factors, anhydrous ammonia fires are infrequent and, where ignition does occur, a combustion explosion is the likely result. In a NFPA fire record analysis of 36 incidents from 1929 through 1969 where released gas or liquid was ignited, 28 incidents, or 78 percent, resulted in a combustion explosion. All occurred indoors.

Emergency Control: At normal temperatures, anhydrous ammonia gas weighs about 0.6 times as much as air. Because of its pronounced solubility in water, the spread of escaping anhydrous ammonia gas can be readily controlled by water spray. If hose streams are used, the toxic and inerting properties of anhydrous ammonia require fire fighters to use protective breathing apparatus. Where liquid contact is likely, full protective clothing should be worn. If release of liquid in cryogenic form occurs, pooling is possible and application of water to pools should be avoided, unless the vapors can be controlled, in order to prevent increasing vaporization rate.

Carbon Dioxide

Classifications: Nonflammable (inert), liquefied, industrial.

Chemical Properties: Carbon dioxide is composed of carbon and oxygen. The ability of carbon to chemically link with as much oxygen as possible prohibits its further oxidation (combustion), making it nonflammable. Although nontoxic, carbon dioxide can cause asphyxiation due to the displacement of air.

Physical Properties: Carbon dioxide exhibits some unique physical properties. At a temperature of −69.9°F (−56.6°C) and a pressure of 60.4 psig (416.4 kPa), it can exist in a container simultaneously as a liquid, solid, and gas (the triple point). At temperatures and pressures above these and below 87.8°F (30.8°C), it exists as both liquid and gas. Above 87.8°F (30.8°C), it exists only in the gaseous phase. At normal temperatures, the gas is about 1½ times heavier than air.

Usage: Carbon dioxide is used primarily to carbonate beverages, to provide an inert atmosphere, and to extinguish fires.

Hazards Inside Containers: Carbon dioxide is shipped in DOT and CTC cylinders and insulated DOT Specification cargo trucks and railroad tank cars. It is stored in cylinders at a pressure of about 850 psi at 70°F (5861 kPa at 21°C) or in insulated ASME Code tanks at pressures of about 200 to 312 psi (1380 to 2150 kPa) and temperatures of −20 to +40°F (−28.8 to +4.4°C).

Cylinders are provided with overpressure protection in the form of frangible (bursting) discs. Insulated tanks are equipped with pressure relief valves. Carbon dioxide cylinders have BLEVE'd as a result of corrosion, with enough energy to nearly demolish small cargo vehicles. The NFPA has no reports of BLEVEs of other types of containers.

Hazards When Released from Containment: In addition to the hazard of asphyxiation, contact with cold vaporizing carbon dioxide can cause frostbite. For these reasons, actuation of automatic room flooding extinguishing system applications should incorporate a time delay if personnel occupy the protected area.

Emergency Control: Carbon dioxide emergencies are "no fire" emergencies involving potential asphyxiation hazards. Structure ventilation indoors and water spray control outdoors are applicable. Self-contained breathing apparatus should be used.

Chlorine

Classifications: Reactive, nonflammable, liquefied, toxic, industrial.

Chemical Properties: Chlorine is a basic chemical element. Although it is nonflammable, it can react with many organic materials corrosively and, in some instances, explosively, particularly with acetylene, turpentine, ether, gaseous ammonia, hydrocarbons, most fuel gases, and finely divided metals.

The reactivity of chlorine necessitates special attention to the container, piping, and equipment materials of construction. Below 230°F (110°C), steel, copper, iron, and lead are widely used. In contact with water, chlorine forms hypochlorous and hypochloric acids which are corrosive to most metals.

Chlorine is toxic enough to be considered as a poison and has been used in warfare as a poison gas. Liquid chlorine can burn skin. Its sharp odor serves as a warning.

Physical Properties: The normal boiling point of chlorine is about −30°F (-34.4°C). At 32°F (0°C), the liquid density is 91.7 lb per cu ft (1468 kg/m^3) and the gas density is 0.2 lb per cu ft (3.2 kg/m^3). The latent heat of vaporization at the normal boiling point is 123.7 Btu/lb (287.7 kJ/kg). At 32°F (0°C), the vaporization of 1 cu ft of liquid will produce about 458 cu ft of gas (1 m^3 of liquid will produce about 458 m^3 of gas). Chlorine has a greenish yellow color.

Usage: Chlorine is used primarily in chemical processing, bleaching, purification of drinking water and swimming pools, and sanitation of industrial and sewage wastes.

Hazards Inside Containers: Chlorine is shipped in DOT and CTC cylinders, in DOT Specification portable tanks (so called "one-ton containers"), and in insulated railroad tank cars.

Chlorine cylinders and 1 ton (1016 kg) containers are provided with overpressure protection by fusible plugs. Tank cars are protected by pressure relief valves. Because the operation of fusible plugs results in complete depressurization, the probability of a BLEVE is low and has not occurred in practice. The NFPA has no record of a BLEVE of a tank car containing chlorine, even though such an occurrence is theoretically possible.

Steel is a suitable material for a chlorine container at normal temperatures; however, it is rapidly attacked at temperatures above approximately 230°F (110°C) and leaks can develop upon fire exposure. For this reason, as well as for BLEVE protection, water should be applied to fire exposed containers. Insofar as possible, application of water directly upon a leak should be avoided because it may enlarge the leak from acid corrosion.

Hazards When Released from Containment: Chlorine presents primarily toxicity and corrosion hazards when released from containment. Since chlorine gas is about 2½ times heavier than air, it will hug the ground.

Emergency Control: The gas can be controlled by water spray. If hose streams are used, fire fighters should wear full protective clothing. Water applied to liquid chlorine will accelerate vaporization.

The chlorine industry has developed an extensive leak control program which includes strategic siting of trained emergency personnel and special equipment kits for stopping the more common leaks in cylinders, ton containers, and tank cars. Information on this program is available from The Chlorine Institute, 342 Madison Avenue, New York, NY, 10017.

Ethylene

Classifications: Flammable, compressed, cryogenic, industrial.

Chemical Properties: Ethylene is comprised of carbon and hydrogen and contains a double chemical bond which imparts a degree of reactivity. Except at very high pressures normally encountered only in chemical processing, it is a stable material. It has a wide flammable range (Table 5-5C) and high burning velocity, reflecting its reactivity. Although it is nontoxic, ethylene is an anesthetic and asphyxiant.

Physical Properties: See Table 5-5A.

Usage: Ethylene is principally used in chemical processing, e.g., manufacture of polyethylene plastic. It is also used to ripen fruit.

Hazards Inside Containers: Ethylene is shipped as a compressed gas in DOT and CTC cylinders and as a cryogenic gas in insulated cargo trucks and railroad tank cars in accordance with DOT Specifications and Regulations. It is stored in cylinders or in insulted ASME Code or API tanks.

Ethylene cylinders (except for medical cylinders which may have fusible plugs or combination safety devices) are protected against overpressure by frangible (bursting) discs. Insulated truck tanks and tank cars are protected by pressure relief valves. Cylinders are subject to failure from fire exposure, but not to BLEVEs because they do not contain liquid. Insulated containers are subject to BLEVEs but actual occurrence of this is rare. One such incident occurred in Spain; however, the cargo tank and insulation system did not comply with North American standards. A few instances of cryogenic truck container overpressure rupture have occurred due to freezing of relief valves which closed due to improper installation of the valves.

Hazards When Released from Containment: Ethylene presents combustion explosion and fire hazards when released from containment. A wide flammable range and high burning rate accentuate these hazards. In a number of instances involving rather large outdoor releases, open air or space explosions have occurred.

Emergency Control: Escaping ethylene presents both "no fire" and "fire" emergency situations. At normal atmospheric temperatures, ethylene gas is very slightly lighter than air. Ethylene gas vaporizing from the cryogenic liquid near its normal boiling point is about 1¼ times heavier than air at 70°F (21°C) and can spread along the ground. Escaping liquid will also pool on the ground. Although the visible fog created is a rough indication of the extent of the hazardous area, the hazard can extend beyond the visible area.

Escaping gas can be controlled by water spray. Contact between water and pooled ethylene should be avoided to prevent increased vaporization unless the vapors can be controlled. Water should be applied to fire exposed containers and the flow of escaping gas should be stopped if possible.

Hydrogen

Classifications: Flammable, compressed, cryogenic, industrial.

Chemical Properties: Hydrogen is a basic chemical element. Hydrogen has an extremely wide flammable range and the highest burning velocity of any gas. Its ignition temperature is reasonably high, but its ignition energy is very low. Because hydrogen contains no carbon, it burns with a nonluminous flame which is often invisible in daylight. Hydrogen is nontoxic.

Physical Properties: See Table 5-5A.

Usage: Hydrogen is principally used in chemical processing, for hydrogenation of edible oils, in welding and cutting, as a metal heat treating special atmosphere, and as a coolant in large electrical generators.

Hazards Inside Containers: Hydrogen is shipped as a compressed gas in uninsulated DOT and CTC cylinders, as a cryogenic gas in insulated DOT and CTC cylinders, and in insulated cargo trucks and railroad tank cars according to DOT Specifications and Regulations. It is stored in cylinders or in ASME Code insulated tanks.

Compressed hydrogen cylinders are protected against overpressure by frangible (bursting) discs or such discs in combination with fusible plug devices. Insulated cylinders, trucks and railcar tanks, and storage tanks are protected by pressure relief valves and frangible discs. Compressed gas cylinders are subject to failure, but not to BLEVEs because they do not contain liquid. Insulated cryogenic containers are theoretically subject to BLEVEs, but NFPA has no record of such an occurrence.

Hazards When Released from Containment: Hydrogen presents both combustion explosion and fire hazards when released from containment. Although its wide flammable range and high burning rate accentuate these hazards, its low ignition energy, low heat of combustion on a volume basis, and its nonluminous (low thermal radiation level) flame exert counteracting influences in many instances.

Because of its low ignition energy, when gaseous hydrogen is released at high pressure, nominally small heat producing sources, e.g., friction and static generation, often result in prompt ignitions. Accordingly, hydrogen is frequently thought of as self-igniting under these circumstances. The record of releases in high pressure applications reveals that fires rather than combustion explosions occur. When hydrogen is released at low pressures, however, self-ignition is unlikely. Rather, hydrogen combustion explosions occur which are characterized by very rapid pressure rises which are extremely difficult to vent effectively. Open air or space explosions have occurred from large releases of gaseous hydrogen.

Because of its very low boiling point, contact between liquid hydrogen and air can result in condensation of air and its oxygen and nitrogen components. A mixture of hydrogen and liquid oxygen is potentially explosive even though the quantities involved are likely to be small. Accidents from this source have been generally restricted to the interiors of liquefaction equipment and to small containers of liquid hydrogen which are handled in the open atmosphere.

At ordinary temperatures, hydrogen is very light, weighing only about 1/15 as much as air. The accordingly high diffusion rate makes it difficult for hydrogen to accumulate in conventional structures unless the escape rate is high. This tends to reduce its combustion explosion hazard.

Emergency Control: Escaping gaseous hydrogen seldom presents a "no fire" emergency situation because it either ignites promptly or rises in the atmosphere rapidly. Hydrogen gas vaporizing from the cryogenic liquid near its normal boiling point is slightly heavier than air at 70°F (21°C); this fact, together with the visible fog of condensed water vapor created, causes it to spread along the ground for sizable distances (depending upon leak size and meteorological conditions). Because of the low gas density of vapors produced from vaporizing cryogenic hydrogen liquid, impounding or diked areas have not been required.

Ignitible mixtures can extend well beyond the visible cloud. Such escapes can be controlled by water spray. Contact between water and pooled hydrogen should be avoided to prevent increased vaporization, unless the vapor can be controlled.

Water should be applied to fire exposed containers and the flow of gas stopped, if possible. Because hydrogen burns with a flame which is often invisible in daylight and because its flames produce low levels of thermal radiation, people have actually walked into its flames. When approaching hydrogen fires, a useful technique is to hold a broom out in front, while progressing slowly and at the same time throwing dirt ahead. The combustibles in the dirt will incandesce and locate the edge of the fire.

Liquefied Natural Gas (LNG)

Classifications: Flammable, cryogenic, fuel.

Chemical Properties: LNG is a mixture of materials all comprised of carbon and hydrogen. The principal component is methane with lesser amounts of ethane, propane, and butane. The composition will vary, depending principally upon whether the source of the natural gas (which has been liquefied) is a transmission pipeline or gas wells. In the former instance, more of the propane and butane is removed prior to introduction into the pipeline. (See Table 5-5D.)

LNG is nontoxic but is an asphyxiant.

Physical Properties: See Table 5-5D.

Usage: LNG is used as a source of natural gas to augment pipeline supplies during periods of extreme demand (peak shaving), to supply gas distribution systems in areas remote from central distribution systems, and as a basic supply of natural gas. To a small extent, it is used as a vehicle propulsion fuel.

Hazard Inside Containers: LNG is shipped as a cryogenic gas in insulated cargo trucks built to DOT Specifications and Regulations and in marine vessels under DOT authorization. It is stored in insulated ASME Code or API tanks. LNG containers are protected against overpressure by pressure relief valves. Such containers are theoretically subject to BLEVEs, but NFPA has no record of such an occurrence.

Hazards When Released from Containment: LNG presents both combustion explosion and fire hazards when released from containment. At the present time, LNG is seldom used indoors and when so used the structure is designed for a combustion explosion hazard in accordance with national standards and regulations. Available test and experience data indicates that escaping LNG is not subject to open air or space explosions.

Emergency Control: Escaping LNG presents both "no fire" and "fire" emergency situations. LNG gas vaporizing from the cryogenic liquid near its normal boiling point is about 1½ times heavier than air is at 70°F (21°C) and will spread along the ground accompanied by the visible fog created by condensed water vapor. This distance will depend upon the leak size and meteorological conditions and upon the geometry of the required liquid impounding area provided. Ignitible areas are roughly defined by the visible cloud, but can extend beyond the visible area. Such escape can be controlled by water spray. Contact between water and pooled LNG should be avoided to prevent increased vaporization unless the vapor can be controlled. Water should be applied to fire exposed containers and the flow of gas stopped, if possible.

TABLE 5-5D. Approximate Properties of LNG

Composition	
Methane	83–99%
Ethane	1–13%
Propane	0.1–3%
Butane	0.2–1.0%
Physical Properties	
Normal boiling point	minus 255 to minus 256°F (minus 160 to minus 164°C)
Density liquid at nbp	3½ to 4 pounds per gallon (0.42 to 0.48 g/cm^3)
Density vapor at nbp (compared with air at 70°F or 21°C)	1.47
Liquid to vapor expansion	600 to 1
Heat of vaporization	220–248 Btu/pound (512–577 kJ/kg) 770–990 Btu/gallon (215–276 MJ/m^3)
Theoretical vaporizing capability of 1 cubic foot of:	
Dry earth	6 gallons LNG
Wet earth	20 gallons LNG
Water	24 gallons LNG (1 gallon water = 3.2 gallons LNG)
Air	0.0005 gallon LNG
Theoretical vaporizing capability of 1 m^3 of:	
Dry earth	0.8 m^3 of LNG
Wet earth	2.7 m^3 of LNG
Water	3.2 m^3 of LNG
Air	0.6 L^3 of LNG
Initial vaporization rate of LNG spill on solid surface—10 cfm vapor per square foot (107.6 m^3/m^2) of LNG surface area	
Initial vaporization rate of LNG spill on water—700 cfm vapor per square foot (7532 m^3/m^2) of LNG surface area	
Steady-state vaporization rate of LNG spill—1 cfm vapor per square foot (134 L^3/m^2) of liquid surface (1-foot-deep (0.30 m) pool evaporates in 10 hours)	
Combustion Properties	
Flammable range	5–14% (methane at normal temperatures) 6–13% (methane near minus 260°F or 127°C)
Heat of combustion	22,000 Btu/pound (51.2 MJ/kg)
Burn rate, steady-state pool	0.2–0.6 inch (5 to 15 mm) per minute
Pool fire flame height—3 times base dimensions of pool (slight wind)	

Liquefied Petroleum Gases (LP-Gas, LPG)

Classifications: Flammable, liquefied (including cryogenic), fuel.

Chemical Properties: LP-Gas is a mixture of materials all comprised of carbon and hydrogen. In commerce, LP-Gas is predominantly either propane or normal butane or mixtures of these with smaller amounts of ethane, ethylene, propylene, iso-butane, and butylene (including isomers). Principal variations in composition occur depending upon whether the source is gas wells or petroleum refineries. (See Tables 5-5C and 5-5E.) LP-Gas is nontoxic but is an asphyxiant.

TABLE 5-5E. Approximate Properties of LP-Gases

	Commercial Propane NLPGA Av.	Commercial Butane NLPGA Av.
Vapor Pressure in psig at:		
70°F	132	17
100°F	205	37
105°F	216	41
130°F	300	69
Vapor Pressure in kPa at:		
20°C	930	103
40°C	1550	285
45°C	1720	345
55°C	2070	462
Specific Gravity Liquid at 60°F (15.5°C)	0.509	0.582
Initial Boiling Point at 14.7 psia	−51°F	15°F
Initial Boiling Point at 101 kPa	−46°C	−9°C
Weight per Gallon of Liquid at 60°F, lb.	4.24	4.81
Weight per m^3 of liquid at 15.5°C	509 kg	582 kg
Specific Heat of Liquid, Btu/lb. at 60°F	0.588	0.549
Specific Heat of Liquid, kJ/kg at 15.5°C	1.37	1.28
Cu ft of Vapor per Gallon at 60°F	36.39	31.26
m^3 of Vapor per L at 15.5°C	0.271	0.235
Cu ft of Vapor per Pound at 60°F	8.58	6.51
m^3 of Vapor per kg at 15.5°C	0.534	0.410
Specific Gravity of Vapor (Air = 1) at 60°F (15.5°C)	1.52	2.01
Ignition Temperature in Air,	920–1120°F (493–604°C)	900–1000°F (482–538°C)
Maximum Flame Temperature in Air,	3,595°F (1979°C)	3,615°F (1990°C)
Limits of Flammability in Air, Percent of Vapor in Air-Gas Mixture:		
(a) Lower	2.15	1.55
(b) Upper	9.60	8.60
Latent Heat of Vaporization at Boiling Point:		
(a) Btu per Pound	185	167
(b) Btu per Gallon	785	808
(c) kJ/kg	430	388
(d) MJ/L	2188	2252
Total Heating Values after Vaporization:		
(a) Btu per Cubic Foot	2,516	3,280
(b) Btu per Pound	21,591	21,221
(c) Btu per Gallon	91,547	102,032
(d) MJ/m^3	93.7	122.2
(e) MJ/kg	50.2	49.4
(f) MJ/m^3	25.5	28.4

Physical Properties: See Table 5-5E.

Usage: LP-Gas is used principally as a domestic, commercial, agricultural, and industrial fuel gas, in chemical processing, and as an engine fuel. In domestic and recreational applications, it is sometimes known as bottled gas.

Hazards Inside Containers: LP-Gas is shipped as a liquefied gas in uninsulated DOT and CTC cylinders and ASME tanks and in DOT Specification cargo trucks, railroad tank cars and marine vessels. It is also shipped at low temperatures near the initial boiling point in insulated marine vessels. It is stored in cylinders, ASME Code tanks, and insulated API tanks.

LP-Gas containers are generally protected against overpressure by pressure relief valves, although some cylinders are protected by fusible plug devices and, occasionally, by a combination of these. Most containers are subject to BLEVEs.

Hazards When Released from Containment: LP-Gas presents both combustion explosion and fire hazards when released from containment. Because most uses of LP-Gas are indoors, the combustion explosion experience dominates. This hazard is accentuated when LP-Gas in the liquid phase is used indoors—reflecting the fact that 1 gal (3.78 L) of liquid propane or butane will produce about 245 to 275 gal (927 to 1041 L) of gas. For this reason, standards and codes severely restrict the uses of liquid-phase LP-Gas indoors. Large releases of liquid-phase LP-Gas outdoors have led to open air or space explosions.

Emergency Control: Escaping LP-Gas presents both "no fire" and "fire" emergency situations. LP-Gas vapor is normally 1½ to 2 times heavier than air, and the vapor produced as LP-Gas vaporizes from the liquid at its normal boiling point is even heavier. Therefore, it will tend to spread along the ground assisted by the created visible fog of condensed water vapor. Ignitible mixtures extend beyond the visible area. Such escape can be controlled by water spray. When propane is stored and handled at atmospheric temperatures, it is unlikely to pool except under very low ambient temperature conditions. Butane stored and handled at atmospheric temperatures and low temperature LP-Gas are likely to pool. Contact between water and pooled LP-Gas should be avoided to prevent increased vaporization, unless the vapor can be controlled. Water should be applied to fire exposed containers and the flow of gas stopped, if possible.

Methylacetylene-Propadiene, Stabilized (MPS)

Classifications: Flammable, liquefied, industrial.

Chemical Properties: Methylacetylene-propadiene, stabilized, is a mixture of materials all comprised of carbon and hydrogen. In commerce, and in compliance with standards covering its use as a metal cutting gas, MPS has the following composition:

Methylacetylene (also known as "propyne") and propadiene (also known as "allene") contain double and triple chemical bonds and are unstable, reactive materials presenting hazards similar to acetylene. The achievement of an adequate degree of stability for this product depends upon the existence of a sufficient quantity of the propane, butane, and iso-butane components of the mixture. MPS is marketed under various trade names, including "MAPP

TABLE 5-5E(a).

1. Methylacetylene-propadiene (in combination, with a maximum ratio of 3.0 moles of methylacetylene per mole of propadiene in the initial liquid phase in a storage container)	—68 mole percent *maximum*
2. Propane, butane, isobutane (in combination)	—24 mole percent *minimum*, of which at least one-third (8 mole percent of total mixture) shall be butane and/or isobutane
3. Propylene	—10 mole percent *maximum*
4. Butadiene	—2 mole percent *maximum*

GAS" (Monsanto Chemical Co.) and "APACHE GAS" (Air Products and Chemicals, Inc.).

MPS has a degree of chemical reactivity with metals similar to acetylene—especially as it pertains to the use of copper and copper alloys. (See "Acetylene" in this chapter.)

MPS is nontoxic but has some anesthetic and narcotic effects. It has a strong odor detectable at concentrations as low as 100 ppm in air. (See Tables 5-5C and 5-5F.)

Physical Properties: See Table 5-5F.

Usage: MPS is principally used as a metal-cutting fuel gas with oxygen.

Hazards Inside Containers: MPS has physical properties very similar to propane and is shipped as a liquefied gas in uninsulated DOT and CTC cylinders and in DOT Specification cargo trucks and tank cars. It is stored in cylinders and ASME Code tanks. Its hazards inside containers are analogous to propane. The pressure of MPS in piping systems is not restricted.

TABLE 5-5F. Selected Properties of MPS*

Molecular Weight (Average-stabilized mixture)	42.3
Heat of Vaporization	227 Btu/lb (528 kJ/kg)
Gross Heat of Combustion	21,700 Btu/lb (50.5 MJ/kg)
Normal Boiling Range, 1 Atm.	−36 to −4°F (−37.8 to −20°C)
Flame Temperature in Oxygen	5301°F (2927°C)
Specific Heat, Liquid C_p—70°F, (21°C) 1 Atm.	0.362 Btu/lb°F [1.5 J/(kg/°K)]
Specific Heat, Liquid C_v—70°F, (21°C) 1 Atm.	0.277 Btu/lb°F [0.9 J/(kg/°K)]
Critical Temperature	245°C (118°C)
Critical Pressure	752 psi (5.2 MPa)
Burning Velocity	15.4 ft/sec (4.7 m/sec)
Explosive Limits in Oxygen	2.5 to 60%
Vapor Pressure at 70°F (21°C)	94 psi (648 kPa)
Specific Gravity of the Liquid (60/60°F)	0.576
Specific Volume of the Gas at 60°F (21°C), 1 Atm.	8.85 cu ft/lb (0.55 m^3/kg)
Heat of Formation	1075 Btu/lb (2.5 MJ)
Minimum Ignition Energy	0.2–0.25 MJ

* MAPP GAS. Dow Chemical Company

Hazards When Released from Containment: MPS presents combustion explosion and fire hazards when released from containment. It is easier to ignite (due to a lower ignition energy) than propane and burns more rapidly; thus it is similar to ethylene in these respects.

MPS is classified as a Group C material for purposes of suitability of electrical equipment installed in classified hazardous areas.

Emergency Control: See "Liquefied Petroleum Gases" in this chapter.

Oxygen

Classifications: Nonflammable (oxidizing), compressed, cryogenic, industrial, medical.

Chemical Properties: Oxygen is a basic chemical element. It reacts with practically all materials and the general reaction is known as oxidation. Combustion is a particular type of oxidation reaction. In most combustion reactions, the oxygen is accompanied by nitrogen, a mixture known as air. Nitrogen contributes nothing to the combustion reaction and actually inhibits it. Therefore, concentrations of oxygen in excess of its concentration in air increase most combustion hazards to a degree directly related to the concentration. This affects all basic combustion parameters except for the heat of combustion. For example, ignition temperatures and energies are lowered, the flammable range is widened, and the burning rate is increased as the oxygen concentration increases—the ultimate effects occurring at an oxygen concentration of 100 percent. In consideration of these properties, the design of systems containing 100 percent oxygen requires particular attention to these factors from a compatibility standpoint.

Physical Properties: See Table 5-5A.

Usage: Oxygen is primarily used in the production of steel, in metal welding and cutting, in medical applications, in life support systems, in aeronautic and aerospace applications, and in chemical processing.

Hazards Inside Containers: Oxygen is shipped as a compressed or cryogenic gas in DOT and CTC cylinders and in tank cars and trucks in accordance with DOT Specifications and Regulations. It is stored in DOT Specification cylinders (insulated or uninsulated) or in ASME Code insulated tanks.

Compressed oxygen cylinders are protected against overpressure by frangible discs or a combination of these with fusible plugs. Insulated cryogenic containers are protected with pressure relief valves. Metals for containers and piping must be carefully selected, depending on service conditions. The various steels are acceptable for many applications, but some service conditions may call for other materials (usually copper or its alloys) because of their greater resistance to ignition and lower rate of combustion.

Similarly, materials that can be ignited in air have lower ignition energies in oxygen. Many such materials may be ignited by friction at a valve seat or stem packing or by adiabatic compression produced when oxygen at high pressure is rapidly introduced into a system initially at low pressure.

Oxygen container failures from fire exposure are rare—probably reflecting standards which require separation of containers from flammable gases and other combustibles. Most oxygen system components have failed due to accumulations of grease, oil, etc., on the surfaces of these components in contact with oxygen, reflecting poor housekeeping practices. Since materials such as grease are rather easily ignited in air, they are especially ignitible in 100 percent oxygen. Their ignition often results in combustion of system components which are noncombustible in air, including metallic components. Such incidents usually involve only small portions of such components, but may occur in a spectacular way, causing property damage and personal injury. They are referred to conventionally as flashes.

Hazards When Released from Containment: Release of compressed or liquefied oxygen is usually observed by acceleration of the fire in whatever is burning. Release of liquid oxygen in the absence of fire presents an increased possibility that an oxygen-fuel mixture may occur prior to ignition. If ignition delay occurs under these circumstances, an explosion may result. Almost any combination of liquid oxygen and a combustible material is potentially explosive due to the rapid combustion circumstances produced. Some commercial explosives have been derived in this manner.

Oxygen is slightly heavier than air at the same temperature.

Emergency Control: Escaping oxygen presents basically a "no fire" situation, but can create a "fire" situation in the vicinity of fired equipment or arcing electrical equipment by causing these to get out of control. For example, metallic components of internal combustion engines have burned in the oxygen rich atmospheres produced by escaping oxygen. Oxygen gas vaporizing from the cryogenic liquid near its normal boiling point is approximately four times heavier than air is at 70°F (21°C) and will spread along the ground assisted by the visible fog of condensed water vapor that has been created. Such escape can be controlled by water spray. Contact between water and pooled liquid oxygen should be avoided to prevent increased vaporization unless the vapor can be controlled. Water should be applied to fire exposed containers and the flow of gas stopped, if possible.

Utility Gases

Classifications: Flammable, fuel.

Chemical Properties: The term "utility gas" is applicable to any flammable gas distributed by gas utilities as a fuel gas. Natural gas dominates this field today. Some LP-Gas is distributed by gas utilities. Most utilities use LP-Gas to augment natural gas supplies during short periods of peak demand which occur in unusually cold weather.

The creation of natural gas is the result of decomposition of organic material by heat, pressure, and bacteriological action in the absence of air, usually below ground. As it is evolved underground, natural gas consists of both flammable and nonflammable gases. The flammable gases are comprised of carbon and hydrogen and are principally methane and ethane with some propane, butane, and pentane. The nonflammable gases are principally nitrogen and carbon dioxide. In commerce, most of the nonflammable gases are removed prior to distribution so that the natural gas consists of about 70 to 90 percent methane with the remainder mostly ethane. (See Table 5-5C.)

Natural gas is nontoxic but is an asphyxiant. This is in marked contrast to the manufactured gas formerly widely distributed as utility gas which contained quantities of poisonous carbon monoxide.

Utility natural gas has no odor of its own and is generally odorized as distributed.

Physical Properties: With the exception of Liquefied Natural Gas (LNG), utility natural gas is distributed in piping as a compressed gas at pressures ranging from approximately ¼ to 1,000 psi (1.72 to 6895 kPa) and in cylinders at pressures up to 3,600 psi (24.8 MPa). The gas has a density of about ⅔ that of air.

Usage: Utility natural gas supplies about one third of the total fuel energy of the United States for domestic, commercial, and industrial heating and power. It is also being used as a motor fuel and is known as "Compressed Natural Gas (CNG)." Natural gas (usually supplied from a gas utility distribution piping system) is compressed to 2,400 to 3,600 psi (16.5 to 24.8 MPa), stored in a fueling station, and dispensed into vehicle containers at the same pressure.

Hazards Inside Containers: Utility natural gas is distributed nearly exclusively by a network of more than 1 million miles (1 million miles equals 1.6 million km) of underground pipeline in the United States and Canada. It is transported from gas wells in producing areas in large diameter transmission pipelines at pressures up to approximately 1,000 psi (6895 kPa). These pipelines are tapped by utility gas companies, and pressures are reduced for distribution—generally to around ¼ to 60 psi (1.72 to 414 kPa). Regulators and safety relief devices are used to control pressures.

Steel and cast iron pipe is used extensively. In recent years, thermoplastic and thermosetting plastic piping have been used extensively in distribution piping at pressures up to approximately 100 psi (689.5 kPa).

Because most gas piping is underground, fire exposure is not a significant problem. Most failures of utility natural gas piping are the result of mechanical damage due to excavation operations and corrosion.

In CNG applications, both the fueling station storage containers and the vehicle containers are either DOT or CTC Specification or Exemption cylinders, or ASME Code containers. DOT/CTC cylinders are protected against overpressure by combination bursting disc/fusible plug devices. ASME Code containers are equipped with spring loaded pressure relief valves.

CNG containers are subject to failure, but not to BLEVEs, because they do not contain liquid.

Hazards When Released from Containment: Natural gas presents both combustion explosion and fire hazards when released from containment. Because most uses of natural gas are indoors, the combustion explosion experience dominates. In approximately three fourths of the combustion explosions, the gas escapes from the underground distribution main in the street or the service connection between the street main and the building piping. In the remainder, the gas escapes from the building piping or gas burning equipment.

Open air or space explosions have resulted from failures of large diameter, high pressure transmission pipelines.

Combustion explosions of biologically produced natural gas are increasing as a result of the use of marshy land and former sanitary landfills for buildings.

The use of CNG in the United States has been limited. Fires in fueling stations have been limited to ignition of the discharge of overpressure relief devices with minor consequences. No vehicle fires have been reported in the U.S. Some fire tests indicate that fires in vehicles operating on their other fuels (gasoline/CNG dual fueled vehicles are common) can be expected to produce short lived flare-ups as the gas is released from the containers through their overpressure relief devices without container failure.

Emergency Control: Escaping natural gas presents both "no fire" and "fire" emergency situations. Although utility gas is considerably lighter than air, most gas escapes underground and can travel underground—usually following underground water, sewer or water piping or electrical conduits—for hundreds of feet (100 ft equals approximately 30.5 m) to enter below grade building spaces or hollow wall spaces. Because of this escape mode, control consists of ventilation rather than use of hose streams.

Fire exposed CNG containers should be kept cool with water.

Because of its lightness, unignited natural gas escaping outdoors is seldom a problem.

Natural gas fires should normally be extinguished by stopping the flow of gas only.

Bibliography

NFPA Codes, Standards, Recommended Practices and Manuals. (See the latest *NFPA Codes and Standards Catalog* for availability of current editions of the following documents.)

NFPA 12, *Standard on Carbon Dioxide Extinguishing Systems.*
NFPA 50, *Standard for Bulk Oxygen Systems at Consumer Sites.*
NFPA 50A, *Standard for Gaseous Hydrogen Systems at Consumer Sites.*
NFPA 50B, *Standard for Liquefied Hydrogen Systems at Consumer Sites.*
NFPA 51, *Standard for the Design and Installation of Oxygen-Fuel Gas Systems for Welding, Cutting and Allied Processes.*
NFPA 51A, *Standard for Acetylene Cylinder Charging Plants.*
NFPA 51B, *Standard for Fire Prevention in Use of Cutting and Welding Processes.*
NFPA 52, *Standard for Compressed Natural Gas (CGN) Vehicular Fuel Systems.*
NFPA 54, *National Fuel Gas Code.*
NFPA 56F, *Standard for Nonflammable Medical Gas Systems.*
NFPA 58, *Standard for the Storage and Handling of Liquefied Petroleum Gases.*
NFPA 59, *Standard for the Storage and Handling of Liquefied Petroleum Gases at Utility Gas Plants.*
NFPA 59A, *Standard for the Production, Storage and Handling of Liquefied Natural Gas.*
NFPA 68, *Guide for Explosion Venting.*
NFPA 99, *Standard for Health Care Facilities. (Chaps 2, 4 and 5)*
NFPA 302, *Fire Protection Standard for Pleasure and Commercial Motor Craft.*
NFPA 328, *Recommended Practices on the Control of Flammable and Combustion Liquids and Gases in Manholes, Sewers, and Similar Underground Structures.*

Additional Readings

"Acetylene," Pamphlet G-1, Compressed Gas Association, Inc., NY.
Allan, D. S., "A Cryogenic Line Leak Detector," *Fire Technology*, Vol. 11, No. 4, Nov. 1975, pp. 270-273.
ANSI B9.1-1971, *Safety Code for Mechanical Refrigeration*, American Society of Mechanical Engineers, NY.
ANSI B31.8-1968, *Gas Transmission and Distribution Piping*, American Society of Mechanical Engineers, NY.
ANSI B138.1-1972, *Design and Construction of LP-Gas Installations at Marine and Pipeline Terminals, Natural Gas Processing Plants, Refineries, and Tank Farms*, American Society of Mechanical Engineers, NY.
ANSI K61.1-1972, *Safety Requirements for the Storage and Handling of Anhydrous Ammonia*, American Society of Mechanical Engineers, NY.
ANSl Z49.1-1973, *Safety in Welding and Cutting*, American Society of Mechanical Engineers, NY.
Atallah, Sami, "U.S. History's Worst LNG Disaster," *Firehouse*, Jan. 1979, p. 29.
Bahme, C. W., *Fire Officer's Guide to Dangerous Chemicals*, National Fire Protection Association, Quincy, MA, 1978.
Benedick, W. B., Kennedy, J. D., and Morosin, D., "Detonation Limits of Unconfined Hydrocarbon/Air Mixtures," *Combustion and Flame*, Vol. 5, 1970, pp. 83-84.
Buckland, I. G., Butlin, R. N., and Annadale, D. J., *Gas Explosions in Buildings*; Part VI, Fire Research Station, Department of the Environment, Borehamwood, Herts, England, 1976.
Burgess, D. S., and Zabetakis, M. G., "Fire and Explosion Hazards Associated with LNG," *Report 6099*, U.S. Bureau of Mines, Washington, DC, 1962.
Cato, R. J., et al., "Effect of Temperature on Upper Flammability Limits of Hydrocarbon Fuel Vapors in Air," *Fire Technology*, Vol. 3, No. 1, Feb. 1967, pp. 14-19.
Clifford, E. A., "A Practical Guide to LP-Gas Utilization," *LP-Gas Magazine*, 5th ed., Duluth, MN, 1973.
Code of Federal Regulations, Title 49, Parts 171-190.
Compressed Gas Association, *Handbook of Compressed Gases*, Van Nostrand Reinhold Co., NY.
Coward, H. F., and Jones, G. W., "Limits of Flammability of Gases and Vapors," *Bulletin No. 503*, USDI, Bureau of Mines, Washington, DC, 1952.
Cryogenic Safety Manual, British Cryogenics Council, London, 1970.
Danakowski, T. D., "Is Liquid Hydrogen Safer than Liquid Methane?" *Fire Technology*, Vol. 17, No. 3, Aug. 1981, pp. 183-188.
Ditzel, Paul C., "BLEVE—Boiling Liquid Expanding Vapor Explosions Can Be Fatal," *Firehouse*, June 1977, p. 28.
Editors of Chemical Engineering Progress, *Safety in Air and Ammonia Plants*, Vol. 2, 1960; Vol. 3, 1961; Vol. 5, 1963; Vol. 7, 1965; Vol. 8, 1966; Vol. 9, 1967, American Institute of Chemical Engineers, NY.
Ellis, Donald L., "Propane Cloud and a Lot of Luck," *Fire Command*, Vol. 41, No. 2, Feb. 1974, p. 18.
"Fire and Explosion Hazards Associated with Liquefied Natural Gas," RI 6099, 1962, USDI, Bureau of Mines, Pittsburgh, PA.
First, M. W., Viles, F. J., and Levin, S., "Control of Toxic and Explosive Hazards in Buildings Erected on Landfills," Public Health Reports, Vol. 81, No. 5, May 1966.
"Flammability and Shock Sensitivity Characteristics of Methylacetylene, Propadiene and Propylene Mixtures," Report No. 3849, 1962, USDI, Bureau of Mines, Pittsburgh, PA.
Gayle, John B., "Explosions Involving Liquid Oxygen and Asphalt," *Fire Journal*, Vol. 67, No. 3, May 1973, pp 12-13.
Handbook of Compressed Gases, Compressed Gas Association, Arlington VA, publ. by Reinhold, NY, 1981.
"Hazards Associated with the Spillage of Liquefied Natural Gas on Water," RI 7448, 1970, USDI, Bureau of Mines, Pittsburgh, PA.

Hilado, Carlos J., and Cumming, Heather J., "The HC Value: A Method for Estimating the Flammability of Mixtures of Combustible Gases," *Fire Technology*, Vol. 13, No. 3, Aug. 1977, p. 195.

Hord, J., "Is Hydrogen Safe?" *NBS Tech Note 690*, National Bureau of Standards, Gaithersburg, MD, 1976.

"Hydrogen," Pamphlet G-5, Compressed Gas Association, NY.

Johnson, D. W., et al, "Control and Extinguishment of LPG Fires," *DOE/EV/06020-T3*, U.S. Department of Energy, Washington, DC, 1980.

Kajino, Higeo, and Noso, Kiyoto, *A Study of Possible Hazards by Leakage of Compressed Hydrogen Gas*, Institution for Safety of High Pressure Gas Engineering, Tokyo, 1979.

Kelly, A. L., "Volunteers Prepared When Butane Tank Explodes," *Fire Command*, Vol. 43, No. 6, June 1976, p. 24.

Kilmartin, John, "Two Liquid Oxygen Explosions," *Fire Journal*, Vol. 65, No. 2, Mar. 1971, pp. 15-22.

Lathrop, James K., "Derailments Underline the Need for Hazardous Materials Training," *Fire Command*, Vol. 45, No. 5, May 1978, p. 26.

———, "LP-Gas Plus Gasoline," *Fire Command*, Vol. 41, No. 9, Sept. 1974, p. 22.

———, "The Terrible Blast of a BLEVE," *Fire Command*, Vol. 41, No. 5, May 1974, p. 14.

Litchfield, E. L., Hay, M. H., and Cohen, D. J., "Initiation of Spherical Detonation in Acetylene-Oxygen Mixtures," RI 7061, Dec. 1967, USDI, Bureau of Mines, Washington, DC.

"LNG Importation and Terminal Safety," Proceedings of Conference of National Academy of Sciences, Boston, June 13-14, 1972.

LNG Safety Research Program, *Report IS 3-1*, American Gas Association, Alexandria, VA, 1974.

"LPG: The Hazards and Precautions," *Fire Protection*, Apr. 1980.

LP-Gas Safety Handbook, National LP-Gas Association, Chicago, IL. Continuous updating.

Lyons, P. R., "What Can We Do About Methane?" *Fire Command*, Feb. 1973, pp. 24-25.

"Mapp® Industrial Gas," *Loss Prevention Data Sheet* 7-94, Factory Mutual System, Norwood, MA.

Matheson Gas Data Book, The Matheson Company Inc., East Rutherford, NJ.

Minzer, G. A. and Eyre, J. A., "Large Scale LNG and LPG Pool Fires," I. Chem. E. Symposium Series No. 71, 1982, pp. 147-163.

National Transportation Safety Board, several reports of investigations of utility natural gas explosions and fires, National Transportation Safety Board, Washington, DC.

Nieuland and Voght, *The Chemistry of Acetylene*, American Chemical Society Monograph No. 99, NY.

"Properties and Essential Information for Safe Handling and Use of Ethylene," SD-100, 1973, Manufacturing Chemists' Association, Washington, DC.

Rasbash, D.J., "Review of Explosion and Fire Hazard of Liquified Petroleum Gas," *Fire Safety Journal*, Vol. 2, No. 4, July 1980, pp. 223-236.

Senesky, J., "Safe Storage and Handling of Compressed Gases," *Plant Engineering*, Vol. 33, No. 10, Oct. 1979, pp. 143-148.

U.S. Dept. of Energy Research Reports, Contract DE-AC06-76 RLO 1830, Pacific Northwest Laboratory, Battelle Memorial Institute (NTIS)

PNL-4401, LNG Annotated Bibliography
PNL-4398, LNG Fire and Vapor Control System Technologies
PNL-3991, Assessment of Research and Development
Needs in LPG Safety and Environmental Control.

Walls, W. L., "LNG: A Fire Service Appraisal," *Fire Journal*, Vol. 66, No. 1, Jan. 1972, Part 1; Vol. 66, No. 2, Mar. 1972, Part 2.

———, "Just What is a BLEVE?" *Fire Journal*, Vol. 72, No. 6, Nov. 1978, p. 46.

———, "Update: Containing the BLEVE Hazard," *Fire Journal*, Vol. 76, No. 5, Sept. 1972, p. 14ff.

Zabetakis, Michael D., "Flammability Characteristics of Combustible Gases and Vapors," *Bulletin 627*, U.S. Bureau of Mines, Washington, DC, 1965.

CHEMICALS

Revised by W. J. Bradford, P.E.

Safe and effective fire control measures when chemicals are involved require a knowledge of the hazardous properties of chemicals. This chapter discusses chemicals that produce hazards due to properties other than or in addition to combustibility. It also contains fire hazard information on combustible chemicals that are not discussed elsewhere in this Section.

For the purposes of this discussion, chemicals are classified according to the following hazardous properties: (1) ability to oxidize other materials, (2) combustibility, (3) instability, (4) reactivity with air or water, (5) corrosiveness, and (6) radioactivity. Although many chemicals possess more than one of these properties, it is customary to classify each by its predominant hazard. The danger of such a procedure is that an unmentioned hazardous property may be overlooked. For this reason it is necessary to refer to chemical dictionaries; manufacturers' data sheets; NFPA 49, *Hazardous Chemicals Data*; NFPA 491M, *Manual of Hazardous Chemical Reactions*; NFPA 325M, *Fire Hazard Properties of Flammable Liquids, Gases, and Volatile Solids*; and similar sources of information when evaluating the hazards of a particular chemical. Benzoyl peroxide, for example, is not only combustible but also very reactive; the degree of reactivity depends on the concentration. Another example is a commercial blasting mixture composed of ammonium nitrate mixed with oxidizable substances. For the purposes of transportation, these mixtures have been classified as oxidizing materials; however, they are also explosives. Thus, NFPA recommendations for the storage and handling of blasting agents closely parallel those for commercial explosives. At the end of each hazard classification, there is a discussion of the properties of the most common chemicals in the class and recommendations for storage and fire fighting.

Later in this Section, Chapter 7, "Explosives and Blasting Agents," and Chapter 9, "Dusts," provide information about chemicals not contained in this chapter. In Section 10, "Process Fire Hazards," two chapters discuss allied topics. Chapter 10, "Chemical Processing Equipment," gives information on chemical processing equipment and describes processing in terms of specific reactions. Chapter 17, "Nuclear Reactors, Radiation Machines and Facilities Handling Radioactive Materials," provides information about the hazards of radioactive materials in processes. Section 11, Chapter 6, "Storage and Handling of Chemicals," details safe storage practices and fire suppression techniques used with the various classes of chemicals.

TOXICITY OF CHEMICALS

The toxicity of chemicals is particularly important from the point of view of fire protection, regardless of the toxic material's fire hazard. A fire or explosion may subject fire fighters to a severe life hazard if the toxic material is accidentally released while they are present. In situations where fire officers are aware of a real or potential toxicity hazard, the decision may be made to forgo effective manual fire fighting.

Before using a chemical, one should obtain and evaluate information about its toxicity. When the toxicity problem appears to be severe, an effort should be made to find a suitable less toxic substitute. If there is no practical way to eliminate the toxic material, protection should be provided for those subject to daily exposure and for fire department personnel. Those who may be exposed during a fire or other emergency should be informed of the potential hazard and advised of the proper protective clothing and breathing apparatus to wear. Automatic fire protection should always be provided when a fire hazard is present in conjunction with a toxicity hazard.

Information on the toxicity of chemicals can be obtained from various sources, including the manufacturers of the chemicals; NFPA 49, *Hazardous Chemicals Data*; Patty's *Industrial Hygiene and Toxicology*; and other references.

There are two ways to protect against the toxic effects of chemicals during handling. First, use the most practical of the available methods for controlling and confining the chemical so that the toxic material cannot be contacted, swallowed, or inhaled in dangerous quantities during

W. J. Bradford, P.E., is a loss prevention consultant. He is a registered professional engineer in Connecticut and Rhode Island, a member of AIChE and SFPE and serves on several NFPA technical committees.

normal operations. Second, in areas where toxic chemicals are handled, educate all persons about the hazards, precautionary procedures, danger signals, and emergency procedures to be followed.

Of the methods used to control and confine toxic chemicals, handling them in a closed system is absolutely necessary. However, a leak in the system is always possible and could subject persons to a severe health hazard. If the gas is not an irritant, personnel must be warned of the exposure by automatically operated toxic gas indicators or other alarm devices. In some installations, it may be possible to maintain a slight negative pressure on the closed system to prevent the escape of toxic materials in case of a minor leak. Generally such systems are ducted to a scrubber where the toxic gas can be neutralized. Wherever possible, processes should be installed outdoors where natural air movement will dilute and dissipate toxic gases or vapors.

OXIDIZING CHEMICALS

Several important groups of chemicals known as oxidizing agents provide oxygen for combustion. Although most oxidizing chemicals are not combustible, they may increase the ease of ignition of combustible materials and they invariably increase the intensity of burning. A few oxidizing agents are susceptible to spontaneous decomposition (instability), and thus possess within themselves all the ingredients for a fire or explosion.

Nitrates

Awareness of the fire hazard properties of the inorganic nitrates is important because these materials are widely used in fertilizers, salt baths, and other industrial applications. Under fire conditions, inorganic nitrates may melt and release oxygen, causing the fire to intensify. Molten nitrates react with organic materials with considerable violence, usually releasing toxic oxides of nitrogen. When solid streams of water are used for fire fighting, they may produce steam explosions upon contact with molten material. Common nitrates are discussed in this section. Other nitrates have somewhat similar properties.

Sodium Nitrate: Noncombustible sodium nitrate promotes combustion of other materials. It evolves oxygen when heated to about 700°F (370°C), thereby increasing the intensity of any fire in its vicinity. It is soluble in water and is hygroscopic. This water solubility (common to most nitrates) is indirectly responsible for many serious fires. Paper, burlap, or cloth bags of nitrates that become moist during shipment or storage retain an impregnation of nitrate after drying and are thus highly combustible. For this reason, nitrates should be transferred from bags or wooden barrels to noncombustible bins for storage. The bags or barrels should be thoroughly washed. For the same reason, storage of bulk sodium nitrate on wood floors or against wood walls or posts is hazardous. An intimate mixture of sodium nitrate and organic material can be exploded by a flame.

Potassium Nitrate: The properties and hazards of potassium nitrate are similar to those of sodium nitrate; however, potassium nitrate is less moisture absorbent.

Ammonium Nitrate: Like other inorganic nitrates, ammonium nitrate is an oxidizing agent and will increase the intensity of a fire. The oxidizing gas it gives off is nitrous oxide rather than oxygen. Chemical and fertilizer grade ammonium nitrate must not be confused with certain ammonium nitrate combustible material mixtures used as explosives. All grades of ammonium nitrate can be detonated if they are in the proper crystalline form, if the initiating source is sufficiently large, or if heated under sufficient confinement (the purest material needing the greatest confinement).

As is evident from the shipboard explosions at Texas City, TX, and Brest, France (both in 1947), and the Red Sea in 1953; the explosion following a freight train wreck at Traskwood, AR, December 1960; and the explosion in a bulk warehouse near Pryor, OK, January 1973, certain conditions other than initiation by explosives can cause ammonium nitrate to detonate. An extensive series of tests to determine the explosion hazards of fertilizer grade ammonium nitrate under fire exposure was conducted by the Bureau of Mines (Van Dolah et al 1966), and from that study it would appear that the following conclusions can be reached:

1. Although detonation initiation in ammonium nitrate as a result of fire exposure cannot be ruled out completely, a direct burning-to-detonation transition in commercial fertilizer grade ammonium nitrate (referred to hereafter as "AN") appears to be possible, if at all, only in a pile of extremely large dimensions with the ignition at the bottom or center of the pile. (The ammonium nitrate involved in the Texas City, Brest, and Red Sea explosions was organic coated, with substantially different burning characteristics from those of the AN manufactured today.)
2. Projectiles derived from nearby explosions can initiate reactions leading to detonations, particularly in hot AN. Ordinary sporting arms bullets are incapable of initiating detonations of AN under normal storage conditions.
3. When AN is intimately mixed with fuel oil, ground polyethylene, or ground paper, transition from burning to detonation is possible, although quite unlikely, in pile sizes that are typical of those found in storage and transportation. Such mixtures are properly classified as blasting agents and should be stored according to the requirements in NFPA 495, *Code for the Manufacture, Transportation, Storage, and Use of Explosive Materials.*
4. Gas detonations are incapable of initiating detonation of AN.
5. Hot AN can be detonated by a high velocity bullet or by projectile impact. (Initiation of detonation of AN in the Traskwood freight train wreck and fire may have been by projectiles derived from a gasoline-nitric acid detonation because tank cars of gasoline and of nitric acid were also in the wreck. Nitric acid may have become mixed with the burning gasoline.)

Tests conducted by Underwriters Laboratories Inc. indicate that mixtures of ammonium nitrate and ammonium sulfate containing not more than 40 percent by weight of ammonium nitrate, and mixtures of ammonium nitrate and calcium carbonate (calcium ammonium nitrate) containing not more than 61 percent ammonium nitrate, are not explosive under conditions met in practice. When exposed to fire, calcium ammonium nitrate forms calcium nitrate and ammonium carbonate which absorbs heat while decomposing to ammonia, carbon dioxide, and water. However, the contamination problem has not been thoroughly investigated.

Ammonium nitrate in water solution is not hazardous unless spilled into combustible material and permitted to dry. Research, however, has shown that solutions containing up to 8 percent water can be detonated (Fukuyuma 1957).

Cellulose Nitrate: For hazards and properties of cellulose nitrate see Chapter 8 of this Section.

Nitric Acid

(See "Corrosive Chemicals," in this chapter.)

Nitrites

Nitrites should not be confused with nitrates. Nitrites contain one less oxygen atom than nitrates but are more active oxidizing agents since they melt and release oxygen at lower temperatures. Because nitrites in mixtures with combustible substances are hazardous, such mixtures should not be subjected to heat or flame. Certain nitrites, notably ammonium nitrite, are by themselves explosive. Nitrites should be treated like nitrates with respect to storage, handling, and fire fighting.

Inorganic Peroxides

Sodium, Potassium, and Strontium Peroxide: Although these chemicals are themselves noncombustible, they react vigorously with water and release oxygen as well as large amounts of heat. Large quantities of sodium and potassium peroxides may react explosively with water, and heat from reaction with just a little water may cause the contents of an entire container to decompose. If organic or other oxidizable material is present when this reaction takes place, fire is likely to occur.

Barium Peroxide: Heat releases oxygen from barium peroxide. Intimate mixtures of barium peroxide and combustible or readily oxidizable materials are explosive and easily ignited by friction or by contact with a small amount of water.

Hydrogen Peroxide: In contrast to the four above mentioned peroxides, which are white powders, hydrogen peroxide is a syrupy liquid. In pure form, it is relatively stable. When heated to and kept at a temperature of 212°F (100°C), 99.2 percent hydrogen peroxide decomposes at the rate of 4 percent per year, and 50 to 90 percent hydrogen peroxide at a rate of something less than 2 percent a day. An increase in temperature increases the decomposition rate of hydrogen peroxide about 1½ times for each 10°F (−12°C). Near the boiling point the rate of decomposition is very rapid, and if adequate venting is not provided, the pressure in the container may cause it to rupture. In general, the dilute material is less stable than the concentrated.

Solutions at concentrations of between 86 and 90.7 percent hydrogen peroxide have been demonstrated to be detonatable (Bureau of Mines 1967). At a concentration above about 92 percent, the liquid can be exploded by shock. Concentrated hydrogen peroxide vapors can be exploded by a spark. At atmospheric pressure, the boiling material must be 74 percent peroxide or higher to produce explosive vapors.

Decomposition of hydrogen peroxide produces water, oxygen, and heat. At concentrations above 35 percent, the heat is sufficient to turn all the water into steam, assuming that the decomposition began at room temperature (72°F) or (22°C). This means that a steam explosion is possible with the sudden decomposition of concentrated material. Decomposition may be caused by contamination with iron, copper, chromium, and many other metals (except aluminum) or their salts. Decomposition can also be caused by combustible dust, or by contact with a rough surface, such as ground glass.

Hydrogen peroxide is a strong oxidizing agent and may cause ignition of combustible material with which it remains in contact. This possibility is remote, except in the case of concentrations greater than 35 percent.

Chlorates

An adequate understanding of the fire hazard properties of the chlorates can be gained from the best known member of the group, potassium chlorate.

Potassium Chlorate: This white crystalline substance is water soluble, noncombustible, and a strong oxidizing agent. When heated, it gives up oxygen even more readily than do nitrates. Mixture with combustible materials, e.g., floor sweepings, should be prevented because under such conditions potassium chlorate may ignite or explode spontaneously. Drums containing chlorates may explode when heated. Sodium chlorate has properties similar to potassium chlorate.

Chlorites

Sodium Chlorite: This is a powerful oxidizing agent that forms explosive mixtures with combustible materials. In contact with strong acids, it releases explosive chlorine dioxide gas. At 347°F (175°C), sodium chlorite decomposes with the evolution of heat.

Dichromates

Among the dichromates, all of which are noncombustible, ammonium dichromate is most readily decomposed. It begins to decompose at 356°F (180°C); above 437°F (225°C) the decomposition becomes self-sustaining and is accompanied by swelling and the release of heat and nitrogen gas. Closed containers rupture at the decomposition temperature.

The other dichromates, such as potassium dichromate, react with readily oxidizable materials and in some cases may cause them to ignite. They release oxygen when heated.

Hypochlorites

Calcium hypochlorite at a concentration above 50 percent by weight may cause combustible or organic materials to ignite on contact. When heated, it gives off oxygen. With acids or moisture, it freely evolves chlorine, chlorine monoxide, and some oxygen at ordinary temperatures. It is sold as bleaching powder or, when concentrated, as a swimming pool disinfectant.

Perchlorates

Perchlorates contain one more oxygen atom than chlorates. They have roughly similar properties, but are more stable than chlorates. They are explosive under some conditions, such as when in contact with concentrated sulfuric acid.

Ammonium Perchlorate: This chemical has great explosive sensitivity when contaminated with such impurities as sulfur, powdered metals, carbonaceous materials, and reducing agents. The pure material in finely divided form can detonate if involved in a fire. A mixture with a chlorate may form spontaneously explosive ammonium chlorate.

Potassium Perchlorate, Sodium Perchlorate, and Magnesium Perchlorate: Each of these chemicals forms explosive mixtures with combustible, organic, or other easily oxidizable materials. Magnesium perchlorate is sometimes used in laboratories in place of calcium chloride as a dessicant. Such use requires vigilance to avoid dangerous contamination.

Permanganates

Mixtures of inorganic permanganates and combustible material are subject to ignition by friction or may ignite spontaneously if an inorganic acid is present. Explosions may occur whether the permanganate is in solution or is dry.

Potassium Permanganate: This chemical reacts violently with finely divided oxidizable substances. On contact with sulfuric acid or hydrogen peroxide, it is explosive.

Persulfates

Persulfates, e.g., potassium persulfate, are strong oxidizing agents which may cause explosions during a fire. Oxygen released by the heat of a fire may cause an explosive rupture of the container, or the explosion may follow an accidental mixture of the persulfate with combustible material.

COMBUSTIBLE CHEMICALS

Carbon Black

Carbon black may be formed by the decomposition of acetylene, by incomplete combustion of natural gas, by a mixture of natural gas and a liquid hydrocarbon, or by cracking hydrocarbon vapor in the absence of air. It is most hazardous immediately following manufacture when bags of the finished product may contain red hot carbon particles. While carbon black adsorbs oxygen, slow smoldering may develop. To prevent this hazard, carbon black is stored in an observation warehouse before shipment or final storage. It is well established that carbon black will not heat spontaneously after thorough cooling and airing, although it may generate heat in the presence of oxidizable oils.

Tests show that a dust explosion hazard does not exist. Red hot metal, electric sparks, and burning magnesium ribbon will not cause carbon black dust clouds to ignite explosively, but ignition has been obtained by using 1 oz (28 g) of gunpowder.

Carbon black comprises 98 percent of the world's production of powdered carbon.

Lamp Black

Lamp black is formed by burning low grade heavy oils or similar carbonaceous materials with insufficient air. It adsorbs gases to a marked degree and often ignites spontaneously when freshly bagged. It has great affinity for liquids, and heats in contact with drying oils. The possibility of lamp black dust explosions is increased by the presence of unconsumed oil that adheres to the carbon. Bureau of Mines' tests indicate that a dust explosion hazard may exist if the oil content exceeds 13 percent. Lamp black should be thoroughly cooled before it is bagged and then stored in a cool, dry area away from oxidizing materials.

Lead Sulfocyanate

Lead sulfocyanate burns slowly. It decomposes when heated, yielding among its decomposition products highly toxic and flammable carbon disulfide and highly toxic but nonflammable sulfur dioxide.

Nitroaniline

This combustible solid melts at 295°F (146°C) and its flash point is 390°F (199°C). In the presence of moisture, it nitrates organic materials, which may result in their spontaneous ignition.

Nitrochlorobenzene

Nitrochlorobenzene is a solid material at ordinary temperatures and gives off flammable vapors when heated.

Sulfides

Antimony Pentasulfide: This is readily ignited and hazardous in contact with oxidizing materials. On contact with strong acids antimony pentasulfide yields hydrogen sulfide.

Phosphorus Pentasulfide: This ignites readily, and in the presence of moisture may heat spontaneously to its ignition temperature (287°F or 142°C). The products of combustion include highly toxic sulfur dioxide and phosphorus pentoxide. Reaction of phosphorus pentasulfide with water yields hydrogen sulfide.

Phosphorus Sesquisulfide: With an ignition temperature of only 212°F (100°C), this chemical is easily ignited and is considered highly flammable. Toxic sulfur dioxide is a product of combustion.

Potassium Sulfide and Sodium Sulfide: These are moderately flammable solids that form toxic sulfur dioxide when burning, and hydrogen sulfide on contact with acids.

Sulfur

Sulfur at ordinary temperatures is a yellow solid or powder consisting of rhombic crystals which melt in the vicinity of 234°F (112°C), depending on their purity. Sulfur boils at about 832°F (444°C). Except in small quantities, it is shipped and stored as a liquid at a temperature below 300°F (149°C). It is combustible, and its vapor forms explosive mixtures with air. (Its flash point is 405°F or 207°C). Finely divided sulfur dust likewise possesses an explosion hazard that requires control during storage and handling. Ignition temperatures of dust clouds vary upward from 374°F (190°C). Sulfur contains varying amounts of hydrocarbons, depending on the source. These hydrocarbons gradually react with the molten material to form combustible and highly toxic hydrogen sulfide. Storage tanks and pits for molten sulfur must be ventilated to prevent accumulation of this gas. Except in the presence of lamp black, carbon black, charcoal, and a few less common substances, spontaneous ignition of sulfur is practically nonexistent. It melts and flows when burning and evolves large quantities of toxic, irritating, and suffocating sulfur dioxide. This gas attacks the eyes and throat, and complicates fire fighting. Sulfur also forms highly explosive and easily detonated mixtures with chlorates and perchlorates, and forms gunpowder when mixed with potassium nitrate and charcoal.

Naphthalene

Naphthalene is combustible both in solid and in liquid form. Naphthalene vapors and dusts form explosive mixtures with air.

UNSTABLE CHEMICALS

Certain chemicals spontaneously polymerize, decompose, or otherwise react with themselves in the presence of a catalytic material or even when pure. Such reactions may become violent.

Acetaldehyde

Acetaldehyde contains the carbonyl group ($C = O$). Like some other carbonyl compounds it undergoes an additional reaction, which can become dangerous in the presence of certain catalysts and at elevated temperatures. Acetaldehyde undergoes a dangerous additional reaction in the presence of an acid catalyst. Caustics can cause an "aldol" condensation which may take place with explosive violence.

Ethyl Acrylate, Methyl Acrylate, Methyl Methacrylate, and Vinylidene Chloride: These are flammable liquids that may polymerize at elevated temperatures, as in fire conditions. If the polymerization takes place in a closed container, the container may rupture violently. These liquids usually contain an inhibitor to prevent polymerization.

Ethylene Oxide

Ethylene oxide may polymerize violently when catalyzed by anhydrous chlorides of iron, tin, or aluminum; oxides of iron (i.e., iron rust) and aluminum; and alkali metal hydroxides. Violent polymerization of ethylene oxide may be initiated by heat or shock. Ethylene oxide reacts with alcohols, organic and inorganic acids, ammonia, and many other compounds. The heat liberated by these exothermic reactions may cause polymerization of the unreacted ethylene oxide. Ethylene oxide vapors may detonate if some initiating heat source is present. Ethylene oxide vapors in a tank exposed by fire may be rapidly heated to their ignition temperature unless the tank is kept wet with water spray. The flammable range of ethylene oxide in air is 3 to 80 percent. Although the upper limit is frequently reported to be 100 percent, explosions of mixtures containing greater than 80 percent ethylene oxide are the result of chemical decomposition.

Hydrogen Cyanide

Hydrogen cyanide is flammable and poisonous. In either liquid or vapor states, it has a tendency to polymerize. The reaction is catalyzed by alkaline materials, and since one of the products of the polymerization reaction is alkaline (ammonia), an explosive reaction will eventually take place. By adding sulfuric, phosphoric, or some other acid to neutralize the ammonia, the rate of polymerization in the liquid can be held down to a safe speed. Potassium cyanide and sodium cyanide on contact with acids release poisonous and flammable hydrogen cyanide vapor.

Methyl Acrylate

(See Ethyl Acrylate)

Methyl Methacrylate

(See Ethyl Acrylate)

Nitromethane

Nitromethane is a combustible liquid. At 599°F (315°C) and 915 psig (6308 kPa), it decomposes explosively. Noteworthy detonations of nitromethane in railroad tank cars occurred at Niagara Falls, NY, and at Mt. Pulaski, IL, in 1958. Although undiluted nitromethane may detonate under certain conditions of heat, pressure, shock, and contamination, the causes of these two tank car explosions were not definitely determined. A possible explanation is contamination of the nitromethane by some material previously carried in the tanks. For a discussion of the hazards of nitromethane and other nitroparaffins, see AIA Research Report No. 12 (AIA 1959).

Organic Peroxides

Organic peroxides are an important group of chemicals widely used in the plastics industry as polymerization reaction initiators, in the milling industry as flour bleaches, and in the chemical and drug industries as catalysts. All organic peroxides are combustible and, as in the case of inorganic peroxides, increase the intensity of a fire. Many organic peroxides can be decomposed by heat, shock, or friction. The rate of decomposition depends on the particular peroxide formulation and the temperature. Some, such as methyl ethyl ketone peroxide, are detonatable. Organic peroxides may be liquid (t-butyl perbenzoate) or solid (benzoyl peroxide) and are often dissolved in flammable or combustible solvents. The peroxides are often found wet with water or diluted with stable liquids. With solutions, sensitive crystals can be formed by freezing.

Benzoyl Peroxide: In undiluted form, benzoyl peroxide ignites very readily and burns with great rapidity, similar to burning an equal amount of black powder. Decomposition by heat is rapid, and if the benzoyl peroxide is confined when heated, explosive decomposition will occur. Decomposition can also be initiated by heavy shock or frictional heat. Most benzoyl peroxide is shipped diluted or water-wet to reduce the fire hazard.

Ether Peroxides: During storage practically all ethers form ether peroxides. When the ether peroxide mixture is heated or concentrated, the peroxide may detonate. With some ethers, the quantities of peroxide formed are too small to be of significance. Peroxide formation is a hazardous property of diethyl ether, ethyl tertiary butyl ether, ethyl tertiary amyl ether, and the isopropyl ethers. Isopropyl ether is said to be considerably more susceptible to peroxide formation than other ethers. Pure dry ether stored under laboratory conditions in a colorless bottle will develop detectable amounts of peroxides in one month. Light seems to be a more important factor than heat, although peroxides have been known to form in amber bottles. Ether sealed in copperplated cans or in otherwise inhibited cans is not likely to form peroxides. Although there is apparently no means yet available to completely eliminate peroxide formation, using any one of numerous patented inhibitors, or copper or iron along with storage in metal or opaque amber glass containers, provides sufficient stability for all practical purposes. Ether should not be dry-distilled unless peroxides have been proved absent.

Styrene

Styrene polymerizes slowly at ordinary temperatures, and the rate increases as the temperature increases. Since the polymerization reaction is exothermic, the reaction will eventually become violent as it is accelerated by its own heat. Inhibitors are added to styrene to prevent dangerous polymerization.

Vinyl Chloride

Vinyl chloride is a toxic and flammable gas that may polymerize at elevated temperatures, as in fire conditions, and cause violent rupture of the container. Vinyl chloride usually contains an inhibitor to prevent polymerization.

Vinylidene Chloride

(See Ethyl Acrylate)

WATER- AND AIR-REACTIVE CHEMICALS

Water-reactive and air-reactive chemicals present significant fire hazards. Significant quantities of heat are liberated during the reactions. If the chemical is combustible, it is capable of self-ignition; if noncombustible, the heat of reaction may be sufficient to ignite nearby combustible materials.

Alkalies (Caustics)

Caustic soda (sodium hydroxide or lye) and caustic potash (potassium hydroxide) are the most common alkalies. Although caustics are noncombustible, they generate heat when mixed with water. In contact with water, dry solid caustics will react. The heat generated (heat of solution) may be sufficient to ignite combustible material. Caustic solutions may generate hydrogen on contact with zinc, galvanized metals, or aluminum.

Aluminum Trialkyls

Most of these organic metal compounds are pyrophoric, i.e., they ignite spontaneously on exposure to air, and react violently with water and certain other chemicals. Triethylaluminum, the most common member of this group, ignites spontaneously in air and on contact with water. When it is mixed with strong oxidizing agents or with halogenated hydrocarbons, violent reactions or detonations may occur.

Anhydrides

Acid anhydrides are compounds of acids from which water has been removed. They react with water, usually violently, to regenerate acids. Organic acid anhydrides are combustible and usually are more hazardous than their corresponding acids, since their flash points are lower. Acetic anhydride has a flash point of 129°F (54°C); propionic anhydride, 165°F (74°C), open cup; butyric anhydride, 190°F (88°C), open cup, and maleic anhydride, 218°F (103°C). Inorganic acid anhydrides, e.g., chromium anhydride and phosphorus anhydride, are not combustible.

Carbides

The carbides of some metals, such as sodium and potassium, may react explosively on contact with water. Many, such as calcium carbide, lithium carbide, potassium carbide, and barium carbide, decompose in water to form acetylene. In addition to the hazard of the formation of flammable gas, another fire hazard of certain carbides is the generation of heat in contact with water. When one-third its weight of water is added to a water-reactive carbide, the temperature may be raised sufficiently to ignite the gas generated. Sodium carbide becomes heated to incandescence when placed in chlorine, carbon dioxide, or sulfur dioxide. The carbides of silicon and tungsten are very stable.

Charcoal

Under certain conditions, charcoal reacts with air at a sufficient rate to cause the charcoal to heat spontaneously and ignite. Charcoal made from hard wood by the retort method appears to be particularly susceptible. Spontaneous heating occurs more readily in fresh charcoal than in old material; the more finely divided it is, the greater the hazard. The principal causes of spontaneous heating of charcoal appear to be: (1) lack of sufficient cooling and airing before shipment; (2) charcoal becoming wet; (3) friction in grinding of finer sizes, particularly of material insufficiently aired before grinding; and (4) carbonizing of wood at too low a temperature, leaving the charcoal in a chemically unstable condition.

Coal

Under some conditions virtually all grades of coal (except high grade anthracite) are subject to spontaneous heating and ignition. Although the basic causes for the spontaneous heating of coal are not well defined, it is believed that adsorption of oxygen or oxidation of finely divided particles is the main cause. While instances of spontaneous heating of coal are numerous, the number of fires from this cause is insignificant when the large number of coal storage piles is considered.

The principal conditions believed to affect the susceptibility of coal to spontaneous heating are: (1) the fineness of the particles, (2) the oxygen adsorptive abilities of the particles, (3) the trapped and confined moisture content of the coal, (4) air trapped in voids in coal piles, (5) the presence of sulfur in the form of pyrites or marcasites, (6) free gases in the pile, (7) foreign substances in the pile, (8) the method and depth of piling, (9) the temperature of the containing walls and floor or of surrounding or surrounded surfaces, and (10) the type and amount of ventilation.

Hydrides

Most hydrides are compounds of hydrogen and metals. Metal hydrides react with water to form hydrogen gas.

Sodium Hydride: A gray-white crystalline, free-flowing powder, sodium hydride will ignite with explosive violence on contact with water. When exposed to air, absorption of moisture may cause ignition.

Lithium Hydride: When this combustible solid reacts vigorously with water, hydrogen gas and heat are evolved.

Lithium hydride dust is likely to explode in humid air. Static electricity may cause the dust to explode in dry air.

Lithium Aluminum Hydride: Like lithium hydride, this chemical is a combustible solid that reacts rapidly with water to form hydrogen gas and heat. The heat will probably cause the hydrogen to ignite. When lithium aluminum hydride is in a solution with ether, a fire involving the solution is essentially an ether fire. Small amounts of water cause the burning to intensify. Combustible materials on which the solution has spilled may ignite spontaneously or be ignited by light friction.

Oxides

Oxides of some metals and nonmetals react with water to form alkalies and acids respectively. This reaction takes place violently with the infrequently used sodium oxide. Calcium oxide, more commonly known as quicklime or unslaked lime, also reacts vigorously with water (slaking) with the evolution of enough heat to ignite paper, wood, or other combustible material under some conditions.

Phosphorus

Two forms of phosphorus, white and red, are in common use.

White (or Yellow) Phosphorus: This type is the more dangerous because of its ready oxidation and spontaneous ignition in air. It is common practice to ship and store white phosphorus under water, usually with the mixture in a hermetically sealed metal container. Periodic checks should be made to be sure that containers do not leak. White phosphorus is very toxic and should not be permitted to come in contact with the skin. On ignition, dense white clouds of toxic fumes are evolved which attack the lungs.

Red Phosphorus: This type is less hazardous than white, does not oxidize and burn spontaneously at ordinary temperatures, and can be shipped and stored without the protection of water, although it should be kept in closed containers away from oxidizing agents. It is formed by heating white phosphorus. The solid form is not toxic but, once vaporized, it takes on all the fire hazards of white phosphorus to which it reverts on condensation. Care should be taken in opening containers of red phosphorus because spontaneous ignition has been known to take place with exposure to air.

Sodium

(See Chapter 10 of this Section.)

Sodium Hydrosulfite

Sodium hydrosulfite burns slowly and produces sulfur dioxide as a combustion product. On contact with moisture and air, sodium hydrosulfite heats spontaneously and may ignite nearby combustible materials.

CORROSIVE CHEMICALS

The term "corrosive" refers to those chemicals that have a destructive effect on living tissues. Although some corrosive chemicals are strong oxidizing agents, they are separately classified to emphasize their injurious effect upon contact or inhalation. It should not be inferred, however, that a chemical is not injurious because it is otherwise classified. For example, caustics, classified as water- and air-reactive chemicals, are also corrosive. *Care should be taken to prevent inhalation, ingestion, and contact with all chemicals unless they are known to be harmless.*

Inorganic Acids

Concentrated aqueous solutions of the inorganic acids are not in themselves combustible. Their chief hazard lies in the danger of leakage and possible mixture with other chemicals or combustible material stored in the vicinity since fire or explosions could occur in some cases.

Hydrochloric Acid: In concentrated solution, hydrochloric acid is hazardous because its reaction with certain metals (including tin, iron, zinc, aluminum, and magnesium) forms hydrogen gas. Strong oxidizing agents mixed with hydrochloric acid causes the release of chlorine gas. A mixture of nitric and hydrochloric acids generates chlorine and nitrous oxide.

Hydrofluoric Acid: Either anhydrous or aqueous hydrofluoric acid is noncombustible and does not cause ignition of combustible materials with which it comes in contact. However, it is highly toxic, irritating to the eyes, and inflicts severe skin burns. When it comes in contact with metals, hydrogen is generated.

Nitric Acid: Under certain conditions nitric acid nitrates cellulose material. Thus, wood that comes in contact with the acid or its vapor may ignite much more easily. Spontaneous heating follows if strong solutions of the acid mix with organic material. In general, concentrated nitric acid nitrates organic materials; dilute acid oxidizes them, giving off oxides of nitrogen during the process. These oxide fumes (colorless to brown) are usually present in fires in buildings where nitric acid is used. A concentration of this gas (actually a mixture of several gases) so small that it is not objectionable at the time of inhalation can result in serious illness or death to the victim, although no effects may be felt for some time. If white fuming nitric acid (more than 97.5 percent nitric acid) is spilled into burning gasoline, it will detonate (Van Dolah et al. 1966).

Perchloric Acid: When misused or used in concentrations greater than 72 percent, perchloric acid can be extremely dangerous. At the normal commercial strength (72 percent), it is a strong oxidizing and dehydrating agent when heated, but a strong nonoxidizing acid at room temperature. Because of this and other advantageous properties, it is widely used in analytical laboratories. The rate of burning of organic substances is greatly increased by contact with perchloric acid. Explosions have occurred at wood and plastic laboratory hoods after long exposure of the hoods to perchloric acid vapors. Strong dehydrating agents, such as concentrated sulfuric acid or phosphorus pentoxide, convert perchloric acid solution to anhydrous perchloric acid, which decomposes even at room temperature and explodes with terrific violence. It also explodes on contact with many organic substances. For these reasons, dehydrating agents should never be mixed with perchloric acid.

Sulfuric Acid: This chemical has the added hazardous property of absorbing water from organic material with which it may come in contact. Charring takes place and sufficient heat may be evolved to cause ignition. Particular care must be taken to avoid painful skin burns inflicted by bodily contact. Dilute sulfuric acid will dissolve metals with the evolution of hydrogen.

The Halogens

The members of the halogen (salt producing) group are all chemically active and have similar chemical properties. The individual elements fluorine, chlorine, bromine, and iodine differ from each other in decreased chemical activity in the order named. Bromine and iodine have the least fire hazard. They are noncombustible, but will support combustion of certain substances. Turpentine, phosphorus, and finely divided metals ignite spontaneously in the presence of the halogens. The fumes are poisonous, as well as corrosive and irritating to the eyes and throat.

Bromine: A dark reddish-brown corrosive liquid which may cause fire in contact with combustible materials.

Chlorine: This is a heavy, greenish-yellow, highly toxic gas given off in some manufacturing processes and by bleaching powder (chloride of lime), especially in the presence of strong acids. It is not flammable itself, but may cause fires or explosions, especially if it comes in contact with acetylene, ammonia, turpentine, hydrocarbons, or finely powdered metals. At 484°F (251°C), mild steel ignites and burns in chlorine. Adequate ventilation should be provided in any process where this gas is generated.

Fluorine: A greenish-yellow gas, fluorine is one of the most reactive elements known. It combines, in most cases spontaneously, with practically all known elements and compounds under suitable conditions. Flourine reacts violently with hydrogen and many organic materials. It explodes in contact with metallic powders, attacks glass and most metals, and reacts explosively in contact with water vapor. It may be safely handled in nickel or monel cylinders. Fluorine reacts with these two metals, but forms a protective nickel fluoride layer that prevents further action. However, moisture or other impurities within the cylinder may cause such violent reaction that the metal will melt and ignite in the fluorine. The tank will then burst and scatter molten metal.

Iodine: This chemical is usually in the form of purplish-black volatile crystals which are corrosive. Reports indicate that iodine is explosive when diffused with ammonia (it forms explosive nitrogen triiodide) and when mixed with turpentine or lead triethyl.

RADIOACTIVE MATERIALS

Radioactive elements and compounds have fire and explosion hazards identical to those of the same material when not radioactive. An additional hazard is introduced by the various types of radiation emitted, all of which are capable of causing damage to living tissue. Under fire conditions, vapors and dusts (smoke) may be formed that could contaminate not only the building of origin but neighboring buildings and outdoor areas. The fire protection engineer's main concern is to prevent the release or loss of control of these materials by fire or during fire extinguishment.

Fire Protection for Radioactive Materials

The life hazard introduced by an escape of radioactive dusts and vapors during a fire makes it vitally important to take all practical steps to prevent a fire from involving these materials. Radioactivity is not detectable by any of the human senses. Special instruments and measuring techniques are required to identify and evaluate it. The hazard is affected by the form of the material, whether solid, liquid, or gas, and by the container in which it is kept or handled.

Radioactivity can cause loss of life, injuries, damage to and extended loss of materials used, equipment, and buildings. Manual fire fighting may be limited by the danger to fire fighters from exposure to radioactivity. Salvage work and resumption of normal operations at a property may be delayed where a fire or explosion causes loss of control over radioactive substances. The need to decontaminate buildings, equipment, and materials presents a serious and complicated problem.

Smoke and products of combustion from fires in places where there are radioactive materials must be controlled. The runoff of water used in fighting fires must also be controlled. Fire fighters require protective clothing and respiratory protection equipment. Fire control must be thoroughly preplanned. With radiation hazards, automatic sprinklers are preferable to measures requiring manual fire fighting. This lessens the amount of radioactive smoke or products of combustion and water runoff to be dealt with.

Handling Radioactive Materials

The possibility of accidental release of radioactive material because of a fire or explosion, with the resultant health hazard to fire fighters and others, is a strong argument for careful attention to methods of fire prevention and control in laboratories and other occupancies handling radioactive materials.

More information may be obtained from NFPA 801, *Recommended Fire Protection Practice for Facilities Handling Radioactive Materials.* This publication calls attention to basic information concerning radiation protection methods and provides some guidance on fire protection practices to those who design and operate such facilities. Most radioactive materials introduce little or no fire or explosion hazard so the fire hazard of facilities handling radiation usually can be determined by a knowledge of the combustibility of the building and its furnishings, and the fire and explosion hazards of the nonradioactive chemicals.

The fire and explosion hazard can be substantially reduced by use of a fire resistive building, noncombustible interior finish and furnishings wherever possible, and enforcement of strict controls to minimize the hazards of flammable liquids and other chemicals that may be necessary for laboratory work or for other occupancies. In any facility handling radioactive materials, one feature that can cause trouble unless the building has been properly designed is the duct system usually required for the safe disposal of contaminated vapors, gases, and dusts. Unless this system is properly arranged, it can spread radioactive contaminants to noninvolved parts of a facility during a fire.

Due to the need to immediately control any fire that might eventually release radioactive materials and the potential health hazard to those who could be exposed to radiation during manual fire control, there can be no question as to the desirability of automatic sprinkler protection for radiation laboratories and other areas where radioactive materials are handled. Special precautions to be followed by fire fighting personnel are described in the book, *Handling Radiation Emergencies* (Purington and Patterson 1977). The procedures stress steps that should be taken to protect fire fighters at all times, including the use of radiation monitoring devices, self-contained breathing apparatus, and regular fire department protective clothing.

MIXTURES OF CHEMICALS

This chapter has presented several examples of dangerous reactions that can occur when certain chemicals are mixed. Examples have been given of chemicals that can increase the ease of ignition or the intensity of burning of the combustible materials with which they are mixed. In order to recognize the innumerable combinations of so called incompatible chemicals, it is necessary to have a knowledge of the potentially dangerous reactions of individual chemicals. NFPA 491M, *Manual of Hazardous Chemical Reactions*, contains more than 3,400 dangerous reactions that have been reported in chemical literature and elsewhere.

Bibliography

References Cited

AIA. 1959. "Nitroparaffins and Their Hazards." NBFU Research Report No 12. American Insurance Association (formerly National Board of Fire Underwriters), NY.

Bureau of Mines. 1967. "Research and Technologic Work on Explosives, Explosions, and Flames: Fiscal Year 1967." IC 8387. Aug 1968. USDI Bureau of Mines, Washington, DC.

Fukuyama, I. 1957. "Sensitive Ammonium Nitrate." *Journal of Industrial Explosives Society*. (Kogyo Kayaku Kyohaishi). Vol 18 No 1. 1957. pp 64-66.

Purington, Robert, and Patterson, Wade. 1977. *Handling Radiation Emergencies*. National Fire Protection Association, Quincy, MA.

Van Dolah, R. W., et al. 1966. "Explosion Hazards of Ammonium Nitrate Under Fire Exposure." RI 6773. USDI Bureau of Mines, Pittsburgh, PA.

NFPA Codes, Standards, Recommended Practices and Manuals. (See the latest *NFPA Codes and Standards Catalog* for availability of current editions of the following documents.)

NFPA 43A, *Code for the Storage of Liquid and Solid Oxidizing Materials*.

NFPA 43C, *Code for the Storage of Gaseous Oxidizing Materials*.

NFPA 49, *Hazardous Chemicals Data*.

NFPA 325M, *Fire Hazard Properties of Flammable Liquids, Gases, and Volatile Solids*.

NFPA 490, *Code for the Storage of Ammonium Nitrate*.

NFPA 491M, *Manual of Hazardous Chemical Reactions*.

NFPA 495, *Code for the Manufacture, Transportation, Storage and Use of Explosive Materials*.

NFPA 655, *Standard for the Prevention of Sulfur Fires and Explosions*.

NFPA 704, *Standard System for the Identification of the Fire Hazard of Materials*.

NFPA 801, *Recommended Fire Protection Practice for Facilities Handling Radioactive Materials*.

Additional Readings

Anderson, W. V., "Hazards of Organic Phosphates," *Fire Journal*, Vol. 60, 6, Nov. 1966, pp. 84-85.

Chemical Hazards Bulletin, (a series) American Insurance Association, NY, (issued and revised at will).

Chemical Safety Data Sheets, (a series) Manufacturing Chemists' Association, Washington, DC (issued and revised at will).

Coffee, R. D., "Evaluation of Chemical Stability," *Fire Technology*, Vol. 7, No. 1, Feb. 1971, pp. 37-45.

Condensed Chemical Dictionary, 9th ed., Van Nostrand Reinhold Co., NY, 1977.

Dean, J. A., ed., *Lange's Handbook of Chemistry*, 13th ed., Handbook Publishers, Inc., Sandusky, OH, 1984.

Doyle, W. H., "Protection in Depth for Increased Chemical Hazards," *Fire Journal*, Vol. 59, No. 5, Sept. 1965, pp. 5-7.

Fawcett, H. H., and Wood, W. S., eds., *Safety and Accident Prevention in Chemical Operations*, Interscience Publishers, NY, 1965.

Fire and Explosion Index Hazard Classification Guide, 5th ed., Dow Chemical Co., Midland, MI, 1980.

Forshey, D. R., et al., "Potential Hazards of Propargyl Halides and Allene," *Fire Technology*, Vol. 5, No. 2, May 1969. pp. 100-111.

Gibson, J. R., *Handbook of Selected Properties of Air- and Water-Reactive Materials*, Library of Congress, Washington, DC, 1968.

Grayson, M., exec. ed., *Kirk-Othmer Encyclopedia of Chemical Technology*, 3rd ed., Interscience Publishers, NY, 1984.

Guide for Safety in the Laboratory, 2nd ed., Van Nostrand Reinhold Co., NY, 1972.

Hygienic Guide Series (reports on individual chemicals), American Industrial Hygiene Association, Westmont, NJ (published periodically).

Kuchta, J. M., Furno, A. L., and Imhof, A. C., "Classification Test Methods for Oxidizing Materials," RI 7594, 1972, USDI Bureau of Mines, Pittsburgh, PA.

Kuchta, J. M., and Smith, A. F., "Classification Test Methods for Flammable Solids," RI 7593, 1972, USDI Bureau of Mines, Pittsburgh, PA.

Laboratory Waste Disposal Manual, Manufacturing Chemists' Association, Inc., Washington, DC, revised Nov. 1972.

Lees, F. P., *Loss Prevention in the Process Industries*, Butterworths & Co., London, England, 1980.

Mandell, N. C., Jr., "A New Calcium Hypochlorite and a Discriminatory Test," *Fire Technology*, Vol. 7, No. 2, May 1971, pp. 157-161.

Nuckolls, A. H., "Fire and Explosion Hazards of Ammonium Nitrate Fertilizer Bases," *Bulletin of Research 20*, Dec. 1940, Underwriters Laboratories Inc., Northbrook, IL.

Pingree, D., "Hay Storage," *Fire Journal*, Vol. 61, No. 4, July 1967, pp. 44-45.

Sax, N. Irving, *Dangerous Properties of Industrial Materials*, 5th ed., Van Nostrand Reinhold, NY, 1979.

Stauffer, E. E., "Extinguishing Toluene Diisocyante Fires with Dry Chemicals," *Fire Technology*, Vol. 11, No. 4, Nov 1975, pp. 255-260.

Steere, N. V., ed., *Handbook of Laboratory Safety*, 2nd ed., Chemical Rubber Co., Cleveland, OH, 1971.

Stull, D. R., "Linking Thermodynamics and Kinetics to Predict Real Chemical Hazards," *Loss Prevention*, Vol. 7, American Institute of Chemical Engineers, NY, 1973, pp. 67-73.

Van Dolah, R. W., Gibson, F. C., and Murphy, J. N., "Further Studies on Sympathetic Detonation," RI 6903, 1966, USDI Bureau of Mines, Pittsburgh, PA.

Van Dolah, R. W., Gibson, F. C., and Murphy, J. N., "Sympathetic Detonation of Ammonium Nitrate and Ammonium Nitrate-Fuel Oil," RI 6746, 1966, USDI Bureau of Mines, Pittsburgh, PA.

Weast, R. C., ed., *Handbook of Chemistry and Physics*, 65th ed., Chemical Rubber Co., Cleveland, OH 1984.

Windholz, M., ed., *The Merck Index*, 10th ed., Merck & Co., Inc., Rahway, NJ, 1983.

Winning, C. H., "Detonation Characteristics of Prilled Ammonium Nitrate," *Fire Technology*, Vol 1, No 1, Feb. 1965, pp 23-31.

Zabetakis, M. D., "Flammability Characteristics of Combustible Gases and Vapors," *Bulletin 627*, U.S. Bureau of Mines, Washington, DC, 1965.

EXPLOSIVES AND BLASTING AGENTS

Revised by Samuel J. Porter

This chapter gives a short description of explosives and explosive materials with emphasis on their fire and explosion hazards. Additional important data concerning the manufacture, transportation, storage and use of explosive materials will be found in NFPA 495, *Code for the Manufacture, Transportation, Storage, and Use of Explosive Materials*, and in other referenced material. NFPA 495, *Code for the Manufacture, Transportation, Storage, and Use of Explosive Materials* is a current and comprehensive guide which should be studied by all who have anything to do with explosives.

The hazardous nature of explosives has long been recognized, and the rapid increase in their production and use makes it necessary to point out those properties that contribute most to the inherent dangers of these very important industrial and military products. In the industrial field alone, approximately 3.7 billion pounds of explosives and blasting agents were used in 1983, compared with 2.7 billion pounds in 1972 and 0.7 billion pounds (1 lb = 0.45 kg) in 1950. Of these totals, approximately 90 percent were used in mining and the remainder principally in construction operations.

Black powder was the first explosive known, although fireworks and pyrotechnics were known and used many centuries earlier (Dickson and Fisher 1958). The preparation and properties of black powder were described by Roger Bacon in the 13th century and first used for military purposes in the 14th century. Black powder was first used for mining operations in the latter part of the 17th century and it remained the dominant commercial explosive until the latter part of the 19th century. In 1846, the compound nitroglycerin was first prepared and its explosive properties studied. The extremely hazardous nature of this compound prevented wide use until 1867 when Alfred Nobel made the discovery that nitroglycerin absorbed in an inert material, such as diatomaceous earth, produced a mixture that was relatively safe to handle. This explosive mixture, with many modifications and improvements, remained the predominant commercial explosive until the introduction of modern blasting agents in the mid 1950s. Meanwhile, the use of black powder decreased to an insignificant quantity largely because of its great fire hazard. In less than 20 years, blasting agents based principally on the sensitization of ammonium nitrate have taken over approximately 90 percent of the market (USDI 1983). This change is based largely on economic considerations and the greatly improved safety of nonnitroglycerin explosive materials.

Samuel Porter is a Consultant in explosives, based in Woodbridge, VA.

NATURE OF EXPLOSIVE MATERIALS

Terminology

An understanding of the nature of explosive materials is essential before one can understand the fire and explosion potential of these admittedly hazardous products. A few general terms are defined below to provide a background for future discussions.

Blasting Agent: Any material designated for blasting which has been tested for blasting cap sensitivity, differential thermal analysis, thermal stability, and fire characteristics according to tests specified in Title 49 of the U.S. Code of Federal Regulations (49 CFR) Part 173 (Fed. Reg. 1972) and found to be so insensitive that there is very little probability of accidental initiation to explosion or of transition from deflagration to detonation.

Detonator: Any device containing a detonating charge that is used for initiating detonation in an explosive. The term includes, but is not limited to, electric blasting caps of instantaneous and delay types, blasting caps for use with a safety fuse, and detonating cord delay connectors.

Explosion: An explosion is an effect produced by the sudden violent production or expansion of gases. It may be accompanied by heat, shock waves, and the disruption or enclosing of nearby structures or materials.

Explosive: An explosive is a substance or a mixture of substances which, when subjected to the proper stimuli, undergoes an exceedingly rapid self-propagating reaction characterized by the formation of more stable products (usually gases), evolution of heat, and the development of a sudden pressure effect through the action of this heat on produced or adjacent gases. A somewhat simpler, but less accurate, definition states that an explosive is any chemical compound, mixture, or device whose primary or common use is to function by explosion. The term includes, but is not limited to, dynamite, black powder, initiating explosives, detonators, safety fuses, squibs, detonating cords, igniter cords, and igniters.

Explosive Material: Includes explosives, blasting agents, water gels (slurries), and detonators.

Propellant: An explosive material which normally functions by deflagration (burning) and is used for propelling purposes.

An explosion may result from: (1) chemical changes, such as those which accompany the detonation of an explosive or the combustion of a flammable gas-air mixture; (2) physical or mechanical changes, such as the bursting of a steam boiler, or a high pressure reaction vessel; and (3) atomic changes, such as those which occur in a nuclear explosion.

Fires and Explosions

A distinction must be made between controlled and unwanted fires and explosions. Controlled combustion reactions are basic to the production of power and necessary for an industrial economy. Likewise, detonation reactions are essential for use by the mining and construction industries. On the other hand, the loss of life and natural resources from unwanted fires and explosions is a matter of national concern. Since fire is a major hazard associated with the utilization of explosives and blasting agents, this chapter discusses the basic nature of these materials and methods for reducing the loss from uncontrolled reactions in these materials.

TYPES OF EXPLOSIVE MATERIALS

Although this chapter mainly concerns commercial explosives, some of the similarities and differences between commercial and military explosives should be considered. Most military explosives have high shattering power and relatively high detonation velocities. Military explosives are often kept in storage for relatively long periods of time and therefore must have good stability. It is essential that these explosives detonate reliably even after storage in a wide variety of conditions. While there are no truly military explosives equivalent to the blasting agents used in commercial operations, the military uses commercial type explosives and blasting agents for operations such as road building, airstrip preparations, and a wide variety of similar operations.

As stated previously, commercial explosives have a wide range of properties. Explosives to be used in underground operations must have relatively good fume characteristics. This requires formulations which are nearly oxygen balanced and therefore produce a minimal amount of carbon monoxide and oxides of nitrogen. The most widely used military explosives, such as trinitrotoluene (TNT) and cyclonite (RDX), are quite oxygen deficient and therefore produce large quantities of toxic gases. Commercial dynamites normally have a fairly long storage life, but blasting agents are more often used soon after manufacture, so storage is not as important a factor. However, prolonged storage of dynamites, slurries, or ammonium nitrate with fuel oil (ANFO) at high or low temperature extremes or high humidity may result in serious changes in sensitivity or effectiveness or both.

Commercial Explosive Types

Explosive materials can be divided into a number of specific types according to their characteristic properties (Cook 1958). The following types are generally recognized in industry.

Primary or Initiating High Explosives: These are quite hazardous materials, but they play a very important role in the utilization of explosive materials. Typical explosives of this type are mercury fulminate, lead styphnate, and lead azide. Because of their hazardous nature they are seldom, if ever, used alone and their principal function is to initiate detonation in less sensitive explosives. These primary explosives are readily detonated by the addition of heat or the application of a mild mechanical shock. They are used almost exclusively as the initiating agent in detonators, and are classified by the Department of Transportation (DOT) as Class A.

Secondary High Explosives: These are materials which are relatively insensitive to mechanical shock and heat and yet are readily detonated by the shock from a primary explosive. They are much more powerful than the primary explosives. They are used for most military purposes and are often used in commercial blasting operations. Secondary high explosives include such products as dynamite, nitroglycerin, TNT, RDX, and pentaerythritol tetranitrate (PETN). Because of its sensitivity, nitroglycerin is never used alone. Secondary high explosives are generally cap sensitive, which distinguishes them from blasting agents, and are classified by DOT as Class A.

Low Explosives or Propellants: These are used primarily for propulsion purposes. They normally function by burning rather than detonation, although many propellants are susceptible to detonation. Black powder, smokeless powder, and solid rocket fuels fall into this category. Fire constitutes the greatest hazard in the handling and use of propellant explosives. Classes A, B, and C may be represented in this group.

Blasting Agents: As stated previously, almost 90 percent of all blasting operations in the United States are carried out using nonnitroglycerin materials. The most commonly used blasting agent is a fuel oxidizer system which consists primarily of ammonium nitrate and a fuel, such as Number 2 diesel fuel. The formulation is termed ANFO and is sold under a variety of trade names. The addition of approximately 6 percent fuel oil to specially prepared ammonium nitrate prills (pellets) produces an amazingly successful blasting material. This combination has the advantage of low cost, satisfactory fume characteristics, and greatly increased safety over the nitroglycerin dynamites. While the combination does not burn readily, it still may change in a well established fire to produce a detonation. ANFO does have a disadvantage in that it is difficult or impossible to use it under very wet conditions. Many variations of the basic ANFO formulation have been developed and successfully used. Powdered aluminum is now added in certain operations to increase the general strength of the material.

Water Gels or Slurries: Operations where a water resistant explosive material is a necessity, and where higher densities than those which can be obtained with ANFO are desirable, have led to the development of water gels or slurries. These may be classed as either blasting agents or explosives, depending on their formulation and sensitivity. Ammonium nitrate is usually the basic oxidizer in this type of explosive, and it may be sensitized by a variety of materials, the most common of which is powdered or flaked aluminum. Other additives include paint grade aluminum, monomethylamine nitrate, TNT, gilsonite, sodium nitrate, and various gelling agents. Approximately 10 percent of the market has been taken over by water gels.

Nuclear Explosives: Nuclear explosives are used for military purposes only.

CLASSES OF EXPLOSIVES

The DOT (Fed Reg. 1972) divides commercial explosive materials into separate classes for transportation purposes. Industry has generally accepted this classification system since it corresponds roughly to the hazards in handling, storing, and transporting these materials. These explosive materials are discussed in the order of decreasing sensitivity.

Class A Explosives

Explosives of this class possess detonating or otherwise maximum hazard. They include dynamite, desensitized nitroglycerin, lead azide, mercury fulminate, black powder, blasting caps, detonators, detonating primers, and certain smokeless propellants.

Class B Explosives

Explosives of this class have a high flammable hazard and include most propellant materials.

Class C Explosives

Explosives of this class include manufactured articles which contain limited quantities of Class A or Class B explosives as one of their components. These explosives include such materials as detonating cord, explosive rivets, fireworks, etc. Class C explosives will not normally mass detonate under fire conditions.

Blasting Agents

Blasting agents are generally considered safer than Class A, B, or C explosives and yet, when properly initiated, they function in the same manner as Class A explosives. Not being cap sensitive, they require a strong primer.

Other Classification Systems

The Bureau of Alcohol, Tobacco, and Firearms (ATF), of the Department of the Treasury, has a classification system based on the type of explosive material involved. In their system, explosive materials are classified as high explosives, low explosives, and blasting agents. Since some low explosives are as hazardous as high explosives, this classification does not indicate relative hazard. While this system is not generally used industrially, it is working satisfactorily from a security standpoint for the control of explosive materials. This classification is employed in the enforcement of Title 11, Regulation of Explosives, of the *Organized Crime Control Act of 1970* (18 U.S.C., Chapter 40). At least annually the ATF director publishes a list of explosives determined to be within the coverage of the law.

Permissible Explosives

While the term "permissible explosive" is not exactly a classification, it does describe a type of explosive which has been tested and approved by the Department of Labor Mine Safety and Health Administration, as meeting the minimum safety requirements for use in underground coal mines (Fed. Reg. 1969). These materials are modified dynamites or water gels which exhibit reduced tendencies to ignite flammable gas-air or gas-air-coal dust mixtures.

Two Component Explosives

Two component explosives, also referred to as plosphoric or binary substances, consist of two or more unmixed, commercially manufactured, prepackaged chemicals, including oxidizing chemicals, flammable liquids, or solids that are not independently classified as explosives. When combined, however, the mixture is classified as an explosive and is stored, transported, and handled as an explosive.

MANUFACTURE OF EXPLOSIVE MATERIALS

With the possible exception of blasting agents, and some water gels or slurries, explosive materials are produced in plants under the close supervision of qualified personnel. While the potential for accidents is relatively high, the safety record is very good, largely because the hazards are well recognized. Fire is a principal cause of accidents, but a normal plant layout keeps this hazard within reasonable limits. Smoking is never allowed near the explosive operations, and the use of flame producing equipment is allowed only in specific locations under strict safety regulations. The relative ease of mixing ANFO mixtures and slurries, in addition to their reduced fire and explosion potential, has led to the construction of many small manufacturing plants which are located in widely scattered areas. Manufacturing methods may include: (1) plant mixed ANFO delivered to site by bulk truck; (2) ANFO mixed by bulk truck at site; (3) plant mixed slurry delivered to site by bulk truck; and (4) slurry mixed by bulk truck at site. These changes in methods have resulted in major variations in safety practices from normal explosive plant safety procedures.

Bulk trucks present a new fire hazard. There have been several fires on the Mesabi Iron Range involving this type of equipment.

TRANSPORTATION OF EXPLOSIVE MATERIALS

The DOT regulates the transportation of all explosive materials in accordance with regulations set forth in *Code of Federal Regulations*, Title 49, Chapter I, Parts 100-199, 1978. Copies of these regulations can be obtained from the U.S. Government Printing Office, or from the Bureau of Explosives of the Association of American Railroads.

Most commercial explosives are now transported over public highways by truck. Military explosives are transported both by rail and by truck. Fire is the most common cause of accidents in truck transportation, and the most probable point of initiation of fire is a truck's tires. Tire fires are quite common and represent a hazard which is difficult to control because the driver of a truck is often unaware of the fire until it has gained considerable headway. Detonators may be transported in limited quantities on the same vehicle with explosives and blasting agents, provided DOT regulations concerning quantity and type of container are observed.

Truck transportation over public highways is of particular concern because there is maximum exposure of the public to fire and explosion hazards. Unfortunately, fires and other types of accident situations tend to draw spectators, which increases the danger of casualties. While rail transportation generally provides a lesser hazard to the public, there is an increased danger resulting from the mixing of different types of cargo in a given train load.

STORAGE OF EXPLOSIVE MATERIALS

A primary objective for the storage of most nonexplosive industrial materials is to provide protection for such materials from their surroundings. Because of the hazardous nature of explosive materials, additional factors must be considered. Safety to the industrial workers involved and to the general public in the vicinity of such storage is of primary importance. Since most explosive materials are utilized by the mining and construction industries, these industries have primary concern in training their employees to provide maximum attainable safety. Employees of the explosive manufacturing firms are well aware of the hazards of their occupation, and safety is an integral part of their work. However, employees in mining and construction industries are less likely to be aware of the precautions required to obtain a satisfactory degree of safety. Another factor connected with storage is indirectly related to safety and involves the element of security. In recent years it has become increasingly important to protect the public from the illicit use of explosives which are often obtained by individuals through illegitimate means. The world wide increase in terrorism has made security an important consideration in the storage of explosives. Most of the explosives used for illegal purposes are obtained from legitimate stores of the materials. To assist in preventing this misuse of explosives, the law now provides that the Bureau of Alcohol, Tobacco, and Firearms of the Department of Treasury (ATF) shall regulate the manufacture, distribution, and storage of explosive materials. Copies of these regulations may be obtained from that agency.

Storage Magazines

To provide proper storage for the wide variety of explosive materials now being produced, five different types of magazines are recognized. The requirements for these five types of magazines are given in the regulations developed by ATF. The specifications for these magazines and the types of materials to be put in them may be obtained from ATF or can be found in NFPA 495, *Code for the Manufacture, Transportation, Storage, and Use of Explosive Materials.*

Magazine storage aside, it is a widespread practice in industry to transport relatively insensitive blasting agents to mining or construction sites in truck trailers. The manufacturer parks the trailer in a designated area, disconnects the cab, and picks up an empty trailer for the return trip. The blasting agents are then transported by small truck to the site of operations. This is considered as providing adequate safety from the fire hazard. When trucks transporting explosives or blasting agents must stop in transit, or the load must be transferred in transit, terminals are now provided in the vicinity of many large cities for motor vehicles carrying explosives. These terminals are well guarded and are operated under strict safety regulations. Requirements for these terminals are contained in NFPA 498, *Standard for Explosives Motor Vehicle Terminals.*

While the probability of the accidental initiation of an explosive while in an approved storage location is slight, accidents still occur. To safeguard the general public, the American Table of Distances for Storage of Explosives (Table 5-7A) has been prepared to indicate the isolation distances for stores of explosives from points of contact with the public. This table is revised periodically by the Institute of Makers of Explosives. While blasting agents are appreciably less sensitive than Class A explosives, once they are initiated to detonation, the damage produced by such an accident is comparable to that for Class A explosives. Therefore, the Table of Recommended Separation Distances of Ammonium Nitrate and Blasting Agents from Explosives or Blasting Agents is used to determine the safe distances for stores of blasting agents from inhabited buildings, railroads, etc. However, because of the lower sensitivity of blasting agents, the separation distances between magazines or stores of explosives and blasting agents need not be as great as that between stores of explosives, as shown in Table 5-7A. Table 5-7B shows these recommended separation distances and the thickness of artificial barricades recommended for these less sensitive materials. When no barricades separate stores, the distances shall be increased as shown in Table 5-7B (Note 2).

FIRE PROTECTION FOR EXPLOSIVE MATERIALS

Fire is a principal cause of accidents involving explosive materials. Although explosives and blasting agents vary in their sensitivity to fire conditions, they are liable to produce a disastrous explosion when subjected to fire. The only effective method for fire protection with explosive materials is to eliminate the sources of fire. This means meticulous housekeeping standards. It is often a small fire in oily rags or wastepaper that triggers a major catastrophe. Smoking or the use of any fire producing equipment should not be permitted where explosive materials are produced, handled, stored, or used. Because of the relative insensitivity of blasting agents, there is always a danger that personnel will become careless around these materials. A fire and subsequent explosion in a blasting agent mixing house near Norton, VA illustrates the potential hazard from the careless use of heat producing equipment (Van Dolah and Malesky 1962).

Some of the materials used in making explosives and blasting agents may be as hazardous as the final product. Materials likely to be found include aluminum powder,

TABLE 5-7A. American Table of Distances for Storage of Explosives

Distances in feet*
As Revised and Approved by The Institute of Makers of Explosives
November 5, 1971.

EXPLOSIVES		Inhabited Buildings		Public Highways Class A to D		Passenger Railways—Public Highways with Traffic Volume of more than 3,000 Vehicles/Day		Separation of Magazines	
Pounds† Over	Pounds† Not Over	Barri-caded	Unbarri-caded	Barri-caded	Unbarri-caded	Barri-caded	Unbarri-caded	Barri-caded	Unbarri-caded
2	5	70	140	30	60	51	102	6	12
5	10	90	180	35	70	64	128	8	16
10	20	110	220	45	90	81	162	10	20
20	30	125	250	50	100	93	186	11	22
30	40	140	280	55	110	103	206	12	24
40	50	150	300	60	120	110	220	14	28
50	75	170	340	70	140	127	254	15	30
75	100	190	380	75	150	139	278	16	32
100	125	200	400	80	160	150	300	18	36
125	150	215	430	85	170	159	318	19	38
150	200	235	470	95	190	175	350	21	42
200	250	255	510	105	210	189	378	23	46
250	300	270	540	110	220	201	402	24	48
300	400	295	590	120	240	221	442	27	54
400	500	320	640	130	260	238	476	29	58
500	600	340	680	135	270	253	506	31	62
600	700	355	710	145	290	266	532	32	64
700	800	375	750	150	300	278	556	33	66
800	900	390	780	155	310	289	578	35	70
900	1,000	400	800	160	320	300	600	36	72
1,000	1,200	425	850	165	330	318	636	39	78
1,200	1,400	450	900	170	340	336	672	41	82
1,400	1,600	470	940	175	350	351	702	43	86
1,600	1,800	490	980	180	360	366	732	44	88
1,800	2,000	505	1,010	185	370	378	756	45	90
2,000	2,500	545	1,090	190	380	408	816	49	98
2,500	3,000	580	1,160	195	390	432	864	52	104
3,000	4,000	635	1,270	210	420	474	948	58	116
4,000	5,000	685	1,370	225	450	513	1,026	61	122
5,000	6,000	730	1,460	235	470	546	1,092	65	130
6,000	7,000	770	1,540	245	490	573	1,146	68	136
7,000	8,000	800	1,600	250	500	600	1,200	72	144
8,000	9,000	835	1,670	255	510	624	1,248	75	150
9,000	10,000	865	1,730	260	520	645	1,290	78	156
10,000	12,000	875	1,750	270	540	687	1,374	82	164
12,000	14,000	885	1,770	275	550	723	1,446	87	174
14,000	16,000	900	1,800	280	560	756	1,512	90	180
16,000	18,000	940	1,880	285	570	786	1,572	94	188
18,000	20,000	975	1,950	290	580	813	1,626	98	196
20,000	25,000	1,055	2,000	315	630	876	1,752	105	210
25,000	30,000	1,130	2,000	340	680	933	1,866	112	224
30,000	35,000	1,205	2,000	360	720	981	1,962	119	238
35,000	40,000	1,275	2,000	380	760	1,026	2,000	124	248
40,000	45,000	1,340	2,000	400	800	1,068	2,000	129	258
45,000	50,000	1,400	2,000	420	840	1,104	2,000	135	270

paint grade aluminum, ammonium nitrate, ammonium perchlorate, sodium nitrate, fuel oil, gilsonite, bagasse, and various types of finely divided gums. NFPA Codes should be consulted for proper storage and protection of these materials.

For all explosive materials, whether in manufacture, storage, transportation, or use, housekeeping should be of the highest order.

FIGHTING FIRES INVOLVING EXPLOSIVES AND BLASTING AGENTS

No attempt should be made to fight a fire involving Class A, B, or C explosives once the fire has actually reached the explosives. Fire fighting should be abandoned and the area evacuated. A distance of 2,000 ft (610 m) is considered reasonable. Remote control equipment may be

TABLE 5-7A. American Table of Distances for Storage of Explosives (Cont.)

EXPLOSIVES		Inhabited Buildings		Public Highways Class A to D		Passenger Railways—Public Highways with Traffic Volume of more than 3,000 Vehicles/Day		Separation of Magazines	
Pounds† Over	Pounds† Not Over	Barricaded	Unbarricaded	Barricaded	Unbarricaded	Barricaded	Unbarricaded	Barricaded	Unbarricaded
50,000	55,000	1,460	2,000	440	880	1,140	2,000	140	280
55,000	60,000	1,515	2,000	455	910	1,173	2,000	145	290
60,000	65,000	1,565	2,000	470	940	1,206	2,000	150	300
65,000	70,000	1,610	2,000	485	970	1,236	2,000	155	310
70,000	75,000	1,655	2,000	500	1,000	1,263	2,000	160	320
75,000	80,000	1,695	2,000	510	1,020	1,293	2,000	165	330
80,000	85,000	1,730	2,000	520	1,040	1,317	2,000	170	340
85,000	90,000	1,760	2,000	530	1,060	1,344	2,000	175	350
90,000	95,000	1,790	2,000	540	1,080	1,368	2,000	180	360
95,000	100,000	1,815	2,000	545	1,090	1,392	2,000	185	370
100,000	110,000	1,835	2,000	550	1,100	1,437	2,000	195	390
110,000	120,000	1,855	2,000	555	1,110	1,479	2,000	205	410
120,000	130,000	1,875	2,000	560	1,120	1,521	2,000	215	430
130,000	140,000	1,890	2,000	565	1,130	1,557	2,000	225	450
140,000	150,000	1,900	2,000	570	1,140	1,593	2,000	235	470
150,000	160,000	1,935	2,000	580	1,160	1,629	2,000	245	490
160,000	170,000	1.965	2,000	590	1,180	1,662	2,000	255	510
170,000	180,000	1,990	2,000	600	1,200	1,695	2,000	265	530
180,000	190,000	2,010	2,010	605	1,210	1,725	2,000	275	550
190,000	200,000	2,030	2,030	610	1,220	1,755	2,000	285	570
200,000	210,000	2,055	2,055	620	1,240	1,782	2,000	295	590
210,000	230,000	2,100	2,100	635	1,270	1,836	2,000	315	630
230,000	250,000	2,155	2,155	650	1,300	1,890	2,000	335	670
250,000	275,000	2,215	2,215	670	1,340	1,950	2,000	360	720
275,000	300,000	2,275	2,275	690	1,380	2,000	2,000	385	770

Note 1: "Explosive materials" means explosives, blasting agents, and detonators.

Note 2: "Explosives" means any chemical compound, mixture or device, the primary or common purpose of which is to function by explosion. A list of explosives determined to be within the coverage of "18 U.S.C. Chapter 40, Importation, Manufacture, Distribution and Storage of Explosive Materials" is issued at least annually by the Director, Bureau of Alcohol, Tobacco, and Firearms, Department of U.S. Treasury.

Note 3: "Blasting Agents" means any material or mixture, consisting of fuel and oxidizer, intended for blasting, not otherwise defined as an explosive: provided that the finished product, as mixed for use or shipment, cannot be detonated by means of a number 8 test blasting cap when unconfined.

Note 4: "Detonator" means any device containing a detonating charge that is used for initiating detonation in an explosive; the term includes, but is not limited to, electric blasting caps of instantaneous and delay types, blasting caps for use with safety fuses and detonating-cord delay connectors.

Note 5: "Magazine" means any building or structure, other than an explosives manufacturing building, used for the permanent storage of explosive materials.

Note 6: "Natural Barricade" means natural features of the ground, such as hills, or timber of sufficient density that the surrounding exposures which require protection cannot be seen from the magazine when the trees are bare of leaves.

Note 7: "Artificial Barricade" means an artificial mound or revetted wall of earth of a minimum thickness of three feet.

Note 8: "Barricaded" means that a building containing explosives is effectually screened from a magazine, building, railway, or highway, either by a natural barricade, or by an artificial barricade of such height that a straight line from the top of any sidewall of the building containing explosives to the eave line of any magazine, or building, or to a point twelve feet above the center of a railway or highway, will pass through such intervening natural or artificial barricade.

Note 9: "Inhabited Building" means a building regularly occupied in whole or in part as a habitation for human beings, or any chruch, schoolhouse, railroad station, store, or other structure where people are accustomed to assemble, except any building or structure occupied in connection with the manufacture, transportation, storage or use of explosives.

Note 10: "Railway" means any steam, electric, or other railroad or railway which carries passengers for hire.

Note 11: "Highway" means any street or public road. "Public Highways Class A to D" are highways with average traffic volume of 3,000 or less vehicles per day as specified in "American Civil Engineering Practice" (Abbett, Vol. 1, Table 46, Sec. 3-74, 1956 Edition, John Wiley and Sons).

Note 12: When two or more storage magazines are located on the same property, each magazine must comply with the minimum distances specified from inhabited buildings, railways, and highways, and, in addition, they should be separated from each other by not less than the distances shown for "Separation of Magazines," except that the quantity of explosives contained in cap magazines shall govern in regard to the spacing of said cap magazines from magazines containing other explosives. If any two or more magazines are separated from each other by less than the specified "Separation of Magazines" distances, then such two or more magazines, as a group, must be considered as one magazine, and the total quantity of explosives stored in such group must be treated as if stored in a single magazine located on the site of any magazine of the group, and must comply with the minimum of distances specified from other magazines, inhabited buildings, railways, and highways.

Note 13: Storage in excess of 300,000 lbs. of explosives in one magazine is generally not required for commercial enterprises.

Note 14: This Table applies only to the manufacture and permanent storage of commercial explosives. It is not applicable to transportation of explosives or any handling or temporary storage necessary or incident thereto. It is not intended to apply to bombs, projectiles, or other heavily encased explosives.

Note 15: All types of blasting caps in strengths through No. 8 cap should be rated at 1½ lbs of explosives per 1,000 caps. For strengths higher than No. 8 cap, consult the manufacturer.

Note 16: For quantity and distance purposes, detonating cord of 50 to 60 grains per foot should be calculated as equivalent to 9 lbs of high explosives per 1,000 feet. Heavier or lighter core loads should be rated proportionately.

* one pound equals 0.4536 kg

† one foot equals 0.3048 m

TABLE 5-7B. Table of Recommended Separation Distances of Ammonium Nitrate and Blasting Agents from Explosives or Blasting Agents[1,6]

Donor Weight		Minimum Separation Distance of Receptor when Barricaded[2] (ft§)		Minimum Thickness of Artificial Barricades[5] (in.)#
Pounds‡ Over	Pounds‡ Not Over	Ammonium Nitrate[3]	Blasting Agent[4]	
	100	3	11	12
100	300	4	14	12
300	600	5	18	12
600	1,000	6	22	12
1,000	1,600	7	25	12
1,600	2,000	8	29	12
2,000	3,000	9	32	15
3,000	4,000	10	36	15
4,000	6,000	11	40	15
6,000	8,000	12	43	20
8,000	10,000	13	47	20
10,000	12,000	14	50	20
12,000	16,000	15	54	25
16,000	20,000	16	58	25
20,000	25,000	18	65	25
25,000	30,000	19	68	30
30,000	35,000	20	72	30
35,000	40,000	21	76	30
40,000	45,000	22	79	35
45,000	50,000	23	83	35
50,000	55,000	24	86	36
55,000	60,000	25	90	35
60,000	70,000	26	94	40
70,000	80,000	28	101	40
80,000	90,000	30	108	40
90,000	100,000	32	115	40
100,000	120,000	34	122	50
120,000	140,000	37	133	50
140,000	160,000	40	144	50
160,000	180,000	44	158	50
180,000	200,000	48	173	50
200,000	220,000	52	187	60
220,000	250,000	56	202	60
250,000	275,000	60	216	60
275,000	300,000	64	230	60

Note 1: Recommended separation distances to prevent explosion of ammonium nitrate and ammonium nitrate-based blasting agents by propagation from nearby stores of high explosives or blasting agents referred to in the Table as the "donor." Ammonium nitrate, by itself, is not considered to be a donor when applying this Table. Ammonium nitrate, ammonium nitrate-fuel oil or combinations thereof are acceptors. If stores of ammonium nitrate are located within the sympathetic detonation distance of explosives or blasting agents, one-half the mass of the ammonium nitrate should be included in the mass of the donor.

These distances apply to the separation of stores only.

Note 2: When the ammonium nitrate and/or blasting agent is not barricaded, the distances shown in the Table shall be multiplied by six. These distances allow for the possibility of high velocity metal fragments from mixers, hoppers, truck bodies, sheet metal structures, metal containers, and the like which may enclose the "donor." Where storage is in bullet-resistant magazines* recommended for explosives or where the storage is protected by a bullet-resistant wall, distances and barricade thicknesses in excess of those prescribed in the American Table of Distances are not required.

Note 3: The distances in the Table apply to ammonium nitrate that passes the insensitivity test prescribed in the definition of ammonium nitrate fertilizer promulgated by the National Plant Food Institute;† and ammonium nitrate failing to pass said test shall be stored at separation distances determined by competent persons and approved by the authority having jurisdiction.

Note 4: These distances apply to nitrocarbonitrates and blasting agents which pass the insensitivity test prescribed in the U.S. Department of Transportation (DOT) and the U.S. Department of the Treasury, Bureau of Alcohol, Tobacco, and Firearms.

Note 5: Earth, or sand dikes, or enclosures filled with the prescribed minimum thickness of earth or sand are acceptable artificial barricades. Natural barricades, such as hills or timber of sufficient density that the surrounding exposures which require protection cannot be seen from the "donor" when the trees are bare of leaves, are also acceptable.

Note 6: For determining the distances to be maintained from inhabited buildings, passenger railways, and public highways, use the Table of Distances for Storage of Explosives in Appendix A of *NFPA, 1973, Code for the Manufacture, Transportation, Storage, and Use of Explosive Materials.*

* For construction of bullet-resistant magazines see Chapter 3 of NFPA No. 495, *Code for the Manufacture, Transportation, Storage, and Use of Explosives and Blasting Agents.*

† Definition and Test Procedures for Ammonium Nitrate Fertilizer, National Plant Food Institute, November 1964.

‡ one pound equals 0.4536 kg

§ one ft equals 0.3048 m

one in equals 25.4 mm

left in position. The rules for blasting agents are the same except that, when not confined, incipient fires in blasting agents can be fought with large amounts of water. The water acts only to cool the burning mass to temperatures below the ignition temperature. When fires develop beyond the incipent stage, the only safe method for handling the situation is to abandon direct fire fighting methods and evacuate the area in anticipation of an explosion. The suggested evacuation distance of 2,000 ft (610 m) should be observed for blasting agents as well.

The Norton, VA fire mentioned previously is a good example of the necessity of retreating from an established fire involving blasting agents. The Luckenback Pier fire in Brooklyn, NY in 1959 shows that even Class C explosives represent a potential hazard. Several tons (one short ton equals 0.91 metric ton) of detonating cord were stored on a pier which became involved in a serious fire. After considerable delay, the entire mass detonated with many lives lost and property damage in the millions of dollars.

Truck fires involving explosive materials pose the greatest threat to the public because the number of trucks in transit is so large. Many of these incidents have resulted in explosions. Immediate evacuation in such incidents is imperative.

DESTRUCTION OF EXPLOSIVE MATERIALS

Surplus, abandoned, or damaged explosives unfit for further use should not be disposed of by burning, detonating, water leaching, chemical decomposing or any other means except under the supervision of the manufacturer or the manufacturer's designated representative. This is especially true in the case of explosives that have been damaged during fire fighting operations. If the manufacturer is not known, the Institute of Makers of Explosives, 420 Lexington Avenue, New York, NY, 10017, should be contacted for assistance.

When bombs or other sabotage devices are suspected, disposal should be under the supervision of trained bomb disposal experts.

Bibliography

References Cited

Cook, Melvin A. 1958. *The Science of High Explosives*. Reinhold Publishing Co., NY.

Dixon, W. T., and Fisher, A. W., eds. 1958. *Chemical Engineering in Industry*. Chapter 12. American Institute of Chemical Engineers, NY.

Fed. Reg. 1972. Title 49, Chapter 1, "Transportation," Parts 170-189. *U.S. Code of Federal Regulations*. U.S. Government Printing Office, Washington, DC.

Fed. Reg. 1969. Title 30, Chapter 1, Subchapter C, Part 15. "Explosives and Related Articles." *U.S. Code of Federal Regulations*. U.S. Government Printing Office. Washington, DC.

USDI. 1983. "Apparent Consumption of Industrial Explosives and Blasting Agents in the United States." *Mineral Industry Surveys*. USDI Bureau of Mines, Pittsburgh, PA.

Van Dolah, R. W., and Malesky, J. S. 1962. "Fire and Explosion in a Blasting Agent Mix House Building, Norton, VA." *RI 6015*, USDI, Bureau of Mines, Pittsburgh, PA.

NFPA Codes, Standards, Recommended Practices and Manuals. (See the latest *NFPA Codes and Standards Catalog* for availability of current editions of the following documents.)

NFPA 490, *Code for the Storage of Ammonium Nitrate.*

NFPA 495, *Code for the Manufacture, Transportation, Storage and Use of Explosive Materials.*

NFPA 498, *Standard for Explosives Motor Vehicle Terminals.*

Additional Readings

"ATF Explosives Laws and Regulations," ATF Publication 5400.7; Bureau of Alcohol Tobacco and Firearms, Washington, DC, Nov., 1982.

Bahme, C. W., *Fire Officer's Guide to Dangerous Chemicals*, National Fire Protection Association, Quincy, 1978.

Blasters Handbook, 75th Anniversary Edition, Explosives Department, E. I. DuPont de Nemours and Co., Wilmington, DE, 1977.

Conkling, John A., "New Federal Standards for Class C Fireworks," *Fire Journal*, Vol. 71, No. 3, May 1977, pp. 27-29, 117.

Construction Guide for Storage Magazines, IME Safety Library Publication No. 1; Institute of Makers of Explosives, Washington, DC, Jan. 1983.

Damon, Glenn H., "Blasting Agents: History, Hazards and Protection," *Fire Journal*, Vol. 59, No. 2, Mar. 1965.

Destruction of Commercial Explosives, IME Safety Library Publication No. 21, Institute of Makers of Explosives, Washington, DC, Dec. 1981.

Dick, R. A., Fletcher, L. R., and D'Andrea, D. V., *Explosives and Blasting Procedures Manual*, Information Circular 8925, U. S. Bureau of Mines, Washington, DC, 1983.

Dick, Richard A., *The Impact of Blasting Agents and Slurries on Explosives Technology*, IC 8560, 1972, USDI Bureau of Mines, Pittsburgh, PA.

Dos and Don'ts, IME Safety Library Publication No. 4, Institute of Makers of Explosives, Washington, DC, June, 1982.

Ellern, H., *Military and Civilian Pyrotechnics*, Chemical Publishing Co., NY, 1968.

Explosion and Explosives: Journal of the Industrial Explosives Society, Japan, Industrial Explosives Society, Japan, publ. periodically.

Explosive Hazard Classification Procedures, DLAR 8220.1; Defense Logistics Agency, Washington, DC, 1981.

"Fireworks Incidents: The 1978 Story," Fire Journal, Vol. 72, No. 6, Nov. 1978, pp. 58-63.

Fordham, S., *High Explosives and Propellants*, Pergammon Press, London, England, 1966.

Glossary of Industry Terms, IME Safety Library Publication No. 12, Institute of Makers of Explosives, Washington, DC, Sept., 1981.

IME Standard for the Safe Transportation of Class C Detonators in a Vehicle with Certain Other Explosives, IME Safety Library Publication No 22, Institute of Makers of Explosives, Washington, DC, Mar. 21, 1979.

Juillerat, Ernest E., "Mail Order Explosives for Children," *Fire Journal*, Vol. 62, No. 5, Sept. 1968, pp. 24-28, 68.

Kinney, Gilbert F., *Explosive Shocks in Air*, Macmillan, NY, 1962.

Litchfield, E. L., Hay, M. H., and Monroe, J. S., *Electrification of Ammonium Nitrate in Pneumatic Loading*, RI 7139, USDI Bureau of Mines, Pittsburgh, PA, 1968.

Meidl, H. H., *Explosive and Toxic Hazardous Materials*, Glencoe Press, Beverly Hills, CA, 1970.

Safety in the Transportation, Storage, Handling, and Use of Explosives, IME Safety Library Publication No. 17, Institute of Makers of Explosives, Washington, DC, April 1974.

Safety Recommendations for Sensitized Ammonium Nitrate Blasting Agents, IC 8179, 1963, USDI, Bureau of Mines, Pittsburgh, PA.

Safety Requirement for the Transportation, Storage, Handling and Use of Commercial Explosives and Blasting Agents in the Construction Industry, ANSI A10.7, American National Standards Institute, NY 1970.

Urbanski, Tadensz, *Chemistry and Technology of Explosives*, 3 vols, Macmillan, NY, 1964.

Van Dolah, R. W., et al., *Explosion Hazards of Ammonium Nitrate Under Fire Exposure*, RI 6773, USDI, Bureau of Mines, Pittsburgh, PA, 1966.

Van Dolah, R. W., Gibson, F. C., and Murphy, J. N., *Sympathetic Detonation of Ammonium Nitrate and Ammonium Nitrate-Fuel Oil*, RI 6746, USDI, Bureau of Mines, Pittsburgh, PA, 1966.

———, *Further Studies on Sympathetic Detonation of Ammonium Nitrate and Ammonium Nitrate-Fuel Oil* RI 6903, 1966, USDI, Bureau of Mines, Pittsburgh, PA.

Van Dolah, R. W., Mason, C. M., and Forshey, D. R., *Development of Slurry Explosives for Use in Potentially Flammable Gas Atmospheres*, RI 7195, USDI, Bureau of Mines, Pittsburgh, PA, 1968.

Winning, C. H., "Detonation Characteristics of Prilled Ammonium Nitrate," *Fire Technology*, Vol. 1, No. 1, Feb. 1965, pp. 23-31.

PLASTICS AND RUBBER

Revised by Bert Cohn, P.E.

The plastics industry is one of the most rapidly developing technologies today. Since World War II, the types, variations, and uses of plastics have proliferated at a tremendous rate, producing a family of materials which is extraordinarily complex and varied. Each of the variations can be considered combustible. Some burn very rapidly, and under certain conditions present extreme hazards to life and property. When such hazardous characteristics are identified, modifications can be made in manufacture and application that can reduce the hazard to acceptable levels. For example, cellulose nitrate motion picture film presents the hazard of flash burning. This hazard was controlled by phasing out the cellulose nitrate film and replacing it with slow burning "safety" film. More recently, the hazards of cellular plastics such as polyurethane and polystyrene, which are used for exposed interior finish in buildings, have been controlled by using flame retardant additives during manufacture combined with thermal barriers installed in the field.

This chapter focuses on the fire behavior of plastics which are used as building materials and on the characteristics of polymers in general. The tests which are used to identify the relative hazards from fire and smoke of plastics used in building applications are also discussed. Plastics in fibers and textiles (including clothing, upholstery fabrics, and mattresses) are discussed in Chapter 3 of this Section. The fire hazards associated with the manufacture, processing, and warehousing of plastics and elastomers are discussed in Section 10, Chapter 12, "Extrusion and Forming Processes."

While being a part of the family of polymers, natural and synthetic rubbers, technically, are not plastics. Yet, because of the many similarities and the formulations that intermix the traditional characteristics of plastics and elastomers, rubbers are often included in discussions of plastics. In this chapter, the general term "plastic" is intended to include rubber, but it should be understood that rubber has a whole complex technology of its own—one that predates the modern plastics industry.

Bert Cohn, P.E., is President of Gage-Babcock and Associates, Inc., Elmhurst, IL.

PLASTICS TERMINOLOGY

The following terms are used in this chapter or are commonly encountered in the application, use, and storage of plastics products:

Additive: Any material mixed with a resin to modify its processing or end use properties. The resulting mix usually is called a "plastic" to distinguish it from the "resin" or principal ingredient. Additives may be dyes, pigments, powdered fillers for stiffening, plasticizers for flexibility, fibers to reinforce, antioxidants, lubricants to aid flow into or release from a mold, or flame retardants.

Binders: The resins in a plastic mixture that hold together all of the other ingredients. They are usually a product of polymerization or a naturally occurring substance of high molecular weight.

Blowing Agent: A material which releases gas upon heating so that the plastic in which it is mixed will expand into foam. The gas may result from boiling of a liquid or from decomposition of the blowing agent (foaming agent).

Copolymers: High molecular weight compounds produced when two or more monomers are involved in a polymerization process.

Crosslinking: The establishing of chemical bonds between adjacent molecules so as to make a network which will reduce solubility or resist flow on heating. The amount of crosslinking may be enough to render the plastic highly resistant to flow at temperatures up to thermal decomposition, in which case the plastic is said to be a thermoset.

Fillers: Materials that modify the strength and working properties of a plastic. They may be used to increase heat resistance or alter the dielectric strength. A wide variety of products are used including wood flour, cotton, sisal, glass, and clay.

Film: A general term for plastic not more than 0.01 in. (0.25 mm) thick, regardless of the process used to make it. "Foil" is used today only to describe metal.

Foam: Plastic with many small gas bubbles; cellular plastic.

Foaming Agent: The same as a blowing agent.

Inhibitor: A material added to prevent or retard an undesired chemical reaction.

Laminate: A composition of two or more layers of plastic firmly adhered by partly melting one or more of the layers, by an adhesive, or by impregnation. High pressure laminates are rigid stock made in presses at pressures above 400 psi (2760 kPa).

Monomers: The small starting molecules, usually gaseous or liquid, used to produce the polymer resins.

Plastic (ASTM definition): A material that contains as an essential ingredient one or more organic polymeric substances of large molecular weight, is solid in its finished state, and at some stage in its manufacture or processing into finished articles can be shaped by flow.

Plasticizers: Organic materials added to plastics to make the finished product more flexible or to facilitate compounding. Some plasticizers increase the combustibility of the plastic while others serve as flame retardants.

Polymerization: The process by which molecules of a monomer are made to add to themselves or to other monomers in a repetitive way, forming a much longer chainlike molecule.

Polymers: Large molecules made by combining smaller molecules by chemical reaction.

Reinforced Plastic: A plastic with a filler which significantly increases flexural, impact, or tensile strength. Additives are usually glass asbestos, cotton, or nylon fibers.

Resin (ASTM definition): A solid or pseudosolid organic material often of high molecular weight, which exhibits a tendency to flow when subjected to stress, usually has a softening or melting range, and usually fractures conchoidally. Note: In a broad sense, the term is used to designate any polymer that is a basic material for plastics.

Sheet: A general term for plastics 0.01 in. (0.25 mm) and thicker, regardless of the method of manufacture.

Stabilizer: A material added to prevent or retard an undesired chemical reaction. When used to prevent polymerization, the preferred term is "inhibitor." Stabilizers are usually present in less than one percent. They may be antioxidants, screening agents for ultraviolet radiation, or chemicals to neutralize decomposition products during hot processing.

Thermoforming: The process of stretching or bending a plastic sheet which has been softened by heating.

Thermoplastic (ASTM definition): A plastic that can be repeatedly softened by heating and hardened by cooling through a temperature range characteristic of the plastic, and that in the softened state can be shaped by flow into articles by molding or extrusion.

Thermoset (ASTM definition): A plastic that, after having been cured by heat or other means, is substantially infusible and insoluble.

THE FAMILY OF POLYMERS

In the classification of engineering materials, polymers are one division under the general category of nonmetals, which also includes ceramics, glasses, and materials such as oils, greases, lubricants, papers, etc. Polymers are further subdivided into three classes—elastomers, thermosets, and thermoplastics—as shown in Figure 5-8A. Basically, all polymers are characterized by their very large molecules, formed in chemical "polymerization" reactions which link two or more molecules into larger molecules. These larger molecules contain repeating structural units of the original molecules. If the linkages result in a rigid "cross-linked" molecular structure, a thermosetting plastic results. If the molecular structure is more flexible, including side chains not rigidly linked to the other molecules, the resulting material is a thermoplastic.

POLYMERS

THERMOSETS	THERMOPLASTICS	ELASTOMERS
Alkyds	ABS	Butadiene
Allyls	Acetals	Butyl
Epoxies	Acrylics	Fluoroelastomers
Melamine	Cellulosics	Hypalon
Phenolics	Fluoroplastics	Isoprene
Polyesters	Ionomers	Natural rubber
Polyurethane	Nylons	Neoprene
Silicones	Olefins	Nitrile
Urea	Polycarbonate	Polysulfide
and others	Polyester	Silicones
	Polyimide	Urethane
	Polystyrene	and others
	Vinyls	
	and others	

FIG. 5-8A. The family of polymers.

The products known as "plastics" include the common types of thermosetting and thermoplastic polymers included in Figure 5-8A, plus many more. In addition to the polymer constituent, most finished plastics contain plasticizers, colorants, fillers, stabilizers, reinforcing agents, lubricants, or other special additives. The variation of individual plastic product formulations runs into the thousands. The ultimate form of each product, whether it appears in solid sections, films and sheets, foams, molded forms, synthetic fibers, pellets, or powders, has a significant effect on its fire properties. Additionally, the fire properties are influenced by the environment, and the resultant hazard by the quantity of the product and the conditions of its use. For these reasons, it is important to conduct fire tests of products in their end use configuration and under conditions which simulate the end use environment.

Elastomers

Elastomers are commonly referred to as "synthetic rubbers" and are characterized by their elastic or rubber-like properties. As a whole, this class of polymers is intended to reproduce or surpass the best properties of natural rubber for particular applications. The fire characteristics of elastomers are usually similar to those of natural rubber. A discussion of both natural and synthetic rubbers is included in this chapter.

Thermosetting Plastics

Thermosetting plastics, or thermosets, are hardened into a permanent shape in the manufacturing process. These materials cannot usually be softened by heating. For those that can be softened, the parts cannot be remelted or restored to their original state before curing. The components of thermosets may be purchased in liquid form as a monomer polymer mixture or as a partially polymerized molding compound. Thermosets are usually formed into their final shape with heat (sometimes under pressure), although some are cured at room temperature.

Thermoplastics

Thermoplastics do not cure or set into a permanent shape under the application of heat since heat will cause them to soften. They can be softened and rehardened repeatedly, although eventual thermal aging will cause a degradation of the material and limit the number of reheat cycles. When heated to a flowable state, thermoplastics can be transferred to a mold and cooled into the shape of the mold. Thermoplastics can be extruded through dies and solidified around other materials, such as insulation over electrical wiring. Generally, no chemical reaction takes place.

There is some overlap between thermosets and thermoplastics. Thermoplastics, after extrusion, can be cross-linked chemically or by irradiation to produce a thermoset product. The molding techniques typical of thermoplastics have been modified to permit injection molding of thermosets, such as phenolics.

Uses of Plastics

Plastics have wide ranging and continuously expanding applications. In building construction applications, plastics can be found in thermal and acoustical insulation, roofing materials, decorative trims and imitation wood parts, ceiling and wall panels, resilient flooring, light fixture diffusers, electrical conduits, insulation on wiring, water and gas piping; bathtubs and sinks, and heating ducts. Plastics are found in the housings and parts of equipment and appliances. Entire pieces of furniture, seat cushions, mattresses, carpeting, wall coverings, and draperies are often plastics, as are countertops, desk tops, and tabletops. Textiles in clothing today typically are partially or totally plastic products. Other uses of plastics include major parts, trim, and interiors of automobiles, buses, airplanes and trains; packaging materials, including plastic wrap; housewares; sports equipment; toys; photographic films, magnetic tapes, phonograph records, and luggage.

Lack of stability under high temperatures and inherent combustibility so far have ruled out the use of plastics for applications where fire resistance ratings are a requirement. One exception is the use of special silicone plastic foams for sealing openings in fire walls; the material retains its sealing and insulating properties while it slowly burns away during fire exposure.

MAJOR GROUPS OF PLASTICS

The major groups of plastics are briefly described in the following summaries. It must be recognized, however, that there are hundreds of variations for each group which are sold under thousands of individual trade names. More detailed information can be obtained from consulting the standard reference works of the industry. (See references at the end of this chapter.) When a plastic is identified only by its trade name, information about it can also be found in these handbooks or in trade name indices, such as those listed at the end of this chapter.

Sometimes it is necessary to identify a particular plastic found in a manufactured product. The constituents are so varied that positive identification is often very difficult, but many of the common plastics have certain, unique characteristics which can be determined from relatively simple tests. These tests are presented in Table 5-8A.

Commonly accepted abbreviations have been developed for many plastic materials and some of these have been standardized by the American Society for Testing and Materials (ASTM D1600). Manufacturers are being urged to offer the abbreviation along with trademark names to aid in the rapid identification of the material. This is particularly important for plastic products used in building construction. For the purposes of this text, the abbreviation is noted either in parentheses in the heading or after the name of the plastic in the descriptions that follow.

ABS

Developed to offset some of the shortcomings of polystyrene, ABS plastics are copolymers of acrylonitrile, butadiene and styrene monomers in varying proportions. They are balanced for mechanical toughness, good service temperature for other than high temperature applications, and ease of fabrication. ABS materials have relatively good electrical insulating properties, good resistance to a wide range of chemicals, and good dimensional stability. Typical applications include automobile dashboards, pump impellers, refrigerator parts, appliance housings, luggage, protective helmets, battery cases, telephones, pipe and fittings, shoe heels, knobs, and tool handles. Typical trade names include Abson, Cycolac, GravoFLEX, Kralastic, Ravikral, Styrophane, and Terluran.

Acetals (AR)

The acetal resins are made primarily by the polymerization of formaldehyde. Copolymers of acetal are also produced. These materials form one of the strongest and stiffest thermoplastics. The products show good resistance to organic solvents; however, they do react with strong acids or bases. These plastics are widely used for equipment gears, automotive equipment, plumbing parts, and appliances. Two basic types are a homopolymer (DuPont's Deltrin®) and a copolymer (Celanese's Celcon®).

Acrylics

The acrylics are basically polymethyl methacrylate (PMMA). Marketed under the common tradenames Lucite and Plexiglas, PMMA is known for its excellent clarity, high transparency, good resistance to sharp blows, strength, and rigidity. It is odorless, tasteless, and nontoxic, but the surface scratches easily, and contact with

TABLE 5-8A. Simple tests for identifying the more common plastics. (Adapted from Briston, J. H. and Grosselin, C. C., *Introduction to Plastics*, Newnes-Butterworth, London, 1968.)

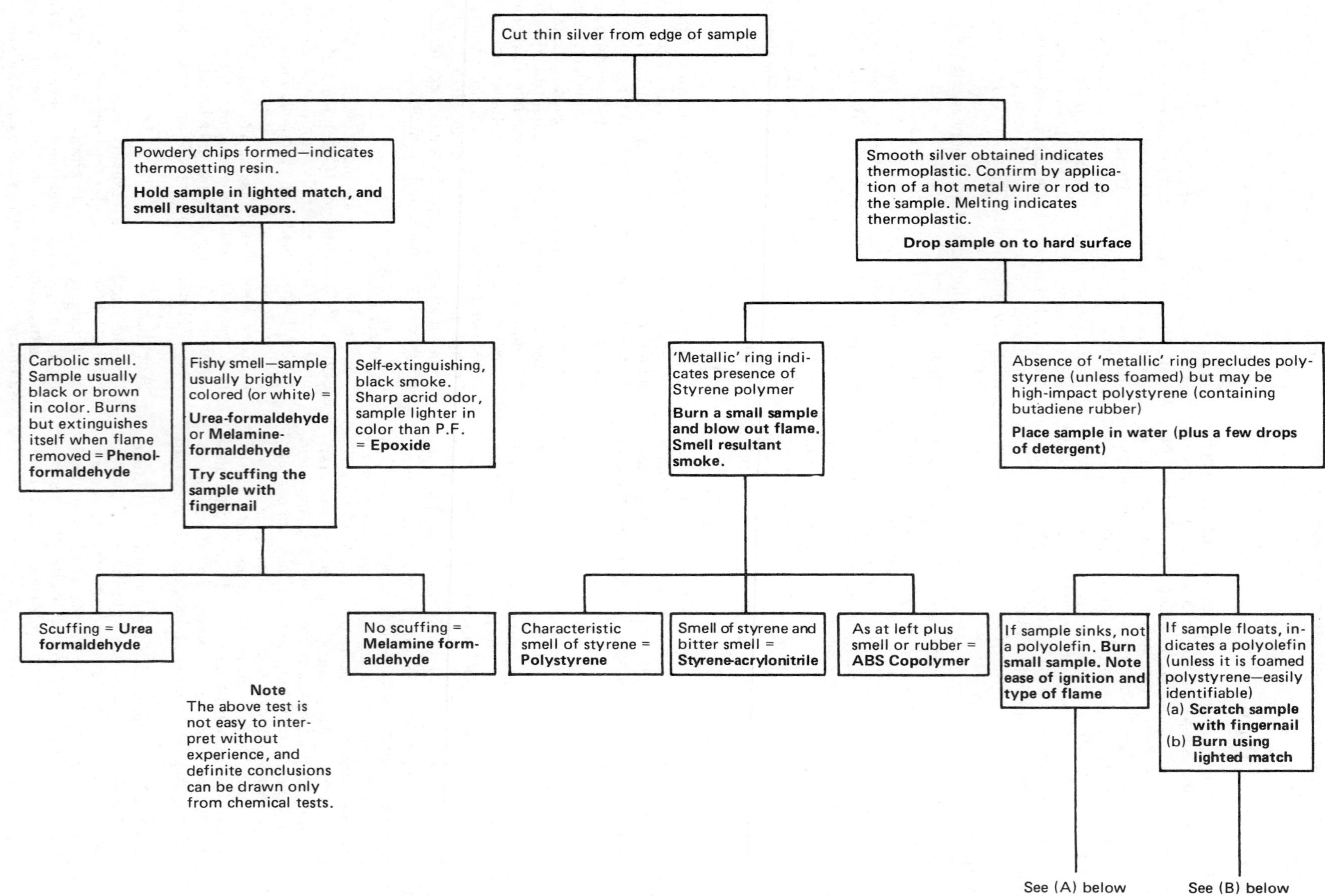

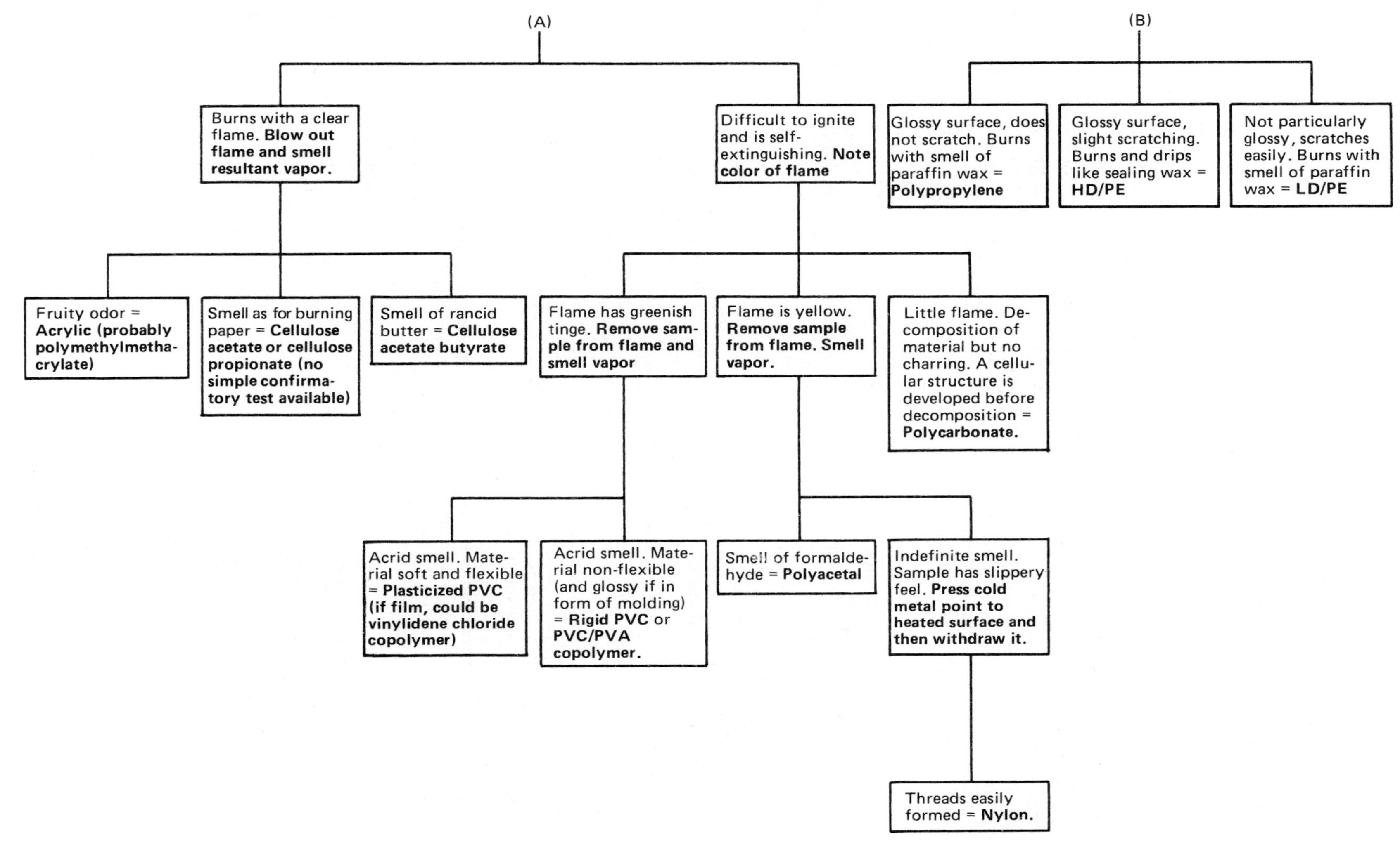
(A)
(B)
Burns with a clear flame. Blow out flame and smell resultant vapor.
Difficult to ignite and is self-extinguishing. Note color of flame
Glossy surface, does not scratch. Burns with smell of paraffin wax = Polypropylene
Glossy surface, slight scratching. Burns and drips like sealing wax = HD/PE
Not particularly glossy, scratches easily. Burns with smell of paraffin wax = LD/PE
Fruity odor = Acrylic (probably polymethylmethacrylate)
Smell as for burning paper = Cellulose acetate or cellulose propionate (no simple confirmatory test available)
Smell of rancid butter = Cellulose acetate butyrate
Flame has greenish tinge. Remove sample from flame and smell vapor
Flame is yellow. Remove sample from flame. Smell vapor.
Little flame. Decomposition of material but no charring. A cellular structure is developed before decomposition = Polycarbonate.
Acrid smell. Material soft and flexible = Plasticized PVC (if film, could be vinylidene chloride copolymer)
Acrid smell. Material non-flexible (and glossy if in form of molding) = Rigid PVC or PVC/PVA copolymer.
Smell of formaldehyde = Polyacetal
Indefinite smell. Sample has slippery feel. Press cold metal point to heated surface and then withdraw it.
Threads easily formed = Nylon.

grit and harsh scouring agents must be avoided. Because it is a thermoplastic, it loses its shape in boiling water, although some cast acrylics have higher heat resistance. In sheets of varying thickness and color, PMMA is used for glazing, building panels, advertising displays, and protective shields. Solutions and emulsions of PMMA are produced for use as surface coatings, adhesives, and finishes. In rods and tubes, it is used for electric lighting and medical illuminating equipment which utilize its capacity to transmit light along its length and around curves. In its molded form, it is used for lenses, dentures, toilet articles, and telephones.

Alkyds

Good insulating properties, excellent moldability, short cure time and low cost are primary reasons for the use of thermosetting alkyds in electrical materials such as housings and cases for switchgear, encapsulation of capacitors and resistors, and parts for fuses, light switches, and insulators. Maximum use temperature of the molded parts is 250 to 300°F (120 to 150°C). Liquid resins are used for enamels and lacquers for cars, stoves, and refrigerators.

Allyls (Allylics)

Allyl resins and monomers are thermosetting plastics which are commercially available as the diallyl esters of phthalic (DAP) and isophthalic (DAIP) acids. Other monomers in this grouping are diallyl maleate (DAM) and diallyl chlorendate (DAC). The latter is used for flame resistant formulations because of the high chlorine content of the resins. They can be polymerized by raising their temperature or by the use of peroxide initiators. The molding and electrical properties of these materials are excellent. They are primarily used for electrical components, decorative laminates, sealants, and coatings.

Cellulosics

Cellulosic plastics are produced by the chemical modification of cellulose, a natural polymer that is a major constituent of plant life. The primary sources of cellulose for this purpose are cotton linters and wood pulp. The term "cellulosic plastics" applies only to plastics whose resinous content is an ester or ether of cellulose. Included in this group are cellulose acetate, cellulose acetate butyrate, methyl cellulose, cellulose triacetate, cellulose propionate, ethyl cellulose, cellulose nitrate, cellulose films, and fibers.

Cellulose Acetate (CA): A thermoplastic made by treating cellulose with acetic acid and acetic anhydride with sulfuric acid as a catalyst. It can be formed into molding powder, foamed by rapid evaporation of a solvent, dissolved for lacquers, cast into film and spun from solution into fibers. It has replaced hazardous cellulose nitrate for photographic use under the name "safety" film. (See also Cellulose Triacetate.) Cellulose acetate molding compositions are tough, of good color and gloss; hardness is controlled by varying the ratio of plasticizer and ranges from rigid to flexible. Primary uses are in electronic components, insulation tapes, food packaging, toys, eyeglass frames, and sound recording and computer tapes. Rigid cellulose acetate foam is used as the core in sandwich panels and marine floats. Variations in composition include the degree of esterification of cellulose and amount and choice of plasticizer, which is usually a phthalate ester. Molding stocks for general purposes ignite easily, give off burning drips, burn with a dark yellow flame, produce moderate smoke, and have an odor of vinegar, unless masked by the odor of the plasticizer. Flame resistant stocks are widely used for molding.

Cellulose Acetate Butyrate (CAB): A plastic made by treating cellulose with a mixture of acetic and butyric acids and anhydrides in the presence of a catalyst. The ratio of acetic and butyric components can be varied to produce the desired flexibility. It absorbs less water than cellulose acetate. Compositions with ultraviolet stabilizers are used for small outdoor and indoor signs. Major uses are for automobile steering wheels and tail lights, telephone hand sets, toothbrush handles, business machine keys, appliance housings, piping, and tubing.

Cellulose Propionate (CP): A thermoplastic molding composition made from cellulose and propionic acid. Molding compositions are usually supplied as granules or pellets, and finished articles can be fabricated by injection molding or extrusion. It has similar uses to CA and CAB, but due to better surface hardness than CAB and less moisture sensitivity than CA, it is also used for mechanical pencils, telephones, and shoe heels.

Cellulose Triacetate: A thermoplastic made by reacting cellulose with acetic anhydride. It differs from cellulose acetate in chemical structure in that it consists of three acetate groups (instead of two) attached to each glucose unit of the cellulose molecule. Because of its high softening temperature, it is not suitable as molding stock. It has lower water absorption than CA and is now the prevalent base for photographic safety film, drawing and printing film, magnetic tape, and transformer and capacitor dielectrics.

Cellulose Nitrate (CN)

Because of the extreme fire hazards associated with cellulose nitrate in its processing, use, and storage, this plastic is discussed in detail later in this chapter.

Epoxies (EP)

Epoxy resins are primarily produced by the reaction of epichlorhydrin and bisphenol-A in the presence of a catalyst. By varying the ratio of the ingredients, the resultant product may be a low viscosity fluid or a high melting solid. Two primary classes are available, i.e., the liquid resins which are combined with curing agents for adhesives, potting and tooling compounds, and the solid resins which are modified with other resins to make coating materials. The epoxy resins also are used as "body solders" for automobile repair, as dies for forming sheet metal, as piping and tubing, and as foamed blocks for construction insulation.

Fluoroplastics

Fluoroplastics or fluorocarbons are commercially available in several major groups under the headings poly (ethylene-tetrafluoroethylene) (ETFE); poly (ethylene-chlorotrifluoroethylene) (ECTFE); polytetrafluoroethylene (TFE); polyfluoroethylene-propylene (FEP); the polychlorotrifluoroethylene (CTFE); polyvinyl fluoride (PVF); and polyvinylidene fluoride (PVF_2). All of these resins are characterized by resistance to solvents and chemical attack and good to excellent weatherability. Six are described below:

Polytetrafluoroethylene (TFE): A representative of a group of completely fluorinated resins. It is handled as general purpose molding powders, fine granules, and aqueous dispersions. The methods of processing include compression molding, extruding, dip coating, and film casting. TFE has a very extended range of serviceable temperatures, from −450 to 500°F (−265 to 260°C). It is used as chemical gasketing, high frequency and high temperature electrical insulation, lubricants, on molded bearings, and coated cookware.

Polyfluoroethylene-propylene (FEP): A copolymer of tetrafluoroethylene and hexafluoropropylene. FEP can be handled through the melt-flow processes involved in extrusion and injection molding. It is widely used in industrial applications, particularly as insulation for process industries involving heat and corrosive atmospheres. It is also used as a lining for vessels, pipes, valves, and fittings handling corrosive fluids.

Poly (ethylene-tetrafluoroethylene) (ETFE): A copolymer of ethylene and tetrafluoroethylene. It is processed by the conventional method of extrusion and injection molding. Its applications include use as gears, pump components, automotive parts, labware, valve linings, electrical connectors, and wire coatings. ETFE can be reinforced with glass fiber.

Polychlorotrifluoroethylene (CTFE): A thermoplastic resin produced in various formulations of fluoroplastics. It differs from the usual fluoroplastic in that its molecular structure contains chlorine. It flows like FEP for injection molding or extrusion but is less thermally stable. It is chemically resistant to many corrosive liquids. Its serviceable temperatures range from −400 to 390°F (−240 to 200°C). Its electrical and thermal properties make it valuable for insulation, cable assemblies, printed circuits, and electronic components. It also is used in the chemical process industries in valves, fittings, and gaskets.

Poly (ethylene-chlorotrifluoroethylene) (E-CTFE): A 1:1 copolymer of ethylene and chlorotrifluoroethylene. E-CTFE is processed by extrusion, molding, rotocast, and powder coating techniques. Applications include chemically resistant linings, coatings, containers, labware, moldings, wire coatings, fibers, and films.

Polyvinyl Fluoride (PVF): A plastic produced through the polymerization of vinyl fluoride. Films are cast from selected hot solvents, and the major use is as a weather resistant surface on building siding made of aluminum, wood, composite boards, or polyester glass fiber laminates.

Furane (Furan)

Furane resins consist primarily of those polymers resulting from polymerization of furfuryl alcohol, of phenol and furfural, and of furfuryl alcohol and phenol. Furfuryl alcohol modified urea formaldehyde resin is also a furane resin. The furanes are thermosetting, and have good bond strength and chemical resistance. Furane resin is used as a bond for sand in foundry molds, as an impregnant for hard board and laboratory bench tops, and as a laminating resin for glass fibers to make tanks and fume ducts.

Ionomers

The term "ionomer" is used to describe the class of thermoplastic polymers in which ionized carboxyl groups create ionic crosslinks in the intermolecular structure. Ionomer resin formulations are available in a wide variety of compositions. They are outstanding plastics with regard to toughness, transparency, and solvent resistance, and resist impact at temperatures as low as −160°F (−107°C). Articles made from these plastics include housewares, toys, containers, safety shields, tool handles, electrical insulation, and packaging. These plastics are also used in the lamination of various building construction materials.

Melamines (MF)

Melamine, one of the members of the amino family, is a thermosetting plastic noted for its extreme hardness, permanent colors, and good electrical insulating characteristics. Alpha-cellulose filled, general purpose grades are used for dishes and kitchenware, where their lack of taste and odor and resistance to food stains make them particularly desirable. Mineral filled grades are used in electrical applications. The resin may be bonded with fabrics, paper, wood veneers, etc., to produce laminates for building applications.

Methylpentene Polymer

Methylpentene polymer is produced by the polymerization of the monomer at atmospheric pressure. The polymer has many properties that make it desirable for specific uses. It has excellent electrical properties and is highly resistant to many corrosive chemicals. This plastic also has good light transmission characteristics. It is used for electrical instrumentation, bottles, chemical process equipment, and food handling equipment.

Nylon (Polyamides)

Nylon is a generic name for a thermoplastic group of linear polyamides. Many different types of nylon are possible; however, six major polymers are commercially available. They are: Nylon 6 (polycaprolactam); Nylon 6/6 (hexamethylenediamine and adipic acid); Nylon 6/10 (hexamethylenediamine and sebacic acid); Nylon 6/12 (hexamethylenediamine and dodecanedioic acid); Nylon 11 (11-aminoundecanoic acid); and Nylon 12 (12-aminododecanoic acid). Copolymers are made from the six types of nylon. Major applications for nylon include monofilaments, wire and cable jacketing, film and extruded rods, and slabs and tubing. Nylon resins are tough (with high tensile and impact strength), have high service temperatures, are resistant to abrasion and fatigue, and resist oils and solvents.

For engineering applications, a number of specialty nylons with improved frictional and wear characteristics have been developed. Typical applications include gears, cams, slide fasteners, gaskets, wire insulation, boat propellers, and appliance parts.

Olefins

See polybutylene, polyethylene, polyallomer, and polypropylene.

Phenolics (PF)

The discovery of phenol formaldehyde plastics (Bakelite) in 1907-1909 marked the beginning of the plastics industry. Many varieties are now available, marked by low cost and good mechanical, electrical, and thermal properties. General purpose, wood filled grade phenolics are suitable for use up to about 300°F (150°C), but glass or mineral filled compounds have long term retention of properties up to approximately 500°F (260°C). Some special, heat resistant types are usable for up to 1 hr at temperatures as high as 1,000°F (540°C).

Paper or fabric impregnated with phenolic resin in a water or organic solvent can be wound into tube form or pressed into flat sheets to make tough thermoset "laminates" with low water absorption and high service temperature. Laminates are used for electrical insulation and bearings for marine propellers and heavy rolling mills. Similar compositions are used as laminating adhesives for exterior grade plywood and as impervious surfacing materials. Paper impregnated with phenolic resin and printed with wood grain is frequently the top surface for wall panels or furniture, with backing of other phenolic laminates or plywood. Phenolics are standard binders for brake shoes, sandpaper, grinding wheels, sand casting foundry molds, and wood chip particle board.

Two types of phenolic foam, reaction foam and syntactic foam, are used in construction or for thermal insulation. Reaction foam is made by curing the phenolic resin with agitation so rapidly that water is produced as a byproduct, and is vaporized and trapped to make the voids in the foam. Syntactic foam is made from hollow spheres of phenolic resin with a phenolic or other resin binder. Each type is used for cores in sandwich panels, and as buoyancy or a stiffening means in marine and aircraft structures. Cast phenolic resin is largely used for decorative applications.

Polyallomers

Polyallomers are made by polymerizing two monomers with catalysts so that blocks of each polymer are formed in the polymer chain. The most common polyallomer contains ethylene and propylene. Applications for polyallomers include films, bottles, appliance parts, automotive parts, toys, and closures.

Polybutylene

Polybutylene is flexible, has excellent creep resistance at room and elevated temperatures, and has higher long term temperature resistance than other polyolefins. Other advantages include good toughness, high tear strength, and good moisture barrier and electrical insulation characteristics. Applications include film and sheeting, hot melt coatings and adhesives, flexible pipe and tubing, and shrink film.

Polycarbonates (PC)

Polycarbonates are thermoplastic polyesters of carbonic acid. The resin is available in several different forms including pellets, powder, film, sheet, rod, plate, and tubing. High impact strength, good electrical properties, clarity, and resistance to creep have caused this material to find many applications, such as in appliances, electrical and electronics equipment, food handling products, and sporting goods. Its transparency makes it useful for lenses, safety shields, and glazing.

Polyester, High Temperature Aromatic

Important characteristics of this material are its dimensional stability at high temperatures (over 600°F or 315°C), self lubricating properties, good stiffness, electrical insulating properties, and solvent resistance. It decomposes rapidly at approximately 1,000°F (540°C). An unusual aspect of this thermoplastic is its very high thermal conductivity, which makes the material useful in applications such as bearings and electrical insulation where failure from localized hot spots is a problem with most plastics. Stickfree coatings for frying pans and encapsulation of electronic parts are among its other uses. Trade names include Ekonol and Ekcel.

Polyesters, Thermoplastic

Unsaturated polyesters show good weathering resistance, light stability, and heat resistance. Their primary application is in the production of reinforced plastics. They are used to fabricate structures ranging in size from electrical components to large boat hulls.

Saturated polyesters do not have the double bond that would permit further polymerization by treatment with initiators, such as organic peroxides. Saturated polyesters of high molecular weight, such as ethylene glycol terephthalate, are used principally in fiber and film production, such as magnetic recording tapes, tough or boil-in-bag food packaging, and satellite balloons. They can be backplated with metals for decorative films and nameplates. The low molecular weight esters are used as plasticizers for vinyl and acrylic resins.

Thermoplastic polyester molding compositions are available. Besides poly (ethylene terephthalate)—PET, there are poly (1,4-butylene terephthalate)—PBT, poly (tetramethylene terephthalate)—PTMT, and poly (cyclohexylenedimethylene terephthalate isophthalate)—PCDT. PBT and PTMT are the same chemically.

Thermoplastic polyesters have engineering applications which include automotive parts, appliance parts, business machine parts, electrical and electronic parts, gears, bearings, pulleys, and pump components. The polyester PET, discovered in Britain during World War II, can be spun into a fiber which is commonly used in textiles (Dacron, etc.).

Polyethylene (PE)

Three major types of polyethylene are marketed: the low, medium, and high density products. In addition, copolymers of ethylene-ethyl acrylate, ethylene-vinyl acetate and ethylene-butylene are produced. The three density ranges are 56.8 to 57.7, 57.8 to 58.7 and 58.8 to 60.2 lb per cu ft (90.9 to 92.4 kg/m^3, 92.5 to 94.0 kg/m^3, and 94.1 to 96.4 kg/m^3).

Low density stocks are waxy, relatively limp, and are the toughest of the polyethylenes with an upper service temperature of approximately 160°F (70°C). As density increases, so does surface hardness, stiffness, resistance to permeation by oils and water, and softening temperatures. For polyethylenes of the highest density, the upper service temperature is 240°F (115°C) under no-load conditions.

The outstanding properties of polyethylenes are: (1) good moldability, (2) good mechanical properties, (3) excellent electrical resistance, (4) low moisture vapor transmission, (5) resistance to solvents and chemicals, and (6) lightness in weight. Polyethylene is widely used as film, sheeting, plastic bottles, piping, tubing, dish pans, garbage containers, laundry baskets, electronic components, insulation, and textiles. Boat and motor vehicle fuel tanks are a recent application for high density polyethylene. Polyethylene can be crosslinked chemically or by irradiation to form a thermoset with high heat resistance.

Polyphenylene Oxide (PPO)

Polyphenylene oxide is a thermoplastic resin formed by the catalyzed oxidation of 2,6-xylenol. It is a specialized, highly engineered plastic with a wide temperature range, good mechanical and electrical properties, and resistance to corrosive materials. It has many electrical and electronic applications, is used for packaging frozen foods, and as a replacement for glass and stainless steel in hospital utensils and medical instruments.

Polyphenylene Sulfide

Excellent chemical resistance and high stiffness at elevated temperatures make this material useful as a coating for various metals, including aluminum, steel, cast iron, brass, etc., and in the petroleum and chemical processing industries. Other uses include cookware, pump impellers, pumps, conveyor rollers, and machine parts.

Polypropylene (PP)

Polypropylene possesses the lowest density of the commercially available thermoplastics. In addition, it exhibits good rigidity, high yield strength, good surface hardness, exceptional flexural properties, resistance to chemicals, and excellent dielectric properties. Like all olefins, it has excellent resistance to water solutions which are highly damaging to many metals.

Primary areas of use include the usual categories of molding as well as in film and piping. It is used for interior trim of automobiles and for many appliance parts. Because of its superior insulating properties, it is used to make electronic components. Carpeting made of polypropylene fibers also is available. Polypropylene filled with asbestos offers high impact strength and stiffness (up to 250°F or 120°C) and low friction.

Polystyrene (PS)

A glassy, transparent thermoplastic, primary characteristics of polystyrene are hardness, rigidity, clarity, and heat and dimensional stability. By varying the polymerization reaction, polystyrenes have been developed with higher heat distortion temperatures. Styrene polymers are marketed in granular and powder form for extrusion or molding. Beads of polystyrene which contain the flammable gas pentane as a blowing agent are sold for the production of polystyrene foam.

General purpose and impact grades of polystyrene are used for refrigerator liners, appliance housings, automotive applications, films for wrapping, sporting goods, toys, and novelties. Heat resistant polystyrene is used for television sets, radios, and illuminated signs. Glass filled polystyrene is used for business machines, electronic equipment, and military hardware. Polystyrene foam is used to make toys, insulation, displays, shipping containers, and molded furniture. Construction panels consisting of cores of polystyrene foam between sheets of plywood, plastic, or aluminum and foamed-in-place insulation are used in the building industry.

Polysulfone

Polysulfone is a thermoplastic resin which contains a diphenylene group. This group imparts certain desirable characteristics to the polymer, including thermal stability (over 300°F or 150°C), oxidation resistance, and rigidity. Glass fibers may be added for improved environmental stress crack resistance. Commercial applications include electronic units, electrical circuit breaker parts, automobile engine parts, aircraft interiors, kitchen range hardware, electroplating equipment, and appliance hardware.

Polyurethane (PUR)

The polyurethanes consist of a group of polymers which are produced in the following general forms: foams (flexible, semi-flexible, and rigid); elastomers (casting compounds and elastoplastic resins, adhesives, coatings, and spandex fiber). The basic reaction used to produce the urethane polymers involves isocyanates and reactive hydrogen bearing materials such as polyethers, castor oils, amines, carboxylic acid, and water. By varying the number of branchings, it is possible to make polyurethanes that are thermoplastic or thermosetting.

The polyurethane foams are widely used in the production of upholstered furniture, bedding, sponges, toys, wearing apparel, and medical dressings. Rigid urethane foams are used for insulation in building construction. Substantial quantities of the polyurethanes are used to produce coatings and adhesives. Its rubberlike qualities aid in the production of specialty fillers and elastomers. Gears, sprockets, and rolls are produced from urethanes. Very thin films are used to produce containers impermeable to gas and moisture.

Silicones (SI)

The silicone resins comprise a large group of polymers consisting of chains of alternating silicone and oxygen atoms with organic groups, e.g., methyl groups attached to the silicone atoms. They are stable at high and low temperatures, have good dielectric properties, resist weathering, do not react with most chemicals, and repel water.

Silicone emulsions are used as mold release agents, defoamers, and waterproofing materials. A compound of liquid silicone and finely divided filler produces a grease suitable for high temperature lubrication applications. Silicone paints, varnishes, and enamels have many uses, including cloth and wire coating. With inorganic fillers, such as glass or asbestos, silicones are molded into such products as connector plugs, coil bobbins, insulators on induction furnaces, and heat barriers in jet engine afterburners. Silicone glass laminate applications include rigid hot air ducts for aircraft, terminal boards, and transformer spools and spacers. Rigid silicone foams can be preformed or foamed in place. High temperature insulation is the principal use of rigid silicone foams.

Ureas (UF)

Urea formaldehyde, as a thermosetting plastic, is especially suitable for any application which must resist heat. The material has excellent resistance to oils and greases, but is not recommended for high humidity environments. Because urea formaldehyde is hard, strong, odorless and tasteless, it has been frequently used in tableware, crockery, and decorative articles for the home. Other uses are for electrical parts, such as switches and outlet covers, bottle tops, toys, and as a laminate in building panels. Plaskon and Beetle are two well known trade names.

Vinyls

The vinyl resins are thermoplastics that exist in a wide variety of polymers and copolymers. Commercial production of the vinyl chloride monomer starts with acetylene or ethylene. Most vinyl plants utilize an oxychlorination process with ethylene to produce vinyl chloride. Copolymers of vinyl chloride and vinyl acetate account for a large part of vinyl production.

The major applications for flexible and rigid vinyls include: automobile seat covers, floor mats, and moldings; shower curtains, window shades, wrappings, bottles, and adhesives; flooring, siding, piping, and wiring; swimming pools, records, and medical appliances.

Polyvinyl Chloride (PVC): This type of plastic is made as a rigid product for a number of building components and as a flexible plasticized stock for upholstery and wearing apparel. It exists in hundreds of individual formulations. It has good abrasion resistance.

The unplasticized PVC softens as it burns, producing white smoke and acrid fumes which can be corrosive. Most of the chlorine content of PVC is released as hydrogen chloride. The fire properties of plasticized PVC are determined to a large extent by the nature of the plasticizer. Building products should be tested individually.

Vinyl Chloride—Vinyl Acetate Copolymers (Polyvinyl Acetate) (PVAC): These contain from 5 to 20 percent vinyl acetate in the polymer. They are used for more flexible stocks than PVC and are more readily softened by plasticizers. Their properties are essentially those of PVC.

Polyvinyl Dichloride: Because this plastic is made by chlorinating PVC, its properties are essentially those of rigid PVC. It is used most where more toughness, stiffness, and heat resistance are needed than is offered by PVC.

Vinyl Acetate Polymer: A thermoplastic that is used in coatings, as a wood primer, and as an adhesive in a milky aqueous dispersion. It is not used as a molding or sheet plastic, as its upper service temperature is below 120°F (50°C).

Vinyl Alcohol Resins (PVA): Because they are made by hydrolysis of polyvinyl acetate, PVA products swell or dissolve completely in water. They are highly impervious to oils and many lacquer solvents. They are used in hoses for spray painting and as water soluble film for dose packaging detergents, etc.

Polyvinyl Butyral (PVB): PVB is produced by condensing butyraldehyde with polyvinyl alcohol. It has remarkable adhesive qualities and is primarily used as plasticized film for laminating automotive safety glass.

Polyvinyl Formal (PVFM): This plastic is produced by condensing formaldehyde with polyvinyl alcohol to make a hard resin. PVFM is then mixed with small amounts of phenolic resin to improve the bond and reduce creep. It is not to be confused with polyvinyl fluoride. (See Fluoroplastics.) Its principal use is as standard insulation enamel for magnet wire in small motors and electronic equipment.

Vinylidene Chloride Polymer: A thermoplastic that as a film has low permeability to water vapor. A principal use is as a coating on cellophane. Filaments are used for webbing on garden furniture and for window screening.

Vinylidene Chloride-Vinyl Chloride Copolymer: This plastic is less rigid than PVC and flows more readily in extrusion and molding. Its major use is as a coating for metals and molded electrical parts.

NATURAL AND SYNTHETIC RUBBER

Like plastics, rubber appears in many shapes and forms, from very soft and pliable products to hard rubber that looks and feels like a molded plastic. Many polymers can be manufactured in elastomeric grades and are classified in that form as synthetic rubbers. Crepe soles fall in that category.

Natural Rubber (NR)

Natural rubber is a constituent of latex, the white milky sap collected from the rubber tree. The processed rubber may be used in sheet form for items like shoe soles, extruded into products like inner tubes, or compression molded with heat to make items like tires and hot water bottles. Liquid latex has a variety of applications. It is often coated on fabrics for waterproofing, mixed with cement for superior bonding qualities, included in water based paints to produce a cohesive film after the paint dries, and expanded into the familiar foam latex, or foam rubber. The density of the foam is controlled by varying the ratio of foam to air and the size of the cavities.

Vulcanization is a curing process that changes the isoprene molecule to a polyisoprene and produces a thermoset polymer that cannot be reshaped or remolded when reheated. By prolonging the vulcanization process, the degree of hardness can be varied, eventually resulting in "hard rubber" (known as ebonite or vulcanite in England). Hard rubber tubes and rods have been popular products, valued for their resiliency, electrical insulating quality, and chemical inertness. Other common products have included automobile battery cases, combs, telephone receivers, casters, wheels, and bushings. In many of these applications, synthetics have replaced natural rubber.

Synthetic Rubber (SR)

These materials are intended to imitate the elasticity of natural rubber, and they can be vulcanized. They have similar long chain molecular structures, built up from copolymerized synthetic monomers, such as butadiene with styrene, or butadiene with acrylonitrile. The former method produces styrene-butadiene rubber (SBR), such as the early Buna and Buna-S rubbers; SBR is used extensively in automobile tires. Nitrile rubber (NBS) is the product of the butadiene-acrylonitrile reaction, while butyl rubber is produced from an isobutylene isoprene copolymer. Neoprene (GR) is a very strong and durable synthetic made from chloroprene.

Synthetic foam rubber usually means a soft flexible product of the polyurethane type and either polyester (AU) or polyether (EU). However, new types are continually being developed, including ethylene-propylene rubber (EPR), polybutadiene (BR), and polyisoprene, the synthetic equivalent of natural rubber.

Silicone rubber (SI) incorporates the inorganic, inert element silicone, which gives it a number of unique properties. It is the only satisfactory soft material for human implants, and it is commonly used for breast implants and many other implants including artificial ears, skull plates, and jawbones. In building applications, selfcuring, water repellant silicone rubber is used as a sealant, and it finds many applications where a nonstick surface is desired. Some formulations have been used successfully as firestops, and to seal openings in fire walls, where its flexibility gives performance superior to grout or plaster.

The publication *Plastics Designs and Materials* (Katz et al 1978) is an excellent source for more information on synthetic rubbers.

CELLULOSE NITRATE (CN)

Cellulose nitrate (also referred to as nitrocellulose and pyroxylin) is made by the action of nitric and sulfuric acids on cellulose materials, such as cotton. Pyroxylin plastic comprises the lower nitrated products and contains from 11 to 12 percent nitrogen.

Cellulose nitrate plastics possess the most unusual and serious burning characteristics of all plastics. Material which has been subjected to heat, such as that salvaged from a fire, may be so altered in composition that it is subject to spontaneous ignition.

When cellulose nitrate products are heated to temperatures above 300°F (150°C), decomposition starts, which generates further heat and soon raises the material to its ignition temperature. Some experimenters have reported decomposition after long continued exposure to temperatures not much above that of boiling water. There are a number of cases on record of ignition from contact with steam pipes and electric light bulbs. Decomposition of cellulose nitrate generates heat and does not depend upon an external air supply.

Some of the gases produced by decomposition are highly toxic. The effect of carbon monoxide is well known. The toxic effect of the oxides of nitrogen is often delayed; persons exposed to these gases may show no immediate ill effects, but fatalities may follow some hours or days after exposure.

Cellulose nitrate, when in solution in acetone or any of the various solvents, has no greater hazard than that of the solvent. If the solvent is lost, the hazard reverts to that of cellulose nitrate.

Most articles formerly made of cellulose nitrate are now made from less hazardous plastics. However, in some countries, common articles such as eyeglass frames are still being made from cellulose nitrate. Wherever cellulose nitrate plastic is still in use, there should be no relaxation of the special fire protection measures that years of experience have shown necessary to safeguard life and property from cellulose nitrate fires.

Processing Cellulose Nitrate

Firesafety in plants processing cellulose nitrate plastic consists essentially of four parts: (1) segregation of hazards, (2) elimination of ignition sources, (3) good housekeeping, and (4) strong fire control facilities. In some plants, finished cellulose nitrate plastic parts may be assembled with other articles which do not involve any special fire hazard; for example, toilet seats. In such cases, the hazard and necessary protection depend upon the quantity of cellulose nitrate used. If the quantities used are large, the same safeguards should be followed as for plants in which the product is made entirely from cellulose nitrate.

Cellulose Nitrate Storage

Storage of cellulose nitrate plastics requires special facilities. Depending on the quantity to be stored, these facilities may be cabinets, vaults, storage rooms, or isolated storage buildings. Storage facilities are described in detail in two NFPA Standards: NFPA 40, *Standard for the Storage and Handling of Cellulose Nitrate Motion Picture Film* (hereinafter referred to as NFPA 40), and NFPA 40E, *Code for the Storage of Pyroxylin Plastic.*

Cellulose Nitrate Photographic Film

All photographic film made in the United States and Canada since 1952 is the safety base type, usually cellulose acetate or triacetate or polyester. An engraving process used in some printing plants engraves on cellulose nitrate plates.

There is comparatively little storage of cellulose nitrate X-ray film, although some hospitals may still store X rays made on cellulose nitrate film many years ago. However, appreciable quantities of old nitrate motion picture and aero film still exist. Small amounts of portrait, industrial, and amateur nitrate film probably exist, but the usual storage of this material (in paper envelopes) poses no serious hazard. Nitrate based film manufactured in the Soviet Union or the German Democratic Republic may occasionally be found.

The existence of old nitrate motion picture and aero film does present a hazard, since persons handling this material may not be familiar with its serious hazards and the elaborate precautions necessary to handle it safely. Although it is unlikely that nitrate film will be projected in tar likea theater handling present day motion pictures, it is quite possible that such material may be shown in small "art" or "film society" theaters. Unfortunately, projectionists may not be aware of the fact that a nitrate print is to be shown. For this reason, theater projection standards for nitrate film should still be kept in effect. Storage of nitrate film should be in accordance with NFPA 40.

The amounts of scrap nitrate motion picture film have greatly decreased, and most of the agencies that formerly handled this material are no longer in existence. This has given rise to the possibility that scrap could accumulate in dangerous quantities. Prior to environmental concern with air pollutants from outdoor burning, the best procedure for disposing of scrap nitrate film was to burn it at a remote location in the open, with all personnel at a safe distance upwind (because toxic oxides of nitrogen are among the products of combustion of this film). If accumulations of scrap nitrate film are a problem, the best course to follow would be to check with environmental authorities to learn if such open burning is permitted locally.

Transportation of Cellulose Nitrate

The U. S. Department of Transportation (DOT) regulates the interstate transportation of cellulose nitrate plastic by rail, highway, or water. These regulations permit dry shipment of cellulose nitrate (pyroxylin) scrap when specially packaged, but if there is evidence, or the possibility of decomposition, the cellulose nitrate must be packaged under water. Shipment of cellulose nitrate sheets, rolls, rods, and tubes is not subject to special DOT regulation, except when shipped by rail express or water.

Articles manufactured entirely of or containing cellulose nitrate plastic do not require special packaging when shipped by land, rail, or water. Special shipping regulations apply to cellulose nitrate wet with alcohol, solvent, or water.

Shipments of cellulose nitrate motion picture and X-ray films are permitted when certain packaging requirements are met, including inside containers made of light metal, cardboard, or fiberboard for individual reels of film. NFPA 40 places restrictions on the transportation of nitrate film in public vehicles so as to protect the public against the consequences of a possible fire involving the film.

Air transportation of cellulose nitrate plastic scrap is not permitted by regulations of the International Air Transport Association. Cellulose nitrate plastic rods, sheets, rolls, tubes, manufactured articles, motion picture and X-ray films are permitted only when specially packaged.

Extinguishing Cellulose Nitrate Fires

Large volumes of water are required to extinguish cellulose nitrate fires. Automatic sprinklers constitute the best general protection for areas where cellulose nitrate materials are stored or handled. Water supplies should be particularly strong. Closer sprinkler spacing than for ordinary occupancies is required in order to supply sufficient water to absorb the heat. The water from sprinklers, in addition to its extinguishing and cooling effects, tends to absorb the poisonous oxides of nitrogen fumes but has no effect on carbon monoxide.

If hose streams are required, the objective should be to get large quantities of water on the fire as soon as possible. Fire fighters should be located upwind or protected by self-contained breathing apparatus to prevent inhalation of the products of combustion.

FIRE BEHAVIOR OF PLASTICS

While there is no special fire hazard for most accepted usages of plastics, some exhibit burning characteristics which are considerably different from those encountered with the more traditional cellulosic building materials.

Special Fire Behavior Problems

Test methods, which in the past have been adequate to indicate the relative hazard of materials under actual use conditions, have failed to predict the fire behavior of some plastics. In addition, different fire conditions cause significantly different burning characteristics. Of principal concern has been fire behavior which poses unusual hazards to life and property, including the following:

Ignitability and Rate of Burning: Although plastics tend to have a higher ignition temperature than wood and other cellulosic products, some are easily ignited with a small flame and burn vigorously. Very high surface flame spread rates have been reported—up to approximately 2 ft per sec (0.6 m/sec), or 10 times the rate of flame spread across most wood surfaces.

Smoke Produced: The burning of some plastics is characterized by the rapid generation of large amounts of very dense, sooty, black smoke. Chemicals added to inhibit flammability may increase smoke production.

Toxic Gases: Any fire will generate lethal products of combustion, principally carbon monoxide. Depending on the plastic and the particular fire conditions, highly toxic gases, such as hydrogen cyanide, hydrogen chloride, and phosgene, may also evolve.

Flaming Drips: Thermoplastic articles tend to melt and flow when heated. In a fire situation, this characteristic may cause the material to melt away from the flame front and inhibit further burning, or it may produce flaming and tarlike dripping which is difficult to extinguish and which may start secondary fires.

Deviation from Test Results: Small scale "Bunsen burner" tests are used for product development and laboratory control purposes. In the past they were used to indicate that certain plastics were "self-extinguishing" or "nonburning" and, presumably, safe for use. Unfortunately, in real life situations the same materials have shown flash burning characteristics. Somewhat larger scale tests, such as the NFPA 255 "tunnel" test (*Standard Method of Test of Surface Burning Characteristics of Building Materials*, hereinafter referred to as NFPA 255), also fail to adequately predict flash burning characteristics under some end use conditions.

Corrosion: Severe corrosion damage to sensitive electronic equipment and to metal surfaces has been reported following fires involving commonly used plastics, such as polyvinyl chloride.

The fire behavior problems summarized above can occur under every conceivable condition of burning, from complete to partial combustion or smoldering and destructive pyrolysis. When the plastics and their constituent modifying agents, including flame retardant additives, burn, they can produce a wide variety of noxious and toxic byproducts in varying concentrations as mentioned previously. In this respect, plastics are similar to most ordinary combustibles, such as wood, leather, wool, silk, etc., in that they are capable of thermal degradation into volatile and gaseous products of combustion able to cause harmful physiological effects when breathed. In general, carbon monoxide is generated more rapidly than other toxic gases and tends to be the principal factor in fire fatalities. Nevertheless, unusually high burning rates, unusually heavy smoke production, and a higher heat content per unit weight are responsible for a greater concern about the fire behavior of certain plastics.

Smoke Generation

Smoke generation from a given polymer may vary widely depending upon the nature of the polymer, the additives present, whether fire exposure was flaming or smoldering, and the type of ventilation which was present. Several comments from Gaskill's comprehensive study on smoke density are especially pertinent (Gaskill 1970):

1. Woods and most polymeric material pyrolytically degrade, yielding smokes that are dense to very dense.

Ventilation tends to clear the smoke, but in most cases it does not reduce the intensity to the point of satisfactory visibility. With fire retardants incorporated into the polymer (at least in the case of solid materials), heat and flame generally produce very dense smokes, and this occurs rapidly.

2. Woods and those polymeric materials that burn cleanly will yield smokes, under conditions of heat and flame, of somewhat less density.
3. Urethane foams under flaming or nonflaming exposures generally yield dense smokes, and with few exceptions, obscuration occurs in a fraction of a minute. Flaming exposure usually causes the smokes to be generated in less than 15 sec and the intensities in this case are usually higher than in the nonflaming case.

Cellular Plastics

Cellular plastics (or foamed or expanded plastics) have been in widespread use since the 1960s. Without adequate flame retardants and without barriers to resist ignition, however, these plastics in flexible form, i.e., mattresses, seat cushions, carpet padding, etc., and in rigid form, i.e., building insulation, have created an unacceptable fire hazard. Many reports tell of fast spreading, high intensity fires and voluminous smoke production. The low resistance of cellular plastics to excessive heat and their ease of ignition contrast sharply with their most valuable property in building construction—their excellent resistance to heat transfer. In a 1969 study of the flammability of cellular plastics (UL 1969), Underwriters Laboratories Inc. reported on an analysis of 34 fires where rigid cellular plastics had been used as building materials. The report indicated that:

1. The fire performance of cellular plastics can vary considerably depending upon whether they are exposed or protected and/or flame resistant, along with the degree of flame resistance. But the reduction in flammability through the use of inhibitors or protective coverings is not conclusively demonstrated by the study.
2. In 8 of 21 fires, a high rate of flame spread was reported and appeared to be especially typical of uninhibited, sprayed-on polyurethane and rigid, polyurethane boards. Flame spread values from NFPA 255 tunnel tests of some of the materials ranged from 5 to 1,375 (red oak 100).
3. The fire behavior of cellular plastics appears to be affected by the extent of foam application, i.e., whether the foam is applied to the walls and ceiling or only to the walls.
4. The fire behavior of the flexible foam plastics appears to be similar to that of the rigid plastic foams. (Peculiar to the flexible foams, though, is the pool of liquid product which can be generated in a fire in the bulk storage of the foam "buns." The liquefied plastic burns like a high flashpoint flammable liquid.)

Whereas untreated rigid polyurethane burns rapidly and is readily ignited by common ignition sources found in the environment, fire retardant grades have substantial resistance to localized, temporary ignition sources under most fire environments. Sustained application of heat and flame causes such foam to burn, but the burning rate tends to be retarded. Often, burning will cease entirely when the external flame source is removed. The production of dense smoke is a characteristic of both treated and untreated urethane.

Untreated rigid polystyrene foam, likewise, ignites easily and burns vigorously with the production of dense, black smoke and a very black, viscous melt product which can burn with the intensity of a flammable liquid. Fire retardant treatments inhibit ignition; the plastic tends to shrink away from fixed heat sources without igniting. The heat of combustion of styrene is about 50 percent higher than that of urethane foam (18,000 Btu/lb vs 12,000 Btu/lb or 2 MJ/g vs 28 MJ/g).

At times, cellular plastics that exhibit good resistance to burning under most exposure conditions have shown a markedly different behavior when surfaces of the exposed plastic are arranged to cause a feedback of radiant heat between the surfaces (Troup 1969). The combination of low thermal conductivity, high rate of heat release, and rapid degradation under heat exposure appear to be responsible for a rapid upward propagation of the flame front occurring in a room corner configuration. When the ceiling is also insulated with the exposed foamed plastic, the corner configuration encourages rapid extension of the fire across the underside of the ceiling. This phenomenon has occurred with very small ignition sources, such as a waste basket fire. The laboratories of Factory Mutual devised the "corner test" to simulate this exposure condition.

Witnesses to fire incidents in buildings which were insulated on the inside with exposed spray on or board foam have reported extremely high flame spread rates across the surface of the insulation immediately after the fire was first noticed. One theory for this phenomenon is that fire had been smoldering unnoticed inside or behind the insulation and generating large quantities of unburned, largely invisible products of combustion which collected under the roof. When the fire eventually broke out into the open, it ignited the accumulation of gases which accounted for the flash fire behavior. Rapid fire spread across the surface of exposed insulation has been reported for fire retardant treated and untreated urethane and styrene.

Rubber

Burning rubber is notable for its smell and dense smoke. The material, as generally used, is not easily ignited and does not burn with unusual rapidity. In normal applications, the most severe hazard appears to be in the use of foam rubber mattresses, particularly in institutional occupancies, such as hospitals, nursing homes, and prisons. Accidental ignition by smoking materials or deliberate ignition with matches or other small sources has caused the rapid evolution of dense, toxic smoke that has resulted in multiple fatality fires.

Foam rubber is subject to spontaneous heating when exposed to a heat source, such as an industrial dryer or a domestic clothes dryer. Cases have been reported of spontaneous ignition of foam rubber in shoes and clothing after the items were removed from clothes dryers. Foam rubber underlayment for carpeting has been reported to have contributed to the extension of fires in buildings; but the U.S. Department of Commerce "pill test" for carpeting appears to weed out the worst offenders.

The bulk storage of rubber items, particularly rubber tires, is subject to hot, difficult to control, and very smoky fires. Surface application of water often is effective only marginally because the fire tends to burrow into the material.

Special sprinkler protection is required to assure control if the storage is more than a few feet high (1 ft = 0.3 m).

Rubber (and plastic) jacketing on grouped electrical cables have provided the fuel for very expensive and potentially very dangerous fires in the electrical and communications industries. Considerable work has been done to raise the operating temperature limits and ignition temperatures of insulation for electric power and control cables. An applicable standard is IEEE 383. Butyl/neoprene and neoprene rubber insulation tend to burn readily and fail such small scale tests, while Hypalon, for instance, is a jacketing material with more resistance to heat and ignition.

No significant differences appear to exist in the ignition or combustion characteristics of natural rubber and the various synthetics. Silicone rubbers can be formulated to reduce combustibility to a minimum under ordinary conditions, but not all silicone rubbers have that quality: the material failed the requirement of no flashover in a quarter scale room fire test conducted at the National Bureau of Standards (NBS 1983).

Test Results

The variation of the fire behavior of plastics when compared to other building materials was further demonstrated in tests conducted at Underwriters Laboratories Inc. (UL 1975). Thirty one materials were subjected to the Steiner tunnel test (ASTM E84, NFPA 255, and UL 723) to establish flame spread and smoke developed classifications. These materials were also subjected to a 20 lb (9.1 kg) wood crib exposure fire in a full size corner configuration. Testing established that the 20 lb (9.1 kg) wood crib presented fire severity conditions about equal to the 5,000 Btu per min (88 kW) burner in the tunnel test, generating a heat flux of about 0.97 Btu per sec/ft^2 (1.1 W/cm^2). Four materials in the test series achieved an identical flame spread classification of 18 (cement asbestos 0; red oak 100), but the smoke that developed and the extent of burning were markedly higher for the flame retardant grade of plastic foam. (See Table 5-8B.)

TABLE 5-8B. Test Results (Cellular Plastics) for a 20 lb (9.1 kg) Wood Crib Fire Test

Code Letter	Material	FSC*	SDC**	Affected Area†
E	3½ in. (89 mm) thick glass fiber blanket insulation, 6% resin by weight, unfaced, 0.65 pcf (10.4 kg/m^3) density	18	3	24%
I	½ in. (12.7 mm) thick wood fiberboard acoustic ceiling panels, integrally flame retardant treated, decorative finished, 18.8 pcf (301.2 kg/m^3) density	18	4	37%
T	½ in. (12.7 mm) thick wood particle board, integrally flame retardant treated 2 in. (51 mm) polyurethane	18	37	25%
Z	2 in. (51 mm) polyurethane spray cellular, plastic on ¼ in. (6 mm) asbestos cement board; 2.2 pcf (35.2 kg/m^3) density, flame retardant	18	935	85%

* Flame spread classification, tunnel test result.

** Smoke developed classification, tunnel test result.

† Percent of 192 sq ft (17.8 m^2) total wall and ceiling area affected in full-scale corner test (7% affected in control fires).

These four building materials were also subjected to exposures in the Heat Release Rate Calorimeter test developed by the National Bureau of Standards (Parker and Long 1972). As reported by UL in the document on the 1973-74 test series, the first three specimens, code letters E, I, and T, yielded identical heat release rates of 0.26 Btu per sec/ft^2 (0.3 W/cm^2) at an exposure energy level of 2.6 Btu per sec/ft^2 (3 W/cm^2). However, a heat release rate 25 times higher (6.6 Btu per sec/ft^2 or 7.5 W/cm^2) was recorded for the flame retardant plastic, code letter Z.

Another cellular plastic, 2 in. (51 mm) extruded polystyrene cellular plastic boardstock (1.8 lb per cu ft or 0.03 g/cc density, flame retardant), obtained a flame spread classification of 3 in the tunnel test, but its smoke developed classification was 785 (red oak 100), and 146 sq ft (13.6 m^2), or 76 percent, of the wall-ceiling area of the corner test structure was adversely affected by the burning.

The fire problem associated with plastics was studied by the Products Research Committee, formed to administer a trust fund established by a number of companies engaged in the manufacture and sale of cellular plastics and their ingredients. The Committee issued its final report in 1980 (PRC 1980) after a five year effort, which included sponsoring a number of research programs. Highlights of conclusions reached by the Committee included the following:

1. Provisions of current building codes, requiring a thermal barrier over cellular plastic insulation in wall cavities, should remain in force until further advances are realized in firesafety engineering.
2. Properly designed interlayers between furniture fabric and padding are an effective means of reducing the fire risk of upholstered furniture.
3. The use of results of many different small scale fire tests to obtain an estimate of behavior of a material in fire is appropriate; however, the present state of the art does not allow regulatory decisions to be made without considerable expert judgment and, preferably, data from full scale fire tests in which the most probable fire scenarios are simulated.

FIRE TESTS FOR PLASTICS

Combustibility characteristics and suitability for use should be determined for any material or assembly on the basis of tests which simulate the end use condition as realistically as possible. The tests should be designed to show how the product will perform when it is subjected to the type of fire exposure which can be anticipated under the conditions of use. The most desirable tests, from the standpoint of fire protection, engineering and codes, are those that are suitable for all products for a particular application and that treat all products equally. Separate tests for plastics or any other group of materials, which do not allow comparison of the fire behavior under like test conditions, are undesirable.

As explained previously, small scale tests are now used mostly for product development and production control. Such tests are also suitable for determining the ignitability of such items as upholstery, where the ignition source is small and of low intensity, but are unsuitable for evaluating the fire characteristics of materials under room fire or other full scale fire conditions.

Somewhat greater intensity fire exposures are produced by standard fire tests, such as the Steiner tunnel test and the radiant panel test, both of which are used to determine the suitability of all kinds of materials as interior finish for walls and ceilings in buildings. As previously indicated, these tests have been proved inadequate to accurately predict the burning characteristics of some plastics under certain conditions. Higher intensity exposures encountered under actual use conditions may cause fire behavior substantially more hazardous than indicated by the results of these tests. The Factory Mutual corner test tends to reproduce these more severe exposure conditions. It is now common to require materials such as plastics to be qualified under more than one test in order to assure their suitability in building construction. Factory Mutual has written about its corner test as follows:

"The characteristic of rapid flame spread across the exposed surface of polyurethane, even with automatic sprinkler protection, prompted the development of the Factory Mutual Corner Test. The test determines the performance of various types of foamed plastic insulation, in combination with various surface treatments, under full scale fire conditions. The test procedure is designed to simulate the exposure that would be expected in an essentially noncombustible occupancy (i.e., an isolated stack of pallets). When sprinklers are provided, the exposure is increased to simulate a fire in a combustible occupancy" (FMRC 1978a).

The full scale corner test is conducted in a 25 ft high (7.6 m) test building and the initiating fire is a 750 lb (340 kg) wood crib. A scaled down version, 1/12 full scale, produces similar results at lower cost.

Other tests which may be required by the authority having jurisdiction are used to determine, separately, ignition temperature, surface flame spread index, and the rate of smoke production. Potential heat (caloric value) of the plastic material may have to be determined, but this parameter is of little significance in fire protection engineering; more important is the rate at which heat is liberated and contributes to the intensity of the fire. A multiplicity of tests to qualify plastics, or any other material, for end use applications may not be required in the future. One standard test method (ASTM E906) uses the Ohio State University Release Rate Apparatus and is capable of producing varying radiant heat fluxes corresponding to the severity of the fire exposure which can be anticipated for the product being tested (ASTM 1983). (See Fig. 5-8B.) Measurements include heat and smoke particles released by the test specimen per unit time and total during the test, toxic gases given off, ease of ignition, and flame spread rate. Research is continuing on this and other heat and smoke release tests, including a "cone calorimeter" developed at the National Bureau of Standards.

The flammability test methods which follow are presently included in the ASTM book of standards specifically for the testing of plastics, Underwriters Laboratories' small scale flammability tests for plastics, and tests for plastics of some federal agencies.

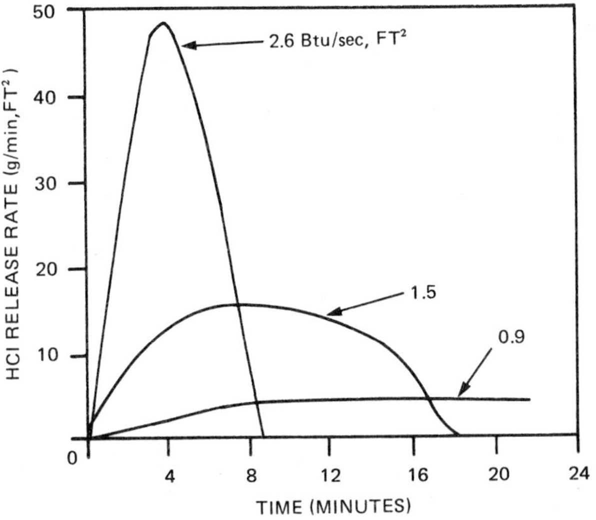

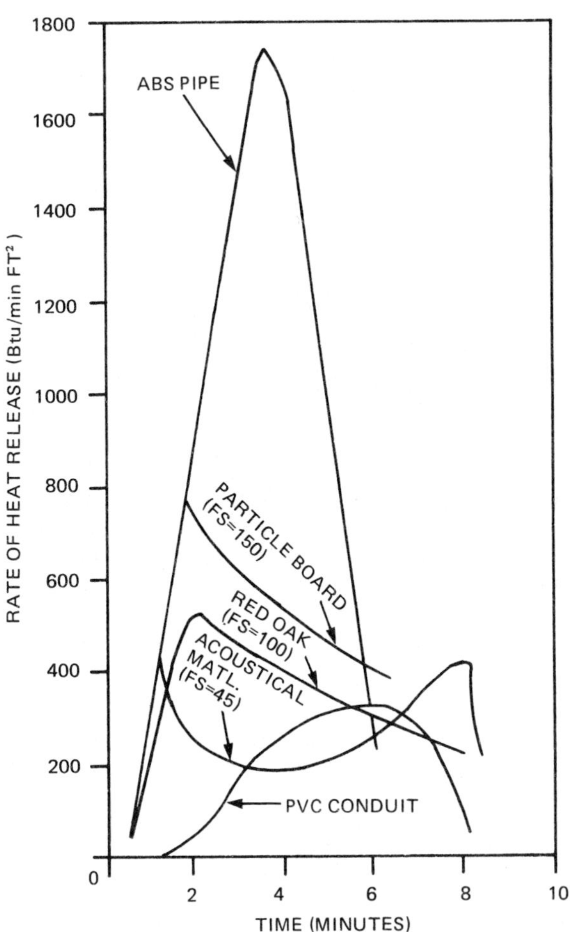

FIG. 5-8B. Typical of the curves generated by the OSU Release Rate apparatus are (a) above, heat release rates for several specimens at an exposure level of 2.6 Btu/sec-ft², and (b) below, hydrogen chloride release rates from a sample of PVC conduit at different exposure levels. FS ASTM E84 flamespread rating. One Btu per sec, ft² equals 0.053 W/m².

American Society for Testing and Materials

ASTM has standardized several small scale test procedures intended for plastics research, product development, and manufacturing control purposes. Results of these tests are not valid for evaluating the fire hazard of plastics under room fire conditions or most other end use conditions. The following standard test methods are included in the 1984 *Book of ASTM Standards*:

D229-82, *Rigid Sheet and Plate Materials Used for Electrical Insulation.* (One of the procedures included is for burning rate and flame resistance.)

D568-77, *Rate of Burning and/or Extent and Time of Burning of Flexible Plastics in a Vertical Position.*

D635-81, *Rate of Burning and/or Extent and Time of Burning of Self-Supporting Plastics in a Horizontal Position.*

D1929-77, *Ignition Properties of Plastics.* (This test uses a Setchkin furnace to heat the specimen prior to ignition by a pilot flame.)

D2843-77, *Density of Smoke from the Burning or Decomposition of Plastics.*

D2863-77, *Measuring the Minimum Oxygen Concentration to Support Candlelike Combustion of Plastics* (Oxygen Index).

D3675-78, *Surface Flammability of Flexible Cellular Materials Using a Radiant Heat Energy Source.*

D3713-78, *Measuring Response of Solid Plastics to Ignition by a Small Flame.*

D3894-81, *Evaluation of Fire Response of Rigid Cellular Plastics Using a Small Corner Configuration* (minicorner test).

Underwriters Laboratories

The UL 746 series of standards recognize the advances made in the safety of electrical products due, in part, to the introduction of plastic materials in their manufacture (Bogue 1979). Plastic handles and enclosures provide electric shock protection, while the moldability and resilience of plastics have made these materials particularly useful for guarding moving parts against accidental contact. Flammability and ignition characteristics of plastics are evaluated by a series of small scale tests under the procedures established in these end product standards. The four "Standards for Safety" in this series are:

UL 746A —*Polymeric Materials—Short-Term Property Tests.*

UL 746B —*Polymeric Materials—Long-Term Property Tests.*

UL 746C —*Polymeric Materials—Use in Electrical Equipment Evaluations.*

UL 746D —*Polymeric Materials—Fabricated Parts.*

Together with UL 94, *Flammability of Plastic Materials for Parts in Devices and Appliances*, these standards are used to establish properties such as resistance to aging, electrical spark resistance, and resistance to heat on molded parts, as well as resistance to ignition and rate of burning. Products which meet the criteria are listed in UL's *Recognized Component Directory*. Four UL originated test methods are used to evaluate the fire hazard as follows:

Hot Wire Ignition Test: Half inch wide, 5 in. long (13 by 127 mm) specimens are wrapped with 5 turns of nichrome wire, which is brought to red heat for up to 5 min. The length of time to cause ignition is recorded as the Hot Wire Ignition Index.

High Current Arc Ignition Test: The specimen is subjected to electrical sparks generated between 2 electrodes at the rate of 40 per min for up to 5 min. Tests are run with the plastic in 4 different positions with respect to the electrodes. The number of arc ruptures which are required before the material ignites is recorded as the High Current Arc Ignition Index.

Vertical Burning Test: Specimens 5 in. long by ½ in. (127 by 13 mm) wide in a variety of thicknesses are supported vertically, and a ¾ in. (19 mm) blue flame is applied to the lower end for 10 sec. The results are used to classify the specimen as 94V-0, 94V-1, or 94V-2, depending on how long it burns after the application of the flame and whether drip particles ignite cotton placed under the specimen.

Horizontal Burning Test: The same size specimen is mounted horizontally, and a 1 in. (25.4 mm) blue flame is applied to the end for 30 sec. (See Fig. 5-8C.) With limited burning, the material will be classed 94HB. More severe testing may be used to establish a 94-5V classification, and classifications 94HBF, 94HF-1 and 94HF-2 are used for foamed materials. The Radiant Panel Test, ASTM E-162, may be used for additional evaluations.

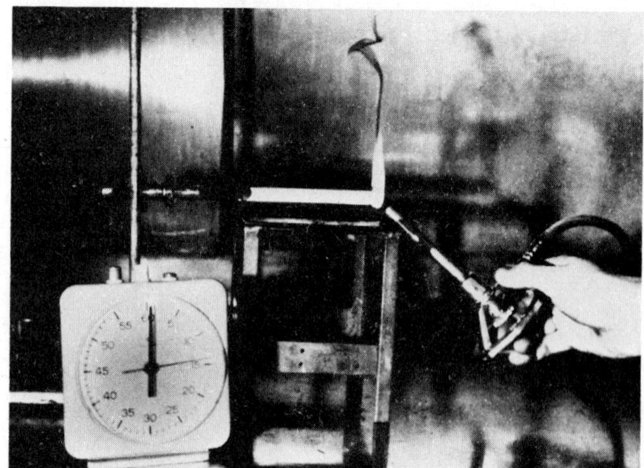

FIG. 5-8C. Specimen being tested under the Horizontal Burning Test of UL Standard 94 for flammability of plastics. (Underwriters Laboratories Inc.)

Federal Agencies

Under MVSS-302, the DOT regulates interior materials for automobiles, using a test for burning rate in the horizontal mode. The U.S. Coast Guard, under Subchapter T, regulates materials used as resins for hulls in boats for hire carrying more than six passengers. The Coast Guard also uses a modified version of ASTM E-136 to establish the "noncombustibility" of materials used in the interior of passenger carrying ships. Guidelines for the flammability of carpets, rugs, children's sleepwear, mattresses, and mattress pads are issued by the U. S. Consumer Product Safety Commission, which is currently studying proposals for low intensity, cigarette ignition tests for upholstered furniture. The U.S. Nuclear Regulatory Commission has issued guidelines for the ignitability and flammability of plastics used in the insulation of electrical cables requiring qualification under test procedures in IEEE 383.

INSTALLATION SAFEGUARDS

In the current state of the art, large areas of cellular plastics left exposed in interior building construction or decoration must be considered to be extremely hazardous. Exposed foam on ceilings or walls, in air handling plenums, in shafts, in other fluelike vertical spaces inside walls, or above hung ceilings should be avoided. A low flame spread classification (e.g., 25 or less) does not assure that a product has limited combustibility under such conditions, nor can it be assumed that automatic sprinklers will control the hazard. Coating such surfaces with a fire retardant paint, intumescent or otherwise, is not likely to produce satisfactory results.

Model building codes require that foamed plastic used as interior wall insulation be covered with a thermal barrier, such as ½ in. (12.7 mm) thick ordinary gypsum board, or other method which reduces the risk of ignition and the subsequent flash fire propensity. The use of foamed plastics in the cavities of hollow masonry walls, such as perimeter insulation around the foundation of a building, insulation below concrete slabs on the ground, and for roof insulation under certain conditions is generally accepted without thermal barrier protection. All such plastics in the thickness and density used must have a smoke developed rating no greater than 450 and a flame spread rating no greater than 75 under ASTM E84 tunnel test (NFPA 255) conditions.

In other than noncombustible or fire resistive building types, foamed plastic insulated, steel or aluminum sheathed building panels are permitted by the model building codes, provided the foam core has a flame spread classification of 25 or less and the space is protected by automatic sprinklers.

When used as interior wall and ceiling finish, plastic materials other than foamed plastic generally are not subject to any special requirements. As for any other materials, plastics are subject to limitations on surface flame spread and often on the smoke generated, as measured by standard test procedures, such as the Steiner tunnel test or radiant panel test. These or special limitations may be applied to plastics used as diffusers in lighting fixtures where it is often acceptable to have a plastic which deforms and drops out of the fixture at an elevated temperature still well below its ignition point. Special limitations may also be set forth by building codes for the use of plastic glazing instead of glass in exterior walls of buildings, where the normal requirement for conformance to the limitations for interior finish may be waived.

Plastic laminates for countertops, kitchen cabinets, tabletops, etc., are not usually included in the definition of interior finish, as regulated by building codes. Even when it is not regulated by local code, the flammability level should be limited to that encountered when natural products are used in these applications. Plastic laminates are commonly available with low surface flame spread classifications, in the range of 25 to 75.

Further guidelines for the installation of plastics in various applications may be found in publications of the Society of the Plastics Industry (SPI various dates) and Factory Mutual Research Corporation (FMRC various dates).

The packaging, storage, and shipping precautions that are customarily observed for ordinary combustible materials apply generally to most plastics. No special shipping requirements for plastics (other than cellulose nitrate) have been adopted by the DOT.

FIRE AND FLAME RETARDANT TREATMENTS

All organic and some inorganic materials burn under proper conditions. Fire and flame retardants can make a given material more difficult to ignite, and when ignited, cause it to burn more slowly. Retardants can be very effective in preventing ignition from low level, short duration flame and heat exposures. Even under moderate exposures, the burning characteristics can be greatly improved by fire retardants, depending upon the product and the additive used. For severe exposures, such as from a fully developed room fire, treated and untreated plastics typically show similar fire behavior.

Fire and flame retardant chemicals are added to plastics during the manufacturing process. Because of the widely varying properties of plastics, no single, universal additive is available, nor is there only one effective mechanism for imparting flame resistance. Fire and flame retardant additives may be blended into the polymer mix during processing, or they may be reactants which change the molecular structure of the polymer. The additives can be chemically reacted throughout the polymer substrate or on the surface of finished or semifinished products. Surface treatments added after the manufacturing process, such as fire retardant paints or coatings, are unreliable without test data to prove the effectiveness of the combination.

Additives blended into the mix may be organic or inorganic materials. Organic additives typically are chlorinated and brominated hydrocarbons or halogenated and nonhalogenated organophosphorous compounds. Among the inorganic flame retardants are the salts of antimony, zinc, molybdenum, and aluminum. Those classed as reactants include brominated aromatics, brominated aliphatic polyols, and phosphorous containing polyols.

Phosphorous additives appear to improve flame resistance by: (1) promoting formation of char to reduce concentration of carbon containing gases; (2) forming a glassy, insulating layer; and (3) either entering into chemical reactions that remove heat from the combustion system, or chemically inhibiting the combustion process, or both. Compounds containing halogens, such as chlorine and bromine, tend to inhibit burning in the gas phase by interrupting chain reactions needed for continuous burning. Other flame retardants involve heat absorption reactions, for example, by providing a heat sink through evaporation of water of hydration (Modern Plastics Encyclopedia 1984-85).

A complicating aspect of treating plastics and other materials is that a fire retardant chemical that works on one substrate system will not necessarily work on other substrates. Thus, fire retardant chemicals for polystyrene will not necessarily work on polyethylene, nylon, wood or cotton. Fire retardant chemicals often impair some other material properties. For example, fire retardant treatments for cotton fabrics generally reduce fabric strength and wear life. Fire retardants in plastics may reduce allowable processing temperatures and impair physical properties. Costs are generally increased.

Smoke generation may or may not be significantly reduced by the treatment. For example, many moderately flammable plastics, such as acrylics, acetate, and polyethylene, emit relatively little smoke when burning. However, chemically changed, they become less flammable, but invariably generate more smoke per unit burned. Two questions arise. Is it preferable to have more flame spread with less smoke, or less flame spread with more smoke? Which condition is likely to result in the greater total amount of smoke?

FIGHTING FIRES IN PLASTICS

Plastics other than cellulose nitrate are classified as ordinary combustibles. Consequently, extinguishing methods suitable for fires involving wood and other ordinary combustibles (Class A fires) should be used to extinguish a burning plastic. Fire protection considerations should include automatic sprinklers, standpipe and hose systems, and water type portable extinguishers. These should be supplemented by fire extinguishers or special automatic systems suitable for flammable liquid and electrical fires where these hazards exist.

The physical form of plastics will greatly influence their fire behavior. Molding pellets which are shipped in bags, drums, or large cartons provide little surface for access to air. These same plastics in such shapes as flashlight cases will have much more access to air and may burn vigorously until heat causes the shapes to melt, thereby reducing the surface area exposed to air. Melting may be a hazard if burning drips carry flame to a lower floor or spread fuel for later ignition. For this reason, during fire fighting some hose streams should be used to cool exposed plastics to keep them from melting and dripping. This is a major advantage of automatic or manual sprinklers in large storage areas.

Foamed Plastics: These plastics have large surface areas but little heat of combustion per unit volume. Under fire conditions, thermoplastic types of foam are readily reduced to small volume. If traces of flammable blowing agent remain in the foamed plastic, a much greater hazard exists. (For firesafety reasons in manufacture and storage, nonflammable fluorinated hydrocarbons should be considered to replace materials as pentane for a blowing agent.)

Toxicity: Possibly because of the long chemical names of some plastics, combustion and thermal decomposition products of plastics have been a cause for concern among fire fighters. For plastics now in commercial use, the hazard of carbon monoxide from partial combustion greatly outweighs the toxic effects of other fire gases, both as to nature and amount. Whether this will remain true for future plastics cannot yet be forecast.

On burning, some plastics, such as polyvinyl chloride or ethylene sulfide rubbery caulking materials, generate hydrogen chloride or sulfur dioxide. Because these are strongly irritating, they ordinarily force evacuation long before their toxic effects become dangerous. These gases also are corrosive to metals and electrical equipment; therefore such equipment should be ventilated, rinsed, or treated with dilute ammonia, and rinsed as soon as possible after exposure. Fighting any but very small peroxide fires at close range is not recommended.

Bibliography

References Cited

Bogue, R. J. 1979. "Spirit of 746." *UL Lab Data*. Underwriters Laboratories. Northbrook, IL.

FMRC. 1978. Rigid Foamed Polyurethane and Polyisocyanurate for Construction. *Loss Prevention Data 1-57*. Factory Mutual System. Norwood, MA.

FMRC. Various dates. Loss Prevention Data of the Factory Mutual System. 1151 Boston-Providence Turnpike. Norwood, MA:

1. 1-57. *Rigid Foamed Polyurethane and Polyisocyanurate for Construction.*
2. 1-58. *Foamed Polystyrene for Construction.*
3. 1-59. *Reinforced Plastic Panels in Construction.*

Gaskill, J. R. 1970. "Smoke Development in Polymers During Pyrolysis or Combustion." *Journal of Fire and Flammability*. Vol 1. July 1970. pp 183-216.

Katz, Cassell and Collier. 1978. *Plastics Designs and Materials.* Macmillan Publishers, London, England.

Modern Plastics Encyclopedia. 1984-85. Vol 56 No 10A. 1979-80 ed. McGraw-Hill, NY.

NBS. 1983. *Fire Hazard Evaluation of Shipboard Hull Insulation and Documentation of a Quarter-Scale Room Fire Test Protocol.* NBSIR 83-2642. National Bureau of Standards. Washington, DC.

Parker, W. J. and Long, M. E. 1972. "Development of a Heat Release Rate Calorimeter at NBS." *ASTM Special Technical Publication 502*. ASTM, Philadelphia. PA.

PRC. 1980. "Fire Research on Plastics: The Final Report of the Products Research Committee." Products Research Committee. John W. Lyons, Chairman. National Bureau of Standards. Washington, DC.

SPI. 1977. *Fire Safety Guidelines for Use of Expanded Polystyrene in Building Construction.* WPS-301. EPS Division. The Society of the Plastics Industry Inc. Des Plaines, IL.

SPI. 1975. *Fact-Finding Reports of the EPS Division.* The Society of the Plastics Industry. Des Plaines, IL:

1. *Flammability of Expanded Polystyrene Insulation Used in Conjunction with Gypsum Wallboard.*
2. *Flammability of Expanded Polystyrene Insulation When Used as Full-Scale Building Enclosure Ceiling Construction Material.*
3. *Flammability of Expanded Polystyrene Insulation Used in Conjunction with Plywood Paneling.*
4. *Flammability of Exposed Polystyrene Insulation (With and Without Wooden Dividers).*

SPI. Various dates. Bulletins of the Urethane Safety Group. Society of the Plastics Industry Inc. 355 Lexington Ave. NY 10017.

1. U-100R. *Fire Safety Guidelines for Use of Rigid Polyurethane Foam Insulation in Building Construction.*
2. U-102R. *An Update Report on Findings of Fire Study of Rigid Cellular Plastic Materials for Wall and Roof Ceiling Insulation.*
3. U-103. *Large-Scale Corner Wall Fire Tests of Spray-On Coatings Over Rigid Polyurethane Foam Insulation.*
4. U-104. *Model Code Provisions Pertaining to Rigid Polyurethane Foam Insulation.*
5. U-105. *Evaluation of the Fire Performance of Carpet Underlayments.*
6. U-106. *Fire Safety Guidelines on Flexible Polyurethane Foams Used in Upholstered Furniture and Bedding.*
7. U-107. *Room-Scale Compartment Corner Tests of Spray-On Coatings Over Rigid Polyurethane Foam Insulation.*

Troup, W. W. J. 1969. *Cellular Plastics and the Building Fire Problem.* Factory Mutual Research Corp. Norwood, MA.

UL. 1969. *Report on Flammability Study of Cellular Plastics, File*

NC522. The Society of the Plastics Industry. Underwriters Laboratories Inc. Northbrook, IL.

UL. 1975. *Flammability Studies of Cellular Plastics and Other Building Materials Used for Interior Finishes.* Subject 723. Underwriters Laboratories Inc. Northbrook, IL.

NFPA Codes, Standards, Recommended Practices and Manuals. (See the latest *NFPA Codes and Standards Catalog* for availability of current editions of the following documents.)

NFPA 40, *Standard for the Storage and Handling of Cellulose Nitrate Motion Picture Film.*

NFPA 43E, *Code for the Storage of Pyroxylin Plastic.*

NFPA 49, *Hazardous Chemicals Data.*

NFPA 231C, *Standard for Rack Storage of Materials.*

NFPA 255, *Standard Method of Test of Surface Burning Characteristics of Building Materials.*

NFPA 701, *Standard Methods of Fire Tests for Flame-Resistant Textiles and Films.*

NFPA 702, *Standard for the Classification of the Flammability of Wearing Apparel.*

Additional Readings

Annamalai, K., and Sibulkin, M., "Ignition and Flame Spread Tests of Cellular Plastics," *Journal of Fire and Flammability*, Vol. 9, 1978, pp. 445-458.

ASTM, "Standard Method of Test for Density of Smoke from the Burning or Decomposition of Plastics," ANSI/ASTM D2843-77, American Society for Testing Materials, 1977.

Barry, Thomas J., and Newman, Bernard, "Some Problems of Synthetic Polymers at Elevated Temperatures," *Fire Technology*, Vol. 12, No. 3, Aug. 1976, pp. 186-192.

Benjamin, I. A., and Parker, W. J., "Fire Spread Potential of ABS Plastic Plumbing," *Fire Technology*, Vol. 8, No. 2, May 1972, pp. 104-119.

Blair, John A., "Fire Research: A Progress Report from the Plastics Industry," *Fire Journal*, Vol. 70, Nov. 1976.

Boettner, E. A., Ball, G. L. and Weiss, B., "Combustion Products from the Incineration of Plastics," *EPA Report 670/2-73-049*, Environmental Protection Agency, Washington, DC, 1973.

Bowen, John E., "Hazards of Burning Plastics," *Fire Chief Magazine*, Dec. 1975.

"Building Insulation: Energy Conservation or Fire Hazard?" *Factory Mutual Record*, Nov.-Dec. 1978.

Cellular Plastics in Construction Applications and Fire Protection, Cellular Plastics Div., The Society of the Plastics Industry, Inc., NY, 1975.

"Combustion Toxicology of Polymers," *NMAB 318-3*, National Materials Advisory Board, Washington, DC, 1978.

Crawford, R. J., *Plastics Engineering*, Pergamon Press, Elmsford, NY, 1981.

Cullis, C. F., and Hirschler, M. M., *The Combustion of Organic Polymers*, Clarendon Press, Oxford, England, 1981.

D'Souza, M. V., and McGuire, J. H., "ASTM E-84 and the flammability of Foamed Thermosetting Plastics," *Fire Technology*, Vol. 13, No. 2, May 1977, pp. 85-94.

Dubois, J. Harry, ed., *Plastics*, 6th ed., Van Nostrand Reinhold Co., NY, 1981.

Einhorn, I., "Physiological and Toxicological Aspects of Smoke Produced during the Combustion of Polymeric Materials," *Environmental Health Perspectives*, Vol. 11, 1975, p. 163.

Ettling, Bruce V., "Enhanced Flammability of Polyurethane Foam," *Fire Technology*, Vol. 9, No. 4, Nov. 1973, pp. 271-274.

"Foamed Urethane, A Solid that Burns Like a Flammable Liquid," *Pamphlet P6805*, Factory Mutual Engineering Corp., Norwood, MA.

"For Architects and Builders: Home Insulation," *Fire Journal*, May 1978.

Frados, Joel, Ed., *Plastics Engineering Handbook of the Society of the Plastics Industry*, Inc., 4th ed., Van Nostrand Reinhold Co., NY, 1976.

Hall, C., *Polymeric Materials: An Introduction for Technologists and Scientists*, Macmillan, London, England, 1981.

Harmathy, T. Z., *Properties of Building Materials at Elevated Temperatures*, DBR Paper No. 1080, Division of Building Research, National Research Council of Canada, Ottawa, Canada, 1983.

Harper, Charles A., ed., *Handbook of Plastics and Elastomers*, McGraw-Hill, NY, 1975.

Hilado, Carlos, *Flammability Handbook for Plastics*, 3rd ed., Technomic Publishing Co., Westport, CT, 1982.

Hilado, Carlos J., "An Overview of the Fire Behavior of Polymers," *Fire Technology*, Vol. 9, No. 3, Aug. 1973, pp. 198-208.

Hilado, Carlos J., *Flammability Handbook for Plastics*, 3rd ed., Technomic, Lancaster, PA, 1982.

Hilado, C. J., Casey, C. J., and Schneider, J. E., "Effect of Pyrolysis Temperature on Relative Toxicity of Some Plastics," *Fire Technology*, Vol. 15, No. 2, May 1979, pp. 122-129.

Ives, J. M., Hughes, E. E., and Taylor, J. K., *Toxic Atmospheres Associated with Real Fire Situations*, National Bureau of Standards, Report 10, 807, Feb. 16, 1972, Washington, DC.

Kanury, A. M., Alvares, N. J., and Martin, S. B., "Flammability Testing of Polymers," *SRI Research Report*, Chemical Manufacturers Association, Washington, D.C., 1977.

Lyons, John W., et al., *Fire Research on Cellular Plastics: The Final Report of the Products Research Committee*, Products Research Committee, 1980.

Madorsky, S. L., *Thermal Degradation of Organic Polymers*, John Wiley Interscience, NY, 1964.

Markstein, G. H., "Radiative Properties of Plastics Fires," *17th Symposium (International) on Combustion*, The Combustion Institute, Pittsburgh, PA, 1979, pp. 1053-1062.

Maroni, W. F., "SLRP Analysis of Recommended Protection for Foamed Plastic Wall Ceiling Building Insulations," *presentation at annual meeting of National Fire Protection Association*, 1974.

Maroni, W. F., "Large-Scale Fire Tests of Rigid Cellular Plastic Wall and Roofing Insulations," *Fire Journal*, Vol. 67, No. 6, Nov. 1973, pp. 24-29.

McCarter, R. J., "Smoldering of Flexible Polyurethane Foam," *Journal of Consumer Product Flammability*, Vol. 3, 1976, pp. 128-140.

Modak, A. T., and Croce, P. A., "Plastic Pool Fires," *Combustion and Flame*, Vol. 30, 1977, pp. 251-265.

Ohlemiller, T. J., and Rogers, F. E., "A Survey of Several Factors Influencing Smolder Combustion in Flexible and Rigid Polymer Foams," *Journal of Fire and Flammability*, Vol. 9, 1978, pp. 489-509.

"Plastic Fire Hazard Classifications," *Pamphlet P6221*, Factory Mutual Engineering Corp., Norwood MA.

"Plastics—General Method of Testing, Nomenclature," *Book of ASTM Standards, Part 35*, American Society for Testing and Materials, Philadelphia, PA, issued annually.

Precautions for the Proper Usage of Polyurethanes, Polyisocyanurates and Related Materials, Technical Bulletin 107, Upjohn Chemical Div., The Upjohn Co., Kalamazoo, MI, 1974.

Rawls, Rebecca L, "Fire Hazards of Plastics Spark Heated Debate," *Chemical and Engineering News*, Vol. 61, pp. 9-16, Jan. 1983.

Reinke, R. E., and Reinhardt, C. F., "Fires, Toxicity, and Plastics," *Modern Plastics*, Vol. 50, No. 2, pp. 94-95, 97-98.

Roberts, A. F., "Polyurethane Foam: Some Studies Relating to Its Behavior in Fire," *Fire Technology*, Vol. 7, No. 3, Aug. 1971.

Salig, Ronald J., "The Smoldering Behavior of Upholstered Polyurethane Cushionings and its Relevance to Home Furnishing Fires," Masters Thesis, Massachusetts Institute of Technology, Cambridge, MA, 1981.

Schaffer, E. L., ed., Behavior of Polymeric Materials in Fire, STP 816, American Society for Testing and Materials, Philadelphia, PA, 1983.

Schafran, Eugene, "Development of Flammability Specifications for Furnishings," *Fire Journal*, Vol. 68, No. 2, March 1974, pp. 36-39.

Schwartz, Seymour, and Goodman, Sidney, *Plastics Materials and Processes*, Van Nostrand Reinhold Co., NY, 1982.

Tewarson, A., and Pion, R. F., "Burning Intensity of Commercial Samples of Plastics," *Fire Technology*, Vol. 11, No. 4, Nov. 1975, pp. 274-281.

Tewarson, A., "Heat Release Rates from Burning Plastics," *Journal of Fire and Flammability*, Vol. 8, 1977, p. 115.

Tewarson, A., "Physico-Chemical and Combustion/Pyrolysis Properties of Polymeric Materials," *NBS-GCR-80-295*, National Bureau of Standards, Gaithersburg, MD, 1980.

Tewarson, A., "Experimental Evaluation of Flammability Parameters of Polymeric Materials," In: *Flame Retardant Polymeric Materials*, Vol. 3, Plenum Press, NY, 1982, pp. 97-153.

Tewarson, A., and Pion, R. F., "Flammability of Plastics: I. Burning Intensity," *Combustion and Flame*, Vol. 26, 1976, pp. 85-103.

Trade Designations of Plastics and Related Materials (rev.), PLASTEC Note N9B, Plastics Technical Evaluation Center, Picatinny Arsenal, Dover, NJ, Oct. 1974. (Available through NTIS, Springfield, VA)

UL 94, *Tests for Flammability of Plastic Materials for Parts in Devices and Appliances*, Underwriters Laboratories Inc., Northbrook, IL, 1982.

Zinn, B. T., et al., "Investigation of the Properties of the Combustion Products Generated by Fire-Retarded Polyurethanes," *Final Report of the Products Research Committee Project No. RP-75-1-15-Revised*, Sept. 1977.

DUSTS

Revised by Richard F. Schwab

Most finely divided combustible materials are hazardous. Deposits of combustible dusts on beams, machinery, and other surfaces are subject to flash fires. When combustible dusts suspended in air are ignited, they can cause severe explosions. If the dusts are oxidizing agents and they accumulate on combustible surfaces, the combustion process would be considerably accelerated in a fire. If an oxidizing agent in the form of a finely divided dust is mixed with other combustible dusts, the violence of the resulting explosion would be much more severe than an explosion without the oxidizing agent dust. On the other hand, inert materials, such as limestone, are sometimes used to quench or arrest combustible dust fires or deflagrations (Palmer 1973).

Although dust explosions have been recorded since 1785, and the principles for controlling them have been published, serious incidents continue to occur. (See Fig. 5-9A.) In less than one week in December 1977, two serious grain elevator explosions resulted in 54 fatalities (Lathrop 1978). Several more explosions were reported within a few months of these, resulting in additional fatalities (Best 1978).

FACTORS INFLUENCING THE EXPLOSIBILITY OF DUSTS

The chance of a dust cloud igniting is governed by the size of its particles, dust concentration, impurities present, oxygen concentration, and the strength of the source of ignition.

Dust explosions usually occur as a series. Frequently, the initial deflagration is rather small in volume but intense enough to jar dust from beams, ledges, etc., or even to rupture small pieces of equipment within buildings, such as dust collectors or bins. Subsequently, a much larger dust cloud develops through which a secondary explosion can propagate. It is not unusual to have a series of explosions propagating from building to building (Palmer 1973).

Richard F. Schwab, P.E., S.F.P.E., AIChE, is Manager of Process Safety and Loss Prevention, Allied Chemical Corporation, Morristown, NJ.

FIG. 5-9A. Smoke from a grain elevator explosion in Westwego, LA. (G. E. Arnold)

Hazards of Dusts

The hazard of any given dust is related to its ease of ignition and the severity of the ensuing explosion. The Bureau of Mines of the U.S. Department of the Interior has developed an arbitrary scale based on small scale tests, which is quite useful for measuring the hazard. The ignition sensitivity is a function of the ignition temperature and the minimum energy of ignition; the explosion severity is a function of maximum explosion pressure and the maximum rate of pressure rise. To facilitate comparisons of explosibility data developed in Bureau of Mines tests, all test results (Jacobson et al 1961) are related to a standard Pittsburgh coal dust taken at a concentration of 0.50 oz per cu ft (500 g/m^3), with the exception of some metal dusts (Jacobson et al 1964).

The ignition sensitivity and explosion severity of a dust are defined as:

Ignition sensitivity =

$$\frac{\text{(Ign. temp.} \times \text{Min. energy} \times \text{Min. conc.) Pgh. coal dust}}{\text{(Ign. temp.} \times \text{Min. energy} \times \text{Min. conc.) Sample dust}}$$

Explosion severity =

$$\frac{\text{(Max. explo. press.} \times \text{Max. rate of press. rise) Sample dust}}{\text{(Max. explo. press.} \times \text{Max. rate of press. rise) Pgh. coal dust}}$$

The Explosibility Index is the product of ignition sensitivity and explosion severity. This method allows one to rate the relative hazard of dusts as follows:

Type of Explosion	Ignition Sensitivity	Explosion Severity	Explosibility Index
Weak	<0.2	<0.5	<0.1
Moderate	0.2–1.0	0.5–1.0	0.1–1.0
Strong	1.0–5.0	1.0–2.0	1.0–10
Severe	>5.0	>2.0	>10.0

Table 5-9A gives the explosibility index, ignition sensitivity, explosion severity, maximum explosion pressure/maximum rate of pressure rise (not necessarily at the concentration of 0.50 oz per cu ft or 500 g/m^3), ignition temperature of both a dust cloud and a layer, the minimum ignition energy of a dust cloud, minimum explosion concentration, and the limiting oxygen concentration in a spark ignition chamber.

Particle Size

The smaller the size of dust particle the easier it is to ignite the dust cloud insofar as the exposed surface area of a unit weight of material increases as the particle size decreases. It is also true that particle size has an effect on the rate of pressure rise. For a given weight concentration of dust, a coarse dust will show a lower rate of pressure rise than a fine dust. The lower explosive limit concentration, ignition temperature, and the energy necessary for ignition will decrease as dust particle size decreases. Numerous studies by various investigators show this effect for a variety of dusts (Hartmann et al 1950).

Decrease in particle size also increases the capacitance of dust clouds, i.e., the size of electrical charges that can accumulate on particles in the cloud (Kunkel 1950). Because capacitance of solids is a function of surface area, the possibility of developing electrostatic discharges of sufficient intensity to ignite a dust cloud increases as average particle size decreases. However, to produce such electrostatic discharges requires, among other things, large quantities of dust in large volumes with relatively high dielectric dust strengths and consequent long relaxation times. Because of the high ignition energies required for dust clouds to ignite in comparison with ignition energies of gases, attributing the cause of dust explosions to static electricity should be held suspect unless definite evidence exists to show that static electricity was a likely cause (Palmer 1973).

Concentration

As with flammable gases and vapors, there is a specific range of dust concentration within which a dust explosion can occur. It is customary to express the concentration figures in terms of weight per unit volume, though without knowledge of the particle size distribution of the sample this expression is meaningless. The results presented in Table 5-9A were obtained with dusts small enough to pass through a 200 mesh screen (74 microns or smaller). Variations in minimum explosive concentrations will occur with changes in particle diameter, i.e., the minimum explosive concentration is lowered as the diameter of particles decreases. Sample purity, oxygen concentration, strength of ignition source, turbulence of dust cloud, and uniformity of dispersion also effect the Lower Explosive Limits (LEL) of dust clouds.

Upper Explosive Limits (UEL) for dust clouds have not been determined mainly because of experimental difficulties. There is also a question of whether a clear cut upper limit exists at all and, from a practical point of view, this information is of questionable use. Curves formed by plotting explosion pressures and rates of pressure rise against concentration show that explosion pressures and rates of pressure rise are at a minimum at the lower explosive limit, then rise to maximum value at a given optimum concentration, then slowly decrease from this point. It is to be noted that the maximum pressure and the maximum rate of rise do not always occur at precisely the same concentration. The destructive effect is determined primarily by the rate of pressure rise.

It appears then that the most violent explosion occurs at a concentration slightly above that required for reaction with all of the oxygen in the atmosphere. At lower dust concentrations, less heat is generated and smaller peak pressures are developed. With dust concentrations greater than those causing the most violent explosions, absorption of heat by unburned dust is apparently the reason for less than maximum explosive pressures.

Moisture

Moisture in dust particles raises the ignition temperature of the dust because of the heat absorbed during heating and vaporization of the moisture. The moisture in the air surrounding a dust particle has no significant effect on the course of a deflagration once ignition has occurred. There is, however, a direct relationship between moisture content and minimum energy required for ignition, minimum explosive concentration, maximum pressure, and maximum rate of pressure rise. For example, the ignition temperature of cornstarch may increase as much as 122°F (50°C) with an increase of moisture content from 1.6 percent to 12.5 percent. As a practical matter, however, moisture cannot be considered an effective explosion preventative since most ignition sources provide more than enough heat to vaporize the moisture and to ignite the dust. In order for moisture to prevent ignition of a dust by common sources, the dust would have to be so damp that a cloud could not be formed.

Inert Material

The presence of an inert solid powder reduces the combustibility of a dust because it absorbs heat, but the amount of inert powder necessary to prevent an explosion is usually considerably higher than concentrations that

TABLE 5-9A. Explosion Characteristics of Various Dusts

(Compiled from the following reports of the U.S. Department of Interior, Bureau of Mines: RI 5753, The Explosibility of Agricultural Dusts; RI 6516, Explosibility of Metal Powders; RI 5971, Explosibility of Dusts Used in the Plastics Industry; RI 6597, Explosibility of Carbonaceous Dusts; RI 7132, Dust Explosibility of Chemicals, Drugs, Dyes and Pesticides; and RI 7208, Explosibility of Miscellaneous Dusts.)

Type of Dust	Explosibility Index	Ignition Sensitivity	Explosion Severity	Maximum Explosion Pressure psig*	Max Rate of Pressure Rise psi/sec*	Ignition Temperature† Cloud °C	Ignition Temperature† Layer °C	Min Cloud Ignition Energy joules	Min Explosion Conc oz/cu ft‡	Limiting Oxygen Percentage§ (Spark Ignition)
Agricultural Dusts										
Cellulose	2.8	1.0	2.8	130	4,500	480	270	0.080	0.055	C13
Cellulose, alpha	>10	2.7	4.0	117	8,000	410	300	0.040	0.045	—
Cocoa, natural 19% fat	0.6	0.5	1.1	68	1,200	510	240	0.10	0.075	—
Coffee, fully roasted	<0.1	0.2	0.1	38	150	720	270	0.16	0.085	C17
Corn	6.9	2.3	3.0	113	6,000	400	250	0.04	0.055	—
Cornstarch commercial product	9.5	2.8	3.4	106	7,500	400	—	0.04	0.045	—
Cork dust	>10	3.6	3.3	96	7,500	460	210	0.035	0.035	—
Cotton linter, raw	<0.1	<0.1	<0.1	73	400	520	—	1.92	0.50	C21
Cube root, South American	6.5	2.7	2.4	69	2,100	470	230	0.04	0.04	—
Grain dust, winter wheat, corn, oats	9.2	2.8	3.3	131	7,000	430	230	0.03	0.055	—
Lycopodium	16.4	4.2	3.9	75	3,100	480	310	0.04	0.025	C13
Milk, skimmed	1.4	1.6	0.9	95	2,300	490	200	0.05	0.05	N15
Rice	0.3	0.5	0.5	47	700	510	450	0.10	0.085	—
Soy flour	0.7	0.6	1.1	94	800	550	340	0.10	0.06	C15
Sugar, powdered	9.6	4.0	2.4	109	5,000	370	400‡	0.03	0.045	—
Wheat flour	4.1	1.5	2.7	97	2,800	440	440	0.06	0.05	—
Wheat starch, edible	17.7	5.2	3.4	100	6,500	430	—	0.025	0.045	C12
Wood flour, white pine	9.9	3.1	3.2	113	5,500	470	260	0.040	0.035	—
Carbonaceous Dusts										
Charcoal, hardwood mixture	1.3	1.4	0.9	83	1,300	530	180	0.020	0.140	—
Charcoal, activated, from lignite	0.1∥	0.1∥	—	41	<100	670	370	#	2.000	—
Pitch, petroleum	4.0	2.8	1.4	82	3,800	630	—	0.025	0.045	—
Carbon black, acetylene	0.1∥	0.1∥	—	—	—	**	900	—	—	—
Coal, Kentucky (Bituminous)	4.1	2.2	1.8	101	4,000	610	180	0.030	0.050	—
Coal, Pennsylvania, Pittsburg (Experimental Mine Coal)	1.0	1.0	1.0	90	2,300	610	170	0.060	0.055	—
Coal, Pennsylvania (Anthracite)	0.1§	0.1§	—	—	—	730	—	0.100††	0.065††	—
Coke, Petroleum	0.1§	0.1§	—	—	200	670	—	∥	1.000	—
Lignite, California	>10	5.0	3.8	94	8,000	450	200	0.030	0.030	—
Chemicals										
Adipic acid	1.9	1.7	1.1	84	2,700	550	—	0.060	0.035	—
Benzoic acid	>10	5.4	2.1	76	5,500	620	Melts	0.020	0.030	—
Bis-phenol A	>10	11.8	2.5	89	8,500	570	—	0.015	0.020	C12
Ethyl hydroxyethyl cellulose	6.0	8.6	0.7	94	2,200	390	—	0.030	0.020	C16
Hexamethylene tetramine	>10	32.7	5.6	98	11,000	410	—	0.010	0.015	C14
Hydroxyethyl cellulose	6.9	4.9	1.4	106	2,600	410	—	0.040	0.025	—
Phthalic anhydride	>10	13.8	1.6	72	4,200	650	—	0.015	0.015	C14
Stearic acid, aluminum salt (aluminum tristearate)	>10	21.3	1.9	87	6,300	420	440	0.015	0.015	—
Stearic acid, zinc salt (zinc stearate)	>10	19.7	2.3	80	>10,000	510	Melts	0.010	0.020	C13
Sulfur	>10	20.2	1.2	78	4,700	190	220	0.015	0.035	C12
Terephthalic acid	6.9	3.0	2.3	84	8,000	680	—	0.020	0.050	C15
Drugs										
Aspirin (Acetylsalicylic Acid) o-$CH_3COOC_6H_4COOH$	>10	2.4	4.3	88	>10,000	660	Melts	0.025	0.050	—
L-Sorbose	1.9	1.0	1.9	76	4,700	370	—	0.080	0.065	—
Vitamin B_1, mononitrate, $C_{12}H_{17}ON_4SNO_3$	8.3	2.7	3.1	101	6,000	360	—	0.060	0.035	—
Vitamin C, ascorbic acid, $C_6H_8O_6$	2.3	1.0	2.2	88	4,800	460	280	0.060	0.070	C15, N12
Dyes, Pigments and Intermediates										
1, 4-Diamino-2, 3-dihydroanthraquinone (90%) 1 methyl-amino-anthraquinone (10%) (Violet 200 dye)	1.0	1.1	.09	64	3,200	880	175	0.060	0.035	—
1, 4-Di-p-toluidineanthra-qui-	1.7	1.7	1.0	73	2,600	770	175	0.050	0.030	—

TABLE 5-9A. Explosion Characteristics of Various Dusts (continued)

none (70%) β naphthalene-azo-dimethyl-aniline (30%) (green base harmon dye)											
1-Methylaminoanthraquinone (red dye intermediate)	1.1	0.9	1.2	71	3,300	830	175	0.050	0.055	—
β-naphthalene-azo-dimethylaniline	3.3	3.9	0.8	70	2,300	510	175	0.050	0.020	—
Metals										
Aluminum flake, A 422 extra fine lining, polished	>10	7.3	10.2	127	20,000+	610	326	0.010	0.045	—
Antimony, milled (96% Sb)	< 0.1	< 0.1	< 0.1	28	300	420	330	1.920	0.420	C16
Boron, amorphous, commercial (85% B)	0.8	0.7	1.1	93	3,300	470	400	0.060	<0.100	—
Chromium, electrolytic, milled (97% Cr)	0.1	0.1	1.2	56	5,000	580	400	0.140	0.230	—
Iron, hydrogen reduced (98% Fe)	0.3	0.7	0.4	61	2,200	320	290	0.080	0.120	C11
Iron, carbonyl (99% Fe)	1.6	3.0	0.5	43	2,400	320	310	0.020	0.105	C10
Magnesium, milled, Grade B	>10	3.0	7.4	116	15,000	560	430	0.040	0.030	—
Titanium (99% Ti)	>10	5.4	2.0	70	6,000	330	510	0.025	0.045	N6, A4, H7‡‡
Uranium	>10	>10	0.9	69	5,000	20	100	0.045	0.060	N1, A2, H2‡‡
Uranium hydride	>10	>10	1.5	74	9,000	20	20	0.005	0.060	N2, A2, H4‡‡
Zinc, condensed (97% Zn, 2% Pb)	<0.1	<0.1	<0.1	50	1,700	690	540	0.960	0.460	N9
Alloys and Compounds										
Aluminum-cobalt alloy (60–40)	0.4	0.1	3.5	92	11,000	950	570	0.100	0.180	—
Aluminum-copper alloy (50–50)	0.3	—	0.9	95	4,000	—	830	—	—	—
Aluminum-lithium alloy (15% Li)	0.6	0.3	1.9	96	6,000	470	400	—	—	—
Aluminum-magnesium alloy (Dowmetal)	>10	2.9	4.5	86	10,000	430	480	0.080	0.020	‡‡
Aluminum-nickel alloy (58–42)	0.6	0.1	4.1	96	10,000	950	540	0.080	0.190	—
Aluminum-silicon alloy (12% Si)	3.6	1.3	2.9	85	7,500	670	—	0.060	0.040	—
Calcium silicide	2.0	0.4	5.0	86	13,000	540	540	0.150	0.060	—
Ferromanganese, medium carbon	0.4	0.4	1.0	62	5,000	450	290	0.080	0.130	—
Ferrosilicon (88% Si, 9% Fe)	<0.1	<0.1	0.2	70	1,000	860	—	0.400	0.425	C19
Ferrotitanium (19% Ti, 74.1% Fe, 0.06% C)	1.3	0.5	2.6	55	9,500	370	400	0.080	0.140	C13
Thermoplastic Resins and Molding Compounds										
Acetal, linear (Polyformaldehyde)	>10	6.5	1.9	113	4,100	440	—	0.020	0.035	C11
Methyl methacrylate polymer	6.3	7.0	0.9	84	2,000	480	—	0.020	0.030	C11
Methyl methacrylate-ethyl acrylate copolymer	>10	14.0	2.7	85	6,000	480	—	0.010	0.030	C11
Methacrylic acid polymer, modified	0.6	1.0	0.6	97	1,800	450	290	0.100	0.045	—
Acrylamide polymer	2.5	4.1	0.6	85	2,500	410	240	0.030	0.040	—
Acrylonitrile polymer	>10	8.1	2.3	89	11,000	500	460	0.020	0.025	C13
Cellulose acetate	>10	8.0	1.6	85	3,600	420	—	0.015	0.040	C14
Cellulose triacetate	7.4	3.9	1.9	107	4,300	430	—	0.030	0.040	C12
Cellulose acetate butyrate	5.6	4.7	1.2	85	2,700	410	—	0.030	0.035	C14
Tetrafluoroethylene polymer (micronized)	0.1‖	0.1‖	—	§§	—	670	570†	§§	‖ ‖	—
Monochlorotrifluoroethylene polymer	0.1‖	0.1‖	—	§§	—	600	720†	§§	‖ ‖	—
Nylon (polyhexamethylene adipamide) polymer	>10	6.7	1.8	95	4,000	500	430	0.020	0.030	C13
Polycarbonate	8.6	4.5	1.9	96	4,700	710	—	0.025	0.025	C15
Melamine formaldehyde, unfilled laminating type, no plasticizer	<0.1	0.1	0.2	81	800	810	—	0.320	0.085	C17
Urea formaldehyde molding compound, Grade II, fine	1.0	0.6	1.7	89	3,600	460	—	0.080	0.085	C17
Epoxy-bisphenol A mixture	1.9	3.8	0.5	85	2,200	510	—	0.035	0.030	—
Phenol furfural	>10	15.2	3.9	88	8,500	530	—	0.010	0.025	C14
Phenol formaldehyde molding compound, wood flour filler	>10	8.9	4.7	94	9,500	500	—	0.015	0.030	C14
Styrene modified polyester-glass fiber mixture (65–35)	5.2	2.0	2.6	91	6,000	440	360	0.050	0.045	—
Polyurethane foam (toluene di-isocyanate-polyhydroxy with	>10	6.6	1.5	87	3,700	510	440	0.020	0.030	—

TABLE 5-9A. Explosion Characteristics of Various Dusts (continued)

fluorocarbon blowing agent), not fire retardant										
Polyurethane foam (toluene di-isocyanate-polyhydroxy with fluorocarbon blowing agent), fire retardant	>10	9.8	1.7	96	3,700	550	390	0.015	0.025	—
Special Resins and Molding Compounds										
Rubber, crude, hard	7.4	4.6	1.6	80	3,800	350	—	0.050	0.025	C15
Rubber, synthetic, hard, contains 33% sulfur	>10	7.0	1.5	93	3,100	320	—	0.030	0.030	C15

* 1 psi = 6.895 kPa.
† °F = 9/5 (°C + 32).
‡ 0.1 oz/cu ft = 100 g/m³.
§ Numbers in this column indicate percentage while the letter prefix indicates the diluent gas. For example, the entry "C13" means dilution to an oxygen content of 13 percent with carbon dioxide as the diluent gas. The letter prefixes are: C = Carbon Dioxide; N = Nitrogen; A = Argon; and H = Helium.
‖ 0.1 designates materials presenting primarily a fire hazard as ignition of the dust cloud is not obtained by the spark or flame source but only by the intense heated surface source.
Guncotton ignition source.
** No ignition.
†† Obtained in an oxygen atmosphere.
‡‡ Reacts with carbon dioxide.
§§ No ignition to 8.32 joules, the highest tried.
‖‖ No ignition to 2 oz per cu ft, the highest tried.
Ignition denoted by flame, all others not so marked (##) denoted by a glow.

would normally be found or could be tolerated as foreign material. The addition of inert material reduces the rate of pressure rise and increases the minimal dust concentration. Rock dusting of coal mines is a practical application of the use of inert dust to prevent explosion of a combustible dust. For general rock dusting of coal mine entries, enough rock dust is usually added to provide an inert dust concentration of at least 65 percent of the total dust (Palmer 1973).

Inert gas is effective in preventing dust explosions because it dilutes the oxygen to a concentration too low to support combustion. In selecting a suitable inert gas, care must be taken to choose one that is not reactive with the dust. Certain metallic dusts, for example, react with carbon dioxide or nitrogen (Jacobson et al 1964). Helium and argon are suitable diluents in such instances.

Oxygen Concentration, Turbulence, and Effect of Flammable Gas

Variations in oxygen concentrations affect the ease of ignition of dust clouds and the explosion pressures. With a decrease in the partial pressure of oxygen, the energy required for ignition increases, ignition temperature increases, and maximum explosion pressures decrease. The type of inert gas used as the diluent for reduction of the oxygen concentration also has an effect apparently related to molar heat capacity.

The combustion of dust takes place at the surfaces of the dust particles. The rate of reaction therefore depends on intimate mixing of the dust and oxygen. It is for this reason that turbulent mixing of dust and air results in more violent explosions than those obtained by ignition in relatively quiescent mixtures (FIA 1966). The data presented in Table 5-9A were gathered by igniting the dust clouds under violently turbulent conditions; thus they cannot be compared to explosion data for flammable vapors which were gathered under comparatively quiescent conditions.

Addition of a small amount of flammable gas to a dust cloud and igniting the resultant aerosol greatly increases the violence of the explosion, particularly at lower dust concentrations. The resultant rates of pressure rise are very much higher than usually anticipated. Without the dust, the remaining fraction of the total combustible in air, represented by the flammable vapor, would be in itself below the LFL. In certain drying operations, involving the evaporation of a flammable vapor from a combustible powder with entrainment of the combustible dust along with the flammable vapor, explosions occurred that were far more violent than anticipated when only the fraction represented by the flammable vapor mixture alone was considered (Cotton 1951). Indeed, explosions occurred in such flammable vapor combustible dust-air mixtures when the flammable vapor-air portion was below the LFL. When encountered, these situations need special safeguards, such as inert gas dilution, explosion suppression, very large explosion vents, and careful static electricity elimination designs.

DUST CLOUD IGNITION SOURCES

Dust clouds have been ignited by open flames, lights, smoking materials, electric arcs, hot filaments of light bulbs, friction sparks, high pressure steam pipes and other hot surfaces, static sparks, spontaneous heating, welding and cutting torches and sparks from these operations, and other common sources of heat for ignition. The dust cloud ignition temperatures given in Table 5-9A for the most part fall between 572 and 1,112°F (300 and 600°C), and a large majority of the reported minimum spark ignition energies were between 10 and 40 millijoules. This can be contrasted to flammable vapor ignition energies with a range for the most part between 0.2 and 10 millijoules. As a general rule, combustible dusts require 20 to 50 times the ignition energy of flammable vapors.

Because ignition temperatures and ignition energies required for dust explosions are much lower than the temperatures and energies of most common sources of ignition, it is not surprising that dust explosions have been caused by all common sources of ignition. For this reason, the elimination of all possible sources of ignition is a basic principle of dust explosion prevention. Ignition sources in dust handling operations are identified and recommendations for their elimination are described in various NFPA Standards for the prevention of dust explosions.

FACTORS INFLUENCING THE DESTRUCTIVENESS OF DUST EXPLOSIONS

While the destructiveness of a dust explosion depends primarily upon the rate of pressure rise, other contributing factors are the maximum pressure developed, the duration of the excess pressure, the degree of confinement of the explosion volume, and the oxygen concentration.

Effect of Rate of Pressure Rise

The rate of pressure rise may be defined as the ratio of the increase in explosion pressure to the time interval of that increase. It is the most important single factor in evaluating the hazard of a dust and principally determines the degree of destructiveness of a deflagration.

The rate of pressure rise is also an important consideration in the design of explosion vents since it largely determines the size of the vent. In many cases, an extremely rapid rate of pressure rise indicates that an explosion vent design is impractical. In referring to Table 5-9A, the empirical term "Explosion Severity" is of a practical value in evaluating this problem. When explosion severities of between 2 and 4 are encountered, very large vent ratios are necessary as well as close attention to building strength and equipment design. Above an explosion severity of 4, most circumstances would preclude an explosion venting design and demand the use of such protective devices as inert gas or explosion suppression systems (FIA 1966).

Effect of Maximum Explosion Pressure

The maximum explosion pressures reported in Table 5-9A are for the most part in excess of 50 psi (345 kPa) and in some cases, even exceed 100 psi (690 kPa). These figures are valid only for test conditions and are subject to changes by variation in particle size, concentration, and other variables. The data in Table 5-9A give some indication of the magnitude of the maximum pressures which can be expected. Considering that an ordinary wall 12 in. (0.3 m) thick can be destroyed by less than 1 psi (6.89 kPa) pressure, it is evident that it is not practical to construct a building strong enough to resist the maximum pressures resulting from a dust deflagration.

One reason why the degree of destruction is not greater in many dust explosions is that the dust is not uniformly dispersed throughout the explosion volume. A dust cloud is rarely ignited under optimum concentration conditions with respect to development of maximum explosion pressures.

Effect of Duration of Excess Pressure

Closely associated with maximum pressure and rate of pressure rise as indications of the destructiveness of dust explosions is the length of time the excess pressure is exerted on the surroundings. The area under the time-pressure curve determines the total impulse exerted. It is this total impulse rather than the force exerted at any one moment that will determine the amount of destruction. The relationship between destructiveness and total impulse partly explains why dust explosions, which generally have slower average rates of pressure rise than gas explosions, may be more destructive than gas explosions.

Effect of Confinement

When a dust explosion occurs, gaseous products are formed and heat is released which raises the temperature of the air in the enclosure. Since gases expand when heated, destructive pressures will be exerted on the surrounding enclosure unless enough vent area is provided to release the hot gases before dangerous pressures are reached.

The Bureau of Mines (Hartman et al 1950) and others (FIA 1966) have examined the effect in some detail. Data from these investigations emphasize the importance of proper vents as a method of reducing damage from explosion pressures. In some instances it may be impractical to provide sufficient vent area to reduce pressures to a safe level. In these situations, the dust producing operation can be conducted in the open, under an inert atmosphere, or protected by an explosion suppression system. The explosion suppression system consists of a flame or pressure detector and a flame quenching agent which is expelled rapidly during the incipient stage of an explosion.

Effect of Inerting on Explosion Pressures and Rates of Pressure Rise

Data published by the Bureau of Mines (Nagy et al 1964) show that reduction in the oxygen concentration in the atmosphere, and mixture of inert powder (fuller's earth) or moisture with the combustible dust, reduces the maximum explosion pressure and rates of pressure rise. These data are shown in Figures 5-9B and 5-9C for explosions of 0.50 oz pcf (500 g/m^3) of cornstarch. A slight reduction in the atmospheric oxygen concentration or slight additions of inert powder or moisture have little effect on the explosion pressure. The effect of the inerting on the dust explosion becomes marked only when limiting values are approached. The maximum rate of pressure rise developed by the explosions appears to decrease almost linearly with use of inerting.

DUST EXPLOSION TEST APPARATUS AND PROCEDURES

There are a number of different types of dust explosion test chambers, and even more procedures for using them, each of which can profoundly affect the test results for any given dust sample. For example, the result of the dust explosion test in a closed chamber is affected by the geometry of the chamber, i.e., interior dimensions (cubical, spherical, cylindrical, etc.), and by the method of creating the dust cloud in the chamber (the method affects the turbulence of the cloud and uniformity of particle dispersion). This situation causes considerable difficulty in attempting to compare

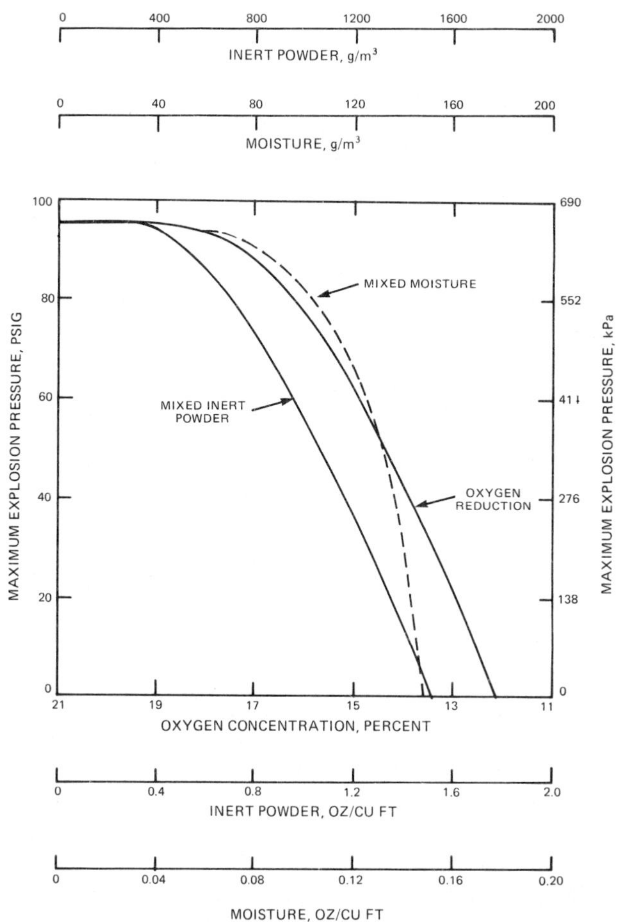

FIG. 5-9B. Effect of inerting on maximum pressure developed by explosion of 0.5 oz per cu ft (500 g/m³) of cornstarch.

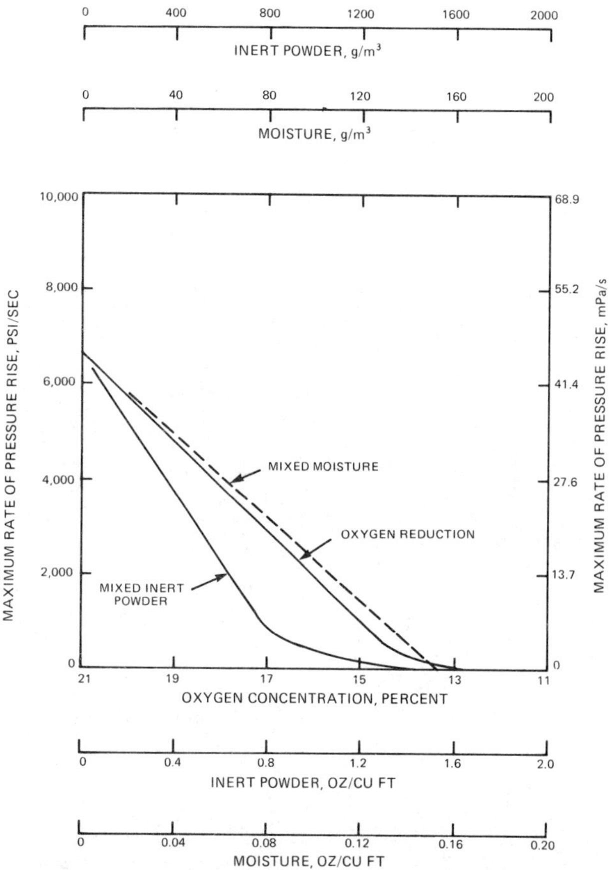

FIG. 5-9C. Effect of inerting on maximum rate of pressure rise developed by explosion of 0.5 oz per cu ft (500 g/m³) of cornstarch.

results obtained by various methods. In fact, unless several test results on the same dust sample are made in different chambers, and a consequent correlation is developed between the test methods, it can be said without reservation that any comparison of data obtained by different tests methods is at the very best only qualitative.

Lack of uniformity in test procedures also rules out any attempt to compare the pressures and rates of pressure rise resulting from tests on flammable vapors and dusts. By far, the largest amount of dust explosion test data has been obtained from tests conducted by the Bureau of Mines. In the field of vapor-gas explosions, there is nothing comparable. The Bureau of Mines test equipment and procedures are described in RI 5624 (Dorsett et al 1960). The test equipment used by the Bureau of Mines is a cylindrical vessel of about 75 cu in. (1200 cm³ or 1.2 liters), standing about 13 in. (0.33 m) high, referred to as the Hartmann Apparatus. (See Fig. 5-9D.) All of the data recorded in Table 5-9A has been collected in this type of test apparatus. Recent work, particularly in regard to generating dust explosion test data for use in designing explosion vents, has shown that it is more desirable to do the test work in special 20 liter (about 5 gal) spherical bombs (Bartnecht 1977; Donat 1977). (See Fig. 5-9E.) Data collected in these containers tends to give more consistent results and follow the "cubic law" described before more accurately.

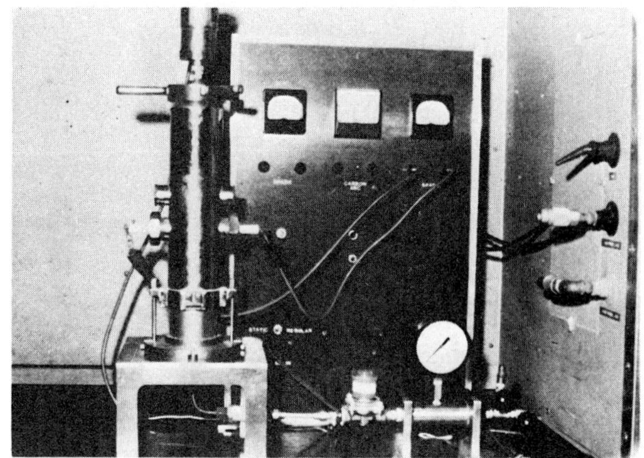

FIG. 5-9D The Hartmann apparatus. (U.S. Bureau of Mines)

Interpretation and Application of Dust Explosion Test Data

Dust explosion test data is gathered in a variety of chambers from a highly turbulent dust cloud to a practically stagnant cloud. This hinders the use of existing test information in the design of process equipment. However, for a constant velocity of propagation or flame speed in a

FIG. 5-9E The Bartknecht apparatus. (W. Bartknecht)

given dust sample suspended in air (or a flammable vapor-air mixture), the following relationship (which is referred to as the cubic law) holds:

$$\frac{R_l}{R_s} = \left(\frac{V_s}{V_l}\right)^{\frac{1}{3}} = \frac{A_l}{A_s}$$

where:

R_l = Rate of pressure rise in the large chamber, psi per sec,

R_s = Rate of pressure rise in the small chamber, psi per sec,

V_l = Volume of large chamber, cu ft,

V_s = Volume of small chamber, cu ft,

A_l = Vent ratio of large chamber, sq ft per cu ft of volume, and

A_s = Vent ratio of small chamber, sq ft per cu ft of volume.

This law indicates that test data gathered in small chambers can be extrapolated to larger chambers with some degree of confidence, though care must be taken to keep the shape of small chambers and large chambers within reasonable limits. If the test is done in small spheres, the results cannot be expected to hold up in the design of explosion relief for a long narrow vessel.

It should be pointed out that the Bureau of Mines test data were gathered under highly turbulent conditions in small chambers, with a high length/diameter ratio and with pressures and rates of pressure rise which cannot be obtained in large buildings and equipment. It has been shown that the cubic law relationship holds up very well to volumes of 100 cubic meters (130 cu yd).

Ignition Temperature

One piece of apparatus developed at the Bureau of Mines consists essentially of an electrically heated vertical cylindrical tube at the top of which is a small container holding a weighed amount of dust. The dust is projected down through the tube as a uniform cloud by a controlled blast of compressed air directed at the dust sample. The ignition temperature is considered to be the lowest temperature at which flame issues from the open bottom end of the tube. Differences in results are due to differences in the size and shape of test containers.

Ignition in Inert Atmospheres

For the determination of the ignitibility of dust in various inert gas-air mixtures, the apparatus consists of a vertical cylindrical tube containing tungsten electrodes to produce a spark for ignition. The tube is designed so that it can be flushed with the inert gas-air mixture prior to the test. The dust sample in a container near the top of the tube is projected downward as a cloud by a controlled blast of the inert gas-air mixture. Flame issuing from the open bottom of the tube signifies ignition.

Minimum Energies for Ignition

Apparatus developed at the Bureau of Mines used to determine minimum energies needed to cause ignition of dust clouds consists of a vertical cylindrical lucite tube. In this test, the dust is dispersed upward in the tube by controlled discharge of compressed air into the dust sample at the bottom of the tube. A spark discharged by an electrical condenser is used as the ignition source and by using condensers with different capacities, spark energy can be varied. Discharge of the condenser is synchronized with the formation of the dust cloud. Visible flame in the tube signifies ignition.

The data presented in Table 5-9A for cloud ignition energies has been gathered in this fashion. Recent work (Eckhoff 1976; Mertzberg 1979) has shown that if the spark is generated in a different fashion with a high resistance in series with the capacitance discharge, it is possible to obtain ignition of the dust cloud considerably below that reported by the Bureau of Mines technique. The results of Eckhoff's work and that of others in the field indicate that ignition is possible at energy levels at several orders of magnitude less than those shown in the table.

Explosion Pressures

Apparatus described by the Bureau of Mines for determining maximum explosion pressures and rates of pressure rise consists of a vertical cylindrical steel bomb into which a weighed sample of dust can be dispersed upward by a jet of compressed air. Pressures are measured by a strain gage transducer mounted in the top of the bomb. A pressure time record is obtained on an oscillograph using a light beam galvanometer system.

Bibliography

References Cited

Bartknecht, W. 1977. "Explosion Pressure Relief." *1977 A.I.ChE. Loss Prevention Symposium*. Vol 11. pp 93-105.

Best, R. 1978. "Two Grain Elevator Explosions Kill Five in Missouri." *Fire Journal*. Vol 72. pp 50-54.

Cotton, P. E. 1951. "New Test Apparatus for Dust Explosions." *NFPA Quarterly*. Vol 45, No 2. pp 157-164.

Donat, C. 1977. "Pressure Relief as Used in Explosion Protection." *1977 A.I.Ch.E. Loss Prevention Symposium*. Vol 11. pp 87-92.

Dorsett, H. G. et al. 1960. "Laboratory Equipment and Test Procedures for Evaluating Explosibility of Dusts." RI 5624. USDI, Bureau of Mines. Pittsburgh, PA.

Eckhoff, R. K. 1976. "A Study of Selected Problems Related to the Assessment of Ignitability and Explosibility of Dust Clouds." *CHR. Michelsens Institute for Vendenskapog Andsfrihet*. Beretninger. XXXVIII, 2. Bergen, Norway.

FIA. 1966. *Dust Explosions Analysis and Control*. Factory Insurance Association (now Industrial Risk Insurers). Hartford, CT (See also: Schwab, R. F. and Othmer, D. F. "Dust Explosions." *Chemical and Process Engineering*. Apr 1964, pp 165-174.)

Hartmann, I. et al. 1950. "Recent Studies on the Explosibility of Corn Starch." RI 4725. USDI, Bureau of Mines. Pittsburgh, PA.

Jacobson, M. et al. 1964. "Explosibility of Metal Powders." RI 6516. USDI, Bureau of Mines. Pittsburgh, PA.

Jacobson, M. et al. 1961. "Explosibility of Agricultural Dusts." RI 5753. USDI, Bureau of Mines. Pittsburgh, PA.

Kunkel, W. B. 1950. "The State of Electrification of Dust Particles on Dispersion into a Cloud." *Journal of Applied Physics*. Vol 21. pp 820-832.

Lathrop, J. K. 1978. "54 Killed in Two Grain Elevator Explosions." *Fire Journal*. Vol 72. No 1. pp 29-35.

Mertzberg, M. et al. 1979. "The Flammability of Coal Dust—Air Mixtures, Lean Limits, Flame Temperatures, Ignition Energies and Particle Size Effects." RI 8360. USDI, Bureau of Mines. Pittsburgh, PA.

Nagy, J. et al. 1964. "Pressure Development in Laboratory Dust Explosions." RI 6561. USDI, Bureau of Mines. Pittsburgh, PA.

Palmer, K. N. 1973. *Dust Explosions and Fires*. Chapman & Hall Ltd. London, England.

NFPA Codes, Standards, Recommended Practices and Manuals. (See the latest *NFPA Codes and Standards Catalog* for availability of current editions of the following documents.)

NFPA 48, *Standard for the Storage, Handling and Processing of Magnesium.*

NFPA 49, *Hazardous Chemicals Data.*

NFPA 61A, *Standard for Prevention of Fire and Dust Explosions in Facilities Manufacturing and Handling Starch.*

NFPA 61B, *Standard for the Prevention of Fires and Explosions in Grain Elevators and Facilities Handling Bulk Raw Agricultural Commodities.*

NFPA 61C, *Standard for the Prevention of Fire and Dust Explosions in Feed Mills.*

NFPA 65, *Standard for the Processing and Finishing of Aluminum.*

NFPA 68, *Guide for Explosion Venting.*

NFPA 69, *Standard on Explosion Prevention Systems.*

NFPA 481, *Standard for the Production, Processing, Handling and Storage of Titanium.*

NFPA 482, *Standard for the Production, Processing, Handling and Storage of Zirconium.*

NFPA 651, *Standard for the Manufacture of Aluminum and Magnesium Powder.*

NFPA 654, *Standard for the Prevention of Fire and Dust Explosions in the Chemical, Dye, Pharmaceutical, and Plastics Industry.*

NFPA 655, *Standard for the Prevention of Sulfur Fires and Explosions.*

NFPA 664, *Code for the Prevention of Fires and Explosions in Wood Processing and Woodworking Facilities.*

Additional Readings

"American Standard Practice for Rock-Dusting Underground Bituminous Coal and Lignite Mines to Prevent Coal Dust Explosions," (ASA Standard M13.1, 1960), IC 8001, 1960, USDI, Bureau of Mines, Pittsburgh, PA.

Atallah, S., "Fumigants and Grain Dust Explosions," *Fire Technology*, Vol. 15, No. 1, Feb. 1979, p. 5.

Bartknecht, W., *Explosionen: Ablauf & Schutzmassnahmen*, Springer Verlag: Berlin, Heidelberg, NY, 1979.

Baumeister, T., and Mark, L., *Mechanical Engineers Handbook*, 7th ed., McGraw-Hill, NY, 1967, pp. 7-38 to 7-45.

Boyle, A. R., and Llewellyn, F. J., "The Electrostatic Ignitability of Dust Clouds and Powders," *Journal of Applied Chemistry*, Vol. 69, June 1950, pp. 173-181.

Bradley, D., and Mitcheson, A., "The Venting of Gaseous Explosions in Spherical Vessels, I and II," *Combustion and Flame*, Vol. 32, 1978, pp. 221-255.

Brown, H. R., "Dust Explosion Hazards in Plants Producing or Handling Aluminum, Magnesium or Zinc Powder," IC 7148, 1941, USDI, Bureau of Mines, Washington, DC.

Brown, H. R., et al., "Fire and Explosion Hazards in Thermal Coal Dying Plants," RI 5198, 1956, USDI, Bureau of Mines, Pittsburgh, PA.

Brown, K. C., and Curzon, G. E., "Dust Explosions in Factories: Field Scale Tests on an Explosion Detector and Two Types of Quick Closing Valves," *Safety in Mines Research Report No. 194*, Dec. 1960, Safety in Mines Research Establishment, Sheffield, England.

Brown, K. C., and James, G. J., "Dust Explosions in Factories: A Review of the Literature," *Safety in Mines Research Report No. 201*, June 1962, Safety in Mines Research Establishment, Sheffield, England.

Bull, B., "Grain Elevator Explosions," *Firehouse*, April 1978, pp. 40-41.

Cardillo, P., and Anthony, E. J., *Dust Explosions & Fires: Guide to Literature*, (1957-1977), Stazione Sperimentale per i Combustibili, San Donato Milanese, Italy, 1979.

Cassel, H. M., "Some Fundamental Aspects of Dust Flames," RI 6551, 1964, USDI, Bureau of Mines, Pittsburgh, PA.

Coffee, R. D., "An Approach to Protection Design," *Fire Technology*, Vol. 4, No. 2, May 1968, pp. 81-87.

"Combustible Dusts," *Loss Prevention Data Sheet 7-76*, Factory Mutual Engineering Corp., Norwood, MA.

Conti, Ronald S., et al., *Thermal and Electrical Ignitability of Dust Clouds*, U.S. Bureau of Mines, Pittsburgh, PA, 1983.

Cross, Jean and Farrer, Donald, *Dust Explosions*, Plenum Press, NY, 1982.

Dawes, J. G., and Maguire, B. A., "Calculation of the Relationship Between Particle Number, Area and Weight Concentration in Coal Mine Dust Clouds," *Safety in Mines Research Report No. 150*, Dec. 1958, Safety in Mines Research Establishment, Sheffield, England.

Donat, C., "Selection and Dimensioning of Pressure Relief Devices for Dust Explosions," *Staub-Reinhaltung der Luft* (English transl.), Vol. 31, No. 4, April 1971, pp. 17-29.

Dufour, R. E., "A New Type of Bomb for Investigation of Pressures Developed by Dust Explosions," *Bulletin of Research No. 30*, March 1944, Underwriters Laboratories, Inc., Chicago, IL.

"Dust Explosions: An Often Unsuspected Potential for Disaster," *CUA Inspector*, Vol. 10, No. 3, 1973, pp. 1-3.

"Dust Explosions in Factories," Ministry of Labour New Series 22, 1963, Her Majesty's Stationery Office, London, England.

"Dust Explosion and Fires: A Manual, National Particle Board Association," Silver Spring, MD, 1977.

"Dust Hazards in the Starch and Dextrine Industries," (Papers given at a conference held September 19-20, 1961 at the Palace Hotel, Buxton England), The British Dextrine Manufacturers' Association, London, England.

Eggleston, L. A., and Pryor, A. J., "The Limits of Dust Explosibility," *Fire Technology*, Vol. 3, No. 2, May 1967, pp. 77-89.

Essenhigh, R. H., "Dust Explosions in Factories: Ignition Testing and Design of a New Inflammator," *Safety in Mine Research Report No. 188*, May 1960, Safety in Mines Research Establishment, Sheffield, England.

———, "Combustion Phenomena in Coal Dusts," *Colliery Engineering*, Dec. 1966, pp. 534-539; Jan. 1962, pp. 23-28; Feb. 1962, pp. 65-72; Mar. 1962, pp. 103-104.

———, "Dust Explosion Research: An Appraisal Conference on Dust Explosions," Pennsylvania State University, Sept. 1962. (See also Fire *Research Abstracts and Reviews*, Vol. 5, No. 1, 1963, p. 55; and Vol. 6, No. 1, 1964, p. 82.)

Essenhigh, R. H., and Brown, K. C., "Dust Explosions in Factories: A New Vertical-Tube Test Apparatus," *Safety in Mines Research Report No. 165*, April 1959, Safety in Mines Research Establishment, Sheffield, England.

Essenhigh, R. H., Froberg, R., and Howard, J. B., "Combustion Behavior of Small Particles," *Industrial & Engineering Chemistry*, Vol. 57, No. 9, Sept. 1965, pp. 33-43.

Essenhigh, R. H., Goldberg, P. M., and Shull, H. E., "Suppression of Dust Explosions in Coal Mines: Use of a Stirred Reactor to Test Mechanisms of Reaction of Coal Dust Flames," Bureau of Mines, OFR-104-78, Combustion Laboratory, Pennsylvania State University, 1978.

Essenhigh, R. H., and Woodhead, D. W., "Dust Explosions in Factories: Speed of Flame in Slowly Moving Clouds of Cork Dust," *Safety in Mines Research Report No. 166*, Sept. 1969, Safety in Mines Research Establishment, Sheffield, England.

Factory Mutual Engineering Division, "Dust Explosions," *Handbook of Industrial Loss Prevention*, 2nd ed., McGraw-Hill, NY, 1967, pp. 66-1-66-16.

Field, Peter, *Dust Explosions*, Vol. 4 of *Handbook of Powder Technology*, (eds. Williams, J. C. and Allen, T.), Elsevier, NY, 1982.

Frank, T. E., "Explosion Venting as a Means of Controlling Dust Explosions," paper presented at the Washington State University Symposium on Particleboard, Washington State University, Pullman, Washington.

Geyerstam, O. et al., "Dust Explosions at Kopingebro," *Socker Handingar*, Vol. 18, No. 3, 1963, pp. 29-83.

Hartmann, I., Jacobson, M., and Williams, R. P., "Laboratory Explosibility Study of American Coals," RI 5052, April 1954, USDI, Bureau of Mines, Pittsburgh, PA.

Hartmann, I., and Nagy, J., "Effect of Relief Vents on Reduction of Pressure Developed by Dust Explosions," RI 3924, May 1946, USDI, Bureau of Mines, Pittsburgh, PA.

Hartmann, I., Nagy, J., and Jacobson, M., "Explosive Characteristics of Titanium, Zirconium, Thorium, Uranium, and Their Hydrides," RI 4835, Dec. 1951, USDI, Bureau of Mines, Pittsburgh, PA.

Heinrich, H. J., and Kowall, Reinhard, "Results of Recent Pressure Relief Experiments in Connection with Dust Explosions," *StaubReinhaltung der Luft* (English transl.), Vol. 31, No. 4, April 1971, pp. 10-17.

Hertzberg, M., Cashdollar, K. L., and Opferman, J. J., *The Flammability of Coal Dust Air Mixtures: Lean Limits, Flame Temperatures, Ignition Energies and Particle Size Effects*, Bureau of Mines, U.S. Department of the Interior, Pittsburgh, PA, 1979.

Jacobson, M., Nagy, J., and Cooper, A. R., "Explosibility of Dusts Used in the Plastics Industry," RI 5971, 1962, USDI, Bureau of Mines, Pittsburgh, PA.

Kunkel, W. B., "Charge Distribution in Coarse Aerosols as a Function of Time," *Journal of Applied Physics*, Vol. 21, 1950, pp. 833-837.

Magison, E. C., *Electrical Instruments in Hazardous Locations*, 3rd ed., Instrument Society of America, Pittsburgh, PA, 1978, pp. 315-334.

Maisey, H. R., "Gaseous & Dust Explosion Venting," *Chemical and Process Engineering*, Part I, Oct. 1965, pp. 527-535; Part II, Dec. 1965, pp. 662-672.

"Metal Dust Ignites," *Fire Journal*, Vol. 90, No. 5, Sept. 1976, p. 47.

Mitchell, D. W., and Nagy, J., "Water as an Inert for Neutralizing the Coal Dust Explosion Hazard," IC 8111, 1962, USDI, Bureau of Mines, Pittsburgh, PA.

Nagy, J., Dorsett, H. G., and Cooper, A. R., "Explosibility of Carbonaceous Dusts," Rl 6597, 1965, USDI, Bureau of Mines, Pittsburgh, PA.

Nagy, J., Dorsett, H. G., and Jacobson, M., "Preventing Ignition of Dust Dispersions by Inerting," RI 6543, 1964, USDI, Bureau of Mines, Pittsburgh, PA.

Nagy, J., and Mitchell, D. W., "Experimental Coal-Dust and Gas Explosions," Rl 6344, 1963, USDI, Bureau of Mines, Pittsburgh, PA.

Nagy, J., Mitchell, D. W., and Kawenski, E. M., "Float Coal Hazard in Mines: A Progress Report," RI 6581, 1965, USDI, Bureau of Mines, Pittsburgh, PA.

Nagy, J., and Portman, W. M., "Explosibility of Coal Dust in an Atmosphere Containing a Low Percentage of Methane," RI 5815, 1961, USDI, Bureau of Mines, Pittsburgh, PA.

Nagy, J., and Surincik, D. J., "Thermal Phenomena During Ignition of a Heated Dust Dispersion," RI 6811, 1966, USDI, Bureau of Mines, Pittsburgh, PA.

Nagy, J., and Verakis, H. C., *Development and Control of Dust Explosions*, Marcel Dekker, New York, NY, 1983.

Nagy, J., Zerlinger, J. E., and Bartmann, I., "Pressure Relieving Capacities of Diaphragms and Other Devices for Venting Dust Explosions," RI 4636, Jan. 1959, USDI, Bureau of Mines, Pittsburgh, PA.

Palmer, K. N., *Dust Explosions and Fires*, Chapman and Hall, London, England, 1973.

Raftery, M. M., "Explosibility Tests for Industrial Dusts," *Fire Research Technical Paper No. 21*, 1968, Ministry of Technology and Fire Offices' Committee, Boreham Wood, Herts, Great Britain. (See also Raftery, M. M., *Staub-Reinhaltung der Luft* (English transl.), Vol. 31, No. 4, April 1971, pp. 1-10.)

Rasbash, D. J., and Rogowski, Z. W., "Relief of Explosions in Duct Systems," Symposium on Chemical Process Hazards with Special Reference to Plant Design, 1961, Institute of Chemical Engineers, pp. 58-68.

Rosenhan, A. K., "How Dust Causes Grain Elevator Explosions," *Firehouse*, April 1978.

Simmonds, W. A., and Cubbage, P. A., "The Design of Explosion Reliefs for Industrial Drying Ovens," Symposium on Chemical Process Hazards with Special Reference to Plant Design, Institute of Chemical Engineers, 1961, pp. 69-77.

Singer, J. M., "Ignition of Coal Dust-Methane-Air Mixture by Hot Turbulent Gas Jets," RI 6369, 1964, USDI, Bureau of Mines, Pittsburgh, PA.

Straumann, W., "Size of Pressure Relieving Explosion Vents in Chemical Plant Equipment," *Chemie-Ingenieur-Technik*, Vol. 3, 1965, pp. 306-316

Verkade, M., and Bluhm, D., "Supplement and Index to Grain-Dust Fire and Explosion Bibliography, November 1976—May 1978," DHEW, NIOSH Publication No. 78-183, 1978, U.S. Department of Health, Education and Welfare, Robert A. Taft Laboratories, Cincinnati, OH.

Verkade, M., and Chiotti, P., "A Bibliography of Topics Related to the Study of Grain-Dust Fire and Explosion," Iowa State University Energy and Mineral Resources Research Institute, Ames, IA, May 1978.

———, "Literature Survey of Dust Explosions in Grain Handling Facilities: Cause and Prevention," IS-EMRRI-2, 1976, Energy and Mineral Resources Research Institute, Iowa State University, Ames, IA.

METALS

Revised by Robert O'Laughlin

This chapter covers the fire hazard properties of combustible metals and storage, handling, and transportation methods to prevent their ignition and to protect against other hazardous properties. Also discussed are the major fire problems that arise while combustible metals are processed in machine shops and foundries.

Nearly all metals will burn in air under certain conditions. Some oxidize rapidly in the presence of air or moisture, generating sufficient heat to reach their ignition temperatures. Others oxidize so slowly that heat generated during oxidation is dissipated before the metals become hot enough to ignite. Certain metals, notably calcium, hafnium, lithium, magnesium, plutonium, potassium, sodium, thorium, titanium, uranium, zinc, and zirconium, are referred to as combustible metals because of the ease of ignition of thin sections, fine particles, or molten metal. However, the same metals in massive solid form are comparatively difficult to ignite.

More information on the hazards of metal dusts is contained in this Section, Chapter 9, "Dusts." Extinguishing agents suitable for use on combustible metal fires are discussed in Section 19, Chapter 5, "Combustible Metal Agents and Application Techniques."

Some metals, such as aluminum, iron, and steel, that are not normally thought of as combustible, may ignite and burn when in finely divided form. Clean, fine steel wool, for example, may be ignited. Particle size, shape, quantity, and alloy are important factors to be considered when evaluating metal combustibility. Dust clouds of most metals in air are explosive. Alloys, consisting of different metals or metallic compounds combined in varying proportions, may differ widely in combustibility from their constituent elements. Metals tend to be most reactive when in finely divided form, and some may require shipment and storage under inert gas or liquid to reduce fire risks.

Hot or burning metals may react violently upon contact with other materials, such as any of the extinguishants used on fires involving ordinary combustibles or flammable liquids. A few metals, such as thorium, uranium, and plutonium, emit, ionizing radiations (e.g., gamma, beta, and alpha rays) that can complicate fire fighting and introduce a contamination problem. The toxicity of certain metals is also an important factor in fire protection.

Temperatures produced by burning metals are generally much higher than temperatures generated by burning flammable liquids. Some hot metals can continue burning in carbon dioxide, nitrogen, or steam atmospheres in which ordinary combustibles or flammable liquids would be incapable of burning.

Properties of hot burning metal fires cover a wide range. Burning titanium produces little smoke, while burning lithium smoke is dense and profuse. Some water moistened metal powders, such as zirconium, burn with near explosive violence, while the same powder wet with oil burns quiescently. Sodium melts and flows while burning; calcium does not. Some metals, e.g., uranium, acquire an increased tendency to burn after prolonged exposure to moist air, while prolonged exposure to dry air may make it more difficult to ignite the metal.

Inasmuch as the extinguishment of fires in combustible metals involves techniques not commonly encountered in conventional fire fighting operations, it is good practice for those responsible for controlling combustible metal fires to gain experience in this area prior to the actual fire emergency. Fire fighters should practice extinguishing fires in those metals in an isolated outdoor location.

Where metals other than those described in this chapter are in use, it is most important that fire fighters gain some experience in extinguishing test fires involving the specific combustible metals.

MAGNESIUM

Properties

The ignition temperature of massive magnesium is very close to its melting point of 1,202°F (650°C). (See Table 5-10A.) However, ignition of magnesium in certain forms may occur at temperatures well below 1,200°F (650°C); magnesium ribbons and shavings can be ignited under certain conditions at about 950°F (510°C), and finely divided magnesium powder can ignite below 900°F (482°C).

Mr. O'Laughlin, P.E., is a Consulting Engineer with Professional Loss Control, Inc., in Oak Ridge, TN.

TABLE 5-10A. Melting, Boiling, and Ignition Temperatures of Pure Metals in Solid Form

	Temperature					
	Melting Point		Boiling Point		Solid Metal Ignition	
Pure Metal	°F	°C	°F	°C	°F	°C
Aluminum	1,220	660	4,445	2452	1832*‡	555*‡
Barium	1,337	725	2,084	1140	347*	175*
Calcium	1,548	842	2,625	1440	1,300	704
Hafnium	4,032	2223	9,750	5399	—	—
Iron	2,795	1535	5,432	3000	1,706*	930*
Lithium	367	186	2,437	1336	356	180
Magnesium	1,202	650	2,030	1110	1,153	623
Plutonium	1,184	640	6,000	3315	1,112	600
Potassium	144	62	1,400	760	156*†	69†*
Sodium	208	98	1,616	880	239‡	115‡
Strontium	1,425	774	2,102	1150	1,328*	720*
Thorium	3,353	1845	8,132	4500	932*	500*
Titanium	3,140	1727	5,900	3260	2,900	1593
Uranium	2,070	1132	6,900	3815	6900*§	3815*§
Zinc	786	419	1,665	907	1,652*	900*
Zirconium	3,326	1830	6,470	3577	2,552*	1400*

* Ignition in oxygen
† Spontaneous ignition in moist air
‡ Above indicated temperature
§ Below indicated temperature

Metal marketed under different trade names and commonly referred to as magnesium may be one of a large number of different alloys containing principally magnesium, but also significant percentages of aluminum, manganese, and zinc. Some of these alloys have ignition temperatures considerably lower than pure magnesium, and certain magnesium alloys will ignite at temperatures as low as 800°F (427°C). Flame temperatures can reach 2,500°F (1371°C), although flame height above burning metal is usually less than 12 in. (305 mm).

As is the case with all combustible metals, the ease of ignition of magnesium depends upon its size and shape. Thin, small pieces, such as ribbons, chips, and shavings, may be ignited by a match flame whereas castings and other large pieces are difficult to ignite even with a torch because of the high thermal conductivity of the metal. In order to ignite a large piece of magnesium, it is usually necessary to raise the entire piece to the ignition temperature.

Scrap magnesium chips or other fines may burn as the result of ignition of waste rags or other contaminants. Chips wet with water, water soluble oils, and oils containing more than 0.2 percent fatty acid may generate hydrogen gas. Chips wet with animal or vegetable oils may burn if the oils ignite spontaneously. Fines from grinding operations generate hydrogen when submerged in water, but they cannot be ignited in this condition. Grinding fines that are slightly wetted with water may generate sufficient heat to ignite spontaneously in air, burning violently as oxygen is extracted from the water with the release of hydrogen.

Storage and Handling (Including Transportation)

The more massive a piece of magnesium, the more difficult it is to ignite, but once ignited, magnesium burns intensely and is difficult to extinguish. The storage recommendations in NFPA 48, *Standard for the Storage, Handling and Processing of Magnesium* (hereinafter referred to as NFPA 48) take these properties into consideration. Recommended maximum quantities of various sizes and forms to be stored in specific locations are covered in this standard. The storage building preferably should be noncombustible, and the magnesium should be segregated from combustible material as a fire prevention measure.

With easily ignited lightweight castings, segregation from combustible materials is especially important. In the case of dry fines (fine magnesium scrap), storage in noncombustible covered containers in separate fire resistive storage buildings or rooms with explosion venting facilities is preferable. For combustible buildings or buildings containing combustible contents, NFPA 48 recommends automatic sprinkler protection to assure prompt control of a fire before the magnesium becomes involved.

Because of the possibility of hydrogen generation and of spontaneous heating of fines wet with coolants (other than neutral mineral oil), it is preferable to store wet scrap fines outdoors. Covered noncombustible containers should be vented.

Magnesium in powder, pellet, or ribbon form is shipped as a flammable solid and must be appropriately labeled according to U.S. Department of Transportation (DOT) requirements. Massive pieces or castings do not require special shipping safeguards.

If dry magnesium scrap in the form of borings, shavings, and turnings is to be shipped in interstate commerce in less than carload lots, the shipper may use closed metal drums. In carload or truckload lots, four-ply paper bags may also be used.

Less than carload lots of scrap in the form of clippings or sheets may be shipped in closed metal drums, wooden barrels, or wooden boxes. For bulk shipments, tight box cars, tightly closed steel covered gondola cars, and closed or completely covered truck bodies may be used.

Process Hazards

In machining operations involving magnesium alloys, sufficient frictional heat to ignite the chips or shavings may be created if the tools are dull or deformed. If cutting fluids are used (machining of magnesium is normally performed dry), they should be of the mineral oil type that have a high flash point. Water or water-oil emulsions are hazardous, since wet magnesium shavings and dust liberate hydrogen gas and burn more violently than dry material when ignited. Machines and the work area should be frequently cleaned and the waste magnesium kept in covered, clean, dry steel or other noncombustible drums which should be removed from the building at regular intervals. Magnesium dust clouds are explosive if an ignition source is present. Grinding equipment should be equipped with a water spray type dust precipitator. (See Fig. 5-10A.) A good grinder installation is designed to operate only if the exhaust blower and water spray are functioning properly. The equipment should be restricted to magnesium processing only.

Molten magnesium in the foundry presents a serious fire problem if not properly handled. Sulfur dioxide or melting fluxes are commonly used to prevent oxidation or ignition of magnesium during foundry operations. The action of sulfur dioxide is to exclude air from the surface of the molten magnesium; it is not an extinguishing agent. Fluxes perform both functions.

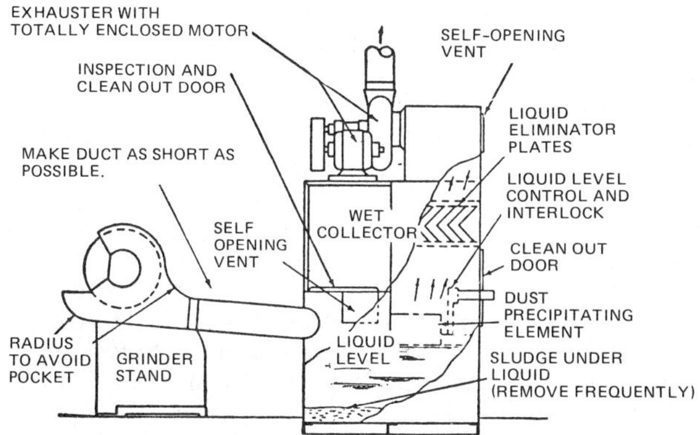

FIG. 5-10A. A schematic diagram of a water precipitation type collector for use in collecting dry combustible metal dust without creating explosive dust clouds or dangerous deposits in ducts or collection chambers. The diagram is intended only to show some of the features which should be incorporated in the design of a collector. It may be used for all combustible metal dusts.

Pots, crucibles, and ladles that may contact molten magnesium must be kept dry to prevent steam formation or a violent metal-water reaction. Containers should be checked regularly for any possibility of leakage or weak points. Steel lined runoff pits or pits with tightly fitting steel pans should be provided, and the pans must be kept free of iron scale. Leaking metal contacting hot iron scale results in a violent thermite reaction. Use of stainless steel pans or linings will eliminate this possibility.

Heat treating ovens or furnaces where magnesium alloy parts are subjected to high temperatures to modify their properties present another special problem. Temperatures for heat treating needed to secure the desired physical properties are often close to the ignition temperatures of the alloys themselves, and careful control of temperatures in all parts of the oven is essential. Hot spots leading to local overheating are a common cause of these fires. Large castings do not ignite readily, but fine fins or projections on the castings, as well as chips or dust, are more readily subject to ignition. For this reason, castings should be thoroughly cleaned before heat treating. Magnesium castings in contact with aluminum in a heat treating oven will ignite at a lower temperature than when they are placed on a steel car or tray.

Magnesium should not be heat treated in nitrate salt baths. Certain commonly used molten mixtures of nitrates and nitrites can react explosively with magnesium alloys, particularly at temperatures over 1,000°F (538°C).

Fighting Magnesium Fires

Magnesium and its alloys present special problems in fire protection. Magnesium combines so readily with oxygen that under some conditions water applied to extinguish magnesium fires may be decomposed into its constituent elements, oxygen and hydrogen. The oxygen combines with the magnesium and the released hydrogen adds to the intensity of the fire. None of the commonly available inert gases are suitable for extinguishing magnesium fires. The affinity of magnesium for oxygen is so great that it will burn in an atmosphere of carbon dioxide. Magnesium may also burn in an atmosphere of nitrogen to form magnesium nitride. For these reasons, any of the common extinguishing methods which depend on water, water solutions, or inert gas are not effective on magnesium chip fires. Halogen containing extinguishing agents (the Halons) react violently with burning magnesium, the chlorine or other halogen combining with the magnesium. However, inerting with noble gases, e.g. helium or argon, will extinguish burning metal.

The method of extinguishing magnesium fires depends largely upon the form of the material. Burning chips, shavings, and small parts must be smothered and cooled with a suitable dry extinguishing agent, e.g., graphite and dry sodium chloride. Where magnesium dust is present, care must be taken to prevent a dust cloud from forming in the air during application of the agent because this may result in a dust explosion.

Fires in solid magnesium can be fought without difficulty if attacked in their early stages. It often may be possible to remove surrounding material, leaving the small quantity of magnesium to burn itself out harmlessly. Considering the importance of prompt attack on magnesium fires, automatic sprinklers are desirable because they provide automatic notification and control of fire. While the water from the sprinklers may have the immediate effect of intensifying magnesium combustion, it will serve to protect the structure and prevent ignition of surrounding combustible material. An excess of water applied to fires in solid magnesium (avoiding puddles of molten metal) cools the metal below the ignition temperature after some initial intensification, and the fire goes out rapidly. By contrast, the fire may be intensified but not controlled with only a small, finely divided water spray.

Magnesium fires in heat treating ovens can best be controlled with powders and gases developed for use on such fires. By using melting fluxes to exclude air from the burning metal, fires in heat treating furnaces have been successfully extinguished. Boron trifluoride gas is an effective extinguishing agent for small fires in heat treating furnaces. Cylinders of boron trifluoride can be permanently connected to the oven or mounted on a suitable cart for use as portable equipment. Boron trifluoride is allowed to flow into the oven until the fire is extinguished, or, where large quantities of magnesium are well involved before discovery or where the furnace is not tight, the boron trifluoride will control the fire until extinguishment can be completed by the application of flux.

TITANIUM

Properties

Titanium, like magnesium, is classified as a combustible metal, but here again the size and shape of the metal determine to a great extent whether or not it will ignite. Castings and other massive pieces of titanium are not combustible under ordinary conditions. Small chips, fine turnings, and dust ignite readily and once ignited burn with the release of large quantities of heat. Tests have shown that very thin chips and fine turnings could be ignited by a match and heavier chips and turnings by a Bunsen burner. Coarse chips and turnings 1/32 by 3/28 in. (0.79 by 2.7 mm) or larger may be considered as difficult to ignite, but unless it is known that smaller particles are not mixed with the coarser material in significant amounts, it is wise to assume easy ignition is possible.

Finely divided titanium in the form of dust clouds or layers does not ignite spontaneously (differing in this respect from zirconium, plutonium, and certain other metals). Ignition temperatures of titanium dust clouds in air range from 630 to 1,090°F, (332 to 588°C) and of titanium dust layers from 720 to 950°F (382 to 510°C). Titanium dust can be ignited in atmospheres of carbon dioxide or nitrogen. Titanium surfaces that have been treated with nitric acid, particularly with red fuming nitric acid containing 10 to 20 percent nitrogen tetroxide, become pyrophoric and may be explosive.

The unusual conditions under which massive titanium shapes will ignite spontaneously include contact with liquid oxygen, in which case it will detonate on impact. It has been found that under static conditions spontaneous ignition will take place in pure oxygen at pressures of at least 350 psi (2413 kPa). If the oxygen was diluted, the required pressure increased, but in no instance did spontaneous heating occur in oxygen concentrations less than 35 percent. Another requirement for spontaneous heating was a fresh surface which oxidizes rapidly and exothermically in an oxygen atmosphere (CEN 1958).

Storage and Handling (Includes Transportation)

Titanium castings and ingots are so difficult to ignite and burn that special storage recommendations for large pieces are not included in NFPA 481, *Standard for the Production, Processing, Handling and Storage of Titanium*. Titanium sponge and scrap fines, on the other hand, do require special precautions, such as storage in covered metal containers and segregation of the containers from combustible materials. Because of the possibility of hydrogen generation in moist scrap and spontaneous heating of scrap wet with animal or vegetable oils, a yard storage area remote from buildings is recommended for scrap that is to be salvaged. Alternate recommended storage locations are detached scrap storage buildings and fire resistive storage rooms. Buildings and rooms for storage of scrap fines should have explosion vents.

There are no special shipping requirements for titanium except when in powder form. When possible, titanium powder is shipped wet (not less than 20 percent water) in tightly closed metal shipping containers, which can be packed in outside wooden boxes. The inside containers are limited to 10 lb (4.54 kg) net weight each, and the gross weight of the outside container cannot exceed 75 lb (34 kg). Cushioning of inside containers by noncombustible material, such as rock wool, is required. Single trip metal barrels or drums with cushioned inside metal drums are permitted by the DOT. Inside containers are flushed with inert gas before filling.

Process Hazards

Contact of molten metal with water is the principal hazard during titanium casting. To minimize this hazard, molds are usually thoroughly predried and vacuum or inert gas protection is provided to retain accidental spills.

The heat generated during machining, grinding, sawing, and drilling of titanium may be sufficient to ignite the small pieces formed by these operations or to ignite mineral oil base cutting lubricants. Consequently, water based coolants should be used in ample quantity to remove heat, and cutting tools should be kept sharp. Fines should be removed regularly from work areas and stored in covered metal containers. To prevent titanium dust explosions, any operation which produces dust should be equipped with a dust collecting system discharging into a water type dust collector. (See Fig. 5-10A.)

Descaling baths of mineral acids and molten alkali salts may cause violent reactions with titanium at abnormally high temperatures. Titanium sheets have ignited upon removal from descaling baths. This hazard can be controlled by careful regulation of bath temperatures.

There have been several very severe explosions in titanium melting furnaces. These utilize an electric arc to melt a consumable electrode inside a water cooled crucible maintained under a high vacuum. Stray arcing between the consumable electrode and crucible, resulting in penetration of the crucible, permits water to enter and react explosively with the molten titanium. Indications are that such explosions approach extreme velocities. The design and operation of these furnaces require special attention in order to prevent explosions and to minimize damage when explosions do occur.

Fighting Titanium Fires

Tests conducted by Industrial Risk Insurers (IRI) on titanium machinings in piles and in open drums showed that water in coarse spray was a safe and effective means of extinguishing fires in relatively small quantities of chips.

Carbon dioxide, foam, and dry chemical extinguishers are not effective on titanium fires, but good results have been obtained with extinguishing agents developed for use on magnesium fires.

The safest procedure to follow with a fire involving small quantities of titanium powder is to ring the fire with a special powder suitable for metal fires and to allow the fire to burn itself out. Care should be taken to prevent formation of a titanium dust cloud.

ALKALI METALS: SODIUM, POTASSIUM, NaK, AND LITHIUM

Properties

Sodium: At room temperature sodium oxidizes rapidly in moist air, but spontaneous ignitions have not been reported except when the sodium is in a finely divided form. When heated in dry air, sodium ignites in the vicinity of its boiling point (1,616°F or 880°C). Sodium in normal room air and at a temperature only slightly above its melting point (208°F or 98°C) has been ignited by placing sodium oxide particles on its surface. This indicates the possibility of ignition at temperatures below the boiling point. Once ignited, hot sodium burns vigorously and forms dense white clouds of caustic sodium oxide fumes. During combustion, sodium generates about the same amount of heat as an equivalent weight of wood.

The principal fire hazard associated with sodium is its rapid reaction with water. It floats on water (density 0.97), reacting vigorously and melting. The hydrogen liberated by this reaction may be ignited by the heat of the reaction. Sodium (like other burning, reactive metals) reacts violently with halogenated hydrocarbons, with halogens such as iodine, and with sulfuric acid.

Potassium: The fire hazard properties of potassium are very similar to those of sodium with the difference that potassium is usually more reactive. For example, the reaction between potassium and the halogens is more violent, and, in the case of bromine, a detonation can occur. There is an explosive reaction with sulfuric acid. Unlike sodium, potassium forms some peroxides during combustion. These peroxides may react violently with organic contaminants.

NaK (Sodium-Potassium Alloys): NaK is the term used when referring to any of several sodium-potassium alloys. The various NaK alloys differ from each other in melting point, but all are liquids or melt near room temperature. NaK alloys possess the same fire hazard properties as those of the component metals except that the reactions are more vigorous. Under pressure, NaK leaks have ignited spontaneously.

Lithium: Lithium, like sodium and potassium, is one of the so called alkali metals. Lithium undergoes many of the same reactions as sodium. For example, both sodium and lithium react with water to form hydrogen, but whereas the sodium-water reaction can generate sufficient heat to ignite the hydrogen, the far less violent lithium-water reaction does not. Lithium ignites and burns vigorously at a temperature of 356°F (180°C), which is near its melting point. Unlike sodium and potassium, it will burn in nitrogen. The caustic (oxide and nitride) fumes accompanying lithium combustion are more profuse and dense than those of other alkali metals burning under similar conditions. Lithium is the lightest of all metals. During combustion, it tends to melt and flow.

Storage and Handling (Includes Transportation)

Because of their reactivity with water, alkali metals require special precautions to prevent contact with moisture. Drums and cases preferably are stored in a dry, fire resistive room or building used exclusively for alkali metal storage. Since sprinkler protection would be undesirable, no combustible materials should be stored in the same area. It is good practice to store empty as well as filled alkali metal containers in the same area, and all containers should be on skids. There should be no water or steam pipes, but sufficient heat should be maintained to prevent moisture condensation due to atmospheric changes. Natural ventilation at a high spot in the room is desirable to vent any hydrogen that may be released by accidental contact of alkali metal with moisture.

Large quantities of alkali metal are often stored outdoors in aboveground tanks. In such installations weatherproof enclosures should cover tank manholes, and the free space within the tank should contain a nitrogen atmosphere. Argon or helium atmospheres should be substituted for nitrogen in the case of lithium.

Dry alkali metal is customarily shipped in watertight, steel barrels or drums. It is also shipped as a solid in tank car quantities. The DOT specifies the type of barrels, drums, and tank cars that are to be used in interstate transportation.

Transfer of an alkali metal to and from tank cars and storage tanks requires heating the metal to its melting point. This is accomplished by circulating hot oil through pipes within the tank. The alkali metal is transferred by a vacuum pump, and as the metal is removed from the tank it is replaced by nitrogen. An alkali metal level indicator should be provided and a low pressure warning device should be part of the nitrogen system.

An alkali metal covered with an organic solvent may be shipped in 1 qt (946 mL) metal cans, each can placed in another can and cushioned with soda ash. The cans are shipped in wooden boxes of 100 lb (45.4 kg) maximum gross weight.

Liquid alkali metal alloys may be shipped in pressure tested metal cans, cushioned with noncombustible material and packed in wooden boxes. Carload and truckload lots are permitted if special steel barrels and drums are used.

For small scale transfer of solid alkali metal from a storeroom to the use area, a metal container with a tight cover is recommended. Alkali metal should be removed from storage in as small quantities as practicable. When stored on work benches, it should be kept under kerosene or oil in a closed container. Alkali metal, with its great affinity for moisture, may react at the time it is sealed in a container with any atmospheric moisture. Due to the possible presence of hydrogen, containers should not be opened by hammering on the lid.

Process Hazards

Liquid alkali metal is valuable as a high temperature heat transfer medium. For example, it is used in hollow exhaust valve stems in some internal combustion engines and in the transfer of heat from one type of nuclear reactor to a steam generator. In the latter process or other large scale use of molten alkali metal, any equipment leak may result in a fire. Where molten alkali metal is used in process equipment, steel pans are located underneath to prevent contact and violent reaction of burning sodium with concrete floors.

Processing of alkali metal is essentially remelting it to form sticks or bricks or to add as a liquid to closed transfer systems. During this handling, contact with moist air, water, halogens, halogenated hydrocarbons, and sulfuric acid must be avoided.

Fighting Fires in Sodium, Lithium, NaK, and Potassium

The common extinguishing agents, such as water, foam, and vaporizing liquids, should never be used because of the violent reactions of alkali metals with them. Special dry powders (essentially graphite) developed for metal fires, dry sand, dry sodium chloride, and dry soda ash are effective. These finely divided materials blanket the fire while the metal cools to below its ignition temperature. Alkali metal burning in an apparatus can usually be extinguished by closing all openings. Blanketing with nitrogen is also effective. In the case of lithium, argon or helium atmospheres should be used.

ZIRCONIUM AND HAFNIUM

Properties

Zirconium: The combustibility of zirconium appears to increase as the average particle size decreases, but other variables, such as moisture content, also affect its ease of ignition. In massive form, zirconium can withstand extremely high temperatures without igniting, whereas clouds of dust in which the average particle size is 3 microns have ignited at room temperature. Dust clouds of larger particle size can be readily ignited if an ignition source is present, and such explosions can occur in atmospheres of carbon dioxide or nitrogen as well as in air. Zirconium dust will ignite in carbon dioxide at approximately 1,150°F (621°C) and nitrogen at approximately 1,450°F (788°C). Tests have also indicated that layers of 3 micron diameter dust are susceptible to spontaneous ignition. The depth of the layer and its moisture content are important variables for ignition. Spontaneous heating and ignition are also possibilities with scrap chips, borings, and turnings if fine dust is present. Layers of 6 micron diameter dust have ignited when heated to 374°F (190°C). Combustion of zirconium dust in air is stimulated by the presence of limited amounts (5 to 10 percent) of water. When very finely divided zirconium powder is completely immersed in water, it is difficult to ignite, but once ignited it burns more violently than in air.

Massive pieces of zirconium do not ignite spontaneously under ordinary conditions, but ignition will occur when an oxide free surface is exposed to sufficiently high oxygen concentrations and pressure. The explanation for this reaction is the same as that cited for a similar titanium reaction. Zirconium fires (like fires involving titanium and hafnium) attain very high temperatures, but generate very little smoke.

Explosions have occurred while zirconium was being dissolved in a mixture of sulfuric acid and potassium acid sulfate. Zirconium has exploded during and following pickling in nitric acid, and also during treatment with carbon tetrachloride or other halogen containing materials. Spontaneous explosions have occurred during handling of moist, very finely divided, contaminated zirconium scrap.

Hafnium: Hafnium has similar fire properties to zirconium. Hafnium burns with very little flame, but it releases large quantities of heat. Unless inactivated, hafnium in sponge form may ignite spontaneously.

Hafnium is generally considered to be somewhat more reactive than titanium or zirconium of similar form. Damp hafnium powder reacts with water to form hydrogen gas, but at ordinary temperatures this reaction is not sufficiently vigorous to cause the hydrogen to ignite. Under some conditions, however, ignition of the hydrogen may be expected to proceed explosively.

Storage and Handling (Including Transportation)

Special storage precautions are not required for zirconium castings because of the very high temperatures massive pieces of the metal can withstand without igniting. Zirconium powder, on the other hand, is highly combustible; consequently, it is customarily stored and shipped in 1 gal (3.78 L) containers with at least 25 percent water by volume. For specific details, refer to NFPA 482, *Standard for the Production, Processing, Handling, and Storage of Zirconium.*

Zirconium powder storerooms should be of fire resistive construction equipped with explosion vents. Cans should be separated from each other to minimize the possibility of a fire at one can involving others and to permit checking of the cans periodically for corrosion. One plant handling zirconium has established the procedure of disposing cans containing powder that have been on the shelf for six months.

Dry zirconium powder or sponge is customarily shipped inside metal cans in wooden or fiberboard boxes. The DOT does not require compliance with these shipping regulations if the particle size of the powder is greater than 20 mesh.

Wet zirconium powder and zirconium sludge may be shipped in glass or polyethylene containers within metal drums or wooden boxes.

Hafnium transportation regulations and storage recommendations are the same as for zirconium.

Process Hazards

In general, processing recommendations for zirconium and hafnium are the same. Handling of zirconium powder, whenever possible, should be under an inert liquid or in an inert atmosphere. If zirconium or hafnium powder is handled in air, extreme care must be used because the small static charges generated may cause ignition.

To prevent dangerous heating during machining operations, a large flow of mineral oil or water base coolant is required. In some machining operations, the cutting surface is completely immersed. Turnings should be collected frequently and stored under water in cans. Where zirconium dust is a byproduct, dust collecting equipment which discharges into a water precipitation type collector is a necessity.

Fighting Fires in Zirconium and Hafnium

Zirconium and hafnium fires can be fought in the same way. Fires exposing massive pieces of zirconium, for example, can be fought with water. Limited tests conducted by Industrial Risk Insurers have indicated that the discharge of water in spray form would have no adverse effect on burning zirconium turnings. When a sprinkler opened directly above an open drum of burning zirconium scrap, there was a brief flareup after which the fire continued to burn quietly in the drum. When a straight stream of water at a high rate of flow was discharged into the drum, water overflowed and the fire went out.

Where small quantities of zirconium powder or fines are burning, the fire can be ringed with a special extinguishing powder to prevent its spread, after which the fire can be allowed to burn out. Special extinguishing powders have been effective in extinguishing zirconium fires. When zirconium dust is present, the extinguishing agent should be applied so that a zirconium dust cloud will not form. If the fire is in an enclosed space, it can be smothered by introducing argon or helium.

CALCIUM AND ZINC

Properties

Calcium: The flammability of calcium depends considerably on the amount of moisture in the air. If ignited in moist air, it burns without flowing at a somewhat lower rate than sodium. It decomposes in water to yield calcium hydroxide and hydrogen, which may burn. Finely divided calcium will ignite spontaneously in air. It should be noted that barium and strontium are very similar to calcium in their fire properties.

Zinc: Zinc does not introduce a serious fire hazard in sheets, castings, or other massive forms because of the difficulty of ignition. Once ignited, however, large pieces burn vigorously. Moist zinc dust reacts slowly with the water to form hydrogen, and, if sufficient heat is released, ignition of the dust can occur. Zinc dust clouds in air ignite at 1,110°F (599°C). Burning zinc generates appreciable smoke.

Storage, Processing, and Fighting Fires in Calcium and Zinc

The storage, handling and processing recommended for magnesium is generally applicable to calcium and zinc.

METALS NOT NORMALLY COMBUSTIBLE

Aluminum

The usual forms of aluminum have a sufficiently high ignition temperature so that its burning is not a factor in most fires. However, very fine chips and shavings are occasionally subject to somewhat the same type of combustion as described for magnesium. Powdered or flaked aluminum, however, can be explosive under certain conditions. (See Table 5-10A.)

When aluminum is substituted for iron, steel, or copper, the lower melting temperature is an important factor if resistance to fire temperatures is essential. Figure 5-10B shows the decrease in strength of an aluminum alloy

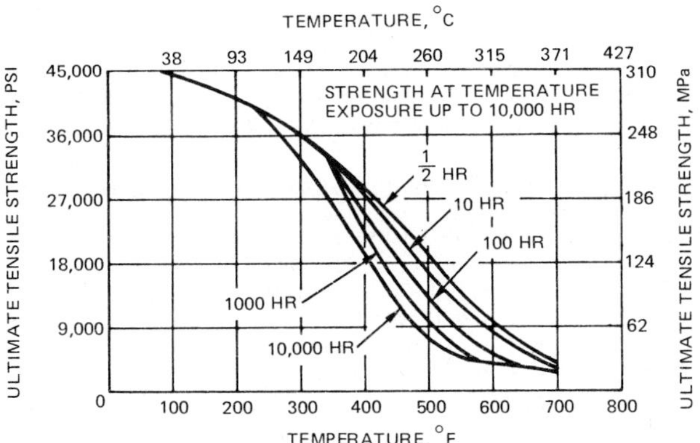

FIG. 5-10B. Effect of temperature on the strength of 6061-T6 wrought aluminum alloy.

(6061-T6) at elevated temperatures. Compare this with Figure 5-10C for steel.

The alloy 6061-T6 is typical of those used for architectural applications. Figure 5-10B also shows the strength of 6061-T6 alloy at various temperatures for five different heating times (DOD undated).

Aluminum roofing laid directly on structural supports without a roof deck may be expected to melt through and vent a fire beneath it, thus permitting fire fighters to enter the building with hose streams or to direct water on the fire

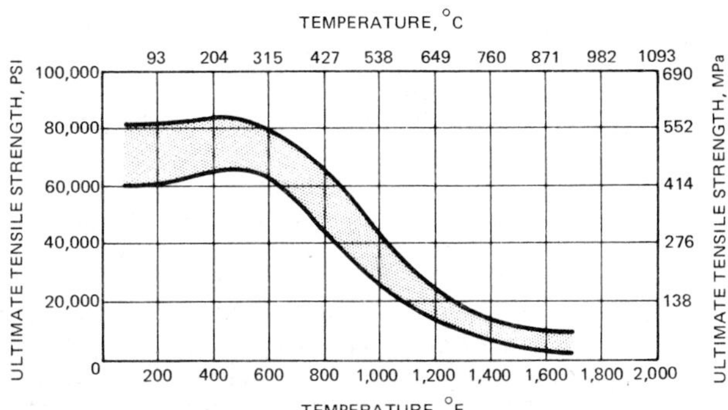

FIG. 5-10C. Strength of A7 and A36 structured steels at elevated temperature. The band indicates performance of a range of structural steels at elevated temperature. Actual performance will vary with the shape, size, and type of assembly exposed to fire conditions. (Data provided by the American Iron and Steel Institute.)

through the melted roof. For fire fighters working on such a roof, however, the hazard is greater than on a steel roof due to the strength loss under fire conditions.

Aluminum tubing is used in place of copper in many applications for flammable liquids and gases. While in many cases aluminum is suitable for tubing, it should be used with respect for its lower melting temperature. A number of aircraft fires have been charged to the failure of thin walled aluminum alloy tubing used in hydraulic lines containing flammable liquids. Under pressure, short circuits caused arcs to the tubing, melting the aluminum, and releasing the flammable liquid.

Iron and Steel

Iron and steel are not usually considered combustible; in a massive form (as in structural steel, cast iron parts, etc.), they do not burn in ordinary fires. Steel in the form of fine steel wool or dust may be ignited in the presence of heat from, for example, a torch, yielding a form of sparking rather than actual flaming in most instances. Fires have been reported in piles of steel turnings and other fine scrap which presumably contained some oil and were perhaps also contaminated by other materials that facilitated combustion. Spontaneous ignition of water wetted borings and turnings in closed areas, such as ship hulls, has also been reported. Pure iron has a specific gravity of 7.86 and a melting point of 2,795°F (1535°C). Ordinary structural steel has a melting point of 2,606°F (1430°C). Its strength when subjected to fire temperatures is shown in Figure 5-10C.

RADIOACTIVE METALS

For all practical purposes, radioactivity does not influence nor is it influenced by fire properties of a metal. Radioactive metals include those few that occur naturally, e.g., uranium and thorium, and those produced artificially, e.g., plutonium and cobalt-60. Any metal can be made radioactive. Since radioactivity cannot be altered by fire, radiation will continue wherever the radioactive metal may be spread during a fire. Smoke from fires involving radioactive materials frequently causes more property damage than the fire. The "damage," however, is not physical but results from radioactive contamination that must be cleaned up to avoid exposure.

Naturally occurring radioactive metals, such as uranium, consist of a mixture of atoms having slightly different mass. These are called isotopes. Since the radioactivity of each of these uranium isotopes is different, it follows that the radiation hazards from any given piece of uranium will change with its isotopic composition, which can be varied by special processing. Change of isotopic content of a piece of uranium alters the nature and magnitude of radiation hazards, but does not alter the chemical or fire properties of the metal. The same is equally true of thorium and plutonium. Prefire planning is particularly important where radioactive materials are involved.

In nuclear reactors, uranium, plutonium, and thorium may build up considerably higher levels of radioactivity. A few nuclear reactors utilize molten metallic sodium as a coolant so the sodium becomes radioactive. Fortunately, the requirements for fire protection are generally consistent with those independently necessitated for operational reasons.

Uranium

Normal uranium is a radioactive metal that is also combustible. Its radioactivity does not affect its combustibility, but can have a bearing on the amount of fire loss. Most metallic uranium is handled in massive forms that do not present a significant fire risk unless exposed to a severe and prolonged external fire. Once ignited, massive metal burns very slowly. A 1 in. (25.4 mm) diameter rod requires about 1 day to burn out after ignition. In the absence of strong drafts, uranium oxide smoke tends to deposit in the immediate area of the burning metal. Unless covered with oil, massive uranium burns with virtually no visible flame. Burning uranium reacts violently with carbon tetrachloride, 1,1,1 trichloroethane, and the Halons. For power reactor purposes, uranium fuel elements are always encased in a metal jacket (usually zirconium or stainless steel).

Uranium in finely divided form is readily ignited, and uranium scrap from machining operations is subject to spontaneous ignition. This reaction can usually be avoided by storage under dry oil. Grinding dust has been known to ignite even under water, and fires have occurred spontaneously in drums of coarser scrap after prolonged exposure to moist air. Larger pieces generally have to be heated entirely to their ignition temperature before igniting. Moist dust, turnings, and chips react slowly with water to form hydrogen. Uranium surfaces treated with concentrated nitric acid are subject to explosion or spontaneous ignition in air.

Thorium

Thorium, like uranium, is a naturally occurring element. Both are known as "source materials," the basic materials from which nuclear reactor fuels are produced. The powdered form of thorium requires special handling techniques because of its low ignition temperature. It is handled dry in a helium or argon atmosphere. The dry metal powder should not be in air because the friction of the particles falling through the air or against the edge of a glass container may produce electrostatic ignition of the powder.

Powdered thorium is usually compacted into solid pellets weighing about 1 oz (28 g) each. In this form it can be safely stored or converted into alloys with other metals. Improperly compacted thorium pellets have been known to slowly generate sufficient heat through absorption of oxygen and nitrogen from the air to raise a steel container to red heat.

Plutonium

Plutonium, like uranium, is radioactive as well as combustible. Plutonium is somewhat more susceptible to ignition than uranium and is normally handled by remote control means in an inert gas or "bone dry" air atmosphere. In finely divided form, such as dust or chips, plutonium is subject to spontaneous ignition in moist air.

Plutonium metal is never intentionally exposed to water, in part because of fire considerations. The massive metal ignites at about 1,112°F (600°C) and burns in a manner quite similar to uranium, except that plutonium oxide smoke damage, i.e., airborne radioactive contamination, is more difficult to control and also more hazardous. Because of certain nonfire hazard considerations, it is necessary to limit the quantity of plutonium kept at one location, thus limiting the maximum size of a fire in this metal. Plutonium, which ignites spontaneously, is normally allowed to burn under conditions limiting both fire and radiological contamination spread.

Bibliography

Referenced Cited

CEN. 1958. "Titanium Does a Fast Burn." *Chemical and Engineering News*. Vol 36, No 31. Aug 1958. pp 36-37.

DOD. undated. "Metallic Materials and Elements for Aerospace Vehicles." *MIL-HDBK-5*. U.S. Department of Defense. Washington, DC.

NFPA Codes, Standards, and Recommended Practices and Manuals. (See the latest *NFPA Codes and Standards Catalog* for availability of current editions of the following documents.)

NFPA 48, *Standard for the Storage, Handling, and Processing of Magnesium*.

NFPA 65, *Standard for the Processing and Finishing of Aluminum*.

NFPA 481, *Standard for the Production, Processing, Handling and Storage of Titanium*.

NFPA 482, *Standard for the Production, Processing, Handling, and Storage of Zirconium*.

NFPA 651, *Standard for the Manufacture of Aluminum and Magnesium Powder*.

Additional Readings

Adams, R. M., ed., *Boron, Metallo-Boron Compounds and Boranes*, Wiley-Interscience, NY, 1964.

Allison, W. W., "Zirconium, Zircaloy, and Hafnium—Safe Practice Guide for Shipping, Storing, Handling, Processing and Scrap Disposal," *AEC Research and Development Report WAPD-TM-17*, U.S. Atomic Energy Commission, Washington, DC, 1960.

Ambrose, P. M., et al., *Investigation of Accident Involving Titanium and Red Fuming Nitric Acid*, IC 7711, Dec. 1963, USDI, Bureau of Mines, Washington, DC.

Broadhurst, V. A., "Processing Titanium Safely," *Metal Industry*, Vol. 96, No. 26, June 24, 1960, pp. 515-517.

Cissel, D. W., et al., *Guidelines for Sodium Fire Prevention,*

Detection, and Control: Report of the Ad Hoc Committee on Sodium Fires, ANL-7691, Argonne National Laboratory, Argonne, IL, 1970.

Conway, J. B., and Grosse, A. V., *High Temperature Project*, Office of Naval Research, Contract N9-ONR-87301, Final Report, 1954.

Douglass, D. L., *The Metallurgy of Zirconium*, Unipub, Inc., NY, 1972.

Eisner, H. S., "Aluminum and the Gas Ignition Risk," *The Engineer* London, England, Vol. 223, No. 5795, Feb. 17, 1967, pp. 259-260.

"Explosibility of Metal Dusts," *Bulletin RI-6515*, U.S. Bureau of Mines, Washington, DC, 1964.

Extinguishment of Alkali Metal Fires, Technical Documents Report APL TDR 64-114, 1964, MSA Research Corporation, Callery, PA.

Factory Mutual Engineering Corporation, "Metals and Alloys," *Handbook of Industrial Loss Prevention*, 2nd ed., McGraw-Hill, NY, 1967, pp. 62-1, 62-6.

Fassell, W. M., Jr., et al., "Ignition Temperatures of Magnesium & Magnesium Alloys," *Journal of Metals*, AIME, July 1951, pp. 522-528.

"Fire Hazards in Aluminum Processing," *The Institution of Fire Engineers Quarterly*, Vol. 27, No. 65, Mar. 1967, pp. 94-97.

"Fire Precautions in Handling Titanium," *American Machinist*, Vol. 99, No. 7, Mar. 28, 1955, pp. 147-9.

"Fire Protection for Combustible Metals," *National Safety News*, Vol. 119, No. 6, June 1979, pp. 75-82.

General Recommendations on Design Features for Titanium and Zirconium Production-Melting Furnaces, DMIC Memorandum 116, July 1961, Defense Metals Information Center, Battelle Memorial Institute, Columbus, OH.

Grosse, A. V., and Conway, J. B., "Combustion of Metals in Oxygen," *Industrial Engineering Chemistry*, Vol. 50, 1958, pp. 663-672.

Guidelines for Handling Molton Aluminum, The Aluminum Association, Washington, DC, 1980.

Handbook of Chemistry & Physics, 60th ed., The Chemical Rubber Co., Cleveland, OH, 1979-1980, published annually.

Handbook on Titanium Metal, Titanium Metals Corp. of America, West Caldwell, NJ.

Harrison, P. L., and Yoffe, A. D., "The Burning of Metals," *Proceedings of the Royal Society*, Vol. 261, 1961, pp. 357-370.

Hartmann, I., Nagy, J. and Jacobson, M., "Explosive Characteristics of Titanium, Zirconium, Thorium, Uranium and Their Hydrides," *Bulletin RI-4835*, U.S. Bureau of Mines, Washington, DC, 1951.

Hill, H. E., "Prevention and Control of Titanium Machining Fires," *Metals Engineering Quarterly*, Aug. 1966, pp. 62, 63.

Jacobson, M., Cooper, A. R., and Nagy, J., *Explosibility of Metal Powders*, RI 6516, USDI, Bureau of Mines, Pittsburgh, PA, 1964.

Keil, A. A., *Radiation Control for Fire and Other Emergency Forces*, National Fire Protection Association, Quincy, MA, 1960.

King, P. W., and Magid, J., *Industrial Hazard and Safety Handbook*, Butterworths, Woburn, MA, 1979.

Lipsett, S. G., "Explosive Reactions Between Molten Materials in Water," *Fire Technology*, Vol. 2, No. 2, May 1966, pp. 118-126.

Long, G., "Explosions of Molten Aluminum in Water—Cause and Prevention," *Metal Progress*, Vol. 75, No. 5, May 1957, pp. 107-112.

Lustman, B., and Kerze, F., Jr., *The Metallurgy of Zirconium*, McGraw-Hill, NY, 1955.

Lyman, Taylor, ed., *Metals Handbook, Vol. 1*, The American Society for Metals, Metals Park, OH, 1961.

McCormick, J. R., and Schmitt, C. R., "Extinguishment of Selected Metal Fires Using Carbon Microspheroids," *Fire Technology*, Vol. 10, No. 3, August 1974, pp. 196-200.

Metals and Ores, Volume 3 of *Chemical Technology, An Encyclopedic Treatment*, Barnes and Noble Books, Imports, Totowa, NJ, 1970.

"Metallic Zirconium Explosions During Dissolution with Sulphuric Acid-Potassium Acid Sulphate," *Accident and Fire Prevention Issue No. 69*, Sept 1957, U.S. Atomic Energy Commission, Washington, DC.

Metals Section, NSC, "Magnesium," Data Sheet 426, National Safety Council, Chicago, IL, 1965.

Ibid.,"Titanium," Data Sheet 485, National Safety Council, Chicago, IL, 1959.

Markstein, G. H., "Combustion of Metals," *Journal of the American Institute of Aeronautics & Astronautics*, Vol. 1, 1963, pp. 550-562.

Mellor, J. W., *Comprehensive Treatise on Inorganic and Theoretical Chemistry*, Longman, Green & Co., London, England, 1946-7, and supplements, 1956-1970.

"Methods Used by an AEC Contractor for Handling Sodium, NaK, and Lithium," *Accident and Fire Prevention Information*, No. 101, Feb. 1960, U.S. Atomic Energy Commission, Washington, DC.

Milich, W., and King, E. C., *Underwater Disposal of Sodium with EB Nozzle*, Technical Report 40, Mine Safety Appliances Co., Pittsburgh, PA, 1955.

Miller, G. S., "Use of Aluminum Wire in Homes Spurs Fears of Fires, Leads Suits," *Wall Street Journal*, Vol. 191, No. 7, Jan. 11, 1978, pp. 1, 25.

Peer, L. H., and Reichling, J. T., "How to Use Pyrophoric Metals Safely," *Mill and Factory*, Vol. 65, No. 2, Aug. 1959, pp. 79-83.

Peloubet, J. A., "Machining Magnesium—A Study of Ignition Factors," *Fire Technology*, Vol. 1, No. 1, Feb. 1965, pp. 5-14.

Peterseim, F. D., *Hazards & Safety Precautions in the Fabrication and Use of Titanium*, TML Report No. 63, Battelle Memorial Institute, Columbus, OH, 1957.

Poster, A. R., *Handbook of Metal Powders*, Van Nostrand Reinhold, NY, 1966.

"Product Information Data Sheet Lithium Metal," *Data Sheet Metal* 110-1061, Lithium Corporation of America, Inc., NY.

Properties and Essential Information for Safe Handling and Use of Sodium, SD-47, Manufacturing Chemists' Association, Washington, DC, 1952.

Purington, R. G., and Patterson, H. W., *Handling Radiation Emergencies*, NFPA, Quincy, MA, 1977.

Republic Steel Corp., et al., "Titanium," *Data Sheet 485*, National Safety Council, Chicago, IL, 1959.

Reynolds, W. C., and Williams, J. J., "An Investigation of the Ignition Temperatures of Solid Metals," NACA Contract NAW-6459, 1957.

Riehl, W. A., et al., "Reactivity of Titanium with Oxygen," R-180, National Aeronautics and Space Administration, Washington, DC, 1963.

Riley, John F., "Na-X, A New Fire Extinguishing Agent for Metal Fires," *Fire Technology*, Vol. 10, No. 4, Nov. 1974, pp. 269-274.

Schmitt, C. R., "Pyrophoric Materials—A Literature Review," *Fire Prevention and Suppression*, ed by Hilado, Carlos J. (Vol. 10, Fire and Flammability Series), Technomic, Westport, CT, 1974, pp. 74-89.

Setting, M., *Sodium—Its Manufacture, Properties and Uses*, Van Nostrand Reinhold Co., NY, 1956.

Smith, R., "Pyrophoric Metals—A Technical Mystery," *NFPA Quarterly*, Vol. 51, No. 2, Oct. 1957.

Smithells, C. J., *Metals Reference Book*, Interscience Publishers, NY, 1962.

Stout, E. L., *Safety Consideration for Handling Plutonium, Uranium, Thorium, the Alkali Metals, Zirconium, Titanium, Magnesium, and Calcium*, LA 2147, Los Alamos Scientific Laboratory, Los Alamos, NM, 1957.

A Study of the Reactions of Metals and Water, AECD 3664, Apr. 1955, Washington, DC.

Thomas, D. E., and Hayes, E. T., *The Metallurgy of Hafnium*, U.S. Atomic Energy Commission, Washington, DC, 1960.

Van Horn, K. R., ed., *Aluminum*, American Society for Metals, Metals Park, OH, 1967.

Warner, J. C., Chipman, J. and Stedding, F., *Metallurgy of Uranium and Its Alloys*, NNES-IV-12A, National Technical Information Service, Springfield, VA.

White, E. L., and Ward, J. J., *Ignition of Metals in Oxygen*, DMIC Report No. 224, Battelle Memorial Institute, Columbus, OH, 1966.

"Zirconium Fire and Explosion Hazard Evaluation," *Accident and Fire Prevention Information*, Issue 45, Aug. 1956, U.S. Atomic Energy Commission, Washington, DC.

"Zirconium Powder," Data Sheet 382 (Revised). National Safety Council, Chicago, IL, 1962. "

"Zirconium Scrap Burning Tests," *NFPA Quarterly*, *Vol. 56, No. 2, Oct. 1960, pp. 110-127.*

"159,000 Lbs. of Zirconium Scrap Involved in Fire," *Serious Accidents Issue* No 84, U.S. Atomic Energy Commission, Washington, DC, 1955.

TABLES AND CHARTS

Revised by Dr. Vytenis Babrauskas

TABLE 5-11A. Heats of Combustion and Related Properties For Pure, Simple Substances*

Material	Composition	W Molecular Weight	Δh_c^u Gross (MJ/kg)	Δh_c^l Net (MJ/kg)	$\Delta h_c^l/r_o$ (MJ/kg O_2)	r_o Oxygen-fuel Mass ratio	T_b Boiling temp. (°C)	Δh_v Latent Heat of Vaporization (kJ/kg)	C_{pl} Liquid Heat Capacity (kJ/kg-°C)	C_{pv} Vapor Heat Capacity (kJ/kg-°C)
acetaldehyde	C_2H_4O	44.05	27.07	25.07	13.81	1.816	20.8	—	1.94	1.24
acetic acid	$C_2H_4O_2$	60.05	14.56	13.09	12.28	1.066	118.1	395		1.11
acetone	C_3H_6O	58.08	30.83	28.56	12.96	2.204	56.5	501	2.12	1.29
acetylene	C_2H_2	26.04	49.91	48.22	15.70	3.072	−84.0	—	—	1.69
acrolein	C_3H_4O	56.06	29.08	27.51	13.77	1.998	52.5	505	—	1.17
acrylonitrile	C_3H_3N	53.06	33.16	31.92	14.11	2.262	77.3	615	2.10	1.20
(allene) → propadiene										
ammonium perchlorate†	NH_4ClO_4	117.49	2.35	2.16	3.97	0.545	—	—	—	
iso-amyl alcohol	$C_5H_{12}O$	88.15	37.48	34.49	12.67	2.723	132.0	501	2.90	1.50
aniline	C_6H_7N	93.12	36.44	34.79	13.06	2.663	184.4	478	2.08	1.16
benzaldehyde	C_7H_6O	106.12	33.25	32.01	13.27	2.412	179.2	385	1.61	
benzene	C_6H_6	78.11	41.83	40.14	13.06	3.073	80.1	389	1.72	1.05
benzoic acid†	$C_7H_6O_2$	122.12	26.43	25.35	12.90	1.965	250.8	415	—	0.85
benzyl alcohol	C_7H_8O	108.13	34.56	32.93	13.09	2.515	205.7	467	2.00	1.19
bicyclohexyl	$C_{12}H_{22}$	166.30	45.35	42.44	12.61	3.367	236.	263		
1,2-butadiene	C_4H_6	54.09	47.95	45.51	13.99	3.254	10.8	—	—	1.48
1,3-butadiene	C_4H_6	54.09	46.99	44.55	13.69	3.254	−4.4	—	—	1.47
(1,3-butadiyne) → diacetylene										
n-butane	C_4H_{10}	58.12	49.50	45.72	12.77	3.579	−0.5	—	2.30	1.68
iso-butane	C_4H_{10}	58.12	48.95	45.17	12.62	3.579	−11.8	—	—	1.67
1-butene	C_4H_8	56.10	48.44	45.31	13.24	3.422	−6.2	—	—	1.53
n-butylamine	$C_4H_{11}N$	73.14	41.75	38.45	12.84	2.994	77.8	372	2.57	1.62
d-camphor†	$C_{10}H_{16}O$	152.23	38.75	36.44	12.84	2.838	203.4	—	—	0.82
carbon†	C	12.01	32.80	32.80	12.31	2.664	4200.	—	—	0.71
carbon disulfide	CS_2	76.13	6.34	6.34	5.03	1.261	46.5	351	1.00	0.60
carbon monoxide	CO	28.01	10.10	10.10	17.69	0.571	−191.3	—	—	1.04
cellulose†	$C_6H_{10}O_5$	162.14	17.47	16.12	13.61	1.184	—	—	1.16	—
(chloroethylene) → vinyl chloride										
(chloroform) → trichloromethane										
chlorotrifluoroethylene	C_2F_3Cl	116.47	2.00	2.00	3.64	0.549	−28.3	188	1.34	0.72
m-cresol	C_7H_8O	108.13	34.26	32.64	12.98	2.515	202.2	399	2.00	1.13
cumene	C_9H_{12}	120.19	43.40	41.20	12.90	3.195	152.3	312	1.77	1.26
cyanogen	C_2N_2	52.04	21.06	21.06	17.12	1.230	−21.2	—	—	1.12
cyclobutane	C_4H_8	56.10	48.91	45.77	13.38	3.422	12.9	—	—	1.29
cyclohexane	C_6H_{12}	84.16	46.58	43.45	12.70	3.422	80.7	357	1.84	1.26
cyclohexene	C_6H_{10}	82.14	45.67	42.99	12.99	3.311	82.8	371	1.80	1.28
cyclohexylamine	$C_6H_{13}N$	99.18	41.05	38.17	12.79	2.984	134.5			
cyclopentane	C_5H_{10}	70.13	46.93	43.80	12.80	3.422	49.3	389	2.23	1.18

Dr. Babrauskas is Head, Flammability and Toxicity Measurement, U.S. National Bureau of Standards, Center for Fire Research.

TABLE 5-11A. Heats of Combustion and Related Properties For Pure, Simple Substances*

Material	Composition	W Molecular Weight	Δh_c^u Gross (MJ/kg)	Δh_c^l Net (MJ/kg)	$\Delta h_c^l/r_o$ (MJ/kg 0_2)	r_o Oxygen-fuel Mass ratio	T_b Boiling temp. (°C)	Δh_v Latent Heat of Vaporization (kJ/kg)	C_{pl} Liquid Heat Capacity (kJ/kg-°C)	C_{pv} Vapor Heat Capacity (kJ/kg-°C)
cyclopropane	C_3H_6	42.08	49.70	46.57	13.61	3.422	−32.9	—	1.92	1.33
(decahydronaphthalene) → *cis*-decalin										
cis-decalin	$C_{10}H_{18}$	138.24	45.49	42.63	12.70	3.356	195.8	309	1.67	1.21
n-decane	$C_{10}H_{22}$	142.28	47.64	44.24	12.69	3.486	174.1	276	2.19	1.65
diacetylene	C_4H_2	50.06	46.60	45.72	15.89	2.877	10.3	—	—	1.47
(diamine) → hydrazine										
diborane	H_6B_2	27.69	79.80	79.80	23.02	3.467	−92.5	—	—	1.75
dichloromethane	CH_2Cl_2	84.94	6.54	6.02	10.65	0.565	39.7	330	1.18	0.60
diethyl cyclohexane	$C_{10}H_{20}$	140.26	46.30	43.17	12.58	3.422	174.		1.87	
diethyl ether	$C_4H_{10}O$	74.12	36.75	33.79	13.04	2.590	34.6	360	2.34	1.52
(2,4 diisocyanotoluene) → toluene diisocyanate										
(diisopropyl ether) → *iso*-propyl ether										
dimethylamine	C_2H_7N	45.08	38.66	35.25	13.24	2.662	6.9	—	—	1.60
(dimethyl aniline) → xylidene										
dimethyldecalin	$C_{12}H_{22}$	166.30	45.70	42.79	13.15	3.254	220.	260		
(dimethyl ether) → methyl ether										
1,1-dimethylhydrazine (UDMH)	$C_2H_8N_2$	60.10	32.95	30.03	14.10	2.130	25.	578	2.73	
dimethyl sulfoxide	C_2H_6SO	78.13	29.88	28.19	15.30	1.843	189.	677	1.89	1.14
1,3 dioxane	$C_4H_8O_2$	88.10	26.57	24.58	9.66	2.543	105.	404		
1,4 dioxane	$C_4H_8O_2$	88.10	26.83	24.84	9.77	2.543	101.1	406	1.74	1.07
ethane	C_2H_6	30.07	51.87	47.49	12.75	3.725	−88.6	—	—	1.75
ethanol	C_2H_6O	46.07	29.67	26.81	12.87	2.084	78.5	837	2.43	1.42
(ethene) → ethylene										
ethyl acetate	$C_4H_8O_2$	88.10	25.41	23.41	12.89	1.816	77.2	367	1.94	1.29
ethyl acrylate	$C_5H_8O_2$	100.12	27.44	25.69	13.39	1.918	100.	290		1.14
ethylamine	C_2H_7N	45.08	38.63	35.22	13.23	2.662	16.5	—	2.89	1.61
ethyl benzene	C_8H_{10}	106.16	43.00	40.93	12.93	3.165	136.1	339	1.75	1.21
ethylene	C_2H_4	28.05	50.30	47.17	13.78	3.422	−103.9	—	2.38	1.56
ethylene glycol	$C_2H_6O_2$	62.07	19.17	17.05	13.22	1.289	197.5	800	2.43	1.56
ethylene oxide	C_2H_4O	44.05	29.65	27.65	15.23	1.816	10.7	—	1.97	1.10
(ethylene trichloride) → trichloroethylene										
(ethyl ether) → diethyl ether										
formaldehyde	CH_2O	30.03	18.76	17.30	16.23	1.066	−19.3	—	—	1.18
formic acid	CH_2O_2	46.03	5.53	4.58	13.15	0.348	100.5	476	2.15	0.98
furan	C_4H_4O	68.07	30.61	29.32	13.86	2.115	31.4	398	1.69	0.96
α-D-glucose†	$C_6H_{12}O_6$	180.16	15.55	14.08	13.21	1.066	—	—	—	—
(glycerine) → glycerol										
glycerol	$C_3H_8O_3$	92.10	17.95	16.04	13.19	1.216	290.0	800	2.42	1.25
(glycerol trinitrate) → nitroglycerin										
n-heptane	C_7H_{16}	100.20	48.07	44.56	12.68	3.513	98.4	316	2.20	1.66
n-heptene	C_7H_{14}	98.18	47.44	44.31	12.95	3.422	93.6	317	2.17	1.58
hexadecane	$C_{16}H_{34}$	226.43	47.25	43.95	12.70	3.462	286.7	226	2.22	1.64
hexamethyldisiloxane	$C_6H_{18}Si_2O$	162.38	38.30	35.80	15.16	2.364	100.1	192	2.01	—
(hexamethylenetetramine) → methenamine										
n-hexane	C_6H_{14}	86.17	48.31	44.74	12.68	3.528	68.7	335	2.24	1.66
n-hexene	C_6H_{12}	84.16	47.57	44.44	12.99	3.422	63.5	333	2.18	1.57
hydrazine	H_4N_2	32.05	52.08	49.34	49.40	0.998	113.5	1180	3.08	1.65
hydrazoic acid	HN_3	43.02	15.28	14.77	79.40	0.186	35.7	690	—	1.02
hydrogen	H_2	2.00	141.79	130.80	16.35	8.000	−252.7	—	—	14.42
(hydrogen azide) → hydrazoic acid										
hydrogen cyanide	HCN	27.03	13.86	13.05	8.82	1.480	25.7	933	2.61	1.33
hydrogen sulfide	H_2S	34.08	48.54	47.25	16.77	2.817	−60.3	548	—	1.00
maleic anhydride†	$C_4H_2O_3$	74.04	18.77	18.17	14.01	1.297	202.0	—	—	—
melamine†	$C_3H_6N_6$	126.13	15.58	14.54	12.73	1.142	—	—	—	—
methane	CH_4	16.04	55.50	50.03	12.51	4.000	−161.5	—	—	2.23
methanol	CH_4O	32.04	22.68	19.94	13.29	1.500	64.8	1101	2.37	1.37
methenamine†	$C_6H_{12}N_4$	140.19	29.97	28.08	13.67	2.054	—	—	—	—
2-methoxyethanol	$C_3H_8O_2$	76.09	24.23	21.92	13.03	1.682	124.4	583	2.23	—
methylamine	CH_5N	31.06	34.16	30.62	13.21	2.318	−6.3	—	—	1.61
(2-methyl 1-butanol) → *iso*-amyl alcohol										
(methyl chloride) → dichloromethane										
methyl ether	C_2H_6O	46.07	31.70	28.84	13.84	2.084	−24.9	—	—	1.43
methyl ethyl ketone	C_4H_8O	72.10	33.90	31.46	12.89	2.441	79.6	434	2.30	1.43
1-methylnaphthalene	$C_{11}H_{10}$	142.19	40.88	39.33	12.95	3.038	244.7	323	1.58	1.12

TABLE 5-11A. Heats of Combustion and Related Properties For Pure, Simple Substances*

Material	Composition	W Molecular Weight	Δh_c^u Gross (MJ/kg)	Δh_c^l Net (MJ/kg)	$\Delta h_c^l/r_o$ (MJ/kg O_2)	r_o Oxygen-fuel Mass ratio	T_b Boiling temp. (°C)	Δh_v Latent Heat of Vaporization (kJ/kg)	C_{pl} Liquid Heat Capacity (kJ/kg-°C)	C_{pv} Vapor Heat Capacity (kJ/kg-°C)
methyl methacrylate	$C_5H_8O_2$	100.11	27.37	25.61	12.33	2.078	101.0	360	1.91	—
methyl nitrate	CH_3NO_3	77.04	8.67	7.81	75.10	0.104	64.6	409	2.04	0.99
(2-methyl propane) → *iso*-butane										
naphthalene†	$C_{10}H_8$	128.16	40.21	38.84	12.96	2.996	217.9	—	1.18	1.03
nitrobenzene	$C_6H_5NO_2$	123.11	25.11	24.22	14.90	1.625	210.7	330	1.52	—
nitroglycerin	$C_3H_5N_3O_9$	227.09	6.82	6.34	—	—	unstable	462	1.49	—
nitromethane	CH_3NO_2	61.04	11.62	10.54	15.08	0.699	101.1	567	1.74	0.94
n-nonane	C_9H_{20}	128.25	47.76	44.33	12.69	3.493	150.6	295	2.10	1.65
octamethyl-cyclotetrasiloxane	$C_8H_{24}Si_4O_4$	296.62	26.90	25.10	14.56	1.725	175.0	127	1.88	—
n-octane	C_8H_{18}	114.22	47.90	44.44	12.69	3.502	125.6	301	2.20	1.65
iso-octane	C_8H_{18}	114.22	47.77	44.31	12.65	3.502	117.7	272	2.15	1.65
1-octene	C_8H_{10}	112.21	47.33	44.20	12.92	3.422	121.3	301	2.19	1.59
(1-octylene) → 1-octene										
1,2-pentadiene	C_5H_8	68.11	47.31	44.71	13.60	3.288	44.9	405	2.21	1.55
n-pentane	C_5H_{12}	72.15	48.64	44.98	12.68	3.548	36.0	357	2.33	1.67
1-pentene	C_5H_{10}	70.13	47.77	44.64	13.04	3.422	30.0	359	2.16	1.56
phenol†	C_6H_6O	94.11	32.45	31.05	13.05	2.380	181.8	433	1.43	1.10
phosgene	$COCl_2$	98.92	1.74	1.74	10.74	0.162	8.3	247	1.02	0.58
propadiene	C_3H_4	40.06	48.54	46.35	14.51	3.195	−34.6	—	—	1.44
propane	C_3H_8	44.09	50.35	46.36	12.78	3.629	−42.2	—	2.23	1.67
n-propanol	C_3H_8O	60.09	33.61	30.68	12.81	2.396	97.2	686	2.50	1.45
iso-propanol	C_3H_8O	60.09	33.38	30.45	12.71	2.396	80.3	663	2.42	1.48
propene	C_3H_6	42.08	48.92	45.79	13.38	3.422	−47.7	—	—	1.52
(*iso*-propylbenzene) → cumene										
(propylene) → propene										
iso-propyl ether	$C_6H_{14}O$	102.17	39.26	36.25	12.86	2.819	67.8	286	2.14	1.55
propyne	C_3H_4	40.06	48.36	46.17	14.45	3.195	−23.3	—	—	1.51
styrene	C_8H_8	104.14	42.21	40.52	13.19	3.073	145.2	356	1.76	1.17
sucrose†	$C_{12}H_{22}O_{11}$	342.30	16.49	15.08	13.44	1.122	—	—	1.24	—
(1,2,3,4-tetrahydronaphthalene) → tetralin										
tetralin	$C_{10}H_{12}$	132.20	42.60	40.60	12.90	3.147	207.0	425	1.64	1.19
tetranitromethane	CN_4O_8	196.04	2.20	2.20	—	—	125.7	196—	—	
toluene	C_7H_8	92.13	42.43	40.52	12.97	3.126	110.4	360	1.67	1.12
toluene diisocyanate	$C_9H_6N_2O_2$	174.16	24.32	23.56	13.50	1.746	120.0	—	1.65	—
triethanolamine	$C_6H_{15}NO_3$	149.19	29.29	27.08	15.30	1.770	360.0	—	—	—
triethylamine	$C_6H_{15}N$	101.19	43.19	39.93	12.95	3.083	89.5	303	2.22	1.59
1,1,2-trichloroethane	$C_2H_3Cl_3$	133.42	7.77	7.28	11.02	0.660	114.0	260	1.11	0.67
trichloroethylene	C_2HCl_3	131.40	6.77	6.60	12.05	0.548	86.9	245	1.07	0.61
trichloromethane	$CHCl_3$	119.39	3.39	3.21	9.60	0.335	61.7	249	0.97	0.55
trinitromethane	CHN_3O_6	151.04	3.41	3.25	—	—	unstable	—	—	—
trinitrotoluene†	$C_7H_5N_3O_6$	227.13	15.12	14.64	19.80	0.740	240.0	322	1.40	—
trioxane	$C_3H_6O_3$	90.08	16.57	15.11	14.17	1.066	114.5	450	—	—
urea†	CH_4ON_2	60.06	10.52	9.06	11.34	0.799	—	—	—	1.55
vinyl acetate	$C_4H_6O_2$	86.09	24.18	22.65	13.54	1.673	72.5	167	2.00	1.05
vinyl acetylene	C_4H_4	52.07	47.05	45.36	14.76	3.073	5.1	—	—	1.41
vinyl bromide	C_2H_3Br	106.96	12.10	11.48	13.95	0.823	15.6	—	2.42	0.53
vinyl chloride	C_2H_3Cl	62.50	20.02	16.86	11.97	1.408	−13.8	—	—	0.86
(vinyl trichloride) → 1,1,2-trichlorolthane										
xylenes	C_8H_{10}	106.16	42.89	40.82	12.90	3.165	138–144	343	1.72	1.21
xylidene	$C_8H_{11}N$	121.22	38.28	36.29	12.79	2.838	192.7	366	1.77	—

† Denotes substance in crystalline solid form; otherwise, liquid if $T_b > 25°C$, gaseous if $T_b < 25°C$.

* Sources: (Domalski 1972; Gray 1972; Rossini 1953; Tinnermanns 1965; Thermodynamics Research Center no date; Stull et al 1969; Landolt et al 1923–61; Reid et al 1977; NACA 1957; Belenyessey et al 1962; Rand Corp 1949; Dreisbach 1955–61; Cox and Pilcher 1970; Kharasch 1929; Lipowitz 1976)

Table 5-11B. Heats of Combustion and Related Properties for Plastics*

Material	Unit Composition	W Molecular Weight	Δh_c^u Gross (MJ/kg)	Δh_c^l Net (MJ/kg)	$\Delta h_c^l/r_o$ (MJ/kg O_2)	r_o Oxygen-fuel Mass ratio	C_{ps} Heat Capacity Solid (kJ/kg-°C)
acrylonitrile-butadiene styrene copolymer	—	—	35.25	33.75			1.41–1.59
bisphenol A epoxy	$C_{11.85}H_{20.37}O_{2.83}N_{0.3}$	212.10	33.53	31.42	13.41	2.343	
butadiene-acrylonitrile 37% copolymer	—	—	39.94				
butadiene/styrene 8.58% copolymer	$C_{4.18}H_{6.09}$	56.30	44.84	42.49	13.11	3.241	1.94
butadiene/styrene 25.5% copolymer	$C_{4.60}H_{6.29}$	61.55	44.19	41.95	13.07	3.209	1.82
cellulose acetate (triacetate)	$C_{12}H_{16}O_8$	288.14	18.88	17.66	13.25	1.333	1.34
cellulose acetate-butyrate	$C_{12}H_{18}O_7$	274.27	23.70	22.3	14.67	1.517	1.70
epoxy, unhardened	$C_{31}H_{36}O_{5.5}$	496.63	32.92	31.32	13.05	2.400	
epoxy, hardened	$C_{39}H_{40}O_{8.5}$	644.74	30.27	28.90	13.01	2.221	
melamine formaldehyde (Formica)	$C_6H_6N_6$	162.08	19.33	18.52	12.51	1.481	1.46
nylon 6	$C_6H_{11}NO$	113.08	30.1 –31.7	28.0 –29.6	12.30	2.335	1.52
nylon 6,6	$C_{12}H_{22}N_2O_2$	226.16	31.6 –31.7	29.5 –29.6	12.30	2.405	1.70
nylon 11 (Rilsan)	$C_{11}H_{21}NO$	183.14	36.99	34.47	12.33	2.796	1.70–2.30
phenol formaldehyde	$C_{15}H_{12}O_2$	224.17	27.9 –31.6	26.7 –30.4	11.80	2.427	1.70
-foam			21.6 –27.4	20.2 –26.2			
polyacenaphthalene	$C_{12}H_8$	152.14	39.23	38.14	12.95	2.945	
polyacrylonitrile	C_3H_3N	53.04	32.22	30.98	13.70	2.262	1.50
polyallylphthalate	$C_{14}H_{14}O$	198.17	27.74	26.19	9.54	2.745	
(polyamides) → nylon							
poly-1,4-butadiene	C_4H_6	54.05	45.19	42.75	13.13	3.256	
poly-1-butene	C_4H_8	56.05	46.48	43.35	12.65	3.426	1.88
polycarbonate	$C_{16}H_{14}O_3$	254.19	30.99	29.78	13.14	2.266	1.26
polycarbon suboxide	C_3O_2	68.03	13.78	13.78	14.64	0.941	
polychlorotrifluoroethylene	C_2F_3Cl	116.47	1.12	1.12	2.04	0.549	0.92
polydiphenylbutadiene	$C_{16}H_{10}$	202.18	39.30	38.2	13.05	2.928	
polyester, unsaturated	$C_{5.77}H_{6.25}O_{1.63}$	101.60	21.6 –29.8	20.3 –28.5	11.90	2.053	1.20–2.30
polyether, chlorinated	$C_5H_8OCl_2$	154.97	17.84	16.71	12.45	1.342	
polyethylene	C_2H_4	28.03	46.2 –46.5	43.1 –43.4	12.63	3.425	1.83–2.30
polyethylene oxide	C_2H_4O	44.02	26.65	24.66	13.57	1.817	
polyethylene terephthalate	$C_{10}H_8O_4$	192.11	22.18	21.27	12.77	1.666	1.00
polyformaldehyde	CH_2O	30.01	16.93	15.86	14.88	1.066	1.46
poly-1-hexene sulfone	$C_6H_{12}SO_2$	148.13	29.78	28.00	14.40	1.944	
polyhydrocyanic acid	HCN	27.02	23.26	22.45	15.17	1.480	
(polyisobutylene) → poly-1-butene							
polyisocyanurate foam	—		26.3	22.2 –26.2			
polyisoprene	C_5H_8	68.06	44.90	42.30	12.90	3.291	
poly-3-methyl-1-butene	C_5H_{10}	70.06	46.55	43.42	12.67	3.426	
polymethyl methacrylate	$C_5H_8O_2$	100.06	26.64	24.88	12.97	1.919	1.44
poly-4-methyl-1-pentene	C_6H_{12}	84.08	46.52	43.39	12.67	3.425	2.18
poly-α-methylstyrene	C_9H_{10}	118.11	42.31	40.45	13.00	3.116	
polynitroethylene	$C_2H_3O_2N$	73.03	15.96	15.06	19.64	0.767	
polyoxymethylene	CH_2O	30.01	16.93	15.65	14.68	1.066	
polyoxytrimethylene	C_3H_6O	58.04	31.52	29.25	13.27	2.205	
poly-1-pentene	C_5H_{10}	70.06	45.58	42.45	12.39	3.426	
polyphenylacetylene	C_8H_6	102.09	40.00	38.70	13.00	2.978	
polyphenylene oxide	C_8H_8O	120.09	34.59	33.13	13.09	2.531	1.34
polypropene sulfone	$C_3H_6SO_2$	106.10	23.82	22.58	16.64	1.357	
poly-β-propiolactone	$C_3H_4O_2$	72.14	19.35	18.13	13.62	1.331	
polypropylene	C_3H_6	42.04	46.37	43.23	12.62	3.824	2.10
polypropylene oxide	C_3H_6O	58.04	31.17	28.90	13.11	2.205	
polystyrene	C_8H_8	104.10	41.4 –42.5	39.7 –39.8	12.93	3.074	1.40
polystyrene-foam	—		39.7	35.6 –40.8			
polystyrene-foam, FR	—		41.2 –42.9				
polysulfones, butene	$C_4H_8SO_2$	120.11	24.04–26.47	22.25–25.01	14.79	1.598	1.30
polysulfur	S	32.06	9.72	9.72	9.74	0.998	
polytetrafluoroethylene	C_2F_4	100.02	5.00	5.00	7.81	0.640	1.02
polytetrahydrofuran	C_4H_8O	72.05	34.39	31.85	13.04	2.443	
polyurea	$C_{15}H_{18}O_4N_4$	318.20	24.91	23.67	13.45	1.760	

Table 5-11B. (continued)

Material	Unit Composition	W Molecular Weight	Δh_c^u Gross (MJ/kg)	Δh_c^l Net (MJ/kg)	$\Delta h_c^l/r_o$ (MJ/kg O_2)	r_o Oxygen-fuel Mass ratio	C_{ps} Heat Capacity Solid (kJ/kg-°C)
polyurethane	$C_{6.3}H_{7.1}NO_{2.1}$	130.30	23.90	22.70	13.16	1.725	1.75–1.84
polyurethane-foam	—		26.1 –31.6	23.2 –28.0			
polyurethane-foram, FR	—		24.0 –25.0				
polyvinyl acetate	$C_4H_6O_2$	86.05	23.04	21.51	12.86	1.673	
polyvinyl alcohol	C_2H_4O	44.03	25.00	23.01	12.66	1.817	1.70
polyvinyl butyral	$C_8H_{14}O_2$	142.10	32.90	30.70	13.00	2.365	
polyvinyl chloride	C_2H_3Cl	62.48	17.95	16.90	12.00	1.408	0.90–1.20
polyvinyl-foam	—		22.83				1.30–2.10
polyvinyl fluoride	C_2H_3F	46.02	21.70	20.27	10.60	1.912	
polyvinylidene chloride	$C_2H_2Cl_2$	96.93	10.52	10.07	12.21	0.825	1.34
polyvinylidene fluoride	$C_2H_2F_2$	64.02	14.77	14.08	11.26	1.250	1.38
urea formaldehyde	$C_3H_6O_2N_2$	102.05	15.90	14.61	13.31	1.098	1.60–2.10
urea formaldehyde-foam	—	—	14.80				

* Sources: (Throne and Grieskey 1972; Krekeler et al 1965; Hogon 1976; NBS no date; Roff and Scott 1971; Joshi 1975; Van Krevelen 1976; Berlin et al 1969; Franz et al 1967)

Notes to Tables 5-11A, 5-11B, 5-11C, and 5-11D

Heats of Combustion: The heat of combustion is, by definition, the enthalpy of reaction when fuel and oxidant at standard conditions are reacted and form products at standard conditions. A unique value for the heat of combustion is possible only if these conditions are fully specified (Rossini 1956; Gray 1972). In normal combustion work the standard conditions are taken as:

1. Fuel and oxidant enter at 1 atmosphere pressure and 25°C (298 K) temperature. An amount of heat, which is equal to the heat of combustion, is extracted, so that the products are also at 25°C and 1 atmosphere.
2. The oxidant is gaseous oxygen.
3. The main products consist of liquid H_2O, gaseous CO_2, and gaseous N_2. There is no CO formed.
4. For fuels containing sulfur, the standard products include liquid $H_2SO_4 \cdot 115H_2O$. For chlorine-containing fuels reference states consisting of either liquid HCl in water solution or gaseous Cl_2 have been used.
5. In the combustion of silicones, the silicon goes to amorphous silica, SiO_2.

The state of the fuel—gaseous, liquid or solid—is not standardized and must be specified. The heat of combustion as defined above is termed the gross or upper value and is customarily determined in an oxygen bomb calorimeter (ASTM undated a). For common materials the value is a negative number; however, customarily a minus sign is included in the definition to make heat of combustion a positive value (ASTM undated b). Heat of combustion, enthalpy of combustion, calorific value and heating value are synonyms, the latter two being used more commonly in the heating industry.

In many cases the products are not cooled down to 25°C. For modest temperature differences the change in the

Table 5-11C. Heat of Combustion of Miscellaneous Substances*

Material	Δh_c^u Gross (MJ/kg)	Δh_c^ℓ Net (MJ/kg)
acetate (see cellulose acetate)		
acrylic fiber	30.6–30.8	
blasting powder	2.1–2.4	
butter	38.5	
celluloid (cellulose nitrate and camphor)	17.5–20.6	16.4–19.2
cellulose acetate fiber, $C_8H_{12}O_6$	17.8–18.4	16.4–17.0
cellulose diacetate fiber, $C_{10}H_{14}O_7$	18.7	
cellulose nitrate, $C_6H_9N_1O_7/C_6H_8N_2O_9/C_6H_7N_3O_{11}$	9.11–13.48	
cellulose triacetate fiber, $C_{12}H_{16}O_8$	18.8	17.6
charcoal	33.7–34.7	33.2–34.2
coal-anthracite	30.9–34.6	30.5–34.2
-bituminous	24.7–36.3	23.6–35.2
coke	28.0–31.0	28.0–31.0
cork	26.1	
cotton	16.5–20.4	
dynamite	5.4	
epoxy, $C_{11.9}H_{20.4}O_{2.8}N_{0.3}/C_{6.064}H_{7.550}O_{1.222}$	32.8–33.5	31.1–31.4
fat, animal	39.8	
flint powder	3.0–3.1	

Table 5-11C. Heat of Combustion of Miscellaneous Substances*

Material	Δh_c^u Gross (MJ/kg)	Δh_c^ℓ Net (MJ/kg)
fuel oil-No. 1	46.1	
-No. 6	42.5	
gasketing-chlorosulfonated polyethylene (Hypalon)	28.5	
-vinylidene fluoride/hexafluoropropylene (Fluorel, Viton A)	14.0–15.1	
gasoline	46.8	43.7
jet fuel-JP1		43.0
-JP3		43.5
-JP4	46.6	43.5
-JP5	45.9	43.0
kerosene (jet fuel A)	46.4	43.3
lanolin (wool fat)	40.8	
lard	40.1	
leather	18.2–19.8	
lignin, $C_{2.6}H_3O$	24.7–26.4	23.4–25.1
lignite	22.4–33.3	
modacrylic fiber	24.7	
naphtha	43.0–47.1	40.9–43.9
neoprene, $C_5H_5C\ell$-gum	24.3	
-foam	9.7–26.8	
Nomex (polymethaphenylene isophthalamide) fiber, $C_{14}H_{10}O_2N_2$	27.0–28.7	
oil-castor	37.1	
-linseed	39.2–39.4	
-mineral	45.8–46.0	
-olive	39.6	
-solar	41.8	
paper-brown	16.3–17.9	
-magazine	12.7	
-newsprint	19.7	
-wax	21.5	
paraffin wax	46.2	43.1
peat	16.7–21.6	
petroleum jelly ($C_{7.118}H_{12.957}O_{0.091}$)	45.9	
rayon fiber	13.6–19.5	
rubber-buna N	34.7–35.6	
-butyl	45.8	
-isoprene (natural) C_5H_8	44.9	42.3
-latex foam	33.9–40.6	
-GRS	44.2	
-tire, auto	32.6	
silicone rubber (SiC_2H_6O)	15.5–16.8	
-foam	14.0–19.5	
sisal	15.9	
spandex fiber	31.4	
starch	17.6	16.2
straw	15.6	
sulfur-rhombic		9.28
-monoclinic		9.29
tobacco	15.8	
wheat	15.0	
wood-beech	20.0	18.7
-birch	20.0	18.7
-douglas fir	21.0	19.6
-maple	19.1	17.8
-red oak	20.2	18.7
-spruce	21.8	20.4
-white pine	19.2	17.8
-hardboard	19.9	
woodflour	19.8	
wool	20.7–26.6	

* Sources: Landolt et al 1927–61; NACA 1957; Throne et al 1972; Moore 1978; Domalski et al 1978; Bostic 1973; Lobanov and Martynovskaya 1972; Ohe et al 1977; Lowrie 1983)

Table 5-11D. Heats of Combustion for Metals*

Material	Δh_c (MJ/kg)
Pure elements	
aluminum	31.04
beryllium	66.43
copper	2.45
iron	7.39
magnesium	24.72
manganese	7.01
molybdenum	6.13
nickel	4.10
tantalum	5.66
tin	3.73
titanium	19.71
zinc	5.37
zirconium	12.07
Copper alloys	
bronze (88 Cu/10 Sb/2 Zn)	2.64
red brass (85 Cu/15 Zn)	2.89
cartridge brass (70 Cu/30 Zn)	3.33
yellow brass (60 Cu/40 Zn)	3.62
Iron alloys	
carbon steels	7.4–7.5
stainless steels	7.7–8.4
Nickel alloys	
Inconel 600	5.40
Monel 400	3.60

* Sources: (Hust and Clark 1973; Stull and Prophet 1971; Brandes 1983)

effective value could be taken as $\approx C_p(T - 25°)$, where C_p is the constant pressure heat capacity of the products. For this relationship to hold, the enthalpy curve has to be smooth and cannot include condensation. Thus it often becomes convenient to define a lower, or net, heat of combustion where the products are as before except that H_2O stays in gaseous form (and sulfur, if any, goes into gaseous SO_2). The relationship between the values for simple fuels can be expressed as:

$$\Delta h_c^l = \Delta h_c^u - 0.2196\ [\%H]$$

where:

Δh_c^l = lower heat of combustion (MJ/kg)

Δh_c^u = upper heat of combustion (MJ/kg)

[%H] = percent, by mass, of hydrogen in fuel

In practice, combustion is rarely complete. For example, some carbon often goes to CO instead of CO_2. Reactions with fuels containing sulfur, or chlorine or other halogens can go into a variety of products, depending on exact reaction conditions. Incomplete combustion acts to lower the heat of combustion. There are no simple general rules to aid in determining the degree of incomplete combustion. Also explosive and propellant materials rarely exhibit reactions where H_2O and CO_2 are the main products; thus, their heats of explosion or reaction will tend to be substantially different from the heat of combustion.

Even though the state of reactants is not described in the definition, most commonly it is taken as the natural state at ambient pressure and temperature. In some references, however, the gaseous state of fuels such as benzene or methanol is assumed, even though they would be normally considered as liquid fuels. The heat of combustion of fuels from the gaseous state is higher than from the liquid state by an amount equal to the fuel's heat of vaporization at 25°C. For combustion of plastic fuels, the most convenient reference state is the solid phase. In such cases, to compute the heat liberated in a given application, it is *not* necessary to subtract the heat of gasification from the heat of combustion.

Heats of combustion (per kg of fuel) are listed in Tables 5-11A through 5-11C. Also listed in Tables 5-11A and 5-11B, is r_o, the stoichiometric oxygen-fuel mass ratio. The heats can be seen to vary quite a bit for different fuels. Recently, however, increasing engineering use is made of the observation that the heat *of combustion per kg of oxygen consumed* is nearly constant for most organic fuels. It can be shown that a value of $\Delta h_c^l/r_o$ = 13.1 MJ/kg O_2 is near-constant (Huggett 1980). An assumption of constant heat of combustion per kg of oxygen is useful in heat release rate measurements and for air-limited combustion problems.

For certain fuels it is more convenient to express heats of combustion by use of equations or charts, rather than tables. Two such instances are woods and certain petroleum products. The gross heat of combustion for dry woods is typically in the vicinity of 18-21 MJ/kg. This value varies, however, according to the species and according to the part of the plant. Furthermore, when wood normally burns, pyrolysates are first released and burned, which are lower in calorific value, followed by burning of the remaining char, which has a higher value. The following empirical correlation can be derived, applicable to all species of dry woods, provided the percent carbon, by weight, is known (Susott et al 1975).

$$\Delta h_c^u = 0.433\ [\%C] - 1.527$$

where Δh_c^u is the gross heat of combustion for dry wood (MJ/kg) and [%C] is the percent of carbon, by weight. This correlation is equally applicable to charred wood and to wood pyrolysates, provided their ultimate analysis is known. For wet fuels, an additional correction should be made,

$$\Delta h_{c,\text{eff}}^u = \Delta h_c\ (1 - M/100) - 0.0257\ M$$

where M is the percent water in the fuel. Lower heats of combustion, if needed, can then be determined as noted above.

For petroleum products a series of empirical rules is available for estimating thermochemical properties, based on specific gravity or on API degrees (Cragoe 1929) (See Table 5-11I.) A determination of the exact heat of combustion by calorimetry is usually not required for typical petroleum products, which include crude oils, fuel oils, kerosenes, and volatile products. These empirical rules are:

Heat of combustion

$$\Delta h_c^u = 52.12 - 8.79\ d^2 - 0.14\ d \quad \text{(MJ/kg)}$$

Percent hydrogen

$$[\%H] = 26 - 15\,d$$

Heat of vaporization, at 25°C

$$\Delta h_v = \frac{240}{d} \quad \text{(kJ/kg)}$$

Specific heat at 25°C, liquid

$$C_{pl} = \frac{1770}{\sqrt{d}} \quad \text{(J/kg°C)}$$

Specific heat at 25°C, vapor

$$C_{pv} = C_{pl} - \frac{380}{d} \quad \text{(J/kg°C)}$$

In all the above relationships, d = specific gravity at 15.6°C (60°F). For conversion from degrees API, see Table 5-11I. The relationship for [%H] can be used to obtain net heat of combustion, according to the general equation given earlier. A more detailed method, requiring additionally the determination of percent aromatics, volatility, and sulfur content, is available as an ASTM method (ASTM undated c).

In some cases the combustion properties, including heat of combustion, are desired for burning metals. The specified combustion product in each case is a stable metal oxide in crystalline form. Table 5-11D lists the heats of combustion for some common metals. For further discussion consult the references (Skinner 1962; Hust and Clark 1973).

Also listed in the tables, where available, are T_b, the boiling point; the heat capacities C_{pl}, C_{pv}, and C_{ps} of the liquid, vapor, and solid phases respectively, taken at constant pressure at T_b; and Δh_v, the latent heat of vaporization at T_b. For purposes where the total heat of vaporization (i.e., heat to bring a liquid at 25°C up to a gas at T_b) is desired, an estimate can be made by adding to Δh_v the quantity $C_{pl} \cdot (T_b - 25)$. More precise values can be obtained by using enthalpy tables (Thermodynamics Research Center no date; Stull et al. 1969).

In some common applications, especially for building materials, a laboratory technique for assessing the possible effect of incomplete combustion is desired. The potential heat test procedure outlined in NFPA 259, *Standard Test Method for Potential Heat of Building Materials* is an appropriate standard technique for this, where only that heat of combustion fraction is determined which can be liberated by heating the specimen at 750°C for two hours in a muffle furnace. In tabulating fuel loads, for instance, it is more appropriate to determine potential heat values than it is to use the full heats of combustion.

Conversion Units: To convert kcal/kg to megajoules/kg multiply by 4.184×10^{-3}
to convert kcal/mol to MJ/kg multiply by 4.184/W
where W = molecular weight (g/mol)
to convert kJ/mol to MJ/kg multiply by 1/W
to convert Btu/lb to MJ/kg multiply by 2.322×10^{-3}

References for Tables 5-11A, 5-11B, 5-11C, 5-11D and 5-11E

ASTM. undated a. *Standard Method of Test for Gross Calorific Value of Solid Fuel by the Isothermal-Jacket Bomb Calorimeter* (D3286). American Society for Testing and Materials.

ASTM. undated b. *Standard Definitions of Terms Relating to Coal and Coke* (ANSI/ASTM D 121). American Society for Testing and Materials.

ASTM. undated c. *Standard Method for Estimation of Heat of Combustion of Aviation Fuels* (ANSI/ASTM D3338). American Society for Testing and Materials.

Belenyessy, L. I. et al. 1962. Heats of Combustion of Complex Saturated Hydrocarbons. *J. Chem. and Engrg. Data.* Vol 7, pp 66–68.

Berlin, A. A. et al. 1969. Intermolecular Interaction in Polyconjugated Systems. *Acad. of Sciences (USSR) Bulletin. Div. of Chemical Science.* pp 1392-1395. July1969.

Bostic, J. E., Jr., Yeh, K-N., and Barker, R. H. 1973. Pyrolysis and Combustion of Polyester. *Journal of Applied Polymer Science.* Vol 17, pp 471–482.

Brandes, E. A. 1983. *Smithells Metals Reference Book.* Butterworths. London, England.

Cox, J. D., and Pilcher, G. 1970. *Thermochemistry of Organic and Organometallic Compounds.* Academic Press. London, England.

Cragoe, C. S. 1929. *Thermal Properties of Petroleum Products* (Miscellaneous Publication 97). National Bureau of Standards.

Domalski, E. S. 1972. Selected Values of Heats of Combustion and Heats of Formation of Organic Compounds Containing the Elements C, H, N, O, P, and S. *Journal of Physical and Chemical Reference Data.* Vol 1. pp 221–227.

Domalski, E. S., Evans, W. H., and Jobe,T. L., Jr. 1978. Thermodynamic Data for Waste Incineration (NBSIR 78-1479). U.S. National Bureau of Standards.

Dreisbach, R. R. 1955-61. *Physical Properties of Chemical Compounds.* v. 1 (1955), v. 2 (1959), v. 3 (1961). American Chemical Society. Washington, DC.

Franz, J., Mische, W., and Kuzay, P. 1967. Zur Bestimmung Von Verbrennungswarmen Von Plasten. *Plaste Und Kautschuk.* Vol 14, pp 472–476.

Gray, D. E., ed. 1972. *American Institute of Physics Handbook.* Section 41. McGraw-Hill. New York.

Hognon, B. 1976. Contribution possible des principaux isolants synthetiques a l'aggravation des dangers en cas d'incendie. *Cahiers du Centre Scientifique et Technique du Batiment.* p 1392 Sept. 1976.

Huggett, C. 1980. "Estimation of Rate of Heat Release by Means of Oxygen Consumption Measurements." *Fire and Materials,* Vol 4.

Hust, J. G., and Clark, A. F. 1973. "A Survey of Compatibility of Materials with High Pressure Oxygen Service." *Cryogenics.* Vol 13, pp 325-336.

Joshi, R. M. 1975. A New Generalized Bond Energy/Group Contribution Scheme for Calculating the Standard Heat of Formation of Monomers and Polymers. Part V. Oxygen Compounds. *J. Macromol. Sci. Chem.* A9, pp 1309-1383.

Kharasch, M. S. 1929. Heats of Combustion of Organic Compounds. *Bur. of Standards J. of Research.* Vol 2, pp 359-430.

Krekeler, K., and Klimke, P. M. 1965. *Kuntstoffe.* Vol 55, pp 758–765.

Landolt-Bornstein Physikalisch-Chemische Tabellen. 1927-61. Fifth series: main work (vol 2, 1923); first supplement (vol 2, 1927); second supplement (vol 2, 1931); third supplement (vol 3, 1936). Sixth series: Part 2 (vol 4, 1961). Springer-Verlag. Berlin, Republic of Germany.

Lipowitz, J. 1976. Flammability of Poly(dimethylsiloxanes). *J. Fire Flamm.* Vol 7, pp 482-503 and pp 504-529.

Lobanov, G. A., and Martynovskaya, L. I. 1972. Estimation of the Enthalpies of Combustion of Resin and Rubber Type Compounds. *Izv. Vyssh. Ucheb. Zaved., Khim. Tekhnol.* Vol 15, pp 1747-1748.

Lowrie, R. 1983. Heat of Combustion and Oxygen Compatibility. *Flammability and Sensitivity of Materials in Oxygen-Enriched Atmospheres* (ASTM STP 812). pp 84-96. Amer. Soc. for Testing and Materials. Philadelphia, PA.

Moore, L. D. 1978. Full-Scale Burning Behavior of Curtains and Draperies (NBSIR 78-1448). U.S. National Bureau of Standards.

NACA. 1957. *Basic Considerations in the Combustion of Hydrocarbon Fuels in Air* (NACA Report 1300). National Advisory Committee on Aeronautics.

NBS. undated. Unpublished tests. U.S. National Bureau of Standards.

Ohe, H., Mastsuura, K., and Sakai, N. 1977. The Heat of Combustion and Oxygen Index of Fibrous Materials. *Textile Research J.* Vol 47, pp 212–216.

Rand Corp. 1949. *Physical Properties and Thermodynamic Functions of Fuels, Oxidizers, and Products of Combustion, I. Fuels.* NTIS No AD 605 967. Battelle Memorial Institute, Columbus, OH.

Reid, R. C., Prausnitz, J. M., and Sherwood, T. K. 1977. *The Properties of Gases and Liquids.* McGraw-Hill. New York.

Roff, W. J., and Scott, J. R. 1971. *Handbook of Common Polymers.* Butterworth. London, England.

Rossini, F. D. 1953. *Selected Values of Physical and Thermodynamic Properties of Hydrocarbons and Related Compounds.* American Petroleum Institute/Carnegie Press. Pittsburgh, PA.

Rossini, F. D., ed. 1956. *Experimental Thermochemistry.* Vol 1. Interscience. New York.

Skinner, H. A., ed. 1962. *Experimental Thermochemistry*, Vol 2. Interscience, New York.

Stull, D. R., and Prophet, H. 1971. *JANAF Thermochemical Tables* (NSRDS-NBS 37). U.S. National Bureau of Standards.

Stull, D. R., Westrom, Jr., E. F. and Sinke, G. C. 1969. *Thermodynamics of Organic Compounds.* Wiley. New York.

Susott, R. A., DeGroot, W. F., and Shafizadeth, F. 1975. Heat Content of Natural Fuels. *J. of Fire and Flammability.* Vol 6. pp 311–325.

Thermodynamics Research Center. undated. *Selected Values of Properties of Chemical Compounds;* and *Selected Values of Properties of Hydrocarbons and Related Compounds.* Texas A & M University, College Station, TX.

Throne, J. L., and Griskey, R. G. 1972. Heating Values and Thermochemical Properties of Plastics, *Modern Plastics.* Vol 49. pp 96-200. November 1972.

Tinnermans, J. 1950-65. *Physio-Chemical Constants of Pure Organic Compounds.* 2 Vols. Elsevier. New York.

Van Krevelen, D. W. 1976. *Properties of Polymers.* Elsevier. Amsterdam, Holland.

Notes to Table 11-5F

In recent years measurement data have started to be tabulated on the engineering performance of plastics. These data are usually measured in heat release rate calorimeters on a bench scale (specimens typically 0.1 m dimension). A number of these data are summarized in Table 5-11F. In many cases wide variations for a property are shown. These can be attributed to (a) specimen size differences, (b) specimen orientation differences, (c) air flow rate differences, (d) heating flux (irradiance) differences, and (e) specimen composition differences. The data for the plastics listed do not represent pure polymers, but rather engineering plastics, which vary in additives, degree of polymerization, etc. When only a single number is

Table 5-11E. Oxygen Index Values*

Material	Oxygen Index Value†
methanol	11–12
benzene	13–16
sulfur	13.6
polyacetal	14.9
polyoxymethylene	15.7
acetone	16.0
kitchen candle	16.0
cotton	16.0–18.5
mineral oil	16.1
polyurethane foam, typical	16.5
cellulose acetate	16.8
styrene-butadiene-rubber foam	16.9
polybutadiene foam	17.1
natural rubber foam	17.2
polymethylmethacrylate	17.3
polyethylene	17.4
polypropylene	17.4
polystyrene	17.6–18.3
polyacrylonitrile	18.0
paper, cellulose filter	18.2
polybutadiene	18.3
ABS	18.3–18.8
cellulose triacetate	18.4
polyisoprene rubber	18.5
rayon	18.7–18.9
polymethylsiloxane (silicone fluid)	18.9
cellulose	19.0
cellulose acetate-butyrate	19.6
epoxy resin	19.8
polyethylene terephthalate	20.0
nylon	20.1–26.0
birch wood	20.5
polyester fibers	20.6
phenolic resin	21.0
phenolic-paper laminate	21.7
polyvinyl alcohol	22.5
polycarbonate	22.5–28.0
polyvinyl fluoride	22.6
red oak	23.0
wool	23.8
neoprene rubber	26.3–40.0
Nomex fibers	26.7–28.5
modacrylic fibers	26.8
polyphenylene oxide	29.0
silicone rubber	30.0
polysulfone	30–32
leather	34.8
carbon powder	35.0
phenol-formaldehyde resin	35.0
polyimide	36.5
polyvinyl chloride	37.1–49
Kynol fibers	38.0
polybenzimidazole	40.6–41.5
polyvinylidene fluoride	43.7
polyvinyl chloride, chlorinated	45–60
carbon black	56–63
polyvinylidene chloride	60
polychlorotrifluoroethylene	95
polytetrafluoroethylene	95

† The values in the table above were determined using the ASTM Oxygen Test (D2863), and represent typical values for materials without added fire retardants. Adding fire retardants to polymers can increase substantially the measured oxygen index value.

* Sources: Cullis, C.F. and Hirschler, M.M. 1981. *The Combustion of Organic Polymers.* Clarendon Press. Oxford, England.

Factory Mutual Research Corp. unpublished data.

Hilado, C.J. 1982. *Flammability Handbook for Plastics.* Technomic. Westport, CT.

Table 5-11F. Miscellaneous Properties or Plastics and Other Combustibles*

Material	Density (kg/m^3)	Min. Radiant Flux for Ignition (kW/m^2)	Min. Mass Loss Rate for Ignition (g/m^2-s)	Energy Required for Ignition (kJ/m^2)	Ignition Temperature (K)	Effective Heat of Gasification (MJ/kg)	Soot Specific Extinction Area (m^2/kg)	Soot Yield (kg soot/ kg specimen)
ABS						3.2		
benzene, liquid							830–900	
cellulose						3.5	10–15	
Douglas Fir				650–770		1.8		
fiberglass, FR	—	7	—	3950	740	1.7–2.2	250–370	
nylon 6/6						2.3	410	0.14
paper, corrugated						2.2		
phenolic, foams, rigid	67	45	5.5	390	944	3.7		
polycarbonate						2.1		
polyethylene	920	19	2.5	1500–5100	761	1.5–2.7	290–510	
polyethylene, foams	36–143	19–22	2.5–2.6	1430–1780	761–807	1.5–2.7	360–810	
polyethylene, foam, chlorinated						2.1–3.1	1010–1360	
polyisocyanurate, foams, ridgid	33–36	23	5.5–6.8	850–930	798	4.5		
polymethylmethacrylate	1170	18	4.4	1300–4200	751	1.6	110–260	0.012
polyoxymethylene	1400	17	4.5	2100–6000	740	2.4	0–10	
polypropylene	905	20	2.7	1000–3500	771	1.4–2.0	380–610	0.07–0.09
polystyrene	1050	29	4.0	1300–6400	846	1.7	730–1170	
polystyrene, foams, rigid	16–34	18–27	4.9–6.3	1200–3200	751–831	1.3–3.1	940–1300	
polytetrafluoroethylene		43		9000	933			0.02
polyurethane, foams, flexible	29–47	16–30	5.3–7.2	150–770	729–852	1.2–2.7	100–500	
polyurethane, foams, rigid	36–321	20–28	6.9–8.4	150–1300	771–838	2.0–4.5	100–1000	
polyvinyl chloride	1305	21	—	3320	780	1.7–2.5	1100–1400	0.23
red oak	740	11	2.5	420–920	730	1.7–5.5	80–130	.0002–0.018
styrene, liquid						0.64	780–920	

indicated, this should be taken as reflecting a paucity of data, rather than a high precision of measurement.

The minimum flux needed for ignition has sometimes been considered to be a constant. Table 5-11F shows that it varies among materials, but only over a range of about 2:1. Ignition temperatures are seen to vary little for common plastics. The total absorbed energy (kJ/m^2) required for ignition has more recently been considered a measure of ignitability.

The effective heat of gasification is a parameter intended to represent the ease of achieving a given heat release rate. The tabulated values, however, do not show the wide range of heat release rate behaviors seen with commercial products. This is partly because for most materials the heat release progress is highly transient, whereas the definition of the effective heat of gasification (average energy required to gasify a unit mass of material) requires assuming steady-state conditions.

Smoke properties can be measured in two ways: (1) Optically, by determining the obscuration along a light beam. This can be expressed as smoke obscuration area per unit specimen mass (m^2/kg). The unit system implied here is logs (Base-e). (2) Gravimetrically, by determining the soot mass evolved, per unit specimen mass.

*Sources for Table 5-11F:

Factory Mutual Research Corp. unpublished data.

National Bureau of Standards. unpublished data.

Tewarson, A. 1982. *Flame-Retardant Polymeric Materials*, Chapter 3. Plenum. New York, NY.

Tewarson, A. 1984. *Particulate Formation in Fires.* Paper presented at Conference on Large Scale Fire Phenomenology. Washington, DC.

Tewarson, A. 1980. *Physico-Chemical and Combustion/Pyrolysis Properties of Polymeric Materials* (NBS-GCR-80-295). U.S. National Bureau of Standards.

Tewarson, A., Khan, M. M., and Steciak, J. 1982. *A Study of Combustibility of Electrical Wires and Cables Used in Rail Rapid Transit Systems.* Transportation Systems Center. U.S. Dept. of Transportation.

Tewarson, A., Lee, J. L., and Pion, R. F. 1979. *Categorization of Cable Flammability.* Part I (EPRI NP-1200, Part 1). Electric Power Research Inst. Palo Alto, CA.

Tewarson, A., Lee, J. L., and Pion, R. F. 1981. *Fuel Parameters for Evaluation of the Fire Hazard of Red Oak* (FMRC J. I.OC6N2.RC). Factory Mutual Research Corp. Norwood, MA.

Table 5-11G. Specific Gravity Equivalents for Degrees Baumé
(Liquids Lighter than Water)

Baumé	Specific Gravity	Lbs per Gallon*	Baumé	Specific Gravity	Lbs per Gallon*	Baumé	Specific Gravity	Lbs per Gallon*
10	1.0000	8.33	45	0.8000	6.66	80	0.6667	5.55
11	0.9929	8.27	46	0.7955	6.63	81	0.6635	5.52
12	0.9859	8.21	47	0.7910	6.59	82	0.6604	5.50
13	0.9790	8.16	48	0.7865	6.55	83	0.6573	5.48
14	0.9722	8.10	49	0.7821	6.52	84	0.6542	5.45
15	0.9655	8.04	50	0.7778	6.48	85	0.6512	5.42
16	0.9589	7.99	51	0.7735	6.44	86	0.6481	5.40
17	0.9524	7.93	52	0.7692	6.41	87	0.6452	5.38
18	0.9459	7.88	53	0.7650	6.37	88	0.6422	5.36
19	0.9396	7.83	54	0.7609	6.34	89	0.6393	5.33
20	0.9333	7.78	55	0.7568	6.30	90	0.6364	5.30
21	0.9272	7.72	56	0.7527	6.27	91	0.6335	5.28
22	0.9211	7.67	57	0.7487	6.24	92	0.6306	5.25
23	0.9150	7.62	58	0.7447	6.20	93	0.6278	5.23
24	0.9091	7.57	59	0.7407	6.17	94	0.6250	5.21
25	0.9032	7.53	60	0.7368	6.14	95	0.6222	5.18
26	0.8974	7.48	61	0.7330	6.11	96	0.6195	5.16
27	0.8917	7.43	62	0.7292	6.07	97	0.6167	5.14
28	0.8861	7.38	63	0.7254	6.04	98	0.6140	5.11
29	0.8805	7.34	64	0.7216	6.01	99	0.6114	5.09
30	0.8750	7.29	65	0.7179	5.98	100	0.6087	5.07
31	0.8696	7.24	66	0.7143	5.95	101	0.6061	5.05
32	0.8642	7.20	67	0.7107	5.92	102	0.6034	5.03
33	0.8589	7.15	68	0.7071	5.89	103	0.6009	5.00
34	0.8537	7.11	69	0.7035	5.86	104	0.5983	4.98
35	0.8485	7.07	70	0.7000	5.83	105	0.5957	4.96
36	0.8434	7.03	71	0.6965	5.80	106	0.5932	4.94
37	0.8383	6.98	72	0.6931	5.78	107	0.5907	4.92
38	0.8333	6.94	73	0.6897	5.75	108	0.5882	4.90
39	0.8284	6.90	74	0.6863	5.72	109	0.5858	4.88
40	0.8235	6.86	75	0.6829	5.69	110	0.5833	4.86
41	0.8187	6.82	76	0.6796	5.66			
42	0.8140	6.78	77	0.6763	5.63			
43	0.8092	6.74	78	0.6731	5.60			
44	0.8046	6.70	79	0.6699	5.58			

* one lb per gallon equals 119.8 kg/m^3

Table 5-11H. Specific Gravity Equivalents for Degrees Baumé
(Liquids Heavier than Water)

Baumé	Specific Gravity	Lbs per Gallon*	Baumé	Specific Gravity	Lbs per Gallon*	Baumé	Specific Gravity	Lbs per Gallon*
0	1.0000	8.33	25	1.2083	10.07	50	1.5263	12.72
1	1.0069	8.38	26	1.2185	10.16	51	1.5426	12.85
2	1.0140	8.46	27	1.2288	10.24	52	1.5591	12.99
3	1.0211	8.51	28	1.2393	10.32	53	1.5761	13.13
4	1.0284	8.56	29	1.2500	10.41	54	1.5934	13.27
5	1.0357	8.63	30	1.2609	10.51	55	1.6111	13.42
6	1.0432	8.69	31	1.2719	10.59	56	1.6292	13.57
7	1.0507	8.75	32	1.2832	10.69	57	1.6477	13.72
8	1.0584	8.81	33	1.2946	10.78	58	1.6667	13.87
9	1.0662	8.88	34	1.3063	10.84	59	1.6860	14.04
10	1.0741	8.94	35	1.3182	10.98	60	1.7059	14.21
11	1.0821	9.01	36	1.3303	11.09	61	1.7262	14.38
12	1.0902	9.09	37	1.3426	11.18	62	1.7470	14.55
13	1.0985	9.15	38	1.3551	11.29	63	1.7683	14.72
14	1.1069	9.21	39	1.3679	11.39	64	1.7901	14.91
15	1.1154	9.29	40	1.3810	11.51	65	1.8125	15.10

Table 5-11H. Specific Gravity Equivalents for Degrees Baumé (continued)

Baumé	Specific Gravity	Lbs per Gallon*	Baumé	Specific Gravity	Lbs per Gallon*	Baumé	Specific Gravity	Lbs per Gallon*
16	1.1240	9.36	41	1.3942	11.61	66	1.8354	15.29
17	1.1328	9.43	42	1.4078	11.72	67	1.8590	15.48
18	1.1417	9.51	43	1.4216	11.84	68	1.8831	15.68
19	1.1508	9.59	44	1.4356	11.96	69	1.9079	15.89
20	1.1600	9.67	45	1.4500	12.08	70	1.9333	16.10
21	1.1694	9.74	46	1.4646	12.21	71	1.9595	16.32
22	1.1789	9.81	47	1.4796	12.33	72	1.9864	16.55
23	1.1885	9.90	48	1.4948	12.46	73	2.0139	16.78
24	1.1983	9.99	49	1.5104	12.58	74	2.0423	17.01
						75	2.0714	17.25

This table and Table 5-11G have been adopted by the U.S. National Bureau of Standards from the formula:

Liquids heavier than water, degrees Baumé $= 145 - \frac{145}{R}$

Liquids lighter than water, degrees Baumé $= \frac{140}{R} - 130$

Where R is the ratio of the density of a liquid to that of distilled water, both densities being taken at 60°F (15.5°C).

NOTE: Degrees API, used to measure the density of petroleum oils, differ slightly from the Baumé scale for liquids lighter than water. API specific gravity is determined by a similar formula, using the figures 141.5 and 131.5 instead of 140 and 130.

* One lb per gallon equals 119.8 kg/m^3.

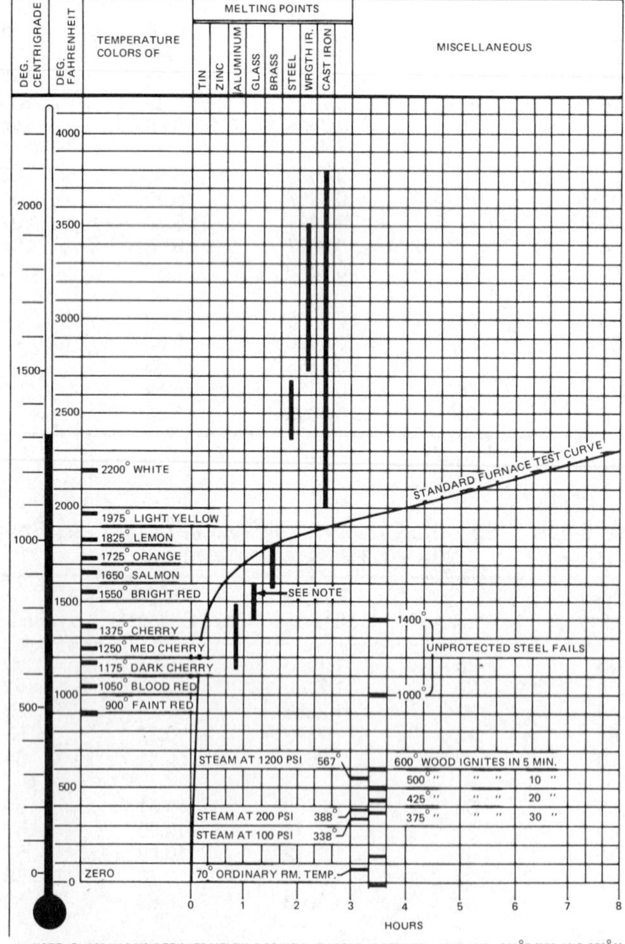

NOTE: GLASS HAS NO DEFINITE MELTING POINT BUT SOFTENS BETWEEN 1400 AND 1600°F (750 AND 850°C)

FIG. 5-1A. Temperature colors, melting points of various substances, and data on the behavior of some materials at elevated temperatures and pressures.

Table 5-11l. Specific Gravity Equivalents, Degrees API

The specific gravity of gasoline and other petroleum products is commonly given in Degrees API (American Petroleum Institute). The API degree is based on the following formula:

$$\text{Degrees API} = \frac{141.5}{\text{specific gravity}} - 131.5.$$

The standard method of test for gravity of petroleum and petroleum products by means of a hydrometer specifies 60°F as the test temperature, and the formula is based on the specific gravity of the liquid tested at 60°F as compared with water at 60°F.

Degrees API	Specific Gravity at 60/60°F	Pounds per Gallon at 60°F	Degrees API	Specific Gravity at 60/60°F	Pounds per Gallon* at 60°F	Degrees API	Specific Gravity at 60/60°F	Pounds per Gallon* at 60°F
0	1.076	8.962	35	0.8498	7.076	70	0.7022	5.845
1	1.066	8.895	36	0.8448	7.034	71	0.6988	5.817
2	1.060	8.828	37	0.8398	6.993	72	0.6953	5.788
3	1.052	8.762	38	0.8348	6.951	73	0.6919	5.759
4	1.044	8.698	39	0.8299	6.910	74	0.6886	5.731
5	1.037	8.634	40	0.8251	6.870	75	0.6852	5.703
6	1.029	8.571	41	0.8203	6.830	76	0.6819	5.676
7	1.022	8.509	42	0.8155	6.790	77	0.6787	5.649
8	1.014	8.448	43	0.8109	6.752	78	0.6754	5.622
9	1.007	8.388	44	0.8063	6.713	79	0.6722	5.595
10	1.0000	8.328	45	0.8017	6.675	80	0.6690	5.568
11	0.9930	8.270	46	0.7972	6.637	81	0.6659	5.542
12	0.9861	8.212	47	0.7927	6.600	82	0.6628	5.516
13	0.9792	8.155	48	0.7883	6.563	83	0.6597	5.491
14	0.9725	8.099	49	0.7839	6.526	84	0.6566	5.465
15	0.9659	8.044	50	0.7796	6.490	85	0.6536	5.440
16	0.9593	7.989	51	0.7753	6.455	86	0.6506	5.415
17	0.9529	7.935	52	0.7711	6.420	87	0.6476	5.390
18	0.9465	7.882	53	0.7669	6.385	88	0.6446	5.365
19	0.9402	7.830	54	0.7628	6.350	89	0.6417	5.341
20	0.9340	7.778	55	0.7587	6.316	90	0.6388	5.316
21	0.9279	7.727	56	0.7547	6.283	91	0.6360	5.293
22	0.9218	7.676	57	0.7507	6.249	92	0.6331	5.269
23	0.9159	7.627	58	0.7467	6.216	93	0.6303	5.246
24	0.9100	7.578	59	0.7428	6.184	94	0.6275	5.222
25	0.9042	7.529	60	0.7389	6.151	95	0.6247	5.199
26	0.8984	7.481	61	0.7351	6.119	96	0.6220	5.176
27	0.8927	7.434	62	0.7313	6.087	97	0.6193	5.154
28	0.8871	7.387	63	0.7275	6.056	98	0.6166	5.131
29	0.8816	7.341	64	0.7238	6.025	99	0.6139	5.109
30	0.8762	7.296	65	0.7201	5.994	100	0.6112	5.086
31	0.8708	7.251	66	0.7165	5.964			
32	0.8654	7.206	67	0.7128	5.934			
33	0.8602	7.163	68	0.7093	5.904			
34	0.8550	7.119	69	0.7057	5.874			

* "Apparent weight," or weight when weighed in air. True weight corrected for weight of air, differs by less than 1/10 of 1 percent. One lb per gallon equals 119.8 kg/m^3. 60°F equals 15.55°C.

TABLE 5-11J. Materials Subject to Spontaneous Heating
(Originally prepared by the NFPA Committee on Spontaneous Heating and Ignition which has been discontinued. Omission of any material does not necessarily indicate that it is not subject to spontaneous heating.)

Name	Tendency to Spontaneous Heating	Usual Shipping Container or Storage Method	Precautions Against Spontaneous Heating	Remarks
Alfalfa Meal	High	Bags, Bulk	Avoid moisture extremes. Tight cars for transportation are essential.	Many fires attributed to spontaneous heating probably caused by sparks, burning embers, or particles of hot metal picked up by the meal during processing. Test fires caused in this manner have smoldered for 72 hours before becoming noticeable.
Burlap Bags "Used"	Possible	Bales	Keep cool and dry.	Tendency to heat dependent on previous use of bags. If oily would be dangerous.
Castor Oil	Very slight	Metal Barrels, Metal Cans in Wooden Boxes	Avoid contact of leakage from containers with rags, cotton, or other fibrous combustible materials.	Possible heating of saturated fabrics in badly ventilated piles.
Charcoal	High	Bulk, Bags	Keep dry. Supply ventilation.	Hardwood charcoal must be carefully prepared and aged. Avoid wetting and subsequent drying.
Coal, Bituminous	Moderate	Bulk	Store in small piles. Avoid high temperatures.	Tendency to heat depends upon origin and nature of coals. High volatile coals are particularly liable to heat.
Cocoa Bean Shell Tankage	Moderate	Burlap Bags, Bulk	Extreme caution must be observed to maintain safe moisture limits.	This material is very hygroscopic and is liable to heating if moisture content is excessive. Precaution should be observed to maintain dry storage, etc.
Cocoanut Oil	Very slight	Drums, Cans, Glass	Avoid contact of leakage from containers with rags, cotton, or other fibrous combustible materials.	Only dangerous if fabrics, etc., are impregnated.
Cod Liver Oil	High	Drums, Cans, Glass	Avoid contact of leakage from containers with rags, cotton, or other fibrous combustible materials.	Impregnated organic materials are extremely dangerous.
Colors in Oil	High	Drums, Cans, Glass	Avoid contact of leakage from containers with rags, cotton, or other fibrous combustible materials.	May be very dangerous if fabrics, etc., are impregnated.
Copra	Slight	Bulk	Keep cool and dry.	Heating possible if wet and hot.
Corn-Meal Feeds	High	Burlap Bags, Paper Bags, Bulk	Material should be processed carefully to maintain safe moisture content and to cure before storage.	Usually contains an appreciable quantity of oil which has rather severe tendency to heat.
Corn Oil	Moderate	Barrels, Tank Cars	Avoid contact of leakage from containers with rags, cotton, or other fibrous combustible materials.	Dangerous heating of meals, etc., unlikely unless stored in large piles while hot.
Cottonseed	Low	Bags, Bulk	Keep cool and dry.	Heating possible if piled wet and hot.
Cottonseed Oil	Moderate	Barrels, Tank Cars	Avoid contact of leakage from containers with rags, cotton, or other fibrous combustible materials.	May cause heating of saturated material in badly ventilated piles.
Distillers' Dried Grains with oil content (Brewers' grains)	Moderate	Bulk	Maintain moisture 7 percent to 10 percent. Cool below 100°F (38°C) before storage.	Very dangerous if moisture content is 5 percent or lower.
No oil content	Moderate	Bulk	Maintain moisture 7 percent to 10 percent. Cool below 100°F (38°C) before storage.	Very dangerous if moisture content is 5 percent or lower.
Feeds, various	Moderate	Bulk, Bags	Avoid extremely low or high moisture content.	Ground feeds must be carefully processed. Avoid loading or storing unless cooled.
Fertilizers Organic, Inorganic, Combination of both	Moderate	Bulk, Bags	Avoid extremely low or high moisture content.	Organic fertilizers containing nitrates must be carefully prepared to avoid combinations that might initiate heating.
Mixed, Synthetic, containing nitrates and organic matter	Moderate	Bulk, Bags	Avoid free acid in preparation.	Insure ventilation in curing process by small piles or artificial drafts. If stored or loaded in bags, provide ventilation space between bags.

TABLE 5-11J. Materials Subject to Spontaneous Heating (Cont.)

Name	Tendency to Spontaneous Heating	Usual Shipping Container or Storage Method	Precautions Against Spontaneous Heating	Remarks
Fish meal	High	Bags, Bulk	Keep moisture 6 percent to 12 percent. Avoid exposure to heat.	Dangerous if overdried or packaged over 100°F (38°C).
Fish Oil	High	Barrels, Drums, Tank Cars	Avoid contact of leakage from containers with rags, cotton, or other fibrous combustible materials.	Impregnated porous or fibrous materials are extremely dangerous. Tendency of various fish oils to heat varies with origin.
Fish Scrap	High	Bulk, Bags	Avoid moisture extremes.	Scrap loaded or stored before cooling is extremely liable to heat.
Foam Rubber in Consumer Products	Moderate		Where possible remove foam rubber pads, etc., from garments to be dried in dryers or over heaters. If garments containing foam rubber parts have been artificially dried, they should be thoroughly cooled before being piled, bundled, or put away. Keep heating pads, hair dryers, other heat sources from contact with foam rubber pillows, etc.	Foam rubber may continue to heat spontaneously after being subjected to forced drying as in home or commercial dryers and after contact with heating pads and other heat sources. Natural drying does not cause spontaneous heating.
Grain (various kinds)	Very slight	Bulk, Bags	Avoid moisture extremes.	Ground grains may heat if wet and warm.
Hay	Moderate	Bulk, Bales	Keep dry and cool.	Wet or improperly cured hay is almost certain to heat in hot weather. Baled hay seldom heats dangerously.
Hides	Very slight	Bales	Keep dry and cool.	Bacteria in untreated hides may initiate heating.
Iron Pyrites	Moderate	Bulk	Avoid large piles. Keep dry and cool.	Moisture accelerates oxidation of finely divided pyrites.
Istle	Very slight	Bulk, Bales	Keep cool and dry.	Heating possible in wet material. Unlikely under ordinary conditions. Partially burned or charred fiber is dangerous.
Jute	Very slight	Bulk	Keep cool and dry.	Avoid storing or loading in hot wet piles. Partially burned or charred material is dangerous.
Lamp Black	Very slight	Wooden Cases	Keep cool and dry.	Fires most likely to result from sparks or included embers, etc., rather than spontaneous heating.
Lanolin	Negligible	Glass, Cans, Metal Drums, Barrels	Avoid contact of leakage from containers with rags, cotton, or other fibrous combustible materials.	Heating possible on contaminated fibrous matter.
Lard Oil	Slight	Wooden Barrels	Avoid contact of leakage from containers with rags, cotton, or other fibrous combustible materials.	Dangerous on fibrous combustible substances.
Lime, unslaked (Calcium Oxide, Pebble Lime, Quicklime)	Moderate	Paper Bags, Wooden Barrels, Bulk	Keep dry. Avoid hot loading.	Wetted lime may heat sufficiently to ignite wood containers, etc.
Linseed	Very slight	Bulk	Keep cool and dry.	Tendency to heat dependent on moisture and oil content.
Linseed Oil	High	Tank Cars, Drums, Cans, Glass	Avoid contact of leakage from containers with rags, cotton, or other fibrous combustible materials.	Rags or fabrics impregnated with this oil are extremely dangerous. Avoid piles, etc. Store in closed containers, preferably metal.
Manure	Moderate	Bulk	Avoid extremes of low or high moisture contents. Ventilate the piles.	Avoid storing or loading uncooled manures.
Menhaden Oil	Moderate to high	Barrels, Drums, Tank Cars	Avoid contact of leakage from containers with rags, cotton, or other fibrous combustible materials.	Dangerous on fibrous product.
Metal Powders*	Moderate	Drums, etc.	Keep in closed containers.	Moisture accelerates oxidation of most metal powders.
Metal Turnings*	Practically none	Bulk	Not likely to heat spontaneously.	Avoid exposure to sparks.
Mineral Wool	None	Pasteboard Boxes, Paper Bags	Noncombustible. If loaded hot may ignite containers and other combustible surroundings.	This material is mentioned in this table only because of general impression that it heats spontaneously.

TABLE 5-11J. Materials Subject to Spontaneous Heating (Cont.)

Name	Tendency to Spontaneous Heating	Usual Shipping Container or Storage Method	Precautions Against Spontaneous Heating	Remarks
Mustard Oil, Black	Low	Barrels	Avoid contact of leakage with rags, cotton or other fibrous combustible materials.	Avoid contamination of fibrous combustible materials.
Oiled Clothing	High	Fiber Boxes	Dry thoroughly before packaging.	Dangerous if wet material is stored in piles without ventilation.
Oiled Fabrics	High	Rolls	Keep ventilated. Dry thoroughly before packing.	Improperly dried fabrics extremely dangerous. Tight rolls are comparatively safe.
Oiled Rags	High	Bales	Avoid storing in bulk in open.	Dangerous if wet with drying oil.
Oiled Silk	High	Fiber Boxes, Rolls	Supply sufficient ventilation.	Improperly dried material is dangerous in form of piece goods. Rolls relatively safe.
Oleic Acid	Very slight	Glass Bottles, Wooden Barrels	Avoid contact of leakage from containers with rags, cotton, or other fibrous combustible materials.	Impregnated fibrous materials may heat unless ventilated.
Oleo Oil	Very slight	Wooden Barrels	Avoid contact of leakage from containers with rags, cotton, or other fibrous combustible materials.	May heat on impregnated fibrous combustible matter.
Olive Oil	Moderate to Low	Tank Cars, Drums, Cans, Glass	Avoid contact of leakage from containers with rags, cotton, or other fibrous combustible materials.	Impregnated fibrous materials may heat unless ventilated. Tendency varies with origin of oil.
Paint containing drying oil	Moderate	Drums, Cans, Glass	Avoid contact of leakage from containers with rags, cotton, or other fibrous combustible materials.	Fabrics, rags, etc., impregnated with paints that contain drying oils and driers are extremely dangerous. Store in closed containers, preferably metal.
Paint Scrapings	Moderate	Barrels, Drums	Avoid large unventilated piles.	Tendency to heat depends on state of dryness of the scrapings.
Palm Oil	Low	Wooden Barrels	Avoid contact of leakage from containers with rags, cotton, or other fibrous combustible materials.	Impregnated fibrous materials may heat unless ventilated. Tendency varies with origin of oil.
Peanut Oil	Low	Wooden Barrels, Tin Cans	Avoid contact of leakage from containers with rags, cotton, or other fibrous combustible materials.	Impregnated fibrous materials may heat unless ventilated. Tendency varies with origin of oil.
Peanuts, "Red Skin"	High	Paper Bags, Cans, Fiber Board Boxes, Burlap Bags	Avoid badly ventilated storage.	This is the part of peanut between outer shell and peanut itself. Provide well ventilated storage.
Peanuts, shelled	Very slight or Negligible	Paper Bags, Cans, Fiber Board Boxes, Burlap Bags	Keep cool and dry.	Avoid contamination of rags, etc., with oil.
Perilla Oil	Moderate to High	Tin Cans, Barrels	Avoid contact of leakage from containers with rags, cotton, or other fibrous combustible materials.	Impregnated fibrous materials may heat unless ventilated. Tendency varies with origin of oil.
Pine Oil	Moderate	Glass, Drums	Avoid contact of leakage from containers with rags, cotton, or other fibrous combustible materials.	Impregnated fibrous materials may heat unless ventilated. Tendency varies with origin of oil.
Powdered Eggs	Very slight	Wooden Barrels	Avoid conditions that promote bacterial growth. Inhibit against decay. Keep cool.	Possible heating of decaying powder in storage.
Powdered Milk	Very slight	Wooden and Fiber Boxes, Metal Cans	Avoid conditions that promote bacterial growth. Inhibit against decay. Keep cool.	Possible heating by decay or fermentation.
Rags	Variable	Bales	Avoid contamination with drying oils. Avoid charring. Keep cool and dry.	Tendency depends on previous use of rags. Partially burned or charred rags are dangerous.
Red Oil	Moderate	Glass Bottles, Wooden Barrels	Avoid contact of leakage from containers with rags, cotton, or other fibrous combustible materials.	Impregnated porous or fibrous materials are extremely dangerous. Tendency varies with origin of oil.
Roofing Felts and Papers	Moderate	Rolls, Bales, Crates	Avoid over-drying the material. Supply ventilation.	Felts, etc., should have controlled moisture content. Packaging or rolling uncooled felts is dangerous.
Sawdust	Possible	Bulk	Avoid contact with drying oils. Avoid hot, humid storage.	Partially burned or charred sawdust may be dangerous.
Scrap Film (Nitrate)	Very slight	Drums and Lined Boxes	Film must be properly stabilized against decomposition.	Nitrocellulose film ignites at low temperature. External ignition more likely than spontaneous heating. Avoid exposure to sparks, etc.

TABLE 5-11J. Materials Subject to Spontaneous Heating (Cont.)

Name	Tendency to Spontaneous Heating	Usual Shipping Container or Storage Method	Precautions Against Spontaneous Heating	Remarks
Scrap Leather	Very slight	Bales, Bulk	Avoid contamination with drying oils.	Oil-treated leather scraps may heat.
Scrap Rubber or Buffings	Moderate	Bulk, Drums	Buffings of high rubber content should be shipped and stored in tight containers.	Sheets, slabs, etc., are comparatively safe unless loaded or stored before cooling thoroughly.
Sisal	Very slight	Bulk, Bales	Keep cool and dry.	Partially burned or charred material is particularly liable to ignite spontaneously.
Soybean Oil	Moderate	Tin Cans, Barrels, Tank Cars	Avoid contact with rags, cotton, or fibrous materials.	Impregnated fibrous materials may heat unless well ventilated.
Sperm Oil—See Whale Oil				
Tankage	Variable	Bulk	Avoid extremes of moisture contents. Avoid loading or storing while hot.	Very dry or moist tankages often heat. Tendency more pronounced if loaded or stored before cooling.
Tung Nut Meals	High	Paper Bags, Bulk	Material must be very carefully processed and cooled thoroughly before storage.	These meals contain residual oil which has high tendency to heat. Material also susceptible to heating if over-dried.
Tung Oil	Moderate	Tin Cans, Barrels, Tank Cars	Avoid contact of leakage from containers with rags, cotton, or other fibrous combustible materials.	Impregnated fibrous materials may heat unless ventilated. Tendency varies with origin of oil.
Turpentine	Low	Tin, Glass, Barrels	Avoid contact of leakage from containers with rags, cotton, or other fibrous combustible materials.	Has some tendency to heat but less so than the drying oils. Chemically active with chlorine compounds and may cause fire.
Varnished Fabrics	High	Boxes	Process carefully. Keep cool and ventilated.	Thoroughly dried varnished fabrics are comparatively safe.
Wallboard	Slight	Wrapped Bundles, Pasteboard Boxes	Maintain safe moisture content. Cool thoroughly before storage.	This material is entirely safe from spontaneous heating if properly processed.
Waste Paper	Moderate	Bales	Keep dry and ventilated.	Wet paper occasionally heats in storage in warm locations.
Whale Oil	Moderate	Barrels and Tank Cars	Avoid contact of leakage from containers with rags, cotton, or other fibrous combustible materials.	Impregnated fibrous materials may heat unless ventilated. Tendency varies with origin of oil.
Wool Wastes	Moderate	Bulk, Bales, etc.	Keep cool and ventilated or store in closed containers. Avoid high moisture.	Most wool wastes contain oil, etc., from the weaving and spinning and are liable to heat in storage. Wet wool wastes are very liable to spontaneous heating and possible ignition.

*Refers to iron, steel, brass, aluminum, and other common metals, for information on magnesium, sodium, zirconium, etc.

SECTION 6
HAZARDOUS WASTES AND MATERIALS

THE ROLE OF THE FIRE PROTECTION COMMUNITY IN HAZARDOUS WASTE CONTROL

Phillip Blye and James Yess

The days are gone when the only activity of the fire service was extinguishing fires. Modern society has become far too complex for fire fighting to be the exclusive activity of fire fighters. Since the scope of the fire fighters' duties has been extended to include the protection of life in emergencies arising from the improper production, storage, transportation, treatment and disposal of hazardous materials it is essential that fire fighters possess information on the role of the fire protection community in the management of these substances. This chapter will present general information on the subject of hazardous waste, direct the reader to applicable federal legislation, and provide general advice regarding emergency preparedness and possible responses to an incident. More detailed and technical accounts are provided in the subsequent chapters of this section. Section 13, "Transportation Fire Hazards" will provide other essential information on handling hazardous materials incidents which may face the Fire Protection Community.

HAZARDOUS WASTE: GENERAL DEFINITION

Every day thousands of dangerous or potentially dangerous substances are used, transported, manufactured, stored, or disposed of in cities and towns throughout the nation. (See Fig. 6-1A.) Hazardous wastes can be called pollutants, toxic substances, hazardous materials, or hazardous substances. Confusion arises because these words are used synonymously to describe all hazardous commodities. For the purpose of this chapter, a hazardous waste may be defined as any material that presents an actual or potential danger to human health and safety or to other living organisms in the environment. The criteria used by the Environmental Protection Agency (EPA) for judging materials to be hazardous are: (1) Ignitability, (2) Reactivity, (3) Corrosivity, and (4) Toxicity. These properties are described in detail in the "Hazardous Waste Control" chapter of this HANDBOOK.

THE HAZARDOUS WASTE PROBLEM

Prior to 1970 the American public was generally unaware of the dangers involved with the transport, storage and disposal of hazardous wastes. The careless disposal of what was merely considered to be trash eventually either found its way into the environment, polluting water supplies and soil, or was encountered by fire fighters attempting to extinguish exploding barrels of unknown liquids in chemical disposal facilities. Before the days of federal regulation, hazardous wastes could be deposited in just about any hole in the ground or body of unprotected water.

FIG. 6-1A. Even small unregulated businesses may store or dump solvents, pesticides and other materials in sufficient quantity to be dangerous to fire fighters responding to an alarm.

According to the EPA there are from thirty to sixty thousand sites across the United States where improper disposal of hazardous waste has taken place. In 1981 alone American industry generated more than 165 million tons

Capt. Blye is Chairperson of the Department of Fire Science Technology at Massasoit Community College in Brockton, MA. He is also currently employed in the fire service. Dr. Yess is an educator and administrator who has developed educational materials for the NFPA.

(150 million metric tons) of hazardous waste. Yet records show that only 53 million tons (58 million metric tons) can be accounted for in legal disposal sites. The U.S. Office of Technology Assessment (OTA) estimates that by 1990 at least $12 billion will be spent on hazardous waste disposal each year in the United States. In 1983 less than $5 billion was spent for this purpose. The high cost of properly disposing of hazardous wastes and the increasing waste produced each year are clear incentives for illegal waste contractors called "midnight dumpers." (See Fig. 6-1B.) It

FIG. 6-1B. Illegal and abandoned dumpsites can pose a serious threat. Hazardous wastes may remain contained for years, and then disperse into the environment without warning.

is important to realize that hazardous wastes need not enter the environment in large quantities in order to have catastrophic effects. Dioxin, for example, which has been the subject of press coverage in both the United States and Great Britain, is at least ten thousand times more dangerous than cyanide. In relatively small concentrations dioxin can cause acute skin disease and is suspected of causing cancer. The safe disposal of hazardous wastes such as dioxin is very expensive because the disposal process requires a specialized incinerator capable of maintaining temperatures as high as 2,372°F (1,300°C). (See Fig. 6-1C.)

It is not unlikely for fire service personnel to encounter old chemical disposal sites in their own communities where corroding barrels leak unidentified flammable solvents into the ground, creating an explosive condition. Dumping sites for industrial byproducts could remain undetected for years; only to be discovered when their presence is an immediate danger to life. Even in the days of strict regulation, hazardous wastes could be stored unlawfully in unsecured containers having no identification. Apart from the ever present risks due to fire from these dangerous substances, fire service personnel responding to an emergency can endanger their health by merely exposing themselves to these materials. It is possible that such exposure could cause chemical burns, asphyxiation, poisoning, or injury from explosion. In addition, more long term, but delayed effects, could result in respiratory disease, cancer, and kidney or liver failure.

Hazardous wastes can find their way into the environment in several ways:

1. Ground water contamination through leaching.
2. Surface water contamination through runoff or overflow.
3. Air pollution by open burning, evaporation, sublimation, or wind erosion.
4. Fire and explosion.
5. Poison through the food chain.

While every hazard is not directly related to a fire emergency, fire service personnel may be the first on the scene to investigate the source of a foul odor or mysterious liquid oozing from the ground. In these circumstances, it is as important to know what not to do as it is to know what to do.

THE ORIGINS OF HAZARDOUS WASTES

Hazardous wastes are usually the byproducts created during a manufacturing process. The production of such common commodities as metals, plastics, paints, jewelry,

FIG. 6-1C. Regulated hazardous wastes are currently disposed of by incineration, chemical neutralization, burial in a secured landfill or by other approved methods.

electronic components, pesticides, clothing, medicines, and fertilizers will yield hazardous wastes as byproducts.

Organizations as diverse as farms, hospitals, and colleges, which are normally considered to be engaged in "clean" activities produce their share of hazardous wastes. Even small unregulated businesses may store or dump solvents, pesticides, and other materials in sufficient quantities to be dangerous to fire fighters responding to an alarm. Therefore, it is extremely important for fire service personnel to realize that there are numerous activities that generate hazardous materials within their communities. It is in their own best interest, as well as in the interests of those they are dedicated to protect, to identify all facilities that produce, store, or serve as a dumping site for hazardous wastes. (See Fig. 6-1D.)

TREATMENT AND DISPOSAL OF HAZARDOUS WASTES

Under the *Resource Conservation and Recovery Act* of 1976 the EPA has established standards for the treatment and disposal of hazardous wastes. The methods used to safely treat, recycle or reduce the dangers of hazardous wastes include those outlined below.

In general there are six disposal treatment methods: (1) disposal at sea, (2) landfill, (3) long term storage, (4) physical treatment, (5) chemical treatment, and (6) biological treatment. Among these are the following:

1. Aqueous treatment — a process which chemically, physically and biologically precipitates metal hydroxides and oxides, neutralizes acidic components, and removes organic contaminants from industrial waste water.
2. Biological treatment — a process which employs living microorganisms that feed on and decompose certain wastes.
3. Carbon absorption — a process involving the use of specifically treated carbon to pickup wastes such as organic compounds.
4. Dechlorination — a technique used to detoxify chlorinated substances by adding nontoxic substances such as hydrogen.

FIG. 6-1D. It is important to inventory and maintain a file of locations where hazardous wastes are likely to be handled or stored. A thorough survey can save lives.

5. Detoxification — a method of incorporating various processes such as coagulation, distillation, oxidation, precipitation, and reduction to render waste liquids less harmful.
6. Incineration — the use of heat at sufficiently high temperatures to convert the waste material into harmless gases and inert solids.
7. Neutralization — a process which renders highly acidic or alkaline wastes inactive by combining proper concentrations of acids with alkalies.
8. Oxidation — the applications of a chemical reaction which transforms primarily organic materials such as cyanides, phenols, and organic sulfur compounds into carbon dioxide and water.
9. Solidification and stabilization — the conversion of wastes into solid masses that are not readily transported or dissolved by water or other liquids in the environment.
10. Recovery and recycling — the process of extracting valuable materials such as metals, acids, alkalies, fuels, and solvents from waste products for reuse.
11. Secure landfill — a facility designed, constructed, and operated in compliance with EPA standards where hazardous wastes can be stored or disposed of in the ground.

Under proper supervision and with appropriate methods, much of the hazardous wastes produced in industrialized nations can be converted to less harmful substances. The major problems for fire service personnel regarding hazardous materials come as the result of improper storage, accidents, deteriorating conditions at unidentified or abandoned dumping and storage sites, and the transportation of these dangerous substances over land and water routes.

SHIPPING OF HAZARDOUS WASTES

Under the *Resource Conservation and Recovery Act* (RCRA) of 1976, any industry generating more than a specified minimum amount of hazardous waste must follow federal guidelines for storage, transportation, and disposal. A "cradle-to-grave" bookkeeping and reporting system is mandated by this legislation, but not all shippers abide by the requirements of the law. Accidents involving hazardous wastes can be dangerous to fire fighters especially when a shipment is being transported illegally. This noncompliance with RCRA guidelines results in deceptive or poorly labeled containers and very little if any documentation, such as a hazardous waste manifest, accompanying the shipment. Regulations of both the Department of Transportation (DOT) and the Environmental Protection Agency (EPA) govern the transport of hazardous wastes. The Hazardous Waste Manifest system, among other things, ensures that the material is properly identified; states the place of origin and destination for treatment, storage, or disposal; classifies the waste; provides the quantity and flashpoint of the substance; and gives special handling instructions. (See Fig. 6-1E.) Each shipper handling the hazardous waste cargo must sign the manifest certifying acceptance.

FEDERAL AND STATE LAWS

Federal Legislation

Resource and Conservation Recovery Act (RCRA): Several laws enacted by Congress and state legislatures were written specifically to protect public health and the environment from accidents or negligence dealing with hazardous wastes. In 1976 Congress passed major legislation establishing a uniform national policy for hazardous and solid waste disposal. This act is called the *Resource and Conservation Recovery Act* (RCRA). Under this act the Environmental Protection Agency (EPA) is authorized to devise management plans for the safe disposal of hazardous wastes and the recovery of energy from solid waste. In addition, EPA, under RCRA, can regulate the treatment, storage, transport, and disposal of hazardous wastes.

Comprehensive guidelines have been established by EPA for tracking the movement, treatment, storage and disposal of these waste products. Among the provisions of the guidelines are the following:

1. Identification and listing of materials classified as hazardous waste.
2. A system of record keeping and labeling.
3. Procedures for providing correct information regard-

ing hazardous waste contents to and by persons transporting, storing or disposing of this waste.

4. A manifest and permit system regulating the transport of wastes to authorized sites.
5. Requirements that transporters of hazardous wastes properly label hazardous waste shipments and carry them only to treatment, storage, or disposal sites licensed under RCRA.
6. Requirements that operators of treatment, storage or disposal facilities maintain records and comply with the manifest system. Only EPA approved methods of treatment, storage, and disposal can be used as indicated in the permit issued by EPA.

HAZARDOUS WASTE MANIFEST
ORIGINAL — NOT NEGOTIABLE

MANIFEST DOCUMENT NUMBER

SHIPPER NUMBER

NAME OF CARRIER (SCAC) CARRIER NUMBER

TO: T /S /D FACILITY	FROM: Generator
E.P.A. ID Code No.	E.P.A. ID Code No.
Address	Address
Destination	Origin
Phone	Phone

WASTE INFORMATION

NO. OF UNITS & CONTAINER TYPE	HM	EPA HAZ. WASTE ID #	DESCRIPTION AND CLASSIFICATION (Proper Shipping Name. Class and Identification Number per 172.101. 172.202. 172.203	UN # or NA #	EXEMPTION OR NO LABELS REQUIRED	FLASH POINT (IN °C) WHEN REQ'D	UNITS WT/VOL	TOTAL QUANTITY	RATE	CHARGES (For Carrier Use Only)

SPECIAL HANDLING INSTRUCTIONS

If an RQ commodity is spilled on a waterway or adjoining land, the incident must be promptly reported to the Federal government at 1-800-424-8802 (toll free) or 202-426-2675 (toll call). If other DOT Hazardous Materials are discharged creating a serious situation, call shipper's telephone number or Chemtrec 1-800-424-9300 immediately.

COMMENTS

On "Collect on Delivery" shipments, the letters "COD" must appear before consignee's name or as otherwise provided in Item 430, Sec. 1

PLACARDS TENDERED Yes ☐ No ☐

REMIT C.O.D. TO: ADDRESS

COD Amt: $

C.O.D. FEE: PREPAID ☐ COLLECT ☐ $

Note—Where the rate is dependent on value, shippers are required to state specifically in writing the agreed or declared value of the property. The agreed or declared value of the property is hereby specifically stated by the shipper to be not exceeding $ ______ per ______

If the shipment moves between two ports by a carrier by water, the law requires that the bill of lading shall state whether it is "carrier's or shipper's weight." ______ Signature

Subject to Section 7 of the conditions, if this shipment is to be delivered to the consignee without recourse on the consignor, the consignor shall sign the following statement: The carrier shall not make delivery of this shipment without payment of freight and all other lawful charges. ______ (Signature of Consignor)

TOTAL CHARGES: $

FREIGHT CHARGES

FREIGHT PREPAID except when box at right is checked ☐ Check box if charges are to be collect

RECEIVED, subject to the classifications and tariffs in effect on the date of the issue of this Bill of Lading, the property described above in apparent good order, except as noted (contents and condition of contents of packages unknown), marked, consigned, and destined as indicated above which said carrier (the word carrier being understood throughout this contract as meaning any person or corporation in possession of the property under the contract) agrees to carry to its usual place of delivery at said destination, if on its route, otherwise to deliver to another carrier on the route to said destination. It is mutually agreed as to each carrier of all or any of, said property over all or any portion of said route to destination and as to each party at any time interested in all or any said property, that every service to be performed hereunder shall be subject to all the bill of lading terms and conditions in the governing classification on the date of shipment.

Shipper hereby certifies that he is familiar with all the bill of lading terms and conditions in the governing classification and the said terms and conditions are hereby agreed to by the shipper and accepted for himself and his assigns.

CERTIFICATION

This is to certify that the above named materials are properly classified, described, packaged, marked and labeled, and are in proper condition for transportation according to the applicable regulations of the Department of Transportation and the E.P.A.

Generator Signature ______ Date ______

TRANSPORTER #1 ______ E.P.A. ID No. ______
Address ______
City ______ State ______ Zip ______ Phone ______
This is to certify acceptance of the hazardous waste shipment.
Transporter No. 1 Signature ______ Date ______

TRANSPORTER #2 ______ E.P.A. ID No. ______
Address ______
City ______ State ______ Zip ______ Phone ______
This is to certify acceptance of the hazardous waste shipment.
Transporter No. 2 Signature ______ Date ______

TREATMENT/STORAGE/DISPOSAL FACILITY
This is to certify acceptance of the hazardous waste for treatment, storage, or disposal.
T/S/D FACILITY Signature ______ Date ______

Printed and Sold by Unz & Co., Division of Scott Printing Corp. 190 Baldwin Ave., Jersey City, N.J. 07306 – N.J. (201) 795-5400/N.Y. (212) 344-2270 Toll Free (800) 631-3098 Form No. 35-666 TRANSPORTER #1 2

FIG. 6-1E. A typical hazardous waste manifest. In a transportation incident, a hazardous waste manifest should accompany the cargo in the cab of the vehicle. If the manifest can be safely recovered, DOT reference numbers can help officials determine the safest way to handle the incident.

7. A mandate for producers and site operators to develop contingency plans in the event of a hazardous waste emergency.
8. Encouragement for states to develop and manage their own hazardous waste programs consistent with RCRA guidelines. In the absence of a state plan, RCRA authorizes EPA to impose a federal plan for the state.

These plans must be the result of consultations with local fire, police, and hospital officials. These organizations must be included as part of the local response team. Contingency plans provide detailed information such as the industrial site plan, the on scene coordinator's phone number, and types of hazards present.

Under this act hazardous waste generators are required to:

1. Consult the list of hazardous and toxic organic and inorganic compounds published by EPA* to determine if a chemical waste is hazardous.
2. Subject all unlisted wastes to chemical analysis to determine if they possess any of the hazardous waste characteristics established in the regulations.
3. Declare a waste hazardous (based on knowledge of the materials or processes used in its production).
4. Obtain an EPA identification number by filing with the EPA's Regional Administrator.
5. Apply for a facility permit if waste is accumulated on the generator's site for more than 90 days. The waste is considered to be stored at the site when it remains on the generator's property for more than 90 days. If this condition continues, the generator must secure a facility permit for storage of hazardous wastes under Section 3005 of the *Resource Conservation and Recovery Act* (RCRA).
6. Treat, store, or dispose of hazardous wastes on their site subject to requirements under RCRA Sections 3004 and 3005.
7. Transport hazardous wastes to other locations for use, storage, treatment, or disposal in properly labeled approved containers, and prepare a Hazardous Waste Manifest.

On November 8, 1984, President Reagan signed *The Hazardous and Solid Waste Amendments of 1984* into law. It calls for changes in waste management practices and closes a number of loopholes in RCRA including:

1. Establishing tighter requirements for hazardous waste land disposal.
2. Regulating underground storage tanks for new or waste products under RCRA.
3. Bringing small quantity generators—those generating between 100 and 1000 kilograms of hazardous waste per month under RCRA's authority.
4. Giving authority to order corrective action at RCRA regulated facilities to the EPA.
5. Setting forth "minimum technological standards" such as double liners, two leachate collection systems, and groundwater monitoring for new land disposal facilities, as well as requiring interim status facilities to either be retrofitted or stop receiving, storing, or treating hazardous waste.
6. Establishing new requirements for delisting hazardous wastes.

Superfund: In 1980 the Congress established a fund of $1.6 billion called the "Superfund" (PL 96-510) to be administered by the EPA. Under this legislation the EPA is authorized to inventory all uncontrolled hazardous waste sites in the nation. If the substances at these sites pose an immediate danger to public health and/or environment, and those responsible for the contamination cannot be

*Title 40, *Code of Federal Regulations*, Part 261, Appendix VIII.

identified or cannot pay for the cleanup, the EPA can use the Superfund to clean up chemical spills or toxic wastes.

Toxic Substances Control Act (TOSCA): TOSCA, passed by Congress in 1976, establishes regulations with respect to the inventory and testing of all chemical substances manufactured or processed in the United States. This act once again makes EPA the central implementation arm of the federal government. TOSCA authorizes EPA to:

1. Develop a uniform listing of all chemical substances.
2. Establish a testing procedure for chemicals already in use and any one of the approximately 1,000 new chemicals developed each year.
3. Determine if these chemicals present an unreasonable risk to health or the environment.
4. Prohibit or limit the manufacture, processing, use, application, and concentration of such chemicals.
5. Recall or seize by civil action hazardous substances which are determined to be imminently harmful to health or the environment.

Federal Water Pollution Control Act: In 1970 the *Federal Water Pollution Control Act* (PL 92-500) was amended in Section 311 to include Oil and Hazardous Spills. Through this legislation both the EPA and the U.S. Coast Guard are mandated to regulate spills of oil and/or other hazardous substances that threaten coastal waters and inland waterways.

State Legislation

Legislation for the control of hazardous wastes varies from state to state. At least 40 states have legislation dealing with the control of hazardous wastes. To receive "full authorization" to regulate hazardous wastes within its borders, a state must institute a regulatory program equivalent to federal legislation and have adequate means of enforcement. For further information on state hazardous waste regulations consult with the appropriate state agency.

PREVENTION AND RESPONSE TO A HAZARDOUS WASTE EMERGENCY

Prevention

Fire service personnel face new challenges with respect to the ever expanding use and development of a broad range of substances for use by industry. Undiscovered multigallon hazardous waste lagoons or a small unlabeled barrel stored in an inconspicious corner of an outbuilding are two of the potential causes of disaster that await fire fighters.

Incidents involving hazardous wastes are complicated by poor identification of hazardous material storage sites, improper labeling of containers, and illegal disposal. The first step for fire service personnel in responding to an emergency involving known or unknown hazardous wastes is planning. These plans should also include preparations for transportation incidents and environmental emergencies which may totally involve local and regional emergency resources. The following list suggests ways in which local fire service personnel can prepare for such emergencies. Accurate record keeping of these activities is vital.

1. Develop a working knowledge of applicable state and federal regulations governing the generation, storage, transportation and disposal of hazardous wastes.
2. Form a working relationship with state and regional coordinators. Record for immediate use all telephone numbers of state and federal emergency response teams.
3. Survey, inventory, and inspect all manufacturers, businesses, educational institutions and health care providers that manufacture, transport, use, store, treat, or dispose of hazardous materials. (See Fig. 6-1F.) Remember that even small unregulated businesses may store, use, and dump hazardous materials. Your municipal dump, if not properly supervised, can be a potential ecological "time bomb."

FIG. 6-1F. Visiting industrial sites, gathering information and planning a response can save lives, jobs and valuable property.

4. Assist industries regulated by RCRA in the development of their plans for handling chemical emergencies. EPA regulations specifically provide for participation in regional and national emergency response plans. Keep copies of these plans on file at the local fire department.
5. If a facility does not fall within the scope of RCRA standards, diplomatically request information about the presence of hazardous materials and where they are located. Uncooperative managers can be reported to the EPA or appropriate state agencies.
6. Know what types of hazardous wastes are likely to be generated in and/or transported in and through your community, by what routes, and the likely disposal areas. Develop procedures to protect against the specific hazard each substance may present.
7. Help to define the roles of local fire, police and hospital emergency personnel in the duties of the local response teams.
8. Develop a comprehensive hazardous waste training program for the fire service. Include sessions designed to familiarize fire service personnel with the operating features of bulk containers such as rail cars, tank trucks and other specially designed vehicles. (See Fig. 6-1G.) Include instant recognition of DOT hazard identification symbols and the correct procedures to be followed in each incident in the training.
9. Conduct a hazardous waste emergency drill to ensure that local response teams know what their assignments are and how to properly execute them.

FIG. 6-1G. Fire service hazardous waste training programs should include sessions utilizing full protective gear such as the chemical resistant encapsulating suits shown above. (Mine Safety Appliances Company)

Immediate Response Guidelines

Planning will not be sufficient if the fire service implementing the plans is not trained in their execution, but even the best plans and training will not cover novel emergency events. There are, however, well-known basic safety precautions that can be taken to protect the lives of first responders and those in the community they seek to protect. Among the necessary actions to take in situations where the hazards from toxic wastes are unknown are the following:

1. Establish an on-site command post.
2. Observe the scene from a safe distance.
3. When approaching an incident, wear appropriate protective gear at all times. (See Fig. 6-1H.) Some hazardous wastes, however, can permeate even the best of protective outerwear.
4. Work in pairs and watch your buddy for signs of giddiness, fatigue, or other forms of abnormal behavior.
5. Try to determine the identity of the hazardous substance as soon as possible. Look for DOT and EPA numbers on container labels.
6. Locate the plant managers and plant safety officer to obtain first hand knowledge of what substances might be involved. They may also have specialized instruments such as a portable photoionizing trace gas analyzer which can be used to detect a wide variety of organic and inorganic vapors. (See Fig. 6-1I.)
7. In a transportation incident, search the cab of the vehicle, if possible, to locate the Hazardous Materials Manifest. DOT reference numbers can help officials determine the safest way to respond.
8. Remove victims and evacuate the area with the assistance of other members of the local response team.

FIG. 6-1H. When approaching an incident, appropriate protective gear should be worn at all times.

FIG. 6-1I. Industrial personnel may have instruments that can be used to identify the hazard. For example, a portable photoionizing trace gas analyzer is routinely used to detect a wide variety of organic and inorganic vapors. This equipment may be available to responding fire fighters.

Bibliography

NFPA Codes, Standards, Recommended Practices and Manuals. (See the latest *NFPA Codes and Standards Catalog* for availability of current editions of the following documents.)

NFPA 49, *Hazardous Chemicals Data.*
NFPA 491M, *Manual of Hazardous Chemical Reactions.*
NFPA 704, *Standard System for the Identification of the Fire Hazards of Materials.*

Additional Readings

Albrecht, A. R., "Handling Spills Involving Hazardous Chemicals Within the Chemical Industry," *Fire Journal*, Vol. 65, No. 3, May 1971, pp. 40-48.

Fire Protection Guide on Hazardous Materials, 8th ed., National Fire Protection Association, Quincy, MA, 1984.

NFPA, *Handling Hazardous Materials Transportation Emergencies*, (slide/tape), National Fire Protection Association, Quincy, MA, 1983.

NFPA, *Hazardous Waste and the Fire Service*, (slide/tape), National Fire Protection Association, Quincy, MA, 1984.

IDENTIFICATION OF THE HAZARDS OF MATERIALS

Revised by Martin F. Henry

This chapter covers several systems that are used to identify the hazards of materials. Such identification systems are designed to minimize the danger to emergency personnel who respond to fires and emergencies involving hazardous materials.

Section 5, Chapter 1, "Fire Hazards of Materials," Chapter 3 of this section, "Handling Hazardous Materials Transportation Emergencies," and Section 12, Chapter 1, "Transportation Fire Hazards," will provide additional information for those concerned with the identification, handling and transport of hazardous materials.

PRESENTATION OF INFORMATION

Hazards are identified in the following ways:

1. Large placards, signs, or decals are posted on storage tanks, in storage areas, and on processing equipment. Information is printed in large type for reading at a distance. These placards usually give only a statement of hazard and a simple warning, such as DO NOT ENTER.
2. Placards are posted in local areas. These are usually more detailed, and may contain instructions for procedures and safety equipment.
3. Department of Transportation (DOT) hazardous materials placards appear on many, but not all, trucks and railway equipment carrying hazardous materials.
4. Containers and shipping packages are labeled.
5. Materials are listed on the shipping papers in the possession of the vehicle operator.

Organization of Information

Hazard information is organized into recognizable parts. These are listed below in order of importance:

1. A flag to attract attention. This can be a symbol, a color, a shape, a word, or a combination of any of these such as the standardized shapes and colors of roadside traffic signs. On a package label it may be a hazard pictograph or a signal word such as DANGER.
2. A statement describing the hazard, such as FLAMMABLE or CAUSES BURNS. There may also be a symbol such as the skull and crossbones that was once used on packages of poisons.
3. Precautionary measures telling what to do or what not to do in order to avoid injury. These precautionary measures may be presented by a symbol, such as a sign on a laboratory door showing a pair of safety glasses, meaning that safety glasses are required in the area. On a container label, precautionary measures are part of the text.
4. Instruction for treatment in case of exposure. This may be symbolic, such as the symbol for a safety shower or eyebath, or it may consist of elaborate first aid instructions on a package label.
5. Special instructions for storage or handling.
6. Instructions for handling fires, leaks, or spills.

Hazard Levels

Hazard information systems recognize four classes, or levels, of hazardous materials:

1. Extremely Dangerous Materials. These materials can cause death or disabling injury on brief exposure, or they are extremely volatile flammable liquids or flammable gases, or detonable materials. A further breakdown is:
   (a) Explosives and explosively unstable materials.
   (b) High-level radioactive materials.
   (c) Highly flammable gases and materials which give off extremely flammable vapors.
   (d) Extremely toxic materials, such as parathion and hydrogen cyanide, which are so poisonous that no bodily exposure should occur.
   (e) Materials which are extremely corrosive to living tissue, such as bromine, which can injure almost instantaneously, or hydrofluoric acid, which can penetrate through the skin to the tissues beneath and cause deep, slow healing burns. Also included are materials that could cause severe eye injury.
   (f) Materials whose combustion products or products

Mr. Henry is Flammable Liquids Field Service Specialist on the staff of NFPA.

of decomposition fit the descriptions given in (a) through (e) above.

2. Dangerous Materials. These are materials that could cause injury from exposure to the detrimental effects of highly flammable or highly self-reactive materials such as:
   (a) Flammable liquids and solids.
   (b) Highly toxic materials which are likely to cause some injury or illness, but not death, from a moderate exposure.
   (c) Materials which could cause destruction of tissue, particularly the eyes, if not removed from the body in a very short time.
   (d) Moderately radioactive materials.

Dangerous materials may, at an extreme, cause some permanent but not disabling injury on exposure.

3. Hazardous Materials. These are materials that could cause temporary disability or injury which presumably would heal without permanent effects. They are moderately combustible or self-reactive, and they include:
   (a) Tear gases.
   (b) Severe irritants.
   (c) Toxic (but not highly or extremely toxic) materials.
   (d) Combustible materials which must be heated before they can be ignited.
4. Nuisance Hazards. These could cause temporary irritation or discomfort that would clear up when the exposure is ended, or they could be materials which are only slightly combustible.

Note that the materials classed as "dangerous" could cause an emergency. The materials classed as "hazardous" would be less likely to cause an emergency, but could add significantly to the seriousness of an emergency. Thus, a "combustible" liquid might not catch fire and cause an emergency. In the presence of a fire arising from some other cause, it could ignite and add to the seriousness of the emergency. In present consumer product labeling practice, materials which could cause an emergency generally use the signal word DANGER. Materials which could add to the seriousness of an emergency use the signal words WARNING and CAUTION. Note, though, that flammable liquids with flash points between 20 and 80°F (7 to 27°C) bear the signal word "WARNING."

A hazard information system is a compromise between two conflicting requirements: immediacy of information and adequacy of information. A hazard symbol, such as the familiar skull and crossbones, can be read at a glance. It has immediacy. Since it doesn't tell how dangerously poisonous the material is, nor whether it enters the body through the skin, the lungs, or the digestive system, it lacks adequacy. A hazard data sheet can give adequate information, but it takes time and good light to read, and it has to be there to be read. A hazard data sheet offers adequacy but lacks immediacy.

DEFINITIONS

The following terms are used in this chapter as defined below.

Flammability: The NFPA definitions of flammable and combustible liquids, the methods of determining the flash point, and the classes of these liquids are given in Chapter 4 of Section 5. However, in some regulations, the classification system is different from the NFPA definitions, not only in the United States but also in other countries. For example, in the United States the labeling for flammability for consumer products is in three classes with the flash point determined by the Tag Open Cup Test Method. Liquids with flash points below 20°F (-7°C) are identified by the signal phrase, "DANGER—EXTREMELY FLAMMABLE;" liquids with flash points from 20 to 80°F (7 to 27°C) are identified by WARNING—FLAMMABLE; and liquids with flash points from 80 to 150°F (27 to 66°C) are identified by CAUTION—COMBUSTIBLE.

Toxicity: In its broadest sense, the term toxicity is defined as the ability of a material to cause bodily harm by chemical action. In common usage, the term toxic substance describes one which can pass through a body surface, i.e., skin, eyes, lungs, or digestive tract, and enter the blood stream. Toxicity is the measure of the injury produced under test conditions. It can affect organs remote from the point of contact. Where injury occurs at the point of contact, two terms are used: "irritation" for a minor though possibly troublesome injury which heals without leaving a scar, and "corrosiveness" for an injury caused by destruction of tissue which leaves a scar when healed.

Several kinds of toxicity are recognized:

Acute Toxicity: This is the effect caused by a single dose or exposure, and may range from simple headache or nausea to disablement or death. Acute toxicity is the kind most important to emergency personnel.

Semichronic and Chronic Toxicity: The effect caused by repeated doses of a material.

Sensitization: A person who has been sensitized may suffer an allergic reaction when exposed again to the same material. For example, in photodynamic sensitization, a person who is sensitized by a material will suffer an allergic reaction when exposed to strong sunlight or ultraviolet radiation.

Carcinogens, Teratogens, and Mutagens: Carcinogens are materials that can cause cancer, teratogens are materials that can cause pregnant females to bear deformed offspring, and mutagens are materials that can cause exposed persons to bear children with traits differing from those of the parents.

LD_{50}, LC_{50}: Acute toxicity is expressed by the term LD_{50}, which means that a dose of a substance was lethal to 50 percent or more of the animals tested. This term is used for liquids and solids which are swallowed or absorbed through the skin. For inhalation hazards of vapors, mists, fumes, and dusts, the term LC_{50} is used, meaning the concentration in air of a substance which is lethal to half or more than half of the test animals used. The term LD_{50} is usually expressed as milligrams of dose per kilogram of body weight of the test animal, and should include the type of test, thus:

$$\text{Oral } LD_{50} \text{ (rat)} = 30 \text{ mg/k.}$$

Note that the smaller the LD_{50}, the more toxic a material is.

Quantitative LD_{50} limits for the levels of toxicity generally increase by a factor of ten for successive classes of toxicity. Thus, for oral toxicity, extremely toxic is applied to materials with an LD_{50} (rat) below 5 mg/kg. For a 70 kg (150 lb) human, this figures out to about seven drops (or less) of a watery liquid. For a solid, it is somewhat less than the weight of a 5 grain aspirin tablet. This means that a very small amount of an extremely toxic material can cause a noticeable illness or injury—the amount, for instance, that might be ingested by swallowing dust trapped in the upper nose if a respirator is not worn. Container labels carry the signal word "DANGER" and a symbol such as the familiar skull and crossbones to indicate the presence of a disabling hazard.

A material highly toxic by oral ingestion is one with an oral LD_{50} (rat) of 5 to 50 mg/kg. For a 150 lb (70 kg) human this is ⅔ of a teaspoonful (3.28 mL) of a watery liquid, or 1/10 oz (2.8 g) of a solid (the weight of eight or nine 5 grain aspirin tablets). Container labels use the signal word DANGER.

A toxic material has an oral LD_{50} (rat) of 50 to 500 mg/kg. It is unlikely that an adult will knowingly swallow enough of a material of this hazard level to cause serious injury. Labels bear the signal word WARNING to indicate a moderate hazard.

Equivalent doses have been worked out for exposure by inhalation (breathing) and absorption through the skin. Rabbits are usually used for skin absorption tests, and rats for inhalation testing.

For materials which cause direct injury at the point of contact, there is an animal test which is used to distinguish between an irritant and a corrosive. So far, there are no quantitative degrees of corrosiveness, but industrial practice does make distinctions in label statements based on human experience and the toxicologists' interpretations of animal tests. In inspecting a warehouse, the following graded phrases on package labels may be found:

WARNING: MAY CAUSE BURNS
WARNING: CAUSES BURNS
DANGER: CAUSES BURNS
DANGER: CAUSES SEVERE BURNS
DANGER: RAPIDLY CAUSES SEVERE BURNS
DANGER: CAUSES SEVERE BURNS WHICH HEAL VERY SLOWLY

Wherever possible, levels of hazard should be assigned on the basis of human experience. When human experience is not available, as in the case of a new chemical or because of a good safety program, animal tests are used. Since no animal is exactly equivalent to a human being, most of the value of such animal tests must come from the skilled interpretations of toxicologists.

Instability (Self-Reactivity) and Chemical Reactivity

Two types of chemical reactions are of interest here: those which absorb energy and those which give off energy. When the energy is in the form of heat, reactions which absorb heat are called endothermic and those which give off heat are called exothermic.

Two kinds of reactivity exist: reactivity and self-reactivity. The term reactivity is used when two or more chemicals react with each other, as when sodium and water combine in a violent and dangerous reaction. Self-reactivity comes from internal activity of a material, as when a stick of dynamite explodes. Radioactivity is a specialized type of self-reactivity which is always considered separately. Explosives are generally considered a specialized area with separate hazard information systems.

Aside from explosives and radioactive materials, there are no formal hazard levels for reactivity. Hazard information is based on experience. NFPA 704, *Standard System for the Identification of the Fire Hazards of Materials*, is the only hazard information system that attempts to cover the whole span of hazardous self-reactivity. The NFPA 704 System also includes information on hazardous reactivity with water.

The terms oxidizer and oxidizing action (or oxidizing reaction) mean different things in different systems. In hazard information systems, the term oxidizer or strong oxidizer defines a chemical which can give off oxygen to increase the burning rate of other materials. Such chemicals are called oxygen donors. Study is being given to use of the term oxidizer for chemicals which can ignite other materials on contact, such as bromine or nitric acid. Currently these are recognized in labeling systems by statements of hazard such as "spillage may cause fire." Although organic peroxides may be oxygen donors under some conditions, they are classed separately because of the severe fire and self-reactivity hazards which many of them present.

THE NFPA 704 SYSTEM OF HAZARD IDENTIFICATION

NFPA 704, *Standard System for the Identification of the Fire Hazards of Materials*, is a symbol system intended for use on fixed installations, such as chemical processing equipment, storage and warehousing rooms, and laboratory entrances. It tells a fire fighter what he must do to protect himself from injury while fighting a fire in the area.

The NFPA 704 Diamond

The information system is based on the "704 diamond" (Fig. 6-2A), which is the vehicle for visually presenting information on health, flammability, and self-reactivity hazards, as well as special information associated with the hazards.

The NFPA 704 diamond symbol is intended to provide immediacy at some sacrifice of adequacy, and there is a tendency to read more into it than it says. The five degrees of hazard, in the order of their descendency, have these general meanings to fire fighters:

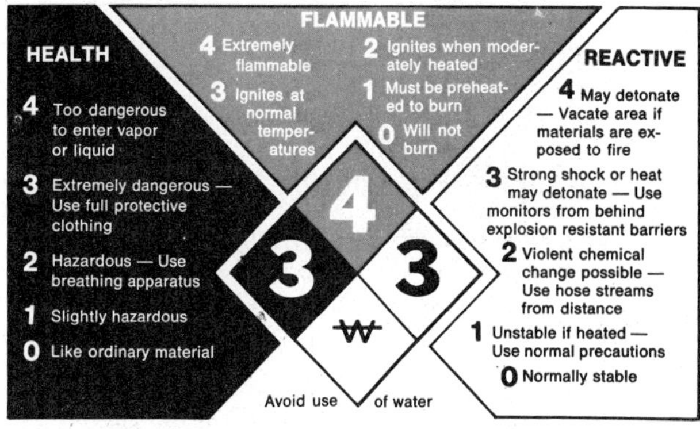

FIG. 6-2A. The NFPA 704 diamond.

4—Too dangerous to approach with standard fire fighting equipment and procedures. Withdraw and obtain expert advice on how to handle.

3—Fire can be fought using methods intended for extremely hazardous situations, such as unmanned monitors or personal protective equipment which prevents all bodily contact.

2—Can be fought with standard procedures, but hazards are present which require certain equipment or procedures to handle safely.

1—Nuisance hazards present which require some care, but standard fire fighting procedures can be used.

0—No special hazards, therefore no special measures.

NFPA 704 describes in detail the hazards and hazard levels which the various numbers indicate for the three hazards. The following, adapted from the appendix of the Standard, summarizes the hazard information and recommends protective actions.

The numbers from 0 through 4 are placed in the three upper squares of the diamond to show the degree of hazard present for each of three hazards. Zero indicates the lowest degree of hazard, and four, the highest. The fourth square, at the bottom, is used for special information. Three symbols for this bottom space are suggested by NFPA 704. (See Fig. 6-2B.) They are:

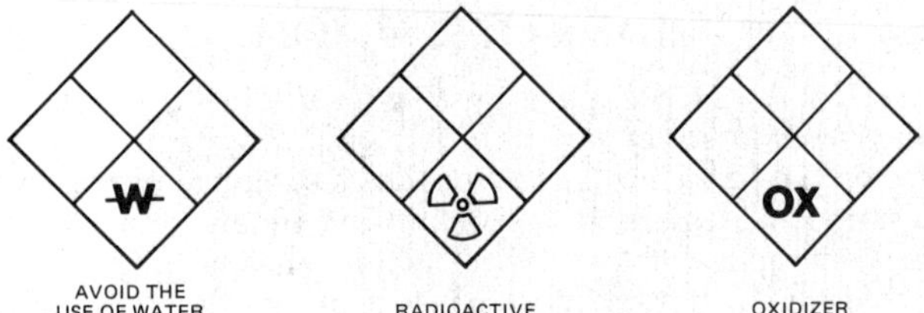

FIG. 6-2B. Special information is presented in the bottom square of the NFPA 704 diamond. This square is color coded white.

1. A letter W with a bar through it (W) to indicate that a material may have a hazardous reaction with water. This does not mean "do not use water," since some forms of water—fog or fine spray—may be used in many cases. What it does say is: "Water may cause a hazard, so use it very cautiously until you have proper information."
2. The "radioactive pinwheel" for radioactive materials.
3. The letters OX to indicate an oxidizer.

Health Hazards

In general, the health hazard in fire fighting is that of a single exposure which may vary from a few seconds up to an hour. The physical exertion demanded in fire fighting or other emergencies may be expected to intensify the effects of any exposure. In assigning degrees of danger, local conditions must be considered. The following explanation is based upon use of the protective equipment normally worn by fire fighters. (See Fig. 6-2C.)

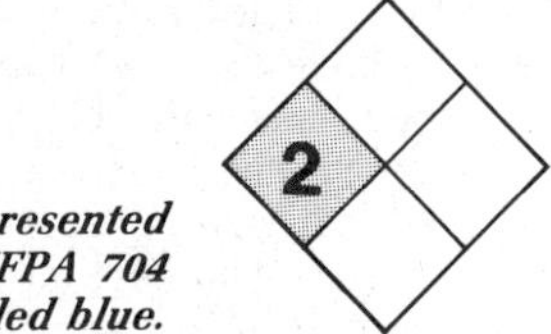

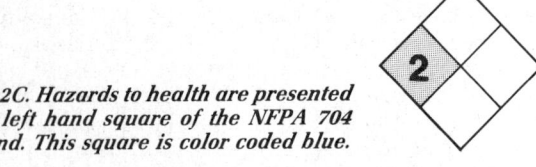

FIG. 6-2C. Hazards to health are presented in the left hand square of the NFPA 704 diamond. This square is color coded blue.

4—Materials which are too dangerous to health for fire fighters to be exposed to. A few whiffs of the vapor could cause death, or the vapor or liquid could be fatal on penetrating the fire fighter's normal protective clothing. Protective clothing and breathing apparatus available to the average fire department will not provide adequate protection against inhalation or skin contact with these materials.

3—Materials which are extremely hazardous to health, but fire areas may be entered with extreme care. Full protective clothing, self-contained breathing apparatus, rubber gloves, boots, and bands around legs, arms, and waist should be provided. No skin surface should be exposed.

2—Materials which are hazardous to health, but fire areas may be entered freely with self-contained breathing apparatus.

1—Materials which are only slightly hazardous to health.

0—Materials which on exposure under fire conditions would offer no health hazard beyond that of ordinary combustible material.

Flammability Hazards

Susceptibility to burning is the basis for assigning degrees within this category. (See Fig. 6-2D.) The method of attacking the fire is influenced by this susceptibility factor.

4—Very flammable gases or very volatile flammable liquids. If possible, shut off flow and keep cooling water streams on exposed tanks or containers. Withdrawal may be necessary.

3—Materials that can be ignited under almost all normal temperature conditions. Water may be ineffective because of the low flash point of the materials.

2—Materials that must be moderately heated before ignition will occur. Water spray may be used to extinguish the fire because the material can be cooled below its flash point.

1—Materials that must be preheated before ignition can occur. Water may cause frothing if it gets below the surface of the liquid and turns to steam. If this is the case, water fog gently applied to the surface will cause a frothing which will extinguish the fire.

0—Materials that will not burn.

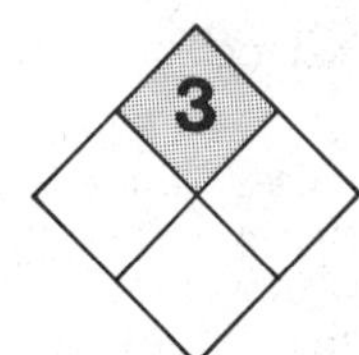

FIG. 6-2D. Flammability hazards are presented in the top square of the NFPA 704 diamond. This square is color coded red.

Reactivity (Stability) Hazards

The assignment of relative degree of hazard in the reactivity category is based upon the susceptibility of materials to release energy either by themselves or in combination with other materials. (See Fig. 6-2E.) Fire exposure was one of the factors considered along with conditions of shock and pressure.

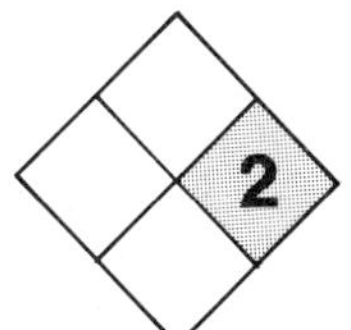

FIG. 6-2E. Reactivity (stability) hazards are presented in the right hand square of the NFPA diamond. This square is color coded yellow.

4—Materials which are readily capable of detonation at normal temperatures and pressures. If they are involved in a massive fire, vacate the area.

3—Materials which when heated and under confinement are capable of detonation and which may react violently with water. Fire fighting should be conducted from behind explosion resistant locations.

2—Materials which will undergo a violent chemical change at elevated temperatures and pressures but do not detonate. Use portable monitors, hoseholders or straight hose streams from a distance to cool the tanks and the material in them. Use caution.

1—Materials which are normally stable but may become unstable in combination with other materials or at elevated temperatures and pressures. Use normal precautions as in approaching any fire.

0—Materials which are normally stable and, therefore, do not produce any reactivity hazard to fire fighters.

Special Information

When W̶ appears at the bottom in the 4th space. (See Fig. 6-2B.)

4—W̶ is not used with reactivity hazard 4.

3—In addition to the hazards above, these materials can react explosively with water. Explosion protection is essential if water in any form is used.

2—In addition to hazards above, these materials may react violently with water or form potentially explosive mixtures with water.

1—In addition to hazards above, these materials may react vigorously but not violently with water.

0—W̶ is not used with reactivity hazard 0.

Methods of Presentation

Considerable leeway is allowed in the presentation of the numbers. The only basic requirement is that numbers be spaced as though they were in the diamond outline. Several methods which have been used are shown in Figure 6-2F. Chapter 5 of NFPA 704 gives recommended layout and sizes for the symbol, and a distance-legibility table, as well as several examples of using the symbol.

Assigning Degrees of Hazard

Numbers (degrees of hazard) for use in the diamond are assigned on the basis of the worst hazard expected in the area, whether it be from hazards of the original material or of its combustion or breakdown products. The effects of local conditions must be considered. For instance, a drum of carbon tetrachloride sitting in a well ventilated storage shed presents a different hazard from a drum sitting in an unventilated basement.

Advantages of the NFPA 704 System

The NFPA 704 System can warn against hazards under fire conditions of materials which other information systems class as nonhazardous. For example, edible tallow produces toxic and irritating combustion products. It would be given a "2" degree of health hazard, indicating the need for air-supplied respiratory equipment.

NFPA 704 also can warn against overall fire hazards in an area. On the door of a laboratory or storage room, it can warn of the worst hazards likely in a fire situation. Such information is useful both in preplanning and in actual fires.

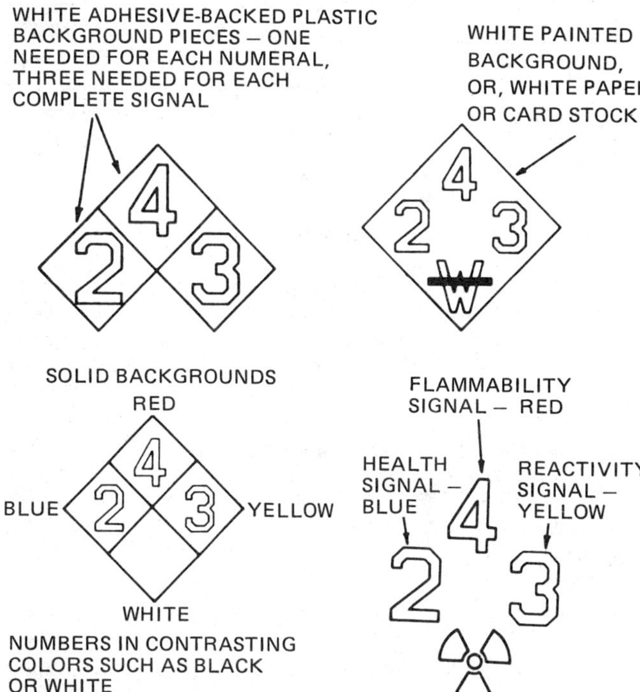

FIG. 6-2F. Methods of presenting the NFPA 704 System Hazard information.

NFPA 704 can be used without a supplementary manual. Because of its simplicity, the general meanings of the numbers can be memorized easily and the whole symbol read and interpreted quickly on the spot and in poor light.

Disadvantages of the NFPA 704 System

The NFPA 704 system gives only minimum information on the hazards themselves. Since the system informs on protective measures, the same number may be used for different types of hazards so that, for instance, a health hazard "3" means "No contact" without saying whether the hazard is corrosiveness to the skin or toxicity by absorption through the skin. Thus, the symbol is most useful only to trained or informed persons.

DEPARTMENT OF TRANSPORTATION (DOT) PLACARDS

The Hazardous Materials Regulations of the Department of Transportation (DOT) require placarding of trucks, trailers, and railway cars carrying dangerous materials. Hazardous materials must be identified on shipping papers regardless of the quantity shipped.

Present DOT Requirements

Over-the-road Equipment: Equipment must be marked on front, rear, and both sides with the hazard name at least 4 in. (102 mm) tall. Explosives, extremely toxic materials, and high strength radioactives in any amount must be marked, along with highly toxic, flammable, and oxidizing materials, corrosive liquids and both flammable and nonflammable compressed gases, in quantities of 1,000 lb (454 kg) or over. A bill of lading indicating mixed hazards totaling over 1,000 lb (454 kg) carries a DANGEROUS marking, and is marked as well for explosives, extremely toxic, and high strength radioactive materials, when they are present. Other than for radioactive materials, a multiple hazard material is marked only with the name of the most severe hazard present. A mixed shipment does not identify hazards.

Railway Equipment: Equipment must be placarded on both sides and ends. Placards for explosives and poison gases are rectangular, the smallest being about 10 by 14 in. (254 by 356 mm). For other hazards, 10¾ in. (273 mm) diamond shaped placards are used. Radioactives show the name on the placard, while all remaining hazards are identified only by a placard with the word DANGEROUS in large red letters. Ladings which have been fumigated in the car and could bear dangerous residues carry a FUMIGATED placard on or near the door. Placards may be attached directly to the car or placed in special holders. Tank cars which have been emptied, but contain residues, must carry DANGEROUS—EMPTY placards. These are the same size as originals, but the right half is solid black.

Air Shipments: These are generally small quantities since many hazardous materials cannot be shipped by air, particularly in passenger aircraft. Placarding of equipment is not required.

Transportation Emergencies—Identification Problems

Certain deficiencies in current DOT placarding and labeling practices contribute to difficulties in identifying the hazardous nature of some commodities found in transit. For example, placards are not required on the following:

1. Class C explosives. These are principally manufactured articles containing explosives, such as fireworks, squibs, and small-arms ammunition.
2. Moderate and low level radioactives.
3. Tear gases.
4. Corrosive solids, including those which could dissolve in water to produce corrosive liquids.
5. Over-the-road equipment containing less than 1,000 lb (454 kg) of highly toxic, flammable, or oxidizing materials; corrosive liquids, or flammable and nonflammable compressed gases, either singly or as a mixed shipment.

The color coding of existing placards and package labels is inconsistent for the various modes of transportation, so the use of colors for quick identification is of doubtful value.

DOT Placard and Labeling System

Effective November 1, 1981, DOT regulations required that each tank truck vehicle must display a designated four digit number. The identification numbering system is based on the system adopted for worldwide use by the United Nations Committee of Experts on the Transport of Dangerous Goods.

The numbers are assigned by governmental authorities under the aegis of the Economic and Social Council of the United Nations, and each number has the same meaning throughout worldwide commerce. Each four digit number identifies a specific hazardous material, and each has no other meaning or use. For example, 1294 will always signify Toluene (Toluol). (See Figs. 6-2G, H and I.)

As an adjunct to the identification system, a manual has been prepared to associate the identification number with a brief, concise instruction that is intended to assist emergency personnel during the first minutes of a hazardous materials accident. This manual is called the *Emer-*

ID No.	Guide No.	Name of Material
1246	27	METHYLPROPENYL KETONE, inhibited
1247	27	METHYL METHACRYLATE, monomer, inhibited
1248	26	METHYL PROPIONATE
1249	26	METHYL PROPYL KETONE
1250	29	METHYL TRICHLOROSILANE
1251	28	METHYL VINYL KETONE
1255	27	NAPHTHA, PETROLEUM
1256	27	NAPHTHA, SOLVENT
1257	27	CASINGHEAD GASOLINE
1257	27	NATURAL GASOLINE
1259	28	NICKEL CARBONYL
1261	26	NITROMETHANE
1262	27	ISOOCTANE
1262	27	OCTANE
1263	26	COMPOUND, PAINT, etc., removing, reducing, or thinning liquid
1263	26	ENAMEL
1263	26	LACQUER
1263	26	LACQUER BASE, liquid
1263	26	PAINT, etc., flammable liquid
1263	26	PAINT RELATED MATERIAL, flammable liquid
1263	26	POLISH, liquid
1263	26	SHELLAC
1263	26	STAIN
1263	26	THINNER
1263	26	VARNISH
1263	26	WOOD FILLER, liquid
1264	26	PARALDEHYDE
1265	27	AMYL HYDRIDE
1265	27	ISOPENTANE
1265	27	PENTANE
1266	26	PERFUMERY PRODUCTS, with flammable solvent
1267	27	PETROLEUM CRUDE OIL
1268	27	NAPHTHA DISTILLATE
1268	27	PETROLEUM DISTILLATE, n.o.s.
1268	27	ROAD OIL
1270	27	OIL, petroleum, n.o.s.
1270	27	PETROLEUM OIL
1271	26	PETROLEUM ETHER
1271	26	PETROLEUM SPIRIT
1272	26	PINE OIL
1274	26	PROPANOL
1274	26	PROPYL ALCOHOL
1275	26	PROPIONALDEHYDE
1276	26	PROPYL ACETATE
1277	68	MONOPROPYLAMINE
1277	68	PROPYLAMINE
1278	26	PROPYL CHLORIDE
1279	27	DICHLOROPROPANE
1279	27	PROPYLENE DICHLORIDE
1280	26	PROPYLENE OXIDE, inhibited
1281	26	PROPYL FORMATE
1282	26	PYRIDINE
1286	26	RESIN OIL
1286	26	ROSIN OIL
1287	26	RUBBER SOLUTION
1288	27	SHALE OIL
1289	26	SODIUM METHYLATE, solutions in alcohol
1292	29	ETHYL SILICATE
1292	29	TETRAETHYL SILICATE
1293	26	TINCTURE, medicinal
1294	27	TOLUENE
1295	30	TRICHLOROSILANE
1296	68	TRIETHYLAMINE

SEE "HOW TO USE THIS GUIDEBOOK" ON THE FIRST PAGE, IF YOU HAVE NOT YET BECOME FAMILIAR WITH THE DETAILS OF USING THESE INDEXES TO THE GUIDES.

FIG. 6-2G. Sample page from the ID Number Index. The four-digit ID Number 1294 indicates that the cargo is toluene, and that the correct Guide Number is 27.

gency *Response Guidebook*, and it is carried by all emergency response vehicles.

To use the *Emergency Response Guidebook*, first identify the material by finding either the four digit ID Number (on a placard or orange panel posted on the vehicle or after the letters UN or NA on the shipping papers) or the name of the material on the shipping papers, placard, label, or package. Next look up either the four digit number or the material name in the Guidebook. The four digit number will identify the material by name and refer you to a two digit Guide Number. (See Fig. 6-2G.) Looking up the material name will refer you to the same two digit Guide Number and give you the four digit ID Number as a cross-reference. (See Fig. 6-2H.) The two digit Guide Number refers you to the proper detailed instruction sheet in the Guidebook. (See Fig. 6-2I.)

Tank trucks carrying hazardous materials must have the Hazardous Identification (HI) number on all four sides. It can be applied in two ways: as an identification number label in addition to the existing placard, with the orange colored label having a black border and black 4 in. (102 mm) high numbers; or the number can be displayed in the existing placard already required by DOT.

However, a tank vehicle transporting gasoline or fuel oil has a third option because special placarding rules

Name of Material	Guide No.	ID No.
TETRAHYDROPHTHALIC ANHYDRIDE	60	2698
TETRAHYDROPYRIDINE	26	2410
TETRAHYDROTHIOPHENE	26	2412
TETRALIN HYDROPEROXIDE, technical pure	48	2136
TETRAMETHYL AMMONIUM HYDROXIDE	60	1835
1,1,3,3-TETRAMETHYLBUTYL HYDROPEROXIDE, technical pure	48	2160
1,1,3,3-TETRAMETHYLBUTYL-PEROXY-2-ETHYL HEXANOATE, technical pure	52	2161
TETRAMETHYL LEAD	56	1649
TETRAMETHYLMETHYLENE-DIAMINE	58	9069
TETRAMETHYL SILANE	29	2749
TETRAPROPYL-ortho-TITANATE	27	2413
TETRANITROMETHANE	47	1510
TEXTILE TREATING COMPOUND	60	1760
TEXTILE WASTE, wet, n.o.s.	32	1857
THALLIUM CHLORATE	42	2573
THALLIUM COMPOUND, n.o.s.	53	1707
THALLIUM NITRATE	42	2727
THALLIUM SALT, n.o.s.	53	1707
THALLIUM SULFATE, solid	53	1707
THIAPENTANAL	55	2785
THINNER	26	1263
THIOACETIC ACID	26	2436
THIOGLYCOL	53	2966
THIOGLYCOLIC ACID	60	1940
THIOLACTIC ACID	59	2936
THIONYL CHLORIDE	39	1836
THIOPHENE	27	2414
THIOPHOSGENE	55	2474
THIOPHOSPHORYL CHLORIDE	60	1837

Name of Material	Guide No.	ID No.
THIOUREA	53	2877
THIRAM	55	2771
THORIUM METAL, pyrophoric	65	2975
THORIUM METAL, pyrophoric	65	9170
THORIUM NITRATE, solid	64	2976
THORIUM NITRATE, solid	64	9171
TIN CHLORIDE, fuming	39	1827
TINCTURE, medicinal	26	1293
TIN TETRACHLORIDE	39	1827
TITANIUM, metal, powder, dry	37	2546
TITANIUM, metal, powder, wet with not less than 20% water	32	1352
TITANIUM HYDRIDE	32	1871
TITANIUM SPONGE, granules or powder	32	2878
TITANIUM SULFATE SOLUTION	60	1760
TITANIUM TETRACHLORIDE *	39	1838
TITANIUM TRICHLORIDE, pyrophoric	37	2441
TITANIUM TRICHLORIDE MIXTURE	60	2869
TITANIUM TRICHLORIDE MIXTURE, pyrophoric	37	2441
TOE PUFFS, nitrocellulose base	32	1353
TOLUENE	27	1294
TOLUENE DI-ISOCYANATE (T.D.I.)	57	2078
TOLUENE SULFONIC ACID, liquid	60	2584
TOLUENE SULFONIC ACID, liquid	60	2586
TOLUENE SULFONIC ACID, solid	60	2583
TOLUENE SULFONIC ACID, solid	60	2585
TOLUIDINES (o-, m-, and p-)	55	1708

* Look for information next to this **NAME** in the TABLE OF EVACUATION DISTANCES in the back of this book. Use this in addition to the Guide Page if there is NO FIRE.

FIG. 6-2H. Sample page from the Materials Name Index. Once again the Guide Number for toluene is 27.

Guide 27

POTENTIAL HAZARDS

FIRE OR EXPLOSION

Flammable/combustible material; may be ignited by heat, sparks or flames.
Vapors may travel to a source of ignition and flash back.
Container may explode in heat of fire.
Vapor explosion hazard indoors, outdoors or in sewers.
Runoff to sewer may create fire or explosion hazard.

HEALTH HAZARDS

May be poisonous if inhaled or absorbed through skin.
Vapors may cause dizziness or suffocation.
Contact may irritate or burn skin and eyes.
Fire may produce irritating or poisonous gases.
Runoff from fire control or dilution water may cause pollution.

EMERGENCY ACTION

Keep unnecessary people away; isolate hazard area and deny entry.
Stay upwind; keep out of low areas.
Wear self-contained (positive pressure if available) breathing apparatus and full protective clothing.
Isolate for 1/2 mile in all directions if tank car or truck is involved in fire.
FOR EMERGENCY ASSISTANCE CALL CHEMTREC **(800) 424-9300.**
If water pollution occurs, notify appropriate authorities.

FIRE

Small Fires: Dry chemical, CO_2, water spray or foam.
Large Fires: Water spray, fog or foam.
Move container from fire area if you can do it without risk.
Cool containers that are exposed to flames with water from the side until well after fire is out.
For massive fire in cargo area, use unmanned hose holder or monitor nozzles; if this is impossible, withdraw from area and let fire burn.
Withdraw immediately in case of rising sound from venting safety device or any discoloration of tank due to fire.

SPILL OR LEAK

Shut off ignition sources; no flares, smoking or flames in hazard area.
Stop leak if you can do it without risk.
Use water spray to reduce vapors.
Small Spills: Take up with sand or other noncombustible absorbent material and place into containers for later disposal.
Large Spills: Dike far ahead of spill for later disposal.

FIRST AID

Move victim to fresh air; call emergency medical care.
If not breathing, give artificial respiration.
If breathing is difficult, give oxygen.
In case of contact with material, immediately flush eyes with running water for at least 15 minutes. Wash skin with soap and water.
Remove and isolate contaminated clothing and shoes at the site.

FIG. 6-2I. Guide Number 27 refers to a set of detailed instructions.

have been established for these products. In order to eliminate the need for changes, a single placard may be used with the HI number associated with the lowest flash point of any distillate fuel ever carried in the tank.

In addition, placards with the words "gasoline" or "fuel oil" on the diamond shaped placard in place of the HI number are also authorized. If a tank contains only gasoline or fuel oir and has the words "gasoline" or "fuel oil" on each side and rear in 2 in. (50 mm) letters, or uses placards having the words "gasoline" or "fuel oil" in place of the HI number, the HI number will not be required.

PACKAGE AND CONTAINER LABELS

A word of warning: Absence of warning labeling on a package or container does not guarantee that the contents are not hazardous. Also, warnings may be printed on a label separate from the label bearing the product name or the manufacturer's signature, and may be on a separate face of the package.

Two separate but related information systems require labels on shipping packages and on product containers.

Department of Transportation (DOT) Labels

Current DOT hazardous materials regulations require diamond shaped labels on shipping containers holding an

OXYGEN placard required for pressurized liquid oxygen.

OXYGEN placard may also be used to identify liquefied oxygen contained in a manner that does not meet the definition of a compressed gas in Sec. 173.300.

CHLORINE placard required only for a packaging having a rated capacity of more than 110 gallons. Use the NON-FLAMMABLE GAS placard for packagings having a rated capacity of 110 gallons or less.

HIGHWAY SHIPMENTS

CARGO TANKS AND PORTABLE TANKS

Above placard may be used in place of FLAMMABLE placard when gasoline is being transported (Sec. 172.542(c)).

Above placard may be used in place of COMBUSTIBLE placard when FUEL OIL that is not classed as a Flammable liquid is being transported (Sec. 172.544(c)).

RAIL SHIPMENTS

EMPTY placard. Each "empty" tank car must be placarded with an EMPTY placard that corresponds to the placard that was required for the material the tank car last contained unless the tank car last contained a Combustible liquid. This placard is required for the following hazardous materials.

Non-Flammable Gas	Flammable Solid
Oxygen	Flammable Solid "W"
Flammable Gas	Oxidizer
Chlorine	Organic Peroxide
Poison Gas	Poison
Flammable	Corrosive

FREIGHT CONTAINERS

FREIGHT CONTAINERS—640 CUBIC FEET OR MORE (Placard each end and each side)

AIR OR WATER—Placard any quantity

HIGHWAY OR RAIL

1. Placard any quantity of hazardous material classes listed in TABLE 1 (Sec. 172.512(a)).
2. Placard 1,000 pounds or more (aggregate gross weight) of hazardous material classes in TABLE 2 (Sec. 172.512(a)).

NOTE: For placarding options for freight containers of less than 640 cubic feet, see Sec. 172.512(b).

CARGO TANKS AND PORTABLE TANKS

1. Cargo tanks containing any quantity of hazardous material must be placarded.
2. Portable tanks having a rated capacity of 1,000 gallons or more must be placarded.
3. Portable tanks having a rated capacity of less than 1,000 gallons need be placarded on only two opposite sides (Sec. 172.514(a)).
4. Cargo tanks and portable tanks must remain placarded when emptied unless reloaded with a material not subject to CFR, Title 49, Parts 100-199 or sufficiently cleaned and purged to remove any potential hazard. (Sec. 172.514(b)).
5. For Combustible liquids, a FLAMMABLE placard may be used on cargo tanks or portable tanks when transported by highway or water.

USE OF THIS CHART DOES NOT RELIEVE PERSONS INVOLVED FROM COMPLYING WITH THE DOT HAZARDOUS MATERIALS REGULATIONS AS PUBLISHED IN TITLE 49, CODE OF FEDERAL REGULATIONS, PARTS 100-199.

FIG. 6-2J. DOT Hazardous Materials Warning Placards. The "explosive" labels are black on orange. The "poison gas" labels are black on white. The "radioactive" label is yellow, black, and white. The "flammable solid" label is red, white, blue, and black. (Department of Transportation Chart 5)

PLACARDING ANY QUANTITY—

TABLE 1

MOTOR VEHICLES, FREIGHT CONTAINERS AND RAIL CARS

Placard motor vehicles, freight containers, and rail cars containing "any quantity" of hazardous materials listed in TABLE 1.

HAZARDOUS MATERIAL CLASSED OR DESCRIBED AS	PLACARDS
Class A explosives	EXPLOSIVES A.
Class B explosives	EXPLOSIVES B.
Poison A	POISON GAS.
Flammable solid (DANGEROUS WHEN WET label only)	FLAMMABLE SOLID W
Radioactive material	RADIOACTIVE.
Radioactive material:	
Uranium hexafluoride, fissile (containing more than 0.7 pct U^{235})	RADIOACTIVE AND CORROSIVE.
Uranium hexafluoride, low specific activity (containing 0.7 pct. or less U^{235})	RADIOACTIVE AND CORROSIVE.

YELLOW III labeled packagings only.

For Uranium Hexafluoride, see Sec. 172.504(a) and TABLE 1.

RAIL PLACARDS

SQUARE BACKGROUND FOR RAIL SHIPMENTS – Each EXPLOSIVE A placard, POISON GAS placard and POISON GAS – EMPTY placard affixed to a rail car must be placed on a square background measuring 14¼ inches on each side with a black border extending to 15½ inches on each side (illustrated in above chart). (See Sec. 172.510(a) and 172.527(a)).

FIG. 6-2J (cont). The "flammable," "flammable gas," and "combustible" labels are white on red. The "poison" and "corrosive" labels are black and white. The "organic peroxide" and "oxidizer" labels are black on yellow. The "nonflammable gas" label is white on green. The "flammable solid" and "dangerous" labels are red, white, and black. (Department of Transportation Chart 5)

OTHER PLACARDING REQUIREMENTS

MOTOR VEHICLES, RAIL CARS AND FREIGHT CONTAINERS

1. Placard motor vehicles and freight containers containing 1,000 pounds or more gross weight of hazardous materials classes listed in TABLE 2.
2. Placard any quantity of hazardous materials classes listed in TABLES 1 and 2 when offered for transportation by air or water.
3. Placard rail cars containing any quantity of hazardous materials classes listed in TABLE 2 except when less than 1,000 pounds gross weight of hazardous materials is transported in TOFC (Trailer on flat car) or COFC (Container on flat car) service.

TABLE 2

HAZARDOUS MATERIAL CLASSED OR DESCRIBED AS	PLACARDS
Class C explosives	FLAMMABLE.
Nonflammable gas	NONFLAMMABLE GAS.
Nonflammable gas (Chlorine)	CHLORINE.
Nonflammable gas (Fluorine)	POISON.
Nonflammable gas (Oxygen, pressurized liquid)	OXYGEN.
Flammable gas	FLAMMABLE GAS.
Combustible liquid	COMBUSTIBLE.
Flammable liquid	FLAMMABLE.
Flammable solid	FLAMMABLE SOLID.
Oxidizer	OXIDIZER.
Organic peroxide	ORGANIC PEROXIDE.
Poison B	POISON.
Corrosive material	CORROSIVE.
Irritating material	DANGEROUS.

Also used for Class C Explosives labeled with an Explosive C label.

Used for materials transported in packaging having a rated capacity of more than 110 gallons.

Other than by water, a FLAMMABLE placard may be displayed in place of a FLAMMABLE SOLID placard except when a DANGEROUS WHEN WET label is specified in Sec. 172.101.

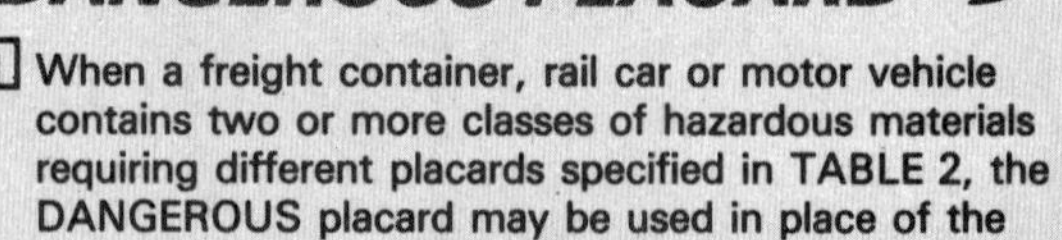

DANGEROUS PLACARD

1. When a freight container, rail car or motor vehicle contains two or more classes of hazardous materials requiring different placards specified in TABLE 2, the DANGEROUS placard may be used in place of the separate placards specified for each class.
2. When 5,000 pounds or more of one class of hazardous material is loaded at one loading facility, the placard for that class in TABLE 2 must be applied.

UNITED NATIONS (UN) HAZARD CLASS NUMBERS

UN hazard class numbers are not required for domestic shipments. The hazard class and division number prescribed for dangerous goods in the United Nations Recommendations entitled "Transport of Dangerous Goods (1970)" may be entered on each placard in the lower corner of the diamond (Sec. 172.519(d)).

FIG. 6-2J (cont). The "oxygen" label is black on yellow. The "chlorine" label is black on white. The "gasoline" and "fuel oil" labels are white on red. The "empty" sticker is black. (Department of Transportation Chart 5)

amount of hazardous material which could cause a hazardous condition in transportation. These color-coded labels bear a hazard symbol, or pictograph, and the name of the hazard. They carry no warning text. Present regulations generally permit only a single label on a multi hazard exposure, that label showing the most dangerous hazard present. Where a material is extremely toxic, explosive, or highly radioactive, the proper label must be used in addition to any other hazard label. In addition the package must be marked with the name of contents as it appears on a Commodity List in DOT regulations. This may be a chemical name or a class name such as "Corrosive Liquid N.O.S.," so is of secondary usefulness. On compressed gas cylinders, the label may be attached to a wire tied tag.

The presence of a DOT label on a package or container indicates a dangerous hazard. Because of exemptions, based mostly on the size of inner containers placed in an outside shipping container, a package without a DOT label may contain significant amounts of fairly dangerous materials, including flammable liquids.

See Figure 6-2J for illustrations of DOT labels.

Warning Labels on Immediate Containers

Containers for hazardous materials in interstate commerce usually bear necessary warning labels, but products of small local distributors may not. Warehouses, particularly those of contract packagers, may contain stocks of filled containers intended for labeling to customers' orders. Where the immediate container is also the shipping container, such as a carboy, a drum, or a compressed gas cylinder, both DOT and warning labels may be present and may be printed on a single label sheet.

Labeling practice is to use the signal word DANGER for corrosive liquids, extremely corrosive solids, poisons, flammable gases, and extremely flammable liquids. The signal word WARNING is used on flammable liquids, less corrosive solids, toxic materials, and similar levels of hazard. The signal word CAUTION is used for the least severe hazards, including combustible liquids and solids, and nonflammable compressed gases. The rest of the warning label is self-explanatory, except to note that warning labels cover only hazards arising from normal use and handling. Hazards under fire conditions are not included in the warning.

Consumer Products: Substances or articles intended or suitable for household use are labeled in accordance with the *Federal Hazardous Substances Act* which vests authority in the Consumer Product Safety Commission (CPSC).

SHIPPING PAPERS

Transport regulations require that a truck driver, a train conductor, or an aircraft pilot have a shipping paper for every shipment of hazardous material in his transport unit. This lists names and quantities of all hazardous materials, including those exempt packages which bear no labels, and should include hazard and emergency handling information.

HAZARD EMERGENCY TEAMS

Emergency assistance in identifying hazards of materials and guidance on handling emergencies involving them is available on quick notice from a variety of sources. The assistance ranges from immediate telephone advice to the dispatching of emergency teams to actually assist in field operation. Among those offering service are:

CHEMTREC.—Phone No. 800-424-9300: CHEMTREC stands for Chemical Transportation Emergency Center, operated by the Chemical Manufacturers Association, 1825 Connecticut Avenue N.W., Washington, DC 20009. It handles only transportation emergencies. When called, a CHEMTREC operator provides immediate advice from CHEMTREC files, then notifies the shipper. CHEMTREC maintains close liaison with the Department of Transportation.

NACA Pesticide Safety Team Network: A service sponsored by The National Agricultural Chemicals Association, 1155 15th St., N.W., Washington, DC 20005, which covers only pesticides. A network of more than 40 teams of specially trained personnel go to the scene of emergencies involving pesticides. Phone answering service is supplied by CHEMTREC.

TEAP: The Canadian Chemical Producers' Association operates a Transportation Emergency Assistance Program through regional teams prepared to give phone and field response.

CHLOREP: The Chlorine Institute, 342 Madison Ave., New York, NY 10017, operates the Chlorine Emergency Plan in which the nearest chlorine producer responds to a problem.

Coast Guard's "React Teams": These teams operate on both the east and west coast, giving skilled help in spills on water.

Environmental Protection Agency: The EPA sponsors teams which go to emergencies. EPA regions and contact addresses are listed in the "Obtaining Technical Assistance in Hazardous Waste Emergencies" chapter of this HANDBOOK.

Many manufacturing companies have organized response capabilities for their own products. In many cases an emergency telephone number is placed on the bill of lading.

Bibliography

NFPA Codes, Standards, Recommended Practices and Manuals. (See the latest *NFPA Codes and Standards Catalog* for availability of current editions of the following documents.)

NFPA 43A, *Code for the Storage of Liquid and Solid Oxidizing Materials.*
NFPA 49, *Hazardous Chemicals Data.*
NFPA 53M, *Manual on Fire Hazards in Oxygen-Enriched Atmospheres.*
NFPA 325M, *Fire Hazard Properties of Flammable Liquids, Gases, and Volatile Solids.*
NFPA 490, *Code for the Storage of Ammonium Nitrate.*
NFPA 491M, *Manual of Hazardous Chemical Reactions.*
NFPA 495, *Code for the Manufacture, Transportation, Storage, and Use of Explosive Materials.*
NFPA 704, *Standard System for the Identification of the Fire Hazards of Materials.*

Additional Readings

ANSI, 2129.1., *Precautionary Labeling of Hazardous Industrial Chemicals*, American National Standards Institute, New York, 1982.

Bahme, C. W., *Fire Officer's Guide to Emergency Action*, National Fire Protection Association, Quincy, MA, 1976.

Bahme, C. W., *Fire Officer's Guide on Dangerous Chemicals*, National Fire Protection Association, Quincy, MA, 1978.

Behrendsen, Darrel J., *Guidelines to the Handling of Hazardous Materials*, Source of Safety, Inc., Denver, CO, 1977.

Cashman, J. R., *Hazardous Materials Engineering—Response and Control*, Technomic Publishing, Lancaster, PA, 1983.

Code of Federal Regulations, Vol. 16, Part 1500, Subchapter C Federal Hazardous Substances Act Regulations, Government Printing Office, Washington, DC.

Coffee, Robert, Evaluation of Chemical Stability, *Fire Technology*, Vol. 7, No. 1, Feb. 1971, pp. 37-45.

Compilation of Labeling Laws and Regulations for Hazardous Substances, Chemical Specialities Manufacturers Association, NY.

DePol, Dennis R. and Cheremisinoff, Paul N., *Emergency Response to Hazardous Materials Incidents*, Technomic, Lancaster, PA, 1984.

Emergency Guide for Selected Hazardous Materials, DOT Office of Hazardous Materials, Washington, DC, 1978.

Fire Protection Guide on Hazardous Materials, 8th ed., National Fire Protection Association, Quincy, MA, 1984.

Handbook of Compressed Gases, Compressed Gas Association, Van Nostrand Reinhold, Arlington, VA, 1981.

"Hazardous Wastes," SW—138, U.S. Environmental Protection Agency, 1975.

Hilado, Carlos J., "Screening Materials for Relative Toxicity in Fire Situations," *Modern Plastics*, Vol. 54, No. 7, July 1977, pp. 64-66, 68.

Isman, W. E. and Carlson, G. P., *Hazardous Materials*, Glencoe Publishing Co., Inc., Encino, CA, 1980.

King, P. W. and Magid, J., *Industrial Hazard and Safety Handbook*, Butterworths, Woburn, MA, 1979.

Kirk, R. E. and Othmer, D. F., eds., *Encyclopedia of Chemical Technology*, 3rd ed., 12 Vols., Wiley—Interscience, NY, 1978-1980.

Lind, C. D. and Whitson, J. C., *Explosion Hazards Associated with Spills of Large Quantities of Hazardous Materials, Phase II*, Department of Transportation, Washington, DC, 1977.

Manual L-1, Guide to Precautionary Labeling of Hazardous Chemicals, Manufacturing Chemists Association, Washington, DC.

Meidl, James H., *Flammable Hazardous Materials*, 2nd Ed., Glencoe Publishing Co., Encino, CA, 1978.

Meidl, James H., *Hazardous Materials Handbook*, Glencoe Press, Beverly Hills, CA, 1972.

Meyer, Eugene, *Chemistry of Hazardous Materials*, Prentice-Hall, Englewood Cliffs, NJ, 1977.

Morton, William I., "Safety Techniques for Workers Handling Hazardous Materials," *Chemical Engineering*, Vol. 83, No. 22, Oct. 18, 1976, pp. 127-132.

NFPA, *Protective Clothing for Hazardous Material Incidents*, (slide/tape), National Fire Protection Association, Quincy, MA, 1984.

NFPA, *Hazardous Waste and the Fire Service*, (slide/tape), National Fire Protection Association, Quincy, MA, 1984.

NFPA, *Hazardous Materials Transportation Emergencies*, 4 Units, (slide/tape), National Fire Protection Association, Quincy, MA, 1983.

Pocket Guide to Chemical Hazards, Pub. No. 78-210, U.S. Department of Health Education and Welfare, Washington, DC, Sept. 1978.

Sax, I., *Dangerous Properties of Industrial Materials*, Van Nostrand Reinhold, NY, 1975.

Schieler, Leroy and Pauze, Denis, *Hazardous Materials*, Delmar Publishers, Albany, NY, 1976.

Turner, Charles F., and McCreey, Joseph W., *The Chemistry of Fire and Hazardous Materials*, Allyn and Bacon, Boston, MA, 1981.

Transport of Dangerous Goods, United Nations Publication, NY, 1977.

Weiss, J., *Hazardous Materials Chemical Data Book*, Noyes, Park Ridge, NY, 1980.

Williams, L. R., et al., *Hazardous Materials Spill Monitoring; Safety Handbook and Chemical Hazard Guide, Part A*, Environmental Monitoring and Support Laboratory, Report No. EPA—600/4—79—008a, U.S. Environmental Protection Agency, Las Vegas, NV, 1979.

Young, M. S., Handling Emergencies Involving Cryogenic Materials, *Fire Command*, Vol. 45, No. 5, May 1978, pp. 20-22.

Zimmerman, O. T., and Lavine, I., *Handbook of Material Trade Names*, Industrial Research Service, Dover, NH, 1953-65 (plus supplements).

HANDLING HAZARDOUS MATERIALS TRANSPORTATION EMERGENCIES

Revised by James V. McKiernan and Callie McDowell

Fire incidents and other transportation of hazardous materials do not happen often enough for fire officers, municipal, county, and state officials, industrial fire chiefs and safety managers, and other trained personnel to gain decision making experience from such emergencies. However, the effects of these materials can be far reaching and subject to rapid changes as an incident progresses. Under such conditions, decision making should be based upon a carefully structured system which leaves few control and communications actions to chance.

General fire problems and fire extinguishment operations for emergencies involving aircraft or airport terminals; motor vehicles; rail transportation; rapid rail transit systems; and marine transportation are defined in the various chapters of Section 13, "Transportation Fire Hazards." Each of these transportation categories has its own unique emergencies, most of which require response from land based fire departments, industrial fire brigades, and other trained groups. For each kind of incident, the fire control and life rescue operations will be influenced by the fire development, or explosion effect, or by the potential for either of these dangers. The assignment of fire control personnel, fire apparatus, and equipment is determined by the individual in charge, usually a chief fire officer. This person must analyze the situation, make quick decisions about the most efficient means of achieving control, and then communicate these decisions to all operating personnel.

Several means of analysis, or models, can simplify the decision making process before an emergency situation occurs. One practical model is called systems analysis, or simply a means of analyzing a system. In the language of modern technology, the word *system* is generally applied to the organization of any human undertaking in order to reach a specific goal (NFPA 1978a).

The individual in command during a hazardous materials emergency is responsible for the functioning of an organized person-machine system. The goal of this system is to return a community, an industrial plant, or other populated location to normal conditions with the least amount of injury to humans and damage to property and the environment (Bahme 1978a).

Mr. McKiernan was Senior Fire Service Specialist on the Staff of NFPA prior to his retirement in 1985. Ms. McDowell is President of Alternative Educational Designs, Ltd, an instructional design consulting firm based in Melrose, MA.

INCIDENT FACTORS

The essentials of a person-machine system include certain key factors or components which can be used for size up of a dangerous incident. They have a direct bearing on the decision for initial action and subsequent decisions. The chart of these incident factors (Fig. 6-3A) does not indicate an order of importance; depending on the situation, any of these factors may be the priority element in decision making.

These factors represent a series of possibilities which must be clarified or identified by the person in command of incident control in order to determine a course of action. The procedure involves a series of questions which must be considered and resolved by decisions and actions (NFPA 1978b).

The incident factors are in four groups—the problem, modifying conditions, potential losses, and possible control measures. After these are identified, the incident commander must specify to the operating forces the control objectives and tactics to be used.

The Incident Problem

Three factors help to define the problem during size up: the stage of the incident; the harmful nature of the material(s); and the type, condition, and behavior of the container or other enclosure of the material.

Stage of the Incident: Emergency personnel may arrive at a transportation emergency incident during any of five stages: when a shipping container is in danger of failing; when the container or transporter has failed or has been damaged, but there has been no ignition or reaction; when ignition or reaction has already occurred; when additional ignition, extension, or reaction is likely; or after the incident has stabilized. Whenever they arrive, their goal should be to bring about stabilization with minimum injuries and loss of life and property.

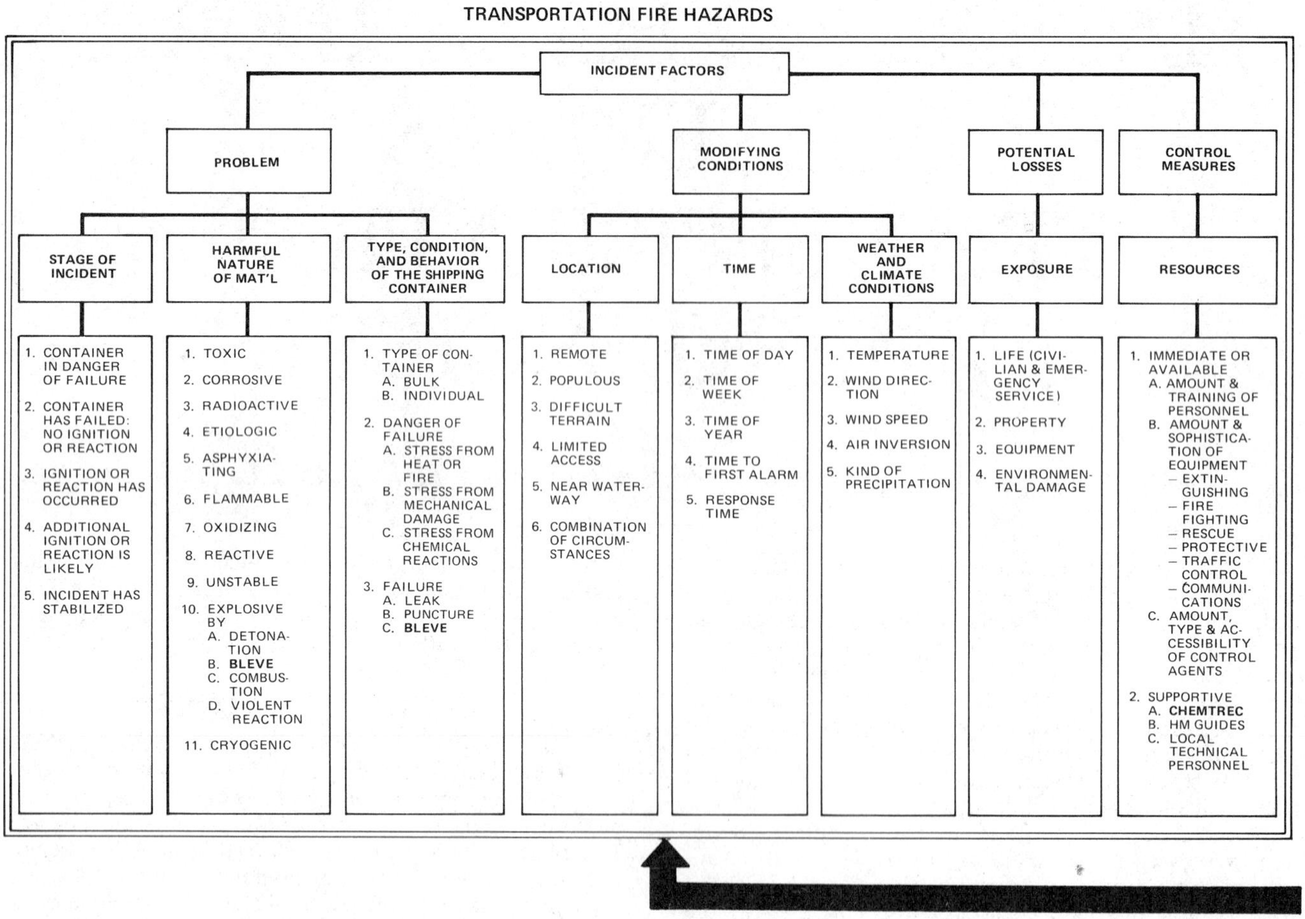

TRANSPORTATION FIRE HAZARDS

INCIDENT FACTORS

- PROBLEM
  - STAGE OF INCIDENT
    1. CONTAINER IN DANGER OF FAILURE
    2. CONTAINER HAS FAILED: NO IGNITION OR REACTION
    3. IGNITION OR REACTION HAS OCCURRED
    4. ADDITIONAL IGNITION OR REACTION IS LIKELY
    5. INCIDENT HAS STABILIZED
  - HARMFUL NATURE OF MAT'L
    1. TOXIC
    2. CORROSIVE
    3. RADIOACTIVE
    4. ETIOLOGIC
    5. ASPHYXIATING
    6. FLAMMABLE
    7. OXIDIZING
    8. REACTIVE
    9. UNSTABLE
    10. EXPLOSIVE BY
        A. DETONATION
        B. BLEVE
        C. COMBUSTION
        D. VIOLENT REACTION
    11. CRYOGENIC
  - TYPE, CONDITION, AND BEHAVIOR OF THE SHIPPING CONTAINER
    1. TYPE OF CONTAINER
       A. BULK
       B. INDIVIDUAL
    2. DANGER OF FAILURE
       A. STRESS FROM HEAT OR FIRE
       B. STRESS FROM MECHANICAL DAMAGE
       C. STRESS FROM CHEMICAL REACTIONS
    3. FAILURE
       A. LEAK
       B. PUNCTURE
       C. BLEVE
- MODIFYING CONDITIONS
  - LOCATION
    1. REMOTE
    2. POPULOUS
    3. DIFFICULT TERRAIN
    4. LIMITED ACCESS
    5. NEAR WATERWAY
    6. COMBINATION OF CIRCUMSTANCES
  - TIME
    1. TIME OF DAY
    2. TIME OF WEEK
    3. TIME OF YEAR
    4. TIME TO FIRST ALARM
    5. RESPONSE TIME
  - WEATHER AND CLIMATE CONDITIONS
    1. TEMPERATURE
    2. WIND DIRECTION
    3. WIND SPEED
    4. AIR INVERSION
    5. KIND OF PRECIPITATION
- POTENTIAL LOSSES
  - EXPOSURE
    1. LIFE (CIVILIAN & EMERGENCY SERVICE)
    2. PROPERTY
    3. EQUIPMENT
    4. ENVIRONMENTAL DAMAGE
- CONTROL MEASURES
  - RESOURCES
    1. IMMEDIATE OR AVAILABLE
       A. AMOUNT & TRAINING OF PERSONNEL
       B. AMOUNT & SOPHISTICATION OF EQUIPMENT
          - EXTINGUISHING
          - FIRE FIGHTING
          - RESCUE
          - PROTECTIVE
          - TRAFFIC CONTROL
          - COMMUNICATIONS
       C. AMOUNT, TYPE & ACCESSIBILITY OF CONTROL AGENTS
    2. SUPPORTIVE
       A. CHEMTREC
       B. HM GUIDES
       C. LOCAL TECHNICAL PERSONNEL

FIG. 6-3A. The Person-Machine System.

It is important to recognize two elements of the emergency, or the problem. First, that there is a relationship between when a fire company or other trained group arrives at the scene of emergency, and when decisions are made to control that emergency. Second, that decision making will continue throughout the emergency.

Characteristics of Materials: The next incident factor which helps to define the problem is the potential harmful effect of materials, and to decide what actions will solve the emergency.

It is imperative to consider all possible hazards, especially when the involved material is not identified. The harmful effect may be toxic, corrosive, radioactive, etiologic, asphyxiating, flammable, oxidizing, reactive, unstable, explosion by detonation, combustion or violent reaction, or cryogenic. While it may only take seconds to identify the material, it may be a time consuming process to determine the characteristics of the material, or to obtain complete and accurate technical information from an outside source, and then to achieve control of the situation. Because decisions regarding control and protective measures may have to be made before obtaining this information, the fire officer or other person in charge may have to rely on other indicators, such as sounds or visual observations, before making a decision that might jeopardize the safety of emergency service personnel and other people (NFPA 1975a).

Many hazardous material transportation incidents which have killed emergency and civilian personnel resulted from fires, explosions, spills, tank ruptures, and BLEVEs (boiling liquid expanding vapor explosions). The possibility of such incidents necessitates identifying the shipping containment system, which is the third incident factor to be considered. Perhaps the most important decision is whether the application of water will decrease or increase the hazard of the situation.

Type of Container: The type, condition, or behavior of the shipping containment system helps define the problem during size up. The system must be evaluated for signs of failure or danger of failure. If the container has not already failed, it can be weakened from thermal, mechanical, or internal chemical stress. Heat or flame can weaken the metal. Impact can cause a rupture or tearing of the metal. Corrosion, exposure to a reaction initiator, or spontaneous ignition of the material can deteriorate the container or weaken it by internal chemical action.

Within limitations, the five senses can be relied upon to detect that shipping containers have failed or are likely to fail within a brief period of time even with intervention. There may be a vapor cloud, a safety relief valve operating,

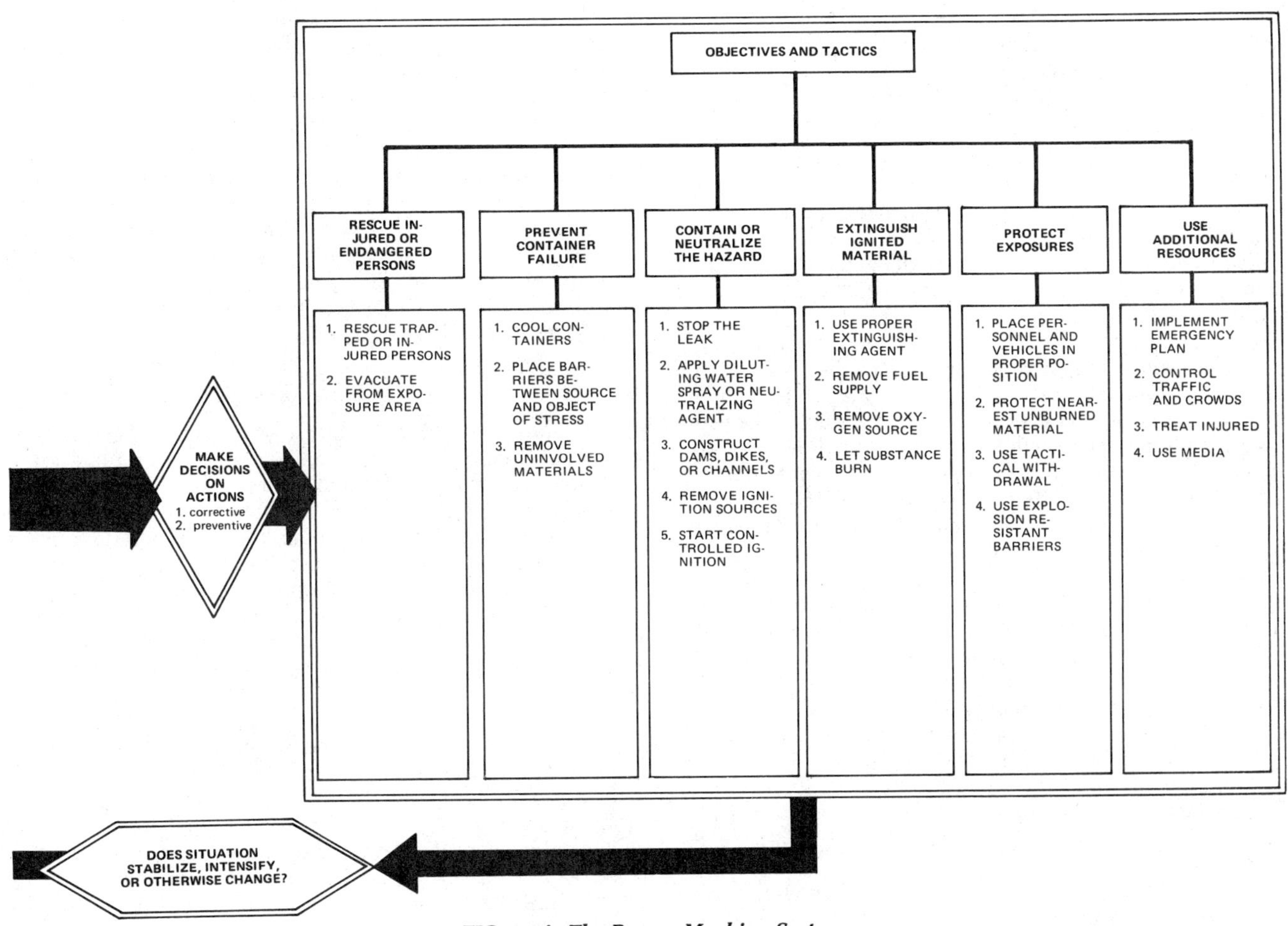

FIG. 6-3A. The Person-Machine System.

or severe flame impingement. It may be possible to taste or smell unusually pungent or offensive odors which indicate product release. However, some gases, such as hydrogen sulfide or hydrogen cyanide, deaden the sense of smell, so it is not always a reliable indicator (NFPA 1975b).

Modifying Conditions

The second major area of concern includes modifying conditions which must be taken into consideration during any emergency—location, time, weather and climate conditions.

Location: The first modifying condition is the location of the emergency. An incident may occur in a remote area, a populous area, on difficult terrain, a limited access highway, near a waterway or any other location where response is affected in some manner. Although sufficient equipment, personnel, and control agents may be available, the location of the incident may severely hamper their access to the scene. A heavily populated area creates additional exposures and possible interference to movement of apparatus and personnel.

If an incident occurs near a body of water, the problem may be complicated by the toxic nature of the material involved, which may endanger the environment.

Time: The second modifying condition is time. Considerations should include: time of day, day of the week, length of time after the incident occurred until first notification, response time, and the time for backup forces to arrive at the scene.

The time and the day of the week may be favorable or unfavorable. A daytime incident may occur when there is good visibility, but vehicle traffic is dense and pedestrians and onlookers are numerous. A weekend incident may mean that exposures are fewer, but if it occurs at night, control efforts may be hindered or it may be difficult to identify the material because of reduced visibility.

Time to first notification and the response time are important, especially when fire impinges on cargo tanks of liquefied gases or flammable liquids. The possibility of a BLEVE or other dangerous rupture increases as time passes. Control efforts may have to be abandoned if too much time has elapsed (Walls 1979).

Weather and Climate: The third incident modifying condition is the combination of weather and climate, including temperature, wind direction, wind speed, air inversion, and kinds of precipitation.

As with location and time, weather and climate conditions can be favorable or unfavorable. They may reduce ignition possibilities, help to dissipate vapor, and carry toxic gases away from a populated area. However, weather can also hamper access of equipment and personnel, reduce visibility, or spread flammable vapor and toxic fumes over a large area.

Potential Losses

Another factor of concern in handling hazardous materials incidents is potential loss, and this is indicated by the kind and number of exposures. These include potential loss of life among civilian and emergency service personnel, property and equipment, and environmental damage. The presence of any of these risks will have direct influence on decision making.

To determine exposure risk, other incident factors must be considered, such as the stage of the incident, the time, location, harmful nature of the materials, and the type of container. The exposure must be defined by the area which may be affected. Understanding how a hazardous material can spread after release from a container will help determine where the exposures will be.

If the material is a gas, it may escape under pressure, as in the operation of a safety relief valve, or through a small tear or other opening. A liquefied gas may form a vapor cloud and fill available space. If the vapor density of a gas is heavier than that of air, the gas may travel along the ground, following the contour to the lowest level. [The vapor densities of some materials are lighter than air at 70°F (21°C) but heavier than air at colder temperatures.] A gas will move according to the wind direction and speed.

If the material escaping is a liquid, it will flow and follow the contour of the ground to the lowest level, or to some place of confinement. A liquid may vaporize and spread like a gas, or seep into the ground or other substances that will absorb it.

A solid can scatter and travel in any direction under the pressure of an explosion. It may become a dust cloud and get carried by the wind.

The lack of information on the escape of material, or the extent of other hazards in transportation emergencies has often presented problems. There are few answers to such qustions as "How far?" and "How wide?" despite the importance of this information for determining control actions and protection of exposures.

Control Measures

The next major category of concern is the means of possible control. To define this potential, the incident commander must consider all resources available to help control the incident. An inventory of resources should include two categories of assistance: resources which are immediate or available, and supportive resources.

Immediate or available resources include the number and level of training of personnel, the amount and sophistication of extinguishing, fire fighting, rescue, protective, traffic control, and communications equipment, and the amount, type, and accessibility of control agents. The latter may include water, foam, sand, or specialized extinguishing agents.

Supportive resources include sources of technical assistance such as CHEMTREC, hazardous materials emergency guides, and technical personnel. These should be determined before an incident occurs. An incident commander should have readily available a written inventory of the kind of assistance that can be expected in order to determine what control measures can be accomplished. It is important to consider personnel, equipment, and control agents from the entire local community and the mutual aid area. This task is best accomplished during the planning of response to emergencies (Bahme 1978b).

In addition to emergency personnel such as fire fighters, law enforcement and emergency medical technicians, rescue, ambulance, and hospital personnel, public works, public utilities, offices of emergency preparedness, chemical manufacturing and transportation industries, military personnel and disaster relief agencies are all potential sources of personnel, equipment and control agents. These groups should be contacted in preemergency planning to determine which resources they can provide.

DECISION MAKING

Once the incident factors have been defined during size up, the next step is to make decisions on the kind of action to be accomplished. For these decisions, it is necessary to weigh certain incident factors against each other. An incident commander must attempt to predict what benefits or losses to people, property, equipment, and the environment will be presented or will occur as a direct result of the control actions.

For example, a derailment involving several tank cars of flammable gas occurs in a remote area. One tank car has failed, ignition has occurred and is threatening failure of another tank car. There is potential for a BLEVE. At the same time, there is a limited water supply. Because the water supply is inadequate and the exposures are limited, it would not be in the best interests of safety to risk the loss of emergency service personnel in trying to prevent an explosion that may not be preventable. Losses would best be kept to a minimum by concentrating on protecting exposures rather than preventing the additional tank car failure.

To assure that the effects of control actions will be in accord with the goal of the person-machine system, that is, a return to normal conditions with the least amount of personnel injury and damage to property and the environment, two kinds of action, corrective and preventive, must be considered, based on size up of the incident factor. Corrective actions are taken to resolve the immediate problem. Preventive actions are taken to prevent the immediate problem from increasing; these actions are intended to keep losses to a minimum.

In order to implement these decisions, it is necessary to specify objectives and tactics.

OBJECTIVES AND TACTICS

Only a limited number of tactics can be used to control any emergency. For a hazardous materials incident, tactics can be grouped in terms of six main objectives: (1) rescue of injured or endangered persons, (2) prevention of container failure, (3) containing or neutralizing the hazard, (4) extinguishing ignited material, (5) protecting exposures, and (6) using additional resources. During an incident, some tactics which relate to each of these objectives may be used.

Rescue and Evacuation

The first objective is to rescue injured or endangered persons. Tactics which can accomplish this include positioning personnel; equipment and vehicles to rescue trapped or injured persons; and evacuating others from the danger area.

Among the difficult problems in a hazardous materials emergency are the rescue and extrication of trapped persons, particularly if this attempt exposes emergency personnel to extreme danger. The best guidelines to follow in life hazard situations are: use only a few personnel to attempt a rescue; make sure that each person is wearing as much protective clothing as required by the hazard; attempt rescue only of persons who obviously are not beyond help; and do not subject a rescuer to needless risk.

Certain actions are usually started prior to extrication of trapped persons. These include: examining unconscious persons for blocked airways; stopping severe bleeding; splinting fractures; covering wounds; and continuing treatment for shock. These practical routines may have to be disregarded if hazardous materials are involved. If they are, the patient should be moved out of danger as soon as possible. Rescuers should not use spark producing equipment in a hazardous atmosphere. Backup personnel can sometimes cover those attempting the rescue with protective water spray. If there is a fire, water spray may be used to clear a path to reach trapped or overcome persons.

If it is known that the material involved in a transportation incident can be dangerous to people in the exposure area, law enforcement officers and other trained persons can evacuate the public. Buildings on densely populated streets should be evacuated from the sides away from the danger. As a minimum safety precaution, all buildings facing the street should be evacuated for at least one block in all directions. For toxic substances, this distance will be much greater and can be obtained from technical sources. After an area is evacuated, roadblocks and patrols should be set up to cover all points of entry into the restricted area.

If evacuation of buildings is not possible because of time, limited personnel, or population density, direct the movement of people into protected areas such as inner corridors, or the sides of the building away from the incident.

Preventing Container Failure

The second tactical objective is aimed at preventing container failure. Tactics which can accomplish this objective include: cooling the container, placing barriers between the container and the source of stress, and removing uninvolved materials (Barr 1978; NFPA 1979; NFPA 1975f; Soros 1979a).

Cooling the container usually is accomplished by applying deluge quantities of water to the vapor space of the tank at the point of flame impingement in order to prevent a BLEVE. This tactic will usually be applied to a tank containing a flammable liquid or liquefied gas.

Extreme care should be exercised when applying water to bulk containers. There must be adequate supply for applying large flows and sustaining this amount until the incident stabilizes or even longer. In addition, the time of fire exposure to the vapor space of the tank should be less than 5 min. BLEVEs have been known to occur within 8 min after flames impinged on the vapor space of uninsulated tanks. The cooling streams should be applied at all points of flame impingement. If these conditions cannot be met, a BLEVE or container rupture should be anticipated. Evacuate all persons, withdraw emergency service personnel, and prepare for the explosion and fireball.

Using fire fighters to operate streams from monitor nozzles exposes them unnecessarily to possible explosions; unattended or fixed position hose streams should be used to cool the containers. Fire fighters can gain valuable time in high exposure areas by placing some kind of barrier or shield between the source of heat and the target hazard.

When two or more materials are involved in an emergency situation, it may be possible to separate them to lessen the hazard. This may involve moving drums or cylinders, or uncoupling tank cars in a railyard and moving them to a safe distance. For safety and efficiency, when possible, transportation personnel should move any bulk containers as directed by the incident commander.

If it is absolutely necessary to move containers, caution should be exercised. Containers may already have been exposed to dangerous amounts of heat or other chemicals. It may be necessary to cool them before moving.

Confining the Danger

The third objective is to confine the danger from the material or from its reaction. Tactics for such confinement include: stopping a leak, applying vapor controlling water spray, damming, diking or channeling, removing ignition sources and starting controlled ignition (Barr 1978).

Stopping Leaks: There are several ways of stopping leaks. A leak at a normally functioning valve or outlet can be stopped by closing the valve. Leaks at punctures or broken valves can be stopped by driving tapered wooden plugs into the opening, or using special kits designed for cylinders or tank cars. If the leak occurs in a tank containing flammable liquid or gas, closing the valve or driving in a plug should be attempted under cover of water spray.

For cryogenic substances, a cautious approach should be used to avoid severe injury from the super cold material. A liquefied gas will cause an icy accumulation at the point of leakage.

If a leaking container is threatening a high exposure area, it might be moved to a safer location before attempts are made to stop the leak.

Leaks can be stopped, or the amount of escaping material can be lessened when a container is turned upright to allow gas instead of liquid to escape, or by turning the container until the leaking portion is above the liquid level. For a tank of flammable liquid with specific gravity less than water, the container can be filled with water to a point above the leak.

Applying Water Spray: Water spray or water fog can be applied in several ways to help contain the hazard. Water spray can be used to direct flammable gas-air mixtures away from ignition sources or toxic vapors away from populated areas. Water spray can also be used to disperse the vapors until a flammable gas-air mixture is no longer possible.

For water soluble materials such as ammonia or sulfur dioxide, spray can be used to clear a path to the shutoff point for the leak. Even if the vapors are not water soluble, the spray should always be used as a protective cover for the personnel trying to stop the leak (NFPA 1975f).

Streams can be applied to flammable and combustible liquids that are water soluble, such as alcohol, to dilute the liquid and thus reduce its hazard.

Spills of corrosive chemicals out of doors should be flushed with large volumes of water to neutralize the acid or dilute the caustic solution to the point where neither will be corrosive. Care must be taken in applying the water because most acids will splash or boil when water is applied.

When water spray is used to knock down vapors from some gases, such as chlorine and hydrogen chloride, a chemical reaction may take place which produces hydrochloric acid. In these circumstances, straight streams should be used with the spray to dilute the acid.

Although water is usually the most effective and inexpensive control agent, there are situations where the use of water or the wrong method of application will increase the hazard. If straight streams are applied to cryogenic pools, they will drastically increase vaporization because of the relative heat of the water compared to the material. However, water spray can control vapors and prevent low temperature damage. Straight streams of water should not be used on concentrated sulfuric acid because of the reactive nature of this material.

Straight streams applied to pools of burning liquids with a flash point above 200°F (93°C), may create a frothing action. If the burning material is in a container, the frothing can cause violent expulsions of burning material. Only under special conditions should water be used on water-reactive materials, such as the alkali metals. Water generates hydrogen and sufficient heat to ignite the hydrogen and cause an explosion.

Damming, Diking, Channeling: Damming, diking, or channeling can direct flammable liquid fires away from high exposure areas, contain toxic or corrosive materials, divert large spills, or channel water runoff.

Public works departments or private contractors should be contacted during the planning stages to assure an accessible supply of smothering agent. Sand, dirt, or dry vermiculite can be used as an absorbent to cover small spills. For corrosive materials, lime or other neutralizers are practical.

Removing Ignition Sources: With flammable liquids and gases that vaporize, steps must always be taken to remove ignition sources. Starting with the downwind areas, all sources of heat, sparks, or friction, indoors and out, should be eliminated.

Starting Controlled Ignition: In certain limited circumstances, it may be advisable to start controlled ignition rather than deal with the hazard in any other way. For instance, if a poisonous flammable gas or liquid is escaping in an inhabited area, it could be better to burn the material under expert technical supervision than let it flow uncontrolled over a large area, or contaminate water supplies.

Extinguishing Ignited Material

The fourth objective in the person-machine system is extinguishing ignited material. Tactics include use of a proper extinguishing agent, removing the supply of fuel, removing the source of oxygen, or letting the substance burn.

Common extinguishing agents are water, foam, dry chemical, dry powder, sand, or earth. Water is the most common because of its cooling effect and its availability in large quantities. However, the decision to use water depends on the physical and chemical properties of the hazardous materials (NFPA 1975d).

Straight streams of water are generally used to cool containers. For other extinguishing or control efforts, water spray should be applied. It is most effective for combustible liquids and flammable solids having flash points between 100 and 212°F (38 and 100°C). However, if applied carefully by properly trained groups, water spray can be used on liquids with a flash point below 100°(38°C) or above 212°F (100°C).

Water spray can be used as a cooling agent to prevent heat or flames from reaching nonignited materials. Direct application on the material of large quantities of water may prevent detonation of explosives if the fire has not reached the cargo, or prevent ignition of air-reactive but not water-reactive flammable solids.

For flammable liquids with low flash points, foam is most effective, but the liquid must be covered with the foam (NFPA 1978c). Applying foam in a fog pattern allows direct application without agitating the liquid, and a rapid covering of the liquid surface. High wind or heavy rain may break up a foam blanket.

Although dry chemicals provide a quick knockdown of flame, in some situations it may be necessary to cool hot metals with foam or water to prevent reignition.

Dry powders are useful for extinguishing combustible metal fires or other water-reactive materials. These materials will also react with foam. The main problem with dry powders is that they usually are limited in amount and availability. Few departments have these materials in sufficient quantities to extinguish large fires (NFPA 1978d). Sand or earth is useful for extinguishing small fires when water would be ineffective and more sophisticated agents are not available.

Removing the fuel which is supplying a gas fire will probably be accomplished by stopping a leak. Stop the leak rather than extinguishing the flame at the leak point. If the flame is extinguished before the leak is stopped, an explosive mixture may be formed with air, which, if ignited, could cause far greater problems. The flame will die out when the leak is stopped (Soros 1979b).

Covering with water spray can be used to extinguish flammable liquids with a specific gravity greater than water. Covering separates the oxygen supply from the burning material. Foam creates the same separation. Water spray can cool the surface of the burning material to a point below its ignition temperature.

Protecting Exposures

The fifth objective is the protection of all exposures. Tactics which may be used include positioning of personnel and equipment, making tactical withdrawal, using explosion resistant barriers, and protecting the nearest unburned material.

Exposures include people, property, equipment, and the environment. Members of the emergency services of any community are a high priority exposure. It is not in anyone's best interests to risk one's own life unreasonably to save other personnel who are beyond help, or equipment, property, and portions of the environment.

One important safety prerequisite for fire fighters and other personnel is the use of proper clothing or equipment. For hazardous materials, such as gasoline or fuel oil in an open air fire, ordinary turnout gear may be sufficient unless it is necessary to rescue someone trapped in a vapor area (NFPA 1975e). However, for materials which are mildly toxic or corrosive, full protective clothing is an absolute necessity for emergency service personnel. This includes a helmet, self-contained breathing apparatus, turnout gear, rubber boots, gloves, bands wrapped around legs, arms, and waist to prevent material from getting in through the large gaps in the clothes, and covering for neck, ears, and other parts of the head or face not protected by the helmet, breathing apparatus, or mask. Toxic or corrosive materials can enter the body through any exposed area, including the eyes and skin.

Substances which require special protective clothing should not be handled unless there is access to and experience with this type of clothing. Acid or ammonia suits are designed to protect against specific hazards, but many departments do not own such suits. For most transportation incidents involving such materials, the best decision to be made is to withdraw, protect exposures, and wait for the experts to arrive. Even special protective clothing will not prevent severe frostbite from skin exposure to cryogenic materials.

Emergency service personnel, such as police, who do not normally use any kind of protective clothing should understand that self-contained breathing apparatus may be a necessity for some rescue situations. A person not trained for wearing this equipment should not attempt to use it during emergency conditions.

The tactical placement of personnel and equipment must be kept in mind during response to an incident. Access routes should be chosen to approach an incident from upwind and avoid flammable, toxic, or asphyxiating vapors. Park vehicles upwind at a safe approach area and stop the automotive engines, which can be a source of ignition.

A horizontal tank fire should be attacked from the upwind side because if the tank ruptures, failure is likely to occur at one end, and the tank may rocket. A cylinder or metal drum may also rocket when an end seam fails.

When operating at a cargo trailer incident, it is best to determine what is inside the van before opening the doors. The result of opening the doors may be explosive. If the doors must be opened, make use of available protection. For a swing-back door, it is best to stand behind the door when opening it. For roll-up doors, there is no shielding capability but personnel can drop down for some protection.

A tactical withdrawal from the danger area may be as effective for keeping losses to a minimum as extinguishing a fire, particularly if the problem cannot be handled with available resources. Withdrawal might also be best for a fire involving a potentially explosive material. Places for possible cover should be noted. Structural shielding, ditches, depressions, or other explosion resistance barriers can protect personnel from flying missiles and radiant heat, although not from overpressure.

When burning material cannot be extinguished, the safest tactic is to keep the fire from spreading. If it is necessary to withdraw, prepare personnel and hose lines for any fires which result from an explosion.

Using Resources

The sixth objective is using additional resources, and this includes implementing an emergency plan, controlling traffic and crowds, treating the injured, and using the media. It is important to consider the use of the person-machine system for reaching an objective. For instance, one of the objectives is rescuing injured or endangered persons. However, during extrication of a trapped person, it may be necessary to evacute a densely populated area. This may be impossible for a small department to accomplish unless additional personnel can do this work.

If a community has an emergency plan, it should be used. A plan can indicate how additional resources can be obtained in the shortest time, and will diminish confusion, inefficiency, and duplication of effort.

In addition to directing evacuation of the public, law enforcement officers can keep crowds and traffic out of exposure areas to allow responding personnel and apparatus to reach the scene and set up operations with minimum delay. Emergency plans should provide specific directions for diverting traffic, especially in areas of limited access. Law enforcement personnel may also be involved in initial rescue, first aid, and the extinguishing of incipient fires.

Emergency medical technicians and rescue and ambulance personnel will have the responsibility for treating the injured and transporting them to a hospital.

Rescue and ambulance personnel must be careful to guard against their own contamination by using proper protective equipment when treating contaminated victims.

During an incident of major proportions, the media can aid in evacuating large areas, controlling panic, and keeping the public informed.

With any discussion of tactics, the list will never be complete. Application of tactics depends on all the incident factors in a specific situation, and the proper direction and capabilities of emergency service personnel. To deal effectively with a hazardous materials emergency, it is important to know those objectives and tactics which will control the problem, and the capabilities of emergency personnel for carrying out these procedures with minimum losses.

EVALUATION OF THE SITUATION

The fourth and last step of the person-machine system is evaluation of information, whether the situation stabilizes, intensifies, or otherwise changes. The system must assure continual evaluation of its output, the making of adjustments, and returning this information into the system with other new input.

In a person-machine system, what happens next during the emergency provides the feedback for adjusting the original input or size up. When actions have been initiated, the incident commander should monitor the situation to detect any indication that the problem may be escalating, intensifying, or stabilizing.

Although it has been emphasized that no major control action should be initiated until a material is identified and all possible hazards are determined, an incident in a populous area may warrant taking some kind of immediate action. In this instance, control measures should take into account the greatest possible hazard or combination of hazards.

Monitoring procedures can provide instant feedback on the success or failure of the initiated control action which may help to eliminate the possibility of certain hazards and emphasize the presence of others. Positive feedback indicates that the tactical measures are working, the incident is moving towards control, and the successful tactical operations should be continued with minor changes. Any negative feedback signals the incident commander to reevaluate the situation, make another size up, make additional decisions on actions, and possibly change objectives and tactics until the situation does stabilize. For instance, if a leak suddenly ignites, a leak-no-fire situation changes to a leak-fire situation. The change is registered and different objectives and tactics are initiated.

SUMMARY

The four steps of the person-machine system—incident factors, decision making, objectives and tactics, and evaluation--combine to form an efficient and effective structure for making decisions during a hazardous materials transportation emergency. It does not purport to be all inclusive since every incident in any community will be unique. However, using the system during planning and training situations will facilitate an automatic thinking process during an actual emergency.

Bibliography

References Cited

Bahme, Charles W. 1978a. *Fire Officer's Guide to Disaster Control.* National Fire Protection Association, Quincy, MA. pp 12-27.

Bahme. 1978b. *Ibid.* pp 18-25.

Barr, Robert C. 1978. *Transportation BLEVEs — Causes, Effects, Guidelines.* Instructor's Manual. National Fire Protection Association, Quincy, MA. Part III.

NFPA. 1979. *Combatting Vehicle Fires.* Instructor's Manual. National Fire Protection Association, Quincy, MA. pp 6 and 7.

NFPA. 1978a. *Handling Hazardous Materials Transportation Emergencies.* Instructor's Manual. Unit 5. National Fire Protection Association, Quincy, MA. pp 20-25.

NFPA. 1978b. *Ibid.* pp 93-109.

NFPA. 1978c. NFPA 10, *Portable Fire Extinguishers.* National Fire Protection Association, Quincy, MA. pp 74-76.

NFPA. 1978d. *Ibid.* pp 83 and 84.

NFPA. 1975a. NFPA 49, *Hazardous Chemicals Data,* National Fire Protection Association, Quincy, MA. pp 8-16.

NFPA. 1975b. *Ibid.* pp 170-174.

NFPA. 1975c. NFPA 49, *Hazardous Chemicals Data.* pp 47 and 275.

NFPA. 1975d. *Ibid.* pp 8-11.

NFPA. 1975e. *Ibid.* pp 15 and 16. (e)

NFPA. 1975f. *Tank Vehicle Fire Fighting.* Instructor's Manual. National Fire Protection Association, Quincy, MA. p 48.

Soros, Charles. 1979a. *Safety in the Fire Service.* National Fire Protection Association, Quincy, MA. pp 149-150.

Soros. 1979b. *Ibid.* pp 145-168.

Walls, Wilber L. 1979. "The BLEVE. Part 1." *Fire Command.* Vol 46, No 5. May 1979. pp. 22-24. Also: Vol 46, No 6, June 1979. pp 35-37.

Additional Readings

Allan, D. S., "A Cryogenic Line Leak Detector," *Fire Technology,* Vol. 11, No. 4, Nov. 1975, pp. 270-273.

Bahme, C. W., *Fire Officer's Guide to Dangerous Chemicals,* National Fire Protection Association, Quincy, MA, 1972.

Basic Training Course for Emergency Medical Technicians, U.S. Government Printing Office, Washington, DC.

Basic Training Course for Emergency Medical Technicians, 2nd ed. (Course Guide, Instructor Lesson Plans, Student Study Guide), U.S. Government Printing Office, Washington, DC, 1977.

Cashman, John R., *Hazardous Materials Emergencies,* Technomic, Lancaster, PA, 1984.

DePol, Dennis R., and Cheremisinoff, Paul N., *Emergency Response to Hazardous Materials Incidents,* Technomic, Lancaster, PA, 1984.

Emergency Medical Services, National Fire Protection Association, Quincy, MA, 1977.

Emergency Response to Hazardous Materials in Transportation: Self-Study Guide, National Fire Protection Association, Quincy, MA, 1982.

Emergency Services Guide for Selected Hazardous Materials, U.S. Department of Transportation, Washington, DC.

Garrard, Charles, "LP —Gas Transport Truck Fire, Aylmer, Ontario," *Fire Journal,* Vol. 59, No. 5., Sept. 1965, pp. 8-10.

Handling Radiation Emergencies, National Fire Protection Association, Quincy, MA, 1977.

Lind, C. D., and Whitson, J. C., *Explosion Hazards Associated with Spills of Large Quantities of Hazardous Materials, Phase II,* U.S. Department of Transportation, Washington, DC, 1977.

Meidl, J. H., *Explosive and Toxic Hazardous Materials,* Glencoe Press, Beverly Hills, CA, 1970.

Meidl, J. H., *Flammable Hazardous Materials,* Glencoe Press, Beverly Hills, CA, 1970.

Meidl, J. H., *Hazardous Materials Handbook,* Glencoe Press, Beverly Hills, CA, 1972.

Meyer, E., *Chemistry of Hazardous Materials,* Prentice-Hall, Inc., Englewood Cliffs, NJ, 1977.

Schieler, L. and Pauze, D., *Hazardous Materials,* Delmar Publishers, Albany, NY, 1976.

Student, Patrick J., ed., *Emergency Handling of Hazardous Materials,* Bureau of Explosives, Association of American Railroads, Washington, DC, 1981.

Terrien, Ernest J., *Hazardous Materials and Natural Disaster Emergencies: Incident Action Guidebook,* Technomic, Lancaster, PA, 1984.

Transport of Dangerous Goods, United Nations Publications, NY, 1977.

Young, M. S., "Handling Emergencies Involving Cryogenic Materials," *Fire Command,* Vol. 45, No. 5, May 1978, pp. 20-22.

Zajic, J. E., and Himmelman, W. A., *Highly Hazardous Materials Spills and Emergency Planning,* Marcel Dekker, NY, 1978.

OBTAINING TECHNICAL ASSISTANCE IN HAZARDOUS MATERIALS EMERGENCIES

Phillip Blye and James Yess

Responding to a hazardous materials incident requires specialized training, equipment and materials, and a more complex coordination with other agencies than other emergency situations. Local emergency service personnel responding to the scene of a hazardous materials incident have the initial responsibility for: (1) determining what materials are involved, (2) the hazards posed by any resulting fire or spill, (3) emergency evacuation of the surrounding areas, (4) rescue of trapped persons, (5) implementing fire and spill control measures, and (6) preventing any further damage to the environment.

When dealing with hazardous materials, technical assistance of one form or another will undoubtedly be required. In order to effectively carry out their responsibilities, local emergency services should first identify and use the equipment, materials, and expertise already available from government agencies and private concerns as well as emergency guidebooks and other printed matter.

Federal, state and local governments, as well as private industry, already have many programs in operation for responding to hazardous materials incidents. These programs may be able to fill the gaps in the local emergency response effort. Knowing the capabilities of other agencies and industry also prevents wasteful duplication. This information is best gathered before any potential incident.

Additional information on dealing with hazardous materials emergencies is contained in the other chapters of this section.

FEDERAL GOVERNMENT AGENCIES

Department of Transportation (DOT)

The Department of Transportation (DOT), through its agencies, has the overall responsibility for regulating different types of transportation within the United States: air, rail, ship, highway, and pipeline. Of its various agencies, several are of particular importance to the emergency services. (See Fig. 6-4A.)

Materials Transportation Bureau (MTB)

In 1974 Congress passed and the President signed the *Hazardous Materials Transportation Act* (HMTA). This act gave the Secretary of Transportation the authority to coordinate the federal regulatory agencies having jurisdiction over the movement of hazardous materials by highway, rail, water, and air. Furthermore, the Act empowers the Secretary to designate which substances are hazardous, how they should be packaged and labeled, the use of placarding, and the routes to be taken. The Secretary can also determine the number of personnel to handle such substances and the minimum training required to handle and transport them. The HMTA also permits the Secretary to set the types, frequency, and procedures to be used in safety inspections.

In 1975 the Materials Transportation Bureau (MTB) was established within DOT to insure uniform implementation of HMTA among its several agencies, such as: Federal Highway Administration (FHWA), Federal Railroad Administration (FRA), Federal Aviation Administration (FAA), and U.S. Coast Guard (USCG). In the following year, 1976, the MTB consolidated all regulations directed at the safe transport of hazardous materials by highway, rail, air, and sea. Enforcement of these regulation is the responsibility of the appropriate agency. These regulations are contained in the *Code of Federal Regulations* (CFR) Title 49, Sections 171-79. The following is a brief summary of each section:

Section 171, General Information Regulations and Definitions: This section defines the purpose and scope of the regulation and includes general information and a list of definitions.

Section 172, Hazardous Materials Table and Communication Regulations: Contained within this section is a list of more than 1,600 materials that have been designated as being hazardous while in transit. It also details requirements for labeling, packaging, and transporting of the materials listed.

Capt. Blye is Chairperson of the Department of Fire Science Technology at Massasoit Community College in Brockton, MA. He is also currently employed in the fire service. Dr. Yess is an educator and administrator who has developed educational materials for the NFPA.

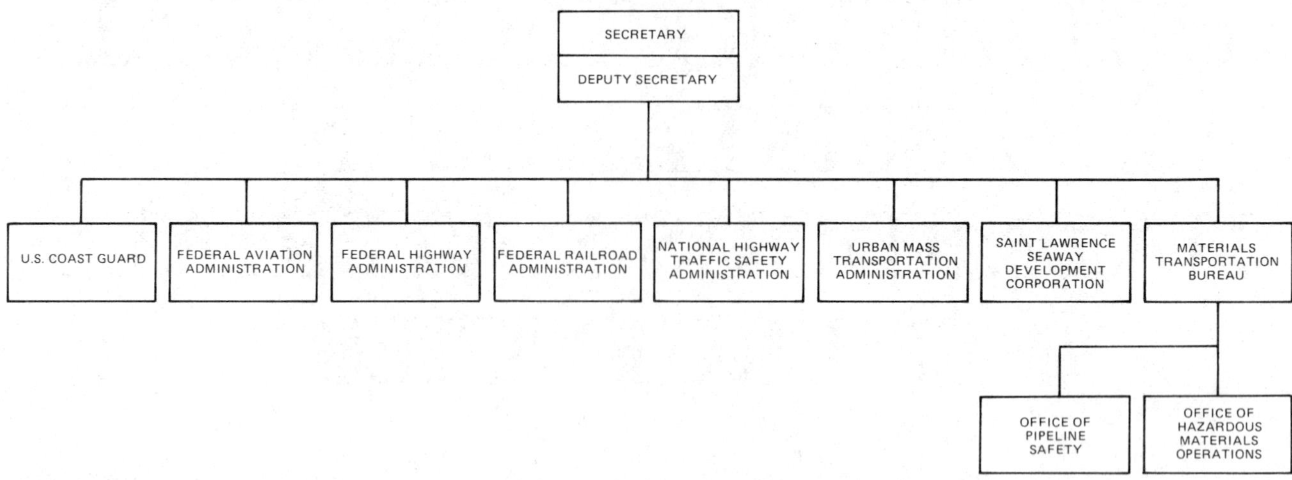

FIG. 6-4A. Organization chart for the Department of Transportation.

Section 173, Shipper's General Requirement for Shipments and Packaging: This section establishes a classification system for hazardous materials. Individual substances are grouped within the system according to the nature and degree of hazard they present.

Section 174, Carriage by Rail: This section specifies rules for handling, loading and placement of hazardous materials transported by rail. It also specifies procedures for inspection of rail tank cars and the proper placarding of rail cars containing hazardous materials.

Section 175, Carriage by Aircraft: Presented within this section are requirements that must be met before hazardous materials can be shipped by civilian aircraft. Limitations on the quantities that may be shipped as well as requirements on cargo compatibility and location within the aircraft are provided.

Section 176, Carriage by Vessel: This section specifies general labeling, handling, storage, and cargo segregation requirements for hazardous material transported by ship.

Section 177, Carriage by Highway: This section applies to all private, common and contract truck carriers. It contains requirements for the loading and proper labeling of both the cargo and vehicle. Procedures are provided for cargo handling during and after an accident, as well as the proper procedure to be followed when reporting an accident.

Section 178, Shipping Containers Specifications: This section details design and construction requirements for containers used to transport hazardous materials. Among the containers regulated under this section are barrels, cylinders, cargo tanks, paper, and plastic bags.

Section 179, Specification for Rail Tank Cars: Design requirements for rail tank cars used to transport specific hazardous substances are detailed with this section.

It also lists specifications for tank car construction materials, welding procedures, valves, and piping. In addition this section details testing procedures to be used in determining tank car integrity.

Federal Highway Administration (FHWA)

Highway safety responsibilities under the jurisdiction of the FHWA are carried out by one of its divisions—the Bureau of Motor Carrier Safety (BMCS). BMCS develops and enforces safety regulations pertaining to interstate transport by commercial carriers. Representatives of the BMCS generally carry out inspections at the facilities of shippers. They may also conduct inspections of interstate trucks while on the road. These various inspections serve to enforce the regulations regarding the proper packaging, labeling, handling, loading, and transporting of hazardous material.

Federal Railroad Administration (FRA)

The FRA is an operating agency of the Department of Transportation and as such oversees the operation of all railroads in the United States. Among the safety regulations it administers are those pertaining to the handling, labeling, placarding, packaging, and shipping of hazardous materials by rail. Offícals from FRA conduct inspections of hazardous materials, tracks, equipment, signals, train controls, and operating practices.

U.S. Coast Guard (USCG)

The safety of ports, waterways, and shipping in waters of the United States comes under the jurisdiction of the USCG. (See Fig. 6-4B.) In addition to its traditional functions of maintaining navigational aids, rescuing vessels and men at sea, and enforcing maritime law, the USCG oversees the shipping of hazardous materials on the waterways. The *Tanker Act* (USC 49, part 39/a) and the *Dangerous Cargo Act* (USC 46, part 170) gave the USCG the authority to regulate the design and construction of tankers operating in waters of the United States.

Foreign tankers which transport hazardous liquid cargoes are regulated through the Letter of Compliance Program (LOC). Under the terms of this program owners of ships transporting hazardous liquids must submit design and construction plans, a description of the cargo containment system, electrical safety, and vapor detection equipment plans for approval. The *Port Safety Act* (USC 50, part 191), the *Ports and Waterways Safety Act* (USC 33, part 1221) and *Executive Order 10173* give the USCG power to regulate piers, wharfs, docks, mooring areas, and the land and buildings adjacent to waterfront areas. Anyone loading, discharging, storing, or transporting hazardous materials must obtain the appropriate permit from USCG.

The USCG has trained a National Strike Force. This Strike Force combines the skilled personnel and high seas equipment necessary for the containment, cleanup, and disposal of hazardous materials spills on the water. Strike teams are located on the East, West, and Gulf Coasts and can respond to an incident within hours. Strike Teams can provide advice on ship salvage, removal techniques, and communications. The Coast Guard also maintains a computerized inventory of oil pollution control equipment by state and region called SKIM.

The Coast Guard can also provide information on the chemical and physical properties of a hazardous material through its Chemical Hazards Response Information System (CHRIS). The system is comprised of a four volume manual, a regional contingency plan, the Hazard Assessment Computer System (HACS), and a response team.

Volume 1, *A Condensed Guide to Chemical Hazards*, contains essential chemical information that can be used during the incipient stage of a hazardous materials incident on the water. Information found in this volume includes: the hazardous nature of the material (assuming the hazardous material has been identified) and immediate actions to be taken to safeguard life, property, and the environment.

Volume 2, *The Hazardous Chemical Data Manual*, provides detailed technical information on the chemical, physical, and toxicological properties of hazardous materials.

Volume 3, *The Hazard Assessment Handbook*, presents methods to calculate the rate and quantity of hazardous materials that may be released from a container under different situations, and the spread of a material in both water and/or air based on time and distance.

Volume 4, *Response Methods Handbook*, provides information on current methodology for handling spills such as containment, material transfer, removal, and disposal. An Appendix to Volume 4 lists spill control equipment manufacturers.

The Hazard Assessment Computer System (HACS) is a computerized version of CHRIS —Volume 3. HACS can provide a more rapid and explicit assessment of the incident than can be done manually.

The USCG also operates the National Response Center, (NRC), 24 hrs, seven days a week in Washington, DC. When notified of an environmental emergency a Regional Response Team is alerted. This team can provide the local emergency services with advice and assistance. The NRC can also provide emergency information on hazardous materials from the CHRIS manuals over the telephone. The number to call for the National Response Center is 800-424-8802.

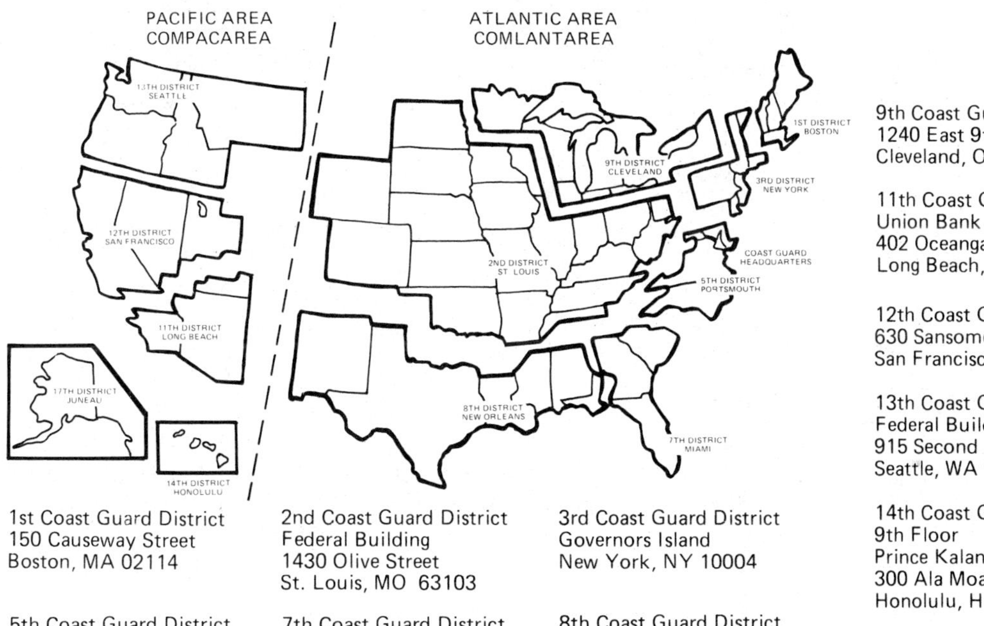

1st Coast Guard District
150 Causeway Street
Boston, MA 02114

2nd Coast Guard District
Federal Building
1430 Olive Street
St. Louis, MO 63103

3rd Coast Guard District
Governors Island
New York, NY 10004

5th Coast Guard District
Federal Building
431 Crawford Street
Portsmouth, VA 23705

7th Coast Guard District
Room 1010
Federal Building
51 SW First Avenue
Miami, FL 33130

8th Coast Guard District
500 Camp Street
Hale Biggs Building
New Orleans, LA 70130

9th Coast Guard District
1240 East 9th Street
Cleveland, OH 44199

11th Coast Guard District
Union Bank Building
402 Oceangate Blvd.
Long Beach, CA 90822

12th Coast Guard District
630 Sansome Street
San Francisco, CA 94126

13th Coast Guard District
Federal Building
915 Second Avenue
Seattle, WA 98174

14th Coast Guard District
9th Floor
Prince Kalanianaole Fed. Bld.
300 Ala Moana Blvd.
Honolulu, HI 96850

17th Coast Guard District
P. O. Box 3-5000
Juneau, Alaska 99802

FIG. 6-4B. U.S. Coast Guard Regions.

National Transportation Safety Board (NTSB)

The National Transportation Safety Board (NTSB) conducts investigations and safety studies on accidents involving the various modes of transportation, including air, rail, ship, highway, and pipeline. The NTSB investigates hazardous materials transportation accidents to determine their causes and makes recommendations to regulatory agencies and parties concerned with improving transportation safety. The NTSB is independent of the DOT and reports directly to the Executive Branch.

Environmental Protection Agency (EPA)

Since its formation in 1970, the Environmental Protection Agency has been concerned with matters of water and air quality, solid waste, and pesticides. The EPA along with the DOT are the primary agencies involved in the regulation of Hazardous Materials. (See Fig. 6-4C.)

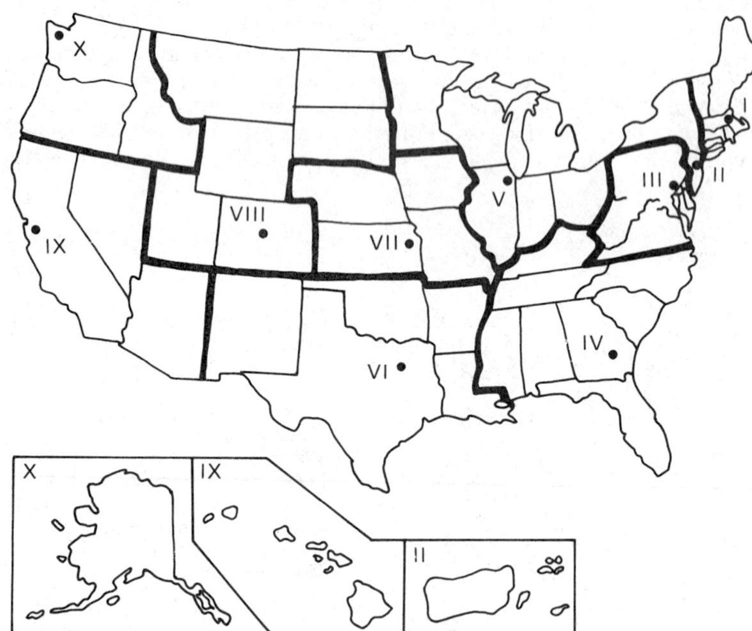

Region I
Chief, Oil & Hazardous Materials Section
Surveillance and Analysis Division
60 Westview Street
Lexington, MA 02173

Region II
Chief, Emergency Response and Hazardous Materials Inspection Branch
Environmental Services Division
Edison, NJ 08837

Region III
Chief, Environmental Emergency Branch
Curtis Building 3ES30
6th and Walnut Streets
Philadelphia, PA 19106

Region IV
Chief, Emergency Remedial & Response Branch
345 Courtland Street, NE
Atlanta, GA 30365

Region V
Chief, Spill Response Section
Environmental Services Division, 5SEES
536 South Clark Street
Chicago, IL 60605

Region VI
Chief, Emergency Response Branch, 6ESE
First International Building
1201 Elm Street
Dallas, TX 75270

Region VII
Chief, Emergency Planning & Response Branch
Environmental Services Division
25 Funston Road
Kansas City, KS 66115

Region VIII
Chief, Emergency Response Branch
Environmental Services Division
1860 Lincoln Street
Denver, CO 80295

Region IX
Chief, Emergency Response Section, T-3-3
Compliance & Response Branch
Toxic & Waste Management Division
215 Fremont Street
San Francisco, CA 94105

Region X
Chief, Environmental Emergency Response Team
Environmental Services Division
1200 6th Avenue
Seattle, WA 98101

FIG. 6-4C. EPA Regional Emergency Response Offices.

In 1970, Section 311, (Oil and Hazardous Substance Spills) was added to Public Law 92-500 *(Federal Water Pollution Control Act)*, mandating the EPA and the Coast Guard to regulate, by prevention and/or enforcement, spills of oil and/or hazardous materials. Regulations promulgated under this law call for: (1) criminal fines and imprisonment for failing to immediately report an incident to the appropriate government agency; (2) civil fines; (3) use of the "Superfund" for cleanup of an incident if those responsible cannot be identified or located; (4) court action to stop the spillage of hazardous materials if there are actual or potential threats to the public welfare; and (5) spill prevention regulations for industry.

Once notified of an incident, the EPA regional office will dispatch an On-Scene Coordinator (OSC) to determine the seriousness of the accident, and the level of technical response required. The On-Scene Coordinator can summon the assistance of a Regional Response Team (RRT). The RRT can provide assistance in planning and preparedness actions to take before an incident occurs, as well as provide coordination and advice during an incident.

At more complex incidents, the EPA On-Scene Coordinator can call upon the EPA's Environmental Response Team (ERT) for assistance in data interpretation of air monitoring, soil sampling, and other aspects of response operation. The ERT will determine the amount of hazardous materials at a site, ascertain how the hazardous materials are spreading, and recommend corrective actions to the OSC. ERT members will also compile a site safety plan and follow up with a report on compliance with the plan.

The OSC can also request aid through the Superfund Emergency Response Program. Superfund emergency responses include sampling of hazardous materials to determine: (1) type, location, and level of contamination; (2) fencing of hazardous sites; (3) removal and disposal of product; (4) lagoon and pond control; and (5) monitoring and provision of alternate water supplies.

If the hazardous material is unknown, the OSC can utilize a computerized information retrieval system to aid in identification through color, smell, or other physical characteristics observed at the scene. Called OHMTADS, for Oil and Hazardous Materials Technical Assistance Data System, the system includes chemical, biological, toxicological, and commercial data on more than 1000 chemicals. The EPA can also be reached through the National Response Center at 800-424-8802.

While the Federal government can provide emergency response resources through the Coast Guard (for coastal incidents) and EPA (for inland releases), these resources are limited and vary regionally. It is therefore important to discuss your needs and their capabilities with local EPA and/or Coast Guard staff. They can advise on:

1. The type and location of the emergency response equipment and personnel they can offer.
2. How to devise a workable emergency response plan.
3. The availability and use of computerized data bases and written manuals on hazardous materials.
4. The availability of contractors to assist in emergency response.
5. The use of the Federal "Superfund" to finance emergency cleanups.
6. How to tap the technical knowledge and experience of federal staff to choose the best response method.

For further information, contact the USCG or EPA On-Scene Coordinator in your region.

Department of Energy (DOE)

The DOE under the Federal Radiological Monitoring and Assistment Plan (FRMAP) assists with the planning for and management of emergencies involving radioactive materials. (See Fig. 6-4D.) FRMAP divides the United States into eight regions, each with a regional coordinating office. In emergency situations, the regional coordinator evaluates the nature of the emergency and provides advice directly over the telephone when appropriate. In serious incidents the DOE will dispatch a technical response team and notify the Nuclear Regulatory Commission.

The Nuclear Regulatory Commission (NRC)

The NRC is authorized to license and regulate the use of radioactive materials. Under an agreement with the Department of Transportation, NRC allows the DOT to regulate the transport of radioactive materials. DOT and NRC have jointly established regulations governing the packaging, labeling, handling, loading, and storing of radioactive materials during transportation. As a general practice, DOT does not regulate the routes used to transport radioactive materials.

STATE GOVERNMENT AGENCIES

Agencies that can provide technical assistance, manpower, materials, and equipment for dealing with a hazardous materials incident vary from state to state. During the planning process, local emergency services are advised to contact their state's information office, Department of Public Safety, Department of Public Health, or Department of Environmental Protection. Ascertaining emergency telephone numbers of concerned agencies, types of technical assistance, manpower, materials, and equipment available, as well as their locations, is vital to the development of a sound emergency.

State agencies that could provide technical assistance and supportive functions for dealing with a hazardous materials incident include:

National Guard: Communications, crowd control, evacuation transportation, radiological monitoring, and helicopters.

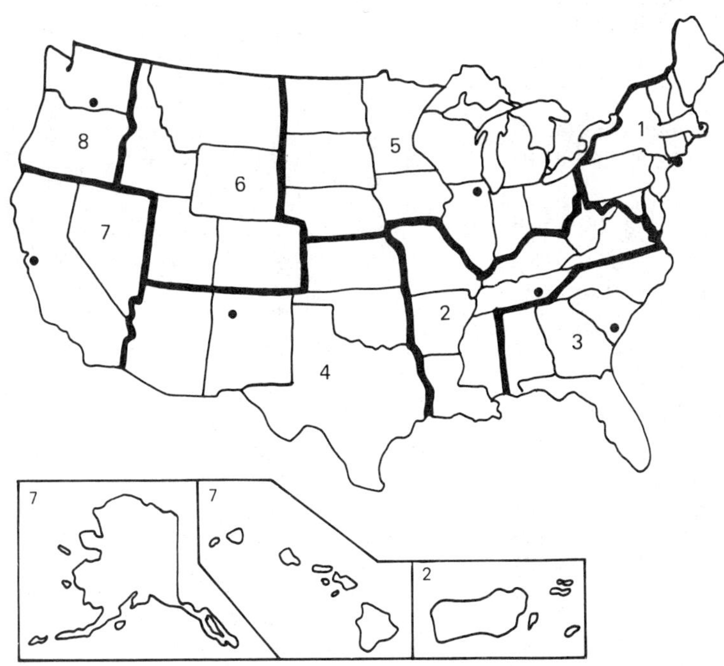

Region 1
Upton, L. I.
New York, NY 11973

Region 2
P. O. Box E
Oak Ridge, TN 37830

Region 3
P. O. Box A
Acken, SC 29801

Region 4
P. O. Box 5400
Albuquerque, NM 87115

Region 5
9300 S. Cass Avenue
Argonne, IL 60349

Region 6
P. O. Box 2108
Idaho Falls, ID 83401

Region 7
2111 Bancroft Way
Berkeley, CA 94704

Region 8
P. O. Box 550
Highland, WA 99352

FIG. 6-4D. FRMAP Regions.

Civil Preparedness: Communications, crowd control, evacuation, and radiological monitoring.

Public Safety: State police and/or sheriffs department for crowd control, traffic control, accident investigation, evacuation, property protection, and communications.

Public Health: Monitoring of water, air, and soil for contamination, treatment of injured personnel, access to hospitals and ambulances, and poison center cooperation.

Public Works: Sand and other containment materials, construction equipment and personnel, and maps of natural watershed areas.

Environmental Protection: Chemists, biologists, and other scientists with specialized knowledge of hazardous materials released into the environment, and knowledge of cleanup contractors.

Agriculture: Soil and water testing, and knowledge of pesticides and fertilizers.

Fire Marshal, Fire Academy, and Institutions of Higher Education: Training and advice in hazardous materials response and suppression, and the use of specialized response equipment.

LOCAL GOVERNMENT AGENCIES

Local government agencies, particularly fire and police departments, are usually the first on the scene of an incident. Federal and state assistance is usually not immediately available. Each segment of local government should know what is expected of it in a hazardous materials emergency. Those who will need to be included in the preincident planning process are:

Fire Departments: Additional manpower, and specialized equipment and materials can be made available through mutual aid programs.

Police Departments: Traffic and crowd control, evacuation, property protection, and accident investigations.

Civil Preparedness: Communications, crowd control, evacuation, and radiological monitoring.

Public Works: Sand and other containment material, as well as construction equipment and manpower, and maps of natural watershed areas.

Public Health: Water, air, and soil monitoring, and local poison center information.

Building Department: Advise on manufacturing locations.

NATIONAL INDUSTRIAL ORGANIZATIONS AND TRADE ASSOCIATIONS

Many organizations that manufacture and transport hazardous materials have joined together and formed national associations. These organizations can provide emergency services and information on safety standards and regulations as well as advice on the proper precautions to take during a hazardous materials incident. Of these national industrial organizations and trade associations, the one that is most important to emergency service personnel is CHEMTREC.

CHEMTREC—Chemical Transportation Emergency Center

CHEMTREC was established in 1971 as a public service of the Chemcial Manufacturers Association (CMA). The Center provides information and/or assistance to those responsible for responding to hazardous material emergencies 24 hrs a day, seven days a week. The CHEMTREC number is 800-424-9300.

When an emergency is reported through this number, the person on duty records the appropriate details and offers advice on what to do in case of spills, fire, or exposure. The person on duty then immediately notifies the shipper. The particulars of the emergency are reported to the shipper who will then offer further assistance to emergency personnel on the scene through direct communications and/or by dispatching personnel to the scene. CHEMTREC maintains a close working relationship with the Department of Transportation and many other public and private organizations involved with responding to hazardous materials emergencies.

Chlorine Emergency Plan—CHLOREP

CHLOREP, administered and coordinated by the Chlorine Institute, is a voluntary program of aid in chlorine emergencies occuring during transportation or at user locations. It operates on a 24-hr, seven day a week basis. Upon receiving an emergency call, CHEMTREC's dispatcher telephones the designated CHLOREP emergency contact who notifies the team leader. The CHLOREP team leader, in turn, telephones the emergency caller at the scene and determines what advice and aid are needed. Figure 6-4E charts the flow of communications during an emergency involving chlorine.

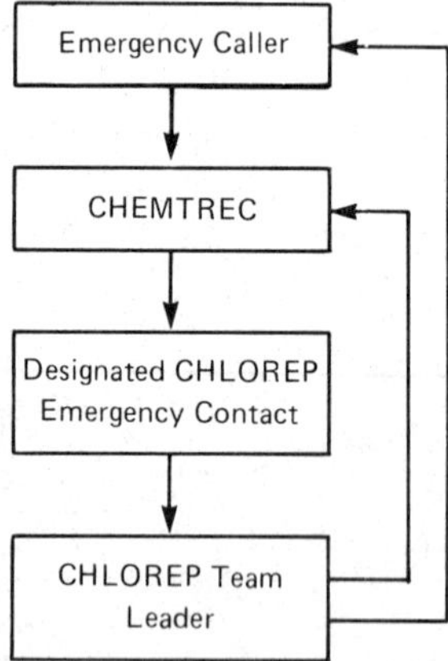

FIG. 6-4E. Flow of communications during an emergency involving chlorine.

Emergency teams, if needed, are dispatched to the site of the incident. Participating chlorine manufacturers sponsor one or more CHLOREP groups consisting of: (1) an emergency contact to whom the request for assistance is directed, (2) a CHLOREP team consisting of a team leader and assistants to handle the emergency, and (3) a home coordinator to provide support at the home location. The personnel are employees of the various chemical producers.

The Pesticide Safety Team Network (PSTN)

In 1970, the National Agricultural Chemicals Association formed a network of safety teams designed to minimize the safety risks arising from the accidental spillage or leakage of Class B poison pesticides. The PSTN is a voluntary association of pesticide manufacturers and is operated as a public service.

Association members participate in the program by cooperatively furnishing personnel, equipment, and expertise for the prompt and efficient cleanup and decontamination of Class B poison pesticides involved in a major accident. More than 40 safety teams currently make up the Network.

Area coordinators of PSTN are called into service by a CHEMTREC operator when a Class B poison pesticide is involved in an accident. The area coordinator will call the producer of the pesticide and jointly work out the details of a response to the emergency. The person reporting the incident is then contacted and provided with advice on what immediate steps to take and that a safety team would be dispatched if needed.

Other Private Sources

In addition to CHEMTREC, CHLOREP, and PSTN there are many other organizations that can provide information, advice, and training materials. They include the following:

American Petroleum Institute
2101 L Street N.W.
Washington, DC 20037

American Trucking Association
1616 P Street N.W.
Washington, DC 20006

American Waterways Operator, Inc.
1600 Wilson Boulevard
Suite 1101
Arlington, VA 22209

Association of American Railroads
1920 L Street N.W.
Washington, DC 20006

Association of Oil Pipelines
Suite 1208
1725 K Street N.W.
Washington, DC 20006

Atomic Industrial Forum, Inc.
Public Affairs and Information Program
7101 Wisconsin Ave.
Washington, DC 20014

Chevron Chemical Company
575 Market Street
San Francisco, CA 94105

Dow Chemical U.S.A.
Bennett Building
2030 Dow Center
Midland, MI 48640

E.I. Dupont de Nemours & Company
Hazardous Materials Section
Wilmington, DE 19898

Hazardous Materials Advisory Committee
Suite 1107
1100 17th Street N.W.
Washington, DC 20036

The Hazmat, Resource Inc.
2100 Raybrook S.E.
Suite 109
Grand Rapids, MI 49506

Information Handling Services
15 Inverness Way East
P.O. Box 1154
Englewood, CO 80150

Interstate Natural Gas Association
1660 L Street N.W.
Washington, DC 20006

National Tank Truck Carriers, Inc.
1616 P Street N.W.
Washington, DC 20006

Occupational Health Services Inc.
400 Plaza Drive, P.O. Box 1505
Seacaucus, NJ 07094

Safety Systems Incorporated
P.O. Box 8463
Jacksonville, FL 32239

SCA Chemical Services, Inc.
Corporate Headquarters
60 State Street
Boston, MA 02109

Water Transportation Association
Suite 2007
60 East 42nd Street
New York, NY 10017

LOCAL INDUSTRY

By identifying, through preemergency planning, those local industries which manufacture, transport, or use hazardous materials, emergency services can secure prompt information and possible technical assistance at the scene of an incident. Local industries such as chemical companies, trucking companies, railroads, oil refineries or storage companies, and licensed pollution spill cleanup contractors may be able to provide emergency services with equipment and personnel. As with any preemergency plan, it is important to keep it continually updated, especially with regard to the names and phone numbers of contact personnel.

Figure 6-4F is a sample hazardous materials preemergency survey form which the authors have found to be most useful. While it is not an official NFPA document, the authors believe that a form such as this should be completed for every major site that handles hazardous materials.

HAZARDOUS MATERIALS PREEMERGENCY SURVEY FORM

1. **NAME OF PREMISES**

2. **LOCATION** (ADDRESS IF POSSIBLE)

3. **CONTACT PERSONS** NAME TELEPHONE

OWNER: __________

MANAGER: __________

ENGINEER: __________

SECURITY: __________

NIGHT/WEEK END CONTACT: __________

4. **TYPE OF PROCESS, USE, STORAGE**

5. **HAZARD OF PROCESS, USE, STORAGE**

6. **TYPE(S) OF HAZARDOUS MATERIAL**
(CHEMICAL NAME, TRADE NAME)

1. __________
2. __________
3. __________
4. __________
5. __________

7. **SPECIFIC HAZARD OF MATERIAL**
(MATCH BY NUMBER) (USE DOT TERMINOLOGY)

1. __________
2. __________
3. __________
4. __________
5. __________

FIG. 6-4F. Hazardous Materials Preemergency Survey Form.

8. SURROUNDING AREA

WEST: OCCUPANCY (MATCH BY NUMBER)

1.
2.
3.

DAY NIGHT
NO. OF OCCUPANTS

1.
2.
3.

TYPE OF HAZARD (LIFE, EXPOSURE)

1.
2.
3.

DEGREE OF HAZARD (SLIGHT, MODERATE, EXTREME)

1.
2.
3.

NORTH: OCCUPANCY (MATCH BY NUMBER)

1.
2.
3.

DAY NIGHT
NO. OF OCCUPANTS

1.
2.
3.

TYPE OF HAZARD (LIFE, EXPOSURE)

1.
2.
3.

DEGREE OF HAZARD (SLIGHT, MODERATE, EXTREME)

1.
2.
3.

W N S E

SOUTH: OCCUPANCY (MATCH BY NUMBER)

1.
2.
3.

DAY NIGHT
NO. OF OCCUPANTS

1.
2.
3.

TYPE OF HAZARD (LIFE, EXPOSURE)

1.
2.
3.

DEGREE OF HAZARD (SLIGHT, MODERATE, EXTREME)

1.
2.
3.

EAST: OCCUPANCY (MATCH BY NUMBER)

1.
2.
3.

DAY NIGHT
NO. OF OCCUPANTS

1.
2.
3.

TYPE OF HAZARD (LIFE, EXPOSURE)

1.
2.
3.

DEGREE OF HAZARD (SLIGHT, MODERATE, EXTREME)

1.
2.
3.

FIG. 6-4F (cont). Hazardous Materials Preemergency Survey Form.

9. FIRE PROTECTION EQUIPMENT

DETECTION/ALARM SYSTEMS ______

______

SPRINKLER SYSTEM ______

______

STANDPIPE SYSTEM ______

______

SPECIAL EXTINGUISHING SYSTEM (CO_2, HALON 1301) ______

______

WATER SUPPLY PUBLIC ______

WATER SUPPLY PRIVATE ______

SPECIAL EXTINGUISHING AGENTS (MET-L-X, G1, ETC.) ______

______

______

BREATHING APPARATUS (TYPE, LOCATION) ______

______

______

ACID SUITS ______

______

______

OTHER ______

______

______

______

10. SPECIAL EQUIPMENT AND MATERIALS

BULLDOZERS ______ TRUCKS ______

CRANES ______ SAND ______

OTHER ______

11. POTENTIAL FOR AN INCIDENT

	YES	NO
IS THERE A CONFLAGRATION POTENTIAL?	___	___
IS THERE AN EARTHQUAKE POTENTIAL?	___	___
IS THERE A WINDSTORM POTENTIAL?	___	___
IS THERE A WINTERSTORM POTENTIAL?	___	___
IS THERE A FLOOD POTENTIAL?	___	___
IS THERE A POTENTIAL FOR WATER CONTAMINATION?	___	___
IS THERE A POTENTIAL FOR STORM AND/OR SANITARY SEWER INVOLVEMENT?	___	___
IS THERE A POTENTIAL FOR AN EXTERNALLY GENERATED ACCIDENT AUTOMOBILE THROUGH FENCE)?	___	___
IS THERE A POTENTIAL FOR SABOTAGE?	___	___

12. IF THERE WAS A MAJOR INCIDENT AT THIS SITE:

HOW FAR WOULD EVACUATION HAVE TO BE EXTENDED (INITIALLY)? ______

______

HOW MANY PEOPLE AT SITE? DAY ______ NIGHT ______

HOW MANY PEOPLE IN EVACUATION ZONE? DAY ______ NIGHT ______

FIG. 6-4F (cont). Hazardous Materials Preemergency Survey Form.

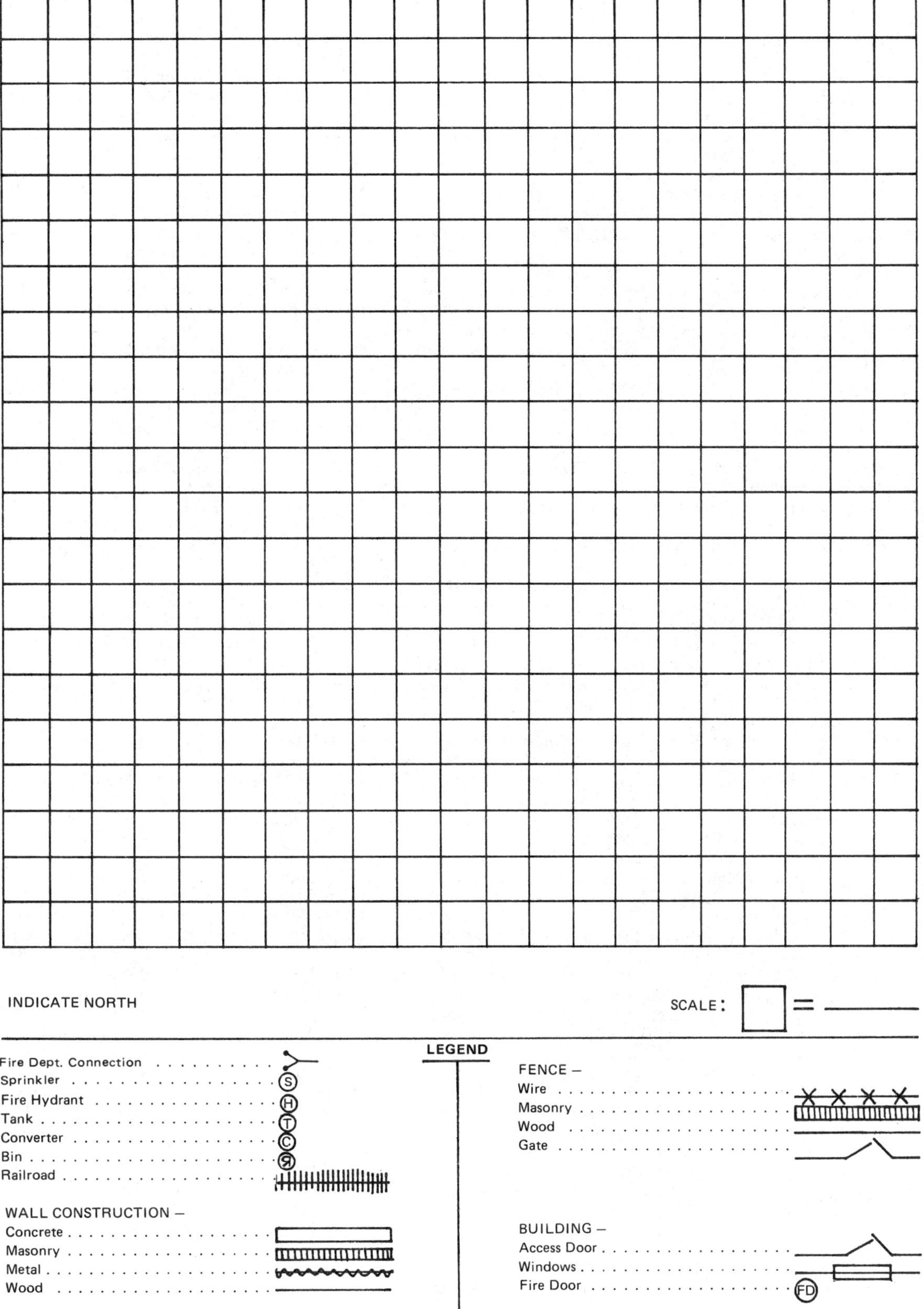

INDICATE NORTH

SCALE: ☐ = ______

LEGEND

Fire Dept. Connection
Sprinkler . . . . . (S)
Fire Hydrant . . . . . (H)
Tank . . . . . (T)
Converter . . . . . (C)
Bin . . . . . (B)
Railroad . . . . .

WALL CONSTRUCTION –
Concrete . . . . .
Masonry . . . . .
Metal . . . . .
Wood . . . . .

FENCE –
Wire . . . . .
Masonry . . . . .
Wood . . . . .
Gate . . . . .

BUILDING –
Access Door . . . . .
Windows . . . . .
Fire Door . . . . . (FD)

FIG. 6-4F (cont). Hazardous Materials Preemergency Survey Form.

Bibliography

Additional Readings

AAI, *Handbook of Hazardous Materials*, Technical Guide No. 7, Alliance of American Insurers, Chicago, IL.

ATA, *Handling Hazardous Materials and Transporting Hazardous Wastes*, American Trucking Association, Safety and Security Dept., Washington, DC.

AAR, Bureau of Explosives, *Emergency Handling of Hazardous Materials in Surface Transportation*, Association of American Railroads, Washington, DC, 1981.

Bennett, G. F., Feates, F., & Wilder, I., *Handbook of Hazardous Materials*, McGraw-Hill, NY.

Cahners Publishing Company, *Hazardous Materials Transportation: A Compliance and Operations Guide*, Cahners, Boston.

Cahners Publishing Company, *1982 Reference Source Handbook on the Transport of Hazardous Materials/Substances/Wastes*, Cahners, Boston.

CMA, *Chem-Cars Manuals*, Chemical Manufacturers Association, Washington, DC.

CMA, *Chemical Safety Data Sheets*, Chemical Manufacturers Association, Washington, DC.

CMA, *Laboratory Waste Disposal Manual*, Chemical Manufacturers Association, Washington, DC.

Chlorine Institute, Inc, *Chlorine Manual*, Chlorine Institute, Inc., NY.

Dennis, A. W., Foley, J. T., Hartman W. F., and Larsen D. W., *Severities of Transportation Accidents Involving Large Packages*, Sandia Laboratories, Albuquerque, NM, 1978.

Ethyl Corporation, *Handling Procedures for Aluminum Alkyl Compounds*, Ethyl Corporation, Industrial Chemicals Division, Baton Rouge, LA.

FEMA, *Guidance for Developing State and Local Radiological Emergency Response Plans and Preparedness for Transportation Accidents*, Federal Emergency Management Agency, Washington, DC.

FEMA, *Disaster Planning Guidelines for Fire Chiefs*. (Prepared by International Association of Fire Chiefs, Inc.: February 1981) Federal Emergency Management Agency, Washington, DC.

FEMA, *Hazardous Materials Management System: A Guide for Local Emergency Managers*, (Prepared by the Multnomah County Office of Emergency Management: July 1981) Federal Emergency Management Agency, Washington, DC.

FEMA/EPA, *Planning Guide and Checklist for Hazardous Materials Contingency Plans*, (Prepared by Rockwell International Corp.: July 1981) Federal Emergency Management Agency, Washington, DC.

FEMA/EPA, *A Training Course on Contingency Planning for The Control of Hazardous Materials Spills*, (Prepared by Rockwell International Corp.) Federal Emergency Management Agency, Washington, DC.

The Fertilizer Institute, *Shipping Papers, Marking and Placarding for Hazardous Materials Transportation*, The Fertilizer Institute, Washington, DC.

General American Corp., *GATX Tank Car Manual*, General American Transportation Corporation, Chicago, IL.

General American Corp., *How to Handle LP-Gas Emergencies*, General American Transportation Corp., Sharon, PA.

Information Transfer, Inc., *Control of Hazardous Materials Spills*, 1976, 1977, 1978, 1979, and 1980 Conference Proceedings.

Isman, W. E. and Carlson, G. P., *Hazardous Materials*, Glencoe Publishing, Encino, CA.

J. J. Keller & Associates, *Hazardous Materials Guide*, J. J. Keller & Associates, Neenah, WI.

Mozingo, Al, *Hazardous Materials and Incident Management*, Tabor Publishing, Portland, OR.

NACA, *Pre-Planning and Guidelines for Handling Agricultural Chemical Fires*, National Agricultural Chemicals Association, Washington, DC.

NCSL, *Hazardous Waste Management: A Survey of State Legislation 1982*, National Conference of State Legislatures, Denver, CO.

NFPA, *Emergency Response to Hazardous Materials in Transportation; Self Study Guide*, National Fire Protection Association, Quincy, MA.

NFPA, *Fire Protection Guide on Hazardous Materials*, 8th Ed. (revised), National Fire Protection Association, Quincy, MA, 1984.

NFPA, *Handling Pipeline Transportation Emergencies*, National Fire Protection Association, Quincy, MA.

NTIS, *Hazardous Materials Incident Response Operations*, National Technical Information Services, Springfield, MA.

NTTC, *Cargo Tank Hazardous Materials Regulations*, National Tank Truck Carriers, Inc., Washington, DC.

Perry, J. H. et. al., eds., *Chemical Engineers Handbook*, McGraw-Hill, NY.

PPG Industries, Inc., *Chlorine Safe Handling*, PPG Industries Inc., Pittsburgh, PA.

PPG Industries, Inc., *Be Careful with Caustics*, PPG Industries Inc., Pittsburgh, PA.

Purington, R. G., and Paterson, W., *Handling Radiation Emergencies*, National Fire Protection Association, Quincy, MA.

RSMA, *Chemical Transportation and Handling Guide*, Railroad Systems and Management Association, Ocean City, NJ.

Robinson, J. S., *Hazardous Chemical Spill Clean-up*, Noyes Data Corporation, Park Ridge, NJ, 1979.

Sax, I. N., *Dangerous Properties of Industrial Materials*, Van Nostrand Reinhold Company, NY.

Smith, A. J., *Managing Hazardous Substances Accidents*, McGraw-Hill, New York.

Stutz, D. R., Ricks, R. C., Olsen, M. F., *Hazardous Materials Injuries: A Handbook for pre-Hospital Care*, Bradford Communications Corporation, Greenbelt, MD.

Union Carbide Corporation, *Precautions and Safe Practices: Liquified Atmospheric Gases*, Union Carbide Corporation, Linde Division, NY.

UPS, *Guide for Shipping Hazardous Materials*, United Parcel Service, Greenwich, CT, 1981.

U.S. Coast Guard, *Chemical Hazards Response Information System* (CHRIS), (Published in 4 volumes), Manual 1, *A Condensed Guide to Chemical Hazards*; Manual 2, *Hazardous Chemical Data*; Manual 3, *Hazard Assessment Handbook*; Manual 4, *Response Methods Handbook*; Superintendant of Documents, Government Printing Office, Washington, DC.

U.S. Department of Commerce, *Recommended Methods of Reduction, Neutralization, Recovery or Disposal of Hazardous Waste, Vols. 1-16*, (Prepared by TRW Systems Group) National Technical Information Service, Springfield, VA.

U.S. Department of Transportation, Federal Highway Administration, *Development of Criteria to Designate Routes for Transporting Hazardous Materials*, (Prepared by Peat, Marwick, Mitchell and Co., September 1980) National Technical Information Service, Springfield, VA.

U.S. Department of Transportation, Office of Hazardous Materials, *Code of Federal Regulations, Title 49; Transportation*, Pts. 170-189, Superintendant of Documents, Government Printing Office, Washington, DC.

U.S. Department of Transportation, Office of the Secretary, *A Guide to the Federal Hazardous Materials Regulatory Program*, (Prepared by the Materials Transportation Bureau), Superintendant of Documents, Government Printing Office, Washington, DC, 1983.

U.S. Department of Transportation, Office of University Research, *A Community Model for Handling Hazardous Materials Transportation Emergencies*, (Prepared by Department of Civil Engineering, Kansas State University: June 1981), DOT, Information Services Division (DMT-11), Washington, DC.

U.S. Department of Transportation, Research and Special Programs Administration, *Community Teamwork: Working Together to Promote Hazardous Materials Transportation Safety*, (Prepared by Cambridge Systematics, Inc: May 1983), DOT, Information Services Division (DMT-11), Washington, DC.

U.S. Department of Transportation, Research and Special Programs Administration, *1984 Emergency Response Guidebook*, DOT, Information Services (DMT-11) or American Trucking Association, Washington, DC.

U.S. Department of Transportation, Research and Special Programs Administration, *Hazardous Materials Transportation Risks in the Puget Sound Region*, (Prepared by Batelle Northwest Laboratories), DOT, Information Services (DMT-11), Washington, DC, 1981.

U.S. Department of Transportation, Research and Special Programs Administration, *Guide for Brokers, Forwarding Agents, Freight Forwarders and Warehousemen*, DOT, Information Services Division (DMT-11), Washington, DC.

U.S. Department of Transportation, Research and Special Programs Administration, *Guide for Carriers*, DOT, Information Services Division (DMT-11), Washington, DC.

U.S. Department of Transportation, Research and Special Programs Administration, *Guide for Hazardous Materials Shipping Papers*, DOT, Information Services Division (DMT-11), Washington, DC.

U.S. Department of Transportation, Research and Special Programs Administration, *Guide for Manufacturers and Vendors of Hazardous Materials*, DOT, Information Services Division (DMT-11), Washington, DC.

U.S. Department of Transportation, Research and Special Programs Administration, *Guide for Hazardous Materials Definition*, DOT, Information Services Division (DMT-11), Washington, DC.

U.S. Department of Transportation, Research and Special Programs Administration, *Guide for Radioactive Materials Definitions*, DOT, Information Services Division (DMT-11), Washington, DC.

U.S. Department of Transportation, Research and Special Programs Administration, *State Hazardous Materials Enforcement Development Program (SHMED)*, DOT, Information Services Division (DMT-11), Washington, DC, 1981.

U.S. Department of Transportation, Research and Special Programs Administration, *Response to Radioactive Materials Transport Accidents*, DOT/RSPA/MTB-79/8, DOT, Information Services Division (DMT-11), Washington, DC.

U.S. Environmental Protection Agency, *Manual for the Control of Hazardous Material Spills: Volume 1-Spill Assessment and Water Treatment Techniques*, National Technical Information Service, Springfield, VA, 1977.

U.S. Environmental Protection Agency, *Manual of Practice for Protection and Cleanup of Shorelines; Volume 1-Decision Guide*, National Technical Information Service, Springfield, VA, 1979.

U.S. Environmental Protection Agency, Office of Solid Waste and Emergency Response, *Hazardous Waste Transportation Interface*, EPA, Office of Solid Waste and Emergency Response, Washington, DC.

U.S. Environmental Protection Agency, *Spill Prevention Techniques for Hazardous Polluting Substances*, (Prepared by Arthur D. Little Company), EPA Office of Public Affairs, Washington, DC.

U.S. Environmental Protection Agency, *EPA Field Detection and Danger Assessment Manual for Oil and Hazardous Material Spills*, EPA Office of Water and Hazardous Materials, Washington, DC.

U.S. Government Printing Office, *Chemical Data Guide for Bulk Shipments by Water*, Superintendent of Documents, Government Printing Office, Washington, DC.

HAZARDOUS WASTE CONTROL

Sami Atallah

Waste is a natural byproduct of many human activities. Although all waste is a nuisance to society, not all waste is necessarily hazardous. Hazardous waste may be defined as that which poses a threat to human health, the environment, and public property if handled or disposed of improperly. The hazard is created by virtue of the toxicity, flammability, explosibility, reactivity, radioactivity, corrosivity, or etiologic (disease causing) potential of the waste.

The varieties, quantities, and hazards of today's waste appear to be directly proportional to the degree of technological development of a nation. It has been estimated that in the United States, over 200 million pounds (91 000 metric tons) of hazardous waste is generated daily, of which about 10 percent is disposed of in an environmentally acceptable manner (Richards 1979). During the course of a year, over 3.5 billion tons (3.3 billion metric tons) of solid waste is produced—roughly 100 lb (45 kg) per day for every man, woman, and child (Reindl 1977a). Solid waste includes animal, mineral, agricultural, and community waste, of which only a small fraction is collected.

In addition to solid waste, many gaseous products such as those produced in combustion processes (e.g., CO_2, NO_x, SO_2, soot, and hydrocarbons) are released in the atmosphere. The pollution of air with such gaseous waste products is not only detrimental to human health, it reduces visibility and contributes to the corrosion and deterioration of clothing, art treasures, historical structures, metals, rubber, and paint. Acid rain, the product of further atmospheric oxidation and dissolution of NO_x and SO_2 in rain water, adversely affects the growth of our forests and agricultural products and the survival of aquatic life in our streams, ponds, and lakes.

Domestic, commercial, and industrial liquid wastes invariably find their way to bodies of water and add to the problem of our dwindling potable water supply. Industrial liquid waste may contain metallic salts that have immediate or long term effects on humans. High concentrations of these salts may also build up in edible fishes, crustaceans, and plants with similar effects on people who consume them. Some organic wastes from commercial, household, and industrial sources reduce the concentration of dissolved oxygen in water supplies. This may lead to the extinction of aquatic life and the growth of disease carrying bacteria and parasites. Other wastes may contain known and suspected carcinogens such as chlorinated hydrocarbon solvents, benzene, and polychlorinated biphenyls (PCBs).

Many federal, state, and local laws and regulations have been enacted to protect the public against exposure to hazardous waste. Federal laws include the *Clean Air Act*, the *Water Pollution Control Act*, the *Solid Waste Disposal Act*, the *Toxic Substances Control Act*, the *Comprehensive Environmental Response, Compensation and Liability Act* (CERCLA) and the *Resource Conservation and Recovery Act* (RCRA). These laws attempt to specify and control the quantities, concentrations, and types of chemicals that may be released and the manner in which they may be disposed of safely. In addition, RCRA encourages the recovery of energy and other resources from waste materials (Glaubinger 1979), while CERCLA, sometimes called the "Superfund Act", makes Federal funds available for the cleanup of old abandoned waste disposal sites and hazardous material spills on land and water.

In addition to these acts, parts of other laws regulate specific waste disposal and handling operations. For example, the *Marine Protection, Research and Sanctuaries Act* regulates ocean dumping and incineration on the high seas; the *Hazardous Materials Transportation Act* regulates the packaging and containerization of hazardous materials; and the *Safe Drinking Water Act* regulates underground injection of hazardous wastes.

The safe disposal of hazardous waste introduces many problems not only to the industry generating the waste, but also to the waste transportation and disposal agency and to fire fighters and other response forces who are called upon to clean up an accidental release. These problems arise because the contents of waste are usually not well defined—sometimes not even known. Waste is inherently a mixture of many components with differing hazardous properties. One component, for example, may be a flammable solvent while another is a highly toxic pesticide.

Mr. Atallah is President, Risk and Industrial Safety Consultants, Inc. Chicago, IL.

Disposing of the solvent by burning may release the pesticide (or its toxic combustion products) to the atmosphere. Additional problems arise in handling and storing waste products. Self-heating and ignition, corrosion of pipe and tank wall materials, and reaction with other noncompatible wastes have been known to cause major catastrophic events. Fire fighters may also be accidentally exposed to a hazardous waste material released as a result of an unrelated fire or an explosion to which they have responded.

This chapter will discuss the sources and characteristics of waste materials with particular emphasis on hazardous waste, summarize waste handling, disposal, and recovery procedures, and examine the various approaches to fire prevention and protection in cases involving hazardous waste. Previous chapters in this section deal with other aspects.

SOURCES OF WASTE

There are many sources of waste materials in our society. To a great extent, the source determines what the composition of the waste may be, its bulk density, quantity, and the type and degree of hazard that it may present. These factors, in turn, determine how the waste should be collected, processed, or disposed of in a practical and safe manner.

Domestic, Commercial, and Industrial Sources

Table 6-5A shows typical breakdowns of the contents of collected solid waste from domestic, commercial, and industrial sources. Obviously, these breakdowns will vary among sources within each category. For example, suburban homes are expected to produce more yard waste than city residences. Restaurants generate more food waste and less paper waste than office buildings. Hospitals dispose of cotton and paper products which may be contaminated with hazardous chemicals and etiologic agents.

Table 6-5A. Typical Contents of Solid Waste (Percent by Weight)

	Domestic	Commercial	Industrial
Paper Products	51.5	69.0	17.4
Glass	15.0	7.0	3.4
Cans and Metals	7.0	10.0	8.0
Plastics	2.0	10.0	
Cloth, Leather, Rags	4.0		0.4
Food	10.0	4.0	8.4
Wood	2.0		17.9
Yard Waste	8.5		
Minerals			36.7
Chemical Waste			3.3
Rubber			1.6
Other			3.1

The nature of an industry determines the content of the waste generated. A furniture plant, for example, discards wood scrap, sawdust, and shavings as well as oily rags, stain, varnish, lacquer, glue, etc., in addition to paper, metal, and other commercial items. The waste products of chemical processing plants differ considerably depending on the raw materials used and the end products processed. These plants may discard substantial amounts of off-grade or reject products, byproducts, tars, spent catalyst, precipitates, etc. On the other hand, nuclear power plants produce radioactive waste.

At any processing plant, some materials will be accidentally spilled in storage and handling. Cleanup operations from filling equipment, fabrication operations, and general plant maintenance will generate some waste. Rejected, aged, or damaged packages; off-specification materials; and overruns will also be discarded.

Transportation Sources

The movement of bulk or package goods may also generate waste. Broken packages, leaking drums, and overheating or freezing of cargo are contributing sources. Washing tank cars or tank trucks produces substantial amounts of waste material which may be emulsified or dissolved in water. A spill of a hazardous liquid or solid chemical cargo as a result of a railway, truck, ship, or pipeline accident contaminates soil and/or nearby bodies of water and constitutes a public hazard and another waste disposal problem. This is particularly true if the accident occurs in a remote area where experienced personnel are not readily available to assess the potential hazard to the public and the environment and to provide guidance in controlling the hazard.

Storage Sources

Warehouse inventory control necessitates that unmarketable or undesirable materials of all kinds, forms, and hazards be occasionally discarded. Tanks must be cleaned as storage needs change. The washing of transfer lines and tanks results in a waste product.

Materials released as a result of plant accidents, such as when packages are damaged by fork trucks, require cleanup and disposal. Water, fire, or explosion damage can also result in substantial amounts of undesired product.

Other Sources

Several other human activities produce waste. Harvesting of agricultural products, mining operations, and demolition of old structures are examples of operations which generate large quantities of waste products with different characteristics.

CHARACTERIZATION OF WASTE

The safe handling, collection, and disposal of hazardous waste can be accomplished only if the physical, chemical, and hazardous properties of its components are known and that information is properly applied. Quite often though, the composition is not known because waste is usually a mixture of many components with differing properties. This section will discuss the physical, chemical, and hazardous properties that may influence the selection of response procedures to accidental releases and the appropriate disposal techniques for wastes.

Physical State

The ease of handling and disposal of waste is highly dependent on its physical state.

Solid Waste: This category includes most domestic, commercial, and industrial trash, metal, wood, plastic scrap, chemicals, food, etc. (See Table 6-5A.) It also includes agricultural waste products, demolition waste, and spent ore from mining operations. The bulk density of these wastes differs considerably depending on the source, mix of materials, and the degree of compactness of the waste. For example, uncompacted household waste has a bulk density of 200 to 250 lb per cu yd (119 to 148 kg/m^3) while waste in a landfill may have a bulk density of 750 to 1,000 lb per cu yd (445 to 594 kg/m^3). Solid waste also includes combustible metal waste (e.g., magnesium shavings) and semisolids such as waxes, soaps, and elastomers.

Liquid Waste: This category includes aqueous and nonaqueous solutions and suspensions generally produced as a result of cleaning operations. Nonaqueous solvents may be flammable (e.g., toluene, and xylene) or noncombustible (e.g., tricholorethylene). Viscous liquids, such as pastes and tars, and suspensions, such as acid sludges, require special handling procedures and equipment.

Gaseous Waste: Most gaseous waste is generated from industrial operations and is disposed of by venting to the atmosphere. Hazardous gaseous waste products are regulated by the Environmental Protection Agency (EPA) and must be properly disposed of such as by flaring or removal of the hazardous components by chemical means. Of particular concern to fire fighting and other response personnel are old unidentifiable gas cylinders with walls that may have become weakened due to age and corrosion. Discarded small butane and LP-Gas cylinders in household trash may also pose a hazard to trash collectors and compactors and to operators of shredding equipment.

Properties of Hazardous Waste

Few discarded materials are so compatible with the environment or so inert as to have no short or long term impact. Hazards that appear minor may have unexpected impacts long after disposal. When two or more hazards pertain to a material, the lesser may not receive the necessary consideration. Mixing of two discarded substances may result in a chemical reaction with severe and unexpected consequences.

Since waste is generally a mixture of many components, its physical and chemical properties cannot be defined with any degree of accuracy. Whenever possible, the approximate composition of a hazardous waste should be ascertained from the originating source or from the manifest accompanying the waste being transported. Generally, when one component predominates, the physical and chemical properties of the waste mixture are nearly those of the major component. This is not true for the hazardous properties of waste mixtures consisting of a relatively harmless major component and small amounts of highly toxic, radioactive, or etiologically active components. The hazard, in this case, is determined by the smaller component.

The Environmental Protection Agency defines hazardous solid waste as that which may: "(i) cause, or significantly contribute to, an increase in mortality or an increase in serious irreversible, or incapacitating reversible, illness; or (ii) pose a substantial present or potential hazard to human health or the environment when it is improperly treated, stored, transported, disposed of or otherwise managed." The determining characteristic of a hazardous solid waste must be "measurable by an available standardized test method . . . or reasonably detectable by generators of solid waste through their knowledge of their waste." (Federal Register 1980).

Basically, EPA standards require that a solid waste be listed and treated as hazardous if it meets the ignitability, corrosivity, reactivity, and/or toxicity characteristics prescribed by the standard. Highlights of the standard are given below.

Ignitability: A solid waste exhibits this characteristic if a representative sample of the waste has any of the following properties:

1. It is a liquid, other than an aqueous solution containing less than 24 percent alcohol by volume, and has a closed cup flash point less than 60°C (140°F), as determined by an approved test method.
2. It is not a liquid and is capable under standard temperature and pressure of causing fire through friction, absorption of moisture, or spontaneous chemical changes and, when ignited, burns so vigorously and persistently that it creates a hazard.
3. It is an ignitable compressed gas as defined by another federal standard (CFR undated a). A flammable compressed gas is defined as one which forms flammable mixtures with air at concentrations less than 13 percent (by volume) or has a flammability range with air which is wider than 12 percent regardless of the lower flammable limit. The standard specifies additional criteria for defining flammable vapors and aerosols.
4. It is an oxidizer such as a chlorate, permanganate, inorganic peroxide, nitro carbo nitrate, or a nitrate, that yields oxygen readily to stimulate the combustion of organic matter (CFR undated b).

Corrosivity: A solid waste exhibits this characteristic if a representative sample of the waste has either of the following properties:

1. It is aqueous and has a pH less than or equal to 2 or greater than or equal to 12.5.
2. It is a liquid and corrodes steel at a rate greater than 6.35 mm (.25 in.) per year at a test temperature of 55°C (130°F).

Reactivity: A solid waste exhibits this characteristic if:

1. It is normally unstable and readily undergoes violent change without detonating.
2. It reacts violently with water.
3. It forms potentially explosive mixtures with water.
4. When mixed with water, it generates toxic gases, vapors, or fumes in a quantity sufficient to present a danger to human health or the environment.
5. It is a cyanide or sulfide bearing waste which, when exposed to pH conditions between 2 and 12.5, can generate toxic gases, vapors, or fumes in a quantity sufficient to present a danger to human health and the environment.
6. It is capable of detonation or explosive reaction if it is subjected to a strong initiating source or if heated under confinement.
7. It is readily capable of detonation or explosive decomposition or reaction at standard temperature and pressure.
8. It is a known forbidden or a Class A or B explosive.

Toxicity: The EPA defines toxic wastes as those that have been found to be fatal to humans in low doses or, in the absence of data on human toxicity, it has been shown in studies to have an oral LD_{50} toxicity (rat) or less than 50 mg/kg, an inhalation LC_{50} toxicity (rat) of less than 2 mg/L, or a dermal LD_{50} toxicity (rabbit) of less than 200 mg/kg or is otherwise capable of causing or significantly contributing to an increase in serious irreversible, or incapacitating reversible, illness. A toxic waste is also one which contains or degrades into a listed toxic material in large enough concentrations to pose a potential hazard to the public and/or the environment. The standard for toxicity provides descriptions of laboratory tests for extracting and analyzing the potentially toxic components of wastes.

COLLECTION, HANDLING, AND DISPOSAL OF WASTE

The disposal of waste may be achieved through a variety of techniques and processes, depending on the physical, chemical, and hazardous properties of the waste. These techniques range from improvements in operations that reduce the total quantity of waste to the introduction of process changes that reduce the concentration of hazardous components to acceptable levels. Recycling and the recovery of energy from waste streams are other approaches which have become economically acceptable as the cost of energy and raw materials continues to increase.

Treatment of Gaseous Wastes

Industrial gaseous waste may be released in the atmosphere provided that the concentrations of the various components meet the allowable levels set by the Environmental Protection Agency and other health organizations. Table 6-5B summarizes the ranges of the hazardous levels of selected components in air as compared with their background level in nature (World Bank 1978).

Gaseous waste streams can be purified by removing undesirable components or by reducing their concentrations to acceptable levels. This may be achieved by chemical conversion, absorption, adsorption, or particulate removal (Reindl 1977b).

TABLE 6-5B. Ranges of Uncontaminated and Hazardous Air Quality Levels*

Contaminant	Uncontaminated	Hazardous to Humans
CO	0.03 ppm	50 ppm (90 min) 10 ppm (8 hr)
NO_2	4 ppb	0.06 ppm (mean 24 hr)
NO	2 ppb	—
HC:		
CH_4	1–1.5 ppm	<500 ppm (aliphatic) (alicyclic)
Other HC	0.1 ppm	<25 ppm (aromatic) <0.06 ppm (HCHO) <0.25 ppm (Acrolein) <50 ppm (Acetaldehyde)
SO_2	<0.002 ppm	<0.04 ppm
O_3	0.01–0.05 ppm	<0.3 ppm
Particulates	10–60 μg/m³	<80–100 μg/m³

* Anon, "Pollution Control Technology," Research and Education Association, New York, NY, 1973.

1. Chemical Conversion: An undesirable component may be allowed to react chemically with other compounds to form a harmless component or one which may be removed easily. Sulfur dioxide, for example, may be removed by reaction with lime or magnesia. Catalytic converters are used to convert nitrogen oxides and hydrocarbons into nitrogen, CO_2 and H_2O.
2. Absorption: Several gaseous pollutants may be removed by absorption. Ammonia and hydrogen chloride, for example, may be removed by water scrubbing. Sulfur dioxide may be absorbed in an alkaline solution.
3. Adsorption: Certain solids with a large pore surface area selectively adsorb small concentrations of specific gases. Activated carbon, silica or alumina gel, fuller's earth and other clays are typical adsorbents. The operator is left with the problem of disposing of the contaminated adsorbent or the concentrated stream of undesirable gas released upon adsorbent regeneration.
4. Particulate Removal: Many industrial gaseous streams contain undesirable dusts, mist, or smoke. These suspensions may be removed by filtration, sedimentation, centrifugal separation, electrostatic precipitation, and wet scrubbing.

Liquid Waste Treatment

The volume and strength of industrial aqueous waste may be reduced by various process modifications. Although the hazard is not completely eliminated, a smaller volume of contaminated water or a stream with a lower concentration of a hazardous component is easier to handle and dispose of. Neutralization of acidic or alkaline waste may be achieved by chemical reaction with the appropriate compounds. Inorganic salts may be removed by precipitation, ion exchange, and adsorption by carbon beds. Holding ponds may be used to retain industrial waste so that is may be disposed of with domestic waste at a constant rate rather than overloading the sewage disposal system at any one time of the day.

Suspended solids may be separated by such processes as sedimentation (allowing heavy particles to sink to the bottom), flotation (skimming floating materials off the surface), drying followed by incineration, centrifuging, and filtration. Colloidal solids are more difficult to remove because their particle size range falls between that of suspended and dissolved solids. Colloidal solids will not settle down unless chemicals are added which help coagulate the solid into larger masses that would sink. Many toxic pesticides, insecticides, and herbicides may also be removed by adsorption in certain clays.

Many food processing operations as well as the manufacture of textiles and paper produce a waste which is high in organic matter. Biological degradation is generally achieved by lagooning, treatment with activated sludge, the use of trickling filters, wet combustion, deep well injection, and other means.

Wastewater from tanneries, slaughterhouses, and canneries may contain etiologic agents. These waters must be sterilized by chlorination, ozonization, or ultraviolet radiation before they are discharged into public water supplies.

Organic liquid waste components may be recovered by fractional distillation. The remaining components may be disposed of by incineration or burial in environmentally acceptable underground locations.

Disposal and Recovery of Solid Waste

Table 6-5A shows that a large fraction of domestic, commercial, and industrial solid waste is combustible. Thus, the burning of scrap and refuse under controlled conditions is one of the most satisfactory methods of disposal. Municipal incinerators operated primarily for domestic and commercial trash will usually accept industrial refuse as long as it does not introduce additional hazards or operating problems.

In recent years, the recovery of glass and metal from these wastes and the use of the combustible component as a fuel supplement has become economically attractive. The basic steps involved in these recovery operations are waste collection, transportation to the disposal site, storage, shredding, separation and recovery of glass and metal, burning of the combustible component, and ash removal. (See Fig. 6-5A.)

Air pollution control equipment is nearly always needed on industrial incinerators. Environmental control devices depend upon the composition of the stack effluent. They may include:

1. Afterburners to ignite combustible particles.
2. Cyclones to remove larger particles.
3. Low energy scrubbers to remove larger particulates.
4. High energy venturi scrubbers to remove fine particulates and/or water soluble acid gases.
5. Electrostatic precipitators to remove fine particulate and mists.
6. Fabric filters (bag houses) to remove fine particulate.

Principal hazards of operating an incinerator are:

1. Overheating due to feeding a large amount of volatiles.
2. Overheating due to an increase in the heat of combustion.
3. Structural failure due to overheating and/or corrosion.
4. Corrosion of scrubber due to acid.
5. Failure of scrubber system.

Compacted trash on transportation vehicles has been known to result in fires. Ignition sources may be smoldering ashes from fireplaces or unextinguished cigarettes. Shredding operations have also been the scene of occasional explosions and fires, in some cases, the result of discarded explosive materials. Small butane and LP-Gas

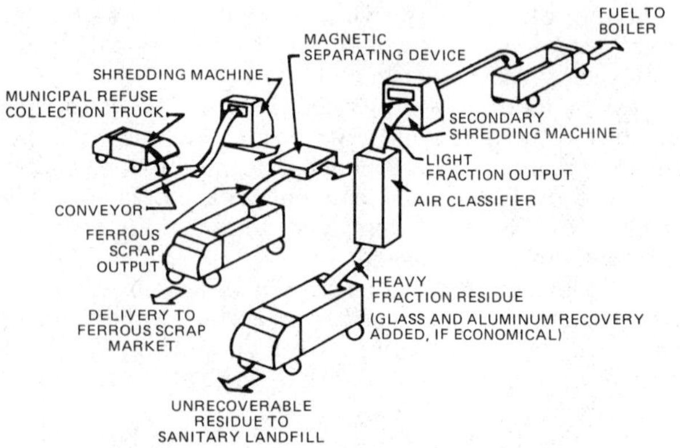

FIG. 6-5A. Typical refuse recovery process.

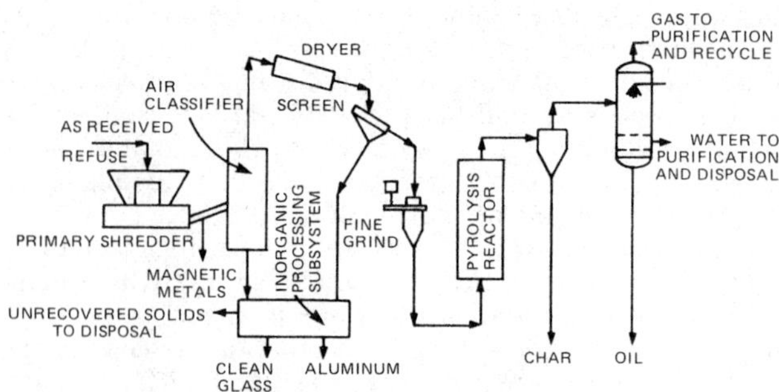

FIG. 6-5B. Flow diagram of a pyrolysis recovery system.

cylinders and gasoline containers with some fluid remaining in them are often the source of the flammable vapor. The ignition sources are frictional sparks between the grinding or shredding teeth, the walls of the shredder, and other solid components of the refuse.

Another approach for recovering the heating content of combustible waste is by pyrolysis. When polymers and organic materials are heated, they yield oils and other recoverable compounds. Although pyrolysis has not yet been widely adopted, it has promise as a conservation measure. A retort is charged with selected scrap and then heated to cause decomposition and distillation of the volatiles and oils. The process is somewhat similar to a coke oven, and similar operating hazards should be expected. (See Fig. 6-5B.)

Most nonhazardous domestic, commercial, and industrial waste today is disposed of in sanitary landfills. In a landfill operation, discarded material is mixed with earth and then compacted in a wide trench or other depression. Layer upon layer of mixture is added and then 2 ft (0.6 m) of earth is spread and compacted on top. This process helps control odor, rodents, and scavenger birds.

Successful landfill operation depends heavily upon biological activity within the fill material. There is some tendency for methane gas to be generated during biological activity. There have been some fires and explosions in buildings constructed over old landfills due to the seepage of biologically generated methane into the building and its eventual ignition. The recovery of natural gas from waste landfills is under way in various parts of the United States.

Hazardous wastes cannot be disposed of in a sanitary landfill. They are either chemically treated or incinerated in properly designed furnaces that ensure the complete degradation of the hazardous components of the waste. Otherwise, they may be disposed of in secure landfills. These are properly designed and engineered containment facilities which ensure that no waste can ever seep into the underground water system to pollute natural bodies of water or the public water supply system. Transporting the hazardous waste from the source to the disposal site may require federal, state, and local permits. The waste must be packaged or containerized in approved containers. A manifest describing the contents, their hazards, and emergency response procedures must accompany the shipment. The operator of the secure landfill generally requires a complete chemical analysis of the hazardous waste so that it may be handled and sited properly.

HAZARD PREVENTION AND CONTROL

Prevention Measures

The probability of exposure to hazardous waste may be reduced through preventive actions which are achieved through preplanning and proper waste handling and storage procedures. Preventive measures include the following:

1. Segregating chemically incompatible wastes. For example, by ensuring that combustible waste does not contain self-heating or oxidizing materials.
2. Disposing of hazardous waste promptly so that only small quantities are allowed to accumulate at any one time.
3. Locating combustible waste piles away from buildings, roadways, and ignition sources.
4. Maintaining flammable and corrosive liquid wastes in approved metal containers.
5. Inspecting hazardous waste containers continuously to ensure their integrity.
6. Properly identifying and marking waste containers. The hazards associated with each waste should be clearly shown on the exterior of each container, storage tank, transport vehicle, or building.
7. Training personnel in proper waste handling procedures, use of personal protective equipment, and emergency response measures to an accidental release of hazardous waste. Training should include realistic drills simulating all credible release scenarios.

A fire department should coordinate and preplan its approach to the handling of hazardous waste emergencies with all waste generating industries or disposal facilities within its jurisdiction. The fire department should be aware of the types, quantities, and hazards of all raw materials, products, and wastes stored at each plant site as well as those that are transported to and from the plant. The fire department should then ensure the availability to its personnel of any necessary specialized handling equipment and materials, detection devices, and personal protective equipment.

Response Procedures

The potential severity of an accident involving the release of a hazardous waste may be reduced through response measures developed and tested under realistic drill conditions. In the absence of such plans, or when the waste components are not known, it is advisable to assume that an industrial waste is hazardous and to treat it with due care until more information becomes available from reliable sources. In general, response procedures include the following:

1. Wearing full protective clothing and breathing apparatus.
2. Using the appropriate detection devices to check for the flammability, toxicity, or radioactivity of the released waste or its products of combustion.
3. Stemming the flow of the waste at the source.
4. Diking the area around a liquid spill to limit its spread to sewage lines and drains and to reduce the vapor release rate.
5. Staying upwind when fumes are being generated.
6. Controlling ignition sources.
7. Evacuating people who are downwind.

The decision to extinguish burning waste should be very carefully made after a full assessment of the consequences. Questions to be considered include:

1. Is the extinguishing agent to be used compatible with the waste?
2. Are sealed drums or containers of flammable liquids or compressed gases in danger of exposure to the fire?
3. Are there any radioactive, explosive, or oxidizing materials in the burning waste or near the fire?
4. Will extinguishment exacerbate the problem by allowing flammable or toxic vapors to be released once the fire is extinguished?
5. Will the use of water result in a severe pollution problem?

SPECIFIC WASTES

Combustible Refuse

Refuse of a highly combustible nature, such as dry waste paper, excelsior, etc., should be collected in metal containers and not allowed to accumulate. Quantity storage of these materials should be separated from buildings, roadways, and ignition sources by a distance of 50 ft (15 m) or more. Transport to an incinerator or landfill on a frequent schedule will minimize the fire hazard.

Drying Oils

Rags or paper that have absorbed drying type oils are subject to spontaneous heating. They should be kept in well covered metal cans and thoroughly dried before collection or transport.

Flammable Liquids and Waste Solvents

Waste solvents have variable flash points, hence varying hazard, depending upon composition. Some may contain solids, tars, waxes, and other materials that impede flow. Chlorinated solvents and water may also be present.

The two most common methods of disposal are by incineration or the use of a disposal contractor. In either case, the solvents are collected in 5 gal (19 L) cans or 55 gal (208 L) drums for disposal. If a contractor is used, the approximate composition of the waste must be provided.

Installations having substantial amounts of waste solvent may install their own incinerator for the purpose. An auxiliary gas or oil burner is needed to maintain adequate fire box temperature. Usually a scrubber section is used to remove acid gases and other pollutants. See NFPA 82, *Standard on Incinerators, Waste and Linen Handling Systems and Equipment*, for further information on industrial incinerator installations.

Polychlorinated Biphenyls (PCBs)

PCBs were manufactured and used in a variety of electrical and mechanical applications from the early 1930s until regulated by the *Toxic Substances Control Act* in 1977. They had been extensively used as a dielectric fluid in capacitors and transformers and as a lubricant in compressors and other machinery. They continue to be used today, but only in enclosed systems.

PCBs have been known to cause temporary skin irritations and nervous system symptoms to humans and are suspected carcinogens. Improper combustion may produce highly toxic and carcinogenic products.

Extreme care must be taken in handling PCBs and fires involving PCBs. Waste PCBs from old electrical equipment must be disposed of by incineration at an approved site or at sea or buried in approved containers in a secure landfill.

Liquid and Solid Oxidizing Materials

Spilled oxidizing materials and leaking or broken containers should be immediately removed to a safe area to await disposal in conformance with applicable regulations and manufacturers' instructions. (See NFPA 43A, *Code for the Storage of Liquid and Solid Oxidizing Materials*.) Most, if not all, oxidizers can be rendered harmless by dilution with water. However, some solutions cause pollution of streams and rivers, and thus require pretreatment.

Combustible and Reactive Metals

Small scraps of sodium from laboratory use can be dissolved in ethyl alcohol (not isopropyl alcohol) and the resulting alkoxide neutralized with acid. Large quantities can be offered to a vendor for purification and recycling. Oil contaminated dispersions or other moderate quantities can be burned in dry pans if provision is made to control the oxide fumes, which are highly alkaline. Disposal of sodium by throwing it directly into water is spectacular but extremely hazardous and will contaminate the water with sodium hydroxide. Equipment contaminated with small amounts of sodium may be cleaned by remotely controlled introduction of water or steam. Potassium, sodium —potassium alloy (NaK) and lithium are handled in the same manner as sodium. Lithium reacts somewhat less violently with water.

It is usually feasible to reclaim magnesium fines and scrap in the form of defective castings, clippings, and coarse chips. If this is to be done, either locally or through a smelting firm, the scrap should be thoroughly dried before remelting. Scrap in the form of fine chips and dust collector sludge can be disposed of in a specially constructed incinerator or by burning in a safely located outdoor area paved with fire brick or hard burned paving brick. By spreading the scrap or sludge in a layer about 4 in. (102 mm) thick, on which ordinary combustible material is placed and ignited, the combustion of magnesium can be safely conducted. Another method of deactivating magnesium sludge is to treat it with a 0.5 percent solution of ferrous chloride. Since hydrogen is generated by this reaction, this method of disposal is conducted in an open container in an outdoor location where the hydrogen can be safely dissipated. (See NFPA 48, *Standard for the Storage, Handling and Processing of Magnesium*.)

Spontaneous combustion of titanium has occurred in fines, chips, and swarf that were coated with water soluble oil. Disposal containers should be tightly closed and segregated so that spontaneous burning would not involve other exposures. Incineration in the open, as for magnesium scrap, may be feasible. (See NFPA 481, *Standard for the Production, Processing, Handling and Storage of Titanium*.)

While small amounts of zirconium fines can be mixed with sand or other inert material and buried, incineration of this combustible metal is generally preferred for larger quantities. Because of the spontaneous ignition potential, personnel carrying scrap should wear flame protective clothing and equipment and carry the scrap in buckets on a yoke between them. A layer may be ignited with excelsior. (See NFPA 482, *Standard for the Production, Processing, Handling, and Storage of Zirconium*.)

Lithium hydride and lithium aluminum hydride can ignite upon contact with water or moist air. Alkaline oxides resulting from incineration may be collected and neutralized.

Radioactive Materials

For most installations a contractor licensed to dispose of radioactive materials can be employed. (See NFPA 801, *Recommended Fire Protection Practice for Facilities Handling Radioactive Materials*.) Radioactive materials that are flammable or combustible must be handled with particular care. Combustible radioactive wastes should not be allowed to accumulate. Storage of such wastes near an air intake is particularly undesirable. If the products of combustion of waste materials containing long-lived radioactive materials are dispersed through air-conditioning or compressed air systems, extensive decontamination may become necessary.

Liquid radioactive waste may be concentrated and retained until the radioactivity has decayed to a safe level. Combustible radioactive waste, such as absorbent paper used to wipe contaminated surfaces, should be placed in approved containers for disposal.

Oil Spills on Water

Booming of spills has been proven effective in containing spills of liquids on relatively calm and current-free waters. Various makeshift designs of booms have been tried with relative ineffectiveness, but commercially available booms which recognize the hydrodynamics and aerodynamics involved in the confinement of spills on water have proven quite effective. Because of ecological considerations, this has become an important means of containing an oil spill, and more effective equipment is now available.

Following confinement of spills on water, various ways of removing the confined liquid have been used, including skimming devices or absorbents. Absorbents such as straw, plastics, sawdust, and peat moss have been spread on the surface of the spill and then collected and burned on shore. Skimming devices operate on several principles, including pumps and separators. Power boats with skimmers on the bow can scoop up oil and water, sending it through an oil separator and rollers to which oil adheres. The oil is then removed by scraping or compression.

Sinking agents to which oil may adhere and sink to the bottom have been used in the past. Several proprietary sinking agents have been developed, but such products as carbonized sand, brick dust, crushed clinkers, and cement also have been utilized. However, sinking agents are objectionable because they contribute to water pollution and are no longer recommended, except possibly for use in the ocean far from land.

Regardless of whether a spill occurs on land or on water, preplanning is imperative and should include prompt notification of all local, state, or federal agencies dealing with the problem of water pollution.

Bibliography

References Cited

CFR. Undated a. "Title 49: Transportation." Subpart G, Section 173.300. *Code of Federal Regulations.*

CFR. Undated b. *Ibid.* Subpart G, Section 173.151.

Federal Register. 1980. Vol 45, No 98, Subpart B. *Criteria for Identifying the Characteristics of Hazardous Waste and Listing Hazardous Waste.* pp 33121-22. May 19, 1980.

Glaubinger, R. S. 1979. "A Guide to the Resource Conservation and Recovery Act." *Chemical Engineering.* Jan 29, 1979. pp 79-81.

Reindl, J. 1977a. "Interrelationships Within the Solid Wastes System." *Solid Wastes Management.* April 1977. pp 22-23, 54-56.

Reindl, J. 1977b. "Examining Disposal and Recycling Techniques for Solid Wastes." *Solid Wastes Management.* May 1977. pp 60, 68, 70, 92, 94.

Richards, W. 1979. "Odd Coalition Scores EPA Over Lack of Waste Rules." *The Washington Post.* Oct 26, 1979. p A3.

World Bank. 1978. *Environmental Considerations for the Industrial Development Sector.* Office of Environmental and Health Affairs. The World Bank. Washington, DC.

NFPA Codes, Standards, Recommended Practices and Manuals. (See the latest *NFPA Codes and Standards Catalog* for availability of current editions of the following documents.)

NFPA 30, *Flammable and Combustible Liquids Code.*

NFPA 40, *Standard for the Storage and Handling of Cellulose Nitrate Motion Picture Film.*

NFPA 40E, *Code for the Storage of Pyroxylin Plastic.*

NFPA 43A, *Code for the Storage of Liquid and Solid Oxidizing Materials.*

NFPA 43C, *Code for the Storage of Gaseous Oxidizing Materials.*

NFPA 43D, *Code for the Storage of Pesticides in Portable Containers.*

NFPA 48, *Standard for the Storage, Handling and Processing of Magnesium.*

NFPA 49, *Hazardous Chemicals Data.*

NFPA 61A, *Standard for Prevention of Fire and Dust Explosions in Facilities Manufacturing and Handling Starch.*

NFPA 61B, *Standard for the Prevention of Fires and Explosions in Grain Elevators and Facilities Handling Bulk Raw Agricultural Commodities.*

NFPA 61C, *Standard for the Prevention of Fire and Dust Explosions in Feed Mills.*

NFPA 61D, *Standard for the Prevention of Fire and Dust Explosions in the Milling of Agricultural Commodities for Human Consumption.*

NFPA 82, *Standard on Incinerators, Waste and Linen Handling Systems and Equipment.*

NFPA 85F, *Standard for the Installation and Operation of Pulverized Fuel Systems.*

NFPA 91, *Standard for the Installation of Blower and Exhaust Systems for Dust, Stock and Vapor Removal or Conveying.*

NFPA 325M, *Fire Hazard Properties of Flammable Liquids, Gases and Volatile Solids.*

NFPA 481, *Standard for the Production, Processing, Handling and Storage of Titanium.*

NFPA 482, *Standard for the Production, Processing, Handling, and Storage of Zirconium.*

NFPA 490, *Code for the Storage of Ammonium Nitrate.*

NFPA 491M, *Manual of Hazardous Chemical Reactions.*

NFPA 495, *Code for the Manufacture, Transportation, Storage, and Use of Explosive Materials.*

NFPA 498, *Standard for Explosives Motor Vehicle Terminals.*

NFPA 651, *Standard for the Manufacture of Aluminum and Magnesium Powder.*

NFPA 654, *Standard for the Prevention of Fire and Dust Explosions in the Chemical, Dye, Pharmaceutical, and Plastics Industries.*

NFPA 655, *Standard for Prevention of Sulfur Fires and Explosions.*

NFPA 664, *Standard for the Prevention of Fires and Dust Explosions in Wood Processing and Woodworking Facilities.*

NFPA 801, *Recommended Fire Protection Practice for Facilities Handling Radioactive Materials.*

Additional Readings

"Controlling Hazardous Wastes —Research Summary," U.S. Environmental Protection Agency, May, 1980.

"Disposing of Small Batches of Hazardous Waste," SW —562, U.S. Environmental Protection Agency, 1976.

Edwards, B. H., and Coghlan, Jordan, K., *Emerging Technologies for the Control of Hazardous Wastes*, Noyes, Park Ridge, NJ, 1983.

Emergency Services Guide for Selected Hazardous Materials, U.S. Department of Transportation, Washington, DC.

"Engineering Handbook for Hazardous Waste Incineration," SW —889, U.S. Environmental Protection Agency, Sept. 1981.

"Firms Avidly Seeking New Hazardous Waste Treatment Routes," *Chemical Engineering*, Sept. 6, 1982, pp. 53-57.

"Flammable and Combustible Waste Disposal in the Plastics Industry," *Bulletin No. 13*, Society of the Plastics Industry, NY.

Guide for Safety in the Chemical Laboratory, 2nd ed., Van Nostrand Co., NY.

Handbook of Key Federal Regulations and Criteria for Multimedia Environmental Control, U.S. Environmental Protection Agency, 1979.

"Hazardous Waste Generation and Commercial Hazardous Waste Management," SW —894, U.S. Environmental Protection Agency, 1980.

Hazardous Waste Management Guide, J. J. Keller and Associates, Neenah, WI, 1983.

Hackman, E., *Toxic Organic Chemical Destruction and Waste Treatment*, Noyes, Park Ridge, NY, 1978.

Hitchcock, D., "Solid Waste Disposal: Incineration," *Chemical Engineering*, May 21, 1979, pp. 185-194.

Linville, J., ed., *Industrial Fire Hazards Handbook*, 2nd, ed., National Fire Protection Association, Quincy, MA, 1984.

Meidl, James, H., *Flammable Hazardous Materials*, 2nd ed., Glencoe, Beverly Hills, CA, 1978.

Metry, Amir A., *The Handbook of Hazardous Waste Management*, Technomic, Lancaster, PA, 1980.

Pocket Guide to Chemical Hazards, U.S. Department of Health, Education and Welfare, DHEW (NIOSH) Publication No 78-210, 1978.

"Report to Congress —Disposal of Hazardous Wastes," SW —115, U.S. Environmental Protection Agency, 1974.

Schieler, Leroy, and Pauze, Denis, *Hazardous Materials*, Demar Publishers, Albany, NY, 1976.

Shen, T. T., Chen, M., and Lauber, J., "Incineration of Toxic Chemical Wastes," Pollution Engineering, Oct. 1978, pp. 45-50.

Sittig, M., *Incineration of Industrial Hazardous Wastes and Sludges*, Noyes, park Ridge, NJ, 1981.

Sittig, M., *Landfill Disposal of Hazardous Waste and Sludges*, Noyes, Park Ridge, NJ, 1979.

"Technologies and Management Strategies for Hazardous Waste Control," OTA —M —197, Congressional Office of Technology Assessment, Washington, DC, March, 1983.

Vance, Mary, *Industrial Waste Disposal: A Bibliography*, Vance Bibliographies, Monticello, IL, 1982.

Wilson, D. G., *Handbook of Solid Waste Management*, Van Nostrand Reinhold, NY, 1977.

Worthy, W., "Hazardous Treatment Technology Grows," *C & EN*, March 1982, pp 10-16.

Zajic, J. E., and Himmelman, W. A., *Highly Hazardous Materials Spills and Emergency Planning*, Marcel Dekker Inc., NY, 1978.

SECTION 7

FIRE SAFETY IN BUILDING DESIGN AND CONSTRUCTION

FUNDAMENTALS OF FIRESAFE BUILDING DESIGN

Revised by Dr. Robert W. Fitzgerald

Building design and construction practices have changed significantly during the past century. One hundred years ago structural steel was unknown and reinforced concrete had not been used in structural framing applications. The first "high rise building" had just been built in the United States.

The design professions have also advanced significantly during the past century. The practice of architecture has changed markedly, and techniques of analysis and design that were unknown a century or even a generation ago are available to engineers today. Building design has become a very complex process, integrating many skills, products, and technologies into its system.

Fire protection engineering has made developmental strides similar to those of other professional disciplines in the building industry. At the turn of the century, conflagrations were a common occurrence in cities. In later years, increased knowledge of fire behavior and building design enabled buildings to be constructed in such a manner that a hostile fire could be confined to the building of origin rather than to the block or larger areas where the fires started. Progress continued in the field of fire protection engineering so that a generation ago a hostile fire could be confined to the floor of origin. At the present time, knowledge is available that enables a hostile fire to be confined to the room of origin or even to smaller spatial subdivisions in a structure.

Much activity is taking place today regarding firesafe building design. The general thrust of some developments appears to be directed toward identification of a rational design methodology to parallel or supplement the traditional 'go or no go' specifications approach. Knowledge in the field of fire protection has undergone and is undergoing development and reorganization that will enable buildings to be more rationally and efficiently designed for firesafety. This section of the HANDBOOK will identify the components of a field that is changing dynamically in its analysis and design capabilities.

The building designer should consult all the chapters in this Section as well as Section 9, "Hazards to Life in Structures." Section 17, "Water and Water Supplies for Fire Protection;" Section 18, "Water Based Extinguishing Systems;" and Section 19, "Special Fire Suppression Agents and Systems," will also supply relevant information.

DESIGN AND FIRESAFETY

The conscious, integrated process of design for building firesafety, if it is to be effective and economical, must be part of the architectural design. All members of the traditional building design team should include, as an integral part of their work, design for emergency fire conditions. The earlier in the design process that firesafety objectives are established, alternative methods of accomplishing those objectives are identified, and engineering design decisions are made, the more effective and economical the final results.

America Burning: *The Report of the National Commission on Fire Prevention and Control* (NCFPC 1973), identifies several areas in which building designers create unnecessary hazards, often unwittingly, for the building occupants. In some cases these unnecessary hazards are the result of oversight or insufficient understanding of the interpretations of test results. In other cases, they are due to a lack of knowledge of firesafety standards.

The Commission's report cites that conscious incorporation of firesafety into buildings is too frequently given minimal attention by the designer, and further that building designers are content, as are their clients, to meet the minimum safety standards of the local building code. Often both assume incorrectly that the codes provide completely adequate measures, rather than minimal ones, as is the case. In other instances, building owners and occupants see fire as something that will never happen to them, as a risk that they will tolerate because firesafety measures can be costly, or as a risk adequately balanced by the provisions of a fire insurance policy or availability of public fire protection.

Fortunately, conditions arising from these attitudes need not exist, much less continue. Information is available for design professionals to incorporate a greater mea-

Dr. Fitzgerald is Professor of Civil Engineering at Worcester Polytechnic Institute, Worcester, MA.

sure of fire protection into their designs. Use of fire protection information requires that the various members of the building design team recognize that fire conditions are a legitimate element of their design responsibilities; as well, they will require a greater understanding of the special loadings that fire causes on building elements and of the countermeasures that can be incorporated into designs.

Objectives of Firesafety Design

Before building designers can make effective decisions relative to firesafety, they must clearly identify the specific needs of the client with regard to the function of the building. After the building functions and client needs are understood, the designer must consciously ascertain both the general and the unique conditions that influence the level of fire risk that can be tolerated in the building. The acceptable levels of risk, and the focus of the firesafety analysis and design process, should be concentrated in the following three areas:

1. Life safety.
2. Property protection.
3. Continuity of building operations.

Since it is difficult to ascertain the level of risk that will be tolerated by the owner, occupants, and community, probing to identify firesafety objectives is an important design function. Consequently, firesafety criteria are often not identified in a clear, concise manner that enables the designer to properly provide for the realization of the design objectives. Unfortunately, it is impossible to provide more than general guidelines that must be considered in building design to assist in the identification of the firesafety objectives in this HANDBOOK.

Life Safety: Adequate life safety design for a building is often related only to compliance with the requirements of local building regulations. This may or may not provide sufficient occupant protection, depending upon the particular building function and occupant activities.

The first step of life safety design is to identify the occupant characteristics of the building. What are the physical and mental capabilities of the occupants? What are the range of their activities and locations during the 24 hour, seven day weekly periods? Are special considerations needed for certain periods of the day or week? In short, the designer must anticipate the special life safety needs of occupants during the entire period in which they inhabit the building.

The identification of life safety objectives is usually not difficult, but it does require a conscious effort. In addition, it requires an appreciation of the time and extent to which the products of combustion can move through the building. It is the interaction of the building response to the fire, and the actions of its occupants during the fire emergency, that determines the level of risk that the building design poses.

The design for life safety may involve one or a combination of three alternatives. These alternatives are: (1) evacuation of the occupants, (2) defending the occupants in place, or (3) providing an effective area of refuge. These alternatives may be evaluated by the likelihood that the building spaces will be tenable for the period of time necessary to achieve the expected level of safety. The criteria for tenability becomes an important part of the design.

The design for building evacuation involves two major components—one is the availability of an acceptable path or paths for escape; the second is the effective alerting of the occupants in sufficient time to allow egress before segments of the path of egress become untenable.

Alerting occupants to the existence of a fire is a vital part of the life safety design. A useful performance objective could be to identify that occupants should have at least x minutes to escape from the time they know of a fire until the escape route is blocked. To accomplish this, the designer must either ensure that the fire and the movement of its products of combustion will be slow enough to provide that time, or special provisions must be incorporated into the building to achieve that objective.

The second life safety design alternative is to defend the individual in place in the building. This may be appropriate for occupancies such as hospitals, nursing homes, prisons, and other institutions. It may be an appropriate alternative to other buildings when the size or design may show that evacuation has an unacceptably low likelihood of success. Defend-in-place design also uses a performance criteria of time and tenability levels.

The performance criteria for time might state that the building space should be tenable for y minutes after the start of the fire. The duration for y could be identified as a period much longer than the duration of any possible fire. The definition of tenability may be quite different from that acceptable for evacuation.

The third alternative is to design for an area of refuge. This involves an occupant movement through the building to specially designed refuge spaces. This type of design is more difficult than either of the other two alternatives because it involves the major design aspects of each. In certain types of buildings this may be a reasonable alternative; however, an evaluation of the effectiveness of the design and its likelihood of success are extremely important.

Life safety design for a building is difficult. It involves more than a provision for emergency egress, it requires attention as to who will be using the building and what they will be doing most of the time. Consideration must then be given to communication, the protection of escape routes, and temporary or permanent areas of refuge for a reasonable period of time for the building occupants to achieve safety.

Property Protection: Specific items of property that have a high monetary or other value must be identified in order to protect them adequately in case of fire. In some cases, specially protected areas are needed. In other cases, a duplicate set of vital records in another location may be adequate; however, the designer should ascertain if the user of the building has property that requires special fire protection.

Continuity of Operations: The maintenance of operational continuity after a fire is the third major design concern. The owner must identify the amount of "downtime" that can be tolerated before revenues begin to be seriously affected. Frequently, certain functions or locations are more essential to the continued operation of the building than are others. The designer must recognize those areas particularly sensitive to building operations, so that adequate protection is provided for the vital business operations conducted in it. Often, these areas need special attention that is not required throughout the building.

In modern buildings the value of the contents of a single room may be extremely high. This value may be due to the cost of equipment or records, or the high cost of business interruption. The sensitivity of equipment and data to the effects of heat, smoke, gases, or water must be addressed. In any event, the designer should protect the specially sensitive rooms from products of a fire either inside or outside of the room.

FIRE HAZARDS IN BUILDINGS

The products of combustion that must be considered by the building designer may be categorized as flame, heat, smoke, and gases, as shown schematically in Figure 7-1A.

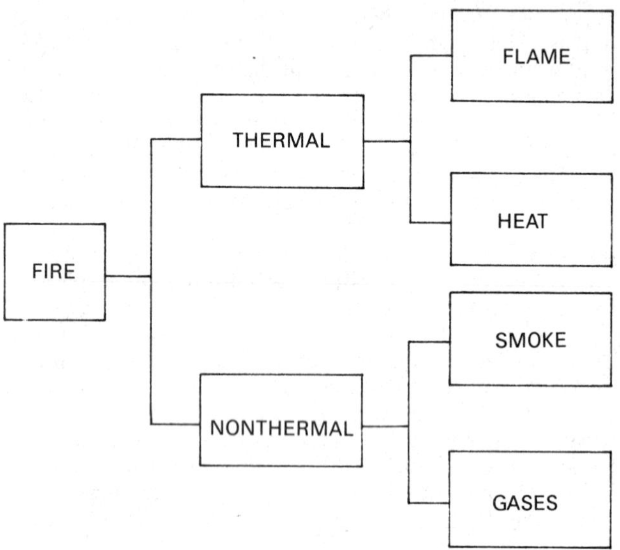

FIG. 7-1A. Schematic representation of products of combustion influencing building design.

The exposed people and property can be protected from the dangers posed by these products through effective building design. The challenge to the designer is to recognize the type of danger posed by each component and to incorporate effective countermeasures into the structure.

Smoke and Gases

Experience has shown that in a building fire the most common hazard to humans is from smoke and toxic gases. Nearly three-fourths of all building-related fire deaths are directly related to these products of combustion. Death often results from oxygen deprivation in the bloodstream, caused by the replacement of oxygen in the blood hemoglobin by carbon monoxide. In addition to the danger of carbon monoxide, many other toxic gases that are present in building fires cause a wide range of symptoms such as headaches, nausea, fatigue, difficult respiration, confusion, and impaired mental functioning.

Smoke, in addition to accompanying toxic and irritant gases, contributes indirectly to a number of deaths. Dense smoke obscures visibility and irritates the eyes and can cause fear and emotional shock to building occupants. Consequently, the occupant may not be able to identify escape routes and utilize them.

Heat and Flames

Heat that results in burns is often incorrectly assumed to be the primary cause of fire death and injury. Although heat injuries do not compare in quantity to those caused by inhalation of smoke and toxic gases, they are painful, serious, and cause shock to victims. Statistically, this thermal product of combustion accounts for nearly 25 percent of fire deaths. In addition to these deaths, the pain and disfigurement caused by burns results in serious, long term complications.

Building Elements and Contents

Property is also affected by the thermal and nonthermal products of combustion, as well as the extinguishing agents. Smoke damage often occurs to goods located long distances from the effects of the heat and flames. Fires that are not extinguished quickly often result in considerable water damage to the contents and the structure, unless special measures are incorporated to prevent that damage.

The collapse of structural building elements can be a serious life safety hazard. Although statistically it has not resulted in many deaths or injuries to building occupants, structural collapse is a particular hazard to fire fighters. A number of deaths and serious injuries to fire fighters occur each year because of structural failure. While some of these failures result from inherent structural weaknesses, many are the result of renovations to existing buildings that materially, though not obviously, affect the structural integrity of the support elements. A building should not contain surprises of this type for fire fighters.

Fast flame spread over finish materials or building contents and vertical propagation of fire are serious concerns. The ability of the fire service to contain or extinguish a fire is significantly diminished if the fire spreads vertically to two or more floors. With a given potential for fire growth, the prevention of vertical fire spread is influenced principally by architectural and structural decisions involving details of compartmentation.

The Fire Growth Hazard

The speed and certainty of fire growth and development in rooms can vary greatly. The contents and interior finish in some rooms are quite safe, and for this type of situation it is unlikely that, once ignited, a fire can grow to full involvement of the room. On the other hand, the interior design of other rooms poses a high hazard which, if an ignition were to occur, could lead to an almost certain full room involvement.

The traditional method of describing the hazard has been through fire loads reflected in use and occupancy classifications. Building types, rather than rooms within buildings, have been grouped with regard to their relative hazard. For example, residential and educational occupancies are considered low hazard because they normally contain relatively low fuel loads in the rooms. Mercantile buildings are normally a moderate hazard, while certain industrial and storage buildings may be considered highly hazardous because they contain a high fuel load.

This type of classification is a basis for building code requirements, and historically it has been quite useful.

However, a more detailed look at the fire growth potential within rooms of building can be a valuable part of a detailed firesafety design. The fire growth hazard, which identifies the speed and relative likelihood of a fire reaching full room involvement, is a useful base from which to design suppression interventions and to evaluate life safety problems. For example, situations where fast, severe fires will occur may call for automatic sprinkler protection, even though it may not be required by a building code.

The basis for the fire growth hazard analysis is the combustion characteristics within the room. The main factors that influence the likelihood and speed with which full room involvement occurs are: (1) fuel load (type of materials and their distribution), (2) interior finish of the room, (3) air supply, and (4) size, shape, and construction of the room.

Fire development in a room is neither uniform nor a certain progression from ignition to full room envolvement. Fires moves through several stages, called realms, in its behavior. Table 7-1A provides guidance on the technical definition of the realms. Within any realm a fire may continue to grow or it may be unable to sustain continued development and die down. Table 7-1A includes a rough guide to the approximate flame sizes that may be used to describe the realms of the fire size. It also describes the major factors that influence growth within the realm. Absence of a significant number of the factors would indicate that the fire would self-terminate, rather than continue to develop.

TABLE 7-1A. Fire Prevention and Emergency Preparedness

1. Ignitors
   a. Equipment and devices
   b. Human accident
   c. Vandalism and arson
2. Ignitable Materials
   a. Fuel type and quantity
   b. Fuel distribution
   c. Housekeeping
3. Emergency Preparedness
   a. Awareness and understanding
   b. Plans for action
      —Evacuation or temporary refuge
      —Self-help extinguishment
   c. Equipment
   d. Maintenance—operating manuals available

Different rooms pose different levels of risk regarding the likelihood of reaching full room involvement and the time in which fire development takes place. The factors in Table 7-1A provide a general guide to the important types of factors.

If one were to focus on a single event that might describe the relative level of risk posed by the contents and interior finish in a room, it would be the ability of flames to reach the ceiling. The arrangement of contents and types of fuels where it would be difficult for a fire to grow to touch the ceiling pose a relatively low fire growth hazard potential. On the other hand, where furniture combustibility and density will allow a fire to develop to ceiling height, or when combustible interior finish is present, the fire growth hazard potential usually is comparatively high.

The fire growth hazard potential is important for several reasons. The first is the obvious concern for the speed and likelihood of full room involvement. However, the factors become the threat to be protected by the automatic and manual suppression activities. The fire growth hazard potential and the speed of fire development become the major factors in designing appropriate fire defenses.

ELEMENTS OF BUILDING FIRESAFETY

Building firesafety may be achieved either by fire (ignition) prevention or, if a fire does occur, by managing its impact through building design. Fire prevention is accomplished by separating ignitors from ignitable materials. While this is certainly an important aspect of the total firesafety picture, experience shows that firesafety cannot rely totally on fire prevention.

Building design and construction features that influence safety are within the decision making authority of the design team, based on the assumption that its firesafety objectives are clearly defined by management, the owners, or other responsible parties, both public and private. The design and construction elements include both active and passive design considerations.

Fire Prevention

The persons responsible for fire prevention are not the same as those responsible for the building design. Table 7-1B describes the elements that comprise firesafety from a prevention consideration. The decisions concerning these elements are predominantly under the control of the building owner or occupant, or both.

Automatic Suppression of Building Fires

For more than a century, automatic sprinklers have been the most important single system for automatic control of hostile fires in buildings. Many desirable aesthetic and functional features of buildings that might offer some concern for firesafety because of the fire growth hazard potential can be protected by the installation of a properly designed sprinkler system.

Among the advantages of automatic sprinklers is the fact that they operate directly over a fire and are not affected by smoke, toxic gases, and reduced visibility. In addition, much less water is used because only those sprinklers fused by the heat of the fire operate, particularly if the building is compartmented.

Other automatic extinguishing systems (carbon dioxide, dry chemical, Halon agents, high expansion foam) also find use in providing protection for certain portions of buildings or type of occupancies for which they are particularly suited.

An automatic sprinkler system has been the most widely used method of automatically controlling a fire. The major elements for determining the effectiveness of an automatic suppression system are (1) the presence or absence of automatic suppression, (2) if present, its reliability (i.e., will water issue from the nozzle?), and (3), if reliable, its design effectiveness.

TABLE 7-1B. Major Factors Influencing Fire Growth.

Realm	Approximate Ranges of Fire Sizes	Major Factors that Influence Growth
1 (Preburning)	Overheat to Ignition.	Amount and Duration of Heat Flux. Surface Area Receiving Heat from Material Ignitability.
2 (Initial Burning)	Ignition to Radiation Point (10 in. [254 mm] high flame).	Fuel Continuity. Material Ignitability. Thickness. Surface Roughness. Thermal Inertia of the Fuel.
3 (Vigorous Burning)	Radiation Point to Enclosure Point (10 in. to 5 ft high flame [254 mm to 1.5 m]).	Interior Finish. Fuel Continuity. Feedback. Material Ignitability. Thermal Inertia of the Fuel. Proximity of Flames to Walls.
4 (Interactive Burning)	Enclosure Point to Ceiling Point (5 ft. [1.5 m] high flame to flame touching ceiling).	Interior Finish. Fuel Arrangement. Feedback. Tallness of Fuels. Proximity of Flames to Walls. Ceiling Height. Room Insulation. Size and Location of Openings. HVAC Operation.
5 (Remote Burning)	Ceiling Point to Full Room Involvement.	Fuel Arrangement. Ceiling Height. Length/Width Ratio. Room Insulation. Size and Location of Openings. HVAC Operations.

Although automatic sprinkler systems have a remarkable record of success, it is possible for the system to fail. Often failure is due to a feature that could have been avoided if appropriate attention had been given at the time of the design, installation, or inspection. Table 7-1C describes common failure modes and their causes. During the design stages these factors should be addressed to increase the probability of successful extinguishment by the sprinkler system.

MANUAL SUPPRESSION OF BUILDING FIRES

The protection offered by a community fire department has an important influence on building fire design. Some buildings are designed in a manner that helps the fire department extinguish fires while they are small; others in a manner that hinders a fire department. Rarely does the designer consciously design the building to enable emergency forces to be able to handle the emergency. The following discussion provides some guidelines for building design to enhance the building's ability to allow the fire department to extinguish a fire with minimal threat to life and property.

Ideally a building is designed so that should a fire occur, it can be attacked before it extends beyond the room of origin. If that is not possible, the building design should retard fire spread so that the fire department will encounter a relatively small, easily controllable fire. The major aspects of building design for manual suppression include: (1) fire department notification, (2) agent application, and (3) fire extinguishment. These aspects are discussed briefly to provide guidance for incorporating features into the building that enable manual fire fighting to be more effective.

Fire Department Notification: The time durations for completing the events through agent application are very dependent upon the speed of the fire spread. Often, buildings have been lost because of insufficient attention to the method of notifying the local fire department. The complete chain of events from detection to correct receipt of an alarm should be a part of every building fire design. It should be consciously designed, rather than left to chance.

The notification process involves four distinct events; (1) detection of the fire, (2) decision to call the fire department, (3) placement of the call, and (4) correct receipt of the notification by the fire department. Frequently, the building design does not consciously incorporate a reliable method for notifying the fire department. Consequently, delayed notification often occurs and the fire may grow to proportions beyond the extinguishment capability of the local fire department. If all four notification events are not consciously addressed in the building design, the owner and occupants should be aware of the possibility of delayed fire department response.

Agent Application: The next critical event is fire department application of agent to the fire. This involves three distinct events for its success. They are: (1) arrival at the site, (2) nozzle enters room, and (3) water discharge from the nozzle.

The fire department response to the site with the appropriate equipment can be estimated by a generally predictable time duration. Alarm handling time, turnout

TABLE 7-1C. Common Automatic Sprinkler Failure Modes.

FAILURE MODE	POTENTIAL CAUSES
Water Supply Valves are Closed When Sprinkler Fuses.	Inadequate Valve Supervision. Company Attitude. Maintenance Policies.
Water Does Not Reach Sprinkler.	Dry Pipe Accelerator or Exhauster Malfunctions. Pre-action System Malfunctions. Maintenance and Inspection Inadequate.
Nozzle Fails to Open When Expected.	Fire Rate of Growth too Fast. Temperature of Link Inappropriate for the Area Protected. Sprinkler Link Protected from Heat. Sprinkler Link Painted, Taped, Bagged or Corroded. Sprinkler Skipping.
Water Cannot Contact Fuel (Note: the intent of this failure mode is to ensure that discharge is not interrupted in a manner that will prevent fire control by a sprinkler).	Fuel is Protected. High Piled Storage is Present. New Construction (walls, ductwork, ceilings) Obstructs Water Spray.
Water Discharge Density is not Sufficient.	Discharge Needs are Insufficient for the Type of Fire and the Rate of Heat Release. Change in Combustible Contents Occurred. Number of Sprinklers Open is too Great for the Water Supply. Water Pressure too Low. Water Droplet Size is Inappropriate for the Fire Size.
Enough Water Does Not Continue to Flow.	Water Supply is Inadequate because of Original Deficiencies, Changes in Water Supply, or Changes in the Combustible Contents. Pumps are Inadequate or Unreliable. Power Supply Malfunctions. System is Disrupted.

time, and travel time can be estimated with reasonable accuracy. During this time period the fire continues to grow. When the cumulative time for response becomes excessive for the fire growth hazard potential, the fire has an opportunity to extend significantly.

Site access then becomes an important part of the building features. Ideal exterior accessibility occurs where a building can be approached from all sides by fire department apparatus. This, unfortunately, is not always possible. In congested areas, only the sides of buildings facing streets may be accessible. In other areas, topography or constructed obstacles can prevent effective use of apparatus in combating the fire.

Some shopping centers and buildings located some distance from the street make the approach of apparatus difficult. If obstructions or topography prevent apparatus from being located close enough to the building for effective use, such equipment as aerial ladders, elevating platforms, and water tower apparatus are rendered useless. Valuable labor must be expended to carry ground ladders or hand carry hose lines long distances.

The matter of access to buildings has become far more complicated in recent years. The building designer must consider this important aspect in the planning stage. Inadequate attention to site details can place the building in an unnecessarily vulnerable position. If its fire defenses are compromised by preventing adequate fire department access, the loss must be compensated for by more complete internal building protection.

The arrival at the site is only a part of the agent application evaluation. The fire fighters must then be able to enter the building, reach the floor of the fire, and find the involved room or rooms. This is often a time consuming, difficult task. Considerable attention must be given to the problem of getting fire fighters and equipment to the fire.

Access to the interior of a building can be greatly hampered where large areas exist and where buildings have blank walls, false facades, solar screens, or signs covering a high percentage of exterior walls. Obstacles that prevent ventilation allow smoke to accumulate and obscure fire fighters' vision. Lack of adequate interior access can also delay or prevent fire department rescue of trapped occupants.

Windowless buildings and basement areas present unique fire fighting problems. The lack of natural ventilation facilities, such as windows, contributes to the buildup of dense smoke and intense heat, which hamper fire fighting operations. Fire fighters must attack fires in these spaces in spite of heat and smoke.

Manual Extinguishment: After the time consuming and sequential events of notification and agent application have transpired, the fire department is ready to fight the fire. The size of fire that is present at the time of agent application determines the fire fighting strategy and likelihood of success of the operation.

Broadly speaking, there are three categories of fire conditions that may be expected. Comparatively small fires may be extinguished by direct application of water. When the fire is larger than the comparatively smaller sizes that can be directly extinguished, the fire is pushed out of the building; i.e., the building is opened (ventilated), and the hose streams drive the fire, heat, and smoke out of the building. Fires that are too large for this operation must be surrounded. All available techniques of ventilation and heat absorption by water evaporation are used; however, the fire area is lost, and the main purpose of this strategy is to protect exposures, both external and internal.

Ventilation

Ventilation is an important fire fighting operation. It involves the removal of smoke, gases, and heat from building spaces. Ventilation of building spaces performs the following important functions:

1. Protection of life by removing or diverting toxic gases and smoke from locations where building occupants must find temporary refuge.
2. Improvement of the environment in the vicinity of the fire by removal of smoke and heat. This enables fire fighters to advance close to the fire to extinguish it with a minimum of time, water, and damage.
3. Control of the spread or direction of fire by setting up air currents that cause the fire to move in a desired direction. In this way occupants or valuable property can be more readily protected.
4. Provision of a release for unburned, combustible gases before they acquire a flammable mixture, thus avoiding a backdraft or smoke explosion.

The building designer should be conscious of these important functions of fire ventilation and provide effective means of facilitating venting practices whenever possible. This may involve access panels, moveable windows, skylights, or other means of readily opened spaces in case of a fire emergency. Emergency controls on the mechanical equipment may also be an effective means of accomplishing the functions of fire ventilation. Each building has unique features, and consequently a unique solution should be incorporated into the design.

Water Supply and Use

Water is the principal agent used to extinguish building fires. Although other agents may occasionally be employed (e.g., carbon dioxide, dry chemical, foams and surfactants, and halons), water remains the primary extinguishing agent of the fire service. Consequently, the building designer should anticipate the needs of both the fire department and automatic extinguishing systems and provide an adequate supply of water at adequate residual pressure.

Water is normally supplied to the building site by mains that are part of the water distribution system. Few cities can supply a sufficient amount of water at required pressures to every part of the city. Consequently, water supplied to hydrants, standpipes, or sprinklers must be boosted by pumps located on fire department apparatus or in the buildings themselves.

Careful attention must be given to water supply, distribution, and pressure for emergency fire conditions. High rise buildings are particularly sensitive in this respect because the water pressures that are required depend upon building height. The water supply needs of large area buildings must be given more careful attention.

Fire conditions that require operation of a large number of sprinklers or use of a large number of hose streams can reduce pressure in standpipe and sprinkler systems to the point where residual pressures in the distribution system are adversely affected. Fire department connections for sprinkler and standpipe systems are important components of building fire defenses. The building designer must carefully consider installation details of fire department connections to make sure they will be easily located, readily accessible, and properly marked. Locations should be approved by the local fire department.

Water Removal: Watertight floors are important in this respect. Salvage efforts can be greatly affected by the integrity of the floors. Of greater importance is the number and location of floor drains. If interior drains and scuppers are available, salvage teams can remove water effectively with a minimum of damage to the structure.

Barriers

Barriers, such as partitions and floors, separate building spaces. These barriers also delay or prevent fire from propagating from one space to another. In addition, barriers are important features in any fire fighting operation.

The effectiveness of a barrier is dependent upon its inherent fire resistance, the details of construction, and the penetrations such as doors, windows, ducts, pipe chases, electrical raceways, and grilles. Although one's confidence in the hourly ratings of fire endurance may be somewhat weak in building fires, unpenetrated rated barriers seem to perform rather well. This may be due to the rather large factor of safety inherent in the codes. On the other hand, it is quite common for rated barriers to fail because of inattention to penetrations. For example, the fire resistance of a 3 hr floor-ceiling assembly can be voided because of large and numerous poke throughs. The fire resistance of a 2 hr partition is lost when a door is left open.

Fire resistance requirements imposed by the regulatory system often have comparatively little value because of inattention to the functional and construction details. In order to predict field performance of barriers, the penetrations and details of construction must be considered.

The major function of barriers is to prevent an ignition in the adjacent room. It is useful to classify ignitions in two categories. One is a massive barrier failure, which would occur when a part of the barrier collapses or when a large penetration, such as a door or a large window, is open. When a massive failure occurs, the adjacent room becomes fully involved in a short period of time. The second type of failure is a localized penetration failure; this occurs when flames or heat penetrates small poke throughs or small windows. A localized penetration causes a hot spot to occur. If ignition occurs, this could lead to a full room involvement by the normal fire development progression.

Structural Collapse

The potential for structural collapse must be determined. Building codes address this aspect through construction classification requirements. The relationship between fire severity and fire resistance to collapse are the principal factors.

Collapse can occur when the fire severity exceeds the fire endurance of the structural frame. However, this is comparatively rare. Structural collapse is more commonly associated with deficiencies in construction. These deficiencies are not evident under normal, everyday use of the building. They become a problem when the fire weakens supporting members, triggering a progressive collapse.

Smoke Movement

Smoke and gases will move through a building much faster and easier than flames and heat. The time duration from ignition until a building space is untenable is an important aspect of firesafety.

Smoke is generated by the fire. When the fire is extinguished, additional smoke development stops. Therefore, if smoke poses a potentially serious protection problem, one solution is to design effective, fast fire suppression into the building. If suppression is not accomplished, the next concern is the time until the building spaces become untenable.

People Protection

The architectural design of a building has a significant influence on its firesafety capabilities. Interior layout, circulation patterns, finish material, and building services all are important factors in firesafety.

During a fire the occupant is exposed to two types of danger. One is exposure to flame and hot products of combustion. This is a problem in the vicinity of burning, but the danger decreases rapidly as the distance from the fire increases. Smoke and toxic gases, on the other hand, present a different type of hazard. The greatest number of fire deaths result from these latter products of combustion, and the danger is present at considerable distances from the location of the fire.

Most fires, even relatively minor ones, may produce large quantities of smoke and gases. These products can obscure vision and irritate the eyes to the extent that visibility is reduced to practically zero. Occupants familiar with their surroundings often experience great difficulty in locating means of egress. The problem is compounded for transients and occasional visitors to the building.

Architectural layout and normal circulation patterns are important elements in emergency evacuation. For example, many large office buildings are a maze of offices, storage areas, and meeting rooms. Even under normal conditions, a visitor can become confused and have difficulty in exiting the building. Clearly marked emergency travel routes can enhance life safety features in all buildings.

Life safety is a major concern in building firesafety. It involves a complex interaction between the fire, the building, and the occupants. The fire conceptually includes its size, propagation, and duration, as well as the smoke movement that will occur. The building design provides for alerting the occupants of a fire emergency, as well as providing temporary refuge until egress is accomplished or until the fire is extinguished. Time is the element that is common to all of these components.

Property and Mission Protection

It is important to reduce property loss from a fire as much as possible. Property includes both contents and the building itself. Many buildings contain rooms whose contents may be valued at many times the cost of the building. It is important to safeguard those spaces, when they are known, by providing special suppression design for fires occurring within the room of origin, and special exposure protection for fires originating outside of these high value spaces.

Exposure Protection

A fire in one building creates an external fire hazard to neighboring structures by exposing them to heat by radiation, and possibly by convective currents, as well as to the danger of flying brands of the fire. Any or all of these sources of heat transfer may be sufficient to ignite the exposed structure or its contents.

When considering protection from exposure fires, there are two basic types of conditions to be considered. They are: (1) exposure to horizontal radiation, and (2) exposure to flames issuing from the roof or top of a burning building in cases where the exposed building is higher than the burning building. Radiation exposure can result from an interior fire where the radiation passes through windows and other openings of the exterior wall. It can also result from the flames issuing from the windows of the burning building or from flames of the burning facade itself. A source for guidelines and data on exposure protection is given in NFPA 80, *Recommended Practice for Protection of Buildings from Exterior Fire Exposures.*

Bibliography

References Cited

NCFPC. 1973. *America Burning: The Report of the National Commission on Fire Prevention and Control.* Superintendent of Documents. U.S. Government Printing Office. Washington, DC.

NFPA Codes, Standards, Recommended Practices and Manuals. (See the latest *NFPA Codes and Standards Catalog* for availability of current editions of the following documents.)

NFPA 13, *Standard for the Installation of Sprinklers.*

NFPA 80A, *Recommended Practice for Protection of Buildings from Exterior Fire Exposures.*

SYSTEMS CONCEPTS FOR BUILDING FIRESAFETY

Revised by Edward M. Connelly and Clifford S. Harvey

Before examining examples of the various systems concepts which are currently being applied to firesafety problems, it is well to establish what is meant by a system. A system is an interdependent set of components; a system description is the description of the components and their interactions. These interactions may be static, i.e., fixed at a point in time. For example, one might look at a bridge, define its components and describe by vectors all of the forces in the members. Similarly, the firesafety of a structure might be evaluated through a rating sheet that lists all of the components affecting firesafety and rate them individually and collectively as to their effects on firesafety. In both examples, however, dynamic factors, i.e., time varying factors, were excluded. For example, the effects of vibration from a truck passage or an earth tremor were not estimated on the bridge; neither was the interaction of the various elements of the firesafety system in the structure with a fire explicitly estimated.

Therefore, there are two types of systems, static and dynamic. The dynamic system differs because time varying interactions are explicitly considered. In both cases, the assumption is made that all parts of the system that are important to the interaction have been identified, and that all other factors are external to the system. The analysis that proceeds in either case is called systems analysis.

SYSTEMS ANALYSIS

The term "systems analysis" is used to describe a careful, methodical, well documented, step by step approach to problem solving. The steps to be taken in systems analysis are:

1. Write an exact statement of the problem to be solved. This also can be in the form of several general statements.
2. Define the system to be considered, including all interacting factors and excluding all others.
3. List all the assumptions made regarding the system and the problem.
4. Document all data and sources of that data.
5. Review the methods and procedures to be used in the analysis.
6. Perform the analysis prescribed.
7. Document the results.
8. Draw conclusions and document those conclusions.

The discipline of systems analysis requires documentation of each step for the review by the person performing the analysis as well as others. Often some steps will be attempted several times. For instance, preparation of accurate statements of the problem to be solved may be attempted more than once before the problem is clearly identified. Further, the methods and procedures used in the analysis may involve construction of models and the use of computer simulation, although it is not absolutely necessary to use those devices.

SYSTEMS APPROACHES APPLIED TO THE FIRE PROBLEM

Firesafety can be incorporated into building design by three different methods. The first (specification codes) is to require design and construction to conform to requirements in specification oriented building codes and standards. The requirements are based on fire experiences and are generally strict.

To overcome the inflexibility of specification codes, interest has centered on performance codes. A major drawback to this approach is that building components are considered on an ad hoc basis. The definition of firesafety performance of the separate components, and the measurement of that performance, are difficult to obtain. Also, when considering the entire building as a system, the satisfactory physical performance of individual components does not assure the desired level of safety for the building as a whole, as some components have a higher probability of successful performance than others.

A third method of achieving firesafe building design is as an integrated subsystem of the building. This integrated method considers the aesthetic functional, structural, electrical, or mechanical subsystems. Buildings can be de-

Mr. Connelly is president, Performance Measurement Associates, Inc., Reston, VA.

Mr. Harvey is president, Firemeasure, Inc., Boulder, CO and the Chief Fire Marshal for that community.

signed for firesafety utilizing an engineering methodology, rather than by strict compliance with codes. This approach to achieving firesafety requires a professional technology in its application, just as do the architectural and engineering disciplines. Moreover, it shows promise of achieving a greater level of cost effectiveness than strict enforcement of codes and most often allows greater flexibility.

In a similar manner, the evaluation of current levels of firesafety and the recommendations for upgrading can only be accomplished properly if the impact of a change in a structure can be evaluated in context. For example, if a building is to be rehabilitated, which requires bringing it up to code level, does this mean strict compliance, or are there alternative means to achieve the same level of firesafety? By logical extension, it is also desirable to estimate the impact of a code or standard change on the level of firesafety considering all other factors bearing on the "firesafety system." So, too, before catastrophic fires it is highly desirable to determine the impact of new processes and materials on the level of firesafety provided through compliance with a given code or standard.

A Qualitative Approach

During the 1960s, there was a growing awareness that modern high rise buildings designed to building codes and standards contained potential firesafety weaknesses. Various researchers and organizations, principally the U.S. General Services Administration and the National Research Council of Canada, identified aspects of weakness. For example, studies identifying the time of evacuation of high rise buildings were published, and the problems relating to air movement in tall buildings were identified. Careful analysis of two fires in New York City, at One New York Plaza (Powers 1971a) and 919 Third Avenue (Powers 1971b), focused attention even more clearly on the problem areas.

About the same time, NFPA formed a committee on high rise structures. After initial meetings, this committee determined that its contribution should not be in developing standards for high rise construction, but rather in identifying a system of firesafety for all buildings. After many modifications, the Firesafety Concepts Tree shown in Figures 7-2A through 7-2E evolved.

FIRESAFETY CONCEPTS TREE

The Firesafety Concepts Tree is a summation of the consensus of the NFPA Committee on Systems Concepts for Fire Protection in Structures. It represents the deliberations and professional judgments of this committee over many years. It is not, and is not represented as, the only manner in which the factors involved in firesafety can be organized.

The Tree as shown in Figures 7-2A through 7-2E shows the elements that must be considered in building firesafety and the interrelationship of those elements. It enables a building to be analyzed or designed by progressively moving through the various levels of events in a logical manner. Its success depends upon the completeness by which each level of events is satisfied. Lower levels on the decision tree, however, do not represent a lower level of importance or performance; they represent a means for achieving the next higher level. In addition to the material in this chapter the reader is refered to NFPA 550, *Guide to the Firesafety Concepts Tree* for detailed guidance in the use of the tree.

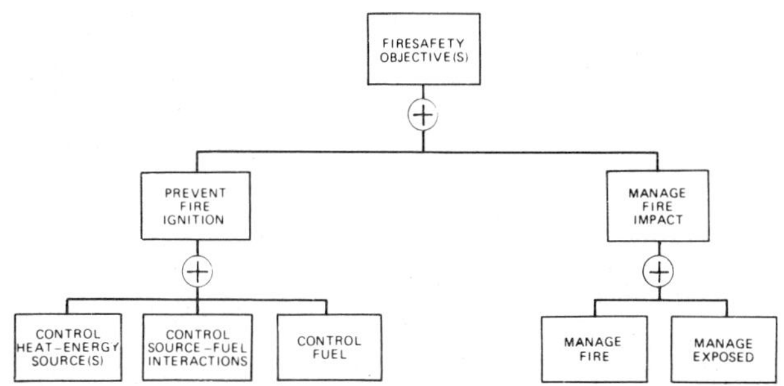

FIG. 7-2A. The principal branches of the firesafety concepts tree as developed by the NFPA Committee on System Concepts for Fire Protection in Structures. Note the symbols at the bottom of the illustration. They apply equally to the tree branches shown in Figures 7-2B to 7-2E.

Firesafety Objectives

The use of the Tree requires that the firesafety objectives (goals) be clearly identified. These objectives describe the degree to which the building should protect its occupants, property contents, continuity of operations, and neighbors. The objectives should be quantified wherever possible, rather than stated in broad or general terms.

The life safety objective, for example, might state that all occupants be safeguarded against the intolerable or untenable effects of the fire. It may be further stated that emergency personnel such as fire fighters, who may be expected to stay in areas considered too dangerous for the occupants, should be protected against unexpected collapse of the building or entrapment. A range of specific life safety objectives may be appropriate for varying types of occupancies. Nursing home requirements, for example, are vastly different from those for offices; and both differ from industrial occupancies or storage facilities.

To identify the property protection objectives, the building designer must ask a number of questions. For example, does a portion of the property have a significantly higher value than the rest of the property? What property cannot be replaced, or if destroyed, would significantly affect continued operations? Are there specific functions at the site that are vital to continuity of operations? Further, are there other functions that may be performed at other sites on an emergency basis? Proper identification of factors such as these enable the designer to tailor the design to reflect the client's needs.

"Prevent Fire Ignition" or "Manage Fire Impact"

The Tree provides the logic required to achieve firesafety, i.e., it provides conditions whereby the firesafety objectives can be satisfied, but it does not provide the minimum condition required to achieve those objectives. Thus, according to the Tree, the firesafety objectives can be met if fire ignition can be prevented or if, given ignition, the fire can be managed. This logical "OR" function is represented by the symbol (+) on the Tree.

The "Prevent Fire Ignition" branch of Figure 7-2B (and associated performance standards) essentially form a fire prevention code. Most of the events described in this

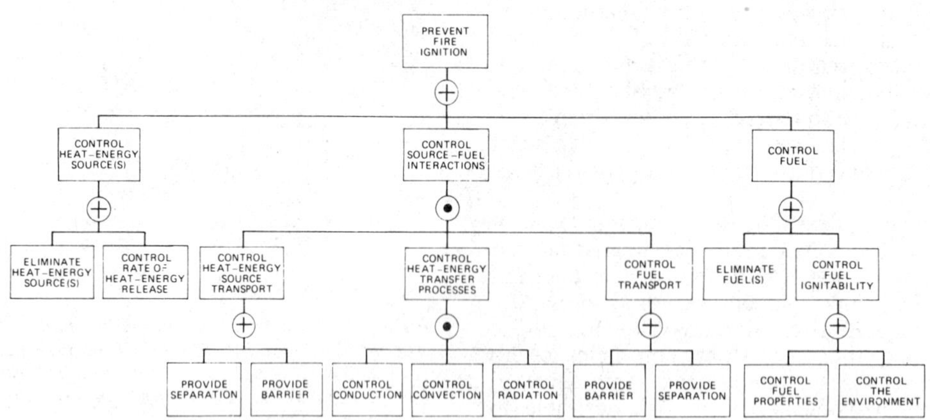

NOTE: ⊕ = "OR" GATE ⊙ = "AND" GATE ▽ = ENTRY POINT

FIG. 7-2B. Components of the "prevent fire ignition" branch of the concepts tree.

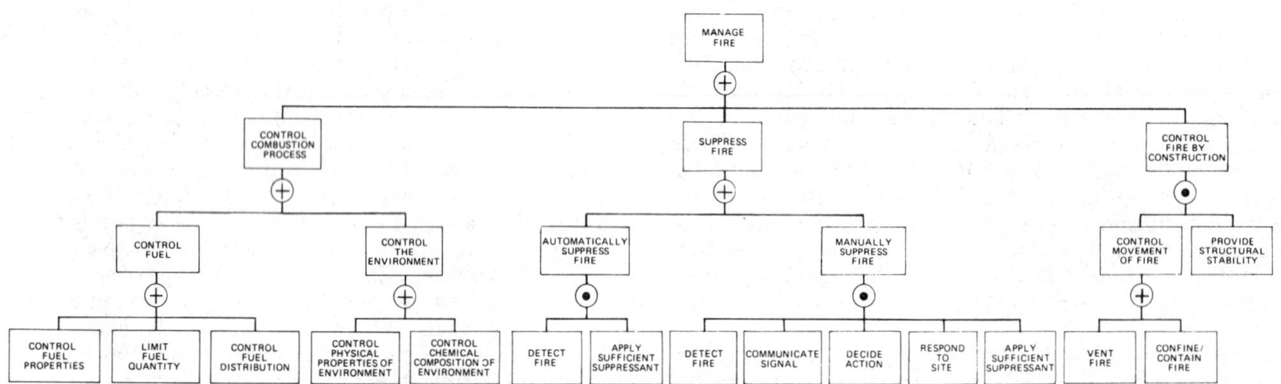

FIG. 7-2C. Components of the "manage fire" branch of the concepts tree.

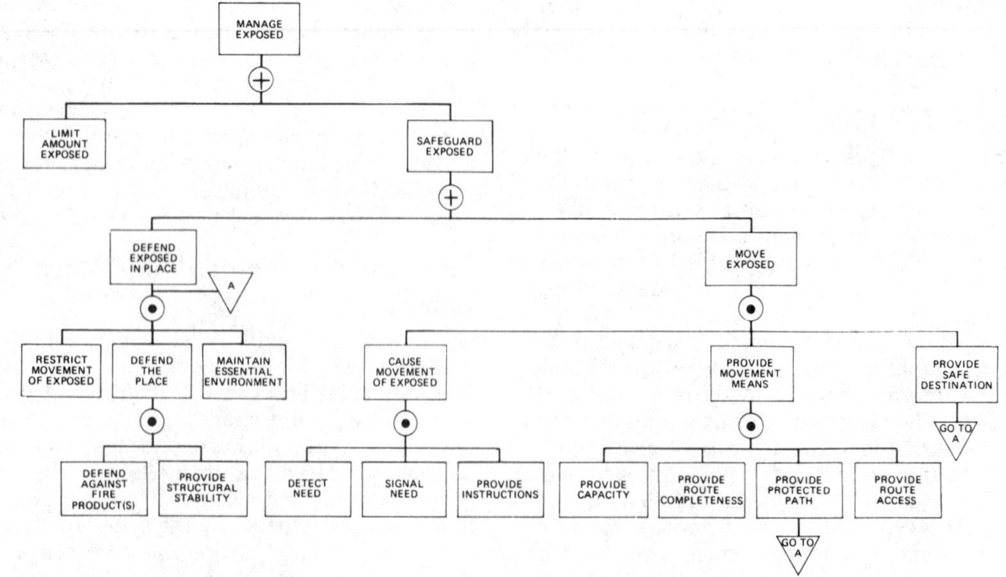

FIG. 7-2D. Components of the "manage exposed" branch of the concepts tree.

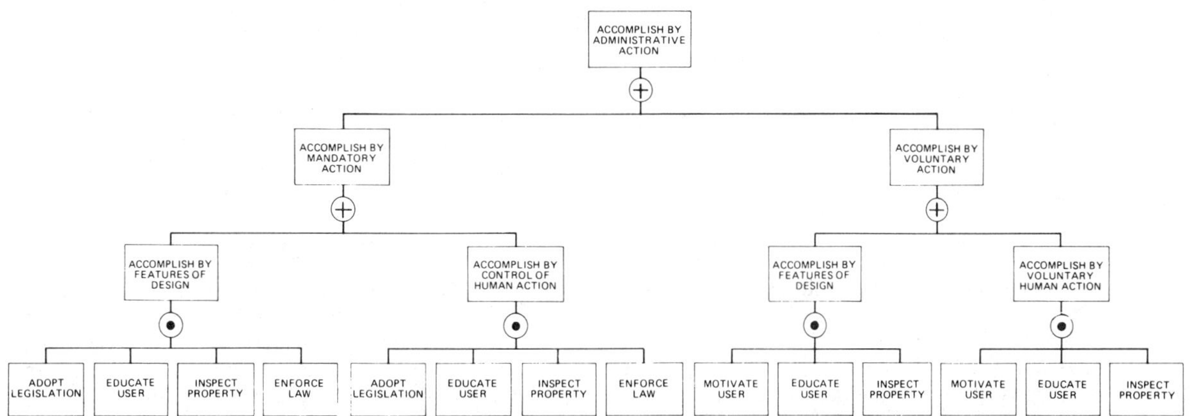

FIG. 7-2E. Components of the "administrative action guide"—a third dimension to the concepts tree.

branch require continuous monitoring for success. Consequently, the responsibility for satisfactorily achieving the goal of fire prevention is essentially an owner/occupant responsibility. The designer, however, may be able to incorporate certain features into the building that may assist the owner/occupant in preventing fires.

It is impossible to prevent completely the ignition of fires in a building. Therefore, in order to reach the overall firesafety objective, from a building design viewpoint, a high degree of success in the "Manage Fire Impact" branch assumes a significant role. Essentially this branch of the Tree (and associated firesafety standards) may be considered as a building code by the design team. After an ignition occurs, all considerations shift to the "Manage Fire Impact" branch to achieve the firesafety objectives.

According to the logic of the Tree, the impact of the fire can be managed either through the "Manage Fire" or "Manage Exposed" branches. (See Fig. 7-2A.) The "OR" gate indicates that the objectives may be reached through either design branch or both, as long as the avenue selected completely satisfies the firesafety objective. Naturally, it is acceptable to do both, which will increase the probability of success over using only one branch.

"Manage Fire" Branch

The firesafety objectives can be achieved by managing the fire itself. Figure 7-2C shows that this can be accomplished by (1) controlling the combustion process, (2) suppressing the fire, or (3) controlling the fire by construction. Here, again, any one of these branches of the Tree will satisfy the "Manage Fire" event. Thus, for instance, in some fires success is achieved where the building construction controlled the fire. And in other fires, success is achieved by controlling the combustion process, either by controlling the fuel or the environment.

The events that control the "Suppress Fire" event are shown in Figure 7-2C. In this Figure the symbol ⊙, (a bold dot in the center of a circle) represents a logical "AND" gate, and signifies that all of the elements in the level immediately below the gate are necessary to achieve the event above the gate. To accomplish the automatic suppression event, for example, all three events—detecting the fire, initiating action, and controlling the fire are necessary. Similarly, to manually suppress the fire, all six events must take place. The omission of any single event is sufficient to break the chain and cause the failure of this automatic suppression event.

In considering the "Control Fire by Construction" event, structural integrity must be provided, and the movement of the fire itself must be controlled. As shown in Figure 7-2C, this can be accomplished either by venting, confining, or containing the fire.

"Manage Exposed" Branch

As shown in Figure 7-2D, the fire impact can be managed by managing the "exposed"—people, property, or functions, depending upon the design aspects being considered. In considering the "Manage Exposed" branch, it can be successful either by limiting the amount exposed or by safeguarding the exposed. For example, the number of people as well as the amount or type of property in a space may be restricted. Often this is impractical. If this is the case, the objectives may still be met by incorporating design features to safeguard the exposed.

The exposed people or property may be safeguarded either by moving them to a safe area of refuge or by defending them in place. For example, people in institutionalized occupancies such as hospitals, nursing homes, or prisons must generally be defended in place. To do this, the "Defend Exposed in Place" branch shown in Figure 7-2D would be considered. On the other hand, alert, mobile individuals, such as those expected in offices or schools, could be moved to safeguard them from fire exposure on either a short term or long range basis dependent upon other key design elements. Figure 7-2D describes the events that must be satisfied if firesafety objectives are to be met by moving the exposed property and people.

Administration Action Guide

The Administration Action Guide (Fig. 7-2E) organizes into a logical structure those actions that might be taken to carry out the elements of the basic Tree. (See Figs. 7-2A through 7-2D.) It can be thought of as the third dimension of the Tree and, as such, a checklist of actions that might be taken. For example, "Provide Safe Destination," (Fig. 7-2D) could be accomplished through code enforcement ("Mandatory Action") or public firesafety education ("Voluntary Action").

USES OF THE FIRESAFETY CONCEPTS TREE

The tree can be adapted for a number of different functions. A particularly useful aspect of the decision Tree is the descriptive feature of important code requirements.

Fire Prevention/Building Codes

The "Prevent Fire Ignition" branch (Fig. 7-2B) is essentially a fire prevention code. It describes and shows the interrelationship between the essential features of such a code.

The "Manage Fire Impact" branch (Figs. 7-2A, 7-2C and 7-2D) is essentially a building code. It enables the code writer to organize the important elements of the building code and to identify their interrelationships. An important feature for building codes, for example, is the subject of alternatives, or "tradeoffs." The only legitimate areas in which alternatives can be established is among those factors below an "OR" gate in the decision tree. The factors below an "AND" gate are necessary and thus cannot be considered as alternatives. It is up to the authority having jurisdiction to decide what levels of protection are adequate to be allowed as tradeoffs for code or Tree mandated items.

Building Analysis

The Concepts Tree provides a framework to support firesafety analyses. It is a fundamental necessity that objectives be established and concurred with before they can be acted upon. Once the fundamental firesafety objectives for a particular building are identified, the designer/architect can analyze the building's design by progressing successively through the various levels of events. Redundancies and deficiencies can be specifically identified and evaluated. Often the weaknesses become so apparent that specific and effective solutions can be economically devised.

The building analysis, supported by the framework of the "decision tree," has among its advantages the capacity of considering separately fire prevention considerations and the requirements involved in managing the fire impact in the event an ignition occurs. This definite separation of events can enable the owner/occupant to take immediate, corrective action for fire prevention, and then take thoughtful, effective action to correct the building deficiencies.

Building Design

The Firesafety Concepts Tree can be employed effectively in building design. If the architect incorporates the Tree during the preliminary planning phase of design, many important decisions and alternatives can be established more effectively. For example, decisions can be made regarding evacuation versus temporary refuge, and their implications on the functions of the building. Specific needs with regard to the decision are then recognized.

The Tree also enables separation of the functions of fire prevention and building design. In this way the responsibilities of the owner/occupant can be differentiated from those of the building design team. Those events that are eventually incorporated into the design can be identified with a specific member of the building design team.

Building codes and NFPA standards are still important factors in building design; the tree should not supercede those documents. Rather, it enables those documents to be interrelated and consequently used more effectively. The Tree helps in design decision making with regard to firesafety; specific minimum design guidelines and standards are still the appropriate domain of the building codes and the standards writers.

QUANTITATIVE MEASURES

The Concepts Tree does not, in its present form, provide a firesafety score for a building. Other techniques have been developed to provide such scores. The major systems, some of which provide static measures while others provide dynamic measures, are described in this section.

Gretener Method

The Gretener Method (Kaiser 1979/80), also known as the Swiss Insurance Method, concentrates on property protection in any type of facility. The method has been widely utilized by Swiss insurance inspectors. This model determines the risk of a facility by dividing the hazard factor by the protection factor. The hazard factor is determined through multiplication of factors for the fuel load, combustibility, smoke generated, etc. The protection factor is determined by multiplication of factors, based on an ordinal scale, for various protection features.

As with other insurance type models, life safety is not stressed. This model might be used to determine the risk to the occupants or in occupancies with a low life hazard; however, in order to select the proper means of protection, a knowledge of fire protection is a prerequisite for the user of this model.

Network Analysis Method

A network is a graphical representation of a system. It shows the logic, sequencing, and relationships of the different events. For example, when the system is a construction project, a network might be used to describe several different ways the sequence of construction activities could be performed. Analysis of the network gives insight into the impact of alternative schedules on such factors as manpower needs, time durations, costs, or probability of completion on schedule. The critical path method (CPM) is a network technique that has become a common tool of construction management.

Network diagrams can be used with fire protection in a number of ways (Fitzgerald 1979). For example, they may be used to describe paths of egress from a building, the movement of flame and heat through a building, the analysis of the likelihood of barrier success, or the comparison of one type of detection system with another. All one needs to construct a network is a clear statement of the logic of the process and attention to a few simple rules of construction and evaluation. Sometimes, though, writing a clear, correct statement of logic is time consuming.

The construction of network diagrams requires that activities or events be stated in a manner that clearly identifies their accomplishment. For example, "notify the fire department," "apply agent to the fire," and "extinguish the fire" are all events which can be clearly recognized at the instant of completion. On the other hand, "manage the fire" or "control the environment" are inappropriate event descriptions in a network, because of the difficulty in specifically identifying their accomplishment.

Another requirement of network construction is that from any given event, future events are described in a manner that includes all possibilities. This requirement can be used to construct networks that may be used to analyze firesafety alternatives. Networks are used by some fire departments to determine relative equivalency to the codes.

Figure 7-2F shows a network that describes fire extinguishment within a room. The starting event is the fire ignition, and the three possible ways the fire can go out are considered: (1) self-termination of the fire itself through some deficiency, such as lack of fuel or discontinuity in the source of fuel; (2) automatic suppression of those fires that do not self terminate; and (3) manual suppression of those fires that do not self-terminate.

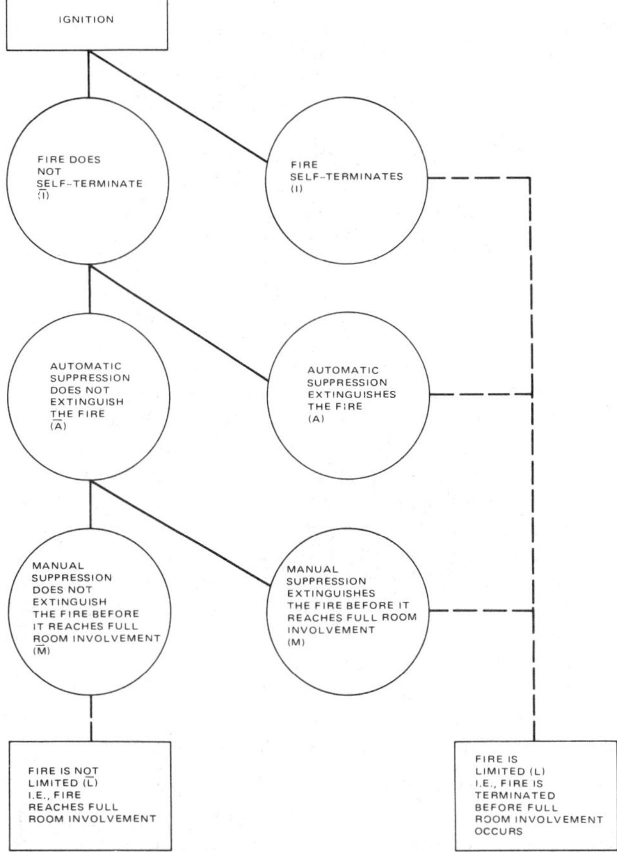

FIG. 7-2F. Network diagram for fire extinguishment within a room.

In Figure 7-2F, all possible occurrences are ensured because each event is described in a binary manner. For example, given ignition, the fire will either self-terminate or continue to burn. There can be no other possible outcome when the events are selected in this manner. For those fires that do not self-terminate, an automatic sprinkler system will or will not put out the fire. If automatic sprinklers do not suppress the fire, the fire department will either extinguish or not extinguish it before it grows to full room involvement. The network diagram shows by the solid lines all connections between possible events.

Figure 7-2F shows all possible occurrences for fire in a room, given an ignition and considering the three methods of fire termination. The path leading to full room involvement can occur only when:

Given ignition
AND
The fire does not self-terminate
AND
Automatic sprinklers do not extinguish the fire
AND
Manual suppression does not extinguish the fire
THEN
The fire reaches full room involvement.

On the other hand, the fire can be terminated before reaching full room involvement by three routes. They are:

(1) Given ignition
The fire self-terminates
OR
(2) Given ignition
AND
The fire does not self-terminate
AND
Automatic suppression controls the fire
OR
(3) Given ignition
AND
The fire does not self-terminate
AND
Automatic suppression does not control the fire
AND
Manual suppression extinguishes the fire.

The fire is extinguished by any one of the three paths. The network shown in Figure 7-2F is a graphic method of describing all of the paths discussed above.

One of the uses of this network is to compare the relative success of two or more alternative fire protection designs. Probability values can be assigned to each of the events, and the value of the probability of fire termination can then be computed.

After probability values for each event are determined, then (because the network is constructed using binary events, i.e., true or not true), only two rules for calculation are needed. They are:

1. The probability of an event is computed by multiplying the probability values for each of the events along the continuous path being considered, provided the events are independent.
2. The probability values of different events having similar outcomes may be added.

Therefore, only multiplication and addition are needed to evaluate a network diagram. Figures 7-2G and 7-2H illustrate evaluation of a network diagram and include probability values assigned to each event. For convenience, symbols rather than word statements have been used to describe the events. A bar over a letter means "NOT." For example, I signifies "the fire self-terminates," while Ī means "the fire DOES NOT self-terminate."

Assume that design alternative "A" resulted in probability values of the events as shown in Figure 7-2G. The symbol P (̄) indicates the probability of an event. Here, the probability that, given ignition, the fire will reach full room involvement is:

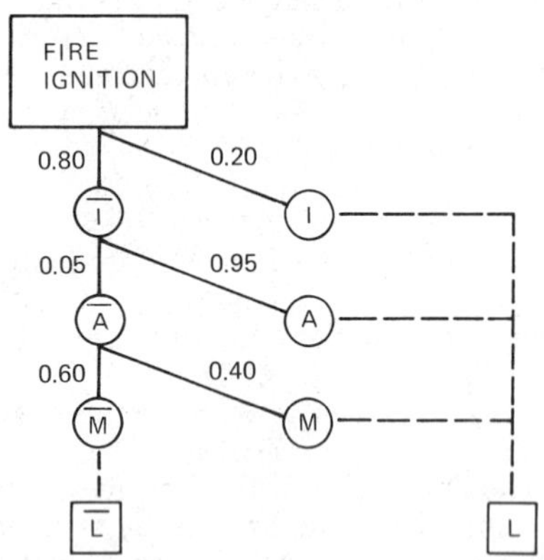

FIG. 7-2G. Network Design System "A."

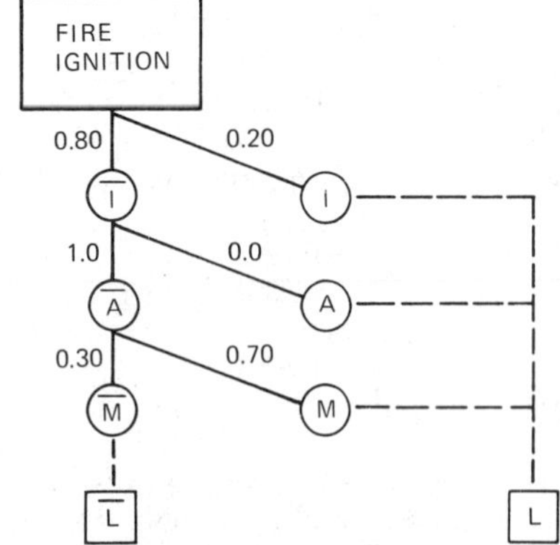

FIG. 7-2H. Network Design System "B."

$$P(\bar{L}) = P(\bar{I})P(\bar{A})P(\bar{M})$$
$$P(\bar{L}) = (0.80)\ (0.05)\ (0.60)$$
$$P(\bar{L}) = 0.024$$

where

$P(\bar{L})$ = Probability fire is not limited
$P(\bar{I})$ = Probability fire does not self-terminate
$P(\bar{A})$ = Probability automatic suppression does not extinguish the fire
$P(\bar{M})$ = Probability manual suppression does not extinguish the fire

That is, this room design will be expected to reach full room involvement only 24 times per 1,000 ignitions.

The likelihood of termination of a fire before full room involvement occurs may be computed by calculating each termination path, and then adding the results. From Figure 7-2G this becomes:

$$P(\bar{L}) = P(\bar{I}) + P(\bar{I})P(\bar{A}) + P(\bar{I})P(\bar{A})P(\bar{M})$$
$$P(\bar{L}) = (0.20) + (0.80)\ (0.95) + (0.80)\ (0.05)\ (0.60)$$
$$P(\bar{L}) = 0.20 + 0.76 + .016$$
$$P(\bar{L}) = 0.976$$

That is, this room design will be expected to cause termination by some means 976 times in 1,000 ignitions.

The sum of the probability values must always be 1.0 for fires that are terminated plus those that are not terminated.

Assume that a different design alternative "B" is proposed, and its analysis resulted in the probability values shown in Figure 7-2H. The probability of successful termination before full room involvement occurs is:

$$P(\bar{L}) = P(\bar{I}) + P(\bar{I})P(\bar{A}) + P(\bar{I})P(\bar{A})P(\bar{M})$$
$$P(\bar{L}) = (0.20) + (0.80)\ (0) + (0.80)\ (1.0)\ (0.70)$$
$$P(\bar{L}) = 0.20 + 0 + 0.56$$
$$P(\bar{L}) = 0.76$$

Design alternative "B" is, therefore, expected to result in termination before full room involvement 760 times in 1,000 ignitions. The numerical results of design alternatives "A" and "B" can be compared, as can the costs, benefits, and the likelihood of each alternative in achieving its design objectives.

The form of network diagram shown in Figure 7-2F can be related to the event logic trees described earlier in this chapter. Since probability values of events along a continuous chain are multiplied, this describes an "AND" gate. Similarly, the addition of like event results is the equivalent of an "OR" gate. The mathematical procedure described above and that of Section C yield identical results.

Because of the direct correspondence between network diagrams and event logic trees (sometimes called decision trees) for the same events, the results are identical. The establishment of independent probability or conditional probability values are also identical. In other words, when the selected events are the same, event logic diagrams and network diagrams are interchangeable, and use the same type of probability values.

The network diagram system is being used by some fire departments to graphically illustrate the strengths and weaknesses of various alternatives to written code requirements. Further, the system is available on microcomputers for quick review of those alternatives.

The National Bureau of Standards (NBS) Firesafety Evaluation System (FSES)

The National Bureau of Standards (NBS) has developed a method for evaluating firesafety of health care occupancies, detention and correctional occupancies, and board and care facilities, that provides broader flexibility than the traditional code approach (Nelson 1979). These Firesafety Evaluation Systems (FSES) have been published and are included in NFPA 101® Life Safety Code® (hereinafter referred to as NFPA 101).

NBS is currently working on similar systems for other occupancies. FSES combines concepts of systems approach with traditional fire risk grading concepts to develop total firesafety performance measurements.

The NBS system evaluates the elements of risk and types of construction in the various health care occupancies by assigning relative numerical values for each of the controlling safety factors involved. Three major factors are taken into account by the system: (1) risk to occupants, (2) the capability of the building and its fire protection system to provide firesafety commensurate with that risk, and (3) the redundant safety capabilities available to assure safety in case any single safeguard fails. The calculated occupancy risk level provides a minimum target to which levels of protection must be provided by the nature of the building design, supplemented by appropriate passive and active fire protection devices.

In FSES, the occupancy risk in a particular type of occupancy is gaged by the number of people who will be affected by a given fire, the kind of fire they are likely to encounter, and their ability to protect themselves. The system requires information on occupant mobility, occupant density, location within the building (for example, in the basement or on upper levels), ratio of occupants to attendants, and the average age of occupants. These individual occupancy risk factors are given numerical values and adjusted to reflect the type of building being considered and whether it is a new design or an existing building. (This variation between new and existing construction is in recognition of the concept in NFPA 101 of demanding different safety requirements for existing buildings than those demanded for new designs or major renovations.)

The next step in the evaluation involves an assessment of the ability of the fire protection system to provide measures of safety commensurate with calculated occupancy risk. The analysis addresses the general safety and three separate safety subsystems.

The evaluation system includes a weighting of individuals factors such as: construction, interior finish (corridors and exit), interior finish (rooms), corridor partitions or walls (separation from exit access), doors to corridors, zone dimensions (exit access), vertical openings, hazardous areas, smoke control, emergency movement routes (exit system), manual fire alarms, smoke detection and alarm, and automatic sprinklers. These factors cover the range of physical materials, arrangements, and safeguards that determine the basic level of safety of the facility. They range from construction and finish, through exit arrangements and distances to detection and extinguishing systems.

The actual evaluation of safety in the facility is determined by the total scores of the safeguards and risks as they apply not only to overall safety but to each of the three subsystems of fire containment, extinguishment, and people movement. The use of the subsystem arrangement of scores recognizes that certain fire protection features influence one or more, but not necessarily all, of the separate subsystems. This arrangement also assures that sufficient safeguards will be present in each of the basic subsystem areas of protection methodologies, and that tradeoff or compensating arrangements will occur only in the areas where appropriate.

FSES sets minimum requirements in the separate subsystems of containment, extinguishment, and movement of people as a basis for measuring safety that is equivalent to applicable provisions of NFPA 101. (These minimum safety levels were derived by using the system to measure NFPA 101, i.e., a building in exact conformance with the Code.)

In order to demonstrate equivalency with NFPA 101, the rating for each of the three subsystems must match or exceed that level which was achieved by the Code itself. The system is therefore directly tied to NFPA 101. It could be adjusted to another code by measuring that code and developing a new set of mandatory requirements. The NBS advises that there are limitations on the evaluation system:

1. Application of FSES is limited to health care occupancies, detention and correctional occupancies, and board and care facilities at this time, because the weighted values have been tailored specifically to the needs of these institutions. Normally, the use of these evaluation systems in other institutions would be excessively restrictive.
2. The evaluation systems can measure safety only relative to NFPA 101 (or another code if used as a comparative base). As such, the system only evaluates absolute firesafety relative to the given code.
3. The system covers the primary physical fire protection features and arrangement of the building but does not cover requirements related to utilities, furnishings, or administrative activities (such as emergency plans and fire drills). These must be considered separately from the evaluation system to assure complete safety. For convenience, an additional checklist has been developed by the NBS to cover all of the items not covered by the evaluation system but required by NFPA 101.

Since the first FSES was developed, others governing different occupancies have been, and are being, developed. Completion of an evaluation system for all occupancies will allow greater flexibility in design of new buildings, and the use of existing buildings. The expectation is that these systems will probably become the standard for decision making when designing or renovating buildings.

Finite State Transition Models

In general terms, a model represents what is known about a subject. A model permits one to ask questions and get answers about real world processes. In this sense models provide an "accounting"—usually the result of interactions of complex factors, which are not easily envisioned without the model. It is the accounting, or manipulation of the data, that makes the model useful as a logic system.

Finite state models provide a way of representing physical processes simply. By constructing a finite state model, a very complex system can be reduced to a simple form to solve practical problems. A familiar example of finite state modeling, are the prospects of promotion in any organization, e.g., a fire department, government agency, etc. The prospects can be represented by identifying the important positions in the organization (these are called states) and determining from experience and other information the possibility of transfer from one position to another. Transfer possibilities include the likelihood of transfer from one position to another as well as the time one is likely to spend in each position before a transfer occurs. Once completed, this finite state model of "Promotion Prospects" can be used to determine the expected time delay and likelihood of reaching each position in the organization.

In contrast to physical models in which physical processes are represented by complex mathematical functions representing those processes, finite state models simply consist of a set of process states and the rules governing transitions from one state to the other. Physical models are frequently used in research projects and might, for example, be used to develop a simpler finite state model. Thus, physical models might be thought of as a "scientist's model." On the other hand, a finite state model, once developed, can be thought of as a "user's model."

Building Firesafety Simulative Model

A finite state model, the "Building Firesafety Simulative Model, (BFSM)" which represents fire growth in a residential building, was developed by NFPA on a HUD sponsored contract (Project H2316) entitled "Firesafety Systems Analysis for Residential Occupancies" (Berlin 1978). In this development effort, seven states of fire development were identified as critical states of interest in a room fire. The states are referred to as realms and are listed in Table 7-2A. In addition to the list of realms (states), the Building Firesafety Simulative Model also has a set of rules which define how the fire development process moves from one realm to another. These rules are statistical relationships since fire development is subject to a certain amount of variability. Two types of rules are used for realm transition descriptors:

TABLE 7-2A. Realm Definitions For Single Family Dwellings*

Realm 1:	Pre-Burning. Situations prior to ignition and following termination of a fire are represented by this realm.
Realm 2:	Established Burning. The fuel exhibits self-sustained burning which includes smoldering.
Realm 3:	Vigorous Burning. The heat release rate exceeds 2 kW. It is expected that the flame height may be approximately 25 cm (9.84 in) and that the upper room air temperature has increased by 15°C (59°F) above ambient temperature.
Realm 4:	Interactive Burning. The upper room air temperature exceeds 150°C (302°F). It is expected that the flame height may be approximately 120 cm (47.25 in) and that the heat release rate may exceed 50 kW.
Realm 5:	Remote Burning. The upper room air temperature exceeds 500°C (932°F). It is expected that the external heat flux returning to the fuel surface exceeds 5 kW/m^2. This realm also includes secondary ignition beyond the room of fire origin with the change in upper room air temperature in this area less than 15°C (59°F).
Realm 6:	Flame Spread. There is flame involvement beyond the room of origin with the change in upper room air temperature in this area greater than 15°C (59°F).
Realm 7:	Major Spread. There is involvement beyond the room of origin with the upper room air temperature exceeding 150°C (302°F).

* Table extracted from NFPA Final Report (Contract H-2316) "Firesafety Systems Analysis for Residential Occupancies."

1. The probability of transfer to each other realm, and
2. The statistical distribution of time the fire will remain in each realm before a transfer occurs.

The specification of the set of realms and the transition descriptors constitute the specification of the finite state Building Firesafety Simulative Model.

Uses of the Building Firesafety Simulative Model: As described above, the Building Firesafety Simulative Model, as an example of a finite state model, provides a flexible representation of fire development as a function of characteristics of the building design and its contents. This model can be used as an adjunct to professional judgment and to quantitative data such as charts and graphs of fire growth in evaluating the firesafety of a specific building design. The fire model should be considered as a very general tool—a kind of fire development "slide rule"—which can be used to assist in solving many problems that require assessment of the properties of fire growth.

The Building Firesafety Simulative Model or other finite state fire models provide a simulation of fire growth given a fire ignition for a specific building design and specific contents. This simulation is implemented on a digital computer so that the user can input specific building design characteristics of interest such as the physical dimensions of rooms, burn characteristics of contents, ventilation, etc., and receive as computer output a description of the fire development as a function of time. This fire development can be described in many forms such as the physical size of the flame, energy output, or temperature of the flame, etc. Also, the generation and spread of the products of combustion can be represented. Since many of the fire characteristics of materials are not known with certainty, they must be represented in the model by their statistical distributions. Thus, for example, the heat of combustion of material used to pad a couch might not be represented by a specific number, e.g., 10,000 Btus per lb (23 MJ/Kg), but rather by distribution of values which represents the distribution found in the real world. In that way distributions representing the uncertainty of each factor important to fire growth can be represented in the model. As a result, each fire simulated by the computer is but one of all the possible fire developments given those distributions. Repeated simulations provide the fire development of likely fires given those distributions. Thus, statistics of fire growth calculated from many replications of the fire simulation can provide an estimate of the distribution of the fire characteristics that would exist in the real world.

In addition to the variation in burn characteristics of building materials and furnishing materials, the variations in building construction and/or building design factors can also be included. When this is done the distributions of the simulated fires represent the effected variations in building construction practice, as well as variation in building design practice.

Use of finite state fire models is facilitated when specific questions regarding fire damage, such as extent of building damage, time of blockage of escape routes, and time available for escape, are included in a model annex, or a set of computer instructions which would continuously test the fire development in a building to determine when the critical conditions occur. For instance, an annex question might be included to ask when one or all fire exits become blocked or when the lethal products of combustion have reached the location of human occupants. Of course each user may want to develop specific questions that reflect his need for information about the fire development. They permit automatic assessment of the frequency of occurrence of risk conditions, and thus permit use of the finite state fire models to answer specific questions about building firesafety.

Limitations

Finite state models provide information based on step changes of fire and products of combustion developments.

As a result, the model does not provide any information about what happens in between realms. This is why it is important to initially select realms that satisfy many purposes.

Furthermore, a finite state model of fire and the products of combustion development processes provides a statistical representation of those processes. Thus, a single fire simulation is simply one of many likely fire developments, just as an actual fire is simply one of many types of fires that could occur. Statistically significant (and therefore useful) results are obtained with a finite state model when many simulations are generated on a computer and their statistics recorded. And, as mentioned earlier, the use of a model annex which asks appropriate questions about the fire development and the fire resistance of the building, provides statistical answers of what might be expected of the general population of buildings and furnishings.

Bibliography

References Cited

Berlin, G. N., et al. 1978. "Firesafety Systems Analysis for Residential Occupancies." Final Report (H-2316). National Fire Protection Association. Quincy, MA.

Fitzgerald, Robert W. 1979. Author's Personal Communications, Worcester Polytechnic Institute. Worcester. MA.

Kaiser, J. 1979/80. "Experiences of the Gretener Method." *Fire Safety Journal*. Vol 2. 1979/80.

Nelson, Harold E. 1979. Author's Personal Communications, National Bureau of Standards. Washington, DC.

Powers, Robert W. 1971a. "New York Office Building Fire." *Fire Journal*. Vol 65, No 1. pp 18-23, 87.

Powers, Robert W. 1971b. "Office Building Fire, 919 Third Avenue, New York City." *Fire Journal*. Vol 65, No 2. pp 5-7, 13.

Additional Readings

Apostolakis, G., *Mathematical Methods of Probabilistic Safety Analysis*, University of California, Los Angeles, CA, 1974.

Barlow, R. E., and Chattergee, Purmendu, *Introduction to Fault Tree Analysis*, Operations Research Center; College of Engineering, University of California, Berkeley, CA, 1973.

Chattergee, Permendu, *Fault Tree Analysis: Reliability Theory and Systems Safety Analysis*, University of California, Berkeley, CA, 1974.

General Services Administration, "Technical Papers Given at the Nov., 1974 International Conference on Firesafety in High-Rise Buildings," GSA, Washington, DC, 1974.

McCormack, John W., "Project Factors and Influences Integral to the Development of Fire Protection Solutions," SFPE TR 84-4, Society of Fire Protection Engineers, Boston, MA, 1984.

Nelson, Harold E., "Directions to Improve Application of Systems Approach to Fire Protection Requirements for Buildings," SFPE TR 77-8, Society of Fire Protection Engineers, Boston, MA, 1977.

Nelson, Harold E., "Credible Engineering Methodologies as a Solution to Bridging the Fire Safety Technology Gap," NBS Conference Paper, National Bureau of Standards, Gaithersburg, MD, Nov. 10, 1983.

Nelson, H. E., and Shibe, A. J., "A System for Fire Safety Evaluation of Health Care Facilities," NBSIR-78-155-1, Center for Fire Research, National Bureau of Standards, Washington, DC, 1980.

Watts, Jack L., and Milke, J. A., "A Study of Fire Safety Effectiveness Statements," USFA Contract No. 7-35563, University of Maryland, College Park, MD, 1979.

CONCEPTS OF EGRESS DESIGN

Revised by Dr. John L. Bryan

Means of egress and their design should be based upon an evaluation of a building's total fire protection system and an analysis of the population characteristics and hazards of the occupants of that building. The means of egress design should be treated as an integral part of the "total" system that provides reasonable safety to life from fire.

This chapter covers the fundamental concepts of good egress design that are the basis for NFPA 101®, *Code for Safety to Life from Fire in Buildings and Structures (the Life Safety Code®)*, (hereinafter referred to as NFPA 101). This standard governs good practices to provide life safety features that can be designed as integral parts of new and existing buildings and structures to provide reasonable safety to occupants in fire emergencies. The components of good means of egress are discussed in some detail, with their functions and relationships one to another in the total concept of proper egress design. Computer modeling and simulation to assist the egress design process are also discussed.

FUNDAMENTALS OF DESIGN

The approach to designing means of egress first requires a familiarity with the reaction of people in fire emergencies. These reactions can differ widely depending upon the physical and mental capabilities and conditions of building occupants. The psychological and physiological factors affecting the use of exits during emergencies are being identified and measured in research studies.

Patterns of movement of people, singularly and in crowded conditions, must also be understood. For example, in buildings used as schools or theaters (housing highly mobile occupants), studies have shown certain reproducible flow characteristics from persons exiting buildings. These predictable flow characteristics have fostered computer simulation and modeling to aid the egress design process; however, no number of practical exit facilities can prevent injury or loss of life to persons if the occupant egress flow is inhibited or prevented by the building itself, personnel, or fire and smoke conditions.

Dr. Bryan is Chairman, Department of Fire Protection Engineering, at the University of Maryland, College Park, MD.

Human Factors

The design and capacity of passageways, stairways, and other components within the total means of egress is related to physical dimensions of the human body. The tendency of people to avoid bodily contact with others should be recognized as a major factor in determining the number of persons who will occupy a given space at any given time. Given a choice, people usually automatically establish "territories" to avoid bodily contact with others.

Studies have shown that the majority of adult men measure less than 20.7 in. (520 mm) at the shoulder with no allowance for additional thicknesses of clothing (Fruin 1977). A "body ellipse" concept is used to develop the design of pedestrian systems. The major axis of the body ellipse measures 24 in. (609 mm), whereas the minor axis is considered 18 in. (457 mm). This ellipse equals 2.3 sq ft (0.21 m^2), which is assumed to help determine the maximum practical standing capacity of a space.

The movement of persons results in a swaying action which varies from male to female and also, depending upon the type of motion, involves stairs, free movement, or dense crowds. Body sway has been observed to range 1½ in. (38 mm) left and right during normal free movement. Where movement is reduced to shuffling in dense crowds and movement on stairs, a total sway range of almost 4 in. (101 mm) has been observed. In theory, this indicates that a total width of 30 in. (762 mm) would be required to accommodate a single file of pedestrians traveling up or down stairs (Pauls 1977).

Crowding people into spaces where less than 3 sq ft of space (0.28 m^2) per person is available under nonemergency conditions may create a hazardous condition. When the average area occupied per person is reduced to 2¾ sq ft (0.25 m^2) or less, contact will be unavoidable. Needless to say, under the psychological stresses imposed during a fire emergency, such crowding and contact could contribute to crowd pressures, resulting in injuries. When a queue occurs because of an artificial, temporary situation or because of some permanent design feature, crowd control becomes difficult and the well being of individuals is threatened.

Factors Affecting Movement of People

There are several factors which determine how quickly people may pass through the means of egress.

Studies in level walkways have shown that an average walking speed of 250 fpm (76 m/min) is attained under free flow conditions with 25 sq ft (2.3 m^2) of space available per person. Speeds below 145 fpm (44 m/min) show shuffling which restricts motion. Figure 7-3A, adapted from Research Report No. 95 of the London Transport Board (LTB 1958), shows the rate of speed reduction for space concentrations of less than 7 sq ft (0.65 m^2) per person. Speeds of less than 145 fpm (44 m/min) result in shuffling, and finally a jam point is reached with one person every 2 sq ft (0.18 m^2). A significant nonadaptive behavior possibility exists whenever egress movement is restricted, and the problem becomes urgent under fire exposure conditions, especially when there is more than one person every 3 sq ft (0.28 m^2).

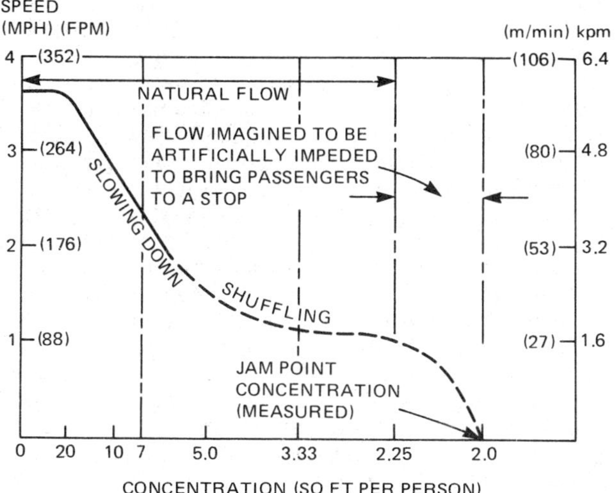

FIG. 7-3A. Speed in level passageways (LTB 1958). SI units: 1 fpm = .305 m/min; 1 sq ft = 0.093 m^2

Calculations of flow rates using velocity (fpm or m/min) and density (people per square foot or square meter) will reveal flow (people per minute per foot or meter of width) increases as the pedestrian area decreases. The flow increases will continue until forward movement becomes restricted to the point that the flow begins to drop. Interestingly, observations of flow rates in one study noted the same flow rate sometimes occurred even though walking speeds of people were significantly different. Investigation revealed that the rate of decrease in speed accompanied with an increase in density results in uniform flow rates over a wide range of conditions.

A study of footways indicates that for passageways over 4 ft (1.2 m) wide, flow rates are directly proportional to width. The London Transport Board, Research Report No. 95 (LTB 1958), determined the flow rate in level passages to be 27 persons per minute per ft (0.30 m) width; travel down stairways was determined at 21 people per minute per ft (0.30 m) width whereas upward travel was reduced to 19 persons per minute per ft (0.30 m) width. Where the width of a footway is less than 4 ft (1.2 m), the flow rate depends upon the number of possible traffic lanes. Absolute maximum flow rates occur when approximately 3 sq ft (0.28 m) is occupied per person, which is applicable to both level walkways and stairs. However, Pauls, in observed and measured evacuations, has empirically determined the maximum flow rates down stairs in high rise buildings to occur when from 4 to 5 sq ft of space (0.37 to 0.46 m^2) is occupied per person as shown in Figure 7-3B (Pauls 1977). An observation confirmed by two separate studies indicates that when flow in opposite directions takes place in a passageway (up to the point where two flows are of equal magnitude in opposite directions), there is no significant reduction in total flow below that which would be predicted on the basis of unidirectional flow in the same passageway.

Further, observations of flow rates at short passageways (less than 10 ft or 3.05 m long) indicate flow can be 50 percent greater than through a long passageway of the same width. Minor obstructions within a passageway do not appear to have a significant effect on flow. Observations made within a 6 ft (1.82 m) wide passageway indicated no effect on flow rates when a 1 ft (0.45 m) projection was introduced. A 2 ft (0.61 m) projection (33 percent reduction in width) reduced the flow rate approximately 10 percent. A major obstruction though, such as occurs at a ticket booth or turnstile may interrupt the movement of people and reduce flow rates.

Corners, bends, and slight grades (up to 6 percent) are apparently not factors in determining flow rates. A slight reduction in speed does occur; however, the flow rate is maintained by an increased concentration of persons.

A center handrail or mullion, which may divide a passageway into narrower sections, can further reduce the capacity of the passageway. In one study, the observed capacity of a 6 ft (1.82 m) wide stairway revealed a reduction from 130 to 105 persons per minute after installation of a center handrail.

Except for the very young and the very old, age does not appear to be a significant factor in determining travel speed. Studies have shown a significant reduction in walking speeds for persons over 65 years of age. Studies have further revealed that a 40 percent increase is possible in the normal walking speed, which tends to discount this factor as a major influence on flow rates (Fruin 1977).

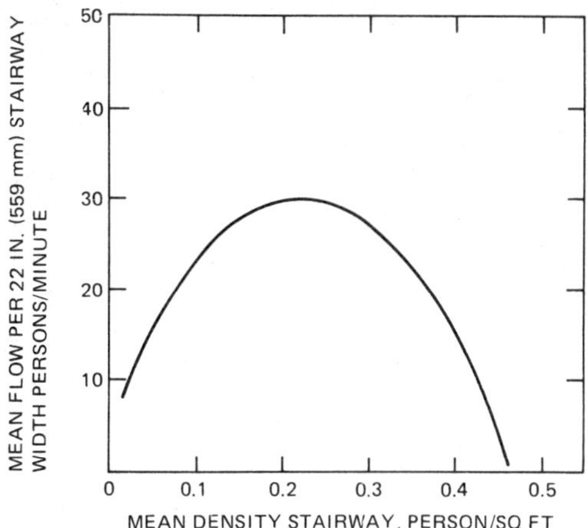

FIG. 7-3B. Effect of density on flow down exit stairways in evacuations of high office buildings (Pauls 1977). SI units: 1 sq ft = 0.93 m^2.

Methods of Calculating Exit Width

Two major principles are used to determine the necessary exit width. They are based on anticipated population characteristics identified with a specific occupancy.

The Flow Method: This method uses the theory of evacuating a building within a specified maximum length of time. Flow rates have traditionally been set at 60 persons per 22 in. (559 mm) width per minute through level passageways and doorways. The flow method may be applied in assembly occupancies (theaters, for example) and educational occupancies where people are alert, awake, and assumed to be in good physical condition. Figure 7-3C illustrates the flow time in seconds relative to the effective stair width per person and the units of width.

Pauls' effective stair width concept advocates the consideration of only the portion of the stair utilized in effective movement by the occupants as observed in functional and practice evacuations. This width is established with 6 in. (150 mm) clearance from each side wall of the stair. (The effective width concept of exit design is included in the Appendix of the 1985 edition of NFPA 101).

The Capacity Method: This method is based on the theory that sufficient numbers of stairways should be provided in a building to adequately house all occupants of the building within the stairways without requiring any movement (flow) out of the stairways. In theory, assuming a stairwell provides a safe and protected area for all occupants within the protective barrier created by the stairway enclosure, evacuation of the building may then be more leisurely, permitting people to travel at a rate within their physical ability. The capacity method recognizes that evacuation from high rise buildings is physically very demanding. Further, evacuation of a health care facility is likely to be slow; thus, design criteria are established to permit holding occupants within exits or areas of refuge.

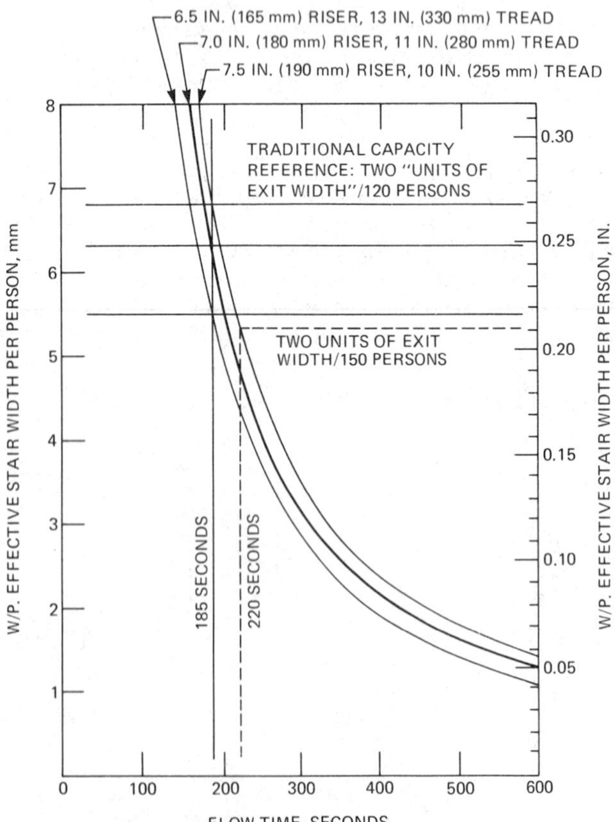

FIG. 7-3C. Relationship between effective stair width and units of exit width per person and flow time for three stair geometries (Pauls 1984a).

Application

The capacity and flow methods may both be applied to efficient egress design depending upon specific circumstances. Where people are expected to be physically or mentally sick, aged, asleep, or incapacitated in any way, evacuation and use of the flow method is unwise. Therefore, the capacity method that provides a place for everyone within an area of refuge is the appropriate method.

There is little time between an alert and the use of an exit in assembly occupancies, and maximum flow rates that cause reductions in the area used by each person may result in reduced traffic flows. On the other hand, the control of children in an educational setting, coupled with their familiarity with the surroundings, their presumed high physical capabilities and their experience with a program of drills, should allow rapid evacuation times. The flow method appears to have its application in those occupancies where people are considered alert, awake, and of normal physical ability. Pauls has reviewed the historical and current literature relative to the principles of people movement, exit width determination, and the design of the means of egress (Pauls 1984b).

Design of Means of Egress

Designing a means of egress involves more than numbers, flow rates, and densities. Safe exit from a building requires a safe path of escape from the fire environment. The path is arranged for ready use in case of emergency and should be sufficient to permit all occupants to reach a safe place before they are endangered by fire, smoke, or heat. Proper exit design permits everyone to leave the fire endangered areas in the shortest possible time with efficient exit use. If a fire is discovered in its incipient stage and the occupants are alerted promptly, effective evacuation may take place.

Evacuation travel distances are related to the content fire hazard; the higher the hazard, the shorter the travel distance to an exit.

Depending upon the physical environment of the structure, the characteristics of the occupants, and the detection and alerting facilities, fire or smoke may prevent the use of one means of egress; therefore, at least one alternate means of egress (remote from the first) is essential. Provision of two separate means of egress is a fundamental safeguard except where a building or room is too small and arranged so that a second exit would not provide an appreciable increase in safety. There are no advantages to separate means of egress if there is travel through a common space or use of common structural features that result in the loss of the two distinct and physically separate means of egress.

One example of a "common" structure is a multistory building where scissors stairs are used (two stairs enclosed within a common shaft, separated by a partition common to both stairs). Scissors stairs are sometimes used to provide the required exit capacity while minimizing the loss of valuable floor space. However, where a set of scissors stairs is provided as the only means of egress when "two remote exits are required," the fundamental principle of two separate means of egress design may be violated. If the common partition between the stairs fails, it would result in the simultaneous loss of both exits during a fire, leaving no alternate means of egress. Therefore, with scissors stairs the validity of the two separate means of egress depends upon the design characteristics and construction of the common partition. (See Fig. 7-3D.)

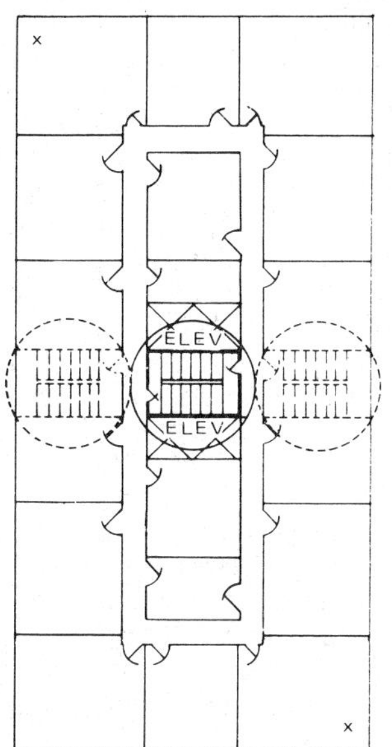

THIS SET OF SCISSORS STAIRS PROVIDES THE SAME DEGREE OF REMOTE EXIT OR ENTRANCE DOORS AS THE CIRCLED STAIRS SHOWN BY DOTTED LINES—TRAVEL DISTANCE FOR ALL OCCUPANTS IS THE SAME, EVEN IF THE DOTTED EXIT STAIRS WERE LOCATED AT OPPOSITE CORNERS DENOTED BY THE CROSS MARK. SPACE IS SAVED, HOWEVER THE INTEGRITY OF THE SEPARATION OF THE 2 SCISSORED STAIRS MAY REMAIN IN QUESTION.

FIG. 7-3D. Advantages and disadvantages of scissors stairs versus conventional stairs.

In some proposed egress designs, it is considered acceptable to discharge all the exits through a single lobby space at the street level, even though this procedure results in egress travel through a common space. This design philosophy further presumes that the lobby may be considered a safe area for all future egress needs during the life of the building. Where two remote means of egress are required, this type of egress design is unsuitable.

NFPA 101 limits openings in exit enclosures to those necessary for access to the enclosure from normally occupied spaces and for egress from the enclosure. Penetration of enclosures by ducts or other utilities constitutes a point of weakness and may result in contamination of the enclosure during a fire. Furthermore, it is not good practice to use exit enclosures for any purpose which could interfere with their value as exits. For example, piping for flammable liquids or gases should not be routed through such spaces.

It has been estimated that 12 million people within the U.S. have limited mobility due to physical disabilities. More than 250,000 persons are confined to wheelchairs. Each year, an estimated 100,000 children are added to the total because of birth defects. The removal of handicapped persons is an important consideration in the design of an emergency means of egress from a building. A 32 in. (813 mm) doorway is considered the minimum width to accommodate a person in a wheelchair. Since handicapped employees or visitors may be found in all types of buildings special life safety considerations are indicated.

LIFE SAFETY CODE®

The *Life Safety Code* (formerly the Building Exits Code), dating from 1927, revised and reissued in successive editions, is developed by the NFPA Committee on Safety to Life, a representative group dedicated to safety of life from fire. NFPA 101 is primarily concerned with the control of conditions which threaten the lives of individuals in building fires. This objective is different from fire protection provisions in building codes, which are concerned with the preservation of property in addition to preservation of life.

Adequate means of egress alone are not a guarantee of life safety from fire. They provide no protection to an individual whose own carelessness, such as setting his or her own clothes on fire, causes a threat to life. Neither do sufficient means of egress alone provide adequate protection in occupancies such as hospitals, nursing homes, prisons, and mental institutions, where occupants are confined, or are physically or mentally unable to escape without effective and immediate assistance. NFPA 101 does recognize such situations and provides life safety measures, including low flame-spread and reduced smoke-producing materials for interior finish. In addition, automatic sprinkler and smoke control systems called for by NFPA 101 are designed to restrain the spread of fire and smoke and thus help to "defend" the occupants within an area of refuge until help comes to assist them in the use of exits or until the fire is extinguished.

In general, saving building occupants from a fire requires the following principles, all of which are identified in NFPA 101:

1. A sufficient number of properly designed, unobstructed means of egress of adequate capacity and arrangement.
2. The protection of the means of egress against fire, heat, and smoke during the egress time determined by the occupant load, travel distance, and exit capacity.
3. The provision of alternate means of egress for use if one means of egress is blocked by fire, heat, or smoke.
4. The subdivision of areas by proper construction to provide areas of refuge in those occupancies where total evacuation is not a primary precaution.
5. The protection of vertical openings to limit the operation of fire protection equipment to a single floor.
6. The provision of detection or alarm systems to alert occupants and notify the fire department in case of fire.
7. The adequate illumination of the means of egress.
8. The proper marking of the means of egress and indication of directions.
9. The protection of equipment or areas of unusual hazard which could produce a fire capable of endangering the egressing occupants.
10. The initiation, organization, and practice of effective drill procedures.
11. The provision of instructional materials and verbal alarm systems in high density and high life hazard occupancies to facilitate adaptive behavior.
12. The use of interior finish materials that prevent a high

flame spread or dense smoke production that could endanger egressing occupants.

Figure 7-3E illustrates some of the principles of exit safety.

NFPA 101 recognizes that full reliance cannot be placed upon any single safeguard since any single protective feature may not function due to mechanical or human failure. For this reason redundant safeguards, any one of which will result in a reasonable level of life safety, should be provided. NFPA 101 also recommends the special protection of hazardous areas and specifies where automatic sprinkler, automatic detection, and other protective systems are required.

The Code is widely used as a guide to good practice and as a basis for local laws or regulations. It differs from building codes, since it generally provides little distinction between the different classes of building construction. However, where total evacuation of a building is not practical, due either to the occupant characteristics or the building environment, the construction type becomes an important variable and should be considered.

The Code also recognizes that all habitable buildings contain sufficient quantities of combustible contents to produce lethal quantities of smoke and heat (Abbott 1971; Lathrop et al 1975). In addition, casualty studies have established the toxic properties of smoke to be the principle hazard to life (Karter 1984), and this hazard is recognized in the Code provisions.

The Code is intended to be applied to both new construction and existing buildings, and is designed to provide a reasonable level of life safety from fire in both types of buildings. The authority having jurisdiction is given considerable latitude in achieving conformance with existing buildings. Each existing building represents a special situation requiring individual attention for the most effective and economical method of a reasonable level of life safety.

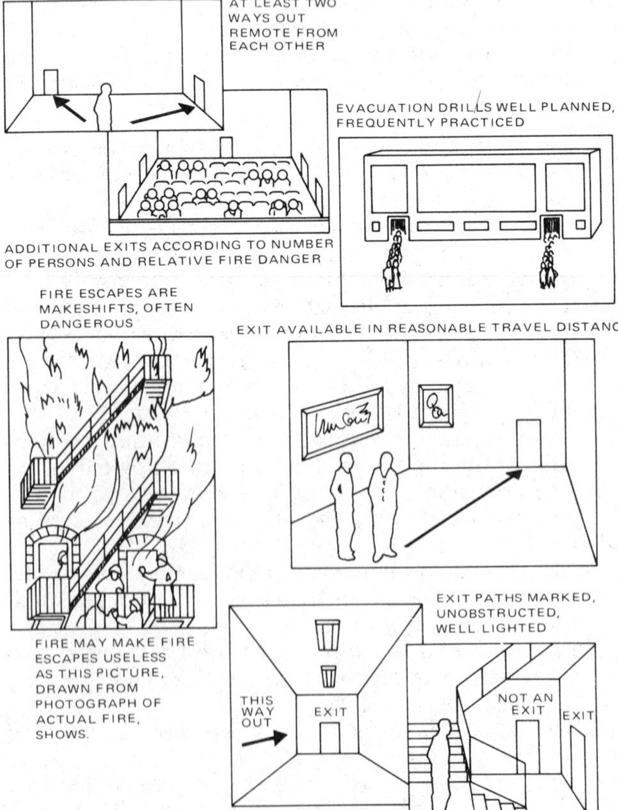

FIG. 7-3E. Principles of exit safety.

The argument that buildings constructed according to all legal requirement many years ago are sufficiently safe now should not be accepted. If the economic cost of reasonable life safety is judged to be prohibitive, the occupancy or the structure should be changed or prohibited because there is no justification for subjecting building occupants to an unreasonable level of peril from a fire.

There may be a variety of opinions and the possibility of disagreement as to what constitutes reasonable life safety from a fire in any given case. It is not possible to guarantee 100 percent life safety to occupants from a fire, but beyond certain conditions the building becomes hazardous to the life of the occupants in a fire. How should the authority having jurisdiction establish the minimum conditions? NFPA 101 provides guidance in such decisions with the help of studies of major loss of life fires (Sharry 1974; Best & Demers 1982), fire development research (Abbott 1971; Powers 1971), personnel evacuation (Keating et al 1978; Pauls 1975 & 1978), and human behavior (Phillips 1974; Bryan 1982).

The Code examines the various occupancy populations according to their perceived life safety hazard, which includes psychological and sociological variables in addition to the physiological and environmental factors. These occupancy classifications are: (1) assembly, (2) residential, (3) health care, (4) educational, (5) detention/correctional occupancies, (6) mercantile, (7) business, (8) industrial, (9) storage, and (10) unusual structures.

Separate and distinct means of egress provisions are made for each occupancy classification, with the various occupancy subgroups included. These classifications, based on the perceived hazard to life safety from a fire, often differ from building code occupancy classifications. For example, mercantile and office occupancies are often grouped together in building codes. However, there appears to be an increased hazard to life in mercantile properties, resulting from the displays of combustible merchandise, the greater density of population, and the transient character of most of the occupants. These factors are not usually found in office and educational buildings, which have a relatively low combustibility content, a lower population density, and normally alert occupants who are in the building daily and presumably have the opportunity to familiarize themselves with the means of egress through functional use and drills.

HAZARDS OF CONTENTS

An evaluation of the hazard of the building contents must take the relative probability of the ignition of combustibles, spread of flames and heat, the probable smoke and gases expected to be generated by the fire, and the possibility of a fire related explosion or other structural failure endangering occupants. The degree of hazard is usually determined by the flammability or toxicity of the contents and the processes or operations conducted in the building. Most NFPA 101 requirements are based on the exposure created by contents with ordinary hazard. Special requirements for buildings with high hazard contents usually consist of special protection systems, construction isolation of the hazard area, reduced travel distances, and additional means of egress.

Influence of Occupancy

To assist in evaluating the content hazards, the Code establishes three classifications of contents: low, ordinary, and high hazard. They are discussed below:

Low Hazard Contents: These are contents of such low combustibility that no self-propagating fire can occur in them; consequently, the only probable danger requiring the use of emergency exits will be from panic, fumes or smoke, or fire from some external source.

Ordinary Hazard Contents: These are contents that are liable to burn with moderate rapidity and to give off a considerable volume of smoke, but from which neither poisonous fumes nor explosions are to be feared in case of fire. This class includes most buildings and is the basis for the general requirements of the Code.

High Hazard Contents: These are contents that are liable to burn with extreme rapidity or from which poisonous fumes or explosions are to be feared in the event of fire. Examples are occupancies where flammable liquids or gases are handled, used, or stored; where combustible dust explosion hazards exist; where hazardous chemicals or explosives are stored; where combustible fibers are processed or handled in a manner to produce combustible flyings; and similar situations.

Influence of Building Construction and Design

A building of fire resistive construction is designed to permit a burnout of contents without structural collapse. Fire resistive design does not assure the life safety of the occupants of such buildings (Sharry 1974; Phillips 1974). However, the ability of a structural frame to maintain building rigidity under fire exposure is important to the maintenance of the fire resistance protection of exit enclosures. Where a two hour fire rated exit enclosure is required, a fire resistive structural frame capable of withstanding stresses imposed by fire for a similar time period is also necessary. It is inconsistent to provide a two hour exit enclosure in a building having a structural frame rated at less than one hour for example, unless special construction precautions are taken to prevent structural failure of the building from adversely affecting the protective construction of the exit enclosures.

The protection of vertical openings is one of the most significant factors in the design of multistory buildings from the standpoint of life safety and exit design. Because of the natural tendency of fire to spread upward within a building, careful attention to details of design and construction are required to minimize this effect. One of the greatest hazards to life safety results from fires which start below the occupants and the means of egress, such as in basements or on the level of exit discharge. Similarly, fires in multistory buildings may result in smoke spread into enclosed exits prior to evacuation (Sharry 1974; Phillips 1974). Conversely, escape is relatively simple from fires which occur above the occupants, provided sufficient warning is given and adequate means of egress are available.

The influence on the life safety of the occupants by the materials used in building construction depends primarily upon whether or not the materials will propagate flame, support combustion, or create dense amounts of smoke when exposed to a fire initially involving the building contents. The use of some materials as insulation, for example, could contribute to rapid flame and dense smoke production spread. Masonry walls enclosing a wood frame interior provide no increased occupant life safety compared with a total wood frame structure.

Exit requirements are based on buildings of conventional design; unusual buildings call for special consideration, for example, windowless buildings or buildings with unopenable windows. Windows provide a number of advantages in a fire. Persons at openable windows have access to fresh air, can see fire department rescue operations in progress, are able to communicate verbally and visually with rescue personnel, and thus may be less subject to stress and anxiety. Windows provide an emergency means of escape and provide accessibility to the building by the fire department for rescue and fire fighting. Automatic sprinklers are considered a primary requirement for life safety in windowless buildings, buildings with unoperable windows, and underground structures.

Influence of Interior Finish, Furnishings, and Decorations

The rapid spread of flame over the surface of walls, ceilings, or floor coverings may prevent occupant use of the means of egress. In general, NFPA 101 limits the flame spread index classification of interior finish materials on walls and ceilings to a maximum of 200, based on results of tests conducted in accordance with NFPA 255, *Method of Test of Surface Burning Characteristics of Building Materials (Tunnel Test)*. Lower ratings are prescribed for the interior finish materials used in exits and the exit access. (Select grade red oak produces a flame spread index classification of 100.) Lower flame spread index classified materials are also required in certain areas in individual occupancies. A fire retardant coating may be used on existing interior finish materials to reduce the rate of flame spread. In areas protected with automatic sprinklers, the use of materials with higher flame spread index classifications is sometimes permitted. Table 7-3A is a summary of the interior finish requirements contained in NFPA 101 for the various occupancy classifications.

The flame spread of floor coverings is evaluated by the Code through the use of NFPA 253, *Standard Method of Test for Critical Radiant Flux of Floor Covering Systems Using Radiant Heat Energy Source*. Two classes of floor coverings are established—Class I finishes with a minimum critical radiant flux of 0.45 W/cm^2, and Class II finishes having a minimum critical radiant flux of 0.22 W/cm^2.

Furnishings and decorations—particularly furnishings—play an increasingly important role in loss of life by fire. Decorations can be treated with a flame retardant. Furnishings, on the other hand, are difficult to control and regulate as a fire hazard, since they are not attached to or part of the building construction or the interior finish materials. Furnishings are moved, refurbished, and replaced; however, there is now a test procedure for measuring the combustibility of upholstered furniture and its susceptibility to ignition (Krasny et al 1981). The U.S. Consumer Products Safety Commission (CPSC) also has a

TABLE 7-3A. Summary of Life Safety Code Requirements for Interior Finish

	Class of Interior Finish*		
Occupancy	Exits	Access to Exits	Other Spaces
Places of assembly—Class A†	A	A	A or B
Places of assembly—Class B‡	A	A	A or B
Places of assembly—Class C§	A	A	A, B, or C
Educational	A	A	A, B, or C
Educational—unsprinklered open plan buildings‖	A	A	A or B
Flexible plan buildings#	A	A	C or low height partitions
Institutional, existing—hospitals, nursing homes, residential-custodial care	A or B	A or B	A or B
Completely Sprinklered	A or B	A or B	A, B, or C
Institutional, new—hospitals, nursing homes, residential-custodial care	A	A	A B in individual room with capacity not more than 4 persons
Residential, new—apartment houses	A or B	A or B	A, B, or C
Residential, existing—apartment houses	A or B	A, B, or C	A, B, or C
Residential—dormitories	A or B	A or B	A, B, or C
Residential, new—1- and 2-family, lodging or rooming houses		A, B, or C	
Residential, existing—1- and 2-family, lodging or rooming houses		A, B, or C	
Residential, new—hotels	A or B	A or B	A, B, or C
Residential, existing—hotels	A or B	A or B	A, B, or C
Mercantile—Class A**	A or B	A or B	ceilings—A or B existing walls—A, B, or C
Mercantile—Class B††	A or B	A or B	ceilings—A or B existing walls—A, B, or C
Mercantile—Class A or B, sprinklered	A, B, or C	A, B, or C	A, B, or C
Mercantile—Class C‡‡	A, B, or C	A, B, or C	A, B, or C
Business	A or B	A or B	A, B, or C
Business-sprinklered	A, B, or C	A, B, or C	A, B, or C
Industrial	A or B	A, B, or C	A, B, or C
Storage	A, B, or C	A, B, or C	A, B, or C
Unusual Structures	A, B, or C	A, B, or C	A, B, or C

* There are three classes of interior finish in the *Life Safety Code*: Class A, flame spread 0-25, smoke developed 0-450: Class B, flame spread 26-75, smoke developed 0-450: Class C, flame spread 76-200, smoke developed 0-450. When a standard system of automatic sprinklers is installed, an interior finish with a flame spread rating not over Class C may be used in any location where Class B is normally specified, and with a rating of Class B in any location where Class A is normally specified, unless specifically prohibited elsewhere in the *Life Safety Code*.

† Class A Places of Assembly—1,000 persons or more

‡ Class B Places of Assembly—300 to 1,000 persons

§ Class C Places of Assembly—100 to 300 persons

‖Open plan buildings—includes all buildings where no permanent partitions are provided between rooms or between rooms and corridors.

\# Flexible plan buildings have movable corridor walls and movable partitions of full height construction with doors leading from rooms to corridors.

** Class A Mercantile Occupancies—stores having aggregate gross area of 30,000 sq ft (2800 m^2) or more, or utilizing more than 3 floor levels for sales purposes.

†† Class B Mercantile Occupancies—stores of less than 30,000 sq ft (2800 m^2) aggregate gross area, but over 3,000 sq ft (2800 m^2) or utilizing any floors above or below street floor level for sales purposes, except that if more than 3 floors are utilized, store shall be Class A.

‡‡ Class C Mercantile Occupancies—stores of 3,000 sq ft (280 m^2) or less gross area, used for sales purposes on street level only. (A single balcony or mezzanine floor with less than half the area of the street level floor and which is used for sales purposes is not counted as another floor.)

standard for the evaluation of the ignitibility of mattresses (CPSC 1972).

The use of some materials in furnishings results in fuels being easily ignited (by a match, for example) and capable of rapid fire spread with accompanying dense amounts of smoke. Some incidents especially illustrate this—a fire principally involving plastic seating in the BOAC Passenger Terminal at New York's Kennedy Airport in August 1970 (Abbott 1971) and a New York office building fire in August 1970 where two persons died (Powers 1971). A fire in December 1974, at the Sac-Osage Hospital, Osceola, MO, resulted in the death of eight persons (Lathrop et al 1975). In the hospital incident, fire was confined to a single patient room and principally involved two plastic mattresses and two chairs with foam cushions. Heavy quantities of black smoke were produced by the fire prior to discovery. As indicated in Table 7-3A, NFPA 101 established a smoke production level of 450 as the maximum value for Class A, B, or C interior finish materials.

Influence of Psychological and Physiological Factors on Egress

Psychological and physiological conditions of the occupant population must be considered in addition to the physical configuration factors of the building in planning means of egress. Recent studies indicate people usually behave adaptively and often altrustically in the stress of fire conditions (Bryan 1982; Keating & Loftus 1981). A heterogenous collective of persons under the influence of alcohol or drugs, as is sometimes present in a place of public assembly, pose the greatest probability of nonadaptive group behavior with a competitive flight, panic type behavior. Historically, this type of nonadaptive behavior has been documented, although recent studies indicate the phenonomenon is rare in occurrence, and depends upon unique predetermined conditions involving both the population and the physical environment of the structure (Bryan 1982; Keating & Loftus 1981; Keating 1982).

In some cases, evacuation procedures and the creation of areas of refuge within high rise buildings encourages occupant movement upward within the building. The effectiveness of this concept has not been completely validated in actual fire emergencies. Because of the orientation of some people toward total evacuation and escape from the building, it is possible, in spite of instruction, that occupants may attempt to evacuate a building in the conventional "down and out" approach (Krasney et al 1981).

The evacuation procedures in federal high rise office buildings, as directed by vocal alarm systems, have continually obtained the selective movement of personnel in both upward and downward directions (Keating et al 1978). In both of the serious high rise office building fires in Sao Paulo, Brazil, there was upward movement of occupants to the roof when the downward movement was inhibited by smoke and heat (Sharry 1974).

In the MGM Grand Hotel fire at Las Vegas, NV, in November 1980, there was upward movement to areas of refuge in the stairways to the roof and the rooms on upper floors when the downward travel was made untenable by smoke and heat (Best & Demers 1982; Bryan 1982).

All exits need to be conspicuously marked since under fire conditions people are likely to be unfamiliar with the various exits from an area and thus neglect alternate means of egress. It is also important that the means of egress from a building be used as a matter of daily routine so the occupants will be familiar with their location and operation. NFPA 101 requires that, in assembly occupancies, the main exit (which also serves as the entrance) be sized to handle at least one-half of the total occupant load of the building.

There are three critical parameters involved for the effective utilization of the zoned evacuation of personnel to areas of refuge within the building (Sharry 1983). They are: (1) proper construction to provide compartmented areas protected from the effects of fire and smoke; (2) an effective verbal alarm system providing clear and comprehensive instructions, with provision for originating on scene instructions from the fire department (Keating et al 1978); and (3) effective evacuation drills to familiarize the occupants with the functioning of the system.

The concept has been advocated that occupants in fire resistive compartmented buildings used for hotels, motels, apartments, condominiums, dormitories, hospitals and other health care facilities should stay in their rooms rather than evacuate, since the rooms are the most adequate area of refuge (Macdonald 1984).

Influence of Fire Protection Equipment

It is unsuitable to rely on manual or automatic fire extinguishing systems in place of adequate means of egress since fire extinguishing systems are subject to both human and mechanical failures. In addition, building areas may become untenable for human occupancy before the fire extinguishing systems are effective. Under no conditions can manual or automatic fire suppression be accepted as a substitute for provision and maintainance of proper means of egress.

Where a complete standard system is installed, automatic sprinklers are sufficiently reliable to provide a major influence on life safety. In addition to providing an automatic alarm of fire, they quickly discharge water on the fire before smoke has spread dangerously. While automatic sprinklers should never be utilized in place of adequate means of egress, they are recognized in various ways by NFPA 101. The Code permits increased travel distance to exits, the use of interior finish of greater combustibility, and, in health care occupancies, the use of combustible construction in situations where it would otherwise be prohibited. Sprinklers are particularly valuable in dealing with travel distance problems in existing buildings.

Automatic fire detection (fire alarm) systems have a valuable function in notifying building occupants of a fire so they may evacuate promptly. However, these systems are not specified in NFPA 101 to any great extent. Automatic fire detection systems only provide warning of a fire and of themselves do nothing to suppress or limit the spread of fire and smoke. An automatic fire detection system is not a substitute for adequate means of egress.

DEFINITION OF THE TERM "EXIT"

NFPA 101 includes the term "exit" in an overall definition of means of egress. A means of egress is a continuous path of travel from any point in a building or structure to the open air outside at ground level and consists of three separate and distinct parts: (1) the way of exit access, (2) the exit, and (3) the means of discharge from the exit.

An exit access is defined as that portion of a means of egress which leads to the entrance to an exit.

An exit is that portion of a means of egress which is separated from the area of the building from which escape is to be made by walls, floors, doors, or other means which provide the protected path necessary for the occupants to proceed with reasonable safety to the exterior of the building. An exit may comprise vertical and horizontal means of travel such as doorways, stairways, ramps, corridors, and passageways.

Exit discharge is that portion of a means of egress between the termination of the exit and a public way.

Figure 7-3F illustrates the relationship in a building of these three areas.

The types of permissible exits are: doors leading directly outside or through a protected passageway to the outside, smokeproof towers, interior and outside stairs, ramps, escalators in existing buildings, and moving walkways. Elevators are not accepted as exits. See Figures 7-3G and 7-3H for illustrations of three common types of exit arrangements.

Exits are measured in units of 22 in. (558 mm) width, taken from the average width of a male at shoulder height. The studies of an occupant's movements on stairs during

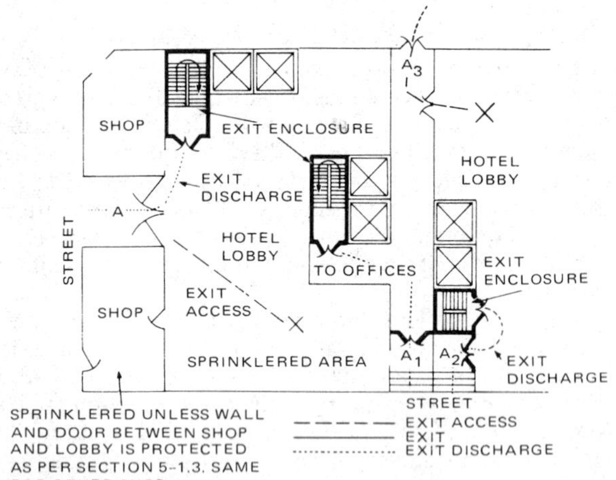

FIG. 7-3F. Examples of exit discharge. To the occupant of the building at the discharge level, the doors at A, A_1, A_2, and A_3 are exits and the path denoted by dashes is the exit access. To the person emerging from the exit enclosure, the same doors and the paths denoted by dotted lines are the exit discharge.

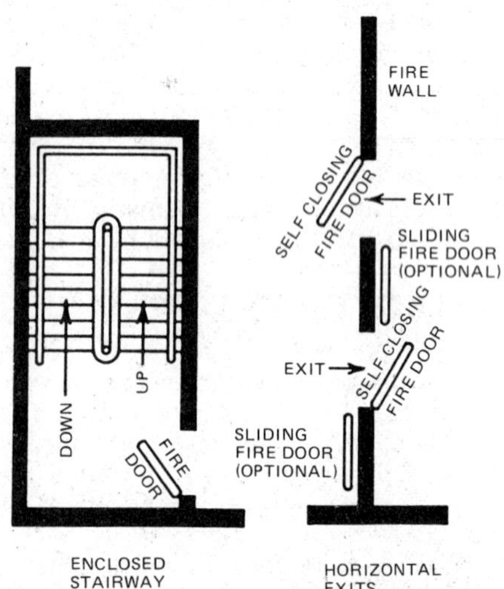

FIG. 7-3G. Plan views of types of exits. Stair enclosure prevents fire on any floor trapping persons above. Smokeproof tower is better, as opening to air at each floor largely prevents chance of smoke in stairway. Horizontal exit provides a quick refuge, lessens need of hasty flight down stairs. Horizontal sliding fire doors provided for safeguarding property values are arranged to close automatically in case of fire when open. Swinging doors are self-closing. Two wall openings are needed for exit in two directions.

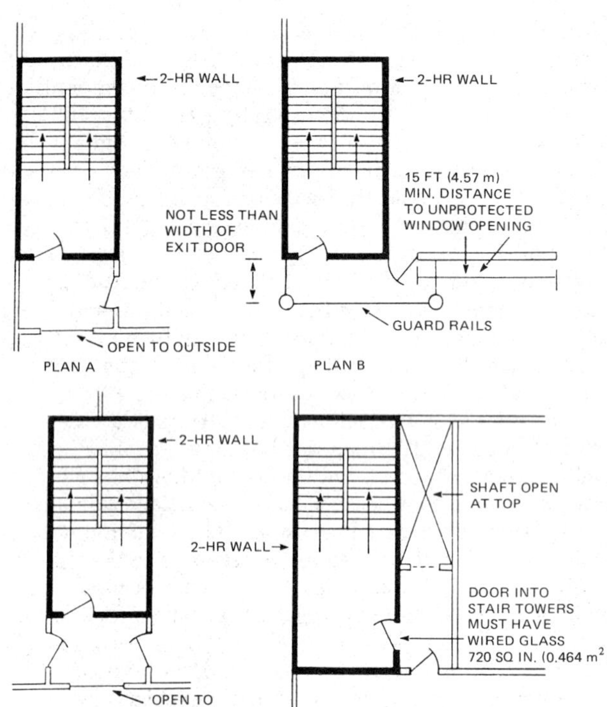

FIG. 7-3H. Four variations of smokeproof towers. Plan A has a vestibute opening from a corridor. Plan B shows an entrance by way of an outside balcony. Plan C could provide a stair tower entrance common to two buildings. In Plan D smoke and gases entering the vestibule would be exhausted by natural or induced draft in the open air shaft. In each case a double entrance to the stair tower with at least one side open or vented is characteristic of this type of construction. Pressurization of the stair tower in the event of fire provides an attractive alternate for tall buildings and is a means of eliminating the entrance vestibule.

both staged and normal evacuations led to advocation of the concept of a 30 in. (762 mm) width for an exit unit (Pauls 1975, 1977 & 1978).

The specific placement of exits is a matter of design judgment, given the specifications of travel distance, allowable dead ends, maximum single path of travel, and exit capacity. NFPA 101 states that exits must be remote from each other. The principle involved is to provide two separate means of egress, so located that occupants can travel in either of two opposite directions to reach an exit. This concept is important when it is necessary for occupants to leave a fire or smoke contaminated area and move toward an exit. If occupants have no choice but to enter the fire environment area to reach an exit, it is doubtful whether they will be able or willing to do so.

The Access to an Exit

The access to an exit is that portion of a means of egress which leads to an entrance to an exit. The access to an exit may be a corridor, aisle, balcony, gallery, porch, or roof. The length of the access establishes the travel distance to an exit—an extremely important feature of a means of egress, since an occupant might be exposed to fire or smoke conditions during the time it takes to reach an exit. The average recommended maximum distance is 100 ft (30.5 m), but this distance varies with the occupancy, depending upon the fire hazard and the physical ability and alertness of the occupants. (See Table 7-3B.) The travel distance may be measured from the door of a room to an exit or from the most remote point in a room or floor area to an exit. In those occupancies where there are large numbers of people in an open floor area or where the nature of the business conducted makes an open floor area desirable (often office buildings), the travel distance is measured from the most remote point in the area to the exits. Conversely, in occupancies where there are only a

TABLE 7-3B. Summary of Life Safety Code Provisions for Travel Distances to Exits

Occupancy	Dead-End Limit, Ft (m)	Travel Limit to an Exit, Ft (m)	
		Unsprinklered	Sprinklered
Places of Assembly	20** (6.1)**	150 (45.7)	200 (70)
Educational	20 (6.1)	150 (45.7)	200 (70)
Open plan	N.R.*	150 (45.7)	200 (70)
Flexible plan	N.R.*	150 (45.7)	200 (70)
Health Care			
New	30	100 (30.5)	150 (45.7)
Existing	N.R.*	100 (30.5)	150 (45.7)
Residential			
Hotels	35 (10.7)	100 (30.5)	150 (45.7)
Apartments	35 (10.7)	100 (30.5)	150 (45.7)
Dormitories	0 0	100 (30.5)	150 (45.7)
Lodging or rooming houses, 1- and 2-family dwellings	N.R.*	N.R.*	N.R.*
Mercantile			
Class A, B, and C	50 (15.2)	100 (30.5)	150 (45.7)
Open Air	0 0	N.R.*	N.R.*
Covered Mall	50 (15.2)	200 (70)	300 (91.4)
Business	50 (15.2)	200 (70)	300 (91.4)
Industrial			
General, and special purpose	50 (15.2)	100 (30.5)	150† (30.5)†
High hazard	0 0	75 (22.9)	75 (22.9)
Open structures	N.R.*	N.R.*	N.R.*
Storage			
Low	N.R.*	N.R.*	N.R.*
Ordinary hazard	N.R.*	200 (70)	400 (121.9)
High hazard	0 0	75 (22.9)	100 (30.5)
Open parking garages	50 (15.2)	200 (70)	300 (91.4)
Enclosed parking garages	50 (15.2)	150 (45.7)	200 (70)
Aircraft hangars, ground floor	20 (6.1)	Varies‡	Varies‡
Aircraft hangars, mezzanine floor	N.R.*	75 (22.9)	75 (22.9)
Grain elevators	N.R.*	N.R.*	N.R.*
Miscellaneous occupancies	N.R.*	100 (30.5)	150 (45.7)

* No requirement or not applicable.
† A special exception is made for one-story, sprinklered, industrial occupancies.
‡ See Paragraph 15-4.2 of *Life Safety Code* for special requirements.
** In Aisles.

few people in small cutoff areas or rooms (such as in hotels and apartments), the travel distance is measured from the door of the room or area to the exits.

In most cases the travel distance can be increased up to 50 percent if the building is completely protected with a standard automatic sprinkler system.

A dead end is an extension of a corridor or aisle beyond an exit or an access to exits that forms a pocket in which occupants may be trapped. Since there is only one access to an exit from a dead end, a fire in a dead end between an exit and an occupant prevents the occupant from reaching the exit. While traveling toward an exit in a smoke filled atmosphere, an occupant may pass by the exit and be trapped in the dead end. In good egress designs, dead end corridors are not utilized. However, the Code permits dead ends in most occupancies, within reasonable limits. Two dead end corridors are illustrated in Figure 7-3I.

The width of an access to exits should be at least sufficient for the number of persons it must accommodate [minimum: new buildings, 36 in. (914 mm), existing buildings, 28 in. (711 mm)]. In some occupancies, the width of the access is governed by the character of activity in the occupancy. One example is a hospital, where patients will be moved in beds; therefore, the corridors in patient areas must be wide enough for a bed to be wheeled out of a room and turned 90 degrees.

A fundamental principle of exit access is the provision of a free and unobstructed way to the exits. If the access passes through a room that can be locked or through an area containing a fire hazard more severe than is typical of the occupancy, the principles of free and unobstructed exit access are violated.

The floor of an exit access should be level. If this is not

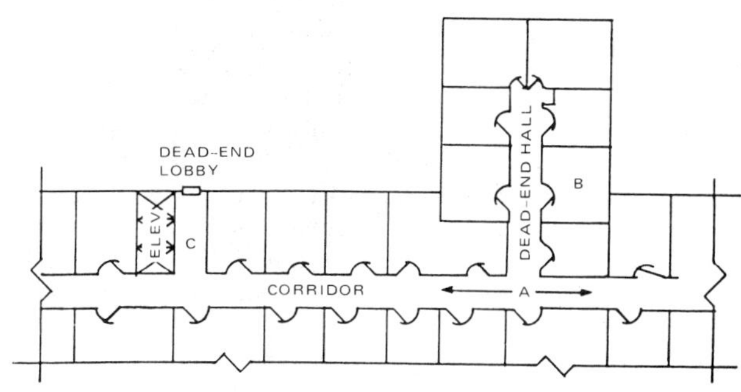

FIG. 7-3I. Examples of two types of dead end corridors.

possible, small differences in elevation may be overcome by a ramp, and large differences by stairs. Where only one or two steps are necessary to overcome differences in level in an exit access, a ramp is preferred because in a crowded corridor people may trip and fall on the stairs if they do not see the steps or notice that persons in front of them have stepped up.

The Discharge of an Exit

Ideally, all exits in a building should discharge directly or through a fire resistive passageway to the outside of the building; but in deference to building owners, operators and designers, NFPA 101 permits a maximum of 50 percent of the exit stairs in some occupancies (such as mercantile and business) to discharge onto the street floor. The obvious disadvantage of this arrangement is that if a fire occurs on the street floor, it is possible for people using the exit stairs discharging to the street floor to be discharged into the fire area. In some occupancies it is impossible to retreat up the exit stairs to a floor above to gain access to another exit because, for security reasons, the doors in the stairway are locked on the inside except at the street floor. Therefore, if any exits discharge to the street floor, the Code requires: (1) that such exits discharge to a free and unobstructed way to the outside of the building, (2) that the entire street floor be protected by automatic sprinklers, and (3) that the street floor be separated from any floors below by construction having a two hour fire resistance rating.

Discharging an exit to the outside is not necessarily discharging to a safe place. If the exit discharges into a courtyard, an exit passageway must be provided from the courtyard. If the exit discharges into a fenced yard, sufficient area must be provided to handle the expected occupant load far enough away from the building so that the occupants will not be exposed to the fire conditions in the building. If the exit discharges into an alley, the alley must be of sufficient width to accommodate the capacity of all the exits discharging into it, and any openings in the building walls bordering the alley should be protected to prevent fire exposure to the occupants proceeding through the alley.

When exit stairs from floors above the street floor continue on to floors below the street floor, occupants evacuating the building may miss the exit discharge door to the street level, continue down the stairway, and enter a floor below the level of exit discharge. Therefore, NFPA 101 requires a physical barrier at the street floor landing to prevent the passing of the level of exit discharge.

Capacity of Unit of Exit Width

The capacity in number of persons per unit of exit width varies with the occupancy—from 30 persons per unit of exit width for health care occupancies to 100 persons per unit of exit width for office buildings and assembly buildings for travel in a horizontal direction. For travel in an inclined direction such as down stairs, the figures vary from 22 persons per unit of exit width in hospitals to 75 persons per unit of exit width in places of assembly. (See Table 7-3C.) The reason for these variations is to establish a consistent total evacuation time in different occupancies, based on the physical ability, mental alertness, age, and sociological roles of the occupants. In occupancies where people sleep or are housed for care, the time taken to reach exits will be greater than in some other occupancies, thus the exits must be of sufficient width to enable egress for nonambulatory occupants and to prevent any waiting to get into the exit. It can be expected that in an assembly occupancy, people will move rapidly to the exits and reach them at about the same time, which requires them to wait before entering the exit. This expected queing time is reduced by the Code provision which requires the main exit to provide at least 50 percent of the total required exit capacity and the provision of a total capacity for all exits of 116 percent of the occupant load.

The capacity of exits is used to establish a consistency of evacuation time on the basis of the rate of travel through a door of 60 persons per minute and down a stairway of 45 persons per minute per unit of exit width, respectively. These figures were established by evacuation counts conducted primarily in federal office buildings (NBS 1935). More recent studies of evacuations in high rise office buildings indicate peak flows of 30 persons per minute and mean flows of 24 persons per minute per unit of exit width down stairways (Pauls 1975, 1977, & 1978).

Occupant Load

Occupant load, or the number of people to be expected in a building at any time for whom exits must be provided, is determined by dividing the gross area of the building or the net area of a specific portion of the building by the area in square feet (or square meters) projected for each person. The amount of floor area projected for each person varies with the occupancy. (See Table 7-3C.) These figures are based on actual counts of people in buildings and reviews of architectural plans. In some situations the actual number of people in a building can be determined at the design stage, in which case this number should be used in the design of the exits. A typical example is an assembly occupancy in which fixed seating is installed. Counting the number of seats provided would obviously give a more accurate figure than multiplying a square foot (or square meter) per person figure by the net floor area.

Computing Required Exit Width

It is necessary to follow these steps to compute the required exit widths from the individual floors of a building: (1) calculate the floor area (net or gross, whichever is applicable); (2) determine from the Code the allowable number of square feet (square meters) per person; (3) divide the number of square feet (square meters) per person into the floor area to determine the number of people for which exits must be provided (occupant load) for that floor; (4) determine from NFPA 101 the capacity of the type of exit(s) to be used for the occupancy being designed; and (5) calculate the number of units of exit width for each type of exit used, based on its capacity.

It should be pointed out that in a multistoried building, if X units of exit width are required from each floor, the stairways serving those floors do not need to be X times the number of floors served in units of exit width. The stairs need only be of sufficient width to serve each floor—but should not be less than the minimum width allowed by the Code.

Street-floor exits may require special treatment, depending upon the occupancy. Some occupancies require street-floor exits to be sized to handle not only the occupant load of the street floor but also a percentage of the load of the exits discharging to the street floor from floors

TABLE 7-3C. Summary of Life Safety Code Provisions for Occupant Load and Capacity of Exits

Occupancy	Occupant Load: Sq Ft (m²) per Person	Capacity of Exits (Number of Persons per Unit of Exit Width): Doors* Outside	Horizontal Exit	Ramp Class A	Ramp Class B	Escalator	Stairs
Places of Assembly	15 (1.39) Net	100	100	100	75	75	75
Areas of concentrated use without fixed seating	7 (2.13) Net						
Standing space	3 (0.91) Net						
Educational		100	100	100	60		60
Classroom area	20 (1.86) Net						
Shops and vocational	50 (4.65) Net						
Day Nurseries with sleeping facilities	35 (3.25) Net						
Health Care		30	30	30	30		22
Sleeping departments	120 (11.15) Gross						
Inpatient departments	240 (22.30) Gross						
Residential	200 (18.58) Gross	100	100	100	75	75	75
Mercantile		100	100			60	60
Street floor and sales basement	30 (2.79) Gross						
Other floors	60 (5.57) Gross						
Storage-shipping	300 (27.87) Gross						
Office areas	100 (9.29) Gross						
Business	100 (9.29) Gross	100	100	100	60	60	60
Industrial	100 (9.29) Gross	100	100	100	60	60	60
Detention and Correctional occupancies	120 (11.15) Gross	100	100	100	100‡ 60†		

* Not more than three risers or 21 in. (533 mm) above or below grade.
† 60 (up).
‡ 100 (down).

above and below. In addition, in those occupancies where floors above and/or below the street floor are permitted to have unenclosed stairs and escalators connecting them with the street floor, the exits must be sufficient to provide simultaneously for all the occupants of all communicating levels and areas. In other words, all communicating levels in the same fire area are considered as a single floor area for purposes of determining the required exit capacity. This identical single fire area factor can have a considerable effect on the sizing of the street-floor exits.

Exits should never decrease in width along their length of travel. Should two or more exits converge into a common exit, the common exit should never be narrower than the sum of the width of the exits converging into it.

Generally, the minimum number of exits is two. However, in certain limited situations, because of a very low occupant load and low fire hazard, one exit may be permitted.

Dynamic Evaluation of Exit Design

Exit design may be evaluated on a dynamic basis using calculated evacuation times. Dynamic evaluation allows for the establishment of goals in terms of evacuation times and certain other key criteria.

For example, the U.S General Services Administration (GSA) has established a goal oriented system approach to building firesafety which utilizes specific sets of goals for normal federal office operations (GSA 1985). The established goals are:

1. All occupants exposed to the fire environment must be able to evacuate to a safe area within 90 seconds of alarm.
2. A portion of this time, not to exceed approximately 15 seconds, can be involved in traveling in a direction toward the fire, for example, in a dead end corridor.
3. All occupants must reach an area of refuge within five minutes of downward vertical travel or within one minute of upward vertical movement.

Further criteria established by the GSA for evaluation uses 3.5 fps (1.07 m/sec) as the rate of horizontal movement. Exit flow rate for level travel is calculated at 60 persons per minute per 24 in. (610 mm) width; 45 persons per minute per 24 in. (610 mm) width down stairs, and 40 persons per minute per 24 in. (610 mm) width up stairs.

Fatigue is judged to become important after five minutes of travel in a downward direction or one minute of travel in an upward direction. Thus, fatigue becomes a "human factor" consideration affecting design.

The time to exit a building can be calculated using information and formulas contained within studies produced by the London Transport Board and the National Research Council of Canada (Ministry of Works 1952; Galbreath 1969).

Calculations are possible based on the number of

persons discharged per unit of exit width per minute. The calculated evacuation time reflects:

1. The time required to fill the stairs with people.
2. The time during which additional persons can enter the stairs from the upper floors.
3. The time required for all persons remaining in the stairs to discharge to the outside.

The following calculation method results in the time for complete evacuation based upon the sum of the three preceding intervals.

$$T = \frac{N + n}{r \times u}$$

where

T = the time in minutes required for complete evacuation by stairs.

N = the number of persons in the building above the first floor.

n = the number of persons who can stand on the stairs at 3 sq ft (0.28 m^2) per person or the number of persons on the floor, whichever is less.

r = the rate of discharge of the stairs in persons per unit exit width per min. (See Table 7-3D.)

u = the number of 22 in. (559 mm) exit units of stair width.

This calculation method assumes occupants will be uniformly distributed over the entire floor area. If, however, all occupants were located 100 ft (30.5 m) from the stairs, additional time would be required. In fact, an additional one-half minute evacuation time would be added to the total.

Further, it is necessary to determine the time required for the person located at the most remote point of the floor under consideration to reach the stairs or exit. If the travel time for this person exceeds the time calculated for all persons to egress by stairs or other exits, this additional time interval becomes a factor in the total evacuation time of the building.

The method of calculation described previously provides guidance in terms of the minimum time necessary for evacuating a building under normal considerations. It is obvious that the loss of an exit stair due to fire or smoke contamination, artificial obstructions to exit paths, etc. could seriously affect evacuation times. From earlier discussions, it was noted that when densities result in less than 7 to 8 sq ft (0.65 to 0.74 m^2) of space per person, forward motion is significantly slowed, as indicated in Table 7-3D.

Table 7-3E records the calculated evacuation times and provides a comparison to actual evacuation times measured as a result of drills conducted during nonemergencies (Galbreath 1968). As noted in Table 7-3E, the actual recorded evacuation times exceeded the calculated times. Similar results have been found using differing formulas for these buildings (Melinek & Booth 1975).

TABLE 7-3D. Relationship Between Concentration of People on Stairs and Forward Movement*

Concentration of People on Stairs sq ft (m^2) per person	Forward Movement, ft per min (m/mm)	Resultant Discharge from Stairs, persons per unit exit width† per min.
2 (0.19)	0 (0.00)	0
2.5 (0.23)	53 (15.2)	39
3.0 (0.28)	75 (22.9)	45
3.5 (0.33)	82 (25.0)	43
4.0 (0.37)	94 (28.6)	43
4.5 (0.42)	106 (32.3)	43
5.0 (0.46)	117 (35.7)	43
5.5 (0.51)	129 (39.3)	43
6.0 (0.56)	139 (42.4)	43
6.5 (0.60)	143 (43.6)	40
7.0 (0.65)	147 (44.8)	39
7.5 (0.70)	150 (45.7)	37
8.0 (0.74)	152 (46.3)	35
8.5 (0.80)	154 (46.9)	33
9.0 (0.84)	156 (47.5)	31
9.5 (0.88)	157 (47.8)	30
10.0 (0.93)	158 (48.2)	29
11.0 (1.02)	158 (48.2)	26
12.0 (1.11)	158 (48.2)	24
13.0 (1.21)	158 (48.2)	22
14.0 (1.30)	158 (48.2)	21
15.0 (1.39)	158 (48.2)	19

* Galbreath 1969.
† unit exit width = 22 in. (559 mm).

COMPUTER SIMULATION AND MODELING OF EGRESS DESIGN

Computer simulation and modeling have become important tools in designing adequate means of egress under a variety of occupancy and structural conditions. The following paragraphs summarize activity to date in applying simulation and modeling techniques to solving egress design problems.

In one study of the critical variables for firesafety in relation to buildings utilized for the housing of the elderly, the problem of fire development and evacuation as a time structured problem was considered (Caravaty & Haviland 1967). This study analyzed the variables related to the occurrence and spread of fire relative to the population in terms of their ability to attain an area of refuge from the effects of the fire. The study established the variables for the fire on a continuum of a "critical time," and the parameters for the survival of the occupants on a continuum of a "reaction time."

The definition adopted for critical time is the elapsed time from the start of the fire to the attainment of the intolerable levels. The definition for reaction time is reported as the amount of time used by the occupant in reacting to the fire situation and achieving safety through evacuating the fire area or by obtaining a place of refuge from the effects of the fire. Thus, within the framework of these definitions, both the parameters of the problem of the human behavior factors involved in a fire incident in a building and the variables of the problem are considered.

This conceptualization of a "critical time" for fire development within a building and the "reaction time" required for occupants to perceive and respond to the fire threat (either by evacuation or movement to an area of refuge) has been thoroughly studied. It has resulted in multiple computer models of this essential egress behavior since Caravaty and Haviland's study in 1967.

A computer model for the simulation of movement of people relative to evacuation through the floor of fire origin has been developed (Stahl 1975). A Markov based model focused on the movement of people through a fire floor from the time of alarm until their safe exit or until

TABLE 7-3E. Time of Evacuation for Certain Existing Office Buildings*

Building No.	Height in Stories	Area of Exit Stairs per Floor, sq ft (m^2)	Width of Exit Stairs, Unit Exit Width	Average Number of Occupants per Floor at Time of Survey	Evacuation Time Calculated by Formula, min	Evacuation Time in Practice Drill, min	Maximum No. of Occupants per Floor By NBC Provisions
1	7	857 (79.6)	10	61	2.1	4½	600
2	7	636 (59.1)	8	108	2.1	5	480
3	9	692 (64.3)	8	133	3.6	4½	480
4	9	408 (37.9)	4	111	6.0	5½	300
5	11	346 (32.1)	4	110	6.7	6½	240
6	11	150 (13.9)	4	100	5.8	7½	240
7	12	314 (29.2)	4	67	4.3	9	240
8	13	319 (29.6)	6	38	1.8	4	360
9	18	260 (24.2)	4	50	5.2	7½	240
10	22	160 (14.9)	2	80	20.0	incomplete	180

* Galbreath 1968.

they became a casualty of the fire. Six variables affecting their movement have been identified: (1) objective location of threatening stimuli in time and space, (2) occupant's prior knowledge of effective egress routes, (3) occupant's perception of the location and severity of the threat, (4) occupant's perception of available alternatives, (5) occupants threat reducing experiences prior to the current movement decision, and (6) the interjection of sudden interpretations to occupant's goal directed behavior.

Stahl has indicated that his decade of effort in modeling the movement of occupants has resulted in an improved computer model and an appreciation of the value of simulation in egress design (Stahl 1982 a&b). Simulation modeling may be appropriate for problems that (1) would otherwise require costly, time consuming and tedious manual effort; (2) cannot be solved through experimentation because of high costs or unacceptable risks to human participants; and (3) for which past experience, intuition, or available data do not provide the proper insight. Figure 7-3J is a comparison of data developed on the effect of variation in density of occupants upon walking speed in two computer simulation programs, and data compiled from two actual pedestrian movement studies.

A modeling technique has been developed which would appear to have potential for comparing the evacuation behavior of occupants assisted by trained personnel, or evacuation without assistance (Berlin 1982a). This model determines the escape potential from an area of a building based on a deterministic approach with a computer algorithm, and evaluates the effects of fire growth and smoke movement on occupants as they try to egress. Occupant evacuation is simulated with evaluation of the effects of the egress design, the density of the occupants in the means of egress, the congestion at door or other restrictions, and the effects of the combustion products on occupants.

The basic components of the original Building Firesafety Model have been applied to the specific problem of the evacuation of board and care homes during a fire occurrence (Berlin et al 1982). This simulation model has added an evaluation of the residents' physical and mental characteristics which affect evacuation. These characteristics are movement variables or impairments and stress behavior, including resistance to assistance, mobility, the need for assistance, and the resident's responsiveness. In addition to residents' characteristics, this model evaluates the staff characteristics critical to providing the required assistance for effective evacuation. Simulations with this model have indicated that residents should be located on the ground floor, near exits. The quickest egress routes should be identified. With increased staff/resident ratios rescuing those farthest away first can reduce evacuation time.

The development and the current applications of the Building Fire Simulation Model (BFSM) and its five submodels have been reviewed. The applications are: (1) the parameter simulation conversion logic, (2) the fire development model, (3) the combustion products model, (4) the escape routing model, and (5) the estimated escape time escape routing model (Fahy 1983). The relationship of these submodels is illustrated in Figure 7-3K.

The concepts of critical time and reaction time as

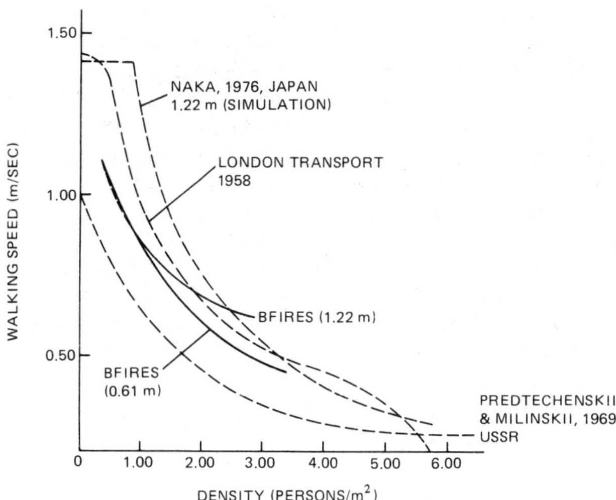

FIG. 7-3J. A comparison of the effect of variation in density of occupants upon walking speed from simulations and studies. The NAKA and BFIRES curves are based on computer simulations. The London Transport and the Predtechenskii & Milinski curves are based on actual pedestrian movement studies. The curves indicate that computer simulation data lies within trends established by real world observation.

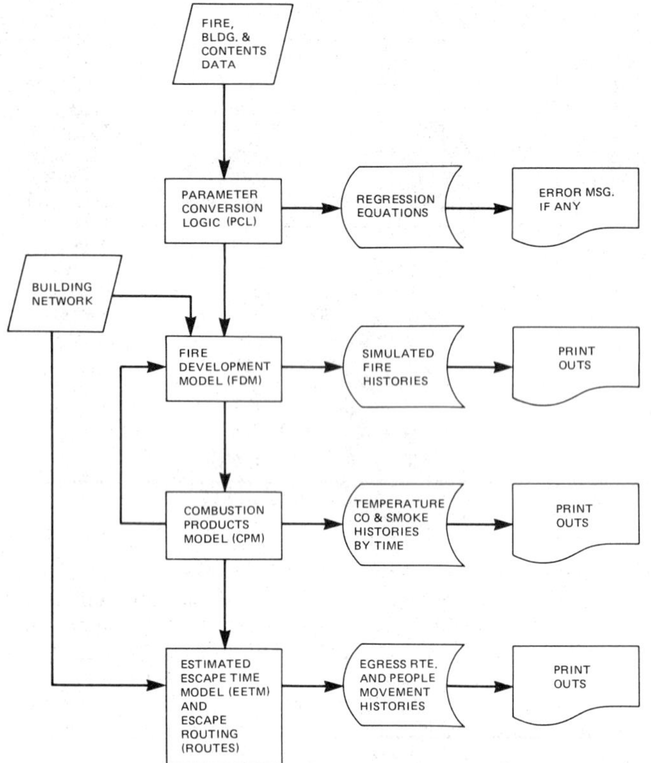

FIG. 7-3K. The Building Fire Simulation Model Design with the Submodels Relationship (Fahy 1983).

originally formulated by Caravaty and Haviland, have been adapted into a determination of the Available Safe Egress Time Model, or ASET (Cooper 1982). This model is a mathematical procedure for the simulation of the conditions which develop between the time of fire ignition and the onset of untenable conditions for human occupancy. Cooper's model thus addresses the fire development and combustion products submodels of the BFSM.

A simulation model has been developed to evaluate the evacuation plans and procedures within a specific building (O'Leary and Gratz 1982). The model requires input data on the physical dimensions of the building areas, the means of egress, and the specified evacuation routes with the number and location of the occupants. The model provides an estimated average evacuation time and an estimated total evacuation time.

Building network evacuation models similar to the model of O'Leary and Gratz have been developed which evaluate the egress environment within the building and the characteristics of the population relative to density and location (Chalmet et al 1982). The model predicts evacuation flows and times as well as identifying queueing problems. The model has been validated with evacuation data from high rise federal office buildings and college dormitories.

An analytical Queing Network Model has also been developed to be used in the analysis of a building design relative to the suitability of the egress system (Smith 1982). This model provides estimates of the evacuation times, the average queue lengths along egress routes, potential impairments, and total egress probability.

EXIT FACILITIES AND ARRANGEMENTS

The following kinds of exit facilities are covered in NFPA 101:

Doors

Doors should swing with exit travel except for small rooms. Vertical or rolling doors are not recognized for use in exits. In places of assembly and in schools panic hardware should be installed on exit doors equipped with latches.

Where doors protect exit facilities, as in stairway enclosures and smoke barriers, they normally must be kept closed to the spread of smoke, or, if open, must be closed immediately in case of fire. Although ordinary fusible link operated devices to close doors in case of fire are designed to close in time to stop the spread of fire, but they do not operate soon enough to stop the spread of smoke. At relatively low temperatures, the smoke accumulation could continue with accumulations reaching untenable levels.

Sometimes people hold self-closing doors open by hooks or by wedges under the door. Doors can also be blocked open to provide ventilation, for the convenience of building maintenance personnel, or to avoid the accident hazard due to swinging doors. The following measures have been provided within the Code to alleviate this undesirable situation:

1. Use of smokeproof towers that protect against smoke even if doors are open.
2. Doors that are normally closed can be equipped to open electrically or pneumatically upon the approach of persons to the door.
3. Doors that are normally closed can be opened and held open manually by monitors, as in schools.
4. Doors that are normally kept open can be equipped with door closers and automatic hold-open devices which release and allow the doors to close on operation of an automatic sprinkler system, automatic fire detection system, and smoke or other products of combustion detection devices.

There are qualifications and limitations applicable to each of these measures, the most important of which is that in the event of electrical failure, the door must close and remain closed unless manually opened for escape purposes.

Another major maintenance difficulty with exit doors is the exterior door which is locked to prevent unauthorized access or for other reasons. NFPA 101 specifies that when the building is occupied all doors must be kept unlocked from the side from which egress is made.

The Code allows a delayed releasing device on exterior exit doors, provided this is permitted by the requirements for the occupancy in question. Where the devices are allowed, the following provisions apply:

1. The building must be protected throughout by an approved and supervised automatic fire alarm or automatic sprinkler system.
2. The release devices are installed only in low or ordinary hazard areas.
3. The devices must unlock upon activation of the fire alarm or automatic sprinkler system.

4. The devices must unlock upon loss of power.
5. The devices must initiate an irreversible process which will free the latch within 15 seconds whenever a force of not more than 15 lb (6.8 kg) is applied to the releasing device, and not relock until the door has been opened. Operation of the releasing device shall actuate a signal in the vicinity of the door.
6. A sign must be placed adjacent to the door that reads: PUSH UNTIL ALARM SOUNDS. DOOR CAN BE OPENED IN 15 SECONDS!
7. Emergency lighting must be provided at the door.

Locks on a door that let people exit but not enter are satisfactory, but even this type may not be satisfactory for security purposes. Possible measures to prevent unauthorized use of exit doors include:

1. Automatic alarm to ring when door is opened.
2. Visual supervision (wired-glass panels, closed circuit television, and mirrors may be used where appropriate).
3. Automatic photographic devices to provide pictures of users.

Of these, only the first two measures have been used to any great extent.

So called exit locks, with a break-glass unit actuated by striking a handle with the hand, are not permitted by NFPA 101 unless installed in conjunction with panic bars. Otherwise they do not comply with the Code provision which reads: "A latch or other fastening device on an exit door shall be provided with a knob, handle, panic bar or other simple device, the method of operation of which is obvious, even in darkness."

Other types of break-glass locks and electrical controls for releasing exits from a central point are not permitted by the Code. The exception is where controls may be necessary as in health care, educational, and detention and correctional occupancies.

A single door in a doorway should not be less than 32 in. (813 mm) wide in new buildings and 28 in. (711 mm) in existing buildings nor more than 48 in. (1.2 m). To prevent tripping, the floor on both sides of the door should have the same elevation.

Panic Hardware

The exit doors in places of assembly and educational occupancies such as schools or motion picture theaters are normally equipped with panic hardware. Basically, panic hardware devices are designed so that they will facilitate the release of the latching device on the door when a pressure not to exceed 15 lb (6.8 kg) is applied to the releasing device in the direction of exit travel. Such releasing devices are bars or panels extending not less than one-half of the width of the door and placed at heights suitable not less than 30 in. (762 mm) nor more than 44 in. (1.1 m) above the floor for the service required.

Fire testing laboratories have tested assemblies which are intended for mounting on or integral with swinging exit doors designed to meet the installation recommendations given in NFPA 101. Panic hardware has also been tested on fire doors to determine the fire resistance of the complete assembly.

Panic hardware is available for use on single and double doors with variations for rim-mounted hardware and mortise or vertical rod devices.

Horizontal Exits

A horizontal exit is a means of egress from one building to an area of refuge in another building on approximately the same level, or a means of egress through or around a fire wall or partition to an area of refuge at approximately the same level in the same building that affords safety from fire and smoke. With a horizontal exit, it is obvious that space must be provided in the area or building of refuge for the people entering the refuge area. NFPA 101 recommends 3 sq ft (0.28 m^2) of space per person with the exception of health care occupancies, where 6 to 30 sq ft (0.56 to 2.79 m^2) of space are recommended. Horizontal exits cannot comprise more than one-half the total required exit capacity except in health care facilities, where horizontal exits can comprise two-thirds of the total required exit capacity. Horizontal exits have been universally applied in health care facilities where the evacuation of patients over stairs is slower and more difficult than taking them through a horizontal exit to a safe area of refuge. A horizontal exit arrangement within a single building and between two buildings is illustrated in Figure 7-3L.

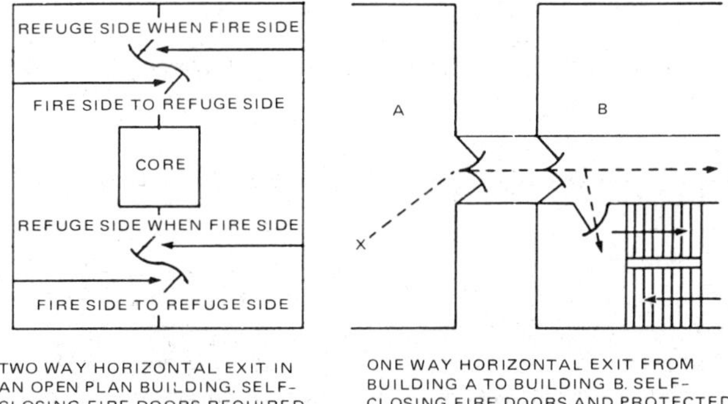

FIG. 7-3L. Types of horizontal exits.

A swinging door in a fire wall provides a horizontal exit in one direction only. Two openings, each with a door swinging in the direction of exit travel, are needed to provide horizontal exits from both sides of the wall. Where property protection requires fire doors on both sides of the wall, a normally open automatic, fusible link-operated horizontally sliding fire door may be used on one side, with a swinging fire door on the other.

Stairs

Exit stairs are arranged to minimize the danger of falling, because one person falling on a stairway may result in the complete blockage of an exit. Stairs must be of sufficient width so two persons can descend side by side; thus a reasonable rate of evacuation may be maintained, even though aged or infirm persons may slow the travel on one side. There must be no decrease in the width of the stair along the path of travel, since this may create congestion.

Steep stairs are dangerous. Stair treads must be wide enough to give good footing. NFPA 101 specifies a minimum 11 in. (279 mm) tread and a maximum 7 in. (178 mm) riser. Landings should be provided to break up any excessively long individual flight. Continuous railings are now recommended for both sides of the stairs. Stairs of unusual width should have one or more center rails.

Two classes of stairs are permitted in the Code for existing buildings, with a single class of stairs for new stairs. There are Class A and Class B stairs for existing buildings; the requirements for each class are given in Table 7-3F.

Stairways may be inside the building where the Code generally specifies protective enclosures. They also may be outside if they comply with the requirements for exterior stairs and are arranged to avoid any handicap to their use by persons having a fear of high places, to avoid exposure to fire conditions originating in the building, and, where necessary, are shielded from snow and ice. Exterior stairs should not be confused with fire escape stairs. (See Fig. 7-3M.)

Construction details of stair enclosures involve the principles of the limitation of fire and smoke spread. Doors on openings from each story are essential to prevent the stairway from serving as a flue. Stairway enclosures should in general include not only the stairs, but also the path of travel from the bottom of the stairs to the exit discharge, so occupants have a protected exclosure passageway all the way out of the building. The stair enclosure should be of 1 hour noncombustible construction when connecting three or less floors, and 2 hour noncombustible construction when connecting four or more floors.

Smokeproof Towers

Smokeproof towers provide the highest protected type of stair enclosure recommended by NFPA 101. Access to the stair tower is only by balconies open to the outside air, vented shafts, or mechanically pressurized vestibules, so that smoke, heat, and flame will not readily spread into the tower even though the doors are accidentally left open.

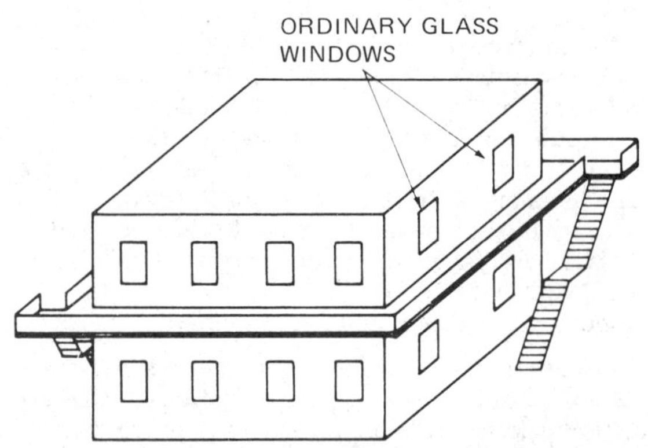

FIG. 7-3M. An example of using outside stairs to provide direct exits to the outside for all rooms in a multistory building. There are no interior corridors through which smoke and flame could spread. This method has application in many types of occupancies, such as schools, motels, small professional buildings, etc. Note that there are two means of egress, remote from each other, from the second story balcony.

Ramps

Ramps, enclosed and otherwise, arranged similar to stairways, are sometimes used instead of stairways where there are large crowds and to provide both access and egress for nonambulatory persons. Ramps are required where differences in floor level would result in less than three stair steps. To be considered safe exits ramps must have a very gradual slope.

Exit Passageways

A hallway, corridor, passage, tunnel, or underfloor or overhead passageway may be designated as an exit com-

TABLE 7-3F. Requirement for New and Existing Building Stairs

	New Stairs	Existing Stairs: Class A	Existing Stairs: Class B
Minimum width clear of all obstructions except projections not exceeding 3½ in. (0.89 mm) at and below handrail height on each side	44 in. (1.12 m) 36 in. (0.91 m) where total occupant load of all floors served by stairways is less than 50.	44 in. (1.12 m) 36 in. (0.91 m), where total occupant load of all floors served by stairways is less than 50.	44 in. (1.12 m)
Maximum height of risers	7 in. (178 mm)	7½ in. (191 mm)	8 in. (203 mm)
Minimum height of risers	4 in. (102 mm)	10 in. (244 mm)	9 in. (229 mm)
Minimum tread depth	11 in. (279 mm)		
Winders	See 5-2.2.2.5.	See 5-2.2.2.5.	See 5-2.2.2.5.
Minimum headroom	6 ft 8 in. (2.03 m)	6 ft 8 in. (2.03 m)	6 ft 8 in. (2.03 m)
Maximum height between landings	12 ft (3.7 m)	12 ft (3.7 m)	12 ft (3.7 m)
Minimum dimension of landings in direction of travel	Stairways and intermediate landings shall continue with no decrease in width along the direction of exit travel. In new buildings every landing shall have a dimension, measured in direction of travel, equal to the width of the stair. Such dimension need not exceed 4 ft (1.22 m) when the stair has a straight run.		
Doors opening immediately on stairs, without landing at least width of door	No	No	No

ponent providing it is separated and arranged according to the requirements for exits.

The use of a hallway or corridor as an exit component introduces some unique considerations. The use of these spaces for purposes other than exiting may violate fundamental design considerations. For example, within an industrial situation, the use of a gasoline powered fork lift within a corridor designated as an exit component would violate the principles of exit design. NFPA 101 specifies that an exit enclosure should not be used for any purpose which could interfere with its value as an exit and specifically prohibits piping for flammable liquids and gases. Furthermore, penetration of the enclosure by ducts and other utilities may violate the protective enclosure.

Each opening in an exit enclosure introduces a point of weakness which could allow fire contaminants to spread into the exit and prevent its use. The typical corridor used as an exit with numerous door openings could result in fire contamination of the enclosure if a door fails to close and latch. The door openings in exit enclosures should be limited to those necessary for access to the enclosure from normally occupied spaces. Therefore, door and other openings to spaces such as boiler rooms, storage spaces, trash rooms, and maintenance closets are not allowed into an exit passageway.

An exit passageway should not be confused with an exit access corridor. Exit access corridors do not have the construction protection requirements of exit passageways, since they provide access to an exit rather than being an extension and component of the exit.

Fire Escape Stairs

Fire escapes should be stairs, not ladders. Fire escapes are at best a poor substitute for standard interior or exterior stairs. The only use of fire escapes permitted by NFPA 101 is to correct exit deficiencies of existing buildings where additional standard stairs cannot be provided.

The same principles of design apply as for interior stairs, though requirements for width, pitch, and other dimensions are generally less strict. The Code gives the following criteria for fire escape stair design:

Fire escape stairs ideally extend to the street or ground level. When sidewalks would be obstructed by permanent stairs, swinging stair sections (designed to swing down with the weight of a person) may be used for the lowest flight of the fire escape stairs. The area below the swinging section must be kept unobstructed so the swinging section can reach the ground. A counterweight of the type balancing about a pivot should be provided for swinging stairs; cables should not be used. Fire escapes terminating on balconies above the ground level, with no way to reach the ground except by portable ladders or jumping, are unsafe.

Many persons having a fear of high places are reluctant to use fire escapes. Design should, as far as possible, provide a sense of security as well as suitable railings and other details actually needed for safety. Fire escapes must be well anchored to building walls and kept painted to prevent rust.

Preferred access to fire escapes is through doors leading from the main building area or from corridors—never through rooms which may have locked doors except where every room or apartment has separate access to a fire escape. Although preferred access to fire escapes is by doors, windows may be used, in which case sills should not be too high above the floor. Windows should be of ample size, and if insect screens are installed they should be of a type that can be quickly and easily opened or removed. Decorative grills or security bars should not be installed across windows which provide access to fire escapes.

In some situations, fire escapes have created a severe fire exposure to people when flames came out windows beneath them. The best location for fire escapes is on exterior masonry walls without exposing windows, with access to fire escape balconies by exterior fire doors. Where window openings expose fire escapes, fixed wired-glass windows in metal sashes should be used. Where there is a complete standard automatic sprinkler system in the building, the fire exposure hazard to personnel on fire escapes is minimized.

In northern climates, outside fire escapes are subject to obstruction by snow and ice. Roofs are sometimes provided as a safe refuge, particularly for fire escapes serving places of assembly.

Escalators, Moving Walkways, and Elevators

Escalators may be recognized as exits in existing buildings if they have enclosures similar to exit stairs and meet the requirements for stairs as to tread width and riser height. They are, however, seldom installed in such a way that they would qualify as exits, and it is common to find installations with the hazard of unprotected floor openings.

Moving walkways may also be used as means of egress if they conform to the general requirements for ramps (if inclined) and passageways (if level).

Elevators are not recognized as exits.

Ropes and Ladders

Ropes and ladders are not generally recognized in codes as a substitute for standard exits from a building. This is proper since there is no excuse for permitting their use except possibly in existing one and two family dwellings where it is economically impractical to add a secondary means of egress. In this case, a suitable rope or chain ladder or a folding metal ladder may be desirable. The homeowner should recognize however, that aged, infirm, very young, and physically handicapped persons cannot use ladders and that if the ladder passes near or over a window in a lower floor, flames from the window can prevent the use of the ladder.

Windows

Windows are not exits. They may be used as access to fire escapes in existing buildings if they meet certain criteria concerning the size of window opening and the distance of the sill from the floor. Windows may be considered as a means of escape from certain residential occupancies.

Windows are required in school rooms subject to student occupancy, unless the building is equipped with a standard automatic sprinkler system, and in bedrooms in one and two family dwellings that do not have two separate means of escape. These windows are for rescue and ventilation and are required to meet criteria for size of opening, method of operation, and height from the floor. NFPA 101 also requires operable bedroom windows in hospitals, nursing homes, and residential custodial care facilities to vent products of combustion and to provide fresh air to the occupants.

EXIT LIGHTING AND SIGNS

Exit Lighting

In buildings where artificial lighting is provided for normal use and occupancy, exit lighting and the illumination of the means of egress is required to assure that occupants can quickly evacuate the building. The intensity of the illumination of the means of egress should be at a value of not less than 1 ft candle (10.77 lu/m^2) measured at the floor. It is desirable that such floor illumination be provided by lights recessed in the wall and located approximately 1 ft (305 mm) above the floor because such lights are then unlikely to be obscured by smoke, which might occur during a fire. In auditoriums and other places of public assembly where motion pictures or other projections are shown, NFPA 101 permits a reduction in this illumination for the period of the projection to values of not less than ⅕ foot candle (2.2 lu/m^2).

Emergency Lighting

The Code requires emergency power for illumination of the means of egress based upon occupancy criteria. For example, emergency lighting is required within places of assembly, certain types of educational buildings, health care facilities, residential buildings with more than 25 rooms (hotels) or living units (apartments), dormitories, Class A and B mercantiles, business buildings subject to occupancy by more than 1,000 persons, parking garages, and underground or windowless structures subject to occupancy by more than 100 persons.

Well designed emergency lighting, using a source of power independent from the normal building service, automatically provides the necessary illumination in the event of interruption of power to normal lighting. Failure of the public utility or other outside electric power supply, opening of a circuit breaker or fuse, or any manual act including accidental opening of a switch controlling normal lighting facilities should result in automatic operation of the emergency lighting system.

Reliability of the exit illumination is most important. NFPA 70, *National Electrical Code®*, details recommended good practices in the installation of emergency lighting equipment. Battery operated electric lights and portable lights or lanterns normally are not used for primary exit illumination, but may be used as an emergency source under the restrictions imposed by NFPA 101. Luminescent, fluorescent, or other reflective materials are not a substitute for required illumination since they do not provide sufficient intensity of illumination to justify recognition as exit floor illumination.

Where electric battery operated emergency lights are used, suitable facilities are needed to keep the batteries properly charged. Automobile type lead storage batteries are not suitable because their relatively short life when not subject to frequent recharge. Likewise, dry batteries have a limited life and there is danger that they may not be replaced when deteriorated.

If normal building lighting fails, well arranged emergency lighting provides necessary exit floor illumination automatically, with no appreciable interruption of illumination during the changeover. Where a generator is provided, a delay of up to ten seconds is considered tolerable. The normal procedure is to provide such emergency lighting for a minimum period of one and a half hours. Most health care occupancies have self-contained electric generating plants for emergency power supplies, not only for exit lighting but also for emergency use in the event of failure of the public utility. Where such emergency electric facilities are provided, they may supply power supply for emergency exit lighting as well as to other critical areas of such buildings.

Exit Signs

All required exits and access ways must be identified by readily visible signs. The character of the occupancy will determine the actual need for such signs. In places of assembly, hotels, department stores, and other buildings with a transient populations, the need for signs will be greater than in a building having a permanent or semipermanent populations. Even in permanent residential occupancy buildings, signs are needed to identify exit facilities, such as stairs, which are not used regularly during the normal occupancy of the building. It is just as important that doors, passageways, or stairs which are not exits (but which are so located or arranged that they may be mistaken for exits), be identified by signs with the words "NOT AN EXIT."

Signs should be so located and of such size, color, and design as to be readily visible. Care should be taken not to have decorations, furnishings, or other building equipment located so the visibility of these signs is impaired. NFPA 101 does not make any specific requirement for sign color, but specifies the size of the sign and dimensions of the letters. The *Code* also specifies the levels of illumination for both externally and internally illuminated signs. The color of exit signs is usually required by local or state ordinances.

Improvement in the physical marking of exits in an office occupancy with point source, red or green strobe lights has been suggested. Placing of corridor illumination on the walls close to the floor to provide effective illumination under smoke conditions, as is the practice in Japan, is a technique worthy of research (Cohn 1978).

Directional "EXIT" signs are frequently required in locations where the direction of travel to reach the nearest exit is not immediately apparent.

ALARM SYSTEMS

Alarm systems to alert occupants to leave the building are normally operated manually. The alarm sounding devices themselves should be distinctive in pitch and quality from all other sounding devices, and the use of these devices should be restricted to evacuation purposes. Vocal alarm systems have been developed and installed in some high rise buildings (Powers 1971). It is, of course, very important that all alarm system devices be distributed throughout a building so as to be effectively heard in every room above all other sounds. Visual, in addition to audible, alarm devices are sometimes used in buildings occupied by deaf persons.

The proper maintenance of alarm systems is most important. Alarm systems should be under the supervision of a responsible person who will make proper tests at specified intervals and have charge of all alterations and additions to the systems.

FIRE EXIT DRILLS

Fire exit drills are essential in schools and are desirable in every type of occupancy to assure familiarity with the exits and their operation. In occupancies such as hospitals, nursing homes, hotels, and department stores, drills are usually limited to employee participation, without alarming patients, guests, or customers. Drills should be planned to get everyone out of the building or to an area of refuge in an orderly manner, as promptly as possible. Fire fighting is always secondary to life safety, and, in general, fire fighting operations should not be started until the evacuation is completed, except in cases where trained fire departments conduct rescue and fire fighting operations simultaneously.

Drills should be held at least once a month or more often, but not at regularly scheduled periods. Drills should occur on all shifts in an occupancy operated 24 hours a day. Drills should be conducted without warning and should simulate typical fire conditions for the occupancy.

School Exit Drills

School fire exit drills are an exercise in descipline, not speed, though reasonably prompt evacuation of the building is important. Students and staff should not be permitted to stop to put on coats. No individuals should be permitted to remain in the building and nobody should be excused from participating in the drill.

The drill should include a roll call by classes at designated assembly areas outside the building to make sure that no one is left behind. There should also be an established routine for a complete check of the entire building, including toilet facilities rooms, to make sure that no one is left behind. All exits should be used in drills, but routes should be varied from drill to drill. Occasional drills should be held that simulate conditions when one exit cannot be used because it is blocked by fire or smoke. All drills should include a provision to simulate the fire department notification procedure.

MAINTENANCE OF THE MEANS OF EGRESS

The provision of a standard means of egress with adequate capacity does not guarantee the safety of the occupants in the event of a need for evacuation of any building. Means of egress which were not properly maintained have been responsible for loss of life in a number of fires. Property managers usually assign definite responsibility for maintenance of mechanical and electrical equipment, but fail to do the same for the means of egress. As a result, it is all too common for inspection authorities to find otherwise safe stairways used for storage of materials during peak sales or manufacturing periods. In apartment buildings rubbish, baby carriages, and other obstructions are sometimes allowed to collect in stairway enclosures. Exit doors are often found locked or the hardware in need of repair. Doors blocked open or removed from openings into stairway enclosures are certain to permit rapid spread of smoke or hot gases throughout the building. Loose handrails and loose or slippery stair treads offer the dangerous probability that persons evacuating a building will fall in the path of others seeking escape. Maintaining the means of egress in safe operating condition at all times is as important to the prevention of loss of life as proper construction of building and the elimination of fire hazards.

Bibliography

References Cited

Abbott, J. C. 1971. "Fire Involving Upholstery Materials." *Fire Journal*. Vol 65, No 4. July 1971. p 88.

Berlin, Geoffrey, N. 1982. "A Simulation Model for Assessing Building Fire Safety." *Fire Technology*. Vol 18, No 1. Feb 1982. pp 66-75.

Berlin, Geoffrey, N., et al. 1982. "Modeling Emergency Evacuation from Group Homes." *Fire Technology*, Vol 1. Feb 1982. pp 38-48, XVIII.

Best, Richard, and Demers, David P. 1982. "Fire at the MGM Grand." *Fire Journal*. Vol 76, No 1. Jan 1982. pp 19-37.

Bryan, John L. 1982. "Human Behavior in the MGM Grand Hotel Fire." *Fire Journal*. Vol 76, No 2. Mar 1982. pp 37-41, 44-48.

Caravaty, Raymond D., and Haviland, David S. 1967. *Life Safety From Fire, A Guide for Housing the Elderly*. Architectural Standards Division. Federal Housing Administration, Washington, DC.

Chalmet, L. G., et al. 1982. "Network Models for Building Evacuation," *Fire Technology*. Vol 18, No 1. Feb 1982. pp 90-113.

Cohn, Bert M. 1978. *Study of Human Engineering Considerations in Emergency Exiting From Secure Spaces*. Gage-Babcock & Assoc. Inc. Chicago, IL.

Cooper, Leonard Y. 1982. "A Mathematical Model for Estimating Available Safety Egress Time in Fires." *Fire and Materials*. Vol 6, Nos 3 & 4, Sept/Dec. 1982. pp 135-144.

CPSC. 1972. FF-4-72, 40 FR 59940, CFR Part 1623. *Standard for the Flammability of Mattresses and Mattress Pads*. U.S. Consumer Product Safety Commission, Washington, DC.

Fahy, Rita. 1983. "Building Fire Simulation Model." *Fire Journal*. Vol 77, No 4. July 1983. pp 93-95, 102-105.

Fruin, J. J. 1977. *Pedestrian Planning and Design*. Metropolitan Association of Urban Designers and Environmental Planners, Inc., New York, NY.

Galbreath, M. 1969. "Time of Evacuation by Stairs in High Building." *Ontario Fire Marshal*. Vol 5, No 2. First Quarter. 1969.

Galbreath, M. 1968. "A Survey of Exit Facilities of High Office Buildings." Building Research Note 64. Division of Building Research. National Research Council, Ottawa, Canada.

GSA. 1985. "Fire Protection Safety and Health." *PBS P5900.2A*. General Services Administration, Washington, DC.

Karter, Michael J. Jr., and Gancarski, Joan L. 1984. "Fire Loss in the United States During 1983." *Fire Journal*. Vol 78, No 5. Sept 1984. pp 48-51, 54-61.

Keating, John P. 1982. "The Myth of Panic." *Fire Journal*. Vol 76, No 3. May 1982. pp 57-61, 147.

Keating, John P., and Loftus, Elizabeth F. 1981. "The Logic of Fire Escape." *Psychology Today*. June 1981. pp 14-19.

Keating, John P., et al. 1978. *An Evaluation of the Federal High Rise Emergency Evacuation Procedures*. University of Washington. Department of Psychology, Seattle, WA.

Krasny, John F., et al. 1981. "Development of a Candidate Test Method for the Measurement of the Propensity of Cigarettes to Cause Smoldering Ignition of Upholstered Furniture and Mattresses." *NBSIR 81-2363*. Center for Fire Research. National Bureau of Standards, Washington, DC.

Lathrop, J. K., et al. 1975. "In Osceola: A Matter of Contents." *Fire Journal*, Vol 69, No 3. May 1975. pp 20-26.

LTB. 1958. "Second Report of the Operational Research Team on the Capacity of Footways." Research Report No. 95. London Transport Board, London, England.

Macdonald, James N. 1984. *Non-Evacuation in Compartmented Fire Resistive Buildings Can Save Lives and Makes Sense*. NFPA 88th Annual Meeting, New Orleans, LA. May 23, 1984.

Melinek, S. J., and Booth, S. 1975. "An Analysis of Evacuation Times and the Movement of Crowds in Buildings." *CP 96175*. Building Research Establishment. Fire Research Station, Boreham Wood, England.

Ministry of Works. 1952. Part III. "Personal Safety and Fire Guarding of Buildings." Post War Building Studies No 29. " Joint Committee on Fire Grading of Buildings. Her Majesty's Stationary Office, London, England.

NBS. 1935. "Design and Construction of Building Exits." Miscellaneous Publication M51. National Bureau of Standards, Washington DC. pp 30-37 (out of print).

O'Leary, Timothy J., and Gratz, Jerre M. 1982. "An Analysis of Fire Evacuation Procedures Using Simulation." *Fire Journal*. Vol 76, No 3. May 1982. pp 119-121.

Pauls, J. L. 1975. "Evacuation and Other Fire Safety Measures in High-Rise Buildings." *Research Paper No. 648*. National Research Council of Canada. Division of Building Research, Ottawa, Canada.

Pauls, J. L. 1977. "Movement of People in Building Evacuations." *Human Response to Tall Buildings*. Dowden Hutchinson and Ross Inc. Stroudsburg, PA.

Pauls, J. L. 1978. "Management and Movement of Building Occupants in Emergencies." *Research Paper No. 788*. National Research Council of Canada. Division of Building Research, Ottawa, Canada.

Pauls, Jake. 1984a. "Development of Knowledge About Means of Egress." *Fire Technology*. Vol 20, No 1. Feb. 1984. pp 27-47.

Pauls, Jake L. 1984b. "The Movement of People in Buildings and Design Solutions for Means of Egress." *Fire Technology*. Vol 20, No 1. Feb. 1984. p 41.

Phillips, A. W. 1974. "You and the High Rise Building Fire." *Technology Report 74-1*. Society of Fire Protection Engineers, Boston, MA.

Powers, W. R. 1971. "New York Office Building Fire." *Fire Journal*. Vol 65, No 1. Jan. 1971. pp 18-23, 87.

Sharry, John A. 1983. "Real-World Problems with Zoned Evaculation." *Fire Journal*. Vol 77, No 2. Mar 1983. pp 32-33, 55.

Sharry, John A. 1974. "South America Burning." *Fire Journal*. Vol 68, No 4. July 1974. pp 23-33.

Smith, J. MacGregor. 1982. "An Analytical Queing Network Computer Program for the Optimal Egress Problem." *Fire Technology*. Vol 18, No 1. Feb. 1982. pp 18-33.

Stahl, Fred I. 1975. Simulating Human Behavior in High-Rise Building Fires: Modeling Occupant Movement Through A Fire Floor From Initial Alert to Safety Egress. *NBS-GCR-77-92*. Center for Fire Research, National Bureau of Standards, Washington, DC.

Stahl, Fred I. 1982a. "B Fires II: A Behavior Based Computer Simulation of Emergency Egress During Fires." *Fire Technology*. Vol 18, No 1. Feb 1982. pp 49-65.

Stahl, Fred I. 1982b. "B Fires II: A Behavior Based Computer Simulation of Emergency Egress During Fires." *Fire Technology*. Vol 18, No 2. May 1982. p 63.

NFPA Codes, Standards, Recommended Practices and Manuals. (See the latest *NFPA Codes and Standards Catalog* for availability of current editions of the following documents.)

NFPA 13, *Standard for the Installation of Sprinkler Systems.*

NFPA 70, *National Electrical Code.*

NFPA 101, *Code for Safety to Life from Fire in Buildings and Structures.*

NFPA 253, *Standard Method of Test for Critical Radiant Flux of Floor Covering Systems Using a Radiant Heat Energy Source.*

NFPA 255, *Standard Method of Test of Surface Burning Characteristics of Materials.*

Additional Readings

Babrauskas, Vytenis, "A Laboratory Flammability Test for Institutional Mattresses," *Fire Journal*, Vol. 75, No. 5, Nov. 1981, pp. 35-40.93.

Behrens, John F., "Horizontal Exits," *Fire Journal*, Vol. 77, No. 2, Mar. 1983, pp. 30-31.

Behrens, John F., "Handrails," *Fire Journal*, Vol. 77, No. 3, May 1983, pp. 14, 132.

Behrens, John F., "Protection From Hazards," *Fire Journal*, Vol. 77, No. 6, Nov. 1983, pp. 5-6.

Bryan, John L., *An Examination and Analysis of the Dynamics of the Human Behavior in the MGM Grand Hotel Fire*, rev. ed., National Fire Protection Association, Quincy, MA, Apr. 1983.

Bryan, John L, *An Examination and Analysis of the Dynamics of the Human Behavior in the Westchase Hilton Hotel Fire*, National Fire Protection Association, Quincy, MA, Mar. 28, 1983.

Bryan, John L., *Smoke as a Determinant of Human Behavior in Fire Situations*, (Project People), Department of Fire Protection Engineering, University of Maryland, College Park, MD, June 30, 1977.

Chalmet, L. G., et al., "Network Models for Building Evacuation," *Management Science*, Vol. 28, No. 1, Jan. 1982, pp. 86-105.

Cooper, L. Y., "A Concept for Estimating Available Safe Egress Time in Fires," *Fire Safety Journal*, Vol. 5, 1983, p. 135.

Francis, Richard L., "A Simple Graphical Procedure to Estimate the Minimum Time to Evacuate a Building," *SFPE TR 79-S*, Society of Fire Protection Engineers, Boston, MA, 1979.

Kisko, T. M., and Francis, R. L., "Network Models of Building Evacuation: Development of Software Systems, " *NBS-GCR-84-457*, National Bureau of Standards, Gaithersburg, MD, 1984.

Klevan, Jacob B., "Modeling of Available Egress Time from Assembly Spaces or Estimating the Advance of the Fire Threat," *SFPE TR 82-2*, Society of Fire Protection Engineers, Boston, MA, 1982.

Klote, J. H., "Elevators as a Means of Fire Escape," *NBSIR 82-2507*, National Bureau of Standards, Gaithersburg, MD, May 1982.

Lathrop, James K., ed., *Life Safety Code Handbook*, 3rd ed., National Fire Protection Association, Quincy, MA, 1985.

Predtechenskii, V. M., and Milinskii, A. I., *Planning for Foot Traffic Flow in Buildings*, Amerind Publishing Co., New Delhi, India, 1978.

Sharry, John A., "Determining Flame Spread," *Fire Journal*, Vol. 70, No. 3, May 1976, pp. 97, 101.

Stevens, R. E., "Designing Life Safety for Places of Public Assembly," *Fire Journal*, Vol. 61, No. 1, Jan. 1967, pp. 22-24.

Stevens, R. E., "Designing Life Safety in Education Occupancies," *Fire Journal*, Vol. 61, No. 2, Mar. 1967, pp. 24-25.

Stevens, R. E., "Designing Life Safety in Institutional Occupancies," *Fire Journal*, Vol. 61, No. 3, May 1967, pp. 50-53.

Stevens, R. E., "Scissors Stairs as Exits," *Fire Journal*, Vol. 59, No. 1, Jan. 1965, p. 40.

Stevens, R. E., "Smokeproof Towers," *Fire Journal*, Vol. 60, No. 1, Jan. 1966, pp. 54-55.

Stevens, R. E., "What is an Exit?" *Fire Journal*, Vol. 59, No. 6, Nov. 1965, pp. 44-45.

Templer, J., *Stair, Shape and Human Movement*, University Microfilms, Ann Arbor, MI, 1977.

BUILDING AND SITE PLANNING FOR FIRESAFETY

Revised by Clifford S. Harvey

Effective, firesafe design begins with conscious analysis and decision making early in the design process. This broad overall approach includes consideration of both interior building functions and layout as well as exterior site planning. This chapter will present some broad guidelines to be considered by building designers in analyzing fire protection measures that must be provided for a firesafe environment for structures and their sites. Detailed information of the various aspects of firesafe site planning and building design will be found elswhere in this HANDBOOK, particularly in the other chapters in this section.

To effectively incorporate the building's fire defenses into the design, the firesafety objectives must first be identified. Chapter 2, "Systems Concept for Building Firesafety," of this section briefly discusses a systematic assessment of firesafety objectives. Decisions must then be made regarding the means to achieve the objectives. The concepts tree in Chapter 2 can be used to obtain an overall perspective of the firesafety system. Venting in conjunction with sprinkler protection continues as a subject of some controversy, and is discussed in greater detail in Chapter 11 of this section.

Concepts of egress design, an extremely important part of the design process, are discussed in Chapter 3, "Concepts of Egress Design," of this section. That chapter, based on NFPA 101, *The Life Safety Code®*, describes the elements that must be considered when designing egress in buildings for life safety from fire.

BUILDING INTERIOR DESIGN CONSIDERATIONS

The architectural design of a building has a significant influence on its firesafety capabilities. Interior layout, circulation patterns, interior finish materials, and building services are all important factors in firesafety. Manual suppression of fires, particularly by fire department operations, is another important consideration. The building design significantly influences the efficiency of fire department operations. Consequently, manual suppression activities should be considered during all phases of architectural design.

A concept of creating firesafety involves the philosophy that the building itself must be designed to either assist the manual suppression of fire or be self-protecting from fire spread. Design features should consider the local fire fighting resources that are available. The fire service cannot be expected to provide complete protection for building occupants and property; it must be assisted by both active and passive building fire defenses, to provide reasonable safety from the effects of fire.

Interior Layout

During a fire the occupants of the entire building are exposed to two types of danger. One is exposure to flame and hot products of combustion – a problem in the vicinity of burning, the danger of which rapidly decreases as the distance from the fire increases. The other is smoke and toxic gases, which presents a different type of hazard. The greatest number of fire deaths result from these latter products of combustion, and the danger is present at considerable distances from the location of the fire.

Most fires, even relatively minor ones, may produce large quantities of smoke and gases. These products can obscure vision and irritate the eyes to the extent that visibility is reduced to practically zero. Even occupants familiar with their surroundings often experience great difficulty in locating means of egress. This problem is compounded for transients and occasional visitors to the building.

Architectural layout and normal circulation patterns are important elements in emergency evacuation. For example, many large office buildings are a maze of offices, storage areas, and meeting rooms. Even under normal conditions, a visitor can become confused and have difficulty in exiting the building. Clearly marked emergency travel routes can enhance life safety features in all buildings.

The building height is also related to the interior layout and firesafety. Under optimum conditions, modern fire department aerial equipment reaches to approximately the seventh floor of a building. Occupants above the

Mr. Harvey is president of Firemeasure, Inc., in Boulder, Colorado, and is the Chief Fire Marshal for that community.

seventh floor cannot be evacuated by exterior aerial equipment. Consequently, interior layout becomes more significant to occupant evacuation and protection, and fire department operations.

BUILDING DESIGN AND FIRE DEPARTMENT OPERATIONS

From the beginning of the construction project, the building designer should consult fire department personnel about fire suppression operations in a building. Such consultation is particularly important in the design phase of a building project. In this way the designer can incorporate features into the building that will aid, rather than hinder, fire suppression and other operations.

Briefly, the operations may be broadly grouped as rescue, fire control, and property conservation. The first priority of any fire fighting operation is rescue, which is defined as the removal of trapped and injured occupants, and the search of all areas to assure that all occupants are out of the building. The larger the building, the greater the number of fire fighting personnel that must be committed to the rescue operation. In some cases, simultaneous fire suppression operations will be necessary to complete the rescue operation.

Accessibility for Fire Fighting

Building design features are important factors in the spread of fire and the ease of suppression. One of the more important design features involves access to the fire. This includes access to the exterior and interior of the building by fire fighters and their equipment.

Access to the building itself requires roads or fire lanes which are designed to withstand the weight of the heaviest apparatus owned by the municipality in which the building is located. These access lanes must be designed to allow fire equipment to get close enough to all sides of the building for rescue and fire suppression functions; lanes must be of sufficient width to insure they will not be obstructed when needed. Street or road widths should not be less than 28 ft (8.5 m) curb face to curb face in low density residential areas, and should be much wider in commercial and industrial areas. Secondary access and emergency fire lanes should be readily visible, at least 12 ft (3.7 m) wide [16 ft (4.9 m) wide at buildings], with adequate turning radii at each corner or turn. Accepted standards dictate a 22 ft (6.7 m) inside turning radius, and a 45 ft (13.7 m) outside turning radius, but larger equipment may need larger radii.

Access to the interior of a building can be greatly hampered where large areas exist and where buildings have blank walls, false facades, solar screens, or signs covering a high percentage of exterior walls. Obstacles that prevent ventilation allow smoke to accumulate and obscure fire fighters' vision. Lack of adequate interior access can also delay or prevent fire department rescue of trapped occupants or the initiation of fire suppression efforts.

Windowless buildings and basement areas present unique fire fighting problems. The lack of natural ventilation facilities such as windows, contributes to the buildup of dense smoke and intense heat, which hampers fire fighting operations. Fire fighters must attack fires in these spaces despite heat and smoke.

Spaces in which adequate fire fighting access and operations are restricted because of architectural, engineering, or functional barriers should be provided with automatic extinguishing systems. A *complete* automatic sprinkler system is probably the best solution to this problem. Other methods that could be incorporated in appropriate design situations include access panels in interior walls and floors, fixed nozzles in floors with fire department connections, automatic smoke venting, and pressurization of certain building areas, although none of these items can take the place of, or be considered equivalent to, a complete automatic sprinkler system.

Ventilation

Ventilation is an important fire fighting operation. It involves the removal of smoke, gases, and heat from building spaces. Ventilation of building spaces performs the following important functions:

1. Protection of life by removing or diverting toxic gases and smoke from locations where building occupants must find temporary refuge.
2. Improvement of the environment in the vicinity of the fire by removal of smoke and heat. This enables fire fighters to advance close to the fire and extinguish it with a minimum of time, water, and damage.
3. Control of the spread or direction of fire by setting up air currents that cause the fire to move in a desired direction. In this way occupants or valuable property can be more readily protected.
4. Provision of a release for unburned, combustible gases before they acquire a flammable mixture, thus avoiding a backdraft or smoke explosion.

The building designer should be conscious of these important fire ventilation functions and provide effective means of facilitating venting practices whenever possible. This may involve access panels, moveable windows, skylights, or other means of readily opened spaces in case of a fire emergency. Emergency controls on mechanical equipment may also be an effective means of accomplishing fire ventilation. Each building has unique features, and consequently a unique solution should be incorporated into the design.

Water Supply and Use

Water is the principal agent used to extinguish building fires. Although other agents may occasionally be employed (e.g., carbon dioxide, dry chemical, foams and surfactants, and halons), water remains the primary extinguishing agent of the fire service. Consequently, the building designer should anticipate the needs of both the fire department and automatic extinguishing systems and provide an adequate supply of water at adequate residual pressure.

Water is normally supplied to the building site by mains that are part of the water distribution system. Few cities can supply a sufficient amount of water at required pressures to every part of the city. Consequently, water supplied to hydrants, standpipes, or sprinklers may need to be boosted by pumps located on fire department apparatus or in the buildings themselves.

Careful attention must be given to water supply, distribution, and pressure for emergency fire conditions. High rise buildings are particularly sensitive because of the water pressures that are required depending upon building height. Large area buildings must also be given more careful attention concerning these water related needs.

Fire conditions that require operation of a large number of sprinklers or use of a large number of hose streams can reduce pressure in sprinkler and standpipe systems to the point where residual pressures in the distribution system are adversely affected. Fire department connections for sprinkler and standpipe systems are therefore important components of the building fire defenses. The building designer must carefully consider installation details of fire department connections to ensure they will be easily located, readily accessible, and properly marked. Locations for the connections should always be approved by the local fire department.

Water Removal: Watertight floors are important in this respect. Salvage efforts can be greatly affected by the integrity of the floors. Of greater importance is the number and location of floor drains — if interior drains and/or scuppers are available, salvage teams can remove water effectively with a minimum of damage.

BUILDING DESIGN AND FIRE SUPPRESSION

The building designer has a significant influence on the relative ease and effectiveness of fire suppression operations. Not only is it important to incorporate the needs of the fire service into the functional layout of the building and its services, but also to properly plan for automatic fire detection and extinguishing capabilities.

Compartmentation

Fire fighting efficiency decreases rapidly as the fire propagates vertically. Fires on two or more floors are extremely difficult to control, and are virtually impossible to extinguish by manual fire fighting methods. Construction methods utilizing such techniques as open stairwells, curtain wall construction, poke through assemblies, and air handling ductwork provide wide avenues for vertical fire spread. It is more difficult to isolate a space effectively in modern construction than it was in pre-World War II construction, yet isolation can be accomplished if attention is paid to the details of design and construction.

Detection, Alarm, and Communication

Time is the most significant factor concerning fire control. If it is during the early development stages of a fire, a matter of minutes is a significant factor both in life safety and in manual fire extinguishment capabilities. Consequently, effective fire detection becomes a vital element in firesafety. After a fire or smoke detector is activated, appropriate action must be initiated. This may involve alarm, communication, fire suppression, or a combination of these. An alarm may alert building occupants to the existence of a fire, but what action should then be taken? Evacuation is the normal procedure in smaller buildings. High rise buildings, on the other hand, cannot reasonably be evacuated without allowance for a greater period of time. Consequently, safe areas should be provided within the building, and instructions should be given concerning the appropriate course of action.

Communications are vital in fire control. Communications help warn occupants, provide sufficient instructions for action, and alert the fire department early about the existence and location of a fire. Too often, fire department notification has been delayed because occupants were under the mistaken impression that a local alarm notified the fire department.

Automatic Fire Suppression

For more than a century, automatic sprinklers have been the most important single system for automatic control of fires in buildings. Many desirable aesthetic and functional features of buildings that might offer some concern for firesafety can be protected by the installation of a properly designed sprinkler system. NFPA currently has no record of a multiple fatality in a fully sprinklered building where the sprinkler system was properly designed and operational at the time of the fire.

Among the advantages of automatic sprinklers is that they operate directly over a fire and are not affected by smoke, toxic gases, and reduced visibility. In addition, much less water is used because only those sprinklers fused by the heat of the fire operate, particularly if the building is compartmented. Other automatic extinguishing systems (carbon dioxide, dry chemical, Halon agents, high expansion foam) also provide protection in certain portions of buildings or type of occupancies for which they are particularly suited.

SITE PLANNING

The architect must utilize a given site in designing the building, and adapt the functional and engineering considerations to the particular site conditions that are present. In a similar manner, the architect should consider site features in arriving at decisions on fire protection. A particular set of site characteristics may significantly influence the type of building fire defenses incorporated by the design team. Among the more significant features are traffic and transportation conditions, fire department access, and water supply. Water supply was discussed briefly earlier in this chapter; a few additional considerations are given here.

Traffic and Transportation

As previously indicated, time is a vital factor in fire control, particularly when response time of the fire department is a major component of the fire defenses. Fire apparatus must respond to a location along streets and at certain times of the day, depending upon traffic conditions, apparatus can be seriously delayed in responding to some locations.

It may or may not be helpful to consider limited access highways for fire department response. Highways may attract motorists, thus improving traffic conditions on other city streets. On the other hand, the limited access highway may divide a community into sections. Means of travel to locations intermediate to the access points can be delayed. This can have a significant effect on response times from fire stations, and must be considered by the design team in selecting the appropriate fire defenses for the building. Lengthy response times dictate greater built in fire protection capability.

Fire Department Access

Ideal exterior accessibility occurs where a building can be approached by fire department apparatus from all sides. This, unfortunately, is not often possible. In congested areas, only the sides of buildings facing streets may be accessible. In other areas, topography or constructed obstacles can prevent effective use of apparatus in combatting the fire.

Some shopping centers and buildings which are located some distance from the street make the approach of apparatus difficult. If obstructions or topography prevent apparatus from being located close enough to the building for effective use, such equipment as aerial ladders, elevating platforms, and water tower apparatus are rendered useless. Valuable manpower must be expended to carry ground ladders or hand carry hose lines long distances.

The matter of access to buildings has become far more complicated in recent years. The building designer must consider this important aspect in the planning stage. Inadequate attention to site details can place the building in an unnecessarily vulnerable position. If its fire defenses are compromised by preventing adequate fire department access, more complete internal protection in the building itself must make up the difference. If fire department access is limited or nonexistent, complete automatic sprinkler protection should be provided throughout the structure.

Location of Hydrants

An adequate water supply delivered with the necessary pressure is important to control a fire. In many new developments such as shopping centers, apartments, and housing projects, inadequate attention is given to the number and location of hydrants. There are examples of areas that present extremely difficult fire fighting problems where mains are so small and hydrants are so poorly spaced that adequate water to control a fire is unavailable. The building designer must consider these conditions when planning for the fire defenses for a development.

FIG. 7-4A. The Burlington Building (right), Chicago, March 15, 1922. Fire in a group of brick, wood joisted buildings (left) spread across the 80 ft (24 m) street to ignite the upper seven floors of the 15 story office building, which was of fire resistive construction but lacked exposure protection for street front windows. Wind carried hot gases against the Burlington Building. Updraft due to convection made exposure most severe above the level of exposing buildings.

FIG. 7-4B. Typical ignition of wood frame dwelling by heat radiated from an exposure fire. Distances of 10 to 20 ft (3 to 6 m) between frame buildings, as commonly specified by building codes, do not eliminate exposure hazard, but usually provide space for fire department operations.

EXPOSURE PROTECTION

A fire in one building creates an external fire hazard to neighboring structures by exposing them to heat by radiation and convection, as well as to the danger of flying brands of the fire. Any or all of these sources of heat transfer may be sufficient to cause an ignition in the exposed structure or its contents. This part of the chapter describes a method for evaluating exposure severity, and suggests methods for protecting buildings from exterior fire exposures. These methods are aimed at protecting combustibles within, and on the exterior of, an exposed building. NFPA 80A, *Recommended Practice for Protection of Buildings from Exterior Fire Exposures* (hereinafter

FIG 7-4C. Fire starting in Wellesley, MA (left), ignited a building in Newton (right) by radiated heat across the Charles River, 100 ft (30 m) distant, against the wind. The river prevented access for fire department operations to protect the exposure.

referred to as NFPA 80A), is a source for further information on exposure protection.

Factors Influencing Severity of Exposure

When discussing protection from exposure fires, there are two conditions to be considered. They are: (1) exposure to horizontal radiation and (2) exposure to flames issuing from the roof or top of a burning building in cases where the exposed building is greater in height than the burning building. Radiation exposure can result from an interior fire where the radiation passes through windows and other exterior wall openings. It can also result from the flames issuing from the windows of the burning building, or from flames of the burning facade itself.

A number of factors significantly influence the danger and intensity of exposing fires to neighboring exposed structures. Among the more important parameters is the severity of the exposing fire (total energy of the fire) itself. It involves both the temperatures developed within the exposing fire and the duration of burning. NFPA 80A describes three levels of exposure severity as light, moderate, or severe. The classifications are based on (1) the average combustible load per unit of floor area, and (2) the characteristics of an average flame spread rating of the interior wall and ceiling finishes. Tables 7-4A and 7-4B serve as a guide in assessing severity on the basis of these properties. The more severe of the two classifications should govern when using these tables.

The duration of the exposing fire and the total heat produced by the fire are related to the guide in Table 7-4A. The speed of fire buildup is influenced by both the nature of the combustibles and by the combustibility of the interior walls and ceiling finish. Table 7-4B relates the flame spread classification with severity.

Besides the temperature and duration of the exposing fire, other variables influence the severity of exposure on buildings. Some of these variables include the following:

1. Exposing fire
   a. Type of construction of exterior walls and roofs.
   b. Width of exposing fire.
   c. Height of exposing fire.
   d. Percent of openings in exposing wall area. Exterior walls that are combustible or which do not have sufficient resistance to contain the fire should be treated as having 100 percent openings.
   e. Ventilation characteristics of the burning room.
   f. The fuel dispersion, or surface to volume ratio of the fuel.
   g. The size, geometry, and surface-to-volume ratio of the room involved.

TABLE 7-4A. Classification of Exposure Severity Based on Fire Loading*

Classification of Severity	Fire Loading in Lbs/sq ft (kg/m^2) of Floor Area
Light	0–7† (0–34)
Moderate	7–15 (34–73)
Severe	15–up (73–up)

† Excluding any appreciable quantities of rapidly burning materials such as certain foamed plastics, excelsior, or flammable liquids. Where these materials are found in substantial quantities, the severity should be classed as moderate or severe.

* Source: NFPA 80A, *Protection of Buildings from Exterior Fire Exposures.*

TABLE 7-4B. Classification of Exposure Severity Based on Flame Spread of Interior Finish*

Classification of Severity	Average Flame Spread Rating of Interior Wall and Ceiling Finish
Light	0–25
Moderate	26–75
Severe	75–up

NOTE: Where only a portion of the exposing building has combustible interior finish (i.e., some rooms only, ceiling only, some walls only, etc.), this factor is considered in judging severity classification in as much as it reduces the average flame spread rating.

* Source: NFPA 80A, *Protection of Buildings from Exterior Fire Exposure.*

TABLE 7-4C. Guide Numbers for Minimum Separation Distances*

Severity			Width/Height or Height/Width Ratio																
Light	Moderate	Severe	1.0	1.3	1.6	2.0	2.5	3.2	4.	5.	6.	8.	10.	13.	16.	20.	25.	32.	40.
Percent Openings			Guide Number (Multiply by Lesser Dimension, Add 5 Feet, to Get Building-To-Building Separation)																
20	10	5	0.36	0.40	0.44	0.46	0.48	0.49	0.50	0.51	0.51	0.51	0.51	0.51	0.51	0.51	0.51	0.51	0.51
30	15	7.5	0.60	0.66	0.73	0.79	0.84	0.88	0.90	0.92	0.93	0.94	0.94	0.95	0.95	0.95	0.95	0.95	0.95
40	20	10	0.76	0.85	0.94	1.02	1.10	1.17	1.23	1.27	1.30	1.32	1.33	1.33	1.34	1.34	1.34	1.34	1.34
50	25	12.5	0.90	1.00	1.11	1.22	1.33	1.42	1.51	1.58	1.63	1.66	1.69	1.70	1.71	1.71	1.71	1.71	1.71
60	30	15	1.02	1.14	1.26	1.39	1.52	1.64	1.76	1.85	1.93	1.99	2.03	2.05	2.07	2.08	2.08	2.08	2.08
80	40	20	1.22	1.37	1.52	1.68	1.85	2.02	2.18	2.34	2.48	2.59	2.67	2.73	2.77	2.79	2.80	2.81	2.81
100	50	25	1.39	1.56	1.74	1.93	2.13	2.34	2.55	2.76	2.95	3.12	3.26	3.36	3.43	3.48	3.51	3.52	3.53
***	60	30	1.55	1.73	1.94	2.15	2.38	2.63	2.88	3.13	3.37	3.60	3.79	3.95	4.07	4.15	4.20	4.22	4.24
***	80	40	1.82	2.04	2.28	2.54	2.82	3.12	3.44	3.77	4.11	4.43	4.74	5.01	5.24	5.41	5.52	5.60	5.64
***	100	50	2.05	2.30	2.57	2.87	3.20	3.55	3.93	4.33	4.74	5.16	5.56	5.95	6.29	6.56	6.77	6.92	7.01
***	***	60	2.26	2.54	2.84	3.17	3.54	3.93	4.36	4.82	5.30	5.80	6.30	6.78	7.23	7.63	7.94	8.18	8.34
***	***	80	2.63	2.95	3.31	3.70	4.13	4.61	5.12	5.68	6.28	6.91	7.57	8.24	8.89	9.51	10.05	10.50	10.84
***	***	100	2.96	3.32	3.72	4.16	4.65	5.19	5.78	6.43	7.13	7.88	8.67	9.50	10.33	11.15	11.91	12.59	13.15

*Source: NFPA 80A, *Protection of Buildings from Exterior Fire Exposure.*

h. The thermal properties, conductivity, specific heat, and density of the interior finish.
2. Exposed building
   a. Type of construction of exterior walls and roofs.
   b. Orientation and surface area of exposed exterior walls.
   c. Percent of openings in exterior wall area.
   d. Protection of openings.
   e. Exposure of interior finish and combustibles to the radiation, convection, and flying brands of the exposing fire.
   f. Thermal properties, conductivity, specific heat, density, and fuel dispersion of the interior finish materials and the building contents.
3. Site and protection features
   a. Separation distance between exposing and exposed building.
   b. Shielding effect of intervening noncombustible construction.
   c. Wind direction and velocity.
   d. Air temperature and humidity.
   e. Accessibility for fire fighting operations.
   f. Extent and character of fire department operations.

Design Guidelines for Exposure Protection

When analyzing or designing a building for exposure protection, two situations are considered. The first occurs where the exposing building is equal or greater in height than the exposed building. In this case, only the thermal radiation from walls or wall openings is considered. The second case arises when the exposing building is of less height than the exposed building.

The criteria for protection is based on the separation distance between the exposing and the exposed buildings for these two cases. Tables 7-4C and 7-4D provide guidelines for determining separation distances that should protect exposed structures from fire spread. They are based on the condition that separation distances are great enough so that ignition of the exposed building or its contents is unlikely, assuming that no means of protection are installed in connection with either building.

TABLE 7-4D. Separation Distance Based on Building Height*

Number of Stories Likely to Contribute to Flaming Through the Roof	Horizontal Separation Distance or Height of Protection Above Exposing Fire.—Ft (m)
1	25 (7.6)
2	32 (9.8)
3	40 (12.2)
4	47 (14.3)

* Source: NFPA 80A, *Protection of Buildings from Exterior Fire Exposures.*

In Table 7-4C, the parameters include severity classification (described in Tables 7-4A and 7-4B), width and height of exposing fire, and percent of openings in exposing wall area. The definitions of the terms are as follows:

Width of Exposing Fire (w): The length in feet or meters of the exposing wall between interior fire separations or between exterior end walls where no fire separations exist. Fire separations such as fire partitions or fire walls should have sufficient resistance to contain the expected fire.

Height of Exposing Fire (h): The height in feet or meters of the number of stories involved in the exposing fire, considering such factors as building construction, closure of vertical openings, and fire resistance of floors. The relevant fire separations must have sufficient fire resistance to contain the expected fire.

Percent of Openings in Exposing Wall Area: This is the percentage of the exposing wall that consists of doors, windows, or other openings within the assumed height and width of the exposing fire. Walls without the ability to withstand fire penetration for the expected duration of the fire should be treated as having 100 percent openings.

In using Table 7-4C, for example, assume that a building has a moderate severity [7 to 15 psf (34 to 73

FIG. 7-4D. Above are two types of freestanding fire walls. At left is a brick fire wall reinforced with concrete abutments to make it freestanding. At right is a reinforced concrete fire wall which protected the dwelling from a fire in an adjacent lumberyard.

FIG. 7-4E. This wired-glass window in a metal frame, though cracked by intense heat of an exposure fire, held in place and combined with an open sprinkler to prevent fire from entering the building. Note the open sprinkler at top outside the window.

kg/m²) of fire load]. Assume further that the width and height of exposing fire is 100 ft (30 m) and 50 ft (15 m) respectively, and that 60 percent of the exposed wall area is open. Using a width to height ratio of 100/50 (30/15) = 2.0, a moderate severity, and 60 percent openings, a guide number of 2.15 is obtained from Table 7-4C. This indicates that the minimum separation of an unprotected building should have from an exposing fire in the building described should be 2.15 multiplied by the smaller dimension plus 5 ft (1.5 m).* In this case the separation should be 2.15 × 50 + 5 or 112.5 ft (2.15 × 30 + 1.5 or 31.5 m).

Naturally, site conditions and economic constraints providing most buildings with the calculated separation distance. When this occurs, the separation may be reduced by protecting the structure. Some of the means of protecting buildings from exposure fires, and thus reducing the desirable separation distance are:

1. Clear space between buildings.
2. Complete automatic sprinkler protection.
3. Blank walls of noncombustible materials.
4. Barrier walls (self-supporting) between building and exposure.
5. Extension of exterior masonry walls to form parapets or wings.
6. Automatic outside water curtains for combustible walls.
7. Elimination of opening by filling it with equivalent construction.
8. Glass block panels in openings.
9. Wired glass in steel sash (fixed or automatic closing) in openings.
10. Automatic or deluge sprinklers outside over openings.

*The 5 ft (1.5 m) is added to the computed values of separation distances partly to account for the horizontal projection of flames from windows and partly to guard against the risk of ignition by direct flame impingement where small separations are involved.

TABLE 7-4E. Adjustments for Reducing Separation Distances*

Means of Protection	Separation Distance Adjustment
Frame or Combustible Exterior Wall:	
1. Replace with blank fire-resistive wall (3 hr minimum)	Reduce to 0 ft (mm)
2. Install automatic deluge water curtain over entire wall with no windows or with wired-glass windows closed by ¾ hr protection	Reduce to 5 ft (1.5 m)
3. Install automatic deluge water curtain over entire wall with ordinary glass windows	Reduce by 50 percent
Noncombustible Exposed Exterior Wall (Fire Resistance Less Than 3 Hrs):	
1. Replace wall with blank fire-resistive wall (3 hr minimum)	Reduce to 0 ft (m)
2. Close all wall openings with material equivalent to wall, or with ¾ hr protection and eliminate combustible projections	Reduce by 50 percent
3. Install automatic deluge water curtain over entire wall with no windows or with glass windows or with windows closed by ¾ hr protection	Reduce 5 ft (1.5 m)
4. Install automatic deluge water curtain on all wall openings equipped with ordinary glass and on combustible projections	Reduce by 50 percent
Veneered Exposed Exterior Wall (Combustible Construction Covered by a Minimum of 4 in. of Masonry):	
1. Replace wall with blank fire-resistive wall (3 hr minimum)	Reduce to 0 ft (m)
2. Close all wall openings with ¾ hr protection and eliminate combustible projections	Reduce by 50 percent
3. Close all wall openings with material equivalent to wall construction and eliminate combustible projections	Reduce to 5 ft (1.5 m)
4. Install automatic deluge water curtain over windows equipped with wired glass or over ¾ hr closed openings and on combustible projections	Reduce to 5 ft (1.5 m)
5. Install automatic deluge water curtain over windows equipped with ordinary glass and on combustible projections	Reduce by 50 percent

TABLE 7-4E. (Continued) Adjustments for Reducing Separation Distances*

Means of Protection	Separation Distance Adjustment
Fire Resistive Exposed Exterior Wall (Minimum 3 hr Rating):	
1. Close all openings with material equivalent to wall or protect all wall openings with 3 hr protection	Reduce to 0 ft (m)
2. Protect all openings with 1½ hr protection	Reduce by 75 percent (maximum required = 10 ft (2.1 m)
3. Protect all wall openings with ¾ hr protection	Reduce by 50 percent (maximum required = 20 ft (6.1 m)
4. Install automatic deluge water curtain on all wall openings with wired glass or with ¾ or 1½ hr protection	Reduce to 5 ft (1.5 m)
5. Install automatic deluge water curtain on all wall openings equipped with ordinary glass	Reduce by 50 percent

* Source: NFPA 80A: SI units: 1 ft = 0.3048 m.

11. Automatic (rolling steel) fire shutters on openings.
12. Automatic fire doors on door openings.
13. Automatic fire dampers on wall openings.

The application of the means of protection described previously will enable the separation distance to be reduced. Guidelines for the distance reduction are given in Table 7-4E.

Bibliography

NFPA Codes, Standards, Recommended Practices and Manuals. (See the latest *NFPA Codes and Standards Catalog* for availability of current editions of the following documents.)

NFPA 80A, *Recommended Practice for Protection of Buildings from Exterior Fire Exposures.*

NFPA 101, *Life Safety Code.*

NFPA 1231, *Standard for Water Supplies for Suburban and Rural Fire Fighting.*

Additional Readings

Brannigan, Francis, L., *Building Construction for the Fire Service*, National Fire Protection Association, Quincy, MA, 1981.

Brunacini, A. V., *Fire Command*, National Fire Protection Association, Quincy, MA, 1985.

Clark, W. E., *Fire Fighting Principles and Practices*, Donnelly Publishing Corporation, New York, NY, 1974.

Fried, E., *Fireground Tactics*, H. M. Ginn Corporation, Chicago, IL, 1972.

Kimball, Warren, Y., *Fire Attack 1*, National Fire Protection Association, Quincy, MA, 1966.

Kimball, Warren, Y., *Fire Attack 2*, National Fire Protection Association, Quincy, MA, 1968.

Law, M., "Heat Radiation from Fires and Building Separation," Technical Paper No. 5, Joint Fire Research Organization, Boreham Wood, Hertfondshire, England, 1963.

Lie, T. T., "Safety Factors the Fire Loads," *Canadian Journal of Civil Engineering*, Vol. 6, 1979, p. 617.

McGuire, J. H., "Spatial Separation of Buildings," *Fire Technology*, Vol. 1, No. 4, Nov. 1965, pp. 278-287.

Shorter, G. W., et al., "The St. Lawrence Burns," *Quarterly*, Vol. 53, No. 4, Apr. 1960, National Fire Protection Association, Quincy, MA, pp. 300-316.

Thomas, P. H., "The Size of Flames from Natural Fires," Nineth Symposium on Combustion, Academic Press, 1963, pp. 844-859.

Williams-Leir, G., "Approximations of Spatial Separation," *Fire Technology*, Vol. 2, No. 2, May 1966, pp. 136-145.

Yokai, S., "Study on the Prevention of Fire Spread Caused by Hot Upward Current," Report No. 34, Nov. 1960, Building Research Institute, Japan.

CLASSIFICATION OF BUILDING CONSTRUCTION

Revised by Richard G. Gewain and David C. Jeans

A well established means of codifying fire protection and firesafety requirements for buildings is to classify them by types of construction based upon the materials used for the structural elements, and the degree of fire resistance afforded by each element.

In early codes, only two classifications of construction were identified: "fireproof" and "nonfireproof." The term "fireproof" was replaced by the term "fire resistive" since it was recognized that no material or building is totally fireproof. It is possible, however, to design buildings which will resist a fire without suffering serious structural damage. Optimum fire resistive design, balanced against anticipated fire severity, is the objective of fire protection requirements in modern codes.

Several distinct types of construction which use combustible framing were originally classified based on the materials—masonry or wood—used in the exterior wall construction and the type and size of the framing members i.e., heavy timber versus conventional framing. As fire resistance ratings for construction assemblies were recognized in building codes, subclassifications of building types were added for both noncombustible and combustible types of construction based on the degree of fire resistance provided.

Code regulations governing the size, area, and height of buildings and their allowable uses are usually predicated on the relative fire load represented by the occupancy and the construction materials used in the building.

This chapter identifies the basic types of building construction that are recognized in NFPA 220, *Standard for Types of Building Construction* (hereinafter referred to as NFPA 220) and in the model building codes. It also reckons the current classification system of building construction equates with the traditional descriptive terms used to identify building types, e.g., "fire resistive," "noncombustible," "ordinary," "frame," etc. that no longer are prime references to construction types.

Mr. Gewain is Director of Fire Research for American Iron and Steel Institute (AISI), Mr. Jeans is Director of Codes and Standards for AISI.

CONSTRUCTION CLASSIFICATIONS

The construction types presently identified in NFPA 220 as well as in the three model building codes, i.e., UBC (Uniform Building Code); B/NBC (Basic/National Building Code); and SBC (Standard Building Code), fall into ten classifications in NFPA 220 and the B/NBC. The UBC and SBC have nine types of construction. All classifications, however, are derived from five fundamental construction types: (1) fire resistive, (2) noncombustible, (3) ordinary (exterior protected), (4) heavy timber, and (5) wood frame. These descriptive names are now being discontinued because they no longer define the construction types as precisely as needed; the names are helpful, however, in tracing the development of building types.

Although the fire resistive construction types differ in detail from code to code, the four model building codes and NFPA 220 all define subtypes of fire resistive construction using as a basis the amount of fire protection (two, three, or four hours) required for the structural members.

Where the interior structural members and floors are of noncombustible materials with fire resistance ratings of one hour or less, the type of construction is generally identified as noncombustible. Most codes subdivide the noncombustible classification into protected and unprotected types.

NFPA 220 and other NFPA Building Construction Standards and the four model building codes employ three broad classifications for combustible types of construction: (1) exterior protected ordinary, (2) heavy timber, and (3) wood frame. Exterior protected ordinary and wood frame each include two subtypes, protected and unprotected. They are almost identical in most respects, except for their exterior wall requirements. Heavy timber construction is unique because it is identified by detailed requirements relating mainly to the size of structural members and their connections. Properties such as combustibility or fire resistance are not specifically included in the requirements for heavy timber construction except that exterior walls are required to be of noncombustible construction.

TABLE 7-5A. Types of Construction And Their Fire Safety Characteristics National Building Code of Canada*

Basic Type of Construction	Group	SubTypes	Characteristics (under fire load conditions)
Combustible Construction	I	Wood frame Wood post and beam Plank Plastic Other unprotected combustible	Fuel contributing and unstable
		Heavy timber construction and other protected combustible construction	Fuel contributing but partially stable to the degree of fire resistance
Noncombustible Construction	II	Unprotected steel construction Ordinary prestressed concrete Thin unprotected reinforced masonry Other unprotected noncombustible construction	Nonfuel contributing but unstable
	III	Steel construction with fire resistance Masonry with fire resistance Reinforced concrete with fire resistance	Nonfuel contributing and stable to the degree of fire resistance

* Table reproduced from "Steel and Fire Safety as required in the National Building Code of Canada—1975," R. V. Hebert, Canadian Steel Industries Construction Council.

Canadian Classifications

Construction type classification in building codes is more of a convenience than a necessity. The National Building Code of Canada (NBCC) does not classify buildings in the traditional manner as do United States codes, but rather specifies fire resistive requirements for the structural components of a building, depending upon its occupancy and its story height and floor area. In this code, two basic types of construction, combustible and noncombustible, are recognized. These are further subdivided by the characteristics of the materials used in construction under fire conditions as shown in Table 7-5A.

The NBCC establishes the areas for the subtypes of construction identified in the table by placing them into three groups which are based upon firesafety characteristics, combustibility or noncombustibility, and stability or instability under fire conditions. These groups are:

Group I—construction is limited to the smallest of buildings.

Group II—construction is limited to small and intermediate buildings (with some variations in treatment).

Group III—construction may be used for all buildings, and is mandatory for the largest and highest buildings, and for some smaller buildings with hazardous occupancies.

Wherever three-quarter hour protected combustible construction is required by the NBCC, unprotected noncombustible construction may be substituted.

New York City Building Code Classifications

The New York City Building Code divides construction types into two groups, noncombustible and combustible; each of these in turn is divided into five types. These subdivisions are distinguished by the fire resistance required for the interior structural members and for the exterior walls.

BCMC (Board for the Coordinating of the Model Codes)

To achieve better uniformity in building code and standard requirements, the MCSC (Model Codes Standardization Council), predecessor to the BCMC (Board for the Coordinating of the Model Codes), established a committee in 1972 to study the classifications and fire resistance requirements for types of construction used in the model building codes and to develop recommendations for the model code organizations.

As a result of its comparative study, the BCMC proposed that the basic types of construction now recognized in the codes be continued but that they be reordered to some degree and be divided into two groups; i.e., "noncombustible" and "combustible." It was also proposed that the identifying names for types of construction, such as "fireproof," "ordinary," "heavy timber," etc., be dropped because current design methods and architecture no longer follow the concepts in vogue when the named building types were established. The classifications proposed are shown in Table 7-5B. (These are the classifications recognized in the current edition of NFPA 220.)

The MCSC also concluded that to rationally compare the various types of construction, a notational system was needed to identify the fire resistance required for three basic elements of the building. These elements are (1) the

TABLE 7-5B. Model Codes Standardization Council Recommended Types of Construction

	Noncombustible	
Type I (433) Type I (332)	Type II (222) Type II (111) Type II (000)	
	Combustible	
Type III (211) Type III (200)	Type IV (2HH)	Type V (111) Type V (000)

exterior wall, (2) the primary structural frame, and (3) the floor construction. A three digit notation was developed as follows:

First Digit—Hourly requirement for exterior bearing wall fronting on a street or lot line.

Second Digit—Hourly requirement for structural frame or columns and girders supporting loads from more than one floor.

Third Digit—Hourly requirement for floor construction.

For heavy timber construction, the notation "H" and not a digit was used for the structural frame and floor construction designations.

Thus, for example, a "332" building would have 3 hr exterior bearing walls, a 3 hr structural frame, and 2 hr floor construction and would correspond to the NFPA 220 Type I (332) building, the Basic Building Code Type IB building, the Uniform Building Code Type I FR (fire resistive) building, the Standard Building Code Type II building.

A comparison of types of construction, based on the MCSC notational system, as found in NFPA 220 and the current editions of the four model building codes is shown in Table 7-5C.

A standard nomenclature was also developed for identifying and defining the structural elements in buildings as they relate to fire resistance. For example, it was found in reviewing various codes and fire protection standards that floor construction was referred to by such terms as "floors," "floor assemblies," "floor and ceiling assemblies," and "floor deck construction." If codes agree, for example, that "floor construction" includes the floor deck and all beams, joists, and other structural elements directly supporting the loads from the floor, as recommended by MCSC, then some misinterpretation of a code's intent would be avoided.

CLASSIFICATION OF BUILDING TYPES

Following the completion of these recommendations in 1974, the model code organizations and NFPA adopted a number of changes to their requirements for types of construction to agree with the MCSC classifications. However, it was recognized that some conflicts between the model codes still remained. In 1975, therefore, the BCMC established a committee to develop more detailed recommendations for types of construction.

In 1980 the committee's recommended definitions of types of construction and fire resistance requirements were finalized. The requirements are based on five basic types of construction. Two are identified as noncombustible construction and three as combustible construction types. Table 7-5D gives the fire resistance requirements for the structural frame, interior bearing walls, floor construction, and roof construction of the five basic types of construction.

The term "structural frame" as used in the table refers to the columns and the girders, beams, trusses, and spandrels having direct connections to the columns, and all other members which are essential to the stability of the building as a whole. The members of floor or roof panels which have no connection to the columns are considered part of the floor or roof construction and are not classified as a part of the structural frame.

Type I Construction (Fire Resistive)

Type I construction (fire resistive) is construction in which the structural members are noncombustible and are fire protected as specified in Table 7-5D. This classification is divided into two subtypes, Type I (443) and Type I (332). The basic difference between the subtypes is in the level of fire protection specified for the structural frame.

For both subtypes, the required fire resistance of those portions of the structural frame and bearing walls supporting roof loads only may be reduced by one hour.

The fire protection requirements for Type I (433 and 332) construction were selected because they provide reasonable firesafety for the structure for occupancies with moderate and low combustible contents. In occupancies with higher fire loads and hazardous uses, fire protection is supplemented by additional protection, usually including an automatic fire extinguishing system. Even in occupancies with moderate fire loads such as in mercantile and in some factory industrial and storage uses, supplementary firesafety precautions are required. These include restrictions on the building size or requirements for automatic fire extinguishing equipment.

In Type I construction, only noncombustible materials are permitted for the structural elements of the building. This is an accepted regulation that appears in practically every modern building code. Obviously, if combustible structural materials were allowed in noncombustible building types, the whole concept of their allowable use (height and area) would become meaningless. However, for practical reasons, the use of some combustible materials in Type I and Type II buildings are permitted for other than structural components. Roof coverings, some types of insulating materials, and limited amounts of wood for interior finish and flooring have been traditionally recognized as not adding significantly to the fire hazard or fire load if these materials are properly regulated and qualified by tests.

Some codes have attempted to regulate combustible materials by using a definition of noncombustible materials which includes two or three alternatives that allow for the acceptance of materials having relatively low fuel content and surface burning characteristics. The purpose

TABLE 7-5C. A Comparison of Construction Types (Based on the MCSC National System

NFPA 220	I (443)	I (332)	II (222)	II (111)	II (000)	III (211)	III (200)	IV (2HH)	V (111)	V (000)
UBC	—	I FR	II FR	II-1 hr	II N	III-1 hr	III N	IV HT	V 1-hr	V-N
B/NBC	1A	1B	2A	2B	2C	3A	3B	4	5A	5B
SBC	I	II	—	IV 1-hr	IV unp	V 1-hr	V unp	III	VI 1-hr	VI unp
BBC	1A	1B	2A	2B	2C	3B	3C	3A	4A	4B

TABLE 7-5D. Fire resistance Requirements for Type I Through Type V Construction (NFPA 220, Standard on Types of Building Construction)

	Type I		Type II			Type III		Type IV	Type V	
	443	332	222	111	000	211	200	2HH	111	000
EXTERIOR BEARING WALLS—										
Supporting more than one floor, columns or other bearing walls	4	3	2	1	0[1]	2	2	2	1	0[1]
Supporting one floor only	4	3	2	1	0[1]	2	2	2	1	0[1]
Supporting a roof only	4	3	1	1	0[1]	2	2	2	1	0[1]
INTERIOR BEARING WALLS—										
Supporting more than one floor, columns or other bearing walls	4	3	2	1	0	1	0	2	1	0
Supporting one floor only	3	2	2	1	0	1	0	1	1	0
Supporting a roof only	3	2	1	1	0	1	0	1	1	0
COLUMNS—										
Supporting more than one floor, bearing walls or other columns	4	3	2	1	0	1	0	H[2]	1	0
Supporting one floor only	3	2	2	1	0	1	0	H[2]	1	0
Supporting a roof only	3	2	1	1	0	1	0	H[2]	1	0
BEAMS, GIRDERS, TRUSSES & ARCHES—										
Supporting more than one floor, bearing walls or columns	4	3	2	1	0	1	0	H[2]	1	0
Supporting one floor only	3	2	2	1	0	1	0	H[2]	1	0
Supporting a roof only	3	2	1	1	0	1	0	H[2]	1	0
FLOOR CONSTRUCTION	3	2	2	1	0	1	0	H[2]	1	0
ROOF CONSTRUCTION	2	1½	1	1	0	1	0	H[2]	1	0
EXTERIOR NONBEARING WALLS	0[1]	0[1]	0[1]	0[1]	0[1]	0[1]	0[1]	0[1]	0[1]	0[1]

(shaded) Those members listed that are permitted to be of approved combustible material.

[1] Requirements for fire resistance of exterior walls, the provision of spandrel wall sections, and the limitation or protection of wall openings are not related to construction type. They need to be specified in other standards and codes, where appropriate, and may be required in addition to the requirements of this Standard for the construction type.

[2] "H" indicates heavy timber members; see text for requirements.

of this definition was to recognize certain materials or nonhomogenous assemblies containing limited amounts of combustible such as gypsum wallboard which, although covered with paper, is used as a fire resistive material. These alternate definitions include limits on surface flame spread rating and on the heat content—the latter being 3,500 Btu per lb (8050 J/kg), somewhat less than one-half that of untreated wood.

Rather than complicate the definition for the accomodation of certain materials, a more fundamental approach is to define limited uses and combustibility characteristics of materials that may be acceptable in buildings of noncombustible construction. This approach is followed in the NBCC and by the BCMC committee in its recommendations for the allowable kinds and extent of use of combustible material in noncombustible buildings.

Type II Construction (Noncombustible)

Type II construction (noncombustible) is a construction type in which the structural elements are entirely of noncombustible or limited combustible materials permitted by the code being applied and either protected to have some degree of fire resistance, either 2 hr [Type II (222)], 1 hr [Type II (111)], or completely unprotected except for exterior walls, Type II (000) construction.

The fire protection required in Type II (222 or 111) construction will afford adequate firesafety for residential, educational, institutional, business, and assembly occupancies without supplementary restrictions. Height limits, however, are commonly prescribed for this type of construction. When used for other occupancies involving a greater fire loading, additional firesafety precautions are usually required, such as more stringent area limitations and automatic fire extinguishing equipment. In occupancies with low combustible contents, the absence of fuel in noncombustible construction not only helps prevent the spread of fire but also reduces potential risk of a fire starting within the structure itself.

The noncombustible feature is valuable because it prevents fire from spreading through concealed spaces or involving the structure itself. Because of this attribute, a fire in a building of noncombustible construction can be controlled more readily. (See Fig. 7-5A.)

Type III Construction (Exterior Protected Combustible)

Type III construction (exterior protected combustible) is a construction type in which all or part of the interior structural elements may be of combustible materials or any other material permitted by the particular building code being applied. The exterior walls are required to be of noncombustible or limited noncombustible materials acceptable to the code and have a degree of fire resistance depending upon the horizontal separation and the fire

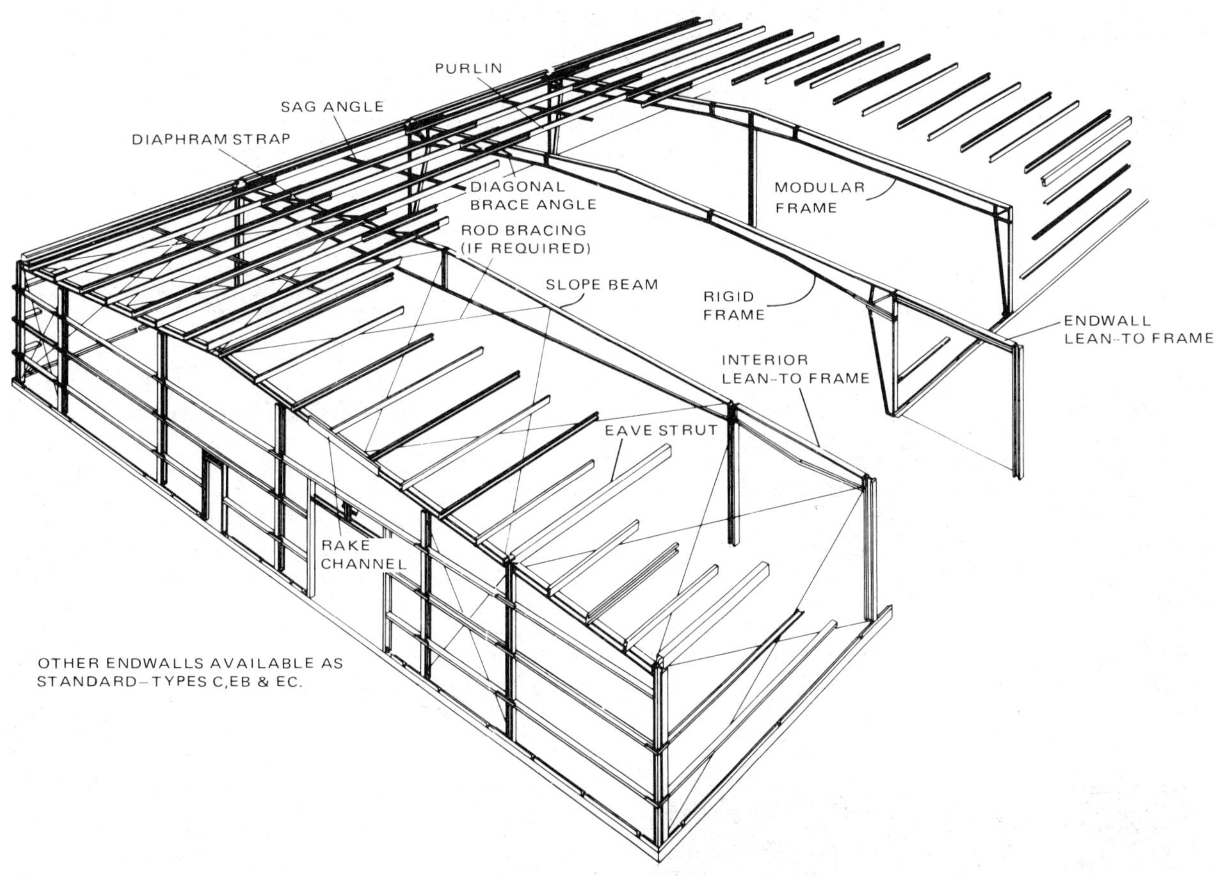

FIG. 7-5A. The framing system of a representative building of Type II (noncombustible) construction. Shown is preengineered pitched roof and lean to framing with structural elements of unprotected steel. Requirements for exterior walls of Type II construction are specified by the applicable building code or standard that is used by the authority having jurisdiction.

load. Type III construction is further divided into protected and unprotected subtypes. Protected construction, Type III (211), has a 1 hr protection rating for the floors and structural elements. Type III (200) construction has no protection for the floors or structural elements. Whether or not fire protection is provided, it is essential that all concealed spaces be properly fire stopped in buildings of combustible construction. This must be done with care in all furred spaces, partitions, ceiling spaces, and attics. Codes are very specific as to the materials to be used for firestopping and the locations where firestopping is required. To be effective, firestopping must completely close off and subdivide the combustible construction into limited areas, thereby restricting the spread of fire and hot gases and allowing additional time for detection and evacuation of the building or area involved.

The 1 hr fire protection provided in Type III (211) construction offers a measure of safety for fire fighting and evacuation before the construction itself becomes involved. It has been well established, however, that combustible parts of any fire rated assembly will be burning actively before the end of the rated time period. For this reason, that portion of the fire load represented by combustible structure must be considered as part of the total potential fire load, whether or not the construction is protected. (Type III construction was once known as "ordinary construction.")

Type IV Construction (Heavy Timber)

Type IV construction (heavy timber) is a construction type in which structural members, i.e., columns, beams, arches, floors, and roofs, are basically of unprotected wood (solid or laminated) with large cross-sectional areas. No concealed spaces are permitted in the floors and roofs or other structural members with minor exceptions. NFPA 220 and most model building codes are specific in the minimum dimensions permitted for the various wood structural members (Table 7-5E) and minimum fire resistive rating required for interior columns, arches, beams, girders, and trusses of materials other than wood that may be permitted as acceptable alternatives to wood members.

Walls, both interior and exterior, including structural members framed into them, can be of noncombustible or limited combustible materials acceptable to the code being applied. Brick and stone were the traditional materials used in early heavy timber, or "mill" construction.

During a fire, heavy timber construction resists failure longer than a conventional wood frame structure because the structural members are larger, have a smaller surface to volume ratio, and take longer to burn. Heavy timber

TABLE 7-5E. Recommended Nominal Dimensional Requirements for BCMC Type IV (2HH) Construction

	Supporting Floors	Supporting Roofs
Columns	8 in. × 8 in.	6 in. × 8 in.
Beams and Girders	6 in. × 10 in.	4 in. × 6 in.
Arches	8 in. × 8 in.	6 in. × 8 in., 6 in. × 6 in., 4 in. × 6 in.
Trusses	8 in. × 8 in.	4 in. × 6 in.
Floors	3 in. T & G or 4 in. on edge w/1 in. flooring	
Roofs		2 in. T & G or 3 in. on edge or 1⅛ in. plywood

SI units: 1 in. = 25.4 mm.

construction is more properly considered as a building system, not just a construction type using large size framing members. It was developed during the mid 1800s by insurance interests for the purpose of reducing fire losses in the many textile factories, paper mills, and storage buildings in the New England states. Through the intelligent use of combustible materials of sufficient mass, the absence of concealed spaces, and by attention to details to avoid sharp corners and ignitable projections, the chance of rapid spread of fire are lessened and the probability of serious structural damage is reduced.

Examples of heavy timber construction and are shown in Figures 7-5B and 7-5C.

Type V Construction (Wood Frame)

Type V construction (wood frame) is a type of construction in which the structural members are entirely of wood or any other material permitted by the code being applied. (See Fig. 7-5D.) Depending upon the exterior horizontal separation, the exterior walls may or may not be required to be fire resistive.

Type V construction is probably more vulnerable to fire, both internally and externally, than any other building type. Accordingly, it is essential that greater attention be given to the details of construction of this basically light wood frame building. Firestopping in exterior and interior

FIG. 7-5B. Elements of a building of Type IV (heavy timber) construction. Note the large size of the columns and beams and the absence of concealed spaces. The exterior wall at far left is of lightweight corrugated steel.

FIG. 7-5C. A variation of Type IV (heavy timber) construction. Shown are haunched arches made of laminated wood (gluelam construction). Beams are anchored to the arches by steel hangers.

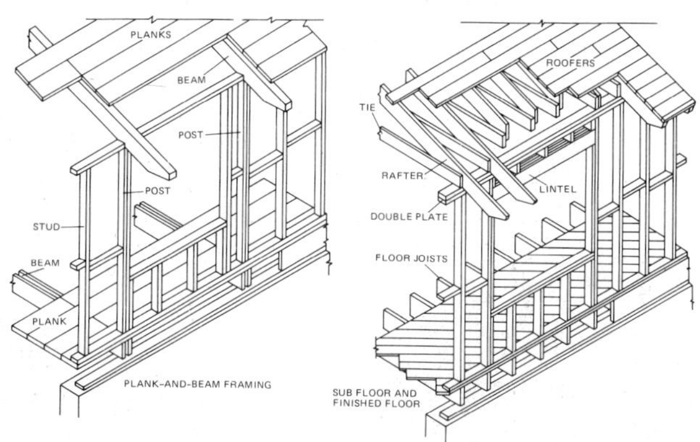

FIG. 7-7D. Two variations on basic Type V (wood frame) construction. At left is plank and beam framing in which a few large members replace many small members used in typical wood framing. At right is conventional wood framing (western or platform construction). Firestopping is essential in concealed spaces. (For details see Chapter 6, "Building Construction," in this section.)

walls at ceiling and floor levels, in furred spaces, and other concealed spaces can retard the spread of fire and hot gases in these vulnerable areas.

Type V construction is subdivided into two subtypes: Type V (111) construction, which has one hour protection throughout, including the exterior walls, and Type V (100) construction, which has no fire protection or fire resistance requirements, except for the exterior walls when horizontal separation is less than 10 ft (3 m).

Mixed Types of Construction

Where two or more types of construction are used in the same building, it is generally recognized that the requirements for occupancy or height and area would apply for the least fire resistive type of construction. However, in cases where each building type is separated by adequate fire walls or area separation walls having appropriate fire resistance, each portion may be considered as a separate building.

Another general limitation included in some codes prohibits construction types of lesser fire resistance to support construction types having higher required fire resistance. In the event of a fire, the risks of a major structural collapse are generally too great to permit this type of design. This limitation does not necessarily apply where construction supports nonbearing separating partitions, which provide protection for exit corridors or tenant spaces.

Bibliography

NFPA Codes, Standards, Recommended Practices and Manuals. (See the latest *NFPA Codes and Standards Catalog* for availability of current editions of the following documents.)

NFPA 203M, *Manual on Roof Coverings and Roof Deck Construction.*

NFPA 220, *Standard for Types of Building Construction.*

NFPA 251, *Standard Methods for Fire Tests of Building Construction and Materials.*

NFPA 255, *Standard Method of Test of Surface Burning Characteristics of Building Materials.*

NFPA 256, *Standard Methods of Tests of Roof Coverings.*

NFPA 258, *Standard Research Test Method for Determining Smoke Generation of Solid Materials.*

NFPA 259, *Standard Test Method for Potential Heat of Building Materials.*

Additional Readings

American Society for Testing and Materials, *Standard Test of Materials in a Vertical Tube Furnace at 750°C,* (ASTM E136-79), Philadelphia, PA, 1979.

ASTM, *Standard Test Method for Surface Burning Characteristics of Building Materials* (ASTM E84-80), Philadelphia, PA, 1980.

Bletzacker, R. W., et al., *Fire Resistance of Construction Assemblies* (EES0245), Engineering Experiment Station, Ohio State University, Columbus, OH, Dec. 1969.

Brannigan, Francis L., *Building Construction for the Fire Service,* National Fire Protection Association, Quincy, MA, 1982.

Fire Structural Use of Timber in Buildings, Ministry of Technology and Fire Officers' Committee, Joint Fire Research Organization, Symposium No. 3, London, England, Oct. 25, 1967.

"Fire-Resistance of Heavy Timber Construction," *Technical News Bulletin* No. 349, National Bureau of Standards, Washington, DC, May 1946.

Fire Resistance Classifications of Building Construction (BMS 92), National Bureau of Standards, Washington, DC, 1942.

Recommended Minimum Requirements for Fire-Resistance in Buildings, Building and Housing, No. 14, National Bureau of Standards, Washington, DC, 1931.

Siegel, L. G., "The Severity of Fires in Steel Frame Buildings," *AISC Engineering Journal,* Vol 4, No 4. Oct 1967, pp 137-142.

BUILDING CONSTRUCTION

Revised by Richard G. Gewain and David C. Jeans

This chapter briefly outlines building construction. It identifies the more common construction assemblies that are used or have been used in building construction in the United States, but is by no means exhaustive. It provides basic information relating how the structural elements of a building relate to one another, and methods by which the final construction is intended to resist the effects of fire. Refer to architectural or engineering design books for more comprehensive information.

FIRE PROTECTION OF BUILDING ELEMENTS

Fire protection for building elements is provided for two reasons. The first is to prevent the spread of fire within or into the building during a protracted or uncontrolled fire exposure and the second is to assure that even under that exposure, the building frame or elements of that frame will not collapse. Such collapse or even the threat of collapse will render fire fighting measures less effective than they might be otherwise.

There are two groups of building elements: load bearing and nonload bearing. Load bearing elements are those which support loads other than their own weight. Nonload bearing elements support only their own weight. Removal of nonload bearing elements would have no effect on the structural behavior of the building as a whole. Table 7-6A lists the structural elements which are common to most buildings separated into these two general categories.

Building codes provide requirements for both structural loads and superimposed live loads. These codes are such that building "failures" are rare; when one does occur, it is apt to be the result of application of unanticipated loadings. In a fire situation, "loads" are induced by heat which may cause thermal stresses, if the members are in any way restrained against expansion, and also at the same time, may cause a loss of strength if not actual consumption of the member due to heat. While restraint in some designs may offset any loss in strength, there is the increased likelihood of collapse in the event of a protracted fire. Almost by definition, properly designed and constructed fire resistive assemblies will retain adequate strength during fire exposure and continue to resist collapse.

TABLE 7-6A. Structural Elements of a Building

Load Bearing Elements
- Compressive Stress
  - Columns
  - Walls, exterior or interior
- Flexurally Stressed
  - Beams, Girders, Trusses
  - Floors, Roofs
- Nonload Bearing Elements
  - Curtain or Panel Walls
  - Partitions
  - Ceilings

Columns

Columns serve to carry building loads to the foundation where the loads are distributed to the supporting earth or rock. The material of the column is established by the type of construction—steel or reinforced concrete if noncombustible, or wood, including heavy timber designs if combustible. In a fire situation, interior columns are the most severely exposed of any structural members, as they may be enveloped in fire on all sides. Exterior columns may not be as severely exposed, and in some designs, the exposure may be minimized by shielding the column by structurally independent nonload bearing shields.

Functions of Walls

Building walls may serve one or more functions and they may be either bearing or nonbearing, and exterior or interior. Beyond that, the following brief descriptions will delineate most functions that walls may serve.

Bearing Wall: A wall that supports vertical loads in addition to its own weight.

Mr. Gewain is Director of Fire Research for American Iron and Steel Institute (AISI), Mr. Jeans is Director of Codes and Standards for AISI.

Curtain Wall: An exterior wall supported by the structural frame of the building. Also called a panel wall.

Enclosure Wall: An interior wall that separates a vertical opening for a stairway, elevator, duct space, etc., that connects two or more floors.

Exterior Wall: A wall that forms a boundary to a building and is usually exposed to the weather.

Fire Partition: An interior wall that serves to restrict the spread of fire, but does not qualify as a fire wall.

Fire Wall: A wall of sufficient durability and stability to withstand the effects of the most severe anticipated fire exposure. Openings in the wall, if allowed, must be protected.

Parapet Wall: A portion of an exterior, fire, or party wall that extends above the roof line.

Nonbearing Wall: An exterior wall that only supports its own weight.

Partition: An interior wall, not more than one story in height, that separates two areas in the same building but is not intended to serve as a fire barrier.

Party Wall: A wall that lies on a common lot line for two buildings and is common to both buildings. Most of these walls may be constructed in a wide range of materials or assemblies. Some of the more special designs include:

Cavity Wall: A wall of two parallel wythes (vertical wall of bricks) with an air space between them. The wythes are connected by metal ties.

Faced Wall: A wall comprised of two different masonry materials; one, the facing wythe; the other, the backup wythe, which are bonded together to act as a single unit.

Hollow Wall: A wall of two parallel wythes, with an air space between them, but without ties to hold the wythes together.

Sandwich Wall: A nonbearing wall whose outer faces enclose an insulating core material.

EXTERIOR WALLS

The primary function of all exterior walls is to protect the inside of the building from the elements such as heat and cold, water, wind, and windblown dust and dirt. In addition, some exterior walls may support floor and roof framing systems and may include portions that support girders or beams.

The fire resistance required for exterior walls is determined not only by the type of construction, but also by the distance the wall is set back from a lot line or an exposing structure. The greater that distance, the less severe the exposure and hence the less fire resistance that would be needed.

Load Bearing Exterior Walls

Load bearing exterior walls are generally constructed of masonry units such as stone, brick, concrete block, or a combination of these materials. Brick veneer is sometimes used as a facing in wood frame construction. In this case, the wood studs support the applied loads, while the veneer provides an attractive, useful exterior surface. Veneers are also used as the exterior face of cavity walls. These latter types are useful only in buildings of very limited height. Figures 7-6A, 7-6B, 7-6C, 7-6D, and 7-6E illustrate some common forms of exterior wall construction.

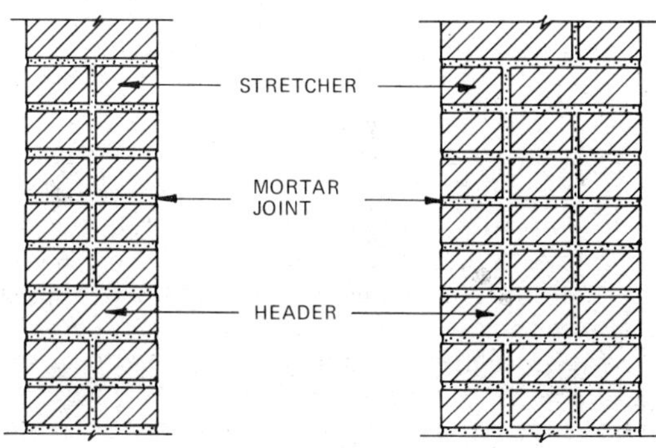

FIG. 7-6A. Typical types of wall and partition assemblies showing an 8 in. (200 mm) brick bearing wall and a 12 in. (300 mm) brick bearing wall.

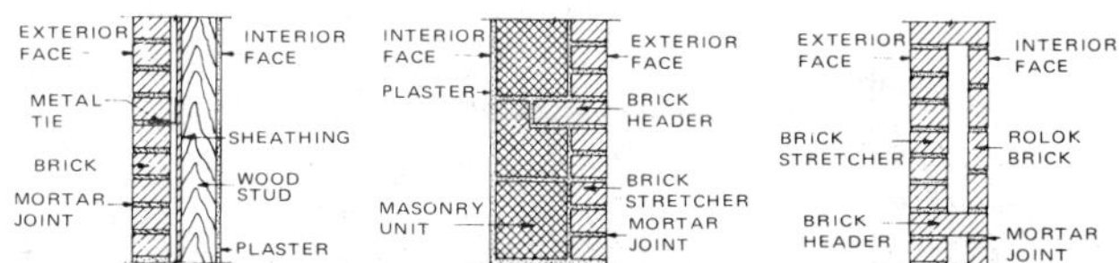

FIG. 7-6B. Typical types of wall assemblies showing (left to right) an exterior brick veneer on wood frame wall, a 12 in. (300 mm) exterior faced or veneered wall, and an 8 in. (200 mm) exterior hollow Rolok Bak brick wall.

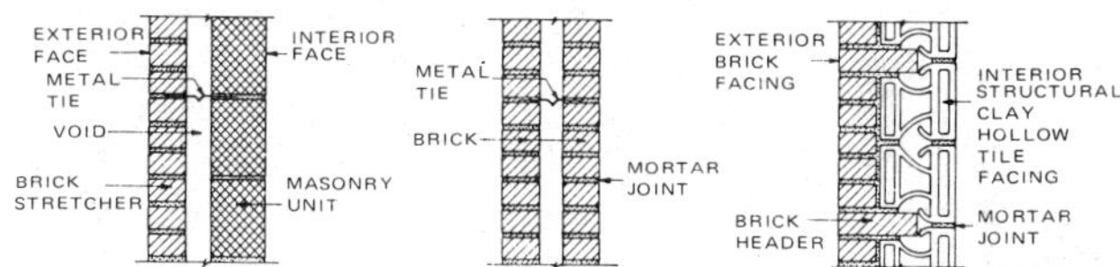

FIG. 7-6C. Typical types of wall assemblies showing (left to right) two examples of exterior nonbearing cavity walls and a 12 in. (300 mm) exterior bearing wall.

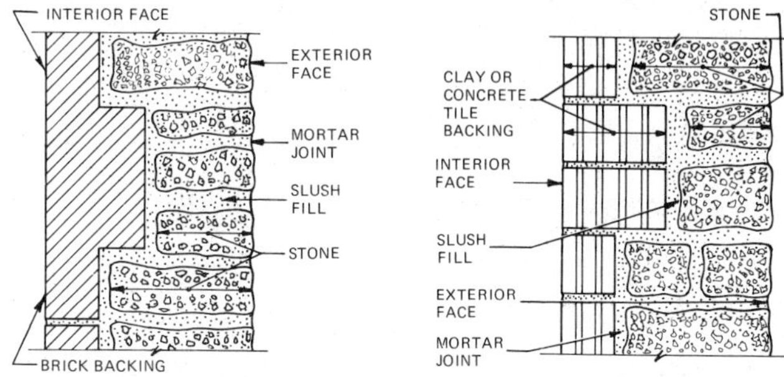

FIG. 7-6D. Stone faced wall assemblies, one with brick backing, the other with tile.

Exterior walls can also be constructed of reinforced concrete which is either poured in place or precast. Masonry veneers are often used as the exposed surface of reinforced concrete walls. The brick veneer is tied to the

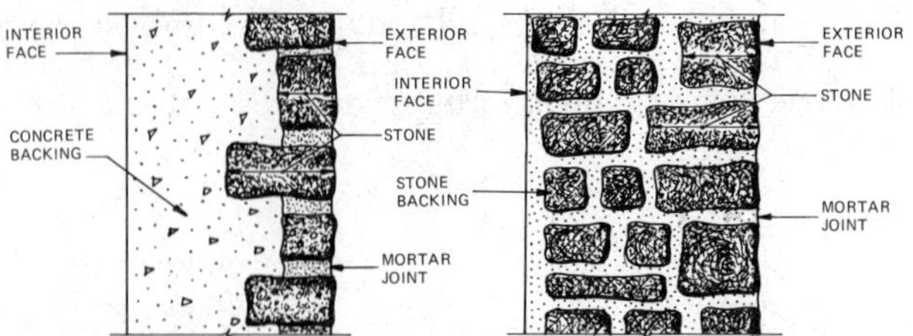

FIG. 7-6E. At left a wall of concrete backing and stone face; at right a solid stone wall.

concrete by metal ties fastened to the concrete and set into the bed mortar joints of the masonry. If the walls are load bearing, the reinforced concrete is designed to support all of the applied loads.

Exterior masonry walls may be nonload bearing, i.e., supporting only their own weight. The floor and roof framing is supported by columns that transfer the loads to the foundation. Usually each story of the exterior wall is supported on spandrel beams, which frame into the columns. Some of the structural difficulties experienced with masonry walls in fires are illustrated in Figure 7-6F.

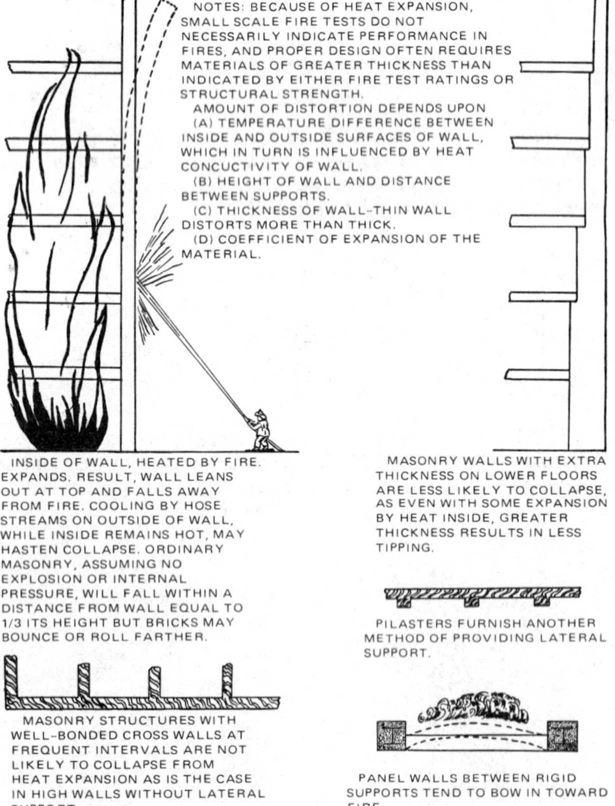

FIG. 7-6F. Heat expansion effects on ordinary masonry walls.

Openings such as doors and windows in masonry walls must have supports to carry the masonry units above the openings. These supports, called lintels, are short beams over the openings to support the masonry. Lintels can be constructed of several materials; steel angles and beams in various combinations are commonly used. Precast concrete lintels or brick arch lintels, have also been common means of supporting masonry units. Figures 7-6G and 7-6H illustrate several types of lintels. Openings may need to be limited in size and number when the exterior wall is close to lot line. Openings that are protected do not have fire resistance as such, but can impede the spread of fire even though they have little resistance to the transmission of heat.

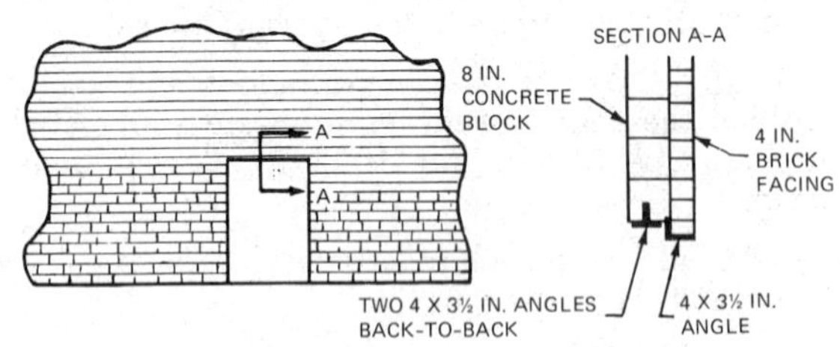

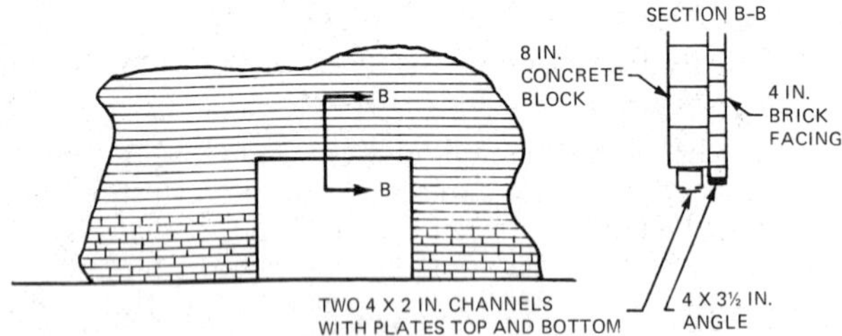

FIG. 7-6G. Lintels formed from steel angles and channels in brick faced, concrete block walls.

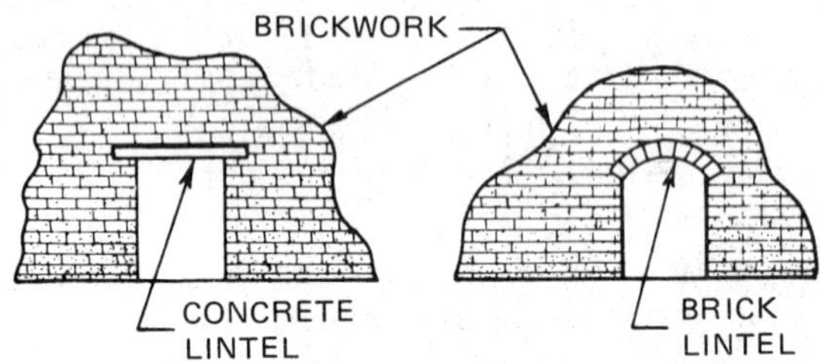

FIG. 7-6H. Use of concrete and brick lintels in masonry walls.

Curtain Walls

Most multistory buildings consist of a steel or reinforced concrete frame of columns, girders, and beams. The building envelope may be enclosed with nonbearing panels supported at each level. The panels are often identified as either curtain or panel walls.

Many materials and types of construction are used for curtain walls. Aluminum and stainless steel curtain walls are, by far, the most common but copper and copper alloys, carbon steel, galvanized metals, porcelain enamel finish, concrete, glass, and plastics are also used. Large window areas are common in this type of wall. Figure 7-6I shows an example of contemporary curtain wall construction.

A complete curtain wall assembly may consist of a panel with finished outside and inside surfaces, insulation, and means of attachment to the building frame. However, this type is not used as frequently as is a metal or

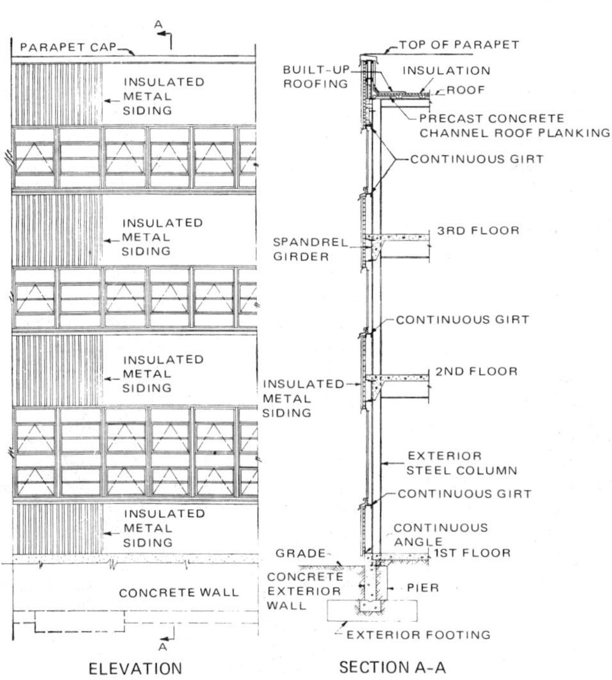

FIG. 7-6I. Elevation of typical bay of steel frame curtain wall construction.

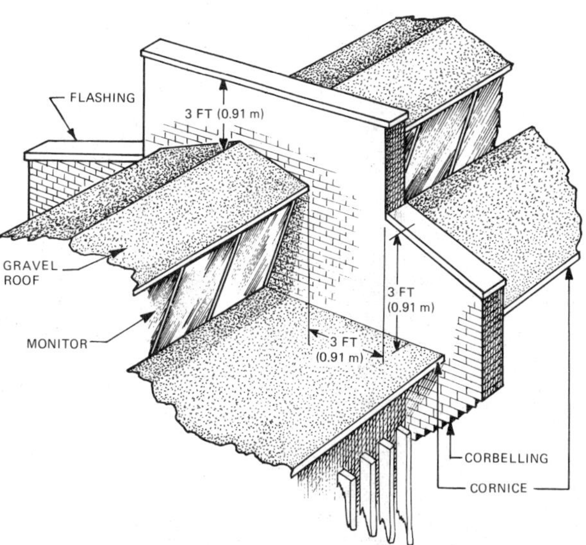

FIG. 7-6J. Fire wall installation on building with combustible roof and monitor.

glass skin assembly reinforced by conventional construction.

Curtain walls are required to have fire resistance ratings up to two hours depending upon the separation distance between the wall and a lot line or exposing structure. Secure fastening of curtain walls is important for many reasons including stability during fire exposure. Curtain walls are generally bolted to clips attached to the columns, spandrel, or floor slab. There is usually a space between the outer end of the floor slab and the inside face of the curtain wall. Unless adequately firestopped, this space may act as a route for vertical fire spread for the entire height of a building.

Parapet Walls

A parapet wall is an extension of fire wall construction above the roof line and prevents a roof fire from spreading to an adjacent building. Where walls extend only a few inches above the roof, ignition of an adjoining roof may occur, depending upon wind direction and velocity and the character of the roof covering. The Factory Mutual System, for example, specifies a parapet height of 30 in. (762 mm). Greater heights may be necessary for safety in some cases, as in walls over 50 or 60 ft (15 or 18 m) in length. Many building codes permit lesser heights, and in large sections of the country 18 in. (457 mm) parapets are considered standard. The higher the parapet, the greater the degree of safety, but considerations of expense and appearance dictate practical compromises. Generally, parapets of intermediate height are used, which perform satisfactorily under ordinary conditions in small buildings.

Fire may also extend around the ends of fire walls where the exterior walls of the building on both sides of the fire wall are combustible. There are two methods of minimizing this danger—one to extend the fire wall several feet out beyond the wall of the building, the other to provide a "T" section at the end of the wall. Where combustible roofs or cornices project beyond the walls of the building, fire walls should be extended to form a break in such combustible construction, as shown in Figure 7-6J. Many otherwise effective fire walls have failed because wood platforms or projecting canopies along the sides of buildings, have spread fire from one fire area to another.

INTERIOR WALLS AND PARTITIONS

Interior walls and partitions used for building corridors or rooms may be either bearing or nonbearing. Bearing partitions are common in the older, wall bearing construction systems and are also employed with industrialized building systems, particularly those of precast concrete. Newer construction tends to use nonbearing partitions.

The need for open space flexibility in modern construction has led to floor and interior partition design that allows partitions to be installed at any location. These moveable partitions are often made of steel studs and gypsum board and they may be installed, then dismantled, and reinstalled at another location when occupancy needs change. Normally, moveable partitions extend only from the floor to the underside of the ceiling.

Interior partitions, particularly nonload bearing partitions, can be installed with a number of other different materials. Wood stud and gypsum board or plaster is another very common type of partition. Partitions constructed of masonry units such as concrete block, structural clay tile, terra cotta, and gypsum block are common in a wide variety of buildings. Figure 7-6K shows typical partition constructions.

In building construction, properly designed interior partitions can act as barriers to the spread of fire. However, to protect certain areas more completely than would be possible with ordinary partitions, fire walls and fire partitions are constructed. Fire partitions are often incorrectly

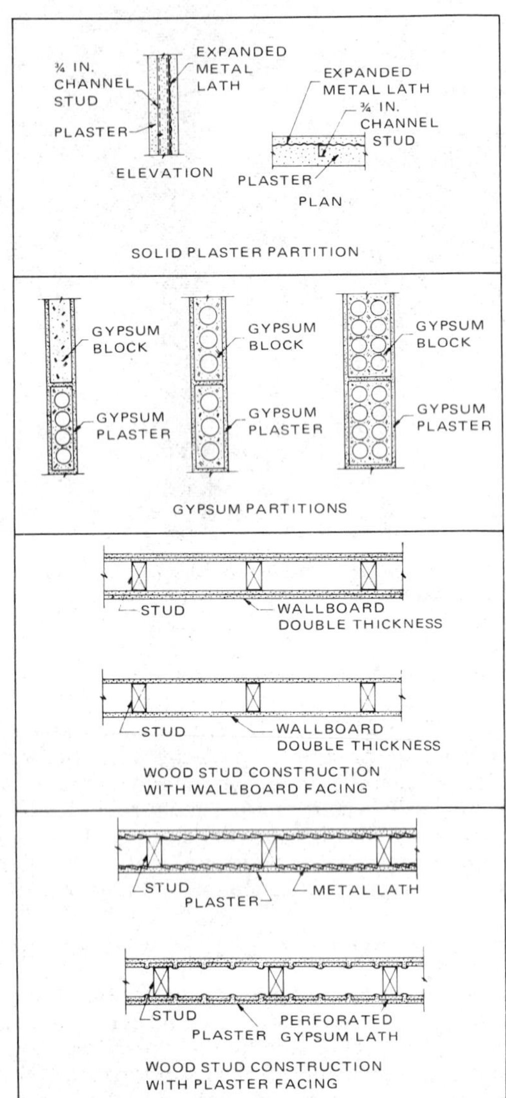

FIG. 7-6K. Typical interior wall and partition constructions.

called fire walls. Fire walls are fire resistive, of superior construction, and cannot be readily modified to meet changing building needs.

Fire Walls

Fire walls are interior walls providing a fire separation between areas of the same building. They should be designed to maintain structural integrity even in cases of complete collapse of the structure on either side of the fire wall. To withstand heat expansion effects, they are commonly made thicker than would be required by normal fire resistance ratings. Also they may be buttressed by cross walls or pilasters if of considerable height or length. In fire resistive buildings, structure supported fire division walls may be used.

Fire Partitions

Usually, a fire partition subdivides a floor or an area and is erected to extend from floor to the underside of the floor above. Fire partitions may be constructed of noncombustible, limited combustible, or protected combustible materials and should be attached to, and supported by, structural members having fire resistance at least equal to that of the partition. A fire partition normally possesses somewhat less fire resistance than a fire wall and does not extend from the basement through the roof, as does a fire wall. The fire resistance ratings for such partitions range from one-half to as much as four hours.

FLOOR FRAMING SYSTEMS

Floor framing systems typically include not only the flooring assembly but also its supporting beams, girders, or trusses. Beams and girders are nearly always an integral part of the floor system, while trusses may serve other purposes (and hence are discussed separately). The range of different types of floor framing systems that include beams and girders and the variety of components that may be included in a given assembly is best portrayed through drawings. Figures 7-6L through 7-6V are typical of many designs that are currently used.

Figures 7-6L and 7-6M are generalized framing plans

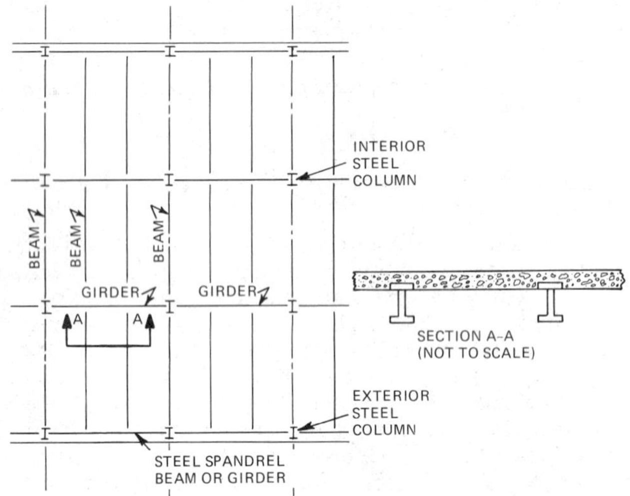

FIG. 7-6L. Portion of floor plan of steel frame structure.

for steel and reinforced concrete floor assemblies. The steel beams shown in Figure 7-6L support the floor assembly and are, in turn, supported by girders. Interior girders support beams on both sides. Girders are often concealed within partitions where their depth is not apparent, and they can directly support the weight of partitions in the space above. The girders frame into steel columns. Figure 7-6M illustrates a slab beam girder column system cast monolithically (no joints) of reinforced concrete. The functions of the various components are the same as those described for the steel frame.

The floor slab supported by the steel beams shown in Figure 7-6L need not be of reinforced concrete. Alternate types of deck are the composite decks shown in Figures 7-6N, 7-6O, and 7-6P. This type of floor system offers many construction advantages. The metal deck acts as a form for the concrete that remains in place after curing. In addition, the cellular panels can be used for energy services raceways where desired. A feature of this floor system is that

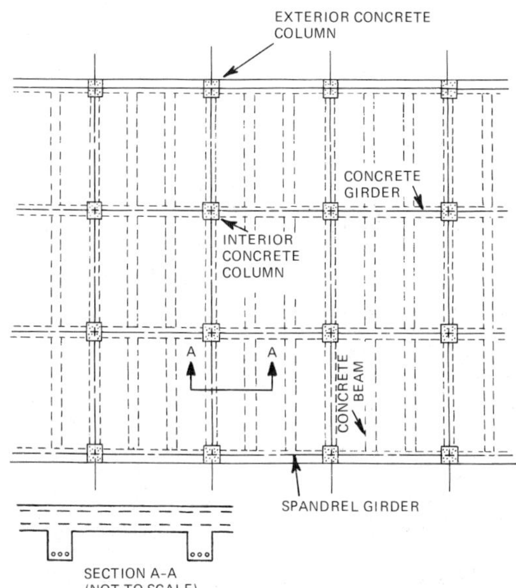

FIG. 7-6M. Portion of floor plan of concrete beam and girder construction.

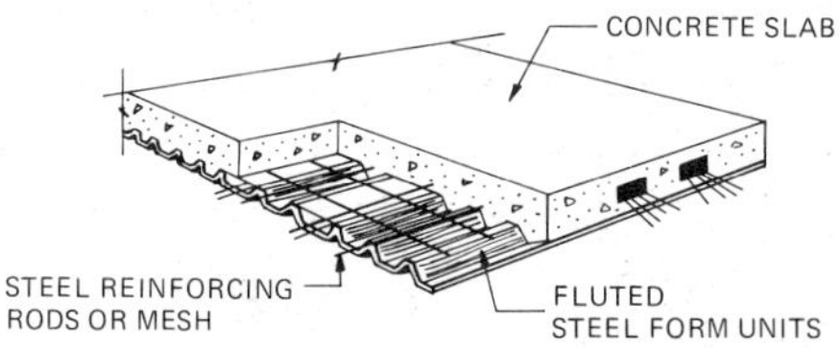

FIG. 7-6N. Example of concrete floor construction showing composite floor assembly.

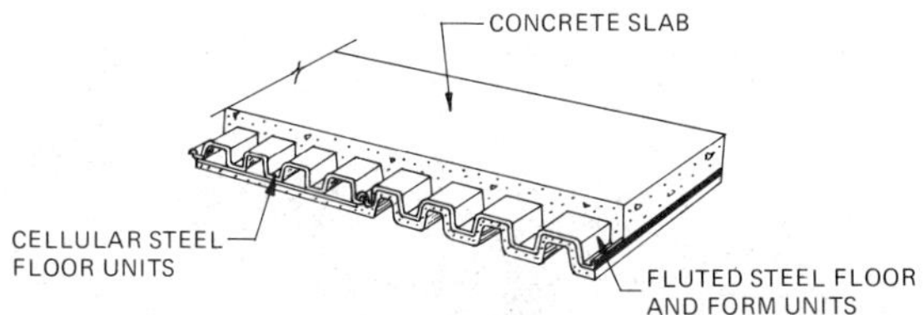

FIG. 7-6O. Example of concrete floor construction showing light gage cellular floor panels alternated with light gage steel forms.

the steel deck also acts as the tensile reinforcement of the concrete floor slab.

Another floor deck modification is to replace the reinforced concrete floor slab with a precast concrete system. Figure 7-6Q illustrates three common precast concrete floor slabs. The precast concrete planks generally have voids through their length which reduce the weight of the planks and act as energy services raceways. The planks may be supported by either steel beams or reinforced concrete beams.

Open web joists which are lightweight prefabricated trusses, may replace steel beams as the intermediate flexural framing and are a very common type of construction both for floors and roofs. The spacing of floor joists is approximately 2 ft (0.61 m), and generally about 4 ft (1.22 m) for roof joists. Figure 7-6R illustrates this type of construction.

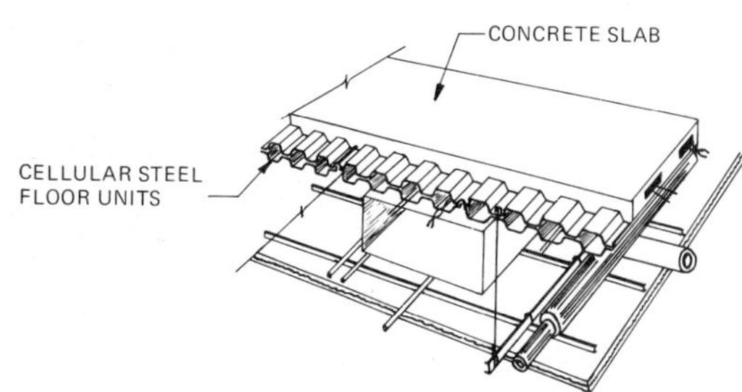

FIG. 7-6P. Example of a concrete floor poured over light gage cellular floor panels. The cellular panels can carry all three basic energy services—power, telephone, and data transmission lines.

Because the spacing of the joists is so close, the floor slab can be thinner and still support the same load. Corrugated steel forms with 2½ in. (63.5 mm) of concrete as a wearing surface is a typical design. The corrugated deck acts as a form for the concrete and contributes to the support of live loads. Lightweight concrete is often used in place of normal weight concrete to reduce the dead load.

Similarly, the reinforced concrete beams shown in Figure 7-6M may be spaced more closely together, permitting both a thinner floor slab and smaller beams. This type of construction is generally called a ribbed reinforced concrete system or a concrete joist system. The ribbed system can be either cast in place or precast. Figure 7-6S illustrates the concrete joist floor, while Figure 7-6DD illustrates two of the more common forms of ribbed precast systems. In each illustration, the ribs are the intermediate flexural framing.

The two way flat slab is an increasingly popular floor system. Two way reinforced concrete slabs, used since the turn of the century, include reinforcement placed in two directions. The slab performs the functions of deck, intermediate flexural framing, and primary flexural framing. The underside of the slab may be either flat or ribbed in two directions. Figures 7-6T and 7-6U illustrate two different forms of this type construction.

Wooden floor systems can be designed to perform the same functions as the basic model in Figure 7-6L. If the reinforced concrete deck were replaced by heavy wood planking 2 in. (51 mm) or more in thickness, and the steel beams, girders, and columns with large wooden members with a minimum dimension of 6 in. (152 mm), the basic form would be that of mill construction. This construction, shown in Figure 7-6V, was quite common for industrial buildings and warehouses during the last century. The general design is still used in locations where wood has aesthetic appeal.

Lightweight, wood frame construction is usually used in buildings of limited size. Floor joists in such construction are normally spaced 16 in. (406 mm) center to center and the vertical supports are often 2 by 4 in. (51 by 102 mm) or 2 by 6 in. (51 by 152 mm) wall bearing studs, again spaced 16 in. (406 mm) center to center. The stud walls may extend only from floor to floor in platform framing (Fig. 7-6W), or they may be extended continuously for two or three floors, a type of framing identified as balloon construction. (See Fig. 7-6X.)

Wood frame construction has little fire resistance because flames and hot gases can penetrate into the spaces between the joists or the studs. Firestopping must be installed at critical locations throughout the building to

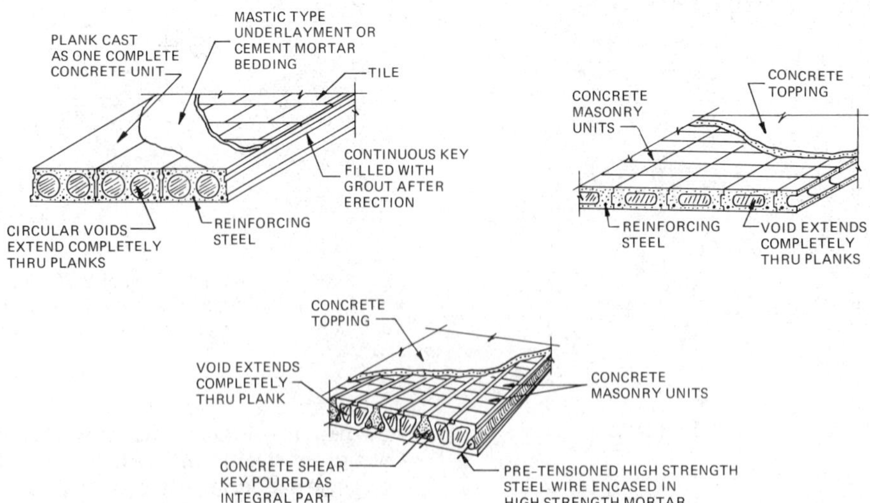

FIG. 7-6Q. Three examples of precast concrete floor slab construction.

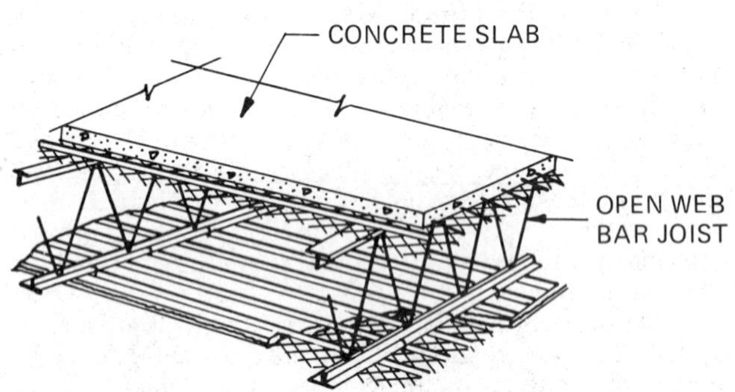

FIG. 7-6R. Example of concrete floor construction showing open web steel joist.

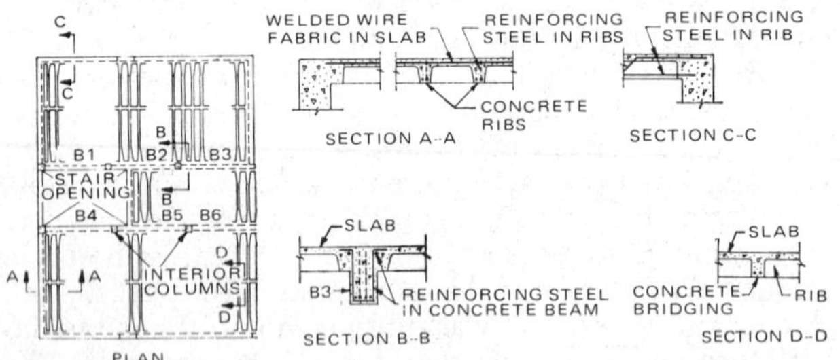

FIG. 7-6S. Partial plan of a concrete joist floor with four sectional views.

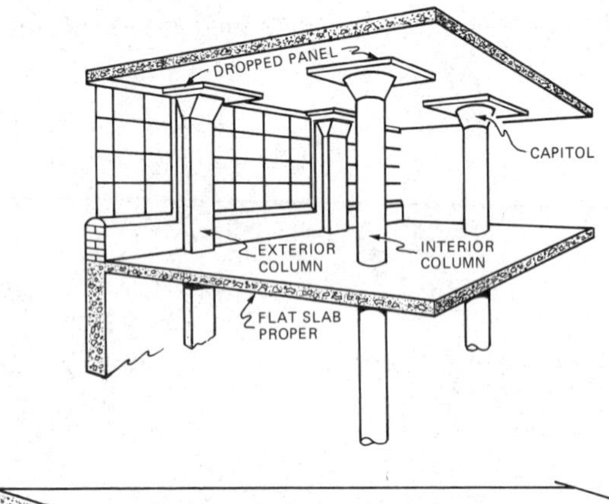

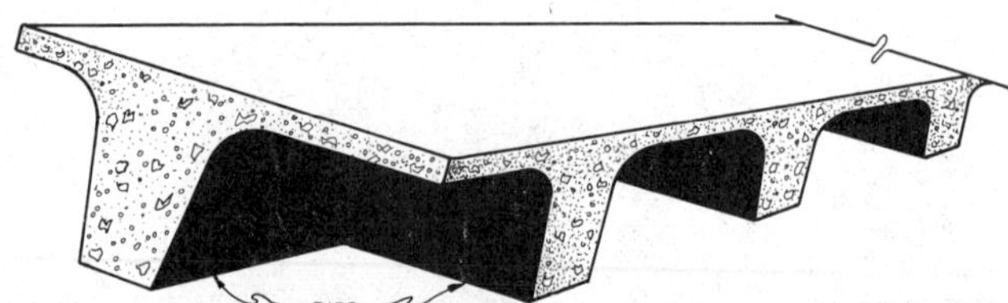

FIG. 7-6T. Two types of two way flat slabs: flat underside and two way ribbed underside.

prevent the rapid vertical and horizontal spread of fire. Firestops are normally made of wooden blocks or noncombustible material. Figures 7-6X and 7-6Y illustrate critical locations for firestopping in balloon frame and in platform frame construction; Figure 7-6Z shows the firestopping for these constructions in greater detail.

Watertightness and Drainage of Floors

Aside from the specifics of the effectiveness of floor framing systems as barriers to fire spread, there is another consideration involving floors in multistory buildings that can have a bearing on their performance in fire situations. This consideration is the degree of watertightness of construction and drainage arrangements for the floors. Massive quantities of water are sometimes needed on fires in unsprinklered buildings of combustible construction or with significant fuel loads. Without good drainage and/or waterproofing arrangements, water could pose the threat of increased water damage on floors below.

The most common and usually most satisfactory type of drainage is to provide scuppers in the walls. Floor drains connected to ample size headers or a combination of scuppers and floor drains may be used. The number of scuppers or drains is dependent upon the hazard and the amount of water likely to be used. For average conditions, the recommendations in Table 7-6B are considered good practice.

Waterproofing techniques such as waterproof membranes, hot mastics, asphalt emulsions, cementitious

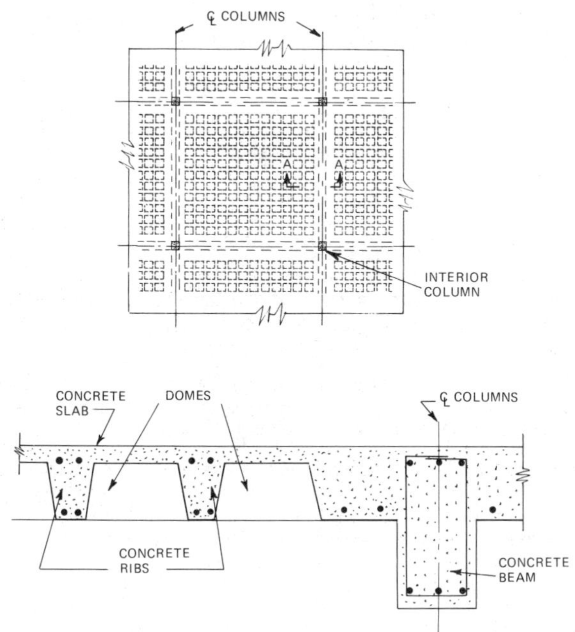

FIG. 7-6U. Partial typical interior two way ribbed floor slab with section view A-A.

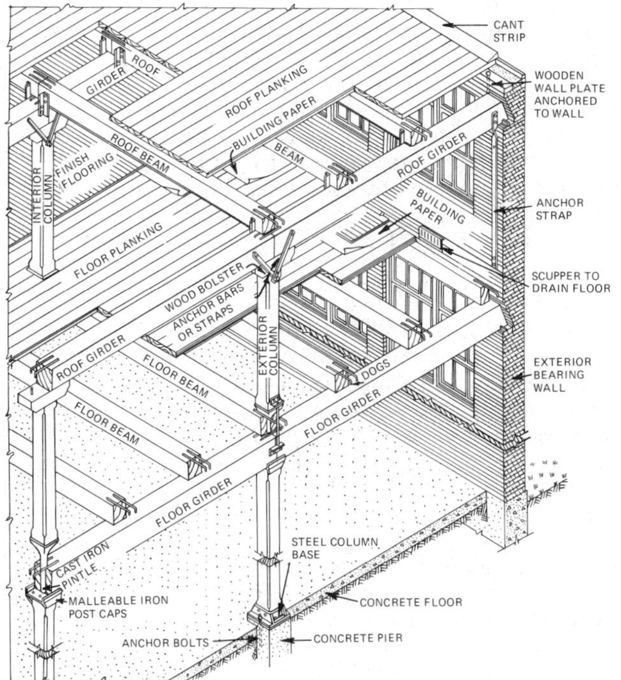

FIG. 7-6V. Components of a heavy timber building showing floor framing and identifying components of a type known as semimill.

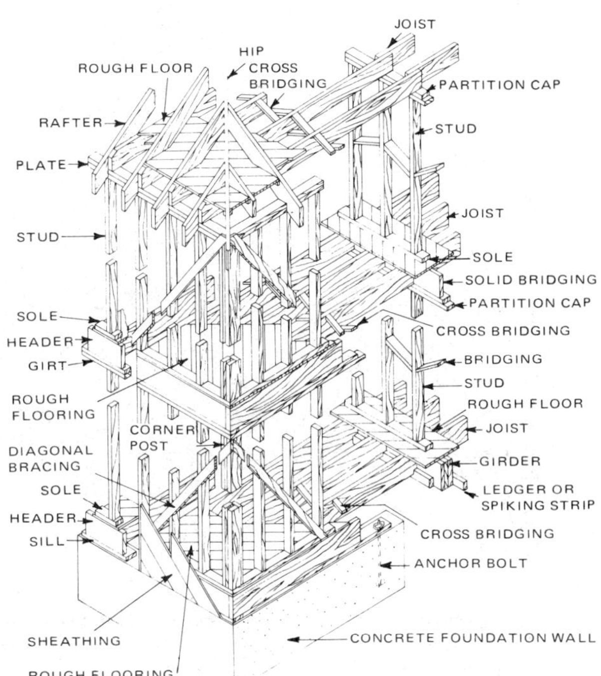

FIG. 7-6W. Example of wood frame platform construction common to dwellings. Structural members are identified.

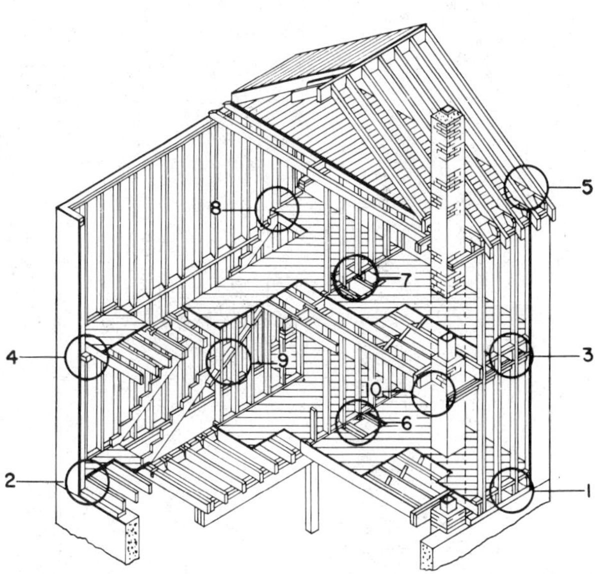

FIG 7-6X. Example of wood ballon frame construction showing points to be fire stopped. Expanded views of points are shown in Figure 7-6Z.

surfacings, etc., are available that can be applied to the floors depending upon the basic type of floor construction in question. The intent of these applications is to provide a watertight barrier on the floor itself which extends up 4 to 6 in. (50 to 150 mm) at walls and columns. Water impounded by these protective measures must be carried off before its weight becomes a loading threat to the building.

Guidance on the cost/benefits of providing waterproofing or drainage systems in building is contained in a Factory Mutual data sheet (FMEC 1977).

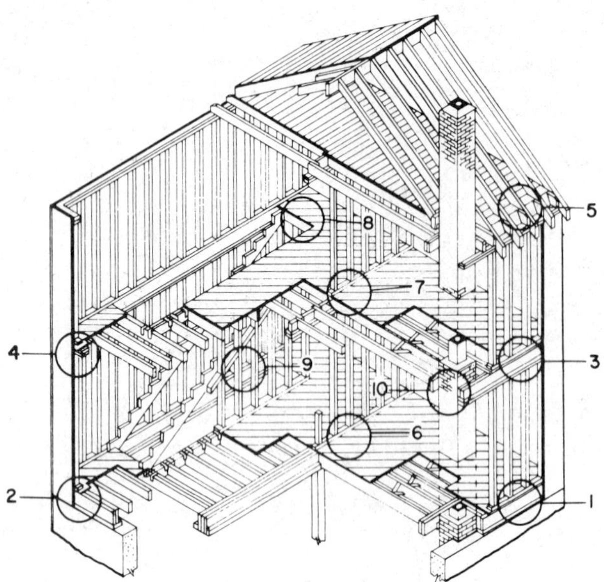

FIG. 7-6Y. Example of wood platform frame construction showing points to be fire stopped. Expanded views of points are shown in Figure 7-6Z.

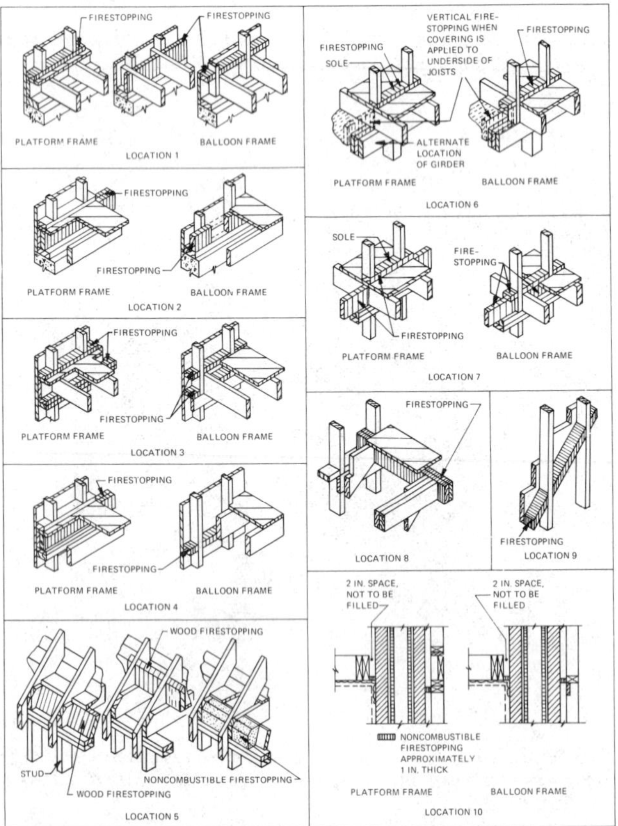

FIG 7-6Z. Details of the application of firestopping to platform and balloon framing. Location numbers coincide with locations circled in Figures 7-6X and 7-6Y.

TABLE 7-6B. Sizes for Scuppers and Drains

Floor Areas	No. of 4-in. (102 mm) scuppers or drains
For floor areas 500 sq ft (46 m²) or less	2
For floor areas 750 sq ft (70 m²)	3
For floor areas 1,000 sq ft (93 m²)	4
Additional scuppers for areas over 1,000 sq ft (93 m²):	
For extra hazard occupancies (quantity and combustibility of contents is very high), or floors of questionable watertightness, or contents especially subject to water damage—one scupper for each additional 500 sq ft (46 m²).	
For moderate hazard occupancies (quantity and combustibility of contents is moderate) with watertight floors—one scupper for each additional 1,000 sq ft (93 m²) or fraction thereof.	
For ordinary hazard occupancies (quantity and combustibility of contents is low) with strictly watertight floors—one scupper for each additional 2,000 sq ft (186 m²) or fraction thereof.	

TRUSSES

Where large areas must be column free or where special occupancy requirements may warrant, trusses may be used for purposes other than roof support. Three such applications are identified as transfer trusses, staggered trusses, and interstitial trusses.

Transfer Trusses

Load transfer trusses create a clear space on a lower floor by directly supporting the loads from columns above the truss or, at times, from tension members below. This design allows large, column free areas for auditoriums, ballrooms, etc. at any level in a high rise building.

Since transfer trusses carry load from more than two floors, building codes generally require that the fire protection be provided by individual protection for each of the truss elements, or by complete enclosure of the truss (for its entire height and length) with envelope protection.

Staggered Trusses

The staggered truss system is primarily intended for high rise residential buildings. Typically, in a staggered truss design, story high trusses span the full building width at alternate column lines on each floor. These trusses are supported only on the two rows of exterior columns. Thus, the interior of the building is column free; at any given elevation, floor construction is alternately supported on the top and bottom chords of adjacent trusses.

It is characteristic of staggered truss applications that these trusses are enclosed in wall construction that separates individual apartments or hotel/motel guest rooms. There may be a control opening in the truss that permits passage space for a corridor.

Since staggered trusses are usually enclosed in walls, the entire wall assembly must have a fire resistance required for this condition.

Interstitial Trusses

The interstitial truss concept was developed primarily to solve the functional needs of hospitals. While steel trusses in interstitial framing systems are quite deep (on the order of 8 ft, or 2.5 m, in height), they do not extend from floor to floor as do staggered trusses. The top chords support the floor above, while the bottom chords support a suspended ceiling system and a walk on surface for maintenance purposes. As such, interstitial trusses are analogous to very deep open web steel joists.

The interstitial space thus created provides a convenient location for the complex mechanical and electrical systems that are necessary components of a modern hospital. Direct access for maintenance, renovation, and replacement of the various system components within these interfloor spaces is provided without significant interference with normal operations.

Since each interstitial truss supports only one floor, all three conventional fire protection methods for trusses are permitted by the model building codes. However, because of practical considerations, individual protection of each truss element or ceiling membrane protection are the most widely used.

FLOOR/CEILING ASSEMBLIES

Ceiling components are important elements in the overall firesafe design of a floor/ceiling assembly. In the event of fire within a room, the ceiling acts as a barrier to protect the structural framing above it. The degree of protection, of course, depends upon the type of material, its installation, and its completeness. Combustible ceilings or ceilings that do not remain in place when subjected to the pressures and temperatures of a fire do not provide a significant degree of protection.

Ceilings may be applied directly to, or they may be suspended from, the underside of the floor framing. Figure 7-6DD illustrates a ceiling connected directly to the floor framing system, while Figure 7-6P shows a ceiling suspended from the floor slab. It is also common to apply a fire protective material such as a plaster or sprayed mixture, directly to the bottom surface of the floor slab. Figure 7-6AA illustrates a floor slab with plaster applied directly to the steel floor deck.

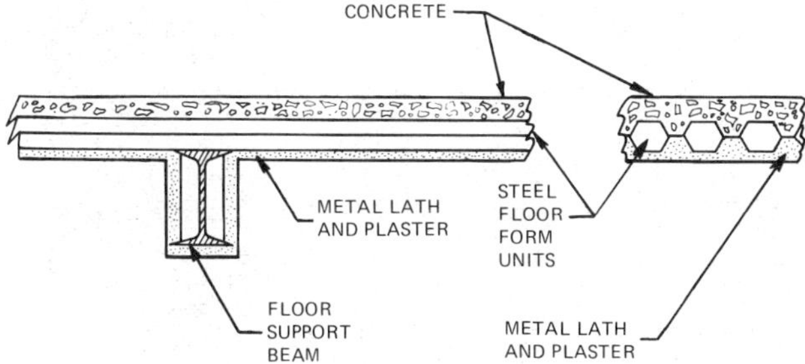

FIG. 7-6AA. Plaster applied directly to underside of floor slab construction.

There are several types of suspended membrane ceilings in common use. One type consists of a lay-in system in which the panels are supported by a grid. Another common type of membrane is installed by fitting slits in the panels into the sections of the suspension system. Suspended membrane ceilings may also consist of traditional lath and plaster or permanently secured gypsum wallboard with finished joists. Both of these latter systems can be either suspended below or attached directly to the basic subflooring system.

In the variety of arrangements currently used, the space above the ceiling may be of combustible construction and may also contain other combustible materials such as duct insulation, wire insulation, and vapor seals. Air handling systems may use ducts or the void space itself as a means of supplying, exhausting, or recirculating air; but associated materials must meet stringent requirements for combustibility and smoke evolution.

The material used for membrane ceiling is important. Gypsum plaster and lath was a common construction of early membranes. Special mineral tile formulations are used for acoustical panel systems that will not warp during fire exposure. The support framing is designed to accomodate thermal expansion with a minimum of distortion.

If lighting fixtures and duct openings are included in membrane ceilings that are part of fire rated assemblies, they both must be of suitable design and properly spaced. Fire test performance is based on specific lighting fixtures and spacing, so indescriminate substitution of lighting fixtures can cause premature failure. If air duct openings are installed in a membrane type ceiling, they should be suitably protected or dampered, as determined by a fire test on the ceiling assembly which incorporates the air handling components.

ROOF FRAMING

The design and construction of roof framing follow the general pattern for floor framing systems—both must support vertical loads and distribute these loads to walls or columns. Roof loads are usually smaller than floor loads. In addition, architectural considerations may demand longer spans than floor framing, and the shape of the roof need not be flat.

The roof covering may be supported by steel or wood joists or by a truss where a longer span is needed. Roof trusses can be made to conform to any roof shape, whether flat, pitched, or curved.

Open web steel or wood joists are often used in flat roof construction. An open web joist is merely a lightweight, parallel chord truss. They are usually spaced about 4 ft (1.22 m) center to center. This close spacing avoids the necessity of purlins, since the roof deck can be designed to span between the joists. Figure 7-6BB illustrates this type of roof framing.

In single story buildings that require a large open space, rigid frames are sometimes used both as columns and roof supports. The portion spanning the roof is rigidly connected to columns on opposite sides of the building to form a single member. The frame itself may provide for a sloping roof. Figure 7-6CC shows a rigid frame structure.

Flat slab and ribbed concrete slab systems may be used for roof construction in those buildings in which that is the predominant type of framing. Other systems utilizing long span prestressed concrete double-tee or channel sections are also common. These units serve both as support and as the deck. (See Fig. 7-6DD.)

Roof decks may also consist of precast reinforced concrete planks fastened to the structural steel supports by galvanized metal clips, as shown in Figure 7-6EE. These planks are manufactured in three general shapes: square edge, channel slab, and tongue and groove. To reduce weight they may be fabricated with aerated concrete using lightweight aggregates.

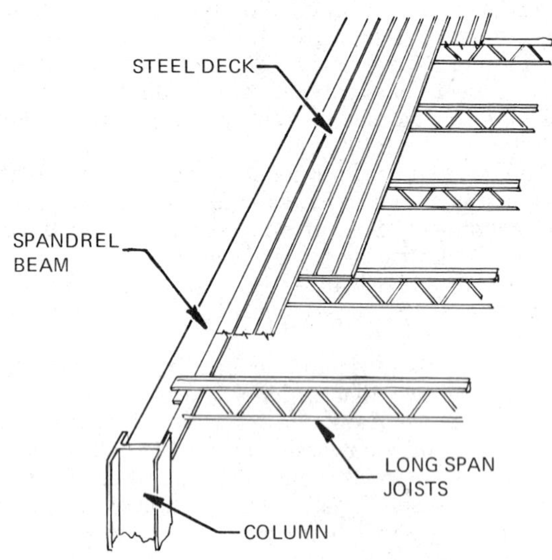

FIG. 7-6BB. Long span joist framing.

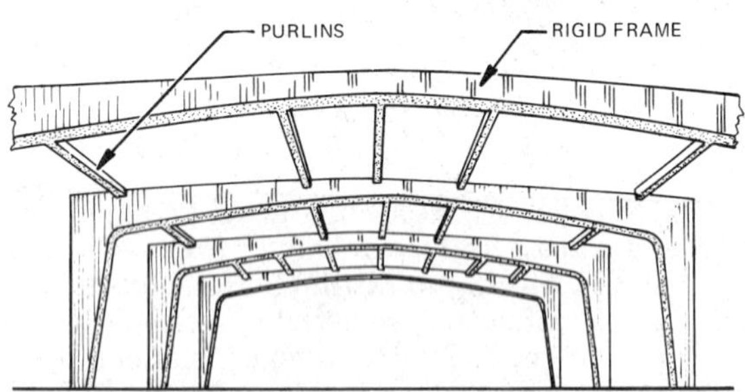

FIG. 7-6CC. A rigid frame.

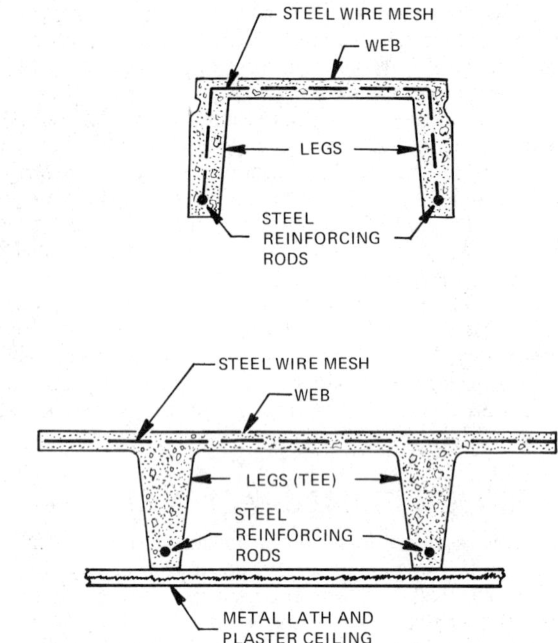

FIG. 7-6DD. Sections through channel and double-tee concrete slabs. The slab at bottom shows a metal lath and plaster ceiling attached directly to the ribs of the slab.

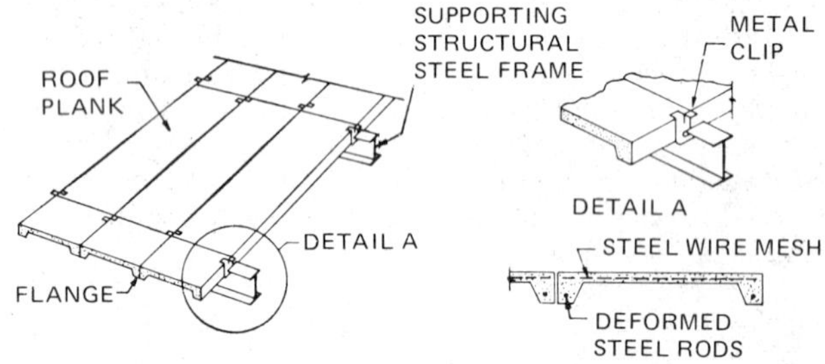

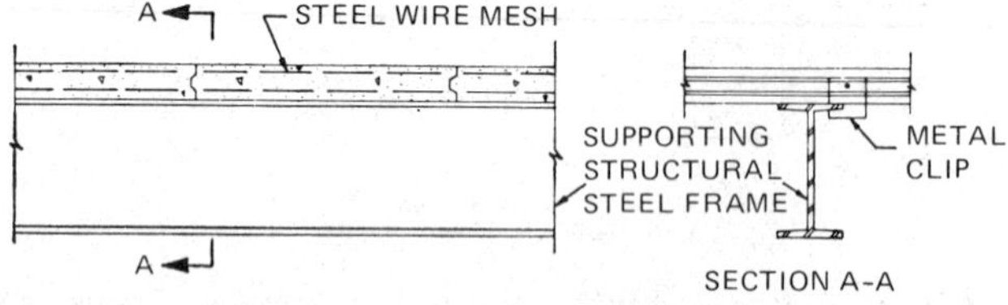

FIG. 7-6EE. Lightweight precast concrete planks.

ROOF COVERINGS

In resistance to ignition and burning, roof coverings range from combustible wood shingles with no fire retardant treatment to built up or prepared coverings which are effective against severe external fire exposure. Insofar as well designed roof coverings can protect buildings from exposure fires, the likelihood of fire spread from one building to another can be reduced considerably.

Fire Retardant Roof Coverings

Definitions of three classes of roof covering that include criteria for resistance to ignition have been prepared by Underwriters Laboratories Inc. (UL). They are:

Class A Coverings: Include roof coverings that are effective against severe fire exposures. These coverings are not readily flammable, do not carry or communicate fire, afford a fairly high degree of fire protection to the roof deck, do not slip from position, possess no flying brand hazard, and do not require frequent repairs to maintain their fire retardant properties.

Class B Coverings: Include roof coverings that are effective against moderate fire exposures. These coverings are not readily flammable, do not readily carry or communicate fire, afford a moderate degree of fire protection to the roof deck, do not slip from position, and possess no flying brand hazard; however, they may require repairs to maintain their fire retardant properties.

Class C Coverings: Include roof coverings which are effective against light fire exposure. These coverings are not readily flammable, do not readily carry or communicate fire, afford at least a slight degree of fire protection to the roof decks, do not slip from position, and possess no flying brand hazard; however, they require repairs or renewals to maintain their fire retardant properties.

Building codes commonly require Class A or B coverings within the fire limits of cities or wherever fire resistive construction is required. Class C roofing is appropriate for other buildings. Many cities specify Class C as the minimum standard for roofing.

Fire testing laboratories classify metal deck roof assemblies by type of construction (FMEC 1955). These classifications are discussed later in this chapter.

Built Up and Prepared Coverings

Fire retardant roof coverings fall within either of two groups. One is the built up covering which, as the name implies, consists of several layers of materials applied or "built up" on the roof decks according to specifications. What is known as a "tar and gravel" roof is actually a built up roof covering consisting of several layers of roofing felts and insulating panels or sheets bonded together by hot or cold cements and topped with roofing gravel. (See Fig. 7-6FF.)

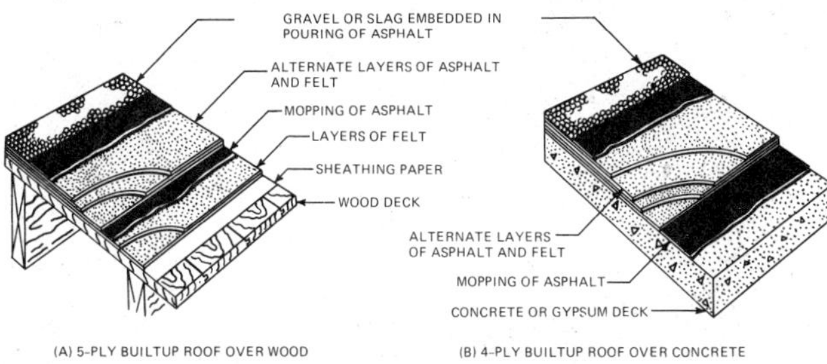

FIG. 7-6FF. Typical built up roof coverings with hot asphalt mopping or coal tar pitch between each two layers.

The second group (prepared coverings) includes fire retardant shingles and sheet coverings, which can be applied only to roof decks capable of receiving and retaining nails and which are sloped sufficiently to permit drainage. In some instances, however, the slope must not exceed a specified maximum in order to qualify for Class A, B, or C fire retardant rating. Asphalt organic felt shingles, i.e., "composition" roofs, are a common example of prepared roof coverings and are frequently used on dwellings. (See Fig. 7-6GG.)

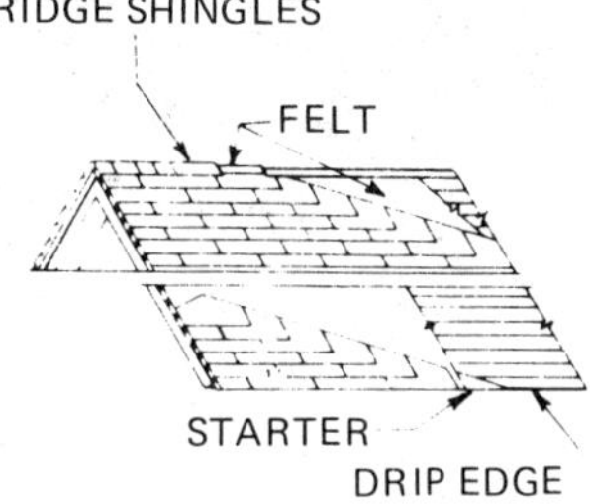

FIG. 7-6GG. Installation of a typical prepared roof covering.

Prepared roof coverings also include such noncombustible coverings as brick, concrete, tile, and slate. Test methods have been developed by UL for measuring the fire retardant characteristics and classifications of roof coverings. These test methods have been adopted in NFPA 256, *Standard Methods of Fire Tests of Roof Coverings*. Tables 7-6C and 7-6D list various types of prepared and built up roof coverings and their fire retardant classifications. Any roof covering can be used for a less severe exposure than the one for which it is listed.

In addition to the built up roof coverings described in Table 7-6D, there are several other types consisting of asbestos, combinations of asbestos and rag felt, glass fiber, aluminum, and tile applied in a special manner with special listed proprietary cements that are listed by testing laboratories.

Roof Deck Insulations and Vapor Barriers

Since the 1953 General Motors fire at Livonia, MI, major attention has been paid to the combustibility of insulated metal roof decks. In that fire, the asphalt mopping in the felt vapor barrier beneath the insulation on the 1,502,500 sq ft (139,582 m^2), flat, steel deck roof melted and vaporized. Asphalt vapor under pressure entered the building through the joints in the steel deck and burned inside. As a result of this and similar fires, research was conducted to find a solution to the problem. The significant findings are summarized in the following paragraphs.

The combustibility of a roof was historically determined by the physical characteristics of the roof deck and its supporting structure, with no consideration given to the combustibility of the roof insulation and covering. Thus, combustible fiberboard insulation and asphalt impregnated felt roof covering normally did not affect the classification of a noncombustible roof deck on noncombustible supports. Large scale fire tests, confirmed by actual fire experience, however, have shown that under certain conditions some metal deck assemblies can contribute to an interior fire.

A series of full scale tests were conducted in 1955 at Factory Mutual Engineering Corp. testing facilities to compare the performance of various metal roof deck constructions with respect to flame spread along the underside of the deck. The tests were run in cooperation with the Metal Roof Deck Technical Institute (now Steel Deck Institute) and the Insulation Board Institute, and the results were published in the pamphlet *Insulated Metal Roof Deck Fire Tests* (FM 1955). Conventional construction, with the insulation fastened to the metal deck with complete moppings of asphalt or with vapor barriers using complete asphalt moppings above and below one or more plies of felt, was shown to contribute significant amounts of fuel when exposed to the heat of an intensive interior fire below the deck. (See Fig. 7-6HH.) The asphalt was vaporized by the heat of the fire below, and unable to escape upwards through the roof covering, was forced down through the joints in the deck, when it ignited and contributed to the spread of fire under the roof deck.

Recognized fire testing laboratories now list metal roof deck construction and components that will not contribute significantly to an interior fire and that meet certain performance standards. A typical built up roof covering of this construction is shown in Figure 7-6II.

Metal roof deck constructions use both combustible and noncombustible insulation fastened with noncombustible adhesives or mechanical fasteners to the upper surface of the deck. Asphalt can also be used as the adhesive if applied in strips (strip mopping), and if the total amount used between the deck and the insulation does not exceed 12 to 15 lb per 100 sq ft (5.4 to 6.8kg/9.3m^2) of roof area. There is no restriction on the use of asphalt in conven-

TABLE 7-6C. Typical Prepared Roof Coverings*†

Description	Minimum Incline In. to Ft	CLASS A	CLASS B	CLASS C
Brick Concrete Tile Slate		Brick, 2¼ in. thick. Reinforced portland cement, 1 in. thick. Concrete or clay floor or deck tile, 1 in. thick. Flat or French type clay or concrete tile, ⅜ in. thick with 1½ in. or more end lap and head lock, spacing body of tile ½ in. or more above roof sheathing, with underlay of one layer of Type 15 asphalt-saturated asbestos felt or one layer of Type 30 or two layers of Type 15 asphalt-saturated organic felt. Clay or concrete roof tile, Spanish or Mission pattern, 7/16 in. thick, 3 in. end lap, same underlay as above. Slate, 3/16 in. thick, laid American method.		
Metal Roofing	12	Sheet roofing of 16 oz copper or of 30-gage steel or iron protected against corrosion. Limited to noncombustible roof decks or noncombustible roof supports when no separate roof deck is provided.	Sheet roofing of 16 oz copper or of 30-gage steel or iron protected against corrosion or shingle-pattern roofings with underlay of one layer of Type 15 saturated asbestos-felt, or one layer of Type 30 or two layers of Type 15 asphalt-saturated organic felt.	Sheet roofing of 16 oz copper or of 30 gage steel or iron, protected against corrosion, or shingle-pattern roofings, either without underlay or with underlay of rosin-sized paper.
				Zinc sheets or shingle roofings with an underlay of one layer of Type 30 or two layers of Type 15 asphalt-saturated organic-felt or one layer of 14 lb unsaturated or one layer of Type 15 asphalt-saturated asbestos felt.
Cement-Asbestos Shingles	Exceeding 4	Laid to provide two or more thickness over one layer of Type 15 asphalt-saturated asbestos felt.	Laid to provide one or more thickness over one layer of Type 15 asphalt-saturated asbestos felt.	

TABLE 7-6C. Typical Prepared Roof Coverings (Cont.)*†

Description	Minimum Incline In. to Ft	CLASS A	CLASS B	CLASS C
Asphalt-Asbestos Felt Sheet Coverings	Not Exceeding 12	Factory-assembled sheets of 4 ply asphalt and asbestos material.	Factory-assembled sheets of 3 ply asphalt and asbestos material or sheet coverings of single thickness with a grit surface.	Single thickness smooth surfaced.
Asphalt-Asbestos Felt Shingle Coverings	Exceeding 4		Asphalt-asbestos felt grit surfaced.	
Organic-Felt (previously referred to as rag felt) Sheet Coverings	Exceeding 4			Sheet coverings of asphalt organic felt either grit surfaced or aluminum surfaced.
Organic-Felt (previously referred to as rag felt) Shingle Covering, with special coating	Sufficient to permit drainage	Grit surfaced, two or more thicknesses.	Grit surfaced, two or more thicknesses.	
Organic-Felt (previously referred to as rag felt) Shingle Coverings	Sufficient to permit drainage	Grit surfaced, two or more thicknesses.	Grit surfaced, two or more thicknesses.	Grit surfaced shingles, one or more thicknesses.
Asphalt Glass Fiber Mat Shingle Coverings	Sufficient to permit drainage	Grit surfaced, two or more thicknesses.	Grit surfaced, one or more thicknesses.	Grit surfaced shingles, one or more thicknesses.
Asphalt Glass Mat Sheet Covering	Sufficient to permit drainage			Grit surfaced.
Fire-retardant treated red cedar wood shingles and shakes	Sufficient to permit drainage			Treated shingles or shakes, one or more thicknesses; shakes require at least one layer of Type 15 felt underlayment.

* Prepared roof coverings as classified as applied over square-edge wood sheathing of 1-in. nominal thickness, or the equivalent, unless otherwise specified. See footnote (Built-Up Roof Coverings) to Table 7-6D. Laid in accordance with instruction sheets accompanying package. Limited to decks capable of receiving and retaining nails.

Where organic-felt is indicated, asbestos felt of equivalent weight can be substituted.

By end lap is meant the overlapping length of the two units, one place over the other. Head lap in shingle-type roofs is the distance a shingle in any course overlaps a shingle in the second course below it. However, with shingles laid by the Dutch-lap method, where no shingle overlaps a shingle in the second course below, the head lap is taken as the distance a shingle overlaps one in the next course below.

Prepared roofings are labeled by Underwriters' Laboratories which indicate the classification when applied in accordance with direction for application included in package.

† 1 in. = 25.4 mm; 1 oz = 29 mL; 1 ft = 0.3 m.; 1 lb = 0.45 kg

tional built up roofing used above the insulation. When a vapor seal is used between the deck and the insulation, a listed, noncombustible, or slow burning material is recommended, fastened to the deck and insulation with a noncombustible adhesive or strip mopping. Testing laboratories list various manufacturers of acceptable vapor seals and adhesives. A roof cover with a slow burning vapor seal is shown in Figure 7-6II.

Factory Mutual Engineering Corp. divides metal deck roof assemblies into two classes. Class I construction is any insulated metal roof deck construction having a special vapor barrier (if any is used) and an adhesive that will not contribute significantly to an interior fire and that will meet or equal FM performance standards. Class II construction is any insulated metal roof deck construction using asphalt in sufficient quantity to provide adequate adhesion for wind resistance, with or without felts between the insulation and the deck, and regardless of the type of insulation used. This class of construction does not meet FM standards and can provide fuel that will contribute significantly to an interior fire. Constructions which are found acceptable are listed by Factory Mutual (FMEC 1983).

Wood Shingles

Untreated wood shingles may be readily ignited by small sparks from chimneys or exposure fires, by radiated heat, or by burning brands. Burning shingles themselves also produce brands, which may be carried by the wind to

TABLE 7-6D. Built up Roof Coverings*†

Description	Minimum Incline In. to Ft	CLASS A	CLASS B	CLASS C
Asphalt organic-felt, bonded with asphalt and surfaced with 400 lbs of roofing gravel or crushed stone, or 300 lbs of crushed slag per 100 sq ft of roof surface, on coating of hot mopping asphalt.	3	4 (plain) or 5 (perforated) layers of Type 15 felt. 1 layer of Type 30 felt and 2 layers of Type 15 felt 1 layer of Type 15 felt and 2 layers of Type 15 or 30 cap or base sheets. 3 layers of Type 15 or 30 cap or base sheets. 3 layers of Type 15 felt. Limited to non-combustible decks.	4 layers of perforated Type 15 felt. 3 layers of Type 15 felt. 2 layers of Type 15 or 30 cap or base sheets.	
Tar-asbestos-felt or organic-felt bonded with tar and surfaced with 400 lb of roofing gravel or crushed stone, or 300 lb of crushed slag per 100 sq ft of roof surface on a coating of hot mopping tar.	3	4 layers of 14 lb asbestos-felt or Type 15 organic-felt. 3 layers of 14 lb asbestos-felt or Type 15 organic-felt.	3 layers of Type 14 lb asbestos-felt or Type 15 organic-felt.	
Steep tar organic-felt	5	4 layers of Type 15 tar-saturated organic-felt, bonded with steep coal-tar pitch, surfaced with 275 lbs of ⅝ in. crushed slag per 100 sq ft of roof surface on steep coal-tar pitch.		
Asphalt organic-felt, plain or perforated, bonded and surfaced with a cold application coating.	12			3 layers of Type 15 felt. 1 layer of Type 30 felt and 1 layer of Type 15 felt. 2 layers of Type 15 or 30 cap or base sheets. 2 layers of Type 15 felt and 1 layer or Type 15 or 30 cap or base sheets.

* Built-up roof coverings are classified as applied over square-edge wood sheathing of 1 in. nominal thickness, or the equivalent, unless otherwise specified.

From the standpoint of relative effectiveness of the different types of wood roof sheathing, the tongue-and-groove boards and ¾ in. moisture-resistant plywood give better results in the brand and flame tests than square-edge sheathing with boards spaced about ¼ in. apart. For classifications based on square-edge sheathing, tongue-and-groove or plywood sheathing can be substituted. Square-edge sheathing boards should be butted together as closely as possible. Reference to ¼ in. spacing is to indicate fire test procedure intended to simulate actual conditions after shrinkage of boards due to age or other reasons.

The minimum weight of cementing material between separate layers of felt is considered to be 25 lbs per 100 sq ft of roof surface.

Types 15 and 30 felts are defined as saturated felts weighing a minimum of 14 lbs and 28 lbs per 100 sq ft of the finished materials, respectively. Where saturated felts are referred to by weight, the weight is minimum and is expressed in pounds per 100 sq ft of the finished material.

Materials intended for built up roof coverings are labeled by Underwriters' Laboratories. The classifications indicated are of generally accepted combinations.

† 1 in. = 25.4 mm; 1 ft = 0.3 m.

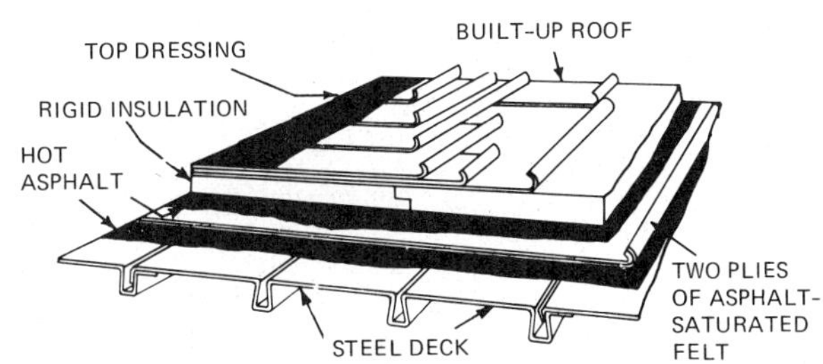

FIG. 7-6HH. Typical built up roof covering with a combustible vapor barrier adhered to the roof deck and to roof insulation by a combustible adhesive.

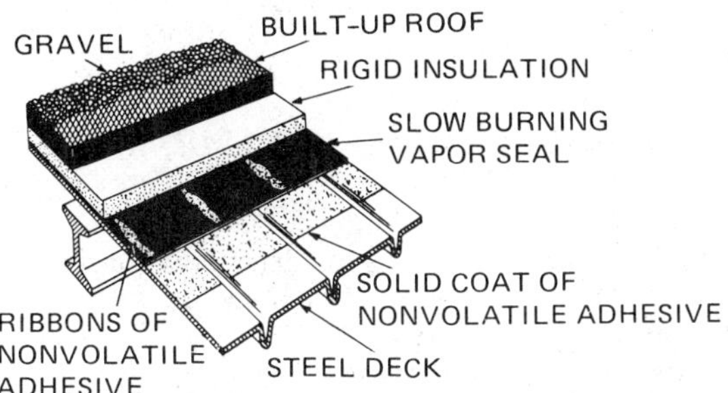

FIG. 7-6II. Typical built up roof covering with a slow burning vapor seal adhered to the roof deck and to the insulation by a nonvolatile adhesive.

start other fires. Age and low humidity increase the susceptibility of wood shingled roofs to ignition.

Treatment of wood shingles with fire retardant coatings has been proposed at various times but has not proved practical to date. Ordinary flame resistant treatments lose their effectiveness with continued exposure to the weather. Wood shingles impregnated with a fire retardant solution are more durable. Shingles of this type have been tested and carry a Class C rating.

Untreated wood shingle roofs are prohibited by law in the congested sections of practically all large cities, and a very large number of cities and towns prohibit their use altogether within municipal limits.

Because of the somewhat greater susceptibility of nonsealing asphalt shingles to windstorm damage, wood shingles have been more widely used than asphalt shingles in certain areas particularly subject to high winds. However, where a recently developed self-sealing type of asphalt shingle with good fire resistance has been used, the incidence of windstorm damage has been materially reduced.

Bibliography

References Cited

FMEC. 1983. *Insulated Steel Deck*. Loss Prevention Data Sheet 1-28. May 1983. Factory Mutual Engineering Corp., Norwood, MA.

FMEC. 1977. *Protection Against Liquid Damage*. Loss Prevention Data Sheet 1-24, Nov 1977. Factory Mutual Engineering Corp., Norwood, MA.

FMEC. 1955. *Insulated Metal Roof Deck Fire Test*. Factory Mutual Engineering Corp., Norwood, MA. May 1955.

NFPA Codes, Standards, Recommended Practices and Manuals. (See the latest *NFPA Codes and Standards Catalog* for availability of current editions of the following documents.)

NFPA 256, *Standard Methods of Fire Tests of Roof Coverings*.

Additional Readings

Bescher, R. H., "A New Class C Treatment for Wooden Shingles and Shakes," *Fire Journal*, Vol. 61, No. 5, Sept. 1967 , pp. 52-56.

Brannigan, Francis L., *Building Construction for the Fire Service*, National Fire Protection Association, Quincy, MA 1982.

Platzker, J., "Regulations on Wood-Shingle Roofing: A Survey," *Fire Journal*, Vol. 60, No. 4, July 1966, pp. 36-39.

UL, *Building Materials Directory and Annual Supplement*, Underwriters Laboratories Inc., Northbrook, IL, (directory annually in January; supplement annually in July).

Wilson, R., "Wood Shingles, 1959," *NFPA Quarterly*, Vol. 53, No. 2, Oct. 1959, pp. 99-110.

Yuill, C. H., "Floor Coverings: What is the Hazard?" *Fire Journal*, Vol. 61, No. 1, Jan. 1967, pp. 11-19.

INTERIOR FINISH

Revised by Norman DeHaan

Three principal elements which determine the fire hazard of a building are the fire resistance of the structure, the contents or process enclosed by the structure, and the characteristics of the interior finish of the structure. The importance of these three elements is frequently misunderstood or underestimated. The elements should be considered separately and as fully as possible in order to appreciate their impact on firesafety.

This chapter will discuss the materials commonly used for wall and ceiling finishes, such as wood, plaster, wallboards, acoustical tile, insulating materials, and decorative materials inside buildings; in addition, interior floor finishes are discussed. Fire resistance and fire resistance ratings, which have no essential relationship to the fire properties of interior finish materials or interior floor finish materials, are fully discussed in Chapter 8, "Structural Integrity During Fire," in this section. (An example of this "nonrelationship": a heavy timber structural member may have a fire resistance rating of 1 hr or more, but it still presents a combustible interior surface; conversely, a bare sheet metal wall has little or no fire resistance rating, since heat penetrates it quickly, but it presents no surface combustibility.)

DEFINITION OF INTERIOR FINISH AND INTERIOR FLOOR FINISH

Interior finish is generally considered to consist of those materials or combinations of materials that form the exposed interior surface of walls and ceilings in a building. Interior floor finishes are considered to mean the exposed floor surfaces of buildings and include floor coverings such as carpets and floor tiles which may be applied over, or in lieu of, a finish floor. Variations of this basic definition of interior finish are found in some building regulations where counter tops, built in cabinets, and even doors are included as interior finish. Many codes such as NFPA 101, *Life Safety Code* (hereinafter referred to as NFPA 101), exclude trim and incidental finish from the requirements for wall and ceiling finish; less rigid requirements are set for trim of a type acceptable to the code in question if they comprise less than ten percent of the aggregate wall and ceiling area.

Free hanging draperies that cover most or all of a wall surface have on occasion been subjected to the requirements for interior finish, as have framed, flexible (folding) door assemblies. The normal tests for interior finish combustibility do not accurately predict fire behavior in these cases. Such materials are more properly tested according to NFPA 701, *Standard Methods of Fire Tests for Flame-Resistant Textiles and Films*. However, when applied to a solid backing such as gypsum wallboard, drapery material can be considered as interior finish and tested accordingly.

Many local building regulations have been expanded to include provisions regulating floorings and floor coverings, either by including them as interior finish or by requiring they meet other criteria based on testing in accordance with one or more test methods.

NFPA 101 does not regard floorings and floor coverings to be interior finish; instead, the Code considers them to be "interior floor finish." Thus, unless the authority having jurisdiction determines that the material used poses an unusual hazard, floorings and floor coverings are excluded from interior finish requirements but are required, where so specified, to comply with special "interior floor finish" provisions. These provisions are based on NFPA 253, *Standard Method of Test for Critical Radiant Flux of Floor Covering Systems Using a Radiant Heat Energy Source* (described later in this chapter). Another test method utilized by some jurisdictions to regulate floorings and floor coverings is the "Chamber Test" developed by Underwriters Laboratories Inc. Further, under the Flammable Fabrics Act, the U. S. Department of Commerce now regulates rugs and carpets for U. S. consumption and for hospitals. Some states also regulate the use of systems office furnishings if the movable partitions exceed a specified height. To meet these criteria, the Business and Institutional Furniture Manufacturers Association (BIFMA) has established voluntary flammability test criteria for their members products. Determination of hazard under the Flammable Fabrics Act is based on the Methanamine "Pill" Test (described later in this chapter).

Mr. DeHaan, FASID and AIA, is president Norman DeHaan Associates, Architects and Interior Designers, Chicago, IL.

TYPES OF INTERIOR FINISH

The types of interior finish materials are numerous and include such commonly used materials as plaster, gypsum wallboard, wood, plywood paneling, fibrous ceiling tiles, plastics, and a variety of wall coverings. Collectively, these finishes serve several functions—aesthetic, acoustical, insulating, as well as protecting against wear and abrasion. Ordinary paint, wallpaper, or other similar wall coverings not exceeding 1/28 in. (0.6 mm) in thickness are not generally included as interior finish, except where deemed to be a hazard by the authority having jurisdiction. Chicago, IL, and Boston, MA, for example, establish stringent wall covering criteria in areas of public assembly, fire exit stairs and passageways, and in lobby egress areas.

The development of certain types of cellular plastics in board, poured in place, and sprayed on form has provided lightweight materials with exceptional thermal insulation. The incorporation of fire retardants into some cellular plastics products made it possible for them to meet building code requirements for interior finishes. As a result, cellular plastics, particularly the sprayed on type, have been widely used as exposed insulation. Rapid fire spread in several widely publicized fires involving exposed cellular polyurethane and polystyrene materials has led to federal action and industry recommendations for protection of such surfaces against ignition and fire spread (FTC 1974, SPI 1974). NFPA 101 states that cellular or foamed plastic materials shall not be used as interior finish, except under certain conditions.

THE ROLE OF INTERIOR FINISH IN FIRES

Most building fires begin when decorative materials, furnishings, or waste accumulations ignite, or when electrical systems or mechanical devices fail. Interior finishes are not usually the first items ignited, except when ignition occurs by overheated electrical circuits, careless use of plumbers' torches, or direct impingement of flame from some other source, e.g., a candle or a match. After the fire has started and intensified, however, the interior finish can become involved and can contribute extensively to the spread of fire.

The combustibility of the interior finish is particularly significant. Combustible interior finish in an enclosure can greatly reduce the rate of heat release necessary for flashover as well as reduce the preburn time before flashover. Full scale room and building fire tests have shown that flashover is caused by thermal radiation feedback from the ceiling and upper walls which have been heated by the fire. This radiation feedback gradually heats the contents of the fire area. When all the combustibles in the space have become heated to their ignition temperatures, simultaneous ignition occurs. Interior finish plays an important role in the occurrence of flashover—an interior finish that absorbs heat readily and holds it, as an insulator would, might reduce the time to flashover. If the finish material is combustible, it will be a source of fuel for the fire. Considering the nature of thermal radiation, the size and shape of the space in which the fire occurs becomes a critical factor (Waterman 1972). More recently, the term "flameover" was coined to denote the rapid spread of flame over one or more surfaces. This reaction was observed in corridor tests conducted by the National Bureau of Standards (NBS 1973).

Once a room fire reaches the stage of full involvement or flashover, and the openings to adjoining spaces allow heat, smoke, and combustion gases to escape, combustible interior finish of any kind or quantity becomes a significant factor in the spread of fire to other areas. Several full scale fire tests have demonstrated that heat, smoke, and noxious combustion gases from burning furnishings may pose a greater threat to the life safety of persons unable to evacuate the room of fire origin than do those from interior finishes (Bruce 1959; Pryor 1969). However, because the life safety of others in the building and potential property damage are important considerations, the nature of the interior finish in and beyond the room of origin is a serious and legitimate concern.

Interior finish relates to a fire in four ways. It can: (1) affect the rate of fire buildup to a flashover condition, (2) contribute to fire extension by flame spread over its surface, (3) add to the intensity of a fire by contributing additional fuel, and (4) produce smoke and toxic gases that can contribute to life hazard and property damage (Christian 1974). An interior finish that would provide the ultimate safety would be made of a relatively dense and noncombustible material; the material would be a good conductor of heat, would not speed up flashover, would not add fuel to the fire, would provide no path for surface flame spread, and would produce little or no smoke or toxic gases. Materials which exhibit high rates of flame spread, contribute substantial quantities of fuel to a fire, or produce hazardous quantities of smoke or toxic gases, would be undesirable.

The effort to develop controls on the use of interior finish began in earnest in 1946 when three hotel fires—the Winecoff in Atlanta, GA; the LaSalle in Chicago, IL; and the Canfield in Dubuque, IA—took the lives of 199 persons. Each of these fires involved delayed discovery of the fire, open stairways, and combustible interior finish. Interior finish materials continue to be a serious life safety factor, as many recent fires indicate; for example, interor finishes contributed to the Beverly Hills Supper Club fire at Southgate, KY, May 28, 1977, in which 164 persons died (NFPA 1978).

METHODS OF APPLICATION

The method of application of an interior finish can have a serious effect on its behavior when the finish is exposed to fire. For this reason, the manufacturer's specifications for application should be strictly followed. Items that require special attention include size and spacing of nails or similar fasteners, type and application of adhesives, and the number of coats and the application rate of fire retardant coatings. Application details are equally important in repair or replacement of interior finishes. Any new material used in retrofit should meet current code requirements. Substandard application or substitution of "equivalent" materials without supporting test data has delayed the occupancy of many new buildings.

Surface finishes should be considered with recognition of the substrate material to which they are attached. A thin combustible finish applied to a noncombustible substrate may present little hazard. The same finish material, however, on a combustible backing presents considerably greater hazards. In the former situation, the substrate will not ignite and will absorb heat during the early stages of fire development; in the latter case, both the surface finish and the backing material become involved.

The adhesive used is also a factor in the fire behavior of interior finishes. Adhesives that soften at moderate temperatures will allow wall or ceiling finishes to drop or peel from place during the growth stage of a fire. This not only increases the susceptibility of the surface material to ignition but also exposes the substrate material which, if combustible, adds fuel to the fire. Some building codes specify that wall and ceiling finishes should not become detached under exposure to elevated temperature [200 to 300°F, (93 to 149°C)] for a specified time interval (usually 30 minutes). Unfortunately, no standard test method has yet been developed for this purpose.

Most building regulations provide that a combustible material in thin sheets less than 1/28 in. (0.6 mm) thick, applied to backings, shall not be subject to the requirements for interior finishes. This exception has been shown by test to be valid for the paper surface on gypsum wallboard and for ordinary paint. This, of course, assumes proper application procedures. Where successive coats have been applied, the fire hazard can be greatly magnified.

Some building regulations further require that combustible interior finish materials of other than the most restrictive class, less than 1/4 in. (6.3 mm) thick and intended to be applied directly to studs or joists, shall be applied over a substrate of a noncombustible material. This requirement followed upon several rapidly burning residential fires where thin combustible paneling that was directly applied to the framing provided fuel for the fire.

The use of sprayed on materials for interior finishes has been proliferating. Some of these materials are coatings that are inherently resistant to attack by fire; other materials, however, especially the sprayed on cellular (foamed) plastics used for insulation or decorative effects, are not resistive to attack by fire. This is particularly true of cellular plastics.

FIRE TESTS FOR INTERIOR FINISHES

The nature of materials and the fire environment vary so widely that the development of a fire test becomes highly complex. Three factors must be taken into account in this development process:

1. The start and growth of fire in a building is affected by the ignition source, space geometry, ventilation, and the nature, amount, and location of other processes and materials in the building.
2. The changing conditions during a fire such as oxygen concentration, rate of heat release, protection systems, etc.
3. Variations in form, composition, density, and application of the materials present.

With these variables in mind, the difficulty in designing a test that will provide a basis for predicting performance under fire exposure becomes obvious. Equally obvious is the impracticality of designing tests to represent all fire conditions. A test designed to represent a "typical" fire situation or to expose materials to one set of "standard test" conditions may not provide a reliable basis for predicting "real life" performance of all materials tested. Thus, there is a constant search for improved test methods having a numerical range of results, and for an adequate array of tests to suitably describe the behavior of the various materials available.

The Steiner Tunnel Test

The 25 ft (7.62 m) Steiner Tunnel Test, also known as ASTM E84, NFPA 255, and UL 723, was developed by A. J. Steiner at Underwriters Laboratories Inc. (UL) (Wilson 1961). After the fatal hotel fires of 1946 mentioned earlier, the need for some method for control of interior finishes was recognized and use of the tunnel test proposed. The method was adopted by the American Society for Testing and Materials (ASTM) as a tentative standard in 1950 and as an official standard in 1958. NFPA adopted the test method tentatively in 1953 and officially in 1958 as NFPA 255, *Standard Method of Test of Surface Burning Characteristics of Building Materials* (hereinafter referred to as NFPA 255).

Figures 7-7A and 7-7B show the general appearance and basic dimensions of the furnace. A detailed description may be found in NFPA 255. Briefly, a 20 in. by 25 ft (508 mm by 7.62 m) specimen is placed on a ledge in the top of the furnace in a face down position. A gas burner is ignited, and flame travel over the bottom face of the test specimen is observed through sealed windows on one side of the furnace, forming the basis of the rating for flame spread. Furnace temperature and smoke density are also recorded, and these figures can provide a basis for calculating fuel contributed by the specimen and smoke developed. All three measurements are relative to asbestos cement board, arbitrarily assigned a 0 rating, and to red oak flooring, arbitrarily assigned a rating of 100. Provision may be made for the measurement and analysis of combustion gases. There is not necessarily a relationship between the flame spread, smoke development, and fuel contribution ratings obtained by this test method.

The tunnel test was designed to provide a moderately severe exposure of approximately 1,400°F (760°C) in the area of flame impingement under controlled conditions. The specimen size is scaled sufficiently large to simulate the effect of joints, lack of uniformity within the material, and the synergistic reaction of composite surface finishes. The draft, fuel input, and flame size are varied to provide a wide range of numerical results.

At least fifteen tunnel furnaces are known to be in use in the U.S. and Canada. While several are reserved for product development purposes by individual companies, the majority are available for product acceptance testing.

Application of Tunnel Test Results

There is a considerable degree of uniformity in the application of tunnel test results with regard to flame spread ratings. Smoke developed ratings are used more extensively today, but with less uniformity. Fuel contributed factors have been applied only in a few scattered instances. Most building regulations base requirements for interior wall and ceiling finish on tunnel test results.

Recognizing that variations do exist, NFPA 101 treats the classification of interior finish materials as follows:

Classification Flame Spread Smoke Developed

Classification	Flame Spread	Smoke Developed
A	0–25	0–450
B	26–75	0–450
C	76–200	0–450

In many building regulations, vertical exits are most severely restricted with a requirement for flame spread

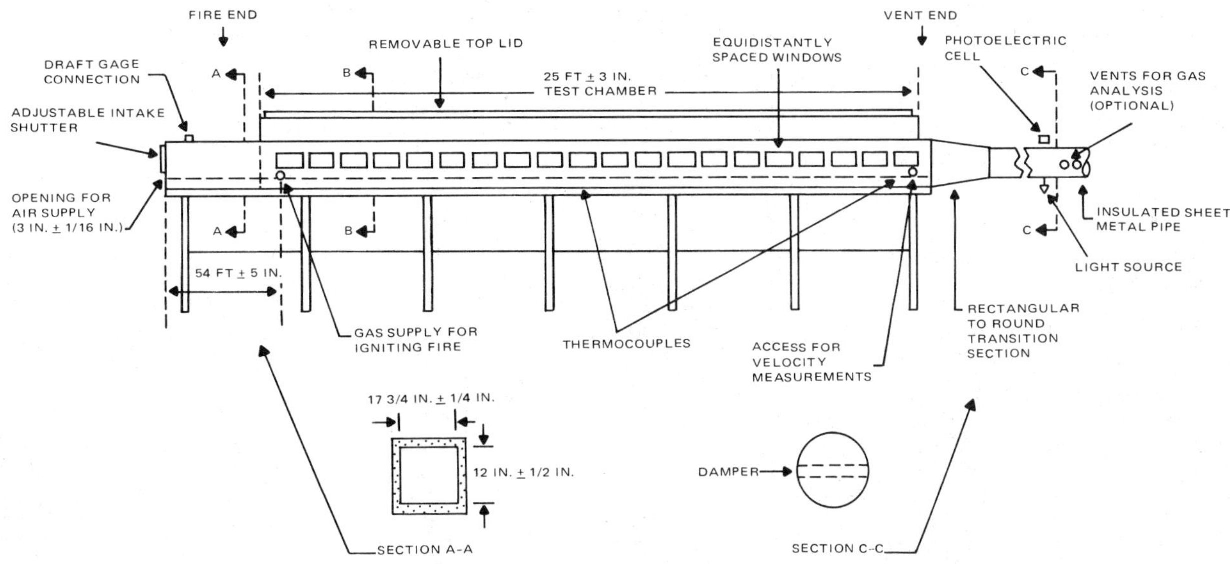

FIG. 7-7A. A schematic diagram of the Steiner Tunnel Test apparatus used for the fire hazard classification of building materials.

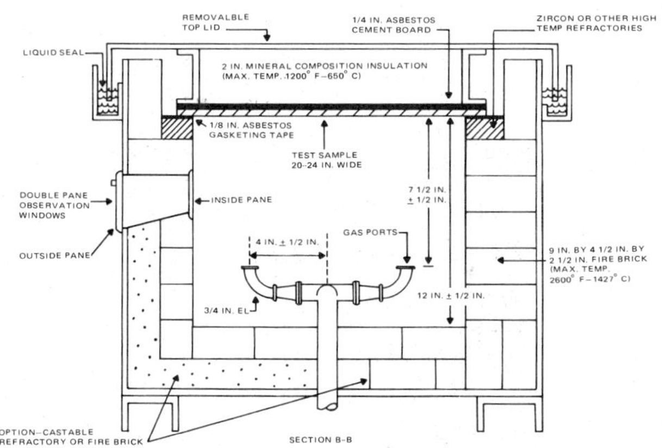

FIG. 7-7B. A cross sectional view of the Steiner Tunnel Test apparatus.

rating not exceeding 25 (or Class A), up to 75 flame spread rating (Class B) in horizontal exitways. The limits on interior wall and ceiling finishes set forth in the 1985 edition of NFPA 101 are given in Table 7-7A, which also includes requirements for interior floor finishes based on two classifications (Class I and II). These classifications are used to evaluate floor covering materials in the critical radiant flux test (as contained in NFPA 258, *Standard Research Test Method for Determining Smoke Generation of Solid Materials*, hereinafter referred to as NFPA 258) described later in this chapter.

The performance of materials in the tunnel test depends upon the quality and composition of the material tested, as well as its application or use. Generally, the higher the numerical rating, the greater the flammability hazard. When automatic fire suppression is provided, NFPA 101 will allow interior finish materials of a higher class in certain occupancies to be used when the structure is so protected.

Eight Foot Tunnel Furnace

Many attempts have been made to develop a smaller scale test that can be used by manufacturers and testing laboratories in place of the 25 ft (7.62 m) tunnel furnace. The advantages of such a facility are obvious—less space required for the equipment, smaller specimens needed, and lower costs.

Such an effort was undertaken at the Forest Products Laboratory in 1951. Initially, a furnace was built following the plan of the 25 ft (7.62 m) furnace, but reduced in scale (Peters and Eichner 1962). The reduction in scale, however, lengthened the tongues of flame from burning specimens and made visual measurement of the flame front difficult. A new approach was devised and a test apparatus developed that was quite different from the 25 ft (7.62 m) tunnel furnace. The test method for using this apparatus was tentatively adopted by ASTM in 1965, and officially adopted for research and development in 1968 (ASTM 1969).

In this test, a 13¾ in. by 8 ft (0.35 by 2.44 m) specimen is supported at a 30° angle from the horizontal by a perforated panel of 12 gage stainless steel. This panel separates the upper part of the furnace from the combustion chamber. A burner at one end of the combustion chamber provides the proper ratio of gas to air to produce a blue flame at approximately 3,400 Btu per min (959.76W). Open view ports on one side of the furnace also allow air to be drawn across the specimen and into a full length stack via a self-induced draft. As in the 25 ft (7.62 m) furnace, asbestos cement board and red oak flooring are used as comparison standards and assigned flame spread ratings of 0 and 100, respectively.

Several manufacturers have built this furnace for product development purposes and some authorities will accept results in lieu of other test data. Good correlations with the large tunnel furnace have been obtained for many materials. Some manufacturers have also constructed scale models of the 25 ft (7.62 m) furnace with good results reported for specific products.

TABLE 7-7A. *Life Safety Code* Requirements for Interior Finish

Occupancy	Exits	Access to Exits	Other Spaces
Assembly—New*			
Class A or B	A	A or B	A or B
Class C	A	A or B	A, B, or C
Assembly—Existing*			
Class A or B	A	A or B	A or B
Class C	A	A or B	A, B, or C
Educational—New*	A	A or B	A or B C on movable partitions
Educational—Existing*	A	A or B	A, B, or C
Day-Care Centers—New	A I or II	A I or II	A or B
Day-Care Centers—Existing	A or B	A or B	A or B
Group Day-Care Homes	A or B	A, B, or C	A, B, or C
Family Day-Care Homes	A or B	A or B	A, B, or C
Health Care—New	A I	A B lower portion of corridor wall I	A B in small individual rooms
Health Care—Existing	A or B	A or B	A or B
Detention & Correctional—New	A I	A I	A, B, or C
Detention & Correctional—Existing	A or B I or II	A or B I or II	A, B, or C
Residential, hotels & Dormitories—New	A I or II	A or B I or II	A, B, or C
Residential, hotels & Dormitories—Existing	A or B I or II	A or B I or II	A, B, or C
Residential, apartment buildings—New	A I or II	A or B I or II	A, B, or C
Residential, apartment buildings—Existing	A or B I or II	A or B I or II	A, B, or C
Residential, Board and Care†			
Residential, 1- and 2-family, lodging or rooming houses	A, B, or C	A, B, or C	A, B, or C
Mercantile—New*	A or B	A or B	A or B
Mercantile—Existing Class A or B*	A or B	A or B	ceilings—A or B existing on walls —A, B, or C
Mercantile—Existing Class C*	A, B, or C	A, B, or C	A, B, or C
Office—New and Existing	A or B	A or B	A, B, or C
Industrial	A or B	A, B, or C	A, B, or C
Storage	A or B	A, B, or C	A, B, or C
Unusual Structures	A or B	A, B, or C	A, B, or C

* Exposed portions of structural members complying with the requirements for heavy timber construction may be permitted.

† See Chapter 21 of NFPA 101.

Notes:

Class I Interior Floor Finish—minimum 0.45 watts per sq cm.

Class II Interior Floor Finish—mimimum 0.22 watts per sq cm.

Automatic Sprinklers—where a complete standard system of automatic sprinklers is installed, interior finish with flame spread rating not over Class C may be used in any location where Class B is normally specified and with rating of Class B in any location where Class A is normally specified; similarly, Class II interior floor finish may be used in any location where Class I is normally specified and no critical radiant flux rating is required where Class II is normally specified.

The Radiant Panel Furnace

While the Forest Products Laboratory developed the 8 ft (2.44 m) tunnel furnace, the National Bureau of Standards (NBS) developed the Radiant Panel Test apparatus. This utilized a much smaller test specimen, 6 by 18 in. (152 by 457 mm), and measured hazard in terms of both flame spread and rate of heat release after ignition. The test method was adopted as a tentative standard by ASTM in 1960 and as a full standard for research and development purposes in 1967 (ASTM 1967).

In this test, the specimen is positioned at an angle of 30° to the horizontal in front of a gas fired, porous refractory vertical panel. The specimen slants toward the radiant panel at the top and a small pilot flame at that location ignites flammable gases developing from the surface of the specimen. Air is drawn over the surface of the specimen at a controlled rate by a fan in the hood under which the equipment is located. The rate of flame travel from top to bottom of the specimen is noted, as well as temperatures that develop in the stack. Smoke generated can be measured by weighing deposits on a filter located in the stack. The test is continued for 15 minutes or until the surface flaming reaches the lower edge of the specimen.

This test apparatus is available commercially and is used in many research, commercial, and industrial laboratories. The numerical results are on a scale similar to that of the 25 ft (7.62 m) and 8 ft (2.44 m) furnaces. In several jurisdictions, radiant panel test results are acceptable in lieu of 25 ft (7.62 m) tunnel test results.

In this test, as with the 8 ft (2.44 m) tunnel furnace, the airflow is against or across the direction of flame travel. This provides a clear definition of the interfacial burning. With the 25 ft (7.62 m) furnace, the flame travel and air flow coincide and with some materials the flame front may be several feet in front of the interfacial burning.

The Corner Test

Seeking a more realistic assessment of the hazard of interior finishes, several laboratories conducted simulated room corner tests some 25 years ago. These usually comprised an 8 ft (2.44 m) high corner construction with 2 to 4 ft (0.69 to 1.22 m) wing walls. A simulated ceiling of the same material was provided. A wood crib placed in the corner at floor level was used for ignition. The degree of flame spread and the rate and amount of smoke developed were the primary observations.

Interest in the rapidity of flame spread across the surface of what was considered low flame spread cellular plastics in actual fires revived use of the corner test. Much larger scale corner tests have been conducted by Factory Mutual Research Corporation (FMRC) and Underwriters Laboratories Inc. (UL). Corners up to 25 ft (7.62 m) in height with wing walls of up to 50 ft (15.29 m) in length have been used to determine the burning characteristics of these cellular plastics and the effectiveness of protective measures such as automatic sprinkler protection or use of a thermal barrier (Christian and Waterman 1970). Results have reinforced the premise that fire hazard cannot be fully judged on the basis of any single fire test method (Maroni 1973).

Work continues on the development of a small scale test that will correlate with the large scale corner test. One such effort consists of a 24 ft (7.21 m) inverted tunnel furnace. Known as the Factory Mutual Wall-Ceiling Channel Test, it is still in the development stage. It is anticipated that this test method will eventually replace the more expensive full scale corner test (Maroni 1974).

Small Scale Tests

Many small scale "bench" tests are used in industry for research and development and quality control purposes. Use of such tests for promotional purposes has been curtailed following action by the Federal Trade Commission (FTC) in 1973. The FTC findings in fact, included some of the larger tests as well, further emphasizing the probability that no single test will predict the reaction of materials when exposed to fire under the wide range of possible uses.

Many of the smaller tests relate to specific types of materials, e.g., wood, plastics, and others. The SS-A-118b Federal Specification Test for flammability of ceiling tiles and the ASTM C209 inclined panel test for cellulosic fiber insulation board generally have been replaced by the 25 ft (7.62 m) tunnel furnace test. Both tests are still used to some extent in their respective industries.

The following is a list of small scale fire tests relating to the flammability of building materials that are primarily useful for product development and experimental laboratory work:

Combustible Properties of Treated Wood by the Crib Test, ASTM E 160-50.

Combustible Properties of Treated Wood by the Fire Tube Apparatus, ASTM E 69-50.

Fire Retardancy of Paints (Cabinet Method), ASTM D 1360-58.

Flammability of Plastics 0.050 in. and Under in Thickness, ASTM D 568-77.

Flammability of Rigid Plastics Over 0.050 in. Thickness, ASTM D 635-77.

Flammability of Flexible Thin Plastic Sheeting, ASTM D 1433-77.

Incandescence Resistance of Rigid Plastics, ASTM D 757-77.

Flammability of Treated Paper and Paperboard, ASTM D 777-74.

Flammability of Rigid Cellular Plastics, ASTM D 3014-76.

Flammability of Plastics using the Oxygen Index Method, ASTM D 2863-77.

Flammability of Finished Textile Floor Covering Materials, ASTM D 2859-76.

There are two tests for surface flammability which are sometimes encountered in addition to the above. Both use a small sample of material to be tested, approximately 4 by 24 in. (102 by 610 mm), placed face down in an inclined position with a small heat source applied at the lower end (Vandersall 1967, Levy 1967).

Smoke Density Chamber

A "bench test" is available for assisting in research, development, and production quality control regarding smoke generation characteristics of materials. Described in NFPA 258, it provides for measurement of the total smoke generated in a chamber from test specimens of solid materials and assemblies up through 1 in. (25.4 mm) in thickness. Despite the specific intent of NFPA 258 to limit the test procedure to use as a research and development tool, some jurisdictions are erroneously requiring it as the basis for accepting smoke development characteristics of some materials.

FIRE TESTS FOR INTERIOR FLOOR FINISHES

Fire tests for interior finishes were developed primarily with wall and ceiling finishes in mind, long before carpeting became popular as an interior floor finish in institutional, commercial, and other occupancies subject to high human occupancy loads. Official attention was directed toward a possible hazard with soft floor coverings in 1960 and 1961 as a result of a series of small fires in the Washington, DC area (Yuill 1967). A well publicized dwelling fire on the west coast in 1967, followed by the Harmer House Nursing Home fire in Ohio in 1970, focused attention on the problem, although carpets were a contributing rather than a causative factor in these incidents (NFPA 1968; Sears 1970).

In 1965, the Public Health Service of the U.S. Department of Health, Education and Welfare issued a directive calling for flame spread limits on floor coverings in federally aided hospitals, citing the 25 ft (7.62 m) tunnel test as the control mechanism. Since the test specimen in this test is held face down, a problem developed for carpets made with synthetic fibers of the type that melt, drip, or delaminate when exposed to elevated temperatures. The mounting of carpets with separate underlays or parts also posed a difficult problem. Thus a need was established for a test specifically designed for interior floor finishes.

Methanamine Pill Test

The Methanamine Pill Test was developed by the NBS as a means of preventing the distribution of highly flammable soft floor coverings. Based on an earlier government purchasing specification, this method was adopted in 1970 as DOC FF-1-70 (Federal Register 1970a) for carpets, and DOC FF-2-70 (Federal Register 1970b) for rugs.

In this test, eight 9 in. (229 mm) square sections of a carpet are oven heated to drive off excess moisture, brought to room temperature in a desiccator, and tested. Each specimen in turn is placed on the bottom of a 1 ft (0.30 m) enclosed cube, which is open at the top, and held in place by a 9 in. (229 mm) square metal plate having an 8 in. (203 mm) circular cutout. The methanamine tablet is placed at the center of the circle and lighted. The specimen fails if the flame advances at any point to within 1 in. (25.4 mm) of the metal ring in the hold down plate. At least seven of the eight specimens must pass the test to meet the established criteria. All carpet made in the U.S. for hospital use must pass this test.

Floor Covering Chamber Test

In 1969 UL started work on a new test for floor coverings with the test specimens mounted in a floor position (UL 1971). The program was sponsored by the U.S. Public Health Service and resulted in what is known as UL-992, the Floor Covering Chamber Test. In essence, it is an 8 ft (2.44 m) version of the 25 ft (7.6^2 m) tunnel furnace, with the test specimen mounted on the floor of the tunnel, and with appropriate modifications in heat input, burner design, airflow, and other test specifications. This test is not as widely used as some others.

Critical Radiant Flux Test

Another test for floor coverings was developed in the laboratories of the Armstrong Cork Company. This is a modification of the radiant panel test for interior finishes with an 8 by 30 in. (203 by 762 mm) specimen in a horizontal position (Hartzell 1974). Thermal energy is supplied by the radiating gas panel, with an ignition source provided by a pilot light at one end. Flame travel is observed through a window in one side of the test chamber. (See Fig. 7-7C.) The critical radiant flux test method

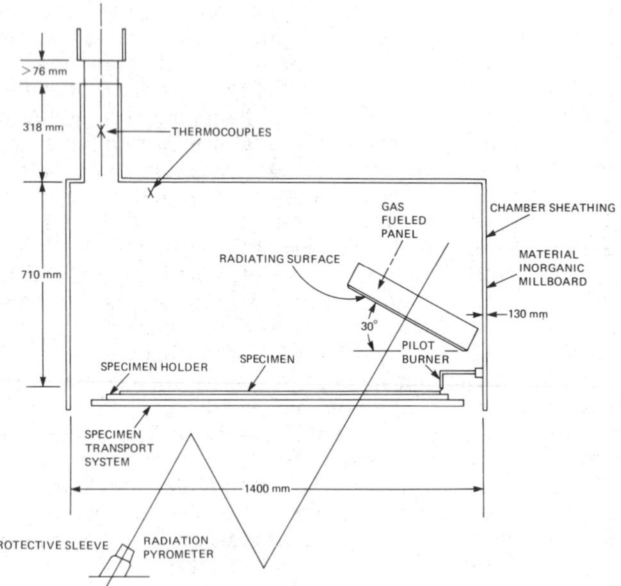

FIG. 7-7C. Flooring Radiant Panel Tester Schematic—Side Elevation.

has been adopted by both NFPA and ASTM as NFPA 253, *Standard Method of Test for Critical Radiant Flux of Floor Covering Systems Using a Radiant Heat Energy Source.*

This test method simulates conditions which have been observed and defined in large-scale corridor experiments. Both the NFPA and ASTM test methods indicate that the method is intended for evaluating floor coverings installed in building corridors having little or no combustible wall or ceiling finish. Corridors with combustible wall and ceiling finishes would be expected to be a greater potential fire hazard. Experimental results suggest that floor coverings would contribute much less to fire spread than wall or ceiling finish in the initial growth and spread of corridor fires (McGuire 1968).

NFPA 101 classifies interior floor finishes in accordance with the critical radiant flux test method. Specifically, floor finishes are grouped into two classes. They are:

Class I: A minimum critical radiant flux of 0.43 Btu per sq ft per sec (0.45 W/cm^2).

Class II: A minimum critical radiant flux of 0.19 Btu per sq ft per sec (0.22 W/cm^2).

See Table 7-7A for assignment of the two classes of floor finishes to the various occupancies identified in NFPA 101.

An advantage of the relatively simple and inexpensive radiant flux test method is that some manufacturers are printing test results on their samples, including test results for various installation methods, e.g., pads, glue down, etc.

While the radiant flux test method is widely used for flammability, some local jurisdictions establish additional criteria for permissible floor finishes based on a test for smoke generation. The Steiner Tunnel Test in NFPA 258 is the test used. Conceived as a laboratory test method, this test provides a procedure for measuring the total smoke generated in a chamber from a specimen of solid materials and assemblies, and uses a photometric system to measure the optical density of the smoke.

FULL SCALE TESTS

Many large scale fire tests have been conducted using existing buildings slated for demolition or available for other reasons. When carefully planned, such tests can and have added significantly to knowledge about the growth and spread of fire in buildings.

One of the first of these tests was conducted in Great Britain in 1949 under the auspices of the Fire Research Station (NFPA 1952). Two 2 story dwellings were used, one with cellulose fiber insulation board lining and the other with plasterboard over fiberboard. One test in each house, both with real or simulated furnishings, was conducted with bedroom doors left open. The other test was conducted with the doors closed. With both types of lining, temperatures in the living room where the fire was started became intolerable in a few minutes. Also, as in many other tests and actual fires, the value of the closed door was clearly demonstrated in minimizing the life hazard in bedrooms.

In 1958, the Division of Building Research of the National Research Council of Canada conducted a series of full scale tests known as the "St. Lawrence Burns," in six dwellings and two larger buildings (Shorter et al 1960). Again, the one objective was to study fire spread as it would affect the life safety of second floor bedroom occupants behind open and closed doors. All dwellings became "smoke logged" within six minutes of the ignition of the wood cribs, regardless of the type of interior finish. Temperatures and gaseous combustion products developed much faster in the dwellings with combustible finishes and in those tests where the bedroom doors were open.

Many large buildings scheduled for demolition were partially burned under the auspices of the IIT Research Institute. In these and in extensive room/corridor tests, much information was developed relative to fire spread, flashover phenomena, and factors affecting life safety in fires.

The Office of the State Fire Marshal in California sponsored a series of tests in unused buildings at Camp Parks. The objective was the development of criteria for egress facilities, particularly corridors. The final report, published in 1974, includes a discussion of the reaction and interaction of wall, ceiling, and floor finishes (WFJ 1974). Precedent for the full scale corridor burns had been established earlier in 1959 and 1960 by the City of Los Angeles Fire Department. Final reports on the Los Angeles tests were published in two volumes by NFPA (LAFD 1959; LAFD 1961).

Furnished room burnout and corridor tests conducted at the Forest Products Laboratories of the U.S. Department of Agriculture compared the relative importance of room furnishings versus room finishes (Bruce 1959). The conclusion that lethal levels of combustion products developed from burning furnishings before the wall finishes were seriously involved was supported by later work at Southwest Research Institute (Pryor 1969). These and other studies recognize that once a fire is well established, the probability is great that all combustibles in the room will become involved. It is also recognized that early detection, containment, and extinguishment in the room of fire origin is important not only for early evacuation from the fire area, but also for the safety of persons elsewhere in the building and for the protection of the building itself.

In 1979-80, the U.S. Fire Administration sponsored a series of 75 tests to evaluate the performance of residential sprinklers in single family and mobile homes. These tests showed the rapid buildup of heat and toxic gases that was possible with ordinary hazard furnishings, particularly when the fire was ignited in the corner of the room or near curtains. There was a significant difference in the levels of combustion products and the times to reach those levels between fires of noncombustible interior finish and combustible interior finish (Cote & Moore 1981).

Bibliography

References Cited

ASTM. 1967. *Standard Method of Test for Surface Flammability of Materials Using a Radiant Heat Energy Source.* ASTM E-162-67. American Society for Testing and Materials, Philadelphia, PA.

ASTM. 1969. *Standard Method of Test for Surface Flammability Using an 8 ft (2.44 m) Tunnel Furnace.* ASTM E 286-69. American Society for Testing and Materials, Philadelphia.

Bruce, H. D. 1959. *Experimental Dwelling Room Fires.* Report No. 1941. U.S. Department of Agriculture. Forest Products Laboratory, Madison, WI, Apr 1959.

Christian, W. J. 1974. "The Effect of Structural Characteristics on Dwelling Fire Fatalities." *Fire Journal.* Vol 68, No 1. Jan. 1974. pp 22-28.

Christian, W. J., and Waterman, J. E. 1970. "Fire Behavior of Interior Finish Materials." *Fire Technology.* Vol 6, No 3. Aug. 1970. pp 165-178.

Cote, Arthur E., and Moore, David. 1981. *Field Test and Evaluation of Residential Sprinkler Systems.* National Fire Protection Association, Quincy, MA.

Dimensions. 1974. "New Fire Research Building." *Dimensions.* Vol 58, No 6. June 1974. pp 123-124.

Federal Register. 1970a. "Carpets and Rugs—Notice of Standard." *Federal Register.* Vol 35, No 74. Apr. 16, 1970.

Federal Register. 1970b. "Small Carpets and Rugs—Notice of Standard." *Federal Register.* Vol 35, No 251. Dec. 29, 1970.

FTC. 1974. *Disclosure Requirements and Prohibitions Concerning the Flammability of Plastics.* Federal Trade Commission, Washington, DC. Aug. 1974.

Hartzel, L. G. 1974. "Development of a Radiant Panel Test for Flooring Materials." *Journal of Fire and Flammability/Consumer Product Flammability Supplement.* Vol 1. Dec. 1974. pp 305-353.

LAFD. 1959. *Operation School Burning: Official Report on a Series of School Fire Tests conducted April 16, 1959 to June 30, 1959, by the Los Angeles Fire Department.* National Fire Protection Association, Quincy, MA, 1959.

LAFD. 1961. *Operation School Burning, No. 2: Official Report on a Series of Fire Tests in an Open Stairway, Multistory School conducted June 30, 1960 to July 30, 1960, and*

February 6, 1961 to February 14, 1961 by the Los Angeles Fire Department. National Fire Protection Association, Quincy, MA.

Levy, M. M. 1967. "A Simplified Method for Determining Flame Spread." *Fire Technology.* Vol 3, No 1. Feb. 1967. pp 38-46.

Maroni, W. F. 1973. "Large Scale Fire Tests of Rigid Cellular Plastic Wall and Roof Insulations." *Fire Journal.* Vol 67, No 6. Nov. 1973. pp 24-30.

McGuire, J. H. 1968. "The Spread of Fire in Corridors." *Fire Technology.* Vol 4, No 2. 1968. pp 103-108.

NBS. 1973. *NBS Corridor Fire Tests: Energy and Radiation Models.* NBS Technical Note 794. National Bureau of Standards, Washington, DC. Oct. 1973.

NFPA. 1952. "British Fire Tests of Fiberboard." *NFPA Quarterly.* Vol 45, No 3. Jan. 1952. pp 218-224.

NFPA. 1968. "Fire in Acrylic Carpeting." *Fire Journal.* Vol 62, No 2. May 1968. pp 13-14.

NFPA 1978. "Reconstruction of a Tragedy: Beverly Hills Supper Club." National Fire Protection Association, Quincy, MA.

Peters, C. C., and Eichner, H. W. 1962. *Surface Flammability as Determined by the FPL 8-Foot Tunnel Method.* FPL Report No. 2257. Forest Products Laboratory, Madison, WI. Nov. 1962..

Pryor, A. J. 1969. "Full Scale Fire Tests of Interior Wall Finish Assemblies." *Fire Journal.* Vol 63, No 2. Mar. 1969. pp 14-20.

Sears, A. B. Jr. 1970. "Nursing Home Fire, Marietta, Ohio." *Fire Journal.* Vol 64, No 2. May 1970. pp 5-9.

Shorter, G. W., et al. 1973. "SLRP Analysis of Recommended Protection for Foamed-Plastic Wall-Ceiling Building Insulations." *Fire Journal.* Vol 68, No 5. Sept. 1974. pp 51-55.

Shorter, G. W., et al. 1960. "The St. Lawrence Burns." *NFPA Quarterly.* Vol 53, No 4. Apr. 1960. pp 300-316.

SPI. 1974. *Fire Safety Guidelines for Use of Rigid Urethane Foam Insulation in Building Construction.* Urethane Safety Group Bulletin. The Society of the Plastics Industry, Inc., New York, NY. May 1974.

UL. 1971. *Standard Method of Test for Flame Propagation Classification of Flooring and Floor Covering Material.* Subject 992. Underwriters Laboratories Inc., Northbrook, IL. Feb. 1971.

Vandersall, H. L. 1967. "The Use of a Small Flame Tunnel for Evaluating Fire Hazard." *Journal of Paint Technology.* Vol 39, No 511. 1967. pp 494-500.

Waterman, T. E. 1972. "Room Flashover—Model Studies." *Fire Technology.* Vol 8, No 4. Nov. 1972. pp 316-325.

WFJ. 1974. "Project Corridor: Fire and Life Safety Research." *Western Fire Journal,* North Highlands, CA.

Wilson, J. A. 1961. *Surface Flammability of Materials: A Survey of Test Methods and Comparison of Results.* ASTM STP No. 301. American Society for Testing and Materials, Philadelphia, PA.

Yuill, C. H. 1967. "Floor Coverings: What is the Hazard?" *Fire Journal.* Vol 61, No 1. Jan. 1967. pp 11-19.

NFPA Codes, Standards, Recommended Practices and Manuals. (See the latest *NFPA Codes and Standards Catalog* for availability of current editions of the following documents.)

NFPA 101, *Code for Safety to Life From Fire In Buildings and Structures.*

NFPA 205M-T, *Tentative Guide for Plastics in Building Construction.*

NFPA 253, *Standard Method of Test for Critical Radiant Flux of Floor Covering Systems Using a Radiant Heat Energy Source.*

NFPA 255, *Standard Method of Test of Surface Burning Characteristics of Building Materials.*

NFPA 258, *Standard Research Test Method for Determining Smoke Generation of Solid Materials.*

Additional Readings

Alderson, Susan, and Bredan, Leslie, *Evaluation of the Fire Performance of Carpet Underlayments,* NFSIR 76-1018, Sept. 1976.

America Burning: The Report of the National Commission on Fire Prevention and Control, Superintendent of Documents, U.S. Government Printing Office, Washington, DC, 1973.

Babrauskas, V., "Estimating Room Flashover Potential," *Fire Technology,* Vol. 16, No. 2, May 1980, pp. 94-104.

Benjamin, I. A., and Davis, S., *Flammability Testing for Carpet,* NFSIR 78-1436, Apr. 1978.

Castino, G. T., et al., "Flammability Studies of Cellular Plastics and Other Building Materials Used for Interior Finish," *U.L. Report,* Underwriters Laboratories, Inc., Northbrook, IL, 1975.

Den Braven, Karen, *Radiative Ignition of Some Typical Floor Covering Materials,* NBSIR 75-967, Dec. 1975.

D'Souza, M. V., and McGuire, J. H., "ASTM E84 and the Flammability of Foamed Thermosetting Plastics," *Fire Technology,* Vol. 13, No. 2, May 1977, pp. 85-94.

Denyes, W., and Quintiere, J., "Experimental and Analytical Studies of Floor Covering Flammability with a Model Corridor," *Journal of Fire and Flammability/Consumer Products Flammability Supplement,* Vol. 1, Mar. 1974, pp. 32-109.

Fang, J. B., "Fire Buildup in a Room and the Role of Interior Finish," *NBS Tech Note 879,* National Bureau of Standards, Gaithersburg, MD, June 1975.

Fung, F. C. W., et al., "The NBS Program on Corridor Fires," *Fire Journal,* Vol. 67, No. 3, May 1973, pp. 41-48.

Groah, W. J., "The ASTM" "Tunnel" Test: "What It Is and How It Works," *The Building Official and Code Administrator,* Vol. 8, No. 6, June 1974, pp. 4-6.

Harmathy, T. Z., "A New Look at Compartment Fires, Part I," *Fire Technology,* Vol. 8, No. 3, Aug. 1972, pp. 196-217; ibid., "Part II," Vol. 8, No. 4, Nov. 1972, pp. 326-351.

Hartzell, L. G., *Development of a Radiant Panel Test for Flooring Materials,* NBSIR 74-495, National Bureau of Standards, Gaithersburg, MD, May 1974.

Klitgaard, P. S., and Williamson, R. B., "The Impact of Contents on Building Fires," *Journal of Fire and Flammability/Consumer Product Flammability Supplement,* Vol. 2, Mar. 1975, pp. 84-113.

Lieberman, P., and Bell, D., "Smoke and Fire Propagation in Compartment Spaces," *Fire Technology,* Vol. 9, No. 2, May 1973, pp. 91-100.

Magnusson, S. E., and Sundstrom, B., "Modeling of Room Fire Growth—Combustible Lining Materials," *Report LUTVDG/(TVBB-3019),* Lund Institute of Technology, Lund, Sweden, 1984.

Magnusson, S. E., and Thelandersson, S., "A Discussion of Compartment Fires," *Fire Technology,* Vol. 10, No. 3, Aug. 1974, pp. 228-246.

McGuire, J. H., and D'Souza, M. V., "The E162 Radiant Panel Flammability Test and Foamed Plastics," *Fire Technology,* Vol. 15, No. 2, May 1979, pp. 102-106.

Moulen, A. W., et al., "The Early Fire Behavior of Combustible Wall Lining Materials," *Fire and Materials,* Vol. 4, No. 4, 1980, p. 165.

Parker, William J., *An Investigation of the Fire Environment in the ASTM E84 Tunnel Test,* NBS Technical Note 945, National Bureau of Standards, Gaithersburg, MD, Aug. 1977.

Pryor, Andrew J., "Full-Scale Fire Tests of Interior Wall Finish Assemblies," *Fire Journal,* Vol. 63, No. 2, Mar. 1979.

Quintiere, James G., *A Characterization and Analysis of NBS Corridor Fire Experiments in Order to Evaluate the Behavior and Performance of Floor Covering Materials,* NBSIR 75-691, National Bureau of Standards, Gaithersburg, MD, June 1975.

Quintiere, James G., and Bromberg, Kevin, *Calculations of Radiant Heat Flux in the Proposed Floor Covering Flame*

Spread Test Apparatus, NBSIR 75-706, National Bureau of Standards, Gaithersburg, MD, Dec. 1975.

Rose, A., *Some Aspects of Carpet Flammability Testing*, DBR Paper No. 712, National Research Council of Canada.

Schaffer, E. L., and Eickner, H. W., "Corridor Wall Linings— Effect on Fire Performance," *Fire Technology*, Vol. 1, No. 4, Nov. 1965, pp. 243-255.

Sherad, Shirley E., ed., *Interior Finish and Fire Spread*, National Fire Protection Association, Quincy, MA, 1977.

"Space Age Contribution to Residential Fire Safety," *Fire Journal*, Vol. 68, No. 2, Mar. 1974, pp. 18-25.

Teller, Harlan, *The Red Carpet Treatment*, UL Lab Data, Underwriters Laboratories, Northbrook, IL, Winter 1976.

Thomas, P. H., "Old and New Look at Compartment Fires," *Fire Technology*, Vol. 11, No. 1, Feb. 1975, pp. 42-47.

Thomm, E. C., "Effect of Carpet Variables on the Methanamine Pill Test," *Journal of Fire and Flammability*, Vol. 4, July 1973, pp. 197-209.

Tu, King-Mon, and Davis, Sanford, *Flame Spread of Carpet Systems Involved in Room Fires*, NBSIR 76-1013, National Bureau of Standards, Gaithersburg, MD, June 1976.

Waksman, D., and Ferguson, J. B., "Fire Tests of Building Interior Covering Systems," *Fire Technology*, Vol. 10, No. 3, Aug. 1974, pp. 211-220.

Waterman, T. E., "Room Flashover—Model Studies," *Fire Technology*, Vol. 8, No. 4, Nov. 1972, pp. 316-325.

Williams-Leir, G., "The Steiner Tunnel Index as a Measure of Flame Spread Rate," *Fire Technology*, Vol. 14, No. 3, Aug. 1978, pp. 223-225.

Williamson, R. B., and Baron, F. M., "A Corner Fire Test to Simulate Residential Fires," *Journal of Fire and Flammability*, Vol. 4, Apr. 1973, pp. 99-105.

Yuill, C. H., *Flame Spread Tests in a Large Tunnel Furnace*, ASTM STP No. 344, 1962, American Society for Testing and Materials, Philadelphia, PA, pp. 3-17.

STRUCTURAL INTEGRITY DURING FIRE

Revised by Robert W. Fitzgerald

The selection of building materials and the design of the details of construction have always played an important role in building firesafety. Two of the important structural fire considerations are the ability of the structural frame to avoid collapse and the ability of the barriers to prevent ignition and resulting flame spread into adjacent spaces.

Three approaches to structural frame and barrier design for fire resistance are in use or under development today. They include:

1. Standard fire resistance testing combined with building code requirements;
2. Analytical calculations to determine the resistance to a standard fire test exposure as a substitute for laboratory testing; and
3. Analytical structural fire engineering design methods based on real fire exposure characteristics.

The traditional method of treating the structural aspects of fire protection is through building code requirements. Codes require structural frames and barrier assemblies to be selected on their ability to pass a laboratory fire resistance test.

Chapter 5 of this Section describes the classification of buildings with regard to fire resistance requirements. This approach has the principal advantage of being easy to administer from a code enforcement viewpoint. It has a major disadvantage of being unable to accurately predict the field performance of a structural frame or barrier. In other words, code compliance without simultaneously translating fire test results into field performance may lead to failure under fire conditions. Studies that correlate building code and fire test results with field performance have not been made.

Although the fire resistance test can be criticized for its shortcomings, it is the only method universally accepted in building codes today. Consequently, the first part of this chapter describes the fire resistance test procedures common in building construction today. Later parts of the chapter discuss methods of calculating fire endurance (resistance) ratings and the behavior of building materials at elevated temperatures. The chapter concludes with a brief discussion of some of the more advanced approaches to designing for structural integrity in a fire.

Dr. Fitzgerald is Professor of Civil Engineering at Worcester Polytechnic Institute, Worcester, MA.

FIRE RESISTANCE TESTING AND STRUCTURAL ASSEMBLIES

The fire endurance of the beams, girders, and columns that comprise the structural frame, and the endurance of the walls, partitions, floor/ceiling assemblies and roof/ceiling assemblies that serve as barriers to flame movement have been an historical basis for classifying buildings and rating frame and barrier capabilities. This part of the chapter will describe the history and procedure for fire testing, as well as discuss the interpretation of laboratory fire tests. In addition, structural behavior of frames and barriers is described.

History of Fire Resistance Tests

One review of fire test methods for building constructions refers to tests on metal and masonry columns conducted in Germany as early as 1884-1886 (Clay 1927). The first large scale fire tests in this country are reported to have been conducted on masonry arches in Denver, CO in 1890. These were followed by tests in New York City in 1896.

Efforts to establish an acceptable test procedure were initiated by Professor Ira H. Woolson of Columbia University and Rudolph P. Miller, chief engineer, Building Bureau, New York City. Preliminary tests by the Bureau, "necessitated by the rapid development of the skyscraper," led to the development of a test furnace (using railroad ties as fuel). This provided a means of establishing hourly ratings for floor constructions.

In 1905, after the Baltimore conflagration, the American Society for Testing Materials (now the American Society for Testing and Materials) established a committee to standardize the test method with Prof. Woolson as chairman and Mr. Miller as secretary. A test method for floor constructions was proposed in 1906 and adopted by ASTM in 1907. A procedure for testing wall and partition constructions was proposed in 1908 and adopted in 1909 (Shaub 1961).

These standards were presented to the NFPA Committee on Fire Resistive Construction for consideration in 1914. In 1916, a joint committee composed of representatives from eleven engineering societies, including NFPA, was organized to revise and update the standard. This was done and the revised standard test method controlled by a standard time temperature curve was adopted subsequently by NFPA, ASTM, and the American Engineering Committee (now the American National Standards Institute). The standard was adopted by NFPA as a tentative standard in 1917 and advanced to official standard status in 1918. Time-temperature curves and testing procedures developed in other countries follow the same general pattern, with minor differences.

Fire Test Procedure

Fire test procedures usually require that columns, floors, partitions, walls, and other structural elements be loaded in a manner calculated to develop, as nearly as practicable, the theoretical working stresses expected in the design. Separate test procedures are provided for load-bearing and nonload-bearing constructions, and for constructions involving restrained and unrestrained beams and girders.

The Appendix to NFPA 251, *Standard Methods of Fire Tests of Building Construction and Materials* (hereinafter referred to as NFPA 251), contains detailed specifications for test procedures, a guide to the determination of restraint required, if any, and a suggested report form. The standard specifies in detail the preparation and conditioning of the test specimens.

Acceptance criteria are specific for the construction or element tested and on predetermined conditions of test (load or no load—restrained or unrestrained). The criteria may include:

1. Failure to support load.
2. Temperature increase on the unexposed surface 250°F (121°C) above ambient.
3. Passage of heat or flame sufficient to ignite cotton waste.
4. Excess temperature (as specified) on steel members.
5. Failure under hose stream (walls and partitions).

The end restraint conditions during the fire test influence significantly the test results, and consequently the ratings. Appendix E of NFPA 251 defines a restrained condition as one in which expansion at the supports of a load-carrying element is prevented during the test. This usually provides some unknown degree of rotational restraint at the ends of the element in addition to prestressing some parts of the assembly. An unrestrained condition is one in which the load-carrying element is free to expand and rotate at its supports.

One reason for incorporating two different conditions of end restraint is to simulate simple (statically determinate) and continuous (statically indeterminate) constructions. Figure 7-8A illustrates simple and continuous construction. In general, continuous construction is inherently stronger than simple construction. However, the amount of increased strength in an indeterminate structure can be quite variable, depending on the type of construction materials used, the location of structural members

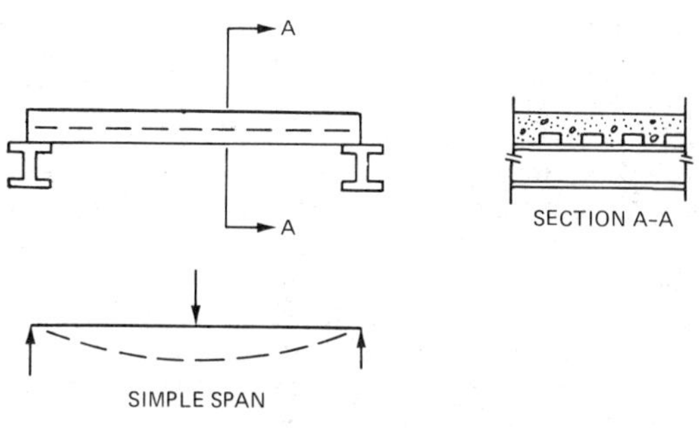

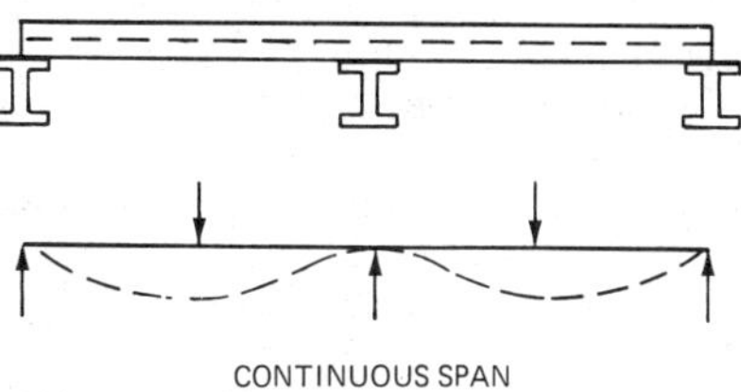

FIG. 7-8A. Influence of structural continuity on collapse mechanisms.

within the structure, the degree of indeterminacy, and the details of construction.

Test results based on the two conditions of support (restrained and unrestrained) were introduced in 1970. Consequently, the degree of restraint for structural frames and assemblies tested and accepted for building construction prior to 1970 is unknown.

Fire Resistance Ratings

The fire resistance is the time period the member or assembly withstood the fire test without failure. While the actual time is recorded to the nearest integral minute, fire resistance ratings are given in standard intervals. The usual fire resistance ratings for all types of members, structural assemblies, doors, and windows are 15 min, 30 min, 45 min, 1 hr, 1½ hr, 2 hr, 3 hr and 4 hr. Therefore, a 1 hr rating indicates that the assembly withstood the standard test for 1 hr or longer. A 2 hr rating indicates that the assembly withstood the standard test for longer than 2 hrs without failure by any one of the failure criteria listed in the fire test protocol.

The principal agencies in the United States that test assemblies of building construction materials for fire resistance are Underwriters Laboratories Inc., and the National Bureau of Standards. There are a number of other agencies that have furnaces or equipment for fire tests of various special assemblies. These include the Factory Mutual System laboratories in Norwood, MA; the Forest Products Laboratory at Madison, WI; the Engineering Experiment Station at Ohio State University, Columbus; University of California at Berkeley; the Portland Cement Association, Skokie, IL; Armstrong Cork Co., Lancaster, PA; the Corbetta Construction Company at College Point, Queens, NY; and the National Gypsum Co., Buffalo, NY. Some of these test furnaces may not conform to the specifications described in NFPA 251. Consequently, those furnaces may be used for research purposes, rather than for rating assemblies.

In England, there are the facilities at the Testing Station of the Joint Fire Research Organization of the Fire Offices' Committee and the Department of the Environment at Boreham Wood, Hertfordshire. In Canada, there are facilities at the Underwriters Laboratories of Canada, at Toronto, and the Building Research Division of the National Research Council of Canada, at Ottawa. Throughout the world, many well equipped fire testing laboratories are making significant contributions to the literature and to the development of international fire test standards.

The results of fire tests conducted by Underwriters Laboratories Inc. are given in the *Fire-Resistance Directory* published annually by UL.

The results of fire tests by governmental agencies have been published in various reports and technical papers. Those issued by the National Bureau of Standards may be consulted in many depository libraries, and lists of those available for purchase may be obtained on request from the NBS. The bulletins and reports of the British Ministry of Technology on the subject are obtainable by purchase from H. M. Stationery Office, or the British Information Services, New York, NY.

The Factory Mutual System publishes annually in its *Approval Guide* illustrations of building assemblies that have met the test requirements for various fire resistance hourly ratings.

Various building codes, which specify fire resistance in terms of fire test results, include tabulations of construction forms that will be accepted as meeting code requirements for specific fire resistance ratings. Such tables are usually based upon fire test data, but some include ratings determined on a judgmental basis or estimated from limited test data.

Not listed are the many useful data sheets and summary tables published by trade associations of the building materials manufacturers. These sources are not cited because they are so numerous and because the information published either pertains to the products of a specific manufacturer or consists of tabulated ratings for a single type of construction material.

As indicated above, the number of test facilities in use today and the variety of technical papers, reports, compilations, etc., reporting on test results that are available would make it an almost impossible task to tabulate systematically in a single source the data on all the various assemblies that have been subjected to test. HANDBOOK users are urged to consult the fire resistance index ratings listing and reports mentioned above for the specifics on design specifications for assemblies that have been tested and for which fire resistance ratings have been assigned.

As a matter of information and to illustrate the scope of testing activities, tabular compilations of test results of representative assemblies will be found later in this chapter. They are included in the HANDBOOK to help give an understanding of basic requirements for test specimens and how the results of tests are recorded. In addition, the information contained in the tables can be useful in identifying, within reasonable limits, the fire resistance capabilities of assemblies as they are found in the field and for which precise data on their fire resistance is not available or no longer available. The tables are not presented as references for design but to familiarize HANDBOOK users with the type of test information that is available.

Variation in Test Results

It should be recognized clearly that the fire resistance rating is the time the member or assembly withstood a standard laboratory fire test. It is not necessarily the time duration the assembly will withstand failure under actual fire conditions. For example, a 2 hr assembly may withstand failure in a standard fire test for greater than 2 hrs. Under actual field conditions, that same assembly may fail in a considerably shorter or a considerably longer time duration. One of the major misconceptions in fire protection is the belief that an assembly rating indicates the time that the assembly will survive an actual fire.

The fact that test time duration and field performance may be vastly different is not intended to demonstrate that fire resistance ratings do not have value or purpose. Over the years this procedure has resulted in improved fire resistance of code complying buildings. However, one must understand the limitations of the procedure so that unthinking reliance is not placed on the construction.

The standard fire test attempts to provide a relative measure of fire performance of comparable assemblies under specified fire exposure conditions. It does not consider its suitability for further use. Many effects from fire tests are observed indirectly, if at all. For example, the temperature gradient through a wall or floor slab results in internal strains and deflections. The strains may cause spalling or other disruptions, and distortions may be severe enough to crack floor slabs and walls, sometimes leading to collapse. The greater the area exposed, the more serious are the results of unequal expansion. General deterioration of the test specimen is not considered, except when it has been involved in failure of the test specimen.

While the test standard specifies the preparation and conditioning of specimens, there are many opportunities for differences between a test specimen and an actual structural element in a building. The test specimen may be superior in both materials and workmanship to those found in a building. It is usually smaller than the construction it represents. Restraints to thermal expansion may be of different magnitudes. Therefore, discretion in the application of fire test results is in order.

Differences in the results of fire tests on apparently equal test specimens of building constructions arise from many factors, such as undetermined differences in the quality of materials, workmanship, moisture content, and test procedures. The standard method of test requires the materials in the specimen and the workmanship to be representative of those in actual buildings. Considerable variation in practice can occur.

The test method permits the intensity of the fire exposure to deviate as much as 10 percent (formerly 15 percent) from that prescribed as standard, and requires adjustment in the reported results to correct for such deviations only for tests of ½ hr or greater duration. The effects of several of the variables encountered in fire testing procedures are found in the technical reports listed in the bibliography at the end of the chapter.

The character and proportions of aggregates and binders have an important influence on results of fire resistance tests. For example, gypsum plaster with lightweight aggregates, mixed with from 2 to 3 cu ft (0.057 to 0.085m^3) to each 100 lb (45 kg) of gypsum, has been shown by test to provide fire resistance from 10 to 70 percent greater than equal thicknesses of sanded gypsum plaster. The use of such aggregate in specific constructions should be adopted only after consulting the lists of ratings based on tests in which these aggregates have been used.

Vermiculite aggregate plaster, if applied too wet, may be subject to shrinkage cracks. Such cracks, if they develop, are likely to occur within a short time. Some expanded perlite aggregates tend to absorb atmospheric moisture and may show destructive expansion after a few years. There may be considerable variation in this effect as perlite is a natural volcanic glass or rock from many sources and varies in composition.

Finishing lime produced from dolomite contains magnesium oxide. That which is designated as "normal" finishing hydrate may not be sufficiently hydrated, and sometimes subsequent gradual hydration results in destructive expansion. Plasters containing such lime are subject to rapid destruction in the event of fire. This difficulty is avoided with lime hydrated under high temperature and pressure and known as "special" or "autoclaved" lime.

So-called stabilized gypsum plaster containing a small percentage of the normal hydrated lime similarly may be subject to deterioration with age and in tests may not provide the fire resistance to be expected from such constructions.

The effects mentioned are not such as to preclude the use of these materials where long life is not a factor or where adequate guarantees are provided. They are merely illustrations of the variability in test results that can occur as a result of differences in materials.

Although standard fire tests are made on fairly large specimens representative of building constructions or assemblies, they do not necessarily produce similar heat-expansion effects to those of fires in buildings having larger wall or floor areas. It may be necessary to specify thickness of construction or lateral support in addition to rated fire resistance in order to guard against the adverse effects of heat expansion or temperature gradients.

In addition to the material variation noted, construction differences may be even more significant. For example, a test does not provide full information on performance of assemblies constructed with components or lengths other than those tested. Performance is often reduced when the spans are increased. Also, great care must be exercised in approving "or equal" clauses in construction specifications. Often substitutions in materials are approved for components that have not been listed with the assembly.

The standard test does not incorporate information regarding the effect on fire endurance of conventional openings in the assembly such as electrical receptacle outlets, pipe chases, etc. Poke-through construction for electrical services or piping often is not patched properly. Transfer grilles, windows, and other penetrations often are incorporated into barriers without considering their impact on the fire resistance.

The factors noted above, in addition to the many deviations to plans and specifications during and after construction, affect significantly the fire resistance ratings of members and assemblies. Often the code calls for, and the architect thinks he is providing, a specified fire resistance, only to have that resistance reduced because of inattention to penetrations and certain important details of construction.

Structural Frame Systems

The fire resistance rating of structural framing members, such as beams, girders, and columns, may be achieved in several ways. Generally, the members are protected by encasing them in a material sufficiently inert to prevent excessive thermal penetration or by providing a membrane protection that delays thermal penetration to the members.

The general behavior of reinforced concrete, structural steel, and barrier assemblies are described in the following paragraphs. Illustrative fire resistance ratings are shown in tabular material later in this chapter in the part entitled, "Illustrative Fire Resistance Ratings for Structural Frame and Barrier Constructions."

Reinforced Concrete Systems

Reinforced concrete construction has had a good experience record with regard to structural collapse. Because concrete has a low thermal conductivity and a low thermal capacity, it provides an effective cover for reinforcing steel. For example, Figure 7-8B shows the temper-

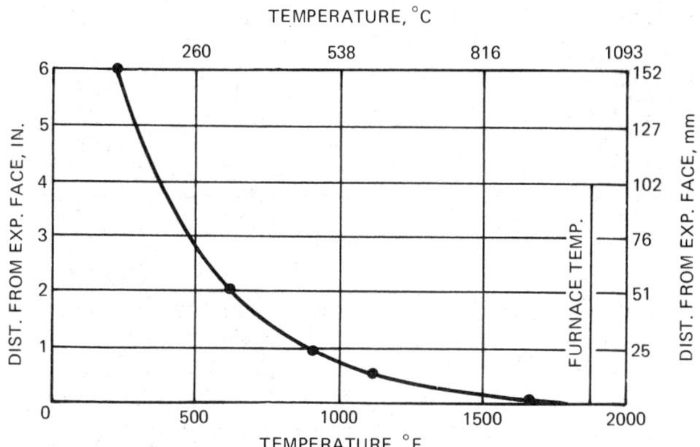

FIG. 7-8B. Thermal gradient in a 6 in slab after 2 hr fire exposure (Benjamin 1961). SI Units: 1 in. = 25.4 mm; °C = 5/9 (°F−32).

ature gradient in a 6 in (152 mm) slab after a 2 hr fire exposure. Although undoubtedly the moisture in the concrete greatly influences the values, the significant feature is the fact that the temperatures vary considerably throughout the thickness, even after a considerable time exposure.

This feature provides some insight into one reason that reinforced concrete systems usually perform comparatively well during fire exposure. Consider, for example, a continuous, monolithic reinforced concrete beam or slab, as shown in Figure 7-8C. Considering the temperature gradient of Figure 5-8B, it will take some time before the tension steel at midspan is affected. Even after it reaches its yield value, the negative steel over the supports has not been seriously affected because of the insulating effect of the concrete and the moisture.

Continuous construction of this type has inherent strength capabilities far greater than statically determinate construction. Considerable stress redistribution can take

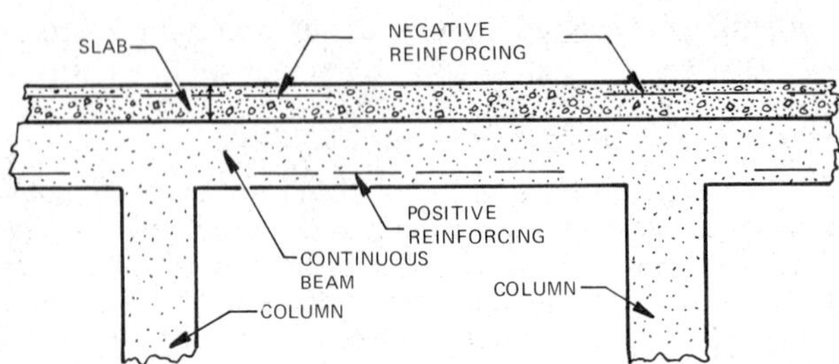

FIG. 7-8C. Monolithic reinforced concrete beam and slab.

place before collapse will occur. It takes time before excessive rotation will develop at all three necessary locations, causing structural collapse of the member. Although the member is weakened by the fire, structural stability against collapse will remain for a considerable period of time.

The level of stress in a reinforced concrete member exposed to the elevated temperatures of a fire has a significant influence on its endurance. Table 7-8A illustrates the effects of stress level on the fire resistance of reinforced concrete columns. The columns used for these tests were 15 by 15 in. (.38 by .38 m) containing four number 9 reinforcing bars (NBS 1961). It can be seen that the magnitude of stress during a fire causes significant reductions in capacity. This is attributed primarily to the reduction in the mechanical properties of steel and concrete at elevated temperatures.

TABLE 7-8A. The Influence of Stress on the Fire Resistance of Concrete Columns

Applied Load, % Design Load	Fire Resistance, Minutes
150	68
100	124
75	198
50	248
30	358

Steel Construction

The popularity of steel frame building construction is due to its high strength, ease of fabrication, and assured uniformity of quality. Exposed structural steel, however, is vulnerable to fire damage. In order to have fire resistance, it must be protected from high temperatures encountered in fires.

Protections for steel beams, girders, and columns, such as encasements of concrete, clay, tile, or gypsum blocks, have been generally superseded by plastered or sprayed-on applications applied either to a furred plaster base, such as expanded metal lath, or to the surface of the member to be protected. The applications may be conventional plasters of portland or gypsum cements combined with appropriate aggregates, or one of the many combinations of mineral fibers with binders. Currently, gypsum board encasement is a popular technique for protecting steel structural elements.

Barrier Systems

Exterior walls, interior partitions, floors, and floor/ceiling assemblies are components that define the architectural layout of rooms in a building. In the normal functional use of a building these components are used to provide privacy, security, protection from the elements, and noise control. They also provide fire protection by delaying or preventing flames from moving from one room to an adjacent room.

The effectiveness of a barrier in preventing flames from moving from one room to another depends upon the fuel load in the room, the fire resistant construction of the barrier, the applied loading, the construction features, and the effect of openings and penetrations in the barrier. The most common cause of fire movement from one room to an adjacent room is through unprotected openings in barriers. Often code requirements and expensive constructions are rendered ineffective because of a lack of attention to the details of opening protection.

Building codes through construction classifications identify the fire endurance requirements of barriers. In addition, special locations, such as around vertical shafts, are required to provide specific fire resistances.

The fire resistance ratings are determined by subjecting the barrier assembly to a standard fire test, as described in NFPA 251. The length of time the assembly withstands the laboratory fire without failure describes the fire endurance. Fire resistance is the endurance time rounded down to the standard durations. Both combustible and noncombustible barriers can obtain fire resistance ratings in the standard fire test.

The cautions expressed earlier in this chapter with regard to the interpretation of the fire resistance time durations also apply here; that is, the time durations reflect laboratory times and are not necessarily the values to be expected in a real fire. For example, one should not expect a 2 hr barrier assembly to withstand a building fire for 2 hrs. Failure could occur much earlier or much later, depending upon construction details.

Floor/Ceiling Assemblies

The fire resistance of floor/ceiling assemblies is important in the prevention of flame movement vertically from floor to floor. Fire resistance durations from the standard fire test is determined from unpenetrated assemblies. The anticipated fire resistance may be reduced greatly or destroyed completely in building construction when holes and poke throughs are not protected adequately, or when details do not conform to the tested assembly. Attention to details is important.

The method of attaching ceilings is a major factor in determining the fire resistance of floors. For example, the nailing of plaster bases of gypsum lath, metal lath, or gypsum wallboards to the soffits of wood joists is often critical. The longer thinner nails, particularly those with cement coatings, conduct less heat to char the wood surrounding them than do the common types of wire nails. Similarly, the integrity of suspended ceilings must be maintained in order to reduce the likelihood of premature failure on some assemblies.

Self-tapping screws, made particularly for the attachment of gypsum boards, offer greater holding power and less damage to the core materials than do nails. Such screws can be used to attach wallboards to either wood or cold-rolled channels without previous drilling.

Consideration must also be given to the character of the plaster base with respect to loosening of plaster mixes from the base on application of heat sufficient to char combustible surfaces. The use of wire, or better yet, wire fabric, to reinforce the plaster mixes applied to such plaster bases assures increased fire resistance. Clearances for longitudinal expansion of metal furring members are required to prevent damage from buckling. The tendency of certain plaster bases and plaster mixes to expand or contract with changes in atmospheric humidity should also be given consideration where resultant cracking might affect the fire resistance of structures incorporating such plaster.

Suspended ceilings with openings for air diffusers and light troughs should be designed so that such openings are not points of vulnerability to fire. Continuous construction above recesses for lighting fixtures and properly designed self-closing dampers for air ducts provide protection.

Roof/Ceiling Assemblies

Roof/ceiling assemblies are tested and rated in a manner similar to the floor/ceiling assemblies. The results are generally comparable. However, it should be noted that roof assemblies often have a given thickness of insulation in place at the test. If additional thicknesses of insulation are desired, the fire resistance rating may be reduced.

Walls and Partitions

The fire resistance of walls and partitions will delay or prevent flames from moving horizontally from one room to an adjacent room. These assemblies are also tested in accordance with NFPA 251. The cautions expressed earlier in this section concerning the relationship between test duration and actual fire performance also apply to these constructions. Walls and partitions are particularly vulnerable to the effects of horizontal communication mechanisms. The fire resistance can be almost completely destroyed by ineffective opening protectives for doors, windows, grilles, ducts, and other openings in these barriers.

The fire resistance ratings are obtained for unpenetrated constructions. When openings are made for doors, windows, transfer grilles, pipe chases, etc., the fire resistance is reduced or destroyed. If the fire resistance of a barrier is important to the fire protection design of the building, attention must be given to the details of design for all openings.

Many fire tests are stopped arbitrarily before an end point criterion has been attained. In some cases the indications are that had the test been continued, the construction would have qualified for a higher fire resistance rating than that assigned to it.

ILLUSTRATIVE FIRE RESISTANCE RATINGS FOR STRUCTURAL FRAME AND BARRIER CONSTRUCTIONS

Fire resistance testing has produced a large amount of experimental data. To illustrate the effect of materials and constructions on the fire resistance rating of assemblies and structural members, several graphs and tables have been included. These graphs and tables can be used both to give guidance on the needs to achieve fire resistance ratings and to provide a basis for judgmental interpretations of the fire resistance of construction encountered in field evaluations.

Reinforced Concrete Members

The details of construction and the kind of aggregates affect the fire resistance of concrete structures. The Building Officials Conference of America (BOCA) divides concrete into two grades on the basis of the fire resistance as affected by aggregates. The two grades are defined as:

"Grade 1 concrete shall mean concrete made with aggregates such as blast-furnace slag, burned clays, and calcareous igneous, and most silicate crushed stones and gravels and shales, as well as any other aggregates performing as required by this code for the appropriate construction when tested in accordance with standard methods of fire tests of building construction and materials listed in Appendix A." (The reference is to NFPA 251, *Fire Test of Building Construction and Materials*.)

"Grade 2 concrete shall mean concrete made with aggregates such as cinders and crushed stones and gravels composed of essentially quartz and quartzite cherts, as well as any other aggregates performing as required by this code for the appropriate construction when tested in accordance with standard methods of fire tests of building construction and materials listed in Appendix A."

The fire resistance ratings for reinforced concrete beams are given in Table 7-8B and for prestressed concrete girders, beams, and slabs in Table 7-8C. As the cold-drawn, high-strength steel tendons used in prestressed concrete are more adversely affected by high temperatures than is normal reinforcement steel, these tendons require a thicker protective cover than is required in conventional reinforced concrete. The fire resistance of reinforced concrete columns is shown in Table 7-8K.

Structural Steel Members

Fire protection generally is provided for structural steel by encasing the members in concrete, lath and plaster, gypsum boards, or sprayed fibers. A second com-

TABLE 7-8B. Reinforced Concrete Beams and Girders of Medium Size

Concrete Grade*	Protective Cover of Reinforcement, Inches	Fire Resistance Rating† Hours,
1	¾	1
	1	2
	1¼	3
	1½	4
2	¾	½–1‡
	1	1–2‡
	1½	2–3‡
	2	2–4‡

* For lightweight concrete having an oven-dried density of 110 pcf or less, the cover shown for Concrete Grade 1 may be reduced 25 percent.

† May be increased if some bars are better protected by being away from corners or in an upper layer, or if beam is large. Should be decreased if beam is small.

‡ Variable depending on spalling characteristics of aggregate. The use of mesh to hold cover in place will give ratings about as high as for concrete of Grade 1.

SI units: 1 in. = 25.4 mm.

mon method of providing fire resistance is by installing a membrane barrier to delay or prevent the heat from raising the temperature of the steel to the critical temperatures. Table 7-8D gives minimum thicknesses of portland cement encasement for steel beams, while Tables 7-8E and 7-8F illustrate the influence on the fire resistance of columns for various methods of encasement. Figures 7-8D and 7-8F illustrate different methods of encasement for steel columns. Figure 7-8I provides a means to estimate the necessary encasing requirements for steel columns.

Floor/Ceiling Assemblies

Fire resistance for floor/ceiling assemblies of reinforced concrete, structural steel, and wood can be achieved in several ways. For example, Figure 7-8E illustrates the influence of floor thickness and aggregate type on the fire resistance of reinforced concrete slabs, while Table 7-8N shows the fire resistance for several types of concrete floor constructions. Table 7-8G shows the test results for structural steel joist floor and roof constructions, and Table 7-8H provides some data on the fire resistance of steel formed concrete floor systems. Illustrative fire resistance ratings for wood floors are provided in Tables 7-8P and 7-8J.

Walls and Partitions

Walls and partitions are commonly constructed of masonry, wood, or metal studs faced with fire resistant materials. Table 7-8I gives the fire resistance of load-bearing brick or clay tile walls. Figures 7-8H shows a graph that allows the fire resistance of burned clay brick walls to be estimated. Tables 7-8M and 7-8O provide fire resistance rating for wood stud walls, while Figures 7-8G gives the fire resistance of stud walls faced with common construction materials to be estimated quickly and conveniently. Finally, Table 7-8L and Figure 7-8J are very useful descriptions of the fire resistance of solid, non load-bearing partition materials.

TABLE 7-8C. Prestressed Concrete Girders, Beams, Joists and Slabs (Grade I Concrete)

Type of Unit	Condition of Restraint	Cross-Sectional Area, sq in.†	Cover in Inches for Fire Rating Shown* 1 Hr	2 Hr	3 Hr	4 Hr
Girders, beams and joists	Unrestrained	40 to 150	2	2.5	—	—
		150 to 300	1.5	2.5	3.5‡	—
		Over 300	1.5	2.25	3‡	4‡
	Axially restrained	40 to 150	1.5	2	—	—
		150 to 300	1	1.5	2	—
		Over 300	1	1.5	1.5	2
Slabs, solid or covered, with flat undersurface	Unrestrained		1	1.5	2	2.5
	Biaxial restraint		0.75	1.25	1.5	2

* Cover for an individual steel tendon is measured to the nearest exposed surface. For several tendons in the same member having different concrete covers, the minimum cover may be reduced slightly. The covers shown may be reduced 25 percent for lightweight concrete having an oven-dried density of 110 pcf or less. For Grade 2 concrete the cover may need to be increased.

† In computing the cross-sectional area of joists, the area of the flange shall be added to the area of the stem, but the total width of the flange so used shall not exceed three times the width of the stem.

‡ Provide against spalling of the cover by means of a light, 2-in., U-shaped mesh, covered about 1 in.

Note: Data in this table are based on 67 Standard ASTM fire tests.

SI units: 1 in. = 25.4 mm.

TABLE 7-8D. Concrete Protection for Steel Beams

(All reentrant portions filled)

Fire Resistance Rating		Thickness of Concrete Protection, Inches: Grade 1 Size of Member (flange width)				Grade 2 Size of Member (flange width)			
Hr	Min	2 to 3¾ in	4 to 5¾ in	6 to 7¾ in.	8 in. and over	2 to 3¾ in	4 to 5¾ in.	6 to 7¾ in.	8 in and over
4	—	4	3¼	2½	2	4¾	3¾	3	2½
3	—	3½	2½	2	1½	4	3	2½	2
2	—	2½	2	1½	1	3	2½	1¾	1¼
1	30	2	1½	1	1	2½	1¾	1¼	1
1	—	1½	1	1	1	1¾	1¼	1	1

Note: Protective concrete having thickness one-fourth of flange width or less shall have steel wire reinforcement spaced not more than four times the thickness of the concrete covering the flange.

SI units: 1 in. 25.4 mm.

TABLE 7-8E. Fire Resistance of Steel Columns with Lath and Plaster Protective Coverings

Metal Column				Protective Covering							Total Area of Materials	Fire-Resistance Rating		Notes
Type of Section	Size In.	Wt per Lin Ft Lb	Area of Metal Sq In.	Design	Type of Covering: Plaster Type	Aggregate	Mix, Volumes	Thickness† In.	Furring† In.	Bond of Covering	Sq In.	Hr	Min	
H	6	44	13	—	Portland cement & lime	Sand	1:1/10:2½	7/8	1	Metal lath	40	—	45	Metal lath furred out
H	6	31	9	—	Two thicknesses of above	Sand	1:1/10:2½	7/8 + 7/8	1 & 1	Metal lath	80	1	30	Metal lath furred out
H	10	49	14.5	A	Gypsum-cement mixture	Lightweight*	Mill mix	1¾	⅜	Wire fabric	125	3	25	
H	10	49	14.5	A	Gypsum-cement mixture	Lightweight*	Mill mix	1 7/8	⅝	Metal lath	125	4	—	½-in. channel behind lath
H	10	49	14.5	A	Gypsum	Lightweight*	1½:2 1½:3	1¾	⅜	Metal lath	125	4	—	Self furring lath
H	10	49	14.5	A	Gypsum	Lightweight*	1½:2 1½:3	1⅜	⅜	Metal lath	102	3	—	Self furring lath
H	10	49	14.5	A	Gypsum	Lightweight*	1½:2 1½:3	1	⅜	Metal lath	78	2	—	Self furring lath
H	10	49	14.5	B	Gypsum	Lightweight*	½:3½ ½:4	2⅛	½	18-gage wire & wire fabric	145	4	—	Gypsum lath
H	10	49	14.5	C	Gypsum	Lightweight*	1½:2 1½:3	1½	1	18-gage wire & wire fabric	140	4 to 4	15 to 40	Gypsum lath
H	10	49	14.5	B	Gypsum	Lightweight*	1½:2½	1½	½	18-gage wire & wire fabric	110	3	40	Gypsum lath
H	10	49	14.5	D	Gypsum	Lightweight*	1:2½	½	⅜	18-gage wire ties	53	1	20	Perforated gypsum lath
H	10	49	14.5	D	Gypsum	Lightweight*	1:2½	⅝	⅜	18-gage wire ties	60	1	30	Perforated gypsum lath
H	10	49	14.5	D	Gypsum	Lightweight*	1½:2 1½:3	1	⅜	18-gage wire ties	80	2	15	Perforated gypsum lath
H	10	49	14.5	D	Gypsum	Lightweight*	1½:2 1½:3	1½	⅜	18-gage wire ties	104	2	30	Perforated gypsum lath
O	7	51	15.5	—	Portland cement & lime	Sand	1:1/10:2½	1¼	7/8	7/8-in. rib lath	70	2	45	Metal lath on cast column

* Lightweight aggregate can be either perlite or vermiculite.
† Dimensions as shown in Figure 7-8D.
SI units: 1 in. = 25.4 mm; 1 lb = 0.43 kg.

TABLE 7-8F. Fire Resistance of Steel Columns Encased with Concrete, Masonry, or Sprayed Fibers

Steel Column			Protective Covering									
Type of Section	Size	Wt per Lin Ft	De-sign	Type of Covering	Thick-ness Out-side Steel t‖	Re-entrant Portion Filled	Plaster Thick-ness P‖	Section Area of Solid Mate-rial	Bond of Covering	Fire Resistance Rating		Notes
	In.	Lb			In.		In.	Sq In.		Hr	Min	
H	8	34		None	0	No	0	8	—	—	10	Bare column
H	6	20	E	Siliceous gravel con-crete, 1:2½:3½ mix	2	Yes	0	100	8-gage wire spiral 8-in. pitch	3	30	NBS test
Plate & angle	6	34	E	Traprock or cinder concrete, 1:6 mix*	2	Yes	0	130	6-gage wire spiral	3	45	UL test
H	8	34	E	Limestone concrete, 1:6 mix*	2	Yes	0	144	6-gage wire spiral	6	30	UL test
H	8	34	E	Limestone concrete, 1:6 mix*	4	Yes	0	256	6-gage wire spiral	7	30	UL test
H	8	34	E	Traprock, granite, cinders, 1:6 mix*	4	Yes	0	256	6-gage wire spiral	7	—	UL test
Plate & angle	6	34	E	Gypsum concrete†	2	Yes	½	114	4-in. mesh fabric	6	30	NBS test gypsum plaster
Plate & angle	6	34	E	Gypsum block	2	No	½	107	1 by ⅛-in. O clamps	4	—	NBS test gypsum plaster
Plate & angel	6	34	E	Cinder block	3¾	Yes	¾	240	Block bond	7	—	NBS test gypsum plaster
H	8	34	E	Common brick	4¼	Yes	0	270	Brick bond	7	—	UL test
H	8	34	E	Semi-fireclay hollow tile	2	No	0	96	Wire ties	1	30	UL test
H	8	34	E	Semi-fireclay hollow tile	4	No	0	158	Wire ties	1	30	UL test
H	10	49	F	Sprayed mineral fiber‡	2¼	Yes	—	164	No special adhesive	5	—	
H	10	49	F	Sprayed mineral fiber‡	3⅜	Yes	—	238	Special adhesive	5	—	
H	8	28	F	Sprayed mineral fiber‡	2	Yes	—	44	Special adhesive	5	—	
I	8	28	F	Sprayed asbestos fiber‡	2	Yes	—	38	No special adhesive	3	—	
	8	35	E	Sprayed asbestos fiber‡	1	Yes	—	90	No special adhesive	2	—	
	8	35	E	Sprayed asbestos fiber‡	1¾	Yes	—	98	No special adhesive	4	—	
	8	35	E	Sprayed asbestos fiber‡	1⅞	Yes	—	120	No special adhesive	4	—	
I	8	28	F	Sprayed asbestos fiber‡	1½	Yes	—	28	No special adhesive	2	—	UL test
O	7.6	24		None (Bare steel pipes filled with concrete)§	0	Yes	0	46	—	—	35	UL test

* Concrete mix—1 part cement to 6 parts total aggregate including sand and coarse aggregate.
† Gypsum concrete—7 parts gypsum stucco to 1 part wood shavings, by weight.
‡ Mineral fibers, with bonding agent as required, sprayed on to all surfaces of column shaft to thicknesses indicated. (Thickness different on account of characteristics of fiber and binder.)
§ Concrete-filled columns require vent holes to prevent explosion in the event of fire.
‖ Dimensions as shown in Fig. 5-8E.
SI units: 1 in. = 25.4 mm; 1 lb = 0.45 kg.

TABLE 7-8G. Steel Joist Floor or Roof Constructions

Joists			Thickness			Thickness	Fire Resistance Rating	
Type	Depth In.	Floor Slab	In.	Flurring	Ceiling Kind	In.	Hr	Min
I or S*	8	T & G wood flooring on 2- by 2-in. wood strips	25/32	3.4-lb metal lath	Gypsum—sand plaster	¾	—	45†
I or S*	8	T & G wood flooring on 2- by 2-in. wood strips	1⅝	3.4-lb metal lath	Gypsum-sand plaster	¾	1†	—
I or S*	8	Reinforced concrete, precast concrete, or gypsum planks	2	3.4-lb metal lath	Gypsum—sand or portland cement-sand plater	¾	1	—
S	8	Reinforced concrete or precast gypsum tile	2¼	3.4-lb metal lath	Gypsum—sand plaster, 1:2; 1:3 mix	¾	2	—
S	10	Reinforced concrete or reinforced gypsum tile or planks	2	3.4-lb metal lath	Neat†† gypsum, or gypsum-vermiculite plaster, 1:2; 1:3	1¾	2	30
S	10	Reinforced concrete	2½	3.4-lb metal lath	Gypsum—sand plaster	⅞	2	30
		Reinforced concrete	2½	3.4-lb metal lath	Gypsum-perlite of gypsum-vermiculite plaster, 1½:2; 1½:3	¾	3	—
S	8	Reinforced concrete perlite, or vermiculite aggregate	2½	3.4-lb metal lath	Gypsum-perlite or gypsum-vermiculite plaster, 1½:2; 1½:3	¾	3	—
S	10	Reinforced concrete, 1:2:4 gravel aggregate	2½	3-lb metal lath	Gypsum-vermiculite or gypsum-perlite plaster, 1½:2; 1½:3	1	4	—
S	10	Reinforced concrete, 1:2:4 gravel aggregate	2	Gypsum lath‡	Gypsum-perlite or gypsum-vermiculite plaster, 1½:2½	⅝	1	—
S	10	Reinforced concrete, 1:2:4 gravel aggregate	2	Gypsum & wires§	Gypsum-perlite or gypsum-vermiculite plaster, 1½:2½	½	2	—
S	10	Reinforced concrete, 1:2:4 gravel aggregate	2	Gypsum‖	Gypsum-vermiculite or perlite plaster, 1½:2; 1½:3	1	4	—
S	10	Reinforced concrete, 1:2:3.4 gravel	2½	Gypsum & wires§	Sprayed-on mineral fiber	¾	3	—
S	10	Reinforced concrete, 1:2.5:3.5 gravel	2	Special Z section#	Special acoustical tiles (see UL list)	⅝	2	—
S	12	Reinforced concrete, gravel aggregate	2	Nailing channels 16 in. o.c.	Type X** wallboard	⅝	1	30
S	12	Reinforced concrete, 1:3:3⅔ gravel aggregate	2	2¾ × ⅞ in. 26-gage channels 14 in. o.c.	Type X** wallboard applied with No. 6 by 1-in. wallboard screws	⅝	1	30
S	10	Reinforced concrete, 1:2:4 gravel aggregate	2	25-gage nailing channels 16 in. o.c.	Gypsum wallboard applied with 1¼-in. long barbed nails ⅜-in. diam. head	⅝	1	—

* I-beam or open web type joists.
† Combustible construction.
‡ All gypsum lath ⅜-in. perforated type.
§ Gypsum lath and No. 20 gage wires attached to nailing channels. Wires attached diagonally to reinforce and support lath and plaster.
‖ One-in. hexagonal mesh wire fabric to reinforce plaster and hold up lath and plater.
\# Special No. 25 gage galvanized steel Z runners 12 in. o.c.
** Type X gypsum wallboard designates gypsum wallboard with a specially formulated core which provides greater fire resistance than regular gypsum wallboard of the same thickness.
†† Unsanded wood-fiber plaster.
SI units: 1 = 25.4 mm; 1 lb = 0.45 kg.

TABLE 7-8H. Floors of Concrete on Steel Floor and Form Units

(Plaster or Sprayed on Fire Protective Covering)

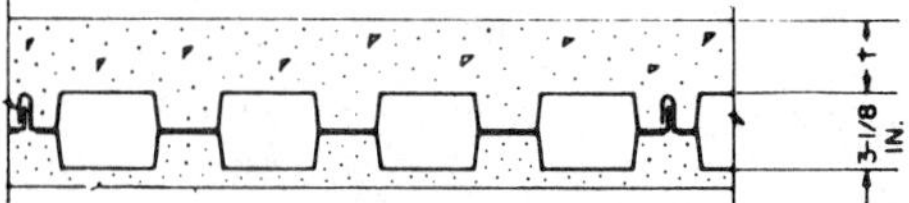

Type A

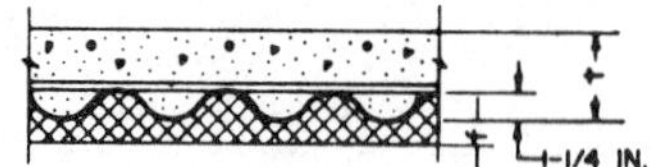

Type B

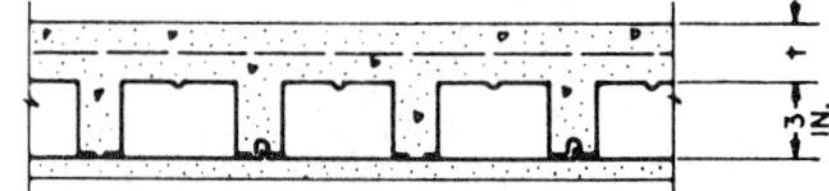

Type C

Type of Floor Unit	Thickness of Floor In.	Furring	Protective Covering Material	Application	Thickness In.	Fire Resistance Rating Hr	Fire Resistance Rating Min
A	5⅝	None	Mineral fibers applied to floor units	Sprayed	½ to 2	5	—
C	5½	None	Same	Sprayed	1½	5	—
A	5⅝	None	Vermiculite or perlite acoustical plastic	Sprayed	1 1/16 to 3 1/16	4	—
B	4½	None	Same	Sprayed	4	—	
B	4½	See Footnote*	Gypsum-vermiculite or perlite plaster	Troweled or Sprayed	⅜ to 1⅝	4	—
C	4	None	Mineral fiber applied to floor units	Sprayed	¾	3	—
A	5⅝	None	Vermiculite acoustical plastic, cellular floor units	Sprayed	½ to 2	2	—
B	5¼	None	None†	—	—	2	—
B	4¼	None	None‡	—	—	1	—
A	8	See Footnote§	Gypsum-vermiculite or perlite plaster, 100 lb gypsum to 2 cu ft for scratch coat and 3 cu ft for browncoat plaster, white finish 1/16 in.	Troweled	⅜	4	—
A	8	See Footnote§	Same	Troweled	1	5	—
A	6⅝	See Footnote‖	Acoustical tiles, T & G edges with saw kerfs	Sheet metal clips	¾	4	4

* Expanded metal lath tack welded or tied to bottom of corrugated steel floor units.
† Floor slab, limestone concrete, 5¼ in. thick.
‡ Floor slab, limestone concrete, 4¼ in. thick.
§ 24-gage, 3.4-lb, ⅜-in. mesh expanded metal lath suspended 2½ to 7½ in. below floor units.
‖ Special furring system to which acoustical tiles are clipped 10¾ in. below floor units.
SI units: 1 in. = 25.4 mm.

TABLE 7-8I. Load-Bearing Brick and Clay Tile Walls

Material	Wall Thickness Inches	Solid Content of Walls Percent	Hollow Units: Number of Cells in Wall Thickness	Hollow Units: Thickness of Shells of Unit Inches	Combustible Members Framed 4 in. into Wall: No Plaster	Combustible Members Framed 4 in. into Wall: Plaster on Two Sides	No Combustible Members Framed into Wall: No Plaster	No Combustible Members Framed into Wall: Plaster on Two Sides
Brick, clay or shale	12	90 to 100	—	—	8	9	10	12
	10*	72	2-in. cavity	—	2	2½	5	7
	8	90 to 100	—	—	2	2½	5	7
	4†	90 to 100	—	—	—	—	1	1½
Load-Bearing Hollow Tile (not partition tile)	12	45	3	0.7	2½	3½	3	6
	12‡	48	4	⅝	2½	4	5	7½
	10*	36	2 + 2-in. cavity	—	—	1¼	—	4
	8	48	3 or 4	—	1	1¾	2½	3½
	8	40	2	—	¾	1½	2	3
	6†	40	2	⅝	—	—	¾	1½

* Cavity wall with metal ties across cavity.
† Nonload-bearing wall restrained on all edges.
‡ Two units, 8 by 12 by 12-in. 6-cell and 3¾ by 12 by 12-in. 3-cell tiles, in wall thickness.
SI units = 1 in. = 25.4 mm.

TABLE 7-8J. Wood Joist Floors with Plaster Ceilings (Combustible)

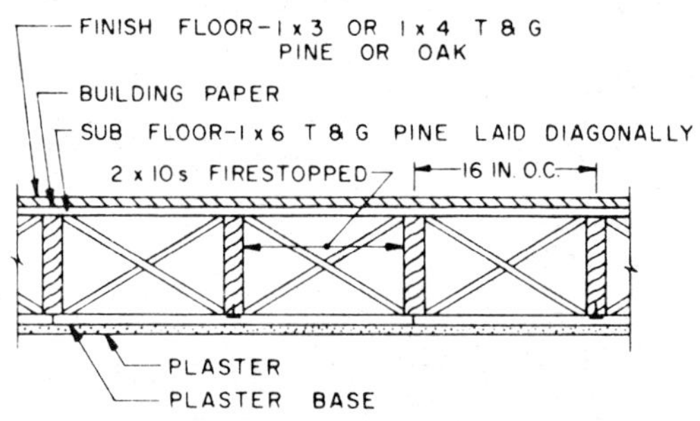

Plaster Base						Plaster						
	Thick-ness in inches or wt.	Nails					Aggregate	Mix	Thick-ness	Ceiling Thick-ness	Fire Resistance Rating	
		Size or Length		Head Diam	Spac-ing							
Type	per sq yd	Inches	Gage	Inches	Inches	Type			Inches	Inches	Hr	Min
Wood lath	3/8	3d	15	11/64	1 7/8	Lime	Sand	1:4½	5/8	1	—	30
Wood lath	3/8	3d	15	11/64	1 7/8	Gypsum	Sand	{1:2 1:3}	½	7/8	—	35
Gypsum, perforated	3/8	1 1/8	13	5/16	3½	Gypsum	Sand	1:2	½	7/8	—	30
Gypsum, perforated	3/8	1 1/8	13	3/8	3½	Gypsum	Sand	1:2	½	7/8	—	45
Gypsum, perforated	3/8*	1¾	12	½	3½	Gypsum	Sand	1:2	½	7/8	1	—
Gypsum, plain	3/8†	{5d 8d	13 12½	5/16 5/16	3½ 8}	Gypsum	Vermiculite or perlite	1½:2½	½	7/8	1	40
Gypsum, perforated	3/8	4d	—	—	3½	Gypsum	Vermiculite or perlite	1½:2½	½	7/8	1	—
Metal lath	3.4 lb	{6d 1¼	11½ 11	¼ 3/8	6 6}	Gypsum	Sand	{1:2 1:3}	¾	¾	—	45
Metal lath	3.4 lb	1½	11	7/16	6	Gypsum	Sand	{1:2 1:3}	¾	¾	1	—
Metal lath	3.0 lb	8d	11½	¼	6‡	Gypsum	Sand	{1:2 1:3}	¾	¾	1	15
Metal lath	3.4 lb	1½	11½	7/16	5	Portland cement§	Sand	{1:2 1:3}	¾	¾	1	—
Metal lath	3.4 lb	1½	11	7/16	5½	Gypsum	Vermiculite or perlite	1½:2 {1⅓:3}	¾	¾	1	30

* Three-in.-wide strips of metal lath with two 1¾-in. 12-gage ½-in. head barbed roofing nails in each joist covering joints of lath to reinforce plaster.
† Plaster reinforced with 1-in. hexagonal mesh wire fabric (1 lb/sq yd) nailed to joists with 8d 2¾-in.-long cement-coated nails spaced 8 in. on centers at each joist.
‡ Additional support for metal lath by ties of 18-gage wire spaced 27 by 32 in. and nailed 2 in. up on sides of joists.
§ Three-lb asbestos fiber and 15-lb hydrated lime per bag of cement added.
SI units: 1 in. = 25.4 mm.

TABLE 7-8K. Fire Resistance of Reinforced Concrete Columns*

Column				Concrete			Reinforcement							
	Section		Load	Aggregates		Mix	Vertical			Lateral		Concrete Cover Thick-ness	Fire Resistance	
No.	In.	Area Sq In.	in 1000 Lbs	Fine	Coarse	Cement Fine Coarse	Num-ber	Bar Size No.	Sq In.	Diam In.	Spac-ing In.	In.	Hr	Min
70	16 × 16	256	101	Fox River Sand	Chicago limestone	1:2:4	4	9	4.00	¼	12	2¼	8 (13+	40 —)
71	16 × 16	256	101	Long Island Sand	New York trap rock	1:2:4	4	9	4.00	¼	12	2¼	7	22
72	17 dia	227	107.5	Fox River Sand	Chicago limestone	1:2:4	6	9	6.00	¼	12	2½	8 (12+	04 —)
73	17 dia	227	107.5	Long Island Sand	New York trap rock	1:2:4	6	9	6.00	¼	12	2½	7	57

Table 7-8K (Cont.)

74	17 dia	227	129	Fox River Sand	Chicago limestone	1:2:4	6	6	2.64	1/4	1 1/2	2 1/4	8 (13+	06 —)
75	17 dia	227	129	Long Island Sand	New York trap rock	1:2:4	6	6	2.64	1/4	1 1/2	2 1/4	8	02
25	16 × 16	256	92	Pittsburgh Sand	Pittsburgh gravel	1:2:4	4	8	3.16	1/4	12	1 1/2	4 (6	— 30)
44	16 × 16	256	92	Long Island Sand	Pure quartz gravel	1:2:4	4	8	3.16	1/4	12	1 1/2	4 (6	— —)
51	16 × 16	256	92	Pittsburgh Sand	Blast furnace slag	1:2:4	4	8	3.16	1/4	12	1 1/2	4 (8	— —)
56	16 × 16	256	92	Pittsburgh Sand	New Jersey trap rock	1:2:4	4	8	3.16	1/2	12	1 1/2	4 (7	— —)
7	18 dia	254	99.75	Pittsburgh Sand	Pittsburgh gravel	1:2:4	8	6	3.52	1/4	12	1 1/2	5 (6	— —)
2	18 dia	254	141	Pittsburgh Sand	Pittsburgh gravel	1:2:4	8	6	3.52	3/16	2	1 1/2	4	—
48	18 dia	254	141	Pittsburgh Sand	Blast furnace slag	1:2:4	8	6	3.52	3/16	2	1 1/2	4 (10+	— —)
85	18 dia	254	141	Elgin, Ill. Sand	Elgin, Ill. gravel	1:2:4	8	6	3.52	3/16	2	1 1/2	4 (12+	— —)
12	18 dia	254	81	Pittsburgh Sand	Pittsburgh gravel	1:2:4	None	—	—	—	—	—	4	—
33	12 dia	113	51	Pittsburgh Sand	Pittsburgh gravel	1:2:4	4	5	1.24	1/4	2 1/8	1 1/2	Avg. of 2 3	—

* Columns Nos. 70 to 75 tested at UL. Test No. 70 of the group was stopped at 8 hr 40 min, others at failure or at 8 hr, and all loaded to failure at end of fire exposure (NBS Tech. Paper 184). All other columns tested at NBS Laboratory at Pittsburgh, Pa. The fire endurance test of Col. No. 7 stopped at 5 hr, all others of series stopped at 4 hr. Figures in parentheses are estimates of fire resistance if tests had continued to failure (NBS Tech. Paper 272).
SI units: 1 in = 25.4 mm; 1 lb = 0.45 kg.

TABLE 7-8L. Solid Partitions: Nonbearing

Materials	Fire Resistance Rating Hr	Fire Resistance Rating Min
Sheathing planks (tongue-and-groove), in 2 layers, each 3/4 in. thick (1 in. nominal) and with joints staggered	—	15*
Same, with layer of 30-lb asbestos felt between planks	—	25*
Planking, pine (tongue-and-groove), 2 in. thick (nominal), set vertically	—	12*
Wallboard, 3/8-in. gypsum, full height facings on two thicknesses 1/2-in. coreboard, full height, cemented with staggered vertical joints to form 1 3/4-in. thick partition, external joints finish taped	1	—
Wallboard, 5/8-in. gypsum, Type X‡, full height facings, cemented and nailed or screwed to ribs made of two thicknesses of 1/2-in. gypsum board 3 1/2 or 6 1/2 in. wide and to 1- by 1 5/8-in. wood runners top and bottom, external joints finished taped	1	—
Wallboard, 1/2-in. gypsum, full height, nailed and cemented to 1-in.-thick coreboard factory laminated from two 1/2-in.-thick by 24-in.-wide coreboards with staggered edges, external joints staggered, butted and finish taped with joint finisher	2	—
Gypsum tile, 3 in. thick, cored	1	—
Gypsum tile, 4 in. thick, cored	1	—
Gypsum tile, 3 in. solid, no cores	3	—
Gypsum tile, 3 in. cored, 1/2 in. of 1:3 gypsum-sand plaster on each side	3	—
Gypsum tile, 4 in. cored, 1/2 in. of 1:3 gypsum-sand plaster on each side	4	—
Gypsum-vermiculite or perlite plaster, 1 1/2:2, 1 1/2:3 by vol., 3/4 in. thick on each side of 1/2-in. gypsum lath, vertical full height, tied to floor and ceiling runners, no studs	1	30
Gypsum-sand plaster, 1:2, 1:3 by wt, 3/4 in. thick on each side of 1/2-in. gypsum lath vertical full height, tied to floor and ceiling runners, no studs	1	—
Partition tile, burned clay, 4 in. thick, 1 cell in thickness	—	10†
Cinder block, 4 in. thick, solid	1†	—
Cinder block, 6 in. thick, 1 cell in thickness	1	15†
Calcareous gravel concrete tile, 4 in. thick, 65 percent solid	—	45‡
Calcareous gravel concrete tile, 8 in. thick, 55 percent solid	2	30

* Combustible.
† When plastered on both sides with 1/2-in. 1:3 gypsum-sand plaster, the tile partition described has 45-min fire resistance and the cinder block and 4-in. concrete tile assemblies described have 2-hr fire resistance.
‡ Type X gypsum wallboard designates gypsum wallboard with a specially formulated core which provides greater fire resistance than regular gypsum wallboard of the same thickness.
SI units: 1 in. = 25.4 mm; 1 lb. = 0.45 kg.

TABLE 7-8M. Wood Stud Walls and Partitions (Combustible)

(Bearing and nonbearing: 2- by 4-in. studs spaced 16 in. on centers, fire stopped)

Material	Fire Resistance Rating: Partition Hollow Hr	Partition Hollow Min	Partition Filled with Mineral Wool† Hr	Partition Filled with Mineral Wool† Min
Plasterless Types of Construction				
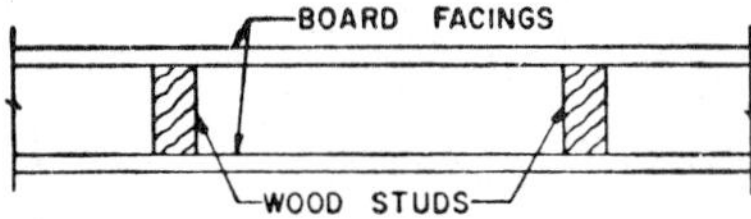				
The following are applied to both sides of studs:				
Sheathing boards (tongue-and-groove) ¾ in. thick	—	20	—	35
Gypsum wallboard, ⅜ in. thick	—	25	—	—
Gypsum wallboard, ½ in. thick (nonload-bearing only for mineral wool filled)	—	40	1	—
Gypsum wallboard, ⅜ in. thick, in two layers each face	1	—	—	—
Gypsum wallboard, ½ in. thick, in two layers each face	1	30	—	—
Gypsum wallboard, ½ in. thick, Type X*, one layer each face	—	45	—	—
Gypsum wallboard, ⅝ in. thick, Type X*, one layer each face	1	—	—	—
Gypsum wallboard, ⅝ in. thick, Type X*, on fire retardant wood fibreboard, ½ in. thick	1	—	—	—
Fir plywood, ¼ in. thick	—	10	—	—
Fir plywood, ⅜ in. thick	—	15	—	—
Plasterless Types of Construction				
Fir plywood, ½ in. thick	—	20	—	—
Fir plywood, ⅝ in. thick	—	25	—	—
Cement-asbestos board, 3/16 in. thick	—	10	—	40
Cement-asbestos board, 3/16 in. thick, on gypsum wallboard, ⅜ in. thick	1	—	—	—
Cement-asbestos board, 3/16 in. thick, on gypsum wallboard, ½ in. thick	1	25	—	—
Plaster and Lath Construction				
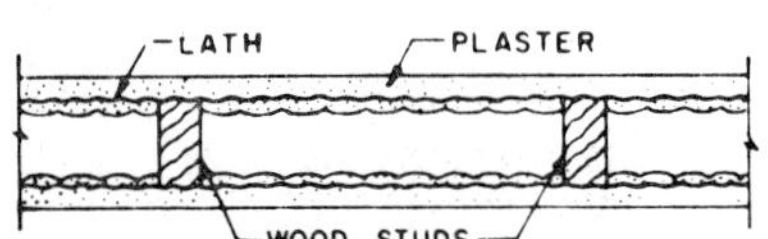				
The following are applied to both sides of studs:‡				
Gypsum-sand plaster, 1:2, 1:3, ½ in. thick on wood lath	—	30	1	—
Lime-sand plaster, 1:5, 1:7.5, ½ in. thick on wood lath	—	30	—	45
Gypsum-sand plaster, 1:2, 1:2, ½ in. thick on ⅜ in. perforated gypsum lath	1	—	—	—
Gypsum-sand plaster, 1:2, 1:2, ¾ in. thick on metal lath	1	—	1	30
Gypsum lath, ½ in. thick, Type X*, and ⅛ in. gypsum-sand plaster, each face	1	—	—	—
Gypsum wallboard, ½ in. thick, Type X*, and 1/16 in. gypsum plaster, each face	1	—	—	—
Portland cement-lime-sand plaster, 1:1/30:2, 1:1/30:3 and asbestos-fiber plaster, 7/8 in. thick on metal lath	1	—	—	—
Gypsum-vermiculite, or perlite plaster, 100 lb gypsum to 2½ cu ft aggregate, ½ in. thick on ⅜ in. perforated gypsum lath	1	—	—	—
Gypsum perlite plaster, 1:2, ¾ in. thick on metal lath	1	—	—	—
Exterior Bearing Wall				
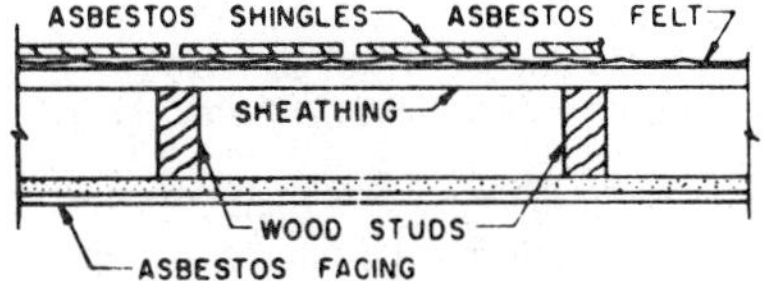				
Outside: Cement-asbestos shingles, 5/32 in. thick, on layer of asbestos felt on wood sheathing, ¾ in. thick, on wood studs. Inside: Cement-asbestos facing, ⅛ in. thick on fiberboard, 7/16 in. thick	—	30	—	—
Outside: Gypsum sheathing, ½ in. thick. Inside 1:2 gypsum-sand plaster, ½ in. thick on ⅜ in. perforated gypsum lath	1	30	—	—

* See footnote ‡ to Table 5-8S.

† Mineral wool fill requires some degree of anchorage so as to be held in place after partition facing has been burned away.

‡ See Table 5-8Q for one side ratings.

SI units: 1 in. = 25.4 mm; 1 lb = 0.45 kg; 1 cu ft = 0.09 m^3.

TABLE 7-8N. Concrete Floor Constructions

Material	Fire Resistance Rating Hr	Min
Reinforced Concrete (Free or partly restrained, 1500–2500 psi)		
3/4 IN. MIN. PROTECTION FOR STEEL REINFORCING; 3/4 IN.		
3 in. thick	—	45
4 in. thick	1	15
6 in. thick; 1-in. minimum protection for steel	2	—
Reinforced Concrete on Precast Joists		
1 IN. MIN. PROTECTION FOR STEEL REINFORCING; 4 IN. MIN.; 30 IN.; 1 IN.; 1 IN.; 8 IN. JOISTS BURNED CLAY OR EXPANDED SLAG AGGREGATE		
Reinforced concrete, 1:3½:4, 3 in. thick, no ceiling	—	45
Reinforced concrete, 1:3½:4, 3 in. thick, ceiling of gypsum wallboard ½ in. thick, nailed to wood, strips wired to joists	1	—
Combination of Tile and Concrete Floors		
4 IN. x 12 IN. x 12 IN. TILE (FIRE CLAY); REINFORCING STEEL		
Concrete 2 in. or 1½ in. thick, and fire clay tile 6 in. or 4 in. thick, no ceiling finish	1	—
Concrete 1½ in. thick, and tile 4 in. thick with gypsum-sand plaster ceiling finish, 1:3 mix, ⅝ in. thick	1	30
Concrete 2 in. thick, and fire clay tile 6 in. thick with gypsum plaster ceiling finish, 1:3 mix, ⅝ in. thick	2	—
Concrete 2½ in. thick, limestone aggregate 4 in. thick, expanded slag concrete tile	3	—
Reinforced Concrete Ribbed Slab		
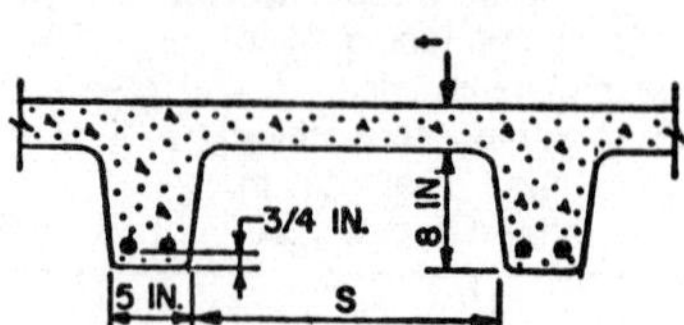		
Concrete, ribbed slab, limestone aggregate, t = 1½ in., S= 20 in.	—	20
Concrete, ribbed slab, limestone aggregate, t = 2½ in., S = 20 in.	—	45
Same with metal lath and ⅞-in. gypsum-sand plaster ceiling, 1:2, 1:3 mix	2	30
Concrete, ribbed slab, limestone aggregate, t = 3 in. S = 30 in.	1	—

SI units: 1 in. = 25.4 mm; 1 psi = 6.89 kPa.

TABLE 7-8O. Various Finishes Over Wood Framing, One Side (Combustible) with Exposure on Finish Side

(See Table 5-8P for 2-side ratings)

Material	Fire Resistance Rating* Min
Fiberboard, ½ in. thick	5
Fiberboard, flameproofed, ½ in. thick	10
Fiberboard, ½ in. thick, with ½ in. 1:2, 1:2 gypsum-sand plaster	15
Gypsum wallboard, ⅜ in. thick	10
Gypsum wallboard, ½ in. thick	15
Gypsum wallboard, ⅝ in. thick	20
Gypsum wallboards, laminated, two ⅜ in.	28
Gypsum wallboards, laminated, one ⅜ in. plus one ½ in. thick	37
Gypsum wallboards, laminated, two ½ in. thick	47
Gypsum wallboards, laminated, two ⅝ in. thick	60
Gypsum lath, plain or indented, ⅜ in. thick, with ½ in. 1:2, 1:2 gypsum-sand plaster	20
Gypsum lath, perforated, ⅜ in. thick, with ½ in. 1:2, 1:2 gypsum-sand plaster	30
Gypsum-sand plaster, 1:2, 1:3, ½ in. thick, on wood lath	15
Lime-sand plaster, 1:5, 1:7.5, ½ in. thick, on wood lath	15
Gypsum-sand plaster, 1:2, 1:2, ¾ in. thick, on metal lath (no paper backing)	15
Neat gypsum plaster, ¾ in. thick on metal lath (no paper backing)†	30
Neat gypsum plaster, 1 in. thick, on metal lath (no paper backing)†	35
Lime-sand plaster, 1:5, 1:7.5, ¾ in. thick, on metal lath (no paper backing)	10
Portland cement plaster, ¾ in. thick, on metal lath (no paper backing)	10
Gypsum-sand plaster, 1:2, 1:3, ¾ in. thick, on paper-backed metal lath	20

* From National Bureau of Standards BMS-92.
† Unsanded wood-fiber plaster.
SI units: 1 in. = 25.4 mm.

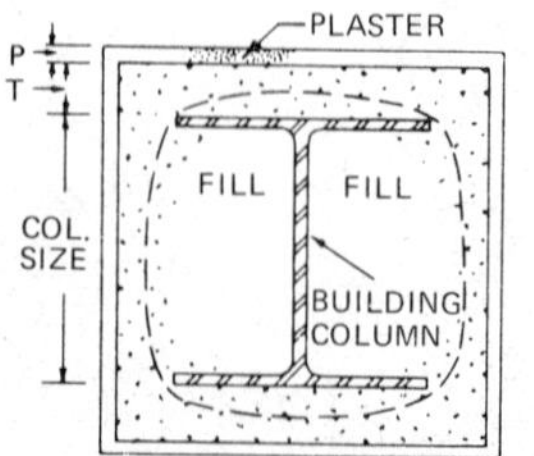

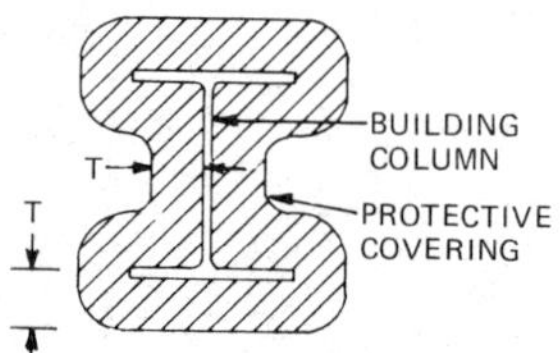

FIG. 7-8D. Typical steel column protections of concrete, masonry, or sprayed fibers.

TABLE 7-8P. Wood Joist Floors with Wallboard Ceilings (Combustible)

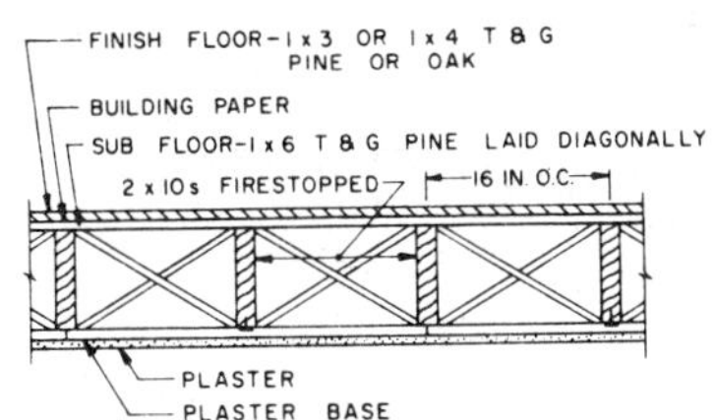

Wallboard Type	Wallboard Thickness Inches	Core Materials	Nails Type	Nails Size	Nails Gage	Nails Length Inches	Nails Spacing Inches	Fire Resistance Rating Hr	Fire Resistance Rating Min
Gypsum	5/8	Type "X"* special fire-retardant gypsum	Cement-coated wire	6d	13	1 7/8	6	1	—
Gypsum	1/2	Type "X"* special fire-retardant gypsum	Cement-coated wire	5d	13½	1 5/8	6	—	45
Gypsum	3/8	Type "X"* special fire-retardant gypsum	Cement-coated wire	4d	14	1 3/8	6	—	30
Two thicknesses of gypsum	1/2 + 1/2	Gypsum	{ Box wire	5d(1) 6d(2)	14 10¼	1¾ 2½	18 6 }	1†	—
Two thicknesses of gypsum	3/8 + 1/2	Gypsum	{ Plasterboard cement-coated	1½ in. 6d	13 13	1½ 1 7/8	7 6 }	—	40
Two thicknesses of gypsum	1/2 + 3/8	Gypsum	{ Plasterboard cement-coated	1½ in. 6d	13 13	1½ 1 7/8	7 6 }	—	35
Two thicknesses of gypsum‡	3/8 + 3/8	Gypsum	Box	{ 4½d(1) 4½d(2)	— —	1½ 1½	6 6 }	—	35
Gypsum	1/2	Gypsum	Box	4½d	15	1½	6	—	25
Gypsum§	3/8	Gypsum	Box	4½d	15	1½	6	—	25
None‖	—	—	—	—	—	—	—	—	14
Acoustical Tile	5/8	12- by 12-in. mineral fiber tiles mounted on special channels						1	—

* Type X gypsum wallboard designates gypsum wallboard with a specially formulated core which provides greater fire resistance than regular gypsum wallboard of the same thickness.

† 1-in. hexagonal mesh 20-gage wire fabric between wallboards nailed with 8d nails 8 in. o.c.

‡ NBS test on floor 4½ by 9 ft; joints of wallboard staggered, but no tape and joint finisher.

§ NBS test; bottom of ceiling covered with 14 lb asbestos paper applied with paperhanger's paste and casein paint.

‖ NBS test on 2 specimens of open-joist floors, each 4½ by 9 ft; fire endurance 15 min and 12 min.

Note: for other similar constructions, see UL *Building Materials List.*

SI units = 1 in. = 25.4 mm.

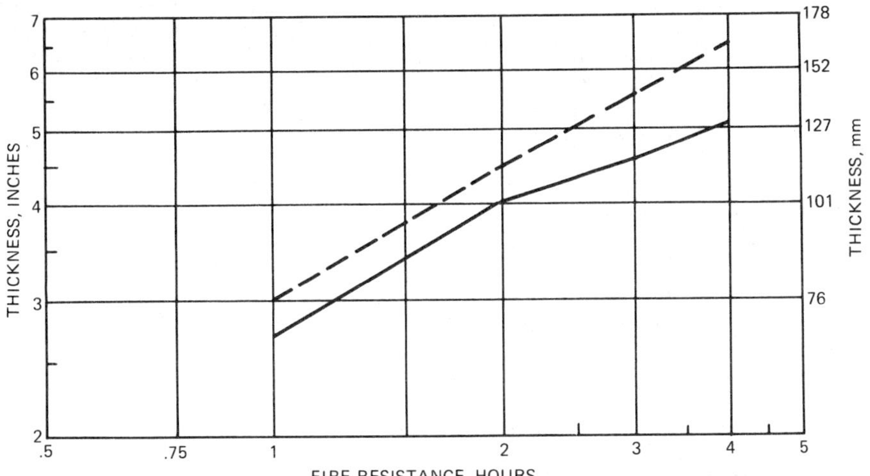

FIG. 7-8E. Fire resistance ratings of reinforced concrete floors of varying thicknesses. The dotted line represents concrete floors made with regular aggregates and the solid line represents lightweight concretes.

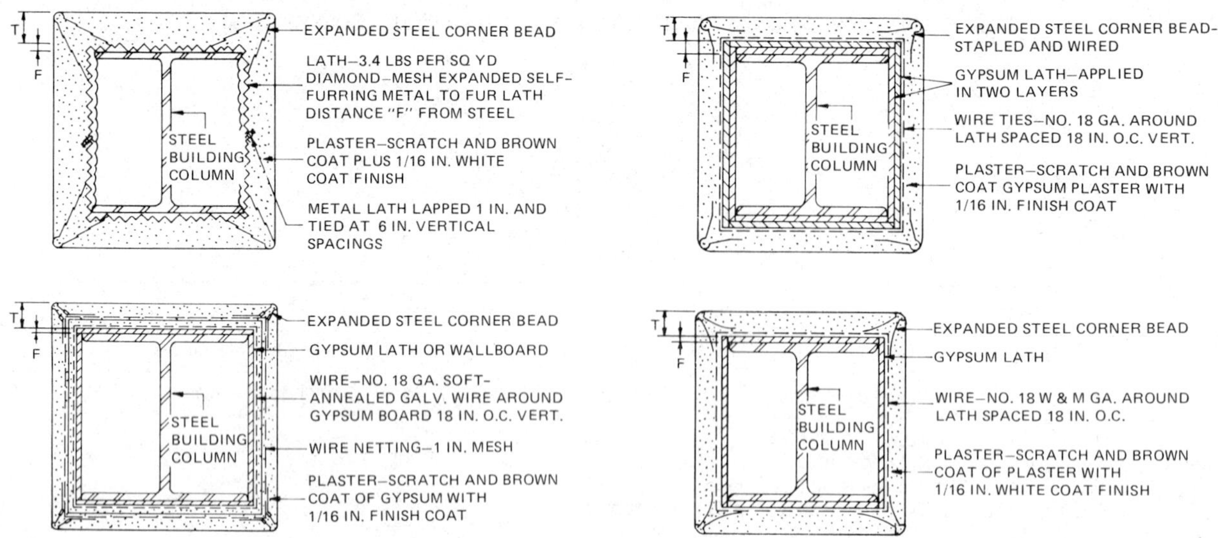

FIG. 7-8F. Typical steel column protection of lath and plaster assemblies.

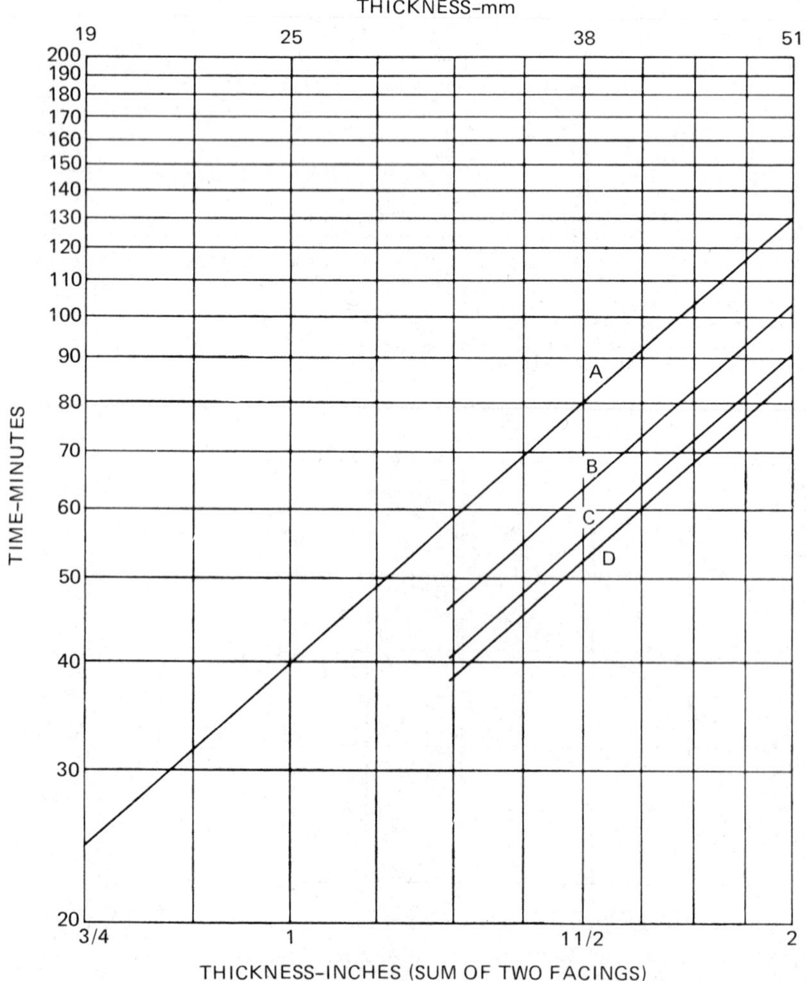

FIG. 7-8G. Fire resistance of wood or metal stud partitions faced with gypsum wallboards or gypsum plaster on metal lath: A, Type X gypsum wallboards or wood fiber gypsum plaster; B, 1:1 gypsum-sand plaster; C, 1:2 gypsum-sand plaster; D, 1:2 and 1:3 gypsum-sand plaster.

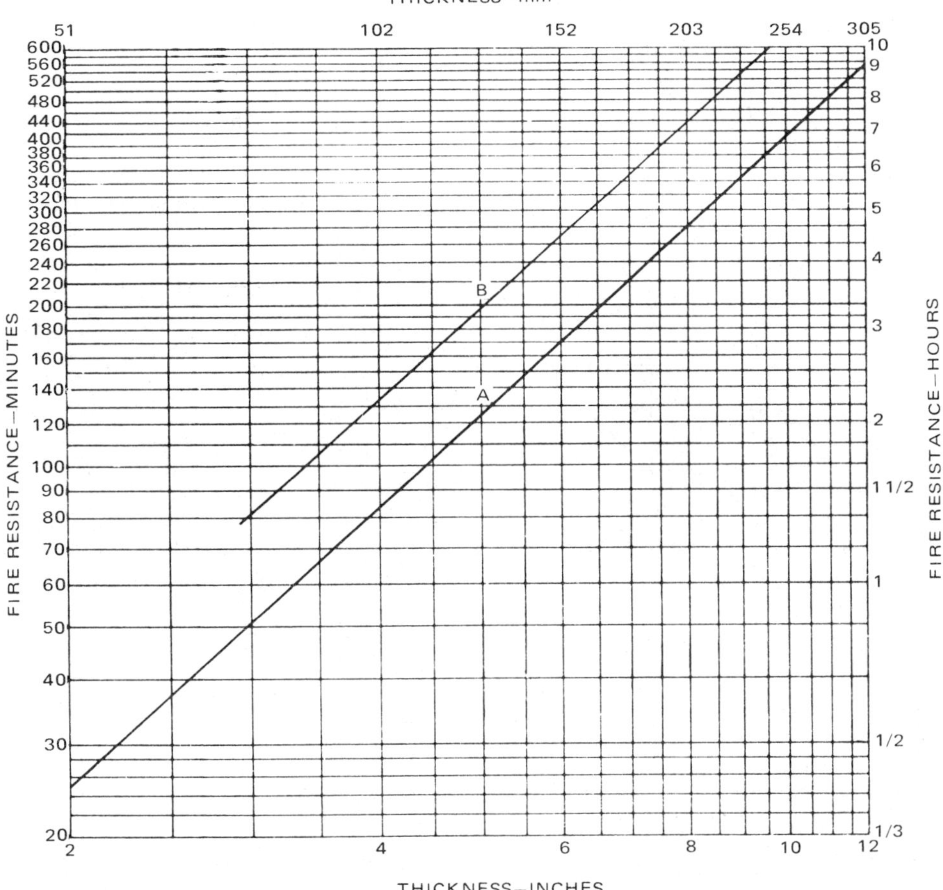

FIG. 7-8H. Fire resistance of burned clay brick walls: A, unplastered; B, plastered both sides.

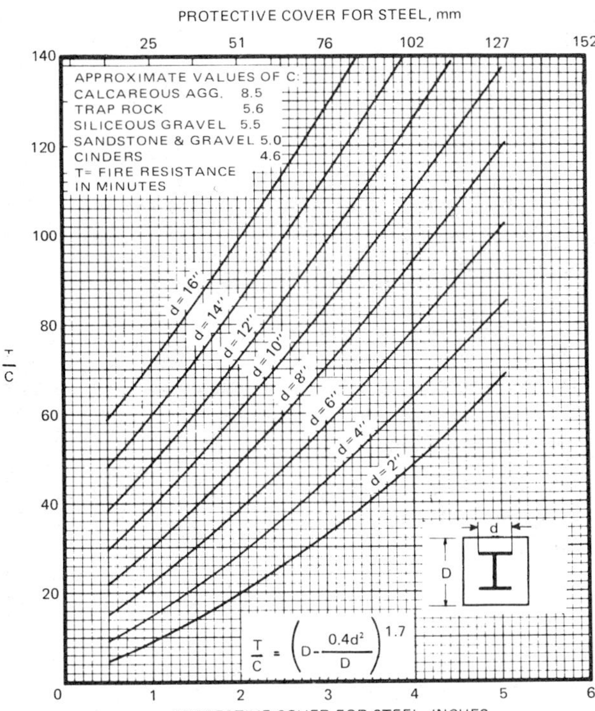

$$\frac{T}{C} = \left(D - \frac{0.4d^2}{D}\right)^{1.7}$$

FIG. 7-8I. Encasing requirements of steel columns and other free standing steel members.

CALCULATING FIRE ENDURANCE

Fire testing has been ongoing for many years, and a large amount of data and experience has been developed during that time. In addition, specific research projects have been undertaken with the express purpose of developing equations by which fire endurance may be determined by calculation rather than by test. Consequently a number of imperial equations have been developed which enables the equivalent fire endurance for a number of types of building elements to be calculated.

The critical temperature for structural steel is normally considered to be 1,000°F (538°C). At that temperature, structural steel will have lost sufficient strength to reduce the yield stress to the value commonly used for the working stress design. Consequently, the load carrying capacity is reduced to about the design load capacity for statically determinate members.

The temperature of 1,000°F (538°C) is, therefore, considered to be the approximate failure temperature for structural steel. The fire resistance of structural steel beams and columns can be extended to acceptable levels by insulating the members. The fire resistance of steel beams and columns may be calculated from theory and empirical testing. Equation 1 has been derived from studies with the objective of calculating the fire resistance of steel beams and columns protected by light insulation (Lie & Stanzak 1973, AISI 1980). The members may be pro-

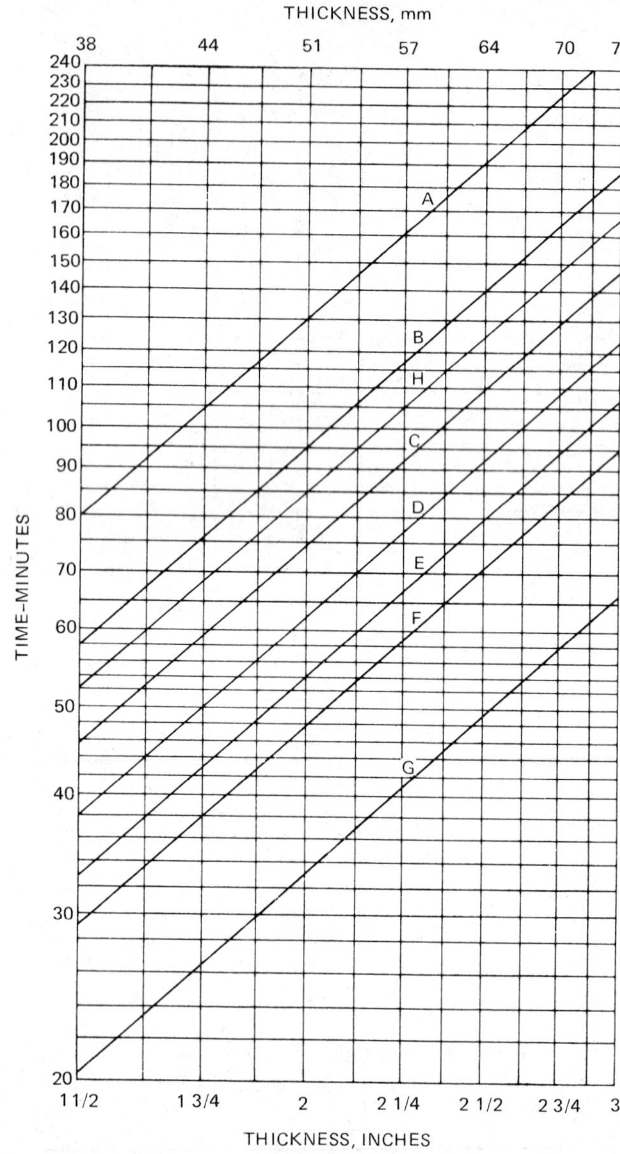

FIG. 7-8J. Fire resistance of solid partitions of metal lath and plaster: A, wood fiber gypsum; B, 1:½ gypsum-sand; C, 1:1 gypsum-sand; D, 1:1½ gypsum-sand; E, 1:2 gypsum-sand; F, 1:2 and 1:3 gypsum-sand; G, 1:2 and 1:3 portland cement + 0.2 lime to cement.

tected by boxing the slope or by a contour protection. The equation is:

$$R = \left(C_1 \frac{M}{D} + C_2\right)l \tag{1}$$

where

R = fire endurance in minutes
M = mass of the member in lb/ft
D = heated perimeter in inches
l = thickness of protection in inches
C_1, C_2 = constants that are empirically derived for the insulating units.

In SI units:

$$R = \left(0.672\ C_1 \frac{M}{D} + 0.039\ C_2\right)l \tag{1m}$$

where

R = fire endurance in minutes
M = mass of the member in kg/m
D = heated perimeter in mm
l = thickness of protection in mm
C_1, C_2 = empirical constants the same as used for U.S. customary units.

For insulating materials, such as mineral fibers, vermiculite, and perlite, having densities (ρ), in the range of 20 to 50 lb per cu ft (32 to 80 kg/m^3), the factors C_1 and C_2 are:

$$C_1 = 1{,}200/\rho \text{ and } C_2 = 30.$$

For insulating materials of the same range of densities, but incorporating cement pastes or gypsum, the factors C_1 and C_2 are:

$$C_1 = 1{,}200/\rho \text{ and } C_2 = 72.$$

Heavy, unprotected steel columns are capable of exhibiting some significant fire resistance. Equations (2) and (3) and (2m) and (3m) can be used to predict the fire resistance of unprotected steel columns (Stanzak & Lie 1973):

$$R = 10.3 \left(\frac{M}{D}\right)^{0.7} \text{ for } \frac{M}{D} < 10 \tag{2}$$

$$R = 8.3 \left(\frac{M}{D}\right)^{0.8} \text{ for } \frac{M}{D} > 10 \tag{3}$$

where

R = fire endurance in minutes
M = mass of the column in lb/ft
D = heated perimeter in inches

In SI units:

$$R = 75.1 \left(\frac{M}{D}\right)^{0.7} \text{ for } \frac{M}{D} < 171 \tag{2m}$$

$$R = 60.5 \left(\frac{M}{D}\right)^{0.7} \text{ for } \frac{M}{D} > 171 \tag{3m}$$

where

R = fire endurance in minutes
M = mass of the column in kg/m
D = heated perimter in mm

Thermal and mechanical properties vary for all materials with the temperature of the material. Because the heat transmission through concrete is quite slow, the variability of properties throughout the member is quite nonuniform. As a result, the behavior of reinforced concrete in fire is very complex. Nevertheless, equations have been developed to predict the fire endurance of reinforced concrete columns.

$$R = \frac{t}{4f} - 1 \tag{4}$$

where

R = fire endurance in hours
T = minimum dimension of the column in inches
F = a factor that considers overdesign, effective length, and percent of steel. (Table 7-8Q provides values for f.)

TABLE 7-8Q. Values for *f*

Overdesign factor	Effective length, kL (kL_{max} = 12 ft)	Effective length, kL 12 ft < kL < 24 ft: $p \leq 0.03$ $t \leq 12$ in.	All other cases
1.00	1.0	1.2	1.0
1.25	0.9	1.1	0.9
1.50	0.8	1.0	0.8

SI units: 1 ft. = 0.30 m; 1 in. = 25.4 mm.

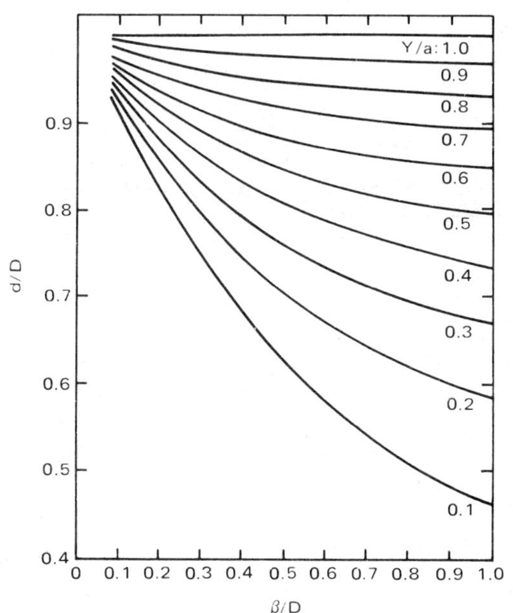

FIG. 7-8K. Critical depth of solid timber beams of rectangular cross section exposed on three sides to fire (Schaffer 1977).

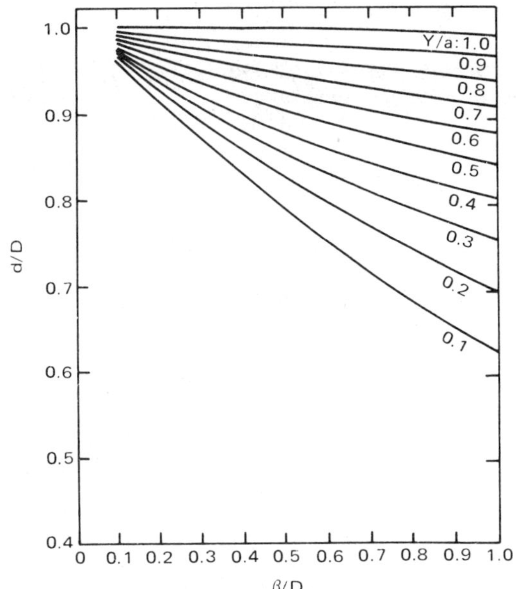

FIG. 7-8L. Critical Depth of solid timber beams of rectangular cross section exposed on four sides to fire (Schaffer 1977).

In SI units:

$$R = \frac{t}{101.6f} - 1 \tag{4m}$$

where
R = fire endurance in hours
T = minimum dimensions of the column in mm
F = the same factor from Table 7-8Q

An equation has been derived for the fire resistance of normal weight concrete slabs (Harmathy 1970, Allen and Harmathy 1972). It is:

$$R = 0.031\ \rho^{1.2}\ L^{1.85} \tag{5}$$

where
R = fire resistance in hours
ρ = concrete density in lbs/cu ft
L = slab thickness in feet

In SI units

$$R = 0.010\ \rho^{1.2}\ L^{1.85} \tag{5m}$$

where
R = fire resistance in hours
ρ = concrete density in kg/m^3
L = slab thickness in meters (m)

When timber members are tested, they burn. As burning continues a char layer is formed on the exposed surfaces. This char layer reduces the useable strength of the timber members. However, the time duration that ensues before the member reaches its critical failure load is the fire endurance.

The fire endurance for wood beams is calculated by determining the breaking strength of the uncharred residual cross section. The time to reach a critical depth may be calculated from the following equations.

$$t_c = (D-d)/\beta \tag{6}$$

$$t_c = (D-d)/2\beta \tag{7}$$

where
t_c = fire endurance time in minutes
D = original depth of the beam in inches (mm)
d = critical residual depth of the beam in inches (mm)
β = charring rate for wood in inches (mm). [An average charring rate for wood is 1/40 in. per min (0.6 mm per min).]

The critical residual depth, d, may be obtained from Figures 7-8K and 7-8L. The family of curves shows ratios of allowable bending strength to ultimate bending strength for the beam.

STRUCTURAL MATERIALS AT ELEVATED TEMPERATURES

All structural materials used in building construction are adversely affected by the elevated temperatures caused by fire. The degree and significance of this adverse behavior depends primarily on the function of the elements and on the degree of protection afforded. The mechanical properties of strength and stiffness decrease as the temperature rises. Other adverse behavior, such as excessive expansion and accelerated creep, also develops with increasing temperatures. In general, however, the design parameters that are of concern at normal temperatures are the same parameters that are of concern at elevated temperatures.

Structural Steel

Steel is the backbone of modern building design. Whether it acts as the reinforcement for concrete or the skeleton framework for buildings, it constitutes the major load carrying material in modern building construction. From the designer's viewpoint steel possesses many qualities, such as high strength and good ductility, that make it an ideal structural material. However, steel, like all other materials, is adversely affected by fire.

Steel is noncombustible and does not contribute fuel to a fire. In the past, these properties have sometimes provided a false sense of security with regard to its durability in a fire because they overshadow the fact that steel loses strength when subjected to temperatures easily attained in a fire. The relative seriousness of the problem depends on several factors, such as the function of the steel element, its level of stress, its surface area and thickness, and the temperature within the steel itself. This temperature can be quite different from the ambient temperature in the compartment.

From a structural viewpoint, the yield stress of steel is the significant parameter in establishing load carrying capacity. The tensile stress-strain diagram for A-36 steel at various temperatures is shown in Figure 7-8M. The compressive stress-strain diagram for this same steel at various temperatures is shown in Figure 7-8N. It can be seen that both the yield stress and the modulus of elasticity decreases with increasing temperatures. Figure 7-8O shows the ratios of modulus of elasticity and yield stress for A-36 steel at various temperatures. As can be seen, both values decrease with an increase in temperature.

The intensity of stress in a steel member influences the load carrying capacity. The higher the load stress, the more quickly a member will fail at elevated temperatures. A temperature of 1,100°F (593°C) is normally considered to be the critical temperature. At this temperature the yield stress in the steel has decreased to about 60 percent of the value at room temperature, which is approximately the level normally used as the design working stress.

This reduction in stress, combined with the reduction in the modulus of elasticity, causes compression members

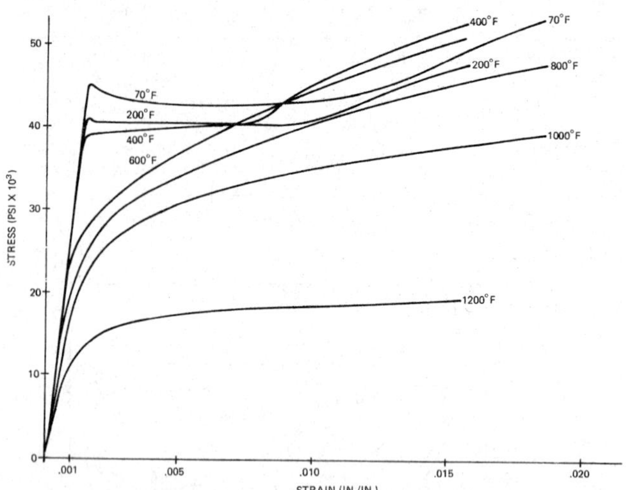

FIG. 7-8N. Stress-strain curves for A36 steel at various temperatures (DeFalco 1974).

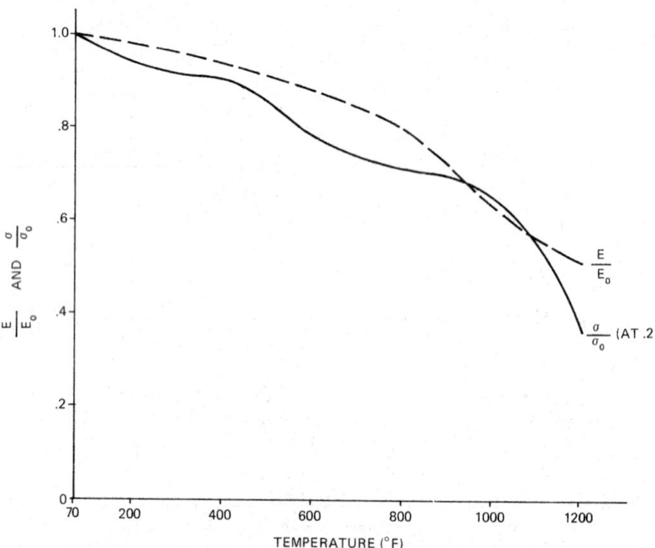

FIG. 7-8O. Ratio of modulus and yield strength with temperature—A36 steel (DeFalco 1974).

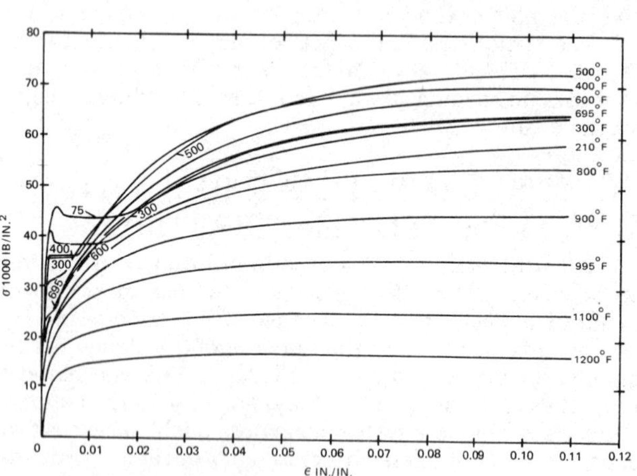

FIG. 7-8M. Stress-strain curves for an ASTM A36 steel (Harmathy & Stanzak 1969).

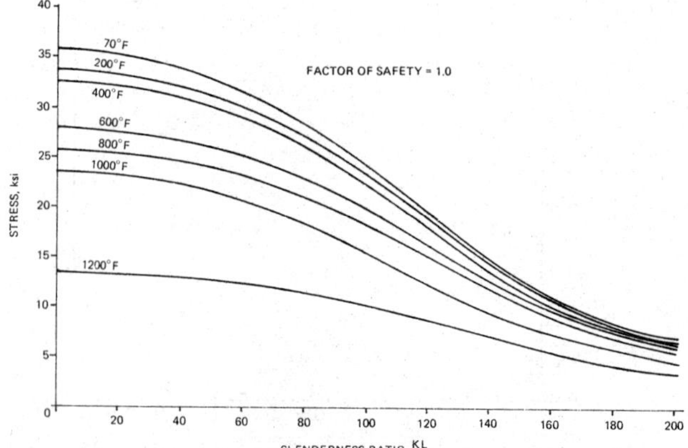

FIG. 7-8P. Critical stress as a function of the slenderness ratio for various temperatures—A36 steel (DeFalco 1974).

to be more sensitive to higher temperatures than tensile or flexural members. Figure 7-8P shows the influence of temperature on the critical stress of compression members of A-36 steel.

It should be recognized that the temperature considered for stress limitations is the temperature within the steel, and not the ambient temperature. Because steel has a high thermal conductivity it can transfer heat away from a localized heat source rather quickly. This property, in conjunction with its thermal capacity, enables steel to act as a heat sink. When the steel has an opportunity to transfer heat to cooler regions, it can take a relatively long time for a member to reach its critical value. On the other hand, an extensive fire that distributes heat simultaneously over a greater area reduces this time considerably.

Related to this thermal activity is the effect of mass and surface area of structural steel members. Heavy, thick sections have a far greater resistance to the effects of building fires than do lighter ones. Unprotected lightweight sections, such as those found in trusses and open web joists, can collapse after 5 or 10 min of exposure.

Another property of steel that has an effect upon its performance at elevated temperatures is its coefficient of expansion. The linear coefficient of expansion of steel at temperatures up to 1,100°F (600°C) is given as:

$$\alpha = 0.0000061 + 0.0000000022\ \Delta t$$

where

α = the coefficient of expansion
Δt = temperature change in degrees Fahrenheit.

In S.I. units:

$$\alpha = 6.1 \times 10^{-6} + 3.96 \times 10^{-9}\ \Delta t$$

where

α = coefficient of expansion
Δt = temperature change in degrees Celsius.

This high coefficient affects the structure in two ways. If the ends of a structural member are axially restrained, the attempted expansion due to the heat causes thermal stresses to be induced in the member. These stresses combine with those of the normal loading causing more rapid collapse. If the structural member is not axially restrained, the increased stresses described do not occur; instead, movement takes place. This movement causes the ends of steel columns to be moved laterally, producing an eccentrically loaded column. In other cases, walls can be moved to the point of collapse by expansion of beams. This creates an extremely hazardous condition both for the building itself and for the fire fighters.

To illustrate the magnitude of movement, consider a 50 ft long steel beam that is heated uniformly over its length from 721° to 972°F. The average value for alpha is 0.0000073, and the increase in length, delta, becomes approximately,

$$\delta = \alpha \angle \Delta t = (7.3 \times 10^{-6})\ (50 \times 12)\ (900);$$

Thus δ = 3.9 inches

Fire Protection of Structural Steel

Because unprotected structural steel loses its strength at high temperatures, it must be protected from exposure to the heat produced by building fires. This protection, often referred to as "fireproofing," insulates the steel from the heat. The more common methods of insulating steel are encasement of the member, application of a surface treatment, or installation of a suspended ceiling as part of a floor-ceiling assembly capable of providing fire resistance. Additional methods, such as sheet steel membrane shields around members and box columns filled with liquid have been introduced.

Over the years, encasement of the structural steel member has been a very common and a very satisfactory method of insulating steel to increase its fire resistance. In floor systems of reinforced concrete slabs supported by structural steel beams, the encasement can be placed monolithically with the floor. Figure 7-8Q illustrates this method. The major disadvantage is the cost, which is related both to the added formwork and concrete and to the increased weight of the supporting members due to the added dead load. To reduce the cost of encasement, systems utilizing lath and plaster or gypsum boards have been developed, as shown in Figure 7-8R.

Because of labor costs and the weight increases of the encasement method, surface treatments applied directly to the member are quite popular. Sprayed-on mineral fiber coatings are widely used for protecting structural steel. While the protection is excellent if applied correctly, it can easily be scraped off the member during construction or renovations. Consequently, sprayed-on mineral fiber coatings are suspect in their effectiveness over long term use. Cementitious materials have been used as sprayed-on coverings. During a fire, however, they can spall. Adhesion problems have also been experienced with this type of sprayed-on coating. Thus effective application, complete

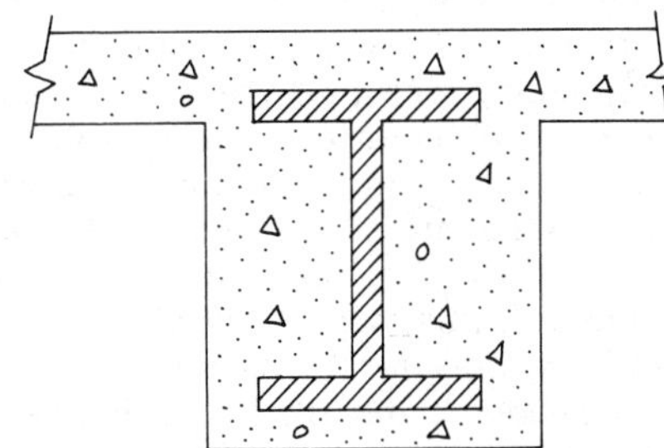

FIG. 7-8Q. Encasement of a steel beam by monolithic casting of concrete around the beam.

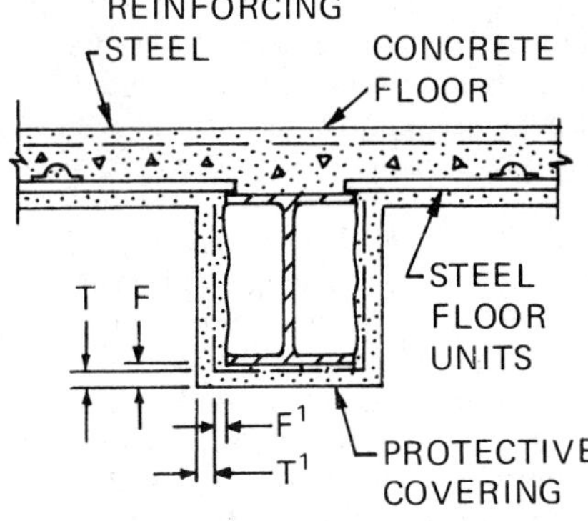

FIG. 7-8R. Furred steel beams with noncombustible protective coverings.

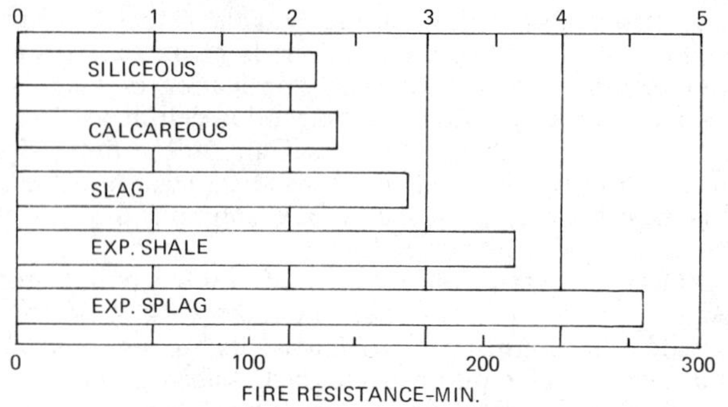

FIG. 7-8S. Effect of various types of aggregate on the fire endurance of 4¾ in. (121 mm) slabs (Benjamin 1961).

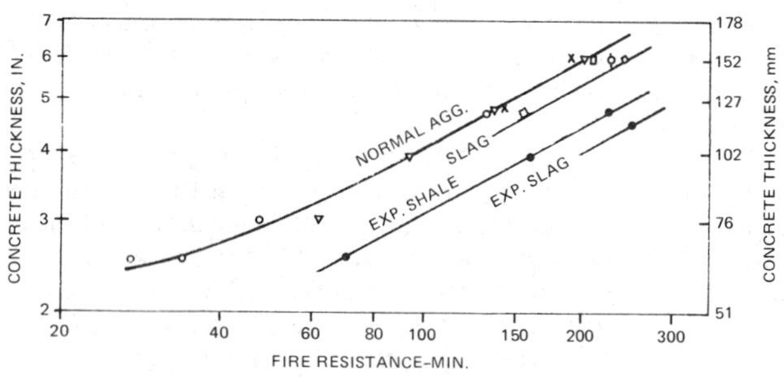

FIG. 7-8T. Relationship of slab thickness and type of aggregate to fire endurance (Benjamin 1961).

coverage, and long term maintenance are attributes that must be evaluated in considering sprayed-on applications.

Intumescent paints and coatings have been utilized to increase the fire endurance of structural steel. These coatings intumesce, or swell, when heated, thus forming an insulation around the steel. They are primarily used for nonexposed steel subjected to elevated temperatures as prolonged exposure to flame can destroy the char coating.

Suspended ceilings consisting of lath and plaster, gypsum panels, or acoustical tile supported on a grid system as part of a floor-ceiling assembly are a popular method of fireproofing. The grid system can be suspended from wire hangers or it can be attached directly to the bottom chord of joists or to the bottom flange of beams. Sometimes ceiling tiles are either mechanically fastened or fitted into splines to prevent the pressures that occur in building fires from lifting them out of place.

The overall effectiveness of this type of barrier protection, often called membrane protection, is questionable. This is due to experiences where a lack of control during construction resulted in improper installation procedures. In addition, maintenance to duct work and fixtures in the plenum area is frequently done by personnel who are not aware of the importance of the integrity of the ceiling to the fire protection system. Removed tiles are not replaced in a manner that will ensure their integrity during a building fire. Consequently, the unprotected steel in the plenum area is exposed to fire and hot gases, which reduce its strength.

Water filled columns have received attention for their fire resistive capabilities. During a fire, heat is transferred away from the location of exposure by convection currents set up in the liquid. Consequently, the heat is transferred before it has an opportunity to raise the temperature of the unprotected steel to its critical value. Although the principle of operation is valid and a few tests have been performed, this type of construction has never been exposed to an actual building fire.

Structural steel members can also be protected by sheet steel membrane shields. The sheet steel holds inexpensive insulation materials in place, thus providing a greater fire resistance. In addition, polished sheet steel has been used in tests to protect spandrel girders. The shield reflects radiated heat and protects the load carrying spandrel.

Reinforced Concrete

Concrete is often used as a protective covering for other materials. Consequently, reinforced concrete buildings give a sense of fire security. However, concrete as a material is also adversely affected by the heat of a fire. Although collapse of reinforced concrete structures is rare, loss in strength, spalling, and other deleterious effects do occur.

When the temperature of a reinforced concrete member is raised, the member loses strength. The amount of strength reduction is influenced by a number of factors. Among the more significant from a structural viewpoint are the type of aggregate, moisture content, type of loading, and level of stress during the fire exposure.

One of the more significant factors in determining the change in strength and the thermal characteristics of concrete is the type of aggregate. Aggregate types can vary widely in different sections of the country. Consequently, numerical values of strength properties are related to a percentage of the original strength, rather than a specific stress value. The qualitative behavior of concrete is generally accurate, however.

Lightweight concrete performs better at elevated temperatures than does normal weight concrete. Not only does it retain more of its strength during heat buildup, it also has a lower coefficient of thermal conductivity. Concrete using aggregates of vermiculite or perlite is particularly good at protecting structural steel from a heated environment.

Figures 7-8S and 7-8T illustrate the effect of aggregates on the fire resistance of reinforced concrete slabs. The lightweight aggregates, such as expanded shale and expanded slag, have considerably more fire endurance than do normal weight concretes made from carbonate and silicious aggregates.

The moisture content of concrete has a significant influence on its thermal performance. A considerable quantity of the heat energy of a fire is expended in vaporizing the absorbed and capillary moisture in concrete. In the case of horizontal members, the water vapor is driven upward and maintains a temperature at the top of the member of 212°F (100°C) until the water has been driven off. This increases the fire endurance, as it keeps the temperature on the unexposed side below that defined as the failure temperature. The voids caused by the evaporation of water contribute to shrinkage and a decrease in concrete strength.

The mechanical properties of concrete are significantly reduced at elevated temperatures. Figure 7-8U shows the effect on the compressive strength and modulus of elasticity by increasing the temperature. There is some question about whether concrete completely regains its strength after exposure to high temperatures.

Prestressed Concrete

The factors that are significant in the behavior of prestressed concrete are similar to those that affect reinforced concrete. In addition to the influences of moisture content and aggregate, the higher strength concrete and the function and type of steel used for prestressing are important considerations.

The concrete used for prestressed concrete is of a higher strength than that used in ordinary reinforced concrete construction. The overall fire resistance of this concrete is somewhat better than the lower strength concrete. The heat transmission is about the same for the two systems. However, there is a somewhat greater tendency for prestressed concrete to spall, thus exposing the prestressing steel.

The greatest problem of prestressed concrete subjected to elevated temperatures involves the prestressing steel. The fact that the steel is put under a high initial stress, coupled with the fact that the reinforcing wires are high carbon, cold-drawn, rather than low carbon, hot-rolled steel, is the root of the problem. Normal prestressing losses due to deformation and creep are considered in prestressed concrete design. At elevated temperatures these losses are accelerated. Creep in the steel increases, and the modulus of elasticity of the prestressing wires is reduced by 20 percent when the temperature reaches 600°F (316°C). These losses reduce the carrying capacity of the member.

The type of steel used for prestressing is more sensitive to elevated temperatures than the steel used in reinforced concrete construction. Not only is its strength reduced at temperatures somewhat less than those for hot-rolled steel, but also that strength is not regained after cooling. Prestressing wires are permanently weakened when they reach a temperature of about 800°F (427°C).

Wood

Depending on its form, wood may or may not provide reasonable structural integrity in a fire. The important factors seem to be the physical size and the moisture content of the members.

Fire retardant treatments delay ignition and retard combustion, thus providing time for extinguishing procedures. However, all wood will burn. The burning of wood produces a charcoal on the surface at the rate of about 1/40 in. per minute (0.6 mm/min). This charcoal provides a protective coating that insulates the unburned wood and isolates it from the flame. Therefore, thicker members provide much more structural integrity over the period of fire exposure than do thin ones.

Heavy timber construction has proven to be an excellent form of construction. It maintains its integrity during a fire for a relatively long time, providing an opportunity for extinguishment. When the fire is extinguished relatively soon, much of the original strength of the members is retained and reconstruction is possible.

Glued laminated frames, arches, and beams have become increasingly popular. These members also provide reserve strength during a fire, and the char can be simply removed by sandblasting to restore the aesthetic appearance.

Wood frame construction utilizes structural members considerably smaller than mill construction. The exposed area is greater, and the fire resistance considerably reduced. When this type of construction is exposed to a fire, it offers relatively little structural integrity.

FIRE BEHAVIOR OF OTHER BUILDING MATERIALS

Many materials other than steel, concrete, and wood are commonly used in modern building construction. They frequently make up a large volume and/or surface area of the structure. Nonbearing partitions, insulation, building services, and finish materials are all important parts of building construction. Some of the nonstructural, thermally inert materials are used as fireproofing. Others contribute significantly to a potential fire.

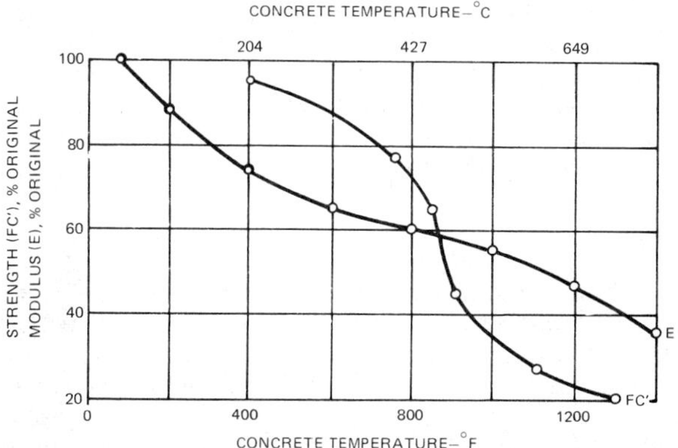

FIG. 7-8U. Effect of temperature on modulus of elasticity and compressive strength. The curves are taken from two different specimens (Benjamin 1961).

Glass

Glass is utilized in three common ways in building construction. The most obvious is glazing for windows and doors. In this capacity the glass has little resistance to fire. It quickly cracks because of the temperature difference between the surfaces. Double glazing does not provide much improvement. Wire reinforced glass is an improvement as it provides somewhat greater integrity in a fire if it is properly installed. However, no glazing should be relied upon to remain intact in a fire.

A second common use of glass in buildings is in fiberglass insulation. The fiberglass does not burn and is an excellent insulator. The fiberglass is often coated with a resin binder, however, which is combustible and which can spread flames, although relatively slowly.

A third form in which glass is found in buildings is as reinforcement for fiberglass reinforced plastic building products. The products include translucent window panels, siding, and prefabricated bathroom units. They have distinct advantages of economy and aesthetic appeal. The fiberglass acts as reinforcement for a thermosetting resin, usually a polyester. The resin, combustible even with fire retardants incorporated in the composition, frequently comprises 50 percent or more of the material. While the fiberglass itself is noncombustible, the products are quite combustible.

Gypsum

Gypsum products, such as plaster and plaster board, are excellent fire protection materials. The gypsum has a high proportion of chemically combined water. Evaporation of this water requires a great deal of heat energy, making gypsum an excellent, inexpensive, fire retardant building material.

Lightweight Concrete

Lightweight concrete, made with noncombustible aggregates resists high temperatures extremely well without degradation. Vermiculite and perlite are the most common types of light concrete used for this purpose. Vermiculite is an inert aggregate made from weathered mica. The mica is crushed and roasted, which causes it to expand to form pellet-like aggregates. Perlite is a volcanic rock which is crushed and heat treated. The heat treatment causes the rock to expand in volume. During the process, water vapor is absorbed by the rock particles, and the perlite takes the form of glasslike, cellular structured particles.

Asbestos

Asbestos is a mineral fiber which is used in several forms in building construction. When it is used with a cementicious binder and sprayed onto structural members, it forms an excellent fireproofing agent. However, the asbestos fibers are a health hazard, and they can be easily scraped off the members, so its use has been significantly curtailed in recent years.

Asbestos is combined with portland cement to make asbestos cement products. Although these products are noncombustible, they often shatter during the temperature buildup of a fire, thus reducing their effectiveness. However, asbestos is also combined with materials other than portland cement to form products, such as asbestos insulation board and asbestos wood. These products behave quite well in fires, providing a great deal of fire resistance and protection.

Masonry

Brick, tile, and concrete masonry products behave well when subjected to the elevated temperatures of a fire. Hollow concrete blocks may crack from the heat, but they generally retain their integrity. Brick can withstand high temperatures without severe damage.

Plastics

There is a wide range of plastic products used by the building industry. They provide numerous aesthetic, physical, and economic advantages in building applications. Their major disadvantage is that all plastics are combustible. Although there are certain treatments which increase ignition temperatures or inhibit flame spread, there is no known additive that will make them noncombustible.

CONTEMPORARY TRENDS IN STRUCTURAL FIRE PROTECTION

There is an increased interest internationally in the development of more rational methods of structural fire design. These methods attempt to establish a theoretical base for the design of structural members and assemblies that will predict their actual fire performance more accurately than present methods. This section briefly addresses some of these modern trends.

Usual Structural Design Procedures

The standard fire resistance test, combined with building code requirements that specify fire endurance ratings for structural frames and barrier assemblies, is the usual method of structural fire design throughout the world today. The code requirements for fire resistance are the subjective judgments of the code-writing groups. A statistical sampling of fuel loads for different types of occupancies is combined with expected population characteristics in those occupancies, building heights and areas, and an anticipated importance of the structure.

The members and assemblies are laboratory tested in accordance with a standard fire test. This test is the equivalent to NFPA 251 for the jurisdiction served by the code. An acceptable design occurs when the test results exceed the code requirements.

Although this procedure is the usual procedure throughout the industrial world today, its validity is subject to increasing doubts. The fire test itself is being severely criticized. In addition to the concerns identified earlier in this chapter, considerable variation can occur for the same assembly when it is tested in different laboratories. The relationship between test results and field performance is unknown. The procedure, however, is relatively easy to administer from a regulatory viewpoint, and many other code requirements are related to the form of construction anticipated by the fire resistance ratings.

Analytical Methods of Design

Research is being conducted along two different, but related, avenues of structural fire design (Pettersson 1980). Both attempt to substitute rational analytical procedures for predicting structural fire behavior.

One procedure is to develop analytical methods by which the fire resistance of a member or assembly may be calculated. This method computes the fire resistance that one would expect from the standard fire test. The method uses structural characteristics of the building, thermal properties of the materials, heat transfer characteristics of the assembly, mechanical properties of structural materials, and design service loads. This method enables the important considerations of structural restraint to be taken into account more easily than the standard fire test. This procedure is identical to the usual design procedure except that calculated fire resistance ratings are substituted for experimental ratings.

Another approach in which considerable research effort has been devoted is an analytical design based on real fire exposure characteristics. There are basically two different methods in this approach. The first relates the time-temperature characteristics of real fires to the structural behavior at elevated temperatures. The second method relates the properties of real fire development to an equivalent standard time-temperature curve. This method relates a theoretically equivalent time to experimental fire test results.

Design Procedures

The Swedish Institute of Steel Construction publishes a design manual entitled *Fire Engineering Design of Steel Structures* (Petersson et al undated). A building designer has the option of designing his structural system in accordance with the traditional code or by the procedures described in the manual.

The manual describes a rational fire engineering design process for steel structures on the basis of performance requirements. The manual consists of four parts. The first part is a relatively detailed description of fire energy impacts and structural behavior at elevated temperatures. This part provides the foundation for the design procedures.

The second part presents the detailed design procedure. Design charts and curves which provide the numerical base are provided. Worked examples comprise Part Three, and an alternate design method based on the concept of equivalent fire duration is given in Part Four.

The American Iron and Steel Institute publishes a design guide, *Fire-Safe Structural Steel* (AISI 1979), for exterior structural members. Exterior members, of course, cannot be tested in the standard laboratory test. The guide provides a design procedure by which exterior members can be designed for fire conditions. The design guide presents step-by-step procedures to arrive at a design.

The basis for this design guide is "Design Guide for Fire Safety of Bare Exterior Structural Steel" (Law 1977). Part 1 of the AISI guide presents the theory and validation of the method, while Part 2 reviews the state of the art of relevant research programs conducted during the past twenty years.

The Portland Cement Association has published a procedure to determine analytically the fire resistance of prestressed concrete structures. This method is described in *PCI Design for Fire Resistance of Precast Prestressed Concrete* (Gustaferro and Martin 1979). The method calculates fire resistance by obtaining temperatures at different locations for the different standard fire test times using tables and diagrams based on test results. This information is then used to compute a fire resistance value.

Bibliography

References Cited

AISI. 1979. *Firesafe Structural Steel, A Design Guide.* American Iron and Steel Institute, NY.

AISI. 1980. "Designing Fire Protection for Steel Columns." American Iron and Steel Institute, Washington, DC.

Allen, L. W., and Harmathy, T. Z. 1972. "Fire Endurance of Selected Concrete Masonry Units." *Journal of the ACI.* Vol 69.

Benjamin, I. A. 1961. "Fire Resistance of Reinforced Concrete." *Symposium on Fire Resistance of Concrete.* American Concrete Institute, Detroit, MI. pp 29 and 31.

Clay, W. 1927. "Standard Fire Tests," *Proceedings of the 13th Annual Meeting of the Building Officials' Conference of America,* Chicago, IL. pp 74-88.

DeFalco, F. D. 1974. "An Investigation of Modern Structural Steels at Fire Temperatures." *Ph.D. Thesis,* University of Connecticut, Storrs, CT.

Gustaferro, A. H., and Martin, L. D. 1977. *PCI Design for Fire Resistance of Precast Prestressed Concrete.* Prestressed Concrete Institute, Chicago, IL.

Harmathy, T. Z. 1970. "Thermal Performance of Concrete Masonry Walls and Fire." *Special Technical Publication 464.* American Society for Testing and Materials, Philadelphia, PA.

Harmathy, T. Z., and Stanzak, T. T. 1969. "Elevated-Temperature Tensile and Creep Properties of Some Structural and Prestressing Steels." *ASTM Special Technical Publication 464.* American Society for Testing and Materials, Philadelphia, PA. pp 186-208.

Law, Margaret. 1977. *Design Guide for Fire Safety of Bare Exterior Structural Steel.* Ove Arup & Partners, London, England.

Lie, T. T., and Stanzak, W. W. 1973. "Fire Resistance of Protected Steel Columns." *Engineering Journal.* Vol 10, No 3.

NBS. 1961. "Fire Tests of Columns Protected with Gypsum." *NBS Research Paper No. 563.* National Bureau of Standards, Washington, DC.

Pettersson, Ove. 1980. "Structural Fire Protection," *Fire and Materials,* Vol 4, No 1.

Pettersson, Ove, et al. undated. "Fire Engineering Design of Steel Structures." *Publication 50,* Swedish Institute of Steel Construction, Stockholm, Sweden.

Schaffer, E. W. 1977. "State of Structural Timber Fire Endurance." *Wood and Fiber.* Vol 9, No 2.

Shaub, H. 1961. "Early History of Fire Endurance Testing in the United States," *ASTM Special Technical Publication 301,* American Society for Testing and Materials, Philadelphia, PA. pp 1-9.

Stanzak, W. W., and Lie, T. T. 1973. "Fire Resistance of Unprotected Steel Columns." *Journal of the Structural Division.* Vol 99, No ST5.

NFPA Codes, Standards, Recommended Practices and Manuals. (See the latest *NFPA Codes and Standards Catalog* for availability of current editions of the following documents.)

NFPA 251, *Standard Methods of Fire Tests of Building Construction and Materials.*

NFPA 252, *Standard Methods of Fire Tests of Door Assemblies.*

NFPA 257, *Standard for Fire Tests of Window Assemblies.*

Additional Readings

Barnett, Jonathan R., "Uses and Limitations of Computer Models in Structural Fire Protection Engineering Applications," *Fire Safety Journal,* Vol. 9, Nos. 1-2, 1985, pp. 137-146.

Brannigan, F. L., *Building Construction for the Fire Service,* National Fire Protection Association, Quincy, MA, 1971.

Bresler, B., "Analytical Prediction of Structural Response to Fire," *Fire Safety Journal,* Vol. 9, 1985, pp. 103-117.

Byrne, S. M., "Fire Resistance of Load-Bearing Masonry Walls," *Fire Technology,* Vol. 15, No 3, Aug. 1979, pp. 180-188.

"Design of Buildings for Fire Safety," *ASTM Special Technical Pub. 685,* American Society for Testing and Materials, Philadelphia, PA, 1979.

ECCS European, *European Recommendations for the Fire Safety of Steel Structures,* Elsevier, NY, 1983.

"Fire Resistance Classifications of Building Constructions," *Building Materials and Structures Report 92,* National Bureau of Standards, Washington, DC, 1942.

Gehri, E., "The Fire Resistance of Steel Structures," *Fire Technology,* Vol. 21, No. 1, Feb. 1985, pp. 22-33.

Harmathy, T. Z., "Thermal Performance of Concrete Masonry Walls in Fire," *ASTM Special Technical Pub. 464,* American Society for Testing and Materials, Philadelphia, PA, 1970, pp. 209-243.

Jeanes, David C., "Predicting Temperature Rise in Fire Protected Structural Steel Beams," *SFPE TR 84-1,* Society of Fire Protection Engineers, Boston, MA, 1984.

Lie, T. T., *Fire and Buildings*, Applied Science, London, England, 1972.

Lie, T. T., "Calculating Resistance to Fire," *Canadian Building Digest 204*, Division of Building Research, National Research Council, Ottawa, Canada, May 1979.

Lie, T. T., and Williams-Leir, G., "Factors Affecting Temperature of Fire-Exposed Concrete Slabs," *Fire and Materials*, Vol. 3, No. 2, June 1979, pp. 74-79.

Milke, J. E., "Overview of Existing Analytical Methods for the Determination of Fire Resistance," *Fire Technology*, Vol. 21, No. 1, Feb. 1985, pp. 59-65.

Odeen, Kai, "The Fire Resistance of Wood Structures," *Fire Technology*, Vol. 21, No. 1, Feb. 1985, pp. 34-40.

Petterson, O., Magnusson, S. E., and Thor, J., "Fire Engineering Design of Steel Structures," *Bulletin 52*, Lund Institute of Technology and Swedish Institute of Steel Construction, Lund, Sweden, 1976.

CONFINEMENT OF FIRE IN BUILDINGS

Revised by John A. Campbell

People and property not directly exposed to a fire can be protected by confining heat and smoke to the area of origin until either the fire is extinguished or burns itself out. People can also be protected by delaying the spread of fire and smoke until the occupants can be relocated to a place of safety.

This chapter discusses barriers that are intended to limit the spread of fire; smoke movement in buildings is discussed in Chapter 10, "Smoke Movement in Buildings," and smoke and heat venting in Chapter 11, "Venting Practices," both in this section. The same barriers used to prevent the spread of fire will almost always be the bulwark of systems designed to prevent the spread of smoke. To be contained, a fire must be bounded by barriers that limit the transmission of heat and hot gases to combustible materials. Such barriers must maintain their continuity and stability under the thermal and physical forces of a fire and during the structural distortions and collapse that sometimes occur within the fire area.

The fire resistance required of a barrier depends upon its intended purpose and on the expected severity of the fire to which it may be exposed. In most cases, the temperature history of the fire provides an approximate but adequate description of fire severity. If the purpose of the barrier is to limit probable fire size independent of any fire suppression actions, the barrier must then withstand the maximum expected fire that would occur during a burnout of combustibles inside the fire zone. However, if the purpose of the barrier is to contain the fire until occupants are evacuated and the fire department gains access, the time necessary for these actions determines the time that the barrier must hold back the fire, unless the fire is expected to largely burn itself out first.

Properly designed and constructed barriers that satisfy fire endurance test criteria are assigned a fire resistance rating that is measured in hours or fractions thereof. Any breach or unprotected opening in a fire barrier will negate its fire resistance rating and may indeed contribute to rapid fire spread through the barrier.

Mr. Campbell is a registered professional engineer and a Principal of Gage Babcock & Associates, Inc., Chicago, IL.

FIRE SEVERITY

The fire severity to which a barrier can be exposed is related to the intensity of a fully developed fire in the space adjacent to the fire barrier. For a room fire, a fully developed fire would be a post flashover fire; however, in large open volumes such as those found in many industrial plants, a fully developed fire may occur in one area without flashover of the entire volume. It is the fully developed fire which first imposes extensive thermal and physical stresses on fire barriers. The initial possibility of barrier penetration occurs prior to full development. A typical developing fire is characterized by a flame front moving over a surface and/or flames localized to a stationary source. Although such a fire does not massively attack fire barriers, it can spread through faults or openings in the barrier and cause local destruction of barrier fire resistance.

A growing fire may or may not continue to flashover or become a fully developed area wide fire. Unless it reaches these stages, it is unlikely to threaten fire resistive barriers except through unprotected openings or serious defects in the barrier system.

Flashover of an enclosure is likely to occur if the temperature of the upper gas layer reaches approximately 1,100°F (600°C). Full scale fire experiments and energy balance analysis of room fires have shown that the temperature of this upper gas layer depends upon the heat released by the burning fire, ventilation of the enclosure, dimensions of the enclosure, and the type of material lining the enclosure (McCaffery et al 1981). These factors also are interdependent with each other to some degree. The larger the enclosure, the higher the heat release necessary to produce the upper gas temperature needed for flashover. Conversely, the greater the combustibility of the enclosure lining, the smaller the fire needed to produce flashover. The combustibility of the lining, or interior finish, is particularly significant. Combustible interior finish in an enclosure can greatly reduce the rate of heat release necessary for flashover as well as reduce the preburn time before flashover. A practical difficulty in estimating the likelihood of room flashover is in determining a value for the heat release rate possible from the room

TABLE 7-9A. Flashover Experience in Full Scale Fire Tests and Some Real Fires in Rooms with Noncombustible Interior Fires*

Furnishing Items Shown as Being Capable of Causing Room Flashover†	Furnish Items Shown as Probably Being Incapable of Producing Flashover
Clothes hanging in a wooden wardrobe.	Cotton inner spring mattress without box springs (some small probability of causing flashover may exist with a wooden headboard).
Clothes hanging in a closet.	Most polyurethane mattresses.‡
Mattress with box springs.	Single upholstered chair.
A few polyurethane mattresses.‡	Padded lounge chair.
Latex foam mattress.	
Overstuffed chair.	
Two upholstered chairs adjacent to each other.	
Sofa.	

* Sources: (Quintiere 1983; Vodvarka & Waterman; AHCA 1975; Davis 1978; Sharry 1974; Babrauskas 1980; Lathrop, 1975).
† Hanging clothes and a latex foam mattress have been shown capable of producing room flashover in as little as 5 min from start of flaming. Other furnishings required an average of about 18 min from the start of flaming to room flashover.
‡ In 17 experimental fires using various types of polyurethane mattresses without box springs, only one developed to sufficient intensity to produce room flashover.

contents. Typical heat release rates from some common furnishings have been summarized from full scale experimental fire data (Quintiere 1983). These data can be used in calculating the likelihood of flashover in a particular room.

It is also possible to qualitatively estimate the likelihood of flashover for some types of rooms. Fire experience and full scale experimental fires have shown that some furnishings can be expected to produce flashover in typical residential and institutional rooms. Examples of such furnishings for rooms with noncombustible interior finish are summarized in Table 7-9A. This data would apply to small office rooms with similar furnishings as well as residential and institutional rooms.

The intensity and duration of a fully developed fire depend upon the quantity of combustibles available, their burning rates, and the air available for combustion. Fire intensity will also be somewhat lower when walls and ceilings absorb significant amounts of energy rather than act primarily as insulation or radiation barriers.

The burning rate of a fully developed fire involving a specific material or grouping of similar materials is determined either by (1) the fuel surface area available to participate in the combustion reaction, or (2) the amount of oxygen available for combustion. These are referred to as fuel surface controlled and ventilation controlled combustion, respectively. The maximum rate of heat transfer into fire separation barriers occurs at the point where the ventilation is just sufficient so that combustion is controlled at the fuel surface (Berry & Minor 1979). At higher ventilation rates, more heat is removed from the fire by the excess air. At lower ventilation rates, the combustion heat release rate is less and more unburned pyrolysis products and fuel particles are vented outside the fire area.

The possibility of failure of fire separation barriers can exist long after the fully developed fire begins to decay. However, in many real fire situations, this threat is mitigated by fire suppression activities. The decay in air temperature in a fire room has been reported as 27 to 36°F (15 to 20°C) per minute after fully developed fires of 10 to 15 minute duration. Other data indicates a decay rate of 18°F (10°C) per minute for fully developed fires under one hour duration and about 13°F (7.2°C) per minute for longer duration fires (Lie 1974). Cooling can be even slower in large debris piles. Fire endurance testing in the U. S. does not simulate a fire decay period, although the decay period is simulated by some European and proposed U.S. testing.

Poorly ventilated fires that are encountered in spaces such as basements, ship holds, or enclosed interior rooms often produce sufficient heat over a long period of time to penetrate separation barriers. These fires typically start with flaming combustion and, as the air in the space is consumed, revert to a state of mixed smoldering and glowing combustion with isolated or intermittent flaming. However, at present there is no adequate experimental data or theoretical approach capable of reliably estimating the effect of these fires on fire resistive barriers.

Standard Time-Temperature Curve

Fire resistive barriers are evaluated in a testing furnace by exposure to a fire whose severity follows a time varying temperature curve known as the standard time-temperature curve. The specified time-temperature history is tabulated in NFPA 251, *Standard Methods of Fire Tests of Building Construction and Materials* and illustrated in Figure 7-9A. The standard time-temperature curve was adopted by the American Society for Testing Materials (ASTM) in 1918 and has been the basis of almost all fire resistance testing ever since.

Following adoption of the curve, the National Bureau of Standards (NBS) conducted a number of full scale fire

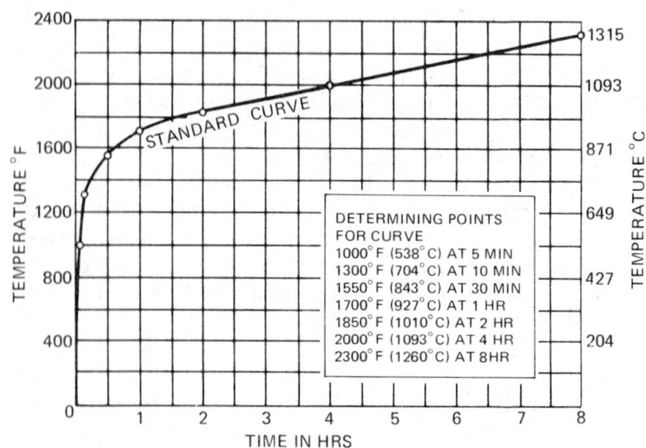

FIG. 7-9A. The standard time-temperature curve.

tests to determine how actual building fires compared with the temperatures represented on the curve (Ingberg 1927, 1928). The tests included two actual buildings that were allowed to burn to destruction and a series of fires in fire resistive test buildings containing contents representative of office, record room, and household occupancies. The principal variable considered in these occupancy fire tests was the amount of combustible materials present, which is defined as the fire load. Although the ventilation in the test buildings was not reported, the windows were equipped with steel shutters that could be adjusted to control ventilation and maximize fire severity. The quantitative importance of ventilation on fire severity was not identified until more than 25 years after these tests. These tests conducted by the NBS provided quantitative data on the temperature history of fires that was representative of various occupancies and fire load at that period of time. Fire load was expressed as the weight of ordinary combustibles in the room divided by the floor area of the room. Loading is the average amount of ordinary combustible material per square foot (m^2) of floor area. The temperature history of the fully developed fires in the three test occupancies was approximately bounded by the standard time-temperature curve.

The NBS developed the concept of equivalent fire severity to define the severity of actual fires that had various temperature histories. This concept states that the area above a base line under the time-temperature curve of a test fire, which is expressed in degree hours, is an approximate representation of the severity of a fire involving ordinary combustibles. The base line used represents the temperature the materials can be exposed to without impairing their fire resistive capabilites. Two fires with differing temperature histories are considered to have equivalent severity when the area under their time temperature curves is similar. This concept permitted comparison of any fire test data to the standard time-temperature curve by relating the area under the test curve to the area under the standard curve.

FIRE LOAD

The original concepts of fire severity and fire load are very important even though they are technically obsolete. These concepts are the basis for many of the fire resistance requirements of building codes and for government agencies. In many cases, use of this original fire severity/fire load relationship was more severe than is indicated by more accurate analysis. Such results are conservative since the resultant error is on the safe side.

Analysis of NBS tests developed an approximate relationship between fire loading and an exposure to a fire severity equivalent to the standard time-temperature curve. The weight per square foot or square meter of ordinary combustibles [wood, paper, and similar materials with a heat of combustion of 7,000 to 8,000 Btu per lb (16 282 to 18 608 J/kg)] was related to hourly fire severity as described in Table 7-9B.

The fire severity/fire load relationship was the first method developed to predict the severity of a fire that would be anticipated in various occupancies. It was used to determine resistance required of fire barriers as well as structural components. Although the technique has its limitations, the fire severity/fire load relationship still provides an approximate but conservative estimate of the probable maximum fire severity in residential, institutional, and some commercial occupancies. Fire load should not be used as an approximate indicator of fire severity with combustibles having a high heat release rate and when fire conditions can produce temperatures significantly higher or lower than the standard time-temperature curve.

TABLE 7-9B. Estimated Fire Severity for Offices and Light Commercial Occupanices

Data applying to fire-resistive buildings with combustible furniture and shelving

Combustible Content Total, including finish, floor, and trim psf	Heat Potential Assumed* Btu per sq ft†	Equivalent Fire Severity Approximately equivalent to that of test under standard curve for the following periods:
5	40,000	30 min
10	80,000	1 hr
15	120,000	1½ hrs
20	160,000	2 hrs
30	240,000	3 hrs
40	320,000	4½ hrs
50	380,000	7 hrs
60	432,000	8 hrs
70	500,000	9 hrs

* Heat of combustion of contents taken at 8,000 Btu per lb up to 40 psf; 7,600 Btu per lb for 50 lb, and 7,200 Btu for 60 lb and more to allow for relatively greater proportion of paper. The weights contemplated by the tables are those of ordinary combustible materials, such as wood, paper, or textiles.

† SI units: 1 psf = 4.9 kg/m^2; 1 Btu/ft^2 = 1.14 J/m^2

Fire load is a measure of the maximum heat that would be released if all the combustibles in a given fire area burned. Maximum heat release is the product of the weight of each combustible multiplied by its heat of combustion. In a normal building, the fire load includes combustible contents, interior finish, floor finish, and structural elements. Fire load is commonly expressed in terms of the average fire load, which is the equivalent combustible weight divided by the fire area in square feet or square meters.

Equivalent combustible weight is defined as the weight of ordinary combustibles having a heat of combustion of 8,000 Btu per lb (18 608 J/kg), that would release the same total heat as the combustibles in the space. For example, the equivalent weight of 10 lb per sq ft (48.8 kg/m^2) of a plastic with a heat of combustion of 12,000 Btu per lb (27 912 J/kg) would be:

$$10 \text{ lb per sq ft} \times 12{,}000 \text{ Btu per lb} = 120{,}000 \text{ Btu per sq ft}$$

$$120{,}000 \text{ Btu per sq ft} \div 8{,}000 \text{ Btu per lb ordinary combustibles} = 15 \text{ lb per sq ft}$$

Technically accurate methods for relating fire severity, fire load, and fire resistance requirements are complex but can be advantageously used in important specific applications. Such methods require consideration of parameters other than the fuel load, such as ventilation, type of enclosure walls, and ceiling. These methods are complex and currently too difficult for general use in design or selection of barrier fire resistance.

TABLE 7-9C. Characteristics of Fire Loads in Office Buildings

	Government Buildings			Private Buildings		
	No. of Rooms	Total Fire Load, psf		No. of Rooms	Total Fire Load, psf	
Room Use	Sampled	Mean	Std. Dev.	Sampled	Mean	Std. Dev.
General	342	7.3	4.4	479	7.7	4.3
Clerical	77	5.8	5.2	146	6.8	4.0
Lobby	15	2.6	1.4	45	5.0	4.2
Conference	39	4.2	6.1	57	5.9	4.6
File	10	17.9	11.9	20	16.2	12.9
Storage	35	11.7	19.2	77	13.2	11.7
Library	2	30.2	7.8	10	23.6	10.8

Notes: Fire load was not reduced to account for combustibles that do not burn completely because they are in steel enclosures. Weight of combustibles was converted to an equivalent weight of combustibles having a heat of combustion of 8,000 Btu/lb.
SI units: 1 psf = 4.88 kg/m^2

Occupancy Fire Load

A number of surveys have identified the fire loads found in various occupancies (Berry and Minor 1979; Culver 1978; Campbell 1978). (See Tables 7-9C, 7-9D, and 7-9E, and Figure 7-9B.)

TABLE 7-9D. Samples of Typical Fire Loads

Type of Room	Contents Fire Load psf	Standard Deviation psf
Living Room	3.9	1.13
Family Room	2.7	.65
Bedroom	4.3	1.15
Dining Room	3.6	1.02
Kitchen	3.2	.77
Hospital Patient Room	1.2	.36
Nursing Home Patient Room	2.6	.62

SI units: 1 psf = 4.68 kg/m^2

Data from some fire load surveys as well as the inherent nature of combustible contents likely to be encountered suggests that the dispersion of fire load within a certain class of rooms can be approximated by either a normal or moderately skewed frequency distribution curve. (See Fig. 7-9C.) The standard deviation, included in Tables 7-9C and 7-9D, can be used to determine the probability that a particular fire load value will not be exceeded in a class of rooms. A fire load which is one standard deviation above the mean value of a normal distribution curve would represent an upper boundary for 84.13 percent of the fire loads in rooms of that class. Two standard deviations above the mean would bound 97.75 percent of the fire loads in that class of rooms, and three standard deviations, 99.86 percent of the fire loads. Thus, if a fire barrier were to be designed on the basis of two standard deviations above the mean, there would be a 97.73 percent probability that this fire load would not be exceeded in a similar room.

The above percentages are exact only if the distribution of fire loads is perfectly normal. If the distribution is more accurately defined by a moderately skewed curve, the percentages only represent close approximations.

TABLE 7-9E. Fire Severity Expected by Occupancy*

Temperature Curve A (Slight)
Well-arranged office, metal furniture, noncombustible building.
Welding areas containing slight combustibles.
Noncombustible power house.
Noncombustible buildings, slight amount of combustible occupancy.

Temperature Curve B (Moderate)
Cotton and waste paper storage (baled) and well-arranged, noncombustible building.
Paper-making processes, noncombustible building.
Noncombustible institutional buildings with combustible occupancy.

Temperature Curve C (Moderately Severe)
Well-arranged combustible storage, e.g., wooden patterns, noncombustible buildings.
Machine shop having noncombustible floors.

Temperature Curve D (Severe)
Manufacturing areas, combustible products, noncombustible building.
Congested combustible storage areas, noncombustible building.

Temperature Curve E (Standard Fire Exposure—Severe)
Flammable liquids.
Woodworking areas.
Office, combustible furniture and buildings.
Paper working, printing, etc.
Furniture manufacturing and finishing.
Machine shop having combustible floors.

* See Fig. 7-9B for the temperature curves identified in this table

Derated Fire Loads

Ordinary combustibles that are completely or largely enclosed in steel containers will not burn completely during a room fire and therefore will not contribute a full 8,000 Btu per lb (16 282 J/kg) to the fire load. The General Services Administration has developed guidelines for determining a derated fire load for office buildings, which can be applied to other occupancies having similar classes of combustibles (GSA 1977). The total contents fire load is divided into three categories: (1) weight of materials completely enclosed in containers such as steel desks or file cabinets, W_E; (2) weight of materials enclosed on five sides

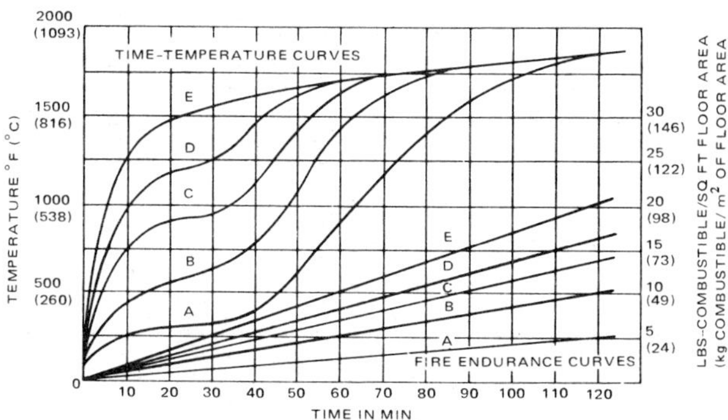

FIG. 7-9B. Possible classification of building contents for fire severity and duration. The straight lines indicate the length of fire endurance based upon amounts of combustibles involved. The curved lines indicate the severity expected for the various occupancies. (See Table 7-9E). There is no direct relationship between the straight and curved lines, but, for example, 10 lb of combustible per sq ft (48.8 kg/m²) will produce a 90 minute fire in a "C" occcupancy, and a fire severity following the time-temperature curve "C" might be expected.

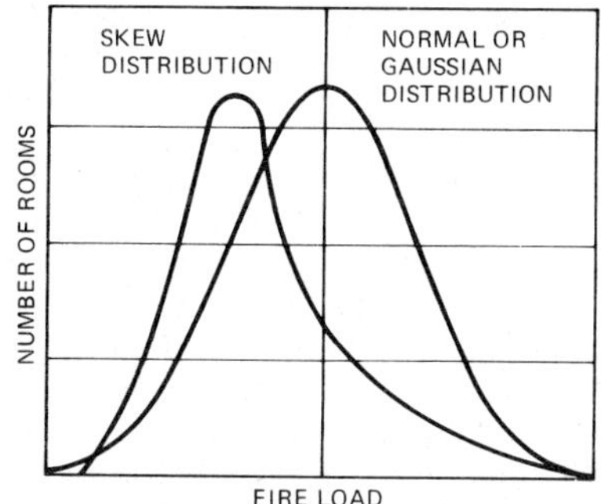

FIG. 7-9C. Expected distributions of sample fire loads.

such as in a steel bookcase, W_{PE}; and (3) weight of free combustibles, W_{FE}.

Completely enclosed combustibles will be heated by the room fire and pyrolyze; and the escaping pyrolysis products will burn. Therefore, the heat that the enclosed combustibles release depends upon the extent to which they are pyrolyzed. This is related to the intensity and duration of the room fire, which in turn is related to the total combustibles in the room. The extent to which the enclosed contents are derated is determined by considering the ratio of the total weight of enclosed combustibles, W_E, to the total weight of all combustibles in the room, F_T.

Thus:

$$F_T = W_E + W_{PE} + W_F$$

A derating factor is assigned to W_E as tabulated below:

Ratio W_E/F_T	Derating Factor, K
under 0.5	0.4
0.5 to 0.8	0.2
over 0.8	0.1

Ordinary combustibles enclosed on five sides by steel, such as in a bookcase, are derated to 75 percent of their weight.

The total derated fire load is given by:

$$F_{DR} = K \times W_E + 0.75\, W_{PE} + W_F$$

The specific fire load is then computed by dividing by the floor area.

This derating procedure can be applied if other than ordinary combustibles are included in the fire load. The total weight of other combustibles must be expressed in terms of an equivalent weight of ordinary combustibles that would release the same total heat in burning. This is accomplished by multiplying the weight of the other combustibles by the ratio of their heat of combustion in Btu per lb, to 8,000 Btu per lb (18 608 J/kg), the heat of combustion of ordinary combustibles.

British Fire Loading Studies

The British have attained a similar objective by grading building occupancies according to hazard. Three classifications, low, moderate, and high fire loads, are defined in terms of Btu per sq ft or J/m² as follows:

Occupancies of Low Fire Load: The fire load of an occupancy is described as low if it does not exceed an average of 100,000 Btu per sq ft (114 000 J/m²) of net floor area of any compartment, nor an average of 200,000 Btu per sq ft (228 000 J/m²) in limited isolated areas, provided that storage of combustible material necessary to the occupancy may be allowed to a limited extent if separated from the remainder and enclosed by appropriate grade fire resistive construction. Examples of occupancies of normal low fire load are offices, restaurants, hotels, hospitals, schools, museums, public libraries, and institutional and administrative buildings.

Occupancies of Moderate Fire Load: The fire load of an occupancy is described as moderate if it exceeds an average of 100,000 Btu per sq ft (114 000 J/m²) of net floor area of any compartment but does not exceed an average of 200,000 Btu per sq ft (228 000 J/m²), nor an average of 400,000 Btu per sq ft (456 000 J/m²) on limited isolated areas, provided that storage of combustible material necessary to the occupancy may be allowed to a limited extent, if separated from the remainder and enclosed by fire resistive construction of an appropriate grade. Examples of occupancies of normal moderate fire load are retail shops, factories, and workshops.

Occupancies of High Fire Load: The fire load of an occupancy is described as high if it exceeds an average of 200,000 Btu per sq ft (228 000 J/m²) of net floor area but does not exceed an average of 400,000 Btu per sq ft (456 000 J/m²) of net floor area, nor an average of 800,000 Btu per sq ft (912 000 J/m²) on limited isolated areas. Examples of occupancies with normal high fire load are warehouses and other buildings used for the bulk storage of commodities of a recognized nonhazardous nature.

The low fire load grading of occupancies used by the British is approximately equivalent to the classification of occupancy represented by the Temperature Curve A; moderate fire load grading by Temperture Curves B, C, and D; and the high fire load grading by Temperature Curve E in Figure 7-9B.

EFFECT OF VENTILATION

The burning rate of fully developed fires may depend upon the available fuel surface area or on the air available for combustion. When ample air is available, the burning rate of a fire depends upon the exposed surface area and on the properties of the combustible itself. Fully developed fires involving ordinary residential furnishings burn at a rate (R) which can be approximated by the equation:

$$R = 0.09\ A_f \qquad \text{(lbs per min)}$$

where A_f is the exposed fuel surface in sq ft or square meters (Waterman et al 1964).
For the general case, the equation becomes:

$$R = \dot{W}_f\ A_f$$

where W_f = surface rate of burning of the fuel, lb per sq ft (kg/m^2) of exposed surface per minute.

Typical furniture found in homes and offices has a surface area to weight ratio between 0.55 and 0.90 sq ft per lb (.113 and .185 kg/m^2) (Harmathy 1976). Loose combustibles, cellular material, hanging clothes, etc., have much higher surface to weight ratios and can produce rapidly developing fires. Packed solid material or heavy furniture items have lower surface to weight ratios and are associated with more slowly developing fires of longer duration.

When a fire cannot get sufficient air to maintain the burning rate associated with fuel surface controlled combustion, it will burn at a ventilation controlled rate. The burning rate for fully developed fires in ordinary combustibles to which air is supplied through broken windows or a doorway is approximated by:

$$R = 0.62\ A_o\ H_o^{1/2} \qquad \text{(lb per min)}$$

where A_o = the opening area in sq ft (m^2) and H_o = the height of the opening in ft (m) (Heselden et al 1970).
Air flow into the fire, through a single opening has been found to be approximately equal to:

$$\dot{m} \simeq 5.06\ C_D\ A_o\ H_o^{1/2} \qquad \text{(lb per min)}$$

where C_D = the orifice coefficient, normally about 0.5 to 0.7 (Babruska & Williamson 1975).
When there are multiple openings or a window wall, the flow is less than predicted by the area and a C_D of about 0.35 is recommended.

The maximum fire intensity, from the standpoint of rate of heat input into barriers, occurs at a ventilation rate just sufficient to sustain a surface burning fire. This can be expressed as

$$\dot{W}_f A_f = 5.06\ C_D\ W_a\ A_o\ H_o^{1/2}$$

where W_a = the weight of fuel each lb (kg) of air will burn completely. Typical values of W_a are presented in Table 7-9F (Babrauskas & Williamson 1975).

The preceding relationships assume that wind is not a factor in ventilation. In addition, if the ventilating opening is an interior door such as from a corridor, adequate combustion air must be available for the fire.

TABLE 7-9F. Fuel Consumed by One Pound of Air at a Stoichiometric Fuel/Air Ratio

Combustible	lb fuel/lb air	g fuel/kg air
wood	0.175	175
polyethylene	0.069	69
polystryene	0.076	76
polyurethane	0.135	135
benzene	0.076	76
methane	0.060	60

Considerable ventilating area is required for a fully developed fire to burn at a fuel surface controlled rate. For example, over one-fourth of the wall area would have to be open in a 20 by 20 ft room (6.1 by 6.1 m) with an 8 ft (2.4 m) ceiling and an exposed combustible surface of 800 sq ft (74.3 m^2) of ordinary combustibles. Many, if not most, building fires will be ventilation controlled at least during the period of time in which containment is a consideration. A fully developed fire can change from ventilation controlled to fuel surface controlled as ventilation changes (such as when windows break out).

A number of mathematical models have been developed to calculate the temperature history of fully developed room fires as a function of ventilation, fuel, and the walls and ceiling of the room (Berry & Minor 1979; Lie 1974; Babrauskas & William 1975; Kawagoe 1967; Harmathy 1976). Calculations also yield the temperature rise on the unexposed wall and ceiling surfaces. In determining fire severity, this technique represents an improvement over the single parameter fire load concept; however, it also is an approximation and has definite limitations. A major disadvantage of these models is their complexity and specificity. At the present time, it is not practical to generally use calculated fire severity for fire barrier design even if it were permitted by codes. However, calculations can be used for specific applications as long as the user recognizes their capabilities and limitations. In addition, qualitative relationships identified by these analyses can provide guidance in the design of fire barriers and assist in comparing advantages of containment and suppression protection.

The principal assumption that must be made in mathematical fire history models is that the temperature is uniform in the fire room. The turbulence associated with a post flashover fire is assumed to mix the hot gases so that no appreciable temperature gradients exist. In fire models, this assumption is used to calculate mass airflow into the room, heat transfer from the room to walls and ceiling, and pyrolysis of the fuel. This assumption is an approximation because temperature differences of between 250 and 500°F (139 and 278°C), have been measured within a room during post flashover fires; larger differences could be encountered in larger compartments (AHCA 1975; Zinn et al 1974; Christian & Waterman 1971).

The maximum intensity of post flashover room fires occurs when the ventilation is just sufficient to permit fuel surface controlled combustion (Berry & Minor 1979). At either higher or lower ventilation levels, the maximum fire temperature is lower (Lie 1974). At higher ventilation levels, the fire duration will be shorter, and at lower levels, it will be longer. The ventilation necessary to sustain fuel surface limited combustion will depend upon the combustibles in the room, their exposed surface area, surface burning rate, and the amount of air each lb (kg) of fuel requires for combustion. More ventilation would be needed to sustain fuel surface controlled combustion if a room contains blocks of foam rubber than if it contains an equivalent fire load of ordinary furnishings. The effect ventilation area has on the temperature of a fully developed room fire is qualitatively shown in Figure 7-9D.

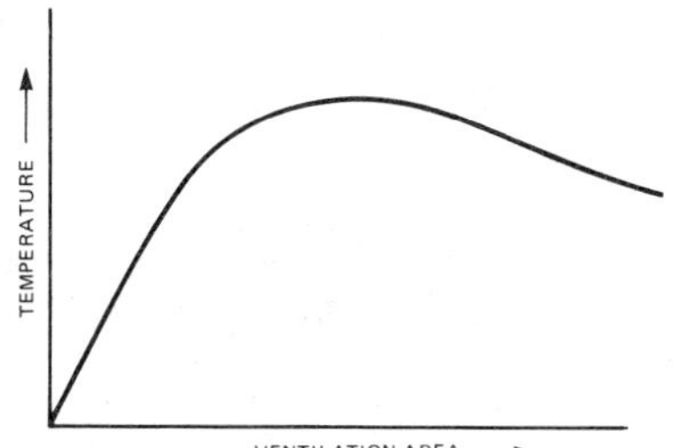

FIG. 7-9D. Typical fully developed fire temperature-ventilation relationship.

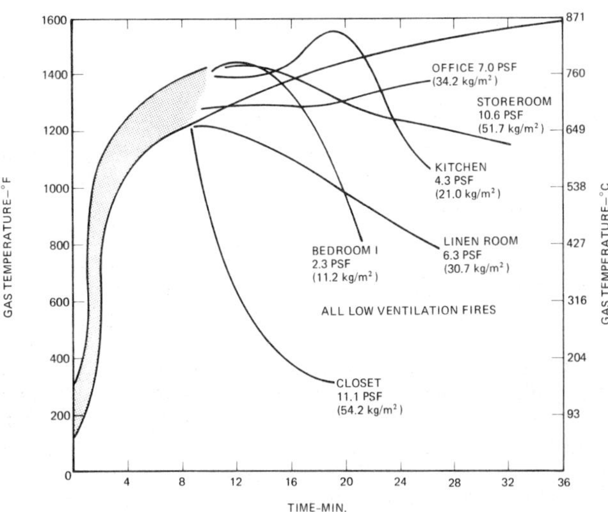

FIG. 7-9E. Gas temperature at ceiling of burning room. (Christian & Waterman 1971).

The majority of experimental fires, which form the basis for the previously described relationships, have emphasized one room fires, fuel consisting of either ordinary household and business furnishings or wood cribs, fuel loads under 10 lb per sq ft (48.8 kg/m^2), at least a level of ventilation provided by an open door or large window representing 10 to 15 percent of the room area. The results of these fires have shown the following:

1. Fire room temperatures may rise much faster than the standard time-temperature curve, remain above the curve for a period of time, and then drop below it.
2. Within a normal ventilation range of about 10 to 15 percent of room area on the low end and 25 to 30 percent on the high end, the maximum fire temperature depends primarily upon fire load, which in turn is related to the exposed fuel surface area.
3. Ordinary combustible materials with a very high exposed fuel surface area 8 sq ft per sq ft (0.74 m^2/m^2) of floor area, can result in maximum fire room temperatures over 2,000°F (1093°C). Maximum fire temperatures involving ordinary furnishings generally ranged between 1,400 to 1,800°F (760 to 982°C).

Temperature histories of some typical small room fires in which ventilation was provided by an open corridor door are illustrated in Figure 7-9E. Higher temperatures associated with the large fuel surface areas of wood cribs are shown in Figure 7-9F; the fuel surface area was about 1,000 sq ft (93 m^2) for the 10 lb per sq ft (48.8 kg/m^2) fire loading.

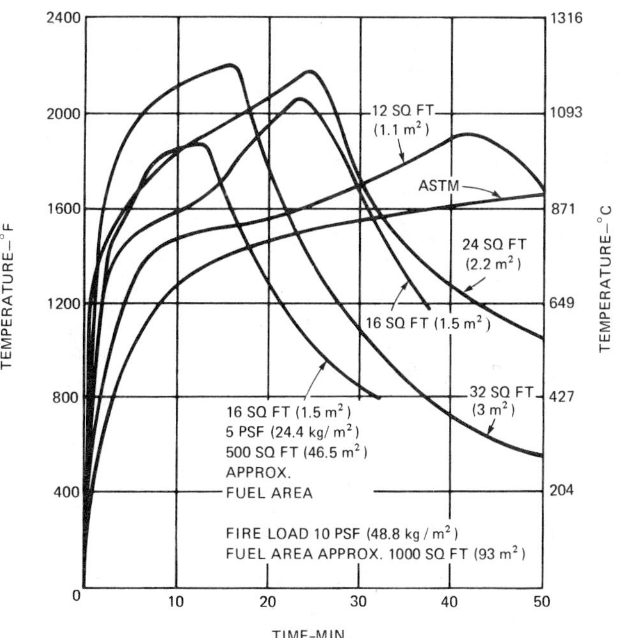

FIG. 7-9F. Effect of window area on average fire room temperature. (AISI 1975).

FIRE SPREAD

Fire spread rarely occurs by heat transfer through, or structural failure of, walls and floor-ceiling assemblies. The common mode of fire spread in a compartmented building is through open doors, unenclosed stairways and shafts, unprotected penetrations of fire barriers, and nonfirestopped combustible concealed spaces. Even in buildings of combustible construction, the common gypsum board or lath on plaster protecting wood stud walls or wood joist floors provides 25 to 30 minute resistance to a fully developed fire as determined by a standard fire test (NBFU 1956). When such barriers are properly constructed and maintained and have protected openings, they will normally contain fires of maximum expected severity in light hazard occupancies. However, no fire barrier will reliably protect against fire spread if it is not properly constructed and maintained and openings in the barrier are not protected.

Although typical wall and floor-ceiling constructions generally have adequate inherent fire resistance in light hazard occupancies, there are practical advantages in requiring rated fire resistive assemblies in new construction. Materials and construction methods required for fire rated assemblies provide quality controls not found in typical construction.

Fire can spread horizontally and vertically beyond the room or area of origin and through compartments or spaces that do not contain combustibles. Heated unburned pyrolysis products from the fire will mix with fresh air and burn as they flow outward. This results in extended flame movement under noncombustible ceilings, up exterior walls, and through noncombustible vertical openings. This is a common way fire spreads down corridors and up open stairways and shafts. Similar phenomena have been observed in flame spread underneath the ceiling in industrial and storage buildings. Combustible interior finish in corridors and vertical openings, which by itself may be incapable of propagating flames, will be heated and may produce pyrolysis products. These products add to those from the main fire, and increase the intensity and length of flames.

Once a room fire becomes fully developed, fire can spread very quickly to adjacent rooms in the absence of any fire separation barriers. When a building contains similar size rooms, the total volume of the rooms involved in fully developed fires tends to increase at an exponential rate (Waterman 1974). This is illustrated in Figure 7-9G,

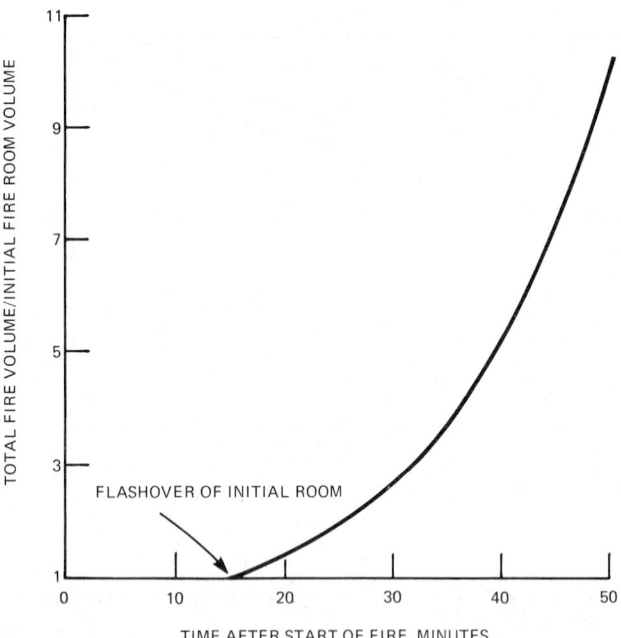

FIG. 7-9G. Exponential fire volume increase for a room flashover time of 15 minutes.

which shows the form of percentage increase in total fire volume based on the initial room volume as a function of time after the initial room flashed over.

Floor to floor fire spread along the outside of a building is not a common occurrence, but it can and does occur, such as in the Hilton Hotel fire in Las Vegas in 1980. Glass panes normally start breaking when heated to 550 to 600°F (288 to 316°C), although some breakage may occur at lower temperatures (Roytman 1979). The hazard is greater in high rise buildings because of fire fighting difficulties and because there are more floors to which fire can spread. Wide spandrels and projections such as balconies will reduce the risk of, although not necessarily prevent, external fire spread.

Whether a barrier will or will not contain a fire can be considered as a probability and reliability problem that parallels reliability problems encountered in the operation of electronic, electrical, and mechanical equipment. Using this analogy, barrier failures can be considered to fall into one of three categories—early failures, random failures, and degradation failures. Early failures would be the result of occurrences such as a door not closing or an opening not being protected. Random failures would be the result of faults in materials or workmanship, a fire of unexpected severity, or design weaknesses not identified by standard testing methods. Random failure probabilities follow an exponential law and would predominate until barrier thermal degradation became a factor. Degradation failures would be similar to wearout failures encountered in electronic, electrical, and mechanical equipment. A probability of a wearout failure follows a normal or Gaussian distribution law, and the probability of barrier degradation failures should be similarly distributed. This technique of considering the three categories of barrier failure probability permits a more realistic comparison to be made between compartmentation and other protective measures. It is also apparent from this analogy that a barrier having a certain fire resistive rating is not necessarily twice as reliable as one having one-half that rating.

Concealed Spaces

Buildings may contain a wide variety of concealed spaces behind walls, above suspended ceilings, in utility chases, behind soffits, under computer room floors, in attics, and elsewhere. The spaces range in size from 1⅝ in. (41.3 mm) stud spaces to 8 ft (2.4 m) and higher interstitial spaces. Fires in concealed spaces burn out of sight of building occupants, and detection is frequently delayed. Manual fighting of concealed space fires can be very difficult because of limited accessibility and inherent venting problems. Vertical concealed spaces can act as flues to spread fire and hot gases.

Fire development and spread in combustible concealed spaces is a classic cause of large losses in fires. The fire may originate in, or spread into, a combustible concealed space and burn for a long time, spreading throughout the concealed volume. The disastrous Beverly Hills Supper Club and MGM Grand Hotel fires both originated in a combustible concealed space (Best 1978, Best & Demers 1982). Fires in horizontal concealed spaces such as cock lofts, attics, interstitial spaces, soffits, etc. have spread undetected until they either burned through the floor or roof above or caused parts of the ceiling below to drop. A fresh air supply, preheated combustible materials, and hot pyrolysis products combine to produce a rapidly developing fire, which may break out of the concealed space over a wide area. Vertical fire spread inside walls and chases can spread fire between floors and into attic spaces where it mushrooms out horizontally. The rapid vertical spread of fire was quantitatively demonstrated during several full scale fire tests in wood frame housing in the Bushwick section of Brooklyn, NY. In one test, temperatures in excess of 1,500°F (816°C) were recorded at the top of a 32 ft (9.75 m) shaft one minute after a fire was started at the bottom (DeCicco 1976).

Fire spread through noncombustible concealed spaces is rarely a hazard unless the space contains combustible materials. Interstitial ceiling spaces frequently contain thermal insulation, communications cable, plastic pipe, etc. If thermal insulation is combustible, fire spread can be similar to that in combustible concealed spaces. Isolated combustible items are not likely to propagate fire horizontally; however, when grouped together, they may. For example, grouped cables in trays spread a fire, relatively slowly, in the Browns Ferry Generating Station fire, although the cables had passed a single cable flame progagation test (NRC 1976). Movement of flame through a noncombustible ceiling plenum space was reported as contributing to the horizontal spread of fire in One New York Plaza (Powers 1970). The interstitial space contained a number of electrical communications cables and exposed foam plastic insulation on the inside of the exterior walls.

Electrical cables that are difficult to ignite are available and being used in nuclear power plants. In tests, they did not propagate fire horizontally except in a multiple stacked tray configuration (Kamerus 1977). Tests have also shown that use of fire retardant coatings can be effective in inhibiting horizontal fire spread; however, there were major performance differences among the products tested (Kamerus 1978). Although some combustibles do not pose a serious threat of horizontal fire spread in noncombustible concealed spaces, they are likely to pose a serious threat of spread in vertical concealed spaces.

Vertical Openings

Unprotected vertical openings have been responsible for many large loss-of-life fires. These openings can act as a chimney—smoke, hot gases, and burning pyrolysis products flow up under stack effect and then mushroom out horizontally. Because of their function, openings such as elevator shafts, stairways, laundry chutes, and air shafts cannot be cut off at floor levels. Vertical openings are enclosed with fire rated walls to prevent fire spread and in the case of vertical exitways, to provide a safe path of egress for occupants and access for fire department personnel. Vertical interior exitways are often the only avenue of escape to the outside or to adjacent floors for building occupants. NFPA 101, *Life Safety Code®* (hereinafter referred to as NFPA 101), provides a comprehensive treatment of exits in general and for various types of occupancies involving both new and existing buildings. Detailed requirements are given for enclosing walls, doors, stairs, ramps, escalators, and other components of an exit system.

Shafts containing power, communication cable, combustible ducts or pipes, and other combustible materials should normally be enclosed and often firestopped at each floor. The reliability of the protection of combustible material penetrations through enclosure walls will normally be lower than for noncombustibles. If fire enters a vertical shaft containing combustible materials, it can spread rapidly upward, generate considerable smoke, and possibly spread to other floors. An eleventh floor storeroom fire in the World Trade Center, New York City, spread into a telephone closet through a louvered door (Lathrop 1975). The closets on each floor were connected through an unfirestopped hole in the slab to the closet on the floor above, up to the forty-first story of the structure. The fire ignited electrical wiring and plywood mounting boards and then spread up and down grouped cables that penetrated each floor. The fire spread up to the sixteenth and down to the tenth floor before it could be controlled.

Room or Suite Compartmentation

The goals of compartmentation in confining a fire to the room or suite of rooms of origin are generally the following:

1. Segregate a space with a higher level of fire hazard than the surrounding area. This is commonly found around rooms or suites used for storage, trash, flammable liquids, furnaces, laboratories, maintenance, painting, etc.
2. Minimize risk of loss to an occupant of one space as a result of a fire in space controlled by another. This is commonly found by separating apartments, office suites, motel/hotel rooms, row houses, etc., from each other.

An added benefit is that compartmentation also limits the size of the fire, which limits the amount of smoke that will be generated and facilitates fire suppression.

Properly designed and installed compartmentation has been successful in limiting many fires to the unit of origin. In a number of high rise fires, occupants who were unable to escape safely through corridors were able to "sit out" a major fire in an apartment adjacent to theirs (Watrous 1969). The most common cause of "failure" of this level of compartmentation has been a door that did not have a closer or was left open. Failure due to construction deficiencies are most commonly reported in buildings of combustible construction.

When a single room is compartmented, there may be insufficient ventilation for a fire to fully develop unless a window is open or happens to break.

Compartmentation is also used to protect locations of high value or critical operations from a fire in the surrounding area. Computer rooms, some control rooms, vaults, record storage, etc., are often protected from a fire involving the rest of the building.

Protection of Corridors

Fire resistive corridor partitions may be required (1) to protect the means of access to an exit for sufficient time to permit occupant evacuation, or (2) to provide a protected means of access to the fire for fire department personnel. In addition, when used in conjunction with room to room or suite to suite fire resistive separation, the corridor wall forms an element in preventing fire spread.

Corridor partitions typically are required to have a one hour fire resistance rating, and codes may require the partition to extend from the floor to the underside of the floor above. Either self or automatic closers are generally required on doors opening into the corridor. Since smoke control is also necessary to protect a corridor as an access to an exit, automatic closers are generally actuated by smoke detectors. Additional information on corridor doors will be found later in this chapter where protection for openings is discussed.

Building Separation

Walls separating two buildings or dividing a building into separate fire areas limit the maximum probable property loss and also can divide one building into what is legally two or more buildings to comply with the height and area limits of building codes. Stability becomes a very important consideration in this class of wall. Although all fire separation walls must maintain their stability under fire exposure for their rated duration, walls separating buildings or dividing a building into two fire areas must maintain stability during a complete burnout on either side, which is often accompanied by structural collapse.

If at least one of the buildings has a fire resistive frame such as reinforced concrete or protected steel, panel walls built in along a column line can provide fire separation. The frame must have sufficient resistance to withstand a fire of the maximum expected severity for that building. Care must be taken so that the floor construction will not adversely affect stability of the wall under fire exposure. In buildings of Type V (wood frame), Type III (ordinary), or Type IV (heavy timber) construction, the combustible framing members can be tied into the fire wall so they are self-releasing in case of collapse. The wall then remains supported by the framing on the unexposed side. The use of self-releasing framing has been generally successful in combustible construction but has questionable reliability in steel framed buildings. Caution must be exercised in the design of these fire walls because some methods of framing combustible members into the wall will lower its fire resistance.

Freestanding fire walls are most commonly used in one or two story industrial and storage buildings of unprotected steel frame construction. The maximum height is generally limited by economic considerations. A freestanding fire wall is entirely self-supporting under vertical loads and is not directly connected to the building framing. The adjacent building framing can provide support under horizontal loading. During a fire the wall must be capable of acting as a cantilever to withstand all horizontal forces directed toward the fire side. Freestanding fire walls should be designed to withstand at least a 5 lb per sq ft (20 kg/m^2) uniform lateral loading without any support from the framing on either side (FMEC 1982).

Tension stress in mortar joints limits the height of unreinforced 12 in. (25.4 mm) concrete block walls to 15 ft (4.6 m). Higher walls either require massive construction, which is rarely practical, or reinforcement. A common reinforcement technique is installation of vertical steel bars, anchored to the foundation within hollow masonry units. Reinforced pilasters can also be built into the wall at periodic points along the perimeter. A typical freestanding fire wall is illustrated in Figure 7-9H.

Freestanding fire walls have very definite limitations under horizontal forces, which may preclude or restrict their use. They are particularly vulnerable to collapse under seismic loads and wind forces received by exterior walls. Additional support must be provided if the walls are temporarily used as an exterior wall or become an exterior wall as a result of a fire. Piping, conveyors, and similar items whose collapse in a fire will exert a horizontal pull should not pass through freestanding fire walls.

Fire walls in steel frame buildings may be tied into the building structure so the forces of collapse on one side are resisted by the framing on the other side. The wall is built along a column line and the columns and horizontal framing are tied into the wall. The columns and framing in the wall line must be protected so they have the same fire resistance as the wall. The walls have to be located in the

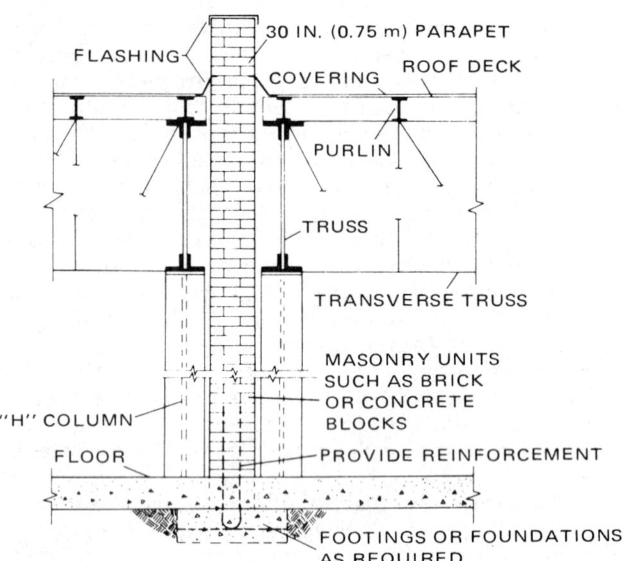

FIG. 7-9H. Typical freestanding wall, used at expansion joints or at joints between buildings. (Factory Mutual Engineering Corporation)

framing so that the strength of framing on one side will be capable of withstanding the pull of collapsing framing on the other side. The construction of a typical tied fire wall is shown in Figure 7-9I.

Fire separation between buildings can also be provided by two blank exterior walls of adequate fire resistance and located very close together. Each wall is tied to the frame in its building; collapse of the frame pulls the wall in, leaving the other wall to resist the fire.

Pipes, conveyors, cable trays, and similar through penetrations of fire walls can present special problems in seismic areas. The need for mobility may conflict with opening protection criteria. The complexity of the problem depends upon whether the systems are to be designed to protect against only one hazard at a time (fire or earthquake) or whether the simultaneous or consecutive occurrence of earthquake and fire should be considered in protection.

Fire Plumes Above Roofs

The configuration factor for plume/roof relationships can be calculated using the formula found in a reference (Hamilton & Morgan 1952). The effect of wind velocity, which deflects the fire plume, must be accounted for in calculation of the configuration factor. At high winds a fire plume can be deflected so it is almost horizontal above a burning roof. However, in high winds, a fire plume also tends to break up and is expected to be cooled by excess air. The angle of the fire plume selected is a matter of professional judgment; use of 45° is considered reasonable by some engineers.

If the roof surface is combustible, the value of the incident thermal radiation calculated will indicate whether ignition of the roof is expected. If the surface is noncombustible or resistive to ignition by radiant heat, it is then necessary to calculate the temperature of the combustible subsurface material to which heat will be transferred from the surface. This is an unsteady state heat

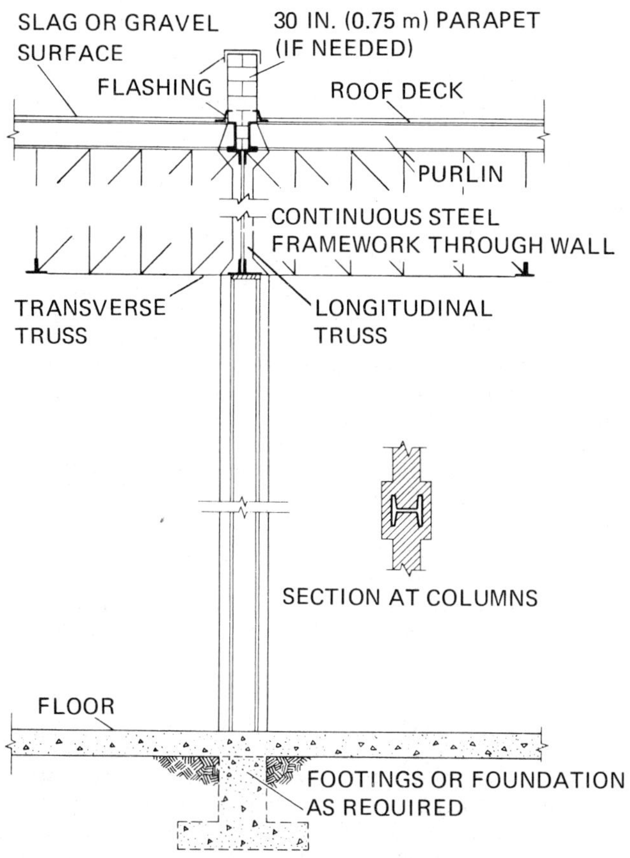

FIG. 7-9I. Typical tied fire wall, used with continuous building framework. (Factory Mutual Engineering Corporation)

transfer problem readily solved by standard numerical procedures.

A simple example of the calculation of incident thermal radiation can be used to illustrate the method of analysis. For the example, consider a 12 ft (3.7 m) high and 60 ft (18.3 m) wide frame building with a wood surfaced roof. The building is subdivided by a masonry fire wall with a 3 ft (0.91 m) parapet. (See Fig. 7-9J.) The expected height of flames above the roof would be about 1.4 times the height of the building, or approximately 17 ft (5.2 m) (Pingree 1968). Assuming a flame temperature of 1,400°F (760°C), the thermal radiation from the fire plume would be about 5.8 Btu per sq ft per sec (6.6 W/m^2). In a moderate wind, a fire plume deflection of 45° is possible. The maximum configuration factor would be about 0.3 and the incident radiation on the wood roof surface about 1.75 Btu per sq ft per sec (2 W/m^2).

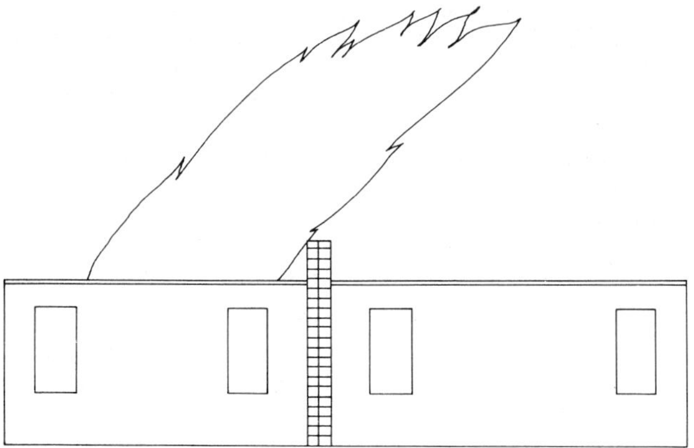

FIG. 7-9J. A fire plume exposing a wood roof over a parapeted fire wall.

At an incident radiation of 1.1 Btu per sq ft per sec (1.25 W/m^2), a wood roof can be sufficiently heated so a small burning brand will ignite it (Law 1963); this is referred to as pilot ignition. Since in the example the incident thermal radiation is well above that required for pilot ignition, flame spread over this fire wall is to be expected if the wind deflects the fire plume toward the exposed roof.

Fire walls must extend through and above a combustible roof to reduce the risk of fire spread over the top of the wall. Codes and standards typically call for fire walls to have a parapet 18 to 36 in. (0.46 to 0.9 m) above a combustible roof. The parapet minimizes the risk of direct flame spread over the fire wall and reduces the risk of fire spread by radiant heat from flames above the roof. The risk of direct flame spread over an 18 to 36 in. (0.46 to 0.9 m) parapet is very low. However, risk of fire spread by radiant heat depends upon the combustibility of the roof surface, the height of the parapet, the dimensions of the fire plume through the roof of the fire zone, the wind velocity, and, if the roof covering is resistive to ignition by radiant heat, its insulating properties and the duration of the exposing fire. If a roof surface is combustible, the risk of fire spread by radiant heat past a parapet can be quite high.

The threat of fire spread by radiant heat over a parapet can be calculated using the thermal radiation values identified in Appendix A of NFPA 80A, *Recommended Practice for Protection of Buildings from Exterior Exposure Fires* (hereinafter referred to as NFPA 80A), and fire plume dimensions (Pingree 1968). Standard heat transfer procedures can then be applied using the formula:

$$I_o = SxT^4x\phi$$

where:

I_o = unit incident radiation
S = Stefan-Boltzman constant
T = absolute temperature
ϕ = configuration factor or view factor

This formula can be used with either customary American or SI units as long as the units are consistent.

Structural Stability of Fire Walls

Fire rated wall requirements may be intended to ensure structural stability rather than to provide a fire barrier. In many cases, for example, hourly rated exterior bearing walls are permitted to have unlimited unprotected window areas; therefore, it is obvious that the rating refers to structural stability rather than provision of a fire barrier. Required fire ratings of interior bearing walls may or may not be intended to provide a fire barrier. Interior bearing walls with an hourly rating for fire separation do require protection of all openings; however, interior bearing walls that have an hourly rating for only structural stability purposes would not need opening protection. Caution must be exercised in selecting the proper hourly rating for interior bearing walls intended for structural stability only. A barrier separation wall is exposed to fire on only one side, whereas an interior bearing wall that would be providing only structural stability could be exposed to fire on both sides. The same hourly rating wall will not provide the same protection in both cases.

Fire resistive walls of a high hourly rating are occasionally improperly specified to separate areas for protection against an explosion hazard rather than a long duration fire. The fire resistance of a wall has no relationship to its ability to withstand explosion forces. A wall with a one hour rating may withstand explosion overpressures better than a four hour wall; no conventional wall will remain in place unless the explosion pressure is limited by venting or suppression.

PROTECTION OF OPENINGS

The protection for openings in a fire rated wall has less resistance to fire spread than is provided by the wall. This is allowable because (1) easily ignited combustibles are not normally found against doors and other opening protection constructions, and (2) fire suppression forces can usually extinguish small localized fires that may spread past the wall. However, because openings do reduce the effectiveness of a fire barrier, fire doors generally cannot occupy more than 25 percent of the perimeter of any fire separation wall.

Factors that reduce the fire resistance of opening protection as compared to that of walls include the following:

1. There is no allowable temperature rise limitation on the unexposed sides of most types of fire doors. The unexposed side of steel doors may actually glow red during fire testing. When there is a limit to or listing of a temperature rise on a fire door such as those used in stairwells, it applies only to the temperature rise during the initial 30 minutes of the fire test. Fire doors also buckle during fire tests and are permitted to move out of their frames in a direction perpendicular to the plane of the door as much as 2⅞ in. (72 mm). Some flaming is also permitted on unexposed sides of fire doors during the test procedures.
2. Requirements for the protection of duct penetrations place no limit on the temperature rise on the duct itself on the unexposed side of the wall.
3. Piping, cable trays, and conduit can penetrate fire walls and although the opening must be firestopped, the temperature rise on the metal penetration itself would normally be above the 250 or 3251°F (139 or 161°C) limits required of the wall.
4. The standard fire test furnace is operated at a negative pressure so that air flows in around protected openings rather than outward. In a real fire there may be a higher pressure on the fire side which is more likely to force flame through indirect paths or firestopped openings.
5. Wired glass vision panels may transmit heat by both conduction and radiation without any restrictions.

Thermally induced distortions of between 4 and 5 in. (102 and 127 mm) were observed in some foreign tests of swinging steel fire doors (Bushev et al 1978). Considerably smaller distortions were noted with wood and insulated metal swinging doors. However, wood doors were noted as being more likely to fail by localized carbonation at hinges, locks, and knobs. Rolling steel fire doors generally do not deform appreciably during a fire. Large distortions of some rated fire doors, both wood and metal, were also reported in "Project Corridor," a series of fire tests conducted by the California State Fire Marshal in a simulated light hazard occupancy (Reagan et al 1974). In some countries, extensible bolts and latches are added to swinging steel fire doors so that the door is held in the frame at five or six points. Distortions of this type of door were reported to be under 1 in. (2.54 mm) during a standard fire test (Bushev et al 1978).

Fire doors, windows, and shutters are the most widely used and accepted means of protecting openings in fire resistive walls. Suitability of these closures is determined through test by recognized testing laboratories, and doors not tested cannot be relied upon for effective protection. Fire door assemblies for the protection of openings depend upon the use of labeled fire doors and frames (where frames are required) and listed or labeled hardware. Hardware for fire doors is referred to as builders hardware (which includes fire exit hardware) and fire door hardware. The doors are tested as they are installed in the field; that is, with the frame, hardware, wired-glass panels, and other accessories necessary to complete the installation.

Classification of Openings and Doors

Fire doors may be classified by an hourly rating designation, an alphabetical letter designation, or a combination. Current practice is to use the hourly rating designation; formerly, the alphabetical letter designation was used. Appendix F of NFPA 80A, refers to openings as A, B, C, D, and E in accordance with the character and location of the wall. The alphabetical classification of the opening does not apply to the closure; however, in actual practice, the distinction between opening classification and door classification was rarely maintained. A three hour door for use in a Class A opening was commonly called a Class A door.

Building codes and NFPA 80A commonly specify a door by its fire protection rating instead of by the classification of opening to be protected. Fire doors are now available with one-half and one-third hour ratings, which do not fall within the specified protection criteria for the classes of opening. The following paragraphs summarize current applications of doors, windows, and shutters for openings in fire resistive walls.

Three Hour Fire Doors: Openings in walls separating buildings or dividing a building into different fire areas are protected by three hour fire doors. This can include the protection of openings in walls enclosing hazardous spaces such as flammable liquid storage rooms. However, many codes only require three hour doors in separation walls required to have a fire resistance of three hours or more. In addition, some authorities require two three-hour fire doors, one on each side of the opening, whenever a wall is required to have a fire resistance of four hours.

One and One-half Hour Fire Doors: Openings in two hour enclosures of vertical openings in buildings are protected by one and one-half hour fire doors. Many codes also permit the use of one and one-half hour fire doors to protect openings in walls separating buildings or dividing buildings into different fire areas when the wall is only required to have a fire resistance of two hours.

One and One-half Hour Fire Doors and Shutters: Openings in exterior walls that can be subjected to severe fire exposure from outside the building are protected by one and one-half hour fire doors and shutters.

One Hour Fire Doors: Openings in one hour enclosures of vertical openings in buildings are protected by one hour fire doors.

Three-quarter Hour Fire Doors: Openings in corridor and room partitions are protected by three-quarter hour fire doors. Sometimes they are also permitted in partitions that subdivide floors of a building. Although three-quarter hour fire rated doors are for use in one hour corridor partitions, many codes permit installation of other types of doors such as one-half or one-third hour fire rated or 1¾ in. (445 mm) solid bonded wood core doors. Either self or automatic closers are generally required on corridor doors. Since smoke control is also necessary to protect a corridor as an access to an exit, automatic closers are generally actuated by smoke detectors.

Three-quarter Hour Fire Doors and Shutters: Openings in exterior walls subject to a moderate or light fire exposure from outside the building are protected by three-quarter hour fire doors and shutters.

Three-quarter Hour Fire Windows: Openings in corridor or room partitions or in exterior walls subject to a moderate or light external fire exposure can be protected by three-quarter hour fire windows.

One-half Hour (30 minute) and One-third Hour (20 minute) Fire Doors: Doors with these ratings are for use where smoke control is a primary consideration. They are also used for the protection of openings in partitions between a habitable room and a corridor when the wall is constructed to have a fire resistance rating of not more than one hour or across corridors where a smoke partition is required. Some codes permit the use of 1¾ in. (445 mm) solid bonded wood core doors in corridor, room, and smoke stop partitions that are required to have a fire resistance of not more than one hour. These doors cannot be used for corridor doors to hazardous areas or to protect openings in enclosures of vertical openings in buildings.

Types of Doors

There are several types of construction for fire doors. They are:

Composite Doors: These are of the flush design and consist of a manufactured core material with chemically impregnated wood edge banding and untreated wood face veneers or laminated plastic faces, or surrounded by and encased in steel.

Hollow Metal Doors: These are of formed steel of the flush and paneled designs of No. 20 gage or heavier steel.

Metalclad (Kalamein) Doors: These are of flush and paneled design consisting of metal covered wood cores or stiles and rails and insulated panels covered with steel of No. 24 gage or lighter.

Sheetmetal Doors: These are of formed No. 22 gage or lighter steel and of the corrugated, flush, and paneled designs.

Rolling Steel Doors: These are of the interlocking steel slat design or plate steel construction.

Tin Clad Doors: These are of two or three ply wood core construction, covered with No. 30 gage galvanized steel or terneplate [maximum size 14 by 20 in. (0.35 by 0.50 m)]; or No. 24 gage galvanized steel sheets not more than 48 in. (1.2 m) wide.

Curtain Type Doors: These consist of interlocking steel blades or a continuous formed spring steel curtain in a steel frame.

Wood Core Doors: These doors consist of wood, hardboard, or plastic face sheets bonded to a wood block or wood particle board core material with untreated wood edges.

Unrated Door Protection

The fire resistance of existing doors can be improved by coating the door with an intumescent paint, although no formal fire resistance rating has been established. This method is not commonly accepted in the U. S., although it has been used in England. Intumescent paint is listed for reducing flame spread not for improving fire resistance. However, tests conducted by the NBS showed that a single layer of intumescent paint, 7 to 8 mil (0.18 to 0.20 mm) thick, significantly delayed char penetration into a soft wood joist (HUD 1976). Ordinary wood panel doors, in which both door and frame were coated with two layers of intumescent, withstood fully developed room fires similarly to 1¾ in. (44.5 mm) thick solid core wood doors in simulated nursing home room fires (AHCA 1975). In another test, a wood panel door was protected with gypsum board on the thin panel and intumescent paint on the remainder of the wood. This upgraded door withstood a standard fire test, modified to simulate the positive pressure of a room fire, longer than a 20 minute rated fire door (Fisher et al 1977). However, there is no fire protection rating assigned to any upgraded fire doors; it is entirely up to the authority having jurisdiction whether or not this is acceptable for improving the firesafety in existing buildings.

Wired Glass

Wired glass is used as vision panels in fire doors, for exterior windows subject to moderate fire exposure, and in one hour rated corridor walls. In addition, it is frequently used in smoke stop barriers and to enclose open stairways in older buildings. Such panels are not allowed in doors having a fire protection rating of three hours or in doors having a one and one-half hour fire protection rating for use in severe exterior fire exposure locations. Vision panels are limited to 100 sq in. (.065 m^2) in one and one and one-half hour rated doors (except as noted previously) and to 1,296 sq in. (0.84 m^2) with a maximum dimension of 54 in. (1.37 m) in three-quarter hour rated doors. However, the total maximum area of glass per door leaf is simply limited to the maximum area tested.

Wired glass is a glass sheet containing an imbedded net of steel which helps distribute heat, lowers thermal stresses, and increases the strength of the assembly. When exposed to a fire, the glass starts cracking after a minute or two but does not open up. The glass begins to weaken when heated to about 1,470°F (799°C) and gradually starts to soften. At about 1,600°F (871°C) it deforms so badly it will drop out (Bushev et al 1978). Its normal 45 minute fire resistance rating corresponds to 1,638°F (892°C) on the standard time-temperature curve. In actual fires, the endurance of wired glass is temperature dependent; if it is never heated above the temperature at which it weakens, it could remain in place for an extended period of time.

Wired glass panels differ from a solid wall primarily in the transmission of radiant heat. Conductive heat transmission is also increased, but high conductive heat transmission is also found in door hardware, frames, etc. The intensity of radiant heat, I, to which a combustible or person would be exposed to through a wired glass panel from a fully developed room fire, is given by:

$$I = k\phi \, I_o$$

where I is the transmissivity of the window, $K\phi$ is the configuration factor, and I_o is the radiant intensity of the exposing fire.

This equation may be used for conventional or SI units as long as their use is consistent.

A radiant intensity of 5 Btu per sq ft per sec (5.7 W/m^2) has been measured for a room fire involving ordinary residential furnishings with average ventilation such as an open door (Battelle Columbus Laboratories 1974). Measurements of well ventilated fires in light hazard occupancies indicate a peak radiant intensity of 7.4 Btu per sq ft per sec (8.4 W/m^2) (Law 1963). Transmissivity of glass windows is reported to range from 0.4 to 0.6 (Law 1963). The configuration factor, ϕ, can be determined from any specific situation from graphs found in most heat transfer texts.

If wood on the unexposed side of the wired glass is exposed to a radiant intensity above 3 Btu per sq ft per sec (3.4 W/m^2), autoignition may occur. Most synthetic materials will require a higher level of radiant heat intensity for autoignition.

When wired glass panels are installed in walls on a deadend corridor or other location in which building occupants have no alternate escape path, the heat radiated through the glass may prevent escape. If persons moving past the glass absorb sufficient heat to cause severe pain, their ability to escape could be impaired. The human threshold of severe pain for exposure to radiant heat depends upon both the intensity and the time over which it is received. Experimental data that has defined this limit (Beuttner 1951) can be approximated by the equation:

$$Q_{max} \simeq 2.3 \, t^{0.25} \quad t \leq 45 \text{ s}$$

where Q_{max} is the total heat absorbed by the skin in Btu per sq ft (W/m^2) over t seconds at the threshold of severe pain. As a person moves past the wired glass panel the view factor, θ, will change. The absorbed heat can be calculated by determining θ for a number of points along the person's path and using a stepwise integration over the time required to move past each point. The normal walking speed used in exit calculations is 3.5 fps (1.07 m/sec) (GSA 1977). If the total heat absorbed exceeds the limits in the equation, the escape of occupants past the wired glass panels may be impaired.

Presently, wired glass is the only glass that has been listed to provide some measure of fire resistance. NFPA 80A refers only to wired glass for vision panels or windows in rated assemblies. However, efforts are being made to develop glazing materials other than wired glass that will qualify for fire resistance ratings. At least a few manufacturers have produced clear, transparent vision panels, most often of glass/plastic sandwich construction which offers resistance to fire. Those that have passed the necessary fire tests can be used instead of wired glass.

Firestopping

The term firestopping is used to describe both barriers to restrict the spread of fire in concealed spaces and materials used to fill gaps around penetrations in walls and ceilings.

In wood frame construction, the normal fire stop inside walls and ceiling spaces is a 2 in. (50.8 mm) thick (nominal) piece of lumber. However, for other applications, firestopping should normally be noncombustible and have a sufficiently high melting point so it will remain in place under fire exposure. Gypsum board, sheet metal, plaster, brick, cement grout, mineral fiber insulation, asbestos cement board, ceramic fiber boards, and even sand are all examples of commonly used firestopping. Synthetic (silicone) materials, developed earlier for nuclear applications, have come into use, especially where an expanding medium is helpful in multiple cable or pipe penetrations.

Protective devices and systems are available for sealing cable, conduit, pipe, and cable tray penetrations through fire rated brick and concrete block walls and concrete slab floors. The fire test acceptance criteria for these exclude an evaluation of temperatures on unexposed surfaces of devices, cables, conduit, trays, etc. Modular devices are sized for the cable, conduit, or pipe, and contain an organic composition that expands when heated and tightly seals the penetration. Construction systems, based on a foamed-in-place fire resistant silicone elastomer, can be used for cable tray, cable, conduit, and pipe penetrations. It expands as it cures and forms a tight but resiliant seal. This material may also be used to close openings. The listings of each system construction include limitations on the size and type of items that may penetrate each assembly and on the size of the openings. There are also specific requirements for supporting the equipment penetrating the barrier so that it will not collapse and pull the sealant out of the opening. Often considerable bracing is required on equipment such as cable trays.

Sand is occasionally used to firestop cable penetrations when need for frequent changes precludes use of permanent firestopping material. The opening in the wall should be located near or even below floor level. A noncombustible trough is built up on each side of the wall with its side and ends extending several inches above the opening in the wall. The cables are run down into the trough, through the wall, and up and out again. The trough is filled with sand to provide firestopping for the cable penetration. This and other unusual firestopping techniques should only be used if and when conventional methods are not suitable and when approved by the authority having jurisdiction.

Considerable attention and research has been directed toward wall and floor penetrations by plastic drain, waste, and vent (DWV) pipes (Attwood 1980; Benjamin & Parker 1972; McGuire 1973; Williamson 1979). Full scale tests have shown that the integrity of a two hour fire resistive chase enclosure could be maintained by encasing the pipe in an 18 in. (0.46 m) steel sleeve and penetrating the chase at a forty five degree downward angle. These results were applicable to 3 in. (76 mm) laterals of both DVC and ABS DWV pipe. Unless otherwise tested and approved, chases should be firestopped at each floor. In many buildings, much of the plumbing is installed in wall cavities. Use of plastic DWV pipe inside fire resistive wall assemblies has been extensively tested, and a number of methods of protection identified, which will not compromise the fire resistance or create a risk of vertical fire spread.

Some designs for curtain walls, particularly in high rise buildings, offer danger of vertical spread of fire behind the curtain walls for the entire height of the building. This can occur when the panels of a curtain wall do not rest tightly for their full length on a floor slab. Rather, there is a continuous relieving angle attachment between the panels and the building that holds the panels away from the floor slab, creating vertical flue spaces between the panels and the end of the floor slab. Where firestopping is not inherent in the design, it is necessary to fill the space with appropriate firestopping materials. This arrangement is generally less satisfactory because of the tendency for both expansion and contraction of the exterior wall and maintenance operations to cause unplanned openings through and around the firestopping.

Atriums

Atriums, which are becoming more common in many new buildings, are a departure from the traditional compartmentation concept of fire containment by enclosing all vertical openings. Alternative arrangements must be provided in atrium buildings to achieve the same level of protection found in buildings using the traditional compartmentation concept. The occupancy should be one in which a rapidly spreading fire would not be encountered, and all occupants should be expected to be aware of any hazard before they are endangered, and in time to move to a place of safety if necessary.

Fire protection in atrium buildings is generally provided by a combination of methods such as compartmentation, ventilation, automatic suppression, and contents and material control. NFPA 101 requires atriums in new construction to be separated from adjacent spaces by fire barriers having at least a one hour fire resistance rating. However, as many as three levels of a building may open directly to the atrium without the need for a fire resistive barrier. As an alternative where one hour barriers are normally required, glass walls may be used provided there are sprinklers spaced 6 ft (1.8 m) apart not more than 1 ft (0.3 m) from the glass along both sides of the glass wall. The sprinklers must be located to ensure that the glass surface is wet when the sprinklers operate. Sprinklers are not required on the atrium side of the glass wall if there is no walkway or other floor area on that side above the main floor level.

Large volume atriums will dilute fire gases and, with proper ventilation, direct their flow safely to the outside. Small atriums may not have the capacity to dilute fire gases, although if they are properly designed and ventilated, fire gases can be channeled to the outside and the risk of horizontal fire spread minimized.

Bibliography

References Cited

AHCA. 1975. "Fire Tests in a Nursing Home Patient Room." *Report HEW*. Contract HSA 105-74-116. American Health Care Association, Washington, DC.

AISI. 1975. "Fire Severity at the Exterior of a Burning Building." *Report 67NK5227A*. American Iron and Steel Institute, Washington, DC.

Attwood, P. C. 1980. "Penetration of Fire Partitions by Plastic Pipe." *Fire Technology*. Vol 16, No 1. pp 37-62.

Babrauskas, V. 1980. "Fire Tests and Hazard Analysis of Upholstered Chairs." *Fire Journal*. Vol 74, No 2. pp 35-39.

Babrauskas, V., and Williamson, R. B. 1975. "Post Flashover Compartment Fires." *Report UCB FRG 75-1*. Fire Research Group. University of California, Berkeley, CA.

Battelle Columbus Laboratories. 1974. "Space Age Contribution to Residential Fire Safety." *Fire Journal*. Vol 68, No 2. pp 18-25.

Benjamin, I. A., and Parker, W. J. 1972. "Fire Spread Potential of ABS Plastic Plumbing." *Fire Technology*. Vol 8, No 2. pp 104-119.

Berry, D. L., and Minor, E. E. 1979. "Nuclear Power Plant Fire Protection—Fire Hazards Analysis (Subsystems Study Task 4)." *Report NUREG/CR-0654, SAND 79-0324*. Sandia National Laboratories, Albuquerque, NM.

Best, R. L. 1978. "Tragedy in Kentucky." *Fire Journal*. Vol 72, No 1. pp 18-35.

Best, R. L., and Demers, D. P. 1982. "Fire at the MGM Grand." *Fire Journal*. Vol 76, No 1. pp 19-37.

Beuttner, K. 1951. "Effects of Extreme Heat on Man, III. Surface Temperature, Pain, and Heat Conductivity of Living Skin in Experiments with Radiant Heat." *Project 21-26-002, Report No. 3*. USAF School of Aviation Medicine.

Bushev, V. P., et al. 1978. *Fire Resistance of Buildings*. Construction Literature Publishers. Moscow, USSR. 1970. National Bureau of Standards. *IT 73-52030*. Amerind Publishing Co. PVT Ltd., New Delhi, India.

Campbell, John A. 1978. "Fire Safety Systems Analysis Data Base." Presented at NFPA Fall Meeting Montreal, Quebec, Canada.

Christian, W. J., and Waterman, T. E. 1971. "Characteristics of Full-Scale Fires in Various Occupancies." *Fire Technology*. Vol 7, No 3. pp 204-218.

Culver, Charles, G. 1978. "Characteristics of Fire Loads in Office Buildings." *Fire Technology*. Vol 14, No 1. pp 51-60.

Davis, S. 1978. "Assessment of Fire Hazards from Furniture." International Fire, Security, and Safety Conference, London, England, Apr. 24-28, 1978.

DeCicco, P. R. 1976. "What to Do with Existing Row-Frame Residential Buildings." *Fire Journal*. Vol 70, No 6. pp 23-31.

Fisher, F. L., et al. 1977. "A Study of Potential Post-Flashover Fires in Wheeler Hall and the Results from a Full-Scale Fire Test of a Modified Wheeler Hall Door Assembly." *Report UCX 77-3*. Fire Research Laboratory, University of California, Berkeley, CA.

FMEC. 1982. "MFL Fire Walls: Barriers Against Destruction." *Factory Mutual Record*. p 13.

GSA. 1977. "Handbook, Building Firesafety Criteria." *PBS P5920.9, Change 6*. General Services Administration, Washington, DC.

Hamilton, D. C., and Morgan, W. R. 1952. "Radiant-Interchange Configuration Factors." *Technical Note TN 2836*. National Advisory Committee for Aeronautics.

Harmathy, T. Z. 1972. "A New Look at Compartment Fires—Parts I and II." *Fire Technology*. Vol 8, Nos 3 and 4.

Harmathy, T. Z. 1976. "Design of Buildings for Fire Safety—Part I." *Fire Technology*. Vol 12, No 2. pp 95-108.

Heselden, A. J. M., et al. 1970. "Burning Rate of Ventilation

Controlled Fires in Compartments." *Fire Technology*. Vol 6, No 2. pp 123-125.

HUD. 1976. "A Compendium of Fire Testing." Feedback, Operation Breakthrough, Vol 5. *HUD-PDR-28-3*. U.S. Department of Housing and Urban Development, Washington, DC.

Ingberg, S. H. 1927. "Fire Tests of Office Occupancies." *NFPA Quarterly*. Vol 20, No 3. pp 243-252.

Ingberg, S. H. 1928. "Tests of the Severity of Building Fires." *NFPA Quarterly*. Vol 22, No 1. pp 43-61.

Kamerus, L. J. 1977. "A Preliminary Report on Fire Protection Research Program (July 6, 1977 Test)." *SAND 77-1424*. Sandia National Laboratories, Albuquerque, NM.

Kamerus, L. J. 1978. "A Preliminary Report on Fire Protection Research Program Fire Barriers and Fire Retardant Coating Tests." *NUREG CR-0381, SAND 78-1456*. Sandia National Laboratories, Albuquerque, NM.

Kawagoe, Kario. 1967. "Estimation of Fire Temperature-Time Curve in Rooms." *Building Research Paper No. 29*. Ministry of Construction, Japanese Government, Tokyo, Japan.

Lathrop, James K. 1975. "World Trade Center Fire." *Fire Journal*. Vol 69, No 4. pp 19-21.

Lathrop, J. K., et al. 1975. "In Osceola a Matter of Contents—Hospital Fire Kills Seven." *Fire Journal*. Vol 69, No 3. pp 20-29.

Law, Margaret. 1963. "Heat Radiation from Fires and Building Separation." *Technical Paper No. 5*. Joint Fire Research Organization. Her Majesty's Stationary Office, London, England

Lie, T. T. 1974. "Characteristic Temperature Curves for Various Fire Severities." *Fire Technology*. Vol 10, No 4. pp 315-326.

McCaffrey, B. J., et al. 1981. "Estimating Room Temperatures and the Likelihood of Flashover Using Fire Data Correlations." *Fire Technology*. Vol 17, No 2. pp 98-119.

McGuire, J. H. 1973. "Penetration of Fire Partitions by Plastic DWV Pipe." *Fire Technology*. Vol 9, No 1. pp 5-14.

NBFU. 1956. "Fire Resistance Ratings of Less than One Hour." National Board of Fire Underwriters, NY.

NRC. 1976. Special Review Group. "Recommendations Related to Browns Ferry Fire." NUREG-0050. U.S. Nuclear Regulatory Commission, Washington, DC.

Pingree, D. 1968. "Looking at Fire Hazards, The Height of Flames Above a Roof." *Fire Journal*. Vol 62, No 3. pp 24-27, 32.

Powers, W. R. 1970. "One New York Plaza Fire." New York Board of Fire Underwriters, NY.

Quintiere, J. G. 1977. "Growth of Fire in Building Compartments." Fire Standards and Safety. *ASTM STP 614*. A. F. Robertson, ed. American Society of Testing Materials. pp 131-167.

Quintiere, J. G. 1983. "A Simple Correlation for Predicting Temperature in a Room Fire." *NBSIR 83-2712*. U.S. Dept. of Commerce. National Bureau of Standards.

Reagan, R., et al. 1974. "Project Corridor, Fire and Life Safety Research." *Western Fire Journal*, Bellflower, CA.

Roytman, M. Ya. 1975. *Principles of Fire Safety Standards for Building Construction*. Construction Literature Publishing House, Moscow, USSR. 1969. National Bureau of Standards TT-71-58002. Amerind Publishing Co., PVT Ltd., New Delhi, India.

Sharry, John A. 1974. "Another Pennsylvania Nursing Home Fire, Wayne, Pennsylvania." *Fire Journal*. Vol 68, No 3. pp 11-14.

Vodvarka, F. J., and Waterman, T. E. 1965. "Fire Behavior, Ignition to Flashover." *Final Report*. Task 2-C, OCD-PS-64-50, OCD Work Unit 2536-C. IIT Research Institute, Chicago, IL.

Waterman, T. E. 1974. "Experimental Structural Fires." *Report J6269*, Contract DAHC 20-72-C-0290, DCPA Work Unit 2562B. IIT Research Institute, Chicago, IL.

Waterman, T. E., et al. 1964. "Prediction of Fire Damage to Installations and Built-Up Areas from Nuclear Weapons." *Final Report—Phase III, Experimental Studies, Appendices A-G*. Contract No. DCA-8. National Military Command System Support Center, Washington, DC.

Watrous, L. D. 1969. "Fire in a High-Rise Apartment Building, Hawthorne House, Chicago." *Fire Journal*. Vol 63, No 3. pp 5-11.

Williamson, R. B. 1979. "Installing ABS and PVC Drain, Waste and Vent Systems in Fire Resistant Buildings." *Fire Journal*. Vol 73, No 2. pp. 36-45.

Zinn, B.T., et al. 1974. "Fire Spread and Smoke Control in High-Rise Fires." *Fire Technology*. Vol 10, No 1. pp 35-53.

NFPA Codes, Standards, Recommended Practices and Manuals. (See the latest *NFPA Codes and Standards Catalog* for availability of current editions of the following documents.)

NFPA 80, *Standard for Fire Doors and Windows*.

NFPA 80A, *Recommended Practice for Protection of Buildings from Exterior Fire Exposures*.

NFPA 101, *Code for Safety to Life from Fire in Buildings and Structures*.

Additional Readings

Babrauskas, V., "Estimating Room Flashover Potential," *Fire Technology*, Vol. 16, No. 2, May 1980, pp. 94-104.

"Fire-Rated Hollow Metal Doors and Frames," National Association of Architectural Metal Manufacturers, 1983.

"Fire Wall Design," *The Sentinel*, Vol. 36, No. 6, Nov.-Dec. 1980, pp. 14-15.

Harmathy, T. Z., "The Fire Resistance Test and Its Relation to Real Word Fires," *Fire and Materials*, Vol. 5, 1981, p. 112.

Quintiere, J., "The Spread of Fire from a Compartment—A Review," *ASTM STP 685*, American Society for Testing and Materials, Philadelphia, PA, 1980.

Tanaka, T., "A Model of Multiroom Fire Spread," *NBSIR 83-2718*, National Bureau of Standards, Gaithersburg, MD, 1983.

SMOKE MOVEMENT IN BUILDINGS

Revised by Harold E. Nelson, P. E.

Smoke and fire gases, inherent in all accidental fires, are dangerous products of combustion that have critical influences on life safety, property protection, and fire suppression practices in buildings. In some fires the volume of smoke is so great that it may fill an entire building and obscure visibility at the street level to such an extent that it is difficult to identify the fire involved building. In other incidents the volume of smoke generation may be considerably less, although the danger to life is not necessarily diminished because of the presence of other airborne products of combustion.

This chapter gives information on the techniques used to evaluate the physical characteristics of smoke movement through buildings, both short and tall, as a basis for designing smoke control systems. It also covers the approaches that can be used to test the effectiveness of designed smoke control systems in the absence of actual performance test involving test fires.

Other chapters in this HANDBOOK of interest in controlling the hazards of smoke are Chapter 11, "Venting Practices," in this section, and Section 8, Chapter 4, "Air Conditioning and Ventilating Systems."

CLASSIFICATION OF SMOKE ZONES

As a fire burns, it:

1. Generates heat.
2. Changes major portions of the burning material (fuel) from its original chemical composition to one or more other compounds, e.g., carbon dioxide, carbon monoxide, water, and/or other compounds.
3. Often (due to less than 100 percent combustion efficiency) transports a portion of the fuel as soot or other material that may or may not have undergone a chemical change.

A major portion of the heat generated as a fuel burns remains in the mass of products liberated by the fire. This mass expands, is lighter than the surrounding air, and rises as a plume. The rising plume is turbulent and because of this, entrains large quantities of air from the surrounding atmosphere into the rising gases. This entrainment:

1. Increases the total mass and volume of the plume.
2. Cools the plume by mixing the cool entrained air with the rising hot gases. Normally the rising plume is hotter at its center and cooler towards the edges where cooler air is entrained.
3. Dilutes the concentration of fire produced products in the plume.

Smoke, as discussed in this chapter, is therefore defined as a mixture of hot vapors and gases produced by the combustion process along with unburned decomposition and condensation matter and the quantity of air that is entrained or otherwise mixed into the mass.

For the purposes of describing smoke movement in buildings, the treatment of smoke movement is divided into two general areas. They are:

1. *Hot Smoke Zone*: This includes those areas in a building where the temperature of the smoke is high enough so that the natural buoyancy of the body of smoke tends to lift the smoke towards the ceiling while clean (or at least less polluted) air is drawn in through the lower portion of the space. Normally this condition exists in the room of fire origin. Also, depending on the level of energy produced by the fire and the size of connecting openings, e.g., open doors, etc., hot smoke zones can readily exist in adjacent rooms or corridors. Industrial and warehouse smoke and heat venting, atria smoke removal, and the movement of smoke in corridors open to flashed-over spaces all involve a hot smoke zone where the smoke is lifted and driven by the buoyant forces produced directly by the fire.
2. *Cool Smoke Zone (Rest of the Building)*: This includes those areas in a building where mixing and other forms of heat transfer have reduced the effect of the driving force of the fire to where buoyant lift in the smoke body is a minor factor. In these areas the movement of the smoke is primarily controlled by

Mr. Nelson is a research fire protection engineer at the Center for Fire Research, National Bureau of Standards (NBS), Gaithersburg, MD. He is also a Fellow and past president of the Society of Fire Protection Engineers. (SFPE).

other forces, such as wind and stack effects and the mechanical heat, ventilating, air conditioning or other air movement systems. In these areas the movement of smoke is essentially the same as the movement of any other pollutant.

SMOKE MOVEMENT IN THE HOT SMOKE ZONE

The volume of combustion products entrained in a rising plume in the hot smoke zone is relatively small, compared with the volume of air in the total mixture. Consequently, the smoke produced by a fire will approximate the volume of air drawn into the rising plume. Figure 7-10A illustrates the process.

In those situations where the height of the plume as measured from the top of the fire to the level of the smoke layer, as shown in Figure 7-10A, is more than about twice the height of the solid body of flame, it is reasonable to estimate the amount of smoke using developed formulas (Thomas et al 1963) and presented in the book, "Smoke Control in Fire Safety Design" (Butcher & Parnell 1979).

The formulas involved were derived from research conducted at the British Fire Research Station more than 20 years ago. This work showed that the amount of smoke could be reasonable estimated solely as a function of the height of the fire plume over a "virtual fire source." The research indicated that the virtual source for a free burning fire having a circular shape would be approximately 0.15 diameters below the burning surface.

The mass of gas drawn into the fire can be estimated as:

$$M = 0.096\ P q_o y^{3/2}\ (g\ T_o/T)^{1/2}$$

where:

M = rate of smoke production (kg/s)
P = perimeter of the fire (m)
q_o = density of ambient air (kg/m^3)
y = distance from floor to bottom of smoke layer (m)
g = acceleration due to gravity (m/s^2)
T_o = absolute temperature of the ambient air (°K)
T = absolute temperature of the flames of the smoke plume (°K)

Typical numerical values of the parameters are:

q_o = 1.22 kg/m^3 @ 17°C
T_o = 290 K
T = 1100 K
g = 9.81 m/s^2

Using these values, the rate of smoke production becomes:

$$M = 0.188\ P\ y^{3/2}$$

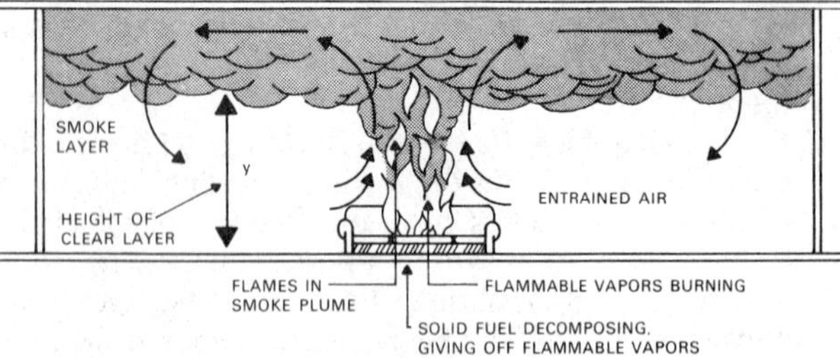

FIG. 7-10A. The production of smoke from a fire.

The simple expression $M = 0.188Py$ includes a series of assumptions, the most important of which are:

1. The tip of the flame is a significant distance below the bottom of the smoke layer. The formula, while useful, is much less accurate in spaces having a low ceiling relative to the height of the fire involved.
2. The fire bed itself covers an area having a length and width that are reasonably approximate to each other. The original formula is based on the assumption of a circular fire. The degree of error in the formula increases as the relationship of length to width increases.

Flame Height

A reasonable estimate or flame height (Alpert and Ward 1963) can be obtained from the expression:

$$H_f = 0.011\ (k\dot{Q})^{0.4}$$

where:

H_f = flame height (m)
k = wall effect factor. The value of k to be used is:
$k = 1$ when there are no nearby walls.
$k = 2$ when the fuel packages near a wall.
$k = 4$ when the fuel package is in a corner.
$\dot{Q}$ = fuel heat release rate (watts).

The results of this formula is shown graphically in Figure 7-10B.

Plume Gas Temperature

Detailed engineering formulas for properties of fire plumes have been presented (Heskestad 1982). In simpler terms, however, (Alpert and Ward 1963) an empirical estimate plume temperature is provided at a given point above the fuel as:

$$\Delta T = \frac{0.222\ (k\dot{Q})^{2/3}}{H^{5/3}}$$

where

ΔT = maximum temperature increase (°C) above ambient (room) temperature.
$\dot{Q}$ = total heat release rate (W).
k = wall factor. The value of k is:

$k = 1$ when there are no nearby walls
$k = 2$ when the fuel package is near a wall
$k = 4$ when the fuel package is in a corner

H = distance (m) above the top of the fuel package (for a pool of flammable liquid, such as gasoline or heptane, H is the distance above the fuel surface minus one pool diameter).

Smoke production is, therefore, dependent upon the perimeter of the fire and the effective height of the gas column above it. As the fire continues to burn, the rate of smoke production will vary as the distance y changes. Figure 7-10C illustrates this variability.

The mass rate of smoke production, M, can be converted into a volumetric rate of production by dividing the expression above by the density of the air (as smoke) at T°C. This factor becomes 1.22 (290/T+273) kg/m^3. Table

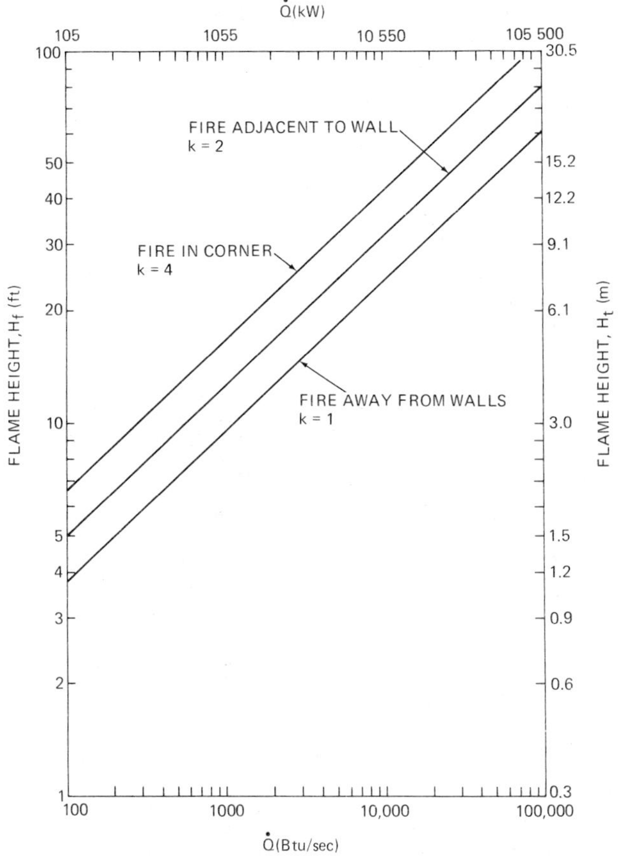

FIG. 7-10B. Flame height versus fire heat release rate.

7-10A illustrates the relationship between mass rates and volumetric rates of flow.

Smoke Filling of a Space

Numerous mathematical models involving complex interrelationships to describe the filling of a space with smoke, the transfer of heat energy from that smoke, and the flow of that smoke through large openings are in advanced stages of development and validation. There are, however, several estimating formulas that have been proposed by various researchers. Several of these are presented here.

The filling time of a space down to the level of the top of the burning item (Cooper 1982) has been estimated to be:

$$t_f = 200A/\dot{q}^{0.6}$$

where

t_f = time to fill the space to the burning surface (sec.)
A = floor area of space $(^2)$
$\dot{q}$ = burning rate (kW)

A modification for a crude estimate of filling time (t_e) to some point other than the fire surface (Nelson 1985) has been proposed as:

$$t_e = t_f\left(\frac{h - d}{h}\right)^{5/2}$$

where

h = distance from top of the fuel to ceiling
d = distance from top of the fuel to bottom of smoke layer

Where there are open windows or doors involved, these proposed formulas can be used only for broad estimates to the point where smoke descends to the height of the window or door soffit. After that, movement through the openings must also be considered in smoke flow. Where the openings are large so that there is an outflow of smoke through the upper portion of the opening and an inflow of makeup air through the lower portion, complex relationships, not easily expressed in Handbook type formulas, are involved.

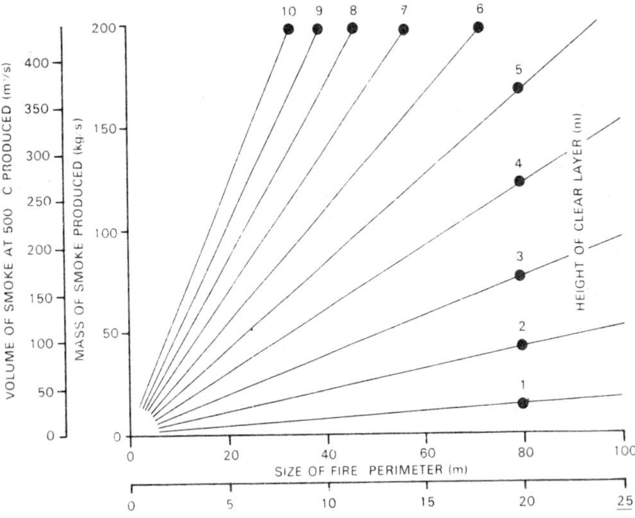

FIG. 7-10C. Rate of smoke production as a fire continues to burn.

Where the flow is through an opening of a relatively small size and is either entirely above the smoke layer, or where the portion of interface is small compared to the full size of the opening, e.g., the space between a door frame and the vertical side of a closed smoke barrier door, it is possible to use adaptations of Bernoulli's equation to estimate the smoke movement. The basic formulas have been assembled (Klote & Fothergill 1983) and adapted to the following statement (Nelson 1985):

$$f = 49A\left(\frac{1}{T_a} - \frac{1}{T_s}\right)^{1/2} h^{1/2}\left(\frac{T_s}{T_a}\right)^{1/2}$$

TABLE 7-10A. Conversion of Mass Rates of Flow of Smoke into Volume Rates of Flow.

Mass rate of flow		Volume rates of flow			
		m^3/sec		ft^3/min	
kg/sec	lb/sec	at 20°C	at 500°C	at 20°C	at 500°C
200	440.8	164.9	436.9	346928	925354
100	220.4	81.9	218.5	173464	462783
90	198.4	73.8	196.6	156308	416399
80	176.6	65.6	174.8	138941	370226
70	154.3	57.4	152.9	121573	323842
60	132.2	49.2	131.1	104205	277670
50	110.2	41.0	109.2	86838	231286
40	88.2	32.8	87.4	69470	185113
30	66.1	24.6	65.4	52103	138517
20	44.1	16.4	43.7	34735	92557
10	22.0	8.2	21.8	17346	46278
9	19.8	7.4	19.7	15631	41640
8	17.7	6.6	17.5	13894	37023
7	15.4	5.7	15.3	12157	32384
6	13.2	4.9	13.1	10420	27767
5	11.0	4.1	10.9	8684	23128
4	8.8	3.3	8.7	6947	18511
3	6.6	2.5	6.5	5210	13852
2	4.4	1.6	4.4	3473	9256
1	2.2	0.8	2.2	1735	4628

where:

f = flow (m^3/min)
A = area of opening (m^2)
T_a = ambient temperature (K)
T_s = smoke temperature (K)
d = depth of smoke from centerline of opening to bottom of smoke layer (m)

In conventional units the formula is:

$$f = 7214A \left(\frac{1}{530} - \frac{1}{T}\right)^{1/2} h^{1/2} \left(\frac{T}{530}\right)^{1/2}$$

$$\Delta P = 7.64 \left(\frac{1}{530} - \frac{1}{T}\right) h$$

where:

f = flow (ft^3/min)
A = area of opening (ft^2)
T_a = ambient temperature (°R)
T_s = smoke temperature (°R)
ΔP = pressure difference (in. of water)

Figure 7-10D shows typical results from this formula.

SMOKE MOVEMENT IN COLD SMOKE ZONES

As smoke is transmitted from the area of fire origin it is cooled by entrainment of air, transfer of the heat from the smoke body to building materials, primarily those in the walls and ceilings, and (to a lesser extent as the smoke

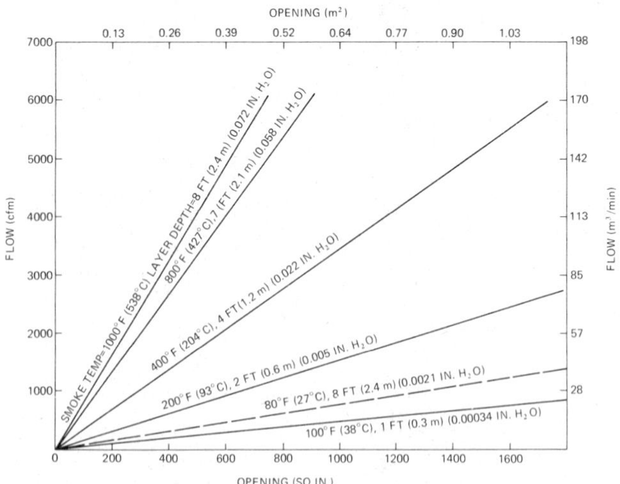

FIG. 7-10D. Volume of smoke flow through an opening.

cools) by radiant energy losses. When smoke from a fire area flows through a relatively small crack, the entrainment of cool air on the unexposed side tends to cool the smoke very quickly. Where the leakage is through larger openings there may be less entrainment relative to the mass of smoke movement at such junctures and therefore, cooling will be slower. However, once the smoke has cooled to a significant degree, the smoke is transported in the same manner as any other pollutant and the primary moving forces are those presented by stack effect, wind effect, and mechanical air movement systems.

SMOKE MOVEMENT IN TALL BUILDINGS

Smoke can behave very differently in tall buildings than in short buildings. In the shorter buildings, the influences of the fire (such as heat, convective movement, and fire pressures) are the major factors which cause smoke movement. Smoke removal and venting practices reflect this behavior. In tall buildings, these same factors are modified by the stack effect, which is the vertical natural air movement through the building caused by the differences in temperatures and densities between the inside and outside air. This stack effect can become an important factor in smoke movement and in building design features used to combat that movement.

The predominant factors that cause smoke movement in tall buildings are (1) the stack effect, (2) the influence of external wind forces, and (3) the forced air movement within the building. The following text describes the theoretical natural air movement which is affected by the first two factors. (Forced air movement caused by the building air handling equipment is presented elsewhere in this HANDBOOK). At this point, however, it should be noted that air movement is considerably influenced by the mechanical systems of the building. Many design solutions to the problem of tenability utilize emergency operation of the mechanical systems.

Stack Effect

The stack effect accounts for most of the natural air movement in buildings under normal conditions. During a fire the stack effect is often responsible for the wide distribution of smoke and toxic gases in high rise buildings.

The stack effect is characterized by a strong draft from the ground floor to the roof of tall buildings. The magnitude of this effect is a function of the building height, the air tightness of exterior walls, air leakage between floors of the building, and the temperature difference between the inside and the outside of the building.

To illustrate the principle of stack effect, consider the schematic of a box with a single opening near the bottom and another near the top, as shown in part *a* of Figure 7-10E. The theoretical natural draft between the two openings is caused by the difference in weight of the column of air within the box and that of a corresponding column of air of equal dimensions outside the box. The magnitude of the theoretical natural draft may be computed using the following formula:

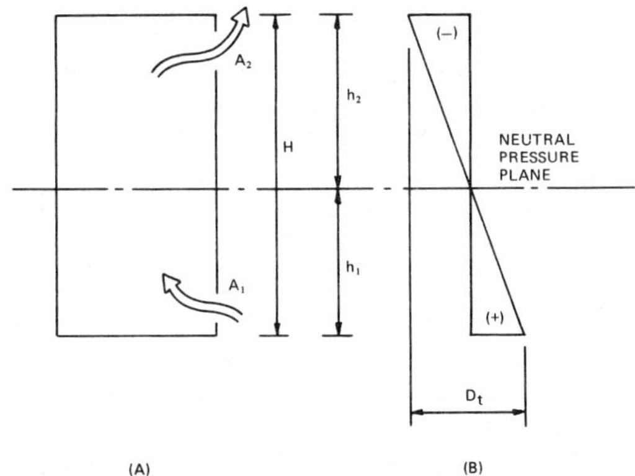

FIG. 7-10E. (a) Air movements caused by pressure, and (b) location of neutral pressure plane in a structure without horizontal barriers and with the two openings shown.

$$D_t = 2.96HB_o\rho\left(\frac{1}{T_o} - \frac{1}{T_i}\right)$$

where:

D_t = theoretical draft in inches of water
H = vertical distance between the inlet and the outlet in feet
B_o = barometric pressure in inches of mercury
T_o = temperature of outside air in degrees Rankine.
T_i = temperature of inside air in degrees Fahrenheit
ρ = density of air at 0°F and 1 atmosphere pressure in pounds per cubic foot

Assuming values of B_o = 29.9 in. and ρ = 0.0862 pcf, this expression reduces to

$$D_t = 7.63H\left(\frac{1}{T_o} - \frac{1}{T_i}\right)$$

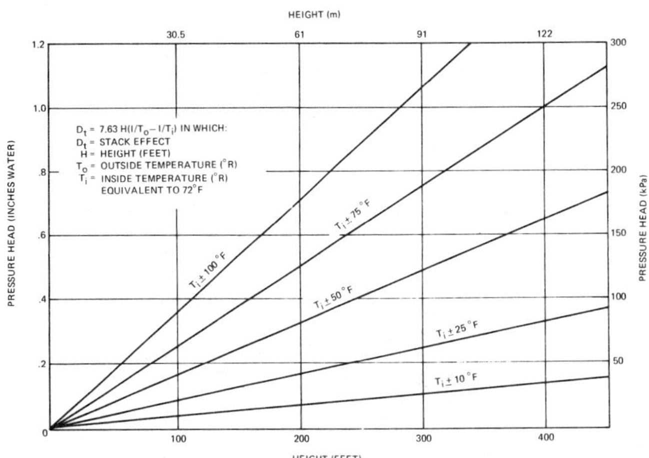

FIG. 7-10F. Stack effect due to height and temperature difference.

Vertical air movement in a building is caused by this natural draft or stack effect. It can be seen that the magnitude of the stack effect is dependent upon both the difference between inside and outside temperatures and the vertical distance between openings. If the inside and outside temperatures are equal, no natural air movement takes place. When $T_o \leq T_i$ the air moves vertically upward, with the lower opening acting as the inlet and the upper opening becoming the outlet. A reverse stack effect occurs when $T \geq T_i$. Under this condition, the upper opening is the inlet and the lower opening becomes the outlet.

Part *b* of Figure 7-10E illustrates the pressures that cause these movements. If it is assumed in this Figure that $T_o \leq T_i$, the exterior pressure will be greater than the interior pressure at the lower opening. This is a positive pressure which forces outside air into the building at that location. The outside pressure at the upper opening is less than the inside pressure; this creates a negative pressure which at that location, forces the inside air to the outside. The pressure distribution between these two locations is assumed to be linear.

If an opening were present in the exterior wall in a region of positive pressure, air would flow into the building. An opening in a region of negative pressure would cause air to flow out of the building. The neutral pressure plane indicates the location where inside and outside pressures are equal. If there were an opening at this level, air would move neither in nor outward. The location of the neutral pressure plane in a structure without horizontal barriers and with the two openings shown in Figure 7-10E can be determined from the following relationship:

$$\frac{h_1}{h_2} = \frac{A_2{}^2T_o}{A_1{}^2T_i}$$

where:

h_1 and h_2 represent the distances from neutral pressure plane to the lower and upper openings respectively

A_1 and A_2 represent the cross-sectional areas of the lower and upper openings respectively

T_i and T_o represent the absolute temperatures of the air inside and outside the building respectively

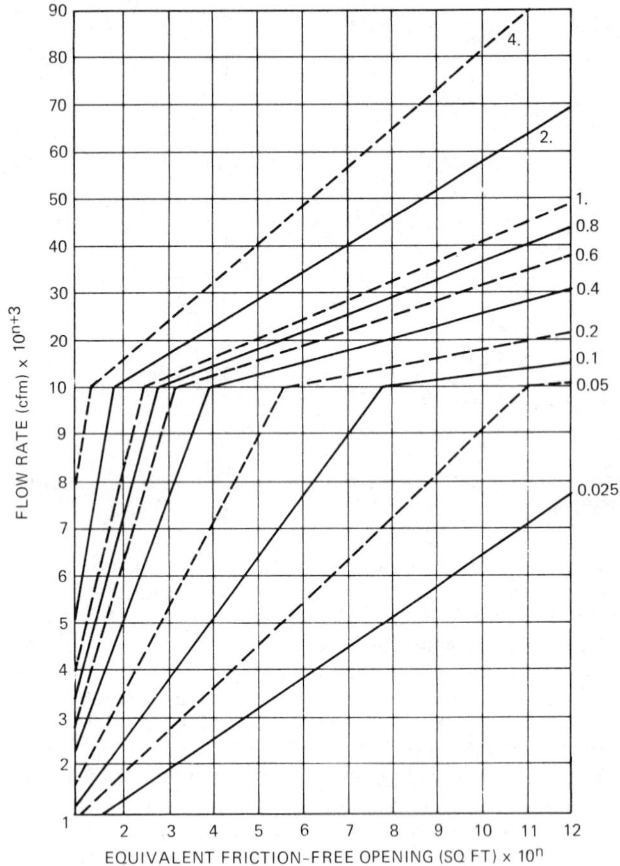

FIG. 7-10G. Airflow required to develop shown pressures in inches of water. 1 in. of water equals 249 Pa.

The magnitude of the pressures created by the stack effect described by the equation $D_t = 7.63H\left(\frac{1}{T_o} - \frac{1}{T_i}\right)$ is shown graphically in Figure 7-10F. These pressures can cause significant air flows, as shown in Figure 7-10G.

An analysis of Figure 7-10F illustrates the significant differences between tall and short buildings regarding air movement. For example, assume that a fire develops a pressure of 0.1 in. (25 Pa) of water in a compartment. Assume further that the outside temperature is 50°F (10°C) lower than the inside temperature, and that the fire occurs at the same level as the lower opening. The curve $T_i \pm 50$°F (10°C) indicates that if the upper outlet were approximately 60 ft (18 m) above the fire, the inlet stack pressure would balance the pressure caused by the fire. A building taller than 60 ft (18 m) would create a greater stack pressure, and theoretically the outside air would move into the building. Rather than venting smoke and heat, this would accelerate burning by driving them into the building.

Influence of Floors and Partitions

The theoretical draft described by Figure 7-10E and the final reduced equation (above) are modified in real buildings by the presence of floors and partitions. These barriers impede free air movement, although a significant flow can take place through openings in the assemblies.

The magnitude and location of the leakage areas in a building naturally varies with its function and type of construction. The National Research Council of Canada conducted studies of air tightness for major separations on four buildings ranging from 9 to 44 stories in height. The measurements were used for computer modeling of the air movement for a 20 story simulated building having a floor plan dimension of 120 by 120 ft (36 by 36 m), and a floor to floor height of 12 ft (3.6 m) (Tamura 1969). The data from the National Research Council of Canada has been established as a table of calculation (Klote and Fothergill 1983). The data are given in Table 7-10B.

TABLE 7-10B. Typical Leakage Areas for Walls and Floors of Commercial Buildings.

Construction Element	Wall Tightness	Area Ratio A/Aw*
Exterior Building Walls (includes construction cracks, cracks around windows and doors)	Tight†	0.70×10^{-4}
	Average†	0.21×10^{-3}
	Loose†	0.42×10^{-3}
	Very Loose‡	0.13×10^{-2}
Stairwell Walls (includes construction cracks but not cracks around windows or doors)	Tight§	0.14×10^{-4}
	Average§	0.11×10^{-3}
	Loose§	0.35×10^{-3}
Elevator Shaft Walls (includes construction cracks but not cracks around doors)	Tight§	0.18×10^{-3}
	Average§	0.84×10^{-3}
	Loose§	0.18×10^{-2}
		A/A_F
Floors (includes construction cracks and cracks around penetrations)	Average#	0.52×10^{-4}

* A = leakage area; A_w = wall area; and A_F = floor area.
† Tamura and Shaw 1976a.
‡ Tamura and Wilson 1966.
§ Tamura and Shaw 1976b.
\# Tamura and Shaw 1978.

These leakage areas are sufficient to allow a substantial air movement throughout the building. Most of the air will flow into the vertical shafts such as stairwells and elevator shafts. Some air will flow vertically from floor to floor through the minor openings in the floor-ceiling assembly. This floor to floor movement is always caused by a pressure differential between the adjacent floors.

Part *a* of Figure 7-10H illustrates the pressure difference characteristics of a building in which stack action causes air movement. The slopes of the pressure lines represent pressure differences between any two regions at the same height. Airflow from one region to another will always be in the direction of the region whose pressure curve is more to the left. This is illustrated by the airflow

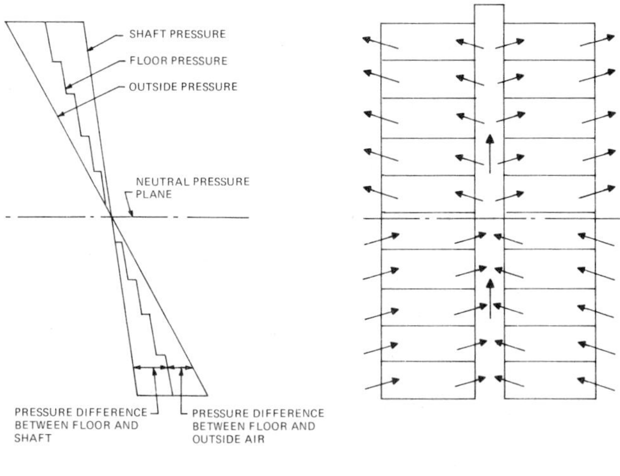

FIG. 7-10H. The pressure difference characteristics of a building in which stack action causes air movement.

directions represented by the arrows in Part b of Figure 7-10H.

Wind Effects

Wind action is another important feature in the behavior of smoke movement. Again, tall and short buildings behave somewhat differently in this regard. Figure 7-10I

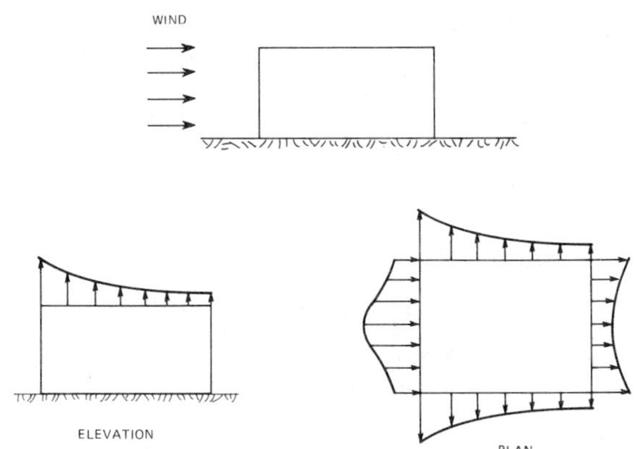

FIG. 7-10I. The air pressure distribution along the four sides and the roof of a building.

illustrates the air pressure distribution along the four sides and the roof of a building. The plan view of the pressures shows that the windward wall is subjected to an inward pressure, while the leeward wall and the two side walls have an outward pressure (suction). The flat roof has an upward pressure, with the maximum amount occurring at the windward edge.

These pressures are caused by the movement of a mass of air around and over the structure. A short, wide building will cause the major volume of air to move over the roof, with correspondingly less air movement around the sides. A narrow, tall building, on the other hand, will cause the major volume of air to follow its path of least resistance around the building, with less movement over the top. The velocities of these movements are the primary cause of the amount and directions of the pressures on the building.

The effect of wind pressures and suctions modify the natural air movement within a building. For example, the negative pressure on the roof of a tall building can have an aspirating effect on a vertical shaft opened at the roof level. This can cause the observed draft to exceed the theoretical draft shown in Figure 7-10F.

Horizontal pressures and suctions cause the neutral pressure planes in exterior walls to move. Positive wind pressure would tend to raise the neutral pressure plane, while negative pressures will lower it. Figure 7-10J illustrates the influence of wind action on air movement in a building.

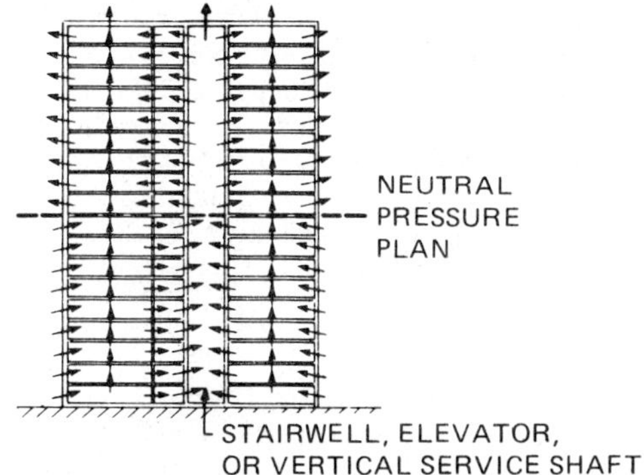

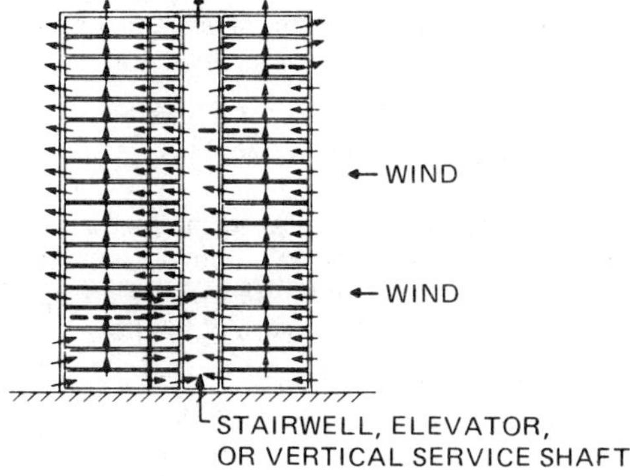

FIG. 7-10J. Influence of wind action on air movement in a building. Note in the presence of significant wind how the neutral pressure plane changes location throughout the building.

DESIGN OF SMOKE CONTROL SYSTEMS FOR BUILDINGS

Working in conjunction with the American Society of Heating Refrigerating and Air Conditioning Engineers

(ASHRAE) and the Center for Fire Research, Klote and Fothergill have published a manual, *Design of Smoke Control Systems for Buildings*. This manual provides guidance in the details of designing smoke control systems. Chapter 2, entitled "Fundamental Smoke Control," should be reviewed by those interested in greater details on cold smoke movement.

The following paragraphs, covering building air flow analysis, are based on the text of the smoke control manual. The information is felt to be of particular worth to anyone making an analysis of smoke movement between space.

Building Airflow Analysis

The movement of cold smoke is determined by the total air flow in the building involved. Measures to control cold smoke movement must therefore include air flow analysis.

The methods of hand calculation presented here are based on the principle of conservation of mass, the hydrostatic equation, and the flow equation (for the latter see "stack effect" earlier in this chapter). There are situations, however, where the airflow path is complicated or where there are several driving forces, so hand calculation is not practical. These cases can be analyzed more readily with the aid of a digital computer.

Effective Flow Areas

The concept of effective flow areas is quite useful for analysis of smoke control systems. The flow paths in the system can be in parallel with one another, in series, or a combination of parallel and series paths. The effective area of a system of flow areas is the area that results in the same flow as the system when it is subjected to the same pressure difference over the total system of flow paths. This is analogous to the flow of electric current through a system of electrical resistances.

Parallel Paths

Three parallel leakage areas from a pressurized space are illustrated in Figure 7-10K. The pressure difference, ΔP, is the same across each of the leakage areas. The total flow, Q_T, from the space is the sum of the flows through the leakage paths:

$$Q_T = Q_1 + Q_2 + Q_3$$

The effective area, A_e, for this situation is that which results in the total flow, Q_T. Therefore, the total flow can be expressed as:

$$Q_T = CA_e \sqrt{\frac{2\Delta P}{\rho}}$$

where:

ΔP = pressure difference
C = flow coefficient (generally in the range of 0.6 to 0.7
ρ = density

The flow through area A_1 can be expressed as:

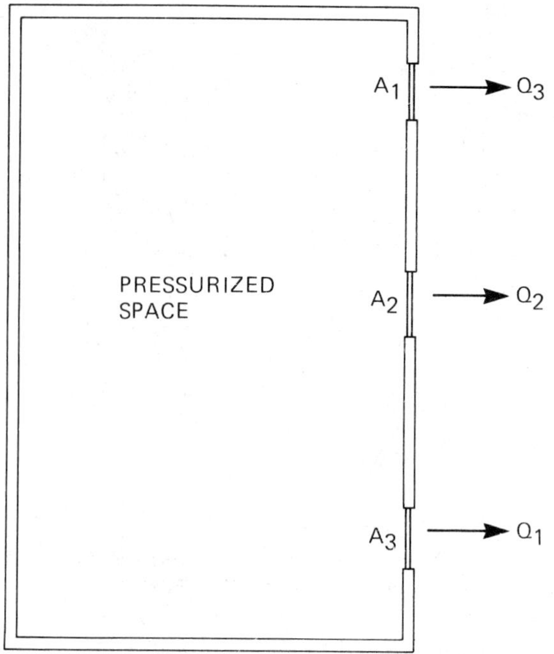

FIG. 7-10K. Leakage paths in parallel.

$$Q_1 = CA_1 \sqrt{\frac{2\Delta P}{\rho}}$$

The flows for Q_2 and Q_3 can be expressed in a similar manner. Substituting the expressions for Q_1, Q_2, and Q_3 into the total flow equation and collecting like terms yields:

$$Q_T = C(A_1 + A_2 + A_3) \sqrt{\frac{2\Delta P}{\rho}}$$

This yields:

$$A_e = A_1 + A_2 + A_3$$

In Figure 7-10K, if A_1 is 1.08 sq ft (0.10 m^2) and A_2 and A_3 are 0.54 sq ft (0.05 m^2) each, then the effective flow area, A_e, is 2.16 sq ft (0.20 m^2).

The above logic can be extended to any number of flow paths in parallel; i.e., it can be stated that the effective area is the sum of the individual leakage paths and can be expressed as:

$$A_e = \sum_{i=1}^{n} A_i$$

where n is the number of flow areas, A_i, in parallel.

Series Paths

Three leakage areas in series from a pressurized space are illustrated in Figure 7-10L. The flow rate, Q, is the same through each of the leakage areas. The total pressure difference, ΔP_T, from the pressurized space to the outside

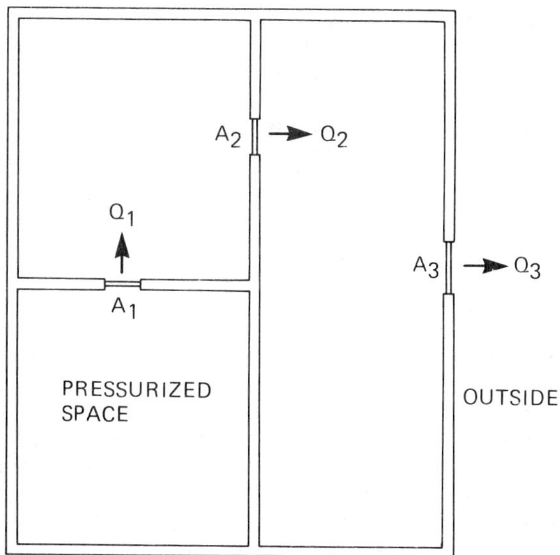

FIG. 7-10L. Leakage paths in series.

is the sum of pressure differences ΔP_1, ΔP_2, and ΔP_3 across each of the respective flow areas, A_1, A_2, and A_3:

$$\Delta P_T = \Delta P_1 + \Delta P_2 + \Delta P_3$$

The effective area for flow paths in series is the flow area that results in the flow, Q, for a total pressure difference of ΔP_T. Therefore, the flow, Q, can be expressed as:

$$Q = CA_e \sqrt{\frac{2\Delta P_T}{\rho}}$$

Solving for ΔP_T yields:

$$\Delta P_T = \frac{\rho}{2}\left(\frac{Q}{CA_e}\right)^2$$

The pressure difference across A_1 can be expressed as:

$$\Delta P_1 = \frac{\rho}{2}\left(\frac{Q}{CA_1}\right)^2$$

The pressure differences, ΔP_T and ΔP_3, can be similarly expressed. Substituting the equation P_T and the expressions for ΔP_1, ΔP_2, and ΔP_3 and cancelling like terms yields the following:

$$\frac{1}{A_e^2} = \frac{1}{A_1^2} + \frac{1}{A_2^2} + \frac{1}{A_3^2}$$

That is:

$$A_e = \left(\frac{1}{A_1^2} + \frac{1}{A_2^2} + \frac{1}{A_3^2}\right)^{-1/2}$$

This same reasoning can be extended to any number of leakage areas in series to yield:

$$A_e = \left[\sum_{i=1}^{n} \frac{1}{A_i^2}\right]^{-1/2}$$

where n is the number of leakage areas, A_i in series. In smoke control analysis, there are frequently only two paths in series. For this case, the effective leakage area is:

$$A_e = \frac{A_1 A_2}{\sqrt{A_1^2 + A_2^2}}$$

Example A:

Calculate the effective leakage area of two equal flow paths of 0.22 sq ft^2 in series.

Let $A = A_1 = A_2 = 0.22$ sq ft (0.02 m^2)

$$A_e = \frac{A^2}{\sqrt{2A^2}} = \frac{A}{\sqrt{2}} = 0.15 \text{ sq ft} \quad (0.014 \text{ m}^2)$$

Example B:

Calculate the effective area of two flow paths in series, where $A_1 = 0.22$ sq ft (0.02 m^2) and $A_2 = 2.2$ sq ft (0.2 m^2).

$$A_e = \frac{A_1 A_2}{\sqrt{A_1 + A_2}} = 0.214 \text{ sq ft} \quad (0.0199 \text{ m}^2)$$

This example illustrates that when two areas are in series and one is much larger than the other, the effective area is approximately equal to the smaller area.

Combination of Paths in Parallel and Series

The method of developing an effective area for a system of both parallel and series paths is to systematically combine groups of parallel paths and series paths. The system illustrated in Figure 7-10M is analyzed as an example.

Figure 7-10M shows that A_2 and A_3 are in parallel; therefore, their effective area is:

$$A_{23_e} = A_2 + A_3$$

Areas A_4, A_5, and A_6 are also in parallel, so their effective area is:

$$A_{456_e} = A_4 + A_5 + A_6$$

These two effective areas are in series with A_1. Therefore, the effective flow area of ther system is given by:

$$A_e = \left[\frac{1}{A_1^2} + \frac{1}{A_{23_e}^2} + \frac{1}{A_{456_e}^2}\right]^{-1/2}$$

Example:

Calculate the effective area of the system in Figure 7-10M if the leakage areas are $A_1 = A_2 = A_3 = 0.22$ sq ft (0.02 m^2) and $A_4 = A_5 = A_6 = 0.11$ sq ft (0.01 m^2).

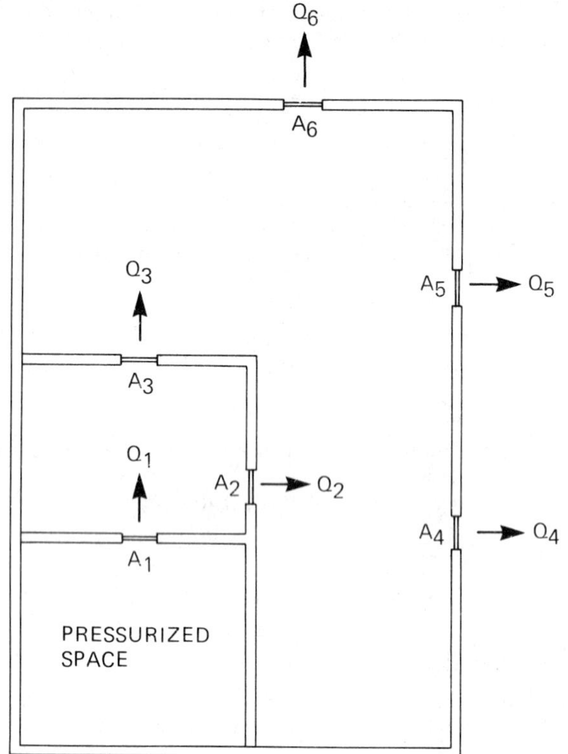

FIG. 7-10M. Combination of leakage paths in parallel and series.

$$A_{23_e} = 0.44 \text{ ft}^2 \ (0.04 \text{ m}^2)$$
$$A_{456_e} = 0.33 \text{ ft}^2 \ (0.03 \text{ m}^2)$$
$$A_e = 0.17 \text{ ft}^2 \ (0.015 \text{ m}^2)$$

Testing Smoke Control Systems

Systems to Control Smoke in Hot Smoke Zone: To be effective, these systems must use the buoyant plume of the fire. For such systems to work, the buoyant plume must be of sufficient force to stratify the smoke. Because of the danger of a fire that produces a significant plume, it is normally impossible to execute a realistic performance test of such a system. The volumes of gas produced, the entrainment of air, and hydraulic lift of the plume can not reasonably be imitated.

The concept of air changes common in HVAC (heating, ventilation and air conditioning) design is a misnomer. The fact that a certain cubic volume of air is entered into a room and that same cubic volume is removed, in no way means that any particular smoke particle has been changed. In the traditional manner of designing both the intake and exhaust at the ceiling levels, it must be expected that a large portion of the molecules of air in the space only casually becomes entrained in the airflow. The majority of the air removed is the same clean air that entered near the ceiling. Where smoke is imposed as an additional factor in the space, it is inappropriate to expect that such air transfer will be meaningful. Buoyant smoke control systems function efficiently only when the entire removal outlet is at the top, above the smoke layer, and the entire intake is below the smoke layer. Any introduction of air above the smoke layer only slows down the process and to some degree thins the smoke as it goes out. Any removal below the smoke layer only removes the clear air. For such systems, the most realistic testing procedure consists of an analysis of the design to determine if it can achieve the desired effect from a rational engineering analysis. Field testing of detectors, fans, dampers, and other hardware is essential to assure that each item individually operates as specified by the design.

Systems to Control Smoke in the Cold Smoke Zone: In these systems, the actual energy forces induced by the fire are normally insignificant. The movement of the smoke is dominated by the movement forces from the HVAC, wind, and stack actions. Essentially, this type of system would have exactly the same effect on controlling either smoke or an unexpected release of a toxic gas. Typical examples are stairwell pressurization systems and zone pressurization systems. In these systems, the primary function of the smoke control devices are to stop or drastically reduce the leakage from the contaminated to the clear area. This is accomplished by: (1) plugging holes which have air movement flowing in the direction of the contaminated area, (2) the dilution of any smoke that may pass into the clear area (due to either steady or episodic inabilities of the smoke control systems to completely stop the passage through openings), (3) the purging of spaces in clear areas where either a brief high volume or a continuing low volume transfer of smoke takes place into a limited volume space, e.g., a stairwell, and (4) assuring a constant movement of air from the clear area towards the contaminated area to prevent a pressure stagnation situation that can defeat the smoke control system. These types of systems can be performance tested using either pressure measuring devices or tracer chemicals. The tracer chemicals can be visible as with smoke bombs, or invisible as with sulfur hexafluoride. When conducting such tests, it is suggested that the leakage condition and the extent of dilution (if dilution is an intended design factor) be checked when:

1. All openings are at the minimum expected level such as when all doors on a protected stairwell are closed. This will demonstrate the maximum build up of pressure.
2. All openings between the protected space and the rest of the building are closed except for the maximum opening between clear areas and the smallest potential fire zone. This will determine if the smoke control system fans can overpower a small area of involvement and bring it to the same pressure as the protected area. Such would create a stagnation condition where smoke can travel into the protected area (stairwell, etc.) or result in pressurizing the fire area pushing smoke in other directions, such as into rooms off a corridor.
3. The number or size of openings from the pressurized area to other areas (clean or contaminated) is progressively increased. This might require a series of tests. Eventually any smoke control system can be defeated by opening enough relief and thereby diverting the flow of air from the pressurized area to some other area not actually requiring protection. It is important for anyone planning to use smoke control as a safeguard to know the system limitations.

Bibliography

References Cited

Alpert, R. L., and Ward, E. J. 1983. "Evaluating Unsprinklered Fire Hazards." *SFPE Technology Report 83-2*. Society of Fire Protection Engineers, Boston, MA.

Butcher, E. G., and Parnell, A. C. 1979. *Smoke Control in Fire Safety Design*. E. and F. N. Spon., London, England.

Cooper, L. Y. 1982. "The Development of Hazardous Conditions in Enclosures with Growing Fires." *NBSIR 82-2622*. National Bureau of Standards, Washington, DC.

Heskestad, G. 1982. "Engineering Relations for Fire Plumes." *SFPE Technology Report 82-8. Society of Fire Protection Engineers, Boston, MA.*

Klote, J. H., and Fothergill, J. W. 1983. "Design of Smoke Control Systems for Buildings." *National Bureau of Standards Handbook 141*. National Bureau of Standards, Washington, DC.

Nelson, H. E. 1985. "Emerging Engineering Methods Applied to Fire Safety Design." *General Proceedings*. Research and Design 85: Architectural Applications of Design and Technology Research. American Institute of Architects Foundation, Washington, DC.

Thomas, P. H., et al. 1963. "Investigations Into the Flow of Hot Gases in Roof Venting." *Fire Research Technical Paper No. 7*. Joint Fire Research Organization, London, England.

Tamura, G. T., and Shaw, C. Y. 1976a. "Studies on Exterior Wall Air Tightness and Air Infiltration of Tall Buildings." *Transactions*. American Society of Heating, Refrigeration, and Air Conditioning Engineers. Vol 82, Part I. pp 122-134.

Tamura, G. T., and Shaw, C. Y. 1976b. "Air Leakage Data for the Design of Elevator and Stair Shaft Pressurization Systems." *Transactions*. American Society of Heating, Refrigeration, and Air Conditioning Engineers. Vol 83, Part II. pp 179-190.

Tamura, G. T., and Shaw, C. Y. 1978. "Experimental Studies of Mechanical Venting for Smoke Control in Tall Office Buildings." *Transactions*. American Society of Heating, Refrigeration, and Air Conditioning Engineers. Vol 86, Part I. pp 54-71.

Tamura, G. T., and Wilson, A. G. 1966. "Pressure Differences for a Nine Story Building as a Result of Chimney Effect and Ventilation System Operations." *Transactions*. American Society of Heating, Refrigeration and Air Conditioning Engineers. Vol 72, Part 1. pp 180-189.

Tamura, G. T., and Wilson, A. G. 1967. "Pressure Differences Caused by Chimney Effect in Three Story High Buildings." *Transactions*. American Society of Heating, Refrigeration and Air Conditioning Engineers. Vol 73, Part II.

Tamura, G. T. 1969. *Computer Analysis of Smoke Movement in Tall Buildings*. Annual Meeting, June 1969. American Society of Heating, Refrigeration and Air Conditioning Engineers.

NFPA Codes, Standards, Recommended Practices and Manuals. (See the latest NFPA *Codes and Standards Catalog* for availability of current editions of the following documents.)

NFPA 204M, *Guide for Smoke and Heat Venting*.

NFPA 258, *Standard Test Method for Measuring the Smoke Generated by Solid Materials*.

Additional Readings

Christian, W. J., and Waterman, T. E., "Characteristics of Full-Scale Fires in Various Occupancies," *Fire Technology*, Vol. 7, No. 3, Aug. 1971, pp. 205-218.

Cooper, Leonard Y., "The Need and Availability of Test Methods for Measuring the Smoke Leakage Characteristics of Door Assemblies," *NBSIR 84-2876*. National Bureau of Standards, Washington, DC., 1984.

Cresci, R. J., "Smoke and Fire Control in High-Rise Office Buildings —Part II: Analysis of Stair Pressurization Systems," *Symposium on Experience and Applications on Smoke and Fire Control*, ASHRAE Annual Meeting, June 1973, Louisville, KY; Atlanta, GA, 1973, pp. 16-23.

Dias, C., "Stairwell Pressurization in a High-Rise Commercial Building," *ASHRAE Journal*, Vol. 20, No. 7, July 1978, pp. 24-26.

Evers, E., and Waterhouse, A., "A Computer Model for Analyzing Smoke Movement in Buildings," *BRECP 69/78*. Building Research Establishment, Boreham Wood, Hertsfordshire, England, Nov. 1978.

Gross, D., "A Review of Measurements, Calculations and Specifications of Air Leakage through Interior Door Assemblies," *NBSIR 81-2214*, National Bureau of Standards, Washington, DC, Feb. 1981.

Heselden, A. J. M., "Studies of Smoke Movement and Control at the Fire Research Station," *CIB Symposium on Control of Smoke Movement in Building Fires*, Fire Research Station, Garston Watford, UK, Vol. 1, 1976, pp. 185-198.

Heselden, A. J. M., "Studies of Fire and Smoke Behaviour Relevant to Tunnels," *BRECP*, Building Research Establishment, Boreham Wood, Hertsfordshire, England, Nov. 1978.

Heselden, A. J. M., and Baldwin, R., "The Movement and Control of Smoke and Escape Routes in Buildings," *BRECP*, Building Research Establishment, Boreham Wood, Hertsfordshire, England.

Heselden, A. J. M., and Baldwin, R., "The Movement and Control of Smoke on Escape Routes in Buildings," *Fire Technology*, Vol. 14, No. 3, Aug. 1978, pp. 206-222.

Hinkley, P. L., "A Preliminary Note on the Movement of Smoke in an Enclosed Shopping Mall," *Fire Research Note 806*, Fire Research Station, Boreham Wood, Hertfordshire, England, Mar. 1970.

Jones, Walter W., "A Model for the Transport of Fire, Smoke and Toxic Gases (FAST)," *NBSIR 84-2934*, National Bureau of Standards, Gaithersburg, MD, 1984.

Jones, W. W., and Quintiere, J. G., "Prediction of Corridor Smoke Filling by Zone Models," *Combustion Science and Technology*, Vol. 35, Nos. 5 and 6, 1984, pp. 239-253.

Klote, J. H., "Smoke Control for Elevators," *ASHRAE Journal*, Vol. 26, No. 4, Apr. 1984, pp. 23-33.

Klote, J. H., "The ASHRAE Design Manual for Smoke Control," *Fire Safety Journal*, Vol. 7, No. 1, 1984, pp. 93-98.

Klote, J. H., "Smoke Movement through a Suspended Ceiling System," *NBSIR 81-2444*, National Bureau of Standards, Washington, DC, 1982.

Klote, John H., "Field Tests of the Smoke Control System at the San Diego V.A. Hospital," *NBSIR 84-2912*, National Bureau of Standards, Gaithersburg, MD, 1984.

Klote, J. H., "Elevators as a Means of Fire Escape," *NBSIR 82-2507*, National Bureau of Standards, Washington, DC, May 1982.

Klote, J. H., and Fothergill, J. W., "Design of Smoke Control Systems for Buildings," *NBS Handbook 141*, National Bureau of Standards, Gaithersburg, MD, 1983.

Lieberman, Paul, and Bell, Diane, "Smoke and Fire Propagation in Compartment Spaces," *Fire Technology*, Vol. 9, No. 2, May 1973, pp. 91-100.

Marchant, E. W., "Effect of Wind on Smoke Movement and Smoke Control Systems," *Fire Safety Journal*, Vol. 7, No. 1, 1984, pp. 55-63.

McGuire, J. H., "Control of Smoke in Buildings," *Fire Technology*, Vol. 3, No. 4, Nov. 1967, pp. 281-290.

McGuire, J. H., and Tamura, G. T., "Simple Analysis of Smoke Flow Problems in High Buildings," *Fire Technology*, Vol. 11, No. 1, Feb. 1975, pp. 15-27.

Morgan, H. P., and Marshall, N. R., "Smoke Hazards in Covered, Multi-Level Shopping Malls: An Experimentally-Based Theory for Smoke Production," Building Research Establishment, Boreham Wood, Hertfordshire, England, May 1975.

Morgan, H. P., and Marshall, N. R., "Smoke Hazards in Covered, Multi-Level Shopping Malls: A Method of Extracting Smoke from Each Level Separately," Building Research Establishment, Boreham Wood, Hertfordshire, England, Jan. 1978.

Morgan, H. P., et al., "Smoke Hazards in Covered Multi-Level Shopping Malls: Smoke Studies Using a Model 2-Story Mall," Building Research Establishment, Boreham Wood, Hertfordshire, England, June 1976.

Phillips, Alan M., "Canada's Fire Research Facility," *Fire Journal*, Vol. 78, No. 2, Mar. 1984, pp. 46-48.

Thomas, P. H., "Movement of Smoke in Horizontal Corridors Against an Air Flow," *Institute of Fire Engineers Quarterly*, Vol. 30, No. 77, 1970, pp. 45-53.

Thornberry, Richard P., "Designing Stair Pressurization Systems," *SFPE TR 82-4*, Society of Fire Protection Engineers, Boston, MA, 1982.

Wakamatsu, T., "Calculation Methods for Predicting Smoke Movement in Buildings and Designing Smoke Control Systems, Fire Standards and Safety," *ASTM STP-614*, A. F. Robertson, ed., American Society for Testing and Materials, Philadelphia, PA, 1977, pp. 168-193.

VENTING PRACTICES

Dr. Gunnar Heskestad

The importance of the buoyant properties of heat and smoke was realized at an early date, as evidenced by an NFPA standard adopted in 1903 that called for smoke vents over theater stages and in the ceilings of theater auditoriums. However, until the advent of effective artificial lighting, buildings were generally small enough so that windows provided adequate smoke venting, other than venting of fire through the roof.

The main body of this chapter covers venting practices as they would be applied to nonsprinklered buildings following the recommendations of NFPA 204M, *Smoke and Heat Venting Guide*. An Addendum at the conclusion of the chapter (in which the several different approaches are incorporated) discusses the current state of the technology involving the sprinkler/vent issue.

HISTORICAL BACKGROUND

Large undivided floor areas present extremely difficult fire fighting problems since fire fighters must enter these areas to fight fires in central portions of the building. If fire fighters are unable to enter because of the accumulation of heat and smoke, fire fighting efforts may be reduced to ineffective application of hose streams to perimeter areas while fire consumes the interior. (See Fig. 7-11A.)

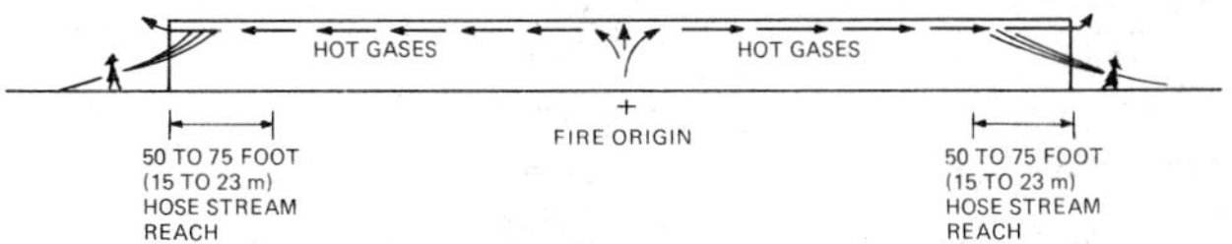

FIG. 7-11A. Behavior of hot gases under a flat-roofed building.

Great impetus to the subject of smoke and heat venting was provided by the General Motors fire at Livonia, Michigan, in 1953 when fire spread horizontally under an unvented metal roof, with 34 acres of undivided area. Fire protection engineers were in general agreement that this fire could have been greatly reduced if there had been effective roof venting. The General Motors fire led to a new approach to the subject by the NFPA Committee on Building Construction, which prepared NFPA 204M, *Guide to Smoke and Heat Venting*, adopted by NFPA in May 1961. In 1968, the venting guide was expanded to include a new section on inspection and maintenance.

A reconfirmation action failed in 1975, as concerns had surfaced over use of NFPA 204M in conjunction with automaticaly sprinklered buildings. Because of this controversy, work on a revision of NFPA 204M continued at a slow pace, but concluded with the 1982 edition.

The 1982 edition of NFPA 204M distinguishes between unsprinklered and sprinklered buildings. The unsprinklered part is considered to represent a major advance in engineered smoke and heat venting. The sprinklered part, limited to a single chapter (Chapter 6), does not recognize that venting is necessarily desirable in a sprinklered building and does not offer a design basis, pending the resolution of some technical questions.

The technical questions which remain unresolved in application of venting to sprinklered buildings are related to (1) the effects of sprinkler discharge on venting effectiveness, and (2) the effects of fresh air introduced into the building on the burning process and the water demand of the sprinkler system. Sprinkler discharge will cool the fire gases, perhaps to the extent that the vent discharge is reduced, hence reducing venting effectiveness. Additionally, the sprinkler sprays will entrain ambient smoke and air, transporting smoke to the floor level and possibly drawing gas from the vent exhaust wherever nozzles are located close to vents, further reducing venting effectiveness. Unless the building is very large, fresh air flowing into the building and replacing smoke escaping through the vents will increase the oxygen concentration in the fire space. The increased oxygen concentrations may cause a more vigorous fire than would otherwise exist, with the possible outcome of increased number of operating sprinklers, in some cases overtaxing the water supply. Against these possible adverse effects must be balanced the likelihood that some improvement in visibility for fire fighting will result in many cases, and the possibility that the number of operating sprinklers may sometimes be reduced

Dr. Heskestad is principal research scientist in the Applied Research Department, Factory Mutual Research Corporation, Norwood, MA.

because of the cooling effect of vent flows. There are as yet no universally accepted conclusions from either fire experience or research.

Essential features specified in NFPA 204M include roof vents operated by fusible links, with curtain boards to confine heat and prevent lateral fire spread, as well as openings for fresh air make-up at low levels. (See Fig. 7-11B.) Two general classes of fires are considered: limited growth fires, which are not expected to grow past a

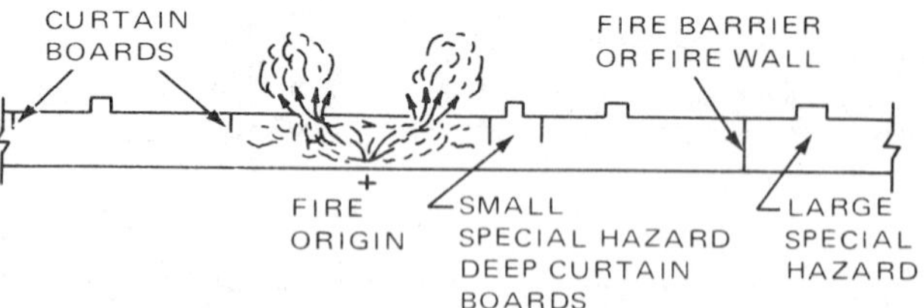

FIG. 7-11B. Behavior of combustion products under vented and curtained roof.

predictable maximum size, and continuous growth fires, which can be expected to grow indefinitely until intervention by fire fighters. Mechanical exhaust at ceiling apertures is treated as an option to natural ventilation at roof vents. The provisions of NFPA 204M may be applied to one story buildings or to the top story of multiple story buildings. With mechanical exhaust, lower stories of multiple story buildings can also be vented.

INDUSTRIAL BUILDING VENTILATION

Although any opening in a roof will reduce some heat and smoke, building designers and fire protection engineers cannot rely on casual inclusion of skylights, windows, or monitors as adequate means of venting. Standards now exist (from Factory Mutual Engineering Corp., Underwriters Laboratories Inc., etc.) that include design criteria and test procedures for unit vents and also call for simulated fire tests as well as engineering analysis.

Unit Vents

Unit vents are relatively small, usually 16 to 100 sq ft (1.49 to 9.29 m^2) in area. Automatic unit vents are of two types, based on manner of operation—fusible link or drop out plastic. (See Fig. 7-11C.) The former type consists of a metal housing with lids which depend upon a temperature rated fusible link to trip the lid mechanism, but alternative modes of operation (by heat, smoke, etc.) are possible. The latter type depends upon a temperature sensitive, transparent or translucent, thermoplastic dome which deforms from its setting and falls out of the roof. Vents designed for manual operation are constructed with metal lids to resist elevated fire temperatures and may be opened from the floor with wires and cables. These units can also be modified for automatic operation.

FIG. 7-11C. Building with roof vents.

Vents may be single unit in which the entire unit opens fully with a single sensor, or multiple units in rows, clusters, groups, or other arrays that will satisfy the venting requirements for the specific hazard.

Mechanical roof ventilators actuated by fire conditions are possible alternatives to roof vents based on natural ventilation, especially in lower stories of multiple story buildings. These ventilators must be capable of functioning under expected high temperature fire conditions.

Curtain Boards

In large area buildings, unless vented areas are subdivided by means of walls or partitions, curtain boards are important for prompt and positive actuation of vents because they bank up heat in the curtained area. They also limit the spread of heat and smoke beneath the ceiling during the design duration of the venting system. Curtain boards can be made from any substantial noncombustible material that will resist the passage of smoke.

The depth of curtain boards would normally be selected to correspond with the design depth of the smoke layer. However, the depth should not be less than 20 percent of the ceiling height to prevent spillage of smoke under the curtain where depth and ceiling height are referenced to the center of the lowest roof vent. It would rarely be desirable to have the curtain extend below 10 ft (3 m) from the floor. Around special hazards, the curtain should preferably extend down to this limit. (See Fig. 7-11D.)

The distance between curtain boards should not exceed eight times the ceiling height to ensure that vents

FIG. 7-11D. Deep curtain board around a special hazard. Equipped with proper venting, a noncombustible curtain board extending down from a ceiling around special hazards will prevent smoke and heat from mushrooming throughout the plant. A curtain board for a heat treating department is shown.

remote from the fire will be effective within the curtained compartment. Smaller curtained areas may be desirable where occupancies are particularly vulnerable to damage. However, it is important that the distance between curtain boards not be less than two times the ceiling height, unless the curtain boards extend down to a depth of at least 40 percent of the ceiling height. The increased curtain depth is needed for these small curtain spacings to prevent significant spillage of fire gases underneath the curtain because of the close proximity of the fire.

Dimensioning and Spacing of Vents

When the area of an individual vent becomes sufficiently large, there is a possibility that a core of clean air from beneath the smoke layer will be included in the exhaust from the vent, which reduces the effectiveness of the vent. To prevent inclusion of clear air, the area of a unit vent or cluster of vents should not exceed $2d^2$, where d is the design depth of the smoke layer or curtain board. In the case of a row of unit vents or a monitor vent, the width of the row or monitor should not exceed d.

A large number of small vents on close spacing is preferable to a small number of large vents on wide spacing. This ensures early operation of the first vents in a fire, reducing the likelihood of initial smoke excursions beyond the curtained area above the fire. In no case should the vent spacing in a rectangular matrix exceed $2H$ where H is the ceiling height measured from the floor to ceiling for flat roofs, and from the floor to the center of the vent for sloped roofs. Alternatively, in a nonrectangular vent matrix or plan, the distance between any point on the floor and the nearest vent should not exceed $2.8H$ (the diagonal of a square whose side is $2H$).

The total vent area for each curtained compartment under the ceiling depends upon the severity of the expected fire and is discussed later in this chapter.

Fresh Air Makeup

Openings must be provided at or near floor level to allow the introduction of makeup air. This is critical in today's tightly constructed and insulated buildings. The total area of these openings must normally be at least as great as the installed vent area for each curtained compartment; otherwise, the inlets will effectively throttle the vent flow. If doors and windows below the designed smoke layer cannot meet the total required inlet area, special air inlet provisions are necessary.

It is essential that a dependable means be provided for admitting inlet air within approximately one minute after the first vent opens. If prompt air inlet is not provided, the entire building may quickly fill with smoke, which will clear only slowly at lower levels (in the design clear layer) in response to the fully developed vent flow after the air inlets have been actuated.

Fire Characterization*

Each curtained compartment, or the ceiling area of buildings requiring no curtain boards, must be furnished with a total installed vent area (or exhaust capacity in case of mechanical ventilation) sufficient to vent fires of the expected severity. In addition to the expected fire severity, the installed vent area (or exhaust capacity) will depend upon the depth of the curtain boards or design depth of the smoke layer. Furthermore, unless the occupancy or hazard is such that the expected fire will peak or level off at a predictable maximum size, the installed vent area (or exhaust capacity) will also depend upon the minimum clear visibility design time as measured from the time the vents first operate.

The fire severity is expressed differently according to the class of the fire, which for purposes of calculating vent areas, are classified as either limited growth or continuous growth fires.

Limited Growth Fires: These are fires that are not expected to grow past a predictable maximum size in terms of heat release rate, expressed in Btu per sec or watts. Special hazard fires can be assigned to this class; so can fires in occupancies with concentrations of combustibles separated by sufficiently wide aisles. The minimum aisle width to prevent lateral spread (by radiation), W_{min}, (Alpert and Ward 1984) can be estimated from an equation for radiant flux from a fire and a (conservatively low) value for the ignition flux of most materials (1.8 Btu per sq ft per sec or 20.4 kW/m^2):

$$W_{min}(\text{ft}) = 0.14\ [Q\ (\text{Btu/sec})]^{1/2} \quad (1)$$

For example, if the predicted maximum fire size is 30,000 Btu/sec (31.6 MW), the aisle width should be at least 24 ft (7.3 m). Table 7-11A contains examples of heat release data, expressed as Btu per sq ft per sec (W/m^2) of floor area. To obtain heat release in Btu per sec (watts), a number in the table is multiplied by the floor area underneath the combustible.

Continuous Growth Fires: These are fires that can be expected to grow indefinitely until intervention by fire fighters. (See Fig. 7-11E.) Starting after an incubation period, the heat release rate of these fires grows continuously, proportional to the square of time. The growth time of a given fire is defined as the interval of time between the effective ignition time and the time when the fire reaches an intermediate energy release rate of 1,000 Btu per sec, (1054 kW). (Any reference heat release rate could have been chosen with a corresponding change in the growth time. The rate 1,000 Btu per sec (1054 kW) was chosen for convenience.) Table 7-11B contains examples of growth times for continuous growth fires.

*The theory and application of venting outlined in this chapter is presented in U.S. customary units. These units reflect the system used in the original research and are consistent with the design recommendations and background materials contained in NFPA 204M. For those more familiar with S.I. units, quantities such as curtain board depth (in meters) and heat release rate (in kW), should be converted to the appropriate conventional units (feet and Btu/sec, respectively) using the conversion factors given at the end of this footnote and at various points throughout the text. After conversion, design data should be inserted in unit dependent formulas to calculate vent areas and mass flow rates. As in the NFPA Venting Guide, 1 ft=0.305 m, 1 Btu/sec=1.054 kW.

TABLE 7-11A. Limited-Growth Fires.

Heat-release rate per unit floor area of fully involved combustibles, assuming 100 percent combustion efficiency.
(PE = polyethylene; PS = polystyrene; PVC = polyvinyl chloride; PP = polypropylene; PU = polyurethane; FRP = Fiberglass-Reinforced Polyester)

	Btu/sec/ft² of Floor Area		Btu/sec/ft² of Floor Area
1. Wood pallets, stack 1½ ft high (6–12% moisture)	120	12. PU insulation board, rigid foam, stacked 15 ft high	170
2. Wood pallets, stack 5 ft high (6–12% moisture)	340	13. PS jars packed in Item 6	1300
3. Wood pallets, stack 10 ft high (6–12% moisture)	600	14. PS tubs nested in cartons, stacked 14 ft high	450
4. Wood pallets, stack 15 ft high (6–12% moisture)	900	15. PS toy parts in cartons, stacked 15 ft high	180
5. Mail bags, filled, stored 5 ft high	35	16. PS insulation board, rigid foam, stacked 14 ft high	280
6. Cartons, compartmented, stacked 15 ft high	200	17. PVC bottles packed in Item 6	300
7. PE letter trays, filled, stacked 5 ft high on cart	750	18. PP tubs packed in Item 6	380
8. PE trash barrels in cartons, stacked 15 ft high	260	19. PP and PE film in rolls, stacked 14 ft high	540
9. FRP shower stalls in cartons, stacked 15 ft high	110	20. Methyl alcohol	65
10. PE bottles packed in Item 6	550	21. Gasoline	200
11. PE bottles in cartons, stacked 15 ft high	170	22. Kerosene	200
		23. Diesel oil	170

SI units: 1 ft = 0.31 m; 1 ft² = .093 m²; 1 Btu = 1.054 kW.

INSTALLED VENT AREA

Limited-Growth Fires

Using theory for air entrainment in fire plumes and gas flow through roof vents, required vent areas can be calculated. The results are presented in Figure 7-11F as recommended vent area per curtained compartment (in square feet) versus expected maximum heat release rate (in Btu per sec) of the combustibles underneath the curtained compartment. Several ceiling heights in the range 15 ft to 60 ft (4.6 to 18.3 m) are represented. The figure pertains to a curtain depth (or design smoke layer depth) which is 20 percent of the ceiling height. It is assumed that the fire is located on the floor, a conservative assumption since the lower the fire source, the greater the air entrainment rate from the clear layer and the required vent area. Moreover, an aerodynamic discharge coefficient of 0.6 has been assumed, which is normal for commercial unit vents. If the discharge coefficient is different from 0.6, the recommended vent areas need to be multiplied by the ratio of 0.6 to the actual discharge coefficient.

For each ceiling height in Figure 7-11F, the respective curve begins at a heat release rate where vents operated by fusible links rated at 100°F (37.8°C) to 220°F (104°C) above ambient are first expected to operate promptly. Furthermore, for each ceiling height, the respective curve terminates near a heat release rate, $Q_{feasible}$, beyond which the feasibility of roof venting, using the present approach, might be questioned. $Q_{feasible}$ can be estimated from the equation:

$$Q_{feasible}\ (\text{Btu/sec}) = 1130\ (H - d)^{5/2} \qquad (2)$$

where H is the ceiling height and d is the curtain depth ($0.2H$) in this case). (At $Q_{feasible}$, the fire may become controlled by the ventilation rate allowed by the roof vents. Ventilation controlled fires do not support a clear layer. Venting at heat release rates greater than $Q_{feasible}$ to maintain a clear layer would require larger vent areas than indicated in Figure 7-11F.)

Along the dashed segments of the curves in Figure 7-11F, gas temperatures in excess of 1,000°F (538°C) may be reached; unprotected structural steel may begin to lose stength and the possibility of flashover exits within the curtained area. The lowest rate of heat release at which this may occur, $Q_{1,000}$, has been estimated from the equation

$$Q_{1000}\ (\text{Btu/sec}) = 69\ (H - d)^{5/2} \qquad (3)$$

where, in the case of Figure 7-11F, $d = 0.2H$.

For curtain depths greater than 20 percent of the ceiling height, the vent areas read from Figure 7-11F may be multiplied by the factors indicated in Table 7-11C. New values for $Q_{feasible}$ and $Q_{1,000}$ can be calculated using equations (2) and (3), respectively, by inserting the appropriate values of H and d.

Continuous Growth Fires

Required vent areas for each curtained compartment for fires in this class depend upon ceiling height (H), growth time (examples in Table 17-11B), spacing of cur-

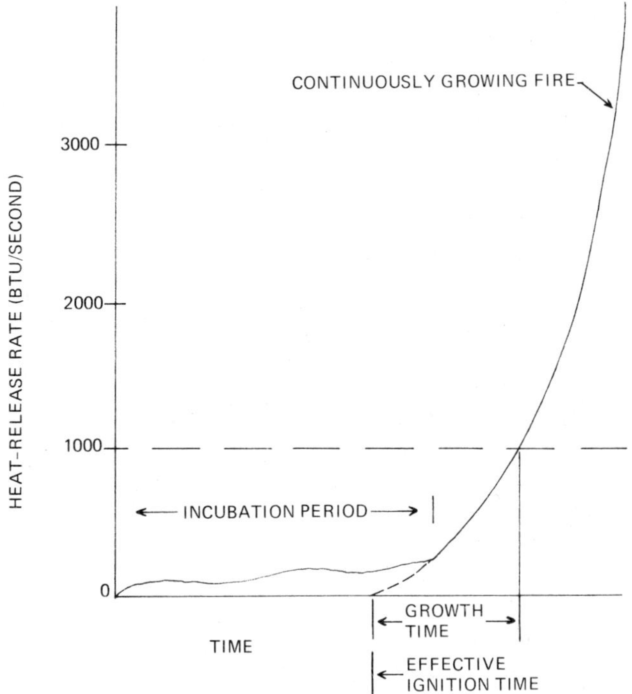

FIG. 7-11E. Conceptual illustration of continuous fire growth.

tain boards (S_c), vent spacing, and means of vent actuation. They also depend upon the desired minimum clear visibility time from the time the first vents operate. The minimum clear visibility design time will facilitate such activities as locating the fire, appraising the severity and extent or the fire, evacuating the building, and making an informed decision on deployment of personnel and equipment to be used for fire fighting.

Table 7-11D lists recommended vent areas per curtained compartment for (1) the minimum recommended curtain depth of 20 percent of ceiling height, and (2) vents spaced at no more than one-half of the curtain board spacing. For other than square curtains, the spacing S_c is interpreted as the largest spacing defined by the curtained area. The tabulated areas are approximate, pertaining to vents that are operated by heat responsive devices of an average thermal inertia, or reponse time index (RTI), of 520 $(ft \cdot sec)^{1/2}$ $[287(m{\cdot}sec)^{1/2}]$ and rated between 100°F (38°C) and 220°F (104°C) above ambient temperature (Heskestad & Smith, 1980). Each entry in the table gives the range of vent areas in 1,000 sq ft (92.9 m^2) associated with the selected range of temperature ratings. Entries boxed in are not possible (since the vent areas exceed the largest possible curtained area of $S_c \times S_c$); however, these entries may be needed for curtain depths greater than 20 percent of ceiling height as treated in a later paragraph. Where values are not given, heat release rates are greater than $Q_{feasible}$ as given in equation (2); the vent area associated with $Q_{feasible}$ is $A_{feasible}$, which has been calculated from:

$$A_{feasible}\ (ft^2) = 8.5\ (H - d)^{5/2}/d^{1/2} \qquad (4)$$

where, in the case of Table 7-11D, $d = 0.2H$. Entries in parenthesis in the table correspond to levels of heat release greater than $Q_{1,000}$ as given in equation (3); the vent area associated with $Q_{1,000}$ is $A_{1,000}$, which has been calculated from:

$$A_{1000}\ (ft^2) = 1.6\ (H - d)^{5/2}/d^{1/2} \qquad (5)$$

where, in the case of Table 7-11D, $d = 0.2H$.

To illustrate the use of Table 7-11D, consider an installation with heat responsive devices rated approximately 100°F (38°C) above ambient, a ceiling height of 20 ft (6.1 m), a growth time of approximately 150 sec, a (square) curtain spacing of 80 ft or 24.4 m ($S_c = 4 \times H$), a curtain depth of 4 ft or 1.2 m (4/20 = 20 percent of ceiling height), and a minimum clear visibility time of 10 minutes. In Table 7-11D, the lower limit 100°F (38°C) of the appropriate entry indicates a vent area per curtained compartment of 0.64 × 1,000 = 640 sq ft (59.5 m^2) for this case.

The recommended vent area for each curtained compartment is reduced if larger curtain depths than minimum (20 percent of ceiling height) are installed. The reduced areas are calculated by multiplying the values listed in Table 7-11D by the appropriate multiplication factor listed in Table 7-11C. To determine if Notes 4 or 6 from Table 7-11D apply to the newly derived values for vent area, the value of $A_{feasible}$ associated with $Q_{feasible}$ is calculated from equation (4) and the value of $A_{1,000}$ associated with $Q_{1,000}$ is calculated from equation (5). Note 4 applies if a vent area is larger than $A_{feasible}$ and Note 6 applies if a vent area is larger than $A_{1,000}$.

Vent areas for each curtained compartment should be distributed among individual vents within the constraints discussed in the section, "Dimensioning and Spacing of Vents," of this chapter. In some cases, the calculated number of vents may be so large that the vent spacing will be considerably smaller than the design spacing for vents assumed in Table 7-11D (½ S_c). The closer vent spacing implies earlier operation of the first vents than is the case for the designs of Table 7-11D. Earlier operation, such as that provided by an auxiliary fire detection system, would provide extra clear visibility times beyond the values of 5, 10, and 15 minutes selected in Table 7-11D. The extra time available with vents spaced at less than ½ S_c may be considered to represent a safety factor.

Selection of Design Basis

The vent area in a curtained compartment need not exceed the vent area recommended for the largest limited growth fire predicted for combustibles beneath the curtained area. Using sufficiently small concentrations of combustibles and minimum aisle widths according to equation (1), it may be possible to satisfy the venting needs using smaller vent areas than required by a continuous growth fire vent design. For example, in the illustration discussed previously for the use of Table 7-11D, the continuous growth fire vent design for a growth time of 150 seconds led to a required vent area for each curtained compartment of 640 sq ft (59.5 m^2). A limited growth fire vent design for a 50 sq ft (4.7 m^2) floor area concentrations at 200 Btu per sec per sq ft (2271 kW/m^2) (if that is the heat-release rate per unit floor area of the combustible considered) involves a total heat release rate per fuel concentration of 10,000 Btu per sec (10 540 kW) and a minimum aisle width of 14 ft (4.3 m), according to equation (1). For the relevant vent parameters of ceiling height

TABLE 7-11B. Continuous-Growth Fires.

Growth times of developing fires in various combustibles, assuming 100 percent combustion efficiency.
(Pe = polyester; PE = polyethylene; PS = polystyrene; PVC = polyvinyl chloride; PP = polypropylene; PU = polyurethane; FRP = Fiberglass-Reinforced Polyester)

	Growth Time (sec)		Growth Time (sec)
1. Wood pallets, stack 1½ ft high (6–12% moisture)	160–320	15. PE bottles in cartons, stacked 15 ft high	75
2. Wood pallets, stack 5 ft high (6–12% moisture)	95–190	16. PE pallets, stack 3 ft high	150
		17. PE pallets, stacks 6–8 ft high	32–56
3. Wood pallets, stack 10 ft high (6–12% moisture)	80–120	18. PU mattress, single, horizontal	120
4. Wood pallets, stack 16 ft high (6–12% moisture)	75–120	19. PU insulation board, rigid foam, stacked 15 ft high	8
5. Mail bags, filled, stored 5 ft high	190	20. PS jars packed in Item 6	55
6. Cartons, compartmented, stacked 15 ft high	60	21. PS tubs nested in cartons, stacked 14 ft high	110
7. Paper, vertical rolls, stacked 20 ft high	17–27	22. PS toy parts in cartons, stacked 15 ft high	120
8. Cotton (also Pe, Pe/Cot Acrylic/Nylon/Pe), garments in 12 ft high rack	22–43	23. PS insulation board, rigid foam, stacked 14 ft high	7
		24. PVC bottles packed in Item 6	9
9. "Ordinary combustibles" rack storage, 15–30 ft high	40–270	25. PP tubs packed in Item 6	10
		26. PP and PE film in rolls, stacked 14 ft high	40
10. Paper products, densely packed in cartons, rack storage, 20 ft high	470	27. Distilled spirits in barrels, stacked 20 ft high	25–40
11. PE letter trays, filled, stacked 5 ft high on cart	190		
12. PE trash barrels in cartons, stacked 15 ft high	55		
13. FRP shower stalls in cartons, stacked 15 ft high	85		
14. PE bottles packed in Item 6	85		

SI unit: 1 ft = 0.31 m.

of 20 ft (16.1 m) curtain depth at 20 percent of ceiling height, heat release rate of 10,000 Btu per sec (10 540 kW), Figure 7-11F indicates a vent area per curtained compartment of 250 sq ft (23.2 m^2), considerably smaller than the 640 sq ft (59.5 m^2) vent area according to the design for continuous growth fire. Of course, the combustible and floor area concentrations must be carefully controlled when limited growth fire vent designs are selected. Furthermore, designs incorporating heat release rates greater than $Q_{1,000}$ are risky since the fire may flash over to all the combustibles under the curtained area. Such designs should be based on the potential heat release rate of all the combustibles under the curtained area and should not be attempted at all if the resulting heat release rate approaches $Q_{feasible}$. For continuous growth fires, designs giving vent areas greater than $A_{1,000}$ cannot be recommended because of the flashover potential.

Mechanical Ventilators

For mechanical venting systems capable of functioning under the expected fire exposure, recommended exhaust capacities per curtained compartments are obtained by simple conversion from the recommended vent areas per curtained compartment discussed in the preceding section. The conversion depends upon the depth of the curtain board or the design depth of the smoke layer, according to Table 7-11E.

ELEMENTS OF VENTING THEORY

Refer to Figure 7-11G—in order to stop the descent of a smoke layer, the mass flow rate of hot gas out of the vents, $\dot{m}_v$, must match the mass injection rate of gas by the fire plume at the interface with the smoke layer, $\dot{m}_p$. Therefore, expressions are needed for the vent flow and for the plume flow.

Vent Flow

The vent flow can be calculated from the following expression (Thomas et al 1963, Thomas & Hinkley 1964):

$$\dot{m}_v = (2\rho_o{}^2 g)^{1/2} \left[\frac{T_o\,(T - T_o)}{T^2}\right]^{1/2} A_v d^{1/2} \qquad (6)$$

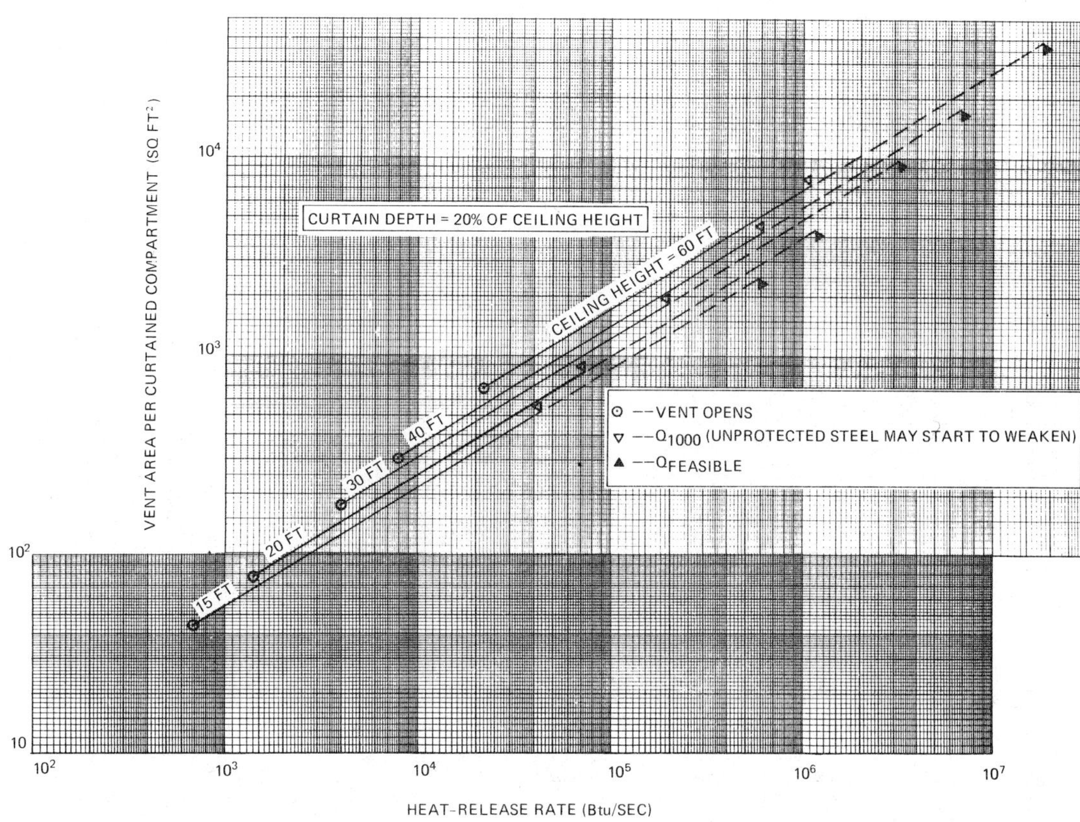

FIG. 7-11F. Limited fire growth; recommended vent areas for each curtained compartment for various maximum heat release rates (Btu per sec or kW/sec).

where $\dot{m}_v$ is the mass flow rate through the vent (lb per sec or kg/sec); ρ_o is the density of the ambient air (lb per cu ft or kg/m^3); g is the acceleration of gravity (ft/sec^2 or m/sec^2); T_o is the ambient temperature (°R or K); T is the smoke layer temperature (°R or K); A_v is the aerodynamic vent area (sq ft or m^2); and d is the depth of the smoke layer (ft or m), generally interpreted as the height of the center of the vent above the smoke interface.

The aerodynamic vent area, A_v, is always smaller than the geometric vent area. For simple apertures, A_v may be taken as 0.6 times the geometric through flow area.

The temperature function $\left[\frac{T_o\,(T - T_o)}{T^2}\right]^{1/2}$ in equation (6) can be approximated at 0.50 for smoke layer temperatures in the range 275 to 1,400°F (135 to 760°C), which covers most vent applications. Then equation (6) becomes simply

$$\dot{m}_v = (\rho_o{}^2\, g/2)^{1/2}\, A_v d^{1/2} \tag{7}$$

To prevent air from the clear layer below from being entrained in the vent exhaust (Thomas and Hinkley 1964), have proposed that the area of the vent not exceed $2d^2$. If the vent is long and narrow, an analogous requirement would be that the width of the vent not exceed d.

TABLE 7-11C. Multiplication Factors for Vent Areas in Figure 7-11F for Curtain Depths other than 20 Percent of Ceiling Height

Curtain Depth in Percent of Ceiling Height	Multiplication Factor
30	0.71
40	0.53
50	0.40
60	0.29
70	0.20
80	0.13

Plume Flow

The mass flow rate injected into the smoke layer by the fire plume is practically equal to the air entrainment rate of the plume from the clear layer; the contribution of the fire source itself is small in comparison. Before the 1982 revision of NFPA 204M, the entrainment theory (Thomas et al 1963) was in existence. However, that theory was considered somewhat unwieldly for the format of NFPA 204M and an alternative, more adaptable and probably more accurate, theory was developed (Appendix A of 1932 edition).

TABLE 7-11D. Vent Areas for Curtained Compartments

Vent Area (in 1000 ft^2) per Curtained Compartment for Heat-Responsive Device Operated Vents With Various Curtain Board Spacings (S_c) and Minimum Clear-Visibility Design Times (5, 10 or 15 minutes).

Ceiling Height, H	Growth Time (Sec)	$S_c = 2 \times H$			$S_c = 4 \times H$			$S_c = 8 \times H$		
		5 min	10 min	15 min	5 min	10 min	15 min	5 min	10 min	15 min
15 ft	20				(1.9–2.1)			(2.2–2.5)		
	40	.87–.97			(.98–1.1)	(1.8–2.0)		(1.1–1.4)	(2.0–2.3)	
	80	(.44–.52)	(.82–.90)		(.52–.64)	(.90–1.0)	(1.3–1.4)	(.64–.83)	(1.0–1.2)	(1.5–1.7)
	150	.25–.32	(.43–.50)	(.63–.70)	.31–.41	(.50–.60)	(.70–.80)	(.41–.58)	(.60–.77)	(.81–.98)
	300	.15–.20	.23–.29	.32–.38	.20–.28	.28–.36	.37–.46	.27–.41	.36–.51	(.46–.61)
	600	.10–.14	.13–.18	.17–.23	.13–.21	.17–.25	.21–.29	.20–.34	.24–.38	.29–.43
20	20	2.2–7.3			(2.5–2.7)			(2.7–3.2)		
	40	(1.1–1.2)	2.1–2.2		(1.2–1.5)	(2.2–2.5)	(3.3–3.6)	(1.5–1.8)	(2.5–2.9)	(3.6–4.0)
	80	.56–.68	(1.0–1.1)	1.5–1.6	.68–.86	(1.1–1.3)	(1.6–1.8)	(.87–1.2)	(1.3–1.6)	(1.8–2.1)
	150	.34–.43	.55–.65	(.77–.88)	.43–.58	.64–.80	(.88–1.0)	.58–.83	(.80–1.1)	(1.0–1.3)
	300	.21–.29	.30–.39	.40–.50	.28–.41	.38–.52	.48–.63	.40–.65	.51–.76	.62–.88
	600	.14–.22	.19–.27	.23–.32	.20–.33	.25–.38	.30–.44	.32–.56	.37–.61	.42–.67
30	20	(2.9–3.2)	5.6–6.0		(3.2–3.8)	(6.0–6.6)		(3.8–4.6)	(6.7–7.6)	
	40	1.5–1.8	(2.7–3.0)	4.0–4.3	(1.8–2.2)	(3.0–3.4)	(4.3–4.8)	(2.2–2.9)	(3.5–4.2)	(4.8–5.6)
	80	.82–1.0	1.4–1.6	(2.0–2.2)	1.0–1.4	1.6–2.0	(2.2–2.6)	1.4–2.0	(2.0–2.6)	(2.6–3.2)
	150	.51–.71	.78–.99	1.1–1.3	.68–.99	.96–1.3	1.3–1.6	.97–1.5	1.3–1.8	(1.6–2.2)
	300	.34–.53	.47–.66	.59–.80	.48–.77	.61–.91	.74–1.1	.74–1.3	.88–1.4	1.0–1.6
	600	.26–.44	.32–.50	.38–.57	.38–.67	.44–.74	.50–.81	.62–1.2	.69–1.2	.76–1.3
40	20	(3.6–4.1)	6.7–7.3	10–11	(4.1–4.9)	(7.4–8.2)	(11–12)	(5.0–6.3)	(8.3–9.7)	(12–13)
	40	1.9–2.3	(3.3–3.8)	(4.8–5.3)	2.3–3.0	(3.8–4.5)	(5.3–6.1)	(3.0–4.1)	(4.5–5.6)	(6.1–7.3)
	80	1.1–1.5	1.7–2.1	2.4–2.8	1.4–2.0	2.1–2.7	2.8–3.4	2.0–3.0	2.7–3.7	(3.4–4.5)
	150	.72–1.1	1.0–1.4	1.4–1.8	.98–1.5	1.3–1.9	1.7–2.2	1.5–2.4	1.8–2.8	2.2–3.2
	300	.51–.84	.66–1.0	.81–1.2	.74–1.3	.89–1.4	1.1–1.6	1.2–2.2	1.4–2.3	1.5–2.5
	600	.41–.74	.48–.81	.55–.89	.61–1.1	.69–1.2	.77–1.3	1.1–2.0	1.1–2.1	1.2–2.2
60	20	4.9–5.9	(8.9–10)	(13–14)	5.9–7.4	(10–12)	(14–16)	(7.5–10)	(12–14)	(16–19)
	40	2.8–3.6	4.6–5.5	6.5–7.4	3.5–4.8	5.4–6.7	(7.3–8.8)	4.8–7.0	(6.7–9.1)	(8.7–11)
	80	1.7–2.5	2.5–3.4	3.4–4.3	2.3–3.5	3.2–4.4	4.1–5.4	3.4–5.6	4.3–6.6	5.2–7.6
	150	1.2–2.0	1.6–2.4	2.1–2.9	1.7–2.9	2.2–3.4	2.6–3.9	2.8–4.9	3.2–5.4	3.7–5.9
	300	.95–1.7	1.2–1.9	1.3–2.1	1.4–2.6	1.6–2.8	1.8–3.1	2.4–4.6	2.6–4.8	2.9–5.1
	600	.82–1.6	.92–1.7	1.0–1.8	1.3–2.4	1.4–2.6	1.5–2.7	2.2–4.4	2.3–4.5	2.5–4.6

For SI Units: 1 ft = .3048 m; 1 ft^2 = 0.093 m^2.

Notes:

1. Vents are assumed to be spaced at one-half of the curtain board spacing.
2. Curtain depth assumed at 20 percent of ceiling height.
3. Each entry is the vent-area range (in 1000 ft^2) associated with heat-responsive devices rated between 100°F and 220°F (38°C and 104°C) above ambient temperature.
4. No Entries: Heat-release rates greater than $Q_{feasible}$.
5. Entries Boxed-In: Not possible, but needed for curtain depths greater than 20 percent.
6. Entries in Parentheses: Correspond to levels of heat release greater than Q_{1000}.

The plume flow, $\dot{m}_p$, from the new theory is calculated according to either of two equations, depending upon whether the convective heat release rate of the fire is larger or smaller than a critical value, Q_c, defined as (fire assumed to be located on floor of building):

$$Q_c \text{ (Btu/sec)} = 11.3\,(H - d)^{1/2} \quad \text{(where } H \text{ and } d \text{ are in ft)} \tag{8}$$

The plume flow equations are:

For $Q \le Q_c$

$$\dot{m}_p \text{ (lb/s)} = 0.022\, Q^{1/3}\,(H - d)^{5/3}\,[1 + 0.19\, Q^{2/3}/(H - d)^{5/3}] \tag{9}$$

For $Q > Q_c$

$$\dot{m}_p \text{ (lb/s)} = 0.097\,(H - d)^{5/2}\,(Q/Q_c)^{3/5} \tag{10}$$

where Q and Q_c are in Btu/sec; H and d in ft.

Required Vent Area

For

$$Q \le Q_c,$$

where Q_c is given by equation (8), the expression for $\dot{m}_p$ in equation (9) is set equal to the expression for $\dot{m}_v$ in equation (7) and solved for A_v, the aerodynamic vent area. For

$$Q > Q_c,$$

The expression for $\dot{m}_p$ in equation (10) is set equal to the expression for $\dot{m}_v$ in equation (7) and solved for A_v.

The resulting relations for A_v can be greatly simplified whenever Q/Q_c is larger than approximately 0.2, which is nearly always the case in vent design. Then the two expressions for A_v can be consolidated into a single relation with the aid of equation (8), valid for heat release rates both smaller and larger than Q_c:

TABLE 7-11E. Conversion from Vent Area of Gravity Vents to Equivalent Mechanical Exhaust Capacity

Curtain Depth (ft)*	Mechanical Exhaust Capacity Per Unit Area of Gravity Vent (SCFM/ft²)†
6	354
8	409
10	457
12	501
16	578
20	647
24	708

SCFM = Standard Cubic Feet per Minute (Standard Temperature and Pressure)
* 1 ft = 0.3 m
† 1 ft² = 0.028 m³

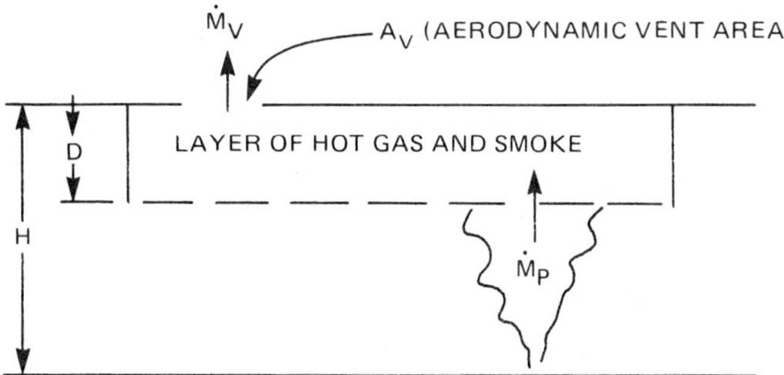

FIG. 7-11G. A schematic of a venting system.

$$A_v\ (\text{ft}^2) = 0.075\ Q^{3/5}\ \frac{H - d}{d^{1/2}} \tag{11}$$

where Q is in Btu per sec, and H and d in ft.

For fires growing with the second power of time (continuous growth fires), equation (11) can be written:

$$A_v\ (\text{ft}^2) = 4.8\ [(t_d + t_r)/t_g]^{6/5}\ (H - d)/d^{1/2} \tag{12}$$

where H and d are in ft; t_d is the detection time, or the time the first vents operate; t_r is the additional, clear visibility time; and t_g is the growth time defined in the section "Fire Characterization."

These are the basic relations used for calculating vent areas in the 1982 edition of NFPA 204M, with adjustments for a discharge coefficient of 0.6 (aerodynamic vent areas, A_v, divided by 0.6). In equation (12), the detection time, t_d, was taken as the time of operation of the first vent in a square matrix (vent farthest possible from the fire location). The vents were assumed to be actuated by heat responsive devices of various temperature ratings above the ambient temperature. Link actuation times (t_d) were calculated from a thermal response equation for heat responsive devices derived previously (Heskestad & Smith 1976), together with generalized data on gas temperatures and velocities under extensive flat ceilings (Heskestad 1972). The RTI value of the heat responsive device was taken as 520 $(\text{ft·sec})^{1/2}$ $[287(\text{m·sec})^{1/2}]$, which is considered to be a conservatively high value for heat responsive devices listed by testing laboratories.

Limiting Heat Release Rates

$Q_{1,000}$ was defined in conjunction with equation (3) as the heat release rate at which gas temperatures in excess of 1,000°F (538°C) may be reached with possible weakening of structural steel and potentials for flashover within the curtained areas. The expression for $Q_{1,000}$ was derived first using equation (10), with substitution for Q_c from equation (8), to obtain $\dot{m}_p$ expressed as:

$$\dot{m}_p\ (\text{lb/sec}) = 0.0226\ (H - d)\ Q^{3/5} \tag{13}$$

where Q is in Btu/sec, and H and d in ft.

The average temperature rise in the smoke layer, ΔT_s (°F), follows from

$$\dot{m}_p \cdot c_p \cdot \Delta T = Q \tag{14}$$

where c_p [Btu per lb per °F or J/(kg)/K] is the specific heat of air and where it has been assumed that no heat is lost to the ceiling and walls. With substitution of equation (13) for m_p, c_p = 0.24 Btu per lb per °F [1004 (J/kg)/K] and ΔT = 1000°F in equation (14), the resulting expression can be solved for Q to give equation (3). Of course, the result is quite conservative since a significant heat loss may occur to the ceiling and the walls of the curtained compartment.

$Q_{feasible}$ was described in conjunction with equation (2) as a heat release rate beyond which the feasibility of roof venting, using the present approach, might be questioned because of the onset of ventilation control of the burning process. Experience has indicated that ventilation controlled fires do not support a clear layer as, for example, in the case of the typical, flashed-over room fire. According to experiments on wood crib fires in enclosures, vented through doors or windows (Croce 1978), ventilation control appears to set in when the so called "ventilation parameter," $A_w\sqrt{h_w}$, in ratio to the mass burning rate R is less than 266 ft $^{5/2}$ per lb per sec), where A_w is the window (or door) area and h_w is the window (or door) height. Generalized to any combustible, the limiting ratio might be expressed as $A_w\sqrt{h_w}$, in ratio to the stoichiometric air requirement associated with the mass burning rate, $R \cdot r$, where r is the stoichiometric mass ratio, air to combustible (assumed at 5.5 for wood). Now, the ventilation parameter $A_w\sqrt{h_w}$, is proportioned to the mass flow rate of air through a window or door opening, m_v (Harmathy 1980):

$$\dot{m}_v\ (\text{lb/sec}) = 0.064\ A_w\sqrt{h_w} \tag{15}$$

(A_w in ft², h_w in ft).

The limiting condition for ventilation control can, therefore, be expressed in terms of the mass venting rate of air in ratio to the stoichiometric air requirement, $\dot{m}_v/R \cdot r$, which turns out to be approximately 3. Using the fact that $R \cdot r$ can be written $Q/(H_c/r)$, where Q is the heat release rate and H_c is the heat of combustion, the limiting mass rate ratio can be converted to a limiting heat release rate:

$$Q = \frac{1}{3}\ \dot{m}_v\ (H_c/r) \tag{16}$$

In application to roof vents, $\dot{m}_v$ can be expressed as in equation (7). Furthermore, H_c/r is quite constant among common combustibles. Substituting equation (7) and taking H_c/r = 1,328 Btu per lb (3,086 kJ/kg), equation (16) becomes:

$$Q \text{ (Btu/sec)} = 133\ A_v d^{1/2} \qquad (17)$$

where A_v is in ft^2, and d is in ft.

In the venting approach used in NFPA 204M, the vent area A_v is given by equation (11), which, divided by 0.6 to convert to geometric vent area and then substituted in equation (17), gives a relation which can be solved for Q. The result is the expression for the limiting heat release rate, $Q_{feasible}$, in equation (2).

Bibliography

References Cited

Alpert, R. L., and Ward, E. J. 1984. "Evaluation of Unsprinklered Fire Hazards." *Fire Safety Journal*. Vol 7. pp 127-143.

Croce, P. A. 1978. "Modeling of Vented Enclosure Fires Part I. Quasi-Steady Wood-Crib Source Fire." *Technical Report, FMRC 7A0R5.GU*. Factory Mutual Research Corporation, Norwood, MA.

Harmathy, T. Z. 1980. "Ventilation of Full-Developed Compartment Fires," *Combustion and Flame*. Vol 37. pp 25-39.

Heskestad, G. 1972. "Similarity Relations for the Initial Convective, Flow Generated by Fire." *Paper No 72-WA/HT-17*, American Society of Mechanical Engineers.

Heskestad, G., and Smith, H. F. 1976. "Investigation of a New Sprinkler Sensitivity Approval Test: The Plunge Test," *Technical Report, FMRC 22485*. Factory Mutual Research Corporation, Norwood, MA.

Heskestad, G., and Smith, H. F. 1980. "Plunge Test for Determination of Sprinkler Sensitivity." *Technical Report, FMRC 3A1E2.RR*. Factory Mutual Research Corporation, Norwood, MA.

Thomas, P. H., and Hinkley, P. L. 1964. "Design of Roof-Venting Systems for Single-Story Buildings." *Fire Research Technical Paper No 10*. Department of Scientific and Industrial Research and Fire Offices' Committee. Joint Fire Research Organization. Her Majesty's Stationery Office, London, England.

Thomas, P. H., et al. 1963. "Investigations Into the Flow of Hot Gases in Roof Venting." *Fire Research Technical Paper No. 7*. Department of Scientific and Industrial Research and Fire Offices' Committee, Joint Fire Research Organization. Her Majesty's Stationery Office, London, England.

NFPA Codes, Standards, Recommended Practices and Manuals. (See the latest *NFPA Codes and Standards Catalog* for availability of current editions of the following documents.)

NFPA 204M, *Guide for Smoke and Heat Venting*.

Additional Readings

Butcher, E. G., and Parnell, A. C., *Smoke Control in Fire Safety Design* E. & F. N. Spon Ltd., London, 1979; also National Fire Protection Association, Quincy, MA.

Carlson, G., and Orton, C., eds, *Fire Ventilation Practices*, 6th ed., International Fire Service Training Association, Oklahoma State University, Stillwater, OK, 1981.

Heskestad, G., "Model Study of Automatic Smoke and Heat Vent Performance in Sprinklered Fires," *Technical Report RC74-T-29*, Factory Mutual Research Corp., Norwood, MA, Sept. 1974.

Langdon-Thomas, G. J., and Hinkley, P. L., "Fire Venting—Small-Story Industrial Buildings," *Fire Note No. 5*, Ministry of Technology, Industrial and Fire Offices Committee, Joint Fire Research Organization, Her Majesty's Stationery Office, London, England, 1965.

Lyons, Robert J., "Automatic Heat and Smoke Venting," *Progessive Architecture*, Apr. 1972.

Miller, E. E., "Position Paper to 204 Subcommittee," *Fire Venting of Sprinklered Property*, National Fire Protection Association, Quincy, MA.

Tewarson, A. "Fire Ventilation," *Combustion and Flame*, Vol. 53, Nos. 1-3, Nov. 1983, pp. 123-134.

Waterman, T. E., *Fire Venting of Sprinklered Buildings, Fire Journal*, Vol. 78, No. 2, Mar. 1984, pp. 30-39, 86.

Waterman T. E., "Fire Venting of Sprinklered Buildings," IIT Research Institute, Chicago, IL, 1982.

ADDENDUM

AUTOMATIC HEAT AND SMOKE VENTING IN SPRINKLERED BUILDINGS

This addendum supplements the venting practices in this chapter as written by Dr. Heskestad. It presents in summary the various views on the controversial and unresolved subject of automatic venting in sprinklered buildings. The history of research testing on the subject is traced. Contributors include Ernest E. Miller, Thomas E. Waterman and Edward J. Ward.*

The Different Views

There has been controversy for approximately two decades over automatic venting for sprinklered buildings.

Proponents of automatic venting in sprinklered buildings claim that venting will aid manual fire fighting by delaying loss of visibility, reduce the risk of structural failure by venting gases with dangerously high temperatures, vent the equivalent of an unsprinklered fire if the installed sprinklers are defeated by human failure or mechanical damage, and substitute for manual roof venting by fire service personnel.

Opponents believe that in some cases, venting may even be detrimental because it draws in fresh makeup air keeping oxygen levels higher than they would be otherwise, causing more vigorous combustion with an increase in fuel consumption and potential for loss of sprinkler control. They also believe sprinkler discharge will cool the fire gases, perhaps to the extent that the vent discharge is reduced, hence reducing venting effectiveness. Further, the sprinkler sprays will entrain ambient smoke and air, transporting smoke to the floor level and possibly drawing gas from the vent exhaust wherever sprinklers are located close to vents and further reducing venting effectiveness.

*Mr. Miller is an Executive Engineer with Industrial Risk Insurers (IRI). Mr. Waterman is Director of Research for the Institute for Advanced Safety Study and while employed by the IIT Research Institute conducted the IITRI test series of 1980-1981 referenced in the section. Mr. Ward is Manager, F&EC Standards Department for Factory Mutual Research Corporation.

Additionally, opponents see little, if any, evidence from research testing of the benefit to fire control obtained from the use of vents in a sprinklered building, and believe that the high cost of installation does not justify their use.

Sprinkler/Vent Research

Research testing on the subject includes:

1. Large Scale Tests: In 1956 FMRC (Factory Mutual Research Corporation) ran a series of large scale tests from which it concluded that the effects of venting in sprinklered buildings are different from unsprinklered buildings. The research project, "Heat Vents and Fire Curtains, Effect on Operation of Sprinklers and Visibility" (FMRC 1956), was conducted in a 120 by 60 ft (36.6 by 18.3 m) test building equipped with 5 ft (1.5 m) draft curtains and vent areas ranging up to 32 sq ft (3 m^2) within a curtained area of 2,280 sq ft (212 m^2). This is a vent ratio of about 1:70—that is 1 sq ft (0.9 m^2) of vent area for every 70 sq ft (6.5 m^2) of floor area. A 5 gal per min (18.9 L/min) gasoline spray fire was used as the exposure, and protection consisted of 160°F (71°C) automatic sprinklers installed on a 10 by 10 ft (3 by 3 m) spacing.

The tests in the series were conducted using various combinations of vents, draft curtains, and sprinklers. Six of these were sprinklered tests. (See Table 7-11F.)

TABLE 7-11F. FMRC Test Results

Test No.	Curtain ft (m)	Vent ft² (m²)	Sprinkler Discharge gpm (L/min)	Number of Sprinklers Operated
1	None	None	15 (57)	48
2	None	32 (3)	15 (57)	44
3	5 (1.5)	32 (3)	15 (57)	24
4	5 (1.5)	16 (3)	15 (57)	24
5	5 (1.5)	None	15 (57)	28
6	5 (1.5)	None	25 (95)	15

Comparison of average temperatures indicates that vents contributed significantly to temperature reductions in the unsprinklered tests in the series; however, they were of marginal value in the sprinklered tests. The greatest reduction in sprinkler operation was attributed to increased sprinkler discharge.

2. FMRC Model Study: In the early 1970s, a FMRC study, "Model Study of Automatic Smoke and Heat Vent Performance in Sprinklered Fires" was conducted (Heskestad 1974). The objective was to experimentally investigate the performance of automatic heat and smoke vents in sprinklered fires in one story buildings, principally in terms of sprinkler water demand, but also in terms of visibility conditions and fuel consumption. The study was performed at FMRC's Norwood, MA, laboratory and involved a 1:12.5 scale model of FMRC's fire test facility at West Glocester, R.I. Automatic, individually fused vents on 50 ft (15.2 m) spacing were employed, often in combination with a draft curtain encompassing a 10,000 sq ft (929 m^2) curtained area.

Of primary interest were venting installations conforming to recommendations of standards setting groups that were current at the time. For a fire in cellulosic material that opened about fifty 212°F (100°C) sprinklers at a density of 0.27 gpm per sq ft [11 (L/min)/m^2], distributed vents alone (without draft curtains) had no effect on the water demand, but delayed loss in visibility from 13.1 to 15.7 min and increased fuel consumption by 31 percent. Vents and draft curtains caused a 35 percent increase in water demand relative to the unvented fire, delayed loss in visibility from 13.1 to 20.2 minutes, and increased fuel consumption by 66 percent.

Favorable effects of venting were observed for a series of large, heptane fires that opened an average of 112 sprinklers at 0.27 gpm per sq ft [11 (L/min)/m^2]. Vents alone reduced the water demand by 8 percent and markedly improved visibility conditions; vents and draft curtains reduced the water demand by 18 percent, but did not improve visibility conditions as much as without draft curtains.

The test results were based on ignition points that were equidistant from the four nearest vents in the vent matrix. Experiments with ignition directly under a vent indicated that some reduction in water demand could be expected for fires that opened 50 sprinklers or less. It was concluded, however, that for randomly starting fires, less than about 13 percent of the fires would benefit, respective to water demand, from closeness to a nearby vent.

Critics counter that the 0.08 scale model experiments, although carefully designed and executed, incorporated numerous uncertainties. Of particular concern was modeling of a full scale fuel bed (rack storage configuration), by using ½ in. (12.7 mm) thick triwall cardboard. Some ten years later, the modeling of room fire development has proceeded to an advanced state in all respects except for the burning fuel; both analytical and experimental models address only the simplest of fuel configurations. Yet the model study assumed that both flame spread and burning rate of the fuel bed were modeled under conditions of depleted oxygen and impacting water droplets.

The FMRC model study was also criticized because of the configuration of the water discharge devices that were used. The study pioneered the use of nondescript miniature nozzles (representing sprinklers) arranged to be operated in small open head deluge systems of groups of 5 to 11 nozzles controlled by individual sensors. Critics claim this arrangement magnified the known imprecision in measuring comparative sprinkler performance. It required multiplying the normal spread experienced in full scale sprinkler operation by a variable factor of 5 to 11. Due to the small deluge systems which operated as groups, sprinkler skipping occurred in units of 5 or more open nozzles, rather than the skipping of individual sprinklers as would be customary in full scale fire research.

Lastly, the critics agreed that although up to three times as much combustibles were consumed in the vented tests as compared to the unvented tests, the implication of detriment was improper. They claimed the explanation for the increased fuel consumption in the vented tests lies in the customary manner of conducting fire tests involving sprinklers. The test fires are allowed to continue far beyond sprinkler control to give the fires every opportunity to escape. During the additional minutes of these tests, the oxygen reduction in the unvented tests limited

TABLE 7-11G. Rack Storage Tests

Test No.	Ventilated	Number of Ceiling Sprinklers Operated
65	No	45
66	No	48
72	Yes	92
73	Yes	30

the fuel consumption, whereas oxygen remained abundant in the vented tests and resulted in greater fuel consumption. Since sprinkler controlled fires continue to smoulder and burn, it was felt that the reported increased fuel consumption was attributable to more complete burning of the rubble and did not represent increased fire damage. Had the fires spread to involve a significantly larger area, additional sprinklers would have operated.

3. Other FMRC Large Scale Testing: More than 80 full scale rack storage tests were conducted between 1968 and 1975. Only three tests employed perimeter ventilation (not to be confused with roof venting directly over a fire) and only two of these three (Nos. 72 and 73) were identical to two other tests (Nos. 65 and 66) except for the ventilation variable. Results of these tests are given in Table 7-11G.

Depending upon which tests are compared, venting can show either a moderate improvement or a dramatic detriment. Factory Mutual's position on the effects of ventilation in these tests is that they are not conclusive.

Venting proponents argue that although the influence of venting was inconclusive because the perimeter windows were too remote, the ventilated tests maintained visibility for 48:40 and 33:00 (minutes:seconds) as compared to 10:30 and 18:00 for the unventilated tests.

Venting opponents claim the effect of ventilation on fire control was best demonstrated during Factory Mutual's large scale fire testing of rubber tires in 1970. This test was made to evaluate sprinkler protection for automobile tires stored in portable steel racks. Storage was located in a single pile 35 ft by 50 ft by 18 ft (11 m by 15 m by 5.4 m) high. Protection consisted of 286°F (141°C), ½ in. (12.7 mm) sprinklers spaced 50 sq ft (4.6 m^2) each with a controlled discharge of 0.6 gal per min per sq ft [24.4 (L/min)/m^2]. There was no ventilation at the start of the test.

The fire was started at the base of a rack. The first sprinkler operated 2 minutes and 15 seconds after ignition. By 8 minutes 20 seconds, forty-three sprinklers had operated and controlled the fire. Only one additional sprinkler operated at 28 minutes. The fire remained under control until 60 minutes into the test when all doors and windows were opened to ventilate the building and sprinklers were left on. The fire then began to spread and grow in intensity. At 117 minutes when it was apparent that sprinklers were failing to control the fire, all doors and windows were closed. A total of 95 sprinklers operated and water was discharged for over 5 hours. Only the 95th sprinkler operated after doors and windows were closed and that was at 118 minutes, 1 minute after closure.

Critics reappraised the ventilated rubber tire fire test (Miller 1982) and contend that the loss of initial sprinkler control was not due to remote window ventilation as reported. Ceiling temperature gradients revealed that during the hour delay before hose extinguishment was attempted of the sprinkler controlled fire, the fire had burrowed to the other end of the pile where sprinklers were not operating and erupted with great intensity. The tightly stacked tires, stored on tread, formed horizontal tunnels that offered avenues of fire spread which were sheltered from sprinkler discharge.

4. IITRI Full-Scale Venting Research Testing: In 1977, the IITRI (IIT Research Institute) was commissioned by the intra-industry Fire Venting Research Committee to review past research and fire experience related to vent/sprinkler interactions in large area single story structures. Based on IITRI's review and other considerations, the Committee ultimately funded some 45 large scale segment experiments in 1980-81 (Waterman 1984). The experiments were conducted in a 75 by 25 by 17 ft (23 by 7.6 by 5.2 m) high portion of IITRI's fire laboratory. Fires were placed in a corner to represent one-quarter of a larger fire in the center of a 150 by 50 ft (46 by 15.2 m) room. To simulate the test area as part of an even larger area, two garage doors on the wall farthest from the fire were partially opened to represent a draft curtain. The area beyond the curtains was also enclosed by vertical air stacks to minimize extraneous wind effects.

To allow variations in vent area, two pairs of automatic roof vents were installed in the laboratory roof. A diagonal sprinkler pattern was chosen to eliminate aisles between sprinklers that were located between the test fires and the vents. Experiments were conducted using propane diffusion burners or stacked pallets as the fuel source. For each fuel source, fire size and sprinkler water supply pressure were varied until a condition of "marginal" sprinkler control was achieved. Then, vent areas were varied to assess effects of venting on sprinkler performance (number of operating sprinklers and temperature levels at various locations). The laboratory size and excessive recirculation of smoke due to the substitution of vertical stacks for added rooms precluded useful measures of smoke obscuration.

Propane source fires produced results suggesting a slight reduction in water demand for vented fires when either 165 or 286°F (74 or 141°C) sprinklers were employed [vents were operated either with 165°F (74°C) links or by simulated smoke detector or waterflow switch actuation].

All pallet fire experiments employed 165°F (74°C) sprinklers and the same vent arrangement previously described. These experiments demonstrated that under marginal sprinkler protection the configuration of the test facility was extremely sensitive to variations in fire intensity from any cause. Once this possibility appeared, a final series of ten "replicate" experiments was conducted, alternating between vented and unvented configurations. These proved to produce widely varying results, apparently independent of the presence of automatic venting. The number of operating sprinklers varied from 7 to 22 per test and the pattern of sprinkler operating times appeared unrelated to the time that vents opened. Other measured factors (temperature, 0_2) varied in the same way. Although the number of operating sprinklers varied widely in these experiments with marginal sprinkler control, a total of 84 sprinklers operated in the 5 unvented tests, 85 in the 5 vented tests. Earlier tests at higher water supply pressures produced good replication independent of venting.

On the basis of the IITRI experiments, the test sponsors concluded that automatic roof venting did not detract from current state of the art automatic sprinkler performance. Likewise, no particularly significant benefit to sprinkler performance was noted. Thus, it was suggested that the role of automatic vents, if used, in sprinklered properties will be to:

1. Vent unsprinklered fires, if installed sprinklers are defeated by human failure or mechanical damage, and
2. Substitute for manual roof venting upon arrival of fire service personnel.

Primarily on the basis of the first listed application, the researchers recommended that where used for sprinklered properties, the design and installation of automatic roof vents for unsprinklered properties follow guidance provided by NFPA 204M, until further research, testing, or experience suggest more beneficial alternate configurations.

Critics of the IITRI research effort do not agree that the experimental results support the main conclusion of the investigation, which is that automatic roof vents do not impair sprinkler control of fires capable of growth (Heskestad 1983).

The critics also claim the finding that automatic roof vents which do not impair the ability of 165°F (74°C) sprinklers to control fires capable of growth contradict the studies described above, which have indicated that the increased availability of oxygen associated with open vents may lead to increased fire intensity (of fires capable of growth), water demand, and consumption of combustibles.

The critics concluded that the reported results do not justify the overall conclusions and recommendations of the IITRI researchers. Their critique suggested the investigation suffered from a number of defects, i.e., (1) situations claimed to be unvented were not; (2) one of the most important principles of venting was violated, i.e., the provision of adequate openings for inlet air; and (3) inadequate control was exercised over the experimental conditions to the extent that major variations observed in fire behavior could not be attributed to the parameter under study.

Conclusions

The proponents of automatic venting in a sprinklered building cite the IITRI research work as substantiation of the merit of vents in sprinklered buildings. On the basis of this data, they believe that automatic roof venting can provide two major contributions to 165°F (74°C) sprinkler protected (large one story) properties:

1. Perform as vents in unsprinklered properties should installed sprinklers be defeated by human failure or mechanical damage.
2. Substitute for manual roof venting upon arrival of the fire services.

The opponents of automatic venting in a sprinklered building believe that the high installation cost of vents is difficult to justify when one considers the limited benefits and possible detriment. They are not cost effective because it is unlikely a large loss will be averted solely due to the presence of vents when automatic sprinkler protection is inadequate or impaired.

Venting may or may not be detrimental, depending upon many factors such as the location of the fire origin in relation to the location of vents. More often than not, vents will increase the water demand (Heskestad 1974). Even when automatic vents result in some improvement, the improvement is not required for sprinklers to gain control. On the other hand, vents can create conditions which will inhibit or prevent sprinkler control (Heskestad 1983).

Installing automatic vents in unsprinklered buildings is acceptable. It would certainly be more beneficial to the building owner, however, to install sprinklers rather than vents to obtain active rather than passive fire control.

Installing manual vents in a sprinklered building is also acceptable, if desired by the building owner. These vents can be used during manual overhaul or at locations which would be expected to produce dense smoke. However, a similar effect can be achieved by venting through windows, by fire fighters cutting holes in the roof, or by ventilation equipment or smoke exhausters.

Through its consensus making standards process, NFPA members agreed to the following statements for the 1985 edition of NFPA 204M:

Paragraph 6-3: A series of tests was conducted to increase the understanding of the role of automatic roof vents simultaneously employed with automatic sprinklers [Section 6-5(c)]. The data submitted did not permit consensus to be developed whether sprinkler control was impaired or enhanced by the presence of automatic (roof) vents of typical spacing and area.

Paragraph 6-4: While the use of automatic venting in sprinklered buildings is still under review, the designer is encourged to use the available tools and data referenced in this document for solving problems perculiar to a particular type of hazard control.

Even though the issue was raised 15 years ago, proponents maintain venting critics have been unable to cite a single fire in which automatic venting was blamed for loss of fire control by sprinklers. Thus to proponents it is paradoxical that fire research, which is typically motivated only by adverse fire experience, should be the sole basis of suspecting venting of detriment. They urge that serious students of venting should obtain, and carefully examine, this research and compare it with related research, both direct and indirect.

Additional research and testing will be needed before definitively establishing the benefits/detriments of automatic venting in a sprinklered building.

REFERENCES

FMRC. 1956. *Heat Vents and Fire Curtains, Effect on Operation of Sprinklers and Visibility*. Factory Mutual Research Corp., Norwood, MA.

Heskestad, Gunnar. 1983. "Review of 'Fire Venting of Sprinklered Buildings' by T.E. Waterman, et al. Interoffice Correspondence to E. J. Ward," Factory Mutual Research Corp., Norwood, MA. (Also see "Letters to the Editor." *Fire Journal*. Vol 78, No 5. Sept 1984. p 6.)

Heskestad, Gunnar. 1974. "Model Study of Automatic Smoke and Heat Vent Performance in Sprinklered Fires." *Technical*

Report FMRC Serial No. 21933 RC74-T-29. Factory Mutual Research Corp., Norwood, MA.

Miller, E. E. 1982. "Reappraisal of Ventilated Rubber Tire Fire Tests." *Fire Venting Mini-Study No. 8*. Industrial Risk Insurers, Chicago, IL.

Miller, E. E. 1980. "A Position Paper to NFPA 204 Subcommittee: Fire Venting of Sprinklered Properties." Improved Risk Insurers, Chicago, IL.

Waterman, T. E. 1984. "Fire Venting of Sprinklered Buildings," *Fire Journal* Vol 78, No 2. Mar 1984. pp 30-39, 86. (Also see "Letters to the Editor," *Fire Journal*, Vol 78, No 5, Sept 1984, p 6.)

STRUCTURAL FIRESAFETY: ONE AND TWO FAMILY DWELLINGS

Bertram M. Vogel, P.E.

It has been estimated that approximately 80 percent of the population of the United States reside in one and two family dwellings. In order to relate to the high losses, i.e., deaths, injuries, and property, that occur in one and two family residential structures, it is desirable to look closely at the various design considerations and field practices that are an integral part of such residential construction.

This chapter is confined to a discussion of the structural elements of one and two family dwellings as they affect the firesafety of the structure. It does not cover the influence of interior finish or heat conservation measures, the hazards of building facilities (heating and electrical installations), or the role of detection and extinguishing systems as life safety features. These subjects are included in other chapters of the Handbook, particulary Chapter 7, "Interior Finish;" and Chapter 13, "Energy Conservation in Buildings;" of this section. See also Chapter 2, "Electrical Systems and Appliances;" and Chapter 3, "Heating Systems and Appliances," in Section 8; Chapter 4, "Residential Sprinkler Systems," in Section 18; and Chapter 3, "Household Warning Systems," in Section 16.

CONVENTIONAL TYPES OF CONSTRUCTION

This part of the chapter will address the various types of construction used in the design and erection of both conventional and manufactured one and two family dwellings (including mobile homes). While specific requirements are dictated by local building codes, and in the case of mobile homes, by federal regulation. In general, most one and two family residential buildings erected are of Type V (frame) or Type III (ordinary) construction as defined in NFPA 220, *Standard Types of Building Construction*. There are, of course, exceptions where such buildings may be erected of Type II, (noncombustible), or Type I (fire resistive), and possibly of Type IV (heavy timber) construction.

Mr. Vogel, is Manager, Eastern Region, Schirmer Engineering Corporation, having retired from his previous position as Codes and Standards Coordinator, Center for Fire Research, National Bureau of Standards.

Type V (Frame) Construction

The principal categories of Type V (frame) construction are: (1) plank and beam, (2) balloon, (3) platform or western, (4) braced, and (5) veneer (a combination of one of the previous four with a facing wythe of brick, concrete masonry units, or stone on the exterior walls). The buildings may be constructed with either basements, cellars, crawl spaces, or slab on grade. Geographical peculiarities or the needs of the owner generally dictate which foundation type is used. Where basements, cellars, or crawl spaces are provided, they play a definite role insofar as residential firesafety is concerned, since in these areas, major heating, air conditioning, and water heater equipment are usually installed. These areas are also used for storage by the building's occupants. Roof framing may be either of the rafter or trussed type. Where trusses are used, interior bearing partitions may not be required.

Plank and Beam Framing: In plank and beam framing, a few large members replace the many small wood members used in typical wood framing, i.e., large dimensional beams more widely spaced (4 by 10 in. or 6 by 12 in.* on 4 or 6 ft † centers) replace the standard floor and/or roof framing of smaller dimensioned members [2 by 8 in.*, 2 by 10 in.* joists or rafters on 16 in.* centers]. The decking for floors and roofs is planking in minimal thicknesses of 2 in.* or more as opposed to ½, ⅝, or ¾ in.* plywood sheeting. Instead of 2 by 4 in.* bearing partitions supporting the floor or roof joist or rafter systems, the beams are supported by posts. Plank and beam framing results in a saving in the number of members and allows for more rapid site assembly.

Balloon, Platform or Western, and Braced Framing: The basic difference between balloon, platform or western, and braced framing is in the manner in which the floor, roof, and bearing partitions are supported and/or erected.

In western or platform framing, stud bearing partitions are set on soles which are set on top of the joists with the

* nominal sizes, for comparison 1 in. =25.4 mm.

† nominal dimensions, for comparison 1 ft =0.3 m.

joists bearing on sills at the foundation level and girts at intermediate levels.

In both balloon and braced framing, the stud bearing partitions are set on sills at the foundation level. In balloon framing the studs in exterior bearing partitions extend for two stories with the joists supported by 1 by 6 in.* ledger strips or ribbon boards at the intermediate level. In braced framing the bearing partition extends for one story only with joists at the second level of bearing partitions bearing on 4 by 6 in.* girts. In both framing types the interior bearing partitions are erected in one story heights.

The use of wood foundation walls has been recently introduced as an alternate construction procedure; and crawl spaces have been utilized as return air plenums for a number of years.

The exterior wall finish of buildings of frame construction, except veneer, generally consists of stucco, wood or plastic siding, or metal siding having either plastic coating, baked enamel coating or in the case of aluminum, an anodized finish.

Veneer exteriors utilize one of the wood framing systems with the addition of a wythe of masonry secured by metal ties to the wood frame exterior walls. Usually an insulating board of some type is provided between the studs and the masonry. Although the exterior exposed surfaces are masonry, veneer construction is considered a combustible frame construction because the load bearing members are wood.

Type III (Ordinary) Construction

The significant difference between Type III (ordinary) and Type V (frame) construction lies mostly with the construction of the exterior walls. While in frame construction the load bearing components of the walls are wood, in ordinary construction the exterior walls are masonry or other noncombustible materials. The interior partitions, floor, and roof framing systems are of wood and utilize, in general, either the platform or braced framing methods previously described.

Manufactured Housing

During the period 1968 to 1977, a total of 3,884,380 mobile homes were manufactured in the United States and an additional 1,546,313 units were produced and shipped from 1978 to 1983 (Bernhard 1980 & MHI 1983). While many of these homes were put into service as construction offices, banking type occupancies, and other nonresidential occupancies, the great majority were utilized as dwelling units.

As more and more manufactured housing units were put into service, the fire service became increasingly aware that although the fire incident rate for manufactured housing was approximately the same as for conventional dwellings, the injury and life hazard and the extent of property damages were three to five times greater (Schaennon & Herrin 1982). While most fires in conventional dwellings are initially localized and small in size, the design characteristics inherent in manufactured housing such as fire load density, room geometry, and combustible finishes, may provide the potential for rapid flame development, resulting in higher temperatures and smoke and toxic gas generation in a relatively short time (Budnick & Klein 1979).

*nominal sizes, for comparison 1 in. = 25.4 mm.

Four major components or subassemblies—the chassis, floor system, wall system and roof system—comprise the structural systems of the manufactured house, which in essence, is designed specifically to meet the requirements of an efficient assembly line production.

The chassis is the structural base of the mobile home, receiving all vertical loads from the walls and floor and transferring them to a substructure of either the wheel assembly (when home is in transit) or to a foundation (when the home is finally erected at permanent site). The chassis generally consists of two longitudinal steel beams, braced by steel cross members. Steel outriggers cantilevered from the outsides of the main beams bring the width of the chassis to the approximate overall width of the superstructure. The running gear used when the unit is in transit is also considered part of the chassis.

The floor system consists of its framing members, generally conventional 2 by 6 in.* or 2 by 8 in.* wood joists, with plywood decking glued and nailed to the joists, fiberglass insulation blankets installed between the joists, and an asphalt impregnated insulation board sealing the bottom of the floor system. Duct work and piping are installed longitudinally within the floor system, often requiring the cutting of the joists. The floor finish is generally carpeting or resilient flooring such as linoleum or tile.

The wall systems, utilizing the stressed skin principle, consist of an interior skin of plywood which is glued and nailed to wood studs, and an exterior skin of aluminum siding. The actual sizing of members is dictated by the structural design analysis. Insulation batts are installed in the stud spaces.

The roof system consists of either the framed wood roof rafter and ceiling joist system or a wood truss system. Roof decking is generally a rigid insulation board attached to the top of the roof rafters or trusses. Finished roofing is galvanized steel or aluminum decking. The ceiling is either acoustical planks or gypsum wallboard attached directly to the bottom of the ceiling joists or the bottom chords of the trusses. Insulation blankets provide the roof insulation.

Steel tie plates reinforce connections between wall and floor systems; diagonal steel strapping binds the floors, walls, and roof into a complete unit.

CONSTRUCTION CONSIDERATIONS FOR FIRESAFETY

Of all the occupancy use group classifications regulated by building codes, one and two family dwellings are the least regulated insofar as firesafety and life safety are concerned. With the exception of fire resistance ratings for exterior walls, and, depending upon location of the wall with regard to distance separation from property lines or other buildings on the same lot and upon the separation between garages and habitable space, building codes do not mandate fire resistance ratings for building elements. The Minimum Property Standards for One- and Two-Family Dwellings (MPS) promulgated by U.S. Department of Housing and Urban Development (HUD) do establish specific fire resistance ratings for bearing walls and parti-

*nominal sizes, for comparison 1 in. = 25.4 mm.

tions, separations between garage and habitable space, and for all floor/ceiling and roof/ceiling assemblies. These regulations, however, are applicable only to those buildings for which Federal Housing Authority (FHA) or Veterans Administration (VA) loan assistance is provided. Even the interior finish requirements of the building codes and MPSs as related to one and two family dwellings are less restrictive than for other occupancy classifications. Thus, the one occupancy that is the source of the greatest number of fire deaths and injuries in the U.S. is probably the least regulated.

In response to the disproportionate fatality, injury, and property damage rate for conventional construction/manufactured housing, in 1976 HUD promulgated the "Mobile Home Construction and Safety Standards," later the "Manufactured Home Construction and Safety Standard," which was based on NFPA 501B, *Standard for Mobile Homes*, which has been withdrawn. These standards have been amended several times to strengthen requirements for interior finish and firestopping and to identify acceptance criteria limiting the use of materials that generate smoke and gases more toxic than those produced by untreated wood (HUD 1984). The number of deaths resulting from fires in manufactured housing decreased by one-third from 1977 to 1980. Although the fatality rate per hundred fires for manufactured housing is still slightly more than twice that for conventional dwellings, there is evidence that the HUD standards are making an impact on this statistic (Gates 1983).

Since the majority of one and two family dwellings are erected of frame or ordinary construction, the physical nature of these types of construction suggest the addition of certain built in features that will inhibit rapid fire spread. The large concentrations of combustible materials in the structural framework of these construction types, compounded by voids in concealed spaces connected to other voids, are basic conditions that contribute to the propagation and migration of fire throughout the structure. These conditions are addressed in the various building regulations used in this country. These regulations are clear in their requirements for the installation of draft stops and firestopping. Each of the three model building codes (BOCA 1984, SBICC 1985, ICBO 1985) and the *One and Two Family Building Code* (CABO 1983) contain, in one form or another, a statement similar to the following: "Firestopping shall be designed and constructed to close all concealed draft openings and to form effectual fire barriers against the spread of fire between stories of every building and in all open structural spaces therein." These code requirements also identify particular locations for installation of firestopping and draft stops. Figures 7-6M and 7-6Y illustrate the points to be firestopped in balloon and platform framing respectively; Figure 7-6Z provides expanded views of these points of firestopping.

The key to structural firesafety is to confine the fire to the room of origin, i.e., compartmentalize the structure so as to preclude the extension of the fire to areas adjacent to the room or area of origin. This is not achievable in the usual one and two family dwelling—it has a frame that is combustible and has interconnecting concealed spaces that can readily expose the entire structure to total involvement with the fire. Unrated gypsum wallboard, the material most commonly used as partition, wall, and ceiling diaphragms in the housing industry, affords a definite degree of fire protection when unpenetrated or when its penetrations are properly firestopped. But, since fire resistance ratings are not generally mandated for one and two family dwellings, the inspection of the installation of this material is not required, and unprotected penetrations are made for heating ducts, electrical wiring and appurtenances, and plumbing. To compound the problem, even though required firestopping is installed, it may be removed, cut or drilled to permit installation of electrical, plumbing, and mechanical work. Chapters 3 and 4 of *Building Construction for the Fire Service* (Brannigan 1982) address wood frame and ordinary construction in depth, pointing out both their attributes and problems regarding firesafety. The U.S. Army Corps of Engineers is in the process of promulgating a technical manual that further addresses firestopping of penetrations of fire resistance rated assemblies (USACE 1985).

Studies of fires in residential buildings of frame or ordinary construction reveal a number of built in contributing factors to the spread of fire (Vogel 1977). Included are:

1. Location of soffit or eave vents directly over exterior wall openings. This provides a direct channel for any fire breaking through the exterior wall opening to spread into the attic or roof space by the flue action of the attic or roof ventilation system. (See Fig. 7-12A.)

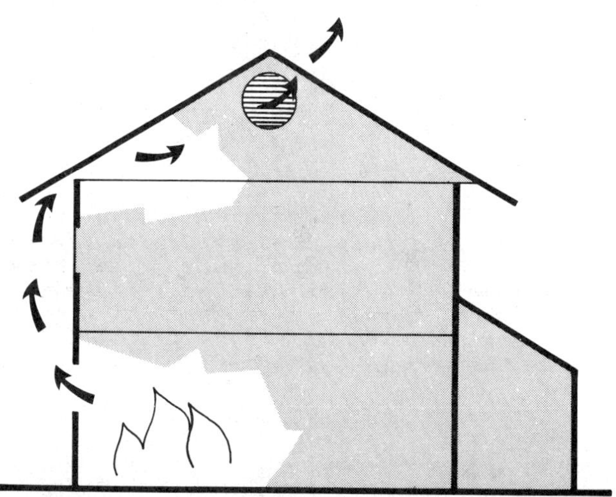

FIG. 7-12A. Attic vents under the eaves have serious potential for outside extension of fire into the attic space. (Francis L. Brannigan)

2. Mansard roofs extending below the top story windows with the windows recessed. The soffit extending from the face of the window to the edge of the roof is often a pressed fiberboard or plywood of a minimum thickness. In addition, the wall construction is not extended in back of the plane of the mansard roof, thus providing an interconnected void. A fire breaking through the window impinges on this soffit, burning through in a relatively short time and involving the attic and roof spaces.
3. Unfirestopped penetrations by plumbing, mechanical, and electrical work of the gypsum wallboard on interior partitions, previously discussed.
4. The widespread use of wood trussed roof framing

precludes the need for the interior bearing partitions normally required where rafter framed roofs are provided. Regulations required joists to be doubled under interior bearing partitions running parallel to joist and bridging to be installed under partitions running perpendicular to joists. This resulted in at least a minimum amount of firestopping of joist spaces. Since trussed roof framing in general does not require interior bearing partitions, nonbearing room division partitions can be located after the floor and roof framing systems are erected. With finished floors now consisting, in the most part, of carpeting laid on a plywood subfloor [versus the old practice of finished wood flooring laid over a nominal ¾ or 1 in. (19 mm or 25 mm) subflooring], there is very little barrier to fire burning through the floor and into the joist spaces and, subsequently, extending to other areas of the structure. (See Fig. 7-12B.)

5. Installing cabinets and other fixtures directly to stud framing without an intervening protective diaphragm presents a situation where the only firestops between a room fire and the concealed stud spaces are the materials that constitute the bottoms and backs of the cabinets or fixtures. (See Fig. 7-12C.) This same practice exists in the installation of unitized bathtubs or shower stalls where ceramic tile is not the water resistant medium.

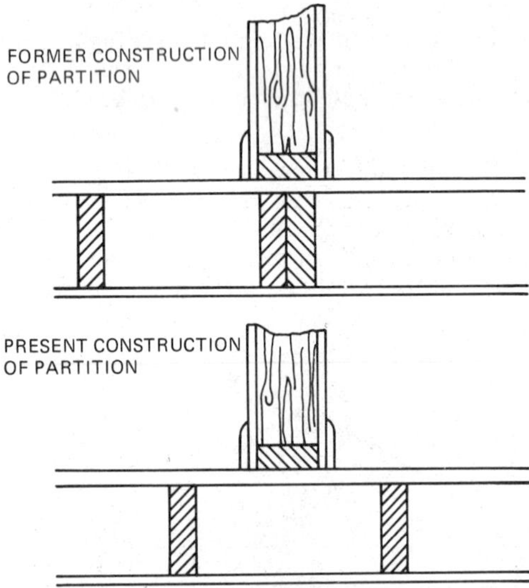

FIG. 7-12B. Conventional practices have been to build partitions over double joists, particularly where the partitions supported upper floors and roof members. New lightweight materials permit partitions to go anywhere on the floor. Fire burning down into the void under the floor can extend from one room to another. (Francis L. Brannigan)

FIG. 7-12C. Soffits (arrow) over kitchen cabinets connect wall voids to ceiling voids permitting fire extension both vertically and horizontally in concealed spaces. (Francis L. Brannigan)

Bibliography

References Cited

Bernhardt, Arthur D. 1980. *Building Tomorrow: The Mobile/Manufactured Housing Industry*. Massachusetts Institute of Technology, Cambridge, MA.

Brannigan, Francis, L. 1982. *Building Construction for the Fire Service*. National Fire Protection Association, Quincy, MA.

Budnick, E. K., and Klein, D. P. 1979. "Mobile Home Fire Studies: Summary and Recommendations." *NBSIR 70-1720*. Center for Fire Research. National Bureau of Standards, Washington, DC.

BOCA. 1984. *Basic/National Building Code*. 1984 ed. Building Officials and Code Administrators International Country Club Hills, IL.

CABO. 1983. *CABO One- and Two-Family Dwelling Code—1983*. Council of American Building Officials, Falls Church, VA.

Gates, Howard. 1983. *Fire Risk in Manufactured Houses-An Update*. Manufactured Housing Institute, Arlington, VA.

HUD. *Minimum Property Standards for One- and Two-Family Dwellings*. U.S. Department of Housing and Urban Development, Washington, DC.

HUD. 1984. *Manufactured Home Construction and Safety Standards*. U.S. Department of Housing and Urban Development. (rev. Oct. 1984.), Washington, DC.

ICBO. 1985. *Uniform Building Code* 1985 ed. International Conference of Building Officials, Inc., Whittier, CA.

MHI. 1983. *Manufacturing Reports*. Manufactured Housing Institute. Arlington, VA. Dec. 1974. Dec. 1981 and Dec. 1983.

SBICC. 1985. *Standard Building Code*. 1985 ed. Southern Building Code Congress International, Inc., Birmingham, AL.

Schaennon, P. S., and Herrin, C. L. 1982. *Evaluation of Mobile Home Fire Safety*. Tri Data., Arlington, VA.

USACE. 1985. "Firestopping." *TM-5-812-2*. U.S. Army Corps of Engineers.

Vogel, B. M. 1977. "A Study of Fire Spread in Multi-Family Residences: The Causes—The Remedies." *NBSIR 76-1194* Center for Fire Research. National Bureau of Standards, Washington, DC.

NFPA Codes, Standards, Recommended Practices and Manuals. (See the latest *NFPA Codes and Standards Catalog* for availability of the current edition of the following documents.)

NFPA 220, *Standard on Types of Building Construction*.

Additional Readings

Brannigan, Francis L., "A Field Study of Non Fire-Resistive Multiple Dwelling Fires," *NBS Special Pub. 411*, National Bureau of Standards, Gaithersburg, MD, Nov. 1974, pp. 178-194.

DeCicco, Paul R., "Fire Hardening of Old Residential Buildings in High-Risk Urban Communities," *SFPE TR 80-5*, Society of Fire Protection Engineers, Boston, MA, 1980.

Fang, J. B., "Fire Endurance Tests of Selected Residential Floor Construction," *NBSIR 82-2488*, National Bureau of Standards, Gaithersburg, MD, Apr. 1982.

Fang, J. B., "Fire Performance of Selected Residential Floor Constructions Under Room Burnout Conditions," *NBSIR 80-2134*, National Bureau of Standards, Gaithersburg, MD, Dec. 1980.

Fire Resistance Design Manual, Gypsum Association, Evanston, IL.

Issen, Lionel A., "Fire Endurance Tests of Residential Walls Containing Branch Circuit Wiring—Preliminary Findings," *NBSIR 78-1415*, National Bureau of Standards, Gaithersburg, MD, Feb. 1978.

ENERGY CONSERVATION IN BUILDINGS

Dr. T. T. Lie

In the wake of energy crises, worldwide efforts are being made to conserve energy. Energy conservation in buildings is an important objective in many countries, and various approaches have been taken to achieve it. Two well known methods for conserving energy in buildings are to increase the amount of insulation used and to seal air leaks. Another method is replacement of energy from nonrenewable sources, such as oil and gas, by energy derived from renewable ones such as the sun and wind.

Although these various measures have contributed to a more efficient use of energy, they have also created problems. Some fire problems, for example, are more ignition sources and an increase in the number of materials that may contribute to spread of fire and the production of smoke and toxic gases. The fraction of air recycled in heated and air conditioned buildings is often increased, which tends to increase the hazard of combustion products because of confinement of these products. Other examples are increases of rate of fire growth and fire temperature (with better insulation), and the reduction of fire resistance of structural elements in specific cases (due to insulation).

This chapter discusses the role of building energy conservation measures, particularly insulation, on the fire behavior of structures. Closely allied with the subject matter of this chapter are Chapter 7, "Interior Finish," of this section, and Chapter 3, "Combustion Products and Their Effect on Life Safety," of Section 3.

INSULATION

Most of the fire problems associated with energy conservation measures in buildings are related to increased use of thermal insulation in the buildings. Insulation affects a fire in all phases of its development from its inception to its decay. These phases will be discussed briefly because knowledge of the characteristics of the various phases of fire that will be considered in this chapter is important for understanding the influence of insulation on firesafety.

Dr. Lie is Research Officer, Division of Building Research, National Research Council of Canada.

Phases of Fire

Normally, a room fire starts as a small fire caused by ignition of a material by a heat source, for example, the flame of a match or the heat of a stove or hot electrical wire. Depending upon various factors such as the nature of the material that is exposed to the ignition source or the size of the ignition source, the fire may self-extinguish or grow to full development. Usually the fully developed stage of a fire is preceded by flashover, which is characterized by instantaneous ignition of materials in all parts of the room.

The course of a fire is usually divided into three periods—a growth period, a fully developed period, and a decay period. (See Fig. 7-13A.) In the growth period, i.e.,

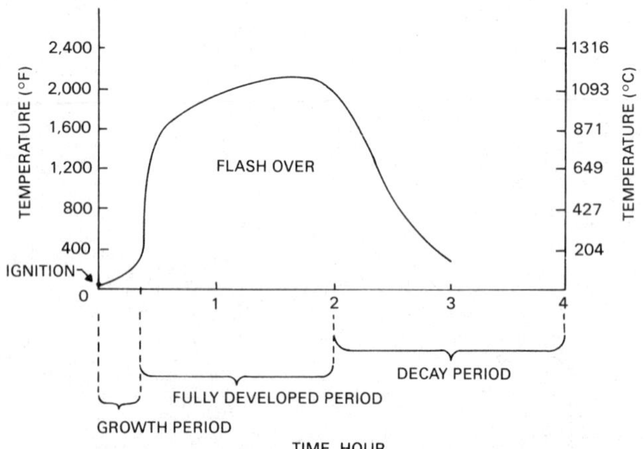

FIG. 7-13A. Temperature course of a typical fire.

the period before flashover, the temperatures in the room are relatively low, even when close to the burning materials, and evacuation from the fire area presents no problems. During flashover, however, the temperature rises very sharply to such a level that survival of the fire by persons still in the room at that stage becomes unlikely. Thus, the time interval between the start of a fire and the occurrence of flashover determines the time that is available for safe evacuation of the fire area.

There is negligible risk of structural failure in the growth period because of the low temperatures. Actual risk of failure of structural elements or of fire separations begins in the fully developed stage of the fire, after flashover. This risk also exists in the decay period, during which remnants of the combustible materials are consumed.

Influence on Ignition of Materials

If thermal insulation is installed to surround heat producing objects such as electrical wires and light fixtures, those objects may become ignition sources. The insulation will impede the dissipation of heat, which will result in heat buildup in the objects. They may become hot enough to ignite combustible materials in contact with them.

Insulation also plays a role in the ignition of a material when it is applied to the back of a thin lining; it reduces heat losses at the back of the lining. The increase of surface temperature at the front side of an insulated interior finish material exposed to an ignition source may be significantly more rapid than that of an uninsulated material. Therefore, the insulated interior finish material may ignite more rapidly than the uninsulated one. The influence of insulation is small for thicker materials, i.e., thicker than ¼ in. (6.3 mm) for cellulosic linings such as wood and board (Lie 1972a).

In addition to the cases mentioned in which insulation indirectly contributes to the ignition of materials, the insulation itself may be a potential ignition hazard. The increased use in recent years of a wide variety of insulating materials, particularly foam plastics, has created concern about this possibility. The plastics industry is now conducting extensive studies to investigate the fire performance of plastics to develop ways in which they can be used with reasonable safety.

Influence on Growth of Fire

During the growth of a fire, unburned material is successively ignited. The rate of growth of a fire depends upon many factors; one of the most important is the nature of the materials that contribute to fire growth. The faster the materials propagate fire and the more heat they develop, the faster the fire will grow. The heat developed by burning materials in the growth period is particularly important in enclosures of rather limited size such as, offices, living rooms, and classrooms. During the growth of the fire, the heat developed by the burning materials accumulates in the room. Because of this, other materials in the room may be heated so severely that after a short time they also ignite and at this stage, flashover occurs.

Adding insulation may significantly affect the rate of growth of a fire. Because of the insulation, heat from fire will be conserved in the room in the same way as heat from heating systems is conserved. Consequently, the accumulation of heat in the room is accelerated, and the flashover stage may be reached much earlier than in a less insulated room. According to estimates, the growth period reduces by a factor of about three by insulating a room of brick-plaster construction with a noncombustible lining of low thermal conductivity (Thomas & Bullen 1980).

In addition to accelerating fire growth by heat conservation, insulation may propagate flames itself, also affecting the growth of a fire.

Propagation of Flames

Once a material, e.g., interior finish in a room, is ignited, the flame may spread across the surface of the material. For flame to spread, heat transferred from the flame to the material must be sufficient to liberate flammable gases from the material near the flame front. Those gases will, in turn, be ignited. The gases are developed initially from the surface of the material; therefore surface temperature is an important factor in determining the attainment of a condition favorable to flame spread. The faster the temperature rise of the surface, the faster the propagation of the flames.

The rate of temperature rise of the surface depends to a high degree upon the thermal properties of the material. In particular, the thermal conductivity (K) and the thermal capacity (ρ c) of the material are important. The lower the thermal conductivity, the less of the heat received at the surface is transferred to the inside of the material. If, in addition, the thermal capacity of the material is small and thus relatively little heat is required to raise the temperature of the material, the temperature rise of the surface may be substantial in a short time. It can be shown for an idealized case that the time to reach a specific critical surface temperature—for example the temperature at which the surface "ignites" is proportional to a quantity, K ρ c, which is often termed the thermal inertia of the material (Carslaw & Jaeger 1959).

In actual practice, however, more factors play a role in the determination of the surface temperature rise. For the purpose of illustrating the role of thermal inertia in flame propagation the idealized case will be assumed in the following discussion (Thomas & Bullen 1979).

In Table 7-13A, typical values are given for the thermal inertia of some materials. The table also gives the time it takes to raise the surface temperature of the material to a level at which wood and plastic produce a significant amount of combustible gases, here assumed to be 400°F (204°C). It is further assumed that heat is supplied at a rate of 1,050 Btu per sq ft per hr (3300 W/m^2). At this rate wood would attain this critical surface temperature in 600 seconds, or 10 minutes. Under the same circumstances, fiber insulation board, which is known to propagate flames rapidly, would attain the critical surface temperature in 23 seconds; for foamed plastics it would take only a few seconds before its surface temperature reaches the critical value.

The very rapid temperature rise of the surfaces of foamed plastics is probably the reason why they may propagate flames with extreme speed; this applies particularly to nonmelting or thermosetting foam plastics. Thermoplastic foam may melt before its temperature reaches a critical value. In that case, the rate of temperature rise of the surface of the molten plastic slows, because the thermal inertia of the plastic surface increases and because heat is used in melting the plastic. Extremely fast propagation of fire is less likely to occur on thermoplastic foams that melt at relatively low temperatures [such as polystyrene and polyvinyl chloride, which melt at temperatures of the order of 250°F (121°C)] than on those that melt at higher temperatures. There are many thermoplastics, however, that do not melt before their surface reaches the critical temperature.

Protected Foam Plastic

Several building codes recognize a material with a flame spread rating of 25 or less [as ascertained by the tunnel test in NFPA 255, *Standard Method of Test of Surface Burning Characteristics of Building Materials* (hereinafter referred to as NFPA 255)] as a low fire hazard material. Tests carried out on urethane foam, however, suggest that foams may accelerate the growth of fire much faster than expected on the basis of tunnel test results.

Concern about this possiblity and the occurrence of a number of large loss fires involving foamed plastic has led to more stringent requirements regarding the use of foams. One of these requirements is that plastic foams should be protected by a thermal barrier capable of preventing excessive heat transfer from heat source to plastic during a specified time.

Methods exist for evaluating the performance of a material as a thermal barrier. One of them is the method described in NFPA 251, *Standard Methods of Fire Tests of Building Construction and Materials* (hereinafter referred to as NFPA 251), in which the area of the test specimen exposed to the test fire is at least 100 sq ft (9.3 m^2). To pass the test, the protection must be capable of preventing excessive temperature rise at the unexposed side for 15 minutes or more when one side of the test specimen is exposed to standard heating. Because the temperatures during standard heating are of the same magnitude as those encountered in the post flashover stage of a fire, the test method, which is intended to evaluate the performance of a protection in the preflashover stage of a fire, is a severe one. Temperatures in the preflashover stage of a house burn test are shown in Figure 7-13B. It can be seen that up to approximately 15 minutes, the fire temperatures are substantially lower than the NFPA 251 standard curve.

Another method, which is less severe but sufficient to evaluate the performance of a protection in the preflashover stage of a fire and which will reasonably ensure that the foamed plastic will not contribute to the growth of a fire for at least the time measured in the test, is described in ULC-S124 (ULC 1976). In this test, the thermal barrier is also exposed on one side to heating of a severity equal to that in the NFPA 251 fire test. The area of the specimen, however, is smaller, about 10 sq ft (0.93m^2), and the test also classifies thermal barriers that provide protection for less than 15 minutes.

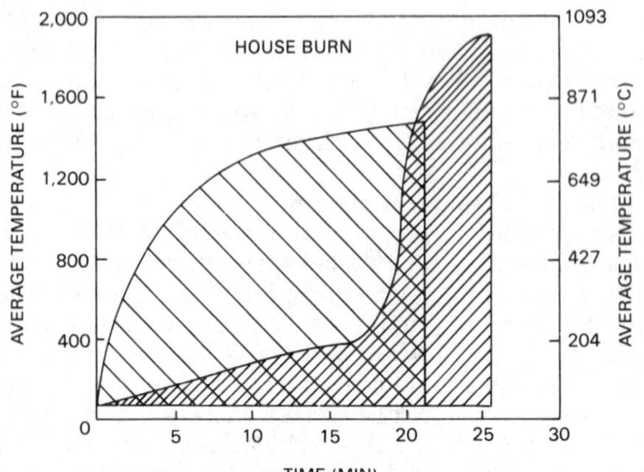

FIG. 7-13B. Temperature-time relationship according to NFPA 251, and for a house burn (Holmes et al 1980).

The times during which materials provide protection against a post flashover fire are given in Table 7-13B. Protection time in this table may be regarded as the approximate time it takes the temperature at the interface of the protective cover and the plastic foam to rise 250°F (121°C) above ambient. The total time before the plastic starts contributing significantly to the fire is thus equal to the time for the fire to grow from ignition to flashover, added to the protection time given in the table. If a combination of protecting materials is used, the total protection time may be assumed to be at least equal to the sum of the protection times of the individual protecting materials.

Because propagation of flame is a process that is determined mainly by the surface characteristics of the exposed materials, provision of a cover will reduce or remove the influence of the foamed plastic on fire growth. In this case, the propagation of flame will be determined by the surface characteristics of the cover material. If it has a low flame spread rating and its behavior is not significantly affected by the foamed plastic insulation, fast growth of a fire due to plastic insulation can be prevented in a room. It is possible, however, for fire to propagate in the cavity behind the cover, e.g., if the cover has a high thermal conductivity or openings are formed in it. It is also possible for fire to originate inside the cavity, e.g., because of overheating of electrical wires due to thermal insulation.

Foamed Plastic in Cavity Walls

Foam plastic in cavity walls is usually sandwiched between two layers of material forming the wall. Sometimes there is an air space between the plastic and the wall, and sometimes the wall material is in direct contact with the plastic on both sides of it.

Depending on the presence of an air space and the type of construction, i.e., whether the building is of noncombustible construction, a number of North American codes require that the insulation has to satisfy specific flame spread ratings determined using the test method described in NFPA 255. Because of doubt whether the contribution of insulation to the propagation of a fire in a cavity is related to its flame spread rating, in particular in those cases where there is no air space, tests were carried out to obtain more insight into the behavior of materials in cavities (Lie 1972). In these cases, the speed of propagation of the fire in the cavity is not determined by the speed of propagation over the surface of the plastic foam, but by the speed with which the plastic at the underside burns away.

The test results indicate that for materials sandwiched between two layers of noncombustible materials without air space, the upward propagation of the fire is slow, even for materials with a high flame spread rating. In Figure 7-13C, a picture is shown of a burned part of a urethane foam test specimen with a high flame spread rating (333 according to NFPA 255) after it has been exposed at the bottom to a standard fire (NFPA 251) for 6 hours. In this

TABLE 7-13A. Thermal Inertia (*Kqc*) and Rate of Surface Temperature Rise of Various Materials

[Rate of Heat Supply: 1050 Btu/(sq ft hr) (3300 w/m^2)]

Material	Density range lb/cu ft (kg/m^3)	Typical values of *Kqc* Btu^2/(sq ft °F)^2 hr [w^2/(m^2 °C)^{2}s]	Time to reach a surface temperature of 400°F (204°C) [sec]
Wood	30–40 (480–640)	1.7 (1.98×10^5)	600
Asbestos board	40–50 (640–800)	1 (1.16×10^5)	353
Gypsum board	40–50 (640–800)	0.6 (7.0×10^4)	212
Fibre insulation board	10–20 (160–320)	0.13 (1.51×10^4)	46
Cork board	5–10 (80–160)	0.06 (7.0×10^3)	21
Foam plastics:	1–3 (16–48)	0.01 (1.16×10^3)	4
Polyurethane			
Phenol-formaldehyde			
Polystyrene			
Cellular PVC			
Isocyanurate			
Fibre batts:	0.5–1.0 (8–16)	0.005 (5.81×10^2)	2
Glass fibre			
Mineral wool			

TABLE 7-13B. Time of Protection Against Postflashover Fire

Material of Protection	Thickness of Protection in. (mm)	Protection Time (min)
Gypsum Wallboard	3/8 (9.5)	11
Gypsum Wallboard	1/2 (12.7)	15
Gypsum Wallboard	5/8 (15.9)	20
Magnesium Oxychloride	1/4–3/8 (6.4–9.5)	10–15
Magnesium Oxychloride	1/2–5/8 (12.7–15.9)	20–25
Plywood	1/8 (4.9)	3
Plywood	1/4 (6.4)	5
Plywood	3/8 (9.5)	8
Plywood	1/2 (12.7)	11
Hardboard	1/8 (4.9)	3
Hardboard	1/4 (6.4)	5
Hardboard	3/8 (9.5)	8
Particleboard	1/4 (6.4)	4
Particleboard	3/8 (9.5)	6
Particleboard	1/2 (12.7)	8

TABLE 7-13C. Smoke Production of Various Unprotected Materials*

Material	Optical density	Time of obscuration (min)
Phenolformaldehyde foam	0.1	100
Wood	1	10
Boards	1-2	5-10
Cork	3	3.3
Polystyrene foam	5	2
Cellular PVC	10	1
Polyurethane foam	15	0.7

*Source (Lie 1968).

FIG. 7-13C. Test specimen after exposure to fire in a cavity wall (no air space, exposure time of 6 hrs). (National Research Council of Canada) (Lie 1972b).

time, the fire had propagated to about 14 in. (356 mm) above the part of the specimen that was exposed to the fire.

The fire may spread rapidly upwards if there is an air space between the wall and foam plastic. The speed of propagation depends not only upon the properties of the foamed plastic but also on several other factors such as the properties of the wall material, width of the air space, and flow of hot gases through the cavity (due to chimney effect). Research is now in progress to examine the influence of a number of important factors on the speed of propagation of fire in cavity walls.

Irrespective of the circumstances, however, spread of fire from story to story in a cavity wall can be prevented by the use of well designed fire stops. The installation of fire stops is prescribed in many codes. The role of fire stops has become very important due to the sudden increase in the use of insulation in cavity walls. Therefore much can be gained by measures that promote the installation of effective fire stops in multistory buildings.

Another consequence of the increasing use of combustible insulation in cavity walls is an increase in the chance of outbreak of fire in the cavity. On the other hand, a fire in a cavity is safer, from the point of view of safety to life, than a fire at the room side.

Insulation in Attics

Whereas fires in cavity walls are relatively low hazard fires, some fires in concealed spaces are potentially hazardous. This is the case if the space is wide and enclosed by combustible materials, as in attics. It is known that in row houses, fire, smoke, and toxic gases may spread quickly through such spaces from house to house.

Increase of building insulation and introduction of new insulating materials may have aggravated this situation. There is concern that widespread installation of thermal insulation may have adverse consequences on the performance of electrical wires, cables, and other electrical devices (Degenkolb 1978). Since such devices generate heat, they may become overheated during operation if they are covered by building insulation. Tests have shown that electrical equipment located in insulated cavities operated at higher temperatures than in uninsulated cavities. Another cause of overheating is the liberation of exothermic heat by insulating materials as they cure.

A type of insulation known as loose fill insulation is frequently used in attics. There are noncombustible insulating materials such as mineral wool, but often combustible insulation made of cellulosic material is used. These cellulosic insulations are generally made of pulverized or finely chopped paper, frequently waste paper (Degenkolb 1978). In this condition, the material is highly flammable and flame retardant compounds have to be added to reduce its flammability. If the materials are properly treated with flame retardants, which can be determined by testing, and fire stops are installed in large attic areas, cellulosic insulation can be utilized without excessive risk.

SMOKE AND TOXIC GASES

Fire statistics show that thermal decomposition products (smoke and toxic gases) are responsible for the majority of fire deaths. Many new materials release harmful decomposition products very rapidly, and some of them produce much more smoke or toxic gas than traditional building materials. With the increased use of these new materials, the problem of smoke and toxic products has become a subject of very real concern.

Regulations to exclude the use of materials with excessive smoke production and methods to determine smoke density are specified in several codes (BOCA 1975, ICBO 1976, NRCC 1977). However, there are as yet no accepted standard methods for evaluating toxicity, although considerable effort is being made towards developing standard procedures for determining fire toxicity of materials.

Smoke

The main danger from smoke is that it reduces visibility. Often it will impede the escape of occupants from a burning building and prolong exposure to the harmful effects of fire such as the smoke itself, toxic products, and heat.

One of the most important factors that determine the reduction in visibility in a given space is the concentration of the smoke generated in that space. A measure of this concentration is the optical density of the smoke, which is a quantity that is directly proportional to the smoke concentration. In Table 7-13C, the second column lists relative optical densities of smoke generated by various materials under similar conditions. The third column lists times required to obscure a given room under the same conditions, assuming that wood smoke would obscure the room in 10 minutes. Although wood is considered here as a single material, there are differences in the smoke production among various species of wood. The figures therefore, should be regarded as approximate and intended to show the order of magnitude of the smoke production of various materials under one specific test condition.

It can be seen that some materials used as insulation, i.e., cork and some foamed plastics, produce significantly more smoke than wood. If the insulation surface area is large in comparison with that of the other combustible materials in the room and they are unprotected, smoke production will be dominated by the insulation. In this case, the time of obscuration, which is a measure of the time available for safe evacuation of the fire area, may drop to levels substantially lower than that if the smoke had been produced solely by wood. If, however, the insulation surface area is a small fraction of the surface area of the combustible materials in the room, not more than ten percent of the surface area of the fire load, smoke contribution of the insulation will not be excessive in comparison with the total smoke produced by all burning materials.

If the insulation is protected, its smoke production may drop to very low values. Table 7-13D shows the influence of protection consisting of ½ in. (12.7 mm) Type X gypsum wallboard on the smoke production of a number of materials.

TABLE 7-13D. Influence of Protection on Smoke Production*

[Protection of ½ in. (12.7 mm) Type X gypsum wallboard]

Material	Optical density	Time of obscuration (min)
Wood (unprotected)	1	10
Cork:		
Unprotected	3	3.3
Protected	0.06	180
Polystyrene foam:		
Unprotected	5	2
Protected	0.8	13
Polyurethane foam:		
Unprotected	16	0.6
Protected	4	2.5

*Source (Lie 1968).

Toxic Gases

Toxic gases can be fatal if they are present for a sufficient time and in sufficient quantities. In recent years, various new materials, especially synthetic polymers, have found increasing use in buildings, and their introduction has heightened the concern of fire authorities over toxic combustion products. Part of this increased concern has arisen because of the lack of information on toxic combustion products and the problem of assessing their potential hazard.

Comprehensive chemical analysis is necessary to identify unusual toxicants that may be produced during fires. For most practical applications, however, this is not necessary since testing for a few of the most important known toxicants often gives sufficient information. Some compounds such as carbon monoxide (CO), hydrogen chloride (HCl), hydrogen cyanide (HCN), sulfur dioxide (SO^2), and oxides of nitrogen, are recognized as harmful products; others such as water vapor and the hydrocarbons, contribute little or no toxic hazard. It is usually sufficient to decompose materials under specified conditions and determine the resulting concentrations of a few of the most important toxicants. From this information, a reasonable indication of the toxicity of the mixture of products can be obtained.

Table 7-13E shows the main harmful products of a number of materials and approximate values of the harmful concentrations for an exposure time of 30 minutes. According to the figures in this table, one of the most toxic products is HCN, which is harmful at concentrations of about 1/30th of the harmful concentration of CO. If the materials that produce HCN are present in small quantities, however, their contribution to toxicity may be less than that of the CO normally produced during a fire. This is also the case if the materials that generate very toxic products are properly protected. According to measurements made during tests in rooms with walls insulated with isocyanurate foam and protected with ¼ in. (6.2 mm) Douglas fir plywood, no HCN was observed in the preflashover period, up to 21 minutes after the ignition (Holmes et al 1980). The results indicate that from the point of view of hampering safe evacuation of people by toxic combustion products, the materials in the room play a far greater role than the insulation in the walls if the insulation is adequately protected, for example, by boards or plywood of not less than ¼ in. (6.4 mm) thick.

HEATING SYSTEMS

Escalating prices of oil and uncertainties in its supply have stimulated the use of heating systems that do not operate on oil. Such heating systems, for example, utilize solar energy, wood, coal, and natural gas. Coal and gas have been used as main fuels for heating of buildings for a long time in several countries. The use of solar energy is relatively new, and wood is enjoying renewed interest as a household heating fuel. In some areas, their use for heating buildings has increased tremendously.

Most of the fire risk introduced by these systems is related to the high temperatures that they may attain. For example, 400°F (204°C) for solar collectors is not abnormal, whereas wood stove encasements may reach temperatures of over 1,000°F (538°C). There is the risk that combustible materials may ignite spontaneously after prolonged exposure to these temperatures.

BUILDING STRUCTURE

Fire Severity

Actual risk of failure of structural elements or fire separations begins after flashover in the fully developed stage of a fire. The temperature of the fire in this stage will be affected by thermal insulation in the boundaries, such as walls and roofs, of the rooms; the greater the amount of insulation, the higher the fire temperature. Less of the heat generated by the burning materials in the room will be absorbed by the boundaries, and as a consequence, the fire gas temperature will increase. The higher temperature in turn will accelerate the rate of decomposition of the materials in the room, which may result in an increase in the heat released in the room. According to existing information, however, this rate is a very weak function of the temperature of the fire gases (Harmathy 1979). Therefore the only significant effect on fire severity caused by adding insulation is expected to be a rise of fire tempera-

TABLE 7-13E. Main Harmful Products of Materials and Harmful Concentrations*

Material	Harmful Product	Harmful concentration in parts per million parts of air (30-min exposure)
Wood and Paper	Carbon Monoxide (CO)	4,000 ppm
Polystyrene	CO—also styrene, but present in smaller quantities	
Polyvinyl Chloride (PVC)	HCl—also corrosive CO	1,000–2,000 ppm
Plexiglas or perspex	CO—also methylmethacrylate, which is as toxic as CO but produced in smaller quantities	
Polyethylene	CO	
Ureaformaldehyde Isocyanurate Polyurethane Acrylic Fibres Wool Nylon	Hydrogen Cyanide (HCN) CO	120–150 ppm

Sources: (Sumi & Tsuchiya 1978; Harmathy 1979; Woolley et al (1979).

tures; whereas, the duration of the fire will be practically unchanged.

In addition to causing higher fire temperatures, insulation in roofs and walls also directly influence the temperatures reached in them by altering the heat transfer through the roof or walls. These temperature changes may affect the fire performance of structural elements and fire separations in various ways.

Roofs

Roof constructions in which fire performance may be affected by the addition of insulation are those supported by steel members such as joists and beams, and those supported by steel wires, such as reinforcing and prestressing steel. Steel loses strength when it is heated to high temperatures; the higher its temperature, the lower its strength. In general, a critical steel temperature can be indicated at which the steel has lost so much strength that it can no longer support the load. Often a temperature of 1,100°F (593°C) is assumed as the critical temperature of structural steel members and reinforcing steel, and a temperature of 800°F (427°C) as the critical temperature of prestressing steel.

Typical roof constructions that are supported by steel members are assemblies consisting of a steel deck supported by steel beams or joists and covered with concrete. Often the steel is protected from attaining excessive temperatures during a fire by a fire resistive ceiling. At present, more roof insulation is placed on top of the concrete than in the past, to conserve energy. Because the insulation reduces heat losses to the air above the roof, all temperatures inside the structure, including the temperatures of the supporting beams, will be higher with increased insulation. If the concrete is sufficiently thick, however, the increase in temperature will be small. According to tests, the addition of insulation on top of the concrete has no significant effect on the fire resistance of the roof assembly if the thickness of the concrete is 2 in. (51 mm) or more (Stanzak & Konicek 1979). If there is no concrete cover and the insulation is placed directly on the steel deck, the fire resistance of the roof assembly may be reduced substantially. The influence on the fire resistance of the assembly depends upon the thermal properties and thickness of the insulation. A maximum thickness can be found beyond which there will be no further reduction in the fire resistance of the assembly. The order of magnitude of this thickness is 1.5 in. (38 mm), which was found during tests with a commonly used noncombustible insulation.

Typical roof constructions that are supported by steel wires are those consisting of reinforced or prestressed concrete slabs. If insulation is added on top of the concrete, the temperature of the reinforcing or prestressing wires may rise more rapidly than if the concrete is uninsulated. The influence of insulation on the temperatures of the wires depends upon various factors such as the properties on the insulation, position of the wires in the concrete, and the thickness of the concrete. If the concrete is not very thin, say not less than 3 in. (76 mm) for normal weight concrete, the influence is small. For lightweight concrete, the influence of adding insulation is less than for normal weight concrete.

Walls

If walls are lined on the room side with an insulating material, the fire temperatures will increase. On the other hand, the temperatures in the wall will be lower, and as a result, the fire resistance of the wall will improve. It is likely that lining a wall will have a favorable effect on its structural fire performance, particularly if the lining is a good insulator.

If the insulation is installed in a cavity, however, the effect on the structural fire resistance of the wall may be unfavorable. This is particularly true for thin walls in which unequal expansion of the wall layers on each side of the cavity may produce large deflections and cause the wall to buckle.

Whereas insulation may reduce the structural fire performance of fire barriers such as roofs, floors and walls, it increases their thermal fire resistance, i.e., their ability to prevent excessive heat penetration through the fire barriers.

Bibliography

References Cited

BOCA. 1975. *The BOCA Basic Building Code*. Building Officials Conference of America, Inc., Chicago, IL.

Carslaw, H. S., and Jaeger, J. C. 1959. *Conduction of Heat in Solids*, 2nd ed. Oxford University Press, London, England.

Degenkolb, J. H. 1978. "Energy Conservation and Its Effect on Buildings from the Fire Protection Viewpoint." *Fire Journal*. Vol 72, No 3. May 1978. pp 86-90 and Vol 72, No 6. Nov. 1978, p 8.

Harmathy, T. Z. 1979. "Effect of the Nature of Fuel on Fully Developed Compartment Fires." *Fire and Material*. No. 3. pp 49-60.

Holmes, Carlton A., et al. (1980). "Fire Development and Wall Endurance In Sandwich and Wood-Frame Structures." *Research Paper FPL 364*. Forest Products Laboratory. U.S. Department of Agriculture.

ICBO. 1982. *The Uniform Building Code*. 1982 ed. International Conference of Building Officials, Pasadena, CA.

Lie, T. T. 1968. "Smoke Contribution of Some Partition Components During a Fire Test." *Technical Note 524*. National Research Council of Canada. Ottawa, Ontario, Canada.

Lie, T. T. 1972a. *Fire and Buildings*. Applied Science Publishers Ltd., London, England. pp 26-27.

Lie, T. T. 1972b. "Contribution of Insulation in Cavity Walls to Propagation of Fire." *Fire Study No. 29*. National Research Council of Canada. Ottawa, Ontario, Canada.

NRCC. 1977. *National Building Code of Canada*. National Research Council of Canada, Ottawa, Ontario, Canada.

Sumi, K., and Tsuchiya, Y. 1973. "Combustion Products of Polymeric Materials Containing Nitrogen in Their Chemical Structure." *Journal of Fire and Flammability*. p 15-22.

Sumi, K., and Tsuchiya, Y. 1978. "Evaluating the Toxic Hazard of Fires." *Canadian Building Digest*. National Research Council of Canada. Ottawa, Ontario, Canada.

Stanzak, W. W., and Konicek, L. 1979. "Effect of Thermal Insulation and Heat Sink on the Structural Fire Endurance of Steel Roof Assemblies." *Canadian Journal of Civil Engineering*. No. 6. pp 32-35.

Thomas, P. H., and Bullen, M. L. 1980. "The Effects of Insulation in Fires." *Fire*. pp 434.

Thomas, P. H., and Bullen, M. L. 1979. "On the Role of K c of Room Lining Materials in the Growth of Room Fires." *Fire and Materials*. No. 3. pp 68-73.

ULC. 1976. *Standard Method of Test for the Evaluation of Protective Coverings for Foamed Plastic*. ULC-S124. Underwriters' Laboratories of Canada. Toronto, Ontario, Canada.

NFPA Codes, Standards, Recommended Practices and Manuals. (See the latest *NFPA Codes and Standards Catalog* for availability of current editions of the following documents.)

NFPA 251, *Standard Methods of Fire Tests of Building Construction Materials*.

NFPA 255, *Standard Method of Test of the Surface Burning Characteristics of Materials*.

Additional Readings

Belles, Donald W., "Fire Hazard Analysis of Foam Plastic Insulation in Exterior Walls of Buildings," *SFPE TR 82-1*, Society of Fire Protection Engineers, Boston, MA, 1982.

Choi, K. K., "Combustion of Insulation in Cavity Walls," *Journal of Fire Sciences*, May/June 1984.

Henderson, Russell L., "Reflective Insulation," Costa Mesa Fire Department, Costa Mesa, CA, Apr. 15, 1980.

Lawson, J. R., "Environmental Cycling of Cellulosic Thermal Insulation and Its Influence on Fire Performance," *NBSIR 84-2917*, National Bureau of Standards, Gaithersburg, MD, Aug. 1984.

Mazzoni, Steve, "Safety Considerations of Energy Savings Materials and Devices," *SFPE TR 78-6*, Society of Fire Protection Engineers, Boston, MA, 1978.

McCarter, R. J., "Combustion Inhibition of Cellulose by Powders —Preliminary Data and Hypotheses," *Fire and Materials*, Vol. 5, No. 2, June 1981, pp. 66-72.

Melott, Ronald K., "Is Energy Conservation Firesafe?" *SFPE TR 78-8*, Society of Fire Protection Engineers, Boston, MA, 1978.

Ohlemiller, T. J., and Rogers, F. E., "Cellulosic Insulation Material —Effect of Additives on Some Smolder Characteristics," *Combustion Science and Technology*, Vol. 24, 1980, pp. 139-152.

Prusaczyk, J. E., "Fire Performance Characteristics in Rooms as the Result of Increased Insulation," *SFPE TR 78-2*, Society of Fire Protection Engineers, Boston, MA.

Schwartz, Kenneth J., "Effects of Thermal Insulation on Fire Resistive Assemblies," *SFPE TR 81-7*, Society of Fire Protection Engineers, Boston, MA, 1981.

SPECIAL STRUCTURES

Revised by John G. Degenkolb

Special structures constitute a variety of constructed items that are designed for specific purposes. They are also different from what is considered typical day to day uses or occurrences for most structures. Typical examples of special structures are towers, tunnels, underground buildings, bomb shelters, highway and railroad bridges, grandstands, tents, membrane structures, signs, mechanical parking garages, marine terminals, piers and water surrounded structures used for other than vessel mooring or cargo handling, and air right structures. Because these structures represent unusual uses or solutions to design problems, they also present unusual or special fire protection design problems.

Additional relevant information may be found in Section 5, Chapter 3, "Fibers and Textiles;" Section 7, Chapter 3, "Concepts of Egress Design;" and Chapter 4, "Building and Site Planning for Building Firesafety;" Section 9, Chapter 2, "Places of Assembly," and Chapter 16, "Occupancies in Unusual Structures;" Section 13, Chapter 4, "Fixed Guideway Transportation Systems," and Section 19, "Special Fire Suppression Agents and Systems."

TOWERS

The major fire problems with towers are: (1) the difficulty of extinguishing fires that may occur high in the tower, (2) limited means of egress for occupants of lookout, air traffic control, and observation towers (this prompts the use of noncombustible materials throughout, and good fire prevention practices), and (3) vulnerability to exposure fires. When towers are built of unprotected steel or wood, a ground fire can cause ignition or structural failure, or both, of the tower. Towers are frequently appended or attached to other structures of lower height. In such cases, egress from the tower should be directly outside where possible. If this is not feasible, the building exits should be designed with consideration to their use by occupants of the tower. Apart from fire damage, but extremely important from a life safety viewpoint, is the possibility of windstorm damage to towers. The Federal Aviation Agency (FAA) has been active in developing building code requirements for the towers which they occupy.

Water Cooling Towers

Functionally, a water cooling tower is a structure designed for the transfer of heat from water to air. This is accomplished by passing the water in small particles through the air where heat is removed from the water partly by an exchange of latent heat (heat required to pass from liquid to a gas) due to the evaporation of some of the water particles, and partly by a transfer of sensible heat (heat required to change the temperature). Figure 7-14A shows a cross section of a typical mechanical induced draft water cooling tower and some of its significant parts.

Water cooling towers may be all metal, metal frame with wooden or plastic fillers and wooden or plastic draft eliminators, all wood (generally redwood), or ceramic construction. Some towers have a noncombustible exterior covering. Wood is used in cooling tower construction primarily because of its durability, the absence of corrosion problems, and economics both in the cost of the original material and for replacement.

Water cooling towers are of atmospheric and mechanical draft types. Mechanical draft towers are of two general types—forced and induced draft. They may be further classified by design as counterflow and crossflow.

Contrary to a popular misconception, fire records of water cooling towers indicate that approximately two-thirds of the fires studied occurred while the tower was in operation. Despite appearance to the contrary, towers do have some relatively dry areas even when operating, particularly the induced draft types.

The majority of cooling tower fires are caused by ignition from outside sources. Most common of these sources are sparks from welding and cutting (either on the tower or on adjacent structures), or sparks from trash fires, incinerators, chimneys, and industrial stacks. Other causes include carelessly discarded smoking materials, lightning, and exposure fires. Another problem is the buildup of windblown leaves and trash that has collected around towers and is easily ignited.

Mr. Degenkolb is a Fire Protection Engineer and code consultant in Carson City, NV.

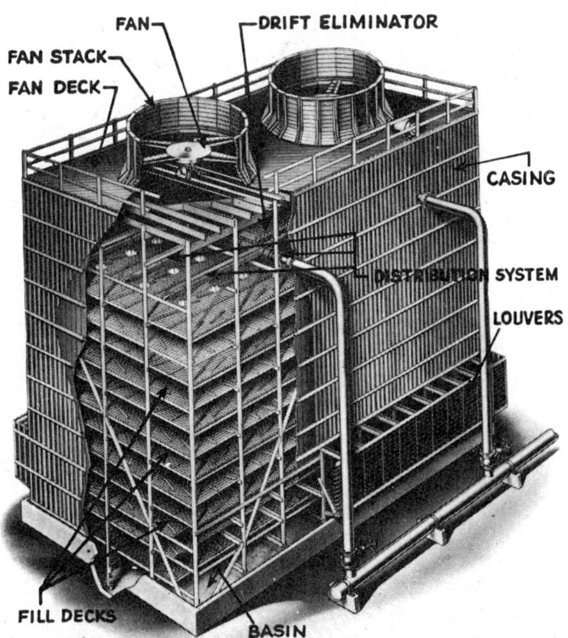

FIG. 7-14A. A typical mechanical, induced draft, counterflow water cooling tower. Water is pumped to the distribution system where it is cascaded over the fill decks and collected in the basin at the base of the tower. The fan draws in air through the louvers and the casing of the tower. Drift eliminators minimize loss of water through the fan stacks.

Fires caused by tower equipment originate from mechanical failures in gear reduction boxes and bearings, misaligned fan blades, or metal fatigue, which can result in localized heating. Electrical breakdowns in motors, short circuits in wiring, and other electrical faults also account for many cooling tower fires.

The construction of the tower is a primary factor in considering the possible fire hazard of a water cooling tower. Small towers [those of main structure not exceeding 2,000 cu ft (57 m^3) in volume] are generally completely noncombustible and therefore are seldom a problem. Larger towers, however, generally contain considerable amounts of combustible materials. Towers on roofs, for example, may be inaccessible to automotive fire fighting equipment; those in yard areas may require fencing and weed control or protection to prevent exposure damage to or from an adjacent building. Once the degree of hazard has been determined, any one or a combination of the following protection measures may be considered when dealing with essentially combustible towers:

1. Adequate separation from exposures.
2. Noncombustible exterior coverings.
3. Division of areas by fire partitions between cells.
4. Automatic fire protection systems (deluge systems or dry pipe or wet pipe automatic sprinkler systems, the latter in warm climates only).
5. Automatic water spray systems (exterior of the tower only).
6. Hydrants and standpipes.
7. Lightning protection.

The key element in a fire protection analysis is to provide a combination of prudent, engineered tower design with sufficient consideration to automatic extinguishment plus good preventive maintenance. This is all directed at providing assurance that the fan deck, fan, and related motor and drive survive intact the effects of a fire incident. This is recommended for two reasons:

1. The entire tower is considerably damaged when the fan deck collapses, dropping it and its substantial weight of equipment down through the tower during a fire.
2. Downtime of the tower is considerably lessened when only the louvers and fill deck material need to be replaced. This is critical since downtime on a cooling tower may cause downtime on a key income producing manufacturing process.

Cooling towers, like other outside structures, must be designed and constructed to prevent damage from windstorms. Care also should be taken that the towers are not under, or adjacent to, high voltage wires or transformers.

Fire prevention involves the usual control of the fire causes previously discussed, plus a regular maintenance program. Scheduled maintenance of mechanical parts for overheating, excessive wear, inadequate lubrication, etc., is an essential part of this program. Preventing fires and fire damage in water cooling towers is covered in detail in NFPA 214, *Standard on Water-Cooling Towers.*

Television, Microwave, and Radio Transmitting Towers

Television and radio transmitting towers frequently have electronic equipment at such heights that fire fighting from the ground is impossible. In such situations an automatic carbon dioxide, halogenated agent, or dry chemical extinguishing system, with an interlock to cut off power to the electronic equipment, should be considered as practical protection. In any case, it is good practice to have a means of shutting off the power to the tower from the ground.

Microwave towers, frequently located in remote areas, sometimes have equipment buildings near their bases for relays, standby generators, and batteries. These buildings should be constructed of noncombustible materials, be protected by automatic detection and extinguishing systems, and have the means for transmitting an alarm to a base station.

Airport and Railroad Control Towers

Airport and some railroad control towers present special fire problems because of the expensive, complicated electrical and electronic equipment normally installed in them. Automatic fire extinguishing equipment is usually justified by the importance of the tower equipment.

In recent years the Federal Aviation Agency (FAA) has proposed amendments for building codes to establish basic guidelines for aviation control towers concerning height and area limitations, limitation on activities, etc. Wood frame buildings, whether protected or not, as well as unprotected noncombustible buildings, are not acceptable. While one means of egress (an enclosed stairway) is accepted, the elevator hoistway and the stairway may be in a common shaft enclosure provided they are separated from each other by an unpierced fire wall and access to each way is well separated from the other. Smoke detection is required for each level. Standby power, independent of that needed for aircraft control purposes, is required for towers over 65 ft (19.8 m) in height. Such towers need not be accessible to the handicapped (ICBO 1985).

Forest Fire and Other Observation Towers

Forest fire and other observation towers should be of noncombustible construction throughout, not only because of the possibility of ignition of the base from a ground fire, but also to avert any fire in the upper structure of the tower. The problem of fire protection in these towers is especially important since they are normally occupied by observation personnel and are remote from fire suppression facilities.

Underground Buildings

The inaccessibility of underground structures results in some unique fire problems. Primary among these are the venting of smoke and gases from fires, and the difficulty in fire fighting and evacuating occupants. The solution of all these fundamental problems in relation to underground structures has only recently been introduced into building codes and standards. The inaccessibility of underground structures eliminates any thought of partial fire protection facilities that might suffice in an aboveground building where dependence is placed upon the public fire service. On the contrary, protecting occupants and property in underground structures prompts maximum utilization of good fire protection.

Regulations for underground buildings are beginning to be included in building code requirements as well as in NFPA 101, *Life Safety Code®* (hereinafter referred to as NFPA 101). The underground buildings regulated are those having an occupied level more than 30 ft (9.1 m) below grade. Generally speaking, they must be provided with automatic sprinkler protection, and in addition when over a certain area per level, each level must be divided into not less than two compartments so egress can be made from an endangered area to a safe area at the same level. To assure that the two levels are well separated from each other, each section is required to be air conditioned separately. Emergency power, mechanical means for evacuating the building, and smoke management are other basic requirements for underground building protection.

Many of these protective measures are also applicable to windowless buildings, since the only advantage such buildings have over underground buildings is that direct access for occupants to the outside can be provided. Windowless buildings provide the same problems of venting, fire fighting, and rescue as underground buildings. They are not intended to be all inclusive nor to eliminate the usual fire protection considerations typical of buildings.

Means of egress from underground buildings are provided for in NFPA 101. Those recommendations take into consideration the panic that may result when there is no direct access to the outside and no windows to permit fire department rescue and ventilation.

Automatic sprinklers must be installed in all such buildings because of combustible contents and concealed spaces, even though construction may be noncombustible. Underground mines and caverns are being converted for use as buildings without specific building construction being incorporated and as such, are to be considered as basic noncombustible construction.

When possible, venting facilities must be provided to exhaust smoke and fire gases from the buildings. Storage buildings may require extensive venting facilities, whereas buildings with low fire loading, noncombustible construction, and automatic sprinklers protection may be cleared of smoke through the ventilation system which provides for normal changes of air.

Emergency lighting for safe evacuation is required since artificial light is a necessity in underground buildings.

The use of combustible construction and other than Class A interior finish should be prohibited. Foam plastics should never be left exposed but should be protected by a thermal barrier acceptable to the authority having jurisdiction.

Compartmentalization (by installation of fire barriers) will aid in limiting the extent and severity of the fire, and provide areas of refuge for occupants.

Manual fire fighting equipment such as standpipe and hose systems and portable fire extinguishers can provide quick extinguishment of fires when handled by trained personnel who may discover a fire in the incipient stage.

Drainage of water from sprinkler discharge or from hose streams can be a serious problem in underground structures and therefore requires early planning. If provisions for drainage are required for other than fire protection reasons, it may be possible to use one drainage system for all purposes, provided it is designed to handle the expected maximum flow.

Tunnels

Tunnels for automotive vehicles and tracked vehicles (trains and subways) present fire problems similar to those of underground buildings, except that the cargo of the vehicles using tunnels may present any number of hazards. The transportation of certain cargoes such as high explosives, through tunnels by automotive vehicles is prohibited in some cities.

Because gasoline as an automotive fuel is commonly involved in automotive fires, handling the hazard of gasoline spill fires is a typical consideration in providing fire protection for tunnels.

Providing means for giving a fire alarm in a tunnel is a primary consideration. This may be satisfied by fire alarm boxes or telephone stations spaced at intervals along the tunnel. This alarm system can also be tied into the traffic control system so that incoming traffic can be stopped. Automatic sprinklers provide good tunnel protection. In addition, an adequate water supply with hose connections should be provided on both sides of the tunnel. Supplemental portable fire extinguishers for handling small fires can be installed in wall cabinets for use by attendants and motorists.

NFPA 130, *Standard for Fixed Guideway Transit Systems* (hereinafter referred to as NFPA 130), provides information relative to fire safety for subways, trains, etc. NFPA 130 provides guidelines for both the vehicles themselves and for the passenger stations.

Fire apparatus equipped to pull disabled vehicles from tunnels, combat vehicle fires, and extricate accident victims should be stationed within a reasonable distance of each end of tunnels. Personnel assigned to such locations should have special training and specialized equipment.

As with underground buildings, ventilation in tunnels is important to dissipate smoke and fire gases. An existing mechanical ventilation system may be satisfactory for emergency use if it is capable of handling the volume required to exhaust smoke and fire gases. Low points in the tunnel should have suitable drainage to handle flammable liquid spills and water used for extinguishment.

Pedestrian Tunnels

Pedestrian tunnels, normally used for entrance to, and egress from, buildings, terminals, subway stations, etc., primarily involve safety to life. Tunnels are commonly used to satisfy exit requirements for large industrial and commercial facilities. Detailed requirements for such tunnels are contained in NFPA 101.

Bomb Shelters

Shelters designed to protect people from the effects of bombing, especially nuclear bombs, are underground structures and therefore subject to the same general fire protection factors as other undergound structures.

Areas of refuge designated as bomb shelters in buildings may be considered as temporary expedients; nevertheless, they should be chosen with care, with consideration to such matters as alternate means of egress, noncombustible construction, sprinklered areas, proximity to gas mains, water pipes, other building services that may be damaged by explosion or fire resulting from bombing, and similar existing factors that could protect or unnecessarily expose people seeking refuge. Storage of emergency supplies should be carefully planned to avoid the problem of exposed combustibles.

BRIDGES

Highway Bridges

Highway bridges are continually exposed to accidents that may cause direct damage from impact or, occasionally, fires that may follow that impact. When flammable cargoes are involved, either on the roadway or on adjacent lands or waters, there is a potential for a serious fire. The record reveals remarkably few incidents however, considering the hundreds of thousands of bridges in service that warrant special fire protection to the structures themselves. Fire resistive or noncombustible construction and means to transmit emergency alarms on longer structures generally suffice. Established response to accidents should include fire fighting capability.

Fire alarm boxes or telephones spaced along a bridge provide means of notifying the fire department or bridge attendants of an emergency condition. As in tunnels, this system can be tied into a traffic control system to stop traffic from driving onto the bridge and to alert vehicles already on the bridge.

Bridges in critical seismic areas may require special anchoring to assure continued use in times of severe earthquake activity.

Since fire fighting is a necessity on bridges, adequate water supply from standpipes is good practice, preferably on both sides of the bridge or on long bridges at intervals along its length.

The location of apparatus to combat bridge fires can be a problem if a bridge is located in a rural area. It is desirable, however, that apparatus be stationed as near the bridge as is practical. In addition to having fire fighting equipment, this apparatus should be designed to move disabled vehicles and should have the equipment needed for rescue.

Railroad Bridges

Railroad bridges are a fire problem because so many of them are constructed of wood, which may have been treated with a flammable wood preservative. There are two kinds of wooden bridges and trestles; open deck and ballast deck. The fire potential is considerably more severe in open deck design. Many railroad bridges are located in remote areas and therefore, if ignited, are generally a complete loss.

Wooden railroad bridges are subject to ignition by such sources as friction sparks and hot metal particles generated when brakes are applied on heavily loaded freight trains on steep downgrades, and burning waste from overheated journal boxes. Exposure fires from burning grass and brush around supporting piles are another hazard.

FIG. 7-14B. Fire fighters were forced to pull hose lines on hand cars across thousands of feet of track to fight a fire in this wooden trestle located in a remote area on Long Island, NY. Water around the trestle was too shallow for fire boats. High winds drove the fire rapidly through the wooden trestle and destroyed about 2,500 ft (762 m) of the structure. (United Press International)

Some satisfactory fire retardant treatments are being used successfully on timbers for bridges; but wooden bridges, even when treated, should not be considered as an adequate substitute for fire resistive or noncombustible bridges. Material treated with a fire retardant may lose its fire retardancy over a period of time, depending upon the environment to which it is exposed.

Fighting railroad bridge fires presents many problems, particularly in areas where bridges are unaccessible to organized fire fighting facilities. One method of providing minimal fire extinguishing capacity is to provide water casks at the approaches to bridges or at intervals along their length for use by train crews. Some railroads keep a magazine of explosives nearby to blow fire breaks in bridges if necessary. In other instances, preemergency planning calls for tearing down sections of platform type bridges to stop the spread of fire.

PLACES OF OUTDOOR ASSEMBLY

The term "places of outdoor assembly" refers to amusement parks, athletic fields and stadiums, grandstands, tents, and similar places where crowds gather to watch an event or take part in recreational activities. It is still in question whether standards cover an "astro" or "super" dome facility which is also used as a place of assembly.

The frequency of fires in these places and the possibility of loss of life from fire and panic in any place where crowds congregate compel the consideration of providing adequate and accessible ways of escape, and of minimum good practice construction features to reduce the probability of fire spread. NFPA 102, *Standard for Assembly Seating, Tents, and Air-Supported Structures* (hereinafter referred to as NFPA 102), provides recommendations on reasonable measures of safety to prevent loss of life.

The primary consideration in these places is the provision for escape of people from a fire area to a place of refuge or to the exterior. This escape may have to be made through gates in a fence, from a grandstand, out of a tent, or from any confinement that restricts the passage of masses of people. Providing sufficient units of exit width to handle the capacity of the area established by a predetermined figure, based on area of floor or ground area per person or on the number of actual seats, is the recommended method of assuring adequate escape means. This assumes that the capacity will not be allowed to be exceeded and that the exits will be within a specified distance from all occupants of the area. It also assumes that the access to exits will be unobstructed. The area of refuge for occupants may be an open space which is well separated from the danger area. Access to the place of assembly for fire apparatus and other emergency vehicles is also necessary.

Exit requirements for open air grandstands and other places of outdoor assembly are less stringent than for places of indoor assembly, because a build up of hazardous smoke is not so much of a problem. Placing a complete roof over a place of assembly may serve to confine smoke, and may increase panic hazard, thereby calling for more exits for quicker evacuation. Some of the new installations, however, finds grandstands within such large structures with such tremendous quantities of air that heavy concentrations of smoke are unlikely. Such exceptionally large structures containing grandstands are the subject of major revisions proposed for NFPA 101 and NFPA 102.

Construction is the second consideration in a place of outdoor assembly. The temporary or portable nature of the structures generally used in such places and the false sense of complacency that may exist due to the open or partially open nature of the structures, whether fixed or portable, may result in construction weaknesses that accelerate fire spread.

Other typical features include temporary wiring, inadequate maintenance and poor housekeeping (particularly when exhibitions involving displays are present), use of considerable combustible materials (including decorations and flammable liquids), and the general lack of proper fire prevention practices.

A good supply of portable fire extinguishers as well as a reliable method of quickly communicating with the closest fire department and other appropriate emergency organizations are necessary for fire protection in places of outdoor assembly.

Tents

Tents became a major concern to fire protection authorities following the disastrous circus fire in Hartford, CT, on July 6, 1944, in which 168 lives were lost. This fire clearly demonstrated the need for flame retardant materials for tents in places of outdoor assembly. Flame resistant materials are now generally required for all tents. Decorative materials used in tents are generally required to be flame retardant. The use of open flame and heating devices is closely regulated or prohibited.

Tents are combustible, even when made of flame resistant material, and it is good practice to limit their number and size in a location, and to keep them separated from each other and from other structures. The area around tents must be kept free of combustible material, including vegetation. Combustible material used inside a tent should be kept to a minimum, and smoking in tents occupied by the public should be prohibited.

Grandstands

Grandstands are typical examples of structures found in places of both outdoor and indoor assembly and which constitute a safety problem. These structures must be built with adequate aisles, proper distance between rows of seats, strong guard railings, and other exit features to assure rapid and orderly evacuation to areas of safety. Without these features, panic situations could develop in emergencies. (See Fig. 7-14C.) They should also be designed to withstand predetermined deadweight and live loads, and anticipated wind loads.

Since many grandstands are constructed entirely or partially of wood, basic fire protection principles of division of areas and separation for exposure protection apply. Long, combustible grandstands are required to be divided by fire partitions or may be built in smaller sections with adequate space between sections.

Frequently, the space underneath a grandstand is an area for storage of materials and for concession booths, and becomes littered with refuse. This is not good practice; the space should be kept clear of such activities. Such refuse was a major factor in the Bradford, England, Soccer Stadium fire on May 11, 1985, which claimed 54 lives.

Amusement Park Structures

Structures supporting amusement rides are generally built of wood, contain considerable electrical wiring, involve gear boxes for pulleys, and similar friction producing machinery, and are usually poorly maintained. Life safety is the primary consideration in evaluating these

FIG. 7-14C. More than 5,000 people escaped without serious injury when fire occurred in this wooden grandstand at Louisville, KY in the early 1950s. The evacuation was not orderly, and several occupants received minor bruises in their haste to escape. A cigarette is believed to have fallen through a crack into a concealed space below the wood flooring where it ignited debris. (Wide World Photos)

structures; therefore, adequate and accessible means of escape in emergency conditions is vitally important. Fire prevention, noncombustible construction, exposure hazards, maintenance, and housekeeping are other factors to be considered.

Recent disasters in amusement park structures have disclosed weaknesses in modern code requirements. There are structures other than the so-called "ride" devices such as the merry-go-round, roller coaster, ferris wheel, etc. Some are for amusement purposes and some are even educational in nature. Others simulate a bobsled run, or a car travelling through a building; still others are those operating in semidarkness such as a mystery house or a haunted castle.

Since most of these structures are windowless (or at least the public spaces are without windows), they should be required to be provided with automatic sprinkler systems using the newly developed quick response sprinklers. Egress provisions almost necessarily are *not* standard and some innovations are needed to meet the intent of NFPA 101 and the model building codes. Of primary importance is a provision for emergency lighting, and good illumination and marking of paths of egress travel. Sprinkler protection is probably the most important requirement in an amusement structure whether it is a permanent type or made up of semi-permanent trucks in place.

A fire in a "haunted castle" attraction which was made up of interconnected commercial trailers in a Jackson Township, NJ, amusement park on May 11, 1984, claimed eight lives. Major factors contributing to the loss of life were lack of properly designed fixed detection and suppression systems, ignition of synthetic foam material and spread of fire to other combustible contents and interior finish, and the difficulty occupants had in trying to escape along a smoke filled convoluted path.

MEMBRANE STRUCTURES

NFPA 102, used for places of assembly, will soon become the *Standard for Assembly Seating, Tents and Membrane Structures* and will not be applicable solely to places of assembly. The term "membrane" replaces "air supported" because it better reflects the various types of structures other than those which are supported by air and in which the occupants of the structure are in a pressurized atmosphere. Some structures are of the air inflated type where the air pressure is within a series of tubes which then form the structure and keep it upright. In this case the occupants are not within a pressurized atmosphere. Cable supported membrane structures, such as in some newer sports stadiums, where the roof element is supported by cables and air pressure may or may not be provided are also included. In other type structures, the membrane is tensioned and weather protection is usually provided without enclosing walls. Finally there are the framework type structures with the membrane stretched over the frame to form enclosing walls and a roof.

By definition a membrane is a thin, flexible, water impervious material capable of being supported by an air pressure of 1.5 in. (3.7 Pa) of water column. The membrane may be of very thin metal, or glass fiber coated with a fluorocarbon and considered to be noncombustible or limited combustible. Others are of a glass, nylon, or similar fibrous material coated with vinyl or silicone. These are considered to be combustible structures even though they are flame resistant treated in accordance with NFPA 701, *Standard Methods of Fire Tests for Flame-Resistant Textiles and Films*.

An air supported structure specifically is a "balloon like" shelter constructed of flexible coated fabrics supported and stabilized against wind loads by a small amount of internal pressure [usually 1 to 1.5 in. of water pressure (2.5 to 3.7 Pa)] supplied by continuously operated centrifugal fans. (See Fig. 7-14D.) There are no beams, columns, girders, or other structural members involved. Variations of the air supported structure include types which have air inflated double walled roof and side panels and a combination structure which is rib supported, but air supported in heavy wind.

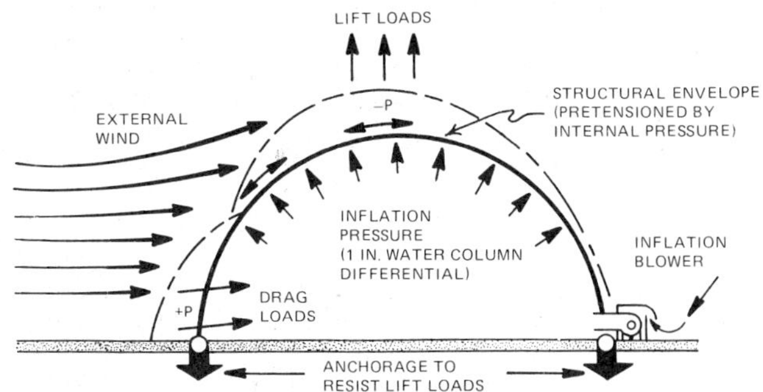

FIG. 7-14D. The aerodynamic and inflation loading of a typical single walled, air supported structure.

A plastic coated synthetic fiber is used in air supported structures. The synthetic fiber will not absorb flame resistive solutions but the plastic coating can contain flame retardants. If the fan serving an air supported structure fails, or if a hole is made in the fabric (by heat from an exposure fire for example), it takes a relatively long time for the structure to deflate. Normal opening and closing of doors will not affect the structure; in fact, breather holes are provided in the skin for air changes.

PIERS AND WATER SURROUNDED STRUCTURES

When piers or related structures surrounded by water are used for places of amusement, restaurants, or passenger terminals, they fall out of the normal use category, i.e., mooring of vessels and handling of cargo, and fall into the "special" structure category. The question of life safety is of primary concern, since the operations listed above all involve gathering or assembling together people in large numbers in or on structures that may have limited egress facilities due to the location or arrangement of the pier or water surrounded structure.

For good design practice for piers falling outside the normal use category, the following recommendations are made when the pier extends more than 150 ft (45.7 m) from the shore. This criterion arises from the recognized maximum travel distance requirement as found in NFPA 101 for the uses listed above. These design approaches are considered equivalent:

1. The pier may be arranged to have two separate ways of travel to shore, as by two well separated walkways or independent structures.
2. The pier deck may be designed to be of open, fire resistive construction, resting on noncombustible supports.
3. If combustible construction is used for the pier deck, superstructure, and/or substructure, they should be protected by an automatic sprinkler system.
4. If the pier is completely open and unobstructed, it may be considered to be of generally safe arrangement if the minimum width is not less than 50 ft (15.2 m) for piers up to 500 ft (152.4 m) in length and, if the length is increased above 500 ft (152.4 m), a 10 to 1 ratio is maintained for the relationship of length to width.

Regardless of whether (1) through (4) above, individually or jointly, are applicable to a particular pier, the minimum requirements of NFPA 101 for providing means of egress apply concurrently. These requirements cover the number, arrangement, location, and protection of means of egress for the type of occupancy involved, plus related questions of hardware, construction, and interior finish for the means of egress. In addition, further guidance for piers is given in NFPA 307, *Standard for the Construction and Fire Protection of Marine Terminals, Piers, and Wharves.*

When a building or structure such as a typical lighthouse, oil drilling platform and related housing facilities is completely surrounded by water, several considerations must be given to life safety from fire. Either sufficient outside ground area such as an island, is needed for refuge, or a fire resistive platform is constructed, for this purpose. This must be coupled with a communication system and a means of transportation for evacuating the occupants from the refuge area. Such transportation will be limited to either boat or helicopter due to the obvious nature and location of these special structures. However, such a system of refuge, communication, and rescue is only as good as the preplanning and ready availability of the transportation on a standby basis. Delays imposed by either poor planning or intolerable weather conditions can lead to a disaster. Thus, such factors must be considered before the event and not retrospectively.

Permanently Moored Vessels

Where a sizable vessel is permanently moored and used as a restaurant, hotel, museum, or convention center, it is for all practical purposes similar to any permanently constructed building occupied for those purposes.

Because the vessels were not built as "buildings" but under completely different standards, it becomes almost impossible to try to make them comply with requirements in NFPA 101 or the model building codes. Passageways are not as wide as would be required for exit access corridors. Doors are usually of smaller dimensions than would be found in a conventional building. The rise and run of stairs, originally known as "ladders" probably do not conform with stair requirements in buildings. As the tide rises and falls, the land connection changes in slope. So, if the vessel is to be accepted for use as other than a ship, some concessions are usually made. A prime means for accepting these nonconforming construction details is to require sprinkler protection for the vessel, limits to the occupant load, strict compliance to interior finish requirements, etc. Generally, the power plant for the vessel is inactivated and shore power substituted after removal of the operating machinery and fuel.

AIR RIGHT STRUCTURES

The economics of the purchase of choice inner urban space, site preparation, demolition of existing structures, and the complex problem of foundation design and caisson construction, all without adversely affecting surrounding structures and without exceeding stringently applied cost limitations, has led to a structure which either totally or partially solves many of those problems—the air right structure. (See Fig. 7-14E.)

The air right structure is generally a facility constructed over existing rail lines, highways, or even other

FIG. 7-14E. An air right hotel and office building shown spanning an expressway and a two track railroad right of way.

buildings. From a firesafety standpoint, its design is not unlike that which would be promulgated if it were isolated on its own site, with one key exception. The air right structure is more likely to be adversely affected by an exposing event on or in the facility above which it is erected. This is due to the fact that the exposure is in the vertical plane, which is ideal for the heat transfer process to take place. Thus, normal good firesafety design practice must be followed, plus allowance for potentially severe fire exposure from the facility straddled by an air right structure. Much of the previous discussion of tunnels and bridges would apply to the air right structure. Normal design guidance is found elsewhere in this text and in the NFPA codes, based on the occupancy of the structure. However, particular emphasis should be centered on NFPA 220, *Standard on Types of Building Construction* in the selection of basic building types, which should be chosen on the basis of hazards involved and exposures encountered.

OTHER SPECIAL STRUCTURES

Signs

In general, outside signs on buildings are not a very serious fire problem because of the limited amount of combustibles involved in such signs and because of their location. However, certain types of signs such as those illuminated by neon tubing, may cause fires because of the wiring and transformers connected to them.

An outside sign may suffer considerable damage from wind and may expose other properties to damage if torn loose by wind. If a sign is of such size, construction, and location as to be considered a fire hazard, it should be protected like any other structure (for example, automatic sprinklers may be installed).

A combustible sign attached to the exterior and covering a large percentage of one side of a fire resistive or noncombustible building may significantly change the firesafety considerations of the building. This is a fire exposure problem not generally contemplated in building design. A large sign can also be hazardous to fire fighters should it fall during fire fighting operations. For these reasons, some building codes limit the size and type of construction of combustible signs that can be installed on buildings.

Mechanical Parking Garages

At one time, parking structures that permitted a driver to operate each vehicle to a parking stall were required to meet fire resistive constrictions far in excess of the fire severity that could exist. Despite the presence of gasoline and combustible tires, the fire load and hazard from private automobiles is very low. Where these structures have permanently open exterior wall areas for the dissipation of smoke and hot gases, the likelihood of a serious fire is likewise extremely low.

Many tests here and abroad have demonstrated the validity of these conclusions. One such test was conducted in 1972 by the American Iron and Steel Institute (AISI) in a multistory open parking structure with exposed steel structural members (Gewain 1973). Three automobiles parked adjacent to each other, each with a fuel tank containing 10 gal (38 L) of gasoline were used as the test ignition and fuel source. The center car was gutted forty-eight minutes after ignition of crumpled newspapers in its rear seat. The contents of its fuel tank were spilled or consumed, but there was no spread of fire to the adjacent cars, and the overhead structural steel was essentially unaffected. Temperatures of the steel remained far below critical levels throughout the test. A similar test conducted by the British at Boreham Wood fire test station yielded similar conclusions (Butcher et al 1968). These conclusions do not necessarily apply to aboveground enclosed garages or to underground garages.

Surveys of actual fire experience in open parking structures have confirmed both that fire severity is low and even the incidence of fire is infrequent.

Building code requirements have recognized this pattern and now permit noncombustible parking structures when used for no other purpose having ample permanent wall openings for the free dissipation of smoke and gases to be of unlimited area and up to 75 ft (22.9 m) in height with no structural fire protection.

Bibliography

References Cited

Butcher, E. G., et al. 1968. "Fire and Car-park Buildings." *Fire Note No. 10*. Ministry of Technology and Fire Offices' Committee/Joint Fire Research Organization. London, England.

Gewain, R. G. 1973. "Fire Experience and Fire Tests in Automobile Parking Structures." *Fire Journal*. Vol 67, No 4.

ICBO. 1985. *Uniform Building Code*. 1982 ed. International Conference of Building Officials. Whittier, CA.

NFPA Codes, Standards, Recommended Practices and Manuals. (See the latest *NFPA Codes and Standards Catalog* for availability of current editions of the following documents.)

NFPA 101, *Code for Safety to Life from Fire in Buildings and Structures.*

NFPA 102, *Standard for Assembly Seating, Tents, and Air-Supported Structures.*

NFPA 214, *Standard on Water-Cooling Towers.*

NFPA 220, *Standard on Types of Building Construction.*

NFPA 307, *Standard for the Construction and Fire Protection of Piers, Marine Terminals, and Wharves.*

NFPA 502, *Recommended Practice on Fire Protection for Limited Access Highways, Tunnels, Bridges, Elevated Roadways, and Air Right Structures.*

Additional Readings

"Automobile Parking Structures," *A Building Code Modernization Bulletin*, 2nd ed., American Iron and Steel Institute, NY, 1971.

Bond, H., "Underground Buildings," *Fire Journal*, Vol. 59, No. 4, July 1965, pp. 52-55.

DeCicco, Paul R., "Life Safety Considerations in Atrium Buildings," *SFPE TR 82-3*, Society of Fire Protection Engineers, Boston, MA, 1982.

Degenkolb, John G., "Atriums," *The Building Official and Code Administrator*, Nov.-Dec. 1983, pp. 18-22.

Harris, Dr. L. M., "Survey of Fire Experience in Automobile Parking Structures in the United States and Canada," Marketing Research Associates, Jan. 1972.

Ivison, J., and Newberry, R., "Fire and Life Safety in Air-Supported Domes," *Fire Engineering*, Vol. 138, No. 5, May 1985, pp. 66-69.

Morgan, H. P., and Savage, N. P., "A Study of a Large Fire in a Covered Shopping Complex: St. John's Centre 1977," *BRE Current Paper CP10/80*, Building Research Establishment, Fire Research Station, Boreham Wood, Hertfordshire, England, Dec. 1980.

Stevens, R. E., "Air-Supported Structures," *Fire Journal*, Vol. 59, No. 10, Sept. 1965, pp. 42-44.

Stevens, R. E., "Water Cooling Towers Will Burn," *NFPA Quarterly*, Vol. 50, No. 9, Oct. 1956, pp. 97-105.

BUILDING CODES AND STANDARDS

Revised by Arthur E. Cote, P.E.

Building codes have been in existence since about 1700 B.C. when King Hammurabi established a law by which a builder could be executed if the house he built collapsed, resulting in the death of the owner. In one form or another, codes have been a part of the regulatory history of most civilized cultures since Hammurabi's age. More frequently than not, building regulations were created as the aftermath of fires and other disasters that claimed many lives injured countless others, caused heavy property damage, and disrupted life in general. The outcry after such calamaties spurred governments and leaders to remedy the conditions they believed caused or contributed to the disasters. Regulating the design, construction methods, and materials used in buildings was perceived as a necessary step in rational defense against the threat of accidental fire and natural disasters involving habitable structures. Thus building codes came into being, and they have been with us ever since.

The discussion in this chapter is limited to modern building codes as they developed through this century. Particular attention is given to the role of model building codes that have been the basis for code enforcement in the United States, and the relationship of those codes to other codes and standards bearing on fire protection and building services.

All chapters in this section of the Handbook have a bearing on the provisions of building codes and their enforcement. Of particular interest would be Chapter 5, "Classification of Building Construction," which contains information on the various types of construction and how they are classified in building codes as a basis for fire protection requirements.

BUILDING CODE DEFINED

A building code is a law that sets forth minimum requirements for design and construction of buildings and structures. These minimum requirements are established to protect the health and safety of society and generally represent a compromise between optimum safety and economic feasibility. Although builders and building owners often establish their own requirements, the minimum code requirements of a jurisdiction must be met. Features covered include structural design, fire protection, means of egress, light, sanitation, and interior finish (Sanderson 1969).

There are two types of building codes. Specification codes spell out in detail what materials can be used, the building size, and how components should be assembled. Performance codes detail the objective to be met and establish criteria for determining if the objective has been met; thus, the designer and builder are free to select construction methods and materials as long as it can be shown that the performance criteria can be met. Performance oriented building codes still embody a fair amount of specification type requirements, but the provision exists for substitution of alternate methods and materials ("trade offs"), if they can be proven adequate.

BUILDING CODES AND FIRE PROTECTION

The requirements contained in building codes are generally based upon the known properties of materials, the hazards presented by various occupancies, and the lessons learned from previous experiences, such as fire and natural disasters (Sanderson 1969). The promulgation of modern building codes in the U.S. began with the disastrous conflagrations and earthquakes which this country experienced at the turn of the century.

For a number of years building codes dealt mainly with structural safety under fire or earthquake conditions. They have since grown into documents prescribing minimum requirements for structural stability, fire resistance, means of egress, sanitation, lighting, ventilation, and built in safety equipment. More than fifty percent of a modern building code now refers in some way or another to fire protection.

Building codes usually establish fire limits or fire districts in certain areas of the municipality. Only certain types of construction are allowed within the fire limits. This restriction is said to reduce the conflagration potential of the more densely populated areas. A type of build-

---

Mr. Cote is Assistant Vice President (Standards), and Secretary to NFPAs Standards Council.

ing construction alone, however, is not necessarily a deterrent to conflagration. Outside the fire limits, the restriction of certain construction types is relaxed, due to decreased building density, (increased spacing between buildings), etc. Unfortunately, as areas outside the fire limits are developed, building density increases and the fire limits are frequently extended. In addition, without construction restrictions, areas outside the fire limits invite the construction of large buildings where public protection is weak or lacking.

Another example of the impact of building codes on fire protection and prevention is the establishment of height and area criteria. The criteria establish size and height of a particular building, based on the building's intended use. Unfortunately, these requirements vary considerably from one area to the next. There is no nationally recognized standard for setting height and area limitations. The types of building construction are important factors in establishing height and area limitations.

Other requirements found in building codes that directly relate to fire protection include: (1) enclosure of vertical openings such as stair shafts, elevator shafts, and pipe chases; (2) provision of exits for evacuation of occupants; (3) requirements for flame spread of interior finish; and (4) provisions for automatic fire suppression systems. Exit requirements found in most building codes are based on those found in NFPA 101, *Life Safety Code®*.

Inasmuch as a building code is actually a law, many state and local jurisdictions write their own codes. But because of the complexities of modern building code development, there are organizations that develop model building codes for use by jurisdictions which can then adopt such model codes into law.

THE FOUR MODEL BUILDING CODE GROUPS

There are four groups that develop or have developed such model building codes. Each organization and its model code are discussed in the following paragraphs.

AIA (American Insurance Association)

The NBFU (National Board of Fire Underwriters), now the AIA (American Insurance Association), first published the NBC (National Building Code) in 1905. The code was used as a model for adoption by cities, as well as a basis to evaluate the building regulations of towns and cities for town grading purposes. The code was periodically reviewed by the NBFU staff, revised as necessary, and republished. The last code revision was in the 1976 edition. The AIA has announced that the National Building Code will no longer be updated and BOCA (Building Officials and Code Administrators) has acquired the right to use the name "National Building Code." The AIA also published a fire prevention code, most recently published in 1976. It too will not be updated.

ICBO (International Conference of Building Officials

The ICBO first published the Uniform Building Code (UBC) in 1927. The UBC is principally used in the western U.S., but has been adopted in municipalities as far east as Michigan.

In addition to publishing the building code, the Conference also publishes a *Mechanical and Plumbing Code*, in association with the International Association of Plumbing and Mechanical Officials, and a *Fire Prevention Code* in association with the Western Fire Chiefs Association. Code changes are made each year and an amended version of the codes is published every three years. Changes made between major reprintings are published in supplements.

The Conference's stated objectives relate to the development and publishing of regulations and educational materials designed to increase the standardization of building construction regulations and the enforcement of these regulations. To encourage this standardization, the Conference maintains a staff of architects and engineers in Whittier, CA, which provides governmental bodies with technical assistance in the administration and enforcement of the ICBO codes.

SBCCI (Southern Building Code Congress International)

The SBCCI (Southern Building Code Congress International) was organized in 1940 and first published the *Southern Standard Building Code* in 1945. The Code is now called the *Standard Building Code* (SBC). The SBC is principally used throughout the southern portion of the U.S.

Like ICBO, SBCCI also publishes mechanical, plumbing, fire prevention, and gas codes to be used in conjunction with the SBC. The code is amended and reprinted every three years with changes made yearly and printed in supplements.

The stated purpose of the SBCCI is to develop, maintain, and promote the use of the series of codes it prepares. In addition, the SBCCI intends to promote standardization in building regulation and enforcement of those regulations. SBCCI maintains a technical staff headquartered in Birmingham, AL.

BOCA (Building Officials and Code Administrators)

The BOCA (Building Officials Conference of America) published the first edition of the *Basic/Building Code* (BBC) in 1950. Since that time, the organization has changed its name to Building Officials and Code Administrators, International. The BBC, now published as the *Basic/National Building Code*, is principally used in the midwest and northeast portions of the U.S.

Like the other code groups, BOCA also publishes mechanical, plumbing, and fire prevention codes. Each of these codes is revised annually and reprinted every three years. Changes approved between reprintings are published in supplements. The stated objective of the organization is to develop and maintain the *Basic/National Building Code*. BOCA maintains a technical staff in Country Club Hills, IL.

TOWARD UNIFORMITY OF BUILDING CODES

Although originally known as regional codes, boundary lines for adoption of the model codes today are much less significant. In fact, much effort was expended towards the elimination of differences between the model codes through such organizations as the Joint Committee on Building Codes (composed of representatives of the organizations sponsoring the model codes and others concerned with the development of codes and standards). Formed in 1949, the Joint Committee later became the Model Code Standardization Council and is now the Board for the Coordination of Model Codes (BCMC). Code uniformity continues to be its prime objective.

OTHER CODE RELATED ORGANIZATIONS

The primary objectives of the model code groups is to provide standardization of construction regulations and enforcement of these regulations. Other organizations with similar goals have also been formed.

NCSBCS (National Conference of States on Building Codes and Standards)

The National Conference of States on Building Codes and Standards (NCSBCS) is a nonprofit corporation founded in 1967 as a result of congressional interest in building code reform. It attempts to foster increased interstate cooperation in the area of building codes and standards and coordinates intergovernmental code administration reforms. NCSBCS is an executive branch organization of the National Governors Association and includes as members governor appointed representatives of each state and territorial government. It has a working relationship with the National Conference of State Legislatures and the Council of State Community Affairs Agencies.

CABO (Council of American Building Officials)

The Council of American Building Officials (CABO) is an organization formed by the three major model code groups in 1972. (The AIA is not a member of this organization.) CABO was formed to represent the model building code process on a national level and to promote the model code process. One of the committees formed by CABO is the BCMC (Board for the Coordination of the Model Codes). Representatives of BOCA, ICBO, SBCC, and, most recently (1981), NFPA, participate in this effort. The purpose of this committee is to develop uniform code language to be included in each of the model codes. Proposals resulting from BCMC are processed in accordance with the code change procedures of each code organization.

Two of the major accomplishments of CABO have been to organize the National Research Board and publish the *One- and Two-Family Dwelling Code*. The National Research Board coordinates the research and evaluation programs of the three model code groups to eliminate the need for a manufacturer working with three different organizations. The *One- and Two-Family Dwelling Code* contains regulations for detached housing which meet the requirements of the three model codes.

NACA (National Academy of Code Administrators)

The National Academy of Code Administrators (NACA) was organized in 1970. The Academy was formed to help educate and train code administrators and building department personnel. The Academy is now defunct.

Association of Major City Building Officials

The Association of Major City Building Officials was formed in 1974. This group recognizes code problems that are peculiar to large cities, defines problem areas, and seeks solutions.

NIBS (National Institute of Building Sciences)

The National Institute of Building Sciences (NIBS) was authorized by Congress in 1974 under Public Law 93-383 as a nongovernmental, nonprofit corporation under the direction of a 21 member board of directors. Fifteen of the board members are elected and six are appointed by the President of the United States, with the advice and consent of the U.S. Senate. In establishing the objectives, powers and, structure of NIBS, Congress also outlined certain functions relating to building regulations in four general areas. These are:

1. Development, promulgation, and maintenance of nationally recognized performance criteria, standards, and other technical provisions for maintenance of life, safety, health, and public welfare that are suitable for adoption by building regulating jurisdictions and agencies. This also includes test methods and other evaluative techniques relating to building systems, subsystems, components, products, and materials with due regard for consumer interests.
2. Evaluation and prequalification of existing and new building technology.
3. Conduct of needed investigations.
4. Assembly, storage, and dissemination of technical data and other related information.

Working under its very broad mandate, NIBS has established a Consultative Council with membership available to representatives of all appropriate private trade, professional, and labor organizations; private and public standards, code, and testing bodies; public regulatory agencies; and consumer groups. The Council's purpose is to insure a direct line of communication between such groups and the Institute and to serve as a vehicle for representative hearings on matters before the Institute.

WOBO (World Organization of Building Officials)

The World Organization of Building Officials (WOBO) was founded in 1984 with the primary objective of advancing education through dissemination of knowledge worldwide in building science, technology, and construction. WOBO was founded because of the increased participation of nations in the global marketplace; the rapid development of new international building technologies and products; and the development of international standards which make it no longer possible for building officials to concern themselves with their own national boundaries.

STATE AND LOCAL GOVERNMENT BUILDING CODES

Within the scope of the police power of state government is the regulation of building construction for the health and safety of the public, a power usually delegated to local governments of the state.

The application of building code requirements is usually for new construction or for major alterations to buildings. Retroactive application of code requirements is very rare. Building code applicability usually ends with the issuance of an occupancy permit. The basic premise that fire legislation should regulate for the safety of current occupants and for current risk is not generally the province of building codes after occupancy. After occupancy fire codes, or more precisely, fire prevention codes, apply. It is also at the point that the authority of the building official usually ends and that of the fire official begins.

This division of authority, however, does not preclude interaction between the two officials during both a building's development and its subsequent use. In practice, many jurisdictions assign responsibilities to fire and/or building officials in both building and fire code application, and the division of authority varies considerably among communities.

The manner in which the states handle the promulgation of building regulations also varies widely. In some states, each local government may have its own code while in others, the local authority may adopt the state building code, but such adoption is optional. In still others, the state code establishes the minimum below which the local regulations cannot go. Finally, in some states, the local government must adopt the state code.

These situations have resulted in a plethora of different local building codes. Some of the local governments adopt one of the model building codes or a code based upon one of the model codes, and others draft their own local code. This lack of uniformity has been criticized by materials producers, building designers, builders, and others; and, some years ago, prompted the appointment of federal commissions to study the situation and make recommendations to the administration. (NCUP 1968; ACIR 1966; PCH 1982)

THE ROLE OF STANDARDS IN BUILDING CODES

Many of the requirements found in building codes are excerpts from, or based on, the standards published by nationally recognized organizations. The most extensive use of the standards is their adoption into the building code by reference, thus keeping the building codes to a workable size and eliminating much duplication of effort. Such standards are also used by specification writers in the design stage of a building to provide guidelines for the bidders and contractors.

Numerous NFPA standards are referenced by the model building codes and, thus, obtain legal status where these model codes are adopted. Notable examples of such referenced NFPA standards are those that deal with extinguishing systems, flammable liquids, hazardous processes, combustible dusts, liquefied petroleum gas, and fire tests.

Three of the model building codes (AIA, BOCA, and SBCCI) contain appendices that list standards published by many organizations including standards making organizations, professional engineering societies, building materials trade associations, federal agencies, and testing agencies. The appendices are prefaced with a statement indicating that the standards are to be used where required by the provisions of the code or where referenced by the code.

One of the model code organizations (ICBO) publishes a book of standards to go with its code. Some of the standards in the book are written by the code making organization; others are standards of other organizations adapted for use by the model code organization. Among the organizations whose standards are referenced by model building codes are the:

Aluminum Association
American Board Products Association
American Concrete Institute
American Institute of Steel Construction
American Institute of Timber Construction
American Iron and Steel Institute
American National Standards Institute
American Plywood Association
American Society for Testing and Materials
American Society of Heating, Refrigeration and Air Conditioning Engineers, Inc.
American Society of Mechanical Engineers
American Welding Society
American Wood Preservers Association
American Wood Preservers Bureau
American Wood Preservers Institute
Architectural Aluminum Manufacturers Association
Brick Institute of America
Gypsum Association
Hardwood Plywood Manufacturer's Association
Illuminating Engineers Society
National Concrete Masonry Association
National Fire Protection Association
National Forest Products Association
National Particleboard Association
Portland Cement Association
Steel Joist Institute
Tile Council of America
Truss Plate Institute
Underwriters Laboratories Inc.
U. S. Department of Commerce

Bibliography

References Cited

ACIR. 1966. *Building Codes: A Program for Intergovernmental Reform*. Advisory Commission on Intergovernmental Relations (ACIR). Superintendent of Documents. U.S. Government Printing Office, Washington, DC.

NCUP. 1968. "Building the American City." Report of the National Commission on Urban Problems (NCUP). Superintendent of Documents. U. S. Government Printing Office, Washington, DC.

Sanderson, R. L. 1969. *Codes and Code Administration*. Building Officials Conference of America, Inc., Chicago, IL.

PCH. 1982. "Report of the President's Commission on Housing." Superintendent of Documents. U.S. Government Printing Office, Washington, DC.

NFPA Codes, Standards, Recommended Practices and Manuals. (See the latest *NFPA Codes and Standards Catalog* for availability of current editions of the following documents).

NFPA 101, *Code for Safety to Life from Fire in Buildings and Structures.*

NFPA 220, *Standard on Types of Building Construction.*

Additional Readings

ASTM Standards in Building Codes, American Society for Testing an Materials, Philadelphia, PA, 1978.

Bihr, J. E., "Building and Fire Codes: The Regulatory Process, Fire Standards and Safety," *ASTM STP 614*, (A. F. Robertson, Ed.), American Society for Testing and Materials, Philadelphia, PA, 1977.

BOCA, *Basic/National Building Code*, 1984 ed., Building Officials and Code Administrators, International, Country Club Hills, IL, 1984.

Curless, M., *Codes, Standards and Fire Protection Engineering*, MP 69-1, National Fire Protection Association, Quincy, MA, 1969.

Fire Protection Through Modern Building Codes, 5th ed., American Iron and Steel Institute, Washington, DC, 1981.

Hansen, A. T., "Applying Building Codes to Existing Buildings," *CBD 230*, Division of Building Research, National Research Council, Ottawa, Canada, Jan. 1984.

ICBO, *Uniform Building Code*, 1985 ed., International Conference of Building Officials, Whittier, CA, 1985.

Harter, Philip J., "Regulatory Use of Standards: The Implications for Standards Writers," *NBS-GCR-79-171*, National Bureau of Standards, Gaithersburg, MD, Nov. 1979.

Kouba Dennis, ed., *Code Administration and Enforcement: Trends and Perspectives*, International City Management Association, Washington, DC, 1982.

Lucht, David A., "Is Our Fire Safety Regulation System Really Working?" *Building Standards*, Nov.-Dec. 1977, pp. 268-273.

McConnaughey, John S., "An Economic Analysis of Building Code Impacts: A Suggested Approach," *NBSIR 78-1528*, National Bureau of Standards, Gaithersburg, MD, Oct. 1978.

Rawie, Carol C., "Estimating Benefits and Costs of Building Regulations: A Step by Step Guide," *NBSIR 2223*, National Bureau of Standards, Gaithersburg, MD, 1981.

Sanderson, R. L., *Readings in Code Administration*, 3 vols. Building Officials and Code Administrators International, Inc., Chicago, IL, 1975.

SBICC, *Standard Building Code*, 1985 ed., Southern Building Code Congress International, Inc., Birmingham, AL, 1985.

Smyrl, Elmira S., "Literature Review: The Building Regulatory System in the United States," *NBS-GCR-80-286*, National Bureau of Standards, Gaithersburg, MD, Oct. 1980.

Standards Activities of Organizations in the United States, National Bureau of Standards, U.S. Dept. of Commerce, Washington, DC., 1984

Sullivan, Charles D., *Standards and Standardization: Basic Principals and Applications*, Marcel Dekker, New York, NY, 1983.

Taylor, D. M., *A Guide for Codes Adoption and Codes Enforcement*, U.S. Department of Housing and Urban Development, Regional Office 4, Atlanta, GA, 1974.

Teague, Paul, E., "Mini-Max Building Codes and their Effect on the Fire Service," *Fire Journal*, Vol. 71, No. 3, 1977.

Vogel, Bertram M., *Standards Referenced in Selected Building Codes, Office of Building Standards and Codes Services*, Center for Building Technology, Institute of Applied Technology, National Bureau of Standards, Washington, DC, 1976.

FIRE HAZARDS OF CONSTRUCTION, ALTERATIONS, AND BUILDING DEMOLITIONS

Revised by John A. Sharry

Buildings and other structures are more vulnerable to fire when they are under construction, alteration, or demolition, regardless of construction type or construction method, than when completed or demolition is finished. The reason for this is that when they occur, fires are likely to spread more rapidly due to the absence or impairment of fire suppression and detection systems, lack of compartmentation, the presence of heavier concentrations of combustibles than would be expected during normal use, and the fact that fires are often not discovered in their incipient stage. In new construction, fires are likely to cause more severe damage because of the lack of all structural members, lack of applied fire resistive materials, and open exposed condition of the structure. Thus fires can cause the destruction of building materials stored on site, lengthy delays in project completion, long term business interruption (in the case of rehabilitation projects), and more importantly, the possible loss of life.

Construction operations can be made reasonably safe in new construction, rehabilitation, or demolition with proper planning, proper allowances in job estimates, and proper on site supervision and control. Although new construction, rehabilitation, and demolition projects differ greatly in the scope of work, they have many firesafety items in common. These common items relate to establishment a.id maintenance of on site fire protection, provisions for temporary fire protection, and maintenance of means of escape for workers and others in the building. NFPA 241, *Standard for Safeguarding Building Construction and Demolition Operations* (hereinafter referred to as NFPA 241M), contains measures that, with prefire planning, will help prevent fire or at least minimize damage when fire occurs.

Guard service on contruction operations is recommended during nonworking hours. Patrols should be planned so that no portion of the job site is overlooked. The use of guard clock and time recording systems is recommended to insure patrols have been made. Guards should be briefed on the fire protection measures available on the job site, on developments during the workday that affect fire hazards and fire protection capabilities, and on emergency procedures. Provision (either a telephone or a fire alarm box) should be made for notifying the fire department in the event of a fire.

NEW CONSTRUCTION

Site Preparation

Site preparation is the first item necessary in any new construction project. This involves the removal of brush, trees, and debris from the site prior to the start of construction. Although such land clearing provides for control of natural cover fuel fires, it mainly provides site accessibility for construction. Today's methods of site preparation differ from the "scorched earth" methods of several years ago in that the preservation of as much natural vegetation as practical is desired. This should present no serious fire problem as long as temporary buildings and materials storage are properly isolated from any natural cover fuel by fire breaks.

Of prime importance in site preparation is the establishment of roadways and the installation of water mains for fire protection. Many jurisdictions now require that permanent all-weather roads and site utilities be provided before construction begins in housing and industrial developments. However, these provisions are not always required for the single building or building complex. In any event, construction site access by temporary or permanent roadways should be provided, as well as water service for fire protection.

While it is usually more economical to provide the permanent water service to serve the construction fire protection needs early in the project, it is sometimes necessary to provide temporary water service in the form of aboveground water mains.

Advance planning can minimize the impact of these requirements by proper construction sequencing and proper site preparation.

Temporary Buildings

Temporary buildings are commonly used on construction sites. They provide for the storage of tools and

Mr. Sharry is Fire Chief of Lawrence Livermore National Laboratory, Livermore, CA.

equipment, and serve as temporary offices for the construction management team. The use of mobile structures such as converted mobile homes, trailers, or manufactured buildings, has largely replaced the construction of temporary buildings.

The hazards are much the same whether a mobile structure or a temporary structure is used. Overloaded temporary wiring, incorrectly installed portable or temporary heating equipment, and inadequate control of open flame devices are examples of the common hazards. It is also important to recognize the high level of fuel load in the form of construction materials and construction plans, specifications, and other documents. In addition, the use of space (sometimes concealed) beneath these temporary structures for miscellaneous storage can be hazardous depending upon the nature of the materials stored. Such spaces and storage can be subject to vandalism and incendiarism. Vegetation growing on, or left in, these spaces can also create hazards.

The protection of these structures, especially mobile structures, is usually possible by the installation of a residential type sprinkler system that can be connected to the site water supply. The cost of this protection is low for the mobile structures, since there is only a one time installation cost and connection costs on the site are minimal.

The exposure protection necessary is often overlooked if several temporary or mobile structures are on the same site. Proper spacing between structures and from the project itself by a minimum of 30 ft (9 m) can eliminate this hazard. Often mobile structures can be attached or installed adjacent to each other, creating a larger building with greater hazards than one structure alone.

Loss Prevention During Construction

Normal building operations require considerable amounts of combustible building materials, forms, scaffolding, etc. Water service at the site should be adequate not only for construction needs but also for fire protection needs in view of the amount of combustibles present. In multistory structures requiring either temporary or permanent standpipes, should be put into service on a floor by floor basis to provide fire fighting capability. Temporary standpipe service is usually a dry standpipe with proper valves at each floor and a fire department siamese connection.

Temporary enclosure of buildings is common and should only be done with some care. Only flame resistive tarpaulins or materials of similar fire retardant characteristics should be used. If temporary enclosure is required, metal or fire retardant treated wood forms should be used and the enclosure materials securely fastened or guarded to prevent contact with heaters or other sources of ignition. If plastic is used, it should be of a type that does not readily ignite and that exhibits slow burning characteristics. (See Fig. 7-16A.)

Concrete should be poured as quickly as possible after combustible forms have been constructed, and the forms removed as soon as possible after the concrete has set properly. During the time the forms are in place, storage and construction operations on that floor should be held to a minimum and ignition sources should be eliminated. Noncombustible forms should be used whenever possible. (See Fig. 7-16B.)

FIG. 7-16A. Large sheets of plastic are sometimes used to enclose buildings under construction to confine heat. (Francis L. Brannigan)

FIG. 7-16B. Combustible formwork is a constant source of danger to concrete buildings under construction. Here fire consumes wood formwork for cast-in-place floor and wall panels in a parking garage.

Where "fireproofing" is specified, it should be applied as soon as possible to afford fire protection to beams, columns, etc. early in the construction phase.

Rubbish and trash accumulations can present a serious fire hazard on the construction site. Daily cleanup of scrap lumber, cardboard containers, and other debris removes the hazard and provides a more orderly working environment. Trash containers should be provided on site for proper storage of debris and should be located at a safe distance from the building or protected to prevent undue exposure of the building.

Cutting and welding operations must be carefully controlled, and proper safeguards utilized. Careless use of welding and cutting torches and plumbing torches have caused numerous fires. NFPA 241 recommends the use of a permit system for welding operations and specifies controls for safe welding operations. Also important is the guidance given in NFPA 51B, *Standard for Fire Prevention in Use of Cutting and Welding Processes.*

Because they were improperly controlled, salamanders have been responsible for many fires. However, since salamanders are often needed, their use should be continuously supervised and the area cleared of combustible materials.

Flammable liquids present obvious fire protection problems. Small quantities can be safely handled in approved safety containers and should be stored in an isolated location. Bulk quantities of flammable or combustible liquids should be avoided, but if bulk quantities are needed, they should be stored in an approved manner. NFPA 395, *Standard for the Storage of Flammable and Combustible Liquids on Farms and Isolated Construction Projects*, provides guidance on flammable liquid storage on construction sites.

Internal combustion engines for pumps, air compressors, etc. should be placed to avoid exhaust discharge near, or in contact with, combustibles. A minimum of a 6 in. (152 mm) clearance to combustibles should be maintained if the exhausts are piped to the exterior. The engine should be shut down during refueling to prevent flash fires. For safety reasons, electric or pneumatic driven equipment, not internal combustion engine driven equipment, should be used in underground or basement locations.

The use of tar kettles should be continuously supervised and only permitted outdoors away from combustibles or on a noncombustible floor or roof in the building. Metal covers should be provided to smother potential fires. Should a fire occur, it is important to remember that water thrown into the tar kettle will cause the tar to froth. Roofing mops should never be left indoors or near ignition sources or combustible materials. Roofing mops soaked with tar have been known to be susceptible to spontaneous ignition and have caused fires. The mops should be thoroughly cleaned, and care should be taken in how they are stored or discarded.

Permanent Safeguards

In most construction operations, the progress of the work toward completion gradually enhances the safety of the structure. As the need for combustible form work and scaffolding, materials storage, temporary buildings and trailers, and other inherent hazards diminishes, they should be removed from the site.

Fire fighter access to the structure is important. Installation of permanent stairways and stairway enclosures should be completed as quickly as possible. Fire walls and other essential features that provide for the confinement of fire should be installed as early as possible. Penetrations of fire resistive construction made for building services should be sealed immediately after the services are installed. Integrity of all fire cutoffs should be maintained.

Permanent standpipe and hose systems should be installed as construction work progresses. In buildings lacking automatic sprinkler systems, or those in which the installation of tanks and other equipment must wait for the completion of structural elements, standpipe and hose systems may provide the only certain source of water for fire fighting throughout the project. Where automatic sprinkler systems are to be a part of the permanent protection for the building, proper correlation of the work will provide water supplies for the system and prompt installation of sprinklers before significant combustibles are present in the building. Blank gaskets used in flanges to permit sprinkler protection by sections as construction work is completed should be provided with lugs that are painted red or are otherwise clearly marked to assure removal when the system is extended.

Permanent wiring systems should replace temporary wiring as rapidly as the removal of form work will permit. Permanent heating plants and temporary nonfired heat exchangers can often be used to supplant the more hazardous salamanders.

Supervision of Fire Protection

All construction, renovation, and demolition projects should have a qualified person in charge of fire protection. This person should be responsible for the maintenance and location of fire protection and suppression equipment, general supervision of safeguards, location of portable heating devices and tar kettles, and the establishment and maintenance of a safe cutting and welding program. On larger projects this person should also devise a written emergency plan for the project. It may be possible to establish an emergency assistance team from among the craftsmen on the project to take initial actions in emergencies. However, care should be taken to ensure that members of emergency action teams are properly trained for the actions they are expected to perform.

On all projects, but particularly on large construction, renovation or demolition projects, there are numerous contractors and subcontractors all contributing to the completion of the project. It is important from a project standpoint, as well as a fire protection standpoint, that someone supervise and coordinate the work done by these various contractors. This person is the key in implementing proper attitudes towards safety and fire protection in the normal course of work. Work guidelines and expectations can be clearly spelled out and enforced in this manner.

DEMOLITION OPERATIONS

If not properly controlled, demolition operations can create more serious problems than construction operations. Demolition operations should include precautions to ensure maintenance of fixed fire protection systems for as long as possible. Automatic sprinkler systems should be modified to allow their removal on a floor by floor basis. The conversion of a wet system to a dry system fed only by a fire department connection is sometimes possible and provides good protection for the building as well as for fire fighting personnel.

Standpipe systems should also be removed floor by floor and should be the last fire protection feature removed. This maintains fire fighting capability to otherwise demolished floors. A wet standpipe can be easily converted to a dry pipe.

It is common practice to provide chutes for the removal of trash and debris from the upper floors. If chutes are used, they should be erected outside the building. Cutting holes through floors to provide an inside chute should be avoided, since this creates a vertical opening through which fire can spread rapidly from floor to floor.

Figure 7-16C shows the results of a demolition operation that violated these principles. The results were disastrous; not only was the building being demolished destroyed by fire, but the fire spread to four other buildings (Sharry 1974).

FIG. 7-16C. Fire originating on the top floor quickly spread throughout this partially demolished building and involved four other buildings. The sprinkler system had been shut down about a week before the fire instead of being shut down in stages as demolition progressed. Contrary to good practice, holes had been made in the floors to facilitate the removal of debris.

REHABILITATION PROJECTS

There has been an increase in the rehabilitation of older buildings in recent years. This effort is partly the result of a desire to preserve our national heritage and partly a result of rising construction costs. From a fire protection standpoint, rehabilitation projects can produce firesafe buildings if the extensive compartmentation usually present in older buildings is at least partially preserved.

From a building code standpoint, existing buildings being rehabilitated usually must meet current code requirements for means of egress, fire protection, and light and ventilation.

The use of equivalency concepts to provide the level of safety intended by a particular code is usually possible when it is practically impossible or undesirable to meet the letter of the code. In buildings with historical significance, it is sometimes undesirable to provide all of the code required safeguards. In such cases, the safety of the occupants takes precedence, and the proper safeguards must be installed to ensure life safety. While improvement of exits to total code compliance may not be possible, providing other compensating features in the building to control fire and smoke may result in equivalent safety. These compensating features may involve the installation of automatic sprinkler protection, smoke detection systems, and smoke control systems. Whatever equivalency is used should be evaluated carefully to ensure that a comparable level of life safety is being provided.

The U.S. Department of Housing and Urban Development (HUD) has produced guidelines for the rehabilitation of buildings. An interesting feature of these guidelines is a volume that provides fire resistance information on archaic building systems (HUD 1980). This information is extremely valuable when attempting to achieve proper fire ratings for building structural systems.

In some rehabilitation projects, the renovations must be made while the building is partially occupied. Special precautions must be taken in these cases. The most important of these is the maintenance of proper exits from the building. Not only must exits be provided for building occupants, but at least one exit must also be provided for building trades workers in the construction areas. The accumulation of trash and other debris must be kept to a minimum, and refuse removed daily.

If fire suppression, detection, or alarm systems are being renovated, their downtime must be kept to a minimum. Proper planning can eliminate unnecessary downtime. Isolation of parts of systems is usually possible to ensure operation in unaffected parts of the building. If it becomes necessary to shut down entire systems, a fire watch should be posted near the main control to restore service in case of fire. Some form of reliable communication should be worked out if the main control is remote from the work area. This precaution is especially important in sprinklered buildings. Removing the entire sprinkler system as part of a rehabilitation project can have disastrous results. (See Fig. 7-16D.)

FIG. 7-16D. Remains of the historic Harrison Building in Philadelphia, PA that was destroyed by fire on May 3, 1984, while the building was undergoing renovation. Although originally equipped with automatic sprinklers and standpipes, the extinguishing systems had been removed from the building as part of the remodeling work. Many openings created by the removal of stairways and elevator shafts from the six floors of the building contributed to the rapid spread of fire.

Means of egress should be readily available to building occupants during normal working hours. Therefore, alterations to means of egress should be accomplished during off hours. In some cases, rerouting of traffic may be a practical alternative. In all cases, exit systems and fire protection systems should be restored at the end of each work day.

In cases where additions to buildings block existing exits, provision must be made for alternate means of egress or for a protected path of travel. The use of combustible materials in temporary exit protection should be avoided.

In any project, whether new construction, rehabilitation, or demolition, the fire department should be kept aware of construction progress. The fire department can thereby provide for command posts and prefire plans and be ready to deal with unusual problems encountered during the course of the project.

Bibliography

References Cited

HUD. 1980. "Guidelines on Fire Ratings of Archaic Materials and assemblies." Vol. 8. *Rehabilitation Guidelines 1980.* U.S. Department of Housing and Urban Development. Washington, DC.

Sharry, John A. 1974. "Group Fire Indianapolis, Indiana," *Fire Journal*, Vol 68, No 4. pp 13-16.

NFPA Codes, Standards, Recommended Practices and Manuals. (See the latest *NFPA Codes and Standards Catalog* for availability of current editions of the following documents.)

NFPA 13D, *Standard for the Installation of Sprinkler Systems in One-and Two-Family Dwellings and Mobile Homes.*

NFPA 30, *Flammable and Combustible Liquids Code.*

NFPA 51, *Standard for the Design and Installation and Operation of Oxygen-Fuel Gas Systems for Welding, Cutting, and Allied Processes.*

NFPA 51B, *Standard for Fire Prevention in Use of Cutting and Welding Processes.*

NFPA 241, *Standard for Safeguarding Building Construction and Demolition Operations.*

NFPA 395, *Standard for the Storage of Flammable and Combustible Liquids on Farms and Isolated Construction Projects.*

NFPA 495, *Code for the Manufacture, Transportation, Storage, and Use of Explosive Materials.*

Additional Readings

Brannigan, Francis L., *Building Construction for the Fire Service*, National Fire Protection Association, Quincy, MA, 1982, pp. 297.

Carney, Jerome P., "Fire Protection During Construction A Critical Time," *SFPE TR 84-7*, Society of Fire Protection Engineers, Boston, MA, 1984.

"Collapse of the Hotel Vendome, Boston, Mass.," *Fire Journal*, Vol. 67, No. 1, Jan. 1973, pp. 33-41.

Feld, Jacob, *Construction Failures*, J. Wiley & Sons, NY, 1968.

Herbstman, Donald, "Fire Protection During Construction," *Fire Journal*, Vol. 64, No. 1, Jan. 1970, pp. 29-32, 89.

Hubitsky, J., "Preventing Fires on Job Sites," *Fire Command*, Vol. 45, No. 7, July 1978.

Johnson, Gerald L., "Trouble Times Three," *Fire Command*, Dec. 1984, pp. 16-19.

Juillerat, Ernest E., "The Menace of Abandoned Buildings," *Fire Journal*, Vol. 59, No. 1, Jan. 1965, pp. 5-10.

McKaig, Thomas H., *Building Failures—Case Studies in Construction and Design*, McGraw Hill Book Company, Inc., NY, 1962.

Peterson, Carl E., "The Occupancy Fire Picture: Apartment Buildings Under Construction," *Fire Journal*, Vol. 63, No. 5, Sept. 1969, pp. 23-25.

Rule, Charles H., "Fire Destroys Philadelphia Building under Construction," *Fire Journal*, Vol. 79, No. 3, May 1985, pp. 42-46, 105-111.

EVALUATING STRUCTURAL DAMAGE

Revised by Herman H. Spaeth

This chapter covers the factors that must be considered in evaluating the integrity of structures for further use and occupancy after exposure to fire. Special attention is directed to the techniques for evaluating the post fire strengths and weaknesses of the basic building materials—steel, concrete, masonry, and wood in structural assemblies.

Other chapters in this HANDBOOK that will be of particular interest concerning evaluating structural fire damage are Chapter 5, "Classification of Building Construction;" Chapter 8, "Structural Integrity During Fire;" and Chapter 9, "Confinement of Fire in Buildings," all in this section.

All structural materials, whether classified as combustible or noncombustible, inherently possess a degree of fire resistance. However, in retrospect, they are adversely affected when exposed to elevated temperatures during a fire. The degree of damage may vary with the basic materials and building configuration. In many cases, damage may be compounded by excessive expansion and instability in unit design.

THE POST FIRE INSPECTION

Immediately after a fire, several interested groups such as the fire department, the building inspection department, and insurance interests, will usually insist on a fact finding inspection. The fire department will be interested in cause of the fire, whether accidental or arson, as well as possible code violations. The building inspection department and the insurance interests will be concerned with the extent of structural and property damage as well as residual stability of any buildings involved.

It is the responsibility of the building department to inspect the damaged structure for evidence of potential collapse. Inspections may be accomplished on a gross basis by inspecting the major load bearing elements to determine prime weakness. Distorted or partially collapsed bearing walls, columns that are misaligned by heat (to the point where loads are no longer supported axially), partial collapse of critical supports, and unsymmetrical loadings are important aspects of post fire inspection. Evidence of removal of lateral supports of flexural members, either by the fire or by subsequent clean up operations, is also of critical concern. Temporary shoring or lateral supports may be required prior to admitting any inspection parties.

The decision to either demolish or reconstruct the damaged building depends upon several factors. Many cities have established statutory requirements and regulations which stipulate a building damaged in excess of 60 percent of its physical proportion must comply with current code regulations if reconstructed. The resulting costs might well discourage rebuilding. New zoning requirements may also prohibit reconstruction.

If the building is to be renovated, the strength of the structural assemblies that were exposed to the fire must be determined. Assessment of damage will involve both evaluation of the structure as a unit, and evaluation of specific structural members and assemblies. The evaluation should be conducted by a qualified engineer, as the influence of distortion and residual stresses may have a significant influence on the load carrying capacity of the structure. The engineer should be provided, if possible, with the building's original plans and specifications. Cooperation with the fire department is essential to establish the path and progress of fire, including method of extinguishment. It may be possible to establish maximum temperatures reached during the fire by examining debris for melted material and by inspecting structural members for spalling, cracking, discoloration, deflection, etc.

The influence of temperature on the mechanical properties of structural materials raises the question of the "use ability" of the material after it has cooled down to ambient temperature. The evaluation of the structural behavior of steel, concrete, masonry, and wood is described below. A more detailed evaluation procedure and a case study of renovation is given in Chapters 12 and 13 of *A Complete Guide to Fire and Buildings* (Marchant 1973).

Mr. Spaeth was Staff Supervisor, Codes and Standards, Insurance Services Offices, San Francisco, CA, prior to his retirement in 1984. He is now a fire protection consultant based in San Francisco.

It is important to determine if the fire resistance rating of specific structural elements or a structural assembly has been reduced to a value less than that required. While a building may remain structurally sound after a fire, i.e., retain its ability to withstand normal loadings, its fire resistance rating may have been impaired to the extent that occupancy cannot be permitted.

Fire resistance requirements for different types of building construction are contained in NFPA 220, *Standard Types of Building Construction*. Construction type is indicated by a Roman numeral. The Arabic numerals following the construction type designation indicate the fire resistance ratings (in hours) of exterior bearing walls, structural frame or columns and girders supporting loads of more than one floor, and floor construction, in that order. In a multistory structure, for example, fire damage may have been confined to a single story with no reduction in the tensile strength of structural members but with damage to the enclosure or protection of a beam or column. If a Type I-443 structure is required in this example and the fire resistance rating of a column supporting more than one floor is reduced to less than four hours, the structure can no longer be defined as a Type I-443. Under these conditions, occupancy of any portion of the structure would violate building code requirements.

STEEL

The use of several grades of steel in buildings and bridges is accepted in modern building codes and design standards. Although different grades of steel have a wide range of strengths, tensile and yield strengths of all steel grades are similarly affected by the temperatures that may be expected in fires in buildings.

The behavior of beams or other members subject to bending stresses under fire conditions is complex. In a building fire, the steel temperature may vary considerably over a single cross section as well as along the length of a structural member. Information relating to the strength properties of steel at elevated temperatures has been derived from the results of tests on small specimens heated so that the entire specimen was at or close to the measured temperature. Conclusions relating to strength characteristics from such small scale tests may not be applicable to steel structural members. Moreover, plastic action results in the redistribution of stresses in steel members that are loaded close to the design limit. This characteristic permits steel members to sustain loads greater than those calculated to be safe on the basis of yield strength alone; therefore, the strength of a given steel member cannot be determined only from isolated temperature data.

In a building fire, parts of the structure may have been exposed to heating followed by abrupt cooling by water from hose streams. This temperature change is usually less severe than what is accepted practice in heat treating of steel during manufacture. Tests performed on structural steel specimens taken from a building that experienced a fire during construction showed no significant loss of either yield or tensile strength. If a steel member shows no distortion due to heat, or if it can be straightened, its physical properties are generally unchanged. Connections between members should be checked, however, for cracks around rivet or bolt holes.

Occasionally, steel exposed to a fire will have a somewhat roughened appearance due to excessive scaling and grain coarsening. The coarsening is caused by exposure of steel to temperatures around 1,600°F (870°C) or higher. The steel will usually have a dark gray color, although other colors may be present if certain chemicals have been involved in the fire. Steel so modified is commonly called "burnt" steel. Members that have become burnt will usually be severely corroded as well, and their suitability for further use is a matter for individual judgment.

Many instances have been reported where straightening of structural steel members distorted due to fire has been both feasible and economical. Following the McChord Air Force Base fire in 1957 near Tacoma, WA, in which two aircraft hangars were seriously damaged, 1,486 structural steel members were either straightened or replaced in the steel frame. Of that number, only 46 members were replaced; the remaining 1,440 members were flame straightened, using welding torches.

The use of cast iron as a structural material has all but ceased. One reason is that cast iron may fracture when heated and suddenly cooled. Fires in buildings may cause heated cast iron columns to collapse if they are struck by water from hose streams. Although rare in modern structures, cast iron columns may still be found in older buildings.

Excluding fire protection features such as extinguishing systems, low fire loads, compartmentation, etc., damage to structural steel members may be minimized by application of protective coatings that provide a period of fire resistance. Generally a fiber or cementitious mixture is applied by spray application; however, the adherence and durability of such coatings, even under normal circumstances, has been the subject of some concern. Too often, questionable application resulted in unreliable fire resistance and, subsequently, considerable structural damage.

Sprayed-on protective coverings can be expected to be easily dislodged during fire fighting and overhaul operations. Usually injury of this kind is easily detected, but a complete examination of the structure must be made to determine the extent of damage. Also, before occupancy is permitted, it is imperative to determine that fire damage to a specific section of a building does not impair the integrity of zones of safety, means of egress, and smoke towers, or the operation of fire doors, fire dampers, or other protective devices or systems in areas not damaged by fire.

At present there is a strong move to assess fire resistance ratings by equating the performance of structural elements to specific fire loadings. While this concept has great merit, it does, however, present a challenge to authorities having jurisdiction. Conceivably, these authorities could be responsibile for assessing the fire resistance rating requirements and maintaining control over fire loads for the life of the building.

Heavy damage to roofs, floors, beams, and trusses have been traced to misinterpreted fire test listings. For fire resistive purposes, a floor, roof, or beam and the ceiling which it supports are considered and tested as a single structural element. The ceiling does not provide a specific fire resistance rating. It should be noted that many floor or roof and ceiling tests provide an assembly rating as well as a beam rating. Untrained persons have assumed the beam rating as protection afforded by the ceiling. Unfortunately, since the ceiling membrane does not provide a fire resistance rating, structural components above the ceiling are subject to damage.

CONCRETE

In a fire lasting one to two hours, concrete will be generally only moderately damaged, and so routine cutting and patching procedures will usually be adequate before repairs. In intense and lasting fires, such as those which may occur in large, heavily stocked warehouses and department stores, severe damage to concrete may be expected. In some cases of record, restoration entailed the removal of severely damaged areas and patching in areas less severely damaged. Experienced engineering judgment is required for evaluating the residual strength of those areas that are somewhere between moderately and severely damaged. An example of one approach to the problem of evaluating the residual strength of apparently marginally damaged concrete structural units will be found in "Prestressed Concrete Resists Fire" (Zollman & Garavaglin 1960).

A waiting period of perhaps several weeks should elapse after the fire has been extinguished before careful study of structural damage is initiated. This delay will allow any damage to the concrete such as cracking, layering, calcination, and discoloration [a change from the natural gray color to pink or brown is indicative of heating to temperatures in excess of 450°F (232°C)] to become more discernible. The thickness of fire damaged concrete in structural members can be determined by chipping with a pick or geologist's hammer. Unsound concrete may be colored and will be more or less soft and friable while sound concrete will give a distinctive ring when struck with a pick or hammer. Cored concrete samples for compressive strength tests and reinforcing steel for tensile strength determinations will enable a closer evaluation of the residual strength of damaged members. Load tests may be applied, but should be conducted only under the supervision of a registered structural engineer.

Two other methods for evaluating fire damage are worthy of mention, but each requires an experienced operator. These make use of the Schmidt hammer and the soniscope. The Schmidt hammer has a spring loaded plunger, which is caused to strike the test surface by the release of a trigger. The rebound of the plunger is a measure of concrete strength. The rebound numbers should not be considered more than a qualitative evaluation of the concrete strength, as there are several variables other than strength which may affect rebound numbers. The soniscope is an electronic device that has been used to gage the soundness of relatively heavy concrete sections such as highway pavement slabs, bridge components, and dams. A high frequency pulse is directed by an electronic sender through the section in question and is picked up by an electronic receiver. The speed at which the pulse travels through the section from sender to receiver, in feet or meters per second, is a gage of the integrity of material in the pulse path. Severe exposure of heavy concrete sections to fire may cause "layering," that is, partial separation of the outer 1 to 2 in. (25 to 50 mm) of concrete of a building member from the interior mass. Such "layering" can absorb the full energy of the pulse, so interpretative experience is necessary. The device has been successfully used in conjunction with other methods in at least one case to evaluate fire damage.

A complete report dealing with a survey of fire damage to a multistory reinforced concrete building and the subsequent repair procedure may be found in the paper "Fire Damage to General Mills Building" (Fruchtbaum 1941).

A severe fire occurred in 1951 in an unsprinklered paper warehouse. This fire lasted for more than 44 hours, and it is estimated that heat intensities of more than 1,600°F (870°C) existed for 3 hours or longer. Even after severe fire, the structure was still standing and the concrete floors prevented the spread of fire and major water damage. An inspection of the structure after the fire showed that the concrete roof and columns appeared to have suffered little damage, except that the roof showed deflection up to a maximum of approximately 2½ in. (64 mm) in a 20 ft (6 in.) span. When holes were cut through the roof, it was found that the concrete was calcined and had a light brown color in varying depths. This damaged concrete had almost no strength, and began to disintegrate after a period of several weeks. As a result, the portions of the roof slab that showed appreciable amounts of calcine were replaced. Many of the columns required removal of the concrete down to the reinforcing steel spiral, and in some cases, calcining appeared inside the spiral. In these cases, the entire column was replaced. The column caps were replaced completely if more than one-half of the column cap showed evidence of calcining. Reinforcing rods showed some reduction in tensile strength but most, were salvaged and reused, and downgraded to about three-quarters of their original strength.

If a decision is made to repair the damage, the usual procedure is to shore up the structure, if necessary, and to remove the fire damaged concrete using a hammer, chisel or a lightweight mechanical hammer. (Use of heavy hammers is discouraged since they might aggravate the damage by inadvertently chipping good concrete.) The steel reinforcements are then cleaned and if necessary, additional reinforcements are added and finally, the concrete is built up with gunite. Durable concrete repairs also have been made with epoxy resins.

The significant difference between conventional reinforced concrete and stressed concrete in fires is the performance of the high tensile steel wire or rods used for pre or post stressing. Under fire conditions, the stressed steel units are liable to rapid loss of strength at temperatures in excess of 752°F (400°C). A load test is recommended in post inspection for significant pronounced sag or an indication of excessive temperature in a range where stress loss may have been sufficient to seriously affect the strength.

BRICK

During production, clay bricks are exposed to temperatures in excess of 2,000°F (1093°C), hence their strength is retained in actual fires. Reinforcing steel embedded in the center of a clay brick wall would normally be protected by a minimum of 3 to 4 in. (75 to 100 mm) and not be affected.

WOOD

Wood is one of the oldest and most widely used building materials. Its behavior in fire conditions varies considerably, depending upon the species of wood and the configuration in design, i.e., solid sawn lumber, glue laminates, plywood, woodchip board, etc. The effect of fire on glue laminates may be considered the same as that on solid sawn lumber (heavy timber), assuming that the adhesives used were not effected by heat. Generally speaking, the Phenol-resorcinal and melamine adhesives are not affected by heat.

Plywood and chipboard are also dependent upon proper adhesives, the difference being slow burning versus a hazardous flash fire caused by delamination.

Redwood, found on the west coast of the U. S., withstands high fire exposures. It is reported that the high resistance is due to the lack of the usual volatile resins and oils found in other woods.

When wood structural members are subjected to fire, the ability to withstand the imposed loads is dependent to a degree upon the amount of remaining cross-sectional area. The average rate of penetration of char when flame is impinged upon an exposed wood member is approximately 1½ in. (38 mm) per hr. Beyond the char area to a point not more than ¼ in. (6 mm) away, the structural properties of wood may be affected by its exposure to high temperatures. The degree of strength loss in this small zone adjacent to the char is not exactly known, but is presumed to be insignificant.

Fire tests made on two solid sawn wood joists, 4 by 14 in. (100 by 350 mm), nominal size, at the Southwest Research Institute showed that after 13 minutes of fire exposure, 80 percent of the original wood section remained undamaged and available to carry the load (Nat. For. Prod. Asso. 1961a). In another test of two 7 by 21 in. (170 by 540 mm) glued laminated beams, after 30 minute fire exposure, 75 percent of the original wood section remained and continued to support the design load (Nat. For. Prod. Asso. 1961b).

The previous tests, as well as actual fire experience, substantiate the fact that wood members will remain in place under fire conditions and continue to support design loads. It is usually in the larger or heavy timber members that char can be scraped clear and an evaluation made by a qualified engineer or architect to determine the remaining load supporting capacity of the wood member. This same process is true of smaller size members, except that they are usually protected by lath and plaster or gypsum wallboard. It can be said then that the strength of a wood member subjected to a prolonged period of fire intensity may be reduced by the loss of cross section, which results in a corresponding change in deformation under a given load. After the char area has been removed, the remaining strength or load supporting capacity can be easily determined through proper design analysis.

Bibliography

References Cited

Fruchtbaum J. 1941. "Fire Damage to General Mills Building." Proceedings of American Concrete Institute, Detroit, MI.

Marchant, E. W., ed. 1973. *A Complete Guide to Fire and Buildings*. Barnes and Noble, NY. p 268.

Nat. For. Prod. Asso. 1961a. "Comparative Fire Tests on Wood and Steel Joists." *Technical Report No. 3*. National Forest Products Association, Washington, DC.

Nat. For. Prod. Asso. 1961b. "Comparative Fire Tests of Timber and Steel Beams." *Technical Report No. 3*. National Forest Products Association, Washington, DC.

Zollman L., and Garavaglin, M. 1960. "Prestressed Concrete Resists Fire." *Civil Engineering*, NY. pp 36-41.

NFPA Codes, Standards, Recommended Practices and Manuals. (See the latest *NFPA Codes and Standards Catalog* for availability of the current edition of the following documents.)

NFPA 220, *Standard on Types of Building Construction.*

NFPA 251, *Standard Methods of Fire Tests of Building Construction and Materials.*

Additional Readings

Davis, Norman H., Jr., "Burning Characteristics of Building Materials," *Proceedings* of the 20th Annual West Virginia Fire School.

Dill, F. H., "Structural Steel After a Fire," *Proceedings* of AISC, National Engineering Conference, Brown Palace Hotel, Denver, CO, May 5 and 6, 1960.

Bessey, G. E., "The Visible Changes in Concrete or Mortar Exposed to High Temperature," Part II, National Building Studies, *Tech. Paper No. 4*, Ministry of Technology and Fire Offices' Committee Joint Fire Research Organization (formerly Department of Scientific and Industrial Research), Building Research Station, Garston, Watford, Hertsfordshire, printed at Her Majesty's Stationery Office, London, England, 1950.

Fire Damage —Civil Liability, *Fire Surveyor*, Oct. 1983.

"Fire Materials of Structures," *Digest No. 106*, June 1958, Building Research Station, Her Majesty's Stationery Office, London, England.

"Fire Performance of Glue-Laminated Timber," *Technical Memo*, Canadian Institute of Timber Construction, Ottawa, Ontario, Canada, Dec. 1958.

"Firemen Fear Floor Collapse," *Fire Engineering*, Nov. 1958.

Gustaferro, A. H., "Experiences from Evaluating Fire-Damaged Concrete Structures," in: *Proceedings* of the International Symposium on Fire Safety of Concrete Structures, ACI Meeting, American Concrete Institute, Detroit, MI, Sept. 1980.

"How Materials Behave in Fire—A Realistic Approach," *Materials Engineering*, Nov. 1976.

"Investigation and Repair of Damage to Concrete Caused by Formwork and Falsework Fire," *Journal of the ACI*, Nov. 1963.

Lie, T. T., "Optimum Fire Resistance of Structures," *Proceedings* of the American Society of Civil Engineers, Jan. 1972.

Lie, T. T., *Fire and Buildings*, Applied Science, London, England, 1972.

"Notes on Repair of Damaged Buildings": Note 2—Repair of Structural Steelwork Damaged by Fire, Nov. 1944, Ministry of Technology and Fire Offices' Committee Joint Fire Research Organization (formerly Department of Scientific and Industrial Research), Building Research Station, Garston, Watford, Herts., Her Majesty's Stationery Office, London, England.

"Notes on Repair of Damaged Buildings," Note No. 13—Reinforced Concrete Columns Damaged by Fire, Sept. 1945, Ministry of Technology and Fire Offices' Committee Joint Fire Research Organization (formerly Department of Scientific and Industrial Research), Building Research Station, Garston, Watford, Herts., Her Majesty's Stationery Office, London, England.

"Notes on Repair of Damaged Buildings," Note No. 19—The Repair of Solid Concrete and Hollow-tile Floors Damaged by Fire, Sept. 1945, Ministry of Technology and Fire Offices' Committee Joint Fire Research Organization (formerly Department of Scientific and Industrial Research), Building Research Station, Garston, Watford, Herts., Her Majesty's Stationery Office, London, England.

"Notes on Repair of Damaged Buildings," Note No. 24 — "Reinforced Concrete Beams Damaged by Fire," July 1946, Ministry of Technology and Fire Offices' Committee Joint Fire Research Organization (formerly Department of Scientific and Industrial Research), Building Research Station, Garston, Watford, Herts., Her Majesty's Stationery Office, London, England.

Parker, T. W., and Nurse, R. W., "Investigation of Building Fires: The Estimation of the Maximum Temperature Attained in Building Fires from Examination of the Debris," Part I, National Building Studies, *Tech. Paper No. 4*, Ministry of Technology and Fire Offices' Committee Joint Fire Research Organization (formerly Department of Scientific and Industrial Research), Building Research Station, Garston, Watford, Hertsfordshire, printed at Her Majesty's Stationery Office, London, England, 1950.

"Repair of Damaged Concrete with Epoxy Resin, Conventional Methods of Repairing Concrete," *No. 57-59*, Aug. 1960, American Concrete Institute Committee 201.

Sharry, John A., et al., "Military Personnel Records Center Fire, Overland, Missouri (Part 1)." *Fire Journal*, Vol. 68, No. 3 (May 1974), pp. 5-9.

Smith, Peter, "Investigation and Repair of Damage to Concrete Caused by Formwork and Falsework Fire," *Journal of the American Concrete Institute*, Nov. 1963, pp. 1535-1565.

"Take the Surprise Out of Building Collapse," *Fire Engineering*, Jan. 1984.

Troxell, G. E., "Prestressed Lift Slabs," *Civil Engineering*, Sept. 1965, New York, pp. 64-66.

Walker, Evans, et al., "Military Personnel Records Center Fire, Overland, Missouri (Part 2)." *Fire Journal*, Vol. 68, No. 4, July 1974, pp. 65-70.

"Why the Increases in Fire Losses —Some Answers," *Fire Surveyor*, Oct. 1973.

SECTION 8

FIRE HAZARDS OF BUILDING SERVICES

HAZARDS OF BUILDING SERVICES

Robert William Ryan

Building services and utilities are those electrical and mechanical systems that provide power, environmental controls, and essential conveniences for the practical use of a structure. These include electrical service and equipment, heating, ventilation, air conditioning, refuse disposal, plumbing, communications, transportation, and conveyance systems. To understand how building services may affect fire behavior, and the fire hazards that could result from these systems and their components, a basic knowledge of the individual systems and their relationship to each other, the structure, and the type of occupancy is needed.

This chapter presents an overview for general discussion of phases of building service system life—design, operation, maintenance, and renovation. For more detailed information about specific systems, see the next five chapters of this section: Chapter 2, "Electrical Systems and Appliances;" Chapter 3, "Heating Systems and Appliances;" Chapter 4, "Air Conditioning and Ventilating Systems;" Chapter 5, "Building Transportation Systems;" and Chapter 6, "Miscellaneous Building Systems." Related information not included in this section may be found in other chapters contained in this HANDBOOK. Section 7, Chapter 2, "Systems Concept for Building Firesafety," discusses the systems approach to building design. Section 7, Chapter 9, "Confinement of Fire in Buildings," covers the design and construction of elevators, escalators, and dumbwaiters, and the protection of floor openings. Section 12 also contains two chapters with pertinent information, Chapter 13, "Air Moving Equipment;" and Chapter 14, "Waste Handling Systems."

DESIGN

It is a challenge to design and install building services and utilities that act in harmony with a building's structural design and use, as well as provide an effective, total fire protection system. The solution of this problem will determine, perhaps for the life of the structure, the effect of fire on the building's occupants and on the fire service.

Mr. Ryan is assistant director, Department of Environmental Safety, at the University of Maryland, College Park, MD.

Although these building systems may be considered separately, it must be understood that each system is only one part of the total building system. Depending upon how they are integrated, the synergism, or combined effect, of these systems on a particular fire scenario may either support fire protection and life safety design objectives, or complicate and even prevent fire control and evacuation efforts.

Systems Approach

Recognizing the potential utility, or possible hazard, of building systems, fire protection engineers stress the need for a systems approach to building design. This approach is particularly effective when building services are under consideration, especially since service and utility systems have become basic elements of the total building fire protection design. In the past, thick stone or masonry walls and floors provided fire zone compartmentation, most areas had operable windows for ventilation, and hazardous operations were segregated in separate structures. Today, fire protection for steel framed, windowless, and large multiple use structures is provided in part by electrical and mechanical systems.

Many building services may be designed to serve a dual role. For example, although heating, ventilation, and air conditioning systems (HVAC) usually function to maintain a controlled environment for the building occupants or equipment, these systems may also contribute actively or passively to the overall building fire protection design. A minimum expectation should be that the HVAC system will not cause fire or smoke to spread. This may be achieved simply by designing the system to shut down during a fire, and by protecting the openings made by the system in structural fire barriers. More effectively, the system can be designed to help control smoke spread by pressurization of exits and areas adjacent to the fire, and by exhausting smoke from the fire area. Recent requirements for smoke control systems have resulted in various design criteria to provide "engineered" systems. There is a wide range of opinion among fire protection engineers as to the effectiveness of many of these systems.

The reliability of such smoke control systems also depends upon the reliability of the electrical systems. Insufficient capacity or inadequate protection of the power supply could prevent the effective transition of the HVAC system from environmental to smoke control. Therefore, critical components such as motor control centers for exhaust fans must be protected from damage during a fire.

Design Team: Each member of a building design team needs to be aware of the performance objectives of the total fire protection system and to integrate their phase of the system accordingly. For example, the electrical engineer must know which control centers are required to provide continuous power to ventilating equipment designed for smoke control, and the mechanical engineer must know which ducts penetrate fire barriers before designing fire and smoke dampers in the system. To ensure that the design of building services will not conflict with fire protection, a qualified fire protection engineer should be retained as a member of the design team from the design stage to the field acceptance inspection and occupancy, and be responsible for coordinating building service systems design with the building fire protection system.

OPERATION

Before buildings are occupied, the building engineering and administrative staff must be prepared to operate the service systems effectively. This means accurate sets of "as built" drawings need to be prepared and provided to the owner and occupants. Representatives of the designer and equipment manufacturer must provide information and training to the building engineering staff so they will be able to operate the systems as designed, and assist the fire service during building emergencies.

Automatic Systems

Building automation systems (BAS) are common in large modern buildings and can be integrated with the fire protection system. Smoke control functions of HVAC systems can be activated by smoke detectors within the HVAC system, or by operation of fire suppression, detection, or alarm systems within the fire area. Elevators can be made to return nonstop to ground level lobbies when heat or smoke is detected. Fire alarm systems in large buildings may activate the buiding's communications system(s) to direct evacuation of fire zones.

Building engineering and maintenance staffs, as well as the local public fire service, must be familiar with the operation of the building systems. When automatic systems are operating according to plan, it could be hazardous, and may make fire suppression more difficult if the systems are shut down or reset prematurely. For this reason, access to system controls needs to be limited to those familiar with the operation of the controls and the possible consequences of early shutdown. It is good practice to provide key-controlled operation for manual override controls or locate them in areas accessible only to authorized personnel.

Training

An ongoing training and orientation program for the building staff and the local fire service is necessary, especially when the fire service is expected to control or utilize a system such as a communications center in a high rise building or a smoke exhaust system in a windowless building. Both the owner and the local fire authorities are responsible for thorough prefire planning and emergency operation of building service systems. The premature shutdown of building services could, in many buildings, cause as much disruption of fire suppression and rescue operations as the premature shutoff of sprinkler systems.

Restoration

After an emergency operation of building systems, it is necessary to check the systems to be certain they have been restored to normal status and are in good operating condition for the next emergency. Some systems are restored simply by resetting or operating remote electromechanical controls such as circuit breakers, elevator reset switches, etc. Other components such as fusible link operated fire dampers in HVAC systems, or any system with damaged component parts, will need to be physically repaired or replaced. Building services that operate as a component of building fire protection should be restored before the building is reoccupied after a fire or emergency. It is good practice for local fire authorities to inspect building services soon after an emergency to ensure that the systems have been restored.

MAINTENANCE

The maintenance of services and utilities is as important as their design and operation. When more complex mechanical and electrical systems are integrated as parts of building fire protection, their long term reliability becomes critical.

The quantity and cost of service system maintenance partially depends upon the design and selection of components and equipment. Maintenance should be one of the considerations of the building design team. In some facilities it may be more appropriate to design simple, relatively maintenance free systems than to overdesign complex systems that will require constant maintenance by skilled mechanics. Systems that are too complex may eventually need to be placed out of service due to a lack of maintenance funds or of competent mechanics. Where building fire protection includes these systems, downtime or shutdown will increase the risk from fire.

Maintenance of building systems has two components: demand (breakdown) maintenance, and preventive maintenance. Elements of good preventive maintenance programs include the availability of manufacturer's equipment data, operating and maintenance instructions, system schematics, qualified mechanics, and assigned, scheduled responsibilities for maintenance personnel.

Unless an adequate maintenance program for building service systems is in place, that system cannot be expected to operate as planned. Poor maintenance is also a major factor in building service failures that cause fires. The author of *Building Mechanical Systems* predicts that the maintenance function is bound to grow in size (Andrews 1977). This is why:

1. More and more complex equipment is being developed for and used in buildings.
2. Competent maintenance people are more difficult to find.
3. Rising materials and equipment costs make repairing the system more economical than replacing it.

4. Soaring energy costs demand that equipment be kept at peak operating efficiency.

To best utilize maintenance dollars, an "operations manual" should be used and a preventive maintenance program established.

RENOVATION

The function of a building will seldom remain static throughout its use. Often building services designed for the original use will be adequate only after structural or occupancy changes have been made. The rehabilitation of older buildings is one of the most obvious examples of this problem. Modern demands for electrical service seldom are met even by the service provided in relatively new buildings. In addition, major renovation projects usually include the revitalization of building services to meet current codes. The provision of three wire grounded circuits and ground fault circuit interrupter devices are examples of this kind of revitalization.

Whenever a change is planned for the use or structure of a building, or for the extension, renovation, or elimination of a building service, the impact this change will have on any other building system and on the total firesafety of the building should be evaluated. Even seemingly harmless changes in building services can jeopardize a building's firesafety. For example, the addition of a laboratory fume hood on an existing exhaust system could reduce volume for existing hoods and allow explosive concentrations of flammable vapors to build up. Perhaps an extension of the communications service could allow fire spread through a fire barrier if the "poke through" holes are not sealed adequately. Of course, all changes in the use of buildings do not increase the risk of fire. Some changes, such as, for example, the elimination of flammable anesthetics from hospital operating rooms, have improved the fire protection effectiveness of existing building services or eliminated the need for some special precautions.

FUTURE DEMANDS

Demands on building service systems will probably increase in the future to meet the changing demands of system size, performance, and critical needs.

Energy

Today, the need for energy conservation has resulted in the application of some ancient design concepts such as solar heating and coverings of earth in new structures. Fire protection demands on building services for these structures are more sophisticated than for those in most existing buildings. Some of the fire protection problems associated with these types of buildings have already been described. Windowless buildings present the same kind of problems as earth covered buildings regarding smoke control systems. Open atrium designs in these types of structures can be adequately protected from fire spread with the imaginative application of existing technology. Energy management systems (EMS) installed primarily for cost effective energy control can and should be used to enhance building fire protection.

Megastructures

Costs of real estate, construction, and transportation, combined with increased population in some areas, will probably encourage the continued construction of very large, multi-occupancy structures. These "megastructures," as they are known, require changes in fire protection philosophy and building service system design. For example, the use of elevators as a means of egress from fires in very tall buildings may be necessary (Kravontka 1976). Such use would require changes both in current elevator design and in public firesafety education. Also, future buildings may require designs that eliminate the need to evacuate during a fire emergency.

Accessibility

The increasing requirements for barrier free design of buildings to allow access by disabled persons has already changed the demand for some building services, such as elevators. In the past, barrier free requirements were concerned primarily with accessibility. The need for designed areas of refuge or special elevators for emergency egress of disabled people from high or deep buildings is obvious. Building service systems can help provide life support or emergency transportation to people who cannot fully utilize standard exit facilities. Parallel and redundant systems for the disabled usually are usually unnecessary; thoughtful design can provide systems that are accessible to everyone.

Summary

The challenge to fire protection authorities is to work closely with other design professionals to develop effective systems that will enhance building firesafety. The importance of building service and utility systems as integral, mutually dependent components of total building fire protection and life safety will remain prominent in the foreseeable future.

Bibliography

References Cited

Andrews, F. T. 1977. *Building Mechanical Systems*. McGraw-Hill, Inc., NY.

Kravontka, Stanley, J. 1976. *Elevator Use During Fires in Megastructures*. Society of Fire Protection Engineers, Boston, MA.

NFPA Codes, Standards, Recommended Practices and Manuals. (See the latest *NFPA Codes and Standards Catalog* for availability of current editions of the following documents).

NFPA 31, *Standard for the Installation of Oil Burning Equipment.*

NFPA 54, *National Fuel Gas Code.*

NFPA 58, *Standard for the Storage and Handling of Liquefied Petroleum Gases.*

NFPA 70, *National Electrical Code.*

NFPA 82, *Standard on Incinerators, Waste and Linen Handling Systems and Equipment.*

NFPA 90A, *Standard for the Installation of Air Conditioning and Ventilating Systems.*

NFPA 90B, *Standard for the Installation of Warm Air Heating and Air Conditioning Systems.*

NFPA 91, *Standard for the Installation of Blower and Exhaust Systems for Dust, Stock and Vapor Removal or Conveying.*

NFPA 96, *Standard for the Installation of Equipment for the Removal of Smoke and Grease-Laden Vapors from Commercial Cooking Equipment.*
NFPA 101, *Code for Safety to Life from Fire in Buildings and Structures.*
NFPA 211, *Standard for Chimneys, Fireplaces, Vents and Solid Fuel Burning Appliances.*

Additional Readings

Anderson, S. A., et al., "Laboratory Fume Hood Control," *Heating/Piping/Air Conditioning*, Vol. 56, No. 2, Feb. 1984.
Baldwin, R., and Heselden, A. J. M., "The Movement and Control of Smoke on Escape Routes in Buildings," *Fire Technology*, Vol. 14, No. 3, Aug. 1978, p. 206.
Bamert, A. E., "Fire Protection in Underground Structures," *Fire Journal*, Vol. 69, No. 1, Jan. 1975, p. 40.
Betz, G. M., "Designing Out Potential Fires in Plant Duct Systems," *Fire Protection, Plant Engineering Magazine.*
Bierwirth, Ed., "When Retrofitting, Build in the Degree of Fire Protection Needed," *ASHRAE Journal*, Vol. 26, No. 4, Apr. 1984.
Bimonthly Fire Record, "Oil-Soaked Air-Conditioning Duct Insulation Ignites," *Fire Journal*, Vol. 76, No. 6, Nov. 1982, p. 30.
Bimonthly Fire Record, "Improperly Installed Exhaust System," *Fire Journal*, Vol. 78, No. 3, May 1984, p. 21.
Bimonthly Fire Record, "Fire In Kitchen Exhaust Hood," *Fire Journal*, Vol. 69, No. 2, Mar. 1975, p. 65.
Bimonthly Fire Record, "Short Circuit in Power Service," *Fire Journal*, Vol. 69, No. 2, Mar. 1975, p. 63.
Borys, K., "Prudential's Enerplex: A Landmark of Energy Efficiency," *Buildings*, Vol. 77, No. 11, Nov. 1983.
Brannigan, F. L., *Building Construction for the Fire Service*, National Fire Protection Association, Quincy, MA, 1983.
Bugbee, Percy, *Principles of Fire Protection*, National Fire Protection Association, Quincy, MA, 1978.
Building Research Advisory Board, *Proceedings of the Conference on Energy Conservation and Firesafety in Buildings*, National Research Council, National Academy of Sciences, Washington, DC, 1982.
Butcher, E. G., and Parnell, A. C., *Smoke Control in Fire Safety Design*, William Clowes & Sons Ltd., London, England, 1979.
Canty, Donald, "Architecture Today: Everything's Happening At Once," *Fire Journal*, Vol. 75, No. 1, Jan. 1981.
Clifford, G. E., *Heating, Ventilating, and Air Conditioning*, Prentice-Hall Inc., West Nyack, NY, 1984.
Clyde, J. E., *Construction Inspection*, John Wiley & Sons, NY, 1983.
Coad, W. J., "Safety Factors in HVAC Design," *Heating/Piping/Air Conditioning*, Vol. 57, No. 1, Jan. 1985.
Degenkolb, J. G., "Atriums," *BOCA Magazine*, Vol. 17, No. 6, Dec. 1983.
Designing Buildings for Firesafety, National Fire Protection Association, Quincy, MA, 1975.
Egan, M. David, *Concepts in Building Firesafety*, John Wiley & Sons, Inc., NY, 1978.
Egan, M. David, *Concepts in Thermal Comfort*, Prentice-Hall, Inc., Englewood Cliffs, NJ, 1975.
Electrical Inspection Illustrated, National Safety Council, Chicago, IL, 1977.
Engineering News–Record, "Buildings on Fire–Are Codes Enough?" Special Report, *The McGraw-Hill Construction Weekly*, Jan. 1981.
The Handbook of Property Conservation, Factory Mutual Engineering Corp., Norwood, MA, 1983.
Featherstone, James J., III, "Use of Computer Technology in Large-Scale Building Fire Protection," *Fire Journal*, Vol. 79, No. 1, Jan. 1985.
Federal Construction Council, *New Approaches to Evaluating Fire Safety in Buildings*, Natonal Academy of Sciences, Washington, DC, 1980.
Fires In High-Rise Buildings, National Fire Protection Association, Quincy, MA, 1974.
Fothergill, J. W., "Computer-Aided Design Technology for Smoke Control and Removal Systems," *Fire Technology*, Vol. 14, No. 2, May 1978, p. 110.
Fung, Francis C. W., "Smoke Control by Systematic Pressurization," *Fire Technology*, Vol. 2, No. 4, Nov. 1975, p. 261.
Fustich, C. D., "Transformer Room Fire Tests," *Technical Report*, U.S. General Services Administration, Washington, DC, 1980.
Galuardi, J. F., *Buildings Maintenance Management*, U.S. General Services Administration, Washington, DC, 1981.
Gewain, Richard G, "Fire Research for Steel HVAC Systems," *Fire Journal*, Vol. 78, No. 6, Nov. 1984.
Glazer, Sarah, ed., "What Has Happened Can Happen – A Delayed Shutdown," *Record*, Vol. 57, No. 1, Feb. 1980, p. 30.
Guide to NFPA National Building Firesafety Standards, National Fire Protection Association, Quincy, MA, 1983.
Harmathy, T. Z., "Design of Buildings for Fire Safety," *Second International Fire Protection Engineering Institute of University of Maryland*, Ottawa, Ontario, Canada, Feb. 1975.
Haviland, D. S., *Toward a Performance Aproach to Life Safety from Fire in Building Codes and Regulations*, U.S. Dept. of Commerce, National Bureau of Standards, Washington, DC, 1978.
High-Rise Building Fires and Fire Safety, National Fire Protection Association, Quincy, MA, 1972.
Jensen, Rolf, "A Technically Sound Basis for Fire Protection Systems Design: The SDS System," *Fire Journal*, Vol. 78, No. 4, July 1984.
Jensen, Rolf, ed., *Fire Protection for the Design Professional*, Cahners Publishing Co., Boston, MA, 1975.
Kaiser, Bruno, "Have They Built a Better Fire-Trap?" *Security World*, Vol. 15, No. 3, Mar. 1978.
Kent, S. R., ed., *The Environmental Services of Buildings*, Ontario Association of Architects, Toronto, Ontario, Canada, 1970.
Klote, J. H., *Smoke Control by Stairwell Pressurization*, Center for Fire Research, National Bureau of Standards, Washington, DC, 1980.
Klote, J. H., and Fothergill, J. N., *Design of Smoke Control Systems for Buildings*, U.S. Dept. of Commerce, National Bureau of Standards, Washington, DC, 1983.
Korte, Robert T., ed., "Where and How to Use Glass Fiber Ductwork," *Heating/Piping/Air Conditioning*, Vol. 49, No. 5, May 1977.
Lathrop, James K., "Atrium Fire Proves Difficult to Ventilate," *Fire Journal*, Vol. 73, No. 1, Jan. 1979.
Lathrop, J. K., ed., *Life Safety Code Handbook*, 2nd ed., National Fire Protection Association, Quincy, MA, 1985.
Lerup, Cronrath, and Chiang, Liu, *Learning From Fire: A Fire Protection Primer for Architects*, College of Environmental Design, Berkeley, CA, 1977.
Levin, B. M., ed., *Fire and Life Safety for the Handicapped*, U.S. Dept. of Commerce, National Bureau of Standards, Washington, DC, 1980.
Linville, Jim, ed., *Industrial Fire Hazards Handbook*, 2nd ed., National Fire Protection Association, Quincy, MA, 1984.
Loss Prevention Data, Factory Mutual System Sheets (published periodically), Norwood, MA.
Magison, E. C., *Electrical Instruments in Hazardous Locations*, The Instrument Society of America, Pittsburgh, PA.
Masters, R. E., "Building Pressurization Approaches," *Proceedings of the Workshop on Engineering Applications of Fire Technology*, U.S. Dept. of Commerce, National Bureau of Standards, Washington, DC, 1980.
McGuiness, W. J., and Stein, B., *Mechanical and Electrical Equipment for Buildings*, John Wiley & Sons, NY, 1971.
McHenry, P. G., Jr., *Adobe and Rammed Earth Buildings*, John Wiley & Sons, NY, 1984.
McLaughlin, David L., "Fire Tests of Fiberglass and PVC Ducts for

Minimum Sprinkler Coverage," *Fire Journal*, Vol. 72, No. 2, Mar. 1978.

McQuiston, F. C., and Parker, J. D., *Heating, Ventilating and Air Conditioning*, John Wiley & Sons, NY, 1982.

Moreland, Frank L., ed., *Alternatives in Energy Conservation – The Use of Earth Covered Buildings*, U.S. Government Printing Office, Washington, DC, 1978.

NFPA, "Trends for 2003," *Fire Journal*, Vol. 77, No. 6, Nov. 1983.

Pauers, William R., "Fire Management via Building Automation," *Heating/Piping/Air Conditioning*, Vol. 51, No. 11, Nov. 1979.

Pearce, Henry A., "Fire Resistant Fluids for Industrial Transformers," *Fire Technology*, Vol. 14, No. 2, May 1978, p. 159.

Peige, J., and Williams, C., eds., *Fire Problems in High Rise Buildings*, International Fire Service Training Association, Oklahoma State University, Stillwater, OK, 1976.

Powers, F. C., ed., "Controlling Plastic Duct Fires," *The Sentinel*, Vol. 35, No. 3, June 1979.

Powers, F. C., ed., "Plastic Ducts: Boon or Bane?" *The Sentinel*, Vol. 35, No. 3, June 1979.

Research Products Corporation, *Design of Grease Filter Equipped Kitchen Exhaust Systems*, Research Products Corporation, Madison, WI, 1973.

Reynolds, K. P., et al. *Innovative Building Design Concepts*, Kelly P. Reynolds, and W. Ted Ritter, Publishers, Chicago, IL, 1980.

Robinson, K. M., ed., "Protecting Department Stores and Shopping Malls from Fire," *The Sentinel*, Industrial Risk Insurers. Third Quarter, 1983.

Rosaler, R. C., and Rice, J. O., ed., *Standard Handbook of Plant Engineering*. McGraw-Hill, Inc., NY, 1983.

Saigal, A., "Computerized Fire and Security System Performance." *Heating/Piping/Air Conditioning*. Vol. 56, No. 11, Nov. 1984.

Sampson, A. F., "Firesafety a Management Concern," *Business Week*, Oct. 6, 1973.

"The General Services Administrations Systems Approach to Life Safety in Structures," *High-Rise Building Fires and Fire Safety*, National Fire Protection Association, Quincy, MA, 1972.

Semple, J. Brooks, "Smoke Control: How HVAC Can Reduce Smoke and Fire Losses," *Heating/Piping/Air Conditioning*, Vol. 49, No. 10, Oct. 1977.

Skelton, R. E., "A Remotely Controlled Fire-Venting System in a Shopping Complex," *Fire Journal*, Vol. 72, No. 1, Jan. 1978, p. 76.

Smith, T., and Post, H., "Fire Safety in Atrium Buildings," *Buildings*, Vol. 77, No. 5, May 1983.

Sotis, L. P., "Plan Today, Build for Tomorrow," *Record*, Factory Mutual System. Vol. 61, No. 3. Fall 1984.

Spinna, R. J. Jr., ed., "Design, Construction and Maintenance of Fire-Safe Structures," *Proceedings of the First Annual Fire Engineering Conference*, Fire Engineering Institute, Riverdale, NY, 1983.

Syska & Hennessey, "Fire Protection: A Total Systems Approach," *Technical Letter*, Syska & Hennessey Engineers, Vol. 29, No. 5, Dec. 1981.

The Systems Approach to Fire Protection, National Fire Protection Association, Quincy, MA, 1975.

Thornberry, Richard P., "Designing Stair Pressurization Systems," *SFPE TR 82-4*, Society of Fire Protection Engineers, Boston, MA.

Yellott, John I., "Solar Energy Update," *Heating/Piping/Air Conditioning*, Vol. 51, No. 1, Jan. 1979.

ELECTRICAL SYSTEMS AND APPLIANCES

Revised by Peter J. Schram

This chapter discusses electrical systems and appliances. Specific topics include: origins of electrical fires; codes and standards; building wiring, design, and protection; electrical household appliances; industrial and commercial equipment; electrical equipment for outdoor use; locations exposed to moisture and dust; communications systems; and special occupancy electrical problems.

The remaining chapters in this section detail the related components of fire hazards of building services; most notably, Chapter 2, "Heating Systems and Appliances;" Chapter 3, "Air Conditioning and Ventilating Systems;" and Chapter 4, "Building Transportation Systems."

Special considerations for the installation of electrical systems may be found elsewhere in this HANDBOOK. See Section 7, Chapter 3, "Concepts of Egress Design;" Section 10, Chapter 5, "Industrial and Commercial Heat Utilization Equipment;" Section 12, Chapter 4, "Medical Gases;" and Section 12, Chapter 12, "Lightning Protection Systems."

If properly designed, installed, and maintained, electrical systems are both convenient and safe; otherwise they may be a source for both fire and personal injury. Electricity may cause a fire if it arcs or overheats electrical equipment and can cause injury or death through shocks and burns.

When an electric circuit carrying a current is interrupted either intentionally, as by a switch, or unintentionally, as when a contact at a terminal becomes loosened, arcing at the switch contacts or heating from the high resistance connection at the terminal is produced. The intensity of the arc and degree of heat depends largely upon the current and voltage of the circuit and the resistance of the contact at the terminal. The temperature may easily be high enough to ignite any combustible material in the vicinity.

An electric arc may not only ignite combustible material in its vicinity, such as the insulation and covering of the conductor; it may also fuse the metal of the conductor. Hot sparks from burning combustible material and hot metal may be thrown about or fall, setting fire to other combustible material.

Mr. Schram is Chief Electrical Engineer, Electrical Field Service, on the staff of NFPA.

When an electrical conductor carries a current, heat is generated in direct proportion to the resistance (ohms) of the conductor and to the square of the current (amperage). The resistance of conductors used to convey current to the location in which it is used, or to convey it through the windings of a piece of apparatus (except resistance devices and heaters), should be as low as practical. Metals such as copper and aluminum are used for this purpose. In other instances, such as in electric heaters, electric cooking equipment, and soldering irons, the heat from the current serves a useful purpose.

As a fire hazard, the heating of electrical conductors is negligible under ordinary circumstances. NFPA 70, *National Electrical Code®* (hereinafter referred to as the NEC) specifies the maximum safe current a conductor may carry without overheating. (See Tables 8-2A through 8-2D.) This depends upon the size of conductor, the conditions of the installation, and the type of insulation. Where these specified currents are exceeded, or where a conductor is overloaded, the generation of heat becomes a hazard through (1) deterioration of the electrical insulation of the conductor, and (2) the excess heat generated. Apparatus or appliances that use electric conductors as heating elements or use an electric arc to generate heat (arc welders, for example) are likely to be fire hazards unless properly installed and used.

One common method of reducing the degree of hazard is to provide sufficient air circulation to prevent unsafe temperatures and premature breakdown of electrical insulation.

All standards governing electric equipment include requirements intended to prevent fires caused by arcing and overheating, and to prevent accidental contact, which may cause an electric shock. These primary sources of electrical hazards should be kept in mind whenever any work is being done on electric equipment.

ORIGINS OF ELECTRICAL FIRES IN BUILDINGS

As stated, electrical fires are caused by arcing and overheating. The International Association of Electrical Inspectors (IAEI) has standardized 19 broad reporting

TABLE 8-2A. Ampacities of Insulated Conductors Rated 0-2000 Volts, 60° to 90°C

Not More Than Three Conductors in Raceway of Cable or Earth (Directly Buried), Based on Ambient Temperature of 30°C (86°F).

Size	Temperature Rating of Conductor, See *National Electrical Code* Table 310-13								Size
	60°C (140°F)	75°C (167°F)	85°C (185°F)	90°C (194°F)	60°C (140°F)	75°C (167°F)	85°C (185°F)	90°C (194°F)	
AWG MCM	TYPES †RUW, †T, †TW, †UF	TYPES †FEPW, †RH, †RHW, †RUH, †THW, †THWN, †XHHW, †USE, †ZW	TYPES V, MI	TYPES TA, TBS, SA, AVB, SIS, †FEP, †FEPB, †RHH, †THHN, †XXHW*	TYPES †RUW, †T, †TW, †UF	TYPES †RH, †RHW, †RUH, †THW, †THWN, †XHHW, †USE	TYPES V, MI	TYPES TA, TBS, SA, AVB, SIS, †RHH, †THHN, †XHHW*	AWG MCM
	Copper				Aluminum or Copper-clad Aluminum				
18	....	....	....	14	....	....	....	....	....
16	....	....	18	18	....	....	....	....	....
14	20†	20†	25	25†	....	....	....	....	....
12	25†	25†	30	30†	20†	20†	25	25†	12
10	30	35†	40	40†	25	30†	30	35†	10
8	40	50	55	55	30	40	40	45	8
6	55	65	70	75	40	50	55	60	6
4	70	85	95	95	55	65	75	75	4
3	85	100	110	110	65	75	85	85	3
2	95	115	125	130	75	90	100	100	2
1	110	130	145	150	85	100	110	115	1
0	125	150	165	170	100	120	130	135	0
00	145	175	190	195	115	135	145	150	00
000	165	200	215	225	130	155	170	175	000
0000	195	230	250	260	150	180	195	205	0000
250	215	255	275	290	170	205	220	230	250
300	240	285	310	320	190	230	250	255	300
350	260	310	340	350	210	250	270	280	350
400	280	335	365	380	225	270	295	305	400
500	320	380	415	430	260	310	335	350	500
600	355	420	460	475	285	340	370	385	600
700	385	460	500	520	310	375	405	420	700
750	400	475	515	535	320	385	420	435	750
800	410	490	535	555	330	395	430	450	800
900	435	520	565	585	355	425	465	480	900
1000	455	545	590	615	375	445	485	500	1000
1250	495	590	640	665	405	485	525	545	1250
1500	520	625	680	705	435	520	565	585	1500
1750	545	650	705	735	455	545	595	615	1750
2000	560	665	725	750	470	560	610	630	2000
	Ampacity Correction Factors								
Ambient Temp. °C	For ambient temperatures other than 30°C, multiply the ampacities shown above by the appropriate factor shown below.								Ambient Temp. °F
31–40	.82	.88	.90	.91	.82	.88	.90	.91	87–104
41–45	.71	.82	.85	.87	.71	.82	.85	.87	105–113
46–50	.58	.75	.80	.82	.58	.75	.80	.82	114–122
51–60	....	.58	.67	.71	....	.58	.67	.71	123–141
61–70	....	.35	.52	.58	....	.35	.52	.58	142–158
71–80	....	....	.30	.41	....	....	.30	.41	159–176

† The overcurrent protection for conductor types marked with a dagger (†) shall not exceed 15 amperes for 14 AWG, 20 amperes for 12 AWG, and 30 amperes for 10 AWG copper; or 15 amperes for 12 AWG and 25 amperes 10 AWG aluminum and copper-clad aluminum after any correction factors for ambient temperature and number of conductors have been applied.

* For dry locations only. See 75°C column for wet locations.

TABLE 8-2B. Ampacities of Insulated Conductors Rated 0-2000 Volts, 60° to 90°C

Single conductors in free air, based on ambient temperature of 30°C (86°F).

Size	Temperature Rating of Conductor, See *National Electrical Code* Table 310-13								Size
	60°C (140°F)	75°C (167°F)	85°C (185°F)	90°C (194°F)	60°C (140°F)	75°C (167°F)	85°C (185°F)	90°C (194°F)	
AWG MCM	TYPES †RUW, †T, †TW	TYPES †FEPW, †RH, †RHW, †RUH, †THW, †THWN, †XHHW, †ZW	TYPES V, MI	TYPES TA, TBS, SA, AVB, SIS, †FEP, †FEPB, †RHH, †THHN, †XXHW*	TYPES †RUW, †T, †TW	TYPES †RH, †RHW, †RUH, †THW, †THWN, †XHHW	TYPES V, MI	TYPES TA, TBS, SA, AVB, SIS, †RHH, †THHN, †XHHW*	AWG MCM
	Copper				Aluminum or Copper-clad Aluminum				
18	....	....	....	18	....	....	....	....	....
16	....	....	23	24	....	....	....	....	....
14	25†	30†	30	35†	....	....	....	....	....
12	30†	35†	40	40†	25†	30†	30	35†	12
10	40†	50†	55	55†	35†	40†	40	40†	10
8	60	70	75	80	45	55	60	60	8
6	80	95	100	105	60	75	80	80	6
4	105	125	135	140	80	100	105	110	4
3	120	145	160	165	95	115	125	130	3
2	140	170	185	190	110	135	145	150	2
1	165	195	215	220	130	155	165	175	1
0	195	230	250	260	150	180	195	205	0
00	225	265	290	300	175	210	225	235	00
000	260	310	335	350	200	240	265	275	000
0000	300	360	390	405	235	280	305	315	0000
250	340	405	440	455	265	315	345	355	250
300	375	445	485	505	290	350	380	395	300
350	420	505	550	570	330	395	430	445	350
400	455	545	595	615	355	425	465	480	400
500	515	620	675	700	405	485	525	545	500
600	575	690	750	780	455	540	595	615	600
700	630	755	825	855	500	595	650	675	700
750	655	785	855	885	515	620	675	700	750
800	680	815	885	920	535	645	700	725	800
900	730	870	950	985	580	700	760	785	900
1000	780	935	1020	1055	625	750	815	845	1000
1250	890	1065	1160	1200	710	855	930	960	1250
1500	980	1175	1275	1325	795	950	1035	1075	1500
1750	1070	1280	1395	1445	875	1050	1145	1185	1750
2000	1155	1385	1505	1560	960	1150	1250	1335	2000

Ampacity Correction Factors

Ambient Temp. °C	For ambient temperatures other than 30°C, multiply the ampacities shown above by the appropriate factor shown below.								Ambient Temp. °F
31–40	.82	.88	.90	.91	.82	.88	.90	.91	87–104
41–45	.71	.82	.85	.87	.71	.82	.85	.87	105–113
46–50	.58	.75	.80	.82	.58	.75	.80	.82	114–122
51–60	....	.58	.67	.71	....	.58	.67	.71	123–141
61–70	....	.35	.52	.58	....	.35	.52	.58	142–158
71–80	....	....	.30	.41	....	....	.30	.41	159–176

† The overcurrent protection for conductor types marked with a dagger (†) shall not exceed 20 amperes for 14 AWG, 25 amperes for 12 AWG, and 40 amperes for 10 AWG copper, or 20 amperes for 12 AWG and 30 amperes for 10 AWG aluminum and copper-clad aluminum after any correction factor for ambient has been applied.

* For dry locations only. See 75°C column for wet locations.

TABLE 8-2C. Ampacities for Insulated Conductors Rated 0-2000 Volts, 110 to 250°C

Not More Than Three Conductors in Raceway or Cable Based on Ambient Temperature of 30°C (86°F).

Size	Temperature Rating of Conductor, See *National Electrical Code* Table 310-13								Size
	110°C (230°F)	125°C (257°F)	150°C (302°F)	200°C (392°F)	250°C (482°F)	110°C (230°F)	125°C (257°F)	200°C (392°F)	
AWG MCM	TYPES AVA, AVL	TYPES AI, AIA	TYPE Z	TYPES A, AA, FEP, FEPB, PFA	TYPES PFAH, TFE	TYPES AVA, AVL	TYPES AI, AIA	TYPES A, AA	AWG MCM
	Copper				Nickel or Nickel-coated Copper	Aluminum or Copper-clad Aluminum			
14	30	30	30	30	40	....	....	....	....
12	35	40	40	40	55	25	30	30	12
10	45	50	50	55	75	35	40	45	10
8	60	65	65	70	95	45	50	55	8
6	80	85	90	95	120	60	65	75	6
4	105	115	115	120	145	80	90	95	4
3	120	130	135	145	170	95	100	115	3
2	135	145	150	165	195	105	115	130	2
1	160	170	180	190	220	125	135	150	1
0	190	200	210	225	250	150	160	180	0
00	215	230	240	250	280	170	180	200	00
000	245	265	275	285	315	195	210	225	000
0000	275	310	325	340	370	215	245	270	0000
250	315	335	....	....	....	250	270	....	250
300	345	380	....	....	....	275	305	....	300
350	390	420	....	....	....	310	335	....	350
400	420	450	....	....	....	335	360	....	400
500	470	500	....	....	....	380	405	....	500
600	525	545	....	....	....	425	440	....	600
700	560	600	....	....	....	455	485	....	700
750	580	620	....	....	....	470	500	....	750
800	600	640	....	....	....	485	520	....	800
1000	680	730	....	....	....	560	600	....	1000
1500	785	....	....	....	....	650	....	....	1500
2000	840	....	....	....	....	705	....	....	2000

Ampacity Correction Factors									
Ambient Temp. °C	For ambient temperatures other than 30°C, multiply the ampacities shown above by the appropriate factor shown below.								Ambient Temp. °F
31–40	.94	.95	.96	....	....	.94	.95	....	87–104
41–45	.90	.92	.94	....	....	.90	.92	....	105–113
46–50	.87	.89	.91	....	....	.87	.89	....	114–122
51–55	.83	.86	.89	....	....	.83	.86	....	123–141
56–60	.79	.83	.87	.91	.95	.79	.83	.91	132–141
61–70	.71	.76	.82	.87	.91	.71	.76	.87	142–158
71–75	.66	.72	.79	.86	.89	.66	.72	.86	159–167
76–80	.61	.68	.76	.84	.87	.61	.69	.84	168–176
81–90	.50	.61	.71	.80	.83	.50	.61	.80	177–194
91–100	....	.51	.65	.77	.80	....	.51	.77	195–212
101–120	....	....	.50	.69	.72	....	....	.69	213–248
121–140	....	....	.29	.59	.59	....	....	.59	249–284
141–160	....	....	....	....	.54	....	....	....	285–320
161–180	....	....	....	....	.50	....	....	....	321–356
181–200	....	....	....	....	.43	....	....	....	357–392
201–225	....	....	....	....	.30	....	....	....	393–437

TABLE 8-2D. Ampacities for Insulated Conductors Rated 0-2000 Volts, 110 to 250°C, and for Bare of Covered Conductors

Single Conductors in Free Air, Based on Ambient Temperature of 30°C (86°F).

Size	Temperature Rating of Conductor, See *National Electrical Code* Table 310-13										Size
	110°C (230°F)	125°C (257°F)	150°C (302°F)	200°C (392°F)		250°C (482°F)	110°C (230°F)	125°C (257°F)	200°C (392°F)		
AWG MCM	TYPES AVA, AVL	TYPES AI, AIA	TYPE Z	TYPES A, AA, FEP, FEPB, PFA	Bare or covered conductors	TYPES PFAH, TFE	TYPES AVA, AVL	TYPES AI, AIA	TYPES A, AA	Bare or covered conductors	AWG MCM
	Copper					Nickel or Nickel-coated Copper	Aluminum or Copper-clad Aluminum				
14	40	40	40	45	30	60	....	....	....	....	....
12	50	50	50	55	40	80	40	40	45	30	12
10	65	70	70	75	55	110	50	55	60	45	10
8	85	90	95	100	70	145	65	70	80	55	8
6	120	125	130	135	100	210	95	100	105	80	6
4	160	170	175	180	130	285	125	135	140	100	4
3	180	195	200	210	150	335	140	150	165	115	3
2	210	225	230	240	175	390	165	175	185	135	2
1	245	265	270	280	205	450	190	205	220	160	1
0	285	305	310	325	235	545	220	240	255	185	0
00	330	355	360	370	275	605	255	275	290	215	00
000	385	410	415	430	320	725	300	320	335	250	000
0000	445	475	490	510	370	850	345	370	400	290	0000
250	495	530	....	....	410	....	385	415	....	320	250
300	555	590	....	....	460	....	435	460	....	360	300
350	610	655	....	....	510	....	475	510	....	400	350
400	665	710	....	....	555	....	520	555	....	435	400
500	765	815	....	....	630	....	595	635	....	490	500
600	855	910	....	....	710	....	675	720	....	560	600
700	940	1005	....	....	780	....	745	795	....	615	700
750	980	1045	....	....	810	....	775	825	....	640	750
800	1020	1085	....	....	845	....	805	855	....	670	800
900	....	....	....	....	905	....	....	....	....	725	900
1000	1165	1240	....	....	965	....	930	990	....	770	1000
1500	1450	....	....	....	1215	....	1175	....	....	985	1500
2000	1715	....	....	....	1405	....	1425	....	....	1165	2000

Ampacity Correction Factors

Ambient Temp. °C	For ambient temperatures other than 30°C, multiply the ampacities shown above by the appropriate factor shown below.										Ambient Temp. °F
31–40	.94	.95	.96	....	.94	.95	....				87–104
41–45	.90	.92	.94	....		....	.90	.92	....		105–113
46–50	.87	.89	.91	....		....	.87	.89	....		114–122
51–55	.83	.86	.89	....		....	.83	.86	....		123–141
56–60	.79	.83	.87	.91		.95	.79	.83	.91		132–141
61–70	.71	.76	.82	.97		.91	.71	.76	.87		142–158
71–75	.66	.72	.79	.86		.89	.66	.72	.86		159–167
76–80	.61	.68	.76	.84		.87	.61	.69	.84		168–176
81–90	.50	.61	.71	.80		.83	.50	.61	.80		177–194
91–100	....	.51	.65	.77		.80	....	.51	.77		195–212
101–120	....	....	.50	.69		.72	....	....	.69		213–248
121–140	....	....	.29	.59		.59	....	....	.59		249–284
141–160	....	....	....	....		.54	....	....	....		285–320
161–180	....	....	....	....		.50	....	....	....		321–356
181–200	....	....	....	....		.43	....	....	....		357–392
201–225	....	....	....	....		.30	....	....	....		393–437

categories for electrical fires on a nationwide basis. These are shown in Table 8-2E.

In even broader terms, the causes of electrical fires could be divided into four categories, as follows:

Worn Out or "Tired" Electrical Equipment Fires: Equipment is actually worn out in service. After a period of years, a wire's electrical insulation deteriorates, thereby resulting in a fire. "Tired" equipment is responsible for the largest percent of the electrical fires of known cause. The leading item within this category is the electric motor; also included in tired equipment fires are those caused by worn out wire insulation, electronic appliances, lamp and other appliance cords, fixtures, and heating appliances.

Aging of electric equipment results in the deterioration of insulation and, in some cases, corrosion or fatigue of the wires.

Improper Use of Approved Equipment: Many fires of known cause are the result of improper use of listed equipment. "Improper use of listed equipment" indicates that although the equipment itself complies with safety

TABLE 8-2E. Analysis of Fires of Reported Electric Origin

Equipment Involved	No. of Fires 1980	1981	1982
AIR CONDITIONERS			
Window or through wall types	71	80	60
Central system types	73	75	102
APPLIANCES—COMMERCIAL, INDUSTRIAL			
Business machines	38	12	15
Food preparation (toasters, mixers, etc.)	20	42	32
Heating (comfort or space)	432	84	74
Medical	10	6	6
Refrigeration, freezers	28	28	8
Tools, hand	4	7	4
Vending machines	7	14	7
Others, specify	33	26	83
Others, specify	35	18	43
APPLIANCES—RESIDENTIAL			
Clothes dryers	163	173	151
Clothes washers	85	88	67
Electric blankets	61	46	43
Food preparation (toasters, mixers, etc.)	101	160	46
Hand tools	4	5	5
Heating (comfort or space)	186	294	224
Irons	18	25	16
Ranges and ovens	411	512	338
Refrigerators and freezers	48	59	61
Radios (tape recorders, reproducing equipment)	38	26	15
Television sets	65	56	53
(a) black and white	75	20	13
(b) color	75	58	55
Others, specify	119	157	69
CABLES			
Armored	21	14	4
Nonmetallic sheathed	54	58	45
Service entrance	60	46	54
Others, specify	121	12	23
CORDS			
On appliances	89	92	113
Circuit extensions	122	130	151
On portable lamps	25	25	35
Others, specify	32	25	32
DISTRIBUTION PANELS			
Fuse	75	75	73
Circuit breaker	90	91	107
FIXTURES—LIGHTING			
Fluorescent (ballast)	408	422	386
Incandescent—Close to ceiling	58	62	48
Incandescent—Drop cord	8	18	4
Incandescent—Chain hung (swag)	3	5	3
Others, specify	11	9	8
LAMPS (bulbs)			
Lighting	60	77	41
Heating	9	10	17
Others, specify	3	5	3
LIGHTING			
MOTORS (not integral with appliance)	24	—	—
Fractional	133	162	62
1 H.P. and over	89	76	60
Elevator	13	7	10
RECEPTACLES	173	136	95
SIGNS	5	—	
Cord connected	24	10	18
Fixed wired	21	47	65
SWITCHES			
Main	18	21	25
Wall type	82	59	54
Industrial control	5	11	3
Others, specify	24	11	6
TERMINATIONS AND SPLICES			
Copper wire	47	31	24
Aluminum wire	29	39	29
Copper to aluminum	10	18	12
Others, specify	49	3	0
TRANSFORMERS			
Control type	24	45	34
Distribution type	80	100	63
Others, specify	27	21	15
WIRES			
In metal raceways	52	40	65
In nonmetallic raceways	63	78	45
Overhead—consumer	242	101	56
Overhead—utility	305	293	534
Christmas decorative	28	15	10
Others, specify	108	45	90
POOLS			
Swimming—wading—decorative			
Lighting (lights)	2	0	0
Equipment	18	3	0
Others, specify	6	0	0
MISCELLANEOUS EQUIPMENT (specify)	2	25	11
REPORTED ELECTRICAL ORIGIN (no other details)	91	357	117
TOTALS	5,360	5,126	4,358

Based on data from the September/October 1981–1983 issues of the *IAEI News*. The IAEI obtains information on the equipment involved in electrical fires from an annual survey of Chief Electrical Inspectors in cities across the country. The IAEI states that the reports received are not complete with regard to full details and should in no way be used in comparison of one material against the other. Statistics from 1980 are based on reports from 174 cities; 176 cities in 1981 and 1982. It should be noted that the data represent reports only from the cities surveyed.

standards of recognized testing laboratories, fires have occurred because such equipment is used under conditions not covered by its listing, i.e., use of No. 18 lamp cord to supply high wattage electric heaters.

Equipment that has been tested and listed by a qualified testing laboratory rarely causes a fire if properly used in accordance with its listing, and if replaced when over age.

The three most prominent offenders in the group of misused equipment fires are: the improper use of (1) heating appliances, (2) electric motors, and (3) extension cords.

Accidental Occurrence: Some electrical fires of known origin are the result of an accidental misuse or oversight on the part of equipment operators. These fires are caused by: clothes left in contact with lamps, materials accidentally dropped into electric equipment, heating appliances left on unintentionally, etc.

Defective Installations: Many electrical fires of known cause result from defective installations. Defective installations are those installed in a manner not acceptable under the NEC, e.g., a fractional horsepower motor automatically started which does not have overload running protection.

The exact causes of a large percent of electrical fires are never determined and this is usually due to the destruction of evidence. Often the cause of the electrical failure cannot definitely be determined, even though the fire is believed to have been, or is obviously the result of, failure in an item of electrical equipment.

CODES AND STANDARDS

All electrical installations in the United States should be made, used, and maintained in accordance with the NEC, the National Electrical Safety Code, and other standards which apply in special situations, e.g., NFPA 70E, *Standard for Electrical Safety Requirements in Employee Workplaces*; NFPA 79, *Electrical Standard for Industrial Machinery* (hereinafter referred to as NFPA 79); NFPA 75, *Standard for the Protection of Electronic Computer/Data Processing Equipment*; and NFPA 99, *Standard for Health Care Facilities* (hereinafter referred to as NFPA 99). There are also special standards for electrical installations on shipboard, aircraft, in hazardous (classified) locations, and other locations.

National Electrical Code (ANSI/NFPA 70)

NFPA 70, *National Electrical Code*, provides for the practical safeguarding of persons and property from hazards arising from the use of electricity. The NEC was first issued under its present name in 1897. It is revised every three years by the National Electrical Code Committee.

National Electrical Safety Code (ANSI Standard C2)

As interest increased in electrical safety in the U.S., a need arose for a code to cover the practices of public utilities and others when installing and maintaining overhead and underground electric supply and communication lines. Accordingly, a National Electrical Safety Code was completed in 1916. Currently this code is published by the Institute of Electrical and Electronic Engineers (IEEE).

Canadian Electrical Code

The Canadian Standards Association currently sponsors and publishes the Canadian Electrical Code. This code establishes essential requirements and minimum standards for the installation and maintenance of electrical equipment. This code can be adopted and enforced by electrical inspection departments throughout Canada and was prepared with due regard for the NEC and the National Electrical Safety Code. Its several parts are the Canadian equivalent of the NEC, the National Electrical Safety Code, and the standards of Underwriters Laboratories Inc., (UL) in the U.S.

The National Electrical Code Handbook

While not in itself a code or standard that can have the force of law, the *National Electrical Code Handbook* is a valuable resource in implementing provisions of the NEC. The NEC Handbook contains the entire text of the NEC supplemented by comments, diagrams, and illustrations that are intended to clarify some of the intricate requirements of the NEC. The NEC Handbook is published by the NFPA; a new edition is published with each new edition of the NEC.

Listed and Labeled Electrical Equipment Manufacturers

Three publications are available that list the names of companies making electric appliances and materials that have been listed or labeled by UL. Devices and materials in the product categories listed have been tested for actual field use in accordance with the NEC and against UL safety standards. The lists are published so that the names of manufacturers of devices that meet test criteria may be readily obtained. The three publications are the *Hazardous Location Equipment Directory*, *Electrical Appliance and Utilization Equipment Directory*, and *Electrical Construction Materials Directory*. They are published annually by UL with six month supplements. Certain types of electrical equipment are covered in other UL directories, such as the *Marine Products Directory*.

There are sections of the NEC that influence the design and construction of electric equipment. UL has complete published standards for most electric fittings, materials, and equipment, including specifications for performance under tests and in service, and further details of design and construction. Other testing laboratories in the U. S. and Canada that test electric equipment include Factory Mutual Research Corporation (FM), ETL Testing Laboratories, Inc. (ETL), and the Canadian Standards Association (CSA). Some laboratories stipulate compliance with the NEC in testing equipment. (Where no published standards are available for special equipment, arrangements usually can be made with testing laboratories for testing on an individual basis.)

BUILDING WIRING, DESIGN, AND PROTECTION

This section contains information on various types of building wiring and equipment, including panelboards and overcurrent protection; types of electric conductors;

identification of conductors, terminals, etc., how to calculate loads, and similar subjects.

Service Entrance

Good practice requires that service entrance conductors comply in all respects with the detailed requirements of the NEC. Figure 8-2A provides typical arrangements of service entrance conductors.

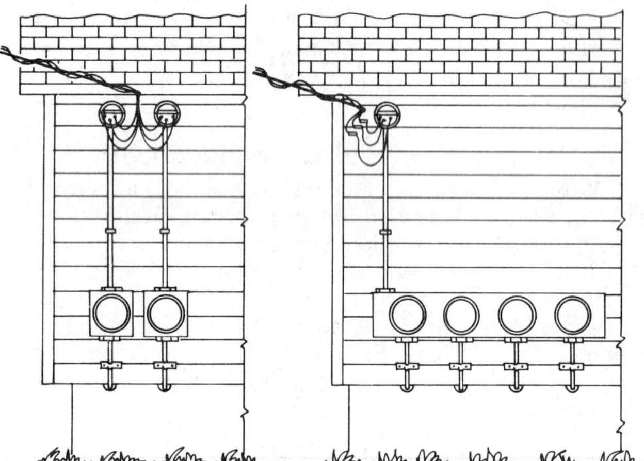

FIG. 8-2A. The diagram at left shows an installation in which two sets of service entrance conductors are tapped from one service drop, with meters mounted on the outside of the building. In the diagram at right, a set of main service entrance conductors is connected to the service drop and is carried through a trough and then through four sets of subservice entrance conductors to each service equipment. Here again, the meters are on the exterior wall.

Means are provided for disconnecting all conductors in the building from the service entrance conductors. The principle is to permit disconnection of all conductors of the service at that location with no more than six operations of the hand. (See Fig. 8-2B.) The limitation is applicable to all of the service disconnecting means located, or grouped, in one place. It does not include the services for fire pumps, emergency electrical systems, etc., which the NEC recognizes as being separate services for specific purposes.

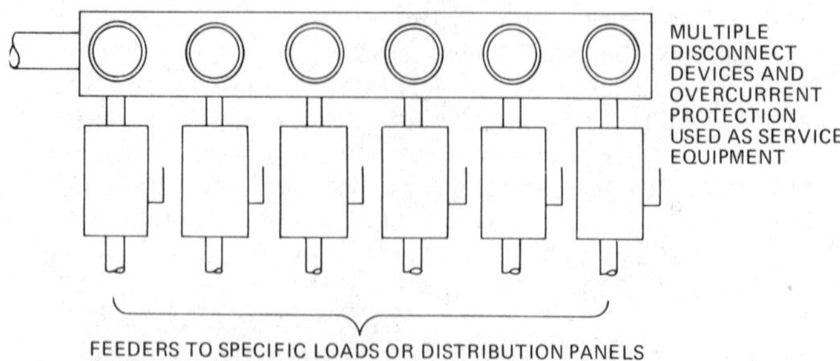

FIG. 8-2B. Typical layout for multiple disconnecting means. As explained in the text, the maximum number of switches is six at any one location and the principle is to permit disconnecting all conductors of the service at that location with no more than six operations of the hand.

The disconnecting means is either an outside or inside installation, at a readily accessible location nearest the point of entrance of the service entrance conductors. Service entrance conductors must be of sufficient size to carry the calculated load. In general, a building or other structure served should be supplied by only one service drop (Fig. 8-2A) or service (underground) lateral, but additional services are permitted for fire pumps, emergency electrical systems, multiple occupancies, large capacities, large areas, and different classes of use.

Underground service conductors must be protected against physical damage. (See Fig. 8-2C.) Tables 8-2F and 8-2G give minimum cover requirements for underground installations. ("Cover" is defined as the distance between the top surface of direct buried cable, conduit, or other approved raceways and the finished grade.)

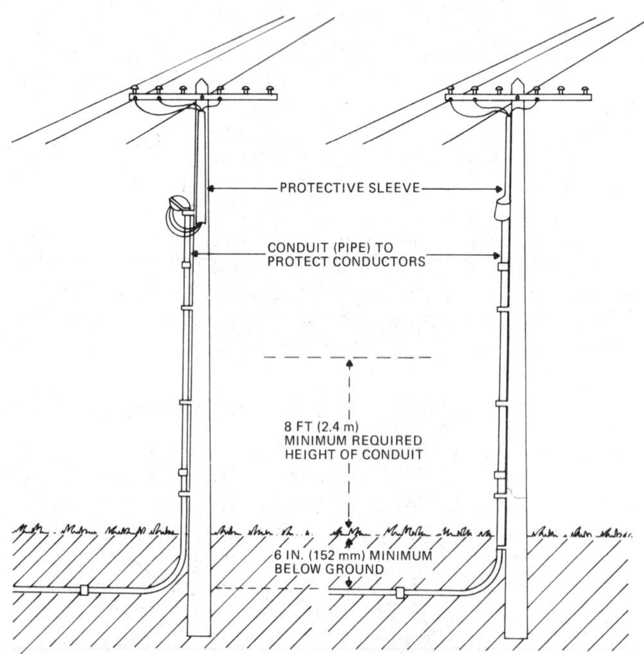

FIG. 8-2C. Service conductors from the outside pole are carried underground from the overhead supply line. Note the service head to protect the conductors against the entrance of water and that the point of connection of the service entrance conductors is below the level of the service head.

TABLE 8-2F. Minimum Cover Requirements, 0 to 600 V
(Table 300-5 of the *NEC*)

Wiring Method	Minimum Burial (Inches)	(mm)
Direct Buried Cables	24	610
Rigid Metal Conduit	6	155
Intermediate Metal Conduit	6	152
Rigid Nonmetallic Conduit Approved for direct burial without concrete encasement	18	457
Other Approved Raceways	18	457

Note: Raceways approved for burial only when concrete encased shall require a concrete envelope not less than 2 in. (51 mm) thick.

TABLE 8-2G. Minimum Cover Requirements, Over to 600 V
(Table 710-3(b) of the *NEC*)

Circuit Voltage	Direct Buried Cables		Rigid Nonmetallic Conduit Approved for Direct Burial*		Rigid Metal Conduit and Intermediate Metal Conduit	
	in.	mm	in.	mm	in.	mm
Over 660-22KV	30	762	18	457	6	152
Over 22KV–40KV	36	914	24	610	6	152
Over 40KV	42	1066	30	762	6	152

Nonshielded cables shall be installed in rigid metal conduit, in intermediate metal conduit, or in rigid nonmetallic conduit encased in not less than 3 in. (76 mm) of concrete.

* Listed by a qualified testing agency as suitable for direct burial without encasement. All other nonmetallic systems shall require 2 in. (51 mm) of concrete or equivalent above conduit in addition to above depth.

Lightning (Surge) Arresters—Building Wiring

Lightning and surge arresters are not required, but may be used, particularly where buildings are supplied by an overhead service and where thunderstorms are prevalent in the area. NFPA 78, *Lightning Protection Code,* (hereinafter referred to as NFPA 78) requires the provision of lightning arresters on electrical service entrance conductors where the structure is provided with a lightning protection system. Surge arresters are also very useful where electronic equipment is used because equipment using solid state electronic components is very sensitive to line voltage surges.

Each lead-in from an outdoor antenna to a radio or television set or to an amateur radio transmitting station needs protection from lightning. Lightning arresters provide good protection, and they are recommended for many other types of service as well, particularly in rural areas.

Grounding Requirements—Building Wiring

Dangerous voltages, which may be a fire hazard and a personal injury (shock) hazard, may be imposed on electric distribution systems and equipment. These voltages may be caused by lightning, inadvertent contact with a high-voltage primary system, breakdown of insulation, surface leakage due to dirt or moisture, or by a wire coming loose from its connection.

By grounding one conductor of the electric circuit and then grounding all exposed metal that may come in contact with a live conductor, a fault to ground occurs if the ungrounded circuit conductor should become accidentally grounded. The fault current finds a path through the equipment grounding conductor and can cause the operation of an overcurrent device (fuse or circuit breaker) in the ungrounded circuit conductor and eliminate the dangerous condition. The impedance of the fault path must be low enough to permit sufficient current flow to cause the overcurrent device to operate quickly.

The metal enclosures of conductors (metal armor of cables, metal raceways, boxes, cabinets, and fittings) must be grounded. (Article 250 of the NEC contains general requirements for grounding electrical installations.)

The NEC also requires grounding of exposed noncurrent-carrying metal parts of cord and plug connected equipment likely to become energized in hazardous (classified) locations, where operated by persons standing on the ground or metal surfaces, or where operated at more than 150 V to ground, except for guarded motors. In some isolated cases, the metal frames of electrically heated appliances should also be grounded. The NEC also specifies grounding of particular cord and plug connected appliances in residential occupancies. These appliances include refrigerators, freezers, air conditioners, clothes washers and dryers, dishwashers, sump pumps, electric aquarium equipment, portable hand-held motor operated tools, and some appliances such as hedge clippers, lawn mowers, snow blowers, and wet scrubbers. Appliances may be excepted from grounding if equivalent protection via double insulation is provided.

A metallic underground water piping system must be used as the grounding electrode, where available, and the buried portion of the pipe is more than 10 ft (3 m) in length. A metal underground water pipe grounding electrode, however, must be supplemented by an additional electrode so the integrity of the grounding electrode system is maintained. In addition, the following grounding electrodes must be used as part of the grounding electrode system, if available on the premises: the metal frame of the building where effectively grounded, a concrete encased electrode, and a ground ring. All grounding electrodes must be bonded together. (Article 250 of the NEC lists various methods of providing grounding electrode systems.) An example of a grounding electrode is shown in Figure 8-2D. Caution is needed to check the piping for insulating joints and nonconducting materials. Effective bonding must be provided around insulated joints and sections, and around any equipment that is likely to be disconnected for repairs or replacement.

Panelboards and Overcurrent Protection—Building Wiring and Equipment

A panelboard has buses (with or without switches) and automatic overcurrent devices for the control and protection of light, heat, or power circuits. The buses are mounted in a cabinet or cutout box which is placed in or on a wall and is accessible only from the front.

No more than 42 overcurrent devices of a lighting and appliance branch circuit panelboard are permitted by the NEC to be installed in any one cabinet or cutout box. Generally, not more than two main circuit breakers or two main sets of fuses are permitted to protect a branch-circuit panelboard. (See Figs. 8-2E and 8-2F.) Figures 8-2G and 8-2H show exceptions to the general rule for individual protection of panelboards.

Panelboards installed in wet locations need weatherproof cabinets and must be mounted so that there is at least 1/4 in. (6 mm) of air space between the cabinet and the wall or other supporting surface. Special requirements of Chapter 5 of the NEC govern installation of panelboards in hazardous (classified) locations.

Overcurrent Protection: Conductors and equipment are provided with overcurrent protection to open a circuit if the current reaches a value that will cause an excessive or dangerous temperature in the conductor or conductor insulation. No feature of an electrical installation deserves

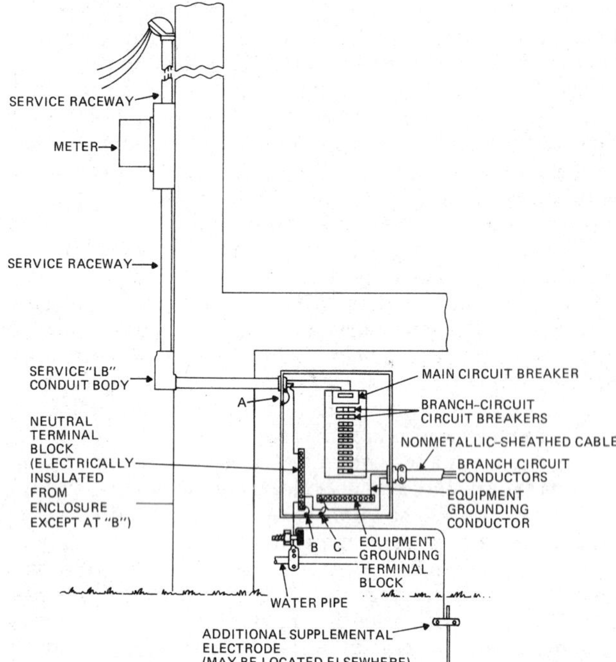

FIG. 8-2D. Grounding at a typical small service [ac, single phase, three wire (120/240 v)]. The service raceway is grounded through the metal of the box; this electrical connection is indicated at "A." The jumper shown is one of the methods employed to assure electrical continuity at service equipment. The service enclosure is itself bonded to the grounded neutral conductor at point "B." This connection is often a removable strap or screw. The meter box and the upper portion of the service raceway or cable are grounded through threaded connectors making threaded connections to the meter box if metal conduit or electrical metallic tubing is used. The meter box on the line side of the service equipment is usually grounded by connection to the grounded neutral and such ground would be required in case of service-entrance cable. The branch circuit equipment grounding conductor is connected to ground through the equipment grounding terminal block which is grounded at point "C." This connection is commonly the mounting screw for this terminal block. The grounding connection shown is to be an underground cold water piping system. The connection should be as required by the NEC.

more careful attention and supervision. In general, an overcurrent device rated or set in accordance with the ampacity, or current carrying capacity, of the NEC should be installed in each ungrounded conductor of each circuit and for each feeder at the point where the conductor to be protected receives its supply. (See Tables 8-2A through 8-2D.) Exceptions for fixture wires, cords, taps, motor circuits, remote control circuits, and fire protective signaling circuits are covered in Articles 240, 430, 725, and 760 of the NEC.

Overcurrent Protective Devices

The most commonly used overcurrent protective devices for feeders, circuits, and equipment are fuses, circuit breakers, and thermal overload units. (Overcurrent and undervoltage relays, etc., are used on high voltage, high current systems.)

Plug Fuses: There are two basic types of plug fuses: (1) the ordinary Edison base type, and (2) the S-type. Either of these may or may not be of the time delay type.

The S-type fuse is designed to prevent tampering or bridging, or at least not without extreme difficulty. Adapters are available that fit the Edison base fuseholders. After an adapter has been properly installed, it cannot be removed without damaging the fuseholder. The adapters are designed to prevent using Edison base fuses in the fuseholder and to prevent using larger (higher rated) S-type fuses in an adapter designed for a lower rating. They also prevent the use of pennies and other common bridging schemes. An Edison base fuseholder will take an Edison base fuse of any size up to a maximum 30 amp rating. Plug fuses of the Edison base type are permitted only for replacement in existing installations where there is no evidence of overfusing or tampering.

Time Delay Plug Fuses: Whether S-type or Edison base design, time delay fuses permit short time current surges such as motor starting currents, without interruption of the circuit. These momentary surges are harmless unless repeated a number of times at short intervals. This makes it

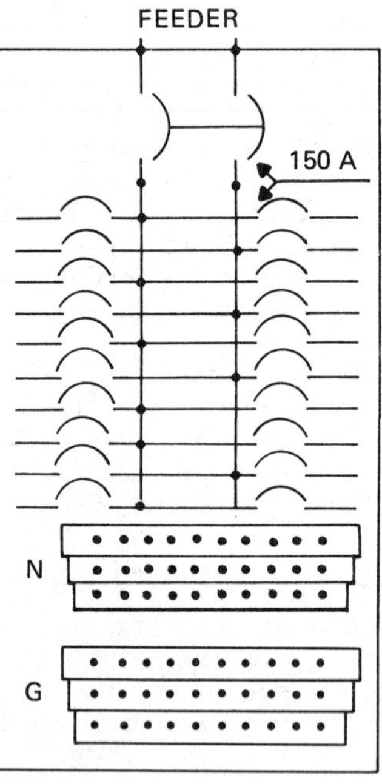

FIG. 8-2E. One main provided (two are permitted). Grounding terminal block may not be required, depending upon wiring system used.

possible to use S-type fuses in sizes small enough to give better protection than a non time delay type, that must be oversized to allow for such surges. In the case of a short circuit or a high current fault, however, the time delay type will operate and clear the circuit as rapidly or even more rapidly than the non time delay type. Some representative plug fuses are shown in Figures 8-2I, 8-2J, and 8-2K.

Cartridge Fuses: These are of both the time delay and the non time delay types. They are also of the one time and the renewable link types. When a one time fuse opens, the entire fuse must be replaced. But when a renewable link fuse opens, the fuse link can be replaced inside the same cartridge unless the cartridge has also been damaged.

Renewable link fuses have two disadvantages: (1) the links can be doubled or tripled, etc., thereby defeating

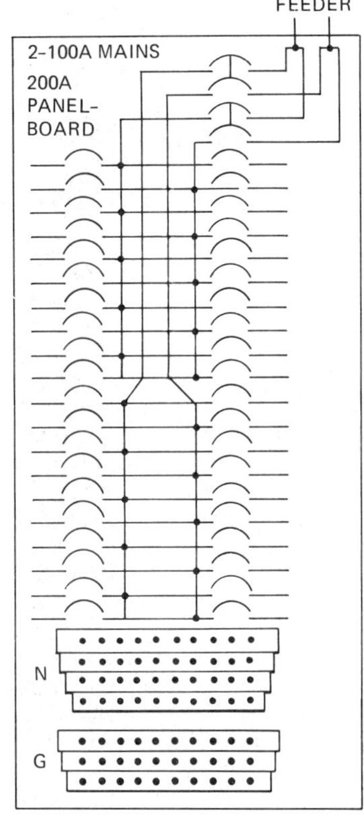

FIG. 8-2F. Two mains required. Grounding terminal block may not be required, depending upon wiring system used.

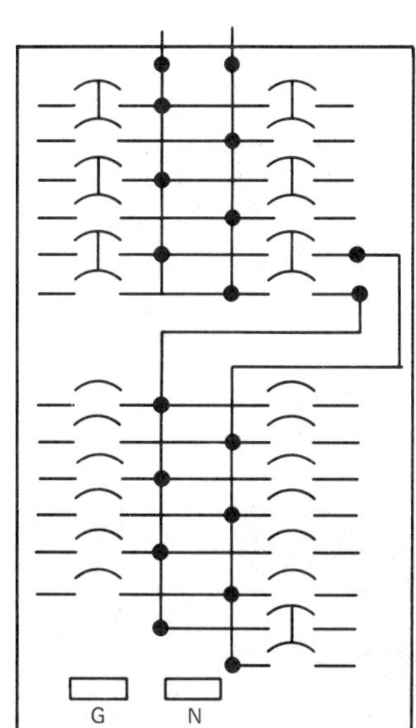

FIG. 8-2H. Circuit breakers in the upper section are the service circuit breakers. One is the main for lower section buses supplying branch circuits. This arrangement is acceptable only for the service to an existing individual residential occupancy where the panelboard was installed prior to application of the 1981 NEC requirements.

their purpose and usefulness, and (2) the links, upon replacement, can be left with loose connections. Some representative cartridge fuses are shown in Figures 8-2L and 8-2M.

Circuit Breakers: There are two basic types of breakers—adjustable trip, and nonadjustable trip. The adjustable trip type may be of either the air or oil immersed type. The setting of the trip point is adjustable between a minimum and a maximum range and is usually used only on large installations having qualified operators and maintenance personnel. They are designed to trip when the current reaches that of the setting. The nonadjustable trip type comes in a molded case, making it extremely difficult or impossible to change its rating. A molded case breaker can be of the inverse time thermal magnetic type, or of the magnetic only inverse time type. Molded case circuit breakers are designed so that the current has to exceed the rating (as is also true with all types of fuses) before it will trip. Unless it is the thermally compensated type, high ambient temperatures can reduce the current required to trip the circuit breaker (or fuse). A representative nonadjustable circuit breaker is shown in Figure 8-2N.

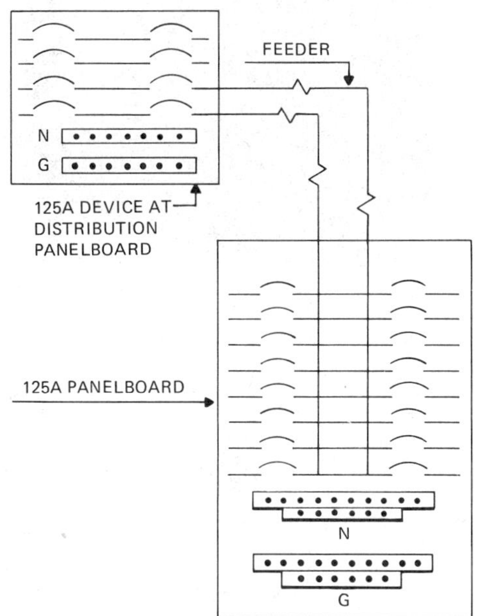

FIG. 8-2G. No mains required. Grounding terminal block may not be required, depending upon wiring system used.

Thermal Devices: These are not intended for protection against short circuits, but only for the protection of overload currents of a comparatively lower magnitude unless otherwise designed. An example is overload protection for a motor where short circuit and ground fault protection is provided by the branch circuit fuses or circuit breakers. For detailed requirements for motors and other equipment, see the applicable provisions of the NEC.

Current Limiting Overcurrent Protective Devices: These devices are provided in some installations where high

FIG. 8-2I. A typical Edison base nonrenewable, single element fuse. (Bussmann Mfg. Div., McGraw-Edison Co.)

FIG. 8-2J. Another Edison base nonrenewable, dual element fuse. (Bussman Mfg. Div., McGraw-Edison Co.)

FIG. 8-2K. A S-type nonrenewable fuse. The time lag type of fuse shown is acceptable but not required by NFPA 70. These fuses have been designed so that tampering or bridging can be done only with difficulty. The National Electrical Code specifies that fuse holders for plug fuses of 30 amp or less shall not be used unless they are designed to use this S-type fuse or are made to accept a S-type through use of an adapter. (Bussmann Mfg. Div., McGraw-Edison Co.)

current (more than 10,000 A) is available under short circuit conditions. They are designed so that when interrupting a specified current, the devices will consistently limit the short circuit current in that circuit to a magnitude substantially less than that obtainable in the same circuit if the device were replaced with a solid conductor having comparable impedance. (See Fig. 8-2O.) Fuseholders for current limiting fuses are designed to make it difficult to install a noncurrent limiting fuse.

Ground Fault Circuit Interrupters (GFCIs)

Circuit breakers and fuses open a circuit and stop the flow of electricity when the flow exceeds the ratings of the circuit breaker or fuse. Lighting and receptacle circuits are

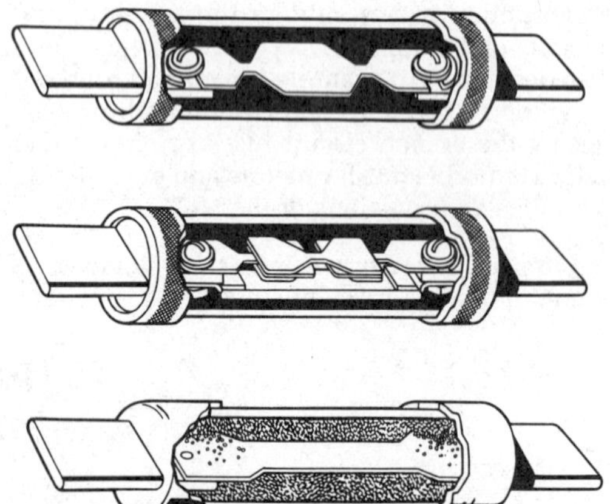

FIC. 8-2L. Three types of cartridge fuses: at top, an ordinary drop out link renewable fuse; at center, a super-lag renewable use; and at bottom, a one time fuse. (Bussmann Mfg. Div., McGraw-Edison Co.)

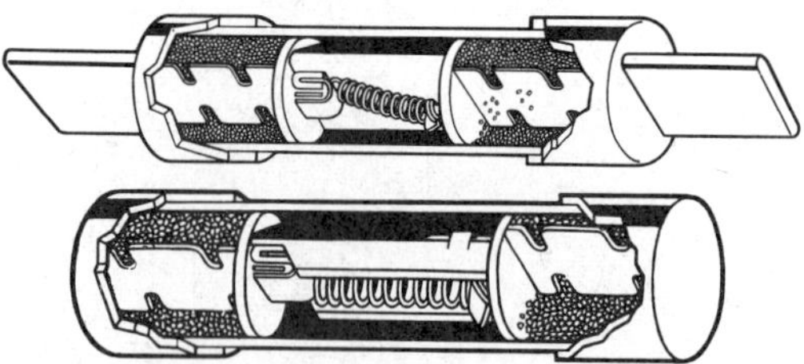

FIG. 8-2M. A dual element cartridge fuse, blade and ferrule type. (Bussman Mfg. Div., McGraw-Edison Co.)

usually rated 15 or 20 A. At times as little as 60 mA (milliamperes) can kill a normal healthy adult; thus the rating of lighting and receptacle circuits is high, relative to the amount of current that could kill a person.

Ground fault circuit interrupters (GFCI) are devices that sense when even a small amount of current passes to ground through any path other than the proper conductor. When this condition exists, the GFCI trips almost instantly, stopping all current flow in the circuit and through the person receiving the ground fault shock.

Simplicity of design is one reason for the reliability of the GFCI. Figure 8-2P shows a typical circuit arrangement of a GFCI for personnel protection. Figure 8-2Q shows types of GFCI units and a GFCI tester.

Present GFCIs are set to operate when line to ground currents exceed 6 mA. Evaluation standards permit a differential of 4 to 6 mA. Even at trips of 5 mA, it should be clearly understood that the instantaneous current will be higher, and any shock during the time the fault is being cleared will be uncomfortable. A shock at 5 mA is also

FIG. 8-2N. A 15 amp, single pole, 120 V, branch-circuit breaker. (Westinghouse Electric Corp.)

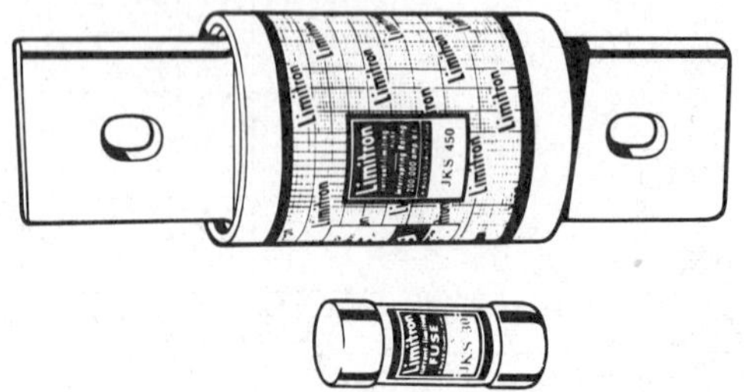

FIG. 8-20. A (rated 0 to 600 A) current limiting over current protection device. (Bussmann Mfg. Div., McGraw-Edison Co.)

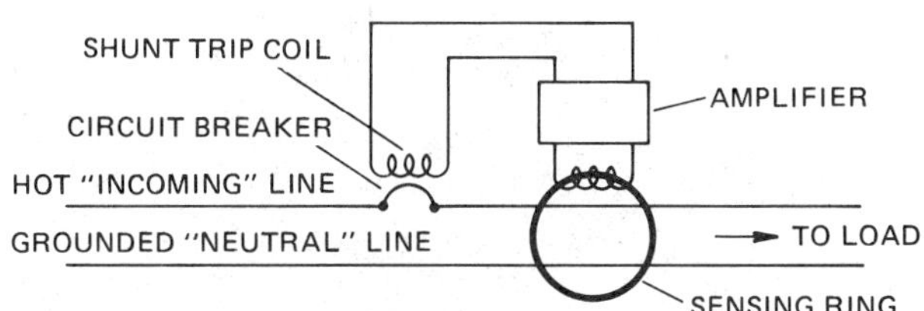

FIG. 8-2P. Circuit arrangement for a typical ground fault circuit interrupter for personnel protection. (I-T-E Imperial)

unpleasant. The key to the ground fault circuit interrupter is the time-current characteristic. Trip out time is about 1/40 of a second (25 milliseconds) when the fault reaches or exceeds 6 mA.

There are two classes of GFCIs—Class A and Class B. The Class A GFCI trips at not more than 6 mA. The Class B GFCI trips when the current exceeds 20 mA, and is for use only with swimming pool underwater lighting fixtures that were installed prior to local adoption of the 1965 edition of the NEC. The Class A GFCI is used where personnel protection against ground fault is required or desired.

GFCIs at Construction Sites: Precautions must be taken to aid the efficient operation of GFCIs on construction sites. Most laboratory tested 120 V appliances have 0.5 mA leakage or less under normal operating conditions. However, moisture and improper maintenance on portable hand held tools, which is common at construction sites, can create conditions under which GFCIs can be expected to trip. Flexible cords with standard attachment plug cap and connector when dropped in water, i.e., puddles, can be expected to cause leakage currents (100 to 300 mA or greater) far in excess of GFCI trip currents. Motors with dirty brushes, carbon tracking on commutators, or moisture in the windings contribute to leakage current. A common sense approach to installing, using, and maintaining GFCI circuits goes a long way toward eliminating nuisance tripping at construction sites. (Actually, tripping under any of the conditions mentioned is not nuisance tripping, but merely a device performing its intended function.)

Moisture is the major culprit in current leakage on wiring and equipment. Panelboards, receptacles, and cord caps and connectors intended for dry locations must not be subjected to wet conditions. Construction receptacles must be centrally located to enable cords 150 ft (46 m) or less in length to be used. There should be sufficient circuits available to minimize the number of tools necessary on each circuit. Receptacles must not be on the same circuit as lighting.

GFCIs in Residential and Commercial Occupancies: The NEC requires GFCIs on all 125 V, single phase 15 and 20 A receptacle outlets in dwelling units installed outdoors with grade level access, in garages (with some exceptions), and in bathrooms of both dwelling units and hotels and motels. Receptacles located within 20 ft (6 m) of the inside walls of swimming pools must be protected by GFCIs—as must receptacles and lighting fixtures near spas, hot tubs, hydromassage bathtubs, and electric equipment serving storable pools and fountains.

Types of Wiring Methods and Materials

The NEC recognizes a number of standard wiring methods, some of which are suitable for general use while others are suitable only for special purposes. In some localities, regulations restrict the use of some of the wiring methods recognized by the NEC. The most widely used

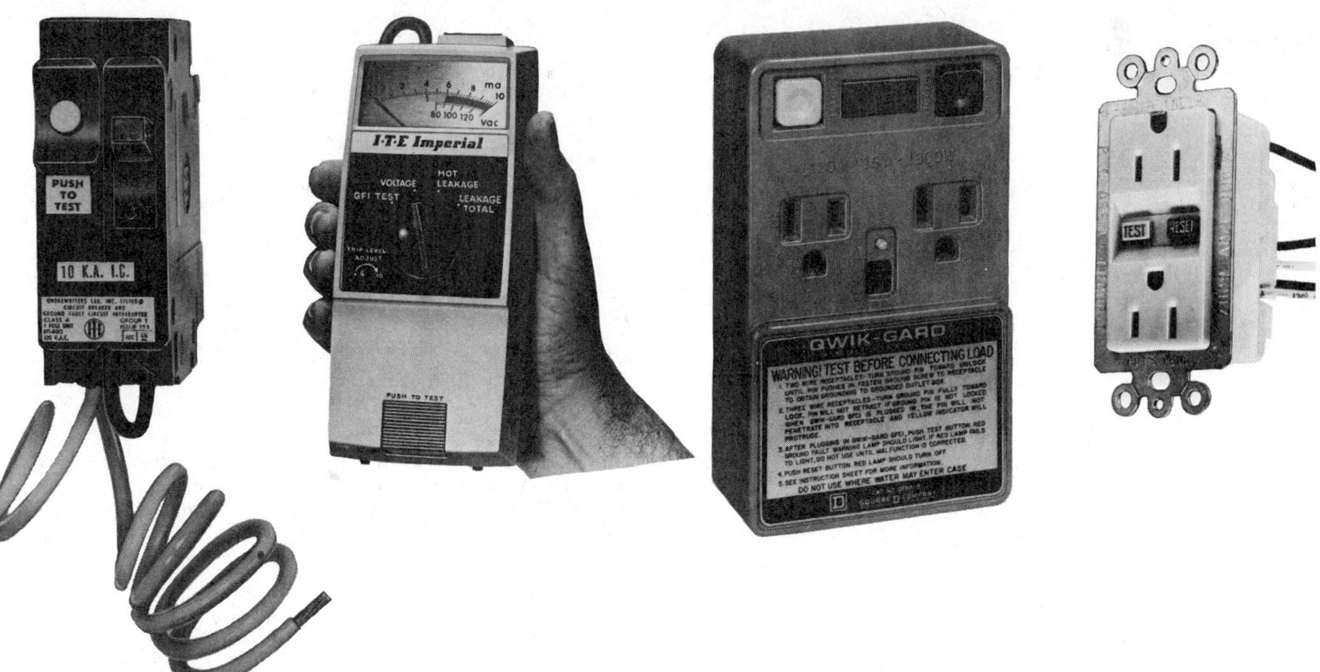

FIG. 8-2Q. A variety of GFCIs are available, including plug in circuit breaker types and duplex receptacle types. These also come in portable units and each has a test switch so that the unit can be checked periodically to ensure continuous proper operation. (I-T-E Imperial, Square D Co.)

wiring methods are rigid metal conduit, rigid nonmetallic conduit, intermediate metal conduit (IMC), electrical metallic tubing (EMT), metal clad cable (MC), armored cable (AC; trade name: "BX"), nonmetallic sheathed cable (NM, NMC; trade named "Romex"), surface metal and nonmetallic raceways, wireways, busways, underfloor raceways, and cellular metal floor raceways. (See Figs. 8-2R, 8-2S, 8-2T, and 8-2U.) In some areas of the country, flexible metal conduit (trade name: "Greenfield") is used for complete wiring systems. None of these wiring methods should be used for either temporary or permanent work where they do not fully conform to NEC safety requirements.

FIG. 8-2R. A Type UF sheathed cable. The insulated conductors are jacketed in parallel and the overall covering shall be flame retardant, moisture resistant, fungus resistant, corrosion resistant, and suitable for direct burial in the earth. (Plastic Wire & Cable Corporation)

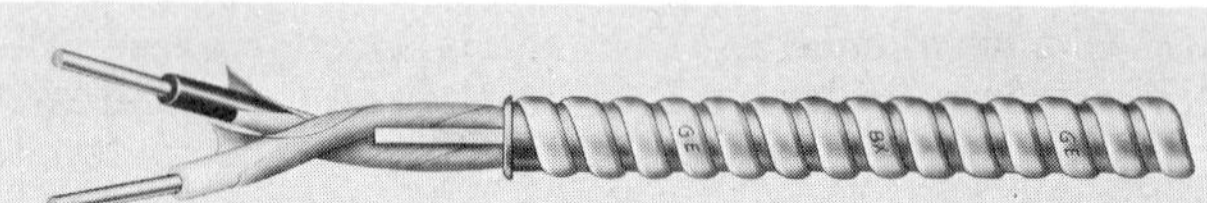

FIG. 8-2S. A typical armored cable (Type AC) that is frequently referred to as "BX." Another type (ACL) has lead covered conductors and is often referred to as "BXL." (Both "BX" and "BXL" are trade names.) (General Electric Company)

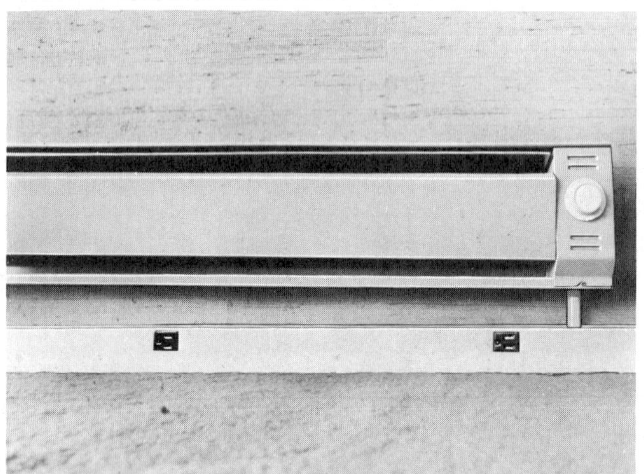

FIG. 8-2T. Installation of a typical metal raceway system showing an all steel baseboard and a multioutlet system. (The Wiremold Company)

FIG. 8-2U. The use of a surface metal raceway in a domestic kitchen installation. Frequent spacing of outlets eliminates interference from cords and reduces risk of injury. (The Wiremold Company)

Identification of Conductors, Terminals, Circuits, Branch Circuits

With few exceptions, all interior wiring systems have a grounded circuit conductor that is continuously identified throughout. The identification for conductors of No. 6 or smaller (except for Type MI cable) should consist of an outer white or natural gray color. Insulated conductors larger than No. 6 should have similar identification or be identified by distinctive white marking at its terminations during installation. The grounded conductors of Type MI cable are also identified by distinctive markings at its terminations during installation.

In general, the terminals of electrical devices to which a grounded conductor is to be connected are identified by being made of metal substantially white in color, have a metallic plating substantially white in color, or the word "white" located adjacent to the terminal. If the terminal is not visible, the conductor entrance hole may be colored white. In the case of screw shell type lampholders, the (white) terminal is the one that is connected to the screw shell.

The equipment grounding conductor of a branch circuit is identified by a continuous green color or a continuous green color with one or more yellow stripes, unless it is bare. Ungrounded conductors of different voltages are permitted to be of any color other than green, white, or gray. In some special applications, such as isolated (ungrounded) circuits in hospital operating rooms, specific colors are required.

Lighting and Appliance Branch Circuits

A branch circuit is that portion of a wiring system extending between the final overcurrent device protecting the circuit and the outlets on the circuit. Branch circuits are further defined as: (1) an appliance branch circuit supplying one or more outlets to which appliances are to be connected and which has no permanently connected lighting fixtures not a part of an appliance, (2) a general purpose branch circuit supplying two or more outlets for lighting and appliances, and (3) an individual branch circuit supplying only one piece of equipment.

Receptacles rated at 15 A connected to 15 or 20 A branch circuits serving two or more outlets must not supply a total load in excess of 12 A for portable appliances. Receptacles rated at 20 A and connected to 20 A branch circuits serving two or more outlets should not supply a total load in excess of 16 A (80 percent of circuit capacity) for cord and plug connected appliances.

The total load must not exceed 80 percent of the branch circuit rating if motor operated appliances are supplied by the circuit or, if in normal operation, the load will continue for three hours or more, as in store lighting and similar loads. Where the current drawn by resistance loads exceeds the motor current rating, as in clothes dryers and similar appliances, branch circuit conductors and overcurrent devices for individual branch circuits supplying such appliances must have a capacity of 125 percent of appliance nameplate ratings.

Lampholders, when connected to circuits having a rating of more than 20 A, must be of the heavy duty type.

Receptacles, when connected to circuits having two or more outlets, must conform to the ratings in Tables 8-2H and 8-2I. Where larger than 50 A, the receptacle rating is not permitted to be less than the branch circuit rating.

TABLE 8-2H. Maximum Cord- and Plus-Connected Load to Receptable
(Table 210-21 (b) of the *NEC*)

Circuit Rating Amperes	Receptacle Rating Amperes	Maximum Load Amperes
15 or 20	15	12
20	20	16
30	30	24

TABLE 8-2I. Receptacle Ratings for Various Size Circuits
[Table 210-21 (b) (3) of the *NEC*]

Circuit Rating Amperes	Receptacle Rating Amperes
15	Not over 15
20	15 or 20
30	30
40	40 or 50
50	50

Individual branch circuits may supply any loads. Branch circuits having two or more outlets may supply only loads as follows:

Branch circuits are for lighting units or appliances rated at 15 and 20 A. The rating of any one cord and plug connected appliance must not exceed 80 percent of the branch circuit ampere rating. The total rating of fixed appliances must not exceed 50 percent of the branch circuit ampere rating when lighting units or other appliances are also supplied.

Branch circuits rated at 40 and 50 A are permitted to supply fixed lighting units with heavy duty lampholders or infrared heating units in other than dwelling occupancies, or fixed cooking appliances in any occupancy. Branch circuits rated larger than 50 A may supply only nonlighting loads.

Receptacle outlets in dwelling occupancies, including guest rooms in hotels and motels, must be installed in every kitchen, family room, dining room, living room, parlor, library, den, sun room, bedroom, recreation room, or similar rooms. Insofar as practical, the outlets should be spaced equal distances apart. (See Fig. 8-2V.) Most appliances and portable lamps are provided with flexible cords at least 6 ft (1.8 m) in length. Countertop appliances for use in the kitchen are usually supplied with shorter cords. It is the intent of the NEC that receptacle outlets be placed around the perimeter of a room and around kitchen counters so that any appliance or lamp placed along the wall or on the counter could be served by an existing outlet without an extension cord. (See Fig. 8-2V.)

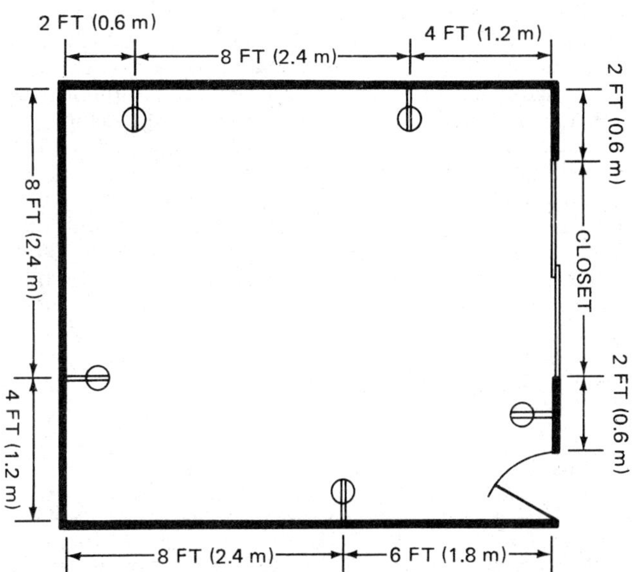

FIG. 8-2V. Illustration of the requirements in the NEC on the placement of receptacles in dwelling occupancies. Receptacles should be installed so that no point along the floor line is more than 6 ft (1.8 m), measured horizontally, from an outlet to permit a lamp or appliance, equipped with 6 ft (1.8 m) cords, to be located anywhere in the room.

Receptacles installed on 15 and 20 A branch circuits must be of the grounding type and must be effectively grounded. This does not mean that all cord and plug connected equipment must be of the grounded type. (See the NEC for specific details on grounding.)

Calculation of Loads

The methods specified in the NEC provide a basis for calculating the expected branch circuit and feeder loads and for determining the number of branch circuits required. Where in normal operation the maximum load of a branch circuit will continue for three hours or more (such as store lighting), the minimum unit loads specified should be increased by 25 percent.

The minimum unit lighting loads in volt-amperes per square foot for various types of occupancies are given in the NEC. In determining the load, open porches and garages in connection with dwelling occupancies are not included. Unfinished and unused spaces in dwellings need not be included unless adaptable for future use.

For other than general illumination lighting, general use receptacles, appliances, and loads other than motor loads, the following minimum unit load should be included for each outlet:

1. Outlets supplying specific appliances and other loads—ampere rating of appliance or load served.
2. Outlets supplying heavy duty lampholders—600 VA.
3. Other outlets—180 VA.*

Feeder Loads

Feeder conductors must have sufficient ampacity to supply the load served. The computed load of a feeder must not be less than the sum of all branch-circuit loads supplied by the feeder subject to the demand factor provisions of the NEC. Tables of demand factors for general illumination, electric ranges and other cooking appliances, and for clothes dryers are found in the NEC. Table 8-2J shows how the demand factors may be applied to that portion of the total branch circuit load computed for general illumination.

An optional method of calculating the load for a one family residence served by a 120/240 V, three wire, 100 A or larger service where the total load is supplied by one feeder or one service is to use the percentages in Table 8-2K.

Examples of branch circuit and feeder calculations for other buildings and occupancies are given in the NEC.

*This provision is not applicable to receptacle outlets as required by the NEC for dwelling units. The 20 A small appliance branch circuits and laundry branch circuit in dwelling units are calculated at 1,500 VA per circuit. Other receptacle outlets rated 20 A or less in dwelling units are considered part of the general lighting load calculated at 3 VA per sq ft. For receptacle outlets in other than dwelling units, each single or multiple receptacle is considered at not less than 180 VA.

An example of how to calculate the load in a single family dwelling is presented to illustrate the principles involved:

For a dwelling that has a floor area of 1,500 sq ft (139 m^2) (exclusive of unoccupied cellar, unfinished attic, and open porches) and a 12 kW range, the load may be computed as follows:

General Lighting Load: 1,500 ft^2 (139 m^2) at 3 volt-amperes per ft^2 (0.09 m^2) = 4,500 volt-amperes
Number of Branch Circiuts Required
General Lighting Load:
4,500 ÷ 120 = 37.5 amp; or three 15 amp, 2-wire circuits; or two 20 amp 2-wire circuits
Small Appliance Load: Two 2-wire 20 amp circuits
Laundry Load: One 2-wire, 20 amp circuit

Minimum Size Feeders Required
Computed Load

General Lighting	4,500 volt-amperes
Small Appliance Load	3,000 volt-amperes
Laundry	1,500 volt-amperes
Total (without range)	9,000 volt-amperes
3,000 volt-amperes at 100 percent	3,000 volt-amperes
9,000 − 3,000 = 6,000 volt-amperes at 35 percent =	2,100 volt-amperes
Net Computed (without range)	5,100 volt-amperes
Range Load	8,000 volt-amperes
Net Computed (with range)	13,100 volt-amperes

For 120/240 V system feeders 13,100 ÷ 240 = 55 amp.
Therefore, feeder size for total load may be seleted on basis of 55 amp load.
Net computed load exceeds 10 kVA so service conductors shall be 100 amperes.

TABLE 8-2J. Lighting Load Feeder Demand Factors
(Table 220-11 of the *NEC*)

Type of Occupancy	Portion of Lighting Load to which Demand Factor Applies (volt-amperes)	Demand Factor Percent
Dwelling Units	First 3,000 or less at	100
	From 3,001 to 120,000 at	35
	Remainder over 120,000 at	25
*Hospitals	First 50,000 or less at	40
	Remainder over 50,000 at	20
*Hotels and Motels—including Apartment Houses without provision for cooking by tenants	First 20,000 or less at	50
	From 20,001 to 100,000 at	40
	Remainder over 100,000 at	30
Warehouses (Storage)	First 12,500 or less at	100
	Remainder over 12,500 at	50
All Others	Total Volt-amperes	100

* The demand factors of this table shall not apply to the computed load of feeders to areas in hospitals, hotels, and motels where the entire lighting is likely to be used at one time; as in operating rooms, ballrooms, or dining rooms.

TABLE 8-2K. Optional Calculation for One-Family Dwelling Unit
(Table 220-30 of the *NEC*)

Load (in kVA)
(1) 100 percent of the nameplate rating(s) of the air conditioning and cooling, including heat pump compressors.
(2) 65 percent of the nameplate rating(s) of the central electric space heating including integral supplemental heating in heat pumps.
(3) 65 percent of the nameplate rating(s) of electric space heating if less than four separately controlled units.
(4) 40 percent of the nameplate rating(s) of electric space heating of four or more separately controlled units.
Plus: 100 percent of the first 10 kVA of all other load. 40 percent of the remainder of all other load.

Flexible Cords and Cables

Flexible cords are made in many types for various kinds of service ranging from wiring for portable lamps to elevator cables. Article 400 of the NEC gives a description, the ampacities, and the intended use of the various flexible cords now available. The ampacity of flexible cords is limited by type and gage of wire used. They must not be overloaded nor the size reduced where the gage is intended for mechanical strength.

Flexible cords are frequently subject to physical damage and rapid wear. Grounds or short circuits may occur if the insulation is damaged, and the resulting arc may ignite the insulation or nearby combustible material. Replacement of flexible cords as soon as they show damage or appreciable wear is of utmost importance.

Under present UL labeling practices, only replacement nondetachable power supply cords, range and dryer power supply cords, and cord sets, such as extension cords, carry the UL label on the product itself. Formerly, the original cords and power supply cords were labeled, and many consumers thought a label on the cord of an electrical appliance signified the appliance itself was investigated for safety. Nondetachable cords on new appliances do not now carry the UL label so the consumer cannot be misled into assuming the appliance itself has been UL tested, which may not be the case.

Flexible cords are not considered by the NEC as a wiring method. They are covered in the NEC chapter on equipment for general use. Flexible cords, including extension cords, are not acceptable as a substitute for the fixed wiring of a structure, but they are acceptable for use in extending the length of a cord on a portable lamp or a cord and plug connected appliance. They should be used in accordance with any instructions included with the listing or labeling, and the flexible cord itself should be used in accordance with Article 400 of the NEC. Flexible cords are not limited solely to "temporary wiring."

Switches

Switches are required for the control of lights and appliances, and as the disconnecting means for motors and their controllers. Switches may be of either the air break or the oil break type. In the oil break type the interrupting device is immersed in oil.

The chief fire hazard in switches is the arcing produced when the switch is opened. This hazard is somewhat greater with oil break switches. If operated much beyond their rated capacity, or if the condition of the oil is poor or its level is not properly maintained, the arc may vaporize the oil, rupture the case, and cause a fire. However, the amount of oil is comparatively small (except in high voltage equipment) and these switches present no hazard if properly used and maintained.

ELECTRICAL HOUSEHOLD APPLIANCES

Electric Heating Equipment: Electric heating equipment is used widely, and it is important that all such devices be tested and listed by qualified electrical testing laboratories. Some fixed heating equipment covered in the NEC is acceptable for installation in direct contact with combustible material. Baseboard heaters listed by UL, for example, have been tested and found to incorporate suitable safeguards against fire hazards that might result from contact with draperies, furniture, carpeting, bedding, etc., although discoloration or scorching (but no glowing embers or flaming) may result on adjacent materials. Other electric air heaters, however, may present fire hazards if they come in contact with combustible materials or if they are covered or blocked in any manner. Space heating systems must not be installed where exposed to severe physical damage unless adequately protected, nor in damp or wet locations unless approved for such locations.

A heater installed in an air duct or plenum must be of a type suitable for that purpose. The NEC covers installation of heaters for ducts and plenums and such items as airflow reliability, problems of condensation, fan circuit interlock, limit controls, and location of disconnecting means. NFPA 90B, *Standard for the Installation of Warm Air Heating and Air Conditioning Systems*, also contains details on central systems.

Electric Ranges, Wall Mounted Ovens, and Counter Mounted Cooking Units: Each of these devices needs a means for disconnection from all ungrounded conductors of the supply circuit, except that a separable connector or a plug and receptacle may serve as the disconnecting means for free standing household ranges. A plug and receptacle connection at the rear base of the range, if accessible from the front by removal of a drawer, is considered satisfactory. Grounding requirements for ranges are covered in Article 250 of the NEC. The NEC gives the methods for calculating feeder loads for household electric ranges and other cooking appliances.

Refrigerators: Fire problems with refrigerators are principally deterioration in service of the fractional horsepower motors used and the possibility of overheating. Refrigerant coils and motors are susceptible to accumulations of lint and oily deposits, so cleanliness is the primary consideration in fire prevention. New devices have sealed motors not requiring oiling or maintenance, but overheating can result from improper use or inefficient cooling of the refrigerant due to coil dirt or damage. Ordinary household or commercial refrigerators must not be used to store flammable liquids. [Refrigerators for use in Class I hazardous (classified) locations and refrigerators for flammable materials storage in laboratories in health related institutions in accordance with Chapter 7 of NFPA 99, are listed by UL.] Exposed noncurrent carrying metal parts of refrigerators and freezers, which are likely to become energized, need to be grounded under provisions of the NEC.

Room Air Conditioning Units: These have the same fire hazards as other motor operated appliances. The exposed noncurrent carrying metal parts must be grounded as required by Article 250 of the NEC.

Incandescent Lamps: Because incandescent lamps produce considerable heat, they inherently possess the hazard of heating and igniting combustible material in contact with them. Under normal conditions, with incandescent lamps in approved lampholders and fixtures where properly guarded, the heating hazard is negligible, but ignition of combustible material may result if lamps are surrounded by or laid on such combustible material. Table 8-2L presents information on surface and base temperatures of standard lamps in open sockets.* The temperature measurements are all taken at an ambient temperature of 77°F (25°C). In most instances where lamps are incorporated in various kinds of lighting equipment, the ambient temperature at which the lamps operate is higher; thus the surface temperatures of the bulbs are also higher. The bulb temperatures may be further increased if the lamp is in other than a vertical, base up position. Figure 8-2W shows surface temperatures of 100 W A-19 and 500 W PS-35 lamps in various positions, but, again, these temperatures were measured in an ambient temperature of 77°F (25°C), so where enclosed in fixtures, the temperatures will be higher. Conventional Christmas tree lamps of U.S. manufacture have approximate average bulb surface temperatures at the hottest spot of approximately 260°F (127°C) for the blue and green colors, and somewhat lower for the white, red, and yellow colors.

In locations where there are flammable vapors or gases, combustible dusts, or readily ignitable fibers or

---

The data in Table 8-2L was extracted from the *IES Handbook*, 5th ed., of the Illuminating Engineering Society and is used here with the permission of the Society. The reader is asked to keep in mind the above comments in the text regarding use of data in Table 8-2L and Figure 8-2W.

TABLE 8-2L. General Service Lamps for 115, 120, and 125 Volt Circuits
(Will Operate in Any Position but Lumen Maintenance is Best for 40 to 1500 Watts When Burned Vertically Base-Up)

Watts	Bulb and Other Description	Base	Fila-ment	Rated Aver-age Life (hours)	Maxi-mum Over-All Length in Inches (mm)	Average Light Center Length in Inches (mm)	Approx-imate Initial Fila-ment Temp. (K)	Max. Bare Bulb Temp. °F (°C)	Base Temp. °F (°C)	Ap-prox-imate Initial Lu-mens	Rated Initial Lu-mens Per Watt‡	Lamp Lumen De-preci-ation** (per-cent)
10	S-14 inside frosted or clear	Med.	C-9	1500	3 1/2 (89)	2 1/2 (63)	2420	106 (41)	106 (41)	80	8.0	89
15	A-15 inside frosted	Med.	C-9	2500	3 1/2 (89)	2 3/8 (60)	—	—	—	126	8.4	83
25	A-19 inside frosted	Med.	C-9	2500	3 7/8 (98)	2 1/2 (63)	2550	110 (43)	108 (42)	230	9.2	79
40	A-19 inside frosted and white¶	Med.	C-9	1500	4 1/4 (108)	2 15/16 (75)	2650	260 (127)	221 (105)	455	11.4	87.5
40	S-11 clear	Inter-med.	CC-2V or C-7A	350 500	3 5/16 (59)	1 5/8 (41)	2800 —	570 (299)	390 (199)	477 —	11.9 —	— —
50	A-19 inside frosted	Med.	CC-6	1000	4 7/16 (113)	3 1/8 (79)	—	—	—	680	13.6	—
60	A-19 inside frosted and white¶	Med.	CC-6	1000	4 7/16 (113)	3 1/8 (79)	2790	255 (124)	200 (93)	860	14.3	93
75	A-19inside frosted and white¶	Med.	CC-6	750	4 7/16 (113)	3 1/8 (79)	2840	275 (135)	205 (96)	1180	15.7	92
100	A-19 inside frosted and white¶	Med.	CC-8	750	4 7/16 (113)	3 1/8 (79)	2905	300 (149)	208 (98)	1740	17.4	90.5
100§	A-19 inside frosted and white	Med.	CC-8	1000	4 7/16 (113)	3 1/8 (79)	—	—	—	1680	16.8	—
100	A-21 inside frosted	Med.	CC-6	750	5 1/4 (133)	3 7/8 (98)	2880	260 (143)	194 (90)	1640	16.9	90
100§	A-23 inside frosted or clear	Med.	C-9	1000	5 15/16 (151)	4 7/16 (113)	—	—	—	1480	14.8	—
150	A-21 inside frosted	Med.	CC-8	750	5 1/2 (140)	4 (102)	2960	—	—	2880	19.2	89
150	A-21 white	Med.	CC-8	750	5 1/2 (140)	4 (102)	2930	—	—	2790	18.6	89
150	A-23 inside frosted or clear or white	Med.	CC-6	750	6 3/16 (157)	4 5/8 (117)	2925	280 (138)	210 (99)	2780	18.5	89
150	PS-25 clear or inside frosted	Med.	C-9	750	6 15/16 (176)	5 1/4 (133)	2910	290 (143)	210 (99)	2660	21.2	87.5
200	A-23 inside frosted or white or clear	Med.	CC-8	750	6 5/16 (160)	4 5/8 (117)	2980	345 (174)	225 (107)	4000	20.0	89.5
200	PS-25 clear or inside frosted	Med.	CC-6	750	6 15/16 (176)	5 1/4 (133)	—	—	—	3800	19.0	—
200	PS-30 clear or inside frosted	Med.	C-9	750	8 1/16 (205)	6 (152)	2925	305 (152)	210 (99)	3700	18.5	85
300	PS-25 clear of inside frosted	Med.	CC-8	750	6 15/16 (110)	5 3/16 (132)	3015	401 (205)	234 (112)	6360	21.2	87.5
300	PS-30 clear or inside frosted	Med.	C-9	750	8 1/16 (205)	6 (152)	3000	275 (135)	175 (79)	6100	20.3	82.5
300	PS-30 clear or inside frosted	Mog.	CC-8	1000	8 5/8 (219)	7 (178)	—	—	—	5960	19.8	—

TABLE 8-2L. General Service Lamps for 115, 120, and 125 Volt Circuits (Continued)
(Will Operate in Any Position but Lumen Maintenance is Best for 40 to 1500 Watts When Burned Vertically Base-Up)

Watts	Bulb and Other Description	Base	Filament	Rated Average Life (hours)	Maximum Over-All Length in Inches (mm)	Average Light Center Length in Inches (mm)	Approximate Initial Filament Temp. (K)	Max. Bare Bulb Temp. °F (°C)	Base Temp. °F (°C)	Approximate Initial Lumens	Rated Initial Lumens Per Watt‡	Lamp Lumen Depreciation** (percent)
300	PS-35 clear or inside frosted	Mog.	C-9	1000	9⅜ (238)	7 (178)	2980	330 (166)	215 (102)	5860	19.6	86
500	PS-35 clear or inside frosted	Mog.	CC-8	1000	9⅜ (238)	7 (178)	3050	415 (213)	175 (79)	10600	21.2	89
500	PS-40 clear or inside frosted	Mog.	C-9	1000	9¾ (248)	7 (178)	2945	390 (199)	215 (102)	10140	20.3	—
750	PS-52 clear or inside frosted	Mog.	C-7A	1000	13 1/16 (332)	9½ (241)	2990	—	—	15660	20.9	—
750	PS-52 clear or inside frosted	Mog.	CC-8 or 2CC-8	1000	13 1/16 (332)	9½ (241)	3090	—	—	17000	22.6	89
1000	PS-52 clear or inside frosted	Mog.	C-7A	1000	13 1/16 (332)	9½ (241)	2995	480 (249)	235 (113)	21800	21.8	—
1000	PS-52 clear or inside frosted	Mog.	CC-8 or 2CC-8	1000	13 1/16 (332)	9½ (241)	3110	—	—	23600	23.6	89
1500	PS-52 clear or inside frosted	Mog.	C-7A	1000	13 1/16 (332)	9½ (241)	3095	510 (265)	265 (129)	34000	22.6	78

* Lamp burning base up in ambient temperature of 77°F.
† At junction of base and bulb.
‡ For 120-volt lamps.
§ Used mainly in Canada.
¶ Lumen and lumen per watt values of white lamps are generally lower than for inside frosted.
** Percent initial light output at 70 percent of rated life.

flyings, etc., lamps in fixtures specially approved for the location must be used.

Electric Discharge (Fluorescent) Lamps: The operating temperature of a fluorescent tube is lower than that of the glass envelope of the incandescent lamp, but high voltage is often used to start the lamp. This requires the use of transformers, reactors, capacitors, and switches; the heat produced by this equipment must be taken into account and the equipment properly safeguarded.

Bulb temperatures on the surface of fluorescent lamps average between 100 and 110°F (38 and 43°C) over most of their length, since this is a requirement for efficient light production. There is a small area of higher bulb temperature directly above the cathode at each end of a fluorescent lamp which ranges from 120 to 250°F (49 to 121°C), depending upon the type involved.

Where fluorescent lamps having an open circuit voltage of more than 300 V are installed in dwelling occupancies, they ideally have no exposed live parts when lamps are being inserted, are in place, or are being removed.

Extreme care is required in mounting fluorescent lamp fixtures containing a ballast on combustible, low density, cellulose fiberboard. Fixtures containing ballasts should not be mounted in contact with low density, cellulose fiberboard unless they are specifically listed by a qualified electrical testing laboratory for mounting in that manner. If not specifically listed for this use, they should be spaced not less than 1½ in. (38 mm) from the surface of the combustible material. The principal hazard is ignition of the low density cellulose fiberboard, which under some conditions can ignite at relatively low temperatures. Integral thermal ballast protection is now provided for fluorescent fixtures installed indoors to protect against ballast overheating as a result of failure of capacitors, lamps, ballast winding shorts, etc.

Electric (High Intensity) Discharge Lamps: The operating temperature of a high intensity discharge lamp such as a mercury vapor lamp is usually much higher than the operating temperature of either Edison base incandescent or fluorescent lamps, exceeding 550°F (288°C) in some sizes. Extreme care is therefore necessary where such lamps are encountered to prevent their contacting combustible material. Although such lamps were at one time used primarily for street lighting and other outdoor lighting applications, their higher lighting efficiency, and new designs providing good color correction and lower wattages, have resulted in increased use indoors. Such lamps usually require ballasts (transformers), sometimes capacitors, and are designed to fit into Edison base lampholders. There are, however, self ballasted lamps that can be

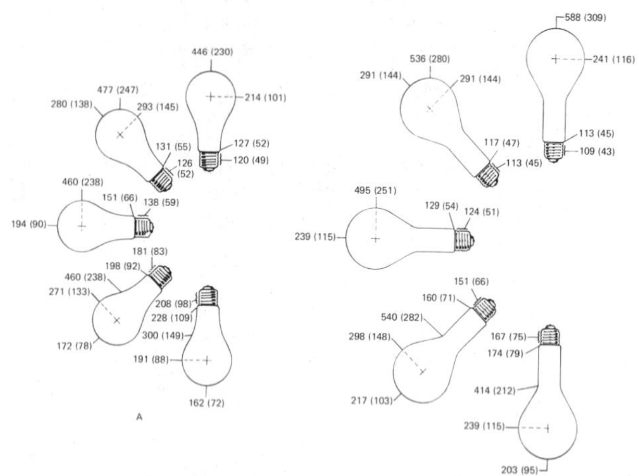

FIG. 8-2W. Surface temperatures of lamps in various positions. At left is a 100 watt, A-19 lamp and at right a 500 watt, PS-35 lamp.

substituted directly for incandescent lamps. High intensity discharge lamps should be used only in fixtures listed by a qualified electrical testing laboratory for their use, as indicated on the fixture.

Portable Hand Lamps: These lamps may constitute a fire and casualty hazard. Metal shell, paper-lined lampholders are not designed to be used as portable lamps. A fire hazard may result from a defective or worn cord or from the breaking of a lamp. A casualty hazard may result from personal contact with bare spots on the cord or by contact with a live metal lamp socket. Portable hand lamps ideally have a handle and, where subject to physical damage or where the lamp may come in contact with combustible material, they have a substantial guard.

Portable hand lamps used in hazardous (classified) locations must be suitable for the purpose. (See Fig. 8-2X.)

FIG. 8-2X. A representative portable hand lamp of a type suitable for use in Class I, Group C and D locations in accordance with Article 501 of the NEC. It has an aluminum guard and globe holder. (Stewart R. Browne Mfg. Co., Inc.)

Portables used on grounded surfaces or in damp or wet locations must be grounded unless supplied through an isolating transformer with an ungrounded secondary of not more than 50 V.

Television and Radio Equipment: Outdoor antennas and lead-in conductors must be of corrosion resistant material and securely supported. (See Fig. 8-2Y.) They must not be attached to poles or similar structures carrying electric

FIG. 8-2Y. A dangerous arrangement of a lightly supported television antenna in the proximity of power lines.

light or power wires or trolley wires, nor should they cross over or be near electric light or power circuits (to avoid accidental contact). Lead-in conductors also must be kept at least 6 ft (1.8 m) from any conductor forming a part of a lightning rod system. The metal sheath of the coaxial cable used for cable television systems must be grounded in accordance with Article 820 of the NEC to avoid electric shock and fire hazards. The sheath must be electrically bonded to the electrical system power ground. Use of a separate ground rod for the cable television ground without such bonding should not be permitted.

Masts and metal structures supporting antennas must be well grounded and each conductor of a lead in from an outdoor antenna must be protected by an approved lightning arrester. Metal masts on buildings are required by NFPA 78 to be bonded to the nearest lightning conductor (where available). This is normally done with standard lightning conductors.

Radio interference eliminators and noise suppressors connected to power supply leads and devices, intended to permit an electric supply circuit to be used in lieu of an antenna, should be listed for the purpose by a qualified electrical testing laboratory.

Operating television sets, even the new solid state sets, can develop considerable heat, so most cabinets housing such equipment are provided with ventilation openings at the rear and bottom. It is important not to install television receivers so the desired ventilation is cut off or significantly reduced (for example, by recessing the set in a wall, bookcase, etc.) unless the receiver is designed for such use.

Washers and Dryers: These devices need a means for disconnecting from all ungrounded conductors of the supply circuit. A separable cord connector or an attachment plug and receptacle may serve as the disconnecting means.

Electric clothes washers and dryers and similar appliances are usually installed within reach of a person who can make contact with a grounded surface or object. Consequently, the exposed, noncurrent carrying metal parts of these machines are grounded to remove the danger of electric shock.

Accumulations of lint in dryers and in lint traps also present a potential fire hazard if not periodically cleaned.

Smoothing Irons: These appliances intended for use in residences are required by the NEC to be equipped with approved means to limit the temperature. The general use of automatic irons has greatly reduced the number of fires resulting from hot nonautomatic irons left in contact with combustible material. However, some hazard still remains because satisfactory ironing of many fabrics requires a degree of heat sufficient to cause ignition if the iron is left in contact with some combustible materials for a considerable period of time.

Electrically Heated Pads and Bedding: These are commonly made of fabrics having certain inherent fire hazards. The user must be warned by means of a marking on the appliance and by instructions packed with it against the common possible abuses which would increase the fire or shock hazard.

Because many fabrics readily absorb moisture, pads listed by UL (unless of the waterproof type) are provided with moisture resistant envelopes. This envelope should be examined frequently for signs of deterioration. Blankets are normally designed so that they may be laundered; some cannot be dry cleaned (as with Stoddard Solvent or perchlorethylene) and are normally labeled "Do Not Dry Clean."

INDUSTRIAL AND COMMERCIAL EQUIPMENT

Furnaces

Industrial furnaces generally employ transformers, which may be either the dry type, askarel insulated type, oil insulated type, or less flammable fluid type, e.g., dimethyl silicone. Although new transformers are no longer made with askarel (a material containing polychlorinated biphenyl, or PCB), there are many such transformers still available. The NEC requires oil insulated transformers of a total rating exceeding 75 kVA to be located in a fire resistive vault.

Oil filled circuit breakers that control arc furnaces are subjected to unusually severe duty and unless frequently inspected and properly maintained, may fail with disastrous results. Circuit breakers on circuits operated at more than 600 V and which are used to control oil filled transformers must be located outside the transformer vaults. Vents on high voltage circuit breakers must be piped outdoors. Electric arc furnace circuits, due to the nature of the operation, are also subject to high surge voltages which can cause failure of the arc circuit breakers used. In these cases, shunt capacitors are installed in the circuit to prevent these high voltage surges.

Inductive and dielectric heat generating equipment employing high frequency alternating currents is used in many heat treating processes. To eliminate both the personal and fire hazards of such equipment, construction and installation should comply with the special requirements of the NEC.

Other hazards of electric furnaces are similar to those of furnaces employing other means of heating.

Motors

Motors cause many fires. Ignition of the motor insulation or nearby combustible material may be caused by sparks or arcs when the motor winding short circuits or grounds, or when brushes operate improperly. Bearings may overheat because of improper lubrication, and sometimes excessive bearing wear allows the rotor to rub on the stator. The individual drives of machines of many different types sometimes makes it necessary to install motors in locations and under conditions that are injurious to motor insulation. Dust that can conduct electricity may be deposited on the insulation, or deposits of textile fibers, etc., may prevent the normal dissipation of heat. Motors should be cleaned and lubricated regularly. All motor installations should comply with the requirements of the NEC, which includes special rules for motors in hazardous (classified) locations.

Machine Tools

The electrical equipment of a modern machine tool or plastics processing machine may vary from a simple, single motor drill press or extruder to a large complicated multimotor, automatic machine involving highly complex control systems and equipment. This latter type is generally custom designed and factory wired. Machine tools incorporate many devices and safeguards to provide safety to life, safety from fire, reduction of machine lost time due to replacement of parts, safety to the machine itself, and safety to the work in process.

NFPA 79, *Electrical Standard for Industrial Machinery*, covers the specific electrical equipment, apparatus, and wiring furnished as a part of an industrial machine, starting at the electrical supply connection, providing the voltage is 600 V or less. The NEC contains requirements for general application, particularly for protection of several motors on one branch circuit.

Electrical equipment is subject to damage by oil, metal chips, coolants, and moving parts. The fixed wiring to the machines specified in the NEC generally should be conductors in conduit, tubing, or Type MI cable. Exceptions are connections to continuously moving parts, which must be extra flexible, multi-conductor cable. For flexible connections where small or infrequent movement is involved, as at motor terminals, flexible metal conduit or liquid tight flexible metal conduit may be used.

Motor Control Center Rooms

Automation in industrial plants has led to the development of what is sometimes called "motor control center rooms." Automation of production machinery has greatly increased the use of motors and their related control circuits. This has resulted in grouping the motor control equipment in relatively large, compartmented, metal enclosures. These are specially designed, factory built equipment, commonly referred to as "motor control centers." They are usually located in large rooms, and electrical failures or fires in the motor control center have often resulted in the destruction of the equipment in many compartments of the control center due to rapid temperature rise.

One of the principal reasons for the large losses has been that multi-conductor circuits have been installed stacked one above the other in cable trays. Heat from arcing or a fire in the control center could raise the temperature in the control center room to a point where normally slow burning cables become readily combustible. A fire in such cables can shut down a process or an entire plant for weeks or months. The NEC covers the use and construction of cable trays, and should be strictly followed where this type of support is used.

The number of motor control centers in one room should be limited or so arranged that exposure of a large number of circuits or equipment to any one fire is avoided. Motor control center rooms should be of noncombustible construction, with a roof or ceiling not readily weakened by a major electrical disturbance or fire. In addition to normal ventilation, the room should be provided with emergency ventilation for removal of heat and smoke in event of a severe fire.

Suggested fire protection for motor control center rooms may consist of preaction sprinkler systems, fixed water spray protection to cover exposed cables, fire detectors to initiate an alarm at a central point in the event of abnormally high temperatures or smoke conditions, a carbon dioxide or halon flooding system, portable extinguishers for Class C fires located at each entrance to the control room, and small hose connections with suitable hose and adjustable spray nozzles readily available near the control room.

Switchboards

Ideally, switchboards are installed in clean, dry locations. They should be under competent supervision and accessible only to qualified persons. Where it is necessary to install a switchboard in a wet location or outside a building, it must be in a weatherproof enclosure. Ample space must be provided for maintenance operations, as stipulated in the NEC.

Insulated conductors grouped within switchboards, as well as the instrument and control wiring, should have a flame retardant outer covering.

Circuit breakers and switches must have ample ratings for the maximum loads and ample interrupting capacity for the maximum short-circuit currents. Contacts of switches and circuit breakers should be kept in good condition, and the oil in oil circuit breakers renewed periodically and kept at the proper level.

Capacitors

Capacitors may be insulated with a combustible or a nonflammable liquid. Capacitors containing more than 3 gal (11 L) of flammable liquid must be enclosed in vaults or outdoor fenced enclosures. These safeguards prevent persons from coming into accidental contact or bringing conducting materials into accidental contact with exposed energized parts, terminals, or buses associated with them.

Capacitors must have a means of draining the stored charge. If no means were provided for draining off the charge stored in a capacitor after it is disconnected from the line, a severe shock might be received by a person servicing the equipment, or the equipment might be damaged by a short circuit.

Resistors and Reactors

Except when installed in connection with switchboards or control panels whose locations are suitably guarded from physical damage and accidental contact with live parts, resistors should always be completely enclosed in properly ventilated metal boxes. A resistor is always a source of heat, and when mounted on a combustible wall, a thermal barrier should be required if the space between resistors and reactors and any combustible material is less than 12 in. (305 mm).

Large reactors are commonly connected in series, with the main leads of large generators or the supply conductors from high capacity network systems assisting in limiting the current delivered on short circuit. Small reactors are used with lightning arresters to offer a high impedance to the passage of a high frequency lightning discharge and aid in directing the discharge to ground. Another type of reactor, having an iron core and closely resembling a transformer, is used as a remote control dimmer for stage lighting. Reactors are sources of heat and therefore should be mounted in the same manner as resistors.

Motion Picture Projectors and Studios

Motion picture projectors of the professional type [employing 35 or 70 mm film having 5.4 perforations per in. (25.4 mm) on each edge] must be located in approved enclosures. Motor driven projector units consist of a listed projector and lamp with motors designed or guarded to prevent ignition of film by sparks or arcs. All professional projectors should be operated by qualified personnel.

Motor-generator sets, transformers, rectifiers, rheostats, and similar equipment for the supply or control of current to projection or spotlight equipment should, if practical, be located in a separate room. If placed in the projection room, the equipment must be so located or guarded that arcs or sparks cannot come in contact with film. Switches, overcurrent devices, or other equipment not normally required or used for projectors, sound reproduction, flood or other special effect lamps, or other equipment (except remote control switches for the auditorium lights, or a switch for the motor operating the curtain and masking of the motion picture screen) must not be installed in projection rooms.

In projection rooms suitable for use with cellulose acetate (safety) film only, a sign reading, "SAFETY FILM ONLY PERMITTED IN THIS ROOM," must be posted on the outside of each projection room door and within the projection room itself in a conspicuous location.

Approved projectors of the nonprofessional or miniature type, when employing cellulose acetate (safety) film, may be operated without a projection room.

Special rules are also given in the NEC to cover electrical installations in motion picture studios, factories, laboratories, stages, or areas of buildings in which work is done on cellulose nitrate film. Provisions include wiring on stages, sets, dressing rooms, viewing, cutting, and patching tables, and film storage vaults.

Cranes and Hoists

Wiring methods and installation of electrical equipment for cranes and hoists are covered in Article 610 of the NEC. Where bare contact conductors are objectionable because of the presence of easily ignitible material, current may be conducted to a crane or hoist via a multiple cable on a cable reel or suitable take up devices.

Where a crane operates over combustible material, the resistors must either be placed in (1) a well ventilated cabinet of noncombustible material that will not emit flames or molten metal, or (2) a cage or cab constructed of noncombustible material that encloses the sides of the cage or cab from the floor to a point at least 6 in. (152 mm) above the top of the resistors.

Collectors must be designed to reduce sparking between them and the contact conductors to a minimum. When operated in rooms where easily ignitible fibers or materials producing combustible flyings are handled, manufactured, used, or stored, the installation must comply with the special requirements of Article 503 of the NEC.

All exposed metal parts of cranes, monorail hoists, hoists, and accessories (including pendant controls) should be metallically joined together into a continuous electrical conductor so the entire crane or hoist will be grounded.

Elevators, Dumbwaiters, Escalators, and Moving Walks

Installation of electric equipment and wiring for elevators, dumbwaiters, escalators, and moving walks are covered in Article 620 of the NEC.

All live parts of electrical apparatus in the hoistways, at landings, in or on cars of elevators and dumbwaiters, or in the wellways or landings of escalators or moving walks must be enclosed to prevent accidental contact. All wiring on panels, in raceways, and in or on cars, including the traveling cables, must have flame retardant, moisture resistant insulation, with the exception of conductors to main circuit resistors. These should be flame retardant and suitable for a temperature of not less than 194°F (90°C).

Traveling cables for operating, control, and signal circuits must be suspended at the car and hoistway in a manner that reduces strain on individual copper conductors to a minimum. Supports for the traveling cables must also be arranged to prevent damage to cables coming in contact with the hoistway or equipment. Where necessary, suitable guards can protect the cables.

To reduce the danger of electric shock, the following equipment must be effectively grounded in accordance with the grounding requirements of the NEC: (1) metal conduit, Type MC cable, or Type AC cable attached to elevator cars, (2) the frames of all motors, elevator machines, and controllers, (3) the metal enclosures for all electric devices in or on the car or hoistway, (4) the frames of nonelectric elevators if accessible to persons and if any electric conductors are attached to the car.

Equipment mounted on members of the structural metal frame of a building is considered to be grounded. Metal car frames supported by metal hoisting cables attached to or running over sheaves or drums of elevator machines are deemed to be grounded when the machines are properly grounded.

Heating Cable

UL has listed a number of heating cables for use within buildings. These devices are primarily designed to be used around water pipes to prevent freezing and facilitate the flow of viscous liquids. The cable is secured to the pipe with straps, and the pipe and heating cable are then encased with thermal insulation. Where the pipe passes vertically through flooring or where the pipe lines run along the floor level, the installation is best protected by sheet steel not less than No. 10 USS gage with the protection extending at least 4 ft (1.22 m) above the floor.

Some units incorporate a thermostat that automatically turns on the heating cable when the temperature drops below a predetermined value. The pipe heating cables are intended to be connected to a permanent supply wiring system. Unless specifically indicated otherwise by marking on the heating cables or in the installation instructions, the heating cables are intended for use only on metallic pipes. Electrical heating cables for outdoor ice and snow melting systems are covered in the NEC.

ELECTRICAL EQUIPMENT FOR OUTDOOR USE

Electric equipment that is installed outside of buildings must be weatherproof in design, i.e., constructed so exposure to the weather will not interfere with its successful operation; otherwise it must be enclosed in weatherproof cabinets. These enclosures are designed to prevent moisture or water from entering and accumulating within them. If mounted on a wall or other supporting surface, there should be at least 1/4 in. (6 mm) air space between the box or cabinet and the supporting surface.

Description of Devices

Electric equipment suitable for outdoor use is tested by qualified electrical testing laboratories. The tests encompass suitability of the materials used and the protective coatings for exposure to sunlight, rain, and snow; effects of heat and cold; and provisions for grounding, etc. The term "raintight," as used in the NEC, is applied to equipment that on exposure to a beating rain will not let water enter. The term "weatherproof" as used in the NEC, is applied to equipment that on exposure to the weather will still operate successfully.

Rainproof, raintight, or watertight equipment can fulfill the requirements for weatherproof equipment where varying weather conditions other than wetness such as snow, ice, dust, or temperature extremes are not a factor.

Electric Signs and Outline Lighting

Except for portable indoor signs, signs and outline lighting equipment are usually constructed of metal or other noncombustible material. Wood may be used for external decoration only if placed not less than 2 in. (50 mm) from the nearest lampholder or current carrying part. Enclosures for outside use should be weatherproof. All steel parts of enclosures must be galvanized or otherwise protected from corrosion. Signs, troughs, tube terminal boxes, and other metal frames must be grounded in the manner specified in the NEC unless they are insulated from ground and from other conducting surfaces and are inaccessible to unauthorized persons.

Each electric sign (other than the portable type) and each outline lighting installation must be controlled by an external operable switch or circuit breaker (handle on outside of switch enclosure), which will open all ungrounded conductors. The switch or breaker should be within sight of the sign or outline lighting installation unless it is capable of being locked in the open position.

Portable outdoor electric signs must be equipped with a GFCI, suitable for use on portable outdoor signs, in the sign, or in the attachment plug cap on the supply cord.

Electric Fences

Wire fences with electrical connections to produce a shock when animals come in contact with them are widely used on farms. To avoid hazard to persons and to stock, the current and the time interval during which the current is on and off must be limited to values which will not cause fatalities or injuries to persons or animals, while still causing an unpleasant sensation of shock. Fatalities and fires have resulted from homemade equipment supplied from ordinary lighting circuits. The open circuit voltage need not be limited if the current is properly limited. UL Research Bulletin No. 14 (UL 1939) and UL 69-1980, *Standard for Electric Fence Controllers*, give detailed information on these installations.

The output characteristics of some controllers are such that combustible material may be readily ignited when a grounded object occupies a position respective to the energized fence to produce an electric arc. To protect against fire from such a cause, users should determine that the controller has been designed and tested with respect to this hazard. Electric fence controllers listed in accordance with the requirements of UL 69 have been so tested.

Marina and Boatyard Wiring

Marina and boatyard wiring presents special outdoor electric equipment wiring problems that are covered in Article 555 of the NEC. Metal raceways and metal boxes must not be depended upon for grounding; a continuous insulated copper conductor, not smaller than No. 12 AWG, must be provided for a grounding conductor from outlet boxes and receptacles to the service ground. Wiring over and under navigable water should be subject to approval by the authority having jurisdiction.

LOCATIONS EXPOSED TO MOISTURE AND NONCOMBUSTIBLE DUSTS

Special electric equipment is required for use where moisture and noncombustible dusts may be present. For years, the NEC referred to such equipment as "vaportight," but this term was dropped from the Code because of confusion in the field between equipment so designated and "explosion proof" equipment. Many users assumed that "vapor tight" equipment was safe to use in atmospheres containing flammable gases or vapors (Class I locations), combustible dusts (Class II locations), or easily ignitible fibers or flyings (Class III locations). To avoid this type of misunderstanding, the term "enclosed and gasketed" was developed by UL for this type of equipment. Some inspection authorities permit enclosed and gasketed lighting fixtures in Class I, Division 2; Class II, Division 2; and in Class III, Divisions 1 and 2 locations when marked to show the operating temperature and maximum wattage of permissible lamps. The NEC requires lighting fixtures to be marked for use in wet and damp locations when so used.

COMMUNICATIONS SYSTEMS

Communications systems, including telephone, telegraph, fire and burglar alarms, watchman and sprinkler supervisory systems, usually operate with low voltages and currents, and if kept free from accidental contacts with higher voltage systems, they present no unusual hazards. There are, however, NEC requirements for the wiring of these systems if the wiring is in a duct, plenum, or other space for environmental air, because of the hazard of spreading products of combustion from one location to another. There are also NEC requirements for separation from other systems and spread of fire in shafts and where fire walls, etc., are penetrated. If the power supply is from storage batteries of appreciable current rating, for example, from the lighting system of a building, or from transformers (other than approved signaling transformers), special precautions should be taken with the wiring and equipment. Because of the importance of these systems, they should be properly designed and installed to give reliable service.

Signaling systems and communication systems must be installed in accordance with the requirements of the NEC and NFPA Standards 71, 72A, 72B, 72C, 72D, 72E, and 74, which cover the proper electrical wiring and equipment for fire alarm services. These standards also cross reference to the NEC. NFPA 72H, *Guide for Testing Procedures for Local, Auxiliary, Remote Station and Proprietary Protective Signaling Systems*, is a guide for testing such systems upon installation and during service. Telephone communication circuits may be used for completing the fire protective circuits between the protected premises and fire alarm headquarters.

EMERGENCY SYSTEMS

Requirements for the installation, operation, and maintenance of emergency system circuits and equipment are given in Article 700 of the NEC. The systems are intended to supply light and power when the normal supply fails. They are generally installed in places of assembly where artificial lighting is required, such as auditoriums, theaters, sports arenas, etc. occupied by large numbers of persons. (See Figs. 8-2Z and 8-2AA.)

Legally required standby systems installed to serve loads, such as heating and refrigeration systems, communication systems, ventilation and smoke removal systems, sewerage disposal, lighting systems and industrial processes, that, when stopped during any interruption of the normal electrical supply, could create hazards or hamper rescue or fire fighting operations, are covered in Article 701 of the NEC.

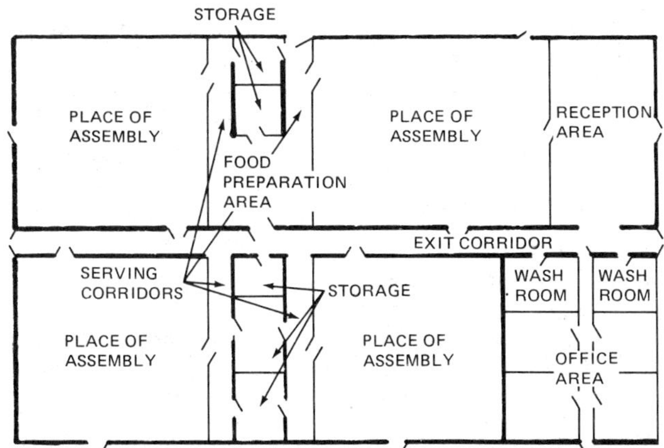

FIG. 8-2Z. The heavy lines represent walls required by typical local building codes to be of fire rated construction. The light lines represent walls not required by typical local building codes to be of fire rated construction.

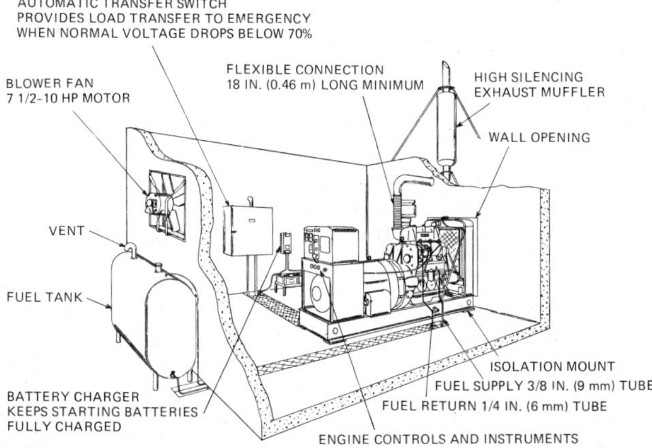

FIG. 8-2AA. Typical installation of a diesel powered standby generator such as might be used for emergency light and power systems. (P & H Harnischfleger Corp.)

NFPA 110, *Standard for Emergency and Standby Power Systems*, covers the performance requirements for such systems.

NFPA 101, *Life Safety Code®* (hereinafter referred to as NFPA 101), specifies where emergency lighting is considered essential for life safety. NFPA 99, *Standard for Health Care Facilities*, Chapter 8, gives minimum factors governing the design, operation, and maintenance of those portions of health care facility electrical systems where any degree of interruption would jeopardize the effective and safe care of hospitalized patients. The provisions in NFPA 99 do not supercede the recommendations in NFPA 101 or the NEC; however, they do limit the type of alternate source of electrical power allowable for use to assure electric power continuity in health care facilities. NFPA 99 recognizes the progressively greater dependence being placed upon electrical apparatus for the preservation of life of hospitalized patients, and is a guide in the selection of the electrical services for emergency supply to all lighting and power equipment considered essential, and for their design and maintenance in health care facilities.

The sources of electric current that can be used for emergency lighting equipment are: (1) a storage battery of suitable capacity, (2) a generator driven by some form of prime mover, (3) a second electric service widely separated electrically and physically from the regular service to minimize the possibility of simultaneous interruption of both services, or (4) connections on the line side of the main service, if sufficiently separated from the main service to prevent simultaneous interruption of supply to both regular and emergency circuits through an occurrence within the building or group of buildings served. NFPA 99 stipulates that the alternate source of power for health care facilities be prime mover driven generators or, where the normal source is an on site generator, an external utility service may be the alternate source. Means must be provided for automatically energizing emergency lights upon failure of the regular lighting system supply. For hospitals, the transition time from the instant of failure of the normal power source to an emergency generator source must not exceed 10 seconds.

Audible and visual signal devices are provided, where practical, to (1) warn of derangement of the emergency source, (2) indicate that the battery or generator is carrying load, (3) indicate when a battery charger is functioning properly, and (4) indicate a ground fault on large 480Y/277-V systems.

No appliances or lamps, other than those required for emergency use, are to be supplied by the emergency lighting circuits. A good emergency lighting system is designed so that the failure of any individual lighting element, such as the burning out of a light bulb, cannot leave any space in total darkness.

The emergency lighting circuit wiring is independent of all other wiring and equipment and must not enter the same raceway, cable, box, or cabinet with other wiring except at transfer switches and exit lighting fixtures supplied from both the normal and the emergency sources.

It is good practice to test the complete system upon installation and periodically thereafter to assure it is maintained in proper operating condition. A written record should be kept of such tests and maintenance.

SPECIAL OCCUPANCY ELECTRICAL PROBLEMS

This section discusses hazardous (classified) locations and other special problems requiring individual attention for which the NEC has specific recommendations.

Hazardous (Classified) Locations—General

Electric lights, motors, and instrumentation are necessary or desirable in many hazardous locations because of the presence of flammable liquids, gases, combustible dusts, or readily ignitable fibers or flyings. In these areas, special electrical equipment is necessary for safety. In Article 500 of the NEC hazardous (classified) locations are divided into three classes depending upon the kind of hazardous material involved; each class is further divided into two divisions according to the likelihood of the potential hazard existing. For the purposes of testing for approval and area classification, various air mixtures (not oxygen enriched) have been grouped on the basis of their characteristics. The groups are given in Tables 8-2M and 8-2N (Section 500-2 of the NEC) and tables from NFPA

TABLE 8-2M. Group Classification and Autoignition Temperature (AIT) of Selected Flammable Gases and Vapors of Liquids having Flash Points below 100°F (37.8°C)

Material	Minimum Cloud or Layer Ignition Temp.[1] °F		°C
Phthalic Anhydride	1202	M	650
Salicylanilide	1130	M	610
Sorbic Acid	860		460
Stearic Acid, Aluminum Salt	572		300
Stearic Acid, Zinc Salt	950	M	510
Sulfur	428		220
Terephthalic Acid	1256	NL	680
DRUGS			
Aspirin	1220	M	660
Gulasonic Acid, Diacetone	788	NL	420
Mannitol	860	M	460
l-Sorbose	698	M	370
Vitamin B1, mononitrate	680	NL	360
Vitamin C (Ascorbic Acid)	536		280
DYES, PIGMENTS, INTERMEDIATES			
Green Base Harmon Dye	347		175
Red Dye Intermediate	347		175
Violet 200 Dye	347		175
PESTICIDES			
Crag No. 974	590	Cl	310
Dieldrin (20%)	1022	NL	550
Dithane	356		180
Ferban	302		150
Manganese Vancide	248		120
Sevin	284		140
THERMOPLASTIC RESINS AND MOLDING COMPOUNDS			
Acetal Resins			
Acetal, Linear (Polyformaldehyde)	824	NL	440
Acrylic Resins			
Acrylamide Polymer	464		240
Acrylonitrile Polymer	860		460
Acrylonitrile-Vinyl Chloride-Vinylidene Chloride Copolymer (70-20-10	410		210
Methyl Methacrylate Polymer	824	NL	440
Methyl Methacrylate-Ethyl Acrylate Copolymer	896	NL	480
Methyl Methacrylate-Ethyl Acrylate-Styrene Copolymer	824	NL	440
Methyl Methacrylate-Styrene-Butadiene-Acrylonitrile Copolymer	896	NL	480
Methacrylic Acid Polymer	554		290
Cellulosic Resins			
Cellulose Acetate	644		340
Cellulose Triacetate	806	NL	430
Cellulose Acetate Butyrate	698	NL	370
Nylon (Polyamide) Resins			
Nylon Polymer (Polyhexa-methylene Adipamide)	806		430
Polycarbonate Resins			
Polycarbonate	1310	NL	710
Polyethylene Resins			
Polyethylene, High Pressure Process	716		380
Polyethylene, Low Pressure Process	788	NL	420
Polyethylene Wax	752	NL	400
Polymethylene Resins			
Carboxypolymethylene	968	NL	520
Polypropylene Resins			
Polypropylene (No Antioxidant)	788	NL	420
Rayon Resins			
Rayon (Viscose) Flock	482		250
Styrene Resins			
Polystyrene Molding Cmpd.	1040	NL	560
Polystyrene Latex	932		500
Styrene-Acrylonitrile (70-30)	932	NL	500
Styrene-Butadiene Latex (>75% Styrene; Alum Coagulated)	824	NL	440
Vinyl Resins			
Polyvinyl Acetate	1022	NL	550
Polyvinyl Acetate/Alcohol	824		440
Vinyl Chloride-Acrylonitrile Copolymer	878		470
Vinyl Toluene-Acrylonitrile Butadiene Copolymer	936	NL	530
THERMOSETTING RESINS AND MOLDING COMPOUNDS			
Allyl Resins			
Allyl Alcohol Derivative (CR-39)	932	NL	500
Amino Resins			
Urea Formaldehyde Molding Compound	860	NL	460
Urea Formaldehyde-Phenol Formaldehyde Molding Compound (Wood Flour Filler)	464		240
Epoxy Resins			
Epoxy	1004	NL	540
Epoxy-Bisphenol A	950	NL	510
Phenolic Resins			
Phenol Formaldehyde	1076	NL	580
Phenol Formaldehyde Molding Cmpd. (Wood Flour Filler)	932	NL	500
Polyester Resins			
Polyethylene Terephthalate	932	NL	500
Styrene Modified Polyester-Glass Fiber Mixture	680		360
Polyurethane Resins			
Polyurethane Foam, No Fire Retardant	824		440
SPECIAL RESINS AND MOLDING COMPOUNDS			
Ethylene Oxide Polymer	662	NL	350
Ethylene-Maleic Anhydride Copolymer	1004	NL	540
Petroleum Resin (Blown Asphalt)	932		500
Rubber, Crude, Hard	662	NL	350
Rubber, Synthetic, Hard (33% S)	608	NL	320

Notes to Table.

[1] Normally, the minimum ignition temperature of a layer of a specific dust is lower than the minimum ignition temperature of a cloud of that dust. Since this is not universally true, the lower of the two minimum ignition temperatures is listed. If no symbol appears between the two temperature columns, then the layer ignition temperature is shown. "Cl" means the cloud ignition temperature is shown. "NL" means that no layer ignition temperature is available and the cloud ignition temperature is shown. "M" signifies that the dust layer melts before it ignites; the cloud ignition temperature is shown. "S" signifies that the dust layer sublimes before it ignites; the cloud ignition temperature is shown.

[2] These materials may be classified in Group E, depending on their resistivity.

TABLE 8-2M. (Continued) Group Classification and Autoignition Temperature (AIT) of Selected Flammable Gases and Vapors of Liquids having Flash Points below 100°F (37.8°C)

Material	Group	AIT °F	AIT °C
Toluene	D*	896	480
Triethylamine	C*	—	—
Tripropylamine	D	—	—
Turpentine	D	488	253
Unsymmetrical Dimethyl Hydrazine (UDMH)	C*	480	249
Valeraldehyde	C	432	222
Vinyl Acetate	D*	756	402
Vinyl Chloride	D*	882	472
Vinylidene Chloride	D	1058	570
Xylenes	D*	867–984	464–529

Notes To Table

* Material has been classified by test.

[1] If equipment is isolated by sealing all conduit 1/2in. or larger, in accordance with Section 501-5(a) of NFPA 70, *National Electric Code,* equipment for the group classification shown in parentheses is permitted.

[2] For classification of areas involving Ammonia, see *Safety Code for Mechanical Refrigeration,* ANSI/ASHRAE 15, and *Safety Requirements for the Storage and Handling of Anhydrous Ammonia,* ANSI/CGA G2.1.

[3] Certain chemicals may have characteristics that require safeguards beyond those required for any of the above groups. Carbon disulfide is one of these chemicals because of its low autoignition temperature and the small joint clearance to arrest its flame propagation.

[4] Petroleum naphtha is a saturated hydrocarbon mixture whose boiling range is 20° to 135°C. It is also known as benzine, ligroin, petroleum ether, and naphtha.

References: Autoignition temperatures listed above are the lowest value for each material as listed in NFPA 325M, *Fire Hazard Properties of Flammable Liquids, Gases, and Volatile Solids,* or as reported in an article by Hilado, C. J. and Clark, S. W., in *Chemical Engineering,* September 4, 1972.

497M, *Manual for Classification of Gases, Vapors and Dusts for Electrical Equipment in Hazardous (Classified) Locations* (hereinafter referred to as NFPA 497M). For Group A, B, C, and D materials with flash points of 100°F (38°C) and higher, see NFPA 497M. For a more complete table of Group G dusts and information on Group E dusts see the following and NFPA 497M.

Group G includes combustible dusts having a resistivity of 10^5 Ω-cm or greater. This group includes grain and food dusts, chemical dusts, plastic dust, and most carbonaceous dusts such as coal dust. Prior to the 1984 NEC, carbonaceous dusts were classified in Group F, but this group was deleted in the 1984 NEC. Equipment approved for Group F, but not for Group E (see below), is not suitable for use in Group E locations where the hazard is caused by the presence of metal dusts. Such equipment is suitable for Group E locations where the hazard is caused by the presence of carbonaceous dusts with a resistivity of less than 10^5 Ω-cm. Group E includes combustible metal dusts regardless of resistivity, and other dusts of similarly hazardous characteristics having a resistivity of less than 10^5 Ω-cm.

Listed equipment is marked to show the Class, Group, and operating temperature, or temperature range, referenced to a 104°F (40°C) ambient. The temperature range, if provided, is indicated by identification numbers, as shown in Table 8-2O. The identification numbers are marked on equipment nameplates.

TABLE 8-2N. Selected Nonconductive Dusts Classified as Group G—Ignition Sensitivity Equal to or Greater than 0.2; Explosion Severity Equal to or Greater than 0.5

Material	Minimum Cloud or Layer Ignition Temp.[1] °F		°C
AGRICULTURAL DUSTS			
Alfalfa Meal	392		200
Cellulose	500		260
Cinnamon	446		230
Cocoa, natural, 19% fat	464		240
Corn	482		250
Corncob Grit	464		240
Corn Dextrine	698		370
Cornstarch, commercial	626		330
Cork	410		210
Cottonseed Meal	392		200
Garlic, dehydrated	680	NL	360
Malt Barley	482		250
Milk, Skimmed	392		200
Potato Starch, Dextrinated	824	NL	440
Rice	428		220
Rice Bran	914	NL	490
Rice Hull	428		220
Safflower Meal	410		210
Soy Flour	374		190
Soy Protein	500		260
Sucrose	662	CI	350
Sugar, Powdered	698	CI	370
Wheat 428			220
Wheat Flour	680		360
Wheat Starch	716	NL	380
Wheat Straw	428		220
Woodbark, Ground	482		250
Wood Flour	500		260
Yeast, Torula	500		260
CARBONACEOUS DUSTS			
Charcoal[2]	356		180
Coal, Kentucky Bituminous[2]	356		180
Coal, Pittsburgh Experimental[2]	338		170
Lignite, California[2]	356		180
Pitch, Coal Tar	1310	NL	710
Pitch, Petroleum	1166	NL	630
CHEMICALS			
Acetoacetanilide	824	M	440
Adipic Acid	1022	M	550
Anthranilic Acid	1076	M	580
Azelaic Acid	1130	M	610
2,2-Azo-bis-butyronitrile	662		350
Benzoic Acid	824		440
Benzotriazole	824	M	440
Bisphenol-A	1058	M	570
Chloroacetoacetanilide	1184	M	640
Diallyl Phthalate	896	M	480
Dihydroacetic Acid	806	NL	430
Dimethyl Isophthalate	1076	M	580
Dimethyl Terephthalate	1058	M	570
3,5-Dinitrobenzoic Acid	860	NL	460
Diphenyl	1166	M	630
Ethyl Hydroxyethyl Cellulose	734	NL	390
Fumaric Acid	968	M	520
Hexamethylene Tetramine	770	S	410
Hydroxyethyl Cellulose	770	NL	410
Isotoic Anhydride	1292	NL	700
Paraphenylene Diamine	1148	M	620
Paratertiary Butyl Benzoic Acid	1040	M	560
Pentaerythritol	752	M	400

TABLE 8-2N. (Continued) Selected Nonconductive Dusts Classified as Group G—Ignition Sensitivity Equal to or Greater than 0.2; Explosion Severity Equal to or Greater than 0.5

Material	Group	AIT °F	AIT °C	Material	Group	AIT °F	AIT °C
Acetaldehyde	C*	347	175	2-Hexanone	D	795	424
Acetone	D*	869	465	Hexenes	D	473	245
Acetonitrile	D	975	524	Hydrogen	B*	752	400
Acetylene	A*	581	305	Hydrogen Cyanide	C*	1000	538
Acrolein (inhibited)[1]	B(C)*	455	235	Hydrogen Selenide	C	—	—
Acrylonitrile	D*	898	481	Hydrogen Sulfide	C*	500	260
Allyl Alcohol	C*	713	378	Isoamyl Acetate	D	680	360
Allyl Chloride	D	905	485	Isoamyl Alcohol	D	662	350
Ammonia	D*[2]	928	498	Isobutyl Acrylate	D	800	427
n-Amyl Acetate	D	680	360	Isobutyraldehyde	C	385	196
sec-Amyl Acetate	D	—	—	Isoprene	D*	743	395
Benzene	D*	1040	560	Isopropyl Acetate	D	860	460
1,3-Butadiene[1]	B(D)*	788	420	Isopropylamine	D	756	402
Butane	D*	550	288	Isopropyl Ether	D*	830	443
1-Butanol	D*	650	343	Isopropyl Glycidyl Ether	C	—	—
2-Butanol	D*	761	405	Liquefied Petroleum Gas	D	761–842	405–450
n-Butyl Acetate	D*	790	421	Manufactured Gas (containing more than 30% H_2 by volume)	B*	—	—
iso-Butyl Acetate	D*	790	421				
sec-Butyl Acetate	D	—	—	Mesityl Oxide	D*	652	344
Butylamine	D	594	312	Methane	D*	999	537
Butylene	D	725	385	Methanol	D*	725	385
Butyl Mercaptan	C	—	—	Methyl Acetate	D	850	454
n-Butyraldehyde	C*	425	218	Methylacetylene	C*	—	—
Carbon Disulfide[3]	—*	194	90	Methylacetylene-Propadiene (stabilized)	C	—	—
Carbon Monoxide	C*	1128	609				
Chlorobenzene	D	1099	593	Methyl Acrylate	D	875	468
Chloroprene	D	—	—	Methylamine	D	806	430
Crotonaldehyde	C*	450	232	Methylcyclohexane	D	482	250
Cyclohexane	D	473	245	Methyl Ether	C*	662	350
Cyclohexene	D	471	244	Methyl Ethyl Ketone	D*	759	404
Cyclopropane	D*	938	503	Methyl Formal	C*	460	238
1,1-Dichloroethane	D	820	438	Methyl Formate	D	840	449
1,2-Dichloroethylene	D	860	460	Methyl Isobutyl Ketone	D*	840	440
1,3-Dichloropropene	D	—	—	Methyl Isocyanate	D	994	534
Dicyclopentadiene	C	937	503	Methyl Mercaptan	C	—	—
Diethyl Ether	C*	320	160	Methyl Methacrylate	D	792	422
Diethylamine	C*	594	312	2-Methyl-1-Propanol	D*	780	416
Di-isobutylene	D*	736	391	2-Methyl-2-Propanol	D*	892	478
Di-isopropylamine	C	600	316	Monomethyl Hydrazine	C	382	194
Dimethylamine	C	752	400	Naphtha (Petroleum)[4]	D*	550	288
1,4-Dioxane	C	356	180	Nitroethane	C	778	414
Di-n-propylamine	C	570	299	Nitromethane	C	785	418
Epichlorohydrin	C*	772	411	Nonane	D	401	205
Ethane	D*	882	472	Nonene	D	—	—
Ethanol	D*	685	363	Octane	D*	403	206
Ethyl Acetate	D*	800	427	Octene	D	446	230
Ethyl Acrylate (inhibited)	D*	702	372	Pentane	D*	470	243
Ethylamine	D*	725	385	1-Pentanol	D*	572	300
Ethyl Benzene	D	810	432	2-Pentanone	D	846	452
Ethyl Chloride	D	966	519	1-Pentene	D	527	275
Ethylene	C*	842	450	Propane	D*	842	450
Ethylenediamine	D*	725	385	1-Propanol	D*	775	413
Ethylene Dichloride	D*	775	413	2-Propanol	D*	750	399
Ethylenimine	C*	608	320	Propionaldehyde	C	405	207
Ethylene Oxide[1]	B(C)*	804	429	n-Propyl Acetate	D	842	450
Ethyl Formate	D	851	455	Propylene	D*	851	455
Ethyl Mercaptan	C*	572	300	Propylene Dichloride	D	1035	557
n-Ethyl Morpholine	C	—	—	Propylene Oxide[1]	B(C)*	840	449
Formaldehyde (Gas)	B	795	429	n-Propyl Ether	C*	419	215
Gasoline	D*	536–880	280–471	Propyl Nitrate	B*	347	175
Heptane	D*	399	204	Pyridine	D*	900	482
Heptene	D	500	260	Styrene	D*	914	490
Hexane	D*	437	225	Tetrahydrofuran	C*	610	321

Since there is no consistent relationship between explosion properties, explosion pressure, maximum experimental safe gap, minimum ignition energy, and ignition temperature, testing requirements for groups and classes are independent of each other.

The operating temperature markings specified in Table 8-2O are not to exceed the ignition temperature of the specific gas or vapor to be encountered.

For information regarding ignition temperatures of gases and vapors, see Table 8-2M, NFPA 497M and NFPA 325M, *Fire-Hazard Properties of Flammable Liquids, Gases, and Volatile Solids.*

Complete definitions of the several classes and divisions of hazardous (classified) locations and the methods of wiring and types of electrical equipment to be used in each are covered in detail in the NEC. Rules applying specifically to commercial garages, aircraft hangars, gasoline dispensing facilities and service stations, bulk storage plants, finishing processes, and areas in health care facilities containing flammable anesthetics are also covered in the NEC.

TABLE 8-2O. Temperature Identification Numbers

Maximum Temperature		Identification Number
Degrees F	Degrees C	
842	450	T1
572	300	T2
536	280	T2A
500	260	T2B
446	230	T2C
419	215	T2D
392	200	T3
356	180	T3A
329	165	T3B
320	160	T3C
275	135	T4
248	120	T4A
212	100	T5
185	85	T6

usually limited to low energy control, signal, and instrumentation systems.

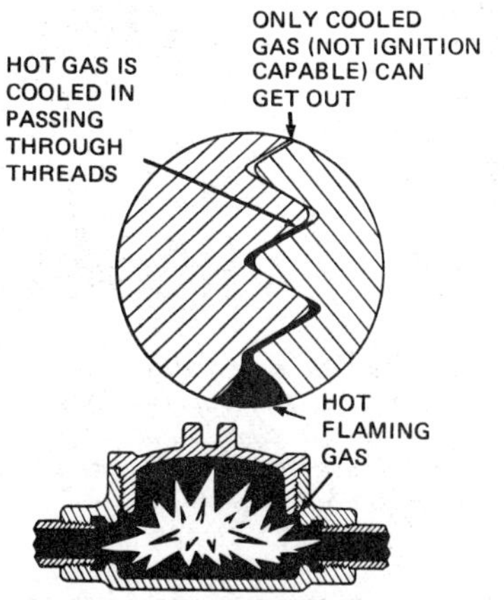

FIG. 8-2BB. Explanation of the principle of "explosionproof" equipment, indicating containment of hot gases within the enclosure. (Crouse-Hinds Company)

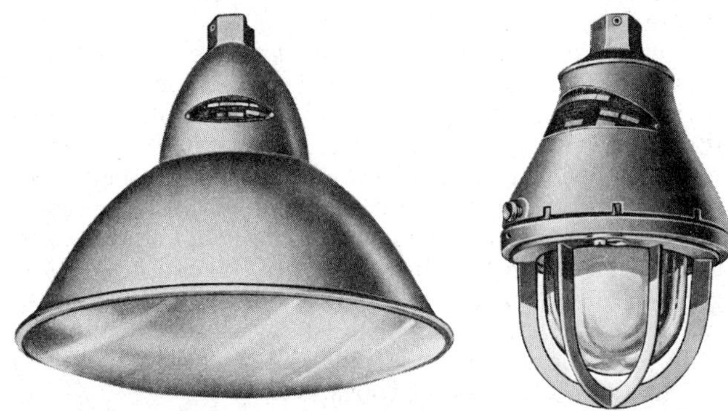

FIG. 8-2CC. Representative lighting fixtures for use in various hazardous locations. (Left) A lighting fixture for use in Class I hazardous locations (60-300 W). (Right) A dust-ignitionproof light fixture for use in Class II hazardous locations. (Crouse-Hinds Company)

Class I, Division 1: Locations in which: (1) ignitible concentrations of flammable gases or vapors can exist under normal operating conditions, or (2) ignitible concentrations of such gases or vapors may exist frequently because of repair or maintenance operations or leakage, or (3) breakdown or faulty operation of equipment or processes might release ignitible concentrations of gases or vapors, and might also cause simultaneous electrical equipment failure. Motors and other rotating electrical machinery, lighting fixtures, most switches, circuit breakers, and similar electrical equipment in these locations must be of the explosionproof type or purged and pressurized type approved for Class I locations of the proper group. (See NFPA 496, *Standard for Purged and Pressurized Enclosures for Electrical Equipment.*) (See Figs. 8-2BB, 8-2CC, and 8-2DD.) Approved intrinsically safe equipment and circuits may also be used. (See NFPA 493, *Standard for Intrinsically Safe Apparatus and Associated Apparatus for Use in Class I, II, III, Division 1 Hazardous Locations,* hereinafter referred to as NFPA 493.) These are

Class I, Division 2: Locations: (1) in which volatile flammable liquids or flammable gases are handled, processed, or used, but in which the liquids, vapors, or gases will normally be confined within closed containers or closed systems from which they can escape only in case of accidental rupture or breakdown of such containers or systems, or in case of abnormal operation of equipment, or (2) in which ignitible concentrations of gases or vapors are normally prevented by positive mechanical ventilation, and which might become hazardous through failure or abnormal operation of the ventilating equipment, or (3) that are adjacent to Class I, Division 1 locations, and to which ignitible concentrations of gases or vapors might occasionally be communicated unless such communication is prevented by adequate, positive pressure ventilation from a clean air source, and effective safeguards against ventilation failure are provided. In general, ordinary types of motors that do not have brushes, switching mechanisms, etc., may be installed in these locations, but motors that do have sliding contacts, switches, etc., must

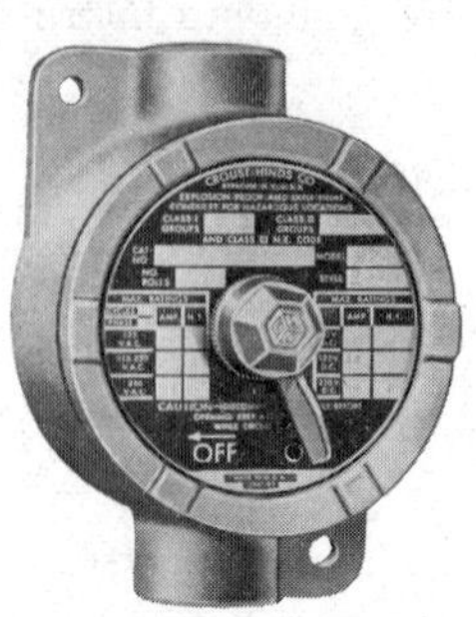

FIG. 8-2DD. Representative switching and panelboard equipment for use in various hazardous locations. (Left) An example of a panelboard for use in Class I hazardous locations. (Right) Tumbler switch for use in Class I and Class II hazardous locations. (Crouse-Hinds Company)

have the switches in explosionproof or other type enclosure approved for Class I, Division 2 locations. Lamps that operate at temperatures above the ignition temperature of the gas or vapor involved must have approved explosionproof or purged and pressurized enclosures. Units located where falling sparks or hot metal from broken lamps might ignite localized concentrations of flammable gases below the Division 2 location must also have suitable enclosures. Switches, circuit breakers, controllers, and other devices that create arcs or sparks and that are intended to interrupt ignition capable energy in normal operation must either be explosionproof, have their contacts in hermetically sealed chambers, or have contacts immersed in oil. Enclosures for electrical equipment that does not interrupt ignition-capable energy or otherwise represent an ignition source under normal conditions may be of the general purpose type. Portable lamps must be of the type satisfactory for use in a Class I, Division 1 location.

Class II, Division 1: Locations: (1) in which combustible dust is in the air under normal operating conditions in sufficient quantities to produce explosive or ignitible mixtures, or (2) where mechanical failure or abnormal operation of machinery or equipment might cause such explosive or ignitible mixtures to be produced, and might also provide a source of ignition through simultaneous failure of electric equipment, operation of protective devices, or from other causes, or (3) in which combustible dusts of an electrically conductive nature may be present in hazardous quantities. Motors must be of an enclosed type approved as dust-ignitionproof for the proper Class II locations. Lighting fixtures, switches, circuit breakers, controllers, and fuses that are intended to interrupt current in normal operation must be provided with dust-ignitionproof enclosures approved for the proper Class II locations. Maximum surface temperatures on equipment in Class II locations under actual operating conditions must not exceed the ignition temperature of the dust. See Table 8-2N for information regarding ignition temperatures of Group G dusts. For additional information, see NFPA 497M. Prior to the 1984 NEC, the surface temperature limits for equipment in Class II, Group G locations were 329°F (165°C) for fixed lighting and similar equipment not subject to overloading, and 248°F (120°C) for equipment such as motors, power transformers, etc., which may be overloaded. Corresponding temperatures for Group F were 329°F (200°C) and 302°F (150°C) and for Group E, 392°F (200°C) for both types of equipment.

Class II, Division 2: Locations: (1) in which combustible dust is not normally in the air in quantities sufficient to produce explosive or ignitible mixtures, and (2) where dust accumulations are normally insufficient to interfere with normal operation of electrical equipment or other apparatus, but combustible dust may be in suspension in the air as a result of infrequent malfunctioning of handling or processing equipment, and (3) where combustible dust accumulations on, in, or near the electrical equipment may be sufficient to interfere with the safe dissipation of heat from the electrical equipment or may be ignitible by abnormal operation or failure of electrical equipment. In general, totally enclosed motors are suitable. Under certain conditions, standard, open type motors without sliding contacts, switches, etc., may be used in many of these locations. Fixed lamps and lampholders must have enclosures designed to minimize the deposit of dust and prevent the escape of sparks, burning material, or hot metal. Switches, circuit breakers, controllers, and fuses should be provided with dusttight enclosures. Division 1 temperature limits also apply to Division 2 locations.

Class III, Division 1: Locations in which easily ignitible fibers or materials producing combustible flyings are handled, manufactured, or used. In general, motors in these locations must be of the enclosed type except that where only moderate accumulations of lint are present and cleaning and maintenance are satisfactory, self-cleaning, squirrel cage, textile motors or standard, open type motors without sliding contacts, switches, etc., may be installed. Lamps and lampholders, switches, circuit breakers, controllers, and fuses must be totally enclosed. Maximum surface temperatures under actual operating conditions must not exceed 329°F (165°C) for equipment that is not subject to overloading and 248°F (120°C) for equipment such as motors, power transformers, etc., which may be overloaded.

Class III, Division 2: Locations in which easily ignitible fibers are stored or handled (except in manufacture). Motors must be of the enclosed type. Lamps, lampholders, switches, circuit breakers, controllers, and fuses must have enclosures similar to those specified for Class III, Division 1 locations. Division 1 temperature limits apply to Division 2 locations.

Equipment for Use in Hazardous Locations

Equipment for use in Class I, Division 1 hazardous (classified) locations as defined in the NEC is sometimes referred to as "explosionproof" equipment. The two basic design criteria for explosionproof apparatus for Class I locations are first, that it withstand internal explosions of flammable gas or vapor-air mixtures and second, that it prevent propagation of the internal explosion to the surrounding flammable atmosphere. In other words, it is recognized that surrounding flammable gas or vapor-air mixtures will enter the enclosure of this equipment and the possibility exists of their ignition within the enclosure. To prevent the propagation of flame to the outside surrounding atmosphere, which may likewise contain flammable vapor-air mixtures, the enclosures of this equipment must: (1) arrest flame at joints or other openings to the outside, (2) be of sufficient strength to resist (without rupture or serious distortion) the internal pressure, and (3) the external surface temperature of the enclosure must not be high enough to ignite the surrounding gas or vapor. The various gas or vapor-air mixtures vary considerably with respect to: (1) the propagation of flames through joints of such assemblies, (2) the pressure developed within the enclosure following ignition, and (3) the ignition temperature of the gas or vapor-air mixture. Thus, equipment is tested and listed for use in the various Groups indicated in Table 8-2M should be used.

Equipment for use in Class I hazardous (classified) locations must also be designed to operate under full load and, in the case of equipment such as motors likely to be overloaded, must operate under overload conditions without developing surface temperatures above the ignition temperature of the flammable gas or vapor with which they are intended to be used.

As an alternative to "explosionproof" motors and generator equipment approved for use in Class I locations, it is permissible to use either of the following two designs:

1. Totally enclosed equipment supplied with positive pressure ventilation from a source of clean air with discharge to a safe area. Controls are arranged to prevent energizing the machine until ventilation has been established and the enclosure purged with at least ten volumes of air, and to automatically de-energize the equipment when the air supply fails.
2. Totally enclosed inert gas-filled equipment supplied with a suitably reliable source of inert gas for pressurizing the enclosure. Devices are provided to assure a positive pressure in the enclosure as well as to automatically de-energize the equipment when the gas supply fails.

When either of these two designs are used, no external surface of the motors, generators, or other rotating electrical machinery must have an operating temperature (in degrees Celsius) in excess of 80 percent of the ignition temperature of the gas or vapor involved, as determined by ASTM test procedure D2155-69. Appropriate devices (heat sensors) must also be provided to detect any increase in temperature of the equipment beyond its design limits and then to automatically de-energize the equipment. The detectors and control equipment used are to be suitable for the atmosphere involved without positive pressure; in other words, they must be explosionproof or intrinsically safe if the location is Class I, Division 1.

NFPA 496, *Standard for Purged and Pressurized Enclosures for Electrical Equipment in Hazardous (Classified) Locations*, provides information for the design of purged enclosures for the purpose of eliminating or reducing within the enclosure a Class I hazardous (classified) location (gases or vapors in air in quantities sufficient to produce explosive or ignitible mixtures). Protective measures include supplying an enclosure with clean air or an inert gas at sufficient flow and positive pressure to achieve and maintain an acceptable safe level of the atmosphere.

Another alternative is the use of "intrinsically safe" equipment and wiring which is defined as equipment and wiring incapable of releasing sufficient electrical energy under normal or abnormal conditions to cause ignition of a specific hazardous atmospheric mixture. NFPA 493, *Standard for Intrinsically Safe Apparatus and Associated Apparatus for Use in Class I, II, and III, Division I Hazardous Locations*, is the standard covering this subject.

In many cases, the amount of "explosionproof," purged and pressurized, or intrinsically safe equipment required can be reduced through the exercise of ingenuity in the layout of electrical installations by locating much of the equipment in nonhazardous areas. The extent of the hazardous areas is normally defined by the codes and standards relating to the storage and handling of the specific flammable liquids, gases, or solids.

Equipment for use in Class II and Class III hazardous (classified) locations presents a somewhat different problem because the equipment is designed to be dust-ignitionproof for Class II, Division 1 and for some Class II, Division 2 locations, and to be dust tight or totally enclosed with telescoping covers for some Class II, Division 2 locations and for Class III locations. It is thus not intended to resist internal explosions of dust-air mixtures. Such equipment is tested in specific dust-air mixtures to determine that the enclosures are dust-ignitionproof for Class II locations and that overheating does not occur when the device is blanketed with dust or lint and flyings.

Garages, Commercial (Repair and Storage)

The NEC requirements apply to locations used for the servicing and repair of passenger automobiles, buses, tractors, trucks, etc., in which flammable liquids or flammable gases are used. NFPA 88A, *Standard for Parking Structures* and NFPA 88B, *Standard for Repair Garages*, should also be consulted in connection with further classifications of, and the fire protection recommendations applying to, garages. The specific hazardous (classified) areas in these garages are defined in the NEC.

Aircraft Hangars

Where aircraft containing gasoline, jet fuels, or other flammable liquids are stored or serviced, the specific (classified) hazardous areas are as defined in the NEC. Reference should also be made to NFPA 409, *Standard on Aircraft Hangars*, for guidance on the construction and protection of these structures.

Service Stations

This occupancy group includes locations where gasoline or other volatile flammable liquids or liquefied petroleum gases are transferred to the fuel tanks or auxiliary fuel tanks of self-propelled vehicles. Reference should be made to NFPA 30, *Flammable and Combustible Liquids Code* and NFPA 30A, *Automotive and Marine Service Station Code*, as well as the NEC for information on the hazardous (classified) areas and other guidance on firesafety in service stations.

Bulk Storage Plants (Flammable Liquid)

Where gasoline or other volatile flammable liquids are stored in tanks having an aggregate capacity of one carload or more and from which such products are distributed, the hazardous (classified) areas are defined in NFPA 30 and the NEC. NFPA 30 also contains many requirements for the safe construction and utilization of bulk storage plants.

Paint Spray Booths and Areas

The NEC hazardous (classified) locations requirements apply to areas where paints, lacquers, or other flammable finishes are regularly or frequently applied by spraying or dipping, or by other means and where volatile flammable solvents or thinners are used, or where readily ignitible deposits or residues from such paints, lacquers, or finishes may occur. Further information regarding safeguards for finishing processes are contained in NFPA 33, *Standard for Spray Applications Using Flammable and Combustible Materials*, and NFPA 34, *Standard for Dipping and Coating Processes Using Flammable and Combustible Liquids*. Hazardous (classified) areas with respect to flammable vapors are defined in the NEC.

Chemical Plants

In chemical plants where flammable liquids or gases are processed or handled, NFPA 497, *Recommended Practice for Classification of Class I Hazardous Locations or Electrical Installations in Chemical Plants*, provides information on the extent of the hazardous (classified) location.

Flammable Anesthetics

Special rules apply in hospital operating rooms and other locations where flammable anesthetics are or may be administered to patients. The NEC defines the anesthetizing (hazardous) area as any area in which any flammable inhalation agent will be administered. The areas include operating rooms, delivery rooms, emergency rooms, and anesthetizing rooms, and other areas when used for induction of anesthesia with flammable anesthetizing agents. In a flammable anesthetizing location the entire area is considered to be a Class I, Division 1 location extending upward to a level 5 ft (1.5 m) above the floor. For further information on the subject of flammable anesthetics, reference should be made to Chapter 3 of NFPA 99.

Places of Assembly, Theaters, and Similar Locations

In places of assembly, theaters, and similar locations where numbers of people congregate, it is important that the electrical equipment be properly designed and installed. The general rules of the NEC as well as its special requirements for these occupancies should be followed in installing the equipment. NFPA 101, *Life Safety Code®*, should be consulted for further information on use of emergency lighting.

Bibliography

References Cited

UL. 1939. *Electric Shock as it Pertains to the Electric Fence.* Research Bulletin No. 14. Underwriters Laboratories Inc., Northbrook, IL.

NFPA Codes, Standards, Recommended Practices and Manuals. (See the latest *NFPA Codes and Standards Catalog* for availability of current editions of the following documents.)

NFPA 12, *Standard on Carbon Dioxide Extinguishing Systems.*
NFPA 30, *Flammable and Combustible Liquids Code.*
NFPA 30A, *Automotive and Marine Service Station Code.*
NFPA 33, *Standard for Spray Application Using Flammable and Combustible Materials.*
NFPA 34, *Standard for Dipping and Coating Processes Using Flammable or Combustible Liquids.*
NFPA 70, *National Electrical Code.*
NFPA 70B, *Recommended Practice for Electrical Equipment Maintenance.*
NFPA 70E, *Standard for Electrical Safety Requirements for Employee Workplaces.*
NFPA 70L, *Model State Law Providing for Inspection of Electrical Installations.*
NFPA 71, *Standard for the Installation, Maintenance, and Use of Central Station Signaling Systems.*
NFPA 72A, *Standard for the Installation, Maintenance and Use of Local Protective Signaling Systems for Guard's Tour, Fire Alarm and Supervisory Service.*
NFPA 72B, *Standard for the Installation, Maintenance and Use of Auxiliary Protective Signaling Systems for Fire Alarm Service.*
NFPA 72C, *Standard for the Installation, Maintenance and Use of Remote Station Protective Signaling Systems.*
NFPA 72D, *Standard for the Installation, Maintenance and Use of Proprietary Protective Signaling Systems.*
NFPA 72E, *Standard on Automatic Fire Detectors.*
NFPA 72H, *Guide for Testing Procedures for Local, Auxiliary, Remote Station and Proprietary Protective Signaling Systems.*
NFPA 74, *Standard for the Installation, Maintenance, and Use of Household Fire Warning Equipment.*
NFPA 75, *Standard for the Protection of Electronic Computer/Data Processing Equipment.*
NFPA 77, *Recommended Practice on Static Electricity.*
NFPA 78, *Lightning Protection Code.*
NFPA 79, *Electrical Standard for Industrial Machinery.*
NFPA 88A, *Standard for Parking Structures.*
NFPA 88B, *Standard for Repair Garages.*
NFPA 90B, *Standard for the Installation of Warm Air Heating and Air Conditioning Systems.*
NFPA 99, *Standard for Health Care Facilities.*
NFPA 101, *Code for Safety to Life from Fire in Buildings and Structures.*
NFPA 110, *Standard for Emergency and Standby Power Systems.*
NFPA 325M, *Fire Hazard Properties of Flammable Liquids, Gases, and Volatile Solids.*
NFPA 409, *Standard on Aircraft Hangars.*
NFPA 493, *Standard for Intrinsically Safe Apparatus and Associated Apparatus for Use in Class I, II, III, Division 1 Hazardous Locations.*
NFPA 496, *Standard for Purged and Pressurized Enclosures for Electrical Equipment in Hazardous (Classified) Locations.*
NFPA 497, *Recommended Practice for Classification of Class I Hazardous Locations for Electrical Installations in Chemical Plants.*
NFPA 497M, *Manual for Classification of Gases, Vapors and Dusts for Electrical Equipment in Hazardous (Classified) Locations.*
NFPA 501A, *Standard for Firesafety Criteria for Mobile Homes Installations, Sites and Communities.*
NFPA 501C, *Standard on Firesafety Criteria for Recreational Vehicles.*
NFPA 501D, *Standard on Firesafety Criteria for Recreational Vehicle Parks and Campgrounds.*
NFPA 513, *Standard for Motor Freight Terminals.*

Additional Readings

Alvares, Norman J., et al., "Thermal Degradation of Cable and Wire Insulations," in ASTM, "Behavior of Polymeric Materials in Fire," *ASTM STP 816*, (E. L. Schaffer, ed.), American Society for Testing and Materials, Philadelphia, PA, 1982.

An Illustrated Guide to Electrical Safety, U.S. Department of Commerce, Occupational, Safety and Health Administration, Washington, DC, 1983.

Beland, Bernard, "Electricity as a Cause of Fires," *SFPE TR 84-5*, Society of Fire Protection Engineers, Boston, MA, 1984

Crouse-Hinds *ECM Code Digest 1984*, Crouse-Hinds Electrical Construction Materials, Division of Cooper Industries, 1984.

Electrical Appliance and Utilization Equipment Directory, Underwriters Laboratories Incorporated, Northbrook, IL.

Electrical Construction on Materials Directory, Underwriters Laboratories Incorporated, Northbrook, IL.

Gomberg, Alan, and Hall, John R., "Analysis of Electrical Fire Investigations in Ten Cities," *NBSIR 83-2677*, National Bureau of Standards, Gaithersburg, MD, Mar. 1983.

Harvey, Clifford S., "Fire in a Code-Complying Electrical Installation," *Fire Journal*, Vol. 73, No. 2, Mar. 1979, pp. 73-78, 90.

Hazardous Location Equipment Directory, Underwriters Laboratories Incorporated, Northbrook, IL.

Magison, E. C., *Electrical Instruments in Hazardous Locations*, 3rd ed., Instrument Society of America, Pittsburgh, PA, 1978.

National Electrical Safety Code, Institute of Electrical and Electronics Engineers, NY, 1984.

Schram, Peter J., ed., *National Electrical Code Handbook*, 3rd ed., National Fire Protection Association, Quincy, MA, 1984.

"Standard for Television Receivers and Video Products," *UL 1410*, Underwriters Laboratories Incorporated, Northbrook, IL.

"Tests for Flammability of Plastic Materials for Parts in Devices and Appliances," *UL 94*, Underwriters Laboratories Incorporated, Northbrook, IL.

Whittington, B. W., "Electrical Installations in Industrial Locations," *Industrial Fire Hazards Handbook*, 2nd ed., National Fire Protection Association, Quincy, MA, 1984, pp. 863-883.

HEATING SYSTEMS AND APPLIANCES

Revised by J. Herbert Witte and Richard L. Stone

Heat producing appliances and associated equipment are among the most prevalent causes of fire because they operate at temperatures above the ignition temperature of many common materials. In addition, combustion type appliances may involve the hazards of an accumulated combustible mixture, the discharge of unburned fuel, and possible exposure of fuel to ignition sources.

Additional information about heating systems and appliances will be found in Section 4, Chapter 1, "The Chemistry and Physics of Fire;" in Section 5, Chapter 5, "Gases;" in Section 8, Chapter 2, "Electrical Systems and Appliances;" and Chapter 4, "Air Conditioning and Ventilating Systems;" in Section 10, Chapter 1, "Boiler Furnaces;" and Chapter 5, "Industrial and Commercial Heat Utilization Equipment;" in Section 11, Chapter 5, "Storage and Handling of Flammable and Combustible Liquids;" and Chapter 6, "Storage and Handling of Gases."

FUELS AND METHODS OF FIRING

Combustion may be defined as a chemical reaction between a fuel and oxygen with evolution of heat and light. In a fuel-burning, heat-producing device this reaction needs to be continuous so a balance will be established between the rate at which oxygen or air and fuel is supplied and the rate at which heat and the products of combustion are removed. Complete combustion takes place when all of the fuel is oxidized by the air supplied to it. Air left over after complete oxidation is called excess air. If only enough air is supplied for complete combustion, perfect combustion results under suitable conditions. This is impractical in most heat producing appliances; thus the appliances operate with some excess air.

Air for combustion is supplied in two ways: (1) primary air is introduced through or with the fuels, and (2) secondary air is supplied to the combustion zone. In some cases, all of the air supplied is primary air, while in others only part is primary air and the balance is secondary air.

Mr. Witte, formerly of Underwriters Laboratories, Northbrook, IL, is a consultant on heating and air conditioning systems. Mr. Stone is Director of Research and Engineering for Metalbestos Systems, a division of Wallace Murray Corporation.

Incomplete combustion in a fuel burning device can produce hazardous carbon monoxide. This condition indicates poor efficiency since the fuel is not completely oxidized, and its total heating value is not obtained. Incomplete combustion usually results from inadequate air supply, insufficient mixing of air and gases, or a temperature too low to produce or sustain combustion.

Solid Fuel—Coal

Coal is one of the principal solid fuels used in heat producing appliances. There are a number of types of coal, each having widely different characteristics. In many cases, there is no distinct line of demarcation between them and the qualities of one type can overlap those of another. Pulverized coal systems, however, require the use of coals having characteristics within the specific range of the coal handling and coal burning equipment.

The seven principal types of coal used in heat producing appliances are:

Anthracite: A clean, dense, hard coal. It burns with a minimum of smoke and has a minimum dust hazard during handling.

Semianthracite: A coal having a higher volatile content than anthracite but not as hard.

Bituminous: This includes many types of coal with different properties depending upon where they are mined. It gives off considerable smoke and soot if improperly fired. Bituminous coal may be subject to spontaneous ignition under some storage conditions and has a dust hazard during handling.

Semibituminous: A dusty soft coal that tends to break up. It may be subject to spontaneous ignition under some storage conditions. It normally produces less smoke than bituminous coal.

Subbituminous: A coal subject to spontaneous ignition under some storage conditions. It burns with very little smoke and soot.

Lignite: A coal having a woody structure and normally a high moisture content. It is subject to spontaneous ignition under some storage conditions. It burns with little smoke and soot.

Coke: A product of the destructive distillation of coal. The type of coke produced is determined by the coal used and the temperatures and time of distillation. Petroleum coke is produced from the destructive distillation of oil.

Combustion of Coal

When coal is burned in a firebox, oxygen in the air passing through the grate (primary air) unites with the carbon in the lower portion of the fuel bed, called the oxidation zone, to form carbon dioxide. Some of this carbon dioxide is then reduced to carbon monoxide in the upper part of the fuel bed, called the reduction zone. Gases liberated from the fuel bed are carbon monoxide, carbon dioxide, nitrogen, and some oxygen. Oxygen from the air admitted over the fuel bed (secondary air) combines with some of the carbon monoxide to form carbon dioxide. When fresh coal is applied to the fire, moisture is driven off as steam and the hydrocarbon gases are distilled, combined with oxygen, and burned above the grate in what is called the distillation zone.

Methods of Firing Coal

Coal is fired either manually or automatically by stokers or pulverized coal burners.

Hand Firing: There are many acceptable hand firing methods. With highly volatile coals, the objective is to leave a suitable bed of glowing fuel to ignite the gases as they are driven off from the fresh coal charge; otherwise the gas may accumulate and explode. Draft regulation is the only automatic feature of a hand fired furnace. Draft can be controlled by a room thermostat, steam pressure, or water temperature. But, as with any hand fired solid fuel, the fire is continuous and overtemperature conditions are difficult to control unless the fire is continually attended.

Mechanical Stokers: These are classified according to their coal burning capacities, and range from those that handle 10 lb (4.5 kg) per hr to those handling over 1,200 lb (540 kg) per hr. A stoker feeds fuel to the combustion chamber and usually supplies air for combustion under automatic control. There are four basic types of stokers: (1) underfed, (2) overfed, (3) traveling and chain grate, and (4) spreader, with fired grate or continuous ash discharge.

Underfed stokers move the coal by screw conveyor or rams. (See Fig. 8-3A.) Overfed stokers feed the coal pneumatically or by rotors. Spreader stokers similarly feed coal by rotors or paddles and discharge onto stationary or traveling grates. Stokers may feed from hoppers or directly from bins. They are equipped with fueling controls that either change the firing rate or stop the fuel altogether. Some stokers are equipped with a control that will stop the feeding of fuel if the fire goes out.

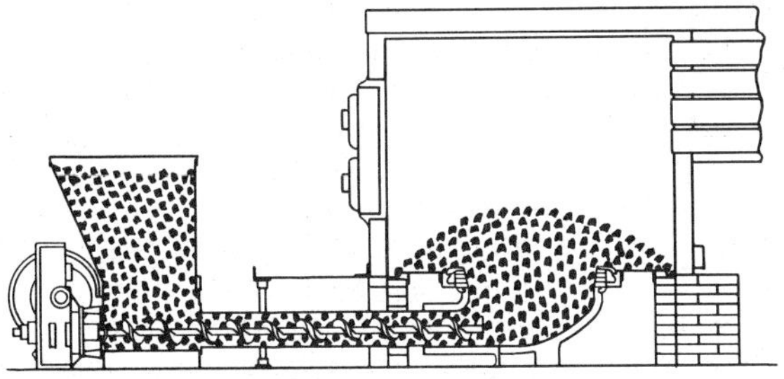

FIG. 8-3A. One type of underfed coal stoker.

Pulverized Coal Systems: These are of two general types: (1) the direct fired, or unit, system and (2) the storage, or bin, system. In the direct fired system, raw coal is fed directly to a pulverizer where it is pulverized, mixed with air, and blown to the burners. In the storage system, coal is delivered to a raw fuel bin and then passed to a pulverizer where it is reduced to a powder and dried for storage in a pulverized fuel bin before delivery to the furnace.

Solid Fuels—Miscellaneous

Miscellaneous solid fuels include: wood (logs, scrap lumber, wood waste from lumber and paper mills (sawdust, shavings, bark), hogged fuel (sawmill refuse run through a disintegrator or "hog" to form uniform chips or shreds), charcoal, briquets, peat, bagasse, sugar cane and a host of other combustibles such as tanbark, wet bark, straw, city refuse, paper, etc. The methods of firing these fuels vary considerably with the type of device and the nature of the particular fuel.

Combustion of wood takes place in two stages. The first is destructive distillation such as that which occurs in a charcoal kiln, where the heat drives off combustible gases from the wood and leaves charcoal. The second stage involves burning of the gases in the air above the wood and burning of the charcoal combined with oxygen, forming an intensely hot and luminous bed of coals. The heat from both burning gases and charcoal distills more volatiles from the wood and increases the temperature and rate of combustion. The acceleration of combustion is limited only by the rate at which air can be brought into contact with the burning wood.

Liquid Fuel—Fuel Oils

This section discusses fuel oil used in heat producing appliances and equipment.

Fuel oil, like other petroleum products, is composed of varying compounds of hydrocarbons. Even the same grade of fuel oil will vary as to its chemical composition, depending upon such factors as the type of crude oil used and the refining process employed.

Types of Fuel Oil: Fuel oils obtained from present day refining processes are known as "cracked" oils, and practically all petroleum products on the market today are obtained by the cracking process. No. 1 and No. 2 fuel oil, as well as those fuels commonly known as kerosene, range oil, furnace oil, star oil, and diesel oil, may still be broadly classed as distillates, and No. 4, No. 5, and No. 6 oil (as well as Bunker C) as residuals.

Oil burners are constructed (and subsequently tested and listed by testing laboratories) for a specific grade or grades of fuel oil. For example, burners listed by Underwriters Laboratories Inc. (UL) have the grade or grades of fuel oil which may be used in each inscribed on the UL listing mark applied to the unit. (See Figs. 8-3B and 8-3C.) It is important that only the proper grade or grades of oil be used in each burner for safety as well as efficiency. Specifications for fuel oil are outlined by the American Society for Testing and Materials in their *Specifications for Fuel Oils*, ASTM D-396, and in Canadian Government Specification 3-GP-28. (See Table 8-3A.)

FIG. 8-3B. Listing mark used on an oil burner listed by Underwriters Laboratories Inc. Note that the grade of oil satisfactory for use in the device appears at the bottom of the marker.

FIG. 8-3C. Listing mark used on an oil fired boiler assembly listed by Underwriters Laboratories Inc. Note the grade of oil satisfactory for use in the device appears at the bottom of the marker.

Methods of Firing Fuel Oil

Two methods, vaporization and atomization, are used to prepare fuel oil for combustion. Air for combustion is supplied by natural or mechanical draft. Ignition is by an electric ignition system, a gas pilot, an oil pilot, or manual means. Operation may be continuous, modulating with high-low flame, or intermittent. While most burners operate from automatic temperature or pressure sensing controls, some simpler types are operated manually.

Oil burners may be classified in several different ways, i.e., by application, by type of vaporizer or atomizer, by firing rates, etc. They are divided into two major groups, residential and commercial-industrial. Vaporizing burners and atomizing burners having capacities of not more than 7 gph (gallons per hour) (25 L/hr) are considered residential types and are intended to be used with oil fuels not heavier than No. 2. The major portion of residential type burners produced are the pressure atomizing type, commonly referred to as the gun type.

Vaporizing Burners: These burners include the sleeve type, the pot type, and the vertical rotary wall flame type. Sleeve type burners (Fig. 8-3D) are used in residential heating and cooking stoves. Natural draft pot burners (Fig. 8-3E) have applications similar to sleeve type burners, while forced draft pot burners may be used in central heating furnaces. Vertical rotary wall flame vaporizing burners are used in residential boilers and furnaces for central heating. While the gun type burner is the most common, many of the burners listed above are still in service.

Atomizing: These burners include high and low pressure types, horizontal rotary cup types, and air and steam atomizing types, the latter used primarily for commercial and industrial applications. High pressure, gun type burners consist of a motor, oil pump (with integral or separate pressure regulating and shutoff valve), strainer, fan, ignition transformer, nozzle, and electrode assembly. (See Fig. 8-3F.) The motor driven pump draws oil from the supply tank and delivers it to the nozzle at pressures of from 100 to 300 psi (690 to 2068 kPa). The nozzle atomizes the oil into fine particles and swirls it into the combustion chamber as a cone shaped spray where it mixes with air and is ignited. They are generally designed to burn No. 2 oil although some of the larger sizes are made for No. 4 oil.

The industrial version of the high pressure burner, known as the mechanical atomizing burner, is a high capacity burner for use with large boilers and industrial furnaces. Figure 8-3G shows a typical piping arrangement for one or more burners supplied by separate pump sets.

The low pressure gun burner is similar to but differs from the high pressure gun burner in two ways. First, the burner includes an air pump to supply compressed air for atomization, and second, the oil and air are delivered to the nozzle at pressures of 15 psi (100 kPa) or less. The size ranges, ignition, and fuel used are as described for the high pressure gun burner.

The horizontal rotary cup oil burner atomizes the oil by spinning it in a thin film from a horizontal rotating cup and injecting high velocity primary air into the oil film through an annular nozzle which surrounds the rim of the cup. Secondary air for combustion is supplied from a separate fan which forces air through the burner wind box. The introduction of secondary air by means of natural draft is not recommended. These burners are used for firing boilers and furnaces and may be used singly or as multiple units on a single appliance. No. 2 fuel may be used with some of the smaller sizes but Nos. 4, 5, or 6 is generally used with the larger sizes. (See Fig. 8-3H.)

In the steam atomizing burner, atomization is accomplished by the impact and expansion of steam. Oil and steam flow in separate channels through the burner gun to the nozzle where the steam and oil mix before being discharged through an orifice into the combustion space. This type of burner is used mainly on large boilers generating steam at 100 psi (690 kPa) and higher and having capacities above 12,000,000 Btu per hr (3.5 MW) input.

Air atomizing burners are designed to use compressed air at either high or low pressure for atomization. The high pressure burner is nearly identical with the steam atomizing burner in design and application. Some burners operate well with either medium.

The low pressure air atomizing burner uses comparatively large volumes of air at low pressure of 5 psi (35 kPa) or less. Air from the blower passes through the burner body and is discharged through annular slots between the body and the nozzle tip, where it meets the film of oil at an angle as it issues from the tip. The impact of the air stream upon the oil film produces a fine mist which is projected into the combustion space. These burners may be fired with light or heavy oil and are most often used to fire industrial heating and processing furnaces. (See Fig. 8-3I.)

Combination oil and gas burners which combine the features of certain oil burners with provisions for burning gas have been developed. These permit the operator to choose either fuel as circumstances may dictate.

Table 8-3A. Detailed Requirements for Fuel Oils*††

Grade of Fuel Oil	Flash Point, deg F (deg C)	Pour Point, deg F (deg C)	Water and Sediment, percent by volume	Carbon Residue on 10 percent Bottoms, percent	Ash, per-cent by weight	Distillation Temperatures, deg F (deg C): 10 percent Point	Distillation Temperatures, deg F (deg C): 90 percent Point		Saybolt Viscosity, sec: Universal at 100°F (38°C)		Saybolt Viscosity, sec: Furol at 122°F (50°C)		Kinematic Viscosity, centistokes: At 100°F (38°C)		Kinematic Viscosity, centistokes: At 122°F (50°C)		Grav-ity, deg. API	Copper Strip Corro-sion	Sul-fur, per-cent
	Min	Max	Max	Max	Max	Max	Min	Max	Min	Max	Min	Max	Min	Max	Min	Max	Min	Max	Max
No. 1: A distillate oil intended for vaporizing pot-type burners and other burners requiring this grade of fuel	100 or legal (38)	0§	trace	0.15	—	420 (215)	—	550 (288)	—	—	—	—	1.4	2.2	—	—	35	No. 3	0.5 or legal
No. 2: A distillate oil for general purpose domestic heating for use in burners not requiring No. 1 fuel oil	100 or legal (38)	20§ (—7)	0.05	0.35	—	—	540§ (282)	640 (338)	(32.6)#	(37.93)	—	—	2.0§	3.6	—	—	30	—	0.5§ or legal
No. 4: Preheating not usually required for handling or burning	130 or legal (55)	20 (—7)	0.50	—	0.10	—	—	—	45	125	—	—	(5.8)	(26.4)	—	—	—	—	‖
No. 5 (Light): Preheating may be required depending on climate and equipment	130 or legal (55)	—	1.00	—	0.10	—	—	—	150	300	—	—	(32)	(65)	—	—	—	—	‖
No. 5 (Heavy): Preheating may be required for burning and, in cold climates, may be required for handling	130 or legal (55)	—	1.00	—	0.10	—	—	—	350	750	(23)	(40)	(75)	(162)	(42)	(81)	—	—	‖
No. 6: Preheating required for burning and handling	150 (65)	—	2.00**	—	—	—	—	—	(900)	(9000)	45	300	—	—	(92)	(638)	—	—	‖

* It is the intent of these classifications that failure to meet any requirement of a given grade does not automatically place an oil in the next lower grade unless in fact it meets all requirements of the lower grade.

† Outside U.S. the sulfur limit for No. 2 shall be 1.0 percent.

‡ Legal requirements to be met.

§ Lower or higher pour points may be specified whenever required by conditions of storage or use. When pour point less than 0°F is specified, the minimum viscosity shall be 1.8 cSt (32.0 sec, Saybolt Universal) and the minimum 90 percent point shall be waived.

‖ The 10 percent distillation temperature point may be specified at 440°F (226°C) maximum for use in other than atomizing burners.

Viscosity values in parentheses are for information only and not necessarily limiting.

** The amount of water by distillation plus the sediment by extraction shall not exceed 2.00 percent. The amount of sediment by extraction shall not exceed 0.50 percent. A deduction in quantity shall be made for all water and sediment in excess of 1.0 percent.

†† This table reprinted from ASTM D-396-69. See complete specification ASTM D-396-69.

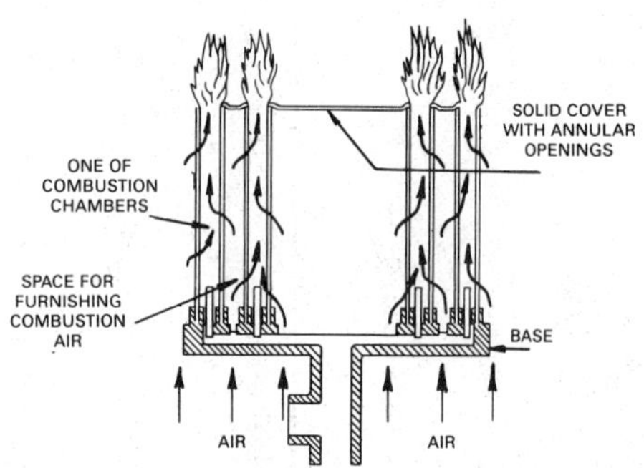

FIG. 8-3D. A sleeve type burner that consists essentially of a flat, cast iron or pressed steel base having two or more interconnecting grooves. Perforated metal cylindrical sleeves are placed in the grooves so that the space between the inner two sleeves and the outer two sleeves becomes a combustion chamber. The space between the second and third sleeves is for the purpose of furnishing combustion air. Covers with annular openings are usually placed on top of the sleeves. Sleeve burners may or may not employ wicks in the annular spaces in the base, depending upon the make and type of burner.

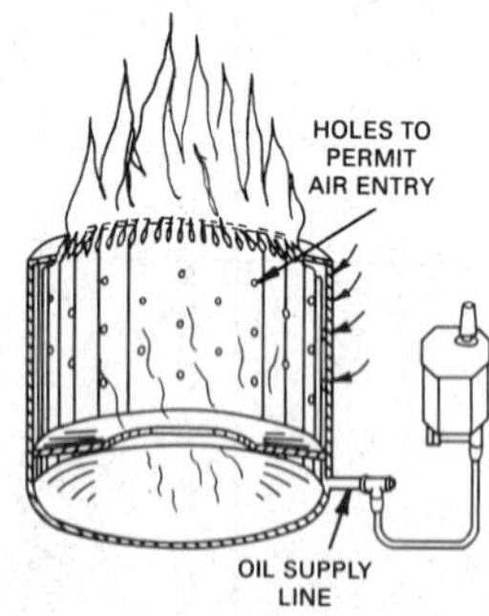

FIG. 8-3E. A vaporizing pot type burner. The oil is introduced into the bottom of the burner and is continually vaporized by the heat of the fire. The holes on the side permit primary and secondary air to enter the burner where they mix with the oil vapors for proper combustion. The primary air is mixed with oil vapors before combustion takes place and the secondary air is supplied to complete the combustion process. At full fire, the velocity of the vapors and the amount of air required result in the flame burning only at the top of the burner.

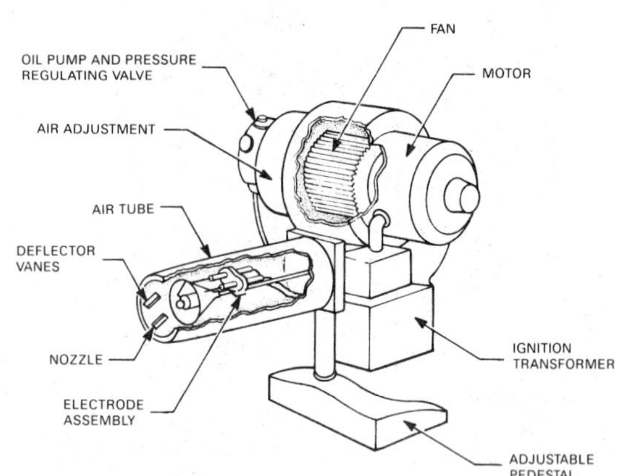

FIG. 8-3F. A high pressure atomizing oil burner. (National Oil Fuel Institute, Inc.)

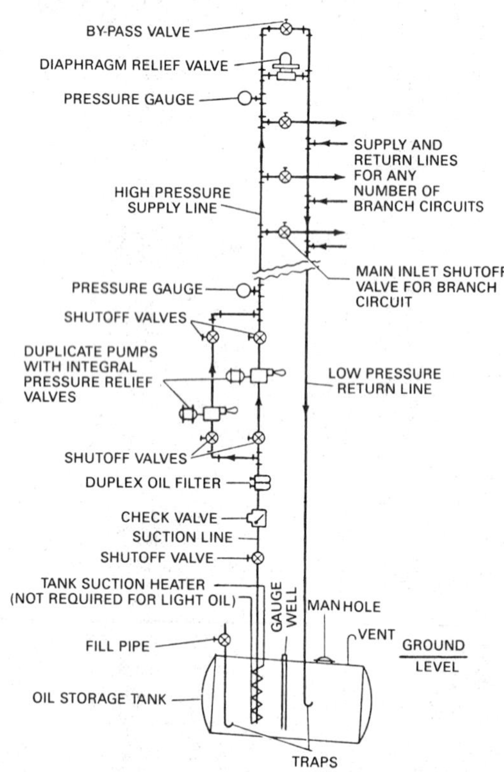

FIG. 8-3G. Typical piping arrangements for burners supplied by separate pump sets. (Combustion Handbook, North American Mfg. Co.)

Fuel Oil Storage

Details on the installation of fuel oil tanks and related piping are described in NFPA 31, *Standard for the Installation of Oil Burning Equipment* (hereinafter referred to as NFPA 31). NFPA 31 covers the installation of tanks underground, unenclosed and enclosed inside buildings, and outside aboveground. The installation requirements in NFPA 31 differ slightly from the requirements for tanks in NFPA 30, *Flammable and Combustible Liquids Code* (hereinafter referred to as NFPA 30), which covers liquids with a wider range of flash points which are used in a variety of locations.

Unenclosed Tanks in Buildings: NFPA 31 currently permits the installation of individual unenclosed tanks of capacities up to 660 gal (2500 L) in the lowest story or basement of buildings. Two such tanks may be installed in any one area, but the aggregate capacity of tankage connected to any one burner may not exceed 660 gal (2500 L). If separation is provided for each tank with capacities of less than 660 gal (2500 L) each, the aggregate capacity may be greater. The separation must be either an unpierced masonry wall or a partition with a fire resistance rating of not less than 2 hr and should extend from the lowest floor area to the ceiling above the tanks. This type of installation is frequently found in multiple unit housing where each unit has its own heating system.

If an unenclosed tank is located above the lowest story or basement of a building, it must not be larger than 60 gal (230 L) capacity. All unenclosed tanks inside buildings must be located not less than 5 ft (1.5 m) from any fire or flame,

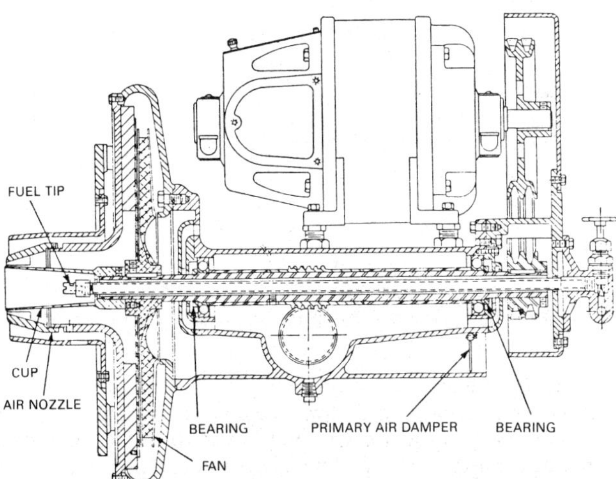

FIG. 8-3H. Horizontal rotary cup oil burner. (National Oil Fuel Institute, Inc.)

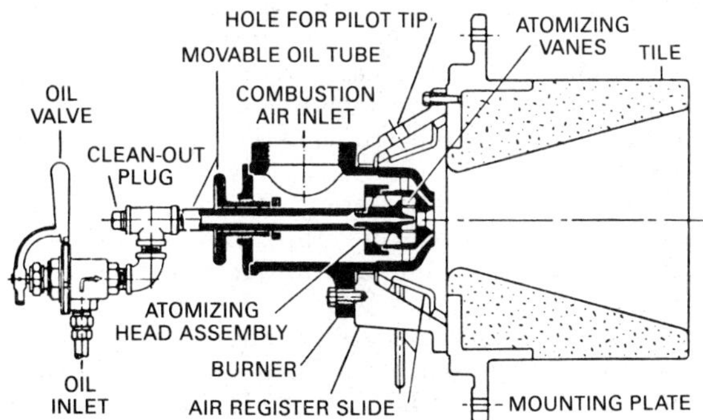

FIG. 8-3I. Low pressure air atomizing oil burner. (Combustion Handbook, **North American Mfg. Co.)**

either in or external to any fuel burning appliance. Furthermore, such tanks must not obstruct quick and safe access to any utility service meter, switch panel, or shutoff valve.

Corrosion of Tanks: Internal corrosion of unenclosed fuel oil tanks, caused by the electrolytic action of water on steel, has been a major problem. Corrosion and subsequent tank leakage have actually been more of a nuisance than a fire hazard. Fire can occur, however, when a burner is located in a pit in the basement and oil from a leaking tank enters the pit. Two generally accepted procedures are recommended to correct the cause of corrosion. One method is to slope the tank to one end and take the burner oil supply line from the bottom of that end. In this way, water is not permitted to accumulate in the bottom of the tank. The quantity of water that comes from condensation inside the tank is so small that it does not present a problem in the safe operation of the burner. The second method is to add a small amount of an alkaline solution to the oil itself or to add a small amount of the solution directly into the tank at periodic intervals.

Enclosed Tanks in Buildings: Tanks larger than 660 gal (2500 L) capacity may be installed in buildings provided the tanks are within an enclosure constructed of walls, floor, and ceiling which has a fire resistance rating of not less than 3 hr. The walls must be bonded to the floor. Any opening into the enclosure is required to have a noncombustible sill or ramp at least 6 in. (152 mm) high and, since it is a Class A (horizontal) opening, it must be provided with a self-closing fire door. The top and walls of the enclosure must be independent of building construction, except that an exterior building wall having a fire resistance rating of not less than 3 hr may also serve as a wall for the tank enclosure. This method of installation is found only in commercial and industrial buildings.

Outside Aboveground Tanks: Tankage not in excess of 1,320 gal (5000 L) may be installed outside aboveground in built up areas. The tanks may be adjacent to buildings, but must be kept away from the line of adjacent property. Not more than one or two tanks having an aggregate capacity of not more than 660 gal (2500 L) are permitted to be connected to one oil burning appliance.

Underground Tanks: These are tanks built to conform to NFPA 31 which also gives special precautions regarding building foundations, property line setbacks, protection against traffic (e.g., vehicle movement over the buried tanks), and for areas subject to flooding. (See NFPA 30 for additional information on protecting tanks containing flammable liquids in locations that may be flooded.)

Centralized Oil Distribution Systems: A system of piping that supplies oil from a central tank or tanks to a number of buildings, mobile homes, recreational vehicles, etc., may be used under certain conditions. For details covering such systems see NFPA 31.

Fill and Vent Pipes: Underground tanks and tanks larger than 10 gal (38 L) capacity inside buildings must have fill and vent pipes terminating outside of buildings. Vent openings and vent pipes must be large enough to prevent abnormal pressure during filling but not smaller than 1¼ in.* pipe size for 550 gal (2082 L) and correspondingly larger for tanks having greater capacity. In addition, outside aboveground tanks must have some form of construction or device to relieve excessive internal pressure that may be caused by an exposure fire.

Gaging devices are required, but on inside tanks they must not allow oil or vapor to be discharged into the building. Test wells are not allowed on inside tanks. Piping to the burner(s) and the required accessories are covered in NFPA 31.

Small Capacity Tanks: There are several types of small tanks used with cooking appliances and room heaters. An auxiliary tank of not over 60 gal (227 L) capacity may be provided between the burner and the main fuel supply tank. An unenclosed supply tank of not more than 10 gal (38 L) capacity for an individual appliance is permitted provided it is placed not less than 2 ft (0.6 m) from a source of heat either inside or outside the appliance being served, and provided that the temperature of the oil in the tank will not exceed 25°F (−4°C) above room temperature at the maximum firing rate.

Integral tanks are furnished by manufacturers as a component part of some oil burning appliances, e.g., kerosene and oil stoves. Stoves with integral tanks [not over 5 gal (20 L) capacity] operate on the gravity or barometric feed principle. The oil in the tank reservoir and in the burner are at the same level. As the oil level drops

*Nominal size; for convenience 1 in. equals 25.4 mm.

due to burning, oil runs into the reservoir from the tank either because a float drops and opens a valve (gravity feed), or because air enters the tank through the cap and allows oil to enter the reservoir until it covers the air opening (barometric feed).

NFPA 31 contains specific provisions on supply tanks for kerosene and oil stoves, portable kerosene heaters, and conversion range oil burners.

Liquid Fuels—Miscellaneous

Alcohol and gasoline are used for some heat producing appliances and equipment to a limited extent. Alcohol stoves and torches, and gasoline torches are examples. NFPA 30 gives additional basic data on the safe storage practices for these liquids.

Gas Fuels

Generally, gas fired, heat producing devices and equipment use natural gas, LP-Gas (liquefied petroleum gas), an LP-Gas/air mixture, or mixtures of these gases. Natural gas is the most commonly used gas fuel in the United States, because gas transmission pipeline systems supply most areas of the country. LP-Gas fired heating devices also are popular, particularly in sparsely settled areas and in recreational vehicles. A number of other flammable gases are used as fuel in many industrial processes for special purposes.

Natural gas consists principally of methane and some ethane, propane, butane, and small amounts of carbon dioxide and nitrogen. Fuel gases can be made in a variety of ways from coal, or by the cracking of oils. Liquefied petroleum gases are largely propane, propylene, butane, and butylene, or mixtures of these gases.

Typical calorific values of various gases as fuel are given in Table 5-5B. Various procedures for sampling, analyzing, measuring, and testing gaseous fuels are outlined in thirteen ASTM standards which have been compiled in the publication, *ASTM Standards on Gaseous Fuel* (ASTM 1974).

Methods of Firing Fuel Gases

Air is either mixed with gas at the burner of a gas fired appliance or premixed with gas. A wide variety of types and sizes of burners are used to mix the gases effectively under different conditions. Every burner is designed to transform the potential energy of the fuel gas into useful heat which can be absorbed in the most effective manner.

Injection (Bunsen) Burners: Practically all residential and commercial gas appliances and some industrial gas fired appliances use injection type gas burners. A jet of gas injects primary air for combustion into the burner and mixes it with the gas. This mixing occurs before the gas reaches the burner ports or point of ignition. Figure 8-3J shows a typical injection type gas burner.

Luminous or Yellow-Flame Burner: In the luminous, or yellow flame, burner, only air externally supplied at the point of combustion is used for burning the gas. The flame is produced without premixing air with the gas.

Catalytic Burners: Another type of burner is the catalytic burner that permits combustion of the gas at temperatures well below the normal ignition temperature of fuel gas air mixtures.

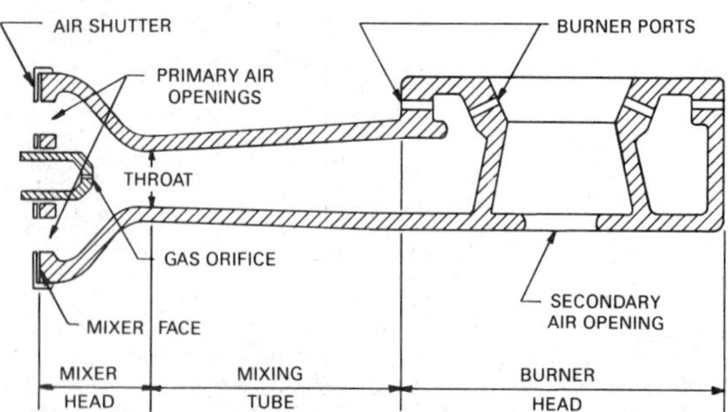

FIG. 8-3J. A schematic view of a typical injection type burner.

Power Burners: In a power burner, either gas or air or both are supplied at pressures exceeding the line pressure of the gas and atmospheric pressure of the air. The added pressure is applied at the burner. When air for combustion is supplied by a fan ahead of the appliance, the appliance is known as a forced draft burner. A premixing burner is a power burner in which all or nearly all of the air for combustion is mixed with the gas as primary air (air that mixes with the gas before it reaches the burner port or ports). A pressure burner is supplied with an air-gas mixture under pressure (usually from 0.5 to 14 in. (0.1 to 3.5 kPa) of water and occasionally higher).

Figure 8-3K illustrates some of the types of industrial appliance gas burners.

Changing from one kind of gas to another may cause serious problems unless the appliance is designed to accept the change. Appliance designs certified by the American Gas Association (AGA) Laboratories are tested for satisfactory performance with one or more gases as requested by the manufacturer. These include natural and mixed gases as well as LP-Gas and LP-Gas-air mixtures. The specific gas or gases that may be used in an appliance are marked on the appliance. Although such appliances may be converted in the field to any one of the marked gases, the conversion may require some interchange of parts and may involve some element of hazard unless carefully performed by experienced, qualified personnel.

Appliance and Piping Installation

Installation, alteration, and repair of gas appliances and gas piping is best done only by qualified agencies fully experienced in this work. Provisions outlined in NFPA 54, *National Fuel Gas Code* (hereinafter referred to as NFPA 54), provide the necessary guidance. Gas should be turned on only after the system has been thoroughly tested to be certain there is no leakage. Any air remaining in piping is a hazard because a slug of air reaching a burner will extinguish the flame. When gas flows again, unburned gas will escape and possibly form an asphyxiating or explosive atmosphere.

The uncontrolled flow of unburned gas into any gas appliance could form an explosive gas-air mixture. Safeguards to prevent this are extremely important, particularly in industrial and other large gas burning devices. Without such safeguards, unburned gas may accumulate in fire boxes or ovens at the time of lighting or when the gas pressure fluctuates. If the pressure is reduced below a certain point, the flame may be extinguished. With a subsequent return of normal pressure, unburned gas might continue to flow. An increase in pressure may increase the

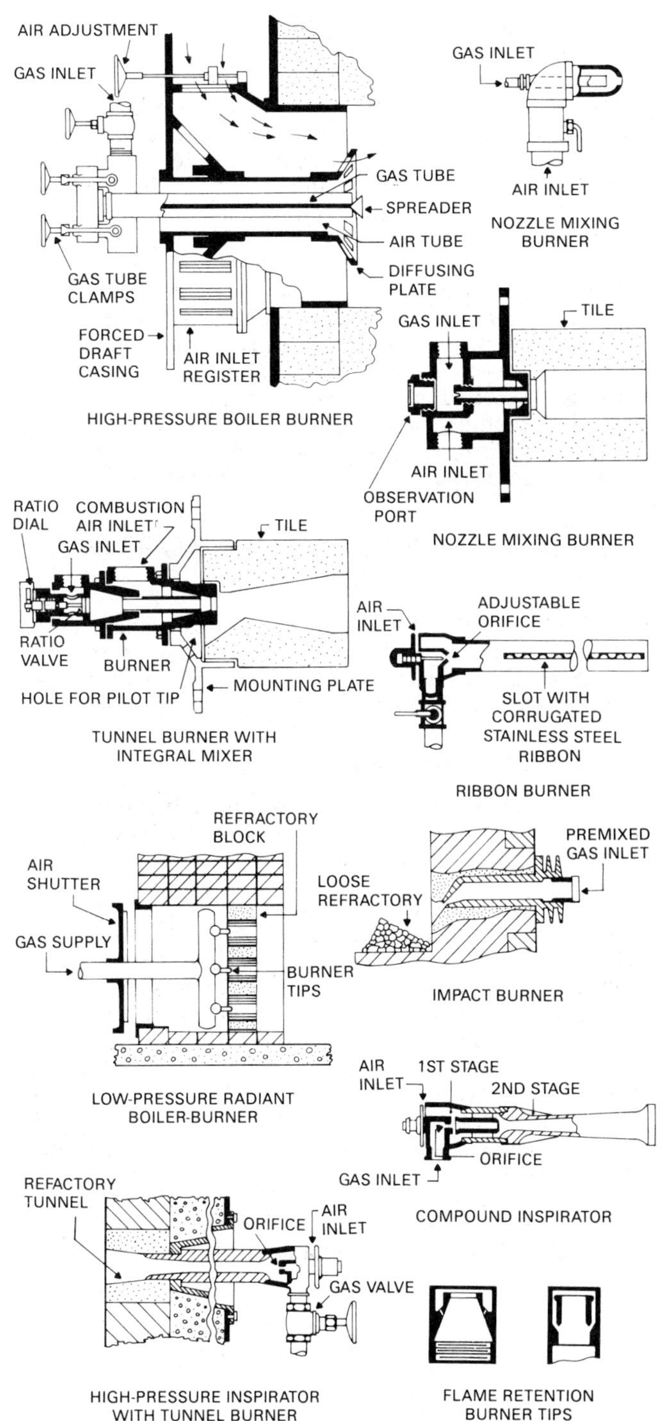

FIG. 8-3K. Types of industrial appliance gas burners.

velocity of flow through the burner in excess of the flame propagation rate. This will force the flame away from the burner. A number of automatic devices are available to safeguard against this hazard, and are commonly found on the larger installations.

Electricity

Heating by electricity involves two basic electrical properties, resistance and induction.

Units of Electrical Heat: Heating calculations for electrical heating are expressed in kilowatt hour (kWh) requirements. One kWh equals 3,412 Btu.

Controls: Automatic regulation of the electrical input (usually through proportional type step controllers) allows heaters to be energized and deenergized in increments small enough to prevent excessive total operation of all heaters during temperature drops.

Method of Installation: Electric heating systems should be installed in accordance with NFPA 70, *National Electrical Code®* (hereinafter referred to as NFPA 70).

Types of Electrical Heating Appliances: Most electrical heating appliances for heating small rooms and for other small heating jobs involve resistor heating elements which are usually one or more metal alloy wires, nonmetallic carbon rods, or printed circuits. Resistor heating is used in radiators, unit heaters, convectors, central hot water systems, central warm air heating systems, and panel type radiant heat installations for walls, floors, and ceilings. Resistor heating is also used for household electrical appliances such as stoves, irons, toasters, etc.

CONTROLS FOR FUEL BURNERS

The uncontrolled flow of unburned fuel into appliances and burners can lead to serious consequences. When fired with pulverized coal, gas, or oil, a combustible mixture could accumulate within the confines of the appliance or equipment. If ignited, it could result in an explosion. To guard against the discharge of unburned fuel or improper mixtures of fuel and air, certain controls are required for all fuel burners.

Primary Safety Controls: These required controls cause the fuel to shut off in the event of ignition failure or flame failure. Except for pot and sleeve type vaporizing burners, shutoff is usually accomplished via a primary safety control. For gas appliances, an automatic gas ignition system is used. Both types of controls sense the presence or absence of flame. Upon ignition or flame failure, they cause the fuel to be shut off in the prescribed period of time. All fuel to the burner should be shut off. Primary safety controls usually provide for starting and stopping the burner in response to changes in demand (room thermostat, process controller, etc.). Controls for commercial and industrial burners may provide a purge period for air to flow through the appliance prior to starting fuel flow. The control also prohibits main burner fuel from being admitted until the control has proved the existence of the ignition flame.

The discharge of unburned oil from pot and sleeve type vaporizing burners should be prevented by a constant level valve or by barometric feed. Each of these provisions maintains a predetermined level of oil in the burner below any point of overflow.

Air Fuel Interlocks: If the safe operation of a burner depends upon a forced or induced draft fan or air compressor to supply combustion air, an interlock is needed to shut off fuel in case the air supply is interrupted.

Atomizer Fuel Interlock: In an oil burning system where air, steam, or other means for atomization can be interrupted without stopping oil delivery to the burner, an interlock must be provided to immediately shut off the oil if atomization fails.

Pressure Regulation and Interlocks: Gas burners and pressure type oil burners require uniform fuel pressure so fuel burns safely and efficiently. Pressure regulators are used to maintain uniform fuel pressures. In situations where fuel pressure fluctuations may be expected, pressure interlocks are provided which shut off the fuel to the main burners when the pressure is too low or too high for safe operation.

Oil Temperature Interlock: Oil burners requiring heated oil for satisfactory operation are equipped with a low oil temperature switch to prevent the burner from starting or to shut it down if the oil temperature falls below the required minimum.

Manual Restart: If a burner is not equipped to provide safe automatic restarting after shutdown, its control system must be arranged to require manual restarting after any control operates to extinguish the burner flame.

Remote Shutoff: It is good practice to provide a remote control for manually stopping the flow of fuel to the burner. An identified switch in the burner supply circuit, placed near the entrance to the room where the burner is located, can be used with electrically powered equipment. A valve in the fuel supply line operable from a location reached without passing near the burner may also be used.

Safety Shutoff Valves: Safety shutoff valves prevent the abnormal discharge of fuel. They should be constructed so they cannot be readily restrained or blocked in the open position. Such valves should automatically close when deenergized, regardless of the position of any damper operating lever or reset handle. Electrically operated valves should not need to be energized to close. Pressure operated valves should close upon failure of the operating pressure.

Safety Control Circuits: Safety control ac circuits are of the two wire type with one side grounded. They must not exceed a nominal 120 V and must be protected with suitable fuses or circuit breakers. All switches should be in the hot ungrounded line. The accidental grounding of such a circuit will not cause a required safety control to be bypassed.

Appliances and equipment fired with fuel burners require some additional controls to avoid excessive pressure, temperature, or other abnormal conditions. These controls are covered later in this chapter.

Appliances listed by testing agencies are equipped with all the required safety controls. The application of controls to fuel burning systems assembled in the field or intended for specific applications should be entrusted only to people specializing in this work. The standards pertaining to fuel burning equipment referenced in the *Bibliography* at the end of this chapter include specific information on controls essential to safe operation of fuel burning equipment.

Safety controls are generally preset by the manufacturer or the installer. Any alteration of these settings or bypassing any safety control could lead to serious consequences. When a faulty control is replaced, a similar model and setting should be used. Figures 8-3L and 8-3M illustrate typical arrangments of safety valves and interlocks for industrial gas and oil burners repectively.

HEATING APPLIANCES AND THEIR APPLICATION

Central heating appliances, room heating and cooking appliances, and miscellaneous heat producing devices, which are fired by the various fuels and methods discussed, are described here.

Central Heating Appliances

Central heating appliances are divided into four broad categories: (1) boilers, (2) central furnaces, (3) floor furnaces, and (4) wall furnaces.

Boilers: These are either of the steam or hot water type and are constructed of cast iron or steel. (See Fig. 8-3N). The *ASME Boiler and Pressure Vessel Code* (ASME 1974) defines low pressure boilers (for low pressure steam heating, hot water heating, and hot water supply) as those steam boilers that operate at pressures not exceeding 15 psi (100 kPa), and hot water boilers as those that operate at pressures not exceeding 160 psi (1100 kPa) and at temperatures not exceeding 250°F (120°C). The *ASME Boiler Code* also covers the construction of power boilers and locomotive boilers.

Boilers are equipped with safety devices to prevent overpressure conditions. Automatically fired boilers also have controls that will shut off the burner or electric

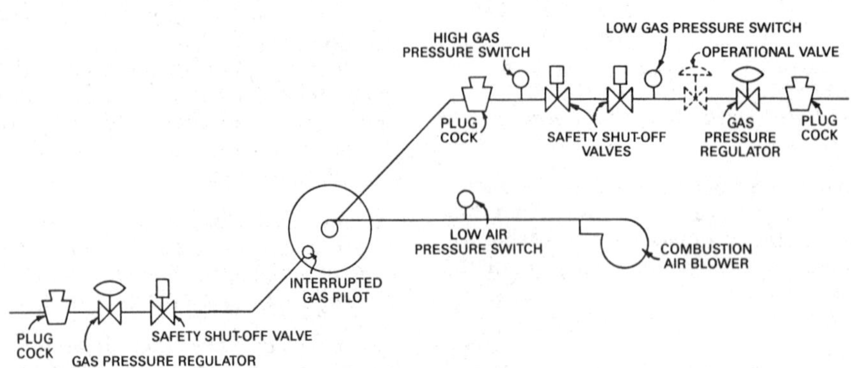

FIG. 8-3L. Typical safety valves and interlock arrangement for industrial gas burner.

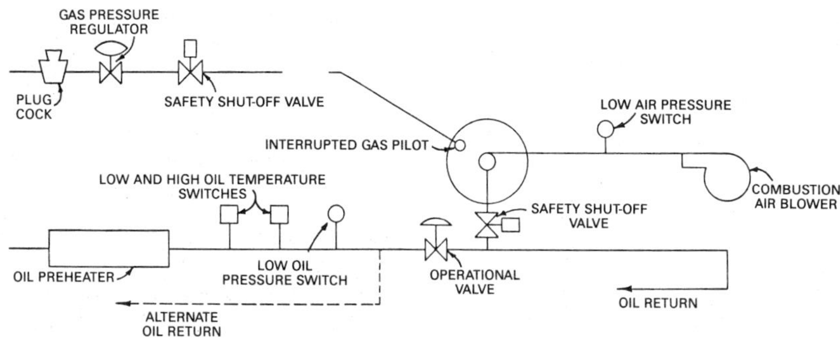

FIG. 8-3M. Typical safety valves and interlock arrangement for industrial oil burner.

heating elements under low water conditions or when a predetermined pressure or temperature has been reached.

The fire problems of boilers involve their mountings and clearances to combustibles. Explosions resulting from unburned fuel accumulations in the firebox or due to overpressure conditions could rupture fuel lines, thus contributing to the potential for subsequent fire or explosion damage.

Warm Air Furnaces: The residential types currently installed in the U. S. are usually equipped with circulating fans, filters, and other features which make them, in effect, air conditioning systems.

Central warm air furnaces are of the following types:

1. Gravity—depends primarily upon the circulation of air by gravity.
2. Gravity with integral fan—fan is an integral part of the furnace construction and is used to overcome internal resistance to airflow.
3. Gravity with booster fan—fan does not restrict flow of air by gravity when fan is not operating.
4. Forced air—equipped with fan that provides the primary means of circulating air. (See Fig. 8-3O.)

Forced air central warm air furnaces may be classified according to the direction of airflow through them as horizontal, upflow, or downflow.

Gravity furnaces are floor mounted and can heat only spaces above them. Forced air furnaces may be floor mounted or suspended and may be found on most any floor of a building, including the attic and roof.

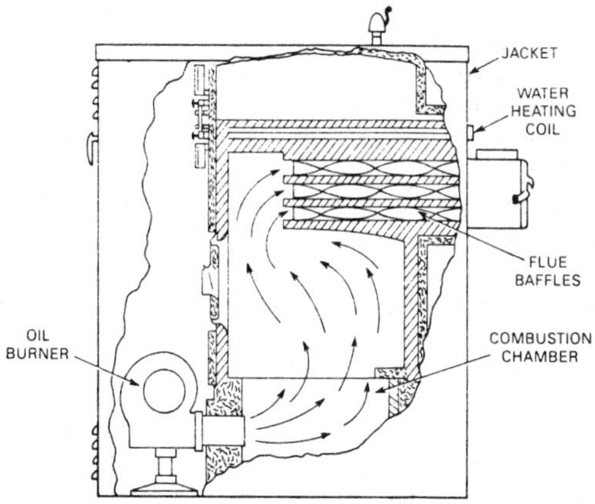

FIG. 8-3N. An oil fired steam boiler. (National Oil Fuel Institute, Inc.)

Under some conditions of operation, plenums of warm air heating furnaces may become hot enough to ignite adjacent unprotected woodwork. Clearances or insulation are necessary in these instances. (Complete requirements for installation of systems in residences may be found in NFPA 90B, *Standard for the Installation of Warm Air Heating and Air Conditioning Systems*, hereinafter referred to as NFPA 90B.)

Automatic controls, also called high limit controls, can shut off the fuel or electric supply whenever the air in the furnace warm air plenum or at the beginning of the main supply duct—at a point not affected by radiated heat—reaches 250°F (120°C) or less, depending upon the type of furnace and its installation. (See Fig. 8-3P.) Automatic controls cannot be set above prescribed temperature limits. Some furnaces used only with special duct systems are factory equipped to permit temperatures above 250°F (120°C) at the inlet of the supply ducts.

Central warm air furnaces also cause fires due to inadequate clearances to combustible construction, lack of proper limit controls, heat exchanger burnouts, and other causes associated with lack of proper installation, servicing, and maintenance.

Pipeless Furnaces: These are essentially gravity warm air furnaces that are not connected to ducts. The furnace is mounted directly under the space to be heated. All the air heated in its outer jacket is discharged through a register above the heat exchanger. They are used for heating small homes and other small one story properties.

Pipeless furnaces may produce dangerously high temperatures, particularly if the air circulation is restricted. For example, if the register in the floor of a small hallway

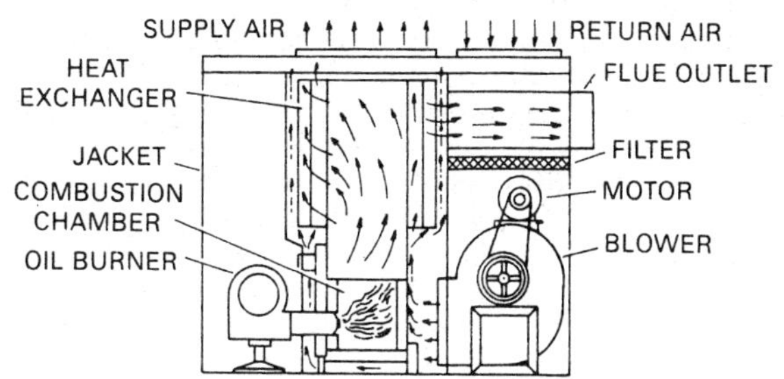

FIG. 8-3O. An oil fired forced air furnace. (National Oil Fuel Institute, Inc.)

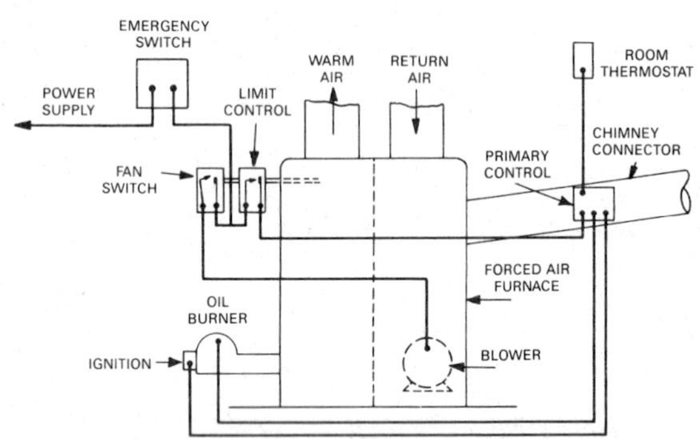

FIG. 8-3P. Control wiring schematic, oil fired forced warm air furnace. (Certain primary controls are mounted on the burner assembly.) (National Oil Fuel Institute, Inc.)

heats adjoining rooms by air circulating through opened doors, the temperature of the entire hall may be raised to a dangerous degree if the doors are closed tightly at a time when the furnace is in full operation. Fires also have occurred when clothing left to dry above a central floor register was ignited. It is good practice to equip automatically fired pipeless furnaces with limit controls.

Floor Furnaces: Most oil and gas fired floor furnaces are designed so they may be installed in combustible floors (Fig. 8-3Q); however, they should not be installed in

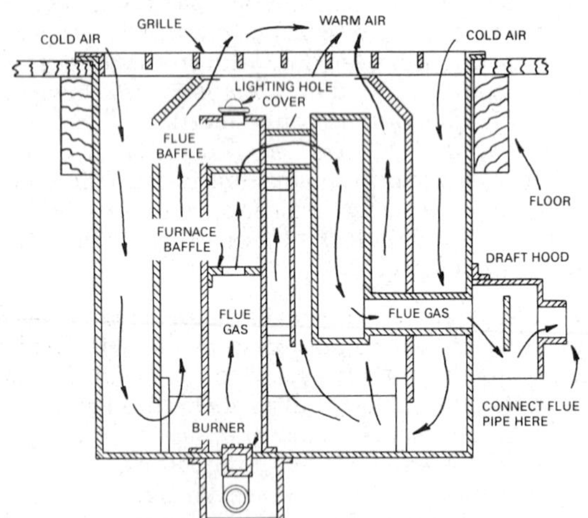

FIG. 8-3Q. A typical floor furnace.

combustible floors if they have not been listed by a testing laboratory for such use. Clearances should be sufficient not only for protection of combustible floors and walls, but also to keep the furnace casing and the piping connected to it out of contact with earth or damp materials.

Temperature limit controls are provided on floor furnaces to shut off the fuel supply when the temperature of the discharged air reaches a predetermined level.

In auditoriums, public halls, or assembly rooms, floor furnaces should not be installed in the floor of any aisle, passageway, or exitway. Floor furnace registers may become hot enough to cause burns under some conditions; therefore, registers in dwellings should not be located in passageways to bathrooms, where persons would be likely to step with bare feet. Users should be warned against covering registers or placing clothing on them to dry. Older types of floor furnaces do not have a temperature limit control, and may become dangerously hot if the air passages are blocked with dust and lint. Inspection and vacuum cleaning or hand removal of this blockage is essential for continued safe operation.

Wall Furnaces: These are self-contained indirect fired gas or oil heaters installed in or on a wall. They supply heated air directly to the space to be heated, either by gravity or a fan through grills on openings or boots in the casing supplied by the manufacturer. The furnaces may be of the direct vent type or are vent- or chimney-connected, depending upon the fuel. Limit controls limit outlet air temperature. The fire problems with recessed wall furnaces are similar to those encountered with most warm air furnaces. A wall furnace installation is shown in Figure 8-3R.

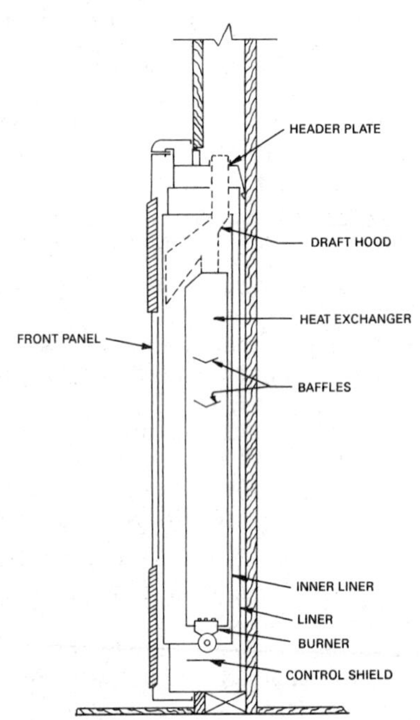

FIG. 8-3R. Typical gas fired wall furnace.

Duct Furnaces: These furnaces are installed directly in ducts of some warm air and air conditioning systems and depend upon air circulation from a blower not furnished as a part of the furnace. They may be oil or gas fired or electrical, and are equipped with a limit control to shut off the fuel supply or electricity at excessive temperatures.

When duct furnaces are used in conjunction with refrigeration coils in a combined heating and cooling system, the furnace should be located upstream from the refrigeration coil or parallel with it to prevent condensation from corroding the furnace. There are, however, furnaces made of corrosion resistant material which may be installed downstream from the refrigeration coil. When the furnace is located upstream from the refrigeration coil, the coil should be designed so that excessive pressures and temperatures will not develop in the coil. A duct furnace installation is shown in Figure 8-3S.

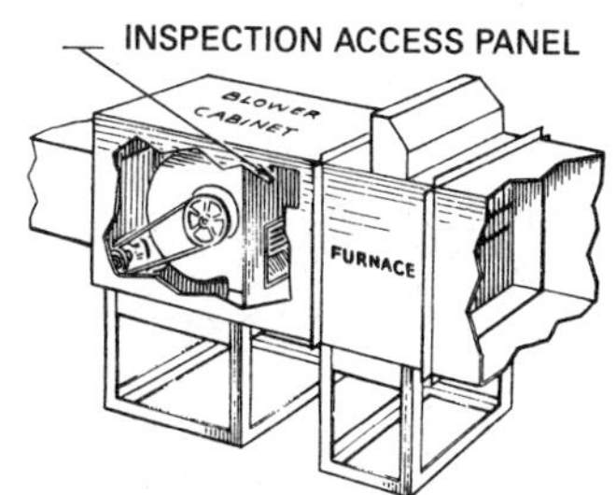

FIG. 8-3S. A duct furnace installation.

Warm Air Heating Panels: These are used in low temperature, forced air systems to circulate warm air through plenums or chambers that have one or more surfaces exposed to the space to be heated. NFPA 90B gives recommendations for the use, construction, and installation of these panels.

Radiant Heating: This type of heating utilizes panels of hot water piping or electric heating elements which usually operate at moderate temperatures. The panels are embedded in plaster walls or ceilings or in cement floors. Hot water pipes or electrical equipment in radiant systems should conform to standard installation practices. NFPA 70 contains specific recommendations on the installation of electrical space heating equipment.

Heat Pump: This term is applied to a type of forced air heating system in which refrigeration equipment is used in such a way that heat is taken from a heat source and given up to the conditioned space when heat service is wanted and is removed from the space and discharged to a heat sink when cooling and dehumidification are desired. These systems frequently have supplemental heating units. In such cases, the units are equipped with an interlock to prevent the unit from operating unless the indoor air circulating fan on the system is running. The units also have temperature limit controls. Hazards are those presented by power, refrigeration equipment, and heat units, if these are part of the system.

Unit Heaters

These are classed as self-contained, automatically controlled, chimney or vent connected air heating appliances with an integral means for air circulation. They may be floor mounted or suspended and are equipped with temperature limit controls. The term unit heater as used in NFPA standards is intended to cover appliances for heating nonresidential properties. Thus it excludes room heaters, floor furnaces, and similar devices. More specifically, a unit heater is an appliance consisting of a heating element and fan housed in a common enclosure and placed within or adjacent to the space to be heated. Unit heaters should not be confused with heat exchangers, which are equipped with fans to circulate heated air. In the latter type, hot water or steam is piped from a heating unit to the heat exchanger in the area to be heated.

Unit heaters that are designed for connection to a duct system may be considered as central heating furnaces and should be provided with the same safeguards. A typical unit heater is shown in Figure 8-3T.

Room Heaters and Cooking Appliances

A room heater differs from a central heating furnace in that it is a self-contained air heating appliance designed

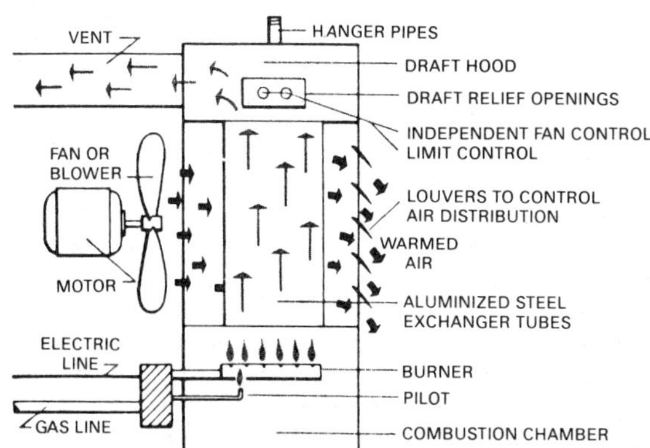

FIG. 8-3T. Typical gas fired unit heater.

for the direct heating of the space around it. External pipes or ducts are not used to distribute the heat. Room heaters fall into two general types, circulating and radiant. A fuel burning circulating heater has an outer jacket surrounding the combustion chamber. It is arranged with openings at the top and bottom so that air circulates between the inner and outer shell. It does not have openings in the sides of the outer jacket in order to permit direct radiation. From a safety point of view, room heaters other than those specifically designed as circulating type are classified as being of the radiant type. They include such appliances as wood and coal stoves, electric and gas logs, open front heaters, wall heaters, gas coal baskets, and fireplace inserts. Care must be taken, especially with heaters that can be moved easily, to place the radiant type so that radiating elements and surfaces are not directed toward walls, drapes, or furniture in close proximity to the heater.

Fuel burning heaters and residential cooking appliances should be connected to a chimney or vent, where appropriate. The exceptions are gas room heaters and gas cooking appliances listed for unvented use.

Fuel burning room heaters are not to be installed in sleeping quarters for use of transients, as in hotels and motels, nor in institutions such as homes for the aged, sanitariums, convalescent homes, orphanages, etc. It is good practice to install direct vent type heating appliances in sleeping quarters where heaters are permitted and in rooms generally kept closed.

Solid Fuel Room Heaters

Coal and wood stoves or room heaters warrant special attention because of their record as a fire cause. The poor record is due to two underlying conditions: (1) inadequate clearances to combustible materials, and (2) improper or inadequate maintenance. Many wood stoves need frequent charging, often involving much regulating of draft controls to prevent overheating. Sometimes, too, weather conditions can affect the chimney draft, causing improper combustion.

A solid fuel heater essentially is a fire chamber surrounded by a decorative enclosure. The chamber may be equipped with manual draft controls or a thermostatic control and provided with a chimney connector and other parts that may be required for installation. Instructions that come with heaters that have been listed by testing laboratories are quite precise on the type of heat shielding and the clearances from combustibles that are required.

The vast majority of solid fuel room heaters are designed to burn only wood, which is easily burned on the bottom of the fire chamber, resting on a bed of firebrick, ashes, or sand. For coal, a grate or other provision to supply air below the fuel bed is needed.

Unless a special chimney is called for by the listing or installation instructions (as in the case of room heaters for use in mobile homes), the typical solid fuel heater can be connected to either a standard masonry chimney or any listed factory built chimney. Laboratory tests for listing heaters establishes the dimensions required for adequate clearances. Because clearances from large heaters may exceed 36 in. (0.9 m), the tests may be performed with supplemental heat protection or shielding for walls. For proper clearances and floor protection see the section of this chapter on installation.

Hazards With Solid Fuel Room Heaters

1. Overfiring. If an appliance is overheated deliberately or through neglect, there is a severe and immediate hazard if the unit or its connector glows even a very dull red. Control is to simply close doors and air dampers until things cool down.
2. Careless handling of ashes containing live coals. Ashes should never be collected in combustible containers, nor should the containers ever be placed on a combustible floor.
3. Inadequate connector clearance. This includes violations of the required 18 in. (457 mm) clearance or lack of adequate protection where a single wall connector must penetrate a combustible wall to join a masonry chimney. Reduced connector clearances are permissible in some instances such as the entry of single wall connectors into listed supports for factory built chimneys. The reduced clearances have been validated by tests and are described in the installation instructions for listed heaters. (See Fig. 8-3U.)
4. Creosote accumulation and fires. Low temperature operation of any wood burning heater for long periods will produce creosote which can deposit in the heater and in its connector and chimney. Airtight woodburning stoves, because of the relatively low temperatures at which they can be fired, can produce an increased rate of creosote deposits in chimneys and connectors. Therefore, frequent inspection and, when required, cleaning of the chimney system is in order when these appliances are used. Creosote deposits will build up even when burning dry seasoned wood of any kind, and enough creosote can accumulate in several days to support a chimney fire. Thin accumulations of creosote can be ignited and burned when the surface temperature reaches the ignition temperature of creosote, which is roughly 900 to 1,000°F (480 to 540°C). However, this could be a dangerous practice because there is no way to guarantee the thickness of the deposit will be uniform over the entire inner surface of the chimney system. Therefore ignition of the creosote deposit could produce temperatures that could result in damage to the chimney system or ignite nearby combustibles. A daily hot fire with a fresh charge of wood may control creosote deposits indefinitely; however, the operator must pay attention and not allow overheating to take place.

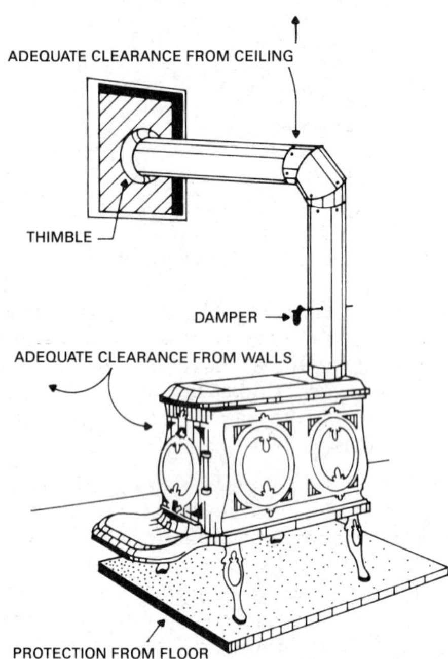

FIG. 8-3U. Adequate clearances should be maintained in order to ensure safe operation of coal and woodstoves.

If creosote is building up excessively, symptoms include drips out of the joints, a dull sound if the connector is tapped, and blockage of the chimney and loss of draft evidenced by smoking. Large accumulations should be removed mechanically, because the excessive amount of fuel deposited in the chimney could be damaging even if the fire is carefully controlled.

Restaurant Cooking Appliances

These include a variety of devices such as ranges, deep fat fryers, steamers, broilers, hot plates and griddles, portable ovens, and others. These devices, like other heat producing devices, require careful installation to avoid overheating of adjacent combustibles. In some of them grease accumulates, adding to the fire hazard. NFPA 54 contains recommendations for the installation of gas-fired restaurant cooking appliances. Ventilation of fixed restaurant cooking equipment is covered in NFPA 96, *Standard for the Installation of Equipment for Removal of Smoke and Grease-Laden Vapors from Commercial Cooking Equipment.*

Miscellaneous Heat Producing Devices

There are so many miscellaneous heat producing devices that it is unrealistic to treat each individually in this text. Certain basic principles apply to all of them, however, including provision for adequate clearances to combustibles, venting products of combustion, adequate air for combustion and ventilation, proper burner controls to safeguard against the hazard of the fuel, proper storage and handling of the fuel, and safety controls to safeguard the heat producing device.

Testing laboratories test and list many miscellaneous types of heat producing appliances such as direct fired heaters, incubators, lanterns, forges, torches, etc.

Kerosene Stoves: These may be defined as self-contained, self-supporting, kerosene burning ranges, room heaters, or water heaters that are not connected to chimneys but are equipped with integral fuel supply tanks not exceeding 2 gal (7.6 L) capacity. Terms often applied to kerosene room heaters are cabinet heaters and space heaters. Since they are not connected to chimneys, they can be moved rather easily although they are not considered portable. This feature of kerosene stoves has frequently resulted in creation of severe fire and life hazards due to improper placement of the stoves in relation to nearby combustibles and paths of exitways. Kerosene stoves listed by testing laboratories incorporate fire protection features which may be missing from those not listed. Among the more important features of listed stoves are the types of construction materials, primary control valves, and drip pans used in them.

Portable Kerosene Heaters: These are similar in hazard to kerosene stoves because they are not connected to a chimney. The hazard in their use, however, is increased by their even greater portability and subsequent misuse and misplacement. Unfortunately, the majority of portable kerosene heaters sold in the U.S. and Canada are not laboratory tested. Heaters that are tested incorporate firesafety features not included in others such as special types of latching devices and integral sheet metal trays under the burners to catch oil drips. They usually employ wick type burners integral with the oil reservoir.

Conversion Range Oil Burners: These consist essentially of a single or double sleeve type burner assembly, regulating valves, and an oil supply assembly with a suitable supporting stand and seamless connecting tubing. A thermal valve located in the burner compartment of the stove adjacent to the burner is installed in the oil supply line. (See Fig. 8-3V).

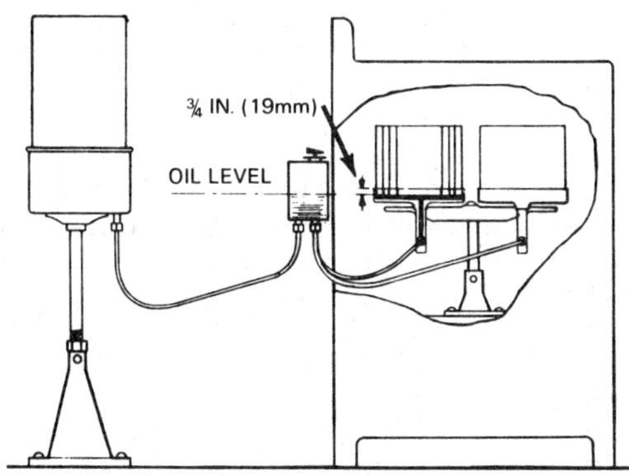

FIG. 8-3V. Typical installation of conversion range oil burner.

Range oil burners, which are found most frequently in the northeastern section of the U.S., are designed to burn kerosene, range oil, or similar fuel. They are primarily installed only in stoves or ranges originally designed to use solid fuel. Range oil burners should not be mistaken for conversion oil burners of the vaporizing pot type designed for conversion of central heating appliances.

Salamanders: These are typical miscellaneous portable heating devices. They are used at construction sites and in unheated buildings. Too frequently, a salamander consists of only a metal drum with some holes punched in it for draft, and wood scrap and other construction waste materials for fuel. These devices present an acute spark hazard, and the hazard of carbon monoxide poisoning because they are not chimney connected. Crude salamanders of this type, and gas and oil fired salamanders that have not been tested and listed by a laboratory, should not be used.

UL and American Gas Association (AGA) laboratories test and list portable gas and oil fired heaters, such as salamanders. One way to use salamanders is to carry heated air indoors through flexible ducts from a heating unit located outdoors. NFPA 58, *Standard for the Storage and Handling of Liquefied Petroleum Gases*, has extensive material on the use of LP-Gas fired portable heaters.

DISTRIBUTION OF HEAT BY DUCTS AND PIPES

Warm Air Distribution Systems

Central warm air heating systems consist basically of a heat exchanger with an outer casing or jacket that is connected to a system of ducts, air passages, or plenums that carry heated air (the supply side) to the spaces to be heated and return air (the return side) from the heated spaces to the heat exchanger.

NFPA 90B contains specific requirements for systems that are installed in one and two family dwellings and other occupancies not exceeding 25,000 cu ft (700 m^3) in volume. Systems that are installed in spaces exceeding 25,000 cu ft (700 m^3) are covered in NFPA 90A, *Standard for the Installation of Air Conditioning and Ventilating Systems.* The text of this portion of the HANDBOOK is confined to a discussion of ducts for warm air heating systems installed in dwellings and other relatively small spaces.

Warm Air Supply Ducts: Warm air ducts must be installed with clearances as given in Table 8-3B. They are substantially constructed of metal or of Class 0 or Class 1 duct materials* and are properly supported and protected against injury.

Supply Air Plenums: It is common practice in some localities to use the crawl space as the supply plenum for warm air heating systems when a one story single family house does not have a basement. These systems utilize a downflow furnace that discharges warm air through directional ducts into the crawl space. An opening cut in the floor of each room of the house and covered with a grill serves as the supply register. The hazards of these systems include: use of combustible construction as a plenum; use of the crawl space for storage; presence of a large enclosed combustible area with no limitations on fire spread with direct openings to every room of the house; and the possibility of reverse flow in the downflow furnace (the cold air return then becomes the supply duct). NFPA 90B gives specific recommendations and limitations applying

*NOTE: Duct materials are tested and listed by Underwriters Laboratories Inc. in accordance with UL Standard 181, *Air Ducts*, and are classified as follows: Class 0—Air duct materials having a fire hazard classification of zero (flame spread and smoke developed). Class 1—Air ducts that have a flame spread rating of not more than 25 without evidence of continued progressive combustion and a smoke developed rating of not more than 50.

Table 8-3B. Installation Clearances for Horizontal Warm Air Ducts

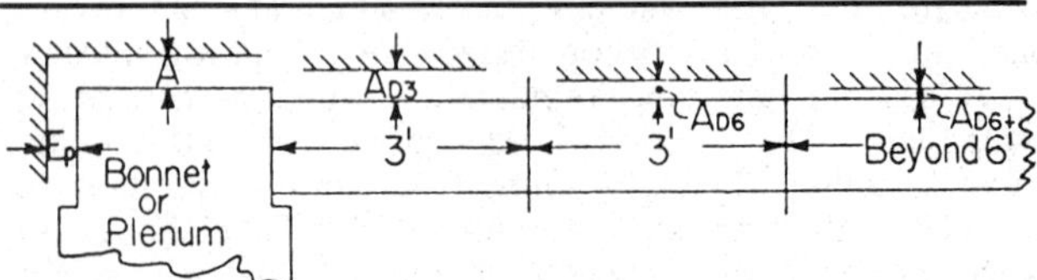

A—Clearance above top of casing, bonnet, plenum, or appliance determined by Table 8-3G.
A_{D3}—Clearance from horizontal warm air duct within 3 ft (0.9 m) of plenum.
A_{D6}—Clearance from horizontal warm air duct between 3 and 6 ft (0.9 and 1.8 m) of plenum.
A_{D6+}—Clearance from horizontal warm air duct beyond 6 ft (1.8 m) of plenum.
E_P—Clearance from any side of bonnet or plenum.

If A Equals	A_{D3} Equals	A_{D6} Equals	A_{D6+} Equals	E_P Equals	Method of Firing
1	1	0	0	1	Automatic oil, comb. gas-oil, or gas
2	2	0	0	2	Automatic oil, comb. gas-oil, or gas
6	6	6	0	6	Automatic oil, comb. gas-oil, or gas
6	6	6	1*	6	Automatic stoker fired†
18	18	6	1	18	Any fuel or control.

* Clearance A_{D6+} to be maintained to a point where there is a change in direction equivalent to 90 degrees or more.

† Furnace must be equipped with limit control that cannot be set higher than 250°F (120°C) and must also have a barometric draft control operated by draft intensity and permanently set to limit draft to a maximum intensity of 0.13 in. water gage (32 kPa) otherwise clearances should be as indicated for any fuel.

to the use of under floor space as supply plenums, and it should be consulted for detailed guidance.

Horizontal supply ducts and vertical ducts, risers, boots, and register boxes within certain limits of the warm air furnace can reach temperatures that can become hazardous; therefore, safe clearances to combustibles must be maintained. The clearances required for the many possible configurations of warm air ductwork are specifically provided in NFPA 90B, which should be consulted.

Registers: To prevent excessive heat from developing in the duct system, one register or grill must be installed without a shutter and without a damper in the duct to it. The exceptions are automatic oil or gas fired systems that have approved temperature limit controls or systems with dampers and shutters designed so that they cannot shut off more than 80 percent of the duct area.

Where registers are installed in the floor over the furnace, as in pipeless furnaces or floor furnaces, the register box should consist of a double wall with an air space of not less than 4 in. (102 mm) between walls. Furnaces with a cold air passage around the warm air passage comply with this requirement.

Return Air Ducts and Plenums: The best way to conduct return air to the furnaces and duct heaters is through continuous ducts. Return ducts do not need to be made of the same materials as supply ducts except for those portions directly over the heating surface or within 2 ft (0.6 m) of the outer casing or jacket of the furnace or duct heater. They should not, however, be made of materials more combustible than 1 in. nominal* wood boards. Under floor spaces may be used for return ducts from rooms directly above, but such spaces are used only if they are not more than 2 ft (0.6 m) in depth from the bottom of floor joists, are clean of all combustible material, and are tightly and substantially enclosed. No vertical stack for return air is connected to registers on more than one floor.

The interior of combustible ducts must be lined with metal at points where there is danger from incandescent particles dropped through a register, such as directly under floor registers and at the bottom of vertical ducts.

Air Filters: Air filters should be the kind that will not burn freely or emit large volumes of smoke or other objectionable products of combustion. Filters qualifying as Class 1 and Class 2 as defined in NFPA 90B are accepted as meeting these requirements. Only liquid adhesive coatings with a flash point of 325°F (163°C) or higher (Pensky-Martens closed cup tester) are acceptable. Filters are not installed in ducts of heating systems unless the system design calls for them. Otherwise the filters may restrict the flow of air and cause dangerous overheating.

Heating Panels: NFPA 90B recommends that heating panels be used only with automatically fired gas or oil burning or electric forced warm air systems that will limit furnace outlet air temperature to 200°F (93°C) or systems equipped with steam or hot water heat exchangers utilizing steam which cannot exceed 15 psig (100 kPa) or hot water which cannot exceed 250°F (120°C).

Panels used with automatically fired forced warm air systems must be of noncombustible material or at least of material with a flame spread rating of not more than 25, as determined in accordance with NFPA 255, *Standard Method of Test of Surface Burning Characteristics of Building Materials*.

Where the warm air supply is from a steam or hot water heat exchanger, the panels may be made of material not more flammable than 1 in. wood boards. No single vertical heating panel should serve more than one story of any building.

Steam and Hot Water Pipes

The temperature of saturated steam varies with the pressure. For example, at atmospheric pressure the temperature of steam is 212°F (100°C), while at 25 psig (172 kPa) its temperature is 266.7°F (130.4°C). Other values of temperature of saturated steam at various pressures are given in Table 8-3C.

Water at high temperature and pressure is also used for heating. The system pressure must always exceed the pressure at the saturation temperature to prevent the water from flashing to steam. This means that the pressure in high temperature hot water systems must exceed the values shown in Table 8-3C at the specific temperature at which the system operates. Most systems operate at temperatures between 250 and 430°F (120 and 220°C).

Recommended clearances for steam and hot water pipes and radiators are given in Table 8-3D.

Radiators and pipes should not be used as racks for drying purposes. Where radiators are placed in window recesses or concealed spaces, such spaces should be lined

*For convenience 1 in. = 25.4 mm.

Table 8-3C. Temperature of Saturated Steam

Gage Pressure		Temperature		Gage Pressure		Temperature	
psi	kPa	°F	°C	psi	kPa	°F	°C
0*		212.0	100.0	35	241	280.6	138.1
1	6.895	215.4	101.8	40	275	286.7	141.5
2	13.79	218.5	103.3	45	310	292.3	144.6
3	20.69	221.5	105.3	50	345	297.7	147.6
4	27.58	224.4	106.9	75	517	320.0	160.0
5	34.48	227.1	108.4	100	690	337.8	169.9
6	41.37	229.6	109.8	125	862	352.9	178.3
7	48.26	232.3	111.3	150	1034	365.9	185.5
8	55.16	234.7	112.6	200	1379	387.9	197.7
9	62.06	237.0	113.9	300	2068	421.7	216.5
10	68.9	239.4	115.2	400	2758	448.2	231.2
11	75.84	241.5	116.4	500	3448	470.1	243.4
12	82.74	243.7	117.6	600	4137	488.8	253.8
13	89.64	245.8	118.8	700	4826	505.6	263.1
14	96.53	247.8	119.9	800	5516	520.3	271.3
15	103.4	249.8	121.0	900	6205	534.0	278.9
20	137.9	258.8	126.0	1000	6895	546.4	285.8
25	172.4	266.7	130.4	1500	10 342	597.6	314.2
30	206.8	274.1	134.5	2000	13 790	636.8	336.0
				3211	22 139	706.1†	374.5

* Zero gage pressure corresponds to an absolute pressure of 14.696 psi (101.329 kPa).

† Critical point, from Keenan's Tables.

Note: Superheated steam has a higher temperature than saturated steam at the same pressure, depending upon the degree of superheat.

Table 8-3D. Installation Clearances for Steam and Hot-Water Pipes and Radiators

Description	Clearances in.	mm
A. Hot-water pipes and radiators supplied by automatically fired gas, gas-oil or oil burning boilers equipped with limit control that cannot be set to permit a water temperature above 150°F (66°C).	None	
B. Hot-water and steam pipes and radiators supplied with hot water at not more than 250°F (121°C) except as permitted in A above, and with steam at not over 15 psig (103 kPa).	1*	25*
C. Steam pipes carrying steam at pressures above 15 psig (103 kPa) but not over 500 psig (3450 kPa)	6	15

* At points where pipes emerge from a floor, wall or ceiling, the clearance at the opening through the finish floor boards or wall or ceiling boards may be not less than ½ in. (13 mm). Each such opening shall be covered with a plate of noncombustible material.

with noncombustible material, have ample air circulation, and be kept clean.

INSTALLATION OF HEATING APPLIANCES

Any source of heat is a potential fire hazard unless it is arranged to prevent the possibility of dangerous temperatures developing on adjacent combustible materials. It is possible for wood and certain other combustible materials to ignite at temperatures far below their usual ignition temperatures if they are continually exposed to relatively moderate heat over long periods of time. To safeguard against this possibility, it is good practice to install heat producing appliances with adequate clearance between the appliance and combustibles. Then, under conditions of maximum heat (long and continued exposure), the temperature of exposed combustibles will not exceed dangerous limits. Minimum clearances for various types of heat producing devices, which should prevent the temperature from exceeding the maximum in exposed combustibles, have been determined by testing laboratories and through field experience. Table 8-3E gives the recommended standard clearances around and above gas and oil fired and electrical residential, commercial, and industrial heat producing appliances. Table 8-3F gives clearances for solid fuel appliances. The term "listed appliance," which appears in the paragraph under the titles of Tables 8-3E and 8-3F, refers to appliances that have been tested by testing laboratories. The proper clearances for these appliances, as determined by tests, are published in a listing. The recommended clearances are also indicated either on the appliance or in the manufacturer's installation manual included with the appliance.

Among laboratories that test and list appliances are the Factory Mutual Engineering Corporation, Underwriters Laboratories Inc., Underwriters Laboratories of Canada, and American Gas Association Laboratories.

Modifications of Clearances

The clearances given in Tables 8-3E and 8-3F will provide reasonable protection for exposed surfaces, including those of considerable area, against assumed conditions of continuous operation of heat producing appliances at maximum temperatures. However, where the appliance is very large, increased clearances to combustible materials may be required. This is a good reason why all appliances should be tested and installed in accordance with the terms of their listing. On the other hand, there are many installations which have not had fires where the clearances are less than that specified. Careful operation of the devices in many such cases has limited temperatures to below the maximum. Yet, every year there are thousands of heating plant fires when cold weather forces long operation at maximum temperatures. An important contributing factor is probably the thorough drying out of exposed wood, which would increase the fuel value under such circumstances. Therefore, where the appliance is used more frequently and for longer periods of time in very cold weather, more frequent checks should be made of the temperature of exposed combustibles and the chimney system.

Installing heating appliances, which under normal operation will not expose adjacent combustible materials to dangerous temperatures, is not sufficient. Clearances and protection also must be designed for reasonable protection when the heating device is operated at its maximum temperature. However, under any conditions of operation the temperature of the exposed combustible materials should not exceed 180°F (82°C). Some provision for ventilation, air circulation, or other method of cooling is necessary to dissipate heat.

In large rooms, a suitable clearance between a heating appliance and combustible material is all that is necessary to prevent ignition. In small rooms or poorly ventilated spaces, particularly where the size of the heating device is large in proportion to the room, dangerous temperatures may be built up no matter how great the clearance unless some provision is made for cooling by air circulation. It is not good practice to put appliances in confined spaces such as alcoves or closets, unless they have been designed and tested for that type of service. The clearances specified

Table 8-3E. Standard Installation Clearances for Heat-Producing Appliances
(See Note 1.)
These clearances apply unless otherwise shown on listed appliances. Appliances should not be installed in alcoves or closets unless so listed. For installation on combustible floors, see Note 2.

		Appliance											
		Above Top of Casing or Appliance in.	mm	From Top and Sides of Warm-Air Bonnet or Plenum in.	mm	From Front See Note 4 in.	mm	From Back in.	mm		From Sides in.	mm	
Boilers and Water Heaters													
Steam Boilers–15 psi (103 kPa) Water Boilers—250°F (120°C)	Automatic Oil or Comb. Gas-Oil	6	152	—		24	610	6	152		6	152	
Water Heaters—200°F (93°C)	Automatic Gas	6	152	—		18	457	6	152		6	152	
All Water Walled or Jacketed	Electric	6	152	—		18	457	6	152		6	152	
Furnaces—Central													
Gravity, Upflow, Downflow, Horizontal and Duct. Warm-Air—250°F (120°C) Max.	Automatic Oil or Comb. Gas-Oil	6[5]	152	6[5]	152	24	610	6	152		6	152	
	Automatic Gas	6[5]	152	6[5]	152	18	457	6	152		6	152	
	Electric	6[5]	152	6[5]	152	18	457	6	152		6	152	
Furnaces—Floor													
For Mounting in Combustile Floors	Automatic Oil or Comb. Gas-Oil	36	914	—		12	305	12	305		12	305	
	Automatic Gas	36	914	—		12	305	12	305		12	305	
	Electric	36	914	—		12	305	12	305		12	305	
Heat Exchanger													
Steam—15 psi (103 kPa) Max Hot Water—250°F (120°C) Max.	—	1	25	1	25	1	25	1	25		1	25	
Room Heaters													
Circulating Type Vented or Unvented	Oil	36	914	—		24	610	12	305		12	305	
	Gas	36	914	—		24	610	12	305		12	305	
Radiant or Other Type Vented or Unvented	Oil	36	914	—		36	914	36	914		36	914	
	Gas	36	914	—				18	457		18	457	
	Gas with double metal or ceramic back	36	914	—		36	914	12	305		18	457	
Radiators													
Steam or Hot Water	Gas	36	914	—		6	152	6			6		
										Firing Side			Opp. Side
Ranges—Cooking Stoves		See Note 7											
Vented or Unvented	Oil	30	762	—		—		9	229	24″*	10	457	18″*
	Gas	30	762	—		—		6	152	6″	6	152	6″
	Electric	30	762	—		—		6	152		6	152	
Clothes Dryers													
Listed Types	Gas	6	152	—		24	610	6	152		6	152	
	Electric	6	152	—		24	610	0			0		
Incinerators		See Note 10											
Domestic Types	—	36	914	—		48	1219	36	914		36	914	

for them are observed whether the enclosing walls are combustible or noncombustible.

Table 8-3G and Figures 8-3W to 8-3 BB show how clearances can be reduced by installing protection between the heat producing device and the combustible material. This information is especially helpful in situations of inadequate clearance where it would be impractical to move the heat producing appliance.

Clearance Reduction Systems: If the air space between heating appliances and combustible material is small, a barrier of metal, metal and insulating material, or masonry heat resistive material should be installed. Metal and masonry shields, or clearance reduction systems, tend also to distribute the heat, preventing in some measure its building up at one location. An air space must be left between the shields and the heat radiating surface on one side and between the barrier and the combustible material on the other. (See Table 8-3G.) The air currents set up by this procedure will prevent the combustible material from

Table 8-3E. (Continued)

Low Heat Appliances Any and All Physical Sizes Except as Noted		Appliance: Above Top of Casing or Appliance See Note 8 in.	mm.	From Top and Sides of Warm-Air Bonnet or Plenum in.	mm.	From Front in.	mm.	From Back See Note 8 in.	mm.	From Sides See Note 8 in.	mm.
Boilers and Water Heaters											
100 ft³ (2.8 m³) or less											
Any psi Steam	All Fuels	18	457	—		48	1219	18	457	18	457
50 psi (345 kPa) or Less											
Any Size	All Fuels	18	457	—		48	1219	18	457	18	457
Unit Heaters											
Floor Mounted or Suspended—Any Size	Steam or Hot Water	1	25	—		—		1	25.4	1	25.4
Suspended—100 ft³ (2.8 m³) or less	Oil or Comb. Gas-Oil	6	152	—		24	610	18	457	18	457
Suspended—100 ft³ (2.8 m³) or less	Gas	6	152	—		18	457	18	457	18	457
Suspended—Over 100 ft³ (2.8 m³)	All Fuels	18	457	—		48	1219	18	457	18	457
Floor Mounted Any Size	All Fuels	18	457	—		48	1219	18	457	18	457
Ranges—Restaurant Type											
Floor Mounted	All Fuels	48	1219	—		48	1219	18	457	18	457
Other Low-Heat Industrial Appliances											
Floor Mounted or Suspended	All Fuels	18	457	18	457	48	1219	18	457	18	457

Commercial-Industrial Type Medium-Heat Appliances		Appliance: Above Top of Casing or Appliance See Note 9 in.	m	From Top and Sides of Warm-Air Bonnet or Plenum in.	m	From Front in.	m	From Back See Note 9 in.	m	From Sides See Note 9 in.	m
Boilers and Water Heaters											
Over 50 psi Over 100 cu ft	All Fuels	48	1.22	—		96	2.44	36	0.91	36	0.91
Other Medium-Heat Industrial Appliances											
All Sizes	All Fuels	48	1.22	36	0.91	96	2.44	36	0.91	36	0.91
Incinerators											
All Sizes	—	48	1.22	—		96	2.44	36	0.91	36	0.91
High-Heat Industrial Appliances											
All Sizes	All Fuels	180	4.57	—		360	9.14	120	0.91	120	0.91

1. Standard clearances may be reduced by affording protection to combustible material in accordance with Table 8-3G.
2. An appliance may be mounted on a combustible floor if the appliance is listed for installation on a combustible floor, or if the floor is protected in an approved manner. For details of protection reference may be made to the Code for the Installation of Heat-Producing Appliances, obtainable from the American Insurance Association, 85 John Street, New York, N.Y. 10038, or Part 6 of the National Building Code of Canada published by the National Research Council, Ottawa, Canada.
3. Rooms which are large in comparison to the size of the appliance are those having a volume equal to at least 12 times the total volume of a furnace and at least 16 times the total volume of a boiler. If the actual ceiling height of a room is greater than 8 ft (2.4 m), the volume of a room shall be figured on the basis of a ceiling height of 8 ft (2.4 m).
4. The minimum dimension should be that necessary for servicing the appliance including access for normal maintenance, care, tube removal, etc.
5. For a listed oil, combination gas-oil, gas, or electric furnace this dimenstion may be 2 in. (51 mm) if the furnace limit control cannot be set higher than 250°F (120°C) or this dimension may be 1 in. (25 mm) if the limit control cannot be set higher than 200°F (93°C).
6. The dimension may be 6 in. (152 mm) for an automatically stoker-fired forced warm-air furnace equipped with 250°F (120°C) limit control and with barometric draft control operated by draft intensity and permanently set to limit draft to a maximum intensity of 0.13 in. (32 Pa) water gage.
7. To combustible material or metal cabinets. If the underside of such combustible material or metal cabinet is protected with asbestos millboard at least ¼ in. (6.4 mm) thick covered with sheet metal of not less than No. 28 gage the distance may be not less than 24 in. (610 mm).
8. If the appliance is encased in brick, the 18 in. (459 mm) clearance above and at sides and rear may be reduced to not less than 12 in. (305 mm).
9. If the appliance is encased in brick the clearance above may be not less than 36 in. (914 mm) and at sides and rear may be not less than 18 in. (457 mm).
10. Clearance above the charging door should be not less than 48 in. (1.2 m).

* 1 in. = 25.4 mm

Table 8-3F. Standard Clearances for Solid Fuel Burning Appliances
For Reduced Clearances, see Table 8-3G

These clearances apply to listed appliances installed in rooms which are large in comparison with the size of the appliances.
Source: NFPA 211, Standard for Chimneys, Fireplaces, Vents and Solid Fuel Burning Appliances.[F]

Kind of Appliance	Above Top of Casing or Appliance AboveTopandSidesofFurnace Planum or Bonnet		From Front		From Back[3]		From Sides[3]	
	in.	mm	in.	mm	in.	mm	in.	mm
Residential Appliances	6	152	48	1219	6	1522[2]	6	1522[2]
Steam Boilers—15 psi (103 kPa) max.								
Water Boilers—250°F (121°C) max.								
Water Boilers—200°F (93°C) max.								
All Water Walled or Jacketed								
Furnaces								
Gravity and Forced Air[4]	18	457	48	1219	18	457	18	457
Room Heaters, Fireplace Stoves, Combinations	36	914	36	914	36	914	36	914
					Firing Side		Opp. Side	
Ranges								
Line Firechamber	30	762[1]	36	914	24	610	18	457
Unlined Firechamber	30	762[1]	36	914	36	914	18	457

[1] To combustible material or metal cabintes. If the underside of such combustible material or metal cabinet is protected with sheet metal of not less than 0.024 in. (0.61 mm) (24 gage) spaced out 1 in. (25.4 mm) the distance may be reduced to not less than 24 in. (610 mm).

[2] Adequate clearance for cleaning and maintenance shall be provided.

[3] Provisions for fuel storage shall be located a least 36 in. (914 mm) from any side of the appliance.

[4] For clearances from air ducts, see NFPA 90B.

[5] Exceptions: Appliances listed for installation with clearances less than specified may be installed in accordance with the terms of their listing and manufacturer's instructions. Heating furnaces and boilers, and water heaters specifically listed for installation in spaces such as alcoves may be installed in accordance with the terms of their listing provided the specified clearance is maintained regardless of whether the enclosure is of combustible or noncombustible material.

attaining a dangerous temperature, even with the smaller clearance.

Limitations of Insulation

To make certain that insulating material will provide adequate protection, one needs to understand its limitations. The insulation may be used on the heating appliance or in combination with sheet metal and air spaces (Table 8-3G) if its value under such circumstances has been verified by tests. Many heat producing appliances have built in insulation, which makes it safe to reduce the required clearances to combustible material.

Insulation alone, however, is not sufficient. Regardless of the thickness of insulation, long continued heat may eventually penetrate it. However, if some method is used to conduct the heat away before it reaches the combustible material, the insulation will provide adequate protection.

Continued high temperatures over long periods of time have caused many fires under apparently safe conditions. Figure 8-3CC shows fires caused under conditions which the layman would consider safe. Solid masses of brick or concrete or plaster finish may not provide any fire protection under some circumstances, especially where masonry coverings are attached directly to the surface of the combustible material. Heat transfer to the combustible material is usually more efficient as the density of the covering material increases. Therefore, without an air space between the masonry and the combustibles, the heat transfer can actually increase resulting in the ignition of the combustibles. However, since the masonry products do have a relatively large mass it will take some time of operation of the appliance to transfer sufficient energy to ignite combustibles. This, however, is entirely dependent upon the weight and dimensions of the masonry and the rate of heat transfer from the appliance. The same rationale would also apply to products of other than masonry attached directly to the surface of combustibles.

Mountings

The limitations of masonry, concrete, metal, and other materials as insulating mediums apply particularly to the underside of stoves, heaters, boilers, furnaces, and other similar heat producing appliances. Tables 8-3H and 8-3I give a list of mountings for various classes of heat producing appliances.

The testing laboratories indicate in their listings whether the appliances tested may be installed on combustible or noncombustible floors. In some cases there is a statement on the appliance itself to the effect that it may be installed on a combustible floor. If no such indication appears, however, it is advisable to check the listing for the particular appliance.

Air for Combustion and Ventilation

In many locations, combustion-type heat-producing appliances have ample sources of air for efficient combustion in addition to the ventilation required to prevent undue temperature rises. In basements of dwellings, for example, sufficient air comes in through the cracks around doors and windows.

In relatively tight rooms such as furnace and boiler rooms, a means to supply air for combustion and ventilation must be provided. Because there are so many variables involved, there is no universally accepted formula for calculating the size of openings necessary to provide adequate air for combustion and ventilation. For example,

Table 8-3G. Reduction of Appliance Clearance with Specified Forms of Protection
Source: NFPA 211,

Clearance reduction system applied to and covering all combustible surfaces within the distance specified as required clearance with no protection (see Table 8-3F).	Maximum allowable reduction in clearance (percent)		When the required clearance with no protection is 36 in. (0.9 m), the clearances below are the minimum allowable clearance. For other required clearances with no protection, calculated minimum allowable clearance from maximum allowable reduction. (See Notes 9 and 10)			
			As Wall Protector		As Ceiling Protector	
	As Wall Protector	As Ceiling Protector	in.	mm	in.	mm
(a) 3½ in. (90 mm) thick masonry wall without ventilated air space.	33%	—	24	610		—
(b) ½ in. (13 mm) thick noncombustible insulation board over 1 in. (25 mm) glass fiber or mineral wool batts without ventilated air space.	50%	33%	18	457	24	610
(c) 0.024 in./0.61 mm (24 gage) sheet metal over 1 in. (25 mm) glass fiber or mineral wool batts reinforced with wire, or equivalent, on rear face with ventilated air space.	66%	50%	12	305	18	457
(d) 3½ in. (90 mm) thick masonry wall with ventilated air space.	66%	—	12	305		—
(e) 0.024 in./0.61 mm (24 gage) sheet metal with ventilated air space.	66%	50%	12	305	18	457
(f) ½ in. (13 mm) thick noncombustibel insulation board with ventilated air space.	66%	50%	12	305	18	457
(g) 0.024 in./0.61 mm (24 gage) sheet metal with ventilated air space over 0.024 in./0.61 mm (24 gage) sheet metal with ventilated air space.	66%	50%	12	305	18	457
(h) 1 in. (25 mm) glass fiber or mineral wool batts sandwiched between two sheets 0.024 in./0.61 mm (24 gage) sheet metal with ventilated air space.	66%	50%	12	305	18	457

Notes to Table 8-3G.
[1] Spacers and ties shall be of noncombustible material. No spacers or ties shall be used directly behind appliance or conductor.
[2] With all clearance reduction systems using a ventilated air space, adequate air circulation shall be provided as described in Fig. 8-3Z. There shall be at least 1 in. (25 mm) between the clearance reduction system and combustible walls and ceilings for clearance reduction systems using a ventilated air space.
[3] Mineral wool batts (blanket or board) shall have a minimum density of 8 lb per ft^3 (128.7 kg/m^3) and have a minimum melting point of 1,500°F (816°C).
[4] Insulation material used as part of clearance reduction system shall have a thermal conductivity of 1.0 (BTU-In.)/(Sq ft-Hr-°F) [(1.730 w/m.K)] or less. Insulation board shall be formed of noncombustible material.
[5] If a single wall connector passes through a masonry wall used as a wall shield, there shall be at least ½ in. (13 mm) or open, ventilated air space between the connector and the masonry.
[6] There shall be at least 1 in. (25 mm) between the appliance and the protector. In no case shall the clearance between the appliance and the wall surface be reduced below that allowed in the table.
[7] Clearances in front of the loading door and/or ash removal door of the appliance shall not be reduced from those in Section 8-6 of NFPA 211.
[8] All clearances and thicknesses are minimums; larger clearances and thicknesses are acceptable. Clearances are not to be less than 12 in. (305 mm) from appliances.
[9] To calculate the minimum allowable clearance, the following formula may be used: $C_{pr} = C_{un} \times (1 - R/100)$. C_{pr} is the minimum allowable clearance, C_{un} is the required clearance with no protection, and R is the maximum allowable reduction in clearance.
[10] Refer to Figures 8-3AA and 8-3BB for other reduced clearances using materials (a) through (h) above.

the tightness and size of the furnace or boiler room, and the operation of exhaust fans and other equipment would affect the static air pressure in the building. Both NFPA 31 and NFPA 54, as well as other standards pertaining to heat producing appliances, include specific recommendations on how to supply the air for combustion and ventilation.

For certain laboratory tested appliances, notably gas and oil burning appliances for installation in closets, the minimum size is specified for an opening to the heater room or space necessary to provide enough air for the operation of the unit.

Clearances for Servicing

Clearances must also be provided for servicing and maintenance of equipment. Lack of proper servicing and good maintenance can result in fires. If lack of space around an appliance makes accessibility difficult, the appliance, or at least some parts of it, will be neglected. This same reasoning applies to appliances located in out of the way places and places difficult to reach. Suspended furnaces and furnaces in attics and underfloor crawl spaces are typical examples of such installations.

CHIMNEY AND VENT CONNECTORS

Chimney and vent connectors are specifically the pipe or breeching used to connect fuel burning appliances with the required chimney or vent unless the chimney or vent is attached directly to the appliance. NFPA 211, *Standard for Chimneys, Fireplaces, Vents and Solid Fuel Burning Ap-*

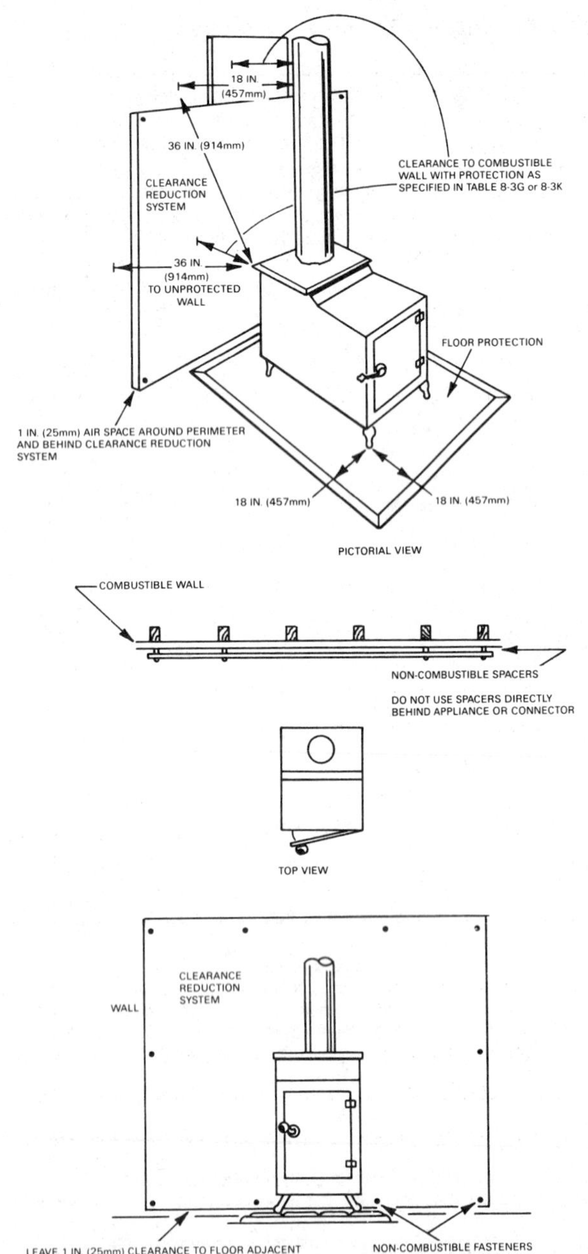

FIG. 8-3W. A clearance reduction system for a heat producing appliance. The system is a barrier of metal, metal and insulating material, or masonry heat resistive material with an air space between the barrier and a combustible surface.

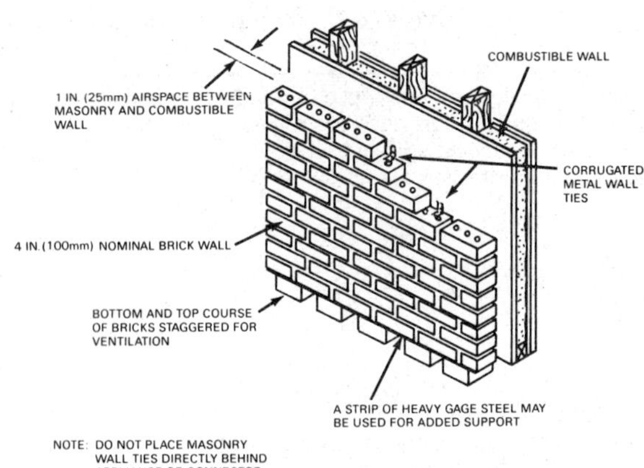

FIG. 8-3X. Details of a masonry clearance reduction system.

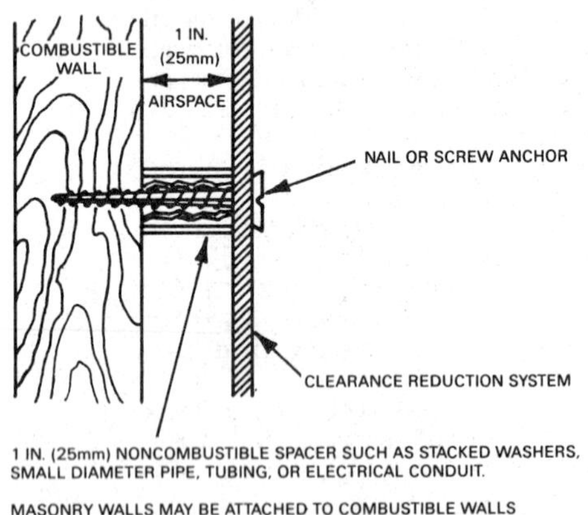

FIG. 8-3Y. Details of anchoring a clearance reduction.

pliances (hereinafter referred to as NFPA 211) includes requirements for chimney and vent connectors and should be consulted for information on materials to be used and the sizing and installation of connectors.

Appliances to be Chimney or Vent Connected

All oil fired appliances are chimney connected except direct fired heaters, listed kerosene stoves, and portable kerosene heaters. All solid fueled appliances are chimney connected.

NFPA 54 is quite specific on the type of gas appliances that are required to be vented through vents or chimneys, and those not requiring vents. Generally, it is the larger type of heating appliances, e.g., boilers and furnaces, that require venting while smaller appliances such as listed cooking stoves and hot plates do not. However, if several smaller appliances that normally do not require venting are located within the same space or room, some may require venting if the aggregate input for all the appliances exceeds 20 Btu per hr per cu ft (0.2 kW/m^3) of the space in which they are installed. NFPA 54 should be consulted for the specifics on the type of appliances requiring venting the conditions under which vents may be omitted or required and the type of chimney, gas vent, or venting system that can be used and their installation.

Listed unvented room heaters may be used, but care should be taken to follow the manufacturer's instructions. Such heaters are equipped with an oxygen depletion safety shutoff system designed to shut off the gas supply to the heater if the oxygen in the surrounding atmosphere is reduced below a safe level. Unvented room heaters are not to be used in sleeping quarters or bathrooms, institutions, convalescent homes, orphanages, etc.

Ventilating hoods and exhaust systems may be used to vent gas utilization equipment installed in commercial applications. They may also be used to vent industrial

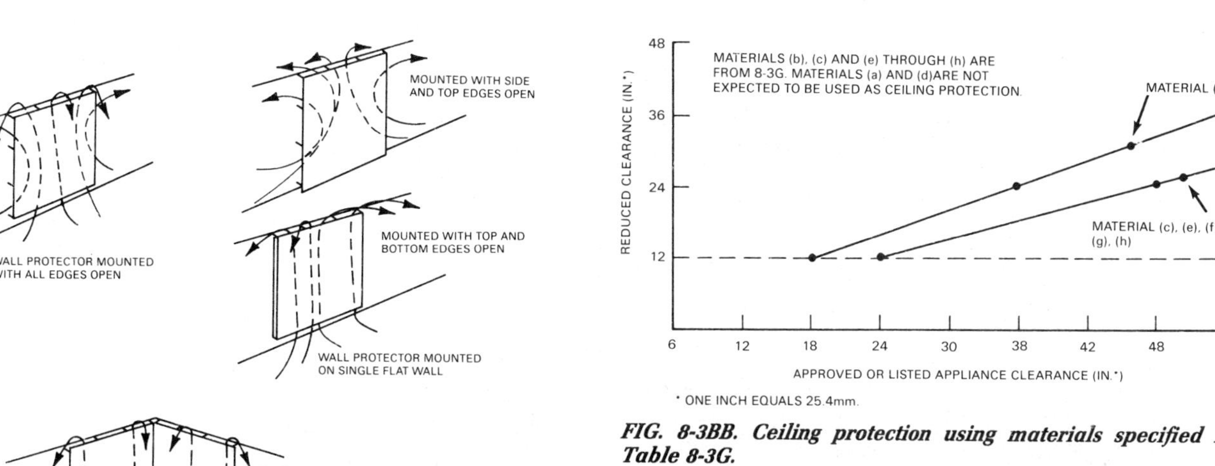

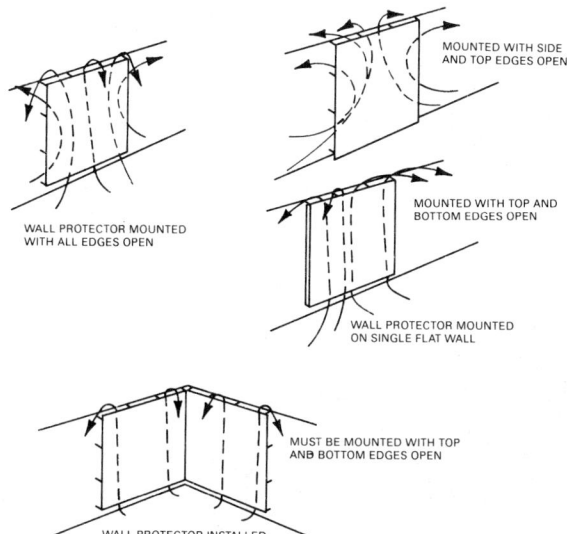

FIG. 8-3Z. Airflow patterns for various configurations of clearance reduction systems having an air space between the barrier and the combustible surface it protects.

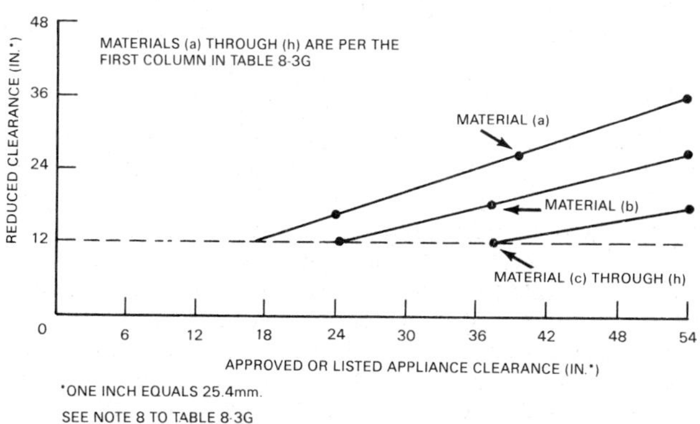

FIG. 8-3AA. Wall protection using materials specified in Table 8-3G.

equipment, particularly when the process itself requires fume disposal. If industrial gas utilization equipment is located in a large and well ventilated space, it may be operated by discharging the products of combustion directly into the atmosphere.

VENTS

Vents are laboratory tested, factory built units used to vent fuel burning heat producing appliances. Specific types of vents have specific uses. (See Table 8-3J.)

Types of Vents

Type B Gas Vents: These vents are used to vent listed gas appliances with draft hoods or which are specifically listed to be used with Type B gas vents.

Type BW Gas Vents: These vents are used with listed gas wall furnaces having capacities not greater than that of Type BW gas vents. (See Fig. 8-3DD.)

FIG. 8-3BB. Ceiling protection using materials specified in Table 8-3G.

Type L Vents: These vents are used to vent oil burning appliances listed for use with Type L vents and gas appliances listed for use with Type B vents.

Single Wall Metal Pipe: Single wall metal pipe, constructed of galvanized sheet steel not lighter than No. 20 galvanized sheet gage (0.020 in. or 0.61 mm) or other noncombustible corrosion resistant material, may be used to vent residential type and low heat gas appliances equipped with draft hoods. They may also be used to vent incinerators used outdoors such as in open sheds, breezeways or carports. (See Table 8-3J.) Single wall metal pipe is not used with solid fuel appliances which require either masonry or factory built chimneys.

Venting High Efficiency Heating Appliances

Manufacturers have developed more efficient heating appliances to conserve energy. To obtain efficiencies approaching 90 percent or better, the flue gases from these appliances have been reduced to temperatures as low as 100°F (38°C). Because these temperatures are too low to allow atmospheric venting, mechanical venting is required to exhaust the flue gases under positive pressure. Furthermore, the temperature of the flue gases are low enough to cause condensation of the water vapor in flue gases. Thus special treatment is needed to provide a venting system that will retain and dispose of the condensate and, in cases of mechanical venting, to prevent leakage of flue gases from the venting system. The information in the 1984 edition of NFPA 54 or venting systems for the types of appliances heretofore available may not be appropriate. High efficiency appliances, if used, should be listed and installed and vented in compliance with the instructions furnished by the manufacturer with the appliance.

Installation of Vents

Types B, BW and L Vents: Installations must be made in compliance with the terms of their laboratory listings and the manufacturer's instructions; making certain that the required clearances are maintained.

Single Wall Metal Pipe: The use of single wall metal pipe is restricted to runs directly from the space in which the gas appliance is located through the roof or exterior wall to the outside. The pipe is not to originate in any attic or concealed space and is not to pass through any attic or inside space, nor through any floor or ceiling. The mini-

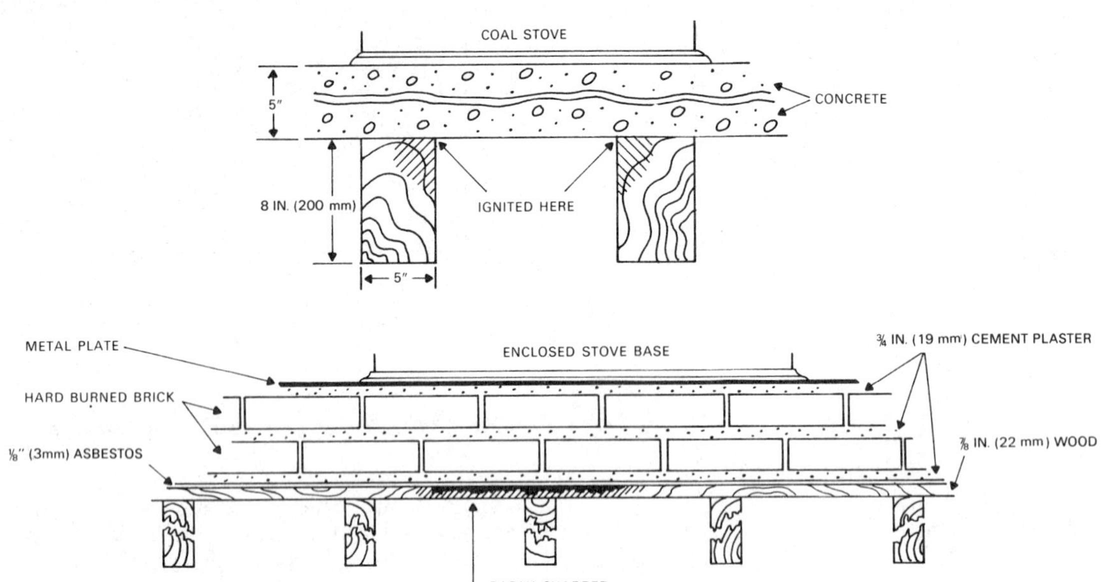

FIG. 8-3CC. Typical fires due to improperly installe heating appliances.

mum clearances from combustible material for single wall metal pipe are as given for connectors in Table 8-3K. (See also Fig. 8-3EE.) Where a single wall metal pipe passes through an exterior wall constructed of combustible material, or the pipe passes through a roof constructed of combustible material, it is protected at the point of passage as specified for connectors in Table 8-3L. If the appliance the pipe serves is a gas appliance for use with a Type B gas vent, protection is by a noncombustible, nonventilating thimble not less than 4 in. (102 mm) larger in diameter than the pipe and extending not less than 18 in. (457 mm) above and 6 in. (152 mm) below the roof, with annular space open at the bottom and closed at the top.

Venting Capacities: The venting capacities of various sizes of Type B gas vents, chimneys, and single wall metal pipe used for venting gas appliances are given in tables in NFPA 54. NFPA 54 also contains data that may be used to calculate the size of vents required under various circumstances and to establish configurations of venting arrangements. Approved engineering methods may also be used.

Firestopping: Vents that pass through floors of buildings requiring the protection of vertical openings are enclosed within walls having a fire resistance rating of not less than 1 hr if the vents are in a building less than four stories in height and not less than 2 hr if the vents are located in a building four stories or more in height.

Draft Hoods

A draft hood is a device built into an appliance, or made a part of the vent connector from an appliance. It is designed to (1) assure the ready escape of the flue gases in the event of no draft, backdraft, or stoppage beyond the draft hood, (2) prevent a backdraft from entering the appliance, and (3) neutralize the effect of stack action of the chimney or gas vent upon the appliance operation.

Draft hoods are an important safety feature of gas appliances and are generally required on all vented appliances except incinerators, dual oven type combination ranges, direct vent appliances, and units designed for power burners or for forced venting by mechanical means.

CHIMNEYS AND FIREPLACES

Chimneys are classified into three major types: (1) factory built, (2) masonry, and (3) metal (smokestacks). In addition to the major chimney classes, there are subclasses, which involve type of building and surroundings. For example, small, well insulated, factory built chimneys may be installed at close clearances in residential frame construction, while larger chimneys for the same type of appliances are suitable mainly for noncombustible surroundings or in well ventilated open room use at much greater clearance to combustibles. Details of installation for chimneys, fireplaces, and vents are provided in NFPA 211, *Standard for Chimneys, Fireplaces, Vents and Solid Fuel Burning Appliances.*

Factory Built Chimneys

A factory built chimney is an assembly of manufactured components that form a completed chimney. Factory built chimneys are tested for compliance with safety standards (ANSI/UL 1976 & 1978).

Factory built chimneys for use with residential type appliances are tested fully enclosed to establish a minimum clearance to combustibles not greater than 2 in. (51 mm). The test is conducted using a flue gas generator which simulates an appliance steadily producing 1,000°F (540°C) gases. A typical factory built chimney for residential appliances is shown in Figure 8-3FF.

For building heating equipment (gases to 1,000°F or 540°C), commercial ovens and furnaces (gases to 1,400°F or 760°C), and medium heat appliances (gases to 1,800°F or 980°C), factory built chimneys are available in sizes from 10 in. (250 mm) to several feet (one foot equals 0.3 m) in diameter. These chimneys are tested for the appropriate temperature and are used mainly in noncombustible surroundings, although tested thimbles are available for penetration of combustible roofs. None of these are suitable for

Table 8-3H. Floor Mountings for Heat-Producing Appliances

Type of Mounting	Required for the Following Types of Heaters and Furnaces
No Floor Protection: Combustible floors.*	Residential- type central furnaces so arranged that the fan chamber occupies the entire area beneath the firing chamber and forms a well-ventilated air space of not less than 18 in. (457 mm) in height between the firing chamber and the floor, with at least one metal baffle between the firing chamber and the floor. Low heat appliances (see Table 7-3E for examples) in which flame and hot gases do not come in contact with the base, on legs which provide not less than 18 in. (457 mm) open space under the base of the appliance, with at least one sheet metal baffle between any burners and the floor. Other appliances for which there is evidence that they are designed for safe operation when installed on combustible floors.
Metal: A sheet of metal not less than No. 24 U.S. Gage (0.024 in./0.61 mm) or other approved noncombustible material, laid over a combustible wood floor.*	Heating and cooking appliances set on legs or simulated legs which provide not less than 4 in. (100 mm) open space under the base. Ordinary residential stoves with legs. Residential ranges with legs. Residential room heaters with legs. Water heaters with legs. Laundry stoves with legs. Room heaters with legs.†
Hollow Masonry: Hollow masonry not less than 4 in. (100 mm) in thickness laid with ends unsealed and joints matched in such a way as to provide free circulation of air through the masonry.	Downflow furnaces.
Hollow Masonry and Metal: Hollow masonry not less than 4 in. (100 mm) in thickness covered with a sheet of metal not less than No. 24 U.S. Gage (0.024 in./0.61 mm), laid over a combustible floor. The masonry will be laid with ends unsealed and joints matched in such a way as to provide a free circulation of air from side to side through the masonry.	Heating furnaces and boilers in which flame and hot gases do not come in contact with the base: Floor mounted heating and cooking appliances. Residential stoves without legs. Residential ranges without legs. Room heaters without legs. Water heaters without legs. Laundry stoves without legs. Residential type incinerators. Restaurant ranges on 4 in. (100 mm) legs. Other low heat appliances on 4 in. (100 mm) legs. Medium heat appliances on legs which provide not less than 24 in. (610 mm) open space under the base.
Two Courses Masonry and Plate: Two courses of 4-in. (100 mm) hollow clay tile covered with steel plate not less than 3/16 in. (4.76 mm) in thickness, laid over a combustible floor. The courses of tile will be laid at right angles with ends unsealed and joints matched in such a way as to provide a free circulation of air through the masonry courses.	Heating furnaces and boilers in which flame and hot gases come in contact with the base. Restaurant ranges. Other low heat appliances.
Fire Resistive Floors Extending 6 in. (152 mm): Floors of fire-resistive construction with noncombustible flooring and surface finish and with no combustible material against the underside thereof, or on fire-resistive slabs or arches having no combustible material against the underside thereof. Such construction will extend not less than 6 in. (152 mm) beyond the appliance on all sides, and where solid fuel is used, it will extend not less than 18 in. (457 mm) at the front or side where ashes are removed.	Floor mounted heating and cooking appliances. Residential type room heaters. Residential type water heaters.
Fire Resistive Floors Extending 12 in. (305 mm): Floors of fire-resistive construction with noncombustible flooring and surface finish and with no combustible material against the underside thereof, or on fire-resistive slabs or arches having no combustible material against the underside thereof. Such construction will extend not less than 12 in. (305 mm) beyond the appliance on all sides, and where solid fuel is used, it will extend not less than 18 in. (457 mm) at the front or side where ashes are removed.	Heating furnaces or boilers. Restaurant-type cooking appliances. Residential-type incinerators. Other low heat appliances.

Table 8-3H. Floor Mountings for Heat-Producing Appliances (Continued)

Type of Mounting	Required for the Following Types of Heaters and Furnaces
Fire Resistive Floors Extending 3 ft (0.9 m): Floors of fire-resistive construction with noncombustible flooring and surface finish and with no combustible material against the underside thereof, or on fire-resistive slabs or arches having no combustible material against the underside thereof. Such construction will extend not less than 3 ft (0.9 m) beyond the appliance on all sides, and where solid fuel is used, it will extend not less than 8 ft (2.4 m) at the front or side where ashes are removed.	Medium heat appliances and furnaces. (See Table 8-3G for examples.)
Fire Resistive Floors Extending 10 ft (3.1 m): Floors of fire-resistive construction with noncombustible flooring and surface finish and with no combustible material against the underside thereof. Such construction will extend not less than 10 ft (3.1 m) beyond the appliance on all sides, and where solid fuel is used, it will extend not less than 30 ft (9.1 m) at the front or side where hot products are removed.	High heat appliances and furnaces. (See Table 8-3G for examples.)

* Where an appliance is mounted on a combustible floor, and solid fuel is used or the appliance is a domestic type incinerator, a sheet of ¼-in. (25 mm) asbestos covered by a sheet of metal not less than No. 24 U.S. Gage (0.024 in./0.61 mm) will be required extending at least 18 in. (457 mm) from the appliance on the front or side where ashes are removed. (The sheet of asbestos may be omitted where the protection required under the appliance is a sheet of metal only.) For residential type incinerators the protection must also extend at least 12 in. (305 mm) beyond all other sides. If the appliance is installed with clearance less than 6 in. (152 mm) the protection for the floor should be carried to the wall.

† Floor protection for radiating type gas burning room heaters which make use of metal, asbestos or ceramic material to direct radiation to the front of the device should extend at least 36 in. (914 mm) in front when the heater is not of a type approved for installation on a combustible floor.

close clearance to combustible enclosures; however, this is seldom a problem in the usual large building of masonry or metal construction. A typical unenclosed type of large factory built chimney for serving building boilers is shown in Figure 8-3GG.

Current information on types, usage, and installation limitations of all UL listed factory built chimneys is given in the UL *Gas and Oil Equipment Directory* published annually with a supplement.

Masonry Chimneys

Field construction requirements for masonry chimneys are detailed in NFPA 211, as well as in building codes. Field erected masonry chimneys, which are subject to the know-how and ability of available labor are not tested, except perhaps for a smoke test. A minimum residential masonry chimney consists of a refractory fire clay tile liner, an air space of roughly ½ in. (13 mm) between the liner and brick, and with all liner joints grouted both to prevent leakage and to center and support the tile liners. One course of common brick around the liner suffices for usual residential chimneys. For higher temperature classifications, firebrick is used as the liner, with additional courses of brick for larger size or greater strength and security. All chimneys should, of course, be supported by footings suitable for the chimney size and weight and provided with cleanouts between the points of entry and the lowest chimney level.

As a general rule, the best and safest chimney operation and appliance control is obtained with only one appliance attached to a chimney flue. Two or three gas appliances, however, may be connected into a masonry chimney flue at one level of a building providing that the connector size, its vertical rise, and chimney size and height are in accordance with tabulated capacity data in NFPA 54. The construction of masonry chimneys for open front fireplaces is the same as for other residential appliances and is specified in NFPA 211.

Metal Chimneys

Metal chimneys are suitable for all the classes of appliances but are not subjected to safety testing of any kind. The major hazard with these chimneys is inadequate clearance to combustibles where they penetrate ceilings and roofs. (See Fig. 8-3HH.)

Metal chimneys may be of single wall metal for low temperature use, such as with gas appliances, or metal lined with firebrick or refractory mortar for medium and high heat service. They may be located inside or outside of buildings but not inside one and two family dwellings or buildings of wood frame construction. The conditions under which metal chimneys may be used are quite limited and are spelled out in detail in NFPA 211. The standard is also quite specific on acceptable materials of construction.

Regardless of permitted clearances, metal chimneys should not be enclosed with combustible construction. It is particularly important that the metal used should be resistant to corrosion if the gases are 350°F (177°C) or below, because exposed outdoor portions of the chimney walls may be below the dew point. The resulting continuous condensation can lead to rapid corrosion and very short chimney life.

Chimney Functions

The purpose of a chimney is to create draft or negative pressure to provide combustion air for the fuel or fuel bed and to remove products of combustion from the appliance and building. For open fireplaces or for gas appliances having draft hoods, chimney draft has very little influence on the combustion process; the chimney serves only as a conduit for carrying away the products of combustion. For forced draft or packaged boilers, the chimney may be under slight positive pressure and again serves only as a duct to carry away combustion products. The advantage of the slight positive pressure, commonly from 0.5 to 1.5 in. of water (120 to 370 Pa) column is that it permits use of

Table 8-3I. Floor Mountings for Solid Fuel Burning Appliances

Kind of Appliance	Allowed Mounting
(1) All forced air & gravity furnaces, steam and water boilers. (2) Residential type ranges, stoves, room heaters, and combination fireplace stove/room heaters, having less than 2 in. (51 mm) of ventilated open space beneath the fire chamber or base of the unit.	Floors of fire resistive construction with noncombustible water heaters, fireplace flooring and surface finish, or fire resistive arches or slabs. These constructions shall have no combustible material against the underside thereof. Such construction shall extend not less than 18 in. (457 mm) beyond the appliance on all sides. These appliances shall not be placed on combustible floors.
(3) Residential-type ranges, water heaters, fireplace stoves, room heaters and combination stove/room heaters having legs or pedestals providing 2 to 6 in. (51 to 152 mm) of ventilated open space beneath the fire chamber or base of the heater.	On combustible floors when such floors are protected by 4 in. (102 mm) of hollow masonry, laid to provide air circulation through the masonry layer. Such masonry shall be covered with 24 U.S. Gage (0.024 in./0.61 mm) sheet metal. The required floor protection shall extend not less than 18 in. (457 mm) on all sides of the appliance. On noncombustible floors, such floors shall extend not less than 18 in. (457 mm) on all sides of the appliance.
(4) Residential-type ranges, water heaters, fireplace stoves, room heaters and combination fireplace stove/room heaters having legs or pedestals providing over 6 in. (152 mm) of ventilated open space beneath the fire chamber or base of the covered heater.	On combustible floors when such floors are protected by closely spaced masonry units of brick, concrete or stone, which provide a thickness of not less than 2 in. (51 mm). Such masonry shall be covered by or placed over a sheet of 24 U.S. Gage (0.024 in./0.61 mm) steel. The required floor protection shall extend not less than 18 in. (457 mm) on all sides of the appliance. On noncombustible floors, such floors shall extend not less than 18 in. (457 mm) on all sides of the appliance.

much smaller chimneys and eliminates some concerns about the draft producing ability or height of the chimney.

The technology of matching the size, height, and configuration of a chimney to the temperature, type, and quantity of fuel for its correct capacity is covered in the *ASHRAE Handbook* (ASHRAE 1979).

Height of Chimneys

A variety of factors determine the height of any chimney, including the type of appliance, need for draft, roof construction, and building height. The minimum height of a chimney (or gas vent) for gas appliances having draft hoods is based on the height specified in the applicable test standard for that appliance. For example, gas furnaces, water heaters, boilers, and room heaters must pass draft hood spillage tests with 5 ft (1.5 m) of chimney height measured from the draft hood to chimney outlet, while wall furnaces require a vent outlet at least 12 ft (3.7 m) above the floor.

In contrast to these low heights for draft hood appliances, some older types of solid fuel boilers require up to 50 or 70 ft (15 to 20 m) of chimney to produce the draft needed to overcome fuel bed and internal flow resistance and to develop rated heat output.

Residential oil furnaces and boilers with pressure atomizing burners are designed to function with negative outlet draft settings in the range of 0.04 to 0.06 in. of water (10 to 15 Pa) column. Manufacturers of factory built residential chimneys have prepared draft chimney height tables which allow for such factors as the number of connector elbows, fuel input rating, and outlet size so that an adequate height can be selected.

Building or dwelling height and configuration also govern chimney height. The generally cited rule for minimum height of the chimney outlet is illustrated in Figure 8-3II. The chimney height selected must be adequate for proper appliance operation and for adequate chimney flow as well as for proper height above the roof. Owing to the aerodynamic complexity of buildings, roofs, and chimneys, the dimensions suggested in Figure 8-3HH sometimes are not adequate for efficient operation, which is attested to by the proliferation of special vent caps, chimney extensions, and other devices used in attempting to cure fireplace or draft problems.

Airflow over the roof must be visualized to analyze a structure for the correct chimney top location and to minimize backdrafts due to wind. Only the pressure or windward side of pitched roofs needs to be considered, because the flow moving up a building wall or toward the ridge separates from the downward side at the ridge or building edge and creates a zone of negative pressure, which circulates and eddies in this protected or cavity zone as shown in Figure 8-3JJ. The building thus may aid or impede chimney flow by using the wind to create a zone of negative or positive pressure at the chimney outlet.

Spark Arresters on Chimneys

Spark arresters are required on refuse burners or incinerators but are desirable for all solid fuels. If sparks are expected from a fireplace or solid fuel appliance, a chimney of any kind should have a spark screen, and/or the roof surface material should be noncombustible. The usual recommendation for an arrester is for a wire mesh cloth or expanded metal having approximately ½ in. (13 mm) square openings. Smaller openings clog rapidly, while larger ones could allow passage of some sparks. Stainless steel material is recommended, with galvanized steel hardware cloth a poor second choice because the coating soon burns off and frequent replacement is necessary. Factory built chimney manufacturers offer a variety of chimney cap options with integral and accessory spark arrester screens.

Selection of Chimneys

As an aid to selecting chimneys, heat producing appliances have been graded by temperatures developed in the heating media or material being heated, and also by

Table 8-3J. Vent Selection Chart

Type of Vent			
Type B—Gas	Type BW—Gas	Type L—Oil	Metal Pipe
Column I All listed gas appliances with draft hoods such as: 1. Central furnaces 2. Duct furnaces 3. Floor furnaces 4. Heating boilers 5. Ranges 6. Built-in ovens 7. Vented wall furnaces listed for use with Type B vents 8. Room heaters 9. Water heaters 10. Horizontal furnaces 11. Unit heaters	Column II 1. Vented wall furnaces listed for use with Type BW vents only	Column III 1. Low temperature flue gas appliances listed for use with Type L vents 2. Gas appliances shown in Column I	Column IV 1. Incinerators used outdoors, such as in open sheds, breezeways or carports. 2. Gas appliances shown in Column I 3. Listed residential and low heat gas appliances without draft hoods and unlisted residential and low heat gas appliances with or without draft hoods

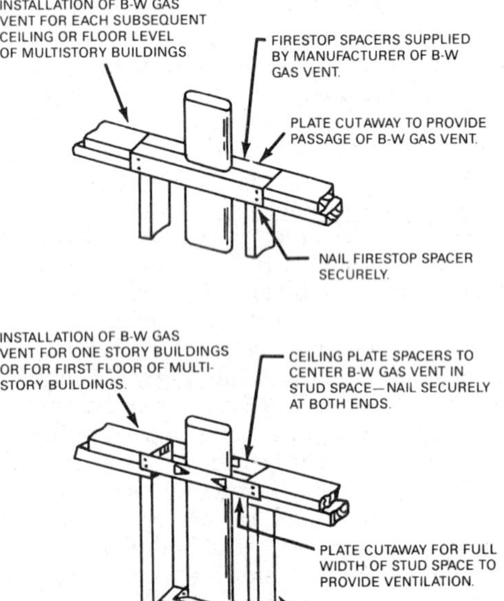
INSTALLATION OF B-W GAS VENT FOR EACH SUBSEQUENT CEILING OR FLOOR LEVEL OF MULTISTORY BUILDINGS.
FIRESTOP SPACERS SUPPLIED BY MANUFACTURER OF B-W GAS VENT.
PLATE CUTAWAY TO PROVIDE PASSAGE OF B-W GAS VENT.
NAIL FIRESTOP SPACER SECURELY.
INSTALLATION OF B-W GAS VENT FOR ONE STORY BUILDINGS OR FOR FIRST FLOOR OF MULTI-STORY BUILDINGS.
CEILING PLATE SPACERS TO CENTER B-W GAS VENT IN STUD SPACE—NAIL SECURELY AT BOTH ENDS.
PLATE CUTAWAY FOR FULL WIDTH OF STUD SPACE TO PROVIDE VENTILATION.
STUDS ON 16 IN. (0.4m) CENTERS
SHEET METAL SCREW BASE PLATE TO HEADER
USE MANUFACTURER'S METHOD OF FASTENING PIPE TO BASE PLATE.
HEADER PLATE OF VENTED WALL FURNACE (ALSO ACTS AS FIRESTOP)

FIG. 8-3DD. Installation of Type B-W gas vents for vented wall furnaces.

size. The grades of low, medium, and high heat appliances (Table 8-3M) can then be used to choose the chimney. This selection method provides help if the appliance outlet gas temperature is consistent with its grade; however, a study of several hundred types of process appliances, ovens, and furnaces has indicated the precepts of the grading system do not always hold true. With the availability of modern temperature instrumentation and controls and the need to conserve energy, the outlet flue gas temperature and other conditions can and should be known more precisely than heretofore for the vast majority of fuel burning equipment. This information, which should be readily obtainable from equipment and appliance manufacturers, can offer a logical basis for chimney selection and for determining safe conditions of installation and use.

As an example of a grading inconsistency, a steam boiler operating at more than 50 psig (350 kPa) is classified in some codes as a medium heat appliance. In selecting a chimney it could be assumed that the outlet temperature was as high as 1,500 to 1,800°F (815 to 980°C). Modern multiple pass boilers, however, operate at controlled outlet gas temperatures only 100 to 200°F (38 to 93°C) above steam temperature. Thus the correct chimney for even a 150 psig (1000 kPa) steam boiler would be selected for an outlet gas temperature of 365.9°F + 200°F or 565.9°F (185°C + 93°C or 278°C). This is well below the 600°F (316°C) which is one minimum temperature at which the medium heat grade begins. Thus, in this instance, neither the grade of appliance based on temperature of the process [2,400°F (1300°C) gas in the flame] or material [365.9°F (185°C) steam] provides a conclusive basis for selecting the boiler's chimney.

Selection of the correct chimney may be aided by the following two checklists. (In those areas with adopted building codes, the code requirements take precedence over the generalities in the checklists; however, if special engineering is needed for a particular application, this approach should generally also meet the intent of any modern performance code.)

Chimney Selection Checklist

Code Factors

Applicable building code
Type or grade of appliance
Appliance manufacturers' instructions
Chimney manufacturers' instructions
Listed, certified appliance
Unlisted appliance
Earthquake zone—for masonry types
Air quality and external discharge factors
Type, occupancy, and use of building
Availability of fire protection service

Engineering Factors

Temperature of gases from appliance
Characteristics of fuel and combustion gases
Corrosives, particulates, deposits, dew point

Table 8-3K. Connector Clearances with Specified Forms of Protection

Type of Protection	Where the required clearance with no protection is:							
	in.	mm	in.	mm	in.	mm	in.	mm
Applied to the combustible material and covering all surfaces within the distance specified as the required clearance with no protection. *(See Figure 8-3DD.)* Thicknesses are minimum.	36	914	18	457	9	229	6	152
	in.	mm	in.	mm	in.	mm	in.	mm
(a) 0.013 in./0.330 mm (28 gage) sheet metal spaced out 1 in. (25.4 mm)	18	457	9	229	4	102	2	51
(b) 3½ in. (88.9 mm) thick masonry wall spaced out 1 in. (25.4 mm) and adequately tied to the wall being protected. *(See Note 4.)*	18	457	9	229	4	102	2	51
(c) 0.027 in./0.686 mm (22 gage) sheet metal on 1 in. (25.4 mm) mineral wool batts reinforced with wire or equivalent spaced out 1 in. (25.4 mm)	12	305	3	76	2	51	2	51

Notes to Table 8-3K:
[1] Spacers and ties shall be of noncombustible material.
[2] All methods of protection require adequate ventilation between protective material and adjacent combustible walls and ceilings.
[3] Mineral wool batts (blanket or board) shall have a minimum density of 8 lb per ft^3 (12.8 kg/m^3) and a minimum melting point of 1,500°F (816°C).
[4] If a single wall connector passes through the masonry wall there shall be at least ½ in. (12.7 mm) open ventilated airspace between the connector and the masonry.

Quantity of combustion products
Operating needs and cycle of appliance
Method of draft control
Location of chimney in or outside of structure
Materials adjacent to chimney
Space available for chimney
Length and insulating ability of chimney

Hazards of Chimneys

For chimneys serving oil, wood, or coal appliances the possibility of fire involves one or more of the following elements: (1) operator error or ignorance, (2) control failure, (3) improper installation, (4) use of a defective or unlisted appliance, (5) ignition of combustible soot or creosote deposits in the chimney, (6) serious cracks or internal collapse of the masonry, (7) defective construction, (8) poor installation, and (9) failure to secure joints of factory built products.

Even masonry chimneys for gas appliances operating at very low temperatures have their problems. Condensation of flue product moisture may eventually disintegrate the mortar, leaving the chimney weak and easily toppled—a hazard to passersby.

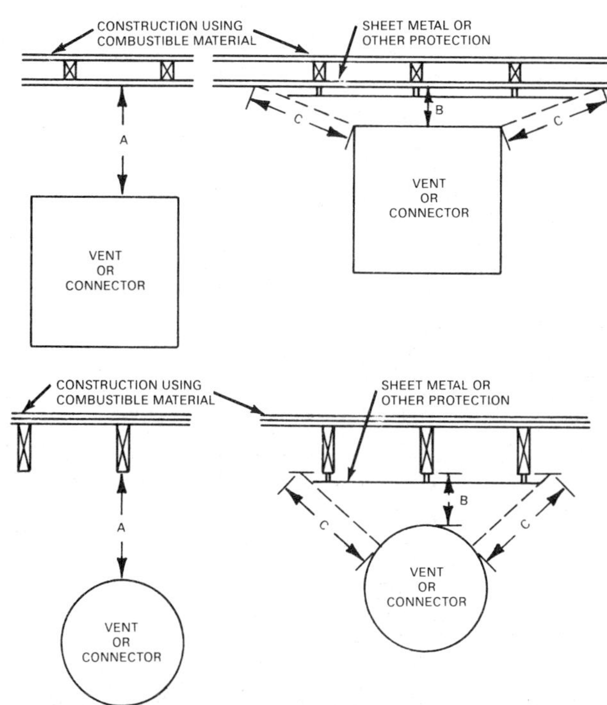

FIG. 8-3EE. Extent of protection required to reduce clearances from combustibles for vents and connectors. A equals the required clearance with no protections as specified in Table 8-3K; B equals the reduced clearance permitted by the types of protection specified in Table 8-3K. The protection applied to construction using combustible material should extend far enough in each direction to make C equal to A.

Table 8-3L. Connectors Through Combustible Partitions

1. Listed Gas Appliances with Draft Hoods (Col. I, Table 8-3J). Oil Appliances listed for Type L Vents (Col. III, Table 8-3J)	Type B or Type L Vent Material for gas appliances, Type L for oil appliances. Installed with listed clearances or single wall metal pipe guarded by ventilated metal thimble at least 4 in. (102 mm) larger in diameter than the pipe.
2. Other low heat appliances	Ventilated metal thimble at least 12 in. (305 mm) larger in diameter than connector, or a metal or burned fire clay thimble in brick or other fireproofing material extending at least 8 in. (203 mm) on all sides beyond thimble.
3. Medium and high heat appliances	Not permitted.

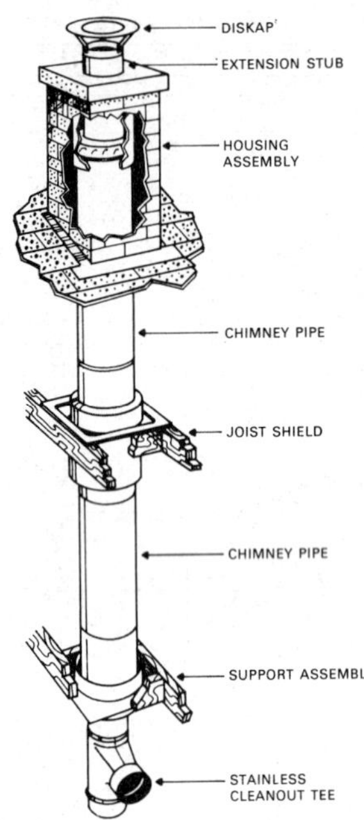

FIG 8-3FF. A typical factory built chimney. (Metalbestos Div. Wallace-Murray Corp.)

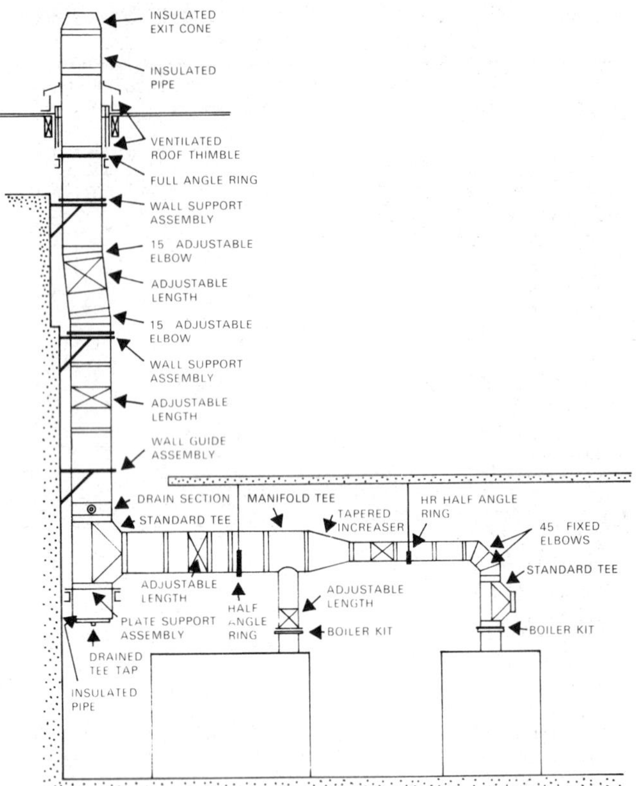

FIG. 8-3GG. A typical unenclosed factory built chimney for serving building boilers.

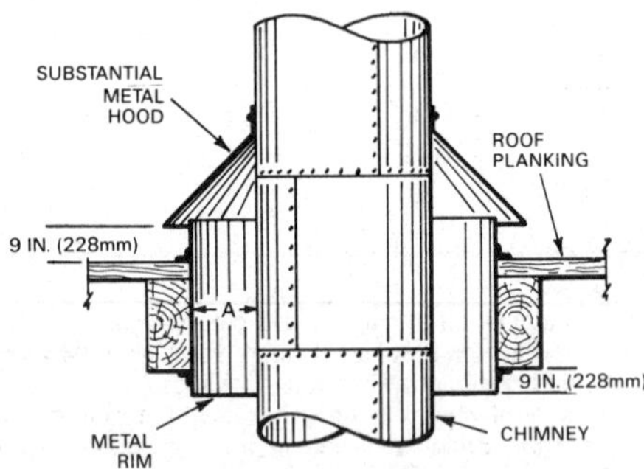

FIG. 8-3HH. Arrangement of a metal chimney through a wooden roof. For chimneys serving residential type or low heat appliances, dimension "a" should be at least 6 in. (152 mm). For chimneys serving medium heat appliances, dimensions "a" should be at least 18 in. (457 mm).

In masonry chimneys, the following conditions may cause a fire in adjacent combustible materials:

1. Filling the space between the chimney and wood framing with combustible insulation, or placing framing into or against the chimney wall.
2. Using a restrictive chimney cap in combination with cracks or failed mortar in a chimney, which may cause excessive leakage of hot gas at roof level, thus endangering rafters and roofing.
3. Prolonged overfiring, which may cause excessive heating on adjacent walls. At temperatures of 500°F (260°C) on the brick exterior surface, 2 in. (51 mm) clearance to nearby combustibles is of little protective value, particularly if there is little ventilation airflow.
4. Creosote or soot fires, insofar as masonry chimneys are susceptible to collecting creosote deposits. When thick or heavy enough, these deposits can burn long enough to cause outer chimney surface temperatures that can ignite the structure around them. Further, upward gas flow velocity may be sufficient to carry sparks up from the fire or burning creosote and to drop them on the roof. After a severe chimney fire, both the tile liner and bricks may have serious cracks due to thermal expansion.

Factory built chimneys provide many opportunities for installation errors and shortcuts which can lead to fire hazards. Among these are:

1. Installation of the chimney ceiling support above the ceiling framing. This necessitates passing the single wall connector through ceiling framing at what could be dangerously close clearance; thus the hazard might more properly be blamed on the connector.
2. Failure to secure joints or align gas carrying parts. Many designs of factory built chimneys allow very little tolerance for misalignment. Failure to make sure that the inner pipes of a multiwall chimney are aligned and connected can allow flames to enter external passages reserved for cooling airflow.
3. Use of mismatched chimney parts. A given design or brand of chimney is intended to be assembled only with identical parts. There is no interchangability and it is extremely bad practice to mix multiwall and insulation filled chimney sections.

The foregoing installation errors may cause fires at relatively low flue gas temperature; however, even a cor-

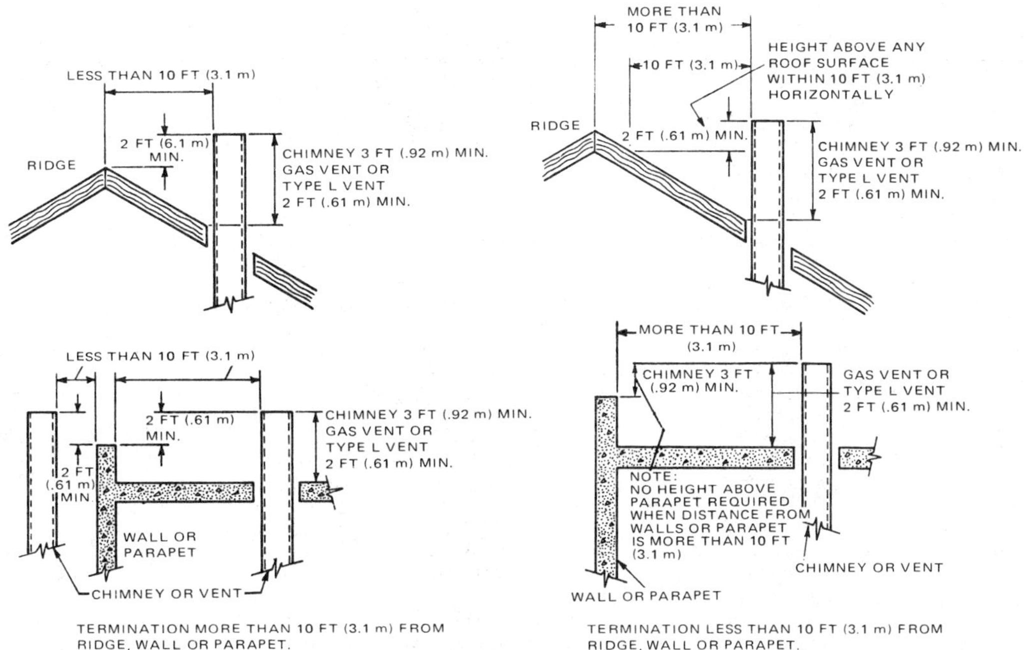

FIG. 8-3II. Termination of vents and chimneys above buildings.

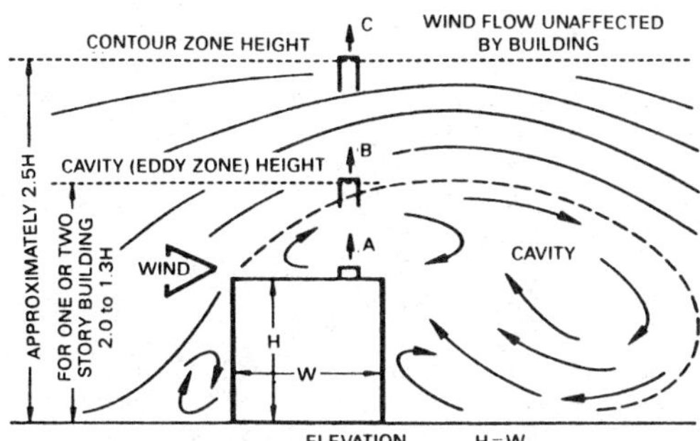

FIG. 8-3JJ. Eddy and contour zones due to airflow for one or two story buildings and their effect on chimneys gas discharge.

rectly assembled and installed chimney lacks the capability of repeatedly acting as a combustion chamber.

Chimney or Creosote Fires

The term creosote as applied to chimney systems refers to the tarry brown or hard black internal deposits produced by burning wood. Combustion of wood produces varying amounts of complex hydrocarbons that can condense and accumulate inside connector and chimney surfaces. These combustible deposits ignite when temperature is sufficiently high and will burn at temperatures in the range of 1,200 to 2,000°F (650 to 1100°C). The temperature and duration of the fire both depend upon the accumulated thickness of creosote, availability of oxygen, type of heating appliance, type of chimney, etc.

In masonry chimneys, intense fires can cause cracking of tiles and brick masonry, as well as excessive heat transfer through chimney walls and hazardous overheated external surface temperatures. In factory built chimneys, as with masonry, the effect of chimney fires varies depending upon attained temperature. Gas temperatures up to 1,700°F (930°C) for periods not in excess of 10 minutes can be endured repeatedly. If it is suspected that an extreme chimney fire has occurred, the liner should be visually inspected. Slight buckling inward [¼ to ½ in. (6 to 13 mm)] is not cause for concern but indicates that temperatures may have reached the 2,000°F (1100°C) level. This should not be allowed to recur. If serious damage of the chimney inner liner is seen, the affected chimney sections should be replaced. In general, if the outer chimney is not discolored and remains shiny, it is unlikely that any building structural damage has occurred since shiny surfaces are unlikely to radiate damaging amounts of heat if the chimney is installed at correct clearance.

Control of Chimney Fires: In woodburning stoves and heaters having air dampers and doors, reducing the air supply by closing them is the best option; then the combustion rate in the chimney depends upon its air supply. When other appliances are connected to the chimney or a barometric damper is used for draft control, these can supply air and must also be closed to control a chimney fire. Open front fireplaces or fireplace stoves without doors may not accumulate creosote rapidly, but under fire conditions air control is difficult. The best policy with fireplaces is routine chimney inspection and mechanical cleaning (brushes and scraping).

FIREPLACES AND FIREPLACE STOVES

Fireplaces of masonry or factory built construction are primarily open fire chamber appliances with no controls on the air supply to the fire. These require a properly sized chimney to carry out the smoke and combustion gases. Fireplace stoves are free standing, factory built appliances operating with an open front, but if these have doors, they may also operate as stoves or heaters with controlled combustion air. The addition of glass doors to a masonry fireplace, or the installation of a closed front fireplace

Table 8-3M. Chimney Selection Chart

Chimney Type (Construction Requirements)				
1. Factory Built—Residential Type and Building Heating Appliance	1. Factory Built—Residential Type and Building Heating Appliance	1. Factory Built—1,400°F (760°C)	1. Factory Built—Medium Heat Appliance	
2. Masonry, Residential Type	2. Masonry, Low Heat Type	2. Masonry, Low Heat Type	2. Masonry, Medium Heat Type	1. Masonry, High Heat Type
	3. Metal, Low Heat Type	3. Metal, Low Heat Type	3. Metal, Medium Heat Type	2. Metal, High Heat Type
Maximum Continuous Appliance Outlet Flue Gas Temperature				
1,000°F (538°C)	1,000°F (538°C)	1,400°F (760°C)	1,800°F (982°C)	Over 1,800°F (982°C)
Types of Appliances to be Used with Each Type Chimney (See Note 2)				
Column I	Column II	Column III	Column IV	Column V
A. Residential-type appliances, such as: 1. Ranges 2. Warm air furnaces 3. Water heaters 4. Hot water heating boilers 5. Low pressure steam heating boilers 6. Incinerators 7. Floor furnaces 8. Wall furnaces 9. Room heaters 10. Fireplace stoves 11. Fireplace stove-room heater B. Fireplaces: 1. Factory built 2. Masonry 3. Freestanding (see Fireplace stove)	A. All appliances shown in Column I. B. Nonresidential type building heating appliances for heating a total volume of space exceeding 25,000 cu ft (708 m^3) C. Steam boilers operating at not over 1,000°F (538°C) flue gas temperature; pressing machine boilers	All appliances shown in Columns I and II, and appliances such as: 1. Class A ovens or furnaces operating at temperatures below 1,400°F (760°C) as defined in NFPA 86 (Note 3) 2. Annealing baths for hard glass (fats, paraffine, salts, or metals) 3. Bake ovens (in bakeries) 4. Candy furnaces 5. Core ovens 6. Feed drying ovens 7. Forge furnaces (solid fuel) 8. Gypsum kilns 9. Hardening furnaces (below dark red) 10. Lead melting furnaces 11. Nickel plate (drying) furnaces 12. Paraffine furnaces 13. Restaurant type cooking appliances using solid or liquid fuel 14. Sulphur furnaces 15. Tripoli kilns (clay, coke and gypsum) 16. Wood drying furnaces 17. Zinc amalgamating furnaces	All appliances shown in Columns I, II and III, and appliances such as: 1. Alabaster gypsum 2. Annealing furnaces (glass or metal) 3. Charcoal furnaces 4. Cold stirring furnaces 5. Feed driers (direct fire heated) 6. Fertilizer dryers (direct fire heated) 7. Galvanizing furnaces 8. Gas producers 9. Hardening furnaces (cherry to pale red) 10. Incinerators, commercial and industrial 11. Lehrs and glory 12. Lime kilns 13. Linseed oil boiling 14. Porcelain biscuit kilns 15. Pulp driers (direct fire heated) 16. Steam boilers operating at over 1,000°F (538°C) flue gas temperature 17. Water-glass kilns 18. Wood-distilling furnaces 19. Wood-gas retorts	All appliances shown in Columns I, II, III, and IV and appliances such as: 1. Bessemer retorts 2. Billet and bloom furnaces 3. Blast furnaces 4. Bone calcining furnaces 5. Brass furnaces 6. Carbon point furnaces 7. Cement brick and tile kilns 8. Ceramic kilns 9. Coal and water gas retorts 10. Cupolas 11. Earthenware kilns 12. Glass blow furnaces 13. Glass furnaces (smelting) 14. Glass kilns 15. Open hearth furnaces 16. Ore roasting furnaces 17. Porcelain baking and glazing kilns 18. Pot-arches 19. Puddling furnaces 20. Regenerative furnaces 21. Reverberatory furnaces 22. Vitreous enameling ovens (ferrous metals)

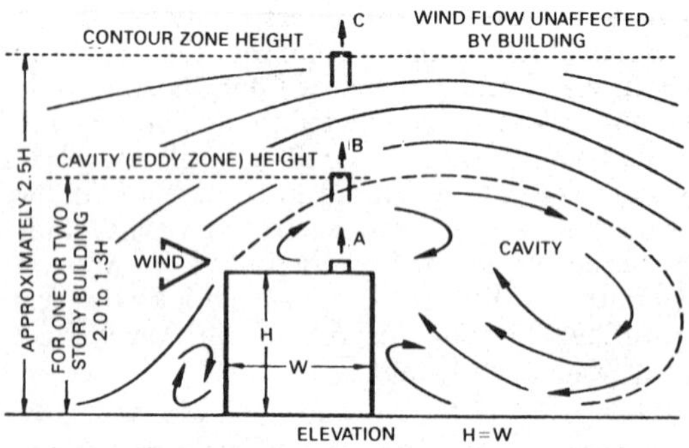

CONTOUR ZONE HEIGHT
C
WIND FLOW UNAFFECTED BY BUILDING
CAVITY (EDDY ZONE) HEIGHT
B
A
WIND
CAVITY
APPROXIMATELY 2.5H
FOR ONE OR TWO STORY BUILDING 2.0 to 1.3H
H
W
ELEVATION
H=W

Notes to Table 8-3M.

1. Single wall metal chimneys or unlisted metal chimneys shall not be used inside one and two family dwellings.
2. For appliance types not listed in Column I through V, the appropriate chimney shall be selected on the basis of the appliance outlet flue gas temperature when appliance is fired at its normal maximum input, and type of surroundings.
3. NFPA 86, *Standard for Ovens and Furnaces: Design, Location, and Equipment.*

insert will transform its combustion characteristics to that of a heater.

Masonry Fireplaces

Field constructed masonry fireplaces must conform to local building codes or provisions of applicable standards (NFPA 211, for example) to assure safe operation. They may be served by a masonry residential chimney or may use a factory built chimney with a suitable factory furnished transition between smoke chamber and chimney. Prefabricated designs of masonry fireplaces and masonry fireplace chimney combinations are available in some localities. When a lining of low duty firebreak (ASTM C64, Type 6) is provided, masonry fireplaces require at least 8 in. (203 mm) of brick thickness or equivalent for the back and sides, with 2 in. (51 mm) clearance to combustibles at the sides and 4 in. (102 mm) minimum clearance from the back. When the lining described above is not used, the thickness of the masonry at the back and sides should not be less than 12 in. (305 mm). The entire fireplace must be supported on a noncombustible footing, with structural floor framing well away from direct heat conduction from the fire zone or external hearth. (See Figure 8-3KK.)

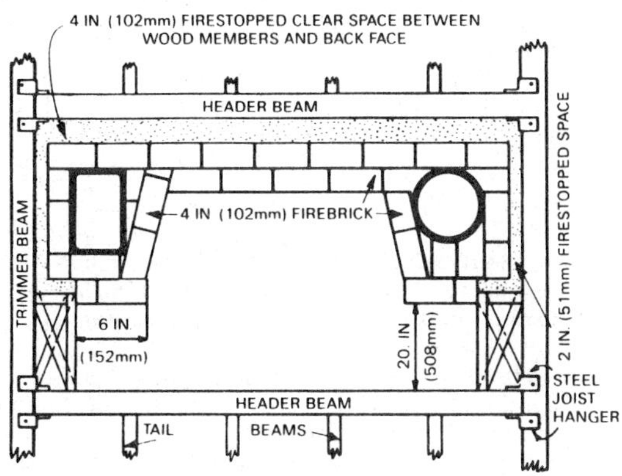

FIG. 8-3KK. Floor framing around a fireplace, when the fireplace opening is 6 sq ft (10.56 m^2) or larger.

Hearth extensions of approved noncombustible material must be provided for all fireplaces. The hearths must extend at least 16 in. (406 mm) in front and 8 in. (203 mm) beyond the sides of fireplaces with an opening of less than 6 sq ft (0.56 m^2). If the opening is 6 sq ft (0.56 m^2) or larger, the hearth must extend at least 20 in. (508 mm) to the sides.

Steel fireplace units or heat circulators incorporating an air chamber may be installed in masonry fireplaces. A total thickness at back and sides of not less than 8 in. (203 mm) must be provided, of which not less than 4 in. (102 mm) is solid masonry. The same clearances to surrounding combustibles apply to fireplaces with circulators.

Factory Built Fireplaces

Factory built fireplaces are mainly of metal construction using multiple air spaces, refractory hearths and liners, and insulation to obtain the required level of safety. Factory built fireplaces are tested and listed to ANSI/UL-127, *Standard for Safety*. Each fireplace must be installed with the specific chimney and other parts called for by the manufacturers' instructions and the terms of its listing.

Chimney Sizes for Fireplaces

For masonry fireplaces, the correct size and height of chimney will assure that all smoke, fire, and gas goes up the chimney. The typical masonry fireplace requires a true chimney flue area of one-eighth of the frontal opening for a minimum height chimney. The nominal size of tile liner must be carefully checked to assure that the internal cross section area is adequate. In addition, the area of opening through the damper throat must be approximately twice the required flue area or greater, or it may reduce flow and possibly cause smoking.

Charts are available for factory built and masonry chimneys in the *ASHRAE Handbook* (ASHRAE 1979) and manufacturers' literature which indicate the correct chimney size and height in relation to frontal opening, based on an average room air velocity of not less than 0.8 ft per sec (0.24 m/s) into the opening. This minimum velocity rule is valid for a wide variety of open fire chamber combustion systems, including masonry fireplaces as well as combination room heaters and fireplace stoves.

Factory Built Fireplace Stoves

These free standing units are usually of metal or combination metal and refractory construction and may require a specific type of chimney and special connectors, or may be connected to any acceptable chimney by conventional means. Fireplace stoves are tested and listed (ANSI/UL 1978), but if the stove has doors it becomes a combination fireplace and room heater, and to be listed for operation with doors closed it must also comply with ANSI/UL 1482. *(ANSI/UL 1982).*

Investigation of many combination fireplace stove heater designs has demonstrated the validity of the geometrical relationship between heater size and required clearance. Simply stated, larger units may require greater than standard clearance. To be sure of correct installation, the manufacturer's installation instruction for listed stoves and heaters should be scrupulously observed. These instructions have been reviewed for consistency with test results and are as essential to the safety of an installation as are the design features of the stove.

While untested fireplace stoves and related heaters can be installed and operated safely, they are seldom accompanied by carefully written instructions. Many imitations of listed designs can be found which are sold through retail outlets having little interest in the final installation. The lack of clear instructions and warnings poses many hazards to the consumer.

Hazards of Fireplaces and Their Chimneys

Open fireplaces, transferring heat primarily by radiation from the fire, can create several kinds of hazards:

1. Ignition by radiation of furnishings or walls located too close to the fire.
2. Ignition of furnishings or carpeting by sparks or from logs rolling out of the fire. For large high fireplaces having a high fuel capacity, the 20 in. (508 mm) hearth width requirement may be grossly inadequate.
3. Ignition of combustible structural materials by excessive heat conducted through hearth or fireplace walls

which can make prolonged overfiring of a masonry fireplace dangerous.

4. Escape of gases from the opening into a construction gap between the fireplace shell and the front facing wall. This is a common installation fault with factory built fireplaces, since a smoking unit—one with insufficient chimney flow—allows very hot gas to escape into this area.
5. Sparks and embers falling onto a combustible floor between the fireplace and its hearth extension. This can be prevented by placing the masonry or factory furnished hearth extension flush against the front of the fireplace, or by using a sheet of metal at the floor surface under the joint between fireplace and its hearth.
6. Installation of glass doors or a fireplace insert. The fireplace may then operate like a high temperature furnace. Fireplace inserts not only can produce higher gas temperatures due to air limiting combustion controls, but may also have insufficient heat dissipation to the room and can increase creosote accumulation in a chimney.
7. Burning highly combustible solid materials such as a dried Christmas tree and wrappings. The high gas temperatures resulting may ignite creosote deposits, or cause damage to the masonry. The high rate of burning may cause the flames, heat, and smoke to billow out of the fireplace opening, possibly igniting nearby combustibles. Regardless of how flammable these materials are, burning them a little at a time obviously will cause no problem.
8. Use of any flammable liquid to start or rekindle a fire, or in any room with or near an open fire.
9. Failure to clean the chimney whenever creosote deposits build up to dangerous thickness. A layer of creosote 1/16 in. (1.6 mm) thick in an 8 in. (200 mm) size chimney 15 ft (4.6 m) high contains as much as 122,000 Btu (128 MJ) of fuel. If this burns off in 15 minutes, the resulting temperature may reach 1,700°F (930°C) at a heat release rate as high as 488,000 Btu per hr (140 kW).
10. Use of an incorrect or mismatched chimney. For safe operation, many types of factory built fireplaces require a chimney that interconnects with its air passages. Failure to connect the fireplace to this specific chimney could be extremely dangerous, and clearly violates installation requirements and the listing.
11. Failure to install firestopping around chimneys, which might be termed an indirect installation hazard. This will not start a fire, but if one starts elsewhere the lack of firestopping may cause severe losses. Because of the numerous possibilities for operator or construction errors in the vicinity of the fireplace itself, every precaution should be taken to prevent rapid involvement of the entire structure in case of fire. Many modern single and multiple residences of frame construction utilize a chase or vertical enclosure inside or external to the dwelling and containing one or more fireplace chimneys. The absence of floors or other internal framing makes it easy to install all the chimneys, but also provides an ideal passage for rapid vertical spread of flame to upper levels and the roof. The shape and size of firestopping for this sort of installation cannot readily be anticipated by the fireplace manufacturer, but carefully fitted sheet metal or lath and plaster may limit the spread and allow time for effective fire fighting.

Bibliography

References Cited

ANSI/UL. 1978a. "Standard for Factory-Built Chimneys." *ANSI/UL 103.* Underwriters Laboratories Inc., Northbrook, IL.

ANSI/UL 1978b. "Standard for Factory-Built Fireplaces." *ANSI/UL 127.* Underwriters Laboratories Inc., Northbrook, IL. 1978.

ANSI/UL 1978c. "Standard for Factory-Built Fireplace Stoves." *ANSI/UL 737.* Underwriters Laboratories Inc., Northbrook, IL.

ANSI/UL 1976. "Chimney for Medium Heat Appliances." *ANSI/UL 959.* Underwriters Laboratories Inc., Northbrook, IL.

ANSI/UL 1982. "Solid Fuel Type Room Heaters." *ANSI/UL 1482.* Underwriters Laboratories Inc., Northbrook, IL.

ASME. 1974. "Heating Boilers." *Section IV. ASME Boiler and Pressure Vessel Code.* American Society of Mechanical Engineers, NY.

ASTM. 1974. *ASTM Standards on Gaseous Fuels (Parts 23, 24, 25, and 26).* American Society for Testing and Materials, Philadelphia, PA.

ASHRAE. 1979. Chapter 26, *ASHRAE Handbook and Product Directory 1979 Equipment.* American Society of Heating, Refrigerating, and Air Conditioning Engineers, Inc., Atlanta, GA.

NFPA Codes, Standards, Recommended Practices and Manuals. (See the latest *NFPA Codes and Standards Catalog* for availability of current editions of the following documents.)

NFPA 30, *Flammable and Combustible Liquids Code.*

NFPA 31, *Standard for the Installation of Oil Burning Equipment.*

NFPA 54, *National Fuel Gas Code.*

NFPA 58, *Standard for the Storage and Handling of Liquefied Petroleum Gases.*

NFPA 70, *National Electrical Code.*

NFPA 85A, *Standard for Prevention of Furnace Explosions in Fuel Oil and Natural Gas-Fired Single Burner Boiler-Furnaces.*

NFPA 85B, *Standard for Prevention of Furnace Explosions in Natural Gas-Fired Multiple Burner Boiler-Furnaces.*

NFPA 85D, *Standard for Prevention of Furnace Explosions in Fuel Oil-Fired Multiple Burner Boiler-Furnaces.*

NFPA 85E, *Standard for Prevention of Furnace Explosions in Pulverized Coal-Fired Multiple Burner Boiler-Furnaces.*

NFPA 86, *Standard for Ovens and Furnaces Design, Location, and Equipment.*

NFPA 86C, *Standard for Industrial Furnaces Using a Special Processing Atmosphere.*

NFPA 90A, *Standard for the Installation of Air Conditioning and Ventilating Systems.*

NFPA 90B, *Standard for the Installation of Warm Air Heating and Air Conditioning Systems.*

NFPA 96, *Standard for the Installation of Equipment for Removal of Smoke and Grease-Laden Vapors from Commercial Cooking Equipment.*

NFPA 97M, *Standard Glossary of Terms Relating to Chimneys, Vents and Heat Producing Appliances.*

NFPA 211, *Standard for Chimneys, Fireplaces, Vents and Solid Fuel Burning Appliances.*

NFPA 255, *Standard Method of Test of Surface Burning Characteristics of Building Materials.*

Additional Readings

AGA Directory of Certified Appliances and Accessories, American Gas Association, Cleveland, OH.

Air Ducts, UL 181, Underwriters Laboratories Inc., Northbrook, IL, 1974.

American National Standard for Gas Utilization Equipment in Large Boilers, ANSl Z83.3, American Gas Association Laboratories, Cleveland, OH, 1971.

"Are Kerosene Heaters Safe?" *Consumer Reports*, Oct. 1982, pp. 499-507.

"ASME Boiler and Pressure Vessel Code, Section 4: Rules for Construction of Heating Boilers," American Society of Mechanical Engineers, NY.

"Clearances and Insulation of Heating Appliances," *UL Bulletin of Research*. Underwriters Laboratories Inc., Northbrook, IL. No. 27, Feb. 1943.

Code for the Installation of Heat-Producing Appliances, American Insurance Association, NY.

Dyer, D., et al., *Residential Consumers Handbook for Wood-Fired Appliances*, Auburn University, AL, Nov. 1981.

"Fireplaces and Chimneys," *Farmers Bulletin No. 1889*, U.S. Department of Agriculture, Washington, DC.

Gas and Oil Equipment Directory, Underwriters Laboratories Inc., Northbrook, IL.

"Gas Vent Tables," *Research Report No. 1319*, American Gas Association Laboratories, Cleveland, OH. Dec. 1960.

Hale, R. A., and Bissell, L. P., "Wood Fuel for Home Burning," *Forestry Notes*, University of Maine Cooperative Extension Service Information Sheet on Forest Conservation, Nov. 1973.

Harris, R. J., *The Investigation and Control of Gas Explosions in Buildings and Heating Plants*, E. and F. N. Spon, London, England, 1983.

Harwood, B., and Kale, D., "Fire Involving Fireplaces, Chimneys & Related Appliances," U.S. Consumer Product Safety Commission, Washington, DC, Sept. 1981.

Havens, David, *The Woodburner's Handbook*, Media House, Portland, ME, 1973.

"Industrial Control Equipment," *UL Bulletin of Research No. 58*, Underwriters Laboratories Inc., Northbrook, IL. Sept. 1964.

Kale, D., "Fires in Woodburning Appliances," U.S. Consumer Product Safety Commission, Washington, DC, Dec. 1982.

Kondritzer, G., and Morrison, J. H., "The Home Structure as a Factor in Burn Injuries," *DHEW Pub. No. (HSM) 72-10026*, U.S. Department of Health, Education and Welfare, Division of Community Injury Control, Cincinnati, OH, May 1972.

Lee, B. T., and Breese, J. N., "Submarine Compartment Fire Study—Fire Performance Evaluation of Hull Insulation," *NBSIR 78-1584*, National Bureau of Standards, Gaithersburg, MD, 1979.

Lee, B. T., and Walton, W. D., "Fire Experiments and Flash Point Criteria for Solar Heat Transfer Liquids," *NBSIR 79-1931*, National Bureau of Standards, Gaithersburg, MD, Nov. 1979.

Loftus, Joseph J., "Evaluation of Wall Protection Systems for Wood Heating Appliances," *NBSIR 82-2506*, National Bureau of Standards, Gaithersburg, MD, May 1982.

Loftus, Joseph J., "An Evaluation of Wall Protection Systems for Wood Heating Appliances," *Fire Journal*, Vol. 77, No. 5, Sept. 1983, pp. 23-25, 20, 39.

Maxwell, T. T., et al., *Design Handbook for Residential Wood-Burning Equipment*, Department of Mechanical Engineering, Auburn University, AL, July 1979.

Maxwell, T. T., et al., "An Investigation of Creosoting and Fireplace Inserts," *NBS-GCR-81-365*, National Bureau of Standards, Gaithersburg, MD, Dec. 1981.

Peacock, Richard D., "A Review of Fire Incidents, Model Building Codes and Standards Related to Wood-Burning Appliances." *NBSIR 79-1731*, National Bureau of Standards, Gaithersburg, MD, May 1979.

Peacock, Richard D., "Intensity and Duration of Chimney Fires in Several Chimneys," *NBSIR 83-2771*, National Bureau of Standards, Gaithersburg, MD, Dec. 1983.

Peacock, R. D., and Ruiz, E., "Fire Safety of Wood-Burning Appliances, Part I: State of the Art Review and Fire Tests, Volumes I and II," *NBSIR 80-2140*, National Bureau of Standards, Gaithersburg, MD, Nov. 1980.

Orton, Vrest, *The Forgotten Art of Building a Good Fireplace*, 2nd ed., Yankee Press, Inc., Dublin, NH, 1974.

Primary Safety Controls for Gas and Oil-Fired Appliances, UL 372, 2nd ed., Underwriters Laboratories Inc., Northbrook, IL, 1971.

"Recommended Rules for Care and Operation of Heating Boilers," Section VI, *ASME Boiler and Pressure Vessel Code*, American Society of Mechanical Engineers, NY, 1974.

"Recommended Requirements to Code Officials for Solar Heating, Cooling and Hot Water Systems," *DOE/CS/34281-01*, U.S. Department of Energy, Washington, DC, June 1980.

Reed, H. L., *A Field Survey of Gas Appliance Venting Conditions*, Research Report No. 1243, American Gas Association Laboratories, Cleveland, OH.

Reineke, L. H., "Wood Fuel Preparation," *U.S. Forest Service Research Note EPL-090*, Jan. 1969, Forest Products Laboratory, Madison, WI, in cooperation with the University of Wisconsin.

Shelton, J., *The Woodburners Encyclopedia*, Vermont Crossroads Press, Waitsfield, VT, 1976.

Shoub, H., "Survey of the Literature on Safety of Residential Chimneys and Fireplaces," *Miscellaneous Publication 252*, National Bureau of Standards, Washington, DC, 1963.

Soderstron, N., "Heating Your Home With Wood," *Popular Science*, 1978.

Specifications for Fuel Oils, D 396-73, American Society for Testing and Materials, Philadelphia, PA, 1974.

Standard for Chimneys, Factory-Built Residential Type and Building Appliance, UL 103, Underwriters Laboratories Inc., Northbrook, IL.

Standard for Room Heaters, Solid Fuel Type, UL 1482, Underwriters Laboratories Inc., Northbrook, IL.

Stone, R. L., "Fireplace Operation Depends Upon Good Chimney Design," *ASHRAE Journal*, Feb. 1969, pp. 63-69.

Using Coal and Wood Stoves Safely, A Hazard Study, National Fire Protection Association, Quincy, MA, 1974.

Vickers, Allan K., "Kitchen Ranges in Fabric Fires," *NBS Tech Note—817*, National Bureau of Standards, Gaithersburg, MD, Apr. 1974.

Walton, William D., "Fire Testing of Roof Mounted Solar Collectors by ASTM E-108," *NBSIR 81-2344*, National Bureau of Standards, Gaithersburg, MD, Aug., 1981.

Walton, William D., "Solar Collector Fire Incident Investigation," *NBSIR 81-2326*, National Bureau of Standards, Gaithersburg, MD, Aug. 1981.

Wood Heating Seminar Proceedings, Documents 1-1977, 2-1977, 3-1978, 4-1979, 5-1979; Wood Energy Institute, Camden, ME.

AIR CONDITIONING AND VENTILATING SYSTEMS

Revised by William A. Schmidt, P.E.

The term "air conditioning" has been defined by the American Society of Heating, Refrigerating, and Air Conditioning Engineers (ASHRAE) as "the process of treating air to simultaneously control its temperature, humidity, cleanliness, and distribution to meet the comfort requirements of the occupants of the conditioned space." Air conditioning and ventilating systems, except for self-contained units, invariably involve the use of ducts for air distribution. The ducts in turn present the possibility of spreading fire, fire gases, and smoke throughout the building or area served.

This and other potential hazards of air conditioning and ventilating systems are the subjects of this chapter. Installation and protection of air heating systems are discussed in Chapter 3 "Air Conditioning and Ventilating Systems," of this Section. Fire confinement, smoke movement and venting practices are considered in Section 7, Chapters 9, "Confinement of Fire in Building," Chapter 10, "Smoke Movement in Buildings," and Chapter 11, "Venting Practices." Refrigeration systems are covered in Section 10, Chapter 15.

Details on safeguarding against these hazards are presented in NFPA 90A, *Standard for the Installation of Air Conditioning and Ventilating System* (hereinafter referred to as NFPA 90A). The Bibliography and additional readings at the end of this chapter contain additional references from which much useful information may be obtained.

SYSTEM TYPES AND OPERATION

The several types of air conditioning systems include: (1) systems in which air is filtered or washed, cooled and dehumidified in summer, and heated and humidified in winter; (2) systems where air is filtered, cooled, and dehumidified; and (3) systems where air is filtered, heated, and humidified. Figure 8-4A depicts a typical arrangement of the various components of a central air conditioning system. A ventilating system simply supplies or removes air by natural or mechanical means to or from a space. The air may or may not be conditioned.

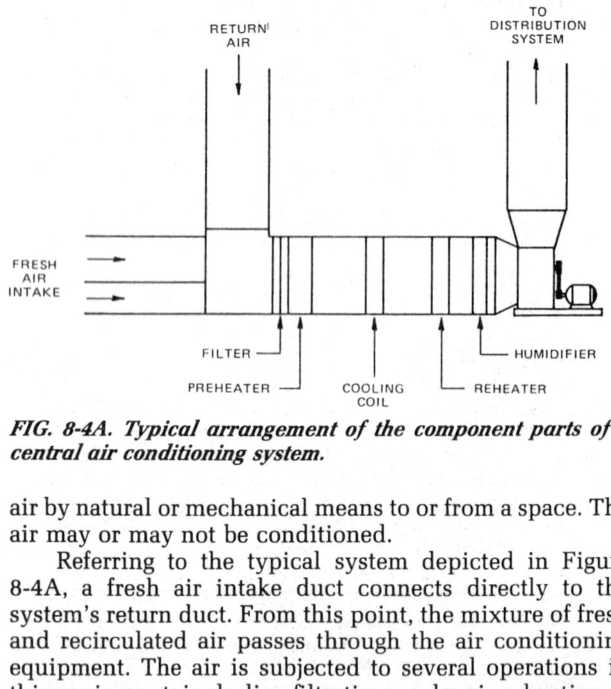

FIG. 8-4A. Typical arrangement of the component parts of a central air conditioning system.

Referring to the typical system depicted in Figure 8-4A, a fresh air intake duct connects directly to the system's return duct. From this point, the mixture of fresh and recirculated air passes through the air conditioning equipment. The air is subjected to several operations in this equipment, including filtration or cleaning, heating or cooling, and humidification or dehumidification. The conditioned air is then circulated continously throughout the area served via the duct system. (See Fig. 8-4B.) Those systems which do not recirculate any air take all of their make up air directly from the outside. A building may contain more than one system, not necessarily of the same type.

LOCATION OF EQUIPMENT

Fans, heaters, filters, and associated equipment that make up a central system air conditioning unit are preferably located in a room separated from the rest of the building area by walls, floor, and floor-ceiling assembly providing a minimum 1 hr fire resistance rating. This arrangement prevents a fire involving the equipment from immediately spreading to adjacent areas of the building.

Mr. Schmidt is a mechanical engineer for the Office of Construction, U.S. Veterans Administration, Washington, DC.

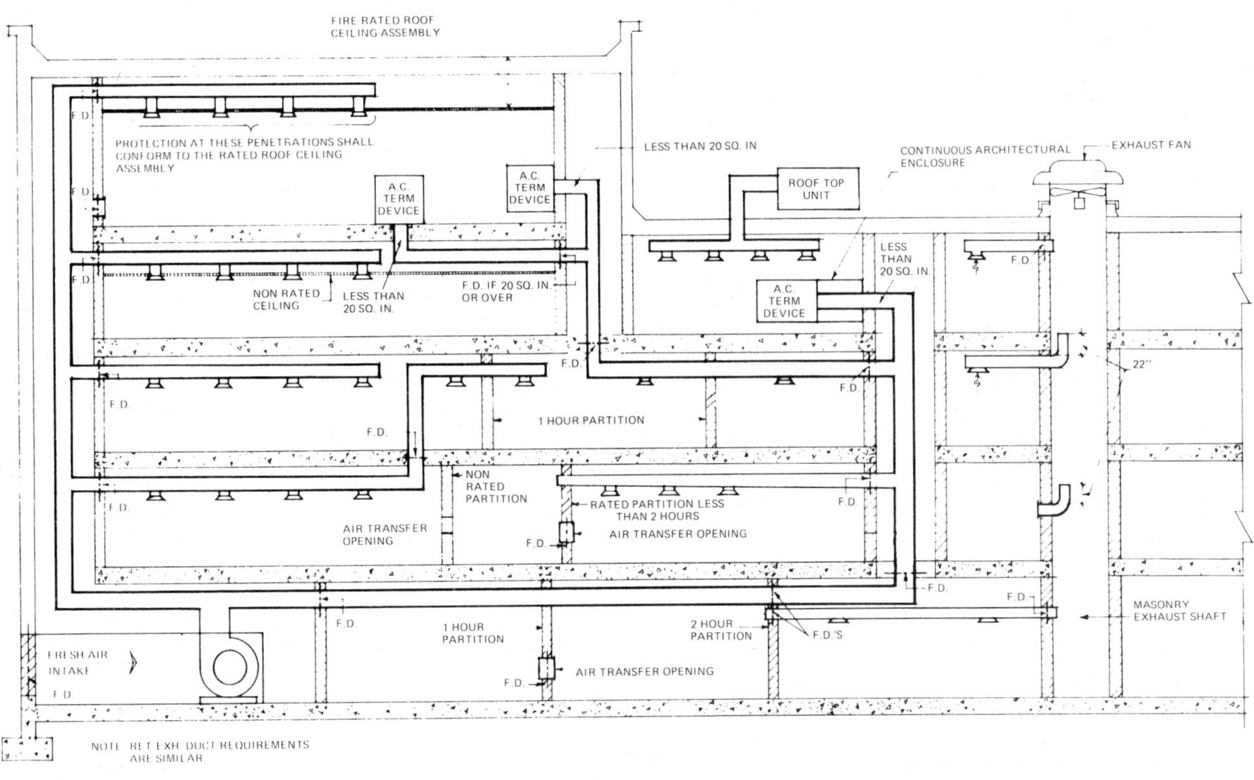

FIG. 8-4B. Typical building duct installation illustrating required protection of walls, ceilings, floors, and shafts.

Such an arrangement also prevents access to the equipment by unauthorized persons. Ideally, service rooms housing air conditioning equipment are protected by automatic sprinklers. At minimum, smoke or heat detectors, or both, should be provided and arranged to initiate an alarm and shut down the air conditioning system. No combustible storage should be allowed in the equipment room.

FRESH AIR INTAKES

The location selected for the fresh air intakes of a system is critical since fire, fire gases, or smoke originating outside the building can easily be drawn in through these intakes and spread throughout the building. Where such a hazard may exist, protection can be provided by the installation at the intakes of fire dampers or smoke dampers that are controlled by fire and smoke detectors.

When considering the location of exterior fresh air intakes, thought must be given to the possibility of sparks or other products of combustion from an exposing fire chimney, or incinerator stack, being drawn into the system. Consideration should not be limited to adjacent buildings or combustible storage. The possibility of fire from a different section of the same building or an adjoining building exposing the intakes is a factor which must also be considered. Since smoke usually rises, the lower the intakes are installed, the less the possibility of drawing in smoke.

All air intakes must be provided with screens of corrosion resistant material not larger than ½ in. (13 mm) mesh to prevent any material from entering the system. Proper maintenance includes periodic removal of any accumulated rubbish or other waste from the immediate vicinity of the intakes. Some fresh air intakes are fitted with a filter which is subject to the same hazards mentioned later in this chapter under "Air Filters and Cleaners." These filters should be protected and maintained accordingly.

AIR COOLING AND HEATING EQUIPMENT

Air cooling equipment presents two basic classes of hazards: those of the electrical equipment and those of the particular refrigerant used. Fire experience is generally good where the cooling equipment is properly installed and maintained. Installation of all electrical equipment should follow both the manufacturer's recommendations and NFPA 70, *National Electrical Code®* (hereinafter referred to as NFPA 70).

In general, the refrigerant poses a toxicity problem and, in a few cases, a combustibility hazard. Even some of the halogenated hydrocarbon refrigerants used in most systems are slightly flammable. The greatest problem associated with refrigeration units is the explosion hazard due to pressurization of the refrigerant. Recommendations for the installation of mechanical refrigerating equipment are contained in the *Safety Code for Mechanical Refrigeration* (ASHRAE 1978).

AIR FILTERS AND CLEANERS

The types of air filters and cleaners used in air conditioning and ventilating systems fall into three general categories: fibrous media unit filters, renewable media filters, and electronic air cleaners. The first two are true filters; the last is a static precipitator. Air filters and cleaners pose a potential hazard due to their function of removing entrained dust and other particulate matter from the air stream. This material builds up on the filter media or precipitator collection plates and if ignited, may burn, producing large volumes of smoke. The smoke and other combustion gases can be circulated throughout the building by the air handling system, thus posing a direct threat to life safety. Not to be overlooked are the possibilities that filter medium may be coated with a combustible adhesive or may itself be combustible.

Underwriters Laboratories Inc. (UL), lists two classes of filter media (UL 1977b):

Class 1: Filters which, when clean, do not contribute fuel when attacked by flame and emit only negligible amounts of smoke.

Class 2: Filters which, when clean, burn moderately when attacked by flame or emit moderate amounts of smoke, or both.

Both of these classes include renewable (washable and reuseable) media and replaceable (disposable) media.

Class 1 filter media are preferred, particularly for systems serving places of assembly such as theaters, auditoriums, department stores, etc. Class 1 filter media are also the obvious choice for systems serving occupancies whose contents are easily susceptible to smoke damage; however, any filter media, if not cleaned or replaced regularly, may become hazardous due to accumulation of combustible dust, etc.

Fibrous Media Unit Filters

Fibrous media unit filters are placed into the air stream and remain there until the pressure drop across the filter reaches some critical point, due to the build up of entrained material. At this point, the filter is either removed, cleaned, and reinstalled or simply discarded and replaced with a new filter. Fibrous media unit filters are of either the viscous impingement type or the dry media type.

The viscous impingement type of filter is characterized by flat or pleated panels of relatively coarse fiber mats. Porosity of the panel is high. The fibers are coated with an adhesive which traps any material entrained in the air stream. The mats are ½ to 4 in. (13 to 102 mm) thick, 1 and 2 in. (25 mm and 51 mm) being most common. This type of filter is most effective when air velocity ranges from 400 to 700 fpm (122 to 213 m/min). Most of these unit filters are ready for replacement when the pressure drop reaches ½ in. of water (124 Pa). The renewable panels must be washed with steam or a hot detergent solution, then recoated with fresh adhesive.

Dry media unit filters are flat or pleated panels of relatively fine fibers, usually ½ to 2 in. (13 to 51 mm) thick. As their name implies, these filters have no adhesive coating. Porosity is not as great as for the viscous impingement type filter; hence particulate removal is greater, depending upon fiber size and porosity. These filters are most effective with an air velocity of 90 to 500 fpm (27 to 152 m/min). Pressure drop may reach as much as 2 in. (500 Pa) of water before replacement is necessary. Disposable dry fibrous media unit filters, particularly of the deep pleated design, are fast becoming the standard for large central air conditioning systems. The dry media classification of air filters also includes membrane and high efficiency types which are used to ultraclean air for clean rooms and similar areas.

Renewable Media Filters

Renewable media air filters can more accurately be termed "moving curtain" air filters, since they operate in exactly such a fashion. The medium, whether viscous coated or dry, is supplied on a large roll and extends across the air stream. Either a timer, light sensitive device, or pressure switch may be used to activate a motor drive which feeds fresh media across the air stream and winds the dirty media onto a takeup spool. Generally, controls are provided that deenergize the drive motor when the media supply is almost exhausted, thus preventing the media from being completely wound onto the takeup spool. A signal indicates that a fresh roll is needed. Operating velocities range from 200 to 500 fpm (70 to 152 m/min), with the dry media types requiring the lower velocities. Filter thicknesses vary from ½ to 2½ in. (13 to 63 mm).

Electronic Air Cleaners

Electronic air cleaners utilize the principle of electrostatic precipitation to remove entrained dust and particulate matter. There are several types of electronic air cleaners, but their basic operation is similar. Entrained particles in the air stream pass through intense, nonuniform electrostatic fields and, due to electric polarization, are collected either on a filter or on charged plates. In some units, equipment is provided to preionize the particulate matter.

Special Industrial Filters and Air Cleaners

There are many industrial applications of air cleaning that involve more than the usual filtering and cleaning methods previously cited. Exhaust systems for flammable vapors, dusts, and other materials in suspension require equipment for recovering the suspended material.

Air cleaning to limit contamination and pollution is of growing concern to industry and public health authorities. Whatever the reason for specialized air cleaning, and whatever the type of equipment used, the selection and protection of the equipment are of concern as the fire hazards are similar to more common types of air cleaning equipment.

One current industrial application of air filtering is in clean rooms where HEPA (high efficiency particulate air) filters are used to filter the atmosphere. These filters must pass an operational efficiency test of 99.97 percent efficiency with 0.3 micron particles. The filters are tested and labeled by UL in accordance with UL 586, *High Efficiency Particulate Air Filter Units*, (UL 1977a). The fire hazards of and protection for clean rooms are discussed in detail in a *Fire Technology* article (Keigher 1967).

Protection for Air Filters and Cleaners

Fire in the filters or air cleaning equipment can release copious quantities of smoke or fire gases that can be distributed by the air handling equipment throughout the area served. Adequate protection to minimize the possibility of such an occurrence must be designed into and around the air conditioning system.

Some means is necessary to prevent smoke and fire gases from being distributed by the main supply fan to all areas served by the system. To accomplish this, detectors are located in the main supply duct, downstream of the filters or air cleaner, to sense smoke or particles of combustion. The detectors are interlocked so as to immediately shut down the entire air conditioning system upon activation. The detectors also control the operation of smoke dampers located in the main supply and return ducts, thus isolating the entire air conditioning section of the system.

With specific regard to viscous impingement filters, the adhesive used should have a flash point of not less than 325°F (163°C) via the Pensky-Martens Closed Tester (ASTM 1980). Also, combustible adhesives should be stored in a safe location, remote from the equipment room housing the air conditioning equipment.

Fire protection should always include facilities for manual fire fighting, usually portable fire extinguishers. Fans, motors, pumps, control circuits, etc., all point to the need for extinguishers suitable for electrical (Class C) fires; however, dry chemical (Class B) extinguishers would be the most suitable for use on filters, especially the viscous impingement type, since the chemical powder would readily adhere to the coated media.

The operational voltage and amperage of electronic air cleaners is lethal; therefore, interlocks should be provided on every access door and panel so that the equipment is immediately shut down if any one of them is opened. An alarm should also be incorporated into this safety interlock circuit.

DUCTS

Ducts are to an air conditioning or ventilating system what pipes are to a water system—a means of distribution. Unfortunately, in fire situations, the ducts may transport deadly smoke and products of combustion instead of breathable air. If proper design and installation precautions are not taken, smoke, fire gases, heat, and even flame can spread throughout the area served by the duct system. Exit corridors used as plenums, lack of smoke detection activated control equipment in the system, and lack of required fire and smoke dampers in appropriate walls, ceilings, or partitions can lead to tragic situations.

Duct Construction

Ducts may be fabricated of metal, masonry, or other noncombustible material. The thickness of materials used in metal ducts of various sizes and methods of bracing, reinforcing, and hanging are covered in the duct manuals of the Sheet Metal and Air Conditioning Contractors National Association (SMACNA 1975-79) and ASHRAE (ASHRAE 1979). SMACNA also publishes manuals on the installation of glass fiber ducts.

UL tests and lists duct materials in accordance with UL 181, *Standard for Factory-Made Air Ducts and Connectors*, (UL 1981a). This standard sets limits on such characteristics as flame spread and flame penetration. Materials are classified as follows:

Class 0: Air Ducts and connectors having surface burning characteristics of zero.

Class 1: Air ducts and connectors having a flame spread rating of not over 25, without evidence of continued progressive combustion, and a smoke developed rating of not over 50.

Class 2: Air ducts and connectors having a flame spread rating of not over 50, without evidence of continued progressive combustion, and a smoke developed rating of not over 50 for the inside surface and not over 100 for the outside surface.

In addition to the above, Class 0 and Class 1 materials must pass a 30 minute flame penetration test and Class 2 materials must pass a 15 minute flame penetration test.

NFPA 90A gives other mandatory provisions for duct linings and coverings, duct tapes, and bands. It also places limitations on the use of flexible duct connectors which must pass through walls, partitions, floors, or ceilings intended to afford fire resistance or smoke control.

Duct Installation

By their very nature, ducts provide means for transferring heat, fire gases, and flame, thus resulting in the spread of fire from one area to another. In addition, at some point or another, a duct probably will pass through a wall, partition, floor, or ceiling which is designed specifically to provide fire resistance. Literally, a hole has been poked through a fire rated design. Theoretically, of course, the duct fills the opening. But if the installation has been made without proper regard for firestopping, this is not valid. Also, under severe fire exposure, the duct will eventually collapse, creating an opening in the fire barrier. An effective method of protecting such penetrations is by the installation of fire dampers.

In the gages commonly used, some sheet metal ducts, if properly hung and adequately firestopped, may protect an opening in a building construction assembly for up to 1 hr. Therefore, ducts passing through fire barriers having a rating of up to 1 hr of fire resistance can possibly present no extraordinary hazard. If the wall, partition, ceiling, or floor is required to have a fire resistance rating of more than 1 hr, a fire damper is required to properly protect the opening. Where it is necessary for a duct to pierce a fire wall, an automatic closing fire damper or fire door assembly having a fire protection rating of not less than 3 hours should be installed at the wall opening. NFPA 90A provides comprehensive recommendations and requirements for the location of fire dampers in rated assemblies.

Fire dampers and ceiling dampers are tested and listed for use in air conditioning and ventilating ducts by UL in accordance with UL 555, *Standard for Fire Dampers and Ceiling Dampers*, (UL 1979). These dampers include single blade, multiblade, and interlocking blade types, all actuated by fusible links. Most fire dampers are designed for vertical installation although some can be installed horizontally as well.

SMOKE CONTROL—PASSIVE

Both NFPA 101, *Life Safety Code®*, and NFPA 90A recognize two approaches to smoke control; passive and active. The passive approach recognizes the long standing compartmentation concept which requires that fans be shut down and fire and smoke dampers in ductwork be closed in fire conditions. The active approach utilizes the building's heating, ventilating, and air conditioning (HVAC) systems to create differential pressures to prevent smoke migration from the fire area, and to exhaust the products of combustion to the outside.

Smoke Dampers

Smoke dampers are required in air conditioning or ventilating ducts which pass through required smoke barrier partitions. NFPA 90A requires smoke dampers to operate automatically upon detection of smoke and must function so that smoke movement through the duct is halted. Basically, they simply interrupt airflow through the duct. NFPA 90A permits the use of fire dampers for smoke control purposes. Obviously a combination fire and smoke damper must meet the requirements for both. Smoke dampers are tested and classified by UL in accordance with UL 555S, *Standard for Leakage Rated Dampers for Use in Smoke Control Systems* (UL 1983).

Smoke Detectors

Smoke detectors must be installed in all air conditioning or ventilating systems over 15,000 cfm (425 m^3/min) capacity. Further, smoke detectors for use in ducts must include features in their design specifically for this purpose. Smoke detectors may shut down the air conditioning or ventilating system, sound alarms, operate smoke control dampers, activate fire suppression equipment, or initiate active smoke control functions. Preference is for detectors with adjustable sensitivity, thus minimizing false alarms due to incidental dust or other particulate matter.

These smoke detectors are located in the main supply ducts, downstream of air filters or air cleaners, and in the main return ducts prior to exhausting from the building or joining the fresh air intake ducts. Smoke detectors at these locations are primarily intended to prevent smoke circulation throughout the entire area served by shutting down the air handling system. Also they prevent circulation of smoke from a fire in the air filters or air cleaners. Due to dilution effects of air picked up from various branch return ducts, these smoke detectors cannot be expected to reliably provide "early warning" smoke detection. For this reason, additional smoke detectors may be provided; for example, in the main branch return duct from each floor or major area division. Such detectors could also be tied into a zoned fire alarm system; however, such an arrangement is not a substitute for a smoke detection system for building protection. For guidance on smoke detector location see NFPA 72E, *Standard on Automatic Fire Detectors.*

SMOKE CONTROL—ACTIVE

The preceding material covered passive control of smoke and fire gases—control designed to prevent products of combustion from moving out of the area of its source. It is an inescapable fact that attempts to completely confine smoke and fire gases are seldom successful. Recognizing this, smoke movement in high rise buildings and its threat to life safety, has been the object of much investigation (McGuire 1967; Hutcheon & Shorter 1968). The discussion of modes of smoke movement, their causes, and effects are outside the scope of this chapter. However, it is worth noting that high rise buildings do present several unique problems, e.g., excessive evacuation time, inability of conventional fire equipment (aerial ladders, snorkels, and monitor nozzles) to reach upper floors, and the reinforcement of existing upward airflow patterns within the building via vertical shafts (stack effect).

In view of the increasing trend towards high rise buildings, it becomes attractive to utilize the air conditioning or ventilating system for smoke control and removal in case of fire. Tests have been conducted on the feasibility of pressurizing emergency exit stairwells and elevator shafts to prevent smoke migration from the fire floor to other parts of a building during evacuation (Kaplan 1973; McGuire & Tamura 1973).

The ASHRAE manual *Design of Smoke Control Systems for Buildings* provides guidelines for designers who wish to provide active smoke control systems for buildings (ASHRAE 1983). This information is generally intended to provide systems that exhaust smoke from the immediate fire area, and provide pressurized outside air to adjacent areas, access corridors, and stairwells. It is fully recognized that this approach would apply more to large HVAC units serving individual floors, or large systems with volume control dampers at each floor.

The smoke control system must maintain safe exit routes with sufficient exiting time for building occupants to either leave or move to designated safe refuge areas.

Exhaust of smoke from the fire area is of primary concern. To replace this exhausted air, the HVAC system should provide 100 percent outside air to adjacent areas, thereby providing a clean atmosphere in exit ways through the area immediately above the fire and preventing smoke from flowing up through any holes in the floor slab and to the area immediately below the fire to provide safe areas for the fire fighters.

It must be fully understood that the smoke control systems mentioned previously are intended to provide a smoke free means of exiting the building. If a fire once gets out of hand, there is danger of ducts burning through and falling from the ceiling, or fans being subjected to intense heat so they may cease to operate. It can then be assumed that initial efforts at extinguishing the fire have failed and that occupants remaining in the building would probably be in serious danger. It is not intended that special electrical generation units, heavy duty fans or heavy ductwork would be used.

FANS, CONTROLS, ETC.

The fan unit on an air conditioning system should not in itself present undue hazard, if properly installed and firmly supported on a rigid foundation. It should also be readily accessible for cleaning, servicing, and lubrication. Fans should be provided with excess vibration switches, wired to sound an alarm and initiate system shutdown when bearing failure is imminent.

Fan motors installed inside ducts or plenums need protective devices to cut off power before temperatures reach a point where smoke may be generated. Most fractional horsepower motors have over temperature protective devices. Thermal overload relays are recommended for fan motors of one horsepower or larger.

Electric wiring to all components should follow NFPA 70. All electrical equipment should be listed by a qualified electrical testing laboratory. Wiring installed in ducts, plenums and other spaces used for environmental air should meet the requirements of UL 910-81.

All air conditioning or ventilating systems need manual shutoffs for use in case of fire or other emergency. This shutoff should be well identified and located where it is readily accessible, such as near building exits. In systems of 2,000 to 15,000 cfm (57 to 425 m^3/min) capacity, it is good practice to provide automatic shutoff controls triggered by thermostatic devices in the same locations as recommended for smoke detectors. A setting of 136°F (58°C) for the thermostatic device in the return air stream and a setting of 50°F (10°C) above maximum operational temperature for the device in the supply stream are recommended.

Energy Conservation

Energy conservation practices require a careful evaluation of the design and control of HVAC systems. To reduce electrical costs, fans are turned off during various portions of the day and during unoccupied hours. Without airflow, the duct fire and smoke detectors will not quickly detect fire and thereby will not provide the protection that may be assumed as inherent in HVAC systems. The use of variable volume air systems is increasing as these systems have an overall lower energy use than most other systems. The variation in duct airflow in such systems can create a problem with the sensitivity of the duct mounted detectors and cause false readings and alarms. Fan cycling and variable volume systems can also cause variations in air pressure relationships of the occupied spaces. An area may have a positive pressure relationship to another area when the fans are operating in one mode, and when the fans are cycled, a completely different relationship can occur which can affect the performance of detection and control systems and calls for careful design evaluation.

UNIT AIR CONDITIONERS

The term "unit air conditioner" may include any factory produced unit serving one room or area. Such units do not involve the hazard of spreading fire from floor to floor or area to area as do ducted systems. However, if improperly designed, installed, or maintained, they may involve the other hazards associated with larger systems. For example, any window sill type unit can overheat through failure of its thermal relay. This could lead to ignition of wire insulation or filter media and the fire could spread to curtains or the wood sash of the window. Defective wiring could lead to the same condition. A common problem involving these units is that they are too frequently plugged into outlets on branch circuits not designed to carry the design load or the circuit is already overloaded. This practice directly violates principles of NFPA 70.

MAINTENANCE

Maintenance and cleaning are of utmost importance to safe operation of any air conditioning or ventilating system. Filters must be changed or cleaned as frequently as necessary. Ducts, particularly on the return side of the system, are cleaned out periodically to prevent hazardous accumulations of combustible dust and lint. Evidence of any defect of wiring or electrical equipment must be checked out immediately and corrected. Repairs on ducts or equipment casings which require welding or cutting should be protected by complete shutdown of the system and thorough cleaning of the area. If possible, the duct section to be repaired should be isolated from the rest of the system.

Proper maintenance includes periodic testing of all fire protection devices including fire suppression equipment, smoke control and fire dampers, alarms, and even vibration switches on fans. A regular testing program should be established. If such testing is outside the function of regular maintenance staff, it should be contracted to an outside agency.

Bibliography

References Cited

ASHRAE. 1979. *ASHRAE Handbook and Product Directory 1979 Equipment*. American Society of Heating, Refrigerating, and Air Conditioning Engineers, Inc., Atlanta, GA.

ASHRAE. 1978. *Safety Code for Mechanical Refrigeration*. No 15-78. American Society of Heating, Refrigerating and Air Conditioning Engineers, Atlanta, GA.

ASHRAE. 1983. *Design of Smoke Control Systems for Buildings*. American Society of Heating, Refrigerating, and Air Conditioning Engineers, Atlanta, GA.

ASTM. 1980. *Standard Method of Test for Flash Point by Pensky-Martens Closed Tester*. D-93. American Society for Testing and Materials, Philadelphia, PA.

Hutcheon, N. B., and Shorter, G. W. 1968. "Smoke Problems in High-Rise Buildings." *ASHRAE Journal*. Vol 10, No 9.

Keigher, Donald J. 1967. "Clean Rooms—Another Fire Protection Problem." Part I. *Fire Technology*. Vol 3, No 4. Nov 1967. pp 261-271.

Koplon, N. A. 1973. *Report of the Henry Grady Fire Tests*. City of Atlanta Building Department, Atlanta, GA.

McGuire, J. H., and Tamura, G. T. 1973. "The Pressurized Building Method of Controlling Smoke in High-Rise Buildings." *Technical Paper No. 394 (NRCC 13365)*. National Research Council of Canada. Division of Building Research, Ontario, Canada, Ottawa

McGuire, J. H. 1967. "Smoke Movement in Buildings." *Fire Technology*. Vol 3, No 3. Aug 1967. pp 163-174.

SMACNA. 1975-79. *Fibrous Glass Duct Construction*. STD 5th ed. 1979; *Low Pressure Duct Construction*. STD 5th ed. 1976; *High Pressure Duct Construction*. STD 5th ed. 1979; Sheet Metal and Air Conditioning Contractors National Association, Inc., Tysons Corner, VA.

UL. 1983. *Standard for Leakage Rated Dampers For Use in Smoke Control Systems*. UL 555S. Underwriters Laboratories Inc. Northbrook, IL.

UL. 1981a. *Standard for Factory-Made Air Ducts and Connectors*. UL 181. Underwriters Laboratories Inc. Northbrook, IL.

UL. 1981b. *Standard for Test Method for Fire and Smoke Characteristics of Cables Used in Air-Handling Spaces*. UL 910-81. Underwriters Laboratories Inc., Northbrook, IL.

UL. 1970. *Standard for Fire Dampers and Ceiling Dampers*. UL-555. Underwriters Laboratories Inc., Northbrook, IL.

UL. 1977a. *Standard for High Efficiency Particulate Air Filter Units*. UL586. Underwriters Laboratories Inc., Northbrook, IL.

UL. 1977b. *Standard for Performance of Air Filter Units*. UL 900. Underwriters Laboratories Inc., Northbrook, IL.

NFPA Codes, Standards, Recommended Practices and Manuals. (See the latest *NFPA Codes and Standards Catalog* for availability of current editions of the following documents.)

NFPA 70, *National Electrical Code*.

NFPA 72E, *Standard on Automatic Fire Detectors*.

NFPA 80, *Standard for Fire Doors and Windows.*
NFPA 90A, *Standard for the Installation of Air Conditioning and Ventilating Systems.*
NFPA 101, *Code for Safety to Life From Fire in Buildings and Structures.*
NFPA 252, *Standard Methods of Fire Tests of Door Assemblies.*

Additional Readings

Associate Committee on the National Building Code, *Measures for Fire Safety in High Buildings*, National Research Council of Canada, Ottawa, Ontario, Canada, 1977.

Barrett, R., and Lachlin, D., "A Computer Technique for Predicting Smoke Movement in Tall Buildings," *Symposium on the Control of Smoke Movement in Escape Routes in Buildings*, Borham Wood, Hertfordshire, UK, Apr. 1969.

Benjamin, I. A. et al, *Control of Smoke Movement in Buildings: A Review, NBSIR 77-1209*, National Bureau of Standards, Gaithersburg, MD, July 1977.

Benjamin, I. A., and Klote, J. H., *Stairwell Pressurization Systems*, National Bureau of Standards, Gaitherburg, MD, June 1979.

Burcher, E. et al, "Smoke Tests in the Pressurized Stairs and Lobbies of a 26-Story Office Building," *Building Services Engineer*, Vol. 39, Dec. 1971. pp. 206-210.

Butcher, E. G., "The Design of Pressurization Systems—A Survey of Current Codes and Discussions of Difficulties," *CIB Symposium on the Control of Smoke Movement in Building Fires*, Vol. 1, Fire Research Station, Building Research Establishment, Garston, Waterford, UK, Nov. 4-5, 1975.

Butcher, E. G. and Parnell, A. C., *Smoke Control in Fire Safety Design*, E & F. N. Spon Ltd., London, England, 1979.

Cresci, R. J., "Smoke and Fire Controls in High-Rise Buildings—Part II: Analysis of Stair Pressurization Systems," *ASHRAE Symposium on Experience and Applications on Smoke and Fire Control*, Louisville, KY, June 1973, pp. 16-23.

DeCicco, P., "Smoke and Fire Control in High-Rise Office Buildings—Part I: Full Scale Tests for Establishing Standards," *ASHRAE Symposium on Experience and Applications on Smoke and Fire Control*, Louisville, KY, June 24-28, 1973.

DeCicco, P. et al, *Fire Tests, Analysis, and Evaluation of Stair Pressurization and Exhaust in High-Rise Office Buildings*, Polytechnic Institute of Brooklyn Center for Urban Environmental Studies, Baywood Publishing Co., New York, NY, 1973.

denOuden, L., "The Effect of Opening and Closing of Doors on the Pressure Levels in a Building in which the Stair-well is Pressurized," *CIB Symposium on the Control of Smoke Movement in Building Fires*, Vol. II, Garston, Waterford, UK, Nov. 1975, pp. 42-49.

Dias, C. "Stairwell Pressurization in a High-Rise Building," *ASHRAE Journal*, July 1978, pp. 24-26.

Dillon, Michael, "Smoke Management System Considerations for Hotel Atriums," *ASHARE Journal*, July 1983.

Erdelyi, B. J., "Test Results: Ducted Stairwell Pressurization System in a High-Rise Building," *ASHRAE Journal*, Feb. 1976, pp. 39-40.

Evers, E., and Waterhouse, A., *A Computer Model for Analyzing Smoke Movement in Buildings*, Scientific Control Systems Ltd, London, England, 1974.

"Exterior Wall Venting in Tall Office Buildings," *ASHRAE Journal*, Aug. 1978.

"Fire and Smoke Technology," *ASHRAE Journal*, July 1978.

"Fire and Smoke Control," *ASHRAE Handbook*, Systems Volume, Chapter 41, 1980.

Fothergill, John, "Smoke Control System Design Tools," *ASHRAE Journal*, July, 1978.

Fung, F., "Evaluation of Smokeproof Stair Towers and Smoke Detector Performance," *NBSIR 75-701*, National Bureau of Standards, Sept. 1975.

Fung, F., "Evaluation of a Pressurized Stairwell Smoke Control System for a 12 Story Apartment Building," *NBSIR 73-277*, National Bureau of Standards, Gaithersburg, MD, June 1973.

Fung, F., "Smoke Control by Systematic Pressurization," *Fire Technology*, Vol. 2, No. 4, Nov. 1975.

Fung, F., and Ferguson, J., "Test and Evaluation of the Smoke Control Features of the Seattle Federal Building," *Proceedings of the Public Buildings International Conference on Fire Safety in High-Rise Buildings*, Nov. 1974.

Fung, F., and Fothergill, W., "Smoke Control Design of Application of Computer Simulation Technology," *ASHRAE Semiannual Meeting*, Chicago, IL, Feb. 1977.

Fung, F., and Zile, R., "San Antonio Veterans Administration Hospital Smoke Movement Study," *NBSIR 75-903*, National Bureau of Standards, Gaithersburg, MD, Nov. 1975.

Fung, F., and Zile, R., "Test and Evaluation of the Smoke Control Capabilities of the San Diego Veterans Administration Hospital," *NBSIR 77-1225*, National Bureau of Standards.

Grupp, David C., "Life Safety Concepts for Today's High Rises," *Heating/Piping/Air Conditioning*, Oct. 1975.

Hedsten, G. C., "Building and Stair Pressurization System," *ASHRAE Journal*, July 1978, pp. 32-35.

Hobson, P., and Stewart, L., *Pressurization of Escape Routes in Buildings*, Heating and Ventilating Research Association, Brachnell, Berkshire, UK, Jan. 1973.

Hutcheson, N., and Shorter, G., "Smoke Problems in High Rise Buildings," *ASHRAE Symposium on Fire Hazards in Buildings and Air Handling Systems*, Lake Placid, NY, June 24-26, 1968.

Imazu, H., "Smoke Evacuation Effect by Forced Air into High Rise Buildings," from *Main Reports on Production, Movement, and Control of Smoke in Buildings*, Occasional Report of Japanese Association of Fire Science and Engineering.

Jones, Walter W., "A Model for the Transport of Fire, Smoke and Toxic Gases (FAST)," *NBSIR 84-2934*, National Bureau of Standards, Gaithersburg, MD, 1984.

Klote, J. H., "Smoke Control by Stairwell Pressurization," *Engineering Applications of Fire Technology Workshop*, Society of Fire Protection Engineers, Boston, MA, 1980, pp. 137-158.

Klote, J. H., "Stairwell Pressurization," *ASHRAE Transactions 1980*, Vol. 86, Part 1, pp. 604-623, 1980.

Klote, J. H., "Smoke Movement through a Suspended Ceiling System," *NBSIR 81-2444*, National Bureau Standards, Gaithersburg, MD, Feb. 1982.

Klote, J. H., "Elevators as a Means of Fire Escape," *ASHRAE Transactions*, Vol. 89, Part 1, 1983.

Klote, J. H., "Designing Effective Zoned Smoke Control Systems," *Building Design and Construction*, Vol. 24, No. 11, pp. 90-93, Nov. 1983.

Klote, J. H., and Fothergill, J. W., "Design of Smoke Control Systems for Buildings," *NBS Handbook 141*, National Bureau of Standards, Gaithersburg, MD, 1983.

Klote, J. H., "Smoke Control for Elevators," *ASHRAE Journal*, Vol. 26, No. 4, pp. 23-33, Apr., 1984.

Klote, J. H., and Bodart X., "Smoke Control by Pressurized Stairwell," *Journal of CIB*, Vol. 12, No. 4, pp. 216-222, 1984.

Klote, John H., "Field Tests of the Smoke Control System at the San Diego V.A. Hospital," *NBSIR 84-2912*, National Bureau of Standards, Gaithersburg, MD, 1984.

Klote J. H., "The ASHRAE Design Manual for Smoke Control," *Fire Safety Journal*, Vol. 7, No. 1, 1984, pp. 93-98.

Klote, J. H., "Field Tests of the Smoke Control System at the Bay Pines VA Hospital," *ASHRAE Transactions*, Vol. 91, Part 1, 1985.

Lie, T., and McGuire, J., "Control of Smoke in High-Rise Buildings," *Fire European Conference of NFPA*, Geneva, Switzerland, Oct. 15-17, 1973.

Maeda, T., et al, "Mixing of Smoke and Air at the Interface of Two Layer Flame," in *Main Reports on Production, Movement, and Control of Smoke in Buildings, from Occasional Report of Japanese Association of Fire Science and Engineering*, No. 1, 1974.

Masters, Richard, "Bankers Trust Plaza," *ASHRAE Journal*, Aug. 1973.

McGuire, J. H., *High Rise Building Fires and Fire Safety*, National Fire Protection Association, Quincy, MA, p. 122.

McGuire, J., and Tamura, G., "Simple Analysis of Smoke-Flow Problems in High Buildings," *Fire Technology*, Vol. 2, No. 1, pp. 15-22.

McGuire, J., and Tamura, G., "Fire and Leakage Tests of Dampers for Smoke Shafts," *DRB Internal Report No. 428*, NRC, Division of Building Research, Apr., 1976.

McGuire, J. et al, "Factors in Controlling Smoke in High Buildings," *ASHRAE Symposium Bulletin, Fire Hazards in Building*, San Francisco, CA, Jan. 1970, pp. 8-13.

Minni, R., "Smoke Infilitration into the Fire-Escape Routes of Tall Buildings," CIB *Symposium on the Control of Smoke Movement in Building Fires*, Vol. I, Watford, UK, Nov. 1975, pp. 245-266.

Moulen, A., *Control of Smoke from Fire in an Air Conditioned Building*, TR 44/153/410 Department of Works, Commonwealth Experimental Building Station, North Hyde, New South Wales 2113, Australia.

Moulen, A., "Fire Precaution in Buildings with Air Handling Systems," Paper 18, *CIB Symposium on the Control of Smoke Movement in Building Fires* held by the Fire Research Station, Garston Watford, UK, Nov. 1975.

Riou, Col. J., "Smoke Extraction in Buildings, A Solution," *Fire International*, Vol. 37, 1972, pp. 41-59.

Sander, D., and Tamura, G., "A Fortram IV Program to Simulate Air Movement in Multi-Story Buildings," *Ottawa DBR Computer Program No. 35*, National Research Council of Canada, Ottawa, Ontario, Canada, Mar. 1973.

Schmidt, William, "Smoke Control System Testing," *Heating/Piping/Air Conditioning*, Apr. 1982.

Schmidt, W. A., "Fire and Life Protection: The Systems Designs Options," *ASHRAE Journal*, July 1978, pp. 21-22.

Schmidt, W. A., "HVAC Systems Can Save Lives," *ASHRAE Journal*, Feb. 1976, pp. 17-19.

Schmidt, W. A., "Smoke Detection: Part of Complete Building System," *Specifying Engineer*, May 1979, pp. 58-62.

Schmidt, W. A., "Smoke Movement Studies in Veterans' Administration Hospitals," *ASHRAE Journal*, Sept. 1975, pp. 21-26.

Semple, Brooks, "Smoke Control: The Retrofit Option," *Heating/Piping/Air Conditioning*, Apr. 1982.

Semple, Brooks, "Smoke Control in High Rise Buildings," *ASHRAE Journal*, Apr. 1971.

Semple, Brooks, "Smoke Control: How HVAC Can Reduce Smoke and Fire Losses," *Heating/Piping/Air Conditioning*, Oct. 1977.

Shavit, Gideon, "Smoke Control for High Rise Life Safety," *Heating/Piping/Air Conditioning*, July 1978.

Shaw, C., and Tamura, G., "A Design of a Stairshaft Pressurization System for Tall Buildings," *ASHRAE Journal*, Vol. 18, pp. 29-33.

Smoke Control in Firesafety Design, National Fire Protection Association, Quincy, MA, 1979.

Tamura, G. T., "Air Leakage Measurements of the Exterior Walls of Tall Buildings," *AHSRAE Spring Conference*, Minneapolis, MN, May 1973.

Tamura, G. T., "Analysis of Smoke Shafts for Control of Smoke Movement in Buildings," *ASHRAE Transactions*, Vol. 76, Part 11, 1970, pp. 290-297.

Tamura, G. T., "Computer Analysis of Smoke Control with Building Air Handling Systems," *ASHRAE Journal*, Vol. 14, No. 8, Aug. 1972, pp. 46-59.

Tamura, G. T., "Experimental Studies on Pressurized Escape," *ASHRAE Transactions*, Vol. 80, Part 2, 1974, pp. 224-237.

Tamura, G. T., "Exterior Wall Venting for Smoke Control in Tall Office Buildings," *ASHRAE Journal*, Aug. 1978, pp. 43-48.

Tamura, G. T., and McGuire, J., *The Pressurized Building Method of Controlling Smoke in High Rise Buildings*, Division of Building Research, Technical Paper No. 394, National Research Council of Canada, Ottawa, Canada, Sept. 1973.

Tamura, G. T. et al, "Air Handling Systems for Control of Smoke Movement," ASHRAE Symposium Bulletin, *Fire Hazards in Buildings*, San Francisco, CA, Jan. 1970, pp. 14-19.

Tamura, G., and Wilson, A., "Natural Venting to Control Smoke Movement in Buildings via Vertical Shafts," *ASHRAE Transactions*, Vol. 76, Part 2, 1970, pp. 279-289.

Tamura, G. T., and Shaw, C. Y., "Air Leakage Data for the Design of Elevator and Stair Shaft Pressurization Systems," *ASHRAE Technical Paper No. 2413*, American Society of Heating, Refrigerating, and Air Conditioning Engineers, Atlanta, GA.

Taylor, R. E., "Air for Smoke Control," *ASHRAE Journal*, July 1978, pp. 27-29.

Taylor, R. E., "The Carlyle Apartment Fire: Study of a Pressurized Corridor," *ASHRAE Journal*, Apr. 1975, pp. 52-55.

Wakamatsu, T., "Calculations Methods for Predicting Smoke Movement in Fires and Designing Smoke Control Systems," *ASTM-NBS Symposium*, Washington, DC, Mar. 1976.

Wakamatsu, T., *Smoke Movement in Buildings —Field Experiments in Welfare Ministry Building and Analysis of Sennichi Building Fire*, Occasional Report of Japanese Association of Fire Science and Engineering, No. 1, 1974.

Wakamatsu, T., "Unsteady Calculation of Smoke Movement in an Actual Fire Building," *CIB Symposium on the Control of Smoke Movement in Building Fires*, Vol. 1, Garston, Waterford, UK, Nov. 4-5, 1975.

Webb, William, "Smoke Control in Buildings, Threat or Promise," *ASHRAE Journal*, Feb. 1976.

BUILDING TRANSPORTATION SYSTEMS

Edward A. Donoghue, CPCA

Because of their impact on firesafety, building transportation systems must be considered during the design stage and should be integrated with the total building fire protection system. The hazards associated with these systems must be included in any prefire planning and fire protection survey.

Building transportation systems, such as elevators and escalators, make high rise buildings feasible. While it would be unrealistic for a megastructure to function without elevators, it should be understood that elevator hoistways can contribute to the spread of smoke or fire. On the other hand, an elevator is a necessary tool required for fire fighting operations in high rise buildings.

The discussion of elevators, dumbwaiters, escalators and moving walks contained in this chapter is intended to complement NFPA 101, *Life Safety Code®* (hereinafter referred to as NFPA 101) requirements that equipment for elevators, etc., be installed in accordance with the *Safety Code for Elevators and Escalators* (ASME 1984b), referred to hereinafter in this chapter as the *Elevator Safety Code*. The three model building codes—*Basic/National Building Code* (BOCA), *Standard Building Code* (SBCC), and *Uniform Building Code* (ICBO)—also refer to the *Elevator Safety Code*.

Section 12, Chapter 7 provides information on material Handling Systems, other than freight elevators and dumbwaiters. Discussions on the "Confinement of Fire in Buildings" (Section 7, Chapter 9) and "Smoke Movement in Buildings" (Section 7, Chapter 10) are pertinent to the subject matter of this chapter. They should be reviewed for a more comprehensive view of the fire problems associated with building transportation systems.

ELEVATORS

An elevator can be defined as a hoisting and lowering mechanism equipped with a car or platform that moves in guide rails and serves two or more landings (ASME 1984a). The two major types of elevators, classified by their driving means, are the electric and the hydraulic electric elevators.

In an electric elevator, the power source is an electric motor. The most common electric elevator is the electric traction elevator (Fig. 8-5A) which employs a grooved traction drive sheave over which pass the suspension ropes that are attached to the car and counterweight. This arrangement is very simple and safe, allowing its use in buildings of any height.

Another type of electric elevator is the winding drum machine found on many older elevators. These elevators employ a winding drum to which the suspension ropes are attached and wind onto. The counterweight ropes may also be fastened to and wind onto the drum. This roping arrangement does not permit a loss of traction as is possible with an electric traction type. It is possible for either the elevator or counterweight to be pulled dangerously into the overhead structure.

Hydraulic Elevators

Hydraulic elevators are powered by a hydraulic plunger or piston. (See Fig. 8-5B.) The rise (travel) of a hydraulic elevator is limited to the length of its piston (usually six floors or less).

Classification of Elevators by Use

Elevators are also classified according to their use, as either passenger or freight elevators. A passenger elevator is used primarily to carry persons other than the operator. A freight elevator is used for freight; only the operator and the persons necessary for unloading and loading the freight are permitted to ride. Service elevator is another term used to describe elevators. A service elevator is officially classified as a passenger elevator that has also been designed for the carrying of freight.

HOISTWAY CONSTRUCTION

Elevators travel in a vertical space or opening (called a hoistway or shaft) extending from the pit floor of a building to the floor or roof above. The *Elevator Safety Code* requires hoistway construction to conform to the applicable building code or, in the absence of a local code,

Mr. Donoghue is Manager, Codes and Standards for the National Elevator Industry, Inc.

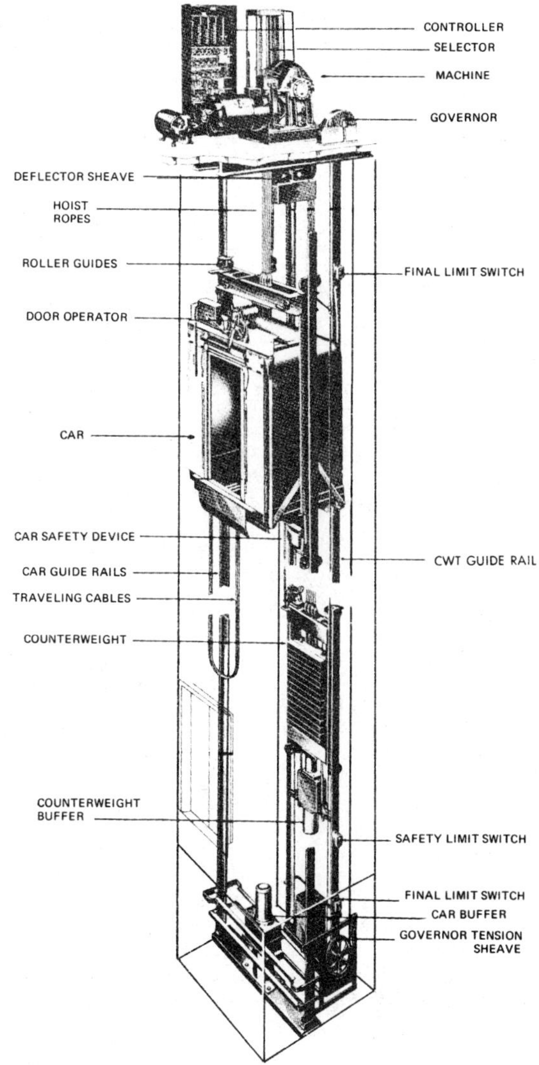

FIG. 8-5A. Electric traction elevator. (National Elevator Industry, Inc.)

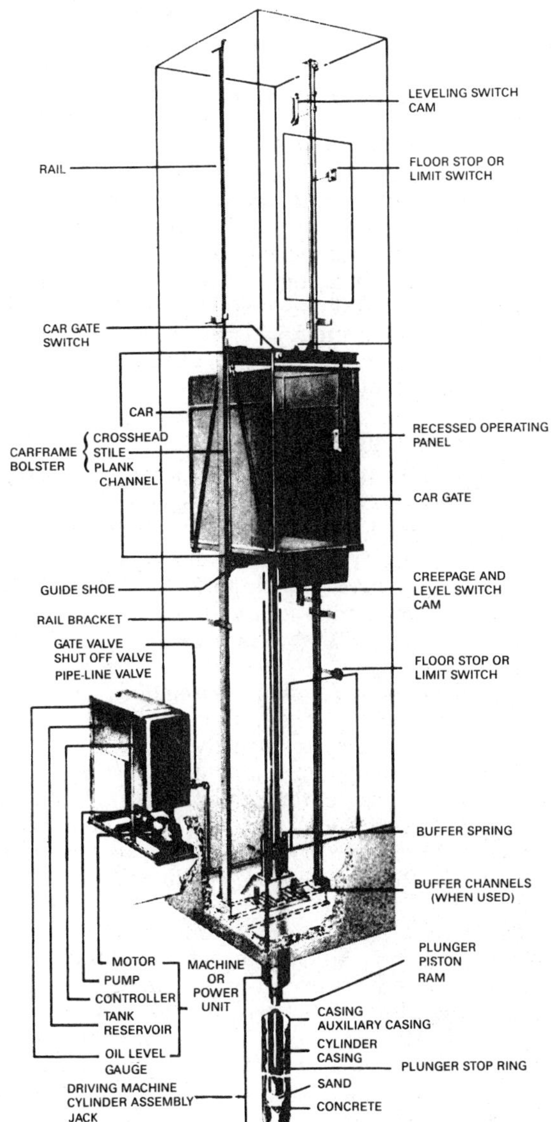

FIG. 8-5B. Hydraulic elevator. (National Elevator Industry, Inc.)

to provisions of one of the three previously listed model codes.

Although elevators have made high rise buildings practical, they have also contributed to dangerous and complicated firesafety problems. Hot smoke and gases from fire, if allowed to accumulate at the top of the elevator hoistway, will result in mushroom fires at the upper floors served by the elevator. This is possible since an enclosed elevator hoistway acts as a chimney or flue (due to a stack effect), which generally results in the movement of smoke and combustion products from lower to upper levels of the building.

Stack effect may also have an adverse effect on elevator door closing even under normal conditions. Pressure differentials must be measured in order to safely regulate the door closing operation.

All of the model building codes require venting of elevator hoistways. Most require venting of the hoistway for elevators serving three or more stories while some permit venting through the machine room with mechanical or natural venting from the machine room to the outside. Other codes prohibit this practice and require cable slots and other openings between the machine room and hoistway to be sleeved from the machine room floor to a point not less than 12 in. (305 mm) below the top of the hoistway vent to inhibit the passage of smoke into the machine room.

Some local codes prohibit venting and require pressurization of the hoistway to a specified minimum positive pressure above the pressure of the elevator lobby. When a pressurized hoistway is provided, it should also contain a vent that can open automatically in case of power failure or a buildup of smoke in the hoistway. Due to the numerous variances and the rapid developments in this area, local code requirements should be checked.

Elevator hoistway enclosure walls are normally fire partitions with a fire resistance rating of two hours. In some situations, no fire resistive rating is required. For example, an elevator that serves only open balconies without penetrating separate fire resistive areas of a building will not require a fire resistive hoistway. In other cases, the fire resistance rating may only be required on one or two walls. An example of this condition is an observation elevator, which permits exterior viewing by passengers, where the entrances to the elevator penetrate fire resistive walls.

The maximum number of elevators permitted in a single hoistway is controlled by building code requirements, which attempt to limit the potential of a fire knocking out all elevator service in a structure. Codes generally require that where there are three or fewer elevators in a building, they can be in the same hoistway enclosure. Four elevators would require at least two separate hoistway enclosures, and where there are more than four elevators, no more than four can be enclosed in the same hoistway enclosure.

Hoistway Entrances

If the hoistway wall is load bearing, the hoistway entrance operating mechanisms and locking devices can be supported by the walls. When the walls are not load bearing, the hoistway entrance operating mechanism and locking devices must be supported by other building structures. Elevator entrance frames are also not designed to carry the weight of the wall above, and a lintel must be provided.

Elevator hoistway entrance assemblies have customarily been rated at 1½ hour, but may be rated at 3 hour, 1 hour, or ¾ hour. They are usually designated as assemblies for Class B openings (openings to enclosures for vertical communication through buildings). A labeled entrance assures that the hoistway entrance door assembly can meet the firesafety requirements specified by building codes and NFPA 80, *Standard for Fire Doors and Windows*.

Passenger elevator doors installed without frames in masonry walls are designed to perform similar to those with frames, under fire exposure conditions, providing the doors overlap the masonry opening the same distance as an assembly incorporating a frame. Masonry type walls are generally considered to consist of brick, concrete block, or reinforced concrete construction. All three of these materials are fire resistant.

Masonry and drywall constructions that have marble or granite veneers are available, but they have not been investigated by testing laboratories under fire conditions, and the performance of such construction cannot be absolutely predicted. Engineering judgment can be applied when comparing granite or marble faced drywalls with solid masonry walls. This judgment can only be made by assuming that the application of the marble or granite facing does not adversely affect the structural integrity of the wall under fire conditions. In the case of drywall construction, the interface between the door and steel frame could be affected by the application of the facing material.

Drywall Enclosure Construction

Drywall construction largely supplanted masonry construction and is the most prevalent hoistway construction used today throughout the United States. Early elevator entrance installations in drywall were similar to masonry installations; however, some design changes were made to assure the integrity of the interface between the entrance assembly and the dry wall construction. Brackets were placed within the frame of the entrance assembly to accept "J" struts that are furnished by the drywall contractor. The "J" strut is of vital importance in assuring a proper entrance to drywall interface. There usually is no difference in the door panels. Figure 8-5C provides a cutaway view of a drywall entrance assembly interface.

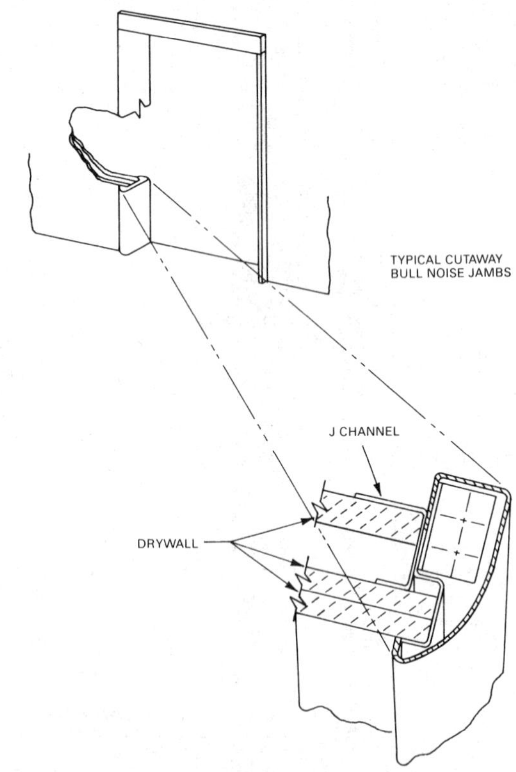

FIG. 8-5C. Details of the interface between drywall construction and bull nose door jamb of an elevator entrance (ASME 1984b).

Based upon a review of past fire performance data of masonry and drywall elevator door entrances, a listing agency determined, that in general, the drywall entrance assembly was more critical with respect to fire performance. As a result of the performance of drywall entrance assemblies, the listing agency could establish coverage for the same assemblies in masonry construction.

The classification marking (label) on passenger elevator doors covers the design and construction of the door panels only. In addition to the door panels, fire door frames for drywall must contain a marking label. Beginning in 1985, the *Elevator Safety Code* required that all passenger elevator frames must be labeled.

The hardware must also be listed to have a completely listed entrance assembly. Passenger elevator door hardware consists of a header, track hangers, pendant bolts, floor sill with guides, sill support plates, and brackets, retaining angles, and closure assemblies. If the hardware is packaged and shipped as a single unit, the *Elevator Safety Code* requires only one label for all components. If not, each component must be individually labeled.

Door Gasketing Materials

Listing agencies recently classified gasketing materials for use on elevator fire door and frames assemblies. The individual gasketing materials are currently investigated for use on particular frame or door types and for specific fire duration periods. The basic standard used to investigate gasketing materials is the Underwriters Laboratories standard *Fire Test of Door Assemblies* (UL 1979). The test, however, does not evaluate whether the gasketing material installed in the door assembly restricts door operation.

Fire departments consider in-car elevator operation crucial in a high rise fire. The question arises as to whether the melting of this gasket material will collect in the door tracks, rollers, etc., thus affecting elevator door operation. The kinetic energy and door closing forces for power operated doors are restricted by the *Elevator Safety Code*. During the last few inches of door travel, the closing speed is reduced. If deposits of gasketing material were to collect in the door tracks, the doors might be unable to move past these deposits, thus preventing complete closing of the door. As a result, the elevator would be immobilized and the elevator hoistway would be exposed to smoke, hot gases, and fire. Another question arises in connection with the effect this seal will have on door closers arranged to close an open door automatically if the car leaves the landing zone for any reason (Donoghue 1983).

Although it may not be as aesthetic as the architect and building owner would prefer, enclosure of elevator lobbies with smoke and draft control doors is the most practical substitute for smoke seals on elevator entrances as a means of controlling the movement of smoke into an elevator hoistway.

Machine Room Construction

When a fire resistive hoistway is required for electric elevators, the machine room must be constructed of equally fire resistive material. When a nonfire resistive hoistway is constructed for an electric elevator or for a hydraulic elevator, and the building code permits, the machine room enclosure must be of noncombustible material extending to a height of not less than 6 ft (1.83 m). Machine rooms are to be kept closed and locked, and should be restricted to authorized personnel who are familiar with the precautions necessary around moving elevator machinery. Machine room floors should be kept free and clean from oil and grease. Material not necessary for the maintenance or operation of the elevator should not be stored in the machine room. Flammable liquids with flash points of less than 110°F (43°C) should not be kept in the machine room. An extinguisher for use on Class C (electrical) fires is required in all electrical machine spaces and should be located convenient to the access door.

When the machine room is not located at the top of the hoistway, equipment not essential to the operation of the elevator is prohibited from being in the machine room because a fire that starts in a basement machine room has a clear unobstructed path to the hoistway. When the machine room is located above the hoistway, it may be in a room with other equipment essential to the operation of the building, provided the elevator equipment is separated from the other equipment by at least a 6 ft (1.83 m) high metal grill or solid partition.

Sprinklers in Hoistways and Machine Rooms

Provisions are outlined in NFPA 13, *Standard for Installation of Sprinkler Systems* (hereinafter referred to as NFPA 13) for sprinklers installed in elevator hoistways and machine rooms. Sprinklers are not required by the *Elevator Safety Code*, but when installed, sprinklers must meet requirements of the safety code in addition to those in NFPA 13. Before the installation of sprinklers in spaces containing elevator control equipment is considered, the following potential problems should be analyzed: (1) any discharge of water on the control equipment that may adversely affect the safe operation of the elevator, (2) the effect of water on safety circuits; water has been known to short out the safety circuits and allow the elevator to run with an open car or hoistway door, etc., and (3) wet brakes, which may be unable to stop and hold a moving elevator.

The *Elevator Safety Code* requires that before a sprinkler in the machine room or hoistway is discharged, the main line power source to the elevator be disconnected. In a fire situation, the sequence of events would probably be as follows: (1) a smoke detector in the machine room or elevator lobby would recall the elevator, (2) the main line switch would automatically be disconnected, and (3) the sprinkler would discharge.

Even if the smoke detector is not activated, or if all elevators have not been returned to the lobby, it is considered safer to stop the car and contain the passengers rather than to allow continued operation with the discharging sprinkler. Fire fighters should be aware that if the fire is in the elevator machine room, the elevators powered from that machine room will not be available for their use.

Other requirements in the *Elevator Safety Code* pertain to the location of sprinkler piping and valving.

Elevator Car Construction

The 1985 edition of the *Elevator Safety Code* requires that all material exposed to the car interior and the hoistway shall be metal or have a flame spread rating of 0 to 75 and smoke development rating of 0 to 450. The materials must be tested in their end use configuration. Padded protective linings for temporary use in passenger cars during the handling of freight are also required to meet these criteria. These parts of the code also provide criteria for materials used in car enclosures, car enclosure linings, and car platforms.

The smoke development rating for materials in cars is equal to NFPA 101 requirements for an exit passageway in a fire resistive building. The smoke development characteristics of materials may be more crucial than flame spread as passengers are enclosed in a confined box and may not be capable of rapid escape.

Napped, tufted, woven, looped, and similar materials must be subjected to a vertical burn test as specified by the *Elevator Safety Code*. This test is based on the applicable portions of *FAA Regulation 25.853*, in addition to the flame spread and smoke development test (ASTM E84). Such material, located on elevator car walls, is extremely susceptible to accidental ignition. Tufted material such as carpeting, may accumulate dust, which could increase the possibility of ignition and flame spread.

Floor covering, underlayment, and its adhesive are required to have a critical radiant flux of not less than 0.45 W/cm^2 as measured by the ASTM E648 test (ASTM 1978). The level of critical radiant flux acceptable for floor systems in elevator cars is based on the assumption that the occupants of an elevator may not be capable of rapid escape. (The flux level equals the requirements recommended for health care facilities.)

Handrails, operating devices, ventilating devices, signal fixtures, audio and visual communications devices, and their housings are not required to meet any of the above criteria as they contribute little to the fire load in relation to the overall elevator car assembly.

Handicapped Considerations

Two national standards address handicapped requirements for elevators. They are the *Minimum Passenger Elevator Requirements for The Handicapped* (NEI 1985) and the *American National Standard Specification for Making Buildings and Facilities Accessible to and Usable by Physically Handicapped People* (ANSI 1985). Both of these standards require that all passenger elevators be made accessible to the handicapped. Planned and practiced procedures are needed to evacuate the handicapped in a fire emergency. (See the discussion on Elevators in Emergencies later in this chapter.)

Emergency Lighting, Alarm, and Communications

The *Elevator Safety Code* requires emergency lighting on all elevators. The power source must originate on the elevator to assure that the passengers in the elevator will not be thrust into total darkness if building power fails. The building emergency power source is not a recognized alternative to this requirement since this source would generally not respond to a branch circuit failure and is usually activated only upon loss of the building power supply. Moreover, it would rely on the elevator's traveling cables to supply the lighting power. This would be useless if the lighting problem was in the traveling cables.

All elevators must be provided with an audible signaling device (alarm) and a means of two way conversation to an accessible point outside the hoistway. (See Figure 8-5D for a diagrammatic description of these requirements.) The two way communications system does not have to be provided with a means of activation from within the car. Most communications systems, excluding telephones, are activated in the main lobby or building emergency control center.

In addition to these requirements, the three model building codes require that all elevators in high rise buildings have a communication system from the elevator lobby, car, and machine rooms to the building's central control station.

ELEVATORS IN FIRE EMERGENCIES

High rise buildings present new and different problems to fire suppression forces, including the use of elevators in emergencies. Elevators are unsafe in a fire because: (1) persons may push a corridor button and have to wait for an elevator that may never respond (valuable time for escape is lost), (2) elevators do not prioritize car and corridor calls (one of the calls may be at the fire floor), (3) elevators cannot start until the car and hoistway doors are closed (panic could lead to overcrowding of an elevator and the blockage of the doors, thus preventing closing), and (4) power failure during a fire can happen at any time and thus lead to entrapment.

Fatal delivery of the elevator to the fire floor can be caused by: (1) an elevator passenger pressing the car button for the fire floor, (2) one or both of the corridor call buttons being pushed on the fire floor, (3) heat that may melt or deform the corridor push button or its wiring at the fire floor, and (4) normal functioning of the elevator, such as high or low call reversal, may occur at the fire floor.

The *Elevator Safety Code* recognizes all the aforementioned unsafe conditions and mandates elevator recall,

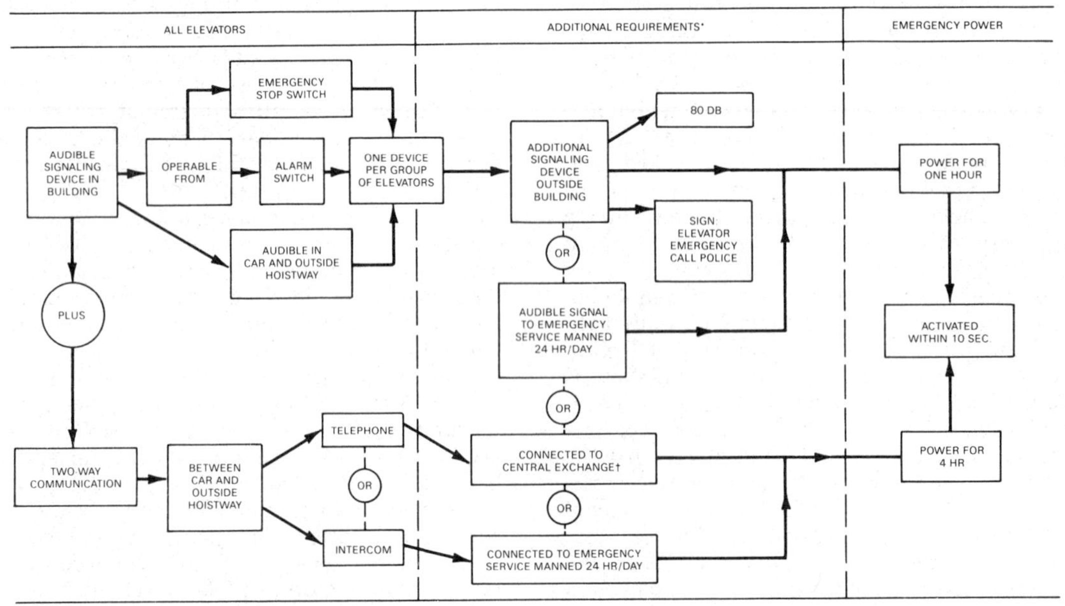

NOTES
* IN A BUILDING IN WHICH A BUILDING ATTENDANT, BUILDING EMPLOYEE OR WATCHSTAFF IS NOT CONTINUOUSLY AVAILABLE TO TAKE ACTION WHEN AN EMERGENCY SIGNAL IS OPERATED
† THE UNIFORM BUILDING CODE REQUIRES A TELEPHONE OR INTERCOM TO AN EMERGENCY SERVICE MANNED 24 HR/DAY FOR ALL ELEVATORS

FIG. 8-5D. Emergency elevator signaling device requirements (ASME 1984b).

more commonly referred to as *Phase I*, Fire Fighters' Service.

Phase I Operations

Phase I is the automatic or manual return or recall of elevators to a designated level or recall floor. Phase I also assures that the elevator is not available to the general public during a fire emergency. (The provision to allow emergency personnel to operate the elevator from within the car on emergency in car operation is commonly referred to as *Phase II*.)

The "designated level" is defined as the main floor or other level that best serves the needs of emergency personnel for fire fighting or rescue purposes. A three position (on, off, bypass) key operated switch in the designated level lobby can be used by emergency personnel to control each single elevator or a group of elevators. The switch in the on position instructs all elevators thus controlled to return nonstop to the recall floor, disregarding all activated call buttons.

Hoistway doors of elevators at the recall floor should remain open. Upon arrival, fire fighters have little time to learn if all elevators are present. Open hoistway doors will reveal which elevators have been recalled and which elevators have to be located to assure that there are no trapped passengers. (Position indicators, which can help emergency personnel locate errant elevators, are required to remain operative during emergencies.)

The open hoistway door at the designated level will also admit air to the hoistway. If the fire is on a floor below the neutral pressure plane, the additional air will reduce the amount of smoke that can enter the hoistway from the fire floor—a beneficial side effect. If the fire is above the neutral pressure plane, there will be an increased flow of air from the hoistway to the floor, which is also beneficial.

Visual and audible signals alert passengers in an automatic operated elevator of the emergency and can help minimize any apprehension that the passengers have when the elevator is returning to the recall floor. In an attendant operated elevator, this signal alerts the attendant of the emergency so that the car can be returned immediately to the recall floor.

More recent elevator recall systems combine the key operated switch with an automatic smoke detection system utilizing detectors at each floor lobby. Elevators are recalled either manually or automatically to the ground floor depending upon which mode initially operates. This recall system is required in many building code provisions for high rise buildings.

To advise the public that elevators may not be available during a fire, the pictograph in Figure 8-5E should be posted above all elevator hall call buttons. Some jurisdictions may require different wording and pictographs, but the message conveyed is similar.

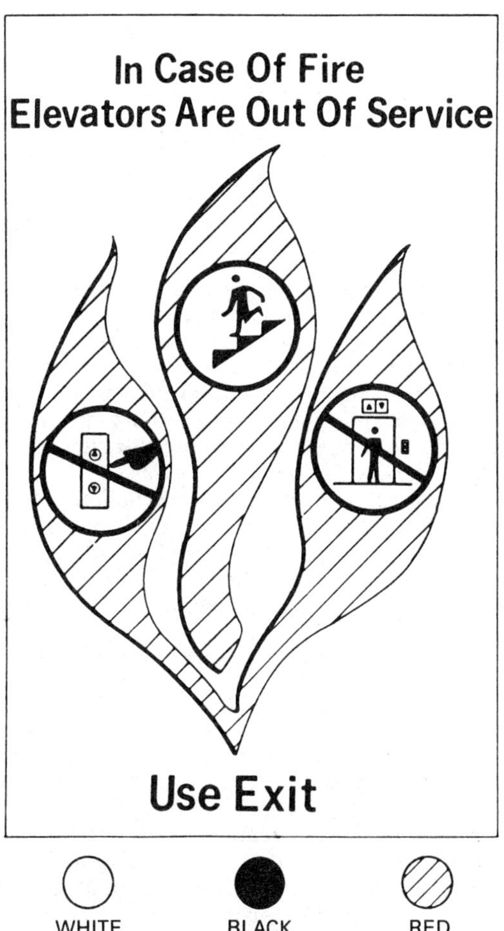

FIG. 8-5E. An example of pictographic warning message for elevator use (ASME 1984b).

Phase II Operations

Phase II fire fighter operations are for the benefit of fire fighters who must have the elevators for their own use. It is standard operating procedure for fire fighters to use elevators not only to carry equipment for fire fighting or evacuation purposes, but also to deliver fire personnel to nonfire floors.

The current *Elevator Safety Code* calls for Phase II operation whenever Phase I operation is required. The rule considers the need for fire fighters to evacuate the handicapped and disabled persons during an emergency. When operating on Phase II, the in-car emergency stop switch remains functional to permit emergency personnel to be selective in guiding car movements.

The required switch keys for Phase I and Phase II operation must be in the same place to assure a speedy response to emergency situations. The use of lock boxes to hold Phase I and Phase II keys is preferred over the use of a jurisdiction wide uniform key as the lock box method provides more assurance that the keys are restricted. The uniform key method has resulted in disruption of normal elevator service, as there is no means of controlling the distribution of elevator keys.

Operating procedures must be incorporated with or be adjacent to the Phase I and Phase II key operated switches so emergency personnel have quick access to instructions during an emergency.

The *Elevator Safety Code* does not require standby power for elevators. This requirement is normally found in the building code; however, the safety code does require that, when standby power is supplied, elevator operation be the same as when normal power is supplied.

For additional information on using fire fighters' service, see the NFPA textbook, *Handling Elevator Emergen-*

cies (NFPA 1979a) and the slide training package, *Using Elevators in High-Rise Fire Fighting* (NFPA 1982).

Evacuation From a Stalled Elevator

Any evacuation of passengers from a stalled elevator car should be performed by a rescue party that is supervised by trained elevator personnel. Their expertise will assure that the evacuation is carried out in a manner that is safe for both the passengers and rescue party. However, waiting for trained elevator personnel may be impossible in a life threatening emergency. Under these emergency conditions, passenger evacuation must be performed by personnel who are carefully selected and trained. Preplanned and practiced procedures are needed for the safety of the rescued passengers and the members of the rescue party. Any rescue party should be cautioned to never try to move an elevator from the controller; this should only be attempted by trained elevator personnel.

Three available references on evacuation procedures are: the ASME guide *Evacuation of Passengers From Stalled Elevator Cars*, the NFPA book *Handling Elevator Emergencies* and the NFPA slide training package, *Elevator Emergency Evacuation Procedures* (NFPA 1979b).

DUMBWAITERS

A dumbwaiter can be defined as a hoisting and lowering mechanism with a car of limited size which moves in guides in a substantially vertical direction and is used exclusively for carrying materials (ASME 1984b). Dumbwaiters are required to conform to requirements in the *Elevator Safety Code*. A dumbwaiter is actually a small elevator which is used for a variety of purposes: mail distribution in office buildings; distribution of medication, food and supplies in hospitals; moving books from floor to floor in a library; etc. A dumbwaiter can either be manually or automatically loaded and unloaded at either the floor or at a counter. Dumbwaiter car size is restricted to discourage riding. As dumbwaiters usually penetrate fire resistive floors, their hoistway construction is critical. The requirements for a dumbwaiter hoistway are the same as those for an elevator. See the section on elevator hoistway construction earlier in this chapter for a detailed discussion on fire resistive hoistway enclosures and entrances.

ESCALATORS

The escalator is another type of vertical people mover. An escalator can be defined as a power driven, inclined, continuous stairway used for raising or lowering passengers (ASME 1984b). In early days, escalators were found principally in department stores and bus and airline terminals. Today they are common in office buildings, hospitals, banks, sports arenas and other locations where many people must be moved quickly from one floor to another.

NFPA 101 requires escalators to have their floor openings enclosed or protected as required for other vertical openings. An important exception allows escalators not to be enclosed in completely sprinklered buildings if the escalators are protected by alternate means. Five enclosure methods that are utilized are:

1. **Sprinkler Draft Curtain Method**: This method is detailed in NFPA 13 and consists of surrounding the escalator opening with an 18 in. (457 mm) deep draft stop located on the underside of the floor to which the escalator ascends. A row of automatic sprinklers located outside the draft stop surrounds the escalator well. A typical installation is shown in Figure 8-5F.
2. **Sprinkler Vent Method**: An automatic fire or smoke detection system, an automatic exhaust system, and an automatic water curtain are combined in this method. When a detector senses a fire, the automatic damper opens and smoke and gases are then vented. Replacement air is drawn through the outdoor air intake above the escalator opening. To control smoke, an exhaust rate of about 60 air changes per hour is required. The sprinkler curtain provides a thermal barrier. A typical installation is shown in Figure 8-5G.
3. **Spray Nozzle Method**: A combination of an automatic fire or smoke detection system and a system of high velocity water spray nozzles comprise this method. The spray nozzles must be of sufficient number with discharge angles such that the escalator opening between the top of the wellway housing and the treadway will be completely filled with a dense water spray on operation of the system.
4. **Rolling Shutter Method**: In this method, escalators above the street floor may be protected by automatic self-closing rolling shutters which completely enclose the top of the escalator wellway. To avoid injury, the escalator must be stopped automatically before the shutter begins to close. The shutter should close slowly and have a pressure sensitive leading edge which will stop the shutter upon contact. A typical installation is shown in Figure 8-5H.

Rolling steel shutters are not recommended at the tops of moving stairways between basements and street floors even though the stairways are not required exits under provisions of NFPA 101. In an emergency, persons seeking egress from basements served by moving stairs could be trapped by fully closed rolling shutters at the street floor level even though other means of egress were available to them. It has also been observed that there is a quite different psychological reaction by those confronted with a closed shutter above their heads as they figuratively try to climb to safety from a basement than by those faced with a closed shutter at an exitway on the same level as the floor from which they are attempting to leave. In the latter case, operation of the shutter would be clearly visible and other means of egress could be readily found and used if the requirements of NFPA 101 are followed.

5. **Partial Enclosure Method**: Two fire rated enclosures, one for the down escalator and one for the up escalator, can be used to prevent the spread of smoke and fire with this method. Automatic fire doors release when a fire is detected. The enclosure should have a fire resistance rating equal to that of the floor ceiling assemblies, and there should be no unprotected escalator penetrations of more than two floors. A typical installation is shown in Figure 8-5I.

In new buildings escalators are not permitted to be used as a means of egress. Modern escalator design requires the removal of steps for normal maintenance, thus making the escalator unusable. A stopped escalator also presents a tripping hazard at the floor entrance and exit, due to the nonuniform riser height of the first and last few steps. In some existing buildings, escalators are part of the

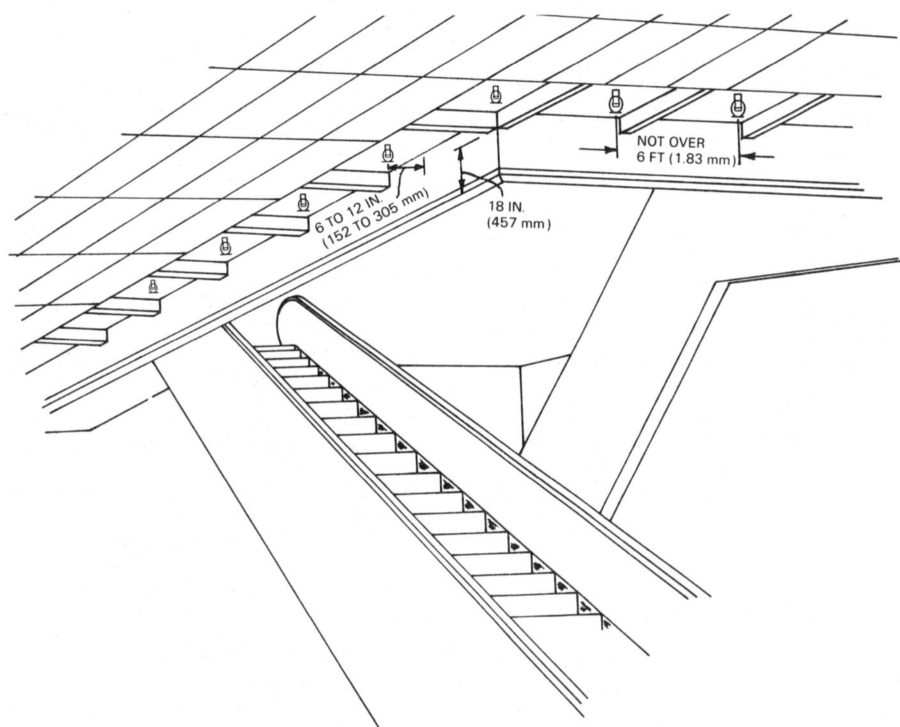

FIG. 8-5F. Sprinklers around an escalator.

means of egress. If they are, they should be arranged so that they can operate only in the direction of egress. If the evacuation plan calls for the use of escalators, efficient crowd control is required to avoid serious crowding and potential panic. NFPA 101 requires the openings for escalators that are used as a required exit to be enclosed in the same manner as exit stairs.

Escalators should not be stopped remotely by any fire detection system. The sudden unexpected stopping of an escalator can cause serious injury to passengers. An escalator should only be stopped in an emergency when there is imminent peril to the riders.

Another important factor in escalator fire prevention is keeping oil, grease, and dust cleaned up in the truss and machine areas. A periodic cleaning schedule should be implemented.

MOVING WALKS

A moving walk is defined as a type of passenger carrying device on which passengers stand or walk, and in which the passenger carrying surface remains parallel to its direction of motion and is uninterrupted. The same fire prevention features described for escalators would apply

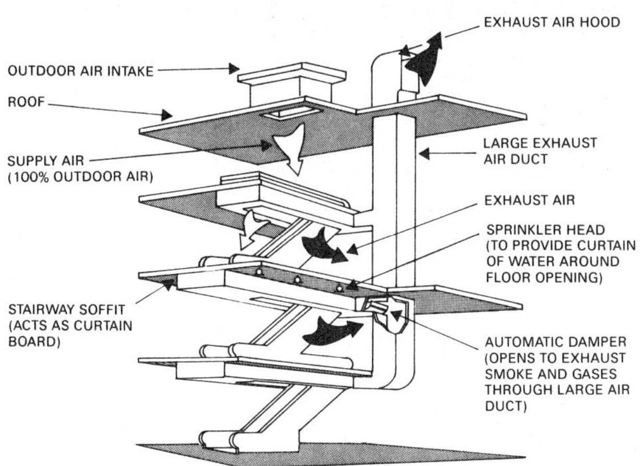

FIG. 8-5G. Sprinkler vent method installation.

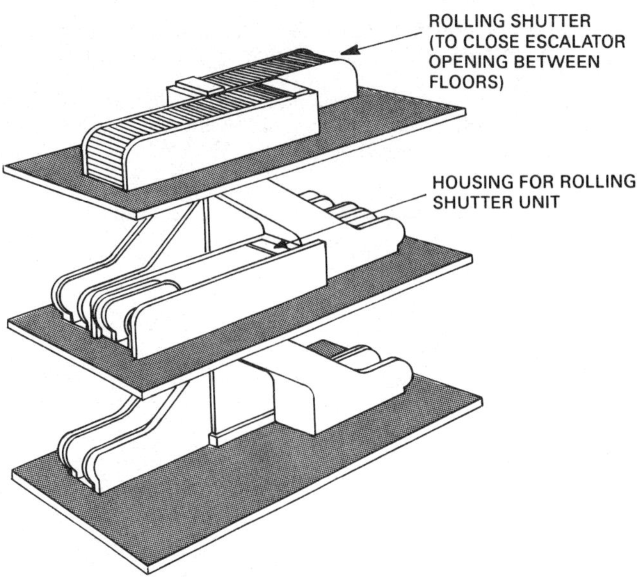

FIG. 8-5H. Rolling shutter escalator installation.

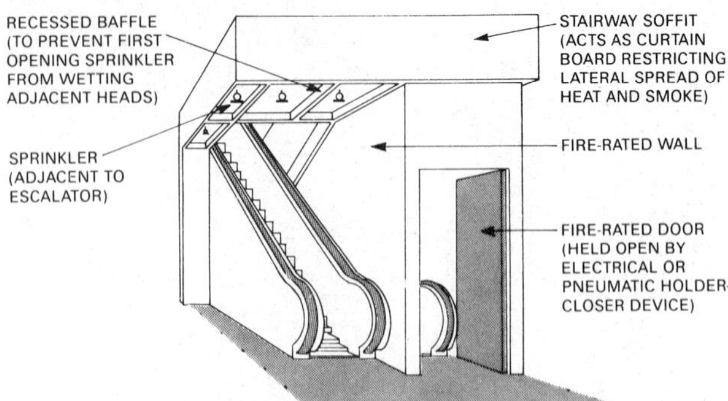

FIG. 8-5I. Partial enclosure escalator installation.

to moving walks. Some moving walks utilize a combustible belt, which is essentially the same as a large conveyor belt. It has the potential to spread a fire either by the belt itself or by movement of burning items along the length of the belt.

TESTS AND INSPECTIONS

Elevators, dumbwaiters, escalators, moving walks and the other passenger conveyance equipment which are required to conform to the requirements of the *Elevator Safety Code* must be tested and inspected periodically. The quality of inspections depends upon the competence of the inspector. The standard for the qualification of elevator inspectors (ANSI/ASME QEI-1) is intended to establish uniform criteria on this purpose, and to serve as a guideline on which certification is based by detailing the expertise necessary to perform inspections.

Bibliography

References Cited

ASME. 1984a. *Evacuation of Passengers from Stalled Elevator Cars, Guide A17.* American Society of Mechanical Engineers.

ASME. 1984b. *Safety Code for Elevators and Escalators.* ANSI/ASME A17.1. American Society of Mechanical Engineers.

ANSI. 1985. *American National Standard Specification for Making Buildings and Facilities Accessible to and Usable by Physically Handicapped People.* ANSI. A117.1. American National Standards Institute, New York, NY.

ASTM. 1978. *Test for Critical Radiant Flux of Floor Covering Systems.* ASTM E648, American Society of Testing Materials, New York, NY.

Donoghue, Edward A. 1983. "Smoke and the Elevator Hoistway." *Building Standards.* May-June 1983; *Elevator World* May 1983.

NEI. 1985. Minimum Passenger Elevator Requirements for the Handicapped. National Elevator Industry, Inc.

NFPA. 1979a. *Handling Elevator Emergencies.* National Fire Protection Association, Quincy, MA.

NFPA. 1979b. *Elevator Emergency Evacuation Procedures* (Slide training package). National Fire Protection Association, Quincy, MA.

NFPA. 1982c. *Using Elevators in High Rise Fire Fighting* (slide training package). National Fire Protection Association, Quincy, MA.

UL 1979. *Fire Test of Door Assemblies.* UL 10B. Underwriters Laboratories Inc. Northbrook, IL.

NFPA, Code, Standards, Recommended Practices and Manuals. (See the latest *NFPA Codes and Standards Catalog* for availability of current editions of the following documents.)

NFPA 13, *Standard for Installation of Sprinkler Systems.*

NFPA 70, *National Electrical Code.*

NFPA 80, *Standard for Fire Doors and Windows.*

NFPA 101, *Code for Safety to Life from Fire in Buildings and Structures.*

NFPA 255, *Standard Method of Test of Surface Burning Characteristcs of Building Materials.*

Additional Readings

BOCA, *Basic National Building Code,* Building Officials and Code Administrators International, Homewood, IL, 1984.

Donoghue, Edward A., *Handbook A17.1 Safety Code for Elevators and Escalators,* American Society of Mechanical Engineers, New York, NY.

Elevator World, Mobile, AL.

ICBO, *Uniform Building Code,* International Conference of Building Officials, Whittier, CA, 1985.

Inspectors Manual for Elevators and Escalators, ANSI/ASME A17.2, American Society of Mechanical Engineers, New York, NY.

Klote, John H., *Smoke Control for Elevators,* US Department of Commerce, National Bureau of Standards, Washington, DC.

SBCC, *Standard Building Code,* Southern Building Code Congress International, Birmingham, AL, 1982.

Strakosch, George, *Vertical Transportation, Elevators and Escalators.* John Wiley and Sons, New York, NY.

Vertical Transportation Standards, National Elevator Industry, Inc., New York, NY.

MISCELLANEOUS BUILDING SERVICES

Revised by Robert William Ryan

This final chapter in Section 8 contains information about building services—communications equipment, plumbing, chutes and utility shafts—that was not presented elsewhere in this section. These systems have no less of an impact on building firesafety than previously described services which should be considered during the design stage and integrated into the total fire protection system. The hazards associated with these systems also must be included in any prefire planning and fire protection survey.

For related information not contained in this chapter, refer to the following chapters. The protection of vertical openings (utility chases and chutes) is discussed in Section 7, Chapter 9, "Confinement of Fire in Buildings." Also of interest are Section 12, Chapter 7, "Materials Handling Equipment" (mechanical and pneumatic conveyors), and Chapter 14, "Waste Handling Systems" (laundry and trash chutes). For more information about central station alarm systems, see Section 16, Chapter 2, "Protective Signaling Systems," and for residential sprinklers using plastic piping see Section 18, Chapter 4, "Residential Sprinkler Installation."

COMMUNICATIONS

Due to the need for immediate and accurate data, communications systems are provided in every type of occupancy. For example, some businesses depend almost entirely upon telephone service, such as a telephone mail order service or a central computer system tied to remote terminals. Loss of access to communication equipment or loss of the equipment itself can ruin a business based on communications. The problems associated with the installation and protection of the communication equipment are: (1) that the equipment often introduces combustible fuel loading into an otherwise firesafe area of the building, and (2) that installation of the equipment can lead to serious reduction of the structural compartmentation originally designed into the facility.

Mr. Ryan is assistant director, Department of Environmental Safety, University of Maryland, College Park, MD.

Various types of communications, such as the public phone system, private phone systems, audible voice page, light or tone page, radio and video, and telecommunications, are used in buildings. The most common system is the telephone. In a large complex, the location, construction, and protection of the cable entrance enclosure in a building must be free of fire hazards and located in a secured area. Unauthorized tampering, deliberate sabotage, or accidental fire could disrupt the entire telephone service. Underground service lines feeding a building should be identified to avoid damage during excavation. Overhead communication lines should not be located above or near hazardous operations subject to fire or explosion.

Wiring

Most communication equipment utilizes electrical service of "low voltage" (50 V or less) which exempts the wiring from being placed in conduit. In addition, the wiring may be polyvinyl chloride (PVC) clad, wrapped in sizeable bundles, and placed in vertical chases that have minimal fire resistance.

Wiring insulation has been shown to contribute significant fuel and smoke generation in fires, even when classed as PVC, "self-extinguishing." The bundling together of the wire in undivided vertical shafts leads to potentially serious fire spread problems and provides fuel in an ideal configuration for continued burning. Usually the wiring enters the floor in communication rooms and is mounted on combustible panels in proximity to one another. This means that the communication room frequently contains an extremely high fuel loading which is an often unrecognized hazard. More fire resistant metal clad inside telephone cables are now available.

Fires originating in communication rooms can penetrate less than substantial partitioning enclosing such rooms, allowing fire to spread to adjacent areas. Fire can also spread rapidly up the open vertical shaft to other communication rooms, causing a multifloor fire. Automatic fire extinguishment, fire resistive compartmentation, and protection of the communication room and the vertical shafts should be considered to adequately safeguard a communication system.

One of the most commonly overlooked hazards of communications systems is the "poke through" penetration of fire barriers to extend service wiring. Holes drilled in floors and walls should be sealed with an approved noncombustible sealant or mechanical device to fill the void space around wiring. (See Fig. 8-6A.)

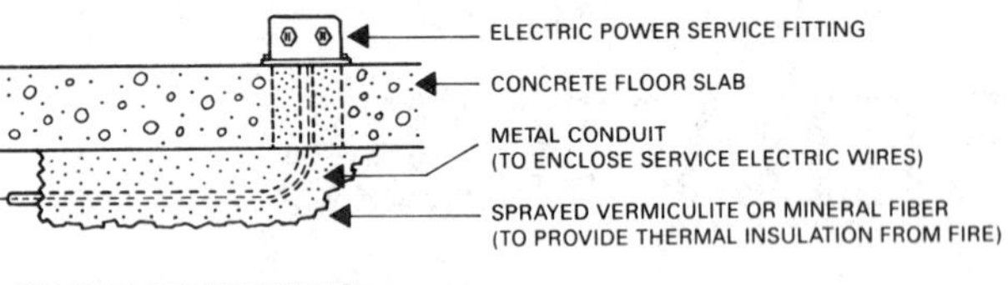

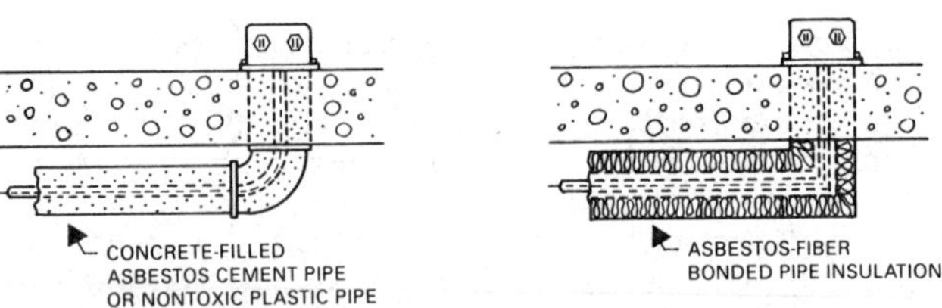

FIG. 8-6A. Fittings for electrical power, telephone service, etc., that are installed through drilled holes in the floors of completed buildings are called "poke through" assemblies. To prevent vertical fire spread, the underside of poke through assemblies can be protected by (1) spray on insulation undercoating, (2) intumescent mastic coating on the conduit, or (3) heat shields.

Central Control Systems

The increased use of computerized central monitoring and control systems in large buildings or multi building facilities provides an opportunity for instant communication of electrical or mechanical system status. Central control systems can be integrated with fire protection and building service systems to provide remote station reporting of fire detection and remote systems control.

Where a central control and monitoring system serves as a central station or remote fire alarm system, it should be protected from fire and power failures in the same way as central station facilities. In addition, the system should be designed to discriminate between emergency and routine signals and to provide an adequate signal to the operator when emergency action is required. Operators of central control systems must be trained to promptly notify fire and security services when an emergency condition is indicated.

Emergency Systems

Communications systems are invaluable during an emergency because they enable those in authority to communicate with people in the buildings from a central communications center. These systems are particularly critical in a high rise structure where the occupants either cannot evacuate the building or must be advised of areas of refuge or alternative egress routes. Emergency communications systems should be supplied from a reliable uninterruptable electrical power supply. Where a building communication center is designed as part of the fire protection system, it is vital that building security and public fire service personnel be trained in its use.

PLUMBING

Plumbing fixtures and piping were not considered a fire hazard, other than during installation, until the widespread use of plastic pipe. Full scale tests have been conducted on typical plumbing installations using ABS plastic (acrylonitrile-butadiene-styrene) for drains, waste, and vent (DWV) pipes, and fittings. These laboratory tests indicated that DWV pipes, and ABS fittings, if separated from other plastic pipe arrays, do not significantly contribute to the spread of fire (Bletzacker and Birle 1973; Troxell 1966). When such pipe is located in a chase, it provides less fire spread potential than when it is located above a suspended ceiling in proximity to other combustible fuel. Past studies of plastic DWV pipe have indicated that:

1. There is no significant fire problem associated with plastic DWV systems in one and two family housing;
2. The contribution of plastic DWV pipe to fire growth in any building is not significant when the pipe is covered with a thermal barrier such as gypsum wallboard; and
3. When tested as part of the rated assembly, plastic DWV pipe can be installed in fire resistant construction without decreasing the rated endurance (Williamson 1979).

In one documented fire, a polypropylene plastic vent spread a fire vertically in both directions as well as horizontally. In addition, the burning plastic dripped onto and ignited combustible material (Attwood 1980).

An advantage of plastic plumbing over copper plumbing with solder joints is that an open flame is not required during installation. However, most of the solvent cements used to join plastic pipe joints are flammable and give off combustible vapors. Precautions are necessary in using these solvents during installation. Care should be used to avoid generation of a flammable concentration of vapor during installation. All materials used to install the pipe, including additional cement and rags used to clean the joints, should be removed from the facility daily.

Plastic pipe has the potential of providing an economical method of "retrofitting" sprinkler systems. The Marriott Corporation, for example, has initiated a program of installing automatic sprinkler protection utilizing polybutylene pipe in all of its existing hotel rooms. The availability of economical piping may provide the incentive necessary for public acceptance of residential sprinklers. The combination of polybutylene plastic pipe and rapid response residential sprinklers has been heralded as the next step in residential occupancy fire protection.

SERVICE CHUTES AND CHASES

Most large or multistory buildings will have at least one vertical shaft or service tunnel constructed for essential building services. The hazards associated with these openings are frequently overlooked and are often given insufficient consideration during design and fire prevention inspection. Because chutes, chases, and barrier penetrations can be a contributing factor in spreading smoke, fire gases, heat, and water damage throughout a building, and in exposing essential building services to change, consideration must be given to arrangement, control of combustible loading, fire detection, and automatic extinguishment (Stevens 1967).

Mail Chutes

Although mail chutes have been in existence for a century, there has been little concern about fire problems with them. This may be attributed to the small floor openings they require, ranging from approximately 21.7 sq in. (140 cm^2) for letter drops, to 9.3 sq ft (0.86 m^2) for bundle chutes. They are also normally capped off at the top, which minimizes any significant stack effect. (The World Trade Center in New York City has more than 430 stories of mail chutes.)

Although fires have been reported in mail chutes, they have been confined to the receiving container. Mail chutes are available that can eject a cigarette dropped into the mail slot back out onto the floor. It is important that the chute is capped or sealed at the top to minimize any flue effect.

UTILITY SHAFTS

Utility chases for piping and wiring are necessary in multistory buildings. Because the utilities must serve each floor, it is more convenient and economical to run them in a common location. These vertical shafts run from the basement up through the building to provide water, sanitation, electricity, and telephone services on each floor.

These utility shafts have contributed to the spread of fire and smoke. Fire studies have shown some of the main weaknesses to be (1) poor shaft construction, (2) non rated access doors, and (3) combustible pipe insulation and jackets (Stevens 1967). One solution for limiting the spread of fire and smoke through chases is to fill in around the shaft solidly with noncombustible material at each floor level. All utilities passing through the fire-smoke stop can be completely filled in around the pipe or sleeve with a noncombustible grouting material. Pipe insulation should not be carried through the chase fire stop.

If barriers at each floor are not possible, fire doors or access doors that have been listed by an independent fire testing laboratory for the protection of openings are recommended for the openings in utility chases.

Bibliography

References Cited

Attwood, P. C. 1980. "Penetration of Fire Partitions by Plastic Pipe." *Fire Technology*. Vol 16, No 1. p 37.

Bletzacker, R. W., and Birle, J. G. 1973. "Standard ASTM Fire Endurance Test and Hose Stream Test on Duplicate Non-Load Bearing Acryionitrile-Butadiene-Styrene (ASB) Plumbing Wall Assemblies." *Building Research Laboratory Report No. 5473*. Ohio State University, Columbus, OH.

Stevens, Richard E. 1967. "Utility Shafts for Piping and Wiring." *Fire Journal*. Vol 61, No 4. pp 22-23.

Troxell, George E. 1966. "Fire Tests of Plastic Vents and Drain Pipes." *Fire Journal*. Vol 60, No 4. pp 52-57.

Williamson, Robert B. 1979. "Installing ABS and PVC Drain, Waste, and Vent Systems in Fire-Resistant Buildings." *Fire Journal*. Vol 73, No 2. pp 36-45.

NFPA Codes, Standards, Recommended Practices and Manuals. (See the latest *NFPA Codes and Standards Catalog* for availabililty of current editions of the following documents).

NPPA 13, *Standard for the Installation of Sprinkler Systems*.

NFPA 54, *National Fuel Gas Code*.

NFPA 58, *Standard for Storage and Handling of Liquefied Petroleum Gases*.

NFPA 71, *Standard for the Installation, Maintenance and Use of Central Station Signaling Systems*.

NFPA 72C, *Standard for the Installation, Maintenance and Use of Remote Station Signaling Systems*.

NFPA 72E, *Standard on Automatic Fire Detectors*.

NFPA 80, *Standard for Fire Doors and Windows*.

NFPA 82, *Standard on Incinerators Waste and Linen Handling Systems and Equipment*.

NPPA 101, *Code for Safety to Life from Fire in Buildings and Structures*.

NFPA 211, *Standard for Chimneys, Fireplaces, Vents and Solid Fuel Burning Appliances*.

NFPA 251, *Standard Methods of Fire Tests of Building Construction and Materials*.

NFPA 252, *Standard Methods of Fire Tests of Door Assemblies*.

Additional Readings

Benjamin, I. A., et al., *Fire Endurance of Gypsum Board Walls and Chases Containing Plastics and Metallic Drain, Waste, and Vent Plumbing Systems*, U.S. Government Printing Office, Washington, DC, 1975.

Bimonthly Fire Record, "Outside Trash Chute Fire," *Fire Journal*, Vol. 78, No. 1, Jan. 1984, p. 16.

Bimonthly Fire Record, "Electrical Fire Spreads Through Store," *Fire Journal*, Vol. 77, No. 2, Mar. 1983, p. 10.

Bimonthly Fire Record, "Short Circuit Results in Loss Over $750,000," *Fire Journal*, Vol. 76, No. 3, May 1982, p. 22.

Bimonthly Fire Record, "Mail Room Fire," *Fire Journal*, Vol. 76, No. 2. Mar. 1982, p. 13.

Bimonthly Fire Record, "Telephone Equipment Destroyed," *Fire Journal*, Vol. 74, No. 2, Mar. 1980, p. 21.

Bimonthly Fire Record, "Short Circuit Ignites PVC Insulation," *Fire Journal*, Vol. 73, No. 4, July 1979, p. 22.

Bimonthly Fire Record, "Welding Spark Ignites Plastic Pipe," *Fire Journal*, Vol. 70, No. 4, Jan. 1976, p. 34.

Bimonthly Fire Record, "Group Cable Fire," *Fire Journal*, Vol. 70, No. 6. Nov. 1976, p. 41.

Church, J. C., *Practical Plumbing Design Guide*, McGraw-Hill Inc., NY, 1979.

Comeford, J. J., and Birky, M., "A Method for the Measurement of Smoke and HCl Evolution from Poly (vinyl chloride)," *Fire Technology*, Vol. 8, No. 72, May 1972, p. 85.

Crane, et al., *Electrical Insulation Fire Characteristics*, U.S. Dept. of Transportation, Oklahoma City, OK, 1979.

Ettling, Bruce V., "Electrical Wiring in Building Fires," *Fire Technology*, Vol. 14, No. 4, Nov. 1978, p. 317.

Frank, Thomas E., "Fire and Explosion Control in Bag Filter Dust Collection Systems," *Fire Journal*, Vol. 75, No. 2, Mar. 1981.

Glass, R. A., and Rubin, A. I., *Fire Safety for High-Rise Buildings: The Role of Communications*, Center for Building Technology, National Bureau of Standards, Washington, DC, 1979.

Lathrop, James K., "Telephone Exchange Fire," *Fire Journal*, Vol. 69, No. 4, July 1975, p. 5.

Lichtenstein, Stan, "Operation Firestop," *Professional Safety*, Vol. 23, No. 6, June 1978.

McGuire, J. H., "Small-Scale Fire Tests of Walls Penetrated by Telephone Cables," *Fire Technology*, Vol. 11, No. 2, May 1975, p. 21.

McGuire, J. H., "Penetration of Fire Partitions by Plastic DWV Pipe," *Fire Technology*, Vol. 9, No. 1, Feb. 1973, p. 5.

McLaughlin, David L., "Fire Tests of Fiberglass and PVC Ducts for Minimum Sprinkler Coverage," *Fire Journal*, Vol. 72, No. 2, Mar. 1978, p. 59.

National Electrical Manufacturers' Association, "NEMA High Rise Fire Alarm Systems Recommendations for State and Local Codes," Washington, DC, 1983.

National Electrical Manufacturers' Association and The Society of the Plastics Industry, Inc., "Characteristics of Polyvinyl Chloride Conduit, Insulated Wire & Piping in Fire Situations," Washington, DC.

Nemeth, Craig W., "Penetration Sealing With Silicone RTV Foams," *Fire Journal*, Vol. 76, No. 1, Jan. 1982.

"Office Building Fire, 919 Third Avenue, New York City," *Fire Journal*, Vol. 65, No. 2, Mar. 1971, p. 17.

Pannkoke, Ted, "Thermoplastic piping: an overview," *Heating/Piping/Air Conditioning*, Vol. 51, No. 4, Apr. 1979.

Plastics and Plastics Products, National Fire Protection Association, Quincy, MA, 1975.

Powers, F. C., ed., "Electric Cables in Trays," *The Sentinel*, Vol. 35, No. 4, Aug. 1979, p. 3.

Powers, F. C., ed., "Why Cable Penetrations Need Fire Stops," *The Sentinel*, Vol. 35, No. 4, Aug. 1979.

Powers, Robert W., "New York Office Building Fire," *Fire Journal*, Vol. 65, No. 1, Jan. 1971, pp. 22-23.

Sawyer, Robert G., and Elsner, James A., "Cable Fire at Browns Ferry Nuclear Plant," *Fire Journal*, Vol. 70, No. 4, July 1976, p. 5.

Schuman, A. R., "Poke-Through Protection," *Fire Technology*, Vol. 11, No. 4, Nov. 1975, p. 294.

Sears, A. B., "Another Home for Aged Fire: Ten Killed," *Fire Journal*, Vol. 65, No. 3, May 1971, p. 7.

Stone, Walter R., "Fire Fatal to Three in Residential Care Facility," *Fire Journal*, Vol. 67, No. 5, Sept. 1973, p. 22.

Watrous, L. D., "High Rise Fire in New Orleans," *Fire Journal*, Vol. 67, No. 3, May 1973, p. 9.

Watrous, L. D., "Fatal Hotel Fire-New Orleans," *Fire Journal*, Vol. 66, No. 1, Jan. 1972, p. 7.

Watrous, L. D., "Tae Yon Kak Hotel Fire-Seoul Korea," *Fire Journal*, Vol. 66, No. 3, May 1972, p.12.

Watrous, L. D., "High Rise Building Fire-Sao Paulo, Brazil," *Fire Journal*, Vol. 66, No. 4, July 1972, p. 107.

SECTION 9
HAZARDS TO LIFE IN STRUCTURES

ASSESSING LIFE SAFETY IN BUILDINGS

Revised by John M. Watts Jr., Ph. D.

Assessment of life safety is the process of estimating the quality of security against fire and its effects. There is no well defined method of assessing life safety from fire in buildings. Life safety is a concept, and no formula can identify or guarantee that a building is safe from fire. First of all, assessment requires an understanding of the fundamentals of the life safety concept. This may be followed by a subjective evaluation of parameters that create risk and those that tend to offset or mitigate a portion of that risk. Checklists may help to ensure consideration of these parameters, but they do not identify relative values or interrelationships of the parameters.

Additional information which can be helpful in assessing life safety in buildings will be found in Section 1, Chapter 2, "Human Behavior and Fire;" Section 2, Chapter 3, "Use of Fire Loss Information"; Section 3, Chapter 3, "Fire Exit Drills;" Section 4, Chapter 1, "The Chemistry and Physics of Fire;" and Section 5, Chapter 2, "Wood and Wood Based Products." Also see Section 7, Chapter 1, "Fundamentals of Firesafe Building Design;" Chapter 2, "Systems Concept for Building Firesafety;" Chapter 3, "Concepts of Egress Design;" Chapter 7, "Interior Finish;" Chapter 9, "Confinement of Fire in Buildings;" and Chapter 10, "Smoke Movement in Buildings." Reference can also be made to Section 16, Chapter 4, "Automatic Fire Detectors," and Section 21, Chapter 5, "Microcomputer Applications in Fire Protection."

LIFE SAFETY FACTORS

Concern for life safety implies avoiding exposure to harmful levels of products of combustion. This goal is usually achieved by controlling the fire process and/or separating endangered individuals from the harmful effects of fire. Details of the fire's development along with characteristics of exposed occupants, determine the magnitude of the risk. Specific safety measures must be employed to reduce the risk. Understanding the interrelationships of these components is the first step in assessment of life safety from fire in buildings.

Time

As a fire develops over time, smoke and heat build to create an environment that is hazardous to life. The rate at which the environment will deteriorate is difficult to predict, since a great many variables are involved. Figure 9-1A is an approximate generalization of the way life hazard increases over time. At ignition (lower left corner) the environment is normal. Most fires develop slowly at first so that, initially, the hazard is small. Eventually, fire intensity will increase more sharply, building up the level of dangerous products of combustion.

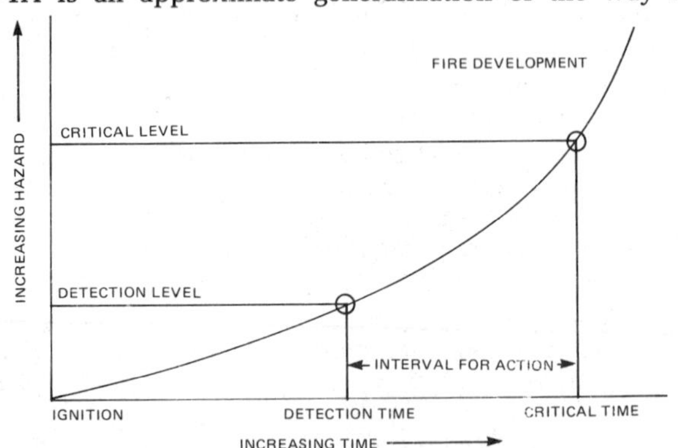

FIG. 9-1A. The approximate rate of deterioration of the environment as a fire progresses.

At some level of accumulation of products of combustion, the fire will be detected—by automatic fire detectors or by personal detection (smell or sight). The level of hazard at which detection occurs corresponds, in Figure 9-1A, to a specific time in the course of the fire—the detection time.

Dr. Watts is the director of the Fire Safety Institute, a Non-profit Information, Research, and Educational Corporation, Middlebury, VT.

The Critical Level

The other important point in fire development is the critical level, i.e., when deterioration of the environment represents danger to life. This will vary according to combustion products emitted and characteristics of exposed persons which make them more or less susceptible to various products of combustion. Because of many unknown and uncontrollable factors in assessment of the effect of fire on people, the critical level is not easy to identify precisely. Corresponding with the critical level of hazard is the time it takes for the fire to produce this situation. This is shown in Figure 9-1A as critical time.

The interval between detection and criticality is the time available to undertake action to prevent occupants from being exposed to the critical hazard level. This action may take various forms such as activation of automatic equipment, evacuation of occupants, or both. Indicated in Figure 9-1A, lowering the level where detection occurs will increase the interval of time available for action. In addition, occupants with greater susceptibility, and hence a lower personal critical level, will have less time to react.

Fire Variations

Fire growth and hazard do not always develop at the same rate. If conditions are insufficient to sustain a fast spreading fire, the rate of deterioration of the environment is reduced, resulting in a decreased slope of the fire development curve. For example, curve B in Figure 9-1B

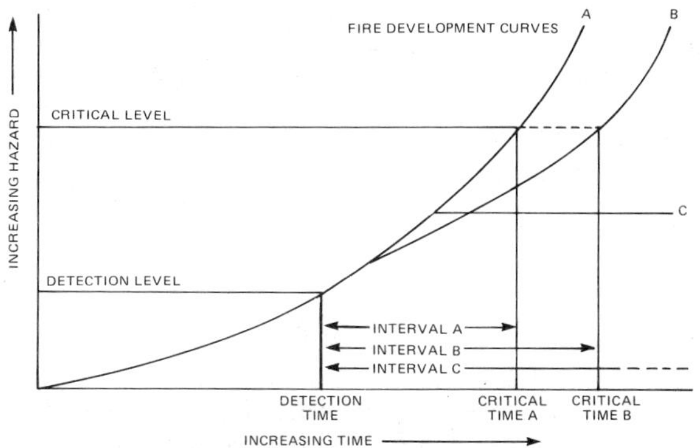

FIG. 9-1B. The difference in deterioration rates between a faster-developing and a slower-developing fire.

represents a slower developing fire than curve A, with a resultant increase in the interval of time available for action. Alternatively, curve C might represent the effect of an automatic smoke control system. Activated after fire detection, the system maintains the atmosphere below the critical level for a longer period of time.

CHARACTERISTICS OF OCCUPANTS

The most difficult component of life safety to evaluate is the population at risk (i.e., the occupants of the building). The difficulty is due to wide variations among building occupants. It is necessary to assess their susceptibility to fire and fire products, and their ability to undertake and follow through with procedures necessary for their safety (i.e., anticipated response to a fire emergency). Indications of these qualities may be found in physical and mental characteristics of the occupants, individually and as a group. Important population parameters are age, mobility, awareness, knowledge, density, and discipline.

Age

Age is the most easily identifiable characteristic of an exposed occupant, and it may directly affect other important characteristics such as mobility, awareness, knowledge, and discipline. Variation of life risk with age is indicated by statistics that show that the very young and the very old suffer higher fatality rates from fire.

Mobility

Mobility is, for most people, a function of age; the very young and very old are less mobile. However, there are many other classes of immobile persons or those with limited mobility. Handicapped persons are present in more and more locations as accessibility to public facilities increases. It is necessary to consider the limitations of people with handicaps in assessment of the life risk. In occupancies such as hospitals, many people are temporarily incapacitated and can be relocated only with assistance. In still other occupancies such as psychiatric institutions or prisons, individuals of average agility may be restrained or incarcerated. Thus, lack of mobility is not necessarily tied to age or even physical capability.

Awareness

Awareness is another characteristic that may be affected by age, as well as by other conditions. A major factor is whether occupants are awake and alert (for example, in offices), or asleep, (for example, in residential occupancies). In addition, limited awareness may be found among those persons to whom drugs have been administered. The influence of alcohol and narcotics is a contributing factor in a significant number of fire deaths (Halpin et al 1975). Most institutional occupancies contain individuals with varying degrees of awareness and a range of decision making capabilities.

Knowledge

Knowledge is perhaps a limited term for the comprehensive concept of self-preservation, which includes factors such as instinct. Some of these characteristics are discernible, if not measurable. Training and drills may increase the level of occupants self-preservation knowledge. This can be extended to include societal self-preservation, whereby certain individuals are trained, or react spontaneously, to assist others. A less direct aspect of knowledge involves occupants' familiarity with the premises. Regular occupants of a particular building are likely to have a better knowledge of the exits than are transient visitors.

Density

Population density, or the number of persons in a given area, is important to the magnitude of risk and to the safe relocation of occupants. The greater the number of people in a given area, the greater the potential loss of life. Studies have shown the relationship of occupant density to speed of movement (Pauls 1977, Melinek and Booth 1975, Peschl 1971). Phenomena, such as formation of arches of people, which obstructs passage through a doorway, are dependent upon occupant density.

Discipline

Like density, discipline is characteristic of occupants as a group more than as individuals. Controllability is evidenced in occupancies such as schools and certain industries. In general, people regularly exposed to disciplinary control and training are apt to respond to authority in a fire emergency with less likelihood of endangering behavior.

THE NATURE OF FIRE IN BUILDINGS

Fire is a chemical reaction. In particular, it is an oxidation process of sufficient magnitude to emit heat and light. Also associated with fire are (1) smoke, which is usually comprised of particulate matter that obscures visibility; and (2) combustion gases, most of which are highly toxic. Although better defined than the human factor, fire is perhaps the least predictable of all common physical phenomenon (Swersey and Ignall 1975). Recently, however, a considerable amount of research has been devoted to understanding the complexities of unwanted combustion (Levine and Pagni 1984). Aspects of fire development that are important to the severity and management of risk to life are ignition potential, fire growth, and smoke spread.

Ignition Potential

The process of ignition involves a heat source coming in contact with or being in sufficient proximity to a fuel. Cooking, heating, and other energy consuming activities involve potential heat sources. Furnishings, clothing, and other building contents provide fuel. It is most often the occupants of a building who bring heat source and fuel together.

Ignition is a function of time as well as of temperature. Fuel subjected to a high temperature for a short period of time may not ignite, but the same fuel may ignite when exposed to a lower temperature for a longer duration. This is one of the limitations of *ad hoc* procedures such as the "match test," for determining ignitability. Information on the frequency of certain ignition patterns is not sufficient to assess ignition potentials because it gives only the number of occurrences and not the number of possible occurrences out of which the given number of ignitions was realized.

Fire Growth

The fire growth stage is the most important to life safety. It is in the growth stage that the room or space of fire origin eventually becomes uninhabitable, i.e., critical hazard levels are reached. Therefore, it is in this stage that detection and action to protect occupants of the room of origin must be effected. Figure 9-1C illustrates the importance of fire growth rate on life hazard.

Many relevant chemical and physical properties of potential fuels manifest themselves in fire hazard tests. The flame spread rating of a material (see NFPA 255, *Standard Method of Test of Surface Burning Characteristics of Building Marerials*) is one observable feature that can give guidance on fire growth characteristics. Flame spread ratings of interior finish vary greatly with the chemical and physical properties of materials.

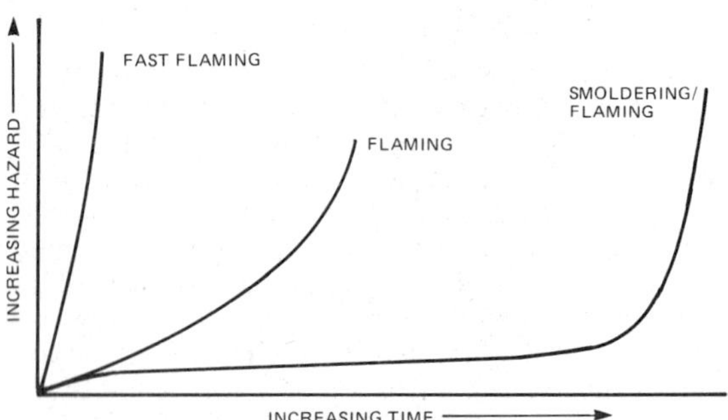

FIG. 9-1C. Effect of fire growth rate on increasing hazard.

While fuel is the most important factor in the very first stages of fire, the surrounding environment becomes significant to fire growth in later development. Materials burning near a wall will have a faster rate of fire growth than materials burning in the center of a room. This is due to the wall heating up and radiating heat to the fuel, reducing heat loss from the combustion system. In the same manner, fires in corners receive radiated heat from two walls and grow even faster. The most important dimension of a room or space is the ceiling height. When a developing flame reaches the ceiling, it flares out, or mushrooms, to produce a large radiating flame surface. This in turn increases the rate of burning and promotes the ignition of additional fuels in the space.

When fire spreads across the ceiling, heat buildup may continue throughout the room until sudden ignition of most other combustible materials in the space occurs. This phenomenon is known as flashover. At this point in fire development, large quantities of high temperature gases are produced, making the area totally untenable. This marks the point beyond which no actions can reasonably be expected to prevent death of any occupants still in the room.

Spread of Smoke and Fire

Smoke and fire spread expose occupants to danger beyond the immediate area of fire origin. Fire developing in a room or compartment generates a positive pressure, which tends to push smoke out of the area of origin to other areas of the building. Even without the pressure of the fire, climatic features and buoyancy can drive smoke and toxic gases to remote parts of a structure. Most commonly, fire products follow the paths designed for use of the occupants—through open doors, along corridors, and up open stairwells. However, there are many other paths by which smoke and gas can travel. Building features such as shafts, ducts, plenums, void spaces, poke throughs, and even exterior windows can all contribute to spread of fire and smoke.

Large complex structures, particularly atrium buildings and multilevel shopping malls, require a thorough analysis of smoke movement features for life safety assessment.

SAFETY STRATEGIES

Ideally, building design considers the risk factors associated with occupants and fire, and includes safety features to mitigate the risks. The discussion in this chapter is intended as a synthesis of the various safety measures applicable to reducing the danger of fire to occupants. Major categories of safety strategies have been identified by the *Firesafety Concepts Tree* (NFPA 1980) as: "Prevent Fire," "Manage Fire," and "Manage Exposed" (occupants).

Fire Prevention

No harm can come from fire if no fire takes place. Fire prevention, therefore, has the potential to eliminate the need for other firesafety measures. However, no satisfactory strategy to completely eliminate unwanted ignition has yet been devised. Fire prevention considerations relate to heat energy sources, fuels, and the mechanisms by which they are brought together.

Major potential ignition energy sources such as electrical power can be largely controlled through regulation, e.g., by following requirements of NFPA 70, *The National Electrical Code®*. In addition, some fuels such as flammable liquids and interior finish may also be addressed by codes and standards. However, not all ignition sources nor all fuels can be regulated. This is also especially true of people—the primary mechanism of ignition.

Assessment of the probability of fire occurrence involves analysis of ignition potential, especially in relation to activities of building occupants, and factors that will decrease the likelihood of ignition.

Fire Management

Since it is functionally impossible to avoid all ignitions, fire control measures are important. Techniques of fire management strategy may be regarded as methods of reducing the slope of the fire development curves portrayed in Figures 9-1A and 9-1B. These approaches strive to: (1) control the rate of production of smoke and heat through alteration of fuel and/or the environment, (2) control the combustion process by manual or automatic suppression, and (3) control the products of combustion through venting and/or containment.

Altering fuel and environment to release fewer harmful products in a fire situation is obviously desirable but difficult to effect. The first obstacle usually is economic considerations. Nevertheless, progress is being made.

Suppression systems must be evaluated in terms of their ability to affect fire development before products of combustion reach critical levels. Venting and smoke control systems should be considered life safety systems. Directing the products of combustion away from occupants is a rational approach where life safety is the primary concern. Design of smoke control systems is the subject of continuing research.

Objectives of the "manage fire" strategy are to reduce risks associated with fire growth, and to reduce fire and smoke spread. Together, these reductions diminish the impact a fire on building occupants.

Occupant Management

This safety strategy is the most complex, dealing with the risk factors of both fire and people. Management, here, involves undertaking emergency action appropriate to expected fire development, as well as to characteristics of the occupants.

To initiate occupant action, there must first be detection and alerting activities. These functions may be performed by automatic equipment or by building occupants. Actions for managing exposed persons involve evacuation, refuge, or rescue. Evacuation is the most common approach where occupants are alert and mobile. In other cases, areas of refuge from fire and smoke within a building are employed, along with removal assisted by emergency personnel.

Emergency egress systems should be assessed in terms of their adequacy and reliability. Adequacy refers to structural components that determine the capacity to evacuate part or all of a building within a safe egress time span. Reliability considers how efficiently the egress capacity will be utilized. Factors of reliability include alerting messages and instructions, signage and emergency lighting, and protection of egress routes from fire, smoke, and toxic gases.

ASSESSMENT TOOLS: EVALUATING THE RISK

The predominant overall guide to life safety from fire in buildings in the United States is NFPA 101, *The Life Safety Code®* (*Code for Safety to Life from Fire in Buildings and Structures*), (hereinafter referred to as NFPA 101). However, the stated purpose of the *Life Safety Code* is to establish minimum requirements. Other tools are required where a higher degree of life safety is desired, or where it is of interest to identify the "worst" life safety risks. NFPA 101 also includes a section on equivalency concepts to foster a performance approach to fire protection. Fire risk assessment is one way of establishing equivalencies that lend themselves to the required technical documentation.

Risk and safety are the components of a generalized approach to assessment of life safety from fire in buildings. Such an approach is summarized in Figure 9-1D. The first

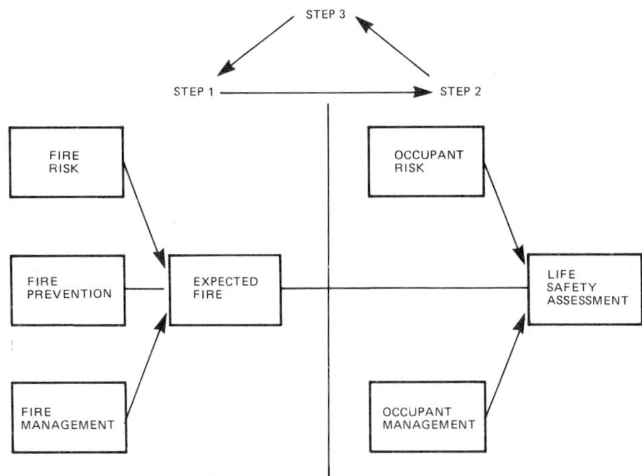

FIG. 9-1D. A general approach to life safety in buildings.

step is analysis of the fire risk parameters discussed and of the safety measures corresponding to fire prevention and fire management. This analysis should yield a concept of the expected fire in the building. The second step is consideration of this expected fire in conjunction with the occupant risk factors and management of exposed occupants. The life safety assessment is then evaluated; if found lacking, step three is a return to the beginning to consider enhanced lifesafety strategies. Thus, assessing life safety in buildings is an iterative process of matching risk and safety factors.

More specific techniques for assessment of life safety from fire in buildings include scenarios, exit analysis, FSES (Fire Safety Evacuation System), BFSM (Building Fire Simulation Model), firesafety effectiveness statements, ASET (Available Safe Egress Time), and a matrix approach.

Scenarios: Fire death scenarios have been devised (Clarke and Ottoson 1976) to help guide national fire research strategy. Scenarios are particularly useful in assessing life safety in buildings. Construction of a fire scenario begins with identification of a particular situation that corresponds to the behavior of fire and occupants within a building. By considering various scenarios, it often is possible to identify and evaluate the potential for life loss. Although they are subjective assessment tools, scenarios can be based on the collective judgment of several experts.

Exit Analysis: This generally refers to calculation methods for determining the time required to evacuate a building or space. The basic engineering approach to exit analysis have been outlined (Predtechinskii and Milinskii 1978). Computer models such as EVACNET (Chalmet et al 1982) are becoming popular for evaluating evacuation capabilities of large buildings. Use of exit analysis in the assessment of life safety is growing, along with application of microcomputers.

Fire Safety Evaluation System (FSES): This was developed for use in health care facilities to identify a level of firesafety in relation to requirements of NFPA 101 (Nelson and Shibe 1978). Thirteen parameters of firesafety are used to evaluate containment safety, extinguishment safety, people movement safety, and general safety. These are compared with a level of risk calculated from other parameters not covered by NFPA 101. FSES has been adopted as Appendix C to NFPA 101 and versions have been developed for certain other institutional occupancies.

Building Fire Simulation Model (BFSM): This was developed to relate firesafety equivalency to the Minimum Property Requirements for residential units (Fahy 1983). Probabilities of fire proceeding through its various stages of development are examined, along with the influence of occupants' actions and effects of the fire on the occupants. These probabilities are then compared to certain firesafety objectives. The BFSM is an important step toward realistic assessment of life safety in buildings. Other computer simulation models are being developed.

Fire Safety Effectiveness Statements: This is the title of Section 13 of the Federal Fire Prevention and Control Act of 1974. A research project explored the concept of firesafety effectiveness statements for existing buildings (Watts et al 1979) and surveyed systematic methods of evaluating firesafety. These methods fell into two categories: schedules, of which the FSES is an example; and analysis, of which the BFSM is an example. It was concluded that the validity of the schedule methods could not be established, and that the analytic methods were too complex for widespread use. Creation of a recognizably valid schedule for Firesafety Effectiveness Statements requires additional research effort.

Available Safe Egress Time (ASET): This is a computer analysis of the time available for action relative to the rate of burning and building dimensions (Cooper and Stroup 1982). ASET synthesizes calculable physical laws with empirical evidence and certain simplifying assumptions to produce the best available approximation of this important life safety parameter. The model is modular and can therefore be enhanced as new information is developed. ASET is another example of the expanding number of microcomputer applications in assessing life safety from fire (Walton 1985).

Matrix Approach to Firesafety: This was developed to evaluate the fire risk in hospitals in England and Wales (Watts 1981). This methodology begins with management policy on firesafety and works through a hierarchical sequence of matrices to identify the most important firesafety components. Relationships of the components are identified by consensus experienced judgment. Determination of the existing level of each component requires application of fire protection engineering principles. The advantage of this approach is that it results in quantitative assessment of fire risk, produced from present methods of evaluating individual firesafety components.

Bibliography

References Cited

Butcher, E. G., and Parnell, A. C. 1979. *Smoke Control in Fire Safety Design*. E. and F. N. Spon, London, England.

Chalmet, L. G., et al. 1982. "Network Models for Building Evacuation." *Fire Technology*. Vol 18, No 1. pp 90-113.

Clarke, Frederic B., III, and Ottoson, John. 1976. "Fire Death Scenarios and Fire Safety Planning." *Fire Journal*. Vol 70, No 3.

Cooper, L. Y., and Stroup, D. W. 1982. *Calculating Available Safe Egress Time (ASET)—A Computer Program and User's Guide*. National Bureau of Standards, Washington, DC.

Fahy, Rita. 1983. "Building Fire Simulation Model." *Fire Journal*. Vol 77. No 4. pp 93-95, 102-105.

Halpin, Byron M., et al. 1975. "A Fire Fatality Study." *Fire Journal*. Vol 69. No 3.

Klote, John H., and Fothergill, John W. Jr. 1983. *Design of Smoke Control Systems in Buildings*. American Society of Heating, Refrigerating and Air-Conditioning Engineers, Atlanta, GA.

Levine R., and Pagni P. 1984. *Fire Science For Fire Safety*, Gordon and Breach Science Publishers, NY.

Melinek, S. J. and Booth, S., *An Analysis of Evacuation Times and the Movement of Crowds in Buildings*, Building Research Establishment, CP 96-75, Borehamwood, UK, October 1975.

Nelson, H. F., and Shibe, A. J. 1978. *A System for Fire Safety Evaluation of Health Care Facilities*. Nov 1978. National Bureau of Standards, Washington, DC.

NFPA. 1985. *Report of the National Fire Research Strategy Conference*. Aug 27-28, 1984. Quincy, MA.

NFPA. 1980. *Firesafety Concepts Tree*. National Fire Protection Association, Quincy, MA.

Pauls, J. L., "Movement of People in Building Evacuations," D. J. Conway, ed., *Human Response in Tall Buildings*, Dowden, Hutchinson & Ross, Stroudsburg, PA, 1977.

Peschl, I. A. S. Z. "Passage Capacity of Door Openings in Panic Situations," *Bouw*, Vol 2, No 9, January 1971.

Predtetschenskii, V. M., and Milinskii, A. I. 1978. *Planning for Foot Traffic Flow in Buildings*. Amerind, New Delhi, India.

Swersey, Arthur J., and Ignall, Edward J. 1975. *Fire Protection and Local Government: An Evaluation of Policy-Related Research*. Rand Institute, Santa Monica, CA.

Walton, D. 1985. "ASET-B, A Room Fire Program for Personal Computers" *Fire Technology*, Vol 21, No 4. pp 293-309.

Watts, John M. 1981. "Matrix Approach to Fire Safety." *SFPE Symposium on Systems Applications for Fire Protection Engineers*. University of Maryland, College Park, MD.

Watts, John M., et al. 1979. *A Study of Fire Safety Effectiveness Statements*. University of Maryland, College Park, MD.

NFPA Codes, Standards, Recommended Practices and Manuals. (See the latest *NFPA Codes and Standards Catalog* for availability of current editions of the following documents.)

NFPA 70, *National Electrical Code.®*

NFPA 101, *Code for Safety to Life from Fire in Buildings and Structures.*

NFPA 255, *Standard Method of Test of Surface Burning Characteristics of Building Materials.*

Additional Readings

Bryan, John L., *Human Behavior in Fire—A Bibliography*, National Bureau of Standards, Washington, DC, Aug. 1978.

Berlin, Geoffrey, N., "A Simulation Model for Assessing Building Fire Safety," *Fire Technology*, Vol. 18, No. 1, Feb. 1982, pp. 66-76.

Canter, D., ed. *Fires and Human Behavior*, John Wiley and Sons, NY, 1980.

Castino, G. T., and Harmathy, T. Z., eds., *Fire Risk Assessment*, American Society for Testing and Materials, Philadelphia, PA, Mar. 1982.

Cooper, Leonard Y., "Calculating Escape Time From Fires," In-Engineering Applications of Fire Technology Workshop, Apr. 16-18, 1980, National Bureau of Standards, Gaithersburg, MD, published by Society of Fire Protection Engineers, Boston, MA.

Clarke, F. B., and Birky, M. M., "Fire Safety in Dwellings and Public Buildings," *Bulletin of the New York Academy of Medicine*, Vol. 57, No. 10, Dec. 1981, pp. 1047-1060.

Friedman, R., "Quantification of Threat from a Rapidly Growing Fire in Terms of Relative Material Properties," *Fire and Materials*, Vol. 2, No. 1, 1978.

Guide to NFPA National Building Firesafety Standards, National Fire Protection Association, Quincy, MA, 1983.

Kisko, T. M., and Francis, R. L., "Network Models of Building Evacuation: Development of Software Systems," *NBS-GCR-85-489*, National Bureau of Standards, Gaithersburg, MD, March 1984.

Lathrop, James K., ed. *Life Safety Code Handbook*, 3rd ed., National Fire Protection Association, Quincy, MA, 1985.

Lerup, Lars, et al., *Learning from Fire: A Fire Protection Primer for Architects*, University of California, Berkeley, CA, 1977.

Pauls, Jake, "Development of Knowledge about Means of Egress," *Fire Technology*, Vol. 20, No. 2, May 1984, pp. 28-40.

Paulsen, R. L., "Human Behavior and Fires: An Introduction," *Fire Technology*, Vol. 20, No. 2, May 1984, pp. 15-27.

Peters, George A., "Why Only a Fool Relies on Safety Standards," *Hazard Prevention*, Vol. 14, No. 2, Nov./Dec. 1977.

Ramachandran, G., "A Review of Mathematical Models for Assessing Fire Risk," *Fire Prevention*, No. 149, May 1982, pp. 28-32.

Rasbash, D. J., "The Definition and Evaluation of Fire Safety," *Fire Prevention Science and Technology*, No. 16, Mar. 1977.

Rowe, William D., *An Anatomy of Risk*, John Wiley and Sons, NY, 1977.

Stahl, Fred I., "BFIRES-II: A Behavior Based Computer Simulation of Emergency Egress During Fires," *Fire Technology*, Vol. 18, No. 1, Feb. 1982, pp. 49-65.

Stahl, F. I., et al., "Time-Based Capabilities of Occupants to Escape Fires in Public Buildings: A Review of Code Provisions and Technical Literature," *NBSIR 82-2480*, National Bureau of Standards, Gaithersburg, MD, Apr. 1982.

Stahl, Fred, I., and Archea, John, "An Assessment of the Technical Literature on Emergency Egress from Buildings," National Bureau of Standards, Washington, DC, Oct. 1977.

Watts, John M., "Systematic Methods of Evaluating Fire Safety—A Review," *Hazard Prevention Journal of the System Safety Society*, Vol. 18, No. 2, Mar./Apr. 1982, pp. 24-27.

ASSEMBLY OCCUPANCIES

Revised by John A. Sharry

Assembly occupancies can be generally defined as structures where groups of people gather for purposes such as deliberation, worship, entertainment, or awaiting transportation. Examples of assembly occupancies are: large meeting rooms, restaurants, nightclubs, dance halls, bars, cocktail lounges, auditoriums (with fixed or loose chair seating), libraries, concert halls, and theaters (both legitimate and motion picture), multipurpose rooms, exhibition halls, convention centers, sports arenas, field houses, and passenger terminals (air or surface transportation).

Because this occupancy is concerned with the safety and hazards of large numbers of people gathered in one place, an established minimum number of occupants must be reached before one of the above examples constitutes an assembly occupancy. Although this minimum number of people can vary, most building codes and NFPA 101, the *Life Safety Code®* (hereinafter referred to as NFPA 101) set this minimum occupant load at 50. Occupancies with similar uses but occupant loads below 50 are in other occupancy categories, usually business.

OCCUPANCY CHARACTERISTICS

Inasmuch as this occupancy involves the safety of a large group of people, the density of the occupant population presents the major safety problem. No other occupancy experiences occupant loads of such density. It is not uncommon for occupant densities to approach 5 sq ft (.46 m^2) per person or greater.

This density produces problems in the physical movement and behavior of the occupants, the capacity of the exits, the maintenance of adequate aisles of proper width, and the method of alerting occupants in case of emergency. Assembly occupancies are usually inhabited by persons who do not use the building frequently and therefore, are not familiar with the location of exits, exit paths, or other safeguards which may be present. In addition, many assembly occupancies such as theaters, concert halls, nightclubs, lounges, and some restaurants are used under conditions of near-total darkness.

All these factors—high occupant density, unfamiliarity with the building, and darkness—are common to the many and varied forms of assembly occupancy. Provisions must be made to deal with all of these factors to avoid panic when an emergency condition occurs.

SPECIAL CONSIDERATIONS

Restaurants and Nightclubs

Both of these assembly uses share common problems, the most prevalent of which is aisles that provide inadequate access to exits. There is a tendency in these assemblies to add tables in an effort to increase profits. While simply uncomfortable under normal conditions, this can be disastrous in an emergency. Proper aisles must be maintained to facilitate exiting. Aisles should be of sufficient width to accommodate movement of people, taking into consideration the position of chairs during normal use and during emergencies. Responsible restaurateurs have found that proper aisle widths actually improve profits by making patrons feel more comfortable, and making table service easier and more efficient. Nightclubs, on the other hand, generally do not value table service ease and efficiency because there is less need for service than in restaurants. Thus, to improve the profit of admission fees, nightclubs tend to squeeze as many people inside the building as possible. This is especially true where dinner service is not offered. Techniques such as using small round tables rather than larger square tables help achieve maximum patron capacity.

In both restaurants and nightclubs, the decor can be elaborate and thematic. Unless properly designed, decor can obstruct or obscure exits, exit marking, emergency lighting, automatic suppression systems, and other safety features.

Methods of alerting patrons to an emergency and methods of notifying the fire department are common problems in these occupancies. There is a reluctance to notify patrons of an emergency before the last possible minute for several reasons:

Mr. Sharry is the Fire Chief at Lawrence Livermore National Laboratory in Livermore, CA.

1. Destroys the "ambience" of the facility.
2. May result in lost revenue from patrons not settling their bills.
3. Gives the establishment a bad reputation (purportedly).

These reasons result in delayed notification and consequent shortened available evacuation time in case of a real emergency. There is reluctance to notify the fire department, for the same reasons, resulting in delayed fire department response.

Fixed Seating

Occupancies utilizing fixed seating present the usual as well as some unique problems. Added to the common problems of adequacy of exits, exit marking, emergency illumination, and notification of patrons in an emergency are the special problems related to seating.

Fixed seating arrangements put a large number of people into a confined environment where there is limited movement. People must move within a row before they can reach an aisle, the aisle then limits their movement in reaching an exit. The design of the seating in occupancies is critical to ensuring rapid egress of all occupants in a fire emergency.

Fixed seating is generally broken into two types, "continental" and "normal." Normal seating involves a limited number of seats (usually a maximum of 14 or 15) between aisles, limited length of dead-end rows, and use of mid aisles and cross aisles to ensure rapid exiting. (See Fig. 9-2A.) The spacing of the seats to provide adequate

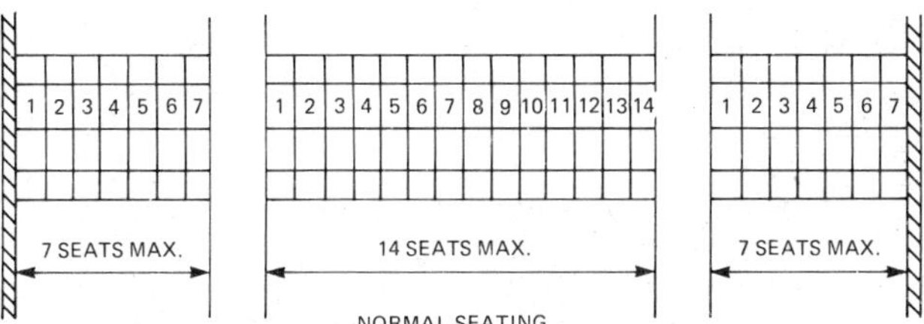

FIG. 9-2A. Arrangement of seats and aisles in an assembly occupancy without continental seating.

width between rows is important for ease of movement. Generally, 12 in (30525 mm) of clear space for movement between rows is considered the minimum. (See Fig. 9-2B.) As patrons leave the rows and enter the aisles, adequate

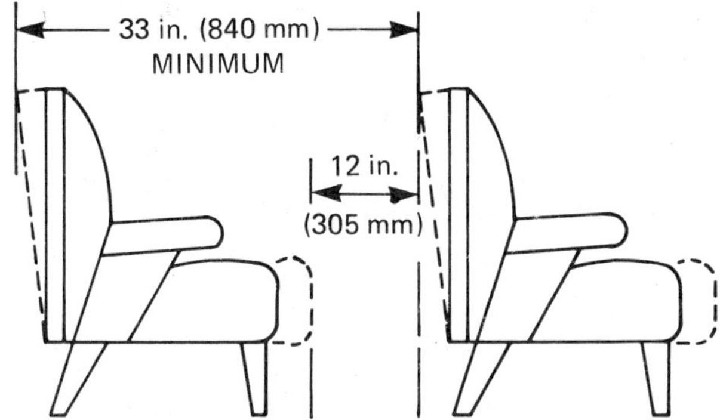

FIG. 9-2B. Correct measurement of minimum spacing between rows of seats without self-rising seats.

aisle width must be available to accommodate the influx. The proper aisle width must continue toward the exit. If cross aisles are used, they must be wide enough to accommodate the flow of people from tributary aisles and to permit movement to the exits. Finally, the exits must be properly sized to accommodate the flow of people from the cross aisles and their tributary aisles. Dead-end aisles should be limited in length or avoided altogether.

Continental seating is characterized by long rows of seats discharging into side aisles which, in turn, discharge directly to the exits. Mid and cross aisles are not used. (See Fig. 9-2C.) Although most codes limit the rows to 100 seats

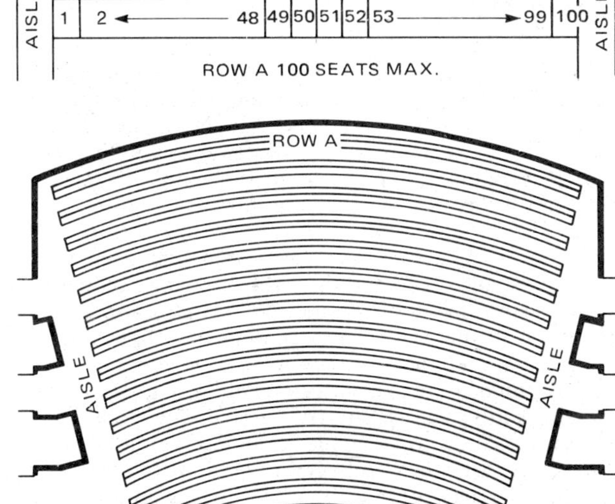

FIG. 9-2C. Arrangement of seats and aisles with continental seating. One hundred seats is the maximum number for one row. The frequency of the exits from side aisles is also depicted in the theater arrangement.

each, the rows are usually functionally limited to fewer seats by the available sight lines. Because the flow of people is less complicated than in normal seating, and because row spacing is greater, the rows function as mini aisles. This results in efficient people movement and superior evacuation times compared with those for normal seating. Continental seating also has the advantage of greater patron comfort and more seats in the prime viewing areas, due to the increased row spacing and longer row length.

Where loose chairs are used in lieu of fixed seating, care must be taken to avoid movement of chairs into a haphazard arrangement which could impair exit access. Generally, loose chairs should be arranged similarly to fixed seating to maintain proper row and aisle widths. To prevent chair movement during use, most codes require that loose chairs be fastened together in groups too large to permit easy repositioning, which might block egress. Many chair manufacturers produce chairs with interlocking features to fulfill this requirement. Usually these features require no tools to join or separate the chairs, which can also be stacked ordinarily for storage.

Multipurpose Rooms

Ballrooms, meeting rooms, cafeterias, and gymnasiums are usually multipurpose rooms and may be used for dining, dancing, sporting events, exhibits, meetings, or receptions. In general, requirements for each use are contained in local codes. Overall, adequate exiting is the key to occupant safety, together with automatic suppression systems if exhibit hall fuel loading is expected.

Where large areas have movable subdividing walls or partitions, each subdivision must provide proper exit accesses and exits—a factor frequently overlooked in planning such facilities. Care must be taken to ensure that proper exiting is designed into the arrangement.

Exhibit Halls

Exhibit halls are large assembly areas which are really multipurpose in nature. Special diligence regarding control of display and chair arrangement must be exercised to ensure unimpeded egress as well as maintenance of proper travel distances to exits. It is recommended that trade shows or exhibitions be required to submit advance layouts for displays and seating, for approval prior to the use date.

Exhibit halls can have very large fire loads unless provisions are made for the proper storage of packing materials needed for transporting displays and surplus quantities of literature. The importance of this cannot be overemphasized, since failure to remove this heavy fuel load to proper storage areas can result in large fires which spread rapidly. The McCormick Place fire in January 1967, in Chicago, IL, is an example of the fire potential in this occupancy. Displays and booths of flimsy construction should be treated with a fire retardant coating to reduce ignition and fire spread.

Because of the fuel load and large number of occupants, moderate and larger size exhibit halls should be provided with an automatic fire suppression system.

Sports Arenas

Sports arenas and field houses usually have fixed seating for large numbers of spectators, as well as a large area. More than 10,000 people in those occupancies at one time is not uncommon. Here, occupant density presents the major problem to life safety from fire. Proper aisles and travel distances are difficult to maintain unless care is taken in the planning stage. On the positive side, these occupancies usually involve very large open areas, permitting smoke to be diluted in the early stages of a fire. By use of smoke control measures, these facilities can be kept reasonably safe for egress. If they are used for exhibitions, the special considerations for exhibit halls must also be followed.

Passenger Terminals

Passenger terminals require special consideration for firesafety because only general rules can be written for their regulation. Densely populated areas are localized, especially at departure and arrival gates. However, most terminals have large open areas with significant fuel loads, which can lead to large losses if a fire is not extinguished in its incipient stage. Here, as in most assembly occupancies, the occupants are generally unfamiliar with the building, its exit routes, and protection features. Proper marking of exits and exit routes along with maintenance of proper travel distances, are particularly important.

OCCUPANCY HAZARDS

Cooking and Open Flames

Many assembly occupancies have cooking associated with their use. The hazard of cooking in the traditional kitchen can be handled by the use of automatic extinguishing systems and proper vapor removal as outlined in NFPA 96, *Standard for the Installation of Equipment for the Removal of Smoke and Grease-Laden Vapors from Commercial Cooking Equipment*. The widespread use of "display cooking" has presented new problems in protecting the cooking surfaces, vapor removal, and humans in case of fire. Display cooking on properly designed and fixed equipment can be made reasonably safe with fixed systems, proper aisles, and trained staff.

The use of open flame devices in tableside and flambe cooking should be carefully reviewed before being permitted. Specially designed equipment and properly trained staff are necessary to ensure safety.

Candles and other open flames require the same degree of awareness as tableside cooking. Several commercial devices are available which make open flame lighting safe, even in the event of a tipover. Coordination between facility operators and the fire department helps maintain an acceptable level of safety.

Stages

Most serious fires in theaters have originated in the stage area. The 1903 Iroquois Theater fire in Chicago, IL, is probably the most vivid example of a stage fire and its tragic effects.

Traditional proscenium stages have: (1) scenery and lighting "flown" above the stage; (2) scenery on the back and sides of the stage; (3) shops located along the back and sides of the stage; and (4) storage, props, trap doors and stage lifts under the stage floor level. The potential for a fire is great with this combination of fuel and potential ignition sources. Since the Iroquois Theater fire, many safeguards have been required to control these hazards. Included among these safeguards are: automatic sprinklers at the ceiling level, below the gridiron, in usable spaces under the stage, and in all auxiliary spaces surrounding the stage; ventilation over the stage, operable manually or automatically by fusible links; and an automatic closing safety curtain to cut off the stage from the audience chamber. All of these features are necessary with a full theatrical stage.

Modern stages are generally modified versions of the classic theatrical stage. More scenery is now moved horizontally, eliminating the need for the high scene loft or fly space. Modern stages also tend to be variations of the traditional proscenium stage, thrusting the stage further out into the audience, or having the stage open to the audience on three sides as in an arena theater—often called "theater in the round."

Whatever the arrangement of scenery on the stage, scenery handling equipment and allied shops need to be protected to prevent a large stage fire from threatening the audience either directly, or indirectly, by inducing panic.

Projection Booths

Projection booths refer to any booth housing equipment for the transmission of light onto a screen, curtain or stage. The hazard is not with the projection of motion pictures or slides made on flammable film, as is commonly assumed, but with the mechanism used to project light. Electric arc, zenon, and other light sources generate hazardous gases, dust, or radiation. They can fail with explosive force under certain circumstances. Therefore, where these light sources are used, a booth around the projection equipment is required to protect the audience. Proper ventilation is also required. Incandescent light sources do not produce the same hazard and therefore need not have the same enclosure.

Modern motion pictures and slides use safety film, which does not introduce the hazards of nitro cellulose found in early movies. Therefore, modern projection rooms are not designed for and cannot safely project nonsafety film.

Storage

In this occupancy, as well as many others, fire and life safety problems often arise from improper storage practices. Storage of tables and chairs in exit ways, storage of scenery so it blocks exits, and storage of supplies in aisles and exit paths are all too common in assembly occupancies. In large measure, this problem is related to improper building design. Little consideration is given to storage needs in the design process. It is essential that space be allocated for proper storage of materials, supplies, props, tables, chairs, and other items associated with assembly occupancies to avoid the creation of unnecessary hazards.

LIFE SAFETY CONCEPTS

Construction

Places of assembly have a high occupant density under variable conditions which, to one degree or another, make alerting and exiting the occupants more difficult than with other occupancies. Because of this, building construction is an important element in the life safety system.

In general, the building should be constructed to allow the structure to remain intact under fire conditions, and not contribute significantly to the fire development for at least the time necessary for evacuation. Obviously, for a one story building, this is relatively easy to accomplish when occupant loads are moderate in size (under 1,000). But even in a one story building with proper exits, a large occupant load (over 1,000) means higher degrees of structural protection are necessary to ensure everyone's safety.

As building heights increase, the level of construction and protection must increase accordingly to permit large places of assembly at higher stories of a building. NFPA 101 specifies types of places of assembly based on construction, height of the assembly occupancy above the level of exit discharge, and fire protection provided, that can be located in buildings.

Alarm Systems

Prompt notification of occupants that there is an emergency condition cannot be overemphasized. However, the sounding of a fire alarm without accompanying verbal instructions might introduce the threat of panic. To overcome this possibility, an appropriate alarm system—perhaps using prerecorded messages—can be designed without undue cost or risk to the occupants.

Exit Hardware

Because of the threat of a large group of occupants rushing to an exit, leaning together on the door, and making the use of traditional knob activated latching devices difficult or impossible, assembly occupancies usually require the use of panic hardware. Security conscious owners in the past have objected to panic hardware because of the potential for slipping the latch from the outside. However, modern panic hardware designs provide for both life safety and security.

There is no excuse for chained or padlocked doors, given today's availability of the right kind of exit hardware.

Interior Finish

Interior finish is a key element to provide for fire control and life safety. Materials used in an assembly occupancy must limit the rate of flame spread and smoke production. Generally, materials are limited to flame spreads of 75 or less (Class A or B) for all but the smallest assembly uses.

Drapes, hangings, tapestries, and other decorative materials must be fire retardant if they constitute a significant part of the decor or if they would provide a continuous path for fire travel. Inadvertent ignition of such materials has been responsible for panic in densely occupied spaces such as the Coconut Grove nightclub fire in Boston, MA, in 1942. Proper treatment to make materials fire retardant is necessary to avoid disaster.

Employee Training

Employees of assembly occupancies are critically important to the life safety of patrons. If the staff is properly trained, they can prevent serious injury or death to many patrons. The Beverly Hills Supper Club fire in 1977 at Southgate, KY, showed the importance of staff assistance in fire survival. Although the building lacked proper exits, staff members saved hundreds of lives by leading patrons to means of escape unknown to the patrons.

It is recommended that management of every assembly occupancy devise an emergency plan in conjunction with its local fire department, to deal with the many contingencies associated with fires and to properly train staff members on safe emergency procedures.

Assembly occupancies can be safely occupied when the following conditions are avoided:

1. Overcrowding.
2. Blocked or impaired exits or means of exit access.
3. Chained or locked exits.
4. Storage of combustibles in other than proper locations.
5. Improper use or control of open flames.
6. Disregard for fire characteristics of materials and decorations.

Bibliography

NFPA Codes, Standards, Recommended Practices and Manuals. (See the latest *NFPA Codes and Standards Catalog* for availability of current editions of the following documents.)

NFPA 101, *Code for Safety to Life from Fire in Buildings and Structures.*

Additional Readings

Klevan, Jacob B. *Modeling of Available Egress Time from Assembly Spaces or Estimating the Advance of the Fire Threat.* Society of Fire Protection Engineers, Boston, MA, 1982.

EDUCATIONAL OCCUPANCIES

Revised by John A. Sharry

Educational occupancies can be generally defined structures used for the gathering of six or more persons for purposes of instruction. Examples of educational occupancies include schools, academies, colleges, and universities as well as nursery schools, day care centers, kindergartens, and other facilities whose purpose is primarily educational even though the students are of preschool age.

Recent actions by model code groups have defined educational occupancies as places where students are present for instruction for more than 12 hours per week or more than four hours per day through the 12th grade. This definition excludes facilities—such as churches or museums—in which instruction is incidental to some other occupancy; in such cases, the requirements of principal occupancy apply.

OCCUPANCY CHARACTERISTICS

Nature of Occupants

People in educational occupancies vary in their ability to deal with an emergency condition, depending upon their age, mental and physical condition, plus the physical characteristics of the facilities. Generally, regulations concerning schools are based on abilities of children in third through eighth grade, with special provisions being made for younger children. High school occupants gain an extra degree of safety because the firesafety design is based on capabilities of younger children.

Occupants in the third through eighth grades should be capable of traversing stairs and acting for their own self-preservation if they are led by an adult. Thus, the aim is to provide sufficient exits and egress paths so that the occupants can be led to an area of safety.

Children, below the third grade level require special consideration because of their limited ability to traverse stairs and move quickly in an emergency. There is a potential for younger children to be overrun on stairs by older students. Most codes require that kindergarten and first grade students be housed on the first or grade level discharge to facilitate exiting. Second grade students are usually limited to one level above the first or grade level, recognizing their improved movement ability.

Mr. Sharry is the Fire Chief at Lawrence Livermore National Laboratory in Livermore, CA.

Design Considerations

Schools having only one story, where each room has a direct exit to the outside, represent the most conservative design for life safety. However, economics often dictate other designs, which can be made safe as well by adhering to certain design principles noted elsewhere in parts of this chapter.

The necessity for accommodating physically and mentally handicapped students within the general school population is a fairly recent development affecting school design. Consideration must be given to access and egress of these students by providing facilities such as elevators, horizontal exits, and areas of refuge in multiple story buildings. Provisions must be made for the use of elevators or other special facilities for evacuation of these students.

Flexible and Open Plan Design

Flexible and open plan design concepts sometimes used in schools differ from conventional designs, in that the walls are demountable and easily moved without major construction. The ability to easily rearrange walls to suit changing educational needs is valuable if proper consideration to firesafety is included in the planning.

Open plan buildings delineate spaces and corridors by use of movable fixtures and low partitions [usually maximum height of 5 ft (1.5 m)]. While this design provides great flexibility, the firesafety of the building is diminished by omission of features that could confine fire and smoke to a single compartment long enough to permit evacuation. The compensating factor is the ability of occupants to observe the entire area over the low partitions, and presumably detect any fire in its incipient stage. Since occupants are awake and alert, their natural faculties serve as fire and smoke detectors. Early detection should make prompt evacuation and extinguishment possible.

Colleges and Universities

Because educational occupancies have requirements based on the behavior of students in the third through eighth grades, these requirements are not applicable to colleges and universities.

College students are generally presumed to be capable of adult behavior because of their age. Therefore, most codes no longer treat colleges and universities as educational occupancies.

Because college and university buildings are actually combinations of other occupancies, codes now consider the actual use of the space and apply the respective occupancy requirements. Classrooms for under 50 occupants are treated as business occupancies; classrooms with an occupant load of 50 and more are considered assembly occupancies, etc. Thus, colleges and universities receive no special fire or life safety treatment because of their educational mission.

Day Care Facilities

Day care facilities have traditionally been associated with the care of children, and thus have been considered as an educational occupancy. Recent changes in society and medical science have indicated a need to expand the scope of day care facilities to also accommodate elderly people. Because many of the physical limitations that apply to children also apply to older adults, this approach seems logical from a fire protection and life safety standpoint. NFPA 101, *Life Safety Code®* (hereinafter referred to as NFPA 101) provides special requirements for adult day care facilities.

It should be recognized that some day care facilities operate twenty-four hours a day to provide care for dependents of people who work at night. These facilities present the greatest hazard because their occupants may be asleep when fire starts. Staff members for day care are required to be awake and alert at all times; however, human nature being what it is, there is the possibility that the staff may be napping when an emergency occurs. Therefore, special provisions are required when day care facilities operate at night. The capabilities of the staff to perform in case of an emergency must be verified regularly to ensure the safety of the occupants.

OCCUPANCY HAZARDS

Mixed Occupancy

Educational occupancies are generally mixed occupancies, with other facilities, such as assembly occupancies often occupying the same building.

With mixed occupancies, care must be taken to ensure that the fire and life safety provision of all occupancies are met. It is important to provide sufficient protection to prevent the educational occupancy from being exposed to the hazards of the other occupancies in the structure.

Hazardous Areas

Areas such as vocational shops, laboratories, home economics laboratories, kitchens, storage rooms, and stages must be separated from educational occupancies. Usually, fire rated walls and doors will provide the necessary degree of protection unless the hazard is severe, in which case automatic suppression systems may also be required.

LIFE SAFETY CONCEPTS

Means of Egress

There is no substitute for properly designed exits, exit access paths, and exit enclosures. Travel distance requirements and dead end restrictions are especially important in educational occupancies because of the nature of the occupants.

Designs involving circuitous exit access paths should be avoided. Proper exit enclosures are important to prevent contamination by smoke and heat. Multistory buildings whose dimensions necessitate stairways away from the building periphery require proper enclosure of the exit path to the exterior. This exit requires that passageways be provided at the level of exit discharge.

Corridor Protection

In standard double loaded corridor designs, precautions must be taken to prevent the corridor from becoming untenable. The use of fire rated corridor construction with proper door controls is necessary to protect this exit access path. Although many codes require self-closing doors in corridors, operational necessities make this impractical in many schools. Therefore, staff training is necessary to insure the proper closing of doors at appropriate times.

Subdivision of any long corridors for smoke control purposes is desirable. By providing smoke doors at 300 ft (90 m) intervals, corridors can be kept reasonably smokefree in areas away from the fire origin.

Exit Hardware

Because of the potential for large numbers of occupants using the same exit doors, and to prevent crushing against unopened doors, panic hardware is required in schools of any substantial size. In no case are chains or padlocks permitted on exits when the building is occupied. If such security devices are used during unoccupied times, provisions for their removal must be made to the satisfaction of local authorities.

Alarm Systems

In general, schools require (as a minimum) a manual fire alarm system. Because of vandalism in public schools in recent years, fire alarm systems sometimes have been turned off or otherwise disabled to prevent malicious false alarms from interrupting the learning process. Recent changes in some codes permit the deletion of manual pull stations if two way communication is maintained in all rooms and if there is a continuously staffed central location where an alarm can be sounded. This allows for an operational fire alarm system without the vandalism problem. NFPA 101 permits such an arrangement.

Staff Training

Educational occupancies should develop emergency plans in conjunction with the local fire department. A plan should include emergency actions, position descriptions for key staff, an evacuation plan, and a fire drill procedure and schedule. The plan should be practiced at various times during the year, with fire drills conducted regularly.

The staff should be well informed as to their responsibilities and practice the plan to ensure smooth functioning during an emergency.

Bibliography

NFPA Codes, Standards, Recommended Practices and Manuals. (See the latest *NFPA Codes and Standards Catalog* for availability of current editions of the following documents.)

NFPA 101, *Code for Safety to Life from Fire in Buildings and Structures.*

Additional Readings

Juillerat, Ernest, E., *Campus Firesafety*, National Fire Protection Association, Quincy, MA, 1977.

Los Angeles Fire Department, *Operation School Burning, No. 12*, Official Report on a Series of School Fire Tests Conducted April 16, 1959 to June 30, 1959, National Fire Protection Association, Quincy, MA, 1961.

Los Angeles Fire Department, *Operation School Burning, No. 22*, Official Report on a Series of Fire Tests in an Open Stairway, Multistory School Conducted June 30, 1960 and Feb. 6, 1961, National Fire Protection Association, Quincy, MA, 1961.

"A Study of School Fires," National Fire Protection Association, Quincy, MA, 1973.

HEALTH CARE FACILITIES

Revised by Donald W. Belles, P. E.

Health care facilities are used for the treatment or care of pesons suffering from physical or mental illness, disease, or infirmity, and for the care of infants, convalescents, or aged persons. These facilities provide sleeping accommodations for occupants who may be incapable of self-preservation because of physical or mental disability or age. Some buildings which house mentally disabled occupants have security measures that limit freedom of movement.

In recent years, facilities—sometimes called ambulatory surgical care facilities—have been developed to provide medical treatment to outpatients. Although patients might be placed under general anesthesia, they are not housed overnight. Occupants exhibit some of the characteristics of people in business occupancies and some characteristics typical of those in health care facilities.

A recent trend is to house persons with physical, mental, or emotional disabilities in a "deinstitutionalized" residential setting. These facilities are variously referred to as board and care facilities, halfway houses, adult congregate living facilities, etc. Residents of board and care homes differ greatly in their ability to respond to a fire threat. Where residents are judged to be incapable of self-preservation or capable of movement only with assistance, consideration should be given to applying safeguards appropriate for health care facilities.

Additional information relating to health care facilities can be found in Section 5, Chapter 6, "Interior Finish;" Section 7, Chapter 9, "Confinement of Fire in Buildings;" Section 9, Chapter 5, "Board and Care Facilities;" and Section 12, Chapter 1, "Electronic Computer/Data Processing."

OCCUPANCY CHARACTERISTICS

Occupants of health care facilities are generally presumed to be incapable of self-preservation. A significant percentage of occupants in hospitals and nursing homes are incapable of self-evacuation or are ambulatory, but incapable of perceiving a fire threat and making a rational choice of response.

There are three types of care in most modern hospitals: ambulatory, general, and intensive care. Given proper directions, and unless smoke or heat is intense, ambulatory patients can make their own way to safety. General care patients may be transported on stretchers or in wheelchairs with some difficulty; horizontal and even some vertical movement is generally possible, although independent evacuation is not. Patients in intensive care are likely to be connected to a variety of life support devices, making movement for even short distances very difficult and evacuation almost impossible without endangering such patients' lives.

Occupants of nursing homes vary from geriatric residents who are capable of evacuation with limited assistance to patients requiring close supervision and skilled nursing care. Those requiring skilled nursing care will generally be incapable of responding to threat of fire. The use of physical restraints and tranquilizing drugs may also prevent a patient from responding to a fire.

The aged present a unique fire problem. There have been fire incidents where elderly people acted contrary to self-interest, either by resisting rescue efforts and reentering burning buildings, or by seeking refuge in their rooms and failing to notify others of a fire. However, the application of health care safeguards should not be made simply on the basis of age, but rather should be based on occupant capability to escape in an emergency.

Because occupants will be incapable of movement or slow to evacuate, a health care facility resembles a ship at sea or a high rise building: it is better to keep the fire from the patient than to remove the patient from the fire; occupants must be defended in place. Thus, health care facility design and operation must incorporate methods by which a fire can be detected early, contained, and fought rapidly and successfully. Accomplishing this requires careful planning of the health care facility and of its day to day operation.

The NFPA *Life Safety Code*® widely used to establish minimum requirement for life safety from fire within health care facilities, sets forth criteria based upon the following general principles:

Mr. Belles is principal, Donald W. Belles and Associates, Inc., Fire Protection Consultants, Madison, KY.

1. Fire resistive construction.
2. Subdivision of spaces (compartmentation).
3. Protection of vertical openings.
4. Provision of adequate means of egress.
5. Provision of exit marking, exit illumination, and emergency power.
6. Limits on the use of interior finish materials.
7. Fire alerting facilites.
8. Smoke control mechanisms.
9. Protection of hazardous areas.
10. Adequate protection of building service equipment.

Total building fire protection for life safety is more necessary in health care facilities than in other occupancies because of the nature of the occupants. At the same time, exits are slightly less important. The first principle of designing a firesafe health care facility must be that safety must not depend wholly upon any single safeguard.

IGNITION SOURCES

A recent NFPA analysis of fires in hospitals and nursing homes indicates that smoking related incidents rank highest or near highest as a leading fire cause. (See Tables 9-4A and 9-4B.) These data reaffirm a study of 938 fires in 75 Massachusetts hospitals during 1973. The earlier study indicated that 29 percent of fires were related to the careless use of smoking materials (Spalding 1975). It should be noted that smoking related fire causes led to approximately 50 percent of the reported fire fatalities.

The NFIRS (National Fire Incident Reporting System) data show that in a majority of the cases where death or injury occurred, the fire began in clothing, a mattress or pillow, bedding, or linen. The data also indicate that the final extent of flame damage was confined to the room of origin in 97 percent of the fires in hospitals and 83 percent of the fires in nursing homes.

FIRE LOADS

Studies show relatively low fuel loads within most spaces in health care facilities. Fire duration varies from approximately 20 minutes for patient areas to several hours, depending upon the space involved.

A study by the National Bureau of Standards (NBS) conducted in 1942 involving three hospitals reveals an average fuel load of 5.7 lb per sq ft (27.8 kg/m^2) (NBS 1942). Fuel loads for typical spaces are indicated in Table 9-4C. A recent study of fuel loads in a U.S. Navy hospital confirmed the relatively low fuel loads for general hospital areas, but indicates higher fuel loads may be anticipated in some spaces such as medical libraries, X-ray file rooms, linen storage, general storage rooms, etc. Further, the increased use of disposables and modern medical record storage practices are likely to result in fuel loads exceeding average. These areas may contain fuel loads sufficient for fires from one to three hours duration. However, such spaces represent a small percentage of the total floor area.

Doctor office areas have been identified as an intermediate fuel load space having combustibles in sufficient quantity to produce fires of 30 to 90 minutes duration.

A major portion of the floor area of health care facilities is used for patient sleeping or treatment rooms. Fuel loads for such spaces are low. For example, fuel loads in nursing homes have been estimated at 2.5 to 3 lb per sq

TABLE 9-4A. NFPA Staff Analyses of Hospital Fire Patterns Based Upon Reports of 5920 Fires During 1982

Cause Categories	Percent of Fires	Comments*
Smoking related	32.0	Also 44% of deaths, 48% of injuries
Incendiary/suspicious	13.8	Also 11% of deaths, 14% of injuries
Cooking equipment	10.0	
(stoves and ovens)	(5.8)	
(hot plates, other portable)	(2.2)	
Electrical distribution system	8.0	
(lighting fixtures, including ballasts)	(2.3)	
Matches, lighters, and other open flame (excluding torches)	6.1	Also 33% of deaths, 12% of injuries
Dryers	3.6	
Air conditioning/refrigerating equipment	2.6	
(refrigerators)	(0.7)	
Torches	2.1	
Heating equipment	2.0	
(fixed local heaters)	(0.8)	
Electronic equipment (X-ray, computer, telephone, etc.)	1.7	
Generators and separate motors	1.3	
Incinerators	1.1	
TVs, radios and phonographs	0.8	
Biomedical equipment	0.5	
Elevators	0.5	
Other appliances	2.5	
Other equipment	2.1	
Unknown and other known**	10.3	

* Note that our sample had only nine fire deaths, so these patterns are subject to considerable statistical uncertainty. The nine deaths translate into an estimated 15–18 civilian fatalities per year in fires reported to the fire department.

** Nearly all of these are unknown cause fires. A few are due to exposure or natural causes.

TABLE 9-4B. NFPA Analyses of Nursing Home Fire Patterns Based Upon 3,059 Fires Reported During 1980–82

Cause Categories	Percent of Fires		Comments
Dryers	20.2		Also 13% of injuries
Smoking related	20.0		Also 52% of deaths, 45% of injuries
Cooking equipment	10.0		
(stoves and ovens)		(7.2)	
Electrical distribution system	8.0		
(lighting fixtures, including ballasts)		(2.8)	
Incendiary/suspicious	7.7		Also 4% of deaths, 20% of injuries
Heating equipment	7.0		
(fixed local heaters)		(2.8)	
Air conditioning/refrigerating equipment	3.4		
Matches, lighters, or other flame (excluding torches)	3.3		Also 22% of deaths, 10% of injuries
TVs, radios and phonographs	1.7		
Generators and separate motors	1.2		
Washers	1.1		
Torches	0.6		
Irons, electric blankets, devices designed to produce heat	0.5		
Other appliances	2.2		
Other equipment	2.3		
Unknown and other known*	10.8		

* Nearly all of these are unknown cause fires. A few are due to children playing, exposure, or natural causes.

TABLE 9-4C. Fuel Loads in Typical Health Care Facilities

	Average Combustible Contents (psf)		
Location	Movable Property	Exposed Woodwork and Floors*	Total
Rooms (single)	0.5	3.2	3.7
Corridors	0	2.6	2.6
Waiting rooms	1.7	1.5	3.2
Janitors' closets and supplies	3.1	3.4	6.5
Doctors' offices	5.7	2.9	8.6
Nurses' offices and rooms	3.1	1.9	5.0
Nurses' infirmary	0.8	2.2	3.0
Diet kitchens and dining rooms	1.2	2.4	3.6
Laundries	4.4	0.6	5.0
Laundries and clothes storage	12.5	0.6	13.1
Dormitories	0.8	2.0	2.8
Pharmacy, dispensary, and stores	5.8	1.9	7.7
Lockers, toilets, and barber shops	0.2	1.2	1.4
Approximate average for entire usable floor area of three hospital buildings surveyed			5.7**

* Combustible floor finish, where present, was ¼ in. linoleum; it was assumed to be the equivalent of 1 lb of combustible material, such as wood, per ft^2 of floor area (4.88 kg/m^2). Doors, windows, trim, mouldings, baseboards, etc., are included.

** This approximate average weight was computed from Table 16 on page 25 of the Bureau Report BMS 92. The value is somewhat high because the highest weight in each bracket of combustible contents was used in figuring it, i.e., in the bracket "0 to 4.9 lb/ft^2 (23.9 kg/m^2)," the value 4.9 lb (2.22 kg) was applied to the indicated area, etc.

ft (9.72 to 14.64 kg/m^2) (HEW 1975). NBS surveys indicate that fuel loads for hospital patient rooms approximate 4 lb per sq ft (19.52 kg/m^2) and that the combustible load in patient sleeping areas ranges from approximately 3 to 4.5 lb per sq ft (14.64 to 22.96 kg/m^2). A recent study of a Navy hospital indicates fuel loads for sleeping rooms average 1 lb per sq ft (4.88 kg/m^2) or less. Assuming standard time/temperature conditions, fire duration for patient areas and a majority of other occupied spaces would be less than 30 minutes.

Disposables

The use of combustible disposable equipment and supplies is on the increase for a variety of reasons, including the hazards of cross infection, personnel shortages, and possible financial savings. Disposables include bedding, gowns, gloves, drapes, collection bags, tubing, dishes, glasses, syringes, needles, and many diagnostic and therapeutic instruments. All such items require packaging, which adds to the combustible load both before and after use. Bulk purchasing is economical, but sometimes results in an overflow of combustibles into areas not designed or protected for such use.

Data Processing and Medical Records

Data processing centers are taking on increasing importance within health care facilities. Used to satisfy general business needs as well as to store patient records, the centers contain quantities of combustibles. These combustibles could expose high value equipment, could result in loss of vital records, and could produce a fire that threatens occupants outside the data processing areas.

Special protection should be considered for data processing centers.

Health care facilities generate and store files in considerable quantity. Files stored in closed steel cabinets do not represent any significant increase in hazard over that typical for most areas. However, files stored on open shelving create a serious fire hazard. Open file storage should be treated as a severe hazard, being both separated by fire rated construction and protected by automatic sprinklers.

FIRE SEVERITY

Determination of relative fire hazard involves considerations beyond total fuel loads. The arrangement of combustibles, their chemical makeup, and their physical state are all factors to be evaluated in addition to room geometry, ventilation rates, fire compartment size, and fire protection facilities. Generally speaking, the faster a fire develops, the greater the threat. Multiple death fires nearly always result from fires which quickly develop to large size, reaching full room involvement or flashover.

Fires which reach flashover will produce acutely lethal atmospheres generating thousands of cubic feet of smoke per minute (1,000 cu ft per min equals 28 m^3/min). Such fires threaten fire rated barriers and produce sufficient energy to drive smoke to remote areas. A fire that grows to full room size in a health care facility represents an unacceptable level of risk; there is a high likelihood of such a fire resulting in injuries or fatalities. Therefore, every effort should be made to recognize and eliminate fuel arrangements that might produce such fires.

Fire tests and actual fire experience have shown that certain arrangements and types of fuels in patient rooms create especially hazardous situations by being able to produce large fires in short periods of time. For example, in January 1976, multideath fires occurring in both the Wincrest and Cermak House nursing homes involved wood wardrobe closets (Best 1976a, 1976b). Tests have shown that combustible wardrobe fires can result in acutely hazardous environments in as little as 120 seconds (O'Neill et al 1980). Full scale tests, conducted by the National Bureau of Standards following the Sac-Osage Hospital fire, which involved a styrene butadiene rubber mattress, resulted in full room involvement approximately 7 ½ minutes, following open flame ignition of bedding (Lathrop 1975).

Any combination of finishes, combustible building materials, or contents and furnishings that could result in full room involvement or flashover in five minutes or less represents a severe fire hazard in a health care facility. Such spaces should always be protected by automatic sprinklers and separated by fire rated construction.

It is a well established fact that furnishings are frequently major contributors to fire growth. Recent developments make it possible to determine whether or not furnishings in a given environment are capable of producing sufficient energy to cause full room involvement.

A simplified equation estimating the rate of heat release required for flashover to occur in a room (Babrauskas 1980) is given by the expression:

$$\dot{Q} \geq 750\, A \sqrt{h}$$

where Q is the rate of heat release (kW), A is the ventilation opening area in sq ft (m^2), and h is its height in ft (m).

Once the heat release rate required to produce flashover is known for a typical room geometry, such information can be compared to actual heat release rates for typical furnishings as determined by tests conducted in a calorimeter such as the NBS furniture calorimeter (Babrauskas et

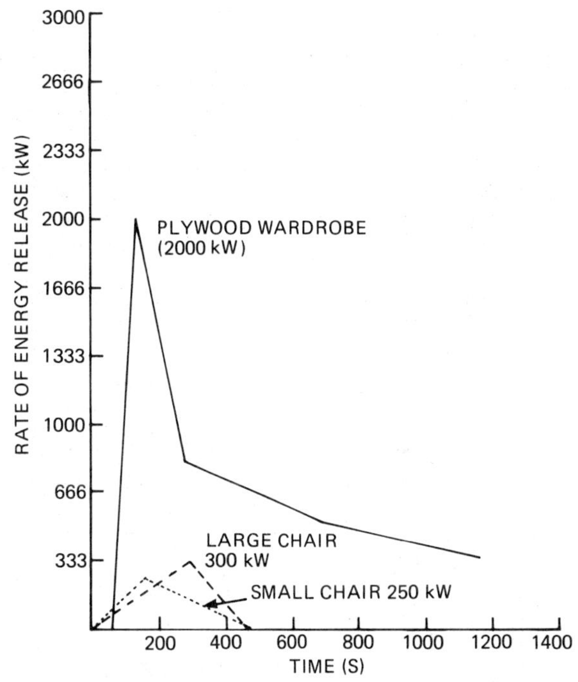

FIG. 9-4A. Heat release rates for miscellaneous furniture.

al 1982). Figures 9-4A and 9-4B are idealized curves developed by the National Bureau of Standards, which

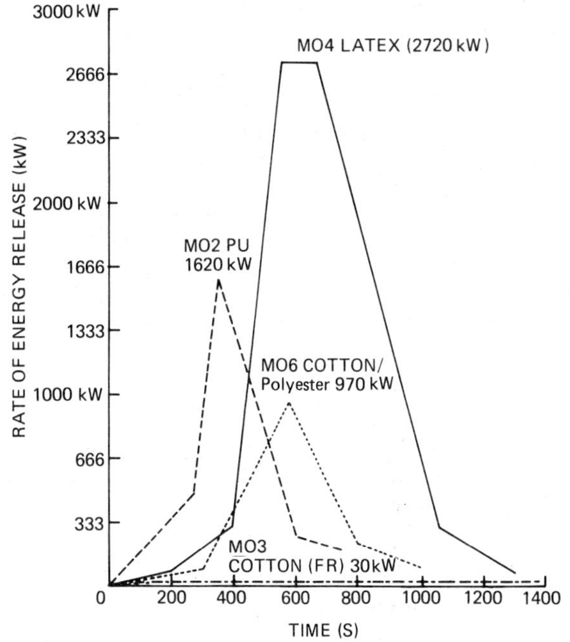

FIG. 9-4B. Heat release rates for beds.

illustrate the rate of heat release for typical furnishings as a function of time. Such information can be used to establish the probability of flashover; it also allows an estimate of the time required to reach critical fire size.

Interior Finish

The initial growth of a fire may be significantly affected by the interior finish of walls, ceilings and floors. Combustible interior finishes such as low density fiberboard ceilings and plastic floor coverings have been found to be significant factors in multideath fires (Juillerat 1962, Sears 1970). Combustible wall and ceiling finishes can act as a fuse, causing a fire to spread to objects remote from the fire origin. Combustible finishes also can cause a fire that would otherwise remain too small to grow large. Any large fire in a confined space creates a potentially lethal atmosphere. Interior finishes, therefore, deserve special attention.

The relative hazard of an interior finish is usually determined by a test conducted in accordance with NFPA 255, *Method of Test of Surface Burning Characteristics of Building Materials*, commonly called the Tunnel Test.

Interior finish on walls and ceilings within the means of egress and any room should be limited to Class A materials (flame spread rating of 25 or less). Class B materials (flame spread rating of 26-75) are considered tolerable, but should be limited to individual rooms with four persons or less. Full scale experiments show that automatic sprinklers are capable of limiting fire growth in rooms with Class C wall and Class D ceiling finishes (HEW 1975; O'Neill et al 1980). Accordingly, in buildings having automatic sprinkler protection, it has been judged acceptable practice to allow the use of materials having higher flame spread classifications than would otherwise be permitted; e.g., Class B materials are sometimes used where Class A is normally specified, and Class C materials where Class B is normally required.

Further, it has been shown that the performance of interior finishes is also related to location (Christian and Waterman 1971). Finishes on the upper portions of walls and ceilings contribute more significantly to flame propagation than finishes on the floor or the lower half of a wall. Therefore, where finishes are limited to the lower half of a wall and are less than 4 ft (1.2 m) above floor level, materials having a higher flame spread rating than would otherwise be allowed might be used without significantly affecting fire growth. For example, where Class A materials are judged necessary, a Class B material might be allowed on the lower portion of a wall.

In the past, floor finish materials were excluded from the requirements for interior finishes. This was based on favorable experience and the assumption that limited exposure would exist at the floor level during actual fires. However, a fire on January 9, 1970, in the Harmer House Convalescent Home in Marietta, OH, caused a significant change in this attitude (Sears 1970). Thirty-two persons died in this fire in which carpeting with foam rubber backing was judged to have played a significant role.

Interior floor finishes in corridors of nonsprinklered health care facilities should be limited to Class I materials—those with a minimum critical radiant flux of 0.45 W/cm^2. Where automatic sprinklers are provided, Class II interior floor finishes—those with a minimum critical radiant flux value of 0.22 W/cm^2—may be used.

BUILDING CONSTRUCTION

Because occupants of health care facilities must be defended in place, construction becomes an especially important factor—especially in multistory buildings. Buildings should be constructed of noncombustible material capable of resisting the effects of fire and maintaining structural integrity.

Buildings of two or more stories should be constructed of noncombustible materials with major structural members having at least two hour fire resistance. Materials that either burn or support combustion, although less desirable, may be used if special precautions are taken. An automatic sprinkler system is an essential part of the total fire defense system for such buildings.

Any evaluation of building materials should include consideration of smoke generating capabilities. When exposed to fire, cellulosic materials with fire retardants sometimes generate unusual amounts of smoke. In addition, plastic construction materials—becoming more common—sometimes generate large quantities of smoke.

Subdivision of Building Spaces

Separation of Patient Sleeping Rooms: Because it may not be possible to remove occupants during a fire, sleeping rooms must sometimes serve as a temporary area of refuge. Therefore, sleeping rooms should be isolated from all other building spaces by fire rated construction. Partitions should be continuous from the floor slab to the floor or roof above through any concealed spaces, such as those above suspended ceilings. (See Fig. 9-4C for typical floor plan for a health care facility.)

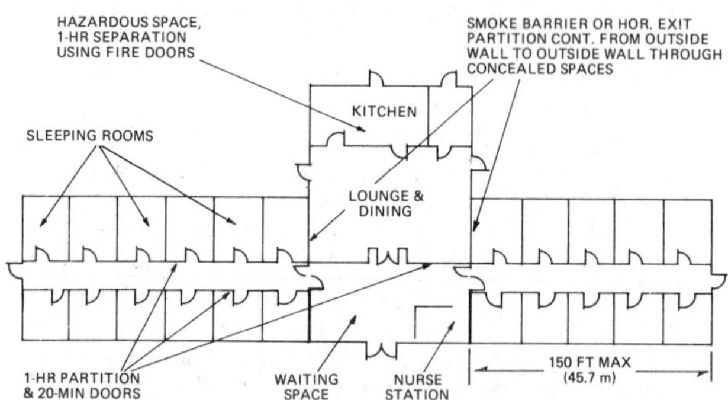

FIG. 9-4C. Typical floor plan for a health care facility.

Considerable controversy exists as to whether or not patient room doors should be equipped with self-closing or automatic closing devices. One school of thought attributes fatality producing fires such as those which occurred at the Harmer House Convalescent Home, Sac-Osage Hospital, Willow Point Nursing Home (Lathrop 1978), and Extendicare Skilled Nursing Facility (Demers 1981) to the fact that the door to the room of origin remained open during the fire. The open door allowed fire effects to spread to the corridor and adjacent rooms. Proponents of door closers argue that these devices would have closed the door to the room of origin and would have confined the fire, thereby preventing many of the fatalities.

Opponents of door closers argue that functional considerations require patient room doors to be open a majority of the time. Further, it is noted that, except for a few isolated instances, trained staff has been able to manually close the door to the room of origin during the many fires that occur annually in health care facilities. Additionally, it might be noted that the overwhelming majority of fire deaths in health care facilities involve persons who are either at the exact point of fire origin or are located in the room of fire origin. Door closers, which cause the doors to patient rooms to be closed at the start of a fire, could decrease the occupants' chance of survival in the room of origin. Closed doors also increase the time required to find the fire and, in the interim, can prevent any dilution of smoke. For these reasons, where patient room doors are equipped with closers, consideration should be given to providing smoke detectors within patient rooms. On detection of fire, the smoke detectors should activate the nurses call system which would annunciate the room of origin and, therefore, minimize search and rescue time.

Automatic sprinkler systems respond automatically to limit fire size. Where such protection is provided, partitions separating spaces occupied by patients need not be fire rated, but should resist the passage of smoke. Partitions may be terminated at suspended ceilings, and glass may be used in corridor doors and partitions without limit as to size or type. Transfer grills should not be used within such doors or partitions.

Any penetration of one hour partitions by building service equipment should be protected in order to maintain the one hour fire rated separation. All spaces around piping and ducts should be tightly sealed with a noncombustible material having adequate fire resistance and capable of retarding smoke transfer.

The effect of duct penetrations on one hour partitions should be carefully evaluated. Considerations should include the effect of metallic versus nonmetallic ducts, the location of openings, the purpose of the duct system, the direction of airflow, interconnected spaces, control features, and the general arrangement of the air handling system. Where steel ducts with a limited number of relatively small openings are installed, protection of the ducts may not be necessary. Where flexible or nonmetallic ducts are employed, consideration should be given to protecting the duct penetration with a fire damper.

Smoke Barriers: Every floor used by inpatients should be subdivided into at least two compartments by one hour partitions capable of retarding smoke. A horizontal exit, when constructed to satisfy the additional criteria imposed upon construction of smokestop barriers, is a desirable alternative to a smokestop partition. To minimize the number of occupants exposed to a single fire, the building should be subdivided into smoke compartments.

Protection of Vertical Openings

Fire and fire produced contaminants tend to spread vertically within a building. Special effort is required to prevent fire on one level from threatening the occupants above; this is especially important in health care facilities.

All shafts should be enclosed with materials that provide a minimum 2 hour fire resistance. Openings in shaft walls should be limited to only those necessary. Any such openings must be protected.

When designing partitions to enclose vertical shafts, consideration should be given to the varying durability of materials. In spaces where partitions may be subject to mechanical injury, materials used to provide floor to floor separation maintain the required fire resistance.

Exit Design

Exits in health care facilities should be limited to doors leading directly to the outside of the building, interior stairs and smokeproof towers, ramps, horizontal exits, outside stairs, and exit passageways.

Vertical evacuation of occupants within a health care facility is, at best, difficult and time consuming. Therefore, horizontal movement of patients is of primary importance.

Horizontal passageways and doors opening into corridors and rooms used for sleeping or treatment should be wide enough to allow the horizontal movement of occupants, even those in beds. Relocation of patients is a slow process, even under favorable staff to patient ratios. Because of the time required to move patients, exit access routes should be protected against fire effects. Spaces open to the corridor should not be used for patient sleeping or treatment rooms, nor should hazardous contents or activities be permitted within them. Such spaces should be arranged to allow direct visual supervision by staff and should be equipped with electrically supervised smoke detectors, which if activated will sound the building fire alarm.

All other spaces containing combustible materials that could prevent use of exit access corridors should be isolated by fire resistive partitions. For example, lounge spaces are typically equipped with combustible furnishings, and any electrical devices (such as lamps and television sets) and smoking in lounges can increase the possibility of ignition.

Horizontal Exits: Horizontal exits are common in health care facilities. Partitions used as horizontal exits and smoke barriers should provide the fire resistance required for exits and, in addition, should satisfy the criteria for smoke barriers. If possible, door openings should be limited to corridors, lobbies, or public spaces. The most desirable arrangement of mechanical systems is one in which the partitions forming the horizontal exit are not penetrated. If penetration by utilities or piping occurs, the space around the piping should be tightly filled with noncombustible materials. If ducts penetrate partitions, fire dampers that close if smoke detectors within the duct activate should be provided.

Since a horizontal exit implies that occupants will be transferred from one side of a partition to the other, adequate space must be available to "store" occupants after movement. At least 30 net sq ft (2.79 m^2) per occupant should be available on each side of the horizontal exit, allowing for the total number of patients in adjoining compartments.

Interior Stairs: Exit stairs should be designed to satisfy the criteria for interior stairs. Stairs should be enclosed with fire resistive materials, with stair openings limited to those necessary for access and discharge purposes. Stairs must be properly protected from the effects of fire.

Exit Features

Since occupant evacuation will move slowly and some occupants may have to be "stored," exit capacities are accorded a relatively low "flow rate". In non-sprinklered health care facilities, the recognized capacity of exits providing travel in stairs is only 22 persons per unit of exit width, and the capacity of doors or horizontal exits where all travel is on one level is 30 persons per unit of exit width. (One unit of exit width equals 22 in. or 559 mm.) Where automatic sprinklers are provided, the exit capacity increases to 35 persons per unit for travel via stairs and 45 persons per unit for travel over level passage-ways.

Limits on the distance of travel reflect the anticipated slow movement. Travel distance should normally not exceed the following:

1. One hundred feet (30 m) between an exit and any room door intended for use as an exit access.
2. One hundred and 50 feet (46 m) between an exit and any point in a room.
3. Fifty feet (15 m) between any point in a sleeping room or suite and the exit access door of that room.

In addition, facilities should be arranged to limit travel in a direction toward the fire to less than 30 ft (9 m). Elevators are not usually counted as required exits because they possess numerous shortcomings which may prevent their use during a fire. However, in the case of critically ill patients, patients in body casts or balkan frames, and others who would be difficult to move, elevators provide the only practical method of evacuation from upper stories of the facility. If separate banks of elevators are located in separate smoke compartments, and staff are well trained, it may be possible to devise a plan in which it will be safe to use elevators during fires.

Exit Marking and Exit Illumination

All exits should be identified by readily visible signs. Where access to exits is not immediately visible, access routes should also be marked.

Illumination of the entire means of egress must be continuous whenever the building is occupied. In some cases, normal street lighting is adequate for illumination of exit discharge. However, consideration should be given to the conditions that would result from a power failure.

Emergency power is also required for the illumination of the means of egress and exit marking. In hospitals, this power should be supplied by the life safety branch of the hospital electrical system. Luminescent, fluorescent, or reflective material should not be substituted for required lighting.

Emergency power supplies should automatically maintain illumination in the event of a power failure, without any appreciable interruption during the changeover from normal to emergency power. Where a generator is provided, the delay should not be more than ten seconds. Where emergency power is supplied by a central system with an engine driven generator, the design should minimize the possibility of any single emergency simultaneously interrupting both normal and emergency power supplies. The switch that transfers power from normal to emergency circuits is one place at which normal and emergency circuits are required to merge. If this switch is exposed to fire, it could simultaneously interrupt power to both normal and emergency circuits. To minimize this possibility, the generator and its attendant fuel hazards should be separated from the remainder of the building by one hour fire rated partitions. Similarly, the transfer switch and other electrical distribution panels and switch gear should be separated from the generator as well as from the remainder of the building.

Fire Alarms

Every building should be equipped with an electrically supervised, manually operated fire alarm system. When actuated, the system should sound alarms that can be heard above ambient noise levels throughout the facility. The fire alarm should also be automatically transmitted to the fire department. Any fire detection or fire suppression system which activates should automatically activate the building alarm system.

Alarm systems, including detection devices, should be provided with an emergency power supply and should be designed according to NFPA 72A, *Standard for the Installation, Maintenance, and Use of Local Protective Signaling Systems for Guard's Tour, Fire Alarm and Supervisory Service.*

Any alarm that is activated should automatically, without delay, provide a general alarm. Presignal systems are not considered suitable for health care occupancies although zoned systems with a coded signal have certain desirable characteristics.

FIRE SUPPRESSION EQUIPMENT

Many authorities believe the most practical and reliable approach to life safety in health care facilities is to use automatic fire suppression—in particular, automatic sprinkler systems. Although persons in the area of origin may still be seriously threatened, persons in adjoining spaces—and, in a number of cases, in the same room—are protected (Boettcher 1967; HEW 1975; O'Neill et al 1980).

The capability of automatic sprinklers to provide a survivable environment for building occupants has been debated at length. Full scale tests have shown that standard sprinklers can extinguish many fires while maintaining a survivable atmosphere outside the room of fire origin (HEW 1975; O'Neill et al 1980). In addition, sprinklers have been shown to be effective in limiting carbon monoxide to nonlethal levels outside the room. However, immediate obscuration results on sprinkler discharge, so the lack of visibility makes rescue efforts difficult. Tests have also shown that privacy curtains may interfere with sprinkler discharge. To obtain full sprinkler effectiveness, the influence of any building design feature or furnishing that would impair sprinkler discharge should be carefully evaluated.

A recent development, the quick response or fast acting sprinkler—which may be up to five times more sensitive than ordinary sprinklers—has been shown in some instances to maintain a survivable atmosphere within the room of fire origin.

Automatic sprinklers should adhere to NFPA 13, *Standard for the Installation of Sprinkler Systems.* Operation of the sprinklers should automatically sound the building fire alarm. The sprinkler system and components should be electrically supervised to ensure reliable operation; this should include gate valve tamper switches with a local alarm at a constantly attended location when the valve is closed. If a single water supply is provided by a connection to city mains, a low pressure monitor should be included. If pressure tanks are the primary source of water, air pressure, water level, and temperature should also be supervised. If fire pumps are provided to boost system pressure, electrical supervision should monitor loss of pump power, pump running indication, low system pressure, and low pump suction pressure.

Portable fire extinguishers should be placed in all buildings, in conjunction with small hose lines, as they may provide an opportunity to control a fire during its early stages. In all cases, however, the fire department should be notified before or at the same time that occupants are attempting to fight the fire. Delayed alarms have resulted in fires growing to large scale which, in turn, threatens occupants prior to fire department notification and arrival.

Smoke Control

Every patient sleeping room should be provided with an outside window or door that can be opened from the inside; this will allow venting of products of combustion if there is a fire. It may also provide fresh air to patients forced to remain in their rooms during a fire.

A specially designed smoke control and exhaust system can be a desirable alternate to the outside window. Special forced air systems, or in some cases adaptation of conventional building air handling systems, can permit venting of products of combustion early in the fire. Such systems may also make it possible to create a pressure differential across physical barriers (floors or partitions) and prevent smoke transfer. The effectiveness of fire partitions is often significantly improved by the use of such systems.

Where adaptation of the building air handling system is contemplated for smoke removal, the design should be as suggested in NFPA 90A, *Standard for the Installation of Air Conditioning and Ventilating Systems.* Consideration should be given to alternate power supplies and electrical supervision of critical system components.

PROTECTION OF HAZARDOUS AREAS

Areas with contents more hazardous than those normally found in health care facilities should be arranged to minimize the exposure to occupants if there is a fire. Hazardous areas such as boiler and heater rooms, laundries, kitchens, and repair shops should be separated by one hour fire rated construction with openings protected by fire doors. As an alternative, the area should be protected by automatic sprinklers and, if the hazard is judged severe, one hour separation should be provided. Spaces requiring both separation and sprinklers are soiled linen rooms, paint shops, and rooms or areas (including repair shops and trash collection rooms) used to store combustibles, supplies, and equipment in hazardous quantities.

Laboratories that contain quantities of flammable, combustible, or hazardous materials should be protected and separated from other building spaces. Pertinent information is found in NFPA 30, *Flammable and Combustible Liquids Code,* which is applicable to repair shops as well as laboratories.

Cooking equipment should be arranged and protected in accordance with NFPA 96, *Standard for the Installation of Equipment for the Removal of Smoke and Grease-Laden Vapors from Commercial Cooking Equipment.*

Nitrous oxide and oxygen are usually piped through hospital spaces from a central distribution point and although these gases will not burn, they do significantly accelerate combustion. Special precautions are necessary and are detailed in NFPA 50, *Standard for Bulk Oxygen Systems at Consumer Sites,* and NFPA 56F, *Standard for Nonflammable Medical Gas Systems.*

Building Service Equipment

Building service equipment should be installed and maintained in accordance with appropriate NFPA standards. Special consideration should be given to the design and installation of heating and air conditioning systems. Because rubbish chutes have been an important factor in a number of fires (Juillerat 1962), due regard should be given to the design of rubbish and linen chutes (including pneumatic systems).

Portable heating devices are unsafe in patient occupied portions of health care facilities; all heating devices should be designed and installed to prevent ignition of combustible materials. Approved suspended unit heaters may be used, except in means of egress and patient sleeping areas, if they are high enough to be out of the reach of persons using the area.

Combustion and ventilation air for boilers, incinerators, or heater rooms should be taken directly from, and discharged directly to, the outside of the buildings.

OPERATING FEATURES

Many fires in health care facilities are caused by careless smokers. Adoption and enforcement of suitable smoking regulations is essential (NFPA 1970). Smoking should be prohibited in any room, ward, or compartment where flammable liquids, combustible gases, or oxygen is being used or stored, and such areas should be posted with suitable signs. Smoking by patients who are under sedation or are not considered responsible should be prohibited. Metal containers with self-closing covers should be available for disposable of smoking materials in all areas where smoking is allowed.

Window draperies and curtains should be made of noncombustible material or material that has been rendered and maintained flame retardant. These window hangings should be capable of passing both the large and small scale tests required by NFPA 701, *Standard Methods of Fire Tests for Flame-Resistant Textiles and Films.* Furnishings, decorations, and other objects should not obstruct exits or exit access routes. Combustible decorations should be prohibited.

Exits and mechanical devices provided to control or limit fire should be maintained to ensure reliable operation. Inspections and tests should be performed as required to verify satisfactory performance.

In facilities with locked exits, adequate staff with keys should always be present to release occupants and direct them away from the fire area to a place of safety during an emergency.

Care should be exercised during construction and repair operations to ensure that such activities do not reduce life safety. Adequate preventive maintenance for mechanical systems, including tests and periodic inspections, are necessary to ensure their reliability.

EMERGENCY PLANNING

Every health care facility must have a fire and evacuation plan, including a disaster plan with which all personnel must be familiar. In addition, personnel should be trained to use fire extinguishers and hose cabinet lines. They must also know how to sound an alarm, move or evacuate patients, and contain the fire.

Each facility should have a safety officer whose prime responsibility is to recognize hazards, act as the liaison with the fire service, and arrange for the training of personnel. While the training of hospital personnel is straightforward and can be accomplished on the job, orienting members of the fire service to hospital problems is more difficult.

Copies of the fire and evacuation plan should be available to all personnel. The plan should contain specific instructions for key (supervisory) personnel if there is a fire. A copy of the plan should also be posted for reference. All employees should have periodic training to ensure readiness.

Emergency drills should include transmission of a fire alarm signal and simulation of emergency conditions insofar as possible without jeopardizing occupants. Drills should be conducted at least quarterly on each shift with at least 12 drills held every year. The drills should be varied to test the alertness of all shifts and if possible, should be unannounced. Use of the building alarm during drills also verifies that it is operating normally.

The fire and evacuation plan should include the following fundamentals:

1. Training personnel in use of the alarm and alarm equipment.
2. Transmission of alarm to the fire department.
3. Details of the fire location.
4. Evacuation practices for all areas.
5. Preparation of building spaces for evacuation.
6. Fire extinguishment.

Emphasis during drills should be placed on immediate notification of the fire department; many fires have spread because of delayed alarms.

Bibliography

References Cited

Babrauskas, Vytenis. 1980. "Estimating Room Flashover Potential." *Fire Technology*. May 1980.

Babrauskas, Vytenis, et al. 1982. "Upholstered Furniture Heat Release Rates Measured with a Furniture Calorimeter." *NBSIR 82-2604*. National Bureau of Standards, Washington, DC.

Best, Richard. 1976a. "The Wincrest Nursing Home Fire." *Fire Journal*, Sept 1976.

Best, Richard. 1976b. "The Cermak House Fire." *Fire Journal*. Sept 1976.

Boettcher, E. N., M.D. 1967. "Hospital Fire Defense: People and Sprinklers." *Fire Journal*. Vol 61. No 4. pp 93-96.

Christian, W. J., and Waterman, T. E. 1971. "Flame Spread in Corridors: Effects of Location and Area of Wall Finish." *Fire Journal*. July 1971.

Demers, David P. 1981. "25 Die in Nursing Home." *Fire Journal*. Jan 1981.

HEW. 1975. "Full-Scale Fire Tests in a Nursing Home Patient Room." *Report 7463*. U. S. Department of Health, Education and Welfare. HEW Contract HSA 105-74-116. Prepared by American Health Care Association.

Juillerat, E. E. Jr. 1962. "The Hartford Hospital Fire." *NFPA Quarterly*. Vol 55. No 3. pp 295-303.

Lathrop, James K., et al. 1975. "In Osceola, A Matter of Contents." *Fire Journal*. May 1975.

Lathrop, James K. 1978. "Nursing Home Fire Causes Two Deaths." *Fire Journal*. Mar 1978.

N.B.S. 1942. "Building Materials & Structures." *Report BMS 92*. National Bureau of Standards, Washington, DC.

NFPA Fire Records Department. 1970. "Hospitals: A Fire Record." *Fire Journal*. Vol 64. No 2. pp 14-27.

O'Neill, John G., et al. 1980. "Full Scale Fire Tests with Automatic Sprinklers in a Patient Room, Phase II." *NBSIR 80-2097*. Center for Fire Research. National Bureau of Standards, Gaithersburg, MD.

Sears, Albert B. Jr. 1970. "Nursing Home Fire", Marietta, OH. *Fire Journal*. May 1970.

Spalding, Charles K. 1975. "The Frequency, Cause and Prevention of Hospital Fires." *Fire Journal*. Mar 1975.

NFPA Codes, Standards and Recommended Practices and Manuals. (See the latest *NFPA Codes and Standards Catalog* for availability of current additions of the following documents.)

NFPA 13, *Standard for the Installation of Sprinkler Systems.*

NFPA 30, *Flammable and Combustible Liquids Code.*

NFPA 50, *Standard for Bulk Oxygen Systems at Consumer Sites.*

NFPA 56F, *Standard for Non Flammable Medical Gas Systems.*

NFPA 72A, *Standard for the Installation, Maintenance, and Use of Local Protective Signalling Systems for Guard's Tour, Fire Alarm and Supervisory Service.*

NFPA 90A, *Standard for the Installation of Air Conditioning and Ventilating Systems.*

NFPA 96, *Standard for the Installation of Equipment for the Removal of Smoke and Grease-Laden Vapors from Commercial Cooking Equipment.*

NFPA 101, *Code for Safety to Life from Fire in Buildings and Structures.*

NFPA 255, *Method of Test of Surface Burning Characteristics of Building Materials.*

NFPA 701, *Standard Method of Fire Tests for Flame-Resistant Textiles and Films.*

Additional Readings

Archea, John, "The Evacuation of Non-Ambulatory Patients from Hospital and Nursing Home Fires: A Framework for a Model," *NBSIR 79-1906*, National Bureau of Standards, Gaithersburg, MD, Nov. 1979.

Bickman, L., et al., "An Evaluation of Planning and Training for Fire Safety in Health Care Facilities—Phase Two," *NBS-GCR-79-179*, National Bureau of Standards, Gaithersburg, MD, Jan. 1979.

Chapman, R. E., et al., "User's Manual for the Fire Safety Evaluation System Cost Minimizer Computer Program," *NBSIR 83-2797*, National Bureau of Standards, Gaithersburg, MD, 1983.

Chapman, R. E., et al., "Economic Aspects of Fire Safety in Health Care Facilities: Guidelines for Cost Effective Retrofits," *NBSIR 79-1902*, National Bureau of Standards, Gaithersburg, MD, Nov. 1979.

Fire Protection for Hospitals, P8404, Factory Mutual System, Norwood, MA.

Firesafety in Hospitals—Employee's Workbook, National Fire Protection Association, Quincy, MA, 1977.

Firesafety Evaluation System for Health Care Facilities—Student Manual, Office of Planning and Education, U.S. Federal Emergency Management Agency, U.S. Fire Administration, Washington, DC, 1980.

"Firesafety Evaluation of Health Care Facilities," *NBSIR 78-1555*, National Bureau of Standards, Gaithersburg, MD, Nov. 1978.

Klein, Burton R., ed. *Health Care Facilities Handbook*, National Fire Protection Association, Quincy, MA, 1984.

Krueger, L. J., and Patton, R. M., "More Fire Safety Can Cost Less," *Hospitals*, Vol. 51, Feb. 1, 1977, pp. 127-132.

O'Neill, J. G., "Fast Response Sprinklers in Patient Rooms," *Fire Technology*, Vol. 17, No. 4, Nov. 1981, pp. 254-274.

O'Neill, J. G., et al., "Full-Scale Fire Tests With Automatic Sprinklers in a Patient Room. Phase III," *NBSIR 80-2097*, National Bureau of Standards, Gaithersburg, MD, July 1980.

Sprague, J. G., "A Common Sense Approach Needed in Dealing with Safety," *Hospitals*, Vol. 51, Feb. 1, 1977, pp. 67-75.

BOARD AND CARE FACILITIES

Revised by Alfred J. Longhitano, P. E.

The residential board and care home, although it has been with us for more than a century, has become an American phenomenon of the 1970s and 1980s. It is estimated that there are more than 300,000 board and care facilities in the United States, and that they may house up to 1,500,000 persons. The numbers are guesses because, in many states, these facilities are unregulated and, indeed, unidentified.

The reasons for the growth of this occupancy are numerous. The population of state hospitals was reduced by two-thirds, or nearly 300,000 patients, in the ten year period from 1969 to 1979. Social, judicial, and economic pressures to remove patients from the restrictive environment of psychiatric hospitals, nursing homes, state schools and other institutional settings helped create a need for board and care facilities. This occupancy provides the deinstitutionalized person with a place to live in a more normal setting and furnishes the support services the resident needs to cope with the requirements of daily life.

Today's young mobile population often leave the elderly living alone and unable to care for themselves in an independent living situation. They might need someone to look after their nutrition, personal hygiene, etc.

Alcohol and drug rehabilitation programs often utilize the halfway house concept, in which the individual is somewhat sheltered from the pressures of being too soon returned to the general population for independent living.

Shelters for battered persons, unwed mothers, and homeless children have become more common in the 1980s.

The fire experience in this type of facility has been horrendous—fires have been numerous and devastating in terms of loss of human life and damage to property. In one month in 1979, four separate fires in board and care facilities claimed 44 lives. The number killed in a single fire has been as high as 33. In addition, total destruction of the facility is not at all unusual.

This chapter will examine what makes this occupancy so susceptible to fire, and what needs to be done if the fire record is to be improved. A partial list of multiple-death board and care facility fires since 1979 appears in Table 9-5A.

Mr. Longhitano is a Principal, Gage-Babcock & Associates, Inc., Mt. Kisco, NY.

TABLE 9-5A. Multiple-Death Board and Care Fires

Date	Occupancy	Civilians Killed	Civilians Injured
4/2/79	Wayside Inn Boarding House Farmington, MO	25	9
4/11/79	1715 Lamont Street Washington, DC	10	4
11/11/79	Coats Boarding Home Pioneer, OH	14	0
7/26/80	Brinley Inn Boarding Home* Bradley Beach, NJ	24	4
8/16/80	Boarding Home Honolulu, HI	3	0
11/30/80	Donahue Foster Home Detroit, MI	5	0
1/9/81	Beachview Rest Home* Keansburg, NJ	31	10
10/28/82	Perrys' Domiciliary Care Home* Pittsburgh, PA	5	1
2/7/83	Silver Leaves Group Home* Eau Claire, WI	6	0
3/13/83	Shannons Foster Care Home* Gladstone, MI	5	3
4/19/83	Central Community Home* Worcester, MA	7	2
8/31/83	Anandale Village* Lawrenceville, GA	8	0

* These are incidents in which NFPA has evidence that elderly occupants or former mental health patients were tenants.

DEFINITION

A residential board and care facility is a residence which always provides shelter and overnight accommodations, often provides meals and, as distinct from an ordinary boarding house, provides the residents with some form of personal care.

This definition could easily describe a hospital or a nursing home. The primary difference is in the definition of personal care. The care given in a hospital or a nursing home is certainly personal, but its primary purpose is to provide for the medical needs of the patient. In the board and care home, the care provided is not medical in nature, nor is it a type that requires the services of doctors and nurses. It is generally a kind of assistance that can be rendered by lay persons who help the residents cope with the rigors of daily living. This care might include looking after the residents' personal hygiene, cleaning their living quarters, assisting those who are physically handicapped to move around, preparing meals, looking after the residents' financial matters, seeing that appointments are kept, and numerous other tasks.

Although there are suspected to be more than 300,000 board and care homes in the U.S., few such facilities call themselves by that name. Some of the kinds of facilities which fit into the definition of this occupancy are:

1. Group homes for mentally retarded persons.
2. Group homes for released psychiatric patients.
3. Rest homes for the aged.
4. Foster care homes.
5. County poor houses.
6. Orphanages.
7. Shelters for battered persons.
8. Shelters for unwed expectant mothers.
9. Halfway houses for rehabilitated alcoholics.
10. Halfway houses for rehabilitated drug abusers.
11. Halfway houses for prison parolees.
12. Shelters for runaways.
13. Rescue missions with overnight accommodations.

Whatever the names, these facilities are all tied together by a common thread of similarity: the facility provides its residents with more than just food and shelter. The residents are usually furnished some form of assistance, support, or care which is not of a medical nature.

OCCUPANCY CHARACTERISTICS

Characteristics of Buildings

A board and care facility falls into the ordinary hazard category by the traditional methods of evaluation under NFPA 101® the Life Safety Code® (low, by comparison with some building codes). There are rarely any massive concentrations of combustible materials and usually no use of flammable liquids or gases. There are no processes to pose "special hazards." Ignition sources are typical of those found in residential occupancies, and fuel loads are generally light. The most significant fire threat in this occupancy is the enormous potential for loss of human life.

The buildings that house residential board and care facilities cover a wide range of structural characteristics. The buildings range from single family dwellings in which a family takes in a few "boarders," to high rise reinforced concrete apartment buildings for the elderly operated by religious organizations, and to commercial rest homes built for the purpose. Construction is likely to be of practically any type.

Characteristics of Residents

A description of the characteristics of the "typical" resident of a board and care facility is just as elusive as a description of a "typical" construction type. Residents of board and care homes do not fit into a stereotype—they may be children or elderly, physically impaired or physically sound, mentally handicapped or brilliant. They may be capable of self-preservation or totally dependent upon others to provide for their safety. There is no one description that fits the residents of all board and care homes.

The variations that can exist in the buildings and the range of capability of residents in these facilities poses the most significant challenge to fire protection professionals.

FIRE PROBLEMS

The problems of board and care facilities generally fall into two categories: structural and behavioral.

Structural Problems

It is probable that the vast majority of the estimated 300,000 board and care facilities are single family dwellings, or single family dwellings to which space has been added to accommodate a number of "boarders." Consequently, it is probable that the vast majority of facilities are in structures built to standards intended for the housing of a normal family having normal capability to detect, respond to, and escape from a fire.

Building Construction: Most facilities are believed to be housed in one, two, and three story wood frame structures. In older buildings, balloon construction and exposed structural members, due to deteriorating plaster, are common structural hazards. Under the best of circumstances, wood frame construction can offer some measure of protection, but fire endurance is limited and the combustible structure can eventually contribute fuel to a fire.

Exit Facilities: Generally there is only a single route that occupants can use from any given room to get to safety outdoors. This escape route frequently consists of a single unenclosed stairway which might take the escaping residents through the area of the house which is on fire. Often, the only alternative escape route is through a window, which in the case of a two or three story building may be more than 20 feet above the ground. Typical of single family dwellings, exit doors are often equipped with locking devices which impede egress, particularly for those who are either unfamiliar with the locks or incapable of operating them.

Compartmentation: There is rarely any substantial protection of vertical openings. Indeed, stairways connecting floors are often wide open.

Separation of rooms from corridors and rooms from each other is either nonexistent or by construction which provides little or no resistance to the passage of fire and smoke. Consequently, a fire in any room can quickly fill the entire building with smoke and toxic combustion products, limiting visibility and making the environment intolerable for human survival.

Interior Finishes: Frequently, walls and ceilings are refinished with paneling and acoustical ceiling tiles without any thought to the combustibility of these materials. The facility owner trying to improve the environment in which his/her charges live by providing attractive decor may be placing them in jeopardy without knowing it.

Highly flammable wall paneling and ceiling tiles contribute to rapid horizontal and vertical fire spread, are frequently a major factor contributing to flashover, and often contribute to the speed with which exit pathways become unusable.

Electrical Wiring and Appliances: Wiring is generally typical of that found in single family dwellings. In older buildings, circuits may be inadequate in number and capacity, insulation may be old and deteriorating, fuses may be of the wrong sizes for the circuits they protect, and wiring modifications are likely to have been made by unqualified people. Use of extension cords, portable heaters, and other residential type electrical appliances contribute to the risk of fire.

Furnishings: The fire hazard posed by furniture and furnishings varies widely. Practically all such products contribute fuel to a fire, with some, of course, being worse than others. There are generally no regulations governing the kinds of furnishings that can be used in a board and care facility. Use of furniture which burns fiercely or which gives off inordinately large quantities of thick smoke or toxic combustion products is a major hazard.

Behavioral Problems

Much of the concern over life safety in board and care facilities arises out of the fact that they house a number of people who, in many cases, have diminished physical and/or mental capacity. Although this is not true in all facilities, or for all residents of a given facility, the very fact that impaired people are residents of a facility which provides them with assistance in coping with daily life suggests that they may have problems fending for themselves.

Behavioral problems generally fall into two categories: evacuation capability and safety consciousness.

Evacuation Capability: Because of the widely differing populations that can be housed in board and care facilities, a broad statement describing the evacuation capability in all such facilities cannot be made.

In the case of homes for the elderly, some residents may be slow moving, unsteady, unable to manage stairs, to see well, or to remember the procedures to be followed in a fire emergency. Some may be sufficiently feeble to make evacuation impossible without much assistance.

In a group home for high functioning mentally retarded persons, evacuation capability could be comparable to that expected of a group of unimpaired teenagers.

Residents of some facilities may be physically handicapped to such an extent that self-evacuation is not possible.

Evacuation capability can be affected by the staffing of the facility. In a facility housing deaf residents, the presence of one staff member who can wake up and alert the residents to the emergency, might be sufficient to normalize the evacuation capability of the group. In facilities housing more debilitated handicapped persons, a large staff might still be incapable of effecting a prompt evacuation of the group.

Safety Consciousness: One of the factors which makes this occupancy highly prone to fires is that the residents are often unaware of or unconcerned about the dangers of their actions. In many of the small facilities, even the management personnel have little awareness of good life safety and fire prevention practices and procedures. Careless smoking habits, improper use of portable heating units, and improper use of electrical extension cords and appliances are fairly common hazards in many of these facilities. In addition some residents of board and care facilities have been known to set fires intentionally.

LIFE SAFETY CONCEPTS

Setting minimum criteria for the protection of life from fire in this broad occupancy type is not a simple matter. As indicated, this is not a homogeneous occupancy. What is minimum protection in one facility, may be over protection or inadequate protection in another.

Perhaps the evacuation capability of the residents is the most significant factor in determining the fire protection features needed to achieve an acceptable level of life safety in a given board and care home. In a facility which houses a group having a fairly normal ability to respond to a fire situation, relatively little protection may be needed. In buildings housing people who can respond to the emergency, but do so slowly, a protection system which provides them a little extra time to react and evacuate may be adequate. For groups that cannot be expected to evacuate to safety unaided, a defense in place protection system may be needed.

One might suspect that over a period of time, all facilities may house residents who are incapable of evacuation; therefore, protection for the worst situation should be provided. This approach is valid in the instance of homes for the aged, where deteriorating physical and mental conditions are predictable.

However, there are many types of board and care facilities which, because of their size and the nature of the services they provide, never deal with people who are incapable of self-preservation. Such facilities should not be required to provide the same degree of built in fire protection as those which house less capable populations. Therefore, it is reasonable to establish fire protection criteria based upon the evacuation capability of the residents of the facility. More or less built in protection, would be required, depending upon the residents' ability to fend for themselves.

For a population having relatively normal evacuation capability, the most basic system of protection consists of:

1. Means to alert the residents to the existence of a fire.
2. At least two routes by which to escape from the endangered area.
3. Prohibition of the use of highly flammable interior finish materials (this limits the rate of fire spread and minimizes the risk of flashover).
4. Some form of enclosure of vertical openings (this limits the vertical spread of fire and smoke).

For a population that is able to evacuate, but does so slowly or with difficulty, a reasonable system of protection would include all the items listed above, plus:

1. Exits that are easier to use than those that might be adequate for a more capable population.
2. Tighter controls on flammability of interior finishes.
3. Inclusion of a degree of fire endurance to maintain building integrity during a longer evacuation period.
4. Inclusion of a degree of compartmentation to confine the fire and smoke, thus maintaining the exit pathways usable during the longer evacuation period.
5. Greater resistance to the vertical spread of fire and smoke.

For a population that is not capable of evacuation, a defense in place protection concept would include all the features listed above, plus:

1. Sufficient structural fire endurance to allow the building and the internal compartmentation to outlast the fire, or
2. A means to automatically suppress the fire before it becomes a serious threat to the structure and the compartmentation.

These protective concepts are the same concepts as those used in various other occupancies in which people sleep overnight. For a detailed set of criteria for life safety in this occupancy, see Chapter 21 of NFPA 101.

Bibliography

NFPA Codes, Standards, Recommended Practices and Manuals. (See the latest *NFPA Codes and Standards Catalog* for availability of current editions of the following documents.)

NFPA 101, Code for Safety to Life From Fire in Buildings and Structures.

Additional Readings

Alvord, D. M., "The Escape and Rescue Model—A Simulation Model for the Emergency Evacuation of Board and Care Homes," *NBS-GCR-83-453*, National Bureau of Standards, Gaithersburg, MD, Dec. 1983.

Berlin, G. N., et al., "Modeling Emergency Evacuation from Group Homes," *Fire Technology*, Vol. 18, No. 1, Feb. 1982, pp. 38-48.

Blye, Philip E. and Yess, James, R., *Fire Safety in Boarding Homes*: A Self-Study Guide for Owners and Operators, National Fire Protection Association, Quincy, MA, 1985.

"Board and Care Homes: A Study of Federal and State Actions to Safeguard the Health and Safety of Board and Care Home Residents," U.S. Department of Health and Human Services, Washington, DC, 1982.

Groner, N. E., "A Matter of Time—A Comprehensive Guide to Fire Emergency Planning for Board and Care Homes," *NBS-GCR-82-408*, National Bureau of Standards, Gaithersburg, MD, Nov. 1982.

Lathrop, J. K., ed., *Life Safety Code Handbook*, 3rd ed, National Fire Protection Association, Quincy, MA, 1984.

Levin, B. M., ed., "Fire and Life Safety for the Handicapped," *NBS-STP-585*, National Bureau of Standards, Gaithersburg, MD, July 1980.

Levin, B. M., and Nelson, H. E., "Firesafety and Disabled Persons," *Fire Journal*, Vol. 75, No. 5, Sept. 1981, pp. 35-40.

Nelson, H. E., et al., "A Fire Safety Evaluation System for Board and Care Homes," *NBSIR 83-2659*, National Bureau of Standards, Gaithersburg, MD, Mar. 1983.

Pearson, R. G., and Joost, M. G., "Egress Behavior Response Times of Handicapped and Elderly Subject to Simulated Residential Fire Situation," *NBS-GCR-83-429*, National Bureau of Standards, Gaithersburg, MD, 1983.

Waller, M. B., and Vreeland, R. G., "Report of a Conference on Fire Emergency Plans in Group Homes for the Developmentally Disabled," *NBS-GCR-81-315*, National Bureau of Standards, Gaithersburg, MD, Mar. 1981.

DETENTION AND CORRECTIONAL FACILITIES

Wayne G. Carson, P. E.

Detention and correctional occupancies include those buildings and facilities where persons are restrained by locks not under their own control. Such occupancies include jails, detention centers, correctional institutions, reformatories, houses of correction, and prerelease centers. Security is a major operational consideration in these facilities and must be included in the fire protection design process.

Large correctional facilities may contain other occupancies such as industrial (vocational shops), business (classrooms and offices), assembly (dining halls, auditoriums and gymnasiums) and storage (warehouses). The principal concern of Chapters 14 and 15 of NFPA 101, *Life Safety Code®* (hereinafter referred to as NFPA 101) is the residential portion of correctional facilities. Except for the locking of doors, the other occupancies are discussed elsewhere in this section.

Additional information relevant to firesafety in correctional institutions may be found in Section 7, Chapter 2, "Systems Concept for Building Firesafety;" Chapter 7, "Interior Finish;" Chapter 10, "Smoke Movement in Buildings;" and Section 20, "Portable Fire Extinguishers."

OCCUPANCY CHARACTERISTICS

The major difference between a detention and correctional occupancy and other residential occupancies is that the occupants are restrained by locks on the doors. Therefore, the occupants are not capable of significant self-preservation actions until someone (staff) unlocks the doors.

The degree of locking can have a significant impact on the risk to the occupants. For example, a facility with only two locks which can be remotely released to allow the occupants out of the building into a controlled exercise yard would generally present less of a risk to the occupants than one that required the manual unlocking of 15 locks with 10 different keys.

There are five types of restraint or locking systems generally found in detention and correctional occupancies. These are referred to as "Use Conditions" by NFPA 101 and the *Standard Building Code* of the Southern Building Code Congress (SBCCI). They are illustrated in Figure 9-6A and defined as follows:

Use Condition I—Free Egress: Free movement is allowed from sleeping areas and other spaces where access or occupancy is permitted to the exterior via means of egress meeting the requirements of NFPA 101. For example, work

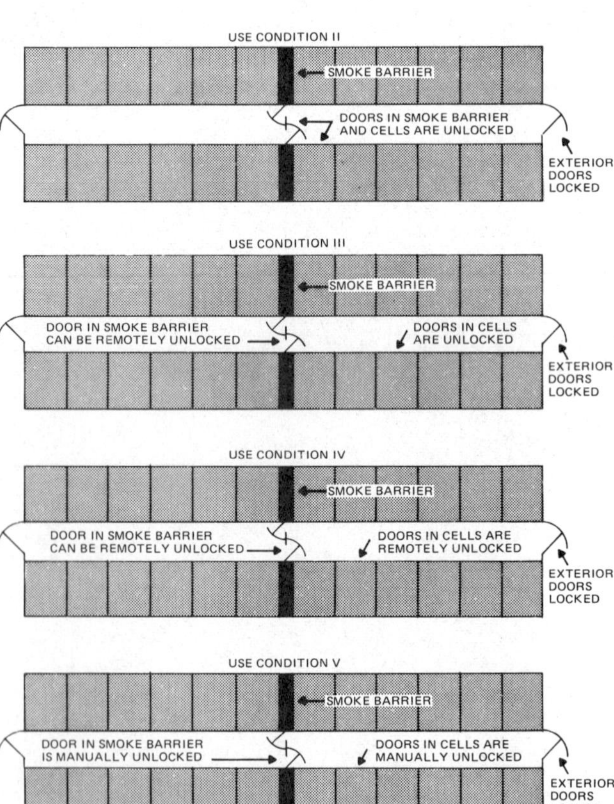

FIG. 9-6A. Classes of restraint, use conditions II, III, IV, and V.

Mr. Carson is Principal, Carson Associates, Inc., Fire Protection Consultants, Warrenton, VA.

release centers where the doors are not locked would be considered Use Condition I and are not considered detention and correction facilities.

Use Condition II—Zoned Egress: Free movement is allowed from sleeping areas and any other occupied smoke compartments to other smoke compartments.

Use Condition III—Zoned Impeded Egress: Free movement is allowed within individual smoke compartments (such as within a residential unit comprised of individual sleeping rooms and within group activity space) with egress from that smoke compartment to another smoke compartment provided by remote control unlocking of doors.

Use Condition IV—Impeded Egress: Free movement is restricted from an occupied space. Remote control release is provided to permit movement from all sleeping rooms, activity spaces, and other occupied areas within the smoke compartment to other smoke compartment(s).

Use Condition V—Contained: Free movement is restricted from an occupied space. Staff controlled manual release at each door is provided to permit movement from all sleeping rooms, activity spaces, and other occupied areas within the smoke compartment to other smoke compartments.

THE FIRE PROBLEM

The fire problem in correctional facilities has been statistically identified and analyzed by NFPA in its report *A Study of Penal Institution Fires* (NFPA 1977a). This study reviewed 52 fires in correctional facilities from January 1967 to July 1977. It identified the significant common characteristics of correctional facility fires and grouped them into several categories as shown in Table 9-6A.

TABLE 9-6A. Statistics of Past Correctional Facility Fires*

Category	Number of Incidents
Type of Act	
Accidental	7
Incendiary	45
Source of Ignition	
Welding	1
Boiler	1
Electrical Equipment	3
Smoking Materials	27
Unspecified	17
Fuel Type	
Mattresses	25
Cell Padding	5
Clothing	11
Paper Products	5
Wood Roof or Floor	2
Flammable Liquid	3
Place of Origin	
Cells	33
Recreation Room	4
Boiler Room	2
Storage Area	4
Cell Blocks	6
Penal Work Buildings	3
Security Aspect (Problem in Evacuation)	
Prompt Evacuation	6
Insufficient Access	2
No Available Key to Locked Door	4
Insufficient Emergency Training	6
No Second Means of Egress	5
Not Indicated	29

* Adapted from Tables 1–6, *A Study of Penal Institution Fires*, National Fire Protection Association, 1977, pp. 1–3.

Forty-five, or 87 percent, of the 52 fires studied were determined to be incendiary. The predominant motives were:

1. To increase chances of escape.
2. To cause malicious damage as a protest against conditions.
3. To show force during a riot.
4. To commit suicide.

In all of the intentionally set fires for which the ignition source could be identified, the ignition source was a match, smoking material, or a cigarette lighter. These are the ignition sources most readily available to inmates.

The data clearly defined the "model" correctional facility fire as incendiary in origin, starting in a cell, involving bedding or clothing, and generating intense heat and dense smoke that quickly spreads to corridors and adjoining areas, significantly hampering the safe evacuation of inmates and staff.

THE SYSTEMS APPROACH TO FIRESAFETY IN CORRECTIONAL FACILITIES

In dealing with the firesafety problem in correctional facilities, two concerns need to be continually addressed: (1) security for the public, and (2) safety from fire for the inmates. The systems approach to firesafety helps to achieve these two objectives through a systematic analysis of each problem area and application of available technologies. Firesafety should be an integrated subsystem of building design, construction, and operation. This method of designing firesafety is known as the systems approach or systems concept of firesafe building design. A system organizes interacting components in such a way that, working together, they can perform a predetermined function or reach a specified objective. The use of the systems approach first requires that firesafety goals be clearly identified. These goals should describe the amount of protection a building should provide its occupants, contents, and operations, and the methods used to achieve these goals should work collectively to reach and maintain a prescribed level of firesafety. These goals should be quantified wherever possible.

The systems approach can be illustrated by the Simplified Firesafety System for Correctional Facilities. (See Fig. 9-6B.) As shown by the box at the top of the diagram, the four objectives of the Simplified Firesafety System are:

Life Safety: The protection of life is paramount in any fire situation. Occupants can be protected by providing a safe area of refuge from the fire within the building or by providing a safe and readily available path of travel to the outside.

Property Protection: The protection of property is also of concern, particularly with today's higher replacement costs. Some areas may be considered more important than others—for example, a control room versus a storage room—and therefore require more protection.

Limited Downtime: This involves providing protection to limit the fire damage so the area can be quickly returned to use. For example, the loss of a storage area may not seriously offset the prison's operation, as trailers or other temporary storage units can be quickly obtained. However, loss of the kitchen area could seriously affect a prison's operation.

Security: The primary function of a correctional facility is to maintain a secure perimeter. Firesafety must be compatible with this function.

These objectives are achieved through a series of goals: ignition control, fuel control, occupant protection, detection and suppression activities, and planning and training operations. For the firesafety objectives to be achieved, each of these goals must be successfully reached. Some methods to attain each goal are listed beneath the goal as shown by Figure 9-6B. For example, for the goal of ignition control the several methods available to reduce ignitions include: (1) control smoking materials, (2) control electrical ignition source, and (3) use alternate grievance procedures to reduce intentionally set fires.

Because different goals do not impact equally on the ultimate level of fire protection, the concept of equivalent protection (also known as trade offs) may be necessary. For example, under certain conditions NFPA 101 permits higher flame spread materials for use as interior finish if automatic sprinklers are provided; or the allowable floor area limit can be increased if fire resistive rather than protected, noncombustible construction is used. The systems approach provides for analysis of all the interacting components of the firesafety system so alternate means of protection can be evaluated. This evaluation may include initial costs, operational and maintenance costs, and impact to operations.

Ignition Control

Ignition control means eliminating unnecessary heat sources so that a fire cannot be ignited either accidentally or intentionally. Unfortunately, total ignition control is nearly impossible to establish and is therefore the least effective fire defense measure. Sources of ignition are nearly always available because cigarettes and matches are, in most cases, available to inmates, and electrical appliances are sometimes allowed in cells. However, an effort must be made to control unwanted or unnecessary ignition sources. The types of lighters or matches can be specified and the overall use of electrical power must be carefully supervised. The number of heat producing appliances such as toasters, hot plates, and space heaters must be controlled by the facility administrator to prevent overloading circuits. Adequate electrical outlets should be provided if electrical devices are allowed in the cell, to discourage the use of extension cords.

Fuel Control

Fuel control means controlling the type, arrangement, and burning characteristics of potential fuels. These fuels may include the building structural system, interior finish materials, and combustible contents such as furnishings and inmates' personal property. Assuming that fires will be ignited, they can be controlled by limiting the type, quantity, and arrangement of the fuels available. By containing the fire and limiting the speed at which fire develops, the impact on life safety from a fire lessens. Slow burning fires give people and suppression equipment time

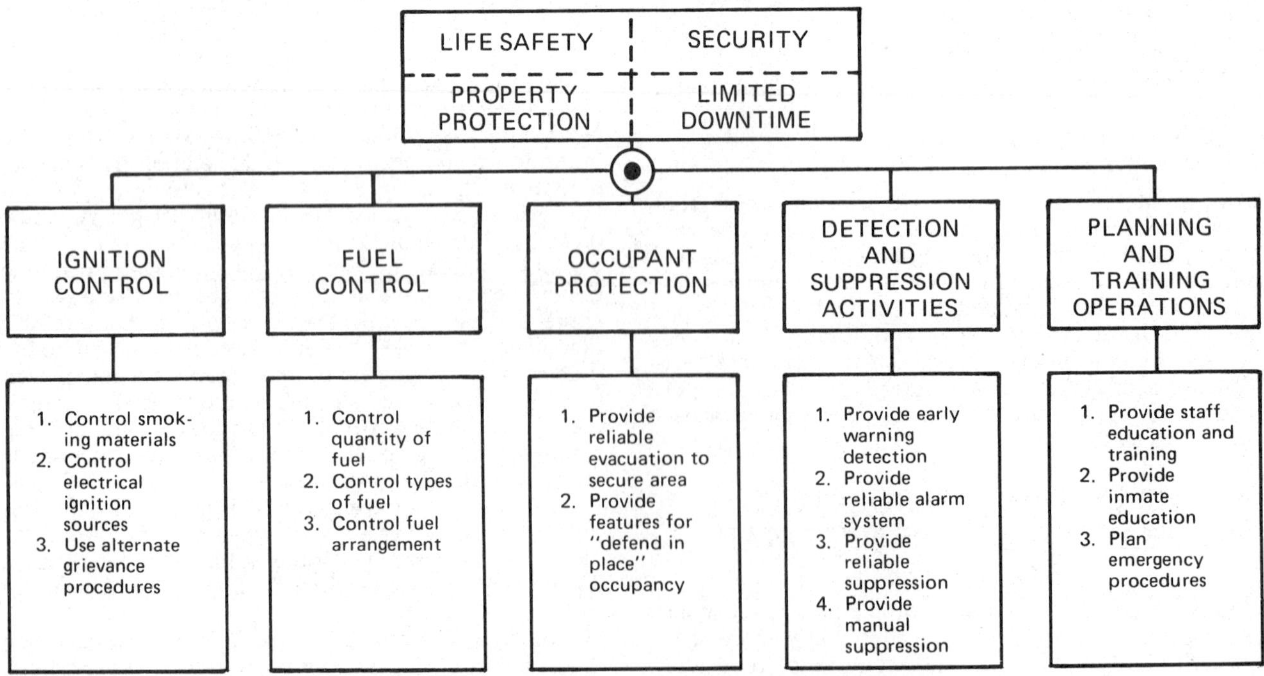

FIG. 9-6B. Simplified firesafety system for correctional facilities.

to react. However, fast fires—those with a rapid rate of smoke development, flame spread, and production of toxic gases—have been the traditional contributor to fire deaths in correctional facilities. Ordinary combustible materials such as low density fiberboard ceilings, thin plywood paneling, and some plastics used in furnishings and interior finish can contribute to fast fire development.

Achieving the goal of fuel control means keeping fuel quantity at a minimum, controlling the type of fuel, eliminating those fuels that develop fast fires, and separating fuels from readily available ignition sources. One part of fuel control is to govern furnishings such as mattresses. Since mattresses and bedding materials are the fuel sources most readily available to inmates, they are the ones most frequently involved in fires. Certain types of polyurethane foam mattresses have been a significant fuel source in many prison fires due to fast fire development, high heat release, and heavy smoke production typical of the mattresses.

Mattresses can be made to resist ignition from cigarettes, which is a slow, low heat ignition source. The federal *Flammability Standard for Mattresses* (FF4-72) requires that the mattress resist ignition by cigarettes placed at 18 specified locations. This test does not include sheets or blankets and does not indicate the resistance to ignition from open flames, nor the intensity of the resulting fire. A series of tests on mattresses by the U.S. Department of Agriculture (JCPF 1977) concluded that "The data show that mattresses containing polyurethane foam easily comply with the standard (FF4-72) where cigarettes are the igniting source, however, such mattresses present a significant hazard where the igniting source is an open flame. Mattresses containing 100 percent polyurethane foam usually burn vigorously until little char remains. Burning in such cases is accompanied by the release of copious amounts of black smoke."

In 1977 the National Bureau of Standards (NBS) conducted fire tests of mattresses and issued a report "Combustion of Mattresses Exposed to Flaming Ignition Sources/Part I—Full Scale Tests and Hazard Analysis." This report categorized mattresses into four groups in order of safety as follows:

Group A: Mattresses that did not exceed any of the tenability criteria for the duration of the 30 minute test. This group included two treated cotton batting mattresses.

Group B: Mattresses that only exceeded the smoke obscuration criterion. Two neoprene mattresses were in this category.

Group C: Mattresses that exceeded all tenability criteria but did not cause full room involvement. This group included three polyurethane foam core mattresses and one of mixed fiber construction. (The best performing of the polyurethane mattresses was associated with a multiple life loss prison fire.)

Group D: Mattresses that exceeded all criteria. Included were one styrene-butadiene latex foam core and one polyurethane foam core mattress. The latex mattress was associated with a multiple life loss fire in a health care institution.

These tests indicate that the best mattresses for use in correctional facilities are those with padding material of cotton treated with boric acid. Mattresses of cotton treated with boric acid pass FF4-72 and perform well when exposed to open flame ignition sources. Such mattresses will burn, but at a much slower rate than urethane foam mattresses (NFPA 1977b). Some promising new materials are entering the market for mattress padding, including treated neoprene. However, since there is no national standard for open flame ignition testing of mattresses, each jurisdiction must evaluate new mattress materials carefully and thoroughly.

Inmate possessions and furnishing in the cell must be controlled in order to limit the size of fire that can be expected. Books, clothing, and other combustible personal property allowed in cells should be stored in closable metal lockers or fire resistant containers. Combustible decorations should be prohibited unless they have been treated with flame retardants. Draperies, curtains, or other decorative and acoustical materials should be noncombustible or rendered and maintained flame resistant according to NFPA 701, *Standard Methods of Fire Tests for Flame-Resistent Textiles and Films.*

Interior finish has been a major factor in several prison fires where multiple deaths have occurred. Interior finish materials contribute fuel to a fire, may accelerate room flashover, and can result in rapid fire spread. Materials that have high flame spread rates, that would contribute substantial quantities of fuel to a fire, or produce hazardous concentrations of smoke or noxious gases should not be used. This would include plastic wall finish material, fiberboard, thin paneling, and untreated plywood. (See Table 9-6B.) Interior finish alternatives with good fire characteristics include paneling treated to provide Class A interior finish (flame spread of less than 25), painted steel, gypsum board, or masonry block.

Occupant Protection

Occupant protection means providing life safety to building occupants in case of fire. This may be accomplished by either evacuation to a secure area or by defending in place. This goal of fire protection is the most controversial one in the firesafety system for correctional facilities because the evacuation aspect is often interpreted by administrators and staff to mean escape. While this is not a valid assumption, means of egress does directly affects security.

With reference to evacuation of inmates, most detention and correctional facilities can be grouped into two general types, based on their physical arrangement. In the first group are larger institutions that have an inner secure area such as a ball field or courtyard. It is possible to evacuate inmates rapidly from the fire area to this area of refuge and still maintain security. In the second category are the smaller city or county facilities where there is no separate area of refuge. In these facilities, which lack an enclosed courtyard or other area in which to evacuate inmates and maintain security, it becomes necessary to defend inmates in place from fire.

A defend in place occupancy is a workable and accepted means of providing firesafety for building occupants and is commonly used to protect persons in hospitals and nursing homes. The technology exists to defend inmates in place so that only a few cells need to be evacuated—possibly only the cell of fire origin.

A defend in place occupancy as well as an occupancy using evacuation to a secure area of refuge needs a reliable means of egress. Although means for defending the occupants in place is provided, it is always considered necessary to provide reliable exits. According to NFPA 101 this means that two distinct paths of travel must be provided from each cellblock or area. Exits should be remote from each other so that both cannot be readily blocked by a single fire. In addition, travel distances must not exceed those defined by NFPA 101; normal distance recommended by this Code is 100 ft (30 m), but the length can vary depending upon the protection provided. Exit access needs to be free and unobstructed, and should not be circuitous or require travel through several rooms, long corridors, or many stairs. The travel distance can be increased up to 50 percent if the building is sprinklered. An exit access that requires travel through a room or area containing a fire hazard higher than usual for the occupancy violates the principles of safe egress. Neither are dead end corridors good practice, since a fire in a dead end between an exit and an occupant can prevent the occupant from reaching the exit. Exitways need to be continuously illuminated, with a provision for emergency lighting. Exits and the path to them should be clearly marked. However, the NFPA 101 permits the elimination of exit signs in inmate housing areas where no visitors are permitted. The rationale for the elimination of exit signs is that the inmates cannot exit until someone unlocks the doors and the inmates are generally familiar with the door locations.

TABLE 9-6B. Flame Spread of Some Building Materials*

Material	Flame Spread
Ceilings	
Glass-fiber sound-absorbing blankets	15 to 30
Mineral-fiber sound-absorbing panels	10 to 25
Shredded wood fiberboard (treated)	20 to 25
Sprayed cellulose fibers (treated)	20
Walls	
Aluminum (with baked enamel finish on one side)	5 to 10
Asbestos cement board	0
Brick or concrete block	0
Cork	175
Gypsum board (with paper surface on both sides)	10 to 25
Northern Pine (treated)	20
Southern Pine (untreated)	130 to 190
Plywood paneling (untreated)	75 to 275
Plywood paneling (treated)	10 to 25
Red Oak (untreated)	100
Red Oak (treated)	35 to 50
Floors	
Carpeting	10 to 600
Concrete	0
Linoleum	190 to 300
Vinyl asbestos tile	10 to 50

* For a comprehensive list of flame spread and smoke developed ratings for proprietary materials, see the current edition of *Building Materials Directory* published by Underwriters Laboratories, Inc.

Compartmentation is an important aspect of occupant protection, whether inmates are evacuated or defended in place. Walls, ceilings, and floors of fire compartment boundaries must be fire resistant. Openings must be protected with self-closing fire rated doors. The degree of fire protection provided by these elements depends on the degree of hazard presented by the occupancy (fuel loading), the type of building construction (i.e., fire resistive, noncombustible, and the function served (such as, a load bearing or nonload bearing wall).

Controlling smoke means controlling the type and quantity of materials used in the building construction and building contents, and providing a means of restricting smoke movement within the building. Dilution, which is reducing smoke concentration by introducing massive quantities of uncontaminated air into a building, is not always satisfactory when used alone, since smoke generation can be of such magnitude that adequate dilution may not be obtained. Prudent use and placement of automatically opening vents can provide some relief from smoke and the products of combustion in small, one story buildings. However, the confinement of smoke can best be achieved (1) by providing a physical barrier such as a wall with self-closing doors and dampers that restrict smoke movement, and (2) by using a pressure differential across the physical barrier to prevent smoke from entering the nonfire area. The building's heating, ventilating, and air-conditioning (HVAC) system can be designed to provide this pressurization. Physical barriers alone are usually not effective in controlling smoke movement due to the many penetrations in these barriers and the uncontrolled pressures created by the fire, the building HVAC, and/or the weather conditions.

Since the primary objective of detention and correction facilities is security, reliable locking systems are essential. Problems with locking systems, however, have played a major role in many correctional fire tragedies. Therefore, whatever their type, locking systems must function reliably in emergency situations, whether a facility is designed to evacuate its inmates or defend them in place. Individual key locks have proven to be the most unreliable for several reasons, including the time required to unlock each door, loss of keys during emergency, key breakage during emergency, heat and smoke prohibiting entry into cell block, and confusion due to number of keys required.

Detection and Suppression Activities

Detection and suppression activities deal directly with the fire—detecting its presence, sounding an alarm to alert occupants, and inhibiting fire growth and ensuring its extinction. For this goal to be achieved, the following need to be provided: (1) an early warning fire detection system, (2) a reliable alarm system, and (3) a reliable fire suppression system.

There are three separate phases of detection and suppression activities, each of which can be performed automatically or manually. (See Table 9-6C.) Automatic and manual systems can also be combined, e.g., detection can be provided by automatic smoke detection (smoke detectors) installed throughout the facility with manual fire fighting by fire brigade personnel using a standpipe hose. However, any system requiring manual intervention requires that someone must make a decision and requires time for human action. The human element is generally the most unreliable part of a firesafety system.

The effectiveness of fire detection and suppression systems is dependent upon the rate of the fire growth and the time from ignition until detection and, eventually, suppression activities are begun. If the fuel is capable of producing a fast fire, as with padded cells for example, then the extinguishing agent, usually water, must be applied to the fire very quickly after ignition. This is very difficult to do with manual fire suppression even if early warning fire detection such as automatic smoke detectors is available. Automatic sprinklers provide the most reliable method of fire detection and suppression without the need for human intervention. With an automatic sprinkler system, the extinguishment phase begins almost simultaneously with detection and alarm, a response difficult to achieve with manual suppression.

Sprinklers have been and are being installed quite successfully in detention and correctional facilities throughout the U.S. and Canada. Sprinklers lessen the likelihood of panic by occupants as they attempt to evacuate, and possibly increase the efficiency of egress. They also provide fire service personnel with a better opportunity to begin rescue and extinguishment operations. Automatic sprinklers can reduce property loss and limit downtime. In addition, sprinklers offer several economic advantages: insurance premiums may be reduced with the installation of sprinkler protection; many building codes offer tradeoffs when sprinklers are used, thereby reducing building costs; and sprinklers can be installed at a discount in correctional facilities by inmates under the supervision of a qualified fire protection engineer. Sprinklers may also result in fewer persons being evacuated, reduce the burden for manual fire fighting, and, in effect, increase security.

TABLE 9-6C. Stages of Detection and Suppression

Stage	Manual	Automatic
Detection	Inmate or guard detects fire	Sprinklers detect fire
Alarm	Guard verbally transmits information to other guards	Activates water flow switch
Suppression	Guards use extinguishers to suppress fire	Sprinklers begin extinguishment immediately

When automatic detection and suppression systems are used, they need to be inspected and tested regularly. In addition, the alarm systems should be connected directly to the local fire department. However, smoke detectors are often not connected to automatically sound the building alarms nor automatically cause the transmission of the alarm to the fire department. Smoke detectors are sensitive devices and are prone to needless or nonfire emergency alarms. Manual suppression ability should include portable fire extinguishers of the proper type and number, readily accessible fire hose stations for use by correctional officers, and a means to stretch fire hoses through double security gates (sallyports) so gates can be closed and security maintained.

Planning and Training Operations

Planning and training means conducting training activities among inmates and staff and planning emergency operating procedures. To achieve this goal, the following methods are used: (1) staff education and training, (2) inmate education, (3) planning emergency procedures, and (4) conducting drills. Planning and training require little capital investment in equipment, but can make a big difference in reducing the impact of a potentially disastrous fire, especially in a system where fire defenses are marginal. In addition, planning and training are important to maintain a high level of security during a fire.

Because firesafety should be everyone's concern, both staff and inmates should be involved in a firesafety program. Each shift should practice the fire emergency plan at least quarterly, and new employees should be routinely briefed on the fire emergency plan. Inmates should be instructed in emergency procedures. Safety information may also be included in an inmate information booklet.

The firesafety program should be designed to meet the needs of the facility; the size and age of the buildings, the security classification of the facility, and its proximity to municipal fire departments should be considered. All programs should include a short description of the background and evolution of the problem; how to recognize, prevent, and reduce fire hazards; information on the fire protection technology available for application at the facility; hands on training for staff and inmate fire brigades; emergency operating procedures; and identification of potential problems that affect firesafety.

Fire brigades should be equipped and trained to deal effectively with a fire emergency. Brigade responsibility and functions will vary with the size of the facility and the size and location of the nearest fire department, but should include calling the fire department, safeguarding lives, providing manual suppression to control the fire until the fire department arrives, and protecting equipment. In large facilities, personnel should be organized into separate fire fighting teams assigned to predetermined areas.

Each building in the correctional facility should have a written fire emergency plan detailing staff action during a fire emergency. The plan provides a guide for evaluating the particular problems and coordinating the response of the fire brigade, fire department, staff, and inmates. Emergency plans should be simple yet comprehensive, specific, flexible, and workable. Preparing the plan can be accomplished in five steps:

1. Define the potential fire protection problems in the particular correctional facility.
2. Set objectives for what can be accomplished with the plan in an emergency.
3. Determine the facility's capability for controlling an emergency situation.
4. Define the roles of the responding agencies, especially the fire department and fire brigade.
5. Put the information into written form.

The local fire department should be involved in the formulation of the fire emergency plan and briefed on building conditions, contents, and fire fighting facilities within the complex. Fire departments with correctional facilities within their jurisdictions should have their own prefire plans ready for a fire emergency.

Bibliography

References Cited

JCPF. 1977. "Resistance of Mattresses Containing Boric Acid Treated Cotton Batting to Open Flame Ignition." *Journal of Consumer Product Flammability*. Vol 4. June 1977. pp 169-188.

NFPA. 1977a. *A Study of Penal Institution Fires*. National Fire Protection Association, Quincy, MA.

NFPA. 1977b. *Fire Safety Bulletin on Penal Institutions*. National Fire Protection Association. Quincy, MA.

NFPA Codes, Standards, Manuals, Recommended Practices and Manuals. (See the latest *NFPA Codes and Standards Catalog* for availability of current editions of the following documents.)

NFPA 13, *Standard for the Installation of Sprinkler Systems*.

NFPA 72A, *Standard for Installation, Maintenance and Use of Local Protective Signaling Systems for Guard's Tour, Fire Alarm and Supervisory Service*.

NFPA 72E, *Standard on Automatic Fire Detectors*.

NFPA 90A, *Standard for the Installation of Air Conditioning and Ventilating Systems*.

NFPA 101® *Code for Safety to Life from Fire in Buildings*.

NFPA 251, *Standard Methods of Fire Tests of Building Construction and Materials*.

NFPA 255, *Standard Method of Test of Surface Burning Characteristics of Building Materials*.

NFPA 701, *Standard Methods of Fire Tests for Flame-Resistant Textiles and Films*.

Additional Readings

Best, Richard, "The Seminole County Jail Fire," *Fire Journal*, Vol. 70, No. 1, Jan. 1976, pp. 5-10.

"Combustion of Mattresses Exposed to Flaming Ignition Sources/Part I—Full Scale Tests and Hazard Analysis," *Report No. NBSIR-77-1290*, National Bureau of Standards, Washington, DC, 1977.

Demers, David P., "Penal Institutions: Firesafety Versus Security," *Fire Journal*, Vol. 73, No. 2, Mar. 1979, pp. 60-62

Demers, David P., "Fire in Prisons—42 Die in Maury County, TN, Jail; 21 Die in Saint John, NB, Detention Center; 5 Die in FCI, Danbury," *Fire Journal*, Vol. 72, No. 2, Mar. 1978, pp. 29-42.

DiMeo, Michael, "Fire in Pennsylvania Prison," *Fire Journal*, Vol. 70, No. 3, May 1976, pp. 70-75.

Federal Bureau of Prisons, "Board of Inquiry into the Danbury Fire," U.S. Department of Justice, Washington, DC, July 7, 1977.

Flammability Standard for Mattresses (FF4-72), U.S. Department of Commerce, Washington, DC, June 1973.

Hartson, C., and Rickey, G., "Ohio's Institutional Firesafety Program," *Fire Command*, Vol. 46, No. 3, Mar. 1979, pp. 34-35.

Knoepfler, N. B., et al., "Mattresses that Resist Open-Flame Ignition," *Fire Journal*, Vol. 72, No. 3, May 1978, pp. 36-40.

Lathrop, James K., "No Freedom for John—2 Die in Youth Correctional Center, Cranston, RI," *Fire Journal*, Vol. 69, No. 3, May 1975, pp. 16-17.

Lee, B. T., "Effect of Ventilation on the Rates of Heat, Smoke, and Carbon Monoxide Production in a Typical Jail Cell Fire," *NBSIR 82-2469*, National Bureau of Standards, Gaithersburg, MD, Mar. 1982.

National Institute of Corrections, *Fire Safety in Correctional Facilities*, available from N.I.C., U.S. Department of Justice, Washington, DC, 1980.

Scala, J. C., *Fire Safety in Correctional Institutions*, National Fire Protection Association, Quincy, MA, 1981.

"Seven Die in Gas Explosion at a Penitentiary Farm Building," *Fire Journal*, Vol. 73, No. 3, May 1979, pp. 25-28.

Tuck, Charles A., *Analysis of Three Multiple Fatality Penal Institution Fires*, National Fire Protection Association, Quincy, MA, 1978.

HOTELS

Gerald R. Kirby

Hotels are defined as buildings in which 15 or more sleeping rooms are rented to transient residents. They may also be identified as motels, inn, clubs, guest quarters, and suite hotels.

These facilities are unique in that they almost always combine several different occupancies under one roof. In addition to sleeping quarters (residential), hotels usually provide space for ballrooms, meeting rooms, exhibition halls and restaurants (assembly); shopping areas (mercantile); and offices and commercial establishments (business). Larger hotels in major cities may have several floors available for sale as condominium apartments. Many large hotels also contain industrial type laundries and dry cleaning facilities, as well as kitchens, which utilize large broilers, ovens, and deep fat fryers.

OCCUPANCY CHARACTERISTICS

Since hotel guests are transients, special consideration must be given to the potential threat to their life safety from fire. For example, the occupants of the residential portion would be sleeping in unfamiliar surroundings and could possibly become disoriented when trying to evacuate under conditions of heavy smoke. Likewise, those persons in the various ballrooms, lounges, and restaurants could become disoriented due to low level lighting, crowd size, and unfamiliarity with escape routes.

Furthermore, it is not unusual to find the residential portion of a hotel full at the same time the banquet halls, ballrooms, lounges and restaurants are operating at capacity. This situation could tax the life safety facilities of a hotel and present a logistical problem for fire fighters trying to locate and extinguish a fire, while simultaneously attempting rescue and evacuation. Added to this are the inherent problems in evacuating high rise buildings with their sealed windows, which are usually inaccessible from fire department ladders, and potential smoke problems in egress halls and stairwells. Moreover, hotels contain a heavy amount of textiles, bedding, carpeting, wall coverings, and lacquered furniture. Even though these items are treated for flammability, they still present a problem due to their large volume.

Presently, building codes are addressing these problems, requiring pressurized stairwells and corridors that provide a smokefree means of egress, as well as fast acting sprinkler heads that will stop a fire in its incipient stage. Thus, the possibility of a large fire is decreased and propagation of smoke is held to a minimum. The codes also mandate smoke removal equipment, as well as smoke and heat detectors in high rise buildings. Smoke detectors automatically shut down air circulation while activating smoke control, stair pressurization, and voice annunciated evacuation messages (BOCA 1984).

In addition to building code requirements, the potential fire and life safety problems that plants of this type present are addressed in the various occupancy sections of NFPA 101, *Life Safety Code®* (hereinafter referred to as NFPA 101). For example, ballrooms, restaurants, lounges, and exhibition halls are dealt with in the chapter of NFPA 101 dealing with assembly occupancies, whereas the portion of the hotel providing sleeping rooms is dealt with in the residential section of the *Code*. Since provisions of NFPA 101 are different for these two major occupancies, the model building codes require that they be separated by construction that will hold a fire in either area for at least two hours. This requirement results in a barrier, which essentially separates the occupancies into different buildings.

LIFE SAFETY FEATURES

Moreover, NFPA 101 provides guidelines for minimum standards regarding various life safety features in the building(s), such as those set forth below:

Means of Egress

This section deals with the type, capacity, number, arrangement, illumination, marking, and measurement of travel distances to exits. These provisions ensure a safe means of egress from the building in emergencies.

Mr. Kirby is fire protection specialist for the Marriott Corporation, based in Washington, DC.

Protection

Enclosures or other means of protection from vertical openings such as stairways, elevator shafts and atriums are required by this section of NFPA 101 to impede the spread of smoke and heat within the building. This section also deals with protection from other potential hazards such as boilers, transformers, and other service equipment subject to explosion.

Interior Finish

Due to the flammability and toxicity of modern day fabrics and furnishings, flame and smoke spread is regulated in this section of NFPA 101.

Detection and Alarm Systems

Inasmuch as hotels are occupied by persons who are usually unfamiliar with the building, provisions are set forth in the *Code* requiring early warning of fire. In order to provide this early detection and warning, the *Code* mandates smoke detection, manual alarm stations, and annunciator panels, as well as central voice command centers in high rise buildings.

Corridor Walls

Another provision of NFPA 101 requires rated corridor walls and self closing doors in guest room areas. (See Figs. 9-7A and 9-7B.) Even though door closing mechanisms are one of the most effective life safety features of the *Code*, they are one of the least expensive to install and maintain. Several recent multiple-death hotel fires have illustrated the importance of self-closing doors (Klem 1982).

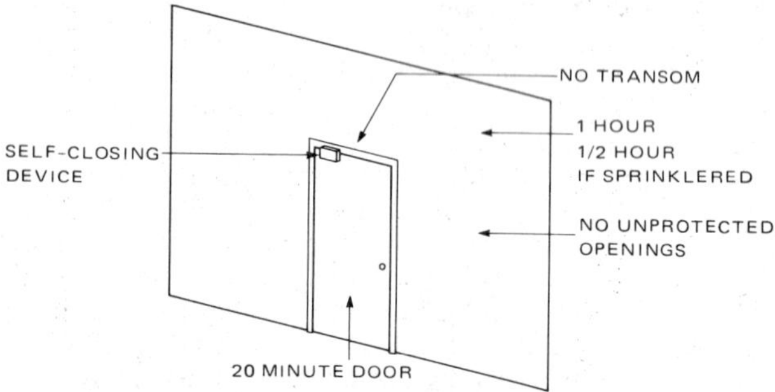

FIG. 9-7A. Provisions for the protection of guest rooms and in new hotels. The need for the self-closing 20 minute door cannot be overemphasized. There are no exceptions or options to the self-closing door.

In most cases, fire and smoke can be contained in the room of origin as long as the door remains closed. This is especially true in sprinklered rooms. Recent tests in San Francisco, CA indicate that fast acting sprinkler heads are not only effective in suppressing fire and minimizing smoke in guest rooms, but they will, in conjunction with door closers, contain smoke and heat in the room of origin, leaving the egress hallways tenable.

Staff Training

Even though these built in life safety features are paramount in the maintenance of a firesafe hotel, a well trained emergency organization is of equal importance.

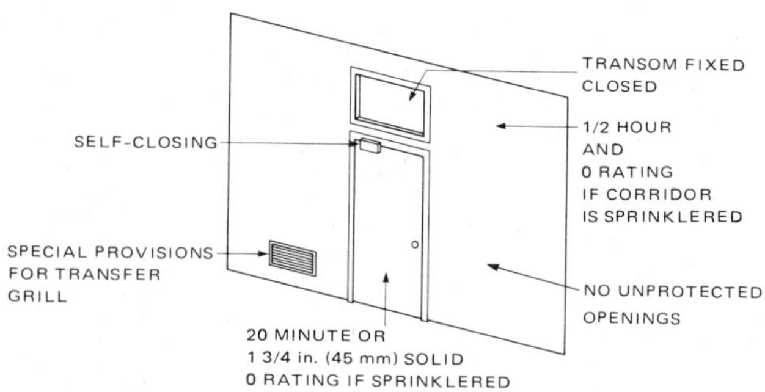

FIG. 9-7B. Provisions for protection of guest rooms and in existing hotels. The need for the self-closing doors cannot be overemphasized. There are no exceptions or options to the self-closing devices. When sprinkler protection is provided, the wall need not be substantial enough to resist the passage of smoke.

Similar to a fire brigade, the emergency organization consists of on duty employees who are trained to conduct fire prevention inspections and tests of life safety systems on a regular basis. In addition, they hold regularly scheduled fire and disaster drills. In the event of an emergency, they are responsible for notification of the fire department, incipient stage fire suppression, evacuation, and sprinkler/fire pump monitoring, as well as salvage and overhaul after extinguishment. All organization members are also trained in first aid and are certified to perform emergency medical treatment, such as cardiopulmonary respiration (CPR) (Marriott undated).

An effective emergency organization, along with automatic sprinklers, compartmentation, effective egress design, and detection and alarm systems will provide hotel guests with a degree of protection that will ensure that the hotel in which they sleep is safe from fire.

Bibliography

References Cited

BOCA. 1984. *The BOCA Basic/National Building Code*. Building Officials and Code Administrators International, Inc., Country Club Hill, IL.

Klem, T. J., and Best, R. L. 1982. "Four Hotel Fires: Lessons Learned." *Fire Journal*. Vol 76, No 6. p 72.

Marriott. undated. *Hotel Fire Emergency Organization*. Marriott Corporation, Washington, DC.

NFPA Codes, Standards and Recommended Practices and Manuals. (See the latest *NFPA Codes and Standards Catalog* for availability of current editions of the following documents.)

NFPA 101, *Code for Safety to Life from Fire in Buildings and Structures*

Additional Readings

Bell et al., *Investigation Report on the Westchose Hilton Hotel Fire*, National Fire Protection Association, Quincy, MA, 1982.

Best, Richard and Demers, David, *Investigation Report on the MGM Grand Hotel Fire*, National Fire Protection Association, Quincy, MA, 1982.

Bryan, John J., *An Examination and Analysis of the Dynamics of*

Bryan, John J., *An Examination and Analysis of the Dynamics of the Human Behavior in the MGM Grand Hotel Fire*, Revised ed., National Fire Protection Association, Quincy, MA, 1983.

Coakley, Deirdre, et al., *The Day the MGM Grand Hotel Burned*, Lyle Stuart, 1982.

Coté, R. M. et al., *Investigation Report of the Ramada Inn Central Fire*, National Fire Protection Association, 1984.

Daly, Thomas G., "Hotel Firesafety—An Update," *Fire Journal*, Vol 79, No 1, Jan. 1985, pp 56-58.

Hotel Fires: Behind the Headlines, National Fire Protection Association, Quincy, MA, 1982.

Operation Life Safety, *The Operation San Francisco Smoke/Sprinkler Test*, International Association of Fire Chiefs, Washington, DC.

APARTMENT BUILDINGS

Kenneth Bush

Apartment buildings are identified as those structures containing three or more living units with independent cooking and bathroom facilities, whether designated as apartment house, tenement, condominium, or garden apartment. Apartments differ from other types of residential occupancies by the provision of individual cooking facilities, the arrangement and number of sleeping rooms, and the number and transient nature of the occupants. In many cases, the definition of apartment building extends to other structures with other outward appearances such as townhouse or condominium arrangements.

OCCUPANCY CHARACTERISTICS

The adequate protection of residential buildings and their occupants from the effects of fire has historically been a major concern to fire protection professionals. This problem is compounded by the construction of multiple dwelling units. Such construction not only adds building size and numbers of occupants, but also adds major complexity due to building design, construction, and fire protection features.

Nearly one-third of the housing units in the United States today consist of multiple dwelling arrangements (Bureau of Census 1983). Of all dwelling units currently within the U.S., 31 percent are renter occupied. These facts, coupled with several other factors, including a higher demand for less expensive housing, the revitalization of large cities, a greater interest in cooperative and condominium type units, and the increasing trend of single person households, clearly indicate an increase in the numbers, types and complexity of the apartment building.

Due to many factors, the fire loss statistics in residential occupancies have reached an alarming rate in all categories of statistical data. Figures have indicated that residential occupancies in the U. S. rank first in the number of civilian fire deaths and injuries, in the amount of property loss due to fires, and in the number of multiple death fires (NFPA 1984). From 1978 to 1982, almost 90 percent of the multiple death fires reported and 85 percent of the associated deaths were in residential properties.

While apartment buildings account for approximately 32 percent of the total housing units, fires in these dwellings account for only 17 percent of the fire deaths and injuries and only 17 percent of the property loss. In addition, these figures indicate that such fire loss statistics have actually decreased from the previous reporting period in 1981. The percentage of civilian deaths in apartment building fires has dropped by 11.3 percent and the property loss has decreased by 15 percent. While the number of civilian injuries has increased by 10.6 percent, this figure indicates that more occupants, although possibly injured, are given more of a chance of escaping an apartment fire. This dramatic turnaround is most likely due to the improvements in the types and methods of building construction, protection of vertical openings, regulation of interior finish, maintenance of required means of egress, adequate building compartmentation, and installation of fixed detection and alarm systems, and, in many cases, fixed suppression systems such as automatic sprinklers. The continued progress in the reduction of fire statistics would seem directly dependent upon the continued efforts of effective code formulation, effective code adoption, and, perhaps most importantly, adequate and complete code enforcement.

OCCUPANCY HAZARDS

As previously indicated, most fire related deaths and injuries, and in particular multiple deaths in the U. S. occur in residential properties. This fact is based upon the high number of buildings classified under this occupancy type, as well as the number of persons who are routinely housed or associated within these types of structures, and the sometimes careless attitude of the occupants. In addition, the basic scenario for a residential fire, i.e., sources of ignition, type and amount of combustible loading, age and condition of the occupants at the time of the fire, and time of fire ignition, add significantly to the potential for life loss and property destruction. Yet, while many studies have demonstrated these facts, such factors have proven to

Mr. Bush is fire protection engineer, Maryland State Fire Marshal's Office.

be very difficult to regulate, and are often not covered by the present codes and standards making process.

Most studies of fatal fires have indicated that cigarettes, particularly when associated with upholstered furnishings in the living area and bedding materials in the sleeping area, continue to be the leading cause of fatal residential fires. Electrical malfunctions or overloaded circuitry follow, but drop almost 15 percent in the number of ignition scenarios. Recent fire experience and the fact that millions of woodstoves and portable electric and kerosene heaters have been sold in the U.S. during the past several years lead to the conclusion that heater related fires will continue to be responsible for hundreds of fire deaths each year. This conclusion is supported by data that show a sharp increase in incidents during the months of November to April, when use of some heating device is required in most areas of the U.S.

More than two-thirds of the fire deaths reported in residential occupancies occurred between the hours of 8:00 p.m. and 8:00 a.m., the time when most people are in their homes and asleep. Many of these deaths occurred when the fire had a significant head start—more than 40 minutes before discovery, for 38 percent of the deaths. In addition, studies of recent fire victims show that these victims have been incapable of escape due to their age, physical or mental impairments such as a high blood alcohol level, or a general lack of knowledge of what to do in the event of a fire.

With the variety of these types of residences, as well as the architectural desire to make buildings esthetically pleasing yet functional, an interesting combination of fire protection demands is developed. In many cases, building designs, space considerations, and parking arrangements create major access problems, particularly for fire fighting apparatus. (See Fig. 9-8A.) Frequently, long hose lays and blocked roadway accesses create new and added problems to fire fighting operations besides existing problems such as weather, personnel deficiencies, and hampered water supplies. (See Fig. 9-8B.) Architectural designs often utilize false partitioning and veneered construction, giving an unrealistic impression of the type and magnitude of the actual building construction. (See Fig. 9-8C.) Although NFPA 101, *Life Safety Code*® (hereinafter referred to as NFPA 101) makes no special requirements for the construction of these buildings, many such requirements are found and enforced through the model building codes, as well as through many local codes and ordinances currently in effect. An additional problem arises in the location and combination of apartments with other occupancy types, particularly above and behind mercantiles and businesses. (See Fig. 9-8D.) Where such multiple classes of occupancy occur in the same building, and are so intermingled that separate safeguards are impractical, means of egress facilities, construction, protection, and other safeguards shall comply with the most restrictive life safety requirements for the occupancies involved.

FIG. 9-8A. Many apartment building designs and parking arrangements create major access problems and hinder fire fighting operations.

FIG. 9-8B. Long hose lays and blocked access routes create major fire apparatus placement and fire fighting operation problems.

FIG. 9-8C. Architectural designs often utilize false partitioning and veneered construction creating an unrealistic impression.

LIFE SAFETY CONCEPTS

The residential fire death problem in all of its aspects is very complex, and several strategies must be presented to help reduce these deaths. The installation and proper maintenance of smoke detectors within the sleeping areas provide an effective, low cost and widely available method of early warning so that evacuation, containment, and

FIG. 9-8D. Additional firesafety problems arise in the location and combination of apartments with other occupancy types, particularly above and behind mercantiles and businesses.

suppression can be initiated during the incipient stages of a fire.

The general public must be taught to protect themselves and others within the residential unit from fire through the safe use and installation of alternative heating devices, the proper storage and handling of flammable liquids, and the identification and correction of home fire hazards. In addition, stringent flammability standards for upholstered furniture, such as those previously issued by the U.S. Consumer Product Safety Commission (CPSC) for children's sleepwear and mattresses, will provide long term firesafety benefits. Specially designed systems that are effective in terms of life safety should be installed to provide a high level of protection to residential occupants, particularly where large numbers of persons are present such as in a high rise apartment building, or where the occupants' ability to escape is otherwise hampered, such as in apartments for the elderly.

Although such firesafety precautions are not always practical, particularly within an individual dwelling unit, a more concentrated effort must be made to confine the fire and its effects to a well defined, limited area of the structure. This task is accomplished through the adoption of accepted firesafety codes, such as NFPA 101. Since many current laws and regulations prevent fire inspection authorities from entering individual living units, adequate fire protection features must be designed and constructed during the initial phases of the project. The regulation of interior furnishings provides only a partial solution to the firesafety problem, because such consumer items are generally in use for many years before being replaced. In addition, individual habits and attitudes of building occupants can never be regulated or otherwise altered.

Because apartment buildings, and residential occupancies in general, present such a diversified collection of building size, configuration, and resident population, many of the regulatory codes pay particular attention to construction designs and fire protection features. NFPA 101 covers the presence of hazards such as cooking and heating equipment, and the degrees of familiarity of the occupant with his/her living space. NFPA 101 has a unique method of regulating the construction and design of apartment buildings and is based upon the type and extent of automatic fire detection and fire suppression equipment provided. Recognizing equivalency provisions found elsewhere in NFPA 101, the residential subcommittee established four equivalent schemes to provide a high degree of flexibility. These designs are considered equal in providing a minimum level of life safety for apartments. Some systems have additional protective capability, permitting higher building heights and larger building areas.

This overall approach provides one of the first system design attempts to be codified. Whereas a total system would have many design approaches, this is a more limited or bounded system in that only four different approaches are available. Yet, a designer can identify an appropriate design from the four approaches based on the building size, height, and arrangement. This allows the designer to put together a safety approach that best fits the building, rather than fitting the building to a single codified design criterion.

FIG. 9-8E. Apartment buildings are classified as ordinary hazard occupancies which include moderately rapid burning of building contents and considerable smoke generation.

Under the special definitions found in Chapter 4 of NFPA 101 the contents of living units in residential occupancies fall into the ordinary hazard classification. Characteristics of these hazards include moderately rapid burning and considerable smoke generation, with the possible presence of toxic combustion products. (See Fig. 9-8E.) Under this degree of hazard classification, neither poisonous fumes nor explosions are to be feared, even though all smoke contains some toxic fire gases. Under normal conditions, no unduly dangerous exposure should develop during the period necessary to escape from the fire area. Generally, this classification represents conditions found in most buildings. Even with the introduction of plastics, foamed materials and other combustible materials in furnishings, the amount and configuration of such materials present a lower degree of hazard than in other occupancies where they may be sold, stored, or displayed. NFPA 13, *Standard for the Installation of Sprinkler Systems*, would classify the contents of apartment buildings as "light" for the design of extinguishing systems. This difference in classifications is based on the threat to life safety assumed in the ordinary classification versus that assumed where the extinguishing capability of the automatic sprinkler system results in the light hazard classification.

In designing the number and types of required exits for apartment buildings, NFPA 101 determines the occupant load on the basis of one person per 200 sq ft (18.6 m^2) of gross floor area, or the maximum probable population of any room or section under consideration, whichever is greater. There are many examples, such as a dormitory type occupancy, that may produce a greater occupant load. However, even though some portions of these buildings are densely populated, the building as a whole may not necessarily exceed one person per 200 sq ft (18.6 m^2) of gross area owing to the space taken for toilet facilities, hallways, closets, and living rooms. These requirements do not preclude the need for providing adequate exit capacities from concentrated use and sleeping areas based on the maximum probable population rather than on the design figure. The capacities of required exits are as specified in NFPA 101.

Although seemingly contrary to the basic code requirements of two distinct, remote means of egress from all locations, several exceptions to this rule are found within the requirements for apartment occupancies. While the traditional interior exit corridor arrangement commonly found in larger apartment buildings and condominiums requires the two remote exit approach, several other options in apartment building configurations permit other arrangements. The common townhouse arrangement, which provides an exit directly to the street or yard at ground level or by way of an outside or enclosed stairway serving that apartment only, permits the front door to be the only required exit; therefore, a rear door, which is often provided, would not be required and could be a sliding door to a porch or patio.

Another exception is a basic design approach for garden apartments where the apartment entrances are enclosed around a single protected stairway. This single stairway is usually open to the exterior, or is glass enclosed. It must be separated from the building by construction of at least 1 hour fire resistance rating. The doors should be similarly treated. Interestingly, the basic design of a typical garden style apartment has influenced other chapters of NFPA 101. Previous editions of the Code have required that all exit stairways continuing beyond the floor of discharge be interrupted at that level by partitions, doors, or other effective means to make clear the direction of egress to the street. However, a recent exception to this requirement has been added allowing these stairways to continue one-half story beyond the level of discharge without such physical interruption. This is a typical exit arrangement found in garden style apartments that are constructed with apartments or service areas on a ground floor or terrace level. Based on this new exception, stairways serving these levels need not be interrupted at the floor of exit discharge for firesafety reasons.

In addition to the number and types of required exits, the arrangement and travel distance to these exits are regulated by the general requirements of Chapter 5 of NFPA 101 with due considerations to the levels of fire protection features within the building. As found in other occupancies, a minimum of 50 percent of the exits from upper floors must discharge directly to the exterior. Appropriate illumination must be provided for public spaces, hallways, stairways and other means of egress. Emergency lighting is required for apartment buildings of more than 12 living units or four or more stories in height, except in townhouse arrangements where every living unit has a direct exit to the outside at grade. Approved exit signs are required in all apartment buildings with more than one exit; however, appropriate discretion must be used in the placement of such signs where exits are visibly obvious, or all occupants are always familiar with the building.

Because a higher degree of transient occupancy of some apartments is a definite possibility, special exiting considerations are made within the individual living units in addition to those found in the public access spaces. All individual living units below the seventh floor must comply with the minimum provisions for operable exterior openings to be used as a secondary means of escape. Not only do these openings require a minimum size, but they also require easy and accessible operation. In apartment buildings higher than six stories, these exterior opening requirements may be waived provided that (1) an approved means of smoke control is provided throughout the building, or (2) the building is sprinklered. A penthouse living unit is permitted in all apartment units except those for the elderly, and it may have regular, curved or spiral staircases connecting up to one level above or below the entrance to the apartment. Additional requirements prohibit any door in an apartment building, including individual unit doors, from being locked against egress while the building is occupied. This permits a door to be locked so that it can be opened from within the building but not from the outside. Ordinarily, dead bolts, double cylinder locks, and chain locks would not meet these provisions. Several tragic multiple death fires have occurred because a key could not be found to unlock the door or the chain was not removed.

In apartment buildings, the protection of vertical openings such as stairways, elevator shafts, and atriums follows the requirements found in NFPA 101. (See Fig. 9-8F.) Except in totally sprinklered buildings, smokeproof towers are required in apartments over six stories. This limits the possibility of smoke contaminating the exits and ensures a safe, protected means of egress. The protection from hazards, as well as the classification of interior finishes, are also based upon the levels of fire detection and/or protection with the building.

Perhaps the best form of fire protection in a residential occupancy is an early warning. All living units within an apartment building must have approved automatic smoke detection. These detectors are usually located in hallways that have access to the sleeping areas. In multiple level units, a detector should also be located at the tops of stairways. Detection systems must be installed in accordance with basic engineering practices. Special consideration should be given to room configuration, air movement, or stagnant air pockets that would reduce detector efficiency; and cooking areas, fireplaces, or other normally smoke producing appliances that would cause

FIG. 9-8F. The vertical spread of fire and other products of combustion is of primary concern to most firesafety codes.

an excess of nuisance alarms. All detectors required for installation in multiple dwelling units must be powered by house current. In addition, where a single apartment unit needs more than one detector, the detectors must be electrically interconnected. Note that this type of detection system is required in addition to the requirements for any other detection or suppression system.

Manual fire alarm systems are required in apartment buildings according to the size and height of the building and the evacuation capabilities of the occupants. As building designs progress in height and complexity, additional requirements are found for special features such as annunciator panels, voice communication systems and automatic transmission of building fire alarms. Voice communication systems are often used as part of a multifaceted system that allows the fire department to contact building occupants during a fire. The extent of automatic sprinkler protection is again governed by the size and height of the building, as well as by which of the four protection options as discussed in chapters 18 and 19 of NFPA 101 is chosen. The Code requires portable fire extinguishers to be located in hazardous areas only, not throughout the entire building. Many building owners place these extinguishers within the individual living units under the direct responsibility of the occupant. The design, installation, and maintenance of all such fire detection and protection systems are based on referenced fire protection standards as well as on sound engineering judgment.

When fires cannot be confined to the area of origin, the means of egress must be sufficiently protected to allow enough time for safe evacuation of all building occupants. This basic concern for providing safety for the occupant in a room during a fire led to a minimum corridor wall construction either to block fire movement into a room from the corridor, or to block fire in a room from entering the corridor. Doors protecting openings in these enclosures must provide a level of protection commensurate with the expected fuel load in the room and the fire resistance of the corridor wall construction. Fuel load studies conducted by the National Bureau of Standards have shown that residential occupancies will have fuel loads in the 20 to 30 minute range. Contrary to the general rule for door assembly considerations, the requirements for doors in residential occupancies apply to the leaf only, with a 1 ¾ in. (45 mm) solid bonded wood core door considered equivalent to a door with a 20 minute fire protection rating. The door rating requirement was meant to establish a minimum quality of construction, not to require a complete listed fire door assembly. However, the requirement that doors between apartments and corridors be self-closing will definitely lead to a significant reduction in fatalities. Studies have shown that fire spread was due mainly to the door being left open as the occupant fled the fire. In other cases, people have died after opening the door to a fully involved room or after causing a room to become fully involved by opening a door and introducing oxygen to the fire. Under both circumstances, a door equipped with spring loaded hinges or a door closer would have prevented smoke or fire from spreading down the corridor and exposing other occupants.

During a fire emergency, the evacuation of all residents from a large apartment building is an inefficient, time consuming process. This problem is compounded with the introduction of larger and taller structures, complex and intricate building designs, and a larger population of residents who are incapable of rapid evacuation, especially by stairways. Realistically, the most expedient movement of a sizable number of residents would be horizontally. The introduction of required smoke barriers and horizontal exits serve three purposes fundamental to the protection of building occupants by: (1) limiting the spread of fire and fire produced contaminants, (2) limiting the number of occupants exposed to a single fire, and (3) providing for horizontal evacuation by creating an area of refuge on the same floor level.

Areas of refuge are required in apartment buildings based upon the height and area of the building and the separation distance of the exit stairways. Since the majority of apartment residents are assumed capable of vertical evacuation through the use of stairways, the location of stairways provides both the required area of refuge as well as the required means of egress. Where travel time to stairways is increased due to excessive horizontal distances and/or where a majority of residents have reduced mobility, specialized areas are required on the same floor level. It is important to remember that the requirement for smoke barriers and horizontal exits is also dependent upon the fire protection features of the building, as well as the building configuration, height, and anticipated mobility of its occupants.

Typically, all building service equipment and utilities in apartment buildings are required to be designed and installed in accordance with sound engineering practices and all applicable NFPA standards. The use of a public corridor as part of the supply, return, or exhaust air system is not permitted; in fact, these corridors must be pressurized against the movement of smoke in certain apartment building designs. The standard requirements for the reliability of the means of egress, fire retardancy for interior furnishings and decorations, and the proper maintenance and testing of fire protection equipment also apply equally to apartment occupancies. However, many of these operating features such as fire evacuation plans and fire exit drills, have been deleted due to the nontransient nature of a majority of the building occupants.

As is clearly evident from the statistical data previously provided, the fire problem with apartment buildings, and within residential occupancies as a whole, is complex, significant, and not easily solved. The ever increasing demand for adequate, yet low cost, functional, and esthetically pleasing housing units has required an ever increasing demand for adequate, safe, and economical fire protection. The problem intensifies with the demand for higher and larger buildings, a higher density of occupants, and a demand for more personal services for building occupants. Although the total solution to these fire problems will undoubtedly require much future study and professional opinion, the obvious short term solution through effective code formulation, adoption, and enforcement is the immediate responsibility of every firesafety professional.

Bibliography

References Cited

Bureau of Census. 1983. *Annual Housing Survey: 1981—General Housing Characteristics.* U.S. Dept. of Commerce, Bureau of Census, Washington, DC.

NFPA Codes, Standards and Recommended Practices and Manuals. (See the latest *NFPA Codes and Standards Catalog* for availability of current editions of the following documents.)

NFPA 13, *Standard for the Installation of Sprinkler Systems.*

NFPA 101, *Code for Safety to Life from Fire in Buildings and Structures.*

Additional Readings

Berl, Walter G., and Halpin, Byron M., "Human Fatalities From Unwanted Fires," *Fire Journal*, Vol. 73, No. 5, Sept. 1979, p. 105.

Derry, Louis, "Fatal Fires In America: How They Happen, Where They Happen, How To Stop Them," *Fire Journal*, Vol. 72, No. 5, Sept. 1979, p. 67.

Fire In The United States, 4th ed., Federal Emergency Management Agency, Washington, DC, 1982, pp. 25-27.

Jones, Jon C., "1982 Multiple-Death Fires In The United States," *Fire Journal*, Vol. 77, No. 4, July 1983, p. 10.

Karter, Michael J., Jr., "Fire Loss in the United States During 1982," *Fire Journal*, Vol. 77, No. 5, Sept. 1983, p. 44.

Lathrop, James K., ed., *Life Safety Code Handbook*, 3rd ed., National Fire Protection Association, Quincy, MA, 1984.

Public Health Service: Vital Statistics of the United States, 1978, National Vital Statistics Division, National Center for Health Services, U.S. Government Printing Office, 1978.

LODGING OR ROOMING HOUSES

Gregory Haley

Lodging or rooming houses include buildings in which separate sleeping rooms are rented, and sleeping accommodations are provided for a total of 15 or fewer persons on either a transient or permanent basis (with or without meals), but without separate cooking facilities for individual occupants. It should be noted that:

1. One- and two-family dwellings may have a room or rooms rented to others not accommodating more than a total of three persons.
2. Sleeping accommodations for more than 15 persons immediately classifies the facility as a hotel.
3. Centralized dining/cooking is permitted, but individual cooking facilities immediately classify the facility as an apartment building.
4. Occupants must be capable of self-preservation. The presence of four or more occupants incapable of self-preservation would immediately reclassify the facility as health care or possibly as a board and care facility.

Based on these restrictions, the classification "lodging or rooming house" is effectively limited in scope to facilities providing sleeping accommodations for four to fifteen persons without separate cooking facilities for individual occupants capable of self-preservation.

OCCUPANCY CHARACTERISTICS

Very few new buildings are being classified as lodging or rooming houses. Most are commonly encountered as existing facilities in converted dwellings, motels, or small apartment houses. Mixed occupancies are generally not encountered except in the case of mercantiles located on the lower floor(s) with sleeping rooms located above. Such an arrangement is not permitted by code unless:

1. The dwelling occupancy and exits are separated from the mercantile occupancy by construction having a fire resistance rating of at least 1 hour.
2. The mercantile occupancy is protected by automatic sprinklers.

Lodging or rooming houses can be in buildings of any construction type permitted by local building codes. There are no minimum construction requirements identified in NFPA 101, *Life Safety Code®* (hereinafter referred to as NFPA 101). With a majority of lodging or rooming houses existing in converted dwellings or sharing space with mercantile operations, construction is generally limited to Type III or V.

Fuel loading is generally considered light, while hazard of contents is considered "ordinary," a trait consistent with a majority of residential occupancies. Unfortunately, extremely fast spreading fires with delayed detection are not uncommon to this type occupancy based on construction characteristics and interior features as noted below.

Occupancy Conversion or Remodeling: In these processes, existing wall and ceiling finishes are removed, exposing combustible structural components. Concealed combustible spaces may be created with the installation of drop ceilings and new wall finishes. These same wall and ceiling finishes, although limited to Class A, B, or C, typically exceed this limitation and significantly affect fire spread and smoke generation characteristics.

Lack of Firestopping: As a result of a lack of building codes or enforcement at the time of construction, many older buildings lack firestopping.

Unprotected Vertical Openings: These penetrations are not limited to stairs. They include shafts, ducts, heat grates or simply old floor openings that were never covered. Some vertical openings exceed three stories in height. Each represents a path for fire and smoke movement and may compromise an exit which is considered protected.

Building Services: In lodging or rooming houses, building services generally are limited to electrical, heating, and possibly air conditioning. Primary consideration should be given to the proper maintenance of heating equipment and the prevention of electrical circuit overloading. Some tenants may be permanent, so consideration should be given to tenant spaces in an attempt to prevent overloading of receptacles or misuse of extension cords.

Mr. Haley is Director of Property Loss Control CIGNA Corporation, Philadelphia PA.

Storage Areas For Personal Belongings: Storage areas are a common problem with apartments and they are occasionally encountered in lodging or rooming houses. These storage bins usually contain tenants' nonvaluable personal items which are often combustible and sometimes flammable. Storage areas are normally located in basements, under stairs, in rooms off egress corridors, etc. and while they are commonly readily accessible, they present a serious hazard.

Supplemental Heating Devices: Inadequate heating can lead to the use of supplemental heating devices including portable electric gas and kerosene space heaters. Each represents a specific hazard and has its own accompanying set of controls. One specific code requirement prohibits locating a stove or heater where its malfunction would block escape in case of fire.

LIFE SAFETY CONSIDERATIONS

Due to the limited number of sleeping accommodations and the lack of separate cooking facilities for individual occupants, lodging or rooming houses are viewed (from a life safety standpoint) as falling somewhere above one- and two-family dwellings and below dormitories, hotels, and apartment buildings. NFPA 101 requirements for this occupancy do not address interior finish, compartmentation, emergency lighting, and other requirements, but rather center on the two key elements essential to life safety.

Means of Escape

For life safety purposes, safe egress from the building is always of paramount concern. In lodging or rooming houses, the egress system consists of "means of escape" which, by definition, is a way out of a building or structure that does not conform to the strict definition of means of egress but does provide an alternate way out.

This term is separate and distinct from means of egress and accepts a reduced level of egress for this specific occupancy in relation to all other occupancy classes, while still maintaining a higher level of egress than required in one- and two-family dwellings.

Sleeping rooms above and below the level of exit discharge are required to have two separate means of escape, one of which should be either an enclosed interior stairway, exterior stairway, or horizontal exit. A fire escape is acceptable for existing buildings only. This primary means of escape must provide a safe path of travel from the building without exposing the occupants to an unprotected vertical opening, except for completely sprinklered buildings three stories or less in height. (See Figs. 9-9A, 9-9B, and 9-9C.)

The specified means of escape are not required to meet the requirements of Chapter 5 of NFPA 101, with the exception of width, stair riser, and tread. Details on the second means of escape are provided under Section 22-2 of NFPA 101, and include either a door or stairway providing unobstructed travel to the outside of the building at street or ground level, or an outside operable window. Note that a second means of escape is not required, provided:

1. The room has a door leading directly outside to grade.
2. The building is completely sprinklered.

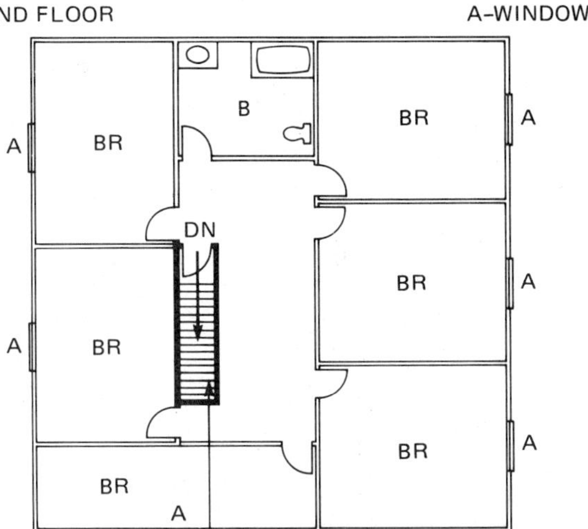

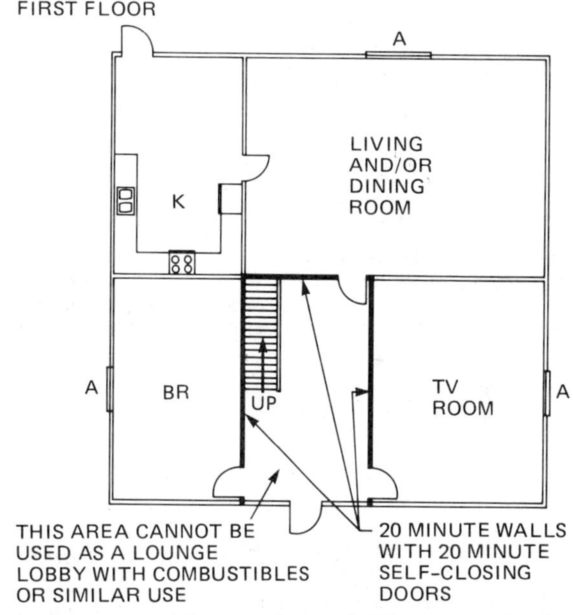

FIG. 9-9A. Means of escape from a lodging or rooming house. This example shows an enclosed interior stairway which discharges directly outside. Access to the interior stairway is via an interior corridor which is not exposed to an unprotected vertical opening. The second means of escape could be via windows. The enclosure at the bottom of the stairway can be done several ways; this is only one example.

Sleeping rooms located on the level of exit discharge are required to have two unspecified means of escape, one of which may be an operable window meeting the requirements of Section 22-2.. One exception is for rooms with doors leading outside to grade, which are considered sufficient themselves.

No required means of escape may be through an area not under the control of the occupant exiting the building or through a door subject to locking.

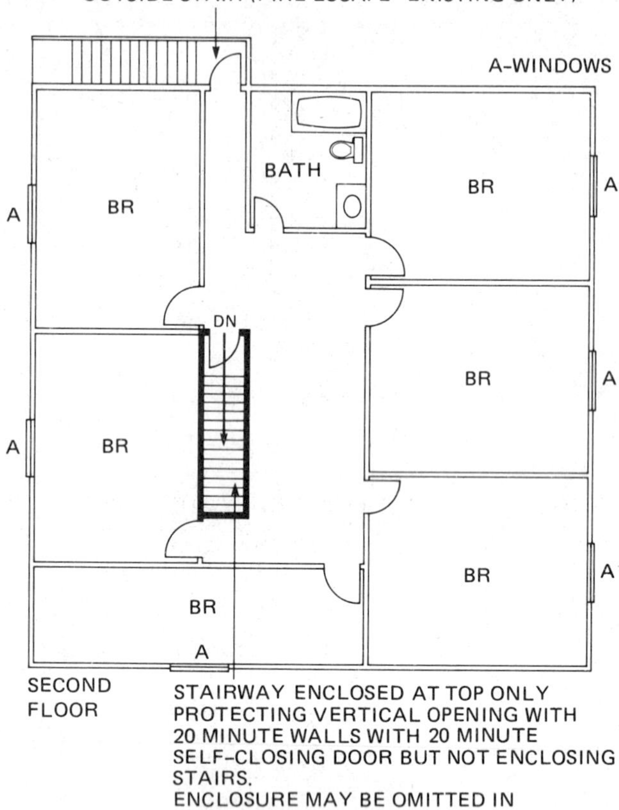

FIG. 9-9B. Means of escape from a lodging or rooming house. This example shows an exterior stairway. The exterior stairway is accessed via an interior corridor which is not exposed to an unprotected vertical opening. The second means of escape is via windows.

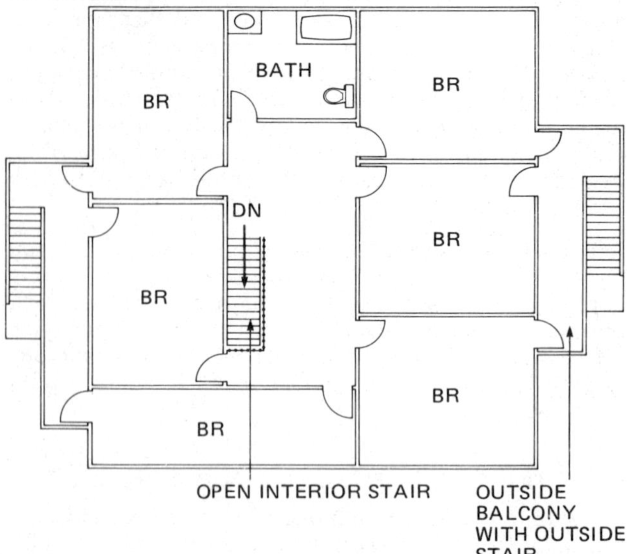

FIG. 9-9C. Means of escape from a lodging or rooming house. This example shows an outside stairway with exterior exit access. Access is not through the interior corridor which is exposed by the open stairway. The second means of escape is via the open interior stairway.

DETECTION AND ALARM SYSTEMS

Detection and alarm systems in residential occupancies play a critical role in the early warning and evacuation of occupants.

A manual fire alarm system installed in accordance with NFPA 72A, *Standard for the Installation, Maintenance or Use of Local Protective Signaling Systems for Guard's Tour, Fire Alarm and Supervisory Service*, is required in all cases.

Smoke detectors, installed in accordance with NFPA 74, *Standard for the Installation, Maintenance and Use of Household Fire Warning Equipment*, and powered by house electrical service are required on each floor level, including basements, but excluding crawl spaces and unfinished attics. When activated, the detectors should initiate an alarm audible in all sleeping areas. This requirement is generally satisfied by interconnecting detectors with the manual fire alarm system.

Past performance of alarm systems in alerting building occupants, particularly in residential occupancies, has generally been less than satisfactory. Assurance that the manual fire alarm system will be audible in all sleeping rooms is critical.

Bibliography

NFPA Codes, Standards and Recommended Practices and Manuals. (See the latest *NFPA Codes and Standards Catalog* for availability of current editions of the following documents.)

NFPA 72A, *Standard for the Installation, Maintenance or Use of Local Protective Signaling Systems for Guard's Tour, Fire Alarm and Supervisory Service.*

NFPA 74, *Standard for the Installation, Maintenance and Use of Household Fire Warning Equipment.*

NFPA 101, *Code for Safety to Life from Fire in Buildings and Structures.*

Additional Readings

Haley, Gregory, "Lodging or Rooming Houses," *Conducting Fire Inspections*, National Fire Protection Association, Quincy, MA, 1982.

Lathrop, James K., ed., *Life Safety Code Handbook*, 3rd ed., National Fire Protection Association, Quincy, MA, 1984.

ONE AND TWO FAMILY DWELLINGS

Harry L. Bradley

One and two family dwellings are defined as residential structures in which not more than two families reside. Dwelling units may be attached or detached from one another. Typical styles would include detached single family dwellings, townhouses, and duplexes. There are many different designs such as a one story ranch, two story colonial, split level, and three level townhouse. Wood frame and masonry are the most common construction types. U.S. Department of Census statistics indicate that approximately 80 percent of the United States population lives in one and two family dwellings (GPO 1981).

Additional information on residential firesafety will be found in Section 16, Chapter 3, "Household Warning Systems;" and Section 18, Chapter 4, "Residential Sprinkler Installation."

Fire and Life Safety Hazards

According to estimates from the Federal Emergency Management Administration (FEMA) in 1981, the leading causes of fire in one and two family dwellings were smoking, heating, cooking, incendiary, electrical distribution systems, appliances, and children playing (NFPA 1983). It is evident that a wide range of fire hazards are associated with residential occupancies.

Heating equipment such as electric, gas, or oil fired furnaces or boilers are found in most homes. It is important that heating equipment be properly installed and maintained. Also, space heaters have become very popular in an effort to conserve energy and fuel costs. Kerosene, electric, and gas fueled space heaters are commonly used and portable space heaters present many firesafety concerns. For these heaters, clearance to combustibles must be maintained, and proper fueling procedures must be performed outside. The correct fuel must always be used and stored properly.

Wood stoves and fireplaces are the most common types of solid fuel burning heating equipment. Many fires in this equipment occur because of improper installation. NFPA 211, *Standard for Chimneys, Fireplaces, Vents and Solid Fuel Burning Appliances*, specifies installation requirements for these types of equipment.

Cooking is the second leading cause of fire, most likely because of the number of people cooking meals each day in one and two family dwellings. Most fire instances involve unattended food cooking or carelessness while using cooking equipment.

Incendiary fires have increased dramatically over the past ten years; the substantial increase is partially due to improved detection and reporting of intentionally set fires. Dwellings are set afire for a number of reasons—homeowners may seek to destroy their own property to collect insurance money; vandalism and attempts to cover up a robbery or other crime are also motives for arson.

Careless smoking accounts for 32 percent of fire fatalities in one and two family dwellings (NFPA 1983) and is the leading cause of fatal fires in the U.S. A common fire scenario involves a cigarette dropped onto upholstered furniture or bedding. The fire sometimes smolders for several hours, then develops into open flames.

The household electric distribution system consists of a fuse or circuit breaker panel, wiring throughout the house, junction boxes, light switches and electric outlets. Fires can occur when an electric circuit is overloaded by placing too many appliances on a single circuit. Frayed wires, broken switches and outlets should be replaced. All electric repairs and alterations should be performed by a qualified homeowner or a licensed electrician.

Washing machines, clothes dryers, dishwashers, hot water heaters and kitchen accessories are examples of home appliances. Improper use, installation, or maintenance of these appliances may cause a fire. Proper precautions should be taken with all home appliances.

With numerous hazards and sources of ignition within the home, unattended children are always a concern—children playing with matches is one of the primary causes of dwelling fires.

Fire statistics for 1983 indicate that 3,825 civilian fire deaths occurred in one and two family dwellings in the U.S. (Karter 1984), accounting for 65 percent of the total fire deaths for the year. Why does this type of occupancy have the highest percentage of fire deaths? As indicated by the leading fire causes described above, many different activities occur daily within the home. Two-thirds of the fire deaths occurred between the hours of 8 p.m. and 8 a.m.

Mr. Bradley is fire protection engineer, State Fire Marshal's Office, Baltimore, MD.

(NFPA 1983). Occupants are most vulnerable while sleeping. Few victims die from burns, most die as a result of smoke inhalation. Ordinary combustibles within the home give off carbon monoxide gas as a product of combustion. Carbon monoxide poisoning causes confusion, unconsciousness, and eventually death.

Life Safety Concept

The philosophy of life safety and fire protection for one and two family dwellings is based on evacuation of the building in the event of a fire. The defend in place concept in other types of occupancies, which utilizes fixed fire protection systems and fire resistant construction, does not lend itself to one and two family dwellings. To allow residents sufficient time to evacuate, a fire must be detected at the incipient stage, the occupants must be alerted and know what action to take, and adequate means of escape must be provided.

Early detection can be accomplished by proper installation and use of residential type smoke detectors. NFPA 74, *Standard for the Installation, Maintenance, and Use of Household Fire Warning Equipment*, recommends smoke detectors be provided on each floor level of the dwelling.

Once the occupants are alerted, prompt action to evacuate must be taken. A written fire plan should be made for each dwelling, showing two means of escape from every room, especially the bedrooms. The plan should be practiced by the entire family. Programs such as NFPA's "EDITH," (Exit Drills In The Home), provide an outline of how to write and practice a home fire plan.

Adequate means of escape from each room should be provided in accordance with the requirements of NFPA 101, Life Safety Code®. Two means of escape are required from each room. One must be a door or stairway, the second can be an operable window. (See Figs. 9-10A and 9-10B.)

Residential type automatic sprinkler systems provide a high degree of protection for one and two family dwellings. These relatively new types of automatic systems are not widely installed at present; however, the technology is gaining in acceptance and use.

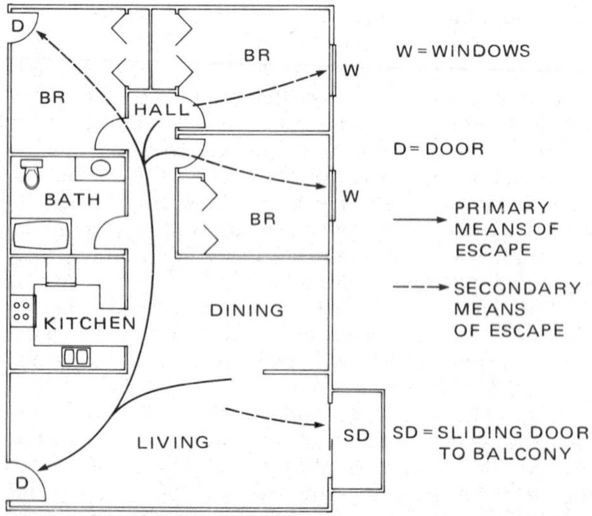

FIG. 9-10A. Means of escape from a three bedroom single family home. The bedroom in the upper left has no window, but has a door directly to the outside. The means of escape from the other bedrooms is through windows which must comply with the requirements of NFPA 101.

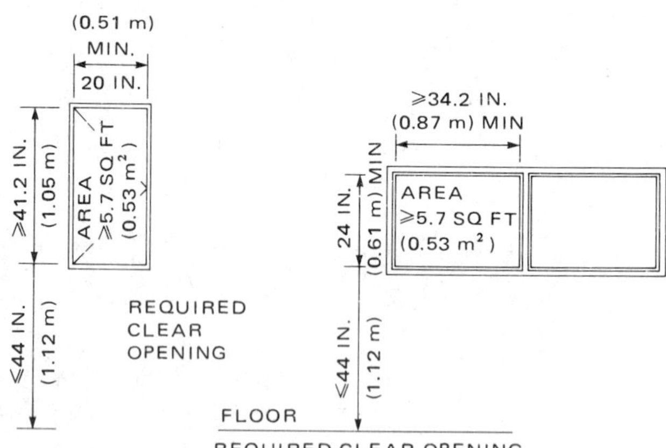

FIG. 9-10B. Minimum size and dimensions of outside windows used as a second means of escape.

Bibliography

References Cited

GPO. 1981. *Annual Housing Survey 1981*. Bureau of Census. U.S. Government Printing Office. p xix.

Karter, Michael J. Jr., with Gancarski, Joan L. 1984. "Fire Loss in the United States." *Fire Journal*. p 50.

NFPA. 1983. *The 1984 Fire Almanac*. National Fire Protection Association, Quincy, MA. p 178.

NFPA Codes, Standards, Recommended Practices and Manuals. (See the latest *NFPA Codes and Standards Catalog* for availability of current editions of the following documents.)

NFPA 74, *Standard for the Installation, Maintenance, and Use of Household Fire Warning Equipment*.

NFPA 101, *Code for Safety to Life from Fire in Buildings and Structures*.

NFPA 211, *Standard for Chimneys, Fireplaces, Vents, and Solid Fuel Burning Appliances*.

Additional Readings

Budnick, E. K., "Estimating Effectiveness of State-of-the-Art Detectors and Automatic Sprinklers on Life Safety in Residential Occupancies," *NBSIR 84-2819*, National Bureau of Standards, Gaithersburg, MD, 1984.

Fang, J. B., and Breese, J. N., "Fire Development in Residential Basement Rooms," *NBSIR 80-2120*, National Bureau of Standards, Gaithersburg, MD, 1980.

Gomberg, A., et al., "Evaluating Alternative Strategies for Reducing Residential Fire Loss—The Fire Loss Model," *NBSIR 82-2551*, National Bureau of Standards, Gaithersburg, MD, 1982.

Home Safe Home, P7838, Factory Mutual System, Norwood, MA.

Issen, Lionel A., "Single-Family Residential Fire and Live Loads Survey," *NBSIR 80-2155*, National Bureau of Standards, Gaithersburg, MD, 1980.

Munson, M. J., and Ohls, J. C., *Indirect Costs of Residential Fires*, Federal Emergency Management Administration, United States Fire Administration, Washington, DC, July 1979.

"A Study of Fatal Residential Fires," National Fire Protection Association, Quincy, MA.

"A Study of One-and Two-Family Dwelling Fires," *NFPA No. FR75-1*, National Fire Protection Association, Quincy, MA.

MOBILE HOMES AND RECREATIONAL VEHICLES

Michael Slifka

In 1983, the Department of Housing and Urban Development (HUD) ordered "mobile" homes be referred to as "manufactured" homes. Since the term "manufactured" has previously been applied to modular housing construction in the residential housing industry, this chapter will use the term "mobile" in its review of life safety and firesafety issues that apply to "mobile" housing. For discussion of the life safety and firesafety aspects of modular residential construction, refer to Chapter 10, of this section, "One and Two Family Dwellings."

MANUFACTURED (MOBILE) HOMES

Mobile homes are factory assembled structures that are transportable in one or more sections. Mobile homes are usually 8 ft (2.5 m) or more in width and range from 32 to 80 ft (10 to 24 m) in length. Each is built on a permanent chassis and can be used as a dwelling unit when placed on stabilizing devices (which may be piers and footings rather than permanent foundations) and connected to the required utilities. Due to the increasing cost of conventional housing and the improved quality of mobile homes, these homes become the choice of residence for almost 300,000 families each year.

Mobile home construction (body and frame design), plumbing systems, heating, cooling and fuel burning systems, and electrical systems are built to a special standard because of their uniqueness.

NFPA 501B, *Standard for Mobile Homes* (or ANSI A119.1) (hereinafter referred to as NFPA 501B), had been the guiding document on firesafety for mobile homes since 1968. Almost all states had adopted the NFPA/ANSI standard for mobile homes, and concerned federal agencies, such as the Federal Housing Administration and the Veterans Administration, have also used it in connection with purchases of mobile homes under loan programs they administered.

However, both the mobile home industry and the federal government were concerned over the lack of uniform enforcement of NFPA 501B between the states, counties, and cities. To assure a safe, low cost form of housing that would be built of uniform quality throughout the U.S., Congress passed the National Mobile Home Construction and Safety Standards Act of 1974 (Title VI of the Housing and Community Development Act of 1974). It established the U.S. Department of Housing and Urban Development (HUD) as the sole source of standards governing the construction of mobile homes in the U.S. HUD was also given the authority to establish the procedural and enforcement regulations necessary to assure that the construction standards were met by manufacturers of mobile homes.

The Act was fully implemented in mid 1976 with the National Conference of States on Building Codes and Standards (NCSBCS) named by HUD as the contract agent for assurring the uniform enforcement of the HUD version of NFPA 501B. (NFPA 501B was withdrawn as an NFPA standard after the *Federal Mobil Home Act* was implemented.) To date, approximately 2.7 million mobile homes have been built under the 1974 Act, adding to the estimated existing mobile home stock of 4.3 million built prior to mid 1976 (Budnick and Klein 1979; Klem 1978). With the infusion of "safer" new mobile homes into the U.S. housing stock, a dramatic effect has been measured with regards to life safety and fire incidents in post 1976 built mobile homes.

Pre 1976 mobile homes were reported to have a death rate of 3.2 per 100 fires (HUD undated). Post 1976 mobile homes built to the federal standard were reported to have a death rate of 1.0 per 100 fires, a greater than two-thirds reduction in the death rate. Similarly, the injury rate in pre 1976 homes per 100 fires was reported to be 5.7 compared to a 3.4 injury rate per 100 fires for post 1976 homes, a 40 percent reduction in the injury rate.

This dramatic downturn in the injury and death rates in fires of mobile home occupants is directly attributable to HUD uniformly applying an updated version of NFPA 501B as the core document for the portion of the HUD standard that applies to firesafety in mobile homes. Some of the pertinent features of the HUD standard that have produced this positive effect on mobile home firesafety are discussed in the following paragraphs.

Mr. Slifka is chief of safety, occupational health, and fire protection for the U.S. Veteran's Administration in Washington, DC.

The HUD standard requires that interior finish be regulated for flame spread characteristics, and in certain areas of the home, for combustibility. Firestopping is required in concealed wall and partition spaces. At least two exterior doors are required, and must be arranged to provide a means of unobstructed travel to the outside. The width of doors is specified, as is the ability to operate locking mechanisms from the inside to facilitate egress. Every room designed expressly for sleeping purposes, unless it has an exterior door to the outside, is required to have at least one outside window large enough to permit emergency egress, with the window openable without the use of special tools. Smoke detectors are mandated in each sleeping area to warn the occupants of any fire condition that might develop within the mobile home.

Utilities

Where LP-Gas is used for heating or cooking, these containers are limited in size, number, and location. They must be mounted on the "A" frame of the mobile home or in a vented compartment vaportight to the inside of the mobile home and accessible only from the outside. Requirements are included to assure secure mounting of the containers so they will not jar loose, slip, or rotate while the home is in transit. Special provisions are included to protect shutoff valves on containers while the mobile home is in transit of storage. LP-Gas regulators must be connected directly to the container shutoff valves or mounted on adjacent support brackets connected by listed high pressure connector to the valve. LP-Gas safety relief devices are required, and the discharge from these devices is regulated as to proximity to any opening in the mobile home. Allowable pressures of gas piping systems are also regulated, as are piping materials, routing, sizing, anchoring, and the provision of shutoff valves. Appliance connectors cannot run through walls, floors, ceilings, or partitions and specific gas piping leakage tests are specified which the manufacturer must perform before delivering.

Where oil is used as the heating fuel, specific requirements are included with regard to oil tanks, fill and vent pipes, liquid-level gages, and shutoff valves. Materials for oil piping systems are specified as to size, type of joints, couplings, grading, hangers, and tests for leakage.

Perhaps one of the most unique features of the HUD standard concerns the installation of heat producing appliances. Especially noteworthy is the requirement that all fuel burning appliances (except ranges, ovens, illuminating appliances, clothes dryers, solid fuel burning fireplaces and solid fuel burning fireplace stoves) be installed to provide for complete separation of the combustion system from the interior atmosphere of the mobile home. Combustion air inlets and flue gas outlets are required to be listed or certified as components of the appliance. The required separation may be obtained by the installation of direct vent systems (sealed combustion systems) or by the installation of the appliance within enclosures so the appliance combustion system and venting system is separated from the interior atmosphere of the mobile home.

Any forced air appliance and its air return system cannot allow a negative pressure to be created which will affect either its combustion air supply or that of any other appliance, nor act to mix products of combustion with circulating air. Other requirements with regard to heat producing appliances include special provisions relating to solid fuel burning fireplaces or fireplace stoves. Circulating air duct materials are specified as to their size and airtightness. Duct materials are regulated as to their design and combustibility. Special provisions are included for registers and grills, and additional special requirements for the use of ducts in expandable or multiple mobile home connections.

Lastly, in the area of heat producing appliances, the HUD standard specifies the maximum construction requirement, thickness, and gage of the metal ventilating hood to be installed over a kitchen range. Additionally, the HUD standard limits the flame spread of the wall finish and cabinet finish in the immediate vicinity of the kitchen range. This combination of requirements effectively isolates or minimizes the direct exposure of the wall surface and the cabinet surface in the nearest vicinity to a kitchen range in the event of a cooking surface fire.

The HUD standard includes specific requirements regarding electrical installations. It concentrates on those provisions which differ from provisions applicable to all buildings under NFPA 70, the *National Electrical Code®* (hereinafter referred to as NFPA 70). Coordination is maintained in both of the documents. Some of the special features relate to the compactness of mobile homes, the fact that they are factory assembled, lighter construction techniques commonly used, under-chassis wiring requirements, wiring of expandable and dual units, need for outdoor outlets, need for bonding of sometimes extensive amounts of noncurrent carrying metal parts, and the grounding of services and appliances.

HUD issued its first update (after nine years) to its standard in 1985. In the area of firesafety, due to the standard's success, HUD produced a "fine tuning" of the standard in lieu of wholesale changes. HUD lowered the egress window latch from 60 to 54 in. (1.5 to 1.4 m) as the maximum height above the floor. More importantly, HUD issued a stricter limit on ceiling interior finish flame spread (to a maximum rating of 75) and expanded its criteria on kitchen cabinet protection. Other minor changes were made including further clarification on firestopping construction techniques in concealed wall, ceiling, and floor construction, the recognition of 5/16 in. (8 mm) thick gypsum board as firestopping material and clarification of smoke detector placement instructions (MHI 1984).

For years, a companion standard to the HUD Mobile Home Construction and Safety Standards was NFPA 501A, *Firesafety Criteria for Mobile Home Installation, Sites and Communities*, ANSI A119.3 (hereinafter referred to as NFPA 501A), which was designed to assure the proper installation of mobile homes whether on private property or in mobile home parks. Special provisions in NFPA 501A cover land utilization, lot facilities, windstorm protection methods, mobile home accessory buildings and structures, permanent buildings in the park, plumbing systems, electrical systems, fuel supply systems, and firesafety considerations.

In 1980, NFPA and NCSBCS jointly formed standards committees under the auspices of the American National Standards Institute (ANSI) and in 1982 cosponsored an update to NFPA 501A. The update is now entitled the NCSBCS *Standard for Mobile Home Installations* (NCSBCS/ANSI A225.1-1982) and the NFPA 501A, *Standard for Firesafety Criteria for Mobile Home Installations,*

Sites and Communities. The new NFPA 501A provides the same scope of coverage as the previously issued standard.

It should be noted that mobile homes have increased in width and length from the minimums of 8 by 32 ft (2.5 by 10 m) to as much as 14 by 80 ft (4 by 24 m), with some designed as units so that they can be joined at the site to form "double wides" or even "triple wides" and thus become sizable homes. Some unintended uses such as employing mobile homes as multi-family occupancies, as dormitories, and as multi-story structures have developed. The basic mobile home also is used as temporary banks or offices, as field offices at construction sites, and for many other purposes. Mobile homes basically are not designed for such uses and need special attention to firesafety where so utilized.

RECREATIONAL VEHICLES

Recreational vehicles are vehicular units primarily designed as temporary living quarters for recreational camping or travel. They either have their own motive power or are mounted on or drawn by another vehicle. The basic entities are the travel trailer, the camping trailer, the truck camper, and the motor home. There are several variations of these recreational vehicles, and new equipment is being constantly developed.

In general, the automotive features of recreational vehicles are covered by federal motor vehicle safety standards. The federal standards do not, however, currently apply to plumbing, heating, and electrical systems in recreational vehicles or to materials of construction relating to interior walls, partitions, ceilings, exit facilities, and fire protection. NFPA 501C, *Standard on Firesafety Criteria for Recreational Vehicles* (and ANSI A119.2) (hereinafter referred to as NFPA 501C), concentrates on plumbing, heating, electrical, and firesafety and life safety considerations. NFPA 501C's requirements for the installation of fuel burning appliances are similar to those found in NFPA 501A insofar as it provides for complete separation of the combustion system from the interior atmosphere. This applies to all types of fuel burning appliances used in recreational vehicles except ranges, ovens, and illuminating appliances. Attention is likewise given to the ventilation and combustion air systems, clearances between heat producing appliances and adjacent surfaces, circulating air system ducts and their sizing, duct supports, registers, and grills.

Electrical systems also are covered in the NFPA 501C as well as in Article 551 of NFPA 70. The bulk of the provisions relate to the electrical equipment and materials required in a recreational vehicle for connection to a wiring system nominally rated 115 V (two wires with ground) or a wiring system nominally rated 115/230 V (three wires with ground). Other special requirements cover low voltage systems (other than circuit supply lines subject to federal or state regulations), combination electrical systems for connection to a battery or direct current supply which may also be connected to a 115 V source, and generator installations.

Fire and life safety requirements in NFPA 501C cover interior walls, partitions and ceilings, exit facilities, and stipulate that each recreational vehicle equipped with fuel burning appliances or an internal combustion engine must be provided with a listed portable fire extinguisher with a specified minimum rating.

In order to facilitate proper use of recreational vehicles, another standard, NFPA 501D, *Standard for Firesafety Criteria for Recreational Vehicle Parks and Campgrounds* (ANSI A119.4), has been developed. The intent of this standard is to provide minimum construction requirements for parks designed primarily for use by owners of recreational vehicles. Attention is devoted to park design and construction, recreational vehicle stand construction, environmental health and sanitation, fuel gas systems, storage of flammable and combustible liquids, electrical systems, and firesafety.

Bibliography

References Cited

HUD. undated. *Fifth Report to Congress on the Manufactured Housing Program*. Department of Housing and Urban Development, Washington, DC.

Klem, T. J. 1978. *Interim Report: The Mobile Home Fire Problem in the United States*. National Fire Prevention and Control Administration (now U.S. Fire Administration), Washington, DC.

MHI. 1984. *Memorandum*. "HUD Standards on Formaldehyde, Firesafety and Ventilation." Manufactured Housing Institute.

Budnick E. K, and Klein D. P. 1979. "Mobile Home Fire Studies, Summary and Recommendations." *NBSIR 79-1720*. National Bureau of Standards, Gaithersburg, MD.

NFPA Codes, Standards, Recommended Practices and Manuals. (See the latest *NFPA Codes and Standards Catalog* for availability of current editions of the following documents.)

NFPA 70, National Electrical Code®

NFPA 501A, *Standard for Firesafety Criteria for Mobile Home Installation, Sites and Communities.*

NFPA 501C, *Standard on Firesafety Criteria for Recreational Vehicles.*

NFPA 501D, *Standard for Firesafety Criteria for Recreational Vehicle Parks and Campgrounds.*

Additional Readings

Budnick, E. K., "Estimating Effectiveness of State-of-the-Art Detectos and Automatic Sprinklers on Life Safety in Residential Occupancies," *NBSIR 84-2819*, National Bureau of Standards, Gaithersburg, MD, 1984.

Budnick, E. K. et al., "Mobile Home Living Room, Fire Studies: The Role of Interior Finish," *NBSIR 78-1530*, National Bureau of Standards, Gaithersburg, MD, 1978.

Budnick, E. K., and Klein, D. P., "Mobile Home Bedroom Fire Studies: Summary and Recommendations," *NBSIR 79-1720*, National Bureau of Standards, Gaithersburg, MD, 1978.

Budnick, E. K., "Fire Spread Along a Mobile Home Corridor", *NBSIR 76-1021*, National Bureau of Standards, Gaithersburg, MD, 1976.

"Fire Performance Evaluation of the Federal Mobile Home Construction and Safety, Standard," U.S. Department of Housing and Urban Development, Washington, DC, 1980.

Gause, W. Peyton, "Mobile Home Flammability," *Journal of Products Liability*, Vol. 3. 1979. pp. 85-106.

Gomberg, A., et al., "Evaluating Alternative Strategies for Reducing Residential Fire Loss —The Fire Loss Model" *NBSIR 82-2551*, National Bureau of Standards, Gaithersburg, MD, 1982.

Herrin, C. L., and Schaenman, P. S., "Data Base for Evaluation of Mobile Home Fire Safety —1979-1980," National Bureau of Standards, Gaithesburg, MD, 1982.

Klein, David P., "Characteristics of Incidental Fires in the Living Room of a Mobile Home," *NBSIR 78-1522*, National Bureau of Standards, Gaithersburg, MD, 1978.

Schaenman, P. S., and Herrin, C. L., "Evaluation of Mobile Home Fire Safety: 1979-1980," prepared for the Department of Housing and Urban Development by Tri Data, Arlington, VA, Mar. 1982.

Consumer Products Safety Commission, *Fact Sheet No. 39*: "Mobile Homes," U.S. Consumer Products Safety Commission, Washington, DC.

MERCANTILE OCCUPANCIES

William Hiotaky, AIA

Mercantile occupancies are those facilities where a wide variety of goods and services are displayed and sold. Examples of such goods and services include clothing and footwear; jewelry; beauty, health, and fitness equipment and instruction; home furnishings, gifts, and luggage; books, cards, and stationery; music, electronic equipment, and photography supplies; sport specialties; toys, hobbies, and pets; drugs, varieties, and tobacco; food and food specialties; and hardware.

Occupancy Contents

Mercantile occupancies' contents are classified in NFPA 101, *Life Safety Code®* (hereinafter referred to as NFPA 101), as ordinary hazard notwithstanding the wide variety of goods with varied degrees of combustibility which might on occasion, suggest a higher or lower classification. On balance, however, the ordinary hazard classification is an appropriate designation assuming that in those instances where a limited higher hazard condition is present, adequate steps are taken to address that particular circumstance.

Major Types of Mercantile Facilities

The variety of facilities in which any or all of the aforementioned goods and services can be found is within the experience of most of the general public. Basically, mercantile facilities can be differentiated as follows:

1. Individual stores with a single category of merchandise such as specialty shops and supermarkets.
2. Individual stores with multiple categories of merchandise such as department stores with or without ancillary nonmercantile uses (such as restaurants or administration offices).
3. Shopping centers that are some combination of the above types in addition to multiple ancillary or accessory nonmercantile uses (such as restaurants and theaters).

Since there are many instances when a mercantile use may be the ancillary or accessory use in connection with a totally different occupancy classification, this chapter covers only those instances where mercantile is the primary occupancy classification.

SINGLE TYPE OF MERCHANDISE STORES

The single category of merchandise retail stores, which for purposes of brevity shall be called Type I stores, comprises the greatest variety of stores with respect to range of floor area as well as to diversity of merchandise.

For example, jewelry stores, tobacco stores, and gift shops are at the lower end of the scale in terms of floor area, whereas furniture stores, supermarkets, and sporting goods stores are at the higher end. The floor area range can vary from a couple of hundred square feet (200 sq ft equals 18.6 m^2), to 30,000 or 40,000 sq ft (2800 or 3700 m^2).

In the past, smaller Type I shops were constructed principally in cities and towns, along business thoroughfares, next to each other, often with basements for stock, and sometimes with an office or residential use immediately above. Many such stores exist today, especially in older business sections of cities; they are sometimes referred to as "Ma and Pa" stores because the proprietors of the store often live in the residential unit directly above. However, these types of stores are slowly fading out of existence.

Because of more restrictive zoning regulations, Type I stores are being constructed without other occupancies above them unless the mercantile occupancy serves an ancillary or accessory use, such as retail shops in a hotel lobby. In addition, basement space, even for stock space, is being phased out since stock space on the same level as the sales area is more efficient and functional.

In previous years, these smaller Type I stores were primarily constructed either entirely of wood or masonry bearing walls with wood joists, floors, and roofs.

The subsequent establishment of fire districts in urban areas plus the adoption of more stringent building code requirements led to the widespread use of noncombustible materials such as masonry walls (either bearing or

Mr. Hiotaky is Senior Vice President of The Taubman Company, Inc., based in Bloomfield Hills, MI.

nonbearing), steel joists, concrete floors, and metal roof decks in the construction of these same Type I stores.

Until recently, smaller Type I stores were almost never sprinklered, nor were larger Type I stores. NFPA 101 may presently require sprinkler protection based on height and area, and, in certain instances, if the stores are located below the level of exit discharge.

Larger Type I stores or chain stores such as supermarkets and "discount" drug stores, did not become prominent until the mid to late 1940s. They replaced the neighborhood grocer and corner druggist who could not compete with the high volume chains.

Larger Type I stores which are approximately 30,000 or 40,000 sq ft (2800 or 3700 m^2) in area are usually one story (with a possible mezzanine area for management offices), and are made of noncombustible unprotected construction. Until recently, they were not required to be sprinklered.

Type I stores are generally free standing buildings, unless they are part of a shopping center complex, which sometimes has a relatively large contiguous parking area surrounding the building.

Type I stores are usually divided into at least two principal areas; the stock space or back of the house employee area and the sales space or customer area. The distribution of areas is approximately one-quarter to one-third stock and two-thirds to three-quarters sales. The larger Type I stores may include nonmercantile accessory uses such as management offices which comprise a small area in relation to the sales and stock spaces. From a retailer's point of view, the distinction between retail stock space and storage space is important. Employees enter the retail stock areas frequently during the day to bring merchandise onto the sales floor. In this sense, the stock space is essentially an extension of the sales space except that the occupant load is considerably less. Stock space is not generally open to the public except as an alternate exit access route in case of fire. Storage areas, on the other hand, are less frequently accessed on the basis that goods are set aside in such areas for extended periods of time. In the retailing industry, storage areas connect with mercantile occupancies that are almost always in separate locations and are essentially distribution warehouses.

Occupant Load —Type I Stores

The occupant load consists of customers and sales employees. In smaller Type I stores, employees double as stock employees; larger Type I stores have a separate workforce employed exclusively in the stock area.

The occupant load will vary with the seasons of the year, such as back to school, Thanksgiving, Christmas, and Easter in addition to occasional clearance sales. NFPA 101 has taken these primary merchandising seasons into account in establishing the occupant load factor for calculating exiting requirements, which seem to work reasonably well.

Fire Load —Type I Stores

As in most other occupancies, contents represent the major potential hazard in Type I stores. Case histories and statistical evidence have proven that sprinklers remain the most effective means of combating fires in mercantile occupancies. In terms of life safety, the issue of cost benefit in installing sprinklers is appropriately addressed in NFPA 101 by the requirements for sprinklers based on height and area limits, which obviate the need for sprinklers in small shops unless, of course, these small shops are accessory uses to an occupancy requiring sprinklers.

MULTIPLE TYPE OF MERCHANDISE STORES

Multiple category of merchandise stores (e.g., a department store), which we shall call Type II stores, are almost as common as the smaller Type I stores.

These types of stores are generally from one to three or four stories in height and can vary in area from 50,000 to 250,000 sq ft (4645 to 23 225 m^2).

Earlier Type II stores, and particularly department stores, were constructed exclusively in business/retail districts in cities. The stores were multistory in height, either of exterior masonry wall bearing construction with wood joists and floors, or reinforced concrete, and had no sprinkler protection. More recently, however, many department stores are constructed of unprotected structural steel and are fully sprinklered.

As the name implies, department stores are divided into departments or sections of different kinds of merchandise, with or without the use of partition walls, and are laid out in a strategic way to influence customer movement patterns vertically and horizontally through the store according to a preconceived merchandising plan.

The vertical movement in multistory Type II stores is more effectively accomplished by means of escalators situated in open wells in highly visible and accessible areas of the store.

While basements were at one time quite common in Type II stores for either sales or stock, this practice generally is no longer followed because basement space is not nearly as desirable for sales or stock as is the same area constructed at or above the ground level.

Unlike the vast majority of Type I stores where the stock space is located in the rear of the store, stock space in Type II stores is often distributed on all four sides of the store. These areas are staffed strictly by stock personnel as opposed to smaller Type I stores where sales personnel often double as stock help.

Another example of a Type II store is the discount store. This store type grew out of the well known "five and dime" variety store. It is usually one story in height and laid out in a manner similar to a retail food supermarket with shopping cart self-service and checkout counters. Merchandise for retail sale in such stores is generally not as expensive or of such quality as in department stores; however, the same sectionalization of merchandise occurs as in a department store.

Accessory Occupancies

To round out the full range of goods and services a customer needs, certain accessory or ancillary occupancies in addition to administrative office areas are often included in Type II stores. Examples of such occupancies are restaurants, beauty shops, travel agencies, and optometry services. Restaurants, and more particularly, the food preparation areas of restaurants warrant special attention because of their potential life safety hazard.

Occupant Load —Type II Stores

As in Type I stores, Type II store occupant load consists of the general public (customers) and sales and stock employees. The occupant load will vary somewhat more dramatically with the primary shopping seasons in Type II stores than in Type I stores. This is especially true during the Thanksgiving/Christmas season when temporary employees are likely to be hired to supplement the permanent workforce, in addition to the heavier volume of customer traffic that can be expected.

Fire Load —Type II Stores

Contents represent the major potential hazard, as in Type I stores. Efforts to control the potential hazard in Type II stores (department stores) are somewhat more complicated because the fire load can vary greatly from one department to another and from floor to floor within the same store. Moreover, accessory occupancies with their own special fire loads must be considered in determining firesafety measures in these stores.

THE COVERED MALL SHOPPING CENTER

A covered mall is commonly defined as a covered or roofed interior area used as a pedestrian way, which connects buildings (or portions of a building) housing single and/or multiple tenants. The covered mall shopping center became very popular in a relatively short period of time, and today, most large shopping centers either planned or under construction are of this type. In addition, many covered malls are now being built in urban areas in conjunction with parking decks, office buildings, hotels, recreational facilities, and public or private transit systems.

Design of Covered Mall Shopping Centers

The covered mall shopping center is generally designed as one building comprised of from one to six or seven sizeable department stores and numerous smaller specialty retail stores, all of which are interconnected by a covered, climate controlled public pedestrian way. The complex may also include other ancillary occupancies such as movie theaters, bowling lanes, ice arenas, project management offices, and other customer service areas with access from the mall.

The mall may be designed as single or multilevel depending upon such factors as topography, the size of the land parcel, and the number and size of the department stores and retail shops. The vast majority of malls in the United States have either one or two levels, with an occasional partial mezzanine.

When a mall has two levels, the parking lot surrounding the building is usually shaped to provide grade access to each level of the department stores and the small shops. This equalizes customer flow so that one level has no advantage over the other as far as sales exposure is concerned. The two level shopping center is distinguished from the two story shopping center in this important respect. (See Fig. 9-12A.) When they are part of a two level covered mall shopping center, department stores are designed and merchandised in a similar manner.

The covered mall proper is designed primarily to serve the requirements for occupant egress, maximization of tenant sales, and promotional activities. The storefronts along the pedestrian "public street" of the mall are usually designed to encourage unrestricted customer flow and traffic control from the mall to the individual shops, as well as to the department store. This system of customer traffic flow and traffic control is the essence of the merchandising concept behind covered mall shopping centers.

Swing doors (whether glass or solid) not only make it less easy to exit, they also promote what is referred to in the retail industry as "threshold resistance," (the reluctance of customers to push, pull, or slide doors to gain access). The storefront closures ideally suited to the requirements for flow and for maintaining security are those that can be fully recessed such as rolling overhead grilles, side coiling grilles, and horizontal sliding doors. Since the ambient temperature in the mall is the same as in the tenant space, the storefront closure need only satisfy a security requirement when the store is unoccupied.

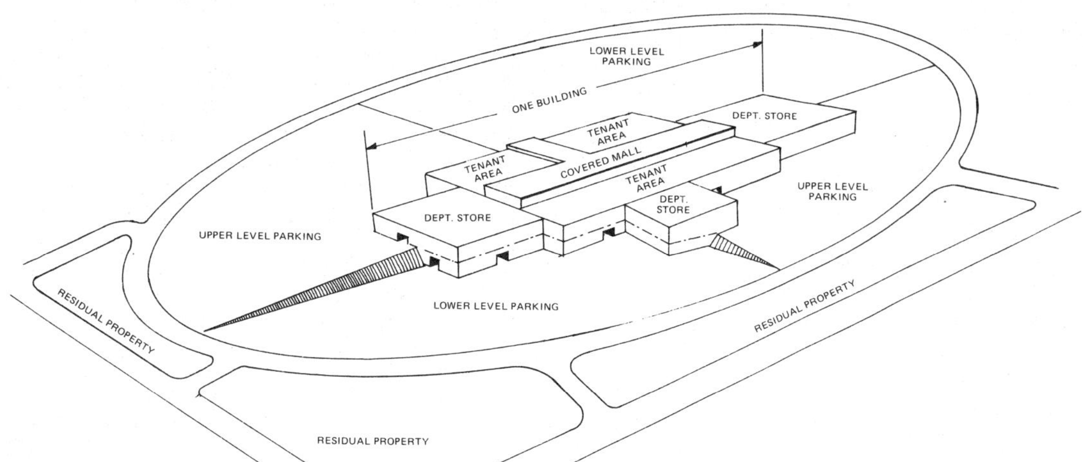

FIG. 9-12A. Many shopping centers today are purposefully designed (and merchandised) to provide an equal distribution of customer flow on all levels.

Occupant Allowance Requirements

Many covered mall shopping centers are constructed by a "fast track" construction procedure, so the ultimate ratio of sales space to storage space for tenant stores is not known when construction begins. In simple terms, fast tracking is a phased construction procedure which allows construction to proceed upon completion of the construction documents for any given phase of the work. Hence, a projection must be made concerning occupant load so the required units of exit width can be determined beforehand.

Many model codes, including NFPA 101, require that the occupant load for Type I and Type II stores be calculated on the basis of a factor of one occupant per 30 sq ft (2.8 m^2) of street floor area and basement sales areas (if planned), plus a factor of one occupant per 60 sq ft (5.6 m^2) of other upper sales floors. The occupant load for stock or storage space which is not open to the public is calculated on the basis of a factor of one occupant per 100 or 300 sq ft (9.3 or 27.9 m^2), depending upon the particular code being used. Whether or not the covered mall proper is assigned an occupant load depends on its designed function.

The Covered Mall Proper — A Systems Approach

In addition to the typical one level covered mall design, the covered mall may be designed as a multilevel space serving multilevel tenant areas and department stores. Although the vast majority of suburban shopping centers built in the U.S. are either one or two level centers with the occasional introduction of a partial mezzanine level, multilevel shopping centers, as a result of urban redevelopment programs, are becoming more commonplace in urban areas.

In most cases, two level covered malls are designed for visual intercommunication of the levels by means of openings in the floor of the upper level. Wherever practicable, visual intercommunication between the levels is important in maximizing exposure of storefronts and merchandise to customer traffic. The floor areas between openings in the upper level serve as bridges that promote cross-mall shopping convenience for the customer. (See Fig. 9-12B and 9-12C.)

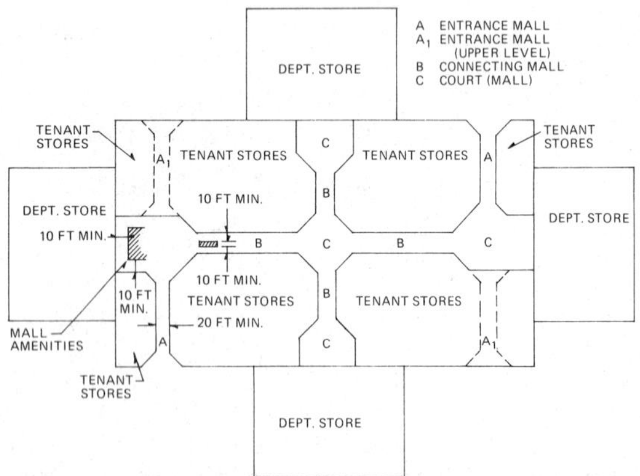

FIG. 9-12B. Typical two level shopping center (lower level).

At the time contruction is started, not everything about the shopping center is known, even to the developer. For example, the final architectural treatment of the exterior and interior spaces may be determined only to a point where the structural elements can be designed. The location of tenant demising walls may be unknown since lease negotiations with prospective tenants may not begin until after the superstructure has been erected. Mechanical, electrical, and fire protection systems are designed for main distribution only, with branch distribution added to the construction documents when the tenant space is defined by lease agreement.

The covered mall proper may contain other amenities for sales promotion and customer service and convenience such as seating areas, art objects, planting areas, pools, specialty retail kiosks, directories, openings in upper levels of multilevel structures, and areas designed for periodic promotional activities. Care must be exercised in the locaton of these amenities so they do not interfere with nor encroach upon required exit access.

Basically, the covered mall may be considered in either of two ways: (1) as a way of exit access from the

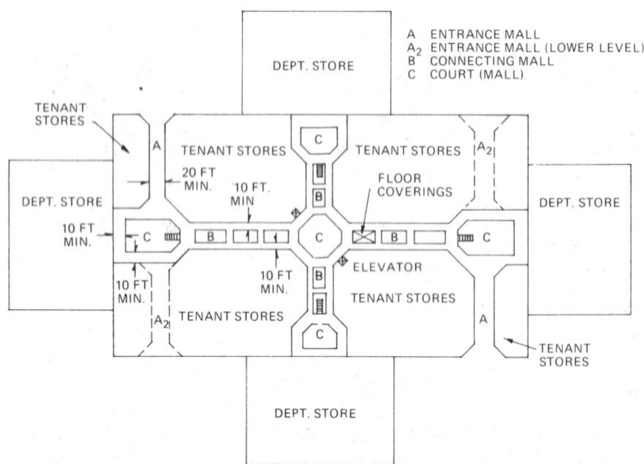

FIG. 9-12C. Typical two level shopping center (upper level).

connected buildings or (2) as a pedestrian way permitting an increase in the distance of travel from each of the tenant stores to a mall exit.

In the first instance, the entire complex is treated as one building or one compartmented department store, where the covered mall is treated simply as an extension of the sales space. In terms of exit access, the covered mall is considered as an aisle common to the various tenant stores that joins the exit access aisles of the tenant stores to mall exits. This treatment, in terms of NFPA 101, leads to the strict application of all design factors as though the shopping center were a Type II store or department store. In the second instance, a systems design is offered where the mall is considered to offer a higher degree of life safety if all of certain minimum conditions of the system are met:

1. The covered mall should be of at least sufficient clear width to accommodate egress requirements for mercantile buildings, but in no case less than 20 ft (6 m) wide in its narrowest dimension.
2. The minimum width dimension of 20 ft (6 m) has particular significance in terms of exit access and

should not be construed as a requirement to provide minimum separation between storefronts on opposite sides of the covered mall.

The need for separation in a covered mall shopping center that is protected by an automatic sprinkler system is not substantiated by the history of recorded accounts of exposure fires involving mercantile as well as various other types of occupancies.

The excellent performance of automatic sprinklers in connection with exposure fires as documented in *Fire; Automatic Performance Sprinklers in Australia and New Zealand, 1886-1968* (Marryatt 1971), together with the sprinkler performance data that was compiled by NFPA, mitigates almost any reasonable doubt regarding an additional requirement for separation over and above the minimum 20 ft (6 m) width requirement for exit access.

The width of the covered mall connecting courts at entrances to department stores, however, is generally found to be well in excess of the 20 ft (6 m) minimum to provide for the amenities described earlier, and at the same time provide for an aggregate of 20 ft (6 m) of exit access. (See Fig. 9-12D.)

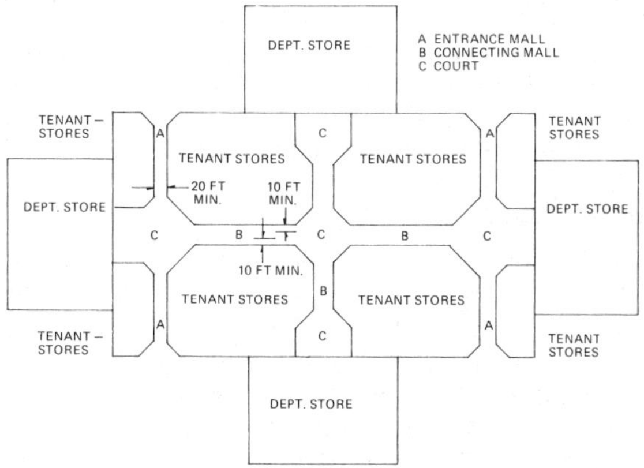

FIG. 9-12D. Typical one level shopping center.

In practice, one is more likely to find entrance malls designed for minimum aisle width. Stores located on either side of entrance malls do not lie along the direct path to department stores, which are commonly called "anchor stores" or "magnet stores" because of their greater customer attraction, do not enjoy as much customer exposure as those located along connecting malls. As a result, stores on opposite sides of the entrance mall are brought closer together by eliminating any mall amenities which in themselves are not likely to generate traffic, thereby maximizing convenient cross-mall customer shopping. The minimum mall width of 20 ft (6 m) is entirely compatible with this merchandising concept and does not impose an unreasonable burden upon developers nor merchants; this minimum width requirement has been successfully applied in shopping centers throughout the U.S.

The covered mall should be provided with an unobstructed exit access of not less than 10 ft (3 m) in clear width on each side of the mall floor area, parallel and adjacent to the mall storefront. Such exit access should lead to an exit having a minimum of three units of exit width. The purpose of the requirement of 10 ft (3 m) of clear exit access is to provide a higher degree of safety in the covered mall commensurate with its use as a continuation of exit access from the tenant stores. Safe, continuous, and unobstructed exit access from each tenant store is provided, while the owner's merchandising and operational requirements for the mall amenities are recognized.

The minimum requirement of 10 ft (3 m) of clear exit access parallel and adjacent to the mall storefronts relates directly to the minimum mall width of 20 ft (6 m). None of the mall amenities mentioned earlier should encroach upon this minimum dimension of 10 ft (3 m). The net effect of this requirement is to preclude the installation of mall amenities in areas of the covered mall that are the minimum 20 ft (6 m) in width. In terms of life safety, this requirement has proved to be reasonable and workable in practice thus far and has the added advantage of not being unduly restrictive nor burdensome to developers.

Occupant Load of Covered Mall Shopping Centers

The units of exit width from the covered mall need only satisfy the calculated occupant load based on all of the connected tenant occupancies. It is not necessary to provide units of exit width from the covered mall commensurate with the minimum mall width, which would necessarily result in wall to wall doors, unless the calculation of the occupant load together with the number and location of mall exits so dictates.

The occupant load factor used in determining exits for the mall proper varies with the aggregate area of the connected (Type I) stores. This sliding scale of occupant load factors is founded on emperical data which indicates that when shopping centers of a certain aggregate area are doubled in size, the resultant occupant load, although increased, does not double; i.e., the relationship between an increase in the size of a shopping center and the increase in the occupant load is not a linear one. (See Fig. 9-12E.)

Sprinkler Protection

The covered mall and all its connected buildings should be provided with an electrically supervised automatic sprinkler system throughout. It is a generally accepted fact that this is the single most effective way to protect life and property from fire. Sprinkler performance data, as referenced earlier, gives rise to objective scrutiny of long standing "accepted" code requirements. For example, accounts of sprinkler performance published in recent years have brought into question the need to provide current established levels of "fire proofing" of steel structural members when the building is protected by supervised sprinkler system.

Considerable data attesting to the effectiveness of sprinkler systems would strongly suggest that perhaps (at least in the context of one and two level covered mall shopping centers) current levels of fire resistance of steel structural members are unjustified in cost benefit terms when the building is protected by an electrically supervised sprinkler system. Another example is the imposition of additional building code restrictions on buildings in fire districts even when such buildings are protected by sprinkler systems.

Originally, fire districts were established to prevent, or at least minimize, the possibility of conflagration primarily in older or more highly congested urban areas. This

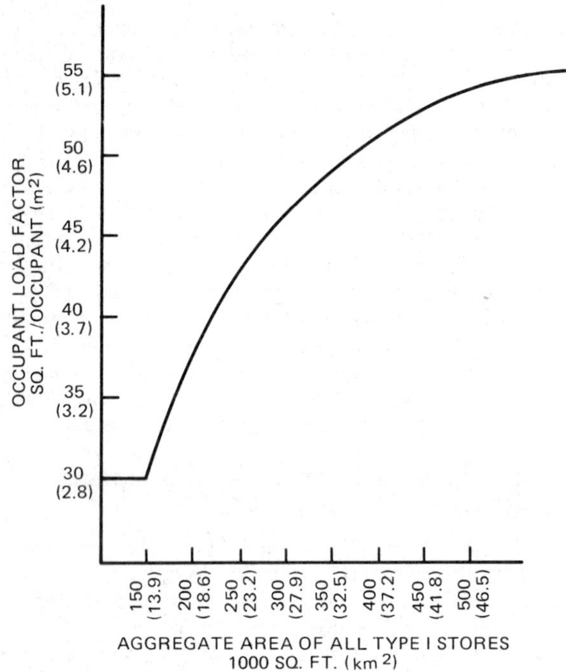

FIG. 9-12E. Relationship of an increase in the size of a shopping center to an increase in the occupant load factor of a shopping center.

resulted in the placement of added restrictions on building area, height, separation, and construction type, without due regard for the effectiveness of sprinkler systems. Recently, model codes have recognized this problem by eliminating many of the unwarranted fire district restrictions.

Horizontal Separation of Type I Stores

Walls dividing stores from each other should extend, to the extent practicable, from the floor to the underside of the roof deck or floor deck above. No separation is required between a store and the covered mall. The purpose of this requirement is to confine fire and smoke spread to the place of origin by inhibiting the passage of fire and smoke through ceilings and over the tops of tenant demising walls into adjacent occupancies. However, the engineering of a smoke evacuation system may require the placement of openings in tenant partitions above the ceilings to effectively prevent the accumlation of smoke. Model codes call for such walls to be constructed of noncombustible materials; in practice, tenant demising walls are usually constructed of metal studs and drywall.

In effect, tenant demising walls represent a form of compartmentation to the extent that no separation is provided at the storefront. A requirement for separation at the storefront would defeat the merchandising purpose of the covered mall, which is to optimize unrestricted customer flow between the covered mall and the tenant store. With this form of compartmentation, in concert with an electrically supervised automatic sprinkler system, the need for any additional fire wall building code requirement becomes highly questionable. Accounts of fires in shopping centers where this configuration was in evidence tend to support the theory that any additional fire wall requirement is unnecessary and in fact burdensome, from an operational point of view in the leasing (and subsequent releasing) of shopping centers where maximum flexibility to rearrange space is important to the continued success of the shopping center.

Smoke Control

The covered mall should be provided with an effective smoke control system. Under fire conditions, and due to the open storefront configuration, any buildup of smoke in a tenant store could ultimately find its way into the covered mall. Effective means of smoke control in the covered mall proper are essential in order to assure its continuous use as a smoke free primary way of exit access.

Sectionalized proprietary heating, ventilation, and air conditioning (HVAC) system service for multiple tenants (as opposed to a separate system for each tenant) is being utilized with increasing frequency to maximize efficiency and control and to minimize energy consumption.

An added benefit in the use of such systems is the ability to utilize them effectively for smoke evacuation purposes in the event of a fire condition occurring within any given tenant space. Thus, the products of combustion can be confined to the area of origin (with the help of the sprinkler system), with the subsequent discharge of smoke to the exterior.

This may necessitate modification of a possible requirement for fire dampers in walls above ceilings between tenant spaces, in order not to negate the smoke removal mode within any given sectionalized area.

General Requirements

In addition to the foregoing, certain conditions of exiting should be present in all covered malls. These conditions of exiting are as follows:

1. Every floor and every store [except stores less than 50 ft (15 m) deep] of a covered mall should have no less than two exits located remote from each other.
2. No less than one-half the required exit widths for each store greater than 3,000 sq ft (279 m^2) in area connected to a covered mall should lead directly outside without passage through the mall.
3. Every covered mall should be provided with unobstructed exit access parallel to and adjacent to the connected buildings. This exit access should extend to each mall exit.
4. Exits from the covered mall should be arranged so that the length of travel from any mall store entrance to an exit should not exceed 200 ft (61 m).

GENERAL LIFE SAFETY FACTORS APPLYING TO MERCANTILE OCCUPANCIES

There are a number of factors related to life safety that are common to all three major types of mercantile facilities.

Prompt discovery is one of the most important factors associated with minimizing any loss due to fire. This is especially important considering that most reported losses in mercantile occupancies occur during nonbusiness shopping hours, from late at night to the very early morning.

The majority of these fires occur in the stock areas, as opposed to the sales areas. Poor housekeeping and smoking appear to be two of the principal causes involving the stock area where, for example, merchandise is removed from cartons for display on the sales floor, and the cartons and packaging are not properly disposed of, thus increasing the hazard.

Sprinkler protection has proven to be the ideal solution to the various hazards described. Properly engineered sprinklers can give warning, as well as attack a fire to the extent of controlling its development, if not extinguishing it altogether.

Every reasonable effort should be made in the design of sprinkler systems to identify and address the causes of unsatisfactory sprinkler performance. This is especially true to Type II and Type III mercantile facilities where the need is greatest.

The need to occasionally renovate existing facilities as a result of new trends in fashion or store design is ever present in mercantile facilities as well as reconstruction of existing facilities due to the releasing of space (especially in shopping centers). Extra care should be taken to mitigate the potential problem of fire during these times, since the hazard is very likely to increase due to the reconstruction and renovation processes.

Certain accessory uses such as the food preparation areas in restaurants, often found in Type II and Type III facilities and more particularly department stores and shopping centers, warrant special attention as very often they are the place of fire origin—much more so than a public eating space. Such areas should be segregated with appropriate construction, if not sprinklered, and appropriate fire supression devices should be installed in food preparation equipment.

CONCLUSION

An important purpose served in the foregoing discussion on the nature of the mercantile occupancy is to form a basis for consideration of code requirements (or reconsideration) that allow the mercantile use to function optimally and yet provide for a reasonable degree of life safety for the occupants of the facility.

All of the requirement related items discussed in this chapter have been practically applied in many stores and covered mall shopping centers throughout the U.S., and have answered the need for life safety without imposing burdensome or unreasonable restrictions on developers and the merchandising needs of covered mall shopping centers.

In addition to NFPA 101, other model codes are presently being revised in an effort to codify requirements for covered mall shopping centers. The success of the effort can depend upon how successfully the life safety concern is in harmony with the retail merchandising purpose of the covered mall shopping center.

There exists the possibility of overreaction in the formulation of code requirements, especially when all the important aspects of the building type being considered are not fully understood. The revision of existing code requirements that have proven to be unreasonably restrictive or unfounded in terms of any real threat to life safety, practicability in application, and cost benefit should be pursued in much the same way as are the efforts being made to rectify potentially dangerous life safety situations in buildings by the adoption of more restrictive and yet reasonable code language.

Bibliography

Referenced Cited

Marryatt, H. W. 1971. *Fire; Automatic Sprinkler Performance in Australia and New Zealand*. Australian Fire Protection Association, Melbourne, Australia.

NFPA Codes, Standards, Recommended Practices and Manuals. (See the latest *NFPA Codes and Standards Catalog* for availability of current editions of the following documents.)

NFPA 101, *Code for Safety to Life from Fire in Buildings and Structures*.

NFPA 220, *Standard on Types of Building Construction*.

BUSINESS OCCUPANCIES

Revised by Clifford S. Harvey

Business occupancies are defined as those occupancies which are used for the transaction of business. They differ from the mercantile occupancy in that, generally speaking, only services are sold in them, while goods are sold in the mercantile occupancy. Examples of a business occupancy might be any office or group of offices, city halls, courthouses, college and university instructional buildings, and ambulatory outpatient clinics. This occupancy is unique in that it generally creates little problem from the fuel load and interior finish standpoint. Except for offices which deal with a great amount of paper, offices can usually be considered a light hazard from a sprinkler design standpoint. Wall coverings tend to be gypsum board coated with paint, or some other low or noncombustible material.

A normal community will find that the predominant occupancy within its corporate limits is the business occupancy, as they tend to take up less overall space than does a clothing store or a restaurant. In one medium sized building, it is not uncommon to find 15 or more different offices, all being used by different clients. Although some business occupancies can be very large [50,000 sq ft (4645 m^2) or more], most are less than 3,000 sq ft (279 m^2) per business.

Information on how to determine life safety equivalency will be found in Section 7, Chapter 2, "System Concepts for Building Firesafety."

BASIC DESIGN PHILOSOPHY AND OCCUPANCY CHARACTERISTICS

Most business occupancies are occupied only during set hours, usually during the daytime. The occupants are expected to be alert, aware of their surroundings, and possess a reasonable degree of familiarity with the facility. With this framework in mind, NFPA 101, *Life Safety Code®* (hereinafter referred to as NFPA 101) views the life safety needs as related to alerting the occupants of an emergency event and providing facilities for total evacuation. NFPA 101 does not require the same degree of structural compartmentation, extinguishing systems, or similar features that would provide for limiting the spread and growth of a fire as may be required for some other occupancies (such as in health care, where occupants are expected to remain for the duration of the fire event). The high rise business occupancy is the only exception to these requirements, where the occupants do not have a reasonable chance of exiting the building upon learning that the fire is in progress.

Historically, business occupancies have been considered to be ordinary hazard types of occupancies. In general, they are relatively free of high potential special hazards and activities. The hazard or risk that is present is normally associated with the common ignition sources arising from heat, light, and power, as well as from a high level of personal activities and support activities such as cafeterias, parking facilities, and small retail operations commonly associated with business occupancies. Business occupancies frequently have a high occupancy load during business hours. While such concentrations are not as dense as the population loads associated with assembly occupancies, they may be heavily distributed throughout a large building at densities approaching those specified by NFPA 101, or 100 sq ft (9 m^2) of gross floor area per person.

Business occupancies can often have high concentrations of value. This occurs particularly with automatic data processing equipment and in financial institutions which actually handle money or other negotiable instruments. Fires in business occupancies can have a critical impact on continuity of services, or the completion of essential missions and services. Special attention should be paid to those items of critical importance to the individual business occupancy such as computers, other automatic data processing equipment, communication networks, and vital or important records or accounts.

EXCEPTIONS TO THE PHILOSOPHY OF TOTAL EVACUATION

There are two exceptions to the total evacuation philosophy for business occupancies. First, both NFPA 101 and the model building codes now call for a very limited degree of compartmentation that enhances traditional built-in features, such as the corridor wall and the

Mr. Harvey is president, Firemeasure, Inc., Boulder, CO, and the Chief Fire Marshal for that community.

office door of most office buildings. The intent is to provide a minimal upgrading in the basic level of construction to retard the fire progress from the space of origin to adjoining areas until after the occupants have left that space and exited to the outside. This compartmentation is supposed to impede the progress of the fire for the first five or ten minutes, during which normal manual rescue or extinguishment can be initiated and critical evacuation can occur. Specifically, to achieve this, NFPA 101 requires partitions having a fire resistance rating of 1 hour to separate corridors from use areas in new construction, where access to exits is limited to corridors. Partitions are not required however, where (1) exits are available from an open floor area, (2) the corridors are within a space occupied by a single tenant, or (3) within a building protected throughout by an approved, automatic sprinkler system. Doors and frames, each with a minimum 20 minute rating (and equipped with a positive latch and closing device), are required to protect openings in 1 hour partitions separating the corridor from use areas. When a glass vision panel is used within that rated partition, that panel is allowed to contain only fixed wired glass, and cannot be larger than 1,296 sq in. (83 613 mm^2) per panel.

The second exception to total evacuation involves recognition of those business occupancies (high rises) where total evacuation has been repeatedly demonstrated as impractical, if not impossible. The high rise business occupancy, has been singled out in the business occupancy chapters of NFPA 101 as requiring special attention to protecting building occupants in place. After extensive review of the model codes, the NFPA Subcommittee on Business Occupancies chose not to recreate a building code for high rise businesses. Rather, they recognized that the automatic sprinkler system was the one dominant feature that assured the greatest life safety for occupants who had to remain in a high rise structure. Hence, the 1976 edition of the *Code*, required all business occupancy buildings over 75 ft (23 m) in height to provide complete automatic sprinkler protection throughout. This height is measured from the lowest level of fire department access, to the floor of the highest occupiable story. This requirement was continued in the 1981 edition, of NFPA 101 and was further clarified in the 1985 edition.

Other equipment or designs providing equivalent life safety protection may be allowed by the authority having jurisdiction (AHJ). This allowance is given because there are always new designs and equipment being developed that would present at least the same level of protection for the occupants as NFPA 101 intended. Generally speaking, although a combination of other items may be deemed equivalent by the AHJ, nothing at this time could be considered as providing equivalent protection to a complete sprinkler system. Before this "equivalent alternative" can be accepted by the AHJ however, documentation must be provided to indicate that the equivalent level of life safety required by NFPA 101 will be met or exceeded by the other requested design approaches.

Ignition Sources

Fires in business occupancies normally result from common hazards, with the most frequent probable sources occurring from smoking and the careless use and disposal of smoking materials, sparks from electrical machines and related office equipment, and torches or other open flame devices used for welding, cutting, or burning in maintenance and alterations operations. The most commonly expected initial fuels are furniture, paper, other working materials used in the conducting of the business, and occasionally the interior finish or insulation. In business occupancies there is frequently a massive flow of paper and similar materials, and a particular concern must be directed to trash and waste paper collection, storage, and disposal.

Types of Construction

Business occupancies can be located in any type of construction. In addition, these occupancies are found in a mix with many other types of occupancies, due to the need for offices in all types of occupancies. Businesses could be located in dwellings, factories, warehouses, laboratories, and all other types of buildings where light, heat, a desk, and working space can be provided. There is however, a common type of building, known as the commercial building, designed specifically for office use. These buildings are generally located in concentrated business districts, and are most frequently located in high value districts or areas. Generally speaking, such buildings are allowed by codes to be of any height, and many are located in high rise buildings.

In recent years, the general style of building construction has changed, and the modification of many older buildings has led to use of the newer designs in terms of style, appearance, and firesafety problems. In the newer buildings, lighter weight types of construction have been used, and instead of the massive heat sink provided by the old style masonry construction, lightweight steel or similar framing has been protected by "fireproofing" material. This materials protects the building frame, but provides no heat sink or other method of assisting in mitigating the impact of fire energy. Various types of block construction and open unpartitioned space have become common, with the loss of the value of inherent compartmentation that was formally traditional to this type of building. Suspended ceilings and additional utilities for air conditioning, power, and illumination, have caused the creation of numerous void spaces between floor and vertical spaces, "poke throughs" and other penetrations of the floor/ceiling assembly.

Modern business and other operations have greatly increased power demands. Building perimeters have been made as tight as possible for energy conservation, and fixed, sealed windows are now common to such occupancies. Even when operable windows are used, they are designed to be difficult to operate, and to be as tight as possible. The result is to create a specifically controlled environment which in normal nonemergency situations are beneficial to the economy and comfort of the building. Under fire conditions however, this can cause large areas to participate in and contribute to a dangerous and possibly lethal fire environment.

Sprinkler systems, and their increasingly common use in buildings of all types have mitigated the problems to some extent. Many buildings which have been renovated and retrofitted with full automatic sprinkler systems actually provide a higher degree of life safety today, than when they were originally built. Extreme care should be taken however, when retrofitting an old building for uses today. Even with the installation of a full sprinkler system, care must be taken to protect all spaces, whether hidden or exposed to view. Some of the more spectacular fires today occur in buildings which were thought to be fully sprinklered, but contained areas that were not fully sprinklered in and of themselves. When the fire occurred in those unprotected spaces, it overpowered the sprinkler system, and the system failed to control the fire before much damage resulted.

MEANS OF EGRESS

Based on the concept of total evacuation by alert occupants of business occupancies, all traditional means of egress are allowed, from doors to stairs and ramps. Even noncomplying existing stairs and fire escapes may be continued in use after review and acceptance by the authority having jurisdiction.

At least two remote means of egress are required for every floor used for business occupants, including floor locations below the street. Any three story business occupancy with less than 30 occupants per floor may have a single exit for the second or third floor if the total travel distance is less than 100 ft (30 m) and the stairway is fully enclosed and does not communicate with the other floors. (See Fig. 9-13A.) Also, any small area occupied by less than 100 persons and having a direct exit to the outside with a total travel distance of less than 100 ft (30 m), including no more than 15 ft (4.8 m) over an enclosed stair, may have a single exit. Other than the above, travel distance to at least one exit is limited to 200 or 300 ft (61 or 91 m) in a fully sprinklered building, with any common path of travel being limited. Pertinent sections of NFPA 101 should be reviewed to determine the exact distances.

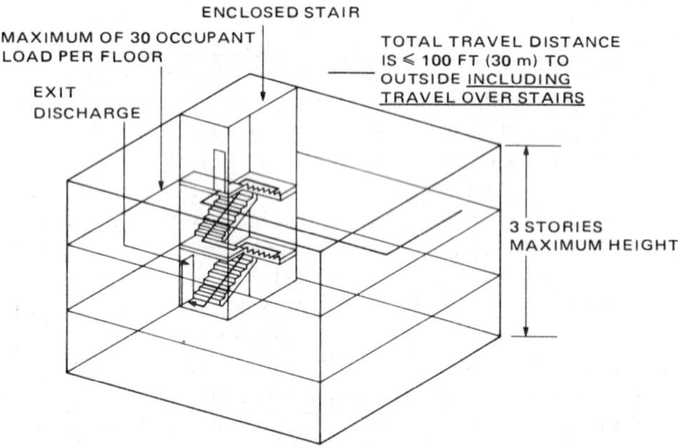

FIG. 9-13A. Single exit from third floor of a business occupancy. Stair is totally enclosed, has opening only at third floor, and discharges directly to street with no communication at second and first floors. A similar arrangement could be provided for the second floor of the same building.

Egress Illumination

Illumination of the exitway must be a minimum of 1 footcandle (1.076 ex) at the floor while the building is in use. Emergency lighting of the exits in case of a power failure is required (1) when the building is two or more stories in height above the level of exit discharge, (2) the occupancy is subject to 100 or more occupants above or below the level of exit discharge, (3) the entire occupancy is subject to 1,000 or more total occupants, or (4) the structure is windowless or underground. For the purpose of these requirements, exit access includes only designated aisles, corridors, and passageways leading to an exit.

Vertical Openings and Interior Finish

Generally speaking, vertical openings such as stairways are required to be enclosed in business occupancies. The exception to this requirement is that if the building is usually protected throughout by an approved sprinkler system, limited unprotected vertical openings are allowed. Interior wall and ceiling finish may be either Class A, B, or C in general office areas, but the exit must be either Class A or B. If the building is sprinklered, all interior finish may be any of the recognized classes.

Alarm Systems

Any business occupancy which is two or more stories above the level of exit discharge, houses 100 or more occupants above or below the level of exit discharge, or is subject to a total building population of 1,000 or more persons, is required to have a manual alarm system. When buildings are protected throughout by an approved automatic sprinkler system, those buildings are not required to have the manual pull stations.

THE LIFE SAFETY CODE AND EQUIVALENCY

Each authority having jurisdiction hears more and more these days about equivalency, or the intent of NFPA 101. They are being faced with numerous decisions each day concerning what to accept and what to reject as "equivalent to the Code." It is relatively easy to determine whether a desired alternative is equivalent to the section of NFPA 101 it is asked to replace through the "Measurement of Building Firesafety" approach used in university courses and by an authority having jurisdiction. Graph the requirements of NFPA 101, and the expected performance of a requested alternative, and compare them. If the requested alternative is equal to or better than the *Code* generated graph, it then can be accepted as an equivalent alternative. With the use of this system, the authority having jurisdiction rarely turns down project requests. The decision as to which direction to proceed is actually up to the developer, and satisfies all.

This concept is extremely easy to use when dealing with business occupancies. Some of the equivalent alternatives are already recognized by NFPA 101, in that the requirement for such items as alarm systems or certain length corridors are waived when the building in question is equipped with a full sprinkler system. The process can even be used to determine what is equivalent to a required sprinkler system, although the results tend to be so inflexible to the user that they are rejected.

The intent of NFPA 101 is to protect people within their chosen living or working environment—either by protecting them in place, or providing the proper number and type of exits so they may expeditiously leave the building. If one part of the system is lacking, (not enough exits or a too long deadend corridor) other parts of the life safety system can be enhanced, strengthened, or increased to compensate for that area of deficiency. For example, when it is impractical or impossible for the occupants to quickly leave the building, some way needs to be found to allow them more time to leave, or to protect them in place for the duration of the fire. Although this in place protection generally takes the form of a full sprinkler system, it could also be accomplished by putting numerous barriers between the fire and the occupants, or by increasing the compartmentation of the building (although these additional barriers to fire spread tend to decrease the usability of the building, in that the users must open and close more doors). If sprinklers are impossible or impractical, and the owner does not want to add additional barriers, a greatly increased number of exits, coupled with a maximum height of the building of two or three stories, might also be considered equivalent.

Each authority having jurisdiction must decide what is acceptable. Using available engineering techniques and the experience gained in the fire service, with the various ways of controlling the spread of fire, it is a relatively easy process to allow alternatives to the Code, and to feel confident about those alternatives.

Bibliography

NFPA Codes, Standards, Recommended Practices and Manuals. (See the latest *NFPA Codes and Standards Catalog* for availability of current editions of the following documents.)

NFPA 13, *Standard for the Installation of Sprinkler Systems.*
NFPA 70, *National Electrical Codelmed.*
NFPA 80A, *Recommended Practice for Protection of Buildings from Exterior Fire Exposures.*
NFPA 101, *Standard for Safety to Life from Fire in Buildings and Structures.*
NFPA 220, *Standard on Types of Building Construction.*
NFPA 232, *Standard for the Protection of Records.*

INDUSTRIAL OCCUPANCIES

Revised by Michael Slifka

The potential for loss of life from fire in an industrial occupancy is directly related to the fire hazard risk of the industrial operation or process within that occupancy. Fire records show that most multiple death fires in these occupancies are the result of: (1) flash fires in highly combustible contents, or (2) explosions in combustible dusts, flammable liquids, or gases.

This chapter is concerned with life safety in industrial occupancies. More information on property protection in industrial occupancies is to be found in Section 10 of this HANDBOOK, "Process Fire Hazards," and Section 18, "Water-Based Extinguishing Systems." Storage facilities, which often are found in industrial occupancies, are discussed in Chapter 15, "Storage Occupancies," of this section."

Every year, the property destroyed by industrial fires constitutes a major amount of the annual fire loss in terms of property damage. Fortunately, a similar number of lives is not lost in factory fires. The low life loss record results from a number of favorable operating features which are commonly found in these occupancies. Continued emphasis on good exit design and daily attention to industrial safety and training programs can help to continue this trend.

LIFE SAFETY RISK

Before proper exits can be designed for an industrial plant, the plant must first be classified for relative degree of fire and life safety risk. This is necessary because the risk varies from plant to plant. Once the building is classified, the exits can be designed according to the requirements set forth in NFPA 101, the *Life Safety Code®* (hereinafter referred to as NFPA 101).

The fire and life safety hazard in an industrial occupancy must be determined carefully. Classification is based on the burning characteristics of the type of *contents* of the building, not on the quantity of materials or type of construction. Contents should be evaluated to determine rate of fire spread, burning rate, toxic fume evolvement, and other factors which affect the amount of time available for evacuation. Once employees have been evacuated, the extent of fire spread becomes a property protection fire problem. Life safety protection and property protection are too often confused in industrial occupancies; life safety is an inappropriate reason for installing extensive property protection. Exit facilities which conform to requirements of NFPA 101 will protect employees, but additional measures may be needed for property protection.

A number of resources should be used to determine the true extent of hazard in any particular plant. Industrial plant operators, NFPA codes and other publications, insurance companies, and previous fire records can all provide useful information. If industrial secrets are reviewed during this evaluation, they should be kept strictly confidential.

Industrial Occupancies—General Design Considerations

The automatic sprinkler system is one of the most important elements of life safety protection in industrial occupancies. Originally developed to protect industrial property, the automatic sprinkler has also been largely responsible for the excellent life safety record in these occupancies. Fire officials and fire protection engineers have recognized this, and in recent years automatic sprinkler systems designed specifically for life safety have been widely adopted for use in buildings with significant life safety hazards.

Automatic sprinklers are particularly effective because they control fire spread and allow employees sufficient time to evacuate the building. In addition, the sprinkler system water flow alarm also alerts occupants of the existence of fire, which improves their reaction time.

Population Characteristics

Occupants of industrial buildings are generally ambulatory and fully capable of responding quickly to fires. To capitalize on this, many industrial plants include life safety measures in their emergency preplanning. The emergency preplan normally includes a means for alerting employees, identifying and posting exit routes, and estab-

Mr. Slifka is Chief of Safety, Occupational Health, and Fire Protection, for the U.S. Veterans Administration in Washington, DC.

lishing group assembly areas for evacuees outside the building where a count of all employees can be taken. Some employees are assigned to ensure that fire protection equipment is operating as designed. It is particularly important that arrangements be made at the preplanning stage to evacuate any physically handicapped persons employed in the plant. The preplan should be routinely evaluated through simulated fire exercises and fire drills; only through such drills can weaknesses in the preplan be recognized and the plan modified.

When determining hazard classification, it is important that the hazard classifications of NFPA 101 be used. A localized high hazard operation in an otherwise low hazard occupancy does not classify the plant as a high hazard industrial occupancy. For example, metalworking plants are normally classified as low hazard occupancies. If a metalworking plant contains a flammable solvent dip tank coater, the classification does not change to high hazard. Adequate means of egress away from the coater is required, but additional exits and a reduction in exit travel distance, as specified for high hazard occupancy, are not required for the entire building. Should the coater be isolated in a separate area, however, the special area would then probably be subject to requirements of a high hazard occupancy.

Operations involving low or ordinary hazard materials and processes are classified for the purposes of life safety as general industrial occupancies. Examples of general industrial occupancies include electronic and metal fabrication operations, textile mills, automobile assembly plants, steel mills, and clothing manufacturing.

High hazard industrial occupancies are those in which flammable liquids are routinely handled, used, or stored, or those in which explosive dusts from grain, wood, flour, plastic, aluminum, magnesium, or other dust generating materials are produced. Other high hazard industrial occupancies are those in which hazardous chemicals or explosives are manufactured, stored, or handled and those in which cotton or other combustible fibers might produce combustible flyings. The classification "high hazard" is limited to those industrial buildings which house extremely hazardous operations and does not include those buildings in which there is only incidental or restricted use of hazardous materials. The storage of limited quantities of flammable liquids (such as paint) in accordance with provisions of NFPA 30, *Flammable and Combustible Liquids Code*, does not require a high hazard occupancy classification. Plants where flammable liquids such as solvents and paints are mixed and blended would, however, have to be classified as a high hazard industrial occupancy.

In some low and ordinary hazard plants, machinery or equipment occupies most of the available floor space and people are restricted to a very small area. In such instances, the entire building need not be evaluated for life safety, because the fewer building occupants require fewer exit facilities. In a special purpose building, exits need only be provided for the actual number of employees in the structure.

Modern Design Practices

Although the life safety record in industry has been relatively good, some major problems may be developing. Modern industrial complexes are much larger, more hazardous, and more complex in operation than the factories and plants of the early 20th century, or even those of 20 years ago. In the past few years, new materials (such as plastics) have been introduced to plants in large quantities. Industrial managers must now pay increased attention to maintaining life safety, especially during daily operations.

Fire Brigades

Most industrial firms train employees to use first aid fire fighting equipment such as in plant standpipes, hose, and fire extinguishers. Thorough training of this type has contributed to a major reduction in fire loss and loss of life in industrial occupancies. Although initial fire fighting measures primarily protect property, they also protect lives. There is no major threat to life if fire spread is restricted.

Requirements of NFPA 101

Most of NFPA 101 was developed through review of past catastrophic fires. There is, as yet, no significant experience of this kind for modern industrial plants. When fully incorporated, the measures in NFPA 101 are sufficient to ensure against major loss of life in industrial plant fires.

To properly arrange the exits of an industrial occupancy, the life safety risk of the plant should be fully evaluated so exits and other protection can be properly designed.

Means of Exit Design Requirements

NFPA 101 requirements for the means of egress in industrial occupancies are similar to those required in any structure. The travel distance to any exit in a general industrial occupancy must be 100 ft (30 m) or less, except where the building is completely protected by automatic sprinklers. With automatic sprinkler protection, travel distance may be increased to 150 ft (45 m). (See Fig. 9-14A.)

It may be impossible to maintain the 150 ft (45 m) travel distance without major renovations, especially in large volume structures. In such cases, exit tunnels, overhead passageways, or travel through fire walls with hori-

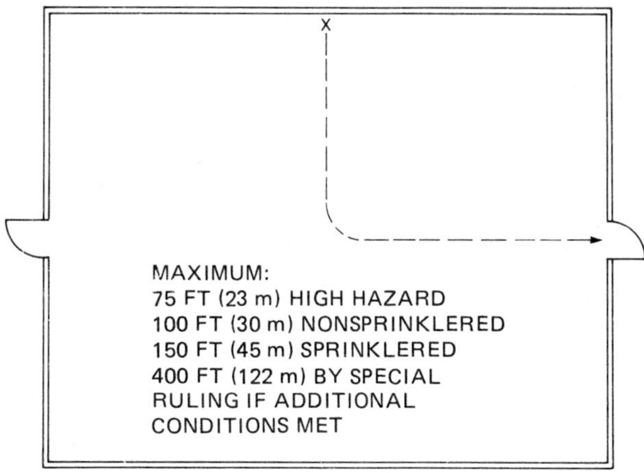

FIG. 9-14A. Summary of travel distance options allowed by NFPA 101.

zontal exits can provide the necessary safeguards. In unusual situations, travel distance to 400 ft (122 m) may be permitted. In such cases, the contents should be limited to low or ordinary hazard in a general industrial or special purpose occupancy. Additional provisions required for extraordinary travel distance should include, as a minimum, the following:

1. Limitation to only one story buildings.
2. Limitation of interior finish to Class A or B.
3. Provision of full emergency lighting in the building.
4. Installation of an automatic sprinkler or other automatic fire extinguishing system. Supervision system for malfunctions, closed valves, and water flow or operational alarm.
5. Provision of smoke and heat venting or other engineered means to limit spread of fire and smoke. The system design must ensure that employees will not be overtaken by heat or smoke within 6 ft (2 m) of floor level before reaching the exits.

In high hazard factories and plants, the travel distance to an exit is reduced to 75 ft (23 m), and no common path of travel is allowed. Regardless of size, every high hazard occupancy is required to have at least two separate and remote exits. In general industrial occupancies, a 50 ft (15 m) common path of travel to two separate exits is allowed.

Illumination of the Means of Egress

As large windowless structures have become more popular, the importance of emergency lighting for the means of egress has increased. Where a building is only occupied in daylight hours and the means of egress is fully illuminated by skylights, windows, or natural light, the normal electrical illumination may be waived. In special purpose industrial occupancies where routine human habitation is not the case, exit illumination and emergency lighting may also be omitted.

Protection for Life Safety in Industrial Occupancies

In general, vertical openings must be fully enclosed in industrial occupancies. The enclosure may be omitted from vertical openings which are not used for exits in buildings with low or ordinary hazard contents protected by an automatic sprinkler system. Another exception to the enclosure requirement is in specially designed industrial buildings that house operations, processes, or equipment requiring openings between floors. Should this be the case, each floor connected by unprotected openings must be provided with exits such as enclosed stairways which are fully protected from obstruction by fire or smoke in the interconnecting floors.

Due to the size and complexity of most industrial structures, a fire alarm system is a necessity. The fire alarm system should alert responsible persons in a continuously manned location so that positive steps can be promptly taken to start fire fighting, employee evacuation, shutdown of hazardous processes and other necessary actions. In high hazard occupancies, the fire alarm system should also sound an alarm to immediately notify employees to evacuate the building. In general industrial occupancies with fewer than 100 persons total, and fewer than 25 persons normally above or below the street level, the fire alarm system may be waived.

High hazard industrial occupancies present a unique fire control problem and a severe life safety hazard. High hazard industrial occupancies, operations, or processes must have automatic extinguishing systems or other equally effective protection such as explosion venting or suppression, to allow occupants to escape before they are exposed to a fire or explosion.

Bibliography

NFPA Codes, Standards, Recommended Practices and Manuals. (See the latest *NFPA Codes and Standards Catalog* for availability of current editions of the following documents.)

NFPA 27, *Standard for Private Fire Brigades.*
NFPA 63, *Standard for the Prevention of Dust Explosions in Industrial Plants.*
NFPA 101, *Life Safety Code.*
NFPA 601A, *Standard for Guard Operations in Fire Loss Prevention.*

Additional Readings

Gold, D. T., *Fire Brigade Training Manual*, National Fire Protection Association, Quincy, MA, 1982.
Linville, J. L., ed., *Industrial Fire Hazards Handbook*, 2nd ed., National Fire Protection Association, Quincy, MA, 1984.

STORAGE OCCUPANCIES

Bruce W. Hisley

Storage occupancies are buildings or structures utilized for the storage or sheltering of goods, merchandise, products, vehicles, or animals. Storage facilities can be separate and distinct facilities or part of a multiple use occupancy. Examples of storage occupancies are warehouses, freight terminals, parking garages, aircraft hangars, grain elevators, barns, and stables.

OCCUPANCY CHARACTERISTICS

Storage occupancies are unique in that they can be classified as low, ordinary or high hazard, or a combination of these where mixed commodities are stored together and not effectively fire separated. Factors that will affect the hazard classification are: (1) burning characteristic of the material stored, (2) combustibility of packaging, (3) method of storage and packaging, and (4) quantity stored. Modern developments in material handling have brought rapid change to storage occupancies, including high rack storage areas that can reach heights of 50 to 100 ft (15 to 30 m).

Computer controlled stacker cranes are now being used to move and place commodities in rack storage areas. Super regional distribution warehouses covering several acres under one roof which may contain two or three level mezzanines are now being developed. Storage occupancies usually have a small number of people in relation to total floor areas. Work patterns usually require employees to move through the structure using industrial trucks. In totally computer controlled warehouses, even smaller numbers of occupants are present, which can complicate the means of egress design.

Ministorage complexes are now being developed which consist of several small rental spaces ranging from 40 to 400 sq ft (8 to 80 m^2) in size in one building. These complexes are unique in that there is very little control over the type and amount of hazardous storage that may be found in a given rental area because of its accessibility to the renter on a 24 hour basis. The rental spaces are usually not separated from each other by fire resistance rated walls and may consist of 50 or more rental spaces in a building.

Storage occupancies can consist of raw materials, finished products, or goods in an intermediate stage of production. These materials can be found in bulk storage, solid piling, palletized storage, and rack storage arrangements. Bins and narrow shelves are also used, usually found in stockrooms, and hold only small amounts of materials for immediate use. The main difference between these storage arrangements having an effect on fire behavior and control is the nature of horizontal and vertical air spaces or "flues" that storage configurations create. Parking garages, grain elevators, and aircraft hangars have other unique and special problems caused by the nature of the operations performed in them. These problems will be addressed later in this chapter.

HAZARDS ASSOCIATED WITH OCCUPANCY

Contents

It is not unusual to find large densely stored quantities of inherently combustible materials, such as textiles and goods stored in combustible packing materials (wood, plastic, and cardboard). Highly flammable materials and gases may also be mixed within the materials in storage. Increasing amounts of plastics are now being used as part of the product as well as the packaging. In determining the hazard classification of contents, consideration should be given to the product, product container, and packaging materials used. Fire behavior will depend upon the ease of ignition, rate of fire spread, and rate of heat release of the commodity.

Commodities are often complex items whose fuel content, arrangement, shape, and form affect their performance in a fire. A packaged commodity must be considered as a whole, since that is the way it burns. Stored commodities are divided into general burning hazard classifications based upon the fire behavior of the typical items in each classification as defined in NFPA 231,

Mr. Hisley is a Training Instructor, Resident Programs Division, National Fire Academy, National Emergency Training Center, Emmitsburg, MD. He was Division Chief, Fire Prevention Bureau of the Anne Arundel County Fire Department, MD, before joining the Academy staff.

Standard for Indoor General Storage (hereinafter referred to as NFPA 231).

In determining the life safety hazard classification, a comparison of the low and ordinary hazard commodities can be made with those classifications found in NFPA 231. Special consideration and study of materials that are considered to be in the high hazard category are required.

The hazard to life safety should be based upon the most severe hazard category in the storage areas where mixed items are stored. The location of the high hazard items may not be limited to a given area or section in the building. This type of condition is common in hardware or automotive parts warehouses. High hazard materials or materials that require special consideration can sometimes be separated to avoid influencing the overall hazard to life safety classification. Examples of these materials are rubber tires, plastic products, combustible fibers, paper and paper products, hanging garments, carpeting, pesticides, flammable liquids and gases, and reactive chemicals.

Aerosol cans which contain flammable products stored at high pressure present a very high hazard risk. When ignited, they can explode to produce fireballs and can rocket throughout a given area to start multiple fires. The bulk storage of these containers should be in specially designed rooms that meet the requirements of NFPA 30, *Flammable and Combustible Liquids Code.*

Arrangement of Contents

The arrangement of the materials being stored has a major impact on fire spread. Fire behavior will depend upon the height of storage, aisle widths, and whether storage is in bulk, solid piles, or palletized piles. Fire development will depend upon the combustible surfaces of the stored goods and the horizontal and vertical flue spaces between the surface of the materials stored. Where combustible fibers are stored, a fast traveling surface fire is possible if the fiber bales have multiple exposed fibers.

Fire usually spreads in a fan shaped pattern—fire at the base preheats the material above which ignites and burns. As the height of the storage increases, the effect on the fire growth can be expected to intensify as the fire moves rapidly upward. Aisles that are at least 8 ft (2.5 m) in width are needed to help restrict the spread of fire across an aisle. Fast spreading fire can be expected in palletized storage caused by the presence of horizontal air spaces formed by the pallets. Rapid fire spread can also be expected in rack storage areas caused by the high heights of the rack and narrow aisles that are usually present when automatic material handling equipment is used.

OPERATIONAL HAZARDS AND FIRE PREVENTION PRACTICES

Storage occupancies can consist of large floor spaces on several levels, occupied by only a few personnel. Because of this factor, when fires start they often go undetected for an extended period of time in spaces that are not protected by automatic fire detection or suppression systems.

From a life safety standpoint, taking these conditions into consideration is very important so that hazardous conditions can be strictly controlled at all times. Common hazards that are generally associated with storage occupancies are careless disposal of smoking materials, poor housekeeping practices, packing and unpacking of goods, storage of idle pallets, and clearances from heat producing equipment. Specially designed materials handling systems can have unique hazards associated with their use.

Smoking

Smoking should be controlled by management at all times. The control of smoking requires a sincere effort by management to enforce observance of permissible and prohibited smoking areas for employees and outside visitors when they enter the shipping and receiving areas.

Industrial (Fork Lift) Trucks

Hazards associated with the use of industrial trucks is found with refueling operations, maintenance, and storage (when not in use). The type of industrial truck being used should be approved for use within the building for the hazardous material being stored. NFPA 505, *Fire Safety Standard for Powered Industrial Trucks Including Type Designations, Areas of Use, Maintenance and Operations,* designates the type of truck that can be used in hazardous areas. All refueling operations should be conducted outside the building. Fuel storage for the trucks requires special attention so that the fuel is properly stored and handled. Areas that are used for maintenance and battery recharging for electrical trucks should be separated from storage areas.

Welding and Cutting Operations

Precautions should be taken at all times during welding and cutting operations. Such operations should not begin until all combustible materials have been removed from the affected area or covered with a fire retardant cover. Portable fire extinguishers or small hoselines should be made ready before the operation begins. A fire watch should be present at all times during the operation and for at least 30 minutes after the operation is completed. Consideration should be given to the use of mechanical fastening devices and saws when repairing or replacing steel racks which would not require this type of operation. All cutting and welding should be conducted in accordance with NFPA 51B, *Standard for Fire Prevention in Use of Cutting and Welding Processes.*

Housekeeping Practices

The importance of good housekeeping cannot be overstressed. The packing or unpacking of goods usually requires that quantities of loose combustibles such as foamed polystyrene beads, cocoons of foamed plastic, shredded paper, excelsior, and straw be present. Other combustibles such as baled fibers, and detached labels or tags can act as kindling to ignite stored materials.

Packing and Unpacking Operations

Areas where packing and unpacking, label sorting, repairing, refinishing, painting, and general maintenance take place are usually associated with the accumulation of several types of combustible materials in many forms. Such areas are comparable to industrial occupancy. It is a good practice to either locate this type of operation away from the storage area or provide a fire resistance rated separation.

Storage of Idle Pallets

Piles of wood or plastic pallets introduce a severe fire condition that is incidental to most storage occupancies. With age they can be ignited easily from a small ignition source. Even in sprinkler protected areas, the undersides of the pallets provide dry areas on which a fire can grow and expand. The process of fire jumping to other pallets continues until the fire spreads through the top of the stack. With the strong updraft of flame very little water from an ordinary hazard designed sprinkler system can reach the seat of the fire. Pallets should be stored in accordance with NFPA 231.

Heat Producing Equipment

The storage of material should be arranged so that adequate clearance is maintained away from heating air ducts, unit heaters, duct furnaces, flue pipes, radiant space heaters, and lighting fixtures. Special care is needed when using temporary or portable heating appliances.

LIFE SAFETY CONSIDERATIONS

In storage occupancies the loss of life potential is dependent upon the fire hazard of the material in storage and the special characteristics of the storage building. Fire spread in high piled or rack storage can be expected to be rapid with some types of materials. In single story buildings where high ceilings are present, the smoke and heat produced by a fire will have a place to accumulate before the means of egress becomes obscured or impassable. In buildings with low ceiling heights, just the opposite will occur. Where multiple level, open grating floor type mezzanines are used, smoke and heat from fires below can be expected to readily block a means of egress.

Particular attention to the means of exit egress and exit locations is required. In older multiple floor warehouses, unprotected vertical openings for freight elevators and other material handling systems are common. These can cause rapid spread of smoke and heat to the upper floors.

Factors that will affect life safety in storage occupancies are (1) the stored commodity, consisting of the product, the packing and the container; (2) the fire behavior of the commodity, which is based on the ease of ignition, rate of fire spread, and rate of heat release; (3) the storage arrangement which includes height of storage and aisle widths; (4) maximum single fire area; (5) smoke and heat venting; (6) exterior fire fighting access, and (7) the form of the stored material.

NFPA 101, the *Life Safety Code®* (hereinafter referred to as NFPA 101) addresses life safety from fire and similar emergencies. Features necessary to minimize danger to life from fire, smoke, fumes, and panic are covered. The *Code* deals with egress and other considerations that are essential to life safety. The minimum requirement of NFPA 101 should be followed in storage occupancies.

Exit Design

The means of egress in storage occupancies are arranged similar to those in other occupancies. The chief difference is that fewer exits are required in storage occupancies. The required exits are based only on travel distance to reach an exit and the minimum number of exits required from a given area. Unlike other occupancies, there is no occupancy load factor to consider. Every type of storage occupancy should have access to at least two separate means of egress, as remote from each other as practicable. These two means of egress should be available from all parts and all levels in the occupancy. In special enclosed areas or rooms not exceeding 10,000 sq ft (929 m^2) in area that do not contain high hazard contents, only one means of egress needs to be provided. In occupancies that only contain low hazard contents there is no limit on the travel distance required to reach an exit. In ordinary and high hazard classified occupancies travel distance to reach an exit can be increased when the occupancy is protected with an automatic sprinkler system. In areas or occupancies that contain high hazard contents, the travel to reach an exit shall be via at least two separate routes remote from each other. In these areas special consideration is needed to make sure that no dead end aisles are present.

Exit doors shall be kept unlocked from the egress side of the door when the occupancy is occupied by employees. Exit doors from high hazard areas should swing in the direction of exit travel and be equipped with panic hardware. Due to the continual movements of stock within a storage occupancy, special attention is needed to keep required means of egress aisles and exit doors from being blocked.

Exit Identification

The means of egress and exit location have to be identified by signs that are readily visible from all directions of exit access. Exit identification can be a problem in high stocked storage areas, so other measures such as painting exit travel paths on the floor may be needed. Consideration to providing adequate illumination for the means of egress is needed, especially in windowless buildings that lack any opening for natural light. Buildings that are occupied at night or lack openings to allow for natural lighting should be equipped with emergency lighting.

Vertical Openings

Vertical openings are permitted in low or ordinary hazard storage occupancies, provided that the openings do not extend more than three floors, the building is protected throughout by an automatic sprinkler system, and the openings do not serve as required exits. Exits from all floor levels shall be protected in accordance with NFPA 101.

Fire Alarm Systems

Due to the large open floor areas and few employees who may be working in many different parts of the building, consideration of a fire alarm system in storage occupancies is needed to alert all occupants and allow for timely exiting. Because of the low risk of fire, occupancies having low hazard contents are not required to have an alarm system. NFPA 101 does require a fire alarm system in ordinary and high hazard content occupancies that are not protected by an automatic sprinkler system, and have a total accumulative floor area exceeding 100,000 sq ft (9290 m^2). In buildings equipped with a fire alarm system, provisions shall be made for the alarm system when activated to sound an alarm in a continuously attended location for the purposes of initiating emergency actions.

BUILDING CONSTRUCTION AFFECTING HAZARDOUS CONDITIONS

Storage occupancies can be located and found in any type of construction. In addition, storage is found in a mix with any other type of use due to the need for storage space that is associated with all types of occupancies. The influence on life safety of materials used in building construction depends largely upon whether or not the material will burn, support combustion, or create unusual amounts of smoke when exposed to fire. Most construction types are suitable for storage. Combustible construction, however, adds to the fire loading and can have combustible concealed spaces contributing to fire spread.

One must consider the type of covering used on steel deck roofs, plastic panels used for insulation within wall panels and roofing, and the flame spread rating of the vapor barrier for interior exposured insulation. Certain expanded plastic insulation core panel walls should be used only in buildings with automatic sprinkler protection. In newer buildings, lightweight steel and metal construction is now commonly used. In some buildings with rack systems, the structural framework of the racks themselves supports the exterior walls and roof of the building.

It is not uncommon to find buildings of up to 100 ft (30 m) in height used for rack storage. Rack systems that are fully automated usually have narrow aisle widths of 4 ft (1.2 m) between the racks. Even in buildings protected by automatic sprinkler systems, when flame spreads to the upper portions of the racks a serious problem can develop in providing complete fire extinguishment within the racks. The use of ground ladders for interior fire fighting is limited to 35 to 40 ft (11 to 12 m) at most. Access to the remaining upper portion for complete extinguishment and overhaul is very limited. The heat produced by the fire may cause structural failure of the rack system or collapse of the stored material into the aisle space. Usually these aisles are only accessible from one direction, and when material is blocking the aisle, access to the fire area is even more difficult.

Buildings or structures with limited exterior access for fire fighting and rescue operations can pose a serious life safety hazard. This type of design can cause large areas to become lethal fire environments. Automatic sprinkler systems are a requirement for life safety in windowless buildings and underground structures.

The installation of heat and smoke vents in storage occupancies should be considered. Some storage occupancies are extremely large areas, which makes dependence upon eaveline windows or doors for smoke and heat removal difficult. In buildings without automatic sprinkler protection, venting can reduce the horizontal spread of smoke and heat along the ceiling, which can aid in the evacuation of the building. In buildings protected by automatic sprinkler systems, venting can aid in smoke removal which in turn can aid fire fighting operations and possible rescue. Smoke and heat venting facilities are essential for life safety in windowless buildings and underground structures.

FIRE PROTECTION SYSTEMS

While automatic sprinklers are considered the main line of fire defense in storage occupancies, they are by no means the only method of protection. Standpipe hose systems, high expansion foam systems, portable fire extinguishers, manual fire fighting operations and alarm systems also have a role in protecting storage occupancies.

Goods stored to great heights, mechanical stacking equipment, and narrow aisles are highly conducive to vertical fire spread. Even with automatic sprinkler protection these types of storage arrangements can pose a high challenge for control. Disastrous fires have occurred in sprinklered warehouses where combustible materials have been stored to heights over 50 ft (15 m). Because sprinkler protection is a critical component in determining life safety risk, it is important that the systems be designed so they are compatible with the commodity stored and the storage heights.

Automatic Sprinklers

Automatic sprinklers are particularly effective for life safety because they warn of the existence of fire and at the same time apply water to the burning area. Properly installed and maintained sprinklers provide a highly effective safeguard against the loss of life and property.

Automatic sprinkler systems are usually required by the model building codes or locally adopted codes when the storage occupancy is classified as a high hazard occupancy. Some codes require sprinkler protection when materials are stored in racks and high piles that exceed 15 ft (5 m) in height. The model building codes usually allow buildings to be constructed with unlimited floor area when protected by sprinkler systems and not used for high hazard storage.

It is important to keep detailed records to determine what type of hazard is located in the storage occupancy. Records indicating the type of material stored, pile arrangement, storage methods, and height of storage should be maintained. Automatic sprinkler systems should be designed and installed in accordance with NFPA 13, *Standard for the Installation of Sprinkler Systems*; NFPA 231; and NFPA 231C, *Standard for Rack Storage of Materials*.

Fire Alarm Systems

Fire alarm systems are of the utmost importance in giving the occupants the needed warning of fire, especially in storage areas that contain only a few employees working on a large floor area in multiple story buildings. Alarm systems should have sounding devices that are distinctive from other sounds. Sounding devices should be distributed so as to be effectively heard in every room or area of a building. It is important to keep the alarm pull stations clear of any storage and well marked for easy identification.

Standpipe Hose Systems

Standpipe hose systems should be installed in storage occupancies that contain rack or high piled storage. Connection for 1½ in.* hoselines should be provided for initial fire fighting and mop up operations. Connections should be so located that all portions of the area can be

*Nominal size. For convenience, 1 in. equals 25.4 mm.

reached. Providing hose connections for rack systems that are over 50 ft (15 m) in height requires special consideration. In addition to the normal floor level connections, additional connections will be needed at different heights to enable the hoselines to reach all areas anticipating the maximum storage height. From a life safety standpoint it is critical that the proper emergency training be preplanned and given to employees in initial fire fighting operations.

Past experience has shown that employees who attempted to fight fire without the proper training or instructions rather than evacuate have been killed or injured. Delay in notifying the fire department has also helped fire to gain considerable headway.

USES REQUIRING SPECIAL CONSIDERATIONS

Parking Garages

Parking garages require special consideration in that the method of operation and use will affect the degree of life safety required. Parking garages can be operated where the unoccupied vehicle is parked exclusively by an attendant or by mechanical means, or where the customers themselves park the vehicle. NFPA 101 addresses exiting requirements when the garage is of a type where the customer parks the vehicle and can occupy all parts of the garage.

Occupancy Characteristics: Vehicle parking garages can be on several levels above or below grade of the enclosed or open type. A garage is considered open when at least 25 percent of the total enclosing walls are open to the atmosphere on two sides of each level. Parking garages can also be found in conjunction with other uses such as hotels, apartments, and office buildings. Garages can be used for both parking and repairs. When this condition occurs, the repair area should be fire separated from the parking area and treated as an industrial occupancy.

Hazards Associated With Garages: Motor vehicles, including automobiles, generally contain fast burning upholstery and gasoline, diesel, or LP-Gas as fuel. Newer model vehicles are now being constructed with major parts made of plastic. These vehicles can create severe fires that produce large amounts of smoke. Garages used for vehicle storage are classified as an ordinary life safety hazard. In some cases, the dispensing of gasoline is conducted in garages. NFPA 30A, *Automotive and Marine Service Station Code*, and NFPA 88A, Standard for Parking Structures list requirements for this type of operation. In addition, NFPA 101 has special life safety requirements when gasoline is dispensed inside. Parking garages are usually only occupied by a few persons at any one time. However, structures that are located near large public assembly complexes such as civic centers or theaters can be occupied by a large number of persons at one time after a show, affair, or event has ended. During this time, several hundred people will be moving throughout the structure either going to their vehicles or waiting in the vehicles to exit the parking garage.

Life Safety Considerations: To assure life safety conditions in a garage, every floor level shall have access to at least two separate exits, arranged so that the path of travel is in two different directions. Vehicle ramps can serve as a means of exit under special conditions. The travel distance allowed to reach an exit depends upon whether the building is open or enclosed and whether the garage is or is not protected by automatic sprinklers.

Special consideration must be given to the location of gasoline dispensing pumps to avoid the possible trapping of occupants if there is a fire or explosion at the pumps. Travel in any direction must lead to an outside exit at the same level, or to properly arranged exit stairs. Where parking areas are located below the gasoline dispensing pumps, each floor must have a direct access to the outside. This arrangement eliminates the possibility of gasoline vapors accumulating within the enclosed exit stair.

The means of egress to exits should be provided with illumination and emergency lighting when the garage is used at night or does not provide natural lighting from windows or other openings during the daytime. Most people using a parking garage are unfamiliar with the location of exits. Special consideration to the marking and identification of exits and access to the exits is needed.

Enclosed parking garages are required to have a fire alarm system when the total aggregate floor area exceeds 100,000 sq ft (9290 m^2) and is not protected by automatic sprinkler systems. Automatic sprinkler protection is required if the garage is enclosed and located below grade and in buildings over 50 ft (15 m) in height having combustible roofs, floor assemblies, or located inside or below buildings used for other occupancies.

Grain and Other Bulk Storage Elevators

These structures are used for the movement, storage, and processing of grain such as wheat, corn, oats, soy beans, and sunflower seeds. (See Fig. 9-15A.) During the handling of the commodity, some dust is generated due to kernel breakage, etc., making the grain stream susceptible to fire and explosion. Storage in bulk form can be found in concrete and steel silos, wooden bins, and steel tanks. Dust explosions are the primary hazard in the grain industry. NFPA 61B, *Standard for the Prevention of Fires and Explosions in Grain Elevators and Facilities Handling Bulk Raw Agricultural Commodities*, lists fire protection and safety requirements for the installation and operation of grain elevators.

Structures of this type are unique and particularly dangerous, posing a serious life safety hazard for all occupants. The hazards of both fire and explosion need special consideration. Past fire experience has shown that grain elevators have accounted for the major portion of life loss in storage occupancies. In most instances, they are used for the storage and handling of high hazard materials with combustible dust present. In many past dust explosions the primary and only means of exiting from the upper parts of an elevator have been damaged or destroyed, trapping the workers on top of the structure.

Life safety considerations that are addressed in NFPA 101 are based on the occurrence of fires and not on the possibility of a grain dust explosion. Basic life safety requirements require that at least two means of egress should be available and located as remote as possible from each other. These means of egress should be available from all working areas and the headhouse, which is located on top of the storage structure. All working areas should have access to at least one stairway that leads to an exit discharge point at the ground level. This stairway should

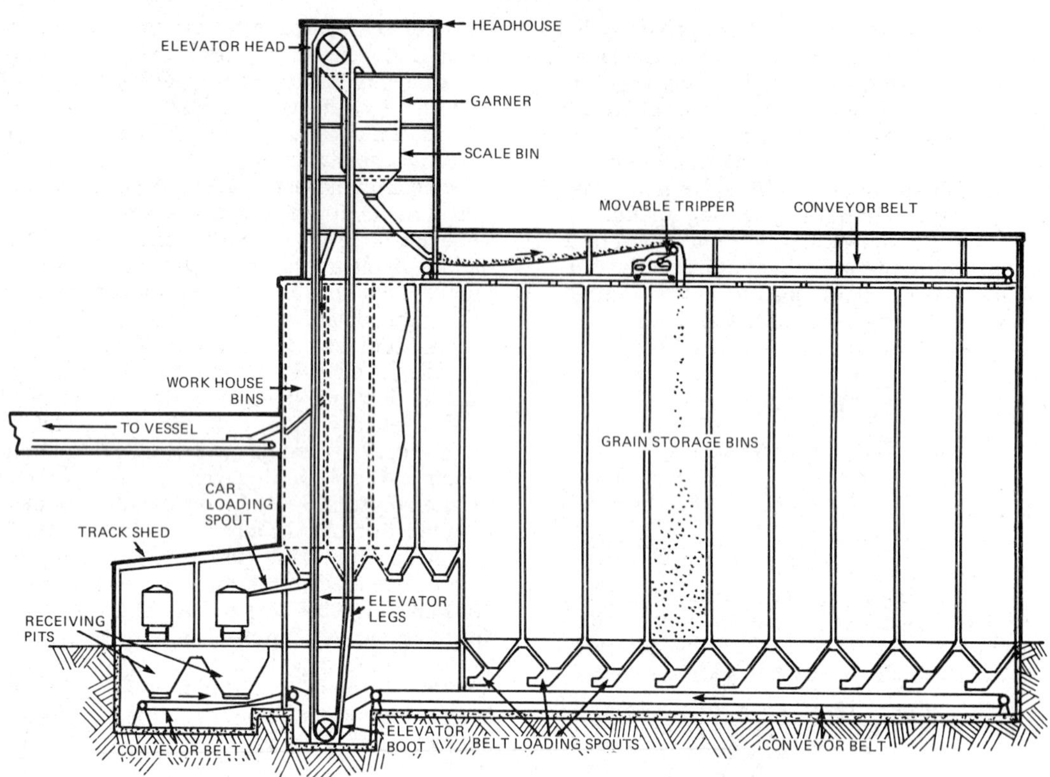

FIG. 9-15A. Cutaway view of a grain elevator.

be constructed within a dust and fire resistive rated enclosure.

In addition to the stairway, an alternate means of escape is required for life safety. This secondary means of escape should provide a continuous path to the ground from the top of the elevator structure, but does not need to meet all the provisions for a means of egress. This secondary escape can consist of exterior stairs or basket type ladders that should be accessible from all working levels of the headhouse. This means of escape will provide passage to the ground or to the top of adjoining structures.

An exterior stair or basket type ladder should be provided from the top of storage structures such as silos, conveyors, galleries or gantries, etc. This secondary escape should be located at the opposite end of the structure away from the headhouse. Most grain elevators have underground passages beneath the silos for conveyors. These spaces should have access to at least two means of egress so located that no deadend condition is present. One of the means of egress can be a ladder or other escape method that leads to a window, hatch, or panel that can be readily opened to allow escape.

Aircraft Hangars

An aircraft hangar is a large structure built to provide weather protection and shop space for aircraft during maintenance and storage. Since it is impractical to remove all fuel from an aircraft prior to moving it into a hangar, there is always the potential of having large quantities of flammable liquids inside the hangar. Fire during maintenance and servicing creates the greatest exposure to personnel in the hangar. Many maintenance procedures involve the use of highly flammable solvents and sometimes unstable toxic chemicals. Special precautions for ventilation and control of ignition sources are needed.

Aircraft hangars provide a particular challenge when designing adequate exits for life safety. The large open area needed for aircraft, especially jumbo jets, requires extensive travel distance to reach exits. This problem is compounded with the hazard of flammable and combustible liquids being present. Exit locations are required to be located at intervals not to exceed 150 ft (46 m) along all exterior walls of the hangar. Each part of the hangar or servicing area shall have access to at least two exit locations. If approved interior horizontal exits (two hour fire walls) are provided, then exit locations at intervals not to exceed 100 ft (30 m) along the wall shall be provided. Since a large hangar door cannot be left open in poor weather, small access doors can be installed within the larger hangar doors. To utilize additional space, mezzanine floors are found in some hangars. Exiting from these levels should be arranged so that an exit can be reached within a maximum travel distance of 75 ft (23 m). These exits shall lead directly to an approved enclosed stairway that discharges directly to the outside, to another cutoff fire separated area, or to outside stairs. Because hangars encompass such large areas, several hazard classifications can be found within the structure at one time. In those areas that are classified as high hazard or where a high hazard operation is being conducted, exits shall be available and located where no deadend conditions are present. For life safety purposes in hangars not protected by automatic sprinkler systems and exceeding 100,000 sq ft (9290 m^2) in total floor area, a fire alarm system is required.

Bibliography

NFPA Codes, Standards, Recommended Practices and Manuals. (See the latest *NFPA Codes and Standards Catalog* for availability of current editions of the following documents.)

NFPA 13, *Standard for the Installation of Sprinkler Systems.*
NFPA 30, *Flammable and Combustible Liquids Code.*
NFPA 30A, *Automotive and Marine Service Station Code*
NFPA 51B, *Standard for Fire Prevention in Use of Cutting and Welding Processes.*
NFPA 61B, *Standard for the Prevention of Fires and Explosions in Grain Elevators and Facilities Handling Bulk Raw Agricultural Commodities.*
NFPA 88A, *Standard for Parking Structures.*
NFPA 101, *Life Safety Code.*
NFPA 231, *Standard for Indoor General Storage.*
NFPA 231C, *Standard for Rack Storage of Materials.*
NFPA 505, *Fire Safety Standard for Powered Industrial Trucks Including Type Designations, Areas of Use, Maintenance and Operations.*

Additional Readings

Linville, J. L., ed., *Industrial Fire Hazards Handbook*, 2nd ed., National Fire Protection Association, Quincy, MA, 1985.
Moulton, Gene A., ed., *Conducting Fire Inspections—A Guidebook for Field Use*, National Fire Protection Association, Quincy, MA, 1982.
Tuck, C. A. Jr., ed., *NFPA Inspection Manual*, 5th ed., National Fire Protection Association, Quincy, MA, 1982.

OCCUPANCIES IN UNUSUAL STRUCTURES

Roger T. Boyce Sr.

Conventional occupancies in unusual structures are broadly defined as occupancies, such as assembly, mercantile, or business, that are located in structures not normally designed for the typical occupancy classification. Some common examples of occupancies in unusual structures would be:

1. A restaurant located on top of a tower (assembly occupancy).
2. A gift shop located in a fixed railroad car, boat, or airplane (mercantile occupancy).

GENERAL

Occupancy Characteristics

Occasionally, an owner will choose to locate a place of business in a structure which is totally different from his/her traditional line of business. The uniqueness of the structure attracts customers and provides a competitive edge for the specific business at hand.

While giving an owner a competitive edge, the uniqueness of a structure can also create confusion and difficult life safety problems not found in more traditional structures. NFPA 101, *Life Safety Code®* (hereinafter referred to as NFPA 101) has detailed provisions for each individual occupancy class, and in addition, provides minimum code requirements for occupancies in unusual structures. It is important to note that an occupancy in an unusual structure does not justify reducing the minimum code standards for the given occupancy classification. As a result, the occupancy must meet the requirements of the given occupancy classification as well as those requirements which may be necessary under the "Occupancies In Unusual Structures" portion of NFPA 101. In addition, the structure should strive to include those life safety features that may be deemed necessary due to the uniqueness of the facility itself.

Mr. Boyce is President of Systems Approach, a fire protection firm based in Newark, DE.

Ignition Sources

Unlike specific occupancy classifications, ignition sources in occupancies in unusual structures will vary tremendously. Once the basic occupancy classification, i.e., business, assembly, educational, has been determined, the ignition sources will be typical for that occupancy class. One must consider the unique features that make the structure unusual and determine what additional ignition sources may be present as a result. As an example, an assembly occupancy located on top of a tower may be more susceptible to lightning as an ignition source than would a traditional structure housing traditional places of assembly. In this case, the uniqueness of the structure requires extra protection that would not normally be indicated.

Fuel Load

Like ignition sources, fuel loads will be dependent upon the primary occupancy classification which is being utilized in the unusual structure. A study of the appropriate chapter for the occupancy classification under consideration will give sufficient insight to basic fuel loads that may be encountered. The fuel load then determines the hazards of contents classification per NFPA 101. When considering the fuel load of any structure, one must take into considertion that it is normally the fuel, or contents, within the building that creates the problem. Typically, the building is a "victim" of the fire, not necessarily the cause of the fire. The few exceptions to this rule would be combustible construction with concealed combustible spaces or combustible interior finishes which constitute a fuel load. Other than the exceptions, the contents should be the primary concern when evaluating the fuel. The following items should be taken into consideration when evaluating the fuel:

1. The amount of fuel, measured in pounds per square foot (kg/m^2).
2. Btu output per pound of fuel per square foot [$(W/kg)/m^2$].
3. Rate of heat release.
4. Configuration of fuel.

5. Burning characteristics of the fuel.
6. Amount of fuel per individual fuel station.
7. Configuration of fuel stations.
8. Burning characteristics of fuel stations.

Once the fuel and its arrangement have been evaluated, then the relationship of the fuel load to the unique features of the unusual structure must be assessed.

Construction

Unusual structures can be of any type of construction classification, and will most often be a combination of two or more construction types. Because a facility is classified as an occupancy in an unusual structure, this does not minimize the need to conform to the minimum construction classification requirements as specified under the traditional occupancy classification requirements.

Because of the uniqueness of the structure and its occupancy classifications, it may be desirable to increase the fire resistive characteristics of the construction classification to compensate for unique hazards that may be created. An example would be increasing the integrity of the egress systems through fire resistive enclosures or passageways, or requiring additionl fire separations and smoke divisions to allow additional time for safe egress. Many times the conceptual design of an occupancy located within an unusual structure will overlook even the most basic elements of good fire protection and prevention. Key components that are commonly forgotten are:

1. Concealed combustible spaces.
2. Protection of vertical openings.
3. Remoteness of exits.
4. Enclosure of exits.
5. Illumination and marking of exits.
6. Controlled interior finishes.
7. Automatic suppression systems.
8. Smoke management systems.
9. Fire alarm systems.
10. Enclosure of hazardous areas.

SPECIFIC UNUSUAL STRUCTURES

Towers

A tower is an independent structure or part of a building used for signaling or some similar limited purpose; towers are not used for routine industrial, business, or assembly purposes. When a tower is used for industrial, business, assembly, or other traditional occupancy purposes, the tower must completely comply with the appropriate occupancy classification and code provisions within NFPA 101 and other applicable model codes.

When a tower is being used in accordance with its definition, the means of egress for a tower is usually a single exit, and in some cases it may be only a ladder. If there is only one means of egress, caution must be taken to ensure that occupants cannot be trapped. No more than 25 persons may be on any one floor of a tower with a single exit. The tower should not be used for sleeping or living purposes. Fire resistive, noncombustible or heavy timber construction should be used, with either Class A or B interior finish.

The amount of combustible materials in, under, or around a tower must be carefully controlled; combustibles should be totally eliminated except for necessary furniture and operating supplies. Careful judgment should be used when determining what combustible materials to allow in the immediate vicinity of the tower. Similar care should also be used when evaluating a tower located near high hazard occupancies. The tower should be separated by a reasonable distance, especially when it has only a single means of egress.

Towers surrounded by water or offshore structures require special consideration and are strictly controlled by the U.S. Coast Guard. The Coast Guard regulations, or those of similar regulatory bodies in foreign countries, normally take precedence over NFPA 101 requirements.

In small towers where there will not be more than three occupants at any one time, the exit may be a ladder. Ladders should be arranged according to the requirements of ANSI A-14.3, *Safety Requirements for Fixed Ladders*, and must provide continuous access to grade at all times. In towers with a single exit ladder, vertical openings need not be protected because such a requirement does not significantly increase protection for occupants. Towers occupied on the top floor only, do not need enclosed stairs since the lack of operations at lower levels minimizes fire exposure. As an alternative, fire escape stairs may be used on open structures.

Windowless Buildings

The hazards of windowless buildings (Fig. 9-16A) are very similar to those of underground structures. Therefore, the same concerns found in underground structures would also be related to the windowless structure. The significant difference from the standpoint of NFPA 101 is that windowless structures are not required to have smoke venting facilities. Obviously, one should look at the feasibility of ventilating the structure and relate this to the classification, nature, and character of the occupancy. In the event that the occupancy is of such a character that a significant occupant load could occur therein, then smoke control and smoke venting capabilities will have a serious impact on the evaluation of the level of life safety.

Typically, a good exit system is not difficult to obtain in a windowless building, and this type of structure lends itself to the use of exterior access panels for fire department and emergency use. It should be kept in mind that access panels not meeting the requirements of the means of egress cannot be considered as an exit.

The most common problem with respect to windowless buildings deals with the identification as to whether the building does or does not have windows. It is very unusual to find a building that does not have at least a few windows. A quick look at the definition of windowless structures and an understanding of its intent will aid in determining whether or not the structure is considered windowless.

The definition from NFPA 101 is actually in two parts or addresses two considerations. The first part is "A building lacking any means for direct access to the outside." This portion of the definition essentially looks at the ability to egress the building in question which is not typically a problem unless the designer intentionally deleted exits along the exterior of the building. Exits are commonly deleted for access control and security purposes; security and access control are obviously important,

FIG. 9-16A. A typical windowless structure.

but cannot be allowed to interfere with a minimum level of life safety in a building.

The second portion of the definition addresses "Outside openings for light or ventilation through windows." This is the most common form of "windowless structure." The problem thus becomes, "How many windows must a building have before it is not considered a windowless structure?"

This determination must be based on the judgment of the authority having jurisdiction, but the definition does give some insight as to the intent of NFPA 101. The *Code* says, "Outside openings for (1) light or (2) ventilation." When NFPA 101 is addressing outside light, it implies that this light would be available to illuminate the means of egress. By the same token, when the *Code* is addressing ventilation, it implies that there will be a sufficient amount of means of ventilation available to allow for a safe period of evacuation. Thus, one must look at the facility and make a judgment as to whether or not (1) there is a sufficient amount of natural light being provided to the interior of the building via windows or opaque panels that would sufficiently illuminate the means of egress; and (2) whether there is adequate capability to ventilate the products of combustion during the period of egress. If there is not a sufficient number of windows or light panels that will allow natural illumination of the means of egress, then the building is most likely considered "windowless." Again, it should be emphasized that judgment must be exercised in determining whether or not the building is truly windowless as compared to the intent of the nature and character of the occupancy and the actual occupancy classification. As an example, a windowless building in a storage occupancy would not be as critical as a windowless building in a place of assembly. One must also keep in mind that an occupant load of 100 or more will be another major deciding factor.

Water Surrounded Structures

This category includes structures fully surrounded by water, including, for example, lighthouse and offshore platforms. Occupant load varies from a limited number of people to offshore oil platforms with 200 to 300 persons. The structures are generally unique. While a water surrounded structure may include areas which contain assembly, industrial, or residential occupancies, egress from the entire structure is limited because of its location.

Other than the fact that the structure is surrounded by water, the traditional occupancy requirements and fire protection features would remain the same with respect to the actual occupancy classification being evaluated. The occupancy should be addressed as if it were not surrounded by water, and the special circumstances that the unusual structure creates should then be applied so the unusual structure features can be evaluated as they relate to the occupancy on hand.

Often the structure is subject to severe environmental conditions including wind, wave action, and heavy rain. For those reasons, NFPA 101 excludes water surrounded structures from provisions of the *Code* and recommends compliance with U.S. Coast Guard regulations.

Open Structures

An open structure is very common in industrial operations. (See Fig. 9-16B.) Therefore, the structure will come

FIG. 9-16B. A typical open industrial structure. This particular building is a solvent extraction plant.

under the guidelines of an industrial occupancy most of the time. The problem common to most inspectors is trying to determine whether the occupancy is general, special purpose, or high hazard industrial. Typically, the high hazard industrial occupancy classification will be determined by the classification of the hazards of the contents which can be present at either general industrial or special purpose. Open air structures are commonly classified under special purpose industrial which is defined in Chapter 28 of NFPA 101.

Chapter 28 of NFPA 101 makes several provisions for special purpose industrial occupancies. The key is that the special purpose industrial occupancy is characterized by a relatively low density of employee population where much of an open air structure is occupied by machinery or equipment. Typically, the open structure facilitates access to the equipment by use of platforms, gratings, stairs, and ladders. The structure may even possess a protective roof to provide some protection from the elements.

Another difficulty with open structures is determining when a structure is open and when it is enclosed. Many open structures will be provided with some walls that are intended to shield the operations from environmental conditions or segregate opertions. This is cause for judgment by the authority having jurisdiction to determine whether or not the open air structure truly meets the definition for open structures or if the facility is a building. A guide that could be used is to determine whether or not the structure would react as an enclosed building in the event of a fire.

Walls acting in conjunction with a roof enclose the combustion process and products of combustion, thereby allowing the fire to spread both horizontally and vertically to other unaffected areas within the building. The advantage of an open structure is to allow the products of combustion to vent to the atmosphere instead of spreading to unaffected areas of the structure. Obviously, this is dependent upon wind and climatic conditions but the overall concept is valid.

One question to consider is whether the walls that are existing during the time of the inspection could cause the products of combustion during a fire to be directed or channeled to other unaffected portions of the structure, thereby preventing their free venting to the atmosphere? If the answer to this question is yes, then one may have to treat that portion of the open structure as a building. It is not uncommon to have buildings constructed within the framework of an open structure, resulting in a typical mixed classification.

Piers

A pier as defined by NFPA 87, *Standard for the Construction and Protection of Piers and Wharves*, (hereinafter referred to as NFPA 87) is a structure, usually of greater length than width, which projects from the shore to navigable water so that vessels may be moored alongside for loading, unloading, or storage. A pier "may be either open deck or provided with a superstructure." Contrary to the exact definition under NFPA 87, piers (sometimes called wharves) can be constructed over land. Occasionally, a designer will construct a pier of earth, pushing its way out from the main body of land into an area which for grade purposes, may not be otherwise previously usable. On this earthen structure the designer will erect a building and occupy it consistent with one of the previous occupancy chapters. Although the pier is not made of traditionally conceived structural elements, it is a pier nevertheless. One must first determine the normal state of events and then assess these features to the unusual condition. Many times a structure is located within what first appears as an unusual circumstance, but it will most likely have the typical fire protection problems.

NFPA 87 deals mostly with property conservation features of piers and wharves and references NFPA 101 for the applicable occupancy chapter. This would include Chapter 30, "Occupancies in Unusual Structures," where piers are addressed. The only exception noted throutout Chapter 30 comes under number of exits, wherein "piers used exclusively to moor cargo vessels and to store materials where provided with proper exit facilities from structures thereon to the pier and a single means of access to the mainland if appropriate with the pier's arrangement." Other than this exception, the appropriate occupancy chapter and the rest of Chapter 30 would apply to piers.

Bibliography

NFPA Codes, Standards, Recommended Practices and Manuals. (See the latest *NFPA Codes and Standards Catalog* for availability of current editions of the following documents.)

NFPA 87, *Standard for the Construction and Protection of Piers and Wharves.*

NFPA 101, *Code for Safety to Life from Fire in Buildings and Structures.*

Additional Readings

Bond H, "Underground Buildings," *Fire Journal*, Vol. 59, No. 4, July 1965, pp. 52-55.

"Fire Hazards of Windowless Buildings." *NFPA Quarterly*, Vol. 58, No. 1, July 1964, pp. 22-30.

"Occupancies in Unusual Structures." *Conducting Fire Inspections.* Chapter 15.

"Safety Requirement for Fixed Ladders," *ANSI A 14.3 −1974*, American National Standards Institute, NY.

SECTION 10
PROCESS FIRE HAZARDS

BOILER FURNACES

Revised by O. W. Durrant

Boiler furnaces use controlled combustion to generate steam to power machinery or to provide heat required by industrial processes. Basically, there are two types of boilers—watertube and firetube. In watertube units, water passes through tubes surrounded by hot combustion gases and is converted to steam. In firetube units, however, the hot combustion gases pass through tubes that are immersed in circulating water, which is converted to steam. Though this chapter discusses watertube boilers, the safety principles outlined are also applicable to firetube boilers. Fuel burning systems and related control equipment for single and multiple burner industrial and public utility boiler furnaces for fuels burned in suspension and on fluidized beds are discussed here. Solid fuel burning units, such as stokers, are not.

Guidance on fire protection for fuel supply systems is found in Section 11, Chapter 4, "Storage and Handling of Flammable and Combustible Liquids," and in Chapter 5, "Storage and Handling of Gases." Additional information on fuel grinders may be found in Section 10, Chapter 11, "Grinding Processes."

THE COMBUSTION PROCESS

Combustion may be defined as the rapid chemical combination of oxygen with the combustible elements of a fuel. There are three combustible elements of significance to this chapter—carbon, hydrogen, and sulfur. Sulfur is usually of minor importance as a source of heat, but can be of major significance in corrosion and pollution problems.

When burned to completion with oxygen, carbon and hydrogen unite according to the following:

$C + O_2 = CO_2$ + 14,100 Btu per lb (or 32 797 kJ/kg) of carbon

$2H_2 + O_2 = 2H_2O$ + 61,100 Btu per lb (or 142 119 kJ/kg) of hydrogen

Air is the usual source of oxygen for boiler-furnaces. These combustion reactions are exothermic, and the heat released is about 14,100 Btu per lb (32 797 kJ/kg) of carbon burned and 61,100 Btu per lb (142 119 kJ/kg) of hydrogen burned. The objective of efficient combustion is the release of all available heat while minimizing losses from combustion imperfections and superfluous air.

The combustion process in a boiler-furnace results from a continual introduction of fuel and air in a flammable mixture. It is necessary to maintain control not only of the flow rates of both fuel and air, but also of the air-fuel ratio (at both the point of ignition and for the entire burner), as well as the source of ignition. If any one of the inputs is interrupted or becomes irregular, ignition can be lost and the controlled flammable mixture can become an uncontrolled explosive mixture.

Figure 10-1A shows schematically how fuel and air

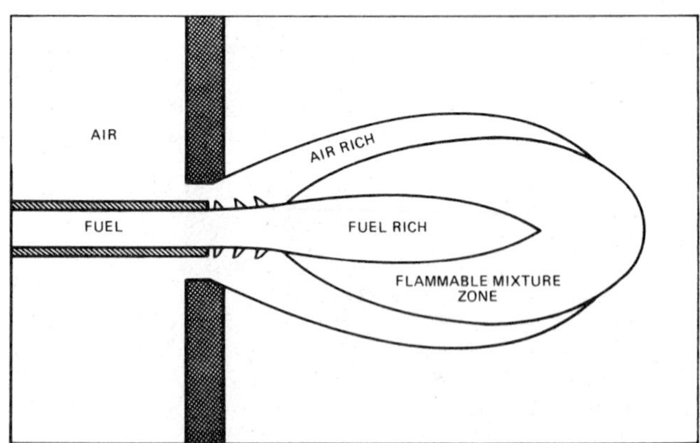

FIG. 10-1A. Schematic of burner flame conditions. (Babcock & Wilcox)

are introduced and burned in a furnace through a modern high capacity burner. The fuel and air are injected in separate impinging streams. Normally the total burner air is limited by some means at the point of introducing the fuel to provide a flammable mixture zone. The zone is often fuel rich, to ensure stable ignition—assuming there is sufficient ignition energy. The mixing and burning of the remaining fuel and air takes place in the furnace. With a turbulent, inherently stable burner, ignition always occurs

Mr. Durrant, retired from Babcock & Wilcox, Barberton, OH, is now a consultant.

at the burner and can be designed to be as stable with a fuel rich ratio as with an air rich ratio. With ignition established at the burner, combustion continues, as the air and fuel continue to mix in the furnace and become flammable until all of the oxygen is consumed (fuel rich ratio) or until all the fuel is consumed (oxygen rich ratio). The combustion process, therefore, proceeds toward completion through a wide range of air-fuel ratios, with the final result dependent on the total ratio of air to fuel over the entire path of the flame.

Fuels

Natural gas, oil, and pulverized coal are the most commonly used fuels to fire boiler-furnaces. However, there are also alternative fuels, such as crushed coal, methanol, or liquefied petroleum gas; and supplementary fuels, such as waste gases or flammable liquids in chemical plants and refineries. Each may demand special precautions in addition to those required for the more common fuels. Sometimes more than one fuel is used; Figure 10-1B shows combination gas and oil burners on a larger public utility boiler furnace.

FIG. 10-1B. Installation for combination oil and gas firing in a large electric utility boiler. (Babcock & Wilcox)

OIL AND GAS BURNING SYSTEMS

The burner is the principal component of oil and gas firing equipment. Its purpose is to introduce fuel and air into the furnace in the proper proportion to sustain the exothermic chemical reactions (combustion) for the most effective release of heat.

Normal use of a steam generator requires operation at different outputs to meet varying load conditions. The specified operating range or load range for a burner is the ratio of full load on the burner to the minimum load at which the burner must be capable of stable ignition and reliable operation. For example, with a boiler of 100,000 lb per hr (45 359 kg/hr) capacity (steam delivered), a load range of four to one on the burners means that the unit can produce stable ignition and complete combustion at any load from 100,000 lb per hr (45 359 kg/hr) down to 25,000 lb per hr (11 340 kg/hr) without necessitating a change in the number of burners in operation.

It is necessary to supply more than the theoretical air quantity to assure complete combustion of the fuel in the combustion chamber. The amount of excess air provided should be just enough to burn the fuel completely in order to minimize the heat loss in stack gases. The excess air normally required with oil and gas, expressed as a percent of theoretical air, is given in Table 10-1A.

TABLE 10-1A. Usual Amount of Excess Air Supplied to Fuel-burning Equipment

Fuel	Type of Furnace or Burners	Excess Air (% by weight)
Pulverized coal	Completely water-cooled furnace for slag-tap or dry-ash removal	15–20
	Partially water-cooled furnace for dry-ash removal	15–40
Crushed coal	Cyclone furnace, pressure or suction	10–15
Fuel oil	Oil burners, register-type	5–10
	Multifuel burners and flat-flame	10–20
Acid sludge	Cone- and flat-flame-type burners, steam-atomized	10–15
Natural, coke-oven, and refinery gas	Register-type burners	5–10
	Multifuel burners	7–12
Blast-furnace gas	Intertube nozzle-type burners	15–18

The most frequently used burner for gas and oil is the circular burner shown in Figure 10-1C. Normally, the capability of an individual circular burner is limited to 190 to 200 million Btu per hr (55 689 to 58 620 kW). The tangentially displaced doors built into the air register provide the turbulence necessary to mix the fuel and air and to produce stable flames. While the fuel is introduced to the burner in a fuel rich mixture in the center, the direction and velocity of the air and the fuel dispersion pattern mix the fuel thoroughly with the combustion air. The circular burner may be modified with an inner register to limit air in the primary ignition zone to limit formation of oxides of nitrogen (NO_x). The principle and performance are similar to those shown for the dual register coal burner. (See Fig. 10-1I.)

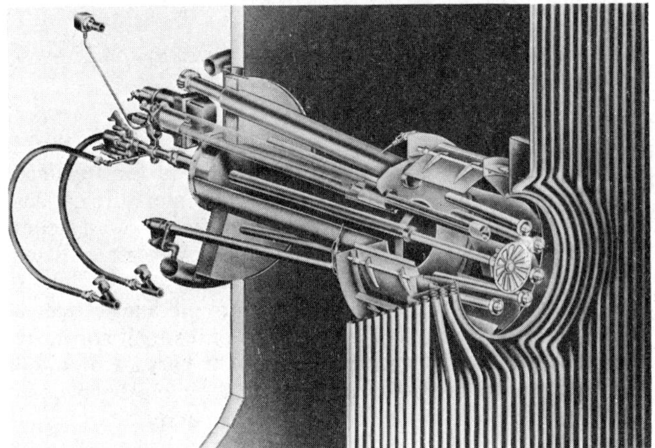

FIG. 10-1C. Circular register for gas and oil firing. (Babcock & Wilcox)

Oil Burners

To burn fuel oil at the high rates demanded of modern boiler units, it is necessary that the oil be atomized; that is, introduced into the furnace as a fine mist. This exposes a large amount of fuel particle surface to air, which ensures prompt ignition and rapid combustion. There are several types of oil atomizers or vaporizers, but the discussion here is limited to the two most popular types—steam or air atomizers, and mechanical atomizers.

For proper atomization, oils with a grade heavier than No. 2 must be heated to reduce their viscosities to between 135 and 150 SUS.* Steam or electric heaters are used to raise the oil temperature to the required level—approximately 135°F (57.2°C) for No. 4 oil, 185°F (85°C) for No. 5, and 200 to 220°F (93.3 to 104.4°C) for No. 6. With certain oils, better combustion is obtained at somewhat higher temperatures than are required for atomization. However, oil temperature must not be raised to the point where vapor binding occurs in the pump that is supplying the oil, since this could cause flow interruptions followed by loss of ignition.

It is also important that oil be free of acid, grit, and other foreign matter that might clog or damage burners or their control valves.

Steam or Air Atomizers: The steam or air atomizer (Fig. 10-1D), in general, operates by producing a steam-fuel (or air-fuel) emulsion that atomizes the oil through the rapid expansion of the steam when released into the furnace. The atomizing steam must be dry because moisture causes pulsations, which can lead to loss of ignition. Where steam is not available, moisture-free compressed air can be substituted.

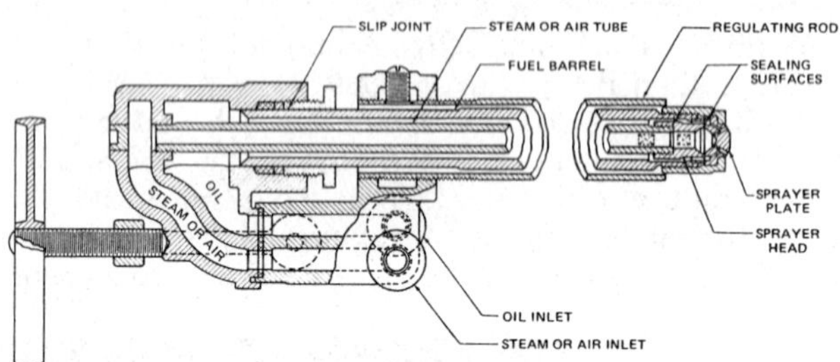

FIG. 10-1D. Steam (or air) oil atomizer assembly. (Babcock & Wilcox)

The steam atomizer performs more efficiently over a wider load range than other types of atomizers. It normally atomizes the fuel properly down to 20 percent of rated capacity; in some instances, steam atomizers have been successfully operated at 5 percent of capacity.

Mechanical Atomizers: In the mechanical atomizer, the pressure of the fuel itself is used as the means for atomization. Mechanical atomizers are available in sizes up to 180 million Btu per hr (52 758 kW) input, about 10,000 lbs (4536 kg) of oil per hr.

Mechanical return-flow atomizers (Fig. 10-1E) are used for many marine installations and some stationary units where the use of steam is inappropriate, impractical or uneconomical. Oil pressure required at these atomizers for maximum capacity ranges from 600 to 1,000 psi (4137 to 6895 kPa), depending on capacity, load range, and fuel. Return-flow atomizers are ideally suited for standard grades of fuel oil where it is desired to meet load variations without changing sprayer plates or cutting burners in and out of service.

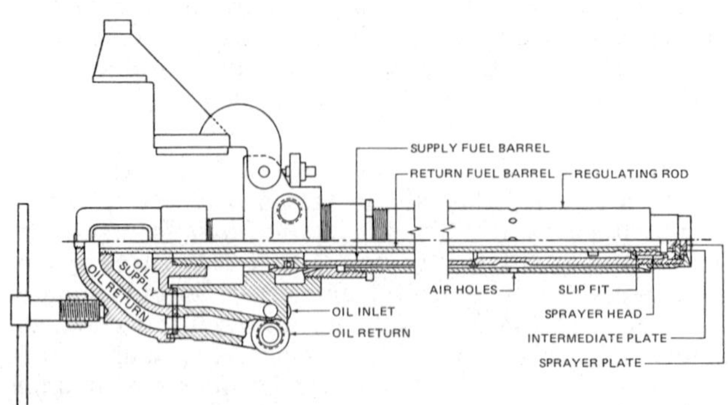

FIG. 10-1E. Mechanical return-flow oil atomizer assembly. (Babcock & Wilcox)

Natural Gas Burners

Natural gas is an ideal fuel for a burner, since it requires no preparation to be suitable for rapid and intimate mixing with combustion air. Basically, gas burners mix fuel and air in either of two ways—premix or external mix.

In premix burners, fuel and a portion of the air are mixed before they are introduced to the burner nozzle input. A common method involves mixing gas and air in the suction side of a mechanical blower. Another method uses the venturi effect. A gas jet creates a negative pressure at the air input orifice and draws air into the mixer. The third method also uses the venturi effect, but in this case, an air jet draws fuel gas into the mixer. The latter system requires a gas regulator to reduce fuel gas pressure to atmospheric before gas and air are mixed.

In external mix gas burners, the fuel and air are mixed external to the nozzle.

In external mix gas burners with individual elements or spuds (Fig. 10-1C), part or all of the gas discharges in front of the impeller, which provides a local fuel rich zone and thus serves as an ignition stabilizer at high loads. Each burner comprises several spuds, each of which is a gas pipe with multiple holes at the end to discharge gas for ignition at the burner throat. Each spud is fitted with a larger diameter flame holder to provide a local fuel rich zone to stabilize ignition at low inputs. Fuel ports are relatively large to minimize plugging during service. Gas spuds are located so that oil will not impinge upon them when provision is made for firing multiple fuels. Because of its unusually good ignition stability, even under severe variations in airflow, the multiple spud type burner is eventually expected to replace most natural gas element designs now used on circular type burners.

With the proper selection of control equipment, a multifuel-fired furnace with a multiple spud type burner is

*Abbrevation for Saybolt Universal Seconds. The efflux time in seconds it takes for a 60 mL sample to flow through a calibrated Universal orifice in a Saybolt viscometer under specified conditions.

capable of changing from one fuel to another without a drop in load or boiler pressure. It is also capable of simultaneous firing of natural gas and oil in the same burner. The primary function of the control system is to balance fuels in the proper relationship with burner airflow to achieve complete and efficient combustion. This type of element is designed for use with natural gas or other gaseous fuels containing at least either 70 percent methane, 70 percent propane, or 25 percent hydrogen by volume. The element is also designed for a maximum input of 173 million Btu per hr (50 706 kW) per burner.

To provide safe operation, ignition of a gas burner should remain close to the burner impeller(s) throughout the full range of allowable gas pressures, not only with normal airflows, but also with abnormally high airflows. Ideally, it should be possible at the minimum load to pass full load airflow through the burner and at full load as much as 25 percent in excess of theoretical air without loss of ignition. With this range of airflow, it is not likely that ignition can be lost even momentarily during some upset in airflow due to improper operation or error.

PIPING AND CONTROL DEVICES

It is essential that boiler-furnace fuels be as free as possible from contamination and be under control to ensure combustion safety. The design and reliability of fuel handling systems are, therefore, important factors in minimizing the risk of explosion and fire.

Oil Fired Systems

Fuel supply equipment should be designed and sized to ensure a continuous, steady flow of fuel that will meet all the operating requirements of the unit. This includes coordinating the burner header pressure regulator (Fig. 10-1F), burner shutoff valves, and associated piping volume to prevent fuel pressure transients that might exceed stable flame burner limits as a result of cutting burners in or out of service.

Fill and recirculation lines should be connected to storage tanks below the liquid level to avoid excessive evaporation and the generation of static electrical charges in free falling fuel.

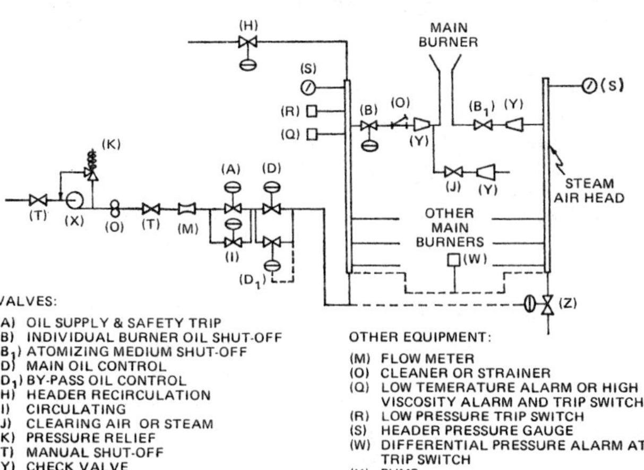

FIG. 10-1F. Schematic of typical main oil burner-steam or air atomizing.

Fuel oils may contain abrasive, corrosive, or waxy contaminants, which may clog filters, produce wear, or otherwise damage oil burning equipment. Strainers, filters, traps, and sumps are some of the devices that may be used to remove harmful contaminants from oil burning systems.

It is good practice to route piping and locate valves to minimize their exposure to physical damage and extreme temperatures that may alter fuel viscosity or pressure.

Burner shutoff valves should be located as close to the burner as possible to minimize the volume of oil that may remain in the burner line downstream of the valve or that may drain into the furnace following an emergency trip or burner shutdown. Positive means should be provided to prevent oil from leaking into an idle burner.

Fuel oil must be delivered to the burners at a specific temperature and pressure to ensure proper atomization. Adequate recirculation provisions must be made for controlling the viscosity of the oil at the burners for initial light-off and subsequent operation. It is important that these systems prevent excessively hot oil from entering the pumps; otherwise, the pumps may become vapor bound and interrupt the supply of fuel to the burners.

Positive means are needed to prevent fuel oil from entering the burner header system through the recirculating valve, particularly from the fuel supply system of another boiler. Check valves have proved unreliable for this function in heavy oil service. Provisions should also be required for clearing or scavenging the passages of an atomizer into the furnace.

Gas Fired Systems

Natural gas fuel supply systems must be able to provide a continuous, steady flow of fuel that is adequate for all operating requirements of the unit and within specified fuel header pressure limits for the burners. The system (Fig. 10-1G) includes a burner header pressure regulating valve and a bypass or minimum pressure regulating valve. The bypass valve comes into use automatically for start up and low flow operation to avoid operating with a header pressure below the minimum pressure prescribed for burner stability.

The portion of the fuel supply system shown in Figure 10-1G which is located outside the boiler room is arranged to prevent excessive fuel gas pressure in the burner supply system, even in the event of failure of the main supply constant gas pressure regulators. Usually this is accomplished by providing full relieving capacity which is vented to a safe location. Where full relieving capacity is not installed, a high supply gas pressure trip may be provided.

The system should have positive means to prevent gas from leaking into an idle furnace. Piping should be vented upstream of the last shutoff valve in any line to a burner or igniter. Provision should also be made in the gas piping to permit testing for leaks and for their subsequent repair. These provisions should include means for making tightness tests of the main gas supply safety shutoff valves and the main burner gas safety shutoff valves.

Vents should be located so there is no possibility of vented gases entering the boiler room, adjacent buildings, or air intakes to ventilating systems. The vents should be placed high enough so escaping gas will not be a fire hazard. Header vent lines should be run independently, and the igniter vent subsystem should be run indepen-

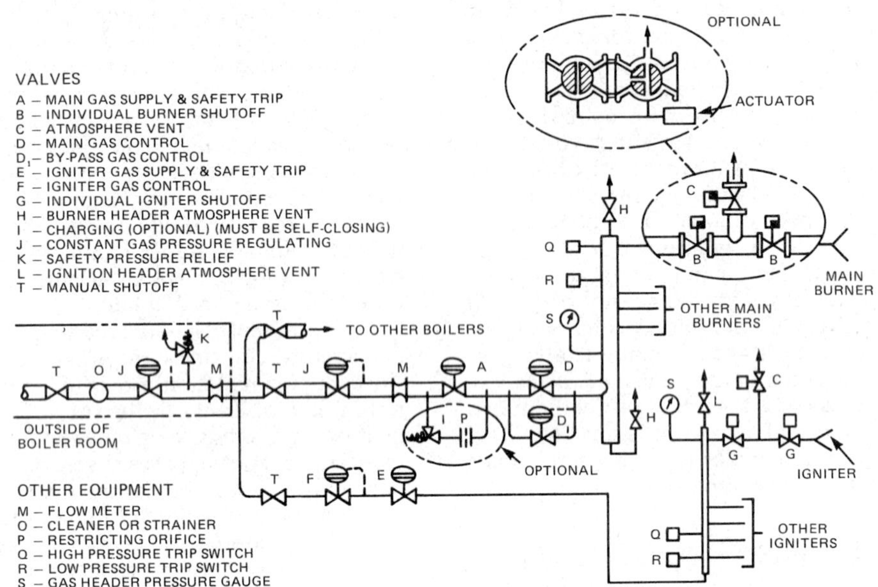

FIG. 10-1G. Typical fuel supply system for natural gas fired multiple burner boiler-furnace.

dently of the burner vent subsystem. There must be no cross connection between venting systems of different boilers.

PULVERIZED COAL SYSTEMS

There are several possible arrangements for pulverized coal fired systems. One of the variables is the location of the primary air fan in relation to the air heater and pulverizer. The combination shown in Figure 10-1H is known as a "hot" system and includes individual primary air fans for each pulverizer located downstream of the air heaters. Often the blowers are located ahead of the pulverizer as shown in Figure 10-1H. Sometimes, the air fans are exhausters located behind the pulverizer.

Coal and primary air are transported from each pulverizer to its associated burners in measured, controlled amounts, well distributed among the burners. The coal-air mixture is combined with a measured and controlled amount of secondary air at the burner to serve several purposes.

The fuel must be volatilized and completely consumed in a continuous process starting with an initial zone of stable ignition at the burner and continuing through the combustion process with no more than a trace of unburned combustibles in the stack gas and ash hoppers.

To control stack emissions, especially oxides of nitrogen, the amount of oxygen in the initial combustion zone is limited. To maintain a low level of nitrogen oxide (NO_x), and to limit stack losses, the total air to each burner and to the burner zone of the furnace must contain a low level of excess air. The limited air should surround the combustion process at each burner to minimize reducing atmosphere in contact with tube surfaces. Stable ignition is necessary at each burner to ensure continuous combustion and a means of flame detection at each burner.

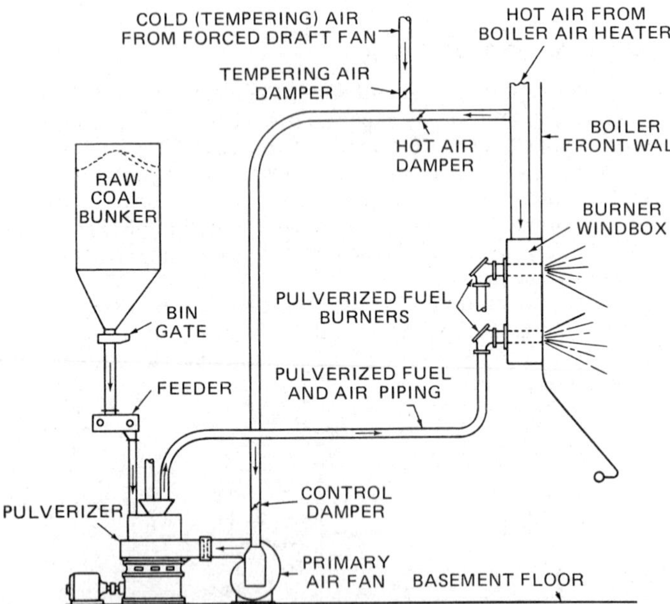

FIG. 10-1H. Direct-firing system for pulverized coal. (Babcock & Wilcox)

Sufficient furnace water wall surface between burners is necessary both to minimize the generation of thermal NO_x by cooling the combustion zone, and to cool ash particles below the temperature that causes slag deposits on furnace or superheater surfaces.

Products of combustion are transported from the furnace through platen and pendant superheater and reheater surfaces, through the convection pass (including the economizer surface), and through the air heater to accomplish the necessary heat transfer before being transported to the air pollution control system(s) by the induced draft (ID).

Achieving the environmental limitation on NO_x levels requires a low turbulence burner with the air and fuel well distributed to minimize the available oxygen in the initial ignition zone and excess air in the furnace. To achieve the conditions discussed in the preceding paragraphs, burner designs similar to that shown in Figure 10-1I, and a

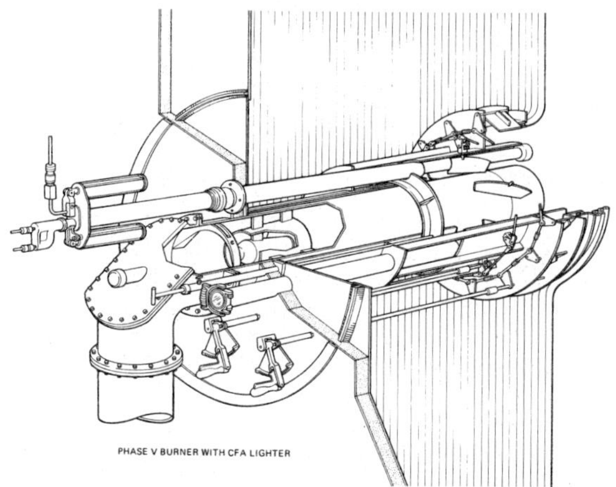

FIG. 10-1I. Dual register burner for NO_x control pulverized coal firing.

compartmented burner system such as that shown in Figure 10-1J were developed.

Figure 10-1I shows the operating principles of a dual register pulverized coal burner used to achieve low levels of nitrogen oxide emissions. The burner, as its name implies, has two registers to proportion the air between the fuel rich, initial ignition zone and the secondary combustion air zone. A third means of adjustment is the position of the swirler vanes to control the turbulence and, thus, the stability of the initial ignition zone. The turbulence and quantity of secondary air to the initial ignition zone is limited to that required for initial ignition and for continuous combustion of the fuel. The remainder of the secondary combustion air is mixed with the fuel within the furnace to provide slow but efficient and continuous combustion until the fuel is completely consumed. A flow distribution device is located in the coal nozzle to ensure a uniform mix of the primary air and coal just prior to emission from the nozzle.

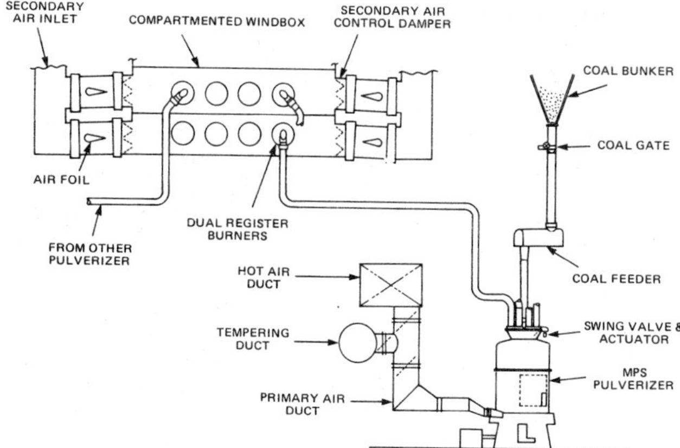

FIG. 10-1J. Pulverized coal firing system. (Babcock & Wilcox)

The compartmented burner system (Fig. 10-1J) provides each burner group served by its pulverizer with a separate windbox with secondary air measured and controlled at each end. Thus, combustion control has been modified from a total boiler basis to a per pulverizer basis to achieve more precise and more flexible control of the combustion process.

Figure 10-1J illustrates another fan-to-pulverizer combination of a pulverized coal system known as a "cold air" or "common primary" fan system. Here, the primary air fans blow cold air through air heaters to supply the common duct for hot primary air, and the common duct for cold primary air that bypasses the air heaters.

The large steam generating capacity of modern boilers, coupled with poorer fuel quality, has created the need for high capacity pulverizers capable of processing 50 tons (45.36 metric tons) or more of coal per hour. The pulverizer shown in Figure 10-1K is a roll and race type that utilizes three large diameter rolls spaced equally around the mill. The rolls are mounted on axles; and, in turn, the roll assemblies are attached by a pivoted connection to a stationary overhead frame that keeps them in their roll path while permitting limited radial freedom of movement. Grinding pressure is supplied by springs that apply force to the axles of the rolls. The grinding ring rotates at low speed and is shaped to form a race in which the rolls run.

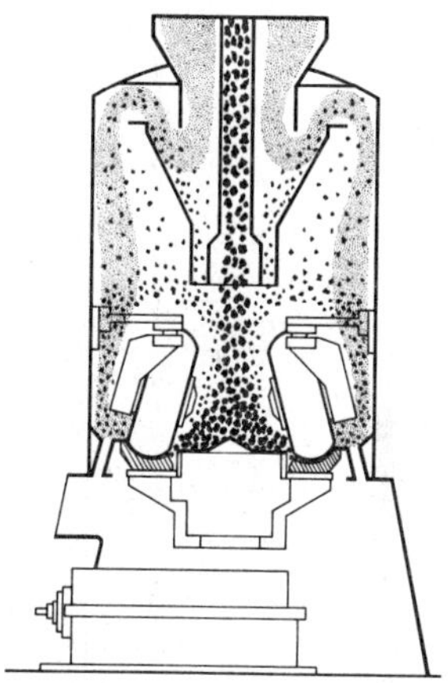

FIG. 10-1K. MPS pulverizer coal recirculation. (Babcock & Wilcox)

Circulation of coal is similar to that shown in Figure 10-1K. Raw coal is fed into the mill either inside or outside of the grinding race. It immediately mixes with the partially ground coal that is circulated within the grinding zone by the airflow through the pulverizer. As the coal is reduced in size, the air carries it to the classifier. Fine coal, along with air, leaves the pulverizer through the outlet pipes. The oversize coal is returned to the grinding zone through a seal at the bottom of the classifier.

FLUIDIZED BED COAL SYSTEMS

The rising cost of fuel, the requirements of air pollution laws regarding sulfur, NO_x; and particulate emission have caused the utility and industrial power industry to develop more economical and efficient boiler systems. A recent technology developed to meet these challenges is the fluidized bed boiler which uses low grade, lower cost energy sources.

Fuel combustion is carried out in a bubbling bed of calcined limestone. Calcium oxide from the limestone reacts with sulfur dioxide produced by combination in a temperature range of 1,500 to 1,600°F (815 to 871°C) to form calcium sulfate in a dry solid state. (See Fig. 10-1L.)

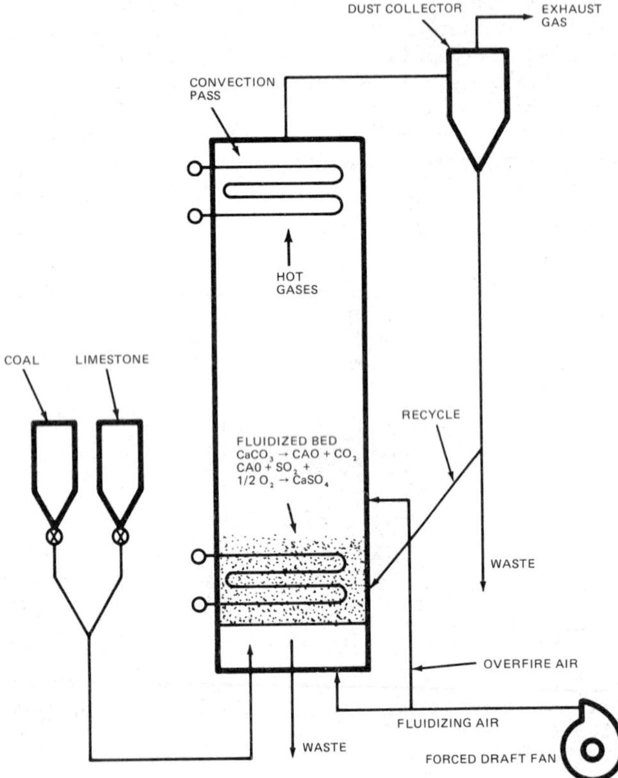

FIG. 10-1L. Fluidized bed system.

In this way, federal emission regulations can be met without installation of flue gas desulfurization equipment.

Cost effective use of low grade fuels is possible because combustion is maintained by the residence time, turbulence, and temperature achieved in the immense, hot mass of the fluidized limestone bed. The temperature of the bed is below the softening temperature of ash so the ash remains in a solid condition throughout the boiler convection pass, superheater, reheater, and economizer. Problems of slagging and fouling are virtually eliminated from the bubbling, fluidized bed boiler.

The major difference between the fluidized bed combustion of coal and pulverized coal combustion is the temperature at which combustion takes place. In Atmospheric Fluidized Bed Combustion (AFBC), the temperature of the combustion environment (usually limestone fluidized by air percolating upward from the bottom) is kept at 1,550°F (843°C). To maintain this desired equilibrium temperature, heat is constantly removed from the bed by a superheater surface or boiler bank within the bed. The 1,550°F (843°C) temperature, compared to 2,500 to 3,000°F (1371 to 1649°C) for pulverized coal boilers, is the source of three major advantages of AFBC over pulverized coal:

1. Sulfur dioxide (SO_2) absorption by a limestone bed is near optimum at this temperature, eliminating the need for a flue gas scrubber.
2. Nitrogen oxides (NO_x) emissions are kept well below federal emissions standards.
3. Slagging and fouling by the coal ash is prevented because the temperature is well below the fusion temperature of the ash.

Bubbles of flue gas are formed in the combustion process and rise through the bed, causing highly turbulent mixing. Smaller limestone particles become entrained in the gas stream and are carried over through the convection pass. These smaller limestone particles are collected in multicone dust collectors for recycling to the fluidized bed. Recyling of the unreacted limestone and unburned carbon for reinjection into the fluidized bed greatly increases efficiency of both sulfur capture and carbon burn up.

The integration of the fluidized bed combustion process with a boiler includes the boiler surface within the bed, superheater surface within the bed, water wall enclosure to contain the bed, and a water cooled air distributor plate to support the bed.

BOILER-FURNACE HAZARDS

Explosion and fire are the principal hazards of boiler-furnaces and their associated fuel supplies, pipes, ducts, and fans. Explosions are the result of the ignition of combustible mixtures of fuel and air that have accumulated in the confined spaces of the equipment. Generally, such accumulations are the result of a malfunction or operator error associated with an inadequate or improper purge or incorrect operation of the burner equipment.

A temporary loss of flame caused by an interruption in fuel or air delivery or in ignition energy may allow a combustible air-fuel mixture to accumulate in the furnace before ignition is reestablished. Leaking fuel may collect in an idle furnace and ignite explosively when the burner is lighted. In multiburner units, the loss of flame at one or more burners may allow an explosive mixture to accumulate, only to be ignited by other burners, either while operating normally or while being lighted. A classic example of operator error is the failure to purge the furnace of combustible mixtures between repeatedly unsuccessful attempts to light off the burner.

Hazards of Oil Firing

Like other petroleum products, fuel oil is a complex blend of hydrocarbons having a wide variety of molecular weights, as well as boiling, freezing, and flash points. When subjected to sufficiently high temperatures, fuel oils will partially decompose or volatize or both, thereby creating new and unpredictable liquid, gaseous, and solid fuels.

Historically, there were two grades of fuel oils—distillates and residuals. Residual fuel oils were left after the distillation of crude oils to obtain distillate fuel oils. Modern refining practices involving the cracking process permit more and different types of finished products to be obtained from a single barrel of crude; thus, a wide range of fuel oils is now available. No. 1 and No. 2 fuel oil, and fuels commonly known as kerosenes, range oil, furnace oil, and diesel oil may still be broadly classified as distillates, while No. 5 and No. 6 fuel oil may be referred to as residuals. Most power boiler fuel systems are designed for the heavier No. 5 or No. 6 oil, although distillate fuels may be used for boilers designed with natural gas as the primary fuel. The use of crude oils, even those that have been given some treatment to remove the gassy light ends, is no longer a common practice. It is necessary to preheat the heavier oils (Nos. 4, 5, and 6) to hold the viscosity of the oil flowing to the burners within acceptable limits that will ensure proper atomization.

If two shipments of fuel having widely different viscosities or specific gravities are stored in the same tank, a significant change in fuel input rate without a corresponding change in airflow may occur and impair the efficiency of combustion.

Combustion efficiency of mechanical atomizing burners may also be affected by a change in orifice size caused by wear. In burners operating with very little excess air, combustibles may collect in the furnace. Periodic flow testing and/or replacement of sprayer plates may be necessary. Unsafe operating conditions can also be created by the failure of installation or maintenance personnel to install a nozzle or sprayer plate in the burner assembly.

Oil flow to individual oil guns can be adversely affected by such conditions as variations in burner elevation, distance from the regulating valve, and pipe size—all of which can be hazardous on low pressure burners.

Hazards of Gas Firing

The primary hazards in gas fired systems are gas leaks and the development of fuel rich mixtures within the furnace or structural enclosures. Potentially hazardous accumulations are most likely to develop within buildings, particularly where gas piping is routed through confined areas that are not adequately ventilated.

Within the furnace it is possible for air-fuel ratios to be altered severely without producing any visible evidence at the burners, furnace, or stack, thus allowing the condition to progressively deteriorate. Combustion systems that respond to reduced boiler steam pressure or steam flow, with an increase in their demand for fuel, are potentially dangerous unless protected or interlocked to prevent the creation of a fuel rich mixture and a loss of ignition.

Natural gas may be either wet or dry. A wet gas implies the presence of distillate, which, if carried into the burners, can result in momentary increases in fuel input and/or a flameout followed by possible reignition and explosion. For this reason, systems using wet gas require special attention.

Gases supplied from one or more sources can introduce unacceptable hazards if they have significant differences in volumetric heating value. With such variable supplies, it is necessary to provide instruments that are responsive to variations in heat value (e.g., specific gravity or heating value meters) and appropriate alarm and combustion compensation devices.

Discharges from relief valves or any other form of atmospheric vents can present a hazard unless special precautions are taken to prevent a source of ignition or reentry into the boiler room.

Burner pulsation is one of the most mystifying problems associated with gas firing and, to a lesser degree, with oil firing. When one or more burners on a large unit begin to pulsate, the action may become alarmingly violent, at times shaking the whole boiler. Adjusting only one burner may start or stop pulsation. At times, only minor burner adjustments eliminate pulsation. In other instances, it may be necessary to make physical alterations to the burners. Alterations may include modifying the gas ports, impinging gas streams on one another, or installing a device that changes the air-fuel mixing characteristics.

Hazards of Coal Firing

Coal varies in size and in the amount of impurities it contains. Wide variations in the size of raw coal may cause erratic or uncontrolled feeding of the coal to the pulverizer. As delivered, coal may contain any number of foreign materials such as scrap iron, wood, rags, excelsior, or rock, which may interrupt coal feed, damage or jam equipment, or become a source of ignition within a pulverizer. Since as little as 0.05 oz (1.4 g) of pulverized coal per cu ft (1 cu ft equals 0.03 m^3) of air forms an explosive mixture and since a large boiler may burn 300 lb (136 kg) or more of coal per second, an explosive mixture will develop quickly if a momentary flameout occurs.

A special hazard is the methane gas that is released from freshly crushed or pulverized coal and which may accumulate in enclosed spaces such as storage bins and within pulverizers and burner piping.

Fuel (finely divided coal) is conveyed through pipes from the pulverizer to the burners entrained in an air stream. To prevent the settling out of coal particles in the burner pipes, and potential preignition, it is necessary to maintain an air velocity that is sufficient to keep the pulverized coal in suspension.

Provisions should be made for cooling down and emptying the pulverizer and/or burner lines when the burners it supplies are shut down. This is to prevent spontaneous ignition and a possible explosion in the pulverizer or burner lines.

Pulverizer fires and explosions are serious hazards. A sudden and considerable increase in the temperature of the air-fuel mixture leaving the pulverizer, or in the temperature of the outer casing, are indications that a pulverizer fire has started. Such a fire may be caused by feeding burning fuel from the raw fuel bin into the pulverizer or by spontaneous combustion of fuel in the pulverizer or piping. If a fire occurs in any part of a pulverized fuel system, it should be looked upon as serious and should be dealt with promptly. The necessary steps for extinguishing this type of fire are the following:

1. Remove the pulverizer from service without emptying it, to avoid creating an air rich condition.
2. Isolate the pulverizer against air infiltration by closing all burner line valves and inlet air dampers, and sealing the air valve.
3. Admit steam or inert gas into the pulverizer through the connection provided for that purpose to smother

the fire and to maintain an inert atmosphere until the coal has been removed and the housing has cooled.
4. Dump the coal stored in the pulverizer through the pyrite gates and into the sluice system.
5. Rotate the pulverizer without operating the primary air fan to complete the swirling and cleaning procedure.

A similar procedure should be followed for pulverizers that have been tripped full of sub-bituminous B, C, or lignite coal without immediate restart to avoid pulverizer explosions while out of service or when started up. Recommended emergency procedures should be reviewed with pulverizer manufacturers to include requirements peculiar to each design.

Meters that detect combustibles, since they measure only gaseous combustibles, are not infallible when used with pulverized coal. Since the generation of combustibles requires both high temperature for volatilization and limited oxygen in the combustion zone, the lack of a meter indication of combustibles does not rule out the presence of unburned coal particles.

The limited operating experience of fluidized bed boilers has not yet identified any unique, hazardous condition. It is expected that there will be less incidence of furnace explosions than for pulverized coal primarily because of the larger size of crushed coal as compared to pulverized coal.

The furnace safety and burner interlock systems for pulverized coal are currently being applied to experimental and pilot plant installations of fluidized bed boilers. This practice will be continued with special emphasis on the requirements for continuous purge procedures and of interlocks for igniters and auxiliary burners. Compartment alarms and trips may be initiated by both high and low bed temperatures. For any installation the interlock requirements should be reviewed with the manufacturer and the consulting engineer for the project.

Open Register Light off or Continuous Purge Procedure

One important consideration is the use of the open register light off or continuous purge procedure for all fuels. This procedure maintains airflows at or above the prescribed minimum of 25 percent of full load volumetric airflow during all operations of the boiler. Volumetric airflow is specified to recognize that weight flow rate must be increased at low air temperature to maintain air and gas velocities.

The open register purge rate or continuous purge procedure is based upon the concept that three basic operating conditions will significantly improve the margin of operating safety, particularly during startup. Those conditions are:

1. A minimum number of required equipment manipulations, thereby minimizing exposure to operating errors or equipment malfunction;
2. A means for establishing the desired fuel rich condition at individual burners during light off; and
3. An air rich furnace atmosphere during light off and warm up by maintaining total furnace airflow at the same rate as that required for the furnace purge.

In its simplest form, the basic procedure that satisfies these three operating objectives is as follows:

1. Place all or most of the burner air registers in a predetermined open position.
2. Purge the furnace and boiler settings with the burner air registers in that same predetermined position. The total airflow for purge must not be less than 25 percent of full load volumetric airflow.
3. Light the first burner or group of burners without any change in airflow setting or in burner air register position.

Each boiler should be tested to determine whether any modifications are required to obtain satisfactory ignition or to satisfy other design limitations during light off and warm up. For example, some boilers will be purged with the registers in the normal operating position.

FIRE AND EXPLOSION PROTECTION

The foundation of fire and explosion protection is prevention. Reliable equipment, good facility design, system monitoring and malfunction alarm instrumentation, operator training, and maintenance are key elements in preventing fires and explosions in boiler-furnace systems.

The entire system—boiler, furnace, fuel supply, air supply, vents, piping, and ducts—should be designed to specific parameters and for specific operating limits. At no time should these limits be exceeded.

Good facility design requires the boiler to be installed in a separate room or structure, preferably of noncombustible construction. Boiler-furnaces should be set on concrete floors or platforms that extend beyond the equipment for a distance of at least 4 ft (1.2 m) in each direction. If they must be set on combustible floors, there should be sufficient air circulation beneath the furnace to keep the temperature of the combustible floor below 160°F (71°C). Metal chimneys or smokestacks should not extend through combustible floors, ceilings, or walls. If such chimneys must be passed through combustible roofs, sufficient clearance and/or insulation should be provided to keep the temperature of the combustible materials below 160°F (71°C). Generally, minimum clearance is considered to be 18 in. (0.46 m).

A system of interlocks should be provided to prevent improper sequencing by operating personnel and to shut down operations if certain critical malfunctions occur. Audible and visual alarms serve to warn operators to take certain corrective steps; other alarms may indicate what automatic functions have been performed to reduce a hazard. New boilers should not be fired until adequate safeguards have been installed and tested.

Safe operation cannot be ensured solely by equipment design and adherence to the manufacturer's operating instructions. Knowledgeable and competent operators who understand the processes involved also are needed. Technical competence should be maintained by a continuing program of retraining so that increasing familiarity with the system does not tempt operating personnel to take "short cuts" in operating procedures or to bypass safety devices.

A program of preventive maintenance is needed to maintain the reliability of the equipment and its control devices. Poor preventive maintenance can lead to frequent corrective maintenance. Cleanliness and good housekeeping, especially where pulverized coal is the fuel, will also contribute to the prevention of fire and explosion. Automatic sprinkler or water spray systems are practical forms of fire protection in all boiler rooms.

Bibliography

NFPA Codes, Standards, Recommended Practices and Manuals. (See the latest *NFPA Publications and Visual Aids Catalog* for availability of current editions of the following documents.)

NFPA 30, *Flammable and Combustible Liquids Code.*
NFPA 68, *Guide for Explosion Venting.*
NFPA 85A, *Standard for Prevention of Furnace Explosions in Fuel Oil- and Natural Gas-Fired Single Burner Boiler Furnaces.*
NFPA 85B, *Standard for Prevention of Furnace Explosions in Natural Gas-Fired Multiple Burner Boiler Furnaces.*
NFPA 85D, *Standard for Prevention of Furnace Explosions in Fuel Oil-Fired Multiple Burner Boiler Furnaces.*
NFPA 85E, *Standard for Prevention of Furnace Explosions in Pulverized Coal-Fired Multiple Burner Boiler Furnaces.*
NFPA 85F, *Pulverized Fuel Systems, Installation and Operation.*
NFPA 85G, *Prevention of Furnace Implosions in Multiple Burner Boiler-Furnaces.*

Additional Readings

Durrant, O. W., "Design, Operation, Control and Modeling of Pulverized Coal Fired Boilers," paper presented to *Boiler-Turbine Modeling and Control Seminar*, University of New South Wales, Sydney, Australia, 1977.
Durrant, O. W., and Krippene, G. C., "Combustion Principles and Processes for NO_x Control Natural Gas-Fired Utility Boilers," presented to the *American Gas Association*, Atlanta, GA, 1972.
Durrant, O. W., and Zadiraka, A. J., "Control of Pulverized Coal-fired Utility Drum Boilers During Load Changes," Presented to *American Power Conference*, Chicago, IL, 1981.
Factory Mutual Engineering Corporation, "Elements of Combustion, Controls, and Safeguards in Industrial Heating Equipment," *Loss Prevention Data 6-0*, Factory Mutual System, Norwood, MA, 1977.
FMEC, "Boiler-Furnaces Oil- or Gas-Fired Single Burner," *Loss Prevention Data*, 6-4, 1976.
FMEC, "Boiler-Furnaces Oil- or Gas-Fired Multiple Burner," *Loss Prevention Data*, 6-5, 1978.
Graham and Trotman, Ltd., eds., *Boiler Operator's Handbook*, State Mutual Book and Periodical Service, Ltd., New York, NY, 1981.
Heil, T. C., and Durrant, O. W., "Designing Boilers for Western Coal," presented to *Joint Power Generation Conference*, Dallas, TX, 1978.
Resource Systems International, *Boiler: Repair and Maintenance*, Reston Publishing Co., Reston, VA, 1982.
Resource Systems International, *Boiler Systems and Components I and II*, Reston Publishing Co., Reston, VA, 1982.
Shields, C., *Boilers*, McGraw-Hill, New York, NY, 1961.
Spring, Harry M., and Kohan, Anthony L., *Boiler Operator's Guide*, 2nd ed., McGraw-Hill, New York, NY, 1981.
Trinks, W., *Industrial Furnaces*, (Vols. 1 and 2), John Wiley and Sons, New York, NY, 1967.
Walker, R. R., and Zadiraka, A. J., "Integrated Fuel and Air Control System for PC Firing," presented to the *American Power Conference*, Chicago, 1975.
Whitney, S. A., and Smith, J. W., "Industrial Fluidized Bed Design and Operation of the TVA Test Facility" presented to *6th International Coal and Lignite Utilization Exhibition and Conference*, Houston, TX, 1983.

STATIONARY COMBUSTION ENGINES

Revised by Frank J. Mapp and Vincent J. Hession

Reciprocating engines and gas turbine engines, the two types of combustion engines in general stationary use, are frequently used as prime movers to drive alternators, pumps, and compressors. Recently there has been a rapid increase in the use of these engines to drive alternators in emergency situations.

Engine driven pumps and compressors handle a variety of fluids, including flammable liquids and gases. This chapter discusses the potential fire hazards associated with such combustion engines. It provides recommendations for the safe location, storage of fuel and installation of the engines.

For a more in-depth discussion of the fire hazards associated with the use and storage of flammable liquids and gases, see Section 5, Chapters 4 and 5, "Flammable and Combustible Liquids" and "Gases," respectively; and Section 11, Chapters 4 and 5, "Storage and Handling of Flammable and Combustible Liquids" and "Storage and Handling of Gases," respectively. Protective devices are covered in Section 16, Section 20, Chapters 1 and 2—"The Role of Extinguishers in Fire Protection" and "Selection, Operation, and Distribution of Fire Extinguishers," which provide detailed information about the use of fire extinguishers.

POTENTIAL HAZARDS

Potential fire hazards associated with combustion engines are: (1) their fuels, lubricants, and in some cases the material being pumped or compressed; (2) the temperature of exposed engine parts, particularly exhaust manifolds and pipes, which can be hot enough to ignite flammable or combustible liquids or other materials; (3) the presence of arcing electrical components, such as starters, generators, distributors, and magnetos, which can ignite flammable liquids and gases; and (4) engine disintegration.

Mr. Mapp and Mr. Hession are District Managers for Bell Communications Research, Inc., Morristown, NJ.

LOCATION

Engines may be located outdoors, inside structures, or on roofs of structures. However, depending on the site, consideration should be given to locating the engine with respect to exterior wall openings, distance to combustible walls and ceilings, and availability of sufficient air for combustion and ventilation. Engine room compartmentation should comply with applicable standards and local codes which specify minimum fire resistance ratings.

Because of the presence of hot surfaces, engines should be isolated from locations where combustible dust or airborne flyings can be produced, and from liquids and gases having unusually low ignition temperatures. Exhaust manifolds and piping can reach temperatures over 1,000°F (538°C) but, in order for ignition to occur, conditions must be favorable for reasonably prolonged contact between the combustible and the hot surface. For this reason, these surfaces seldom ignite flammable liquid vapors or flammable gases. However, dust and flyings can be deposited on these surfaces, and proper precautions should be observed.

FUEL SUPPLY

Many engine applications require on site fuel supplies for reliability. The fuel may be stored outdoors or within a building. Codes and standards restrict fuel tank size and location.

Gasoline, due to its low flash point, presents a greater potential fire hazard than some other fuels. Because of this, gasoline storage in buildings is limited to the capacity of the integral tank on the engine. When larger quantities of liquid fuel are required in buildings, diesel fuel is used.

When gas fueled engines are used, they should be installed in accordance with applicable codes and standards. LP-Gas in the liquid phase should not be stored within a building. Gaseous vapors, e.g., gasoline and propane, are heavier than air and have a tendency to collect in low lying areas. Special precaution should be taken to assure proper ventilation when using these fuels.

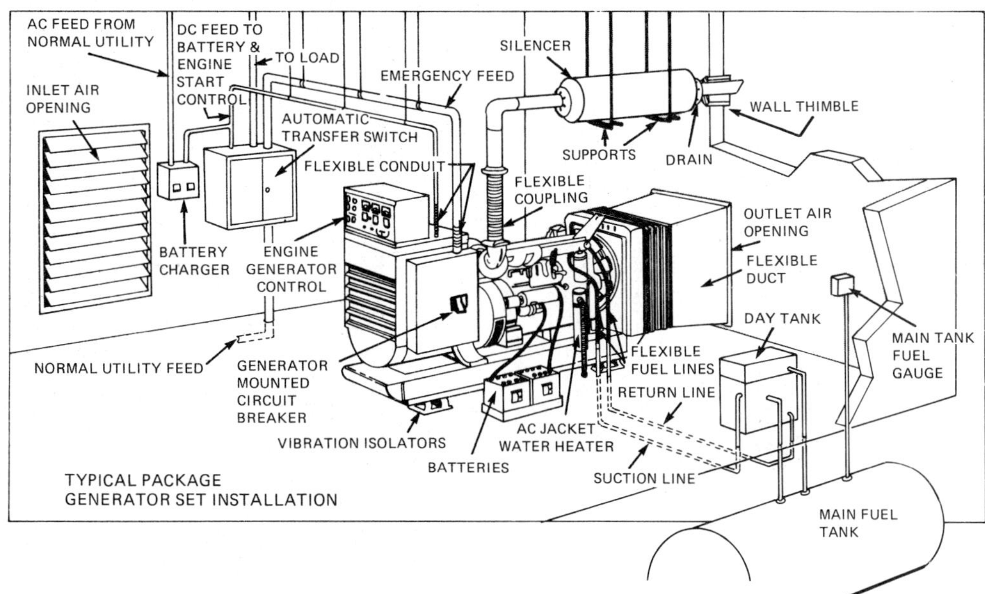

FIG. 10-2A. A typical generator installation used to supply emergency power. (Caterpillar)

INSTALLATION SAFEGUARDS

Codes and standards provide minimum requirements for safe installation of engines and include specifications for engine room compartmentation, fuel supply piping, room ventilation, engine exhaust piping, protective devices, and alarms.

Escaping fuels and lubricants can present hazards regardless of the degree of fuel flammability. Liquid fuel supply systems should be installed to minimize the accidental discharge of fuel into the engine room. Methods or devices such as dikes around fuel storage tanks, drains, float controlled valves, and adequate alarms can be used. Larger installations should have provision for a remote emergency shutdown of the engine. Most installations also have a remote means of shutting down fuel and lubricating oil pumps not directly driven by the engine. Liquid fuel is supplied to engines by pumps; however, gravity feed is permitted from an integral tank on an engine. The fuel supply to tanks above the lowest level of a building should be pumped in a manner acceptable to the authority having jurisdiction.

Backfires through gas-air mixers and carburetor air intakes occur as a result of engine malfunction. These can ignite grease and oil deposits on and around the engine. Flame arresting equipment should be provided at these points if flammable liquids or gases are being pumped or compressed.

A lubricating oil mist is often present in the crankcases of large reciprocating engines which, upon ignition, can result in a combustion explosion in the crankcase. Such engines are often equipped with crankcase explosion vents and sometimes, the crankcase is maintained under a nonflammable atmosphere.

Arcing electrical engine components are safeguarded by flashover shielding, flame arresters, purging, or ventilation if flammable liquids or gases are being pumped or compressed.

Engine exhaust systems, e.g., manifolds, mufflers, exhaust piping, etc., can reach temperatures in excess of 1,000°F (538°C), and special precautions are needed when these items penetrate combustible construction. Many fires have been caused by failure to allow sufficient clearance between exhaust piping and combustible walls, ceilings, or roofs. The exhaust system should terminate outside the structure at a point where the hot gases or sparks will be discharged harmlessly and not come in contact with combustible material.

Engine disintegration is rare. Gas turbines are somewhat more subject to this phenomenon than reciprocating engines because of the turbine's very high rotational speeds. Furthermore, the design of gas turbines for stationary installations usually incorporates construction features to contain the compressor, turbine blades, and rotor parts. The hazard is controlled by providing automatic engine speed governors. An overspeed shutdown device is provided on large engines of both types.

Instructions for normal starting, stopping, operation, and routine maintenance procedures should be posted near the equipment. Appreciable quantities of combustible materials should not be stored in the engine room. Finally, fire extinguishers should be provided in accordance with applicable standards and codes.

Bibliography

NFPA Codes, Standards, Recommended Practices and Manuals. (See the latest *NFPA Codes and Standards Catalog* for availability of current edition of the following documents.)

NFPA 10, *Standard for Portable Fire Extinguishers.*
NFPA 20, *Standard for the Installation of Centrifugal Fire Pumps.*
NFPA 30, *Flammable and Combustible Liquids Code.*
NFPA 37, *Standard for the Installation and Use of Stationary Combustion Engines and Gas Turbines.*
NFPA 54, *National Fuel Gas Code.*

NFPA 58, *Standard for the Storage and Handling of Liquefied Petroleum Gases.*
NFPA 68, *Guide for Explosion Venting.*
NFPA 70, *National Electrical Code.*
NFPA 99, *Standard for Health Care Facilities.*
NFPA 211, *Standard for Chimneys, Fireplaces, Vents and Solid Fuel Burning Appliances.*
NFPA 220, *Standard on Types of Building Construction.*

Additional Readings

ANSI/ASME, *Boiler and Pressure Vessel Code*, American Society of Mechanical Engineers, NY, 1980.
ANSI, B31.2, *Fuel Gas Piping*, American Society of Mechanical Engineers, NY, 1968.
ANSI, *B133.6, Procurement Standard for Gas Turbine Ratings and Performance*, American Society of Mechanical Engineers, NY, 1978.
API, *API 650, Welded Steel Tanks for Oil Storage*, American Petroleum Institute, Washington, DC, 1980.
SAE, *SAE J1349—Dec. 1980, Engine Power Test Code—Spark Ignition and Diesel*, Society of Automotive Engineers, Warrendale, PA, 1980.

HEAT TRANSFER SYSTEMS (Nonwater Media)

Revised by J. G. Coutu

The term "heat transfer fluid" refers to the broad spectrum of either liquid or vapor media used to transfer heat energy at a controlled rate from one place to another. While practically all industrial processes involve the exchange of heat in one form or another, the use of heat transfer fluids is common in the chemical, paint, textile, plastics, food, petroleum, and paper industries. The fluids are used to heat or cool reaction vessels, heat stills, heat rollers for drying paper, calender plastics, and for many other forms of processing.

For additional information relevant to heat transfer systems see Section 5, Chapter 4, "Flammable and Combustible Liquids;" Section 10, Chapter 5, "Industrial and Commercial Heat Utilization Equipment;" Section 10, Chapter 10, "Chemical Processing Equipment;" Section 10, Chapter 13, "Molten Salt Baths;" and Section 11, Chapter 4, "Storage and Handling of Flammable and Combustible Liquids."

TYPES OF TRANSFER FLUIDS

Approximately 95 percent of all heat transfer fluids used are either steam or water. If the temperature of use is above the freezing point of water and below about 350°F (177°C), the choice is usually between these two mediums. On the other hand, if the temperature of use is above or below these two points, it is desirable, if not necessary, to consider other fluids. For temperatures below the freezing point of water, the most common heat transfer fluids are air, refrigerants such as halogenated hydrocarbons, ammonia, brines, or solutions of ethylene glycol and water.

When temperatures increase above 350°F (177°C), the vapor pressure of water increases rapidly, requiring expensive high pressure processing equipment. Therefore, a more practical and inexpensive means of heating or temperature control at higher temperatures is to use a fluid or vaporizing fluid with a high boiling temperature. Care should be taken to use high flash point, thermally stable, and noncorrosive fluids designed for this purpose. The high boiling point liquids that are commonly used can be generally categorized as either heat transfer oils, molten salts, or specially formulated heat transfer mediums.

There are two types of heat transfer systems: liquid and vapor phase. With liquid phase systems (Fig. 10-3A), the temperature of the fluid changes as heat is transferred, resulting in nonuniform temperatures even if large circulation rates are employed. With vapor phase systems, heat is transferred at the saturation temperature of the vapor which affords controlled temperatures. Combinations of these systems may be used. (See Figs. 10-3B and 10-3C.)

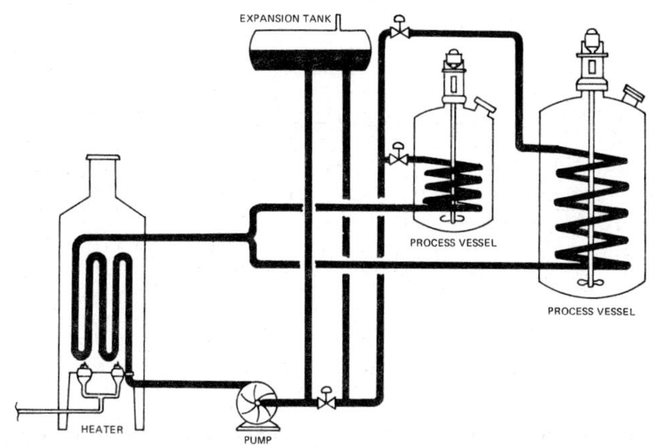

FIG. 10-3A. Liquid phase heat transfer system with forced circulation type heater.

Heat Transfer Oils: These are specially refined petroleum oils for use at temperatures up to approximately 600°F (315°C). At higher temperatures, the oils undergo thermal cracking to produce light hydrocarbons and polymers. Petroleum oils are also subject to oxidation at temperatures above 392°F (200°C) and in contact with air. These oils will provide a long service life and good heat transfer when the manufacturers recommended guidelines are followed.

Mr. Coutu is Executive Engineer, Industrial Risk Insurers (IRI), Hartford, CT.

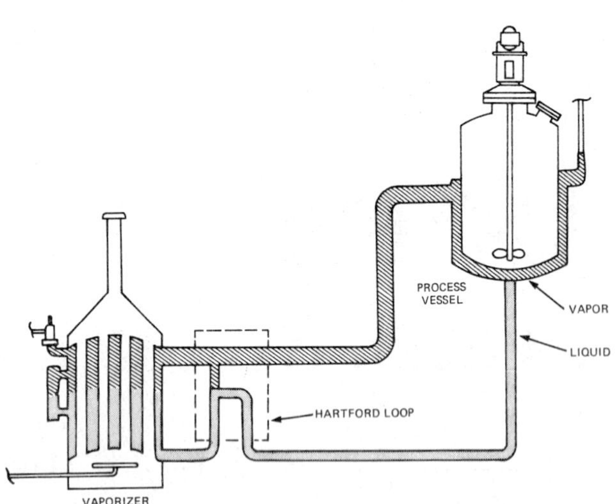

FIG. 10-3B. Vapor phase gravity condensate return heat transfer system. This sytem has a vertical fire-tube type vaporizer with natural circulation.

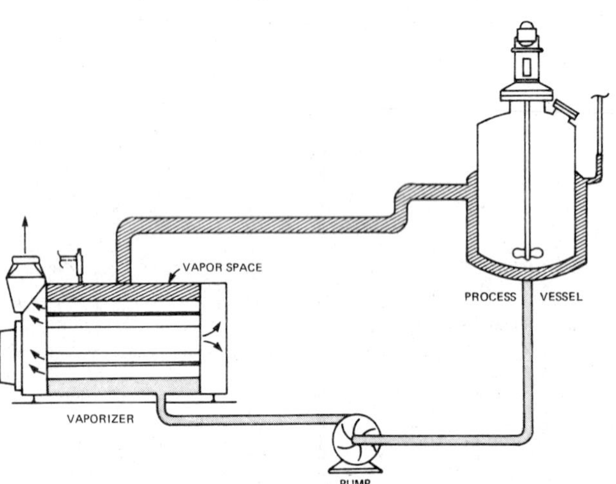

FIG. 10-3C. Vapor phase heat transfer system with pumped condensate return. This system has a horizontal fire-tube vaporizer with natural circulation.

Molten salts: These salts are generally used as heat absorbers or cooling media, and thus serve to control exothermic reactions which occur in the manufacture of certain chemicals. The most frequently used of these salts is a mixture of sodium and potassium salts having a relatively low melting point. Since these salts are solid at room temperature, it is obvious that care must be exercised to keep the temperature of the salt above its melting point to prevent salt solidification in the system. Also care must be taken during start up to control the heat input so the mixture will melt slowly. Many molten salts are strong oxidizers, and this should also be considered where these salts are used.

Synthetic fluids: A variety of specially formulated synthetic fluids are very widely used as a heat transfer fluid. Synthetic fluids are subject to thermal cracking and oxidation; however, they are safe, noncorrosive, essentially nontoxic, and thermally stable when used under the manufacturer's recommended guidelines. The synthetics create a fire hazard if released into the fire box of a fuel fired heater. Table 10-3A gives the physical properties of some common heat transfer fluids and oils.

When these fluids are used in the liquid state, the unit is referred to as a heater, and when the fluid is vaporized the unit is called a vaporizer. Two types of equipment are manufactured: (1) fuel fired heaters which are generally designed for heat duty use in excess of 1 million Btu per hr (29 280 kW), and (2) electric heaters which are usually designed for heat duty use of less than 1 million Btu per hr (29 280 kW). Capacities of vaporizers and heaters range from a few thousand to over 175 million Btu per hr (524 694 kW).

HAZARDS OF ORGANIC TRANSFER FLUIDS

The basic hazards of organic heat transfer liquids are those associated with the heating and transfer of a combustible liquid near or above its fire point in a closed system.

The type of vaporizer shown in Figure 10-3D is designed for relatively low heat release. The lower drum provides a space for the collection of any sludge or foreign substances within the system, preventing such materials from fouling the heating surface. Precautions must be taken to prevent the liquid level from falling below the tops of the tubes and impairing circulation. In direct fired units, impingement of heater burner flames on empty tubes might produce excessive local temperatures, resulting in deposits and ultimate failure of the tubes.

Some organic heat transfer media have very low surface tension and viscosity. To prevent leakage, special care is necessary in the fabrication and erection of equipment. Construction in accordance with the *ASME Boiler and Pressure Vessel Code* is mandatory in nearly all U.S. and Canadian jurisdictions. The wide range in temperature requires provision for expansion of piping. High temperatures may make ordinary relief valve springs unreliable.

Because of the cost of organic fluids, it is common practice to set the blowoff points of safety valves so high that there is very little likelihood of blowing. To allow for this, the ASME Standard requires that reservoirs be designed for a working pressure of at least 40 psi (276 kPa) above the normal operating pressure, but the reservoirs should not be intentionally fired at rates above rated capacity and temperature.

Most corrosive problems result from contamination of the system by foreign chemicals. Contamination by leakage from the process side can lead to attack in the heat transfer loop. The severity and nature of attack in these instances depends on the degree and specific type of contamination. For example, chloride contamination can cause failure in the 300 series of stainless steels because of embrittlement or stress corrosion cracking.

SAFEGUARDS FOR FLAMMABLE TRANSFER SYSTEMS

Where a flammable heat transfer system having a flash point below 300°F (149°C) is heated above its atmospheric boiling point, the vaporizer or heater and the user facility

TABLE 10-3A. Physical Properties of Typical Heat Transfer Fluids

Compound	Freezing Point (°F)	Freezing Point (°C)	Boiling Point (°F)	Boiling Point (°C)	Flash Point (°F) C.O.C.	Flash Point (°C)	Fire Point (°F) C.O.C.	Fire Point (°C)
1,2,4-trichlorobenzene	63	17	417	213	210	99	*	*
tetrachlorobenzene (isomer mixture)	170	77	480	249	None	—	*	*
diphenyl ether—diphenyl eutectic	54	12	495	257	255	124	275	135
biphenylyl phenyl ether (isomer mixture)	99	37	680	360	370	188	410	210
o-biphenylyl phenyl ether	122	50	670	354	370	188	410	210
di- and triaryl ethers	<0	−17	572	300	305	152	315	157
dimethyl-diphenyl ether (isomer mixture)	−40**	−40	554	290	—	—	—	—
tetramethyl diphenyl ether (isomer mixture)	—	—	590	310	—	—	—	—
di-sec-butyl diphenyl ether (isomer mixture)	—	—	705	374	380	193	400	204
dicyclohexyldiphenyl ether (isomer mixture)	—	—	785	418	—		—	—
dodecyldiphenyl ether (isomer mixture)	45**	7	>800	>427	410	210	440	227
ethyldiphenyl (isomer mixture)	<−60**	<−51**	536	280	—	—	—	—
partially hydrogenated terphenyl	−15**	−26**	690	366	335	168	375	191
aliphatic oil	15	−9	720–950	382–510	425	218	475	246
alkylaromatic oil	20	−6	~650	342	350	177	390	199

*—None to boiling point
**—Pour point

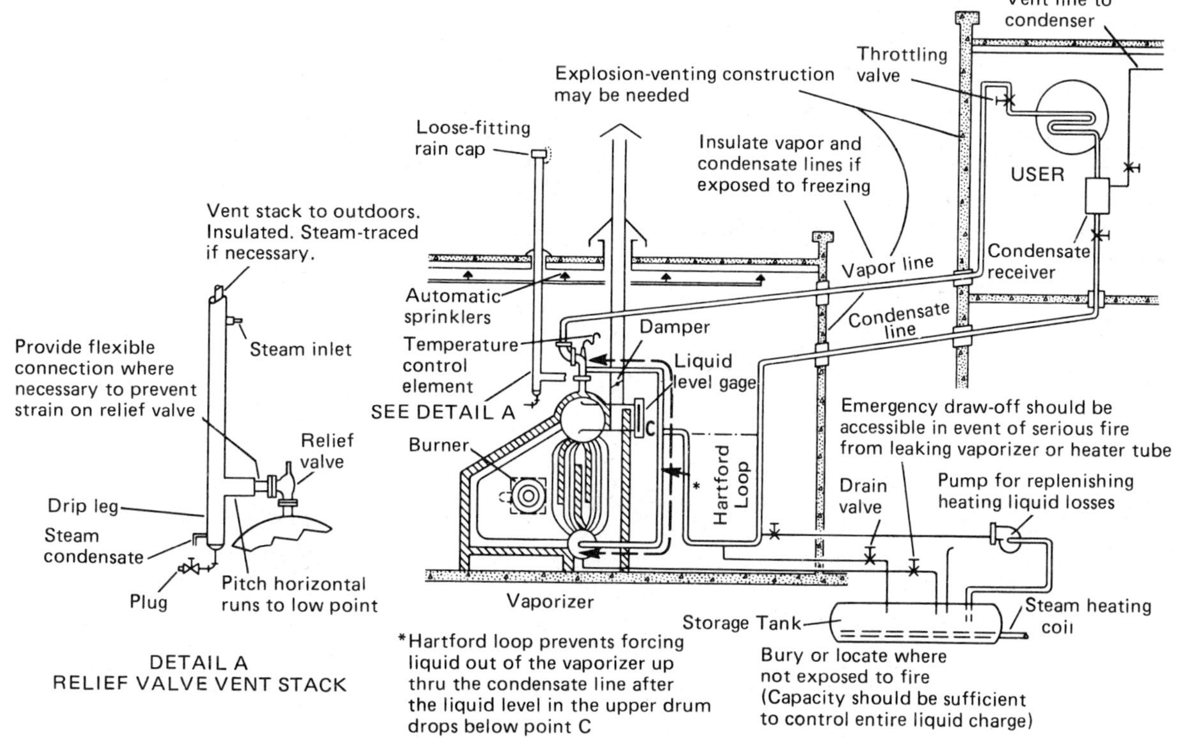

FIG. 10-3D. Gravity return vaporizer arrangement. (Factory Mutual System)

should be located in a detached area or in a cutoff room of damage limiting construction.

If located in a processing building, the vaporizer and heater room should have the following features:

1. Automatic sprinkler protection.
2. Cutoff from other important facilities by construction with a minimum one hour fire resistance rating.
3. Piping securely supported and otherwise protected against mechanical injury with adequate clearance from combustible materials.
4. Floors curbed and drained to protect nearby facilities.

Equipment should always be operated within the temperature and pressure limits specified by the supplier or manufacturer of the heat transfer medium and by the manufacturer of the equipment. Good maintenance is a key factor to any accident free operating system and should include periodic testing of the heat transfer medium, heater fire surfaces, safety valves and control devices.

System vaporizers and heaters should be under automatic control and regulated by temperature and pressure elements in the outlet. On vapor systems, separate safety limit switches should be provided to shut down the heater automatically and sound an alarm either when the vaporizer level becomes dangerously low, or when the vapor pressure or temperature is excessive. On circulating liquid systems, limit switches in addition to the low liquid level cutoff should be provided to shut off the heater automatically if liquid temperature is excessive or if circulation rate is low.

A manually controlled steam or inert gas smothering system for the fire box is advisable where a supply of steam or inert gas is available. If fire occurs in the vaporizer, follow these steps:

1. Shut off the heating units.
2. Open steam or inert gas smothering systems or use portable Class B extinguishers at openings to the fire box.
3. Unless a leak is near the bottom of the vaporizer, do not drain the liquid until after the fire has been extinguished and the unit is cooled down—otherwise, vaporizer drums and tubing may become overheated.
4. If necessary, drain liquid in the vaporizer back to the storage tank. If a storage tank is not available, drain to a safe location.

To minimize the hazard of fires from organic heat transfer media leaking and soaking into pipe insulation, heating spontaneously, and igniting, eliminate the source of leakage when it occurs by repacking valve stems, replacing leaky gaskets, etc. Cover insulation in areas of potential leakage with oil resistant cement or use a nonabsorbent insulation such as cellular glass or reflective aluminum foil.

Bibliography

NFPA Codes, Standards, Recommended Practices and Manuals. (See the latest *NFPA Codes and Standards Catalog* for availability of current editions of the following documents.)

NFPA 10, *Standard for Portable Fire Extinguishers.*
NFPA 12, *Standard on Carbon Dioxide Extinguishing Systems.*
NFPA 13, *Standard for the Installation of Sprinkler Systems.*
NFPA 86, *Standard for Ovens and Furnaces: Design, Location, and Equipment.*

Additional Readings

Albrecht, A. R., and Seifert, W. F., "Accident Prevention in High Temperature Heat Transfer Systems," presented at *AIChE Loss Prevention Symposium,* Atlanta, GA, 1970.

Boiler and Pressure Vessel Code, American Society of Mechanical Engineers, NY.

Factory Mutual Engineering Corporation, "Heat Transfer by Organic Fluids," *Loss Prevention Data Sheet* 7-99/12-19, Factory Mutual System, Norwood, MA, 1975.

"Heat Transfer Media Other Than Water," *Kirk-Othmer Encyclopedia of Chemical Technology,* Vol. 12, 3rd ed., Interscience Publishers, NY, 1973, pp. 171-190.

"Molten Salt for Heat Transfer," *Chemical Engineering,* Vol. 70, 1963, p. 129.

FLUID POWER SYSTEMS

Revised by Paul K. Schacht

Fluid power systems are used for transmitting power or motion to various parts of equipment and machines. The function of the fluid may be power multiplication as in a hydraulic press, actuation of automatic equipment as in die casting, or remote actuation or pressure control of machines or instruments. (See Fig. 10-4A.)

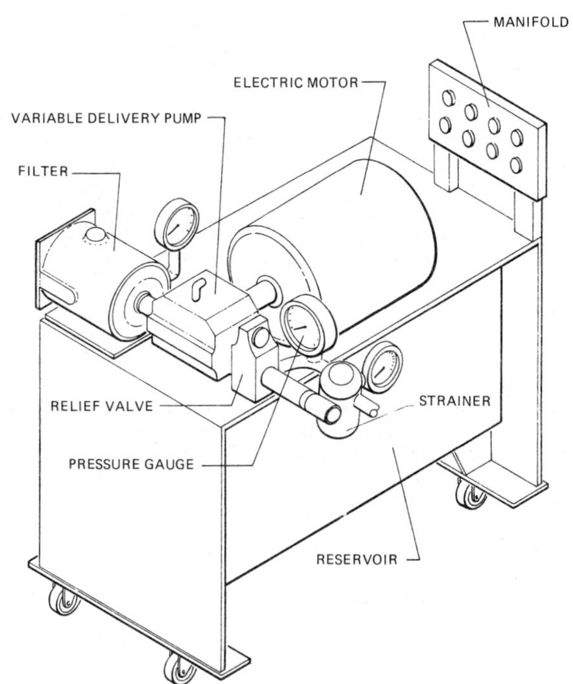

FIG. 10-4A. A nonspecific power unit that might be used to run any of a variety of machines, for example, a drill or press. (National Fluid Power Association)

Mr. Schacht is Manager Research and Development, Racine Hydraulics Division, Dana Corporation and Chairman of the Fluids Coordinating Committee for the National Fluid Power Association.

Hydraulic fluids are generally petroleum based and present fire hazards. For additional information on allied topics consult Section 5, Chapter 4, "Flammable and Combustible Liquids;" and Section 11, Chapter 4, "Storage and Handling of Flammable and Combustible Liquids."

Early fluid power systems used water as the hydraulic medium. Because of its corrosive effect on the metallic parts of the machine and lack of lubricity, it was replaced by petroleum oil. Except for its fire hazard, oil is an ideal hydraulic fluid. It is noncorrosive, is compatible with a wide variety of seals, has good lubricating properties, is readily available, and is relatively inexpensive. Flashpoints range from 300 to 600°F (148 to 316°C) and autoignition temperatures from 500 to 750°F (260 to 399°C).

FLUIDS UNDER PRESSURE

Pressurized oil in fluid power systems presents a considerable fire hazard, particularly where ignition sources are constantly present as in die casting, plastic molding, automatic welding, heat treating of metals, and the engines of mobile equipment.

High pressure pipe with welded and screwed joints, steel and copper tubing, and metal reinforced rubber hose are used to conduct oil at pressures ranging up to 10,000 psi (68 950 kPa). Failure of piping, particularly at threaded sections, failure of components and gaskets or fittings, tubing pulling out of fittings, and rupture of flexible hose have been the principal causes of oil being released from a fluid power system. Lack of adequate supports or anchorage to prevent vibration or movement of piping has often been a factor in these failures. Repeated flexing and abrasion of rubber hose against other hose or parts of machines have created weak spots which eventually failed. Tubing under pressure has failed due to being accidentally cut by torches or stepped on during maintenance procedures.

Guidelines to be followed when selecting hydraulic tubing can be found in the National Fluid Power Association and Society of Automotive Engineers (SAE) standards listed in the Bibliography.

FIRE CHARACTERISTICS

When oil under pressure is released by failure of equipment, the usual result is an atomized spray of mist or oil droplets which, depending on pressure, may encompass large areas. The oil spray is readily ignited and the resulting fire is usually torchlike with a very high rate of heat release.

If systems contain oil, the following recommendations should be met in addition to those recommended for less hazardous fluids:

1. Automatic sprinklers provided in the area.
2. Automatic or remote manual switches to shut off the hydraulic pumps in the event of fire, if over 100 gal (378 L) of fluid are involved.
3. Fire extinguishers provided suitable for Class B fires .
4. Threaded pipe connections avoided wherever possible.
5. Machines isolated from positive ignition sources such as metal casting.
6. Armored hose or enclosed pressure hose in a second tube to contain escaping fluid.
7. Regular inspections of the entire hydraulic system.

LESS HAZARDOUS HYDRAULIC FLUIDS

Less hazardous fluids have been developed to replace petroleum based oils in all types of hydraulic systems. Although these fluids exhibit some degree of combustibility, they represent a much lower fire hazard than ordinary petroleum oils and are preferred when a constant ignition source is present. However, the above recommendations for oil based fluids are nevertheless a reasonable precaution. When converting to less hazardous fluids, and to perform proper maintenance, consult with manufacturers so suitable equipment and procedures can be selected. Although less hazardous, a fire can still occur; accessible and well marked emergency cutoff switches should be provided for shutting down hydraulic pumps if leaks or pipe failures occur. Piping, tubing and hose should be installed in accordance with manufacturer's instructions. Piping and tubing should be anchored or secured as necessary to minimize failure due to vibration, but pipe supports should not prevent normal thermal expansion. Bolted supports should be tightened regularly using torque wrenches.

Precautions to be used in changing from one type of hydraulic fluid to any other are published by the National Fluid Power Association.

Water-Glycol Fluids

Water-glycol fluids normally consist of 35 to 50 percent water, and ethylene or propylene-glycol, thickeners and additives. Recommended temperature limits are 0 to 150°F (−18 to 66°C), with normal operation at 120 to 150°F (49 to 66°C). At higher temperatures, the rate of water evaporation is such that addition of makeup water is frequently required. As the water evaporates, the viscosity increases, and the fire resistance decreases.

Water-glycol fluids are compatible with most types of seals, gaskets and hoses made from nitrite, fluorocarbon, butyl, or ethylene-propylene, but are incompatible with certain types of leather, cork, and paper. These fluids generally have a solvent action on most petroleum compatible paints and coatings.

Synthetic Fluids

Most synthetic hydraulic fluids are phosphate esters or blends of phosphate esters with petroleum oils. Recommended temperature limits are 20 to 150°F (−7 to 66°C). Synthetic fluids are not compatible with seals, gaskets or hoses made from natural rubber, nitrile, or neoprene, and should be replaced with fluorocarbon or other compatible material. Synthetic fluids may also attack paints and electrical insulation.

Water In Oil Emulsions

Water in oil emulsions consist of 35 to 40 percent water, petroleum oil, emulsifiers, and additives. The water is dispersed in fine droplets in the oil phase. Recommended temperature limits for the emulsion are 50 to 150°F (10 to 66°C). At higher temperatures, frequent addition of water is required. Loss of water tends to reduce viscosity and increase flammability. Water in oil emulsions are compatible with most seals and gaskets, and hoses made from nitrile, fluorocarbon or neoprene. Seals or gaskets made from cork, paper, leather, or butyl are unacceptable. These fluids generally have a solvent action on most petroleum compatible paints and coatings.

Bibliography

NFPA Codes, Standards, Recommended Practices and Manuals. (See the latest *NFPA Codes and Standards Catalog* for availability of current editions of the following documents.)

NFPA 10, *Standard for Portable Fire Extinguishers.*
NFPA 13, *Standard for the Installation of Sprinkler Systems.*
NFPA 30, *Flammable and Combustible Liquids Code.*

Additional Readings

American Institute of Physics, *Hydraulic Devices*, McGraw-Hill, New York, NY, 1975.

Anders, James E., *Industrial Hydraulics Troubleshooting*, McGraw-Hill, New York, NY, 1983.

"Development of Nonflammable Hydraulic Fluids and Lubricants," Olin Corporation, Cryogenics Division (U.S. Department of Commerce, Office of Technical Services) Stamford, CT, June 1964.

"Fluid Systems 1978: Looking Ahead Ten Years," *Machine Design*, Vol. 50, No. 22 (Sept. 28, 1978), pp. 2-5.

"Fluids," *Machine Design*, Vol. 50, No. 22 (Sept. 28, 1978), pp. 112, 115-116.

Foitl, R. J., and Kycera, W. J., "Formation and Evaluation of Fire-Resistant Hydraulic Fluids," *Iron and Steel Engineer*, Vol. 41, No. 7, July 1964, pp. 117-120.

Giles, Ronald V., *Fluid Mechanics and Hydraulics*, McGraw-Hill, New York, NY, 1962.

Henrikson, K. H., "Fire Resistant Fluids in Mobile Equipment," 650671, Sept. 1965, Society of Automotive Engineers, New York, NY.

"Hydraulic Fluids," *Data Sheet 7-98*, Factory Mutual System, Norwood, MA, 1974.

"Hydraulic Fluid Power-Fire Resistant Fluids-Information Report on Company Trade Names," T2.13.2R2-1980. National Fluid Power Association, Milwaukee, WI.

Hydraulic Handbook, 8th ed., Gulf Publishing Co., Houston, TX, 1983.

Hydraulic Handbook, 7th ed., State Mutual and Periodical Service, Ltd., New York, NY, 1981.

Johnson, Olaf A., *Fluid Power for Industrial Use: Hydraulics*, 2 Vols., Krieger Publishing Co., Melbourne, FL, 1981.

Merritt, L. C., *Hydraulic Control Systems*, Wiley-Interscience, New York, NY, 1967.

Millett, W. H., "Fire Resistant Hydraulic Fluids," Sept. 1973, E.F. Houghton & Co., Philadelphia, PA. (Paper presented at the Twenty-Ninth National Conference on Fluid Power.)

Pingree, Daniel, "Looking at Fire Hazards: Hydraulic Fluids," *Fire Journal*, Vol. 59, No. 6, Nov. 1965, p. 23.

Pippenger, John, and Hicks, Tyler, *Industrial Hydraulics*, 3rd ed., McGraw-Hill, New York, NY, 1979.

Polack, S. P., "Bureau of Mines Evaluates Fire Resistance of Hydraulic Fluids," *Iron and Steel Engineer*, Vol. 41, No. 8, Aug. 1964, pp. 105-110.

"Pressure Ratings for Hydraulic Tubing and Fittings," SAE J1065, Society of Automotive Engineers, Warrendale, PA.

Protheroe, A. R., "Fire Resistent Hydraulic Fluids: New Capabilities Expand Applications," *Hydraulics and Pneumatics*, Vol. 31, No. 5, (1978), pp. 73-75.

Roberts, A. F., and Brookes, F. R., "Hydraulic Fluids: An Approach to High Pressure Spray Flammability Testing Based on Measurement of Heat Output," *Fire and Materials*, Vol. 5, No. 3, (1981), pp. 87-92.

Sabersky, Rolf H. et. al. *Fluid Flow: First Course in Fluid Mechanics*, 2nd ed., Macmillan, New York, NY, 1971.

"Seamless Low Carbon Steel Hydraulic Line Tubing," ANSI/B93.11M-1981, National Fluid Power Association, Milwaukee, WI.

"Seamless Low Carbon Steel Tubing Annealed for Bending and Flaring," SAE J524, Society of Automotive Engineers, Warrendale, PA.

Snyder, C. E., Krawetz, A. A., Tovrog, T., "Determination of the Flammability Characteristics of Aerospace Hydraulic Fluids," *Lubrication Engineering*, Vol. 37, No. 12, (Dec. 1981), pp. 704-714.

"Standard Practice for the Use of Fire Resistant Fluids in Industrial Hydraulic Fluid Power Systems," ANSI/B93.5M-1979, National Fluid Power Association, Milwaukee, WI.

Stewart, Harry L., *Hydraulic and Pneumatic Power for Production*, 4th ed., Industrial Press Inc., New York, NY, 1977.

"Synthetic Fire-Resistent Hydraulic Fluids for General Industrial Use," *Hydraulic, Pneumatic and Mechanical Power*, April 1979, pp. 145-148.

"Test Method for Steel Separable Tube Fittings for Hydraulic Fluid Power Applications," T3.8.3M R2-1977, National Fluid Power Association, Milwaukee, WI.

Tuve, Richard L., *Principles of Fire Protection Chemistry*, National Fire Protection Association, Boston, MA, 1976.

Warren, S. M., and Kilner, J. R., *Fireproof Brake Hydraulic System*, Aeronautical Laboratories, Wright-Patterson Air Force Base, OH, 1981.

INDUSTRIAL AND COMMERCIAL HEAT UTILIZATION EQUIPMENT

Revised by James M. Simmons

Heat utilization equipment has potential hazards involving both heat generation and the process materials. Fuel fired systems, electrically heated systems, and heat transfer systems all have their individual hazards. The hazards of exposure of adjacent materials and overheating may be common to any type of heating system. The hazards and control methods are discussed in subsequent paragraphs of this chapter.

The process hazards within the equipment may involve combustible materials, flammable liquids, or flammable gases; such hazards would occur in the curing of solvent based coating materials, roasting or drying of combustible agricultural products, or heat treating in a flammable special atmosphere. The principal method of fire or explosion prevention is to prevent overheating or an accumulation of a gas or vapor-air mixture in the explosive range. Subsequent paragraphs in this chapter describe the various classes of ovens, furnaces and dryers, and present general process hazards. The prevention and protection methods which usually are applied to those processes and devices also are described.

Guidelines, rules, and methods applicable to the safe operation of such equipment are covered by NFPA Standards and those of other organizations. Many of these are mentioned and should be referred to for details which could not be included in this text. Additional information is available in this HANDBOOK in Section 5, Chapter 4, "Flammable and Combustible Liquids;" and Section 5, Chapter 5, "Gases." The grading of heat producing applicances is discussed in Section 8, Chapter 3, "Heating Systems and Appliances;" Section 10, Chapter 1, "Boiler Furnaces;" and Section 10, Chapter 3, "Heat Transfer Systems."

Heat utilization equipment is so varied in size, complexity, location, and use that it has been difficult to develop rules that would apply to every type of oven, furnace, or dryer. Users and designers must use engineering and supervisory skills to bring together the proper combination of controls, protective devices, and operator training necessary for proper equipment operation.

Mr. Simmons is Senior Special Hazards Engineer, Factory Mutual Research Corp., Norwood, MA.

Practically all heat utilization equipment failures are caused by people; usually because someone either ignored safety procedures and designs or was unaware they existed. Failures have resulted from inadequate training of operators and maintenance technicians, faulty equipment design, complacency on the part of users, and improper selection of combustion safeguards.

This chapter is not a guide to solving all of these problems. However, if the guidelines and principles it presents are used, a number of problems either will be mitigated or solved.

INDUSTRIAL HEAT UTILIZATION EQUIPMENT

Industrial heat utilization equipment includes a variety of forges, furnaces, kettles, kilns, ovens, and retorts which are heated by gas, oil, solid fuels, or electricity. They may be fired directly, with the products of combustion entering the process space, or indirectly, with radiant tubes or other heat exchanger methods. Heat transfer media such as organic fluids also are used where steam or hot water do not provide the temperature or thermal efficiencies desired for the equipment. Figures 10-5A and 10-5B illustrate typical industrial heat utilization equipment.

Industrial ovens and furnaces, special atmosphere furnaces, vacuum furnaces, after burners and catalytic combustion systems, dehydrators, dryers, and lumber kilns are discussed in this chapter.

Fire and Explosion Problems

When heat utilization equipment is installed, the following factors need to be considered: (1) the proximity and combustibility of the contents of the building where the equipment is located; (2) construction of the building; (3) setting; (4) ventilation; (5) location within the building; (6) heat, gas, and smoke disposal; (7) maximum temperature required; and (8) handling of heated materials in connection with equipment. Fire in combustibles can be prevented by insulation or by separating them from the

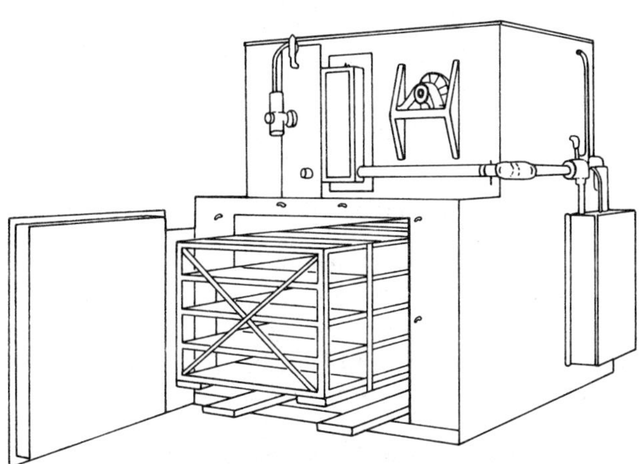

FIG. 10-5A. A typical industrial oven. (Electronics Corp. of America)

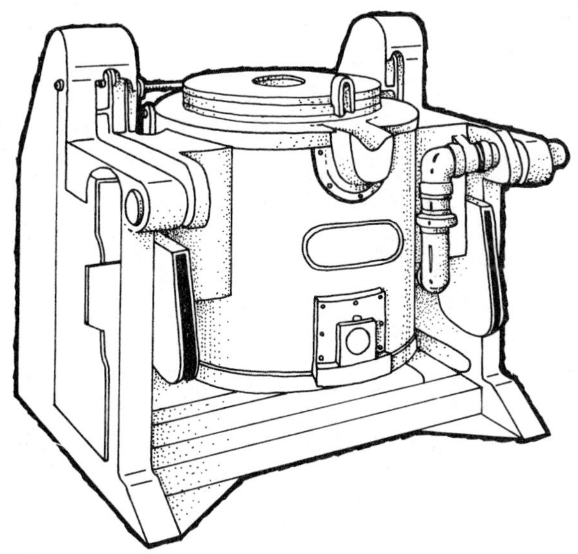

FIG. 10-5B. A fuel fire melting furnace. (Electronics Corp. of America)

source of heat. Overheating can be prevented by temperature controls.

Explosion hazards exist where there are flammable vapor-air mixtures from gas or oil fuel or from the volatiles released from the material being dried. Explosions or fires may be prevented by ventilation and controls that keep the flammable vapor content below 25 percent of the Lower Flammable Limit (LFL) of the vapor-air mixture.

Some special process ovens and furnaces contain hydrogen or other flammable gases for such purposes as annealing copper and heat treating other metal shapes. To operate these devices safely, it is necessary to prevent air from entering the processing enclosure under normal operating conditions. Both normal and unscheduled starts and stops of these devices require special safety procedures that depend heavily upon a skilled operator and adequate control equipment.

OVENS AND FURNACES

This discussion covers the location, design, construction, operation, protection, and maintenance of the industrial heating enclosures known as ovens or furnaces. It does not cover small cabinet or stove type ovens for domestic use. The source of heat for industrial heating may be gas burners, oil burners, electric heaters, infrared lamps, induction heaters, or steam radiation systems. In practically all cases, there are fire or explosion hazards from either the fuel used, volatiles produced by material in the oven, or by a combination of both.

There is no clear distinction between an oven and a furnace. The dictionary definition of an oven is "a compartment or receptacle for heating, baking, or drying by means of heat" and a furnace is "an enclosed chamber or structure in which heat is produced for heating a building, reducing ores and metals, baking pottery, etc." It has been an industry rule of thumb to classify heating devices that do not "indicate color" (operate at temperatures of less than approximately 1,000°F or 540°C) as ovens. This rule does not always apply—coke ovens operate at temperatures in excess of 2,000°F (1098°C), and some furnaces operate at temperatures below 1,000°F (540°C).

NFPA Classification of Ovens and Furnaces

The classification system for heat processing equipment as set forth in NFPA 86, *Standard for Ovens and Furnaces—Design, Location, and Equipment* (hereinafter referred to as NFPA 86), is as follows:

Class A ovens or furnaces are heat utilization equipment operating at approximately atmospheric pressure wherein there is a potential explosion and/or fire hazard which may be occasioned by the presence of flammable volatiles or combustible material processed or heated in the oven. Such flammable volatiles and/or combustible material may, for instance, originate from paints, power, or finishing processes including dipped, coated, sprayed on impregnated materials, or wood, paper and plastic pallets, spacers or packaging materials. Polymerization or similar molecular rearrangements and resin curing are processes which may produce flammable residues and/or volatiles. Potentially flammable materials such as quench oil, waterborne finishes, cooling oil, etc., in sufficient quantities to present a hazard, are ventilated according to Class A standards. Ovens may also utilize a low oxygen atmosphere to evaporate solvent.

Class B ovens or furnaces are heat utilization equipment operating at approximately atmospheric pressure wherein there are no flammable volatiles or combustible material being heated.

Class C furnaces are those in which there is a potential hazard due to a flammable or other special atmosphere being used for treatment of material in process. This type of furnace may use any type of heating system and includes the special atmosphere supply system(s). Also included in the Class C standard are integral quench and molten salt bath furnaces.

Class D furnaces are vacuum furnaces which operate at temperatures above ambient to over 5,000°F (2760°C) and at pressures below atmospheric using any type of heating system. These furnaces may include the use of special processing atmospheres.

Classification by Type of Handling System

Ovens and furnaces may be further classified according to the way a material is handled. The two principal types are: (1) the batch oven or furnace, sometimes referred to as "in and out," "intermittent," or "periodic" types; and (2) the continuous oven or furnace.

Batch Type: In the batch oven or furnace the temperature is practically constant throughout the interior. The material to be heated is placed in a predetermined position and remains there until the process is complete. Material is then removed, generally through the opening by which it entered. Figures 10-5C and 10-5D are examples of a batch type furnace and oven.

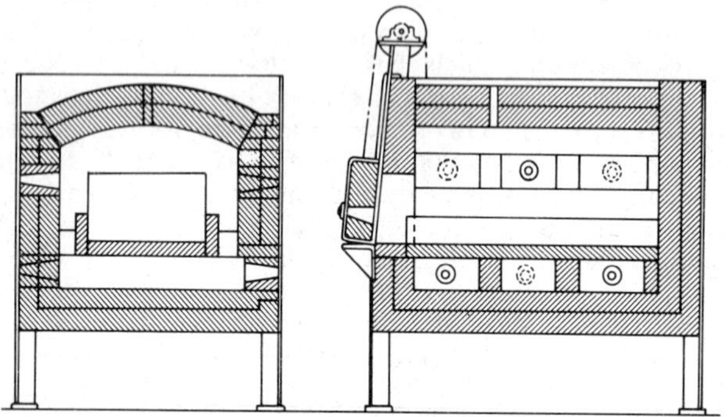

FIG. 10-5C. A large batch type, under and over fired, semi muffle, heat treating furnace.

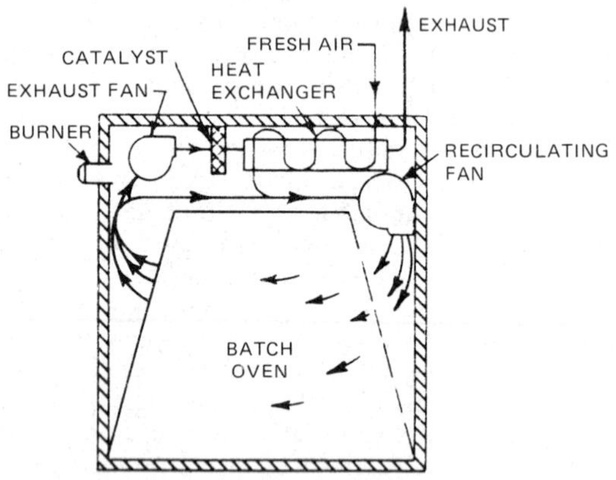

FIG. 10-5D. Batch oven with a catalytic heater for full air pollution control.

Continuous Type: In the continuous furnace or oven, the material moves through the furnace while being heated. The straight line furnace is probably the most common of this type. The material passes through a continuous oven or furnace on a conveyor or on rollers. The oven or furnace may operate at a constant temperature throughout, or it may be divided into zones maintained at different temperatures. Figure 10-5E is an example of a continuous type furnace.

A common variation of the continuous type furnace or oven is one known as a rotating hearth or rotating table furnace. In it, the material is placed on the hearth and is removed after the hearth has completed one revolution. Another design feeds the material being processed through the revolving hearth furnace or tube by means of a stationary internal screw thread.

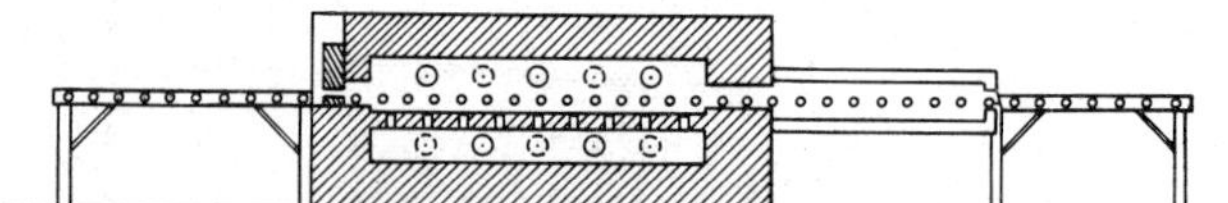

FIG. 10-5E. A continuous roller hearth furnace.

Location and Construction of Ovens and Furnaces

Ovens and furnaces should be located where they will present the least possible hazard to life and property. To prevent or minimize damage from a fire or explosion, they may need to be surrounded by walls or partitions and located either at or above grade, because basements below grade are difficult to ventilate and offer severe obstacles to proper explosion release.

The oven or furnace and the building which houses it need to be of noncombustible construction, and explosion relief venting should be provided where required. Combustibles in the vicinity of the oven should be adequately separated or properly insulated and each oven or furnace should have its own venting facilities. When gas or oil fuel is used, the heater and the oven or furnace should have separate venting unless the products of combustion discharge directly into the oven. Except in special cases, separate mechanical means are needed to provide air for the combustion and ventilation; natural draft is usually inadequate.

Furnaces which exhaust directly outdoors sometimes may be necessary, depending upon the heating process, type of combustion, and the hazard to personnel. If a furnace exhausts directly into a room, the room must have a balanced mechanical ventilation system to bring in fresh air and carry the exhaust outdoors. The supply inlets and exhaust outlets should be arranged to provide a uniform flow of air throughout the area without any dead air pockets. The system should also remove any toxic contaminants at their maximum anticipated rate of release and keep concentrations below the established Maximum Allowable Concentration (MAC) values.

Some furnaces are made of metal with a brick or masonry covering, some are made of metal with metal supports but no covering, and others have an inner lining of metal ceramic fibers or fire clay products. Occasionally furnaces have air spaces or noncombustible fillers between the outer and inner walls.

Ovens and furnaces should be well separated from valuable stock, important power equipment, machinery, and sprinkler risers, so there will be minimum interruption to production and protection if there are accidents to the oven or furnace. They should be readily accessible with adequate space above them for automatic sprinklers, the proper use of hose streams, the proper functioning of explosion vents, and routine inspection and maintenance. Roofs and floors of ovens should be insulated. The space above and below the ovens should be ventilated, to keep temperatures at combustible ceilings and floors below 160°F (71°C).

Oven and Furnace Heating Systems

The three most common methods of transferring heat to materials in ovens or furnaces are: (1) direct contact with the products of combustion, (2) convection and direct radiation from the hot gases, and (3) reradiation from the hot walls of the furnace. In muffle furnaces, the products of combustion are separated from the material being heated by a metal or refractory muffle and heat transfer occurs by radiation. (See Fig. 10-5C.) In liquid bath furnaces, i.e., salt baths or molten metal used for tempering, hardening, galvanizing, tinning, etc., a metal pot containing a liquid is heated and the heat is transferred through the liquid to the material placed in the pot.

Oven Heaters: The two general types of oven heaters are direct fired and indirect fired. In direct fired heaters, the products of combustion enter the work chamber and contact the work in process; this is not the case in indirect fired oven heaters. Instead, heating comes from radiation from tubes or from air passing over tubes and into the oven. Dangerous fuel-air mixtures cannot readily fill the work space of an indirect fired oven. Nevertheless, explosions may still occur from vapors given off by a flammable liquid drying process.

There are several arrangements of these two types of oven heaters—direct fired internal, direct fired external, indirect fired internal, and indirect fired external heaters. Figure 10-5F shows three variations of direct fired external heaters, two of indirect fired internal heaters, and three of indirect fired external heaters, with the advantages and disadvantages for each type. The exhaust ventilation arrangements in these designs provide for air movement through the oven with no recirculation through the exhaust fan. The direct fired external type oven may have a single or relatively small number of burners that simplifies the completing of automatic safety controls.

Furnace Heaters: Furnace heaters are usually arranged like ovens—direct fired internal, indirect fired external, etc., though sometimes the terms are different. If products of combustion are under a hearth and then carried up and into the heating chamber, the furnace is said to be underfired. When the same thing occurs in a chamber at one side of the furnace and passes over a bridge wall into the heating chamber, the furnace is referred to as side fired. A furnace in which the products of combustion are produced in a space above the heating chamber and pass through a perforated arch into the heating chamber is called an over fired furnace. If combustion occurs at some distance above the heating chamber and hearth, and the products of combustion are deflected onto the hearth by an arched roof, the furnace is called a reverberatory furnace. A radiant tube heated furnace is an arrangement for indirect firing. (See Fig. 10-5G.)

Sources of heat: Heat for an oven or furnace may be provided by gas burners, oil burners, electric heaters, infrared lamps, electric induction heaters, steam radiation, or heat transfer medium systems. It is important that flames, heating surfaces, or other possible sources of ignition be located where drippings or dust cannot fall or accumulate on them.

1. *Gas fired*: Gas fuel is any type of gas in common industrial use. It is important that the burner, its adjustment, and the means of combustion be suitable for the type of gas to be burned. The burner may have a single nozzle, with burners located singly or in groups, or it may have multiple nozzles in perforated pipe, ribbon, or slot burners. Burners must light easily, have a stable flame at all ports, and not have the tendency to flash back or blow off over the entire range of turndown under all draft conditions in the oven. A supply of air adequate for complete combustion may be premixed with the gas, nozzle mixed at the burner, or otherwise provided.
2. *Oil fired*: Oven heating systems may be fired with fuel oil. Oil must be vaporized before it can be burned. Most oven and furnace burners atomize and then vaporize oil while mixing it with combustion air. Heat for vaporization is generated by the flame. Atomization may be caused by pressure, mechanical spraying into fine droplets, sonic vibration, low pressure air, by high pressure air, or steam.

Combination gas and fuel oil fired systems use either separate gas and oil burners, or configurations where an oil atomizing nozzle is centered on a gas burner and arranged to use a common combustion air source.

3. *Electric heating systems*: There are five types of electric heating systems for ovens and furnaces: resistance, infrared, induction, arc, and dielectric.

Resistance heat is produced by current flow through a resistive conductor. Resistance heaters may be "open," with bare heating conductors, or "insulated sheath," with heater conductors covered by a protective sheath that may be filled with electrical insulating material. (See Fig. 10-5H.)

Infrared heat is transmitted as electromagnetic waves from incandescent lamps with filaments that operate at temperatures lower than the filament temperature of ordinary incandescent lamps, so that most of the radiation occurs in the infrared part of the spectrum. These waves pass through air and transparent substances but not opaque objects, and release their heat energy to these objects.

Induction heat is developed by currents induced in the charge. Induction heaters have an electric coil surrounding the oven space, and heating is by electric currents induced in the work being processed.

Arc heat is caused by an electric current that passes between either a pair of electrodes or between electrodes and the work, causing an arc that releases energy in the form of heat.

Dielectric heat occurs when dielectric materials are exposed to an alternate electric field. The frequencies are generally 3MHz or more—higher than those in induction heating. This type of heater is useful for heating materials which are commonly considered nonconductive.

Electric systems can be arranged so that processing does not require an oven enclosure. "Ovenless" or unenclosed heating systems can employ lamps, resistance type electric elements, or infrared heaters to vaporize flammable, toxic, or corrosive liquids and their residues. Enclosures around "ovenless" systems are advisable to prevent flammable, toxic, or corrosive vapors from escaping into the general area, and to help provide better ventilation and safeguards for personnel. However, heating systems with energy input of under 100 kW may be

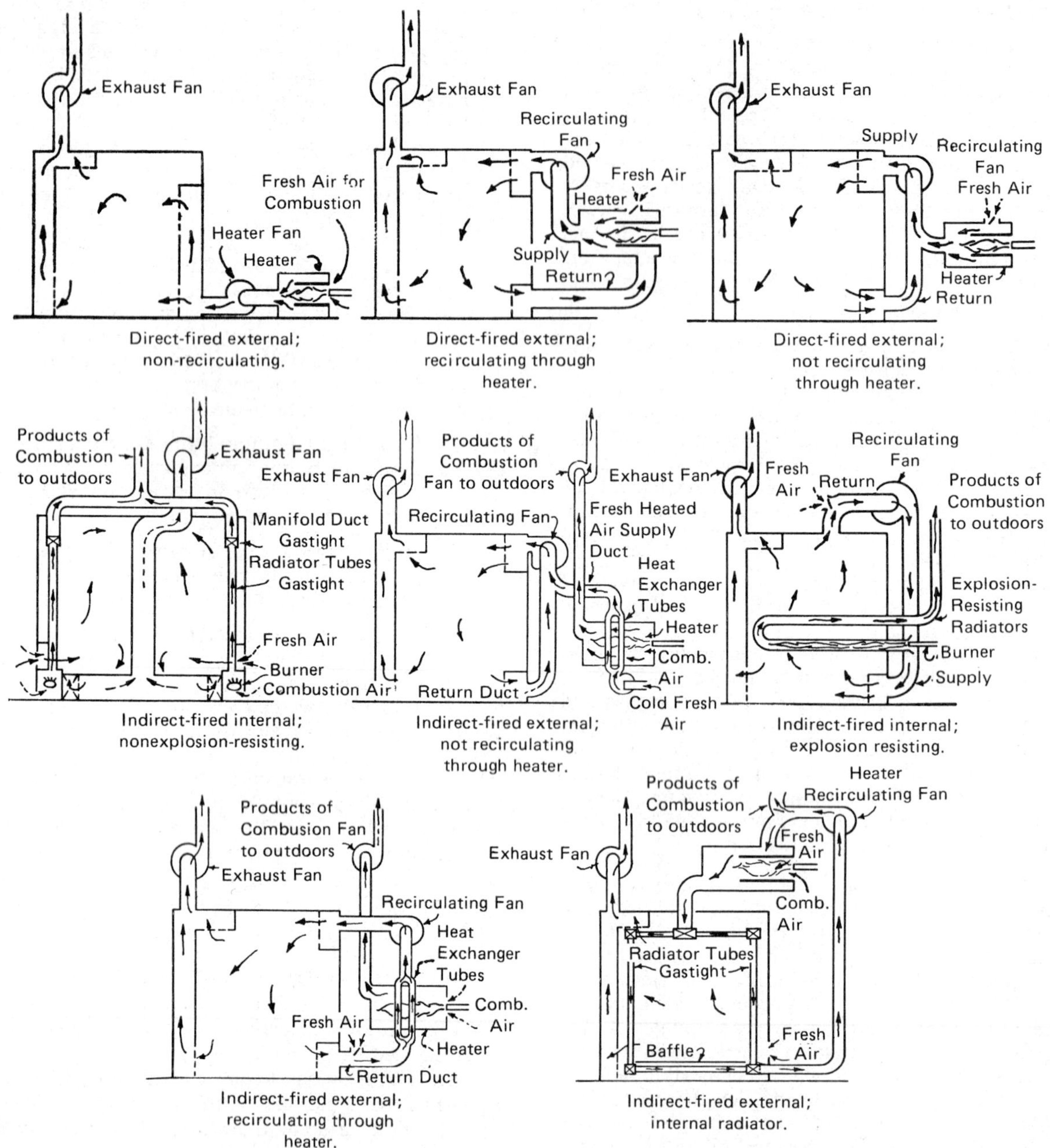

FIG. 10-5F. Types of oven heating systems.

excluded if adequate area ventilation is provided. (NFPA 70, *National Electrical Code®*, gives guidance for electrical installations in hazardous locations.)

All parts of heaters which operate at elevated temperatures within an oven or furnace and all other energized parts must be protected to prevent contact by persons, and to prevent accidental contact with materials being processed and with drippage from the materials.

4. *Steam heating systems*: In steam heating systems, the steam pressure in heat exchanger coils must be regulated at the minimum required to provide the proper drying temperatures. This avoids unnecessarily high temperatures at coil surfaces. The coils must not be located on the floor of the oven or anywhere that paint drippings or other combustibles, such as recirculated lint, may accumulate on them.
5. *Thermal heat transfer fluid systems*: As with steam heating systems, the heat exchanger should be kept free of combustibles such as lint. The heat exchanger may be located external of the work chamber and a recirculating heated air system used.

Fuel Hazards: Gas or vapors from unburned or incompletely burned fuel may, when mixed with air, be within the explosive range. These hazards develop during lighting off, firing, and shutting down the oven or furnace;

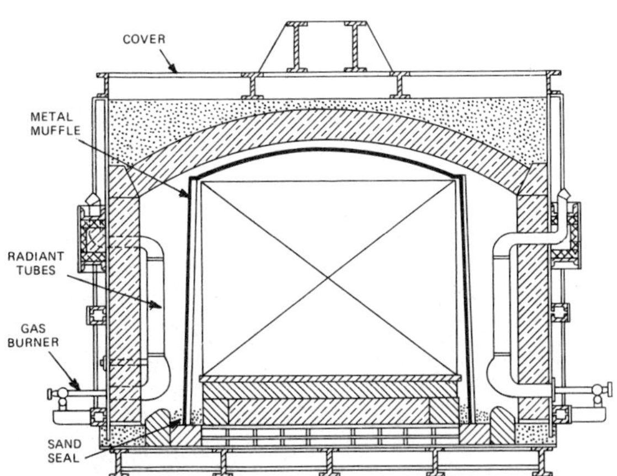

FIG. 10-5G. An annealing muffle furnace.

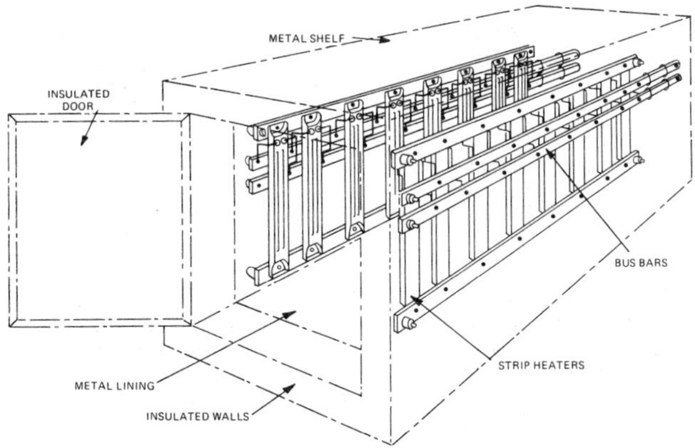

FIG. 10-5H. Strip heaters mounted in a small oven.

therefore, it is necessary to treat each operation as a separate condition requiring certain specific operating procedures to avoid mishaps.

1. *Lighting off*: Before torches, sparks, or other ignition sources are introduced, and until all burners are properly lighted, the operator must take every precaution, using all practical automatic safety controls, to avoid producing dangerous unburned fuel accumulations. The following precautions should be followed:
   (a) Purge possible accumulations of unburned fuel.
   (b) Ignite burners promptly with substantial ignitors.
   (c) If another ignition attempt is necessary, purge before introducing the ignition source and fuel again.
2. *Firing*: In the firing phase, safety requires continuous ignition and complete burning of the fuel before it passes beyond its normal combustion zone. To maintain this, the mixer and burner assembly must proportion fuel and air properly throughout the combustion zone, and the mixture velocity in the combustion zone must be neither too high, causing extinguishment by blowoff, nor too low, causing the flame to flashback or "go out." Thus good burner mixer design will be one of the main factors in safety during firing.

Air for combustion is obtained from the primary and secondary air supplied at the burner. Partial or total failure of the combustion air supply can cause an unstable flame which in turn may lead to flame failure, and the introduction of unburned fuel into the combustion chamber. When too little air is supplied, the result is an over rich mixture and incomplete combustion. The flammable incomplete products of combustion that leave the burner at concentrations too high for prompt ignition may later, within the oven or duct work, become diluted by air into the flammable range and ignite, causing an explosion. Over rich combustion also may produce rapid smothering and extinguishment of the burner flame. The flammable products of incomplete combustion followed by raw fuel likewise may become explosive when later diluted by air in another part of the system. Therefore, precautions must be taken to cut off the fuel and to require manual reset in the event that the air supply for combustion fails.

Liquid fuel, like fuel oil, must be atomized so that it will ignite easily and burn quickly. This can be accomplished by ejecting the liquid fuel at high pressure, or by directing a steam or air jet into the oil stream. Improper oil temperature (high viscosity) which prevents proper flow, partial obstructions in burner tips, and loss of oil or atomizing medium pressures can cause improper atomization. Failure to atomize properly will usually result in an unstable flame which in turn can lead to flame failure. Other things which may cause flame failure are stoppage of fuel supply by an improperly closed fuel valve or other pipe obstruction, and the presence of water in a fuel oil line.

3. *Shutting Down*: Following a shutdown, a dangerous accumulation of unburned fuel may occur in an oven and heating system if any manual fuel valves are left open or are leaking, and/or safety shutoff valves are not tight closing. If the leaking fuel subsequently is ignited by hot refractory material or by an ignition source when starting up, an explosion may occur.

Supervisory Controls for Ovens and Furnaces

It is essential that all ovens and furnaces processing flammable materials, involving flammable vapors, or heated with combustible fuels be provided with adequate supervisory devices that ensure sufficient preventilation, adequate ventilation during operation, and proper operating conditions that will not permit fires or explosions to develop. These devices should be "fail safe." While it is true that a competent operator is essential, it is also true that assistance is needed because the operator cannot continually supervise everything. The operator is, however, responsible for the proper maintenance and testing of the oven's operating and control equipment.

The type of supervisory control equipment used with oven or furnace assemblies depends upon the requirements of the particular operation. A list of the principal types of supervisory controls is given in Table 10-5A. Most of these units are tested by testing laboratories and listed according to the appropriateness of the devices for varying situations.

Full supervisory control of a gas or oil fired oven, especially one that is direct fired, may require the following: (1) determining that all fuel valves are closed and not leaking, (2) establishing and maintaining required ventilation, (3) turning on and igniting the gas pilot only at the conclusion of a required preventilation period, and (4) opening the fuel valves (including the safety shutoff valve) that supply the burner only after the combustion safeguard shows that the pilot flame has been ignited, and that fuel and air (or other atomizing agent) pressures are correct. If the burner flame is not promptly established, the safety shutoff valve will close and the entire cycle must be repeated, starting from the beginning. Once started, operation continues only as long as the supervisory controls indicate normal conditions; failure in any respect shuts down the entire system and the full cycle must begin again.

NFPA 86, *Standard for Ovens and Furnaces; Design, Location, and Equipment* (hereinafter referred to as NFPA 86); NFPA 86C, *Standard for Industrial Furnaces Using a Special Processing Atmosphere*; and NFPA 86D, *Standard for Industrial Furnaces Using Vacuum as an Atmosphere*,

TABLE 10-5A. Supervisory Control Equipment for Ovens and Furnaces

Ventilation controls:
- Air flow switches
- Pressure switches
- Fan shaft rotation detectors
- Dampers
- Position limit switches
- Electrical interlocks
- Preventilation time-delay relays

Fuel supervisory controls:
- Safety shutoff valves
- Supervising cock (FM cock)
- Flame detection units (combustion safeguards)
- Flowmeters
- Firechecks
- Reliable ignition sources
- Pressure switches
- Program relays

Temperature controllers
Temperature limit controls
Continuous vapor concentration indicators and controls.
Conveyor interlocks (with steam and electirc-resistance heating equipment)
Electrical overload protection (with resistance and induction heating equipment)
Low oil-temperature limit controls (on oil-burner equipment using heavy residual fuel oil, such as No. 5 and No. 6, which require preheating)

specify in considerable detail the safeguards needed for different types of ovens and furnaces.

When installations contain a large number of burners, the usual supervisory controls for each pilot and each burner may not be considered practical; in such cases, special devices and arrangements may have to be employed. A special continuous line pilot for multiple burners is illustrated in Figure 10-5I.

Controls for a large number of cup burners with flame propagation between individual burners may be provided as shown in Figure 10-5J.

When burners are so numerous that it is not practical to provide combustion safeguards for each, a system for supervising cocks and gas supervisory controls may be provided. This makes it possible to determine that all fuel valves are closed and that none are leaking. When inter-

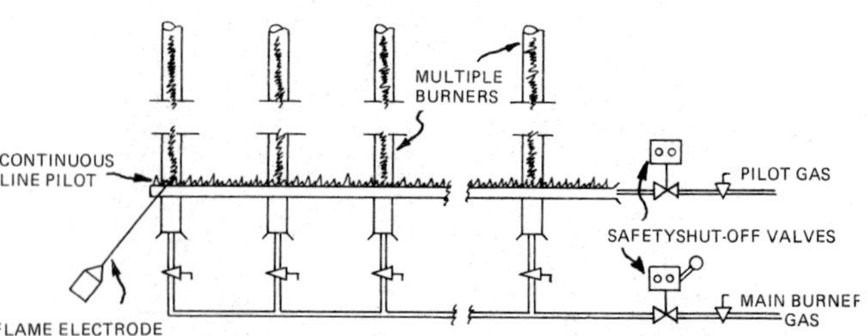

FIG. 10-5I. Arrangement of a continuous line pilot for multiple burners.

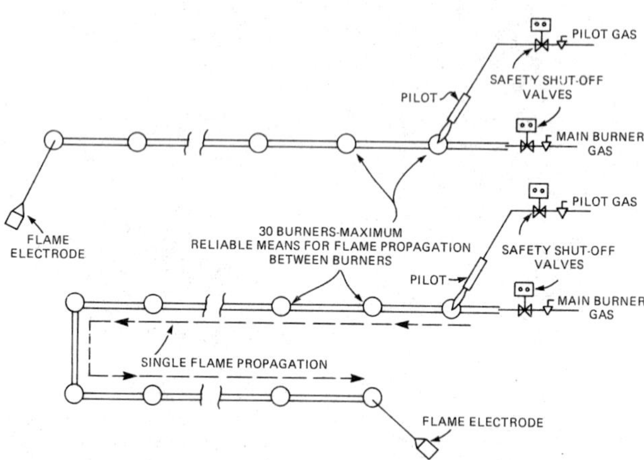

FIG. 10-5J. Typical arrangements for multiple burners of the individual cup type with flame propagation between individual burners.

locking devices indicate that all individual burner valves are closed, the safety shutoff valve may be opened and each burner valve then opened manually and the burner lighted. The supervising cock and two typical layouts are shown in Figures 10-5K, 10-5L, and 10-5M.

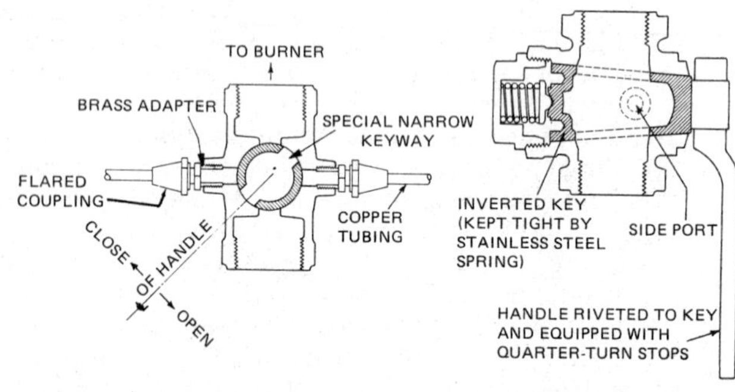

FIG. 10-5K. Details of a supervising cock (FM cock).

Multiburner combustion safeguards are now available, making it possible to install flame supervision for each burner, where previously the supervisory cock system was the only safeguard available.

When the capacity of a gas fired burner system exceeds 400,000 Btu per hr (117 kW), two safety shutoff valves are required. A manually operated plug cock and

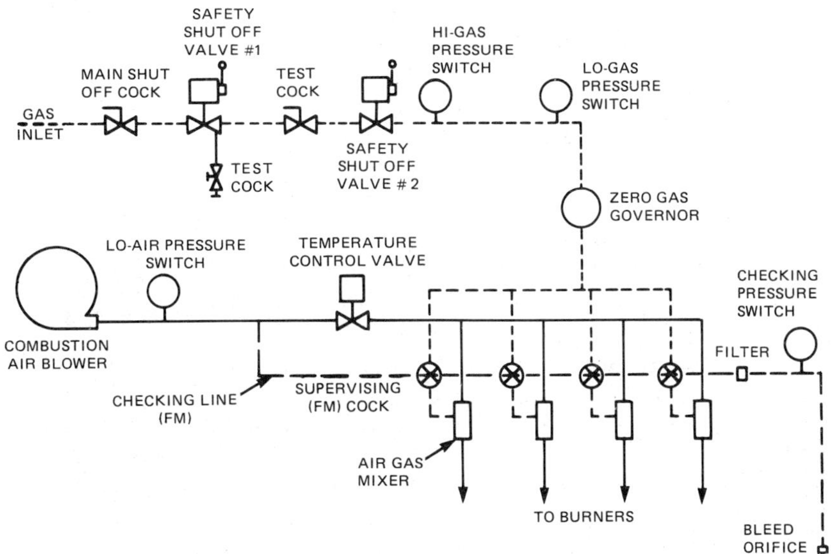

FIG. 10-5L. Example of a supervising cock and fuel gas safety control system.

test tap must be provided downstream of the safety valves. This arrangement assures positive fuel shutoff for testing the valves for leakage.

Safety ventilation is not needed in an oven which never contains flammable or noxious vapors, and it is unnecessary to guard against flammable fuel-air mixtures in an electric or steam heated oven.

Operator Training

Alert and competent operators are essential to safe operations. New operators should be thoroughly trained and tested in the use of the equipment. Regular operators should be retrained at intervals to maintain proficiency and effectiveness. Operators must have access to operating instructions at all times.

Operating instructions should be provided by the equipment manufacturer. These instructions should include schematic piping and wiring diagrams, as well as: (1) light up procedures, (2) shutdown procedures, (3) emergency procedures, and (4) maintenance procedures.

Operator training should include information on: (1) combustion of air-fuel mixtures, (2) explosion hazards, (3) sources of ignition and ignition temperature, (4) atmosphere analysis, (5) handling of flammable atmosphere gases, (6) handling of toxic atmosphere gases, (7) functions of control and safety devices, and (8) purpose and basic principles of the gas atmosphere generators.

Testing and Maintenance for Ovens and Furnaces

The operating and supervisory control equipment of each oven should be checked and tested regularly, preferably once a week. At less frequent intervals, probably annually, a more comprehensive test and check must be performed by an expert. All deficiencies must be corrected promptly, and a regular cleaning program followed to cover all portions of the oven and its attachments. Access openings for cleaning the oven enclosure and the connecting ducts must be provided.

A program for inspecting and maintaining oven safety controls is given in the appendices of NFPA standards for ovens and furnaces.

CLASS A OVENS AND FURNACES

Adequate ventilation must be provided in the operation of Class A ovens and furnaces where there is an explosion potential because of the presence of flammable vapors or fuel-air mixtures. Such fires or explosions may, in general, be prevented by good ventilation and supervisory controls that keep the flammable vapor content well below the LFL.

Ventilation

Ventilation is required while an oven is in operation and flammable vapors are given off. Control devices ensure that the ventilating and preventilating systems are operating. Failure of the ventilating fan causes shutdown of the heating system and the conveyor which carries material into a continuous oven. The following discussion refers only to that ventilation required for safe operation, and not that required for combustion and recirculation. It also does not apply to ovens operating in conjunction with solvent recovery systems which may use a low oxygen atmosphere.

Proper ventilation includes a sufficient supply of fresh air, proper exhaust to outdoors, and properly distributed air circulation sufficient to ensure that the flammable vapor concentration throughout the oven is safely below the LFL at all times. The quantity of fresh air required for safe ventilation is determined by the amount of vapor released during the process.

In general, mechanical ventilation to outdoors is required on all ovens in which flammable or toxic vapors are liberated, as well as for ovens heated by direct fired gas or oil heaters.

Ovens in which flammable or toxic vapors are never released do not require ventilation for safety, if heated by steam or electric energy, or by gas or oil fired indirect

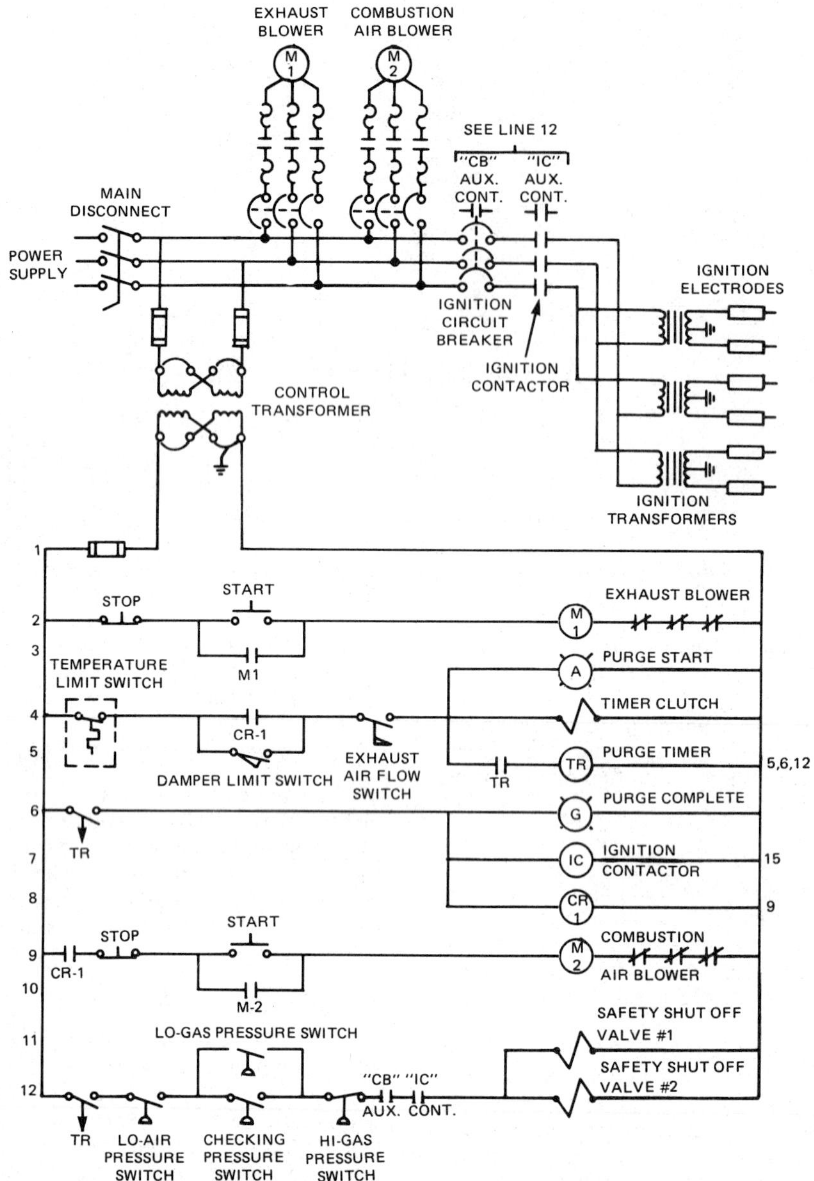

FIG. 10-5M. Electrical circuitry for a supervisory cock and gas safety control system.

heating equipment. On new ovens of every size, ventilation provided by a separate exhauster is advisable whenever appreciable amounts of flammable vapors are given off by the work.

Continuous Conveyor Ovens: The general rule for ventilating continuous conveyor type ovens is to provide not less than 10,000 cfm (283 m^3/min) of fresh air at 70°F (21°C) for each 1 gal (3.79 L) of common solvent introduced into the oven. The basis for this rule is that 1 gal (3.79 L) of common solvent produces a quantity of flammable vapor which will diffuse in air to form roughly 2,500 cu ft (71 m^3) of the leanest explosive mixture. Because a considerable portion of the ventilating air may pass through the oven without completely traversing the zone in which vapors are given off, the ventilation air may be distributed unevenly. To provide a margin of safety, four times this amount of air, or 10,000 cu ft (283 m^3) (referred to 70°F or 21°C), is required for each 1 gal (3.79 L) of solvent evaporated. In certain solvents, however, the volume of air rendered barely explosive exceeds 2,500 cu ft (71 m^3), and the safety factor decreases proportionately.

When a continuous type oven is designed to operate with a particular solvent and the ventilating air can be accurately controlled, the required ventilation can be determined by calculation. (See NFPA 86 for information on how to calculate the volume of vapor produced by 1 gal (3.79 L) of solvent and the volume of air required to provide sufficient dilution to prevent an ignitible mixture.) As with the general rule for oven ventilation, the calculated rate of air change includes a factor of safety four times the volume of air required to prevent an ignitible mixture.

Batch Process (box) Ovens: The nature of the work being processed is the basis for determining the ventilation rate in batch process ovens. Because of the wide variations in the materials, rate of evaporation, and coating thickness, it is preferable to make tests and calculations to figure the proper ventilation rate. However, years of testing and experience have shown that approximately 380 cfm (10.7 m^3/min) (referred to 70°F or 21°C) of ventilation for each 1 gal (3.79 L) of flammable volatiles released from a batch of sheet metal or metal parts being baked after dip coating, is a reasonably safe rate of air change.

For other types of work, the figure of 380 cfm (10.7 m^3/min) (referred to 70°F or 21°C) is also used, unless the required ventilation rates can be calculated from reliable previous experience, or the maximum evaporation rate is determined by tests run under actual operating conditions. In the latter case, a safety margin requires a rate of air change equaling four times the volume of air needed to produce an ignitible mixture. In any event, caution is needed when applying this estimating method to work of low mass such as paper, textiles etc., which will heat quickly, or work coated with materials containing highly volatile solvents. Either condition may give too high a peak evaporation rate for the estimating method.

Temperature Correction: Temperature corrections must be made when using the above rules, since the volume of a gas varies in direct proportion to its absolute temperature (0°F is equivalent to approximately 460° abs and 0°C is equivalent to approximately 273°K).

For example, to supply 10,000 cu ft of fresh air referred to 70°F (530° abs) to an oven operating at 300°F (760° abs), it is necessary to exhaust

$$\frac{760}{530} \times 10{,}000 = 14{,}320 \text{ cu ft of } 300°\text{F air.}$$

The metric equivalent is:

To supply 283 m^3 of fresh air referred to 21°C (294°K) to an oven operating at 149°C (422°K), it is necessary to exhaust:

$$\frac{422}{294} \times 283 = 406 \text{ m}^3 \text{ of } 149°\text{C air.}$$

In some cases, process requirements call for more ventilation than needed to maintain safe conditions in an oven. When this is true, an approximate method of figuring ventilation may be adequate for checking safety requirements. Except in these cases, all factors, including solvent characteristics, type of oven, material being processed, oven temperatures, effect of temperature on the LFL, must be carefully considered so that an adequate safety factor is assured.

Low Oxygen Ovens: Low oxygen ovens are used for evaporation of solvent in an atmosphere where the oxygen concentration is below the flammability level of the solvent. For some solvents this operating value would be approximately 8 percent oxygen, which includes a safety factor of approximately 3 percent. There is a need for oxygen analysis to verify the low oxygen concentrations. Inert gas is used to displace the oxygen (air) during startup, and combustibles during shutdown. These systems are used for solvent recovery from the coating cure process in the oven as shown in Figure 10-5N. The treated product

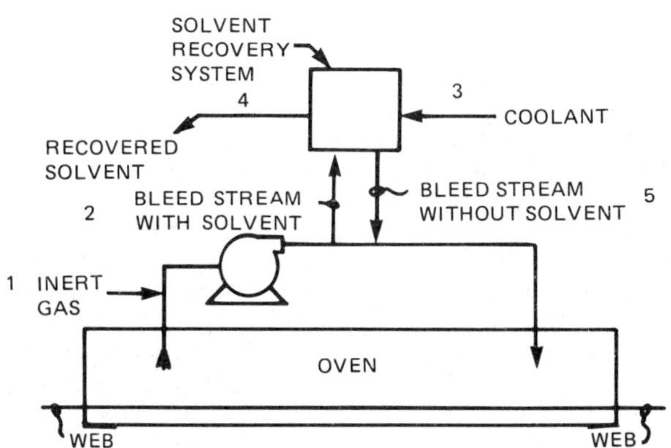

FIG. 10-5N. An example of a low oxygen oven with a solvent recovery system.

passes into and out of the oven enclosure through the oven openings. The oven atmosphere consists of an inert carrier gas which is continuously recirculated through the oven enclosure (line 1). Solvents evolve from the treated product and build up to an equilibrium vapor level which is much higher than the allowable levels in an air atmosphere. A bleed stream (line 2), which is typically 1 percent of the recirculation flow, is processed by a solvent recovery system. A liquid (line 3), which provides cooling through vaporization, acts to condense the solvents which are pumped to storage (line 4). After solvents are stripped from the inert gas stream, it is returned directly to the oven enclosure (line 5) to resume its role as a solvent vapor carrier.

Fire and Explosion Protection

Automatic sprinklers and water spray systems should be considered for heat processing equipment that contains or processes sufficient combustible materials to sustain a fire. The amount of protection required depends upon the construction and arrangement of the oven and the materials handled in it. If combustible material is processed, or if trucks or racks are combustible (or subject to loading with excess finishing material), fixed protection must extend as far as necessary into the enclosure and exhaust ducts. It also should be present where an appreciable amount of flammable drippings from finishing materials accumulates within the oven. If desired, supplementary carbon dioxide, foam, dry chemical, or Halon protection may be permanently installed, but such protection is not a substitute for automatic sprinklers.

The use of steam in ovens and dryers generally is not recommended. However, when there is no alternative, steam smothering systems may be allowed when oven temperatures exceed 225°F (107°C) and large supplies of steam are available at all times. Complete standards have not been developed for the use of steam as an extinguishing agent. Steam is not as dependable as water, carbon dioxide, dry chemical, Halon, or foam.

Portable extinguishers are needed near the oven, oven heater, and related equipment, including dip tanks or other finishing processes operated in conjunction with the oven. Small hose stations with combination nozzles also should be provided so all parts of the oven structure can be reached.

Ovens that may contain flammable gas or vapor mixtures must be equipped with unobstructed relief vents to release internal explosion pressures. These vents, panels, or doors may be secured with explosion relieving hardware or gravity retained panels that provide adequate insulation and possess the necessary structural strength. The weight of the panel should be minimal to permit movement at the lowest practical pressures. Explosion relief panels must be proportioned according to the ratio of area to the explosion containing volume of the oven, with due allowance made for openings or access doors equipped with approved explosion relieving hardware. The preferred ratio is 1 sq ft (0.09 m^2) of relief panel area to every 15 cu ft (0.42 m^3) of oven volume.

CLASS B INDUSTRIAL FURNACES

In many ways, Class B furnaces are similar to Class A ovens and furnaces. But, because no flammable volatiles or residues are present, there is nothing combustible in the construction or contents of the Class B furnace.

In many Class B furnaces, little or no effective explosion relief venting can be provided because of the weight and strength of the walls. Preventilation is important before a source of ignition is introduced into the furnace. In some cases, completely automatic purging may be practical; in others, purging may be partly manual. In either case, supervisory controls are needed to interlock the ventilation, fuel supply, combustion air, safety shutoff valve, and flame failure devices.

In multiburner installations, combustion safeguards with flame supervision should be applied where possible and practical. In some of these installations, furnace design, burner design, and the rigors of the service may preclude the application of flame supervision, and a supervisory cock and gas safety control system may be provided. When furnaces have zones operating at different temperatures, the zones may have to be treated as separate units.

Changes in the usual safety requirements may be permitted in some ovens or furnaces, but if there is any possibility of explosion from unburned fuel, adequate safeguards should be provided. An audible alarm may be installed to indicate unsafe conditions.

CLASS C INDUSTRIAL FURNACES USING A SPECIAL PROCESSING ATMOSPHERE

Special atmosphere furnaces are used to improve the quality of metals and metal alloys by heating them in an atmosphere in which air has been replaced by other gases, some of which are combustible. In most cases, the atmosphere gas is used to prevent oxidation of the metal during heating, but it also may prevent the removal or addition of carbon. Some processes which use atmospheric gases are bright annealing of copper and steel, scale free hardening and annealing of castings, brazing, and sintering. Examples of protective gases used in these processes are hydrogen, charcoal gas, and dissociated ammonia. Also used are various hydrocarbon gases produced by equipment that processes the gas used for firing, generally in the presence of a catalyst. During the past few years, synthetic atmospheres which include methanol and other stored gas components have been used.

Some heat treating furnaces contain an inert atmosphere (carbon dioxide, helium, argon, nitrogen) so they present no special fire hazard; however, these gases could present a health hazard. Other gases produce a flammable atmosphere (hydrogen, dissociated ammonia, incompletely burned hydrocarbon gas, carbon monoxide, and methane) which presents an explosion hazard; therefore their use requires special safeguards. Atmosphere furnaces are not limited to flammable gases but also may contain acid gases such as chlorine and anhydrous hydrochloric. When the latter are used in a special atmosphere, extreme care must be taken to keep air from entering the furnace. Regardless of the type of special atmosphere, Class C furnaces have fuel hazards similar to those of the Class A and B ovens and furnaces discussed earlier (although most Class C units are indirectly fired or electrically heated).

The hazards in Class C furnaces exist chiefly at three times: (1) before the process starts and the flammable atmosphere is replacing air in the furnace; (2) when the process is finished and air is being readmitted; and (3) when, for some reason, the special atmosphere supply is interrupted and air is permitted to enter. If at each of these times sufficient inert gas can be introduced into the furnace to prevent a combustible mixture, there is no danger of an explosion. Automatic introduction of inert gas upon failure of the special atmosphere supply is desirable; at the least, an audible alarm should notify the oven operator of atmosphere supply failure or other upset conditions.

Inert Gas Purge Procedure: Begin by verifying the adequacy of the inert gas supply. After all doors (if any) are closed, make sure that the flammable atmosphere gas, flame curtain, and other valves are closed. The furnace may then be heated to operating temperature. If so, introduce the inert gas at a rate capable of maintaining a positive pressure in the furnace. Sample the furnace atmosphere until two consecutive readings indicate the oxygen content is below one percent. With at least one furnace zone above 1,400°F (760°C), ignite the pilots at the outer doors and effluent vents. The flammable special atmosphere then may be introduced. When the flammable atmosphere is flowing, the inert gas should be turned off immediately. When flames appear at the vestibule effluent ports, the atmosphere introduction has been completed. The curtain burners then may be ignited.

To remove the flammable atmosphere with the furnace at operating temperature, the doors (if any) should remain closed, the inert gas purge actuated and a positive furnace pressure maintained. Shut off the special atmosphere, flame curtain, and other valves; and sample the atmosphere until two consecutive analyses indicate the atmosphere is below 50 percent of its LEL. The furnace is then purged, and the doors may be opened, and the inert gas turned off.

The normal procedure for startup or shutdown of most Class C furnaces when the temperature is above 1,400°F (760°C) is to burn out the air at the start of the process and the flammable atmosphere at the finish. In furnaces with an operating temperature above 1,400°F (760°C), burning may be performed at the start by bringing the temperature up to 1,400°F (760°C) before the special atmosphere is introduced. Automatic means are required to prevent the introduction of flammable fluids into a furnace before the furnace temperature has risen to 1,400°F (760°C). At this temperature the flammable gas will burn in the oven until the oxygen is used up and the hazard removed, allowing the operation to be started. At the end of the process, the heating should be continued so the temperature stays above 1,400°F (760°C); then the flammable gas supply is shut off and air gradually admitted. When burning stops, the flammable gas has been consumed. When an alarm indicates failure of the flammable gas supply or of the heating system, the oven operator must immediately initiate an inert gas purge, or start the admission of air to burn the flammable gas in the furnace. This must be done before the furnace cools to below 1,400°F (760°C). Where operators are not present and flammable gas flow is interrupted because of insufficient temperature inside the furnace, a flow control unit should automatically admit a flow of inert gas that will restore positive pressure without delay.

Special Atmosphere Generators

Special atmosphere generators are a source for the atmospheric gases used in some Class C furnaces. One type of generator (exothermic) produces the atmospheric gas by completely or partially burning fuel gas at a controlled ratio, usually at 60 to 100 percent aeration, while another type (endothermic) produces the atmosphere at a controlled ratio of less than 50 percent. Atmospheres from an exothermic generator may be either inert or flammable, depending upon the generator design and operating range, while those from endothermic generators are always flammable.

Another type of generator is an ammonia dissociator, which, by temperature reaction with a catalyst in an externally heated vessel, produces dissociated ammonia (25 percent nitrogen and 75 percent ammonia) from ammonia.

The special atmosphere generator must be provided with adequate supervisory controls. These would normally include interlocking of raw gas, air (if needed), burners for heating or processing, feed and discharge pressures, etc. Safety shutoff valves are usually provided in feed and discharge piping, and devices also are provided to indicate the pressure or rate of flow of processed gas to the furnace and its analysis. This can vary. The operator is thus assisted in his efforts to supply the furnace with the desired special atmosphere.

The best location for the generator and its auxiliary equipment, such as surge tank, compressor, aftercooler, storage tank, etc., is in a separate, detached building of light, noncombustible construction.

CLASS D VACUUM FURNACES

Vacuum furnaces are used for heat treating metals; they are not, however, limited to this industry. In a vacuum furnace a vacuum pump is used to displace oxygen, and in most cases, to reduce the water vapor content or dew point as well.

Vacuum furnaces are usually batch furnaces. Batch furnaces are further classified into hot wall and cold wall furnaces; the latter are in greater use at the present time. Examples of hot wall and cold wall vacuum furnaces are shown in Figures 10-5O and 10-5P. In the hot wall furnace,

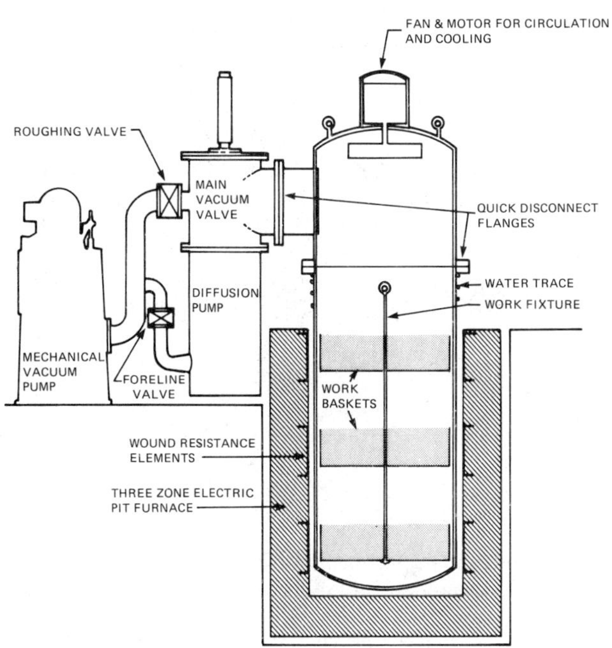

FIG. 10-5O. Example of a hot wall, single pump retort vacuum furnace.

the entire vacuum vessel is heated, though usually not above 1,800°F (982°C) because of the reduction in the strength of materials at elevated temperatures. However, installation of a second vacuum vessel outside the vacuum retort (within which a roughing vacuum is maintained during the heating cycle) permits construction of larger hot wall furnaces with higher operating temperatures.

Cold wall furnaces consist of a water cooled vacuum vessel. Usually the heating elements are inside the vacuum vessel. The walls can be maintained at near ambient temperature during high temperature operations; thus large units operating at high temperatures (4,000 to 5,000°F or 2206 to 2760°C) may be constructed. The two most common methods of heating cold wall furnaces are by resistance and induction, with the heating elements located within the vacuum vessel. The heating elements are usually water cooled, though occasionally they may be air cooled. Insulation may be effected by radiation shields constructed of low emissivity, oxidation resistant metals with proper vapor pressure characteristics. Refractory insulation can be used in some instances; but this is difficult due to outgasing of entrapped air in the refractory.

Mechanical type pumps may achieve vacuum pressures of 10^{-3} to 10^{-2} torr. For greater vacuum, booster oil diffusion pumps are used to achieve pressures in the range of 10^{-5} to 10^{-7} torr; they must be "backed" by a supplementary pump.

Fractionating oil diffusion pumps also are used and are always backed by a rotary pump, or rotary and mechanical booster combination. They produce pressures from below 10^{-3} torr down to about 5×10^{-7} torr. The accidental admission of air into a heated diffusion pump

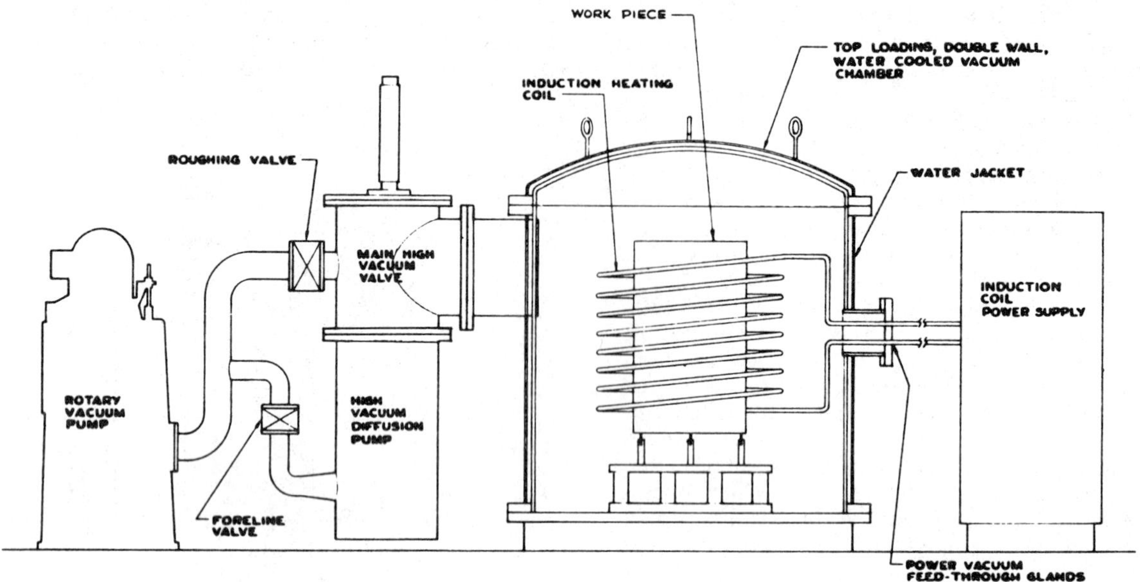

FIG. 10-5P. Example of a cold wall, induction heated vacuum furnace.

has resulted in an explosion of the pump fluid. The explosion forces damaged both the pump and the furnace hot zone.

Vacuum furnaces have the same hazards as most Class C furnaces. Besides those hazards associated with the heat sources, the additional hazards of vacuum furnaces are:

1. Water leaks in either heating elements or vessel jackets can cause explosions. If water enters the furnace at operating temperatures, it will cause more than just a steam explosion.
2. Collapse of the furnace wall if a relief valve on the water jacket fails.
3. Collapse of the vacuum retort in a hot wall furnace, if the vessel uses materials that have inadequate strength at high temperatures. If this occurs in a gas fired unit, the flame can be pulled into the vacuum pump and ignite the oil in the pumps.
4. Vacuum pumps which pull fluids (water or oil) from hydraulic seal pots.
5. The condensed metallic vapors on electrical insulators which can cause short circuiting, because this kind of furnace operates at pressures which can vaporize metals.
6. Short circuiting, if improperly supported heat shields sag at high temperatures and contact heating elements.
7. Hot spots on furnace walls that can weaken the furnace wall, if heat shields sag.
8. Temperature control problems not present in other types of furnaces and ovens. Optical pyrometers must have a line of sight to the work and their accuracy may be seriously impaired by gases, smoke, or discoloration of the sight glass.
9. Heat transfer—unless the thermocouple is actually attached to the part being measured, the heat transfer is based wholly on radiation. In air, a thermocouple receives heat by conduction and convection; therefore, with no air (or gas) in the furnace, the thermocouple response is slower. A gap of as little as 0.001 in. (0.025 mm) between the thermocouple and part or surface being measured can change significantly the response time of the thermocouple. A thermocouple on the heating element could mean that the parts would not reach the desired temperatures because the heating elements would of necessity have a temperature higher than the part (center of furnace), at least until equilibrium has been reached.
10. Induction heating—keeping piping conduits, building columns, beams, etc., out of the induction field, if the furnace is heated by induction. Any one of these items near an improperly shielded furnace can be heated by the induction coil inside the furnace. For instance, a steel bar placed so that it touches the furnace and a metal floor will be visibly hot in a matter of minutes. Of course, this also can happen on a Class B induction furnace.

AFTERBURNER AND CATALYTIC COMBUSTION SYSTEMS

Fume incinerators are combustion oxidation chambers designed to destroy process exhaust vapors or fumes by heat. Both direct flame and catalytic oxidation are used to reduce fumes, odors, vapors, and gases to acceptable exhaust products, such as carbon dioxide and water vapor. Exhaust that contains something besides plain hydrocarbons and oxygenated species may require additional special treatment, scrubbing, and filtration to remove particulate matter as well as halogens, hydroxides, sulfur oxides, and nitrogen oxides.

The advantages of fume incinerators can include reduced cleaning costs, reduced equipment downtime, reduced fire and explosion hazards, compliance with local and state pollution regulations, and savings in plant heating costs. Process fuel consumption can be reduced by the heat recovery of the burned exhaust fumes.

Installations have been damaged by fires in ducts between the process units and the incinerator, explosions of accumulated vapors in the ducts before or during start up, improper operation of gas fired incinerators, and overheating of the catalytic element or combustion chambers. The causes of these fires have been inadequate duct cleaning; inadequate duct design; incomplete prepurging of the ducts, process unit, and incinerator; failure by the operator to follow proper operating procedures; and malfunction of burner and temperature controls.

Afterburner (Direct flame) Incineration

Direct flame fume incinerators can be used for a wide range of organic solvent vapors, organic dusts, and combustible gases. In order to burn, the fumes must be heated to their autoignition temperatures with sufficient oxygen present to complete the chemical reaction. Quenching the burner flame may occur if the burner capacity is insufficient or the flame pattern and mixing are inadequate. The fumes must reach the autoignition temperature and remain there long enough (dwell time) for the chemical reaction to occur (the dwell time is usually 0.4 to 0.8 sec). Ample oxygen, more than 16 percent, is necessary for complete combustion. The step sequence for successful incineration is shown in Figure 10-5Q.

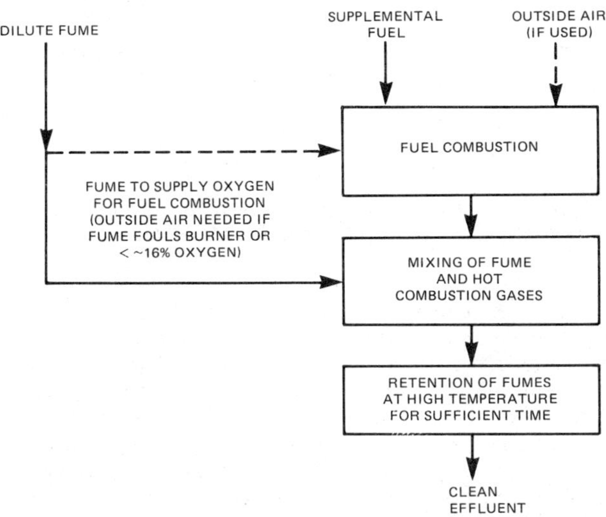

FIG. 10-5Q. Steps required for successful incineration of dilute fumes.

Operating temperatures in the combustion chamber are usually 1,200 to 1,500°F (650 to 816°C). Tests on some units have reported approximately 92 percent conversion efficiency to CO_2 at 1,300°F (704°C) and 96 percent conversion efficiency at 1,450°F (788°C). These conversion percentages of fumes to CO_2 are frequently required by air pollution codes. In any case, complete combustion requires that the fumes have sufficient air, proper mixing with the air, adequate dwell time, and adequate combustion chamber temperature.

Direct flame combustion chambers are usually heavy refractory lined with external burners, like tunnel burners, or light refractory with sectional line burners or line burners with mixing plates. (See Fig. 10-5R.)

When metal construction is used, the design must take into account high thermal stresses and possible overheating of the metals. The combustion chambers, kiln, and boiler burner flames sometimes are used as fume incinerators.

Contaminated process waste streams might be inert gases with a low combustible hydrocarbon content. This mixture may be mixed with sufficient air to ensure combustion, then oxidized in a direct flame or catalytic incinerator.

Special precautions must be taken where concentrated fumes are exhausted from the process. For safety, they are usually diluted with air to below 50 percent LEL for transfer to the incinerator.

Concentrated combustible fumes above the LEL are normally burned in flare stacks or as fuel in various types of heating equipment. The latter requires special burner design and combustion control supervision.

Catalytic Combustion Systems

Air pollution from furnace exhaust often is removed or reduced by catalytic combustion systems. A catalytic heater employs catalysts to accelerate the oxidation or combustion of air-fuel or air-fume mixtures for eventual release of heat to an oven or other process. Catalytic heaters may be used to burn a fuel gas, with substantial portions of the energy released as radiation to the processing zone. Alternately, catalytic heaters may be installed in the oven exhaust stream to release heat from evaporated oven byproducts with available energy returned by a heat exchanger for recirculation through the oven processing zone.

Three types of catalytic combustion elements are available. The first is an all metal mat used either as a fuel fired radiant heater or alternately to oxidize combustible materials in air-fume mixtures. The second type is of ceramic or porcelain construction arranged in various configurations for gas fuel or fume oxidation with catalyst media, including a variety of "rare earth" elements, i.e., platinum, or metallic salts. Both types of these elements are classified as "fixed bed" catalysts since they are normally held rigidly in place by clamps, cement, or other means. A third type element consists of a bed, pellets, or granules supported or retained between screens in a fixed position, but with the individual members free to migrate within the bed.

Heating systems that employ catalysts are widely used to conserve oven fuel and control air pollution emissions. (See Figs. 10-5S and 10-5T.) Catalytic heaters cannot, however, oxidize or consume silicones, chlorine compounds, and metallic vapors as from tin, mercury, and zinc; these elements and various inorganic dusts may retard or paralyze the catalysts.

Installation

All components of the afterburners and catalytic combustion system, related process equipment, and interconnecting ducts must be provided with controls and safeguards to supervise conditions during start up, operation, and shut down. In some installations, many different concentrations of fumes may develop. The fume collection and delivery to the incinerator is an important part of the system. A careful investigation must be made for the incinerator and the associated equipment including the appropriateness of the design and operating procedures.

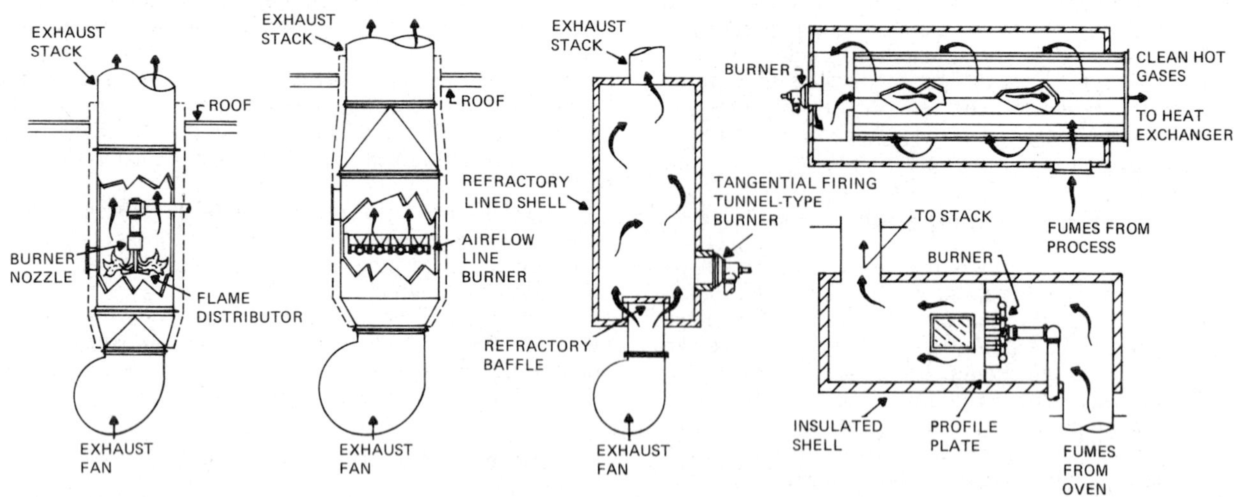

FIG. 10-5R. Typical direct flame fume incinerators. (Maxon Premix Burner Co., Inc.)

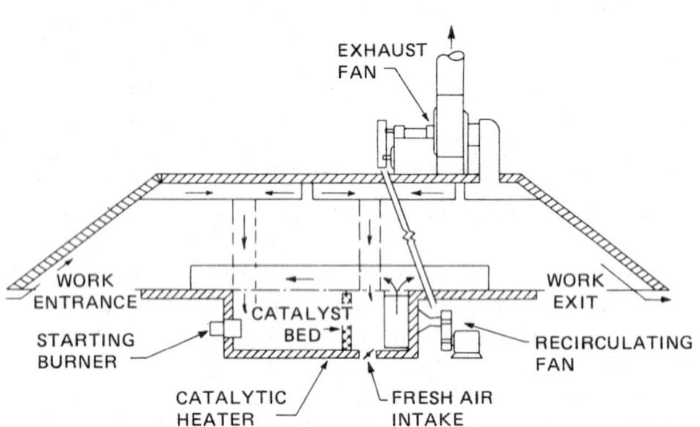

FIG. 10-5S. A direct type catalytic oven heater for partial air pollution control.

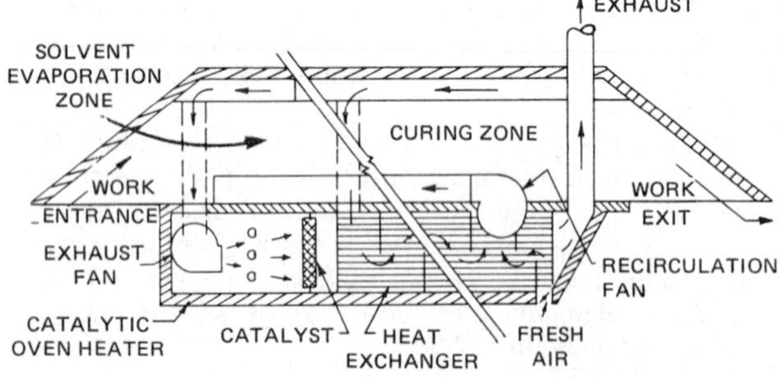

FIG. 10-5T. An indirect type catalytic oven heater for full air pollution control.

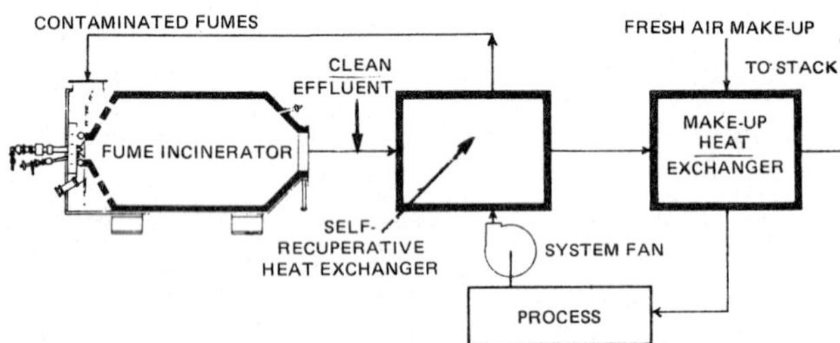

FIG. 10-5U. Typical incineration system incorporating waste heat recovery with fume and process air preheating.

HEAT RECOVERY

Heat exchangers and direct recirculation methods of heat recovery are often used to make process and fume incineration more economical. Some plants have calculated that if heat recovery methods are applied to heat generating processes, including fume incineration, the recovered heat can supply a substantial portion of the entire plant's heat or process demands. (See Fig. 10-5U.)

Recovered heat may be used for: (1) the process as either a sole or supplementary heat source, (2) the process for some zones in a multizone unit, (3) other nearby processes, (4) preheating fumes to incinerators, (5) heating plant makeup air, and (6) a waste heat boiler serving multiple plant services. However, dirty stream deposits in the heat exchanger may make it inoperable. Combustible deposits and flammable liquids heated to high temperatures within the heat exchanger could cause a fire or explosion.

LUMBER KILNS

Drying lumber to a predetermined moisture content is accomplished in a variety of structures called kilns. Kilns make it possible to turn freshly cut, green wood into dry, accurately dimensioned lumber in a much shorter time than seasoning wood in open air. Because large quantities of combustible material are exposed to temperatures that can approximate ignition temperatures, kilns present a high degree of fire hazard.

Although called dry kilns, wood dryers usually employ moisture to maintain a uniform content within the wood during the drying, thus eliminating warping, checking, and splitting. The amount of moisture will vary with the species, and significant variations may be found within trees of the same species. The length of time required to bring the moisture content down to an optimum of about two percent depends upon the species, its original moisture content, the dimensions of the pieces being seasoned, the type of kiln, and the volume of material.

Types of kilns

A dry kiln is basically an oven with controlled heat and humidity. It may operate as a batch dryer or as a progressive dryer. The wood being dried in a batch or compartment kiln remains stationary throughout the process. Temperature and humidity in all parts of the kiln are maintained as uniformly as possible and are adjusted as the wood dries.

Progressive kilns permit green lumber to be introduced in one end while dried lumber is being removed from the other end. Such kilns are designed so that somewhat higher temperatures are maintained at the dry, or discharge end rather than at the loading end.

In natural circulation kilns, heated air rises up through the stacked lumber by convection. (See Fig. 10-5V.) Losing its heat, air travels down to the heating device

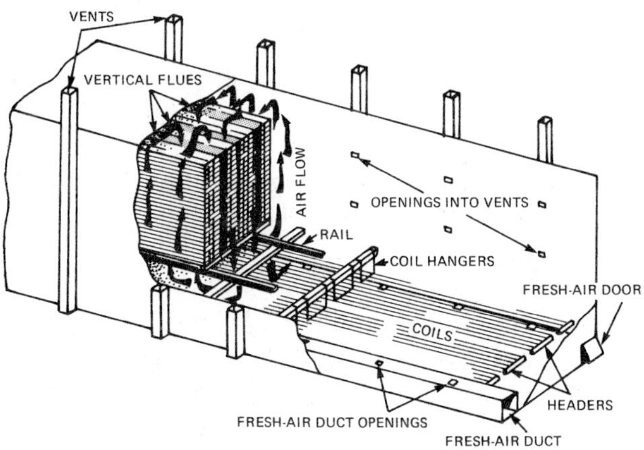

FIG. 10-5V. A natural circulation, steam heated compartment kiln. Arrows indicate air movement during the early stages of the drying cycle.

and is reheated. During the first part of the drying cycle, the air flows up along the sides, over the stacked lumber, and down through air passages in the stack. When the moisture content has been reduced to somewhere between 20 and 10 percent, heated air is directed up through the center of the stack to equalize the drying. Vents in the walls or roof of the kiln exhaust the hot, moisture laden air.

Forced circulation kilns move air through the stacked lumber by either internal or external blowers. (See Fig. 10-5W.) Figure 10-5X shows an internal fan kiln with fans located beneath the floor. Internal fans are reversible, changing the airflow for optimum drying. Internal fan kilns normally require cloth or metal baffles to eliminate turbulence and keep the air flowing in the desired direction.

Much the same airflow systems are used in both batch and progressive kilns. There are some minor differences, however. In progressive kilns, the air flows through the length of the chamber and is discharged at the green end. Because the air has picked up moisture and is cooler by the

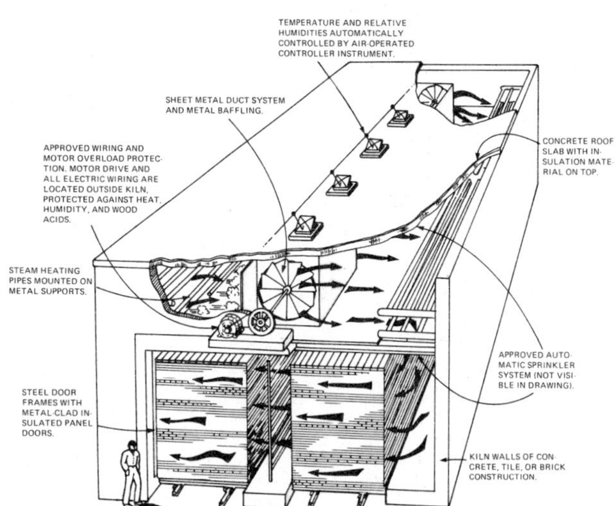

FIG. 10-5W. A forced circulation double track compartment kiln. Note that automatic sprinklers are installed above and below the platform between the kiln and the overhead fan room.

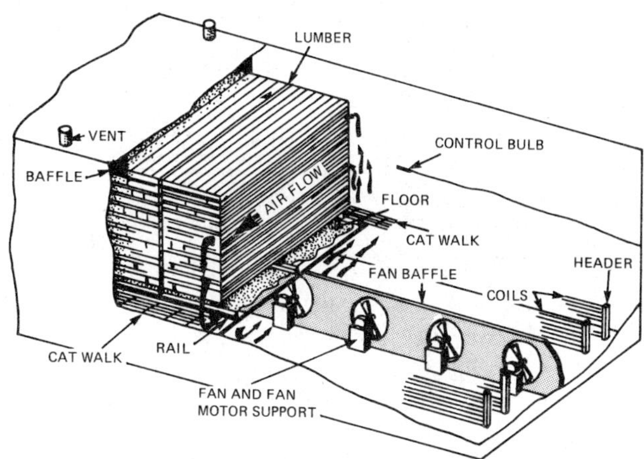

FIG. 10-5X. A compartment kiln with internal fans and stream coils located under the grating at the floor level. Broken pieces of stickers and sawdust can fall through the grates and collect around the coils and fan motors to become a hazard, particularly if high pressure steam coils are used.

time it reaches the loading end of the kiln, the drying rate there is much slower.

Kiln Construction

Unlike most other buildings, kilns are subjected to extreme variations in internal temperature and humidity. Such variations cause unusual expansion and contraction that reduce structural integrity of the kiln and increase heat loss. Untended structural defects also can lead to premature failure of the structure. To reduce the fire hazard, kilns should be of fire resistive or heavy timber construction.

Heat Sources

Dry kilns require a constant source of heat, provided either directly or indirectly, to vaporize the water content of the wood. Direct heating is accomplished by circulating hot gases (produced by burning gas, oil, sawdust, or other fuels) through the stacked lumber. It also can be done by heating large metal surfaces with an open gas or oil flame as in Figure 10-5Y. Air circulated by internal fans passes over the metal which acts as a heat exchanger, then flows over the lumber, carrying off moisture vapor.

Steam is a common source of heat for indirectly heated kilns, but hot gases and electrical resistance heaters also are used. Steam is circulated through pipes and hot gases through ducts.

The Fire Hazards

Lumber kilns present high fire hazards. This is especially true when direct heat systems or high pressure steam systems are used, and in those instances where the structure itself is of combustible materials.

Direct fired kilns present the greatest hazard because open ignition sources are close to the wood. (See Fig. 10-5Y.) The kilns should be considered analagous to Class A ovens in that they should be equipped with all the combustion controls normally required for drying ovens where the heating fuel is introduced into the heating enclosure itself. Indirectly heated kilns utilizing steam coils as heat exchangers and having controlled humidity are usually considered to be of low hazard.

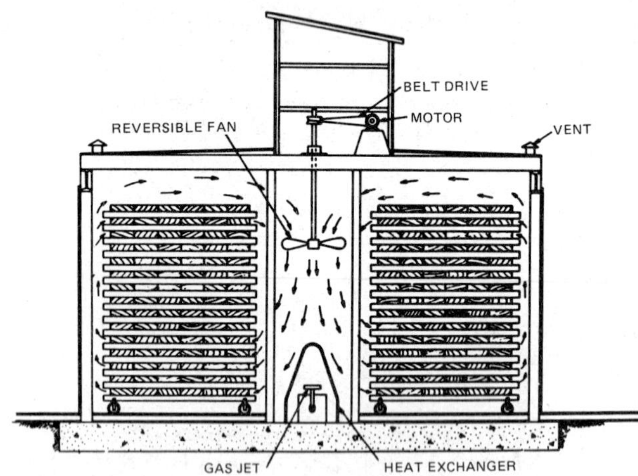

FIG. 10-5Y. A double track, internal fan compartment kiln that is heated directly by a gas burner.

The Safeguards

The important requirements for firesafety are automatic sprinkler protection, sound construction, good housekeeping, automatic humidity control, good air circulation, and proper ventilation. Kilns should be situated at safe distances from storage yards, sheds, and mill buildings. Ideally, they should be of fire resistive construction and equipped with a complete automatic sprinkler system connected to an adequate water supply. The sprinkler protection should extend to fan houses and control rooms. Hydrants or hose connections should be located on the exterior for manual fire fighting.

DEHYDRATORS AND DRYERS

Dehydrators and dryers for agricultural products, commonly referred to as dryers, use heat to reduce the moisture content of products. The hazards of dryers are (1) the possibility of igniting combustible materials near them, (2) the use of fuel or electricity as a heat source, and (3) the ignition of stock being dried.

Types of Dehydrators and Dryers

The three types of agricultural product dryers are: continuous, batch, and bulk. They differ in the arrange-

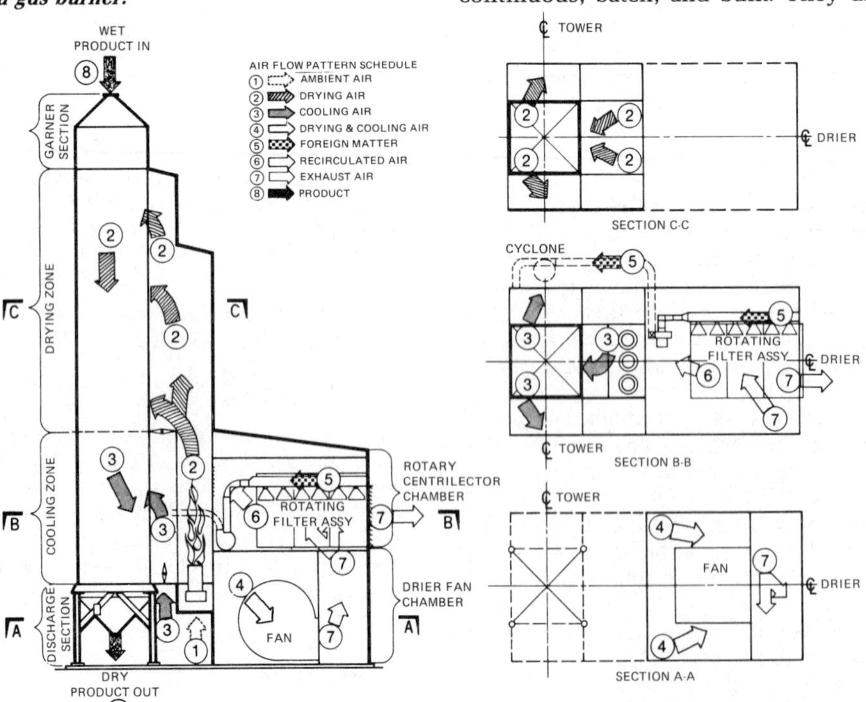

FIG. 10-5Z. A tower type gravity dryer. (Aeroglide Corporation).

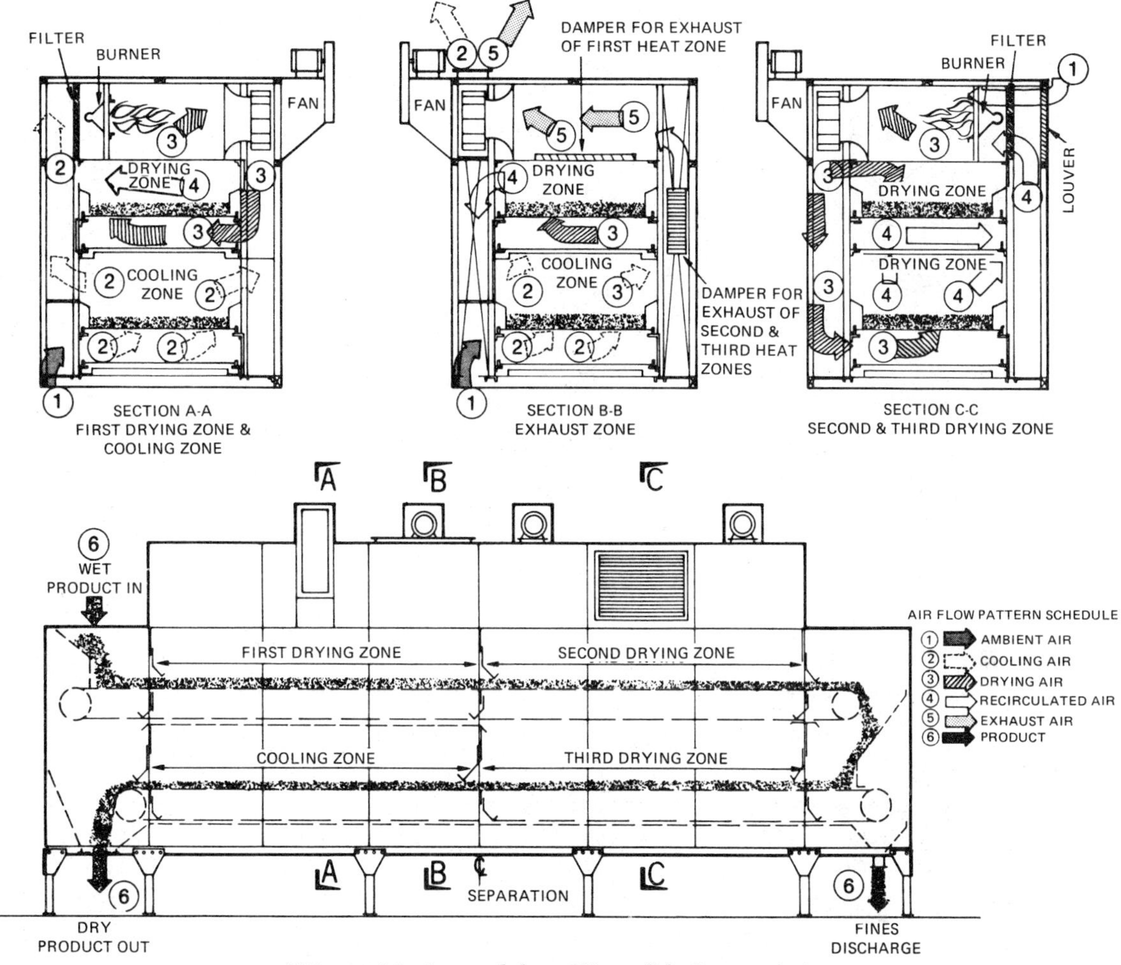

FIG. 10-5AA. A tunnel dryer. (Aeroglide Corporation).

ment and operation of the drying chamber. Continuous dryers include:

1. Drum dryers for milk, puree, and sludges.
2. Spray dryers for milk, eggs, and soup.
3. Flash dryers for chopped forage crops.
4. Gravity dryers (may also be batch type) for small grains, beans, and seeds. (See Fig. 10-5Z.)
5. Tunnel dryers (may also be batch type, and may be further classified according to airflow and whether or not intermediate heating is used) for fruits, vegetables, grains, seeds, nuts, fibers, and forage crops. (See Fig. 10-5AA.)
6. Rotary dryers for milk, puree, and sludges. (See Fig. 10-5BB.)

Batch dryers may be either fixed or portable and include pan dryers for sugar, puree, sludges, and other products. (See Fig. 10-5CC.)

Bulk dryers dry the product in a bin, crib, or compartment in which it is to be stored. They are used to dry seeds, grains, nuts, tobacco, hay, and forage. (See Fig. 10-5DD.)

Methods of Heating

Dryers for agricultural products may be direct fired (where products of combustion contact the material being dried) or indirect fired. The heaters may be oil fired, gas fired, solid fuel fired, electrical, or heated by a heat transfer medium, such as steam. In general, the requirements for burner installation and fuel storage are the same as those for other heat producing devices.

If gas fired infrared heaters or lamps are used, their focal length should be ample so the surface of the drying product does not reach unsafe temperatures. If electrical infrared lamps are used in dryers, the lamps should be located where they cannot collect combustible dust.

Solid fuel furnaces (other than those burning coke and anthracite coal) should not be used where the products of combustion can enter the drying chamber. Indirect solid fuel dryers need temperature controlled heat relief openings to the outside.

Dryer Controls

Some suggested controls for dryers, except those on the heating equipment, include:

1. A method for automatically shutting down the dryer in the event of fire or excessive temperature.
2. A thermostat in the exhaust air when the product is fed automatically from the dryer to a storage building. In the event of excessive temperature, the thermostat (a) shuts off heat to the dryer and stops airflow (except when the product being dried is in suspension), (b)

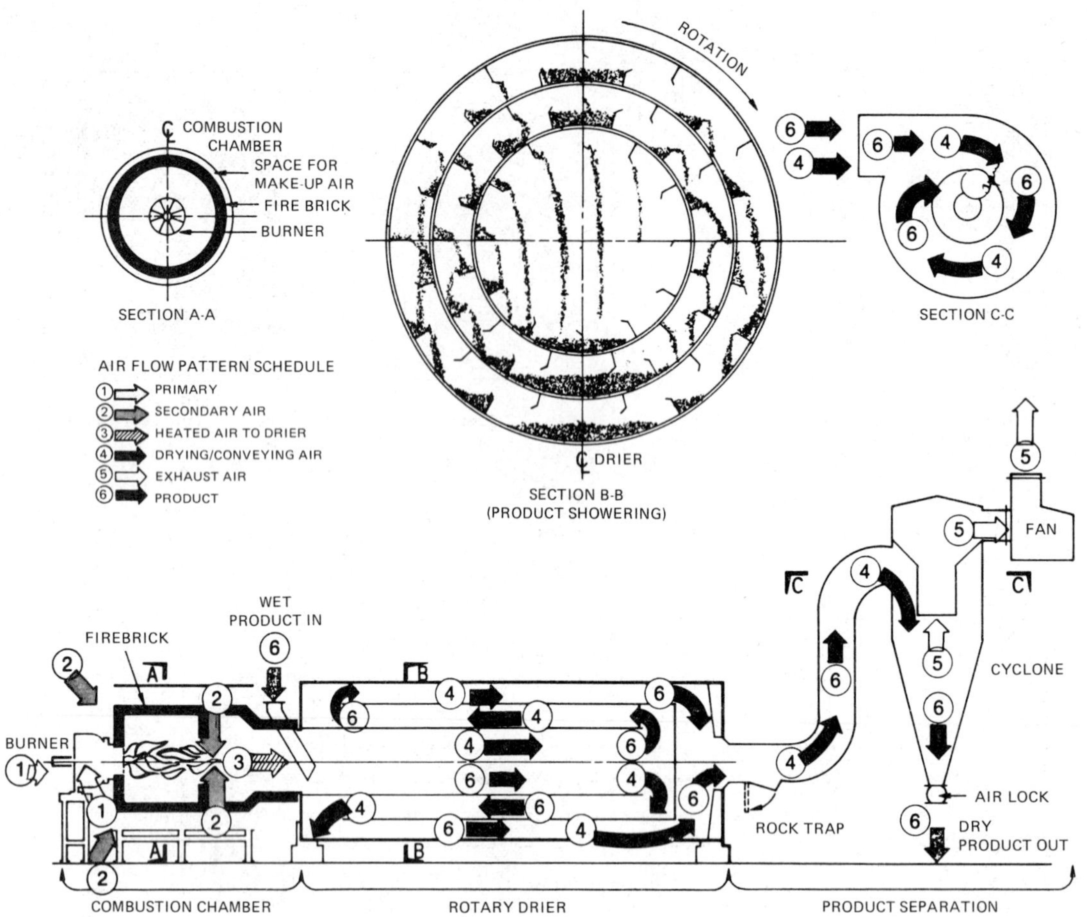

FIG. 10-5BB. A rotary dryer. (Aeroglide Corporation).

stops the flow of the product, and (c) sounds an audible alarm.

3. A thermostat in combustible dryers which, when the temperature of the combustible reaches 165°F (74°C), shuts off heat to the dryer but permits unheated air to pass through, and sounds an audible alarm.
4. A device to shut off heat to the dryer if air movement through the dryer stops.
5. A high limit thermostat located between the heat producing device and the dryer.

Burner Controls

In general, the burner controls for dryers are the same as those for other automatically fired devices. A manual, quick closing shutoff valve should be installed in the

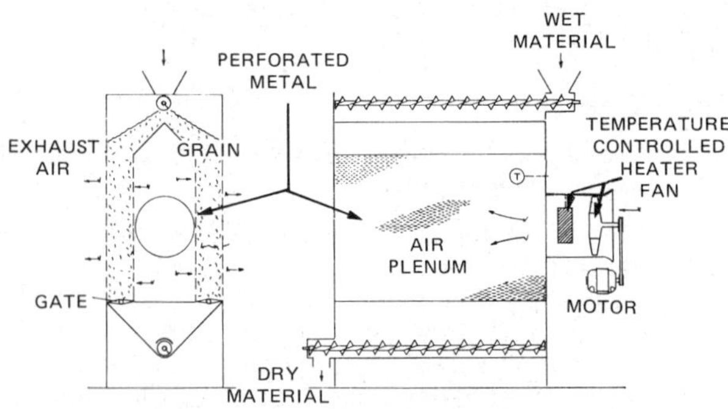

FIG. 10-5CC. A batch type grain dryer.

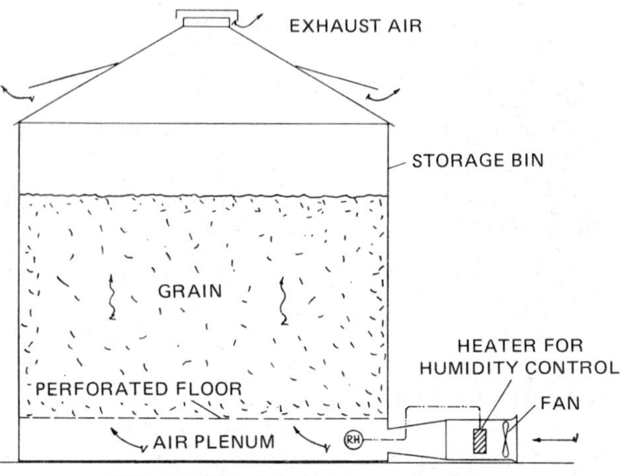

FIG. 10-5DD. A bulk type grain dryer.

supply line of gas and oil fired burners, and controls should be arranged so that following automatic shutdown, manual restart will be necessary. Other control safeguards include flame failure protection, and preventilation of the combustion chamber. All supervisory controls should be arranged to "fail safe."

Construction and Installation of Dryers

Because dryers operate at elevated temperatures, they should be made of fire resistive or noncombustible materials. If combustible materials must be used, dryers must not be subjected to sustained temperatures in excess of approximately 165°F (74°C). Expansion joints should be provided to prevent damage from expansion and contraction.

Secondary air openings for direct fired dryers are screened with ½ in. (12.5 mm) mesh screen so materials cannot enter the combustion chamber. Primary air openings require screens with mesh ¼ in. (6.3 mm) or smaller. An ample supply of easily opened access panels is necessary for inspection, cleaning, and fire fighting.

When stock is moved through the dryers so that it generates static electricity, all conductive parts of the dryers should be electrically bonded and grounded.

Like any heat producing device, a dryer must have adequate clearance from nearby combustibles to prevent overheating. If there is a combustible dust hazard in the same building as the dryer, the heating device and blowers are installed in a dust-free room or area separated from the rest of the building. Ducts to convey heated air to the dryer and exhaust air from the dryer to the outside should be noncombustible.

Extinguishing Equipment

The best way to protect a dryer enclosure is to install water spray heads or automatic sprinklers within the enclosure where possible. An exception to this is the direct fired rotary dryer, which may be damaged by the internal application of water. A carbon dioxide system is satisfactory protection for this type dryer.

To extinguish small fires in and around most dryers, standpipe hoses are most useful, though water type portable fire extinguishers also may be used.

Cooling of Dehydrated Products

A product being dried requires adequate cooling before it is packaged or stored. The amount of cooling required to prevent subsequent ignition will depend upon the properties of each material and how it is to be packaged or stored.

Bibliography

NFPA Codes, Standards, Recommended Practices and Manuals. (See the latest *NFPA Codes and Standards Catalog* for availability of current editions of the following documents.)

NFPA 10L, *Model Enabling Act for the Sale or Leasing and Servicing of Portable Fire Extinguishers (Including Recommended Rules and Regulations for the Administration of the Act).*

NFPA 12, *Standard on Carbon Dioxide Extinguishing Systems.*

NFPA 13, *Standard for the Installation of Sprinkler Systems.*

NFPA 15, *Standard for Water Spray Fixed Systems for Fire Protection.*

NFPA 31, *Standard for the Installation of Oil Burning Equipment.*

NFPA 54, *National Fuel Gas Code.*

NFPA 58, *Standard for the Storage and Handling of Liquefied Petroleum Gases.*

NFPA 61B, *Standard for the Prevention of Fires and Explosions in Grain Elevators and Facilities for Handling Bulk Raw Agricultural Commodities.*

NFPA 70, *National Electrical Code.*

NFPA 86, *Standard for Ovens and Furnaces: Design, Location, and Equipment.*

NFPA 86C, *Standard for Industrial Furnaces Using a Special Processing Atmosphere.*

NFPA 86D, *Standard for Industrial Furnaces Using Vacuum as an Atmosphere.*

NFPA 325M, *Fire Hazard Properties of Flammable Liquids, Gases, and Volatile Solids.*

Additional Readings

Dust Explosion and Fires: A Manual, National Particleboard Association, Silver Spring, MD, 1977.

Factory Mutual Engineering Corp., "Industrial Ovens and Dryers," *Loss Prevention Data Sheet 6-9*, Oct. 1977, Factory Mutual System, Norwood, MA.

Factory Mutual Engineering Corp., "Process Furnaces," *Loss Prevention Data Sheet 6-10*, Nov., 1976, Factory Mutual System, Norwood, MA.

Factory Mutual Engineering Corp., "Elements of Combustion, Controls, and Safeguards in Industrial Heating Equipment," *Loss Prevention Data Sheet 6-0*, Sept. 1977, Factory Mutual System, Norwood, MA.

Feirer, John L., *Wood: Materials and Processes*, Bennett Publishing Co., Peoria, IL, 1980.

Grace, C., "Fluid Choice Takes the Steam Out of Unsafe Process Heaters," *Process Engineering*, (5), 1977, pp. 85, 87-88.

"Hazard and Hazard Prevention in Solvent Evaporating Ovens," *Fire Protection* (127), 1978, pp. 22-25.

Jaffee, H. M., "Grain Elevator Protection, What's Being Done Today?" *Fire Journal*, Vol. 74, No. 3 (May 1980), pp. 131-132.

King, P. W., and Magid, J., *Industrial Hazard and Safety Handbook*, Butterworth Publishers Inc., Woburn, MA, 1979.

Linville, Jim L., ed., *Industrial Fire Hazards Handbook*, National Fire Protection Association, Quincy, MA, 1984.

Rasmussen, E. F., *Dry Kiln Operator's Handbook*, Agricultural Handbook No. 188, Forest Products Laboratory, Forest Service, U.S. Department of Agriculture, Washington, DC, Mar., 1961.

Reed, Robert D., *Furnace Operations*, 3rd ed., Gulf Publishing Co., Houston, TX, 198l.

Standard for Construction, Installation, and Rating of Equipment for Drying Farm Crops, American Society of Agricultural Engineers, St. Joseph, MI, Dec., 1962.

Trinks, W., *Industrial Furnaces*, (Vols. 1 and 2), John Wiley and Sons, New York, NY, (Vol. 1: *Principals of Design and Operation*, 5th ed, 1961; Vol. 2: *Fuels, Furnaces Types and Furnace Equipment: Their Selection and Influence Upon Furnace Operation*, 4th ed., 1967.)

OIL QUENCHING

Revised by Raymond Ostrowski

One process in the heat treatment of metals is a controlled cooling or quenching of heated materials by immersion in a liquid quenching medium. This process hardens and tempers the metal by imparting metallurgical changes in its surface. Due to the combustible nature of quench oils, the process presents serious fire hazard potentials.

Additional hazard and protection information relevant to oil quenching may be found in Chapter 4 of this Section, "Fluid Power Systems." Fire suppression information will be found in Section 18, "Water Based Extinguishing Systems;" and Section 19, "Special Fire Suppression Agents and Systems."

Contributing to the hazards of oil quenching are:

1. Special atmosphere requirements (a gas other than air blanketing the surface of the oil).
2. Temperature requirements of the quench medium.
3. Physical properties of the quench medium.
4. Volume limitations of the quench medium .
5. Size and configuration of process materials.
6. Locations of furnaces and quench tanks.
7. Mutual exposure between quenching and other processing or storage facilities.

Whether the quenching is an automatic and continuous, semiautomatic, or batch operation, it will involve elevators, conveyors, hoists, and cranes, either individually or in combination, to immerse the work in, move it through, and remove it from, the oil bath. (See Fig. 10-6A.) Although all three steps are important to the process, the one that is critical to safety is the entrance of the work into the quench.

Quenching Oils

In most cases, mineral oils are used for quenching, but specific metallurgical requirements may dictate the use of mixtures with animal or vegetable oils. In addition, wetting agents may be blended with certain oils or oil mixtures. In any case, it is important to use a quenching oil with a low viscosity.

Essential quench oil properties include the ability to remain stable over periods of extended usage and to retain fluidity. For unheated quenching, operating temperatures between 100 and 200°F (38 and 93°C) are considered normal. Standard quench oils for use in this temperature range usually have flash points somewhat above 300°F (149°C). For heated quenching, operating temperatures between 200 and 400°F (93 and 204°C) are common. Quench oils used at these temperatures generally have a flash point above 500°F (260°C).

Polymer Quenching

A rather recent development is the use of polymer quenchant in integral quench furnaces. A furnace designed for oil is not designed for a quick change to polymer quenching. Caution should be observed and the original furnace manufacturer should be contacted prior to changing from oil to polymer quenchant.

Quench Tanks

A quench tank should allow proper quenching under normal conditions, and provide for minor variations in equipment control functions and operator error. The design of the tank freeboard, overflow drains, and liquid level control are all critical. A fire that is confined to the liquid surface within a tank is much more readily controlled and extinguished than a fire involving quench oil that has overflowed the tank.

The distance from the quench surface to the top of the tank, with the tank loaded to capacity, is known as the freeboard. Freeboard design should take into account the splashing to be expected when the maximum workload is immersed with maximum speed. The distance between the liquid level with the workload submerged and any openings in the tank wall should be not less than 6 in. (152 mm) below the door or any opening into the furnace.

Adequately sized, fully trapped overflow drains are important safety features for all oil quench tanks. They should direct the overflow to a safe location outside of buildings or into special tanks. As a practical matter, small quench tanks frequently may be installed without overflow drains. However, quench tanks with a liquid capacity

Mr. Ostrowski is Sales Manager with Protection Controls, Inc., Skokie, IL.

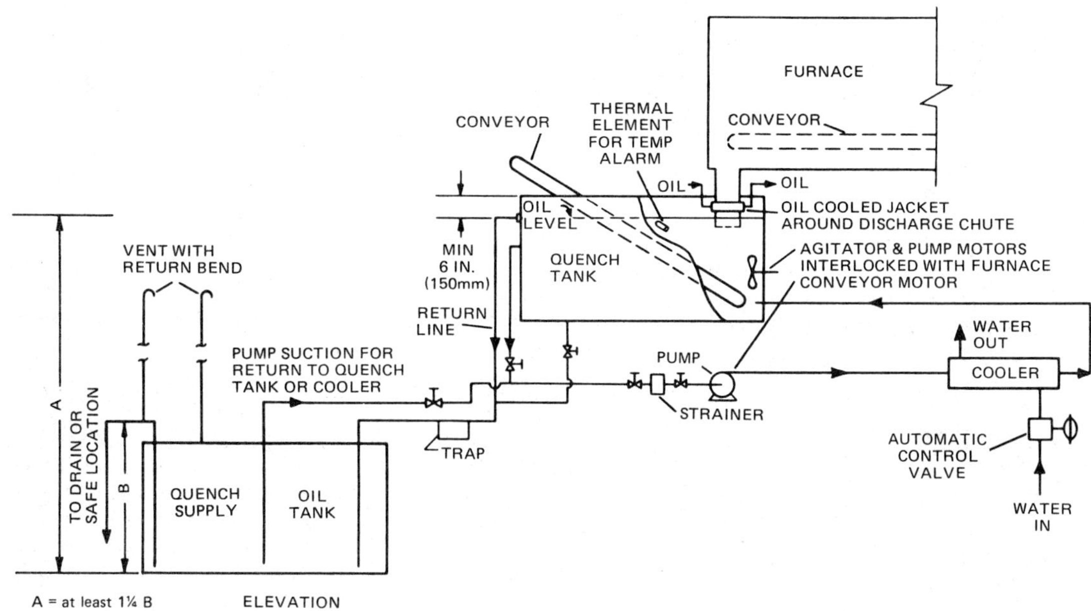

FIG. 10-6A. A schematic of a typical continuous type oil quench tank. (Factory Mutual System)

of 150 gal (0.57 m^3) or a liquid surface area of 10 sq ft (0.9 m^2) or larger should be provided with overflow drains.

While overflow drains should be specifically designed for each tank, certain minimum sizes have been established and accepted. (See Table 10-6A.) For large quench tanks, multiple overflow pipes are preferable to a single large pipe, provided the aggregate cross sectional area is equivalent to that of a single pipe. Piping connections on drains and overflow lines must be designed to permit ready access for inspection and cleaning.

TABLE 10-6A. Quench Tank Overflow Drains, Minimum Pipe Sizes

Liquid Surface Area		Minimum Pipe Diameter (I.D.)	
sq ft	m^2	in.	mm
10–75	0.92–6.7	3	76
75–150	6.7–14	4	101
150–225	14–21	5	127
225–325	21–30.2	6	152
325+	30.2+	8	203

Emergency Drains: Under serious fire conditions, it may be necessary to empty a quench tank in order to reduce the amount of fuel available. This can be performed readily through adequately sized, fully trapped, and valved bottom drains directed to safe locations. Where gravity drainage is not possible, special pumps may be provided for oil removal. Drains and pumps should be sized so that the quench oil can be removed within 5 min. Gravity drains should be sized according to the tank capacities and drain diameters given in Table 10-6B.

Emergency drains must be used only under the guidance and control of well trained and experienced personnel, as improper usage can result in greater hazards. Whenever a flammable gas processing atmosphere is maintained above the quench oil, removal of the oil can create a negative pressure that can result in explosion, an increase in fire severity, or both.

TABLE 10-6B. Gravity Drain Pipe Diameters, I.D.

Tank Capacity		Pipe Diameter (I.D.)	
gal.	m^3	in.	mm
500–750	1.9–2.8	3	76
750–1000	2.8–3.8	4	101
1000–2500	3.8–9.5	5	127
2500–4000	9.5–15	6	152
Over 4000	Over 15	8	203

Tank Location: Heat treating operations involving combustible quench oils should be housed in fire resistive buildings and should be well separated from exit areas, combustible materials, valuable stock, power equipment, and important process equipment.

The safest location for quench tanks is at grade level. Boilover from tanks above grade can be expected to spread fire to floors below, thereby making fire control more difficult and significantly increasing the potential fire loss. Fires in below grade locations will make manual fire fighting difficult and result in a significant increase in fire loss.

Material Transfer

Rapid and complete immersion of the work in process is essential to safe and proper heat treating metallurgical processes. The method of immersion must result in minimal splashing and no overflow of the quench oil outside of the tank. It is also essential that the possibilities for partial immersion be eliminated or minimized. Partial immersion of the work is the most common cause of quench oil fires.

Chutes: Many furnaces are designed so that the work in process drops off the end of a conveyor, through a chute, and into the quench oil. The chute design and construction must allow the work to fall freely under all normal furnace feed conditions. This will involve proper chute sizing to accommodate the work, proper pitch to ensure continued stock motion, and smooth surfaces to prevent the work from being stuck in the chute. (See Fig. 10-6B.)

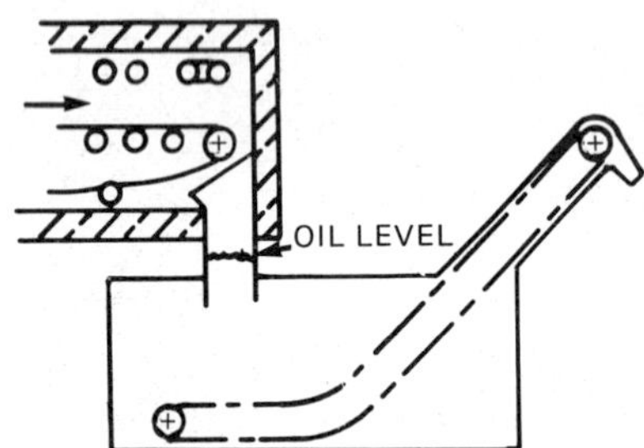

FIG. 10-6B. Bottom Chute Type quench.

Elevators: When stock in process is handled in baskets, elevators are commonly used to immerse the work in the quench oil. (See Figs. 10-6C and 10-6D.) Partial immersions are caused more by baskets and elevators than by any other method. The following are three critical concerns:

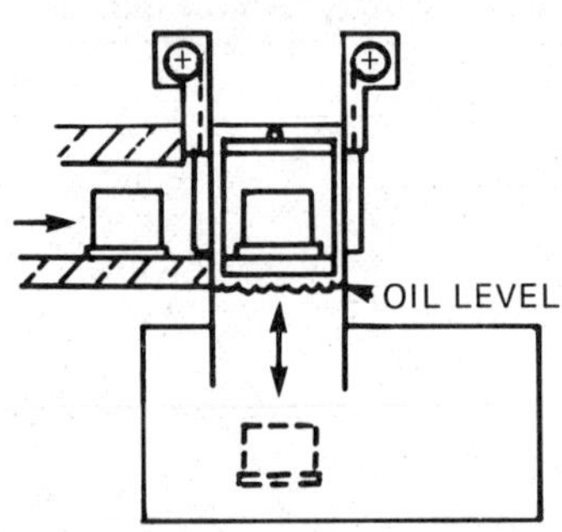

FIG. 10-6C. Dunk Type elevator quench.

1. The elevating mechanism must be adequately supported by structural members to prevent its falling unevenly.
2. Adequate guides must be provided to ensure uniform movement within the quench tank and to prevent an elevator from being wedged, as this could result in partial immersion.
3. Suitable guides and stops must be provided to prevent shifting of the workload as this can cause elevator jamming and partial immersion.

Hoists and Cranes: A hoist or crane is usually required for moving large, specialized workloads into and out of the quench tank. Proper positioning of this equipment can be accomplished by stops and/or limit switches. Mechanical guides may be required to ensure that the work is properly positioned as it enters the quench tank.

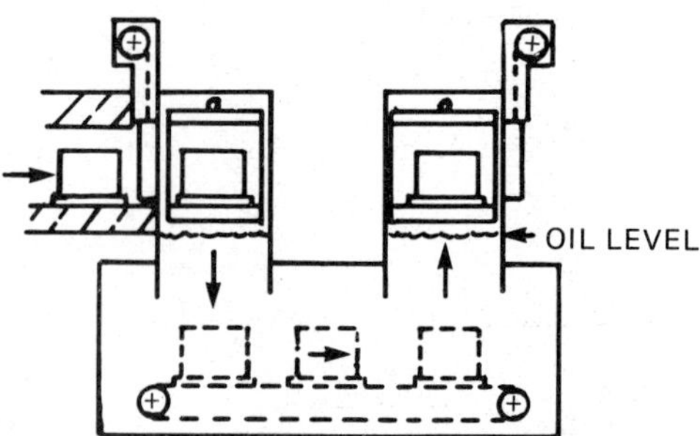

FIG. 10-6D. Dunk Type elevator quench with under oil transfer.

Unless the oil drains off the work at the end of the quench period, an excessive amount of oil will be lost. Usually the work will still be warm at this time and some oil may vaporize. Any ignition source can produce a fire at this point in the process. Oil vapors will condense on cool surfaces above the drain area and thus contribute significantly to the potential fire loss.

Oil Temperature Control

The control of quench oil temperature within specified design limits is essential to the heat treating process as well as for safety. Failure of a cooling system can cause the oil to overheat. What is more hazardous, however, is a cooling system failure that allows water to enter the quenching oil. An excessive amount of water can produce a boilover when a hot workload is immersed in the quench. The exact critical water volume varies somewhat with the specific oil or oil mixtures being used, but if the water content reaches 0.50 percent by volume, the oil is no longer considered safe to use.

Quench oil can be cooled by water circulating through coils in the quench tank, by external heat exchangers, or by water jackets for the quench tank. Water jackets and internal water coils should not be used with combustible quenching oils. In these designs, any mechanical failure of the coil or the quench tank shell will result in water entering the quench medium. Such failures are not readily detectable; the first indication may be a boilover when a hot workload is immersed in the quench.

If an external heat exchanger is used, quench oil must be circulated through the exchanger at a higher pressure than the exchange medium (water). If a leak should develop, oil will enter the water and be wasted. If the water pressure is higher than oil pressure, such a leak would result in water entering the oil, creating a potential for boilover.

The continuous flow of cooling water is essential for temperature control. This can be properly supervised by observing the discharge from an open drain on the water side of the heat exchanger. Waterflow indicators are needed where a completely closed system must be used.

An oil quench tank is built as an integral part of many special atmosphere furnaces. In these cases, the vestibule above the quench tank is water cooled, usually by water jackets. Many failures of the interior jacket walls have released cooling water into the quench tank below. As a result, safety considerations have dictated the use of special materials for vestibule jackets or the use of external coils. With special atmosphere furnaces, the use of a combustible gas for the atmosphere also will result in moisture development when the exit door is opened and the atmosphere is burned out.

Because of the various ways in which water can enter quench oil and become a significant hazard, approved moisture detectors should be used and the oil tested periodically. If moisture content reaches 0.50 percent by volume, quenching operations should be shut down and the oil replaced.

If the oil level is too low and a large workload is immersed in the quench, the oil can overheat. Therefore, low oil level detectors should be used to sound an alarm and shut down quenching operations before overheating occurs.

Agitation is critical to maintaining safe quench oil temperature and uniformity of temperature throughout the bath. If the agitation mechanism fails, localized overheating will occur, which may cause a fire at the surface. Agitation failure and subsequent overheating can cause excessive vaporization, which can raise the pressure inside an enclosed special atmosphere furnace. Gas and oil vapor may then be forced out of the chamber around doors and through vents. Frequently, these escaping gases ignite, damaging the facilities. Agitation systems should be supervised automatically and their failure should result in the safe shutdown of operations.

Where the process requires that the quench be heated, fuel fired, electrically heated, or steam heated, immersion units are used as the heat source. All three methods of heating will create excessive temperatures at the interface of the heating unit and the quenching oil. A quench heating system should be prevented from operating if the oil level is too low, the agitation equipment is not functioning, or the oil temperature is too high. Where combustible oils are used, a temperature controller should be interlocked in the system to prevent quenching and to shut off the oil heating system when oil temperature exceeds a specified maximum limit.

Central Oil System

Many heat treating plants utilize a quenching oil that is common to several operations. Properly designed central oil systems can contribute to the maintenance of reasonably dependable and problem free oil quenching operations. A proper design includes filtering to remove particulate contamination, water removal for safety of operations and prevention of boilover, and cooling to deliver the quenching medium at an acceptable temperature.

Water separation can be accomplished by settling and the use of centrifuges. However, these systems are not dependable enough to eliminate the need for moisture detection devices at the quench tanks or the requirement that oil in quench tanks be tested periodically.

Since central systems involve a constant removal and replacement, of oil from the quench tanks it is essential for the sake of safety that oil flow be supervised. To avoid foaming and boilover, replenishment oil should not be added while the quench temperature is 212°F (100°C) or higher. Regardless of the oil being used, the temperature must never be permitted to rise to a value that is less than 50°F (28°C) below its flash point.

The Safeguards

All automatic shutdowns of quenching operations should result in the workload being completely immersed in, or removed from, the quench. Partial immersion must always be considered hazardous.

Whenever movement of the workload into, or out of, the quench has been stopped by a malfunction, qualified operating personnel must be permitted to override the safety interlocks so that manual attempts can be made to complete immersion or remove the workload. If exit doors must be opened, a fire condition should be anticipated. Extinguishing facilities adequate to protect the operator and prevent property damage should be available.

All safety controls and their interlocking functions should be tested on a regular schedule. The time interval between tests should be adequate for each installation but should not exceed six months. Refer to manufacturers' guidelines.

Hydraulic control systems add to the fire hazard in the vicinity of high temperature equipment. Fire resistant hydraulic fluids should be used. When combustible hydraulic oils must be used, proper maintenance of the hydraulic equipment is vital to safety.

Fire Protection

Fire protection for combustible oil quenching operations requires a detailed evaluation of the inherent fire potentials. One of the most effective forms of area protection is the automatic sprinkler system. Experience has shown that ceiling mounted sprinkler systems will limit building and equipment damage from quench oil fires whether they are confined to the quench tank surface or spread over a large area by a quench oil boilover.

Specific protection for open quench tank surfaces and oil drainage areas is also important. Most oil fires can be extinguished by fixed carbon dioxide or dry chemical systems. In some operations, various foam systems can also be effective. However, the suitability of foam will be determined by the quenching oil used and the temperatures involved. These systems are designed to operate automatically, well in advance of any potential sprinkler system discharge.

Those situations that may require operating personnel to manually release jammed workloads and close furnace doors will dictate the provision of fire control and/or extinguishing facilities for the protection of the operator. In these instances, carbon dioxide or fixed water spray systems are the most effective. The most important fire protection features in every heat treatment shop are the manual fire extinguishing and control equipment and plant personnel properly trained in its use. In addition to an adequate supply of portable, hand carried fire extinguishers, the larger, wheeled extinguishers should also be available.

Appropriately spaced hose connections with water fog or water spray nozzles should be considered an essential part of all heat treating area fire protection. These can provide prolonged periods of fire control and life safety beyond the limited supplies of portable extinguishers.

Bibliography

NFPA Codes, Standards, Recommended Practices and Manuals. (See the latest *NFPA Codes and Standards Catalog* for availability of current editions of the following documents.)

NFPA 10, *Standard for Portable Fire Extinguishers.*
NFPA 11, *Standard for Low Expansion Foam and Combined Agent Systems.*
NFPA 12, *Standard on Carbon Dioxide Extinguishing Systems.*
NFPA 13, *Standard for the Installation of Sprinkler Systems.*
NFPA 14, *Standard for the Installation of Standpipe and Hose Systems.*
NFPA 15, *Standard for Water Spray Fixed Systems for Fire Protection.*
NFPA 17, *Standard for Dry Chemical Extinguishing Systems.*
NFPA 34, *Standard for Dipping and Coating Processes Using Flammable or Combustible Liquids.*
NFPA 70, *National Electrical Code.*
NFPA 86, *Standard for Ovens and Furnaces—Design, Location, and Equipment.*
NFPA 86C, *Standard for Industrial Furnaces Using a Special Processing Atmosphere.*
NFPA 101, *Code for Safety to Life from Fire in Buildings and Structures.*

Additional Readings

Factory Mutual System, "Oil Quenching of Metals," *Handbook of Industrial Loss Prevention 2nd ed.*, McGraw-Hill, NY, 1967.
Flash Point Index of Trade Name Liquids, 9th ed., National Fire Protection Association, Quincy, MA, 1978.
Henry, Martin F., *Flammable and Combustible Liquids Code Handbook*, 2nd, National Fire Protection Association, Quincy, MA, 1984.
Kimura, H., and Maddin, R., *Quench Hardening in Metals*, Elsevier Science Publishing Co., Inc., NY, 1971.
Linville, J. L., ed., "Oil Quenching," *Industrial Fire Hazards Handbook*, 1st ed., National Fire Protection Association, Quincy, MA, 1979.
Trinks, W., *Industrial Furnaces*, (Vols. 1 and 2), John Wiley and Sons, NY, (Vol. 1: Principals of Design and Operation, 5th ed., 1961; Vol. 2: Fuels, Furnace Types, and Furnace Equipment: Their Selection and Influence Upon Furnace Operation, 4th ed., 1967).

SPRAY FINISHING AND POWDER COATING

Don R. Scarbrough

Spray application of coatings is a process basic to the manufacture of a broad variety of fabricated products, and a high percentage of factories operate at least one paint spray booth. Regardless of the purpose for which the coating is applied, flammable or combustible materials are commonly used in the process. Fires in paint operation areas develop very quickly, have high heat release rates, and produce large volumes of toxic smoke. Preventive and protective measures for control of these hazards must be undertaken with special attention to the peculiar hazards of each process.

TYPES OF COATINGS

Fluid Coatings

The most familiar atomizing device for fluid coatings is the air spray gun, which uses jets of high pressure air to break up fluid into a fine mist. Another device commonly seen in high volume production processing is the airless atomizer, which generates a spray of fluid by hydraulic means without using compressed air. Fluid pressures used with this type of atomizer range from 300 psi (2068 kPa) to approximately 3,000 psi (20 685 kPa).

Air and airless atomizers are also used in electrostatic spray operations. For this process, the appropriate atomizer is built into a spray gun that has a high voltage electrical input. Voltages applied to the gun range from approximately 35,000 to somewhat over 100,000 V. There is less overspray with the electrostatic method than with the air or airless spray methods because the charged atomized particles are attracted to the grounded workpiece.

A third type of atomizer depends upon electrostatic forces for its operation. In its most common configuration, a sharp edged disc with a diameter of 6 to 12 in. (152 to 305 mm) is mounted with its axis oriented vertically and is then charged electrically to about 100 kV. The disc is spun about its axis while the coating fluid is poured slowly onto the surface. Centrifugal force spreads the fluid into a thin film and carries it to the sharp edge of the disc, where the film is disrupted by electrostatic forces and sprayed in a 360° pattern. Workpieces are carried to the process zone by a conveyor and collect coating as they pass through. A form of this device that has recently gained popularity has the disc developed into the form of a cup or bell 2 or 3 in. (15 or 76 mm) in diameter that rotates at a high speed—sometimes approaching 60,000 rpm. Bell atomizers are frequently seen mounted in banks of 6 to 12 units at a single spray station.

Powder Coatings

A coating process that has gained broad acceptance is the application of organic coatings in the form of dry powder. In this process, the powder is first suspended in air and then charged electrostatically from a dc power supply operating between 60 and 120 kV. The powder is then directed toward the workpiece and held in place by electrostatic forces. The powder is formed into a continuous coating as it melts during passage through a process oven. The powder coating process differs substantially from the fluid coating processes because no organic solvents are used.

Electrostatic application of powder is most commonly accomplished with spray guns. These spray guns are simple devices to which a mixture of powder and air is fed through a single tube. A separate cable connected to the gun provides the high voltage.

Workpieces having dimensions less than approximately 4 in. (102 mm) can be powder coated with a device known as an electrostatic fluidized bed. In this process, powder is held in a container having an open top and a porous bottom through which an upward flow of air causes the powder mass to levitate or "fluidize." Charging electrodes near the surface impart an electrical charge to the powder. Electrically grounded workpieces pass over the surface of the bed and collect a coating of powder. The coating is then cured in a baking oven.

Although they are not widely used, there are techniques for applying dry powder coatings without the use of electrostatics. These processes typically involve preheating the workpiece to a temperature substantially above

Mr. Scarbrough is Product Safety Staff Consultant for the Nordson Corporation of Amherst, OH.

the melting point of the powder and then applying the powder, either by dipping the workpiece into a fluidized bed or by spraying air suspended powder directly onto the hot surface of the workpiece. The powder melts immediately upon contact and flows to form a film, which is subsequently cured in a bake oven.

SPRAY PROCESS EQUIPMENT AND COMPONENTS

Fluid Supply

Air spray guns draw their coating fluid from either small cups mounted on the guns, or through hoses connected to larger pressurized containers called paint tanks or pressure pots that have capacities varying from 10 to 60 gal (38 to 227 L). In very high production arrangements, the coating may be fed by a pump from a bulk tank.

To ensure satisfactory results, industrial coating mixtures are usually modified shortly before spraying by the addition of solvents to adjust viscosity. A safety tank or container is used to reduce the possibility of a flammable liquid spill while handling or transporting this blended material.

Spray Guns and Devices

Among the various forms of industrial air spray guns, the two most commonly used are the pistol grip hand spray gun and the machine mounted automatic air spray gun. Air is supplied to the hand spray gun through one hose, while fluid is either hose fed or drawn from a gun mounted container. (See Fig. 10-7A.)

Electrical grounding of airless spray guns to drain off static electricity generated during spraying is provided through a wire or conductive layer built into the fluid hose. A manual safety lock is usually provided on hand units to secure the trigger and prevent accidental actuation.

Hand held and automatic electrostatic spray guns have electrically nonconducting extensions at the front end to insulate the energized components from the grounded parts. In addition to being connected to coating fluid and air supplies, these guns are also connected to high voltage power supplies. Voltage at the atomizer is in the 30 to 75 kV range for hand held units and in the 30 to 120 kV range for automatic units. In hand spray guns of this type, the cable carrying the high voltage power to the gun also supplies an electrical ground for the pistol grip and trigger. To prevent electrical sparking from accidental contact of the charged elements of the gun with a grounded object, the cable is usually terminated at the gun in an electrical resistance on the order of 75 to 250 MΩ (megohm). Some automatic electrostatic spray guns are not equipped with a high impedance termination, but rely on the process control approach of maintaining separation between gun and workpiece to prevent electrical sparking.

Figure 10-7B shows an electrostatic disc surrounded by chairs hanging from a conveyor loop. Immediately above the disc is its motor drive. Partially obscured, at the top of the illustration, is a device that moves the disc up and down, enabling it to coat workpieces over the entire height of the carriers. Separate electrical and fluid inputs are remotely controlled. Discs commonly are equipped with variable voltage power supplies that can be adjusted over a range of approximately 50 to 120 kV. Since considerable electrical energy is stored on the disc, adequate distance must be maintained between it and the workpiece to prevent sparking.

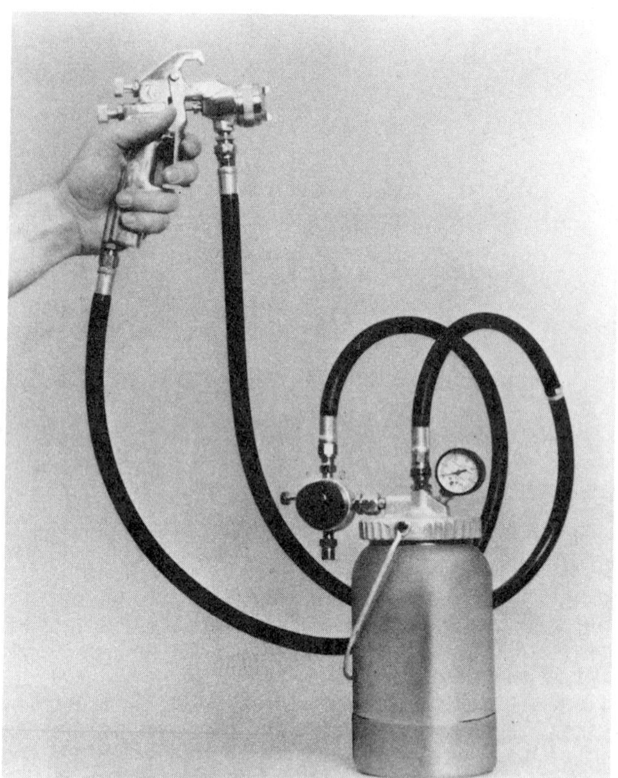

FIG. 10-7B. Electrostatic disc. (Ransburg)

Spray Booths

A spray booth is a power ventilated structure used to enclose a spraying operation. While in most cases the booth structure physically surrounds the spray operation, some configurations leave the spray operation unenclosed and surround the operation with a controlled stream of air drawn into the booth.

The most popular type of spray booth is the "open face" or "open front" arrangement. It is a boxlike structure that has one open side. The ventilation system associated with the booth may provide an airflow horizontal to the floor (cross draft) or vertical to the floor (down draft) as demanded by process requirements.

For operations such as production finishing of automotive bodies, a tunnel spray booth is used. (See Fig. 10-7C.) This structure is virtually always arranged with vertical down draft ventilation and a horizontal floor mounted conveyor running the length of the tunnel. Make-up air is introduced through special ductwork and air diffusers in the ceiling.

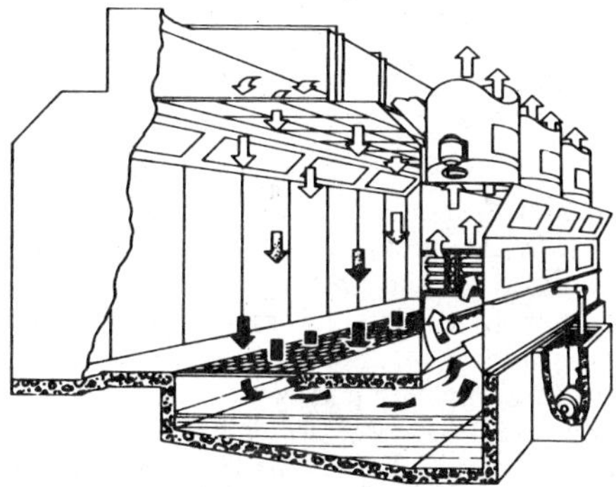

FIG. 10-7C. Tunnel booth. (Binks)

Special Purpose Enclosures

Continuous Coaters: These enclosures are individually engineered for a specific coating process. Within the coater enclosure is an array of spray guns which apply coatings to workpieces that pass through the coater at conveyor speeds between 100 and 600 fpm (30 and 183 m/min). The entry and exit vestibules are equipped with exhaust shrouds to capture any vapors or minor amounts of overspray that drift out from the main enclosure. Most of the overspray is collected in a sump and drawn off through a pumping system, which recycles the coating into the coating process.

Decorating Machines: These machines are used in conjunction with masking devices to paint stripes or other patterns on workpieces similar to automobile side moldings. The mask is mounted in the opening beneath the row of cylinders at the front of the machine, and the workpiece is placed face down on the mask. When the operating cycle is started, air driven pistons in the row of cylinders clamp the work to the machine. Spray guns inside the cabinet automatically tilt in one direction and travel the length of the workpiece. At the end of the workpiece, the spray guns tilt in the opposite direction and travel back to the starting point. The air driven clamps automatically release the work at the end of the operation.

An exhaust system draws fresh air in through the grill at the front of the machine, and internal filters remove overspray from the air vapor stream before it is exhausted. Since the process is totally enclosed, the risk of escaping vapors or overspray and subsequent ignition is significantly diminished.

Spray Rooms

The size, shape, and weight of some workpieces may make the use of spray booths impractical. When this is the case, an entire room dedicated to the spray process is a legitimate alternative. Such rooms, called spray rooms, are classified as hazardous areas. Power ventilation removes the combustible vapors that are released during the process, but velocities are not high enough to capture particulate matter. Therefore, overspray is permitted to settle to the floor where it accumulates as a combustible residue until it is removed by a mechanical cleaning process. Vapor removal systems are most effective when the exhaust intake is located along one wall within 1 ft (0.3 m) of the floor and the make-up air is introduced along the opposite walls.

Open Floor Spraying

A spray operation conducted without a spray booth and in an area not separated by a partition from general factory operations is called open floor spraying. Depending upon the quantity of volatiles to be released in the operation and existing ventilation within the building, forced ventilation may or may not be provided. Whether forced or not, adequate ventilation must be provided to prevent accumulation of flammable concentrations of vapor. Particulate overspray materials are allowed to settle to the floor and their residues are removed by mechanical means. Ignition sources incidental to other factory operations are a great concern whenever open floor spray techniques are used, and it is common to establish a buffer zone surrounding the spray area for about a 20 ft (6 m) radius to separate processes.

Overspray Collectors

Vapor and overspray removal systems typically include a fan to create an airflow, and a collection system which separates and collects particulate matter from the airstream and exhausts vapors to the exterior of the building. Overspray collectors can generally be placed into one of four categories.

Baffle Maze: The baffle maze consists of a series of flat panels arranged in a staggered pattern through which the airstream is directed. A substantial portion of its particulate burden is removed by direct impaction and collection upon the surface of the dry baffle panels where it remains until removed by mechanical cleaning. Such systems are of limited efficiency and permit appreciable amounts of fine particulates to pass through.

Dry Filter: Collectors using paper or fiberglass filter elements are called dry filters and are popular for low to intermediate volume spray operations. The typical filter used in this type of collector is a replaceable element approximately 20 in. (508 mm) square and perhaps 2 in. (51 mm) thick. Particulate residues are permitted to accumulate on the filter until a significant obstruction of the airflow through the filter is noted, then the fouled filters are discarded and replaced.

Waterfall and Cascade Scrubbers: Where high volume spray coating operations are conducted for several hours a day, waterfall or cascade scrubbers are commonly used. The exhaust airstream is either scrubbed directly by sprays of water coming from nozzles or it follows a path that takes it through several stages of waterfall. Figure 10-7D is an example of one variation of this type of scrubber. Particulate matter is accumulated in a water tank from which it is removed either manually or automatically by a sludge removal device.

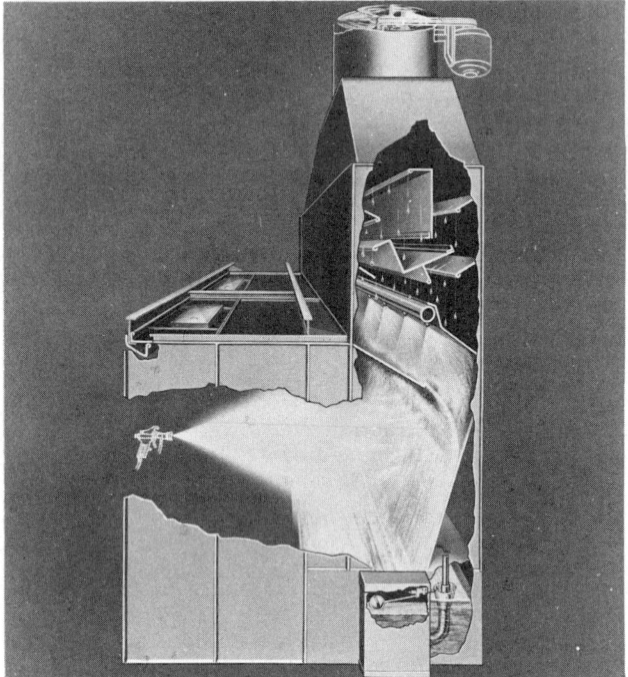

FIG. 10-7D. Water wash booth. (DeVilbiss)

Chemical compounds must be added to the water to prevent paint residues from adhering to the walls or piping thereby creating open channels through the water curtain. Air passing through these open channels will not be adequately cleansed and will carry particulate matter into the exhaust system; the particulate matter may be deposited on the lining of the exhaust stack, on exhaust fan blades, or on the roof of the building.

Venturi Scrubbers: Perhaps the most efficient overspray collector is the venturi scrubber. This device directs the exhaust airflow through a narrow throat (venturi) through which a high velocity spray of water is also directed. Virtually all particulate matter is extracted from the airstream and trapped in the water. The water is processed through a tank where residues are removed by settling and skimming. The same chemical compounds mentioned in the discussion of the waterfall type scrubber must be added to the water used in this system to prevent plugging and blocking of nozzles. Figure 10-7E is a schematic of a venturi scrubber.

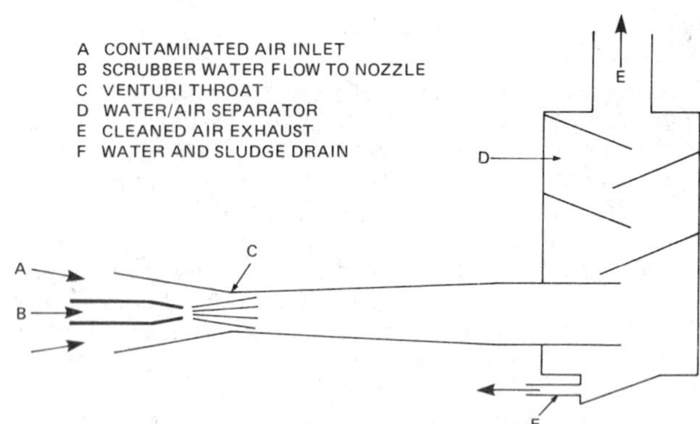

FIG. 10-7E. Venturi scrubber. (Nordson)

POWDER COATING PROCESS EQUIPMENT AND COMPONENTS

Spray Process

The powder coating spray process utilizes a device called a feeder that mixes powder with air and supplies the mixture to the spray guns. The powder may be contained in a fluidized bed or in a hopper having an inverted cone shaped bottom. Feeders may supply a single gun or several guns with a separate ejector for each gun.

The most common coating powder is epoxy powder, but others include acrylic, polyester, vinyl, nylon, butyrate, polyolefin, and alkyd. These powders are shipped from the manufacturer to the user in containers of 25 to 100 lb (11.34 to 45.36 kg) capacity. They may be classified as ordinary combustibles and, as such, may be stored without requirements for extra hazard protection.

Spray guns used for the electrostatic application of dry powders do not differ greatly in appearance from guns used for the electrostatic application of fluids. (See Fig. 10-7F.) Since no atomization is required, the functions of

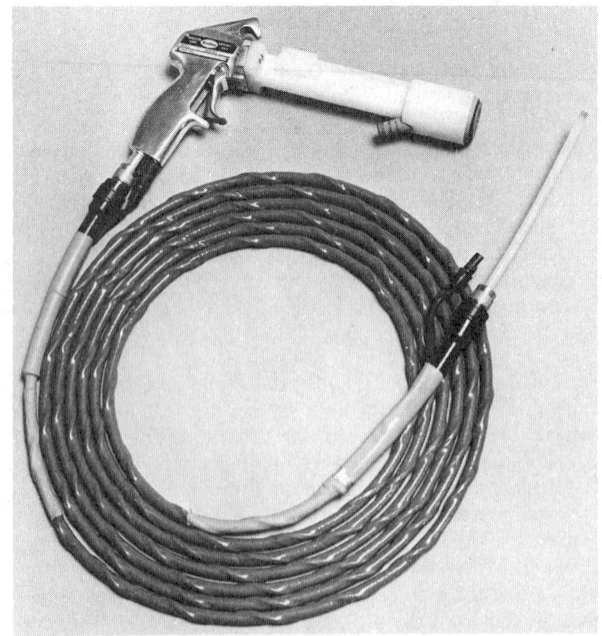

FIG. 10-7F. Hand powder gun. (Nordson

the spray gun are simply to control the shape of the spray pattern and to impress a high voltage charge upon the powder cloud. Powder and electrostatic power input connections may be made at the rear of the barrel or from below to the base of the grip and the forward portion of the barrel. Connections to automatic guns typically are made at the rear.

Electrostatic power supplies rectify stepped up common ac line voltage inputs, either 115 or 230 V, to produce dc output ranging from 30 to 100 kV. Several automatic guns may be connected to a single power supply. For hand gun applications, however, a separate power supply is provided for each gun, and each supply has a control circuit that prevents the pack from being energized unless the trigger is actuated.

The interconnection between power supply and spray gun is usually made through a high voltage coaxial cable. At the termination of the cable within the spray gun, the charging circuit is connected to the gun electrode through a 75 to 250 megohm resistor. In operation, a faint blue glow, called a corona, is developed around the end of the electrode which projects through the front of the gun. Individual powder particles are charged as they pass through the corona.

The most familiar facility in which electrostatic powder spray operations are conducted is a spray booth with a hopper bottom through which exhaust air is drawn into a ductwork that leads to a powder collector. A cloth or fabric filter within the collector separates the powder from the airstream. Powder separated within the collector falls to the bottom of the hopper from which it is extracted and is subsequently reintroduced to the feeder for the spray guns. A basic outline of this arrangement is shown in Figure 10-7G.

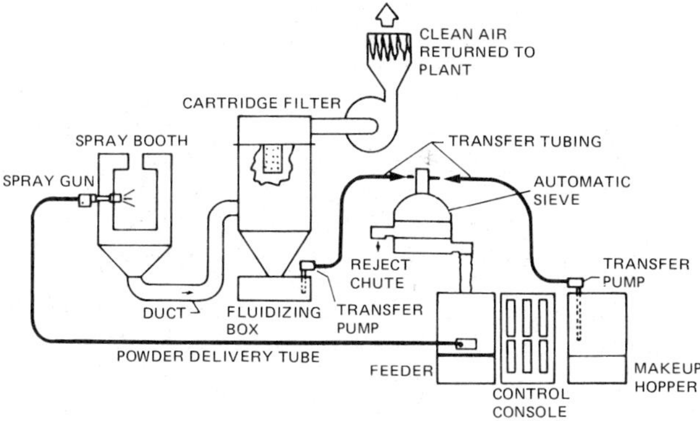

FIG. 10-7G. Diagram of automatic recycle of a single color powder coating system showing the booth, collector, fan, and filter. (Nordson)

A recent development is the integration of the spray booth and powder recovery system into a single structure. (See Fig. 10-7H.) In this equipment, cartridge type filters and the exhaust fan are built into the structure of the spray booth. The ductwork, which conventionally would separate the booth from the powder collector, is eliminated. Collected powder is pneumatically pumped from the hoppers immediately beneath the recovery filters directly back to the spray gun feeders.

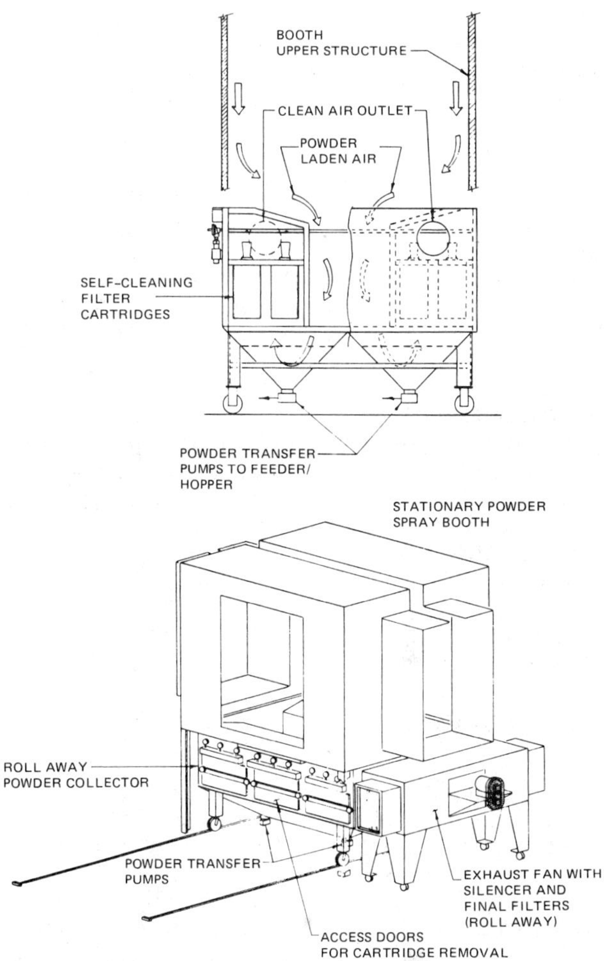

FIG. 10-7H. Integrated powder spray booth/recovery system. (Nordson)

Electrostatic powder coating may also be conducted within an enclosure commonly referred to as a pipe coater. The device consists of a steel enclosure, a ring of automatic electrostatic powder spray guns, and a conveyor. The workpiece enters at one end, passes through a cloud of coating powder, and leaves the coater at the opposite end. The workpiece is preheated to a temperature above the melting point of the powder, so the powder will fuse upon contact. The coater is provided with an exhaust system and powder recovery filter system similar to that used with the booth arrangement described in connection with Figure 10-7G.

Fluidized Bed

Fluidized beds may be constructed in a wide variety of sizes. They have vertical walls (typically steel) and a porous floor (textile or plastic). Air flows upward through the floor and through the powder. When the airstream is adjusted properly, the powder will behave as a liquid, flowing readily around, and contacting all surfaces of any object that is dipped into it. It will immediately fuse to a workpiece that has been preheated above the melting point of the powder. After the workpiece has been removed from the bed, it is placed in an oven to complete the curing of the coating.

An electrostatic fluidized bed has a series of high voltage electrodes near the surface of the fluidized powder mass. Grounded workpieces, which may or may not be preheated, pass over the electrodes and are coated electrostatically, then cured in a bake oven.

Both fluidized bed arrangements usually are provided with peripheral air exhaust systems that collect any powder grains escaping the bed and prevent their distribution into the general factory area.

Cloud Chamber

One other electrostatic powder coating process is the cloud chamber, a box shaped enclosure with entry and exit portals through which workpieces are carried by a conveyor. The enclosure contains an apparatus that generates and maintains a billowing cloud of powder which is charged by high voltage electrodes. Particles of powder are attracted to the grounded workpiece as it passes through the cloud chamber. Powder is electrostatically collected on the workpiece which is then cured in an oven.

FLUID SPRAY PROCESS HAZARDS AND CONTROL

Materials and supplies used in organic spray finishing processes are usually flammable or combustible, are often toxic, and, in some cases, may be highly reactive or unstable. For these reasons, spray finishing operations are considered hazardous, and suitable preventive and protective measures should be taken to minimize the hazards.

Fire Prevention

Hazard Identification: Materials' hazards should be identified and marked upon containers as soon as they arrive in the facility. A detailed method of identification is contained in NFPA 704, *Standard System for the Identification of the Fire Hazards of Materials*. The system provides for the identification of five degrees of health, flammability, and reactivity hazards.

Storage and Handling: Bulk supplies of flammable liquids should be stored outdoors, away from buildings. Smaller quantities are then brought indoors to a mixing room where they are prepared for use. The mixing room should be located adjacent to an outside wall provided with explosion relief vents, and should be isolated from the rest of the building by fire resistive construction. The room should have sufficient mechanical ventilation to prevent the development of flammable vapor concentrations in the explosive range. All flammable liquid containers of greater than 5 gal (19 L) capacity that are kept indoors should be equipped with a special plug that incorporates a pressure relief valve, a vacuum relief valve, and a flame arrester.

Prepared coating materials are transported from the mix room to the spray area in containers or through piping. Containers should be equipped with tightly clamped lids that will retain vapors and liquids in the event the container is upset. Piping should be identified as process pipe containing flammable materials, and a shutoff valve should be installed at each point where a hose is connected to the system. Emergency stop controls and fire alarm interlocks are very important to keep runaway pumps from continuing to feed a fire after hoses have been turned off. All piping, pumps, supply and receiving vessels must be electrically grounded and bonded to prevent accumulation of dangerous static electrical charges during transfer of fluids.

Quantities of flammable or combustible liquids kept in or near process areas should be limited to the amount needed for one working shift at a time. The materials should be put in closed containers equipped with appropriate pressure relief devices.

Flammable or combustible liquids should never be transferred from one container to another by the application of air pressure to the original shipping container. Pressurizing such containers may cause them to rupture, creating a serious flammable liquid spill. Any pressure vessels used should be designed specifically for such use and be manufactured in conformance with the American Society of Mechanical Engineers (ASME) *Code for Unfired Pressure Vessels*.

Fluid hoses connected to some electrostatic equipment must have a special structure to resist both fluid and high voltage electrical stresses that result from use of electrically conductive paint. If ordinary hose or tube is substituted for this use, electrical failure of the hose wall will lead to pinhole failures and ignition of fires.

Control of Vapors and Overspray: Limitation of vapors and overspray to the smallest practicable area is accomplished through the combination of process enclosures and power ventilation systems. Ventilation systems should be checked frequently to assure they are operating properly and that specified flow rates are being maintained.

Ignition Sources: All sources of ignition should be barred from, or controlled in, areas defined as Division 1 or Division 2 in Article 516 of NFPA 70, *National Electrical Code*. The careless use of smoking materials and improperly supervised welding or flame cutting operations pose significant threats to a fluid coating applications system.

To minimize chances of ignition, open flames, spark producing equipment, and any exposed surfaces exceeding ignition temperature of the material being sprayed should be prohibited in either Division 1 or Division 2 areas. No equipment or process capable of producing sparks or hot particles should be located above or adjacent to those areas classified as hazardous unless partitions or other means of separation are provided.

All electric wiring and equipment, unless specially designed and manufactured for use in hazardous areas, may be regarded as potential ignition sources. Only those devices and wiring types listed for this use may be installed or used in areas classified as hazardous. In the event of failure, electric lighting fixtures located above a classified area should be totally enclosed or provided with a guard that will prevent hot particles from dropping into the hazardous area. Exhaust fans should be nonsparking and conform to Air Movement and Control Association (AMCA) Class C requirements.

Static electricity remains the most common source of flammable vapor ignition in electrostatic spray finishing operations. Energized electrodes send generous amounts of electrical charge into the air, and it then collects upon surrounding surfaces. Accumulation of this charge may raise a surface to a high voltage. Electrostatic spray systems should be equipped with electrical interlocks to de-energize the electrostatic power supply any time the spray guns are not actually in use. Most modern electrostatic spray guns that are listed for use with flammables include a charging circuit that limits the energy of a discharge arc to a value below that which is sufficient to cause ignition.

Electrostatic ignition can be prevented by electrically bonding together all the elements of a fluid transfer or spray system along with the workpiece and all other electrically conductive objects located within 10 ft (3 m) of the charged elements, and then grounding the bond. Additional information on bonding and grounding can be found in NFPA 77, *Recommended Practice on Static Electricity*.

Human beings are also conductors of electricity; thus, the spray operator cannot be overlooked in the grounding process. Gripping the grounded spray gun with the bare hand or wearing shoes with electrically conductive soles, provided the floor is grounded, are two workable methods for grounding the operator. Even in the absence of electrostatic equipment, dangerous static electrical charges can be generated by the simple act of walking across a spray area floor that is covered with sticky paint residue. The charging mechanism is identical to that seen in the more familiar act of walking across a wool or nylon carpet on a dry day: a spark and shock results when a doorknob or switch plate is touched. That spark is quite capable of igniting flammable vapors.

Static electrical charge is also generated by the operation of non-electrostatic air and airless spray equipment, resulting in discharge sparks from either the spray apparatus or the object being sprayed. To prevent ignition of fires from such discharges, electrical bonding and grounding of apparatus and the sprayed object is necessary.

Housekeeping: Residues of spray materials are a solid form of fuel. A routine maintenance program should provide for the periodic removal of overspray residue from walls, floor, and ceiling of the spray booth, room, or area, as well as from conveyors and the interior of ventilation ducts. Residue buildup on conveyors may become particularly hazardous by acting as electrical insulation which interferes with workpiece grounding. Contaminated spray booth filters should be either removed from the building as soon as they have been replaced, or kept immersed in water until disposal, as they present a serious spontaneous heating hazard. Only a few hours of "fermenting" is required for a waste container holding paint fouled filters or rags to become so hot that it bursts into flame. Alternately, the spontaneous heating hazard may be eliminated by curing the paint in fouled filters or rags by processing them through a paint bake oven before disposal.

Operator Training: Operating personnel should be exposed to thorough initial training when they begin the job and should be given periodic additional training to maintain their appreciation of the hazards involved, to teach them hazard control procedures, and to educate them in new processes. They should be aware of the inherent hazards of the spray materials being used, particularly if the materials are toxic, chemically unstable, or reactive. They should also be familiar with the system of hazard identification that is employed. Operators should be drilled in proper operating, bonding and grounding, emergency, and maintenance procedures.

Fire Protection

Areas in which fluid spray processes are conducted and areas where coatings and solvents are mixed or stored should be separated from other plant operations by appropriate distance or partitions. Automatic sprinklers inside spray booths, ventilation ducts, and at the ceiling contribute considerably to the control of fire. In open areas, draft curtains extending down from the ceiling around spraying operations, along with smoke and heat vents, will slow the mushrooming of hot combustion gases along the ceiling and thus limit the number of sprinklers that will activate. This will concentrate sprinkler discharge into the area where it is most needed and will minimize smoke and heat interference with fire fighting efforts.

Any fluid transfer or supply system for flammables, whether driven by pump or by pressure vessel, should be arranged with emergency shutdown provisions to interrupt flow in the event of fire or accidental spill, thereby limiting available fuel.

Portable fire extinguishers and standpipe hoses are useful against fires involving small spray operations. The rate of fire development across residue accumulations is very high, however. It must be recognized that in only a few seconds a fire can grow to a size that will be beyond portable extinguisher control, particularly in spray booths of the baffle or the dry filter type or in operations where paint residues are permitted to accumulate on the floor. For enclosed or semi-enclosed coating processes, such as in continuous coaters or automatic decorating machines, automatic fire extinguishing systems may be used to flood the interior of the machines.

A spray booth and its exhaust ductwork should maintain structural integrity (i.e., not collapse) during a fire of nominal proportions so that the major portion of flame, heat, and fire gases produced will vent from the building through the exhaust ductwork, thereby diminishing the chances of fire spread in the building. Spray booths and their exhaust ductwork are required by various codes and standards to be constructed of steel, masonry, or other material of equivalent fire resistance. Survivability of booth structure is considerably enhanced by the cooling effect of water from sprinklers inside the stack and booth. The probability that heat from a fire within a booth and stack will ignite nearby combustible materials can be reduced to an acceptable level by reasonable clearances between any combustible structure or stored materials and the surfaces of the booth or stack (for recommended clearances, refer to NFPA 33, *Standard for Spray Application using Flammable or Combustible Liquids*). Spray rooms are usually required to be separated from other occupancies within the same building by walls, floors, and ceiling structures with a minimum two hour fire resistance rating.

POWDER COATING PROCESS HAZARDS AND CONTROL

Powder coating operations using combustible organic powders are classified as hazardous processes because the powders can burn vigorously when suspended as airborne dusts and can explode when confined. There are no flammable vapors in the powder coating process, and the energy required to ignite a cloud of air suspended coating powder is from 100 to 1,000 times higher than that required to ignite flammable vapors.

Fire Prevention

Storage and Handling: Coating powders that are kept in shipping containers are not customarily classified as hazardous materials and are commonly stored under conditions appropriate for ordinary combustible materials. Reasonable care is required in the handling and movement of the powder containers because a fire hazard could be created should the containers be broken and the powder distributed as an air suspended cloud of dust.

Waste powder is commonly packaged into fiber drums or cardboard cartons lined with plastic bags and shipped to a landfill for disposal in accordance with Environmental Protection Agency (EPA) regulations. If the waste container has been processed through a baking cycle, the waste powder will have fused into a solid block of plastic that is no longer vulnerable to scattering if the container is broken.

Coating Operations: Virtually all spray powder coating processes are conducted in spray booths that are more tightly enclosed and permit considerably less overspray to escape than those booths used for fluid coating processes. All connections in the hoses and piping associated with pneumatic transfer equipment should be routinely checked to confirm that they are secure and that ground connections are in place. Breakage of any such connection during operation could create a substantial hazard by blowing powder into the workspace air outside the process enclosure.

To prevent the escape of powder, the spray booth must have adequate ventilation that provides effective capture velocity at all its openings. Minimum average velocities are considered to be in the vicinity of 60 fpm (18 m/min). Interlocks should be installed to permit operation of the powder application equipment only while the ventilation equipment is in operation and, similarly, to permit operation of the electrostatic power supply only during the time the powder is actually being applied.

Ignition Sources: The control methods for ignition sources for powder coating are similar to those for spray finishing.

Static accumulations on electrically isolated conductive objects are the major cause of fires in electrostatic powder coating installations. Metal workpieces suspended from conveyor hangers that do not provide an effective electrical circuit between the workpiece and ground pass through the coating process zone while discharging hot electrical sparks. This sparking can ignite the powder cloud, and flaming will be sustained by the continuing flow of powder and air. In an electrostatic fluidized bed, the entire volume of the bed can become involved in fire.

Some application equipment will discharge incendiary sparks when approached too closely by a grounded object. Proper use of this type of equipment requires safeguards to prevent any grounded object from approaching the electrically charged high voltage elements closer than twice the sparking distance. Discharge sparks that do occur during use of this type of equipment are usually traceable to improper racking of parts on a conveyor or swinging of the conveyor racks. To reduce the probability of a small fallen part being drawn into a duct and producing mechanical sparking, grill work at the intake of any exhaust duct is highly recommended, and magnetic devices can be added to catch tramp metal.

Fire Protection

Protective measures for powder coating installations are: (1) keep the powder collector from becoming involved in a fire that has originated in the spray operation; (2) equip the collector with pressure relief vents and ductwork in order to prevent its rupture in an explosion; and (3) protect the building from heat and pressure effects that would be generated by an explosion.

To prevent a fire in a powder spray booth from spreading through connecting ductwork, the ventilation system is usually arranged to keep powder concentrations in the ductwork below one-half the Minimum Explosible Concentration (MEC). In those cases where fire has persisted in the spray booth for periods ranging beyond approximately 15 sec, however, the powder collector is threatened by glowing embers that have formed in the spray fire. These glowing embers are independent of powder concentration and are capable of igniting the powder collector.

Powder collectors can be protected by a fast acting flame detector within the spray booth, interlocks to shut down equipment, and a fast acting damper to interrupt airflow in the duct between the spray booth and the powder collector. In operation, this apparatus will recognize any flame that has ignited in the spray booth and respond quickly (within ½ sec) to the flame by shutting down all process equipment energy supplies (including electricity and compressed air) and by closing the ductwork damper. This same equipment can be arranged to extend the response sequence by discharging an inerting gas into the dust collector.

For those application systems that require operation with combustible concentrations of powder in the ductwork between the application area and the powder collector, a potential exists for almost instantaneous involvement of the collector whenever a fire occurs in the application area. Such systems should either be equipped with appropriately engineered suppression systems or designed in such a manner that the powder application and collection equipment will be fire and explosion resistant.

Airborne clouds of powder in a confined space will burn at far higher rates than in unconfined space. Since the interior of a conventional dust collector is a confined space, steps must be taken to accommodate the sudden pressurization and volumes of smoke and fire gases that will evolve in the event of ignition. The usual protective technique used is to install automatic pressure relief vents in the collector housing. These vents will open should an ignition occur and allow the combustion products to escape, thereby limiting pressure development. When the collector is located within a building, ductwork is usually required to direct the vented combustion products to the exterior of the building or to a safe location. Vent duct work must be kept short. Lengths in excess of 10 ft (3 m) are of questionable value.

If powder escapes, it will collect on horizontal surfaces and create a potential for a secondary explosion. The primary explosion may throw loose powder into air in suspension where it could be ignited by flames or embers that persist from the initial combustion. If a substantial quantity of powder is thrown into suspension by the primary explosion, the secondary explosion could be considerably more serious than the primary. Protection is achieved primarily through maintenance of process equipment and procedures and through scrupulous housekeeping to collect any powder that does escape.

The dust explosion hazard is virtually eliminated (through elimination of the confined space necessary to form an explosion) in an integrated spray booth/collector of the type shown in Figure 10-7H and in other apparatus incorporating the same design principles. This class of equipment, since it operates without explosion risk, needs no explosion vent or duct work. However, flame detectors are still needed when automatic spray guns are used, to quickly interrupt powder feed in the event of fire and thereby limit heat damage and formation of toxic smoke.

After ignition of a spray gun powder cloud or a fluidized bed, the flame can usually be extinguished almost instantly by simply turning off the spray gun feeder or the air supply to the fluidized bed. The normal velocity of the feed systems exceeds typical flame front propagation velocities and flame therefore does not backstream in hoses from guns to feeders, even though powder concentrations in hoses may exceed MEC.

Automatic sprinkler systems are commonly required in factory areas where powder coating operations take place. Because sprinklers have rather long operating delays relative to the duration of a powder fire (which is usually measured in seconds), it is common for powder fires to be extinguished by shutting down the equipment before sprinklers can activate. In those cases where small fires have persisted in spray booth residues, portable water fire extinguishers have been found to be adequate.

Bibliography

NFPA Codes, Standards, and Recommended Practices and Manuals. (See the latest *NFPA Codes and Standards Catalog* for availability of current editions of the following documents.)

NFPA 10, *Standard for Portable Fire Extinguishers.*
NFPA 13, *Standard for the Installation of Sprinkler Systems.*
NFPA 30, *Flammable and Combustible Liquids Code.*
NFPA 33, *Standard for Spray Application Using Flammable and Combustible Materials.*
NFPA 68, *Guide for Explosion Venting.*
NFPA 69, *Standard on Explosion Prevention Systems.*
NFPA 70, *National Electrical Code.*
NFPA 77, *Recommended Practice on Static Electricity.*
NFPA 704, *Standard System for the Identification of the Fire Hazards of Materials.*

Additional Readings

Bright, A. W., "Electrostatic Hazards in Liquids and Powders," *Journal of Electrostatics*, Vol. 4, No. 2 (1977-78), pp. 131-147.
Factory Mutual Engineering Corporation, "Spray Applications of Flammable and Combustible Materials," *Loss Prevention Data 7-27*, Factory Mutual System, Norwood, MA.
"Fire Hazards of Electrostatic Paint Spraying," *Fire Prevention*, (124), 1978, pp. 19-20.
Flash Point Index, 9th ed., National Fire Protection Association, Quincy, MA, 1978.
Gillies, M. T., ed., *Powder Coatings: Recent Developments*, Noyes Data Corp., Park Ridge, NJ, 1981.
Henry, Martin F., ed., *Flammable and Combustible Liquids Code Handbook*, 2nd ed., National Fire Protection Association, Quincy, MA, 1984.
Meidl, James H., *Flammable Hazardous Materials*, 2nd ed., Macmillan Publishing Co., Inc., New York, NY, 1978.
Scarbrough, Don R., Chap. 25, "Spray Finishing and Powder Coating," *Industrial Fire Protection Handbook*, J. L. Linville, ed. 2nd ed., National Fire Protection Association, Quincy, MA, 1984.
Talbot, G., "Static Electricity," *Fire*, 70(871), 1978, pp. 397-398.

DIPPING AND COATING PROCESSES

Revised by Nicholas L. Talbot and Paul H. Dobson

This chapter discusses the processes, equipment, hazards, and fire protection of dipping and coating operations that use flammable and combustible liquids. Such processes often involve liquids with tremendous heat energy and heat release capabilities that can cause rapid property destruction when they are involved in a fire, if property is not properly protected. The heat of combustion of flammable liquids is approximately two and one-half times that of an equivalent weight of wood (FMRC 1977). In addition, heavy concentrations of smoke and toxic products of combustion often develop in such fires and make fire fighting extremely difficult and dangerous.

More information on related subjects can be found in Section 5, Chapter 4, "Flammable and Combustible Liquids"; Section 10, Chapter 5, "Industrial and Commercial Heat Utilization Equipment"; and Section 10, Chapter 6, "Oil Quenching."

THE PROCESSES

Applications

Dipping and coating processes include, but are not limited to, finishing, impregnating, priming, cleaning, quenching, and other similar operations during which materials are immersed in, passed through, or coated by flammable or combustible liquids. As is the case with any potentially hazardous operation, a complete evaluation of the process and an understanding of the properties of the liquid(s) in use is essential.

Equipment

Dip tanks, roll coaters, curtain coaters, and flow coaters—the basic process equipment—present similar hazards. Related process equipment often used in conjunction with dipping and coating operations includes conveyors, pumps, piping systems, flammable or combustible liquid storage tanks, ovens, liquid heaters or heat exchangers, agitators, detearing equipment, and ventilation and exhaust systems, including energy recovery and pollution control devices. Support equipment must be selected and installed with full consideration of the inherent process hazards involved.

Dip Tanks: Dip tanks are simply liquid containers of various sizes and shapes designed for the process involved. Tank sizes vary from a few gallons with small sq ft exposed surface area to several thousand gallons with a hundred or more sq ft of surface area (1,000 gal equals 3.78 m^3; 100 sq ft equals 9.29 m^2). Figure 10-8A is an example of a typical dip tank process system. Dip tanks, their drainboards, and covers should be constructed of noncombustible material such as heavy metal, reinforced concrete, or masonry. They should be designed for the process and liquid involved, with consideration given to the static head of the liquid, corrosion, mechanical damage, and ease of maintenance and repair. Supports for large tanks should be made of reinforced concrete or protected steel since spill or exposure fires could weaken tank supports and cause the tank to collapse.

Flow Coaters: In flow coaters, the coating material is applied onto the material being coated from nozzles or slots in an unatomized state. (See Fig. 10-8B.) The excess is collected in a trough or sump below the workpiece, returned to the reservoir, and recirculated to the nozzles by a pump. The workpiece usually enters and leaves through conveyor openings, and a drip tunnel is used in place of the drainboards that are used with dip tanks. The tunnel is enclosed on all sides, except for the conveyor opening, and is sloped toward the trough or sump. Airborne solvent concentration in the tunnel is sometimes used to keep the coating from drying before it gets to the oven.

Curtain Coaters: Curtain coaters are used to apply coating material to flat or slightly curved workpieces. (See Fig. 10-8C.) Coating material is pumped from a reservoir to a coating head which has a small reservoir with a dam or weir forming one side. Coating material pumped to the head overflows the dam and forms a continuous vertical stream which drops down onto the workpiece. Excess coating material drops into a trough and returns to the reservoir to be recirculated.

Mr. Talbot is Senior Field Specialist, IRM Insurance, Clevemont, FL. Mr. Dobson is Senior Engineer with the Factory Mutual Research Corporation, Norwood, MA.

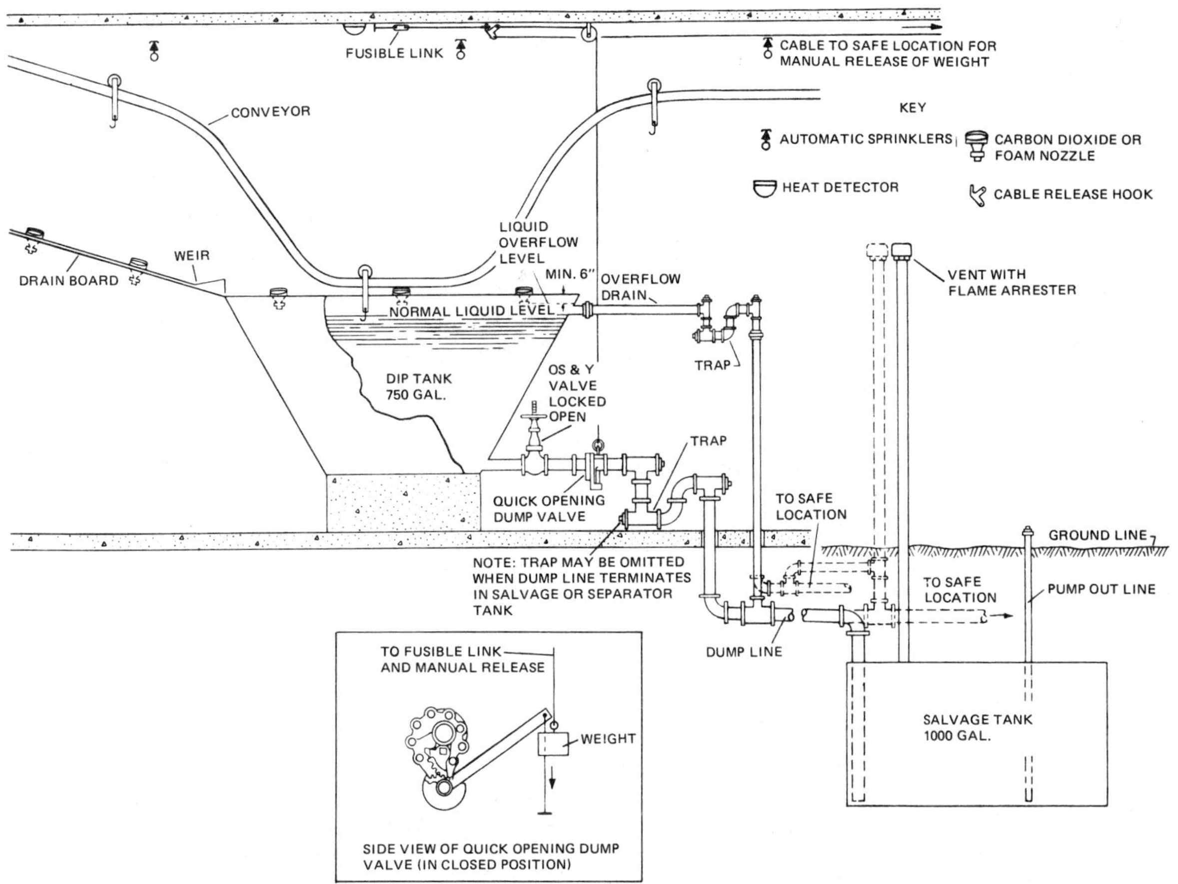

FIG. 10-8A. Dip tank, drainboard, and conveyor system.

Roll Coaters: In a roll coater, coating material is applied by bringing the workpiece into contact with one or more liquid coated rollers. (See Fig. 8-10D.) The coating material may come from an open pan or a nip formed by two rollers. Coating material is supplied from a reservoir by a pump and piping system.

THE PROCESS HAZARDS

Dipping and coating articles or materials by passing them through flammable and combustible liquids involves the danger of fire and explosion of vapor-air mixtures. Generally, the severity of the hazard depends upon the character and flammability of the liquids used, the quantities present, and the rate of vapor generation.

Ease of ignition and the rate of flammable vapor generation from the liquid surface and surfaces of freshly coated articles, floors, drain or drip boards, and other related equipment define the extent of the hazard. The intensity, persistence, and burning characteristics of the flammable vapor evolved and the probability of fire spread from radiated heat or from the flow of burning liquids because of a container rupture, boilover, or overflow complete the hazard picture.

Flammable liquid vapors, which are usually heavier than air, flow to low points. They may travel great distances before being exposed to ignition sources that can cause flashback to the process area.

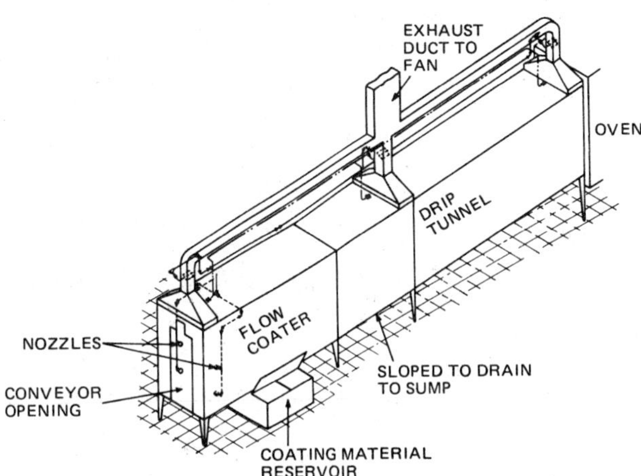

FIG. 10-8B. Flow Coater. (Factory Mutual System)

Flammable liquids are often lighter than, and immiscible in, water. During fire fighting operations, water applied to such liquids may cause overflow and float the burning liquid away from the process to other areas.

HAZARD REDUCTION

Hazards inherent in dipping and coating operations can be reduced by properly locating the processes, install-

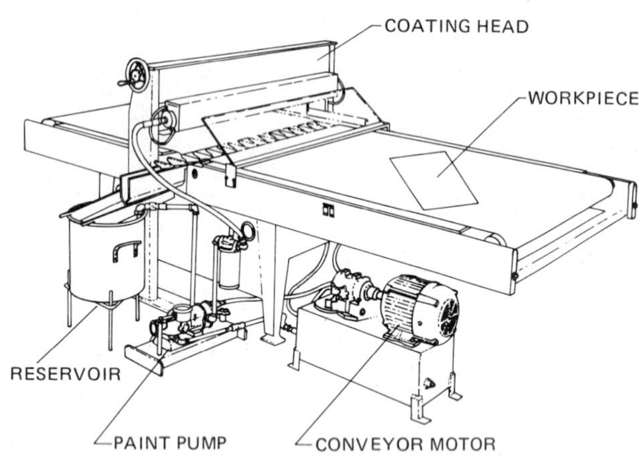

FIG. 10-8C. Curtain Coater. (Factory Mutual System)

ing ventilation and exhaust systems, eliminating ignition sources, and providing proper maintenance and periodic inspections. Special equipment design, employee training,

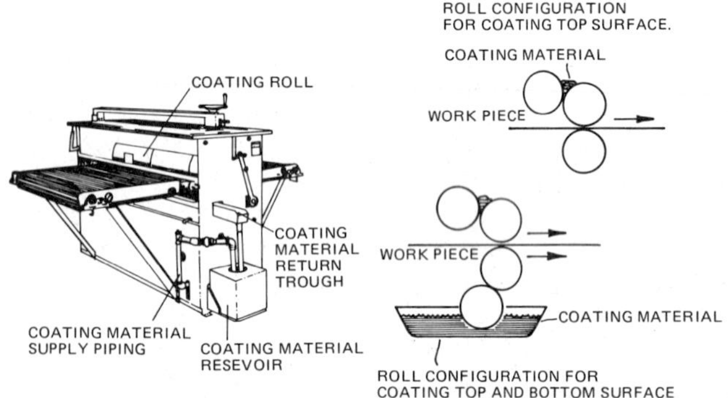

FIG. 8-10D. Roll Coater. (Factory Mutual System)

and proper protection will also help to reduce process hazards.

Process Location

Segregating hazards to confine fires and limit damage is fundamental when selecting locations for dipping and coating operations. A detached or cutoff one-story sprinklered building of fire resistive construction would be one preferred location. When operations are located on upper stories, floors should be made waterproof and equipped with drains. Dipping and coating operations should not be located below grade, directly over basements, or near pits or trenches because liquid drainage and vapor removal will be difficult to achieve. Curbs and trapped drains should be provided to control liquid flow where necessary.

When a process cannot be cut off, it should be located in an area that is clear of combustibles, paths of egress, and other important processes. Combustible floors, ceilings, and surrounding walls should be protected; noncombustible curtain boards should be installed around the perimeter of the process or protected construction area and extended downward as far as practical.

Flammable and combustible liquid fires generally develop rapidly and release large amounts of heat. Therefore, automatic roof vents are preferred, especially in unsprinklered buildings where heat buildup would quickly involve the entire building. Roof vents allow heat and smoke to escape, improving the chances of fire control by automatic sprinklers or the fire department.

When processes are located in confined areas and highly flammable or unstable liquids are used, explosion relief venting may be provided to lessen potential losses. Venting can be in the form of lightweight walls and roofs or explosion relieving wall panels, roof latches, and windows. Explosion relief is discussed in detail in NFPA 68, *Guide for Explosion Venting.*

Figures 10-8E, 10-8F, and 10-8G show common location features of dipping and coating operations.

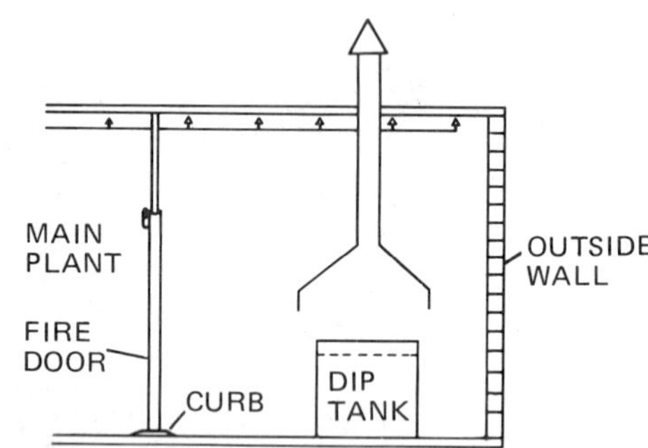

FIG. 10-8E. Most satisfactory arrangement for location of processes. Tank located near an outside wall, cut off from main plant areas by fire resistive construction. (Factory Mutual System)

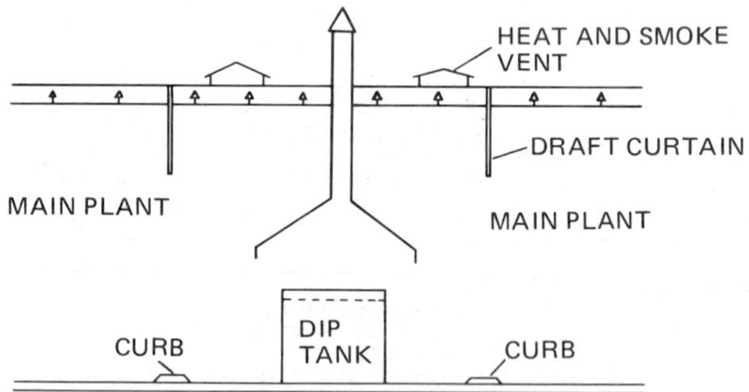

FIG. 10-8F. Satisfactory arrangement. Tank located in main plant area cut off by draft curtains and curbing. Heat and smoke vents should limit opened sprinklers to those near the fire and reduce smoke concentrations. (Factory Mutual System)

Ventilation

When processes involve vapor producing liquids, ventilation is necessary to limit the vapor area to the smallest practical space possible. The vapor area created by dipping and coating processes is defined in NFPA 34, *Standard for Dipping and Coating Processes Using Flammable and Combustible Liquids* as "any area containing vapor at or above 25 percent of the lower flammable limit (LFL) in the vicinity of dipping and coating processes,

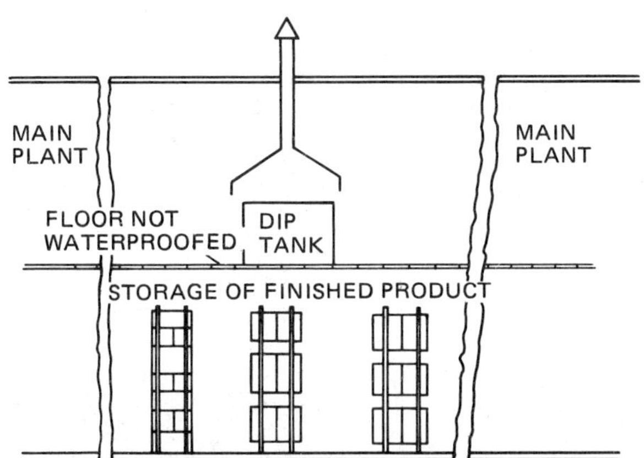

FIG. 10-8G. Unsatisfactory arrangement. Tank not cut off from main plant area. Water from hose streams will leak down through floor and damage finished product. (Factory Mutual System)

their drainboards or associated drying, conveying, or other equipment during operation or shutdown periods."

The extent of the vapor area of a process depends upon the properties of the liquid such as vapor pressure, flash point, boiling point, and evaporation rate, along with the characteristics of the process or wetted surface area exposed. Figure 10-8H clearly shows the difference in exposed wetted surface area between two dip tank operations using the same quantity of liquid. Note that, with all else being the same, the process shown in the upper part of the figure would generate a greater volume of vapor than that in the lower part; thus, it would offer a greater fire potential.

Vapor areas for each process should be kept as small as practical but should not extend more than 5 ft (1.5 m) from the vapor source. The extent of a vapor area can be determined with a combustible gas analyzer. As outlined in NFPA 85, *Standard for Ovens and Furnaces —Design, Location, and Equipment*, theoretical ventilation rate can be determined when the liquid usage rate is known. Since vapor concentrations vary widely from one process to another, a single safety exhaust ventilation system cannot be designed for all operations. The prime objective is to limit the vapor area to the smallest space possible and to eliminate dead air spaces or pockets where vapors can accumulate.

Because flammable liquid vapors are heavier than air, low point peripheral ventilation systems are usually more desirable than overhead hood arrangements. Two types of design are shown in Figure 10-8I. The least efficient system is the overhead hood that allows discharge of some vapors to surrounding work areas; such a system is usually acceptable for unoccupied locations.

When considering exhaust systems, the ratio of air to vapor is important. However, the effectiveness of exhaust air in picking up and diluting the vapor is often lost in design. Entrainment velocity with mixing action becomes the key to efficient use of exhaust air. Exhaust rates per sq ft (1 sq ft equals 0.092 m) of wetted surface area usually range from 50 to 200 cfm (1.4 to 5.7 m^3/min). Effective entrainment of vapor often requires slot velocities of 1,000 to 2,000 cfm (28 to 56 m^3/min).

In designing and evaluating ventilation systems for dipping and coating processes, there are several important considerations:

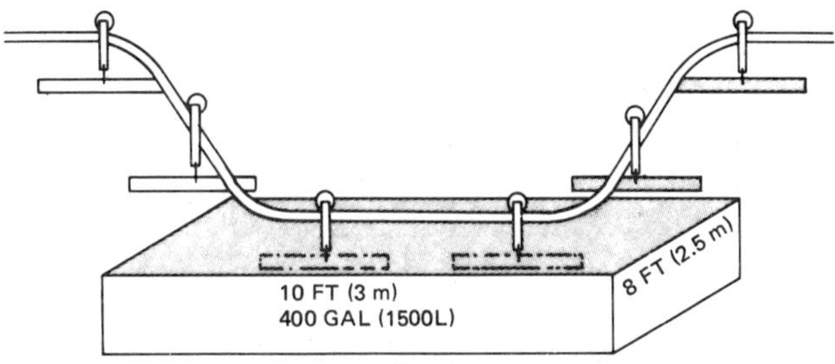

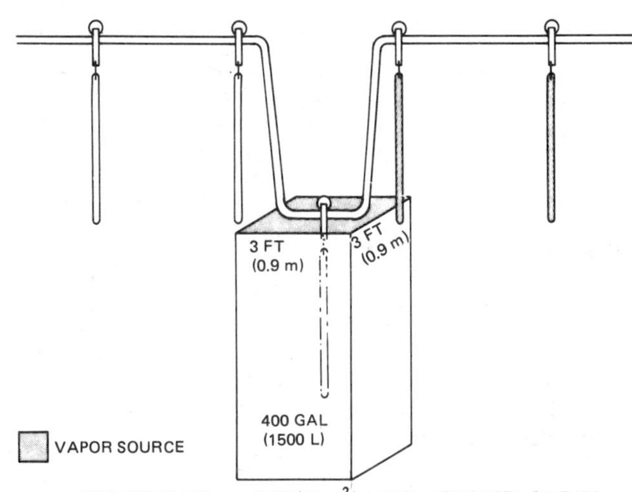

FIG. 10-8H. The difference in vapor source area between two dip tanks with the same coating volume. (Factory Mutual System)

1. Automatic processes should be interlocked to shut down the operation in the event of an exhaust system failure.
2. The supply of make-up air should be sufficient to allow efficient operation of the exhaust fans and to minimize dead air spaces.
3. Ideally, each exhaust system should discharge directly outdoors. When exhaust gases must be treated to satisfy environmental protection requirements or when energy conservation measures are used, individual, direct discharge of each system may not be practical. Manifold systems increase the hazard; therefore, all such systems should be properly designed and strict preventive maintenance measures taken. Reactive materials and coatings should not be used when ducts are connected to a manifold.
4. Exhaust air should not be used for make-up air in occupied spaces. It can be used in unoccupied areas, but only if it has been decontaminated to safe levels.
5. Exhaust ducts should not discharge near air intakes, nor should they discharge less than 6 ft (1.8 m) from a combustible wall or roof or 25 ft (7.6 m) from combustible construction or an unprotected opening in a noncombustible exterior wall.
6. Exhaust ducts should be equipped with ample access doors to facilitate cleaning.

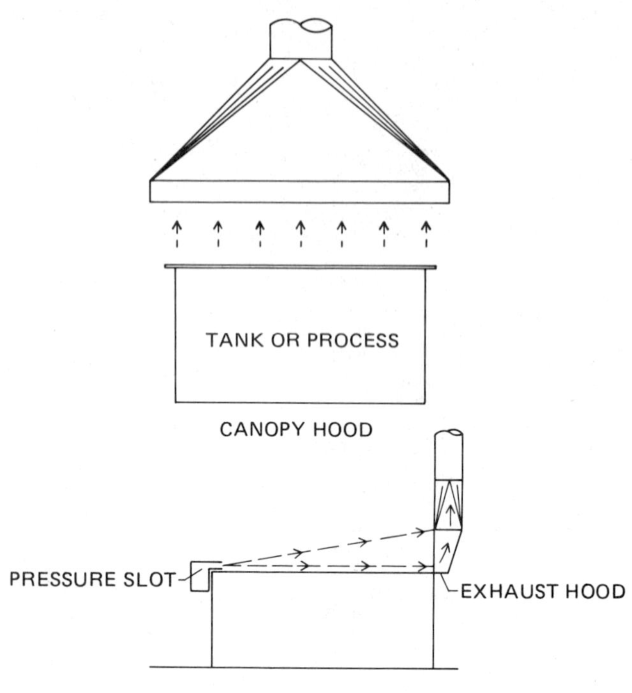

FIG. 10-8I. Ventilation systems used for open process tanks. (Factory Mutual System)

7. Ducts should be constructed of fire resistive material, such as steel or masonry, and should be substantially supported. They should be installed with adequate clearance from combustibles. Dampers should not restrict exhaust ventilation below minimum safe levels.

Ignition Sources

The key to minimizing the hazard of fire or explosion from vapor-air mixtures in dipping and coating processes is to maintain concentrations well below the lower flammable limit (LFL) through properly designed safety exhaust ventilation systems. Without vapor (fuel) and air (oxygen), properly mixed, ignition cannot take place even though sources of ignition are available. As it is almost impossible to keep processes free of flammable vapor-air mixtures in the explosive range, sources of ignition must be eliminated or special equipment must be designed for use in hazardous areas containing flammable vapors and/or combustible residue.

Electrical Equipment: Where flammable liquids are used or where combustible liquids are used at or above their flash point temperatures, electrical equipment should conform to the requirements for hazardous locations as outlined in NFPA 70, *National Electrical Code (NEC)®*. The NEC classifies areas in which special types of electrical equipment must be used. In the case of dipping and coating processes, the classified areas are measured from the vapor source, which may be the liquid surface (open tanks) or wetted surfaces (freshly coated workpieces, drain boards, or floors). The space containing hazardous vapor concentrations normally extends laterally and upward for a short distance from the vapor source and, since the vapors are usually heavier than air, downward to the floor.

Classified areas are shown in Figure 10-8J. Class I,

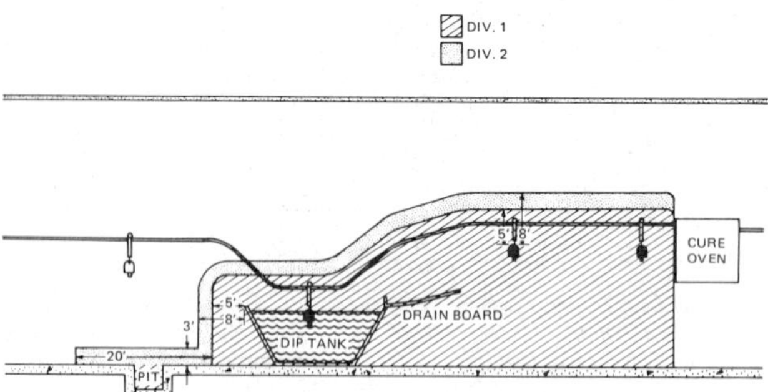

FIG. 10-8J. The Class I, Divisions 1 and 2 hazardous locations for a dipping operation. (Factory Mutual System)

Division 1 electrical equipment is required within a radial distance of 5 ft (1.5 m) from the vapor source and within a horizontal distance of 25 ft (7.6 m) from the vapor source when the equipment is in pits. If a pit is not vapor stopped at a point 25 ft (7.6 m) from the vapor source, the entire pit is considered a classified area. Class I, Division 2 electrical equipment is required 3 ft (0.91 m) beyond Division 1 in all areas down to 3 ft (0.91 m) above the floor. In the space between the floor and 3 ft (0.91 m) above the floor, the Division 2 area extends 20 ft (6.1 m) beyond the Division 1 area. An exception to these electrical requirements is made for tanks containing 5 gal (19 L) or less and having 5 sq ft (0.46 m^2) or less of exposed surface area. These installations generally present less of a hazard and ordinary electrical equipment may be accepted more than 8 ft (2.4 m) from the vapor source.

If electrical equipment must be installed in the vicinity of dipping and coating operations, it should be of a type that can safely be exposed to flammable vapors and combustible residues. Wiring in rigid conduit or in threaded boxes or fittings without taps, splices, or terminal connections is permitted in both residue and vapor areas. Fixed lighting fixtures located outside of, but above, classified areas should be protected against physical damage, which could cause them to ignite flammable vapors.

In areas where the electric service is supplied from overhead wires and lightning is a fairly common occurrence, protection against lightning induced surge voltages should be provided. In the absence of lightning protection, high surge voltages could enter vapor areas and provide a high energy ignition source. Suitable protection includes lightning arresters, interconnection of all grounds, and surge protection capacitors.

Under abnormal conditions, overheating and arcing can occur in electrical circuits and equipment. Therefore, electrical circuits and equipment that service dipping and coating processes should have the proper overcurrent protection, be grounded, and receive regular maintenance.

Frequent changes in dipping and coating processes when new coating materials are introduced, production demand is increased, or new equipment is installed can introduce unsafe electrical installations and poor maintenance habits into the production process. Common problems are overloading of circuits and installation of lighting unsuitable for use in classified hazardous areas.

Other Ignition Sources: The presence of open flames, spark producing processes or devices, and heated surfaces having temperatures high enough to ignite flammable vapors should be prohibited in flammable vapor areas.

Because ovens located directly above or adjacent to dipping and coating operations have been the ignition source for many fires, they should be located as far from dipping and coating operations as practical. When reasonable distances cannot be maintained, noncombustible partitions should be provided. Ventilation exhaust systems of ovens and dryers located directly above or adjacent to dipping and coating processes should be interlocked with process ventilation systems, so that heating and process equipment cannot function unless all ventilation equipment is in operation. Indirect heating systems are usually preferred in drying operations since they do not present ignition sources.

Dipping and coating processes often involve the use of conveyors, rollers, collectors, festoon dryers, and other devices that move the workpiece or liquid through the process. Flowing liquids and materials moving through a process can generate static electricity. A static charge can accumulate on process equipment and a difference in electrical potential can develop which is sufficient to cause a spark containing enough energy to ignite flammable vapor-air concentrations. Static elimination by bonding and grounding process equipment or by humidification or ionization of the local atmosphere is essential in reducing the hazard of fire or explosion.

Of equal importance in hazardous classified areas is the banning of nonessential equipment and processes that are potential ignition sources. Included in this category are cutting and welding operations, portable heaters, and spark producing (ferrous) tools. Smoking should also be banned in these areas.

Special Design Considerations

Some hazards of dipping and coating operations can be reduced by special design features incorporated into the process equipment.

Many flammable and combustible coatings are lighter than, and immiscible with, water. In the event of fire, water from hose streams or sprinkler systems could cause an overflow of a large amount of liquid from the tank unless certain design precautions have been taken.

For example, the top of the tank should be at least 6 in. (152 mm) above the floor, and the liquid surface should be at least 6 in. (152 mm) below the top of the tank. Tanks having a capacity of more than 150 gal (568 L) or a liquid surface area of 10 sq ft (0.93 m^2) or more should be equipped with trapped overflow pipes that discharge to a safe location. Overflow pipes should be capable of handling the maximum delivery of the tank's fill pipes or of the automatic sprinkler discharge. In any case, overflow pipes should never be less than 3 in. (76.2 mm) in diameter.

Larger tanks—those having a capacity of 500 gal (1.9 m^3) or more—should be equipped with a bottom drain of sufficient size to empty the tank in 5 min. The flammable liquid should be drained to a vented salvage tank or other safe location. If gravity flow is not practical, pumps may be used to empty the tank. The drain may be operated manually or automatically, but a manual drain release, if used, should be accessible and at a safe location. If, due to increased hazard or by nature of the operation, bottom drains cannot be provided, the process should be cut off or fully diked, and both an automatic special extinguishing system and sprinklers should be installed.

Conveyor systems used in dipping and coating processes should be arranged to stop automatically in the event of a fire or failure of the exhaust ventilation system.

In processes where there is a possibility of flammable liquids being heated above their boiling points or to within 100°F (37.8°C) of their autoignition temperatures, suitable excess temperature limit controls are needed to prevent rapid vapor buildup and possible autoignition.

Controls should limit the surface temperatures of heated workpieces to at least 100°F (37.8°C) below the autoignition temperature of the coating material used. This limitation does not apply to quenching tanks.

Limit controls, such as liquid level devices, meters, and timers, should be used to prevent overfilling of tanks where automatic filling is designed into the process. If pumps are used, they should be interlocked to shut down in the event of fire.

Maintenance, Training, and Inspection

Location, ventilation, equipment design, and the elimination of ignition sources are primary considerations for reducing the hazards of dipping and coating processes which use flammable and combustible liquids. But maintenance, training, and regular inspection of equipment are just as essential to safe operations. Areas in the vicinity of dipping and coating processes should be kept free of accumulations of combustible residues and unnecessary combustible materials. Spontaneous ignition, due to oxidation or exothermic reaction between various coating components, often occurs when excess residue accumulates in work areas, ducts, duct discharge points, or other adjacent areas. When excess residues accumulate in such locations, operations should be discontinued until conditions are corrected.

Combustible coverings (thin paper or plastic, for example) and strippable coatings are often used to facilitate cleanup of drippings and residues. The increased amount of combustibles which these materials introduce is offset by the improved housekeeping and ease of cleaning they provide. Suitable containers should be provided for the disposal of waste and rags used for cleaning.

Personnel involved in dipping and coating operations should be instructed in the potential safety and health hazards inherent in the particular process. They should understand the nature of the process, the operation and maintenance of equipment, protective measures, and emergency procedures to follow.

Depending upon the size and nature of the process, periodic (usually at least once a month) inspection should be made of dipping and coating processes. Inspections should document the condition of equipment, covers, overflow pipe inlets and outlets, discharge, bottom drains and valves, electrical wiring and equipment, grounding and bonding connections, ventilation equipment, and extinguishing equipment. Defective equipment and unsafe conditions should be corrected promptly.

FIRE PROTECTION

Some form of protection is usually required for all dipping and coating processes. Fire loss experience has shown that damage in small processes is often as severe as that in large operations. Less consideration is usually given to the location of small processes, and they may be located in areas of high value or combustible construction.

By far the simplest device for extinguishing a fire in a process tank is a self-closing cover, held open by a cable and fusible link, which cuts off the fire's air supply. When exposed to the heat of a fire, the solder in the fusible link melts, releasing the cover and allowing it to drop tightly in place over the tank. (See Fig. 10-8K.)

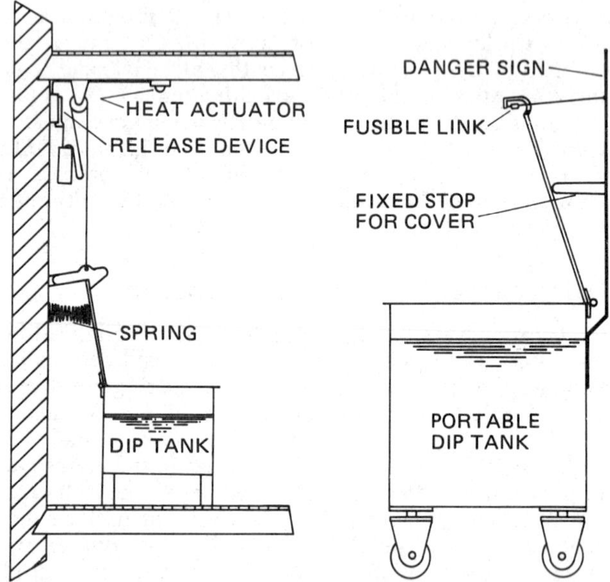

FIG. 10-8K. Automatic closing covers for small fixed and portable dip tanks. (Factory Mutual System)

Automatic sprinklers at ceiling level can provide adequate protection for small processes and for processes where combustible coatings are used with limited liquid surface area exposure. Where processes involve considerable volumes, large liquid surface areas, or large wetted areas of low flash point materials, special extinguishing systems should be provided with automatic sprinklers as backup.

In the case of dip tanks, special systems should be provided for tanks having capacities in excess of 150 gal (569 L) or surface areas greater than 10 sq ft (0.93 m^2). Suitable protection systems use foam, carbon dioxide, a dry chemical, or a halogenated agent as the extinguishing medium. For liquids having flash points above 140°F (60°C), water spray systems may be used. All systems should operate automatically and conform to local fire protection standards. The complexity and value of some processes and the absence of sprinklers as backup often make it desirable to provide redundancy in protection systems.

Dipping and coating process areas should be equipped with portable fire extinguishers suitable for use on flammable and combustible liquids fires.

Due to the flash fire conditions that usually prevail in flammable liquids fires, emergency personnel or fire brigades should be trained to act immediately to bring all protection elements to bear on the fire in hopes of achieving rapid control and limiting possible injury and damage.

Bibliography

References Cited

FMRC. 1977. "Flammable Liquids—General Safeguards." *Loss Prevention Data 7-35*. Factory Mutual System, Norwood, MA.

NFPA Codes, Standards, Recommended Practices and Manuals. (See the latest *NFPA Codes and Standards Catalog* for availability of current editions of the following documents.)

NFPA 10, *Standard for Portable Fire Extinguishers.*
NFPA 11, *Standard for Low Expansion Foam and Combined Agent Systems.*
NFPA 12, *Standard on Carbon Dioxide Extinguishing Systems.*
NFPA 12A, *Standard on Halon 1301 Fire Extinguishing Systems.*
NFPA 12B, *Standard on Halon 1211 Fire Extinguishing Systems.*
NFPA 13, *Standard for the Installation of Sprinkler Systems.*
NFPA 15, *Standard for Water Spray Fixed Systems for Fire Protection.*
NFPA 34, *Standard for Dipping and Coating Processes Using Flammable and Combustible Liquids.*
NFPA 68, *Guide for Explosion Venting.*
NFPA 69, *Standard on Explosion Prevention Systems.*
NFPA 70, *National Electrical Code.*
NFPA 77, *Recommended Practice on Static Electricity.*
NFPA 86, *Standard for Ovens and Furnaces—Design, Location, and Equipment.*
NFPA 91, *Standard for the Installation of Blower and Exhaust Systems for Dust, Stock, and Vapor Removal or Conveying.*

Additional Readings

Meidl, James H., *Flammable Hazardous Materials*, 2nd ed., Macmillan Publishing Co., Inc., NY, 1978.

Talbot, G., "Static Electricity," *Fire*, 70(871), 1978, pp. 397-398.

Weiby, P., and Dickinson, K. R., "Monitoring Work Areas for Explosive and Toxic Hazards," *Chemical Engineering*, Vol. 83, No. 22 (1976), pp. 139-145.

WELDING AND CUTTING

Revised by T. E. Willoughby

In its largest sense, welding includes any materials joining process. As used in this chapter, welding means only fusion welding, i.e., melting together. Similarly, melting is the significant feature in reference to thermal cutting. Both welding and thermal cutting require high intensity energy sources—usually electricity or the heat of combustion of a fuel gas. In this text it is not possible to cover all welding and cutting processes; therefore only basic information about the more common processes is discussed.

For additional information consult Section 5, Chapter 5,"Gases;" Section 10, Chapter 18, "Metal Working Processes;" and Section 11, Chapter 5, "Storage and Handling of Gases."

As a source of ignition, welding and cutting account for about six percent of the fires reported in industrial properties, as well as for many fires in other properties. But education, training, and on the job practice can significantly reduce the potential for welding and cutting fires. There are four factors that must be kept in mind:

1. Two sides of the fire triangle are always present; i.e., a source(s) of ignition and air to support combustion. The other controllable element is the combustible material.
2. The location where the work is being performed. Fires are rare in shops where welding or cutting operations are a regular part of production. However, work carried out in a temporary location using portable equipment is more likely to produce a fire.
3. The kind of process and equipment used and their potential fire effect. Cutting, as well as certain arc welding operations, produces literally thousands of ignition sources in the form of sparks and hot slag. Although arcs and oxy-fuel gas flames are inherent ignition sources, they have rarely caused fires.
4. The supervision and training of the welder. Is the welder trained in the proper use of the equipment and mindful of the exposure? Who has knowledge of and has assessed the risks involved in bringing a torch into the work area and has authorized its use there?

Mr. Willoughby is Associate Director (retired) of the Safety, Health, and Environmental Affairs Department, Linde Division, Union Carbide Corporation.

PROCESSES USING ELECTRICITY

Arc Welding

This term applies to a number of processes that use an electric arc as the heat source for melting and joining metals. The arc is a useful tool because its heat can be concentrated and controlled quite effectively. Frequently, but not always, a filler metal must be used to obtain a good joint. The arc is struck between the metals to be welded and an electrode, which is maneuvered along the joint or which may remain stationary while the work is moved beneath it. The electrode may be consumable or nonconsumable. If the latter, a separate rod or wire may be used as the filler metal. Consumable electrodes supply their own filler metal by melting and, by decomposition of either a covering or a core, may shield the weld zone from unwanted atmospheric effects.

Shielded Metal Arc Welding: This simple well known process is widely used with ferrous based metals. (See Fig. 10-9A.) It produces coalescence* of metals by heating them with an arc between a covered metal electrode and the work. The process requires an alternating or direct current power supply, power cables, and an electrode holder. Shielded metal arc welding can be performed readily in remote, unusual, or confined locations. Consequently, this process is widely used in such industrial applications as construction, shipbuilding, and pipeline erection. Fairly simple, portable units are frequently used in maintenance and field construction work.

Gas Metal Arc Welding: This process uses a continuous, solid wire filler metal, which functions as one terminal of the arc, and a gas to shield the arc and the weld metal. The shielding gas depends on the base metal and process variations. It may be argon (inert), carbon dioxide (oxidizing or oxidizing and carburizing), or mixtures involving these gases with additions of helium or oxygen. Overall,

*This is the preferred American Welding Society term used to describe "the flowing together into one body of the materials being welded."

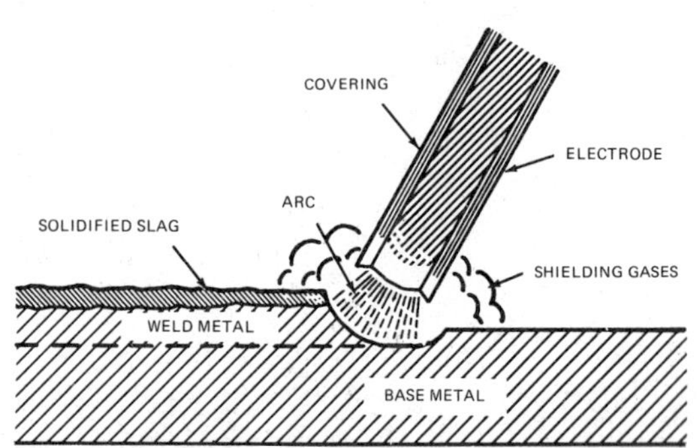

FIG. 10-9A. Schematic diagram of a typical shielded metal arc welding operation. (Lincoln Electric Company)

this process can be used to join just about any metal in any configuration of joint.

Flux Cored Arc Welding: Although this process is similar to gas metal arc welding in equipment used and types of applications, it uses cored rather than solid electrodes. Minerals and alloys in the core assist in weld protection, and many such electrodes are intended for use with carbon dioxide rich gases. Other kinds of cored electrodes are self-shielding and produce their own protective envelope without an auxiliary shielding gas.

Gas Tungsten Arc Welding: This process is similar to gas metal arc welding and flux cored arc welding, but it uses a nonconsumable tungsten electrode for one pole of an inert gas-shielded arc. Filler metal may be used. Shielding gases are argon and helium. Equipment required comprises a power supply, welding torch, source of inert gas, suitable pressure regulators and flowmeters, and connecting hoses.

Plasma Arc Welding: This is characterized by the plasma state, wherein the temperature of argon gas is raised until the gas becomes at least partly ionized, enabling it to conduct an electric current. A plasma arc torch incorporates a tungsten electrode and an orifice through which a small flow of argon forms the arc plasma. Secondary gas, which flows through an outer nozzle cup arrangement that encircles the arc plasma orifice, shields the arc and the weld. The shielding gas may be argon, helium, or mixtures of argon with hydrogen or helium. Filler metal may or may not be used. Equipment required is not unlike that for gas tungsten arc welding with necessary differences in power supply and control, torch design, and gas supply and control.

Submerged Arc Welding: In this process the arc and molten metal are shielded by molten flux and a layer of unmelted flux granules. A continuously fed electrode is submerged in the flux. The arc is invisible, radiation and fumes are minimal, and fire hazard potential is likewise relatively minor compared to other processes.

Resistance Welding

Welding heat for this process is created by resistance to flow of current through the parts being joined. Resistance welding is generally used to join two overlapping metal sheets that may have different thicknesses. Electrodes conduct current through the sheets, which are clamped or rigidly held together to provide good contact and pressure for holding molten metal at the joint.

Flash Welding

Although frequently classed as resistance welding, flash welding is a process unto itself (Weisman 1976). Heat is created by both resistance to current flow and by arcs at the interface. Force is applied after heating, resulting in the expulsion of metal, formation of a flash, and usually a significant shower of sparks. Applications usually involve butt welding rods or bars end to end or edge to edge. The process is almost always automatic. Flash welding machines vary significantly in size, some being large enough to weld sheets in a steel mill.

Electroslag Welding

Electroslag welding is a machine process used primarily for vertical position welding. This process uses a slag, which is conductive while molten, to protect the weld and melt base metal edges and filler metal. An arc is needed to start the process by melting the slag and preheating the work because the unfused, solid slag is nonconductive. Once the process is started, the arc is no longer necessary, and resistance to current flow through the molten slag produces the heat to sustain the process.

Arc Cutting

This term applies to a group of processes that, like the kindred welding processes, melt the metals to be cut by heating with an arc struck between an electrode and the base metal.

Air Carbon Arc Cutting: This process uses electrodes comprising a mixture of graphite and carbon, and coated with a copper layer to improve current carrying capacity. Metal melted by the arc is blown away by a jet of compressed air supplied by conventional compressors, usually at about 80 psi (550 kPa). Current is provided by standard electric welding power supply units.

Plasma Arc Cutting: This cutting process achieves cutting action with an extremely hot jet at high velocity. The hot jet is produced by forcing an arc and inert gas through a small orifice. Arc energy confined to a small area melts the metal, and the jet of highly heated, expanding gases forces the molten metal from the cutting area. High arc voltages, special power supplies, and water cooled torches are required. Argon-hydrogen or nitrogen-hydrogen gas mixtures are commonly used.

OXY-FUEL GAS PROCESSES

Oxy-fuel Gas Welding

This process uses the heat of combustion of a fuel-gas and oxygen flame at high temperatures to melt the workpiece base metal and filler metal, if these are used. The flame (except for oxygen-hydrogen) is readily adjusted to provide a "neutral" environment, as opposed to oxidizing (excess oxygen) and to reducing or carburizing (excess fuel gas). This neutral flame condition is necessary to prevent contamination of the molten metal before it solidifies.

Acetylene remains the preeminent welding fuel because of its unique properties. Recently, other fuel gases or mixtures such as methyl acetylene-propadiene (stabilized), have found limited acceptance. Other hydrocarbons (propane, propylene, butane, natural gas-methane) are not suitable for application to ferrous metals. Hydrogen has a low flame temperature and heat content, and its flame, which is colorless, is difficiult to adjust for oxygen-fuel gas ratio. Oxy-hydrogen welding does have application for welding lead, which is frequently but erroneously called lead burning. Table 10-9A shows maximum and neutral flame temperatures of some fuel gases (Weisman 1976).

With the exception of acetylene, neutral flame temperatures are significantly lower than the maximums, so it is not possible to melt and control the weld, except perhaps for thin sheet metal. At maximum temperatures, the flames are strongly oxidizing and not usable because of deleterious metal oxide formation in the weld metal. Limited use of methyl acetylene-propadiene (stabilized) requires special procedures.

Brazing and Braze Welding

Brazing: This is broadly defined as a welding process in which the base metal is heated but not fused and the joining is accomplished with a filler metal having a melting temperature above 840°F (450°C) (Weisman 1976). Moreover, brazing is characterized by the filler metal being distributed through a closely fitted joint by capillary action.

TABLE 10-9A. Flame Temperatures of Fuel Gases with Oxygen

Gas	Maximum Flame Temperature		Neutral Flame Temperature	
	°F	°C	°F	°C
Acetylene	5615	3102	5612	3100
Methyl acetylene-propadiene (stabilized)	5255	2902	4712	2600
Propylene	5174	2857	4532	2500
Propane	5030	2777	4442	2450
Natural gas-methane	4967	2742	4260	2350
Hydrogen	5200	2871	4334	2390

Braze Welding: Braze welding is of more significance as it relates to fire prevention because of its frequent use in repair and maintenance activities. It differs from brazing in that capillary action in the joint is not a factor; the filler metal is laid down in a groove or fillet at the point of application. The edges of the joint are heated to a dull red color, flux is used, and filler metal is supplied by a bronze rod. The process is widely used on ferrous metals as well as on copper and nickel alloys. Sometimes it provides a solution to the problem of joining dissimilar metals.

Hard Soldering: This common but incorrect term is used to describe brazing with a silver-base filler metal. The operation is often performed with an oxy-fuel gas torch, as in assembling copper pipe joints with silver brazing alloy. Many times, however, the oxygen is not pure gas but is supplied as air, as in the air-acetylene torch. Thus only one small gas supply cylinder may be needed, enhancing portability.*

Surfacing

The oxy-fuel gas flame (usually oxy-acetylene) is used, not in a true welding sense, but to deposit a layer of filler metal on a base metal to obtain certain surface properties or dimensions. Bronze may be used, but more frequently the process is one of hard facing, where the deposited layer is a special alloy that will greatly prolong the life of parts subject to extreme wear or abrasion.

Oxy-fuel Gas Heating Operations

There are numerous industrial and commercial operations that regularly employ an oxy-fuel or air-fuel flame but do not involve a joining or severing operation. The flame and the nature and location of these operations nonetheless may represent an appreciable fire potential. Some such operations are:

Forming: Heating (ironwork, piping, etc.) to facilitate bending, shaping, or straightening.

Annealing, Flame Hardening, Flame Softening: Use of the flame in a controlled manner to obtain specific properties.

Flame Priming: Heating to remove scale and rust in preparing metal surfaces for painting.

Flame Descaling: Heating (generally a steel mill application) to remove scale from bars, billets, slabs, etc., to facilitate machining or inspection.

Other Applications: Paint burning, glass finishing, leather edging, babbitting, or antiquing of wood.

Oxy-fuel Gas Cutting

This term describes a process or group of processes named by the specific fuel gas used (for example, oxy-acetylene cutting, oxy-natural gas cutting) for severing metals by the reaction of high purity oxygen with the metal at elevated temperatures.

Burning iron in oxygen produces iron oxide, normally a solid, but the oxide melts at a temperature below the melting point of iron or steel and runs off as slag. A variation in the process is oxy-fuel gas gouging, wherein a relatively low velocity oxygen jet permits gouging or grooving of a metal surface in a reasonably smooth, well defined manner.

Oxy-fuel Gas Welding and Cutting Equipment

Basic elements are fuel-gas and oxygen supplies, pressure regulators, conduits (hoses or piping) to convey the gases, and a torch to mix and burn the gases in controlled fashion and provide the jet used in cutting operations.

*The air-fuel gas flame, particularly air-acetylene, also finds significant use in soldering, commonly called soft soldering, which by definition involves a filler metal that melts below 840°F (450°C). Plumbing joint work, refrigeration piping, and heating and ventilating duct assembly are applications where the air-fuel gas flame is used. Equipment portability is an advantage—but it may also signify access to places where fire prevention is not easily accomplished and is easily overlooked because a single joint operation can be done in a matter of minutes.

In its simplest form, the equipment comprises a cylinder each of fuel gas and oxygen, a pressure regulator on each cylinder, hoses, and a torch. (See Fig. 10-9B.) A steel mill, however, might have a major installation for cutting and welding operations. (See Fig. 10-9C.) Fuel gas might be supplied from large multicylinder manifolds (possibly truck mounted and replaceable), from storage tanks for liquefied fuel, or from a public utility natural gas main. Oxygen may be supplied from the mill's own on site oxygen generating plant, from storage tanks for liquid oxygen or high pressure gas or both, or from multicylinder manifolds. A major installation may also have an extensive fixed piping distribution network and consuming devices running the gamut from simple hand torches to sophisticated steel conditioning machines.

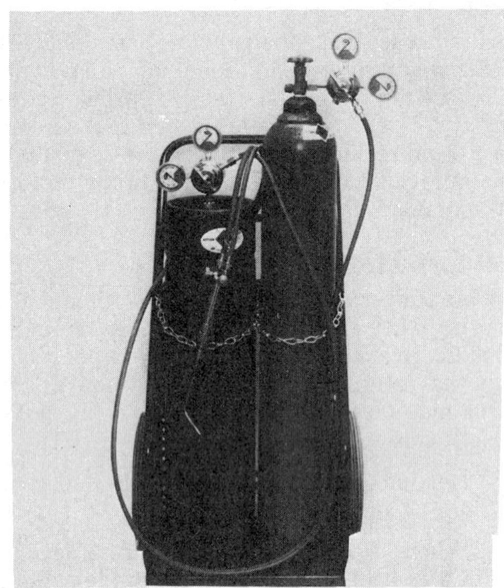

FIG. 10-9B. A portable welding outfit with oxygen and acetylene outfits chained to an easy rolling cylinder truck (cutting attachments not shown). (Linde Div., Union Carbide Corporation)

FIG. 10-9C. A large stationary oxy-fuel gas cutting machine in operation. (Linde Div., Union Carbide Corporation)

Auxiliary but necessary equipment includes standardized (CGA 1976) fuel gas and oxygen hose connections, standardized hose (RMA 1976), protective equipment for service piping systems, and shutoff valves at points (station outlets) where gas is withdrawn from the piping system. Protective equipment for piping, which may be installed at one or more locations, should prevent the backflow of oxygen into the fuel gas supply system, the passage of flashback into the fuel gas supply system, and the development of pressures in excess of system components ratings. In some systems, backflow prevention devices also may be required at station outlets.

Welding and cutting torches have inlet connections and valves for each gas at the rear of the handle. (See Fig. 10-9D.) The gases are controlled by inlet valves and thoroughly mixed before issuing from the torch tip. Cutting torches are similar but provide passageways and separate valving for supplying and controlling the cutting oxygen jet at the center of the tip.

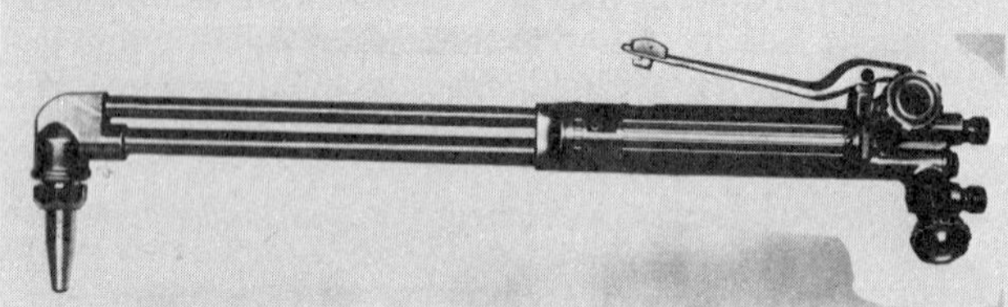

FIG. 10-9D. A typical cutting torch. All control valves are located at the rear of the body. (Linde Div., Union Carbide Corporation)

Mechanized cutting is common. Equipment varies from relatively simple, portable machines, used for straightline work and perhaps circles and some irregular shapes, to highly sophisticated multitorch machines that can trace, (by photocell or other electronic means) intricate shapes and accurately and simultaneously produce a number of parts of the same shape.

THERMAL SPRAYING

Closely allied to welding are several processes, known collectively as thermal spraying,* in which finely divided metallic or nonmetallic materials are deposited in molten or near molten condition to form coatings. Special "guns" are used. (See Fig. 10-9E.) In flame spraying, an oxy-fuel gas flame is used to melt the coating material, and an auxiliary compressed gas may be used to assist in atomizing and propelling coating material to the workpiece.

Electric arc spraying employs an arc between two consumable electrodes of coating material, plus auxiliary compressed gas to atomize and propel the coating particles. Plasma spraying uses a plasma arc as the heat source and for propelling the coating material. In detonation flame spraying, the coating material is melted and propelled by the controlled explosion of fuel gas and oxygen.

THE SAFEGUARDS

Equipment Preparation and Condition

Although it is beyond the scope of this chapter to describe all the fire prevention considerations that are tied

*American Welding Society preferred terminology for "metal spraying" or "metallizing."

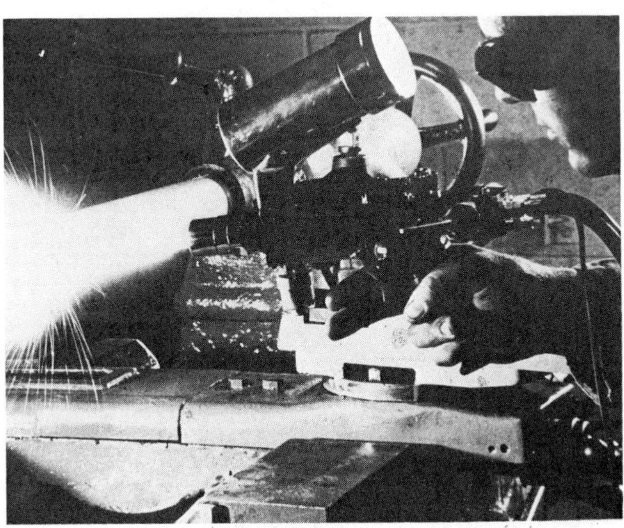

FIG. 10-9E. Thermal spray gun. (American Welding Society)

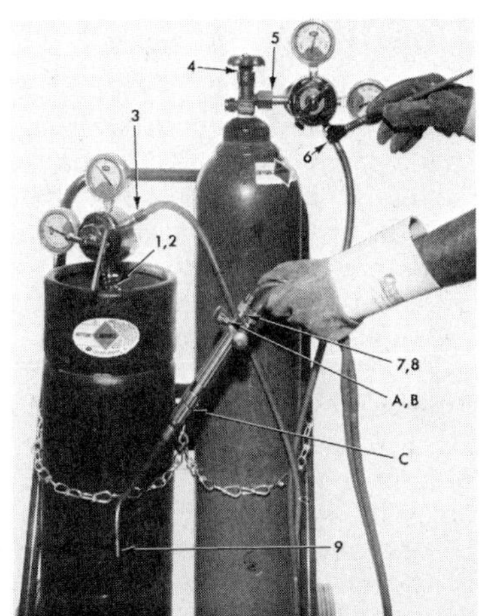

FIG. 10-9F. Before operating an oxy-fuel gas outfit, the equipment should be checked for leaks. After pressurizing both hose lines (with torch valves tightly closed) test for leakage at the following points, using an approved leak-test solution or a thick solution of soap and water: (1,2) Acetylene cylinder connection and acetylene cylinder valve spindle, (3) Acetylene regulator-to-hose connection, (4) Oxygen valve spindle, (5) Oxygen cylinder connection, (6) Oxygen regulator-to-hose connection, (7,8) Hose connections at the torch, (9) Torch tip (for leakage past the torch valves). Later, after lighting the torch, check for leakage at the throttle valve stems (A,B) and at the welding head-to-torch handle connection (C). (Linde Div., Union Carbide Corporation)

into proper equipment design, installation, and maintenance, some significant items have been selected for specific attention.

Oxy-fuel Gas Equipment

Where equipment meeting recognized standards (torches, cylinder manifolds, pressure regulators, pipeline protective devices, etc.) is available, its use is recommended.

Since oxygen is a far more powerful oxidizer than air, oxygen equipment must be kept clean (free of oil, grease, and other combustible contaminants). According to NFPA 51B, *Standards for Fire Protection in Use of Cutting and Welding Processes* (hereinafter referred to as NFPA 51B), "Materials that burn in air will burn violently in pure oxygen at normal pressure and explosively in pressurized oxygen. Also, many materials that do not burn in air will do so in pure oxygen, particularly under pressure." Consequently, it is widely recognized good practice to reserve equipment for oxygen service only.

Proper storage of gas cylinders is important. Details can be found in NFPA 51, *Standard for the Design and Installation of Oxygen-Fuel Gas Systems for Welding and Cutting.*

Cylinders should be moved and handled in accordance with recognized practices (ANSI 1977; CGA 1974). There are many such recognized practices, and their importance may vary with working locations and conditions. However, cylinders should always be supported or located in such a way that they cannot be knocked over.

Oxy-fuel gas cutting and welding equipment should be tested periodically for leaks. (See Fig. 10-9F.) Testing frequency depends upon the specific kind of equipment involved and how often it is used. Hose connections are known possible trouble spots. Also, experience has shown that fires occur at fuel gas cylinder-to-regulator connections simply because someone failed to tighten the joint properly. The ignition of leaking gas at this point may in turn cause the release of cylinder safety devices, especially with acetylene cylinders fitted with fusible plugs, thereby releasing more gas and increasing the size of the fire.

Only standard welding hose (RMA 1976) should be used. It should be frequently inspected for burns, cuts, worn places, abrasions, and similar defects. Taped repairs are unacceptable. Replace the damaged hose, or if feasible, cut out the affected area and insert a proper splice.

When repairs to equipment such as torches and regulators are required (beyond simple operations such as repacking a torch valve), they should be carried out by trained, skilled mechanics. Improper repairs, especially to oxygen regulators, have been responsible for a number of fires and serious personal injuries.

Arc Welding Equipment

1. Where equipment meeting recognized American National Standards (NEMA 1976; UL 1972) is available, it should be used. Installation, including incoming power lines, and grounding of the machine frame or case should comply with NFPA 70, *National Electric Code,* with particular attention to its Article 630, "Electric Welders."
2. Proper storage and handling procedures for cylinders of shielding gases should be observed.
3. At each work location, cylinders should be supported or located in such a way that they cannot be knocked over accidentally. Precautions must be taken that the cylinders are not grounded.
4. Cable sizes should be adequate for current and anticipated duty cycles. Sustained overloading of inadequate cables may burn away insulation. Cables should be inspected frequently for wear and damage, and properly repaired or replaced when necessary.

Precautions for the Work Area

Dangerous sparks—globules of molten, burning metal or hot slag—are produced by both welding and cutting operations. Sparks from cutting, particularly oxy-fuel gas cutting, are generally more hazardous than those from welding because the sparks are more numerous and travel greater distances. In a sense, they are jet propelled by the oxygen or airstreams used in cutting processes. Oxy-fuel gas flames and electric arcs are inherent and obvious ignition sources, as are hot workpieces or sections cut from the base workpiece. However, experience shows these are less frequent ignition sources than sparks.

Either isolation or protection of combustibles is essential, for they may be exposed to sparks that fall through cracks or other openings in floors and partitions. If those sparks are of sufficient mass to retain heat for a time, they may ignite combustibles. The minimum requirements for combustible control in the cutting or welding work area are:

1. Move all combustibles a safe distance away—at least 35 ft (10.6 m) horizontally—and be sure that there are no openings in walls or floors within 35 ft (10.6 m); or
2. Move the work to a safe location; or
3. If neither of the foregoing steps is possible, protect the exposed combustibles with suitable fire resistant guards and provide a trained fire watcher with extinguishing equipment readily available.

These steps are only a partial solution to the problem of preventing cutting and welding fires. There are other important factors to consider. Are there any inconspicuous combustibles in an area proposed for cutting and welding operations? What conditions must be met before cutting and welding operations can take place? Who has the responsibility for authorizing the work to proceed? Are cutters, welders, and their supervisors properly trained in the use of their equipment and in emergency procedures should a fire occur? If an outside firm is engaged to do cutting and welding work, chances are that its employees will be unfamiliar with the premises and its contents. Have they been briefed on the conditions in the areas where they will work?

Based on the fundamental but necessary understanding that welders, their supervisors, and facility management share the responsibility for firesafety, the following abridged version of NFPA 51B should be helpful:

1. Management should establish areas designed and authorized for cutting and welding and/or designate a knowledgeable person to authorize welding or cutting in areas not specifically designed for such processes. This management designee should require trained fire watchers where the potential exists for a significant fire to develop. Fire watchers should also be required where appreciable quantities of shielded combustibles are less than 35 ft (10.6 m) away; where wall or floor openings within 35 ft (10.6 m) expose combustibles in adjacent areas; or where combustibles adjacent to opposite sides of partitions, ceiling, or roofs are likely to be ignited by heat from the work. Management should select contractors with a view to their awareness of risks and the quality of their personnel. Management should also advise contractors of the presence of flammable materials or other hazardous conditions on the property work site.
2. The supervisor of welding and cutting operations in areas not designed for such processes (for example, the plant manager, plant maintenance foreman, contractor, or contractor's foreman) should be assigned the following responsibilities:
   a. Determine what combustible materials are present at the work site.
   b. If necessary, have the work or the combustibles moved, or have the combustibles shielded.
   c. Secure authorization from management, preferably in the form of a written permit.
   d. See that the welder is aware of the authorization and conditions.
   e. See that fire watchers are available when required.
   f. Make a final check for fires one-half hour after completion of welding or cutting operations in cases where a fire watcher was not required.
3. Welders should have their supervisor's approval before starting to cut or weld, should handle their equipment safely, and should continue to work only so long as approval conditions do not change.
4. There are a number of precautions to be observed during welding and cutting operations:
   a. Welding and cutting must not be permitted in flammable (explosive) atmospheres; near large quantities of exposed, readily ignitable materials; in areas not authorized by management; or on metal partitions, walls, or roofs with combustible covering or with combustible sandwich type panel construction.
   b. Floors should be free of combustibles such as wood shavings. If the floor is of combustible material, it should be kept wet or otherwise protected.
   c. If combustibles are closer to the welding or cutting process than 35 ft (10.6 m) and the work cannot be moved or the combustibles relocated at least 35 ft (10.6 m) away, they should be protected with flameproofed covers or metal guards or curtains. This also applies to walls, partitions, ceilings, or roofs of combustible construction.
   d. Openings in walls, floors, or ducts should be covered if within 35 ft (10.6 m) of the work.
   e. Cutting or welding on pipes or other metal in contact with combustible walls, partitions, ceilings, or roofs should not be performed if close enough to cause ignition by heat conduction.
   f. Charged and operable fire extinguishers should be readily available. Trained fire watchers should be posted. In the absence of fire watchers, an important minimum step would be to check the work area and adjacent areas carefully for at least one-half hour after completion of welding or cutting to detect possible smoldering fires.

SPECIAL SITUATIONS AND ADDITIONAL PRECAUTIONS

Containers that Have Held Combustibles

Every year inevitably brings a number of unfortunate and unnecessary accidents involving severe personal injuries or fatalities from explosions or fires resulting from welding or cutting on containers that contain or have contained combustibles. Therefore it is essential that flammable liquids, solids, or vapors be removed from containers by some type of adequate cleaning procedure before welding or cutting.

Depending upon the application, it may be necessary or desirable to supplement the cleaning with inerting, water flooding, or periodic testing (for flammables) of the atmosphere within the container. With some materials, water washing or flushing may be sufficient; others may respond to steam cleaning; some may require alkaline cleaners (trisodium phosphate or caustic soda); some may call for more specialized cleaning methods. Heavy, viscous liquids and solids can be especially difficult to remove completely. Residues are especially dangerous because they may be volatilized or decomposed into flammable products by the heat of the torch or arc. Details concerning these hazards are available (UL 1972; AWS 1965).

Jacketed Containers and Hollow Parts

There are some similarities between the hazards of cutting or welding jacketed containers or other hollow workpieces, and hazards of cutting or welding containers that have held combustible materials. Air confined inside an unvented hollow part will expand when heated, and pressure will increase. Since hot metal rapidly loses its strength, the container or hollow piece may burst with explosive force at the focus of the cutting or welding work. It is well to be suspicious of closed metal parts that seem unusually light. Drilling a vent hole before heating the part may be necessary.

Hot Tapping

Occasionally, there are situations where emergency repair or the complete impracticality of emptying and cleaning demand welding or cutting on a container while it holds flammable gas or liquid (for example, a natural gas transmission pipeline or utility distribution system). Schemes to accomplish such hot tapping with relative safety have been developed (ANSI 1975 and API 1963). Needless to say, any such work should be performed only by specially trained and qualified staff, using recognized and authorized methods.

Public Exhibitions and Demonstrations

Special and enhanced fire safeguarding is necessary when welding or cutting is performed at trade shows and exhibitions. Characteristically, places of public assembly (auditoriums, hotel exhibit halls) and concentrations of people are involved. When welding or cutting work is planned in such exposures, the fire department should have prior notification, operations should be under the control of a qualified person, gas cylinders should be charged to only one-half their maximum permissible content, storage sites for gas cylinders should have special restrictions, and fire extinguishing equipment should be appropriate to the situation.

Personnel Protection and Ventilation

One aspect of welding and cutting fire protection that is sometimes overlooked or inadequately considered is the safety of the operator, helpers, or nearby workers. Flame resistant gloves, woolen clothing, aprons of leather or other durable flame resistant material, cape sleeves or shoulder covers with skull caps under helmets or with goggles for overhead work, leggings for heavy work, and high top safety shoes are generally recommended. Trousers should not be turned up or cuffed on the outside, front pockets on clothing should be eliminated, and sleeves and collars kept buttoned to prevent sparks from entering and lodging in such places. Outer clothing should be free of oil and grease. Cotton instead of woolen clothing may be worn if the cotton is chemically treated to reduce its combustibility.

Adequate ventilation must be provided wherever welding and cutting are performed to protect the operator from inhaling noxious gases and fumes. Potentially hazardous materials may exist in certain fluxes, coatings, and filler metals. In some cases, general natural draft ventilation may be adequate. Other operations may require forced draft ventilation, local exhaust hoods or booths, or personal respirators or air supplied masks.

Oxygen, since it accelerates combustion, must never be used to cool the welder, ventilate a confined space, or dust off clothing. Tests and experience have shown that oxygen saturated clothing or clothing in an oxygen enriched atmosphere will literally burn in a flash, with extremely serious and sometimes fatal results.

Manufacturers' Recommendations

Procedures given in manufacturers' instructions for setting up, connecting, lighting or starting, adjusting, and maintaining equipment have specific, safety oriented purposes. These procedures should be followed. In addition, precautions and safe practices publications with valuable information about fire and personnel safety are available from equipment suppliers.

Bibliography

References Cited

ANSI. 1975. *Gas Transmission and Distribution Piping Systems.* B31.8. Paragraph 841.27. American National Standards Institute, NY.

ANSI. 1977. Z49.1 *Standard for Safety in Welding and Cutting.* American National Standards Institute, NY.

AWS. 1965. *Safe Practices for Welding and Cutting Containers That Have Held Combustibles.* American Welding Society, Miami, FL.

CGA. 1974. Pamphlet P-1. *Safe Handling of Compressed Gases in Containers.* Compressed Gas Association, Arlington, VA.

CGA. 1976. Pamphlet E-1. *Standard Connections for Regulators, Torches, and Fitted Hose for Welding and Cutting Equipment.* Compressed Gas Association, Arlington, VA.

NEMA. 1976. *Electric Arc Welding Apparatus EW-1.* National Electrical Manufacturers Association, Washington, DC.

API. 1963. *Welding or Hot Tapping on Equipment Containing Flammables.* PSD-2201. American Petroleum Institute, Washington, DC.

RMA. 1976. *Specifications for Rubber Welding Hose.* 4th ed. Rubber Manufacturers Association, Washington, DC, and Compressed Gas Association, Arlington, VA.

UL. 1972. *Safety Standard for Transformer Type Arc Welding Machines.* Underwriters Laboratories Inc., Chicago, IL.

Weisman, C. ed. 1976. *Welding Handbook.* 7th ed. Vol 1. American Welding Society, Miami, FL.

NFPA Codes, Standards, Recommended Practices and Manuals. (See the latest *NFPA Codes and Standards Catalog* for availability of current editions of the following documents.)

NFPA 10, *Standard for Portable Fire Extinguishers.*
NFPA 50, *Standard for Bulk Oxygen Systems at Consumer Sites.*
NFPA 51, *Standard for the Design and Installation of Oxygen-Fuel Gas Systems for Welding, Cutting, and Allied Processes.*
NFPA 51B, *Standards for Fire Protection in Use of Cutting and Welding Processes.*
NFPA 70, *National Electrical Code.*
NFPA 306, *Standard for the Control of Gas Hazards on Vessels.*
NFPA 327, *Standard Procedures for Cleaning or Safeguarding Small Tanks and Containers.*

Additional Readings

API, *Publication 2009, Safe Practices in Gas and Electric Cutting and Welding*, 4th ed., American Petroleum Institute, Washington, DC, 1976.
API, *Publication 2013, Cleaning Tank Vehicles Used for Transportation of Flammable Liquids*, 1st ed., American Petroleum Institute, Washington, DC, 1975.
API, *Publication 2015, Cleaning Petroleum Storage Tanks*, 2nd ed., American Petroleum Institute, Washington, DC, 1976.
AWS, *A6.1, Recommended Safe Practices for Gas Shielded Arc Welding*, American Welding Society, Miami, FL, 1966.
AWS, *C5.1, Recommended Practices for Plasma Arc Cutting*, American Welding Society, Miami, FL, 1973.
AWS, *C1.1, Recommended Practices for Resistance Welding*, American Welding Society, Miami, FL, 1966.
AWS, *Safety in Welding and Cutting*, 4th ed., American Welding Society, Miami, FL, 1973.
AWS, *Welding Handbook*, (5 volumes), American Welding Society, Miami, FL.
ANSI, *A13.1, Scheme for Identification of Piping Systems*, American National Standards Institute, NY, 1975.
ANSI, *Z117.1, Safety Requirements for Working in Tanks and Other Confined Spaces*, American National Standards Institute, NY, 1977.
Buehrer, P., "Check List for the Prevention of Fires Arising from Welding and Allied Processes," *Welding in the World*, Vol. 14, No. 5-6 (1976), pp. 122-125.
Compressed Gas Association, *Handbook of Compressed Gases*, 1st ed., Van Nostrand Reinhold Publishing Corp., NY, 1966.
CGA-4, *Method of Marking Portable Compressed Gas Containers to Identify the Material Contained*, Compressed Gas Association, Arlington, VA, 1977.
CGA-V-1, *Compressed Gas Cylinder Valve Outlet and Inlet Connections*, Compressed Gas Association, Arlington, VA, 1977.
CGA, *Hose Line Check Valve Standards for Welding and Cutting*, Pamphlet E-2. Compressed Gas Association, Arlington, VA, 1976.
CGA, *Pipeline Regulator Inlet Connection Standards*, Pamphlet E-3, Compressed Gas Association, Arlington, VA, 1976.
CGA, *Standard for Gas Regulators for Welding and Cutting*, Pamphlet E-4, Compressed Gas Association, Arlington, VA, 1977.
CGA, *Torch Standard for Welding and Cutting*, Pamphlet E-5, Compressed Gas Association, Arlington, VA, 1977.
CGA, G-1, *Acetylene*, Pamphlet Compressed Gas Association, Arlington, VA. 1972.
CGA, *Oxygen*, Pamphlet G-4, Compressed Gas Association, Arlington, VA. 1972.
CGA, *Industrial Practices for Gaseous Oxygen Transmission and Distribution Piping Systems*, Pamphlet G-4.4, Compressed Gas Association, Arlington, VA, 1973.
CGA, *Handling Acetylene Cylinders in Fire Situations*, Bulletin SB-4, Compressed Gas Association, Arlington, VA, 1972.
Lincoln Electric Company, *The Procedure Handbook of Arc Welding*, 12th ed., Cleveland, OH, 1973.
Manz, A. F., *The Welding Power Handbook*, Union Carbide Corporation, Tarrytown, NY, 1973.
NSC, *Accident Prevention Manual for Industrial Operations*, 7th ed., National Safety Council, Chicago, IL, 1974, pp. 947-976.
"Prevent Cutting/Welding Fires," *The Minnesota Fire Chief*, Vol. 16, No. 5 (May/June 1980), pp. 16, 48, 54.
Resource Systems International, *Welding and Cutting Safety*, Reston Publishing Co., Inc., Reston, VA, 1982.
RWMA, *RWMA-16, Resistance Welding Equipment*, Resistance Welder Manufacturers Association, Philadelphia, PA, 1969.
Stewart, John P., *Welder's Handbook*, Reston Publishing Co., Inc., Reston, VA, 1981.
Sullivan, James, *Welding Telchnology*, Reston Publishing Co., Inc., Reston, VA, 1982.
UL, *UL-123, Standard for Oxy-fuel Gas Torches*, 5th ed., Underwriters Laboratories, Inc. Chicago, IL, 1975.
UL, *UL-252, Standard for Gas Pressure Regulators*, 4th ed., Underwriters Laboratories Inc., Chicago, IL, 1973.
UL, *UL-407, Standard for High-Pressure Gas Manifolds*, 3rd ed., Underwriters Laboratories Inc., Chicago, IL, 1972.

CHEMICAL PROCESSING EQUIPMENT

Revised by W. J. Bradford, P.E.

This chapter discusses the chemical reactions, the means of reaction control, and the equipment used to produce chemicals and synthetics such as fibers, plastics, and some drugs. Operations to chemically change the properties of materials, as in the chemical pulping of wood or the tanning of leather are also covered. However, in discussing fire and explosion prevention and loss control, the hazards inherent in each system must be considered as a whole when appropriate separation distances and other means of minimizing damage are examined.

Additional relevant information may be found in this Section in Chapter 3, "Heat Transfer Systems," and Chapter 9, "Welding and Cutting;" in Section 4, Chapter 2, "Explosions," and Section 4, Chapter 4, "Theory of Explosion Control;" in Section 6, Chapter 5, "Hazardous Waste Control;" and Section 12, Chapter 14, "Waste Handling Systems."

PLANT SITING

Loss Prevention in the Process Industries notes that safety is a prime consideration in plant siting (Lee 1980). The most important feature is distance between the site and areas of high population density. Distance always tends to reduce casualties; if a toxic release occurs; the effect of distance is to reduce the concentration of the gas and buy time to permit evacuation. The chemical disaster at Bhopal, India, in December 1984 was undoubtedly exacerbated by the very high population density immediately adjacent to the plant. If the hazard is fire or explosion, distance will reduce the intensity of radiant heat or blast overpressure as well as damage done by missiles. The density of population adjacent to the liquefied petroleum gas tanks which exploded in Mexico City in September 1984 led to very high loss of life. These two incidents resulted in over 3,000 fatalities. They are the worst industrial incidents in history.

A plant with serious toxic gas release potential or major fire or explosion hazards must be designed to high standards of safety and loss prevention. Distance and plant siting cannot be relied upon to prevent disasters but must be considered as a key feature in limiting the potential magnitude of an incident (Melancon 1980).

Separation Distances

Single or allied processes are usually located on individual blocks of land surrounded by access roads. The theory behind the block approach is that fires or unignited spills then can be attacked from all directions. This approach provides maximum latitude to take advantage of both wind direction and the shelter afforded by peripheral structures in the block. If the right of way for these roads is 50 ft (15.25 m) wide, a minimum distance of 50 ft (15.25 m) between structures on adjacent blocks is insured. This distance has proved adequate to prevent the spread of fire through highly protected properties.

Explosions, however, are the cause of about two-thirds of the damage and most of the loss of life in chemical plant disasters. Where an explosion is possible, the 50 ft (15.25 m) distance does not provide sufficient protection. One solution has been to place any unit involving an explosion hazard at least 50 ft (15.25 m) away from the block perimeter. This gives a minimum distance (in the adjacent blocks) of 100 ft (30.5 m) from an explosion hazard to a fire hazard or 150 ft (45.72 m) between explosion hazards. In the case of equipment explosions resulting from an increase in internal pressure, the shock wave would probably be adequately attenuated by that distance. This is true whether the explosion is of the thermal type, represented by an ordinary runaway reaction and failure of the container, or of the shock type, where the pressure increase is caused by a reaction at supersonic speed. In the latter case, the rapid pressure increase could produce shrapnel, so a containment barricade would be needed. If the material that might cause the explosion is a true explosive, whether formed intentionally or accidentally, and whether inside or outside of equipment, the spacing then should be in accordance with Table 10-10A, the American Table of Distances for Storage of Explosives. (See also NFPA 495, *Code for the Manufacture, Transportation, Storage, and Use of Explosive Materials.*)

Mr. Bradford is a loss prevention consultant. He is a registered professional engineer in Connecticut and Rhode Island and is a member of AIChE and SFPE. He serves on several NFPA technical committees.

TABLE 10-10A. General Recommendations for Spacing in Petrochemical Plants

Minimum Distance in Feet*	Process Unit—HH	Process Unit—LH	Tank Farms—HH	Tank Farms—LH	Product Whses—LH	Ship'g. & Rec'g.—HH	Ship'g. & Rec'g.—LH	Service Buildings	Boiler Area	Fire Pumps	Emergency Controls	Water Spray Controls	Turret Nozzles	Emergency Flares	Pilot Plants	Large Cooling Towers	Fire Hydrants	Fired Process Heaters
Process Unit—(High Hazard)[B]	200[A]									250	100	50[5]	50–100 to Center of Target	For 100' Flare that is 25' above Surrounding Equipment, Use 300'	200	150	50 to 250	50 to 100
Process Unit—(Low Hazard)	100[A]	50[A]								150	50	100[5]			200	100		50
Tank Farms—(High Hazard)[C]	250[1]	250[1]	1½ dia. larger							250		100[5]			250	250		200
Tank Farms—(Low Hazard)	200[2]	100[3]	1 dia. larger	½ dia. larger						200					200	200		200
Product Warehouse (Low Hazard)[D]	150	50[4]	250[1]	100[3]	50[4]					200					200	150		100
Shipping & Receiving (High Hazard)[E]	200	200	150[2]	100[3]	150	50				150	100	50[5]			200	200		200
Shipping & Receiving (Low Hazard)	150	100	100	50	20	50	—			100	50				150	150		100
Service Buildings[F]	200	100	200	100	100	150	100	See bldg. chart		100					200	100		100
Boiler Area	200	150	200	150	100	200	100	100	—	—					200	100		100

Recommended Spacing Within Process Units

	React.	Comp.	Tanks	Fract. Equip.	Cont. Rooms
Reactor	25[6]				
Small Compressor House or Pump House	40[6]				
Intermediate Storage Tanks High Hazard Rundown-Feed	100 to 200	100 to 200	17 dia.		
Fractionation Equipment	50	30	100		
Control Rooms[G]	50[6] to 100	50 to 100	100	50 to 100	10

Notes:

(A) Distance between process units is measured from battery limits.

(B) A high hazard process unit includes those with high or extreme explosion hazards.

A process with high explosion hazard is one where there is considerable explosion potential or probability with susceptibility to considerable damage or serious delays in restoration. Examples are: acetylene purification; acrylonitrile formation from hydrogen cyanide and ethylene oxide; acrylonitrile formation from hydrogen cyanide and acetylene; alkylation (using acetylene); partial oxidation of low flash point (below 110°F) (43°C) liquids, such as oxidation of acetaldehyde to acetic acid; vinyl acetate formation from acetylene and acetic acid; vinyl chloride formation from acetylene and hydrogen chloride; and ethylene oxide production, recovery and purification.

A process with an extreme explosion hazard is one employing extreme pressues or is one that is unpredictable and not susceptible to design that will limit the extent of the damage or interruption to minor proportions. These processes, while not susceptible to exact description, can generally be recognized by engineering analysis. A process that uses very high pressures in equipment properly designed for the purpose and in connection with processes of predictable hazard (a synthetic ammonia plant is an example) does not of itself require classification as an extreme explosion hazard. Examples of processes with extreme explosion hazards are: acetylene compression when the partial pressure is 20 psia (139kPa) or higher; storage of organic coated ammonium nitrate or ammonium nitrate with an additive; acrolein drum storage; acrolein recovery and purification and acrolein barreling.

(C) High hazard tanks are those that contain liquids having a flash point below 110°F (43°C) or are tanks or gas holders containing flammable gases other than anhydrous ammonia (for hazard rating purposes, anhydrous ammonia is classified as a liquid having a flash point above 200°F (93°C). Tanks containing unstable liquids or gases, such as acrolein, acrylonitrile, ethylene oxide, hydrogen cyanide, styrene, vinyl acetate and vinylidene chloride require special consideration unless the materials contain stabilizing inhibitors known to be effective.

(D) High hazard product warehouses contain unstable materials, low-flash flammable liquids, or high combustible solids. These require special consideration.

(E) High hazard shipping and receiving denotes stable materials with flash point below 110°F (43°C). High hazard shipping and receiving of unstable materials requires special consideration.

(F) Service buildings include offices, gate houses, change houses, laboratories, shops, garages, maintenance warehouses, cafeterias, hospitals, etc. Experimental laboratories classify as process units.

(G) Control houses serving unusually large or hazardous units and central control houses for multiple units or housing computer equipment, require greater spacing and may require blast-resistant construction.

(1) For specific vertical tank, use 5 diameters.

(2) For specific vertical tank, use 4 diameters.

(3) For specific vertical tank, use 3 diameters.

(4) Standard firewall and sprinklered warehouse acceptable. Limit warehouse to maximum 25,000 sq ft (2322.5 m^2) floor area.

(5) Two stations desirable.

(6) Barricades desirable for hazardous reactors.

(7) Over 100,000 gal (378.5 m^3) requires special consideration.

General Notes:

Keep open flames 100 feet (30.5 m) from vapor hazard area.

Deviation from the distances in this table requires superior construction or special protective installations such as fixed foam systems, water spray systems, automatic sprinklers of a water spray from hose lines capable of supplying about 2500 gpm (9.5 m^3) to the points of need and a trained brigade adequately manned to use the water spray.

In borderline cases, high value requires high hazard classification.

Vertical storage tanks should be individually diked. If not, capacity in a single dike should not exceed 25,000 barrels. For horizontal storage tanks, maximum is 400,000 gal (1514 m^3) per group, with 100 feet (30.4 m) between groups, or other suitable arrangement.

*one ft equals 0.304 m

More conservative views on spacing are taken by the American Oil Company. Its *Safety Guide* says: "Generally, a distance of 250 ft (76.2 m) between units or between units and tankage has been used as a desirable spacing" (AOC 1964). Also, Table 10-10A gives the recommendations of the Industrial Risk Insurers (IRI) with respect to petrochemical plants. Their booklet also suggests distances for separation of units in refineries, gasoline plants, terminals, oil pump stations, and offshore property (IRI no date).

The greater distances suggested by the American Oil Company and IRI are justified by the large amounts of flammables usually present in equipment and piping in oil refineries and petrochemical plants. They are probably adequate for controlling even large flammable liquid fires.

For another type of explosion hazard, no spacing standards exist. This is the explosion which results when a flammable gas or super heated flammable liquid is released into the atmosphere and mixes with air to form a cloud with a volume in the explosive range. Such a cloud can release some 2 to 10 percent of its heat of combustion in the form of explosive energy. In estimating potential losses from vapor clouds, IRI uses the 2 percent figure. This decision was reached after studying the results of 79 cases of vapor cloud formation (Davenport 1977). In 7 cases there was no ignition, in 17 there was ignition but no pressure was generated. The remaining 55 cases formed the data base.

If a hydrocarbon releases 20,000 Btu per lb (46.5 mJ/kg) when burned and 2 percent of that, or 400 Btu per lb (930 kJ/kg) were released explosively by ignition of its mixture with air, then 5 lb (23 kg) of hydrocarbon would release 2,000 Btu per lb (4.65 mJ/kg) explosively—the same as 1 lb (0.45 kg) of TNT. Where such releases are possible, spacing is not a reliable defense against damage since no one can predict where the cloud will drift before it is ignited.

Procedures used to estimate damage from explosion pressure resulting from ignition of a vapor cloud are:

1. Estimate the volume of vapor produced by the release.
2. Assume that a percentage of the heat of combustion of this volume will be released as explosion energy.
3. Equate this energy to an equivalent amount of TNT.
4. Estimate damage, based on the explosion of that amount of TNT.

This is discussed briefly by Davenport (Davenport 1978). More extensive comments, with a table of pressure ranges at which different units of equipment may be expected to fail, have been made by Nelson (Nelson 1977). A discussion by Dr. D. J. Lewis of the Mond Division of Imperial Chemical Industries, appears in Vol. 13 of *Loss Prevention*, a CEP technical manual, published by the American Institute of Chemical Engineers (AIChE). This covers the situation of the drifting vapor cloud and also assigns different degrees of hazard to clouds made up of materials with different degrees of reactivity.

Other Means of Loss Control

As chemical plants process larger quantities of materials, it becomes impractical to provide ever increasing separation of units. Where toxic, flammable, reactive, or otherwise hazardous materials may be spilled, the logical approach is to:

1. Minimize the possibility of uncontrolled spills.
2. Minimize the size of a possible spill (if spills do occur, design features should keep them small).
3. Maximize the spread of a possible spill (if spills are large in volume, design features should keep them confined).
4. Prepare alarm and evacuation plans if toxic releases can occur.
5. Control sources of ignition.
6. Provide protection for exposed property if ignition does occur.

EXPOSURE PROTECTION

Automatic fire control and extinguishing systems are the first line of defense against fire and explosion emergencies. However, if an explosion hazard exists, system designs would require barricades for protection of important components such as deluge valves and dry pipe valves (Rinder and Wachtell 1967). If they might be exposed to explosion pressure, however improbably, process control houses, including areas where people congregate (change houses), should be of explosion resistant design (Bradford 1967; CMA no date; Eggleston et al 1976; Lawrence 1974). The procedure suggested by this author (Bradford and Culberkon 1967) is based on resistance to static pressure; that in the Chemical Manufacturers Association's *Safety Guide* (CMA no date) is based on elastic response to a timed pressure impulse. Lawrence gives a concise exposition of the design theory (Lawrence and Johnson 1974). Reliably available water at pressure and volume adequate to supply the maximum foreseeable fire fighting and exposure protection demand for a minimum of four hours is desirable. Of course, built in or added fire resistance can give excellent exposure protection.

Protection for a chemical plant generally requires a specialized fire department where each fire fighter is equipped with full protective clothing, including self-contained breathing apparatus, and where acid resistant hose and special extinguishing equipment are available. NFPA 49, *Hazardous Chemicals Data* (hereinafter referred to as NFPA 49), is an excellent source for information on personal protection measures needed to fight fires involving different chemicals as well as the tactics that should be used in emergencies. Appropriate symbols from NFPA 704, *Standard System for the Identification of the Fire Hazards of Materials*, posted at locations on the property where chemicals are present, can give helpful guidance on inherent hazards of materials present and planning effective fire fighting operations.

IGNITION SOURCES

The general notes in Table 10-10A recommend that open flames be kept at least 100 ft (30.5 m) from any hazardous area. This note contemplates fixed equipment; however, it could also apply to maintenance work. A better approach is to prohibit open flames or other hot work in all but predetermined specified areas, except where the hot work to be done has been examined and found safe and the work is done under a permit that limits scope and duration. It must be remembered that piping for high pressure steam, heat transfer media, and hot process streams may need insulation to keep it from becoming a source of ignition. When piping operating at high temperatures is insulated, it may be necessary to leave flanges uninsulated. If the bolts joining flanges get too hot, their thermal expansion will relax the compression on the joints and may lead to leakage.

CONTROL OF SPILLS

Large flammable liquid spills should be confined where they may occur. Although retaining dikes are often used, they may not prevent liquid around the spill source from burning. Equipment may be damaged further or the spill aggravated unless efficient, prompt, and preferably automatic fire control measures are taken. A better approach is to provide drainage to an impounding basin located where a fire, if not extinguished, can burn out harmlessly or a nonburning toxic material can be appropriately neutralized. A spill is often ignited at some time during the drainage process, so care must be taken to avoid damage during drainage. If the sewerage system is underground and is not provided with liquid seals at all entrance points, explosions in the system could result. Fire in open drainage ditches could damage anything without adequate protection that is located alongside the ditches or which cross them, such as utility or piping systems. Problems associated with such drainage can be avoided by using the trench system described in NFPA 15, *Standard for Water Spray Fixed Systems for Fire Protection.*

Total confinement of a large flammable gas or vapor spill is not possible, due to the gaseous nature of the spill. Hose lines or monitors with spray nozzles can be used, because of the air entrained by the spray, to dilute the gas cloud and to move it in a desired direction until it dissipates. Fixed water spray nozzles may be used to set up water curtains which prevent a cloud from drifting to a furnace or other ignition source. Units that have been effective in experiments with small, low level vapor releases were discussed in a number of articles in *Loss Prevention* magazine in 1976 (Eggleston 1976; Watts 1976). The use of steam nozzles to dilute and direct vapors upward at an installation in England has been reported (ICI 1971). There are problems with this approach. Large steam capacity must always be available; and there is the need for meticulous grounding of all metallic material that might be charged by steam passing over it or condensing on it, in order to avoid static sparks that might ignite material being dispersed.

Keeping Spills Small

Hazardous spills can be kept small by keeping the amount of hazardous material used in the process small. Where this is impractical, valves to isolate all large quantities of material should be provided. These valves should be installed at each outlet of any large container through which material might escape (except at relief devices). They can also be used to subdivide long runs of large diameter pipe. The valves should be firesafe, i.e., they should not leak appreciably when exposed to fire after being closed, and they should be arranged to operate both remotely and at the valve. If operated by electric motor, both the power wiring to the motor and the signal wiring to the starting switch should be arranged to remain functional during any fire or other emergency requiring valve operation, for at least as long as it would take to discover the emergency, transmit the alarm, and operate the valve. A minimum of 15 minute fire protection for wiring is suggested. If valves are held open by instrument air pressure and closed by spring or gravity when air pressure is lost, plastic tubing can be used in the air line, at least near the valve. The plastic will melt quickly in a fire, and the valve will close before it is damaged. Since it is necessary to detect spills before ignition, flammable vapor detectors of the diffusion head type may be used for that purpose (Johnson 1974).

The distinction between a small spill and a large one must be fixed individually for each plant. A small spill is one that can be easily handled by the exposure protection and confinement means just discussed. All others are large spills. In the case of jets of liquefied flammable gas (Burgess and Zabetakis 1973; Goforth 1970) or hot flammable liquid (Kletz 1975) where spacing is not a reliable defense against explosion damage, various quantities have been proposed as limits that must not be exceeded. One organization suggests 10,000 lb (4536 kg) of gas or vapor as a maximum permissible spill. Another suggests a maximum permissible leakage rate as one that would permit formation of no more than 1,000 lb per min (454 kg/min) of flammable gas or vapor.

Preventing Spills

Because many chemicals cause unusual corrosion or abrasion problems, or their processing requires unusually high or low temperatures, chemical plants should be constructed of appropriate materials and should follow good maintenance practices. Assuming that there are no problems in these areas, the reaction systems themselves may cause trouble.

In general, reactions are used to produce desired commodities that differ chemically from the raw materials from which they are made. The differences are produced by controlled chemical changes. Uncontrolled changes tend to produce both unwanted materials and more or less violent system failures that result in spills. The common denominator in these failures is overheating; whether overheating of a container so that it softens and fails, or overheating of the contents so that the container bursts from too much pressure. Accidents of these types result from a failure to move heat adequately from a heat source to a heat absorbing system or to a coolant from a heat releasing system. Heat is usually produced when materials dissolve, crystallize, condense, or are adsorbed. Heat is also produced mechanically by crushing, grinding, milling, compressing, and pumping, and by exothermic chemical reactions. Heat transfer is also required when heat is needed for evaporation, melting, decreasing viscosity, drying, desorbing, distillation, warming expanding gas, initiating exothermic chemical reactions, and driving endothermic reactions. In any case, loss prevention requires an understanding of heat transfer.

HAZARDS OF HEAT TRANSFER

When a hot fluid, gas, or liquid transfers heat to a cooler fluid through a barrier such as the metal wall of a vessel, the resistance to heat flow, as shown by a temperature drop, is mainly in fluid films on the metal surfaces rather than in the metal. (See Fig. 10-10A.) Flow in the heating and heated fluids is usually turbulent with invisible swirls and eddies; but in the films, it is laminar with thin layers of molecules sliding slowly over each other. The film layer nearest the metal moves slowest and its temperature is nearest that of the metal. This is important because high temperature causes decomposition of most organic and some inorganic fluids. Although a large temperature differential between the fluids moves heat faster than a small one, the hottest part of the heated film must be kept cool enough so that it does not decompose and form a solid film. When a solid film builds up, the temperature of the metal on the heated side may go so high, trying to force the heat through, that the metal is damaged and a "burnout" occurs.

Precautions consist of methods to detect the fluid breakdown such as checking the fluid for tar, solids, or significant change in viscosity or volatility. Where the heater is of a single tube design, an internal buildup will cause an easily observable increased pressure drop as the fluid flows through the tube. For this reason, single pass tube heaters are to be preferred over those of multiple fluid path design.

When the breakdown of the fluid is intentional, as in the case of cracking a petroleum fraction, the internal

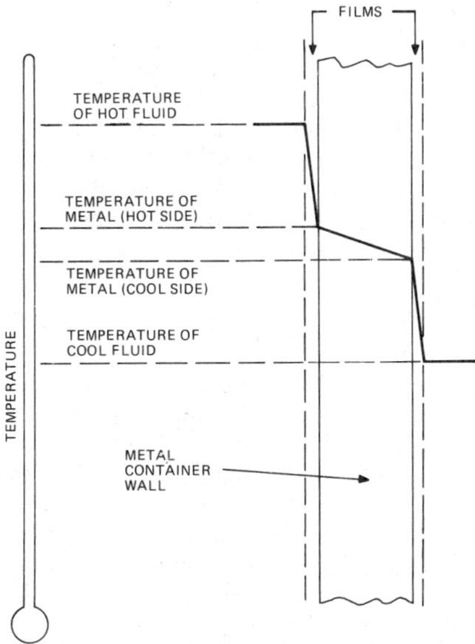

FIG. 10-10A. The effect of fluid films on temperatures at the surfaces of the metal wall of a vessel.

deposits are removed on a scheduled basis before they reach dangerous thickness. This is usually done by burning them out with air or a mixture of air and steam (Armistead 1950). Requirements for safety involve:

1. Purging the unit of flammables, usually with steam, before introducing the oxidizing atmosphere.
2. Purging after burning out the deposits.
3. Safe disposal of the materials purged.
4. A system of valving, usually involving double valves with a purged or vented space between, to insure against mixing air and combustibles.

If the heated fluid is a liquid and starts to boil, heat transfer improves as the film is thinned by the action of bubbles of vapor forming on the metal, growing, and breaking away. This formation of bubbles on the heated surface is called "nucleate boiling." If the temperature of the heat source is then raised in an attempt to increase the rate of boiling, the temperature may become too high. Vapor bubbles then coalesce into a vapor film, instead of breaking away individually. This is called "film boiling." Vapor films offer more resistance to heat flow, so, again, "burnout" may occur.

Endothermic Reactions

Endothermic reactions are most commonly supplied with heat by combustion processes. This energy may be truly direct, i.e., cracking natural gas to acetylene by burning part of it with oxygen and quenching suddenly, concentrating a liquid by submerged combustion (burning a premixed gaseous fuel beneath the surface of a liquid), or roasting ore or cementitious material in a kiln. It may also be first degree indirect (usually and hereinafter called direct since the container, tube, or kettle is directly exposed to the flames); or second degree indirect (usually and hereinafter called indirect) where the flames heat a heat transfer fluid. In this process, the use of tempered flue gas constitutes indirect heating.

The same general hazards exist whether the material being heated is a reacting system or one being decomposed or heated. All cases require consideration of the metallurgical and construction characteristics of the equipment. These problems are accentuated in the case of direct fired equipment. The heater must resist corrosion, pitting, and scaling from the materials being heated as well as from combustion on the firebox side, and must often do so at temperatures much higher than is the case with conventional heat transfer media. Construction must be such that heated and cooled portions can expand and contract freely. This is particularly important if the alloys used enter or pass through a brittle phase at any stage of heating or cooling.

Two more important, but sometimes overlooked, points are:

1. The metal of the heater and heat transfer system and the fluids with which they are in contact must be compatible. The common molten salt heat transfer medium, a mixture of sodium nitrate and sodium nitrite used in indirect fired heaters, will support the combustion of steel if the temperature is high enough and ignition occurs (NBFU 1954); steel and copper in thin sections, as on the fins of heat transfer tubes, can ignite easily in hot chlorine.
2. The heat transfer medium should not create a hazard if it leaks into the material being heated or vice versa. Where leakage one way but not another can be temporarily tolerated (sulfuric acid into water as opposed to water into sulfuric acid, for example), pressures of the coolant and the cooled material must be suitably different. Alternatively, annular tubes with the annulus separating the two fluids may be used with means to detect leakage into the annulus from either direction. To promote heat transfer, the annulus may be filled with a conducting liquid inert to both fluids or, as in "still" tubes, intermittently bridged with heat conducting metal wires or perforated discs.

The preceding discussion applies equally well when the heat transfer is from an exothermic reaction to a fluid used for cooling.

It is impossible to be all inclusive about the hazards of heat transfer. Two examples, however, may indicate the scope of the problem. In the first example, electric induction heat was being used on a cast iron vessel with thermostatic controls. Cast iron growth (caused by precipitation of carbon from the heated cast iron) caused intercrystalline cracking. The cracks interrupted the flow of the induced electric currents throughout the vessel, the thermostats called for more heat, and the vessel eventually failed from localized overheating. In another instance, sodium silicate was condensed on a stainless steel pipe where it acted as a flux to remove one of the alloying elements so that the pipe lost its corrosion resistance and failed.

Exothermic Reactions

Proper application of heat transfer principles is essential to the control of exothermic reactions. Such use requires knowledge of the reaction system. First, it is necessary to know the thermal stability and shock sensitivity of the system and each of its components, raw materials, intermediates, products, and byproducts. Second, the reaction kinetics must be known; that is, the total amount of heat released and how fast it is released. For completely safe process design, it is necessary to have all of this data under conditions that would exist if the reactants were supplied in grossly wrong proportions, in the wrong order, without mixing (both with and without a subsequent late start of the mixing system), at the wrong temperatures, if power failed, if a coolant line broke, or if any other upset condition imaginable should occur.

STABILITY AND SHOCK SENSITIVITY

Most chemicals are reasonably stable and insensitive. Where this is not true, data usually is available from the supplier or in the literature. NFPA 49, discusses specific chemicals, and NFPA 491M, *Manual of Hazardous Chemical Reactions*, covers hazardous systems. Both of the standards provide lists of references. Where data are not available, thermodynamic calculations and laboratory scale stability testing should be used (Coffee 1969, 1973; Davis and Ake 1973; Stull 1973; Trewick et al 1973; Way 1969). Part 41 of the 1979 ASTM Standards contains test procedures developed by Committee E-27 (ASTM annually). Where explosive materials are involved, the hazard may be reduced to an acceptable level by diluting the explosive material in an inert chemical or by providing barricades around the explosive materials.

If the instability hazard is less than explosive, there are various ways to reduce it. Chemical reactions progress faster as the temperature rises, so the methods used involve keeping the material from being heated to an unsafe level. For example, the boiling point of a liquid is lowered by vacuum. Vacuum or freeze drying can avoid overheating. A high boiling material that becomes unstable at or near its boiling point may have steam bubbled through it at a temperature below the boiling point. Vapor from the material goes along with the steam and separation can be made after condensing the steam-vapor mixture. Viscous materials can be concentrated by running them down the heated interior walls of a wiped film evaporator. Mechanically operated blades keep wiping the interior surface and the film cannot thicken, so the film layer nearest the hot wall will not overheat. (See Fig. 10-10A.) Liquid nitrogen or dry ice may be added to material going to a grinder. This not only takes care of the heat produced mechanically but also tends to inert the atmosphere and reduce chances of a dust explosion.

In some cases, pressure may affect stability. Liquids, such as propargyl bromide and nitromethane, will boil away when heated at atmospheric pressure. If heated under pressure, so the boiling point is raised, they sometimes explode. Pure gases such as acetylene, ethylene, and nitrous oxide can be exploded under sufficient pressure. The testing described earlier should be thorough enough to detect such a possibility. Also, when considering systems, one should keep in mind that pressure may cause a phase change. For example, chlorine gas liquefies fairly easily, and a system easy to control when using the gas may be hard to control if the liquid is accidentally formed.

REACTORS

Reaction systems are of two types: continuous and batch. A continuous system may be thought of as a pipe, although it may contain pumps, compressors, bulges that are reactors, etc., with the raw materials and possibly an inert carrier flowing in at one end, or at appropriate intervals along the pipe with products and byproducts flowing out the other end. In a batch process, the chemicals are added to a single unit all at once or in appropriate quantities at appropriate time intervals; then the reaction takes place, the unit is emptied, and the process is repeated. Batch process systems have more flexibility in use than continuous systems.

Continuous Reactors

It is common to have one reactant, pure or diluted, circulating continuously throughout the reactor with another being added, reacting, and its products removed by condensation, washing, filtering, or similar appropriate means. In normal operation, the major safety control is automatically cutting off the appropriate feed stream if the heat transfer system fails. Arrangements are often made for alternate means of supplying the heat transfer fluids. Reliability of the shutoff system may be enhanced by using two valves in series with a vent between them. (This is known as the double block and bleed system.) Depending on the complexity of the system, this same shutoff may be actuated by changes in other process conditions. An example of a process condition change might be too low a temperature, which would indicate that the reaction was not going properly. This signals that concentrations of unreacted materials may be building up to the point where they could suddenly all react with explosive violence and overpower the means of heat removal, or react in a part of the system designed to recover product rather than control the reaction. Also, when starting up or shutting down, the continuous reactor may go through situations that are more hazardous than its normal operation.

Batch Reactors

The essential parts of a batch reactor are:

1. A vessel to contain the reaction at its maximum pressure.
2. A heat transfer system to control the temperature of the reacting mixture.

3. A stirrer or agitor to keep the temperature and composition of the reacting mixture uniform.
4. A relief system to protect the vessel against overpressure.

The Vessel: This may range from an open wood or rubber lined steel tank in which bauxite ore reacts with sulfuric acid to produce a solution of alum to a glass lined or stainless steel pressure vessel. Regardless of size or configuration, the vessel used as a batch reactor will probably have to be entered for cleaning, while the continuous reactor can often be cleaned automatically with a cleaning fluid. Cleaning or entering vessels will require special precautions if flammable cleaning agents are used or if vapors or residues make entry hazardous.

Heat Transfer Methods: Common methods of heat transfer use jackets or coils on the exterior of the vessel or coils on the interior. Others circulate the reacting mixture from the vessel through an external heat exchanger and back to the vessel. A third method allows the reacting mixture to boil. The heat of vaporization then is removed by condensing the vapor externally, which allows the condensed liquid to run back and continue the cycle. This latter method does not provide means to heat the material in the vessel, heat that may be needed to start the reaction even though cooling is needed as the reaction progresses.

Mixing: This is usually performed with an agitator mounted on a shaft driven by an external motor. If an external heat exchanger is used, the mixture leaving it may be jetted back into the vessel to accomplish mixing. Where heat sensitive materials are involved and it is necessary to avoid moving parts because they might rise to a dangerous temperature due to friction, mixing can be done by bubbling in air or inert gas.

Pressure Relief of Batch Reactors

Batch reactors in which an explosive atmosphere might sometimes exist are usually designed to contain the pressure that might result from ignition, or they are purged or inerted to eliminate the hazard. If vessels are constructed according to the Pressure Vessel Code of the American Society of Mechanical Engineers (ASME), they are required to have a relief device adequate to keep the vessel from being overpressured because of overheating from an external source, including an exposure fire, normal heating, or the introduction of fluids at pressures higher than those for which the vessel is designed. The code gives no guidance for the relief of pressure caused by heat of reaction.

Most chemical reactions double in speed with each rise in temperature of 18°F (10°C). This means that a heat removal system would have to remove twice as much heat at 140°F (60°C) as at 122°F (50°C), four times as much at 158°F (70°C), etc. For this reason, it is unsafe to run a batch reaction at a temperature more than 32°F (14°C) above that of the cooling medium (Boynton et al 1959). If the cooling system is overpowered, the reaction keeps increasing in speed, the retained heat builds up pressure, and the pressure must be relieved or the vessel will explode. Austin has given a good explanation of this problem (Austin 1965). The Dow Chemical Company's *Guidelines for Process Scale-up* asks the following questions about runaway reactions (Kline 1974):

Can You Prevent Runaway Reactions?	Yes	No
By adequate heat transfer	____	____
By quenching the reaction	____	____
By stopping feed streams	____	____
By dilution of the reactor contents	____	____

Score: Two or more methods—You're in good shape.
Only one method—Better find a second line of defense.
None of the above—You'd better stop right now and reconsider your design for safety.

However, since even multiple safeguards may fail, emergency venting is supplied as last ditch protection for the vessel. For example, loss of mixing could make all the suggested preventive measures inoperative or dangerously inefficient. Materials that have leaked into the hollow shaft of an agitator have created pressure and spewed materials into the main reaction system, which speeds up the main reaction so that it overpowers the cooling system. Other problems were discussed earlier in this chapter in the section "Exothermic Reactions."

A satisfactory empirical solution to the problem of determining an adequate vent size for a stirred batch reactor ordinarily may be reached by using the diagram in Figure 10-10B (Sestak 1965). Suitable vent sizes are presented as a range of values rather than as one specific figure. Values are selected from the upper part of the range if the vent pipe from the vessel to the outside is long or contains bends, or if the vent device opens at more than one-third in excess of the absolute pressure at which the reaction normally proceeds. Taking atmospheric pressure as 15 psia (103 kPa), a 5 psia (35 kPa) rupture disc is the maximum that should be used with the suggested vent sizes for an atmospheric reaction. A limitation on the data in Figure 10-10B is that the diagram was drawn with a conventional jacketed batch kettle in mind. Such kettles are usually designed for operation from full vacuum to 60 to 70 psig (413 to 483 kPa) internal pressure. Thus the chart may be inadequate and should not be used if the normal reaction pressure exceeds 100 psig (690 kPa).

There is considerable literature on venting batch reactors. Boyle has pointed out that the material vented is not a gas but a multiphase mixture, requiring experimental data for proper vent system design (Boyle 1967). Harmon and Martin have reported valuable work along these lines (Harmon and Martin 1970). Huff has developed a computerized approach to the problem of polymerization reactor venting (Huff 1982). His research establishes the chart in Figure 10-10B as reasonably accurate, within its limitations, and expands on the work of Boyle, Harmon, and Martin. Ogiso, Takagi, and Kitagwa point out that the sudden venting of a low viscosity superheated material (in this case, water) causes an instantaneous pressure reduction to almost atmospheric pressure, followed by the onset of nucleate boiling throughout the system (Ogiso et al 1972). This sudden expansion produces an explosive effect, related to water hammer. A committee for the Design Institute for Emergency Relief Systems (DIERS) has been organized within the framework of AIChE and contributions (dues) are pledged from international sources (manufacturing, insurance, and research) associated with the chemical industry. Funds well in excess of $1 million dollars are being used to conduct research into the problem of adequate venting for batch reactors. As of this

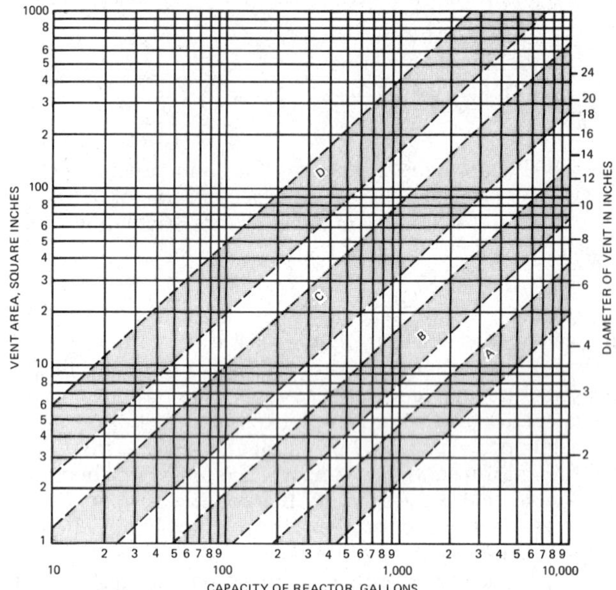

FIG. 10-10B. Vent Sizes for Batch Reaction Vessels. Line A represents endothermic reactions or reactions with a very low exotherm (low heat release). Line B represents reactions with a low heat release per volume of reaction mass. This would include such reactions as suspension polymerizations of vinyl chloride, styrene, butadienestyrene, etc., where 50 percent or more of the material within the reator does not take part in the reaction, being present as a diluent only. Line C represents reactions with moderately high exotherms, such as methyl methacrylate and styrene bulk polymerizations, mono nitration of benzol, toluol, etc., and Friedel-Crafts reactions. Line D represents reactions with extraordinarily high heat release such as oxidations using nitric acid, bulk polymerizations of ethyl acrylate, or methyl acrylate, caustic catalyzed phenol-formaldehyde condensations, etc. In these cases reactors of over 1,000 gal (3.78 m^3) capacity should be avoided.

writing, this work is essentially complete and the results will be made public very soon.

Safety Instrumentation

Safety instrumentation will vary from process to process and plant to plant. There are, however, a few general rules (Doyle 1972):

1. Measure, as directly as possible, the variables of interest. For example, if an agitator is driven by an electric motor, a relay can be used to show that the switch to the motor is closed. An ammeter would be better because that would show current is flowing; a wattmeter would be best because it would show that work is being done and that the impeller has not fallen off the agitator shaft.
2. Reliability and maintenance must be of high quality, e.g., if an external fire could upset the reaction, the warning instrument should not be disabled by the fire. Since safety instruments will not have to operate often, frequent checks must be made to ensure they are in operating condition.
3. Provide redundancy or diversity in the system. If temperature is critical, provision of two thermocouples may lead to two opposite readings. It is best to provide three and trust the majority of the readings. However, since a temperature increase is usually accompanied by a pressure increase, one pressure sensor and one thermocouple may be used, with credence given to the more pessimistic report.

CHEMICAL PLANT OPERATIONS AND EQUIPMENT

The generally accepted unit physical operations require equipment to carry out the following functions:

1. Heat Transfer.
2. Fluid Flow.
3. Crushing and Grinding.
4. Mixing.
5. Mechanical Separation.
6. Distillation.
7. Evaporation.
8. Crystallization.
9. Filtration.
10. Absorption.
11. Adsorption.
12. Drying.

Descriptions of individual pieces of equipment can be found in most engineering handbooks.

Heaters and Coolers (Heat Transfer): These devices are either: (1) direct (flue gas, kilns, spray cooling, vaporization), or (2) indirect (coils and jackets). Heat transfer media usually work by boiling to absorb heat or condensing to give it up, but also may work just by becoming hotter or cooler without phase change. Hazards of heat transfer have been discussed previously in this chapter.

Fluid Flow: This is accomplished by: (1) fans and compressors, (2) pumps, or (3) vacuum jets. Fans, compressors, and pumps heat the materials they move because of mechanical work. Compressors require good aftercoolers. The liquid in pumps can boil, and the pump bearings fail from overheating if a pump operates against a closed discharge valve. If a centrifugal pump suction is inadequate, cavitation may cause the pump to chew itself up. Failure of vacuum devices may permit unwanted air to enter the equipment and cause the temperature in the equipment to rise.

Crushing and Grinding: Equipment for these operations include: (1) mills (impact, ball, hammer, roller, disk), and (2) crushers (cone, gyratory, jaw). All of these devices produce heat and may produce fine material capable of dust explosion. Ball mills may become overpressured as liquid portions of their contents are heated up. If gaskets fail and the liquids are flammable, a dangerous situation will result.

Mixers: Devices used in the mixing process include: (1) tumblers; (2) venturi mixers; (3) kneaders, rolls, and mullers; and (4) propellers, blades, and turbines. These devices also produce heat in varying degrees. When it is necessary to produce a hazardous mixture, like a blend of starch and an oxidizer for use as a flour ager, the mixer is open to avoid an equipment explosion. Any dust collector should be of the water wash type.

Mechanical Separators: These include: (1) cyclones, (2) bag filters (shake or blow back), (3) ore tables, (4) screens, (5) electrostatic precipitators, (6) flotation separators, and (7) centrifuges. The first five are usually dry type and the latter two wet, although there are liquid cyclones and wet screening processes. Devices used in dry service for combustible dusts should have explosion venting or an explosion suppression system. If collectors have internal combustible bags, an internal sprinkler system is desirable. It may not save the bags, but it will save the housing and mechanical devices. When materials wet with flammable liquids are centrifuged, the operation should be conducted under inert gas or with an explosion suppression system.

Stills: The types of stills are: (1) batch, (2) continuous, (3) pressure, (4) vacuum, and (5) steam. All distillations involve heat transfer. The major hazard of this operation is that a flammable vapor may be released to the atmosphere (Doyle 1974). The direct use of steam is discussed in the "Stability and Shock Sensitivity," section of this chapter.

Evaporators: The three types of evaporators are: (1) multiple effect, (2) vacuum, and (3) wiped film. All three use heat transfer. The multiple effect system condenses vapor from the first unit to heat the second, etc. The initial heat transfer medium is used in the first, where the material is most concentrated and boils at the highest temperature. Vacuums and wiped films have been discussed previously in this chapter.

Crystallizers: The two types of crystallizers are: (1) vacuum, and (2) pan. Both involve cooling to remove the heat of crystallization.

Filters and Agglomeraters: Filtration and agglomeration involve: (1) plate and frame filters, (2) Nutsche filters (vacuum), (3) drum type, (4) rotary cell type, (5) agglomerating tables, and (6) pelletizers. No unusual hazards are involved other than possible exposure of flammable liquids to the air.

Adsorbers: The two types of adsorbers are: (1) activated carbon, and (2) zeolites (molecular sieves). Adsorption produces heat just as absorption does. However, when the material being adsorbed is a flammable vapor or gas, the heat is retained in the small pores of the adsorbent. Heat may build up enough to cause fire in the activated carbon or exothermic polymerization of adsorbed reactive materials such as ethylene. This hazard can be minimized by wetting the adsorbent prior to each use.

Dryers: The six general types of dryers are: (1) spray and fluid bed, (2) vacuum, (3) tray, (4) belt, (5) drum, and (6) azeotropic. All these involve heat transfer. In the case of azeotropic dryers, when heat is removed by boiling, condensation, and return of the cool liquid, the unwanted component, usually water, can be discarded if the condensate forms a two phase system. All other types of dryers may present problems of mixing flammable vapors with air. Spray and fluid bed dryers may expose thermally unstable dry materials to hot heat transfer surfaces.

CONCLUSION

A summary of necessary knowledge and procedures to assure safe chemical processing has been suggested as follows (Dow no date):

1. Know the total reaction energy in the system.
2. Know the rate of energy release.
3. Evaluate thermal and shock sensitivity data.
4. Design the process to control the rate of energy release, and in doing so:
   a. Be alert for trace compounds or catalytic impurities which may accelerate reaction rates.
   b. Prevent buildup or concentration of high energy materials in the system. Calculate a material balance.
   c. Design into the process the ability to safely accommodate inadvertent releases.

Bibliography

References Cited

AOC. 1964. *Engineering for Safe Operations*. No. 8. American Oil Company (now Standard Oil of Indiana), Chicago.

Armistead, G. Jr. 1950. *Safety in Petroleum Refining and Related Industries*, 1st ed. John G. Simmonds & Co. Inc., NY. pp 118-119.

ASTM. Annually. *Annual Book of Standards*. American Society for Testing and Materials, Philadelphia, PA.

Austin, G. T. 1965. "Hazards of Commercial Chemical Reactions." *Safety and Accident Prevention in Chemical Operations*. Edited by H. H. Fawcett and W. S. Wood. John Wiley & Sons, NY.

Boyle, W. J. Jr. 1967. "Sizing Relief Area for Polymerization Reactors." *Loss Prevention*. Vol 1. American Institute of Chemical Engineers, NY. pp 78-84.

Boynton, E. D., Nichols, W. B., and Spurlin, H. M. 1959. "Control of Exothermic Reactions." *Industrial & Engineering Chemistry*. Vol 51, No 4. pp 489-494.

Bradford, W. J., and Culbertson, T. L. 1967. "Design of Control Houses to Withstand Explosive Forces." *Loss Prevention*. Vol 1. AIChE, NY. pp 28-30.

Burgess, D. S., and Zabetakis, M. G. 1973. *Detonation of a Flammable Cloud Following a Propane Pipeline Break*. RI 7752. USDI Bureau of Mines, Pittsburgh, PA.

CMA. *Safety Guide SG22, Siting and Construction of New Control Houses for Chemical Manufacturing Plants*. Chemical Manufacturers Association, Washington, DC.

Coffee, R. D. 1969. "Hazard Evaluation Testing." *Loss Prevention*. Vol 3. AIChE, NY. pp 18-21.

Coffee, R. D. 1973. "Hazard Evaluation: The Basis for Chemical Plant Design." *Loss Prevention*. Vol 7. AIChE, NY. pp 58-60.

Davenport, J. A. 1977. "A Survey of Vapor Cloud Incidents." *Loss Prevention*. Vol 11. AIChE, NY. pp 39-49.

Davis, E. J., and Ake, J. A. 1973. "Equilibrium Thermochemistry Computer Programs as Predictors of Energy Hazard Potential." *Loss Prevention*. Vol 7. AIChE, NY. pp 67-73.

Dow. *The ABC's of Reactive Chemical Processing*. Dow Chemical Company. Midland, MI.

Doyle, W. H. 1972. "Instrument Connected Losses in the CPI." *Instrumentation Technology*. pp 38-42.

Doyle, W. H. 1974. "Minimizing Serious Fires and Explosions in the Distillation Process." *Technology Report 74-2*. Society of Fire Protection Engineers, Boston, MA.

Eggleston, L. A., Herrera, W. R., and Pish, M. D. 1976. "Water Spray to Reduce Vapor Cloud Spray." *Loss Prevention*. Vol 10. AIChE, NY. pp 31-42.

Goforth, C. P. 1970. "Functions of a Loss Control Program." *Loss Prevention*. Vol 4. AIChE, NY. pp 1-5.

Harmon, G. W., and Martin, W. A. 1970. "Sizing Rupture Discs for Vessels Containing Monomers." *Loss Prevention*. Vol 4. AIChE, NY. pp 95-102.

Huff, J. E. 1982. "Emergency Venting Requirements." *Plant/Operations Progress*. Vol 1, No 4. pp 211-229.

ICI. 1971. *The Safe Dispersal of Large Clouds of Flammable Heavy Vapors*. Imperial Chemical Industries Ltd. Heavy Organic Chemicals Division, Billingham, England.

IRI. No date. *General Recommendations for Spacing*. Industrial Risk Insurers. Hartford, CT.

Johanson, K. A. 1974. "Gas Detectors by the Acre." *Instrumentation Technology*. Vol 21, No 8. pp 33-37.

Kletz, T. 1975. "Lessons to be Learned from Flixborough." *Loss Prevention*. Vol 9. AIChE, NY.

Kline, P. E., et al. 1974. "Guidelines for Process Scale-up." *Chemical Engineering Progress*. Vol 70, No 10. pp 67-70.

Lawrence, W. E., and Johnson, E. E. 1974. "Design for Limiting Explosion Damage." *Chemical Engineering*. Vol 81, No 1. pp 96-104.

Lees, F. P. 1980. *Loss Prevention in the Process Industries*. Vol 1. Butterworth & Co., Ltd., London, England. pp 210-230.

Melancon, C. L. 1980. "Improving Emergency Control and Response Systems." *Loss Prevention*. Vol 13. AIChE, NY. pp 43-49.

NBFU. 1954. "Potential Hazards in Molten Salt Baths for Heat Treatment of Metals." *NBFU Research Report RR-2*. American Insurance Association, NY.

Nelson, R. W. 1977. "Know Your Insurers Expectations." *Hydrocarbon Processor*. pp 103-108.

Ogiso, C., Takagi, N., and Kitagawa, T. 1972. "On the Mechanism of Vapor Explosion." *Proceedings of the First Pacific Engineering Congress*, Kyoto, Japan.

Rinder, R. M., and Wachtell, S. 1967. "Establishment of Design Criteria for Safe Processing of Hazardous Materials." *Loss Prevention*. Vol 7. AIChe, NY. pp 28-30.

Sestak, E. J. 1965. "Venting of Chemical Plant Equipment." *Engineering Bulletin N-53*. Factory Insurance Association (now IRI), Hartford, CT.

Stull, D. R. 1973. "Linking Thermodynamics and Kinetics to Predict Real Chemical Hazards." *Loss Prevention*. Vol 7. AIChE, NY. pp 67-73.

Treweek, D. N., Claydon, C. R., and Seaton, W. H. 1973. "Appraising Energy Hazard Potentials." *Loss Prevention*. Vol 7. AIChE, NY. pp 21-27.

Watts, J. W. Jr. 1976. "Effects of Water Spray on Unconfined Flammable Gas." *Loss Prevention*. Vol 10. AIChE, NY.

Way, D. 1969. "Fire Protection Engineering." *Loss Prevention*. Vol 3. AIChE, NY. pp 23-25.

NFPA Codes, Standards, Recommended Practices and Manuals. (See the latest *NFPA Codes and Standards Catalog* for availability of current editions of the following documents.)

NFPA 15, *Standard for Water Spray Fixed Systems for Fire Protection*.

NFPA 49, *Hazardous Chemicals Data*.

NFPA 68, *Guide for Explosion Venting*.

NFPA 491M, *Manual for Hazardous Chemical Reactions*.

NFPA 495, *Code for the Manufacture, Transportation, Storage, and Use of Explosive Materials*.

Additional Readings

Atallah, S., and Allan, D. S., "Safe Separation Distances from Liquid Fuel Fires," *Fire Technology*, Vol. 7, No. 1, Feb. 1971, pp. 47-55.

Bahme, Charles W., *Fire Officer's Guide to Dangerous Chemicals*, 2nd ed, National Fire Protection Association, Inc., Boston, MA, 1978.

Bartknecht, W., *Explosions*, Springer-Verlag, Berlin, Heidelberg, NY, 1981.

Bernhardt, Ernest C., ed., *Processing of Thermoplastic Materials*, Krieger Publishing Co., Melbourne, FL, 1974.

Bonyun, M. E., "Protecting Pressure Vessels with Rupture Discs," *Chemical and Metallurgical Engineering*, Vol. 42, May 1945, pp. 260-263.

Chappell, W. G., "Calculating a Pressure-Time Diagram for an Explosion Vented Space," *Loss Prevention*, Vol. 11, AICHe (American Institute of Chemical Engineers), NY, 1977.

Cocks, R. E., and Rogerson, J. E., "Organizing a Process Safety Program," *Chemical Engineering*, Vol. 85, No. 23 (1978), pp. 138-146.

Coffee, R. D., "Dust Explosions: An Approach to Protection Design," *Fire Technology*, Vol. 4, No. 2, May 1968, pp. 81-87.

Coffee, R. D., "Evaluation of Chemical Stability," *Fire Technology*, Vol. 7, No. 1, Feb. 1971, pp. 37-45.

Cousins, E. W., and Cotton, P. E., "The Protection of Closed Vessels Against Internal Explosions," *Paper No. 51-PRI-2*, American Society of Mechanical Engineers, New York, NY, 1951.

Creech, M. D., "Combustion Explosions in Pressure Vessels Protected with Rupture Discs," *Transactions*, Vol. 63, No. 7, American Society of Mechanical Engineers, NY.

Cubbage, P. A., and Marshall, M. R., "Explosion Relief Protection for Industrial Plants of Intermediate Strength," Institution of Chemical Engineers Symposium Series 39, London, England, Apr. 1974.

Davenport, J. A., "Explosion Losses in Industry," *Fire Journal*, Vol. 75, No. 1, Jan. 1981, pp. 52-56, 71-73.

Doyle, W. H., "Protection in Depth for Increased Chemical Hazards," *Fire Journal*, Vol. 59, No. 5, Sept. 1965, pp. 5-7.

"Explosion and Fire Hazards in the Storage and Handling of Organic Peroxides in Plastic Fabricating Plants," *SPI-FPC 19*, June 1964, The Society of the Plastics Industry, NY,

Fawcett, H. H., and Wood, W. S., eds., *Safety and Accident Prevention in Chemical Operations*, 2nd ed., Wiley–Interscience, NY, 1982.

Gibson, A. E., *Processing of Polymer Composite Materials*, Pergamon Press, NY.

Grace, C., "Fluid Choice Takes the Steam Out of Unsafe Process Heaters," *Process Engineering*, (5), 1977, pp. 85, 87, 88.

Halpaap, W., "Special Appliance for the Chemical Industry," *Fire International*, 5(56), 1977, pp. 44-50.

Henry, Martin F., ed., *Flammable and Combustible Liquids Code Handbook*, National Fire Protection Association, Quincy, MA, 1981.

Howard, W. B., "Efficient Time Use to Achieve Safety of Processes," paper presented at EFCE 4th International Symposium on Loss Prevention, Harrogate, England, Sept. 1983.

Joschek, H. I., "Risk Assessment in the Chemical Industries," Plant/Operations Progress, Vol. 2, No. 1, Jan. 1983, pp. 1-5.

King, R., "Plant Hazards," Engineering (London, England), Vol. 216, No. 4 (1976), pp. 277-279.

Kirk, R. E., and Othmer, D. F., eds., *Encyclopedia of Chemical Technology*, 3rd ed, 23 Volumes, Interscience Encyclopedia, Inc., NY, 1978, 1983.

Lewis, B., and Von Elbe, G., Combustion, Flames, and Explosions of Gases, 2nd ed., Academic Press, NY, 1961.

McElroy, Frank E., ed., *Accident Prevention Manual for Industrial Operations*, 7th ed., National Safety Council, Chicago, IL, 1980.

Nostrom, Gail P. II, "Fire/Explosion Losses in the CPI," *Chemical Engineering Progress*, Vol. 78, No. 8, Aug. 1982, pp. 80-87.

Pajgit, O., ed., *Processing of Polyester Fibres*, Elsevier Science Publishing Co., Inc., NY, 1980.

Perry, Robert H., and Chilton, C.H., eds., *Chemical Engineers' Handbook*, 5th ed, McGraw-Hill, NY, 1974.

Pilborough, L., *Inspection of Chemical Plants*, Gulf Publishing Co., Houston, TX, 1977.

"Protection Against Ignitions Arising Out of Static, Lightning, and Stray Currents," *RP-2003*, American Petroleum Institute, Washington, DC, 1967.

Runes, E., "A CEP Technical Manual," *Loss Prevention*, Vol. 6, AICHe, NY, 1972, pp. 63-67.

"Safe and Efficient Plant Operation and Maintenance," Chemical Engineering magazine, McGraw-Hill, NY, 1980.

Sax, N. Irving, *Dangerous Properties of Industrial Materials*, 5th ed., Van Nostrand Reinhold Company, NY, 1979.

Sommer, E. C., "Preventing Electrostatic Ignitions," paper presented at a meeting of the American Petroleum Institute Central Committee on Safety and Fire Protection, Tulsa, OK, Apr. 1967.

Steere, N. V., ed., *Handbook of Laboratory Safety*, 2nd ed., The Chemical Rubber Company, Cleveland, OH, 1971.

Verralin, C. H., ed., *Fire Protection Manual for Hydrocarbon Processing Plants*, 2nd ed., Gulf Publishing Company, Houston, TX, 1973.

Weiby, P., and Dickinson, K. R., "Monitoring Work Areas for Explosive and Toxic Hazards," *Chemical Engineering*, Vol. 83, No. 22 (1976), pp. 139-145.

Windholz, Martha, ed., *The Merck Index*, 9th ed., Merck Company, Rahway, NJ, 1976.

Zabetakis, Michael D., "Flammability Characteristics of Combustible Gases and Vapors," *Bulletin 627*, U.S. Bureau of Mines, PA, 1965.

GRINDING PROCESSES

Revised by Delwyn D. Bluhm, Ph.D., P.E.

This chapter discusses the fire hazards of grinding processes. Grinding (or pulverizing) is an industrial process in which materials are reduced to very small particles. Sometimes the grinding process of combustible and some normally noncombustible materials produces a highly explosive dust that is hazardous when dispersed in critical concentrations in air. A serious dust fire or explosion can begin either in the grinding equipment itself or in the environment surrounding the equipment. Examples of materials having combustible dust hazards are wheat flour, wood flour, sulfur, starch, coal, some plastics, aluminum, and magnesium.

It is important to distinguish the processes used to reduce the size of materials by grinding from those processes which employ abrasive wheels, disks, or drums to surface or shape articles of wood, metal or plastic. The latter processes could produce explosive dusts under certain conditions. However, those operations normally are conducted in environments where there is little likelihood that dust concentrations will reach the Lower Explosive Limit (LEL). This chapter, therefore, is concerned only with the hazards of milling or grinding operations involving large quantities of potentially explosive materials.

Section 4, Chapter 2, "Explosions;" and Chapter 4, "Therory of Fire and Explosion Control;" present guidelines for grinding hazards control. In this Section, Chapter 16, "Food Processing;" and Chapter 19, "Woodworking Processes;" will offer additional insight to the hazards of organic dusts. Section, 11, Chapter 7 comments on "Storage and Handling of Grain Mill Products."

GENERAL

Processes

Grinding operations are performed either wet or dry. Water is an excellent medium for wet grinding, though other liquids are sometimes used. Kerosene, for example, is used for the wet milling of magnesium. Because dust fires or explosions are the major hazards of grinding processes, dry grinding is emphasized in this chapter.

Materials

Agricultural products, such as wheat, corn, and other grains present explosion hazards both during storage and processing into flour or starch. Wood flour, finely divided sawdust, sulfur, coal, and some plastics present the same hazards as magnesium and aluminum. An explosion or fire can originate in the process equipment itself or in the ambient environment into which dust may escape and accumulate.

Actual occurrences, reinforced by laboratory tests, have demonstrated that finely divided particles of almost any readily oxidized material suspended in proper concentration in air will ignite and explode or burn. Table 10-11A classifies 59 different materials into 3 groups according to the range of explosibility. The table includes 9 different plastic dusts but does not cover the entire range of explosive plastic dusts. Table 10-11B gives the explosive characteristics of a variety of agricultural and industrial dusts. Materials other than those mentioned also may be explosive under proper conditions of particle size and oxygen availability.

Table 10-11A cites low, moderate, and high maximum rates of pressure rise for selected explosive materials.

The maximum rate of pressure rise is calculated from the steepest part of the pressure versus time curve. As with the maximum explosion pressure, appropriate adjustments need to be made if the dispersing air effectively increases the ambient pressure. The average rate of pressure rise is also calculated from the pressure versus time curve. An approximate correlation often exists between the average and maximum rates of pressure rise, but the latter is preferred when assessing practical needs. The average rate of pressure rise can be affected by slow development of the explosion and is usually one-half to one-third the value for the maximum rate of pressure rise.

The United States Bureau of Mines has developed three additional measures of the relative explosion hazard: the Ignition Sensitivity, the Explosion Severity, and the Index of Explosibility. Each of these is a dimensionless

Dr. Bluhm is Senior Engineer and Manager of Engineering for Ames Laboratory and Energy & Minerals Resources Research Institute operated by Iowa State University, Ames, IA.

TABLE 10-11A. Classification of Explosive Material According to Rates of Pressure Rise.

Class A Materials (Low maximum rates of pressure rise†)	
Metal Dusts	Miscellaneous Dusts
Antimony	Anthracite
Cadmium	Carbon Black
Chromium	Coffee
Copper	Coke, low volatile
Iron (impure)	Graphite
Lead	Leather
Tungsten	Tea
Class B Materials (Moderate maximum rates of pressure rise‡)	
Metal Dusts, or Powders	Polyethylene
Iron (carbonyl, electrolytic or H_2, reduced)	Polystyrene
	Urea Resins
Manganese	Urea—Melamine
Tin	Vinyl Butryal
Zinc	
	Miscellaneous Dusts
Grains, Spices, etc.—Dusts	Bituminous Coal
Alfalfa	Cork
Cocoa	Calcium Lignosulfonic Acid
Grain Dust and Flour	Coumarone Indene
Mixed Grains	Dextrin
Rice	Lignin
Soy Bean	Lignite
Spices	Peat
Starch	Powdered Drugs
Yeast	Pyrethrum
	Pyrethrum
Plastic Dusts	Shellac
Cellulose Acetate	Silicon
Methyl Methacrylate	Sulfur
Phenolformaldehyde	Tung
Phthalic Anhydride and its resins	Wood Flour
Class C Materials (High maximum rates of pressure rise§)	
Metal Dusts	*Sorbic Acid
Aluminum	*Titanium
*Stamped Aluminum	*Zirconium
Magnesium	Some Metal Hydrides
Magnesium—aluminum alloys	

* These are exceptionally fast.
† ≤7,300 psi/sec (50 000 kPa/sec) measured via the Hartman apparatus.
‡ 7,300 to 22,000 psi/sec (50 000 to 151 000 kPa/sec) measured via the Hartman apparatus.
§ 22,000 psi/sec (151 000 kPA/sec) measured via the Hartman apparatus.

value derived by comparing the measured explosibility parameters for a given dust, as described above, with those of Pittsburgh coal dust. The relative explosion hazards may be derived using the following formulas:

These relative explosion hazards may be derived from the following formulas:

$$IS = \frac{\min T_i' \times \min E_i' \times \min C_e'}{\min T_i \times \min E_i \times \min C_e}$$

$$ES = \frac{\max T_e' \times \max p'}{\max T_e \times \max p}$$

$$EI = IS \times ES$$

where

IS = Ignition Sensitivity
ES = Explosion Severity
EI = Explosibility Index
T_i = Ignition Temperature
E_i = Ignition Energy
C_e = Explosive Concentration
P_e = Explosive Pressure
p = Role of Pressure Rise

(Prime parameters represent Pittsburgh Coal Dust; unprimed parameters represent Dust Samples.)

Known values for these quantities are included in Table 10-11B. In the absence of a sound theoretical basis for predicting explosion hazards of dusts, the Explosibility Index provides a useful evaluation of relative explosibility. An empirical correlation between the U.S. Bureau of Mines indices and a descriptive categorization, the explosion hazard rating, is shown in Table 10-11C.

Materials are ground either in batches or continuously. Continuous processes may be either open or closed circuit. (See Figs. 10-11A and 10-11B.) In air swept mills, air is blown in at one end and the ground material is removed at the other end in air suspension. Batch mills are generally used only where small quantities are processed, due to the high labor costs for charging and discharging the mill.

Grinding equipment is classified according to the manner in which force is applied to the material. The material may be ground (1) between two solid surfaces, (2) by being forced against one solid surface (jet mill method), (3) by the action of the surrounding medium, or (4) by the nonmechanical introduction of energy, such as thermal shock, explosive shattering, or electrohydraulic processes. This chapter is concerned mainly with the first two methods. Also associated with grinding equipment are classifiers which separate out the fine product and return the coarse mill for regrinding.

Fire and Explosion Safety

All mills that produce combustible dusts should be isolated. If possible, they should be in separate, detached, one story, noncombustible buildings entirely above grade. Adequate explosion venting is essential. The surfaces of interior walls and equipment should be smooth to minimize dust accumulations and facilitate cleaning. If a milling operation must be located in a multiple purpose building, it should be in a room with at least one exterior wall which can be arranged to relieve explosion pressures, and the interior walls around the mill area should be of explosion resistant construction.

Although the mill is the source of combustible dust and may also provide an ignition source, explosions generally develop somewhere in the system downstream of the mill. Also, mills are generally constructed substantially enough to withstand explosion pressures. Explosion venting is needed where this is not the case.

As much as practical, nonsparking construction materials should be used to minimize sparks. Magnetic separators should be installed in front of mills to remove foreign ferrous metal, and screens should be used to remove rock and other nonferrous foreign material. Mills should be grounded to minimize the possibility of ignition

Table 10-11B. Explosion Characteristics of Various Dusts

(Compiled from the following reports of the U.S. Department of Interior, Bureau of Mines: RI 5753, The Explosibility of Agricultural Dusts: RI 6516, Explosibility of Metal Powders: RI 5971, Explosibility of Dusts Used in the Plastics Industry: RI 6597. Explosibility of Carbonaceous Dusts: R I 7132, Dust Explosibility of Chemicals, Drugs, Dyes and Pesticides, and RI 1200, Explosibility of Miscellaneous Dusts.)

Type of Dust	E_i Explosibility index	I_s Ignition Sensitivity	E_s Explosion Severity	P_e Maximum Explosion Pressure		p Max Rate of Pressure Rise		T_c Ignition Temperature				E_i Min. Cloud Ignition Energy	C_e Min. Explosion Conc.		Limiting Oxygen Percentage* (Spark Ignition)
								Cloud		Layer					
				psig	kPa	psi/sec	kPa/sec	°C	°F	°C	°F	joules	oz/cu ft	g/m³	
Agricultural Dusts															
Alfalfa meal	0.1	0.1	1.2	66	455	1,100	7585	530	986	—	—	0.320	0.105	105	—
Coca bean shell	13.7	3.6	3.8	77	531	3,300	22754	470	878	370	698	0.030	0.040	40	—
Coffee, raw bean	<0.1	0.1	0.1	33	228	150	1034	650	1202	280	536	0.320	0.150	150	C17
Constarch, commercial product	9.5	2.8	3.4	106	731	7,500	51713	400	752	—	—	0.040	0.045	45	—
Cork dust	>10.0	3.6	3.3	96	662	7,500	51713	460	860	210	410	0.035	0.035	35	—
Grain, dust, winter wheat, corn, oats	9.2	2.8	3.3	131	903	7,000	48265	430	806	230	446	0.030	0.055	55	—
Peat, sphagnum, sun dried	2.0	2.0	1.0	104	717	2,200	15169	460	860	240	464	0.050	0.045	45	—
Pyrethrum, ground flower leaves	0.4	0.6	0.6	95	655	1,500	10344	460	860	210	410	0.080	0.100	100	—
Rice	0.3	0.5	0.5	47	324	700	4827	510	950	450	842	0.100	0.085	85	—
Soy flour	0.7	0.6	1.1	94	648	800	5116	550	1022	340	644	0.100	0.060	60	C15
Wheat flour	4.1	1.5	2.7	97	669	2,800	19306	440	824	440	824	0.060	0.050	50	—
Yeast, torula	2.2	1.6	1.4	123	848	3,500	24133	520	968	260	500	0.050	0.050	50	—
Carbonaceous Dusts															
Charcoal, hardwood mixture	1.3	1.4	0.9	83	572	1,300	8964	530	986	180	356	0.020	0.140	140	—
Coke, petroleum	0.1†	0.1†	—	—	—	200	1379	670	1238	—	—	‡	1.000	1005	—
Graphite	0.1†	0.1†	—	—	—	—	—	§	—	580	1076	—	—	—	—
Lignite, California	>10.0	5.0	3.8	94	648	8,000	55160	450	842	200	392	0.030	0.030	30	—
Metals															
Cadmium, atomized (98% Cd)	—	—	—	7	48	100	690	570	1058	250	482	4.000	—	—	—
Iron carbonyl (99% Fe)	1.6	3.0	0.5	43	296	2,400	16548	320	608	310	590	0.020	0.105	105	C10
Lead, atomized (99%) Pb)	—	—	—	—	—	—	—	710	1310	270	518	—	—	—	—
Magnesium, milled, Grade B	>10.0	3.0	7.4	116	800	15,000	103425	560	1040	430	806	0.040	0.030	30	—
Manganese	0.1	0.1	0.7	53	365	4,900	33786	460	860	240	464	0.305	0.125	125	—
Tantalum	0.1	0.1	0.7	55	379	4,400	30338	630	1166	300	572	0.120	<0.200	<201	—
Thermoplastic Resins and Molding Compounds															
Cellulose acetate	>10.0	8.0	1.6	85	586	3,600	24822	420	788	—	—	0.015	0.040	40	C14
Methyl methacrylate polymer	6.3	7.0	0.9	84	579	2,000	13790	480	896	—	—	0.020	0.030	30	C11
Polyethylene, low-pressure process	>10.0	22.4	2.3	80	552	7,500	51713	450	842	—	—	0.010	0.020	20	—
Polystyrene molding compound	>10.0	6.0	2.0	77	531	5,000	34475	560	1040	—	—	0.040	0.015	15	C14
Thermosetting Resins and Molding Compounds															
Phenolformaldehyde	>10.0	9.3	1.4	77	531	3,500	24133	580	1076	—	—	0.015	0.025	25	C17
Urea formaldehyde molding compound, Grade II, fine	1.0	0.6	1.7	89	614	3,600	24822	460	860	—	—	0.080	0.085	85	C17
Special Resins and Molding Compounds															
Lignin, hydrolized-wood-type, fines	>10.0	5.6	2.7	102	703	5,000	34475	450	842	—	—	0.020	0.040	40	C17
Rubber, synthetic, hard, contains 33% sulfur	>10.0	7.0	1.5	93	641	3,100	21375	320	608	—	—	0.030	0.030	30	C15
Shellac	>10.0	25.2	1.4	73	503	3,600	24822	400	752	—	—	0.010	0.020	20	C14

* Numbers in this column indicate oxygen percentage while the prefix indicates the diluent gas. For example, th e entry "C17" means dilution to an oxygen content of 17 percent with carbon as the diluent gas.
† 0.1 designates materials presenting primarily a fire hazard as ignition of the dust cloud is not obtained by the spa rk or flame source but only by the intense heated surface source.
‡ Guncotton ignition source.
§ No ignition.

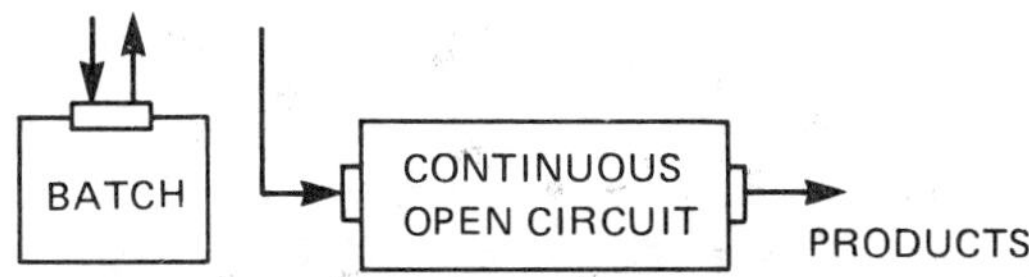

FIG. 10-11A. Batch and continuous grinding systems.

by static sparks. Open flames and smoking should not be permitted, and welding and cutting equipment used only when the mill is shut down and the area has been made entirely dust free.

Good housekeeping is essential. Although well designed grinding mills minimize dust leakage and reduce necessary cleaning, it is not always possible to maintain absolutely tight systems. Vacuum cleaning is the best way to remove any dust which escapes into the room, as it prevents the dust from forming explosive clouds.

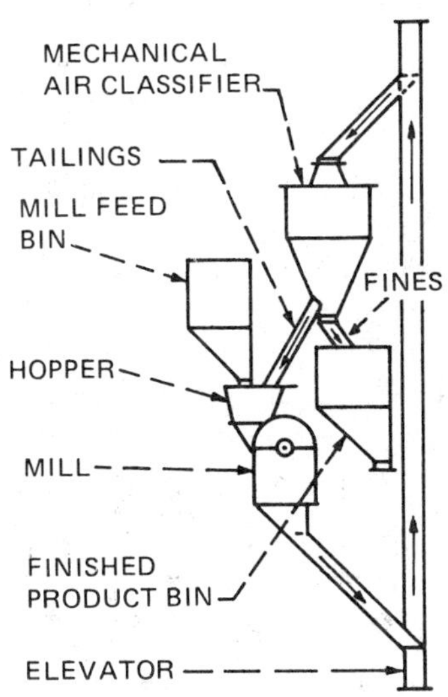

FIG. 10-11B. A hammer mill in a closed circuit with air classifer.

For grinding operations where ignition sources are difficult to control, the equipment may be protected by introducing a continuous flow of inert gas such as carbon dioxide, nitrogen, or flue gas. This will keep the normal oxygen content within the equipment sufficiently low to prevent an explosion.

TABLE 10-11C. Correlation Between Descriptive Categories for Dust Explosions and U.S. Bureau of Mines Indices.*

Type of Explosion	Ignition Sensitivity	Explosion Severity	Index of Explosibility
Weak	<0.2	<0.5	<0.1
Moderate	0.2–1.0	0.5–1.0	0.1–1.0
Strong	1.0–5.0	1.0–2.0	1.0–10
Severe	>5.0	>2.0	>10

* (From: Jacobson et al., U.S. Bureau of Mines, RI5753, "The Explosibility of Agricultural Dusts")

GRINDING HAZARDS

The hazards of grinding operations lie in the fact that the process produces very fine particles of readily oxidized materials that may be mixed with process or environmental air in flammable or explosive concentrations. To separate the properly sized material from the coarser particles being fed into the mill, a system of classifiers is used. For the most part, classifiers are based on the force of gravity and the principles of air drag and particle inertia. A continuous flow of air passes through the mill, regulated so that particles of the desired fineness are carried through to a collecting bin or compartment, while coarser particles fall out and return to the mill for further grinding. This principle is illustrated in Figure 10-11B.

There are several different types of air classifiers, some of which may be external to the mill as in Figure 10-11B. Others may be internal as in the ring-roller mill illustrated in Figure 10-11C. Here the whizzers are rotating blades

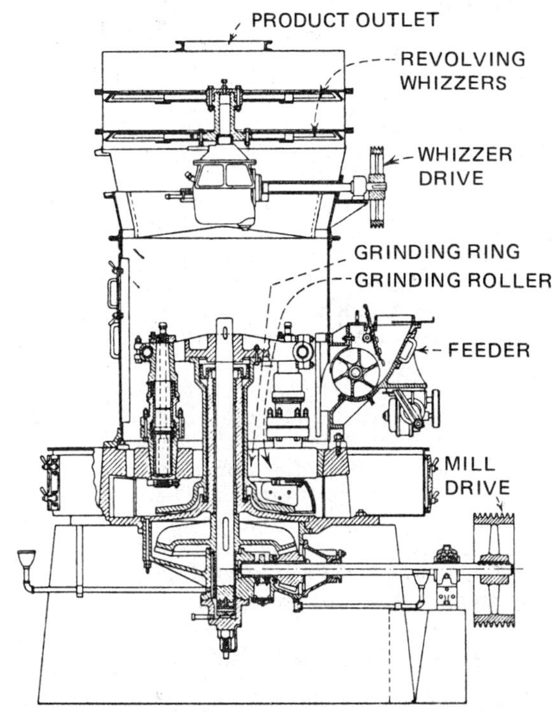

FIG. 10-11C. A Raymond high side mill with internal whizzer classifier.

whose centrifugal action throws the coarser particles outward, permitting them to drop back onto the grinding surfaces. The finished product is carried through an outlet by the airstream.

Theoretically, the process airstream should be fully and tightly enclosed so that it does not carry the finished product or unwanted dust into the atmosphere surrounding the mill. In practice, this is rather difficult to accomplish, and dust builds up in the structure housing the mill. There are, then, two distinct hazards. One is that an explosion may occur within the milling system, and the other is that an explosion may occur in the structure. All that is required is a critical concentration of the material in air and an ignition source.

The lower limit of flammability will vary from one material to another. For some, such as phthalic anhydride, shellac, aluminum stearate, and phenothiazine, it may be as low as 0.015 oz/cu ft (15 g/m^3). For zinc, it may be as much as 0.5 oz/cu ft (500 g/m^3). Upper flammability limits for dusts, i.e., the concentration above which an explosion will not occur, are poorly defined, not usually reproducible, and not yet determined for many dusts. Therefore, any concentration above the lower limit should be considered potentially explosive. The minimum electric spark energy required for ignition of a dust cloud can vary from as little as 10 mJ (millijoule) to as much as 1900 mJ.

Ignition sources, too, can vary widely. Since a bit of metal accidentally entering the mill may cause a spark as it strikes against one of the grinding surfaces, mills should be equipped with magnetic or mechanical separators to remove any tramp metal from the feed system. Sparking may also be caused by tools made of ferrous metals. Other possible ignition sources are static electricity, hot surfaces, friction, open flames, welding arcs, and personnel smoking in the mill.

CHARACTERISTICS OF DUST EXPLOSIONS

Dust explosions in some respects are similar to vapor and gas explosions, but they do differ in some important ways. Like a gas, dust must be mixed with air or another supporter of combustion, and a source of ignition is generally required to cause an explosion. Rarely have dust explosions resulted from spontaneous oxidation and heating. Reaction rates and rates of pressure rise are usually higher in vapor and gas explosions than in dust explosions. However, complete combustion of dust in a given volume of air will frequently develop energy and pressure greater than those developed by the combustion of a gas. Dust explosions, then, are sometimes more disastrous than gas explosions. This, in part, is due to their slower rate of development and longer duration. The slower rate of development results from the fact that the combustion of dust is a surface reaction, and the diffusion of oxygen toward the reacting surface is necessarily slower and less complete than it is in a flammable gas.

Requisite Conditions for Dust Explosions

In order for a dust explosion to occur, four conditions must be satisfied simultaneously:

1. A combustible solid in a finely divided state must be dispersed in an oxidizing medium—usually oxygen in air.
2. The concentration of the dust in air must be within the explosible range.
3. An external source of ignition of sufficient energy and duration to initiate the explosive chain reaction for that particular dust must be present.
4. The chemical reaction must occur in a confined volume.

The rapid chemical reaction, or flash fire, characteristic of explosions, will occur if only the first three conditions are satisfied. However, the rapid buildup of excessive pressures, inherent in the working definition of dust explosions, will result only when the reaction occurs in an enclosed space.

Characterization of Explosion Hazards of Dusts

Parameters employed to describe the relative explosion hazards of various dusts include:

1. Lower and upper limit of dust concentrations within which an explosion is possible;
2. Minimum ignition energy (minimum electric spark energy required for ignition of the dust cloud);
3. Minimum ignition temperature as measured by a furnace apparatus;
4. Maximum oxygen concentration permissible to prevent ignition;
5. Maximum explosion pressure attained during the course of the explosion;
6. Maximum rate of pressure rise (sometimes also the average rate of pressure rise).

Values for these parameters are not fixed but depend on various factors—namely, particle size and shape, ambient temperature and pressure, moisture content of the dust, degree of turbulence in the suspension, and size of the ignition source. (See Table 10-11B.) Experimental work has made possible some qualitative observations of the effects of these factors, but quantitative relations are not available.

Definitions of dust in terms of particle size vary, but normally dust is defined as particles with diameters of 0.1 to 1000 microns (1 micron equals 10^{-6} m).

The minimum explosible concentration and the minimum ignition energy generally tend to decrease with a decrease in average particle size (i.e., the explosion hazard increases with a decrease in particle size). For average particle sizes less than 50 microns, the effect is much less pronounced. According to Palmer, uniform dispersion of the dust in laboratory test equipment may become more difficult as particles become very small (Palmer 1973). Due to the greater cohesiveness of very fine particles, some may exist as agglomerates rather than as individual particles, leading to an apparent reduction in explosibility. A different method of dispersion or a more vigorous ignition source may break up the agglomerations resulting in an increased explosion hazard. Tests performed in large scale coal mines are said to confirm this.

Little experimental evidence is available concerning the effect of particle shape on explosibility. Particles may resemble fibers, needles, or flakes, as well as spheres. Studies employing atomzied spherical particles and flaked aluminum particles indicate that a significant difference in explosibility can occur with changes in particle shape. Size and shape, which are generally measured before ignition, may be altered during preignition stages and subsequently affect propagation of the explosion. Changes may occur as a result of melting, vaporization, expansion to form hollow spheres, and fragmentation.

The effect of ambient temperature on explosibility would be particularly applicable to industrial situations such as dryers, but no information is available on measurements in plant scale units. Theoretically, it would be expected that the minimum ignition energy would decrease as the ambient temperature increased, other factors remaining constant. If the final temperature reached by the combustion products is limited by the molecular dissociation of product gases, then the maximum explosion pressure would be expected to decrease as the ambient temperature increases, and the rate of pressure rise would increase as the ambient temperature increased. Laboratory tests employing coal dust revealed that the minimum ignition energy decreased with increasing ambient temperature as expected, but only over a limited temperature range; thereafter, it increased. Palmer also notes that the minimum ignition energy would depend on previous exposures to high temperatures.

If the pressure in the reaction vessel is atmospheric when the explosion is initiated, rises in pressure during the course of the explosion are not considered to be changes in ambient pressure. Laboratory tests employing methane/air mixtures resulted in proportional increases in maximum explosion pressures with increased initial pressure, but smaller increases were obtained for coal dust air mixtures. The maximum explosible concentration decreased with increasing ambient pressure, while the minimum explosible concentration remained relatively unchanged. Explanations for these effects have not yet been explored.

An increase in moisture content of dust tends to increase the minimum explosible concentration, the minimum ignition energy, and the minimum ignition temperature when measured in small scale apparatus. These effects are apparently due to the absorption of heat in vaporizing water and to the decrease in dispersibility of the dust. Nineteen percent water content (on a weight basis) was found to prevent ignition of starch dust.

Tests with coal dust air suspensions have indicated that explosions tend to increase in severity (i.e., maximum explosion pressure and/or maximum rate of pressure rise increases) with increases in the size of the ignition source. The nature of the ignition source also affects the explosibility of dusts. Organic dusts tend to be ignited more readily by heated coils, but metal dusts react more readily to spark ignition.

Experiments indicate that some dusts produce stronger explosions than others. Metallic dusts, such as stamped aluminum powder, milled and stamped magnesium, and atomized aluminum, produce the most violent dust explosions. Phenolformaldehyde resin, cornstarch, soybean protein, wood flour, and coal dust, respectively, followed the metallic dusts in explosive intensity.

The character and severity of any dust explosion will be affected by several factors. One of these is particle size. For any given material, the finer the particle size, the more violent the explosion. Less energy will be required to ignite the dust, and it will remain in suspension for a longer time, increasing the total force exerted. Figure 10-11D shows the relationship of particle size to the explosibility index.

Turbulence is another factor that contributes to the severity of a dust explosion. Turbulence speeds up the diffusion of oxygen to the reacting surfaces and promotes stronger explosions. The smaller the particle size and the greater the turbulence, the more the dust resembles a gas or vapor in its explosive characteristics.

Relatively little investigation has been undertaken with regard to the effects of turbulence on explosibility. Tests with coal dust air suspensions revealed an increase in maximum explosion pressure and maximum rate of pressure rise with increased turbulence. Both parameters passed through maxima and then decreased, although more slowly in the case of the maximum explosion pressure. Sources of turbulence may include the presence of obstacles and rapid volume expansion due to flame.

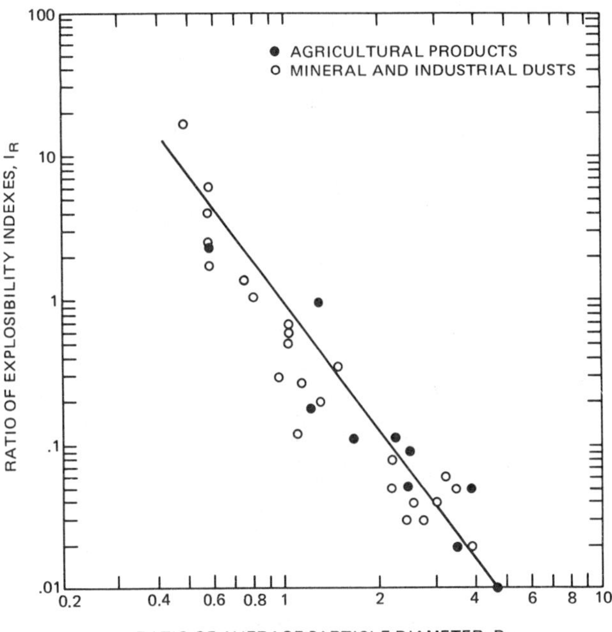

FIG. 10-11D. Effect of particle diameter on relative explosibility index. Relative average particle diameter is the ratio of the mean particle size of the dust to the mean size of the through No. 220 sieve sample. Relative explosibility index is the ratio of the index computed for the dust to that computed for a through No. 200 sieve sample. (From Bureau of Mines Report of Investigation 5753)

Particle Size Classifiers

Classifiers separate out the fine product and return the coarse material (circulating load) to the mill for regrinding along with new material being fed to the mill. If the fines are continuously removed, a mill performs much more efficiently. Closed circuit grinding with size classifiers provides even more uniform size distribution and is also more economical.

Wet classifiers are generally used for large scale wet milling operations, as in cement and ore processing plants. The simplest type of wet classifier is a settling basin arranged so that the fines do not have time to settle out and are drawn off, while the coarse material is raked to a central discharge. Wet classifiers of this type do not present an explosion hazard.

Dry classifiers may be installed external to the mill in a closed circuit (Fig. 10-11B), or they may be internal as an integral part of the mill. (See Fig. 10-11C.)

Air classifiers are used for most dry milling operations. There are a number of different types, but these classifiers all are based upon the principles of air drag and particle inertia. One type directs an airstream across a stream of the particles to be classified. Another has adjustable flow baffles, and still another changes the direction of air flow. (See Fig. 10-11E.) The double cone classifier uses centrifugal action, induced by flow through vanes, which cause coarse particles to move outward and down the wall of the inner cone and thus return to the grinding zone,

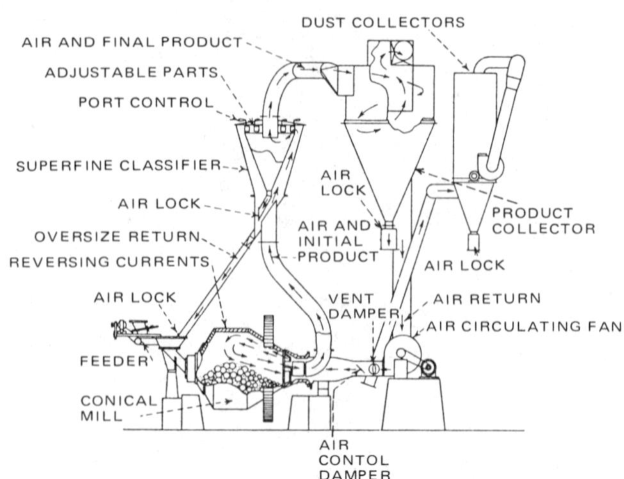

FIG. 10-11E. A Hardinge conical mill with reversed current air classifier.

while the upward moving air stream entrains the fines. (See Fig. 10-11F.)

Rotating blades are the main elements of several types of classifiers. The centrifugal motion established by the rotating blades tends to throw the coarser particles outward and returns them to the grinding zone, while the fines are carried off in the airstream. (See Fig. 10-11C.)

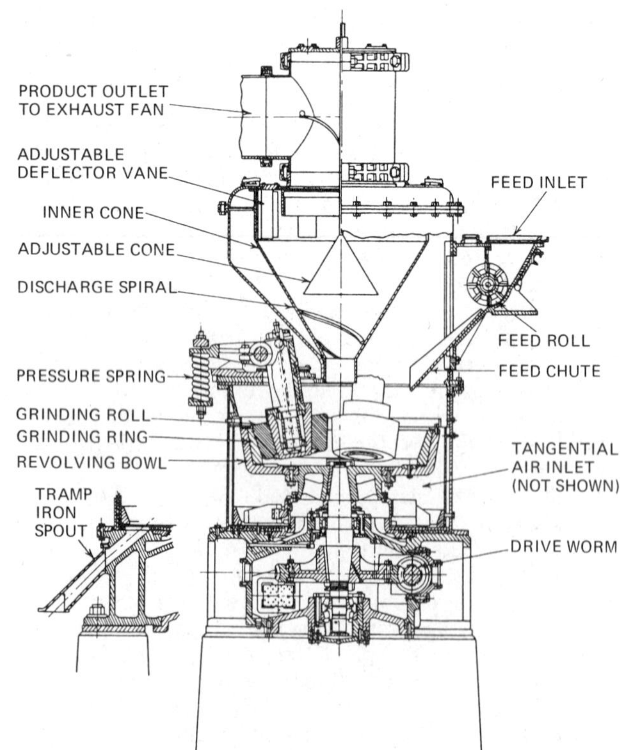

FIG. 10-11F. A bowl mill. (C.E. Raymond, Combustion Engineering Inc.)

GRINDING EQUIPMENT

Equipment for reducing particle size of a given material may be classified into four groups, according to the method of grinding used. A material may be reduced between two solid surfaces; it may be reduced by impact on one solid surface; it may be reduced by action of the surrounding medium; or it may be reduced by such nonmechanical means as thermal shock, explosive shattering, or electrohydraulic processes.

Most grinding is performed by machines that pass the material between two solid surfaces. Such machines can be classified into tumbling mills, ring roller mills, roller mills, hammer mills, attrition mills, and jet mills.

Tumbling Mills: Tumbling mills consist of a cylindrical or conical shell charged with balls of steel, flint, or porcelain, or with steel rods. As the shell revolves about its horizontal axis, the balls or rods tumble about, grinding the material to be reduced against the wall of the shell or between themselves. (See Fig. 10-11E.) The size of the balls or rods and the duration of the operation will determine particle size. Some tumbling mills are compartmented by perforated partitions that allow material to pass from one compartment to another for finer grinding.

Ring Roller Mills: A second type of grinding machine is known as a ring roller mill. Mills of this type consist of a grinding ring or plate which moves between rollers. (See Fig. 10-11C.) The ring may be either horizontal or vertical, and either it or the roller may rotate, grinding the product between the two surfaces. A variation of the ring roller mill is the bowl mill, in which a bowl and ring revolve around stationary rollers to grind the product. (See Fig. 10-11E.) There is no metal to metal contact, and the space between the bowl and rollers can be preset to produce the required particle size.

Roller Mills: Roller mills differ from ring roller mills in that the material to be ground passes between two or more rollers revolving in opposite directions at different speeds. (See Fig. 10-11G.) A scraper blade at the discharge end removes the finely ground material. Most dry materials are ground between rollers having corrugations which determine the final particle size. Corrugations may be either sharp or dull, and they may be used in various combinations to achieve the desired results.

Hammer Mills: Hammer mills have hammers, or beaters, attached to a rotating shaft. (See Fig. 10-11H.) The hammers may be of almost any shape and hinged or fixed to the

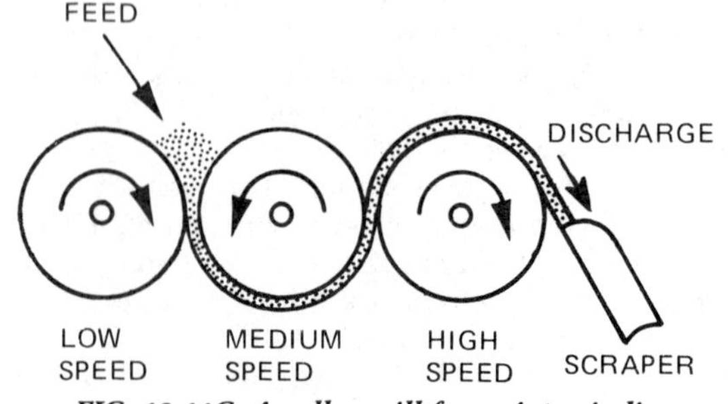

FIG. 10-11G. A roller mill for paint grinding.

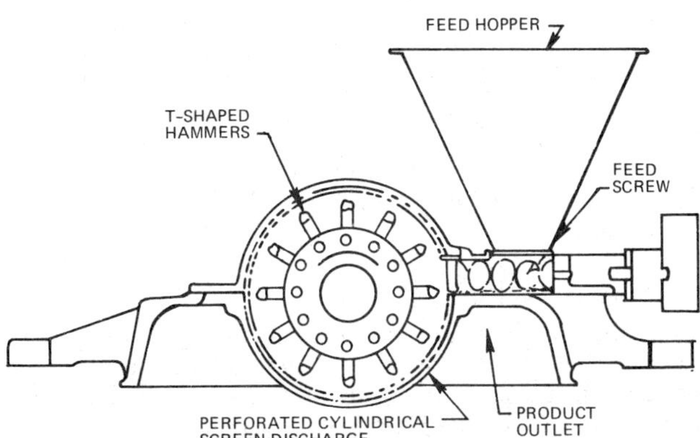

FIG. 10-11H. A Milro-Pulverizer hammer mill. (Pulverizing Machinery Co.)

shaft. The fineness of the finished product is determined by the speed of the rotating shaft, the clearance between the hammers and grinding plates, the number and size of the hammers, the feed rate, and the size of the discharge openings. Two variations of the hammer mill are the disintegrator and the pin mill.

The disintegrator has a vertical rotating shaft with hammers that run at close tolerances to a cylindrical screen. (See Fig. 10-11I.) Material is fed parallel to the

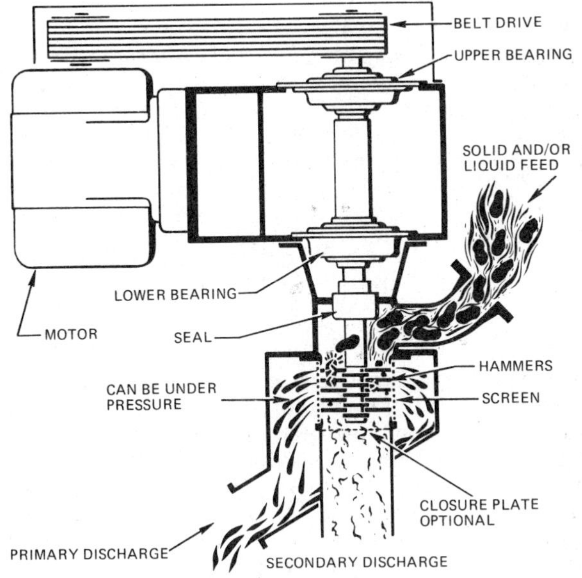

FIG. 10-11I. A Reitz disintegrator. (Reitz Manufacturing Co.)

rotating shaft, and the centrifugal action of the hammers discharges the ground material into a primary chute.

The pin mill is a high speed mill with two disks in which pins are set in alternating circular rows. Either one or both of the disks may rotate. If both rotate, they do so in opposite directions. The material to be ground is broken up between the pins. (See Fig. 10-11J.)

Attrition Mills: Attrition mills use metallic or abrasive grinding plates that rotate at high speed on either a horizontal or vertical plane. One disk may be stationary, or the two may rotate in opposite directions. The material

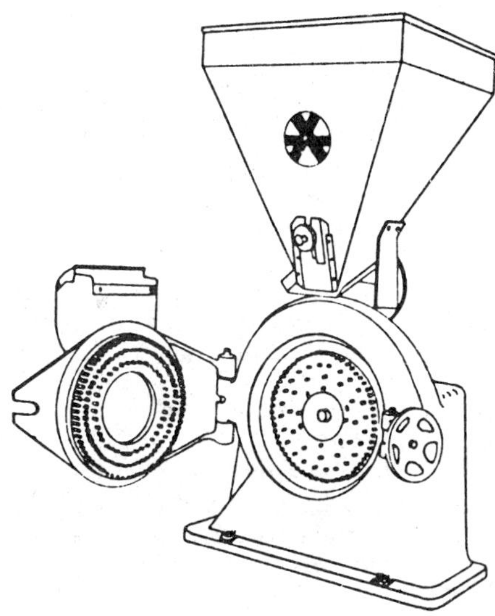

FIG. 10-11J. An Alpine-Kolloplex pin mill. (Alpine-American Corp.)

enters at the axis and is discharged at the periphery of the grinding plates. One type of attrition mill, known as the Buhrstone mill, uses two circular stones instead of metal or abrasive disks between which the material is ground.

Jet Mills: Another type of mill, the jet mill, differs from the others in that the material is not ground against a hard surface. Instead, a gaseous medium is introduced. The gas may convey the feed material at high velocity in opposing streams, or it may move the material around the periphery of the grinding and classifying chamber. The high turbulence causes the particles of feed material to collide and grind upon themselves.

APPLICATION OF GRINDING EQUIPMENT

Agricultural Products

The roller mill is the traditional machine for grinding wheat and rye into high grade flour. Generally, rollers with dull corrugation are used. For very tough wheat, however, a sharp roller is used against a sharp roller, and for other grades, combinations of dull and sharp rollers are used. Rollers with sharp corrugations are used for grinding corn and feed. High speed hammer mills or pin mills are used to produce flour with controlled protein content. Disk attrition mills are also used for grinding wheat.

After the oil has been extracted, soybeans or soybean cake is ground in attrition mills or flour rollers, depending upon whether the product is to be a feed meal or flour. In some cases, a hammer mill may be used as a preliminary disintegrator for pressed cakes, including linseed and cottonseed cake.

Where only medium fineness is required, a hammer mill is used to produce starch, potato flour, tapioca, and similar flours. For finer flour products, a high speed impact mill such as a pin mill is used.

Carbon Products

Bituminous coal and pitch are used as fuel for industrial furnaces, boilers, and rotary kilns. Pulverized coal is either blown directly into the furnace as it is pulverized, or pulverized in a central grinding system and stored in a bin until it is used. Ball, tube, ring-roller, bowl, and ball and ring type mills are used for direct firing of large installations. Ring-roller mills are also used to pulverize coal for bin systems.

Anthracite coal is harder to reduce than bituminous coal. Ball or hammer mills are used to pulverize anthracite coal for foundry facing mixtures. Calcined anthracite, used in the manufacture of electrodes, is generally pulverized in ball and tube mills, or ring-roller mills.

The grinding characteristics of coke vary from petroleum coke, which is relatively easy to grind, to certain foundry and retort coke, which is difficult to grind. Where uniform size of particles with a minimum of fines is required, rod or ball mills are used in a closed circuit with screens.

Natural graphite is classified in three grades: flake, crystalline, and amorphous. Flake is the most difficult to grind to a fine powder, and crystalline is the most abrasive. Ball, tube, ring-roller, and jet mills with or without air classification are used for grinding graphite. For handling large capacities, ball and tube mills are used, especially for the flake and crystalline grades. Graphite for pencils is ground in a jet pulverizer. Ball mills in a closed circuit with air classifiers have been used for grinding artificial graphite. Charcoal and Gilsonite are ground in hammer mills with air classifiers.

Chemicals

Hammer mills are generally used to pulverize dry colors and dyestuffs, with pebble mills used for small lots. Hammer or jet mills with air classifiers for size limitation are used for dyes that are coarsely crystalline. A ring-roller mill is used for fine grinding of sulfur, with inert gas injected into the mill.

Pulverizing of metallic soaps such as stearates requires certain provisions to keep the material cool and in rapid motion. Since these materials tend to cake, batch grinding is not practicable. Stearates are pulverized in multicage mills, screen mills, and hammer mills with air classification.

Organic Polymers

The grinding characteristics of various resins, gums, waxes, hard rubbers, and molding powders are such that when a finely ground product is required, it may be necessary to use a water jacketed mill or a pulverizer with an air classifier in which cooled air is introduced into the system. Hammer mills are generally used for this purpose. Some resins with low softening temperatures can be ground by mixing dry ice with the material before grinding or by introducing refrigerated air into the mill.

Most gums and resins used in the paint, varnish, or plastics industries do not require very fine grinding, so hammer mills or roll crushers will produce a satisfactory product. Some resins used in the phenolic resin industries that require very fine pulverization are ground in a pebble mill and cooled with water or brine in a closed circuit with an air classifier. A ring-roller mill with an internal air classifier is used to pulverize phenolformaldehyde resins.

Hard rubber is ground on heavy steam heated rollers. The materials pass through a series of rollers in a closed circuit with screens and air classifiers. Usually the rollers are of different sizes, and the machines operate at relatively low speeds so as not to generate too much heat.

Molding powders are produced with hammer or attrition mills in closed circuits equipped with either screens or air classifiers.

Cryogenic Grinding

Although cooling is required for some mills because of the material being ground, cryogenic grinding can be applied in any mill to produce a smaller sized particle than could be obtained otherwise. Existing mills can be converted by adding a liquid nitrogen storage tank, a piping system, and a properly designed hopper. In this way, the material can be cooled before it is ground.

The material can be precooled in the hopper by immersing it in a nitrogen bath, it can be cooled in the grinding chamber by spraying it with liquid nitrogen, or both methods can be used simultaneously. Gaseous nitrogen has been used for years to provide an inert atmosphere in mills, particularly in jet mills. Cryogenic grinding can be designed to provide a protective inert atmosphere within the mill. In spice grinding, freezing the spice before grinding gives it a superior appearance and it retains the aroma and flavor usually lost when it is not precooled.

Cryogenic grinding has been applied to the production of powdered coatings used for insulation and protective coatings, and for powders used in the manufacture of bearings with improved wearlife and increased lubrication properties. Other areas where cryogenic grinding may have application are in recycling scrap materials, grinding existing materials for newer applications and processes (rotational molding, spray powder for textile stiffener), and grinding protein concentrations.

PROTECTION AGAINST FIRES OR EXPLOSIONS

Combustion and explosion, or more properly deflagration relative to dusts, are practically identical processes. The difference lies in the speed with which the oxidation reaction takes place. NFPA 68, *Guide for Explosion Venting*, defines deflagration as "burning which takes place at a flame speed below the velocity of sound in the unburned medium." It defines explosion as "the bursting of a building or container as a result of development of internal pressure beyond the confinement capacity of the building or container." Inasmuch as this is usually the result of detonation, the term explosion, as generally used herein, refers to the entire process.

Because an explosion takes place almost instaneously, it cannot be brought under control by containment as can many fires. The effects, however, can be minimized by certain protective measures; these are prevention, venting, inerting, and suppression.

Prevention

The simplest means of preventing an explosion is to keep a critical concentration of dust from developing and to eliminate any potential sources of ignition.

In grinding operations, there are two types of locations where dust accumulation may become critical. One is within the milling or grinding equipment itself. The other is in the surrounding environment, i.e., the structure in which the equipment is housed.

The design and construction of the structure may vary considerably, depending upon the product being ground. It is impractical to construct a building that will withstand the pressure generated by a dust explosion. An ordinary 12 in. (0.3 m) thick brick wall can be destroyed by an internal pressure of less than 1 psi (6.9 kPa). Most dust explosions produce much higher pressures. Typical pressures range from 13 psi (90 kPa) for zinc to 89 psi (614 kPa) for stamped aluminum, with most being above 30 psi (207 kPa).

If a dust explosion hazard exists, the structure should be designed with panels, vents, windows, or other closures that will open at the lowest practical pressures and minimize structural damage.

Materials used should be noncombustible or fire resistive. Any interior walls intended to serve as fire walls should be capable of providing at least three hours of fire resistance under standard fire test methods. Interior stairs, lifts, or elevators should be enclosed in shafts of noncombustible materials and have fire resistive ratings of at least one hour. Such enclosures should be protected by automatically closing fire doors, and any openings in fire walls should be similarly protected.

The number of horizontal surfaces that might collect dust and are difficult to reach or are inaccessible for cleaning should be minimized. Wherever practical, such surfaces should be built up to an angle of at least 60 degrees from the horizontal so that dust will tend to slide off rather than accumulate. Good housekeeping is a necessary part of explosion prevention. It means both equipment and structure must be kept clean so dust cannot accumulate. Dust should be removed by vacuum systems. Brushing or sweeping will disperse the dust and increase turbulence, which in turn increases the possibility of an explosion occurring and increases its potential severity.

Equipment should be made of metal and be as dust tight as possible. It should be designed so that there is continuous suction at openings during grinding, dumping, transfer, and similar operations. The collected dust should be conveyed through tightly constructed ducts or chutes to well designed dust collectors located in a safe place, preferably outside the structure.

Since many dusts can be ignited by low energy sparks, potential ignition sources should be eliminated from or adequately shielded in the area. Welding, cutting, or any other operation that uses an open flame or arc should not be permitted unless the work area is dust free. Smoking should be prohibited. Torque limiting or fluid drive couplings should be designed to dissipate heat readily. Moving equipment, elevators, belts, and conveyors should be grounded or of nonconductive material to eliminate the possibility of static sparks. All electrical wiring should conform to the requirements of NFPA 70, *National Electrical Code®* for hazardous locations containing combustible dusts.

Adequate fire extinguishing equipment, both fixed and portable, should be provided. Hose nozzles should be of the spray type, since solid streams can cause turbulence and dispersion of dust into the air and increase the explosion hazard.

Venting

Vents are openings in the equipment or structure that allow heated explosion gases to escape more readily. Venting does not prevent explosions, but it does serve to limit the maximum pressure resulting from a deflagration. The most effective vents would be free and unrestricted openings; however, these are not always practical.

Vents should be closed in such a manner that they will open under the lowest practical pressure. Typical venting arrangements include rupture diaphragms, hinged or blowout windows and panels, and weakly constructed walls and roofs.

The venting area required, as calculated from empirical formulas, depends upon the expected pressure and its rate of rise, the type of closure, and the volume and strength of the enclosure. Determination of vent area for an enclosure is mostly empirical. Vents should be located where minimum damage will result from the shattering or blowing out of the vent closure.

Inerting

One method of preventing a dust explosion is inerting—the replacement of oxygen in the grinding process with an inert gas such as nitrogen or carbon dioxide. Such equipment as grinders, pulverizers, mixers, driers, conveyors, dust collectors, and filling machines can be protected by this method.

The amount of oxygen that must be replaced to provide a safe concentration depends upon the type of dust, particle size, concentration, turbulence, diluent gas, and intensity of the ignition source. To prevent ignition of carbonaceous dusts by a spark, for example, oxygen should be reduced to 8 percent by nitrogen or to 11 percent by carbon dioxide. However, if a stronger ignition source is present, the oxygen should be reduced to 3 and 4 percent, respectively.

Many factors will affect the use of inerting to prevent explosion. Among them are protection of personnel, the hazard to be protected, the required reduction in oxygen concentration, the availability and cost of the inert gas supply, and the necessary control equipment.

Suppression

Suppression is a technique of stopping an explosion before it develops destructive pressures. The factors involved here are much the same as those used in the extinguishment of fire, namely cooling, limiting the supply of oxygen, and inhibiting flame spread.

Despite the rapidity with which combustion proceeds in an explosion, there is a short period of time before which the destructive force is evolved. During this time, the initial pressure rise can be detected by suitable sensing devices, which automatically trigger the release of the suppressing agent, normally an inert gas or liquid, that inhibits the combustion process.

Suppression systems can be used in confined spaces such as reactors, mixers, pulverizers, mills, driers, storage bins, ovens, bucket elevator transport systems, and pneumatic transport systems. The effective application of such systems requires careful consideration of many factors. Among them are characteristics of the dust, rate of pressure rise, ignition sources, and characteristics of the suppressant. Though the principles of explosion suppression apply to all installations, each must be individually designed to cover the wide range of variables that will be encountered.

Bibliography

References Cited

Johnson, et al. 1961. *Explosibility of Agricultural Dusts.* U.S. Bureau of Mines Report on Investigations 5753, Washington, DC.

Palmer, K. N. 1979. *Dust Explosions and Fires.* Chapman and Hill, London.

NFPA Codes, Standards, Recommended Practices and Manuals. (See the latest *NFPA Codes and Standards Catalog* for availability of current editions of the following documents.)

NFPA 61A, *Standard for the Prevention of Fire and Dust Explosions in Facilities Manufacturing and Handling Starch.*

NFPA 61C, *Standard for the Prevention of Fire and Dust Explosions in Feed Mills.*

NFPA 68, *Guide for Explosion Venting.*

NFPA 69, *Standard on Explosion Prevention Systems.*

NFPA 70, *National Electrical Code.*

NFPA 651, *Standard for the Manufacture of Aluminum or Magnesium Powder.*

NFPA 654, *Standard for the Prevention of Fire and Dust Explosions in the Chemical, Dye, Pharmaceutical, and Plastics Industries.*

NFPA 655, *Standard for Prevention of Sulfur Fires and Explosions.*

NFPA 664, *Standard for the Prevention of Fires and Explosions in Wood Processing and Woodworking Facilities.*

Additional Readings

Allen, J., ed., *Grinding*, Vol. 1, International Ideas, Inc., Philadelphia, PA, 1968.

Barlow, D. W., et al, eds., *Grinding*, Vol. 2, International Ideas, Inc., Philadelphia, PA, 1972.

Bartencht, W., *Explosions: Course, Prevention, Protection*, Springer-Verlag, New York, NY, 1981.

Davenport, J. A., "Explosion Losses in Industry," *Fire Journal*, Vol. 75, No. 1, pp. 52-56, 71-73.

Eckhoff, R. K., *A Study of Selected Problems Related to the Assessment of Ignitability and Explosibility of Dust Clouds*, CHR, Michelsons Institute, Bergen, 1976.

Explosion Hazards and Evaluation, Elsevier, New York, NY.

Factory Mutual Engineering Corporation, "Grain Storage and Milling," *Loss Prevention Data*, 7-75, Norwood, MA, Aug. 1976.

Factory Mutual Engineering Corporation, "Combustible Dusts," *Loss Prevention Data*, 7-76, Norwood, MA, Aug. 1976.

Grain Elevator and Processing Society, *Proceedings: International Symposium on Grain Dust Explosions*, Kansas City, MO, 1977.

Grobel, Edward, "Cryogenic Grinding Gives Process Flexibility," *Cyogenics and Industrial Gases*, Vol. 9, No. 4, July/Aug. 1974, pp. 27-30.

Henderson, S. M., and Perry, R. L. eds., *Agricultural Process Engineering*, 3rd ed., AVI Publishing Co., Westport, CT, 1976.

Jaffe, H. M., "Grain Elevator Protection, What's Being Done Today?" *Fire Journal*, Vol. 74, No. 3, May 1980, pp. 131-132.

Jensen, Rolf, et al., *Fire Protection for the Design Professional*, Cahners Publishing Co., Boston, MA, 1975.

Milling: Methods and Machines, Society of Manufacturing Engineers, Dearborn, MI, 1982.

Milling One and Milling Two, State Mutual Book and Periodical Service, Ltd., NY, 1983.

National Materials Advisory Board, *Prevention of Grain Elevator and Mill Explosions*, Publication NMAB-2, National Academy Press, Washington, DC, 1982.

Perry, Robert H., and Chilton, H., *Chemical Engineers Handbook*, 5th ed., McGraw-Hill, NY, 1973.

Plaster, C., et al, eds., *Milling*, Vol. 1, 2nd ed., International Ideas, Inc., Philadelphia, PA, 1977.

Richey, C. B., et al., *Agricultural Engineer's Handbook*, McGraw-Hill, NY, 1961.

Spencer, A. G., *Milling*, Vol. 2, International Ideas, Inc., Philadelphia, PA, 1969.

Tuve, Richard L., *Principles of Fire Protection Chemistry*, National Fire Protection Association, Quincy, MA, 1976.

Verkade, M., and Chiatti, P., *Literature Survey of Dust Explosions in Grain-Handling Facilities: Cause and Prevention*, Energy and Mineral Resources Research Institute, Iowa State University, Ames, IA, 1976.

Weiby, P., and Dickinson, K. R., "Monitoring Work Areas for Explosive and Toxic Hazards," *Chemical Engineering*, Vol. 83, No. 22, 1976, pp. 139-145.

EXTRUSION AND FORMING PROCESSES

Jerome P. Carroll

There are three broad areas of processing within the plastics industry. The first is the manufacturing, or synthesizing, of the basic plastic or feedstock, which may include compounding with colorants or other additives. The second area is the conversion of the plastic materials into useful articles by molding, extrusion, or casting, processes that heat the plastic so it will flow into a shape that is retained when the plastic is cooled. The plastics industry calls these operations "processing" or "converting," although chemical reactions are not usually significant. The third area in the industry encompasses the largely mechanical operations of bending, machining, cementing, decorating, and polishing, known as "fabricating."

In some cases, plants involved in only the converting and/or fabricating processes may also conduct chemical operations with flammable or reactive materials. This chapter will discuss the fire hazards associated with the processing, or converting, phase of the plastics industry, recognizing that in some instances, hazards associated with all three phases of the industry may be found in a single location.

Additional information on the fire hazards of plastics will be found in Section 5, Chapter 8, "Plastics." Other relevant material will be found in Section 5, Chapter 4, "Flammable and Combustible Liquids;" and in Section 12, Chapter 8, "Housekeeping Practices."

The Basic Hazards

Most plastics are combustible. The combustibility of plastic formulations is influenced by the basic polymers used, the nature of plastic additives, and the form the final product takes. The conversion of feedstock plastics into finished articles may also involve the hazards associated with combustible dusts, flammable solvents, electrical faults, hydraulic fluids, and the storage and handling of large quantities of combustible raw materials and finished products.

Mr. Carroll is Director of Safety, Society for the Plastics Industry (SPI), New York, NY.

Terminology

The reader should become familiar with the following terms before studying the discussion of converting plastic feedstocks into finished products and the hazards associated with the industry.

Additive: Any material used to modify processing or end use properties. The resulting mix usually is called a "plastic" to distinguish it from the "resin" or principal ingredient. Additives may be colorants or stabilizers, powdered fillers for stiffening, plasticizers for flexibility, fibers to reinforce, antioxidants, lubricants to aid flow into or release from a mold, or fire retardants.

Binders: The resins in a plasture that hold together all of the other ingredients.

Blowing Agent: A material which releases gas upon heating so that the plastic in which it is mixed will expand into foam.

Blown Tubing: A thin film made by extruding a tube and simultaneously inflating it with air while hot; distention may be 20 times the diameter of the tube as extruded.

Casting: Flowing material into place with little or no pressure, as contrasted with forcing material into place by molding.

Extrusion: Passing softened plastic under pressure through a die to make an essentially continuous profile; the equipment is called an extruder.

Fabrication: The making of articles by machining, cementing, heat sealing, or thermoforming of preformed sheets, rods, or tubes. The term is used in contrast to "processing."

Fillers: Materials that modify the strength and working properties of a plastic; they may be used to increase heat resistance and alter its dielectric strength. A wide variety of products are used, including wood, cotton, sisal, glass, and clay.

Film: A general term for plastic not more than 0.01 in. (0.25 mm) thick, regardless of the process used to make it. "Foil" is used today only to describe metal.

Finishing: Either: (1) removal of burrs or flash by filing, sanding, or tumbling; (2) buffing or waxing to polish surfaces; or (3) application of decorative or marking treatment, as by painting or metal plating.

Flash: Unwanted projection from molded articles resulting from flow of plastic into space between matching parts of a mold. The term has no fire connotation.

Foam: Plastic with many small gas bubbles, i.e., "cellular plastic." Rigid foams are used as thermal insulation boards for construction, cups for hot and cold drinks, trays for prepackaged meats, and shock resistant packaging. Flexible foams are used for furniture padding, insulation of outer garments, and soft drape upholstery.

Foaming Agent: The same as blowing agent.

Forming: Changing the shape of plastic pieces, such as a sheet, rod, or tube, into a desired configuration.

Gate: (1) that portion of a mold which admits plastic directly to the cavity producing the desired shape, (2) that portion of a plastic which lies in the gate portion of a mold, or (3) a glass, or plastic, or wire screen, which prevents the operator's hands from being caught between the closing parts of a mold.

Laminate: A composition of several layers of plastic firmly adhered together by partly melting one or more of the layers, by an adhesive, or by impregnation.

Lubricant: Material added to improve the feeding of powder or granules into molding or extrusion machines, improve the flow of molten plastic through machines and into molds, or to prevent adhesion of plastic to molds ("mold-release"). Typical lubricants are zinc stearate, carnauba wax, and silicone oil.

Molding: Forcing plastic into a cavity to achieve a desired shape. The term is used in contrast to one form of casting that is taken to mean filling a mold with little or no force.

Monomers: Small starting molecules, usually gaseous or liquid, used to produce the polymer resins.

Plasticizers: Organic materials added to plastics to facilitate compounding or to make the finished product more flexible. Some plasticizers increase the combustibility of the plastic while others serve as flame retardants.

Plastics: Materials that contain as an essential ingredient an organic substance of large molecular weight, are solid in their finished state, and can be shaped by flow at some stage in their manufacture of processing into finished articles.

Polymerization: Adding molecules of a monomer to themselves or other monomers in a repetitive way so as to make a longer chainlike molecule.

Processing: In the plastics industry, converting polymers into useful articles by molding or extruding from granules, depositing film from solvent, or laminating resin and reinforcement. Used in contrast to fabricating. Most often the molding operation uses heat, but is entirely or largely a physical rather than a chemical process.

Reinforced Plastic: A plastic with a filler which significantly increases flexural, impact, or tensile strength. Additives are usually glass, asbestos, cotton, or nylon fibers. Reinforced plastics may be thermoplastic granules for injection molding, or for use in large areas of reinforcement as in layup or pulp molding.

Resin: A solid, semisolid, or pseudosolid organic material which has indefinite and often high molecular weight.

RAW MATERIALS

Many plastic feedstocks are derived from fractions of a petroleum or gas recovered during the refining process. Ethylene monomer (one of the most important feedstocks) is derived in gaseous form from petroleum refinery gas, liquefied petroleum gases, or liquid hydrocarbons.

The monomer is subjected to a chemical reaction, known as polymerization, that causes the small molecules to link together into increasingly longer molecules. Chemically, the polymerization reaction has turned the monomer into a polymer. The polymer, or plastic resin, must next be prepared for use by the processor who will turn it into a finished product. Sometimes it is used as it comes out of the polymerization reaction. More often, however, it is transformed into pellets, granules, flake, or powder. Feedstocks are also available as semisolids, e.g., pastes, or as liquids for casting.

THE PRODUCTION PROCESS

The ways in which plastics can be processed into useful end products are as varied as the plastics themselves. Though the processes differ, there are elements common to many of them.

In the majority of cases, thermoplastic compounds (those resins which may be repeatedly softened by heating and hardened by cooling without effecting a chemical change) must be melted by heat so they can flow. Pressure is often involved in forcing the molten plastic into a mold cavity or though a die, and cooling must be provided to allow the molten plastic to harden. With thermosets (those polymerized plastics which cannot be softened by continued heating without effecting a chemical change, and once hardened cannot be resoftened for useful flow), heat and pressure are most often used. In this case, however, heating (rather than cooling) serves to cure (set) the thermosetting plastic under pressure in the mold. When thermoplastic or thermoset resins are in liquid form, heat and/or pressure need not necessarily be used; although they do play a role in many casting techniques intended for high speed production.

The following descriptions cover the basics of the major manufacturing processes. It should be recognized, however, that there are variations in virtually every process.

Blow Molding: Blow molding is a process generally used only with thermoplastics. It is applicable to the production of hollow plastic products such as bottles, gas tanks, and carboys. Blow molding involves melting the thermoplastics resin, then forming it into a tubelike shape (known as a parison), sealing the ends of the tube, and injecting air (through a needle inserted in the tube) so that the tube, in its softened state, is inflated inside the mold and forced against the walls of the mold. (See Fig. 10-12A.)

Calendering: Calendering is a process used to convert thermoplastics into film and sheeting, and to apply a plastic coating to textiles or other supporting materials. In

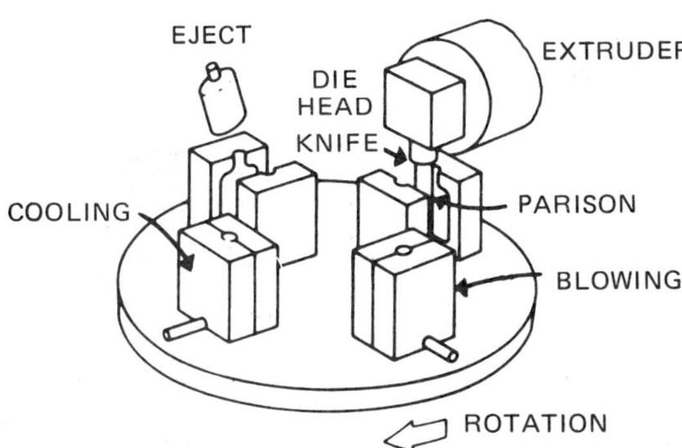

FIG. 10-12A. Diagram of continuous extrusion blow molding set up, using a rotating horizontal table. Plastic parison (in cylindrical shape) is extruded from die into mold which closes on the parison (knife cuts the parison off from the extrudate). Mold then rotates to second station where air is injected into the parison (which is still hot and therefore formable) to blow it out to the shape of the inside mold cavity. At third station, the blown part is allowed to cool and set. Mold finally rotates to last station where finished part (in this case, a bottle) is ejected from mold. (Society of the Plastics Industry)

calendering film and sheeting, the plastic compound is passed between a series of three or four large heated revolving rollers which squeeze the material between them into a sheet or film.

Casting: Casting is used both with thermoplastics and thermosets to make products, shapes, rods, and tubes. A liquid monomer-polymer solution is poured into an open or a closed mold where it finishes polymerizing into a solid. Unlike the molding process, pressure need not be used with casting.

Coating: Coating is accomplished by applying either thermoplastic or thermosetting materials to metal, wood, paper, glass, fabric, leather, or ceramics. (See Fig. 10-12B.)

Compounding: Compounding is the mixing of additives into previously formed resin on masticating rolls or calenders, as for rubber, or in kneading mixers or screw extruders of varied design. Rolls and kneaders are heated by high pressure steam or heat transfer fluids.

NOTE: Screw extruders have largely displaced rolls and kneaders for compounding because they provide better control, continuous output, and less exposure of hot plastic to air.

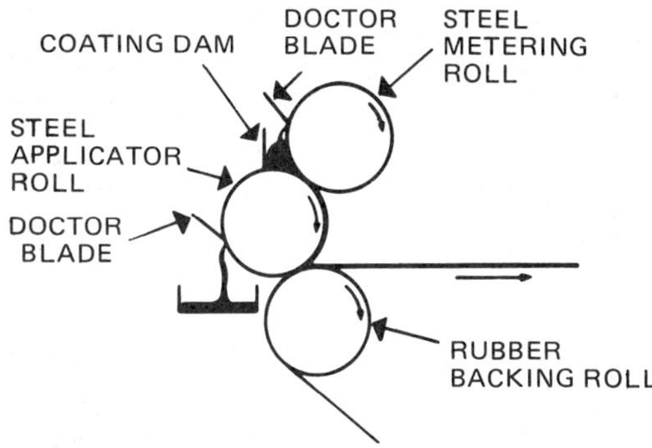

FIG. 10-12B. A typical coating set up, known as a 3 roll nip fed reverse roll coater. Plastic feeds from dam through nip between steel metering and applicator rolls, rotating in the same directions. At bottom of the applicator roll, plastic is laid on top of the substrate (e.g., fabric, paper, etc.) as it comes in contact with the substrate at nip between applicator roll and backing roll (which carries the substrate up from the bottom of set up). Doctor blade is used to scrape off excess plastic from roll. (Society of the Plastics Industry)

Compression Molding: Compression molding is a common method of forming thermosetting materials. In compression molding, the material is squeezed into the desired shape by applying heat and pressure to the material in a mold. (See Fig. 10-12C.)

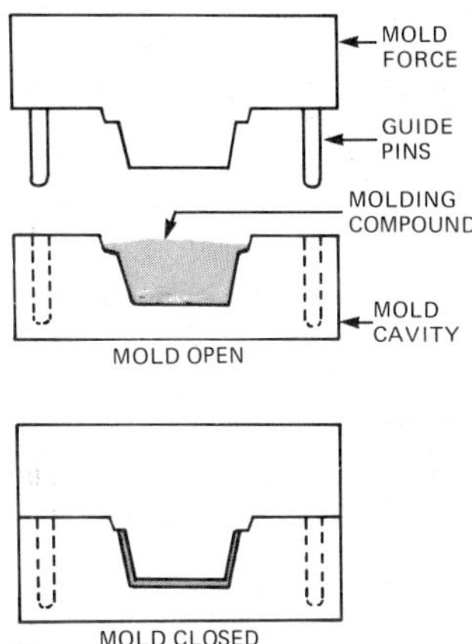

FIG. 10-12C. Basics of a simple two piece compression mold. Plastic molding material is loaded into lower half (cavity) of the heated mold (shown at top). Top half of the mold (mold force) is then lowered and the two halves are brought together under pressure (shown below). The softened molding material is thus formed into the shape of the cavity and allowed to harden with further heating. Mold is then opened and part is removed. (Society of the Plastics Industry)

Extrusion: Extrusion is the forming of thermoplastic materials into continuous sheeting, film, tubes, rods, profile shapes, or filaments, and the coating of wire, cable, and cord. (See Fig. 10-12D.)

Foam Plastics Molding: In this manufacturing process, foams can be used in casting, calendering, coating, rotational molding, flow molding, and even injection molding and extrusion. (See Fig. 10-12E.) The manufacturing of foamed products is discussed in more detail later in this chapter under "Flammable Solvents."

High Pressure Laminating: The high pressure laminating process uses thermosetting plastics to hold together reinforcing materials such as cloth, paper, wood, or glass fibers. Heat and high pressure are used to produce the laminated product.

Injection Molding: A process that uses thermoplastic material which is softened by heat, then allowed to cool and harden. (See Fig. 10-12F.)

Reaction Injection Molding (RIM): RIM is used primarily for molding polyurethane elastomers of foams into end products with solid integral skins and cellular cores. Two

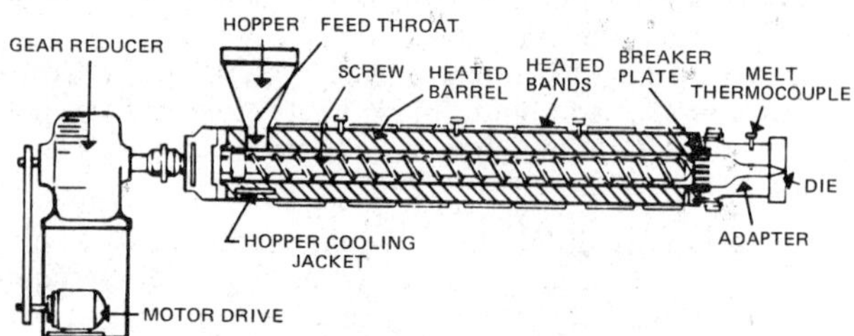

FIG. 10-12D. In a basic single screw extruder, plastic pellets (or powders) are fed through the hopper, through the feed throat, and into a screw that rotates in a heated barrel. The rotation of the screw (which is powered by the drive motor) conveys the plastic forward for melting and delivery through the breaker plate (reducing the rotary motion of the melt), through the adaptor, and into the die which dictates the shape and size of the final extrudate. (Society of the Plastics Industry)

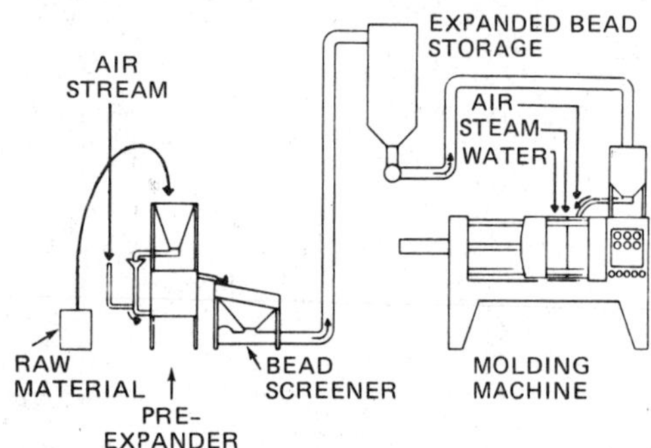

FIG. 10-12E. Among the many variations in molding foamed plastics is this set up for stream chest molding expandable styrene beads into products like foam cups, novelties, building products, etc. In this operation, the expandable beads, containing a blowing agent, are preexpanded with steam, then screened to remove large clumps. The expanded beads next are blown into a storage hopper and allowed to dry and stabilize. From here, they feed into the final mold where steam is again used to complete expansion of the beads so they fill the mold and fuse together. Water is used for cooling, prior to opening the mold and removing the finished foamed styrene part. (Society of the Plastics Industry)

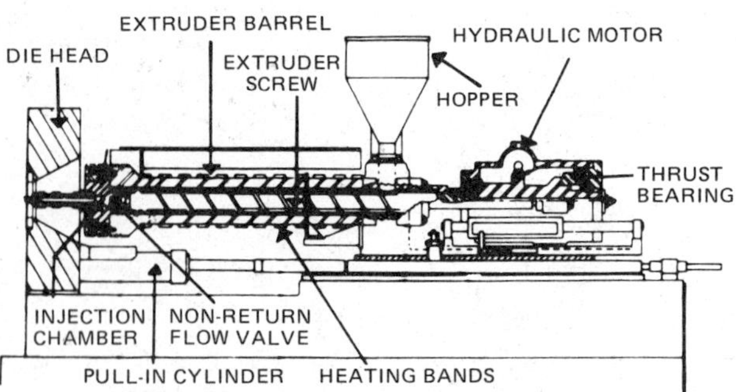

FIG. 10-12F. Diagram of reciprocating screw injection molding machine. Plastic pellets feed through the hopper into the screw (much like the screw in an extruder) where they are compacted, melted, and pumped by the rotation of the screw past the nonreturn flow valve (allowing material to flow right to left, but not from left to right) to the front of the screw where it is allowed to accumulate. At the proper time, the rotation of the screw is stopped and the amount of molten plastic in front of the screw is injected into the mold, using the screw as a plunger activated by the hydraulic injection cylinders. In the mold, the molten plastic flows throughout the cavity, completely filling it. The plastic is then allowed to cool and harden, the mold is opened, and the finished part removed. The back end of the machine shown above contains the motors and drives needed to power the machine. (Society of the Plastics Industry)

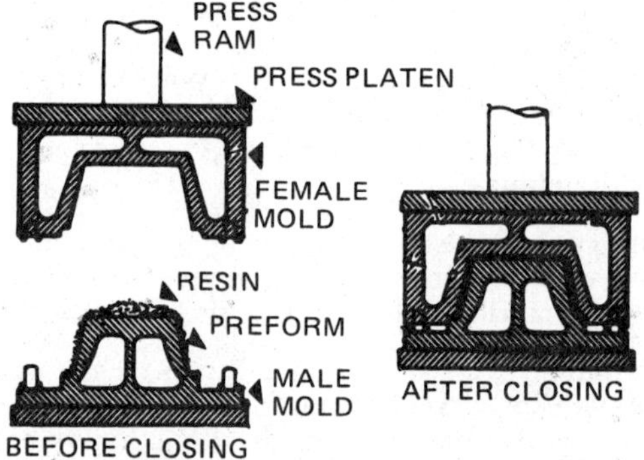

FIG. 10-12G. Matched die molding is one technique for producing reinforced plastic parts. Basically, it is a compression molding process in which resin and glass fibers are shaped into the finished product under heat and pressure between the two halves (male and female) of a mold. The glass fiber reinforcements are laid over the male mold of a "preform," a combination of glass and resin preformed before molding to the basic shape of the part to be molded. Additional liquid resin mix is added before the mold is closed. (Society of the Plastics Industry)

or more pressurized reactive streams are impinged together under high pressure in a mixing chamber. The resulting mixture is then injected, under low pressure, into the mold where the reaction begins, and continues until the liquid mixture has set into a solid or cellular finished product.

Reinforced Plastics Processing: A process in which resins (acting as binder material) are combined with reinforcing materials (usually in fibrous form) to produce composite products having exceptional strength-to-weight ratios and outstanding physical properties. The resins may be either thermosets or thermoplastics. (See Fig. 10-12G.)

Rotational Molding: Like blow molding, rotational molding is used to make hollow one piece parts. Finely powdered plastic or molding granules are placed in a heated cavity that is rotated about two axes to distribute the plastic.

Thermoforming: In thermoforming, thermoplastic sheeting is heated to its softening temperature and forced against the contours of a mold by mechanical means, e.g., tools, plugs, solid molds, etc., or by pneumatic means, e.g., differentials in air pressure which are created by pulling a vacuum between sheet and mold or by using compressed air to force the sheet against the mold. Thermoforming is referred to as pressure forming, vacuum forming, and plug assist forming. (See Fig. 10-12H.)

Transfer Molding: Transfer molding is generally used for thermosetting plastics. It is similar to compression mold-

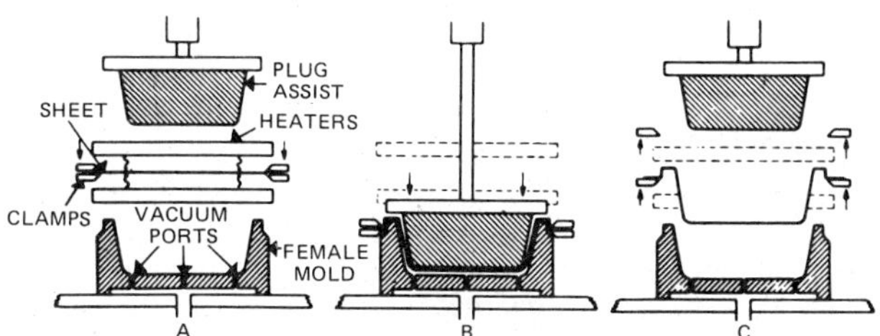

FIG. 10-12H. This variation on the thermoforming of plastic sheet is known as plug assist vacuum forming. In operation, the plastic sheet is clamped in place and heaters move in to heat the sheet top and bottom to soften it (A). Heaters are then withdrawn, and the frame holding the sheet is lowered down to contact the mold. At this point, the plug assist is lowered into the softened sheet, stretching it down to the bottom of the mold cavity (B). After the plug assist has reached its closed position, a vacuum is drawn through the ports to pull the stretched sheet completely into the cavity and finish the forming. Next, the plug assist is withdrawn, the formed sheet is cooled, and the clamps are opened to remove the formed part from the frame (C). (Society of the Plastics Industry)

ing because the plastic is cured into an infusible state in a mold under heat and pressure.

THE FIRE HAZARDS

Plants converting plastic compounds into finished products are subject to a variety of hazards that can result in explosions and fire. The broad area of hazard involves the presence of combustible dusts, flammable and combustible liquids, high heat elements, hydraulic and heat transfer fluids, static electricity, and failure to observe good storage and housekeeping practices.

Dusts

Many, but not all, plastics will burn rapidly when in the form of dust. Some plastics, such as vinyl and Teflon,® do not represent a dust explosion hazard. If dispersed in air, plastic dust can be explosively ignited by a spark, flame, or metal surface above 700°F (371°C). Dust explosions are possible in operations which convey pulverized plastic through pneumatic conveying systems, or produce dust by machining or sanding in finishing work. Additives also require safeguarding while being added to plastics.

Plastic pellets are cubes or cylindrical pellets which have been screened by the maker to remove finer particles. These pellets are free from the hazard of dust explosion; however, dust may be generated by abrasion of these particles when conveyed in a long pneumatic system.

Through regrinding, trimmings from processing operations are cut to small size for reuse. Although regrinding is a shearing and impacting action, it may generate some fine powder. Dust hazards should be considered when much regrinding is performed.

Additives as colorants, fillers, mold-release or flow improving lubricants, plasticizers for flexibility, ultraviolet or heat stabilizers, or modifying resins are also a source of dust hazards. Most of these are charged to mixing equipment as fine powders, making dust explosions possible. Some compounding is done at molding machines in plastics plants of all types.

Generally, the basic chemical structure of the plastic resin governs the explosibility of its dust. Incorporation of combustible fillers may increase the explosibility of dust. Incorporation of low percentages of fire retardants has little effect on the explosibility of dusts.

Dust should be kept to a minimum. Provisions should be made to reduce the possibility of ignition, relieve explosion pressure, and confine and control fire. Guidance in controlling dust explosion hazards in plastics plants is found in NFPA 654, *Standard for the Prevention of Fire and Dust Explosions in the Chemical, Dye, Pharmaceutical, and Plastics Industries*. Other applicable standards are NFPA 68, *Guide for Explosion Venting*, and NFPA 69, *Standard on Explosion Prevention Systems*.

Flammable Solvents

Flammable organic solvents, found in nearly every plastics plant, may be used in very small quantities to apply adhesives, lacquers, or paints to molded or fabricated items; and in large amounts to coat plastic on cloth, paper, leather, or metal, or on metal belts from which a dried film will be stripped. In these uses, the choice of solvents, and hazards in their use, is no different from most lacquering operations. There may be some increase in hazard when solvents are applied to plastics, particularly when printing or coating on fast moving films, because plastics usually have a high electrical resistivity; therefore they generate and retain static charges more readily than paper or cotton fabric.

Use of solvents for the preparation of rigid or flexible foam plastic is increasing. The plastic is moistened with solvent and heated above the boiling point of the solvent in a closed mold or extruder. When the pressure is released, the boiling solvent expands the resin and produces a bubbled structure. Another foaming process heats a resin containing a chemical additive which evolves gas (usually nitrogen), carbon dioxide, or steam. Either the solvent or the chemical agent is called a blowing agent. The most common rigid foam is made from beads of polystyrene moistened with about 10 percent pentane or a similar hydrocarbon. The pentane does not soften the resin at room temperature, but does at the low pressure steam temperature used in molding or extruding. It is imperative that sources of electric spark or flame be avoided. Resin is usually shipped from the manufacturer premixed with the hydrocarbon blowing agent. Containers will have free vapor of pentane, and should be opened outdoors or in an area with good exhaust ventilation. Expanded articles should be aged under forced hot air before shipping to remove nearly all flammable blowing agents.

Flexible urethane foams for upholstery and garments, or rigid foams for pour-in-place construction or refrigeration insulation, are blown by the generation of carbon dioxide. Auxiliary blowing agents, such as methylene chloride for flexible foams and chloro-fluorocarbons for rigid foams, are also used. Hydrocarbons such as pentane are sometimes used for expanding polystyrene foam.

Foams of polyvinyl chloride known as "expanded vinyl," are used in garments and upholstery. These foams usually are made with the chemical blowing agent azodicarbonamide, also called azobisformamide. Above 300°F (149°C) it provides nitrogen gas at a controlled rate, is essentially nonhazardous in the proportions used, and does not give off flammable vapors.

Improper handling of flammable liquids has caused serious fires in plastics plants. The most frequent causes of fires have been failures to recognize the importance of static spark prevention, explosion proof electrical equipment, and vapor removal systems. NFPA standards on flammable and combustible liquids should be consulted.

Static Electricity

Because plastics are such good electrical insulators, static electricity on them can rapidly build up to spark discharge, a hazardous condition if dust or flammable vapors are present. Operations which can generate static are stripping of films from production or printing equipment, or rapid passage of films across rolls or guides. Belts for power transmission are also significant sources of static discharge. Because of their low water absorption and high resistivity, plastics cannot have their static charge dissipated by high ambient humidity as is the practice in cotton, wool, and paper mills. Attention should be given to grounding of equipment and insuring that tinsel conductors firmly contact moving films or filaments. Care should be taken to separate vapor and dust hazards from machines where static electricity ignition sources could develop. NFPA 77, *Recommended Practice on Static Electricity*, is a good source for the safeguards to use in protecting against static electricity hazards.

Heating Elements

Molding and extruding operations for shaping and compounding articles have hazards associated with local overheating of electrical components. Operating temperatures normally range from 300 to 650°F (149 to 343°C), depending upon which plastic is being processed. The upper range is beyond that practical for heat transfer fluids, so electric resistance heating is almost universally employed. Heater bands are required to fuse the resin in the feed section at the upstream end of the extruder barrel. Controllers can stick, permitting resistance heaters to operate above the temperature set for the thermocouple controller. In most cases, the character of extrudate will change markedly well before heater bands get hot enough to be a source of ignition. Cleanliness in molding and extruding areas is vital to reduce the hazard of ignition from overheated bands where flammable vapors may be generated.

Some areas within equipment may not be regularly purged by flow of the plastic feedstock. Material remaining in such areas can be subject to too high a temperature or be kept too long at a normally acceptable temperature. Decomposition may then take place, not often forcefully, but with the release of gases that may be combustible. It is good practice to start heating such equipment first at the downstream end to ensure fluidity of material and hence relief of pressure. (See NFPA 79, *Electrical Standard for Industrial Machinery*.)

Hydraulic Pressure Systems

Hydraulic systems are used to clamp molds and to provide pressure to rams or screws which force molten plastic into molds by compression, transfer, or injection molding. The molten plastic may be at pressures up to 20,000 psi (137 900 kPa), but the hydraulic systems are normally less than 2,000 psi (13 790 kPa). Petroleum fluids have been used in plastic operations where heating elements were generally below 600°F (376°C); the same fluids have a poor record in die casting because pots for the molten metal are at much higher temperatures.

Storage Arrangements

The fire hazards of plastics in storage, whether as feedstocks or as finished articles, are determined by their chemical composition, physical form, and storage arrangement. The storage of plastics generally should not exceed a maximum height of approximately 20 ft (6 m), and venting is desirable in building construction consideration. If fire occurs, large quantitites of smoke are usually generated, making manual fire fighting difficult.

Plastics such as fluorocarbons, rigid or lightly plasticized polyvinylchloride, and phenolics can be protected the same way as any Class III commodity regardless of their physical form or storage arrangement. Pellets and small objects can be protected the same as Class IV commodities (commodity classifications are based on NFPA 231, *Standard for General Storage*).

Thermoplastics such as polyurethane, polyethylene, and plasticized polyvinylchloride, and thermosets such as polyesters, present a severe fire hazard, exceeded only by polystyrene and acrylonitrile-butadiene-styrene (ABS) thermoplastics. These plastic materials will melt and break down (depolymerize), acting and burning like flammable liquids. In the form of foamed material, these plastics represent the most severe fire hazard.

One story buildings without basements are preferable for storage of plastics materials because of greater efficiency for fire fighting, ventilation, and salvage operations.

Housekeeping Practices

Housekeeping is basic to good firesafety. Good housekeeping practices reduce the danger of fire simply by controlling the presence of unwanted fuels, obstructions, and sources of ignition. Listed containers should be provided and properly maintained for the disposal of refuse and rubbish. Spills of flammable liquids or combustible materials should be promptly cleaned up and properly disposed. Removal of combustible dust and lint accumulations from walls, ceilings, and exposed structural members is necessary. Areas containing flammable liquids, vapors, or combustible plastics materials, processing, manufacturing, or storage must be clearly identified to prohibit smoking or the use of open flame devices.

THE SAFEGUARDS

Good fire protection starts with the design of the plant or warehouse, or inspection and modification of the existing facilities, if necessary. Sprinkler protected, noncombustible construction is appropriate in buildings for storage, processing, and manufacturing of combustibles. Automatic sprinklers, as well as standpipe and hose systems, should be supplemented by portable fire extinguishers and special automatic systems suitable for flammable liquid fires and electric fires, where these hazards exist. Sprinkler systems should be designed to provide an initial high sprinkler discharge density over a relatively small area, and a secondary lower discharge density over a large area to protect these types of plastics. Consideration should be given to the provision of roof vents, particularly in large one story warehouses of manufacturing plants.

Building Construction

Long narrow buildings provide greater ease in protection and fire fighting than large square buildings. One story buildings without basements are preferable to multistory buildings which may be subject to the spread of fire between floors.

Storage and manufacturing areas need to be separated from each other by walls with a sufficient fire rating to protect each area and occupancy from the other in case of fire. Preferably, fire walls should be without openings, but if openings are necessary, protection can be provided by self-closing or automatic fire doors suitable for openings in fire walls. Generally, a single fire area should not exceed 50,000 sq ft (4645 m^2).

Fire Control Systems for Processing

An alarm system that alerts building occupants, notifies fire suppression departments, and activates automatic suppression equipment is a desirable protection feature. Rapid extinguishment may be achieved by providing plastics processing equipment, conveyors, and manufacturing machinery subject to ignition or explosion, with automatic fire, smoke, or explosion detecting devices arranged to initiate an alarm and to activate automatic suppression systems (water spray, foam, dry chemical, carbon dioxide, or halogenated extinguishing agents).

Sprinklers are the most important single system for automatic control of fires in plastics plants. Among the advantages of automatic sprinklers are the facts that they operate directly over a fire, and that smoke, toxic gases, and reduced visibility, often associated with fires in plastics, do not affect their operation. Automatic sprinklers, standpipes, and fire hose connections depend upon an adequate water supply delivered with the necessary pressure to control fires.

Bibliography

NFPA Codes, Standards, Recommended Practices and Manuals. (See the latest *NFPA Codes and Standards Catalog* for availability of current editions of the following documents.)

NFPA 10, *Standard for Portable Fire Extinguishers.*
NFPA 11, *Standard for Low Expansion Foam and Combined Agent Systems.*
NFPA 11A, *Standard for Medium and High Expansion Foam Systems.*
NFPA 12, *Standard on Carbon Dioxide Extinguishing Systems.*
NFPA 13, *Standard for the Installation of Sprinkler Systems.*
NFPA 14, *Standard for the Installation of Standpipe and Hose Systems.*
NFPA 15, *Standard for Water Spray Fixed Systems for Fire Protection.*
NFPA 16, *Standard for the Installation of Deluge Foam-Water Sprinkler Systems and Foam-Water Spray Systems.*
NFPA 17, *Standard for Dry Chemical Extinguishing Systems.*
NFPA 30, *Flammable and Combustible Liquids Code.*
NFPA 33, *Standard for Spray Application Using Flammable and Combustible Materials.*
NFPA 35, *Standard for the Manufacture of Organic Coatings.*
NFPA 40E, *Code for the Storage of Pyroxylin Plastic.*
NFPA 68, *Guide for Explosion Venting.*
NFPA 69, *Standard on Explosion Prevention Systems.*
NFPA 70, *National Electrical Code.*
NFPA 77, *Recommended Practice on Static Electricity.*
NFPA 79, *Electrical Standard for Industrial Machinery.*
NFPA 91, *Standard for the Installation of Blower and Exhaust Systems for Dust, Stock, and Vapor Removal.*
NFPA 101, *Code for Safety to Life from Fire in Buildings and Structures (Life Safety Code).*
NFPA 231, *Standard for General Storage.*
NFPA 654, *Standard for the Prevention of Dust Explosions in the Chemical, Dye, Pharmaceutical, and Plastics Industry.*

Additional Readings

Beck, Ronald D., *Plastic Product Design*, 2nd ed., Van Nostrand Reinhold Co., NY, 1980.
Bulletin No. 3, *Fire Hazards of Hydraulic Fluids Used in Processing Plastics*, Society of the Plastics Industry (SPI), NY.
Bulletin No. 6, *Fire and Explosion Due to Electrostatic Charges in the Plastics Industry*, SPI, NY.
Bulletin No. 9, *Planning Good Fire Protection for Plastics Plants*, SPI, NY.
Bulletin No. 10, *Portable Fire Extinguishers for Use in the Plastics Industry*, SPI, NY.
Bulletin No. 12, *Fire Safe Electrical Installations for Hazardous Locations in the Plastics Industry*, SPI, NY.
Bulletin No. 13, *Flammable and Combustible Waste Disposal in the Plastics Industry*, SPI, NY.
Bulletin No. 14, *A Plastics Plant Fire Squad*, SPI, NY.
Bulletin No. 16, *A Basic Fundamental of Fire Protection for Small Plastics Plants*, SPI, NY.
Bulletin No. 21, *Fire Safety in the Storage of Materials in Plastics Plants*, SPI, NY.
Crawford, R. J., *Plastics Engineering*, Pergamon Press, Inc., Elmsford, NY, 1981.
Dubois, J. Harry, ed., *Plastics*, 6th ed., Van Nostrand Reinhold Co., NY, 1981.
Frados, Joel, *Plastics Engineering Handbook*, 4th ed., Van Nostrand Reinhold Co., NY, 1976.
1978 Facts and Figures of the Plastics Industry, SPI, NY, 1978.
Pajgit, O., ed., *Processing of Polyester Fibres*, Textile Science and Technology Series, Elsevier Science Publishing Co., Inc., NY, 1980.
Rosato, D. V., and Lawrence, John R., *Plastics Industry Safety Handbook*, SPI, NY, 1973.
Schwartz, Seymour, and Goodman, Sidney, *Plastics Materials and Processes*, Van Nostrand Reinhold Co., NY, 1982.
The Story of the Plastics Industry, 13th ed. rev., SPI, NY, 1971.

MOLTEN SALT BATHS

Revised by Q. D. Mehrkam

Molten salt baths have become increasingly popular for the heat treatment of metals because they provide rapid and precise heat transfer and the required equipment is relatively inexpensive. A molten salt bath is defined as any heated container that holds a melt or fusion of one or more chemical salts in a fluid state. Metal work to be treated is immersed in the bath. There are many kinds of salt baths that use various salts and salt mixtures and that operate at various temperatures, depending on the results required.

Additional information may be found in Chapter 5 of this Section, "Industrial and Commercial Heat Utilization Equipment." Section 11, Chapter 6 should be consulted for relevant information on "Storage and Handling of Chemicals."

As work is immersed fully into a salt bath, the salt serves as a protective environment. Heat is transferred more rapidly than by any other type of furnace heating. Moreover, because of the constant convection currents in the heated liquid medium, uniform temperature distribution is more easily attained.

TYPES OF SALT BATHS

The various types of salt baths available today can be classified by the heating method, application, or the particular salt mixture utilized.

Salt baths may be electrically heated or fuel fired. Electrically heated baths may have electrodes introduced through the top surface of molten salt or through the sides of the furnace below the salt surface. (See Figs. 10-13A and 10-13B.) In some instances, ribbon type resistance elements are used around a metal pot or container. (See Fig. 10-13C.)

There are two types of fuel fired salt baths. In one, the burners are fired directly outside a retort containing the salt. In most instances the burners are contained in metal tubes, known as radiant tubes, and the tubes are immersed in the molten salt. (See Fig. 10-13D.)

Mr. Mehrkam is President of the Ajax Electric Company, manufacturer of industrial furnaces, based in Huntingdon Valley, PA.

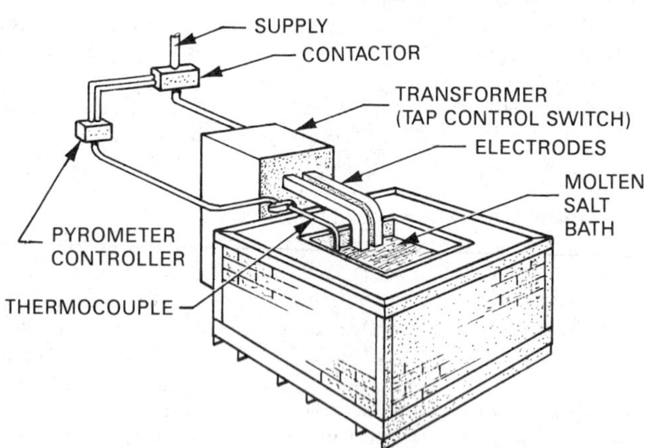

FIG. 10-13A. Typical electrode immersion type electrical heating arrangement.

APPLICATIONS

Major uses of molten salt baths are descaling, liquid carburizing (case hardening), cyaniding and nitriding, neutral hardening, tool steel hardening, annealing, brazing tempering, and isothermal quenching. Each of these operations requires a different salt or salt mixture at different temperatures.

Descaling

For the removal of oxides which are difficult to remove by normal means, i.e., pickling, grit blasting, polishing, etc., there are three different descaling processes. The processes are: the reducing sodium hydride process, the oxidizing sodium hydroxide process, and the electrolytic process. In the reducing sodium hydride process, a fused melt of sodium hydroxide and sodium hydride at temperatures between 650 and 780°F (343 and 415°C) is used. For oxidizing, a melt of sodium carbonate, sodium chloride, and sodium hydroxide is used at temperatures from 700 to 950°F (371 to 510°C). In the electrolytic process, a fused melt of sodium carbonate, sodium chloride, and sodium hydroxide is used in two baths at about 900°F (402°C). In

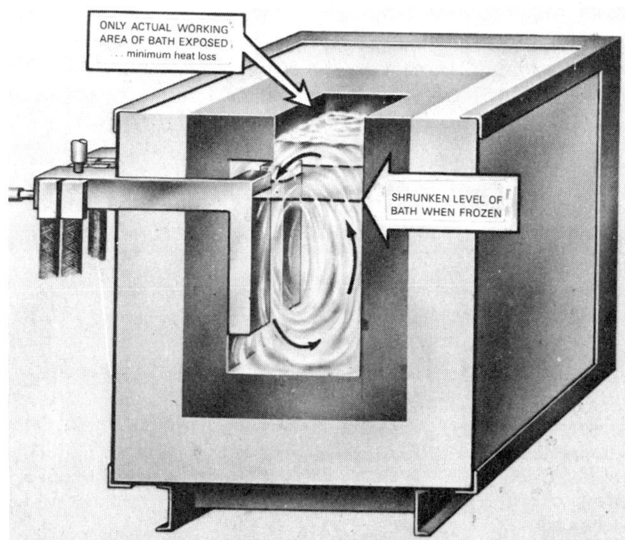

FIG. 10-13B. Submerged electrode type salt bath. (Ajax Electric Co.)

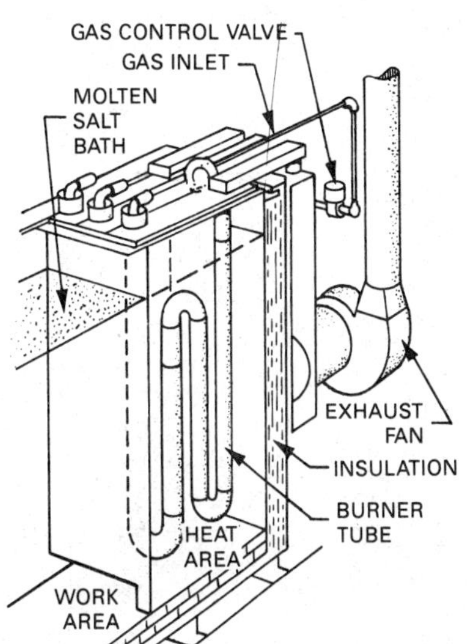

FIG. 10-13D. Gas fired radiant tube salt bath.

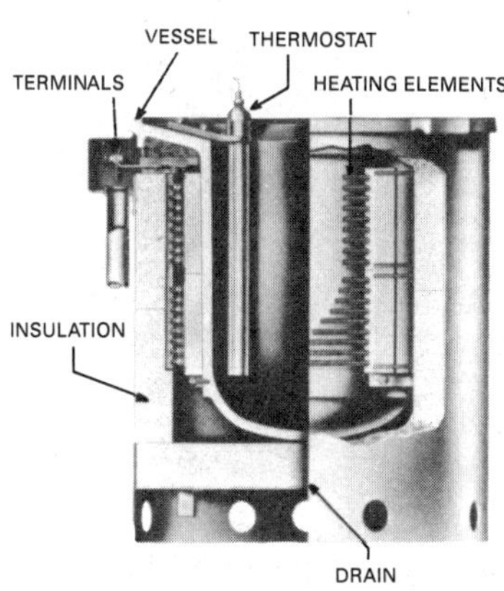

FIG. 10-13C. Resistance heated salt bath with pot. (Sunbeam Equipment Co.)

FIG. 10-13E. Liquid carburizing and isothermal heat treating line. (Ajax Electric Co.)

the first bath, direct current passes from cathode grids to the metal, and in the second bath current passes from the metal to anode grids.

Liquid Carburizing

Case hardening is accomplished by diffusing carbon and small amounts of nitrogen into steel surfaces. The salts used are molten mixtures of sodium cyanide, sodium chloride, and barium chloride at temperatures between 1,450 and 1,750°F (788 and 955°C), depending upon the depth of case required. (See Fig. 10-13E.)

Cyaniding and Nitriding

These two operations are also performed with either sodium or potassium cyanides as the chief ingredient. Usually the operating temperatures are lower, 950 to 1,050°F (371 to 565°C) for nitriding.

Neutral Hardening

The hardening of ferrous alloys without harmful surface effects is performed in mixtures of sodium, potassium, and barium chlorides. Operating temperatures may range between 1,400 to 2,350°F (760 to 1288°C).

Hardening of High Speed Tool Steels

High speed tool steels are hardened in a series of molten salt baths at various temperatures. A preheat bath at approximately 1,400°F (760°C) may be followed by a high heat bath at 2,350°F (1288°C), which in turn is followed by a quench bath at 1,100°F (593°C). All of these baths are mixtures of barium, sodium, and potassium chlorides. (See Fig. 10-13F.)

Isothermal Quenching

To achieve various levels of physical properties in the heat treatment of steels, isothermal quenching rapidly cools metal parts in a molten salt bath. The three major types of isothermal quench are austempering, martempering, and cyclic annealing. Each produces different hard-

FIG. 10-13F. Salt bath for hardening of high speed tool steel. (Ajax Electric Co.)

ness properties in metal alloys. Bath temperatures vary from 400 to 1,300°F (205 to 705°C). The salts used are nitrate/nitrite salts for the lower temperatures and neutral chloride types for the higher temperatures.

Annealing

Process annealing of carbon steels is usually carried out in carbonate and chloride salts at a temperature of 1,250 to 1,300°F (677 to 704°C). For stress relief of carbon steels, the melt is usually a nitrate salt at approximately 1,000°F (538°C). Chloride salts are used to anneal stainless steel products as well as nickel-chrome alloys at temperatures of 1,550 to 2,150°F (844 to 1177°C).

Brazing

Salt bath brazing of ferrous and nonferrous alloys with silver and aluminum alloys, brass, and copper is another popular application for molten salt baths. Depending on the alloy and type of brazing, temperatures and salt mixtures vary considerably. For aluminum dip brazing, the salt is a mixture of chlorides with smaller percentages of sodium and/or aluminum fluorides, which act as fluxes. In copper brazing, barium chloride mixtures are used. Chlorides of sodium and potassium are used for silver alloy and brass brazing applications. Carburizing salts are used for a combination of brass brazing and carburizing.

THE HAZARDS

The hazards common to salt bath furnaces may be divided into the following three types:

1. Fire caused by contact of molten salt with combustibles.
2. Explosion of the salt mixture due to physical or chemical reaction.
3. The danger to operating personnel.

Since salts are used at temperatures from 300 to 2,400°F (149 to 1315°C), the ejection of salt by popping, spattering, or spilling will create a fire hazard if it comes in contact with anything combustible, including wooden floors. Since molten salts have relatively little surface tension and low viscosities, any minor physical disturbance or chemical reaction can result in salt being ejected. When liquids (water, oil, etc.) or materials reactive to the particular salt utilized penetrate the surface of the salt bath, a violent ejection can result. Table 10-13A lists the

TABLE 10-13A. Melting Points of Common Chemical Salts

Salt	Melting point (°F)	Melting point (°C)
Barium chloride	1,764	963
Barium fluoride	2,336	1,280
Boric oxide (Anhydride)	1,071	578
Calcium chloride	1,422	773
Calcium fluoride	2,480	1,360
Calcium oxide	4,662	2,575
Lithium chloride	1,135	613
Lithium nitrate	491	255
Magnesium fluoride	2,545	1,397
Magnesium oxide	5,072	2,802
Potassium carbonate	1,636	892
Potassium chloride	1,429	777
Potassium cyanide	1,174	635
Potassium fluoride	1,616	880
Potassium hydroxide	716	380
Potassium nitrate	631	333
Potassium nitrite	567	297
Sodium carbonate	1,564	852
Sodium chloride	1,479	805
Sodium cyanide	1,047	564
Sodium fluoride	1,796	980
Sodium hydroxide	605	319
Sodium metaborate	1,771	967
Sodium nitrate	586	308
Sodium nitrite	520	271
Sodium tetraborate	1,366	742
Strontium chloride	1,603	873

most common salts and their melting points. Table 10-13B lists the most common salt mixtures and their melting points.

Nitrates are particularly hazardous due to their ability to start and support combustion. When nitrates are overheated to temperatures in excess of 1100°F (593°C) rapid dissociation with the release of nitrous oxide fumes occurs. These fumes are injurious to operating personnel and are corrosive to the adjacent equipment. Explosions may occur if nitrate salt leaks from an externally heated container (pot) into a superheated combustion chamber. The overheating of nitrates can be caused by temperature controls that malfunction, jammed workloads, an accumulation of sludge in the bath, or by operator errors.

If the operating temperature of a nitrate or nitrite bath exceeds the maximum limit, the following emergency action should be taken:

Shut off the heat supply.
Remove all work from the bath.
Move employees to a safe location.

Heated nitrate salts will react strongly with carbonaceous materials such as oil, soot, tar, graphite, and cyanides—all of which may be utilized in some other metal treating process. Accidental mixing of cyanides with molten nitrates can cause explosions of considerable magnitude.

TABLE 10-13B. Melting Points of Common Salt Mixtures*

Mixture and proportion	Melting points (°F)	(°C)
Lithium nitrate 23.3, sodium nitrate 16.3, potassium nitrate 60.4	250	121
Potassium hydroxide 80, potassium nitrate 15, potassium carbonate 5	280	138
Potassium nitrate 53, sodium nitrate 7, sodium nitrite 40	285	141
Potassium nitrate 56, nitrite 44	295	146
Potassium nitrate 51.3, sodium nitrate 48.7	426	219
Sodium nitrate 50, sodium nitrite 50	430	221
Sodium hydroxide 90, sodium nitrate 8, sodium carbonate 2	560	294
Lithium chloride 45, potassium chloride 55	666	353
Barium chloride 31, calcium chloride 48, sodium chloride 21	806	430
Calcium chloride 66.5, potassium chloride 5.2, sodium chloride 28.3	939	504
Calcium chloride 67, sodium chloride 33	941	505
Potassium chloride 35, sodium chloride 35, lithium chloride 25, sodium fluoride 5	960	516
Potassium chloride 40, sodium chloride 35, lithium chloride 20, sodium fluoride 5	990	533
Barium chloride 48.1, potassium chloride 30.7, sodium chloride 21.2	1,026	555
Sodium chloride 27, strontium chloride 73	1,049	565
Potassium chloride 50, sodium carbonate 50	1,086	585
Barium chloride 35.7, calcium chloride 50.7, strontium chloride 13.6	1,110	599
Barium chloride 50.3, calcium chloride 49.7	1,112	600
Potassium chloride 61, potassium fluoride 39	1,121	605
Sodium carbonate 56.3, sodium chloride 43.7	1,177	637
Calcium chloride 81, potassium chloride 19	1,184	640
Barium chloride 70.3, sodium chloride 29.7	1,209	654
Potassium chloride 56, sodium chloride 44	1,220	660
Sodium chloride 72.6, sodium fluoride 27.4	1,247	675
Barium fluoride 70, calcium fluoride 15, magnesium fluoride 15	1,454	790
Barium chloride 83, barium fluoride 17	1,551	845
Calcium fluoride 48, magnesium fluoride 52	1,738	949

* Lowest constant melting points given; proportions are percentages by weight.

The introduction of aluminum or magnesium parts into baths in which the molten salt temperatures approach or exceed the melting points of these metals can cause fire and explosion.

The storage of salts can cause serious problems. Chemicals such as nitrates and cyanides may mix and interact. Many salts are hydroscopic and, when stored in damp areas, will absorb moisture. When these salts are subsequently heated, moisture will be released below the surface and an explosion will ensue.

Intensive external reheating and remelting of a solidified salt bath may result in sufficiently rapid expansion to bulge or rupture the container. Such reheating can generate enough pressure to fracture the surface crust of the salt and scatter hot salts over a wide area.

When exposed to air having even a slight moisture content, the fumes from many salt baths become corrosive because chlorides, fluorides, and other salts will form acids by hydrolysis. These fumes are highly corrosive to adjacent building structures, wiring, and equipment and are, of course, detrimental to the bath operator.

THE SAFEGUARDS

While there appear to be many hazards connected with the operation of molten salt bath furnaces, fire, explosion, and injury can be prevented by the constant observance of ordinary precautions.

Clean dry sand may be used for diking purposes to confine and prevent the spread of the escaped melt. Carbon dioxide or dry chemical extinguishers may be used to extinguish burning carbonaceous material in the immediate vicinity of the salt bath.

The preferred location for a salt bath is a noncombustible area. The bath may be housed in a cement lined pit or curbed area large enough to contain the contents in case of leakage or spill. It should be protected against leakage of liquids from other sources and be provided with baffles to prevent splashover from one tank to another. Salt bath furnaces provided with steel enclosures greatly reduce risks of leakage.

All salts should be shipped and stored in tightly covered containers designed to prevent the absorption of liquids or moisture. Nitrate salts should be stored in a fire resistive, damp free room and separated a reasonable distance from heat, liquids, and reactive chemicals. Nitrate and/or nitrite salts should never be stored in the vicinity of cyanide salts. Only the actual amount required to recharge the bath should be removed from the storage area, and this amount should be melted immediately. Salts should be transported from the storage area to the furnace in suitable containers to prevent loss during transport.

The safety precautions that apply to the heating systems of salt baths are identical to those for the heating systems of other industrial furnaces. Some additional safety measures are necessary. Flame should be tangential to the wall of a salt pot or container when a gas or oil fired heating system is used. The products of combustion should be vented. Radiant tubes and electrodes should be of materials resistant to the corrosive action of the salts used.

The buildup of sludge in any type of salt bath should be avoided, and the sludge periodically removed. Sludge is caused by the deterioration of salt, the introduction of foreign material into the bath, and the dropping of small parts into the bath. Several methods of removing sludge and foreign objects are available. Some equipment has been designed to alleviate the buildup problem. (See Figs. 10-13G and 10-13H.)

Two thermocouples properly located in the bath will provide safe control and protection from overheating. An overtemperature control should be arranged to automatically shut off the heat source and actuate visual and audible alarms in the event of a malfunction of the normal operating controls. All sensing couples must be protected from the corrosive effects of the particular salt used.

For the removal and control of any toxic, corrosive, and heated fumes, all salt bath furnaces should be equipped with hoods made of material that will not be affected by the corrosive nature of such fumes. (See Fig. 10-13I.) In many cases, chemical scrubbers or suitable filters should be installed to remove the particulates. Removal or neutralization of fumes may require special equipment.

All fixtures used for dipping parts in baths, such as hooks, ladles, or baskets, should be of a solid, closed design without corners which could retain water. Closed

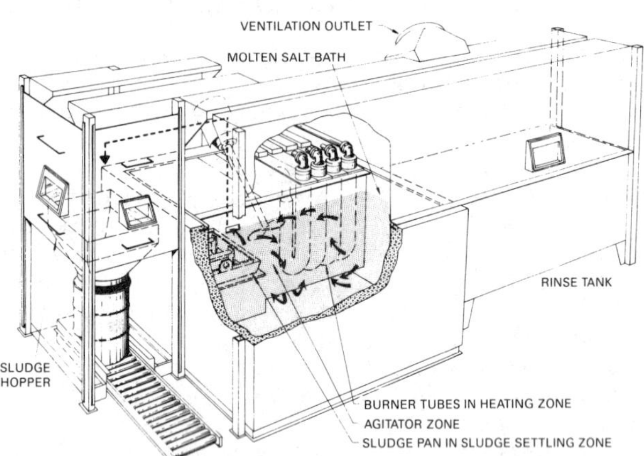

FIG. 10-13G. Descaling salt bath showing sludge pan construction. (Kolene Co.)

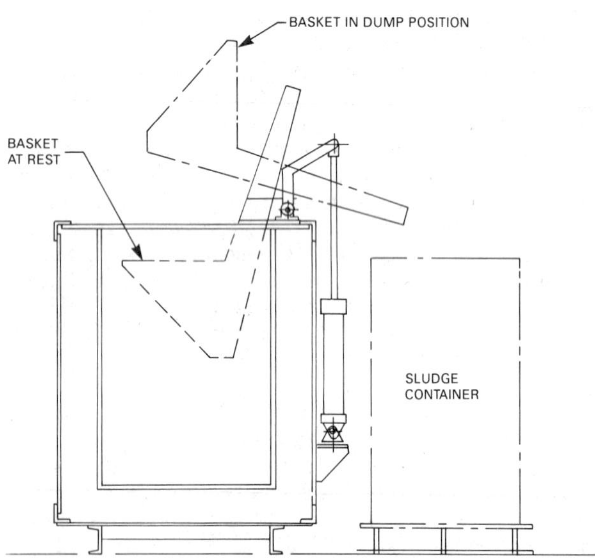

FIG. 10-13H. Automatic basket dumper. (Ajax Electric Co.)

piping or other hollow metallic articles in which air may be trapped should not be inserted into a molten bath without making provisions for venting the heated, expanded air. Immersion of hollow parts in salt baths should be very gradual. Any fixture used in one salt bath should be thoroughly cleaned and dried before it is immersed in another bath.

When electrodes are water cooled, an instrument should be provided to detect failure of the water cooling system.

If a salt bath is used as an internal quench in a furnace with a combustible atmosphere, measures should be provided to prevent carbon precipitation onto the salt surface. In addition, adequate salt circulation should be provided to prevent localized hot spots whenever the salt is exposed to furnace temperatures.

FIG. 10-13I. Typical hooded salt bath construction. (Kolene Co.)

A cold salt bath that has frozen over should be remelted or liquefied by applying initial heat as near to the top as possible. The application of heat to the external surfaces will cause melting near the bottom and leave the top solidified. Gas will then form, causing excessive pressure and failure of the container with the subsequent ejection of hot salts. Breaking the surface crust will result in an explosive ejection of the salt. This hazard can be avoided by inserting a tapered solid bar in the molten salt adjacent to the immersed electrodes. After the salt has frozen, the bar is removed. The cavity created by the removal of the bar will serve as a vent duct for the escape of the gases formed. Gas burners that heat the top layers and gradually heat downward may be used if they are used carefully. For the immersion type of electrodes, a coil set in between electrodes is recommended. Pumps are available to remove salt while it is in the molten state. Considerable care should be exercised in removing any molten salt; it should be pumped directly into steel drums where it can freeze over immediately. Salt and drums may then be properly discarded.

All operators and workers in the vicinity of molten salt baths should be provided with corrosion resistant and heat resistant shoes, gloves, aprons, hard hats, and face shields. Solutions for cleansing eyes and supplies for treating minor burns should be available at all times. Breathing apparatus should also be provided for emergency use against oxides of nitrogen or corrosive fumes such as chlorides and fluorides.

Constant vigilance with safety in mind and good housekeeping by intelligent and well trained personnel can keep a salt bath operation as safe as any other metallurgical process conducted in an industrial furnace.

Bibliography

NFPA Codes, Standards, Recommended Practices and Manuals. (See the latest *NFPA Codes and Standards Catalog* for availability of current editions of the following documents.)

NFPA 10, *Standard for Portable Fire Extinguishers.*

NFPA 70, *National Electrical Code.*

NFPA 86, *Standard for Ovens and Furnaces—Design, Location, and Equipment.*

NFPA 86C, *Standard for Industrial Furnaces Using a Special Processing Atmosphere.*

NFPA 101, *Code for Safety to Life from Fire in Buildings and Structures.*

NFPA 220, *Standard on Types of Building Construction.*

Additional Readings

Bartels, A. L., "Safeguards to Prevent Ignition by Hot Surfaces," *Electrical Review London*, Vol. 203, No. 4, 1978, pp. 29-31.

Janz, George J., *Molten Salts Handbook*, Academic Press, Inc., New York, NY, 1967.

Linville, J. L., ed. *Industrial Fire Hazards Handbook*, 2nd ed., National Fire Protection Association, Quincy, MA, 1984.

Lovering, David G., ed., *Molten Salt Technology*, Plenum Publishing Corp., NY, 1982.

Mehrkam, Q. D., "An Introduction to Salt Bath Heat Treating," Reprint No. 182, Ajax Electric Company, Huntingdon Valley, PA, 1972.

Molten Salt Electrolysis in Metal Production, IMM/North American Publications, Brookfield, VT, 1977.

Trinks, W., *Industrial Furnaces*, (Vol. 1: Principals of Design and Operation, 5th ed., 1961; Vol 2: Fuels, Furnace Types, and Furnace Equipment: Their Selection and Influence Upon Furnace Operation, 4th ed., John Wiley and Sons, NY, 1967.)

Ubbelohde, A. R., *Molten State of Matter: Melting and Crystal Structure*, John Wiley and Sons, NY, 1979.

SOLVENT EXTRACTION

Revised By C. Louis Kingsbaker

Seeds such as soybeans, which contain approximately 20 percent oil by weight, are usually extracted directly. Other seeds, containing more than 20 percent oil, usually are prepared by prepressing to reduce their oil content to 10 to 20 percent. Facilities employing this step are called prepress solvent extraction plants. Some of the major oilseeds processed by this procedure are cottonseed, rapeseed, flaxseed, corngerm, sunflower, safflower, peanuts, and copra.

Additional information relevant to the solvent extraction process may be found in Section 5, Chapter 4, "Flammable and Combustible Liquids;" Section 5, Chapter 9, "Dusts;" Section 6, Chapter 5, "Hazardous Waste Control;" and Section 11, Chapter 7, "Storage and Handling of Grain Mill Products."

THE SOLVENT EXTRACTION PROCESS

Solvents Used

Hexane is the solvent exclusively used in oilseed extraction because it is relatively easy to desolventize from meal and oil products, making them nontoxic; it is essentially immiscible with water; and it can be produced from petroleum at a relatively low cost. Its major disadvantage, the danger of fire and/or explosion, is overcome by careful design, construction, and operation of the solvent extraction plant.

Other flammable solvents, such as mixtures of aromatic hydrocarbons, are used for extracting lignite, Douglas fir bark, and other special raw materials whose end products are not used by humans for internal consumption.

Description of the Process

Unit operations for extraction of all oilseeds are similar, and usually the same equipment can handle the flakes or cakes from these materials. (See Fig. 10-14A.)

For safe storage, oilseeds are brought into the processing plant, checked for moisture content, and dried in a flue gas dryer to about 10 or 11 percent moisture content by weight. To eliminate fire hazards in both the dryer and in the preparation step of the extraction process, the seeds are cleaned to remove trash, dirt, and foreign material before they are dried.

The solvent extraction process is continuous. The rate is usually controlled by measuring at the weigh scale and adjusting the cracking mill feed rolls. The cracking mill splits the hulls and breaks the seeds. Hulls are removed by screens and aspiration. The seeds are flaked between rollers after passing through a rotary steam tube conditioner. Here they are heated to temperatures between 140 and 165°F (60 and 74°C) and the moisture content is adjusted to between 10 to 10½ percent.

Hot flakes are delivered to the extractor, which is usually of the continuous percolation type. The liquid solvent enters above the flake bed and drains or percolates through the bed to a compartment below the flakes to be recycled by a pump to another stage of the extractor. Most percolation extractors have six or more extraction stages.

There are many types of percolation extractors, some of which are the vertical basket, rotary, perforated belt, and rectangular loop extractors. (See Fig. 10-14B.) Fresh, recovered hexane solvent, heated to about 140°F (60°C), contacts flakes near the discharge end of the extractor to remove the remaining oil from the flakes. This solvent, which now contains some oil, is pumped from stage to stage in a direction counter to the flow of flakes, becoming richer in oil with each successive stage. The oil extracted from the flakes and mixed with hexane usually has a concentration of 25 percent when it exits the extractor. The amount of hexane solvent added to the extractor is one weight part of solvent to one weight part of flake. This is defined as solvent ratio and ranges from 0.85 to 1.1. The extractor normally operates at 140°F (60°C) and at either atmospheric pressure or a slight vacuum of ½ in. water gage (124 Pa). Variable speed drive is provided to control the extractor speed for changes of plant rate and also bed height.

After the flakes move beyond the solvent addition point, time is provided for hexane to drain. Extracted and drained flakes are conveyed to a desolventizer-toaster (D-T) in which the solvent is stripped from the flakes with open or sparge steam. Toasting is accomplished by contact

Mr. Kingsbaker is President of C. L. Kingsbaker, Inc., consultants to the oilseed industry.

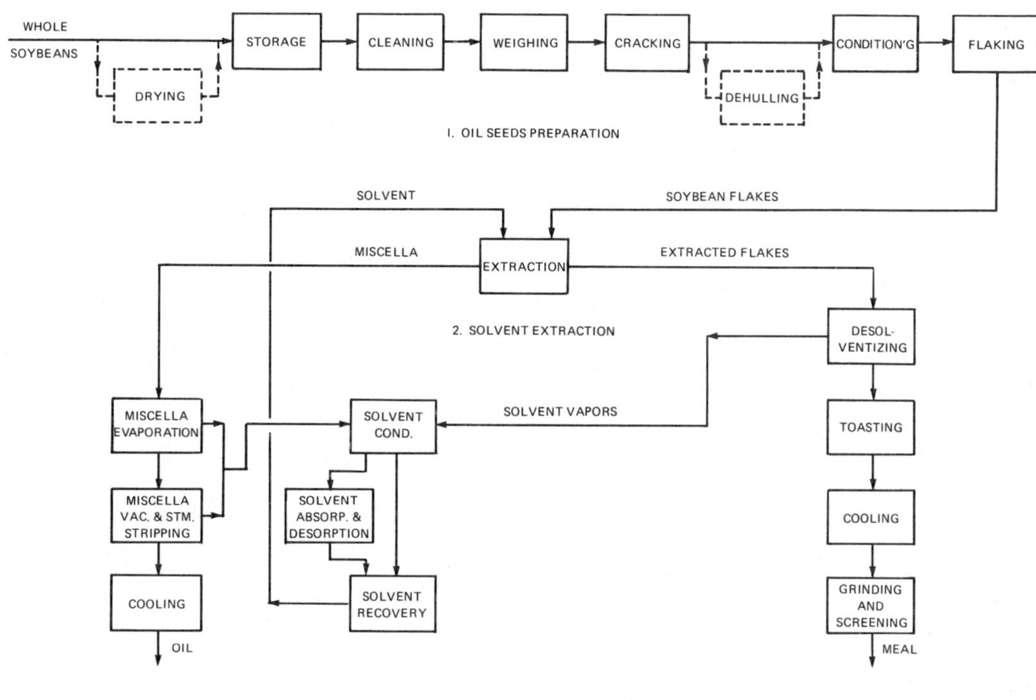

FIG. 10-14A. Flow diagram of the soybean solvent extraction process.

with a series of trays heated indirectly by high pressure steam. The desolventizer-toaster is a vertical, cylindrical unit consisting of multiple trays or decks. (See Fig. 10-14C.) The flakes drop from tray to tray. Solvent is removed by open steam in the top three trays. The steam condenses in the flakes, raising the flake moisture content. Toasting or cooking in the bottom trays reduces the moisture content, cooks the flakes to the desired color, and destroys enzymes that are injurious to animals.

Usually about 15 percent excess sparge steam is used in the desolventizing step. The top of the desolventizer-toaster unit has a normal operating temperature of about 167°F (75°C). The flake outlet temperature from the bottom of this unit must be at least 221°F (105°C). Lower flake outlet temperatures indicate that not all solvent has been removed, which could present a potential safety hazard. The operating pressure at the top of the desolventizer-toaster should be lower than the pressure in the extractor. This will prevent steam vapor from flowing back to the extractor and stop the flow of solvent vapor downward

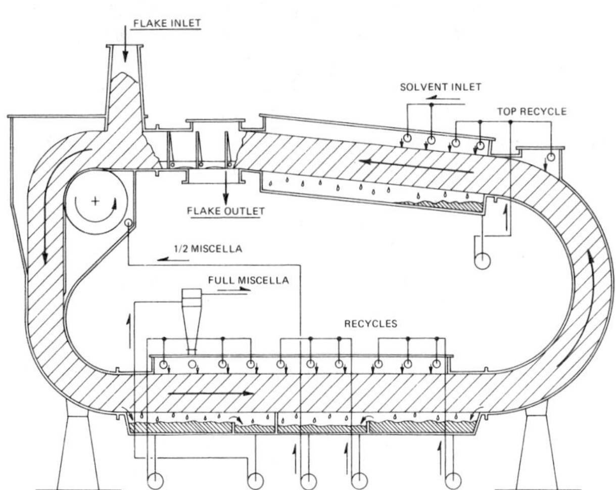

FIG. 10-14B. Rectangular loop extractor.

FIG. 10-14C. Typical desolventizer-toaster (with six trays) used for processing soybeans.

through the desolventizer-toaster unit. Solvent and excess sparge steam vapors leaving the top of the desolventizer-toaster unit are scrubbed with liquid hexane to remove fines, condensed, and the solvent then reused in the extraction step.

From the toaster, the flakes move to a rotary steam tube dryer. After they are dried to 12 percent or less moisture content, and cooled to between 115 and 120°F (46 and 49°C), the flakes are conveyed to the meal finishing system for sizing.

Oil is removed from the solvent in three stages, by two long-tube rising film evaporators and a stripping column. The first stage evaporator operates at a vacuum of about 15 in. of mercury (51 kPa) on the tube side and uses the vapors from the desolventizer-toaster to provide the heat. Miscella from the first stage evaporator has an oil concentration of about 50 to 60 percent. The second stage evaporator operates at atmospheric pressure, using low pressure steam in the shell side. Final removal of solvent from the 92 percent miscella out of the second stage evaporator is carried out in a stripping column operating under vacuum ranging from 22 to 27 in. (74 to 91 kPa) of mercury at a temperature of 210 to 240°F (99 to 115°C) and using open or sparge steam at the bottom of the column. The vacuum dries or removes water from the oil, and the sparge steam strips out the final traces of hexane solvent.

All the solvent is condensed and returned to the solvent separator along with solvent from the flake-desolventizing step. The solvent separator is designed with a 15 minute holding interval to allow water condensed in the process to separate from the solvent. The water flows from the separator to a waste water evaporator, where it is heated to at least 200°F (93°C) with open steam to remove any entrained solvent, before leaving the plant for a waste water trap and finally to a sewer. The separated solvent is returned to the extractor for reuse.

A low temperature alarm which sounds at 180°F (82°C) is required by NFPA 36, *Standard for Solvent Extraction Plants* (hereinafter referred to as NFPA 36) in plants to warn of possible carryover of hexane solvent in process waste water entering public sewer systems. There was no such alarm at a soybean plant in Louisville, KY when a massive quantity of hexane entered the sewers in 1981 and blew up six miles (10 km) of the sewer system.

The extractor and all other tanks and condensers in the process are vented to a separate header and into a vent recovery system. This system, which uses an edible mineral oil to absorb vent gases, consists of two columns filled with ceramic packing. One column absorbs hexane solvent vapors with a flow of cool mineral oil counter to the flow of vapor. The other column is used to strip the absorbed hexane solvent vapors from the mineral oil with open steam. A condensing system is provided to condense these vapors. The mineral oil is recirculated continuously for reuse. A blower is provided on top of the mineral oil absorber to remove the nonabsorbed water and air vapors that discharge into the atmosphere through a flame arrestor.

Efficiency of the Process

Most designers of solvent extraction facilities make efficiency guarantees for the process. The guarantees are important not only for the profitability of the process, but also for the safety of plant operation. Exceeding these guarantees indicates that there are potential safety problems, and steps must be taken to correct them. The guarantees given here are typical for the soybean processing industry and will not vary considerably when other oilseeds are processed. In a process having a capacity that can be measured in U.S. tons per 24 hr day, the following conditions can be expected:

1. Solvent loss of 0.15 percent by weight of beans processed or 0.55 U.S. gal per U.S. ton of beans (2.29 liters per metric ton).
2. Residual oil content of flakes from extractor of 0.5 percent by weight, corrected to 12 percent water by weight on a solvent free basis.
3. Crude oil 0.15 percent of total moisture and volatiles by weight that will not flash at 300°F (150°C) in a closed cup flash tester.
4. Soybean flakes from a meal cooler will not exceed a 0.15 pH rise in urease enzyme activity, measured by the Caskey-Knapp Method.

HAZARDS OF SOLVENT EXTRACTION

The principal hazards in solvent extraction operations are the highly flammable solvents used in the process and combustible dusts associated with the storage and handling of oilseeds. Sources of ignition include arcing electrical equipment, static electricity, and open flame. Ignition can occur during regular operations, maintenance operations, or an emergency. Attempts to operate a facility beyond its design capacity can introduce hazards of abnormal operation.

Normal Operations

Solvent Loss: In order to prevent fires, it is necessary to keep solvent liquid or vapor from leaving the extraction plant and controlled areas. Hexane solvent vapor is about three times heavier than air and, if not dispersed, could possibly enter areas where a source of ignition is present. In order to be ignitable, the vapor concentration in air must be within its flammable limits—1.2 to 6.9 percent by volume.

Solvent loss in an extraction plant is a key indicator to possible hazards during normal operation. A solvent loss of 0.15 percent by weight of the material processed is considered within the bounds of efficient operation. If the loss exceeds 0.30 percent, reasons for the increase must be found and corrective action taken. Table 10-14A shows the five sources of solvent loss from the extraction process and the percentage of loss attributed to each.

TABLE 10-14A. Sources of Solvent Loss from the Solvent Extraction Process

Source	Loss (% by weight)
Flakes from D-T unit	0.04
Leaks	0.04
Vent gas from vent system	0.04
Oil from oil stripper	0.02
Effluent from process waste water	0.01
Total	0.15

It is of primary importance to prevent flammable solvents from leaving the extraction building and entering the preparation building. This can happen if solvent remains in the flakes after they have been removed from the desolventizer-toaster and sent to the preparation building to be ground, sized, and finished. Because the equipment in this building is not of explosion proof design, ignition could occur with disastrous results.

A simple check is to use a "pop" test on a sample from the desolventizer-toaster unit. The flakes are held in a quart (L) size can for 15 minutes. The can then is removed from the controlled area. As the lid is slowly opened, a match at least 8 in. (0.2 m) long is inserted into the can. If there is no flash, the solvent in the flakes is less than 0.04 percent and can be considered safe.

Whenever a transfer of flammable solvents takes place, immediate precautions must be initiated to protect the building until the solvent laden material is removed.

Startup, Shutdown, and Purging: The startup or shutdown of an extraction plant introduces a transient hazard. The equipment, especially the extractor, is full of air before startup. As hexane solvent is added, the concentration of solvent vapor in the atmosphere passes through the flammable range of 1.2 to 6.9 percent by volume. Two methods are available to eliminate any possibility of ignition caused by static electricity when using hot liquid to purge the extractor of air. One method is to design the plant so hot vapors from the second stage evaporator can flow into the extractor to rapidly heat it to 104°F (40°C) and thus replace the air. The advantage of this procedure is to shorten extractor heatup time, which usually takes about 4 hours, to about ½ hour. Stage pumps are not started until the extractor is hot. The second method is to replace the extractor air with carbon dioxide or nitrogen gas before adding hot liquid hexane to the extractor.

Vapor proof slide gates located upstream and downstream of the extractor are closed during the warmup to prevent flow of hexane solvent vapors back to the preparation building.

A normal or controlled shutdown is accomplished by emptying the extraction plant of all oilseed flakes and hexane liquid and purging the empty extractor or desolventizer-toaster unit with air until the unit is cooled. Steam purging is used if normal maintenance is required inside the extractor.

If welding is necessary in the extraction area, personnel must take the following precautions:

1. Empty all solids and liquids.
2. Store hexane from the plant in solvent tanks isolated from the plant.
3. Steam the purge extractor, desolventizer-toaster unit, and all equipment containing hexane or miscella.
4. Fill all vessels with water.
5. Seal off flake connections from the extractor and desolventizer-toaster unit to the preparation building using water.
6. Use a portable gas analyzer to locate possible flammable mixtures.

Inert gas is satisfactory for purging an empty extractor before normal startup procedures. Inert gas can also be used for purging during normal shutdown, but steam purging is a surer method than inert gas purging for removing solvent vapor.

Abnormal Operations

Emergency Breakdowns: The most serious hazard in the operation of a solvent extraction plant is an emergency breakdown in the extractor when it is full of solvent laden flakes and cannot be emptied. Quite often the problem can neither be determined nor corrected without someone having to enter the unit.

The unit should be cooled for at least three hours, during which time a plan for correcting the problem should be devised. Then the unit should be purged with air. Steam purging is impractical and inert gas purging does not ensure a nonflammable mixture. Persons entering the unit must wear air masks and be under constant observation. Ignition sources should be eliminated, since flammable mixtures may be present.

If a desolventizer-toaster unit fails under load, the danger is that the flakes in the cooking trays may overheat and cause a fire. It is important to quickly turn off the heating system in the cooking trays when such a failure occurs. Normally, sparge or open steam can be added to the desolventizer-toaster unit for several hours to desolventize the flakes. The doors on each tray are then opened one by one and the contents carefully raked out until the unit is empty and the problem corrected.

Failures: There should be a sufficient supply of emergency cooling water to operate the condensers long enough for a safe shutdown. If cooling water fails, the process must be stopped, the steam shut off, and the vapor slide gates closed. Similar steps should be taken in the event of steam or electrical failures.

System overloads: Often a solvent extraction plant capacity is increased above its design rating. This situation can create a hazard because normally there is insufficient solvent condensing capacity to recover the solvent. This results in higher vent pressures, with the inherent danger that solvent vapors may be forced into the atmosphere or carried out with the flakes to form flammable mixtures.

FIRE PROTECTION

Fire protection for solvent extraction plants involves building construction and services, plant layout, fire protection equipment and systems, and safe operating procedures. Detailed guidance can be found in NFPA 36, and in other related standards, guides, and recommended practices listed at the end of this chapter.

Layout and Construction

Structural elements of solvent extraction plants should be laid out so that separation distances are adequate to isolate or dissipate flammable concentrations of solvent vapor from ignition sources. (See Fig. 10-14D.)

The extraction and preparation buildings should be of fire resistive or noncombustible construction and equipped with some form of explosion relief. It is important that electrical wiring and equipment be suitable for use in the presence of flammable vapors. All vessels, pipes, tubes, and hoses containing or carrying flammable solvent should be electrically bonded together and grounded to prevent the buildup of a static charge. Portable flammable gas detectors should be available for monitoring the atmosphere.

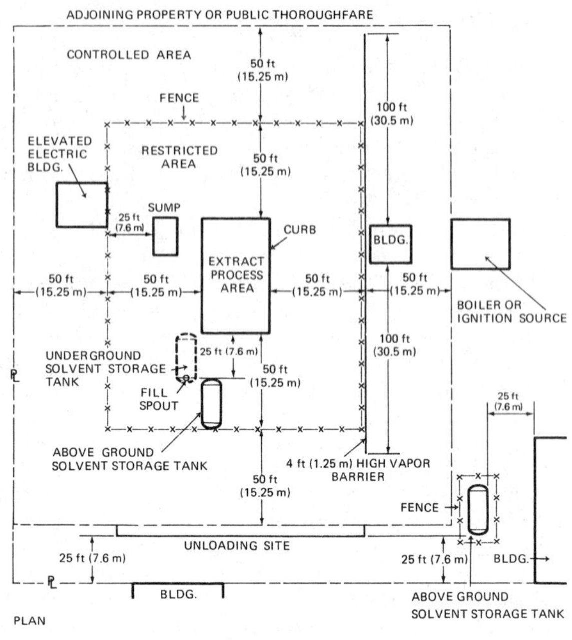

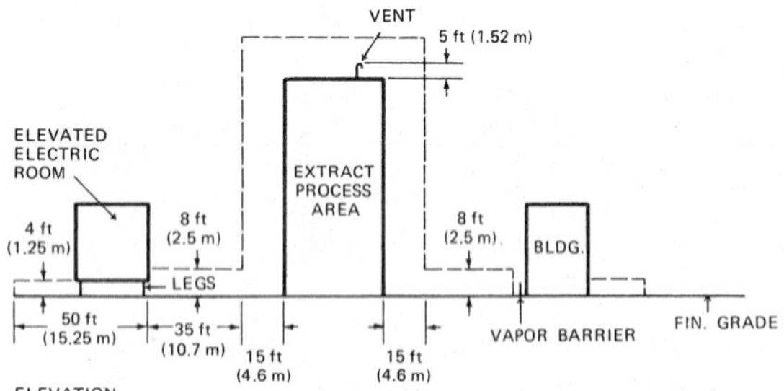

FIG. 10-14D. A typical distance diagram.

For a process involving combustible dust, electrical wiring and equipment should be suitable, a dust collecting system should be provided, and protection against static electricity discharges should be installed.

Fire Protection Equipment

Water spray, deluge, or foam water systems—singly or in combination—are suitable for use in the extractor building. Automatic sprinkler systems are appropriate for use in the preparation building. The solvent unloading and storage area may be adequately protected by portable fire extinguishers of the proper types and sizes if the area is isolated from exposure hazards.

Portable fire extinguishers should be strategically located. The facility may also have standpipe and hose systems equipped with combination nozzles. A system of yard hydrants is an essential part of the protection. Finally, the available water supply must be adequate to meet the needs of the fixed protection and yard hydrant systems.

Policies and Procedures

It is imperative that solvent extraction equipment be operated in a manner prescribed by the manufacturer and that no attempt be made to bypass or defeat safety devices and systems. Company policy should limit the output demanded of the extraction plant so that its design capacity is not overtaxed. Policy should also provide for shutdown in the event of excessive solvent loss and for regular maintenance and cleaning operations. If solvent loss increases from 0.15 to 0.30 percent or 1 gal per ton (4.2 liter per metric ton) of material processed, an abnormal hazard exists and preventive action must be taken.

Fire prevention is a key factor in any fire protection program. This should include the conspicuous posting of safety rules and emergency procedures throughout the plant, a safety education program for new employees, the appointment of a plant emergency organization, and a compulsory monthly safety program. A safety system audit program must be implemented to regularly check the workability of all safety devices and controls.

If a fire should occur, appropriate action would be to activate fire extinguishing systems (if they are not automatically controlled), activate the plant interlock system to shut down all solids conveying equipment in the plant, turn off the main steam valve to the extraction plant, leave the area, and notify the fire department.

Bibliography

NFPA Codes, Standards, Recommended Practices and Manuals. (See the latest *NFPA Codes and Standards Catalog* for availability of current editions of the following documents.)

NFPA 10, *Standard for Portable Fire Extinguishers.*
NFPA 13, *Standard for the Installation of Sprinkler Systems.*
NFPA 14, *Standard for the Installation of Standpipe and Hose Systems.*
NFPA 15, *Standard for Water Spray Fixed Systems for Fire Protection.*
NFPA 16, *Standard for the Installation of Deluge Foam-Water Sprinkler Systems and Foam-Water Spray Systems.*
NFPA 24, *Standard for the Installation of Private Fire Service Mains and Their Appurtenances.*
NFPA 30, *Flammable and Combustible Liquids Code.*
NFPA 36, *Standard for Solvent Extraction Plants.*
NFPA 61B, *Standard for the Prevention of Fires and Explosions in Grain Elevators and Facilities Handling Bulk Raw Agricultural Commodities.*
NFPA 61C, *Standard for the Prevention of Fire and Dust Explosives in Feed Mills.*
NFPA 68, *Guide for Explosion Venting.*
NFPA 70, *National Electrical Code.*
NFPA 385, *Standard for Tank Vehicles for Flammable and Combustible Liquids.*
NFPA 496, *Standard for Purged and Pressurized Enclosures for Electrical Equipment in Hazardous (Classified) Locations.*

Additional Readings

Collings, A. J., and Luxon, S. G., *Safe Use of Solvents*, Academic Press Inc., NY.
"Hazard and Hazard Prevention in Solvent Evaporating Ovens," *Fire Prevention*, (127), 1978, pp. 22-25.
Jaffee, H. M., Grain Elevator Protection, "What's Being Done Today?" *Fire Journal*, Vol. 74, No. 3, May 1980, pp. 131-132.
Marcus, Y., ed., *Solvent Extraction Reviews*, Vol. 1, Marcel Dekker, Inc., NY, 1971.
Sawyer, W., "Relationship of Flash Points of Solvents, Resin Solutions, and Paints," *Journal of Coatings Technology*, 49 (627), pp. 52-55.
Tess, Roy W., ed., *Solvents Theory and Practice*, American Chemical Society, Washington, DC, 1973.

REFRIGERATION SYSTEMS

Revised by Emil E. Roy

Industrial refrigeration systems perform four major functions: (1) provide low temperature storage for food and other perishable products, (2) cool air for building occupants, (3) provide low temperature air or air gas mixtures for critical industrial processes, and (4) convert gases to liquids for more efficient transportation and storage.

HOW THE SYSTEMS WORK

Basically, all refrigeration systems are heat pumps. They take heat from one space or medium and transfer it to another. To accomplish the transfer, a material is needed that will readily absorb, transport, and release heat under controlled conditions. Such materials, or refrigerants, perform these in both the liquid and gaseous states. When the liquid changes to a gas, it absorbs heat from its surroundings. Conversely, when the gas condenses back to a liquid, it releases heat to its surroundings.

At the condenser, where it arrives in a high pressure gaseous state, the refrigerant is condensed back to the liquid state, releasing heat in the process. It then is stored in an accumulator. When temperature controls require refrigeration, the liquid passes through an expansion valve into the evaporator where it vaporizes and absorbs heat. It then is carried back to the condenser.

Refrigeration systems require two pressure levels—low pressure in the evaporator and high pressure in the condenser. Pressure differentials are obtained either by mechanical compression or heat absorption.

Figure 10-15A illustrates the relationship of a mechanical compressor to the evaporator and condenser coils. Mechanical compressors can be of the reciprocating type with horizontal, vertical, or "V" design (Fig. 10-15B) or of the centrifugal type. (See Fig. 10-15C.) Regardless of the type, compressors should be made of materials that are compatible with the refrigerant used and equipped with the proper safeguards against overpressure and leakage. Typical safeguards include relief valves, purge valves, liquid separators, and oil separators.

In heat absorption systems, heat supplied by steam, hot water, or a gas flame raises the pressure of the refrigerant so it will flow through the expansion valve to the evaporator. The refrigerant (lithium bromide for air conditioning applications or ammonia for low temperature process applications) is vaporized and goes into solution with an absorbent, usually water. After it has passed through a heat exchanger, the solution enters a generator in which heat drives the refrigerant from the absorbent. The refrigerant then is purified and returned to the condenser for reuse.

Most of the commercial and industrial refrigeration systems in use today are of the mechanical compressor type. This type permits a wider selection of refrigerants than does the absorbent type.

Refrigeration systems are further differentiated as direct or indirect. In the direct type, the evaporator is in direct contact with the product or space to be cooled, or it consists of air circulating passages connected to the area or product. A household refrigerator is an example of a direct system. An indirect system uses an intermediate heat transfer fluid, which is first cooled by the refrigerant and then circulated to the material or space. The intermediate fluid is called "brine" because early systems used a mixture of sodium or calcium salt and water. Today, organic chemical solutions such as ethylene glycol, as well as brine, are used in indirect systems.

CHARACTERISTICS OF REFRIGERANT GASES

Low temperatures are not usually associated with fire hazards, but refrigeration systems are of concern for two reasons. Some of the products used as refrigerants are flammable, and some are explosive in critical combination with air. Others are toxic and can cause injury or death or interfere with fire fighting operations if they escape during a fire.

Table 10-15A shows refrigerant gases divided into three classes according to their flammability. Those in Group 1 are either very weakly flammable or nonflammable. They also have rather low levels of toxicity and thus present low hazard.

Mr. Roy is Vice President, Loss Prevention Services, Frank B. Hall & Co.

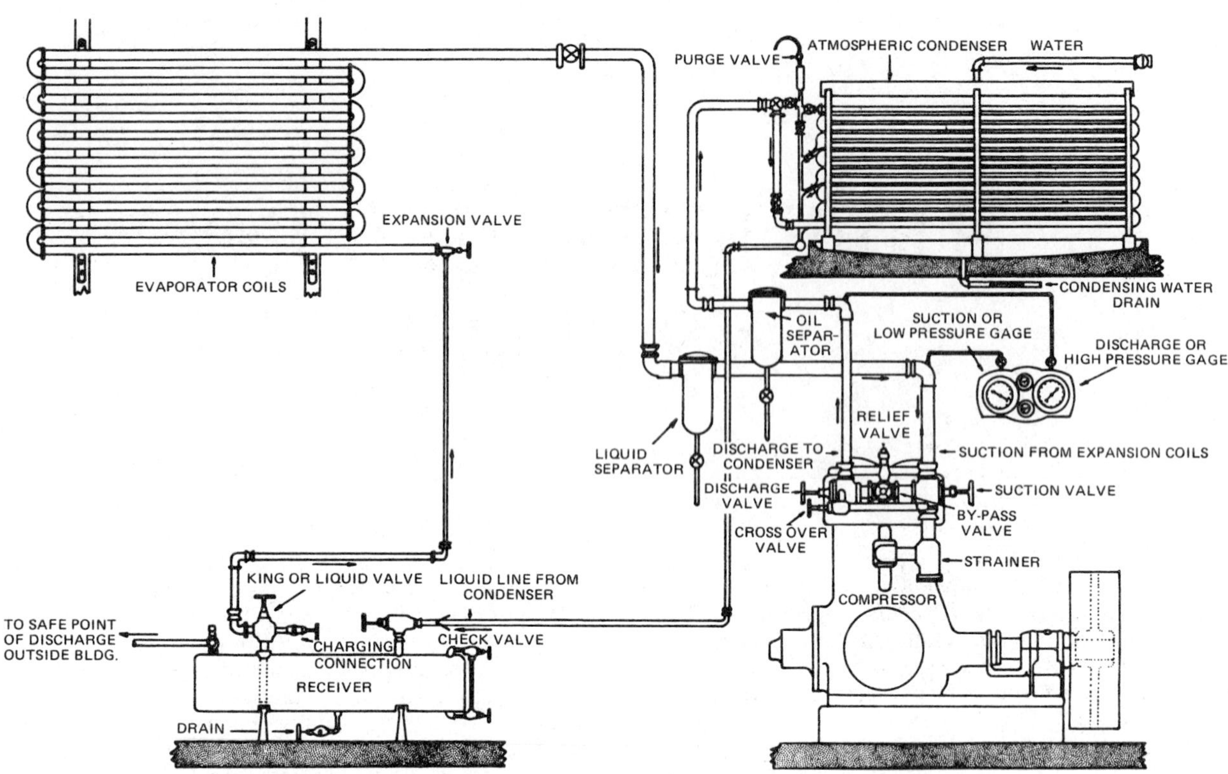

FIG. 10-15A. Typical ammonia refrigeration plant. (Factory Mutual System)

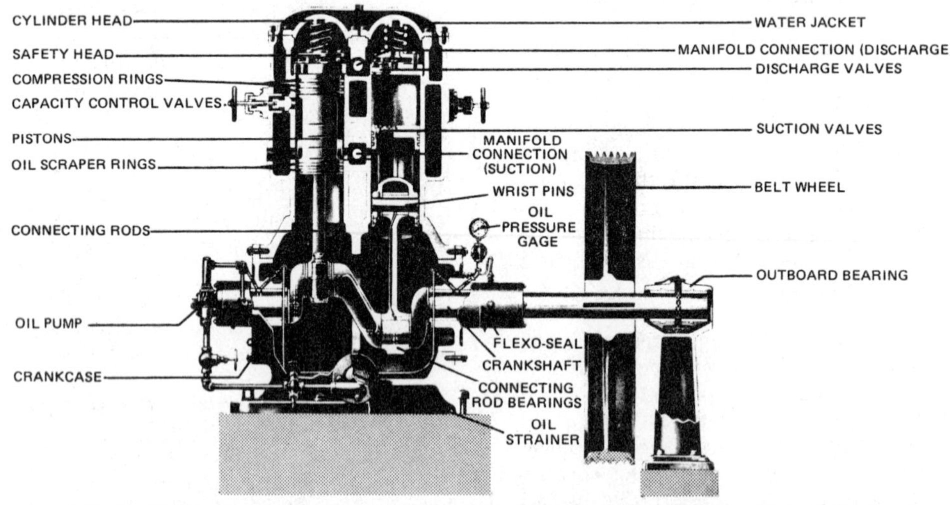

FIG. 10-15B. Reciprocating compressor, vertical single acting. (Factory Mutual System)

The refrigerants in the second group are both flammable and toxic. If they escape from the refrigeration system, they can reach critical concentrations in the surrounding air and produce combustion explosions. They have relatively narrow ranges of flammability, and their upper limits may be reached before combustion is initiated. Group 2 refrigerants in tightly closed refrigerated areas have been involved in combustion explosions. Ammonia systems require careful attention to the compatibility of the materials used. Moist ammonia reacts vigorously with copper, zinc, and many brass and bronze alloys.

Although the refrigerants in Group 3 do not present the same toxicity hazards as those in Group 2, they present much greater fire hazards. They generally have lower flammability limits and a wider flammability range. Methane, ethane, and propane are all components of liquefied natural gas (LNG) and liquefied petroleum gas (LP-Gas), and present relatively the same fire and explosion hazards

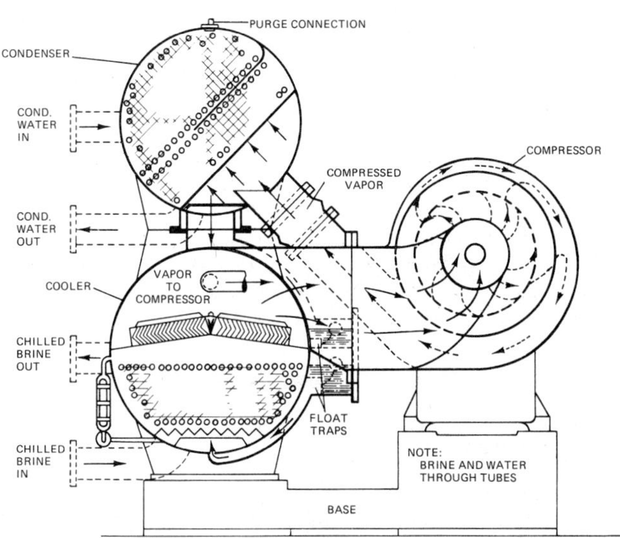

FIG. 10-15C. Cross section of centrifugal refrigeration system showing refrigeration cycle. (Factory Mutual System)

as do LNG and LP-Gas. Ethylene, which is also a component of LP-Gas, has a wide flammability range and a high burning velocity.

HAZARDS OF REFRIGERATION

Since most of the common refrigerants are at least slightly toxic, a leak in the system will be a hazard to health and life safety. The danger to people and products is much greater in a direct system than in an indirect system. However, the degree of hazard is dependent upon how completely the direct portion of the system is isolated from the space or product being cooled.

Leaks can occur for many different reasons. Pipe joints may weaken and fail because of excessive vibration. Dissimilar metals in piping and valves may set up an electrolytic action that destroys the integrity of the junction. Incompatibility between the metal and refrigerant may cause corrosion and failure. Mechanical impurities in the refrigerant can become lodged in valves, causing malfunctions. In compression systems, the introduction of a noncompressible liquid into the suction side of the compressor can cause the compressor to crack or burst. Faulty valves within the system may permit recompression of the gas to pressures high enough to cause failure.

Refrigerants mix readily with air and, if flammable, may reach flammable concentrations ready to be ignited by some ignition source. A sufficient volume of air may dilute the gas, however, diffusion may not take place rapidly enough to avoid the hazard. Experience has shown that most explosions of flammable gas-air mixtures occur with less than 25 percent of the enclosure occupied by the mixture.

TABLE 10-15A. Basic Hazard Data for Some Common Refrigerants

		Boiling point		Calculated density gas	Autoignition temperature		Flammable limits, percent by volume in air		
Classification of refrigerants	Formula	°F	°C	(air = 1)	°F	°C	Lower	Upper	Toxicity
Group 1 (nonflammable, except as noted):									
Carbon dioxide (R-74)	CO_2	−109.0	−78.3	1.52					Slightly toxic
Monochlorodifluoromethane (R-22)	$CHClF_2$	−42.0	−41.1	2.98	1,170	632	Very weakly flammable		Slightly toxic
Dichlorodifluoromethane (R-12)	CCl_2F_2	−22.0	−30.0	4.17					Nontoxic for ordinary exposure
Dichlorofluoromethane (R-21)	$CHCl_2F$	48.0	8.9	3.55	1,026	552	Very weakly flammable		Slightly toxic
Dichlorotetrafluoroethane (R-14)	$C_2Cl_2F_4$	38.4	3.5	5.89					Nontoxic
Trichlorofluoromethane (R-11)	CCl_3F	74.8	23.8	4.79					Slightly toxic
Methylene chloride (R-30)	CH_2Cl_2	105.2	40.7	2.93	1,139	615	Very weakly flammable		Slightly toxic
Trichlorotrifluoroethane (R-113)	$C_2Cl_3F_3$	117.6	47.5	6.46	1,256	680	Very weakly flammable		Slightly toxic
Group 2 (flammable):									
Ammonia (R-717)	NH_3	−28.0	−33.3	0.59	1,204	651	16	25.0	Toxic
Dichloroethylene (R-1130)	$C_2H_2Cl_2$	99–141	37–61	3.35	856	458	9.7	12.8	Moderately toxic
Ethyl chloride (R-160)	C_2H_3Cl	54.0	12.2	2.22	966	519	3.8	15.4	Moderately toxic
Methyl formate R-611)	$C_2H_4O_2$	90.0	32.2	2.07	869	465	5.0	23.0	Toxic
Group 3 (highly flammable):									
Butane (R-600)	C_4H_{10}	31.0	−0.5	2.01	900–1,000	482–538	1.55	8.6	Slightly toxic
Ethane (R-170)	C_2H_6	−128.0	−88.8	1.04	959	515	3.0	12.5	Slightly toxic
Propane (R-240)	C_3H_8	−44.0	−42.2	1.52	920–1,120	493–604	2.2	9.6	Slightly toxic
Ethylene (R-1150)	C_2H_4	−154.8	−103.8	0.98	914	490	2.7	36.0	Slightly toxic

EMERGENCY CONTROLS AND PROCEDURES

Refrigerant gases present two basic hazards—toxicity and flammability—which can create nonfire and fire emergencies, respectively.

Control of the nonfire emergency is accomplished by directing, diluting, and dispersing the gas to prevent it from accumulating in structures or coming in contact with people. Simultaneously, steps should be taken to stop the flow of gas. Air, water, and steam are practical media for accomplishing dilution and dispersion. Water in the form of a spray from hoses or fixed nozzles is the most commonly used carrier for this purpose.

Control of fire emergencies generally consists of reducing the heat of the fire with water and, if possible, shutting off the gas supply. Water should be applied as a spray from hoses or fixed nozzles. Carbon dioxide, dry chemical, and halogenated extinguishing agents are useful in most instances. Ammonia is easily controlled with water, but suitable breathing apparatus and protective clothing should be worn.

With the exception of carbon dioxide, the refrigerants in Group 1 are halogenated hydrocarbons, commonly known as Halons. Halon refrigerants are usually referred to by numbers (R-11, etc.) as in Table 10-15A. They have varying characteristics and, therefore, have varying applications.

R-11 is commonly used to cool water for air conditioning and industrial processing. R-114 is usually employed to cool brines to temperatures between 20 and −20°F (−7 and −29°C). R-12 is used for very low temperatures down to −120°F (−84°C). Though classed as nonflammable, halon refrigerants can present some hazards. R-12, for example, can react vigorously with aluminum at elevated temperatures of approximately 1,200°F (649°C).

These halogenated refrigerants are usually used in centrifugal refrigeration systems. In such systems, the brine is cooled by passing it through tubes immersed in the refrigerant. The vaporized refrigerant is liquefied after it is forced by the compressor into the condenser. At this point, an economizer partially cools the compressed refrigerant by partial evaporation, and the vapor goes to the second stage of the compressor. A purge recovery unit takes air or other noncondensable gas out of the refrigerant and discharges it into the atmosphere.

Refrigeration equipment should be installed with careful regard to the many variables that affect life safety and fire hazards. The type of occupancy should determine whether a direct or indirect system should be installed. Limitations should be imposed upon the quantity of refrigerant available, based upon the concentration that would be attained if the entire quantity were discharged into the area. The quantity, of course, would be determined by the toxicity and flammability of the refrigerant. Materials used in the system should be compatible with the type of refrigerant used, and the system should include the proper safety devices and relief valves. Electrical equipment and wiring should be properly grounded and shielded and all other possible ignition sources eliminated.

Bibliography

Additional Readings

Factory Mutual Engineering Corp., "Refrigeration," *Loss Prevention Data* 7-13, Factory Mutual System, Norwood, MA, Oct. 1975.

Factory Mutual Engineering Corp., "Refrigerated Warehouses," *Loss Prevention Data*, 8-29, May 1976.

"Fire and Explosion Hazards of Compressors," *Fire Prevention*, (122), 1977, pp. 19-21.

Henry, Martin F., *Flammable and Combustible Liquids Code Handbook*, 2nd edition, National Fire Protection Association, Quincy, MA, 1983.

Loader, K., "Fire Protection of Cold Storage Warehouses," *Fire Prevention*, Feb. 1981, pp. 11-15.

McRae, H., "Ammonia Explosion Destroys Ice Cream Plant," *Fire Command*, Apr. 1984, pp. 36-37.

Meacock, M. H., *Refrigeration Processes: A Practical Handbook on the Physical Properties of Refrigerants and their Applications*, Pergamon Press Inc., Elmsford, NY, 1974.

Miller, Rex, *Refrigeration and Air Conditioning Technology*, Bennett Publishing Co., Peoria, IL, 1982.

"Number Designation of Refrigerants," *ASHRAE 34, 1978*, American Society of Heating, Refrigeration and Air Conditioning Engineers, Atlanta, GA.

"Safety Code for Mechanical Refrigeration," *ASHRAE 15, 1978*, American Society of Heating, Refrigeration and Air Conditioning Engineers, Atlanta, GA.

"Sprinkler Protection of Small Cold Stores," *Fire Prevention*, Vol. 143, No. 17, Aug. 1981.

Weiby, P., and Dickinson, K. R., "Monitoring Work Areas for Explosive and Toxic Hazards," *Chemical Engineering*, Vol. 83, No. 22, 1976, pp. 139-145.

FOOD PROCESSING

Jane I. Lataille, P.E.

The food processing industry is especially sensitive to damage from fire, smoke, heat, or water due to their contaminating aspects. Food contamination is the outstanding loss potential common to all food processing and handling facilities. Loss situations that cause relatively minor damage in other industries can cause extremely high amounts of damage to food items since health authorities often condemn items exposed to smoke, heat, or water, even when there is no visible damage.

This chapter discusses the causes of fire losses in the food processing industry, primarily those losses associated with combustible dusts, flammable and combustible liquids, conveyors, refrigeration systems, fuel fired equipment, and general storage. After a discussion of the raw materials that undergo food processing, the fire hazards of process equipment, gases, liquids, combustible dusts, conveyors, and storage are discussed. Finally, the safeguards are presented, including firesafe construction, sprinkler systems, and fixed extinguishing systems.

For the hazards associated with electrical control systems, see Section 8, Chapter 2, "Electrical Systems and Applicances." Section 11, Chapters 4 and 5, "Storage and Handling of Flammable and Combustible Liquids" and "Storage and Handling of Gases," respectively, provide an in depth discussion of the storage practices and hazards of flammable and combustible liquids and gases. For a discussion of the hazards of aerosol food production, see Section 10, Chapter 21, "Aerosol Charging Operations."

RAW MATERIALS

Fruits, vegetables, grains, beef, poultry, fish, and milk are common raw materials in the food processing industry. Other materials are powders such as flour, salt, sugar, cornstarch, and cocoa; liquids such as water, oil, alcohol, molasses, and vinegar; and miscellaneous items such as butter, chocolate, flavorings, colorings, chemicals, and spices. In addition to product ingredients there are cooking oils and fats; various gases used in carbonation, fumigation, and ripening; and fluids associated with heating and cooling systems.

Flour, salt, sugar, and other powdered or granulated raw materials are often pneumatically conveyed (from railroad cars, for example) into storage silos, bins, or hoppers. These materials are highly susceptible to damage by water and other liquids and must be stored in well sealed containers. In multiple story plants they are usually stored on upper floors or in penthouses.

Cooking oils, alcohol, vinegar, and liquids used at room temperature are stored in tanks of various shapes and sizes. Molasses and other liquids requiring heating are stored in steam jacketed tanks. Most liquids storage is found on lower floors, sometimes even in basements. Wherever stored, adequate curbing and drainage must be provided to reduce existing loss potentials, and the chance of damage to other materials, equipment, storage, and the building itself. Generally, storage in basements is not advisable, due to the chance of flooding from surface water runoff, water main breaks, spills from upper floors, reduced fire department accessibility, and the difficulty of venting heat and smoke.

PROCESSES

Dairy Production

The dairy industry depends heavily upon refrigeration. Dairy products must be kept within specified temperature ranges or they may be declared unfit for use. The consequences of lost refrigeration capability are therefore important considerations in planning and protecting refrigeration systems.

In pasturization, milk is heated to kill bacteria and other microorganisms. It then must be cooled within a specified amount of time and kept within acceptable temperature ranges. Milk is also homogenized, or mixed, to give it a uniform consistency. Other dairy products are cream, skim milk, butter, cream cheese, dips, ice cream, buttermilk, yogurt, and cheese. These products also must be kept cooled for acceptable quality.

Jane I. Lataille is an Executive Engineer with Industrial Risk Insurers (IRI), Hartford, CT.

Manufacture of Convenience Foods

Another facet of the food industry is manufacture of convenience food. Examples of convenience products are instant soup, powdered milk, instant coffee, egg and meat substitutes, powdered drinks, instant potatoes, and instant rice. The advantages of these products include economy, ease of preparation, and long shelf life.

The processes most often used in producing convenience foods are flash drying and flash freezing. During these processes, food products are converted into their desired forms by rapid exposure to extreme heat or extreme cold. Some products also contain various chemicals additives for preservation, texture or flavor improvement, coloring, and moisture control.

Preserving

Since most crops can only be grown during limited times, it is necessary to preserve them to achieve year round availability. Meat, poultry, and fish also are preserved. Canning, freezing, and pickling are common means of food preservation. In both canning and freezing, the food is usually cooked in steam jacketed vessels. Salt and vinegar are the primary materials used for pickling.

Baking

Continuous fuel fired ovens are representative of food baking process equipment. The fuel, number of burners, number of zones, temperature, and amount of ventilation will vary for this equipment. Occasionally, electric ovens are used. Baked products include bread, pastry, pies, cakes, cookies, and crackers. In pie and pastry making, fruit must be prepared and cooked before baking. Some ovens are arranged to spray vegetable oil into pans before baking or onto the product during baking.

Candy Making

Most candy making involves cooking or heating, for examples, melting chocolate to pour into molds and cooking flavored sugar mixtures to make hard candy. Process equipment is likely to include kettles, ovens, and cooling tunnels. Many candies are covered with coatings such as chocolate, icings, fruit mixtures, and glazes. The coatings are either melted and mixed or carried in a solvent.

Meat Processing

In processing, meat is cleaned and cut into desired sections. Meat preservation consists mainly of drying, salting (soaking in brine), and smoking. The smoking is performed in smoke chambers. Any other processing is usually done by plants that use meat in their products, such as those making soups, stews, hash, canned meat, meat rolls, or even pet foods.

Warehousing

Some food facilities simply store large quantities of many different food products for distribution to grocery stores. As with non food warehousing facilities, some grocery warehouses are equipped with automatic retrieval systems (automated warehousing). Automatic retrieval systems usually allow the storage of products at a greater density than conventional racks. This increases fuel loading, lessens access to stock, and increases the amount of food exposed to contamination from a single loss. Many other locations may depend on grocery warehouses for food products, consequently, loss prevention is very important for these warehouses.

Miscellaneous Processes

Other food processes, such as frying, do not fall into the previous categories. These processes include the making of potato chips, corn chips, french fries, onion rings, cheese curls, pork rinds, and other snacks. Fryers using hot fat or oil are used to make these products. Nondairy beverages, such as carbonated beverages, fruit drinks and juices, beer, and liquors, represent use of other miscellaneous processes.

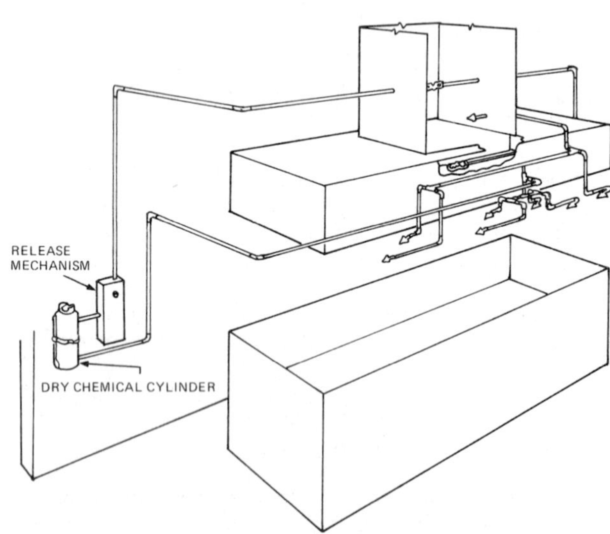

FIG. 10-16A. Layout of a wet chemical extinguishing system for a fat fryer. (Reprinted with permission of Wormald U.S., Inc.)

THE FIRE HAZARDS

Process Equipment

Refrigeration equipment within food processing industry presents hazards which are inherent to compressed gases, potential vapor release, moving mechanical parts, and electrical control systems. Ammonia refrigerant is volatile (Fig. 10-16B) and toxic. Freon is less volatile, but is still considered toxic. Both can contaminate food and otherwise damage equipment and stock. Vessels for cooking or melting are usually heated with electricity or steam. Their fire hazards are usually limited to those of electrical systems and overheating. Boilers that produce the steam for these vessels present the same fire and explosion hazards as those associated with fuel fired equipment.

Baking ovens also present fuel firing hazards. In addition, ovens have interior moving parts and fixed ductwork which become coated with oils and can be involved in a fire. Electrically heated ovens sometimes contain concealed oil-filled transformers which are a concern. Another feature of ovens that requires attention is the presence of combustible insulation. The primary hazards introduced by cooling tunnels are combustible construction (Fig. 10-16C) and/or insulation. Fryers and associated

FIG. 10-16B. Result of an anhydrous ammonia explosion in Houston, TX on December 11, 1983. (Houston Fire Department)

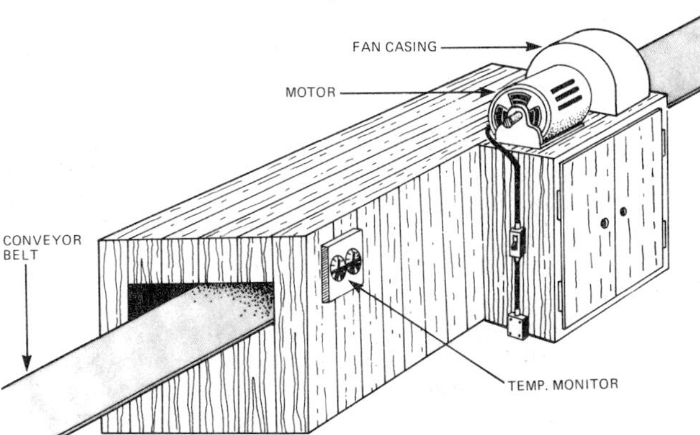

FIG. 10-16C. Cooling tunnels are sometimes made of wood.

ductwork involve the heating of a combustible liquid. The degree of hazard increases as the temperature approaches the flash point of the liquid.

Another characteristic of food processing equipment which contributes to its loss potential is a high level of automation. Process equipment is often custom made and can take a long time to replace. It can also be so much more efficient than other production methods that lost business cannot be made up if the equipment is out of service. For this reason, highly automated processes should be reviewed for duplication, alternate production methods, and outside assistance. Where they exist, duplicate equipment lines should be separated so they will not be exposed to the same potential loss incident.

Gases and Liquids

Some gases used in the food industry are hazardous. Cocoa and other powdered products are treated with gaseous fumigants which, depending upon their composition, could be flammable. Ethylene oxide is a flammable gas used to ripen fruit which has been shipped "green" to distribution centers. Whenever gases are used, their physical characteristics should be checked.

Combustible and flammable liquids are commonly found in food processing plants. Combustible liquids used include lubricating oils, cooking oils, certain alcoholic beverages, and solvent based extracts. Alcohol, high proof alcoholic beverages, and solvents are the most commonly used flammable liquids. In the candy industry, ether is used to dissolve the coating that makes jelly beans shiny. In many types of plants, alcohol is used to dissolve dried, crushed materials or other liquids to make flavorings and colorings. One example of this is the production of vanilla extract. Waxes, which are liquids when heated, also are used in food plants.

Combustible Dusts

Dust hazards are not unusual in the food processing industry. Under the right conditions, any organic dust can explode. Among the many potentially explosive dusts are those from sugar, flour, grains, starches, and cocoa. Since the explosibility of a dust depends partly upon its fineness, the finer dusts (such as confectioners' sugar and cornstarch) present the most severe explosion potential.

Conveyors

Conveyors, which can be used anywhere in a process from raw material transfer to processing, packaging, and storage, pose many hazards. Their drives can jam, misalign, or overheat, and cause friction which can lead to fires. The conveyors themselves can be a means of spreading fire from one area to another; so may the openings made in the walls they pass through. Many belt conveyors have combustible belts. Rubber or plastic belts produce large quantities of smoke when they burn, resulting in a high potential for contamination. Chain conveyors can carry plastic pans. Wide conveyors shield the areas below from ceiling sprinkler protection. Finally, conveyors are part of highly automated systems that cannot function without them.

Storage

Raw materials and food products are stored in various ways. Bins and tanks are often used for powders and liquids. Packaged products are commonly stored on pallets or in racks 80 ft (24.4 m) high or more, including automatic retrieval systems (Fig. 10-16D). Since wood, cardboard, and plastics are used to package food products, the combustible loading in storage areas can become very high.

Some food products, such as vegetable oil spray, are stored in aerosol cans. Under pressure, aerosol propellants can rocket cans during a fire so the fire spreads beyond control. Necessarily, the storage of this type of product must be carefully planned.

THE SAFEGUARDS

Construction Considerations

Since food products have a high susceptibility to smoke contamination, noncombustible construction of buildings is preferred. Fire cutoffs should be provided to separate manufacturing and warehousing areas, to separate duplicate production lines, and to isolate fuel fired equipment, refrigeration systems, and storage of more hazardous materials such as flammable liquids, paper, cardboard, plastics, pallets, and aerosols. Fire cutoffs need to be preserved where conveyors pass through them. To do this, conveyor lines are specially designed to break open at, or hinge away from, the point at which a fire door or

FIG. 10-16D. Erecting storage racks in a grocery warehouse. (Giant Food)

shutter closes. The use of water spray systems is sometimes proposed as an alternative to cutoffs. This is less desirable because water spray by itself cannot prevent the radiation of heat through an opening. Water spray systems also depend upon a water supply and therefore are subject to impairments.

Heating and cooling requirements in the food processing industry call for the widespread use of insulating materials in buildings and process equipment. Noncombustible insulation should be chosen whenever possible. When combustible insulation is used, it should have a low flame spread, as well as fuel-contributed and smoke-developed ratings established by a testing laboratory. Combustible insulation should also be protected by sprinklers and thermal barriers such as gypsum or dry wall. Process equipment should be noncombustible.

Sprinkler Systems

The various combustible materials commonly encountered in the food processing industry suggest the provision of appropriately designed sprinkler protection in all buildings. Sprinklers also should be provided for high storage racks, for process enclosures handling (or built of) combustible materials, under obstructions over 4 ft (1.22 m) wide (such as conveyor belts), and for coolers, freezers, and heating enclosures. Where wood cooling tunnels or other combustible equipment are used, sprinklers should be provided. Hose connections supplied from sprinkler systems should be provided in storage areas and wherever large amounts of combustibles may be present. Water supplies should be sufficient to supply the largest demand area plus hose streams for the amount of time prescribed by the hazard in the protected area under consideration.

Despite low temperatures, it is possible for fires to occur within freezers. Providing sprinkler protection in them, however, requires special design. The usual wet pipe sprinkler system is replaced by a combination dry pipe valve and deluge valve (Fig. 16F). This system reduces the chance of water damage to a minimum by requiring two means of detection before the system operates.

Hose connections are more complicated to install in freezers. They must be kept drained and contain a way to both release the air in the sprinkler piping and open the deluge valve. Maintenance of these complex systems is difficult due to the subfreezing temperatures.

Fixed Extinguishing Systems

A fixed extinguishing system is the most effective protection for some hazards found in the food processing industry. Deep fat fryers and their associated ductwork can be protected with carbon dioxide, or dry chemical extinguishing systems. Large outside tanks of flammable liquids should be protected with water spray or foam systems. Explosion suppression systems are desirable for large or critical equipment handling combustible dusts or for such equipment with insufficient explosion relief. The use of fixed extinguishing systems can be applied similarly to any other hazards involving flammable liquids, heated combustible liquids, and combustible dusts.

Electrical Equipment

Electrical equipment in food processing plants should be suitable for the location in which it is installed (refer to NFPA 70, *National Electrical Code*®). A review should be made of possible vapors or dusts which could be present in each area. Where liquids such as alcohol may be present, electrical equipment suitable for Class I, Group D locations is recommended. Where ether is used, the electrical classification is Class I, Group C. Ethylene oxide requires the use of Class I, Group B equipment. Class II, Group G locations are those where nonconductive dusts are likely to be present.

Other Protection Features

Areas handling flammable liquids should be well ventilated to prevent significant vapor buildup. Enclosures that may contain dusts or vapors should be provided with explosion relief designed to minimize damage. Magnets to extract ferrous debris are used wherever powdered materials enter a storage or distribution system. Portions of systems handling powdered materials from which dust can escape should be provided with permanent dust collection systems. For equipment that handles either dusts or vapors, proper bonding and grounding practices must be observed.

High piled storage areas are a fire fighting challenge even with properly designed sprinkler protection. For these areas, there should be good outside hydrant coverage. Inside hose stations and smoke and heat vents, either manual or automatic, can be used to augment automatic sprinkler protection.

Interlocks should be used on automated processes to run them safely. Fuel fired equipment should be provided with standard combustion safeguards and air and fuel train controls. Ovens and other heated enclosures should have high temperature cut outs. Enclosures which could

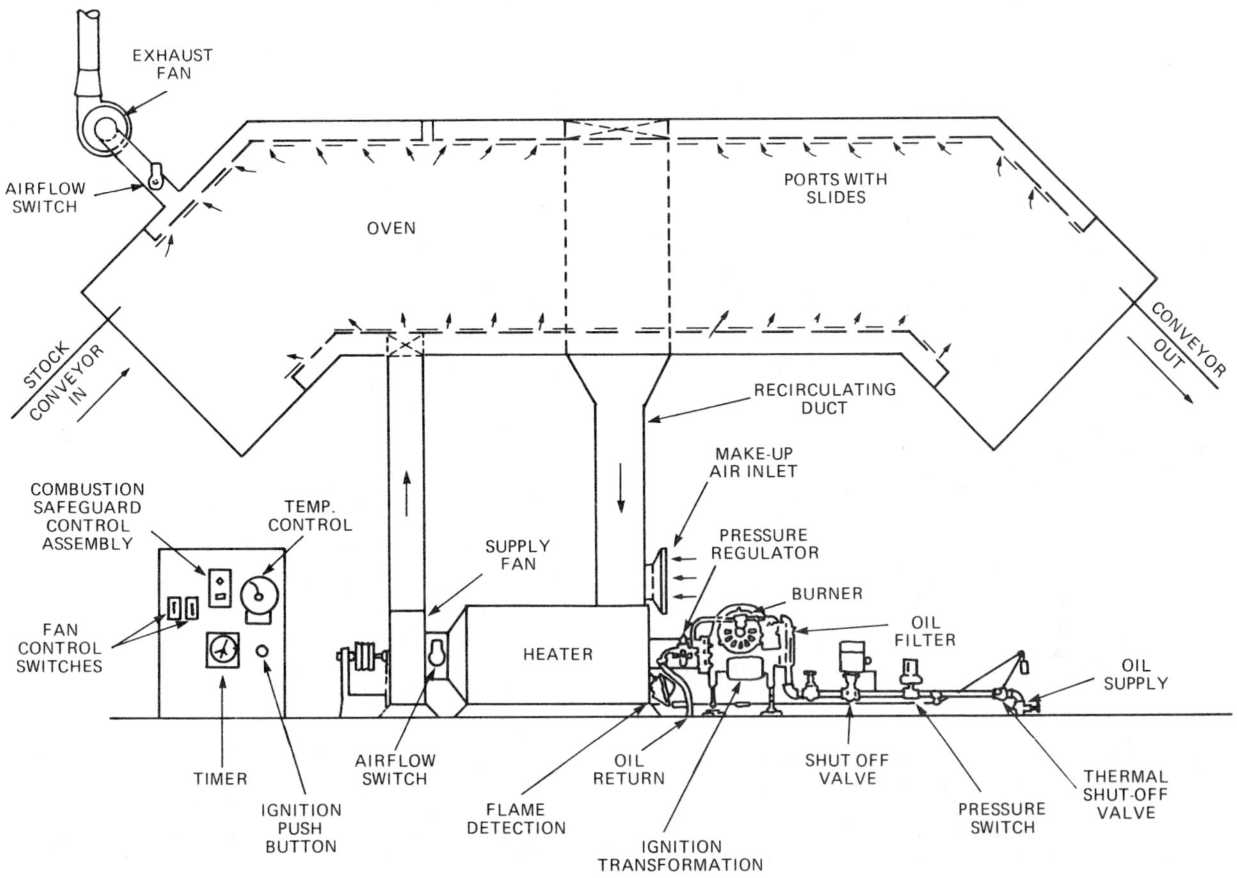

FIG. 10-16E. Conveyor type oven with the controls labeled. (Industrial Risk Insurers)

FIG. 10-16F. A combined dry pipe and deluge system. (Giant Food)

contain hazardous vapors need combustible gas detection systems. Boilers producing steam require excess pressure cut outs, and those producing hot water require high temperature cut outs. Processes that need a flowing liquid to run safely should have provisions to monitor the flow and level of the liquid. Conveyor belts should be monitored for low speed or excessive friction to prevent their malfunction from causing a fire. Continuous feed ovens or dryers requiring ventilation should be interlocked to shut down upon loss of ventilation. Potentially unsafe conditions in any process should be monitored and interlocked to avoid hazardous conditions.

Bibliography

NFPA Codes, Standards, and Recommended Practices and Manuals. (See the latest *NFPA Codes and Standards Catalog* for availability of current editions of the following documents.)

NFPA 10, *Standard for Portable Fire Extinguishers.*
NFPA 11, *Standard for Low Expansion Foam and Combined Agent Systems.*
NFPA 11A, *Standard for Medium and High Expansion Foam Systems.*
NFPA 12, *Standard on Carbon Dioxide Extinguishing Systems.*
NFPA 12A, *Standard on Halon 1301 Fire Extinguishing Systems.*
NFPA 12B, *Standard on Halon 1211 Fire Extinguishing Systems.*
NFPA 13, *Standard for Installation of Sprinkler Systems.*
NFPA 14, *Standard for Installation of Standpipe and Hose Systems.*
NFPA 15, *Standard for Water Spray Fixed Systems for Fire Protection.*
NFPA 16A, *Recommended Practice for Installation of Closed-Head Foam-Water Sprinkler Systems.*
NFPA 30, *Flammable and Combustible Liquids Code.*
NFPA 34, *Standard for Dipping and Coating Processes Using Flammable or Combustible Liquids.*

NFPA 49, *Hazardous Chemicals Data.*
NFPA 61A, *Standard for Prevention of Fire and Dust Explosions in Grain Elevators and Facilities Handling Manufacturing and Handling of Starch.*
NFPA 61B, *Standard for Prevention of Fires and Explosions in Grain Elevators and Facilities Handling Bulk Raw and Agricultural Commodities.*
NFPA 61C, *Standard for Prevention of Fire and Dust Explosions in Feed Mills.*
NFPA 61D, *Standard for Prevention of Fire and Dust Explosions in the Milling of Agricultural Commodities for Human Consumption.*
NFPA 68, *Guide for Explosion Venting.*
NFPA 69, *Standard on Explosion Prevention Systems.*
NFPA 70, *National Electrical Code®.*
NFPA 77, *Recommended Practices on Static Electricity.*
NFPA 85A, *Standard for Prevention of Furnace Explosions in Fuel Oil- and Natural Gas-Fired Single Burner Boiler-Furnaces.*
NFPA 85B, *Standard for Prevention of Furnace Explosions in Natural Gas-Fired Multiple Burner Boiler-Furnaces.*
NFPA 85D, *Standard for Prevention of Furnace Explosions in Fuel Oil-Fired Multiple Burner Boiler-Furnaces.*
NFPA 85G, *Standard for Prevention of Furnace Implosions in Multiple Burner Boiler-Furnaces.*
NFPA 86, *Standard for Ovens and Furnaces—Design, Location and Equipment.*
NFPA 96, *Standard for the Installation of Equipment for the Removal of Smoke and Grease Laden Vapors from Commercial Cooking Equipment.*
NFPA 204M, *Guide for Smoke and Heat Venting.*
NFPA 231, *Standard for General Storage.*
NFPA 231C, *Standard for Rack Storage of Materials.*
NFPA 321, *Standard on Basic Classification of Flammable and Combustible Liquids.*
NFPA 325M, *Fire Hazard Properties of Flammable Liquids, Gases, Volatile Solids.*
NFPA 493, *Standard for Intrinsically Safe Apparatus in Division 1 Hazardous Locations.*
NFPA 496, *Standard for Purged and Pressurized Enclosures for Electrical Equipment.*
NFPA 497, *Recommended Practice for Classification of Class I Hazardous Locations for Electrical Installations in Chemical Plants.*
NFPA 497M, *Manual for Classification of Gases, Vapors and Dusts for Electrical Equipment in Hazardous (Classified) Locations.*
NFPA 650, *Standard for Pneumatic Conveying Systems for Handling Combustible Materials.*

Additional Readings

Factory Mutual Engineering Corp., "Industrial Ovens and Dryers," *Loss Prevention Data 6-9, Oct. 1977.*
Factory Mutual Engineering Corp., "Mechanical Refrigeration," *Loss Prevention Data* 7-13, Oct. 1975.
Factory Mutual Engineering Corp., "Grain Storage and Milling," *Loss Prevention Data* 7-75, Feb. 1982.
Factory Mutual Engineering Corp., "Refrigerated Storage," *Loss Prevention Data* 8-29, May 1976.
Food Processing and Packaging Equipment, 19th ed., Business Trend Analysts, Dix Hill, NY.
Heldman, D. R., and Singh, R. P., *Food Process Engineering*, AVI Publishing Co., Inc., Westport, CT, 1981.
Linko, P., ed., *Food Processing Systems*, (Vol. 1), Elsevier Science Publishing Co., Inc., NY, 1980.
Linville, J., ed., *Industrial Fire Hazards Handbook*, 2nd ed., National Fire Protection Association, Quincy, MA, 1984.
"Moving Fire: Fire Hazards of Belt Conveyors," *Record*, Vol. 54, No. 6, 1977, pp. 18-21.
Schwartzberg, Henry G., and Lund, Daryl, *Food Process Engineering*, American Institute of Chemical Engineers, NY, 1982.

NUCLEAR FACILITIES

Revised by George Weldon, P.E.

In general, radioactive substances and operations involving radioactive materials or devices and equipment which present radiation hazards have the same fire and explosion features as similar materials and operations without radiation hazards. However, due to the hazard to personnel and the possibility of long term contamination of property from exposure to sudden accidental escape of radioactive substances, the subject of protection from radiation hazards deserves special consideration, especially the procedures that must be followed during emergencies.

NUCLEAR REACTORS

A nuclear reactor is a device or assembly for initiating and maintaining a controlled nuclear chain reaction in a fissionable fuel (uranium or plutonium). Nuclear reactors are used to produce energy, to study the fission process, or to produce radioactive materials within the reactor or in a material exposed by the reactor's radiation or radioactive particles. Basically, nuclear reactors may be divided into (1) nuclear power reactors of large size, up to 3,500 MW (megawatts thermal) or 1,100 MW(e) (megawatts electrical), and (2) research reactors that operate at power levels from a few watts to many megawatts. In 1985, there were 95 operating or near operating nuclear power reactors and an additional 30 nuclear power reactors with construction permits in the United States alone. Construction of many others has been delayed—some indefinitely. Dozens of research reactors are now operating, but in general the number of research reactors is not increasing.

Nuclear reactors that include a containment vessel, generating equipment, and heat removal equipment can be as large as the largest fossil fueled electrical generating plants. Most research reactors, however, are so small that they may occupy only one small corner of a room in a typical laboratory building at a research facility or college campus. Therefore, the magnitude of hazard presented by each kind of reactor varies considerably.

Various national and international groups have addressed the need for fire protection requirements for nuclear power reactors and research reactors. These groups include the National Fire Protection Association (NFPA), the American Nuclear Society, the American National Standards Institute (ANSI), the United States Nuclear Regulatory Commission (NRC), the Mutual Atomic Energy Reinsurance Pool, and the American Nuclear Insurers. They are referenced in the Bibliography at the end of this chapter.

The "Defense in Depth" philosophy, which calls for the provisions that follow, is applicable to both power and research reactors. Those provisions are:

1. Fire prevention.
2. Quick detection and suppression of fires that occur.
3. Designing the plant to limit the consequences of fire.

Heat Removal

All nuclear reactors, even very low power training or research reactors, produce heat while operating. This heat either must be dissipated or used, depending upon the amount produced and the purpose for which the reactor is intended.

Reactor Control

Reactor control systems and safety systems are of utmost importance. The control system design is fitted to the technical characteristics of the reactor and is capable of producing power changes at acceptable rates. The control system design also makes it possible to produce and maintain the desired power level within the reactor in such a way that excessive temperatures are avoided. The safety system, which is an addition to the control system, is adopted to the characteristics of the reactor in the instrument and control system. It responds to signals from the instruments by automatic operation, in order to prevent operational variables from exceeding safe limits. On appropriate signals, the safety system warns of incipient performance changes and, if necessary, shuts down the reactor.

A reactor becomes "critical" when the total rate of production of neutrons, under control conditions, is such that self-sustaining reactions occur. Control methods must

Mr. Weldon, P.E., is Engineering Manager for the Mutual Atomic Energy Reinsurance Pool, Norwood, MA.

be rapid and sensitive, and protected from fire. Since the control system is vital to the adequate functioning and safe operation of the reactor, the protection of the control room, cableways, emergency power supply, and electrically or hydraulically operated equipment is of prime importance. Protection for these areas should be fully consistent with that used in computer rooms containing vital records.

Construction Problems

Fire records indicate that one of the more vulnerable periods for fire damage in the lifetime of a large reactor system—such as found in nuclear power plants—exists during the construction stage. (See Fig. 10-17A.) Because the construction of such a plant usually requires many years, the construction hazards acquire a nearly permanent status and should be considered as if they were to be permanent.

FIG. 10-17A. Nuclear facilities are most vulnerable to fire damage during construction, a period of several years, and thus require a good construction firesafety program.

Power reactors present unique fire protection problems during construction since they require a containment vessel as a final protection system. Construction techniques require the containment structure to be built early during the project. This means that much of the subsequent reactor construction takes place inside a large vessel that has limited exits for evacuation and limited fire fighting access. In addition, the vessel confines smoke and other products of combustion, which greatly increases the difficulties of evacuation and manual fire fighting. A good construction firesafety program should ensure that all penetrations of the containment vessel suitable for evacuation remain open and usable during the period that other construction is taking place within the vessel.

Due to the limited access for fire fighting and the lack of normal venting possibilities for smoke and gases, it is imperative to severely limit all combustible materials needed for construction. Metal formwork, scaffolding, platforms, stairways, etc., are preferable to wood. The use of wood in extensive quantities is limited to those kinds of woods appropriately treated to reduce inherent combustibility and flame spread ratings.

Installation of utilities and equipment in the containment vessel requires special care so a low level of combustibles is maintained. Since reactor equipment must meet very high levels of quality assurance, reactor equipment that has been subjected to fire and smoke damage is much more likely to require replacement than similarly exposed equipment in normal industrial installations. Special efforts are necessary to reduce the usual accumulation of packing cases, cartons, insulation, etc., to an acceptable level. This may take the form of conducting all uncrating operations outside the containment vessel and providing special handling devices to transport unpackaged items into the vessel.

Research Reactors

In general, these may present particular problems due to their location in existing buildings (which are sometimes of combustible construction), and the use of much combustible loading in the form of paraffin shielding, wall finishes, and instrumentation. Combustibles should be eliminated wherever possible. For example, water can be used in place of paraffin. Where combustibles cannot be eliminated, automatic sprinkler protection should be provided. Supervision of construction/operating personnel is especially important because a number of different operating groups may be involved with a single reactor.

RADIATION MACHINES

Radiation machines include mechanical and electrical devices that produce or make use of subatomic particles or electromagnetic radiation, or both. X ray machines are used in radiography, therapeutic treatment, and in studies of the behavior of radiation and its effects upon materials. Particle accelerators, while a source of radiation, are primarily devices for imparting extremely high energies to subatomic particles which enter and alter atomic nuclei, thereby providing the means for developing basic information concerning the structure and behavior of matter. Gamma ray sources in the form of radioactive isotopes are employed in radiography equipment. This equipment may produce intense radiation. The use of isotopes as a source of radiation is not discussed in this chapter.

X Ray Machines

Except for the radiation hazard while in use, the hazards of X ray machines are mainly the hazards of high potential and high energy electrical equipment. When shut down, there is seldom any appreciable residual radiation to interfere with fire fighting or salvage operations. Although no flammable gases or hazardous amounts of flammable liquids are ordinarily needed to operate X ray machines, they are frequently encountered in studies of the effects of radiation on a wide variety of substances.

Particle Accelerators

Particle accelerators include Van de Graaff generators, linear accelerators, cyclotrons, synchrotrons, betatrons or bevatrons. The machines are used, as the name implies, to accelerate various charged atomic particles to tremendous speeds and, consequently, to high energy levels. Particle accelerators furnish scientists with atomic particles, in the form of a beam, which may be utilized for fundamental studies of atomic structure. In addition, accelerators furnish high energy radiation which may be utilized for radiography, therapy, or chemical processing.

Particle accelerators emit radiation only while in operation. Attempts to extinguish a fire in the immediate vicinity of the machine should be delayed until the machine's power supply can be disconnected.

Certain "target" materials become radioactive when bombarded by atomic particles, and for this reason monitoring equipment should be used during fire fighting operations to estimate the radiation hazard. The usual hazard presented by particle accelerators is largely that of electrical equipment. There are, however, some important exceptions. Some installations use such hazardous materials as liquid hydrogen or other flammable materials in considerable quantities. Large amounts of paraffin are used for neutron shielding purposes. Another factor is the possible presence of combustible oils used for insulating and cooling.

Industrial applications for particle accelerators include chemical activation, acceleration of polymerization in plastics production, and the sterilization and preservation of packaged drugs and sutures. The general fire protection and prevention measures for these machines should include the use of noncombustible or limited combustible (Type I or Type II) construction housing, noncombustible or slow burning wiring and interior finishing, and the elimination of as much other combustible material as possible. (See NFPA 220, *Standard on Types of Building Construction*.) Automatic sprinkler protection should be provided in areas containing hazardous amounts of combustible material or equipment. Special fire protection should be provided for any high voltage electrical equipment.

FACILITIES HANDLING RADIOACTIVE MATERIALS

The type of equipment used to process radioactive materials depends not only upon the work to be performed, but also upon the degree of hazard associated with the material and the process it is to undergo. Materials with low levels of radioactivity and with little or no inherent fire or explosion hazards require less protective equipment than others. For purposes of personnel protection, the amount and kind of shielding required will depend upon the types of radiation emitted as well as the activity involved. In addition, the chemical and physical nature of the radioactive materials will dictate the degree of containment necessary, as well as the construction materials necessary in the containment system. All equipment to be used for handling and processing radioactive materials should be designed to minimize fire and explosion potentials as well as to protect personnel against harmful radiation exposure and prevent damage to property by contamination. There are many types of equipment and systems for handling radioactive materials, but most may be classified as either benches, hoods, glove boxes, or hot cells.

Benches

Benches are used generally for handling relatively small amounts of alpha or beta emitting materials requiring little or no shielding when handled with gloved hands or tongs. No special ventilation for the bench is provided in most instances and its use is thereby restricted to materials which will not easily become airborne.

Benches should be of noncombustible construction with a nonporous continuous working surface which can be decontaminated easily. Usually, one or two layers of blotting paper on the bench top to absorb small spills will not increase materially the fire hazard.

Hoods

Hoods, sometimes referred to as "fume hoods," are similar to benches except for the addition of an enclosure and exhaust system for removing vapors. The nature of the operations conducted within the hood may require a filter system to prevent the spread of radioactive materials. Filters with a low degree of combustibility are desirable.

Glove Boxes

The term "glove box" refers to a system designed to contain materials, generally alpha radiation emitters, which present little or no external radiation hazard but which can present a serious problem if they become airborne. Glove boxes may be large and used in a wide variety of operations involving flammable liquids and gases, combustible solids, and toxic materials. The sides are fitted with long rubberlike gloves which permit manual operations to be conducted without personal contact with the hazardous materials. Special ventilation and fire protection systems are usually necessary. (See Fig. 10-17B.)

Hot Cells

A hot cell is a heavily shielded enclosure in which gamma emitting radioactive materials can be handled by persons using remote manipulators while viewing the operation through shielded windows or periscopes. Hot cells are constructed preferably of noncombustible materials and contain the minimum amount of combustibles consistent with operational requirements.

In addition to all of the fire and explosion hazards of glove boxes, hot cells also present increased damage potential due to the nature of the high gamma ray producing materials used. The safeguards recommended for glove boxes apply equally to hot cells. Where very high gamma radiation levels are encountered, consideration must also be given to the possible failure of containers as a result of radiation damage.

RADIATION EXPOSURE

Radiation Injury

The harmful effect of radiation is due to its ability to ionize the atoms present in the various compounds which compose the body. How the radiation actually damages the living cells is not exactly known. Unfortunately, the human body has no defense mechanism against radiation; nor can nuclear radiation be detected by any of the five senses. Thus, it is possible for an individual to receive a severe exposure to radiation without knowing it. This

FIG. 10-17B. A typical glove box showing gloves extending from the ports in the viewing window at left. Note the portable fire extinguisher adapted for discharge into the box through a fixed piping arrangement.

danger requires that some form of instrumentation be used to detect radioactivity.

The amount of injury to a person from radiation varies with the type of radiation, how much of the body has been exposed, and whether it is a one time exposure or an accumulation of exposures to small amounts of radiation. Injuries from excessive exposures may not become apparent for days, weeks, months, or years.

Radiation experts report that, similar to many potentially hazardous materials, any exposure to radiation, either external or internal, has some element of risk, which may be too low to be measurable. Certain exposures are believed to be of reasonable risk balanced against the benefits from any activity using radiation exposure, such as medical and dental uses of radiation. Since the human race has evolved on a planet which has always been radioactive from the naturally occurring radioisotopes in the air, soil, and water, it is probable that a low "background" radiation is tolerable. The controversy on radiation exposure concerns the additional amounts that may be tolerable by humans. Standards have been set with the objective of lowering risks to the best practicable levels.

Occupational Exposure from Radiation in Air and Water

Radiological authorities have set very low limits of concentrations of radioisotopes in air and water based on quantities which may be inhaled or ingested. Complete treatment of the subject would require consideration of maximum permissible concentrations in both air and water, but the problem of airborne concentrations is of particular concern in fire situations, as fire fighters and other emergency personnel may be confronted with such material in the fumes, dust, smoke, and gases liberated by a fire in which radioactive materials are involved.

A hard and fast rule should be mandated that self-contained breathing apparatus (SCBA) is to be used whenever exposure to airborne radiation is a possibility. Maximum Permissible Concentrations (MPC) of various radioisotopes in air and water are commonly stated as limits intended to apply to persons who are continuously exposed to the concentrations named. Consequently, the concentrations are far below the exposures which could be tolerated for infrequent exposure, as would most often be the case with fire fighters or emergency workers.

Apparatus is available for taking a sample of an atmosphere. The number of radioactive emissions in the air can be counted with an appropriate counting instrument. From these counts, the extent to which the particular atmosphere is contaminated can be determined. (See Fig. 10-17C.)

Any space normally occupied by persons not primar-

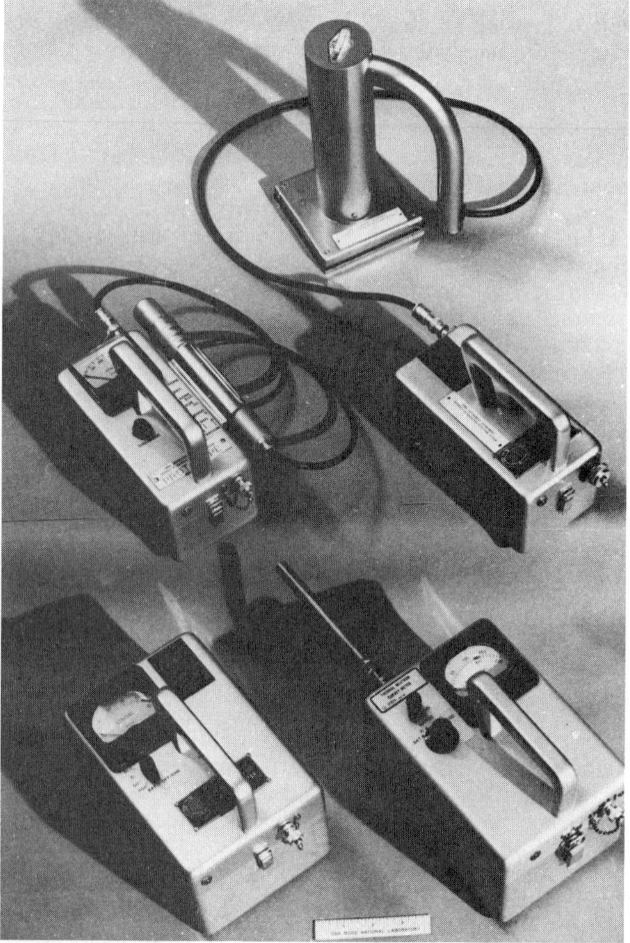

FIG. 10-17C. Portable instruments for measuring developed at Oak Ridge National Laboratory. The meters are for fast neutrons (bottom row at left), thermal neutrons (bottom row at right), beta-gamma (top row left), and for alpha particles (top row right) with an alpha scintillation detector at top of the picture.

ily engaged in radiation work should not be subjected to unreasonable radiation levels; also a person outside the installation should not be subjected to excessive radiation level exposure through contact with radioactive waste or by other means. Air should be monitored continuously for the presence of radiation from fixed sources or from airborne radioactive matter. In emergencies, alarms should sound, and radiation levels be recorded by available commercial instruments.

Fire Department Radiation Exposures

Emergency exposures are usually allowed to exceed those tolerable to persons who work continuously with radioactive materials. In an emergency case, such as a necessary rescue operation, it is considered acceptable for the exposure to be raised, within limits, for single doses. The National Council on Radiation Protection and Measurement has recommended that in a life saving action, such as search for and removal of injured persons or for entry to prevent conditions that would injure or kill numerous persons, the planned dose to the whole body should not exceed 100 rems. During less stressful circumstances, where it is still desirable to enter a hazardous area to protect facilities, eliminate further escape of effluents, or control fires, it is recommended that the planned dose to the whole body should not exceed 25 rems. These rules may be applied to the fire fighter for a single emergency; further exposure is not recommended. Internal radiation exposure may be guarded against by adequate respiratory equipment.

External exposure at the time of a single fire emergency can be judged by the use of commercial radiation survey meters which measure radiation in roentgens or by counted disintegration rates, or by the close observation of the dosimeter indicators carried by individuals. Pocket sized dose rate alarms, which can be carried on the person, are also available. These give an audible signal dependent upon the radiation intensity. Film badges do not provide immediate information.

A rescue procedure which would combine external and internal radiation exposure is usually not attempted. Self-contained breathing apparatus should be used when instruments indicate the presence of any airborne radiation.

FIRE PROTECTION

As noted previously, substances and operations involving radioactive substances or devices presenting radiation hazards have the same fire and explosion features as those of similar materials and operations without radiation hazards. The loss caused by fire, explosion, and accident is affected by the presence of radiation or of radioactive substances in the following ways:

1. Possible interference with manual fire fighting due to the presence of harmful radiation or possible criticality of radiation levels.
2. Possible increased delay in salvage and in normal resumption of operations due to the necessity of decontamination of buildings, equipment, or materials.

Contamination of Property

Entire buildings, land, and important equipment can be rendered unusable for long periods of time because of severe radioactive contamination from the accidental escape of radioactive substances.

Radiological contamination may not stay in one area—it can sift through openings or ventilating systems in the form of dust or vapor, and spread the radioactive material throughout a structure. Careless movement of persons through a contaminated area could also spread contamination to an uncontaminated area.

Once a surface has become contaminated, a decision must be made as to how the particular contaminating material is to be removed, if this is possible. Vacuum cleaning can sometimes be used to remove radioactive dust from building surfaces. If vacuum cleaning is used, however, absolute filters must be used on the exhaust. Hosing with water can be used on some surfaces. Cleaning with soap and detergents is often a hand operation which must be carried out with continuous checks on the amount of exposure that may be tolerated by the persons doing the cleaning. Sand or vacuum blasting can be used on some surfaces and paint may cover alpha contamination.

Plant Fire Protection Organization

In properties where atomic energy is a factor, an in plant fire protection force is recommended. In nuclear reactors and many other such plants, 24 hr/day routines must be maintained for handling fires and emergencies.

Plan for Handling Fires

In plants involving a nuclear reactor, radiation machines, and in other facilities handling radioactive materials, the problems affecting decisions on how best to deal with a fire or other emergency are not those types of problems that can be solved by simply calling the public fire department. As many decisions as possible must be made with respect to the types of fire or emergency to be expected—and these decisions must be made well in advance. The particular fire fighting and personnel safety measures to be taken may involve shutting down or isolating parts of the plant or individual equipment items. The areas where special procedures are necessary must be identified and the procedures for these special areas thoroughly understood by all plant/facility personnel.

Fire/emergency arrangements include provisions for prompt notification of the public fire department, usually through a public fire alarm signal box. However, the plant fire protection department must preplan fire fighting operations with the local fire department so that the local department will be properly coordinated with the plant's own emergency plans. Emergency planning should include measures to prevent the spread of contamination and to promptly decontaminate the area in case of accidental release of radioactive substances.

Fire fighters and other emergency personnel operating in areas where radiation exposure is a danger must be fully trained and provided with suitable protective clothing. Respiratory protective equipment is a must, and competent radiological advisors, equipped with instruments for measuring area and local exposure, are necessary to guide emergency personnel. Dosimeters or other instruments for recording each individual's accumulated radiation exposure are helpful.

A nuclear reactor site must have a generous water supply to facilitate fire control and decontamination operations. Facilities must also be prearranged for safe disposal or storage of water that may be contaminated. The use of noncombustible materials for reactor buildings and equipment will help to avoid complications of fire hazards. For example, all finish materials used for decorative, acoustical, or insulation purposes should both be noncombustible and easy to decontaminate.

The hazard of a reactor structure exposing other buildings to radiation is prevented by appropriate distance separation or fire barriers. To prevent exposure to the reactor, it is always appropriate to separate shops and service spaces from the reactor equipment and structure itself. Wiring ducts in floors introduce an opportunity for the spread of fire or of contaminated liquid or gas from one space to another. Good duct seals separate one space from another. Subassembly or other operations in the preparation of fuel elements for reactors is carried on in work areas separated from the reactor in such a way that fire cannot reach the reactor space.

Equipment for Fighting Fires

Automatic sprinkler systems or specially designed piped water spray systems are the first choice for fire protection in any location where fires may occur in nuclear reactor plants, properties housing radiation machines, and facilities handling radioactive materials. Sprinklers can operate with full effectiveness under radiation or contamination conditions that would make approach by fire fighters impossible.

In spaces where water used in fire fighting would be subject to possible contamination, the collection and disposal of this water must be provided for in the local facilities; this means the facilities should have waterproofed floors and controlled floor drainage. Substantial capacity of such drainage systems would be required if hose streams and manual fire fighting were necessary. By contrast, sprinklers or a specially designed spray system would require relatively modest amounts of water for fire fighting.

If a fire occurs in a containment vessel during construction, the difficulties of access and visability warrant the provision of temporary fixed automatic extinguishing systems when combustibles cannot be effectively controlled. Temporary interior hose stations and an ample supply of portable extinguishing equipment should be within easy reach in all portions of the vessel. Because of the smoke confinement potential, only very fast manual response may be effective, hence the available manual fire fighting equipment should be in excess of normal construction practice to insure the earliest response.

Incompatible Materials

Careful design analysis is required to reduce the fire protection problems inherent in the use of materials that are incompatible in fire situations. As an example, the contemplated use of liquid metal as a reactor coolant/moderator requires special extinguishing systems not compatible with water; in fact, the possibility of a water-liquid metal reaction may justify the exclusion of water systems from the area. If such a decision is made, however, it imposes severe limitations on the presence of flammable oils, plastics, foam insulations, and other materials that generally require copious quantities of water for fire extinguishment. Where such mixed hazards exist, it is imperative that careful consideration be given to the potentials for a failure in one system to cause a failure in the incompatible system. In such cases, either protection systems must be provided that can ensure the extinguishment of fire in either system before it can cause a rupture of the other system, or a single protection system (such as inerting) must be developed that is adequate for either hazard. The difficulties inherent in such problems warrant the most thorough hazards analysis at the earliest design stages.

Bibliography

NFPA Codes, Standards, Recommended Practices and Manuals. (See the latest *NFPA Codes and Standards Catalog* for availability of current editions of the following documents.)

NFPA 10, *Standard for Portable Fire Extinguishers.*

NFPA 48, *Standard for the Storage, Handling and Processing of Magnesium.*

NFPA 72E, *Standard on Automatic Fire Detectors.*

NFPA 220, *Standard on Types of Building Construction.*

NFPA 255, *Standard Method of Test of Surface Burning Characteristics of Building Materials.*

NFPA 259, *Standard Test Method for Potential Heat of Building Materials.*

NFPA 481, *Standard for the Production, Processing, Handling and Storage of Titanium.*

NFPA 482, *Standard for the Production, Processing, Handling and Storage of Zirconium.*

NFPA 801, *Recommended Fire Protection Practice for Facilities Handling Radioactive Materials.*

NFPA 802, *Recommended Fire Protection Practice for Nuclear Research Reactors.*

NFPA 803, *Standard for Fire Protection for Light Water Nuclear Power Plants.*

Additional Readings

UL 586, *Test Performance of High Efficiency Particulate, Air Filter Units,* Underwriters Laboratories Inc., Northbrook, IL.

ASTM E136, *Standard Test Method for Behavior of Material in a Vertical Tube Furnace at 750°C.* American Society for Testing and Materials, Philadelphia, PA.

IEEE 383, *Standard for Type Test of Class IE Electric Cables, Field Splices and Connections for Nuclear Power Generating Stations,* Institute of Electrical and Electronic Engineers, New York, NY.

NCRP 30, *Safe Handling of Radioactive Materials-NBS Handbook 92,* The National Council on Radiation Protection and Measurement, 1964.

NCRP 38, *Protection Against Neutron Radiation,* The National Council on Radiation Protection and Measurement, 1971.

NCRP 39, *Basic Radiation Protection Criteria,* The National Council on Radiation Protection and Measurement, 1971.

Standards of the U.S. Nuclear Regulatory Commission, Code of Federal Regulations, Part 20, Chapter 1, Title 10, U.S. Government Printing Office, Washington, DC.

Nuclear Safety, (bimonthly), U.S. Government Printing Office, Washington, DC.

METALWORKING PROCESSES

Revised by William Atkinson

This chapter describes the various methods used to shape, dimension, and finish metals. These include tool, electrical discharge, and electromechanical machining. The fire and explosion hazards of the various processes, metals, and fluids are indicated and the proper safeguards discussed.

Section 5, Chapter 10 of this HANDBOOK provides additional information on the characteristics of "metals." Extinguishing agents for metals are discussed in Section 19, Chapter 5. For more information, the Bureau of Mines publishes a bulletin on the explosibility of metal dusts (Bureau of Mines 1964).

THE MACHINING OPERATION

Metalworking processes include shaping, dimensioning, or surface finishing of a wide variety of metals. The raw materials, in general, have significant mass and so can be considered as noncombustible unless exposed to massive fire. Coolant/lubricants, varying from alkaline aqueous solutions to light petroleum distillates or compounded oils are necessary adjuncts. The machine tool itself may be a single motor machine, such as a drill press, or it may be a large, multimotored automatic machine with numerical control or a complex computerized numerical control system.

Though the raw materials may be considered noncombustible, the machining operation, by producing chips and cuttings, can present severe fire hazards. The combustibility of these chips depends upon the combustible character of the metal and the surface-to-mass ratio of the chips and fines. Machining produces heat, most of which is absorbed by chips and cuttings, some of which may be subject to spontaneous ignition, and others subject to ignition from outside sources. Table 10-18A shows the combustibility of fine powders and the amount of energy needed to initiate combustion.

Mr. Atkinson is Safety Director, National Machine Tool Builder's Association, McLean, VA.

Tool Machining

It is important to maintain sharp cutting edges on tools. Dull tools tend to produce finer, more readily combustible chips, and, because they are dull, produce more heat during the machining process, increasing the possibilities of ignition.

Chips and cuttings should not be permitted to accumulate at the machine. They should be removed to a noncombustible container and taken to a fire resistive storage area. Where large quantities of chips are produced, a pneumatic system may transport them to a cyclone collector. The system should be designed so that the fine dust concentration will not approach the Lower Explosion Limit (LEL) of the metal dust. The ducts should be so constructed that they will not separate under explosion pressure. Ducts and collectors should be electrically grounded, and the resistance should be checked periodically. Fabric or panel type collectors are unsafe for combustible or explosible metal dusts because the particles are held in an envelope of oxygen and a small static discharge could initiate fire or explosion.

Many machine tools have oil hydraulic control actuators, using petroleum based hydraulic fluids which have flash points between 400 and 450°F (204 to 232°C). Organic substitutes have flash points which are not much higher, but they tend to self extinguish when the ignition source is removed. A failure in a pressurized oil system can result in a jet of oil being sprayed upon a hot surface and igniting. Both local and remote controls should be available to shut off the oil flow.

Alternate Means of Machining

There are also electrical processes used for the shaping and finishing of metals. These require no cutting edge as such.

Electrical Discharge Machining (EDM): EDM sometimes is employed for the machining of conformed cavities using a preshaped electrode. The electrode is spaced 0.0005 to 0.0200 in. (0.013 to 0.5 mm) from the workpiece and is bathed in a continuous flow of dielectric fluid. A direct current of 0.5 to 400 amp at 40 to 400 V is pulsed from

TABLE 10-18A. Ignition and Explosibility of Metal Powders

	Ignition temp. cloud layer		Min. Explosive concentration		Min. ignition energy dust cloud	Maximum pressure		Maximum rate of pressure rise		Explosibility† Index
	°F	°C	oz/ft³	g/m³	millijoules	psig	kPa	psi/sec	Pa/sec	
Aluminum	650	343	0.045	45	50	73	503	20,000	508	>10
Magnesium	620	327	0.040	40	40	90	620	9 000	228	>10
Zirconium	20	7	0.045	45	15*	55	379	6 500	146	>10
Titanium	330	165	0.045	45	25	70	482	5 500	140	>10
Uranium	20	7	0.060	60	45*	53	365	3 400	86	>10
Iron	440	227	0.200	200	72–305	45	310	600	15	0.1
Zinc	680	360	0.500	500	960	48	331	1 800	45	<0.1
Bronze	370	188	1.000	1000	—	44	303	1 300	33	—
Copper	700	371	—	—	—	—	—	—	—	<0.1

NOTE: Data taken from Bureau of Mines. RI 6516, 1964. Table 1, p 4. Data on iron (item 131) from Table A1, p 12. A2, p 16 and A3, p 22.
* In this test 1 gram of powder was used. Larger quantities ignited spontaneously.
† Index of Explosibility. RI 6516, p 3. None 0. Weak 0.1. Moderate 0.1–1.0. Strong 1.0–10. Severe 10.

capacitors at 180 to 260 kHz. Each pulse partially ionizes the dielectric fluid and causes a submerged spark which melts and dislodges a metal particle. Before it is reused, the dielectric fluid must be filtered to remove the particles.

Electromechanical Machining (ECM): ECM is a process in limited specialized use. It operates by anodic dissolution in an electric cell in which the workpiece is the anode and the tool is the cathode. A direct current of 50 to 2,000 amp at 4 to 30 V is imposed. At 1,000 amp the rate of metal removal is approximately 1 cu in. per min (16.4 cm^3/min). Current densities of 100 to 2,000 amp per sq in. (1 sq in. equals 6.45 cm^2) are employed.

Wet collectors for fine metal dusts should be used only after the most careful consideration. Many dusts react with water to produce highly explosive hydrogen. The wetted sludge is also highly explosive unless it is kept submerged under a freeboard of water. (See Fig. 10-18A.)

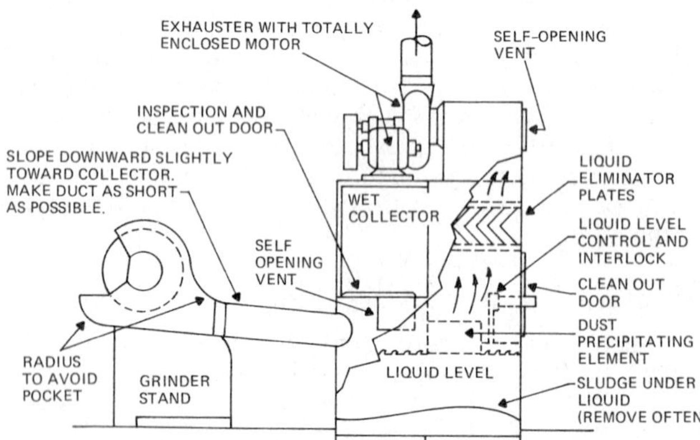

FIG. 10-18A. A schematic diagram of a water precipitation type collector for use in collecting dry combustible metal dust without creating explosive dust clouds or dangerous deposits in ducts or collection chambers. The diagram is intended only to show some of the features which should be incorporated in the collector design. The collector may be used for all combustible metal dusts.

Cutting Fluids

Water based cutting fluids should never be used with machining metals which are water reactive because explosive hydrogen will be evolved and sensible heat will also occur. Essentially all oil based and compounded cutting fluids are flammable to some degree. Compounded oils which are acidic or which contain organic additives that may oxidize and become acidic with use should not be used for the machining of magnesium and certain other metals, because the acid can react with the metal to release hydrogen. Uncompounded mineral oils should be used instead. Drippage of combustible cutting fluid should be absorbed or otherwise collected and then removed from the workroom. Provision should also be made to collect and remove the residual cutting fluid carried by the scrap metal. The waste cutting fluid should not be allowed to accumulate, but should be collected and disposed of in a safe manner.

Cleaning of Machined Components

It is sometimes necessary to clean or degrease components after machining. This is often performed in open tanks or pans using a moderately high flash point naphtha such as Stoddard Solvent. This constitutes a potential fire hazard because the solvent can be easily ignited by a flying spark from a neighboring operation.

Vapor degreasing using inhibited trichlorethylene or tetrachlorethylene (perchlorethylene) is sometimes employed. It has been found that trichlorethylene will react with finely divided aluminum (regardless of the inhibitor) to produce aluminum chloride. The aluminum chloride then catalyzes the breakdown of the trichlorethylene into fine, soft, burning carbon with the evolution of great volumes of hydrochloric acid vapor. The latter corrodes all metallic surfaces with which it comes in contact; the vapor is obviously toxic.

MACHINING CHARACTERISTICS OF COMMERCIAL METALS

Because there are no essentially pure metals used in ordinary commerce and because the commercial metals are alloys to some degree, it is virtually impossible to characterize the machining qualities of any metal except in the most general of terms. The several alloys are often amenable to various types of heat treatment or work hardening, so they have widely varying characteristics. More specific information is available from publications such as the *Machining Data Handbook* (Metcut 1980).

FIRE HAZARDS

The principal hazards to be encountered in a machining operation are:

1. Chip fires at the machine, where ignition is caused by the heat of metalworking, friction of the chip against the tool, or both
2. Spontaneous combustion of cuttings
3. Combustion of oxidizable coolant/lubricants
4. Fine particles which are either combustible or explosible
5. Reaction of certain metals with water or other agents which often results in the evolution of heat and explosible hydrogen
6. Combustion of pressurized hydraulic fluids used for the actuation of machine tools and/or their accessories
7. Combustion of oil vapors deposited on building structures
8. Combustion of oil saturated floors.

The principal sources of ignition are:

1. Smoking materials
2. Heat of cutting
3. Spontaneous oxidation
4. Hot particles from grinding, dressing of grinding wheels, and welding and cutting
5. Hot surfaces such as furnaces, torches, etc.
6. Electrical sparking or arcing
7. Impact ignition of certain pyrophoric surface compounds which sometimes form during the earlier stages of fabrication. (An example of this is magnesium nitride, which sometimes appears on the surface of castings and can explode under the impulse of a very minor impact.)

SAFEGUARDS

Building Construction and Protection

Buildings which house metalworking processes should be of fire resistive construction with a fire resistant roof deck. If the building is large, fire barriers should be installed in the roof support structure, where possible, to limit the spread of fire.

The need for sprinkler protecton should be carefully considered. It will depend largely upon the combustible character of the building structure, especially the floors and roof deck. Small fires at machines can be best controlled with portable extinguishers of the correct classification. If proper housekeeping is practiced, the fire hazard from cuttings and turnings is minimal. Consideration also must be given to the dangers of applying water to certain burning reactive metals such as magnesium.

Buildings where machining operations are performed should be regularly inspected for the accumulation of oily deposits and/or fine combustible metal particles. Cleaning should be undertaken whenever the need becomes apparent. All electrical control equipment, particularly those for numerical control or computer directed control of machines, should be housed in vapor tight enclosures. The equipment itself should be inspected at regular intervals and cleaned when necessary.

Fire Protection

The more combustible metals should never be machined unless a suitable fire extinguisher is immediately available to the machine operator. It is important to remember that extinguishers for Class A, B, and C fires are generally ineffective for burning metal or metal dust. Only extinguishers rated for Class D fires should be used on burning metal. When they are used, extreme care must be exercised to ensure that the emergent force of the extinguishing agent or the action of its application to the fire is not sufficient to create an explosive metal dust cloud.

For their own safety, it is most important that all workers thoroughly understand the hazards involved and be properly instructed so they will fully understand the possible consequences of unconsidered or improper action under stress. In many instances, it is highly desirable that local fire fighters who would normally respond to an alarm be similarly instructed.

Bibliography

References Cited

Bureau of Mines. 1964. *Explosibility of Metal Dusts*. Bulletin RI-6516. U.S. Bureau of Mines. Washington, DC.

Metcut. 1980. *Machining Data Handbook*. Metcut Research Associates Inc. Cincinnati, OH.

NFPA Codes, Standards, Recommended Practices and Manuals. (See the latest *NFPA Codes and Standards Catalog* for availability of current editions of the following documents.)

NFPA 10, *Standard for Portable Fire Extinguishers.*

NFPA 13, *Standard for the Installation of Sprinkler Systems.*

NFPA 30, *Flammable and Combustible Liquids Code.*

NFPA 34, *Standard for Dipping and Coating Processes Using Flammable or Combustible Liquids.*

NFPA 48, *Standard for the Storage, Handling and Processing of Magnesium.*

NFPA 65, *Standard for the Processing and Finishing of Aluminum.*

NFPA 70, *National Electrical Code.*

NFPA 70B, *Recommended Practice for Electrical Equipment Maintenance.*

NFPA 72E, *Standard on Automatic Fire Detectors.*

NFPA 75, *Standard for the Protection of Electronic Computer/Data Processing Equipment.*

NFPA 79, *Standard for Industrial Machinery.*

NFPA 91, *Standard for the Installation of Blower and Exhaust Systems for Dust, Stock, and Vapor Removal or Conveying.*

NFPA 325M, *Fire Hazard Properties of Flammable Liquids, Gases and Volatile Solids.*

NFPA 481, *Standard for the Production, Processing, Handling and Storage of Titanium.*

NFPA 482M, *Standard for the Production, Processing, Handling and Storage of Zirconium.*

NFPA 505, *Fire Safety Standard for Powered Industrial Trucks Including Type Designation, Areas of Use, Maintenance and Operation.*

NFPA 651, *Standard for the Manufacture of Aluminum and Magnesium Powder.*

NFPA 801, *Recommended Fire Protection Practice for Facilities Handling Radioactive Materials.*

Additional Readings

Bartels, A. L., "Safeguards to Prevent Ignition by Hot Surfaces," *Electrical Review London*, Vol. 203, No. 4 (1978), pp. 29-31.

Bellows, Guy, *Machining: A Process Checklist*, 3rd ed., Metcut Research Associates, Cincinnati, OH, 1982.

Feirer, John L., *Machine Tool Metalworking*, 2nd ed., McGraw-Hill, NY, 1973.

Follette, Daniel, ed., by Roy Williams and E. J. Weller; *Machining Fundamentals*, Society of Manufacturing Engineers, Dearborn, MI, 1980.

Heineman, Stephen S., and Genevro, George W., *Machine Tools: Processes and Applications*, Harper and Row, Publishers, New York, NY, 1979.

Heritage, P. ed., *Machining for Toolmaking and Experimental Work*, 2nd ed., (3 vols.), International Ideas, Inc., Philadelphia, PA, 1977.

Holden, A. M., *Physical Metallurgy of Uranium*, Addison-Wesley, Publishers, Reading, MA, 1958.

Jacobsen, M., Cooper, A., and Nagy, J., *Explosibility of Metal Powders, RI-6516*, U.S. Bureau of Mines, Washington, DC, 1964.

Kent's Mechanical Engineers Handbook, John Wiley and Sons, New York, NY.

Kibbe, Richard R., and Neely, John E., *Machine Tool Practices*, 2nd ed., John Wiley and Sons, New York, NY, 1982.

King, P. W., and Magid, J., *Industrial Hazard and Safety Handbook*, Butterworths Publishers, Inc., Woburn, MA, 1979.

Lustman, B., and Kurze, F., *The Metallurgy of Zirconium*, McGraw-Hill, New York, NY, 1955.

Maranchik, *Machining Data for Titanium Alloys*, AFMDC 65-1, Air Force Machinability Data Center, Cincinnati, OH, 1965.

Marks, L. S., *Mechanical Engineers Handbook*, McGraw-Hill, New York, NY.

The Technical Staff of the Machinability Data Center, *Machining Data Handbook*, 3rd ed., (2 vols.), Metcut Research Associates, Cincinnati, OH, 1980.

Van Horn, K. R., ed., *Aluminum*, Vol. 3, American Society for Metals, Metals Park, OH, 1967.

Walker, John R., *Machining Fundamentals*, Goodheart-Willcox Co., Inc., South Holland, IL, 1981.

Warner, J. C., Chipman, J., and Stedding, F., *Metallurgy of Uranium and its Alloys*, NNES-IV-12A, National Technical Information Service, Springfield, VA.

White, Warren T., et al., *Machine Tools and Machining Practices* (2 vols.), John Wiley and Sons, New York, NY, 1977.

Williams, Roy, ed., *Machining Hard Materials*, Society of Manufacturing Engineers, Dearborn, MI, 1982.

WOODWORKING PROCESSES

Edwin P. Bounous

Woodworking plants, except for the newest facilities, have characteristics that make fire control by sprinkler systems and other preventive measures difficult. In general, woodworking facilities are highly combustible structures filled with wood and/or fabric covered finished and semi-finished goods as well as other raw materials. Finishing room exhaust systems plus sawdust and shaving removal systems create high draft conditions throughout most plants.

The two principal problem areas in woodworking facilities are spray finishing and dust collection. (See Fig. 10-19A.) Most finishing room fires start from maintenance workers' torches, saw or grinder sparks, or overheated motors. Other contributing factors to fire are the dirty rags used in finishing operations, accumulation of overspray, and poor ventilation, which leads to a dangerous accumulation of solvent vapors.

Even with the best dust removal systems, the threat of fire and explosion is always present. Primarily, dust fires start at the wood hog (a grinder that turns scrap wood into splinters and chips) where an accumulation of "tramp" metal in the hog trap can become red hot. From the wood hog, fire can spread to cyclones, to the dust house, and often out onto the building roof. Dust fires are hard to control because fires in dust can burrow and smolder even when wet by sprinkler spray.

THE PRODUCTION PROCESS

After lumber has been dried and conditioned to approximately 7 percent moisture content, it is rough milled into long planks and converted to rough blanks which are later converted to finished parts. By the time parts have progressed through rough milling and finish machining, approximately 50 percent of the original material has been discarded. Approximately 35 percent of this discarded material is conveyed to the wood hog to be converted to wood splinters and chips for use as fuel in generating process steam and heat or for manufacturing particleboard, hardboard, or paper.

The process for spray finishing and hot air curing of wood products is much the same as in other industries that involve coating and curing operations. Nitrocellulose coatings, wash coats, and top coats continue to be industry favorites because these materials allow easy refinishing.

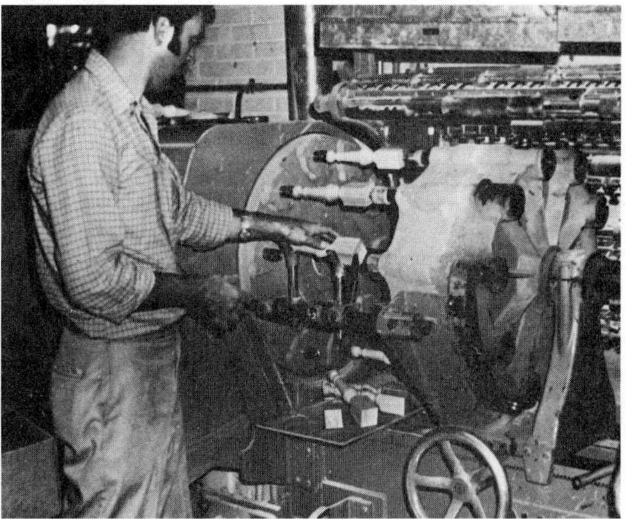

FIG. 10-19A. A sanding machine for applying a smooth finish to furniture legs prior to finishing. The sanding operation is particularly hazardous because of the fineness of the dust created.

The principal flammable solvents used in wood finishing and their approximate flash points are: acetone, −4°F (−20°C); methyl ethyl ketone, 16°F (−9°C); naphtha VM&P, 50°F (10°C); xylene, 81°F (27°C); and toluene, 40°F (4.4°C).

In many modern woodworking plants, the finishing process can employ as many as 20 spray booths in a

Mr. Bounous is recently retired Vice President, Research and Development, Drexel Industries, Drexel, NC. This chapter is a condensation of material that appeared in the *Industrial Fire Hazards Handbook*, 2nd ed, also published by NFPA.

conveyorized operation using 20,000 to 40,000 sq ft (1858 to 3716 m^2) of floor space.

Ovens for curing the finished pieces operate at approximately 250°F (121°C), but even with high temperature, high air velocity (4,000 fpm maximum or 1220 m/min), and conveyor movement through the curing ovens, most multicoat finishes require overnight storage to ensure that all the solvents have been released.

THE FIRE HAZARDS

Insofar as fire origin is concerned, spray finishing and dust collecting are the most hazardous places in most woodworking plants. The most costly fires, other than total burnout fires, are those occurring in the finishing, rubbing, and finished goods storage areas. Fires in lumberyards are also costly and hard to control. Despite the fire hazards presented by the materials and the processes, the worst fires are likely to occur from poor maintenance and repair practices involving welding and similar activities.

In the lumberyard, boards are individually inspected and built into hacks for drying and handling. Lumber hacks are about 14 ft (4.3 m) long, 6 to 8 ft (1.8 to 2.4 m) wide, and vary in height from 4 to 6 ft (1.2 to 1.8 m). Each layer of boards in the hack is separated by spacers known as kiln sticks. Hacks are stacked by forklift trucks back to back and as high as 16 ft (4.9 m). The honeycomb construction and the related easy access to oxygen make it almost impossible to extinguish a fire once it is underway in a lumber hack. Aisle widths should allow fire fighting equipment to maneuver effectively.

Among the main fire hazards around lumber storage areas (which usually have only hydrant protection) are lightning, discarded cigarettes, and boiler room sparks. Portable heaters and bonfires or barrel fires lit for warmth also pose hazards during the winter.

Waste material that ranges from small edgings to fine dust is removed by conveyors, hand trucks, and pneumatic systems. The larger parts, such as slivers, larger edgings, and blocks are conveyed to the wood hog by belt conveyor for reduction into particles small and light enough to be moved to boiler room fuel silos or storage bins. Magnetized pulleys for conveyor systems that carry wood refuse to hogs reduce the danger of fire caused by tramp metal. Trucks take waste material to the hog chute from machines not serviced by the conveyor. Sweepings are a great source of trouble, because metal dropped into the chute can become a source of friction sparks in the pulverizer.

Smaller chips, shavings, and wood dust are carried by air moving equipment connected to cyclones, bag filter systems, or equivalent equipment. In many installations, about 80 percent of the air is returned to the plant. Reinjection of air into the plant atmosphere can cut down on the air make-up requirement and reduce plant heating fuel needs. However, this practice may introduce fire and explosion hazards. NFPA 664, *Standard for the Prevention of Fires and Explosions in Wood Processing and Woodworking Facilities*; and NFPA 91, *Standard for the Installation of Blower and Exhaust Systems for Dust, Stock and Vapor Removal or Conveying*, contain guidance for the installation and protection of air moving systems for wood waste.

The principal hazard of the spray application of finishers involves flammable and combustible liquids and their vapors or mist, and combustible residues in spray booths. Properly constructed booths with adequate mechanical ventilation are the ideal way to discharge vapor to a safe location in order to reduce the possibility of explosion. Minimizing all sources of ignition in the spraying area and constantly supervising the overall finishing process are essential to a safely conducted operation. NFPA 33, *Standard for Spray Application Using Flammable and Combustible Materials*, outlines good practices to be followed in operations involving flammable coatings.

Mechanical ventilation of the spraying area should remove solvent vapors and control overspray. Generally, ventilation provides an average velocity of not less than 100 fpm (30 m/min) over the open face of the booth, sufficient to contain overspray to the booth interior. In a good installation, each spray booth exhausts approximately 25,000 cfm (708 m^3/min). The make up air moves at a significant velocity throughout the plant. These conditions can help a fire to spread and become more intensive.

The curing ovens associated with the finishing process are essentially continuous type Class A ovens as identified by NFPA 86A, *Standard for Ovens and Furnaces: Design, Location and Equipment*. These are enclosures that operate at approximately atmospheric pressure and in which there is an explosion or fire hazard from flammable volatiles and combustible residues. The oven exhaust system may lead directly to the outdoors for removal of the vapors or it may be permitted to exhaust directly into the immediate area. In the latter instance, the area needs balanced ventilation to bring fresh make up air from outside in order to exhaust the vapor laden air. Vapors should be removed before recirculating air into the area.

THE SAFEGUARDS

Automatic sprinkler coverage is the best protection for woodworking facilities. However, in some older buildings, it may be difficult to achieve full protection. NFPA 13, *Standard for the Installation of Sprinkler Systems*, suggests increasing the size of sprinkler supply pipes due to the likelihood that more sprinklers may operate where fire development is rapid. Paint spraying, upholstering, and similar hazardous processes also require larger pipe sizes and closer sprinkler spacing. Whenever possible, these processes should be located in separate buildings or separated from each other by fire rated construction with openings protected by automatic fire doors.

The degree of needed protection can vary from facility to facility, but a basic recommendation is that a yard system of mains and hydrants should be capable of supplying at least 1,000 gpm (3.78 m^3/min), enough to sustain four 2 ½ in. (64 mm) hose streams simultaneously. If conditions warrant, expanded supplies and larger fire stream appliances may be needed for effective fire control. Good unobstructed fire lanes are needed in the yards so that fire equipment can approach the lumber piles. NFPA 46, *Recommended Safe Practice for Storage of Forest Products*, and NFPA 24, *Standard for the Installation of Private Fire Service Mains and Their Appurtenances*, provide guidance on protection for lumber in storage and on supplying sprinklers, hydrants, monitor nozzles, etc. for yard systems

Bibliography

NFPA Codes, Standards, Recommended Practices and Manuals. (See the latest *NFPA Codes and Standards Catalog* for availability of current editions of the following documents.)

NFPA 10, *Standard for Portable Fire Extinguishers.*

NFPA 13, *Standard for the Installation of Sprinkler Systems.*

NFPA 24, *Standard for the Installation of Private Fire Service Mains and Their Appurtenances.*

NFPA 27, *Recommendations for Organization, Training and Equipment of Private Fire Brigades.*

NFPA 30, *Flammable and Combustible Liquids Code.*

NFPA 33, *Standard for Spray Application Using Flammable and Combustible Materials.*

NFPA 46, *Recommended Safe Practice for Storage of Forest Products.*

NFPA 51B, *Standard for Fire Prevention in the Use of Cutting and Welding Processes.*

NFPA 68, *Guide for Explosion Venting.*

NFPA 70, *National Electrical Code®.*

NFPA 86, *Standard for Ovens and Furnaces—Design, Location, and Equipment.*

NFPA 91, *Standard for the Installation of Blower and Exhaust Systems for Dust, Stock, and Vapor Removal or Conveying.*

NFPA 101, *Code for Safety to Life from Fire in Buildings and Structures.*

NFPA 231, *Standard for General Storage.*

NFPA 664, *Standard for the Prevention of Fires and Explosions in Wood Processing and Woodworking Facilities.*

Additional Readings

Buehrer, P., "Check List for the Prevention of Fires Arising from Welding and Allied Processes," *Welding in the World*, Vol. 14, No. 5-6, 1976, pp. 122-125.

Deacon, F. C., "Designing Fire Protection to Limit Monetary Loss," *SFPE Technology Report* No. 80-2, Society of Fire Protection Engineers, Boston, MA, 1980.

"Fires in Furniture Factories," *Fire Protection*, Mar. 1980, pp. 34-35.

"Moving Fire: Fire Hazards of Belt Conveyors," *Record*, Vol. 54, No. 6, 1977, pp. 18-21.

"Prevent Cutting/Welding Fires," *Minnesota Fire Chief*, Vol. 16, No. 5, May/June 1980, pp. 16, 48, 54.

Richards, D., "Fire Risks in the Furniture Industry," *Fire Engineers Journal*, Vol. 38, No. 10, 1978, pp. 23-24.

Trinks, W., *Industrial Furnaces*, Vols. 1 and 2, John Wiley and Sons, NY, 1967.

AUTOMATED PROCESSING EQUIPMENT

John F. Bloodgood, P.E.

Like many other facets of industry, automation has benefitted greatly from the technological developments of the last few decades. Twenty years ago, automation was accomplished by use of large relay control panels. In 1969, programmable controllers (PCs) began to replace relays in the automotive industry. By 1971, other industries were using PCs. In 1977, microprocessor based PCs were controlling automated processes. Today, automation has become very complex and sophisticated. A great deal of care is needed to design safe automated systems.

Programmable controllers have introduced new fire protection problems into the processes which they automate. This is because the PC has been given control over such things as flammable liquids flow, the position of combustible materials, the initiation of sparks and flames, the motion of friction causing parts, and other potential sources of fuel and ignition. As a result, any kind of failure in the PC can cause a fire or explosion. This chapter deals with how to analyze and deal with the increased potential for fire or explosion with automated equipment. The chapter does not cover the hazards and protection of individual pieces of equipment which are included in other chapters of this HANDBOOK, although many of the issues raised also are germane to independently operated machines.

Such automated machinery usually consists of many pieces of equipment acting in a coordinated manner without operating personnel control. These systems are commonly referred to as manufacturing cells or flexible manufacturing systems. Common examples include numerically controlled machine tools (NC), computerized numerically controlled machine tools (CNC), industrial robots, work handling machines or systems, material/workpiece transfer equipment, and associated control and information handling equipment. (See Fig. 10-20A.) Automated systems also control many other types of processes.

Each application of automation must be reviewed for any hazardous situations it may introduce. This is done by reviewing both the PC and the process being controlled.

Mr. Bloodgood is President of JFB Enterprises, Consulting Engineers, Fond du Lac, WI.

FIG. 10-20A. Industrial robots are increasing productivity, lowering manufacturing costs, and improving the product quality. The robots pictured here are spot welding automobile frames. (Robotic Industries Association)

GENERAL CONSIDERATIONS

Automated processing systems consist of two or more machines grouped together to perform a series of operations on materials or parts. Typically, these machines run automatically without an operator, or sometimes one person may run more than one machine. A supervisory controller (usually a computer) directs the interrelations between these machines, transmits information to the individual machine controllers, and monitors the process. The individual machine units (machine and controller) may be supplied by different vendors, thus requiring compatibility between the individual units and the supervisory controller. The issue of compatibility places additional responsibility on the end user for proper installation, maintenance, and training. In addition, the machine controllers and the supervisory controllers utilize sophisticated computer equipment operating in a real time mode that transmits large amounts of information at low levels and high speeds—conditions that require greater attention to the design, installation, operation, and maintenance of these systems.

AUTOMATED PROCESSING EQUIPMENT

Analysis of the Controller

Microprocessor based PCs introduce failures which never had to be dealt with in relay controlled automation. The most common of these failures are described in Table 10-20A.

TABLE 10-20A. Programmable Computer Failures

CPU STALL —	Processor ceases to execute the program.
INPUT/OUTPUT SCAN FAILURE —	Processor fails to scan input/output signals for change in status.
INPUT FAILURE —	Input module locks in "on" position and does not respond to further input changes.
ADDRESSING FAILURE —	Processor fails to correctly consult input/output intelligence.
OUTPUT FAILURE —	Output module locks in "on" position and does not respond to the controller.
MEMORY FAILURE —	Incorrect bit in memory causes an improper instruction to be given.
WATCHDOG TIMER FAILURE —	Timer controlling execution of instructions fails.
PROGRAM CARD FAILURE —	Any problem with the program card.
MOMENTARY POWER OUTAGE —	Temporary loss of power to controller.

The effect these failure modes will have on the process being controlled will then determine the degree of hazard introduced.

Problems of personnel safety, power networks and distribution, grounding, electrical noise interference, software application and maintenance, and failsafe requirements are more evident in these integrated manufacturing systems. Another way PCs introduce fire or explosion potential is by their introduction into areas where relay control panels didn't fit, but which may contain flammable vapors. When this is done, the controller should be intrinsically safe as described in NFPA 493, *Intrinsically Safe Apparatus for use in Division 1 Hazardous Locations.*

Safety

The operation of integrated manufacturing systems is more automated than individual machines in terms of starting motions and the transferring of material or parts without operator intervention. Integrated systems can also cover more complex operations on larger areas of a factory floor. It is therefore necessary to provide additional safety measures over conventional single machine work stations. Typical measures include barriers and gates, presence sensing devices, and visual or audible warning devices to announce pending machine and/or material movement.

The movement of a machine or workpiece and/or a material handling device which occurs at unannounced times could catch personnel in a pinch point and result in a serious injury. Thus the primary safety consideration is to keep personnel out of the work area while the system is operating. Enclosing the system in a cage or fence is one solution to the problem, but it is not always practical since many systems require personnel in the work area during operation. Often, it is not always possible to determine if anyone is in the work area when the equipment is operating. One alternate solution in particularly hazardous work areas is to provide barriers and/or guards, interlocked with the individual controllers or the supervisory controller.

All hazardous motions should be automatically stopped if the interlock is interrupted. Another alternative is to provide presence sensing devices, such as floor mats or light beam curtains, that will either stop all dangerous motions immediately or stop the automatic cycle once the motions in progress have been completed. Visual or audible annunciators that warn of pending axis motions or the cycling of auxiliary equipment also can be used if personnel must be in the work area during automatic operation.

Failsafe travel limits, such as mechanical stops or hard-wired travel switches, always should be used in systems where software controllers for individual machines or system monitoring are utilized. The failsafe travel limits ensure that if the controller should fail or a runaway condition should exist in either an axis motion or an auxiliary cycle, either the motion would be physically restrained or the power would be removed from that motion.

ANALYSIS OF THE PROCESS

For some processes, failure of a PC will result in inconvenience. Usually, more serious problems are introduced. For example, PC controlled robot welders may destroy parts if the arc is not turned off at the right time. They could also ignite combustible materials which mistakenly end up in the area.

PCs controlling hazardous processes such as spray painting, solvent drying, and fuel firing also present many potential fire and explosion hazards. A PC failure could create hazardous conditions by not stopping paint guns, solvent introduction, or fuel flow.

Another complication of automated processing is the potential for interruption of production. In automated processes that cannot be run manually, PC failures will stop production. The length of time needed to repair or replace controllers must be considered. Off-the-shelf controllers are easier to replace than custom made ones, for example. Since off-the-shelf controllers would require reprogramming, software backup should be maintained in a safe location.

Power Networks and Distribution

Careful consideration must be given to the design of the automatic factory power network when large systems are installed. Busbars, which supply electrical power to machines or equipment that are highly inductive or utilize power converters, should be isolated from those which supply power to sensitive electronic equipment. Electrical noise on common power distribution lines can cause malfunctions or permanent damage to electronic equipment. Separate power buses, isolation transformers, and filtering devices are some of the techniques used to minimize the problems associated with power distribution. In some cases, it is necessary to monitor bus loading and automatically schedule the sequencing of particularly heavy electrical loads on the network.

Grounding: The grounding of electrical equipment has always been a consideration for the safety of personnel from electrical shock. In systems using sensitive electronic equipment, grounding is a primary concern to protect the controller (and its input/output connections) from the effects of electrical noise. The grounding of control equipment in industrial environments should be integrated by the equipment user into a coordinated ground system. When control equipment is grounded, it is interconnected into the system ground by building columns or ground rods.

In electrical/electronic controllers, there are usually subsystems that are tied to the ground system. It is essential that the noncurrent carrying metallic enclosures be grounded, including all internal frames and equipment support structures. A specific ground point is provided within the enclosure for the equipment ground to make connection to the ground system. A separate tie point is provided to connect the control common (the zero potential reference point for the electrical and electronic equipment) to the ground system. Having two separate ground reference points separates the two ground systems within the controller, eliminating ground loop problems. Providing two separate ground reference points within the enclosure allows the equipment ground to be connected to the building, while the control ground is connected to a separate ground rod independent of the building structure.

DESIGN SOLUTIONS

For dependability in PC run processes, important controller functions should be supervised. Input circuits, input/output addressing and input/output scanning are verified with a dynamic safety check on the order of several times each second. Output status is monitored. The Central Processing Unit (CPU) is checked by the watchdog timer which itself is provided with a dynamic safety check. Sum and parity checks are used on memories.

Above all else, the entire process control system must be designed to fail safely. The response to detection of PC abnormalities can be an alarm, the shutdown of chosen process functions or both. However, if a PC failure can result in unsafe process conditions, the process should be safely shut down upon this failure. In other words, the system should be failsafe.

Process control devices which may fail in a shorted condition (rectifiers, for example) can be arranged to be disabled manually. Critical controller functions can be independently hard-wired to assure the ability to operate properly in the event of failure.

Electrical Noise Interferences: Because entire books have been written on this complicated subject, only the most important considerations are covered here. The effects of coupling electrical noise into control equipment become more pronounced as the operating speed and equipment complexity increase while the immunity to disturbances decreases. This is certainly the case with controllers applied to industrial automation equipment. The microelectronic circuits operate above 1 MHz with harmonics in the range of 2100 MHz to 1 GHz. Signal levels are typically below 15 volts. Noise coupling from disturbing sources, such as power inverters, motors, contactors and relays, and associated wiring runs can be detrimental to the operating system unless proper techniques are used to reduce the disturbances or minimize coupling into the sensitive circuits. Commonly used methods of reducing the disturbing effects of electrical noise are (1) employing transformers or filters to isolate power wiring from signal and control wiring, or (2) shielding sensitive control and signal wiring by using shielded conductors or coaxial cable. Proper grounding of the shield (usually at only one end) is a must. For extremely sensitive circuits, the use of double shielding (one shield tied to the equipment ground and the other tied to the control common) may be required. Some circuits or subsystems may have to be entirely enclosed in a grounded shield.

Sensitive signal conductors should be separated to the greatest degree practical from noise sources. One example of this problem involves power leads which are run in the same cable or raceway as digital transmission or analog signal lines. Wherever possible, these lines should be in separate cables or raceways. When this is not possible, it is necessary to use shielding or additional conductors that are tied to the ground. Remember that all unused conductors should be grounded at one end to minimize coupling.

The effect of electrostatic discharge is another potentially serious problem that is often overlooked. Microelectronic devices (memory and logic chips) must be handled by personnel who are grounded by either a wrist strap or conductive plates located on the soles of their shoes. However, nonconductive materials (clothing, paper, etc.) in close proximity to the chips also carry an electrostatic charge which can be neutralized by ionized air. The precautions taken for handling microelectronic parts are typically discontinued once the chips are inserted into a printed circuit assembly. Although a completed printed circuit assembly will dissipate some of the charge, the chips are still subject to damage from electrostatic discharge. The grounding or discharging of personnel who are required to work with integrated circuits (including those who assemble, inspect, test, ship, or maintain the equipment) is important to ensure reliable operation of the electronic equipment. An electric charge of thousands of volts can be developed by simply getting up from a chair or paging through a set of schematic drawings. With the present technology, as little as 50 volts of electric charge can permanently damage, degrade the performance, or shorten the life of these components. In industrial systems employing thousands or tens of thousands of these devices, the potential for a malfunction or failure due to electrostatic discharge is very high and should be considered as serious a problem as shock hazard.

Software Application and Maintenance

The use of programmable electronic controllers and software programming has introduced a new series of problems that were not present with hardware systems. Failure modes and malfunctions are harder to detect and correct since the controllers must first verify the proper operation of the hardware and then test with application software. Since the software tends to be unique for each controller (one of the major advantages of programmable electronic controllers), elaborate testing and verification methods must be devised to ensure that the software functions properly under all dynamic conditions. This problem is compounded when the controllers must interact in a manufacturing system. Even if the software is properly certified before shipment to the end user, it is often modified in the field to correct for unexpected conditions of actual use. Many times, the corrected software is put into operation without a complete recertification, or even worse, it is corrected at the job site without being retested. In a large, complex system, the results of this "on the fly" modification method can be disastrous.

The typical failure mode being "fail unsafe" is another problem which may be created with the introduction of electronic circuitry. Silicone controlled rectifiers, triacs, and electronic switching devices usually fail in a shorted condition, resulting in such hazards as axis or auxiliary cycle runaway. Short of removing power to the portion of the system that has failed, the condition cannot be corrected. Typical methods of overcoming this problem are redundant circuits, hard-wired contacts which can break the connection to the drives or motors that control motions or auxiliary cycles, or control circuits which monitor and automatically disable equipment that can create hazardous conditions.

The transmission of large amounts of digital information between system elements presents the potential of system malfunctions if the data are not properly received. Wrong axis commands, machine conditions, or even entirely incorrect part programs may be sent or received. Since there is no operator intervention before machine or controller action is initiated, injury can result to personnel, damage the machine or the work in progress, or initiate a fire hazard. Redundant transmission or error checking and correction are means of reducing the effects of this problem. Proper shielding and protection of these transmission lines is mandatory if these problems are to be kept to a minimum.

Bibliography

Additional Readings

"Computer Technology Can Enchance Industrial Energy Efficiency," U.S. Department of Energy, *Technical Briefing Report*, 1971.

"Digital systems offer flexible, reliable process control operations," *Pulp & Paper*, Feb. 1982.

"Guidelines help you interface field sensors with computers," *Power*, Sept. 1983.

"Modern boiler control and why digital systems are better," *Hydrocarbon Processing*, Aug. 1982.

"Programmable Controller Update," *Plant Engineering*, Mar. 28, 1985.

"Reference File: Programmable Controllers," *Plant Engineering*, Nov. 23, 1983.

"Refinery changeover to microprocessor control," *EC&M*, Aug. 1983.

"Significant process automation changes expected in next decade," *Pulp & Paper*, Feb. 1982.

"Troubleshooting Programmable Controllers," *Plant Engineering*, Oct. 14, 1982.

"Understanding Fiber Optics for Automated Control," *Plant Engineering*, June 23, 1983 and July 21, 1983.

"Understanding Programmable Controllers," *Plant Engineering*, Apr. 17, 1980.

AEROSOL CHARGING OPERATIONS

Henry C. Scuoteguazza

Aerosol products play a large part in the nation's economy. It would be an unusual household or business that did not possess at least one aerosol product at any time. Products so packaged include food (such as whipped toppings), release agents for cooking and industrial processes, deodorants, hair sprays, carburetor cleaners, paints, pesticides, window cleaners, deicers, and engine cleaners, as well as many others.

Due to the nature of the propellants and the pressures under which they are used, aerosol products present some very real fire prevention and protection problems. This chapter is concerned with these problems. Requirements for the storage and handling of the liquefied petroleum gases used as propellants are covered in Section 11, Chapter 5, "Storage and Handling of Gases."

CONTAINERS AND PROPELLANTS

The typical aerosol container is a small, welded-joint, high strength metal can with a capacity of up to one quart containing up to 16 oz (473 mL) of liquid. The top and base of the container are domed to withstand the pressure which may be as high as 240 to 400 psi (1655 to 2758 kPa). A spring-loaded plunger nozzle cap is pressed to release the pressurized product or mixture of liquefied gas and product. (See Fig. 10-21A.)

Current propellants are chiefly hydrocarbons such as propane and butane in various mixtures, depending upon the product and its uses. They are gaseous at ambient temperatures and pressures, and they will condense to liquids at moderate pressures or low temperatures. They maintain constant pressure at a given temperature until all liquid is converted to gas. Other less common propellants are nitrogen, carbon dioxide and nitrous oxide (also methylene chloride). The hydrocarbons are highly flammable, and, although nitrous oxide is nonflammable, it is an oxidizer and can contribute to the fire hazards of aerosols.

The quantity of propellant in an aerosol product ranges from 0.5 to 90 percent of the total weight of the contents. There is some feeling that the flammability of the base product is a major concern and that the propellant adds little to the overall hazard unless the propellant accounts for most of the can's contents (FMEC 1983). Despite this, hydrocarbons are extremely flammable and require careful storage and handling.

The vapor pressure of these materials and the large

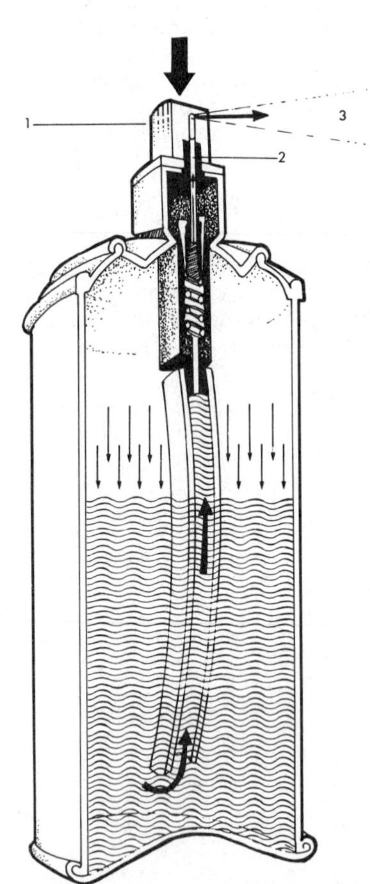

FIG. 10-21A. An aerosol can. When the plunger (1) is pressed, a hole in the valve (2) allows a pressurized mixture of product and propellant (3) to flow through the plunger's exit orifice.

Mr. Scuoteguazza is Standards Coordinator for Factory Mutual Research Corp., Norwood, MA.

quantities of gas they produce when vaporized makes them suitable propellants. The vapor pressure at 70°F (21°C) is 108 ± 6 psig (745 ± 41 kPa) for propane; 31 ± 2 psig (214 ± 13.8 kPa) for isobutane; and 17 ± 2 psig (117 ± 13.8 kPa) for n-butane. They can be mixed to deliver any vapor pressure between the two extremes and thus are suitable for use in a wide variety of products.

The basic fire hazard properties of these hydrocarbon propellants are listed in Table 10-21A.

TABLE 10-21A. Fire Hazard Properties of Hydrocarbon Propellants

	Flammable Limits*		Flash Point		Autoignition Temperature	
	Lower	Upper	°F	°C	°F	°C
Propane	2.1	9.5	−156	−104	940	504
Isobutane	1.8	8.4	−117	−83	890	477
n-Butane	1.8	8.4	−101	−74	860	460

* percent in air by volume

FIRE HAZARDS

Aerosol products have been directly involved in and sometimes responsible for extensive, costly, and occasionally fatal fires. In one instance, an aerosol can which was thrown into an incinerator exploded, blew off the incinerator door, and ignited rubbish. The damage was estimated at $35,000. In another instance, aerosol cans with flammable propellant were stocked close to a heater. Leaking propellant ignited, and bursting containers made fire fighting efforts futile. The damage was estimated at $800,000. At a can filling plant an explosion involving isobutane killed four employees and injured 18 others. Property damage was estimated at $1 million.

Perhaps the most costly fire involving aerosol products was the fire that destroyed the 27 acre (10.9 ha) distribution center of the K Mart Corporation in Falls Township, PA. Palletized petroleum-liquid based aerosol containers were involved in the fires, which overwhelmed the sprinkler protection. The facility was protected with hydraulically designed ceiling sprinkler systems. The extremely fast developing fire spread through the aerosol storage on June 21, 1982; rocketing cans spread the fire throughout this and adjoining areas. The fire overtaxed fire protection systems, resulting in early roof collapse and broken sprinkler piping. There was no loss of life. However, the loss of property was in excess of $100 million (Best 1983).

Classification of Aerosols

To simplify the consideration of the fire hazards involved, aerosols can be divided into two groups:

1. Flammable products with flammable or nonflammable propellants.
2. Nonflammable products with flammable or nonflammable propellants.

Flammable products can be further subdivided as water miscible or water nonmiscible.

FIG. 10-21B. These aerosol cans were located in the area of fire origin of the K Mart fire.

Factory Mutual Engineering Corporation groups aerosols into three levels according to the percentage by weight of flammable substances in the base product:

I. Maximum of 25 percent water miscible or nonmiscible flammable products (i.e., 75 to 100 percent nonflammable products).

II. Twenty five to 100 percent water miscible flammable products, 25 to 55 percent nonmiscible flammable products (remaining 45 to 75 percent is nonflammable product).

III. Greater than 55 percent nonwater miscible flammable products.

Products containing more than 75 percent nonflammable liquids are classified as Level I aerosols. Level I aerosols include such products as shaving cream, spray starch, oven cleaners, rug shampoos, air fresheners, and some insecticides. Their storage hazard is about the same as ordinary combustible goods in cartons. When a can containing Level I aerosol fails, the nonflammable contents tend to quench the flammable contents.

Level II water miscible products include most personal care products such as deodorants (except oil based antiperspirants), hair sprays, antiseptics, and anesthetics, and other products such as some furniture polishes and windshield deicers.

Level III aerosols include automotive products such as undercoaters, engine and carburetor cleaners, furniture polishes, paints, lacquers, lubricants, and some insecticides.

Tests indicate that mixtures containing 55 percent petroleum liquid and 45 percent nonflammable liquid have burning characteristics significantly below those of 100 percent petroleum liquids. If the product has from 25 to 55 percent flammable liquid and the balance is nonflammable, it can be treated as a Level II aerosol.

Fire Behavior of Aerosols

The behavior of filled aerosol cans in fire tests confirms what has often been reported about actual fires. Level II aerosols produce fires so intense that ruptured containers rocketed through the test area, occasionally trailing burning liquid. Sometimes flaming packaging material or plastic caps were propelled beyond the test area.

Level III aerosols burned more intensely and rocketed with trailing burning liquid more frequently than Level II aerosols. In addition, they produced dense black smoke that completely obscured visibility in four to five minutes.

In rack storage tests, the ruptures of both Level II and Level III aerosols spanned the aisle, but ignition occurred only in Level III tests. In both rack and pallet storage, rupturing cans dislodged surrounding containers and exposed others inside the load, increasing the fire severity.

As indicated by the fire incidents cited previously, there are three phases in the production and distribution of aerosol products:

1. Transport, transfer, and storage of bulk propellants.
2. Filling or "gassing" the aerosol containers.
3. Storage of charged containers in warehouses.

Aerosol Filling Plants

Because the hydrocarbon aerosol propellants are highly flammable, container filling plants require special design considerations and operating procedures. Gases present hazards both when they are confined and when they escape. Compressed gases and liquefied gases are affected somewhat differently when heated. When heated, compressed gas simply attempts to expand according to the gas laws, whereas a liquefied gas experiences simultaneous combination of three effects.

First, the gas phase is subject to the same effect as a compressed gas. Second, the liquid attempts to expand and compresses the vapor. Third, the vapor pressure of the liquid increases as the temperature increases. The combined result is a rapid pressure increase when the container is heated. Containers of liquefied gases can release this energy rapidly and violently, resulting in a BLEVE (Boiling Liquid Expanding Vapor Explosion). Also, if some of the propellant is still a liquid at the time of rupture, it will expand immediately and help disperse the base product, especially if the propellant is soluble in the base product.

THE SAFEGUARDS

NFPA 58, *Standard for the Storage and Handling of Liquefied Petroleum Gases* (hereinafter referred to as NFPA 58), sets the standards for storage and handling of liquefied petroleum gases.

Because a release of hydrocarbon propellants within the charging (gassing) room can create an explosive atmosphere, the room should be located and constructed to reflect a combustion explosion hazard. Often the room is a separate building apart from the main production building. The separation should be a minimum of 5 ft (1.5 m) with conveyor openings as small and few as possible. The rooms may be located within the main building, but if so, they should be constructed and vented in accordance with the provisions of NFPA 58 and NFPA 68, *Guide for Explosion Venting*.

If separate from the main building, the filling room should be at least 10 to 25 ft (3 to 7.6 m) from the bulk storage facilities (depending upon building construction); 25 ft (7.6 m) from a property line which can be built upon; and at least 25 ft (7.6 m) from ignition sources. The wall facing the main building should have a minimum pressure rating of 100 psf (488 kg/m^2) (Best 1983.). The venting area should be at least 1 sq ft per 30 cu ft (0.1 m^2/m^3) of enclosed volume. Explosion venting areas should be designed to vent at a minimum static pressure of 20 psf (97.6 kg/m^2).

Even in a well maintained and efficiently operated charging room, enough flammable vapor is released to create a locally explosive atmosphere unless there is a ventilation system which maintains a concentration below the Lower Flammable Limit (LFL). As noted in Table 10-21A, the LFL is about 2 percent by volume in air for hydrocarbon propellants.

Because hydrocarbon propellant vapors are heavier than air, the ventilation system should be designed to let uncontaminated intake air flow over the floor to pick up vapors and discharge them through and out an exhaust stack. It is wise to maintain a slight negative pressure within the charging room to prevent migration of the vapors to other areas.

Equipment failures or other accidents may release quantities of vapor greater than can be handled by the general ventilation. The system, therefore, should be designed to automatically provide a high ventilation level when the monitoring system indicates a vapor concentration of 25 percent of the LFL.

The charging room should be equipped with gas detection devices as part of a system that will shut off the propellant supply, purge the atmosphere, and isolate ignition sources. Such systems are generally electric or electronic. As such, they have the potential of providing an ignition source unless properly selected and installed. NFPA 70, *National Electrical Code*®, and NFPA 493, *Standard for Intrinsically Safe Apparatus and Associated Apparatus for Use in Class I, II and III, Division 1 Hazardous Locations*, describe the selection and installation of the protective system.

Despite the extent and efficiency of fire and explosion preventive measures, a fire or explosion is still possible. As added safety measures, then, fire detection and extinguishing equipment should be installed.

For a charging room, such equipment may be any of the usual systems, i.e, automatic sprinkler, carbon dioxide, foam, dry chemicals, and explosion suppression. The most common extinguishing system is the open head deluge system with rate of rise heat detection (Chem Spec 1979). This system reacts quickly and allows a water application of at least 0.35 gpm per sq ft [14 (L/min)/m^2]. It reacts quickly, is effective regardless of the condition of the room, and involves no physical danger to personnel.

While special fire extinguishing systems (carbon dioxide, foam, dry chemical and Halon) are fast acting and effective extinguishers, they do not replace the need for sprinklers for the following reasons: (1) application must be delayed until personnel can be evacuated; (2) a gas fire that is extinguished before the escape of gas has stopped can lead to an explosion; (3) foam is of value only for fires in flammable products; and (4) such systems have a limited amount of extinguishing material.

Explosion suppression systems have been employed effectively in charging rooms. Such a system provides protection by extinguishing flame before damage occurs. Ultraviolet sensors react in milliseconds, releasing a Halon agent or water which traps the spark or flame before explosive pressures can develop. Halon 1301 is approximately twice as effective as water in an explosion suppression system.

TABLE 10-21B. Arrangement and Protection of Palletized and Solid-Pile Storage of Aerosol Containers

Level	Max Pile Height	Sprinkler Spacing	Temperature Rating & Sprinkler Size	Sprinkler Demand	Hose Stream Demand*	Max. Height Storage to Sprinkler Clearance	Duration Sprinklers & Hose Stream
II	5 ft (1.5 m)	100 ft² (9 m²) max	286°F (141°C) ½ in. (12.5 mm) or 17/32 in. (14 mm)	0.30 gpm/ft² over 2500 ft² (12 mm/min over 230 m²)	500 gpm (1.9 m³/min)	30 ft (9.1 m)	2 hr
II	18 ft (5.5 m)	80–100 ft² (7.4–9 m²)	160°F (71°C) 0.64 in. (16.3 mm) "large drop"	15 heads at 50 psi (344 kPa)	500 gpm (1.9 m³/min)	5 ft or less (1.5 m or less)	2 hr
III	5 ft (1.5 m)	100 ft² (9 m²) max	286°F (141°C) 17/32 in. (14 mm)	0.60 gpm/ft² over 2500 ft² (24 mm/min over 230 m²)	500 gpm (1.9 m³/min)	30 ft (9.1 m)	2 hr
III	10 ft (3 m)	80–100 ft² (7.4–9 m²)	160°F (71°C) 0.64 in. (16.3 mm) "large drop"	15 heads at 75 psi (517 kPa)	500 gpm (1.9 m³/min)	10 ft or less (3 m or less)	2 hr

* For cut-off rooms of less than 2000 ft² (185 m²), hose stream demand is 250 gpm (0.95 m³/min).

TABLE 10-21C. Protection of Rack Storage of Pressurized Containers of Flammable Products

Level	Ceiling Sprinkler Arrangement	In-Rank Sprinkler Arrangement†	Ceiling Sprinkler Demand	In-Rank Sprinkler Demand	Hose Stream Demand*	Duration-Sprinklers and Hose Streams	Clearance Storage to Sprinklers
II	286°F (141°C) rated heads 100 ft² (9.3 m²) max. spacing ½ in. (12.5 mm) or 17/32 in. (14 mm) orifice	165°F (74°C) or less sprinklers 8 ft (2.4 m) apart max. One line at each tier except top. Locate in longitudinal flue spaces of double-row racks.	0.30 gpm/ft² (12 mm/min) over 2500 ft² (230 m²)	30 gpm (0.11 m³/min) discharge per head minimum. Base on operation of hydraulically most remote: (1) 8 sprinklers if one level. (2) 6 sprinklers each of 2 levels if only 2 levels. (3) 6 sprinklers on top 3 levels if 3 or more levels.	500 gpm (1.9 m³/min)	2 hr	15 ft (4.6 m). Need barrier with sprinklers beneath if clearance exceeds 15 ft (4.6 m)
III	286°F (141°C) rated heads 17/32 in. (14 mm) orifice 100 ft² (9.3 m²) max. spacing	165°F (74°C) or less sprinklers 8 ft (2.4 m) apart max. Install in longitudinal flue and on face. Stagger face sprinklers with sprinklers on opposite side of rack.	0.30 gpm/ft² (12 mm/min) over 2500 ft² (230 m²) 0.60 gpm/ft² (24 mm/min) over 1500 ft² (140 m²) to 2500 ft² (230 m²)	Same as Level II	500 gpm (1.9 m³/min)	2 hr	5 ft (1.5 m) or less 5 ft (1.5 m) to 15 ft (4.6 m). If greater than 15 ft (4.6 m), need barrier with sprinklers beneath.

† Provide approved rack storage sprinklers with built-in water shields. Locate longitudinal flue in-rack sprinklers at least 2 ft (0.6 m) from rack uprights. Provide at least 6 in. (150 mm) between sprinkler deflectors and top of storage in tier.
* For cut-off rooms of less than 2000 ft² (185 m²), hose stream demand is 250 gpm (0.95 m³/min).

Disposal of defective cans is an essential part of aerosol product manufacture. After filling, containers are passed through a warm water bath to detect leaks. Leaking containers are removed and crushed in a separate facility. Safety precautions in this area should be on the same order as those in the filling room.

Product development and quality control laboratories associated with aerosol product plants also present hazards of fire and explosion. They should be built, equipped, and operated in accordance with the various safety standards; among these are the NFPA standards listed at the end of this chapter.

Product Warehousing

Wet pipe sprinkler systems are recommended for aerosol storage. Sprinkler spacing, size, and temperature rating will depend upon the classification of the aerosol product and the storage method. Level II aerosols in palletized or solid pile storage should have 1 sprinkler per 100 sq ft (9.3 m^2) delivering 0.30 gpm per sq ft [12.2 (L/min)/m^2] over 2,500 sq ft (232 m^2) as a minimum. Other values for Level II and Level III are shown in Tables 10-21B and 10-21C as developed by Factory Mutual Engineering Corp.

Level I aerosols should be protected as Class III commodities in both palletized and rack storage. Any aerosol products which cannot be identified should also be protected as Level III aerosols.

Whether in rack or palletized storage, Level III aerosols should be in a cut off area with walls or ceiling-high partitions having a 1 hr fire resistance rating. Wet pipe sprinkler protection should be in accordance with the figures in Tables 10-21B and 10-21C.

Rack storage of Level II aerosols also should be in a dedicated area with sprinkler protection as shown in the Tables. It is important to have in-rack sprinklers in every tier to prevent the fire from spreading throughout the rack. For small quantities and picking areas, Level II aerosols may be stored on the first tier with in-rack sprinklers provided specifically for that level and the one above.

Bibliography

References Cited

Best, Richard. 1983. "$100 Million Fire in K Mart Distribution Center." *Fire Journal*. Vol 77, No 2. Mar 1983. pp 36-42.

Chem Spec. 1979. *Hydrocarbon Propellants: Considerations for Effective Handling in the Aerosol Plant and Laboratory*. Chemical Specialties Manufacturers Association, Washington, DC.

FMEC. 1983. "Storage of Aerosol Products." *Loss Prevention Data 7-29S*. Factory Mutual Engineering Corp., Norwood, MA.

NFPA Codes, Standards, Recommended Practices and Manuals. (See the latest *NFPA Codes and Standards Catalog* for availability of current editions of the following documents.)

NFPA 11A, *Standard for Medium and High Expansion Foam Systems*.

NFPA 12, *Standard on Carbon Dioxide Extinguishing Systems*.

NFPA 12A, *Standard on Halon 1301 Fire Extinguishing Systems*.

NFPA 13, *Standard for the Installation of Sprinkler Systems*.

NFPA 15, *Standard for Water Spray Fixed Systems for Fire Protection*.

NFPA 17, *Standard for Dry Chemical Extinguishing Systems*.

NFPA 18, *Standard on Wetting Agents*.

NFPA 30, *Flammable and Combustible Liquids Code*.

NFPA 33, *Standard on Spray Application Using Flammable and Combustible Materials*.

NFPA 45, *Standard on Fire Protection for Laboratories Using Chemicals*.

NFPA 58, *Standard for the Storage and Handling of Liquefied Petroleum Gases*.

NFPA 68, *Guide for Explosion Venting*.

NFPA 69, *Standard on Explosion Prevention Systems*.

NFPA 70, *National Electrical Code*.

NFPA 77, *Recommended Practice on Static Electricity*.

NFPA 91, *Standard for the Installation of Blower and Exhaust Systems for Dust, Stock and Vapor Removal or Conveying*.

NFPA 231C, *Standard for Rack Storage of Materials*.

NFPA 493, *Standard for Intrinsically Safe Apparatus and Associated Apparatus for Use in Class I, II, and III, Division I Hazardous Locations*.

NFPA 704, *Standard System for the Identification of the Fire Hazards of Materials*.

Additional Reading

"Concern Over Warehouse Fires Leads to Tests on Aerosols," *Fire*, Jan. 1981, pp. 410-412.

Davenport, J. A., "Explosion Losses in Industry," *Fire Journal*, Vol. 75, No. 1, Jan. 1981, pp. 52-56, 71-73.

Deacon, F. L., "Designing Fire Protection to Limit Monetary Loss," *SFPE Technology Report No. 80-2*, Society of Fire Protection Engineers, Boston, MA, 1980.

Dennis, Richard, ed., *Handbook on Aerosols*, ERDA Technical Information Center, 1976.

"Fire and Explosion Hazards with Aerosols," *Fire Prevention*, Dec. 1980, pp. 22-24.

Henry, Martin F., *Flammable and Combustible Liquids Code Handbook*, 2nd ed., National Fire Protection Association, Quincy, MA, 1984.

Hinds, Williams C., *Aerosol Technology: Properties, Behavior, and Measurement of Airborne Particles*, John Wiley and Sons, Inc., New York, NY, 1982.

Johnsen, Montfort A., *Aerosol Handbook*, 2nd ed., Dorland Publishing Co., Mendham, NJ, 1982.

Magison, E. C., *Electrical Instruments in Hazardous Locations*, 3rd ed., Instrument Society of America, Pittsburgh, PA, 1978.

Marple, V. A., and Liu, B. Y. H. eds., *Aerosols In the Mining and Industrial Work Environments*, Vols. 1-3, Ann Arbor Science Publications, Woburn, MA, 1983.

"Safe and Efficient Plant Operation and Maintenance," *Chemical Engineering Magazine*, McGraw-Hill, NY, 1980.

"$10 Million Spent for Aerosol Plant Fire Safety," *Fire Engineering*, Vol. 134, No. 12, Dec. 1981, p. 42.

SEMICONDUCTOR MANUFACTURING

Richard P. Bielen and Kathleen Robinson

In the history of technology, nothing parallels the rapid development of the semiconductor industry. In little more than 30 years, the industry has virtually invented itself and become a billion dollar a year business.

Unknown just a few decades ago, semiconductors are now found in a wide variety of goods—from small personal computers to large government information systems, toys to satellites, watches to weapons systems, medical instruments to automatic bank tellers. In 1983 alone, the industry employed almost 200,000 persons and shipments of semiconductor devices reached more than $14 billion.

However, along with its enormous profit and growth potential, the semiconductor manufacturing industry has an enormous loss potential—its manufacturing processes and equipment are often easily damaged and expensive to replace.

This chapter discusses the production process, fire hazards, and the safeguards associated with the semiconductor manufacturing industry. Other chapters within this section cover many of the common industrial processes which also pertain to semiconductor manufacturing; see especially Chapter 3, "Heat Transfer Systems," and Chapter 5, "Industrial and Commercial Heat Utilization Equipment." The hazards associated with flammable and combustible liquids are discussed in Section 5, Chapter 4, and Section 11, Chapter 4. Gases used in the industry are covered in Section 5, Chapter 5, and Section 11, Chapter 5. See Section 12, Chapter 8 for more detail on housekeeping, and Section 12, Chapter 8 for a full discussion of the control of electrostatic ignition sources.

THE PRODUCTION PROCESS

The major product of the semiconductor manufacturing industry is the integrated circuit, which is an array of transistors and other components that are implanted in or deposited on a piece of semiconductor material and fastened to a frame that can be inserted into a printed circuit. The production of such a circuit involves several steps. (See Figs. 10-22A and 10-22B.)

Mr. Bielen is a Fire Protection Engineer and Signaling Systems Specialist with NFPA; Ms. Robinson is Editor of *Fire Journal*.

Crystal Production

The first step in the manufacturing process is making the semiconductor material on to which the circuit information is to be implanted. This material must be electrically neutral so that it does not interfere with the electronic information that will be processed on it. The most common semiconductor material is silicone, although other materials such as gallium arsenide and germanium may be used. Silicone is grown as crystals in water cooled vacuum furnaces that are electrically heated to more than 2,552°F (1400°C). After the crystals are grown, they are sliced into wafers, polished, cleaned, and given a thin surface of silicone dioxide. The wafers are then sent on to the fabrication area for further processing.

Mask Manufacturing

While the semiconductor material is being created, the masks used to produce the circuits on the wafer are also manufactured. A mask is a thin patterned sheet which shields portions of the semiconductor material as the information is imprinted on it. Because it contains only the pattern for a single layer of the circuit element, many masks and layers of patterns may be needed to make a single integrated circuit.

To make a mask, a circuit layout is developed using a computer. The final circuit is then prepared using an optical pattern generator and a step-and-repeat camera or an electron beam lithography machine.

Chip Fabrication

The silicone wafer and the mask now come together during chip fabrication. This process can be compared to that of developing a photograph. In chip fabrication, the "negative" is a mask and the "image" projected is the circuit. The wafer is coated with a light-sensitive chemical called photoresist. To "develop" an image, the wafer is positioned behind the mask and exposed. The unexposed photoresist is then removed in a chemical wash. The photoresist that remains is hardened in an oven, and the oxide that originally coated the wafer is removed in an etching process. This is followed by another cleaning that removes the remaining photoresist. Finally, dopant is

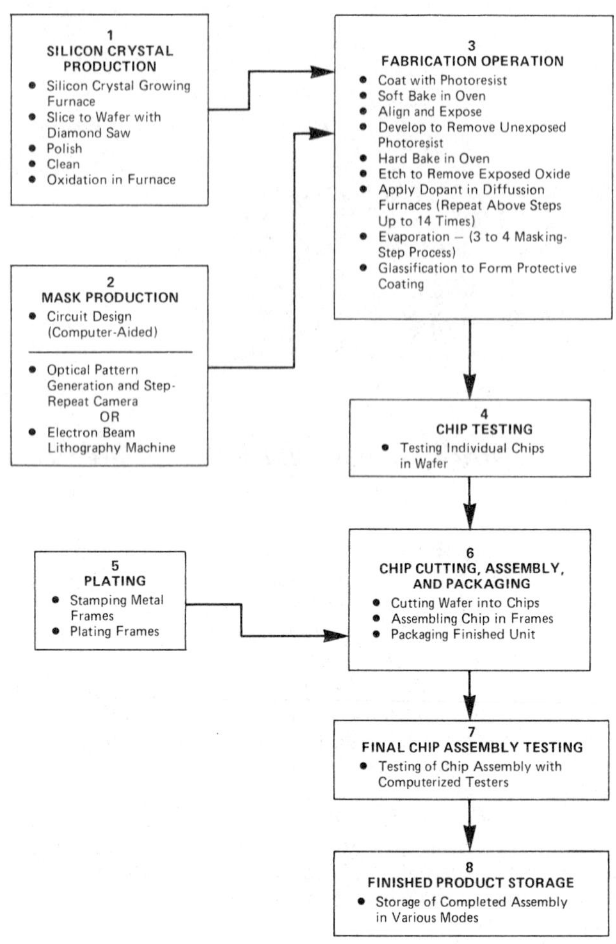

FIG. 10-22A. Flow diagram of the semiconductor manufacturing process.

applied to add the various impurities that give a circuit its desired characteristics.

This process is repeated up to 14 times with different masks until the chip is imprinted with the desired circuit. When the circuit is finished, a coating of metal such as aluminum is applied in another series of masking steps to provide electrical connections to outside circuits.

Chip Testing

Each chip in a wafer must be tested using computer controlled test equipment. Defective chips are marked. The computer maintains statistics on the number and location of good chips per wafer and on the incidence of various defects.

Plating

While the chips are being tested, the frames into which they will be placed are plated using both standard and customized plating equipment. This operation is often performed in a separate building.

Chip Cutting, Assembly, and Packaging

After the chips have been tested and marked, the wafers are cut apart. Defective chips are discarded, and the remainder are fastened into the plated frames. Fine wires connect the chips to the frame leads, and a plastic or ceramic lid is fused onto the frame to protect both the chip and the wires from mechanical damage.

Final Chip Assembly, Testing, and Storage

Various electrical tests with specially designed computerized equipment are run on circuits in the completed chip assemblies. Integrated circuits that must be highly reliable often undergo environmental functional testing. During this type of test, they are subjected to long term temperature variations in specially designed ovens. The finished product is placed in plastic or cardboard containers in cartons. These cartons are then stored in bulk, on pallets, in bins, or on racks.

THE FIRE HAZARDS

A fire during the fabrication of semiconductors can be quite detrimental. During the fabrication and masking operation, semiconductors are susceptible to contamination. Foreign material from smoke can get onto a mask, or water contamination during fabrication can distort a circuit. To avoid contamination, parts of both processes are conducted in clean rooms which are kept as sterile as possible.

Most semiconductor process hazards found in this industry are not unique; they are similar to hazards found in many other industries. For instance, the problems associated with crystal production stem mostly from the furnace in which the crystals are grown. If cooling water comes in contact with the furnace contents, there is potential for a steam explosion. A lack of cooling water could cause a burnout. These hazards may be present wherever a water cooled furnace is used.

Similarly, hazards common to all plating operations exist in semiconductor plating operations. Losses usually occur when electrical devices such as immersion heaters ignite combustible plastic or plastic lined tanks, as well as the exhaust fume ducts and scrubbers.

Combustible plastics also contribute to the fire potential of the semiconductor industry. These plastics are found not only in the tanks, exhaust fume ducts, and scrubbers of plating operations, but in the fabrication and chip testing areas as well. In addition, plastic hoods and containers are used in both wafer cleaning and step-and-repeat camera operations.

Hazardous materials, such as corrosive and flammable liquids and gases, are found in clean rooms and in the ovens in which wafers are baked at various times during chip fabrication. These flammable liquids and gases can be ignited by many ignition sources. When dispensing liquids and when gases move through pipes, static electricity is one source of ignition. The storage of the finished chip assemblies presents the hazards commonly found when combustibles are grouped in bulk, in racks, and on pallets.

SAFEGUARDS

Building Construction

Buildings that house production operations should be of fire resistive, noncombustible construction, or sprinklered and the interior partitions that separate the various processes should also be fire resistive or noncom-

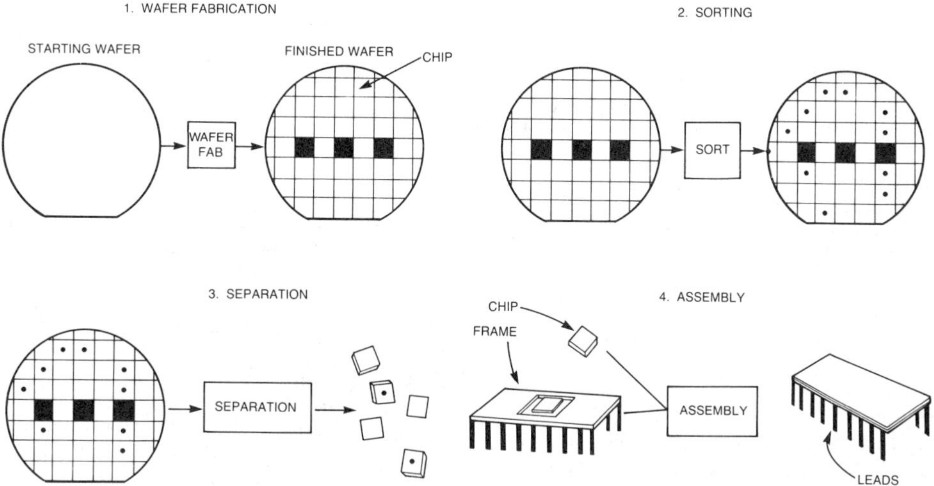

FIG. 10-22B. The basic steps in manufacturing semiconductors.

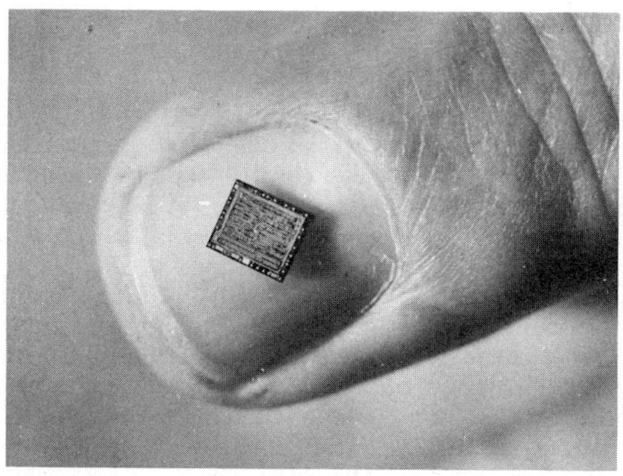

FIG. 10-22C. A completed integrated circuit. (Honeywell)

bustible. The flamespread rating for interior finishes, suspended ceilings, and piping insulation should not exceed 25, and the smoke developed rating should not exceed 50 when tested in accordance with NFPA 255, *Standard Method of Test of Surface Burning Characteristics of Building Materials* (hereinafter referred to as NFPA 255). This test is also referenced in ASTM E-84 and UL-723.

Because of the high costs involved in this industry, consideration should be given to the establishment of multiple production areas. These areas should be separated by fire resistance rated walls, floors, and ceilings with a minimum fire resistance rating of one hour. This will help keep fire, products of combustion and water in one area from contaminating other areas. If the value of a fabrication area is very high, separate fabrication lines should be set up in different buildings or in areas separated by fire walls with a minimum three hour rating. Ventilation, exhaust fume systems, heating and air conditioning, and other utilities should be independent of those in adjacent buildings or areas.

To keep contaminants from spreading throughout the entire building, heating and air conditioning systems should be designed so that they shut down, and dampers close when an alarm is received from a manual box, a water flow alarm, an alarm from an automatic fire alarm system, or other emergency signal. A separate air handling system for clean rooms and fabrication areas should be capable of full exhaust ventilation during a fire or spill. This emergency mode should be capable of either manual or automatic operation.

Ducts for the heating and air conditioning systems should be made of noncombustible materials, such as steel and aluminum, or of Class O or Class I materials having a flamespread rating which does not exceed 25 and a smoke development rating which does not exceed 50. Materials used should be tested in accordance with UL-181, *Standard for Factory-Made Air Duct Material and Air Duct Connectors.*

Semiconductor manufacturing areas should use filters which do not contribute fuel to a fire when clean, and give off only insignificant amounts of smoke when tested in accordance with UL-900, *Standard for High Efficiency Particulate Air Filter Units.*

Static Electricity

Static electricity is a potential source of ignition when filling containers with flammable liquids. Static is generated when liquids move in contact with other materials. The static charge accumulates on an isolated body. If the charge is sufficient, a spark may occur. When pouring liqiuds from one container to another, the fill spout or nozzle should be kept in contact with the container. If it cannot be kept in contact and the two containers are not inherently bonded, a bond wire must be used between them. The containers should be made of a conductive material, preferrably metal.

Housekeeping

All production areas, especially mask manufacturing and fabrication sections, must be clean and free of any materials that are not actually used in the operations. Cartons, paper, or packaging materials should not be stored in these areas. Silicone wafers processed in fabrication areas should be stored in tightly covered containers.

Surveillance

Surveillance of a facility during nonoperating hours will help prevent arson fires and provide early warning of emergencies. Watchclock stations should be installed throughout the facility, and recorded guard service should be maintained in all unattended areas. Television surveillance can be used to supplement guard service. A central station signaling system should be used to monitor sprinkler water flow, control valves, and automatic and manual fire alarms.

Management Support of Loss Prevention Programs

Many fires are the result of failures in the facility's fire protection philosophy rather than weaknesses in the facility's fire protection systems. If management does not organize and train its employees to participate in the plant loss control program, they will not be able to react properly during an emergency and a loss will occur.

Loss prevention and control programs should address such problems as training personnel to respond to fires and other emergencies and to inspect and maintain loss control equipment. In addition, management should draw up procedures for restoring and repairing loss control equipment, and for controlling all phases of construction and renovation. Finally, management should make sure that duplicate master masks or data that can be used to produce master masks are stored in an area that is remote from the mask manufacturing operation. This will help ensure continued production in the event of a loss.

Fire Protection Systems

Because automatic sprinklers are one of the most effective means of controlling and extinguishing fires, they should be installed in most buildings and areas, including interstitial spaces, combustible ductwork, and ventilation hoods. Inside hose connections should be installed in all areas and arranged to remain in service even if sprinkler protection in any one area is shut off.

The sprinkler protection system should be designed to address the particular hazards of the process. Systems protecting bulk storage should be designed according to NFPA 231, *Standard for Indoor General Storage*, while those protecting rack storage should be based on information found in NFPA 231C, *Standard for Rack Storage of Materials*.

Testing areas, step-and-repeat camera enclosures, computer areas, and production areas containing particularly valuable equipment should be protected with automatic total flooding Halon 1301 fire extinguishing systems, as well as sprinkler systems.

Portable fire extinguishers, another protection requirement, should be provided at 25 ft (7.6 m) intervals in fabrication areas and at 50 ft (15.2 m) intervals elsewhere in the building. Additional carbon dioxide or Halon extinguishers should be located in areas that contain small quantities of flammable liquids. Dry chemical extinguishers should not be used in the facility. The powder they discharge can contaminate electronic equipment, circuit masks, and wafers.

Finally, every semiconductor manufacturing facility needs a sufficient, reliable water supply. In fact, most facilities need two independent sources of water, each of which can meet anticipated sprinkler and hose stream demand. Appropriate water distribution systems with underground fire mains should loop the facility, and control valves should divide the loop into sections.

SPECIAL HAZARDS

Flammable Liquids and Gases: Flammable liquids used in production areas should be kept in special enclosures with openings that are curbed and drained. The walls of these enclosures should have a one hour fire resistance rating. The enclosures also should be equipped with automatic sprinklers, exhaust ventilation, and the proper electrical equipment.

Flammable liquids may also be piped directly to the production area through piping that is equipped with easily accessible shut-off valves. Pressurized piping for liquids and gases should be equipped with excess flow control valves that will shut off the flow if a pipe ruptures. Regardless of use, all piping should be color coded or marked in some manner for proper identification of its contents.

Cylinders containing flammable, oxidizing, and corrosive gases should be stored outside the building until used, and secured so they will not tip. Oxidizing and flammable gases should be stored separately.

The number of process gas cylinders brought inside a building should be limited to as few as practical and once inside, kept in ventilated, sprinklered gas cabinets. Electrical equipment associated with these cabinets must comply with NFPA 70, *National Electrical Code®*.

All bulk, gaseous hydrogen, and liquefied hydrogen systems should be designed according to the appropriate NFPA standards. Process areas in which hydrogen is used should be provided with combustible gas detectors that are interlocked to sound an alarm when the hydrogen concentration reaches 15 percent of the Lower Explosive Limit (LEL)*. These detectors should also be arranged to shut off the gas when it reaches 25 percent of its LEL. A method of turning the gas off manually should be provided outside each fabrication area for use during emergencies.

Plating Operations: To prevent losses from immersion heater fires, heat exchangers that use steam or hot water as their heating medium should be installed in place of conventional electrical heaters. Tubing made of Teflon® or of corrosion resistant metals is available to meet any application.

Fires caused by immersion heaters may also damage the fume exhaust systems that remove corrosive and toxic fumes produced by various processing operations. These systems are typically composed of plastic ducts. To prevent fire damage, the ducts should be flame retardant, with a flamespread rating of no more than 25 and a smoke developed rating no higher than 50 when tested in accordance with NFPA 255. Automatic sprinklers should be installed inside ducts that are 8 in. (203 mm) in diameter or larger. Sprinklers also should be installed in fume scrubbers, the plastic exhaust ductwork downstream of the scrubbers, and each vertical connection to a workstation.

*The LEL of hydrogen is 4 percent.

Crystal Furnaces: Furnaces in which crystals are grown should be equipped with a rupture disk and relief vent to prevent overpressurization. Cooling water flow alarms are also needed to warn of interruptions in furnace coolant and emergency cooling water connections. This will help protect the furnaces in the event of a primary cooling water failure.

Environmental Ovens: Depending on the combustibility of the materials being tested, environmental ovens or chambers require special protection devices. These include excess voltage and overcurrent protection, and a maximum temperature cutoff that shuts off power to the oven if the temperature rises above an acceptable level.

Bibliography

References Cited

ASTM E-84, *Standard Test Method for Surface Burning Characteristics of Building Materials.*

UL-723, *Test for Surface Characteristics of Building Materials.*

UL-181, *Standard for Factory-Made Air Duct Material and Air Duct Connectors.*

UL-900, *Standard for High Efficiency Particulate Air Filter Units.*

NFPA Codes, Standards, Recommended Practices, and Manuals. (See the latest *NFPA Codes and Standards Catalog* for availability of current editions of the following documents.)

NFPA 10, *Standard for Portable Fire Extinguishers.*

NFPA 12A, *Standard on Halon 1301 Fire Extinguishing Systems.*

NFPA 13, *Standard for the Installation of Sprinkler Systems.*

NFPA 13A, *Recommended Practice for the Inspection, Testing and Maintenance of Sprinkler Systems.*

NFPA 24, *Standard for the Installation of Private Fire Service Mains and Their Appurtenances.*

NFPA 30, *Flammable and Combustible Liquids Code.*

NFPA 70, *National Electrical Code.*

NFPA 71, *Standard for the Installation, Maintenance, and Use of Central Station Signaling Systems.*

NFPA 72A, *Standard for the Installation, Maintenance, and Use of Local Protective Signaling Systems for Guard's Tour, Fire Alarm and Supervisory Service.*

NFPA 72E, *Standard on Automatic Fire Detectors.*

NFPA 86, *Standard for Ovens and Furnaces—Design, Location and Equipment.*

NFPA 90A, *Standard for the Installation of Air Conditioning and Ventilating Systems.*

NFPA 91, *Standard for the Installation of Blower and Exhaust Systems for Dust Stock and Vapor Removal or Conveying.*

NFPA 231, *Standard for General Storage.*

NFPA 231C, *Standard for Rack Storage of Materials.*

NFPA 255, *Standard Method of Test of Surface Burning Characteristics of Building Materials.*

Additional Readings

Bartels, A. L., "Safeguards to Prevent Ignition of Hot Surfaces," *Electrical Review*, London, England, Vol. 203, No. 4, 1978, pp. 29-31.

Davis, R. H., "Fire Protection Systems (for Computers)," *Data Management*, Vol. 14, No. 11, 1976, pp. 11-14.

Deacon, F. L., "Designing Fire Protection to Limit Monetary Loss," *SFPE Technology Report No. 80-2*, Society of Fire Protection Engineers, Boston, MA, 1980.

Factory Mutual System, "American Experience in the Fire Protection of Computers," *Fire*, Vol. 71, 1979, pp. 498-499.

"Halon Prevents Major Central Processing Unit Fires," *Computer Decisions*, Vol. 10, No. 8, 1978, pp. 56, 58.

Harrington, J. L., and Hopkinson, R. B., *Rack Storage Protection*, Worcester Polytechnical Institute, Worcester, MA, 1977.

King, P. W., and Magid, J., *Industrial Hazard and Safety Handbook*, Butterworths Publishing, Inc., Woburn, MA, 1979.

Nailen, R. C., "Toxic, Flammable Chemicals, Gases, Breed Trouble in Electronic Plants," *Fire Engineering*, Vol. 133, No. 10, Oct. 1980, pp. 54-57.

Perkins, C., and Berenblut, B. J., "Electronic Equipment: What Protection Is Required?" *Fire Surveyor*, Vol. 9, No. 5, Oct. 1980, pp. 28-33.

Philbrick, S. E., "Selecting Cables for Fire-Risk Applications," *Electronics and Power*, Vol. 26, No. 3, Mar. 1980, pp. 232-233.

Wolf, Helmut F., *Semiconductors*, John Wiley and Sons, Inc., NY, 1971.

Yang, Edward J., *Fundamentals of Semiconductor Devices*, McGraw-Hill, NY, 1978.

MINING METHODS AND EQUIPMENT

William H. Pomroy

Mining is an industrial activity characterized by inherent and significant fire and explosion risks. At the turn of the century, mine fires and explosions resulted in loss of life and property damage on a scale unmatched in other industrial sectors. Marked progress has been achieved, however, toward controlling these hazards as evidenced by the steady decline in mine fire and explosion incidents reported in recent years. The purpose of this chapter is to describe basic mining methods and equipment, fire and explosion hazards, and safeguards involved in the mining of coal, metal and nonmetal ores, sand and gravel, and stone.

Additional information relevant to mining processes and hazards may be found in Section 4, Chapter 2, "Explosions;" Section 5, Chapter 7, "Explosives and Blasting Agents;" Section 5, Chapter 9, "Dusts;" Section 10, Chapter 4, "Fluid Power Systems;" Section 11, Chapter 4, "Storage and Handling of Flammable and Combustible Liquids;" and Section 16, Chapter 5, "Gas and Vapor Testing."

Mining can be grouped into two broad categories: surface mining and underground mining. Surface mining requires the removal of the overlying dirt and rock strata (overburden) before the desired minerals are excavated. Underground mining is practiced where the overburden thickness precludes surface mining. It permits the selective extraction of desired minerals with minimal disturbance to overlying strata.

The economics of large scale production and mechanization strongly favor surface mining. Over 95 percent of all mine production is accomplished by surface methods (U.S. Bureau of Mines 1983).

SURFACE MINING METHODS

Numerous surface mining methods have evolved in response to varying surface and subsurface geological conditions, and to innovations in mining equipment.

Methods of surface coal mining, or strip mining, include area strip and various forms of contour mining (Cassidy 1973; U.S. Bureau of Mines 1975). Area strip mining is practiced where the terrain is relatively flat and the coal seams are of moderate thickness (3 to 30 ft or 0.9 to 9 m). It begins with the removal of overburden, from above the coal seam, in a roughly linear strip usually 100 to 170 ft (30 to 52 m) wide and up to several miles long. Overburden depths of 60 to 120 ft (18 to 37 m) are common. As coal is uncovered, it is loaded into trucks for haulage to storage or processing facilities. When an entire strip has been mined, the process is repeated, with the overburden above the adjoining strip of coal cast onto spoil piles in the cut previously mined. (See Fig. 10-23A.) Specialized mining systems resembling metal and nonmetal open pit methods have been developed for multiple, steeply pitched and/or thick seams.

Contour strip mining is most common where flatlying coal seams occur in rolling or mountainous terrain. In contour mining, an initial cut is established along a hillside at the point where the coal seam outcrops or is exposed. Successive cuts then are made into the hillside until an economic stripping limit is reached. Mining proceeds laterally along the hillside, following the outcrop. Contour mining follows the same drill-blast-strip-load-haul cycle as area strip mining, but the equipment employed is generally much smaller.

Most surface metal and nonmetal mining is by the open pit or open cast method (Pfleider 1968). The overburden is first stripped to expose the top of the ore body; stripping and mining then proceed concurrently. The pit expands in depth and area until an economic pit limit is reached. (See Fig. 10-23B.) Pit geometries vary, owing to the irregular shape and orientation of ore bodies, but dimensions of ½ mile (0.8 km) to several miles across and 100 to over 1,000 ft (30 to 300 m) deep are common.

Specialized mining systems have been developed for certain metal and nonmetal ore deposits. Examples include dragline stripping and hydraulicking of phosphates, placer mining, and dredging of alluvial beach sands, gold, and tin. Sand and gravel are generally mined from shallow pits and small surface excavations, or with draglines and dredges from natural or artificial ponds.

Mr. Pomroy is Group Supervisor, Mine Safety Fire Group, Twin Cities Research Center, Bureau of Mines, U.S. Department of the Interior.

FIG. 10-23A. Area strip mining.

Crushed stone and dimension stone are mined from rock quarries. Stone for crushed aggregate generally is mined by conventional open pit methods. Dimension stone is removed in large blocks with the use of wire saws, channeling machines, or by similar means.

Other than maintenance facilities, few fixed structures are involved in the surface mining process. Rather, surface mining is performed out doors by specialized mobile and portable mining equipment. The major equipment types and their primary functions include: blasthole drills to drill holes for explosives; stripping machines such as draglines, stripping shovels, and bucketwheel excavators for overburden removal; loading equipment such as shovels, wheeled loaders, and hydraulic excavators for loading overburden, coal, or ore at the working face; and haulage trucks, trains, or conveyors for transport of materials from the working face to storage or processing facilities (Martin 1982).

FIG. 10-23B. Open pit mine.

A wide variety of miscellaneous equipment is required to support the mining process. This equipment includes scrapers, explosives haulers, dozers, graders, water trucks, pumps, power substations, generators, crushers, cranes, backhoes, and maintenance and utility vehicles.

UNDERGROUND MINING METHODS

Access to underground mine workings from the surface is by shaft, slope, or adit. A shaft is a vertical or nearly vertical opening, usually equipped with hoisting facilities for transporting personnel, supplies, and/or mined products. A slope (also called an incline or decline) is an inclined opening that may be designed to accommodate a belt conveyor for materials haulage, a hoist for personnel and supplies, or diesel trackless haulage. An adit mine (or drift mine) is entered through a horizontal opening in a hillside. An individual mine may employ any or all of these means of access, depending upon the depth and orientation of the coal seam or ore body and the requirements of the mine production system.

Where a single, uniform, relatively flat lying coal seam or ore body is worked, the entire mine is generally confined to a single working level. Where multiple, irregularly shaped, and/or steeply inclined coal seams or ore bodies are worked, mining may be spread over many levels. Large mines may comprise hundreds of miles of drifts or entries (horizontal openings in metal or coal mines, respectively) on many levels and interconnected with shafts, slopes, raises (vertical openings between levels), and ramps (inclined openings between levels).

The predominant constraint influencing the selection of an underground mining method is the amount of support required to prevent the unwanted failure of the roof and/or walls. Factors related to this constraint are the physical character, size, shape, orientation, and depth of the coal seam or ore body, and the characteristics of the surrounding and overlying strata. In addition, planned production capacity, and the presence of water, methane, rock temperatures, radon, or other environmental factors may be important considerations in selecting a mining method.

Operating practices vary widely, since nearly infinite variability exists among these factors from deposit to deposit and even within a single mine. However, the broad spectrum of underground mining methods can be effectively subdivided into three generic classifications. These are (1) methods that produce openings that are naturally supporting (or requiring minimal artificial support), (2) methods that provide for substantial artificial support of the roof, and (3) caving methods where failure of the roof is integral to the mining process.

With few exceptions, coal mining systems are profoundly different from noncoal mining systems, and they are treated separately in the following sections. Where similarities do exist, they are limited to the softer, bedded nonmetallics such as potash, salt, and trona, which are amenable to coal mining systems and equipment.

Underground Coal Mining: The two basic coal mining methods are room-and-pillar, which produces openings that are naturally supporting, and longwall, which requires substantial artificial support (Cassidy 1973). As the name implies, room-and-pillar mining involves the extraction of coal in a regular pattern of rooms separated by pillars of in place coal to support the roof. (See Fig. 10-23C.)

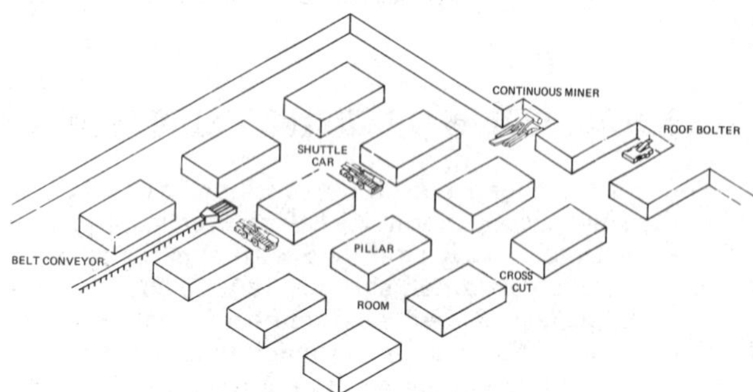

FIG. 10-23C. Room-and-pillar coal mining.

Two systems are used for room-and-pillar coal mining. In conventional mining, the coal is extracted in a sequence of operations, with specialized equipment required to execute each step. Using a cutting machine resembling a large chain saw, a slot is cut into the coal face the entire width of the room along the floor, center, or roof line. Next, a drill is used to bore holes for explosives in the coal face. Coal broken by the explosives is gathered by a loading machine, which either feeds a self propelled shuttle car or may be loaded and hauled by a self propelled scoop. In either case, the coal is discharged onto a conveyor belt or into rail cars for transport to the surface or to hoisting facilities.

In continuous mining, a single machine called a "continuous miner" is used to mechanically cut the coal from the face and load the broken coal into a shuttle car.

After each drill-blast-load cycle in conventional mining, or each machine advance in continuous mining, the mining equipment is withdrawn and the roof is secured by a machine called a "roof bolter." Holes are drilled into the roof and long bolts inserted and anchored to strengthen the roof span and prevent roof failure.

Longwall mining involves the extraction of coal in a nearly continuous operation using a specialized and integrated mining and roof support system. (See Fig. 10-23D.) Using standard room-and-pillar techniques, longwall panels 300 to 1,000 ft (90 to 305 m) wide by 3,000 to 6,000 ft (915 to 1830 m) long are prepared. Each panel is mined in linear slices at full seam height by a mechanical cutting machine or plow that is drawn back and forth across the short face. As the coal is cut, it is forced onto a chain conveyor which extends the entire length of the face. Support of the roof immediately adjacent to the face is accomplished with large, self advancing hydraulic jack units. As the roof support line advances with the coal face, the roof behind it is permitted to cave.

Most coal mining equipment is electrically powered. Undercutting machines, drills, loading machines, continuous miners, shuttle cars, scoops, and roof bolters are generally supplied with power through a trailing cable.

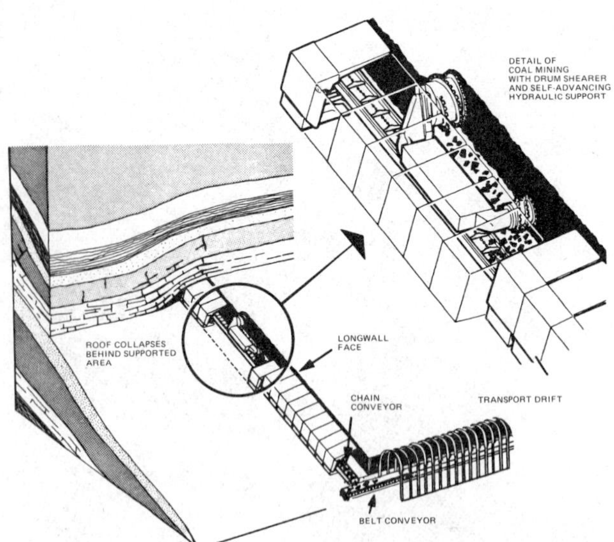

FIG. 10-23D. Longwall coal mining. (Atlas Copco)

Belt conveyors generally are wired directly to a permanent or semipermanent power center, as is longwall equipment. Self propelled rail mounted equipment is generally supplied with power from an overhead trolley wire. Battery powered scoops, personnel transports, and utility vehicles are also common. Some mobile haulage equipment, notably scoops, load-haul-dump (LHD) vehicles (low profile front end loaders or scoop tram), and rail locomotives can be diesel powered. Though diesel usage is growing, it is not yet extensive in underground coal mining.

Numerous ancillary operations must be coordinated with the mining process. The main haulage network, whether rail or belt, must function smoothly to prevent materials handling bottlenecks from interfering with production. The roof in active areas is constantly monitored for failures or signs of impending failure. In heavily traveled areas such as main haulageways, important ventilation routes, etc., heavy timbers or steel arches are often installed to help support the roof. Water supply, power distribution, and ventilation systems require periodic attention, and equipment and facilities inspection, maintenance, and repair are continuous. Dust control also receives high priority, both to protect miner health and to prevent explosive conditions.

Underground Metal and Nonmetal Mining: Underground metal and nonmetal mining methods are more diverse than underground coal mining methods because there is greater variability in the physical character, size, shape, and orientation of metal and nonmetal ore bodies than in most coal seams (Cummins & Given 1973; Hustrulid 1982). Room-and-pillar and sublevel stoping methods are used where minimal artificial roof support is required in the extraction areas. Shrinkage stoping and cut-and-fill are used where the roof and walls in the extraction areas require substantial support. The two most common caving methods are sublevel caving and block caving.

The room-and-pillar method is used for relatively flatlying deposits of moderate thickness with a competent roof. With the exception of the aforementioned softer, bedded nonmetallics, metal and nonmetal mining is generally characterized by larger openings than those found in typical room-and-pillar coal operations, permitting use of larger, more productive mining equipment. (See Fig. 10-23E.)

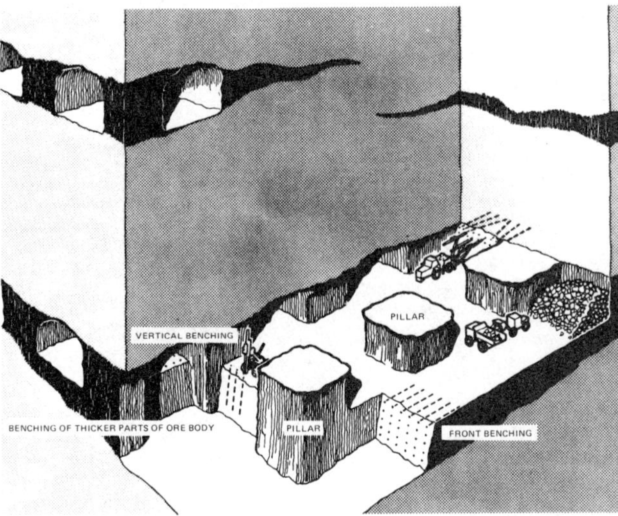

FIG. 10-23E. Room-and-pillar metal and nonmetal mining. (Atlas Copco)

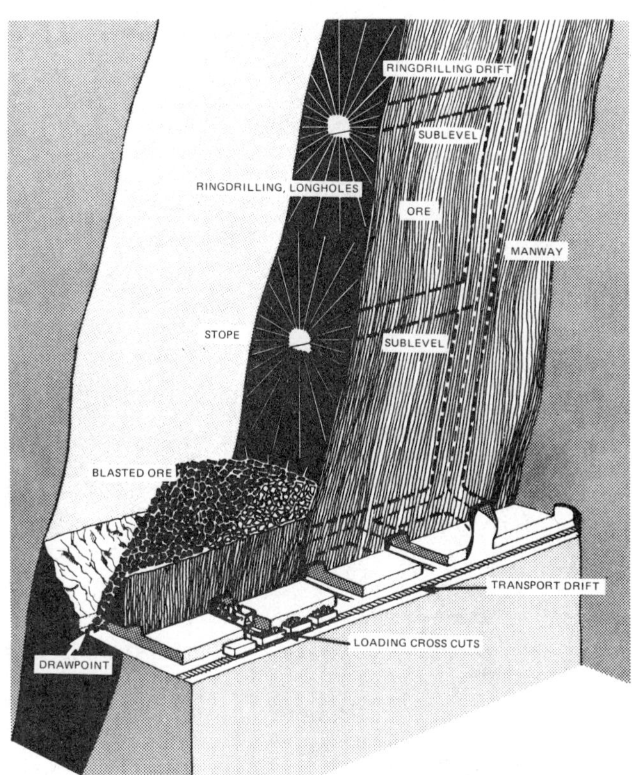

FIG. 10-23F. Sublevel stoping. (Atlas Copco)

Sublevel stoping, used in vertical or nearly vertical ore bodies of highly competent rock, requires considerable premine development to prepare the ore body for production mining.

Sublevel drifts along the longitudinal centerline of the ore body are driven in a vertical column between the main levels of the mine. (See Fig. 10-23F.) On the main haulage levels, drawpoints perpendicular to the sublevel drifts are driven into the ore body. A complete radial pattern of blastholes or parallel rows of vertical blastholes then are drilled at regular intervals along the entire length of each sublevel drift. Each ring or row of holes is individually loaded with explosives and blasted. Every blast results in a vertical slice of ore falling to the bottom of the open cavity (or stope) where it is loaded out through the drawpoints.

If the ore or surrounding rock is too weak or rock pressures too great, open stopes are not feasible. Such ore bodies require artificial support of the stopes during mining. Shrinkage stoping and cut-and-fill are the most common methods employed to extract the ore bodies.

Shrinkage stoping requires very little preproduction development. Drawpoints are driven into the ore body at regular intervals along the main levels, and raises are driven into the ore between levels. (See Fig. 10-23G.) The ore is extracted in horizontal slices, starting at the bottom of the stope and advancing upward. The broken ore in the stope provides a work platform for miners and supports the walls of the stopes. The process of mining horizontal slices is repeated until the upper limit of the stope is reached. The stope is then emptied of ore.

Preproduction development for cut-and-fill mining is similar to that for shrinkage stoping, except only one or two drawpoints are required. (See Fig. 10-23H.) Mining is similar as well, with the ore body mined in horizontal

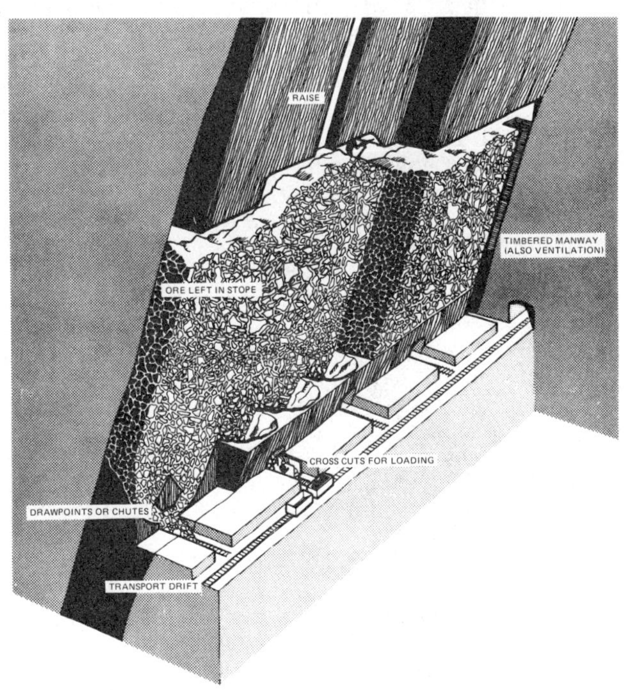

FIG. 10-23G. Shrinkage stoping. (Atlas Copco)

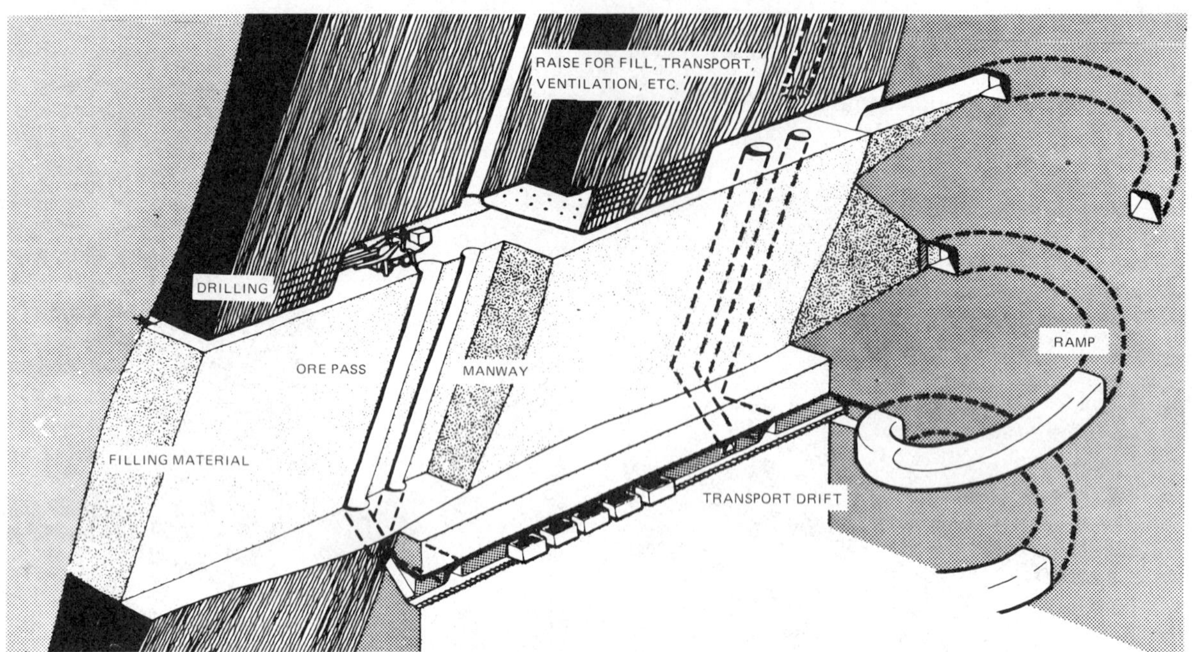

FIG. 10-23H. Cut-and-fill mining. (Atlas Copco)

slices from bottom to top. However, the stope is emptied of broken ore after each slice (cut), and the stope is filled with waste material before mining resumes. The fill provides a work platform for miners and supports the stope walls. Ore is removed from the stope through timber or steel ore chutes imbedded in the fill.

Caving systems are characterized by a high degree of mechanization, high production capacity, and low production costs. In sublevel caving, preproduction development is similar to sublevel stoping, with sublevel drifts driven into the ore at 30 to 50 ft (9 to 15 m) intervals. (See Fig. 10-23I.) Drawpoints on the main levels are not required however, because the ore is retrieved through the sublevel drifts. Fan shaped blasthole patterns are drilled upward at regular intervals along the entire length of each sublevel drift. The drill fans are blasted individually, with the broken ore caving into the sublevel drifts. LHDs are used to load the ore and transport it to the ore pass system.

The principle of block-cave mining is the natural fracturing and caving of ore that has been undercut over a large area. A minimum of drilling and blasting during the ore extraction phase is required, because the ore fractures itself as a result of gravity forces acting on the undercut rock masses.

Preproduction development is undertaken in five steps. (See Fig. 10-23J.) First, a network of haulageways is developed beneath the ore horizon. Next, finger raises are driven up to a set of grizzly drifts. From the grizzly drifts, a second set of finger raises is driven, gradually widening to form funnels at the undercut level. Finally, the entire ore block is undercut.

The weight of the overlying ore and overburden creates enormous stresses in the ore at the bottom of the block, causing it to fracture and cave into the funnels.

A wide variety of equipment is used in underground metal and nonmetal mining. Where space permits, drilling is performed with mobile "drill jumbos" fitted with two or more rock drills. Diesel powered LHDs are used to load

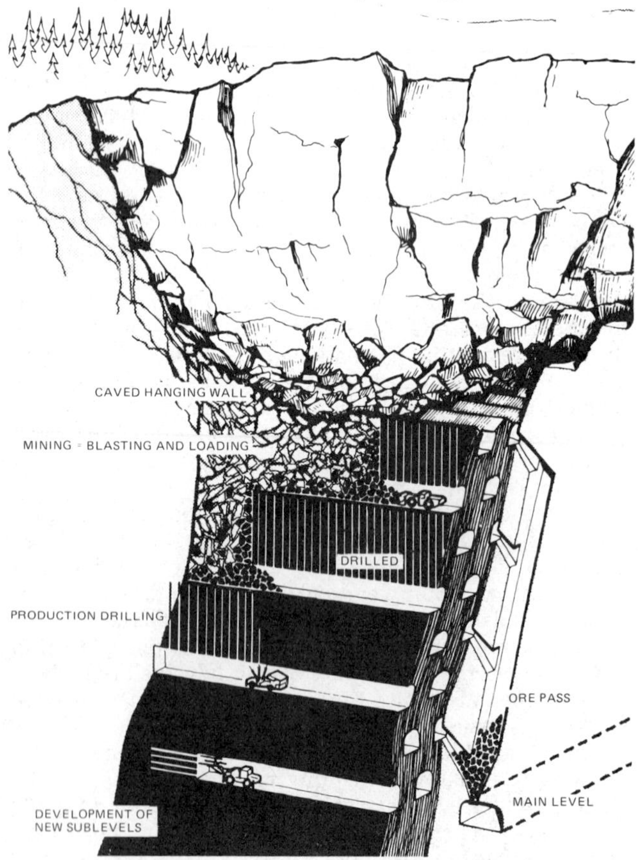

FIG. 10-23I. Sublevel caving. (Atlas Copco)

and haul broken ore from the drawpoint or working face. Where space is limited, drilling is performed with small, portable, manually operated rock drills. Broken ore is

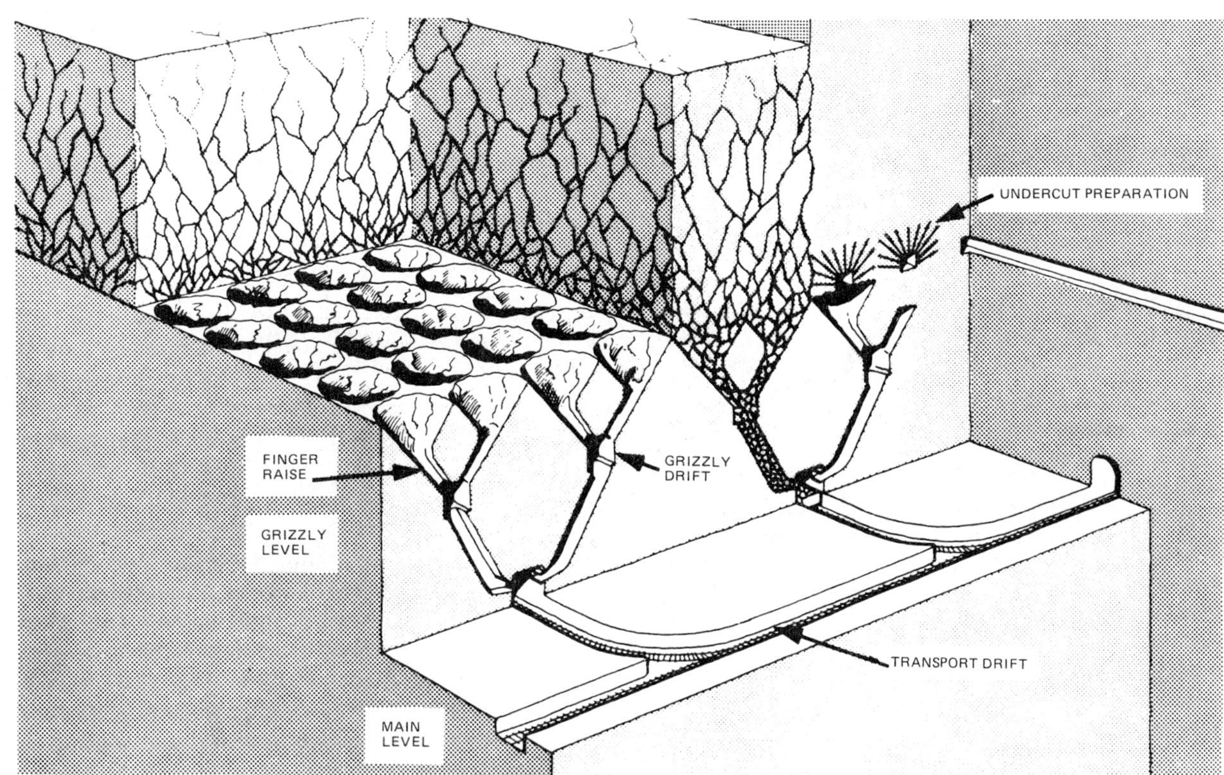

FIG. 10-23J. Block caving. (Atlas Copco)

loaded out by small pneumatic loading machines or by a scraper blade suspended on a cable between a hoist and a pulley, which gathers ore as it is dragged over a pile. Main haulage is usually completed by conveyor belt, rail, or haulage truck.

As with underground coal mining, the ore extraction process is supported by a vast array of ancillary operations. These operations include mining, mine development, and haulage equipment maintenance (which at most mines is performed in large, underground shops). Mines relying on diesel equipment also must have facilities for storage and/or transfer of diesel fuel. In addition, power distribution, ventilation, hoisting, roof control, and dewatering equipment require frequent inspection, maintenance, and repair.

THE FIRE AND EXPLOSION HAZARDS

Fires and explosions pose a constant threat to the safety of miners and to the productive capacity of mines. Mine fires and explosions have historically ranked among our most devastating industrial disasters, with over 10,000 miners killed in coal mine explosions alone from 1880 to 1940 (Nagy 1981).

Although more recent advances in mining and safety technology have reduced losses to a tiny fraction of their former levels, fires and explosions continue to occur, challenging the best efforts of mine safety, production, and engineering personnel to minimize losses. The major fire and explosion hazards in mining are discussed in the following sections.

Mine Fire Hazards

During the 1960s and 1970s about 50 fires per year were reported to federal mine safety authorities. Nonreportable fires (defined as those lasting less than 30 minutes and not resulting in an injury), were estimated at another 200 per year. In the mid 1970s, a steady decline in fire frequency was observed, probably due to the adoption of improved mining and safety technology (roof bolts for roof support, ac powered face equipment in coal mines, etc.). Reported fires now average less than 20 per year (Baker, Nagy & McDonald 1980; McDonald & Baker 1979; McDonald & Pomroy 1980). This downward trend has begun to level off.

Since 1968, mine fires have resulted in over 150 fatalities. Although about one-half of the fires occurred in surface mines, over 80 percent of the fatalities occurred in underground mines where smoke and toxic fire gases are the primary life safety hazards. The major ignition sources, equipment, and burning substances involved in mine fires since 1968 are listed in Tables 10-23A, 10-23B, and 10-23C, respectively.

The largest category of fires in underground mines are of electrical origin. This is especially true in underground coal mines, where nearly half of all fires are electrical. During the 1960s and 1970s, about 40 percent of underground coal mine electrical fires involved face equipment powered through a trailing cable, and about half of those fires were caused by faults in the cable. Face equipment fires have been significantly reduced in recent years by the conversion from dc to ac power.

Except in underground coal mines, more fires originate in diesel equipment than any other source. Combus-

TABLE 10-23A. Percentages of Mine Fire Ignition Sources (1968–1979)*

	Underground Mines		Surface Mines	
	Coal	Noncoal	Coal	Noncoal
Electrical	43.7	18.5	19.0	16.0
Friction	7.8	2.8	9.5	2.0
Spontaneous combustion	10.9	2.4	—	—
Explosives	.8	—	—	—
Hot surface (engines)	—	15.5	38.1	32.0
Welding and cutting	8.5	13.4	9.5	18.0
Other or unknown	28.3	47.4	23.9	32.0

* U.S. Bureau of Mines.

tible liquids (hydraulic fluids, diesel fuel, and lubricants) constitute the greatest hazard. However, wiring hoses, tires, and combustible refuse also are contributing hazards.

Fires on board large surface mining mobile equipment are also a serious hazard. (See Fig. 10-23K.) Fires on diesel powered equipment typically involve leaking high pressure hydraulic lines, which spray a heated mist of highly combustible liquid onto an ignition source such as a hot exhaust manifold or turbocharger. On shovels, blasthole drills, and draglines, faults in electrical equipment are the most frequent cause of fires.

TABLE 10-23B. Percentages of Equipment Involved in Mine Fires (1968–1979)*

	Underground Mines		Surface Mines	
	Coal	Noncoal	Coal	Noncoal
Stationary electrical equipment	27.5	21.6	23.9	18.0
Mobile equipment	35.5	39.1	42.7	32.0
None or unknown	31.2	25.4	6.8	13.5
Other equipment	5.8	13.9	26.6	36.5

* U.S. Bureau of Mines.

Welding and cutting operations are the third leading cause of fires in mines. Welding and cutting operations are the primary cause of shaft fires in both coal and noncoal underground mines. The potential seriousness of shaft fires cannot be overemphasized, because fresh air to underground workings and safe mine evacuation may be impaired by a fire in a shaft.

Exothermic oxidation reactions occur in both coal and metal sulfide ores. When the heat generated by these reactions is not dissipated, the temperature of the rock mass or pile increases. If sufficient temperatures are reached, rapid combustion of coal, sulfide minerals, and other combustibles can result (Ninteman 1978). Although spontaneous ignition fires occur relatively infrequently, they are generally disruptive to mine operations and difficult to extinguish. Spontaneous ignition fires often occur in remote, abandoned sectors of the mine where fire fighting access is difficult. Half of underground coal mine fires lasting more than 24 hours and 57 percent of noncoal fires lasting more than 24 hours are caused by spontaneous ignition.

TABLE 10-23C. Percentages of Burning Substances Involved in Mine Fires (1968–1979)*

	Underground Mines		Surface Mines	
	Coal	Noncoal	Coal	Noncoal
Electrical insulation	25.2	20.2	23.8	20.0
Timber	8.9	22.2	—	—
Coal and combustible ore	28.8	4.0	—	—
Rubber	16.0	14.6	14.3	6.0
Explosives	0.4	1.0	—	—
Flammable and combustible liquids	16.0	26.8	60.9	73.4
Other or unknown	4.7	11.2	1.0	0.6

* U.S. Bureau of Mines.

Heat generated by friction between conveyor belt and conveyor drives or idlers had been a leading cause of fires in underground coal mines until about 1970, when requirements for sequence and slippage switches became effective. Other frictional heat sources include overheated rubber tires and parking brakes on mobile equipment which are engaged during vehicle operation.

For many years, timbering was the most common means of providing both temporary and permanent support of underground openings. The extensive use of timber resulted in a large number of timber fires. Prior to 1967 over half of all underground metal and nonmetal mine fires involved timber. In recent years, the use of roof bolts has reduced significantly the need for timber underground. But older mines contain large amounts of timber from previous years, and timber still is used in varying amounts at most mines to supplement other means of roof control.

About one fire in ten in underground coal mines and one fire in five in underground noncoal mines now involve timber. However, timber fires tend to constitute a greater overall hazard than these figures indicate, because timber fires spread more rapidly, burn longer, and are more difficult to extinguish than fires involving most other extraneous combustible materials. In underground

FIG. 10-23K. Haulage truck fire.

noncoal mines, over 75 percent of the fires lasting longer than 24 hours involve burning wood.

The ultimate objective of underground coal mine fire protection efforts is to prevent the ignition of the coal. Equipment fires are generally limited by the amount of combustible material present. However, once coal is ignited, the entire mine represents the available fuel supply. Over one-third of fires that involve coal as a burning substance last longer than 24 hours, with the average duration being 6 days (McDonald & Baker 1979). Failure to extinguish a fire while it is just developing results in a 50 percent chance that more than 8 hours will be required to bring it under control, and a 5 percent chance that part or all of the mine will be sealed (Mitchell & Burns 1979).

The storage, handling, and use of flammable and combustible liquids pose fire hazards for all sectors of the mining industry. In noncoal underground mines and surface mines, flammable and combustible liquids are involved in more fires than any other combustible material. Most of these fires occur in mobile equipment. Hydraulic fluid at elevated temperature and pressure is a primary hazard.

In underground noncoal mines, the storage of flammable and combustible liquids is an important concern. Many mines store large quantities (up to 30,000 gal or 113 m^3) of flammable and combustible liquids underground. The potential seriousness of a fire in an underground flammable and combustible liquid storage area has prompted extreme care in the design of storage areas, plus the implementation and strict enforcement of safe operating procedures. As a result, few fires have been reported in such storage areas.

Mine Explosion Hazards

Over 100 mine explosions have occurred since 1968, most in underground coal mines, resulting in the deaths of nearly 200 miners (Richmond et al. 1983). Practically all coal mine explosions start with an ignition of methane. However, if the explosion propagates more than a few hundred feet, coal dust becomes the most important fuel source. In underground coal mines, methane explosions outnumber coal dust explosions about six to one. Along with methane explosions in mines are methane ignitions (explosions that do not result in fatalities, injuries, or damage to mine workings or equipment). Methane ignitions outnumber methane explosions about seven to one (Nagy 1981).

The high frequency of methane explosions and ignitions can be attributed to the small amount of methane necessary to produce an explosive mixture in air and to the low energy required to ignite a methane-air mixture. The Lower Explosive Limit (LEL) for methane in air is 5 percent. As little as 0.3 millijoule of electrical energy can ignite methane under the proper conditions (equivalent to 1/50 of the static electricity accumulated by the average person walking across a carpeted floor on a dry day).

The LEL for a suspension of high-volatile bituminous coal dust in air is 0.05 oz per cu ft (0.28 g/m^3). The presence of methane in the atmosphere increases the hazard by producing a linear reduction in the LEL for coal dust. In the absence of methane, coal dust can be ignited by electrical or frictional sparks or by explosives, but the energy required for ignition is much higher if methane is not present. Coal dust explosions are generally triggered by methane ignitions. The flame provides the necessary ignition source and the shock wave causes settled coal dust to become suspended in air.

The destructive potential of ignitions and explosions results from the high temperature flame fronts, and static and dynamic pressures produced. Flame temperatures of 3,600°F (1982°C) and static pressures in excess of 50 psig (345 kPa) are possible. Dynamic ("wind") pressure is directional, and depends upon the air velocity produced by the explosion which, in turn, depends upon the static pressure.

Ignition sources of recent mine explosions and methane ignitions are shown in Table 10-23D.

TABLE 10-23D. Ignition Sources of Mine Explosions and Methane Ignitions (1970–1977)*

	Coal Mines			
	Underground			
	Explosions	Ignitions	Surface	Noncoal Mines
Friction	32	285	0	0
Electric arc	7	13	10	2
Welding	3	24	18	6
Smoking	6	2	3	4
Open flame	0	0	1	10
Battery	2	1	0	1
Lightning	0	2	1	0
Explosives	4	8	2	0
Safety lamp	2	0	0	0
Other	0	0	5	0
Total	56	335	40	23

* Mine Safety and Health Administration, U.S. Department of Labor.

Frictional sparking between cutting bits and rock surfaces (continuous miners, roof bolters, undercutters, face drills, and longwall equipment) causes over 80 percent of the ignitions and explosions in underground coal mines. Other common sources of frictional sparking are drill steel striking machine frames, sandstone impacting sandstone or other hard rock during a roof failure, and mobile vehicle brake linings (Nagy 1981).

Welding and cutting operations are the second leading cause of mine explosions and ignitions. In underground coal mines, the trend since 1970 has been toward increasing numbers of explosions and ignitions caused by welding and cutting operations.

About five percent of underground coal mine ignitions and explosions result from electric arcs (Nagy 1981). This figure is surprisingly low in view of the extensive use of electrical equipment in underground coal mine face areas and the low electrical energies required to ignite methane.

THE SAFEGUARDS

Fire and explosion safety in the mining industry is based on the general principles of fire and explosion prevention—early fire detection and warning, fire suppression, limiting fire propagation, and providing for miner safety during a fire emergency. Implementation of these general principles is discussed in the following sections.

FIRE PREVENTION

Fire prevention practices fall into three categories: limiting fuel sources, limiting ignition sources, and limiting fuel and ignition source contact.

Limiting Fuel Sources: Where there is sufficient frequency and/or severity of fires involving certain combustible materials, and where suitable and less hazardous substitutes for those materials are available, use of the less hazardous materials are preferred, or in some cases, mandated by law. Examples of less hazardous materials include fire resistive hydraulic fluids, conveyor belting, hydraulic hoses, and ventilation tubing, all of which are required in underground mines under certain conditions (U.S. Bureau of Mines 1978). Even more basic is good housekeeping to prevent unsafe accumulations of trash, oily rags, coal dust, and other combustible materials.

Limiting Ignition Sources: Sources of fire that are not essential to the mining process can be banned altogether. Examples are prohibitions on open fires and on smoking in underground coal mines and certain areas of underground noncoal and surface mines. Means to provide the unwanted buildup of heat, such as slippage and sequence switches on conveyors and thermal cutouts on electric motors, are also common.

Limiting Fuel and Ignition Source Contact: Many combustible materials and potential ignition sources are essential to mining operations. Firesafety in these instances depends upon preventing ignition source and fuel source contact. Certain precautions are observed during welding and cutting operations. When welding cannot be performed in firesafe enclosures, areas are wet down and, if practical, nearby combustibles are covered with fire resistive materials or relocated. Fire extinguishers must be readily available and a fire watch posted for as long as necessary to guard against smoldering fires. On mobile equipment, the hydraulic fluid, fuel, and lubricant lines can be rerouted away from hot surfaces, electrical equipment, and other possible ignition sources. Spray shields also can be installed to deflect sprays of combustible liquid from a broken fluid line away from a potential ignition source. Areas with a high loading of combustible materials, both underground and surface, such as timber storage areas, explosives magazines, flammable and combustible liquid storage areas, and shops can be designed to minimize possible ignition sources. Potential sources of ignition such as power substations, battery chargers, and electric motors can be kept clear of extraneous combustible material.

Explosion Prevention

Explosion prevention practices fall into the same three categories as for fire prevention: limiting fuel sources, limiting ignition sources, and limiting fuel and ignition source contact.

Limiting Fuel Sources: Coal dust and methane are the primary fuels involved in underground coal mine explosions. Methane may also be present in noncoal mines (Thimons, Vinson & Krissell 1979). Although every attempt is made to minimize the generation of coal dust in mining processes, the tiny amount of dust sufficient to propagate a coal dust explosion (a layer on the floor as little at 0.005 in. (0.012 mm) thick, if suspended in air, will propagate an explosion, less if methane is present) will be present and is unavoidable. However, when bituminous coal dust is properly mixed with an inert material such as pulverized limestone, dolomite, or gypsum (rock dust), the resulting mixtures will not propagate an explosion.

Methane is treated in two ways. Most commonly, methane is diluted by ventilation air and exhausted from the mine. However, "degasification" in advance of mining may be practiced where unusually high levels of methane are encountered (Hartman, Mutmansky & Wang 1982).

Limiting Ignition Sources: Every effort is made to eliminate extraneous ignition sources from the mining environment. Smoking is prohibited in coal mines, and electrical equipment operating where methane may be present must be "permissible" or "intrinsically" safe. "Permissible" electrical equipment is equipment that has been tested by the federal government and the design, materials of construction, and installation of which will not cause a mine fire or explosion. Use of explosion proof enclosures, specially designed plugs and receptacles, automatic circuit-interrupting devices, and voltage limitations are typical. "Intrinsically safe" equipment is equipment that has been tested by the federal government and is incapable, under normal or abnormal conditions, of releasing sufficient electrical or thermal energy to ignite an explosive gas mixture. Some success has been achieved in reducing frictional sparking at the face by using slower bit rotation, novel bit designs and bit materials, and specially engineered water sprays (Agbde et al 1982; Hanson 1983; Roepke & Hanson 1983). Explosives and diesel powered equipment must also be permissible.

Limiting Fuel and Ignition Source Contact: Because of geologic nonuniformity, varying mining rates, atmospheric pressure, and other factors, the liberation of methane into mine ventilation air from the workings is not uniform. Also, incomplete mixing of methane and ventilation air may occur, with explosive pockets or layers forming, even though the average methane concentration is below the LEL. Thus, dilution and drainage of methane cannot be relied upon to produce nonexplosive atmospheres. Methane checks are made frequently with handheld and machine mounted methanometers. If the methane level reaches or exceeds a predetermined level, all electrical equipment is shut down.

Early Fire Detection and Warning

The elapsed time between the onset of a fire and its detection is critical because fires tend to grow in size and intensity with time. Early fire detection and warning permit the initiation of an emergency plan in the mine (evacuation, fire fighting, etc.) while the fire is still small, or ideally, while it is still in the incipient stage. Over 70 percent of fires detected within 15 minutes cause little or no damage to the mine (McDonald & Pomroy 1980). Advanced fire detection and warning systems using sensitive heat, flame, smoke, and gas analyzers provide the most rapid and reliable indication of fire (Griffin 1979; Kacmar 1982; Johnson & Forshey 1978; Welsh 1982).

Thermal fire detection systems are commonly installed over conveyor belts, belt drives and takeups, and other unattended equipment. In localized, high hazard areas such as flammable and combustible liquid storage areas, refueling areas, and shops, faster acting fire detection devices may be appropriate. Optical flame detectors, which sense either ultraviolet or infrared radiation emitted by a fire, have been used successfully.

Smoke and/or gas detection is the most cost effective approach to providing large area, or even whole mine fire detection coverage (Hertzberg & Litton 1976; Litton 1983a & b; Stevens 1975; Thomas 1983; U.S. Bureau of Mines 1981a; 1977). (See Fig. 10-23L.) Instruments sampling the mine ventilation air streams are capable of detecting increases in particulates (smoke) of less than one micrometer (1 μ m) in size and in concentrations as low as 1 mg/m^3 as well as very small increases in combustion gases. Carbon monoxide (CO), for example, can be reliably detected at levels as low as 2 parts per million (ppm) with instruments intended for permanent underground installation.

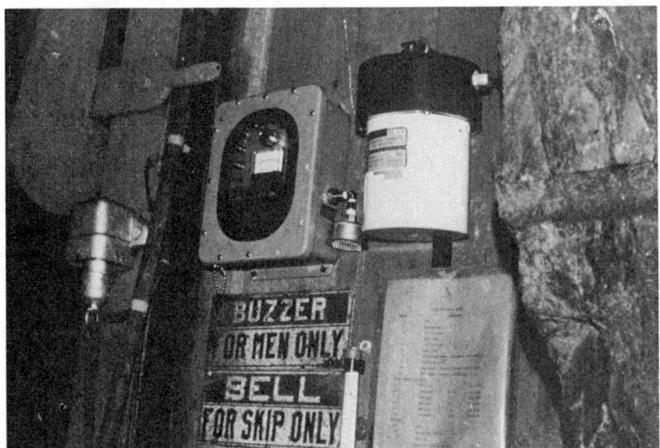

FIG. 10-23L. Fire detection instruments in an underground mine.

Once a fire has been detected and responsible mine officials notified, each miner must be warned of the fire. In coal mines, the most common means of fire warning are shutdown of electric power and/or notification by telephone and messengers. In noncoal mines, shutdown of electric power is generally ineffective because so little equipment is electrically powered. Telephone and messengers are sometimes used, but most miners are too remote from telephones and too widely scattered for either of these means to provide a timely warning. Stench warning is the most common method of emergency communication in noncoal underground mines (Pomroy & Muldoon 1983).

Fire Suppression

The most common types of fire suppression equipment used in underground coal mines are portable hand extinguishers (generally multipurpose dry chemical), water hoselines and sprinkler systems, rock dust (applied manually or from a rock dusting machine), and foam generators. Fire suppression equipment in noncoal underground mines includes portable hand extinguishers and water hoselines. Fire suppression systems, either manual or automatically actuated, are becoming more common for mobile equipment, combustible liquid storage areas, conveyor belt drives, and electrical installations (Johnson 1983; U.S. Bureau of Mines 1982). (See Fig. 10-23M.)

Where large areas of coal or timber are burning and fire fighting is complicated by extensive roof falls, ventilation uncertainties, and accumulations of explosive gas, the only practical alternatives are inerting (with nitrogen, carbon dioxide, or the combustion products of an inert gas generator), flooding, and/or sealing all or parts of the mine (Mitchell & Barns 1979).

All underground miners are trained to use basic fire fighting equipment; however, fires that grow to an advanced stage are fought only by highly trained and specially equipped fire fighting teams.

Mobile equipment is the primary fire hazard in surface mines. Consequently, most surface mining fire suppression equipment is specifically designed for maxium effectiveness on typical mobile equipment fires. Mobile equipment is commonly provided with one or more portable hand extinguishers. Manually or automatically actuated suppression systems may be installed for added protection (Johnson & Forshey 1975; Pomroy, Goodwin & Lynch 1982; Pomroy & Bickel 1980; U.S. Bureau of Mines 1981b).

Limiting Fire Propagation

Means for confining or limiting the rate of propagation of an underground mine fire can help ensure a safer mine evacuation and lessen the hazards of fire fighting. In undergound coal mines, oil and grease must be stored in closed, fire resistant containers and the storage areas must be of fire resistant construction. Transformer stations, battery charging stations, substations, shops, and other installations must be in fire resistant areas or housed in fire resistant structures. Unattended electrical equipment must be mounted on a noncombustible surface and separated from the coal or protected by a fire suppression system. Materials for building bulkheads and seals, including wood, brattice cloth, saws, nails, hammers, plaster or cement, and rock dust must be readily available to each working section. In underground noncoal mines, oil, grease, and diesel fuel must be stored in tightly sealed containers in fire resistive areas at safe distances from explosives magazines, electrical installations, and shaft stations. Ventilation control doors and fire doors are required in certain areas to prevent the spread of fire, smoke, and toxic gas (U.S. Bureau of Mines 1983b).

Providing for Miner Safety During a Fire Emergency

The primary concern during an underground fire emergency is the safety of underground personnel. Smoke and toxic fire gases represent the greatest threat; therefore the mine ventilation system and its relationship to the various working areas, potential fires, primary and secondary escape routes, refuge chambers, etc., are critically important. Detailed emergency preplanning must consider these factors. The use of computer models that simulate mine ventilation systems can be very helpful in developing and optimizing emergency plans and designating escape routes. Advanced models that calculate ventilation changes due to a mine fire (heat-induced buoyant forces causing throttling, reversals, etc.) and can track the distribution and concentration of contaminants are particularly valuable (Edwards & Greuer 1982). To ensure smooth implementation of the emergency plan, miners are provided with comprehensive training and annual retraining in emergency procedures. Fire drills, complete with the

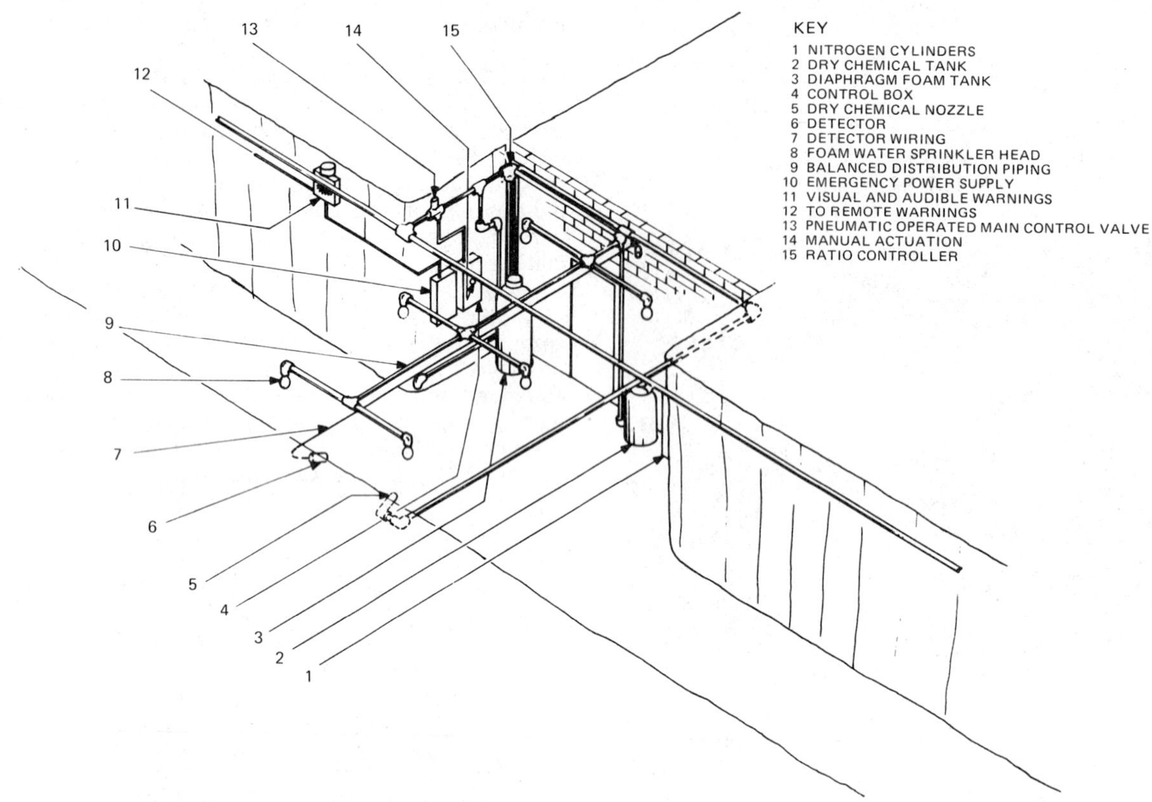

FIG. 10-23M. Fire suppression system for underground diesel refueling area.

activation of the mine warning system and the total evacuation of all mine personnel, are performed frequently to reinforce training and identify weaknesses in the emergency plan.

Although every attempt is made to avoid such situations, evacuating miners are sometimes required to travel through areas contaminted by smoke and toxic fire gases; therefore, appropriate respiratory protection for all miners is required. In coal mines, self-contained breathing apparatus supplying a minimum of one hour of oxygen must be provided. In noncoal mines, one hour rated self-rescue devices that convert carbon monoxide to carbon dioxide must be provided.

Bibliography

References Cited

Agbde, R. O., Whitehead, K. L., Mundell, R. L., and Saltsman, R. D. 1982. *Frictional Ignition Supression by the Use of Cutter Drum Mounted Sprays. Bituminous Coal Research Inc.* Contract Final Report J0395040. U.S. Bureau of Mines, Washington, DC.

Baker, R. M., Nagy, J., and McDonald, L. B. 1980. *An Annotated Bibliography of Metal and Nonmetal Mine Fire Reports.* Contract Final Report J0295035. U.S. Bureau of Mines, Washington, DC.

Cassidy, S. M. ed. 1983. *Elements of Practical Coal Mining.* Society of Mining Engineers of the American Institute of Mining, Metallurgical and Petroleum Engineers Inc., Port City Press, Baltimore, MD

Cummins, A. B., and Given, I. A. ed. 1973. *Mining Engineering Handbook.* Society of Mining Engineers of the American Institute of Mining, Metallurgical and Petroleum Engineers Inc., Port City Press, Baltimore, MD.

Edwards, J. C., and Greuer, R. E. 1982. *Real-Time Calculation of Product-of-Combustion Spread in a Multilevel Mine.* Information Circular 8901. U.S. Bureau of Mines, Washington, DC.

Griffin, R. E. 1979. *In-Mine Evaluation of Underground Fire and Smoke Detectors.* Information Circular 8808. U.S. Bureau of Mines, Washington, DC.

Hanson, B. D. 1983. *Cutting Parameters Affecting the Ignition Potential of Conical Bits.* BuMines RI 8820. U.S. Bureau of Mines, Washington, DC.

Hartman, H. L., Mutmansky, J. M., and Wang, Y. J. 1982. *Mine Ventilation and Air Conditioning.* John Wiley & Sons, NY.

Hertzberg, M., and Litton, C. D. 1976. *Multipoint Detection of Products of Combustion with Tube Bundles.* BuMines RI 8171. U.S. Bureau of Mines, Washington, DC.

Hustrulid, W. A. 1982. *Underground Mining Methods Handbook.* Society of Mining Engineers of the American Institute of Mining, Metallurgical and Petroleum Engineers. Port City Press, Baltimore, MD.

Johnson, G. A. 1983. *Automatic Fire Protection for Mobile Underground Mining Equipment.* Information Circular 8954. U.S. Bureau of Mines, Washington, DC.

Johnson, G. A., and Forshey, D. R. 1975. *Automatic Fire Protection Systems for Large Haulage Vehicles.* Information Circular 8683. U.S. Bureau of Mines, Washington, DC.

Johnson, G. A., and Forshey, D. R. 1978. *In-Mine Fire Tests of Mine Shaft Fire and Smoke Protection Systems.* Information Circular 8788. U.S. Bureau of Mines, Washington, DC.

Kacmar, R. M. 1982. *Reliability of Computerized Mine-Monitoring Systems.* Information Circular 8882. U.S. Bureau of Mines, Washington, DC.

Litton, C. D. 1983a. *Guidelines for Siting Product-of-Combustion Fire Sensors in Underground Mines*. Information Circular 8919. U.S. Bureau of Mines, Washington, DC.

Litton, C. D. 1983b. *Design Criteria for Rapid Response Pneumatic Monitoring Systems*. Information Circular 8912. U.S. Bureau of Mines, Washington, DC.

Martin, J. W., Martin, T. J., Bennett, T. P., and Martin K. M. 1982. *Surface Mining Equipment*. Martin Consultants Inc., Golden, CO.

McDonald, L. B., and Baker, R. M. 1979. *An Annotated Bibliography of Coal Mine Fire Reports*. Contract Final Report J0275008. U.S. Bureau of Mines, Washington, DC.

McDonald, L. B., and Pomroy, W. H. 1980. *A Statistical Analysis of Coal Mine Fire Incidents in the United States from 1950 to 1977*. Information Circular 8830. U.S. Bureau of Mines, Washington, DC.

Mitchell, D. W., and Burns, F. A. 1979. *Interpreting the State of a Mine Fire*. U.S. Mine Safety and Health Administration Informational Report 103.

Nagy, J. 1981. *The Explosion Hazard in Mining*. U.S. Mine Safety and Health Administration Informational Report 1119.

Ninteman, D. J. 1978. *Spontaneous Oxidation and Combustion of Sulfide Ores in Underground Mines*. Information Circular 8775. U.S. Bureau of Mines, Washington, DC.

Pfleider, E. P. ed. 1969. *Surface Mining*. Society of Mining Engineers of the American Institute of Mining, Metallurgical and Petroleum Engineers. Maple Press, York, PA.

Pomroy, W. H., and Bickel, K. L. 1980. *Automatic Fire Protection Systems for Surface Mining Equipment*. Information Circular 8832. U.S. Bureau of Mines, Washington, DC.

Pomroy, W. H., Goodwin, N., and Lynch, F. 1982. *Economic Analysis of Surface Mining Mobile Equipment Fire Protection Systems*. Report of Investigations 8698. U.S. Bureau of Mines, Washington, DC.

Pomroy, W. H., and Muldoon, T. L. 1983. *A New Stench Gas Fire Warning System*. Mines Accident Prevention Association of Ontario, North Bay Ontario, Canada.

Richmond, J. K., Price, G. C. Sapko, M. J., and Kawenski, E. M. 1983. *Historical Summary of Coal Mine Explosions in the United States 1959-81*. Information Circular 8909. U.S. Bureau of Mines, Washington, DC.

Roepke, W. W., and Hanson, B. D. 1983. *Bit Ignition Potential with Worn Carbide Tips*. TPR 121. U.S. Bureau of Mines, Washington, DC.

Stevens, R. B. 1975. *Mine Shaft Fire and Smoke Protection System*. FMC Corp. BuMines Contract Final Report H0242016. U.S. Bureau of Mines, Washington, DC.

Thimons, E. D., Vinson, R. P., and Kissell, F. N. 1979. *Forecasting Methane Hazards in Metal and Nonmetal Mines*. Report of Investigations 8392. U.S. Bureau of Mines, Washington, DC.

Thomas, E. C. 1983. "A Pneumatic Sampling Fire Detection System in an Underground Haulageway." *IEEE Trans. on Industry Applications*. V. IA-19. No. 3. May/June 1983.

U.S. Bureau of Mines. 1975. *Economic Engineering Analysis of U.S. Surface Coal Mines and Effective Land Reclamation*. BuMines Contract Final Report S0241049. Skelly and Loy, Engineers-Consultants.

U.S. Bureau of Mines. 1977. *Metal Mine Fire Protection Research*. Information Circular 8752. U.S. Bureau of Mines, Washington, DC.

U.S. Bureau of Mines. 1978. *Coal Mine Fire and Explosion Prevention*. Information Circular 8768. U.S. Bureau of Mines, Washington, DC.

U.S. Bureau of Mines. 1981a. *Underground Metal and Nonmetal Mine Fire Protection*. Information Circular 8865. U.S. Bureau of Mines, Washington, DC.

U.S. Bureau of Mines. 1981b. "Dragline Fire Protection." *U.S. Bureau of Mines Technology News* 106. U.S. Bureau of Mines, Washington, DC.

U.S. Bureau of Mines. 1982. "Automatic Fire Protection Systems for Underground Fueling Areas." *U.S. Bureau of Mines Technology News* 160. U.S. Bureau of Mines, Washington, DC.

U.S. Bureau of Mines. 1983a. *Mineral Commodity Summaries*. U.S. Bureau of Mines, Washington, DC.

U.S. Bureau of Mines. 1983b. "Improved Fire Doors for Noncoal Underground Mines." *U.S. Bureau of Mines Technology News* 188, Washington, DC.

NFPA Codes, Standards, Recommended Practices and Manuals. (See the latest *NFPA Codes and Standards Catalog* for availability of current editions of the following documents.)

NFPA 10, *Standard for Portable Fire Extinguishers*.

NFPA 11A, *Standard for Medium and High Expansion Foam Systems*.

NFPA 12A, *Standard on Halon 1301 Fire Extinguishing Systems*.

NFPA 12B, *Standard on Halon 1211 Fire Extinguishing Systems*.

NFPA 13, *Standard for the Installation of Sprinkler Systems*.

NFPA 17, *Standard for Dry Chemical Extinguishing Systems*.

NFPA 30, *Flammable and Combustible Liquids Code*.

NFPA 51, *Standard for the Design and Installation of Oxygen-Fuel Gas Systems for Welding, Cutting and Allied Processes*.

NFPA 51B, *Standard for Fire Prevention in the Use of Cutting and Welding Processes*.

NFPA 70, *National Electrical Code®*.

NFPA 120, *Standard for Coal Preparation Plants*.

NFPA 121, *Standard on Fire Protection for Mobile Surface Mining Equipment*.

Additional Readings

"Coal Mine Fire and Explosion Prevention," *Proceedings: U.S. Bureau of Mines Technology Transfer Seminar*, U.S. Bureau of Mines, Washington, DC, 1978.

Cheng, L., Liebman, I., Furno, A. L., and Watson R. W., *Novel Coal-Cutting Bits and their Wear Resistances*, U.S. Bureau of Mines, Report of Investigations 8791, Washington, DC, 1983.

Cybulski, W., *Coal Dust Explosions and Their Suppression* (translated from the Polish), National Technical Information Service, Springfield, VA, 1975.

Hertzberg, M., Cashdollar, K. L., Lazzara, C. P., and Smith, A. C., *Inhibition and Extinction of Coal Dust and Methane Explosions*, U.S. Bureau of Mines, Report of Investigations 8708, Washington, DC, 1982.

Kuchta, J. M., Furno, A. L., Dalverny, L. E., Sapko, M. J., and Litton, C. D., *Diagnostics of Sealed Coal Mine Fires*, U.S. Bureau of Mines, Report of Investigations 8625, Washington, DC, 1982.

Kuchta, J. M., Rowe, V. R., and Burgess, D. S., *Spontaneous Combustion Susceptibility of U.S. Coals*, U.S. Bureau of Mines, Report of Investigations 8474, Washington, DC, 1980.

Liebman, I., and Richmond, J. K., *Suppression of Coal Dust Explosions by Water Barriers in a Conveyor Belt Entry*, U.S. Bureau of Mines, Report of Investigations 8538, Washington, DC, 1981.

Liebman, I., Corry, J., Pro, R, and Richmond, J. K., *Extinguishing Agents for Mine Face Gas Explosions*, U.S. Bureau of Mines, Report of Investigations 8294, Washington, DC, 1978.

Litton, C. D., *Fire Detectors in Underground Mines*, U.S. Bureau of Mines, Information Circular 8786, Washington, DC, 1979.

Sapko, M. J., Mura, K. E., Furno, A. L., and Kuchta, J. M., *Fire Resistance Test Method for Conveyor Belts*, U.S. Bureau of Mines Report of Investigations 8521, Washington, DC, 1981.

"Title 30 Mineral Resources, Chapter I," *Code of Federal Regulations*, U.S. Mine Safety and Health Administration, Washington, DC, 1984.

SECTION 11

STORAGE PRACTICES AND HAZARDS

GENERAL INDOOR STORAGE

Revised by Martin M. Brown, P.E.

This chapter covers the storage of a broad range of combustible commodities. For the purpose of this review, a commodity is defined as a combination of a product and any packing material, container, and storage aids, such as pallets. Information on the storage of specific materials that may present an explosion or fire hazard can be found throughout this HANDBOOK. The reader is referred to the other chapters in this section and the following chapters in Section 5: Chapter 4, "Flammable and Combustible Liquids;" Chapter 5, "Gases;" Chapter 6, "Chemicals;" Chapter 7, "Explosives and Blasting Agents;" and Chapter 9, "Dusts."

FIRE PROTECTION PRINCIPLES

The storage of goods does not involve a high ignition fire risk. However, when warehouse fires do occur, special difficulties often are encountered in controlling and extinguishing them. Automatic sprinkler protection is the practical and preferred means of fire protection, supplemented by manual fire fighting; however, because storage conditions can be unfavorable, sprinkler systems sometimes perform less than satisfactorily. This possibility exists even with complete area coverage and all sprinkler valves fully open.

Modern warehousing has introduced a variety of fast burning substances. For economic reasons, there usually is maximum utilization of space by high piling with minimum aisle distances between piles, large undivided fire areas, and sometimes enormous values of materials in a single fire area. Spaces inside storage configurations provide, to a greater or lesser extent, opportunity for fire to spread beyond the reach of water discharged from sprinklers.

Because of these warehouse realities, experience with automatic sprinklers has in recent years shown the need for great care in the design of the systems, and in maintaining the conditions upon which the design was based. As a result of continuing full scale applied research fire testing, primarily at Factory Mutual Engineering facilities, and the analysis of fire loss experience, principles and practices of good protection have been derived. To a great extent, they comprise the body of standards that are referenced in this chapter.

Mr. Brown is a fire protection consultant based in White Plains, NY.

COMMODITY CLASSIFICATION

Two conditions determine the probable degree of fire intensity and the difficulty facing automatic sprinkler operation. The first is the nature of goods that are stored, i.e., the commodity classification. The other, covered in the next part of this chapter, is the arrangement of the storage.

For general varieties of stored goods, NFPA 231, *Standard for Indoor General Storage* (hereinafter referred to as NFPA 231), and NFPA 231C, *Standard for Rack Storage of Materials* (hereinafter referred to as NFPA 231C), define commodity classifications relative to their heat of combustion, rate of heat release, and rate of flame spread. For guidance in applying fire protection requirements, general goods are divided into Classes I, II, III, and IV, which may include plastics as part of an item. For plastics *per se*, there are additional groupings: A, B, and C.

General Commodity Classifications

Class I: These commodities essentially are noncombustible products arranged on combustible pallets, in ordinary corrugated cartons with or without single thickness dividers, or in ordinary paper wrappings with or without pallets. Such products may have a negligible amount of plastic trim such as knobs or handles.

Class II: These commodities are Class I products placed in slatted wooden crates, solid wooden boxes, multiple thickness paperboard cartons, or equivalent combustible packaging material with or without pallets. In rack storage, only wood pallets are contemplated.

Class III: These commodities are wood, paper, natural fiber cloth, or Group C plastics with or without pallets. In rack storage, only wood pallets are contemplated. The products may contain a limited amount of Group A or B plastics. (See the following section on "Plastics.") A metal bicycle with plastic handles, pedals, tires, and seat; or a wood dresser with plastic drawer glides, handles, and trim, are examples of a commodity with a limited amount of plastic.

Class IV: These commodities are Class I, II, or III products containing an appreciable amount of Group A plastics in corrugated cartons and Class I, II, and III products in ordinary corrugated cartons, with Group A plastic packaging with or without pallets. In rack storage, only wood pallets are contemplated. Group B plastics and free flowing Group A plastics are also included in this class. An example of packing material is a foamed (expanded) plastic cocoon which would hold a typewriter in an ordinary corrugated carton.

In rack storage a distinction is also made, within these four classes, between encapsulated or nonencapsulated loads, the former being pallet loads tightly enclosed on tops and sides by a film of plastic such as polyethylene. This causes increased difficulty for sprinklers, but can be altered to the nonencapsulated status by removing the film from the top.

Plastics

In general, the combustion of a plastic produces about twice the amount of heat per unit of weight as cellulosic materials (wood, paper, cotton cloth). In the four commodity classes, the references to plastics are divided into three groups, designated A, B, and C, which also may be stored independently of Classes I, II, III and IV. Group A includes the fastest burning types, and Group C the slowest.

Plastics Classification

Group A:
ABS (acrylonitrile-butadiene-styrene copolymer)
Acrylic (polymethyl methacrylate)
Acetal (polyformaldehyde)
Butyl rubber
EPDM (ethylene-propylene rubber)
FRP (fiberglass reinforced polyester)
Natural rubber (if expanded)
Nitrile rubber (acrylonitrile-butadiene rubber)
PET (thermoplastic polyester)
Polybutadiene
Polycarbonate
Polyester elastomer
Polyethylene
Polypropylene
Polystyrene
Polyurethane
PVC (polyvinyl chloride —highly plasticized, e.g., coated fabric, unsupported film)
SAN (styrene acrylonitrile)
SBR (styrene —butadiene rubber)

Group B:
Cellulosics (cellulose acetate, cellulose acetate butyrate, ethyl cellulose)
Chloroprene rubber
Fluoroplastics (ECTFE —ethylene-chlorotrifluoroethylene copolymer, ETFE —ethylene-tetrafluoroethylene copolymer, FEP —fluorinated ethylene-propylene copolymer)
Natural rubber (not expanded)
Nylon (nylon 6, nylon 6/6)
Silicone rubber

Group C:
Fluoroplastics (PCTFE —polychlorotrifluoroethylene, PTFE —polytetrafluoroethylene)
Melamine (melamine formaldehyde)
Phenolic
PVC (polyvinyl chloride —rigid or lightly plasticized, e.g., pipe, pipe fittings)
PVDC (polyvinylidene chloride)
PVF (polyvinyl fluoride)
PVDF (polyvinylidene fluoride)
Urea (urea formaldehyde)

NOTE: These categories are based on unmodified plastic materials. The use of flame retarding modifiers or a different physical form of the material, (shredded, for example) may change the classification.

When assigning a Class number to a general commodity, or a Group letter to an essentially plastic commodity, the deciding factor is generally the most hazardous category present in an appreciable amount. If Group A plastic cushioning in a Class IV commodity carton is judged to be of such volume as to dominate the fire behavior of the stored item, the commodity as a whole will fall into Group A plastics and should be protected accordingly. There are two exceptions: (1) if there are multiple layers of corrugation or equivalent outer material that would significantly delay fire involvement of the Group A plastic; or (2) if the amount and arrangement of the Group A plastic within an ordinary carton would not be expected to significantly increase the fire hazard. It frequently is better to assign a next higher class than is present if there is any probability of its being introduced later. The difference in protection cost at the outset would be relatively small.

The classifications already mentioned are for high piled storage, which is over 12 ft (3.65 m) high and typical of modern warehouse practice. When storage is located in stockrooms, multistory buildings, and freight transfer sheds having only low piled goods, it can be classified as Ordinary Hazard-Group 2, or Ordinary Hazard-Group 3, according to NFPA 13, *Standard for the Installation of Sprinkler Systems* (hereinafter referred to as NFPA 13). Typical of Ordinary Hazard-Group 2 would be cold storage warehouses, library stack rooms, and tobacco products storerooms. Typical of Ordinary Hazard-Group 3 would be those having moderate to higher combustibility of content, such as paper, household furniture, whiskey in bottles, and general storage items; the protection requirements in NFPA 13 would apply.

Empty Pallets

Idle pallet storage introduces a severe fire condition that is incidental to most storage practices. Pallets are not classified in the commodity categories listed above because the hazard can be controlled by strictly limiting pile heights or keeping the idle pallets out of the building. When stored indoors, wood and certain solid plastic pallets should be protected as shown in Table 11-1A, unless kept no higher than 6 ft (1.8 m) and in a small area (four stack) detached piles. Idle combustible pallets should not be stored in racks.

Plastic pallets other than the type shown in Table 11-1A should be piled no higher than 4 ft (1.2 m), and be in no more than two stack groups, each at least 8 ft (2.4 m) away from other pallets or 25 ft (7.6 m) from stored

TABLE 11-1A. Protection for Indoor Storage of Wood Idle Pallets or Nonexpanded Polyethylene Solid Deck Idle Pallets

Height of Pallet Storage		Sprinkler Density Requirements		Area of Sprinkler Demand in sq ft and (m^2)			
feet	meters	gpm/sq ft	(L/mm)/m^2	286°F	(141°C)	165°F	(74°C)
Up to 6	Up to 1.8	0.20	8.1	2,000	(186)	3,000	(279)
6 to 8	1.8 to 2.4	0.30	12.2	2,500	(232)	4,000	(372)
8 to 12	2.4 to 3.6	0.60	24.4	3,500	(325)	6,000	(557)
12 to 20	3.6 to 6.1	0.60	24.4	4,500	(413)	—	—

commodities. Otherwise, they are placed outdoors or in cutoff rooms with extra hazard fire protection.

All types of pallets stored outdoors should be kept a sufficient distance from a building, using distances as in Table 11-1B as a guide.

Flammable Aerosols

Flammable liquids in small pressurized spray cans, such as hair sprays and paints, have demonstrated a shocking ability to explode from internal pressure when heated in a fire, creating fireballs and often rocketing projectiles leaving a trail of burning liquid. Any appreciable storage of these aerosol cans has the potential to grossly spread fire, overtaxing sprinkler systems. While technically a flammable liquid problem, these products normally are included in many general product warehouses such as grocery, hardware, and department store warehouses.

Flammable aerosols require fire resistive segregation from other storage, low piling, copious sprinklers within storage racks, and high density ceiling sprinkler protection, such as 0.60 gpm per sq ft over 3,000 sq ft [24(L/min)/m^2 over 279 m^2].

STORAGE ARRANGEMENT

From a fire suppression viewpoint, storage arrangements include: bulk storage, solid piling, palletized pile storage, and rack storage. Bins and narrow shelves also are of importance, but they usually are employed in stockrooms holding moderate quantities of items for direct use. The basic differences between the four categories that affect the fire behavior and difficulty of fire control are in the horizontal and vertical air spaces or "flues" created by the storage configurations.

Bulk Storage

Bulk storage consists of piles of unpackaged materials in a loose, free flowing condition such as powder, granules, pellets, or flakes, or original items such as peanuts. The materials will be found in silos, bins, tanks, or in large piles on the floors of storage buildings.

Conveyor equipment such as belt conveyors, air fluidizing ducts, and bucket conveyors ("legs"), agitate the material during certain stages of its movement. If prone to developing airborne dust, a combustible material presents an explosion hazard, notably in grain storage facilities. Conveyor belts are generally combustible, and together with the commodity, can burn in inaccessible places such as at elevations high above the floor or in tunnels. Automatic sprinklers frequently are needed in the housings around conveying equipment.

Fires in large piles tend to burrow down and require prolonged soaking to be reached. Spontaneous ignition

TABLE 11-1B. Minimum Distances from Pallet to Wall in Outdoor Storage

Wall Construction		Minimum Distance (m) of Wall from Storage of		
Wall Type	Openings	Under 50 Pallets	50 to 200 Pallets	Over 200 Pallets
Masonry	None	0	0	0
	Wired glass with outside sprinklers One Hour Doors	0	10 ft (3 m)	20 ft (6.1 m)
	Wired or Plain glass with outside sprinklers ¾ Hour Doors	10 ft (3 m)	20 ft (6.1 m)	30 ft (9.1 m)
Wood or Metal with outside sprinklers				
Wood, Metal or Other		20 ft (6.1 m)	30 ft (9.1 m)	50 ft (15.2 m)

NOTE 1: Fire resistive protection comparable to that of the wall should also be provided for combustible eave lines, vent openings, etc.
NOTE 2: When pallets are stored close to a building, the height of storage should be restricted to prevent burning pallets from falling on the building.
NOTE 3: Manual outside open sprinklers are generally not a reliable means of protection unless property is attended at all times by plant emergency personnel.
NOTE 4: Open sprinklers controlled by a deluge valve are preferred.

fires may start in the interior of piles, and may be difficult to locate if internal heating has not been continually monitored by heat sensors immersed in the pile. It is questionable whether to install automatic sprinklers over large piles in noncombustible buildings, but good hose stream access around the perimeter is always important.

Solid Piling

Solid piling consists of cartons, boxes, bales, bags, etc., in direct contact with each other. Air spaces, or flues, exist only where there is imperfect contact, or where a pile is close to but not touching another pile. Stacking is by hand or by lift trucks using side clamps, or the usual prongs in the absence of pallets when bales or other packages can be pried up from below without serious damage. (See Fig. 11-1A.)

FIG. 11-1A. A type of solid piling in which spaces are left between cartons to accommodate lift truck prongs. Wide aisles give the lift truck plenty of room to maneuver. The dark pipes at the ceiling are sprinkler lines.

Compared with palletized and rack storage, solid piling gives fire the least opportunity for fire development, and gives water application the greatest chance to be effective. Still, where outer surfaces possess rapid flame spread properties, high piling starting at about 15 ft (4.6 m) can present a severe hazard. Movement of stock during a fire to get at the burning interior is more difficult.

Solid piling of general classes of goods up to 12 ft (3.7 m) high requires sprinkler protection complying with the NFPA 13 sprinkler standards. However, NFPA 231 gives protection requirements for plastics over 5 ft (1.5 m) high, and other piling up to 30 ft (9 m) high.

Palletized Storage

Palletized storage consists of unit loads mounted on pallets. The load usually is an approximate cube, about 3 to 4 ft (0.9 to 1.2 m) in height, consisting of a single package or block of packages presenting a top surface that can sustain the weight of additional pallet loads on top without crushing the packages or forming an unstable pile. Each pallet is approximately 4 in. (102 mm) high and usually composed of wood. (See Fig. 11-1B.) Some are made of metal, plastic, expanded plastic, or cardboard.

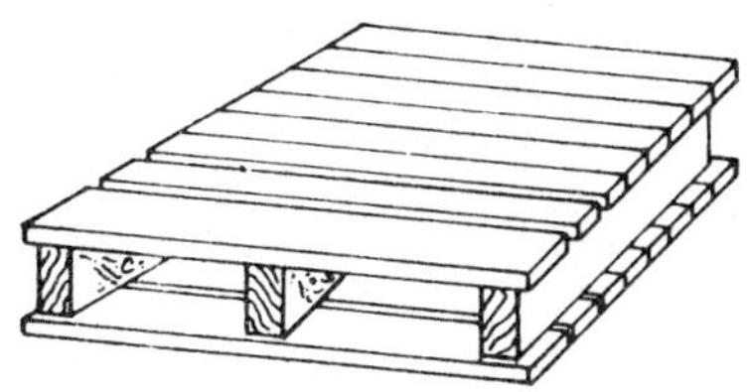

CONVENTIONAL PALLET

FIG. 11-1B. Conventional wood pallet.

The pallet accepts the prongs of a forklift truck, and also can be an integral part of a sizable shipping container. The height of palletized storage, usually 30 ft (9.1 m), is limited by the stackability of the packaged goods.

Horizontal air spaces, formed by the pallets themselves, significantly assist the spread of fire within the pile, because they are out of reach of water from sprinklers. The air spaces usually are continuous in one direction for the entire width of a pile.

Sprinkler system requirements for palletized storage up to 12 ft (3.7 m) high are contained in NFPA 13. However, NFPA 231 gives the requirements for plastics over 5 ft (1.5 m) high, and other goods with heights above 12 ft (3.7 m).

Rack Storage

A storage rack consists of a structural framework into which are placed unit loads, generally on pallets. (See Figs. 11-1C, 11-1D, and 11-1E.) The height of storage racks is potentially limited only by the vertical reach of the materials handling machinery which, like the racks themselves, can be designed for great heights. While most racks are 20 to 25 ft (6.1 to 7.6 m) high, an appreciable number are higher. (See Fig. 11-1F.) In fully automated warehouses, racks are sometimes as high as 100 ft (30 m) or more.

There are numerous variations to open single and double row racks. (See Figs. 11-1G, 11-1H, and 11-1I.) Portable racks approximate pallets except that the upper pallet does not rest on the storage below, but leaves an air space between them.

Automated materials handling machinery for high racks can operate in narrow aisles such as 4 ft (1.2 m) widths, across which fire can readily jump. In racks, horizontal air spaces about 1 ft (0.3 m) in depth are under each tier of supports which provide clearance for han-

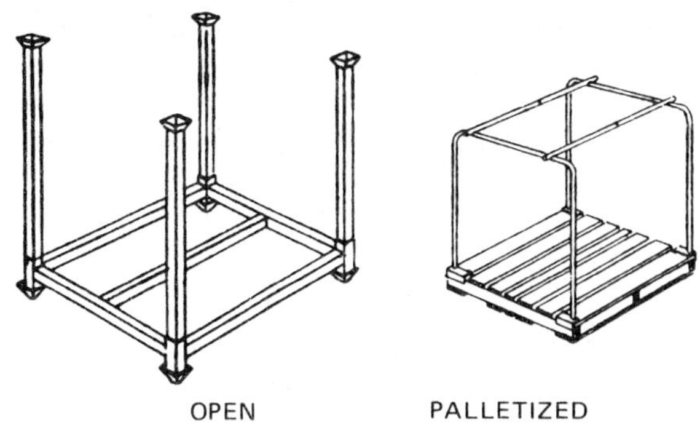

OPEN PALLETIZED

FIG. 11-1C. Two types of portable storage racks.

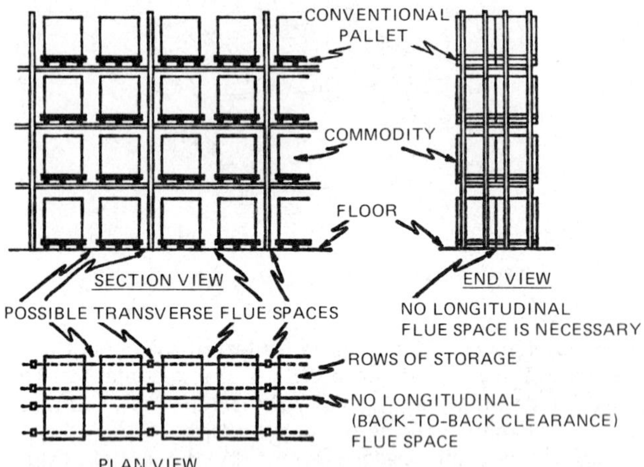

FIG. 11-1D. A typical double row (back to back) storage rack arrangement.

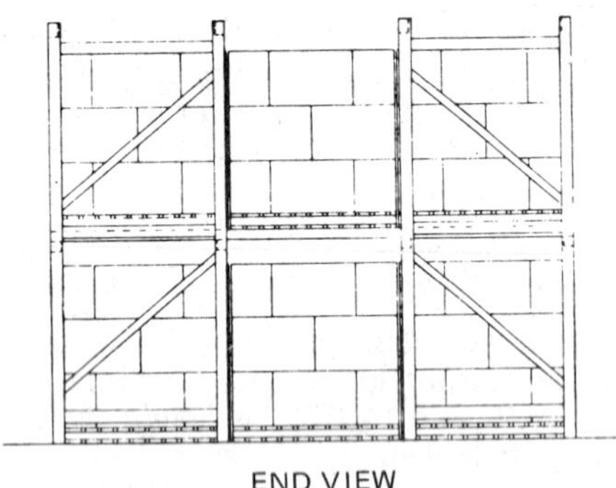

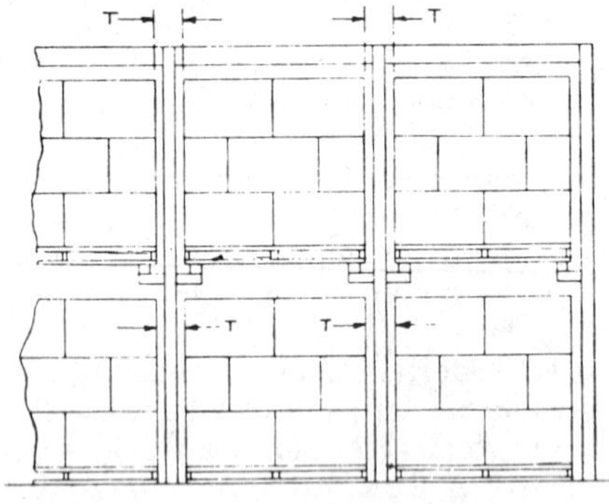

FIG. 11-1E. A multiple row, drive in storage rack that is two or more pallets deep. "T" indicates transverse flue spaces. Fork trucks drive into the rack to deposit and withdraw loads in the depth of the rack.

dling. Since racks are seldom completely filled, the random vacancies also contribute to the air spaces. Vertical flues exist at upright rack members and between unit loads. They allow upward fire spread, but make possible some sprinkler water penetration into the usually open style racks.

In high automated rack storage fires, even if the contents damage is not severe, because materials handling mechanisms must be accurately aligned to function, rack storage is vulnerable to the effects of distortion by heat,

FIG. 11-1F. Fifty-three ft (16 m) high storage racks with narrow aisle. Stacker machine will ride on floor rail, guided from the top. (Unarco Materials Storage)

possibly leading to the impairment of the entire warehouse operation.

Fortunately, racks are a semipermanent type of installation which can support sprinkler piping. Otherwise, a combination of horizontal air spaces sheltered from overhead sprinkler water discharge, great height, and narrow aisles would pose an impossible problem for a conventional sprinkler system installed at the ceiling level.

APPLICATION OF NFPA STORAGE STANDARDS

The NFPA storage standards—NFPA 231, 231C, 231D, *Standard for Storage of Rubber Tires*; 231E, *Recommended Practice for the Storage of Baled Cotton*; and 231F, *Standard for the Storage of Roll Paper*—provide some flexibility in planning protection for storage occupancies instead of merely stipulating minimum requirements for protection. For example, there are curves representing a range of sprinkler discharge densities and the areas they would cover for a variety of storage conditions. (See Figs. 11-1J and 11-1K for examples.) Any point chosen from a curve gives the density area criteria for the particular conditions represented by the curve (higher densities of discharge over smaller areas can result in less damage and are preferred). Often the planner is not bound by one curve, but can alter conditions so that a more favorable curve can be used.

Other choices are available, including different in-rack sprinkler arrangements for rack storage over 25 ft high (7.6 m).

The Influence of Variables

The NFPA storage standards are excellent guides, but they do not cover all situations. It sometimes is necessary

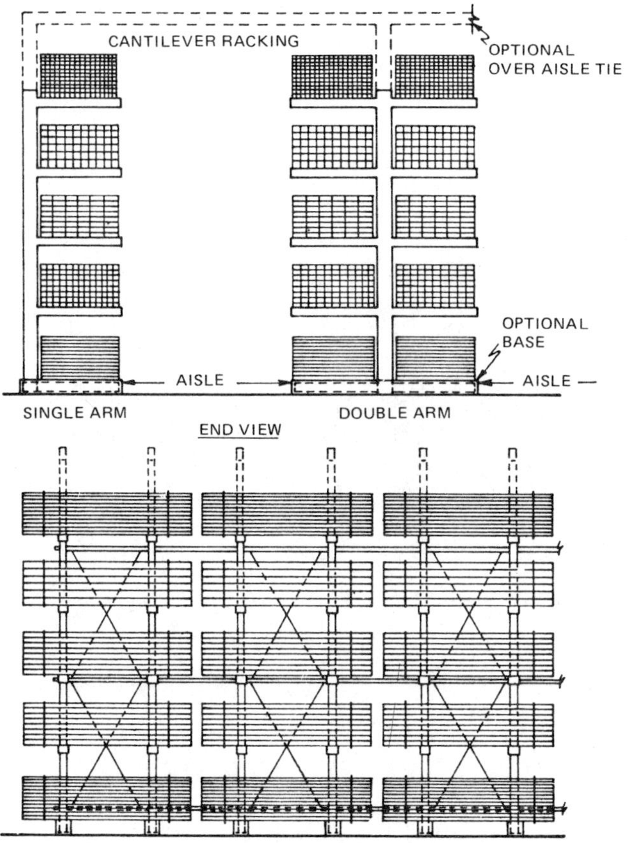

FIG. 11-1G. A cantilever storage rack.

to improvise by extrapolation or interpolation. While intended mostly for planning new installations, the recommendations in the standards assist in upgrading substandard installations and in judging the deficiency of protection so that the cost of an improvement (and options, if any) can be weighed against the seriousness of the deficiency.

Before improvisation upon the basic protection requirements can be attempted, the variables that can be considered must be understood. Armed with an appreciation of these conditions, the evaluation of existing protection and improvisation in original planning or upgrading of protection can be effectively employed. The following variables influence experienced judgment in using the standards and, in fact, play a major role in the requirements of the standards.

Variable No. 1—Storage Height: Other than the fire properties of the commodities themselves, probably no other condition has a more profound influence on the progress of fire in a storage occupancy and difficulty of fire control than storage height. In rack storage tests, fire intensity varied with the square of the storage height. As an example (without in-rack sprinklers), a 10 percent increase in rack storage height from 20 to 22 ft (6.1 to 6.7 m) requires a 35 percent increase in sprinkler discharge density. (See Figs. 11-1L and 11-1M.)

FIG. 11-1H. (Top) Nontypical storage racks with three pallets between uprights, hence three transverse flues. (Clark Industrial Truck Division) (Bottom) Drive-in type multiple row storage racks. Frozen foods are Class 1 commodities. (Merchants Refrigerating Co.)

Although increasing height increases fire intensity, all heights up to 15 ft (4.6 m) in solid piles or 12 ft (3.7 m) in palletized and rack storage are treated the same, i.e.,

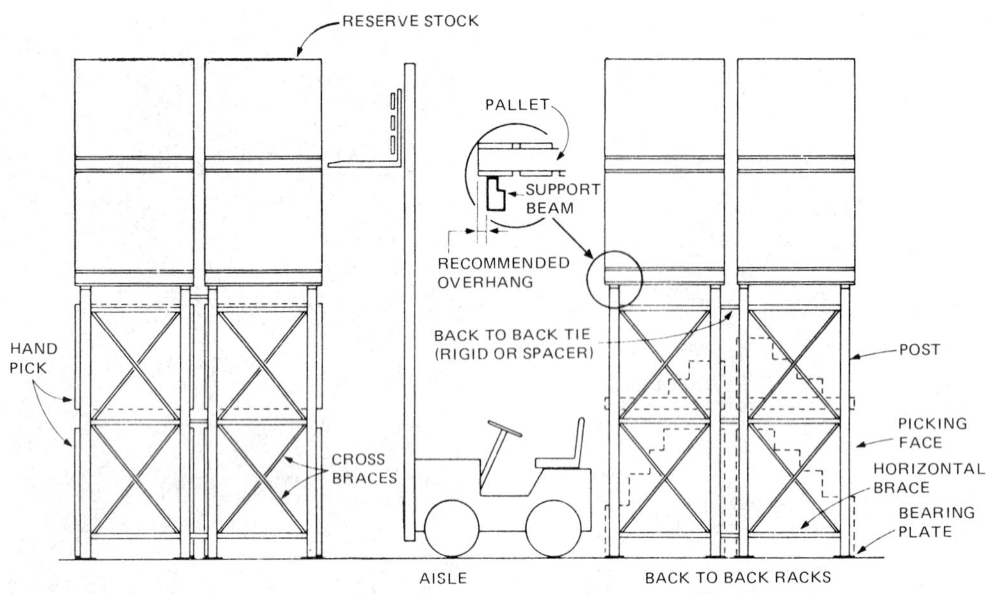

FIG. 11-1I. Common arrangement of double row storage racks with palletized storage atop. (Unarco Storage)

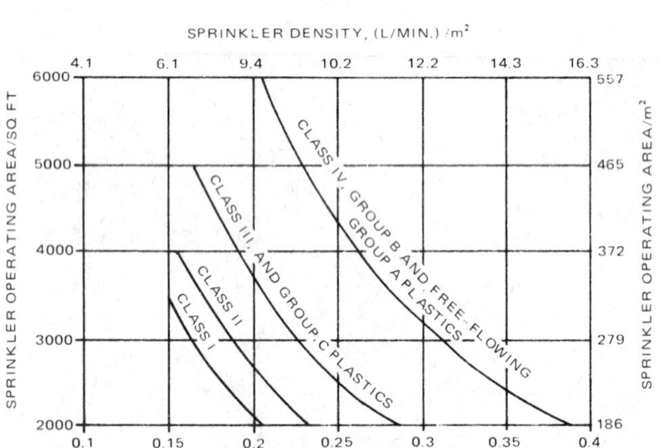

FIG. 11-1J. Wet pipe sprinkler system design curves for 20 ft (6.1 m) high storage and 165°F (74°C) sprinklers. For 286°F (140°C) sprinklers, reduce the sprinkler operating area 40 percent, but to not less than 2,000 sq ft (186 m²).

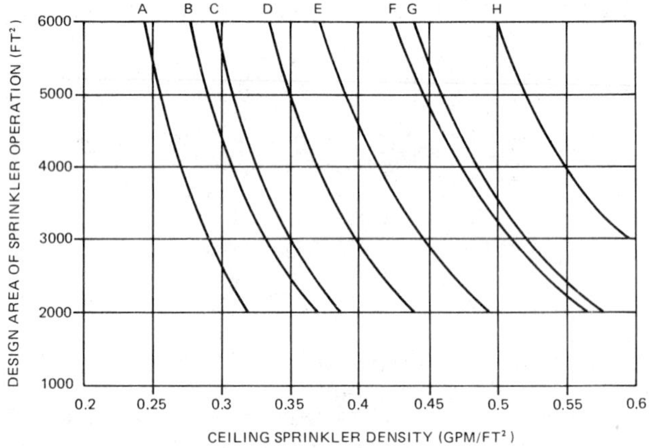

FIG. 11-1K. Typical sprinkler system design curves for rack storage of materials. These curves apply to systems protecting 20 ft (6.1 m) high double row racks holding Class IV nonencapsulated commodities on conventional pallets. Other curves govern other storage conditions and configurations as covered in the NFPA storage standards. [1 ft = 0.31 m; 1 sq ft = 0.093 m²; 1 gpm/ft² = 40.75 (L/min)/m²; 5/9 (°F-32)= °C]

palletized storage 5 ft (1.5 m) high requires the same level of sprinkler discharge as that required for 12 ft (3.7 m) high storage, except for plastics. This illustrates the principle that a safer condition must sometimes be grouped with the more hazardous condition to allow for the effects of future changes that may influence the degree of hazard.

Because of perennial uncertainties in the future use of buildings, it is safe practice to use the maximum height available for storage, for protection calculations, rather than the anticipated maximum storage height. This would be to approximately 18 in. (0.46 m) below the sprinkler deflectors. Of course, with permanent racks, the building height can be disregarded. (Exception: see Variable No. 5.) For racks over 25 ft (7.6 m) high, as height increases there is no change in the ceiling sprinkler requirements, because fire control for all but the highest storage is essentially by in-rack sprinklers. In such high rack storage, ceiling sprinklers primarily protect the ceiling structure and only the top portion of the storage.

Variable No. 2—Aisle Width: Wide aisles help to get water on the fire, retard transfer of fire from one pile to another, and permit access for fire fighting and salvage operations. For rack storage up to 25 ft (7.6 m) high, aisle width is a factor in determining sprinkler density for aisle widths of 4 and 8 ft (1.2 and 2.4 m); for aisle widths

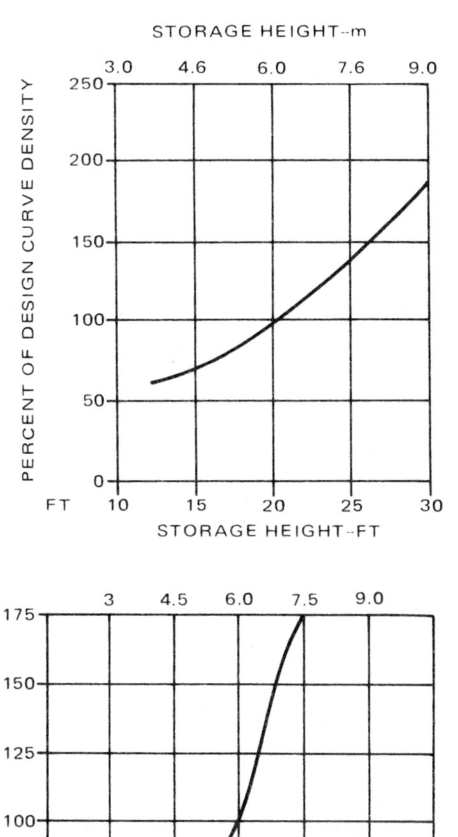

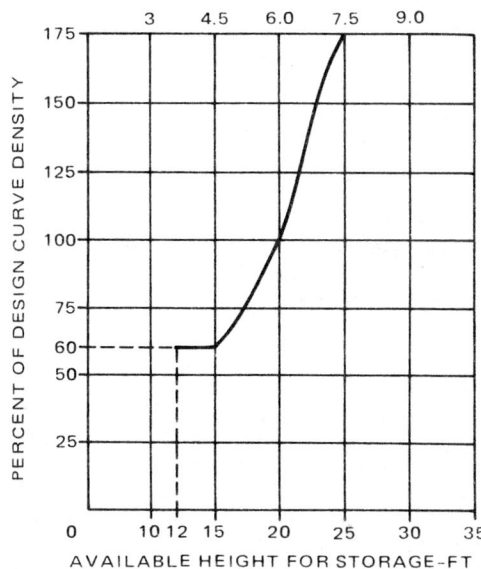

FIG. 11-1L. Ceiling sprinkler density versus storage height curves for both piled (top) and rack (bottom) storage.

between 4 and 8 ft (1.2 and 2.4 m) a direct linear interpolation between curves can be made. (See Fig. 11-1K.).

When space permits, it is desirable for aisle widths to be at least one-half the pile height to restrict fire jumping across aisles. Palletized storage piles should not be more than 50 ft (15.2 m) wide between major aisles, so that small hose streams can penetrate 25 ft (7.6 m) to the center of a pile. Although NFPA 231 does not give "credit" for wide or frequent aisles, the presence of aisles is assumed, and the more and wider the better. Fortunately, aisles usually are necessary for efficient operation in conventional warehouses, with 8 ft (2.4 m) width the normal minimum for maneuvering a lift truck. Where there are frequent aisles, approximately every 10 ft (3.04 m), it is generally agreed that a somewhat lower ceiling sprinkler density is needed over solid or palletized storage than indicated in NFPA 231.

Double row rack storage needs aisles 4 ft (1.2 m) or more in width to achieve the protection intended by NFPA

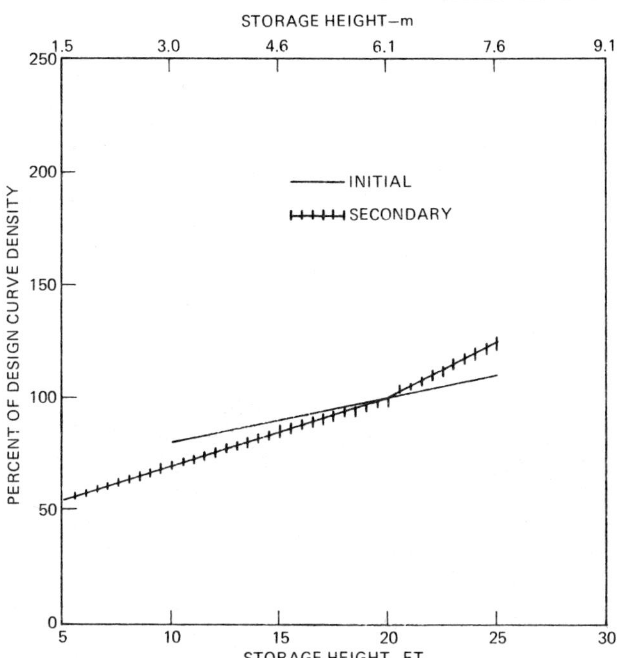

FIG. 11-1M. Height adjustment for Group A plastic.

231C. Should the aisles be, for instance, 3½ ft (1.07 m) or narrower, the storage would have to be classified as "multiple row racks," requiring a higher ceiling sprinkler density. Storage in the aisles in a huge automotive parts warehouse converted it from a double row rack configuration to multiple row racks, thereby making the designed sprinkler coverage inadequate in a disasterous fire.

Variable No. 3—In-Rack Sprinklers: Within racks, there are sizable unsprinklered areas that defy the ability of ceiling sprinklers. In-rack sprinklers were conceived to provide a degree of wetting for those concealed areas, thus interrupting the otherwise free spread (horizontally and vertically) of fire. The higher the rack becomes, the more in-rack sprinklers become essential. In-rack sprinklers are of ordinary temperature—130 to 170°F (54.4 to 76.7°C)—rating, have ½ in. (12.7 mm) orifices, and are either pendent or upright, with water shields to avoid cooling by water spray from any in-rack sprinklers that may be above. (See Fig. 11-1N.)

For rack storage containing solid or slatted shelves of any height, sprinklers are needed beneath each shelf unless transverse flue spaces at least 6 in. (152 mm) wide

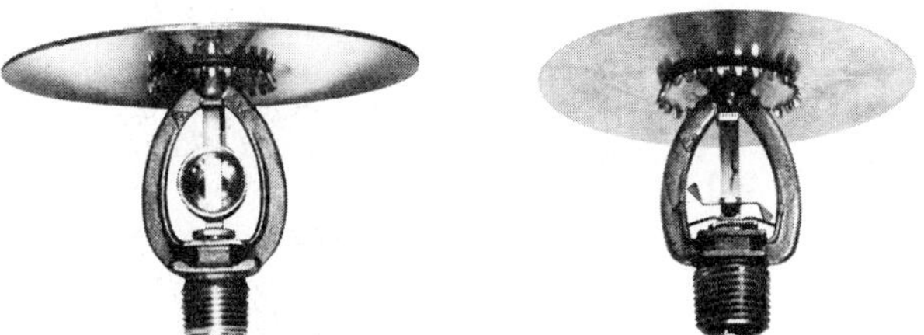

FIG. 11-1N. Typical in-rack sprinklers. Upright standard response (left) and upright fast response (right). Pendent sprinklers can be field equipped with water shields. The fast response type is also appropriate for protection under open gridded catwalks between racks, shelves or bins. (Grinnell Fire Protection Systems Co., Inc.)

are maintained. (An exception is single row racks up to 25 ft or 7.6 m in height.) Otherwise, the rules for open racks apply.

In-rack sprinklers are not needed if an automatic high expansion foam system is used in conjunction with ceiling sprinklers. In-rack sprinklers may also become less important in time if the goals of a new research program are achieved. Early tests indicate that an Early Suppression/Fast Response sprinkler (ESFR) provides faster and better water penetration from ceiling sprinklers in certain moderate height rack storage situations. The ESFR sprinkler, which has a more sensitive sensing element, and an extra large orifice and deflector, will produce larger water drops that quickly pierce the fast rising plume of hot fire gases while the fire is still small, an advantage in certain problem cases. (See also Variable No. 11—Sprinkler Orifice Size.)

Storage Up to 25 ft (7.6 m): For rack storage arrangements without solid shelves up to 20 ft (6.1 m) high, in-rack sprinklers are optional for most conditions, but their presence appreciably reduces the ceiling sprinkler density requirement. If in-rack sprinklers are installed at only one level, they are located at the tier level nearest to one-half to two-thirds of the storage height and are arranged longitudinally in the center of the rack. They become practical when storage height is at least 15 ft (4.6 m).

For the higher range of storage heights under 25 ft (7.6 m), one level of in-rack sprinklers is usually mandatory in double row racks and in multiple row racks, except that two levels are needed when Class IV storage is over 20 ft (6.1 m) high.

Sometimes racks are introduced into an existing building having an ordinary hazard ceiling sprinkler system. Installing one or more levels of in-rack sprinklers may relieve the necessity of upgrading the ceiling sprinkler system.

Storage Over 25 ft (7.6 m): Ceiling sprinkler discharge can effectively wet exposed storage at the aisles only for the upper 15 ft (4.6 m) or so of the rack; thus, for rack storage over 25 ft (7.6 m) high, the lower faces of the rack are protected by "face" sprinklers. They are like other in-rack sprinklers, but located at transverse flues and positioned within 18 in. (0.46 m) of the aisle. They protect against vertical spread of fire over the face of the storage and assist in controlling fire in the racks. Various arrangements of face sprinklers and centrally located in-rack sprinklers have been devised, including the use of horizontal barriers. (See Fig. 11-10 for examples of arrangements for Class I, II, and III commodities.) The arrangements prescribed can be repeated indefinitely as pile height increases.

Horizontal barriers are solid barriers such as sheet metal or wallboard, which cover the entire length and width of the rack, including all flue spaces. Their effect, when equipped with in-rack sprinkler coverage below, is to reduce the number of centrally located in-rack sprinklers needed. The barriers have the same restrictive effect on vertical fire spread as a full complement of face and central in-rack sprinklers at the same level.

Variable No. 4—Temperature Rating of Ceiling Sprinklers: Generally, in storage occupancies of low height, ordinary temperature (135 to 170°F or 57 to 76°C) sprinklers are used, subject to higher temperature ratings

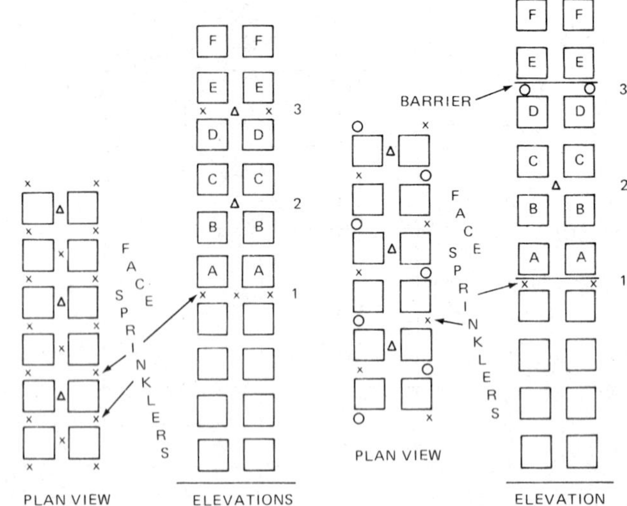

NOTES:
1. SPRINKLERS LABELED 1 REQUIRED WHEN LOADS LABELED A OR B REPRESENT TOP OF STORAGE.
2. SPRINKLERS LABELED 1 AND 2 REQUIRED WHEN LOADS LABELED C OR D REPRESENT TOP OF STORAGE.
3. SPRINKLERS LABELED 1 AND 3 REQUIRED WHEN LOADS LABELED E OR F REPRESENT TOP OF STORAGE.
4. FOR STORAGE HIGHER THAN REPRESENTED BY LOADS LABELED F, THE CYCLE DEFINED BY NOTES 2 AND 3 IS REPEATED, WITH STAGGER AS INDICATED.

FIG. 11-10. In-rack sprinkler arrangement for Class I, II, or III commodities with height of storage over 25 ft (7.6 m). Note how the presence of horizontal barriers within the racks (right) reduces the number of in-rack sprinklers required (left). The symbols O, Δ or X indicate sprinklers installed in a vertical or horizontal alternating or "staggered" arrangement on the sprinkler piping.

for high ambient temperature conditions.* Some engineers recommend intermediate—(175 to 225°F or 79 to 107°C) or high temperature (250 to 300°F or 121 to 149°C)—sprinklers to reduce the number of sprinklers operating unnecessarily beyond the actual fire when the commodity may be expected to produce a rapid rate of heat release, which is often the case.

In high palletized or solid piled storage, or rack storage up to 25 ft (7.6 m), high temperature sprinklers can reduce a sprinkler operating area by 40 percent, as compared with ordinary or intermediate temperature sprinklers. Somewhat surprisingly, the situation is reversed for rack storage over 25 ft (7.6 m). In a series of fire tests, in-rack sprinklers operated prior to ceiling sprinklers, indicating that in the over 25 ft (7.6 m) racks, in-rack sprinklers convert what normally would be a rapidly developing fire, from the standpoint of ceiling sprinklers, to a slower developing fire with a lesser degree of heat release. This result differed from tests involving 20 ft (6.1 m) high storage; in those tests ceiling sprinklers operated before the in-rack sprinklers. Therefore, in over 25 ft (7.6 m) rack storage, ordinary temperature ceiling sprinklers are preferred, as they require a distinctly lower density than high temperature ceiling sprinklers. Example: with

*Automatic sprinklers have six different temperature ratings: ordinary, 135 to 170°F (57 to 76°C); intermediate, 175 to 225°F (79 to 107°C); high, 250 to 300°F (121 to 149°C); extra high, 325 to 375°F (163 to 190°C); very extra high, 400 to 475°F (204 to 246°C); and ultra high, 500 to 575°F (260 to 301°C).

TABLE 11-1C. Storage Over 25 ft (7.6 m) High

Commodity Class	Encapsulated	Double Row Racks Without Solid Shelves, Aisles Wider Than 4 ft (1.2 m): Ceiling Sprinklers Operating Area sq ft	m²	Ceiling Sprinkler Density gpm/sq ft 165°F	286°F	(L/min)/m² 74°C	141°C	Multiple Row Racks: Ceiling Sprinklers Operating Area sq ft	m²	Ceiling Sprinkler Density gpm/sq ft 165°F	286°F	(L/min)/m² 74°C	141°C
	NO	2000	186	0.25	0.35	10.2	14.3	3000	279	0.25	0.35	10.2	14.3
I	YES	2000	186	0.25	0.35	10.2	14.3	3000	279	0.31	0.44	12.6	17.9
	NO	2000	186	0.30	0.40	12.2	16.3	3000	279	0.30	0.40	12.2	16.3
I, II, III*	YES	2000	186	0.30	0.40	12.2	16.3	3000	279	0.37	0.50	15.0	20.4
	NO	2000	186	0.35	0.45	14.3	18.3	3000	279	0.35	0.45	14.3	18.3
I, II, III, IV†	YES	2000	186	0.35	0.45	14.3	18.3	3000	279	0.44	0.56	17.9	22.8

* Good in rack sprinkler arrangement.
† Not as good in rack sprinkler arrangement.

Class IV commodities, a density of 0.35 gpm per sq ft [14.3(L/min)/m²] is required for ordinary temperature sprinklers versus 0.45 gpm per sq ft [18.3(L/min)/m²] for the high temperature type. (See Table 11-1C.)

Variable No. 5—Clearance Below Sprinklers: Although a minimum clearance of 18 in. (0.46 m) is required below sprinkler deflectors to allow a good spread of water discharge, there is a limit to how much clearance is desirable over high piled or rack storage. Beyond approximately 4½ ft (1.37 m), the finer droplets tend to remain suspended in the air or be carried away laterally by the updraft of flames and hot gases. Thus, there are two conflicting variables; under a given roof height, higher storage creates a more severe fire potential, while lower storage creates greater difficulty for sprinklers if the clearance starts to exceed 4½ ft (1.37 m). The two variables may not exactly neutralize each other, so the worst condition of variable storage, including a sloping roof condition, needs to be explored by the designer. The standards reduce the design criteria within the acceptable limits where lesser clearances are provided. (See Fig. 11-1P.)

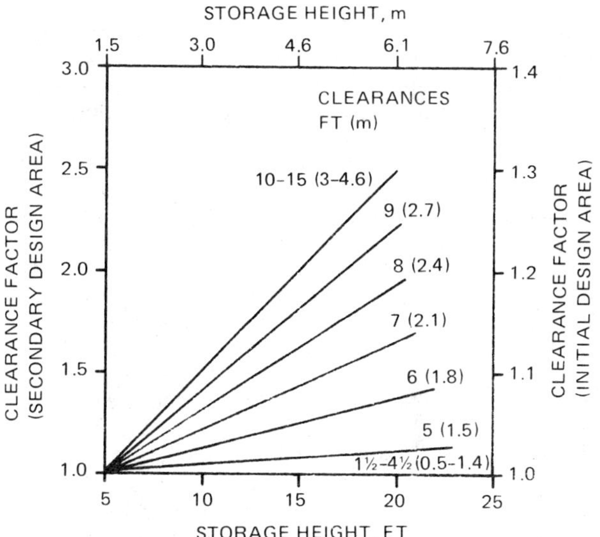

FIG. 11-1P. Clearance adjustment for piled Group A plastic.

Variable No. 6—Dry Pipe versus Wet Pipe Sprinkler Systems*: Where dry pipe systems are used, the delay in water discharge from the earliest operating ceiling sprinklers will allow heat to spread and open a larger number of sprinklers beyond the immediate fire than would be the case with a wet pipe system. To compensate for this, the design area of sprinkler operation is usually increased by 30 percent for dry pipe systems. The density should be selected so that the design area, after the 30 percent increase, does not exceed the upper area limit given in the design curve, such as the curves in Figure 11-1K.

Variable No. 7—Pile Stability: Instability of piles often is undesirable, since it (1) fosters collapse into the aisles, (2) provides a bridge for fire to cross over, (3) impedes fire fighting operations, and (4) possibly endangers building walls. However, tests show that in some cases, sprinklers are more effective after some collapse, spillage of contents, or leaning of stacks across flue spaces has taken place. The favorable effect of a collapse in these situations is evidently in the choking off of the upward flow of fire in flue spaces and exposing some otherwise concealed fire to the water from sprinklers.

Under fire conditions, pile stability is difficult to anticipate, but some guidelines are available. Compartmented cartons containing stiff cardboard dividers were found to be stable under test fire conditions, while noncompartmented cartons tended to be unstable. Storage on pallets within height limits suitable for the commodity to withstand compressive forces, and items held in place by materials which do not deform readily under fire conditions, are both examples of stable storage. If there are leaning stacks, crushed bottom cartons, or reliance on combustible bands for stability, pile instability under fire conditions can be predicted.

The sometimes beneficial effects of pile collapse

*In a wet pipe sprinkler system the piping to which the automatic sprinklers are attached contains water under pressure at all times. In a dry pipe system the piping contains air or nitrogen under pressure; when a sprinkler operates the pressure is reduced, a "dry pipe valve" is opened, and water enters the piping to flow out of any opened sprinklers.

should not, generally, lead one to purposely plan for it. Racks, in particular, should have adequate cross bracing, anchorage at floors, and tight connections, as rack collapses tend to be chain reactions.

Variable No. 8—Encapsulation: A packaging method consisting of a thin plastic sheet such as polyethylene completely enclosing the sides and top of a pallet load containing a commodity package or packages. The plastic covering apparently keeps some of the water from reaching fire that has penetrated a pallet load from below, which is likely to happen in rack storage. Tests show that a higher ceiling sprinkler density is needed for rack storage at all heights, even with in-rack sprinklers, where there is encapsulation. The adverse effect of encapsulation has not yet been appreciably noted in palletized storage pile tests. Standards require no difference in the ceiling sprinkler density, with or without encapsulation, in nonrack storage. To convert an encapsulated pallet load to a non-encapsulated load, the top covering portion of plastic only needs to be removed.

Variable No. 9—Smoke Venting: Smoke removal is important to manual fire fighting and overhaul (mopping up). Ideally, ventilation operations in a well protected warehouse should be deferred until automatic sprinkler operation has reduced all temperatures in the building to ambient, which should occur within 30 minutes of ignition. However, often this is difficult advice for fire fighters who have learned that ideal results are not always attained. To merely pump into the sprinkler system and stand by, waiting for the building to cool down, represents an uncertain situation for the fire department. Hence, fire fighters will most likely make ventilating holes in the roof, and enter the building to put small or large hose streams on the fire.

Automatic Roof Vents: Fire protection engineers still are debating the value of automatic roof vents in sprinklered buildings which, when actuated by temperature, will open and allow smoke and hot gases to escape near the seat of the fire. Tests show they are quite effective in unsprinklered buildings. NFPA 204M, *Guide for Smoke and Heat Venting*, provides guidance. On the other hand, the operation of automatic sprinklers tends to interrupt the operation of the vents, and to spoil the stack effect that causes the hot smoke to escape. In fact, the initial effect of sprinkler operation often is to beat down smoke, obscuring vision within the building.

Because the fire tests upon which they are based did not employ smoke venting, NFPA storage standards do not require smoke removal facilities. However, the standards do comment that venting through eaveline windows, doors, monitors, or gravity or mechanical exhaust systems is essential to smoke removal after the fire is controlled.

Variable No. 10—High Expansion Foam: While automatic high expansion foam extinguishing systems can be an independent means of fire suppression, there is a reluctance to use them as the sole means of automatic fire control. They are expensive, complicated (relative to sprinkler systems), do not protect the roof structure until foam reaches that level, involve the entire contents of a protected area regardless of how small the fire may be, and present a problem of foam residue removal after discharge. However, they have found acceptance used with automatic sprinklers for certain high challenge storage occupancies such as storage facilities for rubber tires, roll paper, and exposed plastics, where excessive foam contact with the stored matter is not a serious problem.

A high expansion foam system uses a series of foam makers at ceiling level. When actuated by the fire detection system throughout the area, the foam system causes a spray of water mixed with special foam concentrate in each foam-maker to strike a screen while a fan blows through it. Many bubbles of uniform size are produced, which cascade down into and gradually fill the warehouse area. The foam has an expansion ratio of up to 1,000 to 1, flashes to steam on contact with burning material, and engulfs other material, keeping it from burning. Building doors automatically close when the system is actuated. NFPA 11A, *Standard for Medium and High Expansion Foam Systems*, contains the requirements for installation of these systems.

When high expansion foam systems are used in combination with ceiling sprinkler systems in general storage facilities, the sprinkler discharge density may be reduced to one-half the density specified otherwise, but not less than 0.15 gpm per sq ft [6.1 (L/min)/m^2] for Class I through IV solid piled or palletized commodities, including idle pallets or plastics, and not less than 0.24 gpm per sq ft [9.8 (L/min)/m^2] for rubber tire storage or 0.25 gpm per sq ft (10 [L/min)/m^2] for roll paper storage. In rack storage, in-rack sprinklers are not required when high expansion foam is installed, and ceiling density can be reduced to 0.2 gpm per sq ft [8.1 (L/min)/m^2] for Class I, II, and III commodities and to 0.25 gpm per sq ft [10.2 (L/min)/m^2] for Class IV commodities.

Variable No. 11—Sprinkler Orifice Size: Theoretically, any required sprinkler discharge density can be achieved with any sized automatic sprinkler that is normally used [½ in. (12.7 mm) or larger orifice] by coordinating the sprinkler spacing and pressure available at the sprinkler. In piping system design, however, when the higher values of density are needed, a larger than ½ in. (12.7 mm) orifice can be beneficial. For example, for densities of 0.40 gpm per sq ft [16.3 (L/min)/m^2] and higher, better results are obtained with large orifice—17/32 in. (13.5 mm)—sprinklers and 70 to 100 sq ft (6.5 to 9.3 m^2) spacing, than when using ½ in. (12.7 mm) orifice sprinklers at 50 sq ft (4.6 m^2) spacing. Substituting the 17/32 in. (13.5 mm) for a ½ in. (12.7 mm) sprinkler results in a delivery of 40 percent more water under the same pressure, with less of the weakly penetrating fine spray.

The NFPA storage standards indicate preferences for sprinkler orifice sizes as follows:

NFPA 231—½ in. (12.7 mm) implied, with economy dictating 17/32 in. (13.5 mm) for higher densities.

NFPA 231C—Densities and areas of discharge are based on tests using ½ in. (12.7 mm) and 17/32 in. (13.5 mm). If "large drop" sprinklers with 0.64 in. (16.25 mm) orifice are used, then instead of densities and areas, numbers of operating sprinklers with minimum sprinkler operating pressure are specified. (See Table 11-1D.) The advantage of large drop sprinklers is primarily in their use for Group A plastics.

NFPA 231D—½ in. (12.7 mm) or 17/32 in (13.5 mm) implied.

NFPA 231E—½ in. (12.7 mm) or 17/32 in (13.5 mm) implied.

TABLE 11.1D. Pressure and Number of Design Sprinklers For Various Hazards (Large Drop Sprinklers)

	Number Design Sprinklers			Sprinkler Temperature Rating
Minimun Operating Pressure (Note 1)	@ 25 psi	@ 50 psi	@ 75 psi	
Hazard (Note 2)				
Double Row Rack Storage with Minimum 5.5 ft aisle width (Note 4) having:				
Class I and II Commodities up to 25 ft with maximum 5 ft clearance to ceiling	20	Note 3	Note 3	High
Class I, II and III Commodities up to 20 ft with maximum 10 ft clearance to ceiling	15	Note 3	Note 3	High
Class IV Commodities up to 20 ft with maximum 10 ft clearance to ceiling	Does Not Apply	20	15	High
Non-expanded Group A plastics in corrugated cartons up to 20 ft with maximum 10 ft clearance to ceiling	Does Not Apply	30	20	High
Non-expanded Group A plastics in corrugated cartons up to 20 ft with maximum 10 ft clearance to ceiling	Does Not Apply	20	Note 3	Ordinary
Non-expanded Group A plastics in corrugated cartons up to 20 ft with maximum 5 ft clearance to ceiling	Does Not Apply	15	Note 3	High

Notes:

1. Open Wood Joist Construction. Testing with open wood joist construction showed that each joist channel should be fully fire-stopped to its full depth at intervals not exceeding 20 ft. In unfirestopped open wood joist construction, or if firestops are installed at intervals exceeding 20 ft, the minimum operating pressures should be increased by 40 percent.

2. Building steel required no special protection for the occupancies listed. Protection requirements are based on rack storage with no solid shelves nor slave pallets.

3. The higher pressure will successfully control the fire, but the required number of design sprinklers should not be reduced from that required for the lower pressure.

4. In addition to the transverse flue spaces required by NFPA 231C, minimum 6 inch longitudinal flue spaces were maintained.

For SI Units: 1 ft = .3048 m; 1 in = 25.4 mm; 1 psi = 6.895 kPa.

NFPA 231F—17⁄32 in. (13.5 mm) required for new installations.

Variable No. 12—Sprinkler Response Time: The sensitivity of a conventional automatic sprinkler to ambient heat varies somewhat with sprinkler height above the fire, closeness to the ceiling, individual design, horizontal air currents, etc. Nevertheless, until recently, sprinkler response time has not been considered in sprinkler system design. It became a critical consideration in residential sprinkler systems in the late 1970s, and now, early tests in the ESFR research program indicate it may be possible to control or even extinguish very difficult storage fires with ESFR sprinklers by achieving adequate water penetration (Actual Delivered Density) early in the fire growth. (See Variable Nos. 3 and 11.)

Requirements have been cited from various NFPA storage standards merely to illustrate some of the statements included under "Variables" and to show how they enter into warehouse fire suppression system design. With an understanding of the variables, the standards themselves, containing comprehensive requirements, graphs, tables, and explanatory indices, should be pursued in detail as needed.

FIRE PROTECTION FOR WAREHOUSES

While automatic sprinklers are considered the main line of fire defense in storage occupancies, they are not the only needed method of protection. Hose systems, portable extinguishers, and manual fire fighting operations also have a role in protecting storage occupancies, and each is discussed in the following pages.

Automatic Sprinkler Systems

For warehouse ceiling sprinkler systems, the most important design consideration is, of course, a required "Sprinkler Discharge Density" at a corresponding "Design Area of Sprinkler Operation." (See Figs. 11-J and 11-1K.)

The storage standards provide charts containing curves from which a density/area point can be selected. For piled Group A plastics, however, *two* points are selected from a curve. (See Fig. 11-1Q.) One is for the "initial density," the other for the "secondary density," when the storage is in solid piles or palletized piles. The

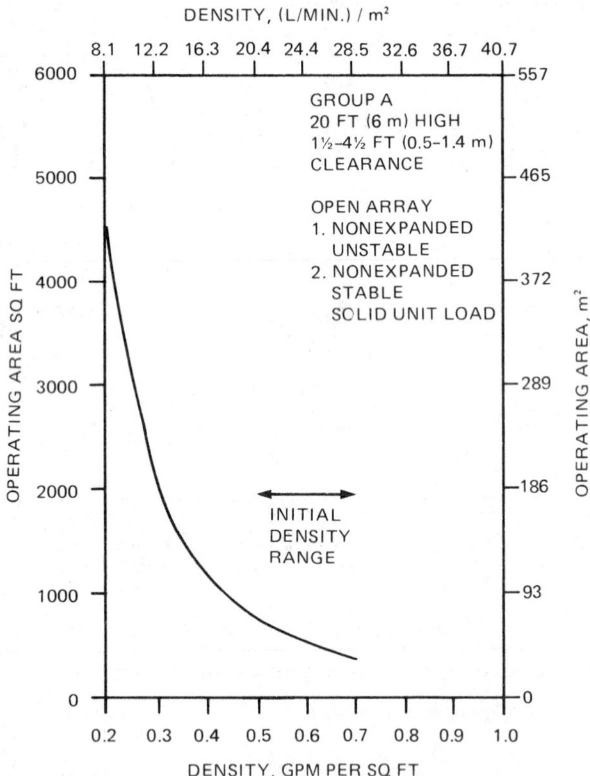

FIG. 11-1Q. An example of a density/area curve indicating initial and secondary densities for piled group A plastics.

initial density is appreciably higher, but only discharges over a smaller area, as would be expected when the first few sprinklers are operating, directly over the fire.

For rack storage of plastics, density/area curves are not used; rather, specific density/areas are stipulated for particular conditions. Deviations from the conditions require well considered deviations from the stipulations. (See Table 11-1E.)

Water Supply

The fire service water supply at a warehouse must satisfy design requirements for ceiling sprinklers, in-rack sprinklers (if any), hydrants, and standpipe and hose systems for indoor use. An adequate water supply will have the capacity to supply these protective system components and still have sufficient remaining pressure in the mains to guarantee sufficient operating pressure at the sprinklers.

Multiplying the needed ceiling sprinkler density by the design area of the sprinkler operation produces a flow requirement in gpm (L/min) to which approximately 10 percent is added to compensate for unavoidable inequalities in the sprinkler system. To the ceiling sprinkler demand is added the in-rack sprinkler demand, if any, based upon operation of a number of the most hydraulically remote in-rack sprinklers. That number depends upon the number of sprinkler levels and the class of commodity involved.

After adding ceiling and in-rack sprinkler water demands, another 500 gpm (1894 L/min) is added for large and small hose streams without altering the residual pressure requirement, under the assumption that they will be used concurrently with sprinkler system operation.

The total water supply estimate can be expressed as a flow rate at a corresponding minimum residual pressure. That pressure is affected by the sprinkler system piping design, but in general at least 30 to 50 psi (207 to 345 kPa) pressure is needed at the ceiling level distribution point or top of the riser. When greater pressure is available, more economic system design is possible, except that for standard orifice sprinklers it is undesirable to have more than 60 psi (414 kPa) at the remote end sprinkler, as it is believed to cause excessive ineffective ultrafine spray in the system discharge when ½ in. (12.7 mm) sprinkler orifices are used.

TABLE 11-1E. Group A Plastics in Cartons in Single and Double Row Racks with 8 ft Aisles, Height of Storage Up Through 25 Ft – 165°F Sprinklers

	Ceiling	In-Rack
Storage Height (ft)/ Clearance (ft)	Density (gpm/ft²)/ Area of Application (ft²)	Sprinklers Needed (Note 2)
15/3	0.45/2000	None
	0.30/2000	One Level
20/3 or 15/10	0.60/3000	None
	0.45/2000	One Level
20/10 or 25/3	0.45/2000	One Level
	0.30/2000	Two Levels
25/10	0.30/2000 (Note 3)	Two Levels

NOTES

1. For 4 ft aisles a density of 0.60 gpm/ft² and an area of application of 1500 ft² should be used. For aisle widths between 4 ft and 8 ft a direct linear interpolation may be made between densities and areas of application.

2. Based on use of 1/2 in. or 17/32 in. orifice sprinklers at 15 psi operating pressure. Sprinklers spaced in the longitudinal flue or an 8 ft spacing with 6 in clearance between the top of stock and the sprinkler deflector.

3. Based on maximum of 5 ft of storage above the top level of in-rack sprinklers. For storage greater than 5 ft up to maximum of 10 ft, increase density to 0.45 gpm/ft².

For SI Units: 1 in = 25.4 mm; 1 ft = .3048 m; 1 psi = 6.895 kPa; 1 gpm = 3.785 L/min; 1 gpm/ft² = 40.74 L/min/m².

The water supply has to be available for a sufficient time (usually two hours), for full fire control by sprinklers, any necessary manual fire fighting, and mop up operations. The reliability requirement for a water supply at a high value warehouse, for example, can lead to a need for a multiple supply, i.e., a good fire pump-suction tank private supply may back up an adequate public supply, or there may be two similar private supplies. Such redundancy is based on good judgment; obviously it is not prudent for a $50 million warehouse to depend entirely upon a connection to a single city water main, no matter how powerful that supply, because of the possibility of its being out of service for some time.

The layout of yard mains, tanks, pump houses, power supply to pumps, hydrants, and valves, warrants qualified fire protection engineering attention in the planning stage. In correcting an existing water supply deficiency, alternate methods can be considered, and the variables discussed earlier in this chapter offer such alternatives. Adding a booster fire pump may be an economical alternative when many sprinkler systems on the property need, and will benefit from, the higher pressure. On the other hand, if just one sprinkler system out of many is deficient, it may be more economical to (1) upgrade the piping layout only in that system, (2) replace ceiling sprinklers for that system with ones of a different temperature rating, (3) install in-rack sprinklers in that area, (4) lower pile heights in that area (not a good long range solution), or (5) a combination of methods.

Large and Small Hose

Warehouses of moderate area close to city hydrants present no special fire suppression problem. But when the smaller dimension of a warehouse exceeds about 200 ft (61 m), hose lays from public hydrants to the far side of the building can be difficult. Private hydrants then become important around the perimeter with access doors to the warehouse near them. Private hydrant spacing of 250 ft (76 m) is common for such situations. Guidance on number and location of hydrants can be found in NFPA 24, *Standard for the Installation of Private Fire Service Mains and Their Appurtenances.* Indoor small hose is recommended in high piled or rack storage buildings for use by occupants on incipient fires. Their relatively good reach and long lasting supply make them more effective than portable fire extinguishers on high storage fires, but extinguishers are also needed.

Portable Fire Extinguishers

NFPA 10, *Standard for Portable Fire Extinguishers,* provides guidance in selecting types and sizes of fire extinguishers and locating them properly. But one caution is in order: There is a tendency to provide dry chemical (ABC type) fire extinguishers everywhere because they are all purpose units. Such extinguishers have a limited range, or reach, and time of discharge. For example, the 20 lb (1 kg) ABC dry chemical unit (about as heavy as can be handled by the average person) has a maximum effective horizontal range of approximately 20 ft (6.1 m) and discharges within 25 seconds. A 2½ gal (9.5 L) water type unit lasts for one minute with an effective range of up to 40 ft (12.2 m), providing a distinct advantage in most warehousing situations, assuming Class BC and units (for flammable liquid and electrical fires) are provided where needed.

Fire Fighting

Even with adequate fire control by sprinklers or other mechanical means, final manual extinguishment of storage fires is not always simple. Sprinkler discharge tends to decrease visibility by the presence of the downpour and the smoke driven down from the ceiling. (See Variable No. 9, Smoke Venting.) In the past, some ill advised fire officers, upon arrival at the scene, ordered sprinklers shut off to assist in getting at the seat of the fire; however, the necessity of working with sprinklers operating, is fortunately now the recognized standard fireground procedure.

Other manual fire extinguishing problems involve getting at residual fire in the upper reaches of high racks, gaining access through aisles blocked by collapsed storage, and removing burning items such as fiber bales or rubber tires from the building, that cannot be fully extinguished while in the original, still smoldering pile. A well planned system of fire protection should simplify the manual fire fighting tasks. Automatic alarms to the fire department from operating sprinkler systems, as well as adequate hydrants, 24 hour emergency access, audible/visual indication of which sprinkler system is operating, and clear identification of sprinkler valves, hydrants, interior small hose stations, and siamese pumper connections all contribute to successful extinguishment. Prefire inspections and planning should be encouraged.

Plant Emergency Organization

A warehouse needs some organization for emergencies. Even if a small building's procedure is merely to designate an individual to notify the public fire department and use a fire extinguisher, a decision must be made on identifying the individual and how the tasks are to be performed. Larger properties frequently warrant inplant assignments for communication, using extinguishers and small hose, checking the open condition of valves, operating the fire pumps, advising responding fire fighters, salvage, and more—all duties of a private fire brigade. (NFPA 27, *Recommendations for Organizing, Training and Equipment of Private Fire Brigades,* gives guidance on the organization, duties, and training of such brigades.)

Fire Prevention

Some commodities such as baled fibers are easily ignited, sometimes merely by friction sparks. Goods in cartons need a substantial or prolonged ignition source, unless inadequate housekeeping permits scrap materials to accumulate where they can easily be ignited and act as kindling for the stored items.

The more common causes of fires in storage are from smoking, electrical causes, industrial trucks (gasoline, LP-Gas, diesel or electrical power) and sparks from flame cutting and welding. Steel racks preferably are assembled by bolts rather than welding, to avoid flame cutting during possible future disassembly. Cutting and welding sparks are prone to start a fire out of range of convenient portable fire fighting equipment.

Arson fires pose a special problem and are combatted by superior automatic sprinkler protection and appropriate security measures.

SPECIAL OCCUPANCIES

A number of commodities have been studied separately because of the fire properties associated with their physical form and method of storage. In addition to the brief descriptions of their fire characteristics which follow, a sampling of sprinkler system density requirements for 15 ft (4.6 m) high storage is shown in Table 11-1F.

Rubber Tires

Fires in rubber tires are infrequent, but are very hot and smoky, and difficult to control and extinguish. High sprinkler densities are needed to tame the fire and protect the building structure; see NFPA 231D.

Tires are piled solid, in compact portable racks referred to as pallets, and in racks. (See Figs. 11-1R and 11-1S.) Since a tire is not packaged, it contributes its own circular air space to the storage, and this forms considerable horizontal or vertical flues. The interiors of tire carcasses can flame vigorously, mostly out of reach of water from sprinklers. Final extinguishment involves the laborious application of water to individual tires, usually by removing them from the building.

In combination with sprinklers, high expansion foam is very effective on rubber tires, and causes minimal water damage or contamination. To obtain full penetration to the inner reaches of the tires, it is advisable to allow an additional hour of foam soaking time after sprinklers are shut off, to maintain the foam level.

Roll Paper

Roll paper is stored on its side or, more commonly, on end. It might be assumed that the latter arrangement would permit easier paths for sprinkler water to penetrate the vertical air spaces and slow the fire; experience with roll paper shows otherwise.

In storage on end, peeling or delamination during a fire is a major problem which can be abated somewhat by metal banding, steel baling wire applied tightly by hand, fire retardant treated tight paper wrappers covering ends as well as sides, or close spacing between columns of rolls. Effective close spacing would occur where stacks or columns are less than 4 in. (101 mm) apart.

NFPA 231F prescribes sprinkler density area requirements according to the arrangement of rolls, whether banded or unbanded, height of piles, and weight of paper, i.e., heavy, medium, or light weight. Light weight paper and tissue are not covered, however.

Carpet Storage

Rugs or carpeting in rolls are generally stored in racks that accommodate 12 to 15 ft (3.7 to 4.6 m) long rolls;

TABLE 11-1F. Protection for Various Specific Materials Stored 15 ft (4.6 m) High

	Sprinkler System Density		Temperature Rating of Head in °F‡	Additional gpm for Hose Streams§	Duration of Flow (hrs)	Area Adjustment for Dry Pipe	Reference Standard	Remarks
	gpm/ft²*	Area ft²†						
Rubber Tires								
Palletized & Fixed	0.48	3550	165	750	3	30%	NFPA 231D	
Rack Storage with Pallets	0.48	2100	286	750	3	30%	NFPA 231D	
Fixed Rack Storage without Pallets or Shelves	0.60	5000	286	750	3	30%	NFPA 231D	Water supply to meet both requirements
	0.90	3000	286	750	3	30%		
Baled Combustible Fibers	0.15	6000	212	500	4	None	FM 8-7	
Hanging Garments	0.60	3000	165	250	1	Avoid DP	FM 8-18	Max. 3 ft. (1.5 m) below sprinklers
Roll Paper								
Rolls on End, Close Spaced	0.30	2000	286	500	2	30%	NFPA 231F	Max. 5 ft (1.5 m) below sprinklers
Banded Rolls, 4 in. (100 mm), or more apart; heavy wt. paper	0.30	2500	286	500	2	30%	NFPA 231F	" "
Banded Rolls, 4 in. (100 mm), or more apart; med. wt. paper	0.45	2500	286	500	2	30%	NFPA 231F	" "
" "	0.60	2300	286	500	4–6	About 75%	FM 8-21	
Baled Waste Paper	0.20	3200	160	500–1000	4–6	35%	FM 8-22	Water supply
Baled Waste Paper	0.20	2700	212–286	500–1000	4–6	40%	FM 8-22	from curves
Carpets								
Cubicle Racks	0.25	3000	286	500	2	None	FM 8-30	
Baled wool	0.15	2500	212	500	4	None	FM 8-7	
Baled cotton	0.25	3000	165	500	2	30%	NFPA 231E	Water supply
Rack storage	0.33	3000	165	500	2	30%	NFPA 231E	from curves
Wood pallets	0.60	4500	286	500	2	None	NFPA 231	

Note: Full requirements are in the referenced standards. Above data is abstracted from the standards for comparison purposes.
* 1 gpm/ft² = 40.746 (L/min)/m²;† 1 ft² = 0.093 m²; ‡ 5/9 (°F−32) = °C; § 1 gpm = 3.78 L/min.

therefore, racks are at least that deep and sometimes placed back to back, forming a total width of 24 to 30 ft (7.3 to 9.1 m) between aisles. Of course, small carpet pieces and rolls may be in cartons and stored in regular solid or palletized piles or conventional racks.

Long carpet rolls would deflect if not supported along their length; hence, solid or slatted shelves are used in the wide racks. An exception would be for carpets individually encased in strong paperboard tubes that have sufficient bending strength to be supported in open style racks without shelving. Note that in-rack sprinklers are needed under each solid or slatted shelf in racks unless there are transverse flue spaces at least 6 in. (152 mm) wide (See Variable No. 3, In-Rack Sprinklers). With carpet racks having as many as 10 tiers, the amount of sprinklers and piping for in-rack sprinklers can be prohibitive. Therefore, protection focuses on providing a minimum 2 in. (51 mm) wide transverse flue at all vertical supports, which are normally 10 to 12 ft (3 to 3.7 m) apart, and providing a level of in-rack sprinklers, including face sprinklers for the wider racks, located about every third tier for storage over 12 ft (3.7 m) high. Tiers vary in height from 24 to 40 in. (0.61 to 1 m), so the sprinkler is installed in a tight area. (See Fig. 11-1T.) Some engineers recommend vertical transverse barriers at every third set of vertical supports instead of relying upon 2 in. (51 mm) wide transverse flues.

FIG. 11-1R. On tread storage of tires in open, portable racks. (Goodyear)

FIG. 11-1S. On side storage of rubber tires in palletized racks. (Goodyear)

With back to back racks, a 12 in. (0.3 m) space should be provided between vertical members, creating a longitudinal flue. If rolls tend to project into this space, toe plates or other means can be fastened to the racks to stop the rolls physically at the rear face. Unless there is a large quantity of carpet with attached expanded (foam) rubber backing, fire does not tend to spread rapidly in normal carpet rolls

FIG. 11-1T. Racks for the storage of carpeting on slatted shelves. The closeness of the tiers to each other plus the depth of the tiers complicate problems of providing in-rack sprinkler protection. (Clark Industrial Truck Div.)

along the length of the rack. Nevertheless, since the racks often extend for several hundred feet (100 ft = 30.5 m), a valuable restriction to fire spread is to provide clear spaces 8 ft (2.4 m) wide about 100 ft (30 m) apart along the length of the rack. This serves as an assist to in-rack sprinklers, or as a fire stop where no in-rack sprinklers are used, usually for storage under 12 ft (3.7 m) high.

Carpet Padding: Padding or backing, usually made of expanded plastic, expanded rubber, or jute generally is stored in carpet warehouses. These materials burn much faster than the carpeting, per se, and should be segregated for special arrangement and protection. Ceiling sprinkler density which is adequate for the warehouse as a whole, probably will not be adequate for the padding unless the storage is kept low, such as less than 5 ft (1.5 m) for expanded rubber or plastic.

Baled Fibers

The main hazard of baled cotton and other fibers of vegetable origin comes from the surfaces of the bales that have a multitude of exposed minute fibers. Fire flashes quickly over these vertical surfaces, as well as over any loose particles on the floor or lint on overhead piping and structure. Fire on one side of an automatic closing fire door can progress through the open doorway on loose floor scraps before the door can close automatically. Housekeeping, therefore, is especially important.

Fire also tends to penetrate between and into bales, requiring removal of burning bales from the building for extinguishment. Considerable smoke is emitted and complicates fire fighting. Certain fibers such as jute, are likely to swell when wet, so they should be piled with regard to stability in a fire, and with proper clearance from walls.

NFPA 231E limits heights of tiered or rack storage to 15 ft (4.6 m) and pile sizes to 700 bales, with a 12 ft (3.7 m) main aisle and 4 ft (1.2 m) cross aisles, in maximum fire divisions of 10,000 bales; the standard calls for large operating areas for sprinklers. Factory Mutual Standard FM 8-7 covers baled combustible fibers generally, including wool which burns much slower (FMEC 1974).

Ordinary dry chemical (BC type) fire extinguishers are very effective in knocking down surface fires on bales; however, they should be backed up by water spray from small hose or garden hose, or water type extinguishers equipped with spray nozzle attachments to extinguish smaller fires that may have penetrated into the bales. "Wet water" (a chemical agent additive to increase water's penetrating and spreading ability) should preferably be used.

Baled Wastes

Paper: Like baled fibers, baled waste paper is stored in solid piles into which fire tends to burrow. When paper is finely shredded, fire can flash over the surface of bales, similar to what happens with baled fiber. Waste paper becomes mushy and difficult to handle when wet; in fact, the integrity of the bales slowly disappears as the hose stream application progresses, making removal of burning bales difficult. Smoke can also be a troublesome problem. Baled paper sometimes is stored at great heights because of its low unit value, but fire protection standards do not apply to storage over 30 ft (9.1 m) high except in racks (FMEC 1977).

Pesticides

Pesticides are not only poisonous to insects and small animals, but are harmful, even fatal, to humans when ingested or breathed as products of combustion. They are stored as solutions in flammable or combustible liquids, as powders or granules in combustible packages, or as compressed gases for fumigation. Special attention must be given to the personnel hazard in addition to the commodity's combustibility. Pesticide fires can endanger persons fighting the fire or standing nearby, and the poisonous runoff from fire fighting can pollute water or soil in the area.

Automatic sprinkler protection is advisable, but provision should be made for safe accumulation and disposal of water runoff. Fire fighting strategy should include plans to use protective clothing and respiratory equipment, spray nozzles rather than straight streams to reduce water runoff, specialized medical back up services, and to avoid container breakup and pesticide dispersal.

NFPA 43D, *Code for Storage of Pesticides in Portable Containers*, gives general requirements for both inside and outside storage of pesticides.

SPECIAL STORAGE FACILITIES

Piers and Wharves

Warehouses often form the superstructure of piers and wharves. They may be referred to as "sheds," which, on piers, can project great distances over the water. They have the same general characteristics as other warehouses, but with more fire protection and suppression problems. Among these problems are reduced accessibility for land based fire department vehicles; the frequently combustible nature of the substructure, which can be exposed to floating burning debris or flammable liquids; the danger of water supply mains freezing; and the hazard of ships colliding with the pier. On the favorable side, commodities tend to be low piled because of the temporary transfer nature of the storage (ship to shore or vice versa).

At the Luckenbach Pier in Brooklyn, NY on December 3, 1956, fire and an explosion in stored Cordeau detonant fuse for explosives killed ten people and caused $7.6 million in damage. It illustrated the need for better limitations on hazardous materials, in this case contiguous with ordinary combustibles.

NFPA 87, *Standard for the Construction and Protection of Piers and Wharves*, requires a water supply for both hose streams and sprinkler systems to be available for at least 4 hours—about double the usual duration. To protect the underside of a combustible substructure, sprinklers should be either standard pendent sprinklers in the upright position, or old style sprinklers in the upright position, to wet the undersurface of the substructure.

It is desirable to limit the area between transverse fire walls in a pier, making these walls continuous with substructure fire walls, which in turn should extend to low water. The walls should be spaced not more than 450 ft (137 m) apart. Below a combustible substructure, transverse fire stops should be spaced not over 150 ft (45 m) apart.

Storage Garages

Motor vehicles, including automobiles and trucks in "dead storage" are housed in storage garages or outdoors. Because of their generally fastburning upholstery and gasoline or diesel oil content, fires in individual vehicles can be severe, but the overall building fire loading is moderate, evidently because the considerable amount of metal in the vehicles absorbs heat, and the average weight of combustibles per square foot (m^2) is not high.

According to NFPA 88A, *Standard for Parking Structures*, storage garages and enclosed parking structures (not open air parking structures) need automatic sprinkler protection if located in basements, in buildings over 50 ft (15.2 m) in height having combustible roof/floor assemblies, or inside or below a building used for another purpose (typically high rise apartments, hotels, or offices). In the latter case, an alternative to sprinklers is a supervised ionization type smoke detection system used together with a mechanical ventilating system capable of exhausting smoke. Dispensing of flammable liquid fuels from underground tanks sometimes is a feature inside such garages, but the pumps should be at street level within 50 ft (15 m) of a vehicle exit or entrance; (See NFPA 30, *Flammable and Combustible Liquids Code*.) Many municipal building codes require automatic sprinkler protection in noncombustible and fire resistive public parking garages that exceed certain areas and/or heights, so applicable codes should be checked.

Refrigerated Storage

Coolers and freezer storage buildings are primarily for foodstuffs, although certain antibiotics, pharmaceuticals, and unstable chemicals are also kept under refrigeration. To date, fire loss experience in refrigerated warehouses has been good during normal occupancy. It is during construction or temporary shutdown for repair that some devastating fires have occurred, usually from cutting and welding operations. The housekeeping is generally excellent, and smoking mostly nonexistent because smoke odor clings to frozen packages. Nevertheless, the potential for large loss is present in the containers, pallets, dunnage, waxy paper containers, and electrical equipment.

Insulation at walls and ceilings can spread fire quickly. Present practice involves prefabricated sandwich panels with insulation cores, lining sheets of expanded polystyrene, or foamed-in-place expanded polyurethane. Older warehouses, and some newer ones, use cork or other cellulose type insulation, or noncombustible insulation such as fibrous glass or expanded glass block. When an expanded plastic lining is used, it should be covered with an approved thermal barrier, which can be cement plaster on lath, gypsum plaster wallboard, fire retardant plywood, or an approved fire resistive inorganic spray material, a few of which are now listed for this purpose by Underwriters Laboratories (UL) and the Factory Mutual System. When a nonplastic combustible lining is used, it too should be covered by a thermal barrier if the occupancy does not require sprinklers.

The same sprinkler system design used for nonrefrigerated buildings is appropriate for this occupancy. However, the commodity hazard classification generally is lowered in freezer areas because of ice in the packages.

Isolated Storage Buildings

Some high value warehouses are needed for only a few years in construction projects in remote areas. At times such a warehouse can contain much electrical and electronic equipment, wire, cables, fixtures, pipe fittings, etc. Also there are experimental or mining sites that may have a large warehouse. These remote areas are often without public water mains or fire departments, making adequate private protection more costly. Factors have to be weighed, such as the criticality of the building relative to the project, and the time needed for replacement of vital contents. Standards and building codes do not rigidly apply in these situations. A suitable decision could be to separate unprotected warehousing into small units, to provide fully adequate private protection, or to compromise and use protective systems with a limited water supply. Any compromised physical protection should be accompanied by increased vigilance regarding guard service, housekeeping, maintenance, and smoking control.

Underground Storage

Large underground warehouses can be a boon to energy conservation, as rock caverns have a nearly constant temperature. In the U. S., some caverns have been used for compressed gas storage and some old mines for records storage (Holdraft 1985). Caves have been developed for warehousing, and include automatic sprinkler protection.

In general, the problems of underground structures include exits, venting of smoke and heat, and access for fire fighting. Automatic sprinkler protection can be installed under constructed floors and ceilings and is vital to protect combustible storage and associated personnel.

Air-Supported Structures

For warehousing in temporary or remote locations such as at construction sites, or as a low cost adjunct to more permanent conventional types of buildings, air-supported structures of considerable size are sometimes used, based on the convenience and economics of the risk involved. (See Figs. 11-1U and 11-1V.) These structures consist of a plastic coated fabric envelope which resembles a balloon and is kept in a rigid condition by low positive air pressure within. When used for warehousing such as palletized or rack storage, the structure may be equipped with loading dock arrangements such as air locks with electric roll up doors. Structures under 150 ft (46 m) in width or diameter may comply with NFPA 102, *Standard for Assembly Seating, Tents, and Air-Supported Struc-*

FIG. 11-1U. A typical air-supported structure that is suitable for warehousing purposes. (Air-Tech Industries, Inc.)

FIG. 11-1V. Interior of a 100 by 300 ft (30 by 91 m) air-supported warehouse, showing rack storage. (Air-Tech Industries, Inc.)

tures, which deals with wind resistance, strength, load distribution, and pressurization of air-supported structures. There is a trend toward using a cable harness net system to encapsulate the envelope in all directions, increasing stability during wind load (ASI 1977).

These structures do not support overhead piping or wiring, and so are unable to satisfy the usual need for automatic fire control systems. Therefore, they are not suitable for long term combustible storage of high value or high hazard. When there are compelling reasons to use these structures, it is preferable to have a number of separate smaller units rather than one large structure.

Construction

Any important building (warehouse or otherwise) merits durable construction that also resists wind and snow loads. Many construction types meeting these criteria are combustible or vulnerable to fire temperatures, but automatic sprinklers are virtually always needed to protect the storage contents and compensate for these hazards.

Buildings used for both manufacturing and warehousing should have a good barrier wall (preferably a true fire wall) between these components because the greater activity (hazard) in the former can expose the high values of the latter to the damaging effects of a fire. Incidental adjoining areas, such as boiler, machinery, or service rooms should be separated by fire partitions.

Steel columns that are within storage racks over 15 ft (4.6 m) high in which there are no in-rack enclosed sprinklers need to be "fireproofed," protected by one or two sidewall sprinklers, or to be under high density ceiling sprinkler discharge, in accordance with NFPA 231C.

Bibliography

References Cited

ASI 1977. "Air Structures Design and Standards Manual, ASI-77." Air Structures Institute, St Paul, MN.

FMEC. 1977. "Storage of Baled Waste Paper." *Loss Prevention Data.* Factory Mutual Engineering Corp., Norwood, MA.

FMEC. 1974. "Baled Fiber Storage." *Loss Prevention Data.* Factory Mutual Engineering Corp., Norwood, MA.

Holdcraft, R. L. 1985. "Fire Protection Criteria for Caves." *Fire Journal.* Vol 79, No 3. pp 35-37, 121.

NFPA Codes, Standards, Recommended Practices and Manuals. (See the latest *NFPA Codes and Standards Catalog* for availability of current editions of the following documents.)

NFPA 10, *Standard for Portable Fire Extinguishers.*

NFPA 11A, *Standard for Medium and High Expansion Foam Systems.*

NFPA 13, *Standard for the Installation of Sprinkler Systems.*

NFPA 13A, *Recommended Practice for the Inspection, Testing and Maintenance of Sprinkler Systems.*

NFPA 14, *Standard for the Installation of Standpipe and Hose Systems.*

NFPA 24, *Standard for the Installation of Private Fire Service Mains and Their Appurtenances.*

NFPA 27, *Recommendations for Organization, Training and Equipment of Private Fire Brigades.*

NFPA 30, *Flammable and Combustible Liquids Code.*

NFPA 43D, *Code for Storage of Pesticides in Portable Containers.*

NFPA 51B, *Standard for Fire Prevention in Use of Cutting and Welding Processes.*

NFPA 71, *Standard for the Installation, Maintenance and Use of Central Station Signaling Systems.*

NFPA 72A, *Standard for the Installation, Maintenance and Use of Local Protective Signaling Systems for Guard's Tour, Fire Alarm and Supervisory Service.*

NFPA 72B, *Standard for the Installation, Maintenance and Use of Auxiliary Protective Signaling Systems for Fire Alarm Service.*

NFPA 72C, *Standard for the Installation, Maintenance and Use of Remote Station Protective Signaling Systems.*

NFPA 72D, *Standard for the Installation, Maintenance and Use of Proprietary Protective Signaling Systems.*

NFPA 72E, *Standard on Automatic Fire Detectors.*

NFPA 87, *Standard for the Construction and Protection of Piers and Wharves.*

NFPA 88A, *Standard for Parking Structures.*

NFPA 204M, *Guide for Smoke and Heat Venting.*

NFPA 231, *Standard for Indoor General Storage.*

NFPA 231C, *Standard for Rack Storage of Materials.*

NFPA 231D, *Standard for Storage of Rubber Tires.*

NFPA 231E, *Recommended Practice for the Storage of Baled Cotton.*

NFPA 231F, *Standard for the Storage of Roll Paper.*

NFPA 505, *Fire Safety Standard for Powered Industrial Trucks Including Type Designations, Areas of Use, Maintenance and Operations.*

NFPA 601, *Recommendations for Guard Service in Fire Loss Prevention.*

NFPA 601A, *Standard for Guard Operations in Fire Loss Prevention.*

Additional Readings

"Aerosol Storage: The Problems and Solutions," *P8207,* Factory Mutual Engineering Corp., Norwood, MA.

Best, Richard, "$100 Million Fire in K-Mart Distribution Center," *Fire Journal,* Vol. 77, No. 2, Mar. 1983, pp. 36-42.

Clarke, Graham, "The Idle Pallet Fire Problem," *Fire Journal,* Vol. 66, No. 4, July 1972, pp. 98-101.

"Concern Over Warehouse Fires Leads to Tests on Aerosols," *Fire,* Jan. 1981, pp. 410-412.

Deacon, F. C., "Designing Fire Protection to Limit Monetary Loss," *SFPE TR 80-2,* Society of Fire Protection Engineers, Boston, MA, 1980.

"Designing to Limit Loss in High-Rise Rack Warehouses," *Kemper Group Report,* Vol. 10, No. 3, Sept. 1981, pp. 2-9.

Goring, G., "Sprinkler Protection for Storage Risks," *Fire Protection,* Vol. 8, No. 2, June 1981, pp. 20-25.

Harrington, J. L., and Hopkinson, R. B., *Rack Storage Protection,* Worcester Polytechnic Institute, Worcester, MA, 1977.

Johnson, R. S., "How to Protect Roll-Paper Storage from Costly Fires," *Inland Printer/American Lithographer,* Mar. 1968.

Linville, J. L., ed., *Industrial Fire Hazards Handbook*, "Industrial Storage Practices," National Fire Protection Association, Quincy, MA. 1984.

Patterson, C. B., "Powered Industrial Trucks: Appraising Their In-Plant Fire Safety," *Fire Journal*, Vol. 66, No. 5, Sept. 1972, pp. 103-104.

"Roll Paper Storage," *Loss Prevention Data Sheet 8-21*, Factory Mutual Engineering Corp., Norwood, MA.

Schirmer, C. W., "Meeting the High-Piled Storage Challenge with Standards," *Fire Journal*. Vol. 65, No. 6, Nov. 1971, pp. 61-66.

"Solid, Palletized, and Rack Storage of Plastics," *Loss Prevention Data Sheet 8-9*, Factory Mutual Engineering Corp., Norwood, MA.

"Stored Plastics—The New High Challenge Risk," *P7422*, Factory Mutual Engineering Corp., Norwood, MA.

"Warehouse Storage: Large-Scale Solutions for Large-Scale Problems," *P8324*, Factory Mutual Engineering Corp., Norwood, MA.

Young, R. A., and Nash, P., "The Fire Protection of Modern High Bay Storages," *Fire Prevention Science and Technology*, No. 18, Dec. 1977, pp. 4-13.

COLD STORAGE WAREHOUSES

Revised by Robert Hodnett

Cold storage warehouses are used primarily for extended storage of food products at temperatures that prevent or retard spoilage. Products such as pharmaceuticals, antibiotics, and unstable chemicals may also require refrigerated storage.

Depending upon the products or processes, warehouse temperatures range from −35°F (−37.2°C) for initial freezing up to 65°F (18.3°C). Despite low temperatures, cold storage warehouses are not immune to fire hazards. In fact, the low temperatures present unusual fire prevention and control problems which may be aggravated when warehouses are located outside municipal water distribution systems. Combustible materials in such warehouses include cork or expanded plastic insulation; wood dunnage, pallets, boxes, and baskets; fiberboard containers; paper and cloth wrapping; and grease impregnated materials. These present the potential for large monetary losses. Fire frequency, however, is relatively low in cold storage warehouses. About half the fires that do occur originate outside the refrigerated areas, damaging unprotected combustible insulation, structural components, and electrical equipment.

CONSTRUCTION OF COLD STORAGE WAREHOUSES

Refrigerated warehouses may be of combustible or noncombustible construction, or a combination of both. The current trend is to build larger and higher buildings in which automated storage racks are structural units supporting the walls and roof of the building. Such construction is acceptable if the rack structure will adequately support wind and snow loads. Although the structural materials themselves are noncombustible, such buildings must be adequately sprinklered. When large amounts of cork and/or foam insulation are ignited, they can burn out of control, generating temperatures high enough to distort steel and cause the structure to collapse.

Polyurethane foam, either in sheets or foamed-in-place, and polystyrene foam are two common insulation materials. When used in walls or ceilings, these materials should be protected by an approved thermal barrier or by a ½ in. (12.7 mm) coat of cement plaster on metal lath attached to the building framing. For polystyrene, the barrier also may be either ½ in. (12.7 mm) Type X gypsum wallboard or ¾ in. (19 mm) fire retardant plywood supported by studs or furring attached to the framing. These thermal barriers are required to be coated with a United States Department of Agriculture (USDA) approved washable finish to meet sanitary standards. If used in the floor, the insulation should be covered with concrete and the joint between floor and wall provided with a cove or curb. Automatic sprinklers should be provided below suspended ceilings, and the space above the ceiling should be protected against fire spread.

STORAGE ARRANGEMENTS

As in other warehouses, storage arrangements in refrigerated warehouses can be bulk, solid piling, palletized, or rack type, with bulk storage minimally used. One exception to this arrangement is the hook storage of meat and poultry products awaiting further processing. This storage method presents minimal fire hazard due to the low temperature and absence of combustible packaging.

Of the three commonly used storage arrangements, solid piling presents the least opportunity for fire development. In this arrangement, cartons, bales, and other packages are in direct contact with each other. Air spaces, or flues, exist only where there is imperfect contact. Palletized storage is limited in height by the ability of the palletized package to sustain the weight of additional packages. The construction of pallets, which are usually made of wood, provides horizontal flues that are usually out of reach of sprinklers. Rack storage is virtually unlimited in height. Vertical flues exist in rack storage at the vertical members and between unit loads. Like palletized storage, rack storage also has horizontal flues.

Mr. Hodnett is a consultant in fire protection and piping systems based in Providence, RI. He was staff liaison to the NFPA Water Extinguishing Systems Committees prior to his retirement from NFPA in 1984.

HOUSEKEEPING PRACTICES

Housekeeping in refrigerated warehouses is usually excellent due to the possibilities of contamination. As a result, fire loss is low during normal occupancy. It usually is during shutdown for repairs or during construction that a large fire loss occurs, most commonly from such hazardous operations as cutting or welding. Short circuits in electrical components that sometimes are damaged by materials handling equipment account for many fires in refrigerated warehouses.

FIRE PROTECTION

A cold storage warehouse should be protected by an adequate sprinkler system. Even though the occupancy might not require sprinkler protection, sprinklers should be installed to protect the insulation and the structure. Room flooding carbon dioxide or high expansion foam systems are not acceptable alternatives to automatic sprinklers. In chill rooms and coolers, sprinkler systems of the electrically operated preaction type are preferred to dry pipe systems. In sharp freezers or holding rooms, systems should combine deluge and dry pipe features; such a system is shown in Figure 11-2A. In this system, water pressure under the dry pipe valve is maintained through a ⅛ in. (3.2 mm) orifice in a bypass. The main water supply is held back by a deluge valve electrically actuated by a separate fire detection system.

Where high racks are used for storage, it is necessary to use in-rack sprinklers designed to reach spaces that cannot be reached by overhead sprinklers. Some storage racks have tubular uprights capable of carrying water which may be incorporated into the sprinkler system if the tube wall has sufficient strength to withstand 700 psi (4827

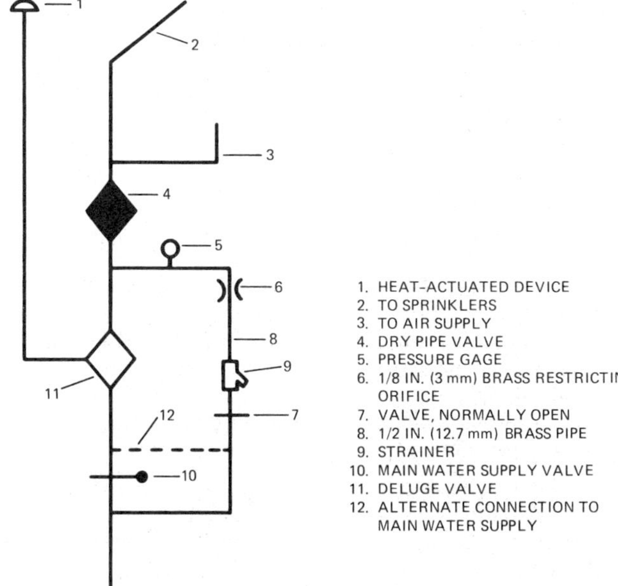

FIG. 11-2A. Piping to chamber between deluge valve and dry pipe valve.

kPa) water pressure or four times the expected water pressure, whichever is greater. It is also necessary that provisions be made for thawing and draining.

All sprinklers in refrigerated warehouses should be pressurized and equipped as described in NFPA 13, *Standard for the Installation of Sprinkler Systems.* They

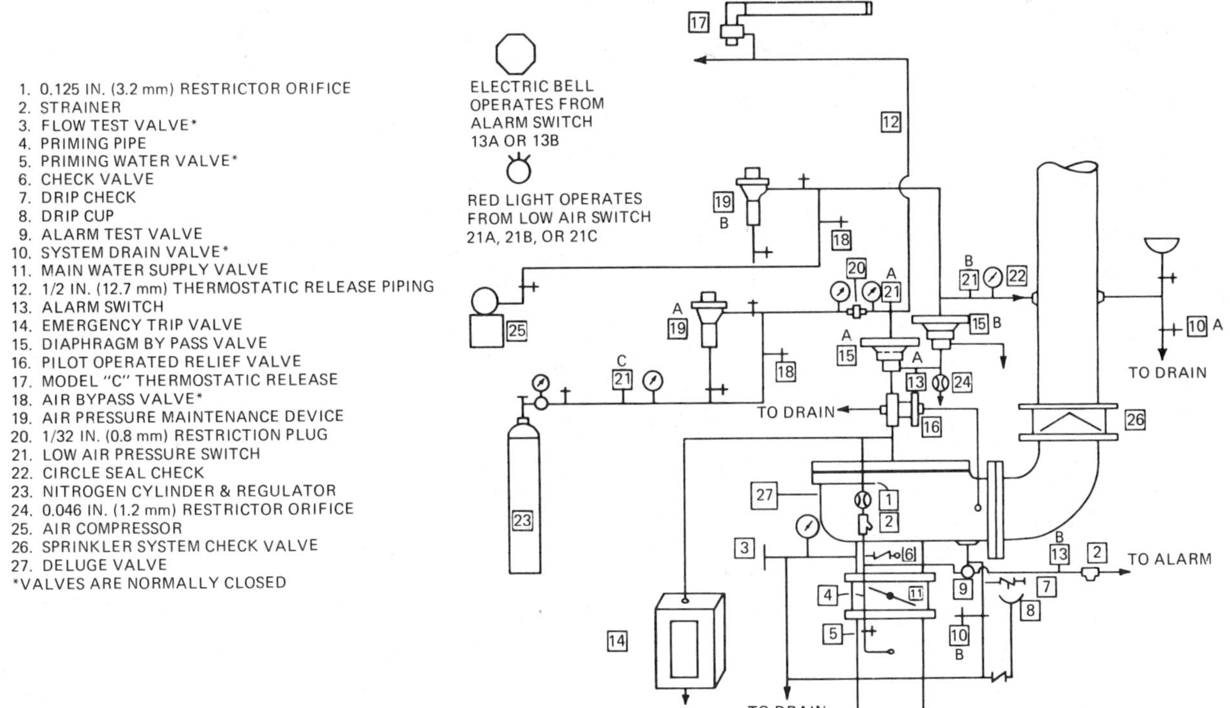

FIG. 11-2B. Pneumatic dual release automatic water control valve.

should be designed so that the systems can be easily inspected, and disassembled for removal of ice plugs.

The combination dry pipe deluge system shown in Figure 11-2A will help to avoid ice formation due to accidental tripping of dry valves. When this combination is used:

1. The dry system should be manifolded to the deluge valve.
2. The protected area should not exceed 40,000 sq ft (3716 m^2).
3. The distance between valves should be as short as possible.
4. The chamber between deluge and dry pipe valves should be pressurized from the supply side of the deluge valve.
5. An approved ball drip valve should be installed on the dry pipe valve.

Properly located heat detecting devices should be provided in the protected area to automatically operate the deluge valve. A deluge valve having electrical actuation is preferred.

An acceptable alternate to the deluge valve/dry pipe valve combination is shown in Figure 11-2B. Approved equipment should be used throughout and all electrical equipment should be compatible. The piping to the heat actuated devices should be supervised. The control piping should be arranged so both loss of sprinkler piping air pressure and operation of a heat actuated device occur before the deluge valve release mechanism operates.

Refrigerant Gases: Refrigerant gases present two basic hazards, toxicity and flammability. Sprinklers are very effective because the spray dilutes, disperses, and cools the escaping gases.

Bibliography

NFPA Codes, Standards, Recommended Practices and Manuals. (See the latest *NFPA Codes and Standards Catalog* for availability of the following documents.).

NFPA 13, *Standard for the Installation of Sprinkler Systems.*

Additional Readings

Factory Mutual Engineering Corporation, "Refrigerated Warehouses," *Loss Prevention Data Sheet 8-29*, Factory Mutual System, Norwood, MA, May 1976.

Linville, J. L. ed., *Industrial Fire Hazards Handbook*, 2nd ed., National Fire Protection Association, Quincy, MA, 1984.

Safety Code for Mechanical Refrigeration, ANSI B9.1, 1971 American National Standards Institute, New York, NY.

OUTDOOR STORAGE PRACTICES

Revised by Martin M. Brown

For outdoor storage, as with indoor unsprinklered storage, truly adequate engineered fire protection is extremely difficult. However, depending upon its hazard, monetary value, area, or a combination of these, a degree of protective safeguards and installations should be provided that is justified by these features. Thus, for small amounts of incidental and transitory outdoor storage, appropriate protection could consist merely of good security fencing and outdoor lighting. On the other hand, elaborate safeguards are warranted for very large, high value storage yards.

Good general practices, developed from fire experience, for sizeable storage areas are outlined in this chapter with references to certain types of storage that have received special study. See also NFPA 231, *Standard for Indoor General Storage*, Appendix C.

Additional information on storage practices involving pallets will be found in Chapter 1 of this section, "General Indoor Storage Practices." Information on water supply for storage facilities is contained in Section 17, Chapter 3, "Water Supply Requirements for Fire Protection." Section 10, Chapter 11, "Welding and Cutting," also should be consulted, as should Chapter 8 of this Section, "Storage and Handling of Solid Fuels."

CHOOSING THE SITE

Size

The area should be of sufficient size to hold the quantity of stored material with allowance for spacing between piles and possible future expansion. Where adequate area is not available, additional sites should be found. Congestion is a major cause of fire spread.

Terrain

The ground should be level if at all possible. Sloping terrain may make piles unstable and present a serious hazard to fire fighters. Refuse or sawdust-filled land, swampy ground, or areas prone to underground fires, are examples of unsuitable terrain.

Exposure

Adequate clearances to adjacent properties should be maintained to minimize mutual fire exposure. Zoning regulations often specify clear distances to a property line. If no such regulations exist, the nature of future development on adjacent property should be considered, for it may reduce the size of the storage area.

Other considerations include features that offer protection from grass, brush, or forest fires, and from potential ignition from sparks given off by railroad rolling equipment, incinerator stacks, electrical transformers on poles, lighted cigarettes flicked from nearby cars etc.

Fire Protection

The manner in which outdoor storage is arranged (including the size of the piles, and the clearances required from adjacent piles, buildings, and property boundary lines) should be influenced by the protection that is available. The local fire protection capabilities in particular should be evaluated before choosing a site (including: the public water supply and its record of reliability, the fire flow available, the type of fire department available and the distance it must travel to the site, and the means available for notifying the fire department of a fire).

Floods and Windstorms

Areas subject to flooding or windstorms are to be avoided. Windstorm is a particular problem in some coastal areas. Fires occurring during high winds will present an exposure problem against which protection is almost impossible. Burning materials and brands will be carried over considerable distances, even though the fire may be confined to a relatively small area.

PREPARING THE SITE

Clearing the Site

Vegetation should be cleared away so that it cannot dry out and become fuel for ignition or fire spread. The site should be leveled as nearly as possible and then paved. If the entire yard is not to be surfaced, care should be taken

Mr. Brown is a fire protection consultant based in White Plains, NY.

to ensure that fire apparatus can easily maneuver throughout the yard. This requires properly developed roadways capable of supporting heavy trucks. Proper drainage of the site is essential.

Many paved outdoor storage yards have painted lines to show aisles and roadways, and the location of yard hydrants, extinguishers, water buckets, and hose houses.

Layout of the Site

A definite plan for the site is advisable. The plan should detail where specific materials are to be piled, the proper separation between piles, and the access to all areas of the site, including fences and their gates.

The density of the packed materials makes a tremendous difference in the way they burn. Lumber stacked in solid, orderly piles perhaps not over 20 ft (6 m) high does not present the same problem as lumber stored in "sticked" piles. In the latter case, fire can spread more rapidly because of the air spaces in the piles. Pulpwood is sometimes stored in piles 200 ft (61 m) in diameter and 100 ft (30 m) high; these piles contain air spaces and pockets, made by the rough and varied shape of the logs, which accelerate burning.

Some materials burn quickly and produce a great deal of heat. Baled cotton, hay, lumber, packing materials, pallets, plywood, pulpwood, and rubber are examples of such materials, and considerable clearance is required between and around these piles. Aisle widths equal to the height of the pile are desirable for hazardous commodities.

Both the yard itself and and the piles within the yard must be easily accessible to fire fighters. Storage close to railroad spurs often hinders access due to boxcars and other railroad equipment left standing on the tracks. Driveways at least 15 ft (4.5 m) in width permit fire equipment to reach all portions of the storage area. Aisles of 10 ft (3 m) or more reduce the danger of fire spread from pile to pile. For unusually large storage yards or moderate sized yards with valuable commodities, main aisles or firebreaks can subdivide the storage in much the same way that fire walls do indoors. Aisle widths will depend upon the commodity, how it is stored, height of piles, normal and abnormal wind conditions, fire fighting forces, and equipment and other factors.

In some cases, packaging and palletizing affect storage methods, particularly if materials are irregularly shaped or small. If the pallets themselves are combustible, they may present an added hazard. Some materials are packed in heavy crates. Usually, this does not present an increased hazard except after severe weathering or where handling has caused breakage or splintering. Tarpaulins and other weather resistant materials used to cover stock should be of fire retardant treated material.

A pile that is stable under normal conditions may present a severe hazard in a fire. It is wise to anticipate the effects of fire and water on the stability of piles when planning their sizes and configurations. Collapsing piles may cause severe fire spread and hamper fire fighting, particularly if there is danger from flying brands.

Areas used to service and maintain equipment, should be designated and properly marked or cordoned off. If flammable liquid fuel is used, it should be stored in an underground tank, if possible. If this is not practical, fuel storage tanks should be located so spills will not flow under or around the storage, and conversely, so that a fire in a storage area will not expose a tank of flammable or combustible liquid.

Installation of Fire Protection

Adequate fire protection must be provided while the site is being prepared. Depending upon the size and location of the site, it may be necessary to install private water mains and hydrants throughout the yard.

In planning for hydrants and yard mains, the water supply, whether from public mains or a private source, should be able to furnish the needed fire flow in gallons or liters per minute.

Based upon accumulated fire experience, the Insurance Services Office has developed formulas to calculate fire flow for a particular large unsprinklered building, based on the area, combustibility of stored goods, and a construction factor that, of course, would not apply to outdoor storage. It is important to realize that fire flows needed for hose stream protection of indoor unsprinklered storage, as well as outdoor storage, are far greater than those for a well designed automatic sprinkler system plus concurrent hose streams. Also, these larger flows will not necessarily minimize the fire damage as well as the lower flows utilized in sprinklered buildings, but they represent reasonable protection.

Table 11-3A contains calculated fire flows for hose stream protection of outdoor storage (assuming early detection, adequate hydrant availability and access).

The four occupancy classes are defined as follows:

1. *Limited combustible*: Low combustibility with limited concentrations of combustible material.
2. *Combustible*: Moderate combustibility.
3. *Free burning*: Examples include cotton bales, furniture stock, wood and wood products.
4. *Rapid burning*: Commodities which either (a) burn with great intensity, (b) spontaneously ignite and are difficult to extinguish, or (c) give off flammable vapors or dusts at ordinary temperatures.

A minimum level of manual fire suppression equipment would be to have rated, properly marked portable fire extinguishers (usually the nonfreezing water type) placed so that the travel distance to the nearest unit is 75 ft (23 m) or less. If the site is active and people are usually present, fully equipped hose houses should be provided and personnel trained to use them in a fire incident.

A means for notifying the fire department is essential. A system that sends a signal directly to the fire department is best, but at minimum, a telephone should be accessible.

Security Measures

Trespassing, vandalism, and theft are problems in many outdoor storage yards. Protection from children (who like to play in yards and are sometimes responsible for setting fires) and vagrants seeking shelter frequently make fencing and lighting a requirement. A fence, however, must not prevent ready access to the yard by the fire department. The fire department should be consulted about incorporating remote gates into the fencing so additional access can be provided under serious fire conditions.

Security guards are very effective in the protection of outdoor storage yards. NFPA 601, *Standard for Guard*

TABLE 11-3A. Fire Flow for Hose Stream Protection of Outdoor Storage*

Area		Limited Combustible		Combustible		Free Burning		Rapid Burning	
sq ft	m^2	gpm	L/min	gpm	L/min	gpm	L/min	gpm	L/min
10,000	929	1,500	5678	1,750	6624	2,000	7571	2,250	8517
20,000	1848	2,000	7571	2,500	9464	3,000	11356	4,000	15142
50,000	4645	3,500	13429	4,000	15142	4,500	17034	6,000	22712
100,000	9290	5,000	18927	5,500	20820	6,500	24606	8,000	30283
200,000	18580	6,750	25551	8,000	30282	9,250	35015	10,000	37854

* Adapted from ISO formulas. Assumes early detection, adequate hydrants, and access.

Service in Fire Loss Prevention, covers the selection, the duties, and the instruction and training of such personnel.

UTILIZING THE SITE

Management

When a yard is selected and laid out, the amount of material it is designed to accommodate is also established. These quantities should not be exceeded since excess material may reduce firesafety. The material and other fire prevention and protection features, including disposal of waste materials, should be monitored by frequent periodic inspections.

In general, the attitude of management toward fire protection in the yard provides an example that employees are likely to follow.

Fire Protection and Prevention

All fire protection equipment, including hydrants, fire pumps, fire extinguishers, and any suppression, detection, or alarm systems on the storage site should be properly maintained and tested.

All materials handling equipment at the site needs to carry a portable fire extinguisher (Class B or C) suitable for any fire. In addition, it is also good practice to carry a portable Class A extinguisher.

The location of all hydrants, hose houses, portable extinguishers, alarm boxes, and other fire protection equipment should be properly marked. Elevated signs are preferable. Arrows and signs painted on the pavement are the minimum markings required.

If material to be torch cut or welded cannot be moved from the storage yard to a remote location, the work should be done only after adequate precautions are taken. (See NPFA 51B, *Standard for Fire Prevention in Use of Cutting and Welding Processes*.)

Depending upon the nature of the material stored, smoking should be prohibited or restricted to specific areas only. In either case, adequate signs should be displayed and the regulations strictly enforced.

A private fire brigade of employees provides at least initial fire fighting. If there are enough employees to organize a fire brigade, the level of organization, training, and equipment provided normally will be governed by the availability and ability of a public fire department. The larger the storage yard and the more valuable its contents, the greater the need for a private fire brigade. NFPA 27, *Recommendations for Organization, Training, and Equipment of Private Fire Brigades* gives helpful information on this subject.

Perhaps the most important duty of a brigade or security guard is to immediately notify the fire department of any fire. The fire department should be called even if brigades can handle the emergency without outside help. In too many cases, notifying the public fire department has been delayed because brigades or security guards unsuccessfully attempted to control fires, resulting in higher losses than necessary.

Maintenance and Housekeeping

Repairs to fencing, lighting, and the yard surface should be made promptly and materials handling equipment should be kept in good condition. Suitable safeguards should be provided to minimize the hazard of sparks from equipment such as refuse burners, boiler stacks, vehicle exhausts, and locomotives.

Good housekeeping also includes controlling weeds and vegetation. Weeds, grass, and other vegetation should be sprayed with an acceptable herbicide or ground sterilizer, or grubbed out. Dead weeds should be removed after destruction, but weed burners should not be used in the removal.

In many instances, scrap lumber, broken pallets and broken containers, bales, or pieces of material stored on the site, are discarded into the clear space around the yard. Daily checks should be made for such discarded materials; where found, they should be properly removed. Care should be taken that wind blown debris and other combustible materials do not accumulate under piles of materials.

If a pile of material falls into an aisle or clear space, the stock should be repiled immediately.

OUTDOOR STORAGE OF SPECIFIC MATERIALS

Discussions of safe storage practices for a few common materials follow. As a source for good practices, see referenced NFPA publications and industry or insurance company standards. Many times, the guidelines for inside storage of a commodity are a good starting point for devising guidelines for outdoor storage.

Wood and Wood Products

Most wood and wood products are stored outdoors. Because wood is moved from the forests only at certain times of the year, large quantities of logs are stored at sawmills, paper mills, and pulp mills. Large circular piles with logs dumped on the top of the pile are called stacked piles. Logs stacked in parallel form, like matches in a box, are called ranked piles. Stacked piles are usually much larger in size, with the height of the pile often approaching 100 ft (30 m). Ranked piles are usually 10 to 15 ft (3 to 5 m) high. Because maintaining separation and aisles is much easier in ranked piles, fire protection is much easier in ranked than in stacked piles. (See Fig. 11-3A.)

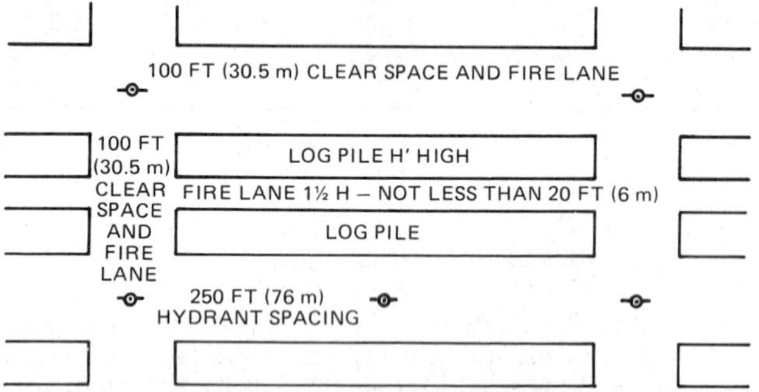

FIG. 11-3A. A good layout for ranked piles of logs in a storage yard.

Large quantities of water are needed to cope with log yard fires. The total volume required may be many thousands of gallons or liters per minute if many portable turrets and monitor nozzles [each discharging about 1,000 gpm (3785 L/min)] are operated simultaneously. A looped underground fire main system with hydrants spaced through the yard is good practice. Monitor nozzles on towers have been common in log yards, especially with high stacked piles. In large stacked piles, burrowing fires are difficult to extinguish even after the surface burning is under control. NFPA 46, *Recommended Safe Practice for Storage of Forest Products* (hereinafter referred to as NFPA 46), provides additional guidelines for the storage of logs.

Chip storage has replaced log storage at many pulp and paper mills. Two completely different types of fires occur in chip piles: surface fires and internal fires. NFPA 46 provides guidelines for fire protection of chips. The amount of water needed to control a chip pile fire will vary substantially, depending upon the size of the pile. Weather conditions, operating methods, geographic location, the type of wood chip, and the degree to which wetting may be employed are all factors influencing fire conditions. Experience indicates that exposure to long periods of hot, dry weather with no regular surface wetting creates conditions under which fast spreading surface fires can occur; and they often require many hose streams.

Likewise, the frequency of pile turnover and operating methods affect the prospects for serious internal fires. Piles built using methods that allow a concentration of fines, and piles stored for long periods of time with no turnover, are subject to internal heating which, if undetected, can create intense internal fires.

Sawdust storage is not common now but once was prevalent at sawmills. Fires in sawdust are not especially serious, but they can smolder and smoke unless the piles are broken open and the fire is completely extinguished.

Lumber storage is quite common in urban areas, particularly in built up neighborhoods. Local ordinances generally prohibit storage of any appreciable quantity in zoned areas, but many lumberyards originally located in unzoned areas now border heavily developed areas. The exposure of stored wood to adjoining areas is a major consideration. Adequate aisles and good housekeeping, along with proper procedures for controlling ignition sources such as salamanders, smoking, etc., will reduce the fire hazard considerably. Large quantities of water are needed to bring a lumberyard fire under control once it has gained headway. Flying brands can spread the fire thoughout the yard and into adjoining areas readily, particularly under windy conditions. NFPA 46 provides guidelines for the safe storage of lumber.

Paper and Paper Products

Some paper and paper products are stored outdoors, but most are stored indoors to protect them from rain and wind. Roll paper is normally stored in large rolls weighing 2 tons (1814 kg) or more. Because the paper is tightly wound, it is very difficult to ignite; however, roll paper stored outdoors is more vulnerable to ignition. Once ignited, fire in roll paper stored outside is difficult to extinguish with manual fire fighting methods.

Baled wastepaper does not produce as intense a fire as does roll paper, but instead tends to produce burrowing fires. Fire in baled wastepaper is difficult to extinguish and generates heavy smoke, which inhibits manual fire fighting. Broken bales are difficult to handle after wetting. Usually a motorized vehicle with a shovel is needed to move the debris to complete extinguishment. Spontaneous ignition is common in wastepaper if foreign materials subject to spontaneous heating become trapped in bales. Fire protection for wastepaper storage is generally the same as for other outdoor storage.

Rubber

Rubber Tires: Tire storage presents a severe hazard. Tires burn rapidly, emitting intense heat and large quantities of dense smoke that hamper manual fire fighting. Large quantities of water at high pressures are usually needed because control is difficult due to the shielded nature of fire within a tire casing. Narrow piles with good aisles are best for outdoor storage.

Baled Crude Rubber: Bales are often stored outdoors. The Factory Mutual System recommends that individual piles be limited to 100 tons (90 720 kg) with a minimum of 30 ft (9 m), but preferably 50 ft (15 m), between piles (FMEC 1971). Piles should be grouped to a maximum of 1,000 tons (90 7200 kg) with 100 ft (30 m) aisles separating them. It is important that burlap used to wrap the bales not be contaminated with oil; if oil gets into a tightly packed pile, spontaneous heating may result.

Baled Combustible Fibers

Although storing baled combustible fibers outdoors is not good practice, much of this material, such as baled cotton in transit from the gin to a warehouse, is temporarily stored outdoors. Factory Mutual recommends that piles be limited to 500 bales per pile with a minimum of 30 ft (9 m) and preferably 50 ft (15 m) clear space between individual piles, from an exterior fence or from buildings (FMEC 1974). Piles of uncleaned baled flax straw should be limited to 300 tons (272 160 kg) and a height of 20 ft (6 m) and should be located 200 ft (60 m) from buildings and 100 ft (30 m) from each other or potential ignition sources. Storing bales of combustible fibers off the ground on pallets provides ventilation and prevents excessive damage from ground moisture. Covering the tops and sides of all piles with a securely fastened tarpaulin of fire retardant material protects them from the weather and mildew.

Coal

The main problem with coal storage is that bituminous coal tends to heat spontaneously. Anthracite coal is not subject to spontaneous heating. Freshly mined coal, however, is more susceptible because it absorbs oxygen more rapidly. Moisture and coal dust also make coal more susceptible to spontaneous ignition.

Bibliography

References Cited

FMEC. 1971. "Storage of Baled Crude Rubber." *Loss Prevention Data Sheet*, 8-1. Factory Mutual Engineering Corp., Norwood, MA.

FMEC. 1974. "Baled Fiber Storage." *Loss Prevention Data Sheet*, 8-7. Factory Mutual Engineering Corp., Norwood, MA.

NFPA Codes, Standards, Recommended Practices and Manuals. (See the latest *NFPA Codes and Standards Catalog* for availability of current editions of the following documents.)

NFPA 10, *Standard for Portable Fire Extinguishers.*

NFPA 24, *Standard for the Installation of Private Fire Service Mains and Their Appurtenances.*

NFPA 27, *Recommendations for Organization, Training and Equipment of Private Fire Brigades.*

NFPA 46, *Recommended Safe Practices for Storage of Forest Products.*

NFPA 51B, *Standard for Fire Prevention in Use of Cutting and Welding Processes.*

NFPA 80A, *Recommended Practice for Protection of Buildings from Exterior Fire Exposures.*

NFPA 231, *Standard for Indoor General Storage.*

NFPA 505, *Fire Safety Standard for Powered Industrial Trucks Including Type Designations, Areas of Use, Maintenance and Operations.*

NFPA 601, *Standard for Guard Service in Fire Loss Prevention.*

NFPA 601A, *Standard for Guard Operations in Fire Loss Prevention.*

Additional Readings

"Coal and Coal Storage," *Loss Prevention Data Sheet*, 8-10, Factory Mutual Engineering Corp., Norwood, MA, Aug. 1975.

"General Storage Safeguards," *Loss Prevention Data Sheet*, 8-0, Factory Mutual Engineering Corp., Norwood, MA, Mar. 1977.

"Outdoor Storage of Wood Chips," *Loss Prevention Data Sheet*, 8-27, Factory Mutual Engineering Corp., Norwood, MA, Dec. 1980.

"Roll Paper Storage," *Loss Prevention Data Sheet*, 8-21, Factory Mutual Engineering Corp., Norwood, MA, Mar. 1978.

"Sawmills and Lumber Yards," *Loss Prevention Data Sheet*, 7-25, Factory Mutual Engineering Corp., Norwood, MA, July 1979.

STORAGE OF FLAMMABLE AND COMBUSTIBLE LIQUIDS

Martin Henry

The first requirement for the storage of flammable and combustible liquids is properly designed containers that are liquid tight and from which release of vapors, if required, is carefully controlled. The containers range from large, upright outdoor storage tanks of millions of gallons (100,000 gals equals 378 m^3) capacity down through drums to small cans containing ounces of liquid.

This chapter discusses the various types of containers used in the storage and transportation of flammable and combustible liquids and the precautions that should be observed in handling the liquids as they are loaded, unloaded, or dispensed.

Additional information relevant to the storage of flammable and combustible liquids will be found in Section 4, Chapter 2, "Explosions," and Section 4, Chapter 11, "Tables and Charts;" in Section 6, Chapter 2, "Identification of the Hazards of Materials;" in Section 11, Chapter 10, "Control of Electrostatic Ignition;" in Section 16, Chapter 5, "Gas and Vapor Testing;" and Section 17, Chapter 5, "Water Storage Facilities and Suction Systems."

TANK STORAGE

Tanks may be installed aboveground, underground, or, under certain conditions, inside buildings. Openings and connections to tanks, for venting, gaging, filling, and withdrawing, can present hazards if they are not properly safeguarded.

Given substantially constructed, properly installed, and well maintained tanks, storage of flammable and combustible liquids presents less of a danger than the transferring of these liquids. The severity of the storage hazard might seem to depend upon the quantity stored. As a practical matter, however, the size of the tank or the number of tanks is less important than such factors as the characteristics of the liquid stored, the design of the tank and its foundation and supports, the size and location of vents, and the piping and its connections.

Flammable liquids expand when heated. Gasolines expand about 0.07 percent in volume for each 10°F (5.5°C) increase in temperature within ordinary atmospheric temperature ranges. The effect of temperature increase on the volume of acetone, ethyl ether, and certain other flammable liquids with higher coefficients of expansion is greater than in the case of gasoline. To avoid danger of overflow, tanks should not be filled completely, particularly where cool liquid is placed in a tank in a warm atmosphere—for example, when filling an automobile tank from an underground gasoline tank on a warm day.

Several methods are used to prevent storage evaporation loss and loss of vapors as the tank is filled. Underground tanks reduce evaporation losses since there is less fluctuation of temperature. Aboveground tanks often are painted with aluminum or white paint to reflect heat, thus decreasing the temperature rise of liquid contents and slowing evaporation of tank contents. Floating roof tanks minimize vapor loss and tend to reduce the fire hazard. Storage of gasoline in pressurized tanks reduces the loss of vapor. In some cases, vapors are conserved by the use of lifter roof or vapor dome tanks, or the vents from several cone roofed tanks may be connected through manifolds to a vapor dome or pressure type tank.

The vapor space in tanks storing flammable liquids with vapor pressures above 4 psia (28 kPa), such as gasoline, is normally too rich to burn. The ratio of vapor to air is above the upper flammable (explosive) limit (LFL). However, if the temperature of the liquid gasoline is in the range of −10 to −50°F (−23 to −46°C), the vapor space may be within the flammable range. When a tank is being off loaded, or when there is a sudden cooling rainstorm on a hot day, there may be a portion of the tank vapor space which will be within flammable limits. This condition may remain for several hours or even days, due to possible stratification of the vapors.

The vapor space in tanks storing low vapor pressure liquids (below approximately 2 psia or 14 kPa) such as kerosene, is normally too lean to burn. The ratio of vapor to air is below the lower flammable (explosive) limit (LFL). However, if the entire body of liquid is heated to its flash point, as may happen during refining processes or by exposure to fire, the vapor space may enter the flammable range. It should be noted that it is the temperature of the liquid and not the temperature of the vapor space that determines the presence of a flammable vapor-air mixture. The oil vapors

Mr. Henry is flammable liquids specialist on the staff of NFPA.

driven off by the heated air in the vapor space are condensed back to liquid by the cooler body of oil. Therefore, the vapors are only flammable for a very short distance above the liquid surface despite the fact that the air within the tank may be considerably above the flash point temperature.

Tank storage of ethyl and methyl alcohol, JP-4 or Jet B turbine fuel, and other liquids of similar vapor pressure (approximately 2 to 4 psia at 100°F or 14 to 28 kPa at 38°C) presents an unusual hazard since the vapors are normally in the flammable range. Storage in floating roof or similar tanks or the addition of inert gas in the vapor space is desirable to reduce the possibility of an explosion in the vapor-air mixture in the tank. Floating roof tanks, cone roof tanks with internal floating roofs, lifter roof tanks, and vapor dome tanks are also used for vapor conservation purposes for Class I liquids.

Aboveground Storage Tanks

Storage tanks come in a variety of designs; however, they may be divided into three general categories of pressure design: (1) atmospheric tanks, for pressures of 0 to 0.5 psig (0 to 4 kPa); (2) low pressure storage tanks, for pressures from 0.5 to 15 psig (4 to 103 kPa); and (3) pressure vessels, for pressures above 15 psig (103 kPa) Some of the more common types of aboveground storage tanks are shown in Figures 11-4A and 11-4B. Pressure tanks and pressure vessels normally are used for vapor conservation purposes, particularly for liquids with high vapor pressures.

Construction: The thickness of the metal used in tank construction is based not only on strength required to hold the weight of the liquid, but also on an added allowance for corrosion. When intended for storing corrosive liquids, the specifications for the thickness of the tank shell are then increased, to provide additional metal and allow for the expected service life of the tank. In some cases, special tank linings are used to reduce corrosion. Periodic inspection should be made to ascertain metal thickness of the tank, and establish safe operating limits, and avoid overstressing the tank. The inspection of tanks for corrosion may be performed by visual inspection, drilling, and calipering, use of sonic devices; provision of weep holes, or experience gained by the storage of similar materials. Sonic devices operate on a principle of the length of time it takes for sound waves to reflect from a surface. Any difference in metal thickness is quickly disclosed by these instruments, which are particularly effective when large areas with many potential corrosion spots are involved. Weep holes, which are very small holes drilled part way into the tank shell, are used occasionally. The principle behind use of these holes is that the lesser metal thickness at the partially drilled hole will show signs of leakage before the strength of the entire tank is endangered by corrosion.

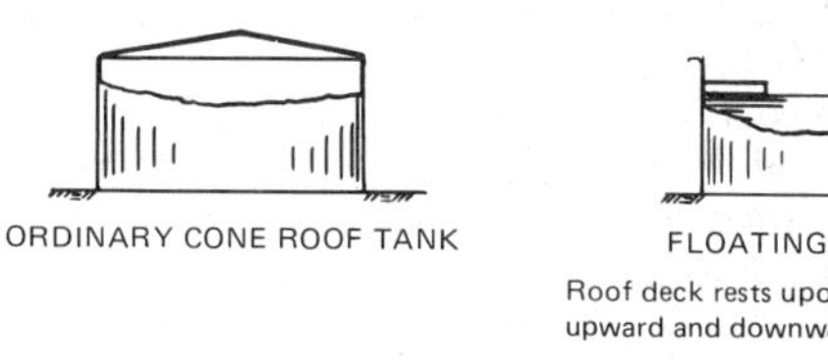
ORDINARY CONE ROOF TANK
FLOATING ROOF TANK
Roof deck rests upon liquid and moves upward and downward with level changes.

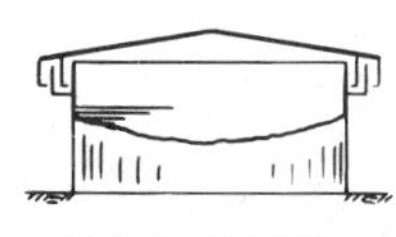
LIFTER ROOF TANK
Liquid sealed roof moves upward and downward with vapor volume changes.

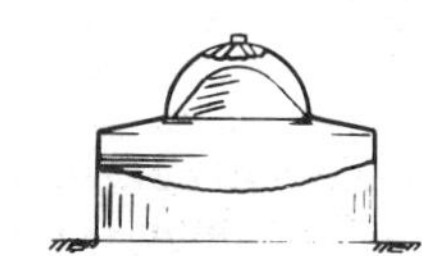
VAPORDOME ROOF TANK
Flexible diaphram in hemispherical roof moves in accordance with vapor volume changes.

FIG. 11-4A. Common types of atmospheric storage tanks.

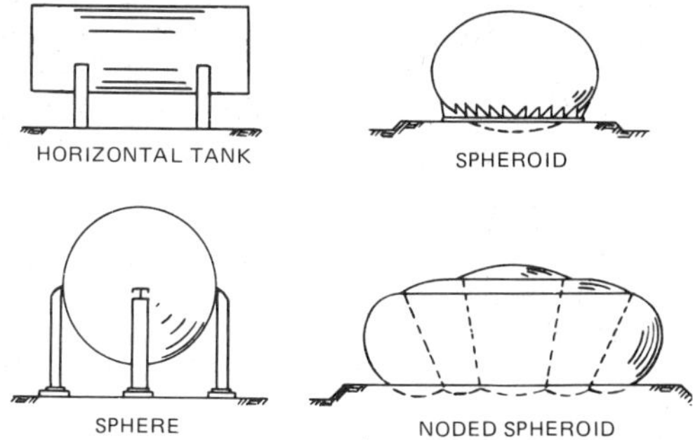
HORIZONTAL TANK
SPHEROID
SPHERE
NODED SPHEROID

FIG. 11-4B. Common types of low pressure tanks or pressure vessels.

All storage tanks should be built of steel or concrete unless the character of the liquid necessitates the use of other materials. Both steel and concrete tanks resist heat from exposure fires. Tanks built of materials less resistant to heat (e.g. low melting point materials) might result in tank failure and fire spread.

Tanks with the label of Underwriters Laboratories Inc. (UL) or those built in accordance with American Petroleum Institute (API) standards meet exacting specifications. For atmospheric, vertical, cylindrical, and aboveground welded tanks, the following formula from API Standard 650, *Welded Oil Storage Tanks*, (API 1979), may be used in calculating the minimum thickness of shell plate:

$$t = 0.0001456 \times D \times (H - 1) \times S,$$

For SI units:

$$t = 0.0398 \times D \times (H - 0.305) \times S,$$

where:

t = minimum thickness, in inches or mm
D = nominal inside diameter of tank, in feet or m
H = height, in feet or m, from the bottom of the course under consideration to the top of the top angle or to the bottom of any overflow which limits the tank filling height
S = specific gravity of liquid to be stored but in no case less than 1.0.

The nominal thickness of shell plates (including shell extensions for floating roofs) is no less than those given in Table 11-4A. The maximum nominal thickness of tank shell plates is ½ in. (18 mm).

Concrete tanks require special engineering, and unlined concrete tanks should be used only for the storage of liquids with a specific gravity of 40° API or heavier. Tanks built of material other than steel should be designed to specifications embodying equivalent steel safety factors.

TABLE 11-4A. Shell Plate Thickness

Nominal Tank Diameter in Feet*	Nominal Thickness in Inches†
Smaller than 50	3/16
50 to, but not including, 120	1/4
120 to 200, incl.	5/16
Over 200	3/8

* 1 ft = 0.304 m
† 1 in. = 25.4 mm

Installation: Aboveground tanks should be installed in accordance with NFPA 30, *Flammable and Combustible Liquids Code* (hereinafter referred to as NFPA 30). NFPA 30 specifies distances from tanks to property lines that can be built upon, and to public ways or to important buildings. These distances will vary depending upon whether the pressures in the tanks (including those attained during fire exposure) can be under or over 2.5 psig (17 kPa) as well as whether the contents are stable or unstable liquids, or liquids having boilover characteristics. Other factors affecting distances are the design of the tank, protection for exposures, fire extinguishing or control systems provided, or other protection features. The spacing between tanks is also specified in NFPA 30.

Municipal ordinances usually prohibit aboveground storage of flammable and combustible liquids in congested business districts, and zoning restrictions commonly exclude them from residential areas and from the vicinity of schools and hospitals. Where a tank is located in an area that may be subjected to flooding, applicable provisions in NFPA 30 should be followed.

Venting and Flame Arresters: An appropriate vent must be provided for normal operation of any tank to permit filling and emptying and for the maximum expansion or contraction of the tank contents with changes in temperature. API Standard 2000, *Venting Atmospheric and Low Pressure Storage Tanks*, provides information on sizing vents for product movement and breathing. Clogged or inadequately sized vents may result in the rupturing of tanks from internal pressure, or collapse due to internal vacuum. During the filling operation, the vents discharge flammable vapors. If the mixture is sufficiently rich or if the vent location is such that the released vapor is a hazard, the vapors should be piped to a place where they can dissipate safely. The vapor release should not occur close to doors, windows, or possible sources of ignition.

Venting devices, which are normally closed when the tank is not under pressure or vacuum, or approved flame arresters, are provided on vent pipes to prevent flashback into tanks. These devices are required when a flammable mixture is present, and for the storage of all Class I liquids. Arresters constructed of banks of parallel metal plates or tubes having a large surface of metal to dissipate heat are more effective than screens for larger openings and are less subject to clogging and corrosion. Heat is absorbed by the metal plates or tubes, which lower the temperature of the vapor below its self-ignition point. However, if the flame arrester is exposed to long burning periods, the metal plates can become sufficiently hot on the bottom side to ignite any flammable vapor-air mixtures in the tank. Figure 11-4D shows average capacities of laboratory tested arresters; Figure 11-4C shows average capacities of flame arrester vent combinations.

Where the liquids stored have flash points in the range of normal summer temperatures, the vapor space above the liquid in the tank will normally contain vapors in the flammable range. Flame arresters have their most important application on such tanks. However, condensation and crystallization of certain liquids and freezing of moisture in winter may make conservation vents and flame arresters impractical. Steam tracing (steam heating) is provided in some cases to prevent freezing or crystallization of contents.

Although a wire screen of 40 mesh* ordinarily will prevent the passage of flame through small openings, it cannot be relied upon as an effective flame arrester because of possible physical damage to the wires or clogging of the mesh by dirt or other residues.

Emergency Venting: In addition to the normal operating vents, emergency relief of internal pressure is required for most aboveground tanks in the event a fire occurs under or around the tank. In the absence of proper provision for such relief, high pressures may be generated when the tank is exposed to external fire, resulting in a Boiling Liquid Expanding Vapor Explosion (BLEVE). Such explosions are infrequent, but when they do occur the results may be disastrous to life and property. BLEVEs can be avoided by providing adequate pressure relief, which permits the vapors to escape and burn at the vents and prevent rupturing of the tank. Emergency relief venting may take the form of loose manhole covers which lift under pressure, weak roof to shell seams, floating roofs, rupture disks, or commonly used emergency relief vents designed for the purpose.

Unless they are adequately vented, horizontal cylindrical tanks under excessive internal pressure commonly fail at the ends. For vertical cone roofed tanks designed with weakened seams at the roof to shell joint, the lifting of the roof or top of the roof to shell seam affords adequate emergency pressure relief. Vertical cone roofed tanks are required to have the roof to shell seam of weaker construction than the bottom to shell seam to prevent failure at the bottom of the tank which would release a wave of liquid. (See Fig. 11-4E.)

The danger of tank failure from internal pressure when exposed to fire depends to a considerable extent upon the characteristics of the liquid, the size and type of the tank, and the intensity and duration of the fire. The smaller the tank or the lesser the volume of the liquid in the tank, the shorter the time under fire exposure before a BLEVE occurs.

Table 11-4B is based upon the required discharge from both normal and emergency vents, derived from a consideration of the probable maximum rate of heat transfer per unit area; size of tank and the percentage of total area likely to be exposed; time required to bring tank contents to boil; time required to heat unwet portions of the tank shell or roof to a temperature where the metal will lose strength; and effect of drainage, insulation and the application of water in reducing fire exposure and heat transfer.

*A woven wire fabric in which there are 40 wires per in. (25.4 mm) in each direction, and 1,600 interstices per sq in. (645 mm^2). The size of the openings depends upon the diameter of the wires as well as upon the mesh of the screens, but the size of wire is approximately uniform for any given mesh, and specifying a screen in terms of its mesh determines the size of the openings with sufficient accuracy for practical purposes. The individual opening in a 40 mesh screen has an area of about 0.00022 sq in. (0.14 mm^2).

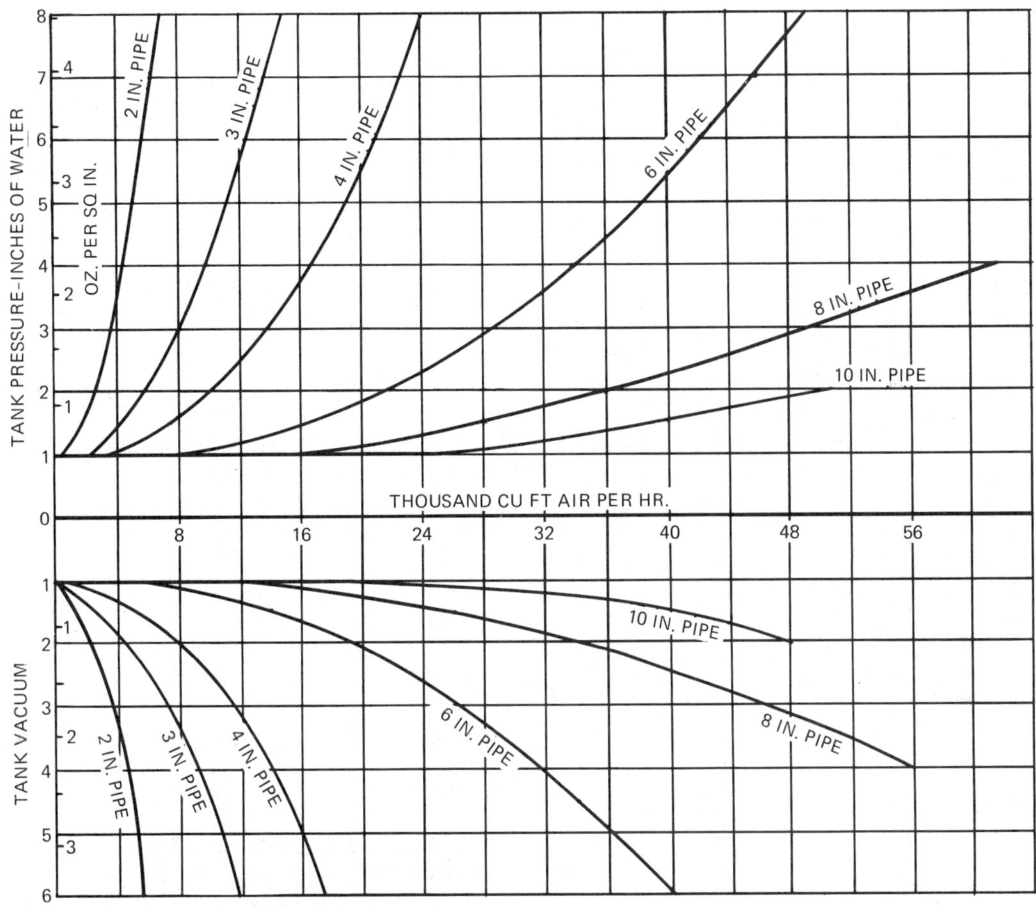

FIG. 11-4C. Average flow capacity of listed flame arrester vent valve combinations for oil storage tanks in thousands of cubic feet of air for positive and negative pressures. (1 in. = 25.4 mm; 1 in. of water = 249 kPa; 1,000 cu ft = 28.3 m³; 1 oz per sq in. = 3 kPa/mm)

Due to the wide variance in flow characteristics of manufactured vents, each vent design under 8 in. (203 mm) should be flow tested rather than relying upon vent size. The flow capacity of larger vents may be flow tested or calculated in accordance with the formula in NFPA 30.

The wetted area referred to in Table 11-4C is the internal portion of the tank in contact with the liquid (assuming a full tank) which is expected to be exposed to the flame of the external ground fire. The requirements for emergency venting are based on the heating of the liquid in the tank. Therefore, the wetted areas (exposed to an external ground fire, such as a spill) vary with different tank designs: 55 percent of the total exposed area of a sphere or spheroid; 75 percent of the total exposed area of a horizontal tank; and, the first 30 ft (9 m) abovegrade of the exposed shell area of a vertical tank.

For tanks and storage vessels designed for pressures over 1 psig (7 kPa), the total rate of venting is given in Table 11-4C. When the exposed wetted area or the surface is greater than 2,800 sq ft (260 m²), the total rate of venting is as given in Table 11-4C, or it can be calculated by the following formula:

$$CFH = 1{,}107A^{0.82}$$

where

CFH = venting requirement, in cubic feet or m³ of free air per hour.

A = exposed wetted surface, in square feet or m².

For SI units:

$$\text{m}^3/\text{hr} = 220\ A^{0.82}$$

The foregoing formula is based on $Q = 21{,}000A^{0.82}$.

The total emergency relief venting capacity for any specific liquid may be determined by the following formula:

$$\text{Cubic ft of free air per hour} = V\frac{1{,}337}{L\sqrt{M}}$$

For SI units

$$\text{m}^3/\text{hr} = V\frac{3{,}107}{L\sqrt{M}}$$

where

V = cubic ft or m³ of free air per hour from Table 11-4B
L = latent heat of vaporization of specific liquid in Btu per lb or kJ/kg
M = molecular weight of specific liquids

The required flow rate may be multiplied by the appropriate factor listed in the following schedule when protection is provided as indicated (only one factor may be used for any one tank):

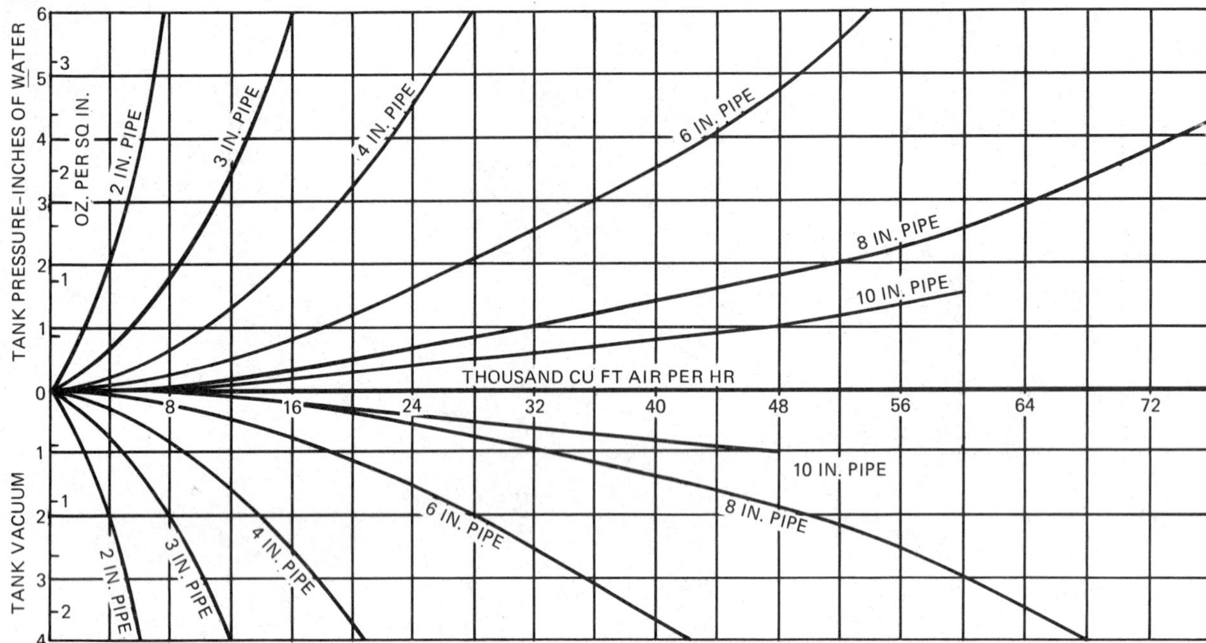

FIG. 11-4D. Average flow capacity of listed flame arresters for oil storage tanks in thousands of cubic feet of air for positive and negative pressures. (1 in. = 25.4 mm; 1 in. of water = 249 kPa; 1,000 cu ft = 28.3 m³; 1 oz per sq in. = 3 kPa/mm)

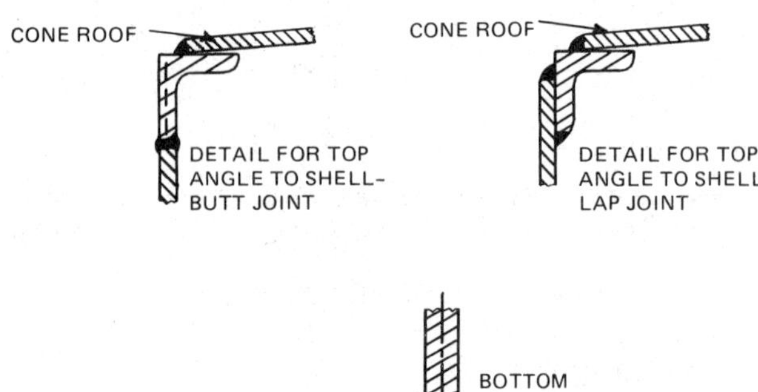

FIG. 11-4E. In cone roofed tanks the roof to shell seams are required to be weaker than the bottom to shell seams (compare top drawings to bottom drawing). Although designed as emergency vents, weakened roof seams, as illustrated, have actually served for explosion relief in some instances as the entire roof has been blown off by an internal explosion.

0.5 for approved drainage for tanks over 200 sq ft (18.6 m^2) of wetted area
0.3 for approved water spray
0.3 for approved insulation
0.15 for approved water spray with approved insulation.

When pressure tanks or vessels are exposed to fire, a violent BLEVE may result if the steel in the vapor space is softened by heat. Localized overheating of the tank shell has been caused by vent fires impinging on tank surfaces. When an exposure fire builds up pressure within the tank, a BLEVE will take place. The outlets of all vents and vent drains on aboveground tanks designed for 2.5 psi (17 kPa) or greater should be arranged to prevent localized overheating of any part of the tank from a vent fire.

Foundations and Supports: Tanks should be set on firm foundations and adequately supported. Vertical tanks are normally set on a slightly elevated pad to provide a sound base, and normally above the adjacent ground level to protect the bottom of the tank from any water in the area. Any exposed piling or steel supports under any flammable liquid tank are to be protected by fire resistive materials to provide a fire resistance rating of not less than two hours.

Drainage and Dikes: Where waterways or properties would be endangered by release of liquids stored in tanks, it is necessary to provide a means to control any spillage. The most desirable method is to locate the tanks on sloping ground. Directional dikes or drainage ditches could then divert spillage away from all tanks and into an impounding basin where the liquid could burn safely without exposing other tanks, property, or waterways. Where impounding basins cannot be used, dikes built around the tanks can prevent the spread of liquid. Dikes may be constructed of earth, concrete, or steel built to withstand the lateral pressure of a full liquid head. When several large tanks are in a single diked enclosure, it may be desirable to place "spill dikes" between tanks. While only a minimum of 18 in. (457 mm) high, these intermediate dikes will prevent any small spills from exposing other tanks in the enclosure. Such spills may result from a leaking valve or connection, or from overfilling the tank.

Diked enclosures are designed to contain the greatest amount of liquid that can be released from the largest tank within the enclosure (assuming a full tank). When calculating the volumetric capacity of the diked enclosure, the volume of tanks within the diked area to the height of the dikes must be considered as unavailable space for the liquid to flow.

TABLE 11-4B. Wetted Area Versus Amount of Free Air per Hour Required for Relief Venting for Aboveground Tanks

(For purposes of calculation the capacity of venting devices, "free air" is defined as air at 14.7 psia and 60°F or 101 kPa and 15.5°C)

ft²	m²	ft³/hr	m³/hr
20	1.86	21,100	597
30	2.78	31,600	895
40	3.71	42,100	1 192
50	4.64	52,700	1 492
60	5.57	63,200	1 790
70	6.50	73,700	2 037
80	7.43	84,200	2 384
90	8.36	94,800	2 675
100	9.29	105,000	2 973
120	11.15	126,000	3 568
140	13.00	147,000	4 163
160	14.86	168,000	4 757
180	16.72	190,000	5 380
200	18.58	211,000	5 975
250	23.23	239,000	6 768
300	27.87	265,000	7 504
350	32.51	288,000	8 155
400	37.16	312,000	8 834
500	46.45	354,000	10 024
600	55.74	392,000	11 100
700	65.03	428,000	12 120
800	74.32	462,000	13 082
900	83.61	493,000	13 960
1000	92.90	524,000	14 838
1200	111.48	557,000	15 772
1400	130.06	587,000	16 622
1600	148.64	614,000	17 387
1800	167.22	639,000	18 095
2000	185.80	662,000	18 746
2400	222.96	704,000	19 935
2800	260.12	742,000	21 011

Note: Interpolate for intermediate values.

Where needed, diked areas are provided with trapped drains to remove rain water or water used for fire fighting. The best practice is to keep drain valves normally closed, and open them at intervals as needed, since permanently opened drains would discharge liquids in case of leakage from the tank. The drain valves should also be accessible under fire conditions, i.e., they should be located outside the dike. Oil separators are effective in skimming off oil flowing on the surface of water, but types of separators now available will not stop the discharge when the entire flow through the drain is oil.

Dikes higher than 6 ft (1.8 m) are not desirable. Certain precautions must be observed where circumstances require high, close dike walls. Means should be provided for access to tanks, valves, and other equipment, and for safe egress from the diked enclosure. Where Class I liquids are stored, normal operation of valves and access to tank roofs must be available without requiring entry below the top of the dike. These provisions reflect a concern for the accumulation of Class I liquid vapors reaching harmful levels when confined in the narrow space between the high dike wall and the tank.

Fire Effects on Aboveground Tanks: Actually, the probability of fire in a modern properly constructed and installed tank is much less than in an ordinary building. Where tanks and buildings are adjacent, building fires

TABLE 11-4C. Rate of Venting For Tanks With An Exposed Wetted Area Larger than 2,800 Sq Ft

ft²	m²	ft³/hr	m³/hr
2,800	260	742,000	21 011
3,000	279	786,000	22 257
3,500	325	892,000	25 258
4,000	372	995,000	28 175
4,500	418	1,100,000	31 149
5,000	464	1,250,000	35 396
6,000	557	1,390,000	39 360
7,000	650	1,570,000	44 457
8,000	743	1,760,000	49 838
9,000	836	1,930,000	54 652
10,000	929	2,110,000	59 749
15,000	1393	2,940,000	83 252
20,000	1858	3,720,000	105 339
25,000	2322	4,470,000	126 577
30,000	2787	5,190,000	146 965
35,000	3251	5,900,000	167 070
40,000	3716	6,570,000	186 043

expose the tanks more often than tank fire expose the buildings. However, experience indicates that if a tank fire occurs, the possibility of fire spread is increased materially when the tank is not built and installed in compliance with NFPA standards.

Unfortunately, a commonly encountered type of bulk plant installation uses tanks on unprotected steel supports, and which loads tank trucks by gravity immediately adjacent to the tanks. The typical fire at such installations originates at the tank vehicle, and in short order the unprotected steel supports fail and drop the tanks to the ground. The tanks may break the piping or rupture on hitting the ground, releasing the entire contents. The steel supports should be protected by two hour fire resistive coverings. The loading rack should be at least 25 ft (7.6 m) from the tanks if Class I flammable liquids are handled, and 15 ft (4.6 m) with Class II and Class III liquids.

Improper and inadequate emergency venting has been a major factor in tank failures during exposure to fire. Since inadequate or improperly designed vents have been the cause of severe BLEVEs, and have resulted in deaths and injuries to fire fighters and spectators, it is of vital importance that the standards on venting be followed. In fighting tank fires it is essential to cool the shell above the liquid level to keep the steel from overheating. Failure to do so could cause the tank to bulge and rupture, regardless of the adequacy of the venting.

A major factor in fires involving storage tanks has been the failure of piping and valves. Such failures have resulted in adding the contents of the tanks to the ground fire. Pipe systems may be located either aboveground or underground. At bulk installations, piping from aboveground tanks is normally placed aboveground to avoid corrosion problems and to aid in detecting pipe leaks. If piping is in open trenches, suitable fire barriers should be placed at certain intervals to prevent the flow of liquid from one section of the plant to another. Underground piping is not subject to fire exposure. All piping must be protected against physical damage and excessive stresses which arise from expansion, contraction, vibration, and settlement. Materials which are subject to failure due to thermal shock, such as cast iron, should not be used. Steel or modular iron pipe and valves should be used for external tank connections through which liquid would normally flow, unless the chemical characteristics of the liquid stored are incompatible with steel. Welded pipe connections or flanged joints are preferred for aboveground piping, particularly for large size pipe. Screwed pipe connections larger than 3 in. (76 mm) in size are subject to disengagement in a long exposure fire unless the connections are back welded. Pipe joints dependent upon the friction characteristics of combustible materials for mechanical continuity of piping are subject to failure under fire exposure conditions.

Most storage tank fires originate with an internal explosion or by a spill fire exposing the tanks. In a typical prolonged tank fire, the shell of a large vertical tank will fold into the tank above the burning liquid level, without splitting the tank shell.

Internal explosions of storage tanks can occur when the vapor space above the liquid reaches the flammable range. Liquids with flash points near the stored temperature are most susceptible to ignition. Liquids with low flash points will have a flammable vapor-air mixture in the vapor space if the temperature is markedly reduced, and high flash point liquids will form a flammable vapor-air mixture when heated. During fire exposures, internal explosions have occurred in tanks containing liquids with flash points above the stored liquid temperature. Such explosions occur because the liquid is heated by the exposure fire and the vapor passes into the flammable range at a time when a part of the tank shell in the vapor space is hot enough to ignite the vapors.

Floating roof tanks provide greater firesafety for aboveground installation, and as a result, are permitted to be located closer to a property line than are other types of tanks. Explosions may occur, however, when floating roof tanks are virtually empty of liquid or the roof has been allowed to rest on the low level supports. (This may create a flammable vapor space.) When fires occur in floating roof tanks, they are normally restricted to the seal space between the roof and the shell and often can be extinguished by portable extinguishers or hand held foam lines. There have been a few cases, however, where a floating roof tank has been overfilled or the roof has sunk.

Where it is judged they are needed, fire protection equipment should be installed on tanks. (See NFPA 11, *Standard for Low Expansion Foam and Combined Agent Systems.*)

Underground Storage Tanks

Underground tanks are designed to safely withstand the service to which they are subjected, including the pressure of the earth, pavement, or possible aboveground vehicle traffic.

Construction: Rigid specifications established by UL assure reasonable safety for those tanks bearing its label. Underground tanks also may be of unlined concrete for the storage of liquids having a specific gravity of 40° API or heavier. Lined concrete tanks may be used for liquids having a lighter specific gravity, providing the lining is suitable for the liquid being stored and has satisfactory adherence to the concrete.

Installation: Underground tanks are generally considered the safest form of storage. Such tanks may be buried outside or under buildings. Tanks that are buried underneath buildings should have fill and vent connections outside the building walls. Tanks should be set on firm foundations and surrounded with soft earth or sand that is well tamped into place.

Underground tanks must be protected against damaging loads imposed by the cover over the tanks and such factors as building foundations and vehicular traffic. Normally, no special protection is needed if the tanks are well supported underneath and buried to a sufficient extent. However, if tanks are located in areas where higher than normal loads may be imposed, paving or additional earth coverage may be necessary. Piping subject to possibly damaging loads or vibrations is frequently protected by sleeves, casings, or flexible connectors to ensure the integrity of the line.

The normal life expectancy of properly installed underground steel tanks is 15 to 20 years. If improperly installed and in a corrosive soil they may leak in less than three years. The soil in which the tank is buried is very important, since some soils may be highly corrosive because of their chemical composition or moisture content. This is particularly true if construction debris, cinders, shale, or other foreign matter is mixed, even in very small quantities, with otherwise "clean" backfill. The use of homogeneous clean backfill and protective coatings prolongs the life of steel tanks and piping. Cathodic protection of buried tanks and piping is often necessary. NFPA 30 requires that tanks and their piping be protected by either using corrosion resistant materials of construction, or by a properly engineered, installed, and maintained cathodic protection system in accordance with recognized standards of design.

Electrolytic corrosion in an electrically conductive tank and piping system may occur at points where metals such as steel and brass, having different electromotive qualities, are connected. Connections of two dissimilar metals should be avoided to prevent galvanic corrosion.

Stray electrical currents may set up corrosive action, but their presence may be difficult to determine until after the action has progressed to the point where damage has been done to the tank or piping. Cathodic protection or insulation is sometimes used to protect underground tanks from stray currents.

Fiberglass reinforced plastic tanks for underground installation eliminate the corrosion problem encountered with steel tanks. However, it is vitally important that the manufacturer's instructions for proper installation of these tanks be followed closely.

Tanks should be anchored or weighted to prevent floating in locations where groundwater level is high or may rise in case of flood. Details of installation and protection are covered in NFPA 30. Typical sizes for underground tanks are shown in Table 11-4D.

The proximity of underground tanks to a building foundation is not a direct measure of the potential danger to the building if a leak should develop in the tanks. Leaking contents from underground tanks have been known to travel several miles underground (1 mi equals 1.6 km) or penetrate 24 in. (610 mm) of waterproofed concrete before appearing in a building. Tanks suspected of leaking should be tested hydrostatically with the same liquid stored in the tank. (See Fig. 11-4F.) Air tests or testing with liquids other than that stored in the tank have been proven to be dangerous and inconclusive for detecting suspected leaks in underground tanks. Additional details of the causes of corrosion, locating leaking tanks,

TABLE 11-4D. Typical Sizes of Underground Flammable Liquid Tanks

Capacity		Diameter			Length		
Gal	L	Ft	In.	m	Ft.	In.	m
300	1 136	3	0	0.91	6	0	1.83
560	2 120	4	0	1.22	6	0	1.83
1,000	3 785	4	0	1.22	11	0	3.35
1,000	3 785	5	4	1.62	6	0	1.83
1,000	3 785	4	0	1.22	11	0	3.35
1,000	3 875	5	4	1.62	6	0	1.83
1,500	5 678	5	4	1.62	9	0	2.74
2,000	7 571	5	4	1.62	12	0	3.66
2,500	9 463	5	4	1.62	15	0	4.57
3,000	11 356	5	4	1.62	18	0	5.49
3,000	11 356	6	0	1.83	14	0	4.27
4,000	15 142	5	4	1.62	24	0	7.32
4,000	15 142	6	0	1.83	19	0	5.79
5,000	18 927	6	0	1.83	24	0	7.32
6,000	22 712	6	0	1.83	29	0	8.84
6,000	22 712	8	0	2.44	16	0	4.88
7,500	28 390	8	0	2.44	20	0	6.10
10,000	37 854	8	0	2.44	27	0	8.23
10,000	37 854	9	0	2.74	21	0	6.40
10,000	37 854	10	0	3.05	17	0	5.18
10,000	37 854	10	6	3.20	15	7	4.75

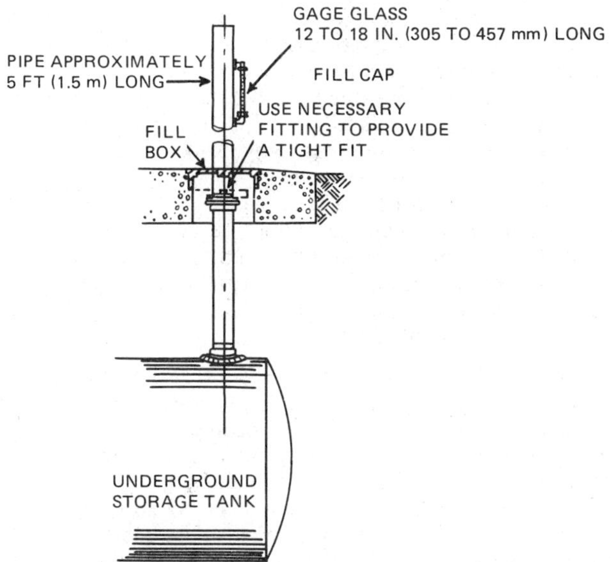

FIG. 11-4F. Simple leak test equipment in a typical underground tank installation.

and removal of liquids in the ground are covered in NFPA 329M, *Recommended Practice for Handling Underground Leakage of Flammable and Combustible Liquids.*

Tanks Inside Buildings

Design: Tanks designed for installation in fire resistive enclosures within buildings have the same metal thicknesses and design features as required for all tanks. Fuel oil tanks designed for use inside buildings without being enclosed in a fire resistive cutoff room are normally restricted to less than 660 gal (2500 L) capacity. Although the metal thicknesses for the tanks designed for differing situations are approximately the same, the location of the pipe connection openings differ. Pipe connections for underground tanks and enclosed fuel oil tanks inside buildings are in the top of the tank only, whereas unenclosed tanks are provided with bottom outlets for gravity feed piping to such installations as oil burning equipment. In certain specialized processing operations, large tanks are required and are designed as low pressure storage tanks or pressure vessels as specified for aboveground storage tanks.

Installation: Storage tanks inside buildings will vary, depending upon the class of liquid and the occupancy of the building. For specific requirements for the installation of tanks inside buildings, see NFPA 30.

Where it is impractical to install a gasoline tank underground, a special liquid tight and vapor tight enclosure without backfill may be used. Such an enclosure is constructed of 6 in. (152 mm) of reinforced concrete with access through the top of the enclosure. Means are also provided for using portable ventilating equipment to discharge flammable vapors to the outside of the building.

Fuel oil tanks over 660 gal (2500 L) capacity may be placed inside buildings in enclosures with a fire resistance rating of not less than three hours. Such tank enclosures may have an opening protected by a self closing fire door designed for Class A openings and a raised, noncombustible, liquid tight sill or ramp. If the sill or ramp is more than 6 in. (152 mm) high, the enclosure walls, up to a height corresponding to the calculated liquid level from a possible leaking tank, should be built to withstand the lateral pressure caused by the liquid head. Provision also should be made for ventilating the enclosure prior to inspection or repair of the tank. Additional storage may be permitted if the fuel oil tank enclosure is located in a room cut off from the rest of the building by walls, floor, and ceiling of two hour fire resistive materials. The installation of fuel oil tanks connected to oil burners is covered in NFPA 31, *Standard for the Installation of Oil Burning Equipment.*

Where the nature of certain types of processing operations calls for tanks to be located inside buildings, particular attention should be given to the desirability of providing automatic sprinkler systems or water spray systems, thermally actuated valves adjacent to the tanks in pipelines, and steel valves and pipe. Vents and fill pipes from tanks inside buildings should terminate on the outside of the building.

Gaging of Tanks

Tank openings for gaging or measuring the quantity of liquid may permit the escape of vapors during the gaging operation. Such openings are particularly undesirable in tanks located in buildings or buried under basements, and are prohibited by NFPA standards unless protected by a spring loaded check valve or other approved device. Substitutes for manual gaging include heavy duty flat gage glasses; magnetic, hydraulic or hydrostatic remote reading devices; and sealed float gages. These devices, however, must be maintained in reliable operating condition. Ordinary gage glasses should not be used since their breakage may permit the escape of liquid.

Cleaning of Tanks

The cleaning of tanks for the purpose of making repairs requires great care. Precautionary measures must be taken to avoid igniting flammable vapors and to protect personnel against toxic vapors (API 1979a; 1979b).

Work on empty tanks must be performed only under the supervision of persons well versed in the fire and explosion hazard, and in the procedures required to properly safeguard the operation. Unless the work can be done with the tank and all connected piping and fittings completely filled with water, all vapors should be removed by cleaning with steam or chemical solutions, or by displacement with water, air, or inert gas. In some instances, the tank vapor space may be filled with inert gas by specially qualified personnel to provide a safe atmosphere. The selection of the method to render a tank safe will depend upon several factors, such as character of liquid, size of tank, flammability and reactivity of residues, and type of work to be performed. With many reactive materials it will be necessary to obtain information from the manufacturer regarding recommended practices for safe cleaning.

Removal of Flammable Vapors by Displacement: This is sometimes referred to as "purging," and may be accomplished by one of several methods.

1. **Displacement with Water**: Where the flammable liquid previously contained is known to be readily displaced by or soluble in water, it can be removed completely by using water to alternate or fill and drain the tank. The operation should be repeated several times until tests with a combustible gas indicator show that the vapors are no longer present. Acetone and ethyl alcohol are examples of water soluble liquids.
2. **Displacement with Air**: Frequently, flammable vapors may be removed by purging with air by the use of venturi type air movers or low pressure blowers, and a safe atmosphere is sustained by continuous ventilation. Air movers should be restricted to those operated by steam, air, or by electric motors approved for use in the atmosphere involved. When steam operates the air mover, the air mover should be bonded or in electrical contact with the tank. When small tank openings cannot accommodate an air mover, the contents can sometimes be purged by compressed air connected to a metallic pipe bonded to the tank. Care should be exercised to prevent overpressuring a small tank if compressed air is used. Irregularly shaped tanks may not be thoroughly purged by this method if the air stream leaves pockets which cannot be reached effectively with the uncontaminated air. In air purging, the concentration of flammable vapors in air in the tank may go through the flammable range before safe atmospheres are obtained. Therefore, all precautions must be taken to minimize the hazards of ignition by static electricity. Air movers should be clamped or bolted to, and thereby inherently bonded to, the vessel being ventilated. (See Fig 11-4G.)
3. **Displacement with Inert Gas**: When nitrogen in cylinders, or carbon dioxide in low pressure containers or in solid form (dry ice) are available in sufficient quantity, they may be used to purge the flammable vapors from tanks without the hazards incident to having the vapor-air mixture in the tank vapor space pass through the flammable range. This procedure should be followed by air ventilation of the tank. High pressure carbon dioxide, such as from a fire extinguisher, should not be used because static electricity is generated. Several explosions or fires have occurred from using CO_2 extinguishers for inerting tanks or vessels.

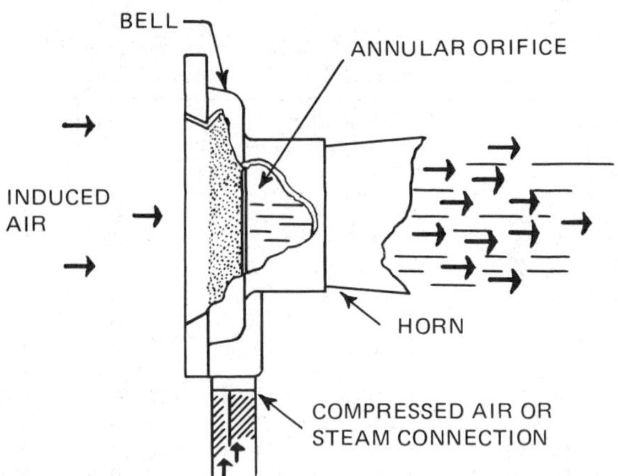

FIG. 11-4G. Schematic drawing of the operation of an air remover. Compressed air or steam is admitted to the bell of the air remover and enters the horn through the annular orifice around the base. As the steam or compressed air passes through the horn its expansion induces a rapid flow of air, equal to approximately ten times the volume of the steam or compressed air.

Inerting of the Vapor Space: Inerting is a means of safeguarding a tank by reducing the oxygen content to the point where combustion cannot take place in the vapor space. However, individuals in direct charge of the work must be thoroughly familiar with the limitations and characteristics of the inert gas being used. Attempting such work without proper knowledge or equipment can be hazardous because of the false sense of security engendered. The oxygen content should be maintained at substantially zero during the entire period when work is in progress. Inerting gases include carbon dioxide and nitrogen. Both may be obtained in tanks, and carbon dioxide may also be obtained in solid form.

Removal of Residues: Liquid or solid residues might release flammable vapors during "hot work." The residues must be removed, and this can be accomplished by means of steam or chemical cleaning or by other recognized methods. In steam cleaning, the rate of steam supply should be sufficient to exceed the rate of condensation, and the steam nozzle should be electrically bonded to the container shell. During the steam cleaning process the entire container is, of course, heated close to the boiling point of water (212°F or 100°C). Chemical cleaning may be necessary to remove some residues, but this may present personnel health hazards which should be guarded against.

Large tanks are continuously ventilated by air movers at the roof manholes. After the atmosphere is sufficiently free for safe entry, more vapor may be released from the residue while it is being removed from the tank. By continuous ventilation during work, the vapor concentration can be kept at a safe limit.

Special care must be exercised to eliminate any source of ignition in the vicinity of the container or in the path of vapors being displaced. Bonding, either with special bond wires or by metal-to-metal contact with clamped or bolted connections providing inherent bonding, must be used where static producing devices are used. All electrical equipment such as inspection lights and motors used in connection with the cleaning operations, must be designed for such use.

Tests for the presence of flammable vapors constitute the most important phase of the cleaning or safeguarding procedure and must be made before commencing any alterations or repairs, immediately after starting any welding, cutting or heating operations, and frequently during the course of such work. The tests made with a combustible gas indicator in good working order will normally produce reliable readings. Considerable care should be exercised to ensure that the indicator is calibrated for the vapors involved, correctly scaled, properly used and read, and the readings properly interpreted and applied. Where an inert gas is used, the oxygen content of the tank is measured to determine whether a hazardous condition exists. This may be determined directly by the use of an oxygen indicator or indirectly, by the use of an indicator showing the concentration of the inert gas being used. All testing must be conducted by persons experienced in the operation and limitations of the indicators used.

When hot work is to be done on small tanks or containers that cannot be entered, the combustible gas indicator should show no appreciable indication of the presence of flammable vapors. Extra precautions are necessary if the container last contained a high flash point liquid, because no vapor reading will show on the indicator. If welding or cutting is done, the heat may vaporize some of the liquid, create a flammable vapor-air mixture in the drum, and explode it. Such containers require special precautions.

Hot work may be performed safely, under qualified supervision, in tank cars, tank trucks, and tanks that can be entered when the flammable vapors are under 20 percent of the lower flammable limit. Additional precautions such as protective clothing and self-contained breathing apparatus, are necessary to protect the health of persons entering tanks that have contained leaded gasoline or other highly toxic residues.

Where toxic materials have been stored in a tank, additional tests may be required to determine if the atmosphere is safe from the standpoint of health. For example, leaded gasoline tanks need to be lead free before entrance can be made without breathing equipment and specialized protective clothing.

More complete instructions for cleaning tanks or containers are covered in the following publications. NFPA 327, *Standard Procedures for Cleaning or Safeguarding Small Tanks and Containers*; NFPA 306, *Control of Gas Hazards on Vessels*. RP 2013; *Cleaning Tank Vehicles Used for Transportation of Flammable Liquids* (API 1975a); and RP 2015, *Cleaning Petroleum Storage Tanks* (API 1975b).

OTHER STORAGE OF FLAMMABLE LIQUIDS

The principal hazard of closed container storage is the possibility of overpressure failure of the container when exposed to fire. This release of liquid adds to the intensity of a fire and may cause the rupture of other containers, resulting in a rapidly spreading fire. Fire tests prove that ordinary automatic sprinkler systems may be inadequate to control a fire involving drums of flammable liquids, or to prevent overpressuring of drums if flammable liquid containers are piled too high. See Table 11-4E for storage limitations in inside storage rooms.

TABLE 11-4E. Storage Limitations for Inside Storage Rooms

Automatic Fire Protection* Provided	Fire Resistance	Maximum Size Ft²	Maximum Size m²	Allowable Loading of Floor Area Gals/sq ft	Allowable Loading of Floor Area L/m²
yes	2 hr	500	46.45	10	407
no	2 hr	500	46.45	4	163
yes	1 hr	150	13.94	5	204
no	1 hr	150	13.94	2	82

Container Storage in Buildings

Special storage buildings or rooms, and other rooms or portions of buildings where containers may be stored, should be designed to protect the containers from exposure to fires in other portions of the building. The life hazard to the occupants of buildings, exposure to other buildings, building construction, and the degree of fire protection provided are factors to be considered when evaluating the amount of container storage in buildings. Details and limitations on closed container storage are given in the NFPA 30.

Specially designed metal storage cabinets are available for storing up to a total of 60 gal (227 L) in small containers. (See Fig. 11-4H.) Specifications for wooden storage cabinets also are given in NFPA 30.

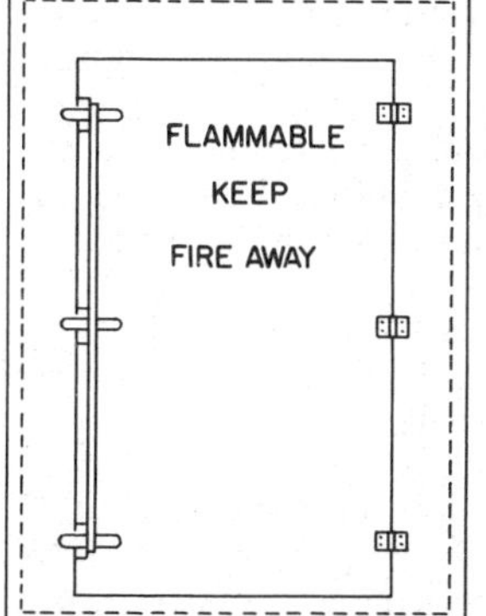

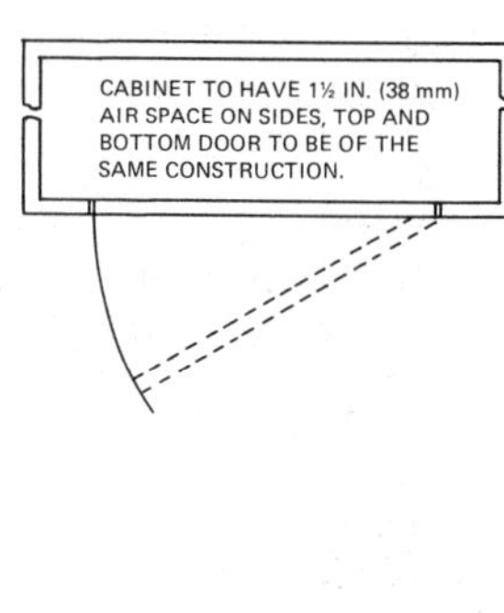

CABINET TO HAVE 1½ IN. (38 mm) AIR SPACE ON SIDES, TOP AND BOTTOM DOOR TO BE OF THE SAME CONSTRUCTION.

FIG. 11-4H. Typical metal storage cabinet recommended by NFPA Committee on Flammable and Combustible Liquids for quantities less than 60 gal (227 L). Material is 18 US gage sheet iron, tight joints. Door has three-point lock with sill raised to at least 2 in. (51 mm) above the bottom of the cabinet.

Drum Storage Outdoors

Outdoor drum storage should be located in such a manner as to reduce the spread of fire to other materials in storage or to other property. (See Table 11-4F.) Areas used for drum storage should be kept free of combustibles and open flames, and smoking should be prohibited.

HANDLING OF FLAMMABLE AND COMBUSTIBLE LIQUIDS

Loading and Unloading

Spills may occur at loading and unloading stations for tank cars and tank vehicles. Tank vehicle and tank car

TABLE 11-4F. Outdoor Liquid Storage in Containers and Portable Tanks

	1				2				3		4		5	
Class	Container Storage Max. per Pile*§				Portable Tank Storage Max. per Pile Gallons*				Distance Between Piles or Racks		Distance to Property Line That Can Be Built Upon†‡		Distance to Street, Alley, or a Public Way‡	
			Height				Height							
	Gal	m³	FT	m	Gal	m³	FT	m	FT	m	FT	m	FT	m
IA	1,100	4.16	10	3.05	2,200	8.32	7	2.1	5	1.5	50	15.2	10	3.0
IB	2,200	8.32	12	3.66	4,400	16.65	14	4.3	5	1.5	50	15.2	10	3.0
IC	4,400	16.65	12	3.66	8,800	33.30	14	4.3	5	1.5	50	15.2	10	3.0
II	8,800	33.30	12	3.66	17,600	66.62	14	4.3	5	1.5	25	7.6	5	1.5
III	27,000	88.28	18	5.49	44,000	166.56	14	4.3	5	1.5	10	3.0	5	1.5

* See 4-8.1.1 of NFPA 30 regarding mixed class storage.
† See 4-8.1.3 of NFPA 30 regarding protection for exposures.
‡ See 4-8.1.4 of NFPA 30 for smaller pile sizes.
§ For storage in racks, the quantity limits per pile do not apply, but the rack arrangement shall be limited to a maximum of 50 (15.2 m) in length and two rows of 9 (2.7 m) in depth.

stations for Class I liquids should be located a minimum of 25 ft (8 m) from storage tanks, other plant buildings, and the nearest line of property that can be developed. Level ground is desirable, and drains, diversionary curbs, or natural ground slope can be utilized to prevent spills from spreading to other parts of the plant or to other property.

Bonding provisions for protection against static sparks must be provided at all loading stations when Class I liquids are loaded. The bonding must be provided between the fill pipe or piping and the tank vehicle. Bonding connections are made before the dome covers are opened. Closed metal piping systems (eliminating exposure to air) for loading and unloading may eliminate the need for special bonding provisions, since the piping system is inherently bonded to the vehicle.

Stray electrical current protection is necessary for ship and tank car loading and unloading in those areas where stray currents may be present.

Tank ships and barges are loaded and unloaded at oil piers which may or may not be used exclusively for this purpose. The location of the oil piers is governed by direction and velocity of the waterway, the range of tides, and the direction and frequency of prevailing high velocity winds. Hazards related to the loading and unloading of vessels are comparable to those for tank vehicles and tank cars, except for the quantities of liquid involved.

Many liquids, including gasoline, jet fuels, toluene, and light fuel oil, can build up dangerous static electrical charges on the surface of the liquid. If a flammable vapor-air mixture is present at the surface of the liquid when high static electrical discharges occur, an explosion or fire can result. Excessive turbulence, pumping two dissimilar materials, the free fall of liquid through the vapor space, and use of filters capable of removing micron sized particles are the most common causes of explosions or fires originating from static electricity in aboveground tanks, tank vehicles, tank barges, and tank ships. Where flammable vapors may exist, and to reduce static caused ignitions, it is recommended that the delivery rate be slowed until the fill pipe is covered. A minimum 30 second relaxation time should be provided downstream of a filter.

Piping and Valves

Substantial piping systems protected against physical damage are preferable to portable containers for conveying quantities of liquids throughout buildings. All piping systems should be designed so that in case of pipe breaks, liquid will not continue to flow by gravity or by siphoning. Valves should be provided at accessible points to control or stop the flow. Emergency remote controls are frequently provided for valves or pumps, particularly at dispensing locations. Other remotely controlled valves installed for normal operating procedures can frequently be used during fire emergency procedures to control or stop the flow of liquids. Valves are available that close automatically if subjected to fire conditions. Dispensing outlets for systems under gas pressure or gravity head should be equipped with self-closing valves.

Pipe materials should be used that (1) are resistant to the corrosive properties of the liquid handled; (2) have adequate design strength to withstand the maximum service pressure (including shock and surge pressures which may be expected) and temperature, and, (3) when possible, are resistant to physical damage and thermal shock. Where low melting point materials such as aluminum and brass, materials that soften on fire exposure such as plastics, or nonductile material such as cast iron, are necessary, special consideration should be given to their behavior on exposure to fire. After installation, piping systems should be tested at 150 percent of the maximum anticipated pressure of the system, or pneumatically tested to 110 percent of the maximum anticipated pressure of the system, but not less than 5 psig (35 kPa) at the highest point of the system.

Since some valves are used infrequently, it is necessary to make periodic maintenance inspections to assure that the valves will operate during emergency conditions. All aboveground piping and connections should be inspected periodically to detect and prevent leakage. Internal corrosion in pipes is a particular problem when liquids are corrosive, or where piping is used for the continuous flow of liquid under high pressure operations common at refineries or chemical plants. Maintenance inspections of the thickness of pipe walls can be performed in several ways similar to the inspection of tanks. Underground piping is also subject to external corrosion and should be protected against such corrosion by suitable coatings and cathodic protection. Tests for possible leaks in underground piping can be performed hydrostatically using the liquid normally handled in the piping system.

Dispensing and Handling Methods

Large quantities of flammable or combustible liquids are best transferred through piping by pumps. Gravity flow is not desirable except as required in process equipment. If positive displacement pumps are used, they should be provided with a pressure relief that discharges back to the tank or to the pump section. The transfer of Class I or Class II liquids by means of air pressure is not good practice. Although inert gas may be used for transferring all classes of liquids, and air pressure may be used for transferring Class III liquids, they are acceptable only if the pressure is controlled, as by use of pressure relief devices, to limit the pressure so it cannot exceed the design pressure of the vessel, tank, or container.

The safest method for handling flammable or combustible liquids is to pump the liquid from buried storage tanks through an adequately designed piping system protected from physical damage, to the dispensing equipment located outdoors or in specially designed inside storage rooms. Such a room should have at least one exterior wall for explosion relief and accessibility for fire fighting, interior walls with 2 hour fire resistance ratings, adequate ventilation and drainage, and be free of sources of ignition.

Where solvents are pumped from storage tanks to the point of use in an industrial building, emergency switches should be located in the dispensing area at the normal exit door or at other safe locations outside the fire area, and at the pumps to shut down all pumps in case of fire.

Where dispensing is by gravity flow, such as in the filling of containers in an industrial operation, a shutoff valve should be installed as close as practical to the vessel being unloaded. A control valve should be located near the end of the discharge pipe. Additionally, in some filling operations, a heat actuated valve is desirable to shut off the flow of liquid.

The preferred method of dispensing flammable and combustible liquids from a drum is by use of a laboratory tested hand operated pump drawing through the top. In certain cases, a laboratory tested drum faucet may be used for dispensing from a drum. However, hand operated pumps are safer than faucets because the hazard of leakage is reduced.

For handling small quantities of flammable and combustible liquids, safety cans are preferred. Safety cans are substantially constructed to avoid the danger of leakage and are designed to minimize the likelihood of spillage or vapor release and of container rupture under fire conditions. (See Fig. 11-4I.) Liquids can also be dispensed from the original shipping containers. Open pails or open buckets are never used for storage.

Flammable liquids should always be handled and dispensed in a well ventilated area free of sources of ignition, and bonding should be provided between the dispensing equipment and the container being filled. (See Fig. 11-4J.)

FIG. 11-4I. Typical safety cans having pouring outlets with tight fitting caps or valves normally closed by springs, except when held open by hand, so that contents will not be spilled if a can is tipped over. The caps also provide an emergency vent when the cans are exposed to fire.

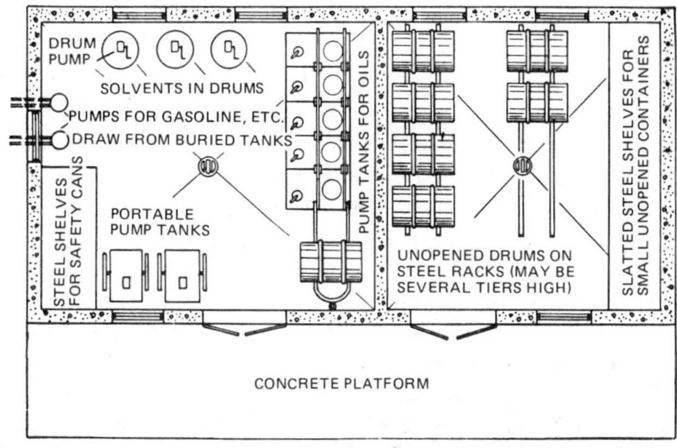

FIG. 11-4J. Layout of a storage and dispensing house.

TRANSPORTATION OF FLAMMABLE AND COMBUSTIBLE LIQUIDS

Tank Vehicles

NFPA 385, *Standard for Tank Vehicles for Flammable and Combustible Liquids*, (hereinafter referred to as NFPA 385), provides for substantially constructed vehicles and tanks that present relatively little danger of fire when involved in minor traffic accidents. Under these recommendations, tanks are constructed to withstand all but the most violent impact without rupturing and releasing liquid. Vents are provided so that if a tank is subjected to fire, vapor will burn at the vent, avoiding the danger of rupture from excessive internal pressure.

For all but viscous liquids, a shutoff valve located inside the shell of the cargo tank is required. It is kept closed except during loading or unloading operations. A shear section is provided in the piping connected to the internal valve. Therefore, in an accident which could damage the piping, the piping breaks at the shear section, leaving the internal valve undamaged and closed.

The operating mechanism for the valve is also required to have a secondary control remotely located which can be used to shut off the valve in case of fire or severe spillage during unloading. In addition, a fusible section is required in the control mechanism for the valve, which permits the valve to close automatically in case of fire. (See Figs. 11-4K and 11-4L.) Other important fire protection features are detailed in NFPA 385.

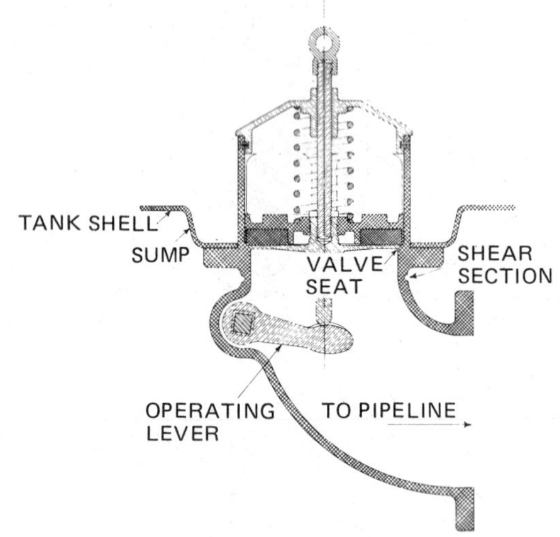

FIG. 11-4K. Shutoff valve located inside tank shell with shear section to leave valve sheet if discharge faucet breaks.

In the past, many cities have attempted to minimize fire problems by specifying the maximum sizes of cargo tanks on trucks. The wisdom of this sort of provision is dubious. Assume, for example, that it is necessary to transport 4,000 gal (15 000 L) of gasoline over city streets to make local deliveries. A legal restriction as to the maximum size of the tank might require the use of four 1,000 gal (3800 L) tank trucks in place of one 4,000 gal (15 000 L) tank. This quadripules the tank truck traffic, with four times the accident potential. The danger from a gasoline fire is not directly proportionate to the quantity of gasoline involved. For this reason, NFPA standards have not recommended any limitation on the maximum size of tank trucks.

Tank truck traffic through congested districts of cities should be avoided as far as possible, since any fire is likely to have more serious consequences in congested districts than in sparsely settled areas. Bypass routes such as those commonly specified for all kinds of through truck traffic, are the obvious solution to this particular problem.

Parking of loaded tank vehicles is another feature that can be appropriately regulated by municipalities. Such parking can be prohibited on city streets. Similarly, tank vehicles should not be parked in public garages. Permissible parking locations for tank vehicles may be specified, if necessary. Any property zoned for aboveground oil storage, for example, the ordinary bulk oil plant, should be a location where tank vehicles can be parked without any undue increase in hazard to the public.

Transportation of flammable liquids in tank vehicles in interstate commerce is governed by regulations of the U.S. Department of Transportation (DOT) as contained in the *Code of Federal Regulations, Title 49—Transportation* (GPO undated). There are comprehensive specifications and labeling requirements for shipping containers for various types of products. (Absence of a label does not necessarily mean that the material is nonhazardous, however.)

Rail, Ship, and Pipeline

The transportation of flammable liquids by railroad tank cars is under the jurisdiction of DOT (GPO undated). The design of tank cars is rigidly controlled and is governed by the nature of the contents being carried. Emergency venting requirements are included in DOT regulations.

The transportation of flammable and combustible liquids in bulk on board vessels in the United States is under the jurisdiction of the Commandant of the U.S. Coast Guard. These transportation requirements include the design of the vessel, whether it be a tank ship or tank barge, as well as requirements for extinguishing systems or portable extinguishers. Further requirements include limitations on container storage of drums or portable tanks, restrictions for passenger carrying vessels, and many other requirements for the safe transportation of flammable and combustible liquids by water. Details of these require-

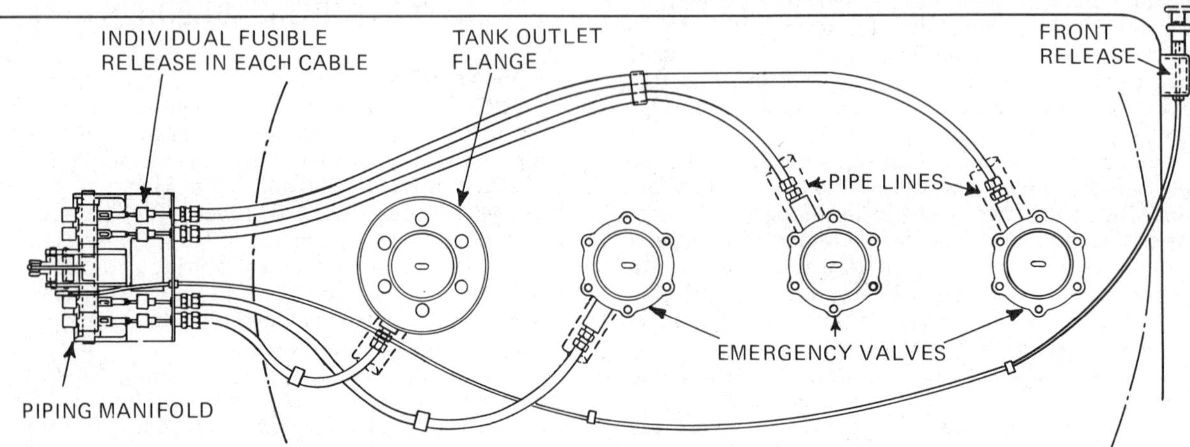

FIG. 11-4L. Bottom view of tank vehicle showing typical installation of emergency valves and controls.

ments can be secured at the nearest U.S. Coast Guard Merchant Marine Inspection Office.

Standards for pipelines transporting liquids are published by the American Society of Mechanical Engineers (ASME). These standards include piping requirements and installation recommendations.

Bibliography

References Cited

API. 1979. *Welded Steel Tanks for Oil Storage*. Standard 650. American Petroleum Institute, Washington, DC.

API. 1975a. *Cleaning Mobile Tanks in Flammable or Combustible Liquid Service*. 5th ed. RP 2013. American Petroleum Institute, Washington, DC.

API. 1975b. *Cleaning Petroleum Storage Tanks*. RP 2015. American Petroleum Institute, Washington, DC.

GPO. Undated. Title 49. "*Transportation*." *Code of Federal Regulations*. U.S. Government Printing Office, Washington, DC.

NFPA Codes, Standards, Recommended Practices and Manuals. (See the latest *NFPA Codes and Standards Catalog* for availability of current editions of the following documents.)

NFPA 11, *Standard for Low Expansion Foam and Combined Agent Systems.*

NFPA 30, *Flammable and Combustible Liquids Code.*

NFPA 30A, *Automotive and Marine Service Station Code.*

NFPA 31, *Standard for the Installation of Oil Burning Equipment.*

NFPA 77, *Recommended Practice on Static Electricity.*

NFPA 306, *Control of Gas Hazards on Vessels.*

NFPA 325M, *Fire Hazard Properties of Flammable Liquids, Gases and Volatile Solids.*

NFPA 327, *Standard Procedures for Cleaning or Safeguarding Small Tanks and Containers.*

NFPA 329M, *Recommended Practice for Handling Underground Leakage of Flammable and Combustible Liquids.*

NFPA 385, *Standard for Tank Vehicles for Flammable and Combustible Liquids.*

Additional Readings

American Conference of Governmental Industrial Hygienists, *Threshold Limit Values*, Cincinnati, OH.

American Industrial Hygiene Association, *Hygienic Guide Series*, Detroit, MI.

Babrauskas, V., "Estimating Large Pool Fire Burning Rates," *Fire Technology*, Nov. 1983.

Bulk Storage of Highly Inflammable Liquids, BP Chemicals Ltd., Technical Booklet SB 122, London, England, 1978.

Campbell, John A., "Combustible Liquids Ignite Below Flash Points in Cloth Dispersion Tests," *Western Fire Journal*, Vol. 27, No. 10, Oct. 1975, p. 9.

Carpenter, R. A., et al., "A System for the Correlation and Physical Properties and Structural Characteristics of Chemical Compounds with their Commerical Uses," Midwest Research Institute, Kansas City, MO. (reprint from *American Documentation*, Vol. X, No. 2).

Claudy, W. D., *Respiratory Hazards of the Fire Service*, National Fire Protection Association, Quincy, MA, 1957, pp. 99-130.

CONCAWE: "Methodologies for Hazard Analysis and Risk Assessment in the Petroleum Refining and Storage Industry," Report No. 10182, Netherlands, 1982.

"Controlling the Power of Flammable Liquids," *Pamphlet No. P7045*, Factory Mutual Engineering Corp., Norwood, MA.

DiMaio, Louis R., et al., "Advances in Protection of Polar-type Flammable Liquid Hazards," *Fire Technology*, Vol. 11, No. 3, Aug. 1975, p. 164.

Factory Mutual Engineering Corporation, *Handbook of Industrial Loss Prevention*, 2nd ed., McGraw-Hill Inc., NY, 1967.

Hawley, G. G., ed., *The Condensed Chemical Dictionary*, 8th ed., Van Nostrand Reinhold, NY, 1971.

Henry, Martin, ed., *Flammable and Combustible Liquids Code Handbook*, 2nd ed, National Fire Protection Association, Quincy, MA, 1984.

Hughes, John R., *Storage and Handling of Petroleum Liquids*, 2nd ed., Chas. Griffin & Co. Ltd., London, England, 1978.

Johnson, Donald M., "New Developments in Bulk Storage of Flammable Liquids," *SFPE TR 81-1*, Society of Fire Protection Engineers, Boston, MA, 1981.

Kirk, R. E., and Othmer, D. F., eds., *Encyclopedia of Chemical Technology*, 2nd ed., 22 Vols., Interscience Encyclopedia, Inc., NY 1963-1969.

Leffler, William L., *Petroleum Refining for the Non-Technical Person*, Pennwell Books, Tulsa, OK, 1979.

Lois, E., and Swithenbank, J., "Fire Hazards in Oil Tank Arrays in a Wind," *Seventeenth Symposium (International) on Combustion*, The Combustion Institute, Pittsburgh, PA, 1978, pp. 1087-1098.

Marks, Alex, *Petroleum Storage Principles*, Pennwell Books, Tulsa, OK, 1983.

Mellan, I., *Industrial Solvents Handbook*, Reinhold, New York, 1970, *Merck Index of Chemicals and Drugs*, 8th ed., Merck & Co., Rahway, NJ, 1968.

National Safety Council, *Accident Prevention Manual for Industrial Operations*, 6th ed., Chicago, IL, 1969.

Perry, J. H., and Chilton, C. H., eds., *Chemical Engineers' Handbook*, 5th ed., McGraw-Hill, NY, 1974.

"Safe Storage and Handling of Flammable and Combustible Liquids," *Professional Safety*, Vol. 20, No. 12, Dec. 1975, pp. 24-27.

"Safety Digest of Lessons Learned," Volumes 3, 6, 7, 8, 9, American Petroleum Institute, NY, 1982.

Safety and Fire Protection Committee, Chemical Manufacturers Association, Inc., *Guide for Safety in the Chemical Laboratory*, 2nd ed., Van Nostrand Reinhold, NY, 1972.

Sax, N. I., *Dangerous Properties of Industrial Materials*, 5th ed., Van Nostrand Reinhold, NY, 1975.

"Shipping," Title 46, Parts 146 to 149; Title 49, "Transportation," Parts 171 to 178, *Code of Federal Regulations*, U.S. Government Printing Office, Washington, DC.

Stevens, A., "Flammable Liquids —Why the Hazard?" *Journal of Chemical Education*, Vol. 56, No. 3, Mar. 1979, pp. 119A-124A.

Stevens, Arthur M., "Handling Flammable Liquids," *Plant Facilities*, Vol. 11, No. 2, Feb. 1979, pp. 30-33.

Van Dolah, R. W., et al., "Flame Propagation, Extinguishment and Environmental Effects on Combustion," *Fire Technology*, Vol. 1, No. 2, May 1965, pp. 138-145.

Vervalin, C. H., ed., *Fire Protection Manual for Hydrocarbon Processing Plants*, Gulf Publishing Co., Houston, TX, 1973.

Walls, W. L., "Just What Is a BLEVE?," *Fire Journal*, Vol. 72, No. 6, Nov. 1978, pp. 46-47.

Weast, R. C., ed., *Handbook of Chemistry and Physics*, Chemical Rubber Co., Cleveland, 1972-1973.

Welker, J. R., and Sliepcevich, C. M., "Bending of Wind-Blown Flames from Liquid Pools," *Fire Technology*, Vol. 2, No. 2, May 1966, pp. 127-135.

Welker, J. R., and Sliepcevich, C. M., "Burning Rates and Heat Transfer from Wind-Blown Flames," *Fire Technology*, Vol. 2, No. 3, Aug. 1966, pp. 211-218.

Welker, J. R., et al., "The Effect of Wind on Flames," *Fire Technology*, Vol. 1, No. 2, May 1965, pp. 122-129.

STORAGE OF GASES

Wilbur L. Walls, PE

This chapter describes essential firesafety characteristics of containers in which gases are stored, as well as some fundamental precepts for storage arrangements.

A gas must be stored in a structure which is gas tight over (1) the range of temperature and pressure conditions present at the storage location, as well as (2) the conditions represented by the transportation environment prior to arrival at a storage site. This necessity usually makes it feasible for the storage container to contain the source of energy needed to remove the gas from storage and convey it to the point of use. This energy is represented by the pressure of the gas in the container.

In the case of containers in which the gas is entirely in the gaseous phase (compressed gases), the pressure is applied at the charging plant or, in the case of pipelines, by compressors spaced along the pipeline. In the case of containers of gas in which the gas is partly in the gaseous phase and partly in the liquid phase (liquefied gases, including cryogenic liquefied gases), the pressure is obtained as a result of heat stored in the liquid, which is directly related to how much the temperature of the liquid is above its normal boiling point.

Therefore, gas containers, whether shaped as tanks or as pipelines, are closed pressure vessels containing considerable energy per unit of volume, and requiring careful design, fabrication, and maintenance. Furthermore, because the containers are closed and excessive pressures can develop when they are exposed to rather nominal heat sources and fire (or by overfilling in the case of liquefied gas containers), overpressure protection is usually needed.

GAS CONTAINERS

Most countries have promulgated regulations for the design, fabrication, and maintenance of gas containers. Although basically similar, the regulations differ in detail. This section concentrates on North American practices.

In North America, there are two types of gas containers—cylinders and tanks. Originally, the distinction between cylinders and tanks was based upon size. Cylinders, being the smaller, were considered portable and the tanks, being the larger, were essentially used in stationary service. Originally, there was also a distinction which reflected the pressures in the containers. Cylinders were thought of as being used for high pressures and tanks for low or moderate pressures. Over the years, these distinctions have lessened so that, today, the only real distinction lies in the regulations or codes under which the container is built.

Practically all gases must be transported from the manufacturer to the user, making the safety of the container in transportation a matter of primary concern. As a result, criteria for many gas containers reflect transportation safety conditions. Because it could be hazardous as well as uneconomical to require gases to be transferred from a shipping container to a separate container for other use, every effort has been made to utilize the same container whenever feasible. This is generally the procedure used for the smaller containers.

Gas Cylinders

Gas cylinders are fabricated in accordance with regulations and specifications of the U.S. Department of Transportation (DOT) in the United States and the Canadian Transport Commission (CTC) in Canada. The requirements are the same in both countries. Prior to April 1, 1967, DOT regulations were promulgated by the Interstate Commerce Commission (ICC), and many cylinders are still in service with ICC markings. Cylinders are generally limited to a maximum water capacity (the capacity when completely filled with water) of 1,000 lb (454 kg), or approximately 120 gal (450 L) of water.

The regulations cover the service pressure the cylinder must be designed for, the gas or group of gases that it can contain, safety devices, and requirements for in service (transportation) testing and requalification. The specifications cover such criteria as metal composition and physical testing, wall thickness, joining methods, nature of openings in the container, heat treatments, proof testing, and marking.

Mr. Walls was engineer, Gases Field Service, from 1962 through 1984 on the staff of NFPA until his recent retirement.

Gas Tanks

Customarily gas tanks are fabricated in accordance with Section VIII (Unfired Pressure Vessels) of the *Boiler and Pressure Vessel Code* published by the American Society of Mechanical Engineers (ASME) or tank fabrication standards of the American Petroleum Institute (API). ASME tanks are usually smaller tanks under moderate pressure, and API tanks are usually very large tanks under low pressure.

Cylinders or tanks that are part of transportation units such as cargo vehicles or railcars, are subject to additional criteria in the regulations, primarily to reflect the fact that the containers are on wheels. Tanks for cargo vehicles are basically ASME code tanks, while tanks for railcars are covered by specific DOT and CTC specifications.

In the late 1960s and early 1970s, a series of serious railcar derailments occurred in the U.S., reflecting the deterioration of tracks and track beds due to inadequate maintenance. These incidents were characterized by both impact and fire-caused Boiling Liquid Expanding Vapor Explosions (BLEVEs) of tank cars containing liquefied flammable gases. As a result, DOT regulations were revised in 1977 to require that these tank cars be insulated and equipped with a type of coupler that prevents vertical disengagement. Unless the insulation is contained in a steel outer jacket, the car also must be equipped with heavy steel plate head shields for impact protection. Existing cars were required to be retrofitted by the end of 1980. In the few derailments involving tank cars outfitted in this fashion, these safeguards have proved effective—a BLEVE either has not occurred at all, or has been delayed for several hours.

A number of such containers are exempted by the regulations. These include certain very small containers and containers with nonflammable cryogenic gases, including oxygen, where the pressure in the container as shipped is below 40 psia (276 kPa).

There is usually a lag in time between the need to ship a gas that is new to commerce and the promulgation of specifications for a transportation container. In such cases, DOT or CTC exemptions, which spell out safety criteria agreed upon by the authorities, are issued. Until recently, all cryogenic flammable gas cargo vehicles were operated under exemptions.

In themselves, the DOT and CTC regulations apply only to cylinders and tanks in transportation in interstate or interprovincial commerce. However, many consensus codes and standards extend these regulations to transportation in intrastate commerce, and also extend applicable criteria to use and storage at consumer sites.

The aforementioned requirements for "in service" reinspection and requalification of DOT specification containers are also extended to use and storage at consumer sites by the consensus standards and codes. However, the ASME Code and API Standards do not contain such provisions. If not covered by consensus standards and codes, in service inspection and requalification become a matter of owner judgment, and such state or local regulations as may apply, or conditions set forth in an insurance contract.

Regardless of the degree of structural integrity incorporated into gas containers, abuse of the container must be avoided. This is especially true of portable cylinders which are subject to mishandling. General handling of safety precautions for gas cylinders are contained in the *Compressed Gas Association Pamphlet P-1.*

Pipelines

Gases used in large volumes are often transported by pipelines. Natural gas is customarily transported by pipeline, as is considerable LP-Gas and some industrial gases such as anhydrous ammonia, oxygen, hydrogen, and ethylene.

Since 1968, most of the transmission and distribution of flammable gases by pipeline has been regulated in the U.S. by the DOT (Office of Pipeline Safety) and covered by federal regulations. These regulations cover such items as pipe materials; design for pressure and other stresses (which, among other criteria, require stronger piping as population density increases); piping components, including valves (emergency flow control valve spacing in the pipeline system is also affected by population density); joining methods; installation of meters, service regulators, and service lines; corrosion control; test requirements; certain operating requirements; and maintenance, including leakage surveys.

Normally, DOT regulations do not apply to piping on consumer premises. Such piping for the more common gases is covered by consensus codes and standards, notably NFPA 54, *National Fuel Gas Code* (ANSI Z223.1), and the *Code for Pressure Piping*, ANSI B31.

STORAGE SAFETY CONSIDERATIONS*

Fire protection safeguards for gas storage reflect the hazards of the container/gas combination and the hazards of the gas when it escapes from the container. Discussions of the basic hazards of gases, gas emergency control, and specific gases are especially pertinent.

During at least some period, any fire incident can manifest both of these hazards at the same time. However, the container/gas combination hazard always exists. The fire hazard of escaping gas, on the other hand, may be negligible if the gas is nonflammable.

Container/Gas Hazard Safeguards

The major container/gas hazard is the BLEVE. The BLEVE hazard is restricted to containers of liquefied gases, and the major cause of such BLEVEs in storage is fire exposure. BLEVEs resulting from corrosion of a container are far less frequent, and impact caused BLEVEs even more so for containers in storage.

Container Insulation: The BLEVE hazard is greatly affected by container features that restrict the opportunity for the container metal to be overheated. Especially notable in this respect is the presence of insulation between an exposing fire and the portion of the container subject to internal pressure. All cryogenic liquefied gas containers are insulated as a matter of functional necessity. The BLEVE hazard can essentially be eliminated by also considering insulation system behavior under fire exposure. North American standards for such systems on storage containers reflect this situation. For example, insulation for flammable cryogenic gas containers is required to be

*The philosophy of this analytical approach is developed in Section 5, Chapter 5, "Gases;" it would be worthwhile to study this before proceeding further in this chapter.

noncombustible and (if in a form such that loss of an enclosing jacket could cause a serious loss in insulating capacity, e.g., powder or granules), the container jacket is required to be steel or concrete rather than aluminum.

Containers of noncryogenic gases are not required to be insulated for operational reasons, and are seldom insulated for firesafety reasons. However, insulation of larger liquefied flammable gas tanks is becoming more frequent in specific installations where other BLEVE prevention measures, e.g., water cooling, are deficient.

A basic BLEVE safeguard is to reduce the chances of fire exposure to the container. This philosophy is also applicable to containers of nonliquefied gases (compressed gases) because, although by definition they are not subject to a BLEVE, they can still fail explosively from fire exposure.

Limiting Combustibles in the Area: Essential to this safeguard is limiting the quantity of combustibles in the vicinity of gas containers. This is valid whether the storage is indoors or outdoors. Except where small quantities of gas are involved, e.g., one or two cylinders of compressed gas, it is highly desirable that storage rooms or docks be of noncombustible or limited combustible construction (see NFPA 220, *Standard on Types of Building Construction*, for characterization). If the building which houses the storage room presents a substantial fire load, the storage room walls should have a suitable fire resistance rating.

Preferably, containers of flammable gases should not be stored with containers of nonflammable gases. This is especially true with respect to gases which, if released, could intensify combustion of the flammable gases. Notable in this respect are oxidizing gases such as oxygen or nitrous oxide. However, it is best to also avoid storage of inert gases near flammable gases because the container hazard reflects only pressure. Storage of nonflammable and oxidizing gases in the same area is safe.

Separation: As a general rule applicable to cylinders and small liquefied gas tanks, separation of flammable gases from nonflammable gases by 20 ft (6 m) is appropriate. Provision of a noncombustible barrier as high as the containers [usually 5 ft (1.5 m)] and having a fire resistance rating of at least ½ hr is an acceptable substitute for this distance where cylinders or small tanks [less than 1,000 lb (454 kg) water capacity] are involved.

The use of barriers around larger containers of flammable gases is questionable unless they permit application of cooling water, prevent pocketing of escaping gas, and allow unrestricted egress of personnel from the area in an emergency.

For restrictions over the proximity of combustibles to oxidizing gases it must be recognized that the degree of combustibility has little practical significance in the presence of these gases. For example, asphalt paving is not normally considered to be combustible, however, in contact with oxygen, especially liquid oxygen, it is not only combustible, but can be explosive.

Application of Water: During a fire exposure the application of water is a basic safeguard to prevent a BLEVE or a compressed gas container failure. In indoor locations, automatic sprinkler protection can greatly limit pressure rise from heat and high metal temperatures from fire exposure. However, for maximum effectiveness, the system should be capable of furnishing a discharge density of at least 0.25 gpm per sq ft [10 (L/min)/m^2] over an area of at least 3,000 sq ft (279 m^2), and the sprinklers should be located not more than 20 ft (6 m) above the floor where the containers are stored. Further characterization of sprinkler protection may be needed if the storage is in racks or in high piles.

A water spray fixed system of similar capacity is also effective and can be installed outdoors. For the protection of larger containers of liquefied flammable gases (chemically not self-reactive), a density in excess of 0.25 gpm of water per sq ft [10 (L/min)/m^2] of container surface is needed to prevent failure, although a 0.25 gpm density [10 (L/min)/m^2] may prevent a crack from proceeding to the point of container rupture.

Overpressure Limiting Devices: The satisfactory operation of container overpressure limiting devices is vital to controlling the BLEVE or compressed gas container failure hazard. Even though these devices by themselves cannot always prevent container failure, they do extend the prefailure time in all cases and often can prevent failure under many fire exposure conditions. It is essential that the device not be blocked closed by corrosion, paint deposits, etc., and not be mechanically damaged. Portable containers should be checked for this every time they are taken into the facility and whenever they are connected to consuming equipment or are filled.

Spring loaded pressure relief valves on the larger stationary storage containers should be tested at five to ten year intervals (more often if in corrosive or reactive gas service).

Relief devices on liquefied gas containers should always directly contact and monitor the vapor space because their relieving capacity is determined under vapor discharge conditions and is severely restricted if liquid is discharged. With the exception of some LP-Gas containers used in mobile engine fuel systems and 1 ton (907 kg) chlorine containers, portable liquefied gas containers should be stored in an upright position to attain this effect.

Care in Handling: Also reflecting the container/gas hazard, it is important that the containers not be subjected to physical abuse. Although quite sturdy as a result of their design as pressure vessels, any dent or gouge can reduce safety factors and, at the least, shorten failure times from fire exposure or lead to impact failures upon subsequent movement. If the valve is broken off on some smaller portable compressed gas containers, the nozzle reaction from escaping gas can be sufficient to propel the cylinder violently. Where the container is designed for a valve protecting cap or collar, these always should be in place during storage or movement.

DOT Requalification: DOT containers must be requalified at intervals specified in the federal regulations. These intervals vary with the kind of gas and type of container. In most cases, this interval is five years, but can be as long as 12 years. A requalified DOT container must be marked with the date of requalification. If such a marking is not present and the manufacture date marking is more than five years old on a compressed gas DOT specification container, it should be brought to the attention of whomever last charged the container.

The container owner is responsibile for requalification. In many instances the user/storer of portable gas containers is not the owner. In others, however, the user/storer is the owner, such as for LP-Gas containers used in industrial truck service and filled on site. Those persons responsible for such containers should be aware of this and the fact that the requalification procedures are rigorous and require considerable technical expertise.

Safeguards for Escaping Gas

The major escaping gas hazard is combustion of flammable gas and is, in turn, manifest as either a fire or a combustion explosion. Fire, of course, can also lead to the previously discussed explosive container failures. Again, the reader is referred to Section 5, Chapter 5, "Gases," for a discussion of the elements of this hazard.

Gases in storage do not present many sources of escaping gas beyond those already discussed under container/gas hazard safeguards. The container itself, its valves, and overpressure protection devices are the only sources of escaping gas, provided that the gas is not being transferred into and out of storage.

Inspection for Leakage: All storage containers should be inspected periodically for leakage from container appurtenances or from the container itself. Portable containers should not be placed into storage if they are leaking. To prevent this, incoming shipments should always be inspected upon arrival.

The senses of sight, sound, and smell are invaluable leak detectors. Although most gases are invisible, some do have color; chlorine, for example. Liquefied gases escaping as liquids can lead to the formation of a visible cloud of condensed water vapor. Because they are stored under pressure, a leak can be accompanied by a hissing sound. Although many gases are odorless, some do possess strong odors—chlorine and anhydrous ammonia are examples. Natural gas and LP-Gas have an odorant added to them. Many gas detection instruments are manufactured. Leak detection solutions are available that, when applied to small leaks, show bubbles.

One of the more common sources of escaping gas in storage locations is the operation of overpressure protection devices. Because the pressure in a gas container is directly related to the gas temperature, and the gas temperature in uninsulated containers reflects the temperature of the surroundings, room temperature and solar radiation will affect the pressure. In general, this pressure will result in operation of overpressure devices on fully charged containers if the temperature reaches 130 to 140°F (54 to 60°C). Therefore, storage areas should not be allowed to reach such temperatures and reflective paint should be maintained on outdoor containers subject to solar radiation.

Leakage from containers themselves is rare and develops slowly, so detection is facilitated prior to the development of a hazardous situation. Leakage from the inner container of an insulated container for a cryogenic gas is indicated by the formation of water condensate or frost on the outside container due to contact with air on the cold surface. While this often may only indicate a void in the insulation rather than leakage, this appearance should be investigated.

Overpressure Protection: Overfilled liquefied gas containers can lead to overpressure device operation at much lower temperatures. (See discussion of Charles' Law in Section 5, Chapter 5.) Portable liquefied gas containers should be checked for proper quantities before being placed into storage. Many such containers may be checked by weighing them. The weight of the empty container (tare weight) must be marked on the container in accordance with DOT regulations. The weight of gas must be determined by the use of data which includes the specific gravity of the gas and its temperature when placed into the container. In other cases, the quantity is checked by the use of a gage which measures the liquid level. The most accurate gage requires that a small amount of gas be released. This should be done outside under carefully controlled conditions. These gages are known as fixed-tube gages or try-cocks. They will release gas only if the level is safe, and will release liquid if there is too much in the container.

Ventilation of Spaces: Indoor storage areas should be ventilated, regardless of the chemical hazard of the gas. In areas used solely for storage (no filling of containers), the amount of ventilation need not be great: one-half to one air change per hour or so. However, even inert gases (because they are often odorless and colorless) can present asphyxiation hazards in unventilated areas.

Controlling Ignition Sources: Ignition sources should be controlled in flammable gas storage areas. The vapor density of the gas, in part, determines the extent of the area in which ignition sources should be eliminated or controlled. Many gases are heavier than air at all times. Others will be temporarily heavier than air when they are released in liquid form and vaporize. In general, flammable gas storage areas (with no container filling) are classified as Division 2 locations for purposes of installing electrical equipment because of the ventilation provided and the nominal leakage potential.

Most flammable gases are Group C or D materials in the context of NFPA 70, *National Electrical Code®*. Acetylene (Group A) and hydrogen (Group B) are common exceptions. Because of the limited variety of Group A and B electrical equipment available, electrical equipment may have to be avoided or be purged and ventilated in accordance with NFPA 496, *Standard for Purged and Pressurized Enclosures for Electrical Equipment in Hazardous (Classified) Locations*, or be intrinsically safe.

Fire protection and control of gas fires is discussed in Section 5, Chapter 5 of this HANDBOOK. Sprinkler and water spray protection has been discussed previously in this section. An essential aspect is that the extinguishment of gas fires by the application of extinguishing agents should be restricted to small leaks where the consequences of reignition can be tolerated.

Bibliography

NFPA Codes, Standards, Recommended Practices and Manuals. (See the latest *NFPA Codes and Standards Catalog* for availability of current editions of the following documents.)

NFPA 50, *Standard for Bulk Oxygen Systems at Consumer Sites.*

NFPA 50B, *Standard for Liquefied Hydrogen Systems at Consumer Sites.*

NFPA 51, *Standard for the Installation and Operation of Oxygen-Fuel Gas Systems for Welding and Cutting.*

NFPA 51B, *Standard for Fire Prevention in Use of Cutting and Welding Processes.*

NFPA 54, *National Fuel Gas Code.*

NFPA 56F, *Standard for Nonflammable Medical Gas Systems.*
NFPA 58, *Standard for the Storage and Handling of Liquefied Petroleum Gases.*
NFPA 59, *Standard for the Storage and Handling of Liquefied Petroleum Gases at Utility Gas Plants.*
NFPA 59A, *Standard for the Production, Storage and Handling of Liquefied Natural Gas (LNG).*
NFPA 70, *National Electrical Codelmed.*
NFPA 220, *Standard on Types of Building Construction.*
NFPA 496, *Standard for Purged and Pressurized Enclosures for Electrical Equipment in Hazardous (Classified) Locations.*

Additional Readings

Compressed Gas Association, *Handbook of Compressed Gases,* published by Van Nostrand Reinhold, NY, 1981.
Coward, H. F., and Jones, G. W., "Limits of Flammability of Gases and Vapors," *Bulletin No. 503,* U.S. Bureau of Mines, Washington, DC, 1965.
LP-Gas Safety Handbook, National LP-Gas Association, Oak Brook, IL, 1978.
"LPG: Storage of Liquified Petroleum Gas," (P7717), Factory Mutual Engineering Corp., Norwood, MA.
Senesky, J., "Safe Storage and Handling of Compressed Gases," *Plant Engineering,* Vol. 33, No. 10, Oct. 1979, pp. 143-148.
"The Design and Construction of Liquified Petroleum Installations at Marine and Pipeline Terminals, Natural Gas Processing Plants, Refineries, Petrochemical Plants and Tank Farms," *API Standard 2510,* American Petroleum Institute, Washington, DC.
Wesson, H. R., et al., "Fire Extinguishment of High Evaporation Rate LNG Spills," *Operating Section Proceedings D-1974,* American Gas Association, 1974.
Zabetakis, Michael D., "Flammability Characteristics of Combustible Gases and Vapors," *Bulletin No. 627,* U.S. Bureau of Mines, Washington, DC, 1965.
Zuber, Kenneth, "LNG Facilities—Engineered Fire Protection Systems," *Fire Technology,* Vol. 12, No. 1, Feb. 1976, pp. 41-48.

STORAGE AND HANDLING OF CHEMICALS

William J. Bradford, P. E.

Safe storage of chemicals requires a knowledge of all of the hazardous properties of the chemicals. Many materials exhibit more than one potential for fire and explosion, and detailed knowledge of hazard potential is best obtained from the manufacturer's detailed literature on the chemical (USCG 1974). In addition, the quantity, size, and nature of the container, as well as its storage arrangement, affects the safety of storage. In addition to knowledge of the hazard potential, it is also necessary for safe handling to have details on the process in which the chemical is to be used. The temperature, pressure, and concentration of all chemicals involved in the process must be known. Other factors to be considered include potentially hazardous byproducts due to normal operating conditions, and byproducts that could be formed by contamination or operation outside the normal parameters. Only with complete information can safe storage and handling of chemicals be achieved.

PRINCIPLES OF GOOD STORAGE

Segregation: The first principle of good storage practice for chemicals is segregation, including separation from other materials in storage, from processing and handling operations, and from incompatible materials. Segregation can take the form of isolation in a separate detached structure, or isolation in the same building (by means of fire walls). Segregation can also be achieved by separating chemicals within the same building by an intervening empty area or by intervening storage of inert or nonhazardous materials. The extent of segregation depends upon the quantity of materials being stored, the physical state of the chemicals, and their degree of incompatibility. In addition, the known behavior of materials under fire conditions will affect the extent and type of segregation needed. Rupture of vessels and possible mixing of incompatible materials under fire conditions must be considered.

Protection Against Physical Damage: Containers for chemicals are designed to be compatible with the materials they contain. However, when containers are subjected to physical abuse, containers may be damaged by the release of chemicals, which can considerably increase fire and explosion hazards. For this reason, protection of the containers against physical damage in shipment, transfer, and storage is very important. In addition, the container may represent a potential hazard under fire conditions. For example, storage of oxidizers in fiber packs or even plastic containers may be safe under normal conditions, but if a fire occurs, the oxidizers and the combustible packages can magnify fire intensity.

Hazard Identification

All storage areas should conspicuously display signs to identify the material stored in the area. The hazard identification system described in NFPA 704, *Standard System for the Identification of the Fire Hazards of Materials* should be used. When materials having different hazard identifications are stored in the same area, the area should be marked to indicate the most severe health, flammability, and reactivity hazard present. Hazard identifications for many materials will be found in NFPA 49, *Hazardous Chemical Data*, and in NFPA 325M, *Fire Hazard Properties of Flammable Liquids, Gases and Volatile Solids*.

Fire Control

The selection of extinguishing agents is determined by the reactivity of the chemical, the physical state of the chemical (solid, liquid, or gas), the toxicity of the chemical, and its expected products of combustion. Water is normally used as the extinguishing agent except on chemicals that may be dangerously water reactive. However, the toxicity of the material in water must be considered as an additional factor. The contamination of potable water supplies by toxic runoff from fire fighting could create a major health hazard.

Automatic protection for storage should be employed wherever possible. The protection must be specifically designed based upon the extent of fire hazard, the arrange-

Mr. Bradford is a loss prevention consultant. He is a registered professional engineer in Connecticut and Rhode Island, a member of AIChE and SFPE and serves on several NFPA technical committees.

ment of material, and the maximum quantity that can be stored.

Manual fire fighting to supplement the automatic protection may be severely limited by the toxicity of the material, obscuration due to smoke, the nature of the products of combustion, or the possibility that fires may lead to explosions in storage areas for reactive chemicals. Fire fighters should wear self-contained breathing apparatus (SCBA) to avoid the hazard of toxic materials. Where explosion is a possibility, manual fire fighting should be conducted from a remote location. In some cases, the hazard will be sufficiently high so that fire fighters should not approach the area to attempt any type of manual fire fighting.

OXIDIZING CHEMICALS

In considering storage facilities for oxidizing chemicals, remember that oxidizing chemicals usually are not themselves combustible, but may provide oxygen to accelerate the burning of other combustible materials. Combustible materials and flammable liquids, therefore, should not be stored in the same storage areas with oxidizing chemicals. Storage buildings should be of noncombustible or fire resistive construction. Combustible packaging and wood pallets may represent a severe hazard and should be eliminated. Because certain oxidizing materials undergo dangerous reactions with specific noncombustible materials, the possibility of dangerous reactions should be considered when deciding on acceptable storage facilities. Chlorates, for example, should not be stored with acids or combustible materials. Inorganic peroxides also react with acids, yielding hydrogen peroxide. A suitable storage facility for inorganic peroxides would be a dry, fire resistive storage room in which there are no combustible contents or acids. When classifying chemicals for storage purposes, one should keep in mind that the label on the container required for interstate shipment classifies the chemical by hazard for transportation purposes only; a knowledge of the fire, health, and reactivity hazards of the chemical should be the guide when arranging storage.

Because of the probability that spilled material will become mixed with combustible refuse, it is important to clean up all spills immediately and thoroughly. Combustible linings of barrels should be removed from the storage area and destroyed as soon as the barrels are empty.

Two NFPA standards are directed specifically toward storage of oxidizing chemicals: NFPA 43A, *Code for the Storage of Liquid and Solid Oxidizing Materials*; and NFPA 43C, *Code for the Storage of Gaseous Oxidizing Materials.*

Ammonium Nitrate: Because of its characteristics and widespread use, ammonium nitrate deserves special mention. Proper storage methods for bagged ammonium nitrate were intensively investigated following the catastrophic Texas City, TX, explosion in 1947 (NFPA 1947) which involved cargoes of ammonium nitrate aboard vessels. Storage recommendations for bagged and bulk ammonium nitrate published in NFPA 490, *Code for the Storage of Ammonium Nitrate*, cover building construction, pile sizes and spacing, separation of ammonium nitrate from contaminating materials that could increase its sensitivity during a fire, elimination of floor drains into which molten nitrate might flow, cleanliness of the storage area, and precautions against ignition sources.

Fire Control Methods

With one or two exceptions, water appears to be the only suitable extinguishing agent for fires involving inorganic oxidizing agents. Water in large quantities should be used to control fires involving nitrates, nitrites, and chlorates. Carbon dioxide and other smothering agents are of little or no extinguishing value because the oxidizing material furnishes its own oxygen for combustion. Although inorganic peroxides decompose when moist and liberate oxygen, water should still be used on fires in combustible materials that are located in the vicinity of the peroxide.

It may be possible to extinguish a fire involving a peroxide spill with a dry chemical extinguisher or by smothering with dry sand or soda ash. If these methods fail, the area should be flooded with water from hose streams.

Self-contained breathing apparatus should be worn by fire fighters. During a fire involving nitrates, one of many dangers is inhalation of oxides of nitrogen. As much ventilation as possible should be provided to permit rapid dissipation of the products of combustion and heat. When water in solid streams strikes molten nitrate, steam explosions may cause a violent eruption of the molten material.

COMBUSTIBLE CHEMICALS

Essentially, all organic chemicals are combustible. Storage practices for such chemicals closely follow those for the more common solid combustible materials discussed in other chapters. Sulfur and sulfides of sodium, potassium, and phosphorus may be considered principally as combustible chemicals. They do have some properties that warrant special mention. Combustion of these materials produces sulfur dioxide, an irritating gas. Storage should be away from oxidizing materials and, in the case of sulfides, should be separated from strong acids.

Water should be used for fighting sulfur and sulfide fires. For fighting fires of sulfur dust, water spray is recommended, principally to avoid stirring up dust clouds and causing a dust explosion. Fires in closed spaces, e.g., tanks of molten sulfur, can best be fought by closing the container and allowing the heavy sulfur dioxide produced by combustion to smother the fire.

UNSTABLE CHEMICALS

Special precautions must be taken for storage of chemicals that are subject to spontaneous decomposition or other dangerous reactions. The precautions should be planned to minimize the possibility of a dangerous reaction and to prevent injuries and extensive property damage if one should occur. Steps to be taken to protect against the hazard will depend upon the conditions that affect the stability of the chemical being stored. Points peculiar to unstable chemicals to be considered are: (1) the catalytic effect of containers, (2) materials in the same storage area that could initiate a dangerous reaction, (3) presence of inhibitors, and (4) the effect of direct sunlight or temperature changes. There is a need for pressure relief vents for containers and explosion venting for the storage area in addition to the usual considerations, such as automatic sprinkler or water spray protection, and elimination of all combustible material from the storage area. Whenever possible, unstable chemicals should be stored in a detached outside location.

Fire fighters should be thoroughly briefed on the proper procedures to be followed if called upon to fight a fire involving or exposing a storage area containing unstable chemicals.

Organic Peroxides: This widely used group of unstable chemicals deserves special mention. Many organic peroxides are diluted to the point where hazardous reactions are impossible; others require special precautions. The manufacturer of the peroxide should be consulted for specific recommendations. The following general recommendations summarize a detailed discussion of organic peroxide storage found in Loss Prevention Data Sheet 7-80, *Organic Peroxides* (FMEC 1972).

1. Develop safe storage procedures using the advice and assistance of the supplier.
2. Limit the occupancy of storage areas to peroxides only.
3. Use storage areas for unopened containers only.
4. Install automatic sprinklers in storage areas and arrange storage to permit ready wetting by sprinkler discharge.
5. Inspect storage areas daily and correct any unsafe conditions.
6. Maintain temperature within the ranges established by the manufacturer.
7. Use explosion proof models if refrigerators or freezers are needed.

Fire Control Methods

Water is the recommended extinguishing agent for fires involving unstable chemicals, including organic peroxides. Automatic sprinklers are the best protection because fires involving such materials can lead to explosions. Manual fire fighting must be conducted from a distance where the fire fighters will be protected from an explosion.

WATER AND AIR REACTIVE CHEMICALS

Water Reactive Chemicals: Anhydrides, carbides, hydrides, alkali metals (lithium, sodium, potassium), and similar chemicals should be stored in dry areas in waterproof and airtight containers that are kept off the floor on skids. The storage area should not contain combustible material and/or incompatible chemicals.

Air Reactive Chemicals: Aluminum hydride, aluminum alkyls, yellow phosphorous, and similar chemicals must be stored to avoid contact with air. Yellow phosphorous, for example, is stored under water. Other materials, such as aluminum alkyls, which react with both water and air, must be stored under inert liquids or inert gas. Storage areas should contain no incompatible materials.

Fire Control Methods

The value of automatic sprinklers depends upon the type of chemical reaction that will occur when the chemical comes in contact with water. Fires involving hydrides, for example, can be smothered with a special graphite base powder or with inert material such as dry, finely divided calcium magnesium carbonate (dolomite). Hydride fires in an enclosed space should not be put out because the continued evolution of hydrogen after extinguishment will create an explosion hazard. For the fires involving air reactive white (or yellow) phosphorus, water will solidify the phosphorus melted by the heat of the fire, after which it can be covered with dry sand or dirt. All of the chemical must be disposed before it can dry out and reignite.

Several of these chemicals are both air and water reactive and the water reaction can be particularly violent, e.g., aluminum alkyls. Fires of such materials can be contained by dry chemical while the material burns out under control. Metallic sodium, which is both water and air reactive, can be extinguished by the use of dry chemical. The residual sodium should then be submerged in oil.

CORROSIVE CHEMICALS

Inorganic Acids, Alkalis: These materials, including sulfuric, nitric, and hydrochloric acids, as well as sodium hydroxide, should be stored in cool, well ventilated places away from incompatable chemicals or combustibles. Acids and alkalis should not be stored in the same area. Spilled acids or alkalis should be diluted at once with large quantities of water. Combustible materials contaminated with nitric, sulfuric, or perchloric acid should be washed thoroughly or removed to a safe location before spontaneous ignition can occur.

Chlorine and Fluorine Cylinders: Preferably, these are stored outside under a noncombustible sun shield. If stored inside, these cylinders should be in a noncombustible, well ventilated, segregated room. Both these gases present a serious inhalation hazard (Chlorine Institute 1969). Areas of suspected gas leakage should not be entered without self-contained breathing apparatus. Fluorine causes severe flesh burns and is also highly toxic by inhalation. Special protective clothing is necessary, in addition to self-contained breathing apparatus.

Fire Control Methods

Water in spray form is recommended for fighting fires in acid and alkali storage areas. Water from straight streams mixing with concentrated acid or a caustic agent will heat and spatter the corrosive chemicals. A fire involving perchloric acid can cause an explosion if the acid becomes mixed with organic material. Ample precautions should be taken to protect fire fighters from possible explosions. Fire fighters should avoid contact with spilled acids and inhalation of their toxic fumes by the use of self-contained breathing apparatus.

Chlorine and fluorine do not represent fire hazards, but the release of dangerous toxic gases should be expected if cylinders are present in a fire. It is mandatory that self-contained breathing apparatus be used when fighting a fire where chlorine or fluorine cylinders are located.

RADIOACTIVE MATERIAL

The direct fire and explosion characteristics of materials are not affected by the radioactivity of its constituents. For example, neither radioactive nor nonradioactive cobalt is a fire hazard, and the frequency or the extent of fires will not be affected. However, because of the radiation hazards during and after a fire, it is extremely important that all practical steps be taken to prevent a fire from involving these materials. Radioactivity is not detectable by any of the human senses; special instruments and measuring techniques are required to identify and evaluate the hazard. Radioactivity can cause injuries, loss of life, and damage to and extended loss of the use of materials, equipment, and buildings. A major problem is the need for decontamination of buildings, equipment, and materials after a fire.

Radioactive materials should be stored with a shielding equivalent to that used in shipping. Storage should be in fire resistant cabinets, safes, or equivalent enclosures, or in wells or pits below the lowest floor level of a building.

Automatic protection is of the greatest importance in areas where radioactive materials are stored, because manual fire fighting exposes fire fighters to radioactivity. There will also be a delay in salvage work and a resumption of normal operations in a location where fire or explosion causes loss of control over radioactive substances. Smoke and products of combustion from these fires especially must be controlled. Fire fighters need protective clothing and respiratory protection equipment. Fire control must be thoroughly planned. Special emphasis must also be placed upon the disposal of water from sprinklers and hose streams to avoid contamination of buildings and equipment.

MATERIAL SUBJECT TO SELF-HEATING

Charcoal: Spontaneous heating hazards with charcoal can be controlled by thorough cooling and ventilation before bagging and storage. It is important to keep the charcoal dry, to prevent contamination with foreign combustibles, and to avoid contact with heat sources. Small quantities of charcoal are normally stored in heavy paper bags. The spontaneous heating hazard of individual small bags, as may be found in a dwelling, is not serious.

When fire occurs in charcoal, water should be used, but directly only on the fire without wetting nonburning material. The damaged and wet materials should be removed from the storage building at once because wet charcoal is even more susceptible to self-heating than when dry.

Agricultural Products: Spontaneous heating in agricultural products can be prevented by control of moisture. Proper curing (as for hay) and adequate aeration will prevent heat buildup. Where the moisture content cannot be controlled, or where any suspicion of spontaneous heating exists, thermocouples may be used in stacks or bales. Pointed hollow metal rods or pipes with holes drilled in the lower ends often are used to permit insertion of temperature instruments into the subsurface areas of the stored material. Regular checks can then be made for the development of any hazardous temperature conditions. Where evidence of dangerous spontaneous heating is noted, the material should quickly be removed from storage.

When spontaneous heating occurs in agricultural products, it is important to have ample fire fighting facilities available. Water hoses are needed to combat a fire that may occur spontaneously on exposure of the hot spots to air. Hay, in and around hot spots, should be wetted thoroughly before complete uncovering or attempted removal.

HANDLING CHEMICALS

Safe handling of hazardous chemicals includes attention to all of the precautions used for their storage. In addition, however, it is necessary to consider other potential problems and employ additional safeguards. Normal handling of the chemicals includes removing them from the manufacturer's shipping container and mixing them with other materials. Therefore, complete details are needed on the process in which the chemical is to be employed. Consideration must be given to the control techniques available to limit potentially hazardous runaway reactions (ASTM undated), which can result in fires or explosions, and to limit the release of flammable and toxic materials from processing.

Reactive and Unstable Chemicals

Particular attention must be given to the handling of reactive or unstable chemicals that can cause fires or explosions. It is necessary to handle all air reactive chemicals in a closed system, which is best kept under an inert gas atmosphere. Use of a vacuum atmosphere is much less desirable because any breach of the integrity of the system will automatically admit air. Dissolving water reactive chemicals in an organic solvent or covering them with a layer of oil will keep moisture from attacking the chemical. For unstable chemicals, an inhibiting catalyst may be employed unless processing of the chemical will be adversely affected. Refrigeration can be used to slow the speed of the chemical reaction. These and other methods should be used to prevent dangerous exothermic reactions involving reactive and unstable chemicals; however, these methods must be supplemented by emergency means such as the use of rupture discs or emergency relief valves for relieving dangerous pressures. In some cases, reactions can be controlled by the introduction of a "short stop" to actually kill the reaction. In other cases, a runaway reacting mass may be dumped into a tank containing a material that will stop the reaction by dilution. Fire fighting personnel should be made aware of the fire and explosion potential in areas where reactive or unstable chemicals are handled.

Toxic Chemicals

Protection against the toxic effects of chemicals during handling can be accomplished in two ways. The first is by using the most practical of the available methods of controlling and confining the chemical so that the toxic material cannot be contacted, ingested, or inhaled during normal operations. The second is by educating all personnel who may be in areas where toxic chemicals are handled about the hazards, danger signals, precautionary procedures to be followed, and proper steps to be taken in case of an emergency.

Toxic chemicals will normally be handled in closed systems. However, leaks can develop in any closed system, thereby subjecting people to a severe health hazard. If the release of a toxic material cannot be seen or smelled, toxic gas indicators will have to be used to alert personnel of the danger. The use of personal protective equipment is always a valuable adjunct to safety in areas where toxic materials are present.

TRANSPORTATION OF CHEMICALS

The safe transportation of chemicals depends upon a knowledge of the hazardous properties of the chemicals, the normal and abnormal conditions to which the chemicals may be exposed during shipment, and the conditions of packing and shipping that will minimize the possibility of accidental release or reaction of chemicals.

In the U.S., shipments of hazardous chemicals in interstate or foreign commerce by land, water, or air must comply with U.S. Department of Transportation (DOT) regulations. Among items covered by these regulations are construction of containers; methods of packing; weight of chemical per package; marking and labeling; loading, placarding, and movement of railroad cars; and regulations for motor vehicle equipment and motor vehicle operation on the highway.

State regulations for intrastate transportation usually agree with DOT interstate commerce regulations. In Canada, shipment of explosives and other dangerous chemicals are regulated by the Canadian Transport Commission (CTC). Chemicals shipped in accordance with Canadian regulations may be shipped to a destination in the U.S. or through the U.S. en route to a point in Canada.

Principles governing the safe transportation of chemicals include:

1. Constructing the container of material that will not react with or be decomposed by the chemical.
2. Excluding chemicals that can react dangerously with each other from the same outside container.
3. Packaging toxic and radioactive chemicals so they will not present a health hazard during normal transportation conditions and will not be released in an accident or under other abnormal conditions.
4. Providing sufficient outage for maximum expansion of liquids under conditions to be expected during transportation.
5. Limiting the amount of chemical that can be released by container breakage or leakage by limiting maximum size of individual containers.
6. Cushioning containers to minimize the possibility of breakage.

WASTE CHEMICAL DISPOSAL

The safe disposal of hazardous chemical waste is controlled by a 1976 federal law, the Resource Conservation and Recovery Act (RCRA- PL 94-590 Subtitle C Hazardous Waste Management). This act provides for cradle to grave regulation of potentially dangerous byproducts spawned by industrial technology. Specifically, the law covers any solid, liquid, or gaseous waste that exceeds federal criteria established for the following characteristics:

1. Toxicity.
2. Persistence and degradability in nature.
3. Potential for accumulation in tissue (bioaccumulation).
4. Reactivity.
5. Flammability.
6. Corrosivity.
7. Radioactivity.

Standards have been published that regulate generators of hazardous waste in several different areas, i.e., recordkeeping practices, labeling and containerization, disclosure of waste composition to haulers and disposers, use of a manifest to monitor the life cycle of a waste stream, and reporting requirements to applicable local, state, and federal agencies. Additionally, standards have been developed to control the storage, transportation, and final disposition to assure that these compounds are handled in properly designed facilities. For example, secured chemical landfills must have a sufficiently impervious base to prevent contaminated leachate from entering adjacent groundwater, and incinerators must achieve a 99.9 percent destruction efficiency.

Recognizing the hazards involved, it is advisable to verify the capabilities of any firm utilized for disposal of hazardous wastes. At the very minumum, operating permits should be reviewed to confirm that a particular waste can be legally processed by any company under consideration for this service.

Bibliography

References Cited

ASTM. undated. *Chemical Thermodynamic Data and Energy Release Computer Program* (CHETAH). American Society for Testing and Materials DS-51, Philadelphia, PA.

Chlorine Institute. 1969. *Chlorine Manual*, 4th Ed. The Chlorine Institute, NY.

FMEC. 1972. "Organic Peroxides." *Loss Prevention Data Sheet 7-80*. Factory Mutual Engineering Corp., Norwood, MA.

NFPA. 1947. "Texas City Disaster." *NFPA Quarterly*. Vol 41, No 1. pp 24-57.

USCG. 1974. *Hazardous Chemical Data*. CG-446-2. U.S. Coast Guard, Washington, DC.

NFPA Codes, Standards, Recommended Practices and Manuals. (See the latest *NFPA Codes and Standards Catalog* for availability of current editions of the following documents.)

NFPA 43A, *Code for the Storage of Liquid and Solid Oxidizing Materials.*

NFPA 43C, *Code for the Storage of Gaseous Oxidizing Materials.*

NFPA 49, *Hazardous Chemicals Data.*

NFPA 325M, *Fire Hazard Properties of Flammable Liquids, Gases, and Volatile Solids.*

NFPA 490, *Code for the Storage of Ammonium Nitrate.*

NFPA 491M, *Manual of Hazardous Chemical Reactions.*

NFPA 655, *Standard for Prevention of Sulfur Fires and Explosions.*

NFPA 704, *Standard System for the Identification of the Fire Hazards of Materials.*

NFPA 801, *Recommended Fire Protection Practice for Facilities Handling Radioactive Materials.*

Additional Readings

Bahme, Charles W., *Fire Officer's Guide to Dangerous Chemicals*, 2nd ed., National Fire Protection Association, Quincy, MA, 1978.

Barron, M. J., "Railroads and Transportation Safety," *Loss Prevention*, Vol. 5, American Institute of Chemical Engineers (AIChE), NY, 1971, pp. 7-12.

Bigelow, C. R., "Chemical Industry Transportation Emergency Information System," *Loss Prevention*, Vol. 5, AIChE, NY, 1971, pp. 21-25.

Botkin, L. A., "Chemical Transport Over the Highway: Equipment Design and Regulations," *Loss Prevention*, Vol. 14, AIChE, NY, 1981, pp. 44-53.

Brasie, W. C., "The Hazard Potential of Chemicals," *Loss Prevention*, Vol. 10, AIChE, NY, 1976, pp. 135-140.

Condensed Chemical Dictionary, 10th ed., Van Nostrand Reinhold Co., NY, 1981.

DeHaven, E. S., "Using Kinetics to Evaluate Reactivity Hazards," *Loss Prevention*, Vol. 12, AIChE, NY, 1979, pp. 41-44.

Doyle, W. H., "Protection in Depth for Increased Chemical Hazards," *Fire Journal*, Vol. 59, No. 5, Sept. 1965, pp. 5-7.

Duch, M. W., et al., "Thermal Stability Evaluation Using Differential Scanning Calorimetry and Accelerative Rate Calorimetry," *Plant/Operations Progress*, Vol. 1, No. 1, 1982, pp. 19-26.

Fawcett, H. H., and Wood, W. S., eds., *Safety and Accident Prevention in Chemical Operations*, Interscience Publishers, NY, 1965.

Fenlon, W. J., "A Comparison of ARC and Other Thermal Stability Test Methods" *Plant/Operations Progress*, Vol. 3, No. 4, 1984, pp. 197-202.

Fire and Explosion Index Hazard Classification Guide, 5th ed., Dow Chemical Co., Midland, MI, 1980.

Grayson, M., exec. ed., Kirk –Othmer, *Encyclopedia of Chemical Technology*, 3rd ed., Interscience Publishers, NY, 1984.

Linville, J. L., ed., *Industrial Fire Hazards Handbook*, 2nd ed., National Fire Protection Association, Quincy, MA, 1984.

Nitromethane: Storage and Handling, NP Series TDS No. 2, 3rd ed., IMC Chemical Group, Inc., Terre Haute, IN.

Novak, R. G., "Hazardous Chemicals Disposal in a Large Chemical Complex," *Loss Prevention*, Vol. 6 AIChE, NY, 1971.

Green, D. W., ed., *Chemical Engineers' Handbook*, 6th ed., McGraw-Hill Inc., NY, 1984.

"The Handling of Explosives," Explosives Division, Department of Energy, Mines and Resources, Ottawa, Canada.

Sax, N. Irving, *Dangerous Properties of Industrial Materials*, 5th ed., Van Nostrand Reinhold Co., NY, 1979.

"Storage and Handling of Organic Peroxides in Reinforced Polyester Fabricating Plant," *Safety and Loss Prevention Bulletin No. 19*, Society of the Plastics Industry, Inc., NY, 1978.

Stull, D. P., *Fundamentals of Fire and Explosion*, M-10, AIChE, NY, 1976.

Stull, D. P., "Linking Thermodynamics and Kinetics to Predict Real Chemical Hazards," *Loss Prevention*, Vol. 7, AIChE, NY, 1973, pp. 67-73.

"Suggested Code of Regulations for the Manufacture, Transportation, Storage, Sale, Possession, and Use of Explosive Materials," *Safety Library Publication No. 3*, Institute of Makers of Explosives, Washington, DC.

Handbook of Chemistry and Physics, 65th ed., Chemical Rubber Company, Cleveland, OH, 1984.

Zabetakis, Michael D., *Flammability Characteristics of Combustible Gases and Vapors*, Bulletin 627, U.S. Bureau of Mines, Pittsburgh, PA, 1965.

Zercher, J. C., "CHEMTREC Up to Date," *Loss Prevention*, Vol. 9, AIChE, NY, pp. 66-70.

STORAGE AND HANDLING OF GRAIN MILL PRODUCTS

Max R. Spencer

Grain handling and storage requires protection to prevent spontaneous combustion and dust explosions. Grain and oil seeds are primarily starch or carbohydrates, protein, fiber, and various vegetable oils. Although these commodities are stable in their raw, whole kernel states and not readily combustible if they are protected from moisture, insects, and fungi, they are capable of flammability under certain conditions.

Microbiological spoilage created by fungi under certain moisture and temperature conditions has caused spontaneous heating. Some higher moisture contents may also produce readily oxidizable hydrocarbon compounds of low ignition temperatures, which, if oxygen is available, can result in fire (Christensen and Kaufmann 1977). The incidence of this source of ignition of fires and explosions in grain elevators is rare, although it has been known to occur. As size reduction of the grain kernel proceeds, the susceptibility to fires and explosions increases dramatically. The frequent handling of grain from farm to consumer progressively creates more broken kernels, hence more dust.

This chapter first identifies and describes the most commonly used grain handlers, including conveyors and bucket elevators. The various grain processes that contribute to fire and explosion potential are presented. The next sections of the chapter explain the specific fire hazards and how to design buildings, devices, and procedures to safeguard against them.

The following chapters in this HANDBOOK will also be of interest to readers of this chapter: Section 5, Chapter 9, "Dusts;" Section 7, Chapter 1, "Fundamentals of Firesafe Building Design;" Section 10, Chapter 11, "Grinding Processes;" and Section 12, Chapter 13, "Air Moving Equipment."

Mr. Spencer is Vice President of Engineering/Operations for Continental Grain Co., in New York, NY. This chapter is a condensation of Chapter 6, "Bulk Grain Handling," in the second edition of the *Industrial Fire Hazards Handbook*, also published by NFPA.

HANDLING

The grain handler basically unloads, stores, conditions for quality, and reloads grain. The process of horizontal and vertical conveying into and out of storage facilities generates the dust at each loading or transfer point throughout the elevator. A high velocity grain stream generates an air movement equivalent to the volume displaced by the grain and to an additional volume known as entrained air. This can be empirically estimated at a ratio often equal to 5 to 10 times the actual displacement volume. Hence, a 60,000 bushel per hr (2114 m^3/hr) falling stream of grain into a silo can generate an air movement of between 6,000 to 11,000 cu ft (170 to 311 m^3) per min [1,000 cu ft (28 m^3) displacement, plus 5,000 to 10,000 cu ft (142 to 283 m^3) of entrained air]. This volume, when forced to exit through constrained openings, readily carries the suspended dust into ambient surroundings unless the dust is captured by an effective aspiration system.

Belt Conveyors

The most common conveyor employed to move grain horizontally from point to point is the troughed-belt conveyor. Generally, this type of conveyance requires shrouded, aspirated loading and discharge points to prevent excessive dust emissions. Totally enclosed belt systems that rely on the enclosure to contain the dust and require little or no aspiration have been developed.

Chain Conveyors

Alternatives to the belt conveyor include the en masse or chain conveyors which are totally encased in an airtight housing that prevents the escape of dust. Normally, only the loading point or discharge point needs to be aspirated.

Screw Conveyors

The helical screw conveyor is a standard for low volume conveying for short distances. The U-shaped housings with removable covers emit little dust and can, in fact, effectively convey pure dust without emissions.

Pneumatic Conveyors

Pneumatic conveying systems are used extensively because they are especially effective for the confinement and movement of finely ground commodities in a processing operation requiring multipoint pickup or distribution. Although capacities are relatively low and energy requirements are proportionately high, this method provides both horizontal and vertical flexibility in product movement and preserves sanitary conditions required of edible products.

Bucket Elevators

For vertical movement of grain and grain products, the bucket elevator remains the principal method of elevating to extreme heights, 100 to 300 ft (30.5 to 91.4 m). The bucket elevator, with its endless belt suspended on a drive pulley, serves as a perfect air "pump" with cups acting as vanes in a constricted housing. Since it is one of the most common sources of dust emission in an elevator facility, it is *the* most common location of primary dust explosions where the cause has been able to be determined (Christensen and Kaufmann 1977). Bucket elevators also have been documented as the principle source of explosions in feed mills and other grain milling industries. (See Fig. 11-7A and 11-7B.)

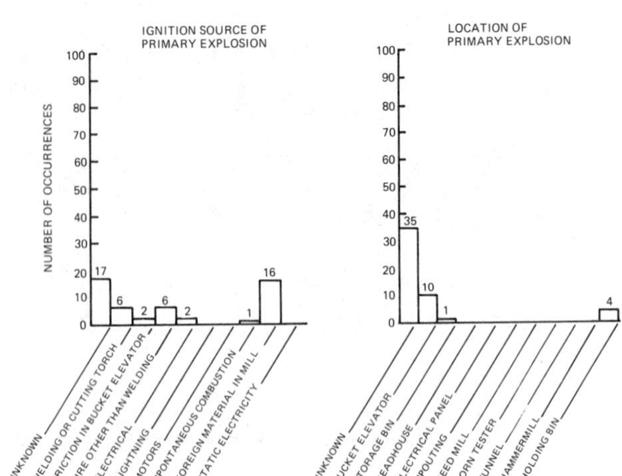

FIG. 11-7B. Causes of U.S. feed mill dust explosions and location of primary explosion, 1958 through 1975. (Continental Grain Co.)

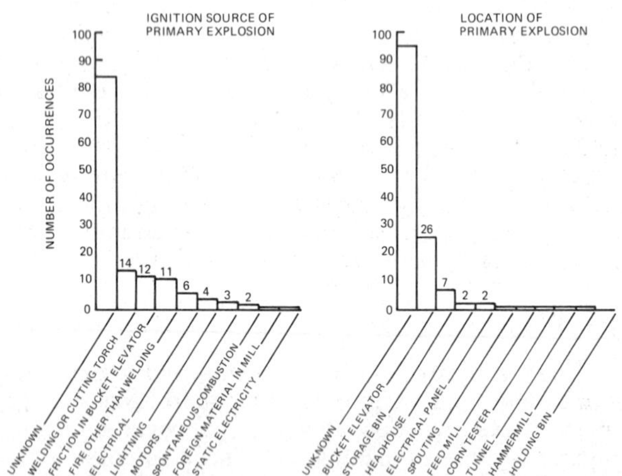

FIG. 11-7A. Causes of U.S. grain elevator dust explosions and location of primary explosion, 1958 through 1975. (Continental Grain Co.)

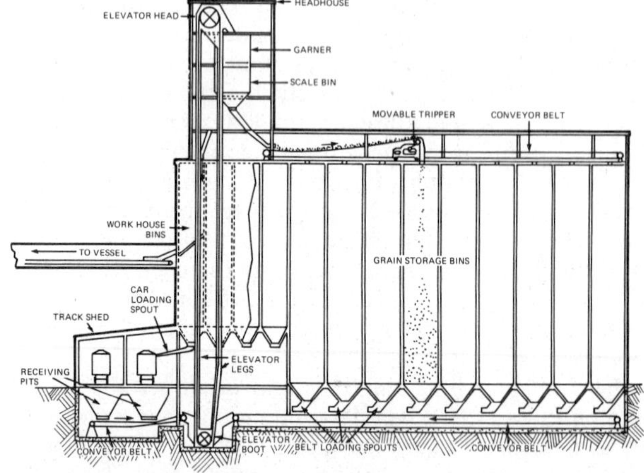

FIG. 11-7C. Diagrammatic section view of a terminal type grain elevator. (Continental Grain Co.)

A common principle in grain elevator design is to elevate the commodity to the highest optimum point, then permit it to flow by gravitational force down through the various garners, weighing scales, cleaners, and finally through spouts into the storage compartment. (See Fig. 11-7C.) Each subsequent rehandling contributes to reducing the quality of the grain and to generating additional quantities of fine particles and dust.

Spouting and Lining

Grain and oilseeds are very abrasive commodities that can rapidly erode steel conveying spouts used to channel the flow through elevators. This problem creates an almost constant demand for maintenance to eliminate leaks and dust emissions. Although patching and repair are adequate temporary measures, they seldom restore a spout to its previous good condition. An alternative has been the wide use of abrasion resistant liners which can be totally replaced without disrupting the outer spout. Lining materials most commonly used include abrasion resistant alloy steel plates, high density synthetic plastics, and certain vitreous ceramics. These can usually be formed or molded to the contour of the spouts and bolted into place without the need for welding or other heating devices.

Where grain impinges or impacts upon a metal surface, the wear characteristics may require alternative lining materials. Synthetic rubber linings, for example, may be more effective in preventing excessive spout wear while at the same time minimizing the breakage of grain kernels themselves upon impact.

Receiving and Shipping

The first and last segments of a grain storage elevator are the machinery and structures needed to unload or load the grain. Whether a railroad or truck receiving system is used, these operations are at ground level, partly or completely in the open, and usually connected to the main storage structure with an underground tunnel beneath the discharge receiving hopper or with an overhead bridge.

The free fall of grain through open spaces into receiving hoppers presents a unique dust control problem which is aggravated by surface winds and the lack of sufficient enclosures to constrain the dust emission. The problem is less of an explosion or fire hazard than a nuisance, since the dust can fly about during the essentially open air operation.

PROCESSES

Grain Drying

Drying the grain to moisture levels low enough to preserve quality is the principal processing operation at most grain elevators. Drier operations are one of the most frequent causes of fire at grain facilities. From 1964 through 1973, the number of fires occurring in grain elevators in the United States exceeded 29,000, or an average of more than 2,900 per year (Chiotti 1977). Although the causes were numerous, many of these undoubtedly were the result of drier operations.

The typical modern grain drier is direct fired, i.e., the heat of the burned fuel is directed into a stream of air that is passed directly through the moist grain. The fuels are principally natural gas, fuel oil, or vaporized liquid propane. Driers are available in a wide array of sizes, from a 50 bushel (1.8 m^3) batch unit for farm use to a continuous flow unit of 6,000 bushels per hr (211 m^3/hr) for the commercial grain handler.

Grain Cleaning

The second most important processing activity in a grain handling facility is usually a screening, cleaning, or scalping system that removes extraneous material such as bits of stalks, stems, seed pods, husks, corncobs, weed seeds, or fine broken grain particles from the grain. Such materials not only affect the quality of the grain, they are also more prone to ignition than the grain itself because of their extremely dry states and high fiber content. The grain is cleaned by passing it over vibrating or gyratory motion screening devices or stationary gravity screens for simple size separation. A positive air aspiration system removes dust generated by the grain movement within the device.

Grinding and Cracking

Some grain facilities serve specialty industries that require grain to be cracked or ground by hammermills or grinders. Both processes are common sources of dust explosions, particularly in feed milling operations, as indicated in Figure 11-7B. The hammermill is frequently used to grind corn and other feed grains for use in rations. Care must be taken to exclude foreign objects from entering the grinding mechanisms, especially stones and metallic objects.

Dust Collecting

Supplementing the conveying, elevating, drying, screening, and storage activity is the dust collection process itself. Each point of handling that produces an emission of dust capable of being suspended in air is normally equipped with a shroud, hood, or suppression device as part of a complete dust control system that isolates, captures, and contains this dust. Many large elevators having complex operations have total aspiration capacity exceeding 500,000 cu ft (14 159 m^3) of air per minute serving hundreds of such pickup points. Newer facilities often require an investment of 10 to 20 percent of the total construction cost for this portion of the fixed assets. With the introduction in the U.S. of federal and state clean air laws, this collected dust no longer can be discharged into the ambient air, but must be collected, stored, and properly disposed.

The dust collector used most commonly is the "bag house" or fabric filter. Dust laden air is passed through filter media and exits virtually 99.95 percent dust free. The dust is recovered in the filter housing, then sold as a grain byproduct for use by the animal feed industry. Such systems are complex processing operations with self-cleaning devices for the bags, automatic discharge mechanisms, continuous conveying to disposal points or storage tanks, and separate load out systems.

The cyclone collector, which was commonly used prior to clean air laws, served as a quasi separator for dust aspirating systems, but was only 75 to 80 percent efficient overall. For fine particles, its efficiency was only 50 to 60 percent, and merely concentrated the collected dust at a central point for discharge into the atmosphere. The cyclone often is still used as a "preskimmer" ahead of filters to remove and recover the very large particles for reentry into the grain stream. The fine dust particles are then collected in the filters for disposal or for reentry into the grain stream, if circumstances and usual operation permit such practice.

The overall design of a grain handling or processing complex must allow for the risk of ignition from the processing equipment itself. Since grain and grain dust are capable of supporting combustion, they should be exposed only to mechanisms that have little or no risk of surface temperatures exceeding the dust ignition temperatures. Any moving piece of machinery will generate heat, and design criteria have been developed to permit safe operation in an atmosphere where dust may be in suspension or settle out on exposed surfaces of the machinery.

Of particular significance to the grain industry is the type of electrical equipment permitted in grain elevators. NFPA 70, the *National Electrical Code®*, (hereinafter referred to as NFPA 70) divides hazardous areas into three broad classifications, depending upon the type of fuel used in each. They are Class I, flammable gases or vapors; Class II, combustible dusts; and Class III, ignitible fibers and flyings. Equipment for electrical installations in grain elevators falls into Class II, more specifically Class II, Group G, which includes grain dusts. This type of equipment is loosely called "explosion proof," but it is more correctly termed "dust-ignition-proof." Where switchgear, starting switches, panelboards, etc., are involved, the housings are sealed to prevent entry of dust or the escape of internal sparking. For motors and other high heat producing electrical devices, the Class II, Group G rating refers to maximum surface temperatures allowed, below which ignition of dust generally will not occur.

THE FIRE HAZARDS

The elements of a grain dust fire or explosion are almost axiomatic; namely, that to be initiated and sustained there must be fuel, oxygen, and an ignition source. To have an explosion, a fourth element is needed: confinement, which contains the rapidly expanding heated gases of combustion within a constraining enclosure until the pressures exceed the ultimate strength of the enclosure.

The precise circumstance under which an explosion of a grain dust will occur is a complex combination of dust particle size, concentration in the air (oz per cu ft, or g/m^3), the energy of the ignition source, and less easily determined factors such as moisture content of the dust (or percent relative humidity of the air) and the actual composition of the dust. Dust from each agricultural commodity has its own explosion characteristics. Although researchers have not agreed precisely on the limits of the various characteristics of a particular dust, their conclusions are generally within an acceptably narrow range. Wheat dust, for example, has been quantified by several sources for minimum concentrations for explosions. (See Table 11-7A.)

Table 11-7A. Minimum Mass Concentrations Required for Wheat Dust Explosions*

Mass Concentration (g/m^3):	20–50†	40‡	50–100§	23#	70‖

* (Lilienfeld 1978); † (Gooijer 1975); ‡ (OSHA 1978); § (Schierwater 1976); # (Weber 1978); ‖ (Marks 1958)

The Fuel

Dust particles emanating from various emission points within a grain elevator are of varying compositions (including silica) and sizes. It is generally agreed by researchers that particle sizes below the 100 μm (micron) range constitute the greatest hazard (Lilienfeld 1978), while the larger particles tend to settle out rapidly. (See Table 11-7B.) Thus, a dust cloud in suspension would most likely be composed primarily of the finer particles (100 μm).

The mechanism of an explosion depends upon the ability of the immediate heat release of a burning particle to ignite and support the burning of adjacent particles (Palmer 1973). As this rapid spread of flame proceeds in a chain reaction from particle to particle, pressure waves and thermal expansion of the air can create an intense shock of sufficient strength to rupture the typical reinforced concrete structure. In studies performed by the U.S. Bureau of Mines (Jacobson et al 1961), the maximum rate of pressure rise ranged from 3,700 psig per sec (26 MPa/sec) for soybean dust to 6,000 psig per sec (41 MPa/sec) for corn dust.

TABLE 11-7B. Setting Rates for Various-sized Particles*

	Rate of Fall			
Size (μm)	in./min	m/min	in./hr	mm/h
100	320.0	8.13	—	
50	160.0	4.06	—	
10	7.0	0.18	—	
5	1.8	0.05	—	
1	—		5.0	127.0
0.5	—		1.4	35.6
0.1	—		0.05	1.3
smaller	—		~0	~0
—		Brownian motion		

NOTE: Specific gravity = 1.0
* (Matkovic 1977)

The dust suspended in ambient air is normally a composite of particle sizes ranging from 1 to 100 μm or more (Matkovic 1977), with the particles 100 μ m or larger settling quickly. (See Table 11-7C). Tests show that the visual opacity in a concentration of 20 g/m^3 (near the lowest required for an explosion) would be near zero at a path length of one meter. While it seems improbable that such a dense cloud would exist within the ambient space of an elevator structure where personnel are present, it is just as probable that such concentrations readily do exist within the confines of bucket elevators, conveyor housings, bins, silos, and connecting spout work, including the +100 μm particles that are constantly generated by the moving grain.

Accumulations of dust on hot surfaces can lead to fires when the dust reaches its ignition point. Although localized fires can readily ignite from this source, the greatest danger is the transmission of such smoldering masses into an area or a conveyance where suspended dust concentrations may exceed the lower limits of explosibility.

Once a dust cloud has ignited, the flame spread speed is nearly instantaneous, depending upon the composition of the dust, the concentration and particle size, and the turbulence of the air. The maximum flame speed seems to occur at or near the stoichiometric balance of dust and oxidant, when there is just sufficient fuel (dust) to consume all the oxygen (Palmer 1973).

Ignition Sources

The second most important element of a grain dust explosion is ignition of the suspended dust cloud by an energy source of sufficient intensity and duration. One ignition source that has been identified in a large percentage of known instances is improper use of welding and cutting equipment. (See Figs. 11-7A and 11-7B.) Other sources may not be identified so easily.

TABLE 11-7C. Size Distribution*

	Dump pit (mostly beeswings)		Belt loading (mostly starch dust)		Main elevator (60:40 mixture; beeswings: starch)		Beans dust		
	Retained on		Retained on		Retained on		Retained on		
Dust size (μm)	%	% cum	%	% cum	%	% cum	%	% cum	Mesh
+ 150	94.8	94.8	—	—	56.0	56.0	16.0	16.0	+ 100
150–100	3.7	98.5	—	—	11.3	67.3	12.1	28.1	100
100–74	1.1	99.6	—	—	7.0	74.0	13.4	41.5	159
74–38	0.4	100.0	—	—	6.0	80.0	9.2	50.7	200
38–21	—	—	31	31	6.0	86.0	16.3	67.0	450
21–16	—	—	28	59	5.0	91.0	16.0	83.0	630
16–8	—	—	22	81	4.0	95.0	11.0	94.0	937
8–6	—	—	10	91	3.0	98.0	4.0	98.0	1875
6–4	—	—	3	94	2.0	100.0	2.0	100.0	2300
4–2	—	—	2	96	—	—	—	—	4500
2–1	—	—	3	99	—	—	—	—	6250
–1	—	—	1	100	—	—	—	—	12500
Total	100.0	—	100.0	—	100.0	—	100.0	—	—

* (Matkovic 1977)

Because the majority of identified explosion locations in grain elevators are in the bucket elevator, it would follow that this piece of operational equipment presents the most serious hazard to the grain handler. This conveyance produces ignition energy in a number of ways. Overloading or stalling of the belt generates intense frictional heat on the revolving drive pulley. This has been known to burn the belting to the point of failure, allowing the severed and flaming pieces to drop within its housing. The introduction of extraneous foreign material such as scrap metal or stones, into the inlet of the bucket elevator where such materials impinge upon the fast moving metal elevating cups, has also been debated as a sparking source of sufficient energy to become the initiating force.

The elevating cups themselves can be torn from the belt and, according to some technicians, be a self-induced source of ignition as can a belt splice failure that allows the entire belting assembly to fall within the housing.

Equally important is the electrical short or sparking from power sources due to failures, overloading, or grounding. Most facilities use motive power voltages of 220 V, 480 V, or even higher. The energy released by such systems is extremely high, but can be minimized as a hazard by strict adherence to the provisions of NFPA 70 for Class II, Group G atmospheres. This applies with equal importance to portable electrical devices, lighting, lower voltage control circuitry, extension drop lights, and communications equipment.

Open flames such as from matches, lighters, cigarettes, and space heating devices, can be obvious sources of ignition, as can direct hits from bolts of lightning.

Mechanically heated surfaces such as bearings on equipment, frictional heat caused when belting rubs against sheet metal, and when dissimilar materials vibrate within equipment housings, can generate sufficient heat to cause the ignition of deposited dust layers. Internal combustion engines, such as tractors, front end loaders, or trucks, have sufficiently high surface temperatures to create ignition when used without precaution in dust laden air.

A variety of opinions exists on the degree of hazard associated with electrostatic discharge in initiating explosions. The presence of this phenomenon and its level of intensity appear to be directly related to the ambient absolute humidity. It is sometimes observed on horizontal belt conveyors and in bucket elevators. Researchers are studying this problem to determine if sufficient energy exists to reliably predict its role in dust ignition. Nevertheless, grounding of all metal housings or frames of equipment is a recommended practice. Where electrically conductive materials are available, their use will preclude much concern with electrostatic discharges.

THE SAFEGUARDS

Building Design

Current design practices require upright silos of reinforced concrete or steel, with a high contiguous structure known as a headhouse, to elevate the grain high enough to pass through necessary scales, samplers, garners, cleaners, and distributors by gravity into the storage bins. Shipping from the silo is a reverse process—that of withdrawing the grain from the silo bottoms, usually in a basement; reelevating high enough to reach outbound scales, cleaners, samplers, and spouting; and then loading into the shipping container.

Enclosures such as silos, tanks, bucket elevator housings, and ancillary structures should be designed to relieve explosion pressure waves as much as possible if such an incident happens. This pressure relief is not always physically possible because of the enclosure's configuration. For example, it is impossible to design a silo with explosion relief panels having a ratio of 1 sq ft (0.09 m^2) of relief surface for each 50 cu ft (1.41 m^3) of volume, as is often recommended. This situation also exists in the explosion venting of bucket elevators, which are often as high as 240 ft (73 m). Where possible, bucket elevators should be located in open air, apart from the structure.

The use of doors, windows, multiple explosion relief panels, and light gage structural coverings on steel structures are practical approaches to an acceptable design. Stairwells and elevator shafts, where they are enclosed, should be protected with listed fire doors. Fire walls should be provided to separate the grain handling function from adjoining grain processing operations, such as flour milling, preparation for oil extraction, feed milling, or grinding. Where tunnels, basements, or other underground structures can be avoided, the alternate use of ground level or aboveground structures can provide maximum openings to the atmosphere. Where sufficient land area is available, structures used for weighing, cleaning, and other operational functions can be located remote from silo storage but interconnected with inclined belt systems. No direct interconnecting openings should be constructed between silos or storage bins. Other methods should be provided for controlling the displaced air from filling and emptying.

All surfaces, bin bottoms, and spout inclinations should be designed to be self-cleaning when the grain or grain products flow. Structural ledges, beams, or other horizontal surfaces should be designed, where practicable, with an inclination of a minimum of 60 degrees from the horizontal to prevent gradual dust buildup of small quantities of suspended dust. Where possible, design should accommodate the flushing down of accumulated dust on vertical walls and overhead structures with water.

NFPA 61B, *Prevention of Fires and Explosions in Grain Elevators and Facilities Handling Bulk Raw Agricultural Commodities*, which has been prepared to assist designers of grain handling facilities, devotes considerable attention to their physical structure.

Mechanical Design

Devices and procedures are available to detect, warn of, and control faulty operation of mechanical equipment or components, including hot bearing sensors, speed indicators, alignment devices, level sensing gages, slowdown detectors, overflow alarms, and pressure gages. These devices can indicate or react to malfunctioning units.

Properly chosen mesh screens, grating, scalpers, or magnets will prevent tramp metal and other extraneous material from entering into the grain handling machinery. These devices can be of the electromagnetic field type, permanent magnets, the grate type magnet, large mesh screening, grizzlies, and specific gravity separators. The use of the devices is particularly important just prior to the point where the grain enters hammermills.

Dust Collection

Dust collection systems must provide sufficient capture velocity at the point of emission, particularly at conveyor loading and discharge points. Design must also provide sufficient air velocity within the ducts to prevent settlement of the particles and subsequent plugging. Blast gates and fresh air inlet dampers are necessary to balance the airflow to all points served by the system so that starving some remote emission points is avoided.

If a bag-house or fabric filter is used, the porosity of the filter media must be maintained to avoid diminishing the airflow. High humidity and fine dust particles often form a cake on the bags and reduce the air passage. In such instances, replacement or laundering of the bags is necessary. (See NFPA 91, *Installation of Blower, Exhaust Systems for Dust, Stock, and Vapor Removal or Conveying.*)

Electrical Design

The development of economical and miniaturized electronic components for use in Class II, Group G atmospheres has made the operation and control of grain elevators more sophisticated and less labor intensive. Remote control of the entire facility from a centralized control center is now a reality in newer installations. This technological advance has dictated the need for redundant detection of malfunctions since people are no longer in an operating area to observe the activity minute by minute.

The more common safeguards include electrical interlocking to shut down simultaneously all activities in a chain reaction fashion when one unit of a sequentially connected operation fails. Enunciator panels can display the exact cause of shutdown to the operator. Ammeters and load indicators can depict the exact level of capacity being carried by conveying equipment. Spring-return pneumatic devices can activate shutdown conditions, even in a total power failure. Fully electronic scales now safely transmit millivolt differentials detected by a load cell or strain gage to a remote amplifier for direct conversion into information on weight.

Grounding of electrical transformers and distribution systems is an important precaution to be followed. Lightning arresting systems as prescribed in NFPA 78, *Lightning Protection Code*, are available for high silo and headhouse structures when these are the highest structures in the area.

Intrinsically safe low voltage communication devices such as portable FM radios, are often used within elevator structures. It is extremely important when the elevator is in operation to use only electrical extension cords, drop lights, and hand tools approved for dusty locations.

Maintenance

A preventive maintenance program is essential to ensure trouble free operation and minimize emergency breakdowns. This is especially important for lubrication of bearings, checking of belt splices, replacement of bent or missing bucket elevator cups, and the early remedy of leaking, worn out spouting.

Dust filters should be checked frequently for torn bags and worn or poorly sealed rotary air locks. Duct work in the aspiration system should be free of plugups, bent ducts, and open blast gates. Grain traps should be emptied frequently. Floor-sweep openings should be kept closed, except when in use.

Electrical junction boxes should be kept closed, lighting fixture protective globes kept in place, and all maintenance work on the electrical systems performed either with the power disconnected or with equally adequate protection in a dust free idle atmosphere.

Housekeeping

Although the design, construction, and operation of a grain facility takes full advantage of all of the safeguards discussed previously, breakdowns, leaks, spills, and other unanticipated instances, which result in dust accumulations or dust suspended in ambient working spaces, may occur despite the best of intentions. Good housekeeping to promptly remove spills or dust accumulations is part of an ongoing management responsibility to prevent the possibility of creating a distinct fire or explosion hazard.

The obvious solution is the use of a broom and shovel. More sophisticated systems are available for situations where they can be installed effectively. These include floor-sweep inlets that are part of the dust control system, portable or permanently installed vacuum cleaning systems approved for Class II, Group G locations or, where permitted, equipment to flush floors and surfaces with hose streams. Ledges, walls, and other surfaces capable of retaining static dust should be vacuumed or swept at frequent intervals.

Grain spills should be promptly recovered, dust ducts frequently checked for plugged condition, blast gates and dampers adjusted for balance to control the aspirated dust, and leaking spout flanges and worn spouting temporarily sealed pending maintenance replacement.

Fumigation

Few if any studies have been conducted on the role that fumigants have played in past fires and explosions. While some authorities seem to think that the proprietary chemicals contained in common fumigants contribute by synergism to the explosibility of dust, no data exists to confirm this.

Fumigants at any elevator facility are used only when infestation occurs. The precautions to follow in handling these chemicals are extremely important for the health hazard to individuals, particularly with respect to toxicity and depletion of oxygen. Scientists have recently begun a more in depth study of fumigant vapors and liquids in grain dust explosions.

Fire Control Systems

Grain storage and handling facilities sometimes contain combustible materials aside from the grains themselves. An automatic sprinkler system can protect all areas containing combustible materials (other than grain). If the facility is located in an area where the water supply is inadequate for an automatic system, a dry standpipe that supplies sprinklers in elevator cupolas and in areas containing combustible materials is an acceptable alternative. Fire departments can hook up to the standpipe and supply the sprinklers from tank truck supplies.

Sprinklers or fixed water spray nozzles are sometimes used for dryer interiors with the degree of protection required governed by the construction and arrangement of the drier and any structure enclosing it, and the product being processed.

Standpipe and hose systems that can service all areas housing combustibles and reach access openings to driers are also effective extinguishing aids. Well supplied hydrants, both on the property and in nearby public ways, are often available for effective manual fire fighting, in addition to tank trucks. Care must be exercised in using hose streams, however, because carelessly directed streams may disperse static dust, cause structural damage to bins, or adversely affect the quality of grain in storage that is not directly involved in a fire or explosion situation.

Early detection is essential for effective fire control. Supervisory and alarm service for automatic extinguishing systems, heat sensing devices, and watch service are the principal methods that can help ensure early detection in those areas where combustibles are stored.

Systems for instantaneous detection and suppression of incipient explosions are available for possible use in specific confined areas such as fuel storage tanks, reaction chambers, etc.; however, the applicability to grain facilities has not yet been established for either the storage areas or the handling machinery. Research projects have been funded to assess the effectiveness of such systems in areas where other means of hazard control are not suitable.

Bibliography

References Cited

Chiotti, Premo. 1977. "An Overview of Grain Dust Explosion Problems." *Proceedings* of the International Symposium on Grain Dust Explosions, Kansas City, MO.

Christensen, C. M., and Kaufmann, H. H. 1977. "Spoilage, Heating, Binburning and Fireburning: Their Nature, Cause and Prevention in Grain." *Feedstuffs.* Vol 49, No 44.

de Gooijer, H., et al. 1975. "Literature Investigation into the Dust Explosion Danger in Industries Storing and Processing Cereals and Flour." *TNO Report 8398,* Rigswijk, Netherlands.

Jacobson, M., et al. 1961. "Explosibility of Agricultural Dust." *Report of Investigations 5753.* U.S. Department of the Interior. Bureau of Mines, Pittsburgh, PA.

Lilienfeld, Pedro. 1978. "Special Report on Dust Explosibility." *GCA-TR-78-17-6,* GCA/Technology Division. EPA Contract No. 68-01-4143. Environmental Protection Agency, Washington, DC.

Matkovic, I. M. 1977. "Dust Composition, Concentration and Its Effects." *Proceedings* of the International Symposium on Grain Dust Explosions, Kansas City, MO.

Marks, L. S., ed. 1958. *Mechanical Engineers Handbook.* 6th Edition. McGraw-Hill Inc., NY.

OSHA. 1978. "Grain Elevator Industry Hazard Alert." Occupational Safety and Health Administration, Washington, DC.

Palmer, K. N. 1973. *Dust Explosions and Fires.* Chapman and Hall, London, England.

Schierwater, F. W. 1976. *Die Konzeption der neuen Explosionsschutz-Richtlinien, insbesondere im Hinblick auf die Schutzmassnahmen gengen Staub explosionen. Staub.* Vol 36, No 43.

Weber, Verh. 1978. *Ver. Beforderung Gewerbefleiss.* p 83.

NFPA Codes, Standards, Recommended Practices and Manuals. (See the latest *NFPA Codes and Standards Catalog* for availability of current editions of the following documents.)

NFPA 10, *Portable Fire Extinguishers.*

NFPA 13, *Installation of Sprinkler Systems.*

NFPA 14, *Installation of Standpipe and Hose Systems.*

NFPA 15, *Water Spray Fixed Systems for Fire Protection.*

NFPA 24, *Installation of Private Fire Service Mains and their Appurtenances.*

NFPA 61B, *Prevention of Fires and Explosions in Grain Elevators and Facilities Handling Bulk Raw Agricultural Commodities.*

NFPA 61C, *Prevention of Fire and Dust Explosions in Feed Mills.*

NFPA 61D, *Prevention of Fire and Dust Explosions in the Milling of Agricultural Commodities for Human Consumption.*

NFPA 69, *Explosion Prevention Systems.*

NFPA 70, *National Electrical Code.*

NFPA 77, *Recommended Practice on Static Electricity.*

NFPA 78, *Lightning Protection Code.*
NFPA 91, *Installation of Blower, Exhaust Systems for Dust, Stock, and Vapor Removal or Conveying.*
NFPA 493, *Intrinsically Safe Apparatus and Associated Apparatus for Use in I, II, and III, Division 1 Hazardous Locations.*

Additional Readings

Christensen, C. M., ed., *Storage of Cereal Grains and Their Products*, American Association of Cereal Chemists, St. Paul, MN, 1982.
"Dust Collectors," *Loss Prevention Data Sheet 7-73*, Factory Mutual Engineering Corp., Norwood, MA.
Dust Control for Grain Elevators, National Grain and Feed Association, Washington, DC, 1981.
Factory Mutual Engineering Corporation, "Grain Storage and Grain Milling," *Loss Prevention Data 7-15*, Aug. 1976, Factory Mutual System, Norwood, MA.
Frank, T. E., "Fire and Explosion Control in Bag Filter Dust Collection Systems," *Fire Journal*, Vol. 75, No. 2, Mar. 1981, pp. 73-80, 94.
"Moving Fire: Fire Hazards of Belt Conveyors," *Record*, Vol. 54, No. 6, 1977, pp. 18-21.
NFGA, *Fire and Explosion Research Council Research Reports* (series), National Grain and Feed Association, Washington, DC.
Spencer, Max R., "Bulk Grain Handling," *Industrial Fire Hazards Handbook*, Second Edition, National Fire Protection Association, Quincy, MA, 1984.
Williams, G. M., "Quantitative Method for the Analysis of Electrostatic Hazards and Risks," Industry Applications Society, *Annual Meeting*, Oct. 2-6, 1977, pp. 1058-1064.

STORAGE AND HANDLING OF SOLID FUELS

Revised by Jack J. Ellis, P.E. and Mel Gould

This chapter discusses the practices and fire hazards associated with the storage and handling of the two principle solid fuels—coal and wood—particularly as they apply to electric generation, industrial, and commercial environments. The hazards associated with the actual use of these fuels in firing boilers and furnaces are covered in Section 10, Chapter 1, "Boiler-Furnaces." This chapter confines itself to the hazards associated with the storage and handling of the fuels to the point of delivery into the firing system.

The following chapters may be of interest to readers of this chapter. Section 11, Chapter 3, "Outdoor Storage Practices," contains information about the outdoor storage of wood products. Section 12 contains three chapters of possible interest: Chapter 7, "Materials Handling Equipment," and Chapter 13, "Air Moving Equipment," both offer information about conveying systems; Chapter 11, "Special Systems for Explosion Damage Control," discusses explosion venting. Section 13, Chapter 2, "Motor Vehicles," contains information about the use of motor vehicles in storage facilities.

COAL AS A FUEL

Coal presents hazards between the time it is mined and its eventual consumption in boilers and furnaces. These hazards are associated with its transportation, storage, and bunkering. They can be classified as: (1) spontaneous heating, (2) dust explosions, and (3) gas generation and explosions.

Spontaneous Oxidation and Heating of Coal

Some coals oxidize far more readily than others. Anthracite, the highest ranked coal, has very low spontaneous heating tendencies. ASTM, D 388, *Specifications for Classification of Coals by Rank*, ranks coal by its tendency to absorb oxygen. The lower the rank, the greater the tendency for coal to absorb oxygen (i.e., oxidize).

Mr. Ellis is director, analytical and technical services for Peabody Coal Company in St. Louis, MO. Mr. Gould is corporate insurance assistant with Great Northern Nekoosa Corporation, Millinocket, ME.

The simple downward classification of coals is: (1) anthracite, (2) bituminous, (3) subbituminous, and (4) lignite.

Generally, the following statements are true of coal:

1. The higher the inherent (equilibrium) moisture, the higher the oxidizing tendency.
2. The lower the moisture and ash free Btu, the higher the oxidizing tendency. The higher the oxygen in the coal, the higher the heating tendency.
3. Sulphur, once considered a major factor, is now thought to be a minor factor in the spontaneous heating of coal. There are many very low sulphur western subbituminous and lignite coals that have very high oxidizing characteristics and there are high sulphur coals that exhibit relatively low oxidizing characteristics.
4. Spontaneous oxidation is a solid-to-gas reaction, which happens initially when air (a gas) scrubs past a coal surface (a solid). Oxygen from the air is absorbed by the coal, raising the temperature of the coal. As the reaction proceeds, the moisture in the coal is liberated as a vapor and then some of the volatile matter that normally has a distinct odor is released. The amount of surface area of the coal that is exposed is a direct factor in its heating tendency. The finer the size of the coal, the more surface is exposed per unit of weight and the greater the oxidizing potential, all other factors being disregarded.
5. Many times, segregation of the coal particle sizes is the major cause of heating. The coarse sizes allow the air to enter the pile at one location and react with the high surface area fires at another location. Coals with a large top size [e.g. ≥ 4 in. (100 mm)], will segregate more in handling than those [≥ 2 in. by O size (50 mm)].
6. It is generally believed that the rate of reaction doubles for every 15 to 20°F (8 to 11°C) increase in temperature.
7. Freshly mined coal has the greatest oxidizing characteristic, but a hot spot in a pile may not appear before one or two months. As the initial oxidization takes place, the temperature gradually increases and the rate of oxidization accelerates.
8. There is a critical amount of airflow through a portion

of a coal pile that maximizes the oxidizing or heating tendencies of coal. If there is no airflow through a pile, there is no oxygen from the air to stimulate oxidation. If there is a plentiful supply of air, any heat generated from oxidation will be carried off and the pile temperature will reach equilibrium with the air temperature, this is considered a ventilated pile.

9. When there is just sufficient airflow for the coal to absorb most of the oxygen from the air and an insufficient airflow to dissipate the heat generated, the reaction rate increases and the temperatures may eventually exceed desirable limits.

The general ranges of densitites of coal are given in Table 11-8A to illustrate the significance of the density of the coal at various stages in its utilization.

TABLE 11-8A. General Ranges of Densities of Coal

Physical Description of Coal	Range of Density lbs. per cubic foot (kg/m³)	Approximate Percentage of Coal Pile Occupied by Air
Coal in the seam before mining	80–85 (12.8–13.6)	0%
Coal size 2″ × 0 normal shipping size as loaded in RR car, truck, etc., *No* Compaction	50–55 (8–8.8)	36%
Coal size 2″ × 0 compacted with track mounted dozers	60–65 (9.6–10.4)	24%
Coal size 2″ × 0 compacted with rubber tired carryall or sheepsfoot roller	70–75 (11.2–12)	12%

STORAGE PRACTICES FOR COAL

The methods used to build outside storage piles and to fill silos, bins, and bunkers are significant in controlling the oxidation hazard of coal. Coal is usually stored either in an uncompacted [50 to 55 pcf (801 to 881 kg/m^3)] or compacted form [70 to 75 pcf, (1121 to 1201 kg/m^3)]. When coal is not compacted, air occupies approximately 35 percent of the volume of the pile, so it is relatively easy for air to move through the pile. Coal is at this density when it is handled on conveyors, or placed in trucks, railroad cars, bunkers or silos, or in any form of live pile. Coal is uncompacted whenever it is moved and falls on itself.

Generally, when coal is compacted it is placed in relatively thin layers of from 6 to 12 in. (15 to 25 mm) thick. Each layer is then rolled over with a device that has a relatively high loading in pounds per square foot or kilogram per square meter of surface (1 psf equals 4.9 kg/m^2). Rubber tired carryalls are normally used for this purpose since they compact the coal while they convey the coal to the desired location.

A coal pile should be compacted on all sides to eliminate any low density areas. When coal is compacted, it has a density of 70 to 75 pcf (1121 to 1201 kg/m^3), and air occupies only 12 percent of the pile. This makes it far less likely that there would be appreciable air movement through the pile. Virtually all types of coal have been stored successfully in this manner, including piles exceeding 1 million tons (1 ton equals 1016 kg) and at depths approaching 100 ft (30 m).

Large users of coal (utilities and large industries) will generally have compacted permanent storage piles and uncompacted "live" storage piles, while the smaller users (small industry, commercial, and residential) normally have uncompacted storage. All coal users have some form of uncompacted storage, and this form of storage may require special fire protection considerations.

Uncompacted Coal Storage

Ideally, all uncompacted coal is stored in properly designed bunkers, silos, bins, and even outside piles so all of the coal moves through the storage system with no dead or unmoving portion of the coal pile. Although the coal is still subject to the slow oxidation, it will move through the system before any appreciable heating can occur. Many new bunkers, silos, and bins are designed for mass flow of coal, and heating problems have been virtually eliminated. Coal piles may be worked in such a manner that the piles are rotated and the first coal in is the first coal out. This achieves the same results as the mass flow conditions in bins where none of the coal remains for prolonged periods of time.

When bunkers, silos, or bins have not been, or cannot be, designed to achieve mass flow, heating of the coal may occur and create problems. As much as practical, sealing off the bottom of the storage system can minimize the airflow through the coal; however, space above the coal in the storage system should not be sealed. This will prevent the accumulation of gases that may become liberated from freshly mined or crushed coal. In these cases, the area above the coal should be ventilated to carry off any accumulation of gases. Some storage systems, particularly ones where pulverized coal is stored, are designed to prevent the formation of a hazardous mixture of gases and coal dust with air.

It may be necessary to routinely draw down the storage system and clean the static deposits of coal from the bunker to minimize problems. Because cleaning bunkers, silos, or bins can be hazardous, special safety precautions should be taken.

If heating occurs in a bunker, silo, or bin, it is generally in an area where static deposits of coal have remained for a period of time. Normally, this condition is resolved by drawing down the bin and removing the heated, or static, deposits. A controlled amount of water, wisely applied, may provide sufficient cooling to allow the user to remove the heated coal and burn it in a normal manner; however, it is very difficult to apply water into stored coal and successfully eliminate further heating. Extreme caution should be practiced when using water on large areas of hot coal in a confined volume to prevent the accumulation of hazardous amounts of water gas that contain varying amounts of hydrogen and carbon monoxide. Generally, moving hot coal in an outside environment will cool the coal substantially.

Care should be exercised whenever using water so that puffs of steam that could impair visibility are not formed. Excessive amounts of water should be avoided because handling and utilization problems may be created.

Selecting an Outside Storage Site: Many plants have uncompacted storage piles on the ground, and some precautions must be taken in selecting and preparing a storage site. First, it should be determined that the proposed site is not over open sewers or other devices that may bring air into the base of the pile. Neither should the site contain any steam lines or systems that could raise the temperature of a portion of the pile beyond normal ambient temperatures.

Once selected, the site should be cleaned of all vegetation and foreign materials, particularly materials that may have heating tendencies. Then it should be graded smoothly and contoured so that water will not drain into the base of the pile. The periphery of a pile should be given special attention in the grading process to make sure that all of the water draining from the pile will be carried away.

Recognizing that uncompacted piles may tend to heat, a good practice is to size and/or shape a pile for easy access to areas that have heat so that they can then be handled appropriately. It is also good practice to limit the height of an uncompacted pile to approximately 15 ft (4.6 m) if the coal is known to have high heating tendencies. Greater pile heights can be tolerated if the coal has low heating characteristics and if the pile can be rotated with ease.

Special Storage Precautions

It is beneficial to handle coal going to storage in a manner that prevents segregation of coarse coal from the fines. Segregation occurs when coal is placed in a conical pile; the coarse coal runs to the edge and outside of the pile and the fines collect in the center. The air can then move readily through the coarse coal to the fines containing the large surface area, greatly encouraging heating of the pile.

Placing coal in storage near a vertical wall presents special problems, because coal tends to be less compacted when it is placed near these vertical surfaces unless special precautions are taken. Preferably, coal should not be stored around supporting beams or similar members. Less dense coal provides a natural flue for air ventilation. When vertical walls or surfaces of any kind cannot be avoided, the coal should be stored along the wall or parallel to the wall, as opposed to pushing the coal up to the vertical surface.

Selecting Coal for Storage

Commercial and domestic users select the size of coal best suited for their burning equipment, and in most cases it is usually a double screened coal with a 1¼ in. (32 mm) top size, a ¼ in. (6 mm) bottom size, and a relatively small percentage of fines. This size of coal, if handled reasonably well, will not segregate and a pile can be considered fully ventilated. There will be a minimum tendency of heating because of fewer fines and because any heat generated will be readily removed from the pile.

Industries and utilities normally use a coal with a larger percentage of fines, and it is advantageous to use a coal that can be handled and stored with ease. Normally, a good shipping size is 2 in. (50 mm) top size or less, but if a larger size is used, extra care should be exercised to minimize segregation and oxidizing in storage.

Detection of Heating

When heating occurs in coal, a vapor is generally emitted that has a distinct odor. It is part of the volatile matter of the coal and may contain some sulphur compounds that are particularly noticeable. In many cases this vapor is visible, and it should alert the operator to start planning a procedure to eliminate a potential hazard. Depending on the materials of construction, and if the heating has occurred in bunkers, bins, or silos, the operator has several choices. One alternative is to select a convenient time and empty the storage vessel to the point where the hot material can be removed, or cooled and removed. In some cases the vessel can be sealed and inerted with an inert gas; this will minimize further heating. Care may still need to be taken in ultimate utilization of the coal.

Heating in a storage pile is also normally detectable by smelling or seeing the vapor. Vapor will travel through the path of least resistance in a storage pile and may emerge from the pile at a location remote from the hot spot. This vapor should not be confused with a moisture vapor that may be noticeable on cool mornings.

Sometimes small diameter pipes of approximately 1 in. (25 mm) in diameter are placed into a pile and temperatures recorded at several elevations; however, it is easy to be mislead as temperatures within a pile can vary, generally within a few feet of each other. Normally, if the pile is less than the temperature the body can tolerate, for example, 125°F (51.6°C), it is reasonable to leave the pile as is, but to continue monitoring. If temperature is over 125°F (51.6°C), plans should be made to remove the hot area. A hot area usually can be dug out and spread out to cool on the surface of the pile if the heated area is not too large and hot. At other times, the coal can be dug out and sprayed with water and mixed with cooler coal and then consumed as soon as possible. When hot coal is fed to pulverized fuel fired equipment, the operator should use extreme caution and be familiar with the potential hazards.

DUST EXPLOSIONS WITH COAL

The conditions that create dust explosions usually occur when dry coal is being transferred from place to place or when an air current disturbs dust that has accumulated on surfaces. An ignition source must also be present at the location of the coal dust and air mixture to create the explosion. Although these types of explosions are rare, when they do occur, injury and loss of life and damage to equipment can be substantial.

There are several precautions that can be taken to help eliminate this problem. Commercial and domestic coals, for example, can be double screened to minimize the number of fines and they may also be treated with oil. Providing sufficient surface moisture to the coal to get the fine dust to stick to the larger sizes can be very effective in many installations. Wetting agents sometimes are used with the water to improve dust control. Many industrial and utility plants use vacuum hoods or other sophisticated systems to minimize the dust problem. Eliminating ignition sources is desirable, but is not always possible; therefore, it is more effective to eliminate coal dust air mixtures that may explode. Some of the ignition sources that can be controlled are smoking, cutting, and welding. Some of the causes that are much more difficult to eliminate are static electricity, occasional electrical sparks, and sparks that may result from dropped tools.

GAS GENERATION AND EXPLOSIONS WITH COAL

Methane and other gases can be liberated from freshly mined and freshly crushed coal. An accumulation of a dangerous proportion of gas from this source is rare, but when explosions do occur, severe and fatal losses may result.

Dangerous accumulations of the gas can occur over the tops of bins, silos, and bunkers, and in underground coal reclaiming facilities. Adequate ventilation is the best solution to prevent dangerous accumulation. The ventilation system should sweep airflow across the tops of bins, silos, and bunkers, and through the conveyor gallery of underground reclaiming systems. Care should be taken in the design and operation of the storage system to minimize any airflow through the stored coal, thus preventing spontaneous heating. In addition, any airflow through the pile should be in a direction that will minimize any dangerous accumulations of gas.

If there is any suspicion of gas accumulations, a methane monitor can be permanently installed and portable detectors can be used by personnel working in these areas.

WOOD AS A FUEL

Wood is becoming widely used because of the cost and occasional shortage of crude oil and because of improved technology for whole tree chipping on site. States such as Maine, New Hampshire, Vermont, and other northern tier states that have reasonably abundant supplies of wood are continuing to research how wood can be converted into usable form to produce energy for heat and power generation. In Maine alone, more than 20 wood to energy projects have been proposed that would have a total capacity of 479 megawatts.

The increasing use of wood as an energy source competes with the use of wood as raw material for pulp, paper, and building products. Consequently, much of the research into wood use today is concentrated upon the less commercially useful wood species and upon wood waste from the forest products industry. The emphasis is upon total use of the tree, either in the production of the product or of the waste as fuel.

Hazards of Wood Fuels

Given a source of ignition, wood in all its forms will burn. As the tree is cut and processed into its eventual use, the hazard of fire increases from two directions: reduction of moisture content from air drying; and reduction of the wood particle size from the standing tree to finished lumber to chips for fuel and chemical pulp, to waste wood particles the size of finely divided dust and flour. As the drying and reduction process proceeds from the relatively large particles down to the dust/flour stage, the hazard changes from that of a Class A combustible material subject to typical burning characteristics to one of such rapid combustion that it is termed an explosion. In considering the storage of wood as a fuel, the hazards encountered will range from those for solid wood to those for wood dust/flour.

Solid Wood: The hazard for solid wood is Class A combustible material not readily ignited but capable of generating a large amount of heat if allowed to burn unimpeded. The rapidity with which fire will spread depends upon pile configuration; moisture of the wood and surrounding atmosphere; physical characteristics of the wood, i.e., size of individual pieces; whether wood is peeled or with bark; and species of wood. Pile size and configuration will determine the amount of radiant heat released.

Chips: This form of wood is being used on an ever increasing scale as fuel for heat and power generation because it can be mechanized on the scale needed. The hazard is again one of a Class A combustible and because of its small size [5/8 to 1¼ in. (16 to 32 mm)], fires in piles tend to be surface type with no more than a few inches (mm) of fire penetration into the pile itself. Internal fires do occur from spontaneous heating as a result of excessive internal heat buildup. These fires are usually from fines that cause too much compaction or humus material within the pile, which is subject to more rapid oxidation than clean wood chips.

The chip fines can present a flash fire or explosion hazard if handled inside and dispersed in a manner which allows the wood dust to collect on structural members or to be suspended in air at concentrations above the Lower Explosive Limit (LEL).

Sawdust and Shavings: Sawdust and shavings, a byproduct of the lumber mill and wood processing plants, are generally used as a fuel on site. If strategically located, the usable material, along with other waste wood, may be shipped to other locations for use in manufacturing pulp, particle board, or pellet fuel. The hazards are Class A combustible material and possible explosion in storage and handling from wood dust buildup.

Other Wood Waste: Wood scrap and edgings are another waste byproduct, particularly of lumber mills and furniture manufacturing plants. This material is generally put through a hog to reduce the size for conveying or blowing to storage bins and boiler rooms for fuel. If the larger pieces of scrap are recoverable for use as chips in pulp manufacturing, they are conveyed to a chipper and pneumatically blown to a loading station for transfer into rail cars or tractor trailers for shipment. The hazards are similar to those for chips and sawdust. The larger wood particles are readily ignited while the fines generated by the sawing, planing, and chipping processes may contain particles small enough to be an explosion hazard.

Pellets: This form of solid wood is in the developmental stage, and is in commercial production only to a limited extent. The process uses pulverized wood waste from the forest products industry which is formed under high temperature and pressure into pellets approximately ¼ by ¾ in. (6 by 19 mm). The finished pellets are stored in bulk form inside to keep them dry; the raw stock of wood waste is stored outside. The finished pellet storage presents little dust hazard but will burn as a Class A combustible material.

Bark: Bark is generated at pulp mills, sawmills, and chip plants as a waste material and used to a limited extent as a soil conditioner and as fuel for limited size boilers (when mixed with other wood scrap from the process), or simply disposed of by burning in teepee type bark burners. Until recent years, bark was basically a disposal problem. Some pulp mills, however, generated too much bark over the years to be disposed of by the above means and it was simply piled at a detached site convenient to the plant. Over many years of operation, these piles have grown into miniature mountains. The sheer height and bulk of the piles has created a fire hazard from internal heating leading to spontaneous ignition, and from external sparks from heavy equipment used to build the pile, particularly during dry periods.

A method has been outlined for eliminating fires in wood bark piles by controlling the flow of air (oxygen) horizontally into the pile (Halsey 1980). Oxygen control is achieved by sloping the pile at a 30 to 40 degree angle from the horizontal. In Figure 11-8A, dimension "A" is the depth at which pressure and moisture allow the chemical reaction to generate enough heat and volatile material to provide two sides of the fire triangle, i.e., heat and fuel. Dimension "B" is the distance air (oxygen) must travel to complete the fire triangle and is approximately three times greater with a 30 degree slope than with a 60 degree slope. The upward flow of heated gases within the pile prevents oxygen from reaching the point of combustion except from the nearly horizontal direction, which means the resistance to oxygen flow increases as dimension "B" increases. Where increasing the slope is not practical, sealing the slope to air penetration is an alternative.

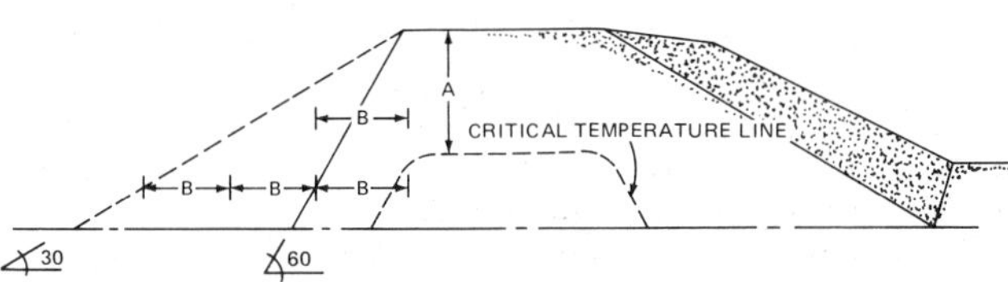

FIG. 11-8A. A method for eliminating fires in wood bark piles by controlling the flow of oxygen horizontally into the pile.

What was formerly a waste disposal problem has become a valuable commodity as the price of fuel oil has gone up and the technology for burning bark on a large scale has improved. For those with such piles, they can be reclaimed as fuel for relatively large bark boilers. The reclaimed bark is mixed with the bark generated daily in the pulp manufacturing process.

STORAGE PRACTICES FOR WOOD FUELS

Logs that are to be brought to the use site for industrial and commercial conversion to chips for fuel should be stored in ranked piles (piled in parallel form), because such piles are lower in height with less volume and less radiant heat exposure than stacked piles (large circular "matchstick" piles). The use of long logs, which is likely, will dictate the ranking method of storage. The piles should be 100 ft (30 m), and preferably more, from the nearest important building. The length and width of piles and pile height should be kept as small as practical to make fire fighting easier and to limit the amount of wood subject to a fire. The storage area should be clean and on solid ground, with good access for fire fighting purposes.

Residential use of cordwood and long logs is generally in the five to fifteen cord range and presents a relatively light hazard. Common sense dictates maintaining as much clear space as possible between the rough storage and dwelling. After cutting and splitting for use, the wood often is moved into the basement or garage for seasoning where there can be a fire potential from heating appliances, electrical wiring, and smoking. Care should be taken to store the wood away from the furnace it is to be burned in, electrical switchboxes, and wiring. A better method is to store the wood in a pile detached from the residence and bring in a week's supply as needed.

The use of wood chips for a residential fuel is still in the developmental stage. Furnaces and the means of automatically stoking them with the chips as well as chip storage are being developed and marketed. Wood chips for fuel, often mixed with other biomass, has forged ahead in industrial and independent energy projects as illustrated by the numerous wood to energy projects proposed in the state of Maine.

Where chips for fuel are used on a large scale, it likely will be in a relatively isolated area close to the wood source and unfortunately away from municipal fire protection facilities and water supplies. Only the larger operations under these conditions can economically justify an independent water supply distribution system, and it will be doubly important to limit outdoor chip pile size and height to dimensions even less than those recommended in NFPA 46, *Recommended Safe Practice for Storage of Forest Products* (hereinafter referred to as NFPA 46). The plant should be located, if possible, near a pond or stream so water is available for hose streams. Any outside storage should be well isolated from processing buildings and surrounding forests.

Storage of chips, sawdust, shavings, and sander wood waste, in silos, bins, and buildings may present fire and explosion hazards. Therefore, storage should be arranged to minimize these hazards by proper location, preferably outside the buildings, and by physical construction of the bin or silo and associated ducts with explosion relief venting and explosion isolating chokes.

Large scale bark burning boilers require a continuous supply of fuel to avoid fluctuations in steam output or the need to fire oil burners to take up the slack. Consequently, it is common practice to provide an enclosed surge storage area capable of providing one or two days supply of fuel. One building design used for such a storage building is the "A" frame style with storage capacities up to 2,000 short tons (1814 metric tons) at 50 to 60 percent moisture. These storage buildings should be designed to minimize the possibility of wood dust buildup on structural building members by minimizing surface areas on which the wood dust can collect. If possible, the walls should be sloped at angles approaching 60 degrees. Structural members that cannot be sloped should be covered with sheet metal designed to eliminate flat surfaces upon which the dust can settle.

HANDLING WOOD FUELS

Logs, either tree length or cordwood, are loaded and piled using mechanical equipment mounted either on the log truck or on a separate vehicle. Logs present minimal fire hazard exposure to the wood handling operation.

Wood chips and wood waste material are generally transferred either pneumatically or by belt conveyor from one place to another. The fans, ducts, cyclones, and collectors associated with air movement of wood fuel constitute a fire and explosion exposure to the process of which they are a part, they should be located so as to minimize this exposure, preferably outside the buildings whenever possible. (See Fig. 11-8B.)

Magnetic separators should be provided to reduce the possibility of a spark igniting the dust. The fans should be downstream of the collectors, if possible, to prevent passage of the wood fuel through the fan.

Belt conveyors present a fire hazard from the belt material as well as from the wood material being conveyed. Sources of ignition are frictional heat from belt slippage, cutting and welding, smoking, and overheated bearings. Belt conveyor fires are sometimes complicated by the need to run them overhead, particularly in large bark burning boilers. Overhead conveyors present an access problem for hose stream use.

Screw conveyors and rotary airlocks should be utilized to provide chokes for fire and explosion isolation where the wood material being handled is fine enough to present an explosion hazard. It is necessary to remove a revolution (helix) of the screw in order to provide the choke. Backdraft dampers can also be used in high velocity systems. (See Fig. 11-8C.)

FIRE PREVENTION FOR WOOD FUELS

Where there is outside bulk storage of wood for fuel in any form, the first line of defense should be to limit the pile size to the absolute minimum required for economical operation. The homeowner and small commercial stick wood consumer will likely limit inventory to one or possibly two years supply, whereas the larger commercial and industrial users who consume wood fuel to generate heat and power will likely restrict inventory to one day or even hours since, in many cases, they are consuming the wood waste as it is generated in the process.

Logs should be ranked rather than stacked and otherwise arranged in accordance with NFPA 46. If large chip piles are necessary, they should be arranged as outlined in NFPA 46. Two or more smaller piles are better than one large pile.

There is indication that foreign material in piles of chips or bark accelerates the buildup of heat, possibly from chemical reaction or simply the increased porosity that allows freer movement of air through the pile. Keeping the piles free of contamination is effective fire prevention.

Pile site selection, whether for chips or bark, should take into consideration the topography of the ground, which should be clean and free of combustibles. Clearance from surrounding forest, brush, or grass should be adequate to prevent an exposure fire from reaching the storage pile. Piles should be detached from buildings and located sufficiently apart from each other to allow fire fighting efforts to control an exposing fire.

Bark piles stored outside, whether rough or hogged bark, create heat which can lead to spontaneous ignition under certain conditions. The piles should be limited in height and contoured to restrict the flow of air (oxygen)

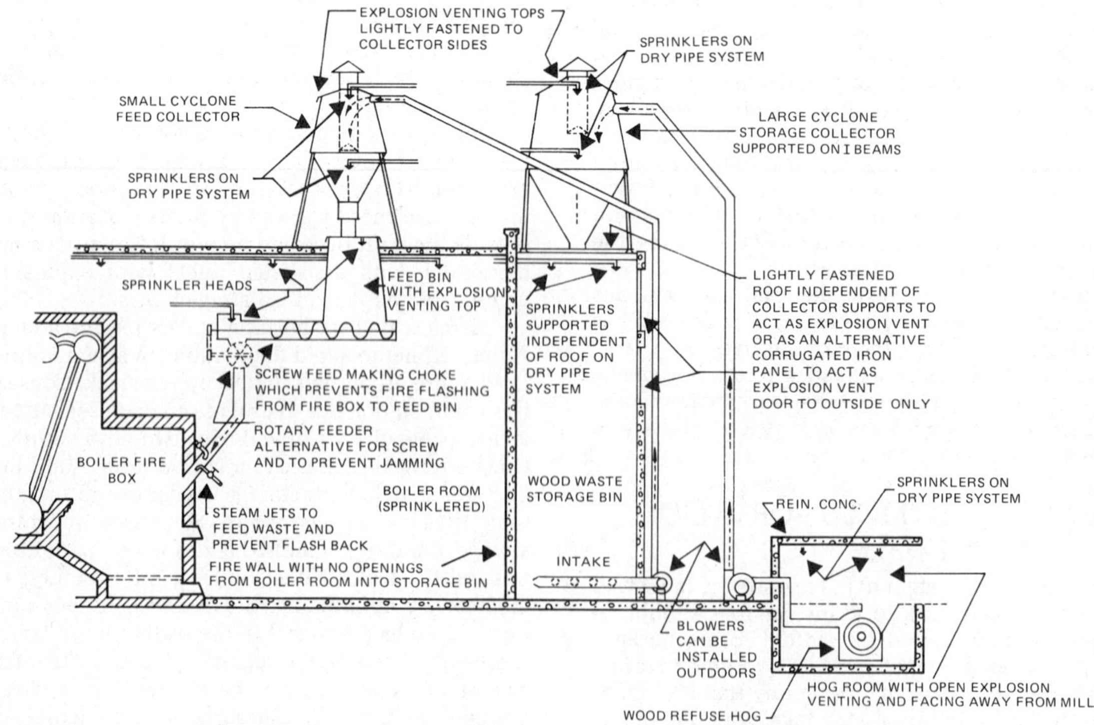

FIG. 11-8B. Suggested arrangement for wood waste firing of boiler to minimize explosion and fire hazard.

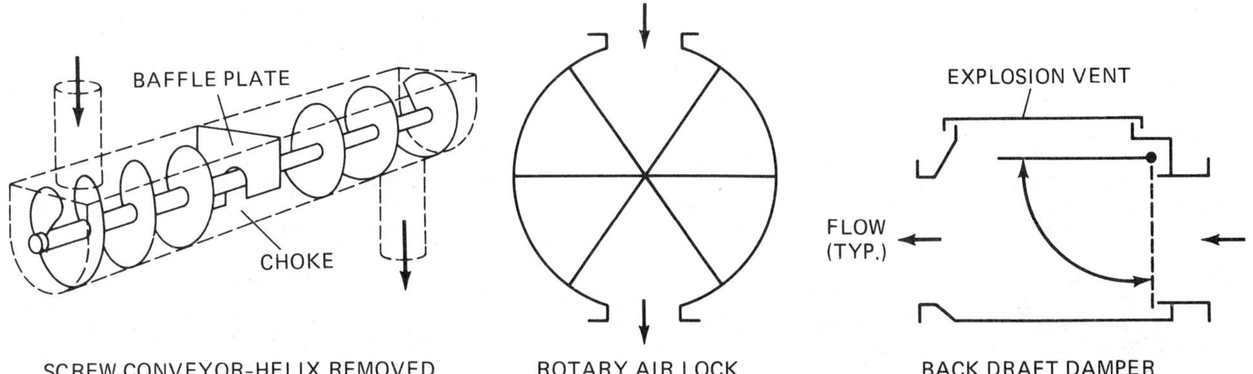

FIG. 11-8C. Types of chokes used to isolate fire and explosion when wood material is fine enough to explode.

through the pile. Heat which is generated within the pile rises to the surface and is dissipated to the surrounding air, which tends to keep the internal temperature below the spontaneous ignition point. However, in northern climates a blanket of ice and snow can impede heat release to the extent that combustion does occur. Thus, restricting the pile height to limit heat from pressure and restricting airflow through the side of the pile becomes doubly important under these conditions. (See Fig. 11-8B.)

Storage bins, silos, and vaults for waste wood fine enough to be an explosion hazard should be outside plant buildings, if possible, or if inside, should be arranged to vent explosions to the outside. For further details, see NFPA 664, *Standard for the Prevention of Fires and Explosions in Wood Processing and Woodworking Facilities.*

The need to control moisture content of wood waste fuel for more efficient combustion may require use of lightly constructed metal buildings on a concrete slab to protect the fuel from the weather. Such buildings should be designed to minimize dust buildup on structural members. Lighting, if required, should conform to applicable provisions of NFPA 70, *National Electrical Code®.*

If the wood waste being stored contains a considerable amount of fines in the wood/flour range (such as sander fines), it may be necessary to incorporate explosion relief in the building design. A partially open structure may be a feasible design.

FIRE PROTECTION FOR WOOD FUELS

The bins, silos, or vaults in which chips or other wood waste are collected should be constructed of metal and located outside the boiler building, where possible. Collector and storage units handling fines that present an explosion hazard should be constructed with explosion relief venting arranged to operate well below the design strength of the equipment. If bulk storage must be in or adjacent to the boiler building, it should be cut off with a fire rated masonry wall.

Ranked log piles, arranged as previously described, should be protected with a yard hydrant system suitable for the storage arrangement and size, as defined in NFPA 46. Where the plant location and size precludes having an adequate water supply and a plant fire brigade, smaller, more segregated piles will be necessary with greater clear space between the wood storage and plant.

Chip and bark piles also require large amounts of water to control and extinguish fires. While it is likely that storage piles of chips and bark to be used as fuel would not be the same size as piles of these materials at pulp mills, conscious effort should be made to keep the piles limited in size in order to reduce the volume of water and the fire fighting effort needed to combat a fire. See NFPA 46 for details of yard fire protection.

Automatic sprinkler protection (on a dry pipe system, where necessary) should be provided in the collecting and storage areas. Where fuel storage buildings are utilized in the system, they should be provided with automatic sprinkler protection in accordance with NFPA 13, *Standard for the Installation of Sprinkler Systems* and NFPA 15, *Standard for Water Spray Fixed Systems for Fire Protection.*

Where bark or chips are transported from the outside storage pile or silo to an inside bin for feeding to the boiler by enclosed conveyors, the structure should be noncombustible and arranged to meet applicable safety codes. The conveyor housing should not exceed 10 ft (3 m) in width to limit the required sprinkler system to one line. The day storage bin should have sprinkler protection inside and, depending upon design, automatic sprinkler protection over it and the conveyor.

Automatic sprinkler protection on a wet pipe system should be provided in the enclosed conveyors and should be hydraulically designed to provide minimum operating pressure on the end sprinklers. Where temperatures in winter will not allow use of wet systems, dry pipe or preaction systems should be used. Regardless of the type system used, the conveyor should be interlocked to shut down upon actuation of the sprinkler or detection system.

Bibliography

References Cited

Halsey, Ken. 1980. "Controlling Bark Pile Fires." *Pulp and Paper Magazine.* Dec. 1980.

NFPA Codes, Standards, Recommended Practices and Manuals. (See the latest *NFPA Codes and Standards Catalog* for availability of current editions of the following documents.)

NFPA 13, *Standard for the Installation of Sprinkler Systems.*
NFPA 15, *Standard for Water Spray Fixed Systems for Fire Protection.*
NFPA 46, *Recommended Safe Practice for Storage of Forest Products.*
NFPA 68, *Guide for Explosion Venting.*
NFPA 70, *National Electrical Code.*
NFPA 91, *Standard for the Installation of Blower and Exhaust Systems for Dust, Stock and Vapor Removal in Conveying.*
NFPA 664, *Standard for the Prevention of Fires and Explosions in Wood Processing and Woodworking Facilities.*

Additional Readings

Bluhm, Delwyn D., "Grinding and Milling Operations," *Industrial Fire Hazards Handbook*, 2nd ed., National Fire Protection Association, Quincy, MA, 1984.
Bounous, Edwin P., "Furniture Manufacturing," *Industrial Fire Hazards Handbook*, 2nd ed., National Fire Protection Association, Quincy, MA, 1984.
Castleberry, Jack C., and Smith, Peter A., "Pulp and Paper Processing," *Industrial Fire Hazards Handbook*, 2nd ed., National Fire Protection Association, Quincy, MA, 1984.
"Coal and Coal Storage," *Loss Prevention Data Sheet 8-10*, Factory Mutual Engineering Corporation, Norwood, MA, Aug. 1975.
"Combustible Dusts," *Loss Prevention Data Sheet 7-76*, Factory Mutual Engineering Corporation, Norwood, MA, Aug. 1976.
Factory Mutual Engineering Corporation, *Handbook of Industrial Loss Prevention*, Chapters 66 and 70, Norwood, MA.
Hall, Keith F., "Wood Pulp," *Scientific American*, Vol. 230, No. 4, April 1974, pp. 52-62.
Industrial Risk Insurers, *The Sentinel*, Jan.-Feb. 1979.
NFPA Industrial Fire Protection Section, Pulp & Paper—Wood Products Industry Committee, "Wood Products," *Industrial Fire Hazards Handbook*, 2nd ed., National Fire Protection Association, Quincy, MA, 1984.
"Outdoor Storage of Wood Chips," *Loss Prevention Data Sheet 8-27*, Factory Mutual Engineering Corporation, Norwood, MA, Dec. 1980.
Reineke, L. H., "Wood Fuel Preparation," *U.S. Forest Service Research Note EPL-090*, Forest Products Laboratory, Madison, WI, Jan. 1969.
Twitchell, Mary, *Wood Energy: A Practical Guide to Heating with Wood*, Garden Way Publishing, Charlotte, VT, 1978.

STORAGE AND HANDLING OF RECORDS

Revised by Thomas Goonan, P. E.

The information explosion brought about by such modern technology as telecommunications and computers has created a massive increase in the total volume of records that are generated. This massive increase has intensified the problems involved in protecting records from fire. Traditional fire resistive containers—insulated file cabinets, safes, and vaults have not lost any of their capabilities to protect paper records. In fact, recent advances have introduced devices that can safeguard magnetic or photographic records. In many cases, however, the volume of records material has become so great that it is impractical to invest in either the devices or the space that would be required to safeguard all the records to the level of protection which is justified. New methods of records storage have been developed that use the maximum cubic capacity of the area assigned to records storage. Some of these storage arrangements not only place the records at risk, but also contain sufficient fire potential to be a danger to the structure and all other operations housed in it.

Additional information on life safety and fire protection as related to records storage will be found in Section 7, Chapter 2, "Concepts of Egress Design," Section 12, Chapter 8, "Housekeeping Practices," in Section 18, "Water Based Extinguishing Systems;" and in Section 19, "Special Fire Suppression Agents and Systems."

This chapter discusses ways of identifying and classifying valuable records to determine the amount of protection, justified by their value, against fire and its associated perils. The relative susceptibility of various records to flame, heat, smoke, and water exposure is discussed. Because water damage is an important byproduct of fire containment efforts, information on salvaging watersoaked documents is provided. The different ways in which the risk of extensive loss of valuable records can be mitigated are considered. Protection against perils not associated with fire is not discussed.

Detailed discussions of the best ways to provide maximum reasonable protection for records are contained in NFPA 232, *Standard for the Protection of Records*, and NFPA 232AM, *Manual for Fire Protection for Archives and Record Centers* (hereinafter referred to as NFPA 232 and 232AM respectively). Further information can be found in the report *Protecting Federal Records Centers and Archives from Fire* (GSA 1977), and the reference book *Protecting the Library and Its Resources* (Gage-Babcock 1963).

Adequate provisions must be maintained for the safety of persons in records storage facilities. This can be a particularly difficult task when the facility is designed to provide maximum security of its contents against illegal entry. However, emergency exits are often necessary to assure life safety for persons in the records area in case of fire, even at some sacrifice in security. Alternate egress from catwalks in records centers often requires a common sense design approach. General good practice recommendations for means of egress should be followed.

Mr. Goonan, P.E., M.S.F.P.E., is a consulting fire protection engineer for the Schirmer Engineering Corporation based in Falls Church, VA.

FIRE RISK ANALYSIS

Maximum possible protection is neither feasible nor desirable for the bulk of what comprises records storage. Most records can be reconstructed, and duplicates are often available in other locations, and their loss may not create any substantial hardship. The value of certain, irreplaceable documents warrants especially sophisticated protective measures to preserve a single sheet of paper. The great majority of records in a particular collection, however, may dictate consideration of less than optimum methods of records protection and anticipation of small fire losses, yet still provide a high degree of assurance against a significant loss.

Records protection programs should be based on an inventory that determines type, volume, rate of acquisition, duplication of data, rate of disposal, and class of importance of the records. Records can be generally classified in one of two ways to assess their value:

Vital Records: These records are irreplaceable; records where reproduction does not have the same value as an original; records that give direct evidence of legal status, or ownership; records needed to sustain a business or avoid delay in restoration of production, sales or service; records of accounts receivable.

Important Records: Records that can be reproduced from original sources only at considerable expense or loss of time.

Other records may be useful, but should be kept well separated from vital and important records because they may constitute a fire exposure in and of themselves.

Fire Risk Evaluation Factors

In considering the protection of valuable records, four basic items must be evaluated. They are:

1. The degree to which the building containing the records, as well as neighboring operations, expose the records to fire, i.e., the possibility of involving the records in a fire originating apart from the records storage activity.*
2. The possibility of fire starting within the records storage activity, including the susceptibility of the records or their container to ignition.
3. The quantity of fuel the records represent, particularly as it relates to available or proposed capability for fire extinguishment and the structural stability of the storage arrangement and the building enclosure.
4. The susceptibility of the records to damage from fire, fire effects (heat, smoke, vapors, etc.), and fire extinguishing efforts (principally water damage and impact damage from hose streams and other extinguishing devices, and physical disruption from manual fire fighting).

When records must be housed in a building that may burn around them, properly rated vaults and containers can give reasonable protection against the external fire exposure. However, the hazard of the records themselves, stored within a vault, file room, segregated floor or section of a fire resistive building, or records center, can be substantial. A major fire in a military personnel records center† is described in the GSA report previously cited (GSA 1977). The protection methods described later in this chapter will provide protection commensurate with the hazard and the sophistication of the systems. The degree of fire risk and the potential for loss in large collections not suitable for cabinet or vault storage may require evaluation by a person knowledgeable in this type of analysis.

Bulk Storage

Bulk storage of records creates a fire hazard in itself. The term bulk storage is used here to describe any sizeable collection of records not contained in vaults, safes, or insulated cabinets. The term includes collections of records ranging from small file rooms to the largest archives or records centers. Storage methods include, but are not limited to, file cabinets; various types of shelving, including open shelves and mobile shelves; palletized cardboard boxes; transport cases; miscellaneous cardboard boxes; and devices for unusually shaped records such as blueprints, magnetic tapes, photographic film, and other media. Locations may range from an area within a general office complex to specially built records facilities. It is not uncommon to house record collections in basements or attics of public buildings, in office buildings, in converted factory or warehouse buildings of various constructions and levels of quality, in public warehouses, underground, or in other facilities protected against wartime disasters.

Open Shelf Storage

The trend toward making the maximum use of available space in buildings has sometimes resulted in using open shelf storage methods, normally with the records held in either file folders or various kinds of cardboard boxes. Typically, the racks of records face each other across 25 to 30 in. (0.6 to 0.75 m) aisles. The aisle exposure presents a wall of paper made up of the sides of boxes or loose ends of paper sticking out of file folders. They can be easily ignited accidentally by any ignition source such as a match, a cigarette, a portable heater, faulty fluorescent ballast, or simply by contact with an exposed incandescent light bulb. Paper ignites at approximately 450°F (230°C). Ignition of a few pieces of paper, on a filing cart, for instance, can readily transmit ignition to the boxes.

Attempts have been made to develop economical methods for increasing the flame resistance of the typical records center cardboard boxes. The most frequently attempted method is coating the box with an intumescent type of fire retardant paint. Tests of records boxes protected by such paint, properly applied, show that the coating will substantially delay actual ignition of the cardboard box material. However, since intumescent paint does not effectively react to heat under about 400°F (204°C), the temperature of any modest exposure fire (such as might occur on a file cart) will weaken the paper in the box to the point where the box will break open under the weight of the paper it contains and expose the ordinary combustible contents of the box to ignition. In a small scale test, conducted as a joint effort by the NFPA Committee on Records Protection and the U.S. General Services Administration (GSA), a fire retardant paint coating on boxes only briefly delayed ignition and the spread of fire up and across the face of the records in storage.

Full scale fire tests were conducted in December 1974 in boxes of paper records stored on open steel shelves 13 ft (4 m) high (GSA 1977). Properly designed sprinklers were adequate to control the fire, but the fire developed very rapidly anyway. From a small ignition source, flames reached the top of the rack in 4½ minutes. Eight seconds later, the facing array of boxes across the 30 in. (0.75 m) aisle flashed into flame from top to bottom. The first sprinkler had a 286°F (141°C) rating and activated in a little more than 5 minutes. By then, the ceiling temperature directly over the fire had reached 1,900°F (1038°C). The direct cause of the intense fire was rupture of the walls of the cardboard cartons and the release of the contents. Paper records exfoliated into the aisle and many burned while falling. A test was conducted with an identical arrangement, except that open files were oriented perpen-

*Determining the extent of the fire hazard to which the records are exposed involves consideration of building construction, contents, possible fire causes, and general features of fire protection. The probable maximum fire severity, rated in hours, should be estimated in accordance with the principles outlined in Section 7, Chapter 8, "Structural Intergrity During Fire."

† A fire involving military records of 19 million U.S. servicemen occurred in an unsprinklered, six story, fire resistive building. The top floor, where the fire started, was completely destroyed. Fire fighting efforts prevented the fire from spreading downward. Had the fire started on a lower floor, it is likely that the entire building would have been lost.

dicular to the aisle, which exposed the edges of paper to easy ignition. In this test, fire spread at a leisurely pace and did not develop into an intense fire. The basic difference in the rate of fire growth between these two fire tests was attributed to whether or not paper records exfoliated into the aisles. The progress of the fire can also be dramatically changed by changing sprinkler temperature ratings and sprinkler drop size characteristics (Chicarello and Troup 1980).

In other tests of high piled storage conducted by Underwriters Laboratories Inc. (UL) (Jensen 1963) and tests of 6 ft (1.5 m) high archival shelving arrangements conducted by the GSA, the fire, at the end of an early and relatively short development stage, preheated a sufficient amount of the exposed boxes so that fire development characteristics changed suddenly, temperatures rose quickly, and the flame enveloped large areas. Neither of the tests involved exposed loose paper typical of systems with open shelf filing.

In open shelf storage, the close proximity of the opposing sides of the aisle can result in increased radiant heat feedback and flaming ignition of evolved gases across the narrow aisle. The severity and rapidity of fire development is accelerated by increased stack height and slowed by increased aisle width.

Properly designed automatic sprinklers will control the fire, but factors which speed fire development and penetration increase the amount of information loss and the amount of irrecoverable damage.

Mobile Shelving

A class of storage devices known variously as mobile shelving, track files, compaction files, moveable files, etc. consists of open shelf units mounted on tracks. In use, all of the shelves are pushed together except that one aisle is left open for access to two units. To access another unit, all of the shelf units are moved on the tracks until access to the desired unit is gained. Units may be moved electrically or manually. The shelves that are pushed together form continuous horizontal tunnels. Mobile shelving containing combustibles should be protected by overhead sprinklers.

A fire in the open aisle of mobile shelving is similar to a fire in open shelf storage and is readily controllable by overhead sprinklers. Because a fire elsewhere in the mobile shelving array is a burrowing fire, it is sheltered from sprinkler application. Full scale fire tests conducted in 1978 indicate that a fire will develop and spread very slowly. When shielded from direct sprinkler discharge, fire is likely to continue spreading through the entire array unless each shelf unit has internal metal dividers (Chicarello et al 1978). Although protection by sprinklers alone may permit fire involvement of the entire array, overhead sprinklers provide adequate protection for a building containing mobile shelving, and prevents fire from jumping to other arrays. In addition to sprinkler protection, a smoke detection system is suggested in mobile shelving areas not continuously occupied to provide a fire warning well in advance of sprinkler operation. Early warning smoke detectors may help limit the extent of fire damage in mobile shelving if skilled forces are quickly available during nonoperating hours.

Plastic Media

The flammability of magnetic media and its containers is a prime concern when such media are stored in bulk quantities. Generally, acetate and polyester base tapes do not present a hazard any more severe than paper. Polystyrene cases and reels, however, present a severe fire hazard condition because they have a high fuel value, a high heat release rate, and shed water. UL tests have shown that storage systems designed to safeguard materials of cardboard or paper composition would not adequately protect materials involving polystyrene. If the containers are made of another plastic, the condition may vary, although most thermoplastics exhibit similar properties. In any case, it is necessary to limit the height and extent of the storage of reels encased in plastic or to design special protection systems for them.

DAMAGEABILITY AND SALVAGE

Ordinary paper records survive reasonably well when exposed to most of the effects of fire, except direct exposure to flaming. With rare exceptions, the burned paper record represents a total loss, while a high recovery rate is practical where the records have been exposed only to water, high humidity, smoke, and moderately high temperatures of about 350°F (177°C). Nonpaper records media tend to be more susceptible to damage than paper.

Photographic Records

Photographic records, whether they are on traditional acetate, glass base, or any special base, consist of an image held in place by an emulsion. The image can be distorted or destroyed under any condition which loosens the emulsion. Tests have demonstrated that these emulsions will not withstand high temperature and humidity conditions, and that they are particularly susceptible to steam (McCrea et al 1956; McCrea and Adelstein 1958). The traditional insulated safe or insulated filing device depends upon the water of crystallization in the insulation material to limit the internal temperature, and is usually vented into the interior of the device. Therefore, it must be expected that a steam atmosphere at 212°F (100°C) or higher will exist within the records protection equipment under severe fire exposure. The tests demonstrated that such exposure would damage or destroy the information on photographic media. A high degree of safety, however, may be achieved by placing the photographic media inside a steel can and sealing it with a moisture resistant tape before storage within a record container.

The fire problem with photographic records that are stored in bulk facilities is similar to that encountered with paper records, except that in salvage efforts it is important to give priority to the photographic records. While the cold water used in extinguishing the fire will not immediately attack the photographic emulsion as would steam, it will tend to cause some softening. If salvage efforts are not undertaken immediately, there is a tendency for the emulsion to stick to adjacent material and damage the image.

Magnetic Media

The more common magnetic records consist of magnetic impulses retained in an iron oxide or similar deposit held to an acetate, polyester, or similar plastic base material. The tapes are usually wound on polystyrene or similar plastic reels and contained in polystyrene cases. This is particularly true of many tapes used with electronic computer systems. The security of the record is primarily related, as with photographic records, to the stability of the emulsion. Even minor distortion is severe in the case of magnetic records since machines can not make subjective judgments regarding distortion. If the emulsion is softened by the heat from fire, for example, there is a tendency for the layers of tape to stick to each other on the reel and to be destroyed in efforts to unreel. A tape can be considered safeguarded only up to temperatures in the range of 150°F (66°C) and in a relative humidity not exceeding 85 percent.

The problem of recovering wet records is the same whether they are damaged by a fire, or some other source, such as flood, hurricane, heavy rainstorm, roof leakage, spillage from operations located above, or a breakdown of any of the numerous water or steam systems in the building. It is generally recognized that virtually any wet paper records can be recovered, provided prompt and proper action is taken.

Good sources for further information on salvaging records are contained in NFPA 910, *Recommended Practice for the Protection of Libraries and Library Collections*; *Salvaging and Restoring Records Damaged by Fire and Water* (Fire Council 1963); *Managing the Library Fire Risk* (Morris 1975), *After the Water Comes* (Spawn 1973), and *Procedures for Salvage of Water-Damage Library Materials* (Waters 1975).

FIRE RISK REDUCTION

Since records media are almost always combustible, 100 percent effective protection is not feasible and efforts should be directed to reducing the risk of fire and its associated effects. One of the most effective means for limiting the disastrous effects of a fire in a records storage facility is to prepare duplicates and store them away from the originals where they will not be subject to the same incident. Often the duplicate copy is on microfilm, which is inexpensive, easily transported, and easily stored at a remote facility. Once such a duplicate has been prepared, the value of the original is reduced considerably unless the original document is required for legal purposes or is of intrinsic or historic value. If the duplication is complete, up to date, and easily accessible, the degree of protection is extremely high, and reduced protective measures may be justified for both the original and the duplicate collection.

If the prior generation of computer records is stored at a remote location, the effect of loss of the present generation of records in use at the computer center would be minimized. Although they are not exact duplicates, the earlier records still greatly simplify the task of reconstructing current records.

Protective Containers

Heavily insulated, massive vaults, safes, and filing cabinets are the traditional way to safeguard valuable records against the effects of fire. The practice of encapsulating the records so that they are fully protected against the greatest potential fire exposure is still viable. Containers are available that will keep the internal temperature and humidity as low as necessary to retain the data on magnetic tape and other temperature sensitive media.

Vaults are usually used where the volume of valuable records is large. They are often the only practical method of protection in nonfire resistive buildings where the probable fire severity exceeds the rated endurance of 4 hr safes. Safes are used for smaller volumes of records where it is necessary to have records near their point of use, where the cost of vault construction would be prohibitive, or where the building does not lend itself to vault construction.

Protective containers for records are rated by tests under standard fire conditions. They are rated according to the time elapsed before the interior of the container reaches 350°F (177°C). This time factor provides a measure of safety, since the ignition temperature of most paper is somewhat higher.

Record containers are also rated for interior temperatures that do not exceed 150°F (66°C) and relative humidity that does not exceed 85 percent at temperatures above 120°F (49°C). Containers rated with these more stringent requirements are intended to provide protection for films, magnetic tapes, disc packs, and similar materials. The development of new and different record materials, however, requires a careful evaluation of their characteristics to assure proper protection. An important limitation on this type of container is that it is not subjected to the traditional drop test while containing magnetic or similar media records. The device should not be considered capable of safeguarding magnetic and similar records in any building that could collapse in a fire and cause debris to fall on the container or the container itself to fall several floors.

Table 11-9A indicates the approximate fire resistance that can be safely expected from various types of record containers. The protection provided varies considerably depending upon design and materials. Thickness of insulation alone does not give a reliable indication of fire resistance, which can be determined accurately only by tests.

Table 11-9A. Fire Resistance of Record Containers

Insulated record vault doors	2, 4, and 6 hrs
Insulated file room doors	½ and 1 hr
Steel plate vault doors (with inner doors)	about 15 min
Steel plate door without inner doors	less than 10 min
Modern safes	1, 2, and 4 hrs
"Old Line," "Iron," or "Cast Iron" safes, 2 to 6 in. wall thickness	Uncertain
Insulated record containers (files and cabinets, etc.)	½, 1, and 2 hrs
Containers with air space or with cellular or solid insulation less than 1 in. in thickness	10 to 20 min
Uninsulated steel files, cabinets—wooden files—wooden or steel desks	about 5 min

An analysis of record container performance based on reports gathered over a five year period was prepared by NFPA (NFPA 1953). The performance of 53 vaults was also analyzed and reported. Table 11-9B summarizes the results of the analysis and shows that the contents of 88 percent of the labeled containers were undamaged whereas only 60 percent of the contents of unlabeled

Table 11-9B. NFPA Analysis: Record Container Performance

	Labeled Containers				Unlabeled Totals
	1 hr	2 hr	4 hr	Totals	
Condition of Container					
Normal After Fire	72	40	14	126	74
Abnormal* After Fire	143	54	22	219	220
Totals	215	94	36	345	294
Cause of Abnormal* Container					
Fell	22	11	5	38	29
Struck	14	0	1	15	10
Exploded	0	0	0	0	6
Previous Damage	2	0	0	2	1
Exposure Judged Excessive†	83	29	5	117	105
Other	10	5	7	22	21
Unknown	12	9	4	25	48
Totals	143	54	22	219	220
Condition of Contents					
Preserved	186	84	35	305	178
Heat/Fire Damage	16	6	1	23	70
Smoke Damage	1	1	0	2	5
Water Damage	12	3	0	15	39
Unknown	0	0	0	0	2
Totals	215	94	36	345	294
Cause of Destruction of Contents					
Door Open	7	2	1	10	8
Exposure Judged Excessive†	14	5	0	19	36
Not Insulated	0	0	0	0	4
Obsolete Design or Construction	0	0	0	0	51
Under Water	7	3	0	10	16
Other	1	0	0	1	1
Totals	29	10	1	40	116

* The condition of a container was classified abnormal if, in the judgment of the inspector, the container was so damaged that it could no longer be safely used to protect records.

† For labeled containers, "exposure judged excessive" means exposure in excess of the labeled rating based on the judgment of the reporting agencies. For unlabeled containers, inspection indicated was excessive for the construction.

containers came through the fires in good condition. Exposure in excess of the labeled rating was the largest single cause of container failure and the principal reason for damage to contents of labeled containers. This analysis confirms the need to consider the probable degree of fire exposure when selecting the proper container. Obsolete design or construction of containers was the most frequent cause of damage to contents in unlabeled containers.

In 90 percent of the vaults equipped with labeled doors, the contents were protected. Contents were protected in only 53 percent of the vaults with unlabeled doors. Deficiencies in door construction were the principal reason for failures of vaults equipped with unlabeled doors.

The Concept of Early Warning

Records administrators usually consider the sprinkler system as an option when they plan records protection. A smoke detection system coupled with manual response, once highly favored by archivists and librarians as a substitute for sprinkler protection, is now generally accepted as a useful supplement to sprinklers under favorable conditions. In a slowly developing fire, a well designed smoke detection system can give an advance warning before heat responsive devices such as automatic sprinklers, operate. If manual fire extinguishment is mobilized by a manned emergency center, it may be able to attack the fire before it gains headway. Whether manual extinguishment causes more or less damage than sprinklers depends upon the skill of the responding force and how the fire progresses while the responding force is mobilizing. In any case, although automatic detection and manual extinguishment can be valuable adjuncts to automatic sprinklers, they would not constitute an acceptable alternative to automatic sprinkler protection under most conditions found in records storage.

Contrary to popular belief, a smoke detection system will not outperform alert occupants, except under the most unusual circumstances. Smoke detectors in records storage are only useful when the area is unoccupied, when rapid mobilization of occupants with hand extinguishers is most unlikely. Detectors on ceilings are unlikely to be activated until smoke from a smoldering fire completely permeates the air, or flaming fire gains sufficient headway to drive a column of smoke and heat to the ceiling.

Smoke detection systems are of little value in open shelf records storage, where fire development is likely to be very rapid. They may provide supplementary protection in high value collections, particularly if measures have been taken to slow down development of a fire (such as storage in totally closed metal containers), or burning material is prevented from falling off shelves. Smoke detection equipment can be useful in unoccupied areas of mobile shelving or other arrangements in which combustible contents are sheltered from sprinklers.

Smoke detection systems are also used to activate extinguishing systems such as carbon dioxide, Halon, or high expansion foam. They are used to activate preaction sprinkler systems (installed to prevent accidental water damage). In this application, failure of the detection system for any reason (e.g., loss of power, loose connection) will prevent operation of the sprinklers, so system reliability is reduced. The reliability of sprinkler systems is much greater than the reliability of smoke detectors because of the design simplicity of sprinkler systems.

Duct detectors are used to shut down fans, activate dampers, or initiate smoke control measures. Detectors are also used to cause held open doors to close, to stop conveyors, and in a variety of single purpose actions.

Heat detectors can perform most of the same duties as smoke detectors, usually with a much longer time lag. An extended time lag is sometimes desirable such as in the opening of a sprinkler, or the release of Halon gas.

Hand Extinguishing Appliances

Even if the facility is equipped with an automatic extinguishing system, manual fire control devices should be provided for staff use. Fire extinguishers are most often encountered. The preferred types are those containing clear water as the extinguishing agent. Although carbon dioxide extinguishers do not leave a residue or dampen documents, they are not effective on the deep seated fires likely to occur in records storage. Halon extinguishers have the same limited extinguishing ability as carbon dioxide, in that they are ineffective on deep seated fires. They also pose a health problem. Although national standards rightly limit human exposure to the various halons, there is no mechanism, except the knowledge and alertness of the extinguisher operator, to manage the extent of his/her exposure to halon. Most extinguisher users, however, are unaware of this hazard and are at risk of overdose. Multipurpose dry chemical extinguishers may be provided, but they leave considerable residue and have only limited effectiveness on deep seated fires.

Water hose lines may supplement, or be used instead of, fire extinguishers. To minimize water damage, rubber lined hose with an adjustable water spray, straight stream, and shutoff nozzle is preferable in records storage areas. Its use greatly expands the fire fighting capabilities of the staff. Hose lines connected to carbon dioxide or Halon storage tanks or cylinders have been used. However, the danger to the fire brigade is much greater than with the use of hand extinguishers containing the same agent and should not be undertaken without extensive training and appropriate protective equipment.

Fire Prevention and Emergency Planning

Although detection and extinguishing systems are important, particularly for the bulk storage of valuable records, the first line of defense remains a good fire prevention and risk reduction program. Good housekeeping, orderliness, maintenance of equipment, and prohibition of smoking in records storage and handling areas are fundamental principles of good records management.

An emergency action plan that is kept current and practiced, is essential to limit damage in case a fire occurs. Staff personnel cannot be expected to approach a fire and use extinguishers or hose lines effectively without adequate training. Fire resistive containers and vaults are of much greater value if there has been training in procedures for fire emergencies. The following suggestions apply:

1. Records should be returned to their places of safety accurately, quickly, and without confusion or oversight. If there is no standard record protection, the best plan is to have the most important records carried out of the building. Drills are valuable to train employees to meet emergencies.
2. Records belonging in vaults or safes should never be left out overnight.
3. Important materials that belong under protection should not be allowed to accumulate on desks.
4. Because records normally safeguarded are often unprotected while temporarily on loan, copies instead of the originals should be loaned, and the originals retained in safe storage.

RECORDS STORAGE

Factory Built Devices

Small quantities of valuable records may be stored in factory built records protection equipment such as insulated records containers, fire resistant safes, and insulated filing devices. These devices are available in varying degrees of resistance to fire, heat, and impact, and the degree of protection to be specified for a particular application will depend upon the severity of the exposure and the items to be stored. UL 72 covers the test procedure applicable to this equipment and the ratings employed (UL 1977) as follows:

Records protection equipment is classified in terms of an interior temperature limit and a time in hours. Two temperature limits are employed; either 350°F (177°C), regarded as a suitable limit for paper records, or 150°F (66°C), regarded as a limiting temperature for most magnetic tape and photographic records. The time limits employed are 1, 2, 3, or 4 hr. The complete rating, consisting of two elements, indicates that the specified interior temperature limit is not exceeded when the record container, safe, or filing device is exposed to a standard test fire as described in UL 72, for the length of time specified.

Ratings are assigned to the various categories as follows:

Insulated Records Containers

Class 150—4 hr
Class 150—3 hr
Class 150—2 hr
Class 150—1 hr
Class 350 (A)—4 hr
Class 350—2 hr
Class 350—1 hr

Fire Resistant Safes

Class 350 (A)—4 hr
Class 350—2 hr
Class 350—1 hr

Insulated Filing Devices

Class 350—1 hr

Insulated File Drawer

Class 350—1 hr

Insulated records containers and fire resistant safes must protect contents from heat to an extent described in the requirements, before and after a fall from 30 ft (9 m). Insulated filing devices, which are not drop tested, are not required to have the strength to endure such an impact.

Insulated record containers, fire resistant safes, insulated filing devices, and insulated file drawers must sustain sudden exposure to high temperatures to an extent described in the requirements without the unit exploding as a result of such exposure.

Ordinary, uninsulated steel files and cabinets provide only a limited measure of protection from an exposure fire, as heat sufficient to char contents is quickly transmitted to the interior. However, they can be very useful where the major fire exposure is from the records themselves. They are commonly used for records having no extraordinary value. They are also used for the organized storage of valuable records in protected facilities including vaults, file rooms, and document buildings. A records vault or file room in which all records are kept in metal file cabinets is much safer than one in which the records are kept in cardboard boxes or on open shelves. If an ignition occurred in an open file drawer, fire development would likely be very slow. An ignition in open shelf files, however, could be beyond manual control within a few minutes.

Files and cabinets made of wood, fiberboard, or other combustible materials add to the fire hazard and release their contents in a fire. They should not be used where they will expose valuable records storage.

Vaults and File Rooms

Standards for the construction of fire resistive vaults and file rooms are contained in NFPA 232. The term "vault" refers to a completely fire resistive enclosure up to 5,000 cu ft (142 m^3) in volume, used exclusively for records storage, with no work to be carried on inside. It is equipped, maintained, and supervised to minimize the possibility of a fire starting within, and to prevent a severe fire of long duration on the outside from entering, provided the vault door is closed. Vaults are designed to fire resistance classifications of 2, 4, and 6 hr, indicating that under standard test conditions, heat will not rise above a specified temperature inside and the construction will withstand both the exposure fire and the application of fire hose streams during that period. Vaults are constructed in the field and do not carry a testing laboratory label. Vault doors, however, are laboratory tested and carry ratings conforming to the vault construction in which they are to be used (GSA 1962). Unlike some other fire doors, vault doors limit the temperature on the interior face to 350°F (177°C) during fire exposure, so that paper could be stored in contact with the door without danger of ignition. Provisions in NFPA 232 require sprinklers for oversize vaults up to 25,000 cu ft (708 m^3) in volume.

Vaults usually contain a substantial fuel load, and by their nature contain only vital and important records. The internal contents of some vaults are more of a fire hazard than any external exposure. In recognition of this situation, vault standards were revised to permit sprinkler systems, lighting, and low energy circuits, with suitable safeguards. This change permits installation of smoke detectors and/or sprinklers to detect and extinguish a vault fire with a minor increase in ignition sources. A fire within an unprotected vault can be disastrous unless it is immediately discovered and extinguished with portable fire fighting devices. Fire extinguishers of the water type or fire hose, or both, should be outside the vault in an accessible location near the door.

A standard fire resistive file room is defined as an enclosure not exceeding 50,000 cu ft (1416 m^3) in volume and 12 ft (3.66 m) in height and designed to a fire resistance classification of 1, 2, 4 or 6 hr. The volume and height limitations restrict the quantity of records exposed to destruction by fire in a single enclosure and reduce the possibility of fire originating within the enclosure.

File room doors are labeled as to their fire resistance following a laboratory test under the procedures of UL 155, *Tests for Fire Resistance of Vault and File Room Doors* (UL 1977), but the ratings are only for ½ or 1 hr. These ratings anticipate that paper and other combustibles will not be stored nearer than 3 ft (0.9 m) from the unexposed face of the door, nor 6 in. (152 mm) to the side from the door jambs.

Lighting, heating, ventilation, etc. are permitted inside file rooms, as are filing cabinets and furniture, provided they are noncombustible. Although it is not to be used as a working space for other than filing purposes, the file room is more susceptible to a fire start than a vault. Due to this vulnerability, installation of automatic fire detection and fire control systems should be considered in relation to the values involved. Automatic sprinklers are required in file rooms having open shelf filing.

Records Centers and Archives

Bulk storage of records in buildings set aside for the specific purpose or in major portions of other buildings [in rooms exceeding the 50,000 cu ft (1416 m^3) fire resistive file room limitation], should comply with the recommendations set forth in NFPA 232AM. For facilities of this size, where the most severe hazard may not be from a fire exterior to the facility but from a fire starting and spreading entirely within the records storage area, the level of fire resistive construction and protection must be individually determined. Within such a facility, secondary operations may have to be segregated from the actual storage by fire resistive construction.* A thorough fire risk analysis by qualified personnel should be conducted due to the high risk of fire from the concentration of combustibles and the probability of total loss of large volumes of valuable documents.

FIRE EXTINGUISHING SYSTEMS

A variety of automatic fire extinguishing systems are suitable for records storage protection; the choice depends upon the economics of a particular situation and the degree of sophistication desired. All of the extinguishing systems described are loss initiated; the system does not

*Full scale fire tests of 14 ft (4.3 m) high open shelf records storage in a sprinklered facility (GSA 1977) indicated the likelihood of ceiling temperatures in excess of 1,200°F (649°C) for periods of up to 8 minutes. Lightweight bar joist roofs often present in records centers are vulnerable to early collapse unless additional precautions are taken, such as might be provided by developments in large drop sprinklers, quick response sprinklers, or fireproof coatings for the steel.

detect and begin extinguishment until a fire is established and a fire loss has occurred. Systems that are more sensitive and reactive are available for highly specialized requirements. A large increase in cost and rate of false operation should be expected. Increased maintenance is usually required in more sophisticated systems to attain the reliability inherent in standard sprinkler systems.

Automatic Sprinklers

Automatic sprinkler systems are the most common type of automatic protection. They have proven effective in records storage fire incidents in limiting both fire and water damage. Water is discharged only in the immediate vicinity of the fire, and salvage techniques have been perfected to the point where recovery of wetted items is no longer unusual. Accidental opening of a sprinkler is rare. The probability of water damage from an accidental discharge can be further reduced using a preaction sprinkler system where water does not enter the piping until a fire detector operates. Preaction systems introduce an additional failure mode, which may be unacceptable. A more sophisticated variation is a system where the control valve cycles on and off, depending upon the actual presence of fire as sensed by a detector in the fire area. Also available are sprinklers which cycle on and off individually, but the piping remains filled with water.

Foam Systems

Tests have shown high expansion foam to be an effective fire control system for records, though few high expansion foam systems are known to have been installed in records centers (AEC 1966). Rapid fire control comparable to or possibly even faster than that achieved by automatic sprinkler protection may be expected. Water damage to any single item will be low, but all items within the room or fire area involved will be affected. (A side effect unrelated to fire was observed in a large scale test, which may be a disaster in itself: The records box labels exposed to foam loosened and slid off.)

Carbon Dioxide Systems

A few automatic total flooding carbon dioxide systems having a high rate of discharge have been installed in record storage or library areas. A properly designed system should promptly control fire with limited damage. To be successful, a high concentration of gas must be maintained for an extended period. The concentration must be maintained until a deep seated fire cools below its ignition temperature. The operation of such systems, however, involves a hazard to life and must be delayed until the area to be flooded is cleared of people.

Halon Systems

Automatic total flooding halon systems utilize flame inhibiting liquefied gas under pressure. Halon 1301 (bromotrifluoromethane) is being marketed to use in specially engineered systems for the protection of valuable records. Primary advantages are that the gaseous agent leaves no residue and does not appear to be toxic in the concentrations in which it is used. Occupants, however, should not be unnecessarily exposed to Halon 1301. Deep seated smoldering fires will not be extinguished by Halon, as it inhibits flaming only.

System reliability

All fire control systems are subject to failure when various conditions are exceeded or not met. Sprinkler failure rates are well kept and publicized. In Australia, where sprinklers are closely supervised, the success rate is more than 98 percent. Other systems, where statistics are very selective or nonexistant, are often assumed to be failure free. Experience indicates this is far from the truth.

Gaseous systems are particularly vulnerable to failure. An independent detection system must work, and must be correctly linked to the gas release. Usually, electric power is required for the detection system, linkage, and release to operate. The gas containers must be filled and connected to distribution piping. If the room has been enlarged since the system was installed, the gas volume applied may be inadequate. If any of the doors fail to close properly, a critical amount of gas may escape. If the ventilating system continues to operate, critical duct dampers fail to close, a window remains open or gets broken, or walls are breached for construction, a critical amount of gas may escape. If fire fighters open doors to assist in extinguishment, a critical amount of gas may escape. If deep seated fires are not extinguished prior to gas dissipation, rekindling of the fire may be beyond the control of manual forces, with no gas supply for reapplication.

High expansion foam systems have similar detection and activation vulnerabilities. They are especially vulnerable to loss of power because substantial power is required to operate the large fans used to generate the foam. Foam is also subject to blockage by doors, walls of stock, and fabric curtains. It is also somewhat vulnerable to loss of containment, although not nearly to the extent of gaseous systems.

Bibliography

References Cited

AEC. 1966. "High Expansion Foam Fire Control for Record Storage Centers." *Atomic Energy Commission Report IDO-12050*, Idaho Falls, ID.

Chicarello, P. J., and Troup, J. M. 1980. *Fire Tests of Records Storage In A Fixed Storage Module Under The Protection of Large-Drop Sprinklers*. Factory Mutual Research Corp. (sponsored by U.S. General Services Administration), West Glocester, RI.

Chicarello, P. J., et al. 1978. *Fire Tests in Mobile Storage Systems for Archival Storage*. Factory Mutual Research Corp. (sponsored by General Services Administration), West Glocester, RI.

Fire Council. 1963. *Federal Fire Council Recommended Practice No. 2, Salvaging and Restoring Records Damaged by Fire and Water*. Federal Fire Council, Washington, DC.

Gage-Babcock & Associates. 1963. *Protecting the Library and Its Resources*. American Library Association, Chicago, IL.

GSA. 1977. General Services Administration. *Protecting Federal Records Centers and Archives from Fire*, Washington, DC.

GSA. 1962. *Safe, Office, Fire Resistant, Burglary Protection*. General Services Administration, Washington, DC.

Jensen, Rolf H. 1963. *Report on Fire Tests in High Piled Combustible Stock*. Underwriters Laboratories, Inc. (sponsored by FIA and NBFU), Chicago, IL.

McCrea, J. L., and Adelstein, P. Z. 1958. "Fire Tests on Microfilm in Insulated Record Containers." *Second Report to the Committee on Protection of Records of the National Fire Protection Association*. Eastman Kodak Co., Rochester, NY.

McCrea, J. L., et al. 1956. *Report to the Committee on Protection of Records of the National Fire Protection Association on Fire Tests on Microfilms Stored in Insulated Records Containers.* Eastman Kodak Co., Rochester, NY.

Morris, John. 1975. *Managing the Library Risk Fire.* University of California, Office of Insurance and Risk Management, Berkeley, CA.

NFPA. 1953. "Record Container Performance." *NFPA Quarterly.* Vol 47, No 2. pp 159-170.

Spawn, William. 1973. "After the Water Comes." *Bulletin of the Pennsylvania Library Association.* Vol 28. No 6. pp 242-251.

UL. 1979. *Tests for Fire Resistance of Vault and Fire Room Doors,* UL 155 Underwriters Laboratories, Inc., Northbrook, IL.

UL. 1977. *Tests for Fire Resistance of Record Protection Equipment,* UL 72 Underwriters Laboratories, Inc., Northbrook, IL.

Waters, Peter. 1975. *Procedures for Salvage of Water-Damaged Library Materials.* Library of Congress., Washington, DC.

NFPA Codes, Standards, Recommended Practices and Manuals. (See the latest *NFPA Codes and Standards Catalog* for availability of current editions of the following documents)

NFPA 10, *Standard for Portable Fire Extinguishers.*

NFPA 11, *Standard for Low Expansion Foam and Combined Agent Systems.*

NFPA 12, *Standard on Carbon Dioxide Extinguishing Systems.*

NFPA 12A, *Standard on Halon 1301 Fire Extinguishing Systems.*

NFPA 13, *Standard for the Installation of Sprinkler Systems.*

NFPA 14, *Standard for the Installation of Standpipe and Hose Systems.*

NFPA 72E, *Standard on Automatic Fire Detectors.*

NFPA 232, *Standard for the Protection of Records.*

NFPA 232AM, *Manual for Fire Protection for Archives and Record Centers.*

NFPA 910, *Recommended Practice for the Protection of Libraries and Library Collections.*

Additional Readings

Bibliography of Micrographics, National Micrographics Association, Silver Springs, MD.

Bibliography on Records Management, Association of Records Managers and Administrators, Bradford, RI, 1980.

Cutler, Harold R., *Engineering Analysis of Compact Storage Fire Tests,* Firepro, Inc., Wellesley Hills, MA, 1979.

Waegemann, C. Peter, *Handbook of Record Storage and Space Management,* Quorum, Westport, CT, 1983.

LIBRARY AND MUSEUM COLLECTIONS

Revised by Stephen E. Bush

Because of their high fuel loads or high value collections, or both, libraries and museums should be looked upon as specialized storage facilities, each with its particular fire protection requirements. While it might appear that libraries and museums are quite dissimilar, most museums contain libraries and many libraries feature exhibits of art or artifacts. The chief differences are in the value, density, and replaceability of the collections; the concentration of combustibles; and the degree of public access. There is a tendency, however, to underrate the fire risk because of the low fuel loads normally housed in such public areas as library reading rooms and museum exhibit galleries. These areas generally have high housekeeping standards; the areas are overseen by staff or by security guards; and in these areas, smoking is controlled, if not prohibited.

Despite their apparently low risk, library reading rooms and museum exhibit galleries have suffered costly fires. A reading of museum and library fire experience as presented by the NFPA Committee on Libraries, Museums and Historic Buildings, in NFPA 910, *Recommended Practice for the Protection of Libraries and Library Collections* and NFPA 911, *Recommended Practice for the Protection of Museums and Museum Collections* (hereinafter referred to as NFPA 911) and others (Morris 1979; Johnson 1963) which illustrates the potential hazard.

Nonpublic areas such as bookstack, storage, and work areas present an even greater hazard. Here the fuel loads are much higher, and housekeeping and smoking are less likely to be rigidly controlled.

FIRE PREVENTION PRINCIPLES

Ignition Control

Fire prevention consists of controlling ignition sources and the supply of fuel. Fuel control must consider all combustible contents, including the collection and its ancillary records, as well as furniture, interior finish, packaging materials, and flammable liquids. Ignition control involves recognizing all sources of heat generation that could cause ignition in combustibles, and isolating these sources from the fuel.

Smoking: Smoking should be permitted only in administrative offices, staff rest areas, and designated visitor lounges.

Electrical Appliances: If not prohibited entirely, space heaters, hot plates, microwave ovens, coffee makers, and other small appliances should be rigidly regulated and closely monitored. Where permitted, each appliance should be:

1. Listed by an independent testing laboratory;
2. Placed on a noncombustible base;
3. Located so there is adequate separation from combustible materials; and
4. Located where air circulation will prevent heat build up.

The use of extension cords to connect heating devices to electric outlets should also be prohibited. Security guards should be furnished with a list showing the location of each heating appliance. During rounds, each appliance should be checked to make sure it is in the "off" position if not being used and attended. This check is especially important during hours when the facility is closed.

Electric Wiring: Overloaded or partially grounded wiring may heat up sufficiently to ignite combustibles without blowing fuses or tripping circuit breakers if the current rating of the fuse is less than that of the wire (Factory Mutual 1967; McElroy 1974). Especially significant is the proliferation of information processing equipment (computer terminals, word processors, microprocessors, minicomputers), and photocopy machines. The use of extension cords as a substitute for fixed wiring is not an acceptable practice because extension cords are subject to mechanical damage and can be easily overloaded. Where permitted, extension cords must be protected from mechanical damage and must never be concealed under carpets (Johnson and Horgan 1979).

Mr. Bush, CSP, ASSE, is safety officer for the Library of Congress, Washington, DC.

Electric Lights: Incandescent lamps near or in contact with combustible materials may ignite them. Surface temperatures on a bulb can vary widely, and can be within the range of the ignition temperature of paper, which is approximately 450°F (232°C). The much cooler surface temperatures of fluorescent lamps do not present the ignition hazards of incandescent bulbs. However, the ballasts in fluorescent light fixtures can be potential sources of ignition, particularly if mounted in heat entrapping installations. They present a potentially severe fire hazard when mounted on combustible, low density, cellulose fiberboard and materials of similar combustibility.

Open Flame Devices: Open flames used in building repairs, laboratory operations, or the interpretation of historic environments, and hot metal sparks or slag from welding and cutting, are severe ignition hazards. Use of open flame devices should require the written approval of the responsible management official. Before signing such an authorization this official should evaluate the risk to life safety and to collection materials, and should prescribe the protective measures to be included as conditions of the authorization. Management conditions for authorization should include as a minimum:

1. Removing exhibit or collection materials a safe distance from the ignition source.
2. Requiring a fire watch of one or more persons with appropriate fire extinguishing equipment through all periods of hazardous activity and continuing beyond any interruption or termination of the activity long enough to assure discovery of any incipient fire conditions (a minimum of 30 minutes when welding and cutting torches have been used).

Heating Equipment: The malfunction of heating equipment is a major potential cause of fires. Boilers and furnaces of central heating systems should be isolated in a fire resistant enclosure, either separate from or attached to the library or museum. Unit space heaters and electric air heaters should not be used in combustible collection storage areas.

Because of the possibility that cellulosic materials may ignite at temperatures far below usual ignition temperatures after extended exposure to relatively moderate heat, combustibles such as paper, wood, and textiles should be kept away from steam or other heat piping and ducts.

Fuel Control

Flammable Liquid Storage and Handling: Flammable liquids permitted inside the library or museum should be contained in safety cans. The quantity permitted outside of storage areas should be limited to the amount required during an eight hour shift. Special attention to adequate ventilation and elimination of ignition sources is required.

Spontaneous Heating: Oil or solvent soaked cloths and rags from oil based paint can heat spontaneously. After use, they should be placed immediately in a metal can with a tight fitting metal top, pending their prompt removal.

Fuel Geometry and Overcrowding: The combustible materials contained in a library or museum may be visualized as arranged in fuel packages—exhibit cases, bookstacks, desks, tables, chairs, etc.—which may also be covered with an additional complement of loose paper or other combustibles. Through the use of distance and low flame spread rated interior finish materials (carpet, wall covering, draperies, etc.), these fuel packages should be arranged to minimize the risk of fire spreading from one fuel package to another. Space limitations resulting from the growth of collections and staff can result in overcrowding; this increases the risk that if a fire occurs, it will grow by igniting adjacent fuel packages rather than burning itself out. Periods of overcrowding therefore require compensating emphasis on other elements of the fire prevention program, especially control of ignition sources.

Loss Potential

Some museum curators, librarians, and archivists believe that water is a greater hazard to their collections than fire. As a result, they place emphasis upon fire prevention almost exclusively and, with some few notable exceptions, limit fire protection to installation of automatic early warning detection systems. However, despite effective fire prevention, fires do occur, and unless there is early detection and quick action to stop fire growth, a large loss can be expected. When one considers that 70 percent of all library fires occur between the hours of 9 pm and 9 am and that an additional 18 percent start between 5 pm and 9 pm, the risk of a total loss is apparent when the institution places complete dependence upon fire prevention. This is the most significant lesson to be derived from a study of fire experience in libraries and museums. The common element in the large loss fires is the absence of automatic suppression systems to prevent small fires from becoming larger. Total dependence upon fire prevention and on manual extinguishment leaves a museum or library vulnerable to large or even catastrophic losses (Harvey 1975).

FIRE PROTECTION

Since fire prevention efforts can lessen but not eliminate the possibility of fire in a museum or library, it is necessary to become familiar with the options available to limit fire spread. Except for the dangers of exposure fires or conflagrations, compliance with local codes should provide for the survival of the building and for the life safety of its occupants. However, sole reliance on code provisions will not always adequately provide for the preservation of building contents, especially high value collections. Fire protection choices must be made within the resources available to the institution.

The Systems Approach to Firesafety

The systems approach to firesafety is a useful management tool for selecting the level of fire protection that should contain fire loss within "acceptable" limits. This approach uses decision tree analysis and other systems techniques to determine the maximum acceptable fire loss and to create a fire control system designed to confine a fire loss within the acceptable limits.

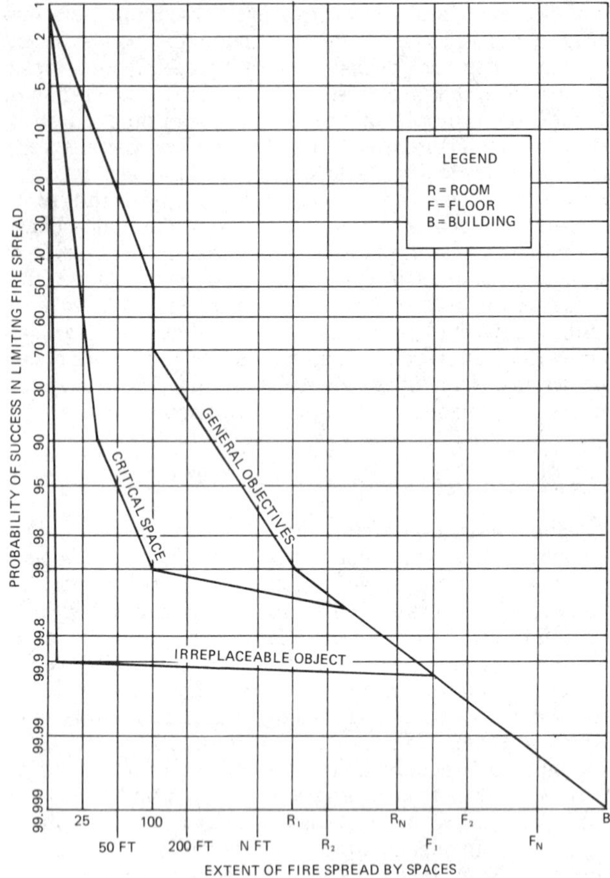

FIG. 11-10A. The extent of fire spread by spaces.

Each institution must develop its own firesafety system objectives based on levels of protection appropriate to the value of collections (many items irreplaceable), the life safety of its patrons and staff, and its obligation for continuity of service to the community or the public. The firesafety objectives adopted by the Library of Congress in Washington, DC may provide a useful point of reference (Fyrepro 1977). The "General Firesafety Objectives" and "Critical Space Objectives" shown graphically in Figure 11-10A are similar to those adopted by the U.S. General Services Administration (GSA) for federal office buildings nationwide (GSA 1979). This graph showing the firesafety objectives for a building represents the minimum probability of success in limiting the growth of a fire from its start to full building involvement.

Designing the Firesafety System

Depending upon the collection values to be protected and the values and characteristics of the building that houses them (including fuel loading), the fire protection system designed to achieve an institution's firesafety objectives should provide for: (1) detection, alarm, and communication; (2) manual extinguishment; (3) automatic sprinklers; and (4) protection of special hazards.

Detection, Alarm, and Communication: Ideally, this system should consist of a property wide automatic smoke and heat detection system giving a local alarm as well as reporting to a central location, preferably to a fire department. At the least, an automatic detection system should be provided in collection and exhibit areas to give early warning of fire; this allows time for trained building occupants to extinguish the fire before automatic suppression equipment is activated. Fire protection systems that fail to provide automatic suppression capability and depend entirely upon manual extinguishment risk disaster when human response is too late or is unavailable.

Manual Extinguishment: Portable extinguishers in the hands of trained users constitute an institution's first line of fire defense. In order to prevent use of the wrong class of extinguisher on a particular fire, some institutions provide only multipurpose portable extinguishers. These should be located appropriately and inspected and maintained regularly so they will be fully charged and operational whenever needed.

Occupant use standpipe hose lines also require training to avoid personal injury and unnecessary property damage.

Manual extinguishment can be provided by the fire department as well as by occupants, especially in buildings without automatic sprinkler systems. Figure 11-10B

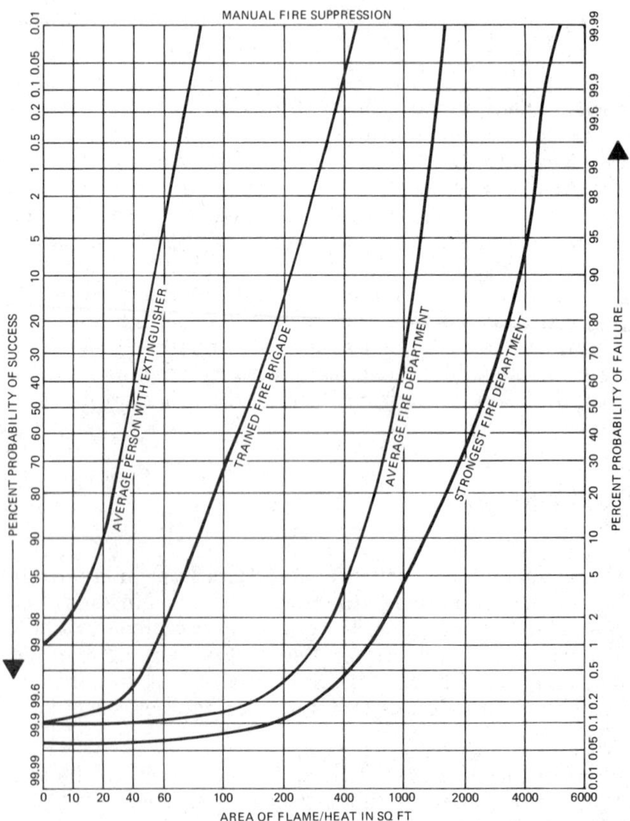

FIG. 11-10B. The probability of success at different levels of manual extinguishment.

plots the probability of success for different levels of manual extinguishment (Fitzgerald 1976). It generally is recognized that if the fire area has reached 2,500 to 3,000 sq ft (232 to 288 m^2) at the time of water application, the fire may be confined, but will not be extinguished quickly. In fact, Figure 11-10B suggests that the probability of fire department success actually begins to drop off rapidly when fire areas exceed 750 to 1,500 sq ft (70 to 140 m^2).

Automatic Sprinklers: For many years, library and museum conservators generally have opposed automatic water sprinkler protection. Fear of water damage (even more than fire damage) is reflected in contemporary library and museum planning references (Myller 1966; Thomson 1978; Brawne 1970). Collections are vulnerable to water damage from flooding conditions associated with problems other than fire, e.g., roof leaks, storm damage, plumbing problems, rivers overflowing, etc. Flooding is not unusual in connection with the use of fire department hose lines to fight fires but unnecessary water damage seldom occurs in facilities completely protected by well designed and maintained automatic sprinkler systems.

Water damage loss experience from sprinklers has been about the same in libraries and museums. Accidental flooding due to defects in sprinklers, water control devices, piping, or associated equipment, has been rare. Based on sprinkler leakage loss reports, the probability that a defective sprinkler may open accidentally is less than one in a million per year.

Several alternatives to the standard wet pipe sprinkler system minimize water damage after the fire has been extinguished. These include pre-action and automatic recycling systems as well as on/off sprinklers. (See NFPA 910, Sec. 5-3.1.) In retrofitted sprinkler installations for library or museum collection areas, preventing leakage can pose a problem when the system is initially charged with water. One solution, used by the architect of the U.S. Capitol in the installation of the Library of Congress wet pipe book stack sprinkler systems, is to require pneumatic testing of each system before proceeding with the normally required hydrostatic testing.

Automatic sprinklers actually minimize the potential for water damage. The fire is stopped by the timely release of 20 to 25 gpm (76 to 95 L/min) in a spray pattern from a sprinkler, in contrast to the large amount of pressurized water from hose streams sweeping materials from shelves and displays into a disorganized mass on a flooded floor. Prospects of salvaging wet materials that have remained in place under the relatively gentle action of the sprinklers are infinitely better than for materials knocked to the floor to be soaked and possibly trampled upon.

Collection items especially vulnerable to water damage are best stored in cabinets, manuscript boxes, or other appropriate containers that shield them from water spray in the event of a sprinkler discharge (Johnson and Horgan 1979).

Since about 1970, a growing number of new and existing libraries and museums have installed automatic sprinkler systems to achieve their firesafety objectives and to keep the fire risk within manageable and insurable limits. The resulting reduction in insurance premiums usually enables the sprinkler system to "pay back" its installation cost (Thomas 1982).

Protection of Special Hazards: Areas housing high value collections or equipment identified as essential for continuity of operations deserve a high level of fire protection. A total flooding Halon 1301 system or other quick action system using an extinguishing agent other than water can provide such protection.* The special system should be installed in addition to, and not as a substitute for, the protection of the general area provided by a standard automatic sprinkler system. The installation cost of a special system in addition to a standard system can be justified: the special system is designed to confine fire growth to an area significantly smaller, i.e., 150 to 300 sq ft (14 to 28 m^2), than would be possible using standard automatic sprinklers alone.

While the same firesafety principles are applicable to the requirements of museum and library collections, several activities common to museums and some libraries require special attention. These are:

1. Storage of art and other valuable objects.
2. Packing and unpacking traveling exhibits.
3. Storage of shipping crates and packing materials.
4. Cleaning and restoration.
5. Construction and renovation of displays.

The areas where these activities take place should be isolated and sprinklered.

CONSTRUCTION CONSIDERATIONS

Intrinsic design features of library and museum buildings should include firesafety aspects of site planning, fire resistant construction and interior finish, smoke control, water supply for fire protection, and special requirements for underground structures, all of which are discussed elsewhere in this HANDBOOK. Topics deserving special attention in this chapter are compartmentation, means of egress, lighting, and gallery flexibility.

Compartmentation

The well protected building is subdivided into compartments whose walls provide barriers to the spread of fire. The compartments should be as small as possible and the walls designed to resist the estimated fire duration, based on the maximum fuel loads in the compartments. Use of automatic sprinkler systems permits larger compartment areas where small compartments would restrict flexible use of space, and compensates for heavy fuel loads in storage areas, including bookstacks.

Areas where quantities of combustible material are in use or storage should be separated by fire walls from the exhibition galleries, reading rooms, other public areas, and from collection storage areas. High concentrations of combustibles are found in carpenter shops; display shops; research, restoration, and conservation laboratories; paint rooms; packing rooms and restaurant kitchens, among other places. The fire walls protecting these facilities should be designed to endure fires in the maximum fuel loads they will enclose, rather than the minimum as required by building codes. Hazardous materials such as

*Among the various special fire suppression systems, carbon dioxide total flooding systems are not generally recommended for normally occupied areas, and Halon 1211 total flooding systems are prohibited in such areas. Both of these gaseous agents can create an atmosphere immediately hazardous to life when used in enclosed areas.

flammable liquids (including some paints), should be housed when possible in a structure separated from the museum or library building. Also, because smoke is damaging to collections as well as harmful to occupants, fire walls should confine smoke to the compartment of fire origin.

Means of Egress

It is common for library and museum security arrangements to require that all departing occupants be channeled through a few exits that can be closely monitored. Unfortunately, this often means that other doors required for exit access or egress from a life safety viewpoint, are kept locked. This problem most commonly occurs in museums and libraries with valuable collections and in libraries with heavily utilized reference collections (e.g., colleges and universities). Use of electromagnetic and electromechanical door locking systems and devices is gaining acceptance as a way to resolve this conflict between life safety and security. Such systems should be designed to include failsafe features designed to provide a level of life safety equivalent to that required by NFPA 101, the *Life Safety Code®*.

Lighting

Art museums generally make use of as much natural light as possible. This often leads to the installation of glass roofs, light diffusing glass ceilings, and window walls, all of which increase the museum's vulnerability to an exposure fire. Conversely, to protect their collections from ultraviolet radiation and achieve complete light control, some museums eliminate most wall and roof openings. Complex wiring systems requiring unusually heavy electrical loads may be needed to compensate for the absence of natural light. Museums also need numerous electrical outlets in floors, walls, and ceilings to provide flexibility for frequent changes in exhibits and avoid the dangerous use of extension cords.

Gallery Flexibility

Movable walls, temporary furring, and lightweight partitions make it easier to rearrange exhibit space and change exhibits. However, such features should not be allowed to impede the movement of occupants in case of an emergency or to increase the fire hazard by adding to the fuel load.

FIRE EMERGENCY MANAGEMENT

Management of a library or museum fire emergency involves decisions and actions that must be taken to mitigate the consequences before, during, and immediately following the incident. Good management includes continuous monitoring to assure that fire protection systems and features are always in operational readiness and are never degraded or defeated by space changes or other building modifications.

A viable fire emergency contingency plan should provide for preemergency training of a fire brigade in evacuation, fire fighting, and salvage operations. In addition to life safety considerations, emphasis should be placed on the need to minimize water damage and to protect or remove endangered high value collections. Prefire planning must also provide for the security of valuable collection materials after they have been removed from the building.

Defense Against Incendiarism: While it may be impossible to eliminate deliberately set fires, emergency management planning must address ways to reduce or deter incendiarism. Some fundamental security measures and procedures are:

1. Uniformed guards on station and patrolling well planned and monitored routes.
2. Surveillance of all areas at irregular intervals.
3. Windows and doors secured with proper locking devices (panic hardware on emergency exit doors allow doors to be opened from the inside while remaining locked from the outside).
4. Windows and doors equipped with intrusion alarms. In some communities, such alarms can be connected directly to police, fire, or central supervisory agency.
5. Closed circuit surveillance by a television system.
6. Thorough security and safety procedures at closing time.
7. Adequate exterior lighting and fences.
8. Regular and close inspection of ash trays, waste receptacles, soiled linen hampers, electric heating appliances, and other places that breed fire. Although not a check against incendiarism alone, this routine can uncover hazards that could involve incendiarism, e.g., a filled wastebasket hiding a time delay igniter.
9. Either elimination of book drops used for after hours return at libraries, or isolation of them from the building by physical separation or by fire protective design. Book drops have been used by arsonists as the portal through which to inflict significant fire losses on libraries (Morris 1980).

SALVAGE

Through the use of freeze/vacuum drying techniques, library preservation specialists have been increasingly successful in salvaging water-soaked library, archival, and other collection materials. However, water is only one of the hazards to which collections are vulnerable during a fire. Others include smoke, acids contained in the combustion products, and chemical agents that may be used in fighting the fire—hazards that are perhaps more dangerous to museum collections than water.

Museum Salvage

With much higher values per item and fewer items than libraries, museums must center their salvage efforts upon removing endangered collections from the fire area to a safe location as quickly and carefully as possible. Conservation oriented fire fighting strategies and techniques enhance salvage efforts and minimize collection losses. Winterthur Museum in Wilmington, DE, has developed appropriate fire fighting techniques that can be adapted by other museums and libraries (Fennelly 1983). Fire fighting efforts are divided into three phases that function concurrently and are directed by the fire marshal and staff. The phases are: (1) collection protection and salvage operations, led by the housekeeping supervisor and staff; (2) security for collections removed from the fire area, the responsibility of the security supervisor and staff; and (3) overall coordination, provided by the building superintendent or a designee.

Library Salvage

The degree of success and expense in salvage of water soaked library collections will depend upon advance planning and the amount of wetting suffered by the materials. Freeze/vacuum drying has been very effective in removing water and smoke odors, while the use of a sterilizing agent (ethylene oxide) during the process arrests the growth of mold. Freeze/vacuum drying cannot, however, restore physical damage inflicted by hose streams, trampling, and mishandling.

Before the Fire

Planning for salvage efforts before the fire must include identification of the following:

1. Collection values. This facilitates decisions under emergency conditions as to which items or collections are to be saved first.
2. Preservation specialists in salvage operations, to provide guidance when needed.
3. Short notice sources of salvage materials, equipment, and facilities.
4. Organized, and trained employees who will be needed in the salvage effort during and after the fire.

During the Fire

1. Remove endangered high value collections.
2. Provide tarpaulin protection for materials remaining in the fire area.

After the Fire

1. Obtain guidance from a preservation specialist experienced in salvage.
2. Wrap and freeze damaged articles.
3. Determine appropriate restoration procedures.

Much less salvage work will be needed in facilities protected by automatic sprinklers. Water absorption is minimized when materials are left on the shelf undisturbed by the gentler action of automatic sprinklers, as opposed to the more forceful action of hose streams. An automatic sprinkler system also greatly reduces the quantity of materials requiring salvage, by limiting the size of the fire area to that covered by the operation of one or just a few sprinkler heads (Fuhlrott and Dewey 1982).

FIRESAFETY IMPLICATIONS OF NEW DEVELOPMENTS

Automation of Library Catalogs

Automation may lead to the replacement of the traditional card catalog. As this occurs, continuity of operation in the library depends increasingly upon the computer and necessitates a high level of fire protection to insure continuity of operation. Duplicates of the magnetic media containing the library's computer based catalog should be maintained in a safe, remote location.

Bookstacks

Multi-tier Bookstacks: Modeled after bookstacks developed for the Library of Congress in 1893, this type of shelving has been used for several decades in many libraries. Typically, decks or tiers are supported from unprotected iron or steel columns and beams at intervals of 7 ft (2.14 m). Vertical openings between decks permit heating of the bookstack areas by convection air currents. Under fire conditions, these same openings function as flues to accelerate vertical fire spread to the tiers above.

Modern versions of these systems are still sold, but without vertical openings in the modules. (See Fig. 11-10C.) The unprotected steel support structure, however, remains vulnerable to collapse in a fire. For this reason each tier should have automatic sprinkler protection.

In new library construction, multi-tier bookstacks are being replaced by "free standing" bookstack ranges installed on structural fire resistant floors.

The difference in fuel load between multi-tier stacks and free standing stacks is substantial. All levels in a multi-tier system are usually considered as one fire compartment, due to the absence of vertical fire barriers between tiers. Fire loads depend upon the number of tiers in the bookstack. With shelf loading at 70 percent of capacity, the expected fire load in each tier would be 55 to 65 psf (268 to 317 kg/m^2). Thus, in a multi-tier bookstack with 10 tiers in a single fire compartment, fire duration would be estimated at 70 to 85 hours. On the other hand, fire in a free standing bookstack that is seven shelves high and set on a fire resistant floor would be limited to a single tier and to a fire duration of 7 to 8 hours.

Compact Storage or Track Files: The practice of mounting ranges of bookstacks or bookcases on carriages which roll on tracks is evident in new library and museum construction and in renovations. This enables one moving aisle to serve several bookstack ranges. The track systems are modular, with two or more mobile ranges between two fixed end ranges. (See Fig. 11-10D.) This compact storage method can result in a fire load in excess of 120 psf (586 kg/m^2) and a fire duration possibly exceeding 15 hours—more than enough to challenge the strongest fire barriers and construction prescribed by building and fire codes.

The NFPA Committee on Libraries, Museums and Historic Buildings recommends automatic sprinkler protection for compact storage installations. To install this storage system without such protection would not only endanger the collections, but might result in building collapse as well. Conclusions from an independent analy-

FIG. 11-10C. A multi-tier installation.

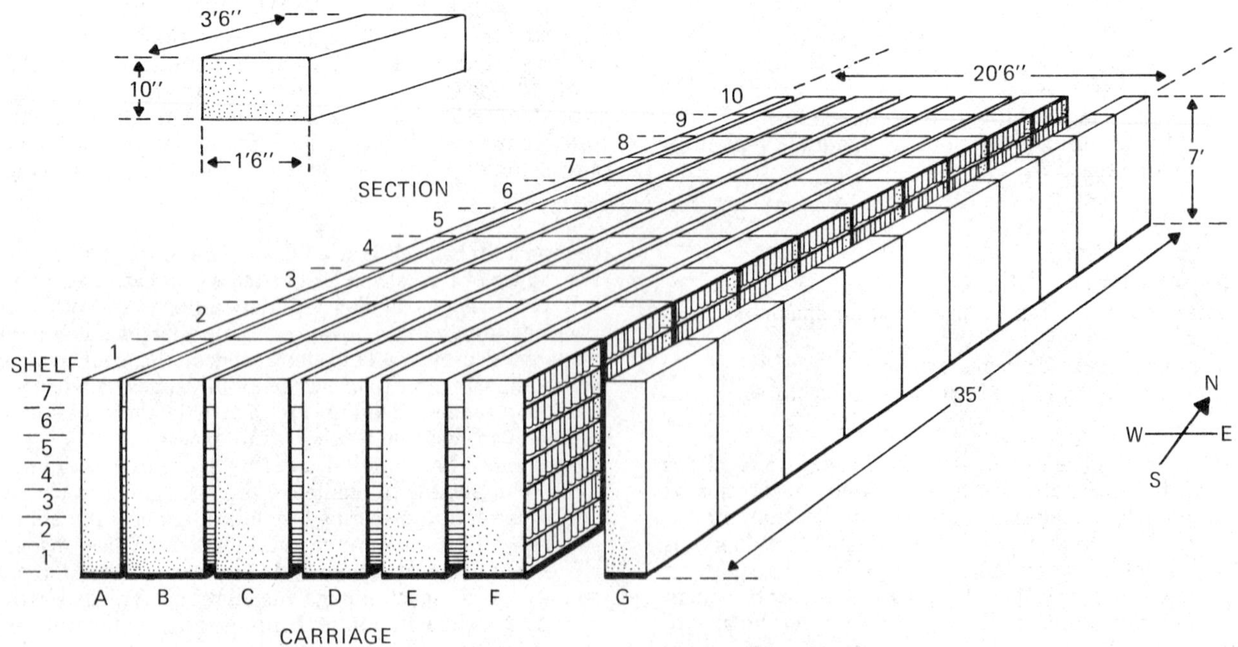

FIG. 11-10D. Compact mobile shelving.

sis (Cutler 1979) of fire tests evaluating the effectiveness of sprinklers on compact storage arrangements (Chicarello et al 1978) indicate a high probability that sprinklers installed over a compact storage module would prevent damage to building structural elements outside the module from fire originating within the module. Further, there is a good probability that sprinklers would prevent fire spread across 4 ft (1.22 m) aisles between compact storage modules and would be effective in preventing fire spread between back to back shelves of adjacent modules when each shelf is backed by a sheet metal panel with a 1 in. (25 mm) air space between adjacent panels. Thus, tests indicate that automatic sprinklers can be expected to limit fire development to that portion of the module of fire origin between the fixed end range and the open aisle. However, total loss of contents within that portion of the storage module is possible. (See Fig. 11-10D.)

Loss of contents within the module of fire origin can be minimized by careful attention to the following considerations in designing the compact storage system:

1. More efficient use of space results in greater density or concentration of values—i.e., increases the number of items subject to fire damage per unit volume of storage space. This risk should be considered in determining the maximum number of ranges to be included in each module.
2. Existing automatic fire detection and fire suppression systems may have to be modified, and Halon 1301 may be ineffective.
3. Compact storage modules may conceal the origin of smoke, compounding the difficulty of locating and extinguishing a fire.
4. Compact storage modules prevent penetration of water from hose streams for fire extinguishment and delay detection by entrapping smoke and heat.
5. The electric motor in each motorized module is a potential ignition source.

Bibliography

References Cited

Brawne, Michael. 1970. *Libraries: Architecture and Equipment*, Praeger, NY.

GSA. 1979. *Building Firesafety Criteria*. General Services Administration, Washington, DC.

Chicarello, P. J., et al. 1978. *Fire Tests in Mobile Storage Systems for Archival Storage*. General Services Administration, Washington, DC.

Cutler, Harold, R. 1979. *Engineering Analysis of Compact Storage Fire Tests*. Library of Congress, Washington, DC.

Factory Mutual Engineering Corp. *Handbook of Industrial Loss Prevention*. 2nd ed. McGraw-Hill, Inc., NY. 1967.

Fennelly, Lawrence J., ed. 1983. *Museum, Archive, and Library Security*. Butterworths, London, England.

Fyrepro, Inc. 1977. *Fire Defense Alternatives: Library of Congress*. Unpublished Report.

Fitzgerald, R. W., and Wilson, Rexford. 1976. *The Systems Technique for Evaluating Building Firesafety*. Worcester Polytechnic Institute, Worcester, MA.

Fuhlrott, Rolf, and Dewe, Michael. 1982. *Library Interior Layout and Design*. Saur, Munich, West Germany.

Harvey, Bruce. 1975. "Fire Hazards in Libraries: Part I—Clearing the Smoke." *Library Security Newsletter*. Vol 1, No 1. Jan. 1975.

Johnson, E. Verner, and Horgan, Joanne C. 1979. *Museum Collection Storage*. United Nations Educational Scientific and Cultural Organization, Paris, France.

Johnson, Edward M., ed. 1963. *Protecting the Library and Its Resources*. American Library Association, Chicago, IL.

McElroy, Frank E., ed. 1974. *Accident Prevention Manual for Industrial Operations*. 7th ed. National Safety Council, Chicago, IL.

Morris, John. 1979. *Managing the Library Fire Risk*. 2nd ed. University of California, Berkeley, CA.

Morris, John. 1980. "Is Your Library Safe from Fire?" *American School and University*. April 1980. p 61.

Myller, Rolf. 1966. *The Design of the Small Public Library*, Bowker, NY.

Thomas, Susan, ed. 1982. "Retrofitting with Sprinklers for Added Protection." *Factory Mutual Record*. May-June 1982. pp 13-19.

Thomson, Garry. 1978. *The Museum Environment*. Butterworths, London, England.

NFPA Codes, Standards, Recommended Practices and Manuals. (See the latest *NFPA Codes and Standards Catalog* for availability of current editions of the following documents.)

NFPA 11, *Standard for Low Expansion Foam and Combined Agent Systems*.

NFPA 12, *Standard on Carbon Dioxide Extinguishing Systems*.

NFPA 12A, *Standard on Halon 1301 Fire Extinguishing Systems*.

NFPA 13, *Standard for the Installation of Sprinkler Systems*.

NFPA 14, *Standard for the Installation of Standpipe and Hose Systems*.

NFPA 70, *National Electrical Code*.

NFPA 101, *Code for Safety to Life from Fire in Buildings and Structures*.

NFPA 232, *Standard for the Protection of Records*.

NFPA 232AM, *Manual for Fire Protection for Archives and Record Centers*.

NFPA 910, *Recommended Practice for the Protection of Libraries and Library Collections*.

NFPA 911, *Recommended Practice for the Protection of Museums and Museum Collections*.

Additional Readings

Cotton, P. E., "Fire Tests of Library Bookstacks," *NFPA Quarterly*, Vol. 53, No. 4, Apr. 1960, pp. 288-295.

Egan, M. David, *Concepts in Building Firesafety*, John E. Wiley, NY, 1978, p. 28.

"Fire Security Problems: Prevention and Cures." *Library Journal*. Dec. 1, 1975, pp. 2203-2204.

Fisher, Walter R., "Protect Our Treasures from Threat of Fire," *Professional Safety*, Vol. 25, No. 2, Feb. 1980, pp. 21-24.

Hodnett, Robert M., ed., *Automatic Sprinkler Systems Handbook*, National Fire Protection Association, Quincy, MA, 1983.

Insurance for Libraries Committee, American Library Association, "Pricing Library Materials for Insurance Purposes," *AIA Bulletin*, July-Aug. 1966, pp. 729-730.

Kallenbach, Wilhelm, et al., *Fire Protection in Historical Buildings and Museums*, Arbeitsgruppe. . .e.v., Hamburg, Germany, 1980 (in German).

Lerup, Lars, et al., *Learning From Fire: A Fire Protection Primer for Architects*, National Fire Prevention and Control Administration, Washington, DC, 1977.

Library Planning: Bookstacks and Shelving, Snead Iron Works, Jersey City, NJ, 1915.

Maddox, J., "Norwood Book Burning," *Wilson Library Journal*, Vol. 34, Feb. 1960, pp. 412-414.

Powers, W. Robert, "Sprinkler Experience in High-Rise Buildings (1969-1979)," *Technology Report 79-1*, Society of Fire Protection Engineers, Boston, MA, 1979.

Schaefer, George L., "Fire!," *Library Journal*, Vol. 85, Feb. 1, 1960, pp. 504-505.
Schell, H. B., "Cornell Starts a Fire," *Library Journal*, Vol. 85, Oct. 1, 1960, pp. 3398-3399.
Schram, Peter J., ed., *National Electrical Code Handbook*, National Fire Protection Association, Quincy, MA, 1984.
Sellers, David Y., and Strassberg, Richard, "Anatomy of a Library Emergency," *Library Journal*, Oct. 1, 1973, pp. 2824-2827.
Swayne, Leo, H., "Fire Proteciton at the National Archives Building, Wahington, DC," *Fire Journal*, Vol. 69, No. 1, Jan. 1975, pp. 65-67.
Tillotson, Robert G., *Museum Security*, International Council of Museums, Paris, France, 1977, p. 46.
Tweedie, A. T., "Freeze-Dried Books," *Library Journal*, Jan. 15, 1974, p. 82.
Waters, Peter, *Procedures for Salvage of Water-Damaged Library Materials*, Library of Congress, Washington, DC, 1975.
Wilson, Rexford, "The L-Curve: Evaluating Fire Protection Tradeoffs," *Specifying Engineer*, May 1977, pp. 110-113.

SECTION 12

SPECIAL FIRE PROTECTION AND PREVENTION

PROTECTION OF COMPUTERS AND ELECTRONIC EQUIPMENT

John H. Uliana, P.E., F.P.E.

With highly accelerated developments in electronics, our society's dependence on computers and other electronic equipment is increasingly magnified. The use of the electronic computer is seen in industry, military, banking, and other businesses of all sizes. These computers are highly sophisticated and are becoming more complex, faster, smaller, and denser as technology increases. Often, the computer's strategic importance in a business is tied to its uniterrupted use and if damaged or destroyed, a vital operation may be totally paralyzed. Because of the immense cost of large computer systems as well as space considerations, it may not be economically feasible to provide backup units or even libraries of all records and data (in the form of magnetic disks or tapes, paper, electronic media, etc.) at remote rooms to assure continued operation in the event of some catastrophe. If a system is custom built for its application, the problem of destruction is compounded because there may be no replacement system available. (See Fig. 12-1A.)

Electronic equipment other than digital and analog computers is also to be considered here. These may be electronic controls for various machines or electronic devices used to monitor or perform production tasks. Many businesses depend on special electronic equipment such as communication systems or test equipment. Electronic laboratory equipment such as production prototypes, simulators, and measuring devices are all vital to a businesses' development of its products. All of the electronic devices mentioned are often more difficult to replace than computer systems and their destruction by fire could have immense impact.

Because of the dependence of businesses on computers and other electronic equipment, the importance of fire protection cannot be overemphasized. This should include protection against arson and sabotage, because computers are prime targets for both.

Additional information relevant to the protection of computers and electronic equipment from fire may be found in Section 11, Chapter 9, "Storage and Handling of Records"; in Section 16, Chapter 2, "Protective Signaling Systems"; Chapter 4, "Automatic Fire Detectors," and in Section 19, Special Fire Suppression Agents and Systems.

FIG. 12-1A. Modern computer room with terminals. (IBM)

PHYSICAL FACILITIES LOCATION AND CONSTRUCTION

Because computers and electronic equipment are valuable, they should be located in a detached building or an area cut off from other portions of the building by walls, floors and ceilings having fire resistance ratings of one hour minimum. These should not be located above, below, or adjacent to areas housing hazardous processes unless there is ample protection for the computer area. These rooms should be made of fire resistive construction with noncombustible interior finish.

Rooms for related necessities such as programmers, maintenance, paper supplies, and recorded data should be separated from the computer or electronic equipment

Mr. Uliana is President and General Manager of Fire Protection Associates, a consulting firm. He is a Life Member of NFPA, and Chairperson of NFPA's Technical Committee on Electronic Computer Systems.

room. Small supervisory or operator offices directly related to the electronic equipment operation may be located in the computer room if combustible materials are kept to a minimum. Only actual computer and electronic equipment and the necessary auxiliary equipment should be permitted within the room itself. Furniture should be made of metal with fire retardant upholstery. Service transformers are not permitted in the room. Supplies of paper and repair or service shops should be strictly limited to the minimum needed for efficient operation and preferably located in another area.

Records that are generated from the electronic computer systems should be protected according to how important and difficult to replace they are. These records fall into five basic categories: (1) input data; (2) memory; (3) programs or software; (4) output data; and (5) engineering plans, specifications, or system architecture sketches. They are stored on hard or flexible magnetic disks, magnetic tapes, paper products (including punch cards where used), photographic material, or electronic hardware. Vital and important records which are irreplacable or very difficult to replace should be stored in separate rooms with fire resistance not less than two hours. When duplicate records are produced these should be stored in a separate room of similar fire resistance rating. (See Fig. 12-1B.)

FIG. 12-1B. Tape storage room. (IBM)

Raised floors are generally used to house electrical wiring, fire protection and extinguishing equipment, and serve as air conditioning plenums. The height must be designed accordingly. The floors, including structural support members, should be concrete, steel, aluminum, resilient tiles, or other noncombustible materials. Fire wall separations must be maintained where openings to another area under the floor need to be made. Carpeting is permitted if it has a flame spread rating of 25 or less and does not restrict lifting of panels to access the underfloor space. Openings in floors for cables or for other uses must be protected and sealed to resist flame penetration, and to prevent debris or other combustibles from collecting beneath the floor. Periodic inspection and maintenance of underfloor space is required to eliminate accumulation of combustibles.

Equipment and Construction Features

Cables needed for power or input/output (I/O) as well as plugs and connectors should be of a listed type and proper size for the application, in accordance with the National Electrical Code®. These should be suitable for installation under raised floors. Openings in the floors should be smooth to prevent damage to cables. No splices or connections should be permitted in the wiring under the floor except within listed junction boxes or approved receptacles or connectors and the number of junction boxes should be kept to a minimum. Cables should be arranged to avoid serious multiple failures because of a fault or overheating in any one conductor.

Wherever possible, equipment should be designed so that no fire will spread beyond the unit it starts in. Where there is floor standing equipment with enough combustible external surface to help spread an external fire, automatic protection is necessary unless the external material has a flame spread rating of 25 or less. Any sound deadening material used inside computer equipment must be noncombustible.

Forced air circulation through each electronic computer is generally needed to remove heat generated. A separate reliable air conditioning system for each computer or other electronic device area is preferred. Air filters used should be a listed type that will not freely burn or emit a large volume of smoke if attacked by flames. As technology advances, computers and other electronic systems become denser and more powerful and as a result, more heat is generated. When plans are made to upgrade some computer systems, the cooling requirements may need to be changed for safe operation. Many modern systems use coolant mediums other than air (water, freon, semiconductor cooling, etc.). The manufacturer should verify that these systems are firesafe. (See Fig. 12-1C.)

FIG. 12-1C. Air conditioning unit. (IBM)

Electrical power supplies and other electrical equipment such as transformers, distribution controls, circuit breakers and disconnect switches should be maintained in a separate cut off room. Where larger batteries and/or alternate generators such as diesel or gas (natural, LPG, or

LNG) are used, they should be kept in a separate, well ventilated, detached room, having adequate air intake supply for the combustion of the engine as well as ventilation of gases such as hydrogen from the batteries and exhaust of engines. It is important not to position these exhausts near where air intakes for heating, ventilation or refrigeration of the building are located. In addition to posing a threat to human safety, this could permit exhaust fumes to enter the building and activate smoke detectors. In the event that the emergency generator exhaust is in the proximity of any building air intake then automatic self closing airtight dampers should be activated when the generator starts. Emergency generator exhausts should not be located near air intakes. (See Fig. 12-1D.)

FIG. 12-1D. Electrical support equipment. (IBM)

FIRE PROTECTION

Computers and other electronic equipment are particularly susceptible to damage from fire and the accompanying heat, steam, and smoke. Equipment and materials may even be damaged by elevated temperatures. Damage to computers and other electronic equipment may begin at a sustained ambient temperature of 175°F (79°C). Damage to magnetic tapes, flexible disks, and similar materials may begin at sustained ambient temperatures above 125°F (52°C). Above 120°F (49°C) the possibility of undoing the damage lessens quickly. Damage to hard discs may begin at sustained ambient temperatures above 150°F (66°C) with the degree of damage increasing rapidly with further elevations in temperature. Damage to paper products may begin at a sustained ambient temperature of 350°F (177°C). Damage to microfilm may begin at a sustained ambient temperature of 225°F (107°C) in the presence of steam, or at 300°F (149°C) in the absence of steam.

Portable Fire Extinguishers

Portable Halon or carbon dioxide fire extinguishers should be provided within the computer or electronic equipment room in accord with NFPA 10, *Standard for Portable Extinguishers*. Occupants should be instructed in their use.

Automatic Sprinkler Systems

Fixed automatic sprinkler systems may be required, especially if more than one computer or electronic system is located in the same room. Automatic sprinklers are usually considered as a back up to special extinguishing systems. The proposed installation and final acceptance should be approved by authorities having jurisdiction.

Special Extinguishing Systems

When the value of the computer equipment and the risk of interruption is deemed critical, a total flooding automatic Halon 1301 system may be used. Total flooding of the entire room should include the under floor plenum. A minimum 5½ percent concentration of the agent will meet standard requirements and permit safe human occupancy, thus permitting orderly shut down and evacuation of personnel. The systems should be designed to operate by manual pull box or push button stations at two or more locations as well as by activation of automatic fire, smoke, and heat detectors placed within the enclosure at ceiling level and under the floor. Alarm signals are also part of the automatic system and should sound in the immediate vicinity and remotely to local emergency personnel or the local fire department. It is important that doors and other openings be kept closed and sealed so that the Halon fire extinguishing agent cannot escape during a fire.

Smoke Detectors

Smoke detectors for detection of fire, transmitting an alarm, and activating a fire extinguishing system should be in accordance with NFPA 72E, *Standard on Automatic Fire Detectors*.

Emergency Plans and Procedures

The primary purpose of the emergency plan is to prepare key personnel to initiate actions during and after any emergency. Plans and procedures should be developed by supervisory personnel, written on paper and reviewed on a periodic basis not to exceed a six-month period. The plan should incorporate contingency factors for emergencies such as power failures, severe weather conditions, tornados, floods, earthquakes, etc. In addition, transportation difficulties such as the impairment of highway and air traffic in the event of a catastrophe, should be evaluated. The initial part of the plan should cover the contingency factor before an incident has occurred. The next part should be the plan of action: who shall do what, when. The plan should include local emergency personnel such as security and/or local fire response organizations and outside fire departments. The local fire departments' staff officers should be invited to visit the computer and electronic equipment rooms to familiarize them with the emergency plan. It is important that they be aware of any fixed automatic fire extinguishing systems, how they operate, and what procedures to follow (such as where to enter the building if summoned by a fire alarm signal).

The second part of the plan is an emergency personnel evacuation procedure. The primary purpose of this plan is safety to life of all persons who may be within the computer room or electronic equipment area. The procedure should be posted and updated periodically (every six months) so that all employees and others entering the area know what to do when the alarm signal is sounded. The alarm signal must be audible throughout the entire area, and when initiated all occupants should immediately exit the area and assemble at a predesignated location. The assembly point could be located in a safe interior part of an adjacent building. The employees should remain in the assembly area until advised otherwise. Several persons should be designated to check their areas to make certain all persons have exited. Laboratories, washrooms, storage rooms, etc. should be checked. The plan should include provisions for the handicapped or those who may need physical assistance. A visible oscillating light should be provided for people who are unable to hear the alarm.

The matter of backup electronic equipment should also be considered in the plan. Alternate equipment should be available, either from with the organization or from an outside organization if operation is to continue during the restoration of damaged equipment. A prearranged plan should consider and test compatibility of backup equipment.

Emergency lighting should also be provided for the entire interior of the building which houses the computers or electronic equipment and offices. Auxiliary battery lighting packs are permitted.

Bibliography

NFPA Codes, Standards, Recommended Practices and Manuals. (See the latest *NFPA Codes and Standards Catalog* for availability of current editions of the following documents.)

NFPA 10, *Standard for Portable Fire Extinguishers.*
NFPA 12, *Standard on Carbon Dioxide Extinguishing Systems.*
NFPA 12A, *Standard on Halon 1310 Fire Extinguishing Systems.*
NFPA 13, *Standard for Installation of Sprinkler Systems.*
NFPA 70, *National Electrical Code.*
NFPA 71, *Standard for the Installation, Maintenance, and Use of Central Station Signaling Systems.*
NFPA 72A, *Standard for the Installation, Maintenance and Use of Local Protective Signaling Systems for Guard's Tour, Fire Alarm and Supervisory Service.*
NFPA 72B, *Standard for the Installation, Maintenance and Use of Auxiliary Protective Signaling Systems for Fire Alarm Service.*
NFPA 72C, *Standard for the Installation, Maintenance and Use of Remote Station Protective Signaling Systems.*
NFPA 71D, *Standard for the Installation, Maintenance and Use of Proprietary Protective Signaling Systems.*
NFPA 75, *Standard for the Protection of Electronic Computer/Data Processing Equipment.*
NFPA 80A, *Recommended Practice for Protection of Buildings from Exterior Fire Exposures.*
NFPA 90A, *Standard for the Installation of Warm Air Heating and Air Condition Systems.*
NFPA 232, *Standard for the Protection of Records.*
NFPA 258, *Standard Research Method for Determining Smoke Generation of Solid Materials.*

Additional Readings

"Computer Centers—Measures to Reduce Losses," *Brand Brandweer*, Vol. 1, No. 3, (1977), pp. 53-56.
Davis, R. H., "Fire Protection Systems (for Computers)," *Data Management*, Vol. 14, No. 11, 1976, pp. 11-14.
"Electronic Computer Systems," *Loss Prevention Data Sheet* 5-32, Factory Mutual Engineering Corporation, Norwood, MA, May 1978.
Factory Mutual Corp., "American Experience in the Fire Protection of Computers," *Fire*, Vol. 7, 1979, pp. 498-499.
Facts About Protecting Electronic Equipment Against Fire, Ansul, Marinette, WI, 1984.
"Halon Prevents Major Central Processing Unit Fires," *Computer Decisions*, Vol. 10, No. 8, 1978, pp. 56, 58.
Harrison, Gregory A., and Lowec, David G., "Computer Fire and Resultant Circuit Card Fire Tests," *Fire Journal*, Vol. 63, No. 2, Mar. 1969, p. 5.
Jacobson, Dan W., "Automatic Sprinkler Protection for Essential Electrical and Electronic Equipment," *Fire Journal*, Vol. 61, No. 1, Jan. 1967, pp. 48-53.
Keigher, Donald J., "Water and Electronics Can Mix," *Fire Journal*, Vol. 62, No. 6, Nov. 1968, pp. 68-72.
Kuechmann, A. M., "On-Site Fire Protection for Computer Media," *Fire Journal*, Vol. 66, No. 6, Nov. 1972, pp. 50-51.
Osborn, Richard W., ed., *Tapping In to the NEC*, National Fire Protection Association, Quincy, MA, 1982.
Perkins, C., and B. J. Berenblut, "Electronic Equipment: What Protection is Required?", *Fire Surveyor*, Vol. 9, No. 5, Oct. 1980, pp. 28-33.
Philbrick, S. E., "Selecting Cables for Fire-Risk Applications," *Electronics and Power*, Vol. 26, No. 3, Mar. 1980, pp. 232-233.
"Standard Practice for the Fire Protection of Essential Electronic Equipment Operations," U.S. Department of Commerce, National Fire Prevention and Control Administration, Washington, DC, Aug. 1978.
Staller, Jack J., and Colling, David A., "The Aftermath of a Computer Fire," *Fire Journal*, Vol. 71, No. 6, Nov. 1977, pp. 29-33.
"Support for Sprinkler Systems in Controversial Settings," *Fire*, February 1981, pp. 455-456.
Williamson, H. V., "New Protection for Electronic Equipment," *Fire Technology*, Vol. 2, No. 4, Nov. 1966, pp. 279-286.

PROTECTION FOR LABORATORIES

Revised by Henry P. Beltramini

This chapter covers the more common laboratory hazard problems that influence the degree of fire protection that may be required. The basic requirements of construction and design of laboratories together with the fire protection systems and procedures for laboratories are essential items of consideration. Standard laboratory equipment should be designed primarily with fire prevention and protection and the diminishment of personnel hazards in mind. The use, storage, handling, and disposal of all commonly used laboratory materials, such as flammable and combustible liquids, compressed gases, biological, and radiological materials must be considered in protection and prevention procedures. To those responsible for laboratory operations to minimize losses from fire or explosion, or both, and to reduce the possibilities of injury and life threatening situations, these concepts should be of primary consideration

This chapter treats the subject of protection for laboratories in a general way. For more information on the hazards of and protection for laboratories consult Section 5, Chapter 4, "Flammable and Combustible Liquids;" Chapter 5, "Gases;" Chapter 6, "Chemicals;" Section 8, Chapter 4, "Air Conditioning and Ventilating Systems," Section 10, Chapter 17, "Nuclear Reactors, Radiation Machines and Facilities Handling Radioactive Materials;" Section 11, Chapter 4, "Storage and Handling of Flammable and Combustible Liquids," Chapter 5, "Storage and Handling of Gases," Chapter 6, "Storage and Handling of Chemicals;" Section 16, "Fire Alarm Systems, Detection Devices and Guard Services;" Section 18, Chapter 6, "Standpipe and Hose Systems," and Section 19, "Special Fire Suppression Systems and Agents."

CONSTRUCTION AND DESIGN

The basic modes of construction and design outlined here are suggested as guides only. Specific details of construction and special facets of consideration change routinely with each type of laboratory. NFPA 45, *Standard on Fire Protection for Laboratories Using Chemicals*, contains detailed information on classifying laboratories based on flammable liquid content, construction, and presence of explosion hazards. NFPA 99, *Standard for Health Care Facilities*, Chapter 7, specialized engineering firms, insurance rating agencies, and the various state and municipal regulatory agencies are also good sources to consult to ensure that all requirements are met. (See Fig. 12-2A.)

FIG. 12-2A. A general view of a hospital laboratory, showing the multitude of chemicals and apparatus found in a typical clinical laboratory. It is imperative that fire fighters be informed of the types of hazardous materials located here. Note the deluge shower on the ceiling, for use if chemicals are spilled. Also note the large number of metal waste baskets available to prevent accumulation of trash on the floor.

Mr. Beltramini is Assistant Vice-President for Administrative Services at Brigham and Women's Hospital, Boston, MA.

Construction

The construction of laboratories depends on a variety of considerations—the type and quantity of potential hazards involved, the area or location with respect to surrounding activities, and the type of fire protection to be provided. Well located laboratories are separated from other laboratories and nonlaboratory space by fire resistive construction with a minimum rating of 1 hr. Interior finish used in laboratory construction and the access to exits from them must be Class A (having a flame spread rating of 0 to 25).

Egress: Means of egress requirements for a laboratory building must agree with the exit requirements specified for the particular occupancy and type of building. (See NFPA 101, *Life Safety Code*®.) Most laboratory buildings are constructed to comply with the exit requirements for general purpose industrial buildings. Exceptions are laboratory buildings or laboratories located in educational, institutional, or other specialized occupancies. They comply with the exit requirements specified for those particular occupancies.

Generally, two means of access opening to an exit access corridor or to the outside of the building are required. These should be located to provide access to exits from all parts of a laboratory unit with an area in excess of 1,000 sq ft (93 m^2), or a laboratory unit containing large quantities of flammable liquids and with an area in excess of 500 sq ft (46 m^2). A second means of egress is also required from a laboratory which has a laboratory hood adjacent to the primary means of egress or which contains substantial quantities of compressed gases. Exit access doors from laboratories should swing in the direction of egress. Furniture and laboratory equipment should be arranged so that the means of access to an exit may be reached easily from any point in the laboratory unit without undue obstruction.

FIRE PROTECTION

All laboratories must have fire protection appropriate to the hazards present. At minimum, they should have portable fire extinguishers (appropriate to the hazard), an adequate fire alarm system, and an emergency evacuation plan.

Automatic Fire Extiguishing/Protection Systems

It is good practice to protect laboratories, including associated storage rooms and enclosures, with an automatic sprinkler system designed and installed according to provisions of NFPA 13, *Standard for the Installation of Sprinkler Systems*. In some instances special hazard laboratories may require nonwater automatic extinguishing and detection systems, such as carbon dioxide, foam, halogenated agents, etc. When such systems are used, they too are installed and maintained in accordance with the appropriate NFPA Standards. Periodic inspection of all automatic extinguishing and detection systems insures that they are cared for properly and maintained in service.

The activation of a fire detection system or discharge of a fire extinguishing system should sound an alarm, preferably at a municipal fire department, to insure prompt response. Additionally, the alarm system should also be arranged to sound alarms both inside the laboratory area and at a constantly attended location, such as a telephone switchboard, communications center, or security office. Activation of the alarm system in turn should activate automatic electromagnetic door closers that affect the fire protection of the laboratory area. Laboratory building air conditioning also should be automatically shut down or switched to total exhaust. The NFPA Signaling Systems Standards give guidance on the various alarm systems and detection equipment that can be used.

Standpipe and Hose Systems

It is also a good practice in laboratory buildings with two or more stories above or below ground level to provide 2½ in. (64 mm) hose connections with hose and combination straight stream and fog nozzle. NFPA 14, *Standard for the Installation of Standpipe and Hose Systems*, is the guide to follow.

Portable Fire Extinguishers

Fire extinguishers suitable to the particular hazards of the laboratory are installed and located so that they will be readily available to laboratory personnel. Water type fire extinguishers suitable for fires in Class A materials are the basic type of unit for laboratories supplemented by sufficient extinguishers suitable for use on Class B (flammable liquid) and Class C (electrical) fires. Special Class D fire extinguishers are advisable where the use of metals or metal hydrides indicates a need for them. NFPA 10, *Standard for Portable Fire Extinguishers*, is the guide to follow.

Fire Prevention Procedures

It is essential to develop good procedures for handling laboratory emergencies. They include manual alarm activation, evacuation procedures, equipment shutdown procedures, and provisions for fire fighting actions. In addition, the following critical areas require special consideration: (1) handling and storage of flammable liquids; (2) handling and storage of other hazardous materials and gases; (3) smoking area controls; (4) electrical hazards, (5) open flame and spark producing work; and (6) radiological and biological hazards.

Evacuation plans must be thoroughly evaluated and then drawn up to include all contingencies. Testing the evacuation plan periodically insures proper and prompt response by all laboratory personnel who must be trained and evaluated periodically in all aspects of the laboratory fire prevention procedures.

EQUIPMENT

All equipment in a laboratory, whether fixed or portable laboratory equipment or fixed building service equipment, must meet specific standards and must be evaluated with respect to safety, fire prevention, damage, and loss in any fire situation. Laboratory and fire brigade personnel, must be familiar with all aspects of laboratory and building service equipment. Municipal fire department personnel must be made aware of all aspects of laboratory and building service equipment before called to fight fires in laboratories. (See Fig. 12-2B.)

FIG. 12-2B. Apparatus used for performing medical tests automatically. Apparatus like this is especially vulnerable to damage by water and smoke. It is essential that electric power and utilities be shut down, and the equipment covered (if possible), to minimize damage during fire fighting.

Ventilation

Duct systems for laboratory heating and ventilating, including warm air heating systems, general ventilating systems, air cooling systems, and laboratory exhaust and hood systems deserve special attention. NFPA 90A, *Standard for the Installation of Air Conditioning and Ventilating Systems*, and NFPA 91, *Standard for the Installation of Blower and Exhaust Systems for Dust, Stock and Vapor Removal or Conveying*, are the good practices to follow.

Fresh air intakes for laboratories deserve special attention. They must not draw in flammable or toxic materials or combustion byproducts emitted either from the laboratory building itself or other related structures nearby. Conversely, exhaust systems for laboratories should not be located near air intakes for related structures, but should be separated and high enough above roof level to afford maximum atmospheric dilution.

Air exhausting from laboratory areas using highly infectious or radioactive materials must pass through high efficiency filters before discharging into the atmosphere. Laboratory units provided with mechanical ventilation throughout or employing fume hoods, or both, as a fixed part of the area exhaust system, should be balanced to provide a negative pressure with respect to the corridors and surrounding nonlaboratory spaces. An exception is those laboratories housing such activities as clean rooms that preclude a negative pressure in relation to surrounding space. A slightly positive pressure is required in these areas and precautions must be taken to prevent escape of atmospheric contaminants in the laboratory unit to surrounding spaces.

Electrical

All electrical installations, both fixed and portable, for laboratories must conform to the requirements of NFPA 70, *National Electrical Code®*. In addition, all electrical appliances and equipment should display a listing with or approval by a testing laboratory for its intended use unless the device is one for which test standards have not been established. All electrical appliances and equipment should have an operator's manual together with a circuit diagram available for maintenance personnel. It is also important that all portable electrical equipment intended for laboratory use be properly grounded by recognized methods.

Installation of a sufficient number of electrical outlets at convenient locations eliminates the use of extension cords. Sufficient circuitry or overcurrent protection devices, which are recommended for all electrical equipment, should be installed to prevent overloading the fuses.

It is good practice to provide emergency power for operating exhaust fans, exit lighting, exit signs, evacuation alarms, and other emergency or fire and safety equipment. In addition, several outlets should be provided with emergency power to prevent loss or interruption of electrical power to critical laboratory equipment or experimentation.

Mechanical

Safe operation of laboratory instruments and equipment depends on purchasing or building apparatus with adequate controls and safeguards, installing the apparatus in safe locations, and using the apparatus within the limitations of the original design or appropriate modifications. Safe design and operation will help to prevent unnecessary accidents, damage or destruction of equipment, and potential fire hazards.

Fume Hoods

Laboratory hoods are the most commonly used means of removing gases, dust, mist, vapors, and fumes from laboratory operations, and of preventing toxic exposures and flammable concentrations. To be effective, a laboratory hood and its associated components must confine contaminants within the hood, remove them through the ductwork, and disperse them so that they do not return to the building through air intake systems. Hood face velocities must be sufficient to assure capture velocity for operating conditions. Recommended face velocities are published by the American Conference of Governmental Industrial Hygienists. Laboratory hoods for hazardous operations, such as perchloric acid or radioactive materials, require special design and are constructed of suitable materials that meet rigid criteria. Fume hood controls should be so arranged that shutting off the ventilation in one hood will not reduce the exhaust capacity or create an imbalance for any other hood. Face velocity must be adequate to prevent backflow of contaminants, especially in the presence of cross drafts. All shutoff valves for services should be located outside the hood enclosure in a readily accessible location. (See Fig. 12-2C.)

Refrigerators

Refrigerators are a valuable asset in most laboratories, but particular care must be taken to insure their proper use. Ordinary domestic refrigerators are completely unsuitable for the storage of flammable liquids and should be so labelled in red letters on the door. Specific laboratory safe or explosionproof refrigerators are available from several manufacturers and should be purchased for storage of flammable liquids. Ordinary domestic refrigeration units can be modified by the removal of all lights, switches, heating units, oven thermostats, and other ignition sources, and by making sure that the unit's entire refrigeration compartment is vapor-tight.

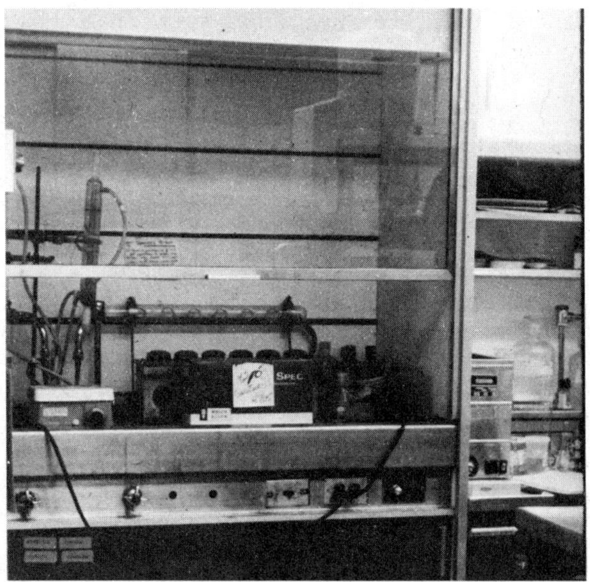

FIG. 12-2C. A laboratory fume hood. The work space is enclosed at the sides, and is covered by a sliding glass panel at the front. Exhaust fans draw in room air from the lower edge of the glass panel and vent it through a plenum to the outside, thus preventing hazardous fumes from entering the room. Special filters are required in the exhaust system if radioactive or hazardous bacteriological materials are used in the hood.

FLAMMABLE AND COMBUSTIBLE LIQUIDS AND HAZARDOUS CHEMICALS

Flammable and Combustible Liquids

Flammable and combustible liquids present a serious fire and explosion hazard in any laboratory. The hazards from a flammable liquid process or the storage of such materials depend on such conditions as the quantity and flammability of the liquid; whether it is in a closed container or in an open system exposed to the air; the probabilities of leakage or overflow; the location in relation to other buildings, equipment, and outside ignition sources; the building construction; and the adequacy of fire protection.

Storage and Use: Flammable and combustible liquids are used from and stored in listed containers. Established laboratory practices are the limiting factors in determining the working quantity of flammable and combustible liquids that are present. All nonworking quantities are stored in well constructed storage cabinets or storage rooms. (See NFPA 30, *Flammable and Combustible Liquids Code* for details of construction.) All laboratory personnel must be thoroughly familiar with the properties and the hazards of flammable and combustible liquids in use.

Handling and Transfer: Only trained personnel should be involved in receiving, transporting, unpacking, and dispensing of flammable and combustible liquids. These activities are carried out in locations in a manner that minimizes the hazards. Good practice dictates that transfer from bulk stock containers to smaller containers is done only in storage rooms or within an adequate fume hood and with adequate grounding.

Disposal: Good safety practices in addition to any applicable governmental regulations must be observed in the disposal of waste flammable and/or combustible liquids. Disposal of these hazardous materials must be accomplished off the premises by a commercial disposal specialist who is competent and possesses knowledge of the basic character and the hazards of the waste.

Chemicals

Hazardous chemicals must not be brought into a laboratory unit unless the design, construction, and fire protection of receiving, using, and storage facilities are commensurate with the quantities and hazards of the chemicals involved. All laboratory personnel must be aware of the hazards of all chemicals in use. Special storage facilities are needed for materials having unique physical or hazardous properties, such as temperature sensitive, water reactive, or explosive materials. A reference source of information is NFPA 49, *Hazardous Chemicals Data*.

Emergency Showers

Safety shower installations are for protection of personnel from acids, caustics, cryogenic fluids, clothing fires, and other emergencies in which volumes of water are needed for diluting, warming or cooling, flushing off chemicals, or putting out clothing fires. Locate emergency showers in conspicuous spots, preferably in usual traffic patterns but not more than 25 ft (8 m) from any laboratory entrance. Avoid locating showers near electrical apparatus and power outlets. Floor drains should be provided and the locations of showers should be plainly marked on the floor. The showers should be tested and flushed at least every six months to ensure they are in operating condition.

Eye Bath

Facilities for quick drenching or flushing of the eyes are needed within the laboratory work area for immediate emergency use by laboratory workers. The location of eye baths should be plainly marked and they should be tested at least every six months to ensure they are in operating condition.

COMPRESSED OR LIQUEFIED GASES

Compressed or liquefied gases and experimental mixtures that have properties which are frequently unfamiliar to laboratory personnel present many problems.

Storage, Handling, and Use

Containers designed, constructed, and tested in accordance with DOT (US Department of Transportation) specifications and regulations are used for storage of compressed or liquefied gases. A good gas container storage room is either a separate room or an enclosure reserved exclusively for that purpose. It should have a fire resistance rating of at least 1 hr and good ventilation. Gas cylinders are secured in position and stored away from heat or other ignition sources in racks. Flammable gases should be separated from other oxidizing gases by construction that has a fire resistance of at least 1 hr. Electrical equipment, if any, in flammable gas storage areas must comply with *National Electrical Code* provisions for Class I, Division 2 hazardous (classified) locations.

Handling and transporting gas cylinders must be consistent with established safety procedures. The hazardous properties of the contents and the hazards of pressurized containers must be considered. Gas cylinders should be secured in place and away from heat or any other ignition source when they are in use. Pressurized gas cylinders are never used without pressure regulators.

Piping Systems: When laboratory equipment is routinely operated with flammable gases supplied from compressed gas cylinders, the cylinders should be located outside the building and connected to the laboratory equipment by a permanently installed piping system. Pressure reducing valves should be connected to each cylinder and adjusted to limit pressure in the piping system to the minimum required gas pressure. Pressure regulators must be compatible with the gas for which they are used. Supply and discharge terminals of piping systems should be permanently marked at both ends with the name of the gas to be piped through them. Never use piping systems for gases other than those for which they were designed and identified. Do not attempt to transfer compressed or liquefied gases from one gas container to another. Additional requirements are covered in NFPA 58, *Standard for the Storage and Handling of Liquefied Petroleum Gases.*

General Safety Rules

The following general rules for handling compressed or liquefied gas should be observed in a laboratory:

1. Know cylinder contents.
2. Know properties of contents.
3. Handle cylinders carefully.
4. Store cylinders in a well ventilated area away from heat.
5. Secure cylinder's position during use, transit, or storage.
6. Never tamper with valves, safety plugs, or packing nuts.
7. Do not strike an electric arc on cylinders.
8. Use equipment suitable for the contents of the cylinder.
9. Do not use cylinders without a regulator.
10. Close cylinder valves when not in use.
11. Never attempt to refill a cylinder.

BIOLOGICAL HAZARDS

Laboratory work done with animals, infectious diseases, or toxic materials presents a variety of safety problems both to laboratory workers and to others whose duties may require their presence in such a laboratory unit. Fire, explosions and other unintended incidents present two types of problems: (1) the dissemination of biological hazards, and (2) the hazards encountered by fire fighting personnel entering the unit. Laboratory fire brigades and local fire fighting personnel must be educated to the hazards they may encounter in laboratories engaged in biological work. Sufficient protective equipment, breathing apparatus, clothing, gloves, etc., must be readily available for use by fire fighting personnel. Laboratories engaged in this type of work must be clearly labeled, and clear and concise safety instructions must be posted for all personnel.

RADIATION HAZARDS

Laboratories using radioactive materials present hazards to both laboratory workers and others required to enter the facility. While rigid safety and health programs have been established for laboratory workers, it is wise to institute a similar program for other personnel such as fire fighters. Good communication between laboratory radiation safety and health officers and public fire fighting personnel can help keep fire fighters informed of all laboratories actively engaged in radioactive work. Such rapport can be the basis for establishing an education program planned to cope with disasters and emergencies. See NFPA 801, *Recommended Fire Protection Practice for Facilities Handling Radioactive Materials* for further discussion of the hazards and safeguards associated with nuclear reactors, radiation machines, and radioactive materials.

PERSONNEL PROTECTION

Personal protective equipment is essential to a sound safety program in any laboratory unit. Personal injuries arising out of a fire or explosion sometimes present serious complications to a fire fighting effort, an effort that is already complicated by the very fact that the disaster involves a laboratory.

Protective Clothing: Personal protective clothing, including head and foot protection, must be made available for use by all laboratory personnel and other personnel required to enter laboratory areas where specific protective clothing is required by the nature of the operations. Specific education programs are useful in instructing personnel in the needs and requirements of protective clothing. Additionally, an eye protection program (safety glasses, goggles or face shields) is beneficial in all laboratory areas.

Shields and Barriers: The hazards of explosion, rupture of apparatus and systems from overpressure, implosion due to vacuum, sprays or emission of toxic or corrosive materials, or flash ignition of escaping vapors require substantial physical protection for exposed personnel. Analysis of the potential force and characteristics of the type of hazards involved can aid in the selection of effective and economical shields or barriers.

Respiratory Protection: Respirators are essential for the protection of laboratory personnel working directly with, or required to enter areas where, infectious diseases, radioactive materials, harmful dusts, mists, fogs, fumes, sprays, or vapor are present, either as a natural result of laboratory work or as the result of an unforeseen disaster. Respirators or masks that are suitable for the specific problem to be dealt with are necessary. Training in the use of respirators is essential for laboratory personnel.

HAZARD IDENTIFICATION

It is good practice to post signs at entrances to laboratories, storage areas, and associated facilities warning personnel, especially emergency personnel, of unusual or severe hazards inside that may or may not be directly related to an emergency situation. Included are signs relating to particularly unstable chemicals, radioactive chemicals, pathogenic or infectious materials, water reactive chemicals, and explosives. NFPA 704, *Standard System for the Identification of the Fire Hazards of Materials*, is a good system to utilize. Periodic review is necessary to ensure that the particular warning label being displayed properly indicates the nature of the material being used within the identified laboratory area. Severe hazards should be discussed with fire fighting personnel to make certain they have sufficient knowledge of what to anticipate.

All individual containers within a specific laboratory area must also be identified as to their contents. Unlabeled containers present a serious potential hazard to not only laboratory workers but also to other personnel required to enter laboratory areas.

Bibliography

NFPA Codes, Standards, Recommended Practices and Manuals. (See the latest *NFPA Codes and Standards Catalog* for availability of current editions of the following documents.)

NFPA 10, *Standard for Portable Fire Extinguishers.*
NFPA 11, *Standard for Low Expansion Foam and Combined Agent Systems.*
NFPA 11A, *Standard for Medium and High Expansion Foam Systems.*
NFPA 12, *Standard on Carbon Dioxide Extinguishing Systems.*
NFPA 12A, *Standard on Halon 1301 Fire Extinguishing Systems.*
NFPA 13, *Standard for the Installation of Sprinkler Systems.*
NFPA 14, *Standard for the Installation of Standpipe and Hose Systems.*
NFPA 15, *Standard for Water Spray Fixed Systems for Fire Protection.*
NFPA 30, *Flammable and Combustible Liquids Code.*
NFPA 45, *Standard on Fire Protection for Laboratories Using Chemicals.*
NFPA 49, *Hazardous Chemicals Data.*
NFPA 50, *Standard for Bulk Oxygen Systems at ConsumerSites.*
NFPA 50A, *Standard for Gaseous Hydrogen Systems at Consumer Sites.*
NFPA 50B, *Standard for Liquefied Hydrogen Systems at Consumer Sites.*
NFPA 53M, *Manual on Fire Hazards in Oxygen-Enriched Atmospheres.*
NFPA 56F, *Standard for Nonflammable Medical Gas Systems.*
NFPA 58, *Standard for the Storage and Handling of Liquefied Petroleum Gases.*
NFPA 70, *National Electrical Code.*
NFPA 72A, *Standard for the Installation, Maintenance and Use of Local Protective Signaling Systems for Guards Tour, Fire Alarm and Supervisory Service.*
NFPA 72B, *Standard for the Installation, Maintenance and Use of Auxiliary Protective Signaling Systems for Fire Alarm Service.*
NFPA 72C, *Standard for the Installation, Maintenance and Use of Remote Station Protective Signaling Systems.*
NFPA 72D, *Standard for the Installation, Maintenance and Use of Proprietary Protective Signaling Systems.*
NFPA 72E, *Standard on Automatic Fire Detectors.*
NFPA 77, *Recommended Practice on Static Electricity.*
NFPA 86, *Standard for Ovens and Furnaces—Design, Location, and Equipment.*
NFPA 91, *Standard for the Installation of Blower and Exhaust Systems for Dust, Stock and Vapor Removal or Conveying.*
NFPA 99, *Standard for Health Care Facilities.*
NFPA 101, *Code for Safety to Life from Fire in Buildings and Structures.*
NFPA 321, *Standard on Basic Classification of Flammable and Combustible Liquids.*
NFPA 325M, *Fire Hazard Properties of Flammable Liquids, Gases and Volatile Solids.*
NFPA 491M, *Manual of Hazardous Chemical Reactions.*
NFPA 801, *Recommended Fire Protection Practice for Facilities Handling Radioactive Material.*

Additional Readings

American Conference of Governmental Hygienists, *Threshold Limit Values for Chemical Substances and Physical Agents in the Workroom Environment with Intended Changes for 1973*, American Conference of Governmental Hygienists, Cincinnati, OH, 1973.
Bond, R. G., Michaelson, G. S., and DeRoos, R. L., *Environmental Health and Safety in Health-Care Facilities*, Macmillan, NY, 1973.
Christensen, H. E., ed., *The Toxic Substances List*, National Institute for Occupational Safety and Health, Rockville, MD.
Christian, F. T., *A Guide to Safety in the Science Laboratory*, Bulletin 74, Florida Department of Education, Tallahassee, FL, 1968.
DeRoo, John L., "The Safe Use of Refrigerator and Freezer Appliances for Storage of Flammable Materials," *Fire Journal*, Vol. 67, No. 2, Mar. 1973, pp. 63-64 .
Factory Mutual Engineering Corp., "Laboratory Fires and Explosions," *Loss Prevention Data Sheet 10-19*, Factory Mutual System, Norwood, MA.
Fawcett, H. H. and Wood, W. S., *Safety and Accident Prevention in Chemical Operation*, Interscience Publishers, NY, 1965.
Fuscaldo, Anthony A., ed., *Laboratory Safety: Theory and Practice*, Academic Press, NY, 1980.
"Glove Box Fire Safety, A Guide for Safe Practices in Design, Protection and Operation," FMRC *Report TID — 24236*, Factory Mutual Research Corp, Norwood, MA, 1967.
Green, Michael F., and Tark, Ames, *Laboratory Safety*, Macmillan, NY, 1978.
Henry, Martin F., *Flammable and Combustible Liquids Code Handbook*, NFPA, Boston, MA, 1978.
Manufacturing Chemists' Association, *Laboratory Waste Disposal Manual*, Washington, DC, 1969.
Manufacturing Chemists' Association, *Guide for Safety in the Chemical Laboratory*, Van Nostrand, NY, 1972.
Meidl, J. H., *Flammable Hazardous Materials*, Glencove Press, Encino, CA, 1970.
Meidl, J. H., *Hazardous Materials Handbook*, Glencove Press, Encino, CA, 1972.
Muir, G. D., *Hazards in the Chemical Laboratory*, Royal Institute of Chemistry, London, 1971.
Pipitone, D., ed., *Safe Storage of Laboratory Chemicals*, John Wiley, NY, 1984.
Quam, G. N., *Safety Practice for Chemical Laboratories*, Villanova Press, Villanova, PA, 1963.
Sittig, M., *Handbook of Toxic and Hazardous Chemicals*, Noyes, Park Ridge, NJ, 1981.
Spindel, W., *Prudent Practices for Handling Hazardous Chemicals in Laboratories*, National Science Foundation, Washington, DC, 1981.
Steere, N., *Handbook of Laboratory Safety*, 2nd ed., CRC Press, West Palm Beach, FL, 1971. Steer, N. V., *Safety in the Chemical Laboratory*, Vol. 1, American Chemical Society, Easton, PA, 1967.
Ibid., Vol. 2, 1971.
Ibid., Vol. 3, 1974.
Welby, P., and Dickinson, K. R., "Monitoring Work Areas for Explosive and Toxic Hazards," *Chemical Engineering*, Vol. 83, No. 22, 1976, pp. 139-145.

OXYGEN-ENRICHED ATMOSPHERES

Revised by George J. Frankel

Oxygen is a clear, colorless, odorless, and tasteless element found commonly in the gaseous state. It comprises about 21 percent of the earth's atmosphere and is by far the most common oxidizing material. An oxygen-enriched atmosphere (OEA) is defined as any atmosphere in which the concentration of oxygen exceeds 21 percent by volume or the partial pressure of oxygen exceeds 160 torr (21.3 kPa), or both. (See Table 12-3A).

The normal concentration of oxygen available in the atmosphere is generally sufficient for man's needs, but oxygen-enriched atmospheres are often needed in medical practice, industry, underwater tunneling and caisson work, space and deep sea exploration, and in commercial and military aviation. They also are inherent to oxygen processing, transport, and storage facilities. In addition, oxygen-enriched atmospheres may develop inadvertently when oxygen or compressed air is transported, stored, or utilized.

TABLE 12-3A. Partial Pressures of Oxygen at Altitudes Above and Below Sea Level.

Total Absolute Pressure				Altitude Above or Depth Below Sea Level — Air or Sea Water		Partial Pressure of Oxygen if Atmosphere is Air	Concentration of Oxygen if Partial Pressure of Oxygen is 160 mm.Hg. or torr
Atmospheres	mm.Hg. or torr	kPa	psia	Feet	Meters	mm.Hg. or torr*	% by Volume
1/5	152	20.0	2.9	38,500	11,735	32	100.0†
1/3	253	33.8	4.9	27,500	8,382	53	62.7†
1/2	380	50.3	7.3	18,000	5,486	80	42.8†
2/3	506	67.6	9.8	11,000	3,353	106	31.3†
1	760	101.4	14.7	Sea Level	0	160	20.9
2	1,520	202.7	29.4	−33	−10	320†	10.5
3	2,280	304.1	44.1	−66	−20	480†	6.9
4	3,040	405.4	58.8	−99	−30	640†	5.2
5	3,800	506.8	73.5	−132	−40	800†	4.2

* This column shows the increased available oxygen in compressed air atmospheres.
† Oxygen-enriched atmosphere.

This chapter discusses the fire hazards which may be associated with oxygen-enriched atmospheres and the methods that can be used to reduce them and to protect against them.

Further information of interest to those dealing with oxygen-enriched atmospheres may be found in Section 5, Chapter 1, "Fire Hazards of Materials;" Chapter 4, "Flammable and Combustible Liquids;" Chapter 5, "Gases;" Section 9, Chapter 4, "Health Care Facilities;" Section 10, Chapter 4, "Fluid Power Systems;" Section 12, Chapter 2, "Protection for Laboratories;" Chapter 4, "Medical Gases;"

Mr. Frankel, Principal Engineer, Engineering Test Operations, Grumman Aerospace Corporation, Bethpage, NY, is Chairman of the NFPA Committee on Fire Hazards in Oxygen-Enriched Atmospheres.

and in Section 19, "Special Fire Suppression Agents and Systems."

FIRE HAZARDS IN OXYGEN-ENRICHED ATMOSPHERES

Fire, as a chemical reaction between a fuel and oxygen, is affected by the concentration of the reactants. The commonly encountered fire involves the 21 percent concentration of oxygen available from the atmosphere.

The degree of fire hazard of an oxygen-enriched atmosphere varies with the concentration of oxygen present, the concentration of any nonflammable (diluent) gas present, and the total pressure. However, an oxygen-enriched atmosphere, as defined, does not necessarily result in an increased fire hazard. Certain oxygen-enriched atmospheres may exhibit combustion supporting properties similar to ambient air; others are incapable of supporting the combustion of normally flammable materials at all. For example, a 4 percent oxygen mixture in nitrogen or helium at a total pressure of 12 atmospheres (1216 kPa) will not support the combustion of paper even though it is an oxygen-enriched atmosphere (because the partial pressure of oxygen is 365 torr (48.7 kPa). However, different concentrations and pressures may accelerate the combustion of these materials, facilitate ignition, and, in general, increase the fire hazard. Most commonly encountered oxygen-enriched atmospheres fall into this latter category.

Fire and explosions have occurred in many diverse circumstances involving both intentional and unintentional oxygen-enriched atmospheres. Among them are:

1. Oxygen production, transportation, and transfer. This may involve truck delivery of liquid oxygen, high pressure oxygen cylinder recharging, oxygen columns and compressors.
2. Medical. This may involve operating rooms and anesthesia machines, oxygen tents and infant incubators, or oxygen respirators such as used by ambulances, rescue squads, fire and police, and near swimming pools.
3. Cutting and welding.
4. Industrial processing. This may involve the manufacture of gasoline, ammonia, methanol, acetylene, nitric acid, and ethylene, etc.
5. Hospital and laboratory supply systems.
6. Space and deep sea activities. This category may involve missile fueling/defueling, rocket engine testing, manned space environment simulation/spacecraft testing, and deep sea diving decompression chambers.
7. Aircraft oxygen systems. Included here are commercial and military aircraft onboard oxygen breathing systems, and aircraft maintenance operations in the presence of oxygen.
8. Inadvertent substitution of oxygen for air or nitrogen.

IGNITION AND COMBUSTION OF MATERIALS IN OEAs

The minimum energy that molecules (including those of fuels and oxygen) must possess to permit chemical interaction is referred to as the activation energy. If the energy released by this chemical reaction is sufficient to impart activation energy to other molecules on a self-sustaining basis, ignition has occurred. The rate of combustion depends upon the chemical nature and physical characteristics of the fuel and oxidant, their relative concentrations, environmental pressure and temperature, and other physical parameters, such as geometry and ventilation.

The likelihood of ignition and the rate of flame propagation of a combustible are greatly influenced by the oxygen content of the environment. In general (although not in every case) the greater the oxygen concentration, the lower the minimum ignition energy required for ignition, and the faster the flame spread rate. This is shown graphically in Figures 12-3A and 12-3B.

Thus, in general, the fire hazard in an oxygen-enriched atmosphere is significantly greater than that in an

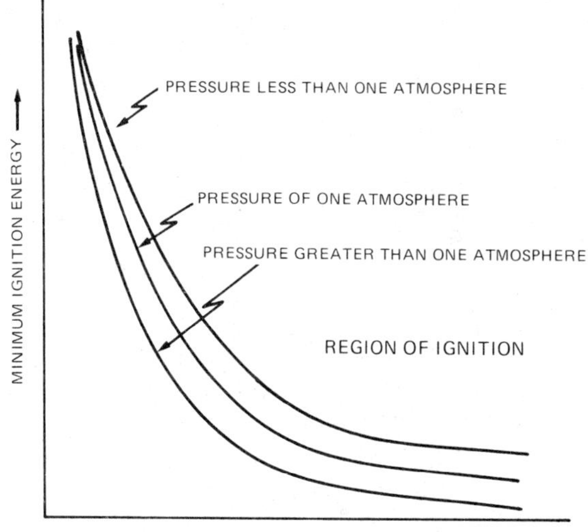

FIG. 12-3A. Minimum ignition energy behavior of combustibles in oxygen-diluent atmospheres at different pressures.

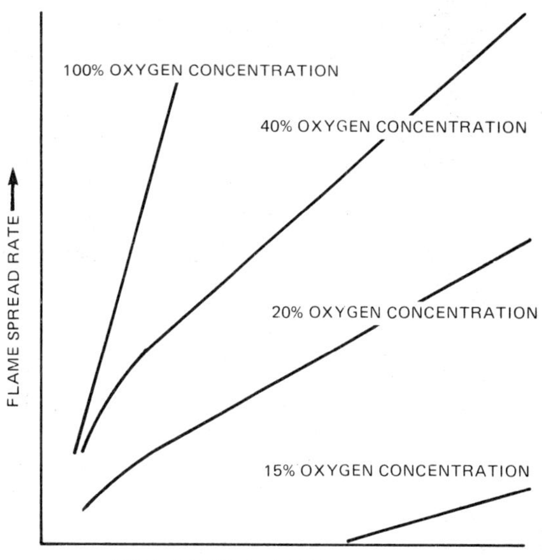

FIG. 12-3B. Effects of atmosphere oxygen content and environmental pressure on flame spread rate.

ordinary atmosphere. Almost all materials are flammable in a pure oxygen environment; therefore, increased oxygen concentration may change the classification of a material from the nonflammable to the flammable category.

Figures 12-3C and 12-3D depict three combustion

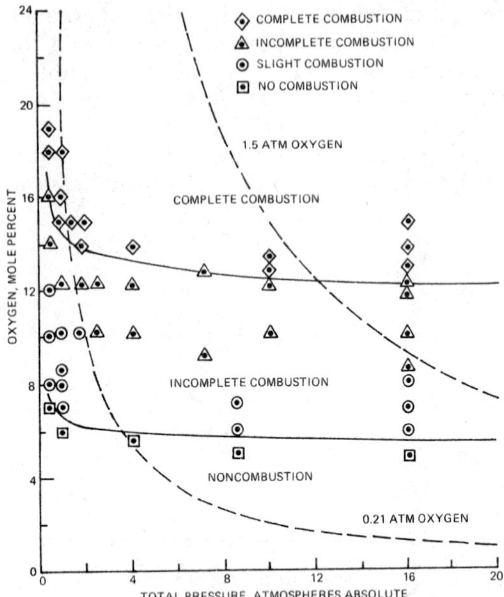

FIG. 12-3C. Illustration of varying degrees of combustion in an oxygen-nitrogen OEA (Dorr 1979).

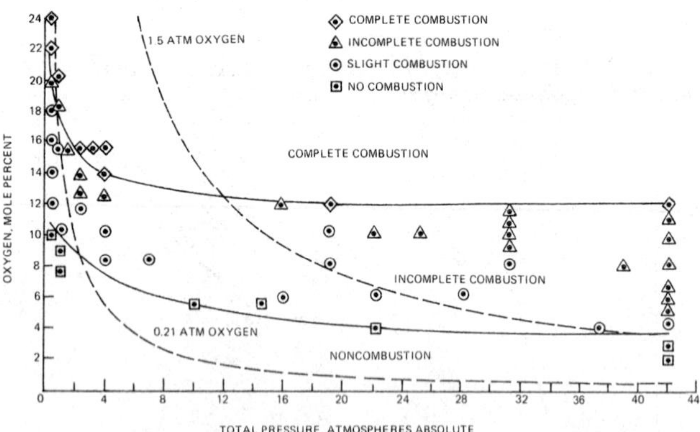

FIG. 12-3D. Illustration of varying degrees of combustion in an oxygen-helium OEA (Dorr 1973).

zones for vertical filter paper strips in mixtures of oxygen-nitrogen and oxygen-helium, respectively. Those combinations of oxygen concentrations and total pressure lying above the 0.21 atmosphere oxygen partial pressure isobar (lower dashed line) are, by definition, oxygen-enriched atmospheres, but they may be located in any of three zones: noncombustion; incomplete combustion; or complete combustion. Note that there are certain oxygen-enriched atmospheres which do not produce an increased fire hazard.

Behavior of Materials in OEAs

Since combustible materials ignite more easily and burn more rapidly in an oxygen-enriched atmosphere than in a normal atmosphere, the careful selection of materials for use in association with an oxygen-enriched atmosphere can do much to reduce the fire hazard.

Combustible Liquids and Gases: For combustible liquids and gases, the potential fire or explosion hazard may be defined in part by the temperature required for the formation of flammable mixtures, the temperature and energy requirements for ignition of the mixture, and the critical fuel concentration (limits) for flame propagation. Such information is presented in Table 12-3B for various representative liquid and gas combustibles in oxygen or air atmospheres, or in both.

The minimum autoignition temperatures (AITs) of most hydrocarbon fuels, solvents, and anesthetic agents fall between 400 and 1,000°F (204 and 538°C) in air at one atmosphere pressure. (See Table 12-3B.) Although the autoignition temperature tends to be lower in oxygen than in air, the differences are not great for many hydrocarbons. In the case of lubricants and hydraulic fluids, however, the effect of oxygen concentration on the AIT tends to be greater than observed for the neat hydrocarbons in Table 12-3B. Figure 12-3E shows that the AITs for five of the hydraulic fluids decrease between 200 and 300°F (93 and 149°C) when the oxygen concentration is increased from 21 to 100 percent.

A correlation of AIT with oxygen partial pressure is shown in Figure 12-3F for several combustible fluids at various pressures and oxygen concentrations.

The limits of flammability are of interest also. Although most lower limits in oxygen do not differ greatly from those in air, the upper limits are usually much higher in oxygen, and tend to be above 50 percent for many materials. Furthermore, some materials, such as the halogenated materials bromochloromethane and dibromodifluoromethane, are flammable in oxygen over a wide range of mixture compositions, whereas they are not reported as being flammable in air. Of the halogenated solvents, trichloroethylene displays the widest range of flammability in both air and oxygen, although elevated temperatures are necessary.

Nonflammable or inert gases are frequently used for explosion prevention. In general, most flammable liquids and gases can be expected to form flammable mixtures over a wide range of oxygen or oxygen-diluent concentrations. Nitrogen is more effective than helium, but not as effective as carbon dioxide or water vapor. Figures 12-3G and 12-3H show the complete range of flammable mixture compositions that may be expected with a hydrocarbon (cyclopropane) in air or oxygen, and various inert gases at atmospheric pressure.

The minimum oxygen percentage below which most hydrocarbon mixtures are not flammable is about 14 percent with carbon dioxide, and 10 to 12 percent with nitrogen. Corresponding values for hydrogen and carbon monoxide are 6 percent and 5 to 5.5 percent respectively.

Combustible Solids: For combustible solids, the ignition and flammability data for any given material are dependent on many variables, e.g., the specimen's physical characteristics, the ignition source, the orientation of the specimen, the environmental characteristics, ventilation

TABLE 12-3B. Ignition and Flammability Properties of Combustible Liquids and Gases in Air and Oxygen at Atmospheric Pressure

Combustible	Flash Point* Air °F	Flash Point* Air °C	Min. Ign. Temperature† Air °F	Min. Ign. Temperature† Air °C	Min. Ign. Temperature† Oxygen °F	Min. Ign. Temperature† Oxygen °C	Min. Ign. Energy‡ Air mJ	Min. Ign. Energy‡ Oxygen mJ	Flammability Limits§ Vol. % Air LFL	Flammability Limits§ Vol. % Air UFL	Flammability Limits§ Vol. % Oxygen LFL	Flammability Limits§ Vol. % Oxygen UFL
				Hydrocarbon Fuels								
Methane	Gas	Gas	1166	630	—	—	0.30	0.003	5.0	15	5.1	61
Ethane	Gas	Gas	959	515	943	506	0.25	0.002	3.0	12.4	3.0	66
n-Butane	−76	−60	550	288	532	278	0.25	0.009	1.8	8.4	1.8	49
n-Hexane	25	−3.8	437	225	424	218	0.288	0.006	1.2	7.4	1.2	52*
n-Octane	56	13.3	428	218	406	208	—	—	0.8	6.5	≤0.8	—
Ethylene	Gas	Gas	914	490	905	485	0.07	0.001	2.7	36	2.9	80
Propylene	Gas	Gas	856	458	793	423	0.28	—	2.4	11	2.1	53
Acetylene	Gas	Gas	581	305	565	296	0.017	0.0002	2.5	100	≤2.5	100
Gasoline (100/130)	−50	−45.5	824	440	600	316	—	—	1.3	7.1	≤1.3	—
Kerosene	100	37.8	440	227	420	216	—	—	0.7	5	0.7	—
				Anesthetic Agents								
Cyclopropane	Gas	Gas	932	500	849	454	0.18	0.001	2.4	10.4	2.5	60
Ethyl Ether	−20	−28.8	380	193	360	182	0.20	0.0013	1.9	36	2.0	82
Vinyl Ether	<−22	−30	680	300	331	166	—	—	1.7	27	1.8	85
Ethylene	Gas	Gas	914	490	905	485	0.07	0.001	2.7	36	2.9	80
Ethyl Chloride	−58	−50	961	516	874	468	—	—	4.0	14.8	4.0	67
Chloroform				Nonflammable								
Nitrous Oxide				Nonflammable								
				Solvents								
Methyl Alcohol	54	12.2	725	385	—	—	0.14	—	6.7	36	≤6.7	93
Ethyl Alcohol	55	12.8	689	365	—	—	—	—	3.3	19	≤3.3	—
n-Propyl Alcohol	59	15	824	440	622	328	—	—	2.2	14	≤2.2	—
Glycol	232	111.1	752	400	—	—	—	—	3.5*	—	≤3.5	—
Glycerol	320	160	698	370	608	320	—	—	—	—	—	—
Ethyl Acetate	24	−4.4	800	427	—	—	0.48	—	2.2	11	≤2.2	—
n-Amyl Acetate	76	24.4	680	360	453	234	—	—	1.0	7.1	≤1.0	
Acetone	0	−17.7	869	465	—	—	1.15	0.0024	2.6	13	≤2.6	60#
Benzene	12	−11.1	1040	560	—	—	0.22	—	1.3	7.9	≤1.3	30
Naphtha (Stoddard)	~100	37.8	~450	232	~420	216	—	—	1.0	6	≤1.0	—
Toluene	40	4.4	896	480	—	—	2.5	—	1.2	7.1	≤1.2	—
Butyl Chloride	20	−6.6	464	240	455	235	0.332	0.007#	1.8	10	1.7	52
Methylene Chloride	—	—	1139	615	1123	606	—	0.137	15.9#	19.1#	11.7*	68
Ethylene Chloride	56	13.3	889	476	878	470	2.37	0.011#	6.2	16	4.0	67.5
Trichloroethane	—	—	856	458	784	418	—	0.092	6.3#	13#	5.5*	57#
Trichloroethylene	90	32.2	788	420	745	396	—	18*	10.5*	41*	7.5	91#
Carbon Tetrachloride				Nonflammable								
				Miscellaneous Combustibles								
Acetaldehyde	−17	−27.2	347	175	318	159	0.38	—	4.0	60	4.0	93
Acetic Acid	104	40	869	465	—	—	—	—	5.4#	—	≤5.4	—
Ammonia	Gas	Gas	1204	651	—	—	>1000	—	15.0	28	15.0	79
Aniline	168	75.5	1139	615	—	—	—	—	1.2#	8.3	≤1.2	—
Carbon Monoxide	Gas	Gas	1128	609	1090	588	—	—	12.5	74	≤12.5	94
Carbon Disulfide	−22	30	194	90	—	—	0.015	—	1.3	50	≤1.3	—
Ethylene Oxide	<0	<−17.8	804	429	—	—	0.062	—	3.6	100	≤3.6	100
Propylene Oxide	−35	−37.2	—	—	—	—	0.14	—	2.8	37	≤2.8	—
Hydrogen	Gas	Gas	968	520	752	400	0.017	0.0012	4.0	75	4.0	95
Hydrogen Sulfide	Gas	Gas	500	260	428	220	0.077	—	4.0	44	≤4.0	—
Bromochloromethane	—	—	842	450	694	368	—	—	NF‖	NF	10.0	85
Bromotrifluoromethane	Gas	Gas	>1100	>593	1215	657	—	—	NF	NF	NF	NF
Dibromodifluoromethane	Gas	Gas	930	499	847	453	—	—	NF	NF	29.0	80

* (NFPA 325m; Humphrey & Morris 1957); open cup method.
† (Litchfield et al 1966; Perlee et al 1966; Scott et al 1948; Zabetakis 1965).
‡ (Lewis & Von Elbe 1961; NACA 1957; Blane et al 1947; Calcote 1952; Litchfield et al 1966).
§ (Fenn 1951; Litchfield et al 1966; Perlee et al 1966; Zabetakis 1965; NFPA 491m).
# Data at 200°F (93.3°C).
‖ NF—No flammable mixtures found in Perlee, 1966.

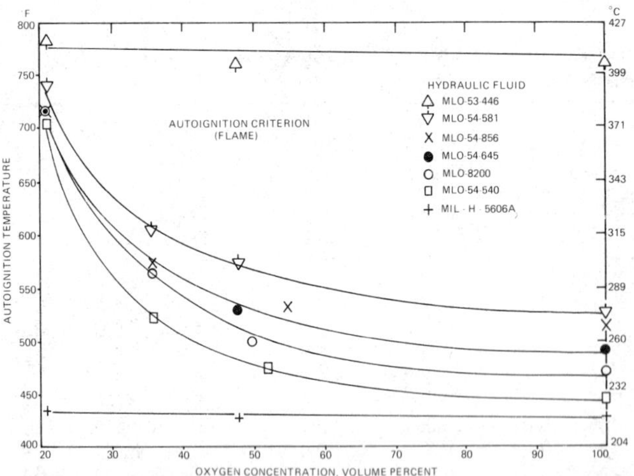

FIG. 12-3E. Minimum autoignition temperatures of seven hydraulic fluids at atmospheric pressure in various oxygen-nitrogen atmospheres (200-cc Pyrex vessel)(Scott et al 1948).

characteristics, inerting diluents, etc. In general, however, ignition temperatures are lower and flame resistance is less in an oxygen-enriched atmosphere than in a normal atmosphere. (See Tables 12-3C and 12-3D.)

Additionally, the rate of flame spread increases with increase in the oxygen concentration at constant pressure or with increase in the total pressure at a constant percentage of oxygen (increased oxygen partial pressure). Various

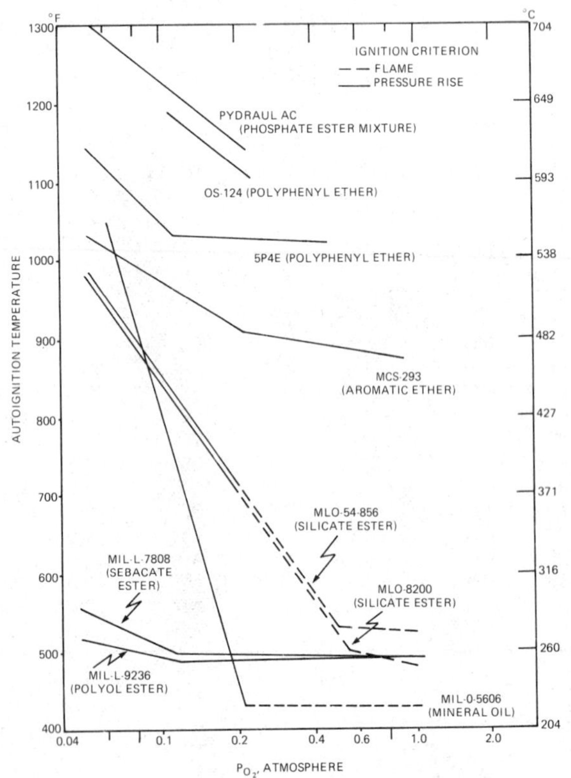

FIG. 12-3F. Variation of minimum autoignition temperature with oxygen partial pressure (PO_2) for various lubricants (Scott et al 1948).

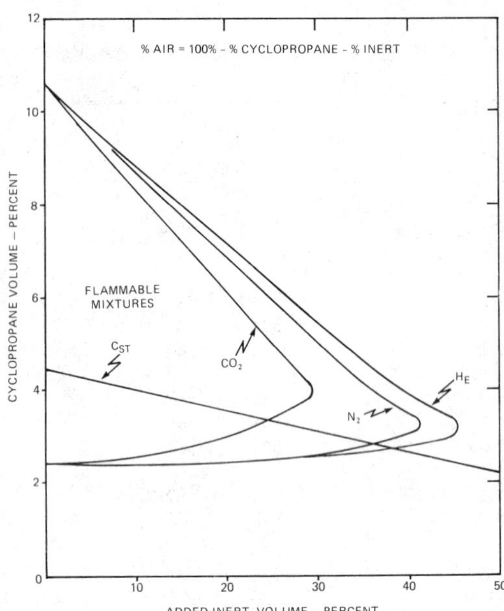

FIG. 12-3G. Limits of flammability of cyclopropane-carbon dioxide-air, cyclopropane-nitrogen-air, and cyclopropane-helium-air mixtures at 25°C and atmospheric pressure (Zebetakis, no date). (Cst = Stoichiometric composition = line defining amount of combustible vapor required for complete combustion.)

flame spread rate data are given in Tables 12-3E, 12-3F and 12-3G.

The rate at which flame spreads under a given set of circumstances in an OEA is the most important single property of a solid material from the fire hazard point of view. Several methods for determining flame spread rate

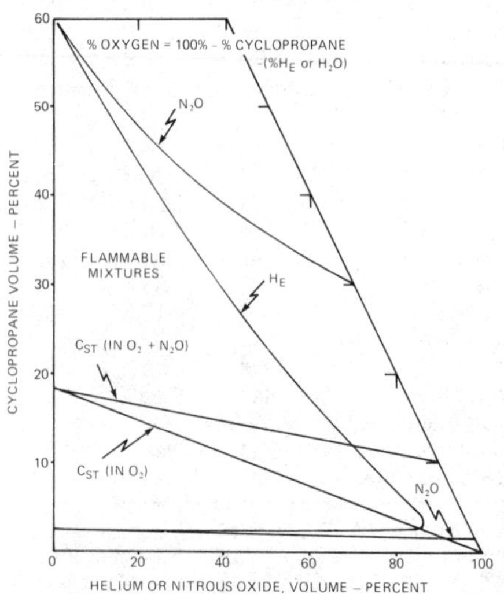

FIG. 12-3H. Limits of flammability of cyclopropane-helium-oxygen and cyclopropane-nitrous oxide-oxygen mixtures at 25°C and atmospheric pressure (Zabetakis, no date). (Cst = Stoichiometric composition = line defining amount of combustible vapor required for complete combustion.)

TABLE 12-3C. Minimum Hot Plate Ignition Temperatures of Six Combustible Materials in Oxygen-Nitrogen Mixtures at Various Total Pressures*

Material	Oxidant	Ignition Temperature, °C§ — Total pressure: 1 Atm	2 Atm	3 Atm	6 Atm
Cotton sheeting	Air	465	440(425)‡	385	365
	42% O_2, 58% N_2	390	370	355	340
	100% O_2	360	345	340	325
Cotton sheeting treated†	Air	575	520(510)	485(350)	370(325)
	42% O_2, 58% N_2	390(350)	335	315	295
	100% O_2	310	—	300	285
Conductive rubber sheeting	Air	480	395	370	375
	42% O_2, 58% N_2	430	365	350	350
	100% O_2	360	—	345	345
Paper drapes	Air	470	455	425	405
	42% O_2, 58% N_2	430	—	400	370
	100% O_2	410	—	365	340
Nomex fabric	Air	>600	>600	>600	560
	42% O_2, 58% N_2	550	540	510	495
	100% O_2	520	505	490	470
Polyvinyl chloride sheet	Air	>600	—	495	490
	42% O_2, 58% N_2	575	—	370	350
	100% O_2	390	—	350	325

* (Kuchta et al 1967).
† Cotton sheeting treated with Du Pont X-12 fire retardant; amount of retardant equal to 12 percent of cotton specimen weight.
‡ Values in parentheses indicate temperature at which material glowed.
§ 1°C = 9/5 × °C + 32

TABLE 12-3D. Flame Resistance of Materials Held Vertically at One Atmosphere Pressure in O_2/N_2 Mixtures*

NRL Sample Number	Material	Combustion in O_2/N_2 Mixtures: 21% O_2	31% O_2	41% O_2
FM-1	Rosin-impregnated paper	Burned	—	—
FM-3	Cotton terry cloth	Burned	—	—
FM-28	Cotton cloth, white duck	Burned	—	—
FM-4	Cotton terry cloth, Roxel-treated	No	No	Burned
FM-5	Fleece-backed cotton cloth, Roxel-treated	Surface only	Burned	Burned
FM-14	Cotton, O.D. Sateen, Roxel-treated	No	Burned	—
FM-15	Cotton, green whipcord, Roxel-treated	No	Burned	—
FM-16	Cotton, white duck, Roxel-treated	No	Burned	—
FM-17	Cotton, King Kord, Roxel-treated	No	Burned	—
FM-29	Cotton white duck, treated with 30% boric acid-70% borax	No	Burned	Burned
FM-30	Cotton terry cloth, treated with 30% boric acid-70% borax	No	Burned	Burned
FM-6	Fire-resistant cotton ticking	No	Burned	—
FM-7	Fire-resistant foam rubber	No	No	Burned
FM-9	Nomex temperature-resistant Nylon	No	Burned	—
FM-10	Teflon fabric	No	No	No
FM-11	Teflon fabric	No	No	No
FM-12	Teflon fabric	No	No	No
FM-13	Teflon fabric	No	No	No
FM-19	Verel fabric	No	Burned	Burned
FM-22	Vinyl-backed fabric	No	Burned	Burned
FM-23	Omnicoated Du Pont high-temperature fabric	No	Burned*	Burned
FM-24	Omnicoated glass fabric	No	No	Burned†
FM-20	Glass fabric, fine weave	No	No	No
FM-21	Glass fabric, knit weave	No	No	No
FM-25	Glass fabric (coarse weave)	No	No	No
FM-26	Glass fabric (coarse weave)	No	No	No
FM-27	Aluminized asbestos fabric	No	No	Burned
FM-32	Rubber from aviator oxygen mask	Burned	Burned	Burned
FM-33	Fluorolube grade 362	No	No	No‡
FM-34	Belco no-flame grease	No	No	No‡

* (Johnson & Woods 1966). † Burned only over igniter. ‡ White smoke only.

TABLE 12-3E. Effect of Oxygen and Storage in Oxygen on Flame Spread Rates over Various Materials (Edges not Inhibited)*

Material†	Flame spread rate (in./sec.#)		
	In air	In 258 mm. Hg. oxygen	
		Before storage	After 30 day storage
Butyl rubber	0.006	0.40 ± 0.04	0.31 ± 0.04
Canvas duck	NP‡	0.25 ± 0.05	—
Cellulose acetate	0.012	0.28 ± 0.12	0.24 ± 0.12
Kel-F	NI	NI	NI
Natural rubber	0.010	0.61 ± 0.05	0.61 ± 0.08
Neoprene rubber	NI	0.32 ± 0.04	0.25 ± 0.05
Nylon 101	NI	0.19 ± 0.05	0.15 ± 0.01
Plexiglas	0.005	0.35 ± 0.01	0.24 ± 0.01
Polyethylene	0.014	0.25 ± 0.05	0.36 ± 0.06
Polypropylene	0.010	0.35 ± 0.01	0.36 ± 0.12
Polystyrene	0.032	0.80 ± 0.20	0.51 ± 0.01
Polyvinyl chloride	NI§	0.10 ± 0.01	0.06 ± 0.01
Silicone rubber	NI	0.14 ± 0.01	0.14 ± 0.01
Teflon	NI	NI	NI
Viton A	NI	0.003 ± 0.002	0.01 ± 0.005

* (Hugget et al 1965).
† All samples except canvas duck, 3 by ½ by ⅕ in (76.2 × 12.7 × 5 mm); canvas duck, 3 by ½ by 1/20 in. (76.2 × 12.7 × 1.3 mm).
‡ NP—No sustained propagation of flame.
§ NI—No ignition of material.
# 1 in./sec = 25.4 mm/sec.

characteristics of materials are in use including NFPA 255, *Standard Method of Test of Surface Burning Characteristics of Building Materials*, and the following American Society for Testing and Materials methods:

1. ASTM E162-67, *Test for Surface Flammability of Materials Using a Radiant Heat Source.*
2. ASTM D568-72, *Test for Flammability of Flexible Plastic.*
3. ASTM D1230-61, *Test for Flammability of Clothing Textiles.*
4. ASTM D2863-74, *Test for Flammability of Plastics Using the Oxygen Index Method.*

Selected materials are tested under conditions of intended use prior to utilization in oxygen-enriched atmospheres.

DESIGN OF SYSTEMS FOR OEAs

Fire hazard considerations in the design of systems associated with oxygen-enriched atmospheres, e.g., heating, ventilating and air conditioning, hydraulic services, gas and compressed air supplies, suction apparatus, both power and electronic electrical systems, etc., include:

1. Characteristics of material of construction (flame spread rate, ignition susceptibility).
2. Risk of fire initiation (expectancy).
3. Presence of potential energy, in the form of compressed gas, etc.
4. Personnel escape paths from occupied areas.

In addition to evaluating the firesafe characteristics of all materials involved in oxygen-enriched environments under end-use performance conditions, oxygen service durability is evaluated by accelerated time tests for deterioration, and by high energy tests for degradation.

The acceptability of candidate materials for oxygen-enriched atmosphere systems applications are, in part, based on data developed by tests such as: (1) flash and fire points; (2) upward/downward combustion propagation rates; (3) calorific fuel values (heat of combustion); (4) electrical wire insulation, coating, and accessory flammability tests; (5) electrical overload and hot-wire ignition tests; and (6) odor toxicity, off-gassing, etc.

Electrical equipment used in oxygen-enriched atmospheres is limited to that which is approved at the maximum anticipated oxygen pressure and concentrations. Since most metals burn freely in oxygen-enriched atmospheres, electrical contacts which do not burn in normal atmospheres could burn away and initiate insulation fires in an oxygen-enriched atmosphere.

Some "fire stopping" techniques which may be considered in the design of a system to minimize ignition potential and fire spread in an oxygen-enriched atmosphere are:

1. Avoidance of mass concentration of combustible materials near potential heat or ignition sources.
2. Spatial separation and configuration to minimize or eliminate flame propagation paths.
3. "Thermal damping" by judicious placement of fire resistant heat sink masses.
4. Flashover barriers.
5. Sealed packaging, e.g., inerted compartmentation, fire resistant encapsulation, etc.
6. Automatic fire monitoring, e.g., infrared thermography, etc.

FIRE EXTINGUISHMENT IN OEAs

Fire extinguishing systems for use in oxygen-enriched atmospheres face new requirements in addition to those imposed on conventional systems because of ignition susceptibility, increased flame spread rate, increased burning intensity, and flammability of normally nonflammable materials. In general, these new requirements cannot be satisfied by the simple extension of classic extinguishment techniques. In addition to special instruction and training of emergency personnel, extinguishing agents and systems must be specially selected.

In view of the increased burning rates of most common materials in oxygen-enriched atmospheres, fire extinguishants should act rapidly to be effective. To protect occupants of affected areas, they should be inherently nontoxic and should not produce significant amounts of toxic decomposition products.

Water has been shown to be an effective extinguishing agent in oxygen-enriched atmospheres when applied in sufficient quantities (Denison & Crosswell 1965; Botteri 1967). Water at a spray density of 50 (L/min)/m^2 (1¼ gpm per sq ft) applied for 2 min will extinguish cloth burning in 100 percent oxygen at a pressure of 1 atm. (101.3 kPa) (Denison & Cresswell 1965). The method of application of the water to the fire is all-important. Extinguishing systems using water must be carefully designed so that all the protected space is covered by the minimum spray density and distributed to a depth sufficient to extinguish stratified fires in nonhomogeneous materials, e.g., layers of cloth in clothing.

TABLE 12-3F. Effect of Oxygen and Storage in Oxygen on Flame Spread Rates over Various Space Cabin Materials (Edges not Inhibited)*

Material†	Flame spread rate (in./sec.)		
	In air	In 258 mm. Hg. oxygen	
		Before storage	After 30 day storage
Aluminized Mylar tape	—	1.95	—
Aluminized vinyl tape	NI†	3.1 ± 0.4	3.0 ± 0.4
Asbestos insulating tape	NI	0.08	0.05
Chapstick	NI	1.82	—
Cotton shirt fabric	NP‡	1.50 ± 0.05	2.10 ± 0.3
Electrical insulating resin	NI	0.27	0.20
Electrical terminal board	NI	0.06 ± 0.01	0.06 ± 0.01
Fiberglas insulating tape	NI	4.2 ± 0.6	2.0 ± 0.8
Foam cushion material	0.19	12.4	11.3
Foamed insulation	0.002	2.2 ± 0.2	3.0 ± 0.3
Food packet, aluminized paper	NI	0.28 ± 0.05	0.26 ± 0.05
Food packet, brown aluminum	NI	0.7 ± 0.30	0.8 ± 0.20
Food packet, plastic	0.33	0.55	0.47
Glass wool	NI	NI	NI
Masking tape	0.17	1.82	—
Paint, 3-M velvet	NI	0.15 ± 0.01	0.31 ± 0.02
Paint, Capon ivory	NI	0.38 ± 0.04	0.35 ± 0.02
Paint, Pratt & Lambert, grey	NI	0.60 ± 0.2	0.24
Pump oil	NI	0.89	—
Refrigeration oil	NI	0.82 ± 0.07	—
Rubber tubing	0.03	0.24	0.25 ± 0.05
Silicone grease	NI	0.92	—
Solder, rosin core	NI	0.18	0.25
Sponge, washing	0.07	8.1 ± 0.1	10 ± 2
Teflon pipe sealing tape	NI	NI	NI
Teflon tubing	NI	NI	NI
Tygon tubing	0.18	0.50 ± 0.05	0.52 ± 0.05
Wire, Mil W76B, orange	NI	0.57 ± 0.05	0.54
Wire, Mil W76B, blue	NI	—	0.57
Wire, Mil W76B, yellow	NI	—	0.54
Wire, Mil W16878, green	NI	NI	NI
Wire, Mil W16878, black	NI	NI	NI
Wire, Mil W16878, yellow	NI	NI	NI
Wire, Mil W16878, white	NI	NI	NI
Wire, misc., white, 3/32	NI	0.33	0.25
Wire, misc., black, 3/16	NI	—	0.40
Wire, misc., brown, 7/32	NI	0.51 ± 0.05	—
Wire, misc., yellow, 7/64	NI	0.89	—
Wire, misc., yellow, 5/32	NI	0.41	—

* (Hugget et al 1965). †NI—No ignition of material. ‡NP—No sustained propagation of flame.

Several halogenated hydrocarbons have been found useful for extinguishing fires in normal atmospheres. Table 12-3H indicates their behavior in pure oxygen as well as their toxicity ratings as given by Underwriters Laboratories Inc. (Toxicity ratings for undecomposed vapor range from 1, the most toxic, to 6, the least hazardous).

Bromotrifluoromethane (Halon 1301) appears to be useful in oxygen-enriched atmospheres. Testing has shown that it will extinguish fires in pure oxygen when used at 50 percent concentration, and in compressed air at 5 percent by volume (Eggleston 1970). Generally, it is effective against few truly deep seated fires in oxygen-enriched atmospheres. NFPA 12A, *Standard on Halon 1301 Fire Extinguishing Systems*, states that concentrations greater than 10 percent are not used in normally occupied areas, and, even then, are used only where evacuation can be accomplished immediately.

Generally, halogenated agents are not recommended against fires that involve highly reactive metals or fuels that contain their own oxidizing agent. Such metals as magnesium and aluminum are known to present an increased fire or explosion hazard with certain halogenated hydrocarbons. The hazard posed by these materials will depend upon their quantity, distribution, and physical form, as well as the composition of the extinguishant; the presence of other combustibles can cause a further complication. All such factors must be weighed by the designer in making any recommendation.

Available data regarding the effectiveness of carbon dioxide is inconclusive.

High expansion foam has been shown to be effective and that respiration in the foam atmosphere is possible (Charno 1969). However, the applicability of this agent to

TABLE 12-3G. Typical Measured Burning Rates for Strips of Filter Paper at 45° Angle*

			Burn Rate, cm/sec					
Total pressure		Atm. abs. / Ft. of sea water	0.21 Atm. —	0.53 Atm. —	1.00 Atm. 0 ft.	4.03 Atm. 100 ft.	7.06 Atm. 200 ft.	10.09 Atm. 300 ft.
Gas Composition (dry basis)								
$\%O_2$	$\%N_2$†	%He						
99.6	0.4	0.0	2.32	3.13	4.19	#	#	#
50.3	49.7	0.0	1.13	1.44	2.36	3.72	5.10	6.34
			1.17			3.77	4.06	
20.95‡	79.05	0.0	§	0.80	1.17	1.82	2.80	3.13
					1.17	1.78	2.28	3.25
					1.10			
49.5	0.0	50.5	1.24	1.87	2.96	4.06	4.90	#
				1.90	2.89		4.82	
					2.89			
20.3	0.0	79.7	§	§	§	2.23	2.61	2.49
47.0	24.6	28.4	#	#	2.74	3.66	4.41	5.53
					2.68		4.64	6.78
20.9	39.6	39.5	#	#	1.38	2.28	2.71	3.72
					1.38	2.28	2.83	3.13
					1.35	1.97	2.74	3.56
					1.27	2.28		3.33
						1.81		3.00
						1.72		

* (Cook et al 1967).
† Includes any argon that was present.
‡ Compressed air.
§ Sample would not burn, even with brightly glowing igniter grid.
# No run was made under these conditions.

TABLE 12-3H. Behavior of Halons in Pure Oxygen

Compound	Halon No.	Toxicity	Behavior in Pure Oxygen*
Carbon tetrachloride, CCl_4	104	3	
Methyl bromide, CH_3Br	1001	2	Flammable
Chlorobromomethane, CH_2BrCl	1011	3	Flammable, decomposes
Dibromodifluoromethane, CBr_2F_2	1202	4	Flammable
Bromochlorodifluoromethane, $CBrClF_2$	1211	5	Nonflammable over tested concentration (14–95%)
Bromotrifluoromethane, CF_3Br	1301	6	Nonflammable over tested range (12–98%)

* (Perlee et al 1966)

each particular candidate system must be evaluated separately with regard to available space and required time.

Little or no data is available on the use of low-expansion foam or dry chemical in oxygen-enriched atmosphere fires. No reliance should be placed on ordinary fire blankets made of wool or asbestos (with an organic binder).

Because of the rapid flame spread in oxygen-enriched atmospheres, fire extinguishing systems must be capable of fast automatic actuation by fire detectors, as well as by manual actuation. Fixed systems utilize an extinguishing agent acceptable for use on fires in oxygen-enriched atmospheres and automatic activation occurs in less than one second of the perception of sensible flame development. In addition to the automatic fixed system in occupied areas, a manually operated water hose not less than ½ in. (12.7 mm) ID, and with an effective nozzle pressure not less than 50 psi (345 kPa) above the ambient pressure, is usually provided.

Bibliography

References Cited

Blanc, M. V., et al. 1947. "Ignition of Explosive Gas Mixtures by Electrical Sparks—Minimum Ignition Energies and Quenching Distances of Methane, Oxygen and Inert Gases." *Journal of Chemical Physics.* Vol 15. p 798.

Botteri, B. P. "Fire Protection in Oxygen Enriched Atmospheres." Fire Journal. Vol 62, No 1. Jan 1967. pp. 48-55.

Calcote, H. F., et al. 1952."Spark Ignition." *Industrial and Engineer Chemistry.* Vol 44, No 11. Nov 1952. p 2656.

Charno, R. J. 1969. *Evaluation of High Expansion Foam for Spacecraft Fire Extinguishment.* N69-20776. Feb 3, 1969. E. W. Bliss Co., Swarthmore, PA, prepared for NASA/JSC (Johnson Space Center).

Cook, G. A., Meierer, R. E., and Shields, B. M. 1967. "Screening of Flame Resistant Materials and Comparison of Helium with Nitrogen for Use in Diving Atmospheres." *First Summary Report on Combustion Safety in Diving Atmosphere.* Mar

31, 1967. U.S. Department of the Navy. Office of Naval Research and Naval Ship Systems Command, Washington, DC.

Denison, D., and Cresswell, A. W. 1965. "The Fire Risks to Man of Oxygen and Rich Gas Environments." *IAM Report No 320*. Apr 1965 and *Report No 343*. Sept 1965. Civil Aeromedical Institute. Aeronautical Center, Oklahoma City, OK.

Dorr, V. A. 1970. "Fire Studies in Oxygen-Enriched Atmospheres." *Journal of Fire and Flammability*. Vol 1. pp 91-106.

Eggleston, L. A. 1970. "Fire Safety in Hyperbaric Chambers." *Fire Technology*. Vol 6, No 4. Nov 1970.

Fenn, J. B. 1951. "Lean Flammability Limit and Minimum Spark Ignition Energy." *Industrial and Engineering Chemistry*. Vol 43, No 12. Dec. 1951. p 2865.

Hugget, C., et al. 1965. "The Effects of 100% Oxygen at Reduced Pressure on the Ignitibility and Combustibility of Materials." *SAM-TR-65-78*. Dec 1965. Brooks Air Force Base, TX.

Humphrey, H. B., and Morgis, G. 1957. "Safety with Solvents." *USDI Bureau of Mines Information Circular 7757*. Washington, DC.

Lewis, B., and Von Elbe, G. 1961. *Combustion Flames and Explosion of Gases*. Academic Press, NY. pp 323-346.

Litchfield, E. L., Kuchta, J. M., and Furno, A. L. 1966. "Flammability of Propellant Combinations." *USDI Bureau of Mines Explosives Research Report 3997*. Oct 30, 1966. Washington, DC. *USDI Bureau of Mines Research Report 3958*. June 30, 1965. Washington, DC.

NACA. 1957. "Basic Considerations in the Combustion of Hydrocarbon Fuels with Air." *NACA Report 1300*. National Advisory Committee for Aeronautics, Washington, DC.

Perlee, H. E., Martindill, G. H., and Zabetakis, M. G. 1966. "Flammability Characteristics of Selected Halogenated Hydrocarbons." *USDI Bureau of Mines RI 6748*, Washington, DC.

Scott, G. S., Jones, G. W., and Scott, F. E. 1948. "Determination of Ignition Temperatures of Combustible Liquids and Gases." *Analytical Chemistry*. Vol 20. Mar 1948. p 238.

Zabetakis, M. G. 1965. "Flammability Characteristics of Combustible Gases and Vapors." *USDI Bureau of Mines Bulletin 627*, Washington, DC.

NFPA Codes, Standards, Recommended Practices and Manuals. (See the latest *NFPA Codes and Standards Catalog* for availability of current editions of the following documents.)

NFPA 12A, *Standard for Halon 1301 Fire Extinguishing System*.

NFPA 50, *Standard for Bulk Oxygen Systems at Consumer Sites*.

NFPA 51, *Standard for the Design and Installation of Oxygen-Fuel Gas Systems for Welding, Cutting and Allied Processes*.

NFPA 53M, *Manual on Fire Hazards in Oxygen-Enriched Atmospheres*.

NFPA 56F, *Standard for Nonflammable Medical Gas Systems*.

NFPA 99, *Standard for Health Care Facilities*.

NFPA 255, *Method of Test of Surface Burning Characteristics of Building Materials*.

NFPA 325M, *Fire Hazard Properties of Flammable Liquids, Gases, and Volatile Solids*.

NFPA 491M, *Manual on Hazardous Chemical Reactions*.

Additional Readings

Aerospace Medical Division, *Proceedings of Fire Hazards and Extinguishment Conference*, Brooks Air Force Base, TX, May 23, 1967.

Attallah, Sami, and DeRis, John N., "Pressure Rise Due to a Fire in an Enclosure," *Fire Technology*, Vol. 5, No. 2, May 1969, pp. 112-121.

Ault, W. E., and Carter, D. L., "The Influence of Hyperbaric Chamber Pressure on Water Spray Patterns," *Fire Journal*, Vol. 61, No. 6, Nov. 1967, p. 48.

Brenneman, J. J., "Oxygen Induced Cabin Fire," *Fire Journal*, Vol. 65, No. 5, Sept. 1971, pp. 26-28.

Cowardin, H. F., and Jones, G. W., "Limits of Flammability of Gases and Vapors," *USDI Bureau of Mines Bulletin 503*, 1952, Washington, DC.

Eggleston, L. A., "Evaluation of Fire Extinguishing Systems for Use in Oxygen Rich Atmospheres," *Final Report prepared for Aerospace Medical Division*, Brooks Air Force Base, May 1967, Southwest Research Institute, San Antonio, TX.

Fitt, P. W., Collings, N. and O'Neill, D., "The Ignition of Solid Materials in Oxygen by Electrical Sparks," in *Fire Prevention and Suppression*, Hilado, C. J., ed., Technomic, Westport, CT, 1974, pp. 62-73.

Harter, J. V., "The Problem of Fire at High Pressure," *Proceedings of the Third Symposium on Underwater Physiology*, ed. by C. J. Lambertsen, The Williams and Wilkins Company, Baltimore, 1967, pp. 76-77.

Johnson, J. E., and Woods, F. J., "Flammability in Unusual Atmospheres, Part I—Preliminary Studies of Materials in Hyperbaric Atmospheres Containing Oxygen, Nitrogen, and/or Helium," *NRL Report 6470*, Oct. 31, 1966, Naval Research Laboratory, Washington, DC.

Kuchta, J. M., and Cato, R. J., "Review of Ignition and Flammability Properties of Lubricants," *Air Force Aero Propulsion Laboratory Technical Report AFAPL-TR-67-126*. Jan. 1968, Wright Patterson Air Force Base, OH.

Kuchta, J. M., et al., "Flammability of Materials in Hyperbaric Atmospheres," *USDI Bureau of Mines Final Report 4016*, Aug. 30, 1967, Explosive Research Center, Pittsburgh.

Kuchta, J. M., Furno, A. L., and Martindill, G. H., "Flammability of Fabrics and other Materias in Oxygen-Enriched Atmospheres, Part I. Ignition Temperature and Flame Spread Rates," *Fire Technology*, Vol. 5, No. 3, Aug. 1969, pp. 203-216.

Litchfield, E. L., and Kubala, T. A., "Flammability and Ignition of Fabrics and Other Materials in Oxygen-Enriched Atmospheres, Part II. Minimum Ignition Energies," *Fire Technology*, Vol. 5, No. 4, Nov 1969, pp. 341-345.

Nakakuki, Atsushi, "Extinction of Fires in Hyperbaric Chambers: Part I, Water Spray Properties and Extinction with Single Spray Nozzles," *Fire Technology*, Vol. 8, No. 1, Feb. 1972, pp. 5-18.

Purser, P. R., "Ignition by Electrostatic Sparks in Hyperbaric Oxygen," *The Lancet*, Dec. 24, 1966.

Segal, L., et al., "Fire Suppression in Hyperbaric Chambers," *Fire Journal*, Vol. 60, No. 3, May 1966, pp. 17-18.

Turner, H. L., and Segal, L., "Fire Behavior and Protection inHyperbaric Chambers," *Fire Technology*, Vol. 1. No. 4, Nov. 1965. pp. 269-277.

Widawsky, Arthur, "Fire Protection System for a Hyperbaric Chamber," *Fire Technology*, Vol. 9, No. 2, May 1973, pp. 85-90.

Wilson, Rexford,, and Ledoux, Edward F., "High-Speed Protection for Personnel in Oxygen-Enriched Atmospheres," *Fire Journal*, Vol. 62, No. 2, Mar. 1968, pp. 23-25.

MEDICAL GASES

Revised by Burton Klein

The use of flammable gases in a hospital represents a fire hazard which is different from that of any other type of occupancy. Many patients cannot be moved easily, even under ideal conditions, because of their dependency upon life-support equipment that is not readily portable. The combination of a flammable gas and an immobile patient can be lethal.

The introduction of nonflammable anesthetic agents shortly after World War II has been offset, in part, by the increased use of oxygen, the proliferation of throwaway items made of paper and plastic, and an ever increasing reliance on electrical appliances. Thus, while practices have changed, the hospital still represents a firesafety problem that is unique in modern society. This section discusses those parts of the hospital firesafety problem related to the storage and use of medical gases, both flammable and nonflammable.

It should also be noted that the hazards and safe practices discussed in this section apply to other health care facilities that use medical gases (e.g. ambulatory health care centers, nursing homes, dental offices, etc.

Other information of interest to those concerned with the problems and hazards of medical gases and equipment will be found in Section 5, Chapter 1, "Fire Hazards of Materials;" Chapter 4, "Flammable and Combustible Liquids;" Chapter 5, "Gases;" Section 11, Chapter 4, "Storage and Handling of Flammable and Combustible Liquids;" Chapter 5, "Storage and Handling of Gases;" in this section, Chapter 2, "Protection for Laboratories;" Chapter 3, "Oxygen-Enriched Atmospheres;" Chapter 10, "Control of Electrostatic Ignition Sources."

ANESTHETIC GASES

The introduction of diethyl ether as an anesthetic in 1846 led to dramatic changes in surgical practice. Its benefits were so obvious that general anesthesia rapidly became a technique thoroughly accepted by all segments of the medical world. Yet diethyl ether had many drawbacks (including flammability) and the search for better agents led to the development and introduction of others.

Initially, these agents were volatile liquids, i.e., agents which are stored as liquids and converted to a gas in a liquid vaporizer. With the advent of more modern, practical methods for compressing, storing, and delivering gases in regulated amounts, the use of the so-called true gases (e.g., nitrous oxide, cyclopropane) became popular.

The enrichment of the anesthetic atmosphere with oxygen (a step nearly simultaneous with the introduction of nitrous oxide) allowed early anesthetists to give patients high concentrations of anesthetic agents while at the same time maintaining adequate levels of oxygenation. The practice of using oxygen-enriched atmospheres as a component of inhalation anesthesia has persisted to this day.

The Flammability of Anesthetics

Initially, most of the anesthetic agents were flammable. The careless use of these agents, through either ignorance or disregard of safety precautions, led to a number of fires, explosions, injuries, and fatalities during the early days of anesthesia. While ether mixed with air will only burn briskly, ether mixed with oxygen generally will explode violently. Cyclopropane-oxygen mixtures create even more violent explosions since this mixture frequently attains stoichiometric proportions.

Although most anesthesiologists and anesthetists have abandoned the use of flammable agents, they are still used by some. The nonflammable agents currently available possess certain toxic properties not shared by cylcopropane or ether, and many anesthetists consider them unusable for certain types of patients. Thus flammable inhalation agents will be used in hospitals until alternatives are acceptable in all situations.

Flammable agents include cyclopropane and ethylene (true gases) and diethyl ether, divinyl ether, and ethyl chloride (which are volatile liquids). Nonflammable anesthetic agents include nitrous oxide (a true gas) and chloroform, halothane, methoxyflurane, ethrane, forane, and trichloroethylene (all liquids).

One method of reducing the flammability of a hydrocarbon (diethyl ether, divinyl ether, and cyclopropane are hydrocarbons) is to substitute halogen atoms (usually fluorine or chlorine) for some of the hydrogen atoms. For

Mr. Klein is Health Care Fire Specialist on the staff of NFPA.

example, chloroform is trichloromethane, a nonflammable but toxic agent, made by substituting chlorine atoms for three of the hydrogen atoms in methane.

The search for a safe, nonflammable agent intensified after World War II. Ultimately fluroxene was developed, the first of a number of fluorine containing inhalation anesthetic agents. Fluroxene is minimally flammable because it is partially halogenated and will not burn when used in the usual concentrations of less than four percent. Yet because of its potential hazard, it is still considered a flammable agent.

Following its introduction into clinical practice, fluroxene was joined by halothane, methoxyflurane, and then ethrane, all halogenated inhalation agents with minimal (or absent) flammability. If methoxyflurane, halothane, or trichloroethylene are heated, they will burn. Under the usual conditions of use and the absence of any active heating well above room temperature, however, these agents are considered to be nonflammable, and are so utilized in anesthetic situations.

Neither nitrous oxide nor oxygen will burn, although both support combustion readily. The oxygen in the nitrous oxide will be released under flame conditions, contributing to the oxidation-reduction reaction.

Safe Practices in Using Anesthetics

A consideration of safe practices in the field of anesthesia and ancillary hospital activities begins with an understanding of the basic chemistry and physics of oxygen and the anesthetic gases, as well as the vapors produced by the volatile liquid agents, such as diethyl ether.

A gas consists of molecules which move individually in a linear fashion at high velocities. The molecular movement increases with a rise in temperature, and decreases as the gas cools. This movement produces a parallel increase or decrease in the pressure exerted by that gas, depending upon its temperature. When a gas is compressed, its individual molecules will be forced closer together, thus increasing molecular movement and the temperature of the gas. Gases possess definite mass, but neither definite shape nor definite volume (i.e., they always assume the shape and volume of their containers, if any).

The relationship between pressure, volume, and temperature is given by the various gas laws. From all of these gas laws, one may establish what is known as the general gas law, i.e., the pressure of any given quantity of gas is proportional to its absolute temperature and inversely proportional to its volume. Algebraically, pressure times volume divided by temperature equals a constant. These gas laws assume the gas is ideal, while in actual practice no gas is ideal.

The gaseous anesthetic agents (cyclopropane, ethylene, nitrous oxide, and oxygen) are stored as compressed gases or as liquids under high pressure. They are dispensed as gases through needle valves and flow meters. In contrast to the true gases, the volatile liquid anesthetic agents (diethyl ether, chloroform, halothane) are stored and dispensed as liquids.

Liquids possess definite volume and mass but no definite shape. At any gas-liquid interface, molecules are continually escaping from the liquid state into the gaseous state. The pressure created by the molecules which have escaped into the gaseous state is called vapor pressure. As the temperature of a liquid is continually raised, the vapor pressure also is raised, approaching atmospheric pressure as the temperature of the liquid reaches its boiling point. When the temperature of the liquid reaches its boiling point, molecules will pass freely into the gaseous state and the liquid will boil.

In the operation of a volatile liquid vaporizer during anesthesia the temperature of the liquid never reaches its boiling point. The carrier gases (nitrous oxide and oxygen) which are used to deliver the inhalation anesthetic agents to the patient pass through the vaporizer in a steady flow. These gases carry away to the anesthesia circuit the vaporized molecules of the volatile liquid. In the absence of applied heat, the temperature of the volatile liquid remaining in the reservoir, and that of the reservoir itself, will fall as the volatile liquid vaporizes (the latent heat of vaporization). As the temperature of the remaining liquid falls so does its volatility. Some anesthetic vaporizers are designed to maintain, by one method or another, the temperature of the liquid at or near room temperature to promote even rates of vaporization.

Safety precautions relating to the use of flammable agents have created other hazards. For example, the conductive flooring necessary to reduce the likelihood of the development of static charges in the members of the operating team and the equipment used in the flammable anesthetizing location, led to a definite shock hazard to personnel and patients. This hazard was met by supplying such locations with electrical power from isolation transformers. Isolated power systems also help prevent sparks (and the corresponding explosion hazard) resulting from certain kinds of electrical appliance failures.

Gas Anesthesia Apparatus

Anesthetic agents are administered by a gas anesthesia apparatus. (See Figs. 12-4A and 12-4B.) This device consists of a number of components. First, there must be a source of gas, which may either be a cylinder or a connection to a central piping system. Next, a regulator may be utilized to reduce the pressure from a higher tank pressure to a lower working pressure. The gases are then fed to needle valves which precisely control the flow of the gas to be delivered to the anesthesia circuit. To measure the flow of these gases, each gas is passed through a flowmeter.

The flowmeter most commonly encountered in modern gas anesthesia apparatus is the rotameter, or Thorpe Tube. This is a tapered glass tube which contains a bobbin or rotor which floats in the gas stream. The space between the outside of the rotor and the inside of the glass tube consists of an annular orifice which increases as the rotor rises in the tube. When the gas passes across this orifice, a fall in pressure occurs. The difference between the upstream and downstream pressures is equal to the weight of the rotor or bobbin when the flowmeter is in equilibrium. Each flowmeter is calibrated by first passing that specific gas through a master flowmeter. Flows are indicated by noting the position of the top of the rotor as related to a card affixed beside the tapered tube.

From the flowmeters the gases enter a common mixing manifold. The mixture of gases then will enter the anesthesia circuit. The anesthetic circuit is the portion of the gas machine from which the patient draws the mixture of anesthetic vapors, gases, and oxygen, and into which the patient exhales. During exhalation, a small amount of

FIG. 12-4A. General view of modern anesthesia apparatus. Nitrous oxide tanks are at left, oxygen tanks at right, with flow-meter and pressure gages at top rear. Carbon dioxide absorber (part of the anesthesia circuit) is glass container at front, draped with hoses. Unit is completely self-contained and requires no electric power. Conductive rubber casters are used to prevent accumulation of static electricity.

FIG. 12-4B. The anesthesia circuit of anesthesia apparatus shown in Fig. 12-4A. Glass cannister (at left) is absorber for carbon dioxide. Smaller glass jar (center) is an ether vaporizer. Two corrugated rubber hoses leading to face mask (at right rear) complete the anesthesia circuit via the rebreathing bag (black rubber balloon behind the vaporizer). Mixed anesthetic-oxygen gas is inserted into anesthetic circuit via small diameter hose at left. Waste anesthetic gases are vented to operating room exhaust air plenum via medium diameter hose at upper left.

carbon dioxide, a product of the body's metabolism, enters the anesthesia circuit. Some method must be utilized to remove this waste gas from the anesthetic atmosphere. The most common method is chemical absolution, employing an absorber which contains one-half to one litre of granular soda lime (soda lime is a mixture of sodium and calcium hydroxides, water and a binder).

Valves in the anesthesia circuit allow the patient to breathe in a unidirectional fashion—from the lungs through the absorber to a rebreathing bag, and then back again to the lungs for inhalation. Sometimes the vaporizer is placed in a breathing circuit. At other times it may be located somewhere between the gas delivery of the anesthesia machine and the breathing circuit.

Storage of Flammable Agents and Nitrous Oxide

In any operating suite in which flammable agents are utilized, provisions must be made for storage of the reserve supplies of gases and flammable liquids. Specifications for storage include a room with a conductive floor which is ventilated to the outside with a minimum of six to eight air changes per hour.

This room with its conductive floor must not communicate directly with a room utilized for the storage of oxidizing gases such as nitrous oxide, oxygen, or compressed air. In addition, these oxidizing gases may not be stored in the same location as the volatile liquid anesthetizing agents.

Waste ether should be stored in a safety can with a spring closing lid and an internal screen to prevent propagation of flame into the can. Waste ether should be disposed off premises by disposal specialists or at a safe location away from the health care facility by competent personnel using procedures established in concurrence with the authority having jurisdiction.

Open cans of ether must not be stored in a refrigerator unless it is of the type acceptable for the storage of flammable liquids, e.g. with a sealed thermostat to prevent accidental ignition of any vapors within the refrigerator cabinet. Such refrigerators usually carry the label of a testing laboratory which specifically indicates that the refrigerator is approved for the storage of flammable liquids.

In some flammable anesthetizing locations, ether is used to prep the skin. This is a dangerous practice and is not recommended. There are better nonflammable agents for preparing the skin, such as inhibited III-trichloroethane or trifluorotrichloroethane. Neither of these solvents will burn, and both are good agents to degrease and prepare the patient's skin for plastic and similar draping.

OXYGEN

Oxygen as a Medical Agent

In the early days of inhalation anesthesia, diethyl ether was administered by the drop technique. Poured or dropped on a gauze covered mask, the ether was volatilized in the air of the operating room atmosphere. No oxygen enrichment was utilized. Today the large majority of inhalation anesthetics are administered with oxygen. In the presence of an oxygen-enriched atmosphere, the lower limit of flammability and the ignition temperature of all flammable agents are reduced. Additionally, such a mixture may approach stoichiometric concentrations, i.e., the proportions of oxygen and fuel may be close to those which will allow complete combustion of the flammable substance with neither reactant being present in any significant amount after the oxidation-reduction reaction.

From a practical standpoint, one cannot eliminate oxygen. Oxygen-enriched atmospheres are commonly utilized in anesthesia apparatus to enhance the administration of the anesthetics and promote a wider margin of safety for the patient. Respiratory therapy also utilizes oxygen-enriched atmospheres to a significant degree. Hyperbaric chambers for various forms of hyperbaric medicine also utilize oxygen-enriched atmospheres either in the form of compressed air or compressed oxygen. Thus, recognizing that oxygen-enriched atmospheres are present to varying degrees, fire or explosion prevention in the hospital must revolve about the control of these fuels and ignition sources.

Oxygen Supplies

The principal method of preparing oxygen for medical use is by the compression and liquefaction of air followed by fractional distillation. Once it has been purified, the oxygen may be transported to, and stored at, the site of consumption as a compressed gas or liquid. Because of the economies in bulk oxygen use, practically all medium size to large hospitals rely on a bulk oxygen storage unit on site. NFPA 50, *Standard for Bulk Oxygen Systems at Consumer Sites*, covers safety standards for the design and operation of such bulk oxygen storage units.

Distribution Systems for Medical Gases

Where bulk oxygen is used at a hospital, a piping distribution system is required. A pipeline system also may be employed for nitrous oxide. Oxygen and nitrous oxide pipeline systems, as installed in hospitals, require the use of copper or brass tubing or pipe with screwed, brazed, or high melting point soldered joints. The system must be pressure tested with oil free dry air or nitrogen before it is put to use.

The piping system terminates at each bedside or operating room in station outlets. These are equipped with a threaded or quick connect terminal keyed to the specific gas in the pipe. Because it is extremely difficult to extinguish a fire being supplied by a continuous source of oxygen or nitrous oxide, all piping systems containing an oxidizing gas for use in hospitals (and similar occupancies) require a remote shutoff valve which can shut off the flow of gas to a station outlet involved in a fire.

Whenever a nitrous oxide piping system is to be installed along with an oxygen system, care must be taken to check all station outlets of both systems for delivery of the correct gas before the systems are put in use. Cross connections have been made with injury or death of patients resulting from anoxia.

All specifications for the design, installation, testing, and operation of oxidizing gas pipeline systems for hospitals are covered in NFPA 56F, *Standard for Nonflammable Medical Gas Systems*. Certain Compressed Gas Association publications also cover these requirements.

Station outlets for medical gas supply and central suction systems must be equipped with noninterchangeable fittings, keyed to the specific supply. Medical gas cylinder outlets are also keyed to the gas supplied. The Diameter Index Safety System is utilized for larger size gas cylinders and mating connectors. For small size cylinders, the Pin Index System is employed. This system uses two holes drilled along a radius beneath the outlet orifice of the valve body, and keyed to that gas. The cylinder yokes are equipped with pins which mate with the holes in the valve body of the cylinder containing the correct gas. The holes and pins are located in such a manner that it is impossible to insert a cylinder in the yoke for a different gas.

Gas cylinders are also color coded. The American standard for color coding includes green for oxygen, blue for nitrous oxide, orange for cyclopropane, brown for helium, red for ethylene, and gray for carbon dioxide. Color combinations are available for gas mixtures of certain percentages. However, color-coding is secondary to labeling *very clearly* the contents of a cylinder.

Cylinder weights, sizes, pressures, and contents are shown in Table 12-4A. from Appendix C of NFPA 99, *Standard for Health Care Facilities* (1984 edition), hereafter referred to as NFPA 99. Data for this Appendix was supplied by the Compressed Gas Association.

In some hospitals, nursing homes and ambulatory health care facilities, oxygen may be supplied to the piping system by a manifold to which cylinders are connected and not by a bulk source as described above. Another common arrangement is individual oxygen cylinders at each site of administration.

Cylinder handling requires certain safety precautions. Cylinders are heavy and mechanical damage to them, as well as injuries to personnel, may result if cylinders are dropped. Cylinders should be transported only by the use of an approved cylinder cart. Cylinders in storage must be chained to a wall, affixed to a cylinder stand, or (in the case of smaller cylinders) securely attached to an item of therapy equipment. Cylinders must not be dropped in such a way to damage the valve. When large cylinders are moved, the cylinder cap must be tightly affixed. If a valve breaks off, the cylinder may be propelled by the reactive release of the high pressure contents. Serious injury or even death of personnel may result.

Oxygen is also utilized in emergency vehicles. Generally small cylinders are employed. Gas anesthesia apparatus also uses cylinders of these sizes.

Transfilling gas from large cylinders to small cylinders is a hazardous procedure, and should only be done by qualified personnel. It should always be done remote from patient areas, in locations acceptable to the authority having jurisdiction. The Compresses Gas Association has published a safety guide to follow when such transfilling is accomplished (Pamphlet P-2.5, *Transfilling of High Pressure Gaseous Oxygen to be used for Respiration*).

RESPIRATORY THERAPY

Respiratory therapy began with the use of oxygen-enriched atmospheres employing oxygen tents. Subsequently, various other methods for application of oxygen in medical care were developed. These included oxygen hoods (especially useful for children), incubators for new-

Table 12-4A. Typical Medical Gas Cylinders—Volume and Weight of Available Contents*‡

All Volumes at 70°F.(21.1°C)

Cylinder Style & Dimensions	Nominal Volume Cu In./Liter	Contents	Name of the Gas: Air	Carbon Dioxide	Cyclo-Propane	Helium	Nitrogen	Nitrous Oxide	Oxygen	Mixtures of Oxygen: Helium	Mixtures of Oxygen: CO_2
		kPa		5578	517				13 500		
B	87/1.43	psig		838	75				1900		
3½″ od × 13″		Liters		370	375				200		
8.89 × 33 cm		Lbs.-Oz.		1-8	1-7¼				—		
		Kilograms		.68	.66				—		
		kPa	13 500	5578	517	11 032	13 500	3275	13 500		
D	176/2.88	psig	1900	838	75	1600	1900	745	1900	**	**
4½″ od × 17″		Liters	375	940	870	300	370	940	400	300	400
10.8 × 43 cm		Lbs.-Oz.	—	3-13	3-5½	—	—	3-13	—	**	**
		Kilograms	—	1.73	1.51	—	—	1.73	—	**	**
		kPa	13 500	5578		11 032	13 500	3275	13 500		
E	293/4.80	psig	1900	898		1600	1900	745	1900	**	**
4¼″ od × 26″		Liters	625	1590		500	610	1590	660	500	660
10.8 × 66 cm		Lbs.-Oz	—	6-7		—	—	6-7	—	**	**
		Kilograms	—	2.92		—	—	2.92	—	**	**
		kPa	13 500	5578		11 032	15 169	3275	15 169		
M	1337/21.9	psig	1900	838		1600	2200	745	2200	**	**
7″ od × 43″		Liters	2850	7570		2260	3200	7570	3450	2260	3000
17.8 × 109 cm		Lbs.-Oz.	—	30-10		—	—	30-10	122 cu ft	**	**
		Kilograms	—	13.9		—	—	13.9	—	**	**
		kPa	13 500	5578		11 032		3275			
G	2370/38.8	psig	1900	838		1600		745		**	**
8½″ od × 51″		Liters	5050	12 300		4000		13 800		4000	5330
21.6 × 130 cm		Lbs.-Oz.	—	50-0		—		56-0		**	**
		Kilograms	—	22.7		—		25.4		**	**
		kPa	15 169			15 169	15 169	3275	15 169		
H or K	2660/43.6	psig	2200			2200	2200	745	2200†		
9¼″ od × 51″		Liters	6550			6000	6400	15 800	6900		
23.5 × 130 cm		Lbs.-Oz.	—			—	—	64	244 cu ft		
		Kilograms	—			—	—	29.1	—		

* Computed contents based on nominal cylinder volumes and rounded to no greater variance than ±1%.
** Pressure and weight of mixed gases will vary according to the composition of the mixture.
† 275 cu. ft. (7800 L) cylinders at 2490 psig (19 237 kPa) are available upon request.
‡ Source: Compressed Gas Association

born infants, oronasal or nasal masks, and the nasal catheter. The most recent development has been devices to mechanically assist or control the patient's breathing while delivering medical gases. This equipment, which intermittently generates pressure, will deliver mixtures of oxygen and therapeutic agents, or air-oxygen mixtures. These devices may be powered by the pressurized oxygen supply or by an electrically operated motor and pump system. One type of respiration unit is shown in Figure 12-4C.

In any oxygen-enriched atmosphere, combustible substances possess a lower ignition temperature and burn much more rapidly than in air. Flame spread is extremely rapid, especially if bedding or clothing are saturated with oxygen. The safe practices for the use of this equipment are detailed in Chapter 5 of NFPA 99.

Respiratory therapy may also employ helium in its pure form to be mixed with oxygen in the dispensing apparatus, or already premixed with oxygen in the cylinder. Helium neither burns nor supports combustion. It does in fact reduce the flammability of the oxygen mixture because it acts as a heat sink, rendering the mixture more difficult to ignite. The helium also tends to quench any flame which may develop.

PRESSURIZED AIR

Air compressed above atmosphere pressure (14.7 psi or 101 kPa) constitutes an oxygen-enriched atmosphere. Although it is not as hazardous as one of pure oxygen, the ease of ignition of combustible substances is nevertheless increased, and flame spreads with much greater rapidity than in atmospheric air at ambient pressures. The fire hazards created by the compressed air atmosphere are significant. It is very difficult to extinguish such a fire unless a water deluge system is employed. Fire blankets

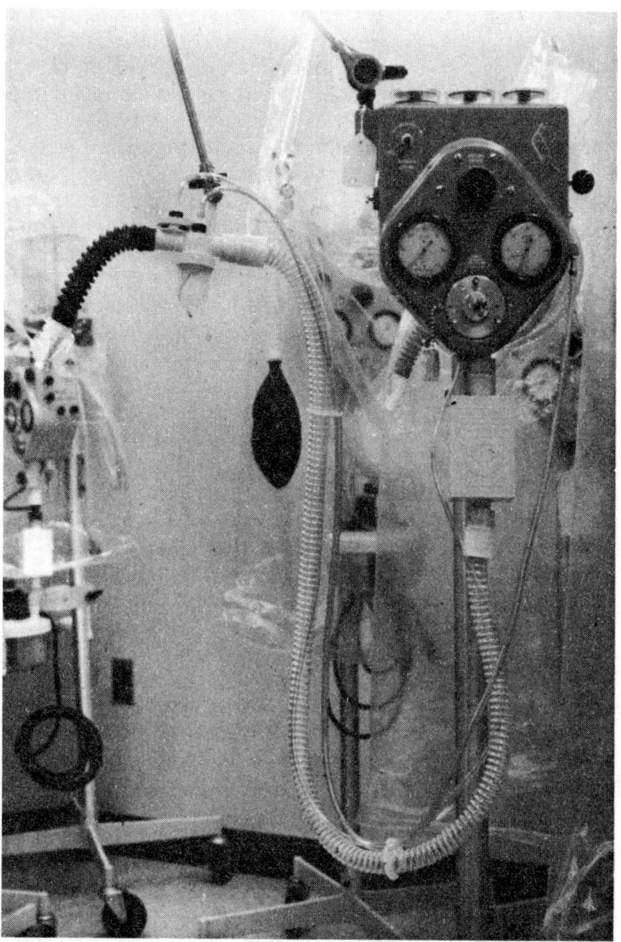

FIG. 12-4C. A respiration unit used for oxygen therapy. The control unit (upper right) is connected to the hospital compressed air and oxygen distribution system via quick-connect couplers on hoses at rear. Desired air-oxygen mixture is fed to patient via bacteria filter (at bottom of control unit) and corrugated hoses at bottom and left. Patient's oxygen mask is attached to black hose at left.

and dry chemical agents are ineffective. The burning material must be cooled by water.

If a fire occurs in an electrically powered item of equipment located in an oxygen-enriched atmosphere, it is first necessary to deenergize the equipment. Then, the fire can be fought with water.

Pressurized air has a wide variety of uses in the modern hospital other than the hyperbaric chamber. If purified, it may be mixed with oxygen to allow ventilation of the patient's lungs without the necessity for giving the patient pure oxygen to breathe. It can be used in laboratory apparatus. NFPA 56F, *Standard for Nonflammable Medical Gas Systems*, includes provisions for the installation of a central piping system for compressed air.

Pressurized Air Supplies

If a hospital elects to utilize a compressor as a source of air, it must be of a type that will not add impurities to the air stream, such as hydrocarbons, carbonaceous particulates, or carbon monoxide, and be able to produce air of the purity recommended by the Compressed Gas Association. To meet these specifications, the air supply to the compressor must be pure. It must not come from the basement of the building in which the compressor is located. Additionally, the air must be free of any lubricants or products resulting from the breakdown of lubricants. Under the pressures and temperatures generated in an air compressor, some lubricants may break down and release carbon monoxide which could be distributed in the compressed air system.

In a hospital using a compressor to furnish air for essential hospital services, including the operating room and as an adjunct to respiratory therapy, the compressor must be connected to the emergency standby power source, a recommended safety feature for all hospitals. The specifications for this power source are set forth in Chapter 8 of NFPA 99.

If the hospital is not located in an area in which the atmosphere is pure, it may be desirable to utilize compressed air in cylinders rather than compressing it on the site. Because the supplier of these cylinders may not have a pure source of atmospheric air, an oxygen-nitrogen mixture may be utilized. This so-called "synthetic air" is much purer than the ambient air because it is made up of pure oxygen and nitrogen in a one to four ratio. It contains no atmospheric pollutants of any kind.

VACUUM SYSTEMS

In addition to central pipeline systems for compressed air, oxygen, and nitrous oxide, central vacuum systems have become a standard feature in modern hospitals. Station inlets for the vacuum system are now commonly installed in emergency rooms, operating rooms, recovery rooms, intensive care facilities, delivery rooms, nurseries, and patient rooms. While portable individual vacuum pump units have been commonly utilized in the past, the trend is to move away from them, including those for low level suction used to remove fluids and air from the patient's thoracic cage.

A central vacuum system requires both collection bottles and trap bottles at the station inlets. The collection bottles collect the fluids. The trap bottles ensure that none of the material will enter the suction line and clog it. A central vacuum system requires adequate piping capacity and pump units of sufficient size to meet peak needs. Two pumps are installed, each sized so that, if one fails, the other can meet the needs of the hospital. The pumps are connected so that they will run alternately to provide even wear. Central vacuum systems are also to be connected to the emergency power system of the facilities since the system provides life support function in many situations.

Modern practice dictates retaining the suction material at the patient care site. This prevents the possibility of contamination and cross infection within the hospital premises. A central vacuum system may also be used in the laboratory and other hospital facilities. If so, it should not be connected to the medical-surgical vacuum system.

In the operation of any central vacuum system, it is vitally important to avoid drawing any flammable gases or liquids into the system. At the opposite end, the discharge of this system is to the outside of the building to avoid contamination back into hospital. Requirements for vacuum systems in hospitals are included in Chapter 6 of NFPA 99.

Waste Anesthetic Gas Disposal (Scavenging)

Central vacuum systems of hospitals are sometimes utilized for waste anesthetic gas disposal (WAGD). For several years it has been recognized that waste anesthetic gases (notably nitrous oxide and halothane) may create an occupational hazard for operating room personnel. Studies have suggested an increased incidence of carcinoma, spontaneous abortion, and minor birth defects in the offspring of such personnel. Anesthesia techniques employed for the use of these agents generally utilize high gas flows. A significant portion of the excess gases are vented to the operating room atmosphere, thus creating the alleged hazard.

If a central vacuum system is to be used for WAGD, it must have a capacity greater than that for conventional operating room requirements (i.e. suctioning only surgical patients). Additionally, a system of this type cannot be employed when flammable anesthetic agents are used unless special precautions are taken to keep the concentration of such agents below their flammable limits.

Workable alternative scavenging systems include a low vacuum-high volume blower system if the operating rooms are equipped with air handling systems and exhaust ducts which afford no recirculation. The "scavenging" inlets from the anesthesia circuit (and ventilator if one is used) are connected to tubing which extends to the plenum of the air exhaust duct. A system of this type could be used in the flammable anesthetizing location to scavenge flammable anesthetic agents. In practice the safe use of flammable anesthetic agents requires low flows and a closed anesthetic circuit. Thus, only small amounts of flammable agents would enter such a scavenging system, and then only at the termination of the anesthesia.

For more details on WAGD, ANSI Z79.11-1982, *Standard for Anesthetic Equipment-Scavenging Systems for Excess Anesthetic Gases* should be reviewed.

Housekeeping Vacuum Systems

In some health care facilities, it may be desirable to have a separate vacuum system for housekeeping purposes. This can be used in conjunction with vacuum equipment for cleaning corridors, drapes, and other related housekeeping chores. This system must, of course, be separate from the central vacuum system used for medical purposes.

HOSPITAL LABORATORIES

Hospital laboratories may utilize a wide variety of gases and gaseous mixtures for chemical analysis, the generation of flames, and other purposes. Chapter 7, "Laboratories in Health-Related Institutions," of NFPA 99 sets forth the design features and practices required for the safe operation of such a laboratory.

GAS STERILIZING

For many years, sterilization by either dry heat or steam heat was standard practice for the eradication of bacteria and other pathogenes from reusable medical equipment. Recent years have seen the introduction of the ethylene oxide sterilizer. This sterilizer employs ethylene oxide, an extremely volatile liquid which vaporizes readily at normal room temperatures, and is also an effective bacteriocidal and virocidal agent.

To sterilize objects, the sterilizer chamber is first evacuated of air, and then filled with the ethylene oxide gas. The gas penetrates the material contained in the chamber and kills all bacteria, viruses, and other pathogenic organisms.

Ethylene oxide is flammable and chemically reactive, and proper operation of the equipment is required in order to avoid the danger of accidental ignition or explosive chemical reaction. The liquid is delivered to the hospital in cylinders which must be stored properly, as is the case with any container of flammable liquid or gas. Ethylene oxide is toxic to man, and proper venting of the ethylene oxide sterilizer to the outside of the building is mandatory.

Other chemical sterilizing and disinfecting agents have been used in the past. These have included chlorine, ozone, sulfur dioxide, and formaldehyde gas. Ozone is generated on site by an ozone generator. The other agents are purchased off site and supplied in cylinders. Chlorine is the most toxic, and leakage from a cylinder will cause a very serious health hazard. Because of the problems in handling these agents, their corrosive nature, and the greater efficacy and safety associated with heat and ethylene oxide sterilization, ozone, formaldehyde, sulfur dioxide, and chlorine are no longer used for medical sterilization purposes.

FIRE PREVENTION AND RESPONSE

Fire and Explosion Prevention

One method of fire and explosion prevention in medically oriented oxygen-enriched atmospheres would be the use of a nonflammable anesthetic agent. Although a large number of useful nonflammable agents are available, under many medical circumstances one or more flammable agents may be indicated. For example, some nonflammable agents may cause serious liver, kidney, or cardiovascular problems. For this reason it is likely that flammable agents will continue to be utilized in some form in clinical anesthesia.

Sources of ignition in flammable anesthetic atmospheres include adiabatic heating of gases, electric sparks, friction sparks, heated objects (including open flame), and static electricity. In view of the hazards involved, fire prevention methods generally strive for redundancy to establish built-in safeguards of a multifaceted nature so that an explosion or fire will not occur even if one element of the safety recommendations is omitted.

To prevent dangerous adiabatic heating of gases, it is recommended that cylinder valves be opened very slowly to allow gradual introduction of the high pressure gas downstream from the cylinder valve. This will allow a slow buildup of pressure and hence temperature, thus allowing the mass of the metal of the regulator or other gas containing component to dissipate the heat rapidly enough to prevent the buildup of dangerously high temperatures. Additionally, safety standards forbid the use of any hydrocarbon substance to lubricate any components containing any oxidizing gas at high pressures.

The elimination of electric sparks, other than static sparks, is directed toward the proper design, application, and maintenance of electrical equipment. Specifications for this electric equipment as well as for line cords and power supplies are set forth in NFPA 70, *National Electrical Code®* and NFPA 99. Recommendations for the proper use of this equipment by hospital personnel are also spelled out in these publications.

The elimination of static electricity from any area designated for the administration of flammable anesthetics, such as an operating room, is another example of the multifaceted approach to hazard control. First, the humidity in such locations is maintained at 50 percent. Second, all floors in the area and immediately adjacent corridors are made conductive (conductive floors are designed to keep all items resting on or moving across the floor at the same potential). Third, personnel entering the area are required to wear either conductive shoes or conductive overshoes that maintain contact with the skin of their feet or legs, to assure an adequately conductive pathway between the body of the person and the conductive floor. Fourth, equipment for use in flammable anesthetizing locations has certain requirements, including specifications for conductive tops as well as conductive casters or feet. This applies to stools, kick basins, mayo stands, operating tables, and gas anesthesia apparatus. Finally, all outer clothing worn by operating room personnel and all patient gowns, drapes, belting, and other components and accessories are of conductive nature. All of these steps preclude the buildup of static electricity in the hospital operating room. The primary documents controlling the use of anesthetic agents are Article 517 of NFPA 70, *National Electrical Code* and Chapter 3 of NFPA 99.

Because there exists the possibility of electrocution in such a conductive environment, both the shoes and the floor have built-in internal resistances which will prevent a direct, low resistance pathway to ground. Additionally, all electrical appliances in the operating room are fed through an isolation transformer equipped with a line isolation monitor. A conductive shoe tester is shown in Figure 12-4D.

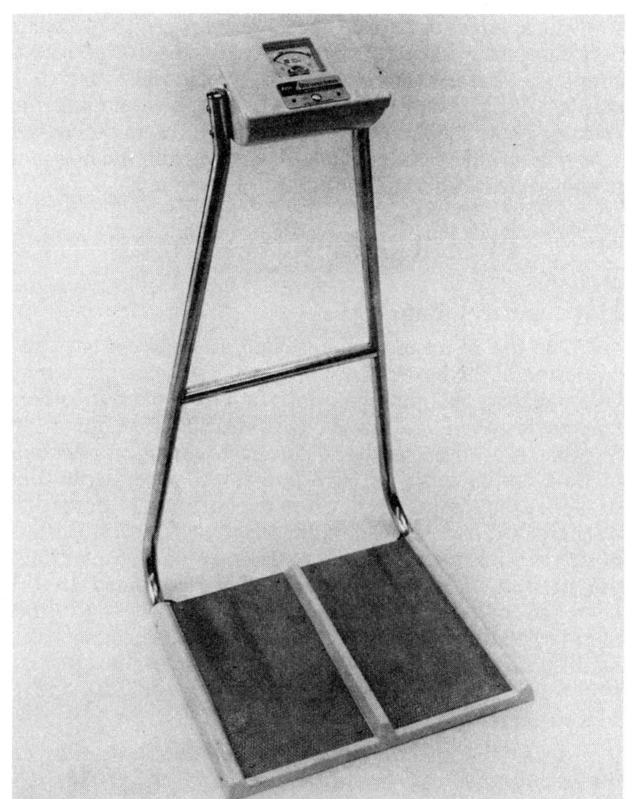

FIG. 12-4D. A conductive shoe tester. Located at entrance to sterile corridor of designated flammable anesthetizing locations, the tester indicates proper function of conductive shoes and "booties." The meter at the top of the frame indicates the resistance between ground and a person standing on the foot plates of the pedestal. (Ohio Medical Products—Airco Division).

Fire Fighting Problems

Oxygen-enriched atmospheres are encountered during practically every inhalation anesthetic administration and all respiratory therapy. When the patient's bedding and clothing become saturated with oxygen, flame spread will be extremely rapid and large quantities of water are required for extinguishment. Dry chemical extinguishers and fire blankets are ineffective. If the fire involves electrical equipment, such equipment must first be deenergized. The fire then may be knocked down with water. Safe practices for the prevention of fires during the use of respiratory therapy are covered in Chapter 5, "Inhalation Therapy," of NFPA 99. Included is a recommended response in the event of a fire involving respiratory therapy equipment.

Storage rooms for flammable anesthetic agents exist in many hospitals. Flammable solvents are present in all hospitals: in pharmacy storage, the hospital laboratory, and possibly housekeeping and other areas. Special fire problems are created if any of these substances are allowed to enter drains, the central suction system, or ventilation ducts. Finally, mechanical hazards are created by the potential energy contained in the compressed gas cylinder.

While high voltage electrical (e.g., radiological) and electronic (e.g., medical monitoring and computer) equipment does not by itself usually create a fire hazard, combustibles stored in the vicinity may be involved in a fire. The free use of water would be inadvisable because of the electric shock hazard to the fire fighter or damage to the equipment. The presence of radioisotopes may also hamper fire suppression methods. Fire service personnel must be made aware of all special problems created by equipment or supplies in any hospital area.

An aid to the identification of hazards associated with medical gases and agents is the NFPA hazards identification symbol system (NFPA 704, *Standard Systems for the Identification of the Fire Hazards of Materials*). This identification scheme, based on the "704 diamond," can visually present information on flammability, health, self-reactivity as well as special information associated with the hazards of the materials being identified. Chapter 7, "Laboratories in Health Related Institutions," of NFPA 99 requires application of the NFPA hazards identification symbols to hospital laboratories. It is also a good prudent policy to apply the system throughout a hospital in such areas as general stores, housekeeping, pharmacy, operating rooms, gas and flammable liquid storage rooms, laboratories, and any other location where volumes of combustible or flammable materials are stored or used, or where certain special fire fighting problems exist. Fire fighting personnel are trained to recognize and decipher the meaning of the "704 diamond."

Adoption by the hospital of the 704 symbol system affords not only enhanced fire protection, but also some measure of prevention and education of employees as well. Fire personnel are instantly apprised of the hazards they face. In addition, hospital employees are educated about the hazardous nature of the materials that they employ daily. Also, it is possible that in deploying the system, personnel may discover hazardous materials no longer in use that can safely be removed from the hospital premises, thus eliminating potential problems.

PRESSURIZED CHAMBERS

Hyperbaric Chambers

One use of an oxygen-enriched atmosphere is in the operation of the hyperbaric chamber. Some chambers may be small and accommodate only a single patient; others may be designed to allow the performance of an operation or other procedure, and are thus large enough to accommodate one or more patients and several attendants. But because the pressure of the air in the chamber is increased, an increased partial pressure of oxygen will result (unless means are taken to limit the oxygen content). Sometimes the percent of oxygen is deliberately increased in the chamber (certain medical problems are treated in this manner). Safety standards for hyperbaric chambers and facilities are detailed in Chapter 10 of NFPA 99.

Hypobaric Chambers

A new therapeutic technique has recently been introduced into medical care: the hypobaric facility. This is a chamber or room in which the pressure is intentionally lowered for therapeutic (generally respiratory) purposes. Hypobaric facilities require, in addition to the means of sealing the doors, a source of suction to lower the atmospheric pressure. The pumps which are used for this purpose are similar to those employed in the central suction system but of much larger capacity. The safe design and operation of the hypobaric facility is covered in Chapter 11 of NFPA 99.

AMBULATORY CARE FACILITIES

Many requirements for the safe use of inhalation anesthetics in the dentist's office and other ambulatory care facilities are identical to those for hospitals. Individual vacuum units are commonly utilized in such a facility, rather than a central system, because of the relatively small amount of suction needed in the dentist's office, as compared to the hospital. However, in larger dental offices, or other ambulatory care facilities in which minor operations are performed under general anesthesia, it may be desirable to incorporate a central vacuum system for the operating rooms and the recovery rooms, as well as the central piping of oxygen and nitrous oxide.

Requirements for inhalation anesthetics in such facilities are contained in Chapter 4, "Use of Inhalation Anesthetics in Ambulatory Care Facilities," of NFPA 99.

Bibliography

NFPA Codes, Standards, Recommended Practices and Manuals. (See the latest *NFPA Codes and Standards Catalog* for availability of current editions of the following documents.)

NFPA 10, *Standard for Portable Fire Extinguishers.*

NFPA 11, *Standard for Low Expansion Foam and Combined Agent Systems.*

NFPA 12, *Standard on Carbon Dioxide Extinguishing Systems.*

NFPA 13, *Standard for the Installation of Sprinkler Systems.*

NFPA 13A, *Recommended Practice for the Inspection, Testing and Maintenance of Sprinkler Systems.*

NFPA 15, *Standard for Water Spray Fixed Systems for Fire Protection.*

NFPA 17, *Standard for Dry Chemical Extinguishing Systems.*

NFPA 18, *Standard on Wetting Agents.*

NFPA 30, *Flammable and Combustible Liquids Code.*

NFPA 43A, *Code for the Storage of Liquid and Solid Oxidizing Materials.*

NFPA 49, *Hazardous Chemicals Data.*

NFPA 50, *Standard for Bulk Oxygen Systems at Consumer Sites.*

NFPA 50A, *Standard for Gaseous Hydrogen Systems at Consumer Sites.*

NFPA 50B, *Standard for Liquefied Hydrogen Systems at Consumer Sites.*

NFPA 53M, *Manual on Fire Hazards in Oxygen-Enriched Atmospheres.*

NFPA 56F, *Standard for Nonflammable Medical Gas Systems.*

NFPA 70, *National Electrical Code.*

NFPA 77, *Recommended Practice on Static Electricity.*

NFPA 99, *Standard for Health Care Facilities.*

NFPA 101, *Standard for Safety to Life from Fire in Buildings and Structures.*

NFPA 325M, *Fire Hazard Properties of Flammable Liquids, Gases, and Volatile Solids.*

NFPA 491M, *Manual of Hazardous Chemical Reactions.*

NFPA 704, *Standard System for the Identification of the Fire Hazards of Materials.*

NFPA 801, *Recommended Fire Protection Practice for Facilities Handling Radioactive Materials.*

Additional Readings

The following pamphlets are published by the Compressed Gas Association, 1235 Jefferson Davis Hwy, Arlington, VA 22202.

CGA No. C-4 (ANSI Z48.). *Method of Marking Portable Compressed Gas Containers to Identify the Material Contained.*

CGA No. C-9. *Standard Color-Marking of Compressed Gas Cylinders Intended for Medical Use in the United States.*

CGA No. G-4. *Oxygen.*

CGA No. G-4.2. *Standard for Bulk Oxygen Systems at Consumer Sites.*

CGA No. G-4.3. *Commodity Specification for Oxygen.*

CGA No. G-6. *Carbon Dioxide.*

CGA No. G-6.2. *Commodity Specification for Carbon Dioxide.*

CGA No. G-7. *Compressed Air for Human Respiration.*

CGA No. G-7.1. *Commodity Specification for Air.*

CGA No. G-8.1. *Standard for the Installation of Nitrous Oxide Systems at Consumer Sites.*

CGA No. G-9.1. *Commodity Specification for Helium.*

CGA No. G-10.1. *Commodity Specification for Nitrogen.*

CGA No. P-2. *Characteristics and Safe Handling of Medical Gases.*

CGA No. P-2.1. *Standard for Medical-Surgical Vacuum Systems in Hospitals.*

CGA No. S-1.1. *Safety Relief Device Standards—Cylinders for Compressed Gases.*

CGA No. V-1. *American National-Canadian Standard Compressed Gas Cylinder Valve Outlet and Inlet Connections.* ANSI-B57.1. CSA-B96.

CGA No. V-5. *Diameter Index Safety System.*

CGA No. V-6. *Standard Cryogenic Liquid Transfer Connections.*

The following document is available from the American National Standards Institute, 1430 Broadway, New York, NY 10018.

ANSI Z79.11. *Anesthetic Equipment-Scavenging Systems for Excess Anesthetic Gases*

PESTICIDES

Revised by Robert Benedetti

Pesticides are extremely diverse in chemical composition and formulation, reflecting the enormous diversity of pests and pest management problems. These tools of pest management are closely regulated under federal law *(Federal Environmental Pesticide Control Act of 1972)*, and are required to be registered with the U.S. Environmental Protection Agency (EPA) with labeling that must include directions for effective use and prominent warnings of at least the primary dangers to the human race, their property, and the environment (including hazards to wild life). Fire hazard warnings are included where appropriate. The label is required to include a statement of the name and percentage amounts of the active pesticidal ingredients.

Additional information relevant to the hazards of storage and use of pesticides may be found in Section 5, Chapter 1, "Fire Hazards of Materials;" Chapter 4, "Flammable and Combustible Liquids;" "Chapter 5, Gases;" Chapter 6, "Chemicals;" Section 11, Chapter 4, "Storage and Handling of Flammable and Combustible Liquids;" Chapter 5, "Storage and Handling of Gases;" Chapter 6, "Storage and Handling of Chemicals;" and in Section 18, "Water-Based Extinguishing Systems."

DEFINITION OF PESTICIDES

The definition of a pesticide as provided in the federal act is comprehensive and includes any chemical or mixture of chemicals or substances used to repel or combat any animal or plant pest, including insects and other invertebrate organisms, all vertebrate organisms, all vertebrate pests such as rodents, fish, pest birds, snakes, and gophers, all plant pests growing where not wanted (such as weeds), and all microorganisms which may or may not produce disease in humans. Household germicides, plant growth regulators and plant root destroyers are also included in the definition.

Pesticides are typed according to their primary or specific control purposes, or to reflect the manner in which they are used. Insecticides (insect control), fungicides (fungi and bacteria control), herbicides (unwanted plant control), nematocides (earth and water worm control), and rodenticides (rodent control) are examples of pesticides classified by control purposes. Fumigants are an example of pesticides that are classified by the manner in which they are used, in that a fumigant is a pesticide that acts in the gaseous, or "fume" state.

Well over one billion pounds of pesticides are produced annually in the United States. In addition to their manufacturing facilities, pesticides can be found stored in agricultural chemical warehouses, farm supply stores, nurseries, farms, supermarkets, hardware stores and other retail outlets, on the premises of commercial pest control operators, and in the home. Large quantities are continually being transported in cargo trucks and railcars.

HAZARDS

In addition to the obvious health hazards of pesticides (because what can kill other forms of life can also kill humans), pesticide formulations are sometimes flammable or combustible. The products of combustion from burning pesticides will most likely be poisonous or toxic. Some formulations may act as oxidizing agents, that can accelerate combustion. Runoff water from fighting fires involving pesticides is likely to be contaminated. It is not uncommon to find pesticides stored in close proximity to other hazardous agricultural chemicals, such as ammonium nitrate fertilizer because the same customer is often involved.

Pesticide formulations are encountered in all three states of matter—liquid, solid, and gas. Flammable or combustible liquids are widely used to dissolve solid pesticides in order to facilitate application. The hazard of flammable or combustible liquids when released from containers can range from a simple spill fire to a combustion explosion of vapors, fogs, or mists during application. Solids are usually stored and handled as granules, dusts, and powders. If they are combustible, they can present a dust explosion hazard when dispersed into the air. Most pesticide powders, however, are specific formulations absorbed onto an inert material such as clay.

Liquids and gases are transported and stored in strong containers which can fail from overpressure resulting from fire exposure.

Mr. Benedetti is Senior Chemical Engineer on the staff of NFPA.

Table 12-5A. Effect of Pesticides Taken Orally*

Name of Chemical	Approximate Lethal Dose for Humans
TEPP, Thimet	3 drops (0.18 mL)
Phosdrin	5 drops (0.30 mL)
Parathion, Systox (Demetron)	9 drops (0.54 mL)
Methyl parathion, EPN	13 drops (0.78 mL)
Endrin, Nicotine	15 drops (0.90 mL)
Guthion	¼ teaspoon (1.2 mL)
Trithion	⅓ teaspoon (1.6 mL)
Aldrin	½ teaspoon (2.4 mL)
Dieldrin, Toxaphene	¾ teaspoon (3.6 mL)
Heptachlor	1 teaspoon (4.9 mL)
Thiodan	1¼ teaspoon (6 mL)
Lindane, Rotenone	1½ teaspoon (7.2 mL)
DDT, Sevin	1 tablespoon (14.8 mL)
Chlordane, Malathion	1 ounce (29.6 mL)
Diazionon, Vapam	2 ounces (59.2 mL)
Sulphenon	3 ounces (88.8 mL)
Phostex	5 ounces (148 mL)
Perthane	1 pound (0.45 kg)
Tedion	2 pounds (0.9 kg)

* Adapted from a chart (No. 4) appearing in the article, "Agricultural Chemicals as a Fire Hazard," by Dr. Robert Jones, in *Fireman*, Vol. 32, No. 4, April 1965, p. 10.

The intrinsic toxicity of pesticides varies considerably although they all must be regarded as dangerous. Variations are taken into consideration in labeling under federal law, and the more toxic materials are required to be labeled with the word "Poison" and a skull and crossbones symbol. Table 12-5A indicates the rather wide range in intrinsic toxicity of several common pesticides.

PESTICIDE STORAGE SAFEGUARDS

Pesticides should be segregated in well identified areas of structures designed to minimize fire exposure to them and in a manner reflecting their own intrinsic fire hazards as flammable, combustible, or reactive chemicals. NFPA 43D, *Code for Storage of Pesticides in Portable Containers*, presents basic guidelines for storing containers up to 660 gal (2.5 m^3) in size.

Where practical, and if large quantities are involved, a detached, noncombustible structure or a first story corner room with direct outside access should be used. Basements are to be avoided as much as possible because of difficulties in fire fighting, salvage, decontamination, and in controlling water runoff. Adequate placarding and security is important.

Because it minimizes both the risk of exposing fire fighters to toxic hazards and the amount of water needed for fire control, automatic sprinkler protection is provided where an adequate water supply is available. More than the usual attention must be paid to disposal of water runoff, regardless of the form of water fire protection provided. Particular attention must be given to providing reasonably watertight floors and runoff controlling drainage patterns around the storage area. Ideally, all runoff around the storage area should drain into an impounding basin.

Safeguards applicable to conventional flammable or combustible liquids or gases are applied to pesticides and their formulations presenting these hazards.

Reactive pesticides and chemicals associated with them should be further segregated. These include ammonium nitrate fertilizer and oxidizing materials for which other NFPA standards are followed. (See NFPA 490, *Code for the Storage of Ammonium Nitrate*, and NFPA 43A, *Code for the Storage of Liquid and Solid Oxidizing Materials*.) Chlorates (oxidizing agents) are often present in herbicides, so herbicides are usually segregated from other pesticides.

USAGE SAFEGUARDS

Many pesticides do not present significant fire hazards in usage. When applied by fogging techniques, fumigants and insecticides can present significant fire hazards. Fumigation is covered in Chapter 10 of NFPA 61B, *Standard for the Prevention of Fires and Explosions in Grain Elevators and Facilities Handling Bulk Raw Agricultural Commodities*. An important provision of this chapter is its utilization of the NFPA 704 hazard identification system for both fire prevention safeguards and emergency personnel protection.

Table 12-5B includes the common fumigants and the hazard signals for them, according to the NFPA 704 system. A list of commercial fumigants under their various

Table 12-5B. Characteristics of Some Common Fumigants

No.	Chemical Name	Chemical Formula	Boiling Point °F	Boiling Point °C	Water Soluble*	Flammable Limits (% Vol. in Air)	Hazard Signal† H	Hazard Signal† F	Hazard Signal† R	Remarks
1	Acrylonitrile‡	CH_2CHCN	171	77.2	No	3-17	4	3	2	May polymerize violently on contact with alkali unless inhibited. Decomposes and releases hydrogen cyanide at high temperatures or in contact with acids.
2	Aluminum Phosphide (formulated)	AlP	See Phosphine							Not flammable in dry state but reacts with moisture to produce phosphine gas.
3	Benzene‡	C_6H_6	176	80	No	1.3-7.1	2	3	0	Can be absorbed through skin.

Table 12-5B. Characteristics of Some Common Fumigants

No.	Chemical Name	Chemical Formula	Boiling Point °F	Boiling Point °C	Water Soluble*	Flammable Limits (% Vol. in Air)	Hazard Signal† H	Hazard Signal† F	Hazard Signal† R	Remarks
4	Calcium Cyanide	$Ca(CN)_2$	See Hydrogen Cyanide							Reacts slowly with moisture to release hydrogen cyanide.
5	Carbon Disulfide‡	CS_2	115	46	No	1.3-44	2	3	0	Low ignition temperature—212F (100°C).
6	Carbon Tetrachloride‡	$C\ Cl_4$	170	77	No	None	2	0	0	Decomposes at elevated temperatures to form phosgene and hydrogen chloride.
7	Chloroform‡	$CH\ Cl_3$	142	51	No	None	2	0	0	Decomposes at elevated temperatures to form phosgene and hydrogen chloride.
8	Chloropicrin	CCl_3NO_2	233.6	112	No	None	4	0	1	Causes severe eye irritation (tear gas). May decompose violently when heated above 390°F (199°C).
9	Ethylene Dibromide	$CH_2Br\ CH_2Br$	268.7	131.5	No	None	3	0	0	
10	Ethylene Dichloride‡	$CH_2Cl\ CH_2Cl$	183	84	No	6.2-16	2	3	0	Decomposes at elevated temperatures to give off phosgene.
11	Ethylene Oxide‡	CH_2OCH_2	51	10.5	Yes	3-100	2	4	3	May polymerize violently in contact with highly reactive catalytic surfaces. Cannot depend upon odor for warning.
12	Hydrogen Cyanide	HCN	79	26	Yes	6-41	4	4	2	Almondlike odor, but do not depend upon odor for warning. May polymerize violently when unstabilized. Can be absorbed through skin.
13	Methyl Bromide	CH_3Br	40	4.4	No	10-16	3	1	0	
14	Methylene Chloride[3]	CH_2Cl_2	104	40	No	15.5-66 (in oxygen only)	2	0	0	May form explosive mixtures with air at high concentrations.
15	Phosphine	PH_3	−125	−87	No	1.79	3	4	1	Decomposes when heated to give phosphorus oxides. Low ignition temperature 212 to 302°F. (100 to 150°C). Explosive under vacuum fumigation conditions.
16	Propylene Oxide‡	CH_2CHOCH_3	95	35	Yes	2.1-21.5	2	4	2	Cannot depend upon odor for warning.
17	Sulfur Dioxode‡	SO_2	14	−10	Slightly	None	3	0	0	Vapors are corrosive to some metals.
18	Sulfuryl fluoride	SO_2F_2	−67	−55	No	None	2	0	0	Cannot depend upon odor for warning.

Notes:
* From standpoint of use of water for extinguishment by dilution.
† H = Health. F = Flammability. R = Reactivity. If the fumigant is a mixture of compounds, use numbers corresponding to the properties of the mixture.
‡ Seldom used singly as fumigant, but are used as components of mixtures.

trade names is published by Underwriters Laboratories Inc. (UL annually). This list specifies the composition of the material and provides a relative fire hazard classification.

Insecticidal fogging presents combustion explosion and flash fire hazards. Safeguards include the use of solvents having flash points well above normal temperatures and limiting the quantities applied to structures in relation to their volume. Some fog applicators are examined and listed by testing agencies.

EMERGENCY HANDLING

Handling an accidental release of pesticides or a fire involving pesticides requires specialized equipment and tactics. These include: (1) thorough knowledge of the specific pesticide formulation involved; (2) the storage arrangement and location of each formulation; and the location of pesticide and fumigation operations to facilitate planning; (3) use of protective clothing and respiratory protection by all personnel involved in emergency operations and clean-up; (4) restriction of manpower to the minimum needed in the working area; (5) minimum use of hose streams to avoid container breakage and pesticide dispersal and to minimize runoff; (6) approach from upwind of the emergency site; (7) evacuation of affected downwind areas; and (8) specialized emergency medical management teams.

The importance of using the least amount of water possible in fire fighting operations cannot be overstressed. Large-scale groundwater pollution by runoff from fire fighting operations is a distinct possibility. In fact, if a storage building is so heavily involved that control is doubtful, efforts should be directed at protecting exposures, thus sacrificing the storage building. This provides two benefits: (1) the fire is allowed to reach higher temperatures, resulting in more efficient and more complete thermal decomposition of the pesticides to the simplest, least toxic combustion products; and (2) the possibility of ground water contamination is almost eliminated. Of course, evacuation of downwind areas is critical in such situations. The bibliographical material contains considerable detail about emergency procedures.

Bibliography

References Cited

UL. Annually. *Classified Products Index*. Underwriters Laboratories Inc. Chicago. Issued annually in July.

NFPA Codes, Standards, and Recommended Practices and Manuals. (See the latest *NFPA Codes and Standards Catalog* for availability of current editions of the following documents.)

NFPA 43D, *Code for Storage of Pesticides in Portable Containers.*

NFPA 61B, *Standard for the Prevention of Fires and Explosions in Grain Elevators and Facilities Handling Bulk Raw Agricultural Commodities.*

Additional Readings

Bahme, C. W., *Fire Officer's Guide to Dangerous Chemicals*, National Fire Protection Association, Quincy, MA, 1972.

"Fire Protection for Pesticide Storage," *NAC News and Pesticide Review* (National Agricultural Chemicals Association), Vol. 28, No. 2, Dec. 1969.

Safety Manual for Handling and Warehousing Class B Poison Pesticides National Agricultural Chemicals Association, Washington, 1969. Hazardous Materials: The Pesticide Challenge, 10-unit audiovisual training program, National Fire Academy, Emmitsburg, MD, 1984.

"Unstable Insecticide," *Fire Journal*, Vol. 62, No. 2, Mar. 1968, pp. 5-9.

FIRE RETARDANT AND FLAME RESISTANT TREATMENTS OF CELLULOSIC MATERIALS

Revised by James R. Shaw, Ph.D.

This chapter discusses the basic principles of fire retardant and flame resistant treatments, their uses and limitations, and methods of treatment for wood, wood-base composite panels, fiberboard, paper, and miscellaneous decorative materials.

Flame retardant plastics are discussed in Section 5, Chapter 8. Information on flame retardant fabrics will be found in Section 5, Chapter 3. Additional information on fire retardants and their uses will be found in Section 5, Chapter 2, "Wood and Wood-based Products," in Section 7, Chapter 7, "Interior Finish," and in Section 12, Chapter 9, "Forest, Brush and Grass Fires."

Cellulosic materials are an integral part of our living environment. Our homes, schools, factories, offices, shops and churches utilize large quantities of cellulosic products in construction, furnishings and decorations. Fire retardant and flame resistant treatments for cellulosic materials can provide additional fire protection where warranted.

Popular usage has created a variety of terms associated with this subject, resulting in frequent misconceptions, misunderstandings, misuses, and an undesirable degree of ambiguity. Terms such as "fire resistant," "fire retardant," "flame resistant," "flame retardant," and "flameproof" are often used indiscriminately and incorrectly.

Of these terms, the term "fire resistant" should not be used in reference to the fire retardant treatment of combustibles. Fire resistant is used properly only to signify the ability of a structure, material, or assembly to resist the effects of a large scale severe fire exposure. ASTM defines fire resistance as "the property of a material or assembly to withstand fire or give protection from it. As applied to elements of buildings, it is characterized by the ability to confine a fire or to continue to perform a given structured function or both." (ASTM 1983). Fire resistance rating is the time, in minutes or hours, that materials or assemblies have withstood a fire exposure as established in accordance with the test procedures of NFPA 251, *Standard Methods of Fire Tests of Building Construction and Materials*. "Flameproof" and its derivatives are also misleading and subject to abuse and misunderstanding; their use should be discouraged.

"Fire retardant" signifies a lesser degree of protection than fire resistant. It should be used in reference to chemicals, paints, or coatings used or intended for the treatment of combustible building materials, and to such treated materials. Again ASTM defines fire retardant as "having or providing comparatively low flammability or flame spread properties." (ASTM 1983). Flame spread rating refers to numbers or classifications obtained according to NFPA 255, *Method of Test of Surface Burning Characteristics of Building Materials*. "Flame retardant" and "flame resistant" may be used more or less interchangeably and denote decorative materials which, due to chemical treatment or inherent properties, do not ignite readily or propagate flaming under small to moderate fire exposure. "Flame retardant" is the preferred term to denote chemicals, processes, paints, or coatings used for the treatment of decorative materials, which include fabrics, foliage, Christmas trees, and similar materials that are decorations or furnishings.

Fire retardants and flame retardant treatments have been frequently misused. They have a widespread appeal due perhaps in part to a general lack of understanding about the limitations of the treatments, and a failure to differentiate between those which are effective and those which are not. One popular misconception is that all fire retardant treatments give a fire resistance rating. This is not true. While some fire resistive coatings may improve the fire endurance of combustible materials and assemblies, other treatments are only effective in retarding the rate of burning and the rate at which fuel is contributed by the treated material. This action, however, can reduce the fire intensity of some materials which otherwise might be very fast burning.

In other words, there are practical limits to fire retardant treatments. They are not all permanent; they affect other properties of the treated material and they add expense. They do, however, provide protection on a competing basis with nonfire retardant treatments and coatings. For example, fire retardant paints have been improved to the point where many provide a colorful,

Dr. Shaw is a scientist in the Fire Technology Unit of Weyerhauser Company.

protective coating with comparable wear resistance and strength of normal oil paints.

BASIC PRINCIPLES

Treatments to reduce the combustibility of wood, fabrics, and other natural materials have a long history reaching back to at least the time of Moses. The early use and development of chemical retardants was limited to materials that were accidentally found to be effective. Research in basic combustion processes and action of retardants was greatly expanded during World War II due to increased military needs for flame resistant fabrics and fire retardant wood; this expansion has continued in recent years. Because cotton textiles and wood have by far the longest history of use and are of great economic importance, research has been directed toward these cellulosic materials.

Burning of Cellulose

The chemical reactions and the mechanisms involved in the combustion of untreated cellulose are, by themselves, highly complex and not yet fully understood. The complexity is magnified by the addition of fire retardant chemicals.

Wood (cellulose) normally does not burn directly. The initial reaction to high heat exposure is pyrolysis, in which some of the products given off are gases, vapors, or mists that can form flammable mixtures with air. On ignition, the combustion of these products is evidenced by visible flames. The solid residues include charcoal, which can undergo a different form of combustion by combining directly with oxygen, as evidenced by glowing. This is an important distinction, since many chemicals which are effective flame inhibitors have little or no effect in retarding flameless combustion, or afterglow. For most applications, flame retardance is more important, but there are some instances where glow resistance is equally necessary.

Theories of Fire Retardant Mechanisms

Four general theories have evolved to explain the mechanisms by which chemicals retard flaming or glowing of cellulosic materials. It is agreed that no single theory is adequate to explain these mechanisms, and that in most cases more than one mechanism is involved. The four theories which offer the most satisfactory explanations for the function of fire retardant chemicals are as follows:

Thermal: There are three possible ways that fire retardant materials work to reduce thermal buildup of the treated combustible: (1) they increase thermal conductivity to dissipate the heat of combustion, (2) they increase thermal absorption to reduce the amount of heat available for pyrolysis, and/or (3) they provide thermal insulation to reduce heat access to the substrate. Thermal insulation is the most important mechanism and is also discussed below under coating theories.

Coating: Some retardants melt or fuse at relatively low temperatures, and it is believed that they may form an insulating coating over the fibers of the treated material, acting to exclude oxygen and inhibit the escape of combustible gases. More effective retardants exhibit a bubbling or foaming action, creating an insulating barrier. Intumescent paints and coatings are good examples of this mechanism.

Gas: Some chemicals function by releasing nonflammable gases such as water, ammonia and carbon dioxide under heat exposure which dilute the combustible gases sufficiently to render the mixture nonflammable in air. Other retardants catalytically inhibit the fire radical chain reactions which occur in the gas phase (flaming combustion). Halogen containing retardants exhibit this type of mechanism and when combined with antimony a synergistic (more than additive) effect is observed.

Chemical: Simplified representations of the chemical mechanism are shown in Figures 12-6A and 12-6B. These

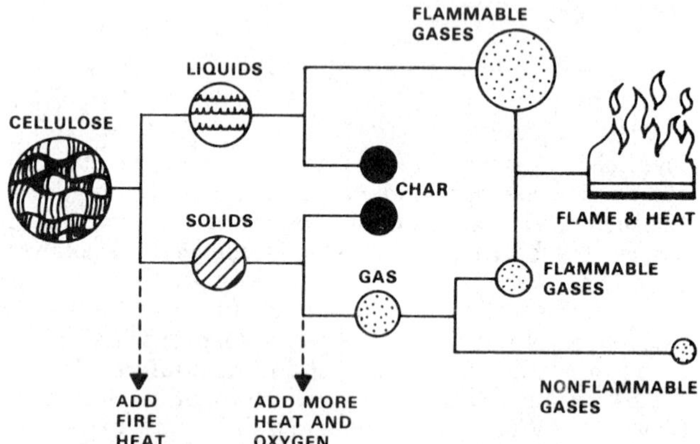

FIG. 12-6A. Schematic diagram of the burning sequence of untreated cellulose. The size of the circle is representative of the volume of product. Note predominance of liquid and flammable gas products.

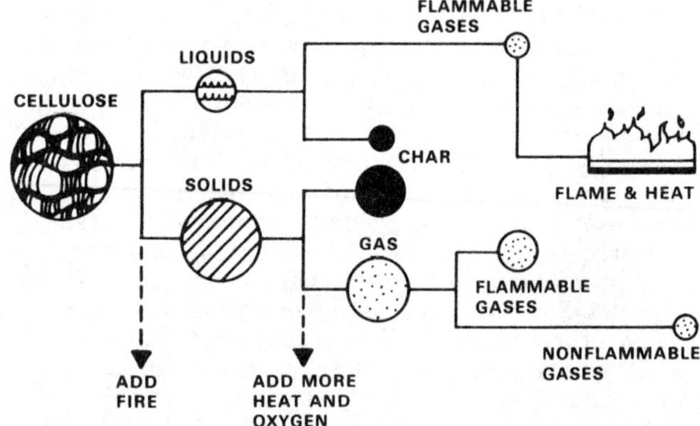

FIG. 12-6B. Schematic diagram of the burning sequence of treated cellulose. The size of the circle is representative of the volume of product. Note the predominance of solids and the severe reduction in liquids and flammable gases emanating from the liquids as compared with the same burning sequence for untreated cellulose as illustrated in Fig. 12-6A.

diagrams illustrate the different effects of fire on untreated and treated cellulose and are diagramatically explained in Flameproofing Textile Fabrics (Little et al 1947). Untreated cellulose breaks down into liquids and a smaller amount of solids. The liquids in turn break down into flammable gases and a little char, and the solids decompose into char and gases, some of which are flammable, some not. The flammable gases volatilized from the liquids combine with those from the solids to cause flaming and the emission of additional heat which continues the reaction.

With treated cellulose, fewer liquids and more solids are formed, and more of the solids become char. The reduced amount of liquids creates some char and some flammable gases. Because the amount of flammable gases from both the liquid and solid phases is now significantly less with treated cellulose, there is less and less heat released. Therefore, flame resistance or at least a reduced burning rate is achieved. Another postulate, an expansion of the chemical theory known as the catalytic dehydration theory, has been advanced (Schuyten 1954).

Studies at the U.S. Forest Products Laboratory (U.S. Forest Products Laboratory 1962-64) have shown that pyrolysis reactions (without the presence of oxygen) for the lignin component of wood start at approximately 220°C (428°F), while that for the cellulose fraction starts near 270°C (518°F). However, once pyrolysis for the cellulose fraction is started, its decomposition is primarily an endothermic reaction and is nearly completed when 400°C (752°F) is reached. The lignin decomposes more slowly, almost in an entirely exothermic reaction, and loses less than half its weight in pyrolysis to 600°C (1112°F).

Fire retardant chemicals greatly influence the pyrolysis reaction of the cellulose fraction, frequently reducing the temperature at which decomposition starts from 270 to 230°C (518 to 446°F). But as a result, the decomposition of the cellulose is promoted directly to form char and water, instead of forming intermediate flammable gases and tar products; and the char yield from the cellulose at 360°C (680°F) is increased from 15 percent up to 30 to 35 percent. The chemical treatment also increases the char yield from the lignin component from 49 percent up to 57 percent, but here the significance is less.

It was further shown that the combustion reactions for wood can be primarily divided into two exothermic stages, one near 330°C (626°F) associated with flaming as the cellulose fraction of the wood undergoes decomposition into flammable gases and tars and becomes ignited, and a second reaction near 430°C (806°F) associated with glowing as the lignin fraction of the wood, which has been changed to charcoal, undergoes oxidation. Fire retardant chemicals greatly reduce the intensity of the first peak, and some chemicals also reduce the intensity of the second peak, thus reducing the heat released by the treated wood over a wide temperature range. Other chemicals reduce the first peak and intensify the second, thus requiring a second chemical to result in maximum retardant performance. Heat release during the pyrolysis reactions for wood was found to be less than 5 percent of that released during the combustion reactions.

All of this data tended to prove that one of the principal reactions of fire retardant chemicals impregnated into wood is to accelerate chemical reactions at certain temperature ranges. This reduces the release of certain intermediate products such as levoglucosan tars and flammable gases, which may contribute to the flaming of wood, and also results in the formation of greater percentages of charcoal and water. Some chemicals are then further effective in reducing the oxidation rate for the charcoal residue. An excellent discussion of these basic principles is contained in the Forest Products Laboratory report, "Theories of the Combustion of Wood and its Control" (Brown 1958).

FIRE RETARDANT TREATING METHODS

There are four basic methods of treatment to produce fire retardance in materials: (1) chemical change (substitutions, admixtures), (2) impregnation (saturation, absorption), (3) pressure impregnation, and (4) coating. Chemical change and pressure impregnation are limited to manufacturing processes and procedures, and are not adaptable for field use. Impregnation and coating methods may be used in the field. Generally, methods employed in manufacturing processes are to be preferred because they provide greater uniformity, permanence, and dependability. Applications in the field, or subsequent to manufacture of the basic material, are more subject to variation in skill and integrity of the processor or applicator. Treatments of the coating and pressure impregnation type are covered in NFPA 703, *Standard for Fire Retardant Impregnated Wood and Fire Retardant Coatings for Building Materials.*

Chemical Changes

Fire retardant chemical changes are primarily effective with plastics and with synthetic fibers because different elements and compounds can be chosen which will alter the burning characteristics of the final product. Since the fire retardant elements are an integral part of the chemical composition, these materials are said to be "inherently" fire retardant. Naturally occurring polymers such as cellulose cannot be made fire retardant by this method. However, another type of chemical fire retardant treatment known as chemical grafting can be applied to natural polymers. It is still in the developmental stage, but shows great promise and utility. In chemical grafting, the material is treated by a multistage process in which the material is chemically activated so that molecules of fire retardant chemical are bonded to molecules of the material being treated. A follow-up curing step ensures permanence.

Impregnation

Impregnation refers to a technique for treating absorbent materials. The flame retardant chemicals are either dissolved or dispersed in a solvent, usually water, and the material to be treated is thoroughly wetted or saturated with the solution. The solution may be applied by spraying or by immersion of the material. The latter process is used for treatment of large volumes of fabric yardage. The excess solution is extracted by passing the saturated material through squeeze rollers before drying.

In the simple water-soluble salt impregnation processes, the result is merely to deposit minute crystals of the salts within or on the surface of the fibers. The process is purely physical, no chemical reaction is involved. With durable treatments, insoluble deposits are formed by chemical reactions or other means.

Certain cellulose based products, such as paper, acoustical tile, and wood-base composition, incorporate a wet pulp stage during the manufacturing process. It is possible to add fire retardant chemicals at the wet pulp stage, resulting in even distribution of the chemicals through the entire mass of the finished product. This practice is prevalent in manufacturing treated papers of many types, and is currently in use by several producers of building materials. Wood fiber acoustical tile and building panels processed by this method have an ASTM E84/ NFPA 255 flame spread rating of less than 25.

Pressure Impregnation

Fire retardant pressure impregnation is used for treating relatively dense nonabsorbent materials, such as wood. This process replaces the air in the wood cells with fire retardant chemical solutions. (The chemicals are deposited as the solutions dry.) The treating solution is usually forced into the wood by the standard vacuum pressure methods used in the wood preserving industry (American Wood Preservers 1977). Compared to simple impregnation techniques, pressure impregnation provided for deeper penetration and much greater chemical retention.

Coating

There are several types of fire retardant coatings that are useful in treating many materials. Coatings may be applied at any stage from manufacture to use. They may either actively inhibit flame spread to some degree or present a noncombustible surface over which flame cannot spread. They are used predominately on nonabsorbent building materials which cannot be treated by any other method, on Christmas trees and similar decorative materials, and to a limited extent on paper and fabrics which for various reasons cannot be treated effectively by impregnation.

The effectiveness of a coating depends on the chemical and physical properties of the material to which the coating is applied, the effectiveness of the coating on this material, the ability of the applier, and the thoroughness of the treatment. Since types of coatings and their effects vary depending on the material treated, they are discussed in the subsections of this chapter that cover materials.

PRACTICAL LIMITATIONS OF TREATMENTS

Ideally, a fire retardant treatment would permanently eliminate all flame spread, fuel contribution, and smoke generation characteristics from a material; the color, feel, strength, weight, and other key characteristics of the basic material would not suffer from the treatment; the method would be inexpensive, easy, and safe. There are no present day treatments meeting all of these qualifications.

WOOD

Wood may be made fire retardant by either pressure impregnation or coatings. While both of these treatments will reduce the flame spread of wood, only some fire retardant coatings are significantly effective in increasing the resistance of wood to degradation under sustained fire exposure or in delaying reduction of its load-bearing capacity.

While the fire retardant treatment which would make wood noncombustible does not exist, the spread of flame from the immediate area of an incipient fire can be retarded and, in some cases, prevented by present-day treatments with a corresponding substantial reduction in fuel and smoke contribution. Also the fire resistance rating of wood structured members in buildings may be improved through the use of fire retardant coatings (White 1983; Lieff 1983, Stumpf 1983).

Fire Retardant Pressure Impregnation

Pressure impregnation treatments deposit chemicals within the fibrous passages of wood. Many chemicals exhibit fire retardant properties, but because of cost limitations or various other objectionable characteristics, comparatively few are considered practical. Chemicals commonly used in the past include monobasic and dibasic ammonium phosphates, ammonium sulfate, borax, boric acid, zinc chloride, and sodium dichromate, usually in various combinations. Newer formulations tend to exclude hygroscopic salts such an ammonium sulfate and zinc chloride and in some cases include organophosphorus compounds and amino resins. The most significant improvements in the new formulations are the high tolerance for high humidity conditions (up to 95 percent R.H.) and exterior application due to their nonleachability. The impregnation operation can be controlled to secure a predetermined absorption of solution. The important considerations are the depth of penetration and the amount of chemical deposited per unit volume of the wood.

Since the fire retarding effect of the treatment is directly related to the amount of the chemical deposited within the wood, it is interesting to compare the effect of an ordinary brush treatment with a pressure-impregnation treatment. A 10-ft (3 m) high fence 50 ft (15.25 m) long of nominal 1-in (25 mm) pine boards would absorb about 4 gals (15 L) of treating solution if applied by brush. The effect on flame spread would be minor. When the same boards are pressure impregnated to refusal (the point at which no more salts are absorbed by the wood), they absorb about 225 gal (852 L) of chemical and the flame spread is cut to one-fourth of the original.

A load or "charge" of wood is locked in a large cylinder, which can be up to 8 ft (2.5 m) in diameter and over 140 ft (43 m) long. The air is then withdrawn and the resulting vacuum draws much of the entrapped air from the wood cells. After a delay to allow the vacuum to stabilize, the retardant chemical solution is bled into the cylinder. After the cylinder has been filled with the treating solution, additional chemicals are forced into it under pressure. As the pressure is raised (up to 200 psi or 1379 kPa), the chemicals are forced into the empty wood cells. The process can be run to refusal and the unabsorbed salt solution drained from the cylinder. A slight vacuum may then be drawn to remove surface chemicals that would cause dripping and slow drying. After the charge is removed and dried, it is ready for shipment and use. Additional details are available in the American Wood Preservers' Association *Book of Standards*.

Advantages: Pressure impregnation permits a large amount of fire retardant chemical to be deposited in the wood under close control and with uniform and predictable results. Samples for test are readily available. Of primary importance, the treatment can be considered permanently effective under normal and proper conditions of use. Since both water soluble and nonleachable fire retardants are available, it is important that nonleachable treatments be chosen for exterior use. Treatments with low hygroscopicity should be used for interior applications where high humidity conditions exist.

Disadvantages: A disadvantage of pressure impregnation is that the wood must be treated before use and after dimensioning. Another is that the strength of pressure treated wood may be reduced slightly (although no reports have attributed structural failures to this cause) while the weight may be increased as much as 4 lbs per cu ft (1.8 kg/m^3). In addition, when only formulations containing hygroscopic salts were available for pressure impregnation, there was a problem with corrosion of metal fasteners. The new fire retardants for wood have eliminated this problem.

Fire Performance Characteristics: The pressure impregnation process can substantially improve the fire performance characteristics of wood, as measured by the Tunnel Test, the room fire test, and heat release rate tests. The effect can be illustrated by a study of a typical wood, Douglas Fir, as shown in Tables 12-6A and 12-6B.

Table 12-6A. Maximum Effect of Pressure Impregnation on Douglas Fir

	Untreated	Pressure-Impregnated
Flame spread	100	As low as 15
Fuel contribution	100	As low as 10
Smoke generation	100	As low as 0

Testing: The effectiveness of fire retardant pressure impregnation treatments is tested by several methods. The Tunnel Test (ASTM E84/NFPA 255) has been used to determine the fire performance characteristics of many pressure impregnation treatments. A standard *Room Fire Test of Wall and Ceiling Materials* (ASTM 1982) is being developed at the National Bureau of Standards Center for Fire Research which will be used to evaluate treated panel products. The test utilizes the oxygen depletion method for calculating the rate of heat release. Data collected by this method will be used in the validation of fire mathematical models. The City of New York originally established three tests—crib (ASTM E160), shavings, and timber—for evaluating fire retardant treated wood. The fire tube test (ASTM E69), developed at the Forest Products Laboratory, has been extensively used both in the United States and in several other countries to measure the effectiveness of fire retardant treatments. Other tests, such as fire penetration, rate of heat release, and ignition tests, have been used by different organizations and individuals.

Recent advances in product standards and quality control have encouraged the use of treated lumber, particularly in some roof deck assemblies and partitions within noncombustible and fire resistive buildings. Lumber for construction can be treated by pressure impregnation to achieve a flame spread rating as low as 15.

Plywood: Fully treated plywood is satisfactory only when surface appearance is unimportant. For decorative purposes where the face veneer must have the beauty of natural wood, untreated face veneers of various species are laminated to treated cores. The flame spread classification of these treated plywood products varies in approximately inverse proportion with the density of the face veneer, and

Table 12-6B. Effect of Pressure Impregnation on Douglas Fir Plywood in the Room Fire Test.*

	Untreated	Pressure Impregnated
Flashover Time	3.3 min	None
Maximum Heat Release Rate	7.3 mW	0.6 mW
Maximum Smoke Generation (O.D.†)	1.11	0.76

* The ignition exposure was 40 kW (net) for 5 min, 160 kW (net) for 5 min and 0 kW (net) for 5 min.
† Optimal Density.

ranges generally from 25 to 75, satisfying all but the most restrictive requirements. Fully treated plywood generally has flame spread ratings less than 25 (Class I).

Relative Toxicity

Combustion toxicity testing of treated and untreated wood suggests that treated wood is "no more toxic than untreated Douglas Fir." This is true even though minor amounts of toxic vapors such as HCN and nitriles are expected to be liberated from some treatments upon degradation (Lieu 1982).

Fire Retardant Coatings

In existing structures of untreated wood, or in cases where fire retardant pressure impregnations are impracticable, fire retardant coatings can be used. Coatings can be applied to any exposed surface of structural members, interior finish, or contents. They may be applied by means of brush, roller, sprayer, or trowel. As in the case of pressure impregnations, the degree of the flame spread reduction secured by a coating is dependent upon the original combustibility of the surface to be protected, the effectiveness of the coating material used, the amount of coating used, the thoroughness of the application, and the size and severity of the exposing fire.

With normal paints, the usual aim is to get maximum coverage with minimum material. The approach is different in the application of fire retardant coatings. Here the objective is to provide the needed amount of coating per unit area required to insure a definite degree of protection. Film thickness for intumescent coatings is a very important factor in their ultimate performance.

Among the many proprietary fire retardant coatings, several have been listed by testing laboratories. Beware of any so-called fire retardant or "fireproof" paint that has not been certified as to its flame spread characteristics by an independent testing laboratory. There are four types of fire retardant coatings for wood. They are:

Intumescent Paints: These coatings, upon the action of heat, expand from a thin, paint like coating to a thick, puffy, burnt marshmallow type coating. The puffy coating results in one or all of the following effects: insulation of fuel from heat, exclusion of oxygen from fuel, production of diluent gases, and reduction of flammable gases. It will retain its effectiveness until broken up either by high heat or by sustained heat.

Mastics: These materials are applied by trowel or by heavy duty spray equipment and form a thick coating on the surface of the combustible material. They vary from hard, ceramic-like materials to soft, tar-like coatings. All withstand significant amounts of heat and inhibit flame spread by the impervious noncombustible membrane they form.

Gas Forming Paints: These coatings when heated release quantities of noncombustible gas which dilutes the oxygen at the protected surface so there is not sufficient oxygen to support combustion.

Cementitious and Mineral Fiber Coatings: Limited research has been conducted on the potential use of cementitious and mineral fiber coatings on wood to improve its fire endurance characteristics. These coatings are commonly used on structural steel to enhance fire endurance by protecting the steel from the intense heat of a fire. Some researchers feel that structural wood systems may be protected in the same manner.

The customary finishing materials for wood, such as ordinary stains, oil paints, enamels, varnishes, shellac, lacquer, and waxes, are of no appreciable value in protecting wood against fire, and in some cases increase combustibility. Paint, however, may be of value in preventing absorption of oil and in promoting cleanliness. It should also be noted that dry, decayed wood is more susceptible to ignition by small sparks from cigarettes than is sound wood, and that paint coatings tend to prevent surface decay.

Advantages: Fire retardant coatings have the following advantages: (1) they can be used on combustible materials already in place, (2) they are relatively inexpensive, and (3) they may be easily applied.

Disadvantages: Fire retardant coatings have the following disadvantages: (1) ease of application can lead to thin and ineffective treatments, (2) unexposed surfaces cannot be treated, (3) limited life and durability, (4) susceptibility to damage, and (5) required maintenance.

Fire Performance Characteristics: The fire performance of wood can be improved through the use of fire retardant coatings. The amount of improvement is determined by the effectiveness of the treatment. Underwriters Laboratories Inc. has established a minimum standard of effectiveness for listed coatings to be a flame spread rating of 50 or less in the Tunnel Test, or a 50 percent reduction in flame spread as compared to the untreated surface, whichever is lower.

Listed coatings, when properly applied, are effective in improving fire performance characteristics, some to a point well below the limits noted above. The variation in the effect of listed fire retardant coatings on Douglas fir is shown as an example in Table 12-6C.

Permanence: Some fire retardant paints and coatings have good permanence when used in interior locations of relatively normal humidity and temperatures. Water resistant, oil based, intumescing fire retardant coatings have been developed (Verburg 1964). But the use of fire retardant coatings in exterior locations, in areas of relatively high humidity, and in areas of high temperature has not been thoroughly tested. One difficulty in testing these exterior paints is the problem of weathering samples. Samples for small-scale tests can be placed in a weather accelerator or weatherometer and the results measured. These small scale tests, however, do not produce significant flame spread results. There is also doubt as to the comparability of weatherometer exposures and actual long term weathering. The samples required for the acceptable large scale tests are too large for accelerated weathering and thus must be exposed to actual weathering, which requires long periods of time. Until this testing problem can be solved, the use of fire retardant coatings in exterior locations is questionable.

Strength and Deterioration: Strength and deterioration of both the materials coated and their fastenings is not a problem with most fire retardant coatings.

Moisture: Only those coatings which use hygroscopic fire retardant chemicals are vulnerable to absorbing moisture, thus reducing their effectiveness.

Toxicity: The reactive toxicity of the total products of combustion of the coated material as compared to the uncoated material should be considered when examining fire retardant coatings.

Preservative Treatments

Where subject to decay or insect attack, or to attacks by marine borers, timber and lumber are commonly pressure treated with coal tar creosote, pentachlors phenol, chromated copper arsenate and ammonical copper arsenate. Such treatments are used in piers, bridges, trestles, and similar structures, and to a lesser extent in buildings.

Oil type preservatives such as creosote cause an increase in fire hazard as compared with untreated wood. Fire retardants utilizing the proper ratio of phosphorous-halogen mixture are effective in reducing the combustibil-

Table 12-6C. Variations in Effect of Different Fire Retardant Coatings on Unprimed Douglas Fir

	Untreated	Painted with Coating X†	Painted with Coating Y†	Painted with Coating Z†
Total coverage (sq ft per gal)	—	100	100	200
(L/m^2)	—	2.45 L/m^2	2.45 L/m^2	1.23 L/m^2
Flame spread*	100	60	10	25
Fuel contributed*	100	35	15	10
Smoke developed*	100	40	5	10
Rate per coat (sq ft per gal)	—	300	200	200
(L/m^2)	—	0.82 L/m^2	1.23 L/m^2	1.23 L/m^2
Number of coats	—	3	2	1

* Where a range of values was obtained, the figure given here is the highest obtained.
† Each coating is listed by Underwriters Laboratories Inc., as meeting the minimum requirements for a fire retardant coating.

ity of the treated wood to that of untreated wood; however, there is an initial flash of "light ends" not observed in untreated wood (Gooch 1959).

Fire is reported to spread more rapidly and generate higher surface temperatures in structures containing newly creosoted wood than in structures of untreated wood, particularly where general fire temperatures are sufficiently high to vaporize the creosote oils. The adverse fire experience in fires in creosoted pier substructures, however, is due primarily to the excessive areas of wooden construction without firebreaks, which are inaccessible for manual fire fighting and lack automatic sprinkler protection.

Water-borne salt preservatives can provide some fire retardant effect but normal preservative concentrations are not recognized as effective fire retardant treatments. Conversely, some fire retardant formulations combine fire retardance and resistance to decay or to attack by insects.

Christmas Trees

Christmas trees become extremely flammable when cut long in advance of use and then brought indoors where heat and low humidity accelerate drying. To reduce the hazard, the tree should be kept indoors only as long as absolutely necessary, the trunk sawed off at an angle at least 1 in. (25 mm) above the original cut end, and the tree placed in water while it is in the house. Water should be added at intervals to the pail, tub, or tree stand in which the tree is placed so that the water level is always above the cut. To achieve a satisfactory degree of flame resistance in any combustible material including Christmas trees, it is essential to get a certain minimum quantity of effective flame retardant chemical either into or on the surface of the material to be treated. Since Christmas trees by their nature are not absorbent, the only way to effectively treat them is to apply a surface coating. Treating Christmas trees with simple solutions of water disolved chemicals such as borox-and-boric acid, diammonium phosphate, or ammonium sulfate is not effective. In order to provide a significant flame retardant value, the testing solution or coating must be thick or syrupy enough to form a fairly heavy coating.

Much misunderstanding in this area results from tests of fresh trees treated with simple water-thin solutions of the type described above. Such tests lead to the erroneous conclusion that the treatment is effective, when the fact is that the tree was naturally flame resistance due to its water content.

Pressurized aerosol containers of flame retardant Christmas tree coatings may be convenient, but their value is questionable due to the very limited quantity of the contents. Many cans would be necessary to adequately treat all parts of larger trees, but the high cost leads the user to apply only a token (and useless) coating.

WOOD BASE COMPOSITE PANELS

The group of materials referred to here as "wood base composite panels" includes products such as insulation boards, hardboards, particleboards, oriented strand boards, and laminated paper boards. They consist of a mixture of fibers, particles or flakes and binder formed under pressure into boardlike panels or tiles. These board or panel materials in final form retain some of the properties of the original wood but because of the manufacturing methods, gain new and different properties from those of the wood.

Wood base composite panels can be fire retardant treated by chemical admixtures, impregnations and coatings. These treatments vary in their effectiveness and are somewhat limited in their use by their effect on the physical properties of the untreated panel.

Admixtures

When mineral fibers, such as glass or rock wood, are added to a mixture of organic fibers, they will reduce the combustibility of the mixture, depending upon the amount of mineral fiber used. If mineral fibers completely replace the organic fibers, then the product becomes noncombustible, provided the binder used to join the fibers is noncombustible or has a sufficiently high ignition temperature. Mixtures of alumina trihydrate, boric acid and wood fibers have been successfully combined into panel products with very low flamespread, fuel contributed, and smoke developed values.

Impregnations

Attempts have been made to reduce the combustibility of organic composite boards by pressure impregnation of the finished product and by treatment of the raw fiber product itself. Treatments have included soluble salts, insoluble complexes, and chemical reactions with the organic materials themselves. The effect of these treatments on combustible composite boards varied with respect to the type of treatment and type and quantity of chemicals used. Treatment by pressure impregnation has not yet proved to be commercially feasible. At least one manufacturer has developed a successful wood pulp treatment process for both composite boards and acoustical tile, and similarly treated wood fiber acoustical tile is produced by other manufacturers.

Coatings

The treatments previously discussed are made during the manufacture of the material. Coatings can be applied during manufacture or to material already in use. Factory applied coatings consist of fire retardant paints; however, many are of questionable effectiveness when exposed to severe fire conditions. Their main effect is to cover the fuzzy surface of the untreated material.

The application of fire retardant paints of the intumescent variety to existing composite board will provide good protection for the exposed side of the board, provided the coatings are maintained. Some composite boards exhibit considerable dimensional instability, expanding when damp and contracting when dry. This movement may expose cracks and gaps in the material which will open untreated areas to the action of heat and rapid flame spread.

Fire Performance Tests of Wood-base Composites

The effectiveness of the treatments of composite boards can be shown with a fire test. Most of the methods listed under the section on wood can be used to evaluate their fire performance characteristics.

PAPER

Paper can be treated by impregnation to make it flame and glow retardant after the ignition source has been removed. In addition to treating various forms of paper, i.e., building paper, wrapping, packing, and decorative paper, these same treatments can be used to preserve foliage, leaves, grass skirts, broom straw, excelsior and similar materials if they are sufficiently absorbent. Adequate wetting may be achieved by dipping or spraying with some materials, while others may require soaking for several hours or even days.

The development of flame retardant products has been given impetus by the requirements for nonflammable display and decorative material in hotels, theaters, schools, hospitals, nursing homes, and other high occupancy buildings. Decorative crepe paper, corrugated display cardboard, textiles, and window shade materials may be made flame resistant during manufacture, which is the most reliable method. Field treating paper is difficult due to variations in paper finishes, sizing, and color fastness. It is preferable to purchase factory treated materials (Segal 1966).

MISUSE OF FIRE RETARDANTS

Fire retardant treatments are used for reduction of flame spread on structural wood members, and combustible interior finish, such as wood paneling, acoustical tile, and wood base composite panels. Flame retardant treatments are used for reducing the flammability of contents, furnishings, and decorative materials, such as curtains, draperies, upholstery, crepe and corrugated papers, Christmas trees, and miscellaneous dry plants and foliage. In all these applications, there are opportunities for misuse or misrepresentation. Some of the more prevalent kinds of misuse are described as follows.

1. Promoting pressure impregnated wood for uses where noncombustible materials are conventionally required. Such substitutions are appropriate in certain limited applications, but even the most effective treatments cannot entirely eliminate some contribution of fuel and smoke.
2. Using pressure impregnated or coated materials which are not intended for use in locations which are exposed to the weather or unusually high humidity conditions.
3. Attempting to treat the newer synthetic fiber fabrics with common water-soluble-salt flame retardant solutions. Such solutions are effective only on cellulose fibers, such as cotton and rayon, and the common animal fibers, wool, and silk. They are not effective for the treatment of nylon, acetate, Orlon, Dacron, Acrilan, or any of the similar synthetics presently in wide use.
4. Marketing flame retardant solutions in small aerosol spray containers for home use. This practice is subject to many abuses, including the lack of product labels to indicate the limitations of the contents and instructions for proper application. Advertising frequently contains wildly exaggerated claims for the product's utility, giving the impression that the contents of one can will "flameproof" virtually everything in the home. The product itself may be excellent and convenient to use, but there is a great risk that the average user will not achieve an effective degree of flame resistance. With few exceptions, flame retardant treatments should be left to skilled persons.
5. Advocating the use of ordinary water-soluble-salt solutions, for the treatment of nonabsorbent materials. Effective treatment of nonabsorbent materials can be achieved only by means of a substantial surface coating, requiring that the treating solution be of a syrupy or paint-like consistency.

The technology of flame retardant treatment is highly useful and can do much to increase firesafety. But because it is also a field vulnerable to entry by unskilled entrepreneurs and charlatans, the buyer should be alert to its limitations: Products which have been tested, rated, and labeled by reputable, independent testing laboratories are the consumers' best safeguard against the pitfalls listed above.

Bibliography

References Cited

American Wood Preservers. 1977. "All Timber Products-Preservative Treatment by pressure Processes." C1-77. American Wood-Preservers Association. *Book of Standards.*

ASTM. 1982. Proposed Method for Room Fire Test of Wall and Ceiling Materials and Assemblies. *1982 Annual Book of ASTM Standards,* Part 18. American Society for Testing and Materials. pp 1618-1638.

ASTM. 1983. Standard Terminology Relating to Fire Standards. E176-83a, *1983 Annual Book of ASTM Standards.* Vol 04.07. American Society for Testing and Materials. pp 416-420.

Browne, F. L. 1958. "Theories of the Combustion of Wood and Its Control." *Report No. 2136.* Forest Products Laboratory, Madison, WI.

Gooch, R. M., Kenaga, D. L., and Tobey, H. M. 1959. "The Development of Fire Retardants for Wood Treated with Oil-Type Preservatives." *Forest Products Journal.* Vol 9, No 10. Oct 1959.

Lieff, M. 1983. "Fire Resistant Coverings." *Fire Resistive Coatings: The Need for Standards.* ASTM STP 826. Morris Lieff and F. M. Stumpf, Eds. American Society for Testing and Materials. pp 3-13.

Leiu, P. J., Magill, J. H., and Alarie, Y. C. 1982. An Evaulation of Some Polyphosphazenes and Commercial Fire Retardants for Wood. *Journal of Combustion Toxicology.* Vol 9. (May 1982) p 65.

Little, R. W., Church, J. M., and Coppick, S. 1947. *Flameproofing Textile Fabrics.* ACS Monography 104. Reinhold, NY.

Schuyten, H. A., Weaver, J. W., and Reid, J. D. 1954. "Some Theoretical Aspects of the Flameproofing of Cellulose." *Advances in Chemistry Series No. 9.* American Chemical Society, Washington, DC.

Segal, L. 1966. "Flameproofing: Facts and Fallacies." *Fire Journal.* Vol. 60, No 6. Nov 1966. pp 41-45.

Stumpf, F. M. 1983. "Spray-Applied Fibrous Material Fire Resistive Coatings." *Fire Resistive Coatings: The Need for Standards.* ASTM STP 826. Morris Lieff and F. M. Stumpf, Eds. American Society for Testing and Materials. pp 14-23.

U.S. Forest Products Laboratory. 1962-64. Eickner, H. W. 1962. "Basic Research on the Pyrolysis and Combustion of Wood." *Forest Products Journal.* Apr 1962; Browne, F. L., and Tang, W. K. 1962. "Thermogravimetric and Differential Thermal Analysis of Wood and of Wood Treated with Inorganic Salts During Pyrolysis." *Fire Research Abstracts and Reviews.* Vol 4, Nos 1 and 2. pp 76-91; Tang, W. K., and

Neill, W. K. 1964. "Effect of Flame Retardants on Pyrolysis and Combustion of Alpha-Cellulose." *Journal of Polymer Sciences*. Part C6. pp 65-81. Brenden, J. 1963. "The Influence of Inorganic Salts on the Products of Pyrolysis of Ponderosa Pine." *M. S. Thesis*. University of Wisconsin. Tang, W. K. 1964. "Effect of Inorganic Salts on the Pyrolysis, Ignition and Combustion of Wood, Cellulose, and Lignin." *Ph.D. Thesis*. University of Wisconsin. Browne, F. L., and Tang, W. K. 1963. "Effect of Various Chemicals on the Thermogravimetric Analysis of Ponderosa Pine." *U.S. Forest Service Research Paper RPL-6*. May 1963. Forest Products Laboratory, Madison, WI.

Verburg, G. B., et al. 1964. "Water-Resistant, Oil-Based, Intumescing Fire Retardant Coatings, I, Developmental Formulations." *Journal of The American Oil Chemists' Society*. Vol 41. Oct 1964.

White, R. H. 1983. "Use of Coatings to Improve Fire Resistance of Wood." *Fire Resistive Coatings: The Need for Standards*. ASTM STP 826. Morris Lieff and F. M. Stumpf, Eds. American Society for Testing and Materials. pp 24-39.

NFPA Codes, Standards, Recommended Practices and Manuals. (See the latest *NFPA Codes and Standards Catalog* for availability of current editions of the following documents.)

NFPA 101, *Life Safety Code®*.

NFPA 220, *Standard Types of Building Construction*.

NFPA 702, *Standard for Classification of the Flammability of Wearing Apparel*.

NFPA 703, *Standard for Fire-Retardant Treatments of Building Materials*.

Additional Readings

Arledter, H. F., Knowles, S. E., and Druck, J. U., "Flame-Retardant Papers and Laminates," *TAPPI* (Technical Association of the Pulps and Paper Industry), Vol. 47, No. 8, Aug. 1964.

ASTM, "Standard Method of Test for Combustible Properties of Treated Wood by the Crib Test," *ASTM E 160-50*, American Society for Testing and Materials, Philadelphia, 1969.

ASTM. "Standard Method of Test for Combustible Properties of Treated Wood by the Fire-Tube Apparatus," *ASTM E 69-50* American Society for Testing and Materials, Philadelphia, 1969.

Bescher, R. H., "Fire Retardant Treated Wood," *The Construction Specifier*, Vol. 23, No. 3, Mar. 1970, pp. 59-62.

Canadian Wood Council, *Wood Fire Behavior and Fire Retardant Treatment: A Review of the Literature, Nov. 1966*, Ottawa, 1966.

Cullis, C. F., and Hirschler, M. M., *The Combustion of Organic Polymers*, Oxford University Press, New York, 1981.

Eichner, H. W., and Schaffer, E. L., "Fire-Retardant Effects of Individual Chemicals on Douglas Fir Plywood," *Fire Technology*, Vol. 3, No. 2, May 1967, pp. 90-104.

George, C. W., and Aylmer, D. B., "Energy Release Rates in Fire Retardant Evaluation," *Fire Technology*, Vol. 6, No. 3, Aug. 1970, pp. 203-210.

Hastie, John W., "Molecular Basis of Flame Inhibition," *Journal of Research of the National Bureau of Standards*, 77A, No. 6, Nov.-Dec. 1973, pp. 733-754.

Hastie, John W., "Chemical Aspects of Flame Inhibition," *NBS Special Pub. 411*, National Bureau of Standards, Washington, DC, Nov. 1974, pp. 30-36.

Jensen, Rolf, ed., *Fire Protection for the Design Professional*, Cahners, Boston, 1975, pp. 181-189.

Kirkpatrick, Allan T., Curtis, Henry, and Adelgren, Alan, "Experimental Measurements of the Thermal Effectiveness of Two Types of Protective Clothing for Fire Fighters," *Fire Technology*, Vol. 18, No. 3, Aug. 1982, pp. 259-267.

Krasny, J. F., Singleton, R. W., and Pettengill, J., "Performance Evaluation of Fabrics Used in Fire Fighters' Turnout Coats," *Fire Technology*, Vol. 18, No. 4, Nov. 1982, pp. 309-318.

Lyons, J. W., *The Chemistry and Uses of Fire Retardants*, Wiley Interscience, New York, 1970.

Montle, J. F., "The Role of Magnesium Oxychloride as a Fire-Resistive Material," *Fire Technology*, Vol. 10, No. 3, Aug. 1974, pp. 201-210.

Rayne, E. T., et al., "Water-Resistant, Oil-Based, Intumescing Fire-Retardant Coatings," *Journal of the American Oil Chemists' Society*, Vol. 41, Oct. 1964, pp. 670-674.

Tryon, G. H. 1966. "Looking at Fire Hazards—Christmas Trees." *Fire Journal*, Vol. 60, No. 6, Nov. 1966, p. 29.

Underwriters Laboratories, Inc., *Building Materials List, Part I*. Underwriters Laboratories Inc., Chicago, published annually.

Vandersall, H. L., "Intumescent Coating Systems, Their Development and Chemistry," in *Fire Prevention and Suppression*, Vol. 10, Hilado, C. J., ed., Technomic, Westport, CT, 1974, pp. 116-159.

Westgate, M. W., et al., *Fire Retardant Paints, Advances in Chemistry Series No. 9*, American Chemical Society, Washington, DC, 1954.

MATERIALS HANDLING EQUIPMENT

Revised by Richard Munson

Materials handling equipment, particularly industrial trucks, mechanical and pneumatic stock conveying systems, and cranes, are essential services in industrial and commercial activity. This chapter describes the basic types of equipment and systems for materials handling that are encountered, and discusses the fire and explosion hazards inherent with them. Protection methods are also discussed.

Readers interested in or concerned with materials handling equipment should also consult Section 5, Chapter 9, "Dusts;" Chapter 10, "Metals;" Section 8, Chapter 2, "Electrical Systems and Appliances;" Section 11, Chapter 7, "Storage and Handling of Grain Mill Products;" Section 12, Chapter 10, "Control of Electrostatic Ignition Sources;" Chapter 11, "Special Systems for Explosion Damage Control;" Chapter 13, "Air Moving Equipment;" and Section 18, Chapter 7, "Water Spray Protection."

INDUSTRIAL TRUCKS

Industrial lift trucks are one of the most common types of materials handling equipment. Although they are available in many special designs to suit the type of load to be handled, the fork and squeeze clamp types are most common. They may be propelled electrically, or by diesel, gasoline, or LP-Gas engines. Unless these vehicles are of a type listed by a testing laboratory, and properly maintained and used, they may introduce serious fire dangers. Industrial trucks are manufactured in many designs to suit the many uses and types of loads to be handled. All of the vehicles should be selected, outfitted, maintained, and operated in accordance with the hazards of the locations.

Type Designations and Areas of Use

NFPA 505, *Fire Safety Standard for Powered Industrial Trucks, Including Type Designations, Areas of Use, Maintenance and Operations*, lists thirteen different type designations of industrial trucks or tractors. These designations are divided into power types: diesel, electric, gasoline, and LP-Gas. In addition, dual-fuel trucks are available. The designation letter reflects the power type, as follows:

1. *Type D* units are diesel powered units having minimal acceptable safeguards against inherent fire hazards.
2. *Type DS* units are diesel powered units that, in addition to all the requirements for the type D units, are provided with additional safeguards to the exhaust, fuel, and electrical systems.
3. *Type DY* units are diesel powered units that have all the safeguards of the type DS units, and in addition, do not have any electrical equipment, including ignition. They are also equipped with temperature limitation features.
4. *Type E* units are electrically powered units having minimum acceptable safeguards against inherent fire and electrical shock hazards.
5. *Type ES* units are electrically powered units that, in addition to all of the requirements for the type E units, are provided with additional safeguards to the electrical system to prevent emission of hazardous sparks and to limit surface temperatures.
6. *Type EE* units are electrically powered units that have, in addition to all of the requirements for the types E and ES units, the electric motors and all other electrical equipment completely enclosed.
7. *Type EX* units are electrically powered units that differ from the type E, ES, or EE units in that the electrical fittings and equipment are so designed, constructed, and assembled that the units may be used in atmospheres containing specifically named flammable vapors, dusts, and under certain conditions, fibers. Type EX units are specifically tested and classified for use in Class I, Group D or for Class II, Group G hazardous locations as defined in NFPA 70, *National Electrical Code®*.
8. *Type G* units are gasoline powered units having minimum acceptable safeguards against inherent fire hazards.
9. *Type GS* units are gasoline powered units that, in addition to all the requirements for the type G units, are provided with additional safeguards to the ex-

Mr. Munson is a Consultant in the Safety and Occupational Health Division of E. I. Dupont de Nemours Company, and Chairman of NFPA's Industrial Trucks Technical Committee.

haust, fuel, and electrical systems.

10. *Type LP* units are liquefied petroleum gas powered units having minimum acceptable safeguards against inherent fire hazards.
11. *Type LPS* units are liquefied petroleum gas powered units that, in addition to the requirements for the type LP units, are provided with additional safeguards to the exhaust, fuel, and electrical systems.
12. *Type G/LP* units operate on either gasoline or liquefied petroleum gas having minimum acceptable safeguards against inherent fire hazards.
13. *Type GS/LPS* units operate on either gasoline or liquefied petroleum gas and, in addition to all requirements for the type G/LP units, are provided with additional safeguards to the exhaust, fuel and electrical systems.

Industrial trucks of the various types are limited to the locations specified in Table 12-7A. Many fires involving industrial trucks that have spread beyond the truck to involve other property have been the result of operating a less than minimum type truck in a hazardous location.

The greatest potential fire source for gasoline, diesel and LP-Gas powered trucks are fuel leaks that are ignited by the hot engine, hot muffler, ignition system, other electrical equipment, or other sparks. This danger is somewhat less for diesel trucks because of the higher flash point of diesel fuel; however, it is especially present in LP-Gas trucks as the vapors are difficult to disperse and tend to gravitate toward lower spots or pits. Care must be exercised with LP-Gas trucks to avoid the high temperatures near ovens, furnaces, and similar sources of heat. Special safeguards in the designs of some types of gasoline, diesel, and LP-Gas trucks help to reduce these fire hazards, but they cannot be completely avoided and areas of use should be rigidly limited. (See Table 12-7A.)

To facilitate identification of truck types and their areas of use, a uniform system of marking has been developed and is described in NFPA 505, *Powered Industrial Trucks.* (See Figs. 12-7A and 12-7B.) Use of these truck and building markers allows easy and quick recognition of a vehicle approved for the location.

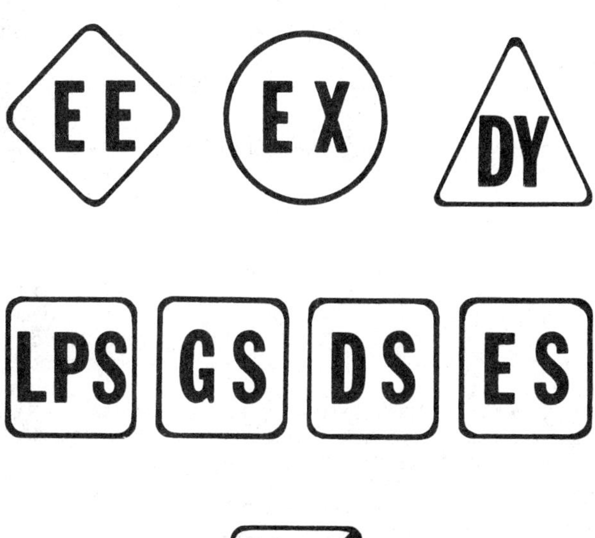

FIG. 12-7A. Markers used to identify the various types of industrial trucks.

FIG. 12-7B. Building signs for posting at entrances to hazardous areas.

Fire Hazards and Prevention

Careless and uninformed operation of industrial trucks has contributed to much property loss. Collision with sprinkler piping, fire doors, and other fire protection equipment; too high tiering of rack storage; and the careless handling of loads, such as containers of flammable liquids, have not only directly contributed to fires but have increased fire fighting difficulties. A complete course of instruction for truck operators can reduce the risk of such accidents. Adequate, clear passageways for truck travel and clear warnings of overhead and exposed piping will also reduce the number of incidents. A good source for safe operating rules is the *American National Standard Safety Code for Powered Industrial Trucks,* ANSI B56.1-1983.

A high number of fires involving industrial trucks are caused by equipment failure due to a lack of maintenance. A system of regularly scheduled maintenance based on engine-hour or motor-hour experience can greatly reduce the incidence of fires of this nature. Trucks are exposed to severe wear and tear, and maintenance programs must be rigidly adhered to. Detecting faulty or damaged fuel connections on gasoline, diesel, and LP-Gas trucks, and removing accumulations of grease and dirt require special attention. Providing portable extinguishers suitable for the hazards presented by the truck and the environment is good practice.

LP-Gas fuel containers require special care to prevent fires. They must not be overfilled. Use a soap solution to check for leaks, never a match or open flame. Means can be provided in the fuel system to minimize the escape of fuel when removable LP-Gas containers are exchanged (i.e., closing the valve on the LP-Gas container and using an approved automatic quickclosing coupling in the fuel

Table 12-7A. Recommended Types of Trucks for Various Occupancies

(Source: Factory Mutual System)

Location	Typical occupancies	Types of trucks†‡ approved and listed
Indoor or outdoor locations containing materials of ordinary fire hazard	Grocery warehouse Cloth storage Paper manufacturing and working Textile processes except opening, blending, bale storage, and other Class III locations Bakery Leather tanning Foundries and forge shops Sheet-metal working Machine-tool occupancies	Electrical—Type E Gasoline—Type G Diesel—Type D LP-Gas—Type LP Dual Fuel—Type G-LP
Class I, Division 1.* Locations in which explosive concentrations of flammable gases or vapors may exist under normal operating conditions or where accidental release of hazardous concentrations of such materials may occur simultaneously with failure of electrical equipment	Few areas in this division in which trucks would be used	Electrical—Type EX# Gasoline, diesel, and LP-Gas—not recommended for this service
Class I, Division 2.** Locations in which flammable liquids or gases are handled in closed systems or containers from which they can escape only by accident or locations in which hazardous concentrations are normally prevented by positive mechanical ventilation	Paint mixing, spraying, or dipping Storage of flammable gases in cylinders Storage of flammable liquids in drums or cans Solvent recovery Chemical processes using flammable liquids Paper and cloth coating using flammable solvents in closed equipment Rubber-cement mixing	Electrical—Types EE, EX** Diesel—Type DY** Gasoline, diesel Types D & DS, and LP-Gas—not recommended for this service
Class II, Division 1.* Locations in which explosive mixtures of combustible dusts may be present in the air under normal operating conditions, or where mechanical failure of equipment might cause such mixtures to be produced simultaneously with arcing or sparking of electrical equipment, or in which electrically conductive dusts may be present	Grain processing Starch processing Starch molding (candy plants) Wood-flour processing	Electrical—Type EX§ Diesel—Type DY Gasoline, diesel Types D & DS, and LP-Gas—not recommended for this service
Class II, Division 2.* Locations in which explosive mixtures of combustible dusts are not normally present or likely to be thrown into suspension through the normal operation of equipment but where deposits of such dust may interfere with the dissipation of heat from electrical equipment or where such deposits may be ignited by arc or sparks from electrical equipment.	Storage and handling of grain, starch, or wood flour in bags or other closed containers Grinding of plastic molding compounds in tight systems Feed mills with tightly enclosed equipment	Electrical—Types EE, EX, ES‖ Gasoline—Type GS‖ Diesel—Type DY, DS‖ LP-Gas—Type LPS‖
Class III, Division 1.* Locations in which easily ignitible fibers or materials producing combustible flyings are handled, manufactured, or used	Opening, blending, or carding of cotton or cotton mixtures Cotton gins Sawing, shaping, or sanding areas in woodworking plants Preliminary processes in cordage plants	Electrical—Types EE, EX Diesel—Type DY Gasoline, diesel Types D & DS, and LP-Gas—not recommended for this service
Class III, Division 2.* Locations in which easily ignitible fibers are stored or handled (except in process of manufacture)	Storage of textile and cordage fibers Storage of excelsior, Kapok, or Spanish moss	Electrical—Types EE, ES, EX, preferred; E‖ Gasoline—Type GS Diesel—Type DS, DY LP-Gas—Type LPS

NOTES:

* Hazardous location as classified in the *National Electrical Code.*

† Type G (gasoline). Type D (diesel), and Type LP (LP-Gas) trucks are considered to have comparable fire hazard.

‡ Type GS (gasoline), Type DS (diesel), and Type LPS (LP-Gas) trucks are considered to have comparable fire hazard.

§ Acceptable for Group G, and for Groups E and F, but subject to special investigation.

‖ Acceptable but subject to special investigation.

\# Class I, Division 1, Group D only; no truck should be used in Groups A, B and C.

** Class I, Division 2, Group D only; may be used in Groups A, B and C areas subject to local authority.

line). Removable LP-Gas containers must be securely mounted to prevent their jarring loose, slipping, or rotating. The safety pressure relief valve opening should always be in contact with the vapor space (top) of the container. When the container is properly installed, an indexing pin and container clamp(s) correctly position the container. It is good practice to examine all LP-Gas containers for defects or damage before refilling.

The number of fires involving battery powered trucks is comparatively small. Nevertheless, electrical short circuits, hot resistors, arcing and fused contacts, and exploding batteries have contributed to the fire occurrence in electric powered industrial trucks.

The two types of batteries in common use are: (1) lead, and (2) nickel-iron. They contain corrosive chemical solutions, either acid or alkali, and therefore present a chemical hazard. On charge, they give off hydrogen and oxygen which, when combined in certain concentrations, can be explosive. Battery charging installations should be located in areas specifically set aside for that purpose. The facilities should include means for flushing and neutralizing spilled electrolyte; barriers for protecting charging apparatus from damage by trucks; adequate ventilation for dispersal of fumes from gassing batteries; and adequate fire protection. A carboy filler or siphon is used in dispensing acid from carboys. Special care is taken to prevent open flame, sparks, or electric arcs in battery charging areas. When charging batteries, the vent caps are kept in place to avoid electrolyte spray. Care also is taken to assure that vent caps are functioning. The battery (or compartment) cover(s) remain open during charging to dissipate heat and gas.

Recharging and Refueling

Refueling and battery charging operations are performed by trained and designated personnel, and only in specified, well ventilated areas (outdoor refueling is recommended where practicable), away from manufacturing and service areas. Smoking is always prohibited in these areas. Scales for weighing LP-Gas containers should be calibrated to insure accurate filling.

Trucks using liquid fuels, such as gasoline and diesel fuel, should only be refueled from approved dispensing pumps in safe locations away from sources of heat and ignition. Care must be taken not to spill fuel or overfill the vehicle fuel tank. Approximately 50 percent of the fires involving trucks using liquid fuels are the result of spillage during refueling.

Maintenance and Repair

Repairs are never made to trucks in Class I, II, and III hazardous locations. Repairs to the fuel and ignition systems of industrial trucks that involve fire hazards should be conducted in locations designated for such repairs. Repairs to the electrical system of battery powered industrial trucks are performed only after the battery has been disconnected. All parts of any industrial truck requiring replacement should be replaced only with parts providing the same degree of firesafety as those used in the original design. Water mufflers should be filled daily or as frequently as is necessary to prevent depletion of the supply of water below 75 percent of the filled capacity. Do not operate vehicles with mufflers having screens or other parts that have become clogged. Immediately remove any vehicle that emits hazardous sparks or flames from the exhaust system from service, and do not return it to service until the cause for the emission of such sparks and flames has been eliminated. When the temperature of any part of any truck is in excess of its normal operating temperature, a hazardous condition is created, and the vehicle must be removed from service until the cause for such overheating has been eliminated.

It is good practice to keep industrial trucks in a clean condition, reasonably free of lint, excess oil, and grease. The conditions under which the truck is operating should govern the frequency of cleaning. Noncombustible agents are preferred for cleaning trucks. Low flash point (below 100°F or 38°C) solvents must not be used. Precautions regarding toxicity, ventilation, and fire hazard should correspond to the agent or solvent used. When antifreeze is required in the cooling system, glycol base material shall be used.

MECHANICAL CONVEYORS AND ELEVATORS

Commonly used equipment in materials handling are mechanical conveyors and elevators which provide an efficient means of transporting bulk and packaged materials. There are many classes and designs of mechanical conveyors. The primary selection considerations are the distance and inclination over which material is to be conveyed; the types of atmosphere and location through which material is to be conveyed; and the lump size, density, flowability, abrasiveness, toxicity, corrosiveness, etc. of the materials to be conveyed. Dust materials or dust atmospheres, or both, require equipment that can be adequately enclosed. Screw and "en masse" conveyors are limited in the lump size they can handle. Temperatures often rule out belt conveyors; long distances usually require the use of belt conveyors. Substantial vertical lifts are best handled by bucket elevators.

From the point of view of fire protection, the considerations of temperature and dust are the most important; in addition, the protection of openings and control of static electrical charges are installation factors that must be given careful attention. In the use of belt and other conveyors, two principal hazards should be considered. These are the materials being handled and the belt itself since both can contribute to fire and the spread of fire.

Temperature

Screw conveyors, vibrating types, and certain pan conveyors are normally the best choice for handling very hot materials or for use in high temperature atmospheres. While the upper limit for most belt conveyors is 200°F (93°C), they are normally restricted to less than 150°F (65°C) (Buffington 1969). Experiments by the Bureau of Mines demonstrate that the highest rate of flame propagation occurs with rubber belts, but that there is little difference in the ignition characteristics among neoprene, rubber, and polyvinyl chloride. Thus, belt conveyors are not suitable for movement of hot or molten materials, and are not used in high temperature atmospheres.

Fire Protection

Loss experience shows that automatic sprinkler or water spray protection is generally needed for important belt conveyors, particularly if they are enclosed. Controls should be provided to shut down the conveyor when sprinklers operate.

Hose stream protection to reach all portions of the conveyor is also needed. This protection can be provided by 1½ in. (38 mm) hose connections to sprinkler lines or from hydrants.

Dust Control

The most frequently occurring factors in fires involving mechanical conveyors are a dusty material, a dust atmosphere, or the dust inevitably created by the materials handling process. Dust will almost certainly be produced in any materials handling systems, especially at inlet and discharge points and the long chutes for free falling materials (Buffington 1969).

Dusty material should be fed to belt conveyors through a choke feed to prevent dust clouds. Where dusty conditions prevail, adequate aspiration should be provided to remove dust to collectors in safe locations. If dust clouds cannot be avoided, it is better to use spiral or enclosed conveyors where the escape of dust can be more readily prevented. Conveyor piping should be of sufficient strength to withstand the maximum pressure produced in explosions of the dust involved. Avoid sharp changes in direction wherever possible, and provide vent pipes to outdoors at any necessary changes in direction and at ends of lines. (This should not be construed as prohibiting the use of explosion relief vents.) Screw conveyors should be fully enclosed in tight noncombustible housings with freelifting covers at discharge end and over each shaft coupling. "En masse" or drag type conveyors should be of substantial metal construction designed to prevent escape of dust. Covers on cleanout, inspection, and other openings should be securely fastened. Conveyors should be designed and constructed to withstand anticipated explosion pressures, considering the pressure release afforded by explosion relief vents. A choke or seal of proper design, or a suppression system, should be installed in a conveyor (other than pneumatic conveyors) to prevent the propagation of an explosion from one building to another or from one portion of a building to another separated by a fire wall.

Dust collectors should be located outdoors or in detached rooms with adequate explosion vents for collectors and rooms. Where it is necessary to use bag-type collectors, they should be enclosed in a metal housing. A single dust collector should not serve several separate processes. Use individual collectors for progressive stages in any one process. Water spray collectors may be located within buildings, but they are not recommended for certain types of dust.

Static Protection

All parts of the machinery and conveyors should be thoroughly bonded and grounded to minimize static discharges. The chances of generating static electricity are increased when heated or dry materials are conveyed or the conveyor belt is operated in a heated or dry atmosphere (FMEC 1976). Static electricity can be controlled by the use of belts made of conductive material, by applying a conductive dressing to the belt surface, or by installing a grounded static collector nearly in contact with the belt just beyond the point where the belt leaves the pulley (FMEC 1976). Pulleys, guards, and other metal bodies should also be grounded.

Protection of Conveyor Openings

Water Spray Method: The protection of openings in walls and floors through which conveyors pass presents difficulties in the installation of ordinary forms of closures due to the objects carried through the opening. Where fire doors or shutters are impractical for conveyor openings, a method of protection incorporating the pressure effect and cooling action of water spray from directed spray nozzles is available. (See Fig. 12-7C.) With proper nozzle design

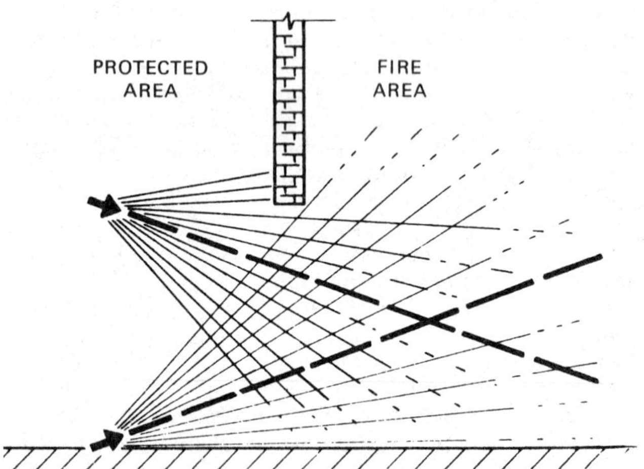

FIG. 12-7C. Spray nozzles protecting a conveyor opening in a fire wall.

and water pressure to provide suitable water velocity and droplet size, the pressure effect from the nozzles will overcome draft due to the temperature difference between one side of the wall and the other and the height of the opening above floor level, unless adverse air currents are prevalent. Since the cooling effect of the spray is directly proportional to the time of exposure of hot gases in the draft to the spray, the effectiveness of the heat absorption may be increased by adding an enclosure to the opening. (See Fig. 12-7D.) Figure 12-7E shows the cooling effect of four ½ in. (12.7 mm) nozzles on an 8 by 8 ft (2.4 by 2.4 m) opening with various draft velocities and nozzle pressures in tests conducted by Factory Mutual Engineering Corporation. The nozzles were discharging water at 28 gpm (106 L/min) each, with an effective angle of about 65°.

The Factory Mutual recommendations for installation include the following:

1. Where fire may be expected to originate on either side of the opening, nozzles are installed on both sides. Nozzles are controlled by an automatic valve actuated by a heat detector. Four nozzles per side are recommended to give complete coverage of the opening. Water discharge rates between 2 and 4 gpm/sq ft [81 and 161 (L/min)/m^2] or more, depending upon the height of the opening and unfavorable draft effects, are considered desirable. Nozzles are located at an angle not more than 30° between the center line of nozzle discharge and a line perpendicular to the plane of the

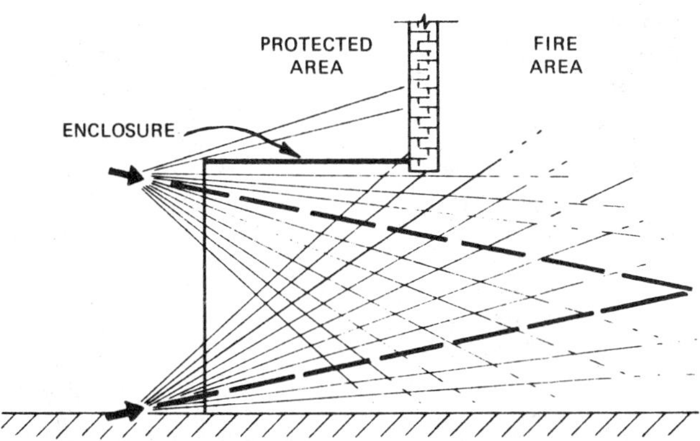

FIG. 12-7D. Greater heat absorption is possible by enclosing the opening for a conveyor.

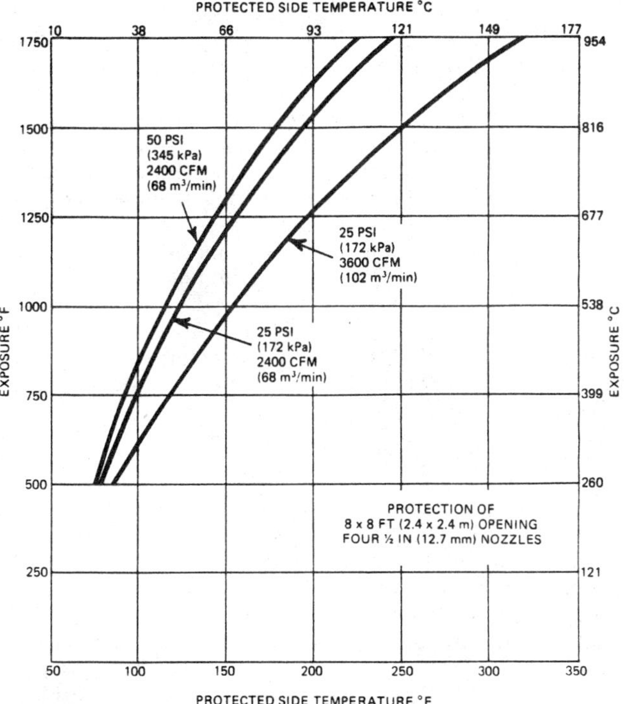

FIG. 12-7E. Exposure temperatures vs. protected side temperatures at various draft velocities and nozzle pressure.

opening. To prevent the nozzle counterdraft from forcing air from the fire area into other areas, all communicating openings to the fire area are protected in a standard manner.

2. Conveyor openings through floors may also be protected by this method, provided an enclosure is constructed around the conveyor from the floor up to or slightly beyond the spray nozzles and draft curtains extending 20 to 30 in. (0.5 to 0.8 m) below and around the floor opening. (See Fig. 12-7F.)
3. The effectiveness of the protection system is, of course, dependent upon rapid detection and appropriate interlocks between the detection system and the machinery.

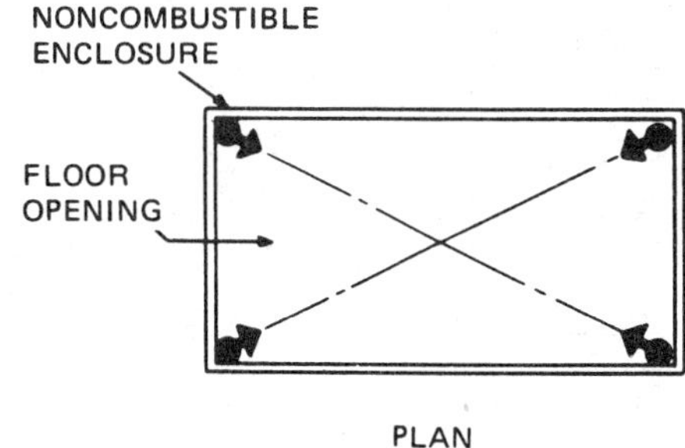

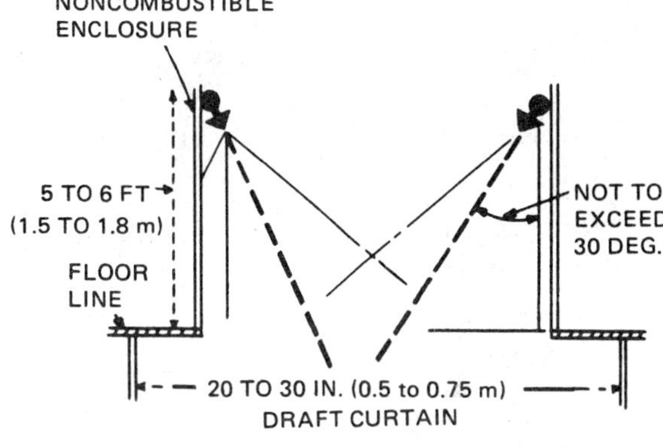

FIG. 12-7F. Spray nozzle protection for floor opening.

Fire Doors: It is a misconception that conveyorized openings cannot be protected by fire doors. Where possible, of course, conveyor penetration of a fire wall is avoided by rerouting, or as is sometimes feasible with a one story building, by running the conveyor through the roof, over the fire wall, and down within an inverted "V" housing arranged to readily vent fire to atmosphere. (See Fig. 12-7G.) Any cutout of a labeled fire door done in the

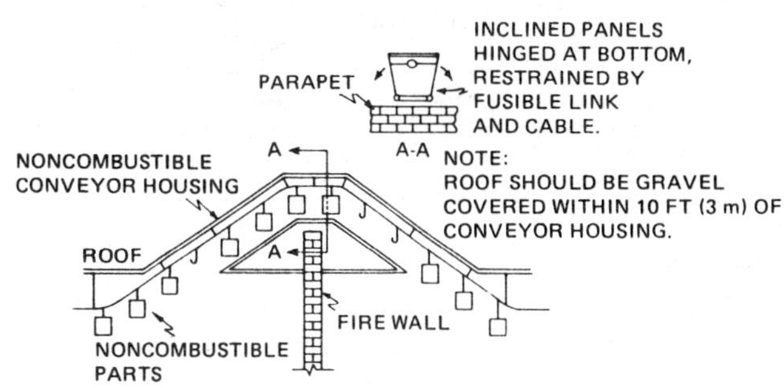

FIG. 12-7G. A conveyor carried over a fire wall.

field to allow for closure about a conveyor track or other components voids its label and this practice should be avoided, if possible. Where notching is distinctly advantageous, a certificate may be furnished by the testing laboratory affixing the label if it is found by inspection that the notched door is in compliance with the laboratory standards in all other aspects.

Figures 12-7H through 12-7J illustrate various conveyor designs and/or programming devices which will

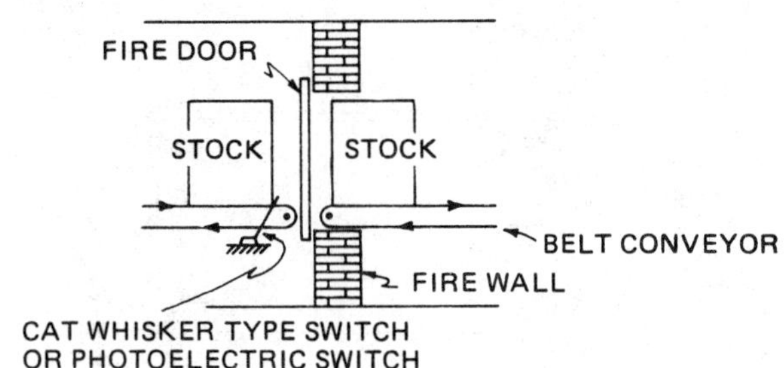

FIG. 12-7H. Protection of openings when a belt conveyor can be interupted.

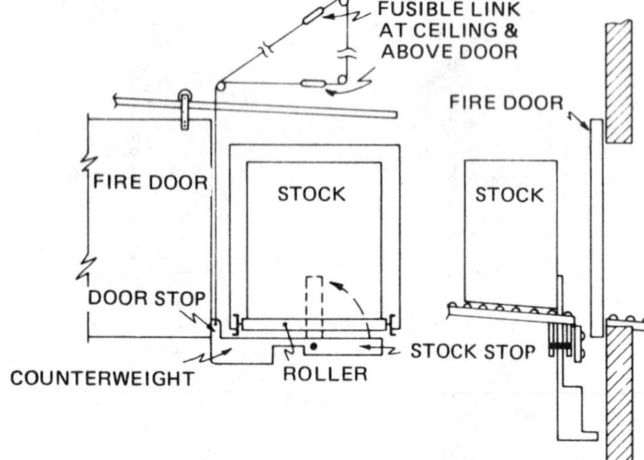

FIG. 12-7I. A method of stopping stock on a gravity roller conveyor.

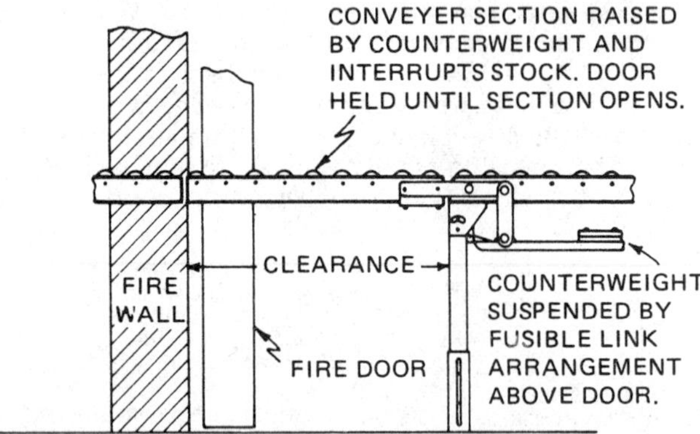

FIG. 12-7J. A counter weighted, hinged section of a roller conveyor.

minimize or eliminate the threat of obstruction to complete fire door closure by the conveyor or conveyed stock.

The illustrations can only show the basic concepts. Proper performance depends equally on conservative design, good workmanship in installation, operating inspection, and maintenance. Guidelines to observe are:

1. Select a design that is as simple and direct acting as possible. Emphasis is on "fail safe" operation.
2. Program the sequence of operating steps and interlocks so that obstruction (conveyor, conveyorized material, etc.) to the door closure is positively and permanently (until manually reset) removed from the door's path before the door is released to close.
3. Design structural and mechanical components, linkages, clearances, etc. in a conservative manner. Counterweights, springs, and other operating forces (uninterruptable by initial fire stages) must have enough reserve strength to handle overload introduced by reasonable anticipated minor changes in configuration and weight of conveyorized material, normal wear, friction, etc. Major changes will necessitate complete re-engineering to ascertain adequacy of the design, with reinforcement as necessary.
4. Incorporate self-releasing features in the design of conveyor components (trolley track, chain, supports, etc.)
5. Maintain ⅜ in. (9.5 mm) clearance between the door and the sill.
6. If it is advisable, provide another fire door on the opposite side of the opening to increase the reliability of the protection of the wall opening in the event of a fire. Similarly, when the property is sprinklered, consideration should be given to the advisability of reinforcing the protection of the opening by a water curtain from automatic sprinklers. Care must be taken that sprinkler discharge does not impede operation of any fusible links on the fire door assembly.
7. Following installation, conduct a number of operating tests that reflect the range of varied adverse conditions which must be anticipated to ascertain that all components operate smoothly, in proper sequence, within specified time interval, and with adequate clearances and tolerances.
8. Close all fire doors during inoperative periods.
9. Routine closure that simulates emergency operation can indicate the continued adequacy of the protection of the opening.

Friction, Overheating

Many fires involving conveyors, particularly those used for such materials as raw cotton, grains, powders, coal, etc., are caused by the heat of friction which results from the accumulations of grease and dirt and the overheating of defective or unlubricated parts, especially rollers. This frequent fire hazard is easily reduced and controlled by patrolling belts, by frequent equipment inspection, and by the removal of dirt and grease build-ups (Mitchell 1967). Early replacement of old and worn parts is also an important item.

Bucket Elevators

These elevators are used in bulk processing plants to convey loads vertically, and they are susceptible to the same fire hazards as the mechanical conveyors. The same precautions need to be taken for temperature, dust control, protection of openings, and elimination of friction and overheating. Elevators are best enclosed in substantial dusttight casings. These casings should be of noncombustible construction, extend without reduction in size through the roof, and be fitted with light weatherproof covers designed to lift readily and relieve explosion pressure within.

Elevators should be installed close to an outside wall of the building and provide short direct vents through the wall at 20 ft (6 m) intervals on tall elevator legs. Doors should be installed to provide access to head and boot pulleys. All doors in elevator leg casings should be dusttight. Ample clearance around elevator boots should be provided for cleaning and oiling. Safeguard elevators against overheating or choking by automatic releases actuated by overloading or reduction below normal operating speed.

PNEUMATIC CONVEYORS

Pneumatic conveying is a common method of transferring dusts from place to place in a building or from one building to another. This process presents an explosion hazard due to the ease and rapidity by which explosive proportions of dust can be introduced into the air (FMEC 1976).

Description of Systems

Pneumatic conveying system consists of an enclosed tubing system in which a material is normally transported by a stream of air having a sufficiently high velocity to keep the conveyed material in motion. Noncombustible gases may be used in place of, or mixed with, air. Such systems are of two principal types, or a combination of the two types.

Pressure Type: Pressure type systems transport material by utilizing air at greater than atmospheric pressure. These systems basically consist of a blower that draws air through a filter; an airlock feeder that introduces materials into the system; tubing or ducts; and a suitable air-material separator.

Suction Type: Suction type systems transport material by utilizing air at less than atmospheric pressure. These systems basically consist of a material and air intake; tubing or ducts; a suitable air-material separator; and a suction fan or blower.

See Figure 12-7K for a schematic drawing of a typical combination pressure type and suction-type system.

Conveyor Ducts

Conveyor ducts are fabricated from nonferrous, minimum-sparking metal, or nonmagnetic, minimum-sparking stainless steel. They are electrically bonded and grounded. Plastic or other nonconductive liners are not desirable.

Inert Atmospheres

Inert gas should be used in conveying systems wherever the concentration of powder or dust may come within the explosive range. The inert gas, such as nitrogen, argon, helium, etc., with a limiting oxygen concentration based on the explosion characteristics of the particles in the system is essential for preventing fires and explosions. Basically, it contains insufficient oxygen to permit combustion, no carbon monoxide, and has a dew point such that no free moisture can condense or accumulate at any point in the system. Further limitations are dictated by the type of material being conveyed, and the character of the inert gas. For example, the inert gas for magnesium dust systems should not contain carbon dioxide. Oxygen limits of three to five percent have been maintained in aluminum powder systems using a controlled type of flue gas. Other limits are applicable where other inert gases and dusts are present (Jacobson 1964).

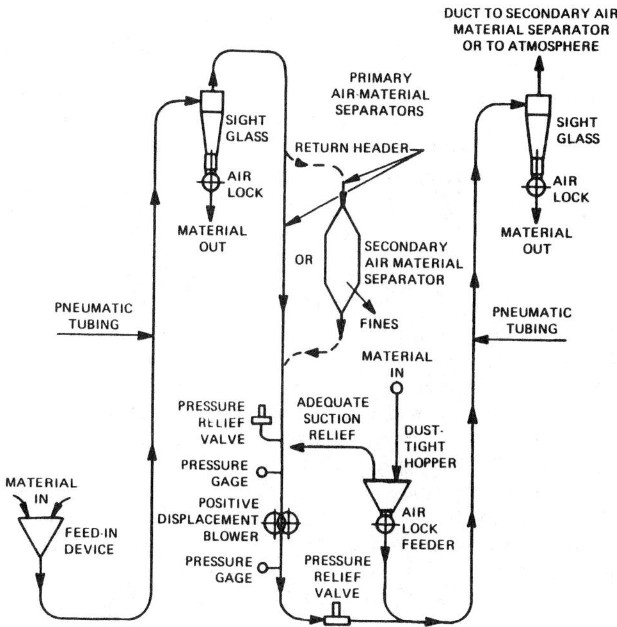

FIG. 12-7K. A typical transfer system of the combination pressure type and suction type with a high capacity.

A continuous monitor is needed to sound an alarm if the oxygen content of the inert gas is not within the established safe range.

Light Metal Conveying: Light metal and light metal alloy powders are produced by various mechanical means of particle size degradation. These processes, as well as certain finishing and transporting operations, have a tendency to expose a continuously increasing area of new metal surface. Most metals immediately experience a surface reaction with available atmospheric oxygen forming a protective oxide coating which then serves as an impervious layer to inhibit further oxidation. This reaction is exothermic, producing sensible heat. If a fine or thin lightweight particle having a large area of new surface is suddenly exposed to the atmosphere, enough heat will be generated to raise its temperature to the ignition point.

Completely inert gas is not used as an envelope to promote operational safety, or for transport of light metal powder in a pneumatic or fluidized transfer device. This would be a very unsafe practice because somewhere in the process of manufacture, packaging, or ultimate use, the powder will eventually be exposed to the atmosphere, where the unreacted surfaces will react suddenly with available oxygen to produce enough heat to cause either a fire or an explosion. To provide maximum safety, a means for the controlled oxidation of newly exposed surfaces is provided as soon as they are exposed by regulating the oxygen content of the inert gas. Tests conducted by the U.S. Bureau of Mines and others have disclosed that an inert gas as described herein is effective for this purpose. This mixture serves to control the rate of oxidation, and at the same time provides an environment which materially reduces the fire and explosion hazard.

Air Conveying

If the conveying gas is air, as is often practiced in atomizing, the dust-air ratio throughout the conveying system is held below the minimum explosive concentration of the metal dust as determined by data from the Bureau of Mines (Jacobson 1964). Although the metal dust-air suspension may be held below the explosive concentration in the conveying system, the suspension will necessarily pass through the explosive range in the collector at the end of the conveying system unless the dust is collected in a liquid, such as in a spray tower. Such wet collection is not always possible or desirable. Any liquid used must be nonflammable, nonreactive with metal dust, or reactive at a controlled minimum rate under favorably controlled conditions; the liquid remaining in or on the product must be compatible with subsequent processing requirements.

In an air conveying system, any dry collector must be considered as an explosion hazard containing a dust-air mixture in the explosive range. It is, therefore, sited in a safe location and is provided with the requisite barricades or other design means for the protection of personnel. Construction is of nonferrous, nonsparking metal, or nonmagnetic, nonsparking stainless steel. The entire system, particularly the collector, is thoroughly and completely bonded and grounded. The entire ground system, when checked with an ohmmeter, should show less than 5 ohms resistance to ground.

Where the conveying duct is exposed to weather or moisture, it should be moisture tight because any moisture entering the system can react with the dust, generating heat and serving as a potential source of ignition.

A minimum conveying velocity is employed throughout the conveying system to prevent the accumulation of dust at any point and to pick up any powder that might drop out during an unscheduled system stoppage.

If the conveying gas is inducted into the system in a relatively warm environment and the duct work and collectors are relatively cold, gas temperature may drop below the dew point, causing condensation of moisture. To avoid this possible condensation, the ducts and collectors are insulated or provided with a heating means.

Relief Vents for Conveyor Ducts

Vents of sufficient area connected to ducts or openings protected with antiflashback swing valves and extending to the outside of the building can provide explosion relief. Care should be taken to limit the inertia of swing valves to the minimum required. Rupture diaphragms can be used in place of swing valves. Wherever damage may result from the rupture of a duct, in case the relief vent fails to offer sufficient pressure relief, the duct is designed for an internal working pressure of 100 psi (690 kPa) minimum. Where the duct is so located that no damage will result from its bursting, it may be of very light construction to intentionally fail as an auxiliary vent for the system.

Fan Construction and Arrangement

Blades and housing of fans that are used to move air or inert gas in conveying ducts are constructed of conductive, nonsparking metal such as bronze, nonsparking stainless steel, or aluminum. In no case should the design be such that the dust is drawn through the fan before entering the final collector. Personnel should not be permitted within 50 ft (15 m) of the fan during operation. This means that the fan and associated equipment are shut down for oiling, inspection, or preventive maintenance. If the area must be approached during operation for pressure test or other technical reasons, it must only be done under the direct supervision of competent technical personnel and with the knowledge and approval of operating management. Ultimately, all fans in dust collector systems accumulate sufficient dust to be a potential hazard; for this reason they are preferably located outside of all manufacturing buildings.

It is good practice to equip fan bearings with suitable instruments for indicating the temperature; such instruments are wired with an alarm device to give notice of overtemperature.

Sight Glass

Avoid sight glasses in pneumatic systems whenever possible. If installed, they should be of noncombustible material that is not readily subject to physical damage. The tubing is supported above and below each sight glass so that the sight glass does not carry any of the system weight and is not, in itself, subject to resulting stresses or strains. The electrical bonding of the system must be continuous around all sight glasses. The strength of the sight glass and its mounting mechanism and its inside diameter are equal to the adjoining tube system. In pressure type systems, connections between sight and tubing are butted squarely and fastened together with rigid, airtight couplings, connectors, or comparable devices. In suction type systems, connections between the sight glass and tubing are butted squarely and sealed with approved sleeves extending a minimum of 3 in. (76 mm) above and below the sight glass and tubing connections. The sleeves are of a material that has elastic properties which provide an air tight seal.

Air-Material Separators

Air-material separators are preferably located outside the building and are provided with lightning protection. Material discharge outlets are provided with a positive choke device. The separator is electrically conductive and bonded. Exhaust air is always discharged to the outside, except where provision is made to recirculate transport air directly back into the pneumatic conveying system. The air-material separators are preferably constructed of noncombustible materials, and the cloth filters are made of low hazard materials. Where it is necessary to use combustible filter media, flame retardant treatment is desirable. In addition, the cloth filters are housed in metal enclosures, and provision is made for cleaning the filters. The separators are constructed so as to eliminate ledges or other points of dust accumulation.

CRANES

Cranes are an essential and necessary part of many industrial or manufacturing operations where heavy materials must be lifted or moved about. Many cranes move along rails such as overhead traveling, gantry, tower, and bridge type. Overhead traveling can be effectively utilized either indoors or outdoors, whereas gantry, tower, and bridge cranes are mainly used outdoors. Other cranes are made mobile by mounting them on wheels. Cranes can be powered by electricity or by combustion engine drive. Normal safeguards and maintenance must be observed with these power systems.

Outdoor cranes are susceptible to damage by high winds unless a positive means of anchorage is provided, in addition to the operating or service brakes. Large cranes which move along rails are generally provided with automatic or manual rail clamps. Other means of anchorage are crane traps, wedges, and cables. Anchorage and support are particularly necessary and advisable when a crane is not in use and unattended.

Crane operators should be thoroughly trained in the proper operation of the equipment, including emergency action which may be necessary in a fire or other hazardous incident. The crane operator's cab should preferably be of noncombustible construction and be provided with an emergency means of egress. It should be kept free of oily waste, rubbish, and other combustibles. All electrical equipment should be in conformance with NFPA 70, *National Electrical Code*, and should be securely mounted and maintained.

Emergency portable extinguisher protection may be desirable in some operator's cabs. In some large equipment, automatic fixed piping extinguishing systems may be advisable.

Bibliography

References Cited

Buffington, M. A. 1969. "Mechanical Conveyors and Elevators." *Chemical Engineering*. Oct 13, 1969. pp 33-49.

FMEC. 1976. *Handbook of Industrial Loss Prevention*. 2nd ed. McGraw-Hill, NY.

Hartmann, I., Nagy, J., and Brown, H. R. 1943 *Inflammability and Explosibility of Metal Powders*. RI 3722. Bureau of Mines, Pittsburgh.

Jacobson, M., Cooper, A. R., and Nagy, J. 1964. *Explosibility of Metal Powders*. RI 6516. Bureau of Mines, Pittsburgh.

Mitchell, D. W. et al. 1967. *Fire Hazard of Conveyor Belts*. RI 7053. Bureau of Mines, Washington, DC.

NFPA Codes, Standards, Recommended Practices and Manuals. (See the latest *NFPA Codes and Standards Catalog* for availability of current editions of the following documents.)

NFPA 15, *Standard for Water Spray Fixed Systems for Fire Protection*.

NFPA 61B, *Standard for Prevention of Fires and Explosions in Grain Elevators and Facilities Handling Bulk Raw Agricultural Commodities*.

NFPA 80, *Standard for Fire Doors and Windows*.

NFPA 505, *Fire Safety Standard for Powered Industrial Trucks Including Type Designations, Areas of Use, Maintenance and Operations*.

NFPA 651, *Standard for the Manufacture of Aluminum and Magnesium Powder*.

Additional Readings

Block, Richard A., *Crane Operation and Preventive Maintenance*, Marine Education Textbooks, Horima, LA, 1983.

"Fire Protection for Belt Conveyors," FMEC Loss Prevention Data Sheet 7-11, Factory Mutual Engineering Corp., Norwood, MA, Aug. 1972.

Johanson, J. R., "Feeding," *Chemical Engineering*, Oct. 13, 1969, pp. 75-83.

Kraus, M. N., "Pneumatic Conveyors," *Chemical Engineering*, Oct. 13, 1969, pp. 59-65.

Kuchta, J. M., Sapko, M. J., and Perzak, F. J., "Improved Fire Resistance Test Method for Belt Materials," *Fire Technology*, Vol. 17, No. 2, May 1981, pp. 120-130.

LaPushin, G., "Transportation and Storage," *Chemical Engineering*, Oct. 13, 1969, pp. 19-21.

"Moving Fire: Fire Hazards of Belt Conveyors," *Record*, Vol. 54, No. 6, 1977, pp. 18-21.

Patterson, C. B., "Looking at Fire Hazards: Powered Industrial Trucks: Appraising Their In-Plant Fire Safety," *Fire Journal*, Vol. 66, No. 5, Sept. 1972, pp. 103-104.

Sapko, M. J., et al., "Fire Resistance Test Method for Conveyor Belts," *Investigative Report 8521*, U.S. Bureau of Mines, Washington, DC, 1981.

HOUSEKEEPING PRACTICES

Kathleen M. Robinson

Good housekeeping is the care and maintenance of property and the provision of equipment and service. It is basic to firesafety and should be a major concern in every type of occupancy, from the simplest dwelling to the most complex industrial facility.

This chapter covers the fundamentals of good housekeeping practices. It discusses in some detail the specifics of practices to be observed in providing firesafe and affective housekeeping in connection with the care and maintenance of buildings, and the cleanliness and order of occupancy processes and outside facilities.

Proper layout of equipment and good storage and handling practices are important and complementary to success in maintaining cleanliness and order. Good storage practices for various materials and the different storage configurations used for them are the subject of chapters in Section 11, "Storage Practices and Hazards," while hazards associated with different types of processes are covered in the chapter's of Section 10, "Process Fire Hazards."

Other chapters in the Handbook that would be helpful in the consideration of good housekeeping practices are Section 6, Chapter 7, "Hazards of Waste Control"; and two other chapters in this section, Chapter 7, "Materials Handling Equipment," and Chapter 14," Waste Handling Systems."

NFPA standards applicable to the subject of good housekeeping will be found in the Bibliography at the end of the chapter.

GOOD HOUSEKEEPING THEORY

A good housekeeping program concerns itself with the less complex aspects of tidiness and order, waste control, and the regulation of such personal practices as smoking, which, without reasonable controls, could lead to hazardous conditions.

Poor housekeeping contributes to loss potential by increasing fire and explosion hazards in several ways:

1. It provides more places for a fire to start.
2. It creates a greater continuity of combustibles that makes it easier for fire to spread.
3. It provides a greater combustible loading for the initial fire to feed upon.
4. It creates the potential for flash fires or dust explosions when layers of lint or dust are allowed to accumulate.
5. It increases the potential for spontaneous ignition.

In addition to the increased hazard, poor housekeeping can have a negative effect on production. Quality proves hard to maintain when the workspace is crowded and messy. Efficiency suffers because people normally tend to work faster and more accurately if their surroundings are clean. Thus, good housekeeping will not only prevent fires but can improve production and employee morale as well.

The Essentials of Good Housekeeping

The degree of effort and attention needed for proper housekeeping is influenced, of course, by the type of buildings and the overall size of the facility involved. But most significant are the specific occupancies of the facility. Some processes produce more waste, leakage, and vapors than others, thus contributing to the extent of housekeeping problems. In addition, the acceptable level of cleanliness varies from occupancy to occupancy. What is satisfactory in a foundry would probably not be tolerable in an office building. And the cleanliness of the average office would hardly be satisfactory for an electronic "clean room."

Proper housekeeping does not just happen. It requires the leadership and wholehearted support of management and the cooperation of all employees. Management cannot merely decree that good housekeeping is a desired goal. It must place the responsibility and authority for achieving that goal with a committee (or individual). The committee, in turn, must inspect, evaluate, and, with management, finally define the desired cleanliness levels for the various sections of the facility. It must also establish the frequency of periodic inspections and devise a report form and distribution list. At the same time, the committee must identify the responsibilities of the individual workers towards both the general area and their immediate workspaces.

Ms. Robinson is Editor of *Fire Journal*, published by the NFPA.

In small locations professional cleaning and maintenance firms may provide the managers and crews for housekeeping. Nevertheless, the building management is still ultimately responsible for auditing the work being performed and making sure that the contractor complies with established housekeeping goals.

Where proper housekeeping does not exist, it is usually because inadequate attention is paid to, or inadequate action taken in, one or more of the following areas.

Communication: Management must obviously publicize its commitment to good housekeeping and the delegation of its authority to the committee or individuals to whom overall responsibility for housekeeping has been assigned. This publicity must be reinforced periodically and recognition given whenever notable improvement or outstanding performance takes place.

If responsibility for housekeeping rests with a housekeeping committee, it must be able to meet with management periodically to review performance, to revise goals, and to offer and substantiate recommendations for major expenditures. The committee should also encourage feedback from employees in the form of suggestions and constructive criticism. Simpler or more efficient housekeeping methods are often more readily identified by the worker who uses them than by management, and action on the proffered suggestions emphasize management's sincere desire to achieve the publicized goals.

Equipment: Housekeeping efforts should not founder from lack of necessary tools or equipment. This includes the tools normally used by maintenance personnel, such as brooms, dust pans, mops, and vacuum cleaners, as well as other items, such as those that encourage the proper disposal of trash. The simple step of putting a sufficient number of easily accessible wastebaskets or trash receptacles at points of need can cut down on the amount of waste deposited on the floor or in the product.

In some production and handling areas, dust, lint, and other waste may be produced constantly. In these cases, vacuum pick up stations tied into an exhaust and collection system may be needed at specific points of waste generation. For area cleaning, powered floor sweepers or rail-mounted traveling cleaners may be warranted. Areas in which large quantities of scrap or discarded packing materials continuously accumulate may need not only large trash containers, but also motorized equipment that can be emptied or replaced frequently.

Layout and Storage: Overcrowding is a major impediment to proper housekeeping. Blocked or restricted aisles limit access and, in so doing, hamper efficient cleaning and trash pick-up. Lack of sufficient workspace and storage capacity leads to inefficient operations, to an inability to create order, and finally to worker frustration. The creative use of racks, shelving, and bins is often a rewarding answer.

With its negative influence on good housekeeping, disorganized and haphazard storage is usually a detriment to effective fire protection, as well. Fire extinguishers, small hose stations, and extinguishing system control valves can become blocked and inaccessible while other fire equipment, such as fire doors, may be made inoperable.

Environment: Equally as important as the natural environment is the artificial environment created in the workplace. The control of process fumes, vapors, dusts, and flyings required today for the well-being of the worker has also had a beneficial impact on overall cleanliness in many work areas.

Adequate lighting, now recognized as a prerequisite for quality and high productivity, also helps improve housekeeping. So do the light colored walls and floors commonly used to make lighting more effective and surroundings more pleasant. They make spills, leaks, and waste accumulations highly visible, and visible waste is more likely to be cleaned up than waste hidden by poor lighting and dark surroundings.

Personnel: Too many housekeeping programs fail because they depend solely on personnel to achieve their goals. They ignore other factors, such as automated equipment and increased employee awareness, that could help keep the facility clean. Nevertheless, adequate personnel is an important part of any housekeeping program and it cannot be overlooked. People will always be needed to make sure that things are being done as they should be; management must recognize this fact if established goals are to be met.

BUILDING CARE AND MAINTENANCE

The three basic requirements for good housekeeping are: (1) proper layout and equipment; (2) correct materials handling and storage; and (3) cleanliness and order. Any facility that implements these basics has laid the foundation for good housekeeping. Using them, the facility can develop special housekeeping practices to deal with its own specific problems.

In pursing cleanliness and order, the care and maintenance of buildings requires special housekeeping practices. These are particularly noteworthy because they either introduce fire hazards into or reduce the fire danger to buildings.

Floors

The general care, treatment, cleaning, and refinishing of floors may present a fire hazard if flammable solvents or finishes are used or if combustible residues are produced in quantity. Many fires have resulted from the use of gasoline to clean floors, for example. In general, cleaning or finishing compounds containing solvents with flash points below room temperature are too dangerous for ordinary use, except in very small quantities. The magnitude of the hazard depends on the conditions of use and the precautions taken. Many cleaning compounds presenting little or no hazard are listed by fire testing laboratories.

Sweeping Compounds: Compounds used for sweeping floors generally consist of sawdust or some other combustible material treated with oil. Such compounds are hazardous, the degree of danger depending upon the characteristics of the oil. The use of sawdust or similar materials to absorb oil spillage increases the fire hazard unnecessarily since noncombustible oil-absorptive materials are available for this purpose.

Floor Oils: Compounds containing oils and low-flash-point solvents are a hazard, particularly when freshly applied. In addition, component oils may be subject to spontaneous heating. To reduce the fire hazard, suitable attention must be given to the safe storage of oily mops, sponges, and wiping rags in metal or other noncombustible containers. Any combustible oil used to excess increases the combustibility of the floor. Oil-soaked floors, the product of years of use, also show increased combustibility.

Floor Waxes: Low-flash-point solvents are hazardous, especially when used with electric polishers. In such instances, ignition might result from friction and sparking. Water emulsion waxes are preferable.

Furniture Polishes: Furniture polishes containing oils subject to spontaneous heating become hazardous when rags that are saturated with these polishes are not disposed of properly. Such oil-soaked rags should be placed in metal or other noncombustible containers.

Flammable Cleaning Solvents: Flammable cleaning solvents need not be used since a number of nonhazardous cleaning agents are available. These relatively safe materials are stable and have high flash points and low toxicity. There are several commercial stable solvents available which have flash points ranging from 140 to 190°F (60 to 88°C) and have a comparatively low degree of toxicity. Safe materials are available for most of the preceding purposes.

Dust and Lint

A necessary procedure in many occupancies is the removal of combustible dust and lint accumulations from walls, ceilings, and exposed structural members. Unless this procedure is performed safely, as by vacuum cleaners or air moving (blower and exhaust) systems, this procedure may present a fire or explosion hazard. In some cases, vacuum cleaning equipment must be equipped with dust-ignition-proof motors to assure safe operation in dust ladened atmospheres.

Care should be taken not to dislodge into the atmosphere any appreciable quantities of combustible dust or lint which might ignite or form an explosive mixture with air. A lot of work can be eliminated by applying suction at locations where dust may escape from processing machinery and conveying the aspirated dust to safely located collectors. Blowing down dust with compressed air may create dangerous dust clouds, and such cleaning should be done only when other methods cannot be used and after all possible sources of ignition have been eliminated. In most localities, it is possible to obtain the services of reliable professional industrial cleaning specialists to remove dust accumulations safely.

Exhaust Ducts and Related Equipment

The exhaust ducts from the hoods over cooking ranges, such as those found in plant cafeterias, present troublesome problems because grease condenses inside the ducts and on exhaust equipment. Grease accumulations may be ignited by sparks from the range or, more often, by small fires in overheated cooking oil or fat. Without these grease accumulations in the hood and duct, stove top fires can often be extinguished or allowed to burn out without causing appreciable damage. Fires occur frequently in frying because cooking oils and fats are heated to their flash points and may reach their self-ignition temperatures when accidentally overheated or spilled on the hot stove top.

Grease Removal Devices: All exhaust systems for kitchen cooking equipment must be equipped with a grease removal device. These include such items as grease extractors, grease filters, or special fans designed to remove grease vapors effectively and provide a fire barrier. Grease filters, including frames, and other grease removal devices should be made of noncombustible materials.

Ducts: There is no practical method for preventing all kitchen duct fires, but the danger can be minimized through a combination of precautions as outlined in NFPA 96, *Standard for the Installation of Equipment for the Removal of Smoke and Grease-Laden Vapors from Commercial Cooking Equipment.* It is good practice to clean hoods, grease removal devices, fans, ducts, and associated equipment frequently. The exhaust system should be inspected daily or weekly, depending on its use, to determine if grease or other residues are accumulating in it.

Clean ducts are essential to firesafety, but they often remain dirty because cleaning them is a difficult and unpleasant job. One source of help is a commercial firm that specializes in this sort of work. In any case, never try burning the grease out; it is a dangerous practice, even though duct systems installed according to NFPA standards are designed to withstand burnout.

In cleaning the exhaust system, avoid using flammable solvents or other flammable cleaning aids. Do not start the cleaning process until all electrical switches, detection devices, and extinguishing system supply cylinders have been turned off or locked in a "shut" position. This will prevent both the exhaust fan and the fire extinguishing system (if the exhaust duct is equipped with one) from actuating accidentally. Once the cleaning process is completed, the switches and other controls should be returned to normal operating position.

Satisfactory cleaning results have been obtained with a powder compound consisting of one part calcium hydroxide and two parts calcium carbonate. This compound saponifies the grease or oily sludge (converts it to soap), thus making it easier to remove and clean. The process requires proper ventilation. Another cleaning method is to loosen the grease with steam and then scrape the residue out of the duct. This has proven to be quite effective.

Spraying duct interiors with hydrated lime after cleaning is a fire prevention method used commercially. This procedure tends to saponify the grease and may facilitate subsequent cleaning, but it does not provide permanent fire retardency.

OCCUPANCY AND PROCESS HOUSEKEEPING

Housekeeping programs must give special consideration to disposal of rubbish, control of smoking habits, housekeeping hazards, and lockers and cupboards, where and as applicable.

Disposal of Rubbish

The proper handling and disposal of rubbish is an integral part of the housekeeping process, and its success depends primarily upon having and observing a satisfactory routine. The proper and regular disposal of combustible waste products is of the utmost importance.

In both industrial and commercial properties, the removal of combustible waste products at the end of each workday or at the end of each work shift is a common practice. In some properties, more frequent waste disposal is necessary. In others, the collection, storage, and disposal routine vary with the nature of the property use. In all cases, however, an adequate program for dealing with this problem is a firesafety essential. Keeping a place tidy also depends on providing enough wastebaskets, bins, cans, and other proper containers so that building users will find tidiness convenient.

Receptacles: Noncombustible containers should be used for the disposal of waste and rubbish. This is true even of such small receptacles as ash trays and wastebaskets and applies, of course, to the larger units found in commercial and industrial properties. Industrial waste barrels should be made of metal and equipped with a fitted cover. Care should be taken to avoid mixing waste materials where such mixing introduces hazards of its own. (See Fig. 12-8A.)

FIG. 12-8A. Waste containers designed to snuff out accidental fires in their contents and to limit external surface temperatures to no more than 175°F (80°C) above room temperature. (Justrite Mfg. Co.)

Plastic wastebaskets of varying sizes are readily available and are popular because they are quiet, attractive, and scratch and dent resistant; however, not all plastic baskets have the same burning characteristics. Some melt and burn readily, adding fuel to the fire and creating a comparatively serious fire exposure problem by collapsing and spilling their burning contents. This is also true of many plastic liners used for wastebaskets and receptacles. Other baskets may contribute relatively little fuel to the fire while maintaining their shape fairly well.

Segregation of Waste: It is not good housekeeping practice to dump all manner of dry waste down refuse chutes or to place it in a common bin or storage receptacle. For example, combustible metal dusts and metal powders dumped into chutes may explode if ignited. Mercury batteries and pressurized containers, such as the ubiquitous aerosol can, may also explode when incinerated or mixed with rubbish which is subsequently burned. Precautions should be taken to keep combustible items separate from each other and from noncombustible items.

Control of Smoking Habits: Smoking may be a difficult problem to handle, particularly because of the personal factors involved. It is a habit that is hard to break, even though it may put the smoker in direct conflict with the firesafety regulations and production standards in effect in various areas. When these areas contain flammable liquids or dusty and linty atmospheres, self-preservation alone makes smoking control relatively easy. However, control of smoking is also required in less obvious places, such as shipping and receiving areas, with their large quantities of loose packing materials, and storage areas, which may have high piled concentrations of combustible materials. In these areas, carefully planned smoking regulations are necessary.

Smoking regulations should be specific as to location and, preferably, time. Areas in which smoking is permissible, as well as those in which it is limited or prohibited entirely, must be clearly marked by appropriate signs that leave no question as to what is allowed where. (See Fig. 12-8B.)

FIG. 12-8B. Examples of signs permitting or forbidding smoking in designated areas. (Factory Mutual System).

In addition to sensible regulations, smoking control also requires adequate receptacles for spent smoking materials. Properly designed ashtrays are essential to safe smoking. They should be made of noncombustible materials, with grooves or snuffers that hold cigarettes securely. Their sides should be steep enough to force smokers to place cigarettes entirely within the ashtray. (See Fig. 12-8C.) In industrial buildings, large containers of sand are often used to conveniently and safely extinguish and dispose of spent smoking materials.

Improperly designed ashtrays may constitute a hazard, particularly if they allow a lit cigarette or cigar to fall or roll away. A lighted butt may too easily come in contact with combustible materials and start a fire under certain circumstances.

The contents of ashtrays must be disposed of carefully because a live butt may well be mixed in with apparently innocuous ashes. If lighted smoking materials were to be dumped into an ordinary wastebasket, they could set paper or some other piece of combustible rubbish on fire. To prevent this from happening, reserve special covered metal containers for discarded smoking materials only.

FIG. 12-8C. Two types of well-designed ash trays.

Industrial Housekeeping Hazards

Some industrial occupancies have special housekeeping problems inherent to the nature of their operations. For these particular problems, specific planning and arrangements are necessary.

Clean Waste and Rags: Clean cotton waste or wiping rags are generally considered to be mildly hazardous, chiefly because they are readily flammable when not baled and there is always the likelihood that dirty waste may become mixed with them. The presence of dirty waste or small amounts of certain oils may lead to spontaneous heating. Reclaimed waste is considered somewhat more hazardous than new waste. It is common practice to handle clean waste in the same manner as dirty waste, although the fire hazard is relatively small.

Large supplies of clean waste are best kept in bins made entirely of metal or of wood lined with metal and provided with covers that are normally kept closed. Several bins may be provided where the supplies are large or where different kinds of waste are kept. The covers on such bins should be counterweighted so that they may be readily raised and lowered. The counterweight ropes can have fusible links to ensure that the covers are closed automatically in the event of fire.

Local supplies of clean waste are usually kept in small, properly marked waste cans. Providing local supply points for clean waste can help eliminate the practice of keeping waste in clothes lockers, drawers, benches, and similar locations. If clean waste is put in such places, workers may mistakenly believe that other, more oily waste is also allowed there when, in fact, the combination of the two may result in fire.

Coatings and Lubricants: Paints, grease, and similar combustibles are widely used at industrial occupancies, and a good housekeeping program will make sure that their combustible residues are collected and disposed of safely. Nonsparking tools are recommended for cleaning spray booths and associated exhaust fan blades and ducts to avoid possible ignition of combustible residues. The discharge of vapors from spray booths should be so arranged that the vapors are conducted directly to the outside and the residues accumulate safely.

Drip Pans: Drip pans are essential at many locations, notably under motors, machines using cutting oils, and bearings. They are also used with borings and turnings that may contain oil. Drip pans should be made of noncombustible material and contain an oil absorbing compound. At many industrial occupancies, commercial oil absorbing compounds consisting largely of diatomaceous earth are used instead of sawdust or sand. The regular removal of oil soaked material is recommended.

Flammable Liquid Spills: Flammable liquid spills may be anticipated wherever such products are handled or used, and some means of coping with these spills must be kept on hand. These include a supply of suitable absorptive material and special tools to help limit the spill. Workers should understand and promptly take the steps needed to cut off sources of ignition, ventilate the area, and safely dissipate any flammable vapors.

Flammable Liquids Waste Disposal: The disposal of combustible liquid waste often presents a troublesome problem. Waste liquids, such as automobile crankcase fluids, must never be drained into sewers but placed in metal drums until they can be disposed of safely. In some cities, there are firms that make a specialty of collecting waste petroleum products and keeping them for further use, such as coating driveways and race tracks. Waste products must not be burned in oil burners unless the burner and accessories have been designed or properly adapted to handle such liquids. Many fire departments like to receive waste oils for use in training fire fighters to handle combustible liquid fires.

Oily Waste: Oily wiping rags, sawdust, lint, clothing, and other items are highly dangerous, particularly if they contain oils subject to spontaneous heating. To dispose of all such materials in ordinary quantities, a standard waste can that has been listed by testing organizations is best. For large amounts, heavy metal barrels with covers are ideal. Good practice calls for cans containing oily waste to be emptied daily and for wiping rags to be kept in covered metal containers until they can be laundered. (See Fig. 12-8D.)

FIG. 12-8D. A portable waste can which is equipped with a self-closing cover. This type of can can be used to store oily waste materials, particularly if they are subject to spontaneous heating. (The Protectoseal Co.)

Oil Puddling: Accumulations of oil can present a housekeeping problem at industrial locations where a considerable amount of oil is used. Poor maintenance of industrial hydraulic elevator installations may result in oil leaks which eventually form puddles on the elevator machine room floors and in the bottoms of hoistway pits. Although most oils used in hydraulic elevator systems have high flash points, any combustible oil can be a source of fire particularly when it is found in puddles that contain accumulations of debris. Puddled oil and materials used to absorb oil spills should be disposed of in metal barrels.

Packing Materials: Almost all packing materials used today are combustible and, consequently, hazardous. Plastic pellets and rigid forms, excelsior, shredded paper, sawdust, burlap, and other such materials should be treated as clean waste. However, large quantities may have to be kept in special vaults or storerooms. Automatic sprinklers are the best protection for areas where considerable quantities of packing materials are stored or handled.

Used or waste packing materials and the crating materials from receiving and shipping rooms must be removed and disposed of as promptly as possible in order to minimize the danger of fire. Ideally, the packing and unpacking processes should be conducted in an orderly manner so that excessive quantities of packing materials do not become strewn about the premises.

Lockers and Cupboards

Many industrial facilities provide their employees with lockers in which to put their personal belongings. These lockers may present a fire hazard if they are untidy or are used as general storerooms for such waste material as oily rags and cloths or paint smeared clothing. These items may ignite spontaneously or they may be accidentally ignited by matches and imperfectly extinguished pipes and cigars that employees inadvertently leave in their lockers.

Wooden lockers can ignite and spread fire. Metal ones are preferred, but they should be inspected regularly. Metal lockers may contain a fire if they are of solid construction, fronts and bottoms, partitions and backs. Backs and partitions made of expanded metal or wire screen may allow fire to spread unchecked and should not be used.

Lockers that are arranged in two tiers, one upon the other, are generally unsatisfactory. They do not hold clothes without mussing them and it often becomes the habit of those using such lockers to keep their clothes outside the locker or to throw them haphazardly into the locker, thus increasing the danger of spontaneous heating when such clothing is spotted with oil or paint.

Some lockers are provided with mechanical exhaust ventilation. Where this is true, NFPA 91, *Standard for the Installation of Blower and Exhaust Systems for Dust, Stock and Vapor Removal or Conveying* should be followed to avoid spreading a fire that originates in a locker.

Industrial plants that furnish and wash their employees' protective clothing may use a system of wire baskets, one per employee, suspended from the ceiling by a small chain running over a pulley instead of lockers. This method has proved successful in maintaining cleanliness and thereby reducing the fire hazard.

Where automatic sprinklers are installed, lockers must have expanded metal or screen tops to enable water from the sprinklers to reach the contents. Paper can be pasted on the tops to keep dust out. Sloping tops are also advisable to help prevent both fires and the accidents. Material cannot be placed on top of a locker designed this way.

Wooden supply cupboards constitute a fire hazard in places like machine and paint shops where woodwork becomes oil or paint soaked and where clothes or oily waste may be left in them. Wooden cupboards should be inspected regularly to make certain that they are always clean. The ideal cupboards for tools and similar items are made entirely of steel.

OUTDOOR HOUSEKEEPING PRACTICES

Good housekeeping practices are as essential out-of-doors as they are indoors. Failure to comply with good housekeeping practices out-of-doors may threaten the security of exposed structures and goods stored outside. The accumulation of rubbish and waste and the growth of tall grass and weeds adjacent to buildings or stored goods are probably the most common hazards. A regular program for policing the grounds is essential.

Grass and Weed Control

Tall grass, dry weeds, and bushes around buildings, along highways, on railroad properties, and along the streets of large industrial and commercial complexes present a definite fire hazard. To reduce this hazard, those responsible for maintaining these properties have always tried to control or destroy such vegetation.

One way to get rid of unwanted vegetation is to apply a chemical solution which poisons the weeds. Among the chemicals used are chlorate compounds, particularly sodium chlorate. Unfortunately, those who use chlorate compounds do not always realize that they are oxidizing agents. When these compounds come in contact with combustible materials, they virtually prime those materials for a fire or explosion. During hot periods in the summer, large numbers of fires have resulted from the use of sodium chlorate solutions on dry grass and weeds. Fires have also been reported in buildings and other structures in which such solutions have been spilled. Chlorate compounds spilled on clothing may present a danger to personnel, too.

There are weed killers that are not toxic and do not pose a fire hazard. Calcium chloride and agricultural borax, applied dry or in solution, are effective nonhazardous weed killers, as are various proprietary solutions. A number of commercial chemical weed killers, such as ammonium sulfamate, have little or no fire hazard and only a slight toxic hazard. Sodium arsenite and other compounds containing arsenic are efficient herbicides, but they are poisonous and not generally recommended.

The amounts of various chemicals needed to effectively kill weeds and the duration of their effect vary depending on the weed-killing agent used, the character of the vegetation, the climate, and the soil. Manufacturers' directions indicate the proper amounts to be used under various conditions.

Another way to remove vegetation is to burn it. This may be done only where environmental regulations permit outdoor burning and must be carefully controlled. Adequate fire extinguishing equipment must be readily available at all times.

Using this method, grass and weeds are usually cut down, collected in piles, and ignited. When the grass is too damp to propagate fire easily, flame-throwing torches may be used on the piles. However, these torches introduce a hazard if not carefully operated.

In fact, all burning introduces a hazard. Grass fires frequently spread out of control and ignite nearby buildings. To avoid this hazard, controlled burning should only be done at certain times of the year and then under the direct supervision of the fire department.

Fire authorities issue fire permits, where permissible, to help them control burning. These permits provide the authorities with an opportunity to educate the public in safe burning. They also help the fire department limit burning to nonhazardous periods of the year and better manage burning done during the hazardous periods.

Outdoor Storage

Goods stored outdoors should be properly separated from buildings of combustible construction and from other combustible storage which might constitute an exposure hazard. These separations should be maintained by the housekeeping staff who must see to it that they are never blocked, even temporarily, by such things as contractors' shacks, discarded crates, pallets, or other combustibles. Obstructed aisles could hamper fire fighting operations if the need for them ever arose. Passageways between storage piles should also be unobstructed and clear of combustibles.

Proper housekeeping also requires that smoking in outdoor storage areas be controlled. Suitable signs should be posted and large noncombustible receptacles should be provided for the disposal of smoking materials before entering a "no smoking" area.

Outdoor Rubbish Disposal

Combustible waste materials stored outdoors to await subsequent disposal as rubbish should be placed not less than 20 ft (6 m), and preferably 50 ft (15 m), from buildings and at least 50 ft (15 m) from public highways and sources of ignition, such as incinerators. It should also be enclosed with a secure noncombustible fence of adequate height.

The most satisfactory solution to the rubbish disposal problem is regular public collection. Burning rubbish is generally unsafe and is not permitted in built up urban areas.

If rubbish must be burned outdoors, it should be done in the early morning or at night because the night moisture reduces the chance that sparks will ignite combustibles in the surrounding area. This is the reasoning behind certain fire department and forestry service regulations that limit outdoor burning to certain days or times of day. Of course, there are some times when outdoor burning is strictly prohibited. Most parts of the United States and Canada experience days when things are so dry that any burning is dangerous.

Even when rubbish is not burned outdoors, it can present a fire hazard. All rubbish dumps, even landfills, are susceptible to fire. And sparks and flying brands from dump fires can carry the fire long distances. This is also true of sparks and brands produced by bonfires and incinerators which lack adequate spark arresters.

Bibliography

NFPA Codes, Standards, Recommended Practices and Manuals. (See the latest *NFPA Codes and Standards Catalog* for availability of current editions of the following documents.)

NFPA 13A, *Recommended Practice for the Inspection, Testing, Care and Maintenance of Sprinkler Systems.*

NFPA 82, *Standard on Incinerators, Waste and Linen Handling Systems and Equipment.*

NFPA 91, *Standard for the Installation of Blower and Exhaust Systems for Dust, Stock, Vapor Removal or Conveying.*

NFPA 96, *Standard for the Installation of Equipment for the Removal of Smoke and Grease-Laden Vapor from Commercial Cooking Equipment.*

Additional Reading

Feldman, Edwin B., *Housekeeping Handbook for Institutions, Business and Industry*, Frederick Fell, NY, 1979.

Higgins, Lindley, R., and Morrow, L. C., eds, "Sanitation and Housekeeping," *Maintenance Engineering Handbook*, 3rd ed., McGraw-Hill, NY, 1977, pp. 14-1 to 14-71.

"How to Apply Good Housekeeping," *The Handbook of Property Conservation*, Ch. 23, Factory Mutual System, Norwood, MA, 1973, pp. 189-193.

Linville, J. L., ed., "Industrial Housekeeping Practices," *Industrial Fire Protection Handbook*, 2nd ed., National Fire Protection Association, Quincy, MA, 1984, pp. 1049-1063.

Phillips, Cushing, *Plant Engineer's Desk Handbook*, Prentice-Hall, Englewood Cliffs, NJ, 1980, pp. 161-173.

Schultz, Joseph F., "Standards for Refuse-Handling in Apartment Houses," *Fire Journal*, Vol. 68, No. 2, Mar. 1974, pp. 82-86.

FOREST, BRUSH, AND GRASS FIRES

C. Bentley Lyon

Each year, on the average, 150,000 fires burn over almost 5.5 million acres (2.2 million hectares) of protected forest, brush, and grass covered lands in Canada and the United States. Protection services cost well over a half billion dollars annually, and the losses approach two billion dollars. The costs do not include the services of thousands of volunteer fire fighters in both countries, nor the expenses of the many city fire departments that fight fires on wildlands (lands that are essentially undeveloped) within or near their jurisdictions.

Most wildfires (any nonprescribed fire burning on wildlands) are suppressed (extinguished) while smaller than one acre (0.4 ha) by a few fire fighters working with hand tools or water handling equipment. But under extremely adverse, conditions some fires spread to well over a 100,000 acres (40 000 ha) and must be fought by 2,000 or 3,000 fire fighters and hundreds of mechanized units during a campaign of one to three weeks.

The wildfire problem is highly variable depending on location. This is because:

1. Fire ignition is dependent on natural (lightning or volcanic) phenomena or human activity, fuel bed characteristics, weather, and the effectiveness of prevention efforts.
2. Fire behavior (how a fire spreads and how intensely it burns) is dependent on local conditions (weather, fuels, and topography).
3. The effectiveness of fire suppression is closely related to accessibility, difficulty of control, and the capability and performance of the local protection agency.
4. Public safety is involved when fires threaten recreation areas, mountain homes, or subdivisions, or when smoke may obscure visibility along highways or near airports.

The above considerations are so variable by location that each region must deal with a unique set of circumstances. In the Mediterranean climate of southern California, a long, dry summer desiccates the chaparral and prepares it for explosive burning during gale force "Santa Ana" winds, posing an enormous threat to human lives and property. In the humid subtropical climate of the southeastern US, early spring and late fall fire seasons threaten extensive forests of pine and hardwood trees. Throughout the boreal forests of central and western Canada, lightning often ignites hundreds of fires during long periods of hot, dry weather causing extensive damage. Wildfires burning in the tundra of Alaska and northern Canada produce vast ecological effects and require special suppression techniques. During droughts, even the marine climate of the Pacific coastal forests extending from northern California to Alaska does not protect these areas from devastating timber fires.

Because of the enormous costs of wildland fire protection and the losses brought about by wildfires, the causes of these fires and the protection programs developed to deal with them are of concern to the public whose interest and well-being are greatly affected.

CAUSES OF FIRES

The leading cause of wildfires in the 1.5 billion acres (0.6 billion ha) of protected wildlands of the United States is incendiarism, accounting for about 30 percent of the fires and about one-third of the burned-over area. Debris burning accounts for about 23 percent of the fires and 16 percent of the burned area, while careless smoking is credited with causing 8 percent of the fires and about 7 percent of the burned area. Lightning typically accounts for about 8 percent of the fires but about 20 percent of the burned area. Other prominent causes of fires include equipment use, children playing with matches, campfires, and railroad use. Approximately 98 percent of these fires are controlled at a size of 100 acres (40 ha) or less, but the other two percent of the fires result in two-thirds of the total burned-over area. All together, in the US, 140,000 wildfires occur each year, burning over about 2.3 million acres (930 000 ha) of forest, brush, and grass covered lands.

The relative importance of the fire causes mentioned above varies considerably among different regions of the country. In terms of both fire numbers and burned area,

---

Mr. Lyon, recently retired, was Staff Fire Control Technologist, Fire and Atmospheric Sciences Research, Forest Service, US Department of Agriculture. He is presently the Vice Chairman of NFPA's Technical Committee on Forests.

lightning is the leading cause in the northwestern part of the country, as well as in Arizona and New Mexico. For the entire eastern half of the United States incendiarism and debris burning are the chief causes. In California, equipment use and incendiarism rank highest while debris burning and lightning are the main problems in the central Rockies. Alaska's wildfires are blamed chiefly on debris burning and lightning.

In Canada, about 9,000 wildfires occur each year, burning over 3.2 million acres (1.3 million ha) of wildlands. About 32 percent of these fires are caused by lightning, but fires started by lightning account for about 86 percent of the burned-over area.

AGENCIES INVOLVED

In the United States, wildland fire protection is handled by federal, state, county, and in some cases, city agencies, and by private corporations and volunteer fire departments. The US Forest Service protects about 200 million acres (80 million ha) of National Forest and other lands. About 487 million acres (197 million ha) of other federal lands (mostly public domain) are protected by the Bureau of Land Management (BLM), National Park Service (NPS), US Fish & Wildlife Service, and the Bureau of Indian Affairs. States, local governments, corporations, and volunteer fire departments protect about 840 million acres (340 million ha) of the country's essentially undeveloped lands. About 158 million acres (64 million ha) or roughly 9½ percent of the wildlands in the United States do not qualify for fire protection.

In Canada, wildland fire protection is handled by ten Provinces, two Territories, and Parks Canada, as well as by private corporations and volunteer fire departments.

COMPONENTS OF WILDLAND FIRE PROTECTION

Fire Prevention

Fire prevention involves all activities concerned with minimizing the incidence of destructive fires. This includes educational, engineering, and law enforcement activities. The most effective educational device in Canada and the US since 1950 has been Smokey Bear, who is credited with having been a major factor in the saving of over $20 billion in the US alone. The Advertising Council (formerly the Wartime Advertising Council) has for many years sponsored public service campaigns, including Smokey's message. Engineering activities include the development of spark arresters for locomotives, chain saws, tractors, and other equipment; the reduction of flammable growth along highways and in areas of heavy public use; and the modification of vegetation to favor less flammable species. Law enforcement activities are aimed mainly at the prevention of incendiarism, currently the leading cause of wildfires nationally.

Fire Presuppression

Fire presuppression activities are undertaken in advance of fire occurrence to help ensure more effective fire suppression. They include overall planning, the recruitment and training of fire personnel, procurement and maintenance of fire equipment and supplies, fuel treatment (hazard reduction), periodic assessment of the fire danger, and developing a system of fuel breaks, roads, water sources, and control lines.

Fire Detection

Discovering, locating, and reporting fires is accomplished in wildland areas by lookouts in fire towers, aircraft patrols, by other employees of the fire agencies, or by private or commercial airline pilots. Thousands of fires are reported annually by travelers, campers, woodcutters, hunters, or other members of the public. In the western United States and parts of Canada, lightning strikes are routinely detected electronically and their locations plotted by computer. Computer programs are being developed to assess the likelihood of ignition at these locations based on current weather and fuel conditions as well as the characteristics of the strikes themselves. Automated fixed point infrared detection systems have been developed, but none are in operational use. Fire detection from satellites is technically possible, but the complications of discriminating between legitimate heat sources and incipient wildfires prohibits the use of this technology.

Fire Suppression

Fire suppression is handled in a variety of ways and with a variety of tools and equipment, depending on the region of the country, local fuels and topography, and what agency or department is taking action. Fires in rural or urban areas which are handled by local fire departments are generally extinguished with a combination of water and hand tools. (See Fig. 12-9A.) Remote fires, out of the reach of water handling equipment, are often suppressed with hand tools by separating the fire from its fuel supply with a "fire line," or trail cut along the edge of the fire. Forest Service and BLM smokejumpers routinely parachute into the back country to extinguish fires in this way using only shovels, Pulaskis (a combination axe-grubbing tool), and chain saws.

FIG. 12-9A. Most wildland fires are suppressed by fire fighters using hand tools and water handling equipment.

Larger fires are fought with bulldozers, tractor-plow units (Fig. 12-9B), engine companies, and crews of fire fighters with lightweight machines and hand tools. In many cases these larger fires are combatted by "indirect attack," a method of suppression in which the control line is located a considerable distance away from the fire's active edge. This is frequently done in the case of a fast

FIG. 12-9B. A tractor-plow unit of the Florida Division of Forestry is unloaded at a fire. (Photo by C. B. Lyon)

spreading or high intensity fire to utilize natural or constructed fire breaks or fuel breaks and favorable breaks in topography. The intervening fuel is usually burned out, but occasionally the main fire is allowed to burn to the line, depending on conditions. The logistics of managing "campaign" fires generally involves establishing a base camp, and in some situations "spike camps," so that fire fighters can be fed and rested near the fire line. Logistical support in the form of qualified fire fighting crews, aircraft, and communications systems is handled in the US by the National Interagency Fire Coordination Center (NIFCC) in Boise, ID, and in Canada by the Interagency Forest Fire Centre (CIFFC) in Winnipeg, Manitoba.

Aircraft play a major role in fire suppression. During the initial action phase, air tankers (Fig. 12-9C) and

FIG. 12-9C. A Boeing B-17 cascades 2,000 gal. (7.5 m³) of fire retardant on a wildfire in Wenatchee National Forest.

"helitankers" are used to cascade water or fire retardants on the fuels immediately adjacent to the fire to slow the fire's advance until ground forces have time to construct fire lines and burn them out if necessary. Aircraft rarely, if ever, extinguish wildfires by themselves. Air tankers carry from about 600 to about 3,000 gal (2.3 to 11.4 m^3) of water or retardant, and either "salvo" their load in a single drop, or make two to four individual ("split") drops.

Most air tanker operations in the United States involve cascading retardant slurries rather than water. "Long-term" retardant formulations are concentrations of salts such as ammonium sulfate or diammonium phosphate in water with the addition of a thickening agent, corrosion inhibitors, rust inhibitors, artificial coloring (for visibility on the fuel), and other ingredients. These products remain effective as retardants even after the water has evaporated, hence their name, long-term. "Short-term" retardants are similar but lack the salt content. While the retardant effect is less durable, the products cost much less. Most retardants are formulated and supplied by commercial producers, but the products must meet performance specifications approved by the agency involved (the US Forest Service maintains specifications for this use). In Canada, air tanker operations involve the Canadair CL-215 and other amphibian aircraft that self-load while skimming the water. Inasmuch as the tanker turnaround time is generally rapid due to the presence of many lakes, retardants are not frequently used.

Helicopters are widely used for firefighting in both Canada and the US. They are used to drop water from either an attached or internal tank or from a suspended "helibucket" (Fig. 12-9D) that can be filled by dipping in a lake, pond, or tank. Helicopters are also used to ferry fire fighters to and from the fire line, to perform reconnaissance, to haul cargo (Fig. 12-9E), to effect rescues, and to

FIG. 12-9D. This "helibucket" is filled by dipping into a lake. The water is then carried to and dropped on the wildfire.

FIG. 12-9E. U.S. Army CH-54A Sky Crane transports a D-4 tractor to an inaccessible Florida wildfire.

serve as airborne command posts. Some helicopters are equipped with special equipment such as "FLIR" (forward looking infrared) for reconnaissance under marginal visibility and NVG (night vision goggles) for operations of all kinds in the dark. Another technology in use involves rappelling fire fighters (called "rappattack" in Canada) by rope into areas where helicopter landing may be impossible or unsafe. A few agencies own and operate their own helicopters, but most are supplied on an hourly or seasonal basis by private contractors.

Some large fires are confined to one jurisdiction and may be suppressed by one agency or department. But frequently wildland fires burn across lands being protected by two or more agencies and require the suppression efforts of from two to more than a dozen agencies and departments. The National Interagency Incident Management System (NIIMS), has been developed to facilitate multiagency collaboration within the US. NIIMS includes an Incident Command System, a training program, a qualifications and certification program, publications management, and supporting technology. According to this system, all participating agencies—federal, state, county and other—adopt a common organization, procedures, and terminology. This makes each agency able to effectively use the closest available qualified personnel and equipment. NIIMS is currently being put into use around the United States.

Overall collaboration in fire operations and planning in the US is effected through the National Wildfire Coordinating Group (NWCG), which includes participation by the US Forest Service, the US Department of Interior fire agencies, the National Association of State Foresters, and the US Fire Administration. In Canada, similar collaboration is achieved via the Canadian Committee on Forest Fire Management, whose purposes are to serve as the national advisory body for the advancement of forest fire research and for the development of improved fire control practices and interagency cooperation in Canada. Standards for the prevention and control of fires in forest, brush, and grass covered lands in the US and Canada are maintained by the National Fire Protection Association through its Forest Committee, which is made up of representatives of government agencies, private companies, and consumer groups. Canada and the US collaborate in the suppression of forest fires through a reciprocal forest fire fighting arrangement. The US and Canadian governments collaborate with friendly foreign countries through consultations and exchanges of personnel and technology in the fire protection business.

Safety on the fire line is stressed by most wildland fire fighting organizations. NWCG agencies, for example, require fire fighters to meet physical fitness requirements, to be outfitted with fire resistant clothing, hard hat and suitable boots, and in many cases, to carry a fire shelter (this is a puptent-like shelter of heat reflective materials). In addition, fire fighters are trained to recognize and avoid high risk situations. Aircraft operations are closely supervised and coordinated to minimize the risk of midair collisions and other accidents. Standard suppression operations call for fire behavior forecasts to facilitate planning and provide for safety in fire strategy and tactics.

Research

Both Canada and the United States maintain fire research programs with the ultimate goal to reduce loss of life, property, and forest resources from wildfires and to better use prescribed fire to achieve forest and range objectives at reduced costs. The Canadian Forestry Service (CFS) of the Canada Department of the Environment is the principal forest research organization in Canada. Its forest fire research program is centered at the Petawawa National Forestry Institute in Chalk River, Ontario. Fire research work is also carried out at Regional Forest Research Centers in Victoria, British Columbia; Edmonton, Alberta; and Sault Ste. Marie, Ontario. The fire research programs address six major areas:

1. *Fire Behavior*: Fuel moisture physics, fire spread physics, prediction of fire danger by forest type, fire/weather interactions, fire danger rating systems, and spatial weather models.
2. *Fire Ecology*: Post fire forest regeneration mechanisms, cyclic forest development from fire to fire, prediction of post fire forest development, and age-class distribution in fire cycled forests.
3. *Fire Suppression*: Performance testing of fire control equipment, air tankers, fire retardants and water additives; aerial ignition devices; backfiring methods; and new suppression methods.

4. *Prescribed Fire*: Tree damage and mortality, use of fire for slash removal, seedbed preparation and vegetation control, design of prescriptions for proper burning conditions, and operational techniques.
5. *Fire Economics*: Estimation of values at risk, effect of fire on timber supply, relation between fire control costs and burned area, allowable cut effect, and ultimate impact of fire on the forest economy.
6. *Fire Management Systems*: Remote sensing applications; computerized systems for integrating weather, fuel type, and terrain into fire spread and growth models; prediction of lighting and human caused fires; air patrol routing; resource deployment; and attack strategy.

Current US fire research is being conducted at six Forest and Range Experiment Stations; much of this work is being done at Forest Fire Laboratories located at Macon, GA; Missoula, MT; and Riverside, CA. Major research areas include the following:

1. *Fire Behavior*: Fundamental studies in fuel chemistry, combustion and ignition processes, fire spread mechanisms, time-temperature heat flux interrelationships, and effects of fuel moisture, wind, and slope; development of systems to aid fire and land managers in dealing with growth of large fires, fuel consumption and energy release, probability of ignition by lightning, and fire spread in nonuniform fuels.
2. *Fire Suppression*: To develop real-time and planning guidelines for individual suppression activities related to primary fuel and fire variables; to integrate production rate knowledge into real-time and planning guidelines for fire suppression strategies and tactics; to develop design criteria for chemical formulations, and aerial and ground delivery systems for primary strategies and tactics.
3. *Fire Effects, Use, and Ecology*: Research to determine how, when, and where prescribed fire may be used to improve tree growth, provide better wildlife habitat, reduce fire hazard, and accomplish other forestry aims; includes studies to ascertain biological responses, impacts on soil, stand structure, etc.
4. *Fire Control Planning*: Studies to relate wildland fires and social benefits desired from the wildlands; to determine the effectiveness of fire prevention activities; to determine the influence of weather and climate on fire occurrence, control, and effects; and to determine the productivity and effectiveness of air tanker systems.
5. *Other*: Additional research projects deal with managing smoke from prescribed fires, reducing residues from forestry activities, evaluating the economics of fire protection systems, and aiding managers with systems for applying knowledge gained through research studies.

During the last two decades, there has been a great increase in use of and interest in prescribed fire as a forestry tool. This is largely due to the enhanced recognition of fire's value in the forest environment as opposed to total fire exclusion, the high cost of mechanical equipment, and environmental objections posed by the use of chemicals. The new emphasis on prescribed burning demands new skills of the forester and new knowledge from research. Fire prevention remains important—fire at the wrong time is still just as disastrous. Fire protection agencies continue to seek out more acceptable and less costly means of protecting resources, lives, and property. Prescribed burning seems to offer at least one opportunity for progress.

Bibliography

NFPA Codes, Standards, Recommended Practices and Manuals. (See the latest *NFPA Codes and Standards Catalog* for availability of current editions of the following documents.)

NFPA 224, *Standard for Homes and Camps in Forest Areas.*
NFPA 295, *Standard for Wildfire Control.*

Additional Readings

Anderson, H. E., "Forest Fuel Ignitibility," *Fire Technology*, Vol. 6, No. 4, Nov. 1970, pp. 312-319.

Bratten, Frederick W., "Containment Tables for Initial Attack on Forest Fires," *Fire Technology*, Vol. 14, No. 4, Nov. 1978, pp. 297-303.

Chandler, C. C., et al., *Fire in Forestry*, John Wiley & Sons Inc., NY, 1983.

Countryman, C. M., "Mass Fire Characteristics in Large-Scale Tests," *Fire Technology*, Vol. 1, No. 4, Nov. 1965, pp. 303-317.

Countryman, C. M., et at., "Fire Weather and Fire Behavior in the 1966 Loop Fire," *Fire Technology*, Vol. 4, No. 2, May 1968, pp. 126-141.

Davis, James B., "A New Fire Management Policy on Forest Service Lands," *Fire Technology*, Vol. 15, No. 1, Feb. 1979, pp. 43-50.

Emori, Richard I., and Saito, Kozo, "Model Experiment of Hazardous Forest Fire Whirl," *Fire Technology*, Vol. 18, No. 4, Nov. 1982, pp. 319-327.

Fahnestock, G. R., "Two Keys for Appraising Forest Fire Fuels," Forest Service Research Paper PNW-99, U.S. Department of Agriculture, Washington, DC, 1970.

Foltz, Jeffrey L., "Disposable Respirators for Forest Fire Fighters," *Fire Technology*, Vol. 17, No. 3, Aug. 1981, pp. 174-176.

Fosberg, Michael A., "Prediction of Prepyrolysis Temperature Rise in Dead Forest Fuels," *Fire Technology*, Vol. 9, No. 3, Aug. 1973, pp. 182-188.

Gaylor, Harry P., *Wildfires, Prevention and Control*, Robert J. Brady Co. Bowie, MD, 1974.

Greenlee, John M., "Wildland Fire Modeling and Strategy Assessment," *Fire Technology*, Vol. 18, No. 3, Aug. 1982, pp. 237-250.

Lindquist, James L., "Building Firelines—How Fast Do Crews Work?" *Fire Technology*, Vol. 6, No. 2, May 1970, pp. 126-134.

Pagni, Patrick J., "Quantitative Analysis of Prescribed Burning," in *Fire Prevention and Suppression*, Hildo, C. J., ed., Technomic, Westport, CT, 1974, pp. 44-61.

Pyne, Stephen J., *Introduction to Wildland Fire; Fire Management in the United States*, John Wiley & Sons Inc, NY, 1984.

Pyne, Stephen J., *Fire in America—A Cultural History of Wildland and Rural Fire*, Princeton University Press, 1982.

Simard, Albert J., et al., "Nondirectional Sampling of Wildland Fire Spread," *Fire Technology*, Vol. 18, No. 3, Aug. 1982, pp. 221-228.

CONTROL OF ELECTROSTATIC IGNITION SOURCES

Revised by Martin F. Henry

Static electricity as a source of ignition is a hazard common to a wide variety of industries and processes. This chapter explains the nature of static electricity and means and methods of eliminating or minimizing it as an ignition source. Information on the hazards of static electricity in specific circumstances and processes will be found in Section 4, Chapter 2, "Explosions;" Section 5, Chapter 4, "Flammable and Combustible Liquids," Chapter 5, "Gases," Chapter 9, "Dusts;" Section 8, Chapter 2, "Electrical Systems and Appliances," and in this section, Chapter 11, "Special Systems for Explosion Damage Control."

The term "static electricity" is used in this chapter to mean the electrification of materials through physical contact and separation, and the effects of the positive and negative charges so formed, particularly where sparks may result which constitute a fire or explosion hazard.

The development of electrical charges in itself may not be a potential fire or explosion hazard. There must be a discharge or sudden recombination of separated positive and negative charges. In order for static to be a source of ignition, four conditions must be fulfilled:

1. There must be an effective means of static generation.
2. There must be a means of accumulating the separate charges and maintaining a suitable difference of electrical potential.
3. There must be a spark discharge of adequate energy.
4. The spark must occur in an ignitible mixture.

Static electricity may appear as the result of motions that involve changes in relative positions of contacting surfaces, usually of dissimilar substances either liquid or solid, one or both of which usually must be a poor conductor of electricity. Examples of such motion that are commonly found in industry are:

1. Motions of all sorts that involve changes in relative positon of contacting surfaces, usually of dissimilar liquids or solids.
2. Steam, air, or gas flowing from any opening in a pipe or hose, when the steam is wet or the air or gas stream contains particulate matter.
3. Pulverized materials passing through chutes or pneumatic conveyors.
4. Nonconductive power or conveyor belts in motion.
5. Moving vehicles.

Less than 100 years ago, electricity was described as a "subtle agent, without weight or form, that appears to be diffused through all nature, existing in all substances," which gives no indication of its presence in the latent state, but is capable of producing sudden and destructive effects in its active state. With the discovery of the electron, this definition now seems prophetic, and the adjective "static" more meaningful. Electricity can move freely through some substances, such as metals, that are called "conductors," but can flow with difficulty or not at all through or over the surface of a class of substances called "nonconductors" or "insulators." This latter group includes gases, glass, amber, resin, sulfur, paraffin, most synthetic plastics, and dry petroleum oils.

When electricity is present on the surface of a nonconductive body, where it is trapped or prevented from escaping, it is termed static electricity. Electricity on a conducting body which is in contact only with nonconductors is also prevented from escaping and is therefore nonmobile or "static." In either case, the body on which this electricity is evident is said to be "charged."

The generation of static electricity cannot be prevented absolutely because its intrinsic origins are present at every interface. The object of most static-corrective measures is to provide a means whereby charges separated by whatever cause may recombine harmlessly before sparking potentials are attained. If hazardous static conditions cannot be avoided in certain operations, means must be taken to assure that there are no ignitible mixtures at points where sparks may occur.

Some electrical terms used in this chapter are defined as follows:

Capacitance: Electrons received by an electrically neutral body of material, such as a person, a car, or an aircraft, raise the voltage at a rate determined by the surface area and shape of the body. The voltage is determined by the

---

Mr. Henry is the Flammable Liquids Specialist on the staff of NFPA.

surface characteristics (capacitance) of the body and the number of electrons on this surface. The larger the body, the more electrons are needed to raise the voltage a specific amount; hence the higher the capacitance of this body.

Capacitance is measured in terms of "farads" or millionths of a farad "microfarads," and millionths of one millionth of a farad "picofarads."

Charge: Measured in coulombs or fractions thereof, the static charge on a body is the number of separated electrons on the body (negative charge) or the number of separated electrons not on the body (positive charge). Electrons cannot be destroyed; so when an electron is removed from one body, it must go to another body. Thus there are always equal and opposite charges produced [leaving behind a positive (+) void]. Since it would be awkward to say there are 6,240,000,000,000,000,000 electrons on a body, we say instead that the body has a charge of one "coulomb." A coulomb is simply a name for this specific quantity of electrons. In electrostatics an even more practical unit is a microcoulomb, representing a charge of 6.24×10^{12} electrons.

Current: Just as water flow is measured in terms of the amount of water that passes a certain point in a specific period of time (gallons or liters per minute), so too is the flow of electrons past a certain point measured by time. The flow of electrons is called current. Current is measured in terms of electrons per second, coulombs per second, or amperes.

Energy: Energy is measured in joules or fractions thereof. A spark is energy being expended. Energy is required to do work. The measure of energy takes several forms. Often it is physical energy, which is measured in foot-pounds or joules or watts. If it is heat energy, it is measured in British thermal units (Btu) or Pascals (P); and if it is electrical energy it is measured in watt-seconds or in joules. A joule is the energy expended in 1 sec by an electric current of one ampere in a resistance of one ohm (approximately 0.738 foot-pound). Static spark energy is usually measured in thousandths of a joule (millijoules or mJ). A static spark needs a minimum amount of energy to cause problems.

Incendive: A spark which has enough energy to ignite an ignitible mixture is said to be incendive. Thus an incendive spark can ignite an ignitible mixture and cause a fire or explosion. A nonincendive spark does not possess the energy required to cause ignition even if it occurs within an ignitible mixture.

Incendivity: The ability of a spark to ignite an ignitible mixture is called incendivity. The energy level required for incendivity varies as described in the text and can be calculated.

Potential: Stored energy is able to do work. In electricity, this ability is expressed in terms of the potential of doing work. Potential in electricity is measured in terms of "volt," "kilovolts," or "millivolts." Potential, or voltage, is measured from a base point. This point can be any voltage but is usually "ground" which is theoretically zero voltage. When one point with a potential of "x" volts to ground is compared with another point with a potential of "y" volts to ground, then we say that a potential difference of "x-y" volts exists between the two. Then, when a point with a potential of 2,500 (+) volts to ground is compared with a point with a potential of 1,500 (−) volts to ground the potential difference is 4,000 volts.

Resistance: Electrical current encounters difficulty in passing through an electrical circuit or conductor. This difficulty can be measured and is called resistance. Resistance can be measured in terms of voltage drop over a part of the circuit but is usually measured in terms of "ohms" or "megohms." The resistance of a circuit in ohms is equal to the ratio of voltage in volts to current in amperes, i.e.,

$$\frac{10\text{ V}}{1\text{ amp}} = 10\text{ ohms}, \frac{1\text{ V}}{1\mu\text{ amp}} = 1\text{ megohm}.$$

STATIC GENERATION

When two bodies are in close physical contact, there is likely to be a transfer of free electrons between them, one giving up electrons to the other, and an attractive force is established. When the bodies are separated, work must be done in opposition to these attractive forces. The expended energy reappears as an increase in electrical tension or voltage between the two surfaces. If the bodies are insulated from their surroundings, both are said to be "charged;" the one having the excess of electrons is said to have a negative charge, while the other is said to have an equal positive charge. If a conductive path is available between them, the charges thus separated will reunite immediately. If no such path is available, as is the case with insulators, the potential increase with separation may easily reach values of several thousand volts.

In many cases, one of the objects has a deliberate or inherently conductive path to the earth, and its charge is immediately lost to the earth, which is considered to have an infinite capacity to absorb or give up electrons. The other (insulated) object now retains its "charge" (often called potential or voltage) with respect to its surroundings. It would hold this charge indefinitely, except for the fact that it must somehow be supported and no supporting insulator (even air) is a perfect nonconductor.

If one of the bodies is itself a nonconductor, the flow of electrons across its surface is inhibited, and the charge tends to remain at the points where electron transfer originally occurred. A highly charged insulating surface can be discharged with the appearance of a spark by bringing an "earthed" conductor close to it, but only a limited area will be so discharged, and sparks so produced seldom release enough energy to cause ignition. Thus, nonconductors, the bodies most directly involved in charge separation, are usually not directly responsible for fires and explosions (an exception is flammable liquids which are discussed later in this chapter). However, such charges can in some cases be the agency for building up or accumulating a charge on a conductive body, which can release all of its stored energy in an incendive spark.

On a conductive body the charge is free to move. Because like charges repel each other, the charge will distribute itself over the surface. If no other body is in close proximity, the concentration will favor the surface having the least radius of curvature; for a point, if the voltage is high enough, the voltage gradient may exceed the breakdown potential of air (about 30,000 volts per centimeter) and the air can be ionized, and a brush discharge may occur.

Like charges repel each other and unlike charges attract because of forces resident in the electrical fields that surround them. These forces have a strong influence on nearby objects. If the neighboring object is a conductor it will experience a separation of charges by induction. Its repelled charge is free to give or receive electrons as the case may be; if another conductor is brought near, the transfer may occur through the agency of a spark, very often an energetic spark. When the inducing charge is moved away from the insulated conductor, there follows a reversed sequence of events, and sparks may result. Thus, in many situations, induced charges are far more dangerous than the initially separated ones upon which they are dependent.

If the object close to the highly charged nonconductor is itself a nonconductor, it will be polarized, that is, its constituent molecules will be oriented to some degree in the direction of the lines of force since the electrons have no true migratory freedom. Because of their polarizable nature, insulators and nonconductors are often called dielectrics. Their presence as separating media enhances the accumulation of charge.

Capacitance

Two conductive bodies separated by an insulator constitute a capacitor or condenser, and where a potential difference is applied between these bodies, electricity can be stored. One body receives a positive and the other an equal negative charge. In many instances involving accumulation of static electricity, one of the bodies is the earth, the insulating medium is the air, and the insulated body is some object to or from which a charge (electrons) has been transferred by one of the mechanisms previously described.

When a conducting path is made available, the stored energy is released, or the condenser is "discharged," possibly producing a spark. The energy so stored and released by the spark is related to the capacitance of the condenser (C) and the voltage (V) in accordance with the following:

$$\text{Energy} = \frac{C}{2} \times V^2$$

or, in practical units;

Energy (millijoules) = $C/2$ (picofarads) × V^2 (volts × 10^{-9}).

Spark Discharge

The ability of a spark to produce ignition of a flammable mixture is governed largely by the energy transferred to the mixture, which will be some fraction of the total stored energy available because some energy is expended in heating the electrodes. Experiments at atmospheric pressure with plane electrodes have shown that the spark breakdown voltage has a minimum value at a critical short gap distance—about 350 volts for the shortest measurable gap (say 0.01 mm). Increased gaps require proportionately higher voltages. At close spacing the heat loss or flame quenching effect virtually precludes ignition of the electrodes.

At most favorable electrode spacing, tests have shown that optimum mixtures of saturated hydrocarbon vapors and gases in air require about 0.25 millijoules of discharge energy to produce ignition. Examples of the capacitance necessary to store the required 0.25 millijoules at various voltages are listed in Table 12-10A.

Table 12-10A. Examples of Conditions Required to Produce Ignition.

	Potential volts	Capacitance pf	Gap Length	
A.	350	4,000	Minimum voltage to jump shortest measurable gap. Quenching effect precludes ignition.	
B.	1,500	222	Gap about 0.5 mm, just exceeding quenching distance. Incendiary sparks.	
C.	5,000	20		1.5 mm
D.	10,000	5	Object the size of a baseball.	3+ mm
E.	20,000	1¼	Object the size of a large marble.	7 mm

For gaps of 1.5 mm or more, substantially longer than the quenching distance, the total energy required to produce ignition is increased somewhat in proportion to the excess of spark length over the diameter of the necessary critical flame volume. This, in turn, may require somewhat greater capacitance than indicated above. This explains why corona discharge from a sharp point at very high voltage may not be incendive.

Ignitible Mixtures

Elimination of ignitible mixtures in the areas where sparks of static electricity can occur is the surest method of preventing static caused fires. This is practical in certain areas, best discussed later in connection with the specific process involved.

Summary

In summary, static electricity will be manifest only where highly insulated bodies or surfaces are found. If a body is "charged" with static electricity, there will always be an equal and opposite charge produced. If a hazard is suspected, the situation should be analyzed to determine the location of both charges and to see what conductive paths are available between them. Tests of the high resistance paths should be made with an applied potential of 500 volts or more, in order that a minor interruption (paint or grease film or air gap) will be broken down and a correct reading of the instrument obtained. Resistances as high as 10,000 megohms will provide an adequate leakage path in many cases; when charges are generated rapidly, however, a resistance as low as 1 megohm (10^6 ohms) might be required. Where bonds are applied, they should connect the bodies on which the two opposite charges are expected to be found.

DISSIPATION OF STATIC ELECTRICITY

A static charge which already exists can be removed or allowed to dissipate itself. The acts or conditions which can accomplish this are the same as those acts or conditions which permit the separation of charges to occur in the first place. Thus, dissipation of static and prevention of its generation are opposite approaches to the same objective.

Humidification

A static charge cannot persist except on a body insulated from its surroundings. Most commonly encountered materials which are not usually thought of as conductors, such as fabrics, paper, wood, concrete or masonry foundations, etc., contain a certain amount of moisture in equilibrium with that in the surrounding atmosphere. This moisture content varies, depending on weather (and to a large measure controls the conductivity of the material) and hence its ability to prevent the escape of static electricity. In an analogous manner, under some conditions water vapor may condense on the surface of some nominally insulating materials, notably glass and porcelain, to render the surface conductive.

The conductivity of the materials under discussion—wood, paper, etc.--is not controlled by the absolute water content of the air but by its relative humidity. This figure, as ordinarily recorded in weather reports and comfort charts, is the ratio of the partial pressure of the moisture in the atmosphere to the partial pressure of water at the prevailing atmospheric temperature. Under conditions of high relative humidity (50 percent or higher) the materials in question will reach equilibrium conditions containing enough moisture to make the conductivity adequate to prevent static accumulations. The generating mechanism may be present, but the generated charge leaks away so fast that no observable accumulation results.

At the opposite extreme, with relative humidities of 30 percent or less, these same materials may dry out, become good insulators, and static manifestations become noticeable. There is no definite boundary line between these two conditions.

It should be emphasized that the conductivity of these materials is a function of relative humidity. At any constant moisture content, the relative humidity of an atmosphere decreases as the temperature is raised. In cold weather, the absolute humidity of the outdoor atmosphere may be low, even though the relative humidity may be high. When this same air is brought indoors and heated, the relative humidity becomes very low. As an example, a saturated atmosphere at an outdoor temperature of 30°F (−1°C) would have a relative humidity of only a little over 20 percent if heated up to a room temperature of 70°F (21°C). This phenomenon is responsible for the common belief that static generation is always more intense during winter months. The static problem is usually more severe during this period because static charges on a material have less ability to dissipate when relative humidities are low.

Where static electricity has introduced operational problems, such as the adhesion or repulsion of sheets of paper, layers of cloth, fibers, and the like, humidifying the atmosphere has proved to be a solution. It is usually stated that a relative humidity of about 50 percent or higher will avoid such difficulties.

Unfortunately, it is not practical to humidify all occupancies in which static might be a hazard. It is necessary to conduct some operations in an atmosphere having a low relative humidity to avoid deleterious effects on the materials handled. High humidity can also cause intolerable comfort conditions in operations where the dry bulb temperature is high. On the other hand, a high humidity may advantageously affect the handling properties of some materials, thus providing an additional advantage. In some cases localized humidification produced by directing a steam jet onto critical areas may provide satisfactory results without the need for increasing the humidity in the whole room. However, it must be remembered that steam which contains droplets of water may itself generate static. Local static can be reduced by providing a low velocity jet of humidified air.

It does not follow that humidification is a cure for all static problems. In particular, it must be remembered that the conductivity of the air is not appreciably increased by the presence in it of water in the form of a gas. If static electricity accumulates on some surface which is heated above normal atmospheric temperature—cloth passing over heated rollers for example—alterations in the relative humidity in the surrounding air may do no good whatsoever.

Another situation where the control of atmospheric humidity appears to accomplish little is in controlling the charge of static electricity that appears on the surface of oils under some circumstances. Such a surface does not absorb water vapor in the same way as paper or wood does, so it can remain an insulating surface capable of accumulating static charges even though the atmosphere above it may have a relative humidity up to 100 percent.

In summary, humidification of the atmosphere to a relative humidity of 50 percent or higher may be a cure for static problems where the surfaces on which the static electricity accumulates are those materials that reach equilibrium with the atmosphere, such as paper or wood, and that are not abnormally heated. For heated surfaces and for static on the surface of oils and some other liquid and solid insulating materials, high humidity will not provide a means for draining off static charges and some other solution must be sought.

Bonding and Grounding

Where natural conditions, including humidity, do not insure a conductive path to prevent static accumulation, artificial conducting paths may be necessary.

"Bonding" is the process of connecting two or more conductive objects together by means of a conductor. "Grounding" (earthing) is the process of connecting one or more conductive objects to the ground, and is a specific form of bonding. A conductive object may also be grounded by bonding it to another conductive object that is already connected to the ground. Some objects are inherently bonded or inherently grounded by their contact with the ground. Examples are underground piping or large storage tanks resting on the ground. Bonding minimizes potential differences between conductive objects. Grounding minimizes potential differences between objects and the ground.

Bond wires and ground wires should have adequate capacity to carry the largest currents that may be anticipated for any particular installation. When currents to be handled are small, the minimum size of wire is dictated by mechanical strength rather than current carrying capacity. The currents encountered in the bond connections used in the protection against accumulations of static electricity are in the order of microamperes (one millionth part of an ampere).

The acceptable resistance in a ground connection depends upon the type of hazard for which it is intended to give protection. To prevent the accumulation of static electricity, the resistance need not be less than 1 megohm and in most cases may be even higher. To protect electrical power circuits, the resistance must be low enough to ensure operation of the fuse or circuit breaker under fault conditions. Any ground that is adequate for power circuits or lightning protection is more than adequate for protection against static electricity.

A bond or ground is composed of suitable conductive materials having adequate mechanical strength, corrosion resistance, and flexibility for the service intended. Since the bond or ground does not need to have low resistance, nearly any conductor size will be satisfactory from an electrical standpoint. Solid conductors are satisfactory for fixed connections. Flexible conductors are used for bonds that are to be connected and disconnected frequently. Conductors may be insulated or uninsulated. Some prefer uninsulated conductors so that defects can be easily spotted by visual inspections. If insulated for mechanical protection, the concealed conductor is checked for continuity at regular intervals, depending on the inspector's experience. Connections may be made with pressure type ground clamps, brazing, or welding. Battery clamps, magnetic, or other special clamps provide metal to metal contact.

A special situation requiring substantial conductors may arise if there is a possibility that a ground wire may be called upon to carry current from power circuits or lightning protection systems. Obviously, any ground that is adequate for power circuits is more than adequate for protection against accumulations of static electricity.

Ionization

Under certain circumstances air may become sufficiently conductive to bleed off static charges.

Static Comb: A static charge on a conducting body is free to flow, and on a spherical body in space it will distribute itself uniformly over the surface. If the body is not spherical, the self-repulsion of the charge will make it concentrate on the surfaces having the least radius of curvature.

If the body is surrounded by air (or other gas) and the radius of curvature is reduced to almost zero, as with a sharp needle point, the charge concentration on the point can produce ionization of the air, rendering it conductive. As a result, whereas a surface of large diameter can receive and hold a high voltage, the same surface equipped with a sharp needle point can reach only a small voltage before the leakage rate equals the rate of generation.

A "static comb" is a metal bar equipped with a series of needle points. Another variation is a metal wire surrounded with metallic tinsel. If a grounded static comb is brought close to an insulated charged body (or a charged insulating surface), ionization of the air at the points will provide enough conductivity to make the charge speedily leak away or be "neutralized." This principle is sometimes employed to remove the charge from fabrics, power belts, and paper.

Electrical Neutralization: The electrical neutralizer is a line-powered high voltage device which is an effective means for removing static charges from materials like cotton, wool, silk, or paper in process, manufacturing, or printing. It produces a conducting ionized atmosphere in the vicinity of the charged surfaces. The charges thereby leak away to some adjacent grounded conducting body. Electrical neutralizers should not be used where flammable vapors, gases, or dust may be present, unless approved specifically for such locations.

Radioactive Neutralizer: Another method for dissipating static electricity involves the ionization of the air by radioactive material. Such installations require no redesign of existing equipment. However, engineering problems are involved, such as elimination of health hazards, dust accumulations, determination of radiation required, and proper positioning in relation to the stock, machine parts, and personnel. These considerations are best worked out in consultation with radiation specialists.

Open Flame: Ionization of the air can also be obtained by an open flame. This method is frequently used in the printing industry to remove static from paper sheets as they come off the press, thus avoiding the mechanical problems involving one sheet of paper adhering to another, but obviously not to avoid an ignition source.

CONTROL OF IGNITIBLE MIXTURES

Despite efforts to prevent accumulation of static charges, which should be the primary aim of good design, there are many operations involving the handling of nonconductive materials or nonconductive equipment which do not lend themselves to this built-in solution. It may then be desirable, or essential, depending on the hazardous nature of the materials involved, to provide other measures to supplement or supplant static dissipation facilities. For example, where a normally ignitible mixture is contained within a small enclosure, such as a processing tank, an inert gas may be used to bring the mixture well below its flammable range. When operations are normally conducted in an atmosphere above the upper flammable limit, it may be practicable to apply the inert gas only during the periods when the mixture passes through its flammable range.

Mechanical ventilation may be applied in many instances to dilute an ignitible mixture to well below its normal flammable range. Also by directing the air movement, it may be practical to prevent the flammable liquids or dusts from approaching an operation where an otherwise uncontrollable static hazard may exist. To be considered reliable, the mechanical ventilation should be interlocked with the equipment to assure its proper operation.

Where a static accumulating piece of equipment is unnecessarily located in a hazardous area, it is preferable to relocate the equipment to a safe location rather than to rely upon prevention of static accumulation.

FLAMMABLE LIQUIDS

Static is generated when liquids move in contact with other materials. This occurs commonly in such operations as flowing through pipes, and in mixing, pouring, pumping, filtering, or agitating. Under certain conditions, particularly with liquid hydrocarbons, static may accumulate. If the accumulation is sufficient, a static spark may occur.

If the spark occurs in the presence of a flammable vapor-air mixture, an ignition may result. Therefore, steps must be taken to prevent the simultaneous occurence of the two conditions.

Standard control measures are designed to prevent incendive sparks, or the formation of ignitible vapor-air mixtures. In many cases, air which may form an ignitible mixture with the vapor can be eliminated or reduced in concentration to render the mixture nonflammable.

Before a container is filled, contact is made between the filling nozzle and the container and the contact should be maintained throughout the filling operation. By this procedure, any difference in potential between the container and nozzle will be dissipated before the filling operation is started, and differences in potential between the nozzle and container are prevented from forming during the filling operation. See Figure 12-10A for methods of bonding containers and nozzles during container filling.

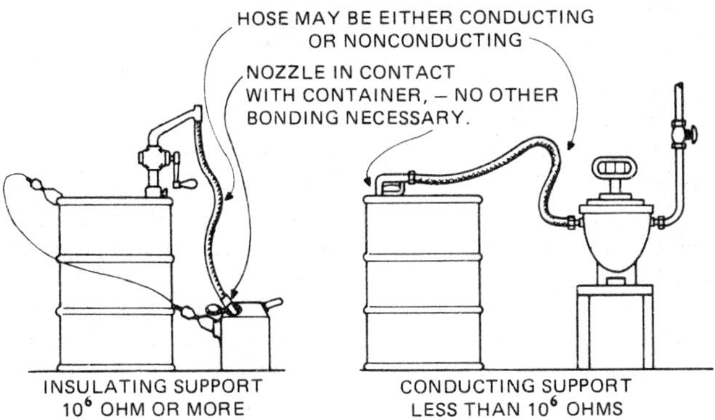
HOSE MAY BE EITHER CONDUCTING OR NONCONDUCTING
NOZZLE IN CONTACT WITH CONTAINER, – NO OTHER BONDING NECESSARY.
INSULATING SUPPORT 10^6 OHM OR MORE
CONDUCTING SUPPORT LESS THAN 10^6 OHMS

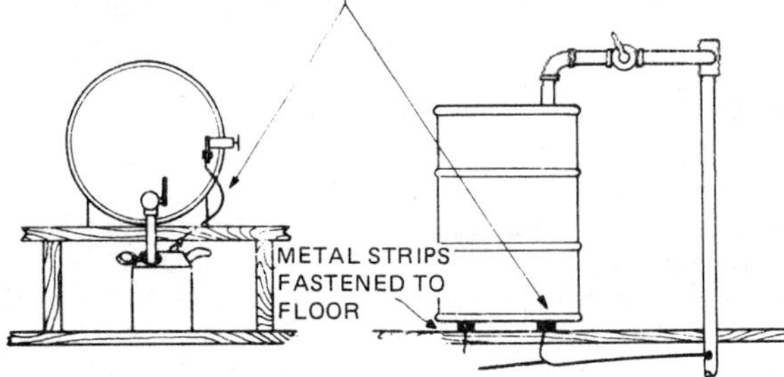
BOND WIRE NECESSARY EXCEPT WHERE CONTAINERS ARE INHERENTLY BONDED TOGETHER, – OR ARRANGEMENT IS SUCH THAT FILL STEM IS ALWAYS IN METALLIC CONTACT WITH RECEIVING CONTAINER DURING TRANSFER.
METAL STRIPS FASTENED TO FLOOR

FIG. 12-10A. Recommended methods of bonding flammable liquid containers during container filling.

The relative static accumulating tendencies of a number of petroleum products have been measured in laboratory tests. In general, aliphatic solvents and lower boiling hydrocarbons exhibited lower charging tendencies than higher boiling products. However, the charging tendency of any given product was found to vary widely from one sample to the next.

The conductivity of a liquid is a measure of its ability to hold a charge. The lower the conductivity, the greater the ability of the liquid to hold a charge. If the conductivity of a liquid is greater than 50 pS/m,* any charges that are generated will dissipate without accumulating to a hazardous potential.

Free Charges on Surface of Liquid

If an electrically charged liquid is poured, pumped, or otherwise transferred into a tank or container, the unit charges of similar sign within the liquid will be repelled from each other toward the outer surfaces of the liquid, including not only the surfaces in contact with the container walls but also the top surface adjacent to the air space, if any. It is this latter charge, often called the "surface charge," that is of most concern in many situations.

In most cases the container is of metal, and hence conducting. Two situations can occur, somewhat different with respect to protective measures, depending on whether the container is in contact with the earth or is insulated from it. These two situations are: (1) an ordinary storage tank resting on earth or concrete or other slightly conducting foundation, and (2) a tank truck on dry rubber tires.

In the first situation, the metal container is connected to ground. The charges that reach the surfaces in contact with the vessel will reunite with charges of opposite sign which have been attracted there. During this process, the tank and its contents, considered as a unit, are electrically neutral, i.e., the total charge in the liquid and on its surface is exactly equal and opposite to the charge on the tank shell. This charge on the tank shell is "bound" there, but gradually disappears as it reunites with the charge migrating through the liquid. The time required for this to occur is called "relaxation time." The relaxation time depends primarily on the conductivity of the liquid. It may be a fraction of a second or several minutes.

During this process, the tank shell is at ground potential. Externally, as already mentioned, the container is electrically neutral. But internally, there may be differences of potential between the container wall and the fluid, lasting until charges on the fluid have gradually leaked off and reunited with the unlike charges on the tank walls.

If the potential difference between any part of the liquid surface and the metal tank shell should become high enough to cause ionization of the air, electrical breakdown may occur and a spark may jump to the shell. A spark across the liquid surface is an ignition hazard where flammable vapor-air mixtures are present. No bonding or grounding of the tank or container can remove this internal surface charge.

In the second situation, when the tank shell is highly insulated from the earth, the charge on the liquid surface attracts an equal and opposite charge to the inside of the container. This leaves a "free" charge on the outside of the tank, of the same sign as that in the liquid, and of the same magnitude. This charge can escape from the tank to the ground in the form of a spark. In filling a tank truck through an open dome, it is this source of sparking which is suspected of having caused some fires; in this case the spark jumps from the edge of the fill opening to the fill pipe which is at ground potential. This hazard can be controlled by grounding the container before filling starts

*picoseimens per meter.

or by bonding the fill pipe to the tank. If grounding the tank is used, the fill pipe must also be grounded.

The foregoing discusses the distribution of charges delivered into a container with a flowing stream. Further generation or separation may occur within the container in several ways to produce a surface charge: (1) flow with splashing or spraying of the incoming stream, (2) disturbance of water bottom by the incoming stream, (3) bubbling of air or gas through a liquid, or (4) jet or propeller blending within the tank.

These charges on the surface of a liquid cannot be prevented by bonding or grounding, but can be rendered harmless by inerting the vapor space, by displacing part of the oxygen with a suitable inert gas, or by increasing the concentration of flammable gas in the vapor space to above the upper flammable limit with a gas, such as natural gas. Use of conductivity additives will rapidly relax the surface charge and prevent the build-up of a hazardous potential.

It must be recognized that mists and foams of flammable and combustible liquids can be ignited by static sparks in much the same way dusts can be ignited. Ignition is possible even though the liquid in the mist is below its flash point.

GASES

Gases not contaminated with solid or liquid particles have been found to generate little, if any, electrification in their flow. When the flowing gas is contaminated with dust, metallic oxides, or scale particles, etc., or with liquid particles or spray, electrification may result. A stream of such particle containing gas directed against a conductive body will charge the latter unless it is grounded or bonded to the discharge pipe.

When any gas is in a closed system of piping and equipment, the system need not be electrically conductive or electrically bonded. Compressed air or steam containing particles of condensed water vapor often manifests strong electrification when escaping.

Carbon dioxide, discharged as a liquid from orifices under high pressure (where it immediately changes to a gas and "snow"), can result in static accumulations on the discharge device and the receiving container. This condition is not unlike the effect from contaminated compressed air or from steam flow where the contact effects at the orifice play a part in the static accumulation. High pressure carbon dioxide should not be discharged into flammable atmospheres because it presents a high risk of ignition due to static spark.

Hydrogen-air and acetylene-air mixtures may be ignited by a spark energy of as little as 0.017 millijoule. In the pure state, no static charges are generated by the flow of hydrogen. However, as gaseous hydrogen is commercially handled in industry, such as flowing through pipelines, discharging through valves at filling racks into pressure containers, or flowing out of containers through nozzles, the hydrogen may be found to contain particles of oxide carried off from the inside of pipes or containers. In this contaminated state, hydrogen may generate static.

The liquefied petroleum gases (LP-Gases) behave in a manner similar to uncontaminated gases in the gas phase and to contaminated gases in the mixed phase. Bonding is not required when LP-Gas vehicles are loaded or unloaded through closed connections so that there is no release of vapor at a point where a spark should occur, irrespective of whether the hose or pipe used is conducting or nonconducting. (A closed connection is one where contact is made before flow starts and is broken after flow has ended.)

DUSTS AND FIBERS

As previously pointed out, the flow of a stream of gas containing small particles can result in a separation of electrons and the accumulation of a static charge on any insulated conductive body with which it comes in contact. Also, dust displaced from a surface on which it rests may develop a considerable charge. The ultimate charge depends on the inherent properties of the substance, size of particle, amount of surface contact, surface conductivity, gaseous breakdown, external field, and leakage resistance in a system. Greater charges develop from smooth than from rough surfaces, probably because of greater initial surface contact. Electrification develops during the first phase of separation. Subsequent impact of airborne particles on obstructions may affect their charge slightly, but if the impact surface becomes coated with the dust, this effect is slight.

Charge generation seldom occurs if both materials are good electrical conductors, but it is likely to occur with a conductor and a nonconductor or two nonconductors. When like materials are separated, as in dispersing quartz dust from a quartz surface, positive and negative charges are developed in the dispersed dust in about equal amounts to give a net zero charge. With materials differing in composition, a charge of one polarity may predominate in the dust. Each of the materials becomes equally charged but with opposite polarity. With a metallic and an insulating material, the former usually assumes a positive and the latter a negative polarity.

Electrostatic charge generation in moving dust normally cannot be prevented. High humidity or grounding of the surface from which dust is dispersed will not eliminate the charge generation. The method of dispersion of the dust, the amount of energy expended in dispersal, the degree of turbulence, and the composition of the atmosphere usually do not affect the magnitude or distribution of the charges.

Not only can dust participate in the generation of static; it may also be the material ignited by static sparks. A suspension of finely divided combustible particles in air has much the same properties as a flammable gas-air mixture, except for a higher ignition energy. It can burn to produce explosive effects. It has a lower flammable limit, but no strictly definable upper limit.

Ignition of Dust by Static Discharge

Dust clouds and layers of many combustible materials (with or without a volatile constituent) have been ignited experimentally by static discharge. In some instances, the charge was generated by movement of dust, in others by a static generator (Wimshurst machine), or by electronic equipment. With dust clouds, it has been shown that a minimum dust concentration exists below which ignition cannot take place regardless of the energy of the spark. At the minimum dust concentration a relatively high energy is required for ignition. At higher dust concentrations (5 to 10 times the minimum), the energy required for ignition is at a minimum.

The energy stored in a capacitative circuit has been previously discussed, where it was tacitly assumed that all of this energy was released in the spark (zero circuit resistance). For the ignition of dusts it has been shown that a circuit resistance of 10,000 to 100,000 ohms may be required for optimum igniting power, indicating that some of the energy is dissipated in the circuit instead of being released in the spark, with a corresponding increase in the total energy required.

A layer of combustible dust can be ignited by static discharge and will burn with a bright flash, glow, or, for some metallic dusts, with flame. Apparently, there is little correlation in the minimum energy required for ignition of dust layers and clouds. Layers of some metallic dusts such as aluminum, magnesium, titanium, and zirconium require less energy for ignition than carbonaceous materials.

Primary explosives, mercury fulminate, and tetryl, for example, are readily detonated by static spark discharge. Steps necessary to prevent accidents caused by static electricity in explosives manufacturing operations and storage areas vary considerably with the static sensitiveness of the material being handled. In all instances in which static electricity was authentically established as the cause of ignition, the spark occurred between an insulated conductor and ground. It has not been verified experimentally that a dust cloud can be ignited by static discharge within itself.

STATIC DETECTORS

The following devices have an application in the measurement and determination of static electricity within the limitations of each device as described:

Electroscopes: The leaf electroscope is a simple but sensitive device that demonstrates the presence or absence of electrical charges by the repulsion of its leaves when the device is charged. Only units intended as portable dosimeters for ionizing, radiation, and one or two classroom demonstration models are available.

Neon Lamps: A small neon lamp or fluorescent tube will light up feebly when one terminal is grounded (or held in the hand) and the other makes contact with any sizeable conductor that carries a charge potential of 70 volts or more. Like the electroscope, it gives but little quantitative information; however, when it passes current, it may give a rough idea of the rate at which charges are being produced in certain operations. Adjustable series-parallel groupings of such lamps and small capacitors can be arranged to give a semblance of quantitative information.

Electrostatic Voltmeters: These meters operate by electrostatic attraction between movable and stationary metal vanes. No current is passed to maintain deflection because one set of vanes (usually the stationary one) is very highly insulated. Small portable, accurately calibrated instruments are available in several ranges from 100 to 10,000 volts. This type of meter may be used for quantitative electrostatic analysis.

Electrometers: Electrometers are frequently used for laboratory and field investigations of static electricity. These instruments employ special input stages designed for high input resistance and low grid current and may be used in several ways. With a small antenna mounted on the grid terminal, they are very suitable for detecting transient electrostatic charge effects; in the presence of a constant field, the charge induced on the grid will leak off and the meter pointer will return to zero. However, as the rate of leakage is not high, the electrometer finds considerable use for "on the spot" static checks. Often electrometers are equipped with high resistance terminal shunts to convert them to current meters in the nanoampere range. They can thus be used in many applications to indicate the rate of charge development. A simple adaptation converts the instrument into a megohm meter as explained in the manufacturer's instruction manual.

Field Mills: A "field mill" is a device that overcomes the serious limitation of the electrometer in that it provides a continuous indication of charge by providing its own transients in the form of continuous grid modulation. It can therefore "look" at a distant charge and determine its potential. The electrical field extending from the charge is chopped by a motor driven variable condenser or window. The resulting pulsations are transformed, amplified, and rectified to give a dc meter deflection proportional to the strength of the field. A similar instrument of suitable design can be immersed in a liquid, as in a pipe line, to measure charge density.

Bibliography

NFPA Codes, Standards, Recommended Practices and Manuals. (See the latest *NFPA Codes and Standards Catalog* for availability of current editions of the following documents.)

NFPA 30, *Flammable and Combustible Liquids Code.*
NFPA 70, *National Electrical Code®.*
NFPA 77, *Recommended Practice on Static Electricity.*
NFPA 385, *Standard for Tank Vehicles for Flammable and Combustible Liquids.*
NFPA 407, *Standard for Aircraft Fuel Servicing.*

Additional Readings

American Petroleum Institute, *Recommended Practice for Protection Against Ignitions Arising Out of Static, Lightning, and Stray Currents.* (RP 20003) (Third Ed.), American Petroleum Institute, Washington, DC, 1974.

Beach, R., "Preventing Static Electricity Fires," *Chemical Engineer* Vol. 88, Feb.d 1, 1965, pp. 85-88.

Bustin, W. M., and Dudek, W. G., *Electrostatic Hazards in the Petroleum Industry*, Research Studies Press, LTD. England, 1983.

Bustin, W. M., Koszman, I., Tobye, I. T., "New Theory for Static Relaxation From High Resistivity Fuel." Presented to a *Special Joint Meeting of the Operating Practices Committee, Division of Refining*, and the *Central Committee on Safety and Fire Protection, Division of Science and Technology*, Esso Research and Engineering Co., Houston, TX, Oct. 1964.

Eichel, F. G., "Electrostatics," *Chemical Engineering*, Mar. 1967, pp. 153-167.

Electrostatic Properties of Materials, (Std. No. 101B--Method 4046), Jan. 1969, General Services Administration, Washington, DC.

"Fire and Explosion Due to Electrostatic Charges in the Plastics Industry," *Bulletin No. 6*, Society of the Plastics Industry, NY.

Guest, P. G., Sikora, V. W., and Lewis, B., *Static Electricity in Hospital Operating Suites: Direct and Related Hazards and Pertinent Remedies*, Bulletin 520, 1953 (reprinted 1962), Bureau of Mines, Washington, DC.

Harper, W. R., *Contact and Frictional Electrification*, Oxford Press, 1967.

Haase, Heinz, *Electrostatic Hazards; Their Evaluation and Control*, 2nd ed., (translated by Michael Wald), Verlay Chemical International, New York, 1977.

Heidelberg, E., "The Ignition of Explosives Mixtures by Static Electricity," *Proceedings of the International Conference on the Safe Use of Electrical Energy in Spaces Where the Danger of an Explosion Exists*, Gottwaldov, West Germany, Paper No. 8, Oct. 1968.

Johnson, O. W., "The Hazard from Static Electricity on Moving Rubber-Tired Vehicles," *Fire Journal*, Vol. 61, No. 1, Jan. 1967, pp. 25-27.

"Let Static Flow," (Part II of article on static sparking), *Record*, Factory Mutual Engineering Corp., May-June 1966.

Leonard, Joseph T., and Carhart, Homer W., *Static Electricity Measurements During Refueler Loading*, (NRL Report 7203) (AD717-347) Naval Research Laboratory, Washington, DC, 1971.

Leonard, J. T., and Clark, R. C., *Electrostatic Hazards Produced by Carbon Dioxide in Inerting and Fire-Extinguishing Systems*, (AD A015 003) (NRL Report No. 7928) U.S. Department of the Navy, Naval Research Laboratory, Washington, DC, 1975.

Loeb, Leonard B., *Fundamentals of Electricity and Magnetism*, Third Edition, Dover Press, 1961.

Martel, C. R., *An Evaluation of the Static Charge Reducer for Reducing Electrostatic Hazards in the Handling of Hydrocarbon Fuels*, Technical Report AFAPLTR-70-22, Air Force Aero Propulsion Laboratory, Wright-Patterson Air Force Base, Ohio, July 1970.

Morris, G., "The Static Hazard in Industry," *Engineering*, Vol. 164, July 18, 1947, pp. 49-51.

Purcer, P. R., "Ignition by Electrostatic Sparks in Hyperbaric Oxygen," *The Lancet*, Dec. 24, 1966.

Rapp, H. W., "Static Electricity," *Safety Newsletter*, National Safety Council, Washington, DC, 1963.

Rogers, George J., and Evans, David D., *Characterizations of Electrical Ignition Sources within Television Receivers*, (NBS Technical Note 1109), National Bureau of Standards, Washington, DC, 1979.

Sommer, E. C., "Preventing Electrostatic Ignitions," Paper presented at *API Central Committee on Safety and Fire Protection*, Tulsa, OK, Apr. 1967.

"Static Electricity," *Bulletin No. 256*, U.S. Department of Labor, Bureau of Labor Standards, Washington, DC, 1963.

SPECIAL SYSTEMS FOR EXPLOSION DAMAGE CONTROL

Revised by Parker Peterson, P.E.

Explosions in industry are accidents caused by a major uncontrolled release of energy. Explosions cause devastating damage, death, and destruction. They may or may not be accompanied by fire damage. Since explosions are probably the most destructive industrial accidents, design and loss prevention engineers devote extensive effort to explosion damage control.

There are at least two fundamental approaches to the problem of control of explosion damage. The first and best one is prevention of the explosion. This gives total damage control by eliminating the cause of damage. Explosion prevention is always the desired objective and the best approach to damage control. There are several factors or elements of explosion prevention that will be covered in this chapter.

Explosion prevention measures for combustion explosions are covered in depth in NFPA 69, *Standard on Explosion Prevention Systems*. Approaches that will be included in this chapter are prevention by elimination of ignition, by oxidant concentration reduction, by combustible concentration reduction, and by suppression of the combustion to prevent pressure rise and resulting explosions.

Flammable or explosive limits of gases or vapors are discussed in Section 4, Chapter 1 of this HANDBOOK, while the various types of explosions are discussed in detail in Chapter 2 of the same Section. Other pertinent references are found in Section 5 of this HANDBOOK, with Chapter 4 covering "Flammable Liquids"; Chapter 5, "Gases"; and Chapter 9, "Dusts". In each of these references in Section 5, information is given on the combustibility of these materials, as well as on the factors influencing the development of combustible mixtures.

Protection Against Damage

Despite the best efforts at prevention, explosions do occur from time to time, and measures to control or limit the damage are essential. The application of protective measures to prevent, limit, or control damage from the explosion is the second fundamental approach to explosion damage control that will be covered in this chapter.

Any flammable dust, vapor, mist, or gas mixed with air or other supporter of combustion will, under certain conditions, burn with sufficient speed to generate high pressures in a confined volume. Consequently, explosion protection systems, in addition to explosion prevention measures, should be considered for equipment, housings, rooms, and buildings associated with the manufacture, handling, processing, and storage of any flammable liquids, dust, vapor, mist, or gas.

The principal factors influencing the generation of pressure from a confined deflagration are: (1) the nature of the material, i.e., chemical structure; (2) the concentration of the material in air or other supporter of combustion; (3) the particle size in the case of dusts; (4) the oxygen or oxidizer concentration; (5) the turbulence of the mixture; (6) the temperature of the mixture; (7) the pressure of the mixture; (8) the effect of moisture and diluting substances, such as inert materials; (9) the nature and location of the ignition source(s); and (10) the pressure relief openings in the enclosure. The violence of the explosion which may result from this confined deflagration is dependent upon the extent and rate of pressure development and the physical construction of the enclosure.

Explosions also occur from rapid release of energy from runaway reactions and decompositions of unstable compounds. Control of damage from accidental discharge of explosives is beyond the scope of this chapter.

CONTROL OF IGNITION SOURCES

If all ignition sources could be eliminated or controlled in any potentially hazardous space, then ignitions, combustions, and resulting explosions could be prevented. It is essential that every effort be made to eliminate all ignition sources.

Open Fires and Hot Work Control

Direct fired equipment should not be located in areas that are, or might become, hazardous. Isolation of fired units from hazardous spaces by vapor tight enclosures fed

---

Mr. Peterson is Manager—Explosion Protections Systems, Protection Systems Division of Fenwal, Inc., and a member of the NFPA Explosion Venting and Explosion Prevention Systems Committees.

with combustion air ducted in from safe locations or isolated by tested and proven flame arrestors and flashback protectors is sometimes resorted to. However, elimination of fired equipment from potentially hazardous areas is preferred.

Strict administrative control of smoking, mobile equipment, and a well administered hot work permit system to prevent inadvertent introduction of an ignition source into hazardous spaces are essential.

Electrical Ignition Sources

NFPA 70, *National Electrical Code®*, must be followed to avoid ignition sources from electrical equipment.

Static Electricity and Grounding

A specially illusive ignition source is static electricity and resulting incendiary spark discharges. Static build-up can occur in many ways. Good grounding of equipment can help reduce static and reduce potential voltage differences between various parts of a system, but does not assure total freedom from static hazards in all circumstances. Static charges can build up in free falling streams of liquids, in dust clouds, in vapor or mist clouds, and in other ways in spite of good grounding. NFPA 77, *Recommended Practice on Static Electricity*, deals with static electricity and its control.

Every effort to control ignition sources is an essential part of explosion prevention. Good control can reduce the probability of ignition and combustion of any given combustible mixture. Most loss prevention engineers assume that, despite the best efforts at ignition control, if a flammable mixture is allowed to exist, it will eventually find a source of ignition and explode. Therefore, even though control of ignition may be vigorously pursued and enforced, it cannot be relied upon to prevent all explosions. Other steps must be taken.

OXIDANT CONCENTRATION REDUCTION

Explosions can be prevented by control of a second element of the fire triangle: the air or oxidant concentration needed to support combustion. NFPA 69, *Standard on Explosion Prevention Systems*, terms this "Oxidant Concentration Reduction." The usual method of oxidant concentration reduction is by purging or inerting the space with a gas that is deficient in oxidant.

Fundamentals of Inerting Systems

The fire and explosion hazard of many materials can be safeguarded during storage and processing operations by the use of a suitable inert gas. This is possible since combustion of most materials will not occur if there is an absence of atmospheric oxygen or if the concentration of atmospheric oxygen is reduced below a certain specific limit.

When inert gas is used as a means of controlling fires and explosions, its principal function is to prevent the formation of explosive vapor-air mixtures, generally in enclosed spaces. Typical examples are its use to inert tanks prior to repair, to empty flammable liquid storage tanks by pressure, to prevent the formation of explosive mixtures in drying ovens, and to blanket flammable products in storage tanks or reaction equipment.

In many cases dust explosions can be prevented by inert gas. Industries that are faced with the explosion hazard of flour, starch, sugar, coal, plastics, and other combustible dusts have found it practical in some cases to install inert gas protection. Equipment such as grinders, pulverizers, mixers, conveying systems, classifiers, dust collectors, sacking machines, and the like can be protected in this manner.

Other applications of inert gas include: (1) use in chemical processes where there is a continuous explosion hazard if oxygen is present; (2) for protection of operations which must be conducted above the ignition temperature of the materials involved; (3) to arrest spontaneous heating of products in storage; and (4) as a means of achieving explosion suppression.

The limitations of inerting methods should be recognized. These include reliability and adequacy of supply, freedom from moisture or any other constituents which could cause contamination or degradation of product, dependability of instrumentation and excess pressure relieving devices, and the suitability of the supply system to insure sufficient quantity under peak demand.

Personnel must be protected by proper respiratory equipment before entering any enclosure where inert gas has been used, and an atmosphere deficient in oxygen may exist. Air supplied or oxygen supplied masks should be used. Self-contained breathing apparatus of short duration has been recognized by the U.S. Bureau of Mines for self-rescue. Filter-canister type masks are inadequate for work involving inert gas because they do not supply any oxygen. The toxicity of certain gases used for inerting must also be recognized.

Basic Installation Criteria

NFPA 69, *Standard on Explosion Prevention Systems*, provides a considerable amount of information on inerting materials, methods for installing inerting systems, and oxidant and combustible concentration reduction, including basic design data for the guidance of engineers.

Hazard Evaluation: Consideration should be given first to the evaluation of all hazards which are to be protected, to the operating equipment involved, and to any abnormal conditions that might be encountered. Such an overall study is essential in order to properly plan the most practical and reliable installation and distribution system.

The need to provide inert gas protection can influence basic plant layout design. It may be desirable, for example, to locate the gas supply near the operation with the largest demand. Likewise, the type, size, and location of process vessels and storage facilities might be quite different if inert gas protection is to be provided.

Type of Inert Gas: Although there are a number of inert gases readily available, the two most common are carbon dioxide and nitrogen. Others, such as argon, helium, and the halogenated hydrocarbons (Halons), or flue gas, may satisfy special needs.

The gas which is most suitable for specific application will depend on a number of factors. In addition to being inert from a fire and explosion standpoint, the gas must be chemically unreactive with the materials to be protected and free from contaminants. If there is a corrosion problem, or if the gas will come in contact with water reactive chemicals, control of moisture must be possible and practical.

A sufficient volume of gas must be available at an adequate pressure to satisfy requirements at periods of peak demand. A completely oxygen free gas may be necessary for process reasons or in order to maintain product quality, though it is not essential for fire protection. However, it is always advisable to maintain as low an oxygen concentration as practicable. The maximum permissible oxygen percentage below which ignition will not occur varies with the flammable materials involved and the inerting medium, as shown in Tables 12-11A, 12-11B, and 12-11C.

Quantity Requirements: The amount of inert gas required and the rate of application will depend upon: (1) the oxygen concentration that is permissible and the amount of oxygen in the inert gas—if any; (2) the "factor of safety" to be observed; (3) the amount of loss through leakage; (4) atmospheric conditions; (5) operating conditions; (6) size and shape of the equipment to be protected; and (7) the method of application.

Methods of Application

There are two ways of applying inert gas to insure the formation of a noncombustible atmosphere within an enclosed tank or space. These may be designated as the (1) batch or fixed volume requirement method, and (2) continuous method. Each has several modes of application which are described in NFPA 69.

Fixed Volume Method: In one variation of this method, the system to be protected is purged and the atmosphere rendered inert by first reducing the pressure and then introducing inert gas. Equipment which normally operates under a vacuum may be inerted for periods of shutdown in this manner. Conversely, inert gas under pressure can be introduced into an enclosure and, after mixing has taken place, the pressure can be reduced to normal by venting the gas to the atmosphere. Several pressurizing cycles may be necessary to sufficiently reduce the oxygen content.

The fixed volume method is useful when containers that are filled with a flammable material are to be emptied and an inert atmosphere provided. In all cases the design limitations of the equipment must be observed. Enclosures may be purged by displacing or diluting the flammable vapors at substantially atmospheric pressure. One or more vents discharging to a safe location are opened and the inert gas added until the oxygen content, as shown by test, has been reduced to the desired limit.

Continuous Methods

Fixed Rate Mode: In this method, the inert gas is added continuously in an amount sufficient to supply peak requirements. The quantity required is based on the maximum inbreathing rate under conditions of sudden cooling, such as that caused by rain, plus the maximum product withdrawal.

This method is relatively simple and has the advantage of requiring no mechanical devices such as regulators or motor valves. The principal disadvantage is wasting inert gas. Since the peak rate is applied continuously, there are vapor losses due to the "sweeping" action of the inert gas passing through the vapor space. An additional objection is the possible plugging by rust, scale, or ice of the orifice plates controlling the flow of inert gas. A personnel hazard also may exist in the immediate area of the vents because of oxygen deficiency. However, where large volumes of inexpensive gas are available, this method may be found most suitable.

TABLE 12-11A. Maximum Permissible Oxygen Percentage to Prevent Ignition of Flammable Gases and Vapors Using Nitrogen and Carbon Dioxide for Inerting

	N_2-Air		CO_2-Air	
	O_2 Percent Above Which Ignition Can Take Place	Maximum Recommended O_2 Percent	O_2 Percent Above Which Ignition Can Take Place	Maximum Recommended O_2 Percent
Acetone	13.5	11	15.5	12.5
Benzene (Benzol)	11	9	14	11
Butadiene	10	8	13	10.5
Butane	12	9.5	14.5	11.5
Butene-1	11.5	9	14	11
Carbon Disulfide	5	4	8	6.5
Carbon Monoxide	5.5	4.5	6	5
Cyclopropane	11.5	9	14	11
Dimethylbutane	12	9.5	14.5	11.5
Ethane	11	9	13.5	11.0
Ether	—	—	13	10.5
Ether (Diethyl)	10.5	8.5	13	10.5
Ethyl Alcohol	10.5	8.5	13	10.5
Ethylene	10	8	11.5	9
Gasoline	11.5	9	14	11
Gasoline				
73–100 Octane	12	9.5	15	12
100–130 Octane	12	9.5	15	12
115–145 Octane	12	9.5	14.5	11.5
Hexane	12	9.5	14.5	11.5
Hydrogen	5	4	6	5
Hydrogen Sulfide	7.5	6	11.5	9
Isobutane	12	9.5	15	12
Isopentane	12	9.5	14.5	11.5
JP-1 Fuel	10.5	8.5	14	11
JP-3 Fuel	12	9.5	14	11
JP-4 Fuel	11.5	9	14	11
Kerosene	11	9	14	11
Methane	12	9.5	14.5	11.5
Methyl Alcohol	10	8	13.5	11
Natural Gas (Pittsburgh)	12	9.5	14	11
Neopentane	12.5	10	15	12
n-Heptane	11.5	9	14	11
Pentane	11.5	9	14.5	11.5
Propane	11.5	9	14	11
Propylene	11.5	9	14	11

Notes:

Data in this Table were obtained from publications of the U.S. Bureau of Mines.

Data were determined by laboratory experiments conducted at atmospheric temperature and pressure. Vapor-air inert-gas samples were placed in explosion tubes and exposed to a small electric spark or open flame.

In the absence of reliable data, the U.S. Bureau of Mines or other recognized authority should be consulted.

TABLE 12-11B. Maximum Permissible Oxygen Concentration to Prevent Ignition of Combustible Dusts Using Carbon Dioxide and Nitrogen for Inerting

	Oxygen Percent by Volume Above Which Ignition Can Occur	
	Carbon Dioxide-Air	Nitrogen-Air
Aluminum (atomized)	2	7
Antimony	16	—
Dowmetal	0	—
Ferrosilicon	16	17
Ferrotitanium	13	—
Iron, Carbonyl	10	—
Iron, Hydrogen reduced	11	—
Magnesium	0	2
Magnesium-Aluminum	0	5
Manganese	14	—
Silicon	12	11
Thorium	0	2
Thorium Hydride	6	5
Tin	15	—
Titanium	0	4
Titanium Hydride	13	10
Uranium	0	1
Uranium Hydride	0	2
Vanadium	13	—
Zinc	9	9
Zirconium	0	0
Zirconium Hydride	8	8

Notes:

Data in this Table were obtained principally from publications of the U.S. Bureau of Mines.

In the furnace test dust clouds of Zr, Th, U, and UH_3 also ignited in CO_2. During heating for several minutes, undispersed layers of samples of the following metal powders ignited (glowed) in CO_2: Stamped Al, Mg, ZnMg-Al, Dowmetal, Ti, TiH_2, Zr, ZrH_2, Th, ThH_2, U, and UH_3. Visible burning of dust layers was also observed in N_2 with powders of Mg, Sn, Mg-Al, Dowmetal, Ti, TiH_2, Zr, Th, ThH_2, U and UH_3.

In the absence of reliable data for combustible dusts, the U.S. Bureau of Mines or other recognized authority should be consulted.

Data were obtained by laboratory experiments conducted at atmospheric temperature and pressure. An electric spark was the ignition source.

Variable Rate Mode: In this method, the inert gas is admitted to the system being protected on a demand basis. As shown in Figure 12-11A, a low volume bleed provides for minor pressure changes in the enclosure while the solenoid valve, which is operated by the pump motor switch, immediately supplies the inert gas flow needed when enclosure contents are being removed. This method has the advantage of reducing vapor losses by maintaining a slightly positive pressure. It is the most efficient procedure since the gas is introduced primarily as needed.

The variable rate method is recommended for maintaining an inert atmosphere in systems that are subject to a wide variation in inert gas demand.

Inert Gases

The inerting media for fire and explosion prevention purposes must be obtained from a dependable source capable of continuously supplying the amount required to maintain an atmosphere that will not support combustion. Distribution should be in conformance with accepted piping standards, with protection provided against physical damage. Main distribution lines should be located so that exposure to fire is minimized.

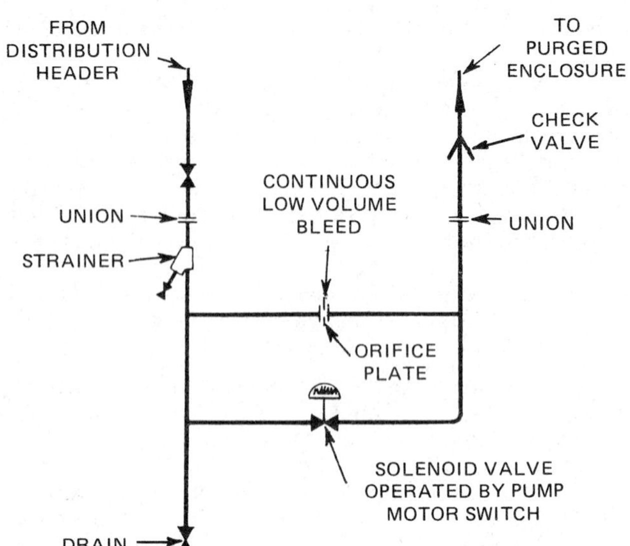

FIG. 12-11A. Schematic sketch showing a method of flow control that can be used with variable rate application.

Carbon dioxide or nitrogen in cylinders is probably the best source of inert gas for small plants, or where the systems to be protected are of small volume and loss through leakage is relatively small. Under some conditions, commercially available carbon dioxide or nitrogen may be the most practical supply—even for large installations. Carbon dioxide fire extinguishers should not be used.

Where no oxygen can be tolerated, carbon dioxide gas or nitrogen as a source of supply has many advantages. Bonding and grounding are important to prevent any static discharge during inerting.

A common method of obtaining inert gas is by the combustion of hydrocarbons as from flame type producers which operate on liquid or gaseous fuels, natural gas, or manufactured gas. The products of combustion generally contain 9 to 12 percent carbon dioxide, some carbon monoxide, and about 85 percent nitrogen. Oxygen will also be present in inert gas, 2 to 8 percent with fuel oil and usually less than 1 percent with natural gas as the fuel. The composition of the gas produced by the combustion of hydrocarbons varies, depending upon the operation of the combustion equipment. If purified inert gas is necessary, the oxygen and traces of carbon monoxide can be removed in one operation. If the carbon dioxide is also undesirable, it can be removed in a second operation. Coolers and scrubbers are needed with flame type producers to remove any particles of glowing soot that might be present, and to make certain the combustion products are not hot enough to ignite any vapor mixtures in the equipment being purged. Inert gas produced in this manner may be stored or piped directly to the point of use. Provision for adequate reserves is important so that continuation of protection can be assured in case of producer failure.

The products of combustion from process furnaces or boiler furnaces find considerable use where large volumes are needed. The flue gas from power plant stacks will contain from less than 9 percent to about 14 percent carbon dioxide, and usually only a fraction of a percent of

Table 12-11C. Maximum Permissible Oxygen Content to Prevent Ignition by Spark of Combustible Dusts Using Carbon Dioxide as the Atmospheric Diluent

Dust	Maximum allowable oxygen concentration, percent
Agricultural	
Clover seed	15
Coffee	17
Cornstarch	11
Dextrin	14
Lycopodium	13
Soy flour	15
Starch	12
Sucrose	14
Chemicals	
Ethylene diamine tetra acetic acid	13
Isatoic anhydride	13
Methionine	15
Ortazol	19
Phenothiazine	17
Phosphorous pentasulfide	12
Salicylic acid	17
Sodium ligno sulfonate	17
Steric acid and metal stearates	13
Carbonaceous	
Charcoal	17
Coal, bituminous	17
Coal, subbituminous	15
Lignite	15
Metals	
Aluminum	2
Antimony	16
Chromium	14
Iron	10
Magnesium	0
Manganese	14
Silicon	12
Thorium	0
Titanium	0
Uranium	0
Vanadium	14
Zinc	10
Zirconium	0
Miscellaneous	
Cellulose	13
Lactalbumin	13
Paper	13
Pitch	11
Sewage sludge	14
Sulfur	12
Wood flour	16
Plastics Ingredients	
Azelaic acid	14
Bisphenol A	12
Casein, rennet	17
Hexamethylenetetramine	14
Isophthalic acid	14
Paraformaldehyde	12
Pentaerythritol	14
Phthalic anhydride	14
Polymer glyoxyl hydrate	12
Terephthalic acid	15
Plastics—Special Resins and Molding Compounds	
Coumarone-indene resin	14
Lignin	17
Phenol, chlorinated	16
Pinewood residue	13
Rosin, DK	14
Rubber, hard	15
Shellac	14
Sodium resinate	14
Plastics—Thermoplastic Resins and Molding Compounds	
Acetal resin	11
Acrylonitrile polymer	13
Butadiene-styrene	13
Carboxymethyl cellulose	16
Cellulose acetate	11
Cellulose triacetate	12
Cellulose acetate butyrate	14
Ethyl cellulose	11
Methyl cellulose	13
Methyl methacrylate	11
Nylon polymer	13
Polycarbonate	15
Polyethylene	12
Polystyrene	14
Polyvinyl acetate	17
Polyvinyl butyral	14
Plastics—Thermosetting Resins and Molding Compounds	
Allyl alcohol	13
Dimethyl isophthalate	13
Dimethyl terephthalate	12
Epoxy	12
Melamine formaldehyde	17
Polyethylene terphthalate	13
Urea formaldehyde	16

Notes:

Data in this table are from U.S. Bureau of Mines Rept. of Inv. 6543. The data were obtained by laboratory experiments conducted at room temperature and pressure, using a 24 watt continuous spark as the ignition source.

For moderately strong igniting sources, such as a low current electrical arc or a heated motor bearing, the maximum permissible oxygen concentration is 2 percentage points less than the corresponding value for ignition by spark.

For strong igniting sources, such as an open fire, flame or glowing furnace wall, the maximum permissible oxygen concentration is 6 percentage points less than the corresponding value for ignition by spark.

The maximum permissible oxygen concentration for ignition by spark, when nitrogen is used as the atmospheric diluent, can be calculated by a "rule of thumb" formula:

$$O_n = 1.3O_c - 6.3$$

where O_n = the maximum permissible oxygen concentration using nitrogen as the atmospheric diluent.

O_c = the maximum permissible oxygen concentration using carbon dioxide as the atmospheric diluent.

Research data on the use of dry powders or water as inerting materials and on the effects of inerting on pressure development in a closed vessel are given in BuMines Repts. of Inv. 6543, 6561 and 6811.

carbon monoxide, although the latter may reach several percent depending upon the efficiency of the combustion process. Likewise, it is not unusual for the oxygen content to reach 15 or 16 percent if dampers are not properly adjusted. So called "blow-off gas" from certain chemical oxidation processes may also be suitable for use in inerting. Purification equipment for these waste gases will vary depending upon composition and specific needs. A means for removing solid particles from the gas stream must be provided, and there will generally be a scrubber, dryer, some means for removing excess oxygen, and a compressor. The exact order of the units will vary with individual installations.

Inert gas can also be obtained by the catalytic oxidation of ammonia with air. This method produces high purity nitrogen with only traces of carbon dioxide, hydrogen, oxygen, and residual ammonia.

Another means of obtaining inerting media is by the liquefaction of air with subsequent fractionation to produce nitrogen. Package units of various capacities using this method are commercially available.

Distribution piping must be properly sized to deliver the required volume of inert gas at adequate pressure, and it must be constructed of suitable material. Moisture traps and strainers may be necessary. Back pressure valves are recommended for each branch from the main distribution line to prevent reduction of the pressure in the main below a predetermined set pressure if there is excessive demand in a branch line. Contamination of the entire inert gas system, due to reversal of flow either through loss of inert gas pressure or by excessive pressure in an inerted unit, may be prevented by installation of check valves. The entire distribution facility should be installed, inspected, and maintained with the same care and attention given to automatic sprinkler systems.

For the special problems of safeguarding aircraft fuel tank atmospheres by the use of inert gas, reference should be made to NFPA 410, *Standard on Aircraft Maintenance.*

The inert gases discussed above are all oxygen deficient and render a space noncombustible by reducing the concentration of oxidant so that the mixture is outside the flammable range. The oxidant deficient gas used is not always inert. A combustible gas, such as natural gas, is sometimes used in the vapor space of a storage tank to keep the space "fuel rich" or above the flammable limit, making the space "oxidant deficient" from an explosion prevention standpoint.

COMBUSTIBLE CONCENTRATION REDUCTION

Containment of the Fuel

The concept of explosion prevention is usually concerned with a specific finite space. Although it can be open air space in the vicinity of certain process operations, it is usually space enclosed by a building, by tank walls, or by other process equipment. Explosions can be prevented by exclusion of the fuel from spaces where it is not wanted, and by good design and maintenance for secure containment of the fuel within the storage tank, process vessels, cyclones, ducts, conveyors, silos, screens, and hoppers where it is supposed to be kept. Proper containment of the fuel will automatically provide for combustible concentration reduction in other spaces.

A second application of the "containment" concept that can sometimes be used is containment of the pressures generated by the combustion. This will be discussed later in this Chapter under the heading "Containment of the Pressure."

Ventilation

General Ventilation: Combustible concentration reduction in a given space can sometimes be attained by general ventilation of the space using clean air to remove dust, gas, or vapor and thus prevent explosions. Properly designed installations may be able to keep the fuel concentration below the LEL (Lower Explosive Limit) at all times.

Even when an abnormal spill raises the fuel concentration perhaps to above the LEL, emergency ventilation may be able to quickly reduce the flammable mixture in the ventilated space to a nonflammable condition. This reduces the length of time an explosion potential exists. It also reduces the overall probability of an accident, but does not eliminate the possibility of an accident. If an ignition source is present in the area during the time an explosive mixture exists, a combustion explosion will occur.

Spot Ventilation: Spot ventilation is often used at points of potential leakage to prevent spread of flammables throughout an area. Pick up velocities, fuel concentrations in duct work and blowers, safe discharge areas, and air pollution regulations need to be considered in designing ventilation systems. Proper design will provide sufficient air flows to keep fuel concentrations well below the LEL throughout the system under the conditions of the largest anticipated peak load that will be incurred. NFPA 69, *Standard for Explosion Prevention Systems,* Appendix A, has useful information and references relative to both general and spot ventilation systems.

Purging

Purging of rooms or enclosures with fuel free air is a specialized utilization of the ventilation principle to prevent explosions by control of concentration of combustibles. Purging, rather than removing fuel from a space, is usually the reverse process of pressurizing the room or enclosure with clean air to prevent the entry of fuel into the space.

Purging of electrical equipment enclosures is covered in detail in NFPA 496, *Standard for Purged and Pressurized Enclosures for Electrical Equipment in Hazardous (Classified) Locations,* hereinafter referred to in this chapter as NFPA 496. Highlights from that standard are included here for reference.

Purging Electrical Equipment

In 1967, the first edition of NFPA 496 was officially adopted. This standard filled the need for a nationally recognized method of using air (or inert gas) to purge and pressurize electrical equipment enclosures in Class I hazardous (classified) locations so that the atmosphere in the enclosure is nonflammable. In such enclosures, the electrical equipment need not be explosion-proof. For example, in chemical plants it is common practice to install as much of the electrical equipment as possible in an air-pressurized control room. As long as a positive pressure is maintained in the control room with clean air, flammable vapors and gases cannot enter from adjoining Class I hazardous (classified) locations of the plant. Motors, instruments, and control equipment can be located in hazardous locations if the enclosure or case is purged and pressurized, and if the enclosure is designed so that external surface temperatures cannot approach the ignition temperature of the gas or vapor in the Class I hazardous (classified) location in which the enclosed equipment is installed.

Several factors have to be considered when designing a purging and pressurizing system, including a source of clean air, safe discharge of air from the enclosure, purging procedures to be followed before equipment in the enclosure is turned on, and construction of the enclosure. Recommendations on these and other considerations are given in NFPA 496 which defines three types of purging with different safeguards for each type. They are:

Type X Purging: Type X purging reduces the classification within an enclosure from Division 1 (normally hazardous) to nonhazardous. Because the probability of a hazardous concentration of gas or vapor external to the enclosure is high and the enclosure normally contains a source of ignition, an essential safeguard in Type X purging is the automatic deenergizing of all equipment in the enclosure if purging is interrupted.

Type Y Purging: Type Y purging reduces the classification within an enclosure from Division 1 to Division 2 (hazardous only under abnormal conditions). Electrical equipment in the enclosure must be suitable for use in Division 2 hazardous (classified) locations (equipment that does not normally contain a source of ignition). Thus, a hazardous condition will exist in the enclosure only if an electrical equipment failure creates an ignition source within the enclosure at the same time that the purging system fails. In Type Y purging, an automatic warning is given when purging fails, but the electrical equipment within the enclosure is not automatically deenergized.

Type Z Purging: Type Z purging reduces the classification within an enclosure from Division 2 to nonhazardous. Since the area surrounding the enclosure is a Division 2 hazardous (classified) location, a hazardous condition will exist within the enclosure only if purging failure occurs simultaneously with some other occurrence that changes the area surrounding the enclosure from a Division 2 to a Division 1 hazardous (classified) location. As in Type Y purging, an automatic warning is required if purging fails so that steps can be taken to reestablish purging, but automatic shutdown of power in the enclosure is not required.

EXPLOSION SUPPRESSION SYSTEMS

Suppression systems are covered in NFPA 69 and are technically explosion prevention systems since they are designed to prevent the "bursting of a . . . container as a result of development of internal pressure beyond the confinement capability of the . . . container" (Rust 1979). Suppression systems do not prevent the combustion from starting, but snuff out the combustion before destructive pressures can develop in the protected equipment.

Fundamentals of Suppression Systems

The suppression of explosions is possible under certain conditions because a short but significant period of time elapses before destructive pressures are developed. If the conditions are right, it is possible to utilize the time available to operate a suppression system. A typical example of the operation of a suppression system is shown in Fig. 12-11B where it can be seen that the explosion pressure is kept to a relatively small value by the actuation of the suppression system during the explosion's incipient growth.

Effective use of the rate of pressure rise to suppress an explosion requires three major considerations in the design of suppression systems. They are:

1. Detection: The explosion must be detected in its incipient stage to allow sufficient time for the operation of the suppression equipment. Due to the relatively short period of time that is available, detection must be automatic, with provisions to discriminate between the explosion and ambient variables that normally exist.
2. Suppression: The mechanism for dispersal of the agent must operate at extremely high speed to fill the enclosure completely within milliseconds after detection of

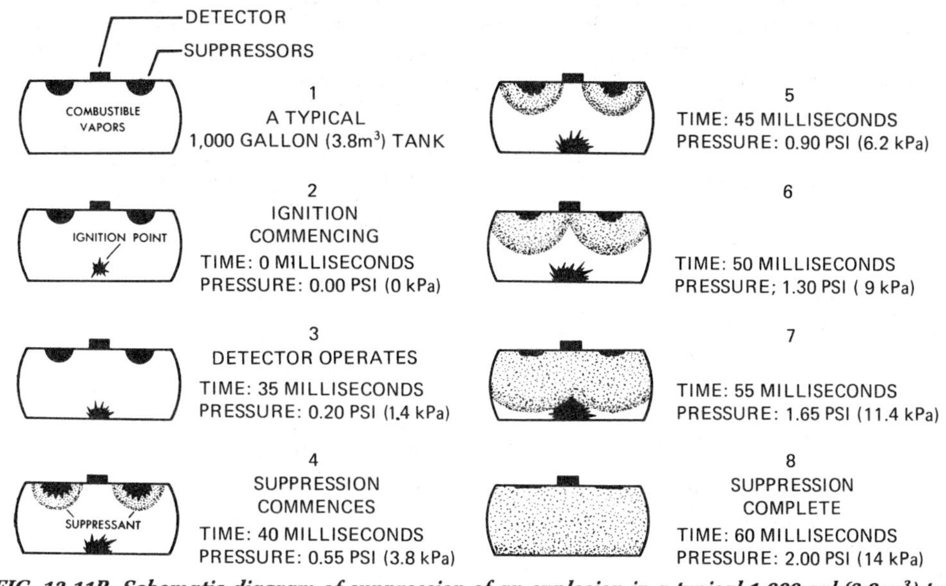

FIG. 12-11B. Schematic diagram of suppression of an explosion in a typical 1,000 gal (3.8 m^3) tank.

the explosion. Actuation of the suppression device must be automatically initiated by the detector to assure no time lag. The extinguishing agent must be dispersed in a very fine mist or dust form at a rapid speed, normally through the use of an explosive force or electroexplosive release of a superpressurized agent.

3. Suppressing Agent: The agent may be a vaporizing liquid compatible with the combustion process to be encountered; a dry chemical or a combination of the two. The mechanism of suppression involves primarily chemical inhibition accompanied by some cooling, inerting, or blanketing. In certain cases, over enrichment of the mixture has been successfully used.

Suppression Equipment

Equipment for use in an explosion suppression system that fulfills the three preceding requirements has been designed as follows.

Detectors: The most commonly used type of detector is a very sensitive and stable type of pressure transducer, designed to close electrical contacts very early during pressure growth. The pressure transducer may be set to sense as low as 0.25 psig (1.72 kPa), depending on plant conditions. Basically, this detector consists of a low inertia diaphragm exposed to the process and coupled to a set of electrical contacts.

If the process being protected operates at negative pressures or undergoes large pressure changes during normal operation, then it may be necessary to use a detector that senses rate of pressure rise rather than a preset static level. Normal settings of a detector of this type are from 5 to 10 psi per second (35-70 kPa/sec.), a range in which most incipient explosions can be detected in time to achieve successful suppression.

Another type of detector is the surveillance detector designed to sense radiation from the incipient explosion. These devices use either photo tubes (UV detectors) or photo cells (infrared detectors). Detectors of this type are necessarily more complex than pressure sensors and require more sophisticated circuitry in translating an alarm condition into a release action. Because radiation is transmitted at the speed of light, and pressure equalizes only at the speed of sound, these detectors will provide somewhat faster response. For most applications, however, this difference is not significant. The principal use for radiation detectors is in applications where the fuel-air mixture is not confined and pressure detection is, therefore, not practicable.

Suppressors: Suppressors are used for storing and releasing an extinguishing agent into the protected volume. Historically, two basically different types have been used. One type utilizes a frangible container and relies on ballistic dispersal of the extinguishing agent provided by a small explosive charge located within. The second, known as "High Rate Discharge" (HRD) extinguishers, contain the agent under pressure and utilize an electroexplosive device to actuate the release mechanism. The trend in recent years has been to standardize on this latter type, namely the HRD units, for most applications.

Frangible suppressors consist of thin wall reservoirs and are generally hemispherical or cylindrical in shape. They are filled with a suppressing agent at the center of which is located an electroexplosive device. The agent is not superpressurized and relies on the hydraulic shock provided by the explosive charge to rupture the walls of the reservoir and disperse the agent throughout the protected volume. The containers are prescored to control the dispersal pattern and minimize the possibility of fragmentation. Because of this these suppressors are relatively weak and are, therefore, limited for use with only the higher boiling agents. These suppressors are expendable and must be replaced following an actuation.

High Rate Discharge extinguishers may be used with any of the suppressing agents, and are superpressurized with dry nitrogen. The release mechanism consists of a rupture disc, actuated by an electroexplosive device. Depending upon the geometry of the protected equipment, these devices may be fitted with spreaders. Although the initial distribution velocity of HRD extinguishers is slightly less than for frangible suppressors, this difference is not considered significant for most applications. Their larger size and consequently greater volume coverage, together with the fact that they are reconditionable, make them more cost effective. This has led to their almost exclusive use in industrial explosion suppression systems.

Design Of Explosion Suppression Systems

For the proper application of an explosion suppression system to a hazardous process or area, the following information is required:

Explosion Characteristics: Inasmuch as effective explosion suppression demands that the entire operating sequence of the suppression system take place during the incipient stage of the explosion, it is important to know the initial rate of pressure rise that the particular fuel-air system can develop under the conditions that prevail within the protected process. In those cases where radiation detectors are going to be used, the spectral characteristics of the combustion must also be known in order to assure that the best combination of sensor and filter is provided. Finally, it is important to know the chemical nature of the fuel-air mixture in order to select the proper suppressing agent and determine the concentration of that agent required to accomplish effective suppression under process operating conditions.

Ignition: In designing the detection subsystem, it is important to diagnose the nature and location of any ignition source. Generally, such sources can be categorized in the following manner:

1. Mechanical
   1.1 Tramp Metal
   1.2 Friction
2. Electrical
   2.1 Electrostatic Discharge
   2.2 Electrical Fault
3. Thermal
   3.1 External Source
   3.2 Spontaneous

Optimum positioning of the detection equipment depends upon advance knowledge of those locations where ignition is likely to occur. If such sources exist only in one small area, it is possible to provide protection with considerably less equipment than if ignition sources are randomly located. If multiple ignition sources are a possibility, then accelerated rates of pressure rise can be anticipated which will in turn put more stringent requirements on the suppression system.

Process Variables: Since the temperature and pressure of a fuel-air mixture prior to an explosion involving it will affect the characteristics of the explosion, it is necessary to have this information in order to determine accurately the factors necessary for system design. These variables will also affect the operation of the equipment, and in cases where exceedingly high temperatures and pressures are encountered, special techniques must be used in order to assure adequate life service for the equipment.

Structural Integrity: Since an explosion is allowed to start and build up to a small pressure before the suppression system operates, it is necessary that the enclosure be able to withstand this pressure.

For those processes operating at normal pressures and temperatures, the maximum pressure that can be expected during a suppression is a transient pressure peak in the order of 3 psi (20 kPa), therefore, ideally the enclosure should be sufficiently strong to withstand this pressure. Actually, depending on the tightness of the process, this pressure may be dissipated in milliseconds and such ponderous pieces of equipment as bag collectors which might not be capable of withstanding steady state pressures of this value would nevertheless be unaffected by the momentary peak due to their inertia. In other cases, equipment might undergo some deformation yet remain intact with the result that a secondary explosion would be avoided.

To illustrate the application of an explosion suppression system, a typical installation for a grinding plant is shown in Figure 12-11C. In plants such as this, the primary suspect ignition source would be a grinder, therefore, a detector would be located as near to this point as possible to achieve early response of the system. Ignition in the grinder may result in initiation of a dust cloud explosion in that region or it may result only in burning material being carried to the cyclone where a greater percentage of fines are in suspension and where a condition more conducive to development of a dust explosion exists. There is also the possibility of electrostatic discharge which would constitute a prime ignition source within the cyclone. Therefore, a detector is also located on this piece of equipment together with high rate discharge (HRD) extinguishers.

High rate discharge extinguishers are deployed to provide optimum coverage within the cyclone itself. An additional container is discharged immediately below the rotary gate valve which will be deenergized upon actuation of the system, but may be sufficiently worn that it does not constitute an effective flame block. Additional high rate discharge extinguishers are installed on the interconnecting ducting between major pieces of equipment and oriented to discharge in opposition to the direction of normal product flow. Their function is to isolate each piece of equipment by intercepting any flame front which

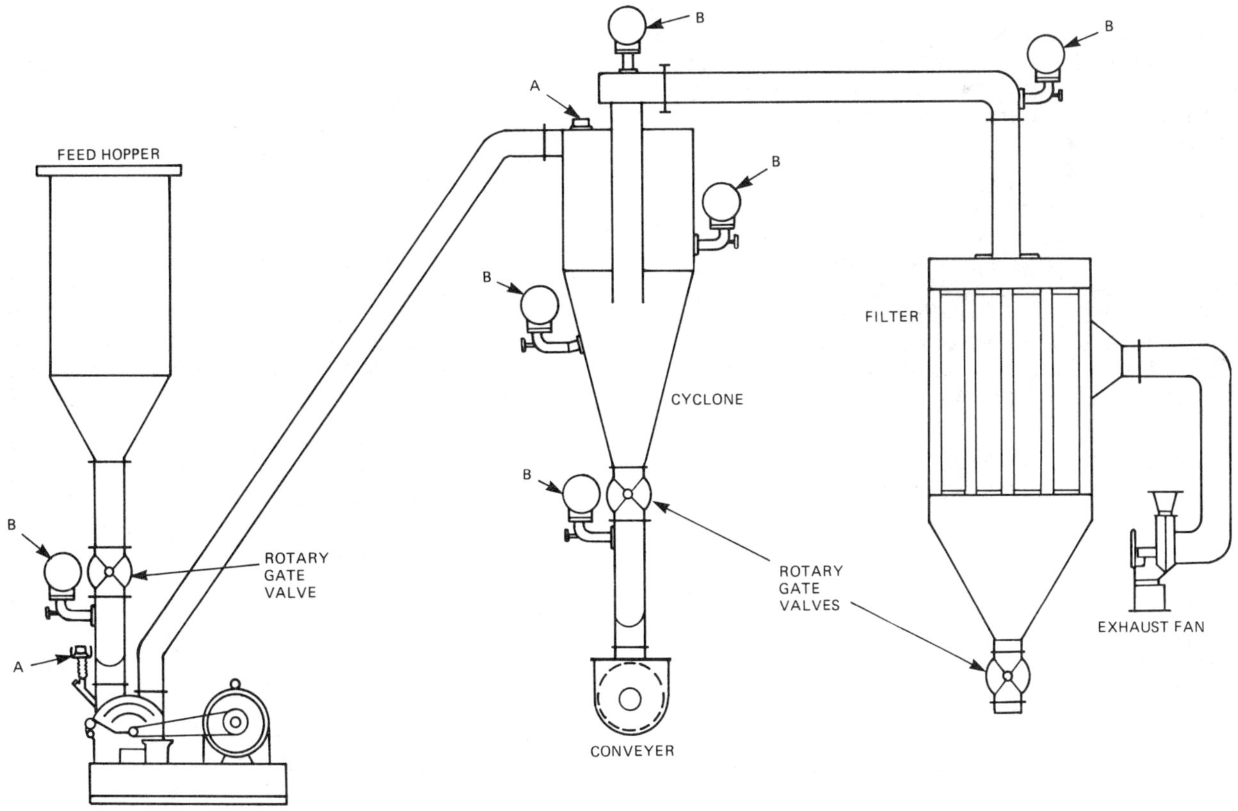

FIG. 12-11C. Explosion protection for a typical grinding plant. The explosion detectors (A) actuate the high rate discharge extinguishers (B).

may be propagating from the piece of equipment within which the explosion originated.

This example assumes that the bag collector is located outdoors which is the recommended practice and that it is suitable vented. Where this is not practicable, explosion suppression would also be applied to the collector hopper with fire detection and suppression in the bag area. Actuation of any detector triggers all high rate discharge extinguishers connected into that particular system via the explosion suppression control panel.

Typical Equipment Applications: Figure 12-11C also illustrates the method of applying an explosion suppression system to a size reduction process. In general, the system is applied by determining the most hazardous area in the process and working from this area to other possible hazards in order to assure that all of these would be successfully suppressed by the equipment. This method has been applied to many other types of industrial processes such as the following:

Storage Tanks: The storage of flammable liquids in many cases presents a severe explosion hazard, and suppression systems have been applied to these tanks to prevent explosions from rupturing and spreading the fire to other tanks in the area.

Coal Pulverizing: Bowl mills, ball mills and other grinding equipment have been protected by this system in order to prevent explosions resulting from the entrance of tramp metal into the grinder, or during startup after a mill trip where glowing particles are suddenly thrown into suspension.

Feed and Grain Processing: Suppression systems have been applied to elevator bucket legs and milling operations encountered in the feed and grain industry where extreme explosion hazards exist due to the dust that is created.

Plastic Grinding: The equipment for grinding plastics is quite similar to that shown in Figure 12-11C and represents an extreme hazard. Several of the suppression installations that are made have operated successfully and have shown the value of this form of protection.

Those processes and materials where explosion suppression systems have found wide usage are listed in Tables 12-11D and 12-11E. It should be noted that in many instances both flammable vapors and dust (hybrid mixtures) may be present. Such mixtures may present an especially critical hazard since the presence of even small amounts of vapor, e.g., one percent or less, can significantly increase the severity of a dust explosion.

TABLE 12-11D. Processes Utilizing Explosion Suppression Systems

Process	System Utilized	
	Dust	Vapor
Size Reduction	X	
Dust Collection	X	
Solvent Recovery		X
Drying	X	X
Conveying		
Mechanical	X	
Pneumatic	X	X
Spray Painting	X	X
Reactors		X
Flaring		X
Storage	X	X

TABLE 12-11E. Industries Utilizing Explosion Suppression Systems

Type	System Utilized	
	Dust	Vapor
Plastics	X	X
Forest Products	X	
Grain and Feed	X	
Pharmaceutical	X	X
Aerosol		X
Waste Disposal	X	X
Power Generation	X	X
Petroleum		X
Propellant and Explosives	X	

CONTAINMENT OF THE PRESSURE

Process Equipment

Maximum pressures from deflagration combustion in a closed vessel reach about 100 to 150 psi (700 to 1000 kPa) when starting from one atmosphere, or about 7 to 10 times the initial pressure.

Much process equipment, built to withstand full vacuum and/or moderate working pressure, may be found on careful strength analysis to have a bursting strength higher than the maximum expected combustion pressure in the system.

Whereas common practice dictates at least a 4:1 safety factor of bursting pressure over working pressure in a vessel, it is not unusual to consider a potential explosion as a "once in a vessel's lifetime" occurrence and depend on vessel yield strength for pressure containment with very little factor of safety. Such a system would limit explosion damage to the vessel itself, which might stretch or deform under worst conditions. The containment approach avoids vessel rupture with catastrophic damage in the area.

If, in fact, pressure containment is the option selected in designing the process equipment, it is imperative that the potential maximum explosion pressures be known. Otherwise, the damage could be catastrophic. For example, if the process contains long duct runs or if any of the process equipment has a high length to diameter ratio, the 7 to 10 multiplier no longer applies. In such configurations the final pressure can exceed the initial pressure by a factor of 100 or more as the result of pressure piling or even detonation with certain combustible vapors and gases.

This approach is practiced only for substantially built vessel systems that can withstand that multiple of the pressure at which a confined combustion deflagration might be initiated, and is not applicable to buildings and lightly built equipment like normal hoppers, dust collectors, tanks, bins, silos, and so forth.

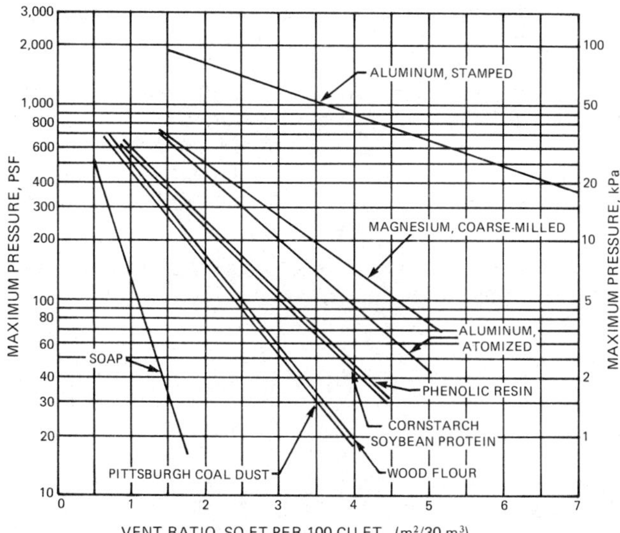

FIG. 12-11D. Effect of unrestricted vents on pressures developed by mild dust explosions. These explosions were produced in a 64 ft (1.8 m³) gallery. (Bureau of Mines, U.S. Department of Interior).

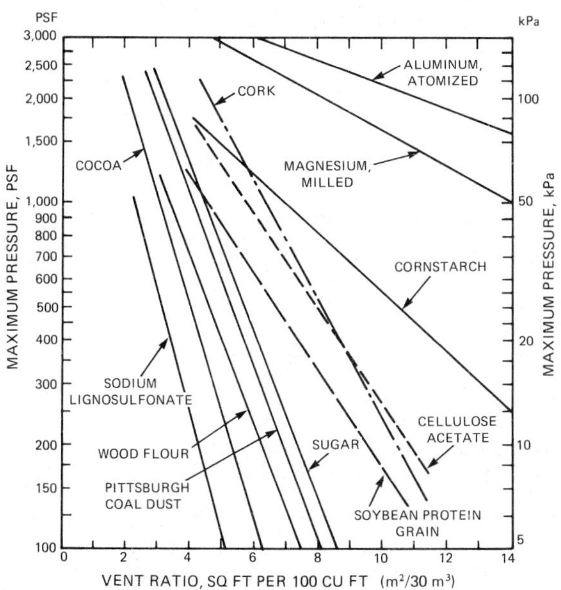

FIG. 12-11E. Effect of unrestricted vents on pressures developed by strong dust explosions. These explosions were produced in a 1 cu ft (.028 m³) gallery. (Bureau of Mines, U.S. Department of Labor).

Area Protection

The containment principle is used in a broader way when hazardous reactions are shielded by containment cell wall and/or barricades to limit the damage from explosion forces.

VENTING

Combustion and/or other exothermic reactions in an enclosed space release heat energy and products of reaction. If the enclosing structure, i.e., vessel, hopper walls, building walls, etc., are not free to expand with the increasing volume of gas contained, the pressure will rise. The expansion ratio, E, for most hydrocarbons is roughly 8 to 1 when burned. If the gases are not confined, the final volume will be eight times the original volume. If they are confined, the pressure will be about eight times the original pressure (Rust 1979). These ratios can be expressed as:

$$E \text{ (expansion value)} = P_2/P_1$$

$$E \text{ (expansion value)} = V_2/V_1$$

where:

P_1 = original pressure
P_2 = final pressure
V_1 = original volume
V_2 = final volume.

Venting a vessel, building, or other enclosed space to prevent or limit explosion damage involves providing an opening large enough to permit the expanding fluids (gases, vapors, and/or liquids) to be pushed out by the rising pressure. The time should be short enough to limit the rising pressure to a value the enclosing structure can withstand without rupture or serious damage.

Explosion Venting Fundamentals

The following five fundamentals of good explosion venting practice should be taken into account whenever the means of minimizing explosion damage are being considered:

1. Explosion damage may be minimized by locating hazardous operations or equipment outdoors, by segregating them in small detached buildings, or by locating them in vented units separated from other portions of a building by pressure resisting walls.
2. Where it is impractical to locate an operation outdoors, the equipment should be placed above grade in a one story building or in the top story of a multistory building.
3. Vents that are properly designed and located will relieve explosion pressures sufficiently in most instances to minimize property damage and prevent injuries.
4. The necessary area of the explosion vents depends primarily upon the expected intensity of an explosion, the strength of the equipment or building, and the type of vent closure.
5. The vent should be located so that upon relieving it will not cause injuries, or structural damage, or allow propagation of flame into another hazardous area.

Basic Considerations for Venting Explosions

The following principles established by test and experience should be considered when applying the fundamentals of explosion venting:

1. Many substances not ordinarily considered combustible will burn and explode under certain conditions of particle size, increased temperature, or increased oxygen concentration.
2. Most ordinary building walls will not withstand a sustained internal pressure as great as 1 psi (7 kPa).

Explosion vents for buildings must be designed to operate at pressures well below those at which the building walls will fail.

3. There is a rise in pressure during an explosion within an enclosure even with open, unrestricted vents, and any delay in opening venting devices increases that pressure. Delay in opening vents may be due to the pressure required to open the vent or to the inertia of the vent closure. Therefore, it is essential that various relieving devices, including devices actuated by detonators, should start to open at as low a pressure as possible, and be made of light construction so that full opening can be quickly attained.
4. Explosion pressures developed in test galleries with vent openings have been determined for a number of combustible dusts. These data, obtained by the Bureau of Mines, are shown graphically in Fig. 12-11D for mild dust explosions, and in Figure 12-11E for strong dust explosions. The mild explosions were produced in the 64 cu ft (1.8 m^3) gallery by dispersing the dust with compressed air; the effective dust concentration was below optimum. The strong explosions were produced in a 1 cu ft (0.028 m^3) gallery when the dust was dispersed more uniformly and during a shorter time interval. The data for the twelve dusts illustrated plot as straight lines on semilog paper as the mathematical relation between the pressure and vent ratio in the range shown is $P = Ae^{kr}$ where P is the maximum explosion pressure, A and k are constants, e is the base of natural logarithm, and r is the vent ratio expressed in square feet (or square meters) of opening per 100 cubic feet (per 30 m^3) of gallery volume.
5. Venting as recommended in NFPA 68, *Guide for Explosion Venting*, will prevent major damage in nearly all instances.

Vent Ratios

Much literature exists which offers recommended ratios of vent area to space volume for various materials. Test results, as shown in Figures 12-11D and 12-11E, often plot maximum pressure rise against vent ratios. Reliances on these old fixed vent ratio recommendations can be misleading, resulting in excessively large and overly expensive vents in some cases and in too small inadequate vents in other cases.

Vessel Venting: More recent work shows that vent area for a given venting pressure for a combustible mixture is proportional to the cube root of the enclosure volume (Bartknecht 1977; Donat 1977; Miller 1978). Such work has been incorporated into NFPA 68.

Various combustible dusts are grouped into three classes according to the violence of the dust explosion generated (Locks 1979). Nomographs are included in NFPA 68 for sizing vents under various conditions. These incorporate provision for taking into account the bursting pressure of the vent closure and the maximum pressure to be attained during venting. Several nomographs for combustible gas explosion venting are also included in NFPA 68.

A more recent publication (Rust 1979) points out the shortcomings of generalizing combustible dust properties into classes and offers a more precise method of calculation based on the specific materials in question and the vessel shape.

NFPA 68 provides a good discussion of the use of test vessels, and the procedures for developing test data and scaling up to larger sizes.

Test vessels of at least 20 liters (0.7 cu ft), or even better, 1 m^3 (35 cu ft), are preferred over the smaller Hartman device. The large amount of Hartman data can be used if nothing else is available. Some allowance should be made for the inaccuracies introduced by the small scale of the Hartmann bomb tests.

Building Venting: A simple approach for building venting is the Runes equation (Runes 1972):

$$A_v = \frac{CL_1L_2}{\sqrt{P}}$$

where:

A_v = vent area in ft^2 or m^2
C = constant for fuel (See Table 12-11F.)
L_1 = smallest room dimension in ft or m
L_2 = next smallest room dimension in ft or m
P = maximum internal overpressure which can be withstood by the weakest structural element in psi or kPa.

This is an important constraint:

$$L_3 \leq 3\sqrt{L_1L_2}$$

where L_3 = longest room dimension, ft or m.

TABLE 12-11F. Recommended Values of *C*

Fuel Identity	*C* for Equation	*C* for Equation
Fuels with flame speed like:	(Customary Units)	(SI Units)
Propane	2.6	6.8
Ethylene	4.0	10.5
Hydrogen	6.4	17.0
Organic Dusts	2.6	6.8
Organic Mists	2.6	6.8
Metal Dusts	4.0	10.5

The Runes Equation has been shown to hold quite well for large values of *A* and low values for *P*, but deviates from actual results above explosion pressures of about 3 psi (20kPa). Since most building walls and roofs, or ceilings will fail below this pressure, the Runes Equation is quite helpful in the range of pressure where it will normally be applied (Howard 1972).

Exotherm and Decomposition

Designing vents to relieve exothermic reactions and decompositions of unstable materials involves many of the same principles used in venting combustion reactions. In exothermic reactions, the controlling factors are the rate of energy release in the vessel, the rate of flow of the vessel contents discharging through the vent system, and hence the time that pressure can continue building up in the system. Boyle, Huff, and others have published procedures for balancing the reaction kinetics against the system fluid flow dynamics to calculate maximum residual vessel pressures (Boyle 1967; Huff 1973). Research and testing continue in this field to refine and further verify the method.

Detonation

If combustion flame front speeds or decomposition pressure wave speeds exceed the speed of sound, by definition the result is a detonation. Systems subject to possible detonation cannot be protected by venting, and damage control must be achieved by isolation and barricading, or in some cases by pressure containment.

Detonation pressures will be far higher than deflagration combustion pressures, and any attempt at designing for containment must consider pressures 100 to 150 times initial pressures. This is sometimes practical and necessary in strong piping systems where deflagration can progress to detonation and pressure piling is likely due to large L/D (length to diameter) ratios in the pipes.

Vent Closures

The most effective vent for the release of explosion pressures is an unobstructed vent opening. However, the fact that very few operations can be conducted in open equipment installed in buildings without walls necessitates the use of various types of vent closures designed to open quickly and automatically under increased pressure

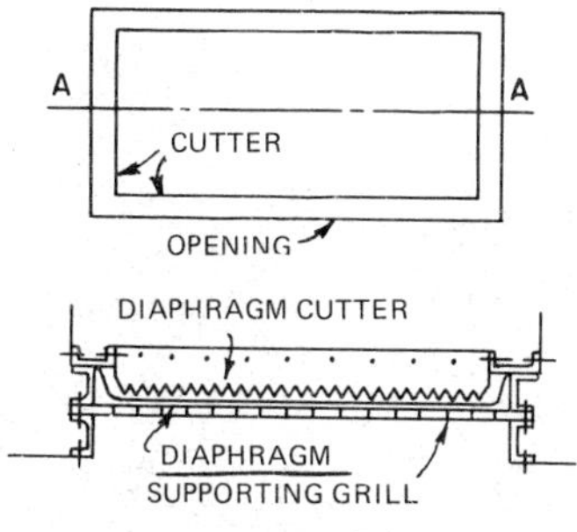

FIG. 12-11F. Explosion vent with diaphragm cutter and supporting grill.

from within. Some types of vent closures are shown in Fig. 12-11F and 12-11G, and further illustrations are given in NFPA 68.

Important fundamental principles of explosion vent design and installation are:

1. In most instances, several small vents will relieve explosion pressures as effectively as one large vent that is equal in area to the combined areas of the small vents. This principle may not hold true in a large structure where the position of the vents in relation to the origin of the explosion is important.
2. Closed vents must be larger in area than open vents in order to provide equivalent explosion pressure relief.
3. Rupture of paper, plastic, and metal diaphragms is facilitated by sawtoothed cutters at the periphery, or by piercing cutters at the center. (See Fig. 12-11F.)
4. Pressure required to rupture diaphragms of the same material and thickness but of different areas decreases with an increase in the area.
5. Pressure required to rupture diaphragms of the same area and material but of different thickness increases with an increase in the thickness.
6. Lightweight, hinged panels are nearly as effective as unrestricted vents for the release of relatively slow explosions (e.g., coal dust). However, this type of vent closure must be designed to remain open after the initial positive pressure because a destructive negative pressure may develop in the explosion area.
7. The nearer a vent is located to the point of origin of an explosion, the greater its effectiveness.
8. Vent closures can be designed to prevent their being accidentally opened by moderate wind pressure. Experimental data showing the relative effectiveness of circular, square, and rectangular openings with and without a vent cover are shown in Figure 12-11H for cellulose acetate explosions in the 1 cu ft (0.028 m^3) gallery.

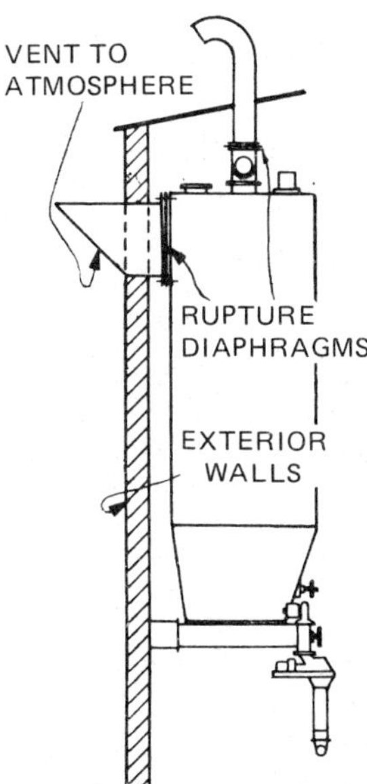

FIG. 12-11G. Dust explosion venting for bulk storage bins or silos, showing explosion vents for tanks.

Vent Panels: Vent panels must be constructed for easy, low pressure release, have low inertia, and be restrained from free fly-away lest they become missiles. If they are exposed in work areas they may need guardrail protection for personnel. A good design guide for building vent panels is to keep the panel mass below 2 lb per sq ft (0.1 kPa), release pressure above 30 lb per sq ft (1.5 kPa) for wind resistance, and below 50 lb per sq ft (2.4 kPa) for release well below normal bursting wall strength. (See Figs. 12-11I, 12-11J, 12-11K.)

Rupture Diaphragms: The pressures at which diaphragms 3 in. (7.6 cm) in diameter or larger will rupture are almost independent of the rate of pressure rise. Since the rupture pressure for a specific thickness of a diaphragm is dependent upon the area of the vent, specific materials can be tested with air pressure for different areas and can be plotted as shown in Figure 12-11L to determine the required vent size which will rupture at a particular pressure.

Maintenance of Vent Closures

Vent closures should be inspected periodically to detect damage, obstructions, corrosion to the working parts, or other conditions which can affect the operation of

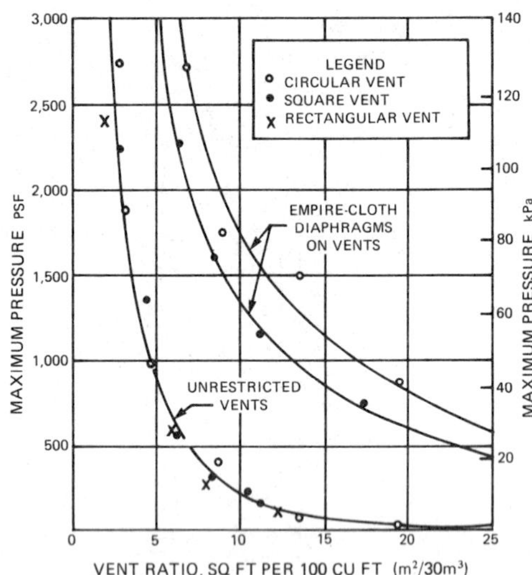

FIG. 12-11H. Relative effectiveness of various types of vent configurations for relieving cellulose acetate dust explosions in a 1 cu ft (0.028 m³) gallery. (Bureau of Mines, U.S. Department of Labor)

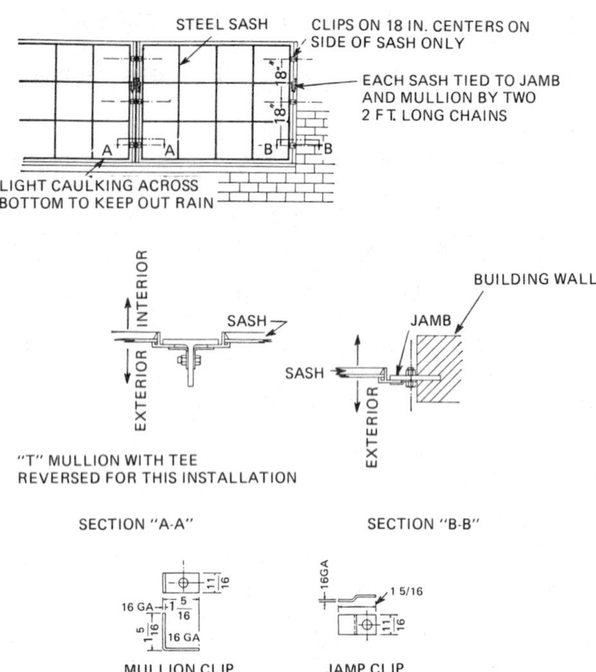

FIG. 12-11J. Method of converting standard steel sash to explosion venting type. (1 in. equals 25.4 mm.)

the vents in time of necessity. Ice crystals have been known to form between the vents and frames due to high humidity when exposed to freezing temperatures. The ice crystals act as a cement on the vent allowing greater pressures to build up before the vent functions. A coating of grease on the adjacent surfaces may prevent the bridging of ice crystals between the members of the vent. Corrosion and paint may also increase the friction in opening vents.

Piping and Duct Systems

Explosions in piping and duct systems can be vented by the use of explosion vents. Ductwork should be as short and straight as possible. Since the explosion wave tends to travel in a straight line, any bends in such ductwork should be vented in the direction of travel of the explosion.

FIG. 12-11K. Almost 100 percent of the wall area of this building is explosion vent area. Each of the lightweight aluminum wall panels is secured by shear pins designed to release the panel at 30 psi (207 kPa). A chain at the top of each panel prevents it from being blown any distance if blown out by an explosion. (Kodak Park Industrial Photo)

FIG. 12-11I. Roof and structural members of a test building remain undamaged after vent panels have relieved overpressure from building during a violent propane explosion.

There should preferably be no duct in the venting system. If a duct must be employed, it should have just as short an effective length (including effective length of bends in producing pressure drop) as possible. It should be noted that combustion can take place in the vent duct itself; i.e., unburned gases may be the first to exit from the vent. This has two implications. The first is that the duct should be capable of withstanding a pressure at least as high as that expected to develop in the vessel during

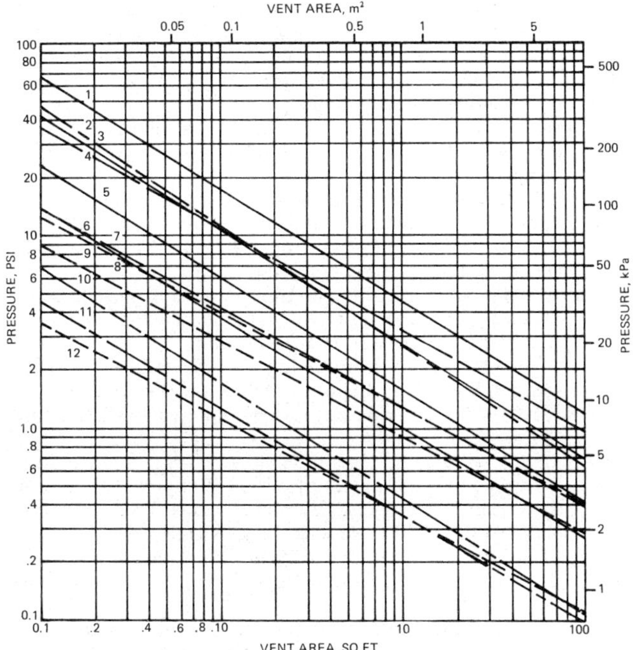

FIG. 12-11L. Rupture pressure vs. vent area of various venting materials. The graph shows that the bursting pressure of venting material is a function of the vent area.

venting. The second is that high turbulence can develop in a duct. In the case of gas burning within the duct, that turbulence could possibly lead to transition of the deflagration to detonation so that a far higher pressure could develop in the duct.

Bibliography

References Cited

Bartknecht, W. 1977. Application of explosion pressure relief to protect apparatus in industrial production facilities, Part II. *Loss Prevention.* Vol 2. American Institute of Chemical Engineers. pp 93-105.

Boyle, W. H., Jr. 1967. Sizing Relief Area for Polymerization Reactors. *AIChE & Loss Prevention Manual.* Vol 1. pp 78-84.

Cocks, Richard E. 1979. Dust Explosions: Prevention and Control. *Chemical Engineering.* Vol 86, No 24. Nov 5, 1979. pp 94-101.

Donat, C. 1977. Application of explosion pressure relief as a protective measure for industrial plant equipment, Part I. *Loss Prevention.* Vol 11. pp 87-92.

Howard, W. B. 1972. Interpretation of a Building Explosion. *Loss Prevention.* Vol 6. American Institute of Chemical Engineers. pp 68-73.

Huff, Jim E. 1973. Computer Simulation of Polymer Pressure Relief, *AIChE Loss Prevention Manual.* Vol 7.

Miller, R. L. 1978 Explosion Pressure Relief. *Proceedings: The User and Fabric Filtration Equipment III.* Niagra Frontier Section, Air Pollution Control Association. Oct 1978. pp 98-112.

NFPA 68, *Guide for Explosion Venting,* National Fire Protection Association, Quincy. 1978. p 48.

Runes, E. 1972. Explosion Venting. *Loss Prevention.* Vol 6. American Institute of Chemical Engineers. pp 63-67.

Rust, Earl A. 1979. Explosion Venting for Low-Pressure Equipment. *Chemical Engineering.* Vol 86, No 24. Nov 5, 1979. pp 102-110.

NFPA Codes, Standards, Recommended Practices and Manuals. (See the latest *NFPA Codes and Standards Catalog* for availability of current editions of the following documents.)

NFPA 68, *Guide for Explosion Venting.*

NFPA 69, *Standard for Explosion Prevention Systems.*

NFPA 410, *Standard on Aircraft Maintenance.*

NFPA 496, *Standard for Purged and Pressurized Enclosures for Electrical Equipment in Hazardous (Classified) Locations.*

Additional Readings

Bartknecht, W., and Kuhnen, G., *Forschungsbericht F45,* Bundesinstitute fur Arbeitsschutz, 1971.

Bartknecht, W., *Explosions —Course Prevention Protection,* Springer-Verlag, NY, 1981.

Bodurtha, F., *Industrial Explosion Prevention and Protection,* McGraw Hill, NY, 1980.

Bonyun, M. E., "Protecting Pressure Vessels with Rupture Discs," Chemical and Metallurgical Engineering, Vol. 42, May 1945, pp. 260-263.

Bourgoyne, J. H., and Wilson, M. J. G., "The Relief of Pentane Vapor-Air Explosions in Vessels," Symposium Chemical Process Hazards, S. 25/29, Institute of Chemical Engineering, 1960.

Brown, Hylton, "Design of Explosion Pressure Vents," *Engineering News-Record,* Oct. 3, 1946.

Brown, K. C., et al., *Transactions of the Institution of Mining Engineers,* Vol. 74, No. 8, 1962, pp. 261-76.

Brown, K. C., and Curzon, G. E., "Dust Explosions in Factories: Explosion Vents in Pulverized Fuel Plants," Research Report 212, 1962, Safety in Mines Research Establishment, Sheffield, England.

Coffee, R. D., "Dust Explosions: An Approach to Protection Design," *Fire Technology,* Vol. 4, No. 2, May 1968, pp. 81-87.

———, *The Testing of Materials for Pressure Relief Vents,* Eastman Kodak Company, Rochester, NY, 1950.

Consion, J., "How to Design a Pressure Relief System," *Chemical Engineering,* Vol. 67, No. 15, July 25, 1960, p. 109.

Cotton, P. E., and Cousins, E. W., "The Protection of Closed Vessels Against Internal Explosions," Paper No. 51-PRI-2, April 1951, The American Society of Mechanical Engineers, NY.

Cousins, E. W., and Cotton, P. E., "The Protection of Closed Vessels Against Internal Explosions," Paper No. 51-PRI-2, 1951, American Society of Mechanical Engineers, NY.

———, "Design Closed Vessels to Withstand Internal Explosions," *Chemical Engineering,* Vol. 58, No. 8, Aug. 1951, pp. 133-137.

Coward, H. F., and Hersey, M. D., "Accuracy of Manometry of Explosions," RI 3274, 1935, USDI Bureau of Mines, Pittsburgh.

Creech, M. D., "Combustion Explosions in Pressure Vessels Protected with Rupture Discs," Transactions of the American Society of Mechanical Engineers, Vol. 63, No. 7, 1941.

Cross, J., Farrer, D., *Dust Explosions,* Plenum Publishing Co., NY, 1982.

Crouch, H. W., et al., "*Maximum Pressures and Rates of Pressure Rise Due to Explosions of Various Solvents,*" Eastman Kodak Company, Rochester, NY, 1952.

Cubbage, P. A., "Flame Traps for Use with Town Gas/Air Mixtures," Gas Council Research Communication GC63, 1959, London.

Cybulski, W., *Coal Dust Explosions and Their Suppression,* (translated from Polish) National Technical Information Service, Springfield, VA, 1981.

Decker, Dick A., "Explosion Venting Guide," *Fire Technology*, Vol. 7, No. 3, Aug. 1971, pp. 219-223.

Donat, C., "Release of the Pressure of an Explosion with Rupture Discs and Explosion Valves," *paper* presented at Achema 73, Frankfurt, West Germany.

Doyle, William H., "Protection in Depth for Increased Chemical Hazards," *Fire Journal*, Vol. 59, No. 5., Sept. 1965, pp. 5-7.

Eckoff, R. et al, *Dust Explosion Experiments in a Vented 500 ft m³ silo*, Christian Michelson Inst., Bergen, Norway, 1982.

Evans, C. H., "Designing Safe Installations for Inert Gas Machines," paper presented at the ASTM 12th Annual Petroleum Mechanical Engineering Conference, Tulsa, OK, Sept. 1957.

Field, P., *Dust Explosions Volume* 4, Elsevier Scientific Publishing Co., NY, 1982.

Fishkin, C. I., and Smith, R. L., "Handling Flammable Dusts," *Chemical Engineering Progress*, April 1964.

Grabowski, G. J., "Progress in Industrial Explosion Protection Systems," *Fire Journal*, Vol. 62, No. 1, Jan. 1968, pp. 21-24.

———, "Explosion Protection Operating Experience," *NFPA Quarterly*, Vol. 52, No. 2, Oct. 1958, pp. 109-19.

———, "Theoretical and Practical Aspects of Explosion Protection," *Fire Protection Manual for Hydrocarbon Processing Plants*, Gulf Publishing Co., Houston, TX, 1964.

Hammond, C. B., "Explosion Suppression—New Safety Tool," *Chemical Engineering*, Dec. 1961, pp. 85-88.

Harris, G. F. P., and Briscoe, P. G., *Combustion and Flame*, Vol. 2, No. 4, Aug. 1967, pp. 329-338.

Hartmann, I., "Bureau of Mines Optical Pressure Manometer," RI 3751, 1935, USDI Bureau of Mines, Pittsburgh.

———, "Dust Explosions," *Marks Mechanical Engineers' Handbook*, 15th ed., McGraw-Hill, 1950, pp. 795-800.

———, "Explosion and Fire Hazards of Combustible Dusts," *Industrial Hygiene and Toxicology*, Vol. 1, Chap. 13, Sec. 2, Interscience Publishers, Inc., New York, 1948, pp. 439-454.

———, "Pressure Release for Dust Explosions," *NFPA Quarterly*, Vol. 40, No. 1, July 1946, pp. 47-53.

———, "Recent Research on the Explosibility of Dust Dispersions," *Industrial and Engineering Chemistry*, Vol. 40, No. 4, April 1948, pp. 752-758.

———, "The Explosion Hazard of Metal Powders and Preventive Measures," *Metals Handbook*, American Society of Metals, Metals Park, Ohio, 1948, pp. 52-54.

Hartmann, I., Cooper, A. R., and Jacobson, M., "Recent Studies of the Explosibility of Corn Starch," RI 4725, 1950, USDI Bureau of Mines, Pittsburgh.

Hartmann, I., and Greenwald, "The Explosibility of Metal Powder Dust Clouds," *Mining and Metallurgy*, Vol. 26, 1945, pp. 331-335.

Hartmann, I., and Nagy, J., "Effect of Relief Vents on Reduction of Pressures Developed by Dust Explosions," RI 3924, 1946, USDI Bureau of Mines, Pittsburgh.

———, "Inflammability and Explosibility of Powder Used in the Plastics Industry," RI 3751, 1944, USDI Bureau of Mines, Pittsburgh.

———, "The Explosibility of Starch Dust," *Chemical Engineering News*, Vol. 27, July 18, 1949, p. 2071.

Hartmann, I., Nagy, J., and Brown, H. R., "Inflammability and Explosibility of Metal Powders," RI 3722, 1943, USDI Bureau of Mines, Pittsburgh.

Hartmann, I., Nagy, J., and Jacobson, M., "Explosive Characteristics of Titanium, Zircorium, Thorium, Uranium and Their Hydrides," RI 4835, 1951, USDI Bureau of Mines, Pittsburgh.

Heinrich, H. J., "Dimensions of Pressure-Release Openings for the Protection of Plants in the Chemical Industry Which Are Endangered by Explosions," *Chemie, Ingenieu, Technik*, Vol. 38, No. 11, Nov. 1966, pp. 1125-33

Howard, W., *Tests of Explosion Venting of Buildings Plant/ Operations Progress*, AIChe, Vol. 1, Jan. 1982, Corrected, Vol. 1, April, 1982.

Howard, W. B., *Loss Prevention*, Vol. 6 (a CEP Technical Manual) American Institute of Chemical Engineers, 1972, pp. 68-73.

———, "An Investigation of Large Electric Motors and Generators of the Explosion Proof Type for Hazardous Locations, Class I, Group D," *Bulletin of Research No. 46*, 1951, Underwriters Laboratories, Inc., Chicago.

Jacobson, M., Cooper, A. R., Nagy, J., "Explosibility of Metal Powders," RI 6516, 1964, USDI Bureau of Mines, Pittsburgh.

Jenett, E., "Design Considerations for Pressure Relieving Systems," Part I, *Chemical Engineering*, July 1963, p. 125; Part II, *Chemical Engineering*, Aug. 1963, p. 151; and Part III, *Chemical Engineering*, Sept. 1963, p. 83.

Jones, W. M., "Determination of Dust Explosion Possibilities," *Special Hazard Study No. 4*, 1940, Factory Insurance Association (Now IRI), Hartford, Conn.

———, "Prevention and Minimizing the Effects of Dust Explosions in Manufacturing Plants," *Special Hazard Study No. 5*, 1940, Factory Insurance Association (Now IRI), Hartford, Conn.

Jones, G. W., Harris, E. S., and Beattie, B. B., "Protection of Equipment Containing Explosive Acetone-Air Mixtures by the Use of Diaphragms," Bureau of Mines Technical Paper 553, 1933, USDI Bureau of Mines, Washington, DC.

Klueg, Eugene P., "Liquid Nitrogen as a Powerplant Fire Extinguishant," *Fire Technology*, Vol. 5, No. 3, Aug. 1969, pp. 197-202.

Le Vine, R. Y., "Electrical Equipment in Outside Chemical and Petroleum Plants," Fire Journal, Vol. 59, No. 3, May 1965, pp. 30-31.

Lunn, G., *Venting Gas and Dust Explosions-A Review*, Institution of Chemical Engineers, Rugby, England, 1985.

Maisey, H. R., "Gaseous and Dust Explosion Venting," Part I, *Chemical and Process Engineering*, Oct. 1965, pp. 526-535; Part II, *Chemical and Process Engineering*, Dec. 1965, pp. 662-672.

Murphy, T. S., "Rupture Diaphragms, Calculations, Characteristics and Uses," *Chemical and Metallurgical Engineering*, Nov. 1944.

Nagy, J., Dorsett, H. G., and Jacobson, M., "Preventing Ignition of Dust Dispersions by Inerting," RI 6543, 1964, USDI Bureau of Mines, Pittsburgh.

Nagy, J., Verakis, H., *Development and Control of Dust Explosions*, Marcel Dekker Inc., NY, 1983.

Nagy, J., Zeilinger, J. E., and Hartmann, I., "Pressure Relieving Capacities of Diaphragms and Other Devices for Venting Dust Explosions," RI 4636, 1950, USDI Bureau of Mines, Pittsburgh.

Palmer, K., *Dust Explosions and Fires*, Chapman and Hall, London, United Kingdom, 1973.

Palmer, K. N., "Dust Explosion Venting—a Reassessment of the Data," *Fire Research Note No. 830*, Aug. 1970, Fire Research Station, Boreham Wood, Herts., England.

Philpott, J. E., *Engineering Materials and Design*, Vol. 6, No. 1, 1963, pp. 24-29.

Pineau, J. P., *Suppression of Gas and Dust Explosions*, Cerchar, Verneuil-en-Halatte, France, 1984.

Potter, A. E., "Flame Quenching," *Progress in Combustion Science and Technology*, Vol. 1, Pergamon Press, Oxford, 1960, pp. 145-81.

Rasbash, D. J., *The Structural Engineer*, Vol. 47, No. 10, Oct. 1969, pp. 404-407.

Rasbash, D. J., and Rogowski, Z. W., "Relief of Explosions in Dust Systems," Symposium on Chemical Process Hazards with Special Reference to Plant Design, Institution of Chemical Engineers, 1961, pp. 58-69.

———, "Relief of Explosions in Propane/Air Mixtures Moving in a Straight Unobstructed Dust," Second *Symposium* on Chemical Process Hazards with Special Reference to Plant Design, Institution of Chemical Engineers, 1964.

Salter, R. L., Fike, L. L., and Hansen, F. A., "How to Size Rupture Discs," *Hydrocarbon Processing and Petroleum Refiner*, Vol. 42, May 1963, pp. 159-160.

Schmidt, H., Haberl, K., and Reckling Hausen, M. K., *Technische Ueberwachurg*, Vol. 7, No. 12, 1955, pp. 423-29.

Schofield, C., *Guide to Dust Explosion Prevention and Protection*, Part 1-Venting, Institution of Chemical Engineers, Rugby, England, 1985.

Schwab, R. F., and Othmer, D. F., "Dust Explosions," *Chemical and Processing Engineering*, April 1964.

Sestak, E. J., "Venting of Chemical Plant Equipment," *Engineering Bulletin No. N-53*, April 1965, Factory Insurance Association (Now IRI), Hartford, CT.

Simmonds, W. A., and Cubbage, P. A., "The Design of Explosion Reliefs for Industrial Drying Ovens," *Symposium* on Chemical Process Hazards with Special Reference to Plant Design, Institution of Chemical Engineers, 1961, pp. 69-77.

Smith, J. B., "Explosion Pressures in Industrial Piping System," Factory Mutual Insurance Association, Norwood, MA, 1949.

Stretch, K. L., "Part I—The Relief of Gas and Vapor Explosions in Domestic Structures," *The Structural Engineer*, Vol. 47, No. 10, Oct. 1969, pp. 408-411.

Thompson, N. J., and Cousins, E., "Explosion Tests on Glass Windows; Effect on Glass Breakage of Varying the Rate of Pressure Application," *Journal of the American Ceramic Society*, Vol. 32, No. 10, Oct. 1949.

———, "Measuring Pressures of Industrial Explosions," *Electronics*, Nov. 1947.

Tonkin, P. S., and Berlemont, C. F. J., "Dust Explosions in a Large Scale Cyclone Plant," *Fire Research Note No. 942*, July 1972, Fire Research Station, Boreham Wood, Herts., England.

Valentine and Merrill, "Dust Control in the Plastics Industry," Transactions of the American Institute of Chemical Engineers, Vol. 38, No. 4, Aug. 1942.

Weldon, George E., "Damage-Limiting Construction," *Fire Technology*, Vol. 9, No. 4, Nov. 1973, pp. 263-270.

Westerberg, William C., "Testing Electrical Equipment for Chemical Atmospheres," *Fire Technology*, Vol. 2, No. 3, Aug. 1966, pp. 226-233.

Wood, L. E., "Rupture Discs," *Chemical Engineering Progress*, Vol. 61, Feb. 1965, p. 93.

LIGHTNING PROTECTION SYSTEMS

Revised by Norman H. Davis III, P.E.

Lightning is a frequent fire cause, and in some areas it is *the* leading fire cause, bringing loss to life and personal injury due to its instantaneous and seemingly unpredictable nature. Each year the National Safety Council estimates that 220 Americans will die (1 per every million) and another 1,500 will be injured from lightning. Nearly 90 percent of these accidents will occur in rural areas.

Unlike many other causes of death and injury that one can run away from, lightning strikes before the warning thunder. The principles of protection and personal safety are well known and are spelled out in NFPA 78, *Lightning Protection Code*, which will be referred to as the NFPA *Lightning Protection Code* in this chapter. It must be remembered, nevertheless, that lightning involves many uncertainties, and that while a given pattern of lightning behavior may be probable, there is no guarantee that a lightning discharge will not deviate from that pattern.

Frequency and Severity of Thunderstorms: The frequency of thunderstorms varies throughout the world. The severity of thunderstorms, as distinguished from their frequency of occurrence, is much greater in some locations than in others. Hence, the need for protection varies geographically, although not necessarily in direct proportion to thunderstorm frequency. A few severe thunderstorms a season may make the need for protection greater than a relatively large number of storms of lighter intensity. (See Fig. 12-12A and Fig. 12-12B for statistical data on the frequency of thunderstorms in the United States and Canada, and Fig. 12-12C for world data.)

Value and Nature of Building and Contents: Buildings often have a historical or sentimental value which is uninsurable. The type of building construction will also influence the degree of protection to be considered. Some buildings do not require supplemental lightning protection due to their construction. The real or intangible value of the contents of the building must also be considered, and in some cases, attention must be given to the nature of the contents and the susceptibility of the contents to damage by induced lightning currents, e.g., storage of explosives.

Appendix I of the NFPA *Lightning Protection Code* contains a "Risk Assessment Guide" which provides guidelines on the need for lightning protection. This Appendix, which is not a part of the requirements of the NFPA *Lightning Protection Code*, was included for information purposes only.

Personal Hazards: The lightning hazard to human beings in a building is a major consideration. Since a stroke of lightning may lead to a considerable degree of discomfort, if not injury or death, lightning protection may be necessary to eliminate possible personal hazards in buildings of any type (other than those constructed of structural steel framing). In places of public assembly, the potential of panic should be considered because a lightning discharge over a structure, even if not accompanied by fire, can have visual and audible effects which may induce panic.

Relative Exposure: In closely built-up towns and cities, the hazard is not as great as in the open country. In the latter, farm barns are most frequently hit. In hilly or mountainous areas, a building located upon high ground is usually subject to greater hazard than one in a valley or otherwise sheltered area.

Indirect Losses: In addition to direct losses due to damage of buildings or their contents by lightning, fire resulting from lightning, killing of livestock, etc., there may be indirect losses. An interruption to business, computer operations or to farming operations, especially at certain times of the year, may involve losses quite distinct from and in addition to those arising from direct property damage. There are also cases where whole communities depend on the integrity of a single structure for a measure of safety and comfort, for instance, the brick chimney of a water pumping plant. A stroke of lightning to the unprotected chimney of a plant of that sort might have serious consequences resulting in lack of sanitary drinking water, irrigating water, water for fire protection, or some similar effect.

Mr. Davis, P. E., Technical Advisor USNC/IEC/TC 81, Secretary NFPA 78, IEEE, IAEI, ASQC, is Associate Managing Engineer, Underwriters Laboratories Inc., Northbrook, IL.

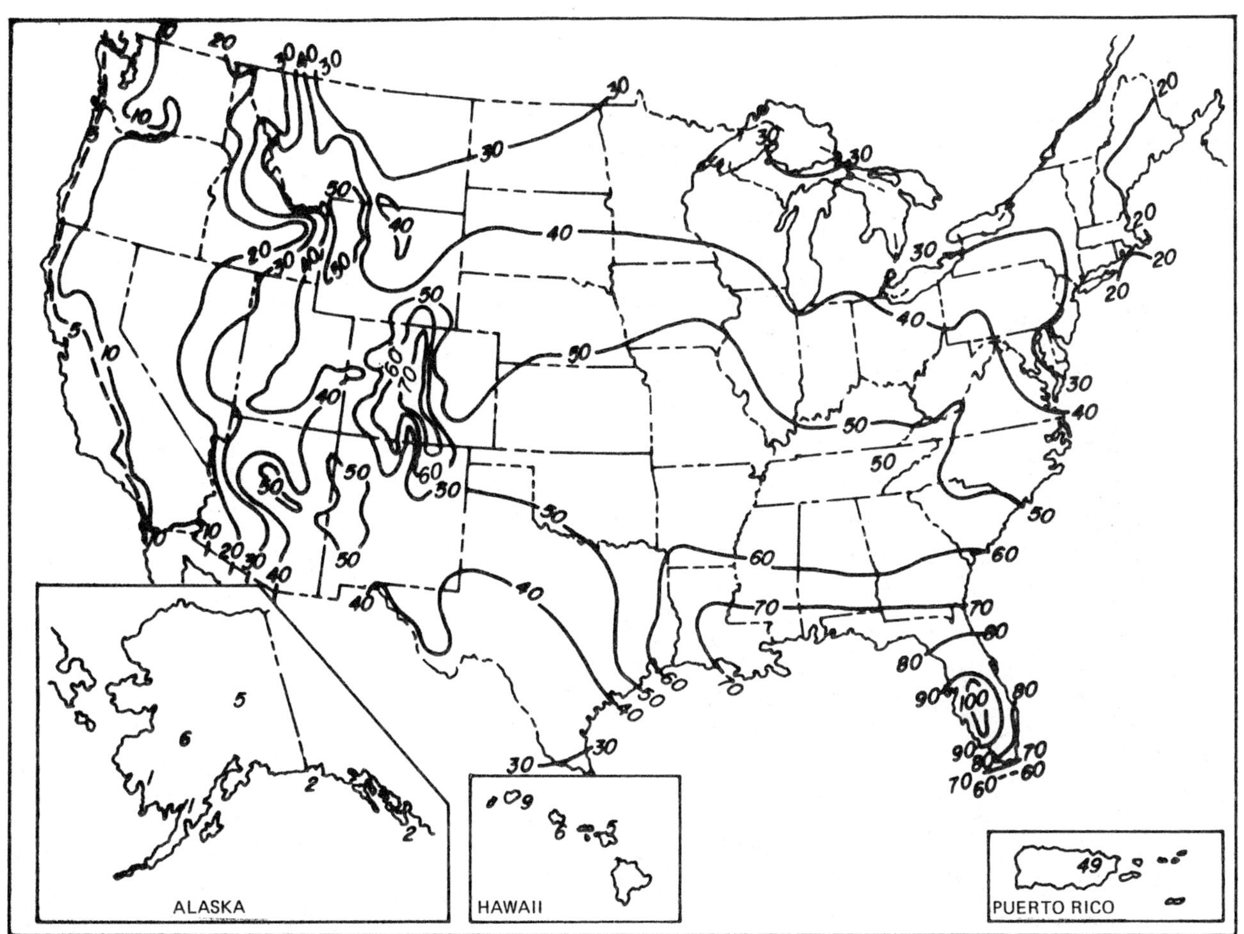

FIG. 12-12A. Statistics for Continental United States showing mean annual number of days with thunderstorms. The highest frequency is encountered in South-Central Florida. Since 1894, the recording of thunderstorms has been defined as the local calendar day during which thunder was heard. A day with thunderstorms is so recorded regardless of the number occurring on that day. The occurrence of lightning without thunder is not recorded as a thunderstorm. (Data supplied by Environmental Science Service Administration, U.S. Department of Commerce)

NATURE OF LIGHTNING

There are four types of lightning strokes: (1) the negative downward stroke; (2) the positive downward stroke; (3) the positive upward stroke; and (4) the negative upward stroke.

Lightning strokes may occur between clouds or between clouds and the earth. In the latter, charges of opposite polarity are generated in the cloud while the charge in the ground below the cloud is induced by the lower cloud charge. In effect, the result is a giant capacitor, and when the charge builds up sufficiently, a discharge occurs.

It is generally agreed that a lightning stroke is initiated by a downward leader, which originates in the charged cloud and progresses toward the earth in discrete steps. It is also generally agreed that the major part of the lightning discharge current is carried in the return stroke, which flows from the earth to the charged cloud along the ionized path created by the stepped leader. However, the path of the lightning stroke and the point struck are determined by the leader. Hence, this discussion is limited to consideration of the initial leader.

The development of a stepped leader is diagrammed in Figure 12-12D. When sufficient charge has built up on a cloud, a "pilot streamer" will develop in a generally downward direction. This streamer ionizes a path in the air. The current associated with it is small, on the order of only a few amperes. The streamer initially extends on the order of 150 ft (45.7 m) below the cloud. After a pause of about 50 microseconds, the streamer will proceed for a second step, again generally downward but usually in a somewhat different direction from the initial step. Subsequent steps occur at intervals of about 50 microseconds. The path of each step is essentially straight, but each new step generally takes a different direction. The change in direction at each junction results in the zigzag path characteristic of lightning. Branches may (and usually do) occur, but for simplification these are not shown in the figure.

As the leader progresses downward, it creates an ionized path of high conductivity in the air. Thus, the tip of the leader essentially remains at cloud potential, and the voltage gradient between the tip and the earth increases as the tip progresses downward. At some critical point (Point 6 in Fig. 12-12D), the voltage gradient be-

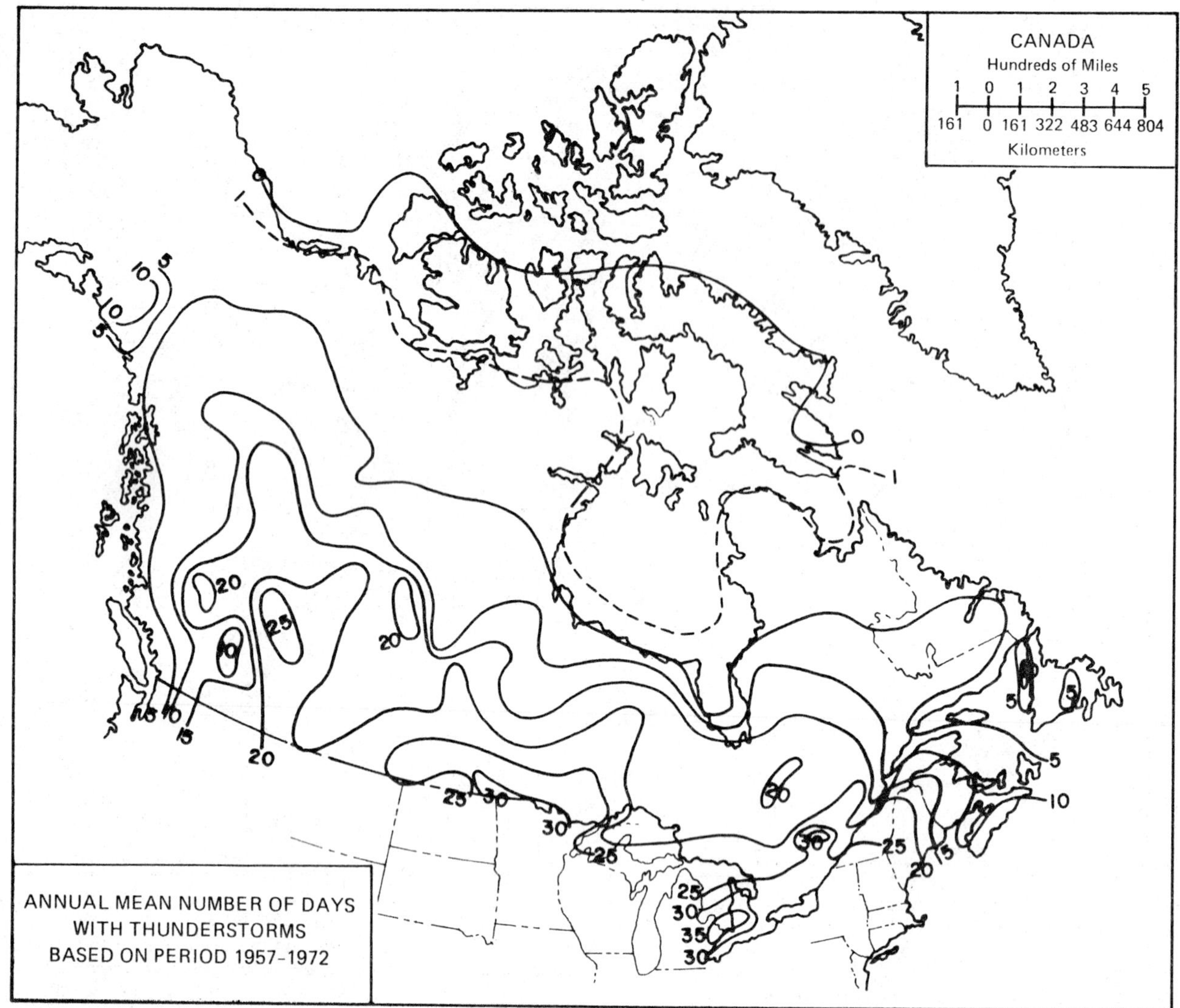

FIG. 12-12B. Canadian statistics showing annual average of days with thunderstorms. Data based on the period 1957-1972. (Meteorological Division, Department of Transportation, Canada.

comes high enough to break down the remaining air gap, and the initial stroke to ground is completed. The point from which the final breakdown occurs is called the point of discrimination, and the distance over which this breakdown occurs is defined as the striking distance.

As many as 40 component strokes have been observed in a single flash. Speeds range from 100 miles per second (161 km/sec) for the first pilot leader stroke to 20,000 miles per second (32,190 km/sec) for the main stroke. Currents range up to 270,000 amperes in extreme cases, lasting for a few millionths of a second, but lesser currents are present for a longer period. Potentials have been estimated as high as 15 million volts.

Because of the high voltage and rapid changes in current flow, induced charges are important. Thus, the NFPA *Lightning Protection Code* requires the interconnection of metallic masses as a part of any lightning protection system.

Lightning causes fire only where sufficient heat is produced to ignite combustible materials. Substantial damage, however, can be produced without resulting in fire. Dry wood beams in houses struck by lightning are often severely splintered and windows are blown outward. Such damage primarily results from pressure generated by the expanding lightning channel. Since some of the effects of lightning can be indirect, it is not necessary for lightning to strike a building to damage it. Because lightning striking overhead wires may be conducted to buildings over the wires, lightning arresters are often provided to minimize such damage. Although there are several types of lightning arresters, all permit the free flow of lightning charges to ground, while preventing the flow of ordinary electric current over the same path.

THEORY OF LIGHTNING PROTECTION

The theory of lightning protection is simple—to provide means by which a lightning discharge may enter or leave the earth without damaging the property protected. There is no evidence that any form of protection can prevent the occurrence of a lightning discharge. A lightning protection system has two functions—to intercept a

FIG. 12-12C. Annual frequency of thunderstorm days as compiled by the World Meteorological Organization, 1956. (World Meteorological Organization)

FIG. 12-12D. Development of stepped leader.

lightning discharge before it can strike the object protected, and to discharge the lightning current harmlessly to earth.

Conventional Concept

The zone protected by a grounded rod or mast is conventionally taken as the space enclosed by a cone which has its apex at the top of the mast. Similarly, the zone protected by a grounded horizontal overhead wire is conventionally taken as a triangular prism with the upper edge along the wire. In either case, the degree of protection is generally assumed to be a function of the shielding angle (the angle between an element of the cone and a vertical line through the apex, or between the side of the prism and the vertical plane through the horizontal wire). This is sometimes expressed as the ratio of the horizontal distance protected to the height of the mast or wire.

Figure 12-12E illustrates this concept and tabulates some generally accepted zones of protection. Previous NFPA *Lightning Protection Codes* stated that a distance equal to the height of a rod or mast in important cases, or up to twice the height of a rod or mast in less important cases, has been found to be substantially immune to direct strokes. The British Standard Code of Practice states that a shielding angle of 45 degrees provides an acceptable level of protection for ordinary structures, but that for structures with explosives or highly flammable contents, the shielding angle should not exceed 30 degrees. Both codes indicate that complete protection cannot be guaranteed in any case. It is assumed, however, that the criteria presented reduces the probability of lightning strikes to a practical minimum. While these criteria provide essentially com-

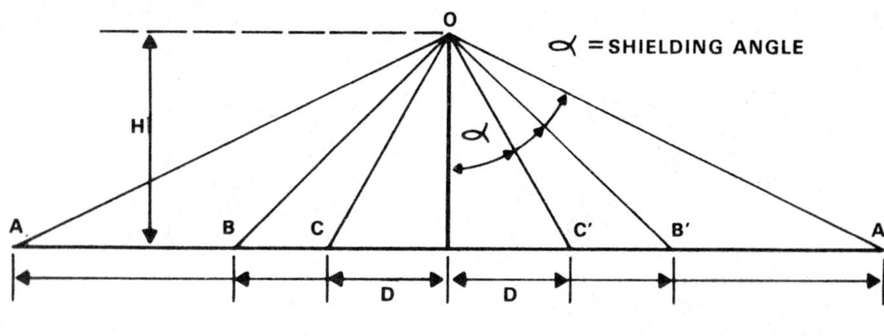

FIG. 12-12E. Conventional concept of protected zones.

ZONE	D/H	α	AUTHORITY	RECOMMENDED FOR
AOA'	2/1	63°	NFPA	ORDINARY CASES
BOB'	1/1	45°	NFPA	IMPORTANT CASES
			BRITISH CODE	ORDINARY STRUCTURES
COC'	0.58/1	30°	BRITISH CODE	DANGER STRUCTURES

plete protection in most cases, theoretical consideration indicates that even the 30-degree shielding angle is not always adequate. Experience with electric power transmission lines corroborates this.

Geometry of Lightning Protection

The geometry of lightning protection is illustrated in Figure 12-12F which shows a hypothetical cross section of an area protected by two vertical masts (or horizontal overhead wires). Consider first a downward leader descending to the right of Mast *A*. Final breakdown will occur when the leader tip approaches within a distance *X* (the striking distance) of any grounded object or surface. If the point of discrimination, *P* (the point at which final breakdown will occur), is equidistant from the tip of the mast and the ground, either the mast or the ground may be struck. If *P* is close to the mast, it will intercept the stroke, and if *P* is farther from the mast, the ground will be struck. If any structure to the right of the mast extends above ground to within a distance less than *X* of Point *P*, the structure will be struck.

It can be seen that the zone protected by a single mast (or by an outside mast of a multimast system, in a direction away from the other masts) is bounded by the arc of a circle of radius *X*, passing through the tip of the mast and tangent to the ground. Geometrically, it can be shown that the protected distance *D* at an elevation *B* above ground is given by the expression shown where *H* is the height of the mast above ground and *X* is the striking distance.

Similarly, for a leader extending downward between the two masts, the protected zone is bounded by the arc of a circle of radius *X* passing through the tips of the two masts. If the masts are close enough together, the space between them is completely protected for an elevation above ground which is a function of mast height *H* and spacing *S*. In this case, the protected distance *R* at elevation *B* and the minimum elevation *G* which is completely in the protected zone, are given by the two expressions shown.

These equations hold only if *H* is less than or equal to *X*, and *S* is less than or equal to 2*X*. Further, *G* must be positive. If *G* is negative, complete protection does not exist between the two masts, and the equation for *D* holds.

The geometric concept shows that the protected zone is not a cone or prism, but is bounded by surfaces generated by concave arcs. This fact is not generally appreciated, although it is at least partially recognized in transmission line shielding. Further, the protected distance is seen to be a function of:

1. Striking distance.
2. Height of mast or overhead wire above ground.
3. The distance between two or more masts or overhead wires.

Figure 12-12F shows why the protected zone between two masts is greater than the total of the zones of the two masts considered individually. Previous NFPA *Lightning Protection Codes* recognized this qualitatively but not quantitatively. The figure also shows why the shielding angle required on transmission lines decreases as tower (and overhead ground wire) height increases.

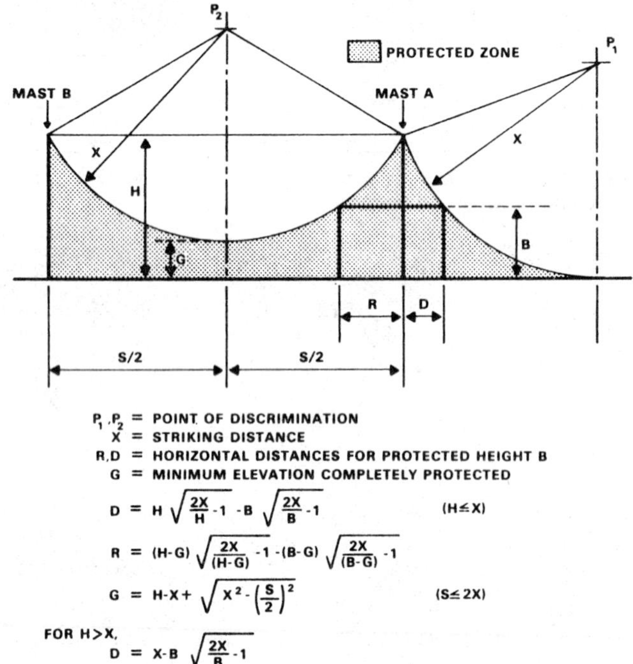

FIG. 12-12F. Geometry of lightning protection.

Striking Distance

The striking distance is related to the potential of the lightning stroke, which is directly related to the charge on the cloud. Since the peak stroke current is also related to the charge on the cloud, it can be shown that the striking distance is directly related to the peak stroke current. For a given mast or ground wire height, the protected distance increases with increasing striking distance. Thus, an adequately designed lightning protection system provides maximum protection against what could be the most damaging strokes. Since minimum protection exists for a minimum striking distance (and minimum stroke current), the lightning protection system should be designed to protect against minimum anticipated striking distances.

A review of lightning stroke current data indicates that the peak stroke current is usually at least 10,000 amp. This corresponds to a striking distance of slightly over 100 ft (30.5 m). It is therefore believed that a protective system based on a minimum striking distance of 100 ft (30.5 m) will provide essentially complete protection. This is considered to be a reasonable design basis.

Design Curves

Using the derived equations and an assumed striking distance, the protection provided by a single mast (or horizontal overhead wire) may be readily calculated. Similarly, the configuration of masts or overhead wires required to protect a given structure may be investigated. However, the calculations are tedious, and it is not convenient to determine directly by simple algebra the mast or overhead wire height required to protect a structure of given dimensions. Hence, curves showing the relationship between protected zones and various mast or horizontal overhead wire configurations are useful. Such curves can be readily constructed graphically. (See Fig. 12-12G and 12-12H.) These curves are based on a striking distance of 100 ft (30.5 m).

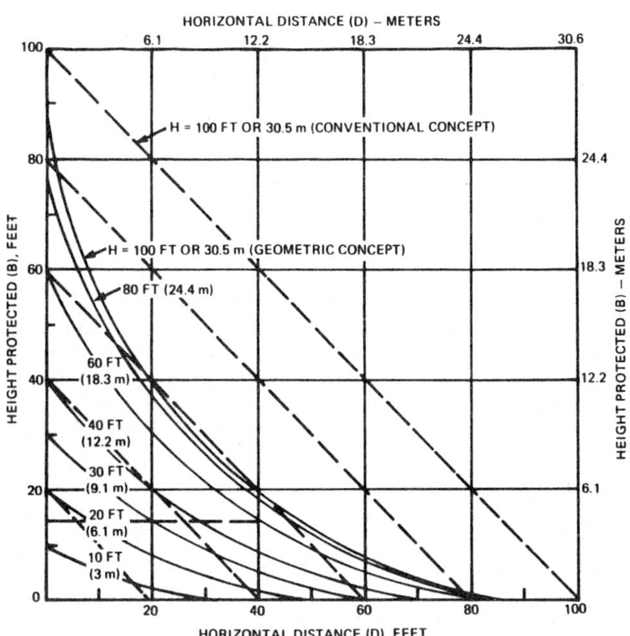

FIG. 12-12G. Zone protected by a single mast or horizontal wire. Solid curves are the geometric concept, with a striking distance of 100 ft. Dashed lines are the conventional concept, with a shielding angle of 45 degrees. "H" equals height of mast or horizontal wire. (One foot equals 0.305 m.)

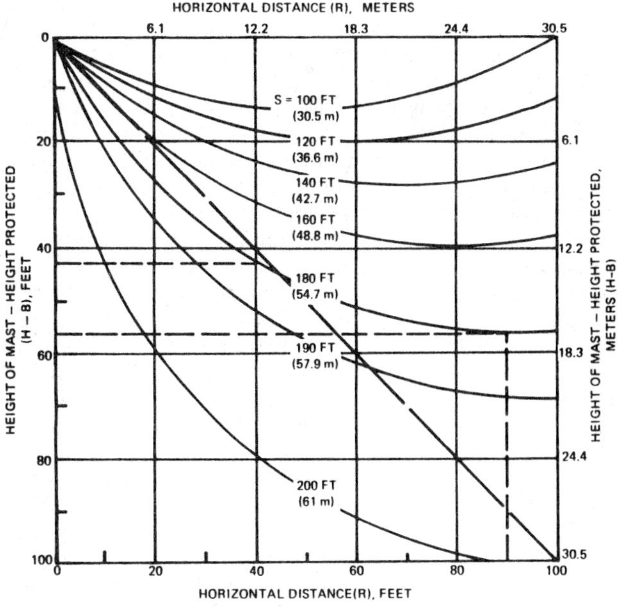

FIG. 12-12H. A zone protected by a single mast or horizontal wire; the striking distance is 100 ft. "H" equals the height of the horizontal wire, "S" is the spacing between masts or wires, and "B" is the height protected above ground. (One foot equals 0.305 m.)

PROPERTY PROTECTION

Components

Air Terminals: According to current United States practice, as indicated in the NFPA *Lightning Protection Code*, conditions required for protection of ordinary buildings are met by placing metal air terminals on the uppermost parts of the building or its projections, with conductors connecting the air terminals with each other and to the ground. By this means a relatively small amount of metal, properly proportioned and distributed, affords a satisfactory degree of protection and it can be installed in a way that interferes only minimally with the contour and appearance of the building. In the current British Standard Code of Practice for "Protection Against Lightning" (CP 326) the use of pointed air terminals (vertical finials) is not considered essential except where dictated by practical considerations.

One theory states that a protective system based on a minimum striking distance of 100 ft (30.5 m) for structures containing flammable liquids and gases will provide essentially complete protection, and a minimum striking distance of 150 ft (45.7 m) will provide a high degree of protection for ordinary structures which are generally capable of withstanding minor direct strokes (Offerman 1969; Lee 1978). This theory supports the use of 10 in. (254 mm) minimum height air terminals above the object to be protected at 20 ft (6.1 m) maximum intervals. A convenient geometric equation:

$$Y = \sqrt{R^2 - (R - h)^2}$$

where:

Y = radius of horizontal protected zone
R = striking distance
h = height of air terminal

illustrates that a striking distance of 100 ft (30.5 m) will provide a 12.9 ft (3.9 m) radius of horizontal protected zone or 25.8 ft (7.9 m) horizontal protected zone between air terminals, and a striking distance of 150 ft (45.7 m) will provide a 15.8 ft (4.8 m) radius of horizontal protected zone or 31.6 ft (9.6m) horizontal protected zone between air terminals.

The subject of sharply pointed versus blunt air terminals is under study by the NFPA Lightning Protection Committee. According to one researcher (Moore 1983), pointed air terminals have a built-in defense against lightning strikes. The strength of the electric field around the tip of an air terminal is limited by a phenomenon called "point discharge." When the electric field reaches a certain strength, but before lightning strikes, the stepped leader current flows from the air through the air terminal to ground. The effect is called point discharge because the point is letting the surrounding air lose charge, or discharge electricity. A pointed air terminal will presumably discharge sooner than a blunt one. Accordingly, Moore concludes that a blunt air terminal is more likely to intercept a lightning strike and complete a circuit.

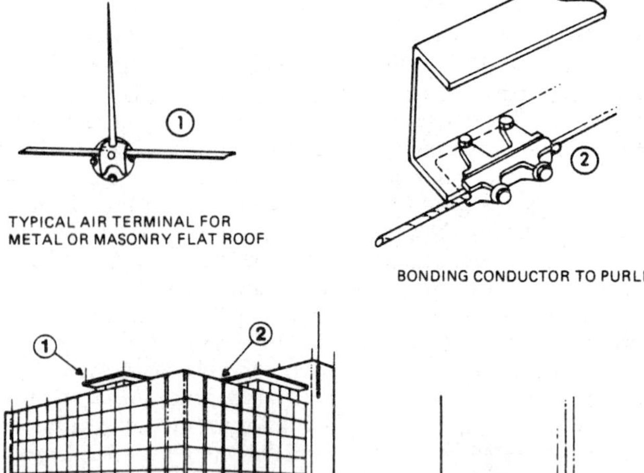

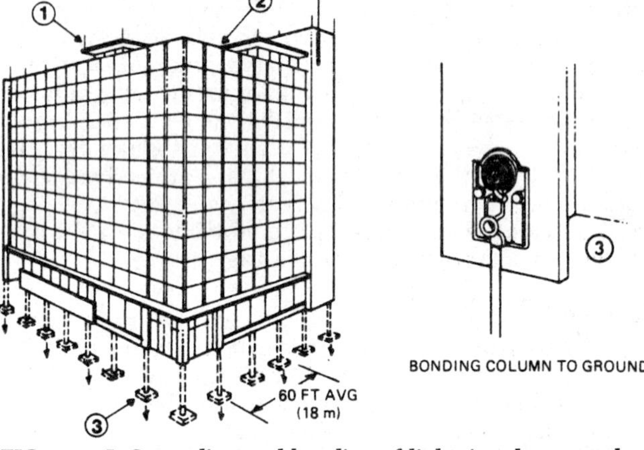

FIG. 12-12I. Grounding and bonding of lightning down conductors. Water pipe grounds (if pipes are metallic) can be made at 1, 2, or 3.

Conductors: The NFPA *Lightning Protection Code* specifies the use of copper or aluminum metals for lightning conductors. Where copper and aluminum are used together, special precautions must be taken against electrolytic corrosion. Conductors are coursed along ridges of sloping roofs, around the edges of flat roofs, and vertically from these and from the air terminals to ground. Vertical runs are designated as "down conductors" and are never less than two on any kind of structure, except for flagpoles, masts, spires, and similar structures. The down conductors are normally placed at diagonally opposite corners of square or rectangular structures, but other factors such as direct coursing, security against displacement, location of metallic elements in the structure, location of water pipes, and favorable ground conditions need to be considered in determining optimum placement. For detailed instructions about the installation of conductors, their form and size, and about the installation of air terminals, refer to the latest edition of the NFPA *Lightning Protection Code.*

The down conductors to ground must be substantial and must provide a reasonably direct path. The blocking effect of electrical induction must be avoided. For example, a lightning conductor that is run through a piece of iron pipe for protection, unless bonded to the pipe at the top and bottom, largely cancels the value of the conductor.

Grounding: Ground connections are essential to the effectiveness of a lightning protection system, and every effort should be made to provide ample contact with the earth. (See Fig. 12-12I.) Resistance of the protective system to ground is not critical. However, resistance and surge impedance must be low enough to prevent side flashes between the down conductor and other grounded objects. Also, the resistance must be low enough to prevent excessive voltage gradients in the soil surrounding the ground electrode. Excessive voltage gradients can be dangerous or fatal to people or animals that may be standing or walking in the vicinity when a lightning stroke occurs. If the down conductor is remote from other grounded objects, and people or animals are not likely to be near the ground electrode during a lightning storm, a relatively high ground resistance can be tolerated.

The NFPA *Lightning Protection Code* does not specify any maximum recommended ground resistance. The British Code recommends a maximum resistance of 10 ohms for a lightning protection system which will normally include two or more ground electrodes. Also 10 ohms is generally considered a reasonably good ground resistance for transmission towers. In this case, a high ground resistance may result in flashover to the line conductors even though the overhead ground wire intercepts the direct lightning stroke. A United States Department of Agriculture Bulletin states that a ground resistance as high as 50 ohms is acceptable for protection of farm structures.

The ground resistance of a ground electrode is directly proportional to soil resistivity. Soil resistivity varies over wide limits; nevertheless 10,000 ohm/cm is a reasonable order of magnitude average for many areas. Figure 12-12J, which is based on derived equations (Sunde 1949), has been prepared to show the variation in resistance to earth of vertical rods or pipes as a function of length in 10,000 ohm/cm soil. For other soil resistivities, the resistance to ground may be obtained by multiplying the indicated resistance by the known soil resistivity divided by 10,000. Vertical rods or pipes driven into the ground are commonly used as ground electrodes. In addition, where buried metallic water pipes are nearby, it is required to

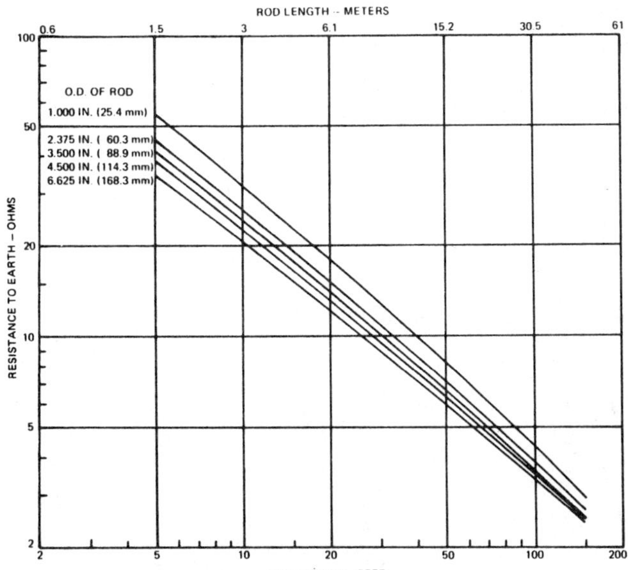

FIG. 12-12J. Resistance of vertical ground rods in 10,000 ohm/cm soil.

connect the lightning protection ground system to the piping system.

If bedrock is on or near the surface, it would be impossible to make a ground connection in the ordinary sense of the term because most kinds of rock are insulating, or at least of high resistivity, and in order to obtain effective grounding, other and more elaborate means are necessary. One effective means is to use an extensive wire network (grid) laid on the surface of the rock surrounding the building, to which the down conductors could be connected. Another is to use ground ring conductors. One ring should be installed at the bottom of the foundation to which the down conductors or structural steelwork are connected, with a second ring conductor provided by blasting a circular trench around the periphery and placing in it a strip conductor. The blasted area should be thoroughly compacted with backfill and the two ring conductors electrically interconnected. Using one of these methods would make the distribution of the electrical potential around the protected building substantially the same as if it were resting on conducting soil, and the resulting protective effect would be equal.

In general, the extent of the grounding arrangements will depend upon the character of the soil, ranging from simple extension of the conductor into the ground where the soil is deep and of high conductivity, to an elaborate buried network where the soil is very dry or of very poor conductivity. Where a network is required, it should be buried if there is soil enough to permit it as this adds to its effectiveness. Its extent will be determined largely by the judgment of the person planning the installation, following the minimum requirements of the NFPA *Lightning Protection Code* and keeping in mind that as a rule, the more extensive the underground metal available, the more effective the protection.

Structural Steel Buildings

Buildings constructed of structural steel framing may be protected by the installation of air terminals at the high parts of the building, connecting such air terminals to the metal framing, and grounding the framing at the bottom end. This assumes that the structural steel framework is electrically continuous, or is made electrically continuous by bonding. Grounding is required from approximately every other steel column around the perimeter, and is arranged so that grounds average not more than 60 ft (18.3 m) apart except as noted below.

Air terminals are not required if the steel columns extend at least 10 in. (254 mm) above the object to be protected. Air terminals are not required on metal roofs where the metal is not less than 3/16 in. (4.8 mm) thick.

Additional grounding is not required if the steel column extends 10 ft (3.05 m) or more into the earth.

Reinforced Concrete Structures

Reinforced concrete buildings, in which the reinforcing rods are electrically bonded together and grounded, are similar to structural steel buildings in regard to lightning protection. In usual building practice, as successive reinforcing bars are fitted, they are made to overlap lengthwise with bars already installed and are then tied together with metal binding wire. With thousands of such connections in a completed framework, the electrical resistance is fully acceptable for the purpose of lightning protection. However, if the reinforcing rods are electrically discontinuous, the building should be treated the same as a building of nonconducting materials. Lightning strokes to reinforced concrete buildings where there are insulating gaps between reinforcing rods are likely to cause cracks at places where beams and floor slabs are connected to their supports. Prestressed concrete buildings are a particular problem. If the wires in precast units are not interconnected (as they frequently are not), individual units are isolated and when a lightning charge is induced, serious damage can result.

Metal Roofed and Metal Clad Buildings

Metal roofing and siding must not be substituted for the main conductors and air terminals of a lightning protection system, unless constructed of 3/16 in. (4.8 mm) minimum sheet metal which has been made electrically continuous by bonding or an approved interlocking contact. This is because there is a likelihood that lightning will puncture thinner sheet metals and ignite wood framing, etc. Acceptable constructions will require ground terminals (ground rods, rings, etc.).

Underwriters Laboratories Inc. Lightning "Master Label Service" for Lightning Protection Systems

UL has had a "Master Label Service" for lightning protection systems since 1923. It provides for both factory inspection and labeling of lightning protection materials (components), as well as field inspections of a substantial number of installations for which Master Labels have been issued. The service covers the installation of labeled lightning protection materials (components) on all types of structures, with the exception of those used for the production, handling, or storage of ammunition, explosives, flammable liquids or gases, and explosive ingredients. Protection of electrical transmission lines and equipment is also not within the scope of the "Master Label Service."

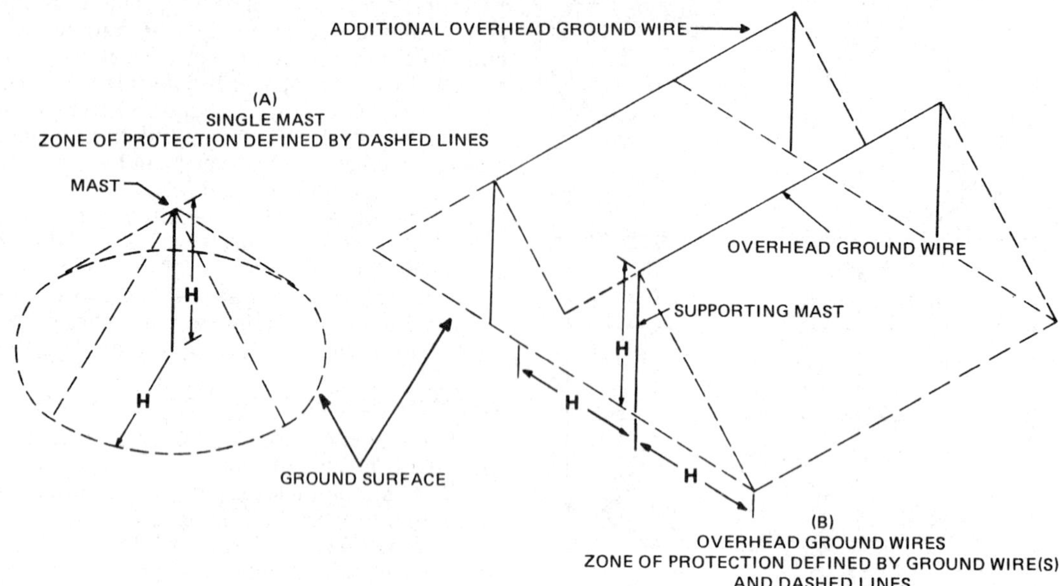

FIG. 12-12K. Zone of protection for mast height "H" not exceeding 50 ft (15 m).

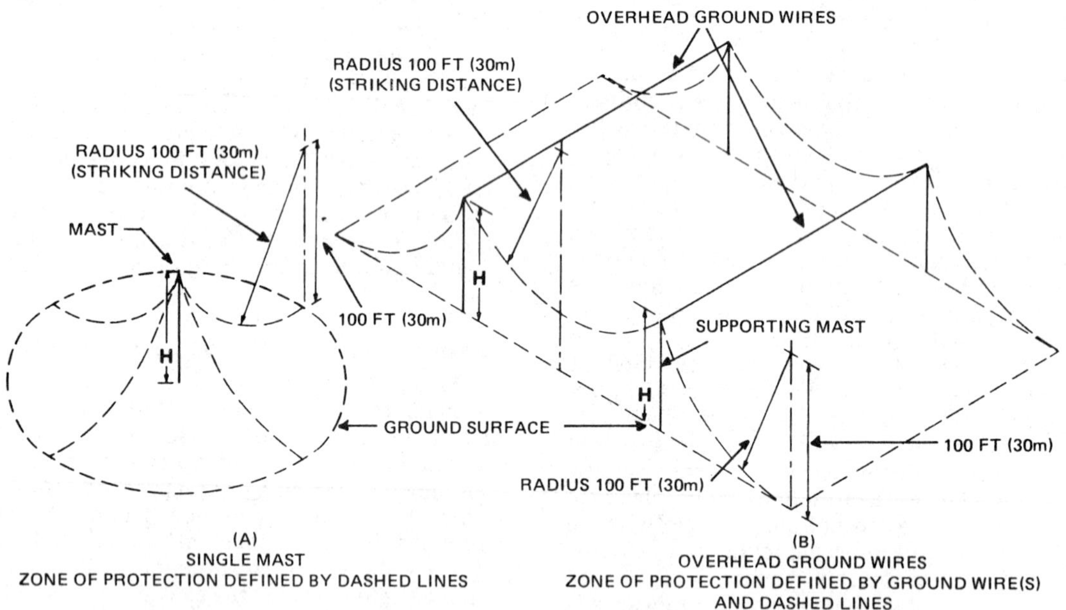

FIG. 12-12L. Zone of protection for mast height "H" exceeding 50 ft (15 m).

Tanks Containing Flammable and Combustible Liquids and Gases

Tanks containing flammable and combustible liquids or flammable gases stored at atmospheric pressures have been set on fire by lightning. Fires may be started by direct hits which ignite vapors escaping from the tank, or may be started on the roofs of wood roofed tanks. A lightning stroke in the vicinity may, by induction, produce sparks which may ignite such vapors. If there are openings in the roof, either intentional or accidental, externally ignited vapors may carry flame inside the tank, possibly resulting in an explosion or fire if there is a flammable or combustible vapor-air mixture inside.

Above ground tanks storing flammable or combustible liquids or flammable gases at atmospheric pressures are considered to be reasonably well protected against lightning if constructed entirely of steel, and if: (1) all joints between steel plates are riveted, bolted, or welded; (2) all pipes entering the tanks are metallically connected to the tank at point of entrance; (3) all vapor openings are closed or are provided with flame arresters (Class I or II liquids as defined in NFPA 321, *Standard on Basic Classification of Flammable and Combustible Liquids*); (4) the tank and roof are constructed of 3/16 in. (4.8 mm) minimum thick sheet metal so holes will not be burned through by lightning strokes; and (5) the roof is continuously welded, bolted, or riveted and caulked to the shell to provide a vapor tight seam and electrical continuity. Where additional protection is desired, the internal supporting members of the roof may be bonded to the roof at 10 ft (3.05 m) intervals, or an overhead ground-wire system or mast protection may be installed in accordance with the NFPA *Lightning Protection Code.* (See Figs. 12-12K and 12-12L.)

Steel tanks in direct contact with the ground or above ground steel tanks connected to extensive properly grounded metallic piping systems are considered to be inherently grounded.

Steel tanks with wooden or other nonmetallic roofs are not considered self-protecting, even if the roofs are essentially vapor tight or sheathed with thin metal. Such tanks should be protected by an overhead groundwire system or mast protection. (See Figs. 12-12K and 12-12L.)

Floating roof tanks with hangers located within a vapor space may be protected by bonding the roof to the shoes of the seal at 10 ft (3.05 m) intervals around the circumference of the tank and by providing insulated joints or installing jumper bonds around each pinned joint of the hanger mechanism. Based upon experience, floating roof tanks without vapor spaces would not appear to need lightning protective measures.

Above ground storage tanks or containers of flammable liquids or liquefied petroleum gas under pressure are considered to be safe from lightning caused explosions since the vapor-air mixture is "too rich" to burn, and the vapor is contained within the tank.

Tall Structures

Spires, masts, and flagpoles of materials other than metal and other slender structures require one air terminal extending 10 in. (254 mm) or more above the uppermost point, a down conductor, and a ground terminal in compliance with the NFPA *Lightning Protection Code.* Church steeples will require one air terminal with a two-way path to ground at the extremities and additional air terminals on the lower level portion that is outside of the steeple zone of protection.

In Britain, brick chimneys, church spires, etc. are protected by using two down conductors connected at the top to an existing metal cap, or to a circular conductor with no air terminals. (See Fig. 12-12M.) Metal smokestacks need no protection against lightning other than that provided by their construction if they are grounded.

Trees, Other Specialized Structures, and Facilities

Trees are sometimes provided with lightning protection where they are especially valuable, are of historical significance, overhang buildings, or provide a shelter for livestock. Guidance on the installation of lightning protection is given in Appendix G of the NFPA *Lightning Protection Code.* other appendices of the NFPA *Lightning Protection Code* cover: (1) sailboats, power boats, small boats, and ships; (2) parked aircraft; (3) livestock in fields; and (4) picnic grounds, playgrounds, ball parks, and other open places.

Grounding Metal Masses

Extensive masses of metal which form part of a building or its appurtenances, such as metal ventilators and television or radio antennas, may need to be bonded to the

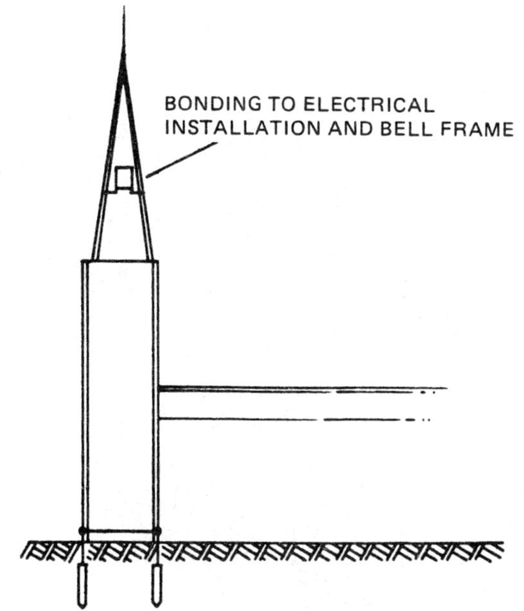

FIG. 12-12M. An installation of a lightning protection system in a church steeple according to British Code requirements. (Note absence of air terminals on lower level portion.)

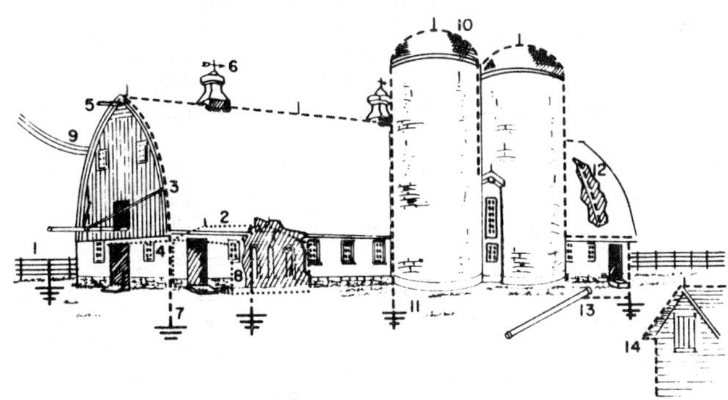

FIG. 12-12N. Lightning protection system on a typical large barn. The numbers indicate the following features: (1) ground attached wire fence; (2) extend protection to any addition; (3), (4), and (5) show bonding conductors to litter, metal door, and hay tracks; (6) shows air terminal on cupolas, ventilators, etc.; (7) indicates ground connection (there should be at least two grounds): (8) illustrates tie-in of metal stanchions; (9) power lines to building need lightning arresters; (10) at least one air terminal should be on each domed silo (at least two on each flat roof or unroofed silo); (11) shows special ground for silo, if required; (12) illustrates connections to vents; and (13) connections to water pipes, and (14) illustrates desirability of protecting adjacent buildings.

lightning protection system. This is dependent upon the ground resistance, the distance from ground to the uppermost point of the object being considered, and the number of down conductors. The grounding of metal windowsills could be a hazard to human life in the event of a direct lightning strike. Details are given in the NFPA *Lightning Protection Code.*

Maintenance

Proper maintenance of lightning protection systems is essential to effective protection. Particular attention

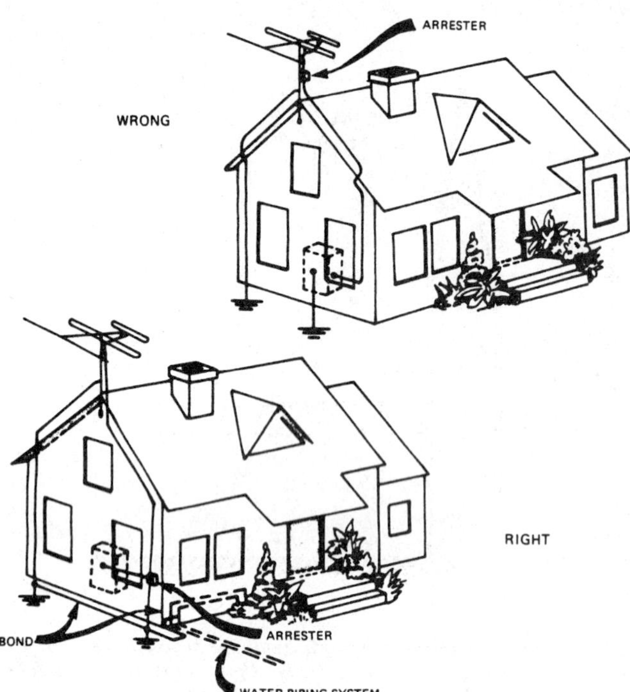

FIG. 12-12O. The right and wrong way of installing a lightning arrester. Installation on the antenna mast leaves the receiver virtually unprotected. If, however, the arrester is mounted approximately as near to earth as the receiver, lightning induced charges may be carried to ground efficiently.

should be given to ground connections, as rods may be broken or corroded at ground level or just below, where the damage is not apparent. Materials (components) may be missing because of storm damage or stolen because of their value.

Lightning Arresters on Electrical Apparatus and Circuits

The installation of lightning (surge) arresters on power and communication lines where they enter structures and at power utility plants is covered in Article 280 of NFPA 70, *National Electrical Code*®. More information on the protection of electrical apparatus and circuits against damage due to lightning appears in the "Additional Readings" section of the bibliography under Surge Arresters. The use of secondary service arresters on electric services is required for buildings equipped with a UL Master Labeled Lightning Protection System. Such arresters may be installed at the yard pole, at the outside electric service entrance, or at the interior service entrance equipment, depending on local regulations. Home lightning protectors are available that are designed for installation at the dwelling weatherhead or indoors at the service entrance equipment. These protectors (arresters) drain lightning surge induced charges harmlessly to ground, and then open to restore electrical service to normal. Before installing a secondary service arrester, it should be determined that the neutral wire is grounded. The proper method of installing lightning arresters on television antennas is shown in Figure 12-12O.

In the United States, except on the Pacific Slope where lightning storms are infrequent, most electric utilities install lightning arresters on the primaries of important transformers on systems of 44,000 volts or less.

PROTECTION OF PERSONS

The lightning hazard is greatest among persons whose occupations keep them outdoors. The probability of injury to the individual from lightning is in general very small, except under certain circumstances of exposure out of doors. Within buildings of considerable size and residential buildings of modern construction, cases of injury from lightning are relatively rare. They are more frequent within small unprotected buildings of the older type. Isolated small schoolhouses and small churches where people may congregate during thunderstorms present a considerable lightning hazard if unprotected.

Guide for Personal Safety During Thunderstorms

Do not go out of doors or remain out during thunderstorms unless it is necessary. Seek shelter inside buildings, vehicles, or other structures or locations which offer protection from lightning.

Seek shelter in the following places which may protect personnel from lightning:

1. Dwellings or other buildings that are protected against lightning.
2. Underground shelters such as subways, tunnels, and caves.
3. Large metal or metal-framed buildings.
4. Large unprotected buildings.
5. Enclosed automobiles, buses, and other vehicles with metal tops and bodies.
6. Enclosed trains and streetcars.
7. Enclosed metal boats or ships.
8. Boats that are protected against lightning.
9. City streets that are shielded by nearby buildings.

If possible, avoid the following places which offer little or no protection from lightning:

1. Small unprotected buildings, barns, sheds, etc.
2. Tents and temporary shelters (not lightning protected).
3. Automobiles (nonmetal top or open).
4. Recreational vehicles (nonmetal or open).
5. Also avoid use of or contact with electrical appliances, telephones, and plumbing fixtures.

Certain locations are extremely hazardous during thunderstorms and should be avoided if at all possible. Approaching thunderstorms should be anticipated and the following locations avoided when storms are in the immediate vicinity:

1. Open fields, athletic fields, and golf courses.
2. Parking lots and tennis courts.
3. Swimming pools, lakes, and seashores.
4. Near wire fences, clotheslines, overhead wires, and railroad tracks.
5. Under isolated trees.

In the above locations, it is especially hazardous to be riding in or on any of the following during lightning storms:

1. Open tractors and other farm machinery operated in open fields.
2. Golf carts, scooters, bicycles, and motorcycles.
3. Open boats (without masts) and hovercraft.
4. Automobiles (nonmetal top or open).

It may not always be possible to choose an outdoor location that offers good protection from lightning. Follow these rules when there is a choice in selecting locations:

1. Seek depressed areas—avoid hilltops and high places.
2. Seek dense woods—avoid isolated trees.
3. Seek buildings, tents and shelters in low areas—avoid unprotected buildings and shelters in high areas.
4. If you are hopelessly isolated in an exposed area and you feel your hair stand on end, indicating that lightning is about to strike, drop to your knees and bend forward putting your hands on your knees. Do not lie flat on the ground or place your hands on the ground.

Bibliography

References Cited

Lee, Ralph H. 1978. "Protection Zone for Buildings Against Lightning Strokes Using Transmission Line Protection Practice." IEEE Transactions on Industry Applications. Vol 1A-14, No 6. Nov/Dec 1978.

Moore, Brook and Krider. 1981. "A Study of Lightning Protection Systems." The Atmospheric Science Program of the Office of Naval Research Under Contract. No. 00014-78M-0090.

Offerman, Paul F. 1969. "Lightning Protection of Structures." IEEE Conference Record of 1969 Fourth Annual Meeting of the IEEE Industry and General Applications Group. 69 C5-IGA.

Sunde, E. D. 1949. *Earth Conduction Effects in Transmission Systems*. D. Van Nostrand Co. Inc. pp 75-80.

NFPA Codes, Standards, Recommended Practices and Manuals. (See the latest NFPA Codes and Standards Catalog for availability of current editions of the following documents.)

NFPA 70, *National Electrical Code.*

NFPA 78, *Lightning Protection Code.*

NFPA 302, *Fire Protection Standard for Pleasure and Commercial Motor Craft.*

NFPA 321, *Standards on Classification of Flammable and Combustible Liquids.*

Additional Readings

"A Feasibility Study of Improved Lightning Protection Systems," Report No. 64PT146, Aug. 6, 1964, available from Clearinghouse for Federal Scientific and Technical Information, Cameron Station, Alexandria, VA.

Electrical Construction Materials Directory, published annually in May with November Supplements, Underwriters Laboratories Inc., Northbrook, IL.

Factory Mutual System Loss Prevention Data Sheets, Lightning Arresters and Grounds, 5-11/14-19.

Golde, R. H., *Lightning Protection*, Edward Arnold (Publishers) Ltd., 25 Hill Street, London, WIX 8LL (ISBN 0 7131 3289 2), 1973.

Golde, R. H., "Protection of Structures Against Lightning," Proceedings, *The Institution of Electrical Engineers*, Vol. 115, No. 10, Oct. 1968.

Golde, R. H. (edited by), *Lightning, Vols. 1 and 2*, Academic Press (Publishers) Inc., 111 Fifth Av., New York, NY 10003 (Lib. Congress No. 77-72088), 1977.

Griscom, S. B., et al., "Five-Year Field Investigation of Lightning Effects on Transmission Lines," IEEE Trans. (*Power Apparatus and Systems*), Vol. 84, April 1965. pp. 257-80.

Gumley, J. R., Invernizzo, C. G., and Khaled, M., "Lightning Protection—A Proven System," *Fire Technology*, Vol. 13, No. 2, May 1977, pp. 114-120.

Hedlund, C. F., "Lightning Protection for Buildings," *IEEE Transactions on Industry and General Applications*, Vol. IGA-3, No. 1, Jan./Feb. 1967, Institute of Electrical and Electronics Engineers, New York.

Hooker, Dan, "Home Lightning Protection Reduces Fire Danger" *Fire Journal*, Vol. 63, No. 3, May 1969.

IEEE Standard 142-1982 (ANSI/IEEE), Chapter 3, *Static and Lightning Protection Grounding.*

Johnson, Erling R., "Accidental Power Cross Results in Improved Alarm System Design," *Fire Journal*, Vol. 68, No. 2, Mar. 1974, p. 7.

Journal of the Franklin Institute, Philadelphia, Vol. 283, No. 6, June 1967, (Special issue on Lightning Research).

Lewis, W. W., *The Protection of Transmission Systems Against Lightning*, Dover Publications, New York, 1965.

The Lightning Flash, published bi-monthly by Lightning Technologies, Inc., 10 Downing Parkway, Pittsfield, MA, 01201.

"Lightning Protection System, General with Regard to Installation," German Standard, DIN 57185 Part 1/VDE 0185 Part 1/11.82 (Engl.), English Translation, 1983.

Malan, D. J., *Thunderstorms and Protection Against Lightning*, The Institution of Certificated Mechanical and Electrical Engineers, South Africa.

Miller-Hillebrand, D., "Lightning Protection," Proceedings of the 3rd International Conference on Problems of Atmospheric and Space Electricity, May 1963.

Peterson, Alvin E., *Lightning Hazards to Mountaineers*, The American Alpine Club, New York, 1962.

"The Philosophy of Lightning Protection," *Fire Journal*, Vol. 61, No. 6, Boston, Nov. 1967.

"The Protection of Structures Against Lightning," British Standard Code of Practice, CP326, 1965, British Standards Institute.

Stone, Walter, "The NEC and You (Protection Against Energized Metal Siding)," *Fire Journal*, Vol. 64, No. 3, May 1970, pp. 33

Timmons, Merrill S., Jr., "Lightning Protection for the Farm." U.S. Department of Agriculture, *Farmers' Bulletin No. 2136.*

Towne, H. M., *Lightning—Its Behavior and What To Do About It*, United Lightning Protection Association, Inc. Webster, NY.

Underwriters Laboratories Inc., *Standard for Lightning Protection Components, UL 96*, 1st ed., Jan. 30, 1981, Northbrook, IL.

Underwriters Laboratories Inc., *Standard for Installation Requirements for Lightning Protection Systems, UL 96A*, Revised Oct. 1983, Northbrook, IL.

Wagner, C. F., and Hileman, A. R., "The Lightning Stroke—II," IEEE Trans. (*Power Apparatus and Systems*), Vol. 80, Oct. 1961. pp. 622-42.

Wagner, C. F., "The Lightning Stroke as Related to Transmission Line Performance," Part I, *Electrical Engineering*, May 1963, pp. 339-47.

Surge Arresters, ANSI Standards

ANSI/IEEE C62.32-1981, Air Gap Surge-Protective Devices (Excluding Valve and Expulsion Type Devices), Test Specifications for the Low-Voltage.

ANSI/IEEE C62.31-1981, Gas Tube Surge-Protective Devices, Test Specifications for.

ANSI/IEEE C62.1-1981, Surge Arresters for Alternating-Current Power Circuits.

ANSI/IEEE C62.2-1981, Valve Type Surge Arresters for Alternating-Current Systems, Guide for Application of.

ANSI C62.33-1982, Varistor Surge-Protective Devices, Test Specifications for.

AIR MOVING EQUIPMENT

Revised by John Bouchard

Many industrial and commercial activities and processes require the removal of smoke, fumes and/or dusts. This necessitates equipment for moving air in and out of the processing area. Both the accumulation and moving of impurities and the substitution of other atmospheres present fire and explosion hazards. Additional information about the elimination of such hazards will be found in Section 5, "Fire Hazards of Materials"; in the following chapters of Section 10: Chapter 5, "Industrial and Commercial Heat Utilization Equipment"; Chapter 10, "Chemical Processing Equipment"; Chapter 11, "Grinding Processes"; Chapter 18, "Metalworking Processes"; Chapter 19, "Woodworking Processes"; in Section 12, Chapter 10, "Control of Electrostatic Ignition Sources"; and in Section 19, "Special Fire Suppression Agents and Systems."

Air moving equipment (AME) includes all mechanical-draft duct systems, of both the pressure and exhaust types, for removal of dusts, vapors, and waste material and for the conveying of materials. The hazards of such systems lie in the possibility of igniting flammable materials or vapors by sources such as sparks generated by fans or foreign material (rocks or tramp metal) or by overheated fan bearings. Then such systems could inadvertantly spread fires through the building.

Air moving systems do, however, play an important role in fire protection. If the materials they remove were allowed to accumulate, the result could be an explosion from vapor-air mixtures, a flash fire due to accumulations of lint or other such things, and generally poor housekeeping conditions conducive to fires in general.

Blower and exhaust installations usually contain centrifugal and axial flow fans; several different designs of each kind are available. (See Fig. 12-13A.) A straight blade centrifugal fan is suitable for exhausting large particles of materials because it resists clogging and the blades can withstand considerable abrasion. Propeller fans will move large volumes of air but only against small resistance. Since fans could not operate efficiently against the friction a duct would introduce, they are not suitable for ducts of any appreciable length.

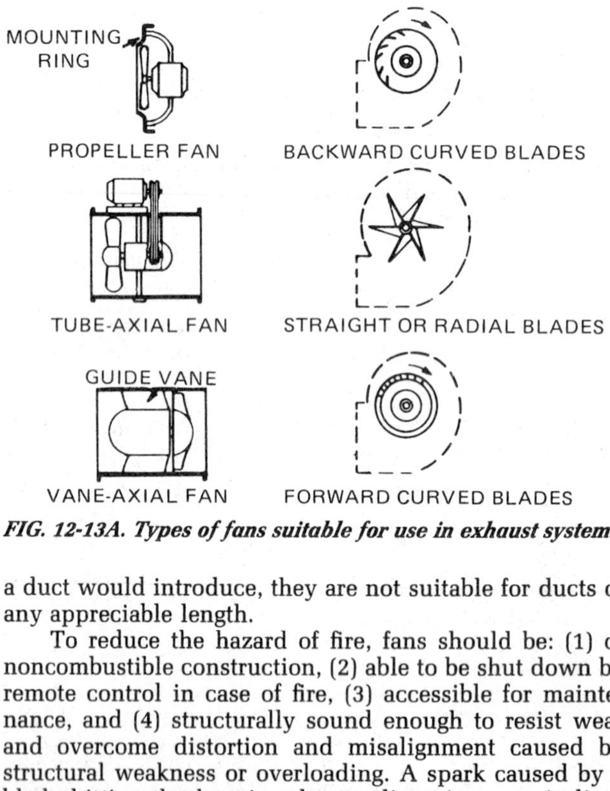

MOUNTING RING
PROPELLER FAN
BACKWARD CURVED BLADES
TUBE-AXIAL FAN
STRAIGHT OR RADIAL BLADES
GUIDE VANE
VANE-AXIAL FAN
FORWARD CURVED BLADES

FIG. 12-13A. Types of fans suitable for use in exhaust systems.

To reduce the hazard of fire, fans should be: (1) of noncombustible construction, (2) able to be shut down by remote control in case of fire, (3) accessible for maintenance, and (4) structurally sound enough to resist wear and overcome distortion and misalignment caused by structural weakness or overloading. A spark caused by a blade hitting the housing due to distortion or misalignment could result in ignition when the exhaust contains flammable solids or vapors. This possibility of friction sparks is minimized if the fan and fan housing are of noncompatible material. Aluminum and magnesium alloys are not compatible with ferrous alloys which may rust.

Details on the installation of air moving equipment are given in NFPA 91, *Standard for the Installation of Blower and Exhaust Systems for Dust, Stock, and Vapor Removal or Conveying*, and NFPA 96, *Standard for the Installation of Equipment for the Removal of Smoke and Grease-Laden Vapors from Commercial Cooking Equipment.*

Mr. Bouchard is Assistant Division Director, General Engineering, on the staff of NFPA.

HAZARDS OF SPECIFIC USES

Ventilation of Flammable Vapors

Because flammable vapors ignite easily, precautions must be taken to eliminate ignition sources in ducts which carry them. Using a single system to exhaust flammable vapors and particles from spark producing processes is obviously dangerous. Similarly, drawing different vapors which individually may not be flammable, but which may be hazardous as a mixture in a single exhaust system, is extremely poor practice. For example, perchloric acid acts as an oxidizing agent at elevated temperatures. If perchloric acid vapors were drawn into a system exhausting organic materials, the combination could cause a fire or explosion.

It is important that vapors be withdrawn from the rooms or equipment in which they are generated and taken directly to the outside of the building. Processes generating flammable vapors are best located along an outside wall of the building to facilitate efficient vapor removal.

When flammable vapors cannot be readily picked up at a specific source, general ventilation through a system of suction ducts with inlets to the room or area may be used. Because suction inlets have minimal directional effect beyond a few inches (1 in. equals 25.4 mm) from the face of the inlet, they should be located so that they produce a sweeping or purging effect that will eliminate pockets in which vapors may accumulate. A make-up air supply, properly located in relation to the point where vapors are generated, and openings for exhaust ducts will aid in the dilution and removal of vapors. The ventilation system should provide sufficient air movement to maintain the vapor concentration below the lower explosive limit in the area where vapors are being liberated. If vapors are toxic, it may be necessary to maintain concentrations well below those required for flammability.

NFPA 86A, *Standard for Ovens and Furnaces: Design, Location, and Equipment*, gives procedures for calculating the volume of air per minute necessary to dilute flammable vapors below their lower flammable limit.

Where heavier-than-air vapors or mixtures are handled, exhaust openings located near the floor line are generally the most effective. Conversely, for vapors or mixtures lighter than air, exhaust openings located near the top of the room, hood, or enclosure are the most effective. Caution in evaluating the specific gravity of vapors is necessary where other than normal temperatures and pressures exist.

Fans: All fans should be suitable for moving vapors and air. If the vapor contains a material which may condense and build up on the fan blades, this possibility should be minimized by the selection of a fan with backward curved blades. Both the fan and the fan housing should be of nonferrous construction.

Ducts: Ducts for systems handling flammable vapors need to be structurally capable of withstanding some fire exposure. Where vapors may condense, continuously welded joints in the ducts will prevent leakage.

Since flash fires or explosions can occur in ducts, ductwork should be installed away from combustibles, but where it will be readily accessible. Ducts should never be installed in a wall or ceiling. Duct systems which exhaust flammable materials should terminate outside in an area where the vapors cannot be ignited and where they will not form deposits, such as overspray residue, on the building or other structures. An enclosure is sometimes used around the discharge end of a duct to prevent deposits on exposed property. These enclosures should be sprinklered.

Frequent cleaning of ducts and any enclosures will minimize both the intensity of a fire in a duct system and any spontaneous heating of deposits subject to this phenomenon.

Electrical Equipment: All equipment which may be exposed to flammable vapors should be that which is specified by NFPA 70, *National Electrical Code®* for such locations.

Fire Extinguishing Equipment: Systems suitable for handling flammable vapors include fixed pipe extinguishing systems and portable extinguishers. The extinguishing agent should be water, dry chemical, or an inert gas such as CO_2 or a halonenated agent.

Ventilation of Corrosive Vapors and Fumes

If corrosive vapors must be exhausted, the degree of expected corrosion is the governing factor in the construction of the ducts. In some cases a heavier gage metal is used, and in others, protective coatings may be sufficient. Occasionally, however, the corrosiveness is so extreme that a special lining is required. Stainless steel has been used very successfully in some cases.

Plastic duct systems are also used, but only if the corrosive vapors are nonflammable and the plastic has a flame spread rating of 25 or less and a smoke developed rating of 50 or less, when tested in accordance with NFPA 255, *Standard Method of Test of Surface Burning Characteristics of Building Materials*. Automatic fire protection is also necessary at the hood, canopy, or intake to plastic duct systems.

A duct system of plastic materials for a typical industrial exhaust system is shown in Figure 12-13B, and a duct system for a typical laboratory hood exhaust system is shown in Figure 12-13C.

Ventilation of Kitchen Cooking Equipment

Exhaust systems for restaurant equipment are troublesome because grease condenses in the interior of the ducts. Grease accumulations may be ignited by sparks from the stove or, more often, by a small fire on the stove caused by overheated cooking oil or fat in a deep fat fryer or on a grill. If the duct did not have a grease accumulation, fires on stove tops could often be extinguished before causing appreciable damage. Fires occur frequently in frying because cooking oils and fats are heated to their flash points. Thus fats and oils can reach their self-ignition temperatures by being accidentally overheated or by being ignited when spilled on the stove top. Details of a typical kitchen range exhaust system are shown in Figure 12-13D.

Fans should be selected according to their ability to exhaust the required quantity of air against all calculated friction losses.

Duct Systems: The following should be considered when designing a good duct system for commercial cooking equipment:

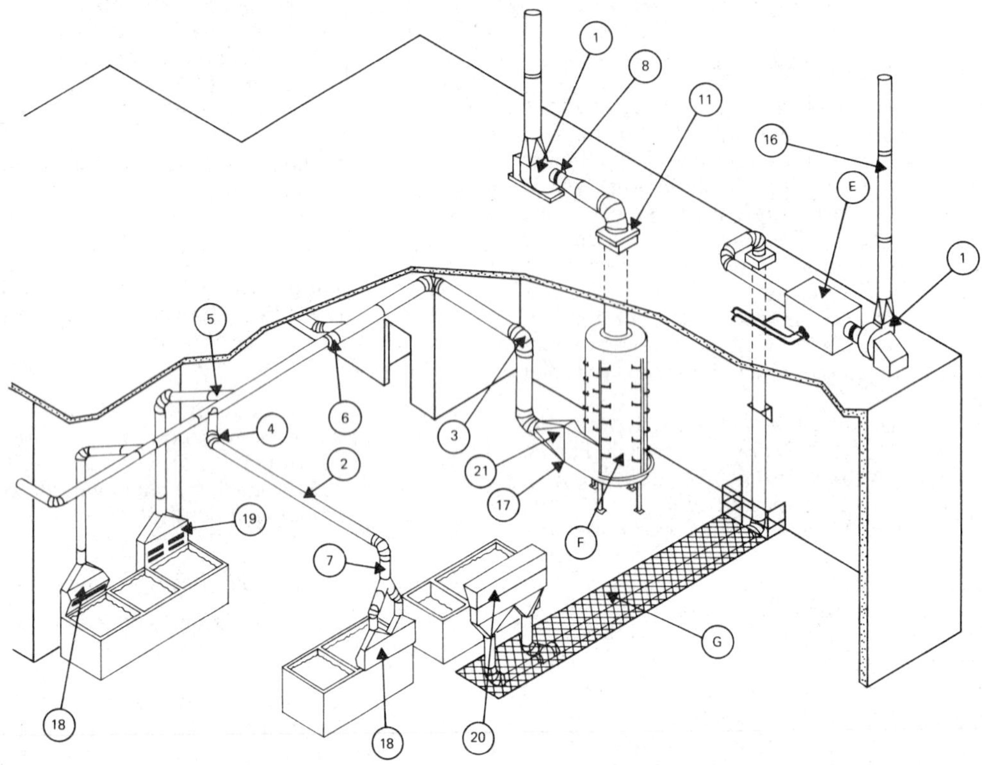

FIG. 12-13B. Exhaust system for one-story building occupied by various type fume hoods with vertical fume scrubber and service trench. (For nomenclature, see below right.)

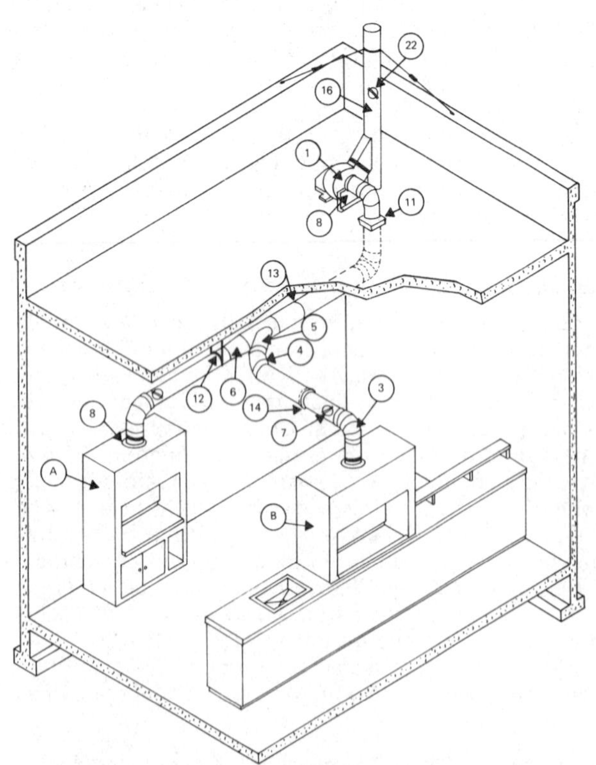

I. EQUIPMENT
- A. CABINET TYPE LABORATORY FUME HOOD
- B. BENCH TYPE LABORATORY FUME HOOD
- C. FILTER BOX FOR SPECIAL OR HIGH EFFICIENCY FILTERS
- D. SHAFT
- E. HORIZONTAL TYPE FUME SCRUBBER
- F. VERTICAL TYPE FUME SCRUBBER
- G. SERVICE PIT OR TRENCH

II. SYSTEM COMPONENTS
1. AIR MOVING EQUIPMENT (CENTRIFUGAL TYPE EXHAUST FAN)
2. HORIZONTAL DUCT SECTION
3. 90 DEGREE ELBOW
4. ELBOW (LESS THAN 90 DEGREE)
5. LATERAL ENTRY
6. TRANSITION
7. MANUAL BALANCING DAMPER
8. FLEXIBLE CONNECTION
9. FIRE DAMPER
10. ACCESS DOOR
11. COUNTERFLASHING
12. DUCT HANGER
13. CIRCUMFERENTIAL GIRTH JOINT (BUTT WELDED)
14. BELL END DUCT SEAM
15. WEATHER CAP
16. FAN DISCHARGE STACK
17. FLANGED DUCT CONNECTION
18. OPEN FACE TANK EXHAUST HOOD (UPDRAFT)
19. SLOTTED FACE TANK EXHAUST HOOD (UPDRAFT)
20. OPEN FACE TANK EXHAUST HOOD (DOWNDRAFT)
21. ROUND TO RECTANGULAR (OR SQUARE) TRANSITIONAL FITTING
22. GRAVITY OPERATED BACKDRAFT DAMPER

FIG. 12-13C. Rooftop exhaust system for one-story building occupied by a cabinet type laboratory fume hood and benchtype laboratory fume hood. (For nomenclature, see above right.)

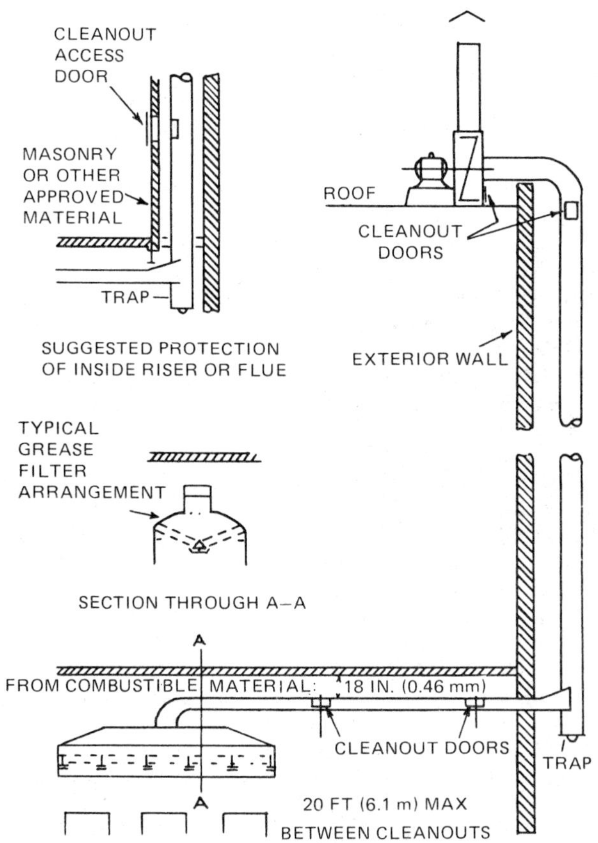

FIG. 12-13D. Typical kitchen range exhaust system arrangement, showing vertical riser outside the building. If necessary to locate riser inside the building, it should be enclosed in a masonry shaft.

1. Design the system to minimize grease accumulations with a minimum air velocity of 1,500 fpm (457 m/min) through any duct.
2. Arrange ducts with ample clearance from combustible materials to minimize the danger of ignition in case of fire in the duct.
3. Use ducts of substantial construction (not lighter than No. 16 Manufacturers Standard Gage steel or No. 18 Manufacturers Standard Gage stainless steel) with all of the seams and the joints having a liquid-tight, continuous external weld.
4. Separate systems to make sure there is no connection with any other ventilating or exhaust system.
5. Lead ducts directly outside the building without dips or traps, unless automatic grease removers are employed at the dips and traps.
6. Provide openings for inspection and cleaning. Do not install dampers in any duct system unless required as part of a grease extractor or extinguishing system.

Although clean ducts are essential, cleaning them is likely to be neglected because it is difficult and unpleasant and can be dangerous if flammable solvents are used. Satisfactory cleaning results have been obtained with a powder compound consisting of one part calcium hydroxide and two parts calcium carbonate. This compound saponifies the grease or oily sludge, thus making it easier to remove and clean. Proper ventilation must be provided and safety precautions taken if cleaning is done inside the duct or fan housings.

Spraying the duct interiors with hydrated lime after cleaning is a fire prevention method in commercial use. Though this tends to saponify the grease and may facilitate subsequent cleaning, it does not provide any permanent fire protection or flame resistance for grease eventually coats the compound.

Grease Removal Devices: All kitchen exhaust systems require a means for removing grease. These may be grease extractors, grease filters, or other grease removal devices, such as water-wash systems and special fans designed to remove grease vapors effectively and provide a fire barrier. Grease filters, including frames, or other grease removal devices should be made of noncombustible materials.

Electrical Equipment: Manual control of the fan motor should be provided near the motor and also near the hood. In addition, it is desirable to have automatic shutdown of the motor by a heat sensitive device in the hood close to the outlet.

All electrical equipment which may be exposed to vapors, grease, or heat should be installed in accordance with NFPA 70, *National Electrical Code* requirements for those conditions. Fixtures should not be installed in ducts or hoods used for removal of cooking smoke or grease laden vapors or located in the path of travel of such exhaust products unless they are specifically designed and tested for such use. Electrical equipment may be placed outside the path of vapor travel by locating it on the outside of the hood with illumination through tight fitting glass panels in the hood.

Fire Extinguishing Equipment: For sizable cooking installations, this equipment may consist of fixed pipe carbon dioxide, dry chemical, foam-water sprinkler, water spray, or sprinkler systems, supplemented by portable alkaline (sodium bicarbonate or potassium bicarbonate) dry chemical extinguishers.

Acidic base extinguishing agents, such as ammonium-phosphate base multipurpose types, impede saponification. Therefore, if the cooking equipment being protected involves exposed liquefied fat or oil in depth, such as deep fat fryers, extinguishers employing these extinguishing agents are not recommended as they may interfere with alkaline-base fixed systems.

Where fixed extinguishing systems are installed, they should: (1) be located so that they easily can be manually activated from the path of egress, (2) be arranged for simultaneous automatic operation of all systems in a single hazard area upon operation of any one system, and (3) automatically shut off all sources of fuel and heat to all equipment protected upon operation of the system.

Dust Collecting and Stock and Refuse Conveying Systems

These systems consist of suction ducts and inlets, air moving equipment, feeders, discharge ducts and outlets, collecting equipment, vaults, and other receptacles designed to collect powdered, ground, or finely divided material.

Collecting and conveying systems are often a very important part of the operations of a plant. When the material they handle is a combustible dust, there is danger of a fire or explosion. These factors make it a prime necessity that the design, construction, and operation of the systems conform to recognized standards. The general recommendations given in the first part of this chapter should be observed, as should the following more specific ones.

Fans: Systems conveying combustible dust should be arranged so that the fan is on the clean air side of the collector, i.e., so that the system operates under suction, and dust collecting equipment removes the dust before the air stream reaches the fan. This arrangement prevents combustible dust from passing through the fan where it may be ignited. If such an arrangement is not possible and the fan must be located between the dust producing equipment and the collector, the blades and spider of the fan and the fan housing should be made of nonsparking material with ample clearance between the blades and the housing. Fan bearings and motors should be outside of casings unless the fan and motor assembly are designed and tested for use in the dust atmosphere present.

Ducts: For conveying dusts or stock, ducts should be constructed of metal. Changes in direction in the duct system should be made with long bends or elbows. This minimizes accumulations of solid particles at the turns.

Following are some general design techniques recommended for ducts conveying dusts and other solid materials:

1. Every duct should be kept open and unobstructed throughout its length.
2. Not more than two branches should connect to any section of the main suction duct.
3. Branch ducts should be connected to the top or side of the main duct at an angle not exceeding 45° inclined in the direction of air flow.
4. Main suction and discharge ducts should be as short as possible.
5. Flexible duct sections should not restrict airflow and should be as short as possible.
6. Systems handling combustible dusts should, as far as possible, be located outside of the building. Branch ducts from each floor should pass through the wall and discharge into the main duct outside.
7. Do not add additional branch ducts without redesigning the system, or blank off disconnected or unused portions of the system without providing orifice plates to maintain required airflow.

Magnetic Separators: As with flammable vapor conveying systems, combustible dust collecting systems should not be connected with processes which may produce sparks. If there is a possibility of particles of ferrous materials ("tramp metal") entering the collecting systems, permanent magnetic separators can be installed at points where ferrous particles may enter. After collection, the dust passes into rooms or bins. These places of collection should be made of noncombustible construction and provided with explosion vents terminating outside of the building.

Explosion Prevention: When ignition sources are difficult to control, inerting may be used to create a safe atmosphere within the system, or an explosion suppression system may be used. There are limits to both systems—inerting is generally only practical for essentially closed systems, while explosion suppression is limited to those materials that can be successfully protected. Efforts to suppress explosions in some materials, such as metal dusts, for example, have not been successful.

Separating and Collecting Equipment

Separating equipment includes cyclones, condensors, wet type collectors, cloth screen and stocking arresters, centrifugal collectors, and other devices used for the purpose of separating solid material from the air stream in which it is carried. Then the material is collected in hoppers, bins, silos, and vaults. See Figures 12-13E, 12-13F, 12-13G, and 12-13H for typical examples of some of this equipment and how sprinklers may be used in some types.

Separating and collecting equipment is designed to withstand anticipated explosion pressures, allowance being made for explosion relief vents. This equipment should be constructed of steel or enclosed in steel and located outside the building. To prevent the equipment from collapsing (which could result in an explosive dust cloud), equipment should be well supported on steel, masonry, or concrete. Equipment and discharge ducts should be well separated from combustible construction and unprotected openings into buildings.

Collectors that must be located indoors, and that cannot be constructed of sufficient strength to withstand anticipated explosion pressures, can be located near outside walls to facilitate explosion venting.

Gravity feed through tightly fitted ducts is the best arrangement for delivering stock from separators, cyclones, or other collection equipment to storage receptacles. Delivery ducts from cyclone collectors should not convey refuse directly into the fireboxes of boilers, furnaces, refuse burners, incinerators, etc.

Where processes produce very fine dusts or where combustible metal dusts are handled, wet collectors provide efficient dust control. Dust is removed by passing the flow of air through a water curtain and is collected as a sludge submerged in water. Except where combustible metal dusts are collected, these wet collectors generally have no fire or explosion hazards.

Where refuse is to be used as fuel, the discharge system from the storage receptacle or intermediate feed bin to the furnace should be designed to prevent a flashback from the furnace. This may be accomplished by means of a choke feeder or choke conveyor so that a positive cutoff is provided. The installation of a steam spray in the duct to the furnace which blows steam in the direction of the fuel flow is recommended because it provides an added safety factor in preventing a flashback. See Figure 12-13E for a typical installation of this type.

The installation of screw conveyors or rotary feeds at appropriate points in stock or dust conveying systems has been used to good advantage in many plants by providing a "choke," which helps stop the spread of a flash fire. The installation of an explosion vent also relieves the pressure of an explosion and prohibits its progress. This combination has often prevented small fires or explosions from becoming large and disastrous.

High Efficiency Air Filter Units: Particulate air filter units are used to remove very fine particulate matter, i.e., to remove not less than 99.97 percent of 0.3-micron diam-

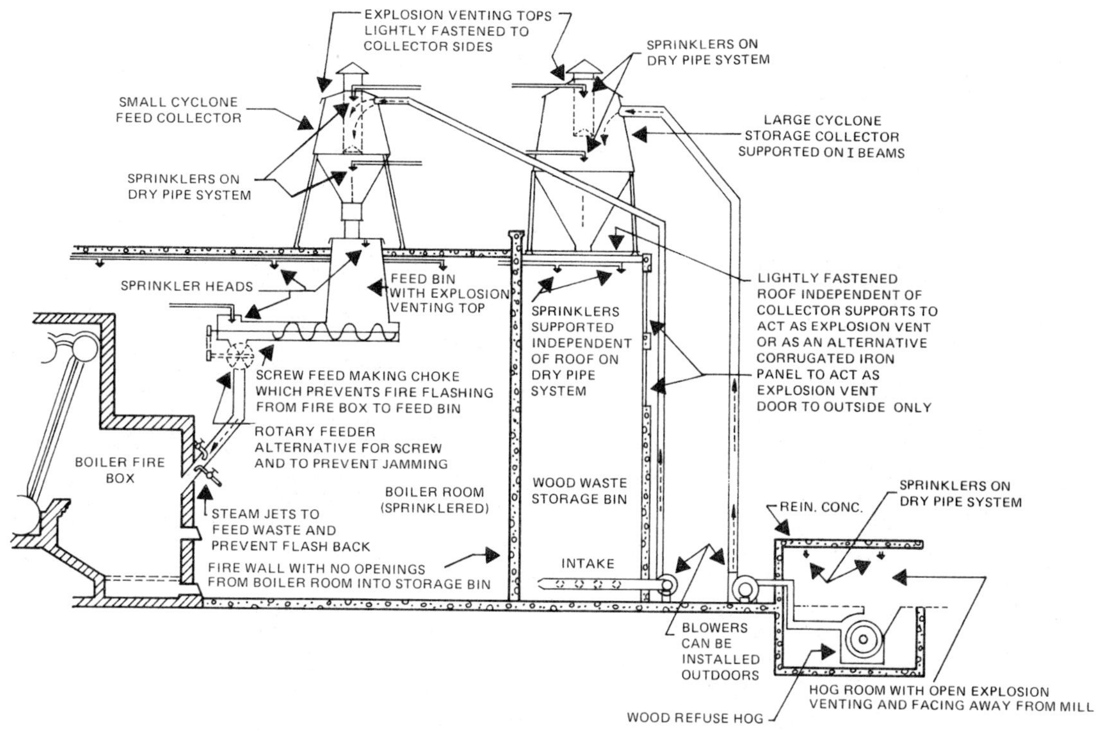

FIG. 12-13E. Suggested arrangement for wood waste firing of boiler to minimize explosion and fire hazard.

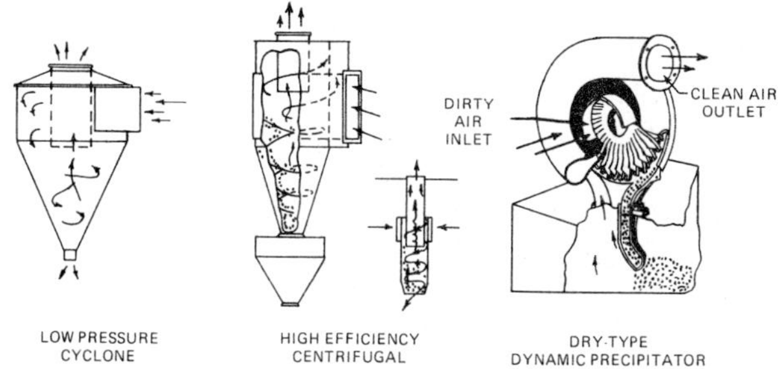

FIG. 12-13F. Examples of typical types of dust collecting equipment.

eter particles, from the air of industrial and laboratory exhaust systems.

The Nuclear Regulatory Commission (NRC) requires high efficiency filters for filtering air exhausted from spaces where radioactive materials are handled. The filter units used are made of a filter medium of glass fiber, or equivalent inorganic material. Although the adhesive used is combustible, it contains a self-extinguishing additive. Fire of sufficient temperature and duration to melt glass fibers can damage the filter at points of contact, but will not stop the exhaust of filtered air through the undamaged area. The NRC installs large banks of these filters, limits the number in a fire area to 100, and protects them with automatic sprinklers.

Fire Extinguishing Systems: Cloth screen or bag type dust collectors used for the collection of fine dusts present a special problem because of the use of a combustible fabric even where the dust is noncombustible. Automatic sprinkler protection may be needed where cloth dust collectors are important for continuity of production or where they provide exposure to other property. Figures 12-13G and 12-13H show typical layouts of sprinkler protection for cloth dust collectors.

Equipment of large volume, such as bins, dust collectors, etc., in which pulverized stock is stored or may accumulate, should be protected by automatic sprinklers. Automatic carbon dioxide, dry chemical, halon, or water spray extinguishing systems can be used effectively in dust collecting systems. Inert gas may be effectively used to create safe atmospheres in conveying systems.

FIRE PROTECTION

Fire Extinguishing Systems

The number of fires originating in and spreading through duct systems, either because of the combustibility of the ducts themselves or because of the combustibility of the material being transported through the system, justifies the installation of automatic extinguishing systems. The systems may be automatically or manually controlled and may utilize whichever of the several extinguishing agents is appropriate or desired.

For example, if the material being transported is a flammable vapor or an easily ignited combustible dust, a deluge system actuated by heat sensitive devices may be most appropriate. If the problem is the combustibility of the duct itself, an automatic sprinkler system may be adequate. In still other systems a special extinguishing agent may be required or deemed more suitable. Sprinklers and discharge nozzles may have to be coated or covered to protect them from the environment inside the duct; they may have to be inspected on a regular basis and cleaned periodically to avoid serious build up of deposits. Simi-

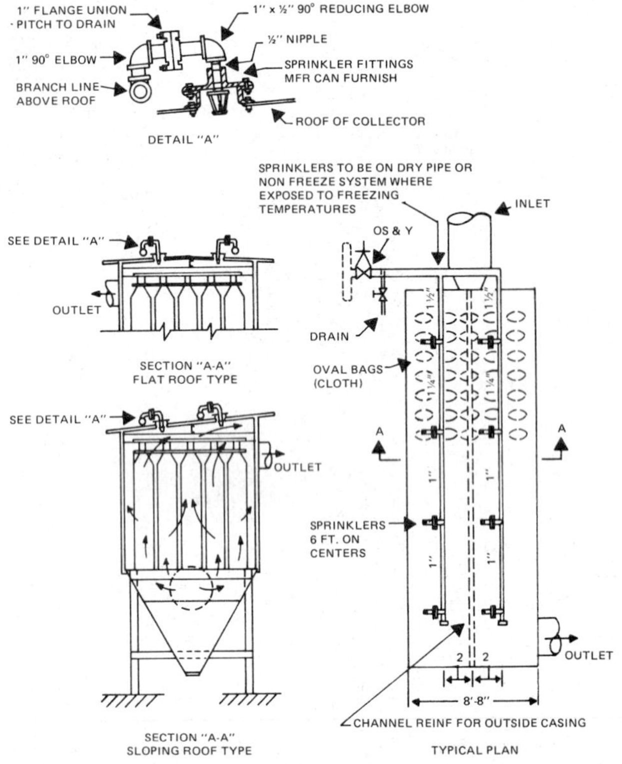

FIG. 12-13G. Typical sprinkler protection for a bag type dust collector.

larly, heat detectors or other types of detectors used to activate the extinguishing system may have to be periodically cleaned of deposits that might hamper their proper operation.

One important problem frequently overlooked is the obstruction of automatic sprinklers or other discharge nozzles by the ductwork. Care should be taken that ductwork does not seriously interfere with the discharge pattern. In many cases, additional sprinklers or nozzles will have to be extended down below the ductwork to provide satisfactory coverage.

Manual Extinguishing Equipment

Portable fire extinguishers of appropriate type or small hose with spray nozzles are helpful where fires may occur in ducts. This is true even where fixed pipe extinguishing systems have been provided. Judicious placement of extinguishers and hoselines with respect to location of access panels will make them more effective.

Explosion Prevention and Venting

Explosion relief vents prevent or minimize damage to duct systems that carry explosive mixtures. The vents should lead by the most direct practical route to the outside of the building. Information in NFPA 68, *Guide for Explosion Venting*, will help when designing explosion vents. The use of suppression and inerting systems is limited by the configuration of the equipment and the physical and chemical properties of the material being conveyed. Details concerning these systems are given in NFPA 69, *Standard on Explosion Prevention Systems.*

Static Electricity

When flammable vapors, dusts or other materials pass through a duct, static charges on the duct are generated. If a charge is allowed to accumulate on an electrically insulated portion of a duct system, it could discharge to an adjoining duct section, at a joint for example, and ignite the material being conveyed. For this reason, exhaust systems carrying flammable vapors, dusts, gases, or other

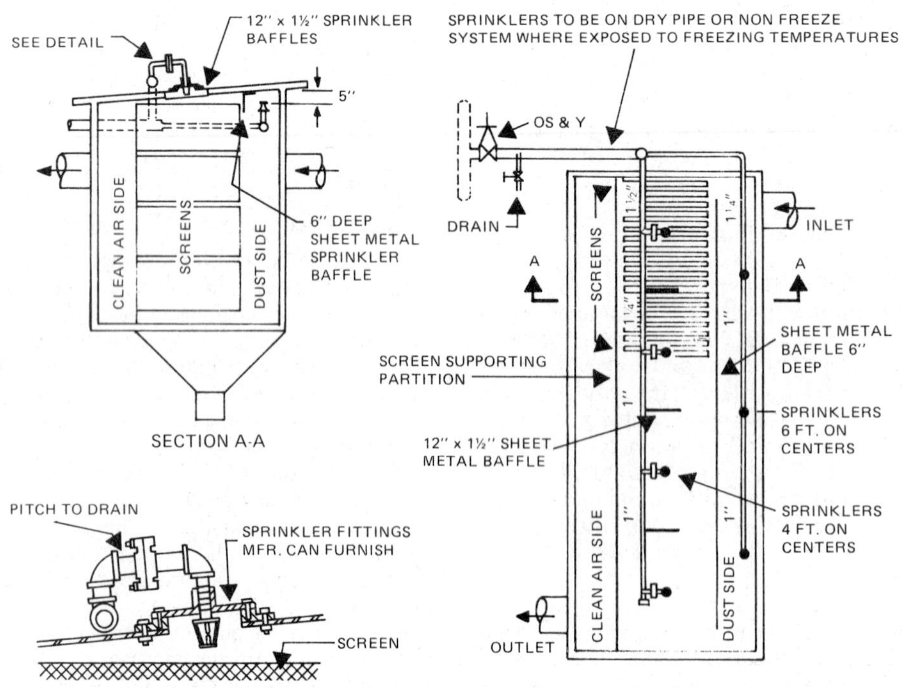

FIG. 12-13H. Typical spinkler protection for a screen type dust collector.

materials should be electrically bonded and grounded. Methods of eliminating dangerous static charges are treated in detail in NFPA 77, *Recommended Practice on Static Electricity*.

Bibliography

NFPA Codes, Standards, Recommended Practices and Manuals. (See the latest *NFPA Codes and Standards Catalog* for availability of current editions of the following documents.)

NFPA 61A, *Standard for Prevention of Fire and Dust Explosions in Facilities Manufacturing and Handling Starch.*

NFPA 61B, *Standard for the Prevention of Fire and Explosions in Grain Elevators and Facilities Handling Bulk and Raw Agricultural Commodies.*

NFPA 61C, *Standard for the Prevention of Fire and Dust Explosions in Feed Mills.*

NFPA 61D, *Standard for the Prevention of Fire and Dust Explosions in the Milling of Agricultural Commodities for Human Consumption.*

NFPA 65, *Standard for the Processing and Finishing of Aluminum.*

NFPA 68, *Guide for Explosion Venting.*

NFPA 69, *Standard for Explosion Prevention Systems.*

NFPA 70, *National Electrical Code.*

NFPA 77, *Recommended Practice on Static Electricity.*

NFPA 86A, *Standard for Ovens and Furnaces: Design, Location, and Equipment.*

NFPA 91, *Standard for the Installation of Blower and Exhaust Systems for Dust, Stock and Vapor Removal or Conveying.*

NFPA 96, *Standard for the Installation of Equipment for the Removal of Smoke and Grease-Laden Vapors from Commercial Cooking Equipment.*

NFPA 255, *Standard Method of Test of Surface Burning Characteristics of Building Materials.*

NFPA 654, *Standard for the Prevention of Fire and Dust Explosions in Chemicals, Dye, Pharmaceutical, and the Plastics Industries.*

Additional Readings

American National Standards Institute Committee on Ventilation, *Fundamentals Governing the Design and Operation of Local Exhaust Systems*, ANSI Z9.2-1971, American National Standards Institute, NY.

American Society of Heating, Refrigerating and Air Conditioning Engineers, Inc., *ASHRAE Handbook and Product Directory—Applications*, NY, 1974.

American Society of Heating, Refrigerating and Air Conditioning Engineers, Inc., *ASHRAE Handbook and Product Directory—Fundamentals*, NY, 1977.

American Society of Heating, Refrigerating and Air Conditioning Engineers, Inc., *ASHRAE Handbook and Product Directory—Systems*, NY, 1976.

Bartknecht, W., *Explosions: Course, Prevention, Protection*, Springer-Verlag, NY, 1981.

Burchsted, C. A., "Basic Requirements for HEPA Filters," *Fire Technology*, Vol. 3, No. 4, Nov. 1967, pp. 271-280.

Chemical Engineering Magazine, *Safe and Efficient Plant Operation and Maintenance*, McGraw-Hill, NY, 1980.

Cocks, R. E., "Dust Explosions: Prevention and Control," *Chemical Engineering*, Vol. 86, No. 24, Nov. 1979, pp. 94-101.

Committee on Industrial Ventilation, *Industrial Ventilation*, 13th ed, American Conference of Governmental Industrial Hygienists, Lansing, MI. 1974.

Frank, T. E., "Fire and Explosion Control in Bag Filter Dust Collection Systems," *Fire Journal*, Vol. 75, No. 2, March 1981, pp. 75-80, 94.

Hutcheon, N. B., "Fire Protection in Air System Installations," *Heating, Piping and Air Conditioning*, Vol. 40, No. 12, Dec. 1968.

Keigher, Donald J., "Fire Protection Considerations in Clean Room Design and Operations," *Fire Technology*, Vol. 3, No. 4, Nov. 1967, pp. 261-271.

Levenback, George, "Grease Fires in Kitchens," *Fire Journal*, Vol. 66, No. 4, July 1972, pp. 69-72.

Roberts, A. F., "Fire in Ducts Under Forced Ventilation Conditions," *Fire Technology*, Vol. 6, No. 1, Feb. 1970, pp. 13-21.

Schmitt, C. R., "Flammable Solvent Cleaning Operations in a Laminar Flow Clean Room," in *Fire Prevention and Suppression*, Hilado, C. J., ed., Technomic, Westport, CT, 1974, pp. 90-96.

Siconolfi, C. A., "Fire Retardancy in Chemical Process Ductwork," *Fire Technology*, Vol. 5. No. 3, Aug. 1969, pp. 217-224.

Smith, Edwin E., "Evaluation of the Fire Hazard of Duct Materials," *Fire Technology*, Vol. 9. No. 3., Aug. 1973, pp. 157-170.

Smith, T., "How Correct Lubrication Can Minimize Risk of Air Compressor Explosions," *Fire*, Nov. 1981, p. 341.

Talbot, G., "Static Electricity," *Fire*, 70(871), 1979, pp. 397-398.

WASTE HANDLING SYSTEMS AND EQUIPMENT

Lawrence G. Doucet

Although typical solid waste materials are rarely a source of ignition, they are a potential source of fuel for fires or ignition from other sources. Therefore, a proper, efficient waste management system incorporating prompt treatment and disposal is essential for good firesafety. Waste management systems include the collection, transport, storage, processing, and final disposal of waste materials. Waste processing systems include such equipment as incinerators, compactors and shredders. This chapter covers the firesafety aspects of waste handling and disposal systems and equipment.

Additional information on other equipment and procedures relating to waste handling systems will be found in Section 4, Chapter 2, "Explosions"; Chapter 4, "Theory of Fire and Explosion Control"; in Section 8, Chapter 3, "Heating Systems and Appliances"; Chapter 6, "Miscellaneous Building Services"; and in this Section, Chapter 11, "Special Systems for Explosion Damage Control."

WASTE STORAGE ROOMS

Rooms in a building or structure that are used for the storage or handling of waste in any form and in amounts exceeding three cubic yards (2.3 m^3), uncompacted measure are subject to the requirements of NFPA 82, *Standard on Incinerators, Waste and Linen Handling Systems and Equipment*. The walls, ceilings and floors of these rooms are required to have fire resistance ratings, and room openings must be protected by self-closing fire doors suitable for Class B openings. In addition, these rooms require automatic sprinklers, in accordance with NFPA 13, *Standard for Installation of Sprinkler Systems*, and the provision of a hand hose suitable for reaching all portions of the room.

WASTE CHUTES AND HANDLING SYSTEMS

Waste chutes are generally fixed systems for transporting waste materials from points of generation or interim storage to centralized areas for processing or disposal. There are basically four types of chute systems:

1. *General access, gravity type*—This type of chute consists of an enclosed vertical passageway through which waste is transferred by gravity alone. All occupants of the building have unlimited access to the chute.
2. *Limited access, gravity type*—This type of chute is similar to a general access type except that access is restricted. Entry is gained through locked chute doors or locked service opening room doors. These chutes are used primarily in hospitals and health care institutions.
3. *Pneumatic chute system*—This type of system uses air flow to transport materials from chute service openings to a central collection area. Such systems have chutes which may run horizontally, vertically or inclined, depending upon blower and design characteristics. High velocity air, which typically transports waste materials at speeds of about 60 mph (97 km/hr), is aspirated through the top of the chute and exhausted at its termination. These systems are usually found in hospitals but they can be used in any size or type of building. Figure 12-14A shows a diagramatic view of a pneumatic chute system.
4. *Gravity-pneumatic chute system*—This type uses a conventional gravity chute to feed a collecting chamber which in turn feeds a pneumatic chute system.

Waste Chute Design and Construction

Waste chute design and construction are critical to firesafety in waste handling systems. Chutes could readily serve as channels capable of carrying smoke and flames throughout a building or facility. They are often considered as waste storage areas since they could become clogged with waste during use and present fire and smoke hazards. Criteria relative to chute construction, chute enclosures, fire dampers, sprinklers, service openings and the like are all directed towards minimizing and controlling potential fire hazards. In this regard, the requirements of NFPA 82 must be strictly adhered to.

Mr. Doucet is a Consulting Engineer with the firm of Doucet & Mainka, P.C., Peekskill, NY.

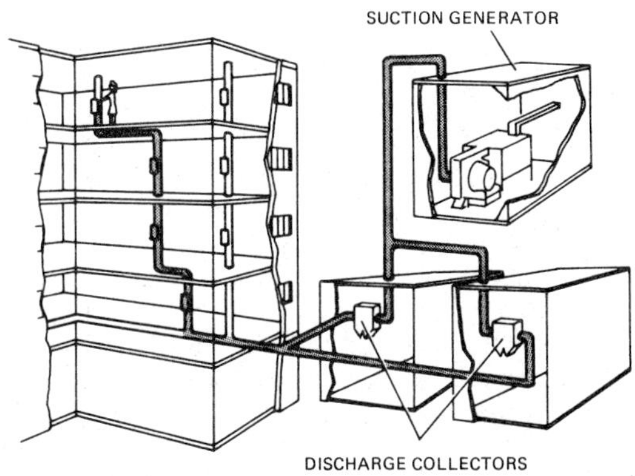

FIG. 12-14A. A diagrammatic representation of a building-wide pneumatic chute system.

Gravity type chutes may be constructed of unlined steel, refractory lined steel or masonry, while pneumatic waste handling systems are only constructed of unlined steel. Unlined chutes are fabricated of stainless, galvanized or aluminized steel for corrosion resistance. Each type of chute construction has separate design and construction criteria, as detailed in NFPA 82, for providing structural integrity combined with adequate firesafety measures.

Chute Size: Gravity type chutes are required to be not less than 22 ½ by 22 ½ in. (570 by 570 mm) or not less than 24 in. (610 mm) diameter. Pneumatic waste handling system chutes are required to be not less than 16 in. (410 mm) diameter but may be smaller if all waste materials are first shredded. However, these sizing criteria are strictly minimum requirements. Waste types and volumes, waste package sizes, user habits and specific applications must be carefully considered before accepting minimum sizing for chutes and service openings. For example, a kitchen compactor may produce a typical package measuring 16 × 16 × 9 in. (570 × 570 × 230 mm), which would have a recommended chute diameter of about 36 in. (915 mm).

Chute Service Openings: Properly sized chute service openings are necessary to prevent chute clogging. Oversized openings increase the probability that overly large or excessive quantities of waste materials will be loaded into the chute and become clogged. Standards for maximum service opening sizes, developed to prevent such conditions, are as follows:

1. General access gravity chute service openings are limited to one third of the cross-sectional area of a square chute or 44 percent of the cross-sectional area of a round chute.
2. Limited access gravity type chutes are limited to two thirds of the cross-sectional area of the chute.
3. Pneumatic chute service openings may be equivalent to or greater than the cross-sectional area of the chute. However, pneumatic system service entrances, or charging stations, comprise a compartment with electrically interlocked inner and outer doors. The capacity of the charging compartment, with only one door being allowed open at a time, limits the volumes or sizes of waste materials which can be loaded into the chute.

Chute Doors: Chute service openings are required to be equipped with fire-rated door enclosures to prevent or minimize the spread of chute fires. Otherwise, natural air flows and pressures could cause flames and smoke to be exfiltrated into other building areas. According to NFPA 82, chute doors must also be self-closing and positive latching types. As noted, pneumatic system service openings are required to have two electrically interlocked doors. For structural integrity, door frames must be securely fastened to both the chute and chute enclosure walls.

Chute Enclosures and Penetrations: In order to insure fire integrity, chute system enclosures, including floor and wall penetrations, require fire-rated protection. Metal chutes, which in themselves offer no fire resistance, must be totally enclosed within walls of masonry or other noncombustible, fire rated material. However, masonry chutes meeting the fire resistance requirements of NFPA 82 are not required to be enclosed. If a masonry chute penetrates a floor without closing the opening entirely, the space between the floor and chute should be filled with material equivalent to the floor in fire resistance.

Wherever a waste chute penetrates a fire rated floor and ceiling assembly, fire wall or other fire rated assembly, a fire damper is required. However, a fire damper may not be needed if the chute is enclosed within a fire rated shaft. Schematic diagrams of typical chute system fire rating requirements are shown in NFPA 82.

Chute Discharge and Terminal Rooms: Waste chutes are required to discharge or terminate in an enclosed room or bin separated from other parts of the building. Since a chute terminal enclosure area is subject to many potential ignition sources, the walls, ceiling and floor of the room or bin must have a fire resistance rating equivalent to that required for the chute. Also, such terminal enclosures are required to have automatic fire doors, automatic sprinklers, a hand hose adequate to reach all areas of the room and a ventilation system designed to exhaust fire and smoke to the outdoors. These requirements are in recognition of the seriousness of fires from waste materials, which are generally difficult to control and typically generate large quantities of smoke.

In gravity-pneumatic type chute systems, the collecting chambers, which can be considered interim discharge or terminal points, are vulnerable to potential fires. They should be adequately sprinklered and provided with ready access for fire fighting measures. Also, the valve room of the system should be sprinklered to aid in cooling the space.

Older gravity chute systems typically discharged directly into an incinerator, and in some systems, the waste chute also served as the incinerator chimney. Because of the tremendous potentials for fire and smoke hazards, most of these older installations have been discontinued or abandoned. Today, waste chutes are not permitted to discharge directly into in incinerator. Furthermore, because of the inherent difficulties in maintaining fire integrity, systems for automatically transferring waste materials from chute terminal enclosures directly into an incinerator are not recommended.

Chute Sprinklers: According to NFPA 82, chute systems must either be sprinklered or capable of containing a fire and venting products of combustion. Metal chutes require automatic sprinklers at the top and at alternate floor levels. Specific requirements for chute sprinklers are in NFPA 13, *Standard for the Installation of Sprinkler Systems*. A good practice is to locate sprinklers where they can be inspected and maintained and yet be out of reach of vandals and beyond the range of falling waste materials. Masonry chutes, or chutes constructed in accordance with requirements for factory built, medium heat appliance chimney sections are not required to be sprinklered.

Chute Service Opening Enclosures: For many years it was general practice to put chute service openings in public corridors for convenience sake. However, aside from the problems of unsightliness and poor sanitation, fire hazards were seriously increased and the fire integrity of a building reduced. Fatalities occurred when chute fires spilled into corridors and cut off means of escape. NFPA 82 requires every service opening to be located in a room or compartment separated from other parts of the building by an enclosure or room having fire rated wall and ceiling assemblies and with the opening into the enclosure protected by a self-locking fire door.

INCINERATORS

Incineration is basically a high temperature combustion process for the treatment and disposal of waste materials. Properly designed and operated systems can reduce the weight and volume of most domestic, commercial, institutional and industrial solid waste by as much as 95 percent or more, thereby greatly reducing off-site haulage of and disposal costs. Incinerators are often used for the destruction and sterilization of biological and pathological wastes at health care facilities, universities and the like. More and more industrial facilities are using on-site incineration for the destruction of hazardous and toxic waste materials. In addition, incineration with waste heat recovery has become an attractive and cost-effective alternative energy source for many facilities. In terms of firesafety, incineration provides prompt and complete waste destruction which substantially reduces interim, on-site waste storage requirements and associated potential fire hazards.

Many early incinerator designs comprised little more than enclosed, single chamber fire boxes in which uncontrolled burning took place. Such systems provided poor waste destruction and were notorious sources of smoke and odor problems. However, modern technology offers systems which are readily capable of meeting demanding waste destruction and performance requirements in compliance with stringent environmental regulations. In addition, properly designed, constructed and operated systems, equipment and support facilities can be completely safe in terms of firesafety.

Incineration Firesafety: The very nature of incineration presents potential fire hazards: storage and handling of combustible, often highly volatile materials; the presence of high temperature, flaming combustion; exhaust and ducting of high temperature combustion gases; handling of hot ashes and the presence of fuel handling and burning equipment.

Incineration firesafety must take into account not only the more obvious measures such as the presence of sprinklers and room fire rating but also less apparent, indirect measures as related to facility planning and design, such as system layout, equipment orientation, waste flow patterns, system operations, and maintenance practices.

Incinerator Categories

Domestic Incinerators: As the name implies, this category of incinerator is used for burning domestic, or household, waste generated in one and two family dwellings. Because of relatively high costs and various operating problems and nuisances associated with these systems, they are generally used only where municipal waste collection services are inadequate or unavailable.

Domestic incinerators are considered gas appliances. They may be either self-contained, factory built units or constructed integral within a masonry wall or chimney enclosure space. A typical self-contained domestic incinerator is shown in Figure 12-14B.

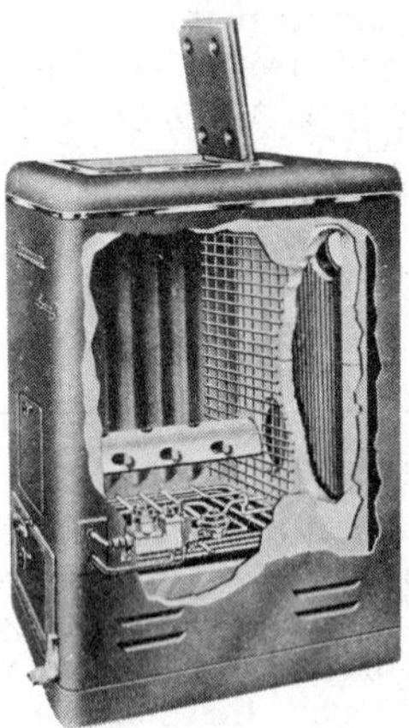

FIG. 12-14B. A typical gas fired domestic incinerator.

Domestic incinerators basically comprise a single combustion chamber and a gas burner for ignition and maintaining temperatures upwards of 1,300°F (705°C). Wastes are typically stored in the incinerator and batch burned when design loading capacity is reached.

Commercial-Industrial Incinerators: This incinerator category applies to commercial and industrial facilities as well as institutional facilities, office buildings and residential complexes. In all, there are more than a dozen different types of commercial-industrial incineration systems for burning hazardous and nonhazardous solids, sludges, liquid and gaseous wastes. However, most of these types are specialized designs for incinerating industrial waste chemicals, byproducts and residues. These special industrial systems are discussed in Chapter 49 of the NFPA *Industrial Fire Hazards Handbook*, 2nd Edition. The most widely and extensively used types of commercial-industrial incinerators are multiple chamber designs, used since the 1940s, and controlled air incineration systems, developed in the early 1960s.

Multiple Chamber Incinerators: As shown in Figure 12-14C, these consist of a primary and one or more secondary combustion chambers. They are designed to operate with high excess air quantities, typically in the 200 to 300 percent range, in order to achieve optimum waste destruction while maintaining burning temperatures in the range of 1,400 to 1,800°F (760 to 980°C). Unfortunately, these

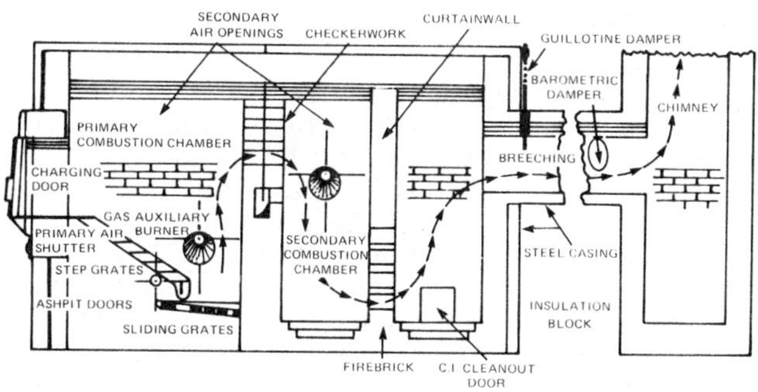

FIG. 12-14C. Typical commercial-industrial, multiple chamber incinerator.

high excess air levels also tend to entrain particulate matter with the flue gases. Multiple chamber incinerators usually require gas cleaning systems, such as scrubbers, for compliance with air pollution control regulations.

Controlled Air Incinerators: These operate basically on a two-stage combustion process. Wastes fed into the first stage, or primary chamber, and burned under starved air conditions are destroyed primarily through a pyrolytic reaction. This process results in dense smoke and pyrolytic products, consisting largely of volatile hydrocarbons, carbon monoxide and other products of combustion which pass to the second stage, or secondary chamber. Here additional air is injected to complete combustion, either spontaneously or through the addition of supplementary fuel. Primary chamber combustion reactions and velocities are maintained at low levels by the starved air conditions in order to minimize particulate entrainment and carryover. Therefore, air pollution control devices are usually not required with controlled air incinerators. A typical controlled incinerator is shown in Figure 12-14D.

Design and Construction

Incineration systems and equipment, including breechings and stacks, are subject to extremely severe operating conditions. These include very high and widely fluctuating temperatures, thermal shocks from wet material, residues which cause slag, clinkers and spalling of refractory linings, explosions from items such as aerosol cans, corrosive attacks from acids and other chemicals, and mechanical scraping and abrasions from metallic wastes as well as operating and cleanout tools. Incineration systems must be suitably designed, constructed, structurally supported and reinforced to resist cracking, warping or distortions which may be caused by these severe conditions. Such failures could allow flames and hot gases to escape into building areas and create dangerous fire hazards.

Outside surface, or skin, temperatures must also be limited by proper incinerator design and construction. Incinerator casings accessible to operators must be insulated or shielded to prevent a temperature rise exceeding 70°F (21°C) over ambient levels, while operating doors and handles must be designed for a maximum 40°F (10°C) rise if metallic and 60°F (15°C) rise if nonmetallic.

Explosion relief protection is also a requirement. Incinerators must be equipped with a suitable relief door or panel with an effective area of 1 sq ft per 100 cu ft (1 $m^2/100\ m^3$) of primary combustion chamber volume.

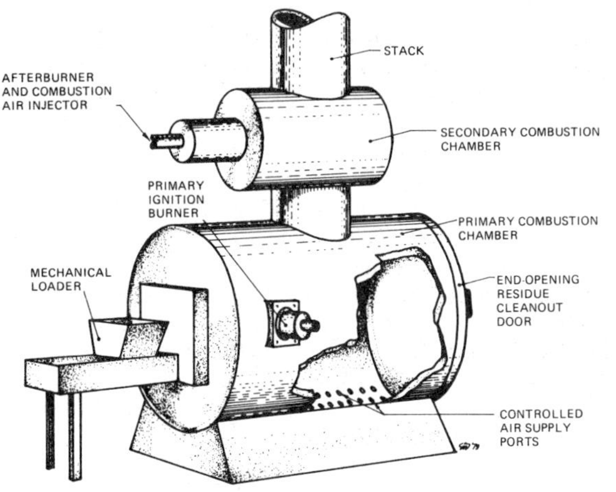

FIG. 12-14D. Cutaway of a controlled air incinerator.

Location and Arrangement

The system should be well planned and implemented to assure that (1) waste and residue containers do not block charging and cleanout operations or access to work areas, or passageways (2) waste material can be charged in a smooth, efficient manner, (3) all parts of the incinerator, including the ash pit, combustion chambers, and breechings are easily and safely accessible for cleaning, repair, and servicing, and (4) clearances above the charging door and between the incinerator top and sides to combustible materials are in compliance with codes and standards.

Charging Systems

The loading of waste materials into an incinerator has a particularly high fire hazard potential in that it provides a direct interface between high temperature, flaming combustion conditions within the incinerator and surrounding building areas. Inadequate designs and improper operations could permit flames and combustion products to escape and ignite nearby waste materials.

Two methods of incinerator charging are manual loading directly through a charging door and mechanical loading with an automatic airlock charging system.

Direct Manual Loading: This charging method is generally employed on domestic incinerators and commercial-industrial incinerators less than about 500 lbs/hr (227 kg/hr) as well as on various commercial-industrial systems which require loading at infrequent intervals. Fire prevention measures include minimizing charging door opening times and eliminating spillage and blockage. Recommendations in this regard include:

1. Charging door openings large enough to accept the largest or bulkiest waste items without blockage or obstruction. Clear opening areas should not be less than 24 by 24 in. (610 by 610 mm). However, excessively oversized openings must be avoided.
2. Charging door openings and frames that have smooth edges and are free of projections which may catch or hang up waste materials and cause spillage.
3. Charging doors that operate easily and which have

positive latching devices and handles that remain relatively cool. Guillotine charging doors should be counterweighted or motor operated.
4. Interlocks to shut off primary chamber burners when the charging door is opened.
5. Loading trays, where appropriate, in front of charging doors to assist the operators in handling and loading heavy and unwieldy waste materials.

Mechanical Loading: Automatic, mechanical charging systems are generally used on incinerators greater than about 500 lb per hr (227 kg/hr) and where controlled charging rates are needed. However, they are sometimes employed on smaller capacity systems where a continuous airlock interface is required. Mechanical loaders are also used to protect against incinerator overcharging conditions.

The most common type of mechanical charging system is the "hopper/ram loader," as shown in Figure 12-14E. Typically, waste is put into the charging hopper

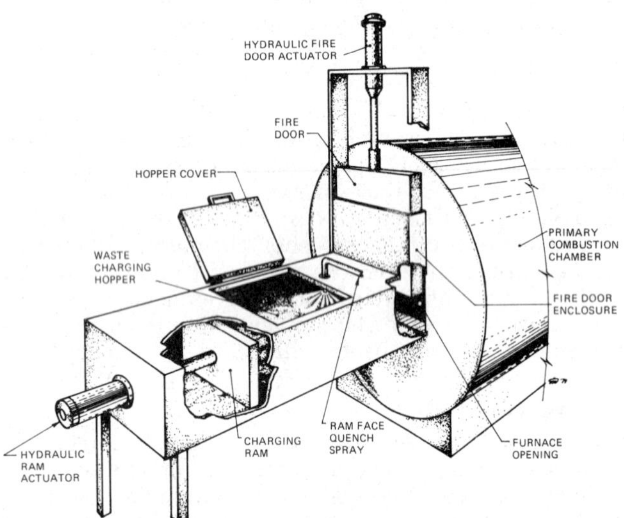

FIG. 12-14E. A typical mechanical loader.

and, upon actuation of a system start switch, the hopper cover closes, the fire door opens, the charging ram injects the waste into the incinerator, the charging ram returns, the fire door closes, and the hopper cover opens for the next loading cycle. Hopper volume and a cycle timer help limit the amount of waste that can be loaded into the incinerator per hour, which helps prevent overcharging. An airlock interface is provided between the incinerator and ambient conditions because the fire door and hopper cover are not both opened simultaneously.

Two potential causes of fire and smoke problems which must be guarded against in hopper/ram loader systems are:

1. Partial closing of fire door: Waste can lodge under the door and prevent it from closing tightly. This waste could ignite and spread fire back into the hopper and beyond. The partially closed door could also significantly affect incinerator draft conditions which in turn could result in heavy smoke escaping under the partially closed door. The fire door should be closed by power, not gravity, in order to seal the opening as tightly as possible should any waste become lodged beneath.
2. Ram ignition hazards: When the charging ram injects waste into the incinerator, its face is directly exposed to high temperatures. Eventually, it could heat up enough to be a potential fire hazard. The hot ram face could cause plastic waste bags and similar materials to melt and adhere to it. If these items do not drop from the ram during its loading cycle, they become ignited and are carried back into the hopper where they will ignite waste remaining in the hopper.

   The charging ram should be cooled either by an internal water circulation or a water spray system which quenches the ram face after every charging stroke. In addition, the loading hopper volume should be such that a single ram stroke completely empties it. This will minimize the risk of waste being in the hopper to catch on fire.

Additional recommended considerations to protect against accidental ignition of wastes in loading hoppers include:

1. Provision of a flame scanner for immediate detection of hopper fires. Actuation of the scanner could sound an alarm, activate a hopper water spray system and/or automatically initiate the loading sequence in order to inject the burning materials into the incinerator.
2. A high-temperature automatic sprinkler directly over the hopper and independent of the room sprinkler system.
3. A manually actuated water spray located above the loading hopper.
4. An emergency switch which would override the normal automatic loading cycle control sequence and cause immediate injection of hopper contents into the incinerator.

Residue Handling and Removal

Residue, or ash, handling and removal systems range from fully manual to fully automatic. Because of high costs, space limitations, operational complexities and technological limitations, most existing older incinerators are cleaned out manually. However, most newer incinerators, and almost all incinerators larger than about 1,000 lb per hr (464 kg/hr), are equipped with an automatic or semiautomatic ash handling and removal system.

Manual Residue Removal: This cleanout method involves raking or shoveling ashes into barrels, bins or tubs. Small capacity incinerators can be cleaned from the outside, but large capacity units may require the operator to enter the primary chamber for cleanout operations. Manual residue removal is objectionable and potentially hazardous in many respects, including: (1) intensive, difficult labor requirements, (2) dangers to personnel from heat, ash, dust and noxious gases, and (3) fire hazards from the handling and containment of unquenched ash residues. Spraying water into the incinerators to facilitate ash removal by quenching is not recommended because the water will damage the furnace refractory due to thermal shock. Water quenching is limited to ash already removed from the furnace.

Automatic Residue Removal: These systems utilize grates, transfer rams or pulse hearths to transfer and discharge ashes from the primary chamber. In semiautomatic type systems, ashes are collected in residue carts. In fully automatic systems, ashes are discharged into a residue conveyor from which they are continuously transferred to a large container for removal. Residue conveyor systems are equipped with a water trough, or sump, which fully saturates the ashes. With cart type systems, the ashes are quenched by water sprays surrounding the discharge chute from the incinerator.

For firesafety protection, it is recommended that ash carts used in both manual and semiautomatic systems be handled with care and, prior to offsite disposal, stored in a fire rated area in a location which is isolated from stored waste materials. Even when quenched by water sprays or hoses, the containers may still be very hot and residues in the ash carts may tend to ignite spontaneously if disturbed and exposed to room air. It is also recommended that considerations be given for dumping ash carts at a landfill rather than within the facility building areas.

Incinerator Chimneys

Incinerator chimneys and high temperature breeching can be of either masonry or refractory lined metal. Design and construction, including clearances, connections, and terminations must be in accordance with applicable codes. Requirements for individual incinerators vary according to secondary combustion temperatures involved. Incinerator chimney terminations should be above roof heights per NFPA 82. In addition they should be equipped with appropriate stack screens.

Domestic incinerators are often connected to conventional chimneys which were designed to withstand temperatures of less than 1000°F (538°C). To protect such chimneys and reduce fire hazards, cooling air must be introduced through a barometric damper to mix with the products of combustion and reduce gas temperatures.

NFPA 82 stipulates that chimneys for commercial-industrial incinerators serve no other purpose. The practice of exhausting incinerators into boiler chimneys or other appliances is not recommended because:

1. Boiler chimneys may not be of suitable construction for the high temperatures associated with incinerators.
2. Boiler chimneys, which may normally operate wet from condensed acidic flue gases, and moisture could rapidly spall and deteriorate when suddenly subjected to high temperature incinerator flue gases.
3. Incinerator flue gases, which vary widely and rapidly, may blow back into boilers or other connected appliances and cause major operating problems.

Auxiliary Fuel Systems

Incinerators require auxiliary, or supplemental, fuel for ignition, warmup and maintaining proper combustion temperatures. Most domestic incinerators fire natural gas, while most commercial-industrial incinerators fire either natural gas or distillate fuel oil. It is imperative that auxiliary fuel systems, including storage and handling systems, burners and controls, be installed and tested in accordance with all applicable requirements of local codes and NFPA standards. Burners should be of a type specifically designed and constructed for use in incineration systems. Burners must also be equipped with automatic ignition systems, and flame safeguard and management systems.

Combustion and Ventilation Air

Depending on the types and heating values of wastes burned, incinerators typically require on the order of 10 to 20 pounds (4.5 to 9 kg) of combustion air per pound (one pound equals approximately 0.45 kg) of waste incinerated. Ventilation air is also required to remove heat which is normally transmitted, or radiated, through incinerator casings in order to maintain comfortable operating conditions and acceptable incinerator room temperature levels. Insufficient supply of combustion and ventilation air may result in operational problems and unsafe conditions. Incinerators located in confined spaces or areas enclosed by tight fitting partitions must be provided an air supply louver or duct in accordance with NFPA 82 requirements. Louvers and ducts providing direct outside air must be sized for one square inch (64.52 mm^2) of free area per 4,000 Btu per hr (1172 kW) of total incinerator heat input. Those providing air from inside building spaces must be sized for one square inch (64.52 mm^2) of free area per 1,000 Btu per hr (293 kW) of total incinerator heat input.

Electrical Service

Power and control wiring, including electrical components and control devices, must be in accordance with applicable local codes and NFPA 70, *National Electrical Code®*. It is also recommended that wiring and conduit not be attached directly to incinerator casings. Wiring within about one foot (0.30 m) of incinerator casings should have high temperature insulation. It is also recommended that incinerator system control panels be located more than 36 in. (0.91 m) from the casing. Wiring, relays, and transformers for auxiliary burners should either be mounted remotely or protected from high temperature damage by the burner-blower fans.

Incinerator Rooms

Domestic incinerators in one or two family dwellings are not required to be in a separated compartment or room provided minimum clearances are maintained. However, all other incinerators, including associated waste and residue handling systems, must be located in a separate, fire rated room or compartment with an automatic fire door suitable for a Class B opening. Such incinerator rooms may also be contained in the same enclosure with the building heating systems and equipment.

Incinerator Operations

To properly manage the severe and complex operating conditions typical of incineration systems, well-trained operating personnel are required. Training programs and comprehensive operating and maintenance manuals and instructions are highly recommended. Such instructions should not only assure proper incinerator operations, but also the continued implementation of safe operating practices.

WASTE COMPACTORS

Waste compactors are devices that use electro-mechanical-hydraulic means to reduce the volume of waste and package it in the reduced condition. The two types of compactors regulated by NFPA 82 are domestic and commercial-industrial.

Domestic Compactors

Commonly called kitchen compactors, domestic compactors are designed for use in dwellings and apartment units for compaction of family developed waste. This appliance reduces the fire hazard of stored waste by retaining it in a metal container under compaction. Units are of the undercounter and moveable types. They must be capable of manual opening such that waste can be easily removed in the event of equipment electrical failure. Figure 12-14F shows a typical domestic compactor.

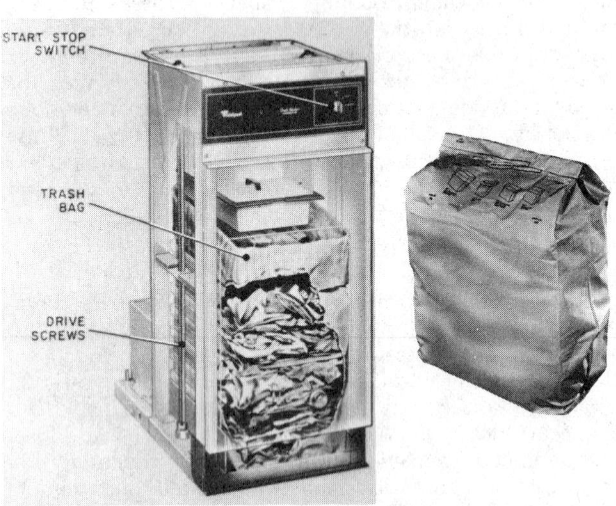

FIG. 12-14F. A domestic compactor unit. Trash is compacted and retained inside the unit in the bag, shown at right, that holds a week's normal household trash for the average family of four. (Whirlpool Corp.)

Commercial-Industrial Compactors

These compactors are used in multiple family dwellings and many other classes of occupancies as the prime waste handling system. They may be located either indoors or outdoors and can be chute or hand fed.

There are four types of compactor systems:

Bulkhead Compactor: Waste is compacted in a chamber against a bulkhead. When a compacted block is ready for removal, a bag is installed and filled with the compacted block.

Extruder: Waste is forced through a cylinder that has a restricted area. Driving forces compact the material and extrude it into a "slug" which is broken off and bagged or placed in a container. The operation of this type of compactor is shown in Figure 12-14G.

Carousel Bag Packer: Waste is compacted into a container in which a bag has been inserted. Various configurations are used. One method utilizes a compaction chamber from which the compacted plug is pushed into the bag. Another uses a cylinder which acts as a compaction chamber that is inserted into the bag. Upon compaction, the cylinder is removed while a compacting head holds the material in place.

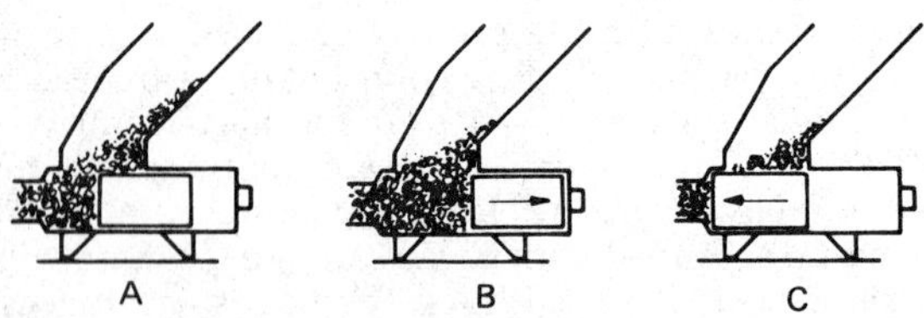

FIG. 12-14G. The sequence of operation of an extruder type refuse compactor. Refuse falling from the chute (A) trips a sensor and starts the compacting cycle. The ram moves backward (B) to let refuse fall into the compactor chamber and then moves forward (C) with more than 40,000 lbs (18 144 kg) of force to force the refuse through a system of restrictors that compacts the refuse and pushes it out of the machine.

Container Packers: Waste is compacted directly into a bin, cart or container. When full, the container can be either manually or mechanically removed from the compactor and compaction area.

Compaction containers must be provided with either a 2 ½ in. (63.5 mm) hose connection suitable for standard fire fighting equipment (and located near the top of the container) or an access door which can be opened without disconnecting the container from the compactor.

Chute Termination Bin

Most compactor waste chutes do not feed directly into the compactor but into a small storage chamber, chute connector or an impact area that is usually large enough to store small quantities of waste. Potential fire hazards can be minimized by installing sprinklers and providing doors large enough to allow access in the event of fire. If a compactor is charged manually from a large bin with an open top, the bin need not be sprinklered. It is sufficient to rely on the sprinklers protecting the compactor room for protection.

The bottom closure of a storage bin (or area) and its ability to be opened under fire conditions deserves attention. If, for instance, there is equipment failure, waste will not only build up in the chute termination chamber but also in the chute itself. In the event of a fire and sprinkler discharge, the weight of soggy refuse could become excessive and jam simple slide devices. Sufficient strength should be built into these closure devices to allow opening them without breakage under such conditions. The most satisfactory situation is one that allows discharge into other receptacles, bins or equipment without excessive chute collection.

Chute fed compactors should be provided with an automatic, special, fine water spray sprinkler in the compaction chamber. However, hand fed compactors do not require such protection.

Compactor Rooms

Compacted material has a wide range of densities and can burn to produce large amounts of smoke. Therefore, storage of compacted materials in buildings should be minimized. Compactor systems are required to be enclosed within fire rated rooms with fire doors suitable for Class B openings. Automatic sprinklers are also required in compactor rooms.

SHREDDERS

Shredders are sometimes used to process waste for volume reduction in order to minimize storage space requirements and to facilitate systems where waste must be mechanically transferred before storage. It is good practice to sprinkler shredder feed bins. If they are serviced by chutes, arrangements must be made to bypass the shredder, since shredders have frequent periods of repair. All bypass areas and storage areas should also be sprinklered and enclosed within fire rated rooms with fire doors suitable for Class B openings. The discharge from shredding devices also requires sprinkler protection.

The possibility of explosion is a particular hazard in connection with shredder operations. This can result from ignition of the dust laden air mixtures which normally surround shredders during operation. A recommended method of protection is to provide an explosion suppressive system within the shredder room and a vent, preferably above the shredder, to safely relieve the impacts of a rapid pressure build-up or explosion.

Bibliography

NFPA Codes, Standards, Recommended Practices and Manuals. (See the latest *NFPA Codes and Standards Catalog* for availability of current editions of the following documents.)

NFPA 13, *Standard for Installation of Sprinkler Systems.*
NFPA 30, *Flammable and Combustible Liquids Code.*
NFPA 31, *Standard for the Installation of Oil Burning Equipment.*
NFPA 54, *Standard for the National Fuel Gas Code.*
NFPA 58, *Standard for the Storage and Handling of Liquefied Petroleum Gases.*
NFPA 68, *Guide for Explosion Venting.*
NFPA 69, *Standard on Explosion Prevention Systems.*
NFPA 82, *Standard on Incinerators, Waste and Linen Handling Systems and Equipment.*
NFPA 86, *Ovens and Furnaces: Design, Location and Equipment.*
NFPA 97M, *Standard Glossary of Terms Relating to Chimneys, Vents and Heat Producing Appliances.*
NFPA 101, *Code for Safety to Life from Fire in Buildings and Structures.*
NFPA 211, *Standard for Chimneys, Fireplaces, Vents and Solid Fuel Burning Appliances.*

Additional Reading

American Gas Association Inc., *Guidebook for Industrial and Commercial Gas-Fired Incineration*, NY, 1963.

Conway, Richard A., and Ross, Richard D., *Handbook of Industrial Waste Disposal*, Van Nostrand Reinhold, NY, 1980.

Doucet, Lawrence, G., *Incineration: State-of-the-Art Design, Procurement and Operations*, paper presented at 71st Annual Meeting of the Association of Physical Plant Administrators of Colleges and Universities, Columbus, OH, June 18, 1984.

Doucet, Lawrence, G., and Knoll, W. G., Jr., *The Craft of Specifying Solid Waste Systems, Actual Specifying Engineer*, May 1974, pp. 107-113.

Engineering Handbook for Hazardous Waste Incineration, *U.S, EPA SW-889*, Environmental Protection Agency, Washington, DC, Sept. 1981.

"Fire Protection for Belt Conveyors," FMEC Loss Prevention Data Sheet 7-11, Factory Mutual Engineering Corp., Norwood, MA.

"Flammable and Combustible Waste Disposal in the Plastics Industry," Bulletin No. 13, Society of the Plastics Industry, NY.

Hitchcock, D., "Solid Waste Disposal: Incineration," *Chemical Engineering*, May 21, 1979, pp. 185-194.

Incinerator Institute of America, *Incinerator Standards*, Falls Church, VA, Nov. 1968.

McColgan, I. J., *Air Pollution Emission and Control Technology: Packaged Incinerators —Economic and Technical Review*, EPS 3-AP-77-3, Canadian Environmental Protection Service, 1977.

Pavoni, J. L., Heer, J. E., Jr., and Hagerty, D. J., *Handbook of Solid Waste Disposal*, Van Nostrand Reinhold, NY, 1975.

Shulz, Joseph F., "Standards for Refuse-Handling in Apartment Houses," *Fire Journal*, Vol. 68, No. 2, Mar. 1974, pp. 82-86.

Solid Waste Handling and Disposal in Multistory Buildings and Hospitals; Volumes I and II, US EPA, SW-34, Environmental Protection Agency, 1972.

Vance, Mary, *Industrial Waste Disposal: A Bibliography*, Vance Bibliographies, Monticello, IL, 1982.

Wilson, D. G., *Handbook of Solid Waste Management*, Van Nostrand Reinhold, NY, 1977.

UPHOLSTERED FURNITURE AND MATTRESSES

Dr. Vytenis Babrauskas and John F. Krasny

Upholstered items (upolstered furniture and mattresses) play a major part in residential fires. It is estimated that there were approximately 80,000 such fires in 1982, causing 1,900 deaths, 10,000 injuries, and $500,000,000 in damage. Approximately 65 percent of these fires were caused by cigarettes. Other major ignition sources were matches, electrical appliances, and hot objects. Cigarette initiated fires lead to approximately 30 percent of residential fire deaths and are the single largest cause of such deaths. However, they account for only 10 percent of the residential fires, and 20 percent of the injuries. Mortality and morbidity in cigarette initiated fires are thus inordinately high. About 40 percent of the victims of cigarette fires are persons other than the smoker.

The need for cigarette ignition resistance of upholstered items is unique—other furnishings such as carpets and curtains, or apparel textile items are rarely if ever, ignited by cigarettes. Unfortunately, cigarette ignition resistance and resistance to flames are not necessarily compatible. Among common upholstery fabrics, medium and heavyweight thermoplastic fabrics (nylon, olefin, polyester) generally resist cigarette ignition but do not resist flame ignition. Similarly, cellulosic materials (cotton, hemp, jute, linen, rayon fabrics and batting) have poor cigarette ignition resistance, but over polyurethane foam they burn at a slower rate than the common combination of thermoplastic fabric and polyurethane foam. Only more expensive fabric/padding combinations, e.g., wool fabric over neoprene padding, combine good cigarette ignition resistance with relatively high resistance to flames.

Many aspects of flammability are similar for upholstered furniture and mattresses. However, the following distinction must be made: mattresses have predominantly flat surfaces which are not as likely to ignite by cigarettes as furniture crevices, i.e., the juncture of seat cushion and armrests and back of upholstered furniture. Similarly, the horizontal flat furniture surfaces are not as easily ignited by flame ignition sources as the vertical surfaces.

Dr. Babrauskas is head of the Furnishings Flammability Program, and Mr. Krasny a Textile Technologist in the Center for Fire Research, National Bureau of Standards, Gaithersburg, MD.

On the other hand, cigarettes on mattresses may be inadvertently covered with sheets, blankets, and/or pillows. This increases the probability of ignition by cigarettes. Also, these intermediary materials may ignite from flames more readily than mattresses, and then expose the mattress to a much more severe fire. Consequently, it is more difficult to develop relevant ignition tests and standards for mattresses than for upholstered furniture.

Upholstered furniture can contain well over a dozen materials, e.g., cover fabrics and fabrics between the cover fabric and the padding (interliners); different padding materials in the seat, sides, and back; weltcords; another fabric and type of padding below the seat cushion (decking); and frames, springs, stiffening for the sides, etc. Mattresses are complex but contain fewer materials. In the ignition process, whether from a cigarette or small flame, the cover fabric and material immediately below it—one or two layers of padding—are important. As the fire progresses, other materials, including the frame and springs become involved, which can affect the manner in which the burning item collapses.

Smoldering of upholstered items, which is the typical consequence of exposure to a burning cigarette, produces enough smoke and toxic gases to incapacitate and kill persons not only in the room of origin but also in adjoining rooms or apartments. Transition from smoldering to flaming does not occur in all cases; when it does, times from placement of the cigarette to flaming of 20 minutes to several hours have been reported (Braun et al 1982; Hafer and Yuill 1970). In a closed room, flaming can revert to smoldering when the oxygen supply is exhausted. With sufficient oxygen, a single upholstered item of sufficient burning rate can lead to room flashover.

Another feature of upholstered item fires is the difficulty of assuring extinction since wetting out or even immersion of the item in water does not assure extinguishment. Items should be removed from the premises, torn apart, and the pieces thoroughly wetted. Many fire victims have wetted down a smoldering area in an upholstered chair or sofa, retired for the night, only to be faced several hours later with impenetrable smoke or a full scale fire.

A recent monograph on upholstered furniture flammability contains an in depth discussion of cigarette and small flame ignition resistance and post ignition behavior (Babrauskas and Krasny 1985). This chapter is essentially a summary of that monograph.

UPHOLSTERED ITEM FLAMMABILITY TESTS

Some cigarette and flame ignition standards and tests for upholstered items are listed in Table 12-15A. They can be classified as: (1) voluntary regulatory standards and tests, (2) federally enforced standards, and (3) state imposed standards.

Voluntary and Regulatory Standards

Voluntary standards are proposed by trade organizations like the Upholstered Furniture Action Council (UFAC) for residential furniture, and the Business and Institutional Furniture Manufacturers Association (BIFMA) for their products. These two standards have also been developed as NFPA 260A, *Standard Methods of Tests and Classification System for Cigarette Ignition Resistance of Components of Upholstered Furniture*, and NFPA 260B, *Standard Method of Test for Determining Resistance of Mock-Up Upholstered Furniture Material Assemblies to Ignition by Smoldering Cigarettes* (hereinafter referred to as NFPA 260A and NFPA 260B. The BIFMA standard was originally developed by the National Bureau of Standards (NBS) for the Consumer Product Safety Commission (CPSC). CPSC decided to hold this standard pending success of the UFAC voluntary approach. A recent report indicates that UFAC has succeeded in improving the cigarette ignition resistance of furniture (Fairall 1984).

Many tests developed by voluntary standards organizations, especially those described in Group C of Table 12-15A, are more rigorous than federal or state standards in that they employ larger ignition sources and/or heating with radiative sources before pilot ignition. These tests often prescribe flame spread and smoke measurements, and are frequently used for research, product development, and regulation of special occupancies by some state and local authorities.

On the other hand, the mattress cigarette ignition standard and regulations are enforced by CPSC on the federal level. The state of California has its own residential and special occupancy regulations; other states may also pass such regulations.

Component or Mockup Tests

Some tests are performed on mockups of furniture or on the actual furniture. Other tests prescribe testing of the individual components, i.e., the cover fabric, padding, weltcord, etc., separately. The component tests do not consider interaction between materials. Sometimes the tests are configured so as to employ "standard materials" for those components not being tested. If that standard material is not the worst case selection, the test results may be difficult to interpret or apply, e.g., upholstered padding in the UFAC –NFPA tests (NFPA 260A and NFPA 260B) is tested while covered with an approximately 480 g/m^2 (14 oz per sq yd) cotton fabric, but many fabrics with higher propensity to ignite from cigarettes are frequently used on furniture.

Cigarette tests specify placement of the cigarette in the most vulnerable configuration, which is the crevice formed by a horizontal and vertical upholstered surface except, of course, for mattress tests. The cigarette is covered with a piece of sheeting. This makes the test slightly more severe and more reproducible, presumably because of exclusion of air currents from the smoldering area.

TABLE 12-15A. Flammability Tests for Upholstered Furniture and Mattresses*

Sponsor	Ignition Source		Status			Test specimen		Occupancy	
	Cigarette	Flame	Voluntary	Regulatory	Proposed	Compon.	Mockup or Actual Item	Res.	Special
A. Upholstered furniture									
UFAC-NFPA 260A	X		X			X		X	
BIFMA-NFPA 260B	X	X†	X				X		X
CPSC	X				X		X	X	
California	X	X		X		X		X	
California		X			X		X		X
British (ISO)	X	X‡		X			X	X	X
B. Mattresses									
CPSC	X			X			X	X	X
Canadian	X			X			X	X	X
California		X		X			X		X
C. Upholstered Furniture and Mattresses (larger flames, radiation)									
ASTM E162, ASTM D3675, ASTM 5903, NFPA 258		X		X§		X	X		X
NY, NJ Port authority		X		X		X			X
MVSS No. 302 (motor vehicles)		X		X			X		X
ASTM E906		X§	X			X	X		X
Cone Cal.		X#			X	X	X		X

* Sources: (UFAC 1984; BIFMA 1980; CPSC 1981; CPSC 1984; CBHF 1980, 1981, 1983, 1984; BSI 1982; CGSB 1977; ASTM 1984a, 1984b, 1984c; GSA 1971; Port Authority 1981; Federal Register 1975; Babrauskas 1982; FAA 1984).

† Small flame test of fabrics and padding.

‡ Seven flame ignition sources of increasing severity, with smallest intended for residential furniture, others for special occupancies (e.g., No. 5, a 17 g wood crib, is specified for special occupancies and children's furniture.

§ Some of these methods are used by state and local authorities for furniture components or assemblies of cover fabric and padding.

Ignitability and rate of heat release rate can be measured.

For flame ignition component tests, ignition is often by a gas flame at the bottom of a vertical specimen, such as foam or fabrics (British test procedure). Flaming ignition in furniture mockups is generally at the crevice location. Ignition sources such as balled up newspaper or three gas flames and four wood cribs (with increasing severity) are used. The smallest British gas flame is intended to simulate a wood match, and furniture which does not pass must be labeled accordingly (however, all British furniture must pass a cigarette test). The same mockup arrangement, with a more severe ignition source, can be used for special occupancies; e.g., children's and contract furniture must pass with ignition source 5, a 17 g (0.6 oz) wood crib. The British test, with some modifications, has been adopted by the International Standards Organization (ISO) and a number of other countries.

Two U.S. flammability tests for transportation seats present a wide range of severity. On the low end of the scale, the Motor Vehicle Safety Standard No. 302 requires a horizontal specimen, and has minimal flame spread requirements. On the other end, a newly adopted Federal Aviation Administration (FAA) test (FAA 1984) specifies a 115 kW, 150 mm (6 in.) diameter kerosene burner directed for two minutes on the side of the seat. The flames must not spread across the seat, no flaming drip must fall, and a maximum weight loss is specified.

Pass/fail criteria in the tests in Group A and B of Table 12-15A are most often expressed in terms of "obvious ignition" and in char length for some cigarette tests and such flame tests as Method 5903. The tests in Group C give quantitative results or ratings: flame spread ratings, ASTM E162 and D3674; smoke density, ASTM 906 and NFPA 258, *Standard Research Test Method for Determining Smoke Generation of Solid Materials* (hereinafter referred to as NFPA 258 and ASTM 906); and heat release rate, ASTM E906 and the cone calorimeter. Like the FAA test, the California institutional test specifies maximum permissible weight loss and temperature, as well as CO and smoke concentrations in the test room.

The cigarette used in all U.S. smoldering ignition tests is a filterless, 85 mm commercial cigarette, (Pall Mall brand). It is representative of many other cigarette brands in its propensity to ignite. However, in recent years it has become widely known that a few commercial cigarette brands have a relatively low propensity to ignite (Krasny et al 1981; Macaluso 1983). This has lead to the submission to the U.S. Congress and about a dozen state legislatures of bills that would require cigarettes to conform to yet to be developed ignitability standards. In 1984, Congress passed H.R. 1880, the "Cigarette Safety Act," which calls for a study of the feasibility of making cigarettes with a low propensity to ignite, without increasing tar and nicotine. As presently understood, this propensity does not relate to the rate of burning or burning cone temperature of cigarettes, but to complex interaction between the tobacco column, the cigarette paper, and the surface on which the cigarette rests. This propensity can apparently be reduced by lowering tobacco packing density and cigarette diameter (both of which reduce fuel loading and tar and nicotine delivery) and, possibly, by changing the relative rate of burning of the tobacco column and cigarette paper, and perhaps other parameters.

CHARACTERISTICS OF IGNITION SOURCES

Table 12-15B shows the characteristics of some ignition sources which have been used in flammability experimentation with upholstered furniture items. The sources are shown ranging four orders of magnitude—5 to 50,000 W. The peak heat fluxes for all the sources, except the methenamine pill, range between 15 and 42 kW/m^2. What differs mainly when ignition source strength is increased is not the peak incident flux, but rather the area over which

TABLE 12-15B. Characteristics of Furniture Ignition Sources*

	Typical Heat Output (W)	Burn Time† (s)	Maximum Flame Height (mm)	Flame Width (mm)	Maximum Heat Flux kW/m^2
Cigarette 1.1 g (not puffed, laid on solid surface)					
bone dry	5	1200	—	—	42
conditioned to 50% R.H.	5	1200	—	—	35
Methenamine pill, 0.15 g	45	90			4
Match, wooden (laid on solid surface)	80	20–30	30	14	18–20
Wood cribs, BS 5852 Part 2					
No. 4 crib, 8.5 g	1000	190			15#
No. 5 crib, 17 g	1900	200			17#
No. 6 crib, 60 g	2600	190			20#
No. 7 crib, 126 g	6400	350			25#
Crumpled brown lunch bag, 6 g	1200	80			
Crumpled wax paper, 4.5 g (tight)	1800	25			
Crumpled wax paper, 4.5 g (loose)	5300	20			
Folded double-sheet newspaper, 22 g (bottom ignition)	4000	100			
Crumpled double-sheet newspaper, 22 g (top ignition)	7400	40			
Crumpled double-sheet newspaper, 22 g (bottom ignition)	17,000	20			
Polyethylene wastebasket, 285 g, filled with 12 milk cartons (390 g)	50,000	200‡	550	200	35§

* Sources: (Behnke 1969; Moulen and Grubits undated; Tu and Davis 1976; Paul and Clevely 1983; Paul 1981; and unpublished NBS data).
† Time duration of significant flaming.
‡ Total burn time in excess of 1800 sec.
§ As measured on simulation burner.
# Measured from 25 mm away.

the flux is applied. In the case of the wastebasket, the area over which fluxes exceed 20 kW/m^2 is about 60 by 700 mm (2.36 by 27.55 in.); for a match this would be approximately 10 by 30 mm (0.39 by 1.18 in.). Higher heat fluxes are required to raise small target areas to ignition temperature than large areas. In addition, one may consider ignition as occurring when the lower flammability limit of pyrolysate gases is reached. This is determined both by the pyrolysate mass flux rate from the surface and by the entrainment and mixing conditions. The latter have not been studied as a function of the heated area size.

FLAME SPREAD

Flame spread has long been the basis of flammability evaluation of building materials; however, little quantitative work has been done on the flame spread over upholstered items. Cotton and rayon cover fabrics by themselves were shown to have flame spread rates inversely proportional to their weight (Lee and Wiltshire 1972); however, when tested over polyurethane padding, the differences between flame spread rates were small. This is another case where component tests can give rise to misleading results, as mentioned earlier.

In another experimental series, the flame spread over seat cushions of chair mockups was measured by counting the number of 100 mm (3.93 in.) squares covered by flames as a function of time from ignition (Krasny and Babrauskas 1984). This flame spread rate was found to correlate fairly well with bench scale measurements on the same fabric/padding composites. There were major differences, however, in the behavior of charring and melting (thermoplastic) fabrics. Cotton and rayon, the cellulosic fabrics, charred ahead of the flame, and on the side of the chair opposite to the one which was burning. Eventually the char burst into flame. The thermoplastic fabrics initially shrank away from the ignition source, but being supported by the foam, the shrunken material bead burned and presented a relatively large ignition source for the foam padding. The fabric on the chair side opposite the burning side ablated, exposing the foam to the radiation. Thus flames spread more rapidly over the thermoplastic fabric covered mockups, even when the foam was neoprene and did not significantly flame during the test.

General flame spread theory has not been much applied to upholstered items, primarily because of the variety of configurations and interaction between fabrics and padding. Dimensions in these items are sufficiently large so flame spread and heat release behavior are dominated by radiative, rather than convective, mechanisms. Some studies of flame spread in the predominant "against the wind" mode, i.e., in the opposite direction to the draft induced by the buoyancy of the flame have been conducted (Rockett 1974).

HEAT RELEASE RATE

In the last few years, the importance of the heat release rate for prediction of flashover potential has been realized. Several methods to measure it have been suggested. One early technique is to directly measure the sensible enthalpy of the fire gas outflow. However, this requires corrections be made for the heat losses to the room walls; these losses vary widely for various wall materials, as well as depending on the heat release rate of the specimen. A method based on this principle, the Ohio State University (OSU) calorimeter, has been used widely.

Since the late 1970s, the oxygen consumption principle for measuring heat release rate has come to the forefront. The principle states that the combustion heat released for most combustible species is proportional to the amount of oxygen consumed to within about ± 5 percent. The proportionality constant is 13.1 × 10^3 kJ per kg of O_2 consumed. By contrast, the heat of combustion of a fuel can vary widely and is given as the heat released per kg of fuel. Furniture calorimeters, large scale instruments in which furniture can be tested, have been built in a number of laboratories. A bench scale instrument, the cone calorimeter (so called because of the cone shaped heater used to irradiate the specimen) has more recently been developed. The OSU method can also be modified to use this principle of heat release measurement.

There are methods to translate, within limitations, small scale heat release rate results on upholstery fabric/padding combinations to flashover potential of furniture in a room. For any one combination, the peak heat release rate measured in an upholstered furniture calorimeter was found to be proportional to cone calorimeter results, as follows:

$$\dot{q}_{fs} = 0.63\ (\dot{q}''_{bs})\ (\text{mass factor})\ (\text{frame factor})\ (\text{style factor})$$

where

$\dot{q}_{fs}$ (kW) = the full scale peak heat release result measured in the furniture calorimeter

$\dot{q}''_{bs}$ (kW/m^2) = the bench scale heat release rate measure in the cone calorimeter under 25 kW irradiance and averaged over 180 s after ignition occurs

mass factor = total combustible mass, kg

frame factor =
- 1.7 for noncombustible frames
- 0.6 for melting plastic frames
- 0.3 for wood frames
- 0.2 for charring plastic frames

style factor =
- 1.0 for plain, primarily rectilinear construction
- 1.5 for ornate, convoluted shapes [with intermediate values for intermediate shapes]

The style factor accounts for certain geometrical flame spread effects on the heat release rate. The frame factor reflects both combustibility and structural integrity effects on fire development.

This rule does not work for fabric constructions with a very low rate of heat release, e.g., heavy cotton fabrics over neoprene foam, since rapid, complete combustion does not occur, and the mass factor overestimates the output for the relatively dense neoprene. Also, the rule is not applicable to types of constructions that lack a conventional structural frame, due to lack of adequate test data. Further experiments could refine this model and extend it to different types of furniture.

Furniture calorimeter results, obtained with unlimited air supply, have been validated against those obtained in room burns with various degrees of air supply limitations. Figure 12-15A shows typical rate of heat release time curves from such experiments, which indicate that no systematic influence of room ventilation conditions was seen. The ability to utilize furniture calorimeter data to represent heat release rates in room fires extends only to the fuel limited regime. Fires with such a small ratio of ventilation to heat release rate as to have no excess oxygen in the combustion gas stream ("ventilation limited") be-

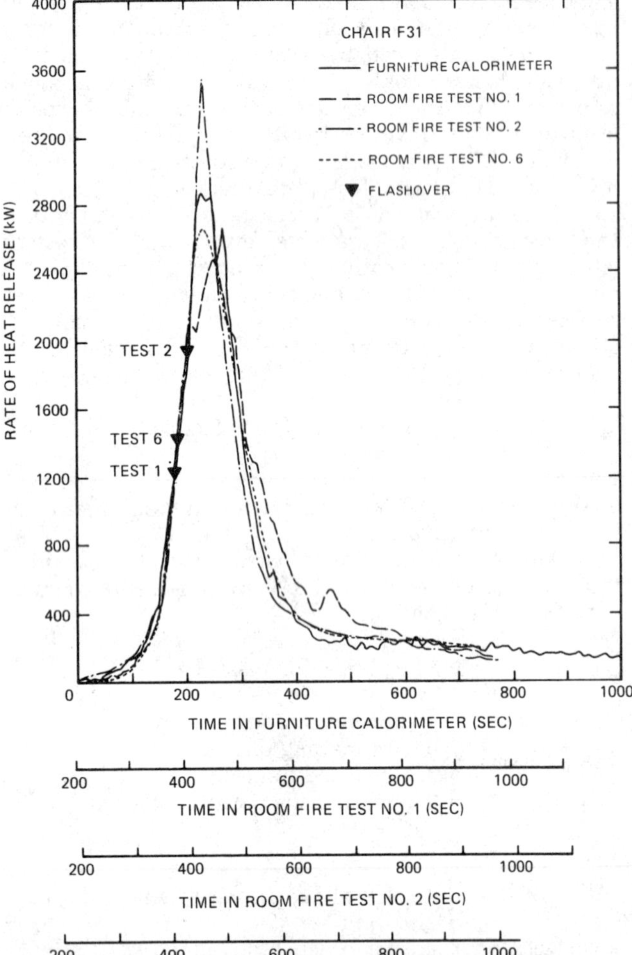

FIG. 12-15A. Comparison of furniture calorimeter and room fire results for a loveseat with Olefin cover fabric, flame retardant polyurethane foam, and a wood frame (three room fires, with varying ventilation openings).

cause of such a small ratio of ventilation to heat release rate, cannot be so simply represented and as yet, have not been quantified for upholstered furniture.

Prediction techniques for computing flashover potential are discussed in Section 21, Chapter 5, "Room Fire Temperature Calculations." In addition to those methods for estimating the potential for flashover caused by a single upholstered furniture item, a few studies have addressed the tendency of one item to ignite other furnishings (Palmer et al 1976; Babrauskas 1982). Irradiances measured at 0.5 m (19 in.) from the edge of chair mockups ranged from 15 to 45 kW/m^2 for combinations of ordinary and flame resistant polyurethane foam covered by four types of fabrics. The upper range is more than sufficient to ignite other upholstered furniture items and many other furnishings, such as wastepaper baskets and wooden chairs placed at that distance. Since distances between furnishings cannot be regulated, flame resistance of all furnishings in special occupancies such as hospital rooms, is important.

SMOKE AND TOXIC GASES

Human incapacitation or fatalities in fires are usually due to smoke and toxic effects rather than to direct burn injuries. Thus, a logical way of characterizing fire hazard would be according to the time for incapacitation. This approach is natural for analyzing standardized room fires, given suitable measurements of toxic gases and a pertinent set of tenability criteria. Such an approach was taken for mattresses (Hafer and Yuill 1970; Babrauskas 1977), where tenability criteria for CO, CO_2, O_2 depletion, heat flux, and smoke were established and applied. This approach remains appealing for cases where a fixed compartment scenario is to be involved. Transportation vehicles are a natural example; recent evaluations of aircraft cabin performance have been assessed in terms of available escape time. In the evaluation of movable furnishings, however, there is no unique possible compartment configuration. However, a recently developed California test for high risk, high density occupancy furniture prescribes pass/fail criteria for smoke and CO concentrations as measured in a fixed room (CBHF 1984).

The most tractable model for toxic gas evolution postulates that a fixed fraction of the specimen mass is realized as any given gas species (or smoke particulates). Thus, the mass rate of production of species is

$$\dot{m}\ (t) = \dot{m}\ (t)\ \Gamma x$$

Here the production of any quantity (x = soot, CO, etc.) is expressed as the fraction, Γx, of specimen mass loss that becomes the species x, multiplied by the specimen mass loss rate in (t). The concentration at any point in a room can then be calculated if airflow rates are known and suitable assumptions on mixing (or stratification) are made. The assumption that the fraction of specimen mass becoming a given gas species is reasonable where fires are not ventilation limited. When fires do become oxygen limited, the most noticeable effect is a rise in CO production, as oxygen is depleted. This has not yet been quantified for upholstered furniture.

The effect of smoke obscuration on lethality is not direct, nonetheless, it is real in that escape can be hindered or precluded where visibility does not exist. The conversion of specimen mass into soot mass is, just as for toxic gas species, dependent upon ventilation and other effects. An examination of the limited available data, however, suggests that the assumption of a constant soot mass ratio is not a bad one. Bench scale measurements of smoke have typically been made in the NBS smoke chamber, where an optical beam attenuation is measured. These optical measurements can be related to the soot particulate mass fraction, Γs, by empirical relationships. Newer tests are being developed that allow dynamic measurements of beam attenuation, along with direct determination of Γs.

TABLE 12-15C. Upholstered Furniture Components Listed in Approximate Order of Descending Ignition Resistance

Ignition Resistance	Cover Fabric	Padding	Interliners	Welt Cords	Construction Parameters
A. CIGARETTE IGNITION RESISTANCE		Specialty foams†	Aluminized fabrics	Aluminized	Flat areas
HIGH	Wool, PVC	Polyester batting	Neoprene sheets	PVC	Flat areas near welt cord
↑	Heavy thermoplastics		Vinyl coated glass fab.		
		SR PU	Novoloid felts	Thermoplastics	Tufts
	Cellulose/thermoplastics blends (depending on thermoplastic percentages)	SR cellulosic batting	Thermoplastic fabrics	SR treated cellulosics	Crevices
		Untreated PU	Cellulosic fabrics	Cellulosics	
		Mixed fiber batting			
	Light thermoplastics	Cellulosic batting			
↓	Light cellulosics				
LOW	Heavy cellulosics				
B. SMALL FLAME IGNITION RESISTANCE AND FIRE GROWTH					
HIGH	FR wool	Specialty foams‡	Aluminized, gas imperme-able fabrics	Effect of welt cord has not been investigated, believed minor	Flat areas
↑	Wool, PVC coated cellulosics‡	FR cellulosic batting			Vertical areas
	Cellulosics‡	FR PU	Neoprene sheets‡		Corner areas
		Cellulosic batting	Novoloid fabrics§		
	Thermoplastics	Polyester batting			
		Untreated PU	Aramid fabrics‡§		
		Latex foam	Vinyl coated glass fabrics‡		
			FR cellulosic fabrics‡§		
↓			Cellulosic fabrics†§		
LOW			Thermoplastic fabrics		

SR—smolder resistant; FR—flame resistant; PU—polyurethane foam.

* Data on the behavior of acrylic are sparse but it seems to act more like cellulosics (smolder) than thermoplastics.

† Neoprene; combustion modified, high resiliency PU.

‡ Heavier materials have higher ignition resistance and generally higher heat release and lower flame spread rate.

§ Fabrics here include woven, knitted and rewoven structures.

MATERIAL RANKING

Table 12-15C summarizes the results from a wide variety of studies on the cigarette ignition and small flame ignition resistance of upholstered items. The materials in the table are listed from top to bottom in the approximate order of decreasing resistance to ignition for cigarette ignition source (Part A) and in order of decreasing resistance to small flames and increasing fire growth after ignition (Part B). There is considerable overlap between these characteristics of materials listed near each other, depending upon such factors as density, amount of smolder resistant or flame retardant agent, fabric finish and backcoating, and other factors.

It should again be emphasized that some materials with good cigarette ignition resistance do not necessarily have good small flame ignition resistance, and vice versa, e.g., thermoplastic fabrics tend to resist cigarette ignition because some of the heat transferred to the fabric is consumed in melting the fibers. On the other hand, cigarettes induce smoldering in medium to heavyweight cellulosic fabrics, with consequent heat transfer to the padding. However, the thermoplastics shrink, melt, and curl away from an open flame and expose the padding, while cellulosic fabrics char and, until the char breaks, protect the padding. Similarly, some flame retardants for polyurethane foam reduce resistance to smoldering, while others enhance both flame and cigarette ignition resistance.

To prevent ignition by cigarettes, it is not necessary to use only the materials listed on top of each column in Table 12-15C, e.g., the combination of medium weight thermoplastic with ordinary polyurethane foam, or the combination of light to medium weight cellulosic fabric with a layer of polyester batting over the polyurethane foam, have a low probability of cigarette ignition. Many wool and medium to heavy PVC-coated fabrics can be used with flame resistant or ordinary polyurethane foam or mixed batting. Material combinations which are, for all practical purposes, cigarette ignition resistant in a crevice configuration, can be chosen on the basis of a few trials in a qualified laboratory. The number of such trials can be held small by using the information from the table.

Unfortunately, no such simple scheme can be recommended for assuring the flame resistance of materials listed in Table 12-15C (Part B). Here even the materials listed on top of each column will ignite if the flame is large enough or applied long enough, or both. However, Part B may be helpful in choosing material combinations with a low probability of ignition from matches, e.g., and which may have lower rates of flame growth when ignited with a larger source.

As examples for use of Table 12-15C (Part B), medium to heavyweight wool and vinyl coated cellulosic fabrics (vinyl coated thermoplastics open up and expose the padding to the flame) have been shown to resist match ignition, regardless of padding material, but ignite with larger flame exposure. Heavy cotton fabric over neoprene padding burned and then smoldered, while the neoprene did not enter the combustion process except for smoldering in the crevice. (Smoldering of neoprene, like that of cotton batting, is hard to extinguish and is best done by immersing torn up pieces in water for prolonged periods.) Total rates of heat release and combustion product formation were negligible after the fabrics stopped burning. Aircraft seats constructed with wool/nylon fabric, an aluminized aramid fabric interliner, and polyurethane foam self-extinguished even after a severe flame exposure for two minutes. This type of interliner is used because it is gas impermeable and prevents pyrolysis gases formed in the urethane foam from reaching the flames (cushion venting is provided in areas not likely to be in the fire path). A polychloroprene type foam sheet interliner (Vonar) was also effective in making these seats self-extinguishing. Thus, relatively firesafe upholstered items are possible, albeit expensive. Aluminized interliners are also effective in increasing cigarette ignition resistance, Vonar perhaps somewhat less so.

Some highly flame resistant treated polyurethane foams have almost as good flame resistance as neoprene but do not smolder. They can be considered for such occupancies as prisons, where cover fabrics and interliners can be expected to be cut open in arson attempts. Cotton batting treated with boric acid is widely used in prison mattresses.

For any combination of fabric and padding material, cigarette ignition resistance is better in flat areas than in crevices. Tufted areas also may have lower cigarette ignition resistance than flat areas, and are usually tested separately, as are areas near the welt edge outside the crevices.

Flames seem to start faster when ignition is on a vertical furniture surface, especially a corner, rather than on a horizontal surface. Some furniture item fires are started by children playing with matches on the underside of furniture items. Making the bottom from thermoplastic fabric which ablates readily, has been suggested as a remedy. All visible furniture surfaces are usually covered by one fabric, but the padding frequently varies greatly in one furniture item: firm foam in the seat; less firm foam or batting in the back, and even less firmness on the sides. On the outside, there may be only light padding, or the fabric may not be in contact with the padding at all. For proper firesafety design, each of these material combinations then needs to be evaluated.

The ranking of materials during fully involved flaming is approximately the same as that given in Table 12-15C. There is little information of the effects of furniture configuration. However, observations suggest that flame spread over curved and convoluted surfaces can be slower initially, but may be appreciably accelerated when larger, radiatively driven flames come to dominate. Peak heat release rate per unit mass, was found to increase when simulated furniture geometry was changed from a single cushion to two, three and four sides of a cube, with ignition in the enclosed space. When cushion thickness was varied, it was found that the peak rate of heat release was not quite proportional to specimen mass, but was somewhat higher, per unit mass, for thinner cushions.

A limited amount of work has been done on the effect of the furniture frames. Near the time of the peak burning rate, the manner in which the frame starts to collapse becomes important. A polyurethane frame produced the lowest burning rate, followed by a wood frame, while a polypropylene frame melted early in the fire (Babrauskas 1983). Wood frames fail at the frame connections, and here metal connections seem to perform worse than glued ones.

Total fuel content of the furniture item can be used to roughly estimate the burning time. A method was recently developed based on this, where the heat release rate curve is approximated as a triangular form (Babrauskas and Krasny 1985).

The amount of smoke data for upholstered furniture items is rather small. In one series of mattress tests, bench scale smoke data were validated against full scale measurements. Because of significant mass differences, these values tend to be determined largely by the padding and not the fabric. Latex foam constituted the worst case, with 20 percent of the specimen mass becoming smoke particulate. Polyurethane foams typically yield 10 to 15 percent smoke, although some specimens yield as low as 5 percent, and, with cotton ticking fabric, 0.5 percent. Neoprene smoke production also depends upon foam formulation; low smoke formulations are commercially available. Polyester and polyester/cotton battings show low values, similar to pure cotton batting. Flame retarded foam generally released more smoke than ordinary polyurethane foam, but at a somewhat lower rate.

Among the fabrics, wool tended to release relatively little smoke at a low rate. Next in order were cellulosic fabrics, thermoplastic fabrics, and worst, PVC cover fabrics. Similarly, the order from best to worst padding was wool, cotton and other vegetable fibers, and polyurethane foams. Fire barrier interliners reduced and delayed smoke release. Modacrylic fabrics produced high smoke release results; aramid fabrics yielded low smoke release results. Acrylic/cellulosic blends seemed to be worse than all-cellulose fabrics.

The major toxic products expected to be found in upholstered item fires are CO, CO_2, HCN, HCl, and NO_x. Depletion of oxygen also has a toxic effect. Perhaps more than in other fields, there is no agreement on relevant test conditions and evaluation of results. It is quite widely accepted that significant differences in toxic effects are expressed by differences in order of magnitude in LC_{50}. On that basis, the differences between most upholstery materials are not significant. However, materials may show a better toxic effect behavior by showing a slower burning rate, not just lower per mass burned toxicity. Based on such a fuller understanding of toxic effects, differences in furniture materials can be seen, even though the LC_{50}s do not differ by an order of magnitude.

CONCLUSIONS

The fire hazard due to upholstered furniture items has to be considered in the context of all relevant parameters. These parameters include:

1. Resistance to smoldering (cigarette) and flame ignitions (not necessarily the same for some typical upholstery materials).
2. Flame spread.
3. Heat release rate (related primarily to flashover propensity in a room) and irradiance (related to ignition of nearby furnishings and walls).
4. Smoke development.
5. Toxic effects due to heat, oxygen depletion, and products such as CO, CO_2, HCl, HCN, NO_x, etc.

Many present test methods for these properties were designed before a total concept approach was seen to be necessary. They need to be reviewed, and modified or replaced by more up to date ones where necessary. The present trend to more intensely look beyond the fire resistance of the building construction and toward that of furnishings will provide major contributions to the overall reduction of fire losses.

Bibliography

References Cited

ASTM. 1984a. "Test For Surface Flammability of Materials Using a Radiant Heat Energy Source." *ASTM E162*. American Society for Testing and Materials, Philadelphia, PA.

ASTM. 1984b. "Test for Surface Flammability of Flexible Cellular Materials Using a Radiant Heat Energy Source." *ASTM D3675*. American Society for Testing and Materials, Philadelphia, PA.

ASTM. 1984c. "Standard Test Method for Heat and Visible Smoke Release Rates for Materials and Products." *ASTM E906*. American Society for Testing and Materials, Philadelphia, PA.

Babrauskas, V., and Krasny, J. F. 1985. "Fire Behavior of Upholstered Furniture," Monograph MN-173. National Bureau of Standards, Gaithersburg, MD.

Babrauskas, V. 1983. "Upholstered Furniture Heat Release Rates; Measurements and Estimation." *Journal of Fire Sciences*. Vol 1. pp 9-32.

Babrauskas, V. 1982a. "Will the Second Item Ignite?" *Fire Safety Journal*. pp 287-292.

Babrauskas, V. 1982b. "Development of the Cone Calorimeter—A Bench Scale Heat Release Rate Apparatus Based on Oxygen Consumption." *NBSIR 82-2611*. National Bureau of Standards, Gaithersburg, MD.

Babrauskas, V. 1977. "Combustion of Mattresses Exposed to Flaming Ignition Sources, Part I. Full-Scale Tests and Hazard Analysis." *NBSIR 77-1290*. National Bureau of Standards, Gaithersburg, MD.

Behnke, W. 1969. "Cigarette Study; Heat Flux and Ember Temperatures." Document, presented to ASTM Committee D-13.

BIFMA. 1980. *Standard and Institutional Furniture Manufacturer's Association First Generation Voluntary Upholstered Furniture Flammability Standard for Business and Institutional Markets*. Business and Institutional Furniture Manufacturer's Association, Grand Rapids, IA.

Braun, E. et al. 1982. "Cigarette Ignition of Upholstered Chairs." *Journal of Consumer Product Flammability*. Vol 9. pp 167-183.

BSI. 1982. "Fire Tests for Furniture, Part I:" "Methods of Test for the Ignitability by Smoker's Materials of Upholstered Composites for Seating, Part II," 1979: "Methods of Test for the Ignitability of Upholstered Composites for Seating by Flaming Sources," 1982: *British Standards BS 5852*. British Standards Institution, London, England.

CBHF. 1984. "Flammability Test Procedure for Seating Furniture for Use in High Risk and Public Occupancies." *Technical Bulletin No. 133*. California Bureau of Home Furnishings, North Highlands, CA.

CBHF. 1983. *Flammability Information Package*. California Bureau of Home Furnishings, North Highlands, CA.

CBHF. 1980. "Flammability Test Procedure for Mattresses for Use in High Risk Occupancies." *Technical Bulletin 121*. California Bureau of Home Furnishings, Sacramento, CA.

CGSB. 1977. *Method of Test for the Combustion Resistance of Mattress; Cigarette Test*. Canadian Government Specification Board, Ottawa, Ontario, Canada.

CPSC. 1984. Code of Federal Regulations, Part 1632. *Standard for the Flammability of Mattresses (And Mattress Pads)*. Final Rule. Oct. 10, 1984. U.S. Consumer Product Safety Commission, Washington, DC.

CPSC. 1981. Code of Federal Regulations, Part 1633. *Proposed Standards for the Flammability (Cigarette Ignition) of Upholstered Furniture*. U.S. Consumer Product Safety Commission, Washington, DC.

FAA. 1984. "Flammability Requirements for Aircraft Seat Cushions." *Code of Federal Regulations 14 CFR Parts 25, 29, 121*. Federal Aviation Administration, Washington, DC.

Fairall, P. 1984. "Analysis of CPSC 40 Chair Test Program." US Consumer Product Safety Commission, Washington, DC.

Federal Register. 1975. "Flammability of Interior Materials—Passenger Cars, Multipurpose Passenger Vehicles, Trucks, and Buses." Motor Vehicle Safety Standard No. 302. *Federal Register*. Vol 35.

GSA. 1971. "Flame Resistance of Cloth; Vertical Methods 5903.2." *Federal Test Standard No. 191*. General Services Administration, Washington, DC.

Hafer, C. A., and Yuill, C. H. 1970. "Characterization of Bedding and Upholstery Fires." Southwest Research Institute, San Antonio, TX.

Krasny, J. F., and Babrauskas, V. 1984. "Burning Behavior of Upholstered Furniture Mockups." *Journal of Fire Sciences*. Vol 2. pp 205-231.

Krasny, J. F. et al. 1981. "Development of a Candidate Test Method for the Measurement of the Propensity of Cigarettes to Cause Smoldering Ignition of Upholstered Furniture and Mattresses." *NBSIR 81-2363*. National Bureau of Standards, Gaithersburg, MD.

Lee, B. T., and Wiltshire, L. W. 1972. "Fire Spread Models of Upholstered Furniture." *Journal of Fire and Flammability*. Vol 3. pp 164-175.

Macaluso, C. 1983. "Cigarette Ignition Studies." Testimony before the Subcommittee on Health and the Environment, Committee on Energy and Commerce on March 21, 1983. US House of Representatives.

Moulen, A. W., and Grubits, S. J. undated. "Paper Ignition Sources Used to Examine the Fire Behavior of Seating and Bedding." *Technical Record 474*. Experimental Building Station. Department of Housing and Construction. Chatswood, N.S.W., Australia.

Palmer, K. N. et al. 1976. "Fire Hazard of Plastics in Furniture and Furnishings; Fires in Furnished Rooms." Building Research Establishment. Boreham Wood, Hertfordshire, England.

Paul, K. T., and Clevely, W. P. 1983. *Unpublished Studies*. Rubber and Plastics Research Association of Great Britain.

Paul, K. T. 1981. "Development and Evaluation of Improved Wooden Cribs for Ignitability Standards in Upholstery and Related Specifications." *Unpublished Report*. Building Research Establishment. Boreham Wood, Hertfordshire, England.

Port Authority. 1981. "Specifications Governing the Flammability of Upholstery Materials and Plastic Furniture." The Port Authority of New York and New Jersey, NY.

Rockett, J. A. 1974. "Mathematical Modeling of Radiant Panel Test Methods." *Fire Safety Research Journal (SP 411)*. National Bureau of Standards, Gaithersburg, MD.

Tu, K. M., and Davis, S. 1976. "Flame Spread of Carpet Systems Involved in Room Fires." *NBSIR 76-1013*. National Bureau of Standards, Gaithersburg, MD.

UFAC. 1984. *Important Consumer Safety Information from UFAC*. Upholstered Furniture Action Council, High Point, NC.

NFPA Codes, Standards, Recommended Practices and Manuals. (See the latest *NFPA Codes and Standards Catalog* for availability of current editions of the following documents.)

NFPA 258, *Standard Research Test Method for Determining Smoke Generation of Solid Materials*.

NFPA 260A, *Standard Methods of Tests and Classification System for Cigarette Ignition Resistance of Components of Upholstered Furniture*.

NFPA 260B, *Standard Method of Test for Determining Resistance of Mock-Up Upholstered Furniture Material Assemblies to Ignition by Smoldering Cigarettes*.

Additional Readings

Alarie, Y., et al., "Toxicity of Smoke During Chair Smoldering Tests and Small Scale Tests Using the Same Materials," *Fundamental and Applied Toxicology*, Vol. 3, 1983, pp. 619-626.

Alarie, Y. C., and Anderson, R. C., "Toxicological and Acute Lethal Hazard Evaluation of Thermal Decomposition Products of Synthetic and Natural Polymers," *Tox. Appl. Pharm.*, 1979, pp. 341-362.

Anderson, B., and Magnusson, S. E., "Fire Behavior of Upholstered Furniture—An Experimental Study," *Report No. R80-4, TVBB-0055*, Lund Institute of Technology, Lund, Sweden, 1982.

Babrauskas, V., et al., "Upholstered Furniture Heat Release Rates Measured With a Furniture Calorimeter," *NBSIR 82-2604*, National Bureau of Standards, Gaithersburg, MD, 1982.

Babrauskas, V., and Krasny, J. F., "Prediction of Upholstered Chair Heat Release Rates From Bench-Scale Measurements," (ASTM STP), to be published.

Babrauskas, V., "Upholstered Furniture Room Fires—Measurements, Comparison With Furniture Calorimeter Data, and Flashover Predictions," *Journal of Fire Sciences*, Vol. 2, 1984, pp. 5-19.

Babrauskas, V., "Performance of the Ohio State University Rate of Heat Release Apparatus Using Polymethylmethacrylate and Gaseous Fuels," *Fire Safety Journal*, Vol. 5, 1982, pp. 9-20.

Babrauskas, V., "Applications of Predictive Smoke Measurements," *Journal of Fire and Flammability*, Vol. 12, Jan. 1981, pp. 51-64.

Babrauskas, V., "Combustion of Mattresses Exposed to Flaming Ignition Sources, Part II, Bench-Scale Tests and Recommended Standard Test," *NBSIR 80-2186*, National Bureau of Standards, Gaithersburg, MD, 1980.

Birky, M., et al., "Study of Biological Samples Obtained From Victims of MGM Grand Hotel Fire," *Journal of Anal. Toxicology*, Vol. 7, 1983, pp. 265-271.

Blankenbaker, S., "Cigarettes and Electric Heating Elements as Ignition Sources," *Proceedings of the Thirteenth Annual Meeting*, Information Council on Fabric Flammability, Atlanta, GA, Dec. 1979.

Clarke, F. II, and Ottoson, J., "Fire Death Scenario and Firesafety Planning," *Fire Journal*, Vol. 70, May 1976, pp. 20-22, 117, 118.

Eighth Annual Flammable Fabrics Report, US Consumer Product Safety Commission, Dec. 1980.

Fang, J. B., "Fire Buildup in a Room and the Role of Interior Finish Materials,"*Tech. Note 879*, National Bureau of Standards, Washington, DC, 1975.

Fitzgerald, W. E., and Kanakia, M., "Calorimetric Evaluation of the Role of Fire Retardants in Selected Polymers," Fire Retardant Chemicals Association *Conference Workshop*, Mar. 12-15, 1978, Technomic, Westport, CT, pp. 21-30.

Fitzgerald, W. E., "Quantification of Fires: 1. Energy Kinetics of Burning in a Dynamic Room Size Calorimeter," *Journal of Fire and Flammability*, Vol. 9, Oct. 1978, pp. 510-527.

Harwood, B., and Kale, D., *Fires in Upholstered Furniture*, US Consumer Product Safety Commission, Washington, DC, May 1980.

Heskestad, G., "A Fire Products Collector for Calorimetrey into the MW Range," *FMRC J. I. OC2E1.RA*, Factory Mutual Research Corp., Norwood, MA, 1981.

Hill, R. G., et al., "Aircraft Seat Fire Blocking Layers: Effectiveness and Benefits Under Various Scenarios," *DOT/FAA/CT-83/43*, Federal Aviation Administration, Atlantic City, NJ, 1984.

Holmstedt, G., and Kaiser, I., "Brand i Vordbaddar," *SP-RAPP 1983:04* Statens Provningsanstalt, Boras, Sweden, 1983.

H. R. 1980. *Cigarette Safety Act of 1984*. Oct. 30, 1984.

Huggett, C., "Estimation of Rate of Heat Release by Means of Oxygen Consumption Measurements," *Fire and Materials*, Vol. 4, 1980, pp. 61-65.

Kaplan, H. L., et al., *Combustion Toxicology: Principles and Methods*, Technomic Publishing Co., Lancaster, PA, 1983.

Krause, R. F., and Gann, R. G., "Rate of Heat Release Measurements Using Oxygen Consumption," *Journal of Fire and Flammability*, Vol. 11, Apr. 1980, pp. 117-130.

Levin, B. C., et al., "Conditions Conducive to the Generation of Hydrogen Cyanide From Flexible Polyurethane Foam," *7th Joint Meeting*, U.S. Japan Panel on Fire Research and Safety, Washington, DC, Oct. 1983.

Levin, B. C., et al., "An Acute Inhalation Toxicological Evaluation of Combustion Products From Fire Retarded and Non-Fire Retarded Flexible Polyurethane Foam and Polyester," *NBSIR 83-2791*, National Bureau of Standards, Gaithersburg, MD, Nov. 1983.

Levin, B. C., et al., "Further Development of a Test Method for the Assessment of the Acute Inhalation Toxicity of Combustion Products," *NBSIR 82-2532*, National Bureau of Standards, Gaithersburg, MD, 1982.

Parker, W. J., "Calculations of the Heat Release Rate by Oxygen Consumption for Various Applications," *NBSIR 81-2427*, National Bureau of Standards, Gaithersburg, MD, 1982.

Prager, F. H., and Wood, J. F., *Full-Scale Investigation of the Fire Performance of Upholstered Furniture*, Part I, (RAPRA 4), International Isocyanate Institute, 1979.

Quintiere, J. G., "An Assessment of Correlations Between Laboratory and Fire-Scale Experiments for the FAA Aircraft Fire Safety Program, Part 1: Smoke," *NBSIR 82-2508*, National Bureau of Standards, Gaithersburg, MD, 1982.

Spears, A. W., "A Technical Analysis of the Problems Relating to Upholstered Furniture and Mattress Fires Relative to Proposed Cigarette Legislation Including a Review of Relevant Patents," *Testimony*, Subcommittee on Health and the Environment, Committee on Energy and Commerce, March 21, 1983, US House of Representatives.

Tewarson, A., "Identification of Fire Properties Relevant to the Prediction of Fire Growth," *Paper RC83-TP-11*, Factory Mutual Research Corp., Norwood, presented at 7th UJNR Panel Meeting, Washington, DC, Oct. 1983.

Thomas, P. H., "Some Conduction Problems in the Heating of Small Areas on Large Solids," *Quart. J. Mech. and Applied Math*, Vol. 10, 1957, pp. 482-493.

Wooley, F. D., and Fardell, P. J., "The Prediction of Combustion Products," *Fire Research*, Vol. 1, 1977, pp. 11-21.

Yockers, J. R., and Segal, L., "Cigarette Fire Mechanisms," *NFPA Quarterly*, Vol. 49, No. 3, Jan. 1956, pp. 213-22.

SECTION 13
TRANSPORTATION FIRE HAZARDS

AVIATION

Revised by James J. Brenneman, P.E.

Among the many factors that contribute to aviation firesafety are the requirements that the aircraft be airworthy and firesafe. The hazards of certain maintenance practices must be recognized and proper control measures applied so operating personnel, aircrews, cabin attendants, and ground mechanical and servicing personnel are well trained in normal and emergency conditions.

Airborne and ground navigational aids, weather, air traffic control, proper design of runways with adequate "overrun" areas free of hazardous obstructions, and proper maintenance of this equipment and facilities are of extreme importance; however, they are beyond the scope of this chapter. This chapter discusses only the problem of fire in aircraft in the operational and ground environment modes.

Additional information on aviation fire hazards may be found in Section 5, Chapter 4, "Flammable and Combustible Liquids"; Section 19, "Special Fire Suppression Agents and Systems"; and Section 20, "Portable Fire Extinguishers."

AIRCRAFT FIRESAFETY

Firesafety in aircraft starts on the drawing board, and aeronautical engineers bear the brunt of responsibility for fire prevention and control. Aircraft require large quantities of fuel, lubricating oils, hydraulic fluids, and Class A combustibles, which are in proximity to potential ignition sources such as power plants, auxiliary power units, electrical systems, and heaters. Many aircraft carry oxygen systems (liquefied and gaseous) and some use oxidizers for auxiliary power units or are equipped with small rocket units for additional takeoff thrust. Consequently, aircraft firesafety requires a skillful blending of reasonable safeguards that are lightweight and do not interfere unduly with the use and mission of the particular aircraft.

Although most nations that have a major aircraft production industry also have extensive regulations governing the use and arrangement of the foregoing items, it is extremely important that the designer be thoroughly versed in the letter and intent of these regulations.

Installed fire detection and extinguishing equipment are normally required for those aircraft areas that possess inherent fire hazards and ignition potentials. Other fire prevention techniques used are separations of systems containing flammable fluids from potential ignition sources, compartmentation of fire hazard areas from critical structural components and flight control systems, and judicious use of other materials that are fire and heat resistive or of very low flammability. Firesafety must receive a high priority because aircraft fires in flight present tragic life hazard potentials. Fires following impact accidents—which can result in the loss of many lives—are also an aircraft design concern.

In commercial, light general aviation, and military reciprocating engine aircraft, the lightweight structure needed to allow the aircraft to perform its mission also permits impact energy to be transmitted to the aircraft occupants. Thus, most deaths occur due to impact trauma than fire.

In modern turbine powered aircraft which operate in a more demanding flight envelope, structures are strengthened by the use of new or heavier gage metals, advanced technology composites, and different design techniques. This aircraft structure absorbs much of the impact energy without transmitting this energy to the occupants. Subsequently, statistics of accidents involving this type of aircraft disclose that the postcrash fire has become the predominant cause of serious injuries and death to the occupants.

As a result of aircraft accident investigations, much research has been conducted, and is still in progress, on methods of minimizing aircraft fatalities. Improved seat belt and buckle design, automatically inflated door mounted evacuation slides, increased numbers of exits and exit system tests and certifications, pretakeoff safety briefings, and removal of potentially dangerous projecting knobs and other items from seat backs are developments which have greatly enhanced the chances of survival in an aircraft accident.

Mr. Brenneman is the Fire Protection Engineer for United Airlines, Inc., San Francisco, CA.

Many methods are being studied to reduce the postcrash fire hazard. They include, but are not limited to, the following:

1. Segregating flammable fluid containers and systems from ignition sources.
2. Improving methods of fuel containment.
3. Modifying the physical characteristics of fuel to reduce probability of ignition in ground impact accidents.
4. Reducing the rate of evaporation and the speed of flame spread over the surface of the spilled fuel.
5. Improving the materials used for interior decor, insulation, sound attenuation, and cushioning to reduce ease of ignition, flame spread, smoke, and toxic gas generation.
6. Compartmentation with lightweight fire walls in the occupied portions of the cabin, blind spaces, lavatories, and trash containers.
7. Onboard cabin fire suppression systems using water or a halogenated extinguishing agent.
8. Early warning fire/smoke detection systems in occupied portions of the fuselage.
9. Emergency lighting systems installed in the lower portion of the cabin to enhance evacuation in a smoke filled environment.
10. Inerting fuel tank vapor space to reduce the possibility of explosion from static electrical discharge and lightning strikes. This can be effective in flight but will do little to contain the postcrash fire, because the inert atmosphere would be lost almost immediately on impact, with the disruption of the tank structure.

Some methods of fuel modification, such as gelling or emulsifying to thicken the fuel, have shown promise; however, they have proven impractical because ground handling of thickened fuel and pumping within the aircraft system created almost insurmountable problems. Fuel modification research, at present, is concentrating on the antimisting additives. A recent large scale crash test disclosed that even this technique may have limitations. Since present antimisting additives must be injected into the fuel at the point of entry into the aircraft, and precise injection ratios are necessary, ground handling of this type fuel is also a problem. In addition, treated fuel is incompatible with untreated fuel; and the fuel reacts with any free water in the fuel or fuel tank pumps to form jellylike lumps.

AIRCRAFT POWER PLANTS

Civil aircraft in the United States are subject to extensive federal regulation through the U.S. Department of Transportation (DOT), Federal Aviation Administration (FAA). The Code of Federal Regulations (CFR), Title 14, "Aeronautics and Space," contains the Federal Air Regulations. Other nations have similar requirements. The following list summarizes the regulations that apply to aircraft power plants:

1. All reciprocating engines, auxiliary power units, fuel burning heaters, or other combustion devices intended for operation in flight must be separated from the aircraft by fire walls, shrouds, or equivalent means so that no hazardous quantities of air, fluid, or flame can pass from these compartments to other portions of the aircraft. These regulations also apply to the combustion, turbine, and tailpipe sections of turbine engines. The fire walls and shrouds must be made of a material that will withstand heat at least as well as steel. When applied to power plants, the material must perform "under the most severe conditions of fire and duration likely to occur in such zones." All openings in fire walls and shrouds must be sealed with close fitting fire resistive grommets, bushings, or fittings. In reciprocating engine nacelles, a fire seal or bulkhead is used to isolate the engine power section and exhaust system from the engine accessory section. In addition, a main fire wall segregates the complete engine assembly (power section, exhaust system, and accessory section) from the remainder of the nacelle section, and as applicable, the wheel well. In gas turbines, the fire seal or bulkhead separates the combustion, turbine, and tailpipe section from the compressor and accessory section. An additional main fire wall isolates the engine assembly from the support pylon or remainder of the aircraft, as applicable.
2. The cowling and nacelle skin are designed to prevent fire from circumventing the fire seals and main fire walls.
3. Tanks containing a flammable fluid cannot be located in a fire zone except when it can be proved that construction, connecting lines, controls, and shutoff means provide protection equivalent to segregation. A specified air gap is required between flammable fluid tanks and a fire wall, and materials that can absorb flammable fluids are prohibited from the area of the tank or any other system containing flammable fluids.
4. Emergency fuel shutoffs are required for each engine, auxiliary power unit, or combustion heater. The emergency shutoffs must be fire resistive or located so the operation will not be affected by a fire in any fire zone. Operation of the emergency shutoff shall not affect other emergency functions, nor the operation of any other engine.
5. Flammable liquid lines in fire zones shall be fire resistive and where required, flexible. This also applies to drain and vent lines for flammable fluids or vapor.
6. Reciprocating and turbine engine air inlets must be arranged safely to prevent backfire flames from entering the fire zone, and designed so discharge from vents and drains cannot enter the air induction system.
7. Exhaust systems must discharge in a safe manner, not expose any portion of a flammable fluid system, and be provided with heat shields wherever they may impinge on other portions of the aircraft.
8. Drains must be provided for all fire zones with the discharge arranged so that drained fluids will not be reingested into any portion of the aircraft.

AVIATION FUELS

The fire hazard properties of aviation fuels are identified according to susceptibility or ease of ignition, flash points, flammability limits, distillation range (initial and end boiling point), and electrostatic susceptibility. (Note: Octane rating has no relation to the degree of fire hazard of a fuel.) Table 13-1A summarizes the characteristics of the more common aviation fuels.

TABLE 13-1A. Summary Data on the Fire Hazard Properties of Aviation Fuels

Characteristics	Gasoline AVGAS	Kerosene Grades JET A JET A-1 JP-5, JP-6 JP-8	Blends of Gasoline and Kerosene JET B and JP-4
Freeze Point*	−76°F	−40°F −58°F	−60°F
Vapor Pressure† (Reid-ASTM D323-58)	5.5 to 7.0 psi	0.1 psi	2.0 to 3.0 psi
Flash Point* (By Closed-Cup Method at Sea Level)	−50°F	+95°F to +145°F	−10°F to +30°F
Flash Point* (By Air Saturation Method)	−75°F to −85°F	None	−60°F
Flammability Limits			
Lower Limit	1.4%	0.74%	1.16%
Upper Limit	7.6%	5.32%	7.63%
Temp. Range for Flam. Mixtures*	−50°F to +30°F	+95°F to +165°F	−10°F to +100°F
Autoignition Temperature*	+825°F to +960°F	+440°F to +475°F	+470°F to +480°F
Boiling Points*			
Initial	110°F	325°F	135°F
End	325°F	450°F	485°F
Pool Rate of Flame Spread‡§	700–800 fpm	100 fpm (or less)	700–800 fpm

Note: Figures vary for some of these values in different data sources. Those shown herein are average figures based on the latest available information.

* $\frac{5}{9}$ (°F−32) = °C

† psi = 6.894 kPa

‡ 1 fpm = 0.3 m/min

§ In mist foam, rate of flame spread in all fuels is very rapid.

Two basic differences which must be considered when evaluating fuel in an operating aircraft and fuel stored in a tank at a refinery or bulk plant are:

1. Because an aircraft travels at different altitudes and through varying ambient temperatures, the conditions of flammability in the tank vapor space can change rapidly. Thus, a fuel type such as Jet A—which is normally too lean in a tank vapor space at sea level with a fuel temperature of 70°F (21°C)—can move into the flammable range as the aircraft gains altitude. Other factors contribute to this change, such as aircraft skin heating from air friction, outgasing of dissolved oxygen, and sloshing of the fuel from air turbulence. Figure 13-1A illustrates changes in the flammability limits of aircraft fuels in the tank vapor space due to altitude and temperature changes.
2. After an impact when major structural damage occurs to aircraft fuel tanks, the fuel may be released as a mist due to forward momentum, splashing, and wind shearing. Regardless of the type of fuel involved, this mist is easily ignitable from disrupted electrical circuits, hot engine surfaces, or ignition sources on the ground. The resulting fireball then acts as the ignition source for other combustibles in the area, including pools of high flashpoint Jet A fuel. If antimisting additives can be successfully developed, they are intended to reduce misting and ease of ignition, or if ignition does occur, to prevent the propagation of flame throughout the mist cloud. In some aircraft accidents where deceleration forces are low, liquid fuel flowing from ruptured fuel tanks or broken fuel

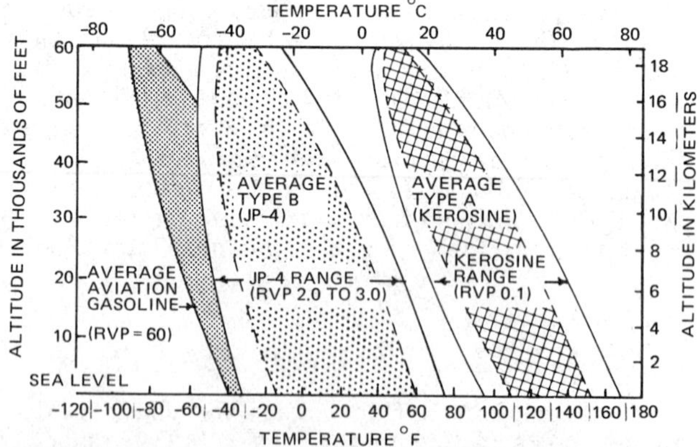

FIG. 13-1A. Flammable ranges of aviation gasoline and Jet A (kerosene) and Jet B (JP-4) turbine fuels, showing variations with altitude. According to laboratory studies, some of the sea level flammable limits shown on this chart for Jet A and Jet B turbine fuel do not agree exactly with Table 13-1A. The test data used to develop this chart were taken from a single source; the table combines and averages data from all available sources.

lines has been vaporized and ignited by hot engine surfaces, hot brakes, heavy electrical arcs, etc. Thus antimisting additives solve only a portion of the postcrash fuel spill fire problem.

Relative Safety of Jet Fuels

Jet A (kerosene grade) turbine fuels apparently offer a safety advantage over other types of aviation fuels, especially during fueling operations and aircraft fuel systems maintenance. Studies indicate there is a measurable reduction of the incidence of fire in impact survivable accidents when the aircraft involved had Jet A fuel in the tanks.

However, once ignition occurs, all fuels exhibit similar behavior, and control measures must be instituted in the shortest time possible to prevent injury and fatalities.

In some portions of the world, Jet A (kerosene grade) fuel is not readily available, and Jet B (JP-4) is used. Jet B is easier and more economical to produce. Some nations also have reduced the minimum flashpoint of Jet A to increase yield of aviation fuel per barrel; the minimum proposed flashpoint has been in the 80°F (27°C) range. Flashpoint testing of fuels as delivered from these nations, has shown the actual flashpoint to be above 100°F (38°C). However, if a worldwide fuel shortage develops, wider use of reduced flashpoint kerosene (RFK) and Jet B can be expected.

Electrostatic Susceptibility

The degree to which a static charge may be acquired by aviation fuels depends upon many factors: the amount and type of residual impurities, dissolved water, the linear velocity through piping systems, and the types of filters and water separators used. Jet A and Jet B are better static generators than AVGAS. While all fuels generate static charges, the electrical conductivity of the fuel bears a direct relationship to the speed of charge relaxation (dissipation).

Antistatic additives have been developed. However, rather than preventing the formation of static charges, they actually increase fuel conductivity, thus considerably shortening the relaxation time. Antistatic additives increase static charge since, by their nature, they are an impurity in the fuel. The charge generation is increased as the fuel passes through filter/water separators and other equipment. If an unbonded object is in the vicinity of the charge generation point before the charge has an opportunity to relax, this object can act as a collector and cause a high energy static discharge.

Although possibly misnamed, antistatic additives have greatly increased the safety of aircraft fueling operations, in all but extenuating circumstances (e.g., the unbonded collector in close proximity to the static generation point).

AIRCRAFT FUEL SYSTEMS

Most aircraft make extensive use of the internal wing volume to store fuel. In larger aircraft, the wing structure is sealed and forms the fuel tank. This is commonly called integral tank, or wet wing construction. Older and light aircraft may incorporate a flexible bladder to contain the fuel within the wing structure, using the wing structure only for support. Separate metal tanks or fiber reinforced nonmetallic tanks are not widely used, and they are mainly located in light aircraft and older transports.

With integral tank construction, and only to a slightly lesser extent with thin wall bladder tanks, it becomes apparent that disruption of the wing structure by ground impact or other damage will result in the release of fuel and the potential for ignition. This may occur even though occupied portions of the aircraft may be only slightly damaged. Thus, fire becomes the prime threat to the occupants. Figure 13-1B illustrates typical fuel occupied spaces in a large turbine powered aircraft.

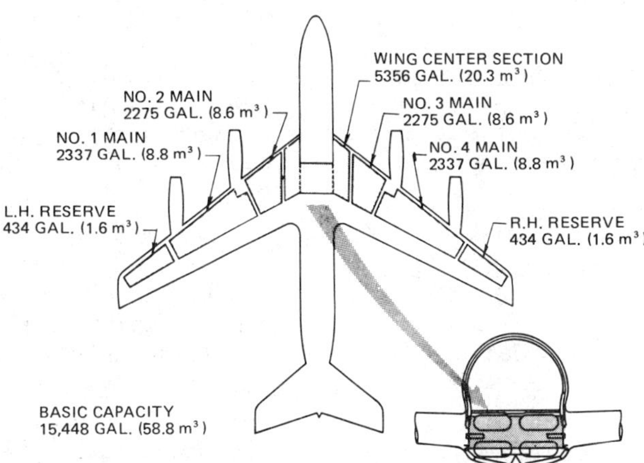

FIG. 13-1B. Arrangement of fuel tanks in a typical turbine powered air carrier aircraft. The reserve tanks and Nos. 1, 2, 3, and 4 are "integral" tanks in this particular airplane, and the wing center section tank is of the "bladder" type.

One solution proposed is the development of crash resistant fuel tank construction with dry break fittings and automatic fuel shutoffs. While the technology to develop this tank has been highly advanced, it has not been widely adopted due to weight limitations and reduction in volume available for fuel storage in long range, military, and commercial aircraft.

However, certain helicopters operated by the U.S. Army are equipped with such crashworthy fuel systems. Statistics compiled after this modification disclosed no thermal fatalities or injuries in almost 800 accidents. Identical Army helicopters without the crashworthy system were involved in over 950 accidents, with a resulting 100 thermal injuries and 86 thermal fatalities.

Open celled foam blocks, cut to fit and placed meticulously in the tank, are used on some military aircraft. While primarily intended for explosion protection after projectile penetration (incendiary bullets, etc.) of the vapor space, they also were tested for improvement of the postcrash fire situation. The tests showed some resulting improvement, but the problems of removal and replacement for tank interior maintenance and loss of available fuel volume have precluded their use in large, long range, civil, and military aircraft.

Some aircraft equipped with these foam blocks have had internal fuel tank fires. Charred foam, discovered when the tank was opened for inspection and repair, has led to the theory that the foam increased the static charging tendency and/or acted as the unbonded collector. However, the presence of the remaining foam in the tank vapor space prevented further propagation of the fire/explosion.

Many aircraft utilize the wing center section (where it passes through the fuselage) for additional fuel storage. In some aircraft, such as the Boeing 727 and DC-9, this is an integral tank with a bladder liner. Newer aircraft designs utilize a double wall tank of metal with a honeycomb core. While such tanks are well protected by the center wing section box structure (the heaviest on the aircraft), they directly expose the fuselage interior. To extend normal operating range, some aircraft utilize fuselage fuel tankage of the double wall type outside of the wing center section box, and thus are deprived of heavy structural protection. As a reasonable compromise, all fuel tanks (at least those within the fuselage) should be of the crash resistant type with dry break fittings and automatic shutoff valves.

Certain models of existing aircraft contain fuel tanks well forward or aft in the fuselage. Fuel storage in the horizontal stabilizer and vertical fin is still being studied.

OTHER DESIGN CONSIDERATIONS

Several design principles which affect basic aircraft firesafety and crashworthiness (excluding cabin furnishings and evacuation systems) are mentioned in the Federal Air Regulations and were developed as a result of accident investigations. They include:

1. When the same structure supports fuel tanks and landing gear, shear pins must be incorporated in the landing gear support structure, allowing the gear to be wiped off without applying structural loads to the fuel tanks.
2. The metal aircraft structure must have electrical continuity to prevent accumulation of static electrical charges, particularly in the fuel tank areas. This is extremely important when designing an all metal aircraft since the aircraft structure acts as a Faraday Cage, shielding all contents from lightning strikes.
3. Static discharge devices and lightning divertors must be located and installed correctly.
4. Fuel lines supplying rear fuselage mounted engines must be designed to ensure proper fire resistance and have the flexibility to resist rupture in a crash situation.
5. Fuel lines and main electrical leads must be segregated.
6. Main electrical power cables in the fuselage must be shrouded in fire resistive flexible conduit.
7. The hydraulic system must be designed properly and use fire resistive fluids. Such fluids presently in use have autoignition temperatures in excess of 1,000°F (538°C) and are only slightly flammable. (Older types of mineral oil based hydraulic fluids are still in widespread use, particularly in the military, and have been responsible for many serious fires.)

AIRCRAFT FUSELAGE COMPARTMENTS

The fire reaction of aircraft cabin interior materials, once ignited, can be crucial from a life safety viewpoint and can result in severe structural damage to the aircraft.

Prior to the mid 1940s, there were few, if any, regulations concerning the flammability of cabin furnishings. Old aircraft at air shows and in museums show the wicker chairs, gauze fabric curtains, and other combustibles used. Following World War II, a qualification test was required that involved igniting small horizontally mounted samples by a short exposure to a Bunsen burner flame. If a sample did not continue to burn after removal of the igniting flame, or burned less than a specified distance in the specified time, it was classified as self extinguishing, or slow burning. Although this test procedure was developed for laboratory comparisons, it created a widespread impression in the aviation industry that cabin furnishings were fire resistive.

After several disastrous fires in flight and on the ground, research was initiated to improve the fire resistive criteria. Various public and private organizations conducted tests which proved the inadequacies of the existing fire resistance criteria and, further, disclosed the problem of heavy smoke and toxic gas generation.

In the late 1960s and early 1970s, the fire resistance criteria were upgraded to require a material sample to be self extinguishing when tested in the vertical position, with exceptions such as floor carpeting. Although aircraft fires have indicated improvement in flame spread of materials qualified by the new vertical burn standards, improvement is still required and the problem of smoke/toxic combustion products remains.

In the mid 1970s, the FAA issued a *Notice of Proposed Rule Making* concerning allowable smoke generation and an *Advance Notice of Proposed Rule Making* concerning toxic gases. After public comment, both these notices were withdrawn temporarily because definitive smoke and toxic gas test procedures had not been developed. (Later, in 1979 the Special Aviation Fire and Explosion Committee was formed. This Committee, after more than one year's study, made recommendations concerning fire blocking layers, emergency lighting, smoke detectors, improved cargo liners, flammability standards on interior cabin panels, etc., which served in part as impetus for further study and future implementation of aviation firesafety improvements.)

One airframe manufacturer subsequently developed smoke emission and toxicity standards for aircraft interior furnishings and materials. Whether these standards are adequate in real life remains to be seen. At least one nation, however, has adopted these standards for all its commercial aircraft.

Some causes of aircraft cabin fires are:

1. Ignition source (usually electrical in nature) in the concealed spaces behind the cabin decorative liner. Such a fire can remain undetected for a considerable time and be beyond the control of hand fire extinguishers by the time it is discovered. Figure 13-1C illustrates an electrical failure which caused a disastrous fire.
2. Ignition of the cabin interior from a postcrash fuel fire or a fire after a fueling operation malfunction. In such a fire, though the interior materials have a certain fire resistance, the heat flux from the exposing fire overwhelms the resistance of the materials.
3. Cigarettes and other carelessly discarded smoking materials on aircraft seats, in trash bins, lavatories, galleys, or receptacles not intended for the purpose.
4. Improper use of flammable liquids during cabin cleaning and refurbishing.

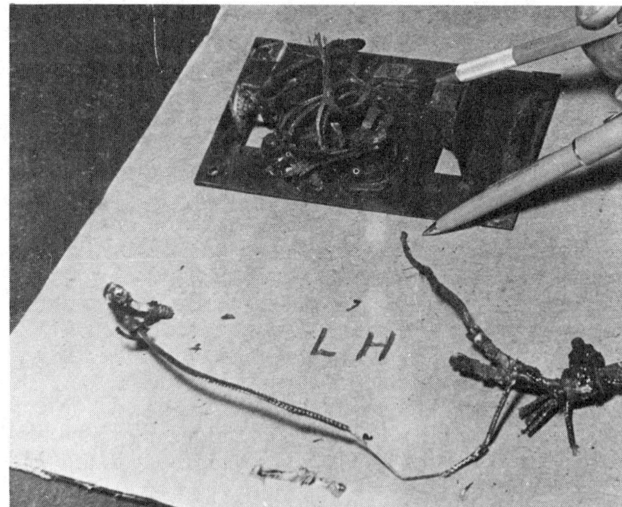

FIG. 13-1C. The shorted wires created a high resistance ground, which ignited extraneous paper towels. The resultant fire burned in the concealed space for 45 minutes before breaking into the open. (Photograph from author's collection)

FIG. 13-1D. Total destruction of the upper portion of the cabin. Seat involvement was moderate and occurred late in the fire. (Photograph from author's collection)

In the late 1970s and early 1980s, researchers developed a method of "fire blocking" aircraft seat cushions, which are usually made of polyurethane foam. This fire blocking layer is inserted between the cushion and decorative upholstery and is intended to delay or prevent the seat cushion involvement in a cabin fire. Under the FAA ruling in 1985 this fire blocking layer became mandatory within a two to three year compliance period.

Fire blocking of seat cushions does reduce the combustibility of the seats. However, unless the fire originates in the seats, fire blocking does not prevent development of serious, often fatal, cabin fires. Flames can propagate fore and/or aft along the ceiling panels, and radiate downward to the seat backs and cushions. By the time the seats become actively involved in the fire, heat and smoke conditions may already be unsurvivable. Figure 13-1D illustrates almost total upper fuselage burnout; seat involvement, however, was moderate. Figure 13-1E is an aircraft totally burned out (no seats were installed at the time of the fire).

In addition to the fire blocking, research is continuing toward the development of an improved, fire resistant aircraft cabin window material to delay or prevent postcrash fire entry into the fuselage via the windows.

Because many aircraft carry supplemental oxygen for the potential failure of cabin pressurization and portable cylinders for first aid purposes, in a crash, localized fire intensification can be expected if the oxygen cylinder fuse plug operates due to fire exposure and releases oxygen in the cabin atmosphere.

Cabin fires in flight are infrequent, and usually are discovered and extinguished in their incipient stage. The potential life loss, however, is severe if fire progresses beyond the point of control by hand fire extinguishers. Research is underway for an onboard cabin fire suppression/inerting system utilizing a halogenated extinguishing agent and/or water carried for other purposes. Some corporate aircraft have been equipped with total flooding extinguishing systems utilizing Halon 1301 and such systems are commercially available. FAA research into total flooding on large commercial transports was somewhat inconclusive; however, the primary aim of the project was to delay penetration of a postcrash fire, rather than control or extinguishment of an inflight cabin fire.

FIG. 13-1E. An electrical spark and use of flammable liquid during maintenance caused this cabin fire. A total burnout occurred with severe damage to the aircraft structure. There were no seats installed in the cabin at the time of the fire. (Photograph from author's collection)

Airline operators and governmental regulations routinely prohibit smoking during aircraft takeoff and landing, while a plane is flying in extreme turbulence, or at any other times the command pilot deems that smoking constitutes a hazard to safety. U.S. regulations and those of most other nations prohibit smoking except in a designated section of the aircraft. Smoking in lavatories, galleys, or while standing in an aisle is strictly prohibited.

Effective in 1985, all lavatories are required to have smoke detectors installed. While the initial rule utilizes domestic type smoke detectors, (which may be of dubious value in an aircraft environment since they were not designed, tested, or approved for this purpose), they can serve as an additional deterrent to smoking in the lavatories. Some aircraft operators have reported a good measure of success in this regard. However, missing batteries and false alarms are a problem. (Smoke detectors specifically designed for aircraft lavatories currently are being developed.)

Trash containers have been redesigned and modified to contain an interior fire, and some aircraft are equipped with a fire suppression system for the trash receptacle. As of 1985, all U.S. aircraft and those registered in most other nations will require such a suppression system for lavatory trash containers regardless of container design.

Hazardous Cargoes

Hazardous cargoes occasionally present a problem to aircraft in flight, on the ground, during ground handling, and in a postcrash situation. This is an international problem, since many shipments cross national boundaries. Hazardous cargoes are regulated by many agencies, among them the U.S. Department of Transportation, the International Air Transport Association, the International Civil Aviation Organization (an agency of the United Nations), and others.

The regulations cover type of cargo that may be shipped, quantities allowed, method of packaging, and the type of aircraft (passenger or cargo only) in which this material may be carried. Flight crews must be informed of the presence of hazardous cargoes on the manifest, and in some cases can refuse to carry the material.

If the rules are followed rigidly, there is little hazard to safety of flight or to ground personnel, whether in normal handling or a postcrash situation. The problem arises when the shipper, either in ignorance or in willful contempt of the regulations, misrepresents the contents of the shipment to a carrier, or does not package the material adequately to ensure containment during transit.

After several serious incidents, the regulations were revised and further restrictions were adopted. Better training is now required for each carrier's receiving agents, standard labels have been adopted, and penalties are assessed against shippers who violate the law.

AIRCRAFT FIRE DETECTION AND EXTINGUISHING SYSTEMS

Fire detection and extinguishing systems are required in certain zones of the aircraft, notably the power plant areas. These systems are defined in design regulations of the nation of aircraft origin.

The systems are similar to those used in ground installations, but there are significant differences. In most aircraft installations, the extinguishing system is activated manually by the flight crew. The detection system must be very sensitive, very stable (free of false alarms), and able to withstand extremes of the environment in which it must function—temperature, vibration, high airflows, and structural flexing. The system must also reset automatically to notify the flight crew when the fire has been extinguished, or when the temperature in the protected zone drops below the set point. System response must be within seconds of the beginning of abnormal conditions.

Extinguishing agent hardware must be lightweight, rugged, and, to overcome the high airflows of flight, be capable of higher rates of agent discharge than would be expected in a similar ground installation.

The most common extinguishing agents used in aircraft are bromotrifluoromethane (Halon 1301) and bromochlorodifluoromethane (Halon 1211). Older aircraft still use carbon dioxide, methyl bromide, and chlorobromomethane. All are effective on flammable liquid and electrical fires.

Cargo Compartment Fire Detection and Extinguishing Systems

Cargo compartments in passenger aircraft often are equipped with fire detection systems. Air starvation is one of the more common techniques used in fire extinguishment in cargo spaces. Upon actuation of fire detection systems, all ventilation to the compartment is sealed off and the fire is allowed to self extinguish due to lack of oxygen. Fire extinguishing systems usually utilizing Halon 1301 or Halon 1211 are also installed for cargo compartments.

Recent regulations have been adopted to improve the fire resistance and fire penetration capabilities of cargo compartment liners. An upper limit has been placed on the volume of a cargo compartment; above this limit, a fire suppression system must be installed regardless of compartment construction.

AIRCRAFT HAND FIRE EXTINGUISHERS

The FAA requires portable hand fire extinguishers in all air transport category aircraft. In some smaller general aviation aircraft, they are recommended but not required.

NFPA 408, *Standard for Aircraft Hand Fire Extinguishers* (hereinafter referred to as NFPA 408), requires portable hand fire extinguishers on all aircraft and describes the type, quantity, minimum capacity, installation, location, and spacing for the extinguishers.

While the FAA recognizes carbon dioxide, water, dry chemical, Halon 1211, Halon 1301, and Halon combinations, at least two of the extinguishers installed on large transport aircraft must be of the Halon 1211 or combination Halon 1211/1301 type.

NFPA 408 no longer recognizes carbon dioxide and dry chemical as suitable for use on aircraft. Carbon dioxide is not recognized as suitable for the following reasons:

1. Ineffectiveness on Class A materials.
2. Chilling effect of solid carbon dioxide (snow) may damage delicate electronic components.
3. Weight of container and agent vs. similar agent capacities of the Halon type.

Dry chemical has long been banned from aircraft in NFPA 408, for the following reasons:

1. Obscuration of the operator's vision, particularly in confined spaces.
2. Possibility of an insulating chemical layer forming on delicate electrical contacts, which could affect continued safety in flight.
3. Dry chemical can be an irritant to the eyes and mucous membranes even though it is not considered toxic.
4. If not immediately cleaned up, certain types of dry chemical can be highly corrosive to aircraft metals.

Some concern has been expressed over the toxicity of both the Halon agents and the products of decomposition in the event of the fire. Figure 13-1F shows the effect of discharging Halon 1211 in a ventilated and unventilated cockpit. These data were generally confirmed by tests conducted by the FAA Technical Center in small aircraft.

While the products of Halon decompostion are indeed toxic, compared to the toxic products of combustion of cabin and flight deck furnishings, the toxicity problem is negligible. (See NFPA 12A, *Standard for Halon 1301 Fire Extinguishing Systems* and 12B, *Standard for Halon 1211 Fire Extinguishing Systems*, and appendices for a full explanation.)

The addition of a flexible discharge device, as recommended by NFPA 408, greatly enhances the capability of the extinguisher to reach hard to get areas behind the instrument panels, under seats, etc. The device enables the operator to hold the unit upright during discharge, gaining full use of the agent charge. Figures 13-1G, 13-1H, 13-1I, and 13-1J illustrate some of the advantages of a nose vs. a short nozzle under simulated conditions of use.

MEANS OF EGRESS FROM AIRCRAFT

In aircraft, emergency exit facilities are particularly important because postcrash fire exposure can be severe and the safe evacuation time for the occupants is limited.

U.S. Federal Air Regulations set the number of exits required in transport category aircraft, depending upon passenger capacity. The manufacturer or operator must demonstrate that evacuation of a full load of passengers can be accomplished in 90 seconds or less, with half the available exits inoperative. These tests require that the test subjects represent an average passenger load, with certain percentages of able bodied males, females, elderly, children, and handicapped. The tests are accomplished without normal cabin lighting; emergency lighting, however, is functional. The exits that are to be blocked are not preannounced to the crew or occupants.

While the data obtained are of value, there are some shortcomings in the tests:

1. Smoke, which could obscure emergency lights or exit signs, is not used.
2. The test subjects know this is an emergency exit certification test and will seek any usable exit. (In real life, most passengers pay little attention to preflight announcements and, consequently, tend to attempt leaving the aircraft by the same door they entered.)
3. Again, many test subjects have taken part in previous demonstrations and, while there is a minimum time limit since last participation, retain a certain knowledge of "what to do in the test."

Escape Systems and Devices

It is acceptable and permissible to utilize the normal entrance and passenger service doors as part of the emergency exit system. If the door is power assisted in its normal mode of operation and operates relatively slowly, it is equipped with a high speed power assist for the emergency mode.

Each emergency exit with a sill height that exceeds a certain distance above the ground, regardless of landing gear configuration (nose gear collapsed, tail high), is required to have an emergency means of allowing the passengers to reach the ground safely. In recent years, this device has usually been an inflatable slide (which sometimes can double as a life raft). Present regulations require the slide to deploy automatically and inflate within a specified number of seconds from the time the exit opens

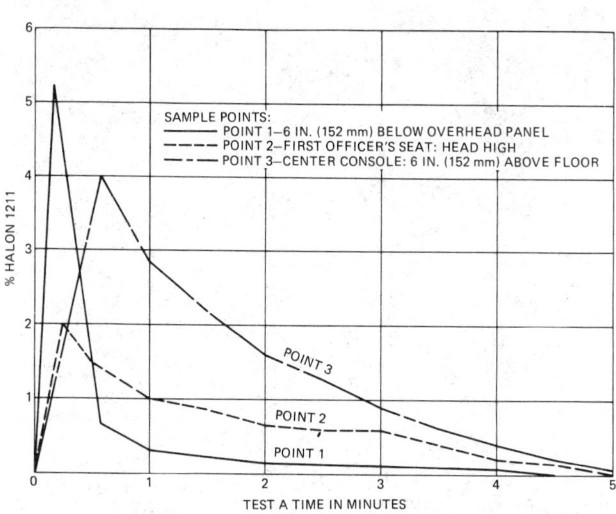

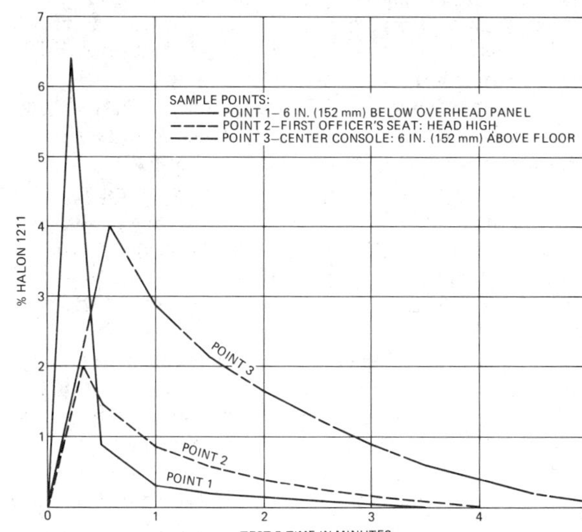

FIG. 13-1F. Halon concentration in a B-737 cockpit (240 cu ft or 7 m³) during and after discharge of a 3 1b (1.4 kg) Halon 1211 extinguisher. Altitude sea level, temperature 68°F (20°C), 11.5 second discharge time, cockpit doors and windows closed. Test A, no forced ventilation. Test B, calculated ventilation rate 1.9 air changes per minute, the minimum for this aircraft. (United Airlines)

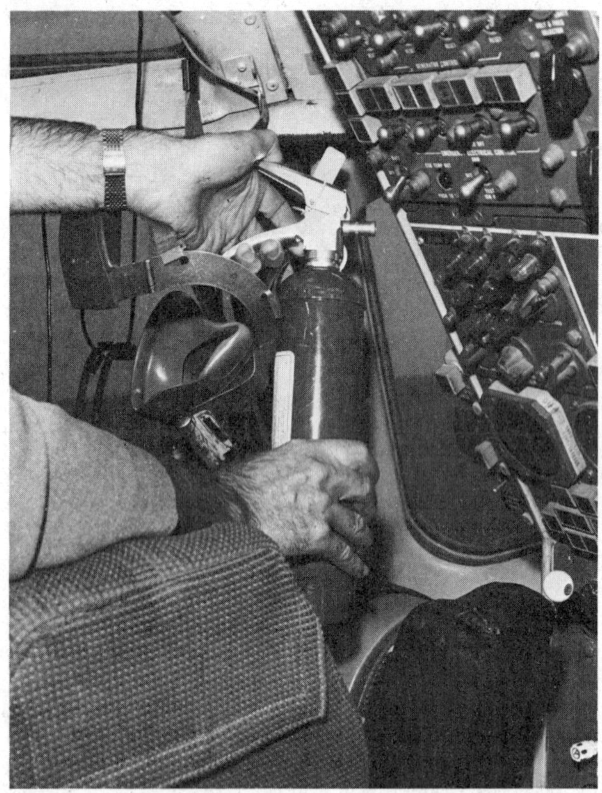

FIG. 13-1G. Effective access to space behind instrument panels is limited due to the short discharge device. (United Airlines)

in the emergency mode. Figure 13-1K shows one type of inflatable escape device.

The openings designated as emergency exits must be operable from the outside and designed to resist jamming due to fuselage distortion. The slides must deploy properly, regardless of wind direction or other influences.

The fire resistance of the slide material is very low, and even slight exposure to flame causes deflation of the slide. Slide material has now been developed which incorporates a reflective coating, reducing the possibility of damage and deflation caused by radiant heat.

In several accidents, obscuration of exit signs and emergency exit lighting by smoke has seriously impeded evacuation of the aircraft. Effective in 1985, all new air transport aircraft are required to have exit signs installed at a lower level and the exit access must be visible from a minimum distance of 35 ft (11 m), assuming a smoke level above the seat back height. In addition, illuminated exit path lights installed in aisle seat armrests or in the floor shall be provided.

Research is being conducted to develop an emergency breathing device for cabin crew members which can be quickly donned, does not impair voice communication or hearing, and is independent of the aircraft oxygen system. Similar devices have been proposed for passengers; however, past tests have indicated that the time necessary in donning such a device, among other factors, can best be used in proceeding to an exit.

Research also is being conducted to develop an effective smoke removal system or procedure for inflight cabin fires. If this research is successful, smoke removal may well obviate the need for a special device for passengers.

Special crash charts for airport fire departments are available from most airframe manufacturers. These show the location of emergency exits on each aircraft type produced by that firm, methods of exit operation from the exterior, location of fuel, oil and hydraulic tanks, oxygen storage, and other hazardous areas. Because none of these items is standardized and may differ in aircraft of the same model series, these charts are an invaluable tool for airport crash crews. Many aircraft operators encourage visits by the fire department for familiarization purposes and when possible, permit hands on operation of the exit devices. At certain locations, mock ups with instructors are available.

SPECIAL MILITARY AIRCRAFT HAZARDS

Among the special hazards of military aircraft are ejection seats and canopy ejectors; armaments, consisting of machine guns and automatic rapid fire cannons, bombs, rockets, nuclear weapons, pyrotechnics, and rocket propellants; and special high energy fuels.

Information concerning fire and rescue for U.S. military aircraft is available from military aviation installations and is given freely to organized fire departments. Information is only limited on subjects which may be sensitive to national military security. Excerpts appear in NFPA 402M, *Manual for Aircraft Rescue and Fire Fighting Operational Procedures* (hereinafter referred to as NFPA 402M).

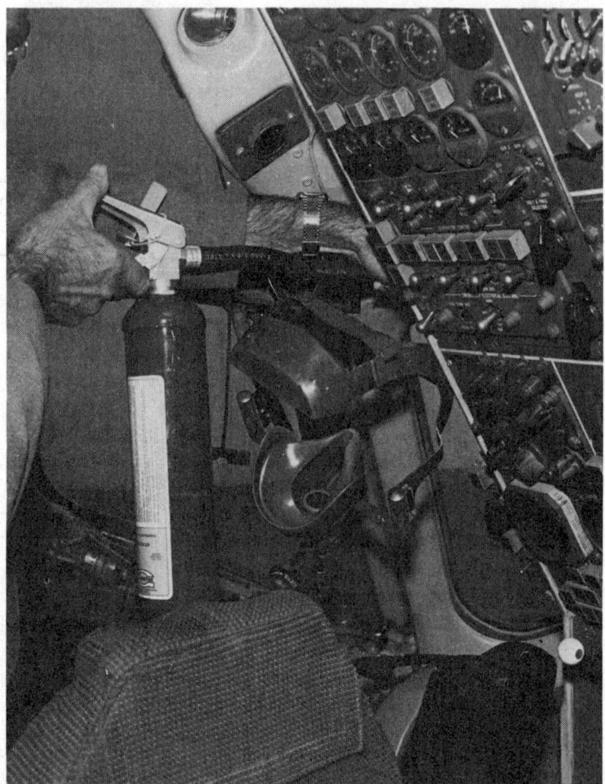

FIG. 13-1H. Improved access to the same space utilizing a short discharge hose. (United Airlines)

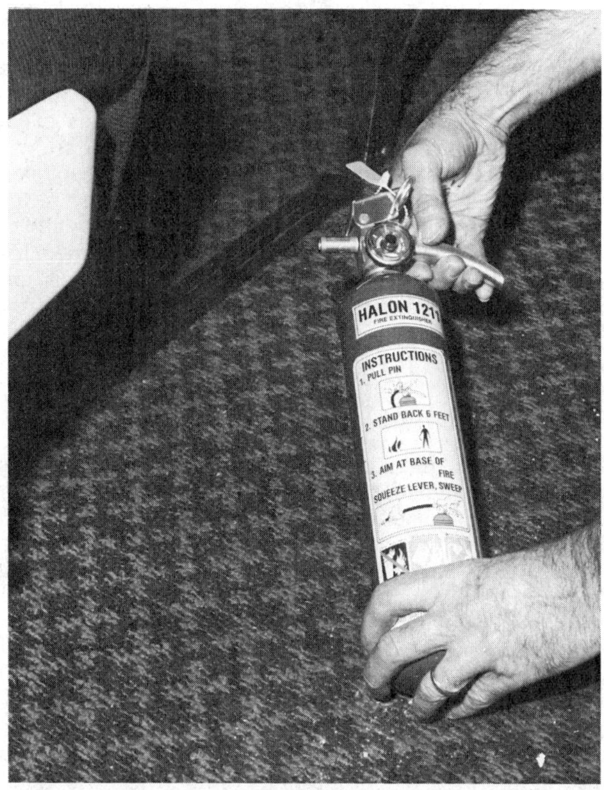

FIG. 13-1I. A considerable amount of agent would be unusable in reaching an underseat fire for final extinguishment. (United Airlines)

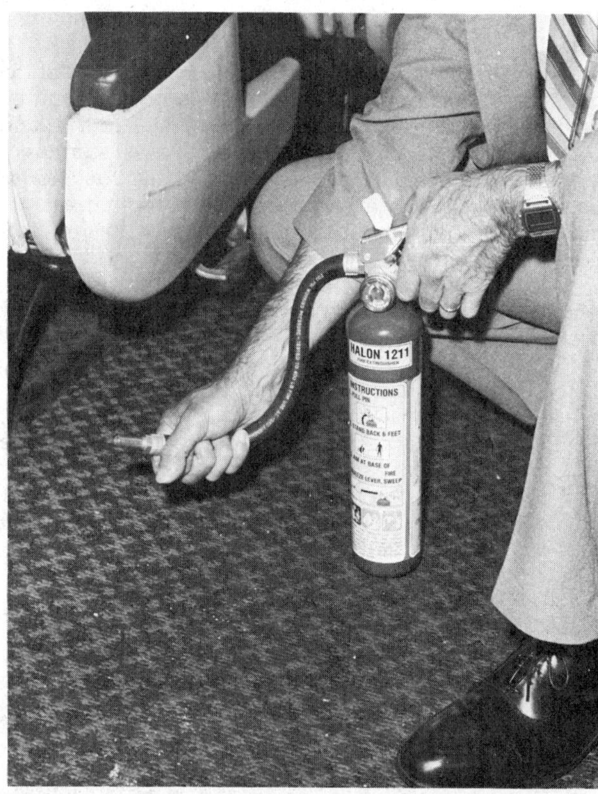

FIG. 13-1J. In the same situation, full use of the contents is obvious. (United Airlines)

AIRCRAFT RESCUE AND FIRE CONTROL

Two distinct types of ground accidents are part of the flight environment of the aircraft (excluding fires from fueling accidents, malfunction of ground servicing equipment, or fires originating during maintenance procedures).

The first type of ground accident is the high speed and/or high angle impact with the ground or other object which results in such structural breakup of the aircraft that survival of the occupants is highly unlikely (fatalities occur due to impact trauma).

The second accident involves relatively low speed and/or shallow ground impact angles. In this type of accident, the occupant survival rate can be very high, especially if there is no postcrash fire, or if the fire is controlled quickly in the area immediately surrounding the occupied portion of the aircraft. An impact survivable accident is illustrated in Figure 13-1L.

Because fire can be anticipated in almost all abnormal aircraft landings plus a few incidental hazards such as overheated brakes or undetected fuel leakage, it is imperative that airports maintain an adequate, well trained, and well equipped fire department (crash crew) at the airport. At smaller airports, the crew often has correlated duties such as security, fueling, or maintenance; they also may be volunteer employees of air carriers. However, all crew members must be available immediately at the sounding of an alarm and be transported to the accident scene immediately. At least one operator should be in the crash house or the immediate vicinity to move the provided equipment rapidly to the scene.

Studies by the International Federation of Airline Pilots Association and the U.S. Airline Pilots Association indicate that the largest percentage of survivable accidents occur within a certain area related to the runway centerline and threshold lights. This "Critical Rescue and

FIG. 13-1K. One type of inflatable, overwing ramp/slide escape device used on wide body aircraft. (United Airlines)

Fire Fighting Access Area" (CRFFAA) is clearly defined and illustrated in NFPA 402M.

To be effective, the demonstrated response time of the first responding fire fighting vehicle to reach any point on the operational runway must be two minutes or less, and to any point remaining in the CRFFAA, in three minutes or less. These conditions must be met when flight operations are in progress; however, NFPA 402M recognizes that mitigating circumstances can make it impossible to meet these recommended times.

Preplanning for anticipated (amber or red alert) emergency landings is also critically important. Prepositioning of equipment and personnel under these circumstances can greatly reduce response time. Illustrated in Figure 13-1M is but one such device used to pre-position equipment for various emergencies.

The amounts of extinguishing agents needed and the type of specialized equipment required in an accident vary, depending on the type of aircraft that use a particular airport. Persons operating a land airport serving any Civil Aeronautics Board Certificated Air Carrier in the U.S. must

FIG. 13-1L. An impact survivable accident followed by fire. (Photograph from author's collection)

have an airport operating certificate. Part of the certification requirements concern the provision of a minimum level of fire fighting and rescue equipment, although under certain conditions a waiver may be granted by the FAA. The minimum level of protection required is determined by the airport *Index* which is based on the frequency and type of aircraft used for air carrier operations. The minimum level was determined, in part, by the use of NFPA standards and private consultant studies; these minimum levels are contained in Part 139 of Title 14 of the *Code of Federal Regulations (CFR)*. In addition, the FAA has published *Advisory Circular* 150/5210-6B, which gives recommended levels of airport protection which are considerably higher than the minimum levels in Part 139.

Other information on the subject of aircraft fire and rescue protection is published in NFPA 403, *Recommended Practice for Aircraft Rescue and Fire Fighting Services at Airports and Heliports*, and the International Civil Aviation Organization in *Annex* 14 (Aerodromes) to the Convention on Civil Aviation.

There are interesting similarities of requirements in the NFPA, ICAO, and the FAA Advisory Circular, as compared to the minimums of CFR Part 139. One major difference in the three is that ICAO and the FAA allow a remission factor so an airport may drop to a lower *Index* if the number of movements (takeoffs and landings) of the largest aircraft to use the airport is below a certain daily or yearly operational level. NFPA does not recognize the remission factor, requiring protection for the largest aircraft that uses the airport, regardless of operational levels.

Other Considerations for Aircraft Rescue and Fire Control

Aside from provision of fire equipment (and assuming adequate personnel for operation), several other important factors directly affect the success or failure of the airport fire rescue crew mission.

The most important factors are training and physical fitness of the personnel. Training, which includes hands on operation of the equipment, frequent "hot drills," in which exposure to radiant heat and unexpected windshifts are experienced, and familiarization with the hazards of the aircraft which regularly utilize the airport, is of the utmost importance. All too frequently after an accident, the airport fire department is criticized for slow response, wasting the extinguishing agent in noncritical fire areas, not knowing how to enter aircraft from the outside, positioning equipment under the escape slide, preventing use of the slide, and other errors. Too often, some of this criticism is true and can be traced to a lack of training.

Funds for training materials are frequently cut back or are not available, and drill fires may be so small as to be meaningless. A hot drill should present a challenge to the crew and severely test its ability to control or extinguish the fire with available agents or equipment, as illustrated in Figure 13-1N.

A sucessful method of training crash vehicle turret operators is currently in use by a major airport operating authority. A scaled down version of the fire fighting "package" of a major crash vehicle is mounted on a small truck chassis. A fire scenario, scaled to the capabilities of the simulated CFR vehicle, is used. The prospective turret operator practices using this simulated arrangement until proficient in its use. Comparative tests and cost records over several years have proved the effectiveness of the simulator in learning turrent operation technique when the operator graduates to the actual CFR vehicles for the final phase of training.

Some of the more obvious benefits of simulated training are reduced extinguishing agent and fuel costs, less wear and tear on the large vehicles, less agent waste practicing with a high capacity turret, and less air pollution. Reduced air pollution is of particular importance in large metropolitan areas.

Crews should exercize regularly and be in top physical condition. The sudden physical and mental stresses of a major aircraft accident are demanding, but can be minimized by good physical conditioning and mental alertness. Airport fire fighters should, as a minimum, meet the qualification standards of NFPA 1003, *Standard for Airport Fire Fighter Professional Qualifications*.

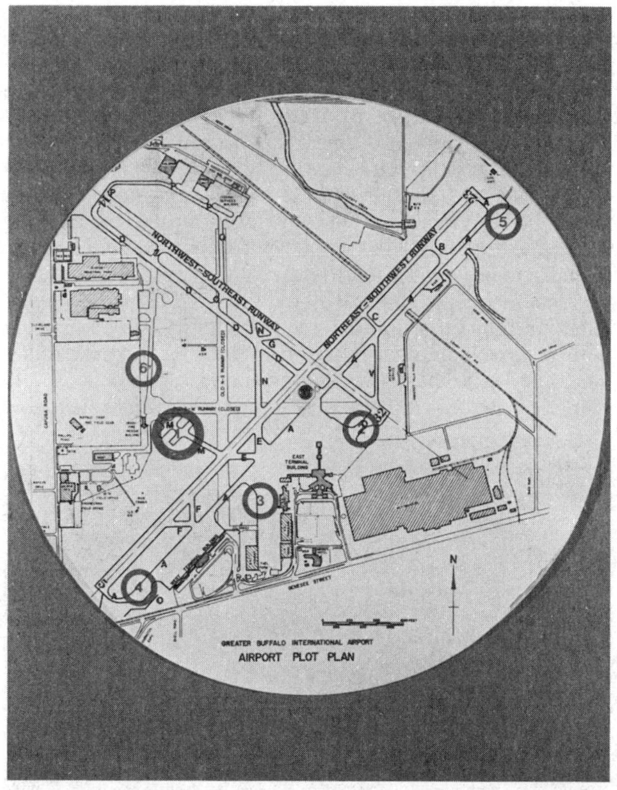

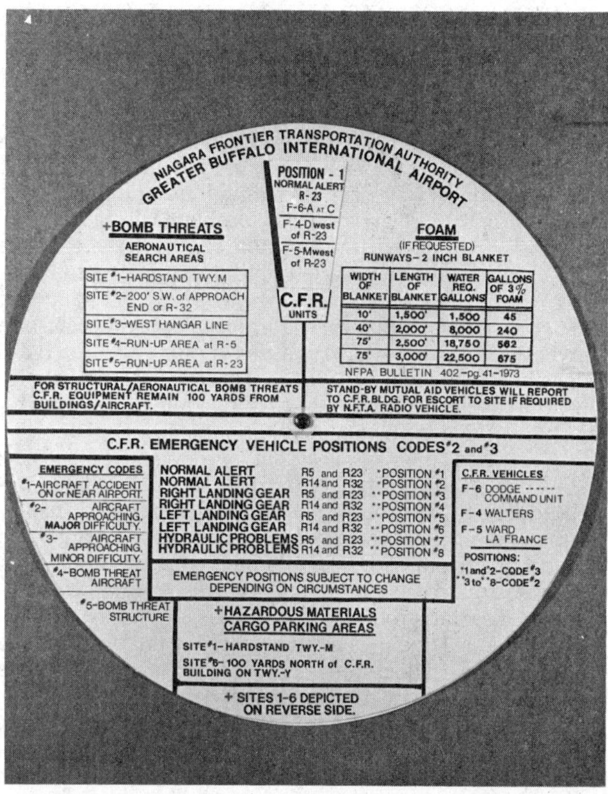

FIG. 13-1M. Right: Airport plot plan indicating preplanned standby position. Left: Rotating the "wheel" gives equipment positions for various types of emergencies for "quick refresher" guidance. (Device from author's collection, courtesy Niagara Frontier Transportation Authority)

FIG. 13-1N. A large scale "hot drill." (Photograph from author's collection).

A second major factor in successful airport fire fighting is the design and operation of the airport fire apparatus. This specialized apparatus must carry all its extinguishing media ready to deliver a massive attack on the postcrash fire. It must have the ability to operate on paved surfaces at acceleration rates and high speeds, but be able to negotiate off pavement terrain at reasonable speeds. Design and performance of the vehicle should be based upon NFPA 414, *Standard for Aircraft Rescue and Fire Fighting Vehicles.*

The vehicle crew should be fully trained in the operation of the vehicle off the paved surfaces of the airport. Regular practice is essential because of the skill required to drive safely over soft ground, since it is impossible to design a vehicle that will negotiate all terrain and retain its other necessary attributes. There have been a few incidents when airport fire vehicles stopped short of an accident scene because the operators thought the terrain too difficult; the vehicle might have reached the scene if these drivers had received adequate training. Practice also will disclose airport areas that are impossible to reach by vehicle. Funding for access roads to these areas should be included in airport improvement plans.

The term "rescue," as used in the context of aircraft rescue and fire fighting, means the control of life-threatening fire in the critical area for the time necessary for all physically able aircraft occupants to escape without help, and continuance of that fire control long enough to assist the severely injured or those pinned in the wreckage. To some people, "rescue" infers physically assisting all occupants out of the aircraft. This is certainly possible for small aircraft; however, in a commercial or military aircraft with more than 10 or 12 occupants, such assistance is impossible with personnel normally available.

AIRPORT AND HELIPORT DESIGN SAFETY

Many factors influence the safety of aircraft operations at airports and heliports: runway lengths and construction, navigational aids, clearances off operational runways below the navigable air space, terrain conditions between runways and taxiways, zoning regulations, and fire protection. Heliports prevent special problems because they frequently are located in congested areas of cities, on the roofs of buildings, on elevated platforms constructed over piers, or at the water's edge. In view of the fact that heliports and helistops that are not an integral part of a conventional airport have special problems, NFPA has recently established a separate technical committee to develop special standards or recommended practices for this type of facility or operation.

A number of serious accidents have occurred because of airport hazards, and their seriousness has been increased due to inadequate zoning regulations and insufficient clearance off the operational runways. Airports are frequently surrounded in whole or in part by water; thus the provision of waterborne crash rescue facilities becomes a special problem, unless the U.S. Coast Guard, Navy, or municipal waterborne rescue and fire control equipment is immediately available to perform rescue services if an aircraft ditches in the water.

The areas off the ends of runways have been the scene of a number of accidents and, where dikes or unimproved terrain conditions exist, create special problems. In some cases, blast fences have been erected at the ends of runways to protect adjacent structures or highways, or to reduce the noise factor during takeoff, particularly for turbine engine aircraft. Under poor visibility conditions, accidents have resulted when aircraft have struck these dikes and blast fences.

During airport construction and repair, especially on the movement areas, the presence of construction and resurfacing equipment has been a factor in accidents. Another principal area where accidents frequently occur is in the overrun portions of runways; a number of the most serious accidents have occurred here when aircraft have been unable to brake effectively, running off the runway, have struck airport boundary fences, or bogged down while traversing unimproved ground surfaces. Unimproved ground surfaces between runways have frequently resulted in aircraft being inaccessible to fire and rescue equipment, especially equipment not specifically designed for off highway use.

SPECIAL AIRPORT FACILITIES AND INSTALLATIONS

Hangars

An aircraft hangar is simply a large structure built to provide weather protection and shop space during aircraft maintenance and storage. However, special fire protection problems associated with a hangar are due entirely to the nature of its occupancy. Since it is impractical and economically unsound to remove all fuel from an aircraft prior to moving it to a hangar, the potential always exists for having large quantities of flammable liquids, mainly aviation fuel, inside the hangar. Quite often, the aircraft contained within the hangar, particularly larger aircraft, are several times more valuable than the hangar, necessitating protection of the aircraft and the hangar structure.

Past hangar designs concentrated primarily on providing excellent protection for the hangar structure. Protection consisted mainly of water type sprinkler systems using the extra hazard pipe schedule, or hydraulically calculated systems, primarily of the deluge type. In certain aircraft hangars, these water only systems are no longer recognized as viable protection by NFPA 409, *Standard for Aircraft Hangars* (hereinafter referred to as NFPA 409). With increased value of the aircraft, it is now common practice to utilize foam water deluge sprinkler systems. Latest designs incorporate foam producing, and oscillating monitor nozzles, in conjunction with all of the previously mentioned systems, to protect the shadow area beneath the wings of large aircraft. (Figure 13-10 illustrates this shadow area.) In nonfire operational tests in actual hangars and large scale tests conducted by the Factory Mutual Research Corporation, this method indicates that considerable progress has been made in protecting aircraft while maintaining the necessary level of building protection.

FIG. 13-10. A typical wide body aircraft maintenance and overhaul facility and structure. Foam/water deluge is the primary protection, with fixed pipe foam/water deluge protection installed beneath the work platforms for the "shadow area." Living joints and flexible connections are provided to allow limited maintenance structure movement to position the aircraft. Structure above the wing and flexible connectors is for fuel tank ventilation. (United Airlines)

A different technique has been more recently developed and used, particularly in older hangars being modernized, with the approval of the authority having jurisdiction. In these cases, overhead closed head type sprinkler systems hydraulically designed to deliver a specified minimum water density over a specified area are installed as a secondary means of protection for the hangar structure. Some of these systems also have the capability of delivering aqueous film-forming foam (AFFF) solution through standard (nonaspirating) sprinkler heads by utilizing wide range proportioning devices.

Primary protection for the hangar, and to a certain extent for the aircraft, is provided by oscillating monitors which provide a complete foam blanket on the floor of the entire hangar within 20 to 30 seconds of actuation. Small roofed enclosures located within the aircraft work areas are provided with sprinklers capable of delivering AFFF solution. Although standard protein or fluoroprotein concentrates may also be used in this type of system, all work to date has utilized AFFF. (Figs. 13-1P and 13-1Q illustrate a test of such a system.)

FIG. 13-1P. Test of a low level foam system in an older, modernized, hangar. This system is designed to cover the entire floor area below the centerline of the aircraft, not just the shadow area. The aircraft is on a hydraulic jack in a normal position for maintenance. Note the dispersed pattern of the foam streams. (United Airlines)

Whether used only for shadow area protection or total floor coverage, foam monitor location, oscillation speed, foam pattern, and barrel centerline height above the floor are critical. Location of the monitors should be such that all areas where coverage is desired can be easily reached. Experience has indicated that, depending upon the arc of

FIG. 13-1Q. Test complete. Note the total floor coverage after only 5 minutes of operation. Systems capacity is 20 plus minutes. (United Airlines)

coverage, no more than 15 seconds should be necessary for a complete cycle. The foam pattern should be dispersed rather than a straight stream to minimize turbulence at the point of stream impact with objects or the floor. Barrel centerline height should be a minimum of 12 ft (4 m) above the floor to extend range and to minimize obstruction by aircraft maintenance equipment, etc. Figure 13-1R illustrates an installation meeting these criteria.

FIG. 13-1R. A foam monitor nozzle. Centerline of this particular nozzle is approximately 12 ft (4 m) above floor level to clear obstructions. (United Airlines)

Due to hangar size, or due to multiple aircraft positions, monitor nozzles can be mounted at the bottom of the truss space to provide coverage for these void spaces. Arrangements should be made to exercise and test the oscillating mechanism from the hangar floor or other easily accessible location. (See Fig. 13-1S.)

The principal means of fire detection for hangars has been rate of rise devices; however, as hangar roof heights increase (sometimes being as much as 150 ft or 46 m from floor to roof deck), the ability of this type of device to perform in the time required to minimize aircraft damage is becoming suspect. In climates with moderate to severe winter conditions, it is becoming increasingly difficult to maintain stability of such systems (due to the opening and closing of hangar doors and the installation of high recovery rate heating systems) and still maintain the sensitivity necessary for rapid fire detection.

Optical type detectors of ultraviolet or infrared types can be used for high speed detection of relatively small fires, while being unaffected by rapid temperature changes.

FIG. 13-1S. A monitor nozzle mounted beneath the roof work to fill void areas unreachable from monitors located near the walls. Centerline of the barrel is approximately 65 ft (20 m) above the floor. Control valves are located at the floor line along the hangar wall. The small diameter lines illustrated are for testing and exercising the water powered oscillating mechanism. (United Airlines)

Care must be taken in selecting the type of detector and control devices for optical detection. They must have demonstrated operational integrity, yet not be overly sensitive and subject to false or unwanted operation. Considerable progress has been made recently in developing optical detection systems that automatically ignore or are insensitive to routine aircraft maintenance procedures involving welding, high intensity lights, boroscopes, radioisotopes, X-ray inspections, etc. Whenever an optical system is installed, its sensitivity to fire should be tested. It is further suggested that a recording device be attached to provide a hard copy record of what the detectors "see" during routine aircraft maintenance operations. During the initial period after installation (usually 60 to 90 days), the detection systems should be in the "alarm only" mode and a careful record made of the recording device printout. Aircraft maintenance records often can determine what operation affected the optical detectors. Figure 13-1T illustrates a detector test fire during extinguishment.

Floor drainage is another overlooked portion of the hangar fire protection system. The drainage system removes excess fire protection water, and also removes large quantities of the liquid fuel from the fire scene. To prove their efficiency, these systems should be flushed thoroughly with high volumes of water at least annually. Oil/water separators should be maintained meticulously to prevent pollution of lakes and streams, and to reduce hazards in the airport drainage system to which the hangar system may be connected.

Aircraft hangars are divided into Types I, II, III, which roughly correspond to the types of aircraft housed in them and the anticipated fuel loadings. NFPA 409 defines these types and outlines the requirements for construction, protection, drainage, fire cutoffs, and other provisions for each type of hangar.

Airport Terminal Buildings

Airport terminal buildings include any fully enclosed extensions which function as passenger concourses (sometimes called fingers or piers), and satellite buildings which serve passenger handling functions. The satellite building may be connected to the main terminal building via tunnels beneath the aircraft operating ramp, or people mover transit systems operating on a fixed guideway which may be above or below ground. Terminals also contain airline related businesses, such as offices, restaurants, and gift shops.

Special attention is given to airport terminal buildings in NFPA 416, *Standard on Construction and Protection of Airport Terminal Buildings*, because these buildings have a severe fire exposure potential from immediately adjacent aircraft fueling and servicing operations and the high occupancy load in terminals.

In recent years, airport security has complicated the fire problems, since preventing access by unauthorized persons to the aircraft operating areas is of prime importance, and directly affects the location and discharge points of the emergency exit facilities.

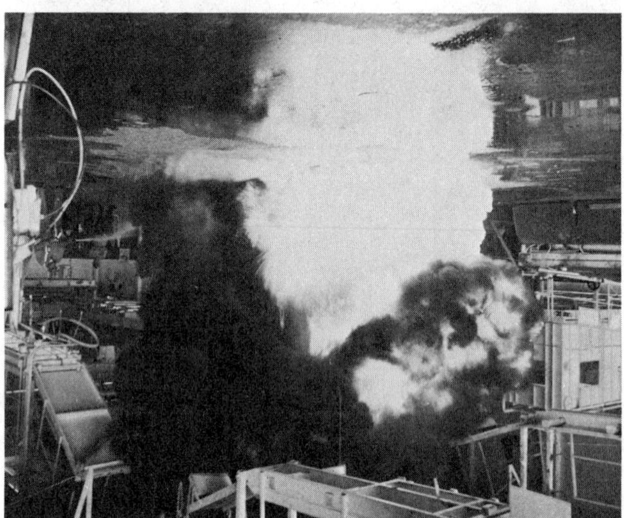

FIG. 13-1T. Extinguishment of a fire following test of the optical detection system. A too "fine" setting of detectors has caused many unwanted operations. In this case a 20 sq ft (2 m^2) circular pan with flame deflector was selected as the minimum size fire which needed to be detected. Actuation occurred in approximately 10 seconds. (United Airlines)

Aircraft Loading Bridges or Walkways

This equipment is frequently installed at the terminal area for passenger convenience and to protect passengers from the weather while they move from the terminal loading gate area to the aircraft. Loading equipment also increases safety by reducing ramp congestion, and segregating passengers from the aircraft servicing equipment and personnel.

In the event of fire caused by a fueling malfunction or other reason, a loading bridge or walkway can serve as a means of egress from the aircraft to a safe refuge, i.e., the terminal building. Most people have the natural tendency to exit an area (aircraft, building) the same way they entered, and since the emergency situation requiring evacuation may occur before the preflight safety and evacuation briefing, the loading bridge may be the only familiar means of exit to passengers. For this reason, NFPA 417, *Standard for Construction and Protection of Aircraft Loading Walkways*, requires five minutes of safe passage from the aircraft to the terminal. This may be accomplished by construction features, special fire protection, or a combination of both. This safe egress must be provided during severe fire exposure to the bridge.

Doors, if any, at the terminal end of the bridge should be equipped with panic hardware and swing in the direction of travel from the aircraft to terminal. In no case should the loading bridge be considered part of the terminal exit system, unless a vestibule and stairs conforming to NFPA 101®, *Code for Safety to Life From Fire in Buildings and Structures*, are provided at the terminal end, even though there may be auxiliary stairs located near the aircraft end. The auxiliary stairs are strictly for the use of aircraft servicing personnel or the flight crew.

Aircraft Fueling Ramp Drainage

Due to the large amounts of flammable liquids (in the form of fuel) handled on aircraft servicing ramps, special attention must be paid to the drainage system provided to prevent exposing the terminal or aircraft unnecessarily in the event of accidental release. NFPA 415, *Standard on Aircraft Fueling Ramp Drainage*, specifies the amount of slope and direction regarding ramp geometry, location of drainage system inlets, water seal traps, and oil/water separators.

Rooftop Heliport Construction and Protection

This is covered in NFPA 418, *Standard for Roof-Top Heliport Construction and Protection*, which deals with the special fire problems existing for this type of facility. The standard covers landing deck construction, drainage, egress, and fire protection.

AIRCRAFT MAINTENANCE AND SERVICING

Second to the postcrash situation, fire during maintenance and servicing (including fuel servicing) creates the greatest hazard exposure to personnel and property.

In order to keep an aircraft airworthy, every part must be inspected, repaired, or overhauled periodically. The frequency of these procedures will vary; operations may be performed all at once in a single maintenance visit, or as part of a progressive overhaul system where a certain portion of a total overhaul is conducted periodically. In addition, periodic inspections are required between major overhauls to repair minor discrepancies, inspect critical structures, and test various components. Nonroutine maintenance can occur at any time to replace or repair a defective or prematurely failed part.

Many maintenance procedures involve the use of high flammable solvents, sometimes unstable or toxic chemicals, and personnel entry into integral fuel tanks.

When situations such as these occur, special procedures must be developed to ensure the safety of the operation. It is sometimes possible to pressurize equipment housings with shop air, using appropriate power/pressure interlock to prevent penetration of flammable vapors. Localized ventilation and curtailment of operations utilizing flammable or combustible liquids are also helpful. In any event, each potentially hazardous procedure should be developed or reviewed by a person who is thoroughly familiar with the basics of firesafety and aircraft maintenance requirements and procedures. NFPA 410, *Standard for Aircraft Maintenance*, should be used to develop local procedures for the more common hazardous aircraft maintenance operations such as welding, spray painting, fuel tank ventilation, etc.

(NFPA 70, *National Electrical Code®*, defines certain areas in hangars and around aircraft as Class I, Group D, Division I or II locations. In this age of sophisticated testing using electronics, eddy currents, X rays, etc., it is often impossible to design or obtain equipment listed or approved for these locations. Welding (usually inert gas shielded arc) is frequently performed on installed engines or aircraft structures.)

Testing of fuel tanks for system leaks is commonly performed in hangars using high flash point Jet A fuel or a special, usually combustible, test fluid. Toward this end, many hangars are provided with a fixed system to board and unload fuel or test fluid. It has also been a practice to park a standard fuel servicing tank truck outside the hangar and run long hoses to the aircraft connection points. This practice is to be discouraged, because the hose is the weak point in any system, and while hoses are necessary for flexible connections, they should be kept as short as possible.

Painting of the interior and exterior of aircraft is also accomplished in hangars. While a few hangars have been constructed specifically for aircraft exterior painting, it is more common to control ignition sources and rely upon the sheer volume of the hangar and natural ventilation to dissipate flammable vapors. In recent years latex interior paints, which contain little if any solvents, are being used for aircraft interiors to permit painting of interior panel partitions and equipment in place rather than to remove it to a standard spray booth. Whenever it is necessary to use flammable paints inside an aircraft, special precautions for ventilation and control of ignition sources must be installed.

Aircraft Fueling

While the potential exists for spills and fires each time an aircraft is fueled, the number of these incidents is rare compared to the total number of aircraft fuelings that take place each day. This excellent safety record is due to the use of NFPA 407, *Standard for Aircraft Fuel Servicing*, a document which is accepted worldwide. The standard outlines specific precautions to control the extent of this hazard and gives recommendations on the design of aircraft fueling, hose, design for aircraft fuel servicing tank vehicles, airport fixed fueling systems utilizing fuel hydrants, hydrant vehicles (to connect the fuel hydrant to the aircraft), and safety procedures to be followed in the event of an accidental spill. The most common type of spill is at the aircraft tank vent point, when an aircraft internal shutoff valve fails or closes slowly, allowing an overfill condition to develop.

MISCELLANEOUS

Figure 13-1U illustrates hazardous intake and exhaust areas of typical modern aircraft turbine engines. These hazards may be encountered by emergency crews responding to a crash situation when, because of damage, an engine has become uncontrollable, or the flight crew may

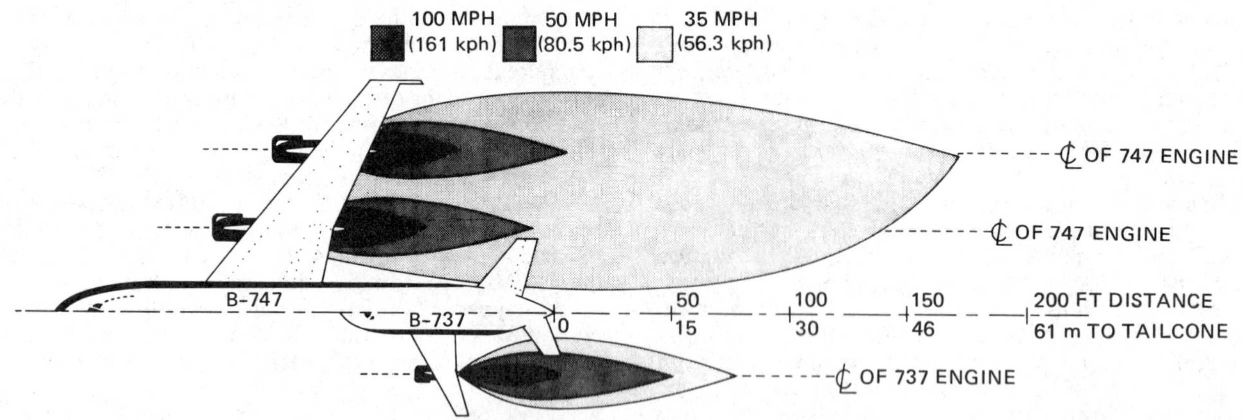

FIG. 13-1U. Comparative blast profiles of typical wide bodied (B-747) and standard bodied (B-737) jets with engines at "idle" power setting. Blast profile increases with power setting to maximum at take off thrust. Most authorities agree that a 25 degree arc area in front of the engine inlet (not shown) should be avoided to prevent ingestion into engine regardless of power setting. (Drawing based on Boeing Commercial Airplane Co. data)

be unaware of the operation of emergency personnel and equipment in the vicinity. Walls and glass areas of terminal buildings and other airport structures which may be exposed to a jet blast or explosion will require design for wind loads in excess of that required by building codes for a particular area.

OTHER ORGANIZATIONS IN THE FIELD OF AVIATION FIRESAFETY

Private organizations other than the NFPA working in various fields of aviation firesafety include:

Aerospace Industries Association of America, Inc., 1725 DeSales St., N.W., Washington, DC 20036

Air Transport Association of America, 1709 New York Ave., N.W., Washington, DC 20006

Air Transport Section and Aerospace Section of the National Safety Council, 444 North Michigan Ave., Chicago, IL 60611

National Transportation Safety Board (U.S. Department of Transportation), Washington, DC 20590. The NTSB has responsibilities for determining the cause of aircraft accidents, except for certain light plane accidents. The NTSB also promotes safety by studies in accident prevention and makes safety recommendations to the FAA.

In Canada, the civil aviation fire protection program is coordinated through the Air Services Fire Marshal of the Canada Department of Transport (Ottawa, Ontario), and the military program through the Fire Marshal, Canadian Forces (Ottawa, Ontario).

International aviation organizations also interested in aviation firesafety include:

International Air Transport Association (IATA), P.O. Box 550, Sucuccursale, Place de L'Aviation Internationale, 1000 Sherbrooke St., W., Montreal, P.Q., Canada H3A 2R2

International Civil Aviation Organization (ICAO), 1000 Sherbrooke St., W., Montreal, P.Q., Canada H3A 2R2 (The Chief of Aerodromes, Air Routes and Ground Aids Division coordinates this program.)

International Federation of Airline Pilots Associations, Office of the Executive Secretary, 1 Hyde Park Place, London W2, England

Airport Operators Council, International, 1700 K St., N.W., Washington, DC 20006

American Association of Airport Executives, 2029 K St., N.W., Washington, DC 20006

Aviation Technical Service Committee of the American Petroleum Institute, 1801 K St., N.W., Washington, DC 20006

Flight Safety Foundation, Inc., 5510 Columbia Pike, Arlington, VA 22204-3194

Society of Automotive Engineers, Inc., 2 Pennsylvania Plaza, New York, NY 10001

U.S. government organizations working in various aviation firesafety fields include:

Department of the Air Force (U.S. Department of Defense): The work of the USAF is coordinated through the Headquarters USAF (PREE), Washington, DC 20230, for operations and administration and through the Air Force Systems Command (Wright-Patterson Air Force Base, OH 45433) for research in the field.

Department of the Army (U.S. Department of Defense): Army aviation firesafety is coordinated primarily through the U.S. Army Agency for Aviation Safety (Fort Rucker, AL 36360) and the U.S. Army Mobile Equipment Research and Development Center (Fort Belvoir, VA 22060).

Department of the Navy (U.S. Department of Defense): Navy aviation firesafety is coordinated primarily through the Naval Facilities Engineering Command, Washington, DC 20390, and the Naval Research Laboratory, Washington, DC 20390.

Federal Aviation Administration (U.S. Department of Transportation): Several branches of the FAA are concerned directly with aviation firesafety, including the Flight Standards Service and Airport Services, which is headquartered in Washington, DC 20590, and the FAA Technical Center, Atlantic City Airport, NJ 08405.

National Aeronautics and Space Administration, 600 Independence Ave., S.W., Washington, DC 20546, is concerned with space vehicle safety problems, and conducts extensive research into conventional subsonic and supersonic aircraft fire problems as well.

Bibliography

NFPA Codes, Standards, Recommended Practices and Manuals. (See the latest *NFPA Codes and Standards Catalog* for availability of current editions of the following documents.)

NFPA 12A, *Standard on Halon 1301 Fire Extinguishing Systems.*
NFPA 12B, *Standard on Halon 1211 Fire Extinguishing Systems.*
NFPA 70, *National Electrical Code.*
NFPA 101, *Code for Safety to Life from Fire in Buildings and Structures.*
NFPA 402M, *Manual for Aircraft Rescue and Fire Fighting Operational Procedures.*
NFPA 403, *Recommended Practice for Aircraft Rescue and Fire Fighting Services at Airports and Heliports.*
NFPA 407, *Standard for Aircraft Fuel Servicing.*
NFPA 408, *Standard for Aircraft Hand Fire Extinguishers.*
NFPA 409, *Standard on Aircraft Hangars.*
NFPA 410, *Standard on Aircraft Maintenance.*
NFPA 412, *Standard for Evaluating Foam Fire Fighting Equipment on Aircraft Rescue and Fire Fighting Vehicles.*
NFPA 414, *Standard for Aircraft Rescue and Fire Fighting Vehicles.*
NFPA 415, *Standard on Aircraft Fueling Ramp Drainage.*
NFPA 416, *Standard on Construction and Protection of Airport Terminal Buildings.*
NFPA 417, *Standard on Construction and Protection of Aircraft Loading Walkways.*
NFPA 418, *Standard on Roof-Top Heliport Construction and Protection.*
NFPA 419, *Guide for Master Planning Airport Water Supply Systems for Fire Protection.*
NFPA 421, *Recommended Practice on Aircraft Interior Fire Protection Systems.*
NFPA 422M, *Manual for Aircraft and Explosion Fire Investigators.*
NFPA 424, *Recommended Practice for Airport/Community Emergency Planning.*
NFPA 1003, *Standard for Airport Fire Fighter Professional Qualifications.*

Additional Readings*

"Aeronautics and Space," *Code of Federal Regulations*, Title 14, Washington, DC (under continous revision).
Federal Aviation Administration Advisory Circulars, U.S. Department of Transportation, Washington, DC (with the following designations and titles):
Aircraft Fire and Rescue Communications, 150/5210-7A, Mar. 16, 1972.
Aircraft Fire and Rescue Facilities and Extinguishing Agents, 150/5210-6B, Jan. 25, 1973.
Airport Fire and Rescue Vehicle Specification Guide, 150/5220-14, Mar. 15, 1979.
Airport Fire Department Operating Procedures During Periods of Low Visibility, 150/5210-0, Oct. 27, 1967.
Airport Operations Manual, 150/5280-1 June 16, 1972.
Fire and Rescue Service for Certificated Airports, 150/5210-12, Mar. 2, 1972.
Fire Department Responsibility in Protecting Evidence at the Scene of an Aircraft Accident, 150/5200-12, Aug. 8, 1969.
Fire Fighting Exemptions under the 1976 Amendment to the Federal Aviation Act, 150/5280-3, Feb. 4, 1977.
Fire Prevention During Aircraft Fueling Operations, 150/5230-3, Apr. 8, 1969.
Hand Fire Extinguishers for Use on Aircraft, 20-42C, Mar. 7, 1984.
Heliport Design Guide, 150/5390-1B, Aug. 22, 1977.
Response to Aircraft Emergencies, 150/5210-11, Apr. 15, 1969.
Water Rescue Plans, Facilities, and Equipment, 150/5210-13, May 4, 1972.
Water Supply Systems for Aircraft Fire and Rescue Protection, 150/5220-4, Dec. 7, 1967.
Williams, C., and Laughlin, J. eds., *Aircraft Fire Protection and Rescue Procedures*, International Fire Service Training Association, Oklahoma State University, Stillwater, OK, 1978.
Published works, including those of many foreign governments and scientists, are available from the National Technical Information Service, Springfield, VA, USA 22161, under the headings "Transportation Safety" and "Air Transportation." In addition, NFPA has, from time to time, published reports of tests, aircraft accidents, etc., which have significant lessons for or impact upon the aviation community. The serious researcher, fire investigator, student, or aviation firesafety enthusiast should contact these organizations for an index of materials available.

*To attempt to list all relevant published work in the limited space available is impossible. Continuous research and testing is being conducted throughout the world in aviation firesafety primarily funded by various governmental agencies. This research, testing, etc., encompasses all aspects of aviation firesafety, emergency evacuation, new concepts and devices in detection and extinguishment, extinguishing agents, etc.

MOTOR VEHICLES

Revised by Hugh McGinley

Many factors influence motor vehicle firesafety: (1) vehicle design and construction; (2) materials from which a vehicle is constructed or which are carried by a vehicle (particularly upholstery, plastic, wood, fuel and hazardous materials cargoes); (3) vehicle maintenance, including safeguards against fire during vehicle repair; (4) vehicle operation, including avoidance of collisions and of other accidents such as those involving smoking materials; and (5) garaging or storage of vehicles.

The degree of fire hazard depends upon: (1) the type of vehicle (e.g., passenger car, motorcycle, motor home, tank truck); (2) the use of the vehicle (pleasure driving, commercial service, off road use); (3) the climate and environment in which it is used (hot weather, high humidity, polluted air); (4) the age of the vehicle; (5) the condition and maintenance of the vehicle (crash damage, leakage of lubricants, underinflated tires, dragging brakes); (6) the material and construction standards to which it is built (particularly the fuel and electrical systems); and (7) the type of fuel used (gasoline, diesel fuel, propane) and the way the fuel is contained in the vehicle.

Research into the causes of motor vehicle fires has not yet yielded complete or reliable statistics or insights into fire hazards. However, individual crash investigations and collections of crash and other loss data indicate the magnitude of the motor vehicle fire problem and some specific causal factors in motor vehicle fires.

Motor vehicle fires represent about 17 percent of all reported fires, and most vehicle fires result from causes other than crashes. Yet serious injury from vehicle fires occurs overwhelmingly from crash related fires—fires that occur at a rate of approximately 1 per 1,000 crashes.

Property loss from motor vehicle fires probably averages several hundred dollars per incident. However, estimates may be skewed by the nonreporting of minor fires and of fires in uninsured vehicles, as well as by ambiguous reporting on police and fire department forms.

Capt. McGinley is Commander, District 15, of the Illinois State Police and a member of the NFPA Technical Committee on Motor Vehicle and Highway Fire Protection.

THE NATURE OF VEHICLE FIRES

As with other fires, motor vehicle fires require a flammable substance, an ignition source, and oxygen. Virtually all motor vehicles carry a flammable fuel—gasoline, gasahol, diesel fuel, or another hydrocarbon compound. Vehicle upholstery, insulating and sound deadening materials, electrical wiring insulation, and plastic body and trim also can fuel a motor vehicle fire. Materials carried as cargo and fare in or on motor vehicles, particularly trucks, also may be flammable.

Ignition sources include: (1) electrical short circuits or other electrical malfunctions that cause excessive heating of conductors or components; (2) sparks from an engine ignition system; (3) hot exhaust system components; (4) engine backfire, (5) overheating of tires, brakes, and bearings; (6) friction generated sparks from a collision or from metal components scraping against the pavement; and (7) careless use of cigarettes and other smoking materials. There is generally sufficient oxygen for a motor vehicle fire.

Table 13-2A summarizes data from a study of the frequency and area of origin of fires in passenger cars involved in noncollision and collision related incidents. Table 13-2B summarizes the locations of the origin of fires on trucks as reported to the Bureau of Motor Carrier Safety (BMCS) of the U.S. Department of Transportation (DOT).

Human injury in motor vehicle fires comes from direct exposure to heat and from inhalation of toxic combustion products. In some cases, injury results from or is exacerbated by the inability of a vehicle occupant to get out of a burning vehicle—the doors or seats are jammed by a crash, or the occupant is injured or stunned by the crash. In one case, a two-door sedan was driven onto the third rail of an electrified railroad track. There were not significant crash forces to cause injury, but a fire was started by the electric current arcing across a short circuit caused by the car body. The front seat passengers escaped, but investigators concluded that the three rear seat passengers were trapped in the burning vehicle probably because of their difficulty in locating and operating the latch to push back the front seat.

TABLE 13-2A. Passenger Car Fires by Type of Incident and Area of Fire Origin*

	Type of Incident			
	Noncollision		Collision-related	
Area of Fire Origin	Number of Fires	Percent	Number of Fires	Percent
Engine	1,085	59	39	54
Passenger	647	35	3	4
Fuel tank	59	3	24	33
Trunk	31	2	3	4
Tire/brake	29	2	3	4
Total	1,851	100†	72	100†

* Source: "1973 National Survey of Motor Vehicle Fires" commissioned by the Insurance Institute for Highway Safety.

† Column does not add to total due to rounding.

Note: Figures do not include fires of unidentified origin. The 1,923 passenger car fires in which the area of origin was identified comprised 83 percent of the 2,325 passenger car fires surveyed, and 73 percent of the 2,637 total motor vehicle fires surveyed.

TABLE 13-2B. Location of Origin of Fire—Property Carriers**

	Moving		Stopped		Total	
Component	No.	Per-cent	No.	Per-cent	No.	Per-cent
Engine	27	5.5	6	8.3	33	4.9
Electrical system	36	7.3	22	12.2	58	8.6
Fuel system*	32	6.4	7	4.0	39	5.7
Service brakes	27	5.5	1	0.6	28	4.1
Parking brakes	3	0.6	0	—	3	0.4
Exhaust system	11	2.2	0	—	11	1.6
Wheel or wheel bearing	26	5.3	3	1.7	29	4.3
Cargo heater	4	0.8	4	2.2	8	1.2
Tires	37	7.5	4	2.2	41	6.1
In cab	10	2.0	13	7.2	23	3.4
Under cab	9	1.8	6	3.3	15	2.2
Around cab	14	2.8	2	1.1	16	2.4
Cargo space	83	16.8	79	43.9	162	24.0
Roof or cargo covering	3	0.6	0	—	3	0.4
Inside mobile home trailer	2	0.4	0	—	2	0.3
Other vehicle or object	90	18.1	20	11.1	110	16.4
Unknown or not reported	81	16.4	13	7.2	94	14.0
	495	100.0	180	100.0	675	100.0

* Broken down as follows:

	Moving		Stopped		Total	
Component	No.	Per-cent	No.	Per-cent	No.	Per-cent
Fuel Tank	19	3.8	3	1.7	22	3.3
Carburetor	2	0.4	1	0.6	3	0.4
Fuel Line	11	2.2	2	1.1	13	1.9
Fuel Pump	0	—	1	0.6	1	0.1

** Source: 1972 Analysis of Accident Reports Involving Fire, U.S. Department of Transportation.

Hazardous materials in motor vehicles present a particular fire problem. The National Transportation Safety Board (NTSB), the BMCS, and the NFPA have all carried out a number of detailed accident investigations into such fires. These investigations have led to changes in the regulations for hazardous materials containers, and other aspects of the transportation of such cargoes. Despite such precautions, hazardous materials cargo fires and explosions occur, primarily when vehicles carrying such cargoes are involved in major collisions.

The NTSB has issued a series of reports on accidents in which hazardous materials exploded or burned following a crash. (These and other NTSB reports are available from the National Technical Information Service, Springfield, VA 22101.) As a result of these and other highway crash investigations, the NTSB makes numerous recommendations to the BMCS as well as to other agencies, concerning modification and enforcement of the BMCS regulations.

FEDERAL MOTOR VEHICLE FIRE STANDARDS

Under the authority of the National Traffic and Motor Vehicle Safety Act of 1966, the National Highway Traffic Safety Administration (NHTSA) of the DOT is authorized to set minimum safety standards applicable to new and used motor vehicles, trailers, and motor vehicle equipment. To date, only two fire related standards have been issued for new vehicles and none has been issued for vehicles in use. The NHTSA also has the authority to investigate motor vehicle defects (some of which may present a fire hazard) and order notification of owners and repair by the manufacturer.

Federal Motor Vehicle Safety Standard (FMVSS) 301 sets requirements for the fuel system integrity of passenger cars, trucks, buses, and multipurpose vehicles having a gross vehicle weight rating (GVWR) of 10,000 lb (4536 kg) or less. The standard also applies to school buses of more than 10,000 lb (4536 kg) GVWR using fuel having a boiling point of more than 32°F (0°C).

FMVSS 301 is intended specifically to reduce the fire hazard resulting from fuel leakage in crashes. It was first made applicable to passenger cars manufactured on and after January 1, 1968. Since that time, the standard has been made increasingly stringent.

Under the present provisions, effective September 1, 1977, all vehicles having a GVWR of 10,000 lbs (4536 kg) or less must pass prescribed front, side, and rear crash barrier tests and a rollover test. Front and rear tests are conducted at impact speeds of 30 mph (48 km/hr), while lateral impact speed is 20 mph (32 km/hr). In the crash barrier tests, fuel leakage cannot exceed 1 oz (28 g) by weight from impact until cessation of motion of the vehicle; 5 oz (142 g) by weight in the 5 min following cessation of motion; or 1 oz (28 g) by weight per min for 25 min.

In the rollover test, the vehicle is turned about its longitudinal axis successively in 90 degree increments and is held for 5 min in each position. Rotation occurs during a time interval of 1 to 3 min. Fuel leakage cannot exceed 5 oz (142 g) by weight in the first 5 min of each segment, nor 1 oz (28 g) per min for the remainder of the segment.

The school bus test procedure for buses over 10,000 lb (4536 kg) GVWR utilizes a special moving contoured barrier striking the test vehicle at the front, rear, and sides.

No rollover test is prescribed for these buses. Maximum permissible fuel leakage is the same as indicated above for the front, side, and rear barrier crash tests.

The other NHTSA firesafety standard, *Federal Motor Vehicle Safety Standard 302*, sets flammability limits for materials used in the interiors of vehicles in the driver and passenger space. Its requirements are directed at reducing the hazards of interior fires caused by smoking and matches. Vehicle appointments specifically covered by the standard include seat cushions, seat backs, seat belts, headlining, arm rests, trim panels, compartment shelves, head restraints, floor covering, engine compartment covering, padding, and the surface of crash protection components.

The standard specifies a horizontal flame test of all materials and incorporates detailed requirements for the test procedure. Material under test must not burn or transmit a flame across the surface at a rate exceeding 4 in. per min (1.7 mm/s).

BMCS, which is a part of the DOT's Federal Highway Administration (FHA), is another federal agency which administers regulations governing the safe operation of commercial vehicles in interstate or foreign commerce. The Bureau's regulations affecting vehicle firesafety include standards for electrical wiring, battery installation and protection, exhaust systems, and heating systems (and restrictions on the use of some types of heaters); a prohibition against the use of flame producing emergency warning devices on vehicles transporting certain classes of hazardous materials; and special regulations governing the driving and parking of vehicles transporting hazardous materials. The Bureau's Federal Motor Carrier Safety Regulations are set forth in *Title 49, Code of Federal Regulations*, Parts 390-397.

The federal regulations governing the transportation of hazardous materials (other than the driving and parking rules which are part of the Federal Motor Carrier Safety Regulations) are promulgated by the Materials Transportation Bureau. This agency is also part of the DOT in the Research and Special Programs Administration. For motor carriers, enforcement of the hazardous materials regulations is handled through BMCS.

The hazardous materials regulations cover matters such as: the proper preparation of shipping papers to facilitate identification of hazardous materials; marking and labeling of hazardous materials packaging; placarding of vehicles transporting materials; regulations applicable to shippers; authorizations of the packaging in which specific materials may be shipped to insure safety; and standards for packaging, including standards for cargo tanks.

Those provisions of the hazardous materials regulations which deal with the identification of the materials are specifically directed to the protection of fire fighters, other emergency service personnel, and the public by warning of the presence of such cargo and by providing information as to the identity of specific materials so emergency service personnel can take proper action in fire prevention, fire fighting, and cleanup.

The full text of the federal hazardous materials regulations is contained in *Title 49, Code of Federal Regulations*, Parts 170-189.

Many states have adopted by reference the provisions of the Federal Motor Carrier Safety Regulations and the DOT hazardous materials regulations. A number of other states have regulations which are similar to the federal requirements. As these trends continue, they will foster increasing uniformity of legal requirements.

The National Transportation Safety Board (NTSB) is an independent federal agency which exercises oversight of safety for all types of transportation. Probably best known for its investigations of aircraft accidents, the agency is also concerned with fire related highway accidents. The NTSB is charged with the responsibility to make recommendations for improved safety, based on its investigations. The majority of its recommendations deal with suggested changes in the regulations and practices of federal and state government agencies, but may be directed to others, such as vehicle manufacturers and users.

In the Congress, the responsibility for legislation and oversight on transportation rests with the Senate Committee on Commerce and the House Subcommittee on Commerce and Finance (Committee on Interstate and Foreign Commerce). These committees have taken a keen interest in the activities of the federal government in the field of motor vehicle safety. They regularly hold legislative and oversight hearings into the activities of the NHTSA and occasionally into the activities of the BMCS and the NTSB. Reports of these hearings are available from the U.S. Superintendent of Documents or from the committees.

DESIGN AND CONSTRUCTION SAFEGUARDS

Many items in the interior of vehicles are flammable, such as upholstery, paneling, trim, carpeting or floor mats. If ignition occurs from any source, and prompt control is not possible, the vehicle can be totally destroyed.

In large truck bodies, and in freight carrying trailers, wood is used as flooring. It withstands the torsional stresses, impacts, and abrasions encountered in this type of service, and replacement of worn or damaged sections is relatively economical. Wood provides safety benefits because blocking and bracing can be nailed in place to prevent shifting of cargo. It offers a high coefficient of friction for safe forklift operation and a slip resistant footing for personnel who handle cargo.

Plywood is used for interior lining material in van bodied trucks and trailers because of its effectiveness in protecting other structural components from damage incidental to the loading, unloading, and transportation of freight. As with wood flooring, the replacement of damaged sections is relatively inexpensive.

Plywood flooring is used in buses for essentially the same reasons. The use of metal flooring would create "tin canning" under torsional stress with resulting noise levels that would be unacceptable to occupants.

These uses of wood have not created any significant fire hazards. In the forms used, the wood is relatively difficult to ignite, and burns slowly. Specific provisions of the Federal Motor Carrier Safety Regulations require that flooring be in good condition. The use of flooring permeated with oil or gasoline is prohibited.

There have been reports of major accidents involving fire, in which the presence of wood has been mentioned as a contributing factor. However, these accidents have been of such great severity that the outcome would appear to be little affected if wood had not been present.

When the cargo must be covered during transportation on flatbed, open top, and other types of nonenclosed vehicles, the use of tarpaulins is the only practicable means of providing the needed protection. Such tarpaulins normally are not given flame retardant treatment, but fires rarely start in them.

Vehicle Fuel Tanks and Systems

The location, construction, and security of fuel tanks are important design features for firesafety in motor vehicles. Liquid fuel tanks for general passenger car use are generally thin-gage steel of various shapes and dimensions, dependent upon other body and chassis characteristics. The vast majority of U.S. built cars, trucks, and buses have liquid fuel tanks that are located at the rear of the vehicle, frequently in a position where they are not entirely enclosed in the body. The most severe casualties in terms of loss of life and property occur from fires following rear end collisions.

The Bureau of Motor Carrier Safety has developed fuel tank regulations applicable to vehicles operating in interstate or foreign commerce. These regulations closely parallel Underwriters Laboratories (UL) standards referred to elsewhere. In 1973, requirements which previously had been limited to gasoline tanks were extended to include tanks for diesel fuel. This action also required protection for diesel fuel crossover lines located below the bottom of a tank or sump. The DOT regulations include requirements for tank construction and installation, and for adequacy of fillpipe closure. Tanks must pass drop, rupture, vent, and spillage tests. Required safety vent systems must limit pressure rise to 50 psig (345 kPa) in a specified fire test.

Tanks must be marked by a manufacturer's certificate that they meet DOT –FHA specifications for sidemounted or nonsidemounted tanks, as applicable.

Efforts have been made to gain DOT approval for plastic fuel tanks for truck service. Despite certain advantages claimed by the manufacturers, doubts about the safety of such tanks in commercial vehicle operations have not been satisfactorily resolved.

Vehicles using flammable compressed gas for a motor fuel must be equipped with fuel systems constructed and installed in accordance with applicable provisions of NFPA 58, *Standard for the Storage and Handling of Liquefied Petroleum Gases (hereinafter referred to as NFPA 58)*, or NFPA 52, *Compressed Natural Gas (CNG) Vehicular Fuel Systems.*

Fuel systems must be maintained so they are free of leaks. Fuel tanks must be capped and connections checked regularly. Metal fuel lines are subject to vibration induced fatigue, and flexible lines may deteriorate due to age and can be damaged by chafing, cutting, or proximity to hot surfaces. In a gasoline system, flooding of the carburetor due to dirt in the float valve may cause an overflow of gasoline onto a hot surface and result in fire.

These problems are somewhat less critical on diesel powered vehicles because of the higher flash point of diesel fuel. In recent years, however, diesel engines have been designed to operate at higher temperatures to control exhaust emissions. This, in turn, has led to higher temperature of the fuel in the return flow line, thus making the fuel more readily subject to ignition if warmed, leaking fuel contacts a hot surface.

NFPA AND RELATED STANDARDS ON MOTOR VEHICLES

Three NFPA standards contain requirements for transportation by truck of specific hazardous materials. NFPA 495, *Code for the Manufacture, Transportation, Storage, and Use of Explosive Materials (hereinafter referred to as NFPA 495)*, details fire and explosion prevention as related to transportation of explosive materials, including driver qualifications, vehicle design, placarding, fire fighting equipment, and vehicle operation. This code is widely used as the basis for state regulations.

Requirements for design and construction of cargo tanks of tank vehicles, for auxiliary equipment, and for operation of tank vehicles are contained in NFPA 385, *Standard for Tank Vehicles for Flammable and Combustible Liquids*, and in NFPA 407, *Standard for Aircraft Fuel Servicing (hereinafter referred to as NFPA 385 and 407)*. NFPA 407 covers only the special requirements for aircraft fuel servicing tank vehicles and aircraft fuel servicing hydrant vehicles.

NFPA 58, mentioned earlier, contains requirements for transportation of liquefied petroleum gases, including requirements for transporting portable containers, requirements for cargo tanks, and for parking and garaging vehicles used to carry LP-Gas cargo. This standard also includes requirements for installation of LP-Gas systems in vehicles fueled by LP-Gas. It has been adopted for regulatory purposes by 47 states, and by the DOT for vehicles in interstate commerce.

NFPA 52, *Standard for Compressed Natural Gas (NG) Vehicular Fuel Systems* covers the design and installation of compressed natural gas (NG) engine fuel systems on vehicles of all types and to their associated fueling (dispensing) systems. NFPA 512, *Standard for Truck Fire Protection (hereinafter referred to as NFPA 512)*, covers property carrying motor vehicles.

As noted above, NFPA 58 covers parking and garaging of LP-Gas cargo vehicles. Two other NFPA standards pertain to parking of trucks: NFPA 513, *Standard for Motor Freight Terminals*, and NFPA 498, *Standard for Explosives Motor Vehicle Terminals (hereinafter referred to as NFPA 513 and NFPA 498)*. Although primarily concerned with fire protection of freight while in a terminal, NFPA 513 contains recommendations for parking vehicles at terminals. Parking of vehicles hauling explosives is the subject of NFPA 498. The seriousness of the problem is evidenced by the explosions of parked explosives trucks in Marshall Creek, PA, in 1964, and in Clyde, TX, in 1974. In the Pennsylvania incident involving the accidental detonation of dynamite, other explosives, and blasting caps, 6 persons were killed, 13 injured, and $600,000 lost in property damaged and destroyed. In the Texas incident, an overheated tire on an oil field truck flashed, causing detonation of explosives on the truck. The accident killed 4 persons and injured 2 others.

The dependence of the fire service upon motor vehicles is well recognized, and it is particularly important that motor fire apparatus be safely constructed. NFPA 1901, *Standard for Automotive Fire Apparatus*, and NFPA 414, *Standard for Aircraft Rescue and Fire Fighting Vehicles*, detail practices that should be followed on these types of apparatus.

Mobile homes are sometimes classified as "vehicles," but modern practice tends to remove them from this category. For 11 years, NFPA issued NFPA 501B, *Standard on Mobile Homes*, that covered body and frame design and construction, and plumbing, heating, and electrical systems for these vehicles. That standard was withdrawn in 1978; authority for the regulation of mobile homes now rests with the federal government through its *Mobile Home Construction and Safety Standards*, Part 280 of the *Code of Federal Regulations*, Title 24.

Recreational vehicles can be divided into four categories: (1) travel trailers, (2) motor homes, (3) camp trailers, and (4) campers. NFPA 501C, *Standard on Firesafety Criteria for Recreational Vehicles*, covers details of plumbing, heating, and electrical systems for travel trailers, motor homes, truck campers, and camping trailers, and includes life safety and exiting requirements for these vehicles when they are used as temporary living quarters.

Requirements for plumbing are developed by the ANSI A119 Committee, of which the Recreational Vehicle Industry Association is Secretariat. Those requirements are published and distributed under one cover as ANSI A119.2, *Recreational Vehicles*.

Underwriters Laboratories Inc.

Certain over-the-road motor vehicle components and signaling appliances which have special fire hazard significance are tested by Underwriters Laboratories Inc. (333 Pfingsten Rd., Northbrook, IL 60662), and listed in UL's *Accident, Automotive, and Burglary Protection Equipment* lists.

The specific items listed include:

1. Electrical equipment (automobile fuses, switches).
2. Fill and vent fittings.
3. Fuel equipment (backfire deflectors, fuel feed systems, electric gasoline gages, automotive type LP-Gas accessories, automotive fuel tanks, tubing).
4. Automotive heaters (combustion types).
5. LP-Gas automotive vehicles (farm and road tractors incorporating fuel systems designed for use of LP-Gas as engine fuel).
6. Mufflers for automobiles.
7. Signals and signaling appliances (highway emergency signals).

UL also has its own standard, UL 395, *Standard on Automotive Fuel Tanks* for liquid fuel tanks mounted outside the frame or in other exposed locations on gasoline or diesel powered trucks, tractors, or trailers. It has issued UL 307(a), *Standard on Liquid Fuel-Burning Heating Appliances for Mobile Homes and Recreational Vehicles*, and UL 307(b), *Standard for Gas Burning Heating Appliances for Mobile Homes and Recreational Vehicles*.

Society of Automotive Engineers, Inc.

The Society of Automotive Engineers, Inc. (SAE; 400 Commonwealth Dr., Warrendale, PA 15096), is concerned with any vehicle that moves under its own power. SAE has a standardization program covering passenger cars, trucks and buses, farm and earth moving machinery, marine propulsion units, aircraft, and space vehicles, as well as the materials and components that go into them. Content of the SAE standards is limited to environmental and operating problems with these vehicles. The *SAE Handbook* contains all of the Society's surface vehicle documents.

OTHER COOPERATING AGENCIES INTERESTED IN HIGHWAY SAFETY

American Petroleum Institute (1220 L St. NW, Washington, DC 20005): Active in the field of truck transportation of petroleum products, API publishes a number of publications of interest and value from the firesafety viewpoint.

American Trucking Associations, Inc. (1616 P St. NW, Washington, DC 20036): Reproduces as a tariff the DOT *Regulations Governing the Transportation of Explosives and Other Dangerous Articles by Motor, Rail and Water*, including specifications for shipping containers. This action (similar to that taken by the Association of American Railroads) makes current DOT regulations available to participating truck carriers. The Association's Safety and Security Department sets trucking industry standards for selection, training and supervision of employees, and develops a variety of safety materials for the use of trucking companies.

Center for Auto Safety (2001 S St. NW, Suite 410, Washington, DC 20009): A consumer advocate organization dedicated to improving the safety and value of automobiles, other motor vehicles, highways, and mobile homes. The Center regularly issues reports, books, comments on federal rulemaking, and other materials.

Chemical Manufacturers Association (2501 M St. NW, Washington, DC 20037): Has an active technical service on transportation and packaging. CMA's *Chemical Safety Data Sheets* (listing available on request) give details on bulk transportation of chemicals.

Compressed Gas Association, Inc. (1235 Jefferson Davis Hgwy., Arlington, VA 22202): CGA provides technical services on products of interest and concern to its membership.

Consumers Union of U.S., Inc. (256 Washington St., Mount Vernon, NY 10553): A consumer information organization with a large technical staff, CU regularly tests automobiles (and occasionally trucks and other motor vehicles) and publishes its findings monthly in *Consumer Reports Magazine*.

Insurance Institute for Highway Safety (600 New Hampshire Ave., NW, Suite 300, Washington, DC 20037): An independent, nonprofit, scientific and educational organization dedicated to reducing the losses (deaths, injuries, and property damage) resulting from crashes on the nation's highways. The Institute is supported by the American Insurances Service Group, the American Mutual Insurance Alliance, the National Association of Independent Insurers, and several individual insurance companies.

Manufactured Housing Institute (1745 Jefferson Davis Hgwy., Suite 511, Arlington, VA 22202): Represents mobile home manufacturers east of the Rocky Mountains.

National LP-Gas Association (1301 W. 22nd St., Oak Brook, IL 60521): Develops LP-Gas standards and provides safety programs for the general public.

National Safety Council (444 N. Michigan Ave., Chicago, IL 60611): A nonprofit organization dedicated to accident prevention with an extensive program on highway safety geared principally to the motoring public. NSC produces a number of publications on transportation and highway safety.

National Tank Truck Carriers, Inc. (2201 Mill Rd., Alexandria, VA 22314): A Conference of the American Trucking Associations, Inc., NTTC assists members (operators of tank truck fleets) in solving problems associated with safe handling of flammable liquids, gases, and chemicals.

Physicians for Automotive Safety (P.O. Box 430, Armonk, NY 10504): This organization of medical doctors who are concerned with preventing death and injury in motor vehicle accidents has taken a particular interest in the safety of children in motor vehicles.

Professional Drivers Council (Suite 612, 2000 P St. NW, Washington, DC 20036): A membership group of truck and bus drivers with an interest in their occupational safety and health.

Recreational Vehicle Institute, Inc. (14650 Lee Rd., Chantilly, VA 22021): Represents the recreational vehicle industry. Manufacturing members produce travel trailers, truck campers, camping trailers, and motor homes.

The Truck Body and Equipment Association, Inc. (4907 Cordell Ave., Bethesda, MD 20814): A trade association of the manufacturers of truck bodies, suppliers of component parts, and distributors. It has a Fire Apparatus Manufacturers Division (formerly a separate organization known as Fire Apparatus Manufacturers Association). The TBEA represents its members in dealing with the DOT on safety legislation.

Truck Trailer Manufacturers Association (1020 Princess St., Alexandria, VA 22314): A trade association of manufacturers of truck trailers. Through its various committees and representatives, it cooperates with NFPA, other allied industry associations, and governmental agencies in matters affecting the design and fabrication of tank transports.

Western Manufactured Housing Institute (3855 E. LaPalma Ave., Anaheim, CA 92803): Represents the manufacturers of mobile homes and recreational vehicles whose equipment is made and/or distributed on the West Coast.

PREVENTION OF MOTOR VEHICLE FIRES

Fire prevention in motor vehicles requires the attention of everyone involved. Vehicle designers must be aware of fire hazards in their products. Sources of heat and ignition must be kept away from flammable materials as much as possible. Consideration should be given to the vulnerability of fuel system components: in their basic integrity, their location away from the perimeter of and from crash collapsible and intrusive parts of the vehicle, and their location away from potentially hot exhaust and electrical components.

Examples of location hazards were found in 1971-1976 Ford Pintos and 1975-1976 Mercury Bobcats. In these cars, a separation between the fuel filler neck and the tank could occur in a rear end collision, resulting in fire. A total 1.4 million of these vehicles were recalled for retrofitting of a longer fuel filler neck and for installation of a protective shield on the front of the fuel tank. Improperly routed fuel lines to the carburetor have been the subject of other recalls because of potential rupture and underhood fires.

Since electrical system failures often cause property damage fires, more attention needs to be paid to the routing of wiring, aging of insulation and components, and failure modes of components. (That is, do they short circuit or open circuit when they fail?)

To minimize the potential for fuel leakage in any crash, extraneous plumbing and openings should be eliminated in fuel and in hazardous liquid and gaseous cargo containers. Necessary plumbing should be designed so it is protected from potential crash damage.

In the construction of vehicles, care must be taken to ensure that the fuel and electrical lines are correctly routed and that connections are secure. In addition, location of sharp or pointed components near fuel or hazardous cargo tanks should be avoided.

The greatest fire hazard during the vehicle repair process is the use of torches, usually in the repair of exhaust systems and crash damage. Some of the items used in vehicle repair are particularly flammable, e.g., paint, solvents, adhesives, and oily rags. Operating an engine with the carburetor air cleaner removed may result in a fire if the engine backfires through the carburetor.

Care must be taken in the maintenance and rebuilding of crashed vehicles to ensure that their fuel and electrical systems are free of hazards not present in the original vehicle.

Vehicle Electrical Systems

Automotive lighting and accessory circuits are usually 6 or 12 V, but vehicle ignition systems have high voltage (though low amperage). Automobile electrical system firesafety is largely a matter of proper installation, fusing, and maintenance. Important check points include: (1) the location and protection afforded the battery; (2) battery cable integrity and protection of exposed battery terminals; (3) adequate flame retardant insulation on all wiring; (4) proper fusing of normal lighting and accessory circuits; (5) proper support, location, and security of all wiring, with adequate protection by rubber insulating bushings where wiring passes through metal; and (6) good ignition system maintenance.

Fires of electrical origin are usually fed by oily deposits in and around the engine, or combustible materials such as interior fabric linings and upholstery are ignited. In the event of collision or upset, electrical short circuits are very apt to occur and may cause ignition of fuel vapors. It is desirable, particularly on buses and trucks, to provide an approved type of manual battery/generator disconnect switch, preferably one incorporating short circuit supervision. However, use of such devices on vehicles is rare.

Miscellaneous Vehicle Hazards

Proper exhaust system installation is important because the hot surfaces of the system may ignite nearby combustible components. Hot carbon particles discharged from the exhaust can ignite flammable liquids, grease, or similar exposed materials.

Overuse of vehicle brakes can result in sufficient overheating to cause a severe smoldering condition, and sometimes fire. Improper adjustment preventing the full release of a brake can lead to similar conditions.

The problem of overheating brakes increases with the size and gross weight of the vehicle. On long downgrades, the driver must rely on the braking effect of the engine for the primary control of the vehicle, saving the brakes for limited use in slowing for sharp turns or unanticipated, localized, hazardous conditions and in case a stop is necessary. The maximum braking effect of the engine is achieved only in the lower gears of the transmission. Particularly on trucks or buses, it is essential for the driver to get into a low gear at the top of the grade before the speed of the vehicle reaches a point at which downshifting is no longer possible. In recent years, engine braking devices have been used widely on diesel powered trucks, especially in mountainous regions. Other types of electrically and hydraulically operated retarders are used in special situations.

Another serious common problem arising from overuse of brakes is the buildup of sufficient heat to cause brake fade. This results in loss of effectiveness of the brakes and the possibility of a major crash, whether or not fire ensues.

If one tire of a dual tire installation is run flat, or is seriously underinflated, flexing may result in sufficient internal heat buildup to cause ignition. This possibility also exists if tires are seriously overloaded, and may be exacerbated by sustained high speed operation and high ambient temperature. Occasionally, fire results if dual tires are spaced so closely together that they rub or chafe against each other when the vehicle is in motion.

Provisions of the Federal Motor Carrier Safety Regulations for all heating systems used on vehicles in interstate commerce have been upgraded. This has led to a significant reduction in the number of fires caused by combustion heaters on vehicles. The reduction of such fires has also been aided by technological advances in the heaters. The few fires that do occur generally result from unforeseen malfunction of a heater, or when freight is loaded too close to a heater or shifts against it in transit.

There is no longer any significant use of alcohol base antifreeze in motor vehicles. Therefore, vehicle fires involving its vapors are seldom a problem.

Farm Tractors and Two-Wheeled Vehicles (Motorcycles, Motorbikes, etc.)

Fire dangers inherent in these vehicles arise from the exceptionally close spacing of such components as the fuel tank, carburetor, ignition system, and exhaust pipe; the lack of protection afforded operators by the vehicles' framings; and the severe vibration forces to which the equipment is normally subjected. Vehicles with air cooled engines are particularly vulnerable; water cooled engines have lower engine operating temperatures. Particular care is required to avoid fuel system leakage or spillage (especially during fueling) because of the high surface temperatures and the close proximity of exhaust piping. Each fuel system should have means, readily accessible to the operator, for quickly stopping the flow of fuel to the tank. Sparks to ground from exposed electrical circuit points also are a major ignition hazard, as are oil and grease accumulations.

Use of Fire Extinguishers

There are no legal requirements for carrying fire extinguishers in private vehicles, although many people do keep them in vans, motor homes, and mobile homes. As a minimum, the vehicle owner should consider purchase of a portable extinguisher with a capability for dealing with Class B (flammable liquids) and Class C (electrical) fires. Also on the market are extinguishers utilizing multipurpose agents for use on Class A fires (ordinary combustibles) as well as Class B and C fires. Anyone equipping a vehicle with a fire extinguisher must understand that the best chance for its successful use lies in: (1) prompt action while the fire is in its early stages, and (2) proper use to avoid wasting the agent and exhausting the extinguisher's contents prematurely. Additional information on the ratings and use of fire extinguishers is found in NFPA 10, *Standard for Portable Fire Extinguishers.*

Commercial vehicles operating in interstate commerce must be equipped with fire extinguishers to comply with Federal Motor Carrier Safety Regulations. These regulations provide for two minimum levels of protection: a 5-B:C extinguisher is required if the power unit is not transporting hazardous materials in sufficient quantity to require marking or placarding of the vehicle; a 10-B:C extinguisher is required on a power unit used for transporting hazardous materials.

Many states have similar requirements for commercial vehicles, through adoption of the applicable provisions of the federal regulations or their own regulations. Some state laws require additional, or larger, extinguishers on tank vehicles transporting flammable liquids or gases, combustible liquids, or Class A and/or Class B explosives.

NFPA 512 references pertinent federal requirements as guidance to truck owners who otherwise may not be subject to regulatory requirements.

Carrying extra gasoline in a can is extremely dangerous because of possible accumulation of fumes. The hazard is even greater if glass or plastic containers are used, because of the greater likelihood of breakage or leakage.

Smoking by a driver or passengers presents the potential for fire if sparks, hot ashes, or incompletely extinguished matches get into upholstery or other interior materials. Such fires can be extinguished by water, or an agent suitable for Class A fires; the burn area then must be watched in case it continues to smolder and eventually rekindles.

While the vehicle is in operation, the driver should be alert for any indication of fire or a fire hazard such as the smell of something hot, of unusual fumes, or of leaking fuel or lubricant. The driver should check for conditions such as exhaust leaks and dragging brakes which can create sufficient heat buildup to cause fire.

In case of an accident, the engines and all electrical accessories of all involved vehicles should be turned off to minimize the danger of fire. At night, headlights can be left on to protect the scene, but persons in the vicinity should be alert for indications of overheated wiring or fire. Smoking and all use of open flames or lights must cease particularly if leaking fuel is detected. If a travel trailer or motor home is involved, the probable presence of liquefied petroleum gas should be kept in mind, smoking should be prohibited, and open flames and lights should be kept well away from the scene.

When a vehicle fire occurs, the following basic steps should be taken regardless of the type of vehicle involved:

1. Shut off the engine and all electrical systems that can be reached.
2. Get everyone out of the burning vehicle and well away from it, preferably uphill and upwind. The potential for fire is one justification for quickly moving injured persons from the scene of an accident.
3. Summon fire department assistance by the best available means.
4. Attempt to fight the fire only if it can be done without endangering anyone's personal safety.

Basic principles to be observed in attempting to fight a vehicle fire are:

1. The fire should be fought from the windward side so that wind, if any, will help carry the fire extinguishing agent toward the fire and help blow smoke and flame away from the person attempting extinguishment.
2. A fire extinguisher should be aimed at the base of the flames.
3. If leaking flammable liquid is involved, work toward the source of the leak in putting out the flames. Ultimate control may depend on the ability to shut off leaking fuel at its source.
4. If flammable gas is burning, allow it to continue to burn until the source of any leak is controlled. Otherwise, extinguishing the fire may permit leaking gas to accumulate with the likelihood of later fire or explosion. If flames are impinging on a flammable gas container, the container should be kept cool to avoid flame induced weakening and possible rupture, or a BLEVE (Boiling Liquid-Expanding Vapor Explosion).
5. If no extinguisher is available, throw sand or dirt on a fire to control flames.
6. If stopping to help at the scene of a vehicle fire, other drivers should park uphill, preferably upwind, and at a safe distance from the burning vehicle.
7. After a fire has been extinguished, a vehicle should not be operated until all burned parts are cool and the cause of the fire has been corrected.
8. During an electrical fire, turn off the ignition switch and lights to try to cut off the flow of current and permit faster extinguishment. If tools are available and it can be done safely, the battery should be disconnected.
9. In fighting an underhood fire, use great care in opening the hood to avoid injury in case of a sudden flareup.
10. All extinguishers used should be inspected and recharged without delay.

The following special considerations apply to fires and potential hazards involving commercial vehicles:

1. Any hazardous situation should be evaluated to avoid endangering lives and wasting extinguishing agent.
2. If a vehicular combination unit is involved, and it can be done safely, the power unit should be unhooked and moved a safe distance away from the vehicle.
3. Drivers must be alert for the hard pulling characteristics that may indicate dragging brakes and readjust the brakes before a fire develops.
4. Drivers should check tire inflation and temperature every 2 hours or 100 miles (161 km), whichever comes first, and at any other stop. For a vehicle placarded for transportation of hazardous materials, this practice is required by BMCS regulations.
5. Drivers should check the rearview mirrors frequently for signs of smoke, and stop and investigate any smoke detected.
6. If a hot or underinflated tire is discovered, it should be moved well away from the vehicle. As an alternative, the driver should remain with the vehicle until the tire is cool to the touch, and then make repairs. If a vehicle is left with a hot tire, the tire may burst into flames and destroy the vehicle and load. With proper tire changing equipment and a pair of heavy leather gloves, a driver can remove a hot tire safely, if it has not burst into flames.
7. Because a tire fire results from internal heat buildup, portable extinguishers normally cannot do more than control open flames. The cooling effect of large quantities of water is required for extinguishment.
8. If fire is suspected inside a closed van body (usually indicated by smoke seeping out around the doors), the doors should be left closed until fire department assistance is at hand. Should the doors be opened, oxygen reaching the fire may cause a sudden flareup. As long as the doors are left closed, the fire can only smolder for lack of oxygen.

In the interest of preventing cargo fires, motor carriers generally prohibit smoking while vehicles are being loaded or unloaded, so sparks or hot ashes will not lodge in the cargo and burst into flame later.

Federal Motor Carrier Safety Regulations prohibit smoking during refueling; while loading, unloading, or transporting explosives, flammable liquids, gases, or solids, or oxidizing materials requiring marking or placarding of the vehicle; and while driving an empty tank vehicle which last contained a flammable liquid or gas and which was required to be marked or placarded.

CARGO TANKS (TANK TRUCKS)

Cargo tanks and portable tanks are large containers for hauling hazardous materials in bulk. Cargo tanks are sometimes called tank motor vehicles or tank trucks. Any person offering a hazardous material in a container for transportation must determine that the container used is authorized for the commodity (per CFR Title 49 Part 172.101). If the container such as a cargo tank is supplied by the carrier, the shipper still must determine that the tank supplied is a proper container for the material being offered. The shipper must either examine the manufacturer's specification marking on the tank, or obtain papers from the carrier certifying the tank is proper for the material (per CFR 49 Part 173.22).

Cargo tanks are the most common transport vehicles used for the movement of combustible, flammable and corrosive materials, as well as for compressed gases (flammable and nonflammable). When these hazardous materials are transported in bulk containers, such as cargo tanks, and an accident occurs, the disaster potential can be enormous.

To ensure that the risks involved in the movement of potentially hazardous commodities are held to a minimum, and to reduce the hazards should an accident occur, the federal government and some state governments specify requirements for the manufacture, operation, maintenance, inspection, testing, and repair of cargo tanks.

However, accident experience has shown a high incidence of leakage resulting from both contact and noncontact accidents, leading to high cost in terms of personal injury and property damage. Data gathered by the Materials Transportation Bureau and the Bureau of Motor Carrier Safety continue to indicate that major causes of unintentional release of hazardous materials are poor maintenance and deviation from tank specifications (NTSB 1983).

Potential hazards to fire fighters dealing with cargo tanks transportating hazardous materials are as follows:

1. Multipurpose tanks holding improper commodities for the specific container.
2. Multicompartment containers hauling materials in more than one hazard class (corrosive/flammable/poison/oxidizer).
3. Defective or missing safety controls.
4. Release of large quantities of hazardous materials.
5. Increase of potential fires (extra equipment required to fight fires and control spills).
6. Potential chances of a BLEVE.
7. Incorrect or missing placarding and identification markings on exterior of cargo tank.
8. Missing, incomplete, or incorrect shipping papers.

Directions for personnel receiving the original notification of a transportation problem should contain the following:

1. Location.
2. Type of container (tank truck, tank trailer, or farm truck).
3. Quantity.
4. Product name (often labeled on container).
5. Current status (fire, vapor, or liquid leakage).
6. Weather and wind direction.
7. Vehicle company and telephone number.
8. Type of area.
9. Number of placard (or papers if available).

Emergency personnel arriving at the scene should approach the vehicle from upwind. If this is not possible, breathing apparatus should be used in a downwind approach. All leaking liquids must be avoided. Many liquids will assimilate through rubber boots and leather shoe soles, enter the blood stream and cause death. Fire and/or hazardous materials units responding to an incident

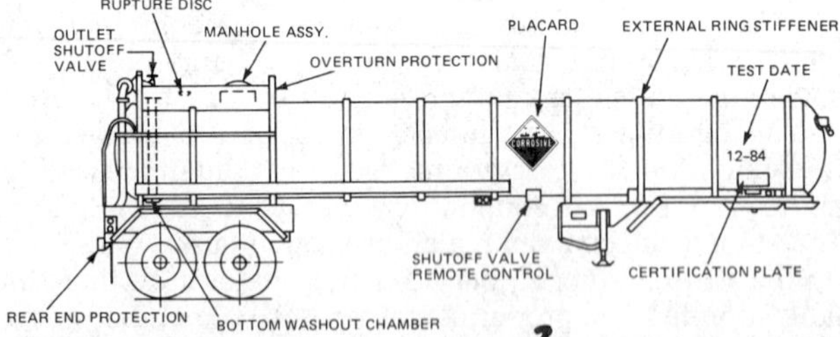

FIG. 13-2A. Corrosive materials cargo tanks must be equipped with internal shutoff valve if unloaded from the bottom.

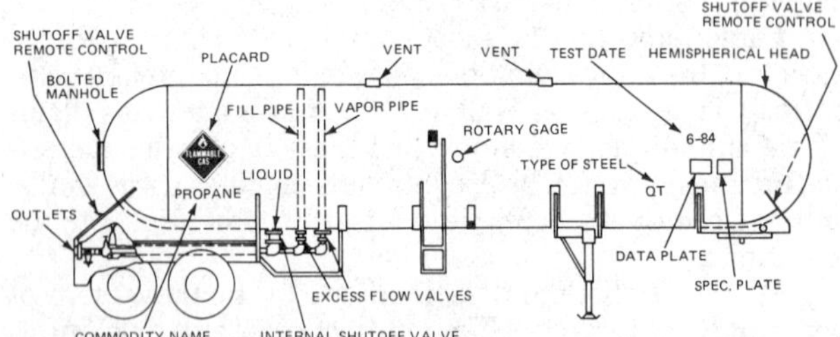

FIG. 13-2B. Compressed gas cargo tanks must be equipped with internal shutoff valves at liquid and vapor discharge openings.

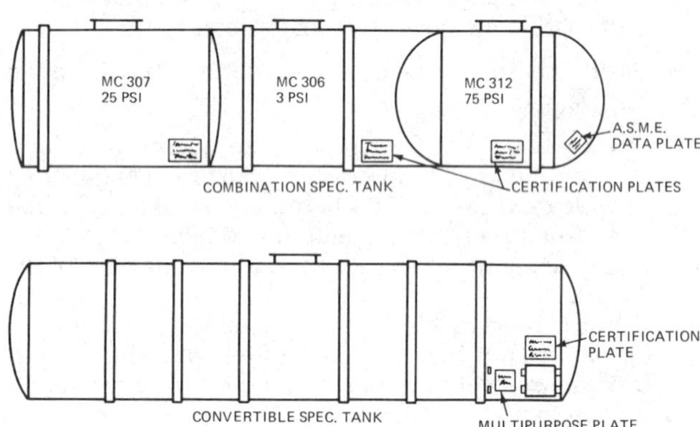

FIG. 13-2C. Profiles of combination (multiple compartment) and convertible (single compartment) cargo tanks.

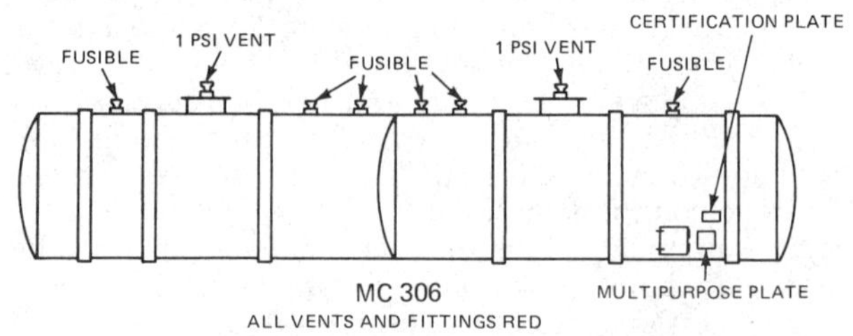

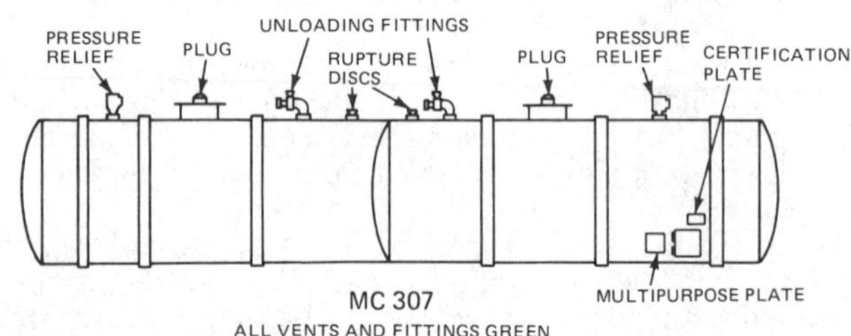

FIG. 13-2D. Profiles of two-compartment convertible cargo tanks.

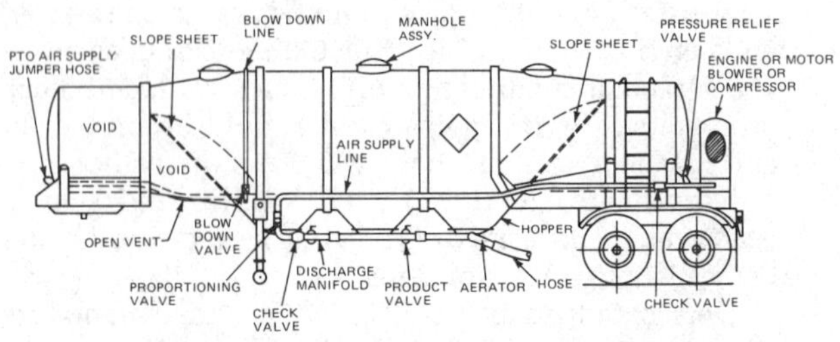

FIG. 13-2E. Profile of a dry bulk cargo tank. Examples of cargo carried are ammonium nitrate (an oxidizer) or corrosive solids.

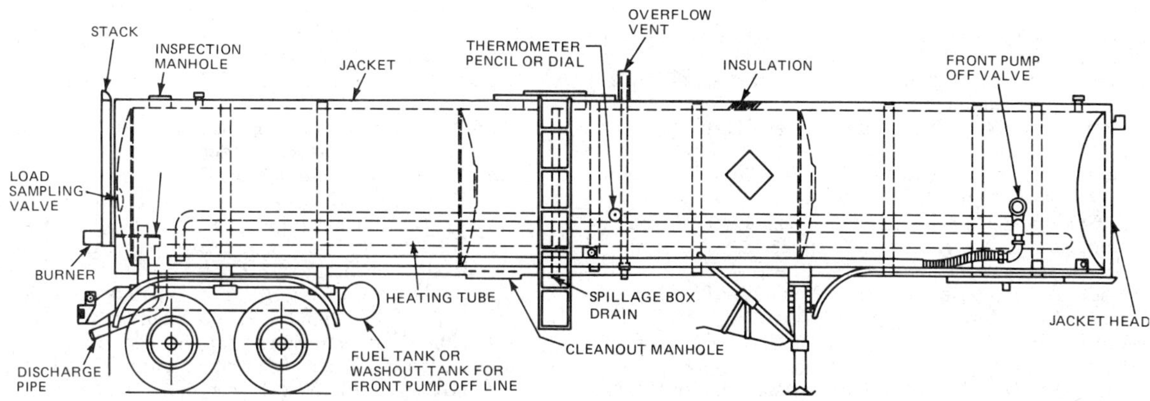

FIG. 13-2F. A nonspecification asphalt cargo tank.

should carry plug and dike material to contain leakage. All leakage must be contained and neutralized so it can be removed in a solid state. Washdowns must be eliminated in order to prevent the leakage from spreading into the environment. Although vaporous gases are usually noncontainable, applying foam or covering the tank with plastic sheets will often curtail the dispersement.

Figures 13-2A through 13-2F display tank construction, denoting areas where leakage is most likely to occur. All of these diagrams should be considered as guidelines rather than constants.

All tank units, with the exception of water tankers, can contain hazardous materials of one or more types. These units, which include farm product tanks, tank trucks, tank trailers and railroad tank cars, must all be treated with equal respect. Emergency personnel responding to an accident scene historically attempt to rescue the truck driver. Unfortunately, rescue often is impossible and the efforts result in serious injury or death of the would-be rescuers. Each emergency situation must be sized up quickly and individually, taking into consideration the safety of every person at the scene.

Bibliography

References Cited

NTSB. 1983. *Safety Recommendation.* H-83-25. National Transportation Safety Board, Washington, DC.

NFPA Codes, Standards, Recommended Practices and Manuals. (See the latest *NFPA Codes and Standards Catalog* for availability of current editions of the following documents.)

NFPA 52, *Standard for Compressed natural Gas (NG) Vehicular Fuel Systems.*

NFPA 58, *Standard for the Storage and Handling of Liquefied Petroleum Gases.*

NFPA 385, *Standard for Tank Vehicles for Flammable and Combustible Liquids.*

NFPA 407, *Standard for Aircraft Fuel Servicing.*

NFPA 414, *Standard for Aircraft Rescue and Fire Fighting Vehicles.*

NFPA 495, *Code for the Manufacture, Transportation, Storage and Use of Explosive Materials.*

NFPA 498, *Standard for Explosives Motor Vehicle Terminals.*

NFPA 512, *Standard for Truck Fire Protection.*

NFPA 1901, *Standard for Automotive Fire Apparatus.*

Additional Readings

Angelo, Phil, "Hazardous Cargo!," *Fire Command*, Vol. 41, No. 8, Aug. 1974, pp. 36-7.

"Bulk Tank Truck, Launceston, Tasmania," *Fire Journal*, Vol. 65, No. 2, Mar. 1971, pp. 22-24.

Cooley, Peter, *Fires in Motor Vehicle Accidents: An HSRI Special Report*, Highway Safety Research Institute, University of Michigan, Ann Arbor, MI, Apr. 1974.

Demers, David P., "Seven Dead in Tank Truck Fire," *Fire Command*, Vol. 45, No. 4, Apr. 1978, pp. 26-7.

Facts for Drivers, Safety and Security Department, American Trucking Associations, Inc., Washington, DC.

"Gasoline Tanker Fires Hit Two Communities," *Fire Command*, Vol. 42, No. 11, Nov. 1975, pp. 19-21.

Handling Hazardous Materials, Safety and Security Department, American Trucking Associations, Inc., Washington, DC.

Johnson, E. F., "Fire Protection Developments in LNG-Fueled Vehicle Operations," *Fire Journal*, Vol. 66, No. 6, Nov. 1972, pp. 11-15.

Lathrop, James K., and Walls, Wilbur L., "Fires Involving LP-Gas Tank Trucks in Repair Garages," *Fire Journal*, Vol. 68, Sept. 1974, pp. 18-20.

McLain, R. F., "Tank Truck Fire in Richmond, VA," *Fire Command*, Vol. 41, No. 4, Apr. 1974, pp. 70-79.

National Transportation Safety Board, *Federal Legislation Affecting the Transportation of Hazardous Materials with Selected Bibliographies and Data*, prepared for National Strategies Conference on Transportation of Hazardous Materials and Hazardous Wastes, Williamsburg, VA, Feb. 1981.

NTSB, *Safety Effectiveness Evaluation—Federal and State Enforcement Efforts in Hazardous Materials Transportation by Truck*, Report NTSB-sEE-81-2, National Transportation

Safety Board, Washington, DC, Feb. 18, 1981.
Schmidt, J. W., and Price, D. L., "Flow of Hazardous Materials on Highways," *Journal of Safety Research*, Vol. 11, No. 3, 1970, pp. 109-114.
"Seven Fire Fighters Injured in LP-Gas Tank 'BLEVE', *Fire Command*, Vol. 41, No. 8, Aug. 1974, pp. 34-35.
Sharry, John A., "Chemical Explosion—Los Angeles, California," *Fire Journal*, Vol. 69, No. 1, Jan. 1975, pp. 62-63.
Sharry, John A., and Walls, Wilbur L., "LP-Gas Distribution Plant Fire," *Fire Journal*, Vol. 68, No. 1, Jan. 1974, pp. 52-57.
Sickles, M. C., "Not Just an Ordinary Accident," *Fire Command*, Vol. 45, No. 6, June 1978, pp. 30-31.
Stillman, Timothy G., "Foam Attack on Tank Fire," *Fire Command*, Vol. 38, No. 10, Oct. 1971, p. 26.
Storrs, C. D., and Lindemann, O. H., "Federal Flammability Standards for Interiors of Motor Vehicles," *Fire Journal*, Vol. 66, No. 4, July 1972, pp. 24-27, 44.
"The Metro Bus Fires," *Fire Journal*, Vol. 70, No. 2, Mar. 1976, pp. 61-71.
Transportation of Hazardous Materials: Toward a National Strategy (Vols. I and II), Special Report 197, Transportation Research Board, National Academy of Sciences, National Research Council, Washington, DC, 1983.
Truck Drivers Handbook, Safety and Security Department, American Trucking Associations, Inc., Washington, DC.
Ward, Will, "Tanker Fire Contained," *Fire Command*, Vol. 45, No. 7, July 1978, pp. 28-29.
"Watch Those Gasoline Tanks!," *Fire Command*, Vol. 38, No. 5, May 1971, p. 23.

The following reports are sponsored by the National Highway Traffic Safety Administration:

An Assessment of Automotive Fuel System Fire Hazards, DOT HS-800 624, Dec. 1971, Dynamic Science, Phoenix, AZ.
Escape Worthiness of Vehicles and Occupant Survival, DOT HS-800 428, Dec. 1970, University of Oklahoma Research Institute, Norman, OK.
Flammability Characteristics of Vehicle Interior Materials, DOT HS-800 205, May 1969, Engineering Mechanics Division, IIT Research Institute, Chicago, IL.
Prevention of Electrical Systems Ignition of Automotive Crash Fire, DOT HS-800 392, Mar. 1970, Dynamic Science, Phoenix, AZ.

RAIL TRANSPORTATION SYSTEMS

Revised by Roger K. Fitch

Railroad trains transport billions of dollars worth of merchandise annually. With approximately 210,000 Class I miles (337 964 km) of main line track operated in freight service in the United States and Canada, the rail network is a vital link in our transportation system. A fire that incapacitates even a portion of the rail system effects much more than transportation.

This chapter identifies the different types of rail rolling equipment used by North American railroads and discusses fire hazards associated with them. Fire problems involving rights of way, specialized railroad facilities, and the hazardous materials transported by rail also are described.

Rapid transit rail systems used to move people in urban environments (mass transit) are covered in Chapter 4 of this Section. Other helpful information will be found in Section 6, Chapter 4, "Obtaining Technical Assistance in Hazardous Materials Emergencies."

Rail rolling equipment, also called "rolling stock," includes locomotives, freight cars, cabooses, and track work (maintenance of way) equipment.

FIRE LOSS STATISTICS

The fire loss statistics in Tables 13-3A, 13-3B, and 13-3C were abstracted from surveys conducted by the NFPA Railroad Section's Committee on Statistics, based on data from U.S. and Canadian railroads.

MOTIVE POWER

Fire Protection for Diesel Electric Locomotives

What is generally referred to as a diesel locomotive is actually a diesel electric locomotive. (See Fig. 13-3A.) The power is developed by the diesel engine which, in turn, drives a main traction generator. The output of this generator is then transmitted to the traction motors (mounted in the trucks and geared to the axles and wheels) which actually drive the locomotive. The numerous support systems of a locomotive include: a battery, which provides power to start the diesel engine; an auxiliary direct current generator, which charges the battery and also supplies low voltage direct current for control and lighting circuits; and an auxiliary alternator (mounted integrally with the main traction generator) which furnishes power for excitation, for the radiator cooling fan motors, and for a blower motor which cleans the inertial carbody filters.

Mainline locomotives have a normal service life exceeding 15 years; thousands of units still in service have operated for 20 years or longer and average between 150,000 and 250,000 miles (241 403 and 402 338 km) per year.

The diesel electric locomotive is an enclosed, self-contained piece of equipment. Fire prevention in the engine room is almost exclusively limited to housekeeping and maintenance. This includes stopping and repairing fuel oil, lubricant oil, or exhaust leaks and assuring that debris and products of leakage do not accumulate. Improved carbody filtration through the development of self-cleaning inertial type carbody filters and carbody pressurization also have helped to improve engine room safety of modern diesel electric locomotives.

Other improvements in recent years have resulted in virtually "fireproof" electrical cabinets on some locomotives. Among these improvements are the use of fire retardant wire and cable insulation and the practice of ventilating electrical cabinets with filtered air. This prevents flammable gases from collecting and keeps dirt out of the electrical cabinet. Other improvements include: better wiring insulation and wiring techniques, increased use of solid state electronics, increased use of circuit breakers and fuses, and the pressurization of electrical cabinets. Design improvements in direct current traction motors and the advent of the ac/dc traction generator (replacing the older dc generator) have also reduced fire hazards.

Fire protection for diesel locomotives includes the following:

Hot Engine Protection: To protect against excessive temperatures in the engine cooling system and hot engine oil,

Mr. Fitch is Fire Protection Engineer for the Union Pacific Railroad Company.

TABLE 13-3A. Railroad Fire Loss Statistics

	1981		1982		1983	
	No.	$	No.	$	No.	$
Group 1—Operations & Transportation						
Brake shoe sparks	97	1,011,003	188	2,597,523	63	292,191
Collision or derailment	15	2,518,276	19	1,728,284	12	2,528,207
Cotton, all causes	4	46,009	3	27,950	2	2,933
Electrical components	80	881,874	53	514,382	42	514,499
Engine exhaust sparks	101	259,990	120	112,645	195	599,973
Freight car or van heaters	6	6,378	3	102,416	17	5,563
Fusees	12	64,859	7	5,300	5	148
Hot journal boxes	34	1,070,538	18	57,627	11	7,217
Hot lading	101	804,136	2	20,800	6	53,120
Internal combustion engines	30	58,260	25	236,624	8	93,394
Other	45	319,605	82	398,403	78	527,985
Group 2—Maintenance & Servicing						
Smoking	17	278,811	11	31,773	19	209,274
Electrical	35	356,280	22	362,641	22	127,930
Flammable liquids	10	143,120	7	2,600	7	1,450
Fueled heaters and appliances	53	864,003	69	1,727,359	29	164,489
Open burning on right-of-way	86	114,257	117	129,956	8	9,875
Radiated heat	10	12,253	1	0	4	50,200
Spontaneous ignition	21	49,927	20	80,465	15	45,622
Welding, cutting or brazing	89	586,721	59	1,640,935	63	354,447
Other	27	35,917	28	74,074	20	366,954
Group 3—Outside Causes						
Exposure fires	60	1,115,647	27	346,171	16	45,034
Lightning and storms	12	200,546	6	13,560	8	100,890
Trespassing including arson	354	8,907,583	269	3,311,590	202	4,856,520
Other	16	364,626	27	428,240	13	62,504
Group 4—Cause Undetermined	262	2,525,644	359	2,612,356	607	1,008,097
Total	1,577	22,596,263	1,542	16,563,674	1,472	12,028,526

devices are installed to prevent buildup of excessive heat in the engine and supporting systems.

Engine Crankcase Protection: A device designed to detect dangerous buildup of pressure in the engine, and to shut down the engine when necessary.

High Voltage Ground Protection: If a high voltage fault develops on the locomotive and goes to ground, the ground relay will detect the fault and disconnect the electrical load to the locomotive.

Emergency Fuel Cutoff Switches: In an emergency, the fuel supply to the engine can be stopped by pressing any one of three emergency fuel cutoff push buttons. Two push buttons are located on the underframe in the vicinity of the fuel filler, one on either side of the locomotive; the third button is on the engine control panel in the cab.

Main Battery Knife Switch: This large, single-throw knife switch is located in the engine cab at the lower portion of the fuse panel. It connects the battery to the locomotive low voltage system.

Fire Protection Equipment: Each diesel electric locomotive is equipped with portable fire extinguisher protection (most commonly, ordinary dry chemical), the quantity and

TABLE 13-3B. Additional Data on Railroad Fire Loss Experience

	1981	1982	1983
Railroads surveyed*	21	23	22
Class I trackage surveyed, miles†	134,418	143,985	145,881
Estimated percentage of total trackage	61.2	67.7	69.5
Total Class I trackage, miles†	219,704	212,698	209,851
Number of fires per mile†	0.01173	0.01071	0.01009
Dollar loss per mile†	$168.10	$115.04	$82.45
Total dollar loss, industrywide (est.)	$36,939,234	$24,448,064	$17,299,171

* The estimated total dollar loss, industrywide, is a simple ratio of reported loss to the percentage of total Class I trackage reported for the U.S. and Canada. As such, it must be considered an approximation. During the three-year time span shown here, most frequent causes of railroad fires are: "Cause undetermined;" "Trespassers, including Vandalism and Arson;" and "Engine Exhaust Sparks." The latter, after years of supposed decline, again became a leading cause of railroad fires in 1981 when the report form was changed to encourage their reporting. On the other hand, the traditional top fire causes, "brake shoe sparks" and "overheated journals," continue to decline in importance.

† 1 mile = 1.6 km.

TABLE 13-3C. Major Causes of Railroad Fires*

	1981		1982		1983	
Cause	$ Loss	Freq.	$ Loss	Freq.	$ Loss	Freq.
Trespasser/Arson	8,907,583	(354)	3,311,590	(269)	4,856,520	(202)
Undetermined	2,575,644	(262)	2,612,356	(359)	1,008,097	(607)
Collision/Derailment	2,518,276	(15)	1,728,284	(19)	2,528,207	(12)
Brake shoe sparks	1,011,003	(97)	2,597,523	(188)	292,191	(63)
Engine exhaust sparks	259,990	(101)	112,645	(120)	599,973	(195)
Welding/Cutting	586,721	(89)	1,640,935	(59)	354,447	(63)

* Compiled by the NFPA Railroad Section from reports representing 60–70 percent of railroad Class I (mainline) track in U.S. and Canada.

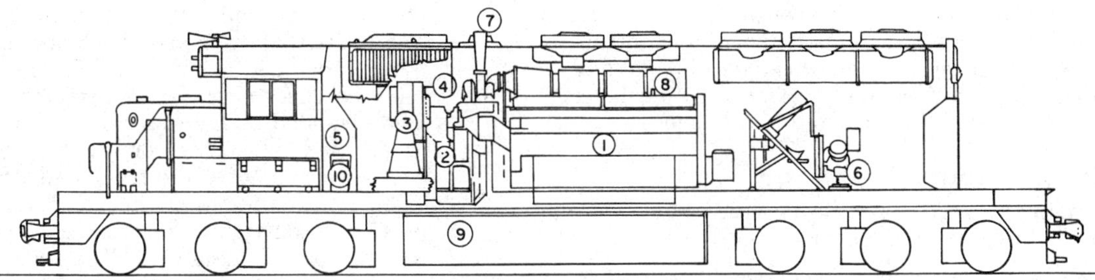

FIG. 13-3A. A diesel electric locomotive showing the location of principal components of the power system. 1, engine; 2, generator-alternator; 3, traction motor-generator blower; 4, auxiliary generator; 5, electrical control cabinet; 6, air compressor; 7, engine exhaust stack; 8, exhaust manifold; 9, fuel tank; and 10, electrical cabinet air filter.

type depending upon the size of the locomotive and its use (yard, local, or mainline service).

Preventive Maintenance and Inspections: Based on lifetime expectancy of component parts, empirical failure data, and Federal Railroad Administration (FRA) regulations, diesel electric locomotives are inspected, tested, and serviced on a regularly scheduled basis.

Fire Protection for Electric Locomotives

Figure 13-3B illustrates the general arrangement of an electric locomotive. This particular unit can deliver 5,100 hp (3.8 MW) at the rail and operates on 25,000 or 50,000 V alternating current from an overhead wire.

Because the electric locomotive derives its power from an electrical circuit, it is necessary to deenergize the circuit to safely extinguish any electrical fire.

Emergency Stop Buttons: These buttons, located one on each side of the locomotive under the operator's cab, can be reached from the ground. Pushing either of the buttons will trip the vacuum circuit breaker on the roofs of all the locomotives in the consist (multiple locomotives in a train) and remove the power.

Manual Pantograph Grounding Switch: Located in the right rear of the locomotive, this switch is used to ground the pantograph before personnel perform maintenance or climb to the roof through the rear hatch. Grounding of the pantograph (catenary), described in special instructions, will vary depending upon the type of locomotive and overhead system employed. Fire fighters should obtain this information from railroad officials and include it in prefire plans.

Emergency Shutdown Switch: This switch, located on the master control housing, is used in emergencies to remove power from the locomotive during single unit operation, or to remove power from all units of the consist during multiple operation.

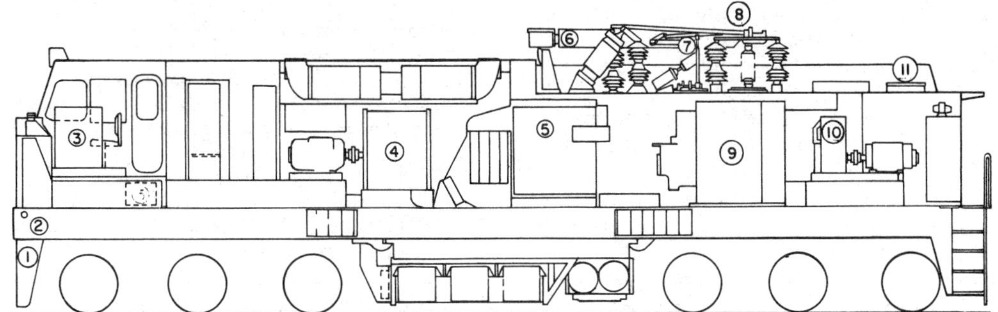

FIG. 13-3B. An electric locomotive showing the location of some essential pieces of apparatus: 1, proximity switch; 2, emergency stop button; 3, operator's console; 4, equipment blower; 5, main rectifier compartment; 6, vacuum circuit breaker; 7, pantograph grounding hook; 8, pantograph (in folded position); 9, transformer; 10, air compressor; and 11, roof hatch.

Electric locomotives, by their very nature, operate with voltages considerably higher than diesel locomotives. Modern electric locomotives have vacuum breakers which trip in cycles, thereby considerably reducing short circuits which could cause fires on older motive power.

Preventive Maintenance: Preventive maintenance is mandatory to preclude downtime caused by fire or equipment failure. Electrical circuits are tested for grounds or other defects when equipment is at major shops, or at intervals of no more than three months. All tests and scheduled maintenance are performed in compliance with U.S. Department of Transportation (DOT) rules, as a minimum.

Fire Protection Equipment: The majority of electric locomotives are equipped with portable fire extinguishing equipment (dry chemical and/or carbon dioxide) and, as with diesel locomotives, the number of extinguishers is based on the size of the locomotive and its use.

Safety Precautions

Certain precautions must be observed when boarding an electric locomotive in an emergency:

1. Never enter any high voltage compartment when the locomotive is energized.
2. Never touch motors, switches, protective barriers, or other electrical apparatus without knowing their exact purpose and function.
3. Never climb on the roof of the locomotive unless the locomotive ground switch is secured and the catenary wire is deenergized.
4. To protect against serious or fatal personal injury, be sure the pantograph is actually latched down before opening the roof hatch.

FREIGHT CARS AND EQUIPMENT

The major causes of fires in freight equipment, including motive equipment, are: (1) collision/derailment, (2) suspected arson, (3) electrical, (4) brake shoe sparks, (5) welding/cutting, (6) hot journals or "boxes," and (7) heating appliances.

Design and preventive maintenance are the keys to reducing or eliminating fires caused by brake shoe sparks, exhaust sparks, hot journals or "boxes," electrical failures, and internal combustion engines. Recent design advances that should lessen the frequency of fires include: increased use of turbochargers, spark retention exhaust manifolds, nonsparking brake shoes, roller bearings, heat detectors, sensors and indication lamps which supervise electrical circuits and components, and the increased use and efficiency of hot box and dragging equipment detectors.

Cotton bales now are wrapped in shrink film, which eliminates the need for metal bale bands. A frequent cause of bale fires has been friction heat, originating when the bale bands moved and ignited loose cotton.

Fires resulting from collisions and derailments are a major concern of the railroad industry. To reduce these incidents, operating rules and safety programs are undergoing major revisions. A program to inventory all grade crossings has been completed recently and the inventory is being used by individual states to establish priorities for crossing projects. This systemized approach to reducing crossing accidents, combined with the Federal Aid Highway Act of 1976 and "Operation Lifesaver," is making the public aware of the dangers associated with railroad crossings.

Refrigerator Cars

Three types of cars are commonly used to transport perishable and semiperishable commodities that require refrigeration, heater service, or protection from the extremes of heat or cold:

1. The standard ice car (RS) is gradually being phased out of service. Few railroads still furnish ice for shipment in such cars; however, the cars may be used to keep commodities from freezing. Portable liquid fuel heaters that burn a mixture of methanol and isopropanol can be installed in the end ice bunkers. Alcohol foam is recommended as an extinguishing agent for these fuels, because water may be ineffective.
2. The insulated bunkerless car (RB) is widely used to protect semiperishable commodities such as canned goods, beer, grocery products, and drugs and medicines from extremes of heat and cold. These cars provide no refrigeration but are equipped for application of portable liquid fuel heaters. The heaters are usually suspended by chains from hooks (four per heater) in the ceiling of the doorway areas. The ceiling above each heater is protected by heat shields.
3. The mechanically refrigerated car (RP) is the most modern and versatile of the freight cars used for perishables. (See Fig. 13-3C.) It can provide automatically controlled temperatures from 0 to 70°F (−18° to 21°C). Foamed in place polyurethane insulation helps to assure consistent and dependable car temperatures. A compressor-evaporator is powered by a self-contained diesel engine generator set within the car. The compressor-evaporator also can be powered directly from the alternator output of a diesel electric unit or from any suitable 220 V standby power supply.

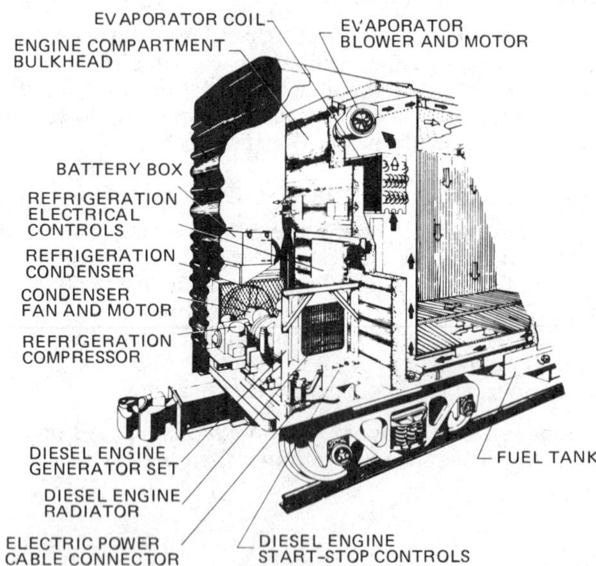

FIG. 13-3C. The engine compartment of a typical mechanically refrigerated car. Note that the start-stop control for the diesel engine is at the lower left, mounted on the side of the car.

The fuel tank for the diesel engine is suspended below the car and normally carries 500 to 550 gal (1.9 to 2.1 m^3) of No. 1 or No. 2 diesel fuel oil. Fire fighting is sometimes complicated by the burning of insulation and because fire fighters have not learned the location of the diesel engine stop control (which is clearly identified). Fire fighters should wear self-contained breathing apparatus when they enter a burning insulated or refrigerator car.

Box Cars

The design of all box cars is essentially the same; only the capacity and length of the cars vary. Additions such as special racks, dunnage devices, and nailable steel floors are made to prevent vibration damage to contents of the car without adding to the combustibility.

Specialized Equipment

Equipment used to meet special shipping needs includes articulated cars, multilevel and enclosed cars for automobiles, "high cube" (large capacity) box cars, 100 ton (90 metric tons) capacity covered hoppers, side loading "all door" box cars, and aerated cars to handle bulk materials which require pneumatic loading and unloading mechanisms to "inhale" cargo at the loading point and "exhale" it at destination.

Cabooses: The chief fire hazards of cabooses are their combustible furnishings and their heating systems. Most cabooses are heated with oil or coal. Fires start because stoves and stove pipes are placed without adequate clearances from combustible materials. Fire protection varies from water extinguishers to portable dry chemical extinguishers, although some carriers have discontinued providing this protection because of continuing problems with extinguisher theft. Cabooses are less likely to be used in the future, as "end of train" electronic devices take over their function.

Intermodal Equipment

This equipment can be a trailer or a container and is loaded on specially designed flat cars. The only difference between the two is that the trailer has wheels. Fire hazards are similar to those of the trucking industry, and the commodities carried range from explosives to frozen foods in refrigerated trailers.

RAIL TANK CARS

Tank cars are bulk railroad transport vehicles designed to carry compressed and liquefied gases, liquids, and solids with relatively low melting points.

Approximately 183,000 tank cars are currently in regular rail service and they comprise approximately 10 percent of the freight cars of the total railroad fleet in the United States. Virtually all tank cars are owned by nonrailroad companies. Tank cars transport nonhazardous as well as hazardous commodities. All pressure tank cars [generally those with tank test pressures of more than 100 psi (689 kPa)] and approximately one-half of the nonpressure tank cars [100 psi (689 kPa) or less tank test pressure] are found in hazardous materials service. An estimated 78 to 82 percent of the annual one million hazardous materials shipments by rail move in tank cars.

In most cases, a tank car is built to transport a single or a limited number of commodities. The number and types of commodities transported depend upon the size of the tank, tank construction, the fittings applied, the type of linings used, and other physical features of the tank.

As a rule, a tank car is built as a single tank permanently mounted on a rail car. However, more than one tank can be mounted on the rail car, as is the case with "ton containers" and high pressure service tank cars.

With ton containers, several multiunit tank car tanks are temporarily mounted on a special rail car for transportation, and are removed for filling and emptying. Ton containers move by highway as well as by rail. With the high pressure service tank car, 25 to 30 horizontal steel cylinders are permanently attached to the rail car. Both of these tank car tanks are discussed in detail below.

Sizes of tank cars vary from a few hundred gallons (100 gal equals 378 L) in the ton containers to a maximum of approximately 45,000 gal (170 m^3) in the jumbo cars. Since 1970, the maximum capacity of a new tank car has been limited by regulation to 34,500 gal (130 m^3) or 263,000 lb (119 295 kg) gross weight on rail (weight of the tank car plus the weight of the commodity being transported). The density of the commodity will determine the maximum tank capacity. Multiple compartments, which may transport different commodities, are found in some nonpressure tank cars.

In a rail accident involving tank cars, it is important to understand the potential hazard, i.e., to identify the contents. Tank cars have certain identifiable physical characteristics that can assist safety personnel in sizing up a situation. The following paragraphs describe some of the types of tank cars that might be encountered; their design, specifications and fittings; and tank car hazards.

It is dangerous for anyone to attempt to identify the contents of a rail tank car by handling its fittings or valves. An emergency responder requiring information on the contents of a rail tank car should get this information directly from the railroad. Any work on a car should be performed only by experienced railroad or shipper representatives. A well intended but incorrect action on a rail tank car can quickly increase the severity of an incident.

Design of Tank Cars

Tank cars are built to the standards common to all freight cars. In addition, they must meet the requirements of both the DOT Hazardous Materials Regulations, found in 49 CFR (*Code of Federal Regulations*) Part 179, and the standards of the Association of American Railroads (AAR), published in their *Specifications for Tank Cars*. Tank car builders must seek approval from the AAR Tank Car Committee before constructing a tank car. Repairs, alterations, or conversions to tank car tanks (at minimum, work requiring welding, riveting, or removal of deformations) must be performed only by AAR certified facilities.

All tank car tanks built since 1960 have two features in common—a circular cross section, and rounded ends, or plates (rounded to more efficiently distribute the internal pressures). Tank car tanks are made up of a shell and two heads. (See Fig. 13-3D.) The shell is constructed of 3 to 7 plates which are formed into cylinders. The heads are steel plates formed into flanged and dished or ellipsoidal pieces. The cylinders are welded together and a head is welded to each end to complete the tank. On some older

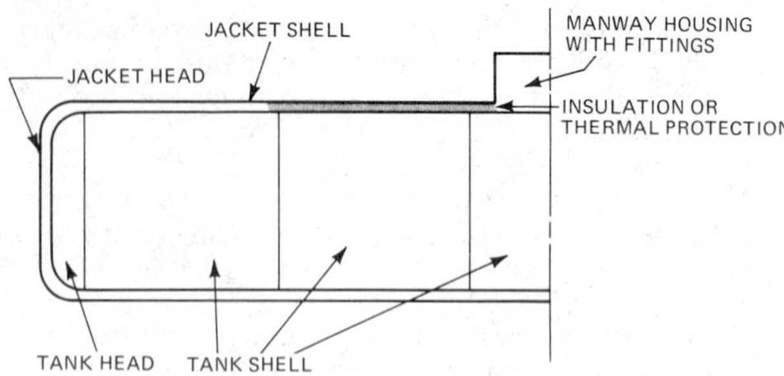

INSULATED OR THERMALLY PROTECTED TANK CAR TANK

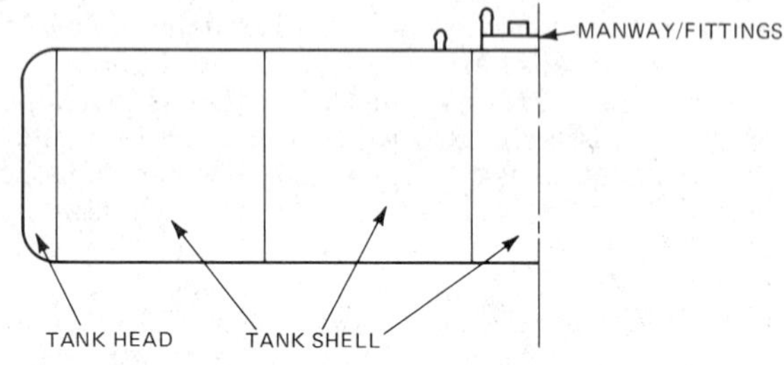

NON-INSULATED TANK CAR TANK

FIG. 13-3D. Tank car tank design and nomenclature.

nonpressure tank cars with expansion domes, the tank parts are riveted together. Fittings and other appurtenances are then added to the tank.

Carbon steel accounts for more than 90 percent of the tank car tanks in use, with aluminum making up most of the remainder. Nickel is used for tanks in selected acid service. Stainless steel, referred to in the regulations as alloy steel, is likewise found in a small number of tanks.

Steel tank cars are built on stub sills because the tank is strong enough to become the structure of the car. Aluminum tank cars, on the other hand, are built on a continuous underframe.

Some tank cars have jackets outside the tank that serve two functions: (1) they protect the insulation or the thermal protection from the weather; and (2) they provide a base for attachment of certain safety appliances—ladders, platforms, etc. Tank car jackets are usually made from carbon steel, although they can be aluminum.

The plate thickness of materials used to construct tank car tanks is specified by regulations. For nonpressure tanks, the plates must be at least 7/16 in. (11.1 mm) steel or 1/2 in. (12.7 mm) aluminum. For pressure tanks, the plates must be at least 9/16 in. (14.3 mm) steel or 5/8 in. (15.9 mm) aluminum. Tank car jackets must be constructed from at least 11 gage [approximately 1/8 in. (3.18 mm)] steel or aluminum plate.

Some tank cars are insulated in order to moderate temperature effects on the commodity being transported. Fiberglass and polyurethane foams are the two most common insulation materials. Cork is required for tank cars in hydrocyanic acid service. Perlite is used in cryogenic liquid tank cars.

Insulated tank cars are recognizable by their squared corners, where the jacket shell and head meet. Other indications include flashing over the bolster or tank bands, and flat sections on the sides of the jacket; also, the visible welds on an insulated tank car may appear rougher than those found on a regular tank.

Certain pressure tank cars may have thermal protection. Thermal protection, not to be confused with insulation, is designed to protect the tank from a pool fire for 100 minutes or a torch fire for 30 minutes.

Two types of thermal protection currently are used:

1. *Jacketed thermal protection*—mineral wool or various ceramic fiber blankets held in place by a jacket.
2. *Sprayed on thermal protection*—a rough textured coating sprayed onto the tank. This coating expands upon exposure to fire and thus protects the tank.

Thermal protection is part of a DOT retrofit program for pressure tank cars which also includes:

1. Top and bottom shelf couplers which are less likely to disengage during derailments.
2. Head puncture resistance (head shields) which protect the lower portion of the heads against punctures (some head shields are built into the jacket, while others are visible as trapezoidal plates of steel mounted on both ends of the tank car).

Some tank car tanks are lined to protect the tank from the commodity being transported. Linings may be materials applied after the tank is constructed, or cladding applied to the base metal before the plates are formed. Rubber is the most commonly used lining in hazardous materials service. Lithcote, (a paintlike material), nickel, stainless steel, and lead are also used as linings. Nickel and stainless steel are both used as claddings.

Some tank cars are equipped with internal or external heater coils. Steam or hot oil from an external source is run through these coils, heating thick or solidified materials (asphalts, heavy fuel oils, phenols, metallic sodium, or petroleum waxes) to make them flow easier while unloading.

Types of Tank Cars

Tank cars are divided into the following groups:

1. Pressure tank cars.
2. Nonpressure tank cars.
3. Cryogenic liquid tank cars.
4. Miscellaneous.
   a. High pressure service tank cars.
   b. Multiunit tank cars.

A tank test pressure of 100 psi (689 kPa) divides pressure from nonpressure tank cars.

Pressure Tank Cars

Generally, pressure tank cars transport nonflammable and flammable gases and Class A poisons. However, pressure tank cars can transport other commodities, depending upon the characteristics of the commodity or the process for filling and emptying the tank. Some of the other commodities transported in pressure tank cars include: ethylene oxide, pyrophoric liquids, N.O.S., sodium metal, motor fuel antiknock compounds, bromine, hydrofluoric acid (anhydrous), and acrolein.

Pressure tank cars range from 4,000 to approximately 45,000 gal (15 to 170 m^3) capacity with test pressures from

100 to 600 psi (689 to 4137 kPa). Pressure tank cars are basically circular in cross-section and are welded. Heads for pressure tank cars are formed convex outward.

Pressure tank cars are provided with:

1. A manway on top of the tank of sufficient size to permit access to the interior of the tank.
2. A manway cover to provide for the mounting of all valves, plus measuring and sampling devices.
3. A protective housing which is approximately 18 to 24 in. (457 to 610 mm) high and 30 to 36 in. (762 to 914 mm) in diameter.

Nonpressure Tank Cars

Nonpressure tank cars transport materials other than nonflammable and flammable gases and Class A poisons. These other materials include flammable and combustible liquids, flammable solids, oxidizers and organic peroxides, Class B poisons, and corrosive materials. The cars also transport many nonhazardous materials such as edible and inedible tallow, fruit and vegetable juices, tomato paste, and caramel. Nonpressure tank cars range from 4,000 to 45,000 gal (15 to 170 m^3) capacities with test pressures from 20 to 100 psi (138 to 689 kPa).

Nonpressure tank cars have circular cross-sections with heads formed convex outward. Nonpressure tank car tanks may be divided into compartments; however, each compartment is treated as a separate tank. Separate compartments can have different capacities.

Nonpressure tank cars are provided with:

1. A minimum of one manway or one expansion dome with a manway, to allow access to the interior of the tank.
2. External projections necessary for the filling and emptying of the tank, in addition to appropriate safety equipment.

Fittings on top of the tank allow personnel to distinguish visually between pressure and nonpressure tank cars. Pressure tank cars have all fittings under a single protective housing, while nonpressure tank cars have visible fittings or an expansion dome.

There are exceptions, however. For example, nitric acid is shipped in nonpressure tank cars with a protective housing, certain pressure tank cars have the safety valve outside the protective housing, and others have an auxiliary pressure manway.

Cryogenic Liquid Tank Cars

These cars are used to transport commodities at temperatures of −130 to −423°F (−90 to 253°C). The commodities transported are liquid hydrogen, liquid ethylene, liquid argon, liquid nitrogen, and liquid oxygen.

Cryogenic tank cars consist of a tank with circular cross-sections supported concentrically within another circular tank section. The space between the tanks is filled with an approved insulating material. The insulation is designed to protect the commodity only for 30 days, making these shipments time sensitive.

On cryogenic liquid tank cars, the tank heads are formed convex outside. These tanks are equipped with piping systems for vapor venting and commodity transfer. They are also equipped with pressure relief devices, controls, gages, and valves on the sides or ends of the tank at ground level. In one series of cryogenic liquid tank cars, the tank is located within a box car.

Miscellaneous Tank Cars

High pressure service tank cars, the rail equivalent of the highway tube trailer, will transport either helium or hydrogen (generally helium). Test pressures range to 4,000 psi (27 579 kPa).

The high pressure service tank car is approximately 40 ft (12.1 m) in length with open frame sides. Inside the frame is a visible cluster of 25 to 30 seamless steel cylinders permanently mounted horizontally on the car. The cylinders are hollow forged or drawn in one piece and are not insulated.

Although listed with tank cars in the regulations, "multiunit tank car tanks," or "ton containers," are also transported by highway. Ton containers carry gases including chlorine, anhydrous ammonia, sulfur dioxide, butadiene, refrigerant or dispersant gases, and phosgene. The tank test pressures range from 500 to 1,000 psi (3447 to 6845 kPa).

Ton containers are designed to be removed from the car structure for filling and emptying; these tanks have a circular cross-section. The heads are welded to the cylinder and may be either concave or convex outside. All fittings are located in the heads. Water capacity for these tanks ranges from 180 to 312 gal (681 to 1181 L). The multiunit tank car tanks are not insulated.

Tank Car Specification Markings

Tank cars are made to precise specifications and are identified by a specification marking stencilled predominently on the right hand side of the car, generally above the trucks. Example: Specification marking DOT-111A-100W4. (See Fig. 13-3E and Table 13-3D.)

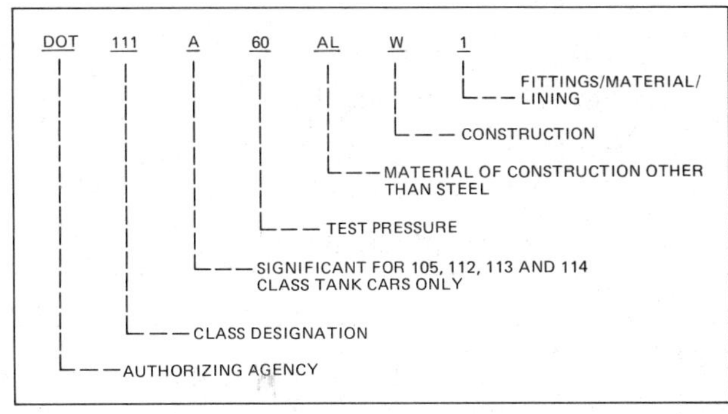

FIG. 13-3E. Sample tank car class and specification marking.

Tank car specifications are prefixed by three letters which represent the authorizing agency as follows:

DOT –Department of Transportation.
AAR –Association of American Railroads.
ICC –Interstate Commerce Commission (regulatory authority assumed by DOT in 1966).
CTC –Canadian Transport Commission.

AAR tank cars can differ from DOT tank cars by as little as the use of a fitting that does not meet DOT requirements, but is necessary for handling the particular commodity being transported. The tank class number is designated by the first three numbers after the authorizing agency designation. (See Table 13-3D.)

Classes for DOT authorized cars are:

103 105 107 110 112 114 120
104 106 109 111 113 115

Classes for AAR authorized cars are:

201 204 207 211
203 206 208

The next set of numbers, if any, indicates the tank test pressure in psi. Except for classes 103, and 104, the two series of numbers (class numbers and test pressure) are separated by a letter—"A," "S," "J," or "T." These letters are only significant for certain tank cars.

The letters indicate the following:

A: equipped with top and bottom shelf couplers.

S: equipped with tank head puncture resistance and top and bottom shelf couplers.

J: equipped with jacketed thermal protection, tank head puncture resistance, and top and bottom shelf couplers.

T: equipped with sprayed on thermal protection, tank head puncture resistance, and top and bottom shelf couplers.

The number and letter combinations (either following the class, where the test pressure is not shown, or following the test pressure) denote:

1. The material used in construction.

   None: Carbon steel

   AL: Aluminum (found in classes 103, 105, 109, 111)

   N: Nickel (found in class 103 only)

   C, D or E: Stainless steel (found in class 103 only) ("C" and "E" are found on cryogenic liquid tank car tanks—DOT-113)

2. The weld construction.

   W: Fusion welding

   F: Forge welding

3. Other features of the car.

TABLE 13-3D. Tank Car Classes

Nonpressure Tank Cars Classes			
DOT-103	DOT-111A	AAR-201A	AAR-206
DOT-104	DOT-115	AAR-203W	AAR-211A
Pressure Tank Car Classes			
DOT-105	DOT-112	DOT-114	DOT-120A
DOT-109	Specification	DOT-111A100W4	
Cryogenic Liquid Tank Car Classes			
DOT-113	AAR-204W	AAR-204X	
Miscellaneous Tank Car Classes			
DOT-106A	DOT-110A	[both multiunit tank car tanks (ton containers)]	
DOT-107A	Seamless steel cylinders in high pressure service		
AAR-207	Hopper car		
AAR-208	Wooden tank		

Tank Car Fittings

Numerous fittings are used with tank cars to allow for filling and emptying the tanks and to provide for the safe transportation of the commodity inside the tank. The following material on tank car fittings is very general. The objective here is to facilitate recognition of the fittings and to give a brief explanation of their normal operation and function.

Manway: The large opening on top of the tank car allowing access to the interior. The manway is closed either with a plate that contains all the valves or with a cover that is opened for loading the tank car.

Liquid Valve: A device used for removing the lading from the tank car from the top by means of the lading's own pressure or by pressurizing the tank car by use of air, nitrogen, or some other gas. The liquid valve reaches into the lading by a pipe called the eduction pipe. (See Fig. 13-F.)

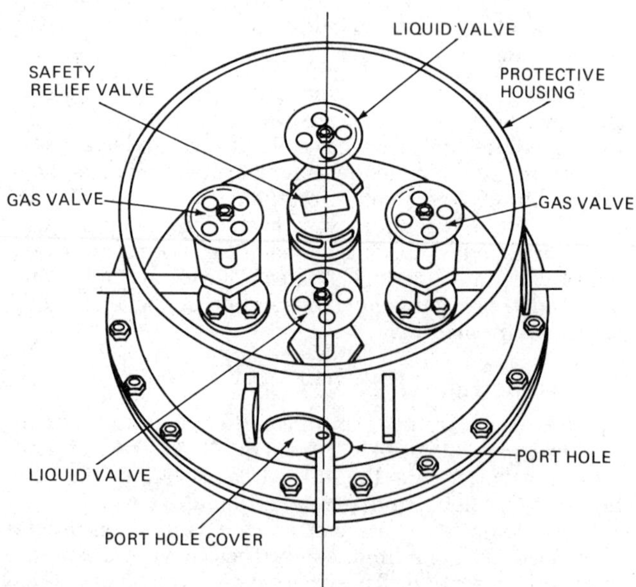

FIG. 13-3F. Details of the protective housing, or dome, on a liquid chlorine tank car.

Vapor/Air Valve: A device used to remove vapor from the tank car or to pressurize the tank car. On a pressure tank car, it is called the vapor valve; on a nonpressure tank car, it is referred to as the air inlet or air valve. (See Fig. 13-3F.)

Safety Relief Device: A device designed to prevent the rise of internal pressure of a container in excess of a specified value, when the container is exposed to abnormal conditions. (See Fig. 13-3F.)

There are three types of safety relief devices:

1. The *Safety Relief Valve*: A safety relief device with an operating part held closed by a spring. The device is intended to open and close at predetermined pressures.
2. The *Safety Vent*: A safety relief device having as the operating part a frangible (breakable) disc installed so as to close the device opening. The frangible disc does not reclose after it has burst.
3. The *Combination Device*: A safety relief device that uses a frangible disc or some other part in conjunction with a spring operated safety relief valve.

Vacuum Relief Valve: A device designed to prevent an excessive internal tank car vacuum by admitting atmospheric air while the tank is being unloaded. The device will reclose after normal conditions have been restored.

Sample Line: A device used to obtain a sample of the lading without the tank car being hooked up for unloading or without otherwise opening the tank car.

Gaging Device: A device used to determine the amount of lading in a tank car. The device can check "innage" (liquid content) on "outage" (vapor space) in the tank car.

Thermometer Well: A closed tube positioned into the tank car, filled with a permanent type antifreeze, that is used to measure the lading temperature.

Bottom Fittings: A fitting on the bottom of the tank car used for unloading or cleaning the car. Such a fitting can be an outlet, sump, or washout.

New Tank Cars

New tank cars meet the new industry requirements. Engineers are designing sloping skids to be welded on bottom outlets of the cars to prevent the outlets from shearing during derailments and releasing the tank car's contents. Some builders are incorporating bottom outlet valves, recessed inside the tank. (See Fig. 13-3G.)

Tank Car Fire Hazards

Fire hazards of tank cars are minimal unless a car is damaged in an accident. If a fire does occur, it is essential to learn the contents of the car before attempting to fight the fire. A tank car exposed to radiant energy from a fire, particularly direct flame impingement, should be approached cautiously. If a commodity absorbs too much heat, vapor pressure will build up and cause the safety valve to operate, or the tank to rupture. If any of the commodity is released, personnel should be prepared to deal with corrosive, toxic, or otherwise harmful effects. Direct contact with vapors discharged from tank cars under emergency conditions should be avoided, unless appropriate protective clothing and self-contained breathing apparatus are worn.

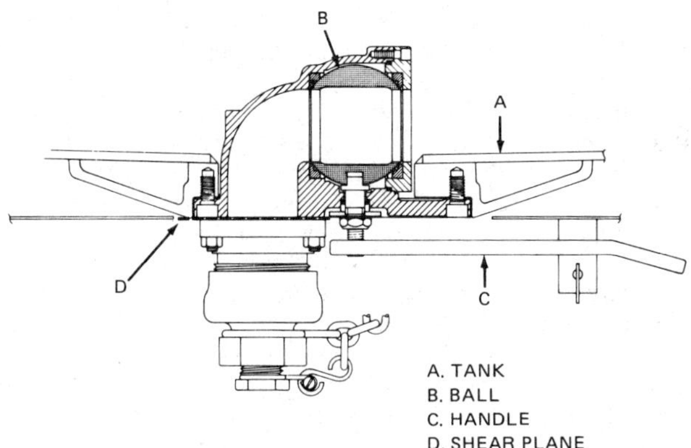

FIG. 13-3G. The internal ball valve is mounted inside the tank flush to the bottom of the tapered saddle. It eliminates steel skids called for in recent regulations for conventional bottom unloading valve protection.

PASSENGER EQUIPMENT

Fires involving railroad passenger equipment are relatively uncommon. Modern steel and aluminum passenger coaches and diners, mail, express, chair, and sleeping cars do not contain many combustibles. The major causes of fires on passenger trains are careless smoking and electrical faults. In diners, hazards are identical to those inherent with stationary restaurants; namely, grease and exhaust system fires. Air conditioning and heating systems must be safeguarded properly, and preventive maintenance is essential.

Self-propelled electrical passenger equipment, such as that used in Amtrak Metroliner service, has safety emergency cutoffs similar to those on electric locomotives. To deenergize the equipment, it is necessary to lower and ground the pantograph. This can be accomplished automatically by operating the pantograph control in the engineer's compartment, or by using the pantograph pole which is inside a tube mounted horizontally on the side of the car. If the car must be mounted, it is necessary to activate a control on the top at the end of each car labeled "Danger High Voltage—Ground Pantograph Before Going on Roof." Each connected car must be deenergized separately by lowering and grounding its pantograph. It is essential to learn from railroad officials the proper operation of the grounding system and to practice its use prior to an emergency.

With the pantograph lowered, the battery operating the emergency lights and doors can be disconnected by using the battery cutoff switch inside the electrical cabinet.

MISCELLANEOUS RAILROAD FACILITIES

Many railroad facilities are similar to those of the manufacturing industry, with familiar hazards. However, other facilities are peculiar to the railroad industry, such as diesel shops, heavy repair shops, paint shops, and fueling facilities. Hazards associated with a diesel shop include cutting, welding, housekeeping, and leakage of fuel.

Fueling facilities need particular protection, not only due to the economic value of the systems, but also due to the value of locomotives and the cost of interruption of service. Fuel pump houses should be of noncombustible construction and, preferably, equipped with automatic fire protection and with light fixtures and devices suitable for hazardous locations. Because leakage is the prime hazard, careful housekeeping and preventive maintenance are essential. Portable and wheeled dry chemical fire extinguishers should be located throughout the fueling area. It is most desirable to have a high volume water supply nearby.

Other facilities are critical to railroad operation and, if involved in fire, could stop the movement of trains in a given territory. These facilities include computer centers, microwave communications facilities, centralized train control equipment, and classification yards. These facilities should be protected by private detection and protection systems. Typical fire suppression systems range from sprinklers to carbon dioxide and Halon 1301.

Bridges and Trestles

Prevention is the key to continued operation of these structures. Susceptibility to fire damage can be reduced by fire retardant treatment of the structure, and by the provision of water barrels, fire detection systems, standby pipe and hose systems, and guard service. At some particularly valuable and critical bridges, fireboats and water buckets (for attachment to helicopters) have been provided. Most railroads have been replacing combustible bridges and other track structures with those of noncombustible construction at every opportunity, thus reducing their exposure to fire related losses.

Exposures to Freight Equipment

Exposure fires that destroy loaded and unloaded freight equipment reflect the need for close analysis of terminal fire hazards. When large classification or switching yards are designed, adequate fire protection—private and public—should be a prime consideration.

Experience in large yards indicates that lack of nearby fire mains and hydrants has been a contributing factor in fire loss. In a disaster, fire fighting activities should be confined to containment and protection of exposed property. A prefire plan should include mapping of fire mains and hydrants near the yard, because experience shows that existing protection in the yard may be rendered useless in a disaster.

Railroad sidings present a similar exposure hazard, particularly at grain elevators, lumber yards, flammable liquid or LP-Gas unloading facilities, and similar industrial properties.

Self-propelled Fire Fighting and Derrick Equipment

Most carriers can clear derailments and rerail freight equipment. They use lifting equipment—steam or diesel derricks—ranging from 150 to 250 ton (136 to 227 metric tons). These are included in a work train which also may include water cars and fire pumps powered by the train air line, or powered independently by an internal combustion engine. Fire hoses, nozzles, foam compound, play pipes, and a quantity of extinguishers are usually provided by the carriers for fire fighting.

Work trains have proved most valuable when an emergency has occurred in a remote and inaccessible area. On occasion, such emergencies have been handled by mounting a fire department pumper on a flat car and taking it to the emergency by train.

Some railroads have self-propelled fire fighting cars that can be used independent of the work train. The availability of such fire fighting equipment, often overlooked in planning, should be considered.

RIGHTS OF WAY

A common, difficult problem is the right of way fire (a fire on land adjacent to the roadbed). Such fires still occur, but their frequency and severity have been reduced by controlled burning, chemical spraying of vegetation, and installation of fire breaks. Cooperation with the U.S. Forest Service has led to development of spark arresters, spark arrester exhaust manifolds, and sparkless brake shoes. Planned vegetation is another solution which can be employed.

Safety

Everyone in the vicinity of a railroad track should assume it is "live" and that a train may appear at any moment. A number of incidents have been recorded in which fire fighting personnel narrowly escaped injury from passing trains whose crews were unaware of the fire fighters presence. At some fires, hose lines have been laid across a railroad track, only to be cut by a passing train.

In case of fire, communications are essential. Railroad personnel should be notified at first opportunity and asked to alert train crews to the location of the fire and fire fighting personnel.

TRANSPORTATION OF HAZARDOUS MATERIALS

Rail transportation of hazardous commodities, which annually represents less than four percent of the carloads of freight moving over the system, is strictly regulated by the DOT through the Materials Transportation Bureau and by the FRA. The FRA Hazardous Materials Branch, which has as its main goal reasonable and adequate safety of the public, regulates shipper preparation and carrier transportation of explosives and other dangerous articles. Canadian railroads are similarly regulated by the Canadian Transport Commission (CTC).

When a rail accident involves hazardous materials, the danger is increased. Product spills, fire, and explosions must be confronted by fire, health, and safety agency personnel.

When accidents, spills, or fires occur in conjuction with the transportation of hazardous materials, the immediate aim of those in charge is to prevent injury and loss of life, and then to prevent property and environmental losses as far as practicable. To do this safely and intelligently, it is necessary to know what hazardous materials are involved in the incident.

Upon notification of an accident or other occurrence involving railroad equipment, it is recommended that the fire chief or other responsible authority take the following course of action:

1. Telephone or otherwise contact the responsible railroad dispatcher. Have this number in advance.
2. Secure from the dispatcher the following information: "WHAT" (what commodities are involved in the accident or fire and the number of cars); "WHERE" (the specific location and best avenues of approach); "WHEN" (when the accident occurred—when access will be permitted—whether or not evacuation of area is necessary.
3. Upon arrival at the scene, establish contact with the senior railroad officer or representative before attempting to deal with the emergency.
4. If there is undue risk to life, retreat to a safe distance and remain there until the incident commander determines that it is safe to reenter the area and take action.

It is essential that firesafety personnel know about the presence, type, quantity, and nature of any hazardous materials in any rail accident or fire. This information is available in two forms.

The first means of detecting the presence of a hazardous materials shipment are the placards affixed to both ends and both sides of the railroad vehicle. Placarding alerts personnel to the presence of a hazardous material and identifies its hazard classification. Sometimes, the commodity name is stenciled on the shell of the tank car. However, due to the nature of the accident, the placard often cannot be read, or it is not prudent to approach the car close enough to read the information. (Binoculars are most helpful here.)

A copy of the shipping papers or waybills, "wheel report" or "consist" is the other method of hazardous materials identification. Upon arrival at the scene and consistent with safety, the first step should be to locate these papers. Regulations require the train conductor to have a piece of paper for each car containing hazardous materials—usually a waybill, but sometimes a switch ticket or other document—which shows the car number, names of the shipper and consignee, name and classification of the chemical(s), placard notation, Standard Transportation Commodity Code "49" series of hazardous materials numbers, and the amount of material in the freight vehicle. Often the emergency telephone number of the shipper also is included.

The wheel report summarizes pertinent information for all train cars in order, and on many roads includes emergency response information for all hazardous materials in the train. In addition to the waybills, the conductor may have a separate list, a "consist," which identifies all the hazardous cars in the train, as well as their location in the train. The train conductor is normally located in the caboose, but may be in the lead locomotive.

Technical assistance on hazardous materials identification at the scene is available from CHEMTREC, toll free at (800)424-9300; (202)483-7616 in Alaska, Hawaii, and the District of Columbia.

In Canada and the United States, the AAR Bureau of Explosives should be notified promptly of accidents, fires or explosions, and of leaking broken containers occurring in connection with the transportation or storage of explosives or other dangerous articles on a carrier's property. The Bureau's telephone number, (202)293-4048, can be reached 24 hours a day, 7 days a week for advice on procedures to follow in an immediate emergency. Bureau representatives are also dispatched to the scene of major emergencies.

When an incident occurring during transportation involves hazardous materials, carriers must forward a hazardous materials incident report to the Secretary, Hazardous Materials Regulations Board, DOT.

Many carriers have prepared guides and recommendations for handling rail emergencies and have made them available to fire departments along their rights of way. In addition, the carriers have assisted the fire service by participating in seminars and providing instructors and training programs at fire department training sessions and fire schools.

Emergency Action Teams

Some carriers have developed Emergency Action Teams that are specially trained to handle emergencies on their lines. Such teams are equipped with special kits and tools to repair leaks and control pollution. This team preparation concept is rapidly spreading throughout the industry.

NFPA Railroad Section

The NFPA Railroad Section was organized in 1963 as a Membership Section of the National Fire Protection Association. It is dedicated to promoting interest in and improving the methods of fire prevention and protection in the railroad industry, and to encouraging proper safeguards and recommended practices against fire through the interchange of ideas and experience.

The NFPA Railroad Section's Annual Meeting is held each May in connection with the NFPA Annual Meeting. This event provides an excellent forum for exchange of ideas between the fire service and railroad firesafety specialists.

Bibliography

Additional Readings

Allender, Peter J., "Specification of Combustible Materials for Passenger Train Design," *Fire and Materials*, Vol. 8, No. 3, Sept. 1984, pp. 113-124.

Brooks, J. L., "Heat Activated Alarm Systems for Railroad Boxcars Carrying Explosives," *TN No. N-1512*, U.S. Navy Civil Engineering Laboratory, Port Hueneme, CA, Dec. 1977.

Bukowski, Richard W., "Fire Protection Systems for Rail Transportation of Class A Explosives: Interim Report," *NBSIR 80-2170*, National Bureau of Standards, Gaithersburg, MD, Nov. 1980.

Bureau of Explosives Tariff No. BOE-6000-E, published by the Hazardous Materials Regulations of the Department of Transportation, Bureau of Explosives, Association of American Railroads, Washington, DC, May 1985.

Comptroller General of the U.S., *The Federal Approach to Rail Safety Inspection and Enforcement: Time for Change*, U.S. General Accounting Office, Washington, DC, 1982.

GATX Tank Car Manual, 4th ed. General American Transportation Corp., Chicago, IL.

Jameel, Haji M., "Rail Rapid Transit Safety Standards: A Feasibility Study," Report No. UMTA-CA-MT-82-106, California Public Utilities Commission, San Francisco, CA, 1982.

Manual of Standards and Recommended Practices, Section C,

Part III, Specifications for Tank Cars, Operations and Maintenance Department, Association of American Railroads, Washington, DC, Jan. 1982.

"Passenger Fire Safety in Transportation Vehicles," *A. D. Little Report No. C-78203*, U.S. Department of Transportation, Washington, DC, May 1975.

Peacock, R. D. and Braun, E., "Fire Tests of Amtrak Passenger Rail Vehicle Interiors", *NBS Tech Note 1193*, National Bureau of Standards, Gaithersburg, MD, May 1984.

"Recommended Fire Safety Practices for Rail Transit Materials Selection," Urban Mass Transit Administration, U.S. Department of Transportation, *Federal Register*, Vol. 47, No. 228, Nov. 26, 1982, p. 53559.

Smith, D. A., "Some Aspects of Fire Safety Design on Railways," *Fire and Materials*, Vol. 8, No. 1, Mar. 1984, pp. 6-9.

Wright, C. J. and Student, P. J., "Understanding Railroad Tank Cars," *Fire Command*, Vol. 52, Nos. 11 & 12, 1985.

FIXED GUIDEWAY TRANSIT SYSTEMS

Revised by Eugene Nunes, P.E. and John J. Troy, P.E.

The environment of a rapid transit system lends itself to some peculiar and difficult firesafety problems. Large numbers of people who move through these subterranean enclosures compound the life safety dangers. Fires within subways have attracted national attention over the years due to the danger facing passengers who must evacuate trains through smoke filled trackways and other emergency exitways.

Among the causes of fire in a subway system are mechanical failure of undercar components, high energy electrical short circuits, accumulation of combustible debris along the trackways and within vent shafts, accumulation of road film and oils on the underside of revenue vehicles, combustible materials within car components, combustible construction of trainways, including cross-ties, and hazards associated with the human factor, e.g., carelessness, arson, or pranks.

This chapter covers areas within a rapid transit system requiring special attention including subways, underground stations, and aboveground stations.

SYSTEM CHARACTERISTICS

Although the most serious threat from fire to transit users is found in subways and underground stations, certain basic considerations must be recognized throughout the transit system.

Traction Power

Traction power, or the electrical energy needed to run the trains, may be carried throughout the system by means of a third rail or overhead wiring. For modern transit systems with traction power provided by a third rail, a coverboard of electrical insulating material is installed throughout the system to prevent people from inadvertently coming in contact with the energized rail. The coverboard should be of sufficient strength to hold a 250 lb (113 kg) weight applied at any point without deflection, and be supported in such a manner as to allow a fire department ladder to rest against it safely. Warning signs, as required, should be affixed to the top of the coverboard in any area accessible to the public access areas.

Disconnect switches should be located strategically throughout the system so traction power can be cut as quickly as possible under emergency conditions. Otherwise, alternate prompt and reliable means for disconnect should be provided.

Access/Egress to Stations

When determining required exit widths, consideration must be given to future occupancy needs. Once a station has been constructed, it becomes very difficult, if not impossible, to place additional stairs or escalators where they will best serve the public. While it is recognized that complying with required exit width may become a major difficulty to designers, a good source of guidance is NFPA 101® *Code for Safety to Life From Fire in Buildings and Structures*, (hereinafter referred to as NFPA 101) as modified by NFPA 130, *Fixed Guideway Transit Systems* (hereinafter referred to as NFPA 130).

To establish required exit widths for any structure, it is first necessary to determine the occupancy classification. The state of California, for example, other jurisdictions, and NFPA 101, have determined that places where people await transportation should be classified as an assembly occupancy. However, effective June 1983, NFPA 130 established requirements for life and firesafety in transit stations, trainways, vehicles, and outdoor vehicle maintenance areas. Transit stations considered in NFPA 130 are only those stations accommodating passengers and employees of the fixed guideway system and incidental occupancies where transit stations are an intregal part of shopping centers, sports arenas, or other similar areas. Where large numbers of people other than transit patrons are involved, NFPA 101 is used to establish exit requirements.

Since rapid transit systems are designed to move people rapidly, most new systems have installed escalators that operate at the rate of 120 fpm (37 m/min). Plans are being formulated to increase this speed. Escalators now are designed to be fully reversible, depending upon com-

Mr. Nunes is Deputy Director of Safety for the Bay Area Rapid Transit (BART) District in Oakland, CA. Mr. Troy is Supervising Engineer for Gage-Babcock & Associates, Inc. in Oakland, CA.

mute conditions. In many instances, escalators will penetrate two or more floor levels to gain access to the intended platform. This can pose special design problems since many designers prefer to maintain an open environment throughout a station complex. Vertical penetration of more than one floor without enclosures has been accomplished in some rapid transit systems by increasing other fire protection components.

When escalators are designed and accepted as required exits, serious study must be given to their usability in evacuating a station under emergency conditions. Consideration also must be given to methods of stopping escalators so as to prevent injury to patrons. One possibility is to notify patrons via a taped public address announcement, which may be activated through the fire alarm system, stating that the escalator will stop within a specified time period. This allows patrons to grasp the handrail and obtain a firm footing or may even alert patrons not to board escalators until they have stopped.

Exit Barriers

Rapid transit stations contain two basic areas to separate those individuals who have paid their fare from those who have not paid. These are commonly called paid areas and free areas. Patrons in the free area are able to move directly to the outside without encountering barriers. Patrons within the paid area must go through fare barriers to gain access to the free area. Under emergency evacuation conditions, either all fare barriers should open automatically, allowing patrons to rapidly evacuate the station, or specially designed access gates for the same purpose should be installed. With increased emphasis being directed to handicapped patrons, the use of access gates becomes more useful in basic station design.

Procedures and methods that allow operators to quickly control or limit the numbers of patrons allowed into a station in the event of delays or a malfunction, are most important. A public address system which will notify and advise patrons during an emergency or a malfunction also is considered essential.

Many stations are completely closed off by grills or steel doors when a station is secured after normal revenue hours. The installation of small doors in the grills can provide easy access for fire department personnel as well as serve as an acceptable means of exit for any employees who may be working in the station.

Emergency Procedures

Due to the variety of hazards involved in a rapid transit system, local fire authorities must establish a direct line of communication with the system's top management. Good inspection techniques must be developed, and coordinated emergency response procedures planned. Access points to all parts of the system must be pinpointed, and engine and truck company response patterns tailored to meet special access problems. Information regarding the system, such as floor plans, electrification, revenue vehicle, structures, fire protection systems, and communications, is needed to properly plan emergency procedures. A wise course for a fire department is to assign one command officer the responsibility of coordinating all the emergency activities required to protect against fires on rapid transit trains, on the right of way, and outside of the right of way. Communication also must extend in the other direction. Fire and other emergencies in adjacent properties may require modified transit operation. In downtown areas, fire suppression activities in nearby buildings can cause water drainage into subways or flammable liquid spills can endanger below grade stations.

The Transit Vehicle

The construction and use of the transit vehicle has become a fire and life safety problem. With a reasonable degree of attention, the other elements of the transit system—guideway, stations, wayside equipment, and maintenance facilities—can be designed with an acceptable level of fire protection. Materials and methods of construction can provide a noncombustible, if not fire resistive, profile. These elements are static and many solutions for achieving a degree of fire resistance are already in use in related structures (buildings, bridges, etc.) and support subsystems (elevators, escalators, electrical and mechanical systems, etc.).

Because of economic and comfort considerations, the modern vehicle has made extensive use of manufactured materials. Included in the vehicle are many attractive and serviceable plastics (foams, fibers, and other synthetic materials) which are often quite combustible and which may emit toxic products of combustion while burning. These problems are not unique to the fixed guideway transit vehicle. Combustible plastics are used in passenger compartments of other modes of transportation, including aircraft, buses, railroad trains, and the family car. The transit vehicle, however, can accommodate several thousand people—far exceeding even jumbo jet capacity, and therefore presents a potential for great loss of life.

A high level of attention must be focused upon materials and finishes of every item that goes into the makeup of each transit vehicle, including insulation for propulsion motors and other electrical wiring. Studies of subway fires have shown the transit vehicle to be the principal cause of the generation of toxic smoke and the contribution of fuel to the fire.

Therefore, the transit vehicle is a primary concern in a fire and life safety program for a transit system. Unfortunately, many of the existing fire test methods used to measure performance of materials in fire test conditions have proven inadequate in predicting actual fire performance. A guideline developed by the Urban Mass Transportation Administration deals with the vehicle fire problem and recommends specific test methods and tests that should be utilized. Also, some transit systems have been able to continue use of a so called "Spartan" vehicle using fewer combustible or toxic materials.

However, a complete fire risk assessment for transit vehicles should be made, including fire propagation resistance, smoke emission, ease of ignition, and rate of heat and smoke release. The total combustible loading of the vehicles should be limited to as low a value as possible. For example, NFPA 130 advises the combustible loading of the vehicle above the floor should not exceed 45,000 Btu per sq ft (511 Mj/m^2). (Values as low as 30,000 Btu per sq ft [341 Mj/m^2] can be achieved.)

Newer individual test methods and the appropriate utilization of test results are part of an overall firesafety effort. One test is detailed in NFPA 253, *Standard Method of Test for Critical Radiant Flux of Floor Covering Systems Using a Radiant Heat Energy Source*, which provides a realistic test to evaluate the flame spread on floor covering systems. Existing tests such as those in NFPA 255, *Standard Method for Test of Surface Burning Characteristics*; NFPA 251, *Standard Method for Fire Tests of Building Construction and Materials*; NFPA 258, *Standard Research Test Method for Determining Smoke Generation of Solid Materials*; ASTM-E-162; ASTM-E-8; and others also can be utilized if results are interpreted in relation to the transit environment. More appropriate tests are being developed to better predict the performance of materials when subjected to elevated temperatures, especially with respect to the release of products of a toxic nature.

Perhaps one of the least appreciated areas of life safety is the severity of toxic products of combustion. The difficulties of removing a person from the fire environment, particularly in subways, demands extreme care in evaluating materials and their use in composite assemblies. To date, the use of test results related to smoke development and generalized statements on avoidance of "recognized" materials generating toxic products of combustion have been the way to address this problem. The development and utilization of tests for toxicity of materials during fire exposure remains an unanswered need for transit safety.

SUBWAYS

Fires within the confining areas of a subway system, especially trainways, are among the most difficult to extinguish due to limited room for fire fighters to operate. Access points for fire personnel, emergency exit locations for passengers, availability of water supply, and ventilation capabilities are among the fundamentals that must be considered in providing adequate safeguards for passengers, and facilities for fire fighters coping with emergencies.

A direct liaison between rapid transit authorities and fire officials is essential to keep emergency forces apprised of current conditions within the underground installations. Floor plans of stations and subways must be reviewed continually. Fire protection included in the original design of the system may have changed, thus requiring updating of fire fighting plans.

Ventilation

Ventilation of underground subways is of primary importance. Serious study must be given to the removal of smoke under fire conditions. Exiting schemes also must be developed which will allow passengers to exit rapidly into fresh air. Some rapid transit systems have installed fully reversible fans within vent shafts. These are capable of moving large quantities of air, by intake or exhaust, thus permitting smoke to be purged from the subway in a predetermined direction so passengers can be evacuated to the nearest station. Ventilation shafts and fans have been placed at each end of an underground station and, in some instances, are located between stations. When vertical shafts are included in the ventilation criteria, the possibility of flammable liquid spills entering the shaft at street level must be considered; adequate sumps to retain such spills must be developed. Electrical equipment which may be installed within shafts and fan rooms should meet the requirements of local authorities for the expected fire exposure.

Access/Egress

Walkways: Primary consideration must be given to the evacuation of large numbers of passengers to safety, which could be a station platform, or directly to the open air via vertical exitways. Many existing systems have evacuated passengers by having them climb from the train to the trackway, and then walk between the rails on crossties until a vertical exit or station platform is reached. Modern transit systems prefer to have passengers evacuate the train to a walkway, which is constructed along the side of the trackway on the side opposite the third rail power source. (See Fig. 13-4A.) This allows passengers to step directly

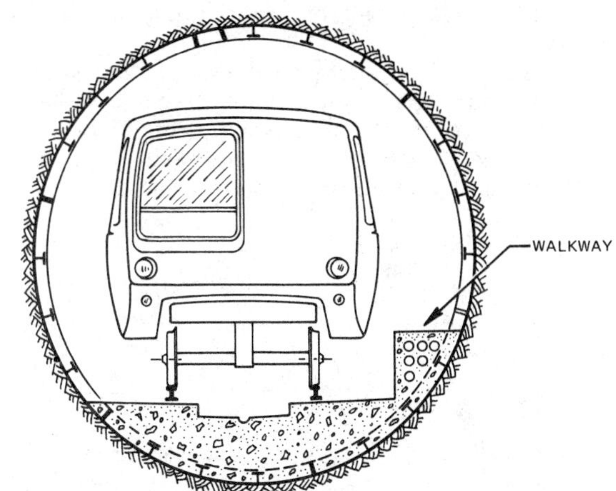

FIG. 13-4A. A sectional view of a single tube subway tunnel, showing the location of the emergency passenger walkway to the floor of the car. The walkway is located on the side opposite the third rail power source.

from the train to the walkway, keeps passengers off the trackway, and allows specially equipped vehicles to respond to train problems. When trackways are separated by concrete walls and fire doors, the walkway system also provides easy access to a train on the opposite trackway and provides an acceptable horizontal exit path that remains relatively free of smoke and heat. Obviously, both adequate and emergency lighting must be provided to allow safe evacuation.

Tunnel or Bore: Many rapid transit systems are presently operating underground on trackways which are completely separated from each other by concrete walls and fire doors. (See Fig. 13-4B.) As previously mentioned, this concept provides a place of refuge for passengers, as long as adequate train control and planning is practiced. Separation walls should have no less than a 2 hour fire resistance rating with openings protected by 1½ hour fire doors. Cross-passageways should not be more than 800 ft (244 m) apart. Hardware normally used in the installation of fire doors and frames has been found inadequate for the pressures generated by a train moving through the tunnel. Special study and consideration must be given to the type of door, frame, hinges, locking device, and method of securing the frame to the opening. Sliding fire doors with counterweights require special attention.

Emergency Exits: When determining exit requirements from an underground subway system, a number of factors must be considered:

1. Length of train.
2. Depth of trackway below grade.
3. Accessibility to exit at grade.
4. Accessibility to exit at trackway.
5. Time and distance needed for people to exit to surface.
6. Available ventilation and controls.
7. Access to trackway for fire services.

Obviously, no transit system can operate safely with unlimited vertical shafts. Confusion would not only develop underground, but also affect required transit security. The intent and basics of NFPA 101, as modified by NFPA 130, should be followed as closely as possible in considering emergency exiting requirements for subway tubes.

Many transit systems have utilized ventilation shafts as emergency exits by constructing conventional stairways in the shaft. Unless the stairway is completely separated from the vent shaft by 2 hour fire resistant construction and adequate control of air currents within the bore is maintained, this concept is extremely risky, since smoke and heat may contaminate the vertical exitway.

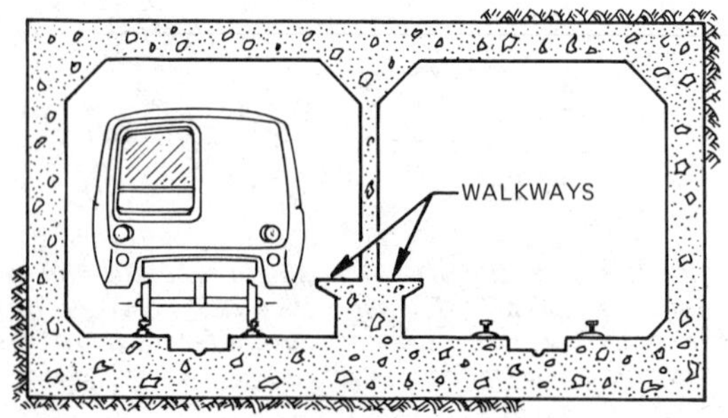

FIG. 13-4B. A sectional view of a subway tunnel having trackways separated from each other by a concrete wall. Openings through the wall, protected by fire doors, can provide an acceptable horizontal exit path from the trackway where an accident has occurred to the other trackway (which remains relatively free of smoke and heat).

Under no circumstances should required exitways contain straight vertical ladders. Internally illuminated exit signs with two sources of power supply should be installed at trackway level to direct passengers, and emergency lighting should be provided within the exit enclosure.

Communication

Probably one of the most serious problems encountered during underground emergencies concerns communications. Emergency telephones are normally identified throughout the subway system by a distinctive light. Most systems designate a blue colored light for emergency phone locations. Communication from trackway direct to a central control room is accomplished by merely picking up the instrument. Some systems utilize the Centrex principle for calls.

Blue light stations identifying the location of emergency telephones can also serve multiple purposes. For example, on the Bay Area Rapid Transit (BART) System headquartered in Oakland, California, the stations include a third rail or power trip button, a maintenance jack, a 110 V outlet, and a 20 lb (9 kg) multipurpose dry chemical extinguisher. The stations are spotted throughout the underground system at 1,000 ft (30 m) intervals or in line of sight, whichever is shortest.

Communication through a subway system is sometimes very difficult. Since fire department portable radios may not function as intended, fire authorities should make every effort to test their communications systems under actual conditions throughout the subway system. Consideration should be given to utilization of hardwired communication systems which may already exist. Portable radios or emergency phones with a direct line to the central headquarters of the transit system or train radios are all useful. Every system should have a radio network providing two way communication connecting personnel on trains and other vehicles, anywhere in the system.

Fire Protection in Underground Trainways

One of the principal fire protection devices in fixed guideway transit systems is the standpipe hose system. NFPA 14, *Standard for the Installation of Standpipe and Hose Systems* (hereinafter referred to as NFPA 14), provides good guidance for installations, and Class I or Class III service should be specified.

Standpipes: Standpipes should be located in stations and throughout the underground trainway system. Standpipes in tunnels should be at least 9 in. (228 mm) in diameter and may be of the dry type. Generally speaking, the locations of standpipe outlets are determined by the local fire authorities with consideration being given to fire department access at tunnel level, available vertical access shafts for emergency forces, availability of street access to the fire department, and siamese connections at grade level.

Some rapid transit systems require that dry standpipe outlets be no more than 200 ft (61 m) apart throughout the underground; others have determined that maximum separation distances of 300, 400, and 500 ft (91, 122, and 152 m) are acceptable.

Some rapid transit systems place bins or boxes containing 2½ in. (53.5 mm) hose in close proximity to dry standpipe outlets, with the inspection and testing of such hose left to rapid transit system personnel. Fire authorities operating under this principle should seriously consider assigning personnel to witness hose inspections and required waterflow testing. In areas where local fire authorities will consider using only their own hose, special consideration must be given to the provisions for transporting hose and other equipment to a fire site.

Special Apparatus: In lieu of the supply of fire hose required at standpipe stations, innovative measures have been taken by some systems to ensure that hose is available when needed. BART, for example, with the approval of fire authorities, has removed all fire hose from its subway system with the exception of 1½ in. (38 mm) hose located within stations. In turn, BART has agreed to furnish five fully equipped, specially designed fire engines capable of traveling by street or rail to previously designated rendezvous points. The responding fire department would then board the apparatus and go to the fire or emergency location within the underground. Each piece of apparatus includes, as part of its inventory, 1,200 ft (366 m) of 2½ in. (63.5 mm) hose, and 400 ft (122 m) of 1½ in. (38 mm) hose, as well as nozzles, a portable emergency generator, a power saw, a cutting torch, extension cords, flood lights, demand breathing apparatus, life lines, miscellaneous tools, a high expansion foam unit and nozzle, a 300 gal (1136 L) water tank, and a small pressure pump.

Fire Extinguishers: Fire extinguishers should be placed on board each vehicle. In addition, fire extinguishers should be placed in proximity to fixed equipment throughout the underground system where ignition sources and/or combustibles exist. Locations should include electrical equipment, sumps, ventilation equipment, etc. If either the blue light or emergency station is suitably located, one or the other would be a preferred placement for extinguishers. The size and type of the extinguisher to be used is dictated by the size and type of exposure expected.

Drainage: Adequate drainage of water accumulated during a fire emergency requires careful consideration. Fire authorities should be made aware of the underground system's drainage capabilities and recognize that large amounts of water could seriously impair other required rescue operations. When fighting structural fires at street level in close proximity to station entrances or street ventilation grills, fire officers should alert the rapid transit officials that water intrusion into their system may be expected.

STATIONS (UNDERGROUND)

Underground rapid transit stations should be constructed of minimum approved noncombustible materials (Type I, Type II, or combinations thereof) as outlined in NFPA 220, *Standard on Types of Building Construction*. Many modern rapid transit systems use reinforced concrete construction for the station's outer shell and protected steel beams, girders, and columns within the shell—which results in a structure with a substantial fire resistive capability. Contrary to general opinion, very few fires of major proportions have occurred within stations. Most station fires are attributed to poor housekeeping or electrical fires occurring in escalator machine rooms. The primary fire hazard exposure to underground stations lies in two areas—concessions and trains. Small concessions within a station where newspapers and magazines are sold would normally have a low risk classification, while major concessions, which could include restaurants or theaters, would have a high potential risk classification for smoke and heat development.

Newspaper and magazine types of concessions should be limited in size and must be placed in areas that do not block required exits. Automatic sprinkler protection should be installed to protect the entire concession area. Possible expansion of the area should be considered when the fire protection system is designed.

Station public areas should be separated from nontransit areas with a fire barrier having at least a 3 hour fire rating. Major concessions or openings from public areas of the station into existing commercial structures should be protected by fire door assemblies having a 3 hour fire protection rating. Even if the commercial activity and the station are owned by the rapid transit system, good separation between the two areas is needed. Fire barriers can be established by installation of fire doors or automatic fire extinguishing systems to prevent fire and smoke spread from the commercial occupancy into the station or from the station into the commercial occupancy. Before any openings into a rapid transit system are made, the existing conditions on both sides of the proposed opening must be reviewed carefully so an equitable fire protection plan is developed.

Protection for Underground Stations

An underground rapid transit station may be compared to a multiple basement structure, except that a rapid transit station generally has a low fire loading factor when compared to a commercial basement structure. Strict application of building or fire code requirements may cause exceptionally high construction costs which are not necessarily justified. If minimal combustible storage or concession areas are involved, large open public areas may be left with little or no fire risk involvement. If such is the case, limiting automatic fire protection systems to those areas containing combustibles and designing the systems for possible expansion may have some merit, although the limitations of partial protection must be clearly understood.

Automatic Sprinkler Systems: All areas in transit stations which are used for concessions, storage and collection, (including other similar areas with combustible loadings, and the steel truss area of all escalators in a single entry station) should be protected by automatic sprinklers. Trainways are excluded. NFPA 13, *Standard for the Installation of Sprinkler Systems* should be used for guidance. Other approved fire extinguishing systems could also be used in lieu of the automatic sprinkler system, but only with the approval of the authority having jurisdiction.

Fire Detection Systems: All nonpublic areas commonly classified as support or ancillary areas, if not protected by an automatic fire suppression system, should be fully protected by a fire detection system. Listed smoke detectors are preferable, although rate-of-rise detectors may be used if authorized by the concerned fire authority.

It is good practice to provide supervision for all fire detection and automatic fire suppression systems. Supervisory signals should be received at the system's command center as well as being annunciated by zones at the station agent's booth or kiosk.

Wet Standpipes: Where separate standpipe and automatic sprinkler systems are provided, wet standpipes are supplied by a piping system usually separate and distinct from that used for automatic sprinkler systems. Fire department siamese connections, properly labeled and installed at grade level, allow the fire department to service either system.

Fire hose cabinets, as recommended in NFPA 14, may contain 1½ in. (38 mm) and 2½ in. (63.5 mm) control valves. A maximum of 100 ft (30 m) of attached 1½ in. (38 mm) hose with nozzle is included within the cabinet. Normally, 2½ in. (63.5 mm) hose is not stored in cabinets because responding fire departments generally prefer to use their own larger hose. A suitable fire extinguisher may be placed in the cabinet together with a hose spanner. To reduce pilferage of fire equipment, cabinets are locked and a breakaway glass panel installed to allow ready access under emergency conditions. Some rapid transit systems have installed intrusion alarms which send a signal to the station agent if a fire hose cabinet has been opened.

SURFACE OR ABOVEGROUND GUIDEWAYS AND STATIONS

Generally speaking, fires occurring in stations at or above grade are no more difficult to combat than any other type of fire at grade. Primary consideration always must be given to the traction power—even though assurance is given that the power is shut off, fire department personnel should operate under the assumption that the third rail or overhead wire is still hot.

Although fires in trains on surface track or aerial guideways present a slightly less serious exposure to the public, certain basic considerations, such as means of escape from elevated trainways, must be addressed in design. As in all other configurations of rapid transit, communication is paramount in any emergency operation. The ability to direct emergency forces to the scene of an incident and to place the system in a safe condition for access is necessary for patrons and emergency personnel.

Access to a trainway also may require special provisions. Locating trainways in a joint corridor with a railroad or freeway may limit access for long distances of the trainway. Trainways extending from population centers through undeveloped areas sometimes extend where surface roads and water mains do not exist. Trainways in congested areas, elevated to unusual heights, and in other sections not accessable to emergency vehicles (such as elevating platforms or paramedic units), could create other problems. Construction of trainways also may cause problems of fire exposure and access to adjacent structures. As an example, an elevated trainway above existing city streets and adjacent to existing buildings may interfere with fire department operations at those buildings.

Similar problems may exist at stations; however, permanent exits provide both patron evacuation and emergency personnel access. Recent systems have incorporated stations in the lower elevations of highrise buildings or within mall complexes having divergent occupancies. In these cases, detailed plans for coordination and compatible design are necessary.

Stations that can be considered to be buildings would normally be reviewed through normal code procedures. If a transit system passes through more than one jurisdiction, particularly if different basic codes apply, development of a common criteria is desirable to achieve uniformity throughout the system.

The subject of trainways, however, is not usually addressed in most existing building codes. Items that should be considered during a plan review of a guideway would include:

1. Designated mileage markers to provide a response location for fire personnel.
2. Locations where overhead trainways may intersect with known street locations.
3. Access gates allowing patrons to exit the system under fire department direction.
4. Availability of wayside communications allowing the fire department to communicate directly to the system's command center.
5. Locations of substations, switching stations, gap breaker stations, or other major electrical installations of concern to local fire authorities.

Bibliography

NFPA Codes, Standards, Recommended Practices and Manuals. (See the latest *NFPA Codes and Standards Catalog* for availability of current editions of the following documents.)

NFPA 10, *Standard for Portable Fire Extinguishers.*
NFPA 13, *Standard for the Installation of Sprinkler Systems.*
NFPA 14, *Standard for the Installation of Standpipe and Hose Systems.*
NFPA 30, *Flammable and Combustible Liquids Code.*
NFPA 72E, *Standard on Automatic Fire Detectors.*
NFPA 101 *Code for Safety to Life from Fire in Buildings and Structures.*
NFPA 130, *Standard for Fixed Guideway Transit Systems.*
NFPA 220, *Standard on Types of Building Construction.*
NFPA 251, *Standard Method of Fire Tests, of Building Construction and Materials.*
NFPA 253, *Standard Method of Test for Critical Radiant Flux of Floor Covering Systems Using a Radiant Heat Energy Source.*
NFPA 255, *Standard Method of Test of Surface Burning Characteristics of Building Materials.*
NFPA 258, *Standard Research Test Method for Determining Smoke Generation of Solid Materials.*
NFPA 259, *Standard Test Method for Potential Heat of Building Materials.*

Additional Readings

The following ASTM publications are available from American Society of Testing Materials, 1916 Race St., Philadelphia, PA:

ASTM E83-81a, *Surface Burning Characteristics of Building Materials.*
ASTM E119-82, *Fire Tests of Building Construction.*
ASTM E136-81, *Standard Method of Test for Noncombustibility of Elementary Materials.*
ASTM C542-78, *Vertical Fire Test Procedures* (FAA-FAR 853).

Birky, M. M., et al., "Measurements and Observations of the Toxicological Hazard of Fire in a Metrorail Interior Mock-Up," *NBSIR 75-966*, National Bureau of Standards, (NBS) Gaithersburg, MD, Feb. 1976.

Braun, E., "A Fire Hazard Evaluation of the Interior of WMATA Metrorail Cars," *NBSIR 75-97*, NBS, Gaithersburg, MD, Dec. 1975.

Braun, E., "Fire Hazard Evaluation of BART Vehicles," *NBSIR 78-1421*, NBS Gaithersburg, MD, Mar. 1978.

Directory of U.S. Department of Transportation Fire Research (DOT), Report No. DOT-TSC-OST-80-1, Department of Transportation, Washington, DC, 1979.

Electrical Insulation Fire Characteristics, Vol. 1, Flammability Tests, Report No. UMTA-MA-06-0025-79-1, I, Urban Mass Transportation Administration, DOT, Washington, DC, 1979.

Electrical Insulation Fire Characteristics, Vol. 2, Toxicity, Urban Report No. UMTA-MA-06-0025-79-1, II, Urban Mass Transportation Administration, DOT, Washington, DC, 1979.

Fire Experience and Exposure in Fixed-Guideway Transit Systems, American Iron and Steel Institute, Washington, DC, 1980.

Fruin, J. J., *Pedestrian Planning and Design*, Metropolitan Association of Urban Designers and Environmental Planners, Inc., NY, 1971.

Gooden, W. E., and Troy, J. J., "Metropolitan Atlanta Rapid Transit Authority-Systems Safety, Fire Protection, and Code Documentation for Stations," *Proceedings Third International System Safety Conference*, Washington, DC, Oct. 1977, pp. 517-527.

Hill, R. G., and Johnson, G. R., "Fire Detection, Extinguishment, and Material Tests for an Automated Guideway Transit Vehicle," *Report No. FAA-NA-76-52*, Federal Aviation Administration. Available through Natioinal Technical Information Service, Springfield, VA, 1976.

Identification of the Fire Threat in Urban Transit Vehicles, Report No. UMTA-MA-06-0051-80-1, Urban Mass Transportation Administration, Department of Transportation, June 1980.

Jenkins, C. E., "BART Experience with New Seat Materials," *Journal of Consumer Product Flammability*, Vol. 9, No. 1, Mar. 1982.

Litant, I., *Guidelines for Flammability and Smoke Emission Specifications*, TSC-76-LFS-6, Transportation Systems Center, Cambridge, MA, July 1976.

Londregan, R. P., "Life Safety in Underground Rail Systems: The Australian Effort," *SFPE TR 83-3*, Society of Fire Protection Engineers, Boston, MA, 1983.

Metsch, W. W., "Meeting Design Need to Control Subway Environment," *Specifying Engineer*, Mar. 1976, pp.86-91.

Moore, William L., Chairman, "Report of the Board of Inquiry on the Fire in the Transbay Tube of the San Francisco Bay Area Rapid Transit District on Jan. 17, 1979," Mar. 5, 1979.

Peacock, Richard D., "Fire Safety Guidelines for Vehicles in a Downtown People Mover System," *NBSIR 78-1586*, National Bureau of Standards, Gaithersburg, MD, Jan. 1979.

Peacock, R. D., and Braun, E., "Fire Tests of Amtrak Passenger Rail Vehicle Interiors," *NBS Tech Note 1193*, National Bureau of Standards, Gaithersburg, MD, May 1984.

Rail Transit Committee, *Guidelines For Design of Rapid Transit Facilities*, American Public Transit Association, Washington, DC, 1979.

"Recommended Fire Safety Practices for Rail Transit Materials Selection," Urban Mass Transit Administration, DOT, *Federal Register*, Vol. 47, No. 228, Nov. 26, 1982, p. 53559.

Subway Environmental Design Handbook, Volume 1, Principles and Applications, 2nd ed., 1976. Associated Engineers—A Joint Venture, Parsons, Brinckerhoff, Duade, Douglas, Inc., Deleuw, Cather and Company; Kaiser Engineers under the direction of Transit Development Corporation, Inc.

Troy, J. J., "Fire Protection Provisions for Rapid Transit Systems," *Fire Journal*, Vol. 70, No. 1, Jan. 1976, pp. 13-17.

Troy, J. J., "Fire Protection for Rapid Transit Systems," *Fire International*, Vol. 5, No. 55, Mar. 1977.

MARINE

Revised by Charles L. Keller, Donald J. Kerlin, P.E., and Robert Loeser

According to United States Coast Guard statistics (1983), more than 15 million pleasure and small commercial boats were registered in the United States. Of that number, 5,569 were involved in accidents that caused the deaths of 1,241 persons. A total of 427 of those accidents, in which 12 persons were killed, were due to fire or explosion. The same 5,569 boating accidents caused 2,913 injuries, 232 of which resulted from fire or explosion. These same accidents resulted in property damage of nearly 16 million dollars.

Marine fire hazards, including those on small pleasure craft, small commercial boats, and commercial vessels are discussed in this chapter. The first four parts cover pleasure and small commercial craft; large commercial vessels are the subject of the balance of the chapter.

Additional information about extinguishing systems suitable for use on vessels will be found in Section 19, Chapter 1, "Carbon Dioxide and Application Systems"; Chapter 2, "Halogenated Agents and Application Systems"; and Section 20, Chapter 2, "Selection, Operation, and Distribution of Fire Extinguishers."

PLEASURE AND SMALL COMMERCIAL BOAT HULLS

Hulls of pleasure and small commercial boats must be arranged so that all compartments are as accessible as practicable. It must be possible to escape from accommodation spaces going forward or aft. Escape hatches are to be unobstructed and adequate for the designed purpose. Engine compartments should be separated from accommodation spaces by bulkheads and/or barriers which will serve as effective fire barriers and minimize the escape of fire extinguishing media discharged in that space. Congestion of engine compartments is unsafe, since it discourages adequate inspection and maintenance. The addition of auxiliary machinery increases ventilation requirements.

Mr. Keller is Marine Field Service Specialist on the staff of NFPA. Mr. Kerlin is Chief, Fire Protection Section, Office of Merchant Marine Safety, U.S. Coast Guard. Mr. Loeser is the Senior Marine Staff Engineer of Underwriters Laboratories, Inc.

Materials

A wide variety of construction materials, such as aluminum, steel, ferrocement, fiberglass reinforced plastic, and wood can be used for the construction of a hull, bulkheads, and superstructures. The choice is usually determined by the hull type, size, intended service, and cost. Where flammable materials such as wood or fiberglass reinforced plastic are used, fire retardant coatings suitable for marine service may be used. In the case of fiberglass materials, fire retardant resins may also be utilized. However, fire retardant resins may reduce the structural strength of the laminate.

Ventilation

Ventilation, which is defined as the positive changing of air by natural or mechanical means within a compartment, may occur (1) by the introduction of fresh air to dilute any contaminated air, or (2) by locally exhausting the contaminated air. Ventilation cannot be relied upon to remove vapors from liquid leakage of gasoline. In fact, in the presence of liquid leakage, ventilation may actually help create an explosive mixture. Natural ventilation only works when there is a wind blowing, and the major factor causing air movement inside the hull is the shape and location of the superstructure or, in the case of runabouts, the position of the windshield.

Liquefied petroleum gas (LP-Gas) and compressed natural gas (CNG) are more likely to jump bulkheads and permeate all spaces in the boat than gasoline, and will fill any space in which released. Compressed natural gas can be released with overhead vents, but, like LP-Gas, will tend to fill any compartment in which it is released. In boats with diesel engines, the ventilation system, in addition to providing sufficient air for proper engine operation, should be designed to help control compartment temperature.

Lightning Protection

Lightning protection can be provided on a boat if the boat has a mast that can be used for the installation of a lightning conductor. For the most effective protection, the mast should be tall enough that an imaginary line drawn at a 45 degree angle from the tip of a pointed lightning rod at the top of the mast will not intersect any part of the boat. From the pointed lightning rod, a number 6 AWG or 8 AWG copper conductor or equivalent conductor should be led as straight as possible to a ground plate at least 1 sq ft (0.09 m^2) in area. Large metallic masses near the lightning ground conductor should be connected to the conductor to prevent side flashes.

On a sailboat, all shrouds and stays should be led directly to ground. Some metal rod type radio telephone antennas may help, if a suitable lightning arrester or gap is provided at all loading coils. The dc bonding ground system and the ac grounding system are not specifically designed to dissipate lightning currents, but can help minimize damage from side flashes.

Illustrations of proper lightning protection appear in Figures 13-5A through 13-5D. In the cabin cruiser illustrated in Figure 13-5B, adequate lightning protection is afforded only by the grounded antenna equipped with a lightning arrester or gap on the loading coil.

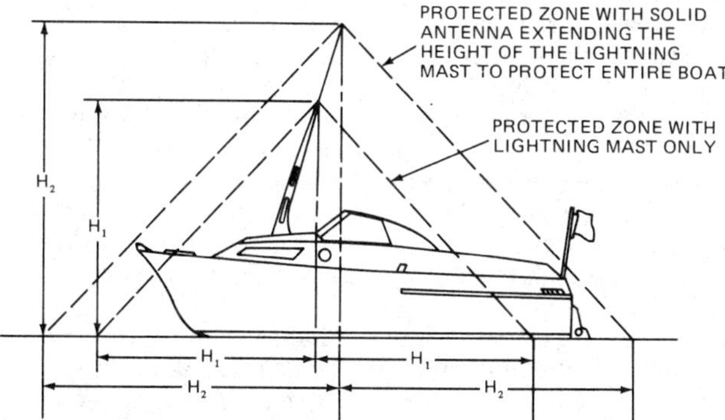

FIG. 13-5A. Proper lightning protection for a boat with a mast not exceeding 50 ft (15 m) above the water.

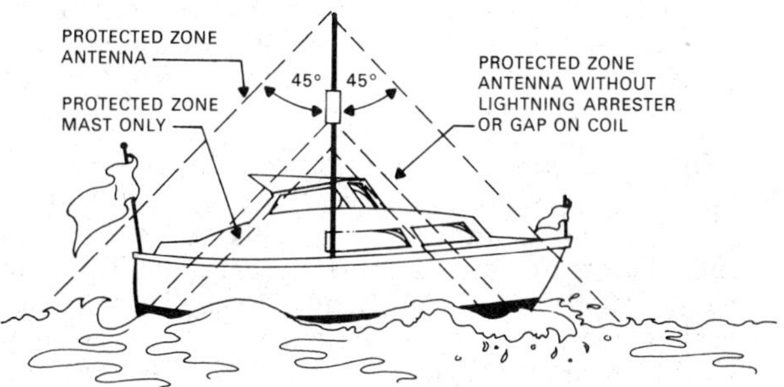

FIG. 13-5B. Diagram of a boat with the radio antenna used as part of the lightning protection system.

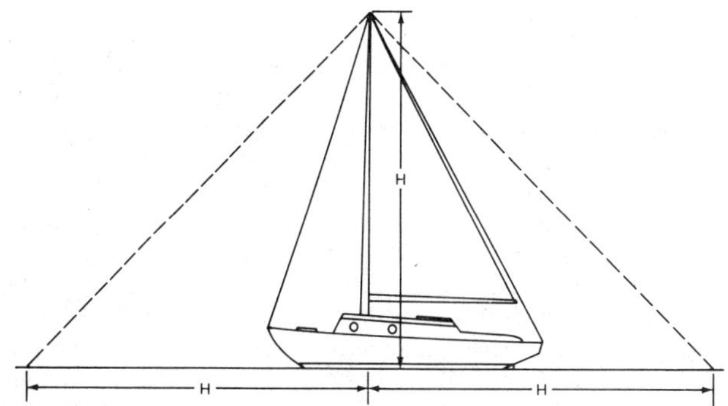

FIG. 13-5C. Proper lightning protection for a small sailboat with a mast not exceeding 50 ft (15 m) above the water.

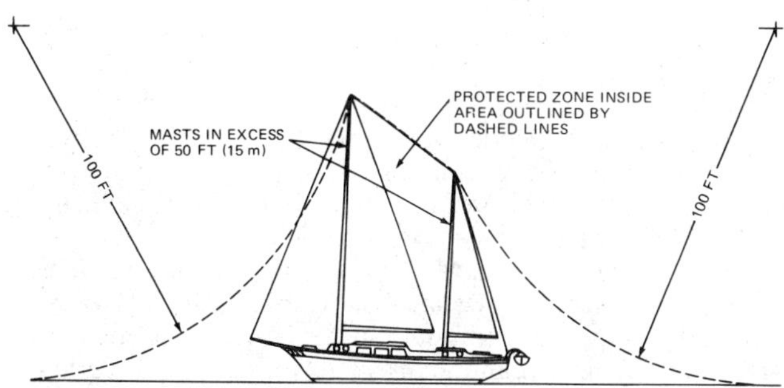

FIG. 13-5D. Proper lightning protection for a boat with masts in excess of 50 ft (15 m) above the water—protection based on a lightning striking distance of 100 ft (30 m).

PLEASURE AND SMALL COMMERCIAL BOAT INSTALLATIONS AND EQUIPMENT

Engines

Marine gasoline and diesel engines may be cooled directly by seawater, with fresh water with a heat exchanger, or by air. Liquid cooled main engines use water jacketed blocks, heads, and exhaust manifolds with the cooling water circulated by an engine drive pump. Fresh water cooled gasoline engines can operate more efficiently at about 180°F (82°C), while salt water engines must be operated at the less efficient 140°F (60°C) to prevent excessive salt precipitation in the block, head, and manifold. For water cooled engines, gages that indicate cooling water discharge temperature and lubricating oil pressure must be located at the helm position(s). It is recommended that a marine strainer be provided at the raw water intake for direct seawater cooled engines and engines using heat exchanger cooling systems. Air cooled engines are not recommended, but if used in an enclosed space, separate ducts must be provided for engine cooling and a warning device must be provided to warn of excessive temperatures.

In addition to the engine cooling and ventilation system, the engine space must meet normal engine space ventilation requirements. Updraft and sidedraft carburetors must be equipped with enclosed drip collectors so no liquid fuel escapes to the bilges. The carburetor must have an approved backfire flame arrester. Diaphragm fuel pumps must be designed and installed to prevent the release of fuel to the bilge in case of diaphragm failure. All electrical equipment in the engine space must be ignition protected. Diesel engines require the use of efficient fuel filters and water separators to prevent damage to the high pressure fuel injection system.

Fuel Systems

In aircraft and land vehicles, any liquid fuel leakage can be drained by gravity to prevent accumulation; this is not possible in a boat. Because both liquid and fuel vapor will accumulate in a boat, the fuel system must be designed to resist vibration, shock, corrosion, and chemical deterioration. Also, the system must be maintained liquid and vapor tight to the hull interior at all times. All parts of the fuel system should be capable of withstanding exposure to free burning gasoline for at least 2½ minutes without fuel leakage, except that fire resistance is not required if the total leakage is less than 5 oz (150 mL) when the fuel line is cut at any point. The fuel system must be electrically bonded to ground, from the fill pipe on deck to the engine to dissipate static electrical charges.

Diesel fuel systems require the installation of efficient fuel filters and water separators to prevent dirt or water from reaching the high pressure injector fuel system. Diesel fuel systems have a fuel return line from the engine to the fuel tank which must meet the same requirements as the fuel feed line to the engine. The fuel in the return line is heated and will gradually raise the temperature of the fuel in the tank. Although diesel fuel does not, at normal temperatures, vaporize and create an explosive air-fuel mixture in the way that gasoline does, leaking diesel fuel has a greater tendency to accumulate and create a fire hazard.

Appliances

Open flame devices are more liable to inadvertent misuse than any other item of boat equipment involving fire risk. It is imperative that such items be specifically designed for marine use and installed to minimize personal and physical hazards. Open flame devices must be labeled for marine use, and printed instructions for their proper installation, operation, and maintenance must be furnished by the manufacturer. A durable and permanently legible instruction sign covering safe operation and maintenance must be provided. The sign should be installed on or adjacent to the appliance where it may be quickly read.

Coal, charcoal, and wood burning stoves should be installed on a hollow tile base or mounted on legs providing a clearance of at least 5 in. (127 mm) to the deck. In addition, the deck must be effectively insulated with a noncombustible material or sheathing.

Continuously burning pilot lights are not permitted on any appliance. Combustion chamber heaters may be used only if the combustion air is completely separated from the atmosphere of the boat.

Liquefied Petroleum Gas (LP-Gas) and Compressed Natural Gas (CNG) Systems

In the interest of safety, it is important that the properties of both liquefied petroleum gas (LP-Gas) and compressed natural gas (CNG) be understood. At ambient temperature, petroleum gas can be liquefied and is stored in containers having a rated working pressure of not less than 250 psig (1724 kPa). Natural gas, on the other hand, cannot be liquefied at ambient temperature. In order to package a sufficient usable quantity for the consumer, it is stored at high pressures in cylinders having a rated working pressure of over 2,000 psig (13 790 kPa). Both LP-Gas and CNG, being free gases, are miscible in air. If released in an enclosed space, they will diffuse throughout the compartment.

The physical properties of each gas are very different. The hydrocarbon constituents of petroleum gas are heavier than air and, if released from the container, will settle. Natural gas, on the other hand, is lighter than air and will rise if released. However, if either gas is released in a confined space aboard a vessel, this difference is of little comfort since in either case it is very possible an explosive mixture could develop; and a source of ignition is all that would be needed to cause an explosion. In recognition of this, NFPA 302, *Fire Protection Standard for Pleasure and Commercial Motor Craft*, (hereinafter referred to as NFPA 302) stipulates that the containers and regulators of either LP-Gas or CNG be located on the open deck, cabin top, outside of cockpits, etc., and protected by a housing vented to open air near the top and bottom. If construction or design of the motor craft prevents this, then the container and regulator shall be mounted in a locker or housing that is vapor tight to the hull interior and above the waterline in an open cockpit. NFPA 302 also covers the method of venting the locker or housing.

NFPA 302 further requires that the distribution lines for either system be continuous lengths of tubing except where a joint is needed to feed each appliance. Lines are required to be secured against vibration, protected against abrasion whenever they pass through decks or bulkheads, protected from physical damage, and accessible for inspection.

Marine gas-consuming appliances should not use pilot lights, but ovens may have flame control, provided the flame is totally off when the oven control is off. It is recommended that all burners have thermal fuel cutoff devices to prevent gas flow when the flame is extinguished for any reason.

It is important that the gas system be checked for leaks at least twice a month, or when the system is serviced or the container changed. NFPA 302 outlines the following procedure to check for leaks:

> "With the appliance valves closed and all other valves open, note pressure on the gage. Close container valve. The pressure should remain constant for at least 10 min. If pressure drops, locate leakage by application of soapy water solution at all connections. Repeat test for each container in multicontainer systems. NEVER USE FLAME TO CHECK FOR LEAKS. NEVER USE SOAP CONTAINING AMMONIA."

If a leak detection device is installed, the above instructions should be modified accordingly.

ELECTRICAL SYSTEMS

Both low voltage direct current (dc) electrical systems and 120/230 alternating current (ac) systems should always be of the two wire type with insulated conductors to and from the source of power and to all accessories.

Direct current electrical systems may be either grounded or ungrounded, but ungrounded systems are rarely used. Although an ungrounded system requires that the cranking motor, alternator, and all accessories (including engine accessories) be of the two wire ungrounded type, such equipment is not readily available. A grounded dc electrical system is actually wired identical to an ungrounded system except at the engine, where the accessories use the engine block as a common ground. Since the engine is also the common ground point for the entire system, using the engine block as a common ground does not result in any stray current flow from the propeller or shaft connected to that engine. To maintain this condition in twin-screw boats, it is necessary to connect the two engines with a heavy conductor capable of carrying the maximum cranking motor current.

To protect against stray current leakage, grounded dc electrical systems are always connected so the engine block is connected to the battery negative terminal, making the propeller shaft and propellers cathodic, if stray current leakage is encountered. If stray current leakage occurs, the underwater parts connected to the cathodic side of the circuit are protected and the anodic items are subject to accelerated corrosion.

Alternating current electrical systems on a boat may be powered from either a shore power outlet on the dock or by an onboard ac generator. A switch prevents both sources from being used simultaneously. The shore power current carrying conductors, including the shore grounded neutral conductor, are never grounded on the boat, but the grounded neutral of the onboard generator may be connected to the boat ground. Only the shore safety grounding wire, which is connected to the noncurrent carrying parts of the system, is grounded on shore and on the boat. Because of the wet environment, shock hazard is substantially more severe on a boat than on shore. Accordingly, the safety grounding wire is very important, but unfortunately can cause accelerated corrosion of any exposed anodic materials, such as aluminum or steel. This problem occurs when the boat is docked at a marina where many boats are connected to the same grounding wire.

Because of interboat corrosion, aluminum, steel, and inboard-outdrive boats with aluminum alloy outdrives must be wired in a special manner to avoid this problem. Such boats must either use an isolation transformer to transfer power from the shore to the boat, or a device called an isolator must be installed in series with the grounding wire from shore. The isolation transformer transfers the electrical power, but current carrying conductors on the boat have no relationship to earth ground. The shore grounding wire is stopped at the transformer core so that no electrical connection exists between the boat with an isolation transformer and any other boat. The isolator is a device that will pass alternating electrical current to allow the grounding wire to function as intended as a protection from electrical shock; the isolator also effectively blocks dc galvanic currents from flowing on the same wire. The isolator, therefore, breaks the dc connection between adjacent boats.

Fire Extinguishers

The fire extinguishment potentials of portable fire extinguishers are normally evaluated by testing laboratories. The U.S. Coast Guard also classifies portable fire extinguishers based upon the UL (Underwriters Laboratories) classification of fires, but uses a different method of indicating extinguishment potentials. These designations are shown in Table 13-5A.

PLEASURE AND SMALL COMMERICAL BOAT OPERATIONS

General Maintenance

Operation and maintenance of pleasure and small commercial boats consists mainly of good housekeeping. The boat must be kept "shipshape" at all times. Clean waste and rags should be kept in covered metal containers; used waste and rags should be kept in separate covered metal containers. Dirty waste and rags should be properly disposed of each time the boat is docked. Flammable paint and varnish removers must be used with caution, and only in well ventilated areas. When repainting interior areas, fire retardant paint should be used. Unprotected electric lights or open flames should not be taken into areas of possible vapor accumulation, and gasoline or other flammable liquid used for cleaning (including paint brushes) should not be used in such areas. Ventilation ducts and exhaust blowers must be maintained at top efficiency.

Good maintenance and sensible fuel system operation are probably the most important fire prevention duties for a boat owner. Gasoline vapors are heavier than air and will not escape from low-lying cockpits such as bilges or tank bottoms unless drawn or forced out. Atmospheric concen-

TABLE 13-5A. Number and Distribution of Fire Extinguishers

Type of Boat	Class of Extinguishers[1]	Minimum Required	Recommended Locations
Open boats under 16 ft (5 m)	B-I	1	Helmsman's position
Open boats over 16 ft (5 m)	B-I	2	Helmsman's position and passenger space
Boats under 26 ft (8 m)	B-I	2	Helmsman's position and cabin
Boats 26–40 ft (8 to 12 m)	B-I	3	Engine compartment, helmsman's position, and galley[3]
Boats 40–65 ft (12 to 20 m)	B-I	4[2]	Engine compartment, helmsman's position, crew quarters, and galley[3]
Boats 65–75 ft (20 to 23 m)	B-I	5[2]	Engine compartment, helmsman's position, crew quarters, and galley [3]
Boats 75–100 ft (23 to 30 m)	B-I	6[2]	Engine compartment, helmsman's position, crew quarters, and galley[3]

[1] One of the required extinguishers shall additionally have the capability of extinguishing Class A fires.

[2] If more than three B-I units are recommended, the extinguishing capacity may be made up of a smaller number of larger units, provided each recommended location is protected with an extinguisher readily accessible, e.g., 3 B-II units may be used in lieu of 4, 5, or 6 of the smaller B-I units.

[3] Extinguishers recommended for "engine compartment" should not be located inside such compartment but near an entrance to the compartment unless someone is normally present in the compartment.

trations of gasoline vapor as low as 1.4 percent and as high as 7.6 percent are flammable. The entire system, including tanks, piping, vent lines, and other accessories must be checked frequently for leaks or evidence of corrosion. Connections must be kept tight. Fuel carried on board outside of the fixed fuel system should be safely stored and stowed in listed or approved containers or in portable tanks.

Fueling

Utmost care must be exercised during fueling operations. Some general guidelines to be observed during fueling operations are:

1. Fueling should never be undertaken at night, except under well lighted conditions.
2. Smoking must be forbidden on board and nearby fueling operations.
3. Before opening tanks, the following precautions should be observed:
   a. All engines, motors, and fans should be shut down.
   b. All open flames should be extinguished.
   c. All ports, windows, doors, and hatches should be closed.
   d. The quantity of fuel to be taken aboard should be determined prior to the start of the fueling operation.
4. The fuel delivery nozzle should be put in contact with the fill pipe before delivery of the fuel is begun; this contact should be continuously maintained until the flow has stopped. (There is a serious hazard from static discharge if this rule is not followed.)
5. Tanks should not be completely filled. A minimum of 2 percent of the tank space should be allowed for expansion. This space allowance should be increased to 6 percent or more if the fuel being taken aboard is below 32°F (0°C). To simplify, do not fill the tank full.
6. After the fuel flow has stopped:
   a. The fill cap should be tightly secured.
   b. Any spillage should be completely wiped up and absorbent material disposed of on shore.
   c. Ventilate all spaces and check for gasoline vapors before starting any engines or operating any appliances.

Storage Between Voyages

Before a boat is stored, even for a short period of time, several precautionary actions should be taken. First, a thorough inspection should be made of the entire vessel. All combustible trash and rags should be removed, as well as painting materials and other nonessential flammable liquids. Fuel lines for both engines and appliances should be secured at both ends. Bilges should be inspected and pumped dry. Shaft logs and rudder bearings should be checked and tightened if necessary. All through-hull fitting valves should be closed, except scuppers which permit drainage of rain water. Batteries should be checked and all electrical services except automatic bilge pumps secured. When maximum ventilation is necessary, locker doors, drawers, and bilge hatches should be secured in a partially opened position.

NFPA BOATING STANDARDS

Requirements relating to powered boats are contained in NFPA 302. These requirements are intended to prevent fire and explosion due to fuel leakage. Among the subjects covered are: hull arrangement, ventilation and lightning protection; engines and their exhaust systems; fuel systems; cooking, heating and auxiliary appliances; electrical systems, and fire protection equipment.

Firesafety requirements for areas used for the construction, repair, storage, launching, berthing, or fueling of small craft are contained in NFPA 303, *Fire Protection Standard for Marinas and Boatyards.* Among the subjects covered are: fire protection; berthing and storage facilities; operational hazards; and electrical wiring and equipment.

OTHER ORGANIZATIONS CONTRIBUTING TO BOATING SAFETY

The American Boat and Yacht Council, Inc.

The Council was established in 1954 so the knowledge, experience, and skills of technicians within the boating industry could be applied to the development of recommended practices and engineering standards for improving and advancing the design, construction, equipage, and maintenance of small craft with reference to their safety. A cooperative working relationship exists between the Council and NFPA insofar as fire protection of motor craft is concerned. Consistency of requirements is a mutual goal.

The National Association of Marine Surveyors

This organization includes professional marine surveyors dedicated to dissemination of information concerning current standards, new materials and their application in the marine field, and control of new hazardous materials within the scope of marine surveyors' operations.

Underwriters Laboratories Inc.

Underwriters Laboratories Inc. (UL), founded in 1894, is chartered as a not for profit, independent organization and tests products for public safety. It develops product standards, including specific tests to determine that a product is manufactured in accordance with UL requirements; UL also operates laboratories for the examination and testing of devices, systems and materials in accordance with those requirements. UL standards are intended for use in the investigation of products to determine their relation to life, fire and casualty hazards, and, in some cases, in relation to crime prevention. The UL Marine Department examines and investigates marine products intended for use on commercial vessels, oil drilling platforms, and pleasure craft. All UL marine standards are designed to indicate compliance with any applicable U.S. Coast Guard regulations. An important factor in UL's listing, classification, or component recognition of products, systems, or materials is that such products are subject to UL's follow-up inspection service at the point of manufacture. This inspection supplements the manufacturer's quality control inspection program and helps determine that the product being manufactured is identical to the product that was tested.

United States Coast Guard

The U.S. Coast Guard publishes regulations for pleasure craft, including fuel and electrical systems requirements. These regulations are found in *33 CFR 183.*

COMMERCIAL VESSELS

Almost any accident involving a vessel presents either a direct or secondary threat of fire. Collision or stranding can rupture tanks or containers of hazardous materials and provide enough frictional heat to ignite the contents. The same hazards can result from the heavy pounding a ship receives in bad weather. Even the leakage of water into a cargo hold can cause accelerated oxidation and eventual combustion of some organic substances. In discussing marine fire hazards, both the transfer of cargo and the operation of the vessel must be considered in addition to the traditional fire problems.

Steadily increasing world population and demands for foreign merchandise have resulted in an ever increasing volume of commercial waterborne tonnage. Several international conventions exist that relate directly or indirectly to fire protection and prevention aboard commercial vessels, and are concerned with construction, operation, manning, training, and inspection. The U.S. is signatory to many of these conventions, whose key points will be stressed in this section. Also, several international codes relate to the carriage of hazardous materials; although not conventions, due to their acceptance by a large number of nations, they carry the weight of a convention.

International Maritime Regulations Standards

The International Maritime Organization (IMO), (formerly IMCO), is a specialized agency of the United Nations and addresses matters relating to maritime activities. IMO is the depository of all treaties dealing with commercial vessel safety, and has developed several recommended standards relating to the same subject. A list of these Conventions and Codes follows, with a brief description of how they relate to fire protection and prevention.

Safety of Life at Sea (SOLAS) '74: Beginning with SOLAS '48 the various International Conventions for the Safety of Life at Sea (SOLAS) address, among other things, fire protection aboard merchant vessels. Although all types of vessels are covered, passenger vessels are addressed in greatest detail, including the construction of bulkheads separating the various compartments. With the inclusion of the first and second set of amendments to SOLAS '74, emphasis is placed upon increased cargo ships and tank vessels, especially regarding structural fire protection details. The reason for this stems from a number of disastrous passenger vessel fires in the 1930s which brought about SOLAS '48, followed by SOLAS '60 and SOLAS '74.

Under the terms of these Conventions, a certificate attesting to a ship's safety issued by a one signatory country is accepted by all other signatory countries. Usually, one of the Classification Societies will perform the survey associated with issuing the certificate. The certificate insures a minimum level of safety and also permits ships to operate freely from one country to another without risk of discrimination due to their safety features. One fire protection feature of SOLAS '74 is the requirement for an inert gas system for new oil tankers over 100,000 deadweight tons (DWT) (95 256 metric tons) and combination oil/bulk solid carriers over 50,000 DWT (47 628 metric tons).

The current SOLAS Treaty, SOLAS '74, was effective internationally on May 25, 1980. Basically, it incorporated resolutions and proposed—but not ratified—amendments that had been developed over the period from 1960 to 1974. Although the amendments to SOLAS '60 never entered into force, the IMO Assembly adopted many of them and recommended that the individual countries incorporate them into their own national maritime safety regulations.

Several changes to SOLAS '74 have been developed since the 1974 Treaty was finalized. These have been expressed in the 1978 Protocol Relating to the International Convention for the Safety of Life at Sea—1974, the 1981 Amendments, and the 1983 Amendments. The majority of the changes in the 1981 Amendments were in Chapter II-2, which deals with fire protection. Also, there were significant changes to Chapter III (Lifesaving Appliances) in the 1983 Amendments.

The 1978 Protocol: The 1978 Protocol came out of the 1978 Tanker Safety and Pollution Prevention (TSPP) Conference. This conference was a direct result of U.S. initiatives in response to a tragic series of tanker casualties that occurred in 1976/77. Since the SOLAS '74 Treaty had not been ratified at that time, the Treaty could not be amended. Therefore, these changes were issued as a Protocol. The 1978 Protocol Relating to the International Convention for the Safety of Life at Sea, 1974, came into force May 1, 1981. It mandates steering gear improvements, collision avoidance aids, dual radars, inert gas systems (IGS) for new and existing tank ships down to 20,000 deadweight tons (DWT) (19 000 metric tons), deck foam systems for new tank ships down to 20,000 DWT (19 000 metric tons), and closed ullage systems for all tank ships fitted with IGS. Finally, it requires IGS for tank ships capable of crude oil washing.

The 1981 Amendments: There are two sets of amendments to SOLAS '74. The first set was developed between 1974 and 1981 by various IMO Subcommittees. In November 1981, the International Maritime Organization's Maritime Safety Committee (MSC) accepted the 1981 Amendments to SOLAS '74; these Amendments came into force September 1, 1984. These include changes to Chapters II-1, II-2, III, IV, V, and VI. The most important changes concern Chapter II-1 (Subdivision and Stability, Machinery and Electrical Installations) and Chapter II-2 (Construction —Fire Protection, Fire Detection and Fire Extinction). Chapters II-1 and II-2 have been updated and virtually rewritten.

The changes to Chapters II-1, II-2 and V contained in the SOLAS '78 Protocol are embodied and partly updated in the first set of amendments. The changes to Chapter II-2 include the suggested standards established in IMO Resolutions A.327(IX) and A.372(X), provisions for halogenated hydrocarbon extinguishing systems, and a new Regulation 62 on inert gas systems. The latter regulation was adopted by the eleventh IMO Assembly as a recommendation in 1979. The changes were so extensive that Chapter II-2 was essentially reorganized and rewritten.

The U.S. Coast Guard is currently incorporating these changes into Title 46 of the *Code of Federal Regulations* (CFR).

The 1983 Amendments: At the June 1983 Maritime Safety Committee meeting a second set of amendments was approved (1983 Amendments). These amendments will become effective July 1, 1986. They include changes to Chapters II-1, II-2, III, IV and VII. The most important changes concern Chapter III (totally rewritten and updated) and Chapter VII (changed to make the International Bulk Chemical Code and the International Gas Code mandatory). The following is a short description of the important changes of interest on a chapter by chapter basis:

Chapter II-2

The changes to Chapter II-2 contain: (1) improvements to the 1981 Amendments dealing with detection systems and inert gas systems; (2) several editorial changes; and (3) a significant rewrite of Regulation 56 (location and separation of spaces). The regulations affected are II-2/3, II-2/11, II-2/12, II-2/15, II-2/27, II-2/36, II-2/37, II-2/40, II-2/42, II-2/49, II-2/52, II-2/54, II-2/56, II-2/59, II-2/61 and II-2/62.

Chapter VII

The major thrust of the changes to Chapter VII involves the incorporation by reference of the International Bulk Chemical Code and the International Gas Carrier Code into SOLAS. These two codes now become mandatory by the 1983 Amendments. In addition, the chapter was divided into three parts—Packaged Goods and Bulk Cargoes, Chemical Tank Ships, and Gas Carriers. The regulations affected include Regulations VII/1 through VII/13.

Convention on Training and Certification of Seafarers, 1978

This Convention, commonly referred to as Standards for Training and Watchkeeping (STW), outlines the training required before obtaining a license. Among the various types of training needed are in fire fighting and fire prevention, as well as familiarity with the cargo systems aboard chemical and gas carriers. Attendance at an approved course or training for an appropriate period under supervised shipboard services is required.

Dangerous Goods Code

As a result of the Safety of Life at Sea Conference of 1960, IMO began work on the development of a code covering the safe transportation by sea of dangerous goods or hazardous materials. Most dangerous goods have as their primary or secondary hazard a potential for fire. If fire is not a hazard, dangerous goods, by their presence, complicate the fire fighting and rescue efforts of emergency forces. Today the IMO *Dangerous Goods Code* is widely recognized by numerous maritime nations and has become the "bible" for seafarers. The IMO *Dangerous Goods Code* is compatible with the method for classifying hazardous materials used by the U.S. for years preceding the code's development. As a result, the code is recognized as a substitute for the safe carriage of hazardous materials aboard foreign and U.S. vessels while in U.S. waters. The *Dangerous Goods Code* is constantly updated by appropriate IMO committees.

Bulk Chemical Code

As a result of the increased demand for chemicals among industrialized trading nations, the method of shipping chemicals has changed from packages to larger quantities. Standard petroleum tankers were modified and shifted to this trade. Several nations, among them the U.S., became convinced that these modified tankers did not have the necessary safety features for protecting the crew, the port, and the environment against chemicals having properties other than flammability. IMO was petitioned to begin development of a code covering the design, construction, and operation of chemical carriers. On October 12, 1971, IMCO adopted a code for bulk chemical carriers. This code includes special fire protection systems using suppressants other than water that will not react with the chemical cargo, yet will be effective when applied. On September 26, 1977, the U.S. Coast Guard adopted the IMO *Bulk Chemical Code* as part of its regulations. In addition to upgrading the various systems aboard these vessels, the code requires an increased level of stability and cargo containment to protect the cargo and ship when involved in an accident.

The *International Bulk Chemical Code*, which is applicable to new ships, will become effective on July 1, 1986. This code, reflecting improvements and previously agreed to amendments to the *Bulk Chemical Code*, becomes mandatory under the second set of amendments to SOLAS '74.

Liquefied Gas Code

In conjunction with development of the *Bulk Chemical Code*, the need became obvious for that code to specifically address the carriage of gases. Gases are transported in other than their natural state, using low temperature, high pressure, or a combination of both. Methane (natural gas), for example, cannot be liquefied by pressure alone; to transport it economically as a liquid, it must be refrigerated to −162°F (108°C). Very specialized containment systems are needed to maintain that temperature for the long sea voyages involved. Also required are specialized fire fighting systems and techniques to put out any fire that may develop aboard one of these bulk gas vessels. IMO now has two bulk gas codes, one for existing and the other for newly built liquefied bulk gas ships. The U.S. adopted the IMO Code for new ships on October 4, 1976. The U.S. Coast Guard requires existing ships to upgrade where the IMO Code also requires it, and maintains the existing U.S. standards when they are more strict than the existing IMO Code.

As noted above, the *International Gas Code*, which is applicable to new ships, will come into force July 1, 1986. This code, containing improvements and previously agreed to amendments to the *Liquefied Gas Code*, becomes mandatory under the second set of amendments to SOLAS '74.

United States Coast Guard (USCG)

The U.S. Coast Guard is the federal agency responsible for enforcing regulations covering the safety of U.S. commercial vessels, both ocean going and inland. The Coast Guard also has authority over foreign vessels operating into and out of U.S. ports (under the authority of the Ports and Waterways Safety Act as modified by the Tanker Safety Act). The Coast Guard has a comprehensive inspection program for foreign tank vessels directed toward prevention of fires and explosions similar to those resulting from the incident aboard the S.S. *Sansanena* in Los Angeles harbor on December 17, 1976. The inspections cover the integrity of the various cargo containment and fire fighting systems aboard a tanker. Included in the cargo containment system are the tanks, piping, and venting system that contain the cargo and its vapors. The Coast Guard also inspects foreign vessels to ensure compliance with the various international treaties and standards described earlier. As the enforcement agency for SOLAS '74, the Coast Guard has the authority and responsibility to ensure that U.S. ships and foreign ships visiting U.S. ports comply with the treaty and its amendments.

In most ports, the USCG Captain of the Port (COTP) now heads the Ports Operations Division of the Marine Safety Office (MSO). The COTP is the U.S. Coast Guard official responsible in a port for ensuring that commercial vessels operate, load and discharge cargo safely, and that they are maintained in accordance with the regulations. The COTP or MSO also is responsible for ensuring that waterfront facilities are maintained and operated in a safe manner, especially regarding fire prevention. Lacking detailed regulations in this area, the COTP or MSO frequently develops local regulations covering waterfront facilities. The COTP or MSO is responsible for the protection of waterfront facilities, and is required to have a detailed contingency plan which is to be implemented in the event of a fire involving a vessel or waterfront facility. This officer is also responsible for responding to pollution incidents involving navigable waters, and should have a contingency plan covering this incident response. Since pollution can involve petroleum products and chemicals, contingency plans for both usually are incorporated into one plan. These plans customarily include the local fire department because the Coast Guard's resources are devoted primarily to fire prevention, limiting its capability to fight fires.

The U.S. Coast Guard Officer in Charge of Marine Inspection (OCMI) is the official responsible for ensuring that U.S. vessels are built and maintained to existing U.S. regulations, most of which are related to the treaties described previously. In nearly every U.S. port, the functions of both the COTP and OCMI are now combined into one office, that of Commanding Officer of the Marine Safety Office. The COTP functions are handled by the Port Operations Division, and most of the OCMI functions are handled by the Inspection Division.

U.S. Coast Guard regulations covering commercial vessels are contained in the following parts of the CFR:

1. 33 CFR Part 157, *Tank Vessels Carrying Oil in Bulk.*
2. 46 CFR Parts 30-40, *Tank Vessels.*
3. 46 CFR Parts 70-80, *Passenger Vessels.*
4. 46 CFR Parts 90-106, *Cargo and Miscellaneous Vessels.*
5. 46 CFR Parts 107-109, *Mobile Offshore Drilling Units.*
6. 46 CFR Part 146, *Carriage of Military Explosives.*
7. 46 CFR Part 147, *Transportation of Bulk Solids.*
8. 46 CFR Parts 150-153, *Carriage of Chemicals in Tank Vessels and Barges.*
9. 46 CFR Part 154, *Carriage of Liquified Gases.*
10. 46 CFR Parts 160-165, *Specifications.*
11. 46 CFR Parts 166-168, *Nautical Schools.*
12. 46 CFR Parts 175-187, *Small Passenger Vessels (under 100 gross tons).*
13. 46 CFR Parts 188-196, *Oceanographic Vessels.*
14. 49 CFR Parts 170-189, *Transportation of Hazardous Materials by All Modes Including Water.*

SHIP CONSTRUCTION

Basically, a ship may be considered as a huge box girder consisting of the hull plating and the main deck. These parts are, in turn, strengthened by such members as the keel, frames, beams, keelsons, stringers, girders, pillars, lower decks, and transverse bulkheads. To appreciate a ship in its entirety, it is well to understand the functions of each of its parts.

The keel is primarily the backbone of the ship. It consists of a rigid fabrication of plates and structural shapes which run fore and aft along the center line of the ship. The stem is connected to the forward end, and at the after end is the stern frame which supports both the rudder and the propeller.

The frames are the ribs of the ship. Their lower ends are attached at intervals along the keel, and their upper ends are attached through brackets to the beams which support the deck. Internal bracing is provided by keelsons and stringers running fore and aft. The frames determine the form of the ship, and support and stiffen the shell plating.

The shell plating, necessary for watertightness, also is one of the principal strength members of the ship. Running continuously from the stem to the stern frame, and from the keel to the weather deck, the shell plating forms three of the sides of the box girder. The plating, aided by the frames, must be able to withstand the exterior water pressure, stress from the buffeting of waves, and rubbing and bumping against docks.

The main deck of the ship forms the fourth side of the girder, and for this reason must be of strong construction. The plating is connected to beams which extend from side to side across the ship. The deck is strengthened by the doubling of plates in regions weakened by openings such as hatchways and companionways, and also under all deck machinery, chocks, and bits. The deck is supported from below by girders and pillars.

The bottom, sides, and main deck of the ship would not be strong enough to withstand the stresses of an ocean voyage without some internal stiffening. In a cargo vessel, this is provided by the lower decks and the main transverse bulkheads. For tank vessels, stiffening comes from transverse and longitudinal bulkheads which separate the vessel into numerous cargo tanks extending from the keel to the main deck. In the newer tank vessels, including all chemical and gas carriers, double bottoms and sides provide added protection from cargo release in case a crack occurs in the shell plating.

In addition to furnishing support for the shell and decks, the main transverse bulkheads are made watertight. The subdividing of the vessel into watertight compartments is achieved so that, in case of damage, water can be confined. All doors through these bulkheads must be fitted with gaskets to make them watertight; also, doors must be kept clear at all times so they can be closed instantly. The bulkheads also are fire resistant and can limit fire spread. This degree of compartmentation can complicate fire fighting efforts unless fire fighters know how a ship is laid out. In an emergency, the general arrangement plan for the vessel should be examined to locate access and egress routes for fire fighting. Some typical general arrangement plans are shown in Figure 13-5E.

The first bulkhead aft of the stem is known as the collision bulkhead, because its purpose is to limit the

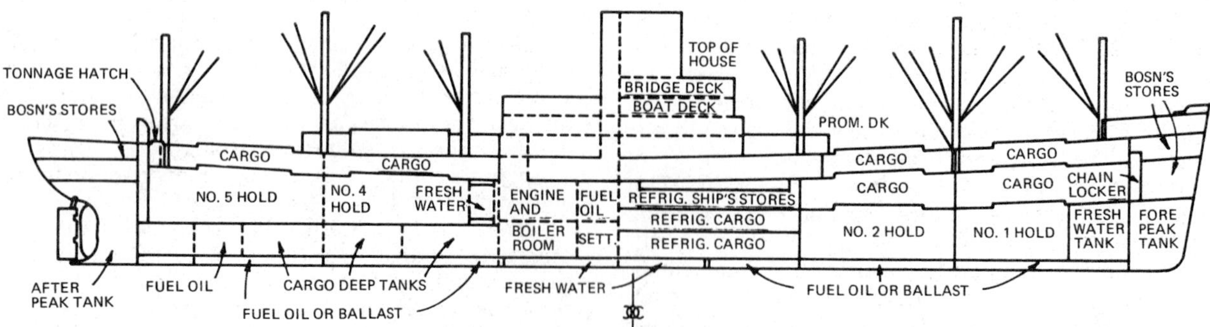

FIG. 13-5E. A cargo passenger vessel showing location of the principal cargo and tank spaces.

flooding that might occur after a collision. No doors or other openings are permitted below the main deck in this bulkhead.

In a cargo vessel, the engine and boiler rooms usually are located amidships. For a tank vessel, the engine room is located in the after section, aft of the cargo tank. Special foundations are necessary to support heavy engines and boilers. In order to provide sufficient headroom for the propelling machinery, it usually is necessary to omit one or more of the decks in this region. To maintain the vessel's strength in the absence of these decks, several extra heavy web frames and transverse beams are fitted.

The propeller shaft extends through the after holds from the engine to the stern gland. Because the propeller shaft must be accessible at all times for inspection and lubrication, it is enclosed in a narrow tunnel known as the shaft alley. The entrance to the shaft alley from the engine room is closed by a watertight door, and the sides are of watertight construction so that a fracture of the tail shaft or similar accident will result in only the tunnel being flooded.

The necessity for good drainage requires special attention to the design and construction of a ship. Free water on the decks, in a hold, or in the bilges is detrimental to the stability of the vessel. Therefore, the drainage system must be as efficient as possible. The decks should be cambered to permit drainage to the scuppers which lead the water either overboard or to the bilges. Sufficient scuppers and suctions must be provided so the drainage will be effective under any condition of list or trim of the ship. Solid bulwarks, where fitted around a deck, should be pierced by large freeing ports to allow any water that is shipped to quickly escape.

Ventilation

The type of ventilation used in a cargo vessel depends upon the nature of the space and the service of the ship. It can be natural, mechanical, or a combination of the two, and may be extended to air heating, cooling, cleaning, humidifying, and dehumidifying. Air movement in natural ventilation systems is created by the difference in density between inside and outside air and depends upon the relative air temperatures. Ventilators dependent upon wind direction and velocity for induction of air currents are used with natural systems. Natural ventilation generally is limited to a few ships, lockers, and storerooms, depending upon their location, and to some dry cargo holds; it also sometimes is used for engine and boiler rooms. In dry cargo holds, ventilation generally is accomplished with vents in the forward end for exhaust, and cowls at the after end for supply. Large ducts are required because of the low velocity necessary for air flow. In a mechanical system, air is moved by various types of fans driven by electric motors. These systems provide positive circulation of air at desired temperature and volume, function regardless of outside atmospheric conditions, and are easily controlled to meet possible variations in requirements.

The ventilation for living accommodations on both cargo and tank vessels is by means of forced air through a duct system similar to that used ashore. All the compartments are interconnected through these ducts. A damper is provided in each duct to isolate one compartment from another in order to prevent the spread of fire. The dampers can operate manually or automatically using a fusible link. Cargo tanks on tank vessels are not ventilated, but utilize a vent system to allow the tank to relieve excess pressure or to allow air ingress if a vacuum develops. This venting arrangement is explained later under "Tank Vessels."

Stability

The stability of a ship, or the property of a floating body to remain at rest in a stable position, is controlled by the interaction of two opposing forces. The center of gravity (Point G in Figs. 13-5F, 13-5G, and 13-5H) is a point

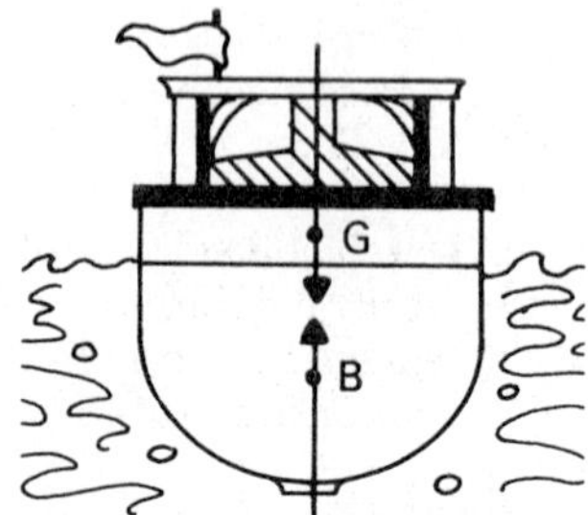

FIG. 13-5F. Ship in a state of vertical equilibrium.

at which the weight of the ship, and any other weight aboard, is concentrated. It remains in a constant position until the weight distribution aboard the vessel is changed, and acts with a downward force. The center of buoyancy may be defined as the center of gravity of the body of water displaced by a floating vessel. It is the center of the immersed part of the ship and acts with an upward force, perpendicular to the surface of the water.

When a ship is in a state of vertical equilibrium, as shown in Figure 13-5F, the downward force of gravity (G) and the upward force of buoyancy (B) lie in the same vertical line. When the ship is inclined by some external force, the center of gravity (G) remains in its original position. The center of buoyancy (B), since it is the center of the immersed part of the ship, shifts toward the lower side as shown in Figure 13-5G. In that situation, the central force of (G) downward, combined with the outer force of (B) upward from the low side, act together to return the vessel to an upright position.

The addition of water into the upper portion of the hull, as shown in Figure 13-5H, produces an entirely different situation. The addition of weight at a high point within the hull raises the center of gravity. The fact that water runs to and stays at the lowest point inclines the vessel and holds it in an inclined position. The center of buoyancy shifts to the lower side, as it did before. If the addition of weight (water) is stopped while the downward force of (G) is inboard (toward the center line) of the upward force of (B), the vessel will settle in a new

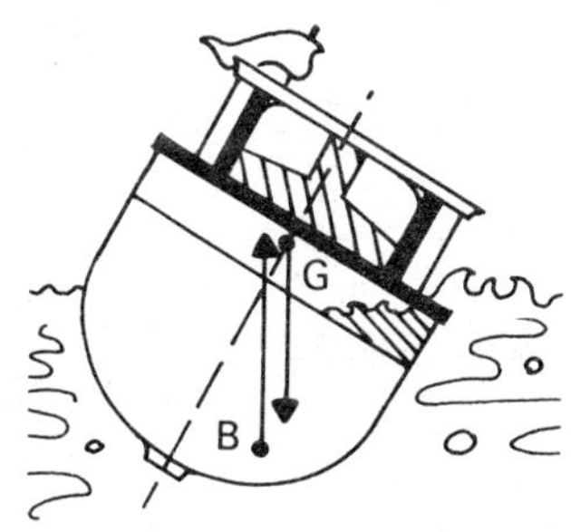

FIG. 13-5G. Ship inclined by an external force.

FIG. 13-5H. The effect of adding water (weight) to the upper portion of a ship.

stabilized position, inclined toward the side of the additional weight. If sufficient water is added to bring the downward force of (G) outside the upward force of (B), as shown in Figure 13-5H, the vessel will continue to roll over and will eventually capsize.

From this explanation of stability, it is easy to see how the addition of water, especially high in the vessel, can dramatically affect the vessel's stability. Whenever a fire is fought on board a vessel, this problem must be recognized. A means of removing the water from a vessel during fire fighting must be established; there is a limit to how much water can be pumped aboard before the vessel will roll over on its side (as the S.S. *Normandie* did in New York Harbor in 1942). To prevent this from happening, a person who understands the principle of stability should be consulted while fighting a shipboard fire. Otherwise, the incident could cause loss of life as well as loss of a vessel.

Tank Vessels

Older tank vessels include pump rooms, one forward and one aft, between the last cargo tanks and the engine room. The cargo and ballast pumps are located within the pump room. All cargo and ballast water passes through the pump rooms into and out of the cargo tanks. A pump room can accumulate flammable vapor from any cargo leakage into the bilges from the various piping and valves within the room. For this reason, the room should be well ventilated before entering, and someone topside should be aware of persons entering a pump room in case of an exposure hazard from an acute concentration of vapors. Tank vessels of newer design, especially the chemical and gas carriers, utilize deep well pumps within each cargo tank, thereby eliminating the need for pump rooms.

Each cargo tank on a tank vessel is fitted with a venting system to regulate the pressure within the cargo tank. On crude carriers and petroleum product carriers, each tank is connected to a common venting system. On a chemical or gas carrier, the vents are independent of each other. This arrangement is required to ensure that vapors from different chemicals do not come in contact with each other, and to prevent incompatible reactions or contamination.

The venting system contains flame screens or flame arrestors. These devices prevent any flame that develops outside the cargo tank from entering the cargo tank and causing an explosion, as happened on the S.S. *Sansinena*. In this case, it was theorized that flame entered the cargo tank through a wasted section of the vent piping.

It is extremely important that the various piping systems aboard a tank vessel, including the vent system, be maintained in good condition and that flame screens are placed over any openings into a cargo tank. During loading of cargo or when ballasting a tank that previously contained cargo, copious amounts of flammable vapors are generated. These vapors are vented to the atmosphere to ensure the cargo tanks are not overpressurized. In the vicinity of a tanker conducting these operations, the air will contain a flammable mixture, especially if low winds are present. Should the flammable mixture be ignited, the flame will be prevented from entering the tank if flame screens or arrestors are in place and all the piping systems that penetrate the cargo tank are intact.

For some toxic chemicals, and all gases, a vapor recovery system is generally employed to return the vapors to shore. With this arrangement, no venting to the atmosphere takes place unless an emergency occurs. The cargo transfer operations and vapor return are balanced to prevent over or underpressurizing of the cargo tank. This procedure eliminates the potential flammable atmosphere around the vessel during loading; however, the various piping systems must be intact to prevent release of any flammable vapors or liquids.

As a result of SOLAS '74 and the TSPP Conference, inert gas systems are now being installed aboard tankers. These systems are intended to ensure that the atmosphere within the cargo tanks does not contain a flammable mixture. Without an inert gas system, all openings into a cargo tank must be protected with flame screens or arrestors and all piping must be in good condition. Otherwise, as discussed previously, a source of ignition can be introduced into the cargo tanks. Experience has shown that precluding entry of all sources of ignition is difficult; consequently, the trend is toward inerting the cargo tanks whenever they contain hydrocarbon vapors. The inert gas system must be designed, constructed, and operated to render and maintain the atmosphere of cargo tanks nonflammable at all times, except when the tanks are required to be gas free. The system must be capable of:

1. Inerting empty cargo tanks by reducing the oxygen content of the atmosphere in each tank to a level at which combustion cannot be supported.
2. Maintaining the atmosphere in any part of any cargo tank with an oxygen content not exceeding eight percent by volume and at a positive pressure at all times in port and at sea, except when it is necessary for such a tank to be gas free.
3. Eliminating the need for air to enter a tank during normal operations, except when it is necessary for such a tank to be gas free.
4. Purging empty cargo tanks of hydrocarbon gas so that subsequent gas release operations will at no time create a flammable atmosphere within the tank.

Figure 13-5I is a simplified illustration of tanker operations using inert gas systems. The rule is never to mix both air and hydrocarbons during operation, but to insert inert gas between the air and hydrocarbon phases of the operation. Specific practices for these systems are described in 46 CFR 32.53, Regulation 62 of SOLAS '74 (describing international requirements which are very similar to U.S. criteria), and the IMO *Guidelines for Inert Gas Systems*. Use of inert gas systems should significantly enhance tanker safety by greatly reducing the risk of fire and explosion within cargo tanks.

BARGES

Virtually any commodity can be shipped by water. The inland waterways industry has developed various types and sizes of barges for the efficient handling of products ranging from coal in open hopper barges to chemicals and "thermos bottle" barges, and from dredged rock in dump scows to railroad cars on car floats. Barging is the only practical mode of marine transportation to move outside machinery, tanks, kilns, and some space vehicles for long distances.

The most versatile, inexpensive and numerous type of barge in the U.S. is the hopper barge. With minor modifications, it can be adapted to the transportation of literally any solid commodity in bulk or package. It is basically a single-skinned, open top box, the inner shell forming a long hopper or cargo hold. The bottom, sides, and ends of the hold are free of appendages. They are generally of welded plate construction with a double bottom for greater safety. They are braced to resist the heaviest of external blows, as well as to absorb the impact of buckets used in loading and unloading. The covered dry cargo barge is used for bulk loading commodities that need protection from weather. Generally, the only difference between these vessels and open hopper barges is that covered dry cargo barges are equipped with watertight covers over the entire cargo hold.

Three basic types of tank barges are used for the transportation of liquid commodities: (1) single-skinned tank barges, (2) double-skinned tank barges, and (3) cylindrical tank barges.

Single-skinned tank barges have bow and stern compartments separated from the midship by transverse collision bulkheads. The entire midship shell of the vessel constitutes the cargo tank. Hydrodynamic considerations require that this huge tank be divided by bulkheads. The usual compartmentation consists of a center line bulkhead with three transverse bulkheads forming four separate cargo compartments on either side. The entire framing structure is inside the cargo tanks.

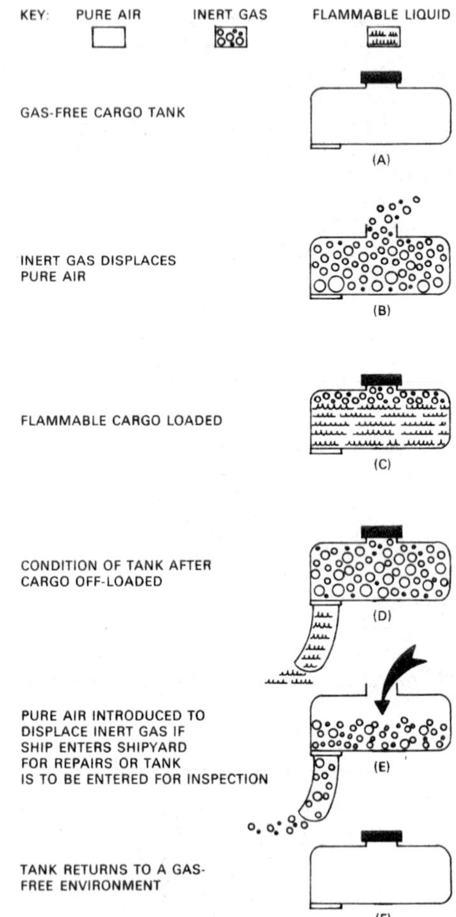

FIG. 13-5I. A tanker inert gas system.

Double-skinned tank barges have, as the term implies, an inner and outer shell. The inner shell forms cargo tanks that are free of appendages and therefore are easier to clean and to line with a protective coating. Poisons and other hazardous liquids require protection of the void compartments between the outer and inner shell.

Barges with independent cylindrical tanks are used to transport liquids under pressure, or, when pressure is in use, to discharge the cargo. Because of the high efficiency of linings and insulation which can be incorporated, cylindrical tank barges can be used to carry cargoes at or near atmospheric pressures. Cylindrical cargo tanks are generally mounted in the barge hopper and are thus free to expand or contract independent of the entire structure. For this reason, they are preferred for high temperature cargoes such as liquid sulphur at 280°F (138°C), or refrigerated cargoes such as anhydrous ammonia at −28°F (−33.3°C).

All barges carrying hazardous materials are required to have a 2 by 3 ft (0.61 m by 0.91 m) warning sign facing outboard without obstruction, on which is printed "WARNING—DANGEROUS CARGO. NO VISITORS, NO SMOKING, NO OPEN LIGHTS." The sign also must specify the names and locations of all cargoes aboard the vessel. In addition to the warning sign, barges carrying hazardous materials must have an information placard giving details about each of the hazardous cargoes aboard. These placards are required to be displayed in a waterproof container as near to the warning sign as practicable.

Information placards for all hazardous materials in an entire tow are required to be maintained in the towboat wheelhouse where they are readily accessible to the operator. In case of any problem, the towboat operator on watch should be able to furnish specific details via radio concerning any hazardous cargo in tow.

Tank Cleaning and Gas Freeing of Cargo Tanks

One of the most hazardous processes in the operation of either tank ships or tank barges is that of tank cleaning and gas freeing. This procedure is carried out in varying degrees for a change of cargo, on or off charter survey, periodic U.S. Coast Guard inspection, and for repairs. Although tank ships are usually cleaned and gas freed at sea, and barges are cleaned in port, the procedure used for both is very similar.

The biggest problem in tank cleaning and gas freeing is the necessity to reduce the flammable vapors in the atmosphere from a point above the Upper Explosive Limit (UEL) to a point below the Lower Explosive Limit (LEL). In other words, the atmosphere of the space must be brought down through the full flammable range of the material of the previous contents. A spark or other source of ignition within the space during the time the atmosphere is within the explosive range can cause an explosion.

The most common and most feared source of ignition during cleaning and gas freeing operations is a spark caused by static electricity discharge. During washing operations, water mist forming in a tank's atmosphere takes on an electrical charge from the cleaning water jet which penetrates the mist. Should any object not properly bonded to the tank be lowered into the tank space, a static discharge—similar to a small lightning bolt—could travel from the mist cloud to the ungrounded object.

Another severe fire and explosion hazard is present while a vessel is being repaired. Even though the tanks may have been thoroughly cleaned and gas freed at sea, there is no guarantee that they will remain in the same condition throughout the duration of repairs. As required by U.S. Coast Guard and Department of Labor regulations, each space is tested and certified to be "safe for hot work" by an NFPA Certificated Marine Chemist before any hot work can begin. The fact that a cargo tank has been found to be completely safe, and so certificated by a marine chemist, does not necessarily mean that it will remain in that condition. Sun beating down on a tank with a rusty interior can cause the release of sufficient flammable vapors inside the tank to bring the atmosphere from a safe condition to well within the explosive range. Therefore, the marine chemist must not only evaluate the atmospheric condition that exists within the space at the time of the inspection, but must also evaluate any weather conditions and vessel interior conditions that might cause regeneration of the flammable vapors. Where there is any doubt, the marine chemist should recommend that additional precautionary measures be taken by the repairer, and should specify the circumstances under which a Marine Chemist should be recalled for further testing. Timely and regular use of the services of NFPA Certificated Marine Chemists has helped to virtually eliminate some of the most severe fire and explosion hazards in the entire marine industry.

SHIPBOARD FIRE EXTINGUISHING EQUIPMENT

Since each ship is different, it follows that fire protection systems will rarely be identical from one ship to another. Under Coast Guard regulations governing U.S. merchant vessels, several alternative fire protection systems may be either permitted or required, depending upon their designed purpose. Foreign flag vessels are likely to have different systems or arrangements. Similar systems manufactured and installed by competing companies can vary widely. Most fire extinguishing and fire detection equipment for use aboard U.S. vessels is approved by the Coast Guard, as are most structural fire protection materials used on U.S. vessels. Other acceptable fire extinguishing equipment, such as portable fire extinguishers, should be listed and labeled by testing laboratories for marine use.

Firemain System

The firemain system is the line that provides fire fighting and flushing water throughout a ship. It is the backbone of all fire fighting systems aboard ships. The quantity of water available for fire fighting is limited only by the capacity of the fire pumps supplying the system. Most vessels are required to have two fire pumps, with all suction inlets, sources of power, etc., for each pump located in separate spaces so that one fire incident will not put both pumps out of operation. An alternative, which must be approved by the Coast Guard, is to install both pumps in the same space and to protect the space with a carbon dioxide extinguishing system. This arrangement is permitted only in unusual circumstances, where separation of pumps would not increase safety—a course usually acceptable for only small vessels. The size of the required pumps depends upon the vessel's size and service, as well as upon the arrangement of the pumps and piping aboard the vessel.

Hydrants on the firemain system are located so that two hose streams may be directed into all portions of the vessel accessible to passengers and crew during navigation. Because one of the streams must flow from a single length of hose, it is essential that the hose be long enough to direct water into all portions of the space—not just long enough to get the nozzle to the door.

International Shore Connection

One particular item of equipment required by SOLAS '74 deserves special mention: the international shore connection. Vessels are required to be fitted with a special bolted plate that can be attached to the firemain. This fitting is designed to eliminate the possible mismatch of hose threads aboard vessels and in port facilities. Provision of such connections is not mandatory for shore installations, but can prove invaluable to organizations having marine fire fighting responsibilities.

Carbon Dioxide Systems Aboard Vessels

Carbon dioxide as an extinguishing agent aboard ship has many desirable properties. It will not, of itself, damage cargo or machinery, and it leaves no residue to be cleaned up after a fire. Even if a ship is without power, a carbon dioxide system, because it is pressurized, can deliver the agent to all parts of a space. Basically, there are two carbon dioxide systems for protecting compartments in a ship—fixed CO_2 systems aboard commercial vessels are classified either as "cargo" systems or "total flooding" systems.

Cargo Systems: Fires in Class A materials (ordinary combustibles) carried in cargo holds usually start with smoldering and production of large quantities of smoke. Rapid burning will occur only when sufficient heat is developed to reach the temperature at which solid combustibles give off sufficient gases to support continued rapid combustion. Until that time, the rate of combustion is relatively slow. After a smoldering fire in a ship's hold is discovered, time to flaming combustion would perhaps be at least 20 minutes, depending upon oxygen available and other circumstances. This would allow time to prepare for fire fighting operations. The cargo hold is sealed and the CO_2 released at a rate normally several bottles at a time, so that the oxygen level falls below the rate that will support combustion.

Total Flooding Systems: Fires in machinery and similar spaces are generally Class B (flammable liquids). In this type of fire, heat buildup is rapid. The safety of a ship depends to a great extent upon the equipment in the machinery space; thus, it is important to introduce the extinguishing gas quickly. Rapid release of extinguisher also prevents heat from causing failure of bulkheads, which would make it impossible to maintain a sufficient concentration of carbon dioxide, and prevents structural members from reaching high temperatures. Quick release of the extinguishing agent also prevents heat updraft from the fire from carrying away the carbon dioxide, and limits the extent of damage to equipment. Discharge of 85 per cent of the required quantity of carbon dioxide in these systems should be completed within 2 minutes (for systems designed to protect roll-on/roll-off cargo spaces, discharge of 100 per cent of the required quantity within 2 minutes is required); slow release as in the case of the "cargo system" might result in no extinguishment. Two separate and deliberate controls are required to avoid unintentional release of the gas. One control releases the minimum required amount of carbon dioxide; the other control operates the stop valve or direction valve.

Liquid cargo tanks aboard cargo and passenger vessels may be protected by a carbon dioxide system that is a modification of the cargo system. A tank system, instead of a steam smothering system or a deck foam system, is most commonly found aboard tank vessels. This system calls for discharge of a specified amount of CO_2 within 5 minutes. The quantity of CO_2 required to protect a given space is based upon a volume factor of 30 (1 lb CO_2/30 ft^3 or 45 kg/m^3). Operating instructions should state the minimum number of carbon dioxide bottles to be released in relation to the amount of cargo in the tank.

Mechanical Foam Systems Aboard Ship

Mechanical foam is produced by introducing foam concentrate in proper proportions into a flowing stream of water and then aspirating with air. Aboard ship, the foam concentrate is normally introduced through proportioning equipment located near the foam concentrate storage container at a central location on the vessel. The foam solution is pumped through fixed piping to foam nozzles, monitors, etc., in the area to be protected. Air is mixed with the foam solution at the discharge nozzle.

Deck Systems: Deck foam systems are required aboard tank vessels constructed after January 1, 1962. This system is intended to protect any deck area, using a predetermined rate of flow from foam stations (monitors or hose stations) aft of the area to be protected. Piping and foam stations are arranged so that ruptured sections of piping in the path of a fire can be isolated. With this arrangement, it should be possible to effectively fight the fire wherever it occurs by working forward from the after house, assuming machinery, foam pumping, and proportioning equipment are located aft.

The concentrate normally supplied for mechanical foam is suitable for use on most, but not all, flammable liquids carried aboard tankers. For example, it is impossible for foam made from ordinary concentrate to form a blanket on alcohols, esters, ketones, or ethers (commonly called water soluble or polar solvents). An alcohol type foam concentrate is available for shipboard fires involving water soluble flammable liquids. An alternate method of combating alcohol type fires is the use of water spray to dilute the flammable liquid and cool the surrounding areas. Also, now available is a foam concentrate that is effective on both polar and nonpolar solvents.

Chemical Carriers

Chemical carriers may be required to have more fire fighting systems than ordinary tank vessels. The systems required for a chemical carrier will be based on the chemical involved; the type of fire protection required is described in 46 CFR 153. A foam system will be specified as polar solvent or nonpolar solvent, depending on whether or not the chemical is water soluble. A water spray system or a dry chemical system may also be required. Each system must be designed to cover each part of the cargo containment system (with exception of the vent riser) on the weather deck. The cargo containment system includes the cargo tank, piping system, venting system, and gaging system.

Gas Carriers

Gas carriers are required to have additional fire fighting systems beyond those of an ordinary tank vessel. An exterior water spray system must be installed to protect all exposed cargo tank surfaces, each boundary of an accommodation space or navigational area that faces the cargo area, and the cargo loading and discharge area. The spray system is activated in the event of a fire to cool down and protect these areas; it is not intended to extinguish a liquefied flammable gas fire. On the other hand, a dry chemical system must be installed to protect the cargo area and cargo piping, and to extinguish a liquefied flammable gas fire that might occur from a broken cargo handling line or a fire in the vent system. Foam systems are not effective against a liquefied gas fire; if installed, they can only augment, not replace, the systems just described.

Passenger Vessels

Like SOLAS '60, SOLAS '74 contains detailed provisions for fire protection aboard passenger vessels during and after their construction. After a number of disastrous passenger ship fires in the 1930s, individual countries developed various approaches to the problem of fire protection aboard these vessels. The International Convention for the Safety of Life at Sea, 1948, predecessor to SOLAS '60, first accommodated these differences of approach by allowing any one of three methods to be employed in the construction of passenger ships. SOLAS '60 retained these three methods with little change. All three of the methods are based upon three basic principles: (1) separation of the passenger accommodation spaces from the remainder of the ship by thermal and structural boundaries; (2) detection, containment, or extinguishment of the fire in the space of origin; and (3) protection of the means of escape. Implementation of these principles takes a different form in each of the three methods, as follows:

Method I: Construction of internal divisional bulkheads using noncombustible, fire resisting materials. Generally, a sprinkler system or fire detection system is not installed. The objective of this method is to confine a fire to the space of origin.

Method II: Installation of a fire alarm system and an automatic sprinkler system for the detection and extinguishment of fire in all spaces in which a fire might be expected to originate. Generally, there are no restrictions on the type or combustibility of internal divisional bulkheads. The objective of this method is to extinguish any fire in the space of origin.

Method III: Use of a system of fire subdivision—each of limited area, dependent upon the size and nature of various compartments—together with the fitting of an automatic detection system and some limit on the combustibility of construction materials. Generally, this method does not include the installation of an automatic sprinkler system. The objective of this method is to detect a fire in the space of origin and to limit its possible growth.

Although SOLAS '60 recognized three methods of construction, SOLAS '74 and regulations governing the construction of U.S. flag vessels only permit use of Method I. Reasons for this are considered in subsequent paragraphs. At this point it is necessary only to note that a number of fires on non U.S. flag passenger ships during the period of 1963-1966 found the Convention lacking on two counts. First, the International Conventions of 1948 and 1960 applied only to ships constructed after the effective date of the Convention. A "grandfather clause" permitted existing passenger ships to continue operation with little or no improvement in firesafety. Second, because of numerous technical advances, the three construction methods permitted by the 1960 Convention did not represent the highest practicable level of firesafety for new passenger ships.

Prompted by disastrous passenger ship fires, IMO undertook an intensive study of passenger ship firesafety in 1966 and 1967. This study resulted in two comprehensive proposals for amendment of SOLAS '60. These amendments have been incorporated into SOLAS '74, which went into effect in 1980.

The first amendment to SOLAS concerned existing passenger ships. It required, in essence, that the level of firesafety of these ships be brought into compliance with one of the three construction methods required by the 1948 convention for new ships. This may be likened to making a building construction code retroactive. The amendment resulted in numerous passenger ships being rebuilt or removed from service.

As important as the first amendment was, it was intended only as an interim measure until many of the older passenger ships became unprofitable to operate. Of greater future importance were the provisions of the second amendment. It eliminated the three construction methods recognized by SOLAS '60 and replaced them with a single system of construction which has several design alternatives. The new method permitted only minimal quantities of combustible material to be used in construction of future passenger ships. Additionally, depending upon the design alternative chosen, an automatic sprinkler system or automatic fire detection system must be fitted throughout the ship. Bulkheads and decks between compartments were required to have a specified fire resistance, determined by the nature of the adjoining spaces and whether or not an automatic sprinkler system was installed. The intended performance of this new construction method is perhaps best summarized by the basic principles underlying the system:

1. Division of the vessel into main vertical zones by thermal and structural boundaries.
2. Separation of accommodation spaces from the remainder of the ship by thermal and structural boundaries.
3. Restricted use of combustible materials.
4. Detection of any fire in the zone of origin.
5. Containment and extinguishment of any fire in the space of origin.
6. Protection of means of escape and means of access for fire fighting.

Other SOLAS '74 Provisions

International requirements for the construction of cargo and tank vessels, although somewhat less detailed than those for passenger vessels, have been incorporated into SOLAS '74 and its initial set of amendments. The current requirements are vastly upgraded from those required in SOLAS '60. Combustible construction is greatly reduced in cargo vessels and completely eliminated in tank vessels. In addition, new tank vessels require deck foam extinguishing systems as well as inert gas systems.

International marine fire protection efforts at IMO have led to firesafety requirements for hovercraft, hydrofoils, and mobile offshore drilling units, as well as for carriage of dangerous goods. Work is continuing on improving fire test methods, fire detection, and machinery space protection.

Sprinkler Systems: Aboard U.S. vessels, manually operated sprinkler systems are employed only in very limited locations. In new SOLAS ship construction, primary dependence is placed upon structural fire protection rather than on automatic sprinkler protection. However, installation of a sprinkler system can be offset by a reduction in the degree of insulation required between adjoining spaces.

For installations on vehicular decks, such as aboard ferry vessels, the sprinkler system should be designed to protect the structural integrity of the vessel, confine the fire to the location of origin, and wash flammable liquid spills to a safe location. Installation on vehicular decks is the primary marine use of sprinkler systems in the U.S.

Bromotrifluoromethane (Halon 1301) Systems: Bromotrifluoromethane (Halon 1301) is one of several halogenated hydrocarbon extinguishing agents introduced in recent years. In general, halogenated agents have high extinguishing efficiency per unit weight. This makes them particularly suitable for installations which are weight critical, e.g., on hydrofoils.

Halon 1301 is the only halogenated agent thus far found satisfactory for marine use in fixed systems. Title 46 of the Code of Federal Regulations does not currently contain any provisions for the design or requirements of Halon 1301 systems aboard commercial vessels. Instead, the approval of these systems is based upon a CFR provision that grants to the Commandant of the Coast Guard the authority to allow new or unique extinguishing systems. It must be demonstrated adequately that a proposed system is equivalent to the existing system required by the regulations before the system can be allowed. Total flooding carbon dioxide fire protection systems are those to which Halon 1301 systems must prove equivalence. Therefore, Coast Guard approved Halon 1301 systems are required by federal law to be designed with an equivalent degree of fire protection capability and reliability. Where basic deviations in design occur due to the use of the Halon 1301 agent, specific adaptations to the regulations apply.

The basic design philosophy of CO_2 systems has been retained for Halon systems:

1. The protected space must be evacuated before the system is discharged. A predischarge alarm is required.
2. Agent storage containers must be located outside of the protected space, except for spaces less than 6,000 ft^3 (170 m^3) and for modular type Halon systems.
3. In order to avoid accidental release, two separate and distinct actions must be performed to discharge the agent.
4. Manual type release devices are required. Automatic release is permitted only for spaces less then 6,000 ft^3 (170 m^3).
5. Detailed instructions must be provided at the remote release station to explain alternate means of discharging the system.

Coast Guard requirements are divided into two areas: (1) machinery space protection of U.S. flag vessels, and (2) fixed systems for pleasure craft. In the first case, the Halon 1301 system has been conceptually approved. Final "type approval" procedures are being developed among the manufacturers, the Coast Guard, and Underwriters Laboratories. Approval is on a ship by ship basis. In the second case, the Coast Guard is in the process of developing a standard for systems aboard pleasure craft. A test program has been developed jointly with UL.

Currently, both Halon 1301 and Halon 1211 portable extinguishers are available, and both carry Coast Guard approval for marine use.

Internationally, IMO has developed detailed requirements—compatible with U.S. requirements—for Halon systems for machinery spaces as well as for roll on/roll off vehicle spaces. A Halon piping diagram for the engine room of a foreign flag 80,000 DWT tanker appears in Figure 13-5J. The diagram appeared in the May 1984 volume of *Chevron's Safety Bulletin.*

Fire Fighting Plan

All U.S. vessels of 1,000 gross tons and over, and U.S. vessels of any tonnage on an international voyage, should permanently display a fire fighting plan of the ship. Although the primary purpose of the plan is to guide shipboard personnel, it should contain the following information that can greatly assist port fire fighting groups:

1. Fire control stations for each deck.
2. Sections enclosed by fire resistive divisions.
3. Fire alarms, detection systems, and sprinkler systems.
4. Fire extinguishing appliances.
5. Access to different compartments, decks, etc.
6. Ventilating systems, including master fan controls, positions of dampers, location of remote controls, and the identification numbers of the ventilating fans serving each section.

Instead of being posted on a plan, the above information can be set forth in a fire fighting booklet. In that case, the booklet is required to be aboard ship at all times.

Whether posted as a fire fighting plan or presented in booklet form, information should also be available on all watertight compartments, openings therein, means of closure, controls, and arrangements for correcting any list due to flooding. In addition, SOLAS '74 requires a duplicate set of fire control plans to be permanently stowed in a prominently marked weathertight enclosure outside the deckhouse for the assistance of shoreside fire fighting personnel. (For illustration of a fire fighting plan for the aft end of a super tanker, see Fig. 13-5K.)

NFPA MARINE STANDARDS

NFPA requirements relating to marine vessels are contained in the following:

NFPA 306, *Standard for the Control of Gas Hazards on Vessels* (hereinafter referred to as NFPA 306). This standard describes the conditions necessary for safety before making repairs on any vessel carrying or having carried, as a fuel or as cargo, combustible or flammable liquids, flammable compressed gases, and chemicals in bulk. It also contains a section on flammable cryogenic liquid carriers.

NFPA 312, *Standard for Fire Protection of Vessels During Construction, Repair and Lay-Up.* This standard covers measures for preventing and controlling fires on vessels while in the building yard, repair yard, or while laid up. Provisions for each of these circumstances appear separately in the standard.

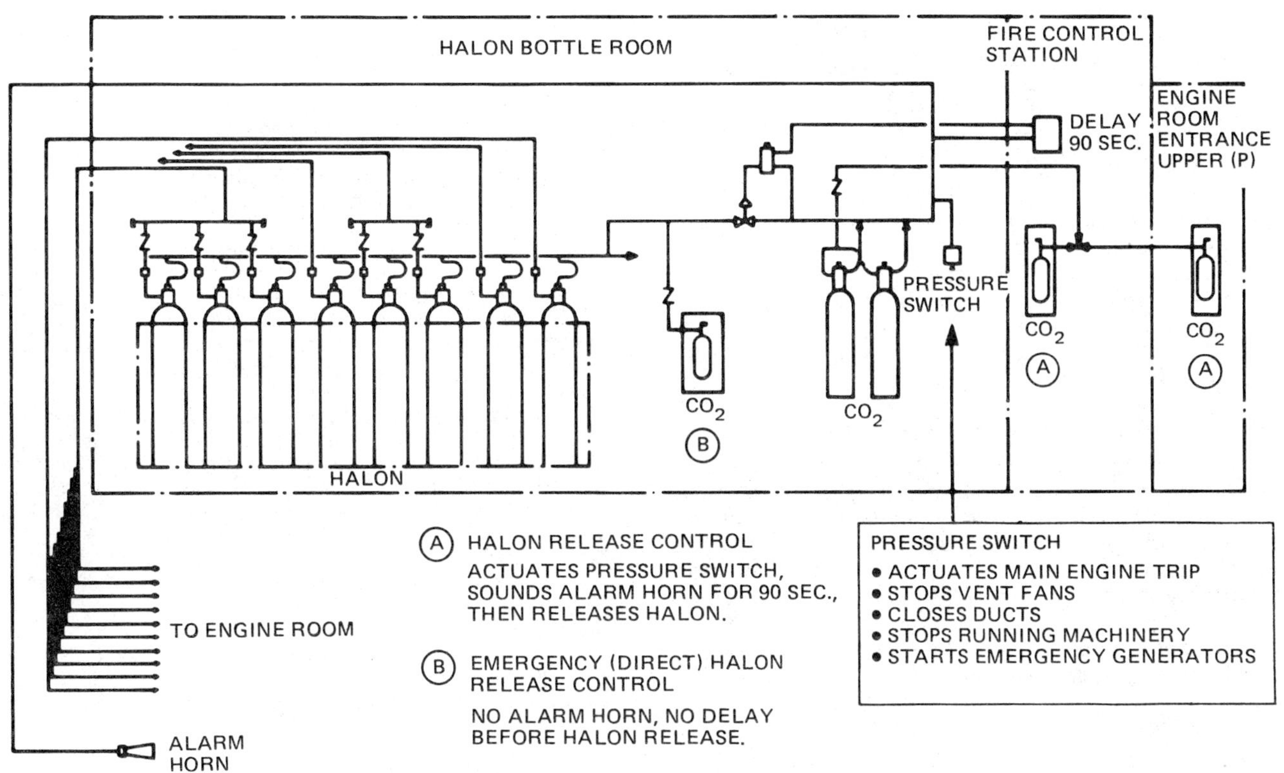

FIG. 13-5J. Halon Piping Diagram for Engine Room (System No. 1).

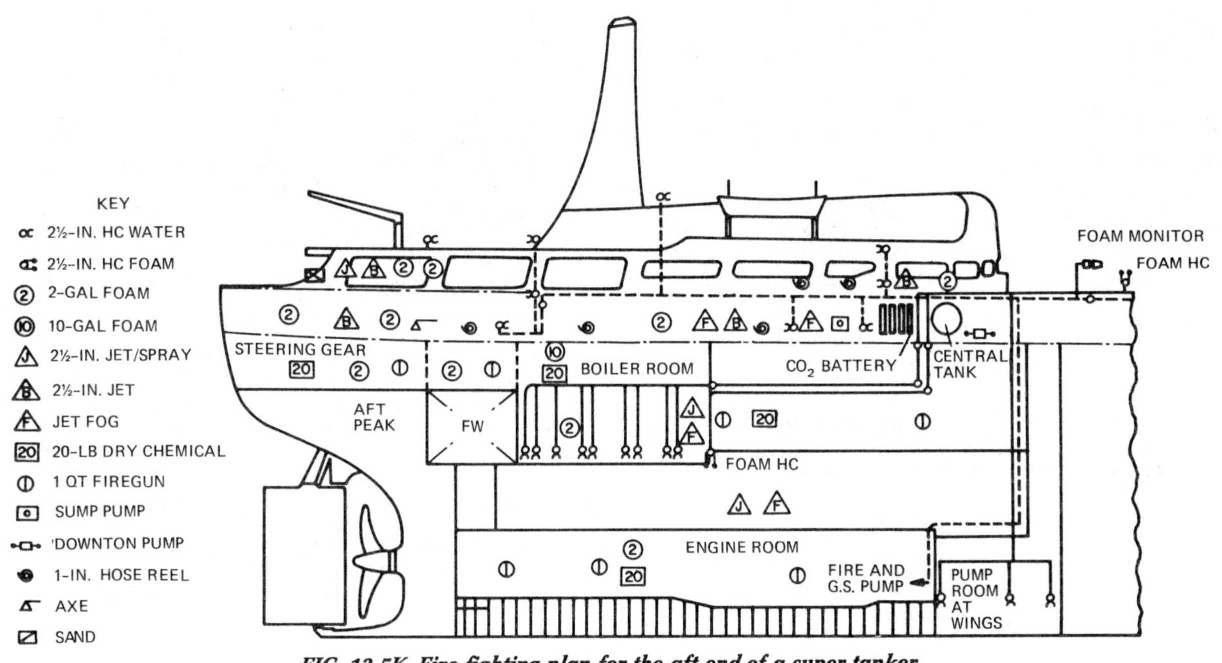

FIG. 13-5K. Fire fighting plan for the aft end of a super tanker.

OTHER ORGANIZATIONS CONTRIBUTING TO MARINE FIRESAFETY

The American Bureau of Shipping

The American Bureau of Shipping (ABS) is the American classification society whose technical committees develop rules for the dimensions and use of materials in the construction and conversion of marine vessels. After a vessel is completed according to the rules and under the constant supervision of an ABS field inspector, it is classed and listed by the Bureau as having met all design and structural safety requirements.

The American Petroleum Institute

In the area of marine fire fighting and fire protection, the American Petroleum Institute (API) is active in the development of domestic and international standards, codes, regulations, etc. Under the policy direction of the API Central Committee on Transportation by Water, most actions are directed by the API Committee on Tank Vessels.

Ad hoc groups are sometimes utilized for special studies of a nonrepetitive nature.

The Marine Chemist Association, Inc.

This is a professional organization of chemists certificated for marine work by NFPA in accordance with provisions of NFPA 306. The association was established in May 1938 as the Marine Chemists' Subsection of the NFPA Marine Section. After the Marine Section was disbanded, the Marine Chemists organized its present association in 1948.

The National Cargo Bureau, Inc.

The Bureau is a nationwide nonprofit membership organization established in 1952 as a private, nongovernmental agency that formulates recommendations to various governments for regulations on the safe stowage of dangerous goods and other cargoes, and on related cargo handling gear.

USCG Advisory Committee on Hazardous Materials

This committee of the National Academy of Sciences—National Research Council is charged with advising the U.S. Coast Guard on scientific and technical matters relating to safe maritime transportation of hazardous materials.

Bibliography

NFPA Codes, Standards, Recommended Practices and Manuals. (See the latest *NFPA Codes and Standards Catalog* for availability of current editions of the following documents.)

NFPA 302, *Fire Protection Standard for Pleasure and Commercial Motor Craft.*

NFPA 303, *Fire Protection Standard for Marinas and Boatyards.*

NFPA 306, *Standard for the Control of Gas Hazards on Vessels.*

NFPA 312, *Standard for Fire Protection of Vessels During Construction, Repair and Lay-Up.*

Additional Readings

Baum, Howard R., and Rockett, John A., "An Investigation of the Forced Ventilation in Containership Holds," *NBSIR 83-2665*, National Bureau of Standards, Gaithersburg, MD, May 1983.

Beatteay, Robert E., *Fire Officer's Guide to Waterfront Fires*, National Fire Protection Association, Quincy, MA, 1975.

Connor, Joseph, *Marine Fire Prevention, Fire Fighting and Fire Safety*, R. J. Brady, Bowie, MD, 1979.

Fire Protection Guide on Hazardous Materials, National Fire Protection Association, Quincy, MA, 1984.

House, W., Frock, W., and Welch, R. O., *Evaluation of Current Packaging Materials Used Aboard Ship and Recommendations for Reducing Their Flammability and Combustion Toxicity; Volume I —Flammability*, prepared for Naval Supply Systems Command, U.S. Department of the Navy by Midwest Research Institute, Kansas City, MO, Mar. 1982.

Kerlin, Donald J., "A Marine View of Fire Protection," *SFPE TR 82-10*, Society of Fire Protection Engineers, Boston, MA, 1982.

Klote, J. H., and Zile, R. H., "Smoke Movement and Smoke Control on Merchant Ships," *NBSIR 81-2433*, National Bureau of Standards, Gaithersburg, MD, Dec. 1981.

Lee, B. T., "Fire Performance Testing of Bulkhead Insulation Systems for High Strength to Weight Ship Structures," *NBSIR 76-1012*, National Bureau of Standards, Gaithersburg, MD, Aug. 1976.

Lee, B. T., and Parker, W. J., "Fire Buildup in Shipboard Compartments—Characterization of Some Vulnerable Spaces and the States of Prediction Analysis," *NBSIR 79-1714*, National Bureau of Standards, Gaithersburg, MD, May 1979.

Lee, B. T., and Parker, W. J., "Fire Performance Guidelines for Shipboard Interior Finish," *NBSIR 79-1700*, National Bureau of Standards, Gaithersburg, MD, June 1979.

Lee, B. T., and Parker, W. J., "Naval Shipboard Fire Risk Criteria—Berthing Compartment Fire Study and Performance Guidelines," *NBSIR 76-1052*, National Bureau of Standards, Gaithersburg, MD, Sept. 1976.

Marine Fire Prevention Firefighting and Firesafety, Maritime Administration, U.S. Department of Commerce, Washington, D.C.

Marine Products Directory, Underwriters Laboratories Inc., Northbrook, IL.

"Navigation and Vessel Inspection Circular 6-72 and 6-72, Change 1", *Guide to Fixed Fire Fighting Equipment Aboard Merchant Vessels*, U.S. Coast Guard, 1972.

"Navigation and Vessel Inspection Circular 6-80", *Guide to Structural Fire Protection Aboard Merchant Vessels*, U.S. Coast Guard, 1980.

"Navigation and Vessel Inspection Circular 2-84", *Amendments to the 1974 Safety of Life at Sea (SOLAS) Treaty*. U.S. Coast Guard, 1984.

Recommended Fire Protection and Fire Prevention Practices for Major Shipyards Engaged in New Vessel Construction, American Hull Insurance Syndicate, NY, May 31, 1977.

Rushbrook, F., *Fire Aboard: The Problems of Prevention and Control in Ships, Port Installations and Offshore Structures*, 2nd ed., Heinman, NY, 1979.

Ships Firefighting Manual, Marine Publications International, State Mutual Bank, London, England, 1983.

Von Iperen, Willen, "A Study of Ventilation of Containership Holds for Carriage of Flammable Liquids," (unpublished) Sea-Land Service, Inc., Elizabeth, NJ, June 18, 1979.

SECTION 14

ORGANIZATION FOR PRIVATE PROTECTION

FIRE LOSS PREVENTION AND CONTROL MANAGEMENT

Revised by Peter K. Schontag, P.E.

Industrial growth and technological developments have vastly increased the complexity of the industrial fire loss prevention process in recent years. Concentration of expensive machinery, equipment, processes and stock, as well as employees, in individual buildings underscores the need to take all possible steps to minimize the possibility of fire and explosion in any workplace.

Industry is obliged by law to provide a reasonably safe working environment. Economically, interruption of production can result in the loss of an important market share. For these and many other reasons, fire prevention and control demands the attention of executive management and provides an important challenge to all management levels. Prudent management involves firesafety measures designed to ensure both the safety of employees and the continuity of operations.

Fire loss prevention and control management is a program of identification, evaluation, and control of fire hazards. While the responsibility for the implementation of such a program is often delegated among the various echelons of management, ultimate accountability rests at the executive level. The firesafe operation of an industrial facility, however, requires the cooperation of all its employees.

MANAGEMENT ECHELONS

Industrial management can be divided into three categories, each of which makes important contributions to the fire loss prevention and control program: (1) corporate or executive management—the board of directors and company officers; (2) middle management—facility managers, department heads, and related staff managers; and (3) line management—the supervisors and section chiefs who direct each phase of the operation.

In a relatively small industrial organization, the owner may perform the functions of all three management levels—executive, middle and line supervisor. Although the owner's loss control problems may be fewer and less complex than those of the management group in a larger organization, some of the same judgments must be exercised. The owner/loss control manager may be at a disadvantage, however, because of the lack of specialized training in loss control and/or the absence of managers with whom to share expertise and creative ideas regarding a sound firesafety program.

Even though nonindustrial organizations such as school systems and banks use different titles to identify levels of management, assigned responsibility and accountability should parallel recommendations for the industrial management hierarchy.

EXECUTIVE MANAGEMENT

Members of executive management establish policy, make major financial decisions, decide what product or service will be provided, and establish production levels. The roles of executive management in fire loss prevention and control management are to stimulate and administer. Strong support from top management is absolutely vital if a loss control program is to be successful. Based on information furnished by middle management, executive management decides the level of protection to be maintained, the amount of risk to be insured, and the acceptable operating risk.

Efficient executive management requires periodic progress reports on the status of the fire loss prevention and control management program. These reports are obtained either through regular management channels or directly from the fire loss prevention and control manager.

MIDDLE MANAGEMENT

Middle management is responsible for collecting adequate information upon which executive and/or middle management can base decisions and establish policy. Middle management makes recommendations that influence company policy, and is accountable for the implementation and success of such recommendations. To best accomplish this, middle management's functions should include:

1. Selecting the site.
2. Planning the facility and its protection systems.

Mr. Schontag is Coordinator of Property Loss Control for the Pacific Gas and Electric Co. He is based in San Francisco, CA.

3. Planning production processes and facility layout.
4. Directing facility construction.
5. Developing a fire prevention and control program.
6. Conducting community relations programs.
7. Providing guidance for line management.

PROGRAM IMPLEMENTATION

This chapter discusses only those parts of management functions which affect the fire loss prevention and control management program. The title "fire loss prevention and control manager" is used here to identify those people in industrial management who are responsible for directing the fire loss prevention and control management program. (Such people may have a variety of other titles, depending upon their organization.) The position of fire loss prevention and control manager may be assigned to any one of an industrial organization's departments. The post should be classified high enough to permit an equitable working relationship with the facility manager, other middle managers, and line managers.

In some industries, fire loss prevention and control managers are responsible for employee safety management, property insurance management, and security management. For example, one large central management organization with a diversified group of component companies and plants assigned an administrative vice president to handle fire protection management along with a number of related functions. Reporting to the administrative vice president was a risk manager who supervised four assistant managers assigned to insurance, fire loss prevention and control, safety, and security.

Generally, in industrial fire protection management situations, the facility manager depends on the fire loss prevention and control manager to see that firesafety policies are carried out in the following areas of shared responsibility:

1. *Selecting the site.* The fire loss prevention and control manager should determine if the following conditions exist at the site under consideration:
   (a) Adequate water supply to fulfill protection requirements in addition to industrial and sanitary requirements;
   (b) Severe exposure to natural hazards; i.e., tornado, flood, earthquake, or heavy snow;
   (c) Extraordinary exposure conditions involving adjacent facilities;
   (d) Acceptable response from support forces;
   (e) Impediments to quick response by support forces; i.e., lift bridges, railroad crossings, or heavily traveled highways.
2. *Planning the facility and its protection systems.* Fire loss prevention and control factors to be considered during the planning stage are:
   (a) Using fire resistant materials;
   (b) Limiting the size of fire areas by the use of fire walls, fire doors, etc.;
   (c) Segregating unusual concentrations of values (i.e., sensitive equipment and highly hazardous operations);
   (d) Making sure emergency exits are adequate in number, size, and location;
   (e) Providing physical protection (i.e., automatic sprinklers, hydrants, and standpipes);
   (f) Providing access routes for fire protection vehicles.
3. *Planning production processes and facility layout.* Fire loss prevention and control management personnel should participate in the consideration of processes which require:
   (a) Segregation to limit exposure to other operations;
   (b) Special protection systems;
   (c) Limited access;
   (d) Special exit facilities.
4. *Directing facility construction.* Fire loss prevention and control management personnel should participate in:
   (a) Advising contractor personnel of fire loss prevention and control management policies; (facility personnel should have authority to enforce compliance);
   (b) Acceptance tests of all fire protection equipment.
5. *Developing a fire prevention and control program.* Fire loss prevention and control management personnel should participate in development of the program by:
   (a) Drawing up a plant self-inspection plan and procedure, along with a systematic way of reporting deficiencies and of eliminating hazardous conditions;
   (b) Participating in the developing of emergency and contingency plans of action;
   (c) Furnishing the risk or insurance manager with information on loss prevention and control management as required by insurance carriers;
   (d) Providing an escort for insurance company representatives making facility inspections;
   (e) Reviewing the fire protection recommendations of insurance company representatives;
   (f) Furnishing the insurance manager with economic justification for protection systems;
   (g) Immediately reporting impairment of protection systems to the insurance manager;
   (h) Furnishing the insurance manager with reports of all fire or explosion incidents.
6. *Conducting community relations programs.* The fire loss prevention and control manager should assist the facility manager and/or the public relations manager in publicizing the facility's fire loss prevention and control management policies to the community and its emergency forces. This type of publicity can help build a favorable company image. The community has an interest in a facility's firesafety planning as it relates to employee safety, to the economic impact of loss of wages due to fire, and to fire exposure of other property.
7. *Providing guidance for line management.* Fire loss prevention and control personnel must provide enough information and guidance to the line management group for it to successfully implement policy and determine proper fire loss prevention and control management procedures.

Implementation of the fire loss prevention and control program does not stop at the middle management level for compliance with company policy. Unless the members of line management adequately perform their function, the program cannot produce satisfactory results.

In regard to employees and work areas under his or her supervision, it is the line manager who:

1. Trains personnel to fulfill their roles in the fire loss prevention and control program;
2. Works with the fire loss prevention and control manager and staff in selecting employees for the private fire brigade or other program assignments;
3. Works with the fire loss prevention and control manager to establish priorities for inspection, testing, and maintenance of fire protection equipment;
4. Inspires employees to develop a fire conscious attitude and respect for the fire loss prevention and control management program;
5. Insists that access areas to all fire protection equipment, as well as all means of egress, be kept open and free from tripping hazards;
6. Works closely with the fire loss prevention and control manager and staff to assure that new equipment or processes are adequately protected before startup;
7. Reports all occupancy changes that may affect any phase of the fire loss prevention and control management program;
8. Works with the fire loss prevention and control manager and staff to ensure that fire exit drills are held regularly;
9. Accounts for the whereabouts of employees following each facility evacuation;
10. Assumes interim command of emergency action when a fire occurs.

Every industry or institution has its own unique problems. The guidelines in this chapter are general and basic. They are intended to emphasize the need for assigning responsibility and planning programs. The variations found in specific operations will become apparent during the fire risk evaluation. Then, after careful review, a good fire loss prevention and control program can be developed.

Bibliography

NFPA Codes, Standards, Recommended Practices and Manuals. (See the latest *NFPA Codes and Standards Catalog* for availability of current editions of the following documents.)

NFPA 1, *Fire Prevention Code.*

NFPA 27, *Recommendations for Organization, Training and Equipment of Private Fire Brigades.*

Additional Reading

Bennett, G., "Investigating the Ins and Outs of Fire Prevention," *Control and Instrumentation*, Vol. 12, No. 2, Feb. 1980, p. 59.

Cooper, D., "Moves Towards Greater Self Regulation," *Fire Engineers Journal*, 40 (120), Dec. 1980, pp. 33-34.

Fire Inspection Management Guidelines, National Fire Protection Association, Quincy, MA, 1982.

Hubitsky, J., "Preventing Fires on Job Sites," *Fire Command*, Vol. 45, No. 7, July 1978, pp. 34-35.

Industrial Fire Protection, IFSTA, Stillwater, OK, 1982.

Introduction to Fire Protection (a training package), National Fire Protection Association, Quincy, MA, 1982.

Linville, J. L., ed., "Industrial Fire Risk Management," *Industrial Fire Hazards Handbook*, 2nd ed., National Fire Protection Association, Quincy, MA, 1984.

Planner, R. G., *Fire Loss Control*, Marcel Dekker, Inc., New York, NY, 1979 (Available from NFPA).

Rutsein, R. and M. B. J. Clarke, "The Probability of Fire in Different Sectors of Industry," *Fire Surveyor*, Vol. 8, No. 1, 1979, pp. 20-23.

"Self-Fire Inspection: A New Concept in Firesafety," *Fire Chief Magazine*, Vol. 24, No. 5, May 1980, pp. 34-35.

Stephens, H. F., "Fire Strategy for Management," *Fire Prevention*, (133), Nov. 1979, pp. 29-30.

Thor, J., G. Sedin, "Principles for Risk Evaluation and Expected Cost to Benefit of Different Fire Protective Measures in Industrial Buildings," *Firesafety Journal*, Vol. 2, No. 3, Mar.1980, pp. 153-166.

Waters, D., "The Fire Protection of Plant and Equipment," Fire, 70 (867), 1977, pp. 185-186, 189.

RESPONSIBILITIES OF THE FIRE LOSS PREVENTION AND CONTROL MANAGER

Revised by Peter K. Schontag, P.E.

Major responsibilities of the fire loss prevention and control manager are to represent the facility manager in all areas of fire risk evaluation, fire prevention, and fire control. Although some of these responsibilities may be delegated to others, the fire loss prevention and control manager is ultimately accountable to the facility manager for the success of the fire loss prevention and control management program. In the absence of a fire loss prevention and control manager, the facility manager must accept these responsibilities.

The fire loss prevention and control manager's primary responsibility is to develop, implement, and manage the fire loss control program. This might include the entire range of subject matter of this HANDBOOK. Some of this manager's major responsibilities are to:

1. Evaluate loss possibilities;
2. Establish procedures for fire risk and loss control;
3. Advise the facility engineering, maintenance, operating and service departments on loss prevention and control matters;
4. Establish standards and specifications for all systems and equipment involved in the fire loss prevention and control management program;
5. Schedule periodic inspections of fire protection systems and equipment;
6. Interpret laws, codes, NFPA standards and related standards applicable to fire loss prevention and control management;
7. Organize the private fire brigade;
8. Assume administrative command of fire control operations;
9. Assist middle management with public agency contact;
10. Provide detailed reports of fire and explosion related emergencies and of any losses incurred.

Mr. Schontag is Coordinator of Property Loss Control for the Pacific Gas and Electric Co. He is based in San Francisco, CA.

EVALUATE LOSS POSSIBILITIES

The fire loss prevention and control manager should make an evaluation of the likelihood of fire occurring in any part of the facility, and of the potential extent of fire spread. The manager may consult with insurance company and other outside advisers, but such advice must not take the place of the manager's own evaluation of the possibilities of fire and other emergencies.

Two specific numbers are useful for the appraisal of the fire problem. The first number is the largest percentage of the property value likely to be affected by fire, heat, smoke, and/or water damage. This number indicates the maximum possibility of loss of physical plant facilities if, for example, normal protection were out of service. The second number is the probable loss. This involves a percentage smaller than the first and takes into account the facility's capabilities for fire fighting and control. These subjective figures should be developed by a qualified fire protection engineer.

Similar calculations can be made not only for fire potential, but also for the effects of other disaster factors. These are physical factors capable of appraisal. With them, management has a working basis for estimating other factors including loss of life, injury, loss of profit, and other losses from interruption of use of the property.

Work Sheets for Loss Evaluation

Data for loss evaluation should be collected on work sheets in a form sufficiently abridged to facilitate consideration by top management. The operating departments (and particularly a loss prevention staff agency) will provide detailed information concerning loss potential; other data will come from the corporation's insurance adviser. Managers of all divisions of the company should be expected to summarize loss possibility information on work sheets for their own use and to furnish the substance of this information to the fire risk manager. From such compilations, the fire loss prevention and control manager should, in turn, present executive management with loss possibility information applicable not only to the facility, but to operation of the corporation or enterprise as a whole.

The starting point for the flow of this management information on loss possibilities should be in the form of a work sheet prepared for each operating department, plant, or property. A suggested form for a work sheet would start with a list all of but the most minor of buildings or operating units. (This list should ignore small buildings and small operating units that could hardly affect major operations of the enterprise.) For each of the listed buildings or units, show the following information:

Approximate Area: This figure can be expressed in thousands or millions of square feet. Its purpose is to show the maximum area that could be involved in a fire, usually between fire walls or within substantial open space separations.

Roof Construction: The type of roof is often the most important factor in estimating the probability of a serious building loss in case of fire. Describe the roof assembly with emphasis on whether the roof is substantially noncombustible or not. In unsprinklered buildings, a wooden roof deck, or any metal roof deck with sufficient asphalt to provide adequate adhesion for wind resistance, and a combustible vapor barrier, produces the likelihood of complete destruction of the building in the event of fire—a risk that management should, in most cases, consider unreasonable.

Occupancy: The loss possibilities are often much greater in buildings which lack automatic sprinklers for protection of combustible contents. Experience with the facility itself will determine the degree to which sprinklers reduce the loss in that particular occupancy. Without such specific experience, it is often fair to assume that, on the average, losses in sprinklered buildings with adequate water supplies and all sprinklers in service will be about 10 percent of losses in unsprinklered buildings.

Approximate Value of Building and Contents: For this purpose, exact figures are not necessary; use figures close enough to indicate the magnitude of loss possible.

Relative Importance of the Occupancy: Use this final part of the work sheet for notations about the building's relative importance to operations and about any special hazards it contains. Report the processes to which the building is devoted and any materials stored in the building, particularly if these present special problems that management must keep in mind (such as the relation of operations in the building to operations elsewhere). Notes under this heading can also refer to salvage and replacement problems in the event the building is lost. List any significant features of the occupancy.

ESTABLISH PROCEDURES FOR FIRE RISK CONTROL

It is understood that the fire loss prevention and control manager will not be responsible for all procedures that affect risk management. Fire risk control must therefore be correlated with other management procedures.

Procedures should be developed to accomplish specific objectives:

1. To lessen loss potential by reducing or controlling hazards;
2. To promote the fastest and most effective reaction by appropriate protective agencies, private and public, in dealing with an emergency;
3. To reestablish normal conditions with as little disruption and delay as possible;
4. To foster good public relations by obviously being prepared for emergencies.

COUNSEL WITH THE FACILITY'S ENGINEERING, MAINTENANCE, AND PLANNING DEPARTMENTS

The facility manager should require close interaction between the property's engineering, maintenance, and planning departments with the fire loss prevention and control manager. Cooperation is especially vital as early as the pre-engineering phase of any proposed structural, equipment, or process change. The loss manager continually needs the expert input of managers of engineering, maintenance, and planning, who in turn should know and apply the fire risk control standards adopted for the facility.

Maintenance department management should be required to integrate facility maintenance procedures with fire risk procedures. This will avoid placing the property in jeopardy by shutting off services that affect fire control equipment. Examples would be closing valves that control protective water supplies, or shutting off electric power to fire pumps. A high priority should be established for maintenance affecting fire control equipment. Advance notice of any impairment should be given to the fire risk manager so alternate protection can be arranged.

Valve Supervision

It is essential to maintain all valves controlling water supplies to automatic sprinklers, hydrants, standpipes, and other fire protection equipment in a wide open position. The fire loss prevention and control manager must know the condition of these valves at all times, and must be notified immediately, or in advance, whenever a valve is closed. Valves are maintained in an open position by one of two methods—sealing or locking—and may be monitored by protective signaling systems.

Sealing: Valves are sealed by placing a wire through a movable part of the valve (wheel or shaft nut) and a fixed part of the valve, then securing the two ends of the wire with a seal. Seals do not prevent the operation of a valve, but are intended to discourage tampering by unauthorized persons. A broken seal is an indication that the valve has been operated or tampered with. Any valve with a broken seal should be checked promptly to be certain it is open, and then resealed. (Sealing and tagging systems are described in NFPA 26, *Recommended Practice for the Supervision and Care of Valves Controlling Water Supplies for Fire Protection.*)

Locking: Valves are locked in the wide open position by padlocks, with or without chains or cables, so they cannot be tampered with or operated maliciously. (See Fig. 14-2A.) These sturdy securing devices are resistant to breakage, except by heavy bolt cutters. The distribution of keys is restricted to the people directly responsible for the fire protection system.

FIG. 14-2A. Methods of locking valves to prevent tampering or malicious operation.

Protective Signaling Systems: Valves are equipped with a device arranged to electrically transmit a signal whenever the valve is operated. (See Fig. 14-2B.) These signals are transmitted through circuits to a location on the premises under the control of the plant proprietary system (see NFPA 72D, *Standard for the Installation, Maintenance and Use of Proprietary Protective Signaling Systems*), or to a location remote from the premises under the control of the local fire department or an organization whose function is to furnish such service (see NFPA 71, *Standard for the Installation, Maintenance and Use of Central Station Signaling Systems*, and NFPA 72C, *Standard for the Installation, Maintenance and Use of Remote Station Protective Signaling Systems*).

Valve Operation: A procedure should be established to notify the fire loss prevention and control manager of the proposed closing of any valve as far in advance of such closing as possible. In this way precautionary measures can be planned. If an emergency valve closing is necessary, the loss manager should be notified immediately. This information should also be transmitted to the proper authority. A red tag should be attached to the valve to indicate that it is shut (Fig. 14-2C), and a card or other device should be conspicuously displayed as a reminder to the person responsible for the fire protection equipment that the valve is closed. When the valve is opened, the red tag should be removed, and the loss manager should notify the proper authority that normal protection has been restored. The red tag system should also be used for other fire protection systems, such as carbon dioxide, dry chemical, and halogenated agent extinguishing systems, and for alarm systems.

Work by Outside Contractors: When work on the property is to be done by an outside contractor, the facility manager should require the engineering department to work with the fire loss prevention and control manager to determine exactly what special requirements must be met by the contractor. These requirements will vary with the job. They might include such items as isolating the contractor's work site; notifying the contractor of special hazards (processes or stored materials at or near the job site); possible hazards the contractor's work might present to the plant equipment, property, and personnel; smoking restrictions; hot work permits; control of contractor's employees and vehicles on the property; fire loss prevention aspects of movement and storage of construction equipment and materials; and waste disposal by the contractor.

Frequent briefing of the contractor's on-site personnel is a desirable way to keep them aware of their responsibilities in the area of firesafety. A brochure containing firesafety information can be developed for distribution to employees of outside contractors, as a tangible reminder of the need for firesafety on the job.

ADVISE OPERATIONS AND SERVICE DEPARTMENTS ON MATTERS OF FIRE RISK MANAGEMENT

Managers of operations and service departments should maintain their equipment and structures in the condition which will best minimize fire loss. These managers must also integrate information on risk management with operating instructions for all standard or repetitive operations, and must make sure that, through training and follow-up, these instructions are observed.

The fire loss prevention and control manager must make sure operations managers have the information needed for employee training in firesafety. Training must be consistent with fire loss prevention and control management policies and procedures. Other responsibilities of the operating managers with which the fire loss prevention and control manager should assist are:

1. Selecting specific people from each department for assignment to duties on fire squads (to fight small or incipient fires).
2. Provide hands-on training in the use of portable fire extinguishers. (It is often convenient to plan such practice when extinguishers are recharged in annual tests.)
3. Instructing employees, and others involved, in fire loss prevention. Covering day-to-day activities as well as emergency procedures for safeguarding processes and machines that operate continuously, important records, raw materials, finished products, etc.

FIG. 14-2B. A typical supervisory switch for an OS&Y (outside screw and yoke) gate valve. (Notifier Company)

4. Preventing overcrowding of stock. Maintain enough aisle space to permit orderly evacuation of people and easy access for fire fighters and their equipment.
5. Working out a detailed plan for evacuation of everyone in each department. Managers should appoint wardens for each floor and room plus monitors, exit guards, room searchers, and other personnel to ensure the successful execution of exit plans.

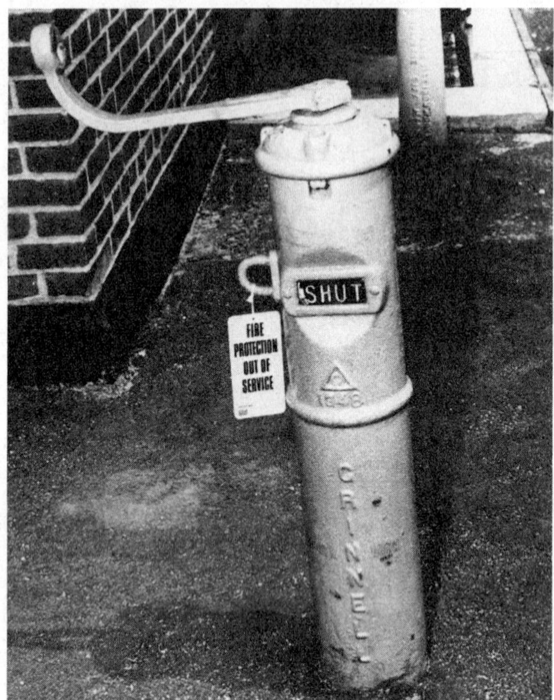

FIG. 14-2C. Shut valve with tag attached.

6. Developing a firesafety orientation program for all new employees.
7. Holding surprise fire exit drills, sometimes having blocked exits and other conditions that might occur during actual fire emergencies.

The fire loss prevention and control manager and the line supervisor of each area share responsibility for assuring that every employee knows what to do when a fire alarm sounds. All occupants should know something about the equipment provided for fire protection, as well as what they can do to help prevent fires. Specifically, this knowledge includes:

1. Their place in the plan for evacuation of the building in case of emergency. Such a plan is essential to ensure occupants' prompt and orderly evacuation, especially where there are materials or processes that would make any fire a quick spreading one.
2. How to use portable fire extinguishers and hand hoses. From prefire planning, occupants should know whether they are to help fight the fire or evacuate the area or building.
3. Some information about fire protective equipment and systems. Occupants should know, for instance, that stock must not be piled too close to sprinklers, since obstruction could prevent distribution of water from sprinklers on a fire in the piled material. Properly keeping fire doors operative, and unobstructed, and keeping all aisles, stairs, and exits clean should be emphasized.
4. How to give a fire alarm; how to operate private fire alarm boxes; and how to operate street boxes in the public alarm system.
5. Where smoking is permitted on the property and where, for safety reasons, it is prohibited.
6. The housekeeping routine. Safe disposal of wiping rags, waste, packing materials, etc., and other measures for orderliness and cleanliness as well as for firesafety throughout the property.
7. Hazards of any special processes in which employees may be engaged.

The fire loss prevention and control manager should work closely with middle management to see that special safeguards against fire are observed whenever a plant is vacated or idle. The term "vacated" refers to a property in which normal operations will not be resumed and from which most or all of the production equipment and furnishings have been removed. An "idle" plant is a property where normal operations have been temporarily interrupted.

Even when a plant is idle, it is essential to maintain its protective system in serviceable condition. Promptly report any impairment, and return the system to normal working order as quickly as possible. The following precautions should be observed:

1. Shutting down hazardous processes and leaving the plant clean and free of rubbish;
2. Providing for necessary plant maintenance and guard service;
3. Periodically inspecting fire protection equipment to ensure that it remains in good operating condition;
4. Notifying the chief of the public fire department that there is a shutdown, permanent or temporary;
5. Reviewing the prearranged plan for receiving and directing the public fire department, and making any alterations necessitated by new conditions resulting from the shutdown.

Establish Standards and Specifications for Equipment

Even though the procurement of equipment may be the responsibility of others, the fire loss prevention and control manager should establish standards and specifications to ensure that such equipment meets the requirements of testing laboratories and is acceptable to the authority having jurisdiction.

Interpret Applicable Laws, Codes, and NFPA and Related Standards

The facility manager should make clear that the fire risk manager has the responsibility for providing interpretations of laws, codes, and standards applicable to fire loss prevention and control management, and should make available the legal guidance required to fulfill this assignment.

Organize and Train the Private Fire Brigade

This subject will be treated in detail in Chapter 3, "Private Emergency Organizations," of this Section.

SCHEDULE PERIODIC INSPECTIONS OF FIRE PROTECTIVE EQUIPMENT

Although some of the equipment inspection and testing may be assigned to others (maintenance or security departments), the fire loss prevention and control manager should establish inspection schedules, determine the type of inspections to be made, and set up the routing procedure for inspection reports. It is also the responsibility of the risk manager to see that inspections are properly carried out on time.

Only general advice can be given regarding the frequency of inspection needed for the various components of the fire protection equipment of a particular property. Figures 14-2D and 14-2E show both sides of a sample inspection sheet designed for general industrial use. The form can be adapted for use in other types of properties. Inspections should also be made to assure good housekeeping, rubbish removal and proper attention to special hazards (Tuck 1982).

The manager or staff of fire loss prevention and control should personally inspect the following equipment:

(1) Every control valve on the piping supplying water for fire protection, particularly valves to automatic sprinkler systems (valve inspections should be recorded, showing whether each valve is open or closed, sealed, or unsealed, and including notes about any conditions needing attention); (2) hydrants; (3) hose house and fire station equipment; (4) fire pumps; (5) water tanks for fire protection; and (6) all other special types of protection.

In addition to these general items of protective equipment, a checklist of items to be inspected should be followed building by building, floor by floor, department by department. On these items, the fire loss prevention and control manager may be able to require department or tenant area managers to make necessary day-to-day inspections and reports. A staff member in fire loss prevention and control management should make periodic supplementary inspections of all items on the checklist. For each building and department, the list should include such items as the following:

(1) fire extinguishers; (2) small hose; (3) fire doors; (4) special hazards; (5) special types of protection, and (6) special routines for firesafety.

Various inspections require a daily, weekly, or other periodic check. A convenient routine is to provide a card for each item to be checked at the location to be examined. Inspectors making the necessary periodic observations record them on the card, along with the date, time, and the name of the person making the inspection.

It is not enough for management to specify periodic checks. The person assigned to the checks must be made to feel that, since the inspection is important enough to be recorded, it must be done correctly. The fire loss prevention and control manager should review these records and periodically summarize them for the facility manager.

ASSIST MIDDLE MANAGEMENT WITH PUBLIC AGENCY CONTACTS AND PUBLIC RELATIONS

During daily operations, the fire loss prevention and control manager often comes in contact with representatives of the local public fire protection and law enforcement agencies. However, routine contacts should not be relied upon to firmly establish the company's relationship with these agencies. The fire loss prevention and control manager, especially if new on the job or representing a new facility, should make an official visit to the headquarters of each of these public agencies. Arrangements should be made for their personnel to tour the facility, whose layout and operation should be explained in detail. Any areas or processes which may present a life hazard or special problem to emergency control personnel should be pointed out and plans for their protection discussed. In order to establish a mutual understanding of operating practices, public agency representatives should meet appropriate members of the facility's middle management.

Generally, public announcements regarding an emergency are made by the facility manager or a member of the public relations staff. It is, therefore, imperative that the fire loss prevention and control manager report the following information to them as early as possible:

1. Probable cause of the emergency;
2. Any injuries or deaths;
3. Estimate of the extent of damage;
4. The number of employees and other people at risk;
5. Expected duration of the emergency situation.

This report should be as complete and factual as possible. The facility manager can then decide how much information should be made public in order to inform the public and dispel rumors.

ASSIST MIDDLE MANAGEMENT WITH INSURANCE CONTACTS

If executive management decides that part of the company's risk is to be shared with insurance carriers, this makes a team of engineers available to the fire loss prevention and control manager for advice and guidance. While no company should expect its insurance carrier to bear the entire responsibility of the fire loss prevention and control management program, generally both the broker and the carrier have on their staffs engineers who are qualified to give advice on loss prevention during any phase of industrial construction or operation.

The property risk or insurance manager is usually a member of the facility's general management group. To ensure the best results from the insurance program, the property risk or insurance manager and the fire loss prevention and control manager should closely coordinate their work. In order to provide adequate coverage at acceptable costs, the insurance carrier will need to make a comprehensive risk evaluation of the facility and maintain a program of periodic inspection. The fire loss prevention and control manager and staff can furnish much of the information required for risk evaluation, and should also provide escort for facility inspection by insurance representatives.

The fire loss prevention and control manager must be aware of what equipment must be test-operated by main-

SAMPLE ONLY

NO ONE BLANK CAN BE DESIGNED TO FIT ALL CONDITIONS. USE THIS BLANK AS A BASIC GUIDE IN DEVELOPING YOUR OWN FORM. ITEMS THAT DO NOT APPLY CAN BE OMITTED. OTHER ITEMS CAN BE EXPANDED AS DESIRED, AND NEW ITEMS CAN BE ADDED. FOR ASSISTANCE, CONSULT THE NEXT FACTORY MUTUAL ENGINEER WHO VISITS YOUR PROPERTY.

FIRE PREVENTION INSPECTION

INSTRUCTIONS TO INSPECTOR: FILL OUT FORM WHILE MAKING INSPECTION. SEND COMPLETED FORM TO YOUR SUPERVISOR FOR NECESSARY ACTION. REPORT SHOULD BE HELD FOR REVIEW BY THE NEXT FACTORY MUTUAL ENGINEER.

PLANT	LOCATION	DATE

VALVE INSPECTIONS

PHYSICALLY TRY LOCKED VALVES AT LEAST MONTHLY AND UNLOCKED VALVES WEEKLY. IN ADDITION VISUALLY INSPECT ALL LOCKED VALVES WEEKLY. RECORD BOTH WEEKLY AND MONTHLY INSPECTIONS.

ALL INSIDE AND OUTSIDE VALVES CONTROLLING SPRINKLERS OR FIRE PROTECTION WATER SUPPLIES ARE LISTED BELOW. CHECK CONDITION OF VALVE AS FOUND. PHYSICALLY "TRY" GATE VALVES INCLUDING NONINDICATING AND INDICATOR POST GATE VALVES. DO NOT REPORT A VALVE OPEN UNLESS YOU PERSONALLY HAVE TRIED IT. FM APPROVED PIVA'S (POST-INDICATOR-VALVE ASSEMBLIES), IBV'S (INDICATING BUTTERFLY VALVES) AND STANDARD OUTSIDE SCREW & YOKE VALVES DO NOT HAVE TO BE TRIED BUT SHOULD BE VISUALLY CHECKED AT CLOSE RANGE.

NO.	VALVE LOCATION	AREA CONTROLLED	OPEN	SHUT	LOCKED	SEALED
1						
2						
3						
4						
5						
6						
7						
8						
9						
10						
11						
12						
13						
14						
15						
16						
17						
18						
19						
20						

THE FACTORY MUTUAL RED TAG ALERT SYSTEM IS USED TO GUARD AGAINST DELAYED REOPENING OF VALVES. FACTORY MUTUAL RED TAGS SHOULD BE USED EVERY TIME A SPRINKLER CONTROL VALVE IS CLOSED. WHEN THE VALVE IS REOPENED THE 2 INCH DRAIN SHOULD BE FLOWED WIDE OPEN TO BE SURE THERE IS NO OBSTRUCTION IN THE PIPING. THE VALVE SHOULD THEN BE RELOCKED.

WERE ANY VALVES OPERATED SINCE THE LAST INSPECTION	☐ Yes	☐ No
WERE FACTORY MUTUAL RED TAGS USED	☐ Yes	☐ No
WAS THE VALVE REOPENED FULLY AND A FULL FLOW 2 IN. DRAIN TEST MADE BEFORE THE VALVE WAS RELOCKED OR RESEALED	☐ Yes	☐ No

COMMENTS:

264 (8-79)ENGINEERING PRINTED IN USA

FIG. 14-2D. Front of Factory Mutual inspection blank.

INSPECT THESE ITEMS AT LEAST WEEKLY

SPRINKLERS — Auto-Matic Sprinklers	ANY HEADS DISCONNECTED OR NEEDED Yes ☐ No ☐	OBSTRUCTED BY HIGH PILING Yes ☐ No ☐	
	HEAT ADEQUATE TO PREVENT FREEZING (NOTE BROKEN WINDOWS, ETC.) Yes ☐ No ☐	Water Pressure	LB. AT YARD LEVEL
COMMENTS			

DRY PIPE VALVES									
	VALVE ROOM PROPERLY HEATED	No. 1 Yes ☐ No ☐	No. 2 Yes ☐ No ☐	No. 3 Yes ☐ No ☐	No. 4 Yes ☐ No ☐	No. 5 Yes ☐ No ☐	No. 6 Yes ☐ No ☐	No. 7 Yes ☐ No ☐	No. 8 Yes ☐ No ☐
	AIR PRESSURE	No. 1 Lbs.	No. 2 Lbs.	No. 3 Lbs.	No. 4 Lbs.	No. 5 Lbs.	No. 6 Lbs.	No. 7 Lbs.	No. 8 Lbs.

WATER SUPPLIES						
FIRE PUMP	TURNED OVER	Yes ☐	No ☐	GOOD CONDITION	Yes ☐	No ☐
	AUTO. CONTROL TESTED	Yes ☐	No ☐	FUEL TANK FULL	Yes ☐	No ☐
	PUMP ROOM PROPERLY HEATED AND VENTILATED	Yes ☐	No ☐	PRIMING TANK FULL	Yes ☐	No ☐
TANK OR RESERVOIR	FULL	Yes ☐	No ☐	HEATING SYSTEM IN USE	Yes ☐	No ☐
	TEMPERATURE AT COLD WATER RETURN (SHOULD BE 42°F MINIMUM)			CIRCULATION GOOD	Yes ☐	No ☐

MFL WALL FIRE DOORS	CONDITION	OBSTRUCTED Yes ☐ No ☐	BLOCKED OPEN Yes ☐ No ☐

OTHER ITEMS

INSPECT THESE ITEMS AT LEAST MONTHLY

MANUAL PROT				
EXTIN-GUISHERS	CHARGED Yes ☐ No ☐	ANY MISSING	ACCESSIBLE Yes ☐ No ☐	ATTENTION NEEDED (Give Location)
INSIDE HOSE	IN GOOD CONDITION Yes ☐ No ☐		ACCESSIBLE Yes ☐ No ☐	
YARD HYDRANTS & HOSE	CONDITION NO. 1 NO. 2		NO. 3 NO. 4	NO. 5 NO. 6
	HYDRANTS DRAINED Yes ☐ No ☐		REMARKS:	

OCCUPANCY			
GENERAL ORDER & NEATNESS	GOOD Yes ☐ No ☐	COMBUSTIBLE WASTE REMOVED ON SCHEDULE (PROMPTLY) Yes ☐ No ☐ COMBUSTIBLE DUST, LINT OR OIL DEPOSITS ON CEILINGS, BEAMS OR MACHINES Yes ☐ No ☐ DESCRIBE AREAS NEEDING ATTENTION INCLUDING YARD:	
ELECT. EQUIP.	DEFECTS NOTED Yes ☐ No ☐		
FLAM. LIQUIDS	SAFETY CANS USED Yes ☐ No ☐ EXCESSIVE IN MFG AREAS Yes ☐ No ☐	DRAINAGE OBSTRUCTED Yes ☐ No ☐ VENT FANS ON Yes ☐ No ☐	
SMOKING REGULA-TIONS	LOCATIONS WHERE VIOLATIONS NOTED		
CUTTING & WELDING	PERMITS ISSUED FOR ALL C&W OPERATIONS Yes ☐ No ☐	LISTED PRECAUTIONS TAKEN Yes ☐ No ☐	
STORAGE	WELL ARRANGED Yes ☐ No ☐	AISLES CLEAR Yes ☐ No ☐	
	ADEQUATE SPACE BELOW SPRINKLERS Yes ☐ No ☐	CLEAR OF LAMPS, HEATERS Yes ☐ No ☐	

DOORS AT CUT OFF WALLS	CONDITION	OBSTRUCTED Yes ☐ No ☐	BLOCKED OPEN Yes ☐ No ☐
Sprinkler Alarms	TESTED Yes ☐ No ☐	OPERATION SATISFACTORY (IF "NO" - COMMENT BELOW) Yes ☐ No ☐	

OTHER ITEMS

INSPECTED BY:		DATE
REVIEWED BY:	TITLE	DATE

FIG. 14-2E. Back of Factory Mutual inspection blank.

tenance personnel to comply with union regulations. By being scheduled in advance, such testing can be accomplished with minimum cost and lost time. Participation by the fire risk manager in insurance related activities gives the insurance representative an opportunity to evaluate the fire loss prevention and control management program. If the program is effective, it can have a beneficial effect on insurance rates. To expedite the internal handling of claims, the facility's risk or insurance manager should be supplied with copies of emergency control and loss reports made by the fire loss prevention and control manager to the facility manager.

ASSUME ADMINISTRATIVE COMMAND OF FIRE CONTROL OPERATIONS

Generally, the local fire chief has command of all fire control activities in the community. However, if the fire risk manager has established good liaison with local fire officials, direction of fire control within the facility can be a joint effort.

Most fire officials realize that they cannot keep themselves as abreast of property and process changes as those spending full time on the premises. In an emergency, the fire service welcomes professional assistance in evaluating the fire and planning attack. The fire loss prevention and control manager should be on hand and constantly available to give the fire ground commander information on construction and occupancy of the fire area and areas adjacent to the fire zone that may be threatened.

Prompt action by fire loss prevention and control personnel can make the difference between a minor incident and a total loss. The fire loss prevention and control manager should assume full command until the local fire department arrives and follows its preplanned fire attack.

PROVIDE DETAILED REPORTS OF LOSS FROM FIRES OR OTHER EMERGENCIES

The fire loss prevention and control manager should keep the facility manager advised during the course of any emergency in order to facilitate administrative decisions and public announcements.

Information should be compiled by on-the-spot assessments and should be as complete as conditions permit. Unless authorized to do so, the fire risk manager should not make statements directly to the news media.

When the emergency is over, the fire loss prevention and control manager should, as quickly as possible, prepare a preliminary damage report. This report should contain:

1. Cause, if known;
2. Major injuries or death of employees and other civilians (non-fire fighters);
3. Estimated damage;
4. Estimated time required for restoration of operations.

The time required for the preparation of a final report will depend on how soon detailed information can be supplied by the responsible departments. This report should amplify or amend the estimates contained in the preliminary report. It should contain recommendations to prevent recurrence of such an incident, and may include a limited number of photographs that clarify or emphasize sections of the text. The report should be considered confidential, with distribution limited to people designated by executive management as having "a need-to-know."

Bibliography

References Cited

Tuck, C. A., Jr. 1982. *NFPA Inspection Manual*. 5th ed. National Fire Protection Association, Quincy, MA.

NFPA Codes, Standards, Recommended Practices and Manuals. (See the latest *NFPA Codes and Standards Catalog* for availability of current editions of the following documents.)

NFPA 1, *Fire Prevention Code.*
NFPA 26, *Recommended Practice for the Supervision of Valves Controlling Water Supplies for Fire Protection.*
NFPA 27, *Recommendations for Organization, Training and Equipment of Private Fire Brigades.*
NFPA 71, *Standard for the Installation, Maintenance, and Use of Central Station Signaling Systems.*
NFPA 72C, *Standard for the Installation, Maintenance and Use of Remote Station Protective Signaling Systems.*
NFPA 72D, *Standard for the Installation, Maintenance and Use of Proprietary Protective Signaling Systems.*
NFPA 601, *Standard for Guard Service in Fire Loss Prevention.*
NFPA 601A, *Standard for Guard Operations in Fire Loss Prevention.*

Additional Reading

Bennett, G., "Investigating the Ins and Outs of Fire Prevention," *Control and Instrumentation*, Vol. 12, No. 2, Feb. 1980, p. 59.
Brannigan, Francis L., *Building Construction for the Fire Service*, 2nd ed., National Fire Protection Association, Quincy, MA, 1983.
Cooper, D., "Moves Towards Greater Self Regulation," *Fire Engineers Journal*, 40 (120), Dec. 1980, pp. 33-34.
Fire Inspection Management Guidelines, National Fire Protection Association, Quincy, MA, 1982.
Hubitsky, J., "Preventing Fires on Job Sites," *Fire Command*, Vol. 45, No. 7, July 1978, pp. 34-35.
Industrial Fire Protection, IFSTA, Stillwater, OK, 1982.
Introduction to Fire Protection (a training package), National Fire Protection Association, Quincy, MA, 1982.
Lees, Frank P., Loss Prevention in the Process Industries, Volumes I and II, Butterworths, London, 1980.
Linville, J. L., ed., "Industrial Fire Risk Management," *Industrial Fire Hazards Handbook*, 2nd ed., National Fire Protection Association, Quincy, MA, 1984.
Planner, R. G., *Fire Loss Control*, Marcel Dekker, Inc., New York, NY, 1979. (Available from NFPA)
Rutsein, R., and Clarke, M. B. J., "The Probability of Fire in Different Sectors of Industry," *Fire Surveyor*, Vol. 8, No. 1, 1979, pp. 20-23.
"Self-Fire Inspection: A New Concept in Firesafety," *Fire Chief Magazine*, Vol. 24, No. 5, May 1980, pp. 34-35.
Smalley, James C., "Managing Today's Volunteer," *Fire Service Today*, Vol. 50, No. 7, July 1983, p. 7.
Stephens, H. F., "Fire Strategy for Management," *Fire Prevention*, (133), November 1979, pp. 29-30.
Thor, J., G. Sedin, "Principles for Risk Evaluation and Expected Cost to Benefit of Different Fire Protective Measures in Industrial Buildings," *Fire Safety Journal*, Vol 2., No. 3, Mar. 1980, pp. 153-166.
Waters, D., "The Fire Protection of Plant and Equipment," *Fire*, 70 (867), 1977, pp. 185-186, 189.
Whitaker, Baron, "Management is Responsible," *Fire Journal*, Vol. 60, No. 1, Jan. 1966, pp. 19-21.

PRIVATE EMERGENCY ORGANIZATIONS

Revised by Peter K. Schontag, P.E.

For many industrial facilities, fire suppression is chiefly a matter of calling the public fire department of the community in which the facility is located. However, a public fire department is organized to protect its entire community rather than to provide specialized services for large institutions or industrial plants. Thus, many industrial facilities find it necessary to establish their own fire brigades.

FACILITY RELATIONS WITH THE PUBLIC FIRE DEPARTMENT

For those facilities dependent upon the public fire department, the facility manager (or the fire loss prevention and control manager) must work out an effective arrangement with the chief of the public fire department. Often, this is merely a matter of the facility having the means to promptly notify the public fire department of a fire or other emergency, and seeing to it that the public fire department has easy access to the property.

Once the fire department is on the scene of a fire, the chief or ranking officer, rather than the company's fire risk manager, is in charge of dealing with the emergency. Among other reasons, this is because a principle of common law gives very broad authority to fire departments in case of fire.

If the property houses operations and processes that are fairly straightforward, no serious complications are likely to develop when the public fire department responds to an alarm.

In many industrial plants, however, the processes are very complicated and potentially dangerous. One plant may contain hazardous materials and processes with which an outside agency could not be expected to be thoroughly familiar. In another plant, only the operating superintendent and staff may understand the process layout at any given time. An industrial occupancy may have extensive quantities of electronic gear that are highly susceptible to water damage; a mercantile occupancy may be filled with fragile merchandise, easily damaged beyond salvaging; and mental hospitals and jails have inherent evacuation and custodial problems. In cases like these, fire fighting benefits from management guidance. Fire loss prevention and control management and the fire department should be in agreement as to their respective functions if an emergency occurs.

In case of an emergency, the fire risk manager should quickly brief the chief of the public fire department on the situation; information from the operating manager on duty can help guide the actions of the public fire department during the fire or emergency. No competent fire officer would try to deal with a complex situation without input from the plant manager or shift superintendent.

The fire loss prevention and control manager or shift fire official must know when the best fire control procedure is not the usual application of water. Alternatives to be reported to the fire chief might involve the operation of valves and equipment to shut down a particular process, transfer of materials from one container to another, or moving materials from one section of the property to another.

Company management should develop a written policy, to be followed by shift personnel, covering response to fires or other emergencies outside the plant. This is particularly important in neighborhoods—usually older ones—where industrial buildings are set close together. Such congestion increases the risk of exposing the company's property to fire that starts elsewhere in the area, making contingency plans essential.

Public Fire Department Mobilization

Some properties are located where not one, but several, public fire departments may respond to a fire. Multiple departments may respond because of an explosion, smoke, or flames visible for a long distance. Most likely, especially in urban areas, notification is by means of a sophisticated electronic communications system that can alert many departments simultaneously.

Anytime the fire departments of several municipalities are working together at one fire, it is important that the respective chief officers have a clear understanding of their relationships with each other. Formal mutual aid agreements usually clarify the responsibilities.

Mr. Schontag is Coordinator of Property Loss Control at Pacific Gas and Electric Company in San Francisco, CA.

The fire loss prevention and control manager should deal with the chief of the public fire department having jurisdiction in the community where the property is located and, through him, with the other public fire departments. If there is no public fire department in the area where the property is located, the management usually has no ready way to deal with haphazard public fire department response to a plant emergency. The problem is complicated if the plant site extends over more than one political subdivision.

If the facility is located outside the limits of the fire department's jurisdiction, an understanding must be reached regarding response procedure. Following are some questions that need to be considered.

1. Are there existing laws or ordinances that give the local fire chief the authority to enter into agreements for extra-territorial protection?
2. What personnel and equipment is available for commitment to service outside the legal jurisdiction?
3. What is the facility's legal liability for fire department personnel and for the personnel of other fire departments responding as part of their mutual aid agreements?

Every facility manager should encourage area communities to have the best fire departments they can afford. Generally, small-unit fire departments are both expensive and inefficient when compared to fire departments organized in a large fire district. Also, the company located in a community with a relatively small fire department cannot expect the depth or expertise and experience found in a large fire department with its many companies, broadly experienced officers, and staff technical services.

Strong, well-staffed fire departments could be organized in county areas by encouraging the establishment of fire districts. Improved fire service can be obtained by many small communities by combining adjoining cities and towns into a fire district large enough to support a better fire department with more qualified officers, technical staff services, standard apparatus and equipment, and trained fire fighters.

PRIVATE FIRE BRIGADE ORGANIZATION

A private fire brigade is a force of fire fighters trained specifically to meet the fire protection and suppression needs of the property. Brigade members should be better acquainted with the property, its hazards, and its problems than the members of any public fire department.

Every property should maintain an organization and preplan to deal with fires and related emergencies. The fire loss prevention and control manager should estimate the potential magnitude of a fire emergency within the property, or of an exposure fire, and the availability of fire fighting assistance from a public fire department. These evaluations help determine the nature of the facility's fire organization.

In its simplest form, this organization would be headed by the manager of the facility, assisted by selected personnel. In properties where enough people are available, they should be organized as a team, or teams, to function as a private fire squad or fire brigade. In the United States, employees required to fight structural fires (as opposed to incipient fires) must have extra training and protective clothing. (See the applicable state and/or federal occupational health and safety regulations.) Availability of fire fighting assistance from a private fire department will affect the size and nature of the private fire fighting organization. The fire department does not necessarily take the place of a private fire brigade in parts of a large property. Any individual fire brigade may respond to alarms in all areas of a property; or each geographical or functional area may have its own fire brigade, according to the needs of the property. A fire brigade should be on duty on each working shift and during periods when the plant is idle.

The equipment likely to be put into service at a fire will determine in part the number of people required for each operating unit or company into which the brigade is organized, as well as the total number needed in the brigade. Operating units or companies may be composed of two or more people to operate a specific item of equipment, or of a larger group to perform more complicated operations. Each brigade should have a chief, and each company within it should have a leader. The fire loss prevention and control manager or an assigned member of the fire risk staff must perform the duties directly associated with the manager's responsibility for loss prevention. From the assigned duties for each, it is clear that the functions of the fire loss prevention and control manager and of the chief of the fire brigade are not the same.

The fire loss prevention and control manager or a member of the fire risk staff should: (1) provide equipment and supplies for the fire brigade or brigades; (2) establish the size and organizational structure of the fire brigades; (3) see that the brigades are suitably staffed and trained; and (4) select the brigade chiefs.

Fire brigade chiefs should have administrative and supervisory abilities. Their duties include:

1. Periodic evaluation of the equipment provided for fire fighting. The chief should set in motion procedures for replacing missing or worn out equipment or repair of inoperative equipment. The chief should call to the immediate attention of the fire loss prevention and control manager, or a fire risk management staff member, any situation likely to reduce the effectiveness of fire fighting operations.
2. Provision of plans of action to meet foreseeable potential fire situations in the plant, subject to approval of the facility manager, the fire loss prevention and control manager and other affected staff (e.g., control production, etc).
3. Periodic review of the brigade roster, and preparation of recommendations that additional members be selected, appointed, trained and made available to keep the roster at full strength.
4. Preparing the plan for training members of the brigade and employees of other departments that could assist the brigade.

Enough assistant chiefs should be appointed to cover the chief's position around the clock. Their order of rank should be established, to provide for succession in the event of absence.

Members of the fire brigade should be people who have met qualifications appropriate for fire brigade work at the particular property. Brigade membership should consist of the personnel necessary for fire fighting teams plus certain operating and maintenance personnel.

To qualify as members of the fire brigade, individuals should be available to answer alarms and to attend required training sessions. A schedule of availability should be established for all members, to prevent conflict of duties and to provide for regular off-duty periods, vacations, sickness, and other absences.

Minimum physical requirements should be established for brigade members. A periodic physical examination is desirable. Employees with heart, lung, back, vision, or hearing disorders or impairments should not be accepted. Members of the brigade should have appropriate identification (such as a card or badge) in order to speed travel (police escort) to the plant in an emergency, and for identification by plant guards to permit movement within the facility to perform fire brigade duties.

Mutual Aid

If the facility is located in an area where there are other industries, the fire loss prevention and control manager may consider entering into a mutual aid agreement with them. While such agreements are usually beneficial to all facilities, some precautions should be observed.

1. Clearance should be received from the company's legal counsel before a written agreement is signed.
2. Only such personnel and equipment should be committed to the program as can be released without jeopardy to the safety of the facility.
3. Assurance should be obtained from the risk or insurance manager that all liability is adequately covered.
4. All fire brigade members should be aware of their obligations under the agreement.
5. Fire brigade members should be trained to handle any special hazards located at the other facilities involved in the agreement.
6. A list of all personnel and equipment should be maintained at a central location, to minimize the number of calls required to request assistance.
7. The local fire department should be aware of the agreement and, if possible, should take part in the mutual aid organization. In many cases, the fire department is willing to serve as the record center, and all assistance calls are handled by its communications system.
8. Activation of the mutual aid system should not delay an alarm report to the public fire department.

Training

A schedule of training should be established for members of the fire brigade. They should be required to complete a specified program of instruction as a condition of membership. Training sessions should be held at least monthly.

Members of the brigade should be instructed in the correct handling of any and all of the fire and rescue apparatus provided. The training program should be adapted to the purpose of the particular brigade. It should include fire fighting with portable fire extinguishers, and hose lines, use of self-contained breathing apparatus, ventilation of buildings, salvage operations, and performing related rescue operations. The training program should meet all state and federal regulations. It also should keep up with problems presented by new fire hazards at the facility as well as with new fire extinguishing equipment.

Assistance in setting up and training the fire brigade can be obtained from outside agencies. Among these are municipal fire departments, state fire schools, state educational extension services, state fire marshals' offices, state insurance inspection bureaus, colleges, and any other agency that offers fire service training. Members of the brigade should be afforded opportunities to improve their knowledge of fire fighting and fire prevention through attendance at outside meetings, seminars, and training sessions, where available.

Where the number of individuals participating in the fire brigade training program warrants such an arrangement, a special space or room in the property should be designated for fire brigade training requiring lectures or classroom instruction. Training aids such as books, literature, and films should be kept at this location. Provision of a special room for the brigade is one way in which membership can be made attractive.

Practice drills should be held to check the ability of brigade members to conduct the operations they are expected to perform with the equipment provided. Occasionally, drills should be held under adverse weather conditions, to work out special procedures needed in wind, heat, rain, etc. Equipment should be operated whenever possible. For example, portable extinguishers should be discharged, respiratory protective gear should be worn and operated, and hose lines should be charged with water. Under the control of the chief and leaders of companies, practice drills should be carried out at a moderate pace, with emphasis on effectiveness rather than speed. This should assure the proper technique and safe operation required at a real fire.

At the conclusion of each practice drill, equipment should be promptly placed in readiness to respond to a fire call.

Sometimes, the number of actual fire and emergency responses is so large that the abilities of fire brigade members are "tested" constantly. Practice drills may not be necessary under such circumstances. Instead, training sessions can be devoted to learning to operate new equipment, observing new hazardous processes in the facility, and similar activities.

An NFPA training course, (NFPA 1981) *Introduction to Fire Protection*, contains guidelines for training industrial fire brigade members and their command officers. NFPA 27, *Recommendations for Organization, Training and Equipment of Private Fire Brigades* and the *Fire Brigade Training Manual* (Gold 1984) provide additional information on the organization and training of industrial fire brigades.

Fire Methods

Employees should be instructed to give an immediate alarm when any fire, abnormal heat, or smoke is detected. Under ordinary circumstances, where there is a public fire department, it should be notified without delay. In situations where a facility maintains a fully equipped professional fire force, the public department may be considered as a reserve or backup for the private department. Normally, the first private fire brigade officers to arrive at the scene should: ascertain whether a public alarm has been turned in; supervise operations pending the arrival of public fire officials; put personnel to work on the fire; and direct previously selected members to attend to salvage operations including covering stock and preventing water damage.

In sprinklered properties, officers of the private fire brigade should assign personnel to the sprinkler valves—to make sure that the valves are open and operating, and to see that the fire pump has been started. In sprinklered buildings, streams from 1½ in. (38 mm) hose will usually be adequate for fire fighting. Caution: Use of too many hose streams should be avoided, because this might deplete the water supply needed for sprinkler systems.

Only the fire chief or person in direct charge of the system should authorize the closing of any sprinkler valve (or other valve) that controls water for fire apparatus.

Procedure After Fire

After a fire, all fused sprinklers should be replaced at once with the proper types of sprinklers, and protection restored as quickly as possible. Spare sprinklers should be readily available for this purpose.

Immediately after any fire, all water main and sprinkler valves should be carefully examined to see that no valves have been closed accidentally. Covering stock and other salvaging operations should be continued.

Fire extinguishers that have been used in fighting a fire should not be rehung on brackets, but should be placed on the floor in order to direct attention to the fact that they need recharging. Patrol of the property should continue for some time after a fire has been extinguished to make sure that it does not rekindle.

The fire loss prevention and control manager should investigate the cause of the fire, and should take steps to prevent similar fires in the future. A written report should be made of each fire.

PRIVATE OR INDUSTRIAL FIRE BRIGADE EQUIPMENT

In properties that have well planned fire protection, principal dependence for fire control is placed on automatic sprinklers and on other special hazard extinguishing systems. The private or industrial fire department's operations usually include spreading waterproof covers, seeing that floor drains are open, and restoring protection after the fire. Hose streams may have to be used on fires in stock stored in yards in some properties, or to protect exposed property from fires in neighboring buildings.

Certain duties in the periodic inspection and maintenance of plant fire equipment, both fixed and portable, may be assigned to members of the fire brigade. However, the fire loss prevention and control manager should establish the necessary schedules for such work, should assign those duties to specific personnel, should see that these inspections and maintenance operations are carried out, and should see that reports are filed with management. In any plant, employees in each work area (not members of the fire brigade) should be encouraged to be constantly on the alert for conditions that would adversely affect fire protection, such as missing fire extinguishers, blocked fire doors, etc., to complement the periodic overall plant inspections made by members of the fire brigade.

Storage space for the fire brigade equipment should be provided so that the equipment can be obtained quickly when needed and can be properly maintained. This may also be a convenient location at which to post the plan of water mains serving the property, with all sectional control valves, sprinkler system valves, hydrants, and fire alarm boxes shown and numbered.

The fire loss prevention and control manager or the fire brigade chief should maintain a list of additional equipment available on the property that might be useful in fire brigade work. This might include such items as portable lighting equipment, power saws and other cutting tools, portable pumps, air-moving equipment, electric motors for replacement purposes, tarpaulins, and roofing material. The list should show where each item of equipment is usually located, and the name of the department, or person, charged with its custody. An up-to-date list of equipment, and service agencies from which equipment or assistance may be obtained, together with telephone numbers, should be maintained.

Equipment Checklist

The brigade, depending on its mission, should be provided with the number and variety of equipment and tools needed to perform effectively. In addition to the fixed or portable equipment provided in buildings and yards, the following principal categories of equipment should be considered by the fire loss prevention and control manager and the fire brigade chief when choosing equipment for the brigade.

1. Portable fire extinguishers.
2. Hose and hose accessories, including hydrant wrenches, hydrant valves, rope tools or hose straps, rope, combination shutoff nozzles, gated wyes, double female hose couplings, and hose spanners.
3. Lighting equipment, including portable electric generators, hand lanterns, and a supply of extra batteries.
4. Forcible entry tools, including axes, saws, plaster hooks and pike poles, claw tools, door openers, and crowbars.
5. Ladders, consisting of a selection of sufficient lengths for the work required.
6. Salvage equipment, including salvage covers, brooms, squeegees, water suction machines, and smoke ejectors.
7. Rescue and first-aid equipment. Specific items provided depend on the extent to which members of the brigade have been trained in their use. The inventory may include first-aid kits and resuscitation equipment—inhalator, resuscitator, or modifications of these devices, with spare cylinders of oxygen or air.
8. Spare and replacement equipment. This should include items for which replacement by members of the fire brigade is practical. Choice of the exact items to be stocked should be determined by their availability and relative importance.
9. Fire brigade personal protective equipment, including helmets, coats, waterproof mittens or gloves, and rubber boots. Self-contained breathing apparatus might also be included. Members must be trained in the safe and effective use of all equipment. The quantity of each item needed should be based on the maximum number of brigade members who might be required to use the items simultaneously.
10. Transportation facilities. The brigade should have transportation equipment as needed for its particular work. In some properties, small trucks for inside use, or motor trucks for outside use, may be desirable. (See

NFPA 1901, *Standard for Automotive Fire Apparatus.*)

Bibliography

References Cited

Gold, David T. 1984. *Fire Brigade Training Manual* (part of training program), National Fire Protection Association, Quincy, MA.

NFPA Codes, Standards, Recommended Practices and Manuals. (See the latest *NFPA Codes and Standards Catalog* for availability of current editions of the following documents.)

NFPA 27, *Recommendations for Organization, Training and Equipment of Private Fire Brigades.*

NFPA 1901, *Standard for Automotive Fire Apparatus.*

Additional Reading

Albrecht, A. Richard, "Handling Spills Involving Hazardous Chemicals Within the Chemical Industry," *Fire Journal*, Vol. 65, No. 3, May 1971, pp. 40-48.

Davis, Larry, and Robinson, Kathleen, "Trends Affecting Fire Brigades," *Fire Service Today*, Vol. 50, No. 7, July 1983, p. 29.

Gold, David T., "A Blueprint for Training a Fire Brigade," *Fire Journal*, Vol. 76, No. 5, Sept. 1982, p. 61.

Lowder, Earl J., "A Better Idea for an Industrial Fire Protection Program" *Fire Journal*, Vol. 62, No. 5, Sept. 1968, pp. 10-13.

NFPA, *Introduction to Fire Protection* (Training Package), National Fire Protection Association, Quincy, MA, 1981.

Rubin, Dennis L., "Industrial Fire Brigades: From Planning to Reality," *Fire Journal*, Vol. 78, No. 3, May 1984, pp. 101-104.

Townley, John P., et al., "Industrial Fire Protection—1967, *Fire Journal*, Vol. 61, No. 5, Sept. 1967, pp. 12-21.

SECTION 15
PUBLIC FIRE PROTECTION

FIRE DEPARTMENT ORGANIZATION

Revised by Carl E. Peterson

This chapter provides an introductory discussion of the various elements of public fire protection. It starts with a brief historical overview of the beginnings of early fire protection and fire regulations in ancient Rome. Discussions follow on fire department objectives, fire service organizations, and fire department personnel and structures.

Consult the remaining chapters in this section for detailed discussions of other aspects of public fire protection. Administration and management, operations, prevention, and arson investigation are described in the following four chapters, while the final three chapters detail facilities and apparatus, planning, and fire department decision making.

Section 14, "Organization for Private Protection," concentrates on fire loss prevention for industry and the organization of fire brigades. Section 3 of this HANDBOOK details fire prevention education outlining the role of the fire service in fire protection and explaining fire prevention practices in commerce and industry.

Today, there are approximately 30,000 fire departments in the United States and there are many types of public fire department organizations as well. Public fire protection is normally (although not necessarily) a function of local government; i.e., a town, city, municipality, county, or organized fire district. State, provincial, or federal properties may also have organized fire departments to protect properties. Likewise, large industries will often provide organized private fire departments at their industrial complexes. The organization and objectives of public fire departments vary according to resources available, and range from simple to complex.

EARLY FIRE PROTECTION AND FIRE REGULATIONS

Generally, public fire departments have evolved from community effort to professional organization in five basic steps:

1. Establishment of a night watch service.
2. Drafting of fire prevention regulations and appointment of fire protection officers.
3. Efforts to salvage building contents from loss by fire.
4. Organization of voluntary fire fighting companies.
5. Appointment of fire fighting officers and personnel.

Possibly the first organized fire protection occurred when Augustus became ruler of Rome in 24 B.C. A "vigile" or watch service was created, and regulations for checking and preventing fires were issued. Night patroling and night watch forces were the principal services, and some of the vigiles had duties more like those of police or soldiers than fire fighters. It is clear from the history of that period, however, that fires were a major problem and that the vigiles were provided with fire fighting tools and equipment (buckets, axes, etc.).

One of the earliest recorded fire protection regulations dates back to the year 872 A.D. in Oxford, England, when a curfew was adopted requiring hearth fires to be extinguished at a fixed hour. Later, William the Conqueror established a general curfew law in England, enforced to prevent both fires and revolt.

Fire Brigades in Great Britain

Until 1830, Great Britain did not provide statutory authority to units of local government for night or other watch services with fire apparatus. The *History of the British Fire Service* (Blackstone 1957) mentions only two cases where this subject was introduced in England before the establishment of civil police forces. One was during the British Civil War in 1643 when a company of 50 women was organized to patrol the town of Nottingham at night.

Fire insurance brigades were formed in England primarily as a result of the London fire in 1666. These brigades were formed by the insurance companies without statutory authority or obligations and the insurance company offices, not the government authorities, decided where the brigades would be located. In London the insurance office fire brigades were consolidated into the London Fire Engine Establishment in 1833, which was taken over by the Metropolitan Fire Brigade in 1865.

Mr. Peterson is Manager, Fire Service Management Systems, at NFPA.

It was not until Edinburgh's 1824 Fire Brigade Establishment that public fire services began to develop modern standards of operation when a surveyor named James Braidwood was appointed chief of the brigade. He selected 80 part time aides between the ages of 17 and 25, and required regular drills and night training. Braidwood wrote the first comprehensive handbook on fire department operation in 1830. His handbook included some 396 standards and explained, for the first time, the kind of service a good fire department should perform.

Fire Protection in Early America

Following a disastrous Boston, MA fire in 1631, the first fire ordinance in the new world was adopted. The ordinance prohibited thatched roofs and wooden chimneys, and was enforced by the "board of selectmen." In 1648 New Amsterdam appointed five municipal "fire wardens" with fire prevention responsibilities; this is often considered to be the origin of the first public fire department in North America.

In Boston, after a 1679 conflagration destroyed 155 principal buildings and a number of ships, laws were adopted requiring stone or brick walls for buildings and slate or "tyle" roofs for houses. This fire also led to the establishment of the first paid municipal fire department in North America, if not in the world. Boston imported a fire engine from England and employed twelve fire fighters and a fire chief. From the start, Massachusetts used paid municipal firemen on a call basis, as contrasted with the unpaid volunteer fire companies that were later organized in other colonies.

Colonial communities required each householder to keep two fire buckets and, upon the ringing of church bells, to report to the scene of fires to form lines for passing water from wells or springs. When hand pumped fire engines where obtained, teams were organized to operate the engines. As late as 1810, Boston citizens were subject to a $10 fine for failure to respond to alarms with their buckets. The laws in a number of states still impose penalties on citizens who refuse to assist in fighting fires upon orders from fire officers.

Fire wardens were appointed in Boston in 1711. With the members of their staffs, fire wardens responded to fires and supervised citizen bucket brigades. By 1715, Boston had six fire companies with engines.

Mutual Fire Societies: The first of a number of mutual fire societies was formed in Boston in 1718 when some of the more affluent citizens organized to assist each other in salvaging goods from fires in their homes or business. Their equipment was a bag in which to collect valuables, a screwdriver, and a bed key to help disassemble and remove beds from burning buildings. About a century later, these mutual fire societies became inactive when fire insurance became available to the more prosperous citizens. The mutual fire societies were forerunners of the salvage corps.

Salvage Corps: After the principal cities of the U.S. organized paid fire departments with steam engines around 1850, insurance interests formed salvage corps to reduce water damage at fires. Gradually, improved fire fighting procedures and the increased expense of operation led to the disbandment of salvage corps. Today, public fire departments handle salvage operations at fires.

Growth of Paid Departments: Lack of discipline of volunteer fire fighters, coupled with their resistance to the introduction of steam pumping engines led to the organization of paid fire departments. Following serious disorders at fires, a paid fire department with horse drawn steam pumpers was placed in service in Cincinnati, Ohio on April 1, 1853. In 1855 two steamers were delivered to New York City, but the volunteer fire fighters would not use them. Ten years later the "Metropolitan Fire Department," using steamers, replaced New York's volunteer force.

Duty Systems and Training

Until World War I, paid members of fire departments worked a continuous duty system with only limited duty off time. By World War II, most paid fire departments had adopted some form of the two platoon system which allowed additional days off and reduced the average number of hours worked per week. The hours worked per week have continued to decline, although in most cities fire fighters work more hours per week than workers in private industry or other municipal employment. Although many fire fighters continue to work 24 hour duty shifts, 1975 federal legislation has had a significant impact upon the working hours of fire fighters.

In 1889 the Boston Fire Department established the first drill school where basic training and uniform company drills were performed. Today most departments provide some degree of training for personnel.

In 1914 New York City established a "Fire College" for advanced officer training, and during that same year the first state fire school was organized in North Carolina. In 1925 Illinois and Iowa started state fire schools for the training of volunteer fire fighters and by 1950 most states were providing systematic fire service training. In 1937 Oklahoma A & M College (now Oklahoma State University) initiated a two year (later four year) college program in fire protection. The initial objective of the program was to provide trained individuals for the fire service. However, the emphasis was later shifted to the industrial and insurance aspects of fire protection. Currently, approximately 200 to 250 institutions in the U. S. offer two year programs in fire protection education. One four year program (the University of Maryland) offers a full curriculum in fire protection engineering. Several colleges offer four year programs of study leading to degrees in fire technology or fire service management, and one (Worcester Polytechnic Institute) offers a masters degree in fire protection engineering.

FIRE DEPARTMENT OBJECTIVES

The foundation of any organization is a set of sound goals and objectives that provide both purpose and direction to the organization. Fire departments, as organizations, are no exception and must establish valid goals and objectives in order to perform effectively. The traditional goals commonly accepted by most fire departments are:

1. To prevent fires from starting.
2. To prevent loss of life and property when fire starts.
3. To confine fire to its place of origin.
4. To extinguish fires.

Whether documented or implied, these are likely the only goals of many fire departments. All four are presented here as broad, general statements that are not definitive in terms of achievement or performance. To be more meaningful, performance oriented objectives should be developed to support the stated goals.

Each fire department, regardless of size, should develop performance objectives that specify the results expected and the time required for achievement. The following material explains a system for developing such objectives and their related enabling objectives. This procedure and the examples given are a brief overview for illustrative purposes. Those desiring to use this concept are advised to consult specific texts on the subject of management by objectives.

Developing Performance Objectives

The first step in developing performance objectives is to determine the purpose of the organization, i.e., why it exists. This should be done not only for the department as a whole, but for each subordinate operating division or section. When determined, these items should show the relationship between the department and each of its operating divisions, and the relationship between the divisions. For example, lists with standard headings such as: "The fire department exists for the purpose of . . ." "The fire suppression operations division exists for the purpose of . . ." "The fire prevention division exists for the purpose of . . ." etc. can be used.

The next step is to develop a list of responsibilities. Each operating division should be asked to develop a list of its current responsibilities and activities. The responsibilities listed should be detailed and specific. Once complete, these lists should be examined to determine where there is overlap or deficiencies in areas of responsibility. This information may be listed in tabular form, as in the following example. Figure 15-1A shows only a partial list for a fire prevention division. Additional lists should be developed for each operating division within the department.

Fire Prevention Division

exists for the purposes of:

- Enforcing the fire prevention code
- Reviewing building construction plans
- Inspecting new construction for compliance with fire protection requirements
- Conducting public fire safety education programs
- Conducting fire investigations whenever a fire is suspicious or cause cannot be determined.
- Issuing Permits
- Inspecting properties for State Health Department

Current Activities are:

- Inspecting institutional and educational properties annually
- Following up on public complaints
- Inspecting properties where permits are due and issuing the permit
- Coordinating with public schools on LNTB curriculum
- Investigating about 30 fires per year

FIG. 15-1A. Example of a "typical" list of responsibilities for an operating division. Each division should develop its own list.

The third step is to write a series of statements that describe desired goals. These statements should describe in detail the definite and measurable goals that a department or division would like to achieve. It is important that the statements be realistic and achievable. Statements that are beyond the current scope or resources of the department may be meaningless unless they are properly labeled as long range goals. Most departments or operating divisions should prepare not one, but a series of statements. Following are two examples of such statements:

1. Fire Suppression Operations Division—Each company will conduct an inspection of all target hazards in their first due area.
2. Fire Prevention Division—Work with the Board of Education to implement the *NFPA Learn Not to Burn Curriculum®* at the elementary school level.

Once the divisions have determined goals, an evaluation should be made by comparing what is currently being done with what has been proposed. This comparison might indicate the need for a decision concerning present status and desired goals. Perhaps the extra effort needed to achieve the desired goals may not be productive, and in many cases there might be gaps between present and desired levels of performance. It is in these areas that priorities should be assessed to establish realistic objectives. In the fire prevention division example, the stated goal is working with the Board of Education to implement the NFPA *Learn Not to Burn Curriculum* at the elementary school level. Assuming that there is currently another fire prevention education program in place at the school system and there are other equally important goals desired by the fire prevention division, it may not be feasible to channel resources into this effort at this time. The implementation of the *Learn Not to Burn Curriculum* might be set aside as a long range goal for the division.

In the example for the fire suppression operations division, the stated goal is that all companies will conduct inspections of all target hazards in their first due area. Are the companies currently conducting inspections? If not, and if it has been decided that all companies will accomplish this goal, then enabling objectives that explain how to achieve this goal must be established.

Enabling objectives must be measurable, and have a standard of performance established, and a definite time period for accomplishment. If the enabling objectives are not expressed in quantitative terms, they will be of little value since there will be no means of measuring the performance needed to achieve the overall objectives or goals.

Management by Objectives

Once specific objectives have been established, consideration should be given to the possibility of developing a more functional operating system. This requires that the specific steps necessary to achieve results be identified, problem areas determined, details of activity planned, and a time sequence programmed. It is important that time deadlines be set at various points so that programs and procedures may be monitored and schedules met.

The establishment of a more efficient operating system necessitates a determination whether or not the present system is adequate. Based upon this decision, changes in the present system may be necessary. Using the previous fire suppression operations division as an example, if it has been decided that all companies are to conduct inspections of target hazards in their first due area, the following questions might be asked:

1. Does the present system have the capacity required?
2. Have the fire fighters been trained in inspection techniques? If not, what training is required?
3. Does the training division have the time and resources to conduct this training?
4. How long will the training take?

The answers to the preceding questions will help fill any existing gaps in the present system, and identify key areas that need improvement.

Next, specific enabling objectives for any changes and improvements should be written. The enabling objectives should be expressed in terms of how these changes and improvements are to be accomplished, and should be used to revise the present operating system or establish a new system.

Before the overall program can be completed, objectives for individual positions within the system must be developed. These objectives must also be specified in terms that are performance oriented and measurable, with a time limit established for their completion. In this way each individual will know what is expected and the amount of time allowed for completion of each activity.

Examples of two individual objectives for the fire suppression inspection project (involving more than one division) are as follows:

1. Fire Suppression—Each company officer will arrange for and ensure that the target hazards in each first due area are inspected starting in January of the next calendar year. Each company will submit weekly progress reports on the inspections completed, and the completed inspection forms. All inspections will be completed prior to the last day of March.
2. Training—The training officer will design and develop a training course for suppression personnel on the techniques of inspecting target hazards. The training course will include all items necessary for the identification of potential hazards and recommendations for their elimination based on the fire prevention code and nationally recognized good practices. Progress reports on course development will be submitted weekly to the chief of training. All work will be completed prior to the last day of October.

Finally, it is necessary to monitor results at timed intervals as specified in the objectives. Monitoring provides an opportunity for evaluation, and documents activity. There may be instances during evaluation when it becomes evident that some of the established objectives are unrealistic (depending upon the department's capacity) and may require change.

Managing an operating system by the use of performance objectives is not a static process. If good objectives are established, the process is dynamic and continuously changing. Constant feedback and realignment, change, and addition of objectives are required.

FIRE SERVICE ORGANIZATIONS

Most public fire service organizations in the states and provinces of North America are usually established by local agencies of government, or special local agencies which provide fire protection. Normally local government is responsible for providing adequate fire protection, and the framework within which the protection operates.

Types of Organizations

The types of public fire protection organizations vary widely. One of the most common types is the *public fire department*, a department of municipal government, with the head of the department directly responsible to the chief administrative officer of the municipality. Most large municipalities as well as numerous small communities operate with this type of organization.

Less common is a *fire bureau*, which is usually a division of a department of public safety. In this type of organization the public safety department head must divide time between several important functions, including police and fire service.

The *county fire department* has gained considerable acceptance in many areas. In this type of organization numerous small suburban municipalities can enjoy the benefits of a large, professionally administered public fire department with staff and service facilities which, ordinarily, few small communities could afford individually. Frequently, this department begins with a county fire prevention office and a fire communications system. The smaller (often volunteer) departments initially remain autonomous for fire suppression purposes, but gradually more functions, including suppression, are assumed by the county organization.

Another type of public fire service organization is the *fire district* which is organized under provisions of state or provincial law. It is in effect a separate unit of government, having its own governing body composed of commissioners or trustees, and is commonly supported by a tax levied through the district. Usually it is organized following a favorable vote of the property owners in the proposed district. The fire district may include portions of one or more townships or other governmental subdivisions. Its fire force is frequently termed a "fire department," although in many cases it is the only department operated by that unit of government.

A fifth common type of fire protection authority is the *fire protection district* which in some states is a legally established, tax supported unit that contracts for fire protection from a nearby fire department or even from a voluntary fire association. This type of organization provides the equivalent of municipal fire protection for rural or suburban areas that might find it difficult to maintain their own experienced and effective fire fighting forces. The fire protection district often provides a source of extra income and special rural fire apparatus for the small municipal fire departments that contract to supply fire protection.

An early type of public fire service organization is the *volunteer fire company or association* that raises its own funds by public activities and subscriptions, frequently with contributions of funds or equipment from interested units of government. Many voluntary fire associations maintain excellent equipment and stations, and also serve as centers for various community activities. Often, volunteer organizations prefer to retain their independence from government, especially when purchasing equipment, although in some instances the activities of independent fire organizations are coordinated through special associations and governmental advisory boards.

FIRE DEPARTMENT PERSONNEL

It is estimated that there are at least 200,000 career fire fighting personnel and more than 1,000,000 on call or volunteer personnel serving in the U.S. The chief distinction between career and on call or volunteer personnel is that career personnel are assigned regular periods of duty and are compensated on a regular basis. On call or volunteer personnel are not normally required to be available except for meetings, training sessions, and fires, and may or may not receive compensation for their services.

While most cities and towns of appreciable size employ only career personnel, some cities utilize auxiliary personnel to supplement their regular force. Other cities and towns may use combinations of career, on call, or volunteer personnel. Some communities maintaining their own fire departments may have a career fire chief, officers, and apparatus operators, but rely upon on call or volunteer personnel to provide the staffing balance necessary for efficient fire fighting operations. Other communities may use career personnel only during normal daytime working hours and rely upon on call or volunteer personnel during the night.

The combinations of personnel utilized—career, on call, or volunteer—are strictly by choice of the community: what might work well for one community might not work well for another. A number of important factors can influence the type of personnel utilized within a fire department. These factors are: (1) the financial resources of the community, (2) the availability of on call or volunteer personnel, (3) the frequency of fire incidents, (4) the range of services expected from the department, and (5) the type of department preferred by the community.

Frequently the financial resources of many smaller communities will dictate that the department be composed entirely of volunteer personnel. Since salaries normally consume a large percentage of the fire department budget, available finances may only be sufficient to purchase and maintain apparatus and equipment. Other communities with adequate financial resources may or may not elect to implement a full career service. This decision may be based on one or a combination of the important factors described below.

On Call or Volunteer Personnel

In some cases where the community wants or needs fire protection, there may not be a sufficient number of persons willing or able to serve in a volunteer capacity. This requires that the community finance the operation of a fire department staffed either fully or partially by career personnel. A situation of this type is probably most evident in the rapidly growing suburban communities surrounding major metropolitan areas. Originally, these outlying areas were composed of small communities that provided a climate suitable for volunteer departments. As times changed, however, these communities experienced rapid growth rates in population, housing, and ancillary services, and demands for fire protection services grew accordingly. The original corps of volunteers was unable to meet the increased demand or to recruit new members since many residents commute out of the community on normal work days and are unavailable for fire department participation. Therefore, many of these once volunteer fire departments have added the services of career personnel.

Effect of Frequency of Incidents on Personnel Selection

The frequency with which the department responds to incidents will determine the type of personnel chosen to staff the department. Large, congested areas, especially those that are heavily commercial or industrial, are likely to produce an increased need for fire department service compared to more sparsely settled, largely residential areas. A department that responds to a large number of incidents each year will tend to inhibit a full volunteer operation unless the department has a large membership and the workload can be apportioned to reduce the commitment required of each individual. There are no set figures on the number of incidents that require career personnel. The determination must be made locally based on the ability of the department to perform at a level acceptable to the community supporting the service. It is reasonable to assume that when the number of incidents increases beyond the level that the volunteers are capable of handling, the members will request a change in department staffing.

Services Expected from the Fire Department

The fire service is often expected to perform many functions beyond strict fire suppression. Many departments today are involved with providing emergency medical service (EMS) for the community which may or may not involve actual transport of victims to medical facilities. However, fire departments involved in EMS can easily double or triple the number of alarms they would usually receive. Other types of emergency or nonemergency service rendered by the fire department will also effect the time demands on the personnel.

Another factor to consider is the involvement of fire department personnel in code enforcement inspection, public education, and fire investigation. Good programs in these areas demand time. If the fire department is expected to assume an active role in these functions, career personnel may be needed for these specific responsibilities. In some areas, state and/or county personnel assume a lot of these responsibilities, leaving the local fire department to handle only the emergencies.

Type of Department Personnel Preferred by the Community

Communities often have preferences for the type of staff members they wish to have serve in their fire departments. The preference may be based on one or several of the previous factors or on entirely different reasons. In some instances, the volunteer fire department serves as a focal point of community activity, and the community is satisfied with the level of service provided by volunteer staff. The community itself must ultimately decide on the type of personnel to be utilized within the department.

There are many fire departments—career, combination, and volunteer—that provide an acceptable level of service to their respective communities. The success of their operations is not dependent upon whether the personnel are paid or unpaid, but upon their individual and collective ability to perform and to accomplish departmental objectives. There are no simple guidelines that set forth the requirements as to the type of personnel to be utilized for fire department operations. The decision is one that must be made at the local level following a careful analysis of all pertinent factors.

FIRE DEPARTMENT STRUCTURE

Fire departments, like other organizations, are comprised of a group of people working together in a coordinated effort to achieve a common set of objectives. For a department to function effectively, it must have an organizational plan that shows the relationship between the operating divisions and the total organization. An organizational plan does not preclude the necessity for active leadership; it merely provides the means by which the organization can be managed effectively. Organizational charts with typical structures of small, medium, and large fire departments are shown in Figures 15-1B, 15-1C, and 15-1D.

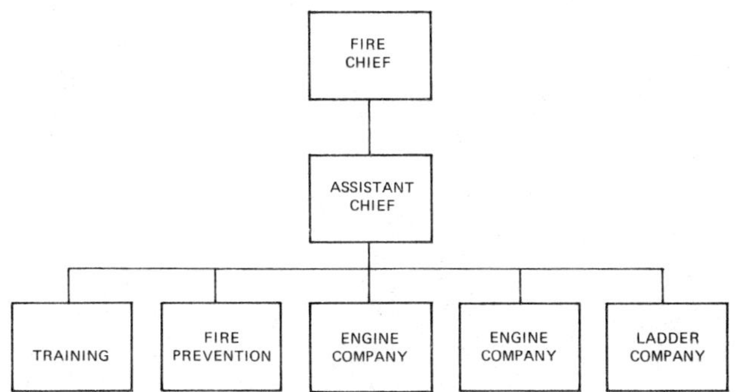

FIG. 15-1B. Typical organizational structure of a small fire department.

Principles of Organization

There has been much written on organization and organizational principles. Although it is impossible to list all of these principles, several which are generally applicable to fire department organizations are briefly described herein.

One of the most basic organizational principles is that the work should be divided among the individuals and operating units according to a plan. The plan should be based on the individual functions that must be performed such as fire prevention, training, communications, etc.

The next principle is that as the department increases in size and complexity, the need for coordination also increases. Small departments are simple organizations that allow frequent personal contact among individuals; thus there is less need for extensive formal coordination. However, as departments increase in size and complexity, they require more extensive coordination of the operating units in order to achieve their objectives.

Another principle of organization is that lines of authority must be established. Individuals should know their relationship to the total organization, and each operational unit or division must know its relationship to the total organization. In many cases individuals are given the responsibility of performing certain tasks, but are not authorized to make some of the decisions necessary to complete the tasks. This restricts the performance of the organization since such individuals must constantly consult their immediate supervisors when decisions are needed. When responsibility is assigned, authority must also be granted.

Unity of command in departmental structure is of prime importance because an individual receiving conflicting orders from several superiors is likely to become confused and inefficient, while an individual receiving orders from only one superior has a better chance to perform more efficiently. Also, the capacity to efficiently supervise becomes limited if too many individuals report to one supervisor. Supervisors directly responsible for too many persons tend to become too involved in supervising subordinates, and leave themselves little time for other important managerial duties.

Line Functions

Line functions in fire departments normally refer to those activities directly involved with fire suppression operations. Fire suppression officers are primarily considered to be line officers. This does not mean, however, that they do not have other functions. As these officers are promoted within the department, their line responsibilities may be equally divided with staff responsibilities. At the highest officer levels within the department, line responsibilities diminish while staff responsibilities increase.

Staff Functions

Staff functions are those activities that do not involve dealing with day to day emergency incidents such as fire suppression, emergency medical service, and various hazardous conditions or hazardous materials spills. Briefly, these staff functions include the following activities:

Fire Prevention: The inspection of all properties for compliance with codes and ordinances; the operation of a public education program; and the investigation of fires to determine the reasons for fire ignition and growth.

Training: Training of all personnel in their job skills; administering continuing education programs in special subject areas; administering the fire department safety program; and organizing and administering prefire planning.

Maintenance: The maintenance of apparatus, equipment, and physical facilities; recommendation of replacement programs for apparatus and equipment; and the development of specifications for purchase of apparatus and equipment.

Communications: Providing and maintaining adequate facilities for the receipt of alarms from the public; and communication with fire companies both in quarters and by radio in the field. Communications functions also include developing and maintaining dispatch policy and response requirements.

Research and Planning: Creating the new knowledge which the fire department needs to provide better service; forecasting long range department goals; and performing the types of analysis necessary for other department functions to assess program effectiveness.

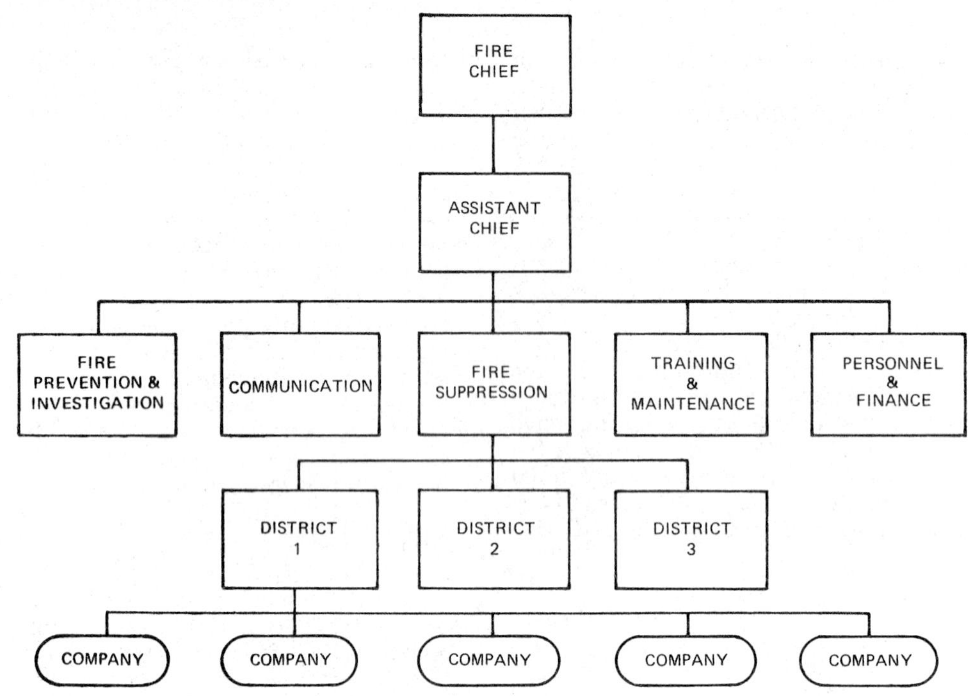

FIG. 15-1C. Typical organizational structure of a medium sized fire department.

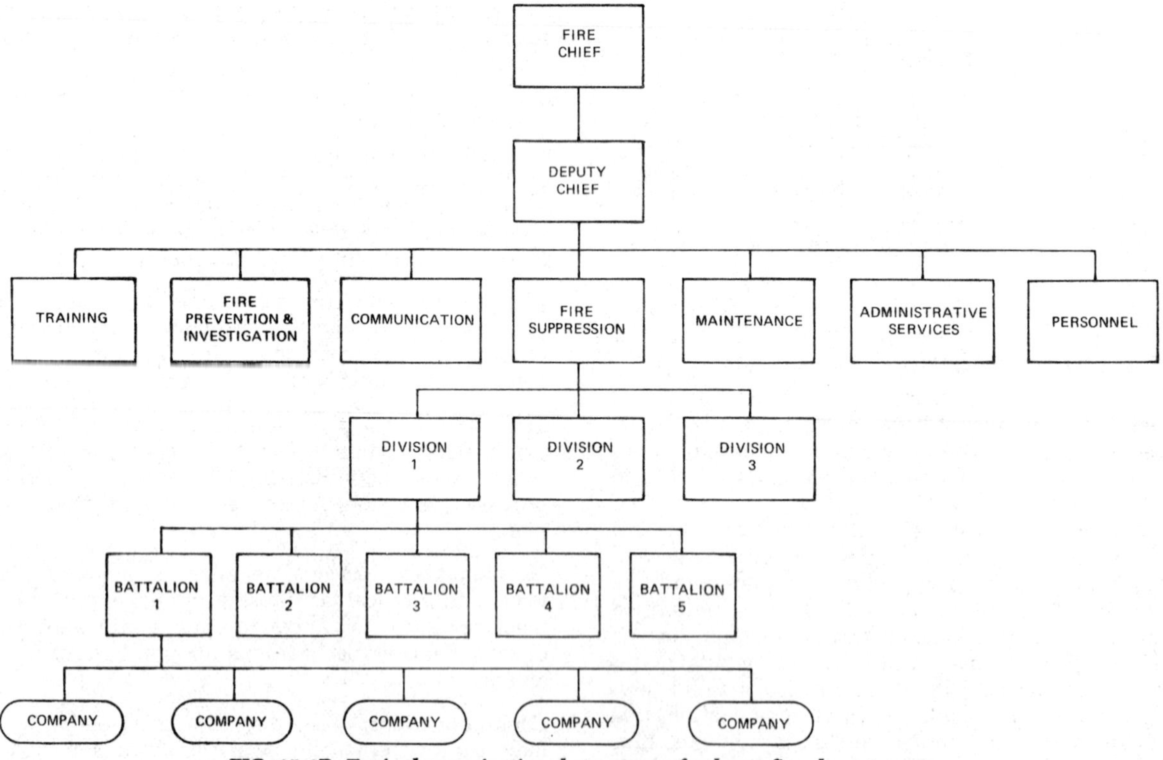

FIG. 15-1D. Typical organizational structure of a large fire department.

Community Relations: Interfacing with the public and the news media to tell the fire department story.

Financial Management: Preparing a budget, monitoring expenses (and income if appropriate) against the budget; planning for capital expenditures; and supervising the purchase and inventory of needed materials.

Personnel Management: Supervising the recruitment, selection, and promotion of personnel; administering the retirement system and the benefits program; and supervising the administration of discipline within the department.

Fire Protection Engineering: Reviewing plans for new construction, major renovation or installation of fire protection systems; developing proposals for code changes;

and assisting in the technical aspects of code enforcement and fire investigations as necessary.

In large fire departments, a staff officer is normally assigned to supervise each of the above functions. Such officers are not normally involved in line functions. The number of personnel assigned to a staff officer will vary with the size of the department and the importance ascribed to the function by the chief and the community. In small departments, line officers may also be assigned staff functions or a single officer may supervise more than one staff function. In the smallest of departments, the fire chief or assistant may directly assume many of the functions.

All of these functions are important to the operation of a fire department. Therefore it is critical within each organization that the person who is to be responsible for the function be aware of the responsibility; all persons within the organization should know who has responsibility for each function as well. Without definite assignments, confusion will arise, important tasks may not get accomplished, and the efforts of personnel within the department will not be utilized in the most effective manner.

Organizational Plans

The manner in which fire departments are organized depends upon the size of the department and the scope of its operations. Organizational plans are designed to show the relationship of each operating division to the total organization. Some departments find projected plans useful for budgetary and planning purposes. It is essential that each fire department have an organizational plan that reflects the current status of the department since a good plan is essentially a blueprint of the organization.

A list of responsibilities or a job description for each position should accompany the organizational plan. In small departments, a single individual may have responsibility for more than one function. For example, a single officer may be responsible for both training and maintenance. This should be detailed in the job description.

The organization chart should show how the various functions that may demand time and support from other personnel or groups will be coordinated within the fire department. Both the personnel within the ranks of the fire department and the public need to see a clear coordinated effort of providing fire protection to the community.

Rules and Regulations

As with any organization, rules and regulations are needed to govern the operations of that organization. This is especially true in the fire service due to the hazardous nature of much of the activity, and the need for clear understanding of expected performance.

Every fire department should have a set of rules and regulations which outline performance expectations for its members, the standard operating procedures for the department, and disciplinary action which can be taken against personnel not following the regulations. These rules and regulations can be, and often are, supplemented by orders from the fire chief who may supplement or clarify the rules or change them for a special event or specific purpose. Both the rules and regulations and subsequent orders from the chief should be written and distributed in such a manner as to ensure all persons are properly made aware of them.

Bibliography

References Cited

Blackstone, G. V. 1957. "The Insurance Fire Brigades." *History of the British Fire Service.* Routledge and Kegan. London, England.

NFPA Codes Standards, Recommended Practices and Manuals. (See the latest *NFPA Codes and Standards Catalog* for availability of current editions of the following documents.)

NFPA 1001, *Standard for Fire Fighter Professional Qualifications.*

NFPA 1002, *Standard for Fire Apparatus Driver/Operator Professional Qualifications.*

NFPA 1021, *Standard for Fire Officer Professional Qualifications.*

NFPA 1031, *Standard for Professional Qualifications for Fire Inspector, Fire Investigator and Fire Prevention Education Officer.*

NFPA 1041, *Standard for Fire Service Instructor Professional Qualifications.*

NFPA 1201, *Recommendations for the Organization for Fire Services.*

NFPA 1202, *Recommendations for Organization of a Fire Department.*

NFPA 1221, *Standard for the Installation, Maintenance and Use of Public Fire Service Communication Systems.*

NFPA 1501, *Standard for Fire Department Safety Officer.*

Additional Readings

Ahlbrandt, R. S., *Municipal Fire Protection Services: Comparison of Alternative Organizational Forms,* Sage Publications, Beverly Hills, CA, 1973.

Bryan, J. L., and Picard, R.C., eds, *Managing Fire Services,* International City Management Association, Washington, DC, 1979.

Campbell, John A., *The Small Business Firm as Provider of Fire Department and Emergency Medical Services in American Communities,* Gage-Babcock and Associates, Elmhurst, IL, 1981.

Caplow, T., *Managing an Organization,* 2nd ed. Holt, Rinehart and Winston, NY, 1983.

Child, J., *Organization: A Guide to Problems and Practice,* Harper and Row Publishers, NY, 1984.

Coleman, R. J., *Management of Fire Service Operations,* Bretton Publishers, North Scituate, MA, 1978.

Didactic Systems, Inc., *Management in the Fire Service,* National Fire Protection Association, Quincy, MA, 1977.

DiNenno, P. J., "Private Contract Fire Protection: An Alternative Management System," *The International Fire Chief,* Jan. 1981, pp. 12-14.

Earnest, Ernest, *The Volunteer Fire Company,* Stein and Day, Briarcliff Manor, 1980.

"Fire Protection through Private Enterprise," *Journal of American Insurance,* Vol. 56, No. 2, 1980, pp. 9-12.

Granito, A. R., *Fire Officers' Guide to Company Leadership and Operations,* National Fire Protection Association, Quincy, MA, 1975.

Hall, J. R., et al., *Fire Code Inspections and Fire Prevention: What Methods Lend to Success,* National Fire Protection Association, Quincy, MA, 1979.

Hickey, Harry E., *Public Firesafety Organization: A Systems Approach,* National Fire Protection Association, Quincy, MA, 1973.

Holzman, Robert S., *The Romance of Firefighting,* Harper & Brothers, NY, 1956.

Mali, Paul W., *How to Manage by Objectives; A Short Course for Managers* Anacom, NY, 1975.

Morrisey, G. L., *Management by Objectives and Results in the Public Sector*, Addison-Wesley Publishers, Reading, MA, 1976.

Office of the Oklahoma State Fire Marshal and Mission Research Corporation, *Basic Guide for Fire Prevention and Control Master Planning*, United States Fire Administration, Emmitsburg, MD, 1978.

Research Triangle Institute, *Evaluating the Organization of Service Delivery: Fire Phase I and IIA*, Prepared for National Science Foundation, Research Applied to National Needs, Washington, DC, 1978.

Research Triangle Institute, International City Management Association, National Fire Protection Association, *Municipal Fire Service Workbook*, Prepared for National Science Foundation, Research Applied to National Needs, Washington, DC, 1977.

Schaenman, P. S., et al., *Procedures for Improving the Measurement of Local Fire Protection Effectiveness*, National Fire Protection Association, Quincy, MA, 1977.

Small Community Fire Departments: Organization and Operation, National Fire Protection Association, Quincy, MA, 1982.

The Fire Safety and Disaster Preparedness Task Group, *Criteria for Master Planning and Resource Allocation*, Public Technology, Inc., Washington, DC, 1978.

The Fire Safety and Disaster Preparedness Task Group, *Urban Guide for Fire Prevention and Control Master Planning*, United States Fire Administration, Washington, DC, 1977.

Thomas, Peter, *Management by Objectives: A Handbook for Governmental Managers and Supervisors*, Masterco Press, Ann Arbor, MI, 1978.

FIRE DEPARTMENT ADMINISTRATION AND MANAGEMENT

Revised by Gary Tokle

The fire service has many unique management needs. It requires: a distinct team spirit; a need for a strong disciplinary influence for concerted and instant reaction on the fireground; a high quality of leadership from its officers; continuous training; an extremely wide range of technical competence; a labor/employer relationship not comparable to that in other occupations; and an ability to deal with the public under both minor and major crisis situations. The fire service is not profit oriented, and it has an obscure productivity pattern. It is a major consumer of tax dollars, uses costly equipment, is heavily dependent upon manpower, and at present has no satisfactory means of measuring the effectiveness of its operation relative to cost. Despite the complexity of these needs, the fire service has generally performed well for many years.

Almost all fire departments were administered by clearly defined organizational structures long before systems techniques were applied to industry and business. A system of task allocation to engine and ladder crews was developed, whereby each person on the apparatus performed certain functions in sequence so the team operated as a coordinated unit without confusion or duplication of effort. Step by step functions were determined many years ago in processes that are now termed "human engineering" and "human factors application."

FUNCTION OF MANAGEMENT

The operation of a fire department is normally a function of local government (in the case of a fire district, possibly the only function) which supports the service and is responsible for the level of service rendered. As with any governmental or business operation, this involves three major areas of responsibility: (1) fiscal management, (2) personnel management, and (3) productivity.

In general, fiscal management practices follow those used by the government agency supporting the department and includes budgeting, cost accounting, personnel costs (including payroll), and purchasing or procurement costs. The degree to which these factors are a direct responsibility of fire department management varies, depending upon the practices of local government.

Fire departments utilize persons with specialized skills who are organized into various operational and staff units. Fire department management is involved to some degree in the recruitment, selection, and promotion of personnel needed to fill various positions in the organization. Largely, these matters are governed by local and/or state law; by personnel agencies, including civil service authorities; and by direct decisions of the governmental agency operating the fire department. The assignment of available personnel to positions provided in the budgeted organizational structure, and supervision of personnel performance, are normally the direct responsibility of the fire department management, although certain assignments are frequently governed by work contract agreements.

Productivity in the fire service is the most difficult ingredient for management to measure. The basic objective of the fire service is the protection of life and property. Modern fire service practice involves two major activities: (1) control of hazards to minimize fire losses and to prevent fires, and (2) dealing with actual fires and emergencies to minimize suffering and losses. It is difficult to assess the number of fires and suffering that have been prevented by fire department activities; however, experience has demonstrated that lack of effective fire prevention and control measures invites disastrous experiences. Likewise, the fact that most fires are suppressed with minimum losses and injuries does not indicate conclusively that an adequate level of fire department service has been provided. Experience shows that major fires and emergencies often arise from combinations of circumstances beyond the immediate control of fire department management, but which must be dealt with effectively to protect the public. It is imperative that fire department management be concerned with maintenance of reasonable standards of organization based upon local and national fire loss experience. Fire department management is responsible for maintaining highly trained and efficient operational units to perform assigned tasks in both the prevention and suppression of fires.

Mr. Tokle is Fire Service Specialist, at NFPA.

In addition to the strictly fire oriented activities many modern fire departments now also provide emergency medical services. These services range from responding as "first responders" to assist ambulance crews, to actually providing paramedic ambulance service. Fire departments that have involved themselves in this aspect of public safety have seen significant increases in the utilization of personnel and equipment.

The hazardous materials response team has been another emerging area for fire department participation. With the ever increasing problems involved with the handling of hazardous materials, fire departments are increasingly being required to step in and develop much greater cababilities in handling such incidents.

PERSONNEL

The recruitment of personnel is seldom the responsibility of a municipal fire department, except in those cases where there is no local governmental personnel agency. Fire districts and volunteer departments recruit their own members.

It is the responsibility of fire department management to notify the personnel agency of existing vacancies in the organization and to request the number of persons needed to fill these vacancies. In connection with recruitment, fire department management has three responsibilities. The first is to recommend appropriate recruitment standards to the personnel agency. The second is to provide the basic training necessary for the new personnel so they can properly perform their assigned duties. The third is to certify, after providing the basic training, that the new members are ready for appointment as permanent fire fighters or, where individuals prove unable to perform satisfactorily, to recommend that their services be terminated before permanent appointment.

Selection of personnel must meet local, state, and federal standards. U.S. courts have ruled previously that there must be no discrimination in hiring practices. Some rulings prohibit residency requirements for recruitment, although fire department rules of employment may stipulate that because of the emergency nature of the work employees must reside within a reasonable distance of the community. One court decision has ruled out examinations that require knowledge of fire department practices and equipment prior to appointment and in service probationary training. Many states have adopted, or are in the process of adopting, minimum fire fighter qualifications standards. Selection practices are a sensitive issue, and knowledgeable counsel should be sought to keep all activities on a sound legal foundation.

In most jurisdictions, applications for employment as a fire fighter are obtained from municipal personnel or a civil service agency. In at least two states recruitment is handled by a state civil service commission. In either system, the fire department is furnished with a certified list of persons eligible for probationary appointment. Age requirements for appointment vary from a minimum of 18 years to a maximum of 35 years of age.

Fire Fighter Qualifications

It is imperative, for obvious reasons, that all persons functioning in the fire service be fully qualified and capable of efficiently performing the wide range of services necessary to protect life and property. Many states have enacted legislation establishing commissions on fire fighter standards which require that all personnel employed by fire departments must satisfactorily complete required basic training before being given permanent employment.

In 1970 the Joint Council of National Fire Service Organizations (JCNFSO) consisting of ten national organizations directly involved in various aspects of the fire service, recommended that national fire service professional qualification standards for fire fighters be developed through NFPA technical committee procedures. The Joint Council established a National Professional Qualifications Board (NPQB) to supervise a nationally coordinated continuing professional development program for the U.S. fire service. The Board has nine members appointed by the Joint Council. The Board agreed that the desired professional qualifications standards should be developed through the NFPA standards making system, and that the secretariat for the committees and the Board would be provided by the NFPA staff. The Board reviews all draft standards before these are submitted to NFPA for final adoption.

In accordance with the objectives outlined, NFPA has established four technical committees composed of peer group representatives, each charged with developing specific areas of the fire service qualification standards. These are: the Fire Fighter Qualifications Committee; the Fire Inspectors and Investigators Qualifications Committee; the Fire Service Instructors Qualifications Committee; and the Fire Service Officers Qualifications Committee. These committees began developing professional qualifications standards in their assigned subject areas early in 1973. Standards have been developed for the following positions: Fire Fighter I, II, and III; Fire Apparatus Driver/Operator; Airport Fire Fighter; Fire Fighter Medical Technician; Fire Officer I through VI; Fire Inspector; Fire Investigation, and Fire Prevention Education Officer; and Fire Service Instructor. It was recognized that NFPA 1001, *Fire Fighter Professional Qualifications Standard* (hereinafter referred to as NFPA 1001), was the first priority because standards for fire officers, training officers, and fire prevention personnel are related to this base. NFPA 1001 covers entrance requirements, including medical examination, and three levels or grades of fire fighter qualifications known as Fire Fighter I, II, and III. Throughout the standard, levels of numerical ascending sequence have been used to denote increasing degrees of responsibility.

At the first level of progression, a Fire Fighter I has demonstrated the knowledge of and the ability to perform the objectives specified for that level, and works under direct supervision. A Fire Fighter II, at the second level of progression, has demonstrated the knowledge of and the ability to perform the objectives specified for that level, and works under minimum direct supervision. A Fire Fighter III, at the third level of progression, has demonstrated knowledge of and ability to perform the objectives specified and works under minimum supervision, but under orders. The standards for officers cover various specified levels of officer qualifications and responsibilities including company level; chief officers of various grades including specified assistant chiefs and departmental management chiefs.

All of the standards are expressed in measurable performance or behavioral objectives covering both required knowledge and demonstrated skills. The standards are prepared for use as a basis for nationally standardized examinations by authorized agencies, and are available for adoption by federal, state, and local authorities in the U.S.

The establishment of standards and testing procedures do not in themselves ensure that all personnel will achieve the required level of competency; NFPA 1001 is a professional qualifications standard and not a training standard. Training programs are necessary to prepare members of the fire service to acquire the skills and knowledge necessary to achieve the terminal performance objectives set forth for each grade. However, training should not be merely random but should be organized to prepare trainees to meet the specified levels of performance which should be demonstrated by performance testing as envisioned in NFPA 1001.

The U.S. Supreme Court decisions based upon constitutional prohibition of sex discrimination in employment has enabled women to apply for fire fighter jobs which has been considerably publicized. In connection with both racial and sexual discrimination, courts have ruled that height and weight requirements are discriminatory. However, candidates may be required to demonstrate their ability to perform the required duties.

Fire fighting requires a major degree of physical strength, and an important factor is the interdependence of fire fighters on each other in fire suppression and rescue operations. If there is a great disparity of physical strength and endurance between crew members, then an unreasonable burden and strain is placed upon those with most strength and stamina. This is not only dangerous to the fire fighter, but also adversely effects the protection of the public.

In some jurisdictions, the services of the fire department training division may be utilized in testing recruits and conducting promotional examinations. In all such cases, recognized standards such as NFPA 1001 should be followed carefully so that the results will not be subject to valid charges of discrimination and qualifications essential to the work will be tested adequately.

Promotional Practices

In the vast majority of fire departments, promotions to various officer ranks are made from personnel serving in the next lower rank or ranks, although more fire departments are recognizing the potential benefits of allowing lateral entry, transfers, and promotions of well qualified personnel from other areas and departments. Promotional procedures are designed to take into account technical qualifications for the particular rank and fire department experience. It is essential that examination procedures in the civil service be fully competitive and nondiscriminatory. In general, as with initial appointments, promotional procedures are administered by personnel departments or by state or local civil service authorities. Usually such authorities employ the advice of persons who are experienced and knowledgeable about the particular job classification, including fire chiefs, fire fighter organizations, and technical consultants. Normally, such advice includes guidance as to the relative weights to be given to experience and to the results of written examinations covering the technical qualifications of the position. In some systems performance ratings are included, but these have tended to be vehicles for various forms of discrimination even though such discrimination may have been quite unintentional. Some supervisors tend to be much more demanding than others when rating performance, and their subordinates often have less favorable performance grades than those of other employees who may actually be less qualified. For this reason, subjective performance ratings should be used sparingly in the promotional process. If performance has not been satisfactory, the matter should be dealt with prior to the promotional process.

Some promotional practices also include an oral interview as part of the process. Again, this can easily become a vehicle for discrimination on the basis of race, color, or mannerisms, and few such oral interviews are scientifically designed or professionally administered. Too often they have been vehicles by which an administration selects and rewards its friends. Therefore, increasing emphasis is being given to requiring candidates for promotion to take a written exam regarding the required technical knowledge.

In recent years some fire departments have relied upon assessment centers as a means of selecting candidates for promotion. Assessment centers have been used for quite some time by industry. The assessment center requires the candidate to demonstrate certain abilities through the use of problem solving exercises, role playing and other simulated exercises. Each person is observed by a trained assessment team and scored according to performance. Some feel that the assessment center is the most realistic means of determining a candidate's suitability for a particular position.

NFPA 1021, *Standard for Fire Officer Professional Qualifications* (hereinafter referred to as NFPA 1021), specifies the levels of performance for fire officers. NFPA 1021, written in performance terms, requires an individual to demonstrate competence by knowledge and performance. More fire departments are including educational requirements such as a community college fire science certificate for all officers, or a bachelors degree for chief officers. It is not uncommon to find fire departments that may seek individuals with graduate level degrees in public administration/management for fire chief positions.

The fire department administration, as an arm of the municipal administration, does have an important role to play in the promotional process. First, it must advise the personnel agency as to the qualifications required in any job or rank to be filled, where such qualifications have not been previously established. Second, when a list of successful candidates for promotion is received from the personnel agency, the agency should advise the promoting authority regarding the promotions to be made. Usual practice is to fill vacancies from the top of the promotional list, except where the head of the fire department specifies in writing valid reasons for rejection of an individual. Such reasons might be a record of serious disciplinary problems including disobedience of written orders, frequent bad judgment when performing assigned duties, a record of conflicts with other employees, and other major personality problems. While such problems might not be serious enough to warrant severance from the present level of employment, they do indicate that the particular candidate, although technically qualified, would be less preferable than another candidate on the list. Personnel records should be available to substantiate any such reasons for rejection.

Personnel Records

A complete personnel record must be maintained for each individual member of the fire department. Such records cover all pertinent facts of each member's fire service career from probationary appointment through retirement, including the original application for employment (or a copy), all assignments, transfers, promotions, commendations, and records of disciplinary action. Duty records must be carefully maintained. It is well for the record to include information on any special skills possessed by the individual that may be useful to the department, as well as education background including fire science courses and other courses completed that may be of value to the department.

In addition to the general record of an individual's service, a training file should be maintained as covered in NFPA 1401, *Recommended Practice for Fire Protection Training Reports and Records.* This will show training periods and subjects in which the individual has received instruction such as apparatus operation, first aid, emergency medical service, and fire inspection. A medical history must be kept of each member, showing absences due to sickness and service connected injuries. A new addition to the individual file is a record of exposure to hazardous materials. This provides a historical record of the fire fighter exposure to toxic materials during employment.

STAFFING PRACTICES OF CAREER FIRE DEPARTMENTS

Staffing levels for fire departments vary considerably and are influenced by such things as population protected, population density, hours per work week of fire fighters, response distances, and fire fighter safety. Generally, fire departments utilize a three or four person platoon system which will accommodate a 56 to 42 hour work week respectively. A four platoon system requires about 25 percent more personnel than a three platoon system. There are a few areas that continue to utilize a two platoon system and in these cases the work week may average 60 hours or more. Staffing levels for major metropolitan cities within the U.S. range from 1 to 3 fire fighters per thousand population with an average of 1.5 per thousand. Communities must assess their needs to determine the level of staffing that meets their requirements. It has been demonstrated however, that when staffing falls below four fire fighters per company, critical fireground operations are not carried out when needed. Tests conducted in 1984 with the Dallas, TX Fire Department indicated that staffing below a crew size of four can overtax the operating force and lead to higher losses.

One study of 25 fire departments working the 42 hour work week in a major metropolitan area showed a median strength of 3.0 fire fighters per 1,000 population. This provides an average of 12 officers and personnel on duty for 20,000 population. When fire departments operate emergency medical service and rescue squads, additional personnel are needed in order to maintain basic fire company strength. In some of the smaller communities, the staffing ratio per population protected may be relatively high because of the need for sufficient on duty personnel for effective initial attack and rescue operations, especially in "bedroom communities" where call personnel are not readily available during the work day and where there are high value properties to be protected that may be more significant than population ratios in determining the number of fire fighters to be provided.

In many core cities, as well as suburbs, the fire departments must protect substantial concentrations of values that exceed the average values related to populations. For example, a core city of 80,000 persons may be the business center for an area of 500,000 persons, and house a high percentage of the low income groups. The number of high rise and large area structures to be protected and the frequency of alarms for fires and emergencies should be considered in determining on duty fire department staffing.

Some very large fire departments may operate with a lower relative strength per 1,000 population than cities of a more average size, because with high population densities these departments have sufficient companies to provide needed coverage while handling working fires. For example, a large city fire department may operate one engine company per 15,000 to 20,000 population and still have a large number of well distributed fire companies, whereas a city of 30,000 persons could not be properly protected with only two engine companies.

Mutual aid plays an important role in providing additional resources. Almost all jurisdictions rely to some extent on mutual aid from surrounding areas to provide fire fighting resources on a routine on major emergency basis. Some departments use automatic mutual aid on initial response. Even large cities are making increased use of both regularly assigned and automatic mutual aid. Often this is practical because companies from neighboring fire departments may be much nearer to a fire location than some of the local fire companies.

Frequently it is impossible for small cities to fully staff all of the fire companies needed for proper distribution of companies throughout the community to handle working fires. In many cases, the population density and the values protected per square mile are relatively low. In such communities, some engine companies may respond with only three persons on duty, and ladder trucks with only two. Such low levels of staffing should be backed up promptly by off shift or call personnel, or by multiple alarm response to assure adequate personnel. Combination fire departments which utilize a mix of career and either paid on call or volunteer fire fighters are found throughout the U.S. In some cases, additional apparatus may be assigned to respond, offsetting deficient company strength. In large geographic area communities with relatively low concentrations of value this may be an acceptable arrangement. However, in general, a minimum of four fire fighters on duty, including an officer, should be provided for each engine company where there is no assigned off shift or call personnel on first alarms.

Fire Department Staffing

The current approach of fire departments is to determine the essential positions in the fire fighting force that must be covered 24 hours a day and 365 days per year, and the total number of duty tours involved. With 24 hour shifts, there is one duty tour per position per day. With 10 to 14 hour or similar tours of duty, there are two per day, or 730 per year.

For example, say that a community has determined that a minimum force of 58 officers and fire fighters of the various ranks should be on duty at all times. With a standard 42 hour nominal average work week, including day and night shifts, each shift or group is on duty 182 or 183 times during a 365 day year (average 182.5 duty shifts per platoon). With 58 persons on duty, this requires 42,340 individual tours of duty per year. A review of fire department records indicates that with vacations, sick and injured leave, and other contractural absences, the individual members of the fire fighting force average not 182.5 but 146.5 tours of duty per year. The required 42,340 tours of duty divided by 146.5 tours worked per person shows that rather than 232 officers and fire fighters (4 shifts × 58 positions), 289 members are actually needed in the fire fighting force, or 72 per platoon to cover the normal anticipated absences. Even with vacations carefully scheduled so that only one person from each shift of each company will be absent at any given time, there may be times when off duty personnel must be used on an overtime basis due to sickness or injury; this is less expensive than carrying more relief fire fighters on the roster than required to maintain normal minimum coverage. Overtime will be required when members are called back for major fires and emergencies; budget allowances should be made for any such overtime.

This example of a fire department requiring 58 fire fighters on duty is based upon minimum effective staffing of eight engine and four ladder companies grouped in two fire districts, commanded by an on duty chief with aides. The minimum staffing for all companies is four persons on duty, including the company officer. Five fire fighters are maintained with two ladder companies in districts of high life hazard and more than average fire duty. To distribute the needed relief personnel, the platoon roster for each company carries one additional fire fighter above the minimum that must be maintained, and one relief officer is assigned to each district headquarters.

In a number of fire departments the minimum levels are not absolute. At times when there is an unusual amount of absence, stations housing more than one company may be allowed to operate one person below the normal minimum rather than pay overtime, unless the company minimum strength is specified by contract or city ordinance.

In general, a nominal 42 hour work week requires not four but five persons per position. Thus, a 20 person company is needed to maintain an average of four on duty. Because this does not provide for two members being away from their shift at one time, personnel swaps between companies or overtime may occasionally be needed.

With a fire department operating on the three platoon system with 24 hour duty tours, each platoon covers 122 tours in 366 days (a leap year). It requires a total of 21,208 tours to have 58 officers and fire fighters on duty. It is found from the records in this example that the average member works 98 tours per year. Thus, 21,208 ÷ 98 requires 216 fire fighters, or 72 for each of the three platoons. However, with a 53 hour maximum work week permitted under federal labor law, each member working more than 53 hours in each 28 day work period would exceed the allowable maximum by 3 hours and be eligible for extra compensation, except possibly when the work cycle is broken by absences.

Where minimum company strength of five persons is desired with a 42 hour work week, most fire companies have 24 persons assigned to the four duty shifts, and 28 assigned for approximately every fourth company. In some fire departments where more vacations are scheduled in summer months, a four person minimum may be maintained in the summer and a five person minimum company strength maintained in the winter. However, most fire departments prefer to maintain minimum staffing year round rather than attempt to adjust for seasonal trends. If additional personnel are needed during severe winter storms, this is provided on a temporary overtime basis and is a part of the overtime account but does not add too significantly to the annual personnel budget, account.

It was customary in the past to allow ten percent for absences from assigned shifts due to vacations and sickness; now the figure commonly is 20 percent or more. Vacations have been increased and work contracts provide for various other compensated absences from duty. An important factor may be an increase in injured leave resulting from reduced on duty staffing per company. With the reduction in work hours, many jurisdictions have both reduced the number of fire companies and the number of on duty fire fighters per company. In some cases, individuals on duty are exposed to much more potential injury than when there were more fire companies with full crews. The injury potential is increased further if chief officers are slow in calling additional fire companies to working fires. Accordingly, in some fire departments where minimum company staffing has been reduced below four persons, operating procedures should require the ordering of additional resources to incidents to provide the personnel necessary to match the size of the emergency.

Minimum Staffing: During recent years an increasing number of fire departments have established "minimum staffing" levels for each fire company or each duty shift. It is a policy in many fire departments not to operate engine or ladder companies with less than four fire fighters including an officer, on duty. In some cases, because of the population and values protected per company and the work load, the minimum established is five persons on duty per company. Where a company member is sick or injured while another member is on vacation and no on duty fire fighter is available to cover the absence, it is the department policy to employ an off duty member of the company on an overtime basis to maintain the essential minimum strength.

Mandatory minimum staffing levels for fire companies are advantageous both to the public and for the safety of personnel. Decisions by labor boards and at least one court have found that minimum staffing agreements or ordinances are reasonable requirements for the protection of the public and personnel. A number of small fire departments which do not attempt to maintain minimum on duty company strength have established a minimum for the duty shift while employing off duty personnel to maintain the predetermined minimum effective strength. Such a plan should account for apparatus that must be operated from the several fire stations before off shift or call fire fighters can arrive to assist.

Even though labor laws or contract requirements may require payment of overtime, managers often find it more economical to use personnel on overtime to cover duty absences which exceed the average allowed in the organization staffing tables, rather than to maintain additional personnel on each fire company duty shift to cover abnormal amounts of absence. However, in the past, there was no allowance in the staffing tables of many fire departments for covering scheduled absences, and fire companies were allowed to run shorthanded, thus seriously compromising their operating efficiency. Also, in a number of cases there has been a failure to allow for the usual amount of sickness and injury. Calculations for the number of personnel needed should also include members on terminal leave and new recruits assigned to a specified period of basic training. Otherwise, overtime costs may exceed the cost of having the needed number of members per shift.

Work Schedules: Work weeks for career fire fighters average from 40 to 56 hours. Most fire departments working an average of more than 50 hours per week have used a 24 hour tour of duty. Most fire departments working 48 hours per week or less have day and night shifts and most popular is the 10 hour day shift and 14 hour night shift. Where the law requires a 40 hour week or payment of overtime, many fire departments work a 42 hour four platoon schedule by paying two hours of overtime. Often this is considerably less expensive than hiring additional personnel, and better teamwork is maintained by keeping crews together on a regular four shift basis.

Occasionally, municipal administrators who are not knowledgeable about fire department operations have suggested that fire fighters be assigned to an 8 hour day, 40 hour per week work schedule similar to the police schedules. This has not proven to be practical nor desirable. On duty police staffing properly varies with time of day and day of week as required by needs for traffic control, patrolling, details, etc. Fire fighting is a team effort with the team including platoon chief officers, company officers, apparatus operators, and fire fighters working together on a regular basis. Serious fires occur at any hour of the day or night, and any day of the week; thus, constant and uniform staffing is essential as is provided by either the 3 or 4 platoon systems. These are readily scheduled in seven 24 hour or fourteen 10 and 14 hour tours of duty per week on either the 42 hour or 56 hour average work week. This may include a 40 hour pay week or a 54 hour pay week with 2 hours of overtime. When the 168 hour calendar week is divided by 8 hour tours, this requires 21 work shifts which cannot be scheduled on a uniform basis with even platoons. In most instances where the 8 hour schedule has been proposed, it has appeared only as a bargaining weapon to counter employee demands for better compensation; invariably it has been rejected. In the entire U.S., there are only a few fire departments using an 8 hour work shift.

The requirement for overtime has resulted in improved mutual aid arrangements because on duty companies from nearby departments can respond much faster than off shift local personnel, and overtime costs are kept down. However, many small fire departments rely heavily upon off duty response on an overtime basis. Usually, a minimum of 2 and often 4 hours overtime pay is guaranteed for each response. Where alarms for structural fires are infrequent, it may be much more economical for a small municipality to contract for overtime response than to provide full on duty fire company staffing around the clock.

Manning of Combination and Volunteer Fire Departments

Part of the population of both Canada and the U.S. lives in communities that cannot support a career or even a combination career/on call fire department. Therefore, volunteer fire departments are essential to providing public fire protection throughout vast areas of both countries. At the last count, approximately 10 percent of the fire departments in the U.S. served communities with populations of 10,000 or more, approximately 18 percent protected communities having populations from 2,500 to 10,000, and over 72 percent of the fire departments served communities having populations under 2,500, or served the fire protection needs of rural areas.

Small community fire departments are often staffed almost entirely by volunteers. Even if a town of 2,500 inhabitants had as many as three career fire fighters per 1,000 population, which is considerably above the average, there would be only two members on each duty shift, so full paid fire departments would not be feasible unless there were a substantial tax base to support a career fire department.

In most small communities, volunteer fire departments are operated on a "neighbor help neighbor" basis in which fire department members (except possibly fire station custodians and sometimes a few career fire fighters on duty), receive no compensation other than possibly some reimbursement for personal expenses and uniforms. To replace this free community labor with minimum staffing by fully salaried personnel would involve an added tax burden on the local population. Various states have recognized the contribution made by their volunteer fire fighters by enacting protective legislation and providing state wide training programs, facilities, and retirement systems.

On receipt of an alarm, members of most volunteer fire departments report to assigned fire stations from which they respond with apparatus. To provide a minimum effective working crew, many such fire departments require that the first piece of apparatus not respond with less than three members. Volunteer fire company response to an alarm should be with a minimum of four members.

Fire department administrators should periodically review response records to determine that sufficient active fire company members are available to respond at all times, and where necessary, should recruit and train additional personnel to provide the required minimum response. All essential staff positions in a well organized all volunteer fire department are covered by assigned volunteer officers. However, in many jurisdictions, career fire prevention training and communication functions are provided by the county or other unit of government.

Where a community can afford on duty career fire fighters, response to alarms is faster and efficiency is increased. The on duty fire fighters take the apparatus to the fire, and the volunteers—notified by radio—go directly to the fire thus saving about three minutes in arrival time on the average. The apparatus operator normally is in charge of the apparatus and of the fire station, but volunteer officers may direct the fire fighting. One difficulty with this arrangement is that there are few, if any, opportunities for advancement for career personnel.

With the increasing legal and technical responsibilities of the fire chief as the principal fire protection officer of a community, the chief should be appointed on the basis of qualifications and experience. Communities should consider adopting NFPA standards for fire officers as a minimum requirement.

An arrangement that works quite successfully is to have the first responding pumper staffed by a career officer, apparatus operator, and, where manpower permits, an additional fire fighter. This force is supplemented by additional personnel assigned to respond on call. The second due engine may be staffed by volunteers, paid call personnel, or may have a career apparatus operator to take the apparatus to the fire, where it is joined by the volunteer or paid call fire fighters. The ladder truck has a career apparatus operator assigned, but is staffed at the fire by volunteer or paid call fire fighters. This arrangement permits a reasonably effective initial fire attack quickly backed by volunteer or paid call members. In all cases there should be but one fire department in any jurisdiction, operating under a clearly defined and unified chain of command.

In a number of combination fire departments, the career fire fighters have complained about being commanded by volunteer officers whom they felt lacked the needed experience and qualifications. All fire officers, whether elected or appointed, should meet the appropriate NFPA qualifications for their rank. Countless volunteer officers have qualified to meet the technical standards for their duties. When a career apparatus operator is assigned to operate volunteer fire company apparatus, he/she should work under the orders of the volunteer officer of that company on the fireground. It is important that administrators make clear the respective roles and duties of all members, career and volunteer.

Call Fire Fighters

Many fire departments in small communities employ fire fighters who have no regular duty shift in the stations, but are paid by the hour or per incident for response to alarms and drills. In some cases such members are loosely termed "volunteers," but under federal labor rules as well as local fire department rules, they are considered paid employees of the fire department and as such, are subject to the requirements of federal wage and hour regulations. The paid on call members may also be employed by other municipal agencies, in which case the time spent on fire department duty may affect their overtime status. In most cases, call fire fighters are local businessmen and tradesmen who are willing to be part time fire fighters. Call fire fighters are expected to meet the same standards of performance as career members of the same rank, but may not be assigned as apparatus operators where there are sufficient career operators on shift duty.

There are various methods used in determining compensation for paid call fire fighters. In many departments they receive the same hourly wage for the rank they hold as do employees who work regular duty shifts. Time is based upon attendance at fires and training sessions, with a minimum hourly rate specified for response to alarms. Often upon reaching a mandatory retirement age they also receive a pro rated pension based on their years of service and hours of duty as call fire fighters. Another method of compensation is a fixed annual salary, based on rank, from which deductions are made for excessive numbers of unexcused failures to respond with their assigned companies. Many fire chiefs arrange to excuse members known to be at their regular employment, except in the case of multiple alarm fires; chiefs should have the authority to dismiss members who frequently fail to respond to fires and assigned training sessions. Still another method of compensation is to make an annual appropriation for call fire service based upon past experience, designed to approximate the hourly wage rate for fire fighters. This is divided on a regular basis among the call members as determined by individual attendance at fires and training sessions, so that members responding most faithfully receive the largest compensations. A fire department should pay for members' insurance, workmen's compensation, and all protective equipment.

It is important that accurate individual service records be kept in the personnel files for all fire fighters as well as for career fire fighters. Response and attendance records are also essential for all members. Members who are habitually late in arriving should be replaced. Volunteer members should be furnished with night turnout suits so that they will lose no time in getting dressed to respond.

In a number of cases, senior fire fighters who cannot respond regularly to first alarms are assigned to operate reserve apparatus or to pilot mutual aid fire companies when serious fires occur. Other senior call members may be assigned to cover the alarm desk when all of the paid apparatus operators are out of quarters.

Keeping accurate response records for all fire fighters has enabled many fire chiefs to hire additional career personnel. In numerous cases municipal officials have believed that because they had the names of several hundred volunteers on the roster, the fire department had ample strength. This has often resulted in serious delays and extension of fires which should have been more readily controlled. Where such situations persist, additional career personnel may be required.

BUDGETING

Fire department budgets are generally prepared and submitted by the fire chief to the elected body that administers the city or district operations. The type of budget used and the manner in which projections are submitted will vary depending upon the jurisdiction. Personnel costs are generally the most significant costs for most municipal fire departments, accounting for approximately 90 percent of the total expenditures of a fully paid fire department. Personnel costs of combination fire departments will comprise approximately 40 to 60 percent of their total budgets.

It is critical for fire department managers to thoroughly understand their jurisdiction's budgeting system. Inadequately prepared budgets can lead to serious monetary problems at the end of the fiscal year. In order to assure smooth operations, all costs must be realistically estimated and expenditures monitored on a regular basis. An effort should be made to develop a long range plan that will project capital replacement costs for items such as staff vehicles, fire apparatus, fire stations, and other major equipment purchases.

Fire apparatus costs normally run from one to two percent of payroll costs. Some fire departments include an apparatus replacement allowance in their budget, but unfortunately this is an item that is too regularly cut, with the result that apparatus replacement may be included in a capital expenditures budget. While this reduces the fire department annual budget, it ultimately results in higher taxes due to interest costs. However, such decisions are generally made above the level of the fire department administration. New fire stations usually are included in a capital improvement budget separate from the fire department budget.

Pension payments to retirees are usually paid out of a separate municipal account and may amount to from 33 to 40 percent of a fire department's current budget.

In large fire departments, there may be separate budget accounts for staff divisions such as fire prevention, maintenance, training, and signal or alarm systems, although the latter may be handled by a different municipal department. Expenditures are charged against specific items in the line budget, and the remaining balance is shown after each expense deduction. Usually the department head or staff division superintendents have authority to make emergency transfers of funds between line categories; transfers between major categories can be made only upon authorization of the municipal management, finance officer, or other governing body.

Fire department administrators are required to submit their budget estimates by a specified time for the coming fiscal year. Usually the budgets are submitted to a finance officer or finance committee, and department heads are interviewed to justify specific items. Although the salary total is likely to be governed by contract with the employees, estimates must be included covering all ranks and including overtime costs. Quite often the actual salary scale is not negotiated prior to submission of the budget, but municipal administrators commonly make a percentage allowance for increases they hope will be accepted in negotiations. When a departmental budget has been approved by the city administration, it must be approved by the city or town council; in some municipalities it must be approved by the financial town meeting. With some municipal charters the council can reduce, but cannot increase, the budget. This is to guard against political pressure on the administration. Once approved, the budget takes effect at the beginning of the fiscal year. If not approved in time, it is customary to permit expenditures at the same rate as the previous year. As indicated, all expenditures must be made against specified items in the budget. Normally there is a set amount above which expenses cannot be incurred without competitive bidding under purchasing department procedures.

PLANNING AND RESEARCH

Planning

Planning for the future needs of a fire department is the most important job of fire department managers. Without adequate planning an administrator will find he/she is handling one crisis after another and can never seem to get ahead. Long range planning has often been neglected by fire department managers, but with budget constraints today it is absolutely essential that this task be accomplished. The U.S. Fire Administration (USFA) has developed a master planning process, available to local communities, that outlines the steps necessary for a local community to determine long range goals for its fire department and how these goals are to be achieved. The NFPA Technical Support Program will assist local governments with the development of a master planning process.

All departments, small or large, need to develop long range plans. These plans need to be flexible and continually updated to reflect the change in the local community as well as developments within the fire service.

Research

In the fire service, the term "research" is commonly used; few fire departments, however, are staffed or financed to support any significant research activity. Limitation in research is due largely to the fact that most fire departments are relatively small organizations which do not have sufficient personnel to meet their on going obligations for furnishing fire protection. True research into efficient equipment design is generally beyond the capability of most fire departments, even for the majority of fire equipment suppliers because this is a relatively small volume competitive business with little profit margin for research and development. In recent years a major part of the engineering effort of fire apparatus builders has been devoted to meeting the increasingly demanding vehicle safety standards of the U.S. Department of Transportation (DOT). Some improvements have been made in various features of fire apparatus design, but not on a uniform or planned research basis.

One area in which fire department research can readily show returns is in fire record analysis, utilizing programs such as the UFIRS (Uniform Fire Incident Reporting System) which was developed by NFPA in cooperation with a selected group of fire departments. Properly utilized fire prevention efforts can be directed against the hazards shown to be most dangerous and significant at any given period, and results of these programs can be effectively analyzed.

Research has been used in a number of instances to help determine optimum locations for fire stations. However, a principal difficulty is that all areas of a community must be covered within reasonable response distances. Considerable judgment and experience is required when considering the many variables such as population densities, valuations at risk, fire frequency and severity, and the number of fire companies required in a given area to apply the required water flow and to maintain coverage during fires. These requirements may be at variance with any optimum location for a given fire station. A number of research studies have placed considerable emphasis upon the arrival time of the nearest fire company, rather than the total fire protection requirements of the area. Arrival times may be of less significance than time required for actually getting to work at a fire, particularly in large high rise structures or shopping centers.

Fortunately, the fire service is not without competent research resources. NFPA conducts research designed to directly help and benefit fire departments. The International Association of Fire Fighters (IAFF) Research Department is available on request to assist its local affiliates with problems. The USFA also conducts research relating to fire service needs.

MANAGEMENT RECORDS AND REPORTS

A records system should be provided to supply the fire chief and other administrative officers with data indicating the effectiveness of the department in preventing and fighting fires, insofar as practicable, to facilitate management of the department. It is essential to maintain complete records of all fires and inspections. The records system should provide data on fire department activities which the fire chief should make available to city officials and the public. The fire chief should specify the records to be kept and methods of gathering data. A records retention and disposal system should be employed. All records should be examined in light of their usefulness since too often there is an accumulation of old records that are no longer needed.

PUBLIC INFORMATION AND COMMUNITY RELATIONS

As a public agency, supported by public funds, the fire department needs public support. The public's understanding and cooperation is required to make fire prevention programs fully effective and is also necessary to enable the fire department to deal effectively within the political environment of its particular community. Without community support it can be difficult for the department to obtain the needed funds to operate at the required level. Accordingly, public information and community relations programs are an important fire department management activity.

In a good public information program it is important to develop and to maintain procedures that keep the public informed of important developments and newsworthy items regarding all types of departmental activities and programs. Press releases should be properly prepared and issued on all such programs and items of general interest. Relations with the news media should be cordial, and representatives of the media should be given all possible cooperation by the fire department. This is not always easy, because hundreds of suburban fire departments are located in communities that depend largely upon a metropolitan press, and items that might be of interest to local citizens may not be newsworthy for the entire region. However, there are usually local weekly papers or local area editions of metropolitan papers which do have local correspondents who can be kept informed of fire department activities. The amount of publicity obtained by fire departments seems to vary greatly in different geographical areas. In some areas, citizens are traditionally interested and involved in local governments and in their fire departments, while in other areas fire departments are not considered to be newsworthy. In the latter case, greater efforts must be made to get fire related information to the public. In general, those communities where citizens regularly participate in town affairs seem to give fire departments better support. Communities with substantial numbers of low income residents subject to fire dangers often are most concerned with the adequacy of fire department services.

Normally, public information programs are of direct concern to the head of the fire department who makes many personal contacts, often holds memberships in various civic groups in the community, and often knows news personnel and editors personally. It is important that news personnel be given information promptly (while it is still timely) when emergencies occur. Most fire departments specify that only the officer in charge at a fire or emergency should give out information. This is to avoid conflicting and inaccurate statements which may be misleading or may compromise the results of subsequent or on going fire investigations.

Some large fire departments have a designated public information officer who is assigned to give the press all possible cooperation and information. In smaller fire departments, this function is usually one of the duties performed by the fire chief or, in cases where fires are under investigation, by the fire marshal.

Community relations should be a year round activity and should embrace various seasonal fire prevention programs such as Fire Prevention Week, clean up week, etc. Often, both the fire prevention bureau staffs and the training division staffs are utilized for such activities. In some fire departments the community relations program is coordinated by the fire prevention bureau as a part of its public education program.

INTERGOVERNMENTAL RELATIONS

Fire departments are but one agency of local government, and much of their success depends upon their working relationships with other local, state, and federal agencies. Some of the more important contacts are mentioned here.

Building Department

Proper construction and arrangement of buildings is essential to a sound fire protection program.

State laws and local ordinances, or agreements between fire and building departments, increasingly require written approval by the head of the fire department concerning specified fire protection features before building permits can be issued. Also, close cooperation is needed between these departments to control serious fire hazards which commonly are present while buildings are under construction, before the required fire resistance or protection features have been installed. In small communities the fire chief generally must handle this assignment, but in many fire departments it is one of the responsibilities delegated to the fire prevention bureau.

Law Enforcement

Cooperation between the fire department and law enforcement officials is essential. Regular law enforcement response to fire alarms is necessary to control traffic and crowds. It is also important for fire and law enforcement agencies to develop coordinated plans in the event of an incident which requires the evacuation or closure of areas. Law enforcement and fire officials also need to develop working relationships to deal with fire investigations. In many areas a combination of fire/law enforcement fire investigation teams have been developed. In these cases, both agencies work together and bring their special expertise to a coordinated effort in cases where arson is suspected.

Water Department

Adequate water supplies, including hydrant service, are essential for fire fighting and are the responsibility of the water department. A knowledgeable fire officer should be assigned to maintain liaison with the water authority. All too often water authorities have little knowledge of the water flow requirements of the fire department for various areas and types of property. Thousands of fire hydrants have been improperly set because water crews did not understand the proper location and setting of hydrants required for efficient fire fighting. In some communities the fire department is responsible, by ordinance, for determining the location and setting of hydrants. It is important that all hydrants be serviced regularly and after each use, especially in cold weather; the fire department should promptly report all hydrants it has used in a particular incident to the water department. Alert fire departments maintain a list of hydrants in each fire company inspection district with flow data on each hydrant. Hydrants should be properly marked for flows and painted for night time visibility.

Personnel Department

Members of career fire departments are public employees, and as such their recruitment and promotion may involve cooperation with the personnel agency, which in some jurisdictions is responsible for conducting entrance and promotional examinations. A fire department officer should be assigned to maintain liaison with the personnel office.

Finance Department

Fire departments need to work closely with finance officials when developing and administering budgets. Budgeting is generally very complex and it is important that fire officials prepare budget documents correctly. After budgets have been prepared and adopted it is necessary to assure that proper records of expenditures are maintained and that spending is kept within the adopted budget. Although some fire departments may have a finance officer, liaison with the finance department may be handled by designated staff officers.

Purchasing

All purchases exceeding stipulated amounts must be made according to specifications, and usually with competitive bidding by the purchasing department. Close liaison is necessary to assure that specifications are properly drawn to meet fire department needs, and that when bids are opened any proposals that deviate from specifications are rejected. This liaison may be handled directly by a staff officer.

Data Processing

Increasingly, fire departments are utilizing electronic data processing for keeping fire records, payroll records, and for statistical analysis. Each fire department should have persons knowledgeable in the use of data processing. It may be desirable to appoint one officer as coordinator of this activity, but commonly there would be an administrative committee in the fire department which would include representatives of plans and research (where provided), administration, fire prevention, and fire suppression. This same committee may also be involved in long range planning for the department.

Planning

Fire departments, particularly those in rapidly growing areas, should maintain a close working relationship with local and regional planning groups. These agencies can provide valuable information on growth patterns that will affect the resources of the department. The plans, studies, and reports prepared by the planning agencies can be used to determine the need to increase personnel, additional equipment and station facilities.

PROCUREMENT OF EQUIPMENT AND SUPPLIES

In most municipalities, fire department equipment and supplies are procured through purchasing departments. Items that are common to all departments may be requisitioned from the purchasing agency and charged to the appropriate fire department account. Where items are of a specialized nature such as fire apparatus, or fire fighting tools and equipment, purchasing specifications must be prepared by the fire department, approved by the purchasing department, and advertised for bids. In preparing specifications for such items as fire apparatus and fire hose, current NFPA standards should be followed. The fire chief, with advice from the apparatus and equipment superintendent, should determine whether or not proposals submitted by bidders adequately meet specifications. In many jurisdictions, the law requires that a contract be awarded to the lowest responsible bidder. However, quite frequently bidders take exceptions to various details in the specifications or offer substitutes, this requires that a fire chief exercise judgment as to whether or not such proposals meet the intent of the specifications. If they do not, the bids should be rejected; but if the proposals conform to the specifications and are within the appropriation or budget item, the contract should be awarded. Upon delivery, new equipment should be tested in accordance with the provisions of NFPA standards.

When emergency purchases must be made, it is customary to require bids from several suppliers. If the amount involved is small and funds are available in the appropriate budget item, the fire chief can authorize the expenditure. If funds are not available in the fire department budget, authorization and funds must be obtained from the municipal management or finance officer.

RESOURCE ALLOCATION AND UTILIZATION

The principal resource of a fire department is its highly trained personnel. The vast majority of personnel are assigned to the fire fighting division, and possibly two percent to three percent of the personnel are assigned full time to the fire prevention bureau. Thus, for most effective resource allocation, maximum use must be made of the fire fighting personnel through careful time utilization schedules assigning appropriate allocations of work periods to apparatus and equipment maintenance, fire service training, and scheduled inspections. Such programs require close coordination between the chiefs of the fire fighting, training, and fire prevention bureaus or divisions. In well managed fire departments, most of the routine fire prevention inspections are conducted by fire companies in their assigned inspection districts. Time utilization studies in some fire departments have shown that undue amounts of time were being wasted on common janitorial duties that could be done more cheaply under contract or by less skilled labor. In many cases only the fire chief is available on a regular basis to handle all staff duties, and accordingly, many of the management functions outlined in this text tend to be neglected. Personnel are not fully utilized when they are not given management responsibilities for the various programs within the fire department. One difficulty in the past, has been that many fire chiefs had no training or experience in staff and business management and did not know how to delegate responsibilities.

Currently, changes in attitudes plus newly acquired management skills are rapidly compensating for past deficiencies, and many fire chiefs have acquired the management skills necessary to manage a complex organization. Considering that 20 percent or more of all fire department personnel are officers, there should be ample, technically qualified, managerial help available even in a small fire department who can contribute to maximum efficiency and productivity.

Bibliography

NFPA Codes, Standards, Recommended Practices and Manuals. (See the latest *NFPA Codes and Standards Catalog* for availability of current editions of the following documents.)

NFPA 901, *Uniform Coding for Fire Protection.*

NFPA 902M, *Fire Reporting Field Incident Manual.*

NFPA 903M, *Fire Reporting Property Survey Manual.*

NFPA 1001, *Standard for Fire Fighter Professional Qualifications.*

NFPA 1002, *Standard for Fire Apparatus Driver/Operator Professional Qualifications.*

NFPA 1003, *Standard for Airport Fire Fighter Professional Qualifications.*

NFPA 1004, *Standard on Fire Fighter Medical Technicians Professional Qualifications.*

NFPA 1021, *Standard for Fire Officer Professional Qualifications.*

NFPA 1031, *Standard for Professional Qualifications for Fire Inspector, Fire Investigator and Fire Prevention Education Officer.*

NFPA 1041, *Standard for Fire Service Instructor Professional Qualifications.*

NFPA 1201, *Organization for Fire Services.*

NFPA 1202, *Organization of a Fire Department.*

NFPA 1401, *Recommended Practice for Fire Protection Training Reports and Records.*

Additional Readings

Anderson, D. L., ed., *Municipal Public Relations*, International City Management Association, Washington, DC, 1966.

Aronson, J. Richard, and Schwartz, Eli, eds., *Management Policies in Local Government Finance*, International City Management Association, Washington, DC, 1975.

Bahme, C. W., *Fireman's Law Book*, 14th ed., National Fire Protection Association, Quincy, MA, 1967.

———, *Fire Service and the Law*, National Fire Protection Association, Quincy, MA, 1976.

Beck, A. D. Jr., and Hillmar, E. D., *A Practical Approach to Organization Development through MBO-Selected Readings*, Addison-Wesley Publishing Co., Reading, MA, 1972.

Bennis, W. G., *Changing Organizations*, McGraw-Hill, Inc., NY, 1966.

Berry, D. J., Fire Litigation Handbook, National Fire Protection Association, Quincy, MA, 1984.

Bormann, E. G., et al., *Interpersonal Communications in the Modern Organization*, Prentice-Hall Publishers, Englewood Cliffs, NJ, 1969.

Brown, F. G., and Murphy, T. P., eds., *Emerging Patterns in Urban Administration*, Lexington Books, Lexington, MA, 1970.

Bryan, J. L., and Picard, R. C., eds., *Managing Fire Services*, International City Management Association, Washington, DC, 1979.

Caplow, A., *How to Run Any Organization: A Manual of Practical Sociology*, Dryden Press, Hinsdale, IL, 1977.

Casey, James F., ed., *The Fire Chief's Handbook*, Fire Engineering Book Service, NY, 1978.

Chaiken, Jan M., et al., *Fire Department Deployment Analysis*, The Rand Fire Project, NY, 1979.

Child, J., *Organization: A Guide for Managers and Administrators*, Harper and Row, NY, 1977.

Clark, R. T., "Compulsory Arbitration in Public Employment," Public Employee Relations Library, No. 37, Public Personnel Association, Chicago, IL, 1963.

Coe, Charles K., *Cost Cutting Guide for Volunteer Fire Departments*, Institute of Government, University of Georgia, Athens, GA, 1982.

Coe, Charles K., et al., *Changing Local Government Fire Rates: The Costs and Benefits*, Institute of Government, University of Georgia, Athens, GA, 1979.

Coleman, R. J., *Management of Fire Service Operations*, Duxbury Press, North Scituate, MA, 1978.

Cutlip, S. M., and Center, A. H., *Effective Public Relations*, 14th ed., Prentice-Hall Publishers, Englewood Cliffs, NJ, 1971.

Czamanski, Daniel Z., *The Cost of Prevention Services: The Case of Fire Departments*, D. C. Heath & Company, Lexington, MA, 1975.

Davis, Paul O., *The Firefighter's Survival Manual—A Guide to Physical Fitness*, National Fire Protection Association, Quincy, MA, 1983.

Didactic Systems, *Management in the Fire Service*, National Fire Protection Association, Quincy, MA, 1977.

Drucker, P. F., *The Effective Executive*, Harper and Row Publishers, NY, 1967.

——— *Managing in Turbulent Times*, Harper and Row Publishers, NY, 1980.

Erven, Lawrence W., *Managing Fire Services Study Guide*, Davis Publishers, Santa Cruz, CA, 1980.

Favreau, Donald F., *Fire Service Management*, Reuben Donnelley Corp., NY, 1969.

Fire Department Personnel Management Handbook: Managing the Entry of Women and Minorities, Federal Emergency Management Agency, U.S. Fire Administration, Washington, DC, 1979.

Fire Labor Relations, International City Management Association, Washington, DC, 1983.

Fire Officer's Guide to NFPA 1021, National Fire Protection Association, Quincy, MA, 1983.

Fire Personnel Practices, International City Management Association, Washington, DC, 1983.

Fire Protection Administration for Small Communities and Fire Protection Districts, International Fire Service Training Association, Oklahoma State University, Stillwater, OK, 1980.

Fordyce, J. K., and Weil, Raymond, *Managing with People: A Manager's Handbook of Organization Development Methods*, Addison-Wesley Publishing Co., Reading, MA, 1971.

Fyffe, David E., and Rardin, Ronald L., "An Evaluation of Policy Related Research in Fire Protection Service Management," Georgia Institute of Technology, School of Industrial and Systems Engineering, Atlanta, GA, 1974.

Getz, Melvin, *The Economics of Urban Fire Departments*, Johns Hopkins University Press, Baltimore, MD, 1977.

Guide to Firefighter Qualifications Training Programs, National Fire Protection Association, Quincy, MA, 1978.

Hatry, Harry P., et al., "How Effective Are Your Community Services? Procedures for Monitoring the Effectiveness of Municipal Services," The Urban Institute and International City Management Association, Washington, DC, 1977.

Hatry, Harry, and Cotton, John, "Program Planning for State, County, City, George Washington University," State-Local Finances Project, Washington, DC, 1967.

Havlick, J. Robert, and Bodnar, Ernest B., "Centralized Municpal Communications Center," Management Information Service Report No. 262, International City Management Association, Washington, DC, Nov. 1975.

Heisel, W. D., et al., *Line-Staff Relationships in Employee Training*, International City Management Association, Chicago, IL, 1967.

Hickey, H. E., *A Minimum Statistical Data Base for Statewide Public Fire Department Management Information Systems*, Johns Hopkins University, Silver Springs, MD, June 1972.

Hickey, H. E., *Successful Public Relations*, National Fire Protection Association, Quincy, MA, 1974.

Hoetmer, Gerard J., "Police, Fire and Refuse Collection and Disposal Departments: Personnel, Compensation and Expenditures," International City Management Association, Washington, DC, Aug. 1982.

Hughes, C. L., *Goal Setting: Key to Individual and Organizational Effectiveness*, American Management Association, NY, 1973.

Jenaway, W. F., "Personnel Management in the Volunteer Fire Service," *Fire Command*, Vol. 45, No. 8, Aug. 1978.

Kimball, W. Y., *Fire Attack 1 and Fire Attack 2*, National Fire Protection Association, Quincy, MA, 1966.

Koontz, H. O., *Appraising Managers as Managers*, McGraw-Hill, Inc., NY, 1971.

Kraemer, K. L., *Policy Analysis in Local Government: A Systems Approach to Decision Making*, International City Management Association, Washington, DC, 1973.

Kraemer, Kenneth L., et al., *The Integrated Municipal Information System: The Use of the Computer in Local Government*, Praeger Publishers, NY, 1974.

La Patra, J. W., *Applying the Systems Approach to Urban Development*, Dowden, Hutchinson & Ross, Inc., Stroudsburg, PA, 1973.

Lee, Robert D., Jr., and Johnson, Ronald W., *Public Budgeting Systems*, 2nd ed., University Park Press, Baltimore, MD, 1977.

Managing People: Fire Service Personnel Strategies, National Fire Protection Association, Quincy, MA, 1984.

McGregor, Douglas, *The Professional Manager*, McGraw-Hill, Inc., NY, 1967.

Miller, E. C., *Objectives and Standards: An Approach to Planning and Control*, American Management Association, NY, 1966.

Morgan, J. S., *Practical Guide to Conference Leadership*, McGraw-Hill, Inc., NY, 1966.

National Fire Prevention and Control Administration, National Fire Safety and Research Office, *A Basic Guide for Fire Prevention and Control Master Planning*, National Fire Prevention and Control Administration, Washington, DC, 1977.

Nigro, F. A., *Management Employee Relations in the Public Service*, Public Personnel Association, Chicago, IL, 1969.

Odiorne, G. S., *Management by Objectives: A System of Managerial Leadership*, Pitman Publishing Corp., NY, 1965.

Page, James O., *Effective Company Command for Company Officers in the Professional Fire Service*, Borden Publishing Company, Alhambra, CA, 1973.

Parker, J. K., *Introduction to Systems Analysis, Management Information Service Reports*, No. 298, International City Management Association, Washington, DC, 1968.

Research Triangle Institute, et al., *Research Applied to National Needs*, prepared for the National Science Foundation, U.S. Government Printing Office, Washington, DC, 1977.

Rosenbauer, D. L., *Introduction to Fire Protection Law*, National Fire Protection Association, Quincy, MA, 1978.

Sagoria, Sam, ed., *Public Workers and Public Unions*, Prentice-Hall, Inc., Englewood Cliffs, NJ, 1972.

Sasso, C. O., *Coping with Public Employee Strikes: A Guide for Public Officials*, Public Personnel Association, Chicago, IL, 1970.

Smalley, James C., *Funding Sources for Fire Departments*, National Fire Protection Association, Quincy, MA, 1983.

Snook, Jack W., and Olsen, Dan C., *Recruiting, Training and Maintaining Volunteer Firefighters*, MDI, Lake Oswego, OR, 1982.

Starrett, P. F., *Mass Communications for Local Officials*, Arizona State University, Institute of Public Administration, Tempe, AZ, 1970.

Stern, James L., et al., *Final-Offer Arbitration: The Effects on Public Safety Employee Bargaining*, D. C. Heath & Company, Lexington, MA, 1975.

Stieber, Jack, *Public Employee Unionism: Structure, Growth, Policy*, Brookings Institution, Washington, DC, 1973.

Stoner, James A. F., *Management*, Prentice-Hall, Englewood Cliffs, NJ, 1978.

Terry, G. R., *Principles of Management*, 6th ed., Richard D. Irwin, Inc., Homewood, IL, 1972.

Tien, J. M., and Chiu, S. S., "Impact Analysis of Some Cost Reduction Options for a Metropolitan Fire Department," *Fire Technology*, Vol. 21, No. 1, Feb. 1985, pp. 5-21.

"Training and Education in the Fire Services," *Proceedings* of a Symposium conducted by the Committee on Fire Research, National Research Council, National Academy of Sciences, Washington, DC, April 8-9, 1970.

Urban Guide for Fire Prevention and Control Master Planning, National Fire Prevention and Control Administration, Washington, DC, 1977.

Vance, J. E., *Information Communication Handbook: Policies for Working with the Media for Public Officials, Citizens, Business and Community Groups*, published by the author, St. Paul, MN, 1973.

Vougioukles, Carol L., *Management of the Entry of Women in the Fire Service: A Directory of Resources*, International Association of Fire Fighters, Washington, DC, 1981.

Webb, Walter C., *The Volunteer Fire Service: A Management Perspective*, International City Management Association, Washington, DC, 1974.

Webber, Ross A., *Management: Basic Elements of Managing Organizations*, Richard D. Irwin, Inc., Homewood, IL, 1979.

Wellington, H. H., and Winter, R. K., Jr., *The Unions and the Cities*, Brookings Institution, Washington, DC, 1972.

FIRE DEPARTMENT OPERATIONS

Revised by J. Gordon Routley

The basic organization and orientation of almost all public fire departments is primarily directed toward fire suppression and emergency service delivery. The most realistic appraisals of such orientation, in almost every case, reaffirms the need for the fire department to respond to and control fires as they occur. In many cases this role has been extended to include the delivery of additional emergency services to deal with situations presenting an immediate threat to lives and property in the community. While the fire service may place an emphasis on fire prevention, public education, risk reduction and hazard abatement programs, its ability to respond and to control fires is an overriding operational priority.

Fire suppression is generally organized around a system of decentralized fire stations, providing the capability to respond quickly with personnel and equipment to control and extinguish fires. This organization may be staffed by career, part time or volunteer personnel; may reflect a variety of characteristics derived from local needs, structure, and tradition; and may be involved in the delivery of other emergency and nonemergency services. The ability to respond to the life safety and property protection needs of the local community is the common denominator in fire department emergency operations.

This chapter discusses various basic components of fire department operations: fire suppression, fire prevention, emergency medical services, and fire ignition sequence investigation. It illustrates the typical fire department organization and how the fire suppression operation relates to the total organization. Next, the chapter details communications, maintenance of equipment, and training; finally, hazardous materials are discussed.

Other relevant chapters in this HANDBOOK that specifically discuss fire department organization, fire prevention and code enforcement, and fire and arson investigation can be found elsewhere in this section. Additional information concerning related topics are located in Section 2, "Fire Loss Information;" and Section 6, "Hazardous Wastes and Materials."

Mr. Routley is Assistant to the Fire Chief, Phoenix, AZ.

ORGANIZATION FOR FIRE SUPPRESSION

The company is the basic organizational unit of the fire department involved in fire suppression. A company is a complement of personnel operating one or more pieces of apparatus under the supervision of a company officer. A number of different types of tactical companies are used by fire departments depending upon local needs. Engine and ladder companies are the most numerous, although a variety of others are often provided to perform specialized functions, including rescue companies and personnel squads, as well as companies operating special tactical or support function apparatus.

Engine Company: Engine companies normally comprise the largest number of companies within any fire department. The basic unit of apparatus is the pumper, which carries hose, nozzles, an on board water tank, and a pump. The engine company's basic role in tactical operations is to deliver water through hose lines to control fires, although a variety of additional functions and equipment generally is assigned to them. In most cases, at least one engine company is based at each fire station to respond quickly and initiate fire control operations at a fire scene. The engine company is considered the basic unit of a fire department and is supplemented by other types of companies.

Ladder Company: The basic ladder company apparatus is an aerial ladder or elevating platform device, which provides access at higher levels or directs elevated master streams on a fire. Ladder trucks also carry a complement of ground ladders and a selection of hand and power tools. Ladder companies perform a supporting role in fireground operations, which includes search and rescue, forcible entry, ventilation, salvage, overhaul, and the use of ladders to gain access and rescue persons above ground level.

Ladder companies are provided in relationship to the degree of urban development and the need for aerial apparatus. In a densely developed city, one ladder company may be provided for every two or three engine companies, while in a rural area no ladder companies may be established. Where there are no ladder companies, their supporting functions must be assumed by other types of companies.

Rescue Company: Many fire departments utilize separate rescue companies for both fire fighting and nonfire related rescue incidents. In many departments, rescue companies are primarily involved in the delivery of emergency medical services and physical rescues, such as extricating victims from vehicle accidents, removing injured persons from perilous locations, and assisting victims of industrial accidents.

Rescue company vehicles range from small vehicles designed for emergency medical service delivery to heavy squad vehicles carrying a large variety of tools and equipment.

In fire fighting operations, rescue companies are usually assigned primarily to search and rescue, along with responsibility for medical treatment. Additional duties often involve activities similar to a ladder company, particularly forcible entry, ventilation, and the use of power tools.

Squad Companies: Various types of squad companies are used by many fire departments. Squads usually supplement engine and ladder companies with additional personnel or highly specialized apparatus. The term "squad" is occasionally used interchangeably with rescue company.

Special Apparatus: In addition to the apparatus normally assigned to engine and ladder companies, fire departments often employ a selection of specialized vehicles including off road vehicles for brush fires, water tankers, hose wagons, foam pumpers, hazardous materials units, lighting trucks, breathing air supply trucks, and command vehicles. These may be organized as individual companies or they may be assigned to regular companies.

Fire apparatus may be purchased with a variety of options and configurations to suit the needs of a particular community or fire department. These options include aerial devices, water towers, and foam systems on engine company apparatus; ladder trucks with high volume pumps; remote control nozzles; or large electrical generators and air supply systems.

Personnel Requirements

Effective fireground operations are highly dependent upon the individual and collective capabilities of the companies responding to an incident. The apparatus assigned to a company basically defines the role which that company will be expected to assume in a fire fighting operation. In order to operate effectively, each company must have skilled and trained personnel who can perform basic functions in a standard manner. An untrained, unskilled, or understaffed crew will not perform effectively, regardless of the apparatus and equipment provided.

The crew size assigned to a company is a local decision and varies with the fire risk in a particular area, the number and types of companies that can respond to incidents, and the fiscal capabilities of the organization. While engine and ladder companies with a minimum of six crew members are generally considered desirable, many fire departments work very effectively with crews of four on each apparatus. In urban areas, crews with less than four personnel are often considered deficient, although many fire departments have geared their operations to smaller crews with reduced capabilities.

Command Personnel: Individual fire companies, particularly engine companies, routinely handle a large proportion of small fires and other incidents. Larger incidents involving several companies require overall command and coordination and begin to involve command officers in a supervisory role. In most cases, the officers providing routine administrative supervision at higher levels also are responsible for fireground command. In a small fire department, there may be only one or two chief officer levels over the company officer level; however, additional levels are provided in most larger departments.

Battalions/Districts: Battalion or district chiefs are generally responsible for supervising a number of fire companies assigned to a particular area of a city. Six to ten companies is a normal span of battalion control. Chief officers at this level normally command incidents at a first alarm level with up to six companies in operation.

Divisions: Division chiefs are normally responsible for the management of three or four battalions and are often the highest ranking on duty officer in a larger department. In addition to routine management responsibilities, a division chief would normally respond and assume command of a two alarm fire or major working incident.

Higher level command officers with overall management responsibilities for the entire fire department would normally respond and assume command responsibility for major incidents involving large numbers of companies.

FIREGROUND OPERATIONS

Fire suppression operations involve the utilization of a fire department's resources to combat a fire. The success of a fire fighting operation depends upon the ability of a fire department to effectively and efficiently use the available resources to protect lives and property. The fireground commander is responsible for managing the available personnel and equipment to achieve maximum results, depending upon the situation, associated conditions, and resources available. The same principles apply to other types of incidents in addition to fires which the department may be expected to handle.

A fireground commander is responsible for the direction and control of operations in every incident. From the arrival of the first unit at the fire scene there should be one identified person in command, with the responsibility and authority to direct all phases of the operation. The officer in charge of the first arriving company assumes the role of fireground commander until relieved by a chief officer. In complex situations, command may be transferred one or more times as higher ranking chief officers arrive and assume command.

The fireground commander is responsible for strategic decision making and the translation of strategic goals into tactical objectives and task assignments. (See Fig. 15-3A.) Intermediate levels of command are responsible to the fireground commander for geographical portions of the operation (sectors) or for the supervision of particular functions. These sector officers coordinate the operations of a group of companies under the overall command of the fireground commander.

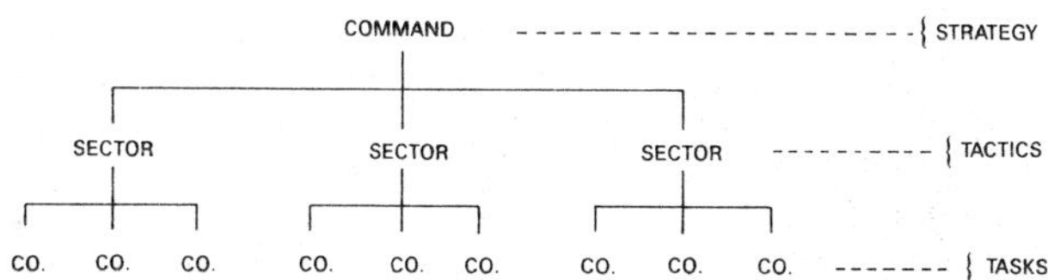

FIG. 15-3A. A common type of fireground organization. The fireground commander develops a strategic plan. The sector officers use tactical operations to achieve the goals of the plan, and assign specific tasks to individual fire companies within each sector.

A major incident may require a fairly complex command staff with a number of sector officers reporting to the fireground commander. Several officers may be assigned to directly support the fireground commander at a command post, providing information or further subdividing the span of control into manageable units.

Standard operating procedures are an important aspect of fireground command. Every fire department should have a set of procedures that outlines the basic operating principles to be employed for any situation, from the most simple to the most complex. The procedures should be flexible enough to allow fire fighters to react to different situations and incremental to permit adjustment to the scale of the incident. Standard operating procedures provide a menu of basic functions that can be employed as needed and establish consistent approaches to fire control situations.

The fireground commander must utilize strategy and tactics in the management of a fire suppression or similar incident.

Strategy: Strategy involves the development of a basic plan to most effectively deal with a situation. The plan must identify major goals and prioritize objectives for the tactical elements. Strategic decisions are based upon an evaluation of the situation, the risk potential, and the capabilities of the available resources.

The strategic options available to the fireground commander involve some very important, but basic decisions. The most fundamental decision is the choice between offensive and defensive modes of operation, based on the capability of available resources and the risk to personnel. For offensive operations, companies extend hose lines into the interior of an involved fire area and extinguish the fire where they find it. In defensive operations, heavy streams are applied from the exterior to confine or control a fire, conceding the loss of the involved area. Offensive and defensive modes of operation must not be mixed in the same place at the same time. The fireground commander must make a conscious decision to identify what can and what cannot feasibly be saved without undue risk to personnel.

Strategic decisions also identify the priorities for committing resources to various tactical positions and activities based on standard approaches and prevailing conditions. The generally accepted priorities for strategic decisions are: (1) rescue, (2) fire control, and (3) property conservation. The strategic plan identifies where and when the forces will attempt to control the fire and how their activities will be combined and prioritized.

Tactics: Tactics are the methods selected by the fireground commander to implement the strategic plan. The tactical objectives define specific functions that are assigned to groups of companies operating under sector officers. The achievement of these objectives contributes to the strategic goals and must be compatible with the overall strategic plan.

Tactical activities must be conducted in relation to three distinct phases or priorities. The first priority is to provide for the safety of the public by extending search and rescue efforts to all areas where potential victims may be located. The second priority is to control the fire and the third priority is to conserve property. Fireground tactics usually involve a coordinated mixture of tasks directed toward these objectives in the above order.

Tasks: The translation of tactical objectives to task assignments results in assignments to individual companies. A company generally is generally involved in the actions relating to one or two specific tasks at any particular time. These tasks must be coordinated and combined to achieve tactical objectives.

Tactical Functions

Several tactical operations may be employed at each fire incident, and several tactical operations may be carried out simultaneously during multicompany operations. Every company must be trained to carry out all basic operations and be prepared to contribute to tactical objectives when possible.

Search and Rescue: Rescue is the first and most important consideration at any fire incident and until completed, may preclude any fire control efforts. The fireground commander may have to initiate fire control activities to protect the rescue operation or to keep the fire away from potential victims. Rescue operations may be simple, requiring only one or two fire fighters, or may require resources beyond the capabilities of the entire first alarm assignment. All involved or threatened occupancies should be thoroughly searched for occupants without delay, with companies specifically assigned to this task. Every company must be prepared to perform search and rescue as a first priority. Rescue operations may be compounded by the time of the incident, the occupancy, and the height and construction of the structure. Rescue is the only acceptable reason for exposing fire fighters to otherwise unnecessary risks.

Exposure Protection: The second fire suppression priority is to control the fire. This begins with confinement of the fire to the property initially involved. The most basic responsibility of fire departments, with respect to property, is to protect the community from large-loss fires. Failure to adequately protect exposed structures may allow a fire to extend beyond the building of fire origin. The problem of exposure protection may be compounded by closely spaced buildings, combustible construction, the type of occupancy, the lack of fire department access to the fire, and the lack of fire department resources. Exposure protection is a vital and necessary tactical consideration, and should be the major objective in defensive fire control situations.

Confinement: The confinement of a fire to its area of origin is often a complex problem. Fire control is achieved when the fire is successfully confined to a manageable area. All avenues of possible fire travel must be secured. The concept of surrounding the fire (over, under, and around) is necessary for successful confinement. Additional factors that influence the success or failure of confinement operations are the type of fuel involved, the location of the fire, building construction features, the presence of built-in fire suppression systems, and the availability of fire department resources.

Extinguishment: Offensive fire control strategies are aimed at controlling and extinguishing the fire where encountered by attack forces. Successful offensive operations are regulated by the type of fuel involved, the location of the fire and the degree of involvement, and the ability of fire control forces to apply sufficient extinguishing agents directly on the fire. In some instances, the use of special extinguishing agents may be required.

In defensive operations, final extinguishment may be achieved only when the fire burns down to a size which can be extinguished by the fire department. Defensive tactics depend upon the department's capability to apply large volumes of water or other agents to confine and eventually extinguish the fire.

Ventilation: Ventilation operations are the planned and systematic removal of heat, smoke, and fire gases from the structure. In some cases, it may be necessary to initiate ventilation with rescue in order to protect occupants from combustion products and heat and to provide visibility and tenability during rescue operations. Ventilation is also necessary during confinement and extinguishment to aid in locating the fire and to provide safer working conditions for fire suppression personnel, as well as to reduce overall damage to structure contents.

Property Conservation: Salvage operations are conducted by fire suppression personnel to conserve property by minimizing damage to the structure and contents due to heat, smoke, and water. This involves, in part, covering contents and removing excess water. Salvage is an integral part of tactical operations and should commence as soon as possible to prevent additional damage to the structure and its contents.

Overhaul: Overhaul operations are required to completely extinguish the fire, place the structure in a safe condition, and aid in determining the fire ignition sequence. Overhaul may involve only a few personnel for a short period of time, or large numbers of personnel over an extended period. Extensive overhaul may require the use of special pieces of equipment beyond those normally provided by the fire department. It is important that extensive overhaul not be commenced prior to a thorough investigation to determine the cause of the fire. Once the investigation is complete, overhaul should continue to ensure that the premises are left as safe as possible, and that all fires are extinguished.

Fire Fighting Safety

Fire fighting is considered to be the most dangerous occupation in North America. Fire fighter deaths and injuries occur at a rate exceeding that for all other categorized labor activities. While inherently dangerous, it is possible and essential to improve the safety of fire department operations. Safety should be a primary concern for all personnel, involving everyone from the fireground commander to the individual fire fighters.

All fire fighters must be provided with complete turnout clothing and protective equipment, including self-contained breathing apparatus, which comply with the appropriate NFPA standards. Regulations must be implemented and enforced to require use of the proper protective equipment in every situation. All fire departments should implement a complete safety program, including a physical fitness program for personnel, health monitoring and medical support, the provision and use of protective equipment, ongoing training, and the management of situations which present an unusual hazard to personnel. A comprehensive breathing apparatus program for all fire incidents is an essential element of a successful fire department safety program.

Safety must be a primary concern of the fireground commander at all times. In complex situations, safety officers should be assigned specifically to monitor conditions and advise the fireground commander and sector officers on hazardous conditions and safety concerns. Sector officers must be particularly concerned with the safety of personnel under their direction. Every tactical option must be evaluated in terms of its risk to fire fighters.

Transportation Incidents

All public fire departments need to be as well prepared for transportation incidents as they are for structure fires. These incidents may range from automobile fires to aircraft accidents and train derailments, and may occur anywhere and at any time. Preincident planning is essential for large scale incidents such as those involving mass transit systems, airports, and railways.

Fires and accidents that occur on bridges, freeways, railroads, tunnels, mass transit systems, and other locations, must be a part of the planning process for every fire department since these incidents combine difficult access, limited water supplies, and special requirements for rescue and fire suppression. These situations may require specialized apparatus, equipment, training, and procedures so personnel can deal effectively with the variety of problems that can occur. Incidents ranging from a gasoline tanker burning on a bridge to a subway train collision and fire in an underground tunnel must be considered, and effective plans must be in place to deal with them. Some of the most disastrous fire incidents have occurred where unanticipated events and circumstances exceeded the capabilities of emergency response organizations.

Large quantities of flammable liquids and other hazardous materials are constantly in transit by road and rail. Any incident involving trucks or freight trains should be considered as a potential hazardous materials incident. Personnel must be familiar with labeling and placarding systems for hazardous materials, the use of cargo manifests and waybills to identify cargos, and resources such as CHEMTREC, a 24 hour emergency service provided by the Chemical Manufacturers Association through their toll free telephone number (1-800-424-9300).

Air: Aircraft crashes may occur on or off airfields and often present a major fire and rescue problem. For effective rescue, fire department action must be rapid and effective. Many public fire departments also provide crash/rescue protection at airports. There are several good sources of information on dealing with aircraft accidents; consult the Bibliography at the end of this chapter.

Marine: Waterborne vessels range from small pleasure boats to barges, cargo carriers, warships, cruise ships, and supertankers. All vessel categories present differing problems that require a thorough knowledge of ship construction, design, stability, and the special techniques of shipboard fire fighting.

In the case of cargo ships, the nature of the cargo may vary from inert materials to general combustibles and highly flammable and explosive materials. As with any other transportation fire, restraint is required until positive identification is made of the materials involved. Materials can be identified from the ship's manifest, which is usually available from the ship's master.

Fire departments with port facilities should be trained in fighting shipboard and waterfront fires. Marine fires may require the use of fireboats and often cannot be handled in the same manner as shore fires.

Some basic factors to consider in fighting marine fires are: stability of the vessel, which may be affected by application of water; ventilation; water pollution and the possible obstruction of waterways or port facilities should the vessel capsize or sink; ability to deal with fire away from wharfside; maximum use of ship's built-in extinguishing systems; and fire spread through vessel bulkheads. Another important factor is the need for the fire department to interact with port authorities, the U.S. Coast Guard, and the master of the vessel concerned. The often unclear divisions of authority and responsibility should be clarified before an incident occurs.

Pipelines: Underground pipelines carry highly flammable products at high pressures through many populated areas. The risk involved in a rupture of one of these lines may be extremely critical. Fire departments should be aware of the locations of underground pipelines, the products carried, and how to shut down the pipelines in an emergency.

Mutual Aid and Major Emergencies

The possibility of fire and disaster problems that exceed the capacity of the local fire fighting forces must always be considered. For this reason, most fire departments traditionally have rendered mutual assistance to other departments in times of need. Mutual aid plans establish procedures for requesting and dispatching help between fire departments so that each party will know what is expected. Mutual aid plans may include the following functions: (1) immediate joint response of several fire departments to high risk properties, (2) joint response to alarms adjacent to the boundaries between fire department areas (automatic aid), (3) coverage of vacated territories by outside departments when the resources of the local department are engaged, (4) provision of additional units to assist at major fires that may be too large for the local department to handle, and (5) provision of specialized types of fire fighting equipment not available locally in adequate quantity for the particular incident.

Mutual aid plans also should include provisions for standard operating procedures, interdepartmental communications, common terminology, maps, adaptors, and other considerations that directly affect the department's ability to operate effectively. Command responsibility, jurisdictional questions, insurance coverage, and legal constraints should be covered in written agreements supported by enabling legislation to properly establish mutual aid systems for the particpating departments.

Some jurisdictions have extended the mutual aid concept to multijurisdictional agreements in which fire department resources are pooled or merged into an integrated system, with standardized training procedures and communications. These networks may include shared facilities, joint purchase of specialized apparatus and equipment, and a coordinated approach to long range planning.

True mutual aid is a relationship in which each member is prepared to render assistance to the parties of the agreement. Many larger fire departments are very willing to provide assistance to smaller jurisdictions when major incidents occur that obviously exceed local capabilities. In many places, there are programs of outside aid whereby communities or individual properties known to be deficient in fire fighting resources contract in advance to pay another jurisdiction for certain fire fighting assistance. In some instances, the contract covers basic first alarm response; in other cases additional assistance is provided for fighting major fires. In the cases of both mutual aid and outside aid, definite agreements should be made in advance according to the legal requirements governing fire department operations outside normal jurisdictional areas.

It should be recognized that some existing mutual aid agreements contain deficiencies. Each local department may have its own operating methods and types of equipment, so that maximum coordination may be hampered. The parties to the plan may render assistance only to the extent that they feel they can do so without seriously reducing the local protection, although the better plans provide effective coverage for all districts dispatching apparatus. These weaknesses should be identified and corrected in advance. State fire training programs, large scale exercises, and fire officer command schools and conferences will help to resolve these problems.

Experience with natural disasters and large scale incidents has focused attention upon the importance of plans and organizational procedures for systematically mobilizing fire forces for large scale operations. Some states and provinces have established large area disaster plans involving all emergency services, coordinated under standard procedures such as the Integrated Command System (ICS). For successful disaster operations, it is imperative that such plans integrate with the normal organizational and command procedures used by fire departments. These large scale mutual aid networks form a natural basis for smaller scale, more routine mutual assistance plans.

NONEMERGENCY ACTIVITIES

Fire suppression personnel perform several standard activities while they are not involved in tactical operations in order to support effective fireground performances. These nonemergency activities include prefire planning, fire prevention activities, and training.

Prefire Planning: Prefire planning should involve all fire suppression personnel on a continuing basis. Prefire plans should complement standard operating procedures by increasing the fireground commander's knowledge of complex occupancies and special risks. Plans should be sufficiently flexible to allow for varying conditions and should utilize the framework of standard operating procedures. Plans that are excessively rigid or complex may handicap more than benefit an operation.

The following steps are part of the prefire planning process:

1. Information gathering—Pertinent information such as building construction features, occupancy, exposures, utility disconnects, fire hydrant locations, water main sizes, etc. that might significantly affect fire fighting operations must be collected at the selected site.
2. Information analysis—The information gathered must be analyzed for what is pertinent and vital to fire suppression operations, so a plan can be formulated into a format that can be used on the fireground. A prefire plan usually includes a site plan or map; floor plans and diagrams identifying pertinent features, hazards and fire control equipment; and additional text outlining special problems, specific tactics, hazardous contents, and information on parties responsible for specific areas.
3. Information dissemination—Prefire plans should be assembled and distributed in a standard format to support their use on the fireground. This may include plans maintained on paper and carried in fire apparatus, microfilmed plans used with viewers in command vehicles, or information stored in computer systems accessed by mobile data terminals.
4. Class review and drill—Any company that might be involved at the prefire plan location should review the plan on a regular schedule. If possible, periodic drills and familiarization tours with all companies involved should be scheduled on the property.

Prefire plans are desirable for all target hazards, special risks, and large complexes. Standard operating procedures are usually sufficient for single family dwellings and smaller occupancies. Prefire planning is a necessary adjunct to tactical operations and should aid in efficient operations, reduced fire losses, and an optimum level of fire protection.

Fire Prevention: The participation of fire suppression personnel in fire prevention and public education activities should be as important as their participation in fire suppression operations. Fire department resources are generally deployed throughout the protected area to provide rapid emergency response—those companies usually spend only a small percentage of their time actually fighting fires. Fire departments committed to a comprehensive fire prevention and inspection program use fire suppression personnel on a regular basis for routine inspections within their first due response areas. They also utilize staff support from full time fire prevention personnel for follow up inspections, legal enforcement actions, and special technical inspections. The involvement of fire suppression personnel in fire prevention and public education activities is a good utilization of resources; it decreases the risk of fire and familiarizes personnel with occupancies in their response areas.

Training: It is essential that a portion of a fire suppression forces' duty time be devoted to training. Training activities provide the opportunity to develop the skills and knowledge necessary to implement tactical operations. The elements of a good training program are discussed in detail later in this chapter.

Risk Analysis

A detailed knowledge of the potential fire problem in any area protected is not only necessary in planning resources, responses, and locating facilities, but also for dealing with incidents when they occur. The community's fire suppression capability should be directly related to the level of fire risk in the community. This information is also critical when a fire officer is called upon to make rapid, often irrevocable decisions affecting the commitment of personnel and apparatus at the fire scene. The fire officer needs an appreciation of the way fire can be expected to behave in any given situation. The following factors should be considered in determining risk potential:

Life Risk: How many occupants are present? Are the occupants likely to be active, physically capable and able to help themselves, or are they likely to be infirm, aged, bedridden, handicapped, or otherwise unable to take any action towards self rescue? Are the occupants likely to be asleep and, therefore, more prone to being trapped? Does the building provide early warning, fire suppression systems and adequate exits, or are the occupants highly dependent upon the fire department to be found and rescued?

Contents: Does the building have a high fire load? Are the contents highly flammable, producing rapid fire spread and high heat release rates? Are they liable to explode, produce dense smoke, or give off highly toxic products of combustion? Are the contents extremely valuable or highly susceptible to heat, smoke, or water damage?

Construction: Is the construction fire resistive and likely to maintain structural integrity, or is it unprotected and likely to fail early in the fire suppression operation? Are the interior finishes liable to cause rapid flame spread and extension of fire? Is the building compartmented with fire walls and doors, or are there openings in floors, walls, ducts, and shafts that will allow a fire to spread? Is the building old, or has it been remodeled with possible resultant weaknesses? Are there very large structures or groups of buildings that present a situation in which a developed fire is virtually impossible to control? Are there unsprinklered high rise buildings that present an excessive demand for personnel and equipment to support aboveground fire suppression operations?

Built-in Protection: Are there sprinklers which might restrict growth and spread of fire? Are there fire doors, compartments, or other fire protection arrangements which would help to confine the fire?

Time: Are there occupants who are alert and would discover a fire soon after its inception, are occupants asleep, or is there no one present to call the fire department? If so, is there a smoke or fire detection system? Is the system connected to the fire department or a central station system? Is the building at risk some distance from a fire station? Are there railroad crossings, heavy traffic patterns, or other restrictions on speed of fire department response? Are there any other factors that would delay discovery, alarm, or response?

Suppression Resources: What is the availability of water or other extinguishing agents? Is special apparatus required? Will there be problems regarding entry or access? Will there be rescue problems? Will there be a need for long duration breathing apparatus? Are there any other factors affecting the fire department response or action? Are adequate fire suppression resources available, including mutual aid and reserve personnel and equipment? Are additional resources, beyond the capability of the fire department, needed?

In summary, to ascertain risk potential it is essential to gain an in depth understanding of the dynamics and behavior of fire, building construction, human factors, and the properties and hazards of materials.

Assessment of fire hazards in the area, together with the technical knowledge of fire, provides an understanding of the risk potential which can take some of the uncertainties out of fire suppression activities and fireground decision making. In assessing risk potential, whether for prefire planning or incidents, all of the factors in Table 15-3A should be considered individually and collectively.

This analytical approach can be expanded and used for almost any risk and for planning resources, response levels, and preparing action plans. It also can be used to develop training exercises with group reaction techniques to prepare officers and officer candidates to think in terms of a systematic assessment for fireground decision making.

In addition to structural fires, fire departments must be prepared to face a wide variety of incidents. These include fires involving outside storage, brush and wildlands, transportation systems, and industrial installations. Most fire departments also are responsible for hazardous materials incidents and rescue of persons trapped by floods, vehicle accidents, building collapses, cave ins, and numerous other situations. Many fire departments also have assumed a major role in delivery of emergency medical services, which often involves more emergency than fire suppression activity.

TABLE 15-3A. Interrelated Factors Affecting Risk Potential

1. Life Risk	Handicapped, sleeping, concentrated, etc.
2. Contents	Readily combustible, extremely hazardous, quantities, disposition, etc.
3. Construction	Type, age, condition, compartmentation, height, extent, exposures, interior finishes, etc.
4. Built in Protection	Sprinklers, fire doors and walls, smoke detection, etc.
5. Time	Delayed alarms, long distances, traffic, etc.
6. Suppression Resources	Water, apparatus, equipment, manpower, etc.

The fire department has become an all purpose emergency service and requires a diversified array of equipment and training to fulfill this diversified mission. Many of the standard procedures and much of the equipment used for structural fire fighting is transferable to these additional roles, but additional specialized equipment and methods often are needed to deal with the identifiable risks and hazards in a community. Preparedness for these situations is an important element of fire department planning.

FIRE PREVENTION

Fire prevention encompasses all the means used by fire departments to decrease the incidence of uncontrolled fire. The fire prevention methods employed by fire department personnel involve engineering, education, and enforcement. Good engineering practices can do much to provide built-in safeguards to help prevent fires from starting and to limit the spread of fire should it occur. Education can instruct and inform groups and individuals of the dangers of fire and its possible effects, and of appropriate behaviors and reactions when faced with a fire situation. Enforcement is the legal means of correcting deficiencies that pose a threat to life and property. In addition, fire investigation aids fire prevention efforts by indicating problem areas that may require additional educational efforts or legislation to correct deficiencies.

In the United States, fire prevention has historically taken a secondary role to fire suppression as the principal activity of the public fire department. In 1973, fire prevention received its greatest endorsement when the National Commission of Fire Prevention and Control reported on the fire problem in America. Throughout the report, top priority was given to the necessity for increased fire prevention activities to reduce fire loss. The results of such efforts are being more clearly defined every year as progressive departments initiate more effective fire prevention efforts, in addition to maintaining their fire fighting forces.

FIRE IGNITION SEQUENCE INVESTIGATION AND REPORTING

Fire ignition sequence investigation and fire reporting are important functions in the operation of a fire department. While it has long been recognized that fire prevention is one of the major concerns of all fire departments, it has not been as well recognized that comprehensive investigation of fires and all the factors influencing or contributing to the fire ignition sequence is the very foundation on which fire prevention is built. Without the extensive and detailed information obtained from these investigations, it is impossible to develop effective regulatory codes, standards, inspection and suppression procedures, and similar actions designed to prevent or control fire. This investigative process should examine fire ignition sequence and causative factors, occupancy, area of origin, time of day, climatic conditions, and other factors that directly relate to fire cause and origin. This data should be compiled on an individual case basis and analyzed for community trends on a geographic, time of day, seasonal, or circumstantial basis.

A duty equally important to comprehensive investigation is accurate reporting of such investigations. This has seldom been recognized as an essential part of fire protection, with a result that fire departments and fire protection interests in the past, have often depended upon unreliable projections and reports to identify fire problems and make judgments involving building and fire prevention codes and standards, special fire protection problems, and corrective approaches.

There is increasing interest and recognition of the need for improvements in both fire ignition sequence investigating and fire reporting by all public agencies, from the national level down to the smallest volunteer department. There are informational guidelines and standards available to assist in the development of fire investigating techniques and reporting systems.

NFPA 901, *Uniform Coding for Fire Protection*, establishes uniform language, methods, and procedures for fire reporting and coding. In this system, all information can be coded for electronic or manual data processing using the NFPA Uniform Fire Incident Reporting System (UFIRS) or the Federal Government adaptation known as the National Fire Incident Reporting System (NFIRS).

The use of electronic data processing is essential if meaningful and readily retrievable fire experience data is to be collected. Data processing capability allows for the input and analysis of great volumes of individual reports relating to the factors and combinations of factors influencing the ignition sequence and communication of fires. This makes it possible to effectively find, evaluate, and compile fire information on a local, regional, or national basis. The information can be used to identify problems and evaluate possible solutions that may have a major effect on fire suppression procedures, structural and fire protection codes, public education programs, fire protection equipment, and all other fire protection areas.

COMMUNICATIONS

An effective communications system is a key factor in fire department operations. The communications system is responsible for:

1. Receiving notification of emergencies from the public.
2. Alerting and dispatching personnel and equipment.
3. Coordinating the activities of units engaged in emergency incidents.
4. Providing nonemergency communications for coordination of fire department units.

Receiving Alarms

Most alarms are received from the public via the public telephone system or municipal alarm systems. In many areas combined emergency service answering centers receive telephone calls for police, fire, and emergency medical services through a "911" emergency number, while in other jurisdictions the fire department receives calls at a separate communications office. Advanced "911" telephone systems may include the capability to automatically identify the location where the call originates for the dispatcher. Small departments with low activity levels may receive alarms through regional or countywide dispatch systems, or through a variety of systems that provide 24 hour coverage.

As a result of high false alarm rates, many cities have removed municipal alarm systems or converted from telegraph to systems providing voice contact between the dispatcher and the caller. The vast majority of fire alarms in most areas are received from private and public telephones. Municipal fire alarm systems are covered in NFPA 1221, *Standard for the Installation, Maintenance and Use of Public Fire Service Communication Systems.*

The communications center should provide recording equipment for all telephone lines and radio channels to provide both immediate playback capability (to verify information) and a complete record of all activity. In addition to telephones and municipal alarm systems, many fire departments receive automatic fire alarms from systems installed in buildings connected directly to the fire department or through a private alarm company.

Dispatch Procedures

The second phase of fire department communications is the actual alerting and dispatching of personnel and equipment to an incident. The complexity of this phase varies greatly with the size and population of the area served and the number of units under the control of the dispatch center.

The dispatcher must identify the units to respond to an incident based on the geographic location and the type of situation indicated. The selection criteria for response units must be determined in advance, based on distance from or response time to the reported location. The type and number of units due to respond to each type of incident is also determined in advance, based on risk criteria and different unit capabilities. The dispatcher must know the status of each unit in the system and have sufficient information to dispatch substitute units if initial response units are engaged at another incident or otherwise unavailable. This information is also needed if the dispatched units request additional assistance or a multiple alarm. Many larger fire departments and regional systems have installed computer aided dispatch systems (CADS) that combine the complex set of information required to manage these functions with the speed necessary for emergency dispatch. Smaller systems generally rely on printed "running cards" or policies to provide the dispatcher with response information for each zone.

The response units may actually be alerted by radio, microwave, telephone, telegraph, and vocal circuits and in some cases, outside sirens or horns for volunteer or on call personnel. Most departments use a voice message from the dispatcher to the responding unit, which may be carried over the radio or wired circuits to the fire station. There should be at least two separate means of communication between the alarm center and each fire station for backup in case of equipment failure. In addition to voice messages some fire departments use printers to provide a written dispatch message, or telegraph systems as a backup to the vocal message. Units out of quarters are normally alerted by radio if available to respond.

Radio Communications

Units responding to or engaged at incidents should have frequent radio contact with other units and with the alarm center. For larger departments, this may require several radio channels to provide sufficient communications capacity. The alarm center must be able to contact responding units to provide additional information or directions while en route to an incident, while units at the scene must be able to request or return additional resources. The fireground commander should be able to contact the dispatcher at all times, providing progress reports, advising on the need for assistance, or releasing units from the scene.

For effective coordination, units engaged at the scene should have radio contact with the fireground commander and with each other. Each company officer and sector officer should be provided with a portable two way radio on a fireground channel.

Units released from an incident by the fireground commander should advise the dispatcher when they are available to respond to further incidents.

The radio systems used for emergency activity also are used to maintain radio communications with units which are engaged in activities out of quarters, but available for dispatch. When viewed as a highly decentralized and mobile organization, the fire department's need for an effective radio communications system is evident. The system must reach units in every geographical area of the jurisdiction and provide sufficient capacity and channels to handle the volume of communications generated by major emergency situations. Regional mutual aid channels to coordinate activities involving units from multiple jurisdictions are essential for these situations.

Staffing

Fire dispatching requires specially trained operators who are familiar with fire department operations and equipment. In very small communities, the fire telephone number may be arranged to ring in a number of locations where appropriate action can be taken to dispatch fire apparatus. This situation may exist in volunteer fire departments without regular watch service at a fire station; dispatching may be handled by the police department, a town office, or other locations that provide 24 hour coverage. Increasingly, small fire departments are being served by regional fire communications centers, which can be properly equipped and staffed.

Larger communities and regional communications centers require trained, capable personnel to be on duty at all times. The number of personnel on duty depends upon the workload, but must include sufficient operators to handle the volume of communications required by busy activity periods and working incidents. Personnel and capabilities should be added to the communications center when needed, due to major emergencies and high activity situations. Off duty personnel or additional trained dispatchers can be called in for such incidents.

Maintenance and repair personnel for the system should be on duty or on call at all times. All equipment must be properly maintained and tested for maximum reliability, and backup equipment or systems should be provided in the event of any key component failure. Every fire department should have a backup facility where basic communications capability can be transferred in the event of a critical situation that makes the primary center unusable.

Information Retrieval and Storage

An important part of communications center operation is to have immediately available any information that may be needed in dispatching assistance to fires and emergencies. The data files must include a complete geofile index of all streets, intersections, and related numbering systems in the area. The geofile should also include all target hazards such as schools, hospitals, and major buildings.

The system must allow the dispatcher to determine the appropriate response zone and map location for any reported emergency so the proper units can be dispatched. The continuing maintenance and updating of these files is an important ongoing function, whether the information is maintained in hard copy form (cards or index systems) or in a computer system. The same approach is necessary to keep prefire plan information and maps up to date in the communications center, as well as on responding fire apparatus.

Additional information that should be included in the communications center data files includes telephone numbers for responsible parties for major buildings and businesses, utility companies, and other agencies to contact during fires, as well as a variety of other individuals who may have to be notified or requested to respond in the event of certain incidents.

Electronic data processing equipment speeds information retrieval, and makes it possible to program and retrieve considerably more data than is quickly available by more conventional means. Computer aided dispatch systems automate many or all of the dispatch and information management functions, including the provision of specific data on individual occupancies. Information available on a particular location may include data on building arrangements, construction, hazards, code inspection access, water supplies, and even prior fire experience. The system may provide for direct transmission of this information to terminals in responding vehicles or storage in small retrieval systems carried in command vehicles.

Station/Personnel Alerting Systems

Some fire departments maintain a 24 hour watch at a console or control room in each fire station. The person on watch is responsible for receiving dispatch messages and controlling alerting devices within the station. Controls for lights, electrically operated doors, traffic control devices and similar equipment are usually located at the watch desk. Instead of maintaining a watch in each station, other fire departments control some or all of these devices from the communications center . Radio signals or hard wired circuits may be used to control these devices.

Where stations are staffed entirely by volunteer or on call personnel, the communications center may alert personnel by activating sirens or horns, or by tone activation of radio pagers carried by personnel or maintained in their homes.

MAINTENANCE OF FIRE DEPARTMENT EQUIPMENT

The maintenance of fire department apparatus and equipment at peak operating efficiency is a primary fire department responsibility. The safety of the public and fire department personnel, as well as fire fighting efficiency, depend considerably upon the effectiveness of the maintenance program. When properly cared for, fire apparatus is expected to give years of reliable service. Each fire department needs a clearly established policy concerning maintenance responsibilities and routines to avoid possible equipment failures during operation.

Every fire department should assign an officer responsibility for apparatus and equipment. (This may be a part time job in small fire departments.) This officer should be responsible for the condition, status, and department needs for both apparatus and equipment, and should assist in preparing specifications in accordance with accepted standards for the procurement of new items.

Complete records should be maintained for every piece of apparatus and equipment, showing all repairs and service performed. Service tests of pumping engines should be conducted annually with test records maintained and compared with original equipment performance. Service tests and structural examinations also should be conducted regularly on ground ladders, aerial ladders, elevating platforms and other equipment. Tests should be conducted in accordance with NFPA 1931, *Standard on Design, and Design Verification Tests for Fire Department Ground Ladders*, and NFPA 1904, *Standard for Testing Fire Department Aerial Ladders and Elevating Platforms*.

Some fire departments maintain service vehicles and employ mechanics to check and service apparatus in quarters and respond to fires to oversee mechanical equipment operation and perform emergency repairs. Other departments have maintenance personnel on 24 hour call. Some small fire departments without full time maintenance personnel, have one or more fire fighters who fill in as part time mechanics in addition to their fire fighter duties.

A preventive maintenance program should be set up for fire apparatus and equipment, with all minor repairs made promptly. Since fire apparatus is subject to frequent starts and relatively high speed operation with cold engines, this approach becomes critical to maintain dependable apparatus.

Much routine apparatus maintenance is the responsibility of the assigned apparatus operators under the direction of company officers. This maintenance should be carried out on a daily and weekly basis following a set checklist. Among the items for which apparatus operators are responsible are: (1) keeping the fuel tank full, (2) checking the engine oil level, (3) checking batteries, (4) maintaining tire pressure at the specified level, (5) maintaining the water level in the radiator, (6) keeping the apparatus water tanks full, and (7) checking all lights. All items needing repair should be reported immediately and the work completed without avoidable delay. Frequently-used parts should be in stock or readily available. The maintenance or service manuals delivered with each piece of apparatus should be on file and followed.

Apparatus taken out of service for repairs or maintenance should be replaced by reliable, properly equipped reserve apparatus. Every fire department should have at least one reserve pumping engine in good useable condition. Some fire departments have a reserve fleet ranging from 25 to 33 percent of their first line equipment, including reserve ladder trucks, rescue trucks, and ambulances. This reserve apparatus also may be placed in service by off shift personnel for major fires or during periods of peak activity.

Minor equipment and power tools also need routine maintenance and servicing. All equipment items should be inspected periodically and repaired or replaced as needed.

Every fire department should have a repair and servicing facility commensurate with its apparatus and equipment needs, including fire hose. Due to the risk of fire, a repair shop area should be separated from the operating areas of a fire station, if it is not located in a separate building.

The fire department repair shop should have necessary work benches, ample power outlets, and jacks and hoists to handle heavy equipment. The shop should meet all fire protection standards for a repair garage, and areas used for painting should be properly protected. A shop facility should include an equipment repair section and a section or area where fire hose and couplings can be properly repaired. In some cases, pump testing equipment is provided at the shop; in others, these facilities are at the fire department training center.

In some communities, apparatus repair is assigned to a municipal service garage. This arrangement may be satisfactory if a regular fire apparatus repair and service section is provided and staffed by persons who are qualified for the specialized nature of this work and available on a 24 hour basis. Often, general mechanics do not have the necessary skills to carry out a proper fire department maintenance program, and fire apparatus may receive a low priority in a municipal repair garage. In some cases, it is advisable to send apparatus to the manufacturer for major repairs or rebuilding.

TRAINING

The level of performance demonstrated by a fire department is usually a good indication of the type, quantity, and quality of the training provided. In addition to departmental training programs, most states provide some type of fire service training that, depending upon the state, ranges from basic fire fighter training to fire department management programs.

Large city and county fire departments usually have their own training facilities and full time training staffs. In these larger fire departments, training takes several forms, starting with continuing in-service training conducted daily by company officers. This may be supplemented by scheduled programs on various subjects at a fire training center, and specialized training for drivers, apparatus operators, and other personnel. Additional officer training may be conducted monthly or annually at a fire college or a staff and command school. New fire fighters normally undergo an intensive recruit training class before assignment to regular duty. Recruit training normally requires two to three months of full time participation.

Training Objectives

The goal of any fire department training program should be the provision of the best possible training so that each person within the department will operate at acceptable performance levels relative to their rank and assignment. Ideally, training courses should have their own instructional objectives, a list of enabling objectives showing how the instructional objectives can be reached, and stated methods that explain how anticipated or desired behavioral changes can be measured.

State Fire Training Programs

Many state fire training programs operate under the umbrella of an educational agency within the state (usually vocational or occupational education) or under the auspices of a state university or college. In other cases, the director of fire training operates under the state fire marshal's office. In Delaware, for example, the state fire training program operates under a commission appointed by, and reporting directly to, the governor.

Some state fire training directors have no full time instructors, while others have a small cadre of full time senior instructors or supervisors who, in addition to their teaching functions, may be responsible for: (1) assigning and supervising part time field instructors, (2) developing instructional outlines, (3) developing and producing audio-visual training aids and props, and (4) monitoring the required records associated with a state program. State programs also may provide training for recruit fire fighters for small departments. Additionally, state programs provide training for specialists within the service, conduct seminars on material of current interest and importance, and serve as a valuable informational reference.

National Fire Academy

The National Fire Academy (NFA) is a center for fire protection related education, information, and expertise, aimed at improving the professional development of fire service personnel and allied professionals. The Academy, which is operated by the Federal Emergency Management Agency (FEMA), offers a wide variety of courses in the general areas of incident management, fire technology, and fire prevention. These courses are delivered through an on campus program and an extensive outreach program.

The resident student program consists of courses conducted at the National Emergency Training Center (NETC) in Emmitsburg, MD. Students, mainly from the fire service, attend these courses at minimal cost through a stipend program, which covers the costs of tuition, travel, and accommodations in campus dormitories. Most of the individual courses last for two weeks and are linked so students can either advance through different levels in a particular subject area or develop management expertise in a variety of subjects.

The outreach program provides for the delivery of short term (generally two days) courses, to students in locations all over the U.S. The delivery of these courses is coordinated through state and local training agencies and is concentrated on weekends to allow maximum participation from both career and volunteer fire service personnel. These courses are designed to be transferred to state and local delivery systems for wider distribution through an increasing network of instructors.

A full time professional faculty and staff at the Academy are supplemented by a large network of adjunct faculty members who teach courses in their particular areas of expertise. Both resident and adjunct faculty include individuals from the fire service and other backgrounds. The main campus facility provides classroom space, laboratories, educational resources, and living quarters for several hundred students.

The Open Learning Fire Service Program is an additional aspect of the National Fire Academy educational system. Through a network of eight accredited colleges and universities this program coordinates independent study programs for fire service personnel. Students may earn degrees through these educational institutions by combining a series of upper division fire related courses with credits in related subject areas. The open learning concept provides a high degree of flexibility for individual students with correspondence courses and seminars as well as opportunities to transfer credits earned at other educational institutions.

Fire Service Organization Training Activities

Several fire service organizations sponsor conferences and seminars as a part of an annual meeting, or separate seminars on materials of current importance. Examples of these organizations include the National Fire Protection Association, (NFPA) the International Society of Fire Service Instructors (ISFSO), the International Association of Fire Chiefs (IAFC), and local and state fire service organizations. Numerous training manuals are published by the International Fire Service Training Association (IFSTA).

Utility Company and Private Agency Training Programs

At the local and regional levels, training programs are available from major utility companies (gas and electrical companies), state agencies (forestry, public safety, and civil defense), and private agencies that interact in various ways with the fire service.

College Training Programs

There are hundreds of college fire science programs in the U.S., most are offered by community and junior colleges on either a full or part time basis. In most cases, an associate degree is awarded upon successful completion of the program, and some institutions award certificates for partial completion. A few four year degree programs with emphasis on fire service management are also offered.

College fire science programs are usually designed to supplement (but not replace) training provided by fire departments and state training agencies. The curriculum design for many such programs emphasizes the arts and sciences, with special courses in fire protection subjects that are beyond the scope of basic fire training programs. A majority of programs emphasize supervisory and management skills along with fire science.

Certification Standards

In an effort to lend uniformity to fire service training, the National Professional Qualifications Board for the Fire Service has developed a set of standards based on performance objectives. The standards, developed through NFPA, are:

NFPA 1001, *Standard for Fire Fighter Professional Qualifications.*

NFPA 1002, *Standard for Fire Apparatus Driver/Operator Professional Qualifications.*

NFPA 1003, *Standard for Airport Fire Fighter Professional Qualifications.*

NFPA 1004, *Standard on Fire Fighter Medical Technicians Professional Qualification.*

NFPA 1021, *Standard for Fire Officer Professional Qualifications.*

NFPA 1031, *Standard for Professional Qualifications for Fire Inspector, Fire Investigator, and Fire Prevention Education Officer.*

NFPA 1041, *Standard for Fire Service Instructor Professional Qualifications.*

These standards are being applied by several states and municipalities. Some states offer formal certification. The certification program may be mandatory or voluntary, depending upon the authority having jurisdiction over the process, and may provide for certification in a number of different areas and levels. The formal certification of individuals provides a standard of preparation and performance for different positions and may be applied to hiring, promoting, retaining, and transferring personnel.

Training Records and Reports

Recent court decisions regarding job related examinations have placed additional importance on training records and personnel evaluations. Training records and reports are particularly important at the entry or recruit level. Training records should be clear and concise and should document a fire fighter's advancement in the department. The content of training records and reports is particularly important when personnel are evaluated for pay increases and promotions. In larger departments, training records may be integrated into a larger management information system along with other personnel and management data.

EMERGENCY MEDICAL SERVICE

Emergency medical service (EMS) has become an important function of many fire departments. While ambulance services have been operated for decades by many fire departments, the major growth in this area has occurred since 1970, stimulated by higher standards for patient care, training, and equipment. Increased interest from the public, the medical community, and local government in emergency medical service delivery has prompted many fire departments to increase their participation in EMS, and the quality of EMS delivery consequently has improved greatly.

The addition of EMS delivery to a fire department is a natural extension of responsibilities and often can increase organizational productivity without compromising fire protection. Fire service personnel are oriented toward delivering emergency service and assisting citizens. Fire stations are decentralized to provide rapid response to all areas, while fire communications systems are designed to receive emergency requests from the public and dispatch assistance without delay, both of which are required to provide effective EMS delivery.

Fire departments have taken several different approaches to EMS, depending upon state and local regulations, medical systems, and funding sources. In some systems the fire department operates ambulances that provide both treatment and transportation to medical facilities, while in other systems the fire department provides first response and on scene patient care, with transportation provided by another agency or private ambulances. When ambulances are operated by the fire department, the personnel may or may not be fire fighters, although both ambulance personnel and fire fighters generally operate under the direction of fire department command officers.

Emergency medical training for fire service personnel generally falls into the categories of first responder, Emergency Medical Technician (EMT), or Paramedic (EMT-P). Training and certification standards include written and practical examinations and usually require continuing education to qualify an individual for recertification. Paramedics often receive 1,000 or more hours of advanced medical training and are certified to perform specific tasks including the administration of drugs and intravenous solutions under the direction of a base hospital physician. Communications systems provide voice and telemetry contact between paramedics at the scene and the base hospital. Specific certification requirements and treatment authorizations for paramedics depend upon individual state requirements.

The frequency of emergency medical service calls in a community generally outnumbers fire alarms by a factor of 2 to 1. Although implementation of EMS response by fire fighting companies places an increased workload on these units, it has proven very efficient in some jurisdictions. Providing basic EMS equipment on engine and ladder companies with one or more EMT certified fire fighters allows these units to provide initial response to most life threatening situations. First responders and EMTs are generally backed up by paramedic personnel responding on ambulances or rescue vehicles. Some fire departments have all personnel trained and certified at the EMT level, and have dual role paramedics assigned to engine and ladder companies.

There is some concern that increased commitment to emergency medical service will have a detrimental effect on the fire department's ability to provide effective fire suppression. This has not proven to be a problem in most cases, although fire department administrators must be constantly aware of this possible conflict. In extremely busy areas, it may be necessary to provide fire protection and EMS with separate resources.

HAZARDOUS MATERIALS

For many years, fire departments have responded to incidents involving hazardous materials as an unofficial extension of their responsibilities for life safety and property protection. This role resulted from the basic reality that no other agency could provide the technical orientation, protective equipment and breathing apparatus, combined with emergency response capability that was necessary to deal with these incidents. The entire subject of hazardous materials, including prevention, regulation, and incident response, has become a major responsibility for many fire departments. This is reflected by the increasing number and severity of incidents involving hazardous materials and the realization of the massive fire and life safety risks that they may present.

Hazardous materials incidents are primarily related to the escape of various substances from their normal containment systems. These events may occur as a result of leaks, spills, ruptures, container failures, fires, explosions, chemical reactions, transportation accidents, and other circumstances that could allow a substance to be released from containment. The specific hazards may include toxicity, corrosive properties, reactivity with other chemicals, fire or explosions, biological agents, and radioactive properties and may range from immediate risks to life and property to long term environmental consequences. Tens of thousands of substances, many of which are used, stored, and transported routinely in every community, present varying degrees of these hazards, either alone or in combination with each other. In most cases, the fire department is the first agency that will be called upon to take action if a hazardous materials incident occurs.

Every fire fighter should have the basic training to recognize a hazardous materials incident and to take actions that will reduce the immediate risk to the public and emergency personnel. In many cases, the only appropriate action may be to evacuate the public from the area and avoid exposure to the hazardous material until qualified personnel with appropriate protective equipment and control devices can respond and take action. In other cases, simple actions taken by initial responding crews may be sufficient to control the situation.

In many situations, the key element is the ability to identify the materials involved and their specific hazardous properties. This requires an in depth approach to pre-incident planning for fixed facilities and the use of reference materials to evaluate risks and develop action plans. The NFPA 704 Diamond system is designed to be used with properties containing hazardous materials to identify the nature of the hazards that may be involved. In the case of transportation incidents, placards required by U.S. Department of Transportation (DOT) regulations and shipping papers can help to identify the materials involved. Basic reference materials should be carried on every emergency vehicle; command vehicles should carry several more detailed references. There should be a standard procedure established for calling CHEMTREC (previously referenced in this chapter) when information is needed to help identify or deal with materials involved in transportation incidents. Many fire departments also have established procedures to call upon computer data bases, toxicology centers, universities, local suppliers, and specific individuals for assistance in dealing with hazardous materials emergencies.

Hazardous materials response teams have been established by many fire departments to provide rapid intervention for leaks, spills, transportation accidents, and other situations involving hazardous substances. These teams consist of specially trained personnel who are equipped with advanced protective clothing, breathing apparatus, test equipment, reference materials, absorbents, neutralizing agents, leak and spill control devices, and a variety of other items that may be needed for these incidents. Team members are usually fire fighters who respond to hazardous materials incidents with specially equipped apparatus, in addition to their regular duties. In some areas, the number and severity of incidents justify the establishment of companies that are fully staffed for hazardous materials incidents exclusively. The establishment and operation of a hazardous materials team requires extensive study, preparation, and training relative to the specific hazards that may be anticipated in a particular area.

The hazardous materials response team should be prepared to deal with a wide variety of possible incidents, with potential consequences ranging from localized risks to major disasters. Contingency planning should include consideration of large scale incidents, which may require evacuating residents, decontaminating exposed persons, detaining specialized medical resources, and effective interaction with state and federal agencies.

The prevention of hazardous materials incidents and the control of associated risks introduces a whole new area of regulation and control for the fire service. In some areas, local ordinances provide for the control of hazardous materials risks through codes applied and enforced by the fire service. In other areas, this enforcement authority is placed with health and environmental protection agencies, requiring the fire service to establish a cooperative working relationship with the authoirities having jurisdiction. The control of hazardous materials risks is an area of critical concern for the safety of fire service personnel, who usually respond to emergency situations involving these materials.

Bibliography

NFPA Codes, Standards, Recommended Practices and Manuals. (See the latest *NFPA Codes and Standards Catalog* for availability of current editions of the following documents.)

NFPA 295, *Standard for Wildfire Control.*

NFPA 402M, *Manual for Aircraft Rescue and Fire Fighting Operational Procedures.*

NFPA 403, *Recommended Practice for Aircraft Rescue and Fire Fighting Services at Airports and Heliports.*

NFPA 704, *Standard System for the Identification of the Fire Hazards of Materials.*

NFPA 901, *Uniform Coding for Fire Protection.*

NFPA 1001, *Standard for Fire Fighter Professional Qualifications.*

NFPA 1002, *Standard for Fire Apparatus Driver/Operator Professional Qualifications.*

NFPA 1003, *Standard for Airport Fire Fighter Professional Qualifications.*

NFPA 1004, *Standard on Fire Fighter Medical Technicians Professional Qualifications.*

NFPA 1021, *Standard for Fire Officer Professional Qualifications.*

NFPA 1031, *Standard for Professional Qualifications for Fire Inspector.*

NFPA 1041, *Standard for Fire Service Instructor Professional Qualifications.*

NFPA 1221, *Standard for the Installation, Maintenance and Use of Public Fire Service Communications Systems.*

NFPA 1410, *A Training Standard on Initial Fire Attack.*

NFPA 1501, *Standard for Fire Department Safety Officer.*

NFPA 1904, *Standard for Testing Fire Department Aerial Ladders and Elevating Platforms.*

NFPA 1931, *Standard on Design, and Design Verification Tests for Fire Department Ground Ladders.*

NFPA 1971, *Standard on Protective Clothing for Structural Fire Fighting.*

NFPA 1972, *Standard on Structural Fire Fighters' Helmets.*

NFPA 1973, *Standard on Gloves for Structural Fire Fighters.*

NFPA 1975, *Standard on Station/Work Uniforms.*

NFPA 1981, *Standard on Self-Contained Breathing Apparatus for Fire Fighters.*

NFPA 1982, *Standard on Personal Alert Safety Systems (PASS) for Fire Fighters.*

NFPA 1983, *Standard on Fire Service Life Safety Rope, Harnesses, and Hardware.*

Additional Readings

Air Operations for Forest, Brush and Grass Fires, National Fire Protection Association, Quincy, MA, 1975.

Bahme, C. W., *Fire Service and the Law*, National Fire Protection Association, Boston, MA, 1976.

Brannigan, F. L., *Building Construction for the Fire Service*, 2nd ed., National Fire Protection Association, Boston, MA, 1983.

Brunacini, Alan V., *Fire Command*, National Fire Protection Association, Quincy, MA, 1985.

Clarke, William E., *Fire Fighting Principles and Practices*, Fire Engineering Book Service, NY, 1974.

Ennis, A. Morrison, "Getting Water to Rural Fires," *Fire Engineering*, July 1970.

Erven, Lawrence W., *Fire Chief's Handbook Study Guide*, Davis Pub., Santa Cruz, CA, 1979.

Fire Protection Guide on Hazardous Materials, 8th ed., National Fire Protection Association, Quincy, MA, 1983.

Gaines, Glen A., *Firefighting Operations in Garden Apartments and Townhouses*, R. J. Brady, Bowie, MD, 1978.

Grant, R. W., *Public Fire Safety Inspections*, National Fire Protection Association, Quincy, MA, 1967.

Jacobs, D. T., *Physical Fitness and the Public Safety*, National Fire Protection Association, Quincy, MA, 1980.

Lyons, Paul R., ed., *Fire Officer's Guide to Extinguishing Systems*, National Fire Protection Association, Quincy, MA, 1977.

Management in the Fire Service, National Fire Protection Association, Quincy, MA 1978.

Mendes, Robert F., *Fighting High-Rise Building Fires: Tactics and Logistics*, National Fire Protection Association, Quincy, MA, 1975.

Page, J. O., *Emergency Medical Services*, National Fire Protection Association, Quincy, MA, 1978.

Richman, Harold, *Engine Company Fireground Operations*, R. J. Brady, Bowie, MD, 1975.

Soros, Charles C., and Lyons, Paul R., *Safety in the Fire Service*, National Fire Protection Association, Quincy, MA, 1979.

Storey, T. G., "FOCUS: A Computer Simulation Model for Fire Control Planning," *Fire Technology*, Vol. 8, No. 2, May 1972, pp. 91-103.

Tobin, E. T., Davis, J. B., and Mandt, C., "Automated Forest Fire Dispatching—a Progress Report," *Fire Technology*, Vol. 5, No. 2, May 1969, pp. 122-129.

Tuck, C., ed., *NFPA Inspection Manual*, 5th ed., National Fire Protection Association, Quincy, MA, 1983.

Walsh, Charles V., and Marks, Leonard, *Firefighting Strategy and Leadership*, McGraw-Hill, NY, 1976.

The following additional readings are published by the International Fire Service Training Association, Oklahoma State University:

Essentials of Fire Fighting.
Forcible Entry.
Fire Fighter Study Guide.
Ground Ladder Practices.
Public Fire Education.
Fire Hose Practices.
Salvage and Overhaul Practices.
Fire Stream Practices.
Fire Apparatus Practices.
Fire Ventilation Practices.
Fire Service Rescue and Protective Breathing Practices.
Fire Service First Aid Practices.
Fire Prevention and Inspection Practices.
Fire Service Practices for Volunteer Fire Departments.
Fire Service Orientation and Indoctrination.
Fire Service Training Programs.
Photography for the Fire Service.
Water Supplies for Fire Protection.
Firefighter Safety.
Aircraft Fire Protection and Rescue Procedures.
Ground Cover Fire Fighting Practices.
The Fire Department Officer.
Fire Department Facilities, Planning, and Procedures.
Fire Service Instructor Training.
Leadership in the Fire Service.
IFSTA Source Material Reference Guide.

FIRE PREVENTION AND CODE ENFORCEMENT

Revised by James R. Bell

Fire prevention includes all fire service activity that decreases the incidence of uncontrolled fire. Usually, fire prevention methods utilized by the fire service focus on inspection, which includes engineering and code enforcement, public firesafety education, and fire investigation.

Inspection, including enforcement, is the legal means of discovering and correcting deficiencies that pose a threat to life and property from fire. Enforcement is implemented when other methods fail. Education informs and instructs the general public about the dangers of fire and about firesafe behavior. Fire investigation aids fire prevention efforts by indicating problem areas that may require corrective educational efforts or legislation. Good engineering practices—another fire prevention method—can provide built in safeguards that help prevent fires from starting and limit the spread of fire should it occur.

In the United States, the full value of fire prevention was not realized until fire departments and agencies began to compile meaningful information concerning the causes and circumstances of fires. Such information highlighted problem areas and led progressive departments to initiate more effective fire prevention efforts. The results of fire prevention efforts are borne out every year statistically, confirming the effectiveness of fire prevention programs. The fire prevention effort, as the most important priority in cutting fire losses, received a endorsement in 1973, when the National Commission on Fire Prevention and Control published the results of its in depth study on the fire problem in America. In its report "America Burning," the Commission emphasized the necessity for increased fire prevention activities to reduce fire. It was the Commission's position that increasing emphasis on fire prevention would measurably affect the fire loss picture in the U.S. Statistics validate this position for areas where fire prevention activities are implemented.

FIRE PREVENTION PERSONNEL

Effective fire prevention depends upon a personnel network dedicated to enforcing fire codes, educating the general public, and investigating fire causes. This network of people is usually organized at the state, county, and/or local level.

State and Provincial Fire Marshal/Commissioner

Most states have offices to over see certain phases of fire protection. (See Table 15-4A.) The state's chief fire protection administrator usually is called the state fire marshal. The organization of state fire marshal offices differs from state to state. (See Table 15-4B.) Most fire marshals receive their authority from the state legislature and are answerable to the governor, a high state officer, or a fire commission. In some states, the fire marshal's office may be a division of the state insurance department, state police, state building department, state commerce division, or other state agency. A few offices are organized as separate agencies.

State fire marshals' offices normally function in areas beyond the scope of municipal, county, or fire district organizations. Depending upon the authority outlined in enabling legislation or administrative regulations, state fire marshal offices may have combination responsibilities in the following areas: old code enforcement, fire prevention inspections, plans review, fire and arson investigation, fire data collection, and analysis; fire service training, public fire education, fire legislation development, explosives and other hazardous materials regulation, manufactured housing regulations, mobile home regulations, electrical inspections, and state agency and public fire protection consulting.

State fire marshals may their share responsibility with local fire officials, or each may have specifically defined areas of concern. Local fire protection officials sometimes are granted the authority to act as agents for the state in stipulated areas of inspection, enforcement, and investigation.

At the national level in Canada, the Fire Commissioner is responsible for firesafety enforcement in all nonmilitary dominion properties and for the gathering of national fire statistics. The Canadian Forces Fire Marshal oversees properties of the armed forces, where fire protec-

Mr. Bell is a fire protection consultant based in Annandale, VA.

TABLE 15-4A. Functions of the State Fire Marshal's Office

STATE	Code Enforcement	Fire & Arson Investigation	Plans Review	Fire Prevention Inspections	Fire Data Collection	Fire Data Analysis	Fire Service Training	Public Fire Education	Fire Legislation Development	Manufactured Housing Regulations	Mobile Home Regulations	Electrical Inspections	Boiler Inspections	Distribution of Funds to Fire Dept.
Alabama	X	X	X	X	X	X	X	X		X	X	X		
Alaska	X	X	X	X	X	X		X						
Arizona	X	X	X	X	X		X	X	X					
Arkansas	X	X	X	X			X	X	X					
California	X	X	X	X	X	X	X	X	X					
Colorado*										X				
Connecticut	X	X	X	X	X	X	X	X	X				X	
Delaware	X	X	X	X	X	X			X					
Florida	X	X		X	X	X	X	X	X					
Georgia	X	X	X	X				X			X	X		
Hawaii†			NONE											
Idaho	X	X		X	X	X			X					
Illinois	X	X	X	X	X	X			X				X	
Indiana	X	X	X	X	X	X	X	X	X					
Iowa	X	X	X	X	X	X		X	X	X	X			
Kansas	X	X	X	X	X	X		X	X					
Kentucky	X	X	X	X	X	X	X	X	X	X	X	X	X	X
Louisiana	X	X	X	X	X	X		X	X	X	X	X	X	
Maine	X	X	X	X	X	X		X	X					
Maryland	X	X	X	X	X	X		X	X			X		
Massachusetts	X	X	X	X	X	X		X	X					
Michigan	X	X	X	X	X	X	X		X					
Minnesota	X	X	X	X	X	X	X	X	X		X			
Mississippi		X									X			
Missouri		X							X					
Montana	X	X	X	X	X	X		X	X					
Nebraska	X	X	X	X	X	X		X	X	X	X	X		
Nevada	X	X	X	X	X	X	X	X	X			X		X
New Hampshire	X	X	X	X	X	X		X	X			X		
New Jersey‡	X			X	X	X	X	X	X					
New Mexico	X	X		X	X	X	X	X	X					X
New York#	X	X	X	X	X	X	X	X	X					
North Carolina	X		X				X	X	X	X	X	X		
North Dakota	X	X	X	X	X	X		X	X					
Ohio	X	X	X	X	X	X	X	X	X					
Oklahoma	X	X	X	X	X	X		X	X					
Oregon	X	X	X	X	X	X	X	X	X					
Pennsylvania	X	X	X		X									
Rhode Island	X	X	X	X	X	X		X	X					
South Carolina	X	X	X	X		X	X	X	X					
South Dakota	X	X	X	X	X	X		X	X					
Tennessee	X	X	X	X	X	X		X	X	X	X	X	X	
Texas	X	X	X	X	X	X		X						
Utah	X	X	X	X	X	X	X	X	X					
Vermont		X			X									
Virginia	X		X											
Washington	X	X	X		X	X		X	X					
West Virginia	X	X	X	X	X	X	X	X	X			X		X
Wisconsin		X												
Wyoming	X	X	X	X	X	X	X	X	X			X		

Footnotes:
* Colorado—Division of Fire Safety established July 1, 1984.
† Hawaii—State fire marshal's office disbanded April 1979, all functions transferred to county fire marshals offices.
‡New Jersey—State Bureau of Fire Safety established May 10, 1984.
New York—State fire administrator appointed Sept. 1, 1979.

TABLE 15-4B. State Fire Marshal Organizational Patterns

Under Department of Insurance
- Alabama
- Florida
- Idaho
- Mississippi
- New Mexico
- North Carolina
- Tennessee
- Texas
- Washington

Under Department of Public Safety
- Alaska
- Colorado
- Connecticut
- Iowa
- Louisiana
- Maine
- Maryland
- Massachusetts
- Minnesota
- Missouri
- New Hampshire
- Utah

Under a Separate Government Department
- California
- Illinois
- Kansas
- Nebraska
- Nevada
- Oklahoma
- Rhode Island
- South Carolina
- Wyoming

Under a Regulatory Agency
- Indiana
- Kentucky
- New Jersey
- South Dakota
- Virginia

Under Division of Emergency and Military Affairs
- Arizona

Under State Police
- Arkansas
- Michigan
- Pennsylvania

Under Cabinet Level Official
- Georgia (Controller General)
- Montana (Attorney General)
- New York (Secretary of State)
- North Dakota (Attorney General)
- Ohio (Director of Commerce)
- Oregon (Secretary of Commerce)
- Vermont (Controller General)
- Wisconsin (Attorney General)

Under Fire Commissioner
- Delaware

tion facilities include a number of operating fire departments.

The Association of Canadian Fire Marshals and Fire Commissioners consists of the previously mentioned dominion and armed forces fire officials, plus all provincial and territorial fire commissioners and fire marshals. The Association exerts a major influence on firesafety policy in Canada.

In Canada, each province has a provincial fire commissioner or fire marshal; likewise, the territories have territorial fire marshals. In most cases, the provincial/territorial fire officer has a wide variety of responsibilities including code enforcement, fire service training, operation of fire reporting systems, and fire investigation as well as support of local fire departments.

Chief of Fire Prevention or Local Fire Marshal

Because of the importance of fire prevention activities, certain members of the department should specialize in fire prevention functions. Staff size of a fire prevention division, bureau, or fire marshal's office will depend upon the community's needs and the department size. The division may be divided into sub departments for inspections, public education, construction plans review, and fire investigations. (See Table 15-4C.) These sub departments are headed by subordinate officers.

The fire prevention division personnel should consist of those department members best qualified for this work. In addition, qualified technical specialists should be available when possible—the department's effectiveness depends upon the technical skills possessed by the fire prevention personnel.

The laws of the state, county, municipality, or fire district often delegate responsibility and authority for fire prevention to the fire chief or fire department head who may then delegate the authority to an individual or division, depending upon the size of the department. In some states, the local fire marshal's authority and area of enforcement responsibility is derived from state laws, and independent of the powers and duties of the department chief.

The fire chief may directly supervise fire prevention activities in smaller departments. In larger departments, the division head should be a high ranking chief officer, functioning as a staff officer to the fire chief. The chief officer for the fire prevention activities may be described as chief of the fire prevention division, chief inspector, or fire marshal.

Fire Inspector or Fire Prevention Officer

Fire inspectors or designated fire prevention officers should be selected for these tasks based on their technical training and their ability to motivate people. Individuals in this position are responsible for conducting fire inspections and also may be responsible for other division duties including fire investigations, public education, or plans review.

An inspector conducting an inspection should be able to persuade property owners or occupants to maintain firesafe conditions. The inspector who relies only on police power cannot accomplish as the much as inspector who relies on salesmanship. Enforcement of the fire law, however, may be necessary to achieve compliance with the laws or codes of the jurisdiction and to assure firesafe conditions.

Plans Review Specialist

The review of plans and specifications is a code enforcement process intended to ensure compliance with provisions of both the building code and the fire preven-

TABLE 15-4C. Organization of Local Fire Prevention Functions in the Fire Marshal's Office

	Fire Prevention, Code Inspections, and Enforcement	Construction Plans and Specification Review	Public Fire Education	Fire Investigation	Records, Statistics, and Evaluation
Personnel:	Fire Inspector Fire Officer Fire Company Personnel	Fire Protection Engineer Plans Review Specialist	Public Fire Education Officer Fire Officer Fire Inspector Fire Fighter	Fire Investigator Fire Officer Fire Fighter	Fire Records Specialist Programmers Statisticians
Functions and Activities:	Inspections Permits Violations Summons Court Cases Complaints Night Inspections Red Tags Hazardous Materials	Plans Review Specification Review Consultations On-Site Inspections Certificate of Occupancy Sign-Off Systems Testing Condemnations Use Permits Liaison with: Public Works Building Electrical Mechanical Zoning Housing	Fire Prevention Behavior Education Fire Safety Behavior Education Special Groups Seasonal Programs Juvenile FIre Setters Smoke Detector Programs Operation EDITH Clean-up, Paint-up, Fix-up	Fire Cause Detection Fire Investigation Case Preparation Courtroom Testimony Prosecutor Liaison Arson Task Force Local, State, and Federal Investigative Agencies	Inspections Violations Found Violations Cleared Violations Pending Reports: Monthly Quarterly Annual Special Fire Experience Statistics

tion code. Individuals in plans review must be technically proficient in both the intent and the letter of the applicable codes. The fire department plans review specialist examines site plans, building and fire protection system plans, and specifications, and may prepare recommendations or consult with individuals involved in building construction, remodeling, or renovations. The plans review specialist may regularly interact with other public safety officials and agencies at the local, state, and federal level (including the building inspectors; the public works, zoning, and health departments, and others) to ensure the construction of safe, code conforming buildings.

Fire Protection Engineer (F.P.E.)

Due to the complexity and magnitude of fire protection problems, the services of fire protection engineers can be beneficial. Although most fire protection engineers are employed on a consulting basis, some public fire protection agencies employ full time staff engineers. The staff fire protection engineer contributes a high level of technical ability to the plans review process, interpretations, consultations, and the preparation of recommendations for a broad range of fire protection problems. The fire protection engineer's credentials, based on professional training, are recognized by architects, engineers, builders, and other professionals involved in the building process.

Fire Prevention Education Officer

The public education officer creates public awareness of fire as a personal, family, business, and community threat, and then tries to motivate the general public of all ages to do something about fire risks based on proper firesafe behavior. Personnel selected for this function should be knowledgeable in a wide area of fire technology and know how to develop and present effective fire prevention and firesafety programs. This job requires someone with the imagination, creativity, motivation, communication skills, and adaptability needed to convey firesafety messages to all population segments of the community.

Depending upon the size of the department and the type of fire prevention and firesafety program it undertakes, all levels of fire department personnel can and should participate in presenting the firesafety and prevention message to the community. Nonuniformed personnel, including senior citizens and citizens with education backgrounds, also are being used successfully by the fire department in this area of fire prevention.

Fire Investigators

Fire cause determination and subsequent investigation can be the responsibility of the fire department or outside agencies. When a fire department has this responsibility, fire cause detection and investigation can be assigned to the fire chief, chief of the fire prevention bureau, fire marshal, battalion chiefs, company officers, fire inspectors, or personnel specified as fire investigators.

Investigative personnel should be perceptive, inquisitive, and thorough in determining facts and conducting an investigation. Investigation experience should be supported by training in areas such as chemistry, criminal law, forensics, and criminal investigation. An increasing number of progressive departments have fire investigation personnel who have been given criminal justice training and police powers.

NFPA 1031

NFPA 1031, *Standard for Professional Qualifications for Fire Inspector, Fire Investigator and Fire Prevention Educator Officer*, specifies the minimum requirements of professional competence, in terms of performance objectives, for these positions. This standard may be used by all organizations employing or assigning personnel to these positions such as police agencies involved in fire investigation, but it is specifically oriented to individuals with a fire service background.

FIRE PREVENTION INSPECTIONS

Fire prevention inspections are conducted by state or local fire department personnel in compliance with laws and ordinances which usually require that specific fire inspections be conducted. Occupancies normally inspected include places of public assembly, educational, institutional, residential (except the interior of dwellings), mercantile, business, industrial, manufacturing, storage, and special hazards structures. In addition to such mandatory inspections, the fire department also may conduct voluntary fire inspections such as home firesafety surveys.

The organization and operation of fire prevention inspection programs varies. Inspections required under the fire prevention code traditionally have been conducted by personnel from the fire prevention division. However, this responsibility is being shared increasingly with fire company personnel. The department's training program should provide all its members with training in fire prevention.

Fire Prevention Inspection Objectives

Inspections conducted as part of code enforcement help to ensure satisfactory life safety conditions within a structure. The condition of exits, interior finish, operation of exit doors, emergency lighting, exit signs, and all fire doors should be inspected. Inspection of exiting facilities should include inspection of the exit discharge area.

Inspections, which are intended to prevent fires from occurring, are effective because the inspector identifies fire hazards which could cause a fire, allow a fire to develop, or allow a fire to spread. In addition to locating and correcting possible fire causes, the fire inspector should check any accumulation of combustible trash and debris, storage practices, maintenance procedures, and safe operation of building utilities.

Inspections assure the proper installation, operation, and maintenance of fire protection features, systems, and appliances within the building. The inspection process should ascertain regular testing of each fire protection system, whether by the fire department or by others.

Fire detection equipment, alarms, annunciation and notification systems, sprinkler valve operation, supervisory switches, and fire pumps all should be regularly tested as part of the inspection process. Other fire protection features, including standpipes and fire escapes, should be tested or closely examined to detect possible malfunctions due to deterioration from weather and corrosion. Portable fire extinguishers should be checked as to proper type, placement, maintenance, testing, and distribution in the structure.

Technical information on the building and the processes therein should be collected during the inspection. When used in prefire planning, such information can be valuable to the fire department in case of a fire at this property. The type of construction, vertical openings, utility type and placement, fire protection systems, fire department access, tenancy, hazardous materials, or special life hazard conditions are the kinds of information that should be noted during inspections and used to develop fire fighting plans.

Inspections provide an opportunity to educate the owners or occupants of a building about firesafe behavior and the need for adequate fire and life safety conditions in the areas under their control. "Selling" fire prevention is the key to success in obtaining code compliance and how fire prevention is "sold" should be an important consideration in training programs for inspection personnel. When inspection programs are properly designed, inspectors may achieve more through public education and persuasion than through exercising their enforcement authority. The persuasive effect of the inspector's presence coupled with the inspector's ability to spot and directly remove hazards enhances the value of the inspection program.

Fire Company Inspections

In many communities, in service fire suppression personnel conduct most or all regular inspections. Company fire inspection procedures may include conducting building surveys, correcting common problems concerning life safety conditions, locating fire hazards, and testing fire protection systems. Inspections also may include checking use and storage of hazardous materials, which may require issuing a hazardous use permit. The inspection process helps familiarize fire company personnel with individual buildings and locations in their jurisdiction, at which they may fight fires or perform other emergency duties.

Company inspections usually are conducted in the fire company's first due area. Inspections and reinspections may be scheduled by the company officer, battalion or district officer, or the fire prevention bureau, according to requirements of the fire prevention code or the department. Before fire fighters perform inspections, they should receive proper training and be authorized to conduct inspections as fire prevention officers.

The inspection may uncover some situations that are easy to correct or others that require more technical skill. In either case, follow up responsibilities should be assigned to the fire companies, fire company officers, and district chiefs according to the department's resources. The fire prevention division may help to obtain compliance by building owners after violations are located by fire companies so it is essential for fire prevention personnel to closely coordinate their activities with those of fire company personnel.

Fire Prevention Division Inspections

In some departments, inspection personnel from the fire prevention division have the sole responsibility to conduct fire prevention inspections. The objectives of their inspections, like those of the general inspections performed by fire companies, are to locate conditions that violate the fire codes and that may cause fire or endanger life and property. Each inspection must be thorough. The owner or occupants of the property being inspected must be notified when unsafe conditions are found. Because inspections usually are required by law in order to locate any violations and to see that they are corrected, legal means are available for enforcement. To perform such work competently, the inspector should have appropriate technical education and specialized training, or equivalent experience.

Where fire companies perform some or all regular inspections, fire prevention division personnel may be concerned with the initial inspection, follow up inspections, or enforcement actions necessary to correct fire code violations. Fire prevention division personnel may perform inspections to issue permits as required by the fire prevention code, and also may be concerned with buildings and premises that require a high level of code application. The inspector may need to confer with the property management or with fire department officers to prepare comprehensive reports, recommendations, or orders.

Both fire company and fire prevention division inspections may reveal conditions hazardous to the health and safety of owners, occupants, and the general public. Even when these hazardous situations are not specifically covered in the fire laws, such conditions should be brought to the attention of the owners or occupants; other inspection authorities should be so notified for the record. This mutual assistance in the inspection of buildings or premises can greatly enhance the level of life safety for the general public. Close cooperation among responsible inspection authorities should be encouraged and where possible, recognized in municipal ordinances or interdepartmental administrative agreements.

Night inspections of public assembly and business occupancies should be part of routine inspection procedures to assure compliance with occupancy load requirements and that the means of egress are maintained in good operating order.

Inspections based on citizen complaints should be conducted by the fire prevention division in a timely manner. An informed and interested public can be an important asset in uncovering unsafe conditions which might have developed after routine inspections were conducted. Where possible, the complaintant should be informed of the results of the inspections and of corrective actions required for building owners to comply with the law or to ensure firesafety.

Dwelling Inspection Process

Residential inspections are not considered mandatory because the Constitutional rights of citizens ensure the sanctity of a person's home. For many years however, many fire departments have inspected dwellings on a voluntary or by invitation basis and have been successful in reducing the residential fire loss experience. Inspections of this nature, whether conducted by career, volunteer, or other fire department personnel, should be supported by a strong publicity program. Adequate training must be provided for all personnel involved. Voluntary inspections can be conducted either as a separate part of the fire inspection program or as part of fire prevention programs which may include fire escape planning and practices such as Operation EDITH (the acronym for Exit Drills In The Home), smoke detector drives, spring clean up, burn prevention programs, heating firesafety, etc.

CODE ENFORCEMENT

Fire Prevention Codes

The fire laws of the state, county, fire district, or community delegate general responsibility and authority to the fire officials involved in fire prevention activities. A fire prevention code adopted into law that outlines specific fire prevention requirements and enforcement procedures is essential for an effective fire prevention program.

State law or local ordinances may outline how a fire prevention code is written or adopted. The fire prevention official may have authority to write a local fire prevention code or to adopt a nationally developed code by reference. Using a model fire prevention code is preferable to using one developed locally—because nationally developed consensus codes are based on a broad spectrum of fire prevention experience, they may protect a fire prevention official from being accused of developing a fire code that is too stringent or biased.

There are additional advantages to adopting a model fire prevention code. A model code: (1) provides a document which may be recognized as authoritative by architects, engineers, and builders who already are familiar with code requirements; (2) provides an interpretation process by which the local official, if a question arises, may determine the intent of the code developing body; and (3) allows for periodic revision of the fire prevention code to reflect new technology and current thinking.

Several organizations publish model fire prevention codes that localities may adopt by reference for enforcement use. NFPA publishes NFPA 1, *Fire Prevention Code*, which incorporates many NFPA standards by reference. Each of the three major model building code organizations also has developed its own model fire prevention codes, correlated with other codes in its series such as model building and mechanical codes. Building Officials and Code Administrators International (BOCA) publishes the *Basic Fire Prevention Code*; the International Conference of Building Officials (ICBO) publishes the *Uniform Fire Code*; and the Southern Building Code Congress International (SBCCI) provides the *Standard Fire Prevention Code*. Each code references codes, standards and other official NFPA documents. These documents constitute the *National Fire Codes®*, considered the most authoritative set of firesafety regulations in the U.S.

Code Administration

Most fire prevention codes specify similar enforcement procedures and contain similar requirements. A typical fire prevention code contains an administrative section which establishes the legal framework and organization for the fire prevention program. This section usually outlines the code's applicability to occupancy types and specifies whether it applies to new or existing buildings, or both.

The code may define the enforcement authority of the fire prevention official and assistants, including lawful right of entry for inspections, the right to issue orders to correct hazardous conditions within a building, and the right to order evacuation of an unsafe building or premises. The code may further assign duties which require the investigation of fires and maintenance of fire records. Permit requirements, conditions, and procedures to issue notices of violation may be included.

The language of the administrative section should be written to provide the fire prevention official with as broad a scope of authority as possible. This facilitates enforcement of the fire prevention code, other county or city ordinances, and state laws that pertain to the firesafety of buildings and premises, as well as the control of hazardous materials within the jurisdiction. This section also may outline owner or occupant responsibility to properly maintain a structure and its fire protection features.

Subsequent code sections may define important code related terminology, outline general fire precautions, and establish the proper installation, operation, testing, and maintenance procedures for fire protection systems and appliances within a building. Final sections of the code may outline the use and maintenance of specific types of equipment, processes, and occupancies, or describe handling requirements for various types of hazardous or explosive materials. Appendices may reference supplemental standards or other helpful data in the administration or application of the fire prevention code.

Under the fire prevention code or other ordinances of the jurisdiction, a Fire Prevention Code Appeals Board or similar body may be empowered to hear appeals of orders made by the fire official under the fire prevention code. This board should consist of technically knowledgeable and unbiased members of the community.

Enforcement Procedures

The fire prevention process includes a number of enforcement procedures used to establish compliance with the fire code. They are described in the following paragraphs.

Permits: The permit is an official document issued by the fire prevention division to authorize the performance of a specific activity. Permits are issued in the name of the fire official for the use, handling, storage, manufacturing, occupancy, or control of specific hazardous operations and conditions. The permit should be issued only if the condition meets code requirements.

The permit process provides the fire prevention official with information on what, where, how, or when specific hazards are being installed, stored, or used within the jurisdiction. The process allows cross-checking with building, zoning, public health, or other departments' requirements for the use outlined on the permit. Further, it allows the fire official to review and approve devices, safeguards, and procedures that may be needed to assure the safe use of hazardous materials or operations. The lack of a required permit constitutes a misdemeanor in legal terms and is grounds for stopping an operation in or use of a structure.

The permit is the property of the issuing agency, not the permit holder. A license or permit authorizes, by law, the right of entry for inspection purposes to ensure compliance with the permit requirements. If an inspector operating under a fire code permit is refused entrance to perform a regulatory inspection, this refusal constitutes grounds to halt operation in or use of the structure involved.

Permits usually are issued for the place, process, and operator, and are nontransferable so that the notification function of the permit process is maintained.

Certificates: A certificate, which is a written document issued by the authority of the fire official to any person or business, grants permission to conduct or engage in any operation or act for which certification is required. Depending upon the needs of a jurisdiction, certificates of approval may be required before smoke or heat detectors, fire extinguishers, fireproofing materials, or other fire protection devices are offered for sale to the general public. This procedure is designed to ensure that only those fire protection appliances which will function in a satisfactory manner are offered for sale to the public.

Certificates of fitness are issued to individuals or businesses with demonstrated proficiency in skills, training, and testing in areas that affect firesafety. This includes pyrotechnics and persons who handle explosives; persons who install fire protection equipment and systems, including sprinklers; and persons who perform maintenance on fire extinguishers and fire detection systems. Certification requirements may include a financial bond or liability insurance where conduct of the regulated activity is inherently hazardous.

Licenses: A license is a permission granted by competent authority to those individuals about to engage in a business, occupation, or otherwise lawful activity. Licenses are issued to provide knowledge of specific business locations, ensure compliance with particular standards, and add a source of revenue to the community. The fire prevention division may issue some licenses directly and may be involved in the check off process for licenses issued by other municipal departments.

Enforcement Notices

When violations of the fire laws or ordinances are discovered during regular or spot inspections, the violations must be called to the attention of the owner or occupants. When the situation is not corrected during the inspection, or is an uncorrected recurring violation, several types of enforcement procedures may be available to the fire inspection officer.

Warnings or Notices of Violation: The inspection official may issue these notices to the owner or occupant, stating that a specific violation of fire regulations and/or a fire prevention code has been identified during the inspection. Using discretion and judgment, the fire official may designate an appropriate period of time for correction of the violation cited. Recommendations for correction of the violation also may be noted on the form. The owner or occupant must sign the form and keep a copy of the document. Reinspection after the allotted time period is essential to ensure that the correction has been made. If it has not, further legal action should be taken.

Red Tag or Condemnation Notices: These notices usually are attached to appliances, systems, or equipment that would be unsafe and dangerous if allowed to remain in operation. The red tag process requires a competent service technician repair or replace the item prior to removal of the tag and return of the equipment to operation. This process also may require that the fire official inspect the completed work before granting approval to resume the operation of the equipment in question.

Citations or Summonses: Violation of the fire prevention code requirements or other fire regulations usually are considered misdemeanors. An authorized fire official may issue a citation or summons to an individual who is in violation of the law. These documents constitutes a notice for the violator to appear before the appropriate court.

Warrants: A warrant is an order issued by a magistrate or agent of the court which directs a law officer to arrest a violator of the law and bring the violator before the court to respond to the charges specified in the document. Arrest warrants usually are issued in felony cases. The person requesting the warrant must provide factual information to the magistrate or court officer issuing the warrant, showing the existence of sufficient "probable cause" of a violation of a law.

When authorized by state law or local ordinances, fire marshals, fire investigators, and other fire officials may be empowered to issue summons, serve warrants, and make arrests. Police or peace officer training, in accordance with the standards of the state, is required by most states before a fire official may serve warrants and make arrests for felony violations of state laws.

Where violations represent a clear hazard to life or property, it may be necessary for the fire official to take immediate action to correct unsafe conditions. Entrance into private property, shutdown of an operation, evacuation of a building, or withdrawal of permits are actions that the fire official may find necessary to ensure public safety. These powers may be explicitly stated or may be implied in discretionary powers and "duty to act" requirements of appointing laws or ordinances. In all cases, actions must be based on clearly demonstrable threats to public safety, showing that delay would provide an unreasonable danger to residents, occupants, guests, or the public, and that this judgment is based on accepted standards or concepts of safety.

RECORD KEEPING

Keeping records and files on all actions taken by the fire prevention division or the fire marshal's office is an essential part of code enforcement and administration. All instruments and records of code enforcement, all documents relating to inspections, violation notices, summonses, plans review comments and approvals, fire reports, investigations, permits issued, certificates issued, etc., are to be handled as legal documents. Well organized and well maintained inspection files and building records are essential foundations for enforcement actions. Complete and accurate records also are needed to measure fire department effectiveness in accomplishing fire prevention goals, and to provide department management with information for budgetary and administrative purposes.

A file should be maintained for each inspected property which summarizes information about the inspected premises and contains copies of all inspection reports (Schaenman and Swartz 1974). Records and reports of fire prevention activities should be clear, concise, and complete. Every time an inspector or fire prevention officer inspects a location, information about that location should be included in a record or a report. The occupancy file of each building visited should include a complete history of the building site; building plans and specifications (where possible); fire protection system plans; information on and permits issued for the use, storage, and handling of hazardous materials; correspondence; inspection reports; and records of fire incidents at that location.

Files are needed for properties where a certificate of occupancy, a license, or a permit has been issued. Files also should be maintained on properties with automatic sprinklers, standpipe systems, private hydrants, or other private means of fire protection. A special notation or key should identify any system or equipment that is legally required. With a well maintained system of files, established data need not be recompiled for each inspection. Inspection work planning is expedited, and time is more effectively used in the field for the inspection itself.

Computer Applications

In recent years, computer technology has helped fire prevention officials handle record keeping and manage fire prevention programs. Many communities have found their records are more accurate and complete when computers are used for processing and/or organizing inspection information. Fire prevention reporting systems should be designed for the collection, storage, and processing of inspection data, and to schedule periodic inspections according to local policy. All data (occupancy description, activity, violations) are collected on a standard form during an inspection. After the inspection forms are completed, they are processed according to local needs.

Computers can produce management reports related to inspection activity and fire prevention personnel resource allocation. In addition, separate master file listings can be provided of occupancies currently under inspection and all occupancies to be inspected. Inspection forms or schedule sheets can be computer printed to show when periodic inspections are required and indicate violations to be checked for correction during follow up inspections. These management reports should document:

1. The occupancy inspected, locating time of day/date, and name of inspector;
2. The code violations that are corrected (filed by type, number, and kind of occupancy);
3. How the fire department utilized its resources to accomplish fire prevention program objectives.*

These systems save inspector time and provide data in a form useful for decision making by operational management. Because they contain a current inventory of occupancies, the systems can provide information for fire protection fiscal and long range master planning.

---

*NFPA has developed a Fire Inspection Reporting System (FIRS) to assist communities in this area. The FIRS system can be used in any fire prevention program. See also: NFPA 903M, *Fire Reporting Property Survey Manual*, and survey forms 903SR, "Basic Structure Report," and 903TR, "Basic Occupancy Report," for collecting basic information about a property. By using NFPA 901, *Uniform Coding for Fire Protection* and its data classifications and definitions, the data collected can be maintained in a uniform manner. Prefire records using NFPA 901 data classifications and definitions are consistent with postfire reporting information coding contained in NFPA 902M, *Fire Reporting Field Incident Manual*, the National Fire Incident Reporting System (NFIRS), and other reporting programs based on NFPA 901.

PLANS REVIEW PRACTICES

Traditionally, the activity of the building department has involved the design, construction, and final occupancy inspection of the building. The fire department's role usually has begun upon occupancy of the building and has concerned the maintenance of life safety conditions and fire protection systems, as well as firesafe storage and handling of contents.

Today, the fire department's role in the building construction process is changing. An increasingly important fire prevention function involves participation of the fire official and the fire marshal (or a designated representative) in the review of building plans and specifications and in the construction process. In most cases, plans review is conducted in close cooperation with the local building, zoning, and public works departments, or with state agencies which may have this review authority.

The review of building plans and specifications provides the fire service with its best opportunity to see that fire protection standards are met prior to the completion of construction and occupancy of the building. The type and depth of the review process depend upon the community's needs and the functions of other local departments or state agencies with review authority.

Whenever possible, fire officials should participate in preconstruction conferences; address questions relating to fire protection features in the planned building, building code or fire prevention code requirements; and comment during the plans review process. If questions and issues concerning the effect of construction on firesafety are discussed at this time with the architect, engineers, contractors, and other code officials, misunderstandings and conflicts that may arise during the construction or final finish phase can be prevented. The fire official can emphasize concerns and coordinate responsibilities with other code enforcement officials. Design professionals and contractors benefit from this procedure as well, because problems which otherwise would cost them time and money are eliminated before construction begins.

Site Plan Reviews: The site plan review provides the fire official with the first look at new construction and additions to existing buildings. The site plan provides an overview of the intended construction in relation to existing conditions and includes information such as building placement, exposures, size, type of construction, occupancy, water supply (public and private mains), hydrant placement, and access. The site plan also may provide information about existing conditions that must be modified such as abandoned flammable liquid tanks or pipelines in the area. The contour of the land may be significant. Also, present and projected uses of adjoining properties, including zoning, should be reviewed at this time.

Preliminary Building Plans: The preliminary building plan gives the fire official an opportunity to comment on those features of the building which significantly affect life safety and protection of the building from fire. The depth and scope of the review and comments will depend upon local conditions.

The provision of required fire protection systems is a primary concern in plans review. Because 75 percent of the building code may relate directly to provision for life safety in case of a fire within the building, the review may include the type of occupancy, allowable areas and heights, fire separations, fire resistance of construction, interior finish, occupancy loading, number and location of exitways, protection of vertical openings, and special hazards.

Final Building Plans and Specifications: When the final building plans are submitted, they should include modifications required by the review official and agreed to by the design professional submitting the plans. Plans may be approved if they agree with the applicable code requirements. A building permit can then be issued and construction may begin.

Plans review and approval must be followed by on site inspections to ensure that the fire protection features and systems are constructed and installed as planned and approved. All deviations from the approval plans should be documented. All agreements reached on site or by telephone should be documented for all parties in follow up correspondence or file memos. This correspondence and a copy of the approved plans should be retained in an organized filing system or on microfilm or microfiche so a permanent record of building construction is available for future reference.

As practical, building construction information should be provided to fire companies responsible for fire suppression and/or fire inspection of the building. Information provided in all of the plans discussed in this section can aid in preplanning for fire operations.

Certificates of Occupancy: If the fire marshal's office is not the primary issuing agency for occupancy certificates, then either the fire marshal's office or the fire prevention division should be involved in the final inspection process. Each agency should certify the building before the Certificate of Occupancy is issued. This certificate indicates that all requirements under the building and other applicable codes have been met and that the building is safe and habitable. All fire protection systems should be tested and placed in service before occupancy is allowed. The building should not be occupied until inspection agencies have approved it. This final review helps to ensure the life safety of the occupants and to verify that any required corrections will be made in the occupancy.

CONSULTATION

The public looks to the fire service for answers and advice concerning fire problems. Due to its unique ability to provide this information, the fire department should offer consulting services to the community. Fire prevention officers must be capable of explaining fire codes, fire related sections of building codes, and the application of specific standards, to help design professionals, contractors, and members of the building trades in their duties. Consultation also should be available directly to property owners, managers, occupants, and members of the general public, who may not be as familiar with fire problems and their solutions.

To maximize this service, the fire department should inform and instruct citizens about the best ways to deal with fire hazard situations. When the fire prevention division receives frequent calls about the same problem, it may wish to develop standard recommendations for specific situations. The fire department should maintain a library of up to date reference materials. In addition, the chief and other officers should be acquainted with individuals outside the fire department who have specialized knowledge or experience and who could act as resource persons. These and other consulting services usually are based in the fire prevention division or the fire marshal's office.

FIRE INVESTIGATION

The thorough investigation of fires is an integral part of the fire department's commitment to public safety. Fire investigation includes two areas: (1) fire cause determination and (2) investigation of criminal actions which may have contributed to a fire.

Fire cause determination is of major importance to a fire prevention program. Analysis of the causes of fires within a community is the basis for establishing fire prevention program priorities and providing firesafety information to the public. Maximum utilization of resources must be based on firm information about the local fire problem. As information is accumulated from fire investigations, the data become available on which to base corrective programs. Over a period of time, this data will indicate trends in the area's fire problem; a fire prevention program designed to tackle priority areas then can be implemented.

If it is determined that a fire is caused by arson or other unlawful burning of property, a full criminal investigation should be conducted. Fires classified as "incendiary or suspicious" should be investigated. Assertive fire investigations by skilled, trained specialists can have a positive deterrent effect on arsonists.

In communities or states where fire department investigators have police powers, fire department personnel may conduct an investigation to the point of arrest and incarceration of fire perpetrators. Where fire department personnel do not have police powers, fire cause determination may be made by the fire department with the followup investigation conducted by the police, sheriff, state police, or investigators from the state fire marshal's office.

PUBLIC FIRE EDUCATION

Firesafety education is an increasingly valuable area of public fire protection. This function has undergone rapid change and growth, with the focus on the importance of prevention in public fire protection planning.

Public firesafety education has two facets: fire prevention education and fire reaction education. Both facets are necessary to fundamentally change the way the general public views fire, and to encourage people to act in a firesafe manner. The public needs to be motivated and instructed in actions that minimize the chances and dangerous effects of a fire, should one occur. The basic method of instruction is changing from "preaching" to "teaching." Information is best presented in short, interesting, even humorous messages which require thought and place responsibility for action on the message recipients. Recipients relate most effectively to the firesafety message when messages are presented in a dramatic context. Cultural and language factors should be considered when firesafety messages are prepared for communication throughout the community.

National firesafety campaigns can be matched to local efforts. NFPA's Learn Not to Burn® effort is one such national campaign, with messages designed to supplement, not supplant, local fire prevention education programs. For example, the 30 second television spots with actor Dick Van Dyke, which are popular and effective in teaching firesafe behavior to the public, are available free of charge to all television stations. Community programs can be developed using the "Learn Not to Burn" theme and applying it to priority local fire problems.

Community awareness and participation at the grassroots level are the keys to encouraging people to adopt firesafe behaviors. Civic and service clubs, youth and fraternal organizations, neighborhood action groups, the business community, schools, and other groups are contributing to the growth of firesafety education. Highly effective programs in such areas as escape planning, smoke detector installation, fire hazard inspections, burn prevention, juvenile firesetting problems, firesafety for babysitters, and firesafety for the elderly have been conducted by community groups in conjunction with local fire prevention personnel.

Thousands of teachers are giving classroom instruction on firesafe behaviors. NFPA developed the *Learn Not to Burn Curriculum®* to assist in this effort. The Curriculum designed to teach 25 basic firesafety and fire prevention behaviors presents comprehensive teaching strategies for teachers and fire departments to convey key life safety messages to children from kindergarten through the eighth grade. Because the curriculum is designed to integrate easily into existing school subjects, it does not represent an additional course fro instructors.

The Public Education Office of the U.S. Fire Administration (USFA) has furnished extensive assistance to help communities and states develop comprehensive public fire education programs. This federal agency has provided funding, technical support, and regional and national workshops to facilitate the exchange of information. State exchange and resource centers for public firesafety education material have been established through the efforts of the Public Education Office. These efforts have greatly increased the amount of information exchanged among fire educators and encouraged the growth of their educational programs.

Many fire departments and fire prevention divisions have appointed full time firesafety education officers. As more people have assumed these positions, and with the development of NFPA 1031, *Standard for Professional Qualifications for Fire Inspector, Fire Investigator and Fire Prevention Education Officer*, the level of talent and professionalism of firesafety educators has increased. Many fire departments no longer require firesafety education officers to be trained fire fighters; rather, these departments recognize the value of hiring educators with teaching experience and skills to develop and present educational programs geared to all ages and groups.

Groups with Special Information Needs

Groups with information needs similar to those of the general public, but different enough to require specialized programs include educational, industrial, institutional, high rise, civic, service, elderly, professional, handicapped, and commercial groups. Many innovative ways are used to reach such groups with firesafety information.

Seasonal Activities

Informative educational programs of interest to the general public often are built around the four seasons of the year. Public education programs include National Fire Prevention Week (always the week that includes the date of October 9) and Operation EDITH. Many materials are available from NFPA for use in public education programs, including some featuring Sparky the fire dog—symbol of personal and property firesafety.

Bibliography

References Cited

Schaenman, Phillip S., and Swartz, Joseph. 1974. *Measuring Fire Protection Productivity in Local Government.* National Fire Protection Association, Quincy, MA.

NFPA Codes, Standards, Recommended Practices and Manuals. (See the latest *NFPA Codes and Standards Catalog* for availability of current editions of the following documents.)

NFPA 1, *Fire Prevention Code.*

NFPA 1031, *Standard for Professional Qualifications for Fire Inspector, Fire Investigator and Fire Prevention Education Officer.*

NFPA 1202, *Recommendations for Organization of a Fire Department.*

NFPA 1301, *Guide to Public Fire Prevention Criteria.*

Additional Readings

Adams, Robert C., ed., *Firesafety Educator's Handbook*, National Fire Protection Association, Quincy, MA, 1982.

Bahme, Charles W., *Fire Service and the Law*, National Fire Protection Association, Quincy, MA, 1976.

Bare, William K., *Fundamentals of Fire Prevention*, Wiley and Sons, NY, 1977.

Berry, Dennis J., Fire Litigation Handbook, National Fire Protection Association, Quincy, MA, 1984.

Bryan, John L., and Picard, Raymond C., eds., *Managing Fire Services*, International City Management Association, Washington, DC, 1979.

Bugbee, Percy, *Principles of Fire Protection*, National Fire Protection Association, Quincy, MA, 1978.

Building Officials and Code Administrators International, Inc., *The BOCA Basic/National Fire Prevention Code*, Country Club Hills, IL, 1984.

Building Officials and Code Administrators International, Inc., *Legal Aspects of Code Administration*, Country Club Hills, IL, 1980.

Clet, Vince H., *Fire-Related Codes, Laws, and Ordinances*, Glencoe Press, Encino, CA, 1978.

Directory of State Building Codes and Regulations, National Conference of States on Building Codes and Standards, Herndon, VA, 1982.

Fire and Emergency Resource Directory, National Fire Protection Association, Quincy, MA, 1984.

Fire Inspection Management Guidelines, National Fire Protection Association, Quincy, MA, 1982. (See especially Guideline No. 11, "Resource Directory," for extensive bibliographic material on fire prevention and code enforcement.)

Fire Marshals Association of North America, *National Conference of State Fire Marshals Report*, National Fire Protection Association, Quincy, MA, 1985.

Hall, John R. Jr., et al., *Fire Code Inspections and Fire Prevention: What Methods Lead to Success?*, National Fire Protection Association, Quincy, MA, 1979.

International Fire Service Training Association, *Fire Prevention and Inspection Practices*, IFSTA 110, Oklahoma State University, Stillwater, OK, 1974.

——— *Public Fire Education*, IFSTA 606, Oklahoma State University, Stillwater, OK, 1979.

Lucht, David A., *Fire Prevention Planning and Leadership for Small Communities*, National Fire Protection Association, Quincy, MA, 1980.

Morse, H. Newcomb, *Legal Insight*, 2nd ed., National Fire Protection Association, Quincy, MA, 1975.

Moulton, Gene A. ed., *Conducting Fire Inspections: A Guidebook for Field Use*, National Fire Protection Association, Quincy, MA, 1982.

National Commission on Fire Prevention and Control, *America Burning, The Report of the National Commission on Fire Prevention and Control*, U.S. Government Printing Office, Washington, DC, 1973.

Public Fire Education Planning, United States Fire Administration, U.S. Government Printing Office, Washington, DC, 1979.

Robertson, James C., *Introduction to Fire Prevention*, 2nd ed. MacMillan, NY, 1979.

Rosenbauer, Donna L., *Introduction to Fire Protection Law*, National Fire Protection Association, Quincy, MA, 1978.

Schaenman, Phillip S., et. al., *Procedures for Improving the Measurement of Local Fire Protection Effectiveness*, National Fire Protection Association, Quincy, MA, 1977.

———, *Small Community Fire Departments: Organization and Operation*, National Fire Protection Association, Quincy, MA, 1982.

Standard Fire Prevention Code, Southern Building Code Congress International Inc., Birmingham, AL, 1985.

Tuck, Charles A., *NFPA Inspection Manual*, 5th ed., National Fire Protection Association, Inc., Quincy, MA, 1982.

Uniform Fire Code, International Conference of Building Officials, Whittier, CA, 1984.

FIRE AND ARSON INVESTIGATION

Robert E. Carter

While this chapter is entitled "Fire and Arson Investigation," its principle thrust is arson fires. However, without fire investigation, there will normally be no arson investigation, so the two are closely related.

Although there has been a significant decrease in the number of incendiary and suspicious structure fires in recent years, fires thus classified in 1983 accounted for 12.9 percent of all structure and vehicle fires and for 23.3 percent of all property damage in structure fires. An estimated 970 civilian fire deaths, representing 19.1 percent of total fire deaths in structures, occurred as a result of incendiary or suspicious fires. The direct monetary loss in 1983 from these fires was $1,421,000,000. Note that this is direct loss and does not consider effects such as blighted or devastated neighborhoods, loss of tax base as a result of buildings destroyed, loss of jobs and income from industrial facilities deliberately burned, increased insurance rates, possibly increased taxes for fire protection, and the threat of injuries and deaths. Further, these figures cover only fires actually classified as incendiary or suspicious. It has been estimated by fire officials that at least one-half of those fires classified as being of undetermined origin are in fact incendiary.

This chapter is not intended to provide instruction on the technical aspects of fire and arson investigation. Rather, its purpose is to provide information to those individuals and/or organizations on how to best manage and organize fire and arson investigation.

Who is responsible for fire and arson investigations? Determining fire cause has traditionally been a function of fire service personnel. The fixing of investigative responsibility becomes confusing when the cause of the fire is determined to be suspicious. Some fire chiefs consider investigation of fires a fire service function, but the police chief, on the other hand, may argue that arson is a crime and should be investigated by law enforcement personnel. When these conflicting philosophies exist in the same jurisdiction, obvious problems arise; in extreme cases two separate, uncoordinated investigations may be in progress simultaneously.

Mr. Carter was Chief Fire and Arson Investigation Specialist on the staff of NFPA until he recently retired.

An equally frustrating situation occurs when the fire chief believes that arson investigations should be a police function while the police chief maintains that they are a fire service responsibility. The result here may be that no investigations are conducted.

When fire/arson investigations are assigned to the fire department, the function often is assigned to the fire prevention bureau. The adopted code may read: "The fire prevention bureau within the department of fire shall determine the origin and cause of all fires." Problems develop in this procedure when fire prevention personnel or fire inspectors assigned this responsibility are not given the authority, including police powers, to carry out the assignment. In some jurisdictions these individuals may be instructed that when the incendiary origin of the fire is established, they are to request the services of police personnel. This procedure creates several potential problems. First, how many indicators are required before police assistance is justified and requested? Second, morale among those who are doing much of the search at the fire scene—a dirty, unrewarding job—reaches rock bottom when headlines announce, "Police Arrest Arsonist," followed by a news story that does not even mention participation by fire service personnel. Third, detectives assigned to arson cases may lack arson training and the required investigative expertise. Fourth, work schedules of fire and police personnel are often not compatible.

Similar problems arise when arson investigation is assigned to the police department. Police personnel overloaded with other cases often cannot devote the hours, days, or weeks necessary to conduct a thorough, comprehensive arson investigation. Furthermore, a detective may be pulled out of the middle of the investigation to respond to a police emergency.

Is there a procedure for fire/arson investigation that can eliminate or reduce the impact of the aforementioned problems? Experience in jurisdictions across the country which have developed a fire/arson investigation unit consisting of both fire and police department representatives indicates that this is a logical and workable arrangement. Details of this type of unit will be discussed in this chapter.

ORGANIZING AN INVESTIGATION UNIT

Size of the Unit

The number of personnel assigned to the arson unit depends upon several factors including population, the identified arson problem, and financial constraints of the department. Initially, size must be based on a common sense evaluation of the need for personnel. Some departments determine personnel needs by population alone; for example, one arson unit member for each 20,000 population. This approach is not always workable because some jurisdictions are more prone than others to high arson rates.

In determining the most effective functioning size of the unit individuals must realize that a thorough investigation of fires is a time consuming process. If individuals do not have adequate time, fire cause determination and follow up investigations will not be accomplished.

As the activities of the unit develop, investigative activities and their results must be carefully evaluated. If pressured investigators are attempting to conduct five or six cases simultaneously, working an excessive number of hours to keep up, ignoring or glossing over other suspicious fires, or closing out investigations not actually completed, the squad then needs additional personnel.

Unfortunately, members of the arson unit often work in an atmosphere of semi-isolation, discussing the workload and the need for personnel only with each other. Supervisory personnel frequently must monitor and evaluate the performance of the arson squad. Records should be kept which identify the results of their efforts as well as work hours, number of investigations, and response time.

Selection

Many of the most effective arson units consist of both fire and police department representatives. This structure utilizes fire behavior and legal information as they both apply to investigative procedures and techniques.

The process of selecting members of the arson unit should include a review of each prospective member's departmental records, including evaluation reports. This review will indicate not only appropriate experience and performance levels, but also how the individual performs as a team member. A review board with people knowledgeable in fire/arson investigation should interview applicants for the arson squad. Professional competence should not be the only consideration in selecting members; interest and motivation also must be evaluated. The review board should ask applicants why they want to become arson investigators, what they can contribute to the unit, and similar questions. Applicants should be asked how they would react psychologically to the relatively low success rate in arson investigations. Applicants may be subject to some rather harsh questioning, which tests their reaction to pressure and ability to control their emotions in a stressful environment.

Review board members should make every effort to evaluate how applicants cooperate and work with others in the same areas and across departmental lines. There is no place on the team for a loner who continually withholds information on investigations from other members and refuses to share expertise. In substance, to work well on the arson unit, members must be unselfish, strongly motivated to contribute to the overall success of the unit, and not seek individual recognition.

Educational and academic criteria must be established for all applicants. Training courses and self-study programs attended should be considered in evaluating applicants. Duties performed in arson investigations involve review of legal and technical documents, insurance policies, claims forms, and financial reports and statements. Additionally, arson investigators must prepare satisfactory reports at the conclusion of their investigation. Furthermore, testifying in court requires investigators to express themself clearly, and concisely.

Fire investigation is a physically demanding job, a fact that often is not considered in selecting arson squad members. Examination of the fire scene can involve many hours of what is actually manual labor—restoring furniture, fixtures, doors, and other items to prefire position; crawling through small spaces in attics and basements; and digging through many thick layers of debris. The physical demands of this activity are obvious. Members of the arson unit must be healthy, active, physically able individuals.

In the past, there has been a tendency to consider the fire investigation, prevention, inspection, and training areas as logical assignments for individuals who can no longer carry out other duties in the organization. These assignments are made occasionally because a disabling injury incurred in the line of duty in either the fire or police departments. This trend must not be continued or tolerated, except under those conditions where the individual involved can carry out investigative duties effectively despite any disabling injury. It must be possible to remove personnel from the arson team for unsatisfactory performance or reassign them at their own request. The arson team must perform as a team. A dissident member in the group not only will perform unsatisfactorily, but also will disrupt the total team operation.

Police Department Personnel: Experience indicates that detectives should be police department members selected for the arson unit. Uniformed patrol personnel do not have the experience and expertise required to conduct the highly specialized investigations needed for arson. Members of the arson squad will benefit from on-the-job training, but they should possess basic investigative skills prior to joining the squad and be able to conduct interviews, process evidence, interrogate suspects, prepare cases for and testify in court.

Fire Department Personnel: Within fire departments in most metropolitan areas there are individuals who are experienced in fire cause determination, but not in basic investigation. Expertise in this area will be attained through extensive training and work with experienced police officers for an appropriate period of time.

Location

The location of the arson unit is a decision which may significant impact on the success of the operation. There are three possible locations for the unit: (1) in the local prosecutor's office, (2) in the fire department, and (3) in the police department. Each of these alternatives is discussed in the following paragraphs.

Local Prosecutor's Office: Historically, arson investigators have complained that prosecutors were inaccessible during critical stages of an investigation. On occasion investigators needed legal advice immediately, for example, about whether evidence collected during the investigation warranted arrest of a suspect or suspects. During an investigation, with the unit under the direction of the prosecutor's office, a prosecutor might have recommended action to enhance and expand the investigation.

Investigators believe that a pretrial conference with the prosecutor is necessary to review the facts of the case, examine physical evidence, review anticipated testimony by experts and other witnesses, and evaluate the potential value of circumstantial evidence. This approach answers the criticism that, because of a lack of regular involvement in arson investigations and prosecutions, prosecutors did not possess the expertise required to effectively prepare and present a case in court.

Until recently, overall participation by prosecutors in arson investigation has been limited or nonexistent. Now in some jurisdictions, however, prosecutors organize, direct, and supervise the entire arson investigation and prosecution program. A special prosecutor is appointed to head up the program because of severe arson losses, disputes between fire and police personnel over investigative duties and responsibilities, and the lack of an effective arson control effort. This provides legal advice during various phases of the investigation, ensures that prosecutors will acquire the necessary expertise to competently prosecute cases, and gives a high priority to arson.

Perhaps a more logical arrangement establishes liaison between the arson unit and the prosecutor's office, with one or more members of the prosecutor's office designated to work on a continuing basis with the arson unit. This leaves the arson unit under the operational control of either the fire or police department while providing for involvement of the prosecutor's office in appropriate phases of the investigation.

Fire Department: In many cities the arson unit is located in the fire department under the operational control of a fire officer. Although the fire department is usually concerned with fires regardless of cause, arson fires pose a serious threat to the safety and lives of fire fighters. The fire department also has a responsibility to identify suspicious fires, collect fire data, and conduct effective arson prevention and public information programs.

Police personnel assigned to the arson squad may remain under the administrative control of the police department. The fire department supervisor of the unit directs the operation, assigns cases, develops work schedules, coordinates training, and provides the necessary investigative equipment. The police department maintains administrative control covering payroll, promotions, and similar matters for their personnel in the arson squad.

When the arson unit is in the fire department, an important consideration is the specific location of the unit within the department organization. One option is to place the squad directly under the supervision of the department chief; this obviously provides a high degree of visibility and recognition for the unit, and a direct line of communication to the department head. A concern is that arson investigation may be given such a high priority that other departmental operations such as suppression, inspection, and prevention are neglected.

On the other hand, the arson squad can be placed in the fire prevention bureau. Under these circumstances it is probable that the unit will lose its identity among the other activities of the bureau, including inspections, code enforcement, education, and other necessary but non arson related functions.

A compromise location for the unit must be found high enough in the organizational structure to command respect, yet not so high as to interfere with the unit's operational efficiency. Probably the most acceptable option is to place the unit under a fire officer who reports directly to the chief of the department. Regardless of whether this individual is a lieutenant, captain, or fire marshal, the duties and responsibilities of this position must be clearly defined, including relationships with other supervisory personnel. The direct supervision of the arson unit must be vested in a knowledgeable individual with a complete understanding of the functions, responsibilities, and role of the squad relative to the entire fire department operation.

Police Department: When the arson squad is located in the police department, fire service personnel may remain under the administrative control of the fire department. Several potential problems are created when the police department assumes operational control of the unit. Of primary concern is the fact that the police department may be preoccupied with other criminal investigations. Arson investigation, may then become a low priority, with members of the arson unit sometimes being assigned to other investigative activities. Additionally, the chief of the department or of the detective division may be unable to adequately supervise the arson unit because of other criminal matters.

At the same time, locating the arson unit in the police department presents certain advantages. Police department records are readily available and could be valuable during the investigation of questionable fires because some suspects may have criminal records not related to arson. Some departments also have laboratories for testing arson evidence, and arson intelligence files. Those jurisdictions which locate the unit in the police department consider arson a criminal problem requiring access to criminal investigative facilities. While these same facilities are available when the unit is in the fire department, use of the facilities is easier when the unit is located in the police department.

UNIT OPERATIONS

Work Schedule

Depending upon the size of the jurisdiction involved, at least one member of the arson squad should be on duty or on call at all times to promptly respond to a request for investigation. Ideally, the investigator should reach the scene while the fire is still in progress to observe the behavior of the fire, how it spreads, etc. and also to look for potentially important witnesses.

Communication is critical to the success of any work schedule. The members of the unit must be familiar with the work schedule. The individual responsible for dispatching members of the arson unit must not only be able to locate unit members, but also must clearly understand priorities for investigation. For example, if simultaneous requests are received to investigate a small one car garage fire and an apartment building fire which may involve fatalities, judgment must be exercised based on established standard operating procedures. At the same time, investigators should not be called off one case to respond to a request for investigation of a minor fire. If requests for investigation of two serious suspicious fires are received simultaneously, back up members of the unit must be available for response.

The need for adequate radio communication between unit members is obvious; however, if dispatchers are not familiar with the arson squad operation, members of the unit may waste time rushing from fire scene to fire scene with no significant results.

Screening Fires

Ideally, the arson squad should respond to all fires to determine origin and cause, thus identifying accidental as well as suspicious or incendiary fires. However, in actual practice there are few jurisdictions which have the personnel to operate in this manner so various systems must be developed to select fires for investigation.

Probably the most widely used practice involves identifying the suspicious nature of the fire by the fire officer in charge. This would appear to provide an effective screening process utilizing this officer's fire knowledge and on site observations. However, experience in some jurisdictions seems to indicate that various factors, including fireground responsibilities and lack of training, make this individual ineffective for this particular task, resulting in failure to identify actual incendiary fires. Nevertheless, this system does offer direct communication with fire fighters who can convey their observations promptly about the scene. The efficiency of this operational concept can be improved by appropriately training the fire fighters in detecting arson and the fire officers in identifying incendiary or suspicious fires.

In other jurisdictions the officer responsible for arson investigation concentrates on those fires where the investigation will contribute most significantly to the reduction of the arson rate. Initially, the determination must be made as to the number of fires the arson unit can effectively investigate each year. After this determination is made, the most viable means of aligning the number of investigations to this figure must be identified.

One means is to set a minimum loss figure, with losses above this figure involving investigation by the arson unit. While this procedure may establish a reasonable workload for the unit, it also produces some problems. For example, a community or a section may be subject to an epidemic of minor fires set by a lone "firebug" or by a gang of juveniles. Unless some investigation is made of these fires, the arson frequency and the seriousness of the fires will increase significantly. In substance, a series of so called lesser fires over a long period of time can be as serious as a major fire. A basically similar screening process consists of limiting arson investigations to those fires involving large loss, fatalities, and injuries. Another criteria for determining whether investigations will be performed consists of identifying those fires which most probably involve arson. Both of these arrangements result in problems similar to those cited previously, and there will be exceptions to any screening plan.

Some localities call for an intermediate investigation following up on the initial cause determination by the fire officer. This needless duplication of effort and questionable utilization of personnel can be eliminated if fire officers are properly trained in origin and cause determination. (Of the various current screening processes, this appears to be the least efficient.)

As stated previously, maximum efficiency can be attained if the arson squad, can respond to all fires. Accepting that this procedure is beyond the available resources of most communities, a viable alternative is for the arson unit to conduct an initial investigation of fires identified as suspicious by the fire officer. If the arson squad does not have sufficient personnel to accomplish this, additional personnel should be assigned to the unit.

Vehicles and Equipment

Proper investigative tools for an arson unit are needed and, equally important, must be available at all times. Each member of the arson unit should be assigned an automobile which is kept on a 24 hour basis both on and off duty. This provides for prompt response by individuals and when the entire unit is called on for emergency.

Except for special purposes (which will be discussed later), these vehicles should be unmarked and of various makes, models, and colors, with "blind" tags so they will not be immediately identifiable.

An increasing number of jurisdictions include vans and mobile crime laboratories among the vehicles available to the arson squad. A midsized van for example, can be used to privately interview witnesses or question suspects without arson squad members having to leave the fire scene; and serve as a storage facility for fire investigation equipment and tools; and provide a mobile office for the investigators. This type of vehicle may be particularly useful in "stake out" and surveillance activities, providing its use and purpose are not readily apparent. In order for this unit to be practical, it should be of reasonable size, and highly maneuverable. In short, investigators should be able to bring such a vehicle to the fire scene as quickly as possible, and the vehicle must be useful after arrival.

The mobile crime lab serves a different purpose. Because of the equipment aboard the vehicle, the mobile lab is necessarily larger, bulkier, less maneuverable, and thus subject to delays before reaching the fire scene. The mobile lab serves a definite purpose and is quite useful for prompt, on the scene examination of samples, especially at major fire losses. Of course, such tests normally provide only preliminary identification of accelerants. Complete laboratory testing with more sophisticated equipment is required to provide positive and specific data about the samples tested.

Problems arise when an arson squad uses only one vehicle for both purposes. When outfitted with a laboratory and the necessary equipment, the unit becomes so bulky that its maneuverability is substantially reduced, thus largely eliminating its immediate usefulness to the arson squad at the scene.

Tools of the Trade

Investigation of a fire scene to determine origin and cause can be challenging even under favorable conditions. The difficulty of the job can be compounded if the investigator lacks the basic tools to accomplish the task.

For those individuals charged with either initial fire cause determination or complete fire scene investigation, the following are some essential tools of the trade, purpose, and use. Some of the items described may not be required in initial fire cause determination.

Accelerant or Hydrocarbon Detector: For determining the presence of an accelerant during fire scene examination, several detection models are available currently and their usefulness and cost varies noticeably. The L.I.S. combustible gas detector (cost approximately $5,000) receives high ratings by forensic experts.

Axe (hand or fire department, pickhead type): For tearing down plaster board, walls and partitions.

Brooms and Brushes: Of various sizes, needed to sweep floors and rugs after heavier material has been removed.

Camera: Good quality 35 mm camera with a compatible flash unit, which the investigator can use effectively. Camera batteries must be fresh. Arson squad members will photograph the fire scene before and during various stages of the fire scene examination.

Cardboard Boxes: For consolidating or transporting of "dry" evidence in which the presence of an accelerant is not suspected, such as a window or door lock, a padlock, or similar item found at the scene.

Carpenter's Level: To check the slope of floors, which may indicate the direction of flow of flammable liquids that resulted in unusual char patterns.

Char Depth Rule or Gauge: To measure the depth of char in wooden surfaces. This can be accomplished by sharpening the penetrating end of a small diameter rod such as a "pipe smoker's knife," and marking it off in eighths of an inch (3 mm). A gauge used for measuring tread depth on tires also is suitable.

Claw Hammer or Tool: To tear out molding, etc., pull out nails. For uses similiar to the axe.

Compass, Small: To establish magnetic directions from the involved structure. Accurate determination assists in sketching, diagramming, report writing, and court testimony.

Cultivator (small hand type): To dig through debris without disturbing potential evidence.

Film: An essential part of an investigation. There are few investigators who have not at some time in their career been caught at the scene without any film, or the wrong type of film for the job at hand.

Glass Jars and Vials, (various sizes): To collect and transport suspected flammable liquid samples or samples with a high liquid content to a laboratory. Jars and lids must produce an airtight seal. While it is preferrable to hand carry such evidence to a laboratory, it can be shipped with careful wrapping and packing.

Hoe: To remove debris in greater quantity than with a small shovel or trowel.

Knife: Scraping, cutting, and for a variety of uses.

Labels and Tags: To identify evidence; may be attached to item or material with string, but stick on type of labels and tags are preferable.

Lights: Probably the most neglected and yet most essential tool of the arson investigator's trade. Conducting a fire scene examination in a pitch dark building by the light of a two cell flashlight with weak batteries is an exercise in futility. A preliminary inspection of a fire scene can be made with a strong flashlight or a battery powered fluorescent hand lamp, sometimes referred to as a camping or "safari" light. To effectively examine an entire room for char and burn patterns, however, the entire room should be lit up with floodlights and spotlights powered by portable generators, or by electricity brought in from adjacent buildings. Proper illumination is essential for an arson investigation.

Magnifying Glass: For examining details such as separated wiring to determine it was cut, shorted circuits or burns to the wire.

Marking Pencils: To mark evidence.

Mason's Trowel: A small trowel to move levels of debris. A small sand shovel, such as a child might use at the beach, is also suitable.

Metal Cans: Unused clean metal paint cans of various sizes, usually quart (1 L) and gallon (4 L) size are the best containers readily available to package fire debris suspected of containing flammable liquids. These cans are sturdy, and easily sealed and resealed. Larger cans of 5 to 30 gal (19 to 113 L) are useful for collecting and preserving larger samples of evidence.

The metal lid must be firmly in place on this type of container, sealed with a strip of pressure-sensitive tape running around the container and adhering to both the lid and the container. Metal containers do not present a problem for laboratory analysis of the sample because, after a small hole is drilled in the top and the sample withdrawn, the container can be sealed with aluminum foil and tape.

Some cans currently used for water based emulsion paints contain a liner of yellowish transparent film or an opaque gray finish. When tested, this liner does not in itself contribute significant hydrocarbons and is impermeable to gasoline and other corrosives. The liner protects the metallic surface of the interior of the container only as long as the liner is intact. If the surface of the liner is broken and the metallic surface is exposed, wet debris in the can may rust the container from the inside.

Coffee cans with plastic tops are not recommended as evidence containers. Although convenient and easily obtained, the fault with these containers lies in their plastic (polyethylene) cover, which will leak the hydrocarbon vapors. If these containers are used, a sheet of aluminum foil should be put over the top before the plastic cover is pressed on top. (Obtaining an airtight seal may be a problem.)

Polystyrene containers are totally unsuitable for collecting and preserving liquids such as gasoline, since they break down the polystyrene and render the evidence useless. Likewise, polyethylene envelopes and bags should not be used for collecting evidence in arson cases; significant accelerant loss and the possibility of sample contamination cannot be avoided.

Paper Towels and Wipe Cloths: To wipe clean surfaces on which numbers or letters are illegible because of soot, ash, or discoloration.

Plastic Bags: Bags made of polyester, polyolefin or nylon can be used as containers for evidence suspected of containing hydrocarbons. Polyethylene bags are not acceptable for containing this type of evidence, as leakage of the hydrocarbon vapors occurs. These and paper bags can be used for collecting and transporting "dry" evidence that does not contain a flammable liquid.

Pliers: For handling evidence without contamination, and for other miscellaneous purposes.

Protective Clothing: The same emphasis placed on protective clothing during fire fighting and salvage operations should apply during fire scene examination. A hard hat or fire helmet should be worn when the possibility exists that either the investigator may bump into hanging objects or be struck by falling structural materials. Waterproof boots with steel toe and arch protection are necessary against penetration by nails and other sharp articles. Heavy duty gloves are required, particularly when moving metal roofing or other objects with sharp edges. An easily washable outer garment such as coveralls, can be worn over other clothing or alone. Turnout coats and trousers may be used, but these can become very uncomfortable because of bulk and weight. Safety glasses are appropriate where wiring and other sharp objects are hanging down in the structure.

Pry Bar: Useful for purposes similar to those of the axe and claw hammer.

Rake: A small hand rake with a short handle is useful for sifting through debris.

Reporting Tools: Useful items for taking notes and for sketching and diagramming the fire scene include clipboards, pads, grid paper, pencils, ballpoint and colored marking pens, rulers, and notebooks. It is recommended that a permanent bound notebook be used (to prevent allegations that pages have been removed).

Saws, (various sizes and types): A handsaw, preferably an electrical circular type, is essential to cut structural members. This tool is particularly useful for removing sections of flooring for submission to a laboratory. A hacksaw for cutting metal is frequently required.

Scissors, Shears: To cut and remove sections of upholstered furniture, mattress covers, drapes, curtains, and miscellaneous cloth items. Heavy-duty shears may be required to cut carpeting, rugs, and rug pads.

Screwdrivers, (various sizes, both straight blade and phillips): To remove electrical fixtures and mechanical controls on various household appliances. Large screwdrivers can be used for prying.

Shovel, (short handle, square point, flat end): To remove and uncover debris at floor level.

Sifting Screen: To sift for minute evidence in debris; screens can be attached to a portable frame. May be sand or ¼ in. (6 mm) mesh screen, depending on type of material being processed and the evidence being sought.

Sponges or Cotton Swabs: To sop up a suspect flammable liquid for transfer to a container. Each item can be used only once.

Stapler: For a variety of purposes.

String or Rope: Heavy string or light rope (clothesline weight) for marking off security, work, and search areas.

Tape Measures and Other Measuring Devices: For obtaining dimensions of both the interior and exterior of the burned structure. May consist of 50 or 100 ft (15 or 30 m) steel tape measure or 6 to 12 ft (2 to 4 m) tape and folding rule.

Tape Recorder: To record observations made during the fire scene examination.

Tin Snips: To cut metals of various thicknesses.

Tweezers and Tongs: To pick up small items which may be used as evidence without damaging or contaminating the same.

Wire Cutters: To facilitate removal of electrical wiring.

Wrenches, (various sizes): To remove electrical or mechanical controls from furnaces, sprinkler systems, and similar equipment.

Training

An appointment to the arson squad does not automatically make a person a full fledged investigator; training is necessary. A training program developed for members of the arson unit must take into consideration the professional background of the individuals. Arson investigation training for fire and police oriented individuals will differ noticeably in content.

Personnel whose investigative activities in the fire department have been confined to origin and cause determination must receive basic and advanced law enforcement training. Subjects covered should include interviewing witnesses, questioning suspects, taking statements, writing reports, and testifying in court. While these investigative activities are important, it is critical for the fire oriented person to understand legal and judicial matters, including rules of evidence, search and seizure, and arrest procedures. It is a fallacy to assume that fire investigation experience, as opposed to arson investigation experience, will provide this vital knowledge. Illegally seized evidence and possible false arrests can result from a lack of understanding of legal requirements in both areas. The experience of work with police members of the arson unit supplements but does not replace training, education, and study in arson investigation.

The training provided for police and fire department personnel will differ greatly. In those jurisdictions where the responsibility for fire cause determination is vested in the fire department, police personnel who continue the criminal part of the investigation develop little expertise in fire scene examination. Detectives newly assigned to the arson unit obviously have no knowledge of fire behavior nor expertise in fire scene examination. Therefore, training must be provided on subjects such as the chemistry of fire, fire behavior, origin and cause determination, and fire scene examination. Additionally, to function effectively, police personnel must acquire a basic understanding of fire department equipment, terminology, and operations.

Upon completion of the training outlined above, advanced training should be conducted for the entire unit as a group. This training should be intensive and is most effective if held under realistic conditions. Training in fire scene examination should take place at a fire scene;

interviews and interrogations should be practiced on live subjects; testifying in court should be as realistic as possible, with a mock trial based on facts developed during the simulated investigations. Prosecutors and defense council should participate in this trial, bringing the investigator face to face with problems encountered in the courtroom. This latter activity will emphasize to the investigator the need for a thorough arson investigation with absolute attention to detail.

Appropriate sections of NFPA 1031, *Standard for Professional Qualifications for Fire Inspector, Fire Investigator and Fire Prevention Education Officer* may serve as a measure of the level of knowledge and competence acquired by members of the arson unit at the completion of training. While this standard is specifically oriented to individuals with a fire service background, it may be used by all organizations to establish levels of professional competence for individuals assigned to arson investigation.

This training enables all members of the arson unit to function effectively throughout the entire investigation. Jurisdictions in which fire cause determination is assigned only to selected members of the unit, with criminal investigation assigned to others, may not find such intensive training necessary. Experience has shown that this separation of duties is probably not the most effective system concerning results attained, utilization of personnel, and development of skills by members in the unit.

A training activity not directed to members of the arson unit but which has a significant impact on the success of the operation, consists of courses of instruction on arson detection for fire fighters. Such training is not intended to turn fire fighting personnel into investigators or fire cause determiners. At this level, programs should have one basic objective: to train fire fighters to observe various conditions and circumstances which may indicate arson.

ARSON DETECTION AWARENESS BY FIRE FIGHTERS

Because of their knowledge of fire behavior, fire fighters are in a unique position to note conditions and events which indicate that a fire is of suspicious origin. The average citizen at a fire scene cannot normally assess the scene with any degree of expertise. Fire department personnel responding to the fire may be the only ones who then have the opportunity to make pertinent observations, because the structure later may be totally destroyed. If fire fighters do not note conditions and report their observations to the fire officer in charge, fire cause determination will be difficult. If fire fighters do not report their observations of suspicious circumstances, it is highly probable that no investigation will follow. Fire fighters also may be valuable witnesses in an arson trial.

Historically, fire department involvement has been confined to saving life and property. While the majority of fire fighters now accept additional roles, there are those who do not consider arson detection among their responsibilities. Perhaps the most effective method of motivating fire department personnel to become involved in arson detection is to point out that incendiary fires threaten their own personal safety.

The typical arsonist has little concern for the safety of those responding to fires. Arsonists, particularly the so called "torch," find it necessary to use flammable liquids in quantity, particularly if a large or complex structure is the target. Additionally, dynamite charges connected with primer cord may be part of the set fire. In one incendiary fire, three fire fighters were killed when numerous white plastic containers of various sizes filled with gasoline exploded as the fire fighters entered the building. In another case, simultaneous explosions caused by connected dynamite charges and several 55 gal (208 L) drums of gasoline occurred prior to fire department arrival, with no multiple fatalities and injuries.

In order to detect arson, there are a number of specific observations each fire fighter should be aware of when responding to *every* fire. These observations can be organizaed into a logical sequence.

When the Fire Call Arrives

The time of the first fire call is important information, as it will assist the investigator in tracing the development of the fire. The reporter probably discovers the fire and can describe its appearance at the time of discovery. When a fire described as "only a small fire" at the time of discovery develops into a major conflagration in the short time between discovery and fire department arrival, this information is important to the investigator because it indicates the likely presence of an accelerant.

The means of receiving alarms varies from those localities with sophisticated equipment that record the entire operation, to those where the fire calls are received by telephone in a 24 hour diner or simultaneously in the home of several fire department members. By whatever means calls are received, investigators will want to obtain all the information possible about the fire.

In some cases investigators will inquire concerning other fires called in at about the same time as the suspected arson. This may divert fire equipment from the target structure shortly before the arson is committed. In jurisdictions where fire suppression resources are limited, such as in a rural community, equipment and personnel may be removed many miles (one mile equals 1.6 km) from the scene of a set fire. Obviously these circumstances give the fire a good start before fire apparatus arrives.

The identity of the person reporting the fire is important for several reasons. As stated previously, the person probably discovered the fire and can describe its appearance. This individual can be questioned by the investigators on what he/she saw, where the fire was burning, how it spread, any individuals or vehicles that were noticed in the area, and any other pertinent observations.

The identity of the discoverer and reporter can be important because this person may also be the firesetter, especially in the event of a series of suspicious fires. Experience has shown that the discoverer and reporter of an unusual number of fires should be considered a suspect regardless of circumstances. This is particularly applicable when the firesetter falls into the "hero" category, such as a security guard who seeks recognition for discovering and fighting a fire. Unfortunately, fire department members on occasion have set and then "discovered" fires, following up with heroic deeds of rescue and fire fighting.

En Route to the Fire

There are several important observations fire fighters can make en route to the fire scene. General weather conditions should be noted, including temperature, wind direction, and whether the sky is clear, cloudy, or stormy. Cold temperatures would normally dictate that all windows and doors of a structure would be closed. If windows and doors were open in cold weather, this may indicate that the arsonist wanted to assure a sufficient supply of oxygen to the fire. Warm temperatures would eliminate the need for any heating equipment, so if fires are blazing in a fireplace or wood burning stove, it is probable that the firesetter planned to blame the fire on flying sparks or on an overheated chimney connector.

The direction and approximate velocity of the wind is important to the arson investigator. For example, if the initial call indicates a small fire on the exterior of a large structure, a favorable (to the arsonist) wind of 10 to 15 mph (16 to 24 km/h) might involve the entire structure in fire in a relatively short period of time. However, if such heavy involvement occurred under similar conditions (except for a wind blowing in the opposite direction), the presence of an accelerant to cause rapid fire spread is likely. The arsonist may try to blame lightning for the fire, but if no storm occurred in the area, lightning is obviously not the cause of the fire.

Weather conditions such as heavy snow, ice, or flooding which might impede the arrival of fire apparatus should be noted. The arsonist may select the time when such conditions exist, knowing that any delay in fire department response will increase the probability of total destruction of the target structure. This rationale is particularly applicable where the perpetrator has planned a fire in advance, waiting for the appropriate time to carry out the scheme.

Fire fighters en route to the fire should note any situation which delays arrival of the apparatus. It may be that the arsonist will create obstacles, such as dragging trees across the highway or stretching cables across a road or street. The arsonist may take advantage of a street closed for repair, a bridge stuck in the open position, or similar conditions. Normal early morning or late afternoon rush hour traffic also slows fire department response.

Approaching the Fire Scene

As fire fighters get close to the fire scene, there are several observations which may indicate a fire of suspicious origin. Vehicles quickly leaving the scene or heading away from the fire should be noted when practical. In urban communities, the volume of traffic makes such observations impractical. In less congested environments such as small towns or rural areas, a fire creates more excitement and attracts people; thus vehicles heading away from the fire scene should be noted and a description of the vehicle and its occupants given to the investigators. People leaving the scene on foot should be noted, especially if they seem to act in an unusual manner. A general description of individuals leaving the scene can be very important, especially when a community is plagued by a series of suspicious fires.

Arrival on the Fire Scene

An important item that should be noted is the time of arrival at the scene. Along with a description of how the fire appeared to arriving fire fighters, this will enable the investigator to determine whether the spread of the fire was abnormally fast.

Arsonists want to conceal their preparations for burning a building from outside observation, so windows and doors may be covered with blankets, plywood, corrugated cardboard, or other materials. This not only enables the firesetter to carry out plans in secret, but also prevents early discovery of the fire. An additional benefit (to the arsonist) from the use of corrugated cardboard is that when exposed to fire in single sheets, it will burn fiercely and accelerate the intensity and spread of the fire.

On arrival, fire fighters should note any individuals suffering from burns. It might not be unusual for the occupants of a residential structure to be burned, but a burned individual who has no reason to be on the premises should be identified for further interview by the investigators. This person may well be the arsonist who got careless when setting the fire, allowed gasoline on other flammable liquids vapors to accumulate which caused the building to explode when the match was ignited.

Another observation which may indicate preparation for arson includes hydrants which have been tampered with or blocked. There are reported incidents where the hydrant threads have been damaged to such an extent that hook up to the hydrant is impossible. Pieces of wood, rocks, bricks, etc., that are dropped or stuffed into the hydrant may render it completely inoperable. It is the responsibility of the investigators to determine whether these are acts of vandalism typical in that neighborhood, or preparation by the arsonist to ensure complete destruction of a target structure.

The exact location of the fire is an important observation of arriving fire fighters. This determination is not always as easy as it sounds. Fire fighting operations may be complex and involve multiple points of attack and entry. In the case of a large structure which eventually burns completely, the task of an investigator is greatly simplified if the location of the fire can be pinpointed in its early stages, because the investigator can then concentrate origin and cause determination efforts in that area. In general, any information fire fighters can provide on fire origin, behavior, and spread will assist the investigator in the fire scene examination phase of the investigation.

The color of smoke and flames should be noted by the arriving fire fighters. Heavy black smoke and deep red flame indicates burning of a petroleum based product. These conditions become significant when it can be established that no products such as rubber or synthetics were present.

While Fighting the Fire

The means by which fire department personnel gained entry to the involved structure is important information to the investigators. As part of the investigative process, every effort will be made to determine whether the building was secure, with windows and doors locked at the time of the fire. If at the time of the fire department's arrival, the building is locked and secured with no signs of forced entry prior to the fire, persons having keys might be suspected. On the other hand, if the investigators locate multiple points of entry, the entry point or points made by fire department personnel (as opposed to those made by an unknown individual prior to the fire) must be identified.

The arsonist/owner may create some conditions to make it appear a breaking and entering occurred in the structure prior to the fire. Buildings which are open to entry pose special problems to investigators because the fire could have been set by vagrants or trespassers who were accustomed to freely entering and exiting the involved structure.

During fire fighting, unusual odors such as gasoline, kerosene, or fuel oil should be noted. Since most personnel entering the structure will be wearing self-contained breathing apparatus (which limits detection of odors), the pump operator or other individuals working outside the involved structure may best be able to detect unusual odors.

A critical observation for arson detection is the presence of several separate, unconnected fires in the target building. Such multiple points of origin cause rapid spread and increase the intensity of a fire.

Other indications of arson are "streamers" or "trailers" in the structure, used to spread the fire from room to room and perhaps from floor to floor. Streamers may consist of a flammable liquid poured in a path, but more often other material, such as rags, sheets, newspapers, etc., are used, with some accelerant being poured on this material to assure burning. Streamers composed only of cloth often do not burn as the arsonist anticipates, and in this event substantial portions of the material will remain. Even when some flammable liquid is poured on such materials, a residue may remain due primarily to a lack of oxygen in the area next to the floor. When a liquid is poured directly on the fire floor, more intense burning will occur and the flammable liquid pattern normally will be obvious.

Other conditions which may indicate arson are attempts to expose wooden surfaces such as wood strips in ceilings and walls, and dressers and dresser drawers. Additionally, interior doors may be propped open and holes punched in walls and ceilings to assist the spread of the fire. Unusually rapid fire spread may indicate the presence of an accelerant and thus arson. An illogical location of a fire, with no reasonable explanation for its starting at that particular point, is suspicious. For example, a fire in a closet, or in the center of a room with no furniture or fixtures nearby, or up the outside wall of a structure with no electrical installation or other accidental fire causes in the area, are all conditions which should be noted by fire fighters and reported to investigators.

Investigators should question fire fighters on their observations concerning contents of the involved structure. In an accidental fire, if the structure involved is a dwelling occupied by a family of four or five persons, the normal quantities of furniture, kitchen accessories, appliances, clothes, and personal effects should be present. In cases where such items are conspicuously absent, there is a strong possibility that they were removed either well prior to or immediately before the fire. At the time of fire department arrival, these items may be sitting in the yard of the house, having been saved by neighbors and passersby. If no people removing the items are observed, the questions must be asked on how, when and by whom these items were removed.

In mercantile losses, fire fighters should note the absence of stock, fixtures, and display cases. When the merchandise has been removed, there are several possible explanations—the owner could not stock inventory because income from the business was down; or merchandise was removed prior to the fire either by the owner/arsonist in preparation for submitting a fraudulent claim or by burglars who set the fire in an attempt to cover up the crime.

Other signs of preparation for arson in mercantile and industrial losses include fire doors propped open, firestops tampered with or damaged, and sprinkler systems shut down. Efforts also may be made to interfere with fire fighter entry and movement; such efforts include doors barred, heavy items placed to block entry, and large holes cut in the floor at logical fire fighter entry points.

While it may be difficult for fire department personnel to devote attention to people at the fire scene, some observations can be made which are significant to the investigators. If members of a family supposedly were aroused from a sound sleep by smoke detectors and were lucky to escape, it would be unusual for them to be fully clothed in what might be considered normal daytime dress. When a community is suffering a series of fires, fire fighters should look for familiar faces at the scene; this type of firesetter will normally stay at the scene to observe the excitement. The behavior of this individual can vary from simply watching the fire and the fire fighting activities to laughing and talking with anyone, apparently enjoying the entire incident. This person may be waiting at the fire scene, anxious to lead fire fighters to the exact location of the fire, and then participate in the fire fighting operations, giving directions and shouting orders.

The observations of fire fighters must be reported to the fire officer in charge. Training fire fighters to make observations relative to the entire fire fighting operation is essential to the success of the arson unit, because fire department personnel may provide on scene information which can lead to the identification, apprehension, and conviction of the arsonist.

Fire department functions which are valuable to the investigators relate to postfire activities. It is essential that the premises be secured from unauthorized and unnecessary entry to prevent destruction, whether deliberate or accidental, to potential evidence for an arson investigation.

At some point in the investigation, the investigator must restore the fire scene to prefire condition as far as possible. This involves placing furniture, appliances, doors, etc., in original positions in the structure, even though these items may be damaged or partially destroyed by the fire. The task of restoration becomes much easier if the fire department modifies overhaul and clean up operations, minimizing the use of high pressure water streams, and postponing removal of contents.

INDUSTRIAL ARSON

While the extent of industrial arson may be difficult to identify, the federal government estimates that 23.8 percent of fires in industrial occupancies are incendiary/suspicious. Industrial arson is, in fact, a major problem. The following discussion examines the who, why, where, how, and when of industrial incendiarism, as well as defenses against such fires.

Perpetrators of and Reasons for Industrial Arson

Current and former employees, competitors and customers, special interest social or political groups, security personnel, employees or outsiders during labor disputes, and intruders are among the people who set industrial fires.

Present and past employees are particularly dangerous as firesetters because they are familiar with plant layout and operations; know the locations of critical areas, inadequately secured entrances, and combustibles and accelerants; and are familiar with the operation of protective systems and general security procedures. To some degree, past and present employees can move with relative freedom about the plant and may have keys to locked areas.

Some employees use arson to cover a crime such as embezzlement, theft, or looting. Through arson, other employees may express a grievance against plant management or against another employee. Former employees may use arson to protest being laid off or furloughed, or to retaliate against the individual responsible for the arsonist's loss of employment.

Competitors may turn to arson to put an industry out of business or to disrupt production, providing the competitor with the opportunity to secure a contract. For dissatisfied customers, arson can be a way to protest shabby treatment or express anger over a faulty product.

For social and political interest groups, arson is a way to demonstrate opposition to a product being manufactured or a test being conducted. Terrorism—demonstrating the group's ability to destroy the facility—as well as sabotage to reduce or halt production of items used by the military, also are among the motives for industrial arson. Although social and political interest groups may seem disorganized or lacking in leadership, their persistent dedication to a cause makes them extremely dangerous. Members of these groups will risk physical harm and arrest to accomplish their goals. Individuals committing acts of sabotage are often well trained, well equipped, and organized to carry out their assignment.

Security personnel hired to prevent arson and other disruptions may become firesetters. Working conditions for night security personnel can be far from ideal. The temptation to set, discover, and extinguish a small fire (and then to be praised as a hero) may be great. Other security personnel set fires to justify their position, especially when rumors of cutbacks, reductions in personnel, or contract changes are circulating. Firesetting also can be seen as a way to convince management that security personnel are necessary and effective.

During labor disputes, employees or outsiders can use arson to disrupt plant operations, frighten management into settling the dispute, or intimidate employees who remain on the job.

Each of the arsonist types mentioned, and their motives, are somehow related to the industry involved. In contrast, intruders have no relationship to the industry they target for arson. The motives of some intruders who commit industrial arson are related to motivation for other crimes. For example, intruders may use arson to divert security personnel to one area of the plant while they commit a theft in another area. Arson may be a way of covering another crime such as burglary, theft, or pilferage.

The sake of destruction is another possible motive for arson. Malicious vandalism is often committed without any justifiable reason. Vandalism has been identified as the motive for as much as 40 percent of all incendiary fires, and industrial facilities are logical targets. Plants in urban areas are especially susceptible, since abandoned and vacant buildings offer easy access, minimal security protection, and little chance of apprehending firesetters.

Locations of Industrial Arson

Storage areas, vacant buildings, and docks and platforms (including trucks and railroad cars) are likely locations for industrial arson. Office and sales areas are less likely to be targets for incendiarism since they are too often occupied.

Few employees are present in storage areas, even during working hours. Stacked or piled materials provide an adequate fuel supply. Storage areas are often hidden from view, minimizing the arsonist's chances of discovery and interference while easy, safe access provides means for quick entry and exit from the area.

Vacant buildings offer the arsonist many advantages similiar to those in storage areas. In addition, vacant buildings offer easy access for juveniles; vandals, or vagrants who loiter in the buildings or use them as temporary sleeping quarters. Vacant buildings are particularly vulnerable to arson if the exterior has deteriorated.

Items being loaded or unloaded on docks and platforms provide a fuel source for the arsonist. The number of people moving in and out of these areas makes access control difficult. The discovery of smouldering fires in trucks or railroad cars may be delayed until the vehicles are some distance from the loading point. Office and sales areas are less likely targets for industrial arson, partially because the areas are normally occupied during working hours and access to these buildings after business hours is more limited than in other areas. The limited fuel supply of all the facilities discourages incendiarism.

How and When Industrial Incendiary Fires Occur

Motive is probably the most important factor for an arsonist. Security personnel and investigators must carefully consider the method used to set a fire, as this may enable them to establish a motive and then take appropriate action to apprehend the arsonist and prevent another fire.

The determined arsonist may set two or more fires simultaneously, or may set several fires over a period of time. If the first attempt at firesetting was unsuccessful, the arsonist may make another attempt shortly thereafter and may use an accelerant.

Approximately 70 percent of the arson losses for a leading insurer occur during non daylight hours. Statistics like this emphasize the need for additional security during evening hours.

Prevention and Control of Arson Losses

Arson prevention and control measures for industry fall into two general categories: (1) stable inanimate measures such as fences, lighting, alarm systems, and locked doors and windows; and (2) human, animate measures. The later include controlling personnel movement, screening and escorting visitors, and recognizing security problems. These actions are not confined to security personnel, and should involve various levels of plant personnel.

Plant management has significant responsibility in arson prevention. For example, since statistics have shown that a significant number of incendiary arson fires are set by past or present employees, supervisory personnel at all levels must attempt to deal effectively with the disgruntled employee. Plant management also should develop and maintain rapport with the community, simply because hostile individuals in a community are more likely to set fire to the facility.

Security personnel are also extremely important to arson prevention. The duties and responsibilities of security personnel include employee control, visitor control, and general security services. Identification is the key to personnel control. Security personnel should only permit someone to enter the plant (or restricted areas of the plant) when the person is known to security or displays a badge or pass. Personal recognition is generally practical only when there are 25 or fewer employees. Since checking passes can become perfunctory, security personnel must be reminded frequently to check photographs, expiration dates, and signatures on badges, and also to check any alterations of the badge.

Employees should be restricted to their own work areas, if possible. Security personnel should check bundles and briefcases that workers carry into the building; this check should be within reasonable limits and respecting the rights of individuals.

Visitor Control

All visitors should furnish positive identification and sign a register when entering and leaving the plant. Security personnel should check packages and issue identification badges to visitors.

Security personnel should be especially careful about identifying vendors and service people. For example, arsonists posing as a telephone repair crew had complete access to an industrial facility for two weeks while preparing for a large incendiary industrial fire.

Whenever possible, all visitors should be escorted while they are on the building premises. When this is not possible, employees should be asked to meet their visitors in the security area. A somewhat less effective procedure is to alert employees that visitors are in the plant en route to their area, and ask them to notify security when the visitors arrive at their destination.

All plant personnel, not just security forces, should be alert to strangers wandering around the property. Employees should note strangers and promptly report this to security personnel.

General Security Services

Around the clock security patrols are recommended. It is important to vary the pattern of these patrols, since intruders may study the timing and route of patrols in planning their intrusion. Entrances and exits also should be guarded. All security systems such as alarms, locks, fencing, and lighting, must be checked periodically. Janitors can be an important element in the security network, particularly if they are asked to notice changes in the normal appearance of the plant's public areas.

Postfire Actions

No facility is completely immune to the threat of arson. A standard operating procedure should be developed in anticipation of a fire. The procedures should specify the person(s) who will be responsible for coordinating the investigation and members of the investigation team.

Two actions should be included whatever procedures are developed. The first is that all fires, no matter how small, should be investigated. So called "minor fires" may indicate the presence of an arsonist and signal more serious problems. If the fires persist, information on minor fires may prove essential in discerning a pattern and identifying the activity of the firesetter.

Second, the area of fire origin should be left undisturbed until a thorough examination has been conducted even though there may be pressure to downplay the fire, to clean up, or to get back into production. Unfortunately, these actions make a successful investigation, apprehension of the arsonist, and prosecution less likely.

Bibliography

NFPA Codes, Standards, Recommended Practices and Manuals. (See the latest *NFPA Codes and Standards Catalog* for the availability of current editions of the following documents.)

NFPA 1031, *Standard for Professional Qualifications for Fire Inspector, Fire Investigator, and Fire Prevention Education Officer.*

Additional Readings

Bates, Edward B., *Elements of Fire and Arson Investigation*, Davis Publishing Company, Santa Cruz, CA, 1975.

Battle, Brendan P., and Weston, Paul B., *Arson: Detection and Investigation*, Arco Publishing Company, NY, 1978.

Bouquard, Thomas J., *Arson Investigation*, Charles C. Thomas Publishers, Springfield, IL, 1983.

Brannigan, Francis L., Bright, Richard G., and Jason, Nora H., eds., *Fire Investigation Handbook*, National Bureau of Standards Handbook 134, U.S. Government Printing Office, Washington, DC.

Carroll, John R., *Physical and Technical Aspects of Fire and Arson Investigation*, Charles C. Thomas Publishers, Springfield, IL, 1979.

Carter, Robert E., *Arson Investigation*, Glencoe Publishing Company, Encino, CA, 1978.

DeHaan, John D., *Kirk's Fire Investigation*, 2nd ed, John Wiley & Sons, Inc., NY, 1983.

Dennett, Michael F., *4ire Investigation*, Pergamon Press Inc., Elmsford, NY, 1980.

International Fire Service Training Association, *Fire Cause Determination*, Fire Protection Publications, Oklahoma State University, Stillwater, OK.

NFPA, Arson Mini-Guide Series, (five books), I. *The Arson Epidemic*; II. *The Fire Department's Role in Arson Investigations*; III. *The Arson Investigator's Responsibility*; IV. *Expanding the Arson Investigation*; V. *The Arson Trial: Courtroom Procedures*, National Fire Protection Association, Quincy, MA, 1982.

National Fire Protection Association, "Who Pays for Arson?" (Six panel consumer folder), National Fire Protection Association, Quincy, MA, 1982.

Phillipps, Calvin C., and McFadden, David A., *Investigating the Fireground*, Robert J. Brady Co., Bowie, MD, 1982.

FIRE DEPARTMENT APPARATUS, EQUIPMENT, AND FACILITIES

Revised by A. K. Rosenhan

Just as fire requires the three legs of the fire triangle to exist, so does the fire fighter require three sets of resources to combat fire—apparatus, equipment, and physical facilities. This chapter covers various aspects of these three resource areas and explains their function, organization, layout, and usage in the fire department.

FACILITIES

Any organization must have a physical location to base its activities. The location may be a simple storage building or space, or may comprise a complex of buildings, space, and specialized structures in one or more locations.

Administrative Offices

Administrative offices may be located in an office building of local government such as a city hall or county courthouse, at a fire headquarters building, or a central fire station. Administrative offices should include space for the administrative head of the organization plus staff personnel, including offices for such activities as fire prevention, fire investigation, planning and research, budgeting, and personnel. The size and space requirements will be dictated by the size of the organization.

All offices and facilities should be clearly marked, with appropriate doorbells, buzzers, or communication systems for both overall security and usage after hours. The public should be constrained to use the designated public areas of the facility unless escorted by fire department personnel.

When planning new administrative offices or expansion of existing offices, the activities that occur within each office should be considered. Offices that are frequented by the public should be easily accessible and arranged so that the general public does not have to pass through other offices or quarters. Separate entrances may be desirable for staff personnel. Ample storage space should be provided for the filing and storage of records and reports.

When fire investigators are assigned there should be separate rooms for interviewing witnesses and suspects, as well as secure storage rooms for the preservation of evidence. If frequent staff meetings are held or the department operates under a board or commission, it may be desirable to have a large meeting room. This room, if properly equipped with communication equipment, can also serve as an operations center for the department during major emergencies. There should also be ample parking space around the building for the public, assigned personnel, and department vehicles.

Fire Stations

Determining the location for a fire department facility is an important decision. Many aspects of initial investment, annual operating costs, practicality, and public reaction and support come into play, as well as the most important aspect of location—providing the best possible fire protection to the area.

Determining Location: The area to be protected by the proposed station is the most important factor in determining station location. Some areas in the community such as those containing primarily business, industrial, mercantile, institutional, and multifamily residential occupancies will contain higher risk potentials than others. In some instances it may be undesirable to locate a station in a high risk area. The station should therefore be located within a reasonable response distance on the perimeter of this area. Distribution should provide a concentration of companies for response into high risk areas without depleting other areas of the community should a second or major fire occur. In other areas with an equal risk throughout, such as residential areas, stations should have equal response time from all parts of the area. Where stations are staffed by volunteer fire fighters, consideration must be given to locating them in or near areas where the volunteers live and work to facilitate quick response to a fire.

Topographical features of a community may also affect fire station locations and the total number of stations required. Natural and artificial barriers that divide communities, such as rivers, mountains, limited access highways, and railroads, may limit response routes and require

Mr. Rosenhan is president of A. K. Rosenhan Consulting Engineers, Ltd., in Mississippi State, MS.

additional stations. Heavily traveled and one way streets are undesirable for fire station locations because of access problems and limited response routes. When a station is located close to a community boundary, its response area is reduced. If the community is anticipating annexation this may still be desirable. If it is anticipated that communities or governmental entities will be establishing mutual aid areas, an overall plan of fire department facilities will provide enhanced protection with a minimum of investment. Coordination with community or regional planning agencies can provide valuable information on growth potential and land use when planning new fire station locations, or in cooperative fire response operations.

Fire department records can also provide valuable information for plotting fire station locations. Fire experience may indicate specific areas of high rate of incidence. Response times may also prove useful. Although computers are currently used in some communities to aid in plotting fire station locations, this is not the answer for all communities, since a valid data base must be established before computers can be utilized.

Several computer program packages used to analyze fire department run records, response times, fire incident locations, etc., are readily available but can be expensive, complicated, and may not satisfy all local needs. A common sense approach involving fire officials, city engineers, insurance rating officials, and citizen input will prove effective. Figure 15-6A shows a plot plan for a typical district station for urban and suburban services, and Figure 15-6B a plot plan for a typical rural station. See Figure 15-6C for elevation and plan view of a typical urban fire station, and Figure 15-6D for elevation and plan view of a typical rural fire station.

Site of Structure: Once a general location has been determined, the site and configuration of the structure may be determined. The station should be set back far enough from the street to permit a paved ramp in front of the building so apparatus can be cleaned prior to parking inside. The site should also allow sufficient parking for personnel. Parking and traffic patterns on the site will differ for volunteer and paid personnel. Additional space should be provided for outside training activities and other activities which may be held at this location.

Design Features of Fire Stations: The size of fire stations should be compatible with the maximum anticipated number of personnel and pieces of equipment that will be assigned. Construction details should emphasize three important aspects: (1) maintenance, (2) traffic patterns within the station, and (3) fire protection features. All portions of the station should be designed so they are easy to clean and maintain. Stations that require extensive amounts of time for cleaning and maintenance reduce departmental productivity in terms of prefire planning, fire prevention, and training. Stations that require frequent painting are also costly in terms of maintenance. Traffic patterns within the station should be carefully considered—all personnel must be able to reach the apparatus room from all parts of the building with a minimum of confusion.

Fire protection features of the stations themselves must not be overlooked. Since fire stations contain all the hazards normally found in a garage, assembly hall, dormitory, and school, protection from hazards found in such occupancies should be included in the building design. This includes fire detection devices, fire suppression devices, and proper exits, plus other items in local building code requirements. Storage and dispensing of motor vehicle fuel require strict adherence to applicable codes.

Based on safety considerations, it is preferable that fire stations be only one story high but, because of lot size, function, and economics, two stories may be necessary. Basement areas should be avoided, particularly under apparatus floors, since this tends to add materially to the cost of the station because of the weight that must be supported. Apparatus room floors should be of concrete slab construction with care taken to avoid a slick finish. Painted concrete floors are generally unsatisfactory, but if color is desired it would be preferable to add it to the finish layer of the floor. Some departments utilize steel treads or runners under the apparatus, but these are expensive. Air entrained concrete should be used to avoid salt damage. The floor should be pitched for adequate drainage, but not so steeply that the apparatus will roll toward the doors when the brakes are off. Drains across the front of the apparatus room are sometimes installed to permit road salt and dirt to be washed off the apparatus.

Apparatus doors should be as wide and high as possible. Provision should be made to accommodate future apparatus such as aerials and platforms, which may require more height than a standard pumper. Maximum allowable height on roadways is 13½ ft (4.1 m); width will vary by state, but a minimum doorway width of 106 in. (2.7 m) is recommended. Doors should generally open vertically and, if power operated, be equipped with a manual release for operation during a power failure. Controls for apparatus doors should be provided at each door and duplicated at a central watch desk (if provided). Any door controls at a watch desk should be clearly marked as to function and location. The use of wireless controls, operable by the company officer or driver, is a convenient method of controlling doors if a watch desk is not continuously manned. This remote operation capability will conserve apparatus bay heat in winter as well as provide security for the station and any remaining apparatus. Lanyards hung from the ceiling and located near the driver's or officer's door on the fire apparatus may also be used for apparatus room door control. A predetermined procedure is essential for the timely and safe operation of apparatus doors. While doors with large expanses of glass may look attractive and provide clear views, glass needs cleaning, is subject to damage, and may be a security problem in some areas.

Heating and Ventilating: The heating and ventilating system should be adequate and installed in accordance with appropriate NFPA standards. The system must have the ability to recover rapidly after the apparatus responds to a winter alarm and the station doors are not immediately closed.

The inclusion of solar heating, semiunderground construction, and other energy conservation construction features may cause additional maintenance problems and compromise simplicity of building operation. The long range economics are of interest in building operation, but basic functional reliability should be of utmost importance in designing and equipping a fire department facility.

Adequate ventilation of the apparatus room should be provided to avoid concentrations of carbon monoxide and

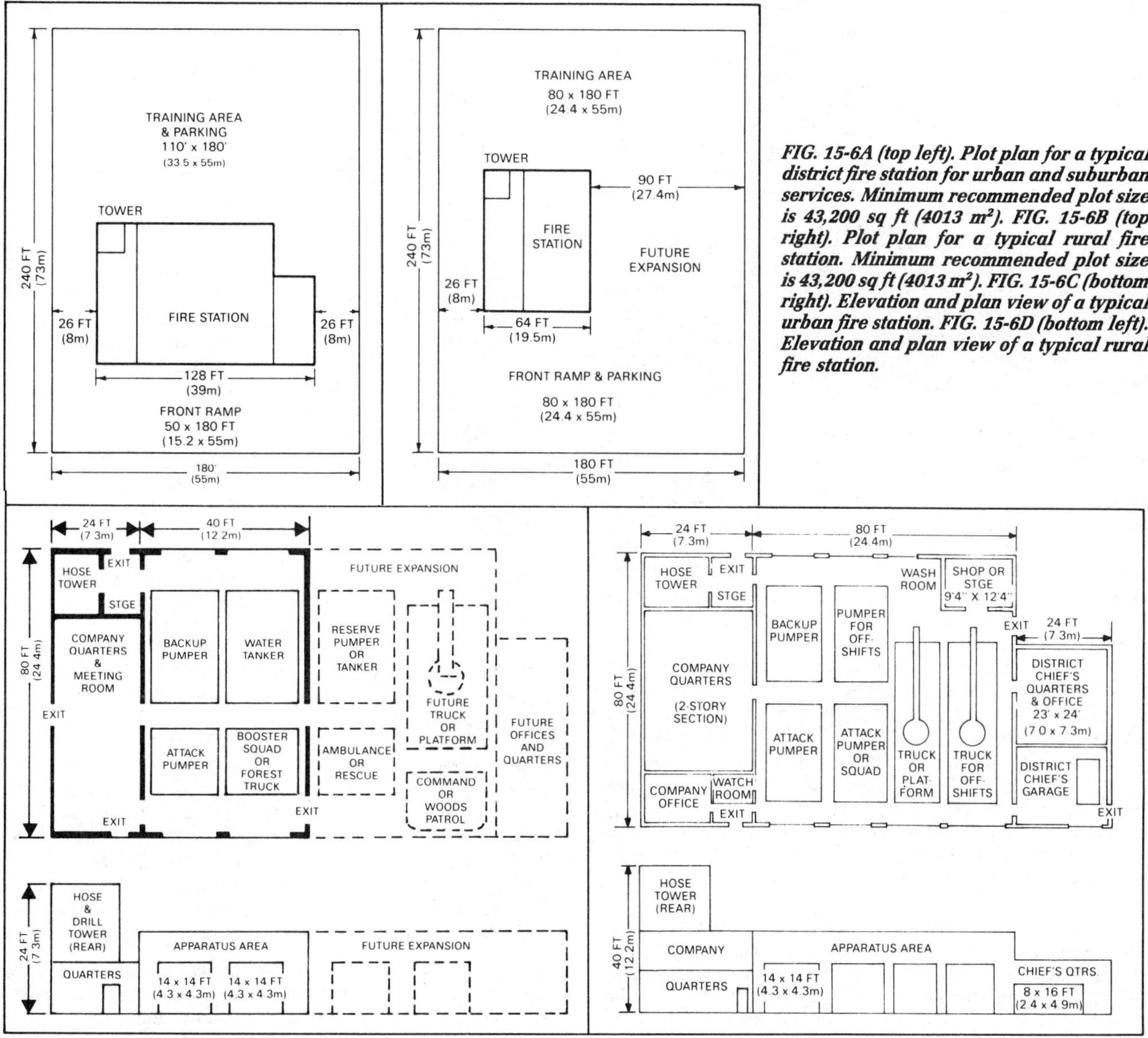

FIG. 15-6A (top left). Plot plan for a typical district fire station for urban and suburban services. Minimum recommended plot size is 43,200 sq ft (4013 m²). FIG. 15-6B (top right). Plot plan for a typical rural fire station. Minimum recommended plot size is 43,200 sq ft (4013 m²). FIG. 15-6C (bottom right). Elevation and plan view of a typical urban fire station. FIG. 15-6D (bottom left). Elevation and plan view of a typical rural fire station.

exhaust fumes during any engine warmup, drilling, or servicing which may be conducted. Exhaust hoses or mechanical ventilation equipment should be provided for prolonged engine use or, if no other provisions are made, apparatus doors should be opened for ventilation.

An adequate electrical system should be installed in accordance with NFPA 70® the *National Electrical Code®* with adequate numbers and sizes of receptacles provided for any anticipated need. Consideration should be given to GFI (ground fault interruptor) circuits on receptacles in or around the apparatus floor and in bathroom/kitchen areas. Overhead drop cords on reels to serve each apparatus location are convenient methods of supplying power for built in battery chargers and other uses. These reels must be located out of the way of overhead door tracks and other items.

Battery charger leads should be handy to the apparatus position, preferably with the cable on self-retracting reels. Battery chargers should be so located that when used, the leads do not present a hazard to fire fighters attempting to board the apparatus to respond to an alarm.

Emergency power should be provided to a station that is used for any dispatch or communication function. This emergency power may consist of a simple battery arrangement with a minimum of 12 hours of capacity, to an engine driven generator. The fuel supply to any engine driven device is usually natural gas (where available) due to its reliability and cleanliness. Oil and gasoline driven engines, while reliable and satisfactory, do require regular maintenance and checking of the fuel supply to eliminate problems with the fuel itself. Any emergency power supply should be inspected and tested on a regular basis to ensure operational ability and capability.

In addition to the water services usually provided for a public building of this type, consideration should be given to facilities for refilling department tankers and apparatus booster tanks. While this can be done outside in good weather, it may be desirable to have water hose leads suspended over each apparatus position for quick refilling during cold weather. In rural areas, water cisterns or tanks near the station may be desirable.

A locally made fitting [1½ in. (38 mm) female to 2½ in. (63 mm) male], will allow a booster tank to be filled using an auxiliary suction and valve, thus eliminating the need for personnel to climb onto apparatus.

Training Area: A separate classroom for use as a training area is preferable, but if that is not feasible, the meeting hall should be planned with training in mind. Training facilities should include a wall mounted chalkboard, projection screen, room darkening shades, bookshelves, etc. Ample electrical outlets, as well as speaker wiring, should be provided.

Watch Desk: The point at which alarms are received within the station should be in the form of a desk arrangement with wall space for maps, schedules, and instructions, and with ample space for the necessary radio equipment, alarm control devices, door controls, and traffic signal controls. The area should be as soundproof as possible, and should allow clear visibility of the entire apparatus room. A desirable location is near the front entrance to the station, the point where visitors may enter and seek information.

If the watch desk is not located near an entrance, there should be a doorbell or remote speaker to communicate with visitors. An electric door latch may be appropriate for after hours visitors. If a watch desk is used for monitoring or dispatching of other emergency services, there should be consideration given to protection in the event of tornado, high winds, etc. This protection may be in the form of a basement location (with adequate provision for drainage), reinforced walls and roofs, and shutters for any windows or other openings.

Apparatus Room: It is preferable that the apparatus room floor area be unobstructed by columns or projection of other rooms into this space. Ample space must be provided to permit work around apparatus, hoses to be changed, and free movement when responding to alarms. Additional space is needed for clothing racks and hose storage if not provided in separate rooms. Multitruck stations require a minimum width of 20 ft (6 m) per track. Single truck stations should be at least 24 ft (7.3 m) wide. Depth is dependent upon the number of pieces of apparatus to be housed, but ample space must be provided front and rear to permit routine maintenance, ease of response, and repacking of hose.

Consideration should be given to drive-through apparatus bays if permitted by the lot size and configuration. This would allow apparatus to reenter the station without backing up, which may damage apparatus and buildings as well as creating a hazard to personnel.

Stations should have adequate office facilities for all officers assigned including both company officers and battalion or district chiefs. Dormitories, locker rooms, washrooms, kitchen, and recreation rooms are also required when full time personnel are assigned. The storage of equipment, supplies, and fuel for apparatus and hose drying are additional considerations for fire stations. Fuel storage should be outside the structure, and filling fuel tanks inside the station should be avoided. Utility shutoffs should be clearly marked and easily accessible.

Training Facilities

Training facilities are a definite prerequisite to the training and education of fire service personnel. Much training can be done without formal facilities. However, certain areas of training become more meaningful and productive when carried out at especially designed facilities. Training facilities may range from the low cost, self designed, and self constructed facility to those designed by architects and constructed by professional contractors. The sizes, shapes, and varieties range from a three story, pitched roof ladder tower constructed of telephone poles and lumber for a cost of several thousand dollars, to a complete fire academy facility which can cost several million dollars. Constraints imposed by environmental protection laws that dictate air and water pollution standards must also be considered.

Many existing training facilities are underused and understaffed because some fire departments are constrained by budget limitations. Those interested in constructing a facility are advised to explore the use of existing training facilities within the immediate area, or to enter into an agreement with other contiguous communities and share both the expense and use of such a facility.

Those considering the construction of a facility should also visit existing facilities prior to the selection of, or in the company of, the architect who will design the facility. A utilization plan is an important factor in the budget justification and must be considered at a very early stage. County, regional, and state training facilities are considerably more difficult to plan than a local or municipal facility. In planning utilization, it is desirable to know the exact length and content of the curriculum and the frequency with which the courses will be attended. Deducting periods when the facility will not be used such as during the winter months in areas affected by snow and ice, and during holiday and vacation periods, can help establish a utilization formula.

Location: Careful and detailed consideration concerning the location of a training facility is necessary, because it is almost impossible to change locations once the commitment is made and ground is broken. Factors involved when considering locations for training facilities, regardless of jurisdiction, include:

1. Land selected for the construction of a training facility should be remote from all types of occupancy, although accessibility must be a careful consideration for state, regional, and county facilities.
2. Topography is important, and flat land is most desirable for economic reasons. Soil conditions must be satisfactory for both drainage and stability.
3. Water in adequate quantities is an absolute necessity for training purposes, and drainage considerations are important. Availability of sewage disposal is also important.
4. Electricity should be available in adequate quantity.
5. Availability of fuel supplies for practice fires should be considered.

Design Features of Training Facilities: There are numerous and varied designs of training facilities. The intended use of the facility must be projected in detail before the actual design is finalized. Experience is important, and before final decisions are made, assistance should be sought from persons experienced in the scheduling and operation of training facilities. Various organizations, both state and national, have specific publications on training facility features and designs.

There are two basic design concepts for training centers. One concept is that all essential features of a training facility, i.e., ladder tower, forcible entry area, smoke house, burn building, and draft pits be incorporated into one building. The second concept that has been advanced in recent years is for separate buildings or areas to be designated for each of the major functions carried out at the training facility. The disadvantage of the second concept is that the cost is greater than for a building that contains all features under one roof. One advantage is that activities in a specific building or area do not interfere with activities being carried out in other buildings or areas. Another advantage is that the facility may be constructed in steps, over a period of time, as funds and priorities dictate. If adequate land is available when the facility is being designed, it is easy to incorporate a buffer zone of trees or other natural barriers between each operational area or building.

Essential Features of Training Facilities: There are many essential features that must be included when planning a training facility. Again, intended use dictates which features (and space requirements) will be incorporated. The size of municipal fire departments is usually stable when training facilities are being considered. Communities experiencing growth should anticipate their future needs during the planning stage. Communities that are annexing land or protecting contiguous unincorporated areas should consider future increased demands which may be placed on such a facility. Some of the essential features common to most training centers include:

1. An adequate amount of space for the administration area.
2. Storage space for paper and office supplies, as well as clean rack storage for printed material used by instructors and students.
3. Library facilities.
4. Audiovisual and training aid facilities, as well as an audiovisual laboratory, viewing and listening areas, and space for the storage and maintenance of all audiovisual equipment.
5. Classrooms with adequate lighting.
6. Apparatus storage, even if only on a temporary basis. The minimum amount of space should accommodate at least one pumper with adequate room for an instructor and students. An exhaust system is mandatory to carry the exhaust from the apparatus to the outside.
7. Towers, preferably constructed of masonry, with enclosed interior stairways and exterior fire escapes are desired including a dry standpipe system with an outlet at each floor; windows of varied styles and sizes; window openings; doors; flat and pitched roofs; skylights, and cutout panels within the roofs; floors and partitions; safety railings at any openings, and elevator shafts to simulate high rise occupancies.
8. Fire buildings, frequently called burn buildings, with at least two stories and a basement, provisions for movable partitions to change the configuration of the interior, windows with steel casings and wired glass, exterior doors, and stairways and bulkheads of adequate size.

   Fire buildings should be constructed of proper materials to withstand fire, high pressure fire hoses, thermal shock, etc. Two basic methods of construction are used: one is fire resistive masonry placed on the surface of standard concrete, steel, or other construction. The other utilizes replaceable fire resistive panels. The services of an architect or engineer should be utilized when designing a fire building due to the need for specialized construction and maintenance techniques.
9. Adequate space for storage, maintenance, and recharging of breathing apparatus is essential.
10. Areas for flammable liquid fire fighting training, consisting of both props and pits.
11. Underground storage space for fuel supplies for training operations, large enough to accommodate from 1,000 to 10,000 gal (3785 to 37 850 L of fuel).
12. Flammable gas training facilities that simulate actual fire conditions, with specially constructed tanks having extra safety features. These tanks are equipped with relief valves and other safety and warning features that confront students with a real fireground experience.
13. Drafting pits, both for training purposes and for the annual testing of pumpers. Design features should include proper baffles to prevent swirling and air entrapment when water is returned to the pit. Also, provisions must be made for filling and draining the pit.
14. Driver training facilities.
15. Areas for rescue training.
16. Control tower, from which all fire operations can be observed. The tower should be of sufficient height to provide a clear view of the entire facility. It should also have controls for lighting and power to the individual facilities on the training ground. Two way selective communications to each individual building or area on the site is desirable along with an adequate public address system.
17. Personnel needs such as cafeterias, dining facilities, and locker and shower facilities for instructors and students. Most training centers have the administration, classroom, storage, and apparatus housing area in one complex, and the remaining buildings and areas located throughout the site. In these cases, the central facility should be designed so students can enter directly into the locker and shower area to clean or change before entering other sections of the building. Dormitory accommodations must also be considered in regional and state training facilities.

Communications Center

The most important aspect of fire department communications is the facility where calls for assistance are received and action is taken. Large communities may require one large center and possibly satellite centers to handle message loads, while smaller communities may require only a watch desk located in a fire station. In addition to receiving calls for assistance, the alarm center must be capable of handling all radio communications for the department, keeping a continuous record on the status of all companies, handling communications to each station, and maintaining current files and maps of streets. It must also have the personnel resources for the required record keeping.

The design of communication centers should focus on the operations room where all calls and alarms are received, and from which all alarms are transmitted to companies or stations. A desk or console may be provided which is occupied by the dispatcher or dispatchers on duty. Console design must be functional so that all necessary operations can be performed efficiently. The operations room may contain all equipment necessary for the receipt of calls from the public, receipt of alarms from street boxes, a department telephone switchboard if necessary, radio transmitting and receiving equipment, tone alerting equipment, and computer terminals and video display units, if used.

Equipment requirements are dictated by the size of the operation and the type of services that are provided. Centers in communities that have communitywide alarm systems require space for alarm equipment and standby batteries. Space is also required for radio and telephone equipment. If computer assisted dispatching is used or anticipated, additional space is required.

Maintenance Facilities

Facilities for maintenance and repair work should be provided for all types of fire department equipment if cost effectiveness can be demonstrated. However, it may be more economical for some departments that seldom use these services to contract with outside agencies, thereby reducing overhead costs. If outside agencies are used, there should be explicit agreements that fire department needs will receive top priority.

Some communities share maintenance and repair facilities with various city departments due to cost considerations, but again priorities should be established with regard to emergency vehicle repair and service.

Some larger departments maintain specialized shops for all types of maintenance and repair such as apparatus, ladder, alarm equipment, breathing apparatus, and radio equipment. The need for such facilities is dictated by the demand for service. New or additional facilities should be cost effective.

Some departments use mobile vans that are equipped to perform a specified level of maintenance at the station, requiring less time for apparatus to be out of its response area.

APPARATUS

The tools for fighting fire and providing emergency services are apparatus and equipment. Without the proper tools, used correctly, a fire department is useless and cannot do its job. This section contains information on the standards used and procurement policies that should be followed in specifying and acquiring apparatus and equipment.

The basic fire fighting vehicle in North America is a diesel or gasoline driven truck that carries a rather extensive assortment of tools and equipment for fighting fires. Such equipment may include a pump, hose, water tank, ladders, and various portable tools and appliances. The amount and capacities of the various fire fighting components carried vary in accordance with the intended service of the particular vehicle, generally taking into account performance standards which can be measured by tests. NFPA 1901, *Standard for Automotive Fire Apparatus*, (hereinafter referred to as NFPA 1901), which also has been adopted by the Insurance Services Office and the International Association of Fire Chiefs (IAFC), is designed to facilitate procurement in accordance with the various components required and specified by the purchaser. In Canada, ULC Standard S-515 is generally similar and certification and testing is conducted by Underwriters Laboratories of Canada.

NFPA 1901 deals with the various designs, functions, and components of apparatus; the standard also gives details of procurement, provisions applicable to all types of apparatus, and the performance requirements for each type of apparatus.

The size of the pump and the amount of water, hose, and equipment carried will vary with the intended type of service such as municipal, suburban, or rural. In general, the largest pumps are provided to protect areas such as industrial districts requiring application of large fire flows. Pumps of 1,000, 1,250, or 1,500 gpm (3785, 4732, or 5678 L/min) capacity are popular for general municipal service, and 1,000 gpm (3785 L) pumps are common in rural and suburban service. A 500 gpm (1893 L/min) pump is the minimum size recognized by most insurance rating groups.

In many fire departments, it has been found desirable to supplement units equipped primarily for hose stream service with other vehicles that provide a large variety of tools and equipment such as ladders, forcible entry tools, generators, lights, and rescue equipment. Such units are variously termed ladder, rescue, or squad trucks, depending upon the primary type of equipment carried.

On the fireground, pumping apparatus must be positioned for the efficient application of hose streams, and should be positioned just beyond or just short of the front of the building, leaving the front of the building open for ladder trucks. For effective master stream use, pumpers must be well positioned. Other apparatus may be placed for convenient use of the equipment. Ladder, squad, and rescue trucks should have their own regularly assigned crews who can utilize their equipment effectively while pumper companies are getting streams into service.

About one-third of the work at an average fire involves the use of tools and equipment in duties classed as rescue or truck work. Apparatus suitably equipped for such service should be available at all structural fires and at emergencies such as highway fires and accidents. While pumping engines, squad trucks, and rescue trucks may carry considerable equipment, the larger aerial ladder and elevating platform trucks commonly transport additional equipment not usually found on other apparatus. Both these apparatus provide vehicle supported, power operated equipment for access above the normal reach of manual ladders and effective elevated stream service.

A wide variety of combinations of fire fighting equipment is provided on automotive fire apparatus. Besides combinations of water pump, hose, and water tank, commonly termed a "triple combination," it is not uncommon to provide a booster pump and small stream equipment on ladder trucks or elevating platforms, or a water tower on a pumping engine. Various desired combinations can be obtained by specifying the desired provisions from NFPA 1901. When fire apparatus is equipped with long ground ladders and other usual ladder truck equipment in addition to the usual pumping engine equipment, the apparatus is referred to as a "quadruple combination" or "quad." If a power operated aerial ladder or elevating platform is added, such apparatus is termed a "quint." However, operating efficiency and capability tends to decrease if too many functions are to be performed by one unit of apparatus, and the unit cost of such apparatus substantially increases. In addition, if one feature of the apparatus requires repair, the entire apparatus must be taken out of service.

Increasingly, apparatus is being designed for functional performance as desired by individual fire departments. Initial fire attack capability for pumpers is being improved not only with the addition of preconnected hose lines but frequently with the addition of elevated stream equipment. Pumpers are often built with considerable compartment space for emergency equipment and may serve as combination rescue pumpers.

Provisions Applying to All Types of Apparatus

The latest NFPA automotive fire apparatus standards are: NFPA 1901; NFPA 1904, *Standard for Testing Fire Department Aerial Ladders and Elevating Platforms* (hereinafter referred to as NFPA 1904); and NFPA 1921, *Standard for Fire Department Portable Pumping Units.*

Each of these documents has a general section entitled "administration" which defines the purpose and scope of the standard, defines the various terms employed in the standard, and discusses the various components specific to the standard type covered. The procedures for bidding, writing specifications, and acceptance are also listed. A prospective purchaser should obtain the latest editions of these standards to have all available information on a specific type of apparatus.

Altogether, NFPA 1901 calls attention to many options that should be considered to make the apparatus suitable for local needs when writing fire apparatus specifications. For example, requirements for fighting large brush fires in southern California are quite different from those occasioned by fire fighting in below zero weather in northern Minnesota. The standard provides for desirable standardization of many items for which a uniform performance (measurable by tests) is desirable, but allows the necessary flexibility to meet local needs that could not be supplied by one assembly line production. Such differences include rated pump capacity, water tank capacity, length of aerial ladders, height of elevating platforms, and more than 100 other important design features.

Besides the functional characteristics of a piece of motorized fire apparatus, considerable thought should be given to personnel safety and operations. The design of the vehicle cab, riding spaces for fire fighters, and the location of various components should be studied carefully. Only fully enclosed cabs with sufficient seating for all fire fighters who are expected to ride on the apparatus should be specified. Besides the aspects of weather and social climate, the overriding consideration is that of personnel safety. Fire fighters should have assigned seating and not be allowed to ride in exposed positions on the rear or sides of apparatus. The use of seat belts must be mandatory in all cases. The increased use of four door, fully enclosed cabs provides a safe environment from weather as well as in case of a motor vehicle accident. NFPA 1901 covers the basic aspects of driver compartment and cab design.

The positioning of the pump operator's panel may be to the side or located in a top midship position. Likewise, the positioning of valve controls and other equipment will affect the overall, efficient, and safe operation of the apparatus.

Fire Apparatus Engines

Engines may be either diesel or gasoline types, but there are few gasoline engines capable of supplying sufficient power for modern apparatus. Fuels other than diesel or gasoline, such as LP-Gas, are acceptable if they meet the intent of NFPA 1901. The standard does not contain any minimum engine size because engine size and horsepower must be chosen to correspond with the conditions of service and apparatus design.

Local availability of service is important with regard to engine maintenance. Since there are few gasoline engines available in the sizes necessary for general fire service usage, midrange diesels are rapidly capturing this market.

Carrying capacity is one of the most important, but least understood, features of a vehicle. Under NFPA 1901, it is the responsibility of the apparatus manufacturer to provide gross axle weight rating (GAWR) and gross vehicle weight rating (GVWR) adequate to carry a 1,200 lb (545 kg) personnel weight, a full water tank, the specified hose load, plus a miscellaneous equipment allowance of 2,000 lb (909 kg) for pumpers, 1,500 lb (682 kg) for water towers, and 2,500 lb (1136 kg), not including ladders, for aerials and elevating platforms, and 900 lb (409 kg) for light attack vehicles. GAWR is the value specified by the vehicle manufacturer as the loaded weight on a single axle system, and GVWR is the value specified by the manufacturer as the loaded weight of a combination vehicle such as a tractor trailer unit. GVWR is the value specified by the manufacturer as the loaded weight of a single vehicle, and is the sum of the weights of the chassis, body, cab, equipment, water, fuel, crew, and all other loads. All vehicles are designed for "rated GVWR" or maximum total weight, which should not be exceeded by the apparatus manufacturer or by the fire department after the vehicle is placed in service.

Too many fire departments have seriously overloaded apparatus by adding more equipment than the vehicle was designed to carry. There are a number of factors included in the GVWR, such as the springs or suspension system, the rated axle capacity, the rated tire loading, and the weight distribution between the front and rear axles. One of the critical factors is the size of the water tank. Water weighs about 8⅓ lb per U.S. gal (1 kg/L). A value of 10 lb (4.54 kg) is often used when estimating the weight of a full tank, making 2½ tons (2.27 metric tons) for a full 500 gal (1892 L) tank. The improper distribution of weight between front and rear wheels can make a vehicle difficult to control, and may require tires of different sizes to carry the load. Overloading not only affects the handling characteristics but undoubtedly results in increased maintenance problems with transmissions, clutches, and brakes.

Compliance with Federal Standards

The federal government has adopted motor vehicle safety standards applicable to all manufacturers of vehicles, including fire trucks. This legislation is enforced by the U.S. Department of Transportation (DOT). Under these standards it is unlawful to sell a truck not in compliance with the current federal standards. Manufacturers cannot accept specifications which would make the vehicle perform unlawfully, delete required items, or include items that are illegal. Because fire apparatus is complex and may require considerable lead time between the signing of a contract and delivery, the federal regulations provide that standards applicable at the time of contract shall be those complied with, provided that the delivery of the apparatus takes place within two years.

Additional requirements are placed upon apparatus and engine manufacturers based upon the Clean Air Act which is enforced by the Environmental Protection Agency (EPA). Engines cannot be modified once approved by EPA. These standards have resulted in changes in engine performance, and often require the use of larger engines than used previously to obtain the same vehicle performance. Likewise, the standards often call for more frequent maintenance checks and operations due to the pollution control equipment now required.

Electrical Power for Apparatus

Sufficient electrical capacity for the vehicle and its components is a major requirement for fire apparatus. Electricity is essential to reliable starting of the apparatus under all temperature and weather conditions. Emergency warning lights and sirens impose a heavy electrical demand, with radios and speakers as an added load. Hose reels, pump controls, and other items of equipment may also be operated electrically. An electromechanical siren alone may require about 300 amps to start, and 100 amps to continue operating. The radio takes additional amperage depending upon equipment, and the standard light load is about 40 amps. Much of the equipment must perform when the apparatus is at a fire with the engine idling or turned off.

NFPA 1901 permits either dual or single battery systems, but dual batteries are recommended. Minimum capacities required are 120 amp hr rating at a 20 hr discharge rate (520 cold cranking amps) for gasoline engines, and 200 amp hr rating at 20 hr discharge rate (900 cold cranking amps) for diesel engines using 12 V starting systems. It is unacceptable to attempt to meet battery requirements with two small batteries of dissimilar size and age because these will not charge or discharge at an equal rate, nor have equivalent reliability and power. The standard stipulates that when a dual battery system is supplied, each battery shall be of the capacity required for a single battery system.

NFPA 1901 requires provision of an electric alternator (alternating current generator) and rectifier equipment of sufficient output to meet the vehicle's electrical requirements. If 110 V power is required, a separate three phase transformer or rectifier unit should be provided in connection with the alternator to produce 110 V dc at special outlets. These outlets can supply power to floodlights of suitable size. Portable tools and other electrically driven equipment supplied from these outlets must be designed for dc operation or have universal motors (because alternators produce current at frequencies varying from 80 to 160 cycles per second); ac motors designed only for 60 cycle current will soon burn out. Separate 110 V ac power supplies may be specified.

Where a dual battery system is provided, it is important that each battery be maintained at peak efficiency and that polarized receptacles be provided for charging.

Apparatus Equipped with a Fire Pump

NFPA 1901 contains provisions applying to apparatus equipped with a fire pump. The most common basic fire fighting vehicle is termed a pumper or pumping engine. This standard includes sections on design and performance of the pump, construction requirements, operating controls and devices, and suction and booster hose which may be supplied with the pumper. The design requirements include the rated capacity, the pump suction capability, and the engine capability to be specified. Construction requirements cover the construction of the pump and the pump suction inlets and discharge outlets. At least one gated suction inlet is required, normally a 2½ in. (63 mm) inlet unless a larger size is specified. This inlet is an auxiliary suction inlet and is provided so an additional feeder line can be taken into the pump without shutting down when operating with marginal incoming water supply. This is a restricted piping inlet and should not be used for normal supply to the pump. Whenever possible, the large suction inlet should be used. If it is desirable to supply the pump with feeder lines, a pumper suction siamese connected to the large suction intake should be used. A pumper suction siamese has two, three, or four gated 2½ in. (63 mm) female swivel intakes, and a large female swivel connection sized for the large pumper supply inlet. A suction may also consist of a large [4 in. (100 mm) or greater] gated suction for large diameter hose. The section of the standard on operating controls covers the pump operator's position and various controls for the pump including engine controls, priming device, and pressure gages and tachometer.

Generally a fire pump is powered by the vehicle engine through a device which transfers power from the rear wheels to the pump itself. Power must be transferred with the vehicle in a stationary position and only after a parking brake has been set. Other designs allow for power to be transferred to the pump as the vehicle continues to move, with power being transferred through a power takeoff (PTO) device. Interlocks may provide for the safe operation of pump shift controls. Some designs have the pump mounted at the front of the chassis where it is driven through a clutch arrangement from the front of the engine. There are advantages and disadvantages to the front mount configuration.

A section in the standard on suction hose and booster hose sets forth their requirements, if specified. Hard suction hose is not required to be carried unless specified or if suitable suction hose is available for testing, because many pumpers serve in areas where high pressure water supply systems exist and there is no opportunity for drafting at the fireground. Some fire departments use different types and sizes of preconnected hose lines in lieu of booster hose. Pumpers designed for use as water supply engines at hydrants or at draft do not need booster hose. However, provision must be made for some apparatus to carry small hand line equipment for initial fire attack and mopping up operations.

Standard sizes of fire pumps for installation on apparatus are: 500, 750, 1,000, 1,250, 1,500, 1,750, and 2,000 gpm (1893, 2839, 3785, 4732, 5678, 6624, and 7571 L/min, respectively). Normal truck chassis and power train generally provide ample power for 1,000 gpm (3785 L/min) and larger pumps; thus, a decreasing percentage of smaller size pumps are currently sold. Relatively few are of the 500 gpm (1893 L/min) rating, which was popular years ago when vehicle engines were less powerful. A 500 gpm (1893 L/min) capacity is the minimum capacity recognized for a pumper. However, the minimum size for a pump should be 1,000 gpm (3785 L/min) for most fire department service due to the increased volume available at higher (above 150 psi or 1034 kPa) pressures.

Rated pumping capacity is determined at 150 psi (1034 kPa) net pump pressure. The pump must be capable of drafting rated capacity at a 10 ft (3 m) lift at altitudes up to 2,000 ft (610 m) above sea level. Where the pumper is to serve at a location at an altitude above 2,000 ft (610 m), NFPA 1901 calls for the purchaser to specify the altitude and whether a 10 ft (3 m) lift suction capability is to be furnished, so that the proper engine and pump can be provided.

The performance requirements for acceptance which a fire pump must meet are covered in detail in NFPA 1901. These tests include a hydrostatic test to be conducted by the manufacturer, and certification that the vehicle has been operated for a minimum of 2 hours before delivery. A 3 hour certification test is also required. UL Bulletin "Subject 822" covers the test procedures in greater detail. The purchaser may specify that the certification test be conducted by engineers of Underwriters Laboratories Inc. (UL), or other capable testing agencies. The certification test includes a 3 hour test at draft during which the pump must deliver its rated capacity for 2 hours at 150 psi (1034 kPa) net pump pressure, followed by 2 one-half hour periods during which 70 percent of rated capacity is delivered at 200 psi (1379 kPa), and 50 percent of rated capacity at 250 psi (1724 kPa). The apparatus is then given a 10 minute overload test discharging rated pump capacity at 165 psi (1138 kPa) to demonstrate reserve engine power. Automatic pump pressure controls are to be tested with the pump discharging at 90, 150, and 250 psi (621, 1034, and 1724 kPa). A vacuum test is conducted to prove the ability of the pump to hold a vacuum with the primer off and discharge outlets uncapped. Where the apparatus has a water tank, the specified rate of flow from tank to pump must be maintained for at least 80 percent of the rated tank capacity for tanks of 300 gal (1135 L) capacity or larger. The minimum specified tank to pump flow rate is 250 gpm (946 L/min), but is increased to 500 gpm (1892 L/min) when a pumper has a water tank in excess of 750 gal (2839 L) capacity.

Service Tests

All pumpers should undergo service tests at least annually and after any major repairs. These tests chiefly demonstrate that the pumper/engine combination is capable of meeting the performance requirements of the original certification or acceptance tests. Records of service tests are important evidence of proper apparatus maintenance. Some rating organizations still conduct services tests on a periodic basis while others require the tests be performed by the fire department and satisfactory records kept.

A satisfactory service test should consist of:

1. Twenty minutes of pumping 100 percent rated capacity, preferably at draft, at 150 psi (1039 kPa) net pump pressure.
2. Ten minutes of pumping 70 percent rated capacity at 200 psi (1379 kPa) net pump pressure, and
3. Ten minutes of pumping 50 percent rated capacity at 250 psi (1724 kPa) net pump pressure.

Engine speed should be recorded for each condition. A "spurt" test need not be conducted, but if care is taken to ensure that the pump does not cavitate, running the pumper with wide open throttle at 165 psi (1138 kPa) net pump pressure may give a good indication of engine condition.

The necessary instrumentation is simple. Smooth bore nozzles of appropriate size and a Pitot gage, to be used to measure capacity, are the only special equipment required. A test gage may be substituted for the discharge pressure gage on the pumper panel, or a hand revolution counter may be used instead of the tachometer to measure engine speed; however, the gages and tachometer on the pumper are usually accurate enough to be used for service tests.

Pump capacity may be determined by connecting one or more lines to a nozzle or nozzles discharging into the reservoir when drafting (or other suitable area if pumping from a hydrant), and measuring the flow by means of the Pitot gage held at the nozzle(s). The latter must be carefully secured. Table 15-6A shows selected nozzle sizes and pressures for various flow rates.

Note: The nozzle pressure is given to the nearest psi or kPa. This may provide a flow of 1 or 2 gpm (3.78 or 7.57 L/min) above that required. In general, the minimum number of lines of hose should be used that will provide desired nozzle pressure with available pump discharge pressure. Pump discharge valves may be throttled to control the flow and increase pressure losses between the pump and nozzle.

TABLE 15-6A. Selected Nozzle Sizes and Pressures for Various Flow Rates

Flow		Nozzle Size		Pressure	
(gpm)	(L/min)	(in.)	(mm)	(psi)	kPa
250	946	1	25	72	496
350	1325	1⅛	29	89	614
375	1420	1¼	32	66	455
500	1893	1⅜	35	80	552
525	1987	1⅜	35	80	552
625	2366	1½	38	88	607
700	2650	1⅝	41	80	552
750	2839	1⅝	41	92	634
825	3123	1¾	44	93	641
1000	3785	2	50	71	490
1050	3975	2	50	78	538
1250	4732	2	50	110	458
1500	5678	2¼	57	99	683

When drafting, net pump pressure is determined by adding an allowance for the lift, friction, and entrance loss in the suction hose to the discharge pressure. Pumping tests should be conducted at a vertical lift of 10 ft (3 m) or less (the vertical lift is the vertical distance from the surface of the water to the center of the pump suction inlet). The net pump pressure is the sum of the discharge pressure (corrected for any gage error), and the dynamic suction lift (the dynamic suction lift is the vertical lift plus friction and entrance loss in the suction strainer and hose). The dynamic suction lift may be determined either by measuring the negative pressure (vacuum) in the pump suction manifold with a manometer (or other suitable test gage which measures vacuum accurately), or by adding the vertical lift and the value of friction and entrance loss taken from Table 15-6B. (To be most accurate, gage readings should be corrected for the difference between the height of the gage and the centerline of the pump suction inlet, but usually this is not a significant amount.) Thus the net pump pressure can be calculated by either of the following formulas:

$$P = D + (H \times 0.5), \text{ or } P = D + 0.43\,(L + F)$$

where

P = net pump pressure, psi (kPa),
D = discharge pressure, psig (kPa),
H = manometer reading, in. Hg (Pa),
L = vertical lift, ft (m),
F = friction and entrance loss, ft of water (Pa).

TABLE 15-6B. Friction and Entrance Loss in 20 ft (6.1 m) of Suction Hose, Including Strainers

	Suction Hose Size (Inside Diameter)													
	4 in.*		4½ in.*		5 in.*		6 in.*		2–4½ in.*		2–5 in.*		2–6 in.*	
Flow Rate GPM#	ft† water	in.‡ Hg	ft† water	in.‡ Hg	ft† water	in.‡ Hg	ft† water	in.‡ Hg	ft† water	in.‡ Hg	ft† water	in.‡ Hg	ft† water	in.‡ Hg
500	5.0	4.4	3.6	3.2	2.1	1.9	0.9	0.8						
350	2.5	2.1	1.8	1.6	1.0	1.0	0.4	0.4						
250	1.3	1.1	0.9	0.8	0.5	0.5	0.2	0.2						
750	11.4	9.8	8.0	7.1	4.7	4.2	1.9	1.7						
525	5.5	4.9	3.9	3.4	2.3	2.0	0.9	0.8						
375	2.8	2.5	2.0	1.8	1.2	1.1	0.5	0.5						
1000			14.5	2.5	8.4	7.4	3.4	3.0						
700			7.0	6.2	4.1	3.7	1.7	1.5						
500			3.6	3.2	2.1	1.9	0.9	0.8						
1250					13.0	11.5	5.2	4.7	5.5	4.9				
875					6.5	5.7	2.6	2.3	2.8	2.5				
625					3.3	2.9	1.3	1.1	1.4	1.2				
1500							7.6	6.7	8.0	7.1	4.7	4.2	1.9	1.7
1050							3.7	3.3	3.9	3.4	2.3	2.0	0.9	0.8
750							1.9	1.7	2.0	1.8	1.2	1.1	0.5	0.5
1750							10.4	9.3	11.0	9.7	6.5	5.7	2.6	2.3
1225							5.0	4.6	5.3	4.7	3.1	2.7	1.2	1.1
875							2.6	2.3	2.8	2.5	1.6	1.4	0.7	0.6
2000									14.5	12.5	8.4	7.4	3.4	3.0
1400									7.0	6.2	4.1	3.7	1.7	1.5
1000									3.6	3.2	2.1	1.9	0.9	0.8

* 1 in. = 25.4 mm; † 1 ft of water = 2988 Pa; ‡ 1 in. of Hg = 3377 Pa; # 1 gpm = 3.78 L/min

Where a suitable drafting site is not available, a service test may be conducted using a hydrant capable of providing the required flow. In this case, the net pressure is determined by subtracting the intake pressure at the pump, as indicated by the compound gage on the pumper panel, from the discharge pressure.

The results should be compared to the data from the original certification and/or acceptance tests and from previous service tests, if any. An example of such a comparison is shown in Figure 15-6E.

If rated capacity cannot be obtained at 150 psi (1034 kPa) and the pump is not cavitating, either the pump or the engine needs attention. If the engine speeds have increased excessively, the pump is probably worn and needs to be repaired. If a spurt test is performed at the same net pressure as in previous tests, e.g., 165 psi (1138 kPa), and significantly less flow can be obtained although engine speed has not increased markedly, the engine is out of tune or for other reasons is not developing as much power as before.

CAPACITY		NET PRESSURE		ENGINE SPEEDS (RPM) CERT. TEST	SERVICE TESTS		
(GPM)	(L/min)	(PSI)	(kPa)	(1983)	(1984)	(1985)	(1986)
1000	3785	150	1034	1850	1880	1920	1940
700	2650	200	1379	1670	1680	1690	1700
500	1893	250	1724	1790	1800	1820	1840

FIG. 15-6E. Data comparison of pump pressure acceptance and service tests.

A secondary purpose of the test is to ensure that the

pumper is generally in good condition, pump casing and various fittings are tight, and that the transfer valve is operating properly (if pump is series parallel type). If for any reason it appears that the pump, engine, pump accessories, or other parts of the power train and pumping equipment are not in good condition, the apparatus manufacturer or an authorized representative should be contacted for advice so the condition can be corrected.

Hose Bodies

NFPA 1901 contains provisions applying to apparatus equipped with a hose body. A standard hose body for pumpers has at least 55 cu ft (1.56 m^3) of space and is designed to carry a minimum of 1,500 ft (457 m) of 2½ in. (63 mm) hose, and 400 ft (122 m) of 1½ in. (38 mm) hose. Standard hose capacity for a pumper/ladder truck is 40 cu ft (1.13 m^3) and the truck is designed to carry 1,000 ft (305 m) of 2½ in. (63 mm) hose, plus 400 ft (122 m) of 1½ in. (38 mm) hose.

Many fire departments specify a minimum of 2,000 ft (610 m) carrying capacity for 2½ in. (63 mm) or larger hose, and a capacity for 800 to 1,000 ft (244 to 305 m) of 1½ in. (38 mm) hose. Fire departments should specify that the hose body be divided to permit laying of two or more lines of 2½ in. (63 mm) or larger diameter hose, and two or more lines of 1½ in. (38 mm) hose.

With the larger capacity pumpers now commonplace, additional hose capacity is required to move the larger volumes of water. A convenient way to relate needed hose capacity to pumper discharge requirements is on the relative water carrying capacity of various sizes of hose at normal operating pressures, as follows: 250 gpm (946 L/min) for 2½ in. (63 mm) hose, 350 gpm (1325 L/min) for 3 in. (76 mm) hose, 500 gpm (1893 L/min) for 3½ in. (89 mm) hose, 750 gpm (2839 L/min) for 4 in. (102 mm) hose, and 1,300 gpm (4921 L/min) for 5 in. (127 mm) hose. Thus, to move 500 gpm (1893 L/min) 1,000 ft (305 m) requires 2,000 ft (610 m) of 2½ in. (63 mm) hose or 1,000 ft (305 m) of 3½ in. (89 mm) hose. (Some increases in flow may be obtained by higher discharge pressures, especially in short lines.) Many pumpers in rural and suburban areas carry at least 2,000 ft (610 m) of big hose line in a split load. By splitting a hose load into two bed sections, either one long lay or a shorter double line lay may be made quickly. Unless hose laying apparatus is arranged to lay multiple lines, or large diameter hose is provided, considerable time and personnel may be required when flows above 250 gpm (946 L/min) per pumper are needed.

Increasingly, fire departments are using hose of diameters larger than the nominal 2½ in. (63 mm) size. While larger diameter hose costs more, it may actually be more economical as more water can flow through one large diameter line than multiple smaller ones. Care should be taken to specify the hose carrying capacity needed since larger hose requires more space on the apparatus. Also, the size of hose reels, whether for booster hose or for larger hose, must be specified. Normally, hose reels for large diameter hose are equipped with a power rewind.

Several types of fire hose are now available on the market, including double jacketed woven hose and the newer lighter weight synthetic type. All hose is pressure rated and should comply with NFPA 1961, *Standard for Fire Hose.*

The relatively large flows available with large diameter hose require increased attention to the use of dump valves on the suction side of pumps, pressure relief capability on various fittings, and fireground safety. Large diameter hose cannot be moved easily once charged, and vehicles should not drive over it.

Booster Pumps

NFPA 1901 applies to apparatus equipped with a booster pump. The term "booster pump" applies to pumps of less than 500 gpm (1893 L/min) rated capacity. Normally, these pumps are not subject to UL certification tests. When equipment was developed to supply small hose on fire apparatus, replacing the former chemical tanks, the equipment consisted of a small tank of water at zero psi (kPa) pressure and a permanently connected pump of about 30 gpm (114 L/min) capacity which took suction from the tank and supplied ¾ in. (19 mm) hose. There were no other discharge connections, but commonly a 2½ in. (63 mm) hose inlet was provided to replenish the tank supply as was the case with many chemical engines. Additional suction and discharge connections were added later.

Booster pumps, used predominantly for fighting grass fires and other small fires, are usually capable of supplying one or two reels of 1 in. (25 mm) booster line, and the pumps have a capacity of up to 100 gpm (379 L/min) at pressures not exceeding 250 psi (1724 kPa). Often pumps for use on grass fires are designed to be operated when the apparatus is in motion. This feature is called "pump and roll."

Some light attack vehicles will have separate engines to drive a small pump so "pump and roll" may be accomplished without mechanical complexity. Likewise, the use of skid mounted fire fighting packages complete with pump and engine would allow the use of temporary, nonfire fighting chassis to be used as necessary.

When one or two lines of 1½ in. (38 mm) hose are to be supplied, a capacity of 250 gpm (946 L/min) at 150 psi (1034 kPa) is desirable. A standard fire pump is required for predominantly structural fire fighting. There is little excuse for failure to provide at least the 500 gpm (1893 L/min) minimum for structural fire fighting, because the apparatus engine generally has ample power for the job.

When working pressures exceeding 250 psi (1724 kPa) are desired in a small pump, the pump discharge capacity should be at least 50 to 60 gpm (189 to 227 L/min) at the specified pressure. In some cases this small volume/high pressure characteristic is specified as an additional capability for a fire pump meeting standard pressure/volume characteristics.

Some fire apparatus will have a separate pump for small volume/high pressure operations. Other designs call for the use of a third stage in the pump to develop the higher pressures. Suitable valving and controls should be provided to preclude the introduction of these higher pressures into ordinary fire hoses.

Years ago when most booster lines discharged approximately 15 gpm (57 L/min) using ¾ in. (19 mm) hose, very high pressures for small stream service became popular. Poorly informed fire fighters mistakenly thought that high pressure contributed to fire extinguishment. Tests demonstrated that the heat absorption necessary for the cooling and extinguishment of fires was due entirely to the volume of water discharged in spray form, and that excessive pressures were not only unnecessary, but much of the high pump pressure was lost due to high friction losses in the small diameter hose used.

Likewise the use of very high pressure lines can be a safety hazard unless fire fighters have been properly trained. Streams from such units can injure personnel as well as cause damage to surroundings.

Small booster pumps can be powered by PTO units attached to SAE six bolt openings on the chassis transmission. Larger booster pumps usually are driven by eight bolt PTOs, flywheel PTOs, or by separate engines. Purchasers should indicate the volume and pressure desired of booster pumps and the tests to be conducted. Pumping tests should demonstrate the pump capacity and pressure, engine capability, suitability of pump controls, engine controls, and suction capability if applicable. All tests should be conducted in accordance with NFPA 1901.

Water Tanks

The majority of fire trucks equipped with a pump also carry a water tank. The tank supplies water to the pump for initial hose streams before hydrants or suction sources may be brought into use. NFPA 1901 requires a minimum water tank capacity of 300 gal (1136 L) for pumpers and 150 gal (568 L) for ladder trucks or elevating platforms for which water tanks are specified. Most new pumpers have water tanks larger than 300 gal (1136 L), with a 500 gal (1893 L) tank being the norm.

NFPA 1901 deals with water tank construction, tank connections including a capped fill opening of not less than 5 in. (127 mm) diameter, tank overflow, and venting. It also covers special provisions for tanks used for mechanical foam or water additive agents, and specifies flow rates from tanks to fire pumps and to booster pumps. For tanks of 300 to 750 gal (1136 to 2839 L) capacity, a flow rate of 250 gpm (946 L/min) is specified (up to the capacity of the booster pump when that is smaller) to permit immediate fire attack with one 2½ in. (63 mm) hose line or two smaller streams. For tanks of over 750 gal (2839 L) capacity, a flow rate to the pump of 500 gpm (1893 L/min) minimum is required. All rates of flow between the water tank and the pump shall be maintained for 80 percent of the tank capacity, unless otherwise specified.

Tanks may be constructed of steel, aluminum, fiberglass, or a plastic material. Further, they may be coated with various nonfouling and corrosion resistant materials to improve service life. Some designs have anodic protection to prevent corrosion. All tanks must have baffles to minimize water surging when the vehicle is in motion. The basic fittings are specified in NFPA 1901.

Mobile Water Supply Apparatus

In rural districts and in outlying districts of cities where hydrant distribution is not complete, supplemental water tank apparatus, defined as "mobile water supply apparatus" is commonly provided. Such apparatus (covered in NFPA 1901) has a permanently installed fire pump or booster pump. Suitable hose and fire fighting tools for accomplishing intended tasks should be carried on the vehicle.

Many tankers carry portable or collapsible water tanks into which the water supply is immediately dumped so the tanker may go for another load. The addition of a large dump valve facilitates the quick discharge of the tank water into the portable tank. The portable tank is positioned adjacent to a pumper which then takes suction from it. Multiple portable tanks may be used with large diameter, lightweight plastic syphons used to connect the tanks.

Normally, mobile water supply apparatus is supplemental to fire department pumpers. A minimum 1,500 gal (5678 L) tank capacity is specified. Larger tanks mean heavier vehicles and may limit the mobility of apparatus on rural roads and bridges. The efficiency of mobile water supply apparatus in transporting water to fires depends upon the over the road mobility of the vehicle, the ability of fire fighters to quickly unload water at a fire, and the ability to quickly refill the tank to transport additional water. Under the standard, the tank must be able to be dumped and filled through a single connection at a rate of at least 1,000 gpm (3785 L).

Many rural fire departments use large "nurse" tankers as a portable reservoir to supply pumping engines at a fire. In some cases, the large tankers are kept supplied by a shuttle of the smaller mobile tankers. The capacity of the large tankers may be a maximum of 5,000 gal (18 927 L). To meet the requirements of NFPA 1901 and DOT regulations, where tankers of 1,500 gal (5678 L) capacity or larger are specified, either tandem rear axles, semitrailer construction, or both, may be needed. A truck chassis with tandem rear axles is suggested for mobile water supply apparatus from 1,500 to 2,000 gal (5678 to 7571 L); for tanks over 2,500 gal (9464 L) a semitrailer chassis with tandem trailer axles is needed. A maximum water tank capacity of 4,800 gal (18 170 L) or 20 tons (18 metric tons) of water is recommended. Power brakes and power steering are important considerations in design of mobile water supply apparatus, and experienced truck drivers are needed for safe operation of large tank vehicles.

Aerial Ladders

Aerial ladders have been used by fire departments for more than a century to gain access to upper floors and roofs of buildings. The first aerials were manually operated before spring assist and air hoists were introduced. These were followed by the present hydraulic hoists and ladder controls powered by the truck engine. More reliable steel or aluminum truss construction has replaced the wooden beams used in the earlier aerial ladders.

NFPA 1901 covers provisions applying to apparatus equipped with an aerial ladder. The chapter on aerial ladders is divided into sections dealing with the aerial ladder and its equipment, control devices, and the type of vehicle chassis to be used. Once the vehicle has been positioned and stabilized, one person has full control of all operations of the ladder; the ladder is independent of the integrity of the building on fire.

Common sizes of aerial ladders are 65, 75, 85, 100, 110, and 135 ft (19.8, 22.8, 25.9, 30.5, 33.5, and 41.1 m, respectively), and the user must specify the length desired and whether a single chassis vehicle or a tractor drawn vehicle is to be provided. In recent years the 65 ft (19.8 m) aerial has been specified less frequently for ladder trucks. The 85 and 100 ft (25.9 and 30.5 m) aerials have become increasingly popular due to increased height of buildings and because of the advantages of greater horizontal reach when fighting fires. Aerial ladders up to 206 ft (62.8 m) can be purchased.

In the last several years, the rear mount aerial ladder has become common. In some cases the shorter aerial ladders are provided as a useful addition to a combination attack pumper. The length of an aerial is measured by a plumb line from the top ladder rung to the ground, with the ladder fully extended at its maximum elevation. NFPA 1901 requires a minimum width of 18 in. (457 mm) at the narrowest point of the ladder.

Aerial ladders may be equipped with a platform or fixed waterway to the tip of the fly section or platform, and can be equipped with electric power, intercoms, etc. NFPA 1901 as well as manufacturer's literature and specifications should be consulted when assembling specifications for aerial ladders.

A detachable ladder pipe which provides elevated fire stream service is standard equipment for an aerial ladder. When supplied by 3 in. (76 mm) hose up the ladder, 600 gpm (2271 L/min) is the practical maximum which can be supplied in normal fireground operations. Tips provided with the ladder pipe are 1¼, 1⅜, and 1½ in. (32, 35, and 38 mm), and a 500 gpm (1893 L/min) spray nozzle.

A fixed waterway on the ladder may supply a greater flow and care must be taken to minimize ladder whipping when the supply is turned off and on. Ladder pipes may be attached at either the tip of the fly section, on a platform, or permanently attached on the end of the bed section.

NFPA 1901 deals with control devices for aerial ladders. An operator's position is required from which controls are operable. Controls are usually grouped at a pedestal. The operator's position must be located so the operator's line of sight coincides with the axis of the ladder when it is in any position. Lighted and clearly marked control devices are provided to: immobilize the vehicle; transfer power to the ladder mechanism; stabilize the vehicle; lock in and release the ladder from its bed; control elevation, rotation, and extension of the ladder; indicate the angle of elevation and the load limit; and control the engine for ladder operation. Details on testing aerial ladders and elevating platforms are covered in NFPA 1904.

Those aerial ladders which are equipped with platforms may have a duplicate set of controls mounted in the platform, thus allowing a platform operator to maneuver and control the ladder. Another type of design provides for the remote control of a ladder pipe nozzle via either electric wires or radio control. Some ladder pipes may be controlled from the ground by lanyards.

Elevating Platforms and Aerial Towers

Fire trucks with hydraulically operated elevating platforms are widely used and are covered in NFPA 1901. Some of these are on trucks that also carry ground ladders and other equipment for ladder company service, while the smaller size platforms may be mounted on attack pumpers.

Platform apparatus are of three principal designs. In one, the platform is mounted on an articulated boom which travels in an arc as desired by the operator. In the second, the platform is mounted on an extendable or telescopic boom much in the fashion of an aerial ladder; the third design has the platform mounted on booms that are both articulated and telescopic.

Platform equipment is available with booms designed for maximum elevation from approximately 50 to 100 ft (15.2 m to 30.4 m). Under the standard, the nominal height of a platform assembly is measured by a plumb line from the top surface of the platform to the ground, with the platform raised to its maximum elevation. It is important in selecting platforms for reach and elevation under fireground conditions to obtain charts from the manufacturer showing reach under recommended operating conditions. The articulated boom design provides its maximum horizontal reach at approximately half of its maximum elevation. The telescopic extendable boom of the same nominal length generally permits somewhat greater horizontal reach at higher and lower elevations. The telescopic type often has the added feature of an extending ladder attached to the boom, which provides access to and from the platform while elevated, which is not the case with the articulated boom.

All designs provide stable platforms with dual controls located both on the platform and at the turntable (which is the overriding control). NFPA 1901 requires that the platform handle a payload of at least 700 lb (318 kg) in all recommended operating positions. An advantage in operation is that the platform can be moved quickly from window to window for fire fighting or rescue. The standard specifies that the platform must be capable of reaching its maximum elevation, extension, and a rotation of 90° within 150 seconds.

Platforms are designed to provide water tower service superior to that provided by ladder pipes. NFPA 1901 requires that the platform turret or turrets be capable of rotation through at least 45°, while ladder pipes are limited to a horizontal rotation of 15°. Also, the mobility of the platform permits the operator to quickly change the turret location as desired. The apparatus must be designed so that regardless of the position of the platform and the direction of the stream, the equipment can be operated safely while discharging 750 gpm (2838 L/min) through a 1¾ in. (44 mm) tip; many platforms are built to handle 1,000 gpm (3785 L/min) discharge. Standard nozzle tips for platform turrets are 1⅜, 1½, and 1¾ in. (32, 35 and 38 mm, respectively), and a 500 gpm (1893 L/min) spray nozzle. Some platforms have two turret nozzles.

As with aerial ladders, NFPA 1901 also covers operating mechanisms; platforms and equipment; control devices, including platform and engine controls; load limitations; and turret nozzle operations.

Water Towers

The success of elevating platforms on the fireground during the 1960s was a prelude to the use of hydraulically operated water towers designed to apply large flows on a fire from effective heights and positions. Such water towers have proven to be very useful and have been widely accepted in both large and small communities as an important type of equipment. Modern water towers are designed to discharge from 300 to 1,000 gpm (1135 to 3785 L/min) or more in either a straight stream or spray, to move the boom and nozzle under the control of the operator from lowest position to maximum height, and to extend horizontally for the best application of water on the fire.

Many of the newer water towers are mounted on attack pumpers, permitting the pump operator to supply the pressure and volume needed for the tower stream. In many cases the initial operation is from the apparatus water tank, while hose lines are being placed in service to supply the apparatus. Some towers are mounted on hose trucks, permitting the equipment to lay its own supply lines from pumpers or from high pressure fire mains.

Like elevating platforms, water towers may be of either articulated or telescopic boom design. Heights specified may be from 50 to 75 ft (15.2 to 22.9 m), with the smaller size designed to be mounted on standard pumpers or hose trucks. The telescopic type may be equipped with a ladder extending the full length of the boom.

NFPA 1901 covers water tower apparatus. Except for the lack of the platform feature, the water tower capability specified is similar to that of platform turret nozzles. The tower is mounted on a turntable or pedestal to permit rotation in either direction through 360 degrees, and can rotate with the nozzle in operation at rated capacity. All controls are grouped at the control station adjacent to the turntable so the operator has full control of the pressure, volume, and stream pattern.

Chief Officers' Vehicles

Each career fire department and many volunteer fire departments find it essential to provide automotive transportation for staff command officers above the fire company level. In the larger fire departments, sizeable fleets of automobiles are maintained. Automobile transportation is necessary for chief officers, fire prevention officers and inspectors, training officers, communications officers, arson squads, etc. Some fire departments pay a mileage allowance for the use of private cars. This is not a very good arrangement, however, because efficient public service is better performed by publicly owned vehicles, especially where response to emergencies may be involved. Inadequate transportation in staff agencies often results in wasted staff time.

All fire department vehicles, including automobiles, that respond to fires and emergencies must be equipped with warning lights, sirens, and radio communications facilities with the emergency networks (including mutual aid frequencies) on which the department operates. The vehicles should be clearly identified as to department or governmental ownership. The vehicles should carry protective clothing and equipment for the officers and their aides, directories containing prefire plans on properties in the area served, water distribution plans, reference books to hazardous materials, portable fire extinguishers, first aid equipment, and hand lights.

Many cities provide district or battalion fire chiefs with station wagons in which the necessary equipment is mounted for ready access. Reserve cars are needed not only to cover vehicles undergoing mechanical repair, but for offshift officers to use when recalled to duty to cover vacated districts.

Generally, these cars respond to more alarms and acquire more mileage than individual fire company apparatus, and a replacement program for the cars should be followed. In busy areas, cars may be used for fire duty for one or two years before being placed in reserve or assigned to staff work.

Floodlight Trucks or Trailers

These units have a generator of 5,000 W or larger, generally driven by its own engine, and may carry floodlights and various power tools. These units provide extra generating capacity that can be used at fires and at emergencies involving interruption of normal power sources, supplementing the emergency lighting and power equipment carried on fire apparatus. Pigtail grounded adapters for both two and three wire services are required as part of the emergency electrical equipment.

Much apparatus has permanently mounted elevating floodlights which are supplied with electrical power from the engine alternator or a built in generator. These lights illuminate both the immediate area of the apparatus and the fireground. This promotes personnel safety and helps to identify the fire apparatus in areas where traffic may be a problem.

Because fire department emergency electrical equipment must operate under conditions of extreme wetness, it is important that properly installed and grounded three wire services be provided for generator and lighting units.

Specialized Apparatus

There is a wide diversity of fire apparatus designed for special types of service. These units should be designed with the rigors of fire duty in mind and should conform to the general vehicle standards set forth in NFPA 1901.

Many fire departments operate squad or rescue trucks carrying emergency equipment to supplement that carried on pumpers and ladder trucks. Many rescue trucks are enclosed to provide shelter for first aid and emergency work during inclement weather.

A SCBA (self-contained breathing apparatus) service unit should be designed for field servicing of SCBAs and recharging of depleted air cylinders on the fireground. Small fire departments may not need a separate vehicle, but any fire department committed to using SCBA will find it necessary to provide field support capability.

Another specialized type is the forest fire truck. Generally these are smaller vehicles, frequently with all wheel drive and equipped with a small water tank, booster hose, forestry hose, a small capacity pump, a number of portable water type extinguishers and various hand tools for fighting wildland fires. When operated by municipal fire departments, the forest or brush fire trucks may have a 500 or 750 gpm (1893 or 2839 L/min) front mounted fire pump, and a 500 gal (1893 L) water tank. The apparatus may respond to structural fires for initial fire attack in the district served and may also be equipped to respond to highway fires.

A few cities still have salvage trucks equipped with 20 or more waterproof covers and other equipment used to reduce water damage. This supplements the salvage equipment carried on pumpers, ladder trucks, and squad trucks. Salvage trucks also carry dewatering pumps, water vacuums, and other specialized salvage equipment.

Fireboats

Fireboats are available in various sizes and types in accordance with local needs, and vary from large tugs to fast jet propelled fire and rescue craft. Some cities maintain fleets of fire fighting vessels. Some of these are operated by fire departments and some by port authorities. U.S. Navy and Coast Guard vessels also are equipped for fire fighting service.

Fireboats are principally used to: (1) protect vessels in the harbor; (2) protect piers, pier sheds, and cargoes along the waterfront; (3) protect yachts and houseboats in basins and marinas; (4) assist in marine rescue from all types of water accidents; and (5) serve as pumping stations for providing large flows at fires within reach of hose supplied by the boats. Fireboats also may provide a valuable emergency source of water for fire protection should an earthquake or other accident interrupt the normal supply. Some fire departments operate hose trucks or "fireboat tenders" out of fireboat stations. Such trucks carry 1 mi.(1610 m) or so of large diameter hose to permit utilization of fireboat pumping capacity at shore fires along the waterfront. In harbors subject to freezing, fireboats should have ice breaking equipment.

Selecting the size, type, pumping capacity, and equipment carried by fireboats should depend upon the type of service expected of the vessels. Fireboats must carry much of the same equipment as pumpers, ladder trucks, and rescue vehicles on land. They need foam making capability, and should carry quantities of special extinguishing agents such as carbon dioxide.

Pumping capacity for individual fireboats varies from 500 gpm (1893 L/min) for very small craft to 10,000 gpm (37 854 L/min) or more for larger vessels. Pumping capacity is rated at 150 psi (1034 kPa) discharge pressure, as with motor pumpers. Some small jet powered fireboats rely upon their jet pumps to supply water for fire fighting. These pumps may develop substantial volumes at pressures adequate to supply turret nozzles attached to the pump, but it may be difficult to provide the necessary pressure to supply the hose streams to fight ship fires or fires ashore, unless additional pumps are provided to meet standard pressure requirements.

Where fire protection is required for yacht basins and where water rescue service must be provided, time is of the essence and boats should be berthed as near the area of anticipated need as practicable. Boat fires and explosions are not only life safety hazards, but also can cause considerable financial loss, with closely berthed yachts.

In some cities, waterfront areas present some of the most serious fire hazards because of substandard construction, large undivided areas, and hazardous storage. Lumber, packaged materials, chemicals, and flammable liquids often are stored and handled in bulk. Wharf sheds may lack fire divisions, and fires may extend between wharf decks and the water, presenting troublesome problems of access, particularly in tidewater areas. In some fire departments, fireboat crews include scuba divers trained in underwater recovery and in fighting fires under piers. The best fireboat protection may be at a disadvantage where piers are structurally weak, lack automatic sprinkler protection, are without draft curtains or underpier firestops, and lack skylights or other means for venting roofs.

Personnel to staff fireboats varies from three fire fighters on the smaller fire rescue craft to a full fire company of five or six fire fighters plus a licensed marine crew for larger vessels. A licensed marine crew consists of a pilot with the required papers for the waters served by the vessel, a marine engineer, and one or two assistants. The assigned fire company should have the same officers and personnel provided for an engine company ashore in the same district.

Smaller fireboats do not require a licensed marine crew, provided that qualified fire fighters are assigned as boat operators. It is essential that these operators complete the U.S. Coast Guard course for powerboat operators and know the rules that must be observed for all vessels. It is also essential that each fireboat carry proper papers for navigation of the waters in which it is to serve. Fireboats need radar as well as radio communications for safe operation under all weather conditions and, at times, in heavy smoke. Self-contained breathing apparatus is essential for fireboat crews.

Where fireboats are used infrequently, the cost of providing personnel for these vessels has been a problem for fire department administrators. In some instances this has been partially resolved by berthing the fireboat adjacent to a land fire company. The problem with this arrangement is that many waterfront fires require simultaneous operations from both land and water, and it is necessary to train personnel from covering companies to work from the vessel when the assigned company is out of quarters. The most practical solution, where proper crew strength and maintenance of a large fireboat is not economically feasible, has been replacement of the vessel with a more modern type requiring fewer crew members and lower maintenance costs.

At some ports fireboats are owned and operated by port authorities. This has sometimes resulted in poor coordination of fire fighting operations between the fireboat crew and land based fire companies. It is important that control of fire fighting operations be under the direction of the chief officer at a fire so that personnel will not be endangered by powerful fireboat streams directed into areas where hose crews and truck crews are working.

In a number of locations fireboats also furnish valuable fire protection for various island and resort communities not readily accessible to land fire apparatus.

Airport Crash Trucks

Specialized apparatus is required for aircraft rescue and fire fighting service at airports. In many cities such equipment is staffed by fire department companies; in other cities, it is maintained and operated by an airport authority.

The NFPA Technical Committee on Aircraft Rescue and Fire Fighting has issued NFPA 402M, *Manual for Aircraft Rescue and Fire Fighting Operational Procedures*; NFPA 403, *Recommended Practice for Aircraft Rescue and Fire Fighting Services at Airports and Heliports*; and NFPA 414, *Standard for Aircraft Rescue and Fire Fighting Vehicles* (hereinafter called NFPA 402M, NFPA 403, and NFPA 414). Because of off highway performance needs, the vehicle weight of crash trucks should be distributed equally over all wheels. Greater axle and chassis clearances are needed than for standard fire apparatus, as well as high acceleration characteristics. NFPA 414 calls for a drive that provides multiplication of torque from the engine flywheel to the wheels of the vehicle. Positive drive to each wheel is required to negotiate soft ground, snow, and ice. The positive drive can be provided by torque proportioning or no spin differentials, or by other automatic devices that assure that each wheel (rather than the axle) is driven independently.

Three types of vehicles are covered in NFPA 414: (1) major fire fighting vehicles, (2) Rapid Intervention Vehicles (RIV), and (3) combined agent vehicles. RIVs are intended to reach the emergency site quickly so rescue operations are started before the major vehicles arrive. Airport crash apparatus should include a small rescue vehicle (which can also serve as a command car), and a nurse tanker carrying additional water and extinguishing agents where needed. Crash trucks can be modified for fighting fires in hangars and other airport structures, but this should not detract from the primary function of these vehicles.

The amount of personnel assigned to operate with airport crash trucks vary according to the apparatus provided, the aircraft using the facility, and the distribution of air traffic over a 24 hour period. The federal government sets standards for such operations. Apparatus should be located and staffed so crash trucks can respond to any area of the airport and extinguishing agent application initiated within three minutes of the sounding of the alarm. All vehicles should have qualified operators assigned, but may be staffed by trained airport personnel who respond to an emergency upon call.

Apparatus Procurement Policies

The responsibility for procurement of fire apparatus and equipment should rest with the agency operating the fire department, whether a municipality, fire district, or private industry. In some cases, procurement is a cooperative measure in which a larger unit of government contributes to the cost of providing fire apparatus for fire department organizations serving in its territory. In a few cases, state governments contribute to local fire apparatus costs. In some areas, several fire departments have banded together to generate standard apparatus specifications to buy apparatus in quantity and realize significant cost savings. Likewise, various fire department groups will cooperate in volume purchases of supplies.

When the fire department is an agency of municipal government, it is the responsibility of the latter to provide for all public fire fighting equipment. Providing the necessary monies is the responsibility of the appropriating fiscal authorities, but the actual selection of the equipment should be the responsibility of the fire department management. The chief administrative officer of the fire department (aided by a staff of technical specialists, including a master mechanic) should keep the municipal administrator informed regarding the age and condition of department equipment, and should prepare current specifications consistent with applicable national standards covering items ready for replacement. In autonomous volunteer fire departments there may be a purchasing committee appointed to handle procurement of apparatus. In municipalities, actual purchase is usually handled through a municipal purchasing department. The purchasing department should not try to tell the fire department what type of fire apparatus it should use, but should see that required procurement procedures such as open competitive bidding are followed.

The Appendix of NFPA 1901 contains helpful suggestions covering specifications, proposals, contracts, and acceptance concerning apparatus purchase.

In general, purchase and replacement costs of fire apparatus should be a regular item of the fire department capital budget. In most cases, except for accidents, the requirements can be planned and funded on a long range basis. Systematic apparatus replacement provides the fire department with reliable apparatus at all times. Improvements in automotive and fire apparatus design can be introduced, maintenance costs become more favorable, and increased operating efficiency and equipment reliability will be sustained.

Lease or lease/purchase of fire apparatus can be an alternative to outright purchase. Income tax benefits and depreciation accrue to the actual, commercial owner of the apparatus with the fire department (which is tax exempt) realizing savings in overall life cycle cost. In addition, the department does not have to expend a large purchase price at one time, but can spread the cost of apparatus out over several years. In addition there is no capital tied up in the event the department wishes to trade, exchange, or dispose of the apparatus. Closed and open end leases are available, and the finance department of the municipality should carefully evaluate these alternatives. In a smaller or volunteer department a local banker can advise the purchasers of the advantages of a lease versus purchase arrangement.

Any such lease or lease/purchase contract should clearly spell out who is responsible for maintenance, repairs, and liability for the apparatus. Likewise, the conditions under which such an arrangement may be terminated, by either party, should be clearly stated.

The normal life expectancy for first line fire apparatus will vary from city to city, depending upon the amount of use of the equipment and the adequacy of the maintenance program. In general, a 10 to 15 year life expectancy is considered normal for first line pumping engines. First line ladder trucks have a normal life expectancy of 15 years. In fire departments where ladder trucks make substantially fewer responses to alarms than engines, a planned first line service of 20 years may be warranted for ladder trucks. Some smaller fire departments that have infrequent alarms operate pumping engines up to 20 years with reasonable efficiency, although obsolescence will make the older apparatus less desirable even if it is mechanically functional. In some types of service, including areas of high fire frequency, a limit of only 10 years may be reasonable for first line service. The older apparatus may be maintained as part of the required reserve as long as it is in good condition, but in almost no case should much reliance be placed on any apparatus more than 25 years of age.

The various state insurance rating organizations may put specific limitations on the age of first line apparatus. In addition the rebuilding or rehabilitation of apparatus calls for a decision as to its actual "age." Apparatus rehabilitation is a growing business and a viable alternative to purchasing new apparatus.

Some fire departments follow a policy of assigning new apparatus to the busiest fire companies for a few years, and then reassign it to less busy companies to average out the work load over the life of the apparatus. This practice is sometimes questionable, as it affects morale and may merely overuse new apparatus. With two piece engine companies, a newer pumper may be assigned as the initial attack apparatus, with an older pumper used mainly for water supply operations.

To achieve these objectives, a fire department with 15 first line pieces of apparatus may schedule replacement of one unit per year, with five or six of the older apparatus kept as reserve. A fire department with five first line engines may schedule replacement of an engine each three years with three older engines kept as reserve. A department with two ladder companies plus one reserve ladder truck may schedule replacement of ladder trucks on an eight year cycle so none of the trucks, including the reserve, will exceed 24 years of age.

The cost of new fire apparatus varies according to the type, size, and special equipment provided. At this writing, properly equipped pumpers meeting NFPA standards and federal regulations cost from $80,000 to $165,000 or more, plus cost of special features provided. Aerial ladder and platform trucks commonly cost from $175,000 to over $450,000, depending upon the equipment carried. While costs have increased due to inflation and requirements of federal highway safety standards, fire protection needs are an authorized use for federal revenue sharing funds. Many municipalities are utilizing part of these funds to procure needed fire apparatus and to meet part of their fire department personnel budgets.

All purchase specifications for municipal fire apparatus should be based upon the requirements of the latest edition of NFPA 1901, as applicable to the type of apparatus desired. NFPA 1901 includes basic provisions that will provide a complete, well engineered fire fighting vehicle. Apparatus manufacturers are geared to provide equipment that meets this standard.

Acceptance Tests and Requirements

Acceptance tests are designed to demonstrate that apparatus will perform as specified in the purchase contract. Tests should be performed prior to delivery or within ten days after delivery and the tests should be conducted by the manufacturer's representative in the presence of such person(s) as the purchaser may have designated in the delivery requirements. Normally, the fire chief or a designated representative is the acceptance authority, exercising this authority following satisfactory completion of tests and inspections to assure compliance with the purchase specifications.

The acceptance tests and requirements for fire apparatus and its various components are given in NFPA 1901. Applicable to all types of apparatus are the road tests designed to prove that the apparatus has ample power for the intended service. Road tests are conducted with apparatus fully loaded with personnel, hose, water, and the equipment specified to be carried. From a standing start, through the gears, the vehicle must attain a true speed of 35 mph (56 km/h) within 25 seconds for pumpers, and within 30 seconds for heavier apparatus such as tank trucks, aerials, and elevating platforms. The vehicle must also accelerate from 15 to 35 mph (25 to 56 km/h) in 30 seconds without the operator moving the gear selector. This demonstrates that the vehicle has the necessary "pick up" to operate safely and efficiently in traffic. The vehicle must attain a top speed of 50 mph (83 km/h), or a higher speed where specified. The required brake performance must also be tested. While the tests mentioned demonstrate power needed to negotiate grades found in most communities, any special ability needed to climb very steep grades may be tested if specified.

The pump tests including certification tests have been outlined earlier in this chapter. If a pump test conducted by the manufacturer is desired at point of delivery, this must be conducted as the purchaser has specified. In all cases with a pump, the manufacturer must supply with delivery a copy of: (1) the engine manufacturer's certified brake horsepower curve, showing the maximum no load governed speed, (2) a record of pumper construction details as indicated on a form in NFPA 1901, (3) the pump manufacturer's certification of suction capability, (4) the pump manufacturer's certification of hydrostatic test, and (5) the certification of inspection by a testing agency.

A test plate, required at the pump operator's position, gives the rated discharge and pressures, together with engine speed determined by the manufacturer's test for the unit. The no load governed speed of the engine as stated on a certified brake horsepower curve is also given. Tests are conducted to see that the water tank has the minimum capacity specified and also gives the flow rate specified.

Structural strength of an aerial ladder is tested by its ability to lift 200 lb (91 kg) on the free end of the main section of the ladder. An operations test demonstrates ability to raise and extend the ladder to full height with 90° rotation in 60 seconds after the truck is set for operation.

The stability test for elevation platform apparatus includes rotating a complete 360° with 150 percent of the manufacturer's rated payload on the platform with the platform at maximum horizontal reach. The operational test consists of extending the platform to full height and a 90° turn within 150 seconds. All such tests must be completed smoothly without undue vibration. The apparatus must also comply with the turret nozzle discharge requirements. For water towers the full extension with 90° rotation must be completed within 120 seconds after the vehicle is set. Tests also are required to demonstrate required water delivery and performance of other specified features.

EQUIPMENT CARRIED ON APPARATUS

NFPA 1901 includes listings of equipment and appliances needed with various categories of fire apparatus. The standard lists equipment which the apparatus manufacturer is required to furnish with each type of apparatus, including lists of equipment required to be carried on each type of apparatus.

Apparatus must be equipped with the tools necessary to accomplish fireground operations. Where apparatus is delivered with only the minimum items of equipment, other equipment must be supplied as needed. The latest edition of NFPA 1901 should be consulted for up to date equipment lists. These lists are generally used when evaluating fire department equipment.

Equipment Carried on Pumpers

Pumping engines carry a wide assortment of tools and equipment needed to make them self-sustaining fire fighting units. An important feature on pumpers is fire hose of various sizes. This includes large diameter supply hose for supplying the pump from a hydrant and, where needed, hard suction hose for drafting. Pumpers normally carry from 1,500 to 2,000 ft (457 to 610 m) or more of 2½ to 5 in. (63 to 127 mm) hose that is used for supply lines and for supplying water to nozzles. Small hose is needed either for streams supplied from the apparatus water tank, or for reducing the larger lines when small streams are needed. Normally, 400 to 800 ft (122 to 244 m) of 1½ in. or 1¾ in. (38 or 44 mm) hose is carried for this service. If hose reels are provided, 200 to 300 ft (61 to 91 m) of 1 in. (25 mm) or larger hose may be carried on each reel.

Pumpers carry a wide selection of nozzles for use with various hose sizes. Nozzles come with 1½ in. (38 mm) and 2½ in. (63 mm) fittings and may be a combination spray and straight stream type, or separate spray and solid stream tips may be carried. A portable deck gun with suitable tips enables the pumper to be operated with a minimum number of fire fighters. The gun also provides a large volume fire stream with greater reach than that obtained from smaller nozzles. Such equipment should be carried on each engine. A pumper also should carry a distributor nozzle for use in cellars or on fires in concealed spaces, and possibly a bayonet nozzle for partition or ceiling fires.

Various hose equipment, tools, and fittings are available for pumpers, including: hydrant adapters; a suction hose strainer when suction hose is carried, including siamese connections for combining flows from two lines into one; wye connections for supplying two hose lines from one; gate valves; adapters for any nonstandard threads; double male and double female connections to permit reversing of hose lays; spanners for hose connections, hose clamps, hose straps or rope hose tools; hose jackets for burst hose; and a hose hoist.

Depending upon standard operations, fire departments may not require all of the previous items. Some fire departments having numerous preconnected 1½ in. (38 mm) hose lines may have little occasion to use a reducing wye connection. Where wyes are used, each outlet should be gated to permit control of individual lines. Siamese connections either should be gated or have clapper valves. In departments where large diameter and Storz coupled hose is used, a suitable quantity of adapters should be provided for the swift and easy hookup of different hose types. Likewise, provision should be made for the use of relief valves at distribution points due to the large flows afforded by large diameter hoses.

The pumper should carry an assortment of minor ladder company tools including a 14 ft (4.27 m) roof ladder, a 24 ft (7.38 m) extension ladder, a short pike pole or ceiling hook, fire axes, a claw tool, a crowbar, rope, and bolt cutters. Where a pumper serves a district having buildings more than two stories high and there is no ladder truck in the station to which the pumper is assigned, a 35 ft (10.7 m) extension ladder should be carried in lieu of the 24 ft (7.38 m) extension ladder.

It is recommended that a pumper carry a number of salvage tools including at least two salvage covers, sprinkler stoppers, brooms, shovels, squeeges, buckets, and a smoke ejector. Rescue and emergency equipment carried on pumpers include a standard first aid kit, self-contained breathing apparatus with spare cylinders for each fire fighter assigned (but a minimum of four SCBA on each pumper), and blankets. Pumper lighting and power equipment includes electric hand lanterns, floodlights or spotlights, and preferably a portable electric generator or an alternator serving power outlets. Electric hand lights may be of the rechargeable type. On board battery chargers and power supplies, supplied by 110 V ac power from the firehouse supply, will ensure fully charged hand light and fire apparatus batteries. In addition to the preceding hose and nozzle equipment, the pumper carries portable extinguishers and often other special extinguishing equipment. For rural service, a portable fire pump should be carried plus possibly rakes, hay forks, long handled pointed shovels, and approximately six backpack water type extinguishers.

Equipment Carried on Ladder Trucks

The amount of space available on pumpers is necessarily limited by space requirements for hose and water tanks. Most fire departments generally provide a number of ladder trucks, and operate these as separate fire companies. They are termed ladder companies because of the prominence of the long ladders, but the ladders form only a part of the large inventory of equipment carried.

Among the tools carried on a ladder truck are: flat head and pick axes, crowbars, door openers, pike poles of varying lengths, scoop and plain type shovels, ½ in. (12.7 mm) and ¾ in. (19 mm) rope, pitchforks, sledges, a battering ram, bolt cutters, wire cutters, hydraulic rescue tools, 10 and 20 ton (9 and 18 metric ton) hydraulic jacks, hydraulic or air powered spreaders, a power saw and blades, a block and tackle, pull down hooks, gas and water shutoff wrenches, an oxyacetylene cutting outfit, compressed air or hydraulic cutters, bale hooks, and a tool box with an assortment of hand tools.

Ladder company crews are normally responsible for search, rescue, and ventilation while engine companies are responsible for getting water on a fire. Ladder company equipment should include at least six sets of self-contained breathing apparatus (plus spare cylinders for each SCBA), a 24 unit first aid kit, four blankets, two litters or stretchers, a resuscitator (with spare oxygen cylinders), a body bag, and a life net. Pompier or ladder belts should be carried to permit members of ladder crews to tie in when working on a ladder. Some ladder trucks carry single spar pompier or scaling ladders. In a few cases, a gun used for projecting a lifeline is carried for areas where it may be needed in outdoor rescues from places of difficult access. An inflatable boat and other water rescue equipment may be carried on the truck.

A ladder company should also be equipped for salvage service, with: ten salvage covers each 12 by 18 ft (3.7 by 5.5 m), two roof covers, a syphon or small portable dewatering pump, four squeegees, four mops, mop wringers, four brooms, heavy plastic tape, rolls of tarpaper or heavy plastic sheeting for "boarding up a structure," claw hammer and nails, heavy duty staples, a smoke ejector with a minimum 5,000 cfm (142 m^3/min) capacity, smoke deodorant, twelve assorted sprinkler heads, sprinkler wrenches for upright and pendant sprinkler heads, and sprinkler stoppers or wedges. It also is very useful for ladder company personnel to carry a two wheeled hand truck for moving heavy objects.

Ladders: A standard complement of ladders for ladder trucks consists of 163 ft (50 m) of ground or portable ladders such as one 40 ft (12 m) extension, one 35 ft (11 m) extension, one 28 ft (8.5 m) extension, one 20 ft (6 m) single ladder with roof hooks, one 16 ft (5 m) single ladder with roof hooks, one 14 ft (4.25 m) extension, and one 10 ft (3 m) collapsible ladder. Ladders must be constructed to meet standard performance requirements in NFPA 1931, *Standard on Design, and Design Verification Tests for Fire Department Ground Ladders* (hereinafter referred to as NFPA 1931). In ordering a ladder truck, careful consideration should be given to the ladder needs of the area served. Manually raised extension ladders more than 40 ft (12 m) long are generally not required, because aerial ladders and elevating platforms provide adequate emergency access to upper floors and roofs. Other considerations are that additional personnel are required to raise longer ladders, and space available for ladders on combination ladder trucks and on elevating platform trucks is limited.

However, some fire departments do need more and longer ladders, especially for apartment and tenement structures. Such fire departments often specify and use from 300 to 400 ft (91 to 122 m) of ladders for ladder trucks. This usually involves one or two 50 or 55 ft (14 or 17 m) extension ladders, one or two 40 or 45 ft (12 and 18 m) extension ladders, two additional 35 ft (11 m) extension ladders, and 25 and 30 ft (8 and 9 m) single or wall ladders in addition to other single ladders. At a building where numerous rescues must be made from upper floors, it is unconscionable to have a ladder company arrive without an ample number of ground ladders capable of reaching above the third floor. Ladders may be constructed of wood or aluminum, with some fiberglass ladders available. In some states the local interpretation and enforcement of Occupational Safety and Health Administration (OSHA) standards will affect the type of ladder used. Folding ladders, step ladders, and multiple configuration ladders are available and can be chosen for local practice and conditions.

Ladder Loads and Safety Tests: Loads that fire ladders are designed to support vary according to the ladder length and type; loads are specified in NFPA 1931. Load limitations are based on use of the ladder supported at the top end against a building, and with the ladder set at a 75½° angle. This angle is measured from the horizontal, with the ladder in the raised position. At angles less than this, the ladder may be unable to support the design load.

NFPA 1932, *Standard on Use, Maintenance, and Service Testing of Fire Department Ground Ladders*, calls for service testing of all ladders at least annually and at any time the ladder is suspected of being unsafe. The tests are designed to show that the ladder is safe for continued use under the conditions for which it was originally designed. The service test procedures must be carefully observed to ensure that ladder damage does not occur. Ladders also should be visually inspected as outlined in NFPA 1932 to detect loose rungs, loose belts and rivets, weld defects, cracks, splintering, breaks, discoloration, and other signs of possible weakness that might warrant testing to determine whether the ladder is safe to use or needs repair.

Combination Apparatus

In many communities combination apparatus is useful for both engine and ladder company service. It may be desirable to provide small stream equipment on ladder trucks, especially in districts where the trucks may arrive at fires before an engine company. This situation may occur when the first due engine is at another fire and the ladder company must wait for a covering engine to arrive. In some cities with low pressure water systems, the engines must always connect directly to hydrants, and small stream equipment on ladder trucks is used for initial fire attack. Even where ladder companies do not normally operate hose streams, a large capacity pump on an aerial ladder or platform truck may be useful for boosting elevated stream pressure.

It is impractical to expect a single vehicle to provide all of the operational capabilities needed on the fireground. However, a pumper ladder combination may be used as the initial fire attack apparatus. Positioned close to the building on fire, the apparatus provides preconnected attack lines plus essential ladder company tools, while other pumping engines supply water. Often an aerial ladder or elevating platform is provided on such combination apparatus. Good staffing is needed for effective use of such combination apparatus even if personnel are assigned to supporting apparatus. Where a single vehicle is provided for both types of service, it must carry a minimum amount of both types of equipment because of space and load limitations. An overloaded piece of apparatus, or one with inadequate hose carrying capability must be avoided.

Fire Extinguishers

At least two approved portable fire extinguishers are required as basic equipment on all fire apparatus. The variety selected must be suitable for use on Class A, B, and C fires. The minimum sizes called for are extinguishers with 20-B:C rating in dry chemicals, 10-B:C rating in carbon dioxide, and 2-A rating in water. Generally, dry chemical fire extinguishers are more effective than carbon dioxide types on outside fires. Portable fire extinguishers can often be used effectively in conjunction with water spray nozzles on flammable liquid fires.

The use of Halon fire extinguishers may be appropriate for certain types of fires, especially those involving electronic equipment. In addition, Class D extinguishers may be used on certain types of combustible metal fires. It is recommended that fire apparatus lacking a water tank and pump be provided with several water type extinguishers in addition to Class B and C extinguishers. The stored pressure Class A extinguisher is generally used because it is economical and easily rechargeable. Most fire departments carry water pump extinguishers for grass and brush fires.

Foam and Water Additive Equipment

The potential for serious flammable liquid fires has caused many fire departments to provide, on the apparatus, various types of foam making equipment or other agents which are more effective than plain water. Built in proportioners and foam tanks are often specified for new pumpers; in other cases, foam making nozzles, portable foam making or generating equipment, and foam supplies are carried on the apparatus. Increasingly, new pumpers are being equipped to use an aqueous film-forming foam (AFFF) additive to seal off the surface of flammable liquid fires and smother the fire. Special foams are available for certain types of polar solvent/alcohol and unusual liquid fires.

Forcible Entry Tools

Ordinarily, structural fire fighting cannot be fully successful unless fire fighters enter the building or the part of the structure where the fire occurs. Fire apparatus carry a variety of tools to permit access to locked or closed areas. This equipment allows fire fighters to gain entrance through doorways and windows and to open walls, partitions, ceilings, or roofs to uncover hidden fire, ventilate the structure, or rescue occupants.

The use of hydraulic spreaders on a building is common because more leverage is obtained than can be supplied manually. Frequently a heavy door can be sprung with these tools without permanent damage caused. Such tools also are useful in freeing persons trapped in elevators, wrecked vehicles, and by machinery. Inflatable bags in various sizes and configurations are also available for gently lifting large weights; the air supply from a SCBA cylinder is the usual source for inflating such equipment. An air hose also may be used.

An important aspect of forcible entry is access through costly tempered glass doors where there is no alternate path for entry. Often, to drive out the lock cylinder is an effective method of entry causing the least damage. Special tools are available for this purpose.

Radio Equipment

Effective communications is essential for any fire department operation. Every fire department vehicle should be equipped with a two way radio that operates on multiple frequencies for dispatch and fireground communications. NFPA 1901 provides specifications for a weatherproof radio equipment compartment on a fire department vehicle. It is highly desirable to provide radio speakers both inside and outside the vehicle cab, that can be heard above the usual noise on the fireground so messages can be received by personnel a distance away from the apparatus. A speaker and microphone should be provided at the pump operator's position, as well as in the cab.

Every fire company vehicle, as well as chief officers' vehicles, should carry portable radios for effective fireground communications. This is important in any fireground command system and provides control over operating units. The radios in fire chiefs' vehicles should be equipped to monitor various frequencies as well as to lock in on a frequency specified by the communication center for particular operations.

Fire chiefs' vehicles also may be equipped to send and receive hard copy messages (which may be associated with computerized dispatching and recording) and also may have facilities for receiving visual displays of information, including maps and fire hazard data that may be transmitted from the communication center.

Electric Lights and Generators

Lights facilitate operations and help to reduce injuries. Each fire apparatus should carry a minimum of two hand lights or lanterns. In addition, each fire fighter needs a hand light for work in any unlighted area. Batteries can be kept fully charged with battery charging equipment on the fire apparatus.

The fire service is often faced with the need to provide electric power at fires and emergencies. Power demands may include: (1) power to operate fire apparatus and its appurtenances, (2) power to operate tools and lights on the fireground, (3) emergency power to maintain essential services such as communications, and (4) emergency power for temporarily replacing essential community services. This demand may be met in several ways: (1) by the provision of generating equipment, transformers, and outlets as part of a vehicle's electric system; (2) by portable generators on fire apparatus; (3) by special "lighting trucks" or mobile generator units; and (4) by standby generators at locations such as fire stations and communication centers.

A portable generator with a minimum 2,500 W rating with three portable floodlights and six 50 ft (15 m) lengths of electric cable on a reels is required for every ladder truck. Such equipment should be carried on pumpers and squad trucks where ladder trucks do not regularly respond. Portable generators capable of generating either direct or alternating current should be investigated. Where smoke ejector fans and equipment other than lights are to be served, a 3,500 W or larger generator is desirable.

Generators may be mounted on apparatus and arranged for automatic starting, but should also be capable of being removed from the apparatus. Some fire apparatus carries two generators, so one can be dropped off where needed while the other provides the required electrical capacity on the unit. Sometimes one portable generator and one generator driven by the apparatus engine are provided. Small, self contained portable light/generator sets are available which may be easily transported to various locations. These units eliminate the need for long extension cords and also eliminate the safety hazards associated with electricity in water and debris.

When power tools, such as heavy duty cutting tools, are supplied with power by generators, individual circuits should be protected by circuit breakers so that if a tool should stall it will not consequently stall or damage the generator while other equipment is operating. Electric cables, fittings, and lights should be of a heavy duty waterproof type that can be used safely without danger of shock or damage when immersed. A common error is made by using conductors that are too small to carry the load required to operate appliances efficiently. Many fire departments provide large capacity conductors serving a multiple outlet, to which individual appliances can be attached near the point of use. The inclusion of ground fault interrupter (GFI) devices will enhance operator safety.

Portable Pumps

NFPA 1901 recommends that both pumpers and water tank apparatus serving rural areas carry a portable pump. In selecting portable pumps, fire departments should be careful to obtain models that will give the needed flow characteristics at a safe, continuous engine speed. The user should obtain from the manufacturer or supplier the discharge-pressure curve for the model selected.

Portable pumps for fire department service are covered in NFPA 1921, *Standard for Fire Department Portable Pumping Units*, except for special types for forest fire service, which are covered in NFPA 295, *Standard for Wildfire Control*. Such portable pumps are too small in capacity to be of much value either for fighting structural fires or for filling and discharging fire department tankers.

Portable pumps for fire department service are of the centrifugal type. They are grouped in categories based upon the pressure-volume characteristics that make them suitable for various classes of work. Small streams at high pressure are intended mainly for grass and brush fire operations. Pumps delivering relatively large volumes at low pressures can serve as a supply pump for fire trucks where the water supply source is beyond the reach of suction hose. These pumps also may be used as a dewatering pump. Fairly large volume flows at higher pressures are considered valuable in hilly areas, where ordinary portable pumps do not develop sufficient pressure to overcome elevation or "head."

Smoke Ejectors

Smoke ejectors are required equipment on ladder trucks and other apparatus used for performing ventilation service. The most commonly used smoke ejector has a 16 in. (406 mm) diameter and is rated at 5,000 cfm (142 m^3/min) at 1,750 rpm. There are also 24 in. (610 mm) ejectors which are rated up to 11,750 cfm (333 m^3/min). Both ac and dc motors are available with explosion proof features; operating ability on 110 V or 220 V is optional. Other units are gasoline engine driven. In fire department operations, smoke ejectors are often used for pulling out smoke and gases as well as for blowing in fresh air. At serious fires, it is not unusual to use a number of smoke ejectors at various locations to move the desired quantity of smoke and air.

An important use of smoke ejectors is to reduce smoke damage from minor fires. Prompt use of ejectors helps prevent soot particles from settling. Some fire departments also use ejector fans to spray deodorants after a fire.

When ejectors are used to remove concentrations of combustible gases, explosion proof equipment will be required. Suggestions for successful use include: (1) having an adequate power supply and conductors to operate ejector motors efficiently, (2) placing fans at openings to blow out, so fumes will travel the shortest possible distance, (3) placing fans as high as possible in openings using secure hooks and other supports as needed, (4) using prevailing wind when possible, (5) closing openings around the equipment so smoke will be drawn through the fan (some departments have folding covers designed to block openings around the equipment), (6) opening doors and windows to establish cross ventilation, and (7) removing screens and other obstructions likely to block the movement of air.

Life Guns

A life gun designed to shoot a rope line to persons in distress is equipment occasionally carried on ladder or rescue trucks. One type of gun, powered by a .30-06 cartridge, has a launcher that permits variations in distance from 100 to 650 ft (30 to 198 m) using a 1/16 in. (1.59 mm) nylon line with a 180 lb (82 kg) tensile strength. Primarily, the life gun is used in water rescues (such as for persons stranded on rocks from overturned boats) and to rescue persons from cliffs or canyons. Fire fighters need extensive practice to use is equipment efficiently.

PROTECTIVE EQUIPMENT FOR FIRE FIGHTERS

Each year one out of every two fire fighters is injured, and many die, in the line of duty. Often these mishaps occur because fire fighters are not sufficiently equipped with, or do not properly utilize, protective clothing and equipment. No equipment can guarantee a fire fighter's safety, but SCBA and full protective clothing offer fire fighters a better chance to carry out their jobs without suffering injury or death. The safety odds are further improved if that equipment is carefully selected, properly maintained, and used at all appropriate times. In addition to protecting fire fighters, SCBA and protective equipment will make fireground operations more effective.

Injuries to fire fighters are costly to the community. Though the initial cost of good protective clothing may seem high, it is small compared to the expense of hospitalization or replacement of an injured fire fighter. Recruitment and training expense, medical expense, disability payment, and early retirements are costly. The loss of a volunteer fire fighter disrupts a community's fire service effort.

The protective equipment that fire fighters wear while combating structural fires must be viewed as a system. That system must include SCBA and full protective clothing. Together, these must protect fire fighters from toxic fumes and gases, heat, moisture, puncture, impact, and electrical shock.

Self-Contained Breathing Apparatus

NFPA 1981, *Standard on Self-Contained Breathing Apparatus for Fire Fighters*, (hereinafter referred to as NFPA 1981) states that "all fire fighters exposed to hazardous atmospheres from fires and other emergencies, or where the potential for such exposure exists, shall be provided with self contained breathing apparatus. Only SCBA approved by the National Institute for Occupational Safety and Health and the Mine Safety and Health Administration (NIOSH/MSHA) shall be considered to meet the provisions of the Standard."

The following are the types of SCBA available:

1. Open Circuit Apparatus. An apparatus of the following two types from which exhalation is vented to the atmosphere and not rebreathed:

Positive Pressure (Pressure Demand) Type: An apparatus in which the pressure inside the facepiece, relative to the immediate environment, is positive during both inhalation and exhalation.

Demand Type: An apparatus in which the pressure inside the facepiece, relative to the immediate environment, is positive during exhalation and negative during inhalation.

Of the above two types, only positive pressure (pressure demand) SCBA is allowed by NFPA 1981 for use by fire fighters. Also, OSHA requires that positive pressure SCBA only be used for structural fire fighting.

2. Closed Circuit Apparatus. An apparatus of the type in which the exhalation is rebreathed by the wearer after the carbon dioxide has been effectively removed and a suitable oxygen concentration restored.

Open circuit SCBA is by far the most common SCBA in use by the fire service. Positive pressure SCBA provides protection against inward leakage that could be caused by improper facepiece fit or a small tear in the breathing tube. For open circuit apparatus, positive pressure SCBA provides the best respiratory protection to the fire fighter. NFPA standards and OSHA regulations specify only positive pressure (pressure demand) for open circuit SCBA used by the fire service.

Self-contained breathing apparatus (SCBA) for fire service use must have a minimum rated service life of 30 minutes. Approvals from NIOSH/MSHA for duration of use are based on tests conducted by NIOSH. Open circuit SCBA are tested with a breathing machine at a use rate of 40 L/min (10 gal/min) and must be able to supply air for the rated service life or longer. Closed circuit SCBA are tested for service life with test subjects wearing the apparatus and performing prescribed activity, because the service life of such apparatus is based on metabolic use of oxygen and no mechanical test is currently available which can simulate this type of use.

However, because work performed by the user may be more or less strenuous than the work level of the test procedure, actual service life of the SCBA may be affected. During extreme exertion, for example, service life may be reduced as much as 50 percent.

Service duration of each unit depends upon such factors as:

1. The degree of physical activity by the user.
2. The physical condition of the user.
3. Emotional conditions of the user such as stress or excitement (which may increase the user's breathing rate).
4. The degree of training or experience the user has had with such equipment.
5. Whether or not the cylinder is fully charged at the beginning of use.
6. Possible presence of carbon dioxide (CO_2) in the compressed air supply at levels greater than the 0.4 percent found in normal air.
7. Atmospheric pressure, e.g., use in a pressurized tunnel or caisson. At two atmospheres the duration will be one-half as long as the rated service life; at three atmospheres the duration will be one-third as long as the rated service life.
8. Condition of the breathing apparatus.

Fire departments must not use SCBA with shorter than 30 minutes rated service life because these units can allow fire fighters to extend into dangerous positions without sufficient time to conduct operations or safely escape. Short duration SCBA are not approved for entry protection for fire fighters.

The closed circuit SCBA used by fire departments is an oxygen rebreathing type. This SCBA has a cylinder supplying oxygen to the wearer through a pressure reducing valve (regulator) and a container of chemicals to remove carbon dioxide from the exhaled air. In this apparatus, the opening of the cylinder valve supplies oxygen to a breathing chamber. When the user inhales, air flows from the breathing chamber through an inhalation hose to the facepiece. Exhaled air passes through an exhalation hose to a carbon dioxide absorber where CO_2 is removed and the air then enters the breathing chamber where it is enriched by fresh oxygen and then rebreathed. Closed circuit SCBA is available in the positive pressure design.

Breathing Air Quality: Air supplies for SCBA should be at least Class D breathing air as specified by the Compressed Gas Association (CGA) in Pamphlet G-7.1 Specification for Air (ANSI A86.1-1973). Class D breathing air has less than 20 parts per million (ppm) of carbon monoxide. The moisture content of the air supply (expressed in dew point) should be appropriate to the conditions of use. This varies according to climatic conditions in which the SCBA is used.

Fire departments should have the quality of breathing air checked at least monthly whether or not they fill their own recharging cylinders. Air samples can be tested by a public health laboratory equipped for such tests, or by a private testing laboratory.

Use of Breathing Apparatus: Without SCBA in a burning building, fire fighters are no better off than victims trapped by the fire.

In a recent NFPA study, one in ten fatalities suffered by fire fighters in the line of duty was due to inhaling smoke or toxic gases. More than half of the remaining fire fighters who died did so as a result of heart attacks, and inhalation of carbon monoxide is a significant contributing factor to heart attacks. Heart cell death can occur in less than ten minutes of unprotected exposure to fire products. Fire fighters must understand how lethal fire produced atmospheres can be, even when most of the smoke seems to have dissipated.

Carbon monoxide, which is produced by all fires, is a major killer. Two or three breaths in an atmosphere that contains only five percent carbon monoxide can be fatal. When combined with other toxic products of combustion, even smaller amounts of carbon monoxide may prove deadly.

Self-contained breathing apparatus should not be used without thorough training with the specific type of equipment available for use, and practice under restricted breathing and visibility conditions. Fire fighters wearing SCBA should not work alone and should be under supervision of officers. Self-contained breathing apparatus should not be donned by those who have already been subjected to exertion and smoke. The use of SCBA does not protect an individual against excessive heat, gases, and poisons that attack the body through the skin.

Prior to use, the minimum pressure in the compressed air cylinder of open circuit SCBA should not be less than ten percent of the pressure used in rating the service life of the unit, e.g., 2,015 psig (13 893 kPa) for a cylinder rated at 2,216 psig (15 279 kPa). Self-contained breathing apparatus must be equipped with an audible low air warning signal to warn the user when the air or oxygen supply is low. When this alarm sounds, wearers should leave the contaminated area immediately. "Buddy breathing" operations should not be used, as both persons are likely to succumb to the contaminated atmosphere.

It should be the written policy of every fire department that all personnel expected or likely to respond to and function in, areas of atmospheric contamination be equipped with and trained in the proper use and maintenance of self-contained breathing apparatus. Each line fire fighter should be accountable for one SCBA and responsible for checking the condition of that SCBA at the beginning of each shift and after each use, or at any other time it may be necessary to have the equipment in a ready state. In volunteer fire departments, the SCBA should be checked for a ready condition at least weekly. Company officers should assign responsibility for the proper use and function of a specific SCBA to each crew member. If a SCBA is found to be functioning improperly, it should immediately be taken out of service, tagged, reported, and replaced as soon as possible. Replacement SCBAs should be available within the fire department.

Each fire department should adopt a written policy which requires SCBA to be worn whenever fire fighters are operating below ground level, above ground level, in a contaminated atmosphere, in situations where the atmosphere may become contaminated, or in an oxygen deficient atmosphere.

Fire service personnel should resist the tendency to prematurely remove SCBA during routine fire situations. All must be aware of the respiratory hazards that exist in ordinary as well as extraordinary fire situations. It is generally true that carbon monoxide levels increase during overhaul, due to the incomplete combustion of smoldering materials. SCBA should not be removed until the atmosphere has been checked with a carbon monoxide indicator and determined to be safe. A reading of greater than 50 ppm of carbon monoxide in the atmosphere should require the use of SCBA. Either continue to use SCBA or change the atmosphere.

Special caution is necessary in the use of breathing apparatus in pressurized atmospheres, not only because of the increased hazard of fire and explosions but also because the period of protection may be reduced by the higher pressure. Another problem exists when fires are difficult to reach and the time required to get to the fire and return to outside air utilizes most of the respirable air of the equipment. Long duration SCBA are available for such situations.

A number of fire departments have scuba (self-contained underwater breathing apparatus) designed for their own purposes. Fire department SCBA should not be used for underwater work.

Protective Clothing for Fire Fighters

Fire fighters should not enter burning buildings or buildings that have been subjected to appreciable fire damage unless they are wearing full protective clothing, including SCBA. It is the responsibility of the authority operating the fire department to provide protective clothing and equipment for all fire fighting personnel, preferably through direct procurement by the fire department.

At a minimum, protective equipment should include: helmets and hoods to safeguard the head, neck, and ears; face shields for face and eye protection; protective coats and protective trousers for the torso, legs, and arms; gloves to protect the hands and wrists; boots to protect feet and ankles; and SCBA to protect the respiratory system.

Special situations may call for approach, proximity, or entry suits. Approach and proximity garments protect against radiant heat while near fire, and entry suits are designed to allow entry into total flame. Various types of special protective clothing are available for hazardous materials incidents, but must be chosen for protection from a specific exposure. Specific decontamination or disposal procedures must be followed for the chemical to which the garment was exposed.

Protective clothing should be properly sized for the individual wearer, allowing the person to perform duties without unnecessary hindrance. All protective equipment should be supported by a manufacturer who offers good service and maintenance. Most important, its use should be mandated and supported by the fire department and all of its members.

Helmets: A fire fighter's helmet should meet the minimum performance requirements of NFPA 1972, *Standard on Structural Fire Fighter's Helmets*, for exposure to convected and radiated heat; resistance to flame; resistance to penetration; resistance to top, front, side, and back impact; and insulation from electrical shock. A fire fighter's helmet should have flame resistive flaps that protect the ears and the neck; the flaps are intended to be worn in the full down position during all fireground operations. A retention system must provide a secure fit. The helmet should be waterproof and divert falling liquids and debris but not hinder the use of SCBA, and it should have an attached protective face shield. The helmet shell, suspension system, and an energy absorbing liner will protect the head from impact and penetration.

An important accessory to the helmet is a lightweight fabric protective hood such as those worn by race car drivers. These hoods are made from a flame resistive fabric, such as aramid. The hood is worn over the head and neck and under the coat and helmet, protecting the head, forehead, neck, and throat area.

Face shields provide partial eye and face protection from heat, sparks, liquids, and flying debris. The SCBA facepiece will also provide protection from these hazards. Although the polycarbonate type face shields are susceptible to scratching, replacement face shields are inexpensive and each could be replaced twice a year for about $10 per year.

Coats and Trousers: NFPA 1971, *Standard on Protective Clothing for Structural Fire Fighting*, covers minimum requirements for fire fighters' protective coats and trousers, and outlines a layered protection system. The outer layer is the shell. The shell of both coat and trousers must be made of a flame resistive fabric that will not be destroyed through charring, separating, or melting when exposed to 500°F (260°C) in a forced air laboratory oven for a 5 minute period. The outer shell may be of a light color for better visibility and have adequate (a minimum of 352 sq in. or 0.28 m^2) retro reflective fluorescent markings on the coat of lime yellow or red orange.

The second layer is the vapor barrier that prevents moisture from penetrating through to the wearer. It should be made of a coated, flame resistive material and must be sewn in at the neck or waist. The third layer is a thermal liner that must be of sufficient thickness for insulation and also be sewn at the neck or waist. This liner provides thermal protection from radiant, conducted, and convective heat. In cold weather an additional flame resistive, snap in liner may help to provide warmth.

The vapor barrier and thermal liner are stitched in at the neck or waist area to prevent the removal of these components. The shell of the coat and trousers must never be worn without the thermal liner and vapor barrier because the garment would offer little, if any, protection against conducted, convected, and radiated heat. Since fire fighters normally work in a high temperature environment, protection from heat transfer is critical and necessary.

The protective coat should have a collar with a throat enclosure that will protect the neck. The collar should be worn up and closed during all fireground operations. The coat should have a storm flap closure in front to protect against heat, steam, and water. To keep from snagging objects, the coat should be secured by a means such as reverse hooks and dee rings, or hook and pile closure.

The length of the turnout coat is particularly important. It must overlap protective trousers by a sufficient margin to adequately protect the wearer in any working position such as kneeling or crawling. Coats worn without protective trousers do not provide adequate torso and leg protection. Protective coats and trousers should be worn as an ensemble for a more complete system of protection. Protective trousers are worn with knee boots.

Tests have proven that thicker liners of flame resistive, nonmelting materials give greater protection. Therefore, station uniforms worn under turnouts can give additional protection. These garments should be made of flame resistive fabrics meeting the requirements of NFPA 1975, *Standard on Station/Work Uniforms*. Station uniforms made of synthetic fabrics that melt and drip (such as many "wash and wear" fabrics) should be avoided. Volunteers should be alert for flame resistive fabrics when purchasing street clothes, or have flame resistive jump suits or coveralls readily available to don before the coat and trousers.

Boots: Boots must be resistive to heat, puncture, impact, and water. Good fit is important because it will lessen foot and leg fatigue and can prevent blisters. Boots should have ladder shanks that spread weight across the sole of the boot to provide comfort while working from ladders. Safety toe caps can protect toes from heavy impact even though the outer shell of the boot is damaged. Shin padding will help to prevent lower leg injury. Thermal insulation and pull loops should be standard in any fire fighting boot. When protective trousers are not worn, ¾ length boots must be worn fully turned up. Boots will last longer when kept free of grease and debris that deteriorates rubber. Three-quarter length boots should not be kept rolled when in storage, and boots should not be stored in cold areas.

Gloves: Hand protection is essential for fire fighters. NFPA 1973, *Standard on Gloves for Structural Fire Fighters*, covers hand protection by gloves. Gloves come in many fabrics and styles. Coated fabric gloves can keep hands dry but must be flame resistive and have thermal, flame resistive liners to help prevent burns. Flame resistive leather gloves can provide good puncture and tear resistance but some may hamper hand dexterity. Flame resistive thermal liners to prevent heat transfer are also required. NFPA 1973 provides performance criteria for heat and flame resistance, cut and puncture resistance, water penetration, dexterity, and sizing.

Other Protective Equipment: Life belts should be used by fire fighters engaged in ladder work. They are worn over the protective coat and provide a sturdy, quick connect clamp to prevent fire fighters from falling.

Personal Alert Safety Systems (PASS) are devices that sound an audible alarm for help if the wearer is unable to do so. PASS monitors fire fighter motion and signals an audible alarm when motion is undetected for more than 30 seconds. Also, the fire fighter can actuate the audible alarm if in need of assistance. NFPA 1982, *Standard on Personal Alert Safety Systems (PASS) For Fire Fighters*, is the standard covering the design and testing of PASS devices. PASS should be worn by fire fighters whenever operating in any hazardous area.

Rope used to raise or lower persons is termed life safety rope and is covered by NFPA 1983, *Standard on Fire Service Life Safety Rope, Harnesses, and Hardware*. The standard provides design, construction, and testing criteria for the rope and associated harnesses and hardware. Rope is classified as one person or two person.

Fire service administrators and members must take notice of the distressing statistics regarding injury and death. It is the responsibility of the fire department to provide all members who engage in emergency operations with complete protective clothing and SCBA that meets at least current minimum standards, and to assure use of these protective measures. It is the fire fighter's responsibility to wear SCBA and complete protective equipment at all appropriate times on the fireground or emergency scene.

HOSE, COUPLINGS, AND NOZZLES

Fire hose is the vital link between the water supply and nozzles used to project streams on the fire. Hose must be rugged, dependable, capable of carrying water under substantial pressures, and yet flexible and sufficiently easy to handle. Fire department management should be concerned with selection of the proper grades and types of hose, and maintenance to assure maximum useful life. These two chief factors are covered in detail in NFPA 1961, *Standard for Fire Hose*, and NFPA 1962, *Standard for the Care, Use, and Maintenance of Fire Hose Including Connections and Nozzles* (hereinafter referred to as NFPA 1962). Aside from unavoidable mechanical injury at fires, dependability and length of life of hose rest on three factors: (1) the quality and suitability of the hose purchased, (2) the care with which it is handled at fires, and (3) the maintenance and care of hose in quarters. Under average field conditions and with proper care, hose should be serviceable for a minimum of ten years unless subject to damage in use or training. There is no reason for discarding hose at the end of ten years if it is in good condition and passes annual service tests.

Types of Fire Hose

Fire hose is manufactured in the following principal sizes (measured by the internal diameter): ¾, 1, 1½, 1¾, 2½, 3, 3½, 4, 4½, 5, and 6 in. (19, 25, 38, 44, 63, 76, 89, 102, 114, 127 and 152 mm, respectively). The internal diameter of the hose waterway principally determines the flow delivered at a given expenditure of pressure energy. For example, two lines of 3 in. (76 mm) hose carry as much water as three lines of the nominal 2½ in. (63 mm) size; one line of 3½ in. (89 mm) hose carries approximately as much water as two 2½ in. (63 mm) lines.

Fire hose is constructed with a number of different forms in accordance with the intended type of service. Most fire hose has an interior rubber tube or lining to provide a smooth, watertight conductor and one or more jackets to provide the strength to withstand the intended pressure and protect the tube against abrasion damage. Certain types of hose also have a protective cover.

Woven Jacket, Rubber Lined Fire Hose: Fire departments, both public and industrial (brigades), are most familiar with woven jacket, rubber lined fire hose. The rubber lining may be natural or synthetic, or a combination of these materials. Jackets may be single or multiple woven. Single jacket hose is intended for use at industrial yard hydrants and at standpipes where it will not get substantial use. Multiple jacket hose is used by public fire departments because service conditions require the additional protection and greater working pressures provided by the extra jacket or jackets.

For many years the woven jacket was made entirely of cotton fibers. In recent years much manufactured hose has had cotton warp (lengthwise) threads and synthetic fiber filler circumferential cords. The synthetic filler thread adds strength, reduces weight, and results in a more flexible hose that is easier to fold into a hose bed. Today, much of the municipal double jacket hose is made of all synthetic yarns. This hose is lighter weight than cotton jacket hose, considerably more flexible, and a greater amount can be carried on apparatus or on standpipe racks. It is immune to damage from mildew although it may be more susceptible to heat damage. Both the all cotton and cotton synthetic jacket hose are subject to mildew, although both are available from manufacturers in treated form that retards the formation of mildew.

The 1½ in. (38 mm) hose is particularly applicable to inside fire fighting for ease in handling and for initial fire attack where appropriate. A larger backup line is always recommended. The 1¾ in. (44 mm) hose may be used in place of 1½ in. (38 mm) hose and can provide approximately 165 gpm (625 L/min) in relatively short hose lays.

More 2½ in. (63 mm) fire hose is carried on fire apparatus than any other size. It is used to provide 250 to 350 gpm (946 to 1325 L/min) hose streams from hand lines, and also in multiple lines to supply master streams. Dual 2½ in. (63 mm) lines are used to supply flows from 500 to 600 gpm (1893 to 2271 L/min). The 3 in. (76 mm) and 3½ in. (89 mm) hose are used as supply lines to supply large stream devices including ladder pipes.

Many fire department pumpers carry short lengths of large diameter woven jacket, rubber lined hose (called soft suction) for supplying the pump from hydrants. Such hose is frequently preconnected to a pump inlet. Sizes used include 4, 4½, 5, and 6 in. diameters (102, 114, 127, and 152 mm, respectively).

Supply Hose: In addition to the double jacket, rubber lined fire hose designed to withstand rugged service (including high pump discharge pressures), many fire department pumpers carry a lighter weight large diameter fire hose having a single synthetic jacket. This hose supplies pumps from hydrants, or is used in relay between pumpers. Such large diameter hose is available in 4, 4½, 5, and 6 in. diameters (102, 114, 127, and 152 mm, respectively). It is furnished with a rubber or synthetic lining and frequently has a protective synthetic cover.

Suction Hose: Hard suction hose has a rubber lining and layers of fabric reinforced with a spiral wire set in rubber between the reinforcement layers to prevent collapse of the hose when drafting water with pressure in the hose below atmospheric pressure. Usually the hose is furnished in 10 ft (3.05 m) lengths, and the common diameters are 4 to 6 in. (102 to 152 mm). Smaller sizes are available for special applications such as for supplying portable fire pumps. Hard suction hose should not be used for supplying pumps from hydrants. Large diameter soft suction hose should be used for pumper supply, and the hose connected to the largest discharge on the hydrant (pumper supply connection or steamer connection) and to the large pumper suction intake.

Hard suction hose should be of a design that has a low friction loss and will not collapse under a vacuum of 23 in. of mercury (78 kPa), and will also withstand a hydrostatic pressure test of 200 psi (1379 kPa).

Booster Hose: Many pumping engines and other fire apparatus are equipped with one or two reels, each carrying 200 to 300 ft (61 to 91 m) of this hose which has an inner rubber tube, a reinforcement consisting of fabric wrapping(s) or braid(s), and an outer rubber cover. Sizes ordinarily used by fire departments are ¾ and 1 in. (19 and 25 mm), with the latter size preferable because it carries approximately twice as much water at a given pressure.

Special Types of Hose: Unlined linen hose for standpipes or hose houses is being phased out, and is NOT allowed in new installations or as a replacement hose. Where unlined linen hose is found unsuitable for service, or in new installations, single jacket lined hose is installed instead.

Hose used by forest service fire fighters is mostly 1 to 1½ in. (25 to 38 mm) with a single woven jacket, either lined or unlined. Lined hose is used extensively. The lower friction loss with a given size is an advantage of the lined hose; it takes much less time for water to reach the nozzle through a long line of lined hose.

Covered, woven jacket, rubber lined hose in various sizes is used extensively in locations where the extra protection afforded by the coating or cover is required.

Examination of Hose for Compliance

Purchasing fire hose by specification alone does not ensure the obtaining of the desired quality and hydrostatic pressures. To assure this, definite tests must be performed under each specification item, conducted in a standard manner by qualified persons. Visual inspection alone does not detect inferior or defective lengths of hose. The cost for the testing service is nominal compared with the investment involved in purchasing hose and the importance of reliability and suitability of intended service. Many fire departments and other purchasers specify that a test report be furnished for each individual length of hose purchased, and that the test report include notation regarding the attachment and condition of couplings. In many cases money is saved by purchasing tested and listed fire hose that is constructed to meet the specified performance tests, thus helping to assure a long reliable service under anticipated conditions of use.

Care, Maintenance, and Use of Fire Hose

To be reliable, fire hose should always be cared for properly. It should not be used for other than training or fire fighting service, except in emergencies and with the approval of fire department officials. Burst hose at a fire may cause serious injury to fire fighters and other persons, and may mean loss of time in bringing a fire under control. The proper use, care, and maintenance of various types of fire hose is the subject of NFPA 1962.

Care of Woven Jacket, Rubber Lined Hose: With the exception of booster and hard suction hose, most of the fire hose carried on fire department apparatus is of woven jacket, rubber lined construction. Hose carried on fire apparatus may be flat loaded to reduce edge wear, and loaded so that air can circulate around it. Where apparatus serves in areas subject to frequent rain and snow, the hose compartments should be protected by removable decks fitted with waterproof tarpaulins or covers extending downward over the rear opening of the hose body. Such covers provide air space over the hose and also protect the hose from the direct rays of the sun when fire apparatus is out of quarters during inspection and training periods.

Where hose is installed at yard hydrants at industrial plants, it should be kept in well ventilated hose houses in such a way that air can circulate and excessive heat can be avoided.

To prevent damage and permanent set to the rubber tube, hose should be removed from the apparatus or hose house at least quarterly and reloaded in a different position. It is considered desirable to also run water through the hose at least quarterly. When hose has been charged with water, whether at fires or servicing drills, it should be replaced on the apparatus by spare hose so the required amount will be available for fire fighting while used hose is being cleaned and dried. Whether hose jackets are of cotton or synthetic construction, the hose should be cleaned and dried after use to remove possible abrasive or contaminating materials, and to protect the hose compartment of the apparatus against rust or water damage.

Care of Hose at Fires: When used for fighting fires, fire hose is subject to severe strains, pressure surges, and mechanical injury. Care should be taken to lay hose so that injury will not result from contact with sharp or rough objects at fires. Too often hose is treated as though it were rugged water pipe instead of a flexible tube protected only by the fabric jacket. Vehicles should not be driven over hose lines. Where it is necessary for fire department vehicles to cross, hose bridges should be used where possible; however, it is desirable to detour all nonemergency traffic from the fire area. When it is necessary to hoist lines, hose rollers can make the task easier and prevent mechanical injury. When hose lines are extended up ladders, the hose should be supported by hose rope tools placed to take the strain off couplings.

Pressure surges are a principal cause of damage to fire hose. Shutoff nozzles should be opened and closed slowly because sudden closure of nozzles can cause severe pressure surges or shock waves which are unpredictable and can be extremely damaging both to hose and pumping apparatus. Pressure relief devices on pumping engines should always be used to control sudden increases in pressure. In pump operation it is preferable to reduce pressure at the pump when convenient to do so before shutting nozzles because this avoids pressure surges which may occur even where the engine governor or relief valve is functioning properly.

The usual required relief valves or pressure governors are designed to protect the discharge side of the pump. When water is relayed from a pump at a water source to a pump at a fire, special precautions should be taken to prevent damaging pressure surges. If not provided as part of the pumping apparatus, a relay relief valve should be attached to the inlet or suction side of the receiving pump to which the relay hose line is to be attached. The lower the setting of this relief valve on the receiving pump inlet, the greater the protection to the pump and to the hose supplying the relay. Some pumpers are equipped with large diameter, powered valves for suction control. These valves should be controlled so as to not actuate too quickly as "water hammer" will result which can damage hoses, equipment, etc., and cause a safety hazard.

In cold climates, care should be taken to prevent water from freezing in or on the hose. Once water is turned on, some water should be left running through the hose until the line is no longer needed. During freezing weather it is common practice to keep water moving through the hose. This is because moving water does not freeze readily. Sharp bends should be avoided in any hose in or on which ice has formed; frozen hose can be damaged by such bends. Care must be used in chopping ice from hose after a fire. Hose which has been frozen must be service tested.

Care must be taken to avoid the burning of the hose at forest and grass fires. As fire is knocked down and the nozzle is advanced, the hose often comes in contact with hot spots or embers unless care is exercised. An advantage of unlined hose in forestry service is that water seeping from the hose helps protect it against such damage.

Care of Hose after Fires: When hose has been returned to quarters it should be laid out where it can be swept and washed as needed. Some fire departments have machines for this purpose. It is important to remove dirt and other foreign material from the jackets. A scrub brush and mild soap and water may be used, but frequently a small hose is used for washdown purposes. Clean dry hose should be loaded on the apparatus to replace the used hose, except for rubber covered booster hose which merely needs to be wiped clean as it is rerolled on the reel.

Hose should be thoroughly dried after cleaning to remove grime and contaminants. Hanging in hose towers or laying on racks are the best methods for drying hose. Where drying cabinets are used, a sufficient number should be provided to properly service the hose. Some fire departments have a central hose depot that cares for all hose and supplies clean hose to the fire stations as needed. When hose has been thoroughly dried, it should be removed from the drying equipment, rolled for storage, and placed on storage racks ready for use.

Hose Records

Good hose records are necessary to keep accurate data on hose performance. These records may be kept in a book or on printed card forms. Records for hose on racks, reels, or in enclosures may be kept at the hose location or at a central location on the premises where the hose is located. Fire department records should include a complete hose inventory, and a record of use of hose by the individual fire fighting units to which it is assigned. Upon delivery and acceptance, each length of hose is given an identification number. This is used to record its history throughout its service life, and ultimately shows the reason the hose was condemned and removed from service. Such records enable fire department administrators to determine the cost effectiveness of the various sizes and types of hose in service, the work to which hose is subjected, service tests, repairs, and other pertinent data. If hose fails within the guarantee period, this will also be indicated.

Each length of hose should be identified by an assigned code number corresponding to the fire department hose record. The number may be stenciled near the coupling on the hose with an ink or paint used that will not be deleterious to the hose, or the number may be stamped on the bowl or swivel of the coupling. Coupling bowls may be damaged by improper number stamping. The proper procedure is to insert a special steel plug with rounded edges into the end of the expansion ring. One sharp blow from a sharp steel numbering die should then stamp the coupling.

Some fire departments color code the couplings, as well as various other tools, to identify the fire company to which the equipment is assigned. This enables each company to readily identify and pick up its own hose and equipment used at a fire. Where mutual aid operations are frequent, each length of hose should be stenciled with an identification of the fire department owning the hose.

Each fire company may be required to keep a hose record book. This shows when each length of hose is placed on particular apparatus, when it is removed after use, and whether it is used for fires or drills. When hose has been removed from the apparatus for use at fires or drills and so recorded, it is not necessary to make a scheduled hose change except for the required service tests. Where apparatus has several hose compartments, the compartment from which hose was last removed may be marked with chalk indicating that change of hose is not required for that compartment. However, if hose has not been changed within three months and a compartment has been partly unloaded at a fire, it may be desirable to reload the entire compartment after a fire so that hose at the bottom of the load will not be neglected.

Responsibility for maintaining the required hose records should be spelled out in departmental rules or orders. The hose shop or maintenance division should keep a hose work report. Records should indicate the age of the hose, the vendor, the fire company to which it was assigned, and also the cause of failure or reason for condemning an individual length of hose. (See Fig. 15-6F for typical hose record card.)

HOSE RECORD CARD

ID NO. ______ CITY ______ ENGINE CO. ______

SIZE (DIA.) ______ LENGTH ______ TYPE HOSE ______ CONST. ______ COST ______

MFG. ______ PART NO. ______ DATE RCD. ______ DATE IN SERVICE ______

VENDOR ______ CPLG. MFG. ______ TYPE CPLGS. ______

REPAIRS

DATE	KIND	COST	NEW LENGTH	NEW ID NO.

REMARKS: ______

FIG. 15-6F. One type of hose record card.

Service Tests for Fire Hose

Reliability of fire hose is essential to good fire department service. Service tests should be conducted on each hose length at least annually and after repairs, or at any time that hose has had hard usage and its condition is suspect. NFPA 1962 includes required service test procedures and includes a typical annual hose test record form. The standard also specifies service test pressures.

The development of the test pressure introduces a serious accident potential even when stated procedures are followed. Extra care must be taken to see that all air is bled from the hose before allowing pressure to rise. The hose must be tested on a surface that permits the hose to elongate freely, lie flat, and not be twisted.

NFPA 1962 details service test methods using either a hose testing machine or fire department pumpers. In most fire departments, it is more practical and convenient to have each company test its own hose, using its pumper to supply the pressure. It is important to use a hose test gate valve with a ¼ in. (6 mm) opening to permit the test pressure to be maintained after the hose has been filled and the valve closed, but which will not permit a pressure surge should the hose burst while testing.

The specific service test procedure specified in NFPA 1962 must be followed.

Unlined Fire Hose

Unlined fire hose, usually called linen fire hose, was designed for standpipe use and is intended for first aid fire protection only. New installations now require lined or unlined hose to be replaced by lined hose when the former is found to be in doubtful condition. It should be service tested but must be thoroughly dried before being placed back in service. Single jacket lined hose is available for interior standpipe use. It has lower friction loss, does not cause water damage from seepage, works well with a shutoff nozzle, is usually made of 100 percent synthetic yarns that eliminate problems from mildew, and can be more easily service tested, cleaned, and dried.

Moisture can cause rapid deterioration unless the hose has been treated to resist rot and mildew. Valves must be kept in good condition so there will be no leakage. Hose must be protected from condensation on the standpipes.

Hose stored on a rack or reel should be carefully examined at least once a year. This annual physical inspection consists of a visual examination to see that the hose has not been vandalized, the couplings and nozzle are attached, and there is no evidence of mildew, rot, chemical, vermin, and abrasion damage. If the hose is in doubtful condition, it should be service tested. Also, the hose and nozzle should be checked to ensure they are free of debris.

The hose yarns will lose effectiveness if impregnated with liquids other than water. The hose must not come in contact with oil, grease, or corrosive chemicals. Care must be taken to keep polish which may be used on racks or couplings from coming in contact with the hose. (For further details on unlined hose for standpipe use, see NFPA 14, *Standard for the Installation of Standpipe and Hose Systems.*)

Unlined and lined forest fire hose is treated to inhibit mildew. The hose may be cleaned and dried thoroughly after use and after service tests. In the absence of a hose washing machine, hose should be washed with a scrubbing brush using water and mild soap, followed by a thorough rinsing. When thoroughly dry, hose should be stored in a cool, dry room where air will circulate, and the hose will be out of contact with damp floors or walls.

Fire Hose Connections

NFPA 1963, *Standard for Screw Threads and Gaskets for Fire Hose Connections,* covers ten sizes of threaded connections, from ¾ to 6 in. (19 to 152 mm), used in fire protection. The standard gives the dimensions for screw thread connections, gages, gaskets, gasket seats, and the size of the threaded connections. These standards apply to fire hose couplings, suction hose couplings, relay supply hose couplings, fire pump suctions, discharge valves, fire hydrants, nozzles, adaptors, reducers, caps, plugs, wyes, siamese connections, standpipe connections, sprinkler connections and all other hose fittings, connections, and appliances that connect to or with fire pumps, hose, and or hydrants.

These threaded connections are defined as the "American National Fire Hose Connection Screw Thread," abbreviated throughout NFPA 1963 as "NH" (also known as NST and NS). Each of the ten standard sizes is designated by specifying in sequence the nominal size of the connection, the number of threads per inch, and the thread symbol as follows:

0.75–8 NH	3.5–6 NH
1–8 NH	4.0–4 NH
1.5–9 NH	4.5–4 NH
2.5–7.5 NH	5.0–4 NH
3–6 NH	6.0–4 NH

The need for standard fire service connections is most apparent during mutual aid operations. When hose couplings, nozzles, and other equipment do not conform to standards, it becomes necessary to use adaptors, thus adding complications to fire fighting operations. Equally important, standard hose fittings are more readily available and can generally be obtained at a more favorable price than where nonstandard fittings are specified. This is important to business and industry as well as to public agencies such as the fire department and water utility. Moreover, most nonstandard threads are very poorly defined, and often new purchases are not fully compatible with earlier equipment. The Appendix of NFPA 1963 gives suggestions for the use of adapters to connect NH standard and nonstandard threads.

There are a number of screw threads which give the appearance of compatibility with NFPA 1963; however, what often appears to be a good connection when lines are not pressurized may be loose, cross threaded, or contain improperly mated connections that have a tendency to fail when hose lines are charged. This both impairs fire fighting operations and presents a serious hazard to personnel.

Brass couplings of standard alloy, machined in a normal manner, weigh approximately 5 lb (2.25 kg) for the 2½ in. (63 mm) size and 1⅜ lb (0.68 kg) for the 1½ in. (38 mm) size.

Aluminum couplings have mostly replaced brass on new hose orders. They are almost three times stronger than cast brass, and weigh approximately 2.3 lb (1.04 kg) for the 2½ in. (63 mm) size and 0.8 lb (0.36 kg) for the 1½ in. (38 mm) size. Many of the problems associated with brass couplings do not present themselves with aluminum; i.e., swivels out of round, and bowls overstressed because of too much expansion. Aluminum and brass couplings should not be left connected to each other over long periods of time without inspection because of bimetal corrosion.

Attaching couplings requires a considerable degree of skill and experience. Couplings can be cracked and damaged by applying too much pressure with the expander, however, the coupling may not hold if insufficient pressure is applied. Also, the correct size expansion ring must be used for the type of hose and coupling bowl diameter. Unless a fire department uses a large amount of hose and has a mechanic skilled and experienced in attaching couplings, this work should be performed by the manufacturer of the hose or by a local coupling supplier who has the trained personnel and equipment necessary. All hose must be service tested after being recoupled.

It is desirable to use new couplings with new hose. Where hose must be recoupled, couplings should be checked for the number of times they have been previously expanded before reusing them. Reattaching couplings too many times may cause them to be overstressed where they will not hold properly when water pressure is applied. When a coupling has been expanded several times, the diameter of the bowl may become so enlarged that a standard expansion ring does not hold.

Care should be taken by fire fighting personnel to avoid dropping couplings on hard surfaces because this may damage the couplings. It is important that vehicles not be driven over couplings. This can cause couplings to be forced out of round. Couplings out of round should be repaired by a person experienced in this work with the needed equipment; otherwise, it is better to replace the coupling. Likewise, threads may be damaged by dropping hose or by dragging the hose ends.

After use, coupling threads and swivels should be examined and repaired as necessary and maintained so they easily can be made up by hand. A good practice when loading hose on apparatus is to examine each coupling before it is connected to the load. Any hose with defective couplings should be set aside for repair. As a general rule, couplings should have no oil or grease applied, because oil may damage the hose. If dirty, couplings should be cleaned in a pail of water.

Some fire departments are equipped with dies and other tools for chasing and refurbishing couplings. Coupling gaskets should be renewed as needed. This is one of the items to be checked when hose is being reloaded or changed. Gaskets that project into the waterway should be replaced.

Suction Hose Connections: NFPA 1901 specifies the size of suction hose normally needed to permit pumpers to pass capacity tests at draft. In some fire departments, prescribed suction hose sizes are kept only at the maintenance shop. Where hard suction hose is carried on the apparatus, a standard 5 or 6 in. (127 or 152 mm) size may be provided for all pumpers.

Until the standard for pumper supply hose threads was developed in 1955, each pump manufacturer furnished thread dimensions of their own choosing so that different makes of pumps could not use the same suction hose. Purchasers should specify that all threaded connections on apparatus and equipment comply with NFPA 1963.

Hydrant Outlets and Outlet Threads: A standard fire hydrant for fire department service should have at least one large outlet for pumper supply and two outlets for 2½ in. (63 mm) hose. Because most pumping engines in municipal service are rated to deliver at least 1,000 gpm (3785 L/min), it is desirable that water mains be capable of delivering at least this quantity to individual hydrants at a residual pressure of 20 psi (138 kPa) while a group of hydrants in the vicinity of the fire are supplying the water flow required for the area. Otherwise, it is often necessary to run supplemental supply lines from pumpers at other hydrants to properly supply pumps nearest a fire. Cities using pumpers of more than 1,000 gpm (3785 L) rated capacity may need stronger hydrant supplies in areas requiring relatively high water flows.

The hose threads provided on hydrant outlets are properly of concern to the fire department. Whatever the size of supply hose used, each pumping engine is required to carry adapters to permit the connection of the engine supply hose thread to the local hydrant outlet threads where hydrant threads and fire department hose threads are different. Adapters are needed to permit connection of large pumper supply hose (soft suction) to a 2½ in. (63 mm) outlet where hydrants do not have steamer connections. In most cities, the large hydrant outlets are equipped with 4½ in. NH threads, although some areas, including California, use a 4 in. NH hydrant outlet.

Some cities in the U.S. still have separate, high pressure water systems which are used specifically for fire fighting. These hydrants are of a special design and can have fittings usable on the high pressure system only.

A single 2½ in. (63 mm) outlet cannot provide sufficient water for capacity operation of pumpers unless hydrant residual pressures remain relatively high. Even where two 2½ in. (63 mm) outlets are used to supply a pumper, the flow is considerably less than that provided from a single 4½ in. (114 mm) outlet at the same pressure. Cities make a mistake when they fail to provide pumper supply outlets on hydrants on low pressure water systems. The more critical the water supply problem, the greater the importance of an efficient waterway.

Nozzles

The minimum number of hand nozzles to be carried on various pieces of fire apparatus is indicated in the equipment lists of NFPA 1901. Choice of nozzles for each class of service is left to the individual fire department. Spray and straight stream nozzles for 1, 1½, 1¾, and 2½ in. hose (25, 38, 44, and 68 mm, respectively), and for large streams, are available in types with various flow characteristics.

Other than variable gallonage, automatically adjusting constant pressure nozzles, the size of nozzle orifice, whether solid stream tip or spray type, determines the discharge at a given nozzle pressure. Fire department pumping engines have rated capacities based upon 250 gpm (946 L/min) increments, and a flow of 250 gpm (946 L/min) is designated as a standard fire stream. A standard stream is the flow from a 1⅛ in. (29 mm) nozzle tip used on a 2½ in. (68 mm) hand line at 50 psi (345 kPa) nozzle pressure. Years ago, with small capacity underpowered pumpers, it was often necessary to use small nozzle tips on long lines of hose. Now, a 1 in. (25 mm) tip flowing approximately 200 gpm (757 L/min) is the smallest size recommended for a 2½ in. (63 mm) hand line. Some fire departments regularly use a 1¼ in. (32 mm) tip on hand lines. These tips can be supplied by 2½ in. (63 mm) hose and deliver 325 to 350 gpm (1230 to 1325 L/min).

Fire department nozzles for 1½ in. (38 mm) hose normally are designed to discharge approximately 95 to 125 gpm (360 to 473 L/min). Nozzles for 1¾ in. (44 mm) hose normally are designed to discharge approximately 95 to 200 gpm (360 to 757 L/min). Nozzles for 1 in. (25 mm) hose generally discharge in the 20 to 30 gpm (76 to 114 L/min) range.

A wide variety of types and flows are available with spray nozzles. It is important that the user select nozzles having the desired characteristics for the intended class of service and desired gallonage (L). *Caution*: Some early types of spray or "fog" nozzles for 1½ and 2½ in. (38 and 63 mm) hose provided relatively small flows; where still in service, they should be replaced with modern types.

Spray nozzles for hose streams are of six general types:

1. Open nozzle—fixed spray pattern: Provides a fixed spray pattern discharge at the rated discharge of the nozzle; usually attached to shutoff valve.
2. Combination nozzle: Provides either a solid stream through a smooth bore orifice or a fixed spray pattern discharge at the rated discharge for each pattern. Pattern selection is provided through a control valve that includes shutoff.
3. Variable gallonage—adjustable stream pattern: Provides discharge patterns that can be selected from shutoff to straight stream and from narrow to wide angle spray at varying discharge rates. The discharge rate can vary considerably and is directly affected by the pattern setting. There is no control of the discharge rate.
4. Constant gallonage—adjustable stream pattern: Provides discharge patterns that can be selected from straight stream to narrow and wide angle spray, all at the rated discharge of the nozzle. The stream pattern selector may include shut off. The discharge pattern setting does not alter the discharge rate.
5. Adjustible gallonage—adjustable stream pattern: Provides discharge patterns that can be selected from shutoff to straight stream and from narrow to wide angle spray. The discharge rate is also adjustable and can be selected through a predetermined range. For 1½ in. (38 mm) nozzles the range is usually 65 to 95 to 125 gpm (246 to 360 to 473 L/min), and for 2½ in. (63 mm) nozzles, 200 to 250 to 300 gpm (757 to 946 to 1136 L/min).
6. Variable gallonage—constant pressure—Adjustable stream pattern: Provides discharge patterns that can be selected from shutoff to straight stream and from narrow to wide angle spray. A device in the nozzle provides automatic adjustment to the orifice size to provide a constant nozzle pressure of approximately 100 psi (690 kPa). The discharge rate varies greatly depending upon the volumn of water being supplied. There is no control of the discharge rate other than through operating the shutoff valve.

The nozzles as described in numbers 3 through 6 above should be equipped with shutoff valves either as an integral part of the nozzle or as an attached item. Fixed spray pattern nozzles are recommended for use on electrical fires to avoid the possibility of inadvertently applying a straight stream.

Combination spray nozzles usually consist of nonadjustable spray nozzle and a separate smooth bore nozzle opening for a solid stream with a valve operable by the user to give the desired type of stream or to shut off the nozzle. Adjustable pattern spray nozzles make it possible for the user to adjust the cone of spray from a straight stream to wide angle while the stream is in use. Some of these nozzles have predetermined settings for angles of spray. With these nozzles the discharge can vary considerably as the spray angle is changed.

Other adjustable spray nozzles are designed to give a practically constant discharge rate for straight stream and spray angles. These types of nozzles are popular with fire departments because the friction loss characteristics in the hose and nozzle pressures remain constant with the selected orifice setting.

Also popular are nozzles with adjustable gallonage selections. These nozzles allow fire fighters to scale the flow up or down depending upon the size of the fire and the success of attack operations. Communications with the pumper supplying the line is very desirable so the pump operator may adjust engine pressures accordingly. Otherwise, careful monitoring of pump gages is necessary.

Variable gallonage, constant pressure nozzles are designed to maintain a constant nozzle pressure. This produces streams of considerably varying discharge, yet with about the same visual characteristics and stream reach. This can be very misleading, as all streams look "good." Caution must be exercised as fire fighters can have confidence in a good looking stream which lacks the adequate flow for proper fire attack. It must be remembered volume, not pressure alone, extinguishes fire,.

Special nozzles such as those used in the aspiration and delivery of foaming agents are manufactured by several firms. Sizes on these special nozzles vary from 1 and 1½ to 2½ in. (25 to 63 mm) in size.

Very high pressure spray nozzles are designed to deliver small flows at pressures in the 400 to 800 psi (2756 to 5516 kPa) range at the pump. This equipment, designed for fire department use, is supplemental to standard apparatus. Either separate high pressure pumps or extra pressure stages on volume pumps are required. Water is delivered to these nozzles through ¾ in. (19 mm) high pressure hose carried on reels. The quantity of water delivered is usually 12 to 30 gpm (45 to 114 L/min). Caution: These very high pressure and very limited gallonage streams have an extremely limited use at most fires and should not be used to combat structural or outdoor fires.

Spray nozzles are widely used on standpipe hose for first aid fire protection, and are available in both 1 and 1½ in. (25 and 38 mm) sizes.

Use of Hose Spray Nozzles: In the majority of fire departments, nozzles used for initial fire attack with 1½, 1¾, and 2½ in. (38, 44, and 63 mm) hand lines are of the adjustable spray type. Such nozzles are suitable for fires in ordinary combustibles, on flammable liquids, and both inside and outside of buildings, including grass and brush fires. They can also be used on fires in transformers and other electrical equipment. In general, the discharge ratings for fire department spray nozzles are based upon discharge at 100 psi (690 kPa) at the base of the nozzle.

Spray nozzles are effective in more quickly absorbing heat than straight streams. The fog or spray generates large volumes of steam when exposed to high heat conditions within confined areas. Steam generation can cause discomfort for fire fighters within highly heated areas. The spray can also provide protection for fire fighters from convective and radiant heat.

When using spray nozzles for interior fire fighting operations, extreme care must be taken not to drive fire and the products of combustion into uninvolved or unaffected portions of the building. Spray nozzles entrain a tremendous amount of air in the direction of discharge and act as a large blower. Discharge patterns from 1½ in. (38 mm) spray nozzles can displace over 17,000 cu ft per minute (480 m^3 per min) of air, with highly negative results if applied improperly. The practice of attacking structural fires from the interior should always be followed whenever conditions permit. Attack should be undertaken from the unburned side driving fire, heat, and products of combustion back into the area already burned, and either horizontal or vertical coordinated ventilation should be provided in the fire area to allow the spray stream to drive heat and products of combustion out of the building. Spray streams can induce enough air into an oxygen deficient but otherwise flammable atmosphere to cause a flashover or back draft. Attacking from the burned side, or into a window where fire is showing, will only drive the fire, heat, and products of combustion throughout uninvolved portions of the building and increase the damage and dollar loss, as well as endanger the safety of other fire fighters.

Where range, penetration, and force of streams are desired, solid stream nozzle tips are generally used. Solid stream nozzles may also be used where debris in water being drafted may clog a spray nozzle, or when extremely cold temperatures may cause a freezing problem with spray equipment.

Master Stream Equipment: The rated capacity of pumping apparatus cannot be utilized to best advantage unless nozzles are available with which to apply the desired fire flow. Unfortunately, it is not uncommon to find pumpers of 1,000 gpm (3785 L/min) and larger capacity carrying only hand line nozzles. Often, the tips carried are too small to provide standard streams from hand lines; thus, only a small percentage of the available pumping capacity can be utilized in applying needed water flow.

Full pump capacity can be most effectively employed through master stream appliances. Preferably, every pumper should carry at least a portable master stream appliance. In many fire departments such nozzles are preconnected or prepiped on pumpers for fast initial attack on serious fires.

Master stream appliances, whether portable or permanently connected to fire apparatus, should have nozzle tips and hose thread connections as specified in NFPA 1963. Smooth bore solid stream tips are the type needed in the majority of cases when master streams are necessary. However, in addition to the solid stream tips, each such device should be provided with a large capacity spray nozzle. Master stream appliance spray nozzles are available in types similar to types 3 through 6 of hand line spray nozzles indicated earlier in this chapter. The standard covers two types of master stream devices—those rated under 1,250 gpm (4731 L/min), and those rated over 1,250 gpm (4731 L/min) but less than 3,000 gpm (11 356 L/min). For both classes, all inlet connections (other than connections piped permanently to a pump) must be fitted with internal swivel connections having standard thread, at least one of which shall be the 2.5-7.5 NH thread.

The discharge ends of master stream appliances designed to discharge from 350 to 1,250 gpm (1325 to 4731 L/min) must have the 2.5-7.5 NH thread for attaching smooth bore nozzle tips or spray nozzles. If stacked tips are used, one of these tips may have the 1.5-9 NH standard thread. Discharge ends of stream devices designed to discharge in excess of 1,250 gpm (4731 L/min), but less than 3,000 gpm (11 356 L/min), must have the 3.5-6 NH standard thread for attaching nozzle tips or spray nozzles. However, such large capacity appliances shall be provided with a reducer fitting, 3.5-6 NH female by 2.5-7.5 NH male.

Master stream appliances designed to discharge flows between 350 and 1,250 gpm (1325 to 4731 L/min) must have 2.5-7.5 NH standard thread for internal entrance thread. Master stream appliances designed for flows above 1,250 gpm (4731 L/min) but less than 3,000 gpm (11 356 L/min) at standard operating pressures must have the 3.5-6 NH standard fire hose thread for the internal entrance thread. Discharging ends and nozzles for large stream devices over 3,000 gpm (11 356 L/min) must be designed so that all inlet and outlet threads are the NH standard thread.

Standard tip sizes (diameters of orifices in inches/millimeters) of smooth bore tips for master stream appliances rated under 1,250 gpm (4732 L/min) are: 1⅜ (500 gpm), 1½

(600 gpm), and 1¾ (800 gpm) [35 mm (1893 L/min), 38 mm (2271 L/min), and 44 mm (3028 L/min), respectively]. For master stream appliances rated over 1,250 gpm (4732 L/min) but less than 3,000 gpm (11 356 L/min), standard tip sizes are: 1¾ (800 gpm), 2 (1,060 gpm), and 2¼ (1,350 gpm) [44 mm (3028 L/min), 51 mm (4013 L/min), and 57 mm (5110 L/min), respectively]. The flows indicated are for approximately 80 psi (552 kPa) nozzle pressure, and vary with the square root of nozzle pressure. The maximum size tip recommended for a ladder pipe supplied by 3 in. (76 mm) hose on the ladder is 1½ in. (38 mm), which discharges 600 gpm (2271 L/min) at 80 psi (552 kPa).

Fire departments should exercise caution in purchasing small master stream appliances that are supplied with only two 2½ in. (68 mm) inlets. Although these appliances are usually lighter in weight, they do not produce very effective master streams. The efficient maximum discharge of such appliances is frequently only 500 gpm (1893 L/min). Fire departments should also avoid purchasing master stream appliances that do not have waterways of adequate size, and stream straighteners (a straight, smooth bore section of waterway) between the inlets and the swivel portion of the appliance, and between the swivel portion and where the nozzle tips are attached. Without these stream straighteners, turbulent flow occurs and results in very high friction losses [45 psi (310 kPa) and over] in the appliance. These conditions create very unsatisfactory and inefficent streams.

Bibliography

NFPA Codes, Standards, Recommended Practices and Manuals. (See the latest *NFPA Codes and Standards Catalog* for availability of current editions of the following documents.)

NFPA 13E, *Recommendations for Fire Department Operations in Properties Protected by Sprinkler and Standpipe Systems.*

NFPA 14, *Standard for Installation of Standpipe and Hose Systems.*

NFPA 24, *Standard for the Installation of Private Fire Service Mains and Their Appurtenances.*

NFPA 295, *Standard for Wildfire Control.*

NFPA 402M, *Manual for Aircraft Rescue and Fire Fighting Operational Procedures.*

NFPA 403, *Recommended Practices for Aircraft Rescue and Fire Fighting Services at Airports and Heliports.*

NFPA 412, *Standard for Evaluating Foam Fire Fighting Equipment on Aircraft Rescue and Fire Fighting Vehicles.*

NFPA 414, *Standard for Aircraft Rescue and Fire Fighting Vehicles.*

NFPA 1001, *Standard for Fire Fighter Professional Qualifications.*

NFPA 1002, *Standard for Fire Apparatus Driver/Operator Professional Qualifications.*

NFPA 1003, *Standard for Airport Fire Fighter Professional Qualifications.*

NFPA 1021, *Standard for Fire Officer Professional Qualifications.*

NFPA 1031, *Standard for Professional Qualifications for Fire Inspector, Fire Investigator and Fire Prevention Education Officer.*

NFPA 1041, *Standard for Fire Service Instructor Professional Qualifications.*

NFPA 1201, *Recommendation for the Organization for Fire Services.*

NFPA 1202, *Recommendation for Organization of a Fire Department.*

NFPA 1221, *Standard for the Installation, Maintenance and Use of Public Fire Service Communications Systems.*

NFPA 1231, *Standard on Water Supplies for Suburban and Rural Fire Fighting.*

NFPA 1402, *Guide to Building Training Centers.*

NFPA 1410, *Training Standard on Initial Fire Attack.*

NFPA 1501, *Standard for Fire Department Safety Officer.*

NFPA 1901, *Standard for Automotive Fire Apparatus.*

NFPA 1904, *Standard for Testing Fire Department Aerial Ladders and Elevating Platforms.*

NFPA 1921, *Standard for Fire Department Portable Pumping Units.*

NFPA 1931, *Standard for Fire Department Ground Ladders.*

NFPA 1932, *Standard on Use, Maintenance, and Service Testing of Fire Department Ground Ladders.*

NFPA 1961, *Standard for Fire Hose.*

NFPA 1962, *Standard for the Care, Use, and Maintenance of Fire Hose, Including Connections and Nozzles.*

NFPA 1963, *Standard for Screw Threads and Gaskets for Fire Hose Connections.*

NFPA 1971, *Standard on Protective Clothing for Structural Fire Fighting.*

NFPA 1972, *Standard on Structural Fire Fighters' Helmets.*

NFPA 1973, *Standard on Gloves for Structural Fire Fighters.*

NFPA 1975, *Standard on Station/Work Uniforms.*

NFPA 1981, *Standard on Self-Contained Breathing Apparatus for Fire Fighters.*

NFPA 1982, *Standard on Personal Alert Safety Systems (PASS) for Fire Fighters.*

NFPA 1983, *Standard on Fire Service Life Safety Rope, Harnesses, and Hardware.*

Additional Readings

Arthur D. Little, Inc., *Guidelines for the Selection of Chemical Protective Clothing*, prepared for U. S. Environmental Protection Agency, Los Alamos National Laboratory, Los Alamos, NM, Mar. 1983.

Bailey, Milton, "Insulated Boots—Field and Laboratory Evaluation," *Fire Technology*, Vol. 15, No. 4, Nov. 1979, pp. 283-290.

Brunacini, Alan V., *Fire Command*, National Fire Protection Association, Quincy, MA, 1985.

Crouch, Keith G., "Fire Fighter's SCBA: Belt Radiant Heat Resistance," *Fire Technology*, Vol. 17, No. 1, Feb. 1981, pp. 39-53.

Fire Apparatus Practices, International Fire Service Training Association, Oklahoma State University, Stillwater, OK, 1980.

Foltz, Jeffrey L., "Disposable Respirators for Forest Fire Fighters," *Fire Technology*, Vol. 17, No. 3, Aug. 1981, pp. 174-176.

Mission Research Corporation, "Guide for Preparing Fire Pumper Apparatus Specifications; Part 1: Executive Summary," *FA-29-1*, prepared for the Federal Emergency Management Agency, U.S. Fire Administration, Washington, DC, 1980.

Perkins, R. M., "Insulative Values of Single-Layer Fabrics for Thermal Protective Clothing," *Textile Research Journal*, Vol. 49, No. 4, 1979, pp. 202-212.

Saunders, P. B., and Ku, R., "Sequencing the Purchase and Retirement of Fire Engines," *NBS Special Pub. 411*, U.S. National Bureau of Standards, Gaithersburg, MD, Nov. 1974, pp. 201-214.

SCBA: A Fire Service Guide to the Selection, Use and Care of Self-Contained Breathing Apparatus, National Fire Protection Association, Quincy, MA, 1981.

Schwope, A. D., et al., *Guidelines for the Selection of Chemical Protective Clothing*. Volume 1: Field Guide.

Veghte, James H., *Design Criteria for Fire Fighter's Protective Clothing*, Janesville Apparel, Dayton, OH, 1981.

Worsham, J. P., *Rural Fire Protection: A Selected Bibliography*, Vance Bibliographies, Monticello, IL, Aug. 1978.

EVALUATION AND PLANNING OF PUBLIC FIRE PROTECTION

Revised by Dr. John Granito

The purpose of this chapter is to demonstrate that adequate, cost effective public fire protection is essential for a community. Public fire protection needs to be carefully planned and requires that certain logical steps be taken to achieve a comprehensive, acceptable, and workable plan. There are two important aspects to any good plan: (1) the plan itself, which must be feasible and directed toward clear goals, and (2) the process by which the plan is developed, which must ensure that all major goals are considered and every constituency to be affected by the plan is reasonably involved in the planning process.

Without adequate involvement of the necessary constituencies, implementation of the plan may likely fail due to a lack of cooperation. For a satisfactory plan to evolve, the planners must decide the end results they wish to achieve (goals), determine the status of the community in relation to those goals (evaluation), and calculate how much and what kinds of progress actually can take place over a certain period of time (objectives, tactics, timeframe). Each of these three steps—setting goals, evaluation, working out the details—requires the collection and analysis of relevant information, usually called data collection. Broad goals are achieved through planned strategies and precise objectives are achieved through implemented tactics. Each must be relevant to the other.

Because some degree of public fire protection is almost always in place, it is common for the entire process to begin with an evaluation of fire protection that is already available. The information obtained from the evaluation, when analyzed in terms of broad, generally recognized public protection goals, identifies needs and provides fire protection officials with the approximate parameters of the plan to be developed. As already noted, that plan cannot be developed without the involvement of a wide variety of community groups. Even if the various constituencies seem willing to allow fire protection officials to develop and write the comprehensive plan without their consultation, the fire officials should be cautious; the citizens eventually must be willing to accommodate and pay for the implementation. Since a comprehensive plan envisions a larger group or system of integrated parts, a number of organizations and agencies outside the fire department will need to play important roles in implementation of the plan.

One basic aspect of a comprehensive public fire protection plan is the concept that it is infinitely better for fires not to occur than to occur, even if they are extinguished quickly and at relatively low loss rates. The goal of reducing the incidence of fire, then, involves all aspects of fire prevention. Historically, much more energy and many more resources have been devoted to evaluating, planning, and implementing fire suppression/fire fighting capabilities than fire prevention capabilities. Simply stated, the United States as a whole has focused on putting out fires rather than on preventing them. That focus is still very necessary and exceedingly important, but additional emphasis on preventing fires is necessary as well. Planning groups have obvious difficulty, however, in evaluating the degree of effectiveness of fire prevention programs.

A reasonably effective method exists for conducting the evaluation, involving two kinds of analysis. The first kind of analysis requires the community to maintain consistent and carefully recorded information each year concerning the number of fires which occur, and the cost of those fires in lives and dollars. When those human and physical costs are added to the expense of maintaining both fire prevention and suppression systems, a standardized total cost of fire to the community can be compared with that same cost in earlier years, or over an average of three or more years. Necessary adjustments for inflation, community growth or decline, and for other important variables can be made by local officials, thus permitting a community to compare its present to its prior fire performance.

The second kind of analysis involves communities identifying other communities which are similar in ways important to fire protection (such as size, construction, hazards, and geography) and comparing their own total performance to the total performance of the similar communities. In the first analysis, the community uses an internal data source (itself) as a yardstick, and in the second analysis it uses an external yardstick (other communities). As more data are collected concerning the total

Dr. Granito is Vice President for Public Service and External Affairs, State University of New York, Binghamton, NY.

cost of fires in various communities operating with various public fire protection plans, any given community will be able to benefit relatively quickly not only from its own experience, but from the experiences of others. The growing focus on reducing the incidence of arson is one example.

Important to this data collection, evaluation, and planning process is the degree of ability and willingness of the community to finance the total level of fire protection required by the plan's goals. The ability and willingness to implement segments of the plan which may be limited by existing legislation (or lack of it), or a lack of innovative approaches to solution identification must be considered. While fire protection officials must always be concerned with reducing the total cost of fire (fire loss plus costs of prevention and suppression), citizens living in tight economic times will ultimately reserve the right to make decisions, or "trade offs," concerning the level of protection they wish their tax dollars to purchase.

To assist citizens in making decisions concerning budgets, fire protection officials must accurately describe the effect on total cost if additional or fewer resources are applied to particular prevention or suppression efforts. This kind of technical knowledge and analytical ability by officials provides a crucial element to comprehensive planning and evaluation. It is one of the most important responsibilities of the public fire protection officer.

EVALUATION

In addition to assessing the capabilities of the fire department and related agencies, fire officials evaluating suppression capabilities must take a number of factors into account. Examples include known combustibles, the life hazard that exists, fire frequency, climatic conditions, demographic and geographic factors, and a basic consideration of the specific role of a public fire department in providing fire protection to the community. Failure to adequately consider each of these factors can lead to a large-loss fire. A fire department's suppression capabilities can never be expected to compensate for the deficiency or lack of built in fire protection systems.

Rural Fire Protection

One principal difference between rural and urban fire departments is that rural departments must pay more attention to water supply problems, even though this is an important factor in any department's operations. Rural fire department operations and apparatus emphasize not only fire fighting requirements, but also the provision of water for fire fighting. Rural fire apparatus must have large water tanks to permit effective initial attack on fires while supplementary water supplies are being brought into action. Supplementary water supplies include suction sources on or adjacent to rural properties, and mobile water tanker vehicles for transporting water from more distant sources. Rural fire departments often use apparatus and hose to relay water from sources several thousand feet (1,000 ft equals 304.8 m) from the emergency. Initial response of pumpers, tankers, and auxiliary apparatus should be adequate for a quick attack on the burning property. With adequate highways and well designed apparatus, it is possible to bring substantial fire fighting forces to an emergency in rural areas in sufficient time for a properly planned and executed initial fire attack operation to be effective. Most rural properties are now located in areas which enjoy some level of fire protection. Some properties, of course, may have to depend entirely upon their own private fire protection and whatever help they may obtain from forestry agencies or distant fire departments.

Minimum protection for a rural area would include a pumper with a large water tank, plus a water tank vehicle responding on an initial alarm. Properly designed tanks should be able to transport water from a source one mile (1.6 km) from the scene so a minimum of 100 gpm (378 L/min) can be pumped at the fire scene by the pumper. Since a larger flow often is required to provide adequate fire protection services, additional tankers must be used, or suction sources within reasonable distance of the fire scene must be identified. Rural apparatus should carry 3½ in. (89 mm) or larger lay-in hose to provide adequate water supply at the fire scene. It is always advisable to lay large diameter fire hose from the water supply source to as near the fire scene as possible in order to avoid extensive friction loss. At the emergency, large diameter hose is sometimes connected into smaller handlines, or used more often to supply another pumper from which handlines are extended. Other pieces of equipment such as rescue and aerial ladder vehicles should be provided as needed to carry out the mission. Generally, elevated master streams are not needed extensively in rural operations, and normal ladder truck equipment for rescue, forcible entry, ventilation, and salvage operations is carried on the pumper and equipment vehicles.

To be even minimally effective in controlling a fire, the initial responding apparatus must reach the emergency scene within approximately ten minutes of the sounding of the alarm, which is sometimes delayed itself. This apparatus should be capable of extinguishing small fires, preventing "flashover" or very rapid fire spread, or possibly preventing the expansion of fires to other exposed structures. When there is no public or private fire protection agency, any fire that is beyond the control of portable extinguishing equipment (extinguishers, garden hose, etc.) may be expected to burn until all fuel in a given fire area has been consumed. Minimum levels of fire protection, while serving to confine fires to isolated properties, leave much to be desired by the property owner who suffers the loss and the fire department whose morale is often affected by its inability to successfully control the average fire. Because of distances, lack of water, and a general resource deficiency, small rural communities frequently suffer heavy fire loss.

Urban Fire Protection

In urban areas, inadequate fire department response to initial alarms can be a major factor in fire losses due to high population densities and increased exposure of structures adjacent to the fire. The number of simultaneous fire fighting operations that may need to be conducted also dictates the total amount of personnel and equipment needed to provide effective fire fighting operations. In all but the smallest structural fire, several operations must be carried on simultaneously and the fire attack must be made from several points. This cannot be accomplished by the crew of a single fire apparatus. Multiple apparatus must be positioned properly, and adequate waterflow made available to cope with the amount of fuel (fire load) involved or exposed.

In simplest terms, structural fire fighting involves simultaneous operation of three units under a chief officer: (1) a pumper company to undertake a fast initial fire attack; (2) a pumper to provide adequate water supply for a continuing operation; and (3) a company to handle rescue, ventilation, salvage, and various other services not related to hose lines. At large-structure fires, additional fire fighting personnel are needed to cover the various points of fire attack, and in some cases various functions can be handled more efficiently by specially trained crews such as rescue companies and hazardous material teams operating from specially equipped apparatus.

In a light hazard residential district, the minimum effective initial fire alarm response should consist of three pieces of apparatus, two of which should be equipped to conduct pumping and water supply operations with the remaining vehicle equipped for the other operations. Each apparatus should carry the necessary tools and appliances to perform the designated operation. Twelve fire fighters and a chief officer would be the personnel required for reasonably satisfactory operation of this equipment. Not only degrees of fire loss but also civilian and fire fighter deaths and injuries as they are related to the size of the responding crew, type of fire, life hazard and the time required to receive the alarm, respond to the location, and conduct the initial attack, must be considered in initial response.

Since research indicates that fire "flashover" often will occur in typical wood frame structures, the location of fire stations and the size of responding crews should be carefully calculated so an adequate number of fire fighters and supervisors are quickly on the scene.

The rationale for a 12 person response team is based on two fire fighters driving the two pumpers and remaining to operate the pumps, distribute equipment, help other personnel don protective masks, and operate radios; four additional fire fighters positioning and operating two hose lines, advancing and repositioning the lines as necessary; four fire fighters performing laddering, forcible entry, ventilation, rescue, salvage, and other "ladder company" duties; two fire fighters handling hydrants, water supply lines, and the advancing and operating of a third hose line. One person also functions as direct supervisor (company officer) of the hose lines, and another as supervisor of laddering and other operations. The chief officer directs overall operations of the working fire. Response crews with fewer personnel and equipment risk being unable to perform rescue and extinguishment rapidly enough to do much good at a working fire. While some case studies indicate that fewer personnel are sufficient to handle a percentage of fire calls, there is a dramatic history of small fires which have extended to large-loss fires with possibly related fire fighter and civilian injuries or deaths because of inadequate initial response.

In most smaller and medium sized communities, all initial response (first alarm) apparatus will not arrive at the fire scene simultaneously. In many departments with on duty personnel, apparatus has to respond from more than one station, and some apparatus have longer travel times to the fire scene. In volunteer departments, personnel must travel varying distances to get to the fire station or the fire scene, and thus all apparatus cannot go into operation at the same time. Those fire fighters and vehicles which cannot arrive at the fire scene within the first critical time period can do no good in the initial attack, regardless of the department's response assignment ("running card"). Communities often have a false sense of security in this regard, until actual response times are tested and working initial attack personnel counted.

The minimum fire force recommended for any small community is two pumpers, a ladder or squad truck, an auxiliary pumper or brush fire fighting vehicle, and possibly a tanker for use in areas where hydrant water supplies are limited. Where necessitated by fire frequency or response distances, additional pumpers, ladder trucks, and tankers may be needed. Reserve apparatus is desirable not only to permit the repair of first line equipment without reducing fireground forces, but to provide additional fire fighting units during major emergencies.

Commercial, industrial, and mercantile areas generally require an additional piece of pumping apparatus in response to the initial alarm. (See Table 15-7A.) If properties with considerable life hazard are involved (schools, hospitals, nursing homes, etc.) at least four pumping apparatus, two aerial ladder trucks, and two chief officers should be considered a minimum response on initial

TABLE 15-7A. Evaluation of Fire Department Response Capability

High Hazard Occupancies (Schools, hospitals, nursing homes, explosive plants, refineries, high rise buildings, and other high life hazard or large fire potential occupancies)
At least 4 pumpers, 2 ladder trucks, 2 chief officers, and other specialized apparatus as may be needed to cope with the combustible involved; not less than 24 fire fighters and 2 chief officers.
Medium Hazard Occupancies (Apartments, offices, mercantile and industrial occupancies not normally requiring extensive rescue or fire fighting forces)
At least 3 pumpers, 1 ladder truck, 1 chief officer, and other specialized apparatus as may be needed or available; not less than 16 fire fighters and 1 chief officer.
Low Hazard Occupancies (One, two or three family dwellings and scattered small businesses and industrial occupancies)
At least 2 pumpers, 1 ladder truck, 1 cheif officer, and other specialized apparatus as may be needed or available; not less than 12 fire fighters and 1 chief officer.
Rural Operations (Scattered dwellings, small businesses and farm buildings)
At least 1 pumper with a large water tank (500 gal (1.9 m^3) or more), one mobile water supply apparatus (1,000 gal (3.78 m^3) or larger), and such other specialized apparatus as may be necessary to perform effective initial fire fighting operations; at least 6 fire fighters and 1 chief officer.
Additioinal Alarms
At least the equivalent of that required for Rural Operations for second alarms; equipment as may be needed according to the type of emergency and capabilities of the fire department. This may involve the immediate use of mutual aid companies until local forces can be supplemented with additional off-duty personnel.

alarms. Especially large numbers of personnel are needed for search and rescue operations in these properties, with several fire fighters needed to "sweep and search" each floor.

The required fire fighting units should arrive on scene close enough in time after the initial alarm to operate as an effective fire fighting unit following planned tactical procedures. Often, the "task force" concept, where vehicles are housed and respond together as a tactical unit, may prove to be a most efficient fire fighting tactic, even though a slight increase in response time is necessary in some areas of the first alarm district. Increased efficiency can often outweigh this slight increase in response time. A minimum task force unit should consist of two pumping engines, a ladder truck, and a chief officer. Any plans for fire station consolidation or relocation should consider the possibility of a task force.

In evaluating the adequacy of fire protection in any given area, major consideration must be given to the ability of the fire department to efficiently handle any reasonably anticipated work load. This requires an evaluation of the possibility of several simultaneous working fires, weather factors that may contribute to the spread of fire, the delay in response or the possibility of slow operations at the scene, and other demographic or geographic conditions that might affect the frequency of fire occurrence and the response time of initial fire fighting units. Where fire frequency is such that any fire company may expect two or three working fires per day, or where structures to be protected require a heavy initial response, closer geographic spacing of or increased personnel assigned to individual fire companies may be necessary. The number of other fire fighting or related operations such as grass, brush, rubbish, and automobile fires, and emergency rescue operations may also require greater than normal staffing of equipment and closer spacing of fire companies. Major structural fires may result when the normal first alarm coverage in a district is depleted through coverage of these other emergencies which makes remaining fire fighting forces inadequate. Staffing fire apparatus at a level far below minimum requirements will usually result in less effective fire fighting performance. This factor also has an adverse effect on the number of required fire companies for various alarms, since additional fire companies must be dispatched to the scene of an emergency to provide adequate coverage.

The highly desirable practice of assigning emergency medical responsibility to the fire department must be calculated into the staffing formula. It is also difficult to obtain effective teamwork and coordination with understrength crews. Some fire departments have attempted to solve this problem by supplementing their crews with part time or volunteer fire fighters, or by providing off duty fire fighters with tone activated radio receivers and paying them for overtime when they respond to a fire. The on duty personnel make the initial fire attack and holding action while off duty personnel provide the additional assistance needed for continuing fire fighting operations, although off duty personnel do not respond immediately. Efficiency is definitely lost, and increased fire losses can be expected with this arrangement. Such protection should not be relied upon to replace adequately the required staffing and equipment needed immediately at the scene.

Personnel requirements are not merely a matter of numerical strength, but are based on the establishment of a well trained and coordinated team necessary to utilize complicated and specialized equipment under the stress of emergency conditions. A general practice should be to avoid attempting to operate more fire companies than can be effectively staffed, even if some response distances must be somewhat increased. The effectiveness of pumper companies must be measured by their ability to get required hose streams into service quickly and efficiently. NFPA 1410, *Training Standard on Initial Fire Attack*, should be used as a guide in measuring this ability. Often a crew of fewer than four fire fighters may be unable to apply half as much water in a given time with their equipment as a company of four or five fire fighters. Seriously understaffed fire companies generally are limited to the use of small hose streams until additional help arrives. Often this action may be totally ineffective in containing a small fire and in conducting effective rescue operations. Research indicates that a crew of four is only 65 percent as effective as a crew of five, and that a crew of three is only 38 percent as effective as a crew of five.

Consideration must also be given to maintaining an adequate concentration of additional forces to handle multiple alarms at the same fire, while still providing minimum fire protection coverage for the other areas under fire department protection. If available personnel prove adequate for routine fires but inadequate for major emergencies, arrangements should be made to supplement the fire protection coverage by calling back the off shift personnel and by promptly calling nearby fire departments for mutual aid. Off shift personnel may operate reserve apparatus or relieve or supplement personnel on the fireground. Fire companies not dispatched or utilized on the fire scene should be repositioned throughout the remaining area of the jurisdiction to assure minimum response times to other alarms.

Reserve apparatus should be properly maintained and equipped, and when placed in service should be staffed to a degree commensurate with standard fire apparatus requirements. Since it may take up to 30 minutes or more to place reserve units in service with personnel recalled in an emergency, these reserve units should not be completely relied upon to provide an adequate level of fire protection services.

In cases where several fire departments occupy adjacent or contiguous territories, arrangements should be made for joint response along common boundaries to high-risk hazards and for assistance in covering vacant fire stations at times of major fires. Mutual aid or mutual response should not be relied upon to always provide assistance in major emergencies, since there could be times when local commitments will preclude the anticipated assistance. Mutual aid agreements do not reduce the responsibility of each jurisdiction to maintain adequate facilities to handle normal fire protection needs. It also must be assumed that teamwork and tactical efficiency at a fire will be somewhat less than that expected of equal units from the same department under a united command. Often, however, specialized units (such as hazardous materials response teams) are organized to protect larger areas encompassing several fire departments.

In the past it was a common practice to relate the number of pumping engines and their pumping capacity, and other apparatus personnel requirements, to the population to be protected. With the industrialization of many areas and the construction of commercial shopping centers, hospitals, schools, and nursing homes in residential areas, it is possible that concentrations of life hazard and property value in areas of small or large populations may require substantial fire fighting forces. Those with the responsibility for providing public fire protection must be prepared to cope with fire potential in any location in the jurisdiction. Fire department response requirements are now based on the water flow in gpm (L/min) that may have to be applied. A good rule of thumb is to provide one pumper company (plus supporting units) for each 500 gpm (1.9 m^3/min) that may be needed, plus personnel for rescue and other operations which need to be performed simultaneously with the advancing of hose lines.

Some may argue that it is not the public's responsibility to provide adequate fire protection to high hazard risks that should have built in fire protection systems. However, failure to attempt to provide fire protection for large taxable values on which the economy of a community may be based would place some municipal officials in an untenable position, and subject them to extreme criticism in the event of a serious fire loss.

As already described, time is another critical factor in the evaluation of public fire protection. It is generally considered that the first arriving piece of apparatus should be at the emergency scene within five minutes of the sounding of the alarm, since additional minutes are needed to size up the situation, deploy hose lines, initiate search and rescue, etc. The old adage says, "The first five minutes of any fire is the determining factor as to whether that fire will remain a small fire or become a large fire." While this may not always be true, delays in sounding an alarm obviously must be minimized or eliminated, as well as delays in responding and initiating rescue and attack. Time, however, cannot become the all important factor at the expense of safety.

There are increasing instances where specialized apparatus and equipment must be available to municipalities. One category of specialization concerns apparatus designed to handle hazardous materials, including spills of petroleum products and other chemicals which require special extinguishing agents such as foams or dry powders, and special equipment to apply these agents. These dangerous substances may be present because of manufacturing or storage facilities, or transportation routes in the district. Another category of specialization includes apparatus and equipment needed because of particular structures or facilities such as airports, research laboratories, hospitals, high rise buildings, oil and gas wells, and marine ports. In some communities fire departments are expected to conduct specialized functions such as extricating at automobile wrecks, performing water and mountain rescue, as well as providing emergency medical services. These services may also require special equipment and apparatus.

As with standard fire fighting equipment and apparatus, specialized tools cannot be used effectively and safely unless personnel are highly trained in their use under a wide variety of circumstances. Whether personnel are volunteer or career, in rural or urban areas, no plan can be implemented and no reasonable level of fire protection afforded to the community unless well designed and managed training programs are carried out.

Fire Prevention

As noted previously, fire prevention activities are somewhat difficult to evaluate. In a real sense, if prevention activities are effective, fires and fire related tragedies occur with less frequency. There is a reduction or absence of fire activity, and these results are evident statistically although they do not appear in dramatic news clips and photographs. Without careful and systematic long term recordkeeping concerning the incidence of fires and fire losses, the effect of prevention programs cannot be documented. Inability of fire officials to demonstrate the value of committing some additional community resources into the broad range of possible prevention activities may well result in a withdrawal from prevention programs and a subsequent increase in the need for a much larger suppression budget. Rational decisions and sound recommendations concerning evaluation and planning cannot be made unless fire officials learn what changes there can be to total fire cost by reallocating resources applied to the total fire defense system.

Both evaluation and planning require recognition of the component and integrated parts of a fire prevention system. Until recent years prevention was seen as a narrow band of activities, often greatly limited or nonexistent in most smaller communities. In urban areas it was limited frequently to the periodic inspection of certain types of buildings. More modern approaches to fire prevention recognize that a comprehensive program includes all organized activities, other than suppression, which reduce the incidence of fire and fire related losses. Ideally, these activities would be carried out in communities of every size, whether rural or urban, with appropriate adjustments made for community size, type, location, and fire history.

Prevention activities may be categorized in several ways, but it is usually helpful to group them as follows:

1. Activities which relate to construction such as building codes, the approval of building and facility plans, and occupancy certification.
2. Activities which relate to the enforcment of codes and regulations such as inspections of different occupancies, the licensure of certain hazardous facilities, and the design of new regulations and codes, or legislation to adopt existing model codes.
3. Activities which relate to the reduction of arson such as fire investigation and the collection of information and data related to setting fires.
4. Activities which relate to the collection of data helpful in fire protection such as standardized fire reporting, case histories, and fire research.
5. Activities which relate to public education and training, including fire prevention safeguards, evacuation and personal safety steps, plant protection training for industrial and other work groups, hazardous materials and devices safeguards, and encouragement to install early warning and other built in signaling and extinguishing devices.

An analysis of the community's fire history, conducted during the evaluation phase of the fire prevention plan, will usually indicate to fire experts and citizen groups which categories need strengthening. Comparing the number of fires and fire related incidents, plus fire loss (property, life, injury) statistics over several years as more prevention activities are phased in, provides an assessment of program effectiveness. Calculating the total cost of fire to a community (fire loss plus prevention costs plus suppression costs) will enable the fire department and the community to estimate the efficiency or cost effectiveness of a proposed prevention program.

PUBLIC PROTECTION CLASSIFICATIONS

The "Grading Schedule for Municipal Fire Protection," developed originally by the National Board of Fire Underwriters (NBFU) and continued by its successor, the American Insurance Association and then by the Insurance Services Office (ISO), has provided a guideline for municipalities to classify their fire defenses and physical conditions. The gradings obtained under the schedule were used in establishing base rates for fire insurance purposes. The schedule has been subject to change with the state of the art, and sweeping changes were made in the 1980 edition with the development of a revised "Fire Suppression Rating Schedule."

The Fire Suppression Rating Schedule (NCFPC 1973) produces ten different Public Protection Classifications, with Class 1 receiving the most rate recognition and Class 10 receiving no recognition. The Fire Suppression Rating Schedule simply defines different levels of public fire suppression capabilities which are credited in the individual property fire insurance rate relativities.

Starting in 1975, on a state by state basis following Insurance Department approval, ISO implemented the Commercial Fire Rating Schedule (CFRS) which was a major revision to the method used to develop individual property rate relativities. The CFRS reviews and correlates the construction, occupancy, exposures, and private and public fire protection (represented by the Public Protection Classification number). This correlation allows development of an equitable rate relativity applicable to the individual property. To this rate relativity, statistical experience adjustments are applied either by ISO or by affiliated companies to produce the applicable fire insurance rate. The Fire Suppression Rating Schedule (FSRS) represents a revision in the method used to derive the Public Protection Classification number used in the CFRS. The Public Protection Classification number is also used as a rate relativity variable for most class related properties, in addition to construction and occupancy variables.

The Grading Schedule for Municipal Fire Protection (ISO 1974), although a much improved system from previous editions, was not developed as an integral part of the individual property rating system. The previous schedules were somewhat independent primarily due to their historical development by the NBFU, which was not an insurance rating organization. The previous schedules were used more to quantify underwriting information, but did define different levels of public fire protection that could be used for a specific rating.

The Fire Suppression Rating Schedule is designed to assist in an objective review of those features of available public fire protection that have a significant influence on minimizing damage once a fire has occurred. This revision ties very logically to the review of contributive and causative hazards which can be performed with the Commercial Fire Rating Schedule.

Comparison of Details

The following material outlines the specific changes that are incorporated into the Fire Suppression Rating Schedule as compared with the 1974 *Grading Schedule for Municipal Fire Protection*.

Building, electrical, and fire prevention laws, and climatic conditions are not included in the FSRS (Table 15-7B) for the following reasons:

TABLE 15-7B. Comparison of Schedules: Major Items

1974 Schedule		1980 FSR Schedule	
	Relative Weight (Percent)		Relative Weight (Percent)
Water Supply	39	Water Supply	40
Fire Department	39	Fire Department	50
Fire Alarm	9	Fire Alarm	10
Building Laws	1.7		
Electrical Laws	.8		100
Fire Prevention Laws	10.5		
Climate Conditions	(variable)		
	100		

Comparison of Schedules: Major Items.

1. Most communities have adopted one of the available model codes, thus differences were not really being measured.
2. Evaluation of codes enforcement and climatic conditions is subjective when trying to review on a citywide basis.
3. Results of the law and its enforcement manifest themselves in the actual conditions found in individual properties when surveyed for application of the Commercial Fire Rating Schedule. (See Table 15-7C.)

Comparison of Schedules: Water Supply.

4. The 1974 schedule and its predecessor schedules contained many standards or benchmarks of evaluation that were detailed specifications. If a condition in a community did not meet the specifications, deficiency points were assigned. Specifications decrease a schedule's flexibility in use to properly evaluate changing conditions and technology. Because the FSRS has eliminated essentially all of the specifications that were in the 1974 schedule, it is referred to as a performance schedule. The only criteria in the FSRS that could be considered as specifications are the descriptions of the minimum conditions to be reviewed under the schedule.

As an example of the performance orientation of the FSR schedule with the 1974 schedule, it was essentially impossible for a community to receive other than a Class 9 rating if it did not have a water supply with conventional underground water mains. However, the FSR schedule is based upon delivery performance only, without regard to the delivery method used, and is adaptable to crediting the delivery of water by other means. Specific reference is made to water carried and delivered to the fire location by fire department apparatus.

TABLE 15-7C. Comparison of Schedules: Water Supply

1974 Schedule	Relative Weight (Percent)	1980 FSR Schedule	Relative Weight (Percent)
Adequacy of Supply Work	6	Supply Works	
Reliability of Source of Supply	6	Fire Flow Delivery	
Reliability of Pumping Capacity	3	Distribution of Hydrants	35
Reliability of Power Supply	4	Hydrants—Size, Type and installation	2
Condition, Arrangement, Operation and Reliability of System Components	4	Hydrants—Inspection and Condition	3
Adequacy of Mains	16		40
Reliability of Mains	2		
Installation of Mains	2		
Arrangement of Distribution System	2		
Additional Factors and Conditions Relating to Supply and Distribution	4		
Distribution of Hydrants	5		
Hydrants—Size, Type and Installation	2		
Hydrants—Inspection and Condition	2		
Miscellaneous Factors and Conditions	6		
	64 (Max=39)		

5. As previously mentioned, the FSRS has been designed to directly interface with individual property insurance rate development. As a result, numerous items that were evaluated both in the individual property review and in the previous public fire protection review have been removed from the FSRS.

The review and enforcement of building, electrical, and fire prevention codes have been deleted from the FSRS because the presence of hazards due to poor codes or lack of enforcement are recognized in an individual property survey and rate review.

Section I of the FSRS can be used to develop a public protection classification which reflects the city's ability to handle fires in small to moderate sized buildings; these are defined as those buildings having a needed fire flow of 3,500 gpm (13 m^3/min) or less. Section II of the FSRS can be used to develop public protection classifications for large individual properties having needed fire flows greater than 3,500 gpm (13 m^3/min).

Most communities design their fire protection needs around normally expected fires. This design is recognized in the differing concept between these two sections. The Public Protection Classification in Section I is applicable to average buildings, and the influence of the fire protection demands for larger buildings has been removed from that analysis. Section II can be applied individually to each large building to develop an individual public protection classification that reflects the available protection to that specific property.

The 1974 schedule and its predecessors did not provide significant recognition of the decreased need for public fire suppression when large properties are protected by automatic sprinkler systems. The FSRS does recognize this important fact by excluding all properties with standard automatic sprinkler systems from the development of needed fire flow figures.

The water supply items dealing with reliability in the 1974 schedule are not included in the FSRS because recent history has indicated that the dependability and replacement capability of water supply equipment is such that measurement of these features has little bearing on the water supply performance. The remaining items of the 1974 schedule that are not included in the FSRS are omitted because they either require subjective review or are not considered significant enough to measure the performance of the water system for fire insurance rating purposes. The emphasis of the FSRS review regarding a water system is the consideration of the actual water supply that is available for fire suppression at representative locations throughout the city.

As in the water supply feature, items that require subjective review and consider minor features are not included in the FSRS. (See Table 15-7D.) The FSRS places emphasis on the review of the first alarm response by the fire department because of the importance of initial attack to minimize potential losses. Additionally, the entire fire department item has been given increased weight in the overall review in recognition of the value of fire department operations in early stages of fire suppression and the fact that some operations are possible with less than optimum water supplies.

As in the water supply and fire department sections, features that are subjective or are minor items dealing with the fire alarm system are not included in the FSRS. (See Table 15-7E.) The emphasis has been placed on the performance of handling and dispatching fire alarms rather than the method of notification.

Rate Level Effect

A testing program was conducted to determine any potential rate level effect resulting from the introduction of the FSRS. The test results indicate that the introduction of the FSRS will have a negligible rate level impact.

A sample of 500 communities was selected so distribution by protection class and by population group within the protection classes corresponded with the average premium distribution across the U.S. Such a selection process was used so the test results would properly reflect any premium change caused by the introduction of the FSRS.

The communities actually tested were randomly selected from across the country. The Fire Suppression Rating Schedule was applied to recent ISO file data. To calculate the expected rate level impact, the redistribution of protection classes was correlated to the current nationwide average protection class rate differentials. The test results indicate a 95 percent probability that the rate level

TABLE 15-7D. Comparison of Schedules: Fire Department

1974 Schedule	Relative Weight (Percent)	1980 FSR Schedule	Relative Weight (Percent)
Pumpers	4.8	Engine Companies	11
Ladder Trucks	3.4	Ladder Service Companies	6
Distribution of Companies and Types of Apparatus	4	Distribution of Companies	4
Pumper Capacity	4.4	Pumper Capacity	5
Design, Maintenance, and Condition of Apparatus	3	Department Manning	15
Number of Officers	2	Training	9
Department Manning	8		50
Engine and Ladder Company Unit Manning	6.4		
Master and Special Stream Devices	1		
Equipment for Pumpers and Ladders	2		
Hose	2.8		
Condition of Hose	1.6		
Training	6		
Response to Alarms	2		
Fire Operations	8		
Special Protection	6		
Miscellaneous Factors and Conditions	6		
	71.4 (Max=39)		

TABLE 15-7E. Comparison of Schedules: Fire Alarm System

1974 Schedule	Relative Weight (Percent)	1980 FSR Schedule	Relative Weight (Percent)
Communication Center	0.8	Receipt of Fire Alarms	2
Communication Center, Equipment and Current Supply	2.8	Operators	3
Boxes	1.2	Alarm Dispatch Circuit Facilities	5
Alarm Circuits and Alarm Facilities, Including Current Supply at Fire Stations	2		10
Material, Construction, Conditioin and Protection of Circuit	1		
Radio	0.8		
Fire Department Telephone Service	1.7		
Fire Alarm Operators	0.8		
Conditions Adversely Affecting Use	1.3		
Credit for Boxes Installed in Residential Districts	−.4		
	12.4 (Max=9)		

effect will be minus 0.08 percent, plus or minus 1.11 percent.

Not considered in the test was the fact that ISO normally resurveys each community only once every ten years unless requested sooner by the community. Therefore, the actual rate level impact would be somewhat lower than the calculated impact.

Monitoring

The Fire Suppression Rating Schedule involves many significant changes in the approach to establishing Public Protection Classifications. As in any comprehensive new rating program, the possibility of unforseen rating situations cannot be overlooked. Consequently, following implementation, it is intended that the results will be monitored very carefully to insure that the program is working as intended with a minimal number of changes needed. Any required changes to the schedule that are identified through the monitoring program will be forwarded immediately.

Implementation

The revised schedule will be applied to communities during the normal resurvey program or when significant changes occur in a community that would require a resurvey. Section II of the Fire Suppression Rating Schedule pertaining to individual nonsprinklered properties will not be applied unless Section I has been applied to the community; (Section II will also not be applied in cities that are statistically rated).

When application of the schedule results in an improvement in the Public Protection Classification, the new classification will be implemented as soon as insurers and producers are notified through ISO's manual page distribution procedures. Should a community retrogress, the community officials will be notified of the change and the possible effect on insurance rates. If within 90 days the community decides to engage in a program that would improve the classification, the implementation of the poorer classification will be delayed indefinitely as long as an improvement program begins within one year and continues to completion.

Transition Rule

The usual impact of a change in public protection classification, for those properties which have such classifications as a rating variable, is approximately five percent for each change in class. Even though it is highly unlikely that there would be any significant rate impact for individual property by a change in the Public Protection Classification, the following transition rule is used for any unusual situations that may arise.

When used solely as a result of the application of the Fire Suppression Rating Schedule, and the applicable Public Protection Classification number increases by one or more, the following shall apply to any resulting individual property rate increases:

1. The rate increase shall be limited to 25 percent for the first publication.
2. If the rate increase is between 25 and 50 percent, the balance of the increase shall be included when the rate is next published for the risk, but not sooner than one year after the effective date in Item 1 above.
3. If the rate increase exceeds 50 percent, Items 1 and 2 above shall apply, except that an increase over 50 percent shall not be published sooner than two years from the effective date in Item 1 above.

PLANNING

Whenever a community—rural, suburban, or urban—considers its fire defenses, it must scrutinize the past and present and make predictions or forecasts for the future. Reviewing the past is called "data analysis" and depends upon good recordkeeping. "Evaluation," which is looking at the present, requires the ability to examine a situation objectively. The process of forecasting future conditions requires that a "planning" process be followed. This planning process results in a plan and its implementation, so that future challenges to the community are met. As the plan is implemented, the process must include the establishment of a "feedback loop," providing a continuing assessment of how well the plan is contributing to successful completion of goals and objectives, and feeding revised data back into the plan so continuing redesign occurs.

Fire protection organizations, and especially fire departments, need to develop several kinds of plans related to fire prevention and fire suppression. These plans are quite specific, directed at one clearly defined goal, and operational over a relatively brief time period (usually from one to five years). Typically, the plans are internal to the department and do not involve broad based planning groups from the outside. Examples of these types of plans, which are most often technical in nature, are apparatus replacement plans, training program plans, revised initial response plans, plans for a special hazardous materials attack unit, and plans for adapting fireground procedures to incorporate the use of larger diameter hose. However, once department planning begins to consider aspects of fire protection which will have an impact on external groups, those groups will need to be consulted and incorporated into the planning process. Fire department planning, for example, must dovetail with new, broad based, emergency management planning.

Examples of fire department planning which require the early involvement of other groups are station relocations or closings, building inspection programs, public education programs, and changes in the scheduling of work platoons. These plans, while involving some other groups, are still fairly narrow in scope and usually can be formulated over a relatively short time.

A third type of planning, called comprehensive or master planning, addresses the total community fire protection problem, incorporating both prevention and suppression, and obviously involves many community agencies and organizations, perhaps even county, state, and federal agencies. Master planning is a necessity for communities and is aimed at integrating all community efforts at prevention and suppression, and improving efficiency and cost effectiveness of those activities. Improved total community performance is the goal of this planning. Its degree of success must be measured by figures relating to the total cost of fire to the community, and not just in gains for one subsystem. Master plans often consist of a number of subplans from various agencies which are developed at the same time as part of a larger, total process, and which fit together to make a comprehensive and integrated plan.

Comprehensive plans have clearly stated goals with agreed upon ways of measuring their attainment. These overall goals are reached through overall strategies acceptable to all involved agencies and to the citizens who must pay for public fire protection. Each goal is composed of some number of subgoals or objectives and for each objective there is a tactic designed to reach that objective. All objectives make sense in terms of the overall strategies. When the objectives and tactics are laid out on a time line, the overall time required to implement the comprehensive plan is then known, and the timing for attaining each objective is apparent.

Fire protection has been largely a local responsibility, and for good reasons it seems destined to remain so.* Each community has a set of conditions unique to itself; it cannot be assumed a system of fire protection that works well for one community will work equally well for other communities. To be adequate, the fire protection system must respond to local conditions, especially to changing conditions. Planning is the key: without local level planning, the fire protection system is apt to be ill suited to local needs and not adapt to the changing needs of the community.

Excellent fire protection (for example, in the form of automatic extinguishing systems such as residential sprinklers) is technically available and certainly can be provided with the resources of most communities. Even with considerable public support, however, this protection may require several years to attain. In the meantime, in every fire jurisdiction (whether a municipality, county, or region) standards aiming at significantly increasing fire protection must be set. The sections below discuss some of the concepts to be defined in setting these standards.

Adequate Level of Fire Protection: The question of "adequacy" is addressed not only in day to day needs, but in major contingencies that can be anticipated and for future needs as well. A definition of "optimal" protection is needed—in contrast to "minimal" protection which fails to meet contingencies and future needs—and "maximal" protection which is usually more expensive than a community can afford.

Comprehensive planning must include contingencies drawn from an analysis of community hazards. This process of hazard identification and analysis, a first step in emergency management, is crucial to fire department planning since fire departments are called to respond to almost all types of emergencies and disasters.

Reasonable Community Costs: Fire, both as threat and reality, has its costs, including deaths, injuries, property losses, hospital bills and lost tax revenues, plus the costs of maintaining fire departments, paying fire insurance

*Some of the following information has been extracted in whole and in part from *America Burning* (NCFPC 1973).

premiums, and providing built in fire protection. Each community must decide upon an appropriate level of investment in fire protection. Some costs which are beyond the public's willingness to bear should be transferred to the private sector (as when buildings over a certain size or height or with a certain occupancy are required to have automatic extinguishing systems). Service and use fees should also be considered (Coleman 1980).

Acceptable Risk: A certain level of fire loss must be accepted as tolerable simply because of limited resources of a community. Conditions that endanger the safety of citizens and fire fighters beyond the acceptable risk must be identified as targets for reduction.

Consideration of these matters helps to determine what functions and emphasis should be assigned to the fire department, other municipal departments, and the private sector both now and in the future. It helps to define new policies, laws, or regulations that may be needed. Most importantly, consideration of these matters makes it clear that firesafety is a responsibility shared by the public and private sectors. Because the fire department cannot prevent all fire losses, formal obligations to have built in fire protection fall on owners of certain kinds of buildings. For the same reason, private citizens have an obligation to exercise prudence with regard to fire in their daily lives. But prudence also requires education in firesafety, and the obligation to provide that education appropriately falls in the public sector, chiefly the fire department. The public sector (again, chiefly to the fire department) also has an obligation to see that requirements for built in protection in the private sector are being met.

A fire department, then, has more than one responsibility—nor are the aforementioned responsibilities exhaustive. At least nine important functions for fire departments can be identified:

Fire Suppression: Fire fighters need proper training and adequate equipment to save lives, extinguish fires quickly, and also to ensure their own safety.

Specialized Emergency and Disaster Services: These include hazardous materials incidents, floods, earthquakes, multiple vehicle accidents, cave ins, collapsed buildings, volcanic eruptions, searches for lost persons, attempted suicides, etc.

Life Safety Paramedic Services: Capabilities needed during fires and other emergencies include first aid, resuscitation, and possibly paramedical services. ("Paramedic services" means emergency treatment beyond ordinary first aid, performed by fire service personnel under supervision—through radio communication, for example—of a physician.)

Fire Prevention: This includes approval of building plans and actual construction; inspection of buildings, their contents, and their fire protection equipment; public education; and investigation of causes of fires to guide future fire prevention priorities.

Firesafety Education: Fire departments have an obligation to bring firesafety education not only into schools and private homes, but also into occupancies such as restaurants, hotels, hospitals, and nursing homes, with a greater than average fire potential or life safety hazard to people.

Deteriorated Building Hazards: In coordination with other municipal departments, fire departments can work to abate serious hazards to health and safety caused by deteriorated structures or abandoned buildings.

Regional Coordination: Major emergencies can exceed the capabilities of a single fire department, and neighboring fire jurisdictions should have detailed plans for coping with such emergencies. But effectiveness can also be improved through sharing of day to day operations—as, for example, an areawide communication and dispatch network.

Data Development: Knowledge of fire department performance and how practices should change to improve performance depends upon adequate recordkeeping. Microcomputers and minicomputers now play an important role in fire protection and emergency management planning.

Community Relations: Fire departments are representative of the local community that supports them. The impression they make on citizens affects how citizens view their government. Volunteer departments dependent upon private donations must, of course, also be concerned with community relations. Moreover, since fire stations are strategically located throughout the community, they can serve as referral or dispensing agencies for a wide range of municipal services.

As communities set out to improve their fire protection, they must not consider the fire department alone. The police have a role in reporting fires and in handling traffic and crowds during fires. The cooperation of the building department is needed to enforce the firesafety provisions of building codes. The work of the water department in maintaining the water system is vital to fire suppression. In firesafety education, the public schools, the department of recreation, and the public library can augment the work of the fire department. Future community development and planning will influence the location of new fire stations and how they will be equipped.

The nine functions above are just the obvious examples of interdependence. Although it may seem trivial, the manner in which house numbers are assigned and posted, for example, can affect the ability of a fire department to respond quickly and effectively to emergencies.

Master Planning

Fire protection is only one of many community services. Not only must it compete for dollars with other municipal needs such as the education system and the police department, but in planning for future growth, the fire protection system must account for the changes in progress elsewhere in the community. For example, if a slum area is to be torn down and replaced with high rise apartment buildings, the fire protection needs of that area will change. Changes in zoning maps will also change the fire protection needs in different parts of the community.

To cope with future growth, local administrators are turning increasingly to the concept of master planning of municipal functions. Such plans include an examination of existing programs, projection of future needs of the community, and a determination of methods to fill those needs. They seek the most cost effective allocations of resources to help assure that the needs will be met.

A major section of a community general plan of land use should be a master plan for fire protection, which should be written chiefly by fire department managers. This plan should, first of all, be consistent with and reinforce the goals of a city's overall general plan and its timeframe. For example, managers should plan the deployment of manpower and equipment according to the kind of growth and the specific areas of growth that the community foresees. Goals and priorities for the fire department should be set in the master plan. Not only is it important to set objectives in terms of lives and property to be saved, but also to decide allocations among fire prevention, inspection, firesafety education, and fire suppression as the best way to accomplish the objectives. A master plan must be an integrated part of the overall emergency management plan for the community.

Having established goals, the department officers should use the plan to establish "management by objectives (MBO)" within the fire department. Management is most effective when each person is aware of how tasks fit into the overall goals, and is committed to getting specific jobs done in a specified time.

Because fire departments exist in a real world where a variety of purposes must be served with a limited amount of money, it is important that every dollar be invested for maximum return on investment. The fire protection master plan should not only seek to provide the maximum cost benefit ratio for fire protection expenditures, but should also establish a framework for measuring the effectiveness of these expenditures. Lastly, the plan should clarify the fire protection responsibility for other groups, both governmental and private, in the community.

Devising a Fire Protection Plan

Key questions to be asked by those planning for fire protection are:

1. Why is planning necessary for us at this time?
2. What do we need to start the process, and are the necessary groups committed to the process?
3. What are the necessary steps in the planning process?
4. How will the plan be implemented?
5. Are all aspects of the plan legally possible and enforceable?
6. How will the plan be evaluated? Will it be a part of the Integrated Emergency Management Plan of the community?
7. How will feedback be gathered and the plan modified and updated?

In its *Introductory Summary: Fire Prevention and Control Master Planning*, the U.S. Fire Administration, (NFPCA no date), points out the following in providing its overview of the planning process:

> "Master planning is a participative process which should result in the establishment of a fire prevention and control system which is goal oriented, long term, comprehensive, provides known cost/loss performance, and adapts continually to the changing needs of your community."
>
> "Master planning should consider all community elements . . . related to fire prevention and control system elements."
>
> "Master planning involves the participation of all parties interested in the development of a defined cost/loss relationship . . ."
>
> "Master planning allows you . . . to systematically analyze fire prevention and control through common sense procedures . . . master planning has three phases: preplanning, planning, and implementation. The preplanning phase gets necessary commitments, committees, estimates and schedules, and go ahead approvals. The planning phase gathers and analyzes data, sets goals and objectives, determines anacceptable level of fire protection service, identifies alternatives, and constructs the plan. The implementation phase never ends, because the plan is on going and always being revised and updated."

The following can serve as guidelines to fire department administrators for developing and presenting a master fire protection plan as part of the comprehensive master plan outlined earlier:

Phase I

1. Identify the fire protection problems of the jurisdiction.
2. Identify the best combination of public resources and built in protection required to manage the fire problem, within acceptable limits:
   a. Specify current capabilities of and future needs for public resources;
   b. Specify current capabilities and future requirements for built in protection.
3. Develop alternative methods that will result in trade offs between benefits and risks.
4. Establish a system of goals, programs, and cost estimates to implement the plan:
   a. Develop department goals and programs, including maximum possible participation of fire department personnel of all ranks;
   b. Provide goals and objectives for all divisions, supportive of the overall goals of the department;
   c. Strive to develop management development programs that increase acceptance of authority and responsibility by all fire officers as they strive to accomplish established objectives and programs.

Phase II

1. Develop a definition of the roles of other government agencies in the fire protection process.
2. Present the proposed municipal fire protection system to the city administration for review.
3. Present the proposed system for adoption as the fire protection element of the jurisdiction's general plan. The standard process for development of a general plan provides the fire department administrator an opportunity to inform the community leaders of the fire protection goals and system, and to obtain their support.

Phase III

In considering the fire protection element of the general plan, the governing body of the jurisdiction will have to pay special attention to:

1. Short and long range goals.
2. Long range staffing and capital improvement plans.

3. Code revisions required to provide fire loss management.

Phase IV

The fire loss management system must be reviewed and updated as budget allocations, capital improvement plans, and code revisions occur. Continuing review of results should concentrate on these areas:

1. Did fires remain within estimated limits?
2. Should limits be changed?
3. Did losses prove to be acceptable?
4. Could resources be decreased, or should they be increased?

Bibliography

References Cited

Coleman, R. J. 1980. "Service and Use Fees Place the Burden on Users," *The International Fire Chief*. Vol 46, No 7. July 1980. pp 14-16.

ISO. 1974. *Grading Schedule for Municipal Fire Protection*. Insurance Services Office, NY.

NCFPC. 1973. *America Burning: The Report of the National Commission on Fire Prevention and Control*. U.S. Government Printing Office, Washington, DC.

NFPCA. no date. *Introductory Summary: Fire Prevention and Control Master Planning*. U.S. Department of Commerce. National Fire Safety and Research Office, Washington, DC.

NFPA Codes, Standards, Recommended Practices and Manuals. (See the latest *NFPA Codes and Standards Catalog* for availability of current editions of the following documents.)

NFPA 1201, *Recommendations for the Organization for Fire Services*.

NFPA 1202, *Recommendations for Organization of a Fire Department*.

NFPA 1221, *Standard for the Installation, Maintenance and Use of Public Fire Service Communications*.

NFPA 1401, *Recommended Practice for Fire Protection Training Reports and Records*.

Additional Readings

Backoff, Robert W., *Measuring Firefighter Effectiveness: A Preliminary Report*, School of Public Administration, Ohio State University, Columbus, OH.

Boswell, C. R., *Standards for Rural Fire Protection*, Department of Extension Forrestry, Kansas State University, Manhattan, KA, 1978.

Chrissis, James W., "Locating Emergency Service Facilities in a Developing Area," *Fire Technology*, Vol. 16, No. 1, Feb. 1980, pp. 63-69.

Didactic Systems, Inc., "Fire Prevention Activities," *Management in the Fire Service*, National Fire Protection Association, Quincy, MA, 1977, pp. 168-199.

Federal Emergency Management Agency, *Hazard Identification, Capability Assessments, and Multi-Year Development Plan*, CPG1-34 (Overview); CPG 1-35 (Workbook); CPG1-35a (Response Book); Washington, DC, Jan. 1985.

Federal Emergency Management Agency, *Planning Guide and Checklist for Hazardous Materials Contingency Plans*, FEMH-10, Washington, DC, July 1981.

Federal Emergency Management Agency, *The Integrated Emergency Management System: Process Overview*, U.S. Government Printing Office, 1984-445 004/18624, Washington, DC, 1984.

Fire Suppression Rating Schedule, Insurance Services Organization, NY, 1980.

"First Responder": An Issue to Be Addressed in Master Planning, Federal Emergency Management Agency, U.S. Fire Administration, Emmitsberg, MD, Oct. 1980.

Granito, J. A., "The Chief's Job, circa 2006," *Fire Command*, Aug. 1981.

Granito, J. A., "Holding the Horses," *Fire Service Today*, Sept. 1983.

Granito, J. A., "Trends in Fire Service Management," *Fire Service Today*, Aug. 1983.

Hickey, Harry E., "A Comparative Analysis of Resource Allocation Plans for Urban Fire Safety," *FPP-B 77-1*, Johns Hopkins University, Laurel, MD, Aug. 1977.

Hickey, Harry E., *Public Fire Safety Organization*, National Fire Protection Association, Quincy, MA, 1973.

Houlihan, John C., *Alternatives to Traditional Public Safety Delivery Systems: A Tale of Two Cities*, Institute for Local Self-Government, Berkeley, CA, 1977.

Huse, Edgar F., *The Modern Manager*, West Publishing Company, St. Paul, MN, 1979.

International City Management Association, *Managing Fire Services*, ICMA, Washington, DC, 1986.

International City Management Association, MIS Report 16/I, *Hazardous Materials Incidents: Improving Community Response*, (J. Granito), Washington, DC. 1984.

International City Management Association, Baseline Data Report 16/4, *Emergency Management*, (G. Hoetmer), Washington, DC, Apr. 1983.

International City Management Association, Baseline Data Report 15/2, *Fire Personnel Practices*, (J. Granito), Washington, DC, Feb. 1983.

International City Management Association, *Assessment of the Transferability of the Community Fire Master Planning Program*, ICMA, Washington, DC, 1977.

Master Planning for Fire Protection, Federal Emergency Management Agency, U.S. Fire Administration, Washington, DC, Mar. 1980.

"National Association of Counties, Multi-Jurisdictional Fire Protection Planning," prepared for the Federal Emergency Management Agency, U.S. Fire Administration, Washington, DC.

National Fire Prevention and Control Administration (now U. S. Fire Administration), *A Basic Guide for Fire Prevention and Central Master Planning*, U.S. Dept. of Commerce, Washington, DC, 1978.

Research Triangle Institute, et al., "Evaluating the Organization of Service Delivery," Fire, Research Triangle Institute, Center for Population and Urban-Rural Studies, Durham, NC, (no date).

Research Triangle Institute, et al., *Municipal Fire Service Workbook*, Government Printing Office, Washington, DC, (no date).

Schaenman, Philip S., and Seits, Edward F., *International Concepts in Fire Protection—Practices from Japan, Hong Kong, Australia and New Zealand*, TriData, Arlington, VA, 1985.

Urban Consortium on Fire Safety and Disaster Preparedness Task Force, *Criteria for Master Planning and Resoures Allocation*, Public Technology, Inc., Washington, DC, 1978.

Webber, Ross A., *Management: Basic Elements of Managing Organizations*, Richard D. Irwin, Inc., Homewood, IL, 1979.

Worsham, J. P., *Rural Fire Protection: A Selected Bibliography*, Vance Bibliographies, Monticello, IL, 1978.

ANALYSIS OF FIRE DEPARTMENT DATA AND DECISION MAKING

By Carl Peterson

Smaller fire departments face many of the same decisions and need much of the same data as large fire departments. With the availability of low cost computer power, many of the applications discussed in this chapter are achievable in the smaller department by using microcomputers. On the other hand, some of the applications may be adequately handled by a good manual record keeping system with periodic analysis of the data to support the information needs of the department.*

Today, effective management is more critical than ever to the fire service. Economic conditions have reduced local government tax revenue, leading to diminished resources, yet the fire service, like all government units, is being asked to maintain acceptable service levels.

The contradiction of "doing more with less" can be resolved only with improved management practices that result in higher productivity. There are many paths to this goal; one such path is information technology.

While one might argue that the computer is not a panacea, information technology has become a pervasive management tool in many fire departments. Specifically, the computer is being used to help:

1. Strategically position capital resources such as fire stations.
2. Dispatch and manage fire fighting resources.
3. Manage personnel resources.
4. Control fire department performance.
5. Plan and control budgets.
6. Manage fleet maintenance.
7. Combat arson.
8. Improve organizational communications.
9. Manage energy consumption.
10. Manage fire prevention, inspections, and permits.
11. Manage emergency medical operations.
12. Manage financial resources.

Mr. Peterson is manager, Fire Service Management Systems, NFPA.

*This chapter was condensed from *Fire Department Information Technology in Nine Western Communities*, prepared for the Federal Emergency Management Agency by the Consortium of Western Fire Agencies on Integrated Information Systems.

Probably no fire department has automated all of these applications. However, almost all large fire departments use a computer in one or more of these areas.

MANAGEMENT ISSUES

The successful function of a fire department's information system depends upon how that function is organized and managed. There is, of course, no "single best way" to manage the information resource, since every situation is unique.

There is growing recognition that the information resources of an organization are crucial to that organization's overall effectiveness. Consequently, the information resource is a valuable asset that must be well managed. While there is little controversy over the truth of this statement, there is some confusion over what constitutes the information resource. Mostly, the argument is a matter of emphasis of various professional specialists whether the data administrator, the communications expert, or the records manager.

The broad definition of the information resource would include everything connected with information: data bases, reports, libraries, word processing, computers, etc. The more restrictive view used in this chapter defines the information resource as computers, software, data, and personnel.

A balanced information resource fulfills three needs: (1) strategic planning, (2) management control, and (3) support of daily operations. Each of these is being met by computerized information systems in some fire departments, but it is rare that a department addresses all three needs.

Strategic planning is the process by which management allocates long range resources such as fire stations, fire fighting equipment, and fire fighters. Some fire departments have developed sophisticated models for this purpose while others have contracted for the models and services of outside resource groups. Most fire departments, however, depend upon the statistical information from fire incident and emergency medical reporting systems.

Management control is necessary in several areas: (1) finance, (2) personnel, (3) fire prevention, (4) arson prevention, (5) fire suppression, and (6) energy management. Control information is often provided as a byproduct of systems designed to support day to day operations. For example, personnel control information may be secondary to the real reason the personnel system was developed—to increase the efficiency of the personnel and the record keeping system.

Most fire service information systems are designed to support day to day operations. The most common example of this type of system is computer aided dispatch (CAD). Other systems support financial management, arson investigations, equipment inventory and maintenance, inspections, permits, organizational communications, and personnel management.

While systems may be classified as suggested above, they are rarely developed explicitly in that fashion. Rather, the packaging of information is a function of priorities, systems planning, organizational needs, and other situational factors. There are, however, certain recurring patterns to the way information is packaged into application systems.

COMPUTER AIDED DISPATCH

Quite literally, the dispatching operation is the nerve center of the fire department. It is here that resource status is maintained, alarms are received, fire stations are alerted, and situations are managed. Because of its importance and high visibility, the dispatching function is usually one of the first fire department operations to be computerized.

Fundamentally, computer aided dispatch (CAD) is oriented toward operations. Its purpose is to support current activity rather than provide managerial control or strategic planning information. However, the functions of CAD systems often go beyond operational support. In an integrated information systems environment, the CAD system may produce nonoperational information (e.g., incident reporting data) and may use nonreal time inputs (e.g., occupancy data).

CAD systems support several functions of the typical dispatching operation:

1. *Alarm Receipt*: Telecommunications management, box decoding, central station communications, etc.
2. *Communications and Signaling Management*: Alerting (unit/station, two way exchange of data, audit switching, frequency selection, etc.).
3. *Status Maintenance*: Resources, equipment, personnel, system supervisor, etc.
4. *Response Assignment*: Dynamic assignment based on current availability and relative location of fire fighting resources to the incident.

Although CAD is usually credited with reducing response time, the payoff is not necessarily in this area. Some departments even report a *drop* in response time after implementing a CAD system. The real benefits of CAD occur in the management of fire suppression activity. Thus, rapid recall of information such as apparatus status and occupancy data is at least as important as response time.

The need for a CAD system must of course be balanced against the cost of entry, which can be considerable. San Francisco, CA, for example, paid more than $2 million while the Phoenix, AZ, CAD cost approximately $5 million.

Choosing a correct approach to implementing a CAD system is a multifaceted problem. The first facet of the problem is related to policy. In some cases dispatching is a shared police-fire responsibility, in others, dispatch serves multiple jurisdictions. The organizational situation can have an important influence on other aspects of the approach as well as the ultimate usefulness of the system.

A major starting point in CAD development is a decision on the scope of the system. Is the system to be developed only for CAD or will it include other applications? Most departments elect to develop a standalone CAD system with the necessary interfaces to other applications (most commonly the sole interface is with the fire incident reporting system).

Closely related to the issue of scope is the degree of modularity built into the system. A highly modular system permits the system to be expanded to include enhanced CAD functions as well as other applications. It also allows new software to be tested in a nonreal time mode and operators to be trained without interfering with operations. Modularity is important in all applications, but it is especially critical in CAD systems because the systems are subject to almost constant change.

Another facet of the approach is the decision on outside versus in-house development. Most fire departments choose to have an outside contractor develop, test and implement the system, but the city of Dallas, TX, successfully developed its system in-house.

Communications is the final aspect of approach. Standard telephone service is typically used for communications between the dispatch center and the fire stations. Increasingly, other ways are being found to transmit data. The 800 mHz radio frequencies have been used successfully, as have microwave links.

PERSONNEL APPLICATIONS

Personnel systems are among the most complex of all fire department applications. Consisting of a number of subapplications, these systems are important in large departments to support the myriad functions of personnel administration: basic personnel records management, time and attendance reporting, payroll (sometimes part of financial management), duty, schedules, physical fitness and medical records, employee evaluation, training records, position control, badge history and control, and disability/injury analysis.

In the absence of computer support for at least some of the personnel administration functions, large fire departments would find it difficult, if not impossible, to maintain all the necessary paperwork without a large clerical staff. Nevertheless, the complexity of personnel systems has discouraged widespread implementation, although most fire agencies have installed some of the subapplications. Also, personnel systems often are citywide in scope thus precluding (or at least discouraging) the development of specific fire applications.

The need for personnel information systems exists primarily at the operational level. Although output from these systems is used for both strategic planning and managerial control, the greatest near term benefit is in making personnel administration more efficient.

The operational benefits are obvious. In a 1,000 person fire department, a large volume of information is needed to carry out standard personnel functions. Fire fighters must be scheduled, and personnel balanced daily. Detailed training records are required for each department member. Time and attendance records are necessary for payroll and record keeping. Employees must be evaluated and considered for promotion. These functions require an efficient way of storing and retrieving information.

The information that is useful for day to day operations also has residual value for strategic planning and managerial control. For example, the roster data may be aggregated in such a way that is useful for position control or personnel resource planning. Time and attendance data may be reviewed by management in an effort to control excessive sick leave or overtime. Much of the information needed for strategic planning and managerial control cannot be defined prior to when it is needed. Therefore, it is useful to have the facility for answering ad hoc inquiries using the operational data base.

At the highest level of design, the approach taken will depend upon the relationship of fire department personnel functions to city (or county) personnel functions. For example, the payroll may be prepared by the city from prescribed time and attendance records. In cases like this, it may be difficult for the fire department to develop some of the components of a personnel system.

A second aspect of the design approach is the degree to which the components of the personnel system are integrated. In most cases, the personnel system has been built up over several years, with one component implemented at a time. There has been no overall design that provides for integration of files and input. This piecemeal approach is gradually being replaced by comprehensively planned systems, and although the components may still be implemented separately, the plan provides for eventual integration.

The final aspect of the design approach relates to the extent to which the data base is on-line. Batch systems have been heavily favored but these are losing favor to on-line updating of data bases as well as on-line retrieval. The on-line approach, although more expensive, permits direct involvement of the users and thus increases the benefits of the systems.

FLEET MANAGEMENT

After the personnel costs, capital and operational expenditures for vehicles and equipment represent the greatest use of fire department funds. Fleet management systems provide the information needed for budgeting and controlling those expenditures as well as for ensuring that the fleet is maintained in a high state of readiness. The need for information about the fire department fleet exists at all levels: strategic, managerial, and operational.

To accomplish these purposes, fleet management systems may involve several components: fleet inventory, maintenance records, vehicle replacement, fuel consumption, internal billing for vehicle use (e.g., in the case of citywide carpools), tax reporting, and performance analysis. Some or all of these components may be present in any given system.

At the strategic level, information about the existing fleet is needed to plan and budget for vehicle acquisition. To make sound replacement decisions, it is important to have accurate information on maintenance frequency, operating costs, age, etc., for each major piece of equipment. Indeed, those decisions are often so complicated and have such political overtones that decision support software (operations research models) is needed to help analyze the data.

At the managerial control level, fleet managers need computer support to control preventive maintenance intervals, work in progress, gasoline mileage, and staffing of the maintenance facility. Other managers in the department also need control information. For example, administrative services managers need financial data for budgetary control and line managers need to control the availability of fire fighting apparatus.

Personnel at the operational level also need output from a fleet management system. Most important, of course, is the ability to have instant access to vehicle availability information. While this requirement is not always provided by the fleet management system per se, dispatchers need to know what equipment is ready to respond to a given type of emergency. Maintenance personnel also need ready access to certain information, e.g., vehicle performance, previous repair records, interchangeable parts data, and standard repair times for all vehicles.

Few fire departments develop their own fleet management systems. The preferred approaches involve using a commercially available system (either a software package or a time sharing service) and a citywide fleet management system. Both approaches are viable alternatives to custom designed software, but both have some problems.

The time sharing service is a stand alone service dedicated to fleet management. It is difficult, if not impossible, to interface such systems directly with other department or city systems. Also, added functions such as vehicle replacement models, cannot be easily integrated with this type service. Yet, this approach can be implemented easily and is probably cost effective, especially for fire departments that do not have highly integrated systems.

The citywide approach also suffers from some of the same deficiencies. Since citywide systems are typically designed for standard street vehicles, they may ignore certain features that would be useful to fire departments. For example, the city system may not have a provision for capturing and maintaining elapsed time meters on pumps, or may not provide for helicopter maintenance records. If these deficiencies are to be eliminated, the fire department must take an active part in defining information requirements. When this is done, the citywide approach can be quite worthwhile.

FIRE INCIDENT REPORTING SYSTEMS

Incident reporting systems are used to maintain statistics on fire, emergency medical, and other emergencies to which a fire department responds. This application is probably the most common use of information technology.

The methods of capturing data are worthy of mention. The most common is batch data entry, where a form or forms are completed after each incident and the data is keyed in batches at a terminal or other data entry device by someone other than the person completing the form. Departments with a CAD System usually collect at least some of the Fire Incident Reporting System data as a byproduct of a CAD operation. This data can then be merged with incident report data.

Departments are now moving in the direction of on line data entry from terminals in the fire station. The process sometimes consists of entering a complete fire report from previously completed forms or field notes. In other cases, the data already captured by a CAD system is displayed on the terminal and this data is supplemented by data from the officer in charge of the fire company or the incident. Allowing the officer in charge to directly enter the data has the advantage of ensuring completeness and correctness because the data can be edited as it is entered and mistakes corrected immediately.

FIRE PREVENTION APPLICATIONS

Fire prevention management information systems are inventory and tracking systems for storing fire protection related information on a city's buildings and information on the fire department's contacts with those buildings. The application is useful in planning and managing fire prevention activities, controlling hazardous contents, and prefire planning for fire suppression. The heart of a fire prevention system is a flexible, address based file of city properties.

Fire prevention records in most cities are in very primitive condition. Long often handwritten narratives, are the rule rather than the exception. This makes it virtually impossible to develop any information on patterns in hazards of violations across the city or for particular property classes. Even within a file on a single property, it can be difficult to ascertain whether the fire department has been providing the kind of frequency of contacts called for by the department's policies. This difficulty exists because such files are often only dumping grounds for any and all papers bearing the property's address. Minor, unofficial correspondence and newsclips may share space with basic inspection reports and detailed, lengthy construction plans. Files on properties not covered by the fire prevention code (such as single family dwellings), on fire education contacts, and on contacts by other agencies (such as the building and housing departments), are generally nonexistent or wholly separate and incompatible with the fire inspection files. Fire incident records are rarely tied to files on the violation and hazard histories that may have caused the fires. In fact, it would be difficult to identify any aspect of a fire department's record keeping that is more in need of refinement and automation than fire prevention.

The first requirement for an adequate fire prevention system is a comprehensive address based inventory of city properties. This is a very difficult file to create. A 1979 study (Hall et al 1979) of fire inspection practices in eleven cities, including two with computerized inspection records, found that every city reported fires in properties that should have been subject to fire inspections but were not shown in the inspection records. These were not new businesses or businesses operating illegally out of homes—they were legitimate businesses of some years standing that simply had been overlooked.

It may be impossible to guarantee file completeness, but this can be improved. Address files should be linked to the fire incident records, preferably through a link to the city's CAD so that emergency runs to corresponding unlisted addresses will lead to the creation of an inspection file.

Sources for both the initial building inventory and periodic updates are the city's tax records, the housing department, the building department, any inspection agency, and city owned utilities such as the water and light departments. On the other hand, some lists (e.g., the telephone company 'Yellow Pages," cross indexes, and mail carrier routes) tend to be unwieldy or poor sources for finding missing addresses.

The fire inspection procedure itself can be a source of effective property list updates. Inspections conducted on a 'block by block, always go around the block" basis allows inspectors to see properties they might miss if they were to organize their routes with strict attention to those properties identified only by the computer.

Once the file is created, it can be used as a basis for organizing and scheduling inspections. The system can be set to print inspection reports for properties scheduled for inspection in any given week (thereby helping inspectors plan their work), and for reinspections to check hazard abatement. Inspector workloads can be balanced either by the number of inspections or by the total time spent on inspections if statistics are kept on the typical duration of inspections by type and size of property.

The NFPA Fire Inspection Reporting System (FIRS)* works in this manner. A basic property address file is built with each property assigned an inspection date and inspection interval. An actual fire inspection form showing the data about that property that is contained on the computer file is then computer printed when a property is due for its periodic inspection. A fire inspection report is also generated when there are outstanding violations and the inspector needs to return to the property to check for compliance. A series of reports allows the fire prevention manager to track outstanding inspections, to monitor the workload of inspectors, and to evaluate the types of violations by property use.

Fire prevention systems can also be used in the control of hazardous materials. In addition to supporting license and permit controls on hazardous materials and processes, a fire prevention file can be expanded to include information on the type, quantity, and location of hazardous materials and the type, frequency, and location of hazardous activities. The prime difficulty in making such a system work is that the situation is so changeable. Types, quantities, and locations of materials may fluctuate widely, according to the demands of the business. Annual or even semiannual updating of a central file may be insufficient to give a picture of the city's true hazard profile.

Ownership information on properties can be among the most difficult to obtain. Current ownership probably is not needed until legal proceedings have to be instituted. It is easier and more reliable, to check legal ownership in those relatively few cases as they occur than to try to develop and maintain a file on legal ownership for all properties.

EMERGENCY MEDICAL SERVICE

Emergency medical service (EMS) is now provided by many fire departments. EMS activities include rescue operations, first aid, and transportation to a medical facility. Ambulances are sometimes based at fire stations and dispatched by the fire department. Fire equipment and fire fighters may be dispatched for EMS activity if necessary.

*More information on FIRS is available from Service Management Systems, NFPA.

In many jurisdictions, the cost of emergency medical service is billed to the patient. Since this represents a substantial revenue source for the city or county, the billing process is the focus of most EMS systems. Two other areas are often also supported—EMS statistics and computerized radio frequency control.

EMS systems requirements are based on the volume of EMS activity. With several thousand incidents per month, computerized processing of bills is necessary to prevent long billing delays and resulting cash flow problems for a city or town.

The statistics generated from EMS systems are used in both long range plans and operational planning and control. Response times, locations of incidents, and units responding are required to plan and budget for resources and determine if the resources are being deployed properly.

Billing and statistics subsystems are typically batch oriented. The ambulance attendants complete a form which is keyed and processed to create the basic incident record. Payment of the bill or other additional data may cause the basic record to be updated either in batches or on-line mode.

In some fire departments, EMS statistics are generated directly from the billing system and the EMS incident record is part of that system. In other departments, data from the billing system is transferred to the fire incident reporting system from which statistics are generated. Having all incident data on the same data base is helpful in studying fire department emergency activity.

Many large jurisdictions are having increasing difficulties in managing the allocation of radio frequencies and eliminating radio interference. Computer control of channel assignments helps ensure that a paramedic can communicate with hospital personnel in life threatening situations.

The Coordinated Paramedic Communications System in Los Angeles County (CA) was designed to provide improved radio communications between paramedics in the field and paramedic base hospitals. The system serves both public and private providers of paramedic services.

The system design is based on dynamic channel assignments, i.e., biomedical communications channels (MED 1-8) are assigned as needed. With the aid of a computer, a radio telephone operator at the Communications Coordination Center assigns channels to paramedic squads and notifies corresponding paramedic base hospitals of ensuing "radio runs."

The system has:

1. Significantly reduced radio frequency (RF) interference, which often causes communications problems. Disruptions of communications had hampered the operations of paramedics and hospital personnel during the treatment of patients in the field, sometimes in life threatening situations.
2. Increased RF coverage and extended it into areas not previously covered.
3. Decreased the average elapsed time between a paramedic squad's decision to talk to a hospital by radio and the corresponding initial response by hospital personnel.
4. Increased the radio communications capability of base hospitals.

MATHEMATICAL MODELS

Most management information systems are quite simple conceptually and do not involve any mathematical modeling, not even simple arithmetic operations such as addition and subtraction. Systems in personnel, fire prevention, etc., are fundamentally automated list making and list matching programs. As such, they are primarily useful in helping fire service managers track large volumes of information, but not in assessing trade offs or aiding in resource allocation decisions.

In this context, a mathematical model based Management Information System refers to systems that use mathematical formulas in providing management information suitable for comparison of alternative courses of action. The most widely used systems are those that address issues of fire station location and fire company deployment at those stations. The models used in these systems generate statistics which indicate the level of service that would be provided to each part of the city if a certain number of fire stations and companies were in force and deployed at certain locations. Other mathematical models address other deployment related issues such as variance of responses by time of day or when and where to move up fire companies for temporary coverage of areas whose assigned companies are already fighting other fires (relocations/move ups).

Deployment of fire stations, companies, and personnel has traditionally been a matter of intuition and judgment, with few rules to follow except a generalized objective of placing stations on or near main thoroughfares and in areas that are far from existing stations. As fire department budgets have become more constrained, however, fire chiefs have needed more precise tools to address deployment. Fire station configurations that look similar on a map may produce substantially different patterns of fire suppression service. Adding or taking away an additional company or station, or providing units to respond may make a slight difference in one city and a tremendous difference in another. A fire chief who can produce detailed statistical descriptions of the implications of different decisions is much more likely to be able to obtain the decision that is desired.

The need for modeling help on deployment decisions is particularly great for departments serving rapidly growing or rapidly shrinking cities, including cities that are rapidly growing in some areas and rapidly shrinking in others. In a city where the geographical distribution of fire demand is changing dramatically, it is likely that major changes may be needed in numbers and locations of fire stations and companies or staffing by company.

Such changes involve millions of dollars in construction costs and more millions each year to staff and equip new companies. Even if a city is shrinking, millions of dollars in potential cost reductions are involved in decision on how much to shrink, and there are potentially costly implications in the decisions on where to cut back. With millions at stake, the thousands of dollars required to set up and conduct a systematic deployment analysis can be seen as a sound investment, especially when one considers the millions of dollars in fire loss that may be implied in the changing response patterns produced by a deployment change. Once the analytic system is set up, it can be used repeatedly at relatively little additional cost whenever new deployment decisions or questions arise.

There are dozens of published models for use in fire service deployment analysis, but the two widely used approaches are those developed by Public Technology, Incorporated (PTI)* and the Rand Corporation†. Both approaches generate statistics on travel time from proposed fire station locations to the parts of the city assigned to the stations. The statistics include: (1) average travel time by region of the city, reflecting differences in fire frequency from one block to the next, (2) maximum travel times, as one indication of whether or not all parts of the city will receive adequate services, and (3) percentage of the city reachable in less than some benchmark travel time (for example, two minutes), where the benchmark may be lower for high priority, high hazard properties.

The differences between the two approaches are less in the kind of information they produce than in the ways that information is produced. The Rand approach, for example, takes advantage of the fact that most cities have streets laid out in a grid network and so uses a mathematical formula, with parameters developed for the particular city, to estimate travel times based on the grid locations of fire stations and fire involved property. A city that already has geocoding will find this approach especially easy to implement if the geocoding includes the estabishment of X-Y coordinates. PTI, on the other hand, calculates travel times over a large number of individual route segments through actual apparatus runs, then calculates station to station travel time on a sequence of route segments that form a full route from the station to the fire. This approach requires more time and data collection expense, but can improve accuracy in cases where a city is split by a river, railroad tracks, or other major obstruction that can be crossed in only a few places.

Another difference lies in the organization and documentation of the two systems. PTI is organized to provide technical assistance, and the company has procedures and personnel able to handle all aspects of the analysis process from setting up the study team to presenting and selling the results to all concerned parties. Accordingly, PTI's documentation has a step by step, simplified format, designed to be used by an end user assisted by a PTI field person. Rand, on the other hand, is primarily a research organization that supplies reports, programs, and specifications off the shelf. Rand's documentation provides far more detail on case studies and on the reasoning behind their models, including information on why the models work as they do and how the organizational process of doing a study has operated in a few locations. But as an organization, Rand is not designed to help install systems and shepherd studies in all the cities that want to pursue deployment analysis.

Other differences between the two approaches are really differences in emphasis. Both approaches, for example, permit special attention to target hazards, with travel time profiles and requirements for only those properties. PTI puts much more emphasis on this aspect of the study, however.

The Seattle, WA Fire Department has developed an approach to deployment analysis that includes the additional variables of staffing per unit and total response staffing. Seattle undertook a comprehensive analysis of the relationship between fire loss/spread and response time. Based on the parameters of response time experienced, minimal correlation was found. These results agreed very closely with a study done in Great Britain. More recently, the Portland, OR Fire Bureau conducted a similar analysis, and again, virtually no correlation was observed. It should be noted that in the case of both Portland and Seattle, mean response times fell between three and four minutes, with a maximum of ten minutes and relatively few times exceeding seven minutes. It may be that studies conducted of areas with significantly larger mean and maximum response times will reveal some correlation between response time and loss/spread. However, the majority of municipalities in the United States report mean response time of less than six minutes. Within these parameters, virtually no correlation has yet been observed.

These analyses encouraged Seattle to look more closely at the other major variables of deployment, staffing per unit, and total response staffing. Many departments have constant staffing per unit, but overall response staffing is normally a matter of policy and, therefore, a definable variable. As budgetary pressure increases, per unit staffing necessarily will be considered. The general objective of cost effectiveness necessitates the analysis of staffing deployment at all levels of the fire department.

Seattle developed a deployment model that considers the three variables of response time, staffing per unit, and total response staffing. Seattle's model is intended to provide a comparative index of the level of service (LOS) provided to every intersection in the city. The output of the model can be used to determine optimum station locations as well as staffing deployment. However, the primary purpose of the model is to provide comparative indices of level of service provided at all intersections for any alternative deployment.

Seattle is currently developing a level of need (LON) model which produces an index of *need* for each intersection stated in the same units of LOS.

A LOS model for emergency medical services has also been developed. Analyses of EMS incidents indicated that, unlike fire, response time is overwhelmingly the most critical variable in EMS LOS. As a result, the EMS and fire LOS models are complementary in utility.

Seattle has also developed a response assignment system that determines the units for all first and second alarm response assignments by type and number for every intersection. A second element of this system generates relocations (move ups) for all levels of alarms 2 through 5 and all response assignments for alarms 3 through 5. This system is called the Vehicle Allocation and Update System (VAUS) and is designed to be used in either an on-line, real time environment (CAD) or off line to produce and maintain assignment files (i.e., run cards).

ARSON INFORMATION SYSTEMS

The principal advances in the area of arson information systems have been carried out through cooperative neighborhood or community block organizations working with the fire service in their communities. A number of microcomputer data bases have been developed to profile the properties in the neighborhood and assist in identifying arson prone buildings. The community group or public

*For more information on Public Technology Systems and Support contact Public Technology, Inc., Decision Support System Group, 1140 Connecticut Ave., NW, Washington, DC 20036.

† For more information on Rand Corporation Systems contact Rand Corporation, 1700 Main Street, Santa Monica, CA 90406.

officials were then able to implement intervention strategies appropriate for the situation.

Much of the early work sponsored by the Federal Emergency Management Agency has led to a developed and tested microcomputer based Arson Information Management System (AIMS) which is now being supported and distributed to the International Association of Arson Investigators.* The system provides both identification of arson prone buildings and a case management system for fire investigators.

NFPA is testing a neighborhood AIMS which allows for the development of a data base with information about neighborhood properties. This data base can then be culled for lists of properties meeting a user defined profile of those traits that exist in arson fires in that neighborhood. Intervention strategies can then be initiated to prevent future arson fires.

There are many municipal and private data bases or record systems that are typically used to support an arson information management system. These include data bases maintained by data from the fire department incident reporting system and code enforcement program, the building department, the assessor's office, the police department crime reporting system, the prosecutor's office, the tax collector's office, the registry of deeds, mortgage companies, and the insurance industry.

Data collection and verification from all these sources is expensive and time consuming. However, automatic integration of existing computer data bases has not proved successful, due to the number of data bases and the fact that they were never established with all the data elements needed as keys to ensure a proper matching of records. As such, most communities get listings of the data needed from each data base and manually transfer it to the AIMS data base.

OFFICE AUTOMATION

Office automation is a relatively new term used in connection with a group of applications that support clerical and managerial activities. Sometimes called the "office of the future," these applications include word processing, electronic filing, electronic mail, "tickler" systems, teleconferencing and video conferencing, and personal calendar management. Depending upon one's perspective, it may also include more traditional data processing functions such as data base access.

The purpose of these applications is to make the executive and the clerical worker more productive. While one might say the same about traditional information systems, the focus here is different. Traditional information system applications are process oriented; office automation, on the other hand, is more people oriented. The objective is to make office personnel more productive by allowing them to use computers as a natural extension of their minds and bodies. Hence, one often hears terms like "ergonomics" or "user friendly" when discussing office automation.

Executives and clerical workers need office automation to cope with the complexities of the modern office. Any fair sized fire agency must contend with an explosion of information in the form of reports, forms, memoranda, and computer files. The efficient production and effective management of this information resource strains even the most dedicated staff. The computer, through automation applications, extends human intellectual capabilities.

The most common office automation application is word processing. Most fire departments employ word processing to replace typewriters, and thereby obtain more productivity from clerical workers. Reports can be corrected and printed without extensive retyping, form letters can be personalized with the addressee names, and hard copy output can even be made to look like typeset copy.

Electronic mail extends the capabilities of word processing. Documents prepared on interconnected terminals can be transmitted immediately to any other terminal in the network of terminals. Since documents so prepared can be stored on a disc, electronic filing and retrieval is also possible. An index of key words reduces the possibility of "misfiling" and enhances the ability to retrieve *all* information relating to a particular point that needs researching.

Executives (or their secretaries) also perform two other tasks that are adaptable to office automation: tracking tasks assigned to subordinates, and arranging meetings. In the first instance, the executive usually maintains some kind of tickler file that shows the date work is due. By maintaining the tickler file on the computer, the executive can review a list of work due today, next week, or anytime in the future. Also, before assigning new work, he or she can identify a staff person who is underloaded or overloaded. In the second instance, executives may maintain their calendars on computer files. If the calendar files are made public, then it would be easy to arrange a time for a meeting that would be convenient to a group of executives.

The various approaches to office automation may be classified as to the degree of centralization involved. Historically, office automation has meant the decentralized use of stand alone word processing machines. Word processors are often brought in on a case by case basis to improve clerical productivity in selected instances.

Some organizations use the centralized approach of implementing office automation applications on the central computer. This approach is usually less than satisfactory. The central computer and the data communications network are not designed to cope with the unstructured formats and communications volumes associated with word processing and electronic mail.

To provide the needed integration while retaining the flexibility for stand alone units, some organizations are turning to so called "shared logic systems." These systems are full fledged minicomputers specifically designed (both in terms of hardware and software) to handle office automation, although the more advanced systems also support standard data processing chores. As user needs grow, more systems can be added and networked together. In addition, the office computer can be connected into the central computer network. This allows the user to have a single work station on a desk top that can perform all the office automation functions and still have access to the data bases retained on the central computer.

There are several variations on these themes. Depending upon how intelligence is distributed throughout the network, various functions can be performed at the work station, the minicomputer, or the central computer. For example, a local work station might handle word process-

*For more information on the FEMA/USFA Arson Investigation Management System contact International Association of Arson Investigators, P.O. Box 600, Marlboro, MA 01752.

ing as a stand alone task, use the minicomputer to transmit electronic mail, and make inquiries into the data bases at the central computer site. The network provides the mechanism to integrate all these activities.

External Data Bases

Much of the technical and legal information needed by fire departments exists on commercial data bases. These data bases are created for general use, and users are charged for access on the basis of a flat monthly rate and/or the time actually used. In a very real sense, commercial data bases are a public utility.

The source for the information stored in these data bases is primarily current technical literature and legal materials. In some cases, abstracts of documents and books are stored while in others, the complete text is stored.

Data bases are provided by numerous firms. Literally hundreds of data bases covering various fields are available today through scores of companies such as *Reader's Digest*, *The Source*, and *The New York Times*, and more are being offered daily.

The data bases of greatest utility to the fire service are probably those dealing with hazardous materials and legal information. Chemical and hazardous materials data bases are now available which allow a user to search for a hazardous material by chemical structure, trade name, generic name, CAS registry number, chemical or physical properties, and textual description. These data bases can be powerful tools in handling chemical spills, fires involving hazardous materials, and some emergency medical situations.

Legal information systems are generally of two types: full text storage and, search and abstracts (headnotes) only. Both types cover case law, statutory law, and some administrative rulings. Mead Data Central's LEXIS* is representative of the full text approach. LEXIS users enter key words, and documents containing the proper sequence, count, and/or combination of those words are displayed. A service of West Publishing Company (WESTLAW†) allows the same type of search but the object of the search is the West headnotes, rather than the full text.

With the rapid introduction of two way cable television systems and videotext services, there will likely be more data bases and quick access methods available to fire departments.

Financial Applications

Financial applications include those systems that support budgeting, budget control, general ledger, accounts payable, and billing. Payroll is sometimes included, although more often it is a part of the personnel system. Billing may also be included in specific systems, e.g., emergency medical for patient billing, and brush clearance to bill property owners for the removal of illegal brush. The implementation of financial systems in fire departments has been slow. Partly because the support of staff functions such as finance, has been secondary to the support of line activities. In addition, municipality wide financial systems often preempt those in the individual departments.

*For more information on LEXIS contact Mead Data Central, 200 Park Avenue, New York, NY 10166.

† For more information on WESTLAW contact West Publishing Company, 50 W. Kellogg Blvd., St. Paul, MN 55165.

Perhaps the most critical need for financial information in the fire department occurs during the yearly budget cycle. The arduous and politically risky process of allocating resources requires considerable analysis. Decisions must be supported by the kinds of facts best stored and manipulated by a computerized financial system.

Budget control is also a high priority for fire agencies. Management needs to know if the department is staying within budget. Even if central systems are available for this purpose, they may not provide data in sufficient detail or in a form that is useful to the fire department.

The remaining needs addressed by financial systems are primarily operational oriented: general ledger, billing, accounts payable, and payroll. The last three requirements more often than not are a part of associated applications, e.g., payroll in the personnel system, billing in the emergency medical system, and accounts payable in the procurement system. Manual or automated "bridges" are then provided to general ledger and budgeting systems.

The most common design approach to financial systems is a strongly centralized one. This is not surprising in view of the fiduciary responsibility of the chief financial officer in a city or county. Since these systems typically represent the city or county's first experience with computers, they are usually quite old and, therefore, batch oriented. While some fire departments have on-line access, that situation is not common.

Energy Management

Energy management systems provide information for control of energy costs. These systems provide information about the consumption of electric power, natural gas, and heating oil. In some cases, the use of motor vehicle fuel may be monitored, although this function is more properly a function of fleet management systems.

Beginning with the oil embargo of 1973, energy costs have represented a major portion of fire department budgets. In addition to this, energy costs in some parts of the nation have been exacerbated by other factors. In the Pacific Northwest, for example, cost overruns on new nuclear power plants have contributed to soaring energy costs. In this kind of environment, the cost of implementing a computerized energy management system can be recovered quickly through lower energy costs.

The typical approach to the design of energy management systems is to capture energy billing data, update master files in batch mode, and print various reports for administrators. More sophisticated approaches include real time monitoring systems and even real time energy control. Computer based evaluation models are also used to assess the impact of conservation measures.

SYSTEMS INTEGRATION

The typical organization does not have a systematic approach to information systems development. The approach is more often piecemeal without an overall plan for systems integration. By analogy, an airplane designed this way might have three wings, an engine with insufficient power for takeoff, and a fuel tank where there should be cargo or passengers. In short, while each subsystem might fulfill its intended function, the overall system is a failure.

The answer to this problem is information resource planning. Like other organizations, the fire department needs an overall plan for integrating the various component subsystems. This does *not* mean that all components have to be implemented at one time to have a workable system. But it does require a strategic plan for eventual integration and an action plan indicating how to get there.

Normally there are two levels of integration to be considered. The first of these is integration of fire department systems with citywide systems. Considerable benefits can arise from, say, having a central data base of occupancy data that can be used by the assessor, tax department, and fire department. However, information resource managers in the fire department should be generally wary of data bases over which they have no control. On the other hand, the potential benefits are well worth the necessary effort in working toward a well developed citywide plan. At a minimum, the fire department should think through potential interfaces with city systems and provide the "hooks" for future connections.

The second level of integration is within the fire department itself. Here, there is significant opportunity for controlling the design of systems and data bases and their interfaces. The remainder of this chapter discuss some techniques for planning such an integrated systems environment.

An organization uses information systems to support strategic planning, managerial control, and operational activities. These activities may be classified into a number of business functions, e.g., scheduling personnel, dispatching, and budgeting. In turn, each function requires a certain subset of the organizational data base. Obviously, different functions may require overlapping data subsets.

This type of analysis suggests two ways to integrate systems. The first approach is related to the way the business functions are "packaged" into application systems. For example, a typical entry level application in a fire department is computer aided dispatch (CAD). In most cases, CAD is implemented on a stand alone minicomputer and supports only basic dispatching functions. However, the boundary of the system could be expanded to include other functions such as inspections and incident reporting. The rationale for this would be based on a broader definition of the dispatch function. In the case of inspections, it could be assumed that the fire fighters need information about hazardous materials, occupancy, guard dogs on premises, etc. In the case of incidence data, CAD collects much of the data needed to report fire incident statistics. Thus, while the focus is still on CAD, the systems boundary would be enlarged to include "subordinate" or supporting functions. The supporting functions would be developed as subsystems with their own data bases (or files) and "hooks" provided to tie them together.

The other approach is based on a "data model" of the organization. The focus here is on the data and their interrelationships. Whereas the data tradionally have been owned by the application, now the application has become almost insignificant, replaced by "transaction sets" which update and query the corporate data base.

In reality, one does not define a single physical data base but rather groups the data bases according to some coherent scheme. Such a scheme should ensure that like entities are grouped together. Although other arrangements are possible, this line of thinking suggests the following groupings:

1. *Personnel Data Base*: Includes data on scheduling of personnel, training, payroll (if not part of the financial system), accidents, assignments, etc.
2. *Financial Data Base*: Includes data on budgets, flow of funds, etc. This data base is likely to be citywide in scope.
3. *Incidents Data Base*: Includes historical data on closed incidents, both fire and emergency medical.
4. *Fire Prevention Data Base*: Includes a building inventory and data on occupancy, ownership, hazardous materials, etc. Arson data may also be included in this data base.
5. *Operational Data Base*: Includes data about incidents in progress. This data is maintained for relatively short durations. At the close of an incident, or periodically, it may be moved to the Incidents Data Base. This prevents overloading the real time CAD system with nonreal time data.
6. *Supplies and Equipment Data Base*: Includes supplies inventory and equipment inventory and may include fleet maintenance. This is typically a citywide data base.

The data base planning approach has the potential of providing a higher level of integration and a more stable systems environment than the applications planning approach. It is not without problems, however. Users may not want to relinquish control over their data to a central data administration function. Data modeling requires specialized skills often not available to fire departments. Perhaps most importantly, it can lead to "analysis paralysis," the reluctance to implement anything until the last detail has been planned. While this is true of any planning approach, the tightly coupled nature of data base planning increases the danger of overplanning. Yet, the concept of data models is too powerful to be ignored.

There are numerous methodologies for defining both application and data base boundaries, developing data models, and planning the integrated systems environment. Three methodologies are mentioned here, but there are numerous others.

The Business Systems Planning (BSP) methodology was developed by IBM and has been used in a variety of organizations. It is extensive in scope and involves users at all levels to define the business functions and data classes. It is especially useful in providing an overview of the systems environment, but is usually not updated on a regular basis because of the resources required.

Another approach uses a comprehensive planning methodology where one year and five year plans are updated annually. The planning process should originate with the central information systems department even if the fire department has its own data processing capability. The fire department can then develop its plan to coordinate with the central information system plan. This will allow for better future integration with citywide systems.

A third approach is to use structured methodology on a project level (Gane and Sarson 1979). Strictly speaking, this is a project development rather than a strategic planning methodology. The hierarchical nature of the approach permits adaptation for planning purposes, although the level of planning is too detailed for pure long range planning; however, it can be a useful adjunct to such planning.

These three approaches are just some of the many methodologies that are available to the systems planner.

There are many others, some of which are fully automated. Whatever the approach taken, the information system manager is advised to select one that addresses the problems of systems and/or data base integration. The piecemeal approach is more costly and frustrating in the long run than a carefully planned strategy.

Bibliography

References Cited

Hall, John R. Jr., et. al. 1979. *Fire Code Inspections and Fire Prevention: What Methods Lead to Success.* National Fire Protection Association, Quincy, MA.

NFPA Codes, Standards and Recommended Practices and Manuals. (See the latest *NFPA Codes and Standards Catalog* for availability of current editions of the following documents.)

NFPA 901, *Uniform Coding for Fire Protection.*
NFPA 902, *Fire Reporting Field Incident Manual.*

Additional Readings

Bryan, John L., and Picard, Raymond C., eds., *Managing Fire Services,* International City Management Association, Washington, DC, 1979.

Centaur Associates, Inc., "Fire Suppression Crew Size: Literature Review," prepared for the Federal Emergency Management Agency, U.S. Fire Administration, Washington, DC, 1982.

Gane, C., and Sarson T., *Structured Systems Analysis: Tools and Techniques,* Prentice-Hall, Inc., Englewood Cliffs, NJ 1979.

Ignall, E., et al., "Fire Severity and Response Distance: Initial Findings," *Fire Technology,* Vol. 15, No. 4, Nov. 1979, pp. 245-261.

Jackson, Durward, P., ed., *Fire Department Information Technologies in Nine Western Communities,* Center for Information Resource Management, California State University, Los Angeles, CA, 1982.

Kelesar, P., and Blum, E. H., "Square Root Laws for Fire Engine Response Distances," *Management Science,* Aug. 1973.

Larson, D. A., and Small, R. D., "Analysis of Large Urban Fire Departments—Part I: Theory. Part II: Parametric Analysis and Model City Simulations," Pacific-Sierra Research Corp., Los Angeles, CA, 1982.

MacGillivray, Lois, "Decision-Related Research on the Organization of Service Delivery Systems in Metropolitan Areas: Fire Protection," Inter-university Consortium for Political and Social Research, Ann Arbor, MI, 1979.

Nickey, B. B., and Chapman, C. B., "Evaluating Fire Prevention Effectiveness Through a Probability Model," *Fire Technology,* Vol. 15, No. 4, Nov. 1979, pp. 291-306.

Pietrzak, Lawrence M., "The Effect of Fire Engine Road Performance on Alarm Response Travel Times," *Fire Technology,* Vol. 15, No. 2, May 1979, pp. 114-121, 129.

Uniform Fire Incident Reporting System, National Fire Protection Association, Batterymarch Park, Quincy, MA.

Walker, Warren E., "Changing Fire Company Locations: Fire Implementation Case Studies," U.S. Dept. of Housing and Urban Development, Washington, DC, 1978.

Walker, W. E., et al., *Fire Department Deployment Analysis,* Elsevier, NY, 1979.

EMERGENCY MEDICAL SERVICES

James O. Page

Since the late 1960s, emergency medical services (EMS) have become an integral part of fire service operations in approximately 75 percent of all U.S. fire departments. Today, two-thirds or more of all alarms commonly received by the typical fire department are requests for emergency medical care. Conservative estimates are that fire service personnel provide emergency medical care to more than 15 million persons annually in the United States.

Many factors have contributed to this trend. Federal and state legislation has produced more stringent training and equipment requirements for EMS providers. In many locales, the traditional ambulance services (operated by funeral homes, hospitals, or private ambulance companies) abandoned this activity when faced with the training and equipment mandates. Frequently, by choice or mandate of elected officials, fire departments assumed responsibility for the service.

An EMS system generally can be defined as a coordinated arrangement of public safety and medical resources to provide the victim of a sudden and unexpected illness or injury with the greatest potential for survival and recovery. The use of fire department and other public safety resources in commencing rescue and medical care in the field, at the earliest possible time, has been shown to improve the statistical chance of recovery from life threatening medical emergencies.

There are various categories of prehospital emergency care personnel within the EMS system.

First Responder: An individual who is a member or employee of a public safety agency and who has taken a First Responder course, Crash Injury Management course, Advanced First Aid and CPR (cardio pulmonary resuscitation) course, or been trained as an EMT (emergency medical technician), and who serves on an emergency unit which normally is dispatched first and/or is the first emergency unit to arrive at the scene of reported medical emergencies.

EMT (Emergency Medical Technician): An individual who has been trained and certified in one or more categories of prehospital life support and patient transportation. Among the 50 states there are many official types of EMT training and certification. Generally, however, "EMT" refers to a basic life support technician, trained in a program consisting of 80 to 100 hours of instruction and observation. "EMT-Intermediate" generally refers to an EMT who has received additional training and certification to provide certain limited advanced life support functions. Within this category would also be "EMT-D" (EMT-Defibrillation), an individual who has been trained in recognition of certain cardiac abnormalities and treatment of them with electrical defibrillation. The "EMT-P" (EMT-Paramedic) presently is the most thoroughly trained of the EMTs. In keeping with that training, most EMT-Ps are authorized to provide advanced life support services to patients while operating under medical control from their program's medical director.

ROLES AND RESPONSIBILITIES

Published medical reports during the 1970s clearly showed that trained first responders, operating in conjunction with an ambulance service, could have a very positive effect on several types of life threatening illness or injury. First responders, trained in basic life support, could maintain a patient's circulatory and neurological status while advanced life support personnel (paramedics) responded from longer distances. Due to the availability and placement of fire company personnel throughout many communities, the fire service has been selected by medical and elected officials to serve in the first responder role.

Some of the fire service movement into EMS has been a matter of replicating successful programs from other locales. This change in the traditional fire service role has coincided with changes in the fire protection environment. Urban renewal projects eliminated entire blocks of fire prone occupancies in cities throughout the country. Improved fire and building codes and stricter enforcement have reduced the number and seriousness of building fires in most areas. Faster, more mobile fire apparatus, lighter and more versatile fire fighting tools and equipment,

Mr. Page is Publisher, Jems (Journal of emergency medical services), Solana Beach, CA. He also serves as Chief of EMS and Training, Carlsbad, CA, Fire Department.

FIG. 15-9A. Paramedics initially treat victim at the fire scene (Engine & Ladder Company Operations).

universal emergency telephone numbers, and improved training of fire fighters all have changed the nature of fire protection in most communities.

ECONOMIC IMPLICATIONS

At the same time, national and international economics have affected municipal governments. Budgets for fire protection are under intense cutback pressure, while costs for fire protection have greatly increased. Reductions in fire company personnel have been common, along with demands for increased productivity in the fire department.

In many cases, the fire service evolution into EMS has been beneficial both for the fire service and for the public. The ever growing public demand for prehospital emergency care and transportation coincided with efforts to balance municipal budgets by trimming fire department staffing. Overwhelming public support for the fire department's EMS activities in many cases offset the budget cutting and, in essence, preserved fire fighter positions.

THE IMPACT OF EMS ON A FIRE DEPARTMENT

Despite the obvious benefits, the widespread expansion of fire departments into EMS has not been without problems. From the beginning, opinion has been divided as to whether fire departments should participate in medical aid services. Conflicts within fire departments have been common, as have conflicts between fire departments and other organizations or agencies vying for the opportunity to perform EMS roles.

Compared to the single function fire service, involvement in EMS requires a fire department to interact with numerous unrelated people and institutions in the community. The success of a fire department's EMS program often is determined by the department's ability to meet the expectations of the medical, hospital, and health planning entities. The sheer volume of EMS calls, the demands placed on fire department EMS personnel, the life and death nature of medical aid, and the medical/legal environment result in major management challenges.

THE COMPETITIVE ENVIRONMENT

A new wave of competition for the right to provide prehospital EMS services is developing in many areas. Sophistication and improved management in the private ambulance sector will produce pressure on fire service EMS programs and local elected officials. At the same time, many hospitals are viewing control or operation of ambulance services as a means of filling empty beds. Where fire departments tail to meet the management challenges of EMS, there is the potential that their EMS activity will be transferred to another provider.

IMPROVEMENTS, CHANGES, AND EVOLUTION

While techniques and procedures for prehospital patient care are generally standardized throughout the U.S., improvements and innovations threaten to make those standards obsolete. Procedures which once were considered advanced life support increasingly are being adopted by basic life support EMS services; for example, application of pneumatic antishock garments, and use of adjunctive ventilation devices. This constant evolution imposes upon the fire department a responsibility to keep up with the state of the art and to provide for continual training in standard as well as new techniques.

Many fire departments have accepted a first responder role—providing basic life support with engine company personnel trained as emergency medical technicians (EMTs). This is a valuable adjunct to the EMS system and in many cases can have a positive effect on medical outcomes. But even this role is likely to change in the next few years. In a few locales, first responders have already been equipped with cardiac defibrillators and have been trained to function as EMT-Ds (EMT-Defibrillation).

A pilot study using fire department personnel in several communities of Washington state, demonstrated that the EMT-D concept can save a significant percentage of cardiac arrest patients. Though not permitted to administer medications or use adjunctive ventilation devices, EMT-D personnel are trained to recognize certain lethal cardiac arrhythmias and apply electrical countershock (defibrillation).

Among the requirements for implementing the EMT-D concept are appointment of a physician medical director, voice and ECG tape recording of all calls, and frequent retraining and retesting of all EMT-D personnel. Despite these requirements, there is growing interest in the concept among fire departments and other EMS providers. Recent articles in medical journals have indicated that this concept may be the next major evolutionary change in prehospital EMS.

A BETTER FIRE SERVICE

Controversy is likely to continue over EMS in the fire service. Regardless, this expansion of traditional fire department services already has had major impact on the fire service nationally. It has forced participating departments to relate to other professions and disciplines and has altered the public's view of the fire service. It has created complex management challenges for fire service administrators. In the final analysis, this introduction of expanded roles and services probably has enhanced the typical fire department's ability to protect the community from fire and other emergencies.

LEGISLATION AFFECTING EMS

Federal legislation governing the delivery of emergency medical services, and providing for funding grants, was rescinded by the Omnibus Act of 1981. Block grants to the states were intended to replace some of the EMS funding which was deleted by the Omnibus Act. However, internal competition for health care funds in most states has further diminished federal funds for EMS.

Most states have recently increased their expenditures to their state EMS agencies. However, availability of those funds to individual EMS providers, and the process for application, varies greatly among the states. Similarly, state laws regarding training, licensure, continuing education, equipment, and authorized emergency procedures vary widely.

Research of individual state EMS laws and regulations is necessary to assure compliance. Copies of applicable laws and regulations are available from state EMS agencies, which are most often a division or branch of the state health department.

FUTURE OF THE EMS-TSS PROGRAM

In the early 1980s the U.S. Fire Administration (Federal Emergency Management Agency) created the Emergency Medical Services Technical Support Services Program (EMS-TSS). Under this program, experienced fire service EMS administrators were selected to serve as facilitators and consultants to fire departments throughout the U.S. Their services were made available without charge.

The EMS-TSS Program ended in August 1984. However, because the program has been very successful, it may be extended or renewed by the U.S. Fire Administration.

Bibliography

Additional Readings

Committee on Trauma, American College of Surgeons. "Essential Equipment for Ambulances." *Bulletin of the American College of Surgeons*. Sept. 1977.

FEMA/USFA, *Fire Service EMS Planning Guide*, North Central Texas Council of Governments, Nov. 1982.

FEMA/USFA, Emergency Medical Services Technical Support Services (EMT-TSS), Report 1984, U.S. Fire Administration, Emmitsburg, MD.

Grant, Harvey, and Murray, Robert. *Emergency Care*. 3rd ed. Brady, Englewood Cliffs, NJ, 1983.

Page, James O., and Smith, Bradley H. "Fire Service/EMS Program Management Guide," Federal Emergency Management Agency, Washington, DC., 1981.

Page, James O. *The Paramedics*, Backdraft Publications, 1979.

Page, James O. *Emergency Medical Services*. Second Edition. National Fire Protection Association, Quincy, MA, 1978.

U.S. Department of Health, Education and Welfare. *Emergency Medical Services Systems Program Guidelines*, U.S. Government Printing Office, Washington, DC, 1979.

U.S. Department of Health, Education and Welfare. *EMS Handbook for Patient Recordkeeping and List of Minimum Data*, DHEW Pub No HSA 77-2034, U.S. Government Printing Office, Washington, DC, Aug. 1977.

Weed, Susan L. American Trauma Society. *Emergency Medical Services*. Division of Emergency Medical Services, Department of Health, Education and Welfare, Washington, DC. Sept. 1978.

SECTION 16

FIRE ALARM SYSTEMS, DETECTION SYSTEMS, AND GUARD SERVICES

FIRE DEPARTMENT EMERGENCY COMMUNICATION SYSTEMS

Max R. Schulman and Thomas F. Sawyer

This chapter is intended to provide the HANDBOOK user with an overview of the total emergency communications responsibilities of the fire service, together with some of the principal management and operational systems and/or techniques through which the respective responsibilities may be achieved. The fire service relies upon four major elements for success: intelligence, personnel, supplies, and good communications.

Intelligent and well trained personnel are of no value if there are no reliable means to direct them to the right place at the right time. The fire department's communications divisions are entrusted to provide this direction.

For additional information, consult the NFPA documents referenced in this chapter and review the experiences of other similar organizations. Make an individual needs assessment of your own community and fire department, and from that, determine the adequacy of the emergency communications services provided. In all cases, the requirements of NFPA 1221, *Standard for the Installation, Maintenance and Use of Public Fire Service Communication Systems* (hereinafter referred to as NFPA 1221), should be followed. Also, Chapter 2 in this Section provides detailed information on protective signaling systems.

COMMUNICATIONS CENTERS

In creating any communications network, you must first establish a center point (communications center) through which all related process and control functions will be directed. (See Fig. 16-1A.) A communications center may serve a single agency or area, multiple agencies under the control of a single political subdivision, several similiar agencies from adjacent political subdivisions, an entire county, or any other large political jurisdiction. Combining fire and police communications within a specific political jurisdiction (and fire communications for several adjacent jurisdictions) has become an increasingly common management concept. If properly planned, such consolidations can provide positive results.

FIG. 16-1A. A communications center equipped with an automated status board (top left). The status board is controlled by a fully computerized, multiple operator dispatching network. (Phoenix, AZ Fire Department)

The communications center is where all requests for emergency assistance under the jurisdiction of a fire department are received and translated into an appropriate response. Further, the communications center should be structured to provide logistical support to the responding units until the situation is under control. Accordingly, as an operating entity, the communications center must ensure that the equipment, personnel, and established procedures are adequate to satisfactorily perform in emergencies.

Even the smallest fire department must have some means of receiving notice of a fire. In small communities, the communications center may consist of a telephone in a private residence or business establishment where someone is always available and where there is a switch to

Mr. Schulman is Chairman of the NFPA Committee on Public Fire Service Communications, and president of Schulman Associates, Ltd., Buena Park, CA, a consulting engineering firm specializing in fire service electronics. Mr. Sawyer is Assistant Chief of the Phoenix Fire Department, AZ.

sound a siren or other device to alert fire fighters. Quite frequently, the police station serves as the communications center. As fire departments grow, a room should be provided to house the fire alarm dispatcher and related facilities in a fire station or other suitable structure.

Site Selection and Configuration Considerations

The following is a list of items that should be considered in the development of a communications center complex:

1. Location.
2. Seismic stability.
3. Security.
4. Emergency electrical power.
5. Wiring access (computer floor).
6. Lighting.
7. Air conditioning (computers/dispatch).
8. Air conditioning back-up.
9. Console layout (per Occupational Safety and Health Administration Regulations).
10. Console arrangement.
11. Acoustics (background noise suppression).
12. Restroom facilities.
13. Kitchen facilities.
14. Decoration.
15. Rest areas (lounges).
16. Dormitories (sleeping quarters).
17. Emergency rations.
18. Alternate locations.

The communications center should be designed and constructed so the probability of interruption to operations due to fire or other causes will be minimized. A building in a park or other open space is most desirable and can often be designed in harmony with the surroundings. When a separate building is not feasible and the communications center must be housed with other operations such as city hall offices or in a fire station, the center should be properly separated from the remainder of the building by vertical and horizontal separations with a fire resistance rating of at least two hours and preferably with entrance only from the outside. Protection should be provided against fire exposures and unauthorized entry.

Alternate Dispatch Location

Serious consideration should be given to an alternate dispatch location. A bomb threat, or an actual fire in the center requires developing an alternate dispatch center and planning when to evacuate the communications center. This location should be totally removed from the primary location, and should have all the necessary requirements for long term usage.

Communications Center Security

Because of the types of activities conducted and the records maintained in any modern communications center, building and systems security must be addressed. Building security begins at the entrance to the communications center, and can be as simple as a dead bolt lock or as complex as a computerized entry system with closed circuit television monitoring equipment. Fire and smoke detector systems, and automatic and hand held extinguishing systems should be provided. Because the communications center will be manned for twenty-four hour periods, an emergency lighting system should also be provided. The information gathered and maintained by a communications center must be properly secured. Tape and disk storage and access to files should be limited to authorized personnel only. Off premise storage of backup software, etc., is essential.

Emergency Power

The communications center must be able to operate in the event of a primary power failure. If the primary power is provided from a utility distribution system, the secondary power may be either an engine driven generator with a four hour battery, a circuit from an independently different utility, or two standby generators. Standby generators should be tested weekly for at least one hour and should be able to power the system for at least four hours.

Uninterruptable Power Source

A Computer Aided Dispatch system (CAD) cannot tolerate the power surges and interruptions normally associated with commercial power systems. A momentary power surge or outage could blank the computer's memory or damage its many sensitive circuits. To prevent this, an Uninterruptable Power Source (UPS) is used as an isolator element between the commercial ac power lines and the computer system. This UPS consists of a set of batteries, a battery charger, and a dc to ac converter. The inverter is phased to the commercial power frequency and produces the same power. Thus, any fluctuations or interruptions in commercial power would not affect the computer.

FIRE DEPARTMENT RADIO COMMUNICATIONS SYSTEMS

Radio System Selection

Considerable thought should be given to selecting the proper radio system and professional personnel should be used to assure the proper functioning of each system.

Radio Frequency Selection

Several bands of radio frequencies are available to the fire radio service. These bands are VHF low band, VHF high band, UHF 45 MHz, and UHF 800 MHz band. Each band of frequencies has its advantages and disadvantages, and the selection of a particular band will depend upon factors such as frequency availability, area to be covered, type of terrain, number of radio units required, frequencies used by bordering fire districts, mutual aid agreements, type of operation, and use of emergency medical radios.

Due to the lack of available frequency bands, it may be necessary for fire departments or districts to share radio frequencies. For small fire districts, sharing radio channels with neighboring departments could be advantageous.

Type of Radio System Operation

The several types of radio systems available are:

Simplex: This system utilizes a single radio frequency for both transmit and receive in all radios for each channel. With a simplex system, only one radio can transmit at any time and all others must receive.

Two Frequency Half Duplex: This system utilizes separate frequencies for transmit and receive. As in the simplex system above, half duplex allows only one radio to transmit at any one time.

Two Frequency Full Duplex: The two frequency full duplex system utilizes separate transmit and receive frequencies and permits simultaneous conversations in two directions.

Two Frequency Repeater System: With this system, a high powered base station "repeater" is centrally located at a favorable site overlooking the area to be covered, preferably on a mountain top, a tall building, or a tower. The repeater receives a transmission from any radio in the system on one radio frequency and instantly retransmits or "repeats" the message on a second frequency which is received by all other radios in the system. A repeater system greatly increases the range of all of the radios in the system. It is especially useful in hilly or mountainous terrain where it might be impossible to talk directly from unit to unit. Repeater systems are two frequency half duplex systems as described above.

Tone Coded Squelch: In a tone coded squelch system, each radio transmits a subaudible tone along with the voice signal. This audio tone is used to "turn on" the receiver so the conversation will be heard. Only the proper tone will activate the receiver, which eliminates the reception of radio interference.

Radio Paging Systems: Contacting fire department personnel who do not normally listen to the radio system can be easily accomplished by a radio paging system. This system consists of a paging "encoder," which is located in the communications center, and individual personnel pagers. Pagers are also available that will respond to both an individual and group call.

Pagers are available for tone-only or tone-and-voice operation. When utilizing tone-only pagers, each individual is given a telephone number to call or location to respond to when the tone activates. With tone-and-voice pagers, the tone alerts the individual and turns on the receive audio. At that time, the dispatcher can relay a verbal message to the individual.

Multichannel Use of Radio Frequencies

The size and complexity of each fire department will determine the number of radio channels necessary for adequate communications. Very small departments should require only one radio channel for their private use, while departments in large metropolitan areas may require several. The Federal Communications Commission (FCC) has definite rules regarding channel loading, and excessive radio congestion must be proven before another radio channel will be licensed. In a large department, it is customary to have one or more dispatch radio channels and one or more tactical radio channels for controlling units in the field. In this case, the jurisdiction is divided into districts and each district is given a tactical radio channel. In addition, all departments adjacent to one another should share a common mutual aid radio channel. This is provided for by the FCC and is usually managed on a statewide basis.

Base Station or Repeater Location

Care should be taken in selecting the main base station or repeater location for the radio system. Each radio coverage area is unique and must be dealt with on an individual basis. A high, central location such as a mountain top or large hill overlooking the area to be covered, a tall building, or similar structure is generally the ideal site for the main base station location. (This depends upon coverage requirements, frequency availability, and requirements for channel sharing.) Where neither of the above are present, a high antenna tower can be utilized. Towers are required, even on mountain tops, to assure proper antenna system operation.

The availability of commercial power and telephone access is also desirable when locating the main base station site. Standby power supplied by a generator or batteries (or both) must be provided to assure the site remains "on the air" in the event of a commercial power outage.

In some locations, where commercial power is unavailable, solar power (solar cells) and batteries, or generators may be used as the primary source of power for the site. If the base station or repeater location is remote and unattended, provisions should be made to secure the area against weather, vandals, etc. A chain link fence should surround the area and all radio equipment should be located inside a proper building, out of sight and protected from the weather.

Standby Base Station or Repeater Location: A separate location should be provided for a standby base station or repeater. This standby location should have the same complement of radios as the main station. A good high central location is also desirable for this site. These standby radios will be used if one or all of the radios at the main site fail. A separate commercial power source should be provided, if possible, and standby power should also be available.

The communications center should have the capability of selecting either the main or standby radios. The standby radios should be used periodically to ensure they are operable and ready for use in case of an emergency.

Radio Site Antenna Configuration

When more than one radio is installed at a single site, an antenna system may be required. It may be impractical for each radio to have its own separate antenna because considerable interference could be generated using separate antennas. When several compatible radios are used, it is customary to provide one or more transmit antenna systems and a separate receive antenna system. A base station repeater using a single antenna requires a duplexer in order to transmit and receive simultaneously. Several transmitters can be connected to a single antenna by use of an antenna combiner; several receivers can be connected to a single antenna through the use of a multicoupler. Professional assistance should be utilized to design the antenna system for all radio sites.

The antenna design should provide maximum radio frequency (RF) isolation between transmitting and receiving antennas. This is usually accomplished by mounting them one above the other on the antenna structure. Coaxial cables should be used for connections between radios and antennas. These cables should be high quality, low pass with length as short as practical. Good grounding of the antenna system is required to prevent damage due to lightning.

Access from Communications Center to Radio Site

If base station radios are located at the communications center, individual wires or "house cables" can be used to connect the dispatcher's consoles to the proper base station. If the base station radios are located remotely from the communications center, radios can be connected by leased telephone lines, a microwave radio system, or a point-to-point radio.

Mobile Radios

Mobile radios are ideal for use in fire apparatus or sedans. They have high RF power output (30 to 50 watts) and, therefore, considerable range. Their use, however, is restricted to vehicles, and they cannot be carried into buildings or structures.

Portable Radios

Portable radios are much lower in RF power than mobile radios (0.1 to 7 watts), and, therefore, have a considerably shorter operating range. Their main advantage is portability. They can be hand carried and allow communications outside the vehicle, giving the user considerable flexibility. The reduced RF power, however, can cause problems in communications reliability inside structures in some areas.

Diversity Receiver-Voter System: Since portable radios have a very low RF power output, this low power creates problems for communications over large areas. Portables can normally receive the high power base station, but they do not always have sufficient power to transmit back to the main radio site. This problem can be eliminated by a diversity receiver-voter system in which radio receivers are installed in a uniform manner at various locations throughout the coverage area. This provides a receiver for each subarea in the system. In this manner, a portable radio should not be more than a mile or two from one of the receivers in its subarea.

The signal from each of these receivers is fed to the communications center. The voter is capable of switching from receiver to receiver in midsyllable, so, as the RF signal rises and fades at each receiver, the dispatcher hears only the best signal received. A diversity receiver-voter system can also be incorporated into a repeater system to greatly exceed portable-to-portable communications. A block diagram of a simple diversity (satellite) receiver-voter system is shown in Figure 16-1B.

Mobile Repeaters or Extenders: An extender is simply a radio which receives on one frequency and immediately retransmits the signal on another frequency, usually in a different band. For example, in a rural area with a low band simplex radio system, high band portable radios could be used near a fire engine equipped with a mobile repeater operating at 50 to 100 watts to increase the range of the portable radios.

Mobile Data Terminals: Mobile data terminals are small computer terminals that may be connected to a radio system, allowing that unit to access another mobile terminal, base station terminal, or computer. These units generally consist of a small computer system that converts digital information into a signal that a radio can transmit or receive. These units may have preprogrammed display capabilities where by pressing one key, a "canned" message such as unit status, condition, operator name, or even the time of day can be sent. These units are so reduced in size they can be hand held (including the radio).

Screen formats vary from simple light indicators to complex visual images including text and graphic characters. Some systems even have city maps or building drawings showing specific items relative to access, hydrant location, standpipes, and hazardous materials storage. One major consideration, however, is that these units require dedicated, conventional radio channels.

Station Dispatch Notification: There are three items which should be included as part of a fire department dispatch/alerting system: (1) a voice announce or public address system to alert the company(s) to an emergency response requirement; (2) a remote control, either wire or radio controlled for station lights/annunciators; (3) failsafe communications to ensure the proper sending of the dispatch message. Vehicles may even be included in this alerting process by the use of digital decoders or other suitable audible alarms.

Mobile Communications

There are several factors to consider when establishing a mobile communications system. This chapter cannot specify the exact number of channels or frequencies needed; a fire department should be aware of the various requirements so a system can be selected more knowledgeably.

An analysis of service area and system configuration will help determine the wattage rating needed and the number of channels or frequencies required to support current usage and provide for future expansion. Small departments may be able to operate with one or two frequencies. However, if you consider the possible need of a mutual or automatic aid frequency, a police or interagency frequency, a disaster communications frequency, and the required individual departmental frequencies, it is quickly obvious why careful planning is needed.

To ensure quality communications, headsets should be considered for the driver, company officer, and fire fighters. To further support the suppression personnel as much as possible, each apparatus should be equipped with at least two radio units, one of which should be a portable radio and charger unit for the company officer. A portable unit will allow the officer to leave the truck while maintaining communications with the apparatus, other units, and the communications center without returning repeatedly to the truck.

Emergency Medical Service Communication

From the early 1970s, the fire service has become more involved in providing emergency medical services (EMS), both basic and advanced life support, to the citizens they serve. Each state has an EMS communications plan on file with the FCC and these plans identify how the EMS communication system will be implemented in that state. The EMS system differs from the normal fire communications system in the approach the FCC has taken to frequency allocation. The FCC has set aside ten UHF frequency pairs for communications between field EMS personnel and base hospital personnel. These ten channels are shared and used in all fifty states. Therefore, to assure quality, noninterfering communications on these ten channels during critical life threatening incidents, it is imperative that local coordination prevail. The ten frequency pairs are outlined in Table 16-1A.

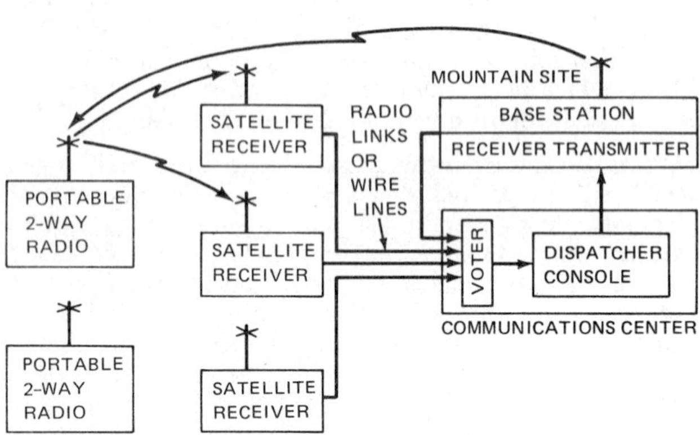

FIG. 16-1B. Block diagram of a simple diversity (satellite) receiver-voter system.

Hazardous Materials Team Communications

The development of hazardous materials response teams in the fire service has resulted in new communications requirements for safety. These new requirements are in two major areas: (1) hazardous materials information, and (2) HAZMAT team communications from within encapsulating suits.

A major element in the successful outcome of hazardous materials incidents is the ability to rapidly obtain information about the material involved. This generally involves contacting outside agencies for a variety of information. Mobile radio telephone patch(es) may be considered, but this usually ties up the only available radio channel. Mobile cellular telephones should be installed in the hazardous materials team apparatus to facilitate this need for information. This approach also allows a printer to be installed along with the telephone so that one of the hazardous materials computerized information services can be used for on scene, rapid printed information.

Mobile telephone also allows the hazardous materials team members direct access to HAZMAT information services, reducing the potential for errors. The medical component of the HAZMAT team should be equipped with a UHF MED radio for medical control or to obtain toxicological information from base hospitals.

Many hazardous materials team members must work inside encapsulating suits. They require special communication systems designed to meet their requirements for operational safety. Two way wireless communications, preferably hands free, should be considered. A variety of radio systems which use bone, ear, or throat microphones for voice transmission are available to meet these needs. These radio systems come in simplex, half duplex, or full duplex. They use push-to-talk or voice actuated circuits to transmit, and mobile repeaters may be integrated into them. The fire department should thoroughly examine its needs in this area prior to purchase.

PERSONNEL

No matter how well designed, installed, and maintained, no system is better than the people who operate it. Communications systems should be supervised by responsible and competent people whose duties include directing, testing, and inspection programs. Fire alarm dispatchers should be healthy, temperamentally suited to the position, and free of physical and/or mental handicaps that could affect their ability to efficiently handle assigned duties. They should have the ability to remain calm, take decisive action during emergencies, remain alert during periods of inactivity and while carrying out normal repetitive operations, and to work harmoniously with other people, and should be familiar with general fire department operations.

Work schedules should be structured to provide for maximum efficiency. Stress and high activity leads to dispatcher fatigue. Therefore, dispatchers should be provided with a place to take work breaks away from the communications center.

Communications personnel should not become isolated from the fire department. Their structured training should be directed toward maintaining and improving their dispatching skills, but it must also include a full understanding of all department operations, the most important being fire suppression. All standard operating procedures should be in writing to assure standard performance.

CONSOLIDATION OF COMMUNICATIONS CENTERS

Many small to medium sized fire jurisdictions have found that consolidating communications functions with other agencies, they can provide a higher level of service to their citizens at a lower overall cost. One of the major benefits of consolidation for these agencies is the ability to provide highly trained, alert radio dispatchers who are on duty 24 hours a day. A larger communications center will also have adequate backup to handle multiple, simultaneous incoming alarms, while the smaller center may not have the personnel to cope with them. The major benefit of consolidation is to provide an increased level of service at reduced costs.

A fire communications center is labor intensive. If a combined communications center can handle the same workload with fewer employees, the result can be a substantial savings to the jurisdictions involved. Suppose there are four medium sized fire departments operating independently in the same valley, each with its own communications center. Typical staffing for the individual centers might be three full time dispatchers, each working a 24 hour shift. Their time would be supplemented by overtime pay and/or the assignment of fire suppression personnel to cover vacations, sick leave, etc. In this case, the four individual fire departments would be served by

TABLE 16-1A. EMS Frequency Pairs

Channel #	Base Frequency	Mobile Frequency	Channel Name
1	463.000	468.000	MED-ONE
2	463.025	468.025	MED-TWO
3	463.050	468.050	MED-THREE
4	463.075	468.075	MED-FOUR
5	463.100	468.100	MED-FIVE
6	463.125	468.125	MED-SIX
7	463.150	468.150	MED-SEVEN
8	463.175	468.175	MED-EIGHT
9	462.950	467.950	MED-NINE
10	462.975	467.975	MED—TEN

twelve dispatchers. By combining the four communications operations into one central communications center, the same workload could be handled by nine dispatchers working 24 hour shifts.

Methods of consolidating fire communications centers vary, but the two most popular are: (1) forming a Joint Powers Authority (JPA) to provide this service as an independent agency, or (2) one or more departments contracting with another fire department to provide this service. Either system can work well and provide the same benefits to the participants.

Joint Powers Authority

A Joint Powers Authority is an independent government entity, established to provide specific services. There are several benefits of a JPA arrangement for combined fire communications versus other arrangements but the major benefit is shared administration. With a JPA operating with an independent Board of Directors, no one agency which might set policy unacceptable to another agency is "in control." In this way, each participant has equal representation and equal responsibility to make the operation a success.

REPORTING A FIRE OR EMERGENCY

Reports of fires and other emergencies under the jurisdiction of the fire service originate from three principal sources: the general public, the business community (industrial, institutional, commercial and mercantile) and other public service/safety agencies. The reporting process may utilize any of the following means, individually or in combination: public commercial telephone systems, public emergency reporting/fire alarm systems, privately operated automatic alarm systems, and two way radio communications systems.

Commercial Telephone Facilities

The most common method, and no doubt the least costly to the community involved, is the conventional commercial telephone network. However, the important factor of automatic location identification is not obtainable in a majority of the areas presently served.

Only the "911" system provides both the calling number and the location of the related telephone instrument directly to the dispatcher. This indication is presented visually on a cathode ray tube (CRT), and can be automatically recorded if the communications center includes CAD or computerized logging support systems.

Although there are some disadvantages to total dependency upon a utility owned, operated, and maintained communications network, specifically with regard to emergency reporting, telephones are often the only means of communication available in urban centers as well as suburban and rural areas. The installation of outdoor telephone booths increases the availability of telephones for emergency reporting. In general, the public telephone system, widely used for reporting fires and other emergencies, provides a very viable function.

Municipal Fire Alarm Systems

Scope, installation, maintenance, and use of all municipal fire alarm systems, regardless of the principles of operation, are covered in detail in NFPA 1221. This standard makes no distinction between a coded, voice, or code-voice alarm system, because each must perform the same function, providing a means by which an alarm can be transmitted from a street alarm box to the communications center.

From the standpoint of the dispatcher responsible for receiving the alarm, the use of a public emergency reporting station (fire alarm box) eliminates the difficulty of determining the location from which the alarm is being transmitted. When actuated, each device must transmit a distinct numerical code in addition to any other functions or capabilities provided.

A general abandonment of jurisdictionally owned municipal fire alarm systems or more aptly, public emergency reporting systems, began to take place in many major metropolitan centers in the mid 1970s. The movement, based on economics and the efficiency of the service provided, has since spread to smaller suburban jurisdictions.

The average metropolitan system consisted of a mix of public reporting stations (street boxes) and auxilliarized fire alarm systems interconnected to master boxes. In most cases, the operational jurisdictions reduced their obligations toward auxilliarized interconnects by referring all risks exclusive of schools, hospitals, churches, government buildings, and major hotels, to privately operated central stations. Public reporting stations were eliminated shortly thereafter. Those agencies who still maintain and operate such systems have essentially directed their applications towards auxilliarized systems.

The expanding implementation of "911" systems has eliminated the necessity of public reporting stations (street boxes), at least in those areas where publicly accessible phones are well distributed and functionally usable. Public emergency reporting systems are electrically operated networks divided into three basic categories: (1) coded, (2) voice, and (3) coded voice combination.

A public emergency reporting system may be used for the transmission of other emergency signals or calls, provided such transmission does not interfere with the proper handling of fire alarms. For example, systems employing voice communications between the street alarm box and communications center can be used to transmit alarms of nonfire emergency nature. Fire alarm boxes in a radio type system can be provided with push buttons for calling the police department, the ambulance service, or other emergency services, and signals can be transmitted directly to the service called. With parallel type telephone systems

FIG. 16-1C. (Left) A coded telegraphic type box. (Right) An interior view of the telegraphic alarm box, showing the actuating lever (arrow) that normally protrudes through the hole shown on the inner door at right. When depressed, the handle releases the spring wound mechanism and sends a coded signal from the code wheel shown at center. (Gamewell Corp.)

FIG. 16-1D. (Left and Center) Typical battery/dc powered coded radio type alarm boxes equipped for multiple messages (fire, police, ambulance, etc.). (King Fisher Co.) (Right) A user-power-coded radio box. This type of devise must incorporate an automatic test feature to meet Standard requirements. (Signal Communications Corp.)

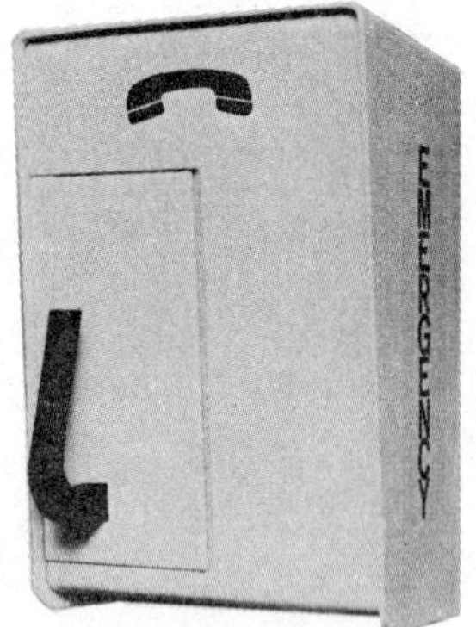

FIG. 16-1E. A voice-telephone alarm box, series type, for handset operation only. (Gamewell Corp.)

FIG. 16-1F. A combination telephone and coded telegraphic type alarm box. (Gamewell Corp.)

(each box served by a separate circuit), the dispatcher can crossconnect to the proper emergency service, such as the police department, or systems can be arranged so authorized persons, such as law enforcement personnel, can be connected directly to their own departments.

Coded Systems

The coded category is itself divided into two classifications: (1) the wired telegraphic system, and (2) the radio (wireless) system.

The wired telegraph type box is actuated by depressing a lever which is accessible through a small door. (See Fig. 16-1C.) This starts a spring wound clockwork mechanism and transmits a code number by the rotation of a code wheel that opens and closes the circuit. The transmitted code number of each box is different so the location of the box transmitting the alarm can be determined. If a circuit wire is broken, the telegraph box system usually is designed to transmit the coded signal through a ground connection.

Radio boxes are similar to wired telegraphic boxes. (See Fig. 16-1D.) When activated, they transmit a coded signal, representative of their geographical location; however, in addition to the basic numerical identification code, each box must be equipped to send at least three specific messages identifying the actuation cause. A "test" message must be transmitted at least once in each 24 hour period, assuring the monitoring point (communications center) that the box is working. If a box is stuck or physically battered, a specific "tamper" message must be transmitted. If the box is used for public reporting only, the third required message would be "fire." Radio boxes are available that allow transmission of up to 50 messages actuated either publicly (fire, police, medical aid, etc.), privately (through auxilliarized connections), or combination of both.

Voice Systems

Voice systems are wired telephone networks. The fire alarm box is essentially a telephone handset installed in a specially designed housing. (See Fig. 16-1E.) The location from which the alarm is transmitted is definitely established without dependence upon voice transmission and without interference from other boxes. Removing the handset from its cradle in the box lights a lamp on the communications center switchboard where the number of the box and the time of message can be recorded.

Like the coded system category, voice system technology is itself divided into parallel and series classifications:

Coded Voice Combination: Boxes which also contain a telegraph mechanism are called combination telephone-telegraph type, and both voice and coded signals can be transmitted over the same circuit to the communications center. (See Fig. 16-1F.) The series telephone box may be intermixed with telegraph street boxes and master boxes.

Privately Operated Automatic Alarm System

There are two sources of alarms that constitute a private agency notification to a governmentally operated communications center, central station and remote station systems. Both operations are described in Chapter 2 of this Section.

Two Way Radio Communications

A commonly overlooked alarm/emergency notification source is the two way radio network of the responding service(s). Notification of incidents in watershed areas are typically reported by mobile patrols. Similarly, requests for supplemental dispatches to multiple alarm fires usually originate from units in the field.

Contact between agencies in adjacent jurisdictions as well as contact between agencies of a single jurisdiction may often utilize radio to expedite communications.

PROCESSING COMMUNICATIONS WITHIN THE DEPARTMENT

Regardless of the method by which an alarm is received, the location of the emergency is the minimum information required to respond to an emergency. The nature of the incident being reported is also obviously essential. However, one must bear in mind that thousands of responses to numerically coded (location) nonvoice notifications are made each year.

Data Collection Systems

To efficiently operate a fire department communications center, reliable data must be available. To be useful, this data must be both accurate and readily available. The data may take a wide variety of forms, from card indexes to sophisticated computer files, and may include such categories as the status of vehicles, the availability of resources, geographic information, and personnel information. The following categories outline the various types and uses of data that can be collected and maintained by a communications center:

Incident Related Data: It is important (and often a legal responsibility) to keep accurate records of fire and emergency medical incidents to which the department responds. The type of data worth collecting include:

1. Time of dispatch.
2. Address.
3. Identifications of responding units.
4. Arrival times at the scene.
5. Departure times from the scene.
6. Radio traffic tape logs.
7. Telephone tape logs.
8. Times of significant events at the scene.

Operational Related Data: This data consists of information which must be rapidly available at all times to maintain efficient operation of the communications center. The data include:

1. Geographic data.
2. Equipment data.
3. Response order data.
4. Additional resource data.

Administrative/Management Data: This category includes data which may be required for the general fire department management:

1. Personnel records.
2. Personnel scheduling data.
3. Personnel specialty skills information.
4. Occupancy information.

Reports: Many types of reports are necessary for overall fire department management. Some of these reports will be required at regular intervals, while others will be required on demand to fill a specific departmental information need. A list of probable reports are:

1. Response time reports.
2. Emergency activity reports.
3. Occupancy activity reports.
4. Specific incident reports.
5. Personnel staffing reports.
6. Reports for outside agencies.
7. Fire loss/injury reports.
8. Equipment related reports.

These reports can be used to project future needs of the department such as location of new fire stations, hiring of additional personnel, and relocation of existing equipment.

Receipt of Alarms

Most alarms or emergency notifications are received by telephone over a conventional commercial telephone network. Automatic recording devices must be interconnected to all types of voice reporting circuits, including radio systems. However, recording devices cannot assure the credibility or accuracy of the information received.

Correct telephone numbers are the only means of positively identifying the location of an informant. The cross matching of telephone numbers to addresses is a telephone company service, provided on a select basis, to public safety agencies.

The 911 systems are designed to provide the answering dispatcher with an immediate reference to the number of the telephone in use and its specific location. The data is usually displayed on a CRT device. The data input is formulated for electronic data processing (EDP) usage and can be fed concurrently to a departmental CAD system.

All public emergency reporting systems (municipal fire alarm systems) provide an individual numeric code for each unit installed, and the proper numeric code is transmitted when use of a specific unit is initiated. All incoming coded signals are permanently recorded on printers integral to the specific systems, and in some cases are also displayed on direct readout (LED, CRT, etc.) devices.

The translation of the numerically coded data into a usable address is performed with the aid of a running card file, which, incidentally, is a necessary factor in the processing of all alarms and requests for emergency assistance.

Voice Recording and Reproducing Systems

The voice recording and reproducing system should consist of a multichannel recording device which provides a permanent, time referable, unalterable, and court admissible recording of all telephonic/radio communications within a fire department.

The watch commander and one of the dispatchers should be totally familiar with the operation and maintenance of the multichannel recording device. Familiarizing other department personnel is advisable, but responsibility should lie with no more than two individuals per watch/shift.

A multichannel recording system will provide:

1. Verification of emergency messages.
2. A log of all communications within the department.
3. Protection of the department against civilian or other department complaints.
4. Substantiation or negation of claims against the department.
5. Added protection to the civilian population.
6. A training tool for dispatch personnel.
7. Validation of response time.

The importance of these various benefits will differ from department to department.

Running Card Files

A running card file contains a dispatch plan for every area of the jurisdiction served. The number of running cards in a given file is dependent upon the size of the jurisdiction and the manner in which it has been indexed for responses. The most common method of indexing is to assign numeric designations to all intersections or cadasteral networks of horizontal and vertical crosspoints. There is a specific running card to each intersection or crosspoint. Public emergency reporting stations (alarm boxes) are normally assigned an identity code corresponding to the intersection or crosspoint site closest to their physical location.

When an emergency notification is received over conventional commercial telephone network, through a 911 system, or by two way radio, the dispatcher utilizes the addresses and/or intersection data obtained from the informant to find the correct running card.

Status Keeping Systems

The ability to maintain an accurate record of the current status of all emergency units in the system is essential to a successful fire dispatch system and will aid in: (1) minimizing response times, (2) ensuring appropriate response, and (3) collecting nonemergency unit activity information.

Status keeping may involve the tracking of various pieces of information such as:

1. Availability of the unit for emergency response.
2. Service capabilities of the unit (paramedic service, extrication equipment, all-terrain capability, etc.).
3. Location of the unit.
4. Means of contacting the unit (radio, phone, etc.).
5. Miscellaneous items based on the particular department's operating philosophy.

The mechanism of a status keeping system will vary, depending upon the number of units on which status must be kept and the actual dispatch process utilized by a department. A small department might utilize a manual system with a small "clip board," while a large department may require a computer.

Computer Aided Dispatch (CAD)

Computers are being used in all areas of the fire service. These areas include personnel management, budgeting, fire prevention, training, inventory, etc. One major use is the dispatching of fire units. The degree which a CAD system can be implemented depends upon the size of the department. A small fire department with one to three stations does not require a complex CAD system with mobile digital terminals, station terminals, and printers. Dispatch software packages are available for personal computers, and small departments can use them to rapidly recover the information needed to expeditously dispatch fire units, i.e., preplans, hazardous material information, lists of other responsible public safety agencies, names, addresses, and phone numbers of responsible parties—the list is endless. (See Fig. 16-1G.)

Caution must be used in both the selection and use of personal computers. Computer professionals should match the application to the proper computer. The dispatch computer should not be used for other department applications. It should be kept free of time consuming processes to allow for rapid retrieval of dispatch information. CAD systems for medium to large departments are more complex and must be designed for the needs of an individual department.

Dispatch Circuits and Equipment

A dispatch circuit is the means by which the fire alarm dispatcher notifies fire companies to respond to an alarm. The location from which the alarm was received is the minimum information that must be transmitted.

NFPA 1221 provides requirements, some of which are:

1. Two separate means of transmitting alarms to fire stations shall be provided at the communications center. However, only one means of transmittal is required when less than 600 alarms per year are received.
2. Each alarm transmitted, date, and time shall be automatically recorded.
3. Devices for transmitting coded or other types of signals shall be arranged for manual setting and operation. (The automatic dispatching of fire companies by the use of computers is under development).

Note: CAD, unlike fully automated dispatching systems, is not a circuit. CAD is a method used to assess related response information. CADs identify the companies normally assigned, and the current status of each of the companies.

Types of Dispatch Circuits

Telegraph Circuits: Telegraph dispatch circuits are wired circuits, arranged for transmitting and receiving coded alarm signals. Signals are received at fire stations on a variety of instruments including gongs, bells or tappers, and punch or printing registers that register the coded signal with holes in, or marks on, a paper tape.

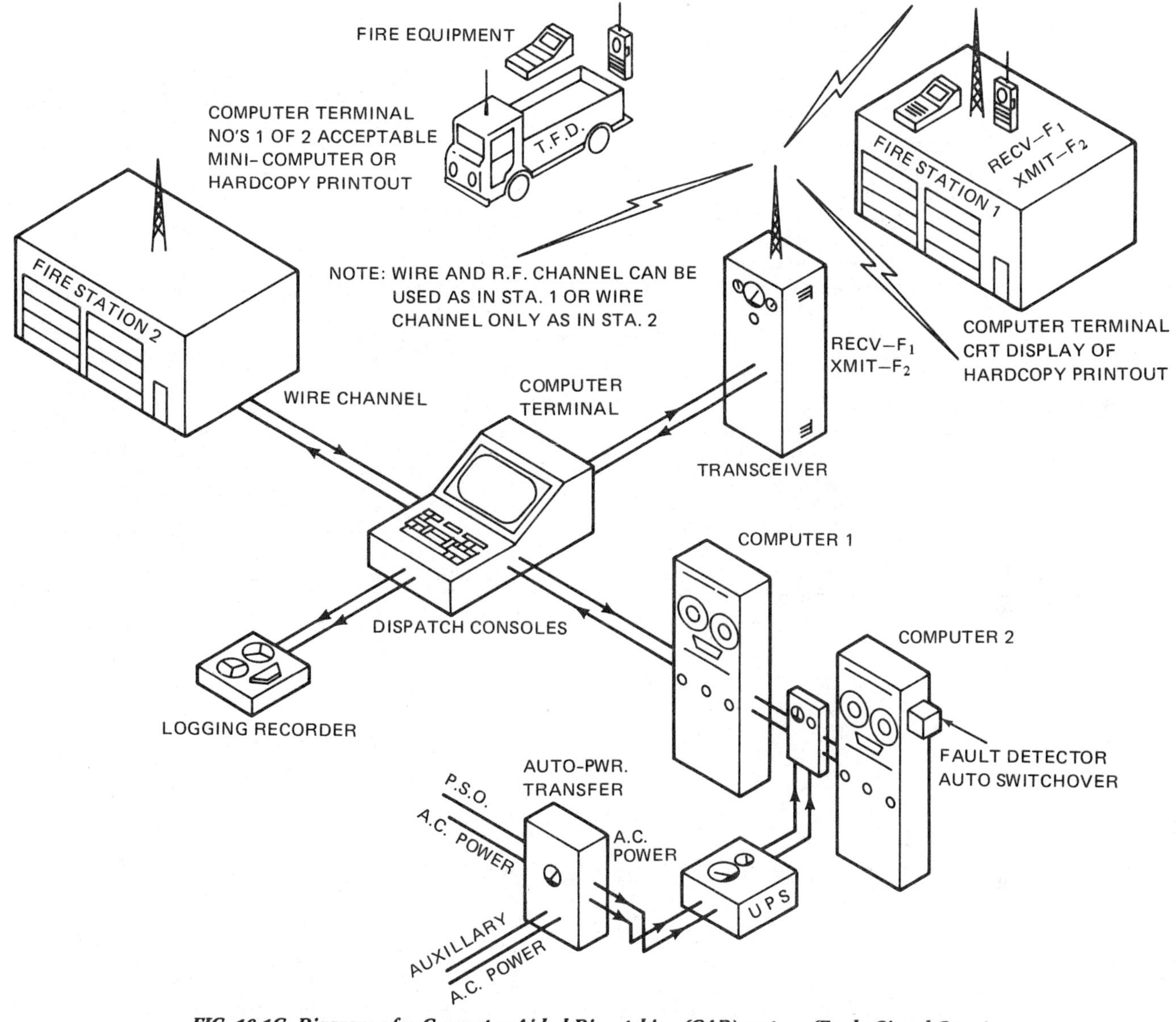

FIG. 16-1G. Diagram of a Computer Aided Dispatching (CAD) system. (Eagle Signal Corp.)

Teletype Circuits: The teletypewriter dispatch circuit (either wired or microwave) is another form of dispatch circuit where a fire alarm dispatcher transmits alarm information to various fire stations by means of a teletypewriter. The alarms are received by the teletypewriter which reproduces the information on a roll of paper. Their major advantage is that more information, if available, can be transmitted to the fire station.

Facsimile Circuits: Facsimile dispatch circuits are wired circuits that incorporate the *telewriter*, a device formerly used by hotels and railroads. The dispatcher writes out instructions on pressure sensitive paper tape with a special stylus, and the instructions are reproduced at the fire stations on a roll of paper tape.

The *teleprinter* is a facsimile device which reproduces written information, including drawings. Receiving units are generally installed in fire stations, and also installed in fire department vehicles when radio is used.

Voice Amplification Circuits: Voice amplification dispatch circuits are wired circuits. As the name implies, this is an arrangement whereby the dispatcher talks into a microphone at the communications center and the information is amplified and received on loudspeakers located in the fire stations.

Radio Channel: A radio channel is another form of dispatch circuit for the transmission of fire alarm signals. Transmitting equipment is located at the communications center with receivers in each fire station. Transceivers on fire apparatus provide communications with companies in the field. A radio system has the advantage of not being connected to or dependent upon wires, unless the transmitter is located at a point remote from the dispatcher.

Telephone Alerting Systems: Various types of telephone alerting systems are available from telephone companies.

Telephone systems are flexible and can be zoned to alert specific people or areas, rather than alerting the general public.

Other Alerting Facilites

Because they are manned entirely by volunteer fire fighters, many fire departments have no one on duty at a fire station to receive an alarm. In these instances, bells, air horns, whistles, or sirens are commonly used to alert the volunteers when an alarm has been received. A satisfactory arrangement is to sound coded signals which communicate the location of the fire. Radio receivers in volunteer fire fighter residences or places of business, and radio pagers (portable receivers carried on the person) are frequently used for receiving alarms.

Supervision and Testing Facilities

The communications center is where trouble signals are received and testing facilities are generally provided. In a leased telephone system, some test facilities are located in telephone buildings. In the event of trouble, both an audible and a visual signal should be provided. An audible signal may be shut off, but it should be designed to operate again in the event of trouble in any of the other circuits connected to the same supervisory device.

Bibliography

NFPA Codes, Standards, Recommended Practices and Manuals. (See the latest *NFPA Codes and Standards Catalog* for availability of current editions of the following documents.)

NFPA 70, *National Electrical Code®*.

NFPA 71, *Standard for the Installation, Maintenance, and Use of Central Station Signaling Systems.*

NFPA 72B, *Standard for the Installation, Maintenance, and Use of Auxiliary Protective Signaling Systems for Fire Alarm Service.*

NFPA 72C, *Standard for the Installation, Maintenance, and Use of Remote Station Protective Signaling Systems.*

NFPA 1221, *Standard for the Installation, Maintenance, and Use of Public Fire Service Communication Systems.*

Additional Readings

Bahme, Charles W., "The Watts Fires and Their Lessons," *Fire Journal*, Vol. 60, No. 2, Mar. 1966, pp. 10-14.

Glicksman, Mark, "Voice Alarms," *Specifying Engineer*, Vol. 49, No. 5, May 1983, pp. 67-73.

Nielsen, D. J., and Ryland, H. G., "Computer Command-Control," *Fire Journal*, Vol. 62, No. 3, May 1968, pp. 43-48.

Singer, Edward N., "Radio Communications in High-Rise Buildings," *With NY Firefighters*, Vol. 42, No. 1, 1091, pp. 10-12.

PROTECTIVE SIGNALING SYSTEMS

Revised by Charles E. Zimmerman, P.E.

A fire detection and alarm system is one of the key elements among the overall fire protection features of any building. Properly designed, installed, operated, and maintained, a fire alarm system can help limit fire losses in buildings, regardless of occupancy. Also, since many of the fire deaths in the United States result from building fires, the use of detection and alarm systems in buildings can reduce the loss of life from fire. This chapter will discuss the operational characteristics of the various types of protective signaling systems currently in use.

TYPES OF SYSTEMS

Fire protective signaling systems are classified according to the functions they are expected to perform. Their installation, maintenance, and use are specified in the several NFPA standards listed in the bibliography at the end of this chapter.

The basic features of each system are:

1. A control unit.
2. A primary (main) power supply which usually is the local light and power service.
3. A secondary (standby) power supply.
4. One or more initiating device circuits to which automatic fire detectors, manual fire alarm boxes, waterflow alarm devices, and other alarm initiating devices are connected.
5. One or more alarm indicating device circuits to which alarm indicating signals such as bells, horns, speakers, etc., are connected, or to which an off premises alarm is connected, or both.

The secondary (standby) power (number 3 above) is required to automatically supply the energy to the system within 30 seconds whenever the primary (main) power supply is incapable of providing the minimum voltage required for proper system operation. Since fire protective signaling systems are required to operate properly when the applied voltage rises to 110 percent of normal, as well as when the voltage falls as low as 80 percent of normal, the secondary supply must not supply energy as long as the primary voltage remains above 85 percent of the rated voltage. The size of the secondary supply usually is measured in the amount of time that the secondary supply will operate the system. Each system type has its own standby time requirement which is explained in the appropriate sections of this chapter.

The secondary supply can be a storage battery, generators, or a combination of a battery and a generator. The secondary supply may be used to power the trouble signals on the system.

Many control units have more than one initiating device circuit, so that the fire location can be indicated on an annunciator panel by floor, wing, subsection, or room. The annunciator can be built into the control unit or located in a lobby, maintenance area, telephone switchboard room, or other space where it is accessible to building and fire service personnel.

Local System

The main purpose of a local protective signaling system is to sound local alarm signals for evacuation of the protected building. (See Fig. 16-2A.) A system could be limited to the basic features indicated above. A local system standby power supply must operate the system for a period of 24 hours under normal load and then operate the alarm system for 5 minutes. This would cover a normal day of occupancy and then a 5 minute alarm to evacuate the building.

In a local system, the alarm is not relayed automatically to a fire department; instead, when the alarm sounds, someone must notify the fire department. If the building is unoccupied at the time of the alarm, fire department response would depend upon a neighbor or passerby hearing it and notifying the department.

Auxiliary System

An auxiliary protective signaling system has circuitry connecting the alarm initiating devices to the municipal fire alarm system, either through a nearby master fire alarm

---

Mr. Zimmerman is a consultant in fire alarm signaling systems based in Norfolk, VA. He was NFPA staff liaison to the NFPA Signaling Systems Technical Committees prior to his retirement in 1983.

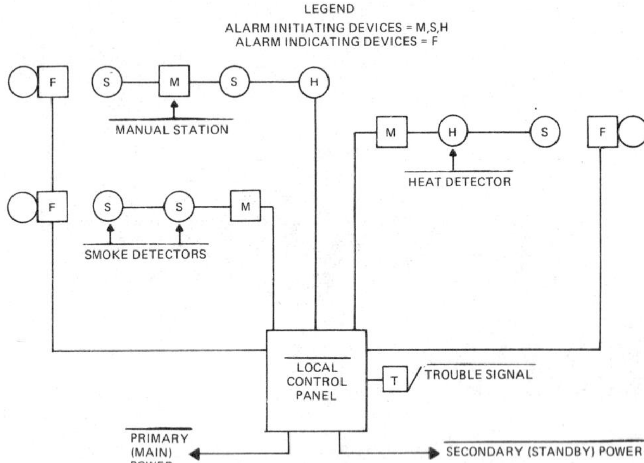

FIG. 16-2A. Typical local fire alarm system.

box (Fig. 16-2B) or through a dedicated telephone line run directly to the municipal communication center switchboard. The signal received by the fire department is the same received when someone manually activates the municipal fire alarm box. Because fire department personnel know which municipal boxes are auxiliarized, responding fire fighters can check for an alarm originating within the premises.

The standby power supply for an auxiliary system is required to operate the system for 60 hours of normal operation followed by 5 minutes of alarm. Since this system has been installed primarily for building protection, the 60 hour span is needed to carry the system over a weekend period. In an auxiliarized system, the protected building may or may not have an evacuation alarm.

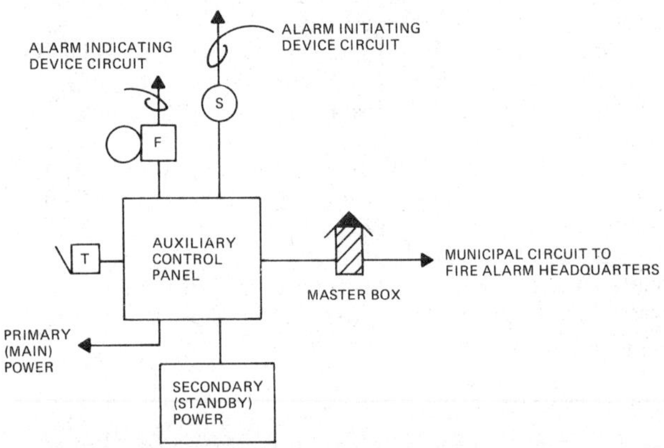

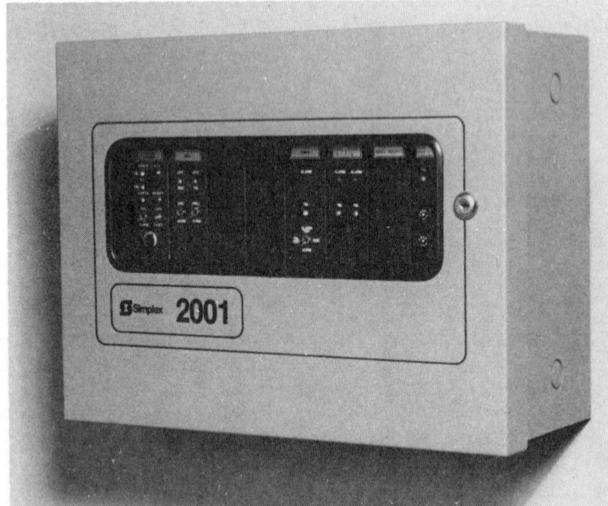

FIG. 16-2B. (Top) Typical auxiliary fire alarm system. (Bottom) Auxiliary system panel. (Simplex Time Recorder Co.)

Remote Station System

A remote station signaling system, while similar to an auxiliary system, has an alarm signal that is received at a remote location which is attended by trained personnel 24 hours a day. (See Fig. 16-2C.) The receiving equipment usually is located at a fire department facility, police station, or telephone answering service. The signal is transmitted over a leased telephone line and is indicated audibly and visually at the remote station. If the remote station is not at the fire department, the remote station personnel notify the fire department of the alarm. System

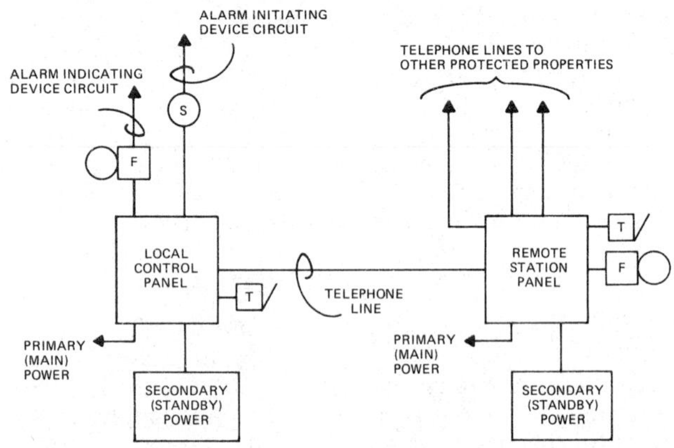

FIG. 16-2C. (Top) Typical remote station fire alarm system. (Bottom) Remote station system panel. (Potter Electric Signal Co.)

trouble signals usually are transmitted automatically to the remote receiving station. The control unit at the remote station, and, if needed, at the protected premises, also is required to have an independent secondary power supply that will operate the system for 60 hours followed by 5 minutes of alarm. As for an auxiliarized system, the protected premises of a remote station system may or may not have an evacuation system.

Proprietary System

The proprietary system is a widely used type of control unit in large commercial occupancies. (See Fig. 16-2D.)

Proprietary and central station systems are somewhat similar, except that the central supervisory station receiving the fire alarm signal in a proprietary system is operated by an individual with a proprietary interest in the protected buildings. The central supervisory station of a proprietary system is generally a guard office within, or near, the building or group of buildings protected by the system. On the other hand, in a central station system, the central station is manned by operators who perform the service for a fee and have no proprietary interest in the protected buildings.

Many existing proprietary systems have separate initiating device circuits for each building zone or subsection, similar to the local, auxiliary, and remote station systems. However, due to the increasing use of electronics, newer proprietary systems for larger buildings often have signal multiplexing and built in minicomputer systems. These systems receive all signals from the building over one or more pairs of wires, and determine the exact location of the fire by use of different frequencies or digitally coded information transmitted over the wires. Figure 16-2E shows the complexity and interrelationship among the many facets of a modern computer controlled proprietary/high rise communication system.

While a proprietary transmitting unit is similar to the other types of transmitting units, the receiving console could be quite different. A proprietary receiving console consists of individual lights, a digital display, or a CRT visual display indicating the alarm point, along with an audible alarm to alert the console operator, and a hard copy printer.

Large proprietary multiplex and computer controlled systems usually do much more than indicate fire alarms to the operator and sound an alarm. These systems often provide for smoke control within the building by automatically closing and opening dampers in heating, ventilating, and air conditioning systems, and turning on exhaust fans. They also may adjust elevator controls so that elevators bypass fire floors, and are automatically routed to the lobby floor for fire department use. In addition to increasing flexibility, use of multiplexing signals greatly reduces the amount of wire used in a building. Computer based proprietary systems often include energy management capabilities resulting in energy savings—a major factor in the recent sizeable increase in the use of these systems in large buildings.

Proprietary control units are required to transmit trouble and alarm signals to the central supervising station. The control unit at the central supervising station as well as remotely located control equipment must have a secondary power supply that will operate the system for

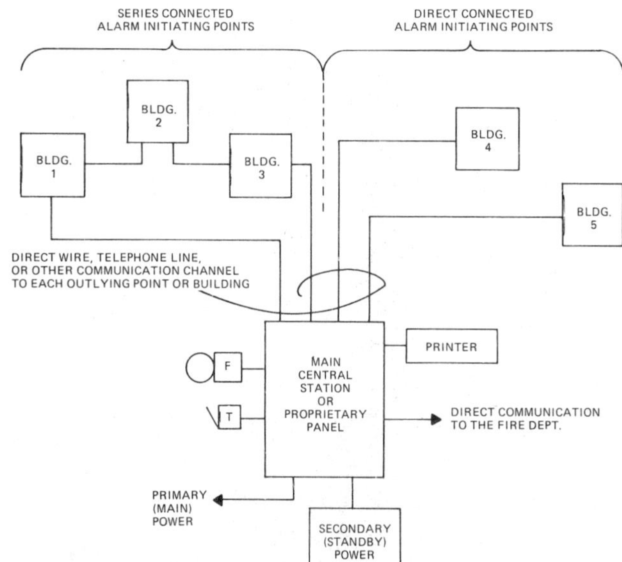

FIG. 16-2D. (Top) Typical proprietary or central station fire alarm system. (Middle) Proprietary system panel. (Autocall Division, Federal Signal Corporation) (Bottom) Proprietary system panel. (Pyrotonics)

24 hours and then be capable of operating for alarm. Since operators are constantly on duty with a proprietary system, 24 hours of standby is considered sufficient.

Emergency Voice/Alarm Communication System

This system is used to supplement a local, auxiliary, remote station, or a propriatary protective signaling sys-

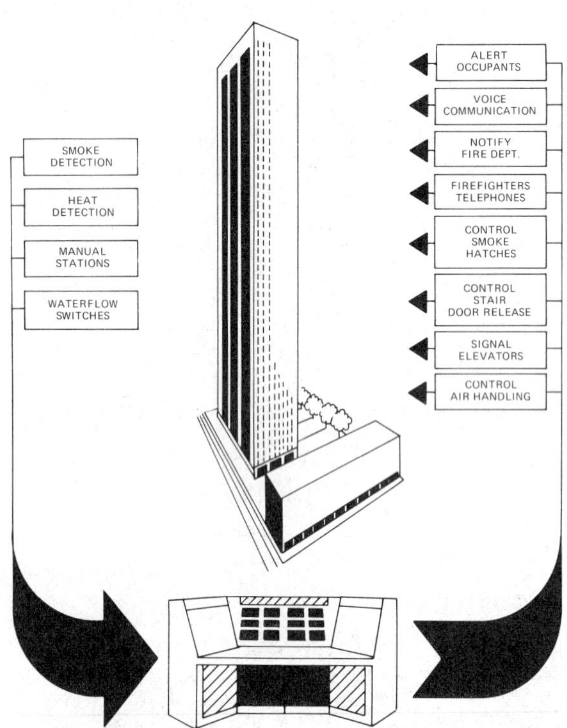

FIG. 16-2E. Typical high rise fire alarm and communication system.

tem. Its standby power supply must operate the voice/alarm signaling service for a fire or other emergency condition, for two hours preceded by 24 hours without the primary operating power. Also, because of the emergency nature of a voice communication system, special requirements in NFPA 72F, *Standard for the Installation, Maintenance and Use of Emergency Voice/Alarm Communication Systems*, cover the survivability of the system so that fire damage to one paging zone will not result in loss of communication to another. (See Figs. 16-2F, 16-2G, and 16-2H.)

FIG. 16-2F. A voice communication system panel. (Autocall Division, Federal Signal Corporation)

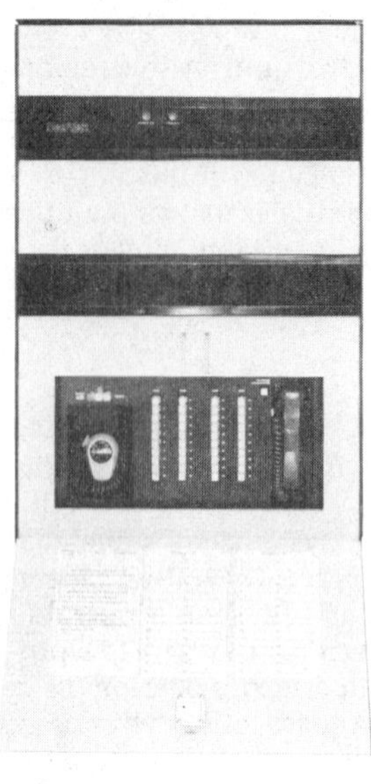

FIG. 16-2G. A voice communication system panel. (The Gamewell Corporation)

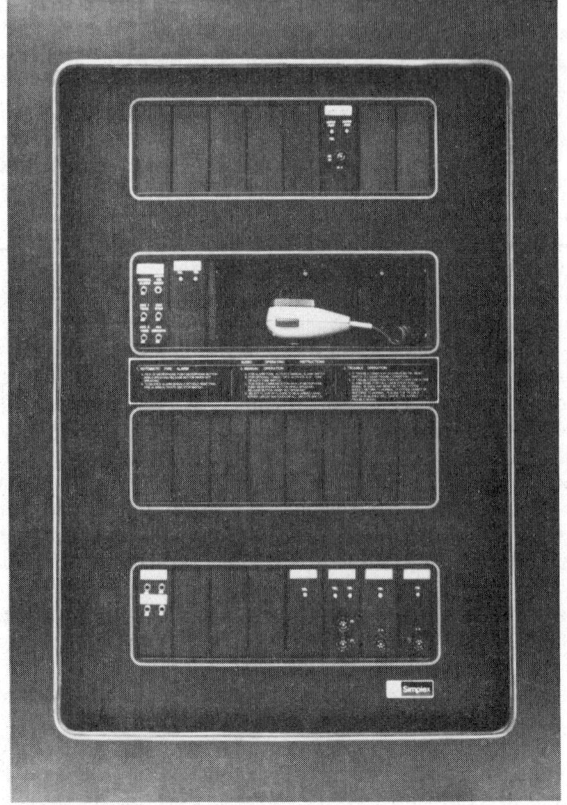

FIG. 16-2H. A voice communication system panel. (Simplex Time Recorder Co.)

The voice/alarm system consists of a series of high reliability speakers located throughout the building. They are connected to and controlled from the fire alarm communication console located in an area designated as a fire command headquarters. From the fire command headquarters, individual speaker zones or the entire building can be selected to receive voice messages which give specific instructions to the occupants. Some systems have fire warden stations on each floor, or fire zones, to which a fire warden would go to assume local command and pass on specific evacuation instructions. Operation of the command headquarters usually is assumed by a trained building employee until the fire department arrives, at which time the officer in charge takes over. The system also may be used during fire fighting operations for communication with the fire fighters.

FIG. 16-2I. A central station office. (ADT)

One important aspect of a voice communication system is that since complete building evacuation is not always feasible, occupants can be instructed to relocate to "safe" areas in which to wait out the fire. In such cases, communication with these people can be maintained to prevent panic and to facilitate further relocation if necessary.

Central Station System

The central station system is similar to a proprietary system. (See Figs. 16-2D and 16-2I.) However, there are differences in the means of signal transmission between the protected building and the receiving station.

The McCulloh circuit, the oldest transmitting means in central station use, normally transmits over two wires but can be switched manually or automatically to transmit over one wire and ground. With this capability, the system will not be rendered inoperative by a break or ground fault in a single wire. It is customary to connect the plants of several subscribers to a single transmission circuit. Each such circuit terminates in a recording instrument, and each subscriber has one or more coded signal numbers not repeated on that particular circuit.

A second means of signal transmission in central station systems is called the direct wire circuit. As its name implies, this is a direct dedicated (unswitched) pair of leased telephone lines running between the building and the central station panel. In such cases, codes are unnecessary because each building is on a separate circuit.

Limitations on circuit loadings and other means of transmission are specified in NFPA 71, *Standard for the Installation, Maintenance, and Use of Central Station Signaling Systems.*

When a signal is received at a central station, the appropriate authorities are informed. The central station dispatches someone to the protected building when equipment must be reset manually. Central station systems require secondary (standby) power supplies that will operate the system for a period of 24 hours.

TYPES OF SIGNALS

Fire alarm systems provide several distinct types of audible signals. (To eliminate confusion between the term "supervised" and "supervisory," the NFPA signaling standards are being revised to change the term "a supervised circuit" to "monitoring integrity of installation conductors." The requirements still are that a "trouble signal" must sound upon the occurence of a single open or ground which would prevent normal operation of the system. On the other hand, a "supervisory signal" indicates the off normal position, or condition, of some part of a sprinkler or other extinguishing system.)

Trouble Signal

A "trouble signal" is given when a fault occurs in a supervised (monitored) device or circuit of a protective signaling system. Circuits that are normally supervised include main power, alarm initiating, and alarm indicating circuits. Trouble signals for remote station, proprietary, and central station systems are usually received at a central supervisory station. In local and auxiliary systems, trouble signals are usually sounded in areas where maintenance personnel are normally present.

Supervisory Signal

In sprinklered occupancies, a sprinkler "supervisory signal" is given when a critical component in the sprinkler system is in an abnormal condition. These conditions include such factors as low water service pressure, loss of power to a fire pump, closing of a water supply valve, low water level of a water supply tank, or near freezing temperature in an outdoor water supply tank. Local and auxiliary system supervisory signals are usually sounded in areas where maintenance personnel are normally present. Remote station, proprietary, and central station system supervisory signals are usually received at the remote station or central supervisory station.

Alarm Signal

When a fire is detected, an alarm signal is given upon operation of either a manual or automatic initiating device (manual box or fire detector). Although alarm signals generally involve the sounding of audible signals throughout the building, in large buildings they may be sounded initially only in the vicinity of the immediate fire area. In some systems, the alarm signal may be a taped or live voice message broadcast over a fire alarm speaker system.

Requirements for alarm/notification signals can be found in NFPA 72G, *Standard for the Installation, Maintenance and Use of Notification Appliances for Protective Signaling Systems.*

Noncoded Signal

The alarm signals produced by a fire alarm system may be the continuous sounding of audible signal devices distributed throughout the protected area. When these devices are sounded continuously, the system is "noncoded."

Coded Signal

When the devices are sounded intermittently in a prescribed pattern, the system is referred to as a "coded"system. Various types of coded systems are employed, depending upon the size and particular needs of the installation.

Visual Signals

Because public buildings should be accessible to handicapped people, fire alarm systems in such buildings often include visual alarm signals to alert occupants with impaired hearing. A combination horn/light unit has become popular for this purpose; such a unit consists of an alarm horn and a high intensity strobe light. When the unit is energized, the horn sounds and the light flashes. Individual strobe light units can be retrofitted into existing buildings or installed where alarm horns are not a desirable fire alarm evacuation signal. When visual signals are included in the fire alarm system, the signals must be carefully located so they can be seen from every point in the building.

Evacuation

The manner of sounding alarms should be standardized throughout as large a geographical area as possible. Uniformity prevents people from being misled or confused by different alarm sounds as they move from one locality to another.

When a distinctive fire alarm signal is used to notify building occupants to evacuate, the use of a well known fire alarm evacuation signal is recommended to facilitate quick and positive recognition of the meaning of the alarm.

The recommended fire alarm evacuation signal is a uniform Code 3 temporal pattern, using any appropriate sound, keyed ½ to 1 sec "ON," ½ sec "OFF," ½ to 1 sec "ON," ½ sec "OFF," ½ to 1 sec "ON," and 2½ sec "OFF," with timing tolerances of ±25 percent, repeated for not less than 3 minutes. The minimum repetition time, however, may be interrupted manually for voice communication.

The recommended standard fire alarm evacuation signal is intended only as an evacuation signal. Its use should be restricted to situations where occupants hearing the signal should evacuate the building immediately. It should not be used when (with the approval of the authority having jurisdiction), the planned action during a fire emergency is not evacuation. Another signal should indicate relocation of the occupants from the affected area to a safe area within the building, or their protection in place (e.g., in high rise buildings, health care facilities, penal institutions, etc.).

PROTECTIVE SIGNALING SYSTEM CIRCUITS

Each protective signaling system has a number of circuits connected to the system control unit. Each circuit provides a general function and may be arranged in a number of different ways. Major circuits found in protective signaling systems are:

Basic Initiating Circuit: A typical basic alarm initiating circuit (formerly called a "class B" circuit), consists of a two wire circuit with an end-of-line resistor. Initiating devices, with normally open contacts, are connected in parallel. (See Fig. 16-2J.) A small supervisory current normally flows through both relays (or an equivalent solid state circuit), the field installation, wires, and resistor. The current is sufficient to energize T_1 but not A_1. Operation of an initiating device (detector) shunts the resistor. This increases the current in the circuit, and energizes A_1 to initiate the alarm. A short between the wires or a ground fault on both wires also will cause an alarm.

A single open on the circuit would deenergize T_1 and initiate a trouble signal. Also, it would render inoperable all initiating devices which are electrically beyond the open.

When an automatic fire detector has integral trouble contacts, they must be wired between the last detector and the end of line resistor. This prevents the opening of a trouble contact from impairing alarm operation from other fire detectors. (See Fig. 16-2K).

Operation with Open Fault Condition: To operate the system with an open circuit fault (formerly called a "class A" circuit) on the initiating circuit, a typical wiring arrangement would be as shown in Figure 16-2L. The circuit would be connected to a control panel designed to receive the signal in spite of an open on the field wiring circuit. In this case, since the end of the line also is at the

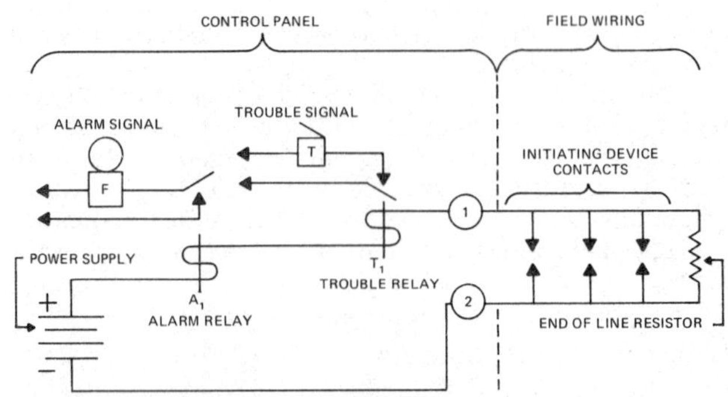

FIG. 16-2J. Typical basic initiating device circuit shown in supervisory (monitored) condition (power on).

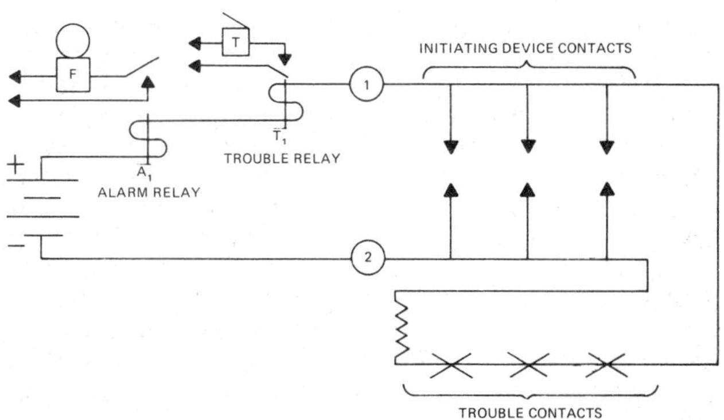

FIG. 16-2K. Typical basic initiating device circuit, with trouble contacts, shown in supervisory (monitored) condition (power on).

fire alarm control panel, relay T_1 deenergizes, when an open on the wire occurs, sounding the trouble signal "T." The panel then would be conditioned manually or automatically by the operation of S_1 and S_2 to connect terminals 1 and 4, and 2 and 3 together. Then the system could operate for an alarm despite an open circuit on either or both of the initiating circuit wires. However, should two opens occur on a single wire, such as at point "A" and "B," the device to the right of point "A" would not be able to operate the system.

Many variations of this circuit provide for operation on one or more fault conditions. In the past, each was placed in the so called "class A" category. However, it was difficult to determine the capability of a system by this brief title. To address the need for an expanded description, the NFPA 72D, *Standard for the Installation, Maintenance and Use of Proprietary Protective Signaling Systems*, now contains two tables which further classify

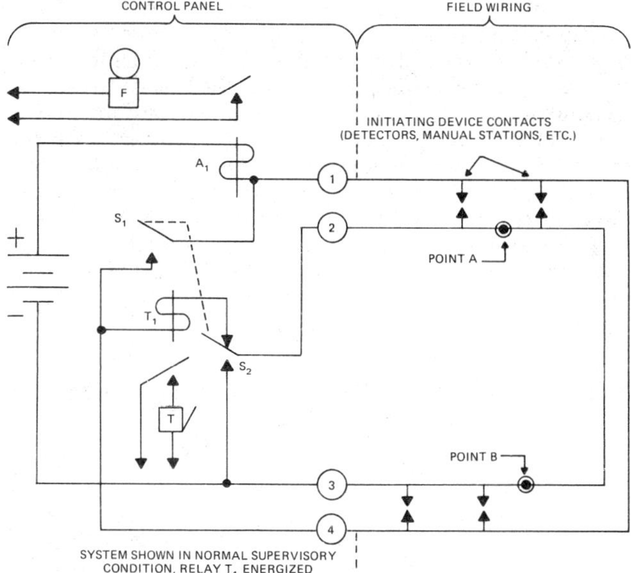

FIG. 16-2L. Typical initiating device circuit for operation with open fault condition. Shown in normal supervisory (monitored) condition (power on)

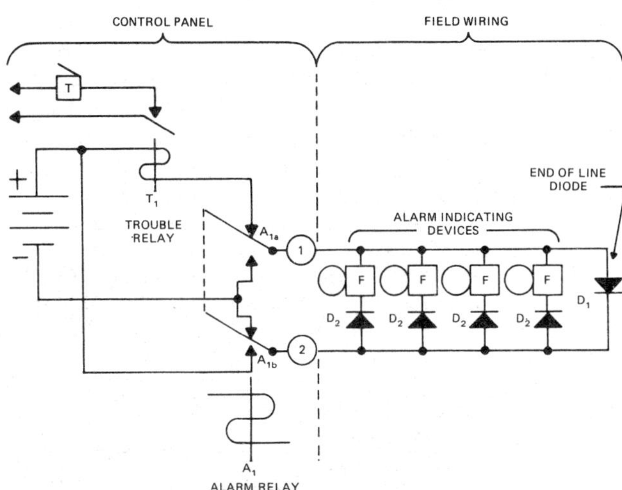

FIG. 16-2M. Typical supervised indicating device circuit shown in supervisory (monitored) condition (power on).

the performance characteristics of initiating device and signaling line circuits. The circuits now are called "styles," with five (A-E) for initiating device circuits and seven (1-7) for signaling line circuits.

Indicating Device (Appliance) Circuits

Indicating device circuits also may be basic supervised circuits or be capable of being conditioned to operate despite an open circuit on the installation wiring.

A typical basic indicating device circuit is shown in Figure 16-2M. In the normal supervisory condition, current would flow from the negative side of the power supply through the normally closed contact (A_{Ib}) of the alarm relay A_1, out terminal 2 and through the end of line diode D_1, back through terminal 1, through the normally closed contact (A_{1a}) through the supervisory relay coil T_1, to the positive side of the power supply.

In this supervisory condition, diodes D_2 block the flow of current through the alarm devices "F," preventing their operation. Sufficient current flows through relay T_1 to energize it as shown. Should an open occur in the wiring to the alarm devices, or in D_1, relay T_1 will deenergize, sounding the trouble signal "T."

When an alarm occurs, and alarm relay A_1 is energized, contacts A_{1a} and A_{Ib} transfer from the upper, normally closed position, to the lower, normally open position. This changes the potential at terminal l from plus (+) to minus (−) and at terminal 2 from minus to plus. Diode D_1 now blocks the flow of current through the end of line, and diodes D_2 conduct current to the indicating devices "F," causing them to operate.

An arrangement which functions similarly to the initiating device circuit also can be provided to condition the indicating device circuit to operate despite of an open circuit.

A variation on this arrangement is the parallel supervision circuit. It is similar to the end of line device circuit, except that the panel contains additional circuitry in the form of a bridge circuit. After all the devices are connected, the bridge circuit is balanced at the panel. Once the circuit is balanced for all connected devices, an open circuit in the wiring, or the coil or contacts of a single device, unbalances the circuit and causes a trouble signal.

The advantage of the parallel supervision circuit is that in addition to supervising the installation wiring, all of the indicating device's coils and contacts also are supervised.

Distinctive Signals

All fire alarm systems produce distinctive alarm and trouble signals. Combination systems also may produce sprinkler supervisory, burglar alarm, or process monitoring signals in addition to the fire alarm signal. When combination systems are used, a fire alarm signal must be clearly recognizable and take precedence over all other signals, even though a nonfire alarm signal may be initiated first.

Signaling Line Circuits

In all except local systems, signals are transmitted from the protected premises to a central supervisory station. The circuits over which these signals are transmitted are referred to as signaling line circuits. Although these circuits may take different forms, depending upon the type of system, all are required to be supervised (monitored) in some manner.

In auxiliary systems, a hardwire pair connects the control panel to the trip coil (or its equivalent) in the municipal master box.

In a remote station system, the signaling line circuit is typically either an end of line circuit similar to Figure 16-2J or a reverse polarity circuit similar to Figure 16-2M. In a very basic way, the remote station in Figure 16-2J would be the "control panel" and the signaling line circuit the "field wiring" between terminals 1 and 2 and the first initiating device. The initiating devices would be found in the protected building. In Figure 16-2M, the "control panel" would be the one in the protected building, the signal line circuit would be the "field wiring" between terminals 1 and 2 and the first alarm indicating device, and the remote station would be one of the alarm indicating devices.

Proprietary and central station systems, particularly the multiplexing types, may contain numerous signaling line circuits. Any circuit extending from a protected building to the central supervisory station would be considered a signaling line circuit.

Multiplexing

The term multiplexing means the transmission of multiple signals over a single transmission path. In multiplexing fire alarm systems, signals from many initiating devices or groups of initiating devices are multiplexed over a single wire pair. The signals can be reproduced and identified by a device or group of devices at the receiving unit. The primary advantage of multiplexing of signals is that in large installations the amount of wiring needed to connect all devices to the panel is minimized, while still allowing devices or zones to be identified individually.

In general, there are two basic types of multiplexing. These are referred to as "time division" multiplexing and "frequency division" multiplexing.

Time Division Multiplexing: Each device or group of devices is assigned an address code when installed. The control panel then scans each device sequentially to determine its current status. For example, the system will address Device 1 by transmitting a specific digital code over the signaling line circuit. If Device 1 reports back the code for normal, the system then will address Device 2. Each time the addressed device reports, the report is compared with the status of that device on the previous scan. As long as the device status on two subsequent scans is the same, no signals are generated. If, however, a "change of state" is reported (a different status on two subsequent scans), the receiving unit then notifies the operator by audible and visual signaling that a change of state has occurred. Thus, all changes of state are indicated to the operator. In this way, the system operator is notified of every status change of every device or group of devices in the system. These status changes might include fire alarm, device trouble, sprinkler supervisory, abnormal process conditions, burglar alarm, or many other types of signals.

While the alerting tone generated to get the attention of the operator may be common to all signals, signal differentiation is required by a visual means. The operator can tell the difference between a fire alarm and an abnormal process condition by the color of a light, an alphanumeric code, or by a message displayed on a CRT or printer. In addition, the system may incorporate a signal priority system. In such a case, where simultaneous signals are obtained from several devices, the system displays the signals in a predetermined order, depending upon the type of signal and its assigned priority in the system.

Fire alarm signals always take precedence over all other signals. Following fire alarm signals, the order might be sprinkler supervisory, burglar alarm, processing monitoring, and trouble signals. The order of priority is determined by the importance of the signal received. For example, where certain abnormal process signals may be indicative of impending explosion or fire, these signals may be given a higher priority than burglar alarm signals.

Frequency Division Multiplexing: In a system using this second major type of multiplexing, each device or group of devices transmits a specific frequency tone over the signaling line whenever it changes state. Rather than addressing each device sequentially, the system can transmit multiple signals simultaneously.

As in time division multiplexing, frequency division multiplexing equipment generally also includes a prioritizing system. Regardless of the type of multiplexing used, the systems must be designed so that the time between the sensing of a fire alarm at an intiating device and the time it is recorded or displayed at the receiving station does not exceed 90 seconds. In addition, if multiple simultaneous status changes of any type occur in the system, the systems are required to record these changes at a rate not slower than either 50 signals or 10 percent of the total number of initiating device circuits connected (whichever is smaller), within the 90 second time period without the loss of any signals.

COMPATIBILITY AMONG SYSTEMS

One frequent question from building designers and equipment owners pertains to compatibility among equipment produced by various manufacturers. That is, if the control unit is manufactured by one company, how is compatibility determined with regard to other manufacturers' detectors, indicating devices, annunciator panels, etc.? The key to all system compatibility is the installation wiring diagram provided with the basic control unit. If a given circuit extending from the control panel is compatible only with specific devices, those devices will be specified by manufacturer and model number. If a circuit contains only an electrical rating and does not specify specific devices, then it may be compatible with any device having similar electrical ratings. If there is any question about compatibility of equipment, the laboratory listing the equipment should be consulted.

CODE REQUIREMENTS

Code requirements applicable to fire alarm systems can often be confusing, because a number of different codes all apply to a given circumstance.

First, it is important that all equipment be "listed" by a testing laboratory. In many cases, use of listed equipment is mandated by local codes; but even if not mandated, listing is one way to be sure that an independent agency has evaluated the equipment.

To determine the type system required for a given occupancy, consult a code such as NFPA 101® *Code for Safety to Life from Fire in Buildings and Structures* (commonly called the Life Safety Code®). The occupancy code will specify the type of system and type of detection that is required. For example, a nursing home would be required to have a fire alarm system that provides evacuation signals, contains smoke detectors in the corridors as a minimum, and is connected to the fire department. Once the type of system and the type of detection required by the occupancy has been determined, guidelines for the installation of fire detectors found in NFPA 72E, *Standard for Automatic Fire Detectors* should be followed.

In addition to the occupancy and device standards, the applicable system standard should be consulted for the installation requirements of the system (i.e., control unit, indicating devices, manual boxes, etc.). Using the nursing home example cited above, if an auxiliary system were the type system selected, the standard to be consulted is NFPA 72B, *Standard for the Installation, Maintenance and Use of Auxiliary Protective Signaling Systems for Fire Alarm Service*. The standard provides guidelines necessary to install the control unit and its associated subsystems, as well as information on system maintenance and testing.

Maintenance and Testing: After a protective signaling system is installed, it is extremely important to conduct a proper acceptance test and regular periodic tests on all parts of the system. Guidelines for these tests can be found in NFPA 72H, *Guide for Testing Procedures for Local, Auxiliary, Remote Station and Proprietary Protective Signaling Systems*. Maintenance of signaling equipment, along with tests, is as important as the installation of the system in the first place. Maintenance must not be overlooked in the overall fire protection plan.

Bibliography

NFPA Codes, Standards, Recommended Practices and Manuals. (See the latest *NFPA Codes and Standards Catalog* for availability of current editions of the following documents.)

NFPA 13, *Standard for the Installation of Sprinkler Systems.*

NFPA 70, *National Electrical Code.*

NFPA 71, *Standard for the Installation, Maintenance and Use of Central Station Signaling Systems.*

NFPA 72A, *Standard for the Installation, Maintenance and Use of Local Protective Signaling Systems for Guard's Tour, Fire Alarm and Supervisory Service.*

NFPA 72B, *Standard for the Installation, Maintenance and Use of Auxiliary Protective Signaling Systems for Fire Alarm Service.*

NFPA 72C, *Standard for the Installation, Maintenance and Use of Remote Station Protective Signaling Systems.*

NFPA 72D, *Standard for the Installation, Maintenance and Use of Proprietary Protective Signaling Systems.*

NFPA 72E, *Standard on Automatic Fire Detectors.*

NFPA 72F, *Standard for the Installation, Maintenance and Use of Emergency Voice/Alarm Communication Systems.*

NFPA 72G, *Standard for the Installation, Maintenance and Use of Notification Appliances for Protective Signaling Systems.*

NFPA 72H, *Guide for Testing Procedures for Local, Auxiliary, Remote Station and Proprietary Protective Signaling Systems.*

NFPA 101® *Code for Safety to Life from Fire in Buildings and Structures.*

Additional Readings

"A Proposed Standard Fire Alarm Signal," *Fire Journal*, Vol. 69, No. 5, Sept. 1973, p. 24.

"Assessment of the Potential Impact of Fire Protection Systems on Actual Fire Incidents," by the Johns Hopkins University, prepared for the Federal Emergency Management Agency, U.S. Fire Administration, May, 1980.

Cellantani, E. N., "A Co-ordinated Detection/Communication Approach to Fire Protection," *Specifying Engineer*, Vol. 48, No. 5, May 1982, pp. 58-62.

"High-Rise Fire Alarm Systems—Recommendations for State and Local Codes," National Electrical Manufacturers Association, Washington, DC, Aug. 1984.

Humphreys, W. L., "The Alarming Problem," *Fire Journal*, Vol. 67, No. 5, Sept. 1973, pp. 15ff.

"Implementing a Community-Wide Automatic Residential Remote Alarm System: The Westland Plan," prepared for the Federal Emergency Management Agency, U.S. Fire Administration, Nov. 1980.

Johnson, Erling R., "Accidental Power Cross Results in Improved Alarm System Design," *Fire Journal*, Vol. 68, No. 2, Mar. 1974, p. 7.

Johnson, Joseph E., "Engineering Early Warning Fire Detection," *Fire Technology*, Vol. 5, No. 1, Feb. 1969, pp.5-15.

Kravontka, Stanley J., "False Fire Alarms in Urban Public Schools," *Fire Technology*, Vol. 10, No. 3, Aug. 1974, pp. 220-227.

Kravontka, Stanley J., "A Fire Signal for Deaf School Children," *Fire Technology*, Vol. 11, No. 1, Feb. 1975.

Letts, J. B., "Fire Alarm Control and Indicating Equipment," *Fire Surveyor*, (U.K.), Aug. 1982.

McPherson, Robert W., "A New Concept of Evacuation Alarm Signaling in High-Rise Buildings," *Fire Journal*, Vol. 65, No. 6, Nov. 1971, pp. 8, 9.

Pezoldt, V. J., and Van Cott, H. P., "Arousal from Sleep by Emergency Alarms: Implications from the Scientific Literature," NBSIR 78-1484, National Bureau of Standards, Washington, DC.

"Remote Detection and Alarm for Residences: The Woodlands System," prepared for Federal Emergency Management Agency, U.S. Fire Administration, May 1980, Reprint Mar. 1981.

Stimson, Donald L., "Designing a Fire Detection System for a Large Hospital," *Fire Journal*, Vol. 67, No. 2, Mar. 1973, p. 65.

HOUSEHOLD WARNING SYSTEMS

Revised by Charles E. Zimmerman, P.E.

Why are household warning systems needed? Statistics show that approximately 71 percent of structural fires occur in residential occupancies; residential fires are responsible for approximately 78 percent of all fire deaths; and approximately 57 percent of the total structural dollar loss results from residential fires. These appalling statistics could be reduced significantly with the widespread use of smoke detection devices in residential occupancies.

Why smoke detectors? Actual fire tests in residential occupancies have shown that measurable amounts of smoke have preceded measurable amounts of heat in almost all cases. In the other cases, the smoke and heat appeared almost simultaneously.

This chapter will discuss the reasons household fire warning systems are needed, the history of residential fire warning systems, research into system effectiveness, the importance of having a system in the home and how to make the best use of it, and the importance of an effective escape plan. The types of systems and detectors will also be covered, along with guidelines for installing and testing household fire warning equipment.

Although heat detectors for residential use have been available since 1921, field tests have shown that they are not as effective as smoke detectors in detecting fires in the home.

Residential smoke detectors were relatively expensive when they began to appear in the marketplace in the late 1960s. In 1970, however, the introduction of a battery operated smoke detector, along with several new line powered smoke detectors, initiated a period of public acceptance of single station residential smoke detectors.

RESIDENTIAL FIRE DETECTOR RESEARCH

Current performance requirements and installation practices for residential fire detectors are the result of knowledge and experience gained from several test programs. The first major test programs using heat and smoke detectors were "Operation School Burning" and "Operation School Burning No. 2," conducted by the Los Angeles Fire Department in 1959 and 1961. These tests had a major impact on fire protection and illustrated two significant points: first, the tests proved that smoke detectors could provide a high level of life safety due to their fast response to a fire; and second, detectable quantities of smoke usually preceded detectable levels of heat.

The next significant test series was conducted by the Bloomington (MN) Fire Department in May 1969. Although all detectors employed in the Bloomington tests were system connected units, the results of these tests—such as time of detector response, order of response as a function of detector and fire type, and conditions at time of response—correlated well with subsequent tests.

In 1974, NFPA members adopted a major change to NFPA 74, *Standard for the Installation, Maintenance, and Use of Household Fire Warning Equipment* (hereinafter referred to as NFPA 74). Prior to this, NFPA 74 required that in addition to a smoke detector, a home be protected with a heat detector in every major room of the dwelling. The 1974 edition recognized two facts: (1) the mandatory total system approach was impractical for the majority of the population, and (2) heat detectors were not the most effective detectors for the home. Therefore, the responsible NFPA Sectional Committee developed a matrix of four "levels of protection" from which the homeowner could choose. The four levels of protection concept was finally adopted by the NFPA membership, but not without a fight. The opposition (principally some fire chiefs and fire marshals) felt that they could not recommend anything less than complete protection, and that effectiveness of Level 4, which provided for one or two smoke detectors in the home, had not been proven by test data.

Based on this latter argument, the National Bureau of Standards (NBS) contracted with IIT Research Institute and Underwriters Laboratories (UL) to provide just such test data. The results of the first phase of this test program were published in August 1975 under the title "Detector Sensitivity and Siting Requirements for Dwellings." The study, more commonly known as the "Indiana Dunes Report," showed, in fact, that Level 4 coverage was inad-

Mr. Zimmerman is a consultant in fire alarm signaling systems based in Norfolk, VA. He was NFPA staff liaison to the NFPA Signaling Systems Technical Committees prior to his retirement in 1983.

equate under certain conditions. The report recommended that at least one smoke detector on each story of the residence be used as a minimum requirement. This was referred to as the "every level" smoke detection system. Based primarily on the results of this report, NFPA 74 was revised again in 1978 by dropping the four levels of protection and adopting a minimum required system of one smoke detector on each story (floor level) of a home. Also, because of the slow response of heat detectors when compared to smoke detectors, NFPA 74 no longer permitted the use of heat detectors as the prime method of fire detection.

Shortly after publication of the first phase of the Indiana Dunes Report, (unknown to them at the time), the Minneapolis Fire Department conducted a similar independent study on smoke detector performance. The Minneapolis and Dunes tests were very similar in that each was conducted in a two story house with basement, with detectors installed on each floor. Likewise, the conclusions of the Minneapolis and Dunes tests were essentially identical; thus the Minneapolis tests provided an independent verification of many of the conclusions of the Dunes tests.

ELEMENTS OF RESIDENTIAL FIRE PROTECTION

Three key elements are necessary to the overall protection of residential occupancies from the loss of life. These elements are: (1) minimization of fire hazards, (2) installation of smoke detectors, and (3) escape planning. Each element is important individually, and all three are interdependent in any successful residential fire protection scheme.

Minimizing Fire Hazards

It is always easier to prevent a fire before it occurs than to detect it once it has occurred; therefore, minimizing fire hazards in and around the home is the first key element to minimizing hazards. This includes removing all unessential flammable liquids and other highly combustible items from the home, properly storing essential combustibles away from ignition sources, and not overloading electrical circuits. Further, it should also mean strict adherence to and enforcement of building and fire codes in the construction of residential occupancies, use of only low flame spread materials for interior finish, and use of low flammability or fire retardant treated materials in residential furnishings. It must be realized that if a fire develops rapidly enough, residential smoke detectors may not provide enough time for the occupants of a dwelling to escape before their exit routes are cut off. Therefore, it is highly dangerous for people to be careless, thinking smoke detectors will protect them.

Installation of Smoke Detectors

The second key element in minimizing hazards is the installation and proper maintenance of the minimum acceptable number of smoke detectors. As required by NFPA 74, the acceptable minimum consists of one smoke detector outside each sleeping area, one on each habitable story, and one in the basement of the home. It is also recommended that the minimum package be supplemented by additional smoke detectors in other areas as needed. For example, a smoke detector could be located inside each bedroom (especially if the occupants sleep with their bedroom doors closed), in living and family rooms, and smoke or heat detectors in the kitchen, attached garage, attic, and furnace room. Each additional detector could provide more time for the family to escape. When more than one smoke detector is installed, the detectors should be arranged so all of the alarm devices sound when any one device detects smoke. This can be done either by installing multiple station smoke detectors or a complete system with smoke detectors, panel, and separate alarm sounding devices.

Equally as important as the installation is proper care and maintenance of the detectors. Each residential smoke detector has an owner's booklet describing the necessary maintenance procedures. A smoke detector with a dead battery will be useless in a fire. In fact, analysis of fire reports involving death or serious injury where smoke detectors have been installed shows that, in almost all cases, the detectors were inoperative at the time of the fire because the homeowner had failed to replace a worn out battery, failed to install the detector properly, had intentionally disconnected the power due to false alarms, or failed to properly evacuate the dwelling. Because of the battery problem, NFPA 74 mandates use of an ac primary power source in all new construction. It would follow that even in existing households smoke detectors should be powered by the ac house current, not only by batteries.

Escape Planning

The third key element in residential fire protection is the development and practice of a family escape plan. Remember, residential smoke detectors can do no more than warn occupants of a fire in the home. Detectors will not extinguish the fire. Also, depending upon the speed at which the fire is developing, the amount of time available for escape may be limited. It is critical that all occupants leave the home immediately and *call the fire department from a neighbor's house*. Many deaths have been reported in fires where people have unwisely taken extra time to get dressed, gather valuables, or look for pets. Also, it is important that everyone have an alternate way out of each room in case the primary exit is blocked by fire. There must be a prearranged outside meeting place so everyone will know that the entire family has escaped the fire. Deaths have resulted because people reluctant to climb out of windows were overcome by smoke, and, equally tragic, many have died when they reentered a burning home to look for people who had in fact already escaped.

Information on how to formulate a family escape plan is contained in the homeowner's booklet provided with each residential smoke detector and by NFPA in its "E.D.I.T.H." (Exit Drills In the Home) materials.

TYPES OF SYSTEMS

A residential fire detection system may range in size from one single station smoke detector in an apartment, mobile home, or recreational vehicle to a centrally wired system containing numerous detectors and separate alarm signaling devices. The fire detection system may include a burglar alarm system and an emergency medical alert system. It may be connected to a central receiving headquarters by leased telephone lines or a two way cable television system. The number of devices and the complexity of the system are determined, on the low end, by the minimum acceptable for a given dwelling, and, on the high end, by the extent of the occupant's needs and finances.

Single and Multiple Station Detectors

A single station detector (Figs. 16-3A and 16-3B) is a self-contained device which consists of a detecting element, electrical control components, an alarm sounding appliance, and provision for connection to a separate power supply source or an integral power source, or both. The detector is connected to the house power system by a line cord and plug, a Class 2 plug in transformer is connected to the detector by power limited wire, or directly wired to an outlet box like an electrical fixture.

FIG. 16-3A. Single station photoelectric smoke detectors. (Chloride Pyrotector, left, and Electro Signal Lab, Inc., right)

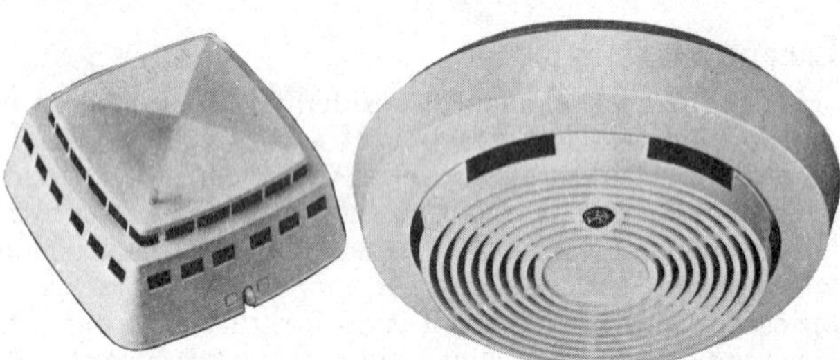

FIG. 16-3B. Single station ionization smoke detectors. (Statitrol Division, Emerson Electric—left, and BRK Electronics—right)

A direct wired type, when provided with one or more additional wires for interconnection with other similar detectors, is referred to as a multiple station detector. (See Fig. 16-3C.) When interconnected according to the instructions provided, an alarm from any one of the interconnected detectors sounds the alarm in all the detectors. This is important because the detector near the bedrooms should sound when a detector located in another area senses smoke. The number of detectors which can be interconnected varies from one model to another. In addition, some interconnection methods allow the connection of heat detectors or manual fire alarm boxes to the interconnecting wire. These detectors may be used to form a household system without the need for a separate control panel.

Some smoke detectors are equipped with a transmitter which sends a signal to a receiver unit, which is usually located in a bedroom. If the detector is an ac plug in model, a line carrier transmitter impresses a high frequency signal on the house wiring. The receiver unit detects the signal as long as both are on the same power company transformer. If the detector is battery powered, it may contain a radio transmitter (similar to those used in garage door openers) with a range up to 200 ft. The receiver unit is usually ac powered. Some models also send a periodic signal to indicate to the receiver that the smoke detector is still functioning. The detector will also send a signal when the battery needs replacement.

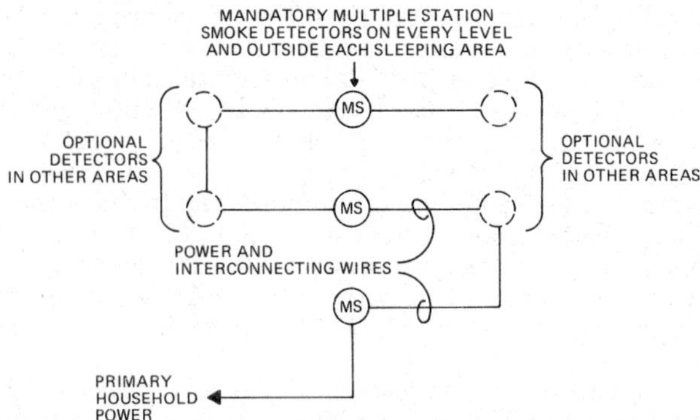

FIG. 16-3C. Typical household multiple station smoke detector system.

Household Fire Warning System

In addition to single and multiple station detectors, there are household fire warning systems similar in makeup and operation to the local fire alarm systems described in NFPA 72A, *Standard for the Installation, Maintenance, and Use of Local Protective Signaling Systems for Guard's Tour, Fire Alarm and Supervisory Service.* Such a system typically consists of a household system control unit (Fig. 16-3D) which derives main power from the ac house wiring, and usually contains a rechargeable standby battery capable of operating the system for at least 24 hours. In addition, this system uses either smoke detectors of the system connected type or single station smoke detectors that contain alarm contacts. The fire

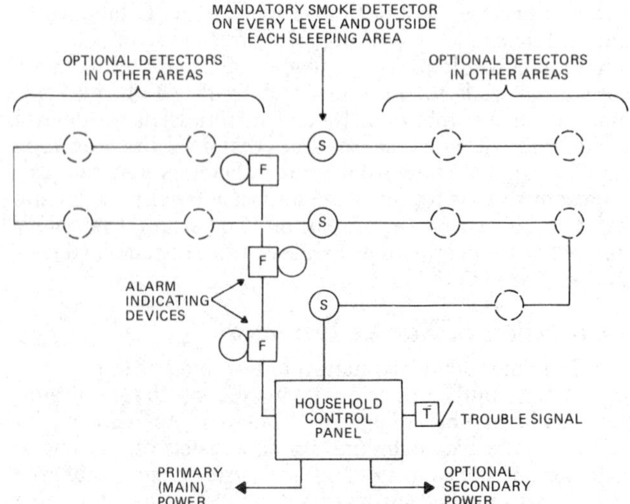

FIG. 16-3D. Typical household fire alarm system, with separate control panel.

warning system also could use heat detectors and manual fire alarm boxes, and has separate alarm indicating devices such as bells, horns, or electronic sirens. The control unit also might have provision for connection to an automatic dialer, central station, or television cable to transmit the alarm to a point beyond the household. However, when a residential system is connected to some other alarm point, the requirements of the other signaling system standard also must be met.

Such standards include NFPA 71, *Standard for the Installation, Maintenance, and Use of Central Station Signaling Systems*; NFPA 72B, *Standard for the Installation, Maintenance and Use of Auxiliary Protective Signaling Systems for Fire Alarm Service*; NFPA 72C, *Standard for the Installation, Maintenance and Use of Remote Station Protective Signaling Systems*; and NFPA 72D, *Standard for the Installation, Maintenance and Use of Proprietary Protective Signaling Systems*.

Combination Systems

Both wired and wireless household systems often include burglar alarm in addition to fire alarm functions. This could make the system more cost effective. In combined systems, the fire alarm signal must take precedence over the burglar alarm signal. The audible alarms are required to be distinctive so the homeowner can immediately discern the difference between a fire and a burglary. In addition, a system which provides for transmission of alarms to a point beyond the household by means of leased lines or television cables will sometimes have a medical alert feature which can be used to summon medical help.

DETECTION SYSTEM COMPONENTS

Smoke Detectors

A single station smoke detector is a self-contained fire alarm device which consists of an assembly of electrical components including a smoke sensing chamber, an alarm sounding appliance, and either integral batteries or provision for connection to a separate power supply source. Single and multiple station residential smoke detectors may be either of the ionization or photoelectric type or may combine sensors of both types within the same unit.

Tests have shown that listed smoke detectors of either the ionization or photoelectric type should provide adequate warning to the occupants for most residential fires. It might be noted that, as a class, ionization detectors respond slightly faster to open flaming fires than photoelectric detectors. Conversely, the photoelectric detectors respond faster to smoldering fires. Since residential occupancies experience both flaming and smoldering fires, an extra margin of occupant safety might be achieved if the home is equipped with one detector of each type. However, both types of smoke detectors are subjected to, and must pass, the same test fires for listing by a testing laboratory. Typical test fires are described in ANSI/UL Standards 217 and 268.

It is well known that compliance with strong smoke detector laws offers the potential of significant reduction in loss of life and property from fire. For uniformity, laws adopted statewide are preferable to local ordinances. Many communities now require residential units from single family dwellings to hotels and motels to be equipped with smoke detectors.

Sounding Devices (Appliances)

A single station smoke detector sounds an audible alarm at the device itself. The sound level output must be at least 85 dBA at a distance of 10 ft (3.05 m). This should be sufficient to alert most sleeping people as long as the detector is located on the same floor and near those who are sleeping. A general rule of thumb is that a device should provide at least 15 dBA above the maximum ambient sound level at the ear of the sleeping person. (For comparison, conversational speech has been shown to register 65 dBA and traffic on an "average" street corner 75 dBA). In addition to background noise, the stage of sleep affects the volume necessary to wake a person.

Long distances between the detector and the bedroom, closed bedroom doors, a noisy appliance in the bedroom such as an air conditioner or humidifier, or impaired hearing could prevent an occupant from being awakened by the detector. In such cases, an alarm should sound in the bedroom and detectors should be interconnected so at least one member of the household will be awakened.

A wired or wireless household system normally has an alarm indicating device circuit for connection of various alarm signals such as fire alarm bells, horns, or electronic sirens. These devices can produce higher sound levels and can be located independently of the detectors. They can be clearly audible throughout the house or even outside, where neighbors might hear the alarm.

The sound frequencies produced by the alarm horn should be considered when selecting a device. Regardless of the type of detector installed in the home, each alarm should be tested with someone in each bedroom. This can ensure that alarms can be heard from all bedrooms.

INSTALLATION PRACTICES

Detector Location Guidelines

Detectors should be located on the ceiling at least 4 in. (100 mm) from the wall; or on the side wall, 4 to 12 in. (100 to 300 mm) from the ceiling to the top of the detector. (See Figs. 16-3E, 16-3F, and 16-3G.) On floors containing bedrooms, the ceiling of the hallway serving the bedrooms is the usual detector location. On other floors, detectors should be located near the stairways to intercept smoke rising from lower floors.

Detectors should be located no closer than 3 ft (0.9 m) from heating vents so that air issuing from the vent will not blow the smoke away from the detector. Some smoke detectors are not suitable for location within kitchens, because of false alarms from cooking vapors. Also, some smoke detectors are not recommended for garages, where automobile exhaust might cause alarms, or for attics or other unheated spaces where extremes of temperatures or humidity might affect their operation. Before a smoke detector is installed in any of these locations, its specifications should be checked to ensure it is appropriate for the intended area.

Another consideration is the temperature of the mounting surface. If a smoke detector is installed on a surface which can be even a few degrees warmer or cooler than the room temperature, its response may be delayed by the cold or warm air layer at the mounting surface. Smoke detectors should not be installed on uninsulated exterior

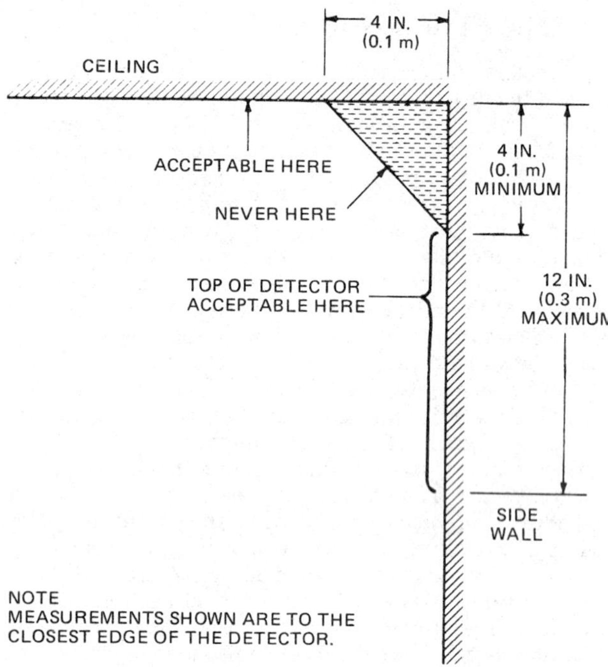

FIG. 16-3E. Proper mounting of detectors.

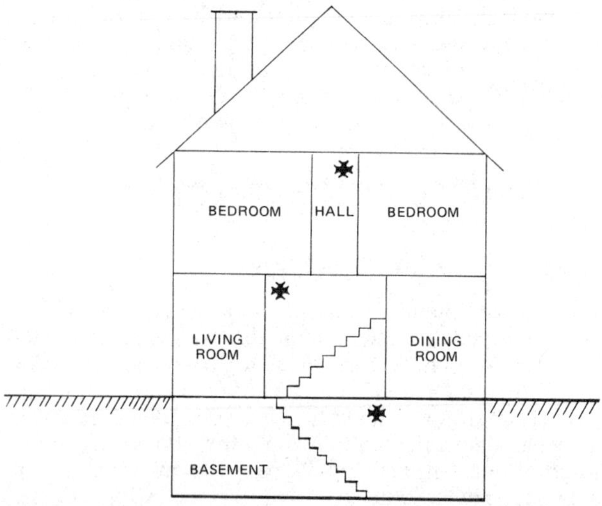

FIG. 16-3F. A smoke detector (indicated by cross) should be located on each story.

walls, ceilings below uninsulated attics, or ceilings containing radiant heating coils.

Special Considerations for the Handicapped

In general, households with handicapped or elderly occupants need a higher level of protection to provide additional escape time. If the handicapped person could not escape without assistance, provisions should be made for someone to provide help. Detectors which signal to remote receivers could be used, for instance, to sound the alarm in a neighbor's home. For hearing impaired occupants, some detectors can be connected to bed vibrators, flashing lights, or fans to provide tactile as well as aural stimulation.

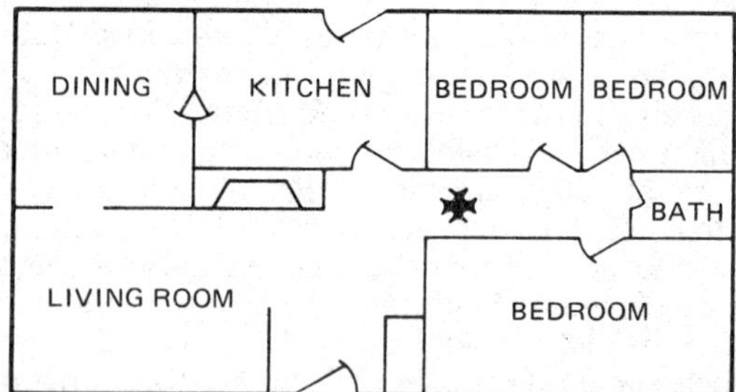

FIG. 16-3G. A smoke detector (indicated by cross) should be located between the sleeping area and the rest of the family living unit.

MAINTENANCE

Household fire warning equipment should be maintained in accordance with the recommendations of the equipment manufacturer. In general, this means little more than keeping the equipment clean and free of dust, and replacing the batteries when needed. Detectors should never be disconnected due to nuisance alarms. If a detector is removed from its mounting bracket to stop a nuisance alarm caused by cooking, place the detector in an obvious location as a reminder to remount it. Most such nuisance alarms can be corrected either by moving the detector to another location, farther from the cooking source, or by replacing it with a smoke detector that is not as sensitive to cooking smoke. Also, unless the detector has a built in sensitivity adjustment, an individual should never tamper with the factory set sensitivity of a detector.

Testing

The importance of testing smoke detectors cannot be emphasized too strongly. Smoke detectors are basically electronic devices, and since all devices will fail at some time, the failure rate projection of a smoke detector can be calculated. From these projections, the life expectancy or the average failure rate of the smoke detector can be determined.

To reduce the period of time that a smoke detector could be out of service because of failure, the detector that has failed can best be identified by periodic testing and then replaced. Table 16-3A shows that over a 30 year period, if a detector with a calculated failure rate of 3.5 failures per million hours is not tested, the probability of it failing is 60 percent. If so, the owner could be unprotected for a period of almost 900 weeks, or 19 years.

Notice that over the same 30 year period, with a test only once a year, the unprotected time would fall from almost 900 weeks to approximately 43 weeks. Referring to Figure 16-3H, notice that if the frequency of testing is increased from one year to one month, the out of service time would drop from 43 weeks to about 6 weeks. All of these figures have a built in factor of two weeks so a defective detector can be removed and a replacement can be obtained and installed. This dramatically shows how important it is to test a smoke detector frequently.

Guidelines for testing smoke detectors can be found in NFPA 74; in NFPA 72H, *Guide For Testing Procedures for Local, Auxiliary, Remote Station and Proprietary Detective Signaling Systems*; and in the manufacturers' instructions.

TABLE 16-3A. Probability of Failure and Average Unprotected Time for Typical Ranges of Detector Failure Rates and Test Intervals for Various Service Life Periods

Service period	1 Year				10 Years				20 Years				30 Years			
Failures per million hours	2.0	3.0	3.5	4.0	2.0	3.0	3.5	4.0	2.0	3.0	3.5	4.0	2.0	3.0	3.5	4.0
Unprotected time (weeks)																
Test interval																
1 week	2.5	2.5	2.5	2.5	2.7	2.8	2.9	3.0	3.0	3.2	3.3	3.5	3.2	3.6	3.8	4.0
2 weeks	3.0	3.0	3.0	3.1	3.3	3.4	3.5	3.6	3.6	3.9	4.0	4.2	3.9	4.3	4.6	4.9
1 month	4.2	4.2	4.2	4.2	4.5	4.7	4.8	5.0	4.9	5.4	5.6	5.8	5.4	6.0	6.4	6.7
3 months	8.6	8.6	8.7	8.7	9.3	9.7	9.9	10.1	10.1	10.9	11.4	11.9	10.9	12.3	13.0	13.8
6 months	15.2	15.2	15.3	15.3	16.4	17.1	17.5	17.9	17.8	19.3	20.1	21.0	19.3	21.7	23.0	24.4
1 year	28.2	28.6	28.6	28.7	30.6	32.0	32.7	33.5	33.3	36.1	37.6	39.3	36.1	40.6	43.0	45.6
No test	28.2	28.6	28.6	28.7	268	271	273	276	550	565	573	581	848	881	898	914
Probability of failure during service period (%)	1.7	2.6	3.0	3.4	16.1	23.1	26.4	29.6	29.6	40.1	45.8	50.4	40.1	54.4	60.1	65.0

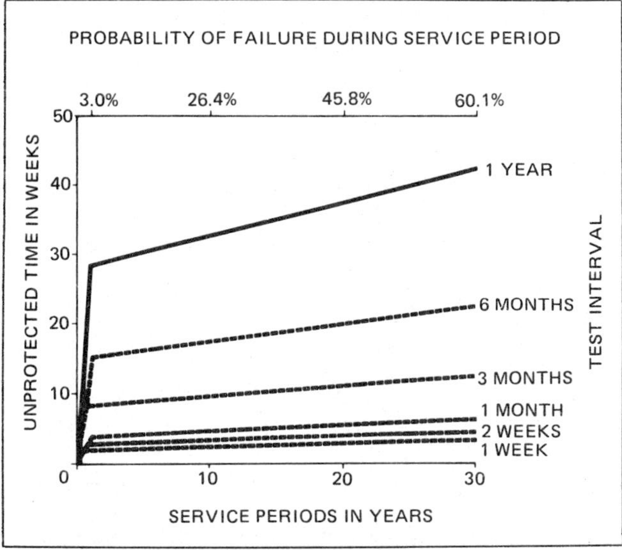

FIG. 16-3H. A plot of unprotected time versus service life with different test intervals. The typical case of 3.5 failures per million hours is used with the expected percent failure also indicated.

In summary, fire prevention in the home involves good planning, good housekeeping, installation of an adequate fire warning system, and ongoing maintenance and testing for the system.

Bibliography

NFPA Codes, Standards, Recommended Practices and Manuals. (See the latest *NFPA Codes and Standards Catalog* for availability of current editions of the following documents.)

NFPA 71, *Standard for the Installation, Maintenance, and Use of Central Station Signaling Systems.*

NFPA 72A, *Standard for the Installation, Maintenance, and Use of Local Protective Signaling Systems for Guard's Tour, Fire Alarm and Supervisory Service.*

NFPA 72B, *Standard for the Installation, Maintenance and Use of Auxiliary Protective Signaling Systems for Fire Alarm Service.*

NFPA 72C, *Standard for the Installation, Maintenance and Use of Remote Station Protective Signaling Systems.*

NFPA 72D, *Standard for the Installation, Maintenance and Use of Proprietary Protective Signaling Systems.*

NFPA 72E, *Standard on Automatic Fire Detectors.*

NFPA 72H, *Guide for Testing Procedures for Local, Auxiliary, Remote Station and Proprietary Protective Signaling Systems.*

NFPA 74, *Standard for the Installation, Maintenance, and Use of Household Fire Warning Equipment.*

Additional Readings

ANSI/UL 217, Safety Standard for Single and Multiple Station Smoke Detectors.

ANSI/UL 268, Safety Standard for Smoke Detectors for Fire Protective Signaling Systems.

Berry, C. H., "Will Your Smoke Detector Wake You?" *Fire Journal,* July 1978, pp. 105-108.

Brannigan, Vincent, "The Legal Implications of Mandatory Home Fire Detection," *Fire Journal,* Vol. 71, No. 2, Mar. 1977, pp. 59-65.

Bright, R. G., "NBS Answers FEMA's Criticisms of the Indiana Dunes Test of Residential Smoke Detectors," *Fire Journal,* Vol. 71, No. 5, Sept. 1977, p. 47.

Bukowski, R. W., et al., "Detector Sensitivity and Siting Requirements for Dwellings," *NBSGCR 75-51,* National Bureau of Standards (NBS), Washington, DC.

Bukowski, R. W., "Testing Residential Fire Detectors: Field Investigation of Residential Smoke Detectors," *Fire Journal,* Vol. 71, No. 2, Mar. 1977, p. 18.

Bukowski, R. W., "Investigation of the Effects of Heating and Air Conditioning on the Performance of Smoke Detectors in Mobile Homes," *NBSIR 79-1915,* NBS, Washington, DC, 1979.

Christian, W. J., and Dubivsky, P. M., "Household Fire Detection and Warning: The Key to Improved Life Safety," *Fire Journal,* Vol. 67, No. 1, Jan. 1973, p. 55.

Custer, R., and Bright, R. G., "Fire Detection, the State-of-the-Art," *NBS Technical Note 839,* NBS, Washington, DC.

"Fire Detection Systems in Dwellings," Los Angeles Fire Department Tests, *NFPA Quarterly,* Jan. 1963.

Gancarski, J. L., and Timoney, T., "Home Smoke Detector Effectiveness," *Fire Technology*, Vol. 20, No. 4, Nov. 1984, pp. 57-62.

Garner, B. W., and LiCalsi, J., "Conceptual Design for an Automatic Residential Remote Fire Alarm System," prepared for Federal Emergency Management Agency, U.S. Fire Administration, Mar. 1980.

Gawin, W. M., and Bright, R. G., "Mobile Home Smoke Detector Siting Study," *NBSIR 76-1016*, NBS, Washington, DC, May 1976.

Harpe, S. W., Waterman, T. E., and Christian, W. J., "Detector Sensitivity and Siting Requirements for Dwellings," *NFPA SPP-43A*, National Fire Protection Association, Quincy, July 1976.

Hirsch, S. N., "Application of Infrared Scanners to Forest Fire Detection," International Workshop Earth Resources Survey Systems *Proceedings 2*, 1971, pp. 153-169.

Kahn, M. J., "Detection Times to Fire-Related Stimuli by Sleeping Subjects," *NBS-GCR-83-435*, NBS, Washington, DC, June 1983.

McGuire, J. H., and Ruscoe, B. E., "The Value of a Fire Detector in the Home," *Fire Study No. 9*, Division of Building Research, National Research Council of Canada, Dec. 1962.

Moore, D. A., "Remote Detection and Alarm for Residences: The Woodlands System," *Fire Journal*, Vol. 74, No. 1, Jan. 1980.

Nober, E. H., et al., "Waking Effectiveness of Household Smoke and Fire Detection Devices," *NBS-GCR-83-439*, NBS, Washington, DC, July 1983.

Nelson, H. N. Jr., "The Need for Full Function Test Features In Smoke Detectors," *Fire Technology*, Vol. 15, No. 1, Feb. 1979.

Operation School Burning and Operation School Burning No. 2, Los Angeles Fire Department, 1959, 1962.

Pezoldt, V. J., and Van Cott, H. P., "Arousal from Sleep by Emergency Alarms: Implications from the Scientific Literature," *NBSIR-78-1484 (HEW)*, NBS, Washington, DC, 1978.

"Residential Smoke Alarm Report," *The International Fire Chief*, Sept. 1980, pp. 62-67.

"Survey and Analysis of Occupant Installable Smoke Detectors: A Summary Report," U.S. Dept. of Commerce, June 1978.

"U.S. Army Aircraft In-Flight Fire Detection and Automatic Suppression Systems," prepared by Walter Kidde and Co., Belleville, NJ, for the U.S. Army Air Mobility Research and Development, July 1972.

Waterman, T. E., "Fire Detector Response Versus Available Escape Time in Residences," ITT Research Institute, Chicago, IL.

"What You Should Know About Smoke Detectors," U.S. Consumer Product Safety Commission, Apr. 1982.

Wilson, R. A., et al., "Airborne Infrared Forest Fire Detection System: Final Report," *Research Paper INT-93*, U.S. Dept. of Agriculture Forest Service, 1971.

AUTOMATIC FIRE DETECTORS

Revised by Charles E. Zimmerman, P.E.

As soon as a fire starts, it produces a variety of environmental changes that help make its presence known. Human beings are excellent fire detectors—they possess the senses of smell, sight, hearing, taste, and touch. However, man's senses can also be unreliable. Beginning in the mid 19th century, a number of mechanical, electrical, and electronic devices have been developed to mimic human senses in detecting the environmental changes created by fire.

The most common elements of a fire that can be detected are heat, smoke (aerosol particulate), and light radiation. Complicating the matter are two facts: (1) not all fires produce all of the elements; and (2) nonfire conditions can produce similar ambient conditions. The fire protection engineer must decide which of the elements produced by a fire might be expected from hostile fires and which similar ambient conditions might result from nonfire situations.

Even if all of these elements—heat, smoke, and light—are present in a given fire, the magnitude of the different elements must exceed some theoretical basic level during fire development. It is also helpful to determine which element will appear first. This is especially true if life safety is involved. This chapter will discuss various automatic fire detectors and their principles of operation.

HEAT DETECTORS

Heat detectors are the oldest type of automatic fire detection device. They began with the development of automatic sprinklers in the 1860s and have continued to the present with a proliferation of various types of devices. A sprinkler can be considered a combined heat activated fire detector and extinguishing device when the sprinkler system is provided with waterflow indicators connected to the fire alarm control system. Waterflow indicators detect either the flow of water through the pipes or the subsequent pressure change upon actuation of the system.

Mr. Zimmerman is a consultant in fire alarm signaling systems based in Norfolk, VA. He was NFPA staff liaison to the NFPA Signaling Systems Technical Committees prior to his retirement in 1983.

Heat detectors which only initiate an alarm and have no extinguishing function are still in use. Although they are the least expensive fire detectors and have the lowest false alarm rate of all automatic fire detector devices, they also are the slowest in detecting fires. A heat detector is best suited for fire detection in a small confined space where rapidly building high heat output fires are expected, in areas where ambient conditions would not allow the use of other fire detection devices, or where speed of detection is not a prime consideration.

Heat detectors are generally located on or near the ceiling and respond to the convected thermal energy of a fire. They respond either when the detecting element reaches a predetermined fixed temperature or to a specified rate of temperature change. In general, heat detectors are designed to operate when heat causes a prescribed change in a physical or electrical property of a material or gas.

Operating Principles of Fixed Temperature Heat Detectors

Fixed temperature detectors are designed to alarm when the temperature of the operating element reaches a specified point. The air temperature at the time of alarm is usually considerably higher than the rated temperature because it takes time for the air to raise the temperature of the operating element to its set point. This condition is called thermal lag. Fixed temperature heat detectors are available to cover a wide range of operating temperatures ranging from about 135°F (57°C) and higher. Higher temperature detectors are necessary so that detection can be provided in areas normally subjected to high ambient (nonfire) temperatures, or in areas zoned so that only detectors in the immediate fire area operate.

Fusible Element Type: Eutectic metals—alloys of bismuth, lead, tin, and cadmium that melt rapidly at a predetermined temperature—can be used as operating elements for heat detection. The most common such use is the fusible element in an automatic sprinkler. Fusing of the element allows the cover on the orifice to fall away, water to flow in the system, and the alarm to be initiated.

A eutectic metal may also be used to actuate an electrical heat detector. The eutectic metal is often used as a solder to secure a spring under tension. When the element fuses, the spring action closes contacts and initiates an alarm. (See Fig. 16-4A, items D, F, and G.) Devices using eutectic metals cannot be restored; either the device or its operating element must be replaced following operation.

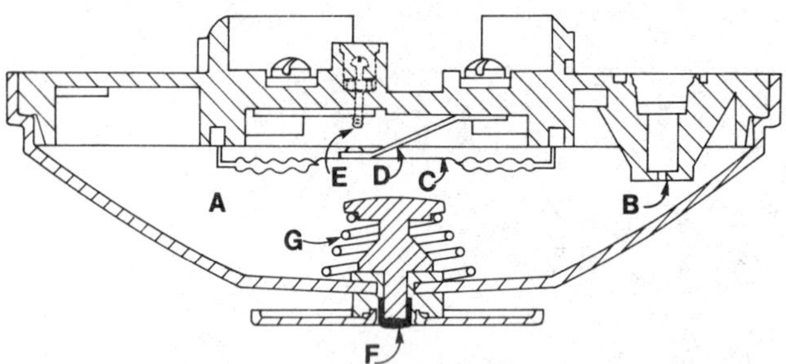

FIG. 16-4A. A spot type combination rate-of-rise, fixed temperature device. The air in chamber A expands more rapidly than it can escape from vent B. This causes pressure to close electrical contact D between diaphragm C and contact screw E. Fixed temperature operation occurs when fusible alloy F melts releasing spring G which depresses the diaphragm closing contact points. (Edwards Company)

Continuous Line Type: As alternatives to spot type fixed temperature detection, various methods of line type detection have been developed. The detector shown in Figure 16-4B uses a pair of steel wires in a normally open circuit.

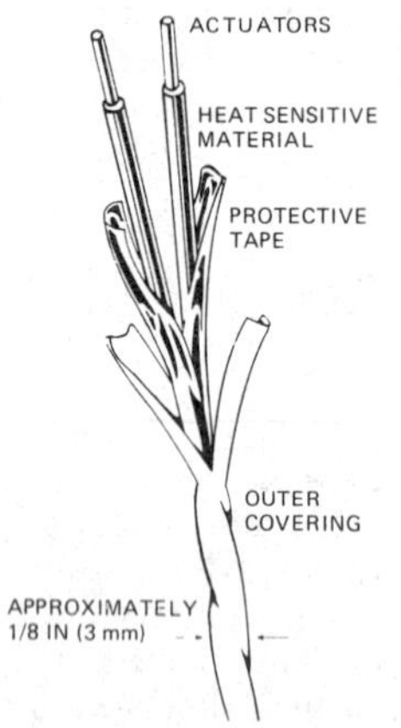

FIG. 16-4B. Line type heat detector. (The Protectowire Company)

The conductors are held apart by a heat sensitive insulation. The wires, under tension, are enclosed in a braided sheath to form a single cable assembly. When the design temperature is reached, the insulation melts, the two wires contact, and an alarm is initiated. Following an alarm, the fused section of the cable must be replaced to restore the system.

A similar alarm device utilizing a semiconductor material and a stainless steel capillary tube has been used where mechanical stability is also a factor. (See Fig. 16-4C.) The capillary tube contains a coaxial center conductor separated from the tube wall by a temperature sensitive glass semiconductor material. Under normal conditions, a small current (i.e., below alarm threshold) flows in the circuit. As the temperature rises, the resistance of the semiconductor decreases, allows more current flow, and initiates the alarm.

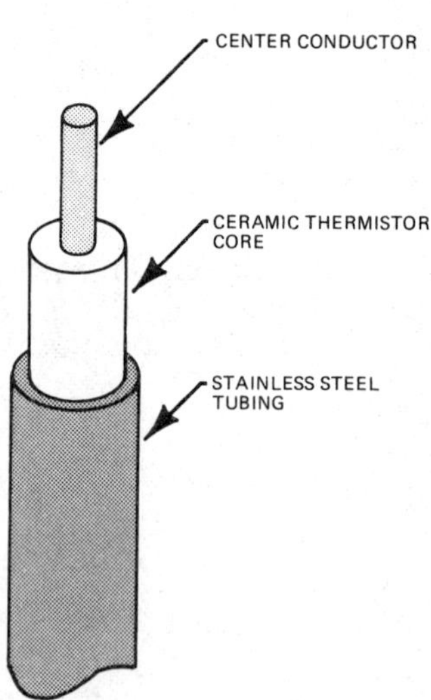

FIG. 16-4C. View of the construction of the continuous thermal sensor showing outer tubing, ceramic thermistor core, and center wire. (Alison Central Inc.)

Bimetallic Type: When two metals with different coefficients of thermal expansion are bonded together and then heated, differential expansion causes bending or flexing toward the metal having the lower expansion rate. This action closes a normally open circuit. The low expansion metal commonly used is invar, an alloy of 36 percent nickel and 64 percent iron. Several alloys of manganese-copper-nickel, nickel-chromium-iron, or stainless steel may also be used for the high expansion component of a bimetal assembly. Bimetals are used for the operating elements of a variety of fixed temperature detectors. These detectors are generally of two types—the bimetal strip and the bimetal snap disc.

As it is heated, a bimetal strip deforms in the direction of the contact point. With a given bimetal, the width of the gap between the contacts determines the operating temperature; the wider the gap, the higher the operating point. The operating element of a snap disc device is a bimetal disc formed into a concave shape in its unstressed condition. (See Fig. 16-4D.) Generally, a heat collector is attached to the detector frame to speed the transfer of heat from the room air to the bimetal. As the disc is heated, the stresses developed cause it to suddenly reverse curvature and become convex. This provides a rapid positive action that closes the alarm contacts. The disc itself is not usually part of the electrical circuit.

All heat detectors using bimetal elements are automatically self-restoring after operation, when the ambient temperature drops below the operating point.

Rate Compensation Detectors

A rate compensation detector is a device that responds when the temperature of the surrounding air reaches a predetermined level, regardless of the rate of temperature rise. (See Fig. 16-4E.)

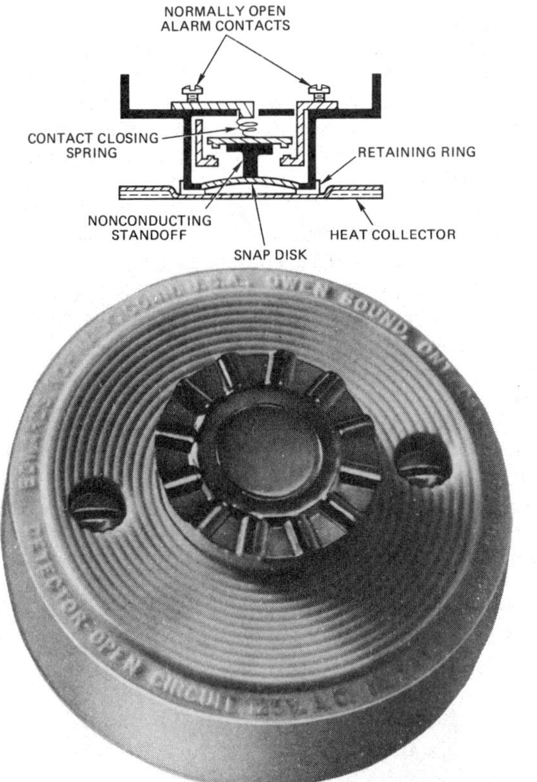

FIG. 16-4D. Spot type fixed temperature snap disc detector. (Edwards Company—bottom)

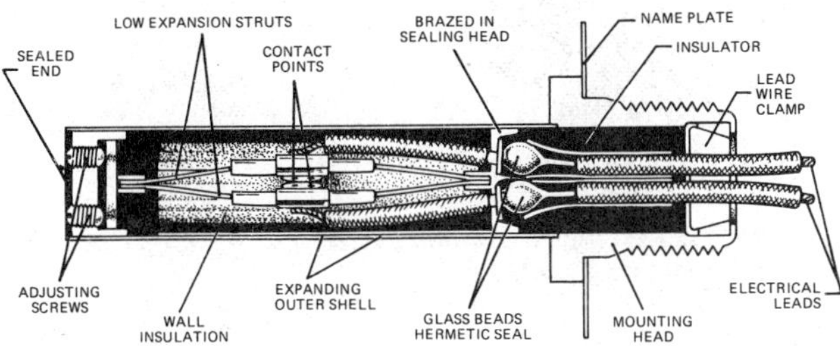

FIG. 16-4E. Section of spot type rate compensation detector. (Fenwal, Inc.)

A typical example is a spot type detector with a tubular casing of metal that tends to expand lengthwise as it is heated, and an associated contact mechanism that will close at a certain point in the elongation. A second metallic element inside the tube exerts an opposing force on the contacts, tending to hold them open. The forces are balanced so that with a slow rate of temperature rise, there is more time for heat to penetrate to the inner element. This inhibits contact closure until the total device has been heated to its rated temperature level. However, with a fast rate of temperature rise, there is less time for heat to penetrate to the inner element. The element therefore exerts less of an inhibiting effect, so contact closure is obtained when the total device has been heated to a lower level. This, in effect, compensates for thermal lag.

Thermal detectors using expanding metal elements are also automatically self-restoring after operation, when the ambient temperature drops to some point below the operating point.

Rate-of-Rise Detectors

One effect that a flaming fire has on the surrounding area is to rapidly increase air temperature in the space above the fire. Fixed temperature heat detectors will not initiate an alarm until the air temperature near the ceiling exceeds the design operating point. The rate-of-rise detector, however, will function when the rate of temperature increase exceeds a predetermined value, typically around 12 to 15°F (7 to 8°C) per minute. Rate-of-rise detectors are designed to compensate for the normal changes in ambient temperature [less than 12°F (6.7°C) per minute] which are expected under nonfire conditions.

In a pneumatic fire detector, air heated in a tube or chamber expands, increasing the pressure in the tube or chamber. This exerts a mechanical force on a diaphragm that closes the alarm contacts. If the tube or chamber were hermetically sealed, slow increases in ambient temperature, a drop in the barometric pressure, or both, would cause the detector to initiate an alarm regardless of the rate of temperature change. To overcome this, pneumatic detectors have a small orifice to vent the higher pressure that builds up during slow increases in temperature or during a drop in barometric pressure. The vents are sized so that when the temperature changes rapidly, as in a fire situation, the rate of expansion exceeds the venting rate and the pressure rises. When the temperature rise exceeds 12 to 15°F (7 to 8°C) per minute, the pressure is converted to mechanical action by a flexible diaphragm. Pneumatic heat detectors are available for both line and spot type detectors. A schematic of a spot type pneumatic heat detector is shown in Figure 16-4A, and a line type is shown in Figure 16-4F.

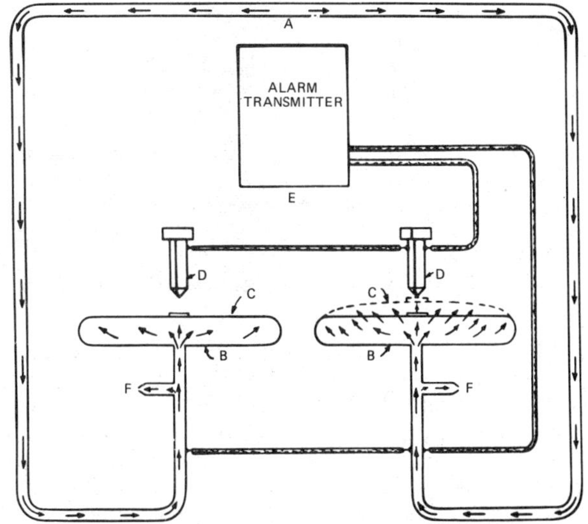

FIG. 16-4F. Line type rate-of-rise heat detector. The copper tubing A is fastened in a continuous loop to ceilings or walls and terminates at both ends in chambers B having flexible diaphragms C which control electrical contacts D. When air in the tubing expands under the influence of heat, pressure builds within the chambers causing the diaphragms to move and close a circuit to alarm transmitter E. Vents F compensate for small pressure changes in the tubing brought about by small changes in temperature in the protected spaces. (American District Telegraph Co., and its associated companies)

Line Type: The line type (Fig. 16-4F) consists of metal tubing, in a loop configuration, attached to the ceiling or side wall near the ceiling of the area to be protected. Lines of tubing are normally spaced not more than 30 ft (9.1 m) apart, not more than 15 ft (4.5 m) from a wall, and with no more than 1,000 ft (305 m) of tubing on each circuit. Also, a minimum of at least 5 percent of each tube circuit or 25 ft (7.6 m) of tube, whichever is greater, must be in each protected area. Without this minimum amount of tubing exposed to a fire condition, insufficient pressure would build up to achieve proper response.

In small areas where the line type tube detectors might have insufficient tubing exposed to generate sufficient pressures to close the alarm contacts, air chambers or rosettes of tubing are often used. These units act like a spot type detector by providing the volume of air required to meet the 5 percent or 25 ft (7.6 m) requirement. Since a line type rate-of-rise detector is an integrating detector, it will actuate either when a rapid heat rise occurs in one area of exposed tubing, or when a slightly less rapid heat rise takes place in several areas where tubing on the same loop is exposed.

Spot Type: The pneumatic principle is also used to close contacts within spot detectors. (See Fig. 16-4A.) The difference between the line and spot type detectors is that the spot type contains all of the air in a single container rather than in a tube that extends from the detector assembly to the protected area(s).

Sealed Pneumatic Line Type Detectors

A pneumatic line type heat detector that is not vented to the air is shown in Figure 16-4G. The unit consists of a

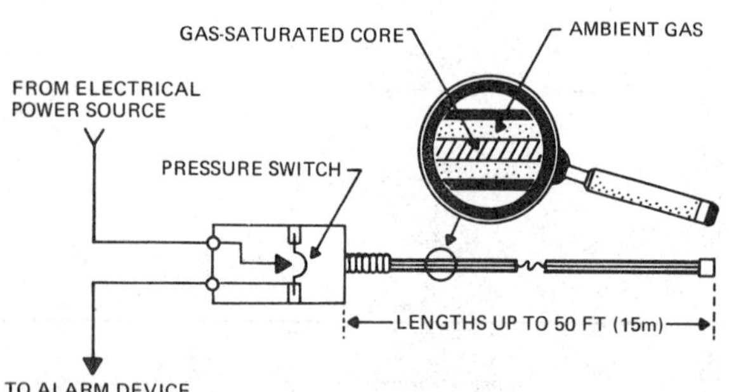

FIG. 16-4G. Line type overheat detector. (Systron-Donner)

capillary tube containing a special salt that is saturated with hydrogen gas. At normal temperatures, most of the hydrogen is held in the porous salt, and the pressure in the tube is low. As the temperature increases, at any point along the tubing, hydrogen is released from the salt, increases the internal pressure, and eventually trips a diaphragm pressure switch. This system supervises the integrity of the capillary tube with a second pressure switch, which monitors the low pressure present at normal temperatures.

Combination Detectors

Combination detectors contain more than one element which responds to a fire. These detectors may be designed to respond from either element, or from the combined partial or complete response of both elements. An example of the former is a heat detector that operates on both the rate-of-rise and fixed temperature principles. Its advantage is that the rate-of-rise element will respond quickly to a rapidly developing fire, while the fixed temperature element will respond to a slowly developing fire when the detecting element reaches its set point temperature. The most common combination detector uses a vented air chamber and a flexible diaphragm for the rate-of-rise function, while the fixed temperature element is usually leaf spring restrained by a eutectic metal. (See Fig. 16-4A.) When the fixed temperature element reaches its design operating temperature, the eutectic metal fuses and releases the spring, which closes the contacts.

Thermoelectric Effect Detectors

A device whose sensing element comprises a thermocouple or thermopile unit produces an increase in electric voltage in response to an increase in temperature. This potential is monitored by associated control equipment, and an alarm is initiated when the voltage increases at an abnormal rate.

Thermopile devices, which operate in the voltage generating mode, use two sets of thermocouples. (See Fig. 16-4H.) One set is exposed to changes in the atmospheric

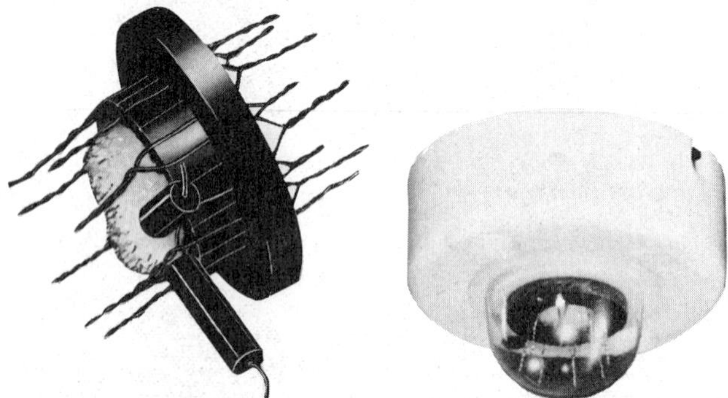

FIG. 16-4H. Spot type thermoelectric effect detector. (American District Telegraph Co.)

temperature. During periods of rapid temperature change that are associated with a fire, the temperature of the exposed set increases faster than the temperature of the unexposed set, generating a net voltage. The voltage increase associated with this potential is used to operate the alarm circuit. This voltage need not be within only one detector, because the thermopile units are connected in series and thus the small voltages produced at each unit on a circuit can cumulatively produce an alarm. The sensitivity of a thermopile detector is related directly to the number of thermocouple junctions within the device. The more junctions present, the greater the sensitivity. Sensitivity also be increased by designs that focus radiative energy on the exposed junctions.

SMOKE DETECTORS

A smoke detector will detect most fires much more rapidly than a heat detector. This section will describe the various principles of smoke detector operation and their applications.

Smoke detectors are identified by their operating principle. Two of the operating principles are ionization and photoelectric. As a class smoke detectors using the ionization principle provide somewhat faster response to high energy (open flaming) fires, since these fires produce large numbers of the smaller smoke particles. As a class smoke detectors operating on the photoelectric principle respond faster to the smoke generated by low energy (smoldering) fires, as these fires generally produce more of the larger smoke particles. However, each type of smoke detector is subjected to, and must pass, the same test fires at testing laboratories in order to be listed.

Ionization Smoke Detectors

Smoke detectors utilizing the ionization principle are usually of the spot type. An ionization smoke detector has a small amount of radioactive material that ionizes the air in the sensing chamber, rendering the air conductive and permitting a current flow through the air between two charged electrodes. This gives the sensing chamber an effective electrical conductance. When smoke particles enter the ionization area, they decrease the conductance of the air by attaching themselves to the ions, causing a reduction in ion mobility. When the conductance is below a predetermined level, the detector responds. (See Fig. 16-4I.)

Photoelectric Smoke Detector

The presence of suspended smoke particles generated during the combustion process affects the propagation of a light beam passing through the air. The effect can be utilized to detect the presence of a fire in two ways: (1) obscuration of light intensity over the beam path, or (2) scattering of the light beam.

Light Obscuration Principle: Smoke detectors that operate on the principle of light obscuration consist of a light source, a light beam collimating system, and a photosensitive device. When dense smoke obscures part of the light beam, or less dense smoke obscures more of the beam, the light reaching the photosensitive device is reduced, and this initiates the alarm. (See Fig. 16-4J.) The light source is usually a light emitting diode (LED), a reliable long life source of illumination having a low current requirement. Pulsed LEDs can generate sufficient light intensity for use in detection equipment while operating at even lower overall power levels.

Most light obscuration smoke detectors are the beam type and are used to protect large open areas. They are installed with the light source at one end of the area to be protected, and the photosensitive device at the other. In some applications, mirrors direct the beam over the desired path and this determines the area of coverage. For each mirror used, the rated beam length of the device must

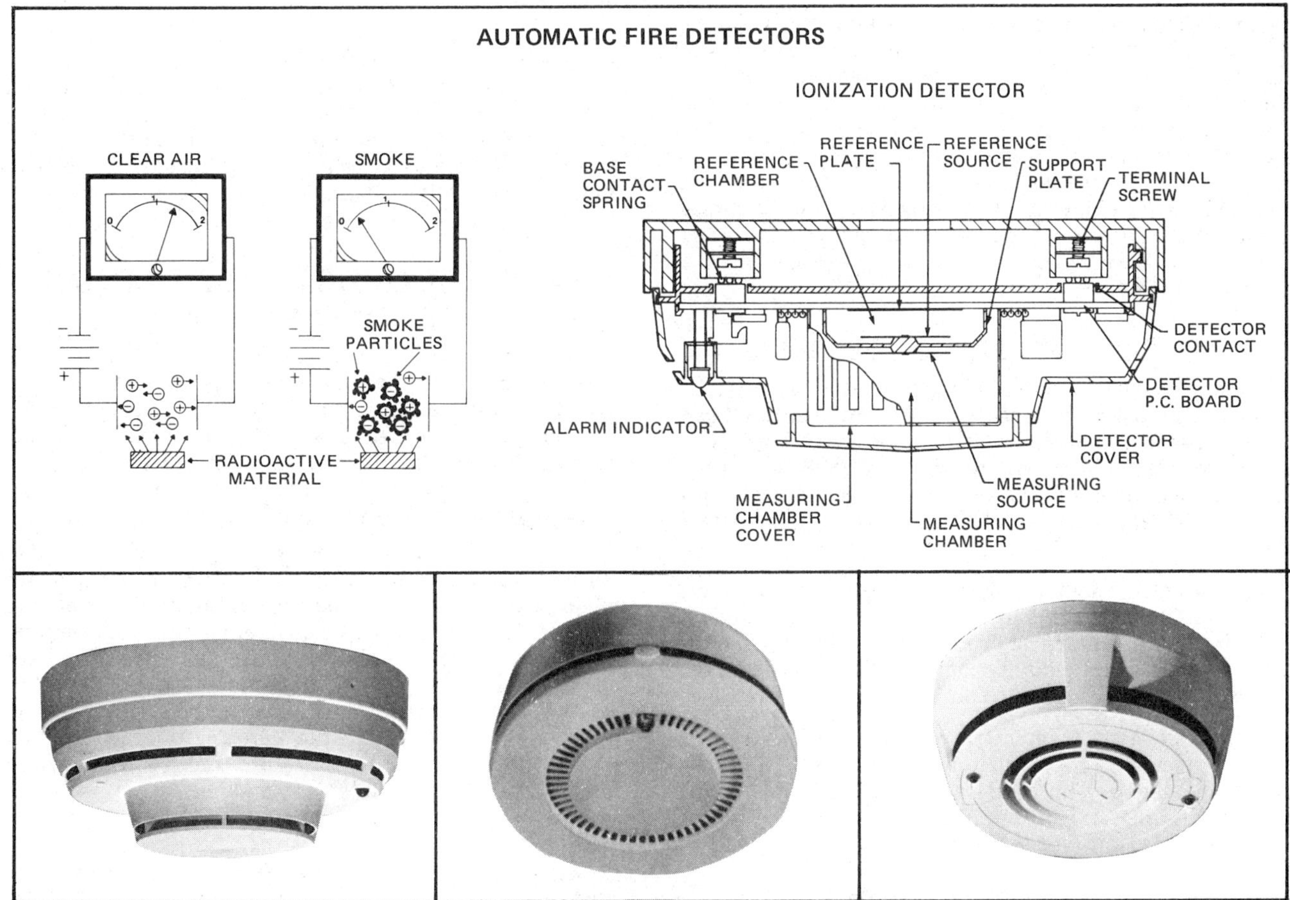

FIG. 16-4I. (Top left) Principle of operation for ionization smoke detector; (Top right) Cross-section view of an ionization smoke detector, (Pyrotronics); (Lower left) Ionization smoke detector, (Pyrotronics); (Lower center) Ionization smoke detector, (Honeywell Inc.); (Lower right) Ionization smoke detector, (The Gamewell Corporation)

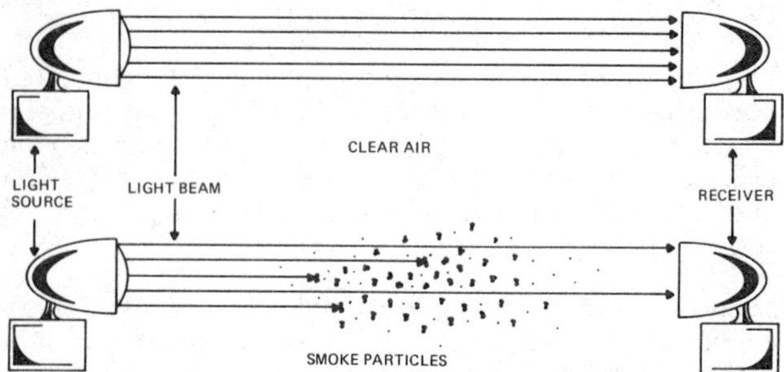

FIG. 16-4J. Principle of operation for photoelectric obscuration smoke detector.

be progressively reduced by one-third. Projected beam detectors are generally installed close to the ceiling.

Light Scattering Principle: When smoke particles enter a light path, scattering results. Smoke detectors utilizing the photoelectric light scattering principle are usually of the spot type. They contain a light source and a photosensitive device arranged so the light rays normally do not fall onto the device. When smoke particles enter the light path, light strikes the particles and is scattered onto the photosensitive device, causing the detector to respond. (See Fig. 16-4K.) The photosensitive device used in scattering detectors usually is a photodiode or phototransistor.

Cloud Chamber Smoke Detection Principle: A smoke detector utilizing the cloud chamber principle is usually of the sampling type. An air pump draws a sample of air from the protected area(s) into a high humidity chamber within the detector. After the air sample has been raised to a high humidity, the pressure is lowered slightly. If smoke particles are present, the moisture in the air condenses on them, forming a cloud in the chamber. The density of this cloud is then measured by a photoelectric principle. The detector responds when the density is greater than a predetermined level.

GAS SENSING FIRE DETECTORS

Many changes occur in the gas content of the environment during a fire. In large scale fire tests, it has been observed that detectable levels of gases are reached after detectable smoke levels and before detectable heat levels. One of two operating principles—semiconductor and catalyic element—may be used in a gas sensing fire detector.

Semiconductor Principle: Fire-gas detectors of the semiconductor type respond to either oxidizing or reducing gases by creating electrical changes in the semiconductor. The subsequent conductivity change of the semiconductor causes actuation of the detector.

Catalytic Element Principle: Fire-gas detectors of the catalytic element type contain a material which in itself remains unchanged, but which accelerates the oxidation of combustible gases. The resulting temperature rise of the element causes detector actuation.

FLAME DETECTOR

A flame detector responds to radiant energy visible to the human eye (approximately 4,000 to 7,700 angstroms) or outside the range of human vision. Such a detector is sensitive to glowing embers, coals, or flames which radiate energy of sufficient intensity and spectral quality to actuate the alarm.

Due to their fast detection capabilities, flame detectors are generally used only in high hazard areas such as fuel loading platforms, industrial process areas, hyperbaric chambers, high ceiling areas, and atmospheres in which explosions or very rapid fires may occur. Because flame detectors must be able to "see" the fire, they must not be blocked by objects placed in front of them. The infrared type of flame detector, however, has some capability for detecting radiation reflected from walls.

Infrared Flame Detectors

An infrared (IR) detector is basically composed of of a filter and lens system used to screen out unwanted wavelengths and focus the incoming energy on a photovoltaic or photoresistive cell sensitive to infrared energy. (See Fig. 16-4L.) IR flame detectors can respond to the total IR component of the flame alone, or in combination with flame flicker in the frequency range of 5 to 30 Hz.

A major problem in the use of infrared detectors receiving total IR radiation is the possible interference of solar radiation in the infrared region. When detectors are located in places shielded from the sun, such as in vaults, filtering or shielding the unit from the sun's rays is unnecessary.

Ultraviolet Flame Detectors

Ultraviolet (UV) detectors generally use either a solid state device such as silicone carbide or aluminum nitride, or a gas filled tube as the sensing element. (See Fig. 16-4M.) UV detectors are essentially insensitive to both sunlight and artificial light.

AMBIENT CONDITIONS AFFECTING DETECTOR RESPONSE

Ambient conditions need to be considered in the selection, placement, and response capability of detectors. The improper selection of a class of detector or the improper placement of a detector can create problems ranging from no alarm to excessive false alarms.

Ambient Background Level

When selecting a detector for a specific location, consider the ambient background to which the detector might be exposed under nonfire conditions. For example, an IR or UV detector used where there is gas or arc welding can generate false alarms due to the presence of radiant energy which the unit was designed to detect. These responses, however, are considered false alarms since they do not result from a "hostile" fire. Also, detectors that respond to smoke particles are especially prone to false alarms from such sources as cooking fumes, cigarette smoke, and automobile exhaust fumes.

Heating, Ventilating, and Air Conditioning (HVAC)

In buildings or portions of buildings where forced ventilation is present, smoke detectors should not be

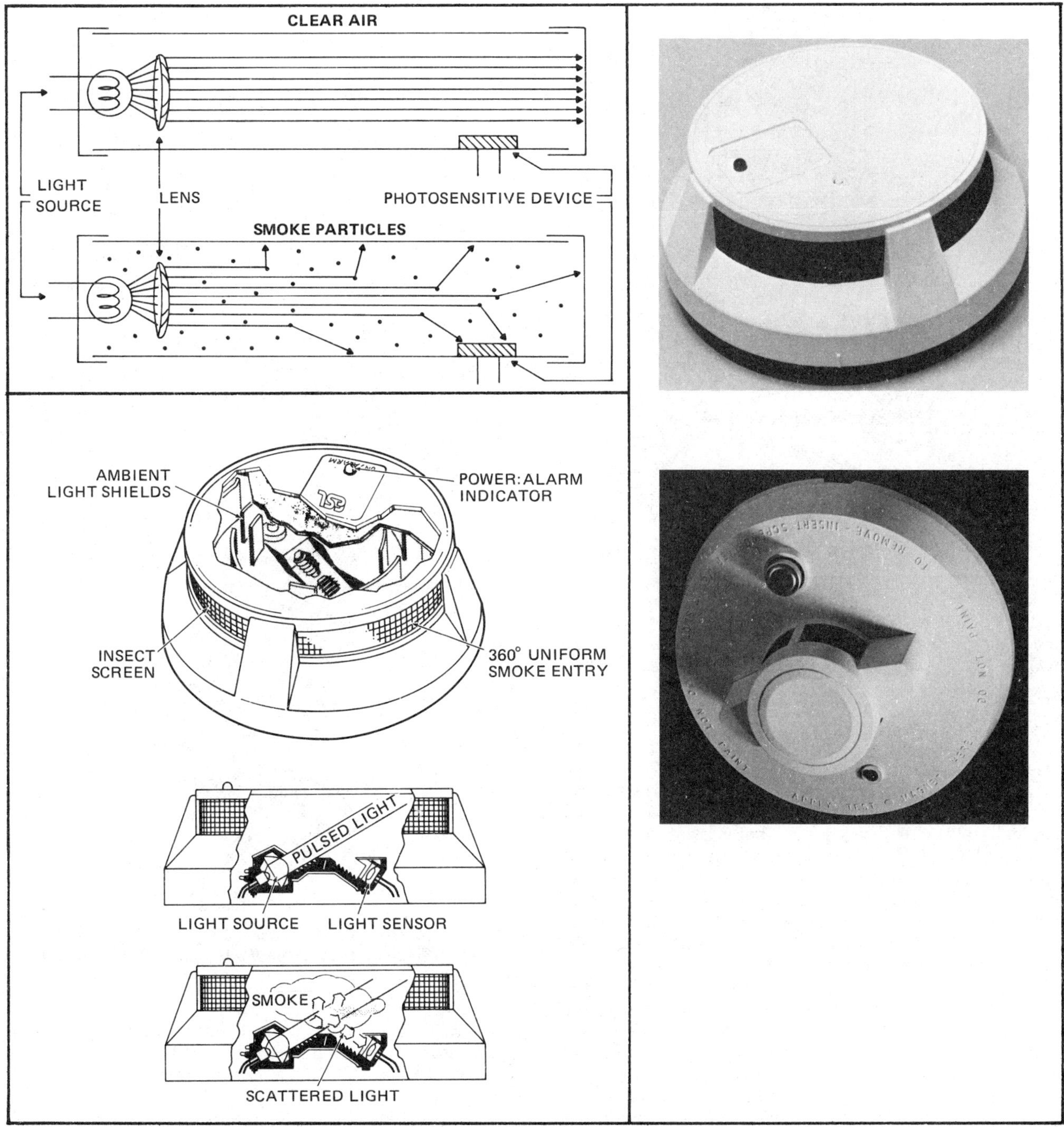

FIG. 16-4K. (Top left) Principle of operation for photoelectric scattering smoke detector; (Lower left) Cross-section view of a photoelectric scattering smoke detector. (Electro Signal Lab, Inc.); (Top right) Photoelectric smoke detector. (Electro Signal Lab, Inc.); (Lower right) Photoelectric smoke detector. (Chloride Pyrotector)

located where air from supply diffusers could dilute smoke before it reaches the detector. Detectors should be located to favor the airflow toward return openings. This may require additional detectors, since placing detectors only near return air openings may leave the balance of the area with inadequate protection when the air handling system is shut down.

SELECTION OF DETECTORS

When planning a fire detection system, choice of detectors should be based on the kinds of potential fires expected. The type and quantity of fuel, possible ignition sources, ranges of ambient conditions, and the value of the protected property should all be considered.

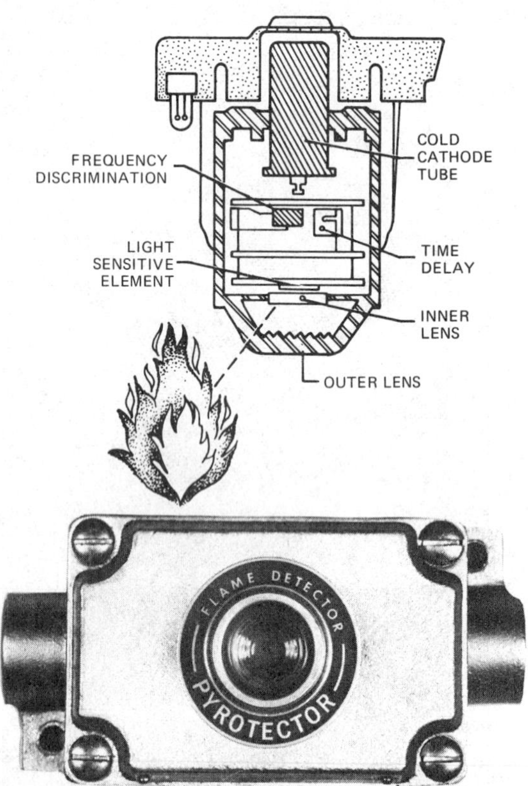

FIG. 16-4L. (Top) Cross-section view of an infrared flame detector. (Pyrotronics). (Bottom) Infrared flame detector. (Chloride Pyrotector)

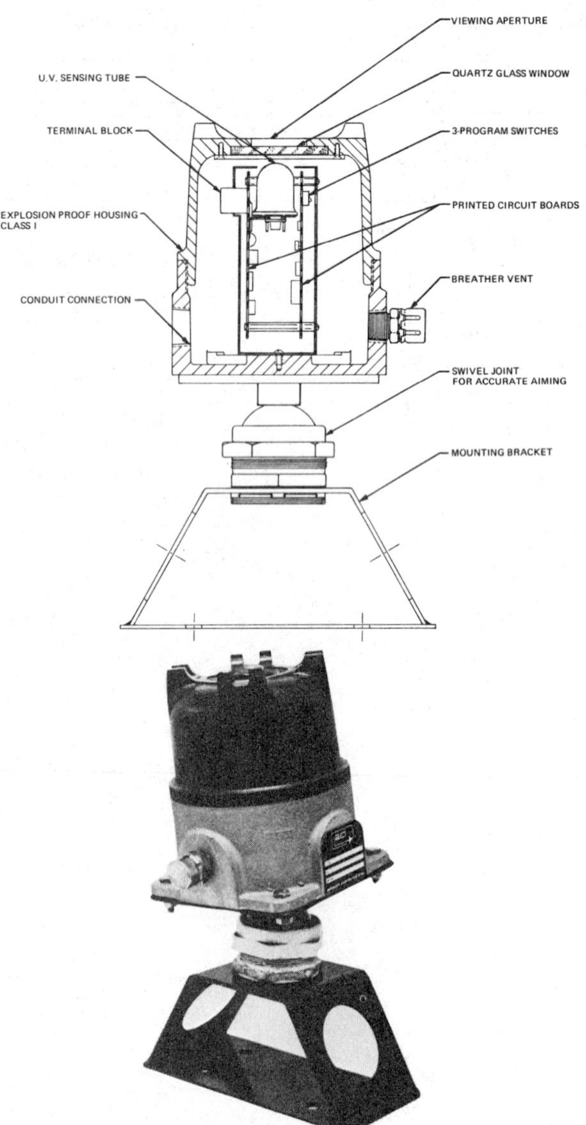

FIG. 16-4M. (Top) Cross-section view of an ultraviolet flame detector. (Bottom) Ultraviolet flame detector. (Alison Control Inc.)

In general, heat detectors have the lowest cost and false alarm rate but also the slowest response time. Since the heat generated by small fires tends to dissipate fairly rapidly, heat detectors are best used either to protect confined spaces or directly over hazards where flaming fires could be expected. They are usually installed in a grid pattern according to their recommended spacing schedule or with reduced spacing for faster response. The operating temperature of a heat detector should be a minimum of 25°F (14°C) above the maximum expected ambient temperature in the area protected.

Smoke detectors are more expensive than heat detectors, but have a faster response time. They are better suited than heat detectors to protect large open spaces because smoke does not dissipate as rapidly as heat does in the same size space. Smoke detectors are installed either according to prevailing air current conditions or in a grid layout.

Ionization smoke detectors are useful where flaming fires are a possibility. Photoelectric smoke detectors are best used in places with potential for smoldering fires or fires involving low temperature pyrolysis polyvinylchloride (PVC) wire insulation.

Flame detectors offer extremely fast response, but will warn of any source of radiation within their sensitivity range. False alarm rates can be high if this kind of detector is improperly applied. Because flame detectors are "line of sight" devices, care must be taken to ensure they can "see" the entire protected area and they will not be accidentally blocked by stacked material or equipment. Sensitivity of these units is determined by flame size and flame distance from the detector. Although fairly expensive, flame detectors are well suited to protect areas where explosive or flammable vapors or dusts may be present, because they are usually available in "explosion proof" housings.

DETECTOR INSTALLATION

After selection of the most suitable detectors for the job, the next step is to determine where to locate the detectors within the space to be protected.

Spot type detectors are usually installed on the ceiling or side wall with the edge of the detector located no closer than 4 in. (100 mm) from the wall or ceiling. When heat detectors are installed at their listed spacing, detection times will approximately equal the operating times of standard 165°F (74°C) link-and-lever sprinklers. If faster response is desired, detector spacing should be reduced. Also, where ceilings are high or ceiling construction is not smooth, spacing should be reduced accordingly. Specific information on the treatment of joisted, beamed, and sloped ceilings can be found in Appendix A of NFPA 72E, *Standard on Automatic Fire Detectors* (hereinafter referred to as NFPA 72E).

Spacing of Heat Detectors on High Ceilings

Because hot air is diluted by colder air as it rises from a fire, it has always been assumed that heat detectors should be spaced closer together on a high ceiling to achieve the same response time that the detectors would provide on an 8 to 10 ft (2.5 to 3 m) ceiling. To determine how much closer the detectors should be, a series of test fires were conducted using fires of various sizes and taking into consideration ceiling height, ambient temperature, and fire development. The data clearly shows that heat detectors must be spaced closer together on a high ceiling to achieve the same response time as on a 10 ft (3 m) ceiling.

A typical chart (Fig. 16-4N) from Appendix C of NFPA 72E, shows a heat detector with a listed spacing of 30 ft (9 m) responding to a slowly developing fire. The example shown indicates that to detect a fire with a heat release rate of 500 Btu per second (527 kW) on a 10 ft (3 m) ceiling, the detector with a listing of 30 ft (9.1 m) should be placed on 17 ft (5.2 m) centers. If the ceiling is 20 ft (6.1 m) high, the same detector should be placed on 8 ft (2.4 m) centers to achieve the same response time as on a 10 ft (3 m) ceiling.

One might ask why a detector with a listing of 30 ft (9.1 m) is not spaced on 30 ft (9.1 m) centers when mounted on a 10 ft (3 m) ceiling. The answer is that the above example considered detecting a smaller fire. The test fires used by the testing laboratories more nearly approximated the 1,000 Btu/sec (1055 kW) fire line on the chart. To detect a slowly developing fire with a heat release rate of 1,000 Btu/sec (1055 kW), the chart shows that on a 10 ft (3 m) ceiling the recommended spacing would be 30 ft (9 m), and on a 20 ft (6 m) ceiling, the recommended spacing would be 19 ft (5.8 m).

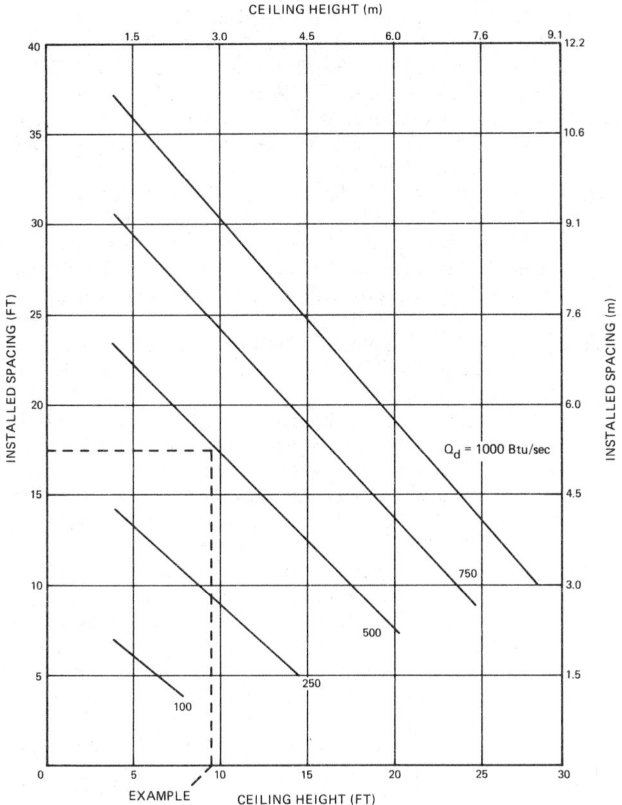

FIG. 16-4N. Example of a design curve for a fixed temperature heat detector with a listed spacing of 30 ft (9.1 m) for slow fires.

Although Appendix C of NFPA 72E also contains charts for smoke detectors, they are only theoretical and do not apply to a smoldering fire.

When installing any type of heat detector, consideration should be given to sources of heat within the protected space which might cause false alarms. For example, heat detectors should be located away from unit heaters and ovens where surges of hot air might be expected.

Proper installation of smoke detectors is more critical than for heat detectors because in a smoldering fire, smoke transport is strongly influenced by the convective airflow patterns within the protected area. For this reason, smoke detectors are not assigned a listed spacing by the testing laboratories. Although a grid pattern can be used as a starting point, care must be taken to appropriately locate the heating supply registers and return air registers. Smoke detectors should be located away from turbulence caused by hot air outlets. Their location should favor return air, because the return air will draw smoke toward the detector, and air velocity at the return tends to be lower.

Special Applications

Air duct smoke detectors are installed either in or on return air ducts or at the return air openings of HVAC systems in buildings. This prevents recirculation of smoke from a fire through the HVAC system within the building. Upon detection of a fire, the associated control system initiates an alarm and either shuts down the circulating blowers or switches them to a smoke exhaust mode. Detectors installed in air duct systems must never be considered substitutes for open area protection. This is because of dilution of smoke laden air by clean air from other parts of the building, and because the smoke may not be drawn into the HVAC system when the system is shut down.

Smoke activated devices are also used to automatically close smoke doors in buildings to limit the spread of smoke in case of fire. Separate corridor ceiling mounted smoke detectors can be connected to electrically operated hold open devices on the doors, or smoke detectors can be built into the door closure units themselves.

Smoke stratification should also be considered when smoke detectors are installed. Smoke may stratify below a ceiling due to temperature gradients or airflow along the ceiling. Where stratification is possible, additional smoke detectors can be installed at alternate levels.

Installation of gas sensing detectors is similar to that of smoke detectors. Fire gases tend to flow with smoke and are similarly affected by convected flows within the protected space. Gas detectors must also be located away from sources of oxidizable gases or vapors such as aerosol sprays or hydrocarbon solvents, as these could cause false alarms.

The requirements of flame detectors are unlike those of heat or smoke detectors because spacings are not relevant for line of sight devices. Rather, flame detectors must be located so they can "see" light radiation emanating from any point within the protected space. Because the "cone of vision" for a given flame detector varies with the detector design, manufacturer's recommendations for area coverage must be followed. Flame detectors need to be shielded or located so they will not "see" radiant energy from nonfire sources that might cause false alarms.

DETECTOR MAINTENANCE AND TESTING

A thorough maintenance and testing program of all fire detectors is essential to provide continuous detector operation. A regular test of every smoke detector is of utmost importance because testing is the best way to determine that the detector has not failed.

Bibliography

NFPA Codes, Standards, Recommended Practices and Manuals. (See the latest *NFPA Codes and Standards Catalog* for availability of current editions of the following documents.)

NFPA 72E, *Standard on Automatic Fire Detectors.*
NFPA 72H, *Guide for Testing Procedures for Local, Auxiliary, Remote Station and Proprietary Protective Signaling Systems.*
NFPA 74, *Standard for the Installation, Maintenance, and Use of Household Fire Warning Equipment.*

Additional Readings

Alpert, R. L., "Response Time of Ceiling Mounted Fire Detectors," *Fire Technology*, Vol. 8, No. 3, Aug. 1972, pp. 181-195.
"Automatic Fire Detection: False Alarms," *FPA Journal*, No. 88, pp. 146-148.
Bankston, C. P., et al., "Review of Smoke Particulate Properties," *NBS-GCR-78-147*, National Bureau of Standards (NBS), Washington, DC, 1978.
Benjamin, I. A., "Detector Response in Large Buildings," *Engineering Applications of Fire Technology Workshop Proceedings*, Apr. 16-18, 1980, Gaithersburg, MD, published by Society for Fire Protection Engineers, Boston, 1983.
Bright, R. G., "A New Test Method for Automatic Fire Detection Devices," *Fire Technology*, Vol. 13, No. 2, May 1977, pp. 105-113.
Bright, R. G., "Report of Fire Tests on Eight TGS Semiconductor Gas Sensor Residential Fire/Smoke Detectors," *NBSIR-76-990*, NBS, Washington, DC, 1976.
Bright, R. G., and Custer, R. L. P., "Fire Detection, The State of the Art," *Technical Note 839*, NBS, Washington, DC, 1974.
Bryan, John L., *Fire Suppression and Detection Systems*, Glencoe Press, Beverly Hills, CA, 1974.
Bukowski, R. W., "An Evaluation of Light Emitting Diodes as Source Lamps in Photoelectric Smoke Detectors," *Fire Technology*, Vol. 11, No. 3, Aug. 1975, pp. 157-163.
Bukowski, R. W., and Bright, R. G., "Taguchi Semiconductor Sensors as Residential Fire/Smoke Detectors," *Fire Journal*, Vol. 69, No. 3, May 1975, p. 30.
Bukowski, R. W., and Mulholland, G. W., "Smoke Detector Design and Smoke Properties," *Technical Note 973*, NBS, Washington, DC, Nov. 1978.
Butler, H., Bowyer, A., and Kew, J., "Locating Fire Alarm Sounders for Audibility," Building Services Research and Information Association, United Kingdom, Aug. 1981.
Drouin, J. A., and Cote, A. E., "Smoke and Heat Detector Performance: Field Demonstration Test Result," *Fire Journal*, Vol 78, No. 1, Jan. 1984, pp. 34-38, 69.
"Fire Detector Systems: The Range of Choice," *FPA Journal*, No. 91, pp. 18-23.
Fry, J. F., "The Behavior of Automatic Fire Detection Systems," *Fire Research Note 310* (U.K.), 1970.
Hertzberg, M., et al., "The Spectral Growth of Expanding Flames—The Infrared Radiance of Methane-Air Ignitions and Coal Dust-Air Explosions," *RI 7779*, U.S. Bureau of Mines, 1973.
Hertzberg, M., and Litton, C. D., "The Optical Detection of Underground Fires and Explosions," Storch Award Symposium, 164th American Chemical Society Meeting, New York, NY, Aug. 1972.
Hertzberg, M., and Litton, C. D., "An Optical Infrared Detection System for Methane-Air Explosions," U.S. Bureau of Mines, 1973.
Heskestad, G., and Delichatsios, M. A., "Environments of Fire Detectors-Phase II-Effect of Ceiling Configuration. Vol. I. Measurements." *NBS-GCR-78-128*, NBS, Washington, DC, 1978.
Heskestad, G., and Delichatsios, M. A., "Environments of Fire Detectors. Vol. II. Analysis." *NBS-GCR-78-129*, NBS, Washington, DC, 1978.
Katzel, Jeanine, "An Overview of Automatic Fire Detectors," *Plant Engineering*, Sept. 27, 1984.
Lawson, D. I., "A Laser Beam Fire Detection System," *Fire Technology*, Vol. 6, No. 4, Nov. 1970, pp. 305-311.
Lee, T. G. K., "An Instrument to Evaluate Installed Smoke Detectors," *NBSIR-78-1430*, NBS, Washington, DC, 1978.
Lee, T. G. K., and Mulholland, G. W., "Physical Properties of Smokes Pertinent to Smoke Detector Technology," *NBSIR 77-1312*, NBS, Washington, DC, 1977.
Luck, H., "Economics of Fire Protection with Fire Detectors," *Fire Technology*, Vol. 9, No. 1, Feb. 1973, pp. 56-64.
Mniszewski, K. R., et al., "Analysis of Fire Detectors Test Methods/Performance: A Summary Report," Federal Emergency Management Agency, U.S. Fire Administration, June 1980.
Nelson, Hjalmar N., "The Need for Full-Function Test Features in Smoke Detectors," *Fire Technology*, Vol. 15, No. 1, Feb. 1979, pp. 10-19.
O'Sullivan, E. F., Gohosh, B. K., and Turner, J., "Experiments on the Use of a Laser Beam for Heat Detection," *Fire Technology*, Vol. 7, No. 2, May 1971, pp. 133-134.
Raber, Samuel, and Ellner, Irving, "Battery-Operated Smoke Detectors, The Need for Effective Battery Supervision," *Fire Journal*, Vol. 72, No. 5, Sept. 1978, pp. 104-108.
Ramachandran, G., "Economic Value of Automatic Fire Detectors," Building Research Establishment Information Paper 27/80, Fire Research Station, Borehamwood, Hertfordshire, UK, 1980.
"Results of Tests Made from Six Separate Fires in a Three Bedroom Single Family Dwelling," Bloomington Fire Department, Bloomington, MN, May 1979.
Rickers, H. C., "Residential Smoke Detector Reliability Handbook," Reliability Analysis Center, Griffiss Air Force Base, New York, NY, 1979. Also available as *NBS-GCR-79-161* and *NBS-GCR-79-162*, NBS, Washington, DC.
"Smoke Detectors," *Consumer Reports*, Aug. 1980, pp. 475-479.

Waterman, T. E., Mniszewski, K. R., and Spadoni, D. J., "Cost/Benefit Analysis of Fire Detectors," IIT Research Institute for the U.S. Fire Administration, Sept. 1978.

Wickham, Robert T., "Detectors," *Specifying Engineer*, Vol. 49, No. 5, May 1983, pp. 86-88.

Zimmerman, Charles E., "Development of Light Emitting Diodes for Photoelectric Smoke Detectors," *Fire Technology*, Vol. 11, No. 3, Aug. 1975, pp. 153-156.

GAS AND VAPOR TESTING

Richard L. Swift

Prevention of accidental fire involving combustible gases requires an accurate knowledge of where and when ignition of a mixture within the flammable range can occur, so additional precautions can be taken to prevent ignition. A combustible gas is defined as any flammable or combustible gas or vapor that can, in sufficient concentrations by volume in air, become the fuel for an explosion or fire. Measurement of the oxygen concentration is important (1) as a means of evaluating the effectiveness of inerting processes, (2) as a means of protecting personnel from the effects of oxygen deficiency, and (3) as a means of determining that sufficient oxygen is present in the atmosphere to ensure reliable combustible gas readings. Of increasing importance is the need for an awareness of the toxic properties of gases or vapors which may also be present in the atmosphere.

Enclosed areas where combustible gases or vapors can be present are found in public utility operations; petroleum production, refining, and marketing; chemical and petrochemical plants; metallurgical industries; distilleries; paint and varnish making; marine operations; and many other industrial activities. Combustible gases may be detected by instruments such as combustible gas indicators, flammable vapor detectors, combustible gas analyzers, and the like. All of these instruments operate by sensing the characteristics of a sample and translating these characteristics into appropriate data.

In this chapter, portable instruments are defined as those requiring no external power source. Such devices may operate continuously for several days or longer. Stationary instruments are those which are permanently installed and, in general, operate on line power.

COMBUSTIBLE GAS INSTRUMENTS

Operating Principles

Many of the instruments in current use utilize the principle of "catalytic combustion." Mixtures of combustible gas and air cannot be ignited to cause self-sustaining flame unless the concentration of gas exceeds a minimum value called the Lower Flammable Limit (LFL) or Lower Explosive Limit (LEL). (The two terms are synonymous and are used interchangeably. For uniformity only "LFL" will be used in this chapter.) Mixtures containing much lower concentrations, approaching zero, can "burn" on the surface of heated platinum, yielding heat in direct proportion to the gas or vapor concentration. This is called "surface" or "catalytic" combustion.

If the heated surface is an electrically heated platinum wire connected in an appropriate circuit (the Wheatstone bridge circuit, Fig. 16-5A), the heat released by catalytic combustion can further increase the temperature of the wire, resulting in a change in electrical resistance, and a corresponding indication on an electric meter. The meter may be analog (scale with needle) or digital. The usual combustible gas indicator (hot wire type) operates in this manner. Other somewhat similar devices may employ a solid or porous catalytic mass instead of a wire and may sense the temperature by means other than the increase in electrical resistance.

A more recent development is a type of instrument using solid state sensors (semiconductors). Generally, these sensors are less specific to combustible gases than the catalytic hot wire filament. However, this type of instrument can be useful as "go-no-go" type of detector.

In operation, gas diffuses through a porous metal disc and comes into contact with the heated surface of a silicone chip. As gas concentrations increase, the resultant chemical reaction reduces the semiconductor resistance logarithmically, producing an output voltage that can be linearized by the instrument's microelectronic circuits.

The combustible gas instrument must be set to read "zero" when uncontaminated air is drawn into it. When making tests, the amount of gas present in the sample is read on a scale that shows the amount in terms of the fraction or percent of the lower flammable limit concentration, usually abbreviated as "percent LFL." This is practical, since the heat of combustion at the LFL concentration of most combustible gases is approximately the same. Some vapors, e.g., carbon disulfide, give somewhat low readings in terms of true explosibility; others, such as natural gas, may give readings on the high side. Special

Mr. Swift is Product Line Manager, Portable Instruments, Mine Safety Appliances Company, Pittsburgh, PA.

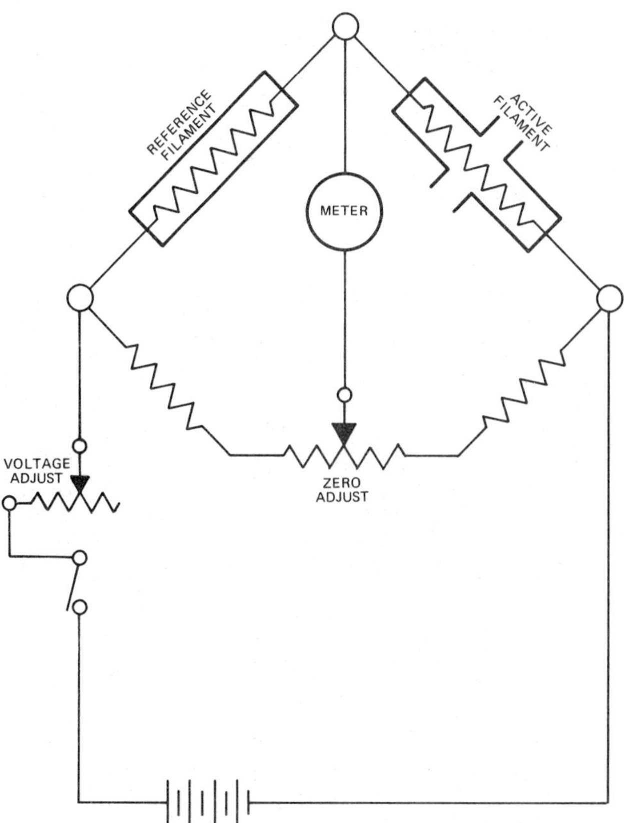

FIG. 16-5A. The Wheatstone bridge circuit in which the "Active Filament" is a platinum wire on which catalytic combustion takes place when a sample of the atmosphere being tested is passed across it. The resultant change in resistance in that arm of the circuit is translated into a corresponding deflection of an electric meter.

calibrations are used where exact readings at the upper part of the scale are required. For most purposes, where the sought after condition is a gas free atmosphere, these differences disappear as the reading approaches the previously set zero.

The fact that the hot wire type of instrument responds to all combustible gases or mixtures of combustible gases, irrespective of chemical composition, is usually advantageous in that it is unnecessary to know the exact identity of a gas to evaluate its fire and explosion risk. Where it is desirable to selectively measure the presence of some specific combustible gas in the possible presence of other combustibles, special or more refined techniques are necessary. Infrared absorption, gas chromotography, and flame ionization are some of the principles which may be used.

Other instruments, some generally similar in appearance to the ordinary combustible gas indicator, depend for their operation upon properties of a gas or vapor other than combustibility, such as refractive index, density, diffusion, or thermal conductivity of a gas. Calibrated upon a specific gas, these instruments can be designed to give readings on higher concentrations up to pure gas. The instruments are not selective, however, and will respond to almost any gas or vapor, regardless of combustibility.

In some instruments, the catalytic combustion principle and the thermal conductivity principle are combined in a single device to permit extending the useful range into and beyond the flammable range. However, each section of the instrument retains the limitations of the particular method employed.

Operation and Limitations

As actual ignition of flammable mixtures may occur within the instrument, flame arresters are provided in the inlet and outlet connections to the chamber housing the catalytic filament to prevent combustion within the chamber from traveling outside and igniting the atmosphere being sampled. Flame arresters are suitable for conventional gas-air mixtures. If the atmosphere is oxygen enriched, i.e., the atmosphere contains more than 21 percent oxygen by volume or the partial pressure of oxygen exceeds 160 mm of mercury (21.27 kPa), special arresters may be necessary.

Tests may be made remotely by drawing a sample from a suspect location through a hose to an indicator, or on site by carrying the instrument (previously set to zero in fresh air) into the suspect area. The former is preferred as a preliminary test because the operator is not exposed to possible concentrations of flammable or toxic gas.

Instruments of the hot platinum filament type are designed specifically for measuring combustibles in air. They depend upon an oxidation reaction to operate. The oxygen in the air is necessary for proper instrument function. They will give a reasonably accurate indication of the presence of combustible gas even when the oxygen concentration has been substantially reduced by oxidation, biological action, or displacement. As a rule of thumb, if the oxygen content of the atmosphere tested is less than ten percent, a catalytic filament type instrument may not give a true reading. Without special adaptation, they cannot be relied upon to demonstrate whether or not the objective of an inerting procedure, i.e., the elimination of an explosive mixture, has actually been achieved.

Sampling steamy atmospheres is not recommended, since readings normally will be erroneously high and condensation of water within the instrument is likely to cause problems. Sampling from ovens or driers at substantially elevated temperatures should be avoided unless it is known that the vapor is so dilute that condensation will not occur at the temperature of the instruments. However, if the solvent involved has a flash point below atmospheric temperature, explosibility readings can be relied upon irrespective of the oven temperature.

Some catalytic filaments lose their catalytic properties if exposed to substances such as silicones, dust, and the vapors of tetraethyl lead. The listings of testing laboratories specify which instruments utilize filaments with sufficiently high temperatures to make them suitable for use on the vapors of gasoline containing tetraethyl lead. Some manufacturers supply an accessory chemical filter whose vapors will combine with the lead to form a compound that will prevent poisoning of the filament. Recent research and development has produced filaments which are less susceptible to poisoning. Manufacturer's literature and instruction manuals should be checked for the latest information about specific instruments.

CALIBRATION CHECKS

The response of these instruments to combustible gases must be verified regularly. Some authorities recommend verification be performed daily before use. Such verification, usually called a calibration check, is accomplished by passing a known concentration of a combustible gas in air through the instrument. The response is then compared with performance data supplied by the manufacturer, and adjustments are made in accordance with that manufacturer's instructions. Most suppliers of combustible gas detection equipment have available pressurized cylinders that contain calibration gas-air mixtures. Such cylinders are equipped with regulators to deliver the mixture to the instrument at the proper pressure. A typical calibration gas-air supply is shown connected to a portable combustible gas indicator in Figure 16-5B. Similar equipment can be used to check a remote sampling head.

FIG. 16-5B. A calibration gas-air supply connected to a portable combustible gas indicator through a pressure regulator and adapter hose. (Mine Safety Appliances Co.)

Some instruments are supplied with calibration curves showing the response of the particular unit to combustible gases other than that on which the device was calibrated. Calibration curves for one particular model of combustible gas indicator are shown in Figure 16-5C. Since all manufacturers do not use the same gas for base factory calibrations, the curves for one instrument cannot be used with curves of another type.

CALIBRATION CURVES FOR FIELD REFERENCE

GAS OR VAPOR	L.E.L. % BY VOL.	CURVE NO.	GAS OR VAPOR	L.E.L. % BY VOL.	CURVE NO.
ACETONE	2.5	5	HYDROGEN	4.0	1
ACETYLENE	2.3	3	METHYL ALCOHOL	6.7	2
BENZENE	1.4	5	METHYL ETHYL KETONE	1.8	6
CARBON DISULFIDE	1.0	10	NATURAL GAS	4.8	3
CARBON MONOXIDE	12.5	1	OCTANE	1.0	9
ETHYL ACETATE	2.2	7	PENTANE	1.4	5
ETHYL ETHER	1.7	7	PROPANE	2.2	4
GASOLINE	1.3	8	TOLUENE	1.3	4
HEXANE	1.2	7	XYLENE	1.0	7

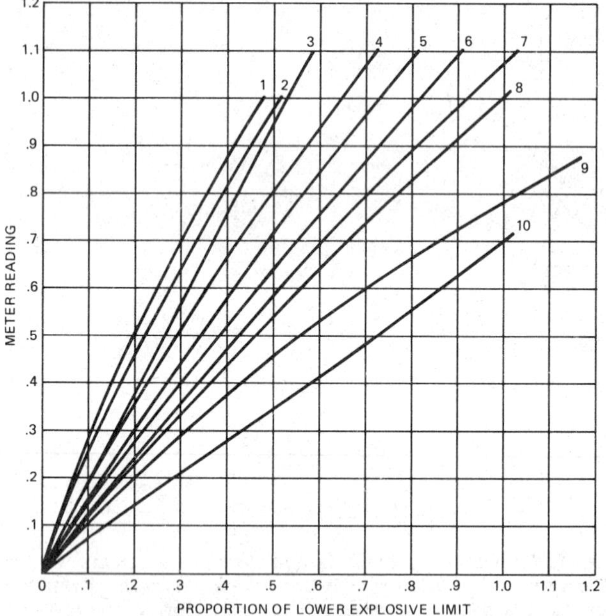

FIG. 16-5C. Calibration curves for one particular make of combustible gas indicator. Calibration curves for manufactured indicators may be different because of physical arrangement of the catalytic unit. Note that despite wide variations at higher concentrations, the error becomes negligible as the zero concentration is approached.

TYPES OF INSTRUMENTS

Portable Indicators

Portable indicators are generally lightweight, battery powered devices intended for field use such as testing sewers, manholes, basements, ducts, containers, or tanks where the presence of a combustible mixture is suspected. The batteries used may be carbon-zinc, alkaline, mercury, nickel-cadmium, or lead-acid. Samples are drawn through a hose or tube, which is frequently equipped with a rigid extension or "probe" for ease in reaching inaccessible points. Suction is provided by a hand operated rubber aspirator bulb or a battery powered built-in motor driven pump.

There are portable combustible gas indicators equipped with audible and visual alarms. Usually powered by rechargeable batteries, they are designed to run continuously for eight to ten hours on a full charge. The alarm circuits are energized when a preset concentration of combustible gas is reached. These instruments usually depend upon diffusion and convection to bring the sample to the filament chamber. (See Fig. 16-5D.)

A number of portable indicators are available for use in hazardous locations. (See Fig. 16-5E.) The electrical circuits have been designed to meet the requirements of Article 500 of NFPA 70, *National Electrical Code®*,

FIG. 16-5D. A portable combustible gas alarm device.

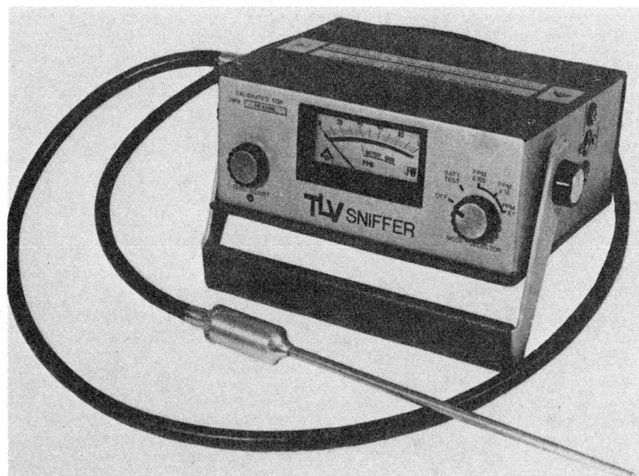

FIG. 16-5E. A portable flammable and combustible gas detector. (Bacharach Instrument Co.)

and have been tested and listed by testing agencies. The manufacturers' label and instructions indicate the type of approval granted for the particular instrument.

Stationary Analyzers

Stationary analyzers are permanently installed line powered devices for continuously analyzing air samples from one to as many as twenty or more points. The analyzers may be either of the remote detection (diffusion) type (Fig. 16-5F) or may draw samples through tubing to a

FIG. 16-5F. A continuous type of combustible gas analyzer for multiple location sampling. Inset shows a typical sensing for remote mounting. (Mine Safety Appliances Co.)

central location by means of a suction pump or equivalent. Functions may include audible or visible alarms in addition to continuous recording of data, two level alarms, e.g., a warning light at 40 percent of LFL and an audible alarm at 50 percent of LFL, automatic shutdown or startup of equipment, and similar features as required by individual applications.

The central equipment is available either for installation in nonhazardous locations such as control rooms, or in listed explosion proof enclosures for hazardous areas.

Stationary analyzers are frequently tailored to meet the exact requirements of the particular installation, and may also employ operating principles other than catalytic

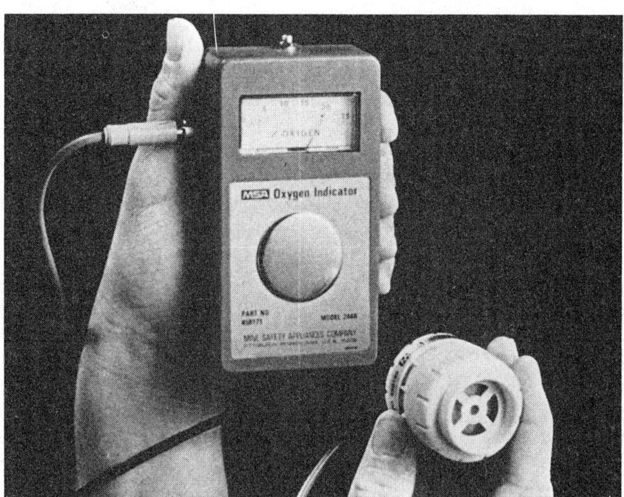

FIG. 16-5G. An individual oxygen indicator for remote sampling. (Mine Safety Appliances Co.)

combustion. The output signals from such instruments can also be incorporated into computer controlled data acquisition networks.

Other Types of Indicators

Numerous other devices are available, some reasonably portable, that are suitable for determining the presence of contaminating gaseous substances in air. Previously mentioned are those depending upon refractive index, density, diffusion, or thermal conductivity of gases. Other operating principles include glass tubes filled with granular material, where a color change indicates the presence of some specific substance; gas chromatography; infrared absorption; flame ionization (specific for hydrogen containing compounds) and others. Generally these indicators are not specifically designed for combustible gases or vapors, but serve as special purpose instruments

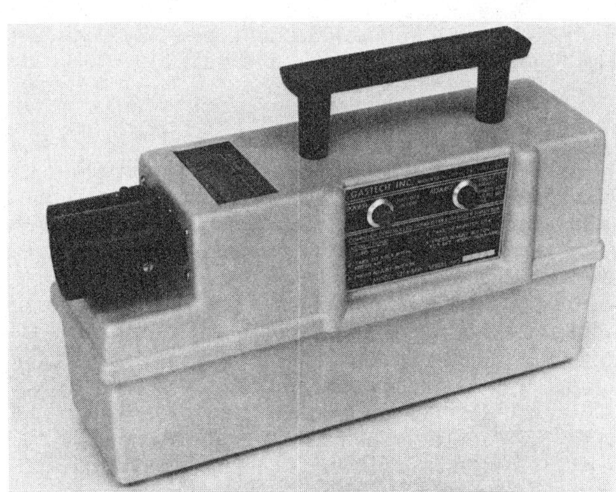

FIG. 16-5H. A microprocessor controlled dual range combustible gas oxygen monitor that uses synthesized speech to produce vocal instructions and alarm warning statements and sounds. Computerization has simplified operation to two pushbutton controls. (Gas Tech Inc.)

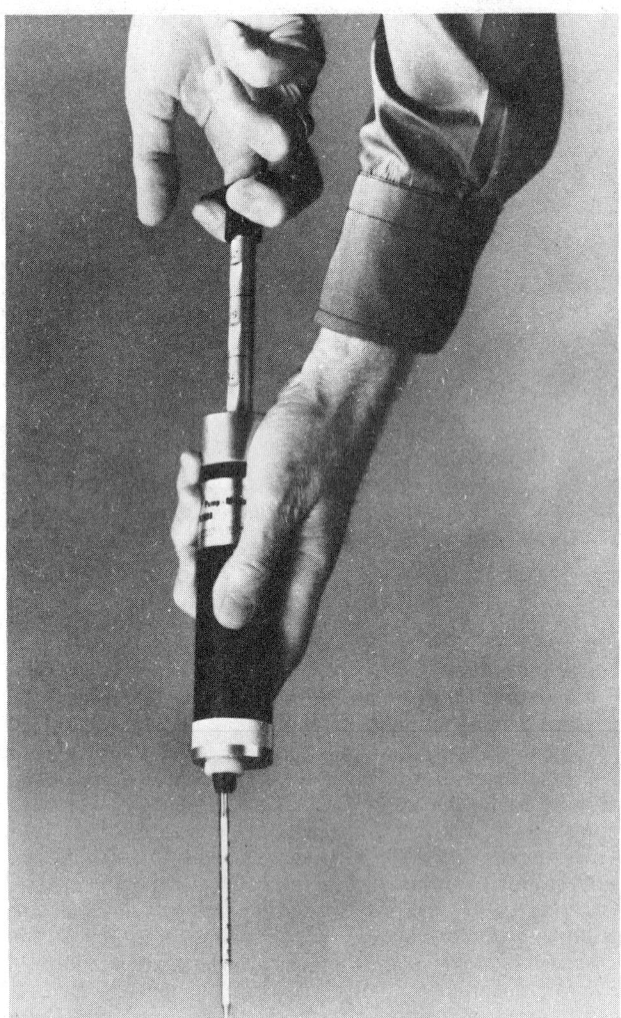

FIG. 16-5I. Length-of-stain toxic gas detector tube and sampling pump.

in process control applications not necessarily associated with fire protection.

Oxygen Indicators

Devices similar in appearance to combustible gas indicators are available for measuring the oxygen content of the atmosphere in a closed space. The devices operate either as an individual unit or as a combination unit also incorporating a combustible gas indicator.

The portable oxygen indicator (Fig. 16-5G) operates on an electrochemical principle, the partial pressure of the oxygen in the atmosphere controlling the rate of diffusion through a porous membrane and into an electrochemical cell with suitable electrolytes and electrodes. The oxygen enters into an electrochemical reaction, generating a current which in turn, reads the percentage of oxygen directly on a scale graduated from 0 to 25 percent.

Except for setting the meter to read 20.8 percent while aspirating fresh air at the elevation (atmospheric pressure) where tests are to be made, no other calibration is necessary for accurate reading in the range of oxygen concentrations in which it is permissible for people to work.

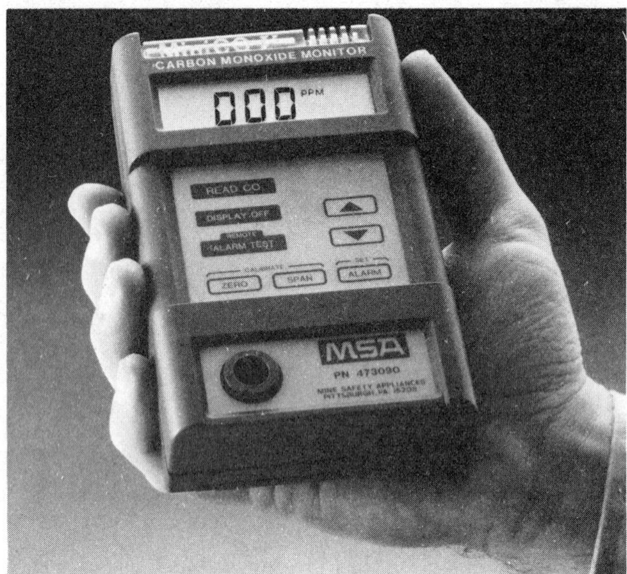

FIG. 16-5J. Electrochemical type toxic gas indicator and alarm for carbon monoxide monitoring.

(Current regulations require a minimum of 19.5 percent oxygen, otherwise air supplied or self-contained breathing equipment must be used.) Enclosed spaces should also be tested to ensure sufficient oxygen is present to permit reliable gas measurements. Where readings are intended to be significant in the low range, i.e., with inerting systems,

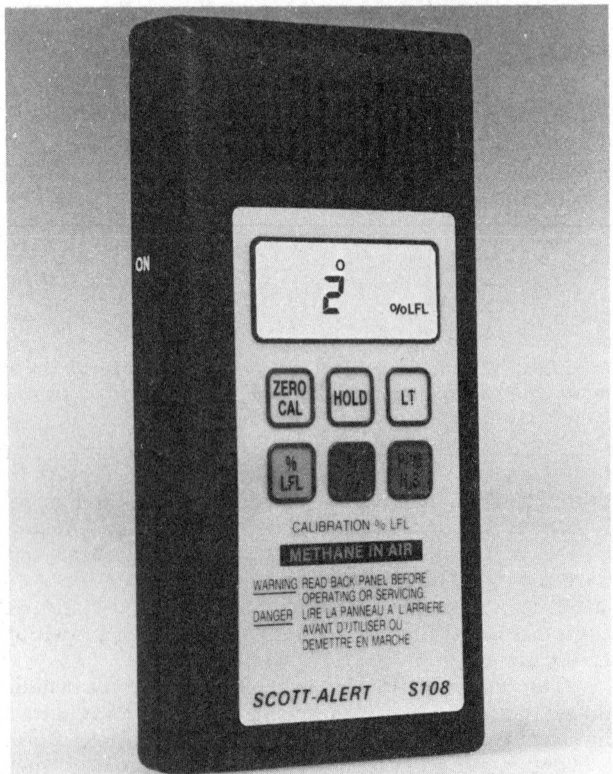

FIG. 16-5K. Diffusion sampling combustible gas, oxygen, and toxic gas indicator. (Scott Aviation)

FIG. 16-5L. Sample-draw combustible gas, oxygen, and toxic gas indicator. (Mine Safety Appliances Co.)

the zero setting can be confirmed by sampling an oxygen free gas such as propane or nitrogen.

Some models use an aspirator bulb or pump to draw the sample into the instrument, and have an appearance and configuration very similar to the standard combustible gas indicator. Other models utilize diffusion sampling, with the oxygen cell suspended at the end of a flexible cable. In either case, the electrical energy levels available from the cell are far below those which could constitute a source of ignition. Figure 16-5H shows a combination combustible gas and oxygen detector.

Toxic Gases and Vapors

Many combustible substances also present a health hazard. A concentration of a gas in air might not be flammable but could be a very serious toxic (health) hazard. As previously stated, combustible hazards are expressed as percent of LFL or a percent by volume. Toxicity is expressed in parts per million (ppm) where one percent by volume equals 10,000 ppm. Many toxic gases are health hazards in concentrations of less than 100 ppm.

The presence of toxic gases in confined spaces must be considered, either because of normal manufacturing procedures or as products of combustion resulting from a fire. Carbon monoxide is one byproduct which is almost always present.

A wide variety of instruments, including both portable indicators and stationary analyzers, exist for the measurement of toxic gases. Ranging from simple chemical filled detector tubes (Fig. 16-5I) to complicated electrochemical systems (Fig. 16-5J), they are relatively specific to individual chemicals or families of chemicals. Figure 16-5K shows a portable diffusion type indicator which simultaneously tests for combustible gas, oxygen, and a toxic gas. Figure 16-5L depicts a sample-draw type portable instrument for similar testing.

Bibliography

NFPA Codes, Standards, Recommended Practices, and Manuals. (See the latest *NFPA Codes and Standards Catalog* for availability of current editions of the following documents.)

NFPA 70, *National Electrical Code®*.

NFPA 325M, *Fire Hazard Properties of Flammable Liquids, Gases, and Volatile Solids*.

Additional Readings

Air Sampling Instruments, American Conference of Governmental Industrial Hygienists Latest Edition. Canadian Standards Association-Standard C22.2 No. 152-M1984 Combustible Gas Detection Instruments.

Bukowski, Richard W., and Bright, Richard G., "Taguchi Semiconductor Sensors as Residential Fire/Smoke Detectors," *Fire Journal*, Vol. 69, No. 3, May 1975, pp. 30-33.

Donohue, Michael L. "Working Safely in Confined Spaces," *Fire Chief Magazine*, Oct. 1984, pp. 44-46.

Hartz, N. W. "Use of Combustible Gas Indicators," *NFPA Quarterly*, Vol. 52, Apr. 1959, pp. 357-365.

Holtzberg, J. T., "Carbon Monoxide Detectors Revolutionize Fire Protection," *Coal Mining and Processing*, Vol. 18, No. 4, Apr. 1981, pp. 118-120.

Riley, J. F., "Detection and Analysis of Binary Gas Phase Mixtures," *Fire Technology*, Vol. 9, Feb. 1973, pp. 15-23.

Swift, R. L., "Testing Equipment for Air Contaminants in the Marine Industry," *Proceedings of the Marine Chemists Association*, National Fire Protection Association, Quincy, MA, 1978.

Welby, P., and Dickinson, K. R., "Monitoring Work Areas for Explosive and Toxic Hazards," *Chemical Engineering*, Vol. 83, No. 22, 1976, pp. 139-145.

Yuill, Calvin H., "Smoke: What's In It?" *Fire Journal*, Vol. 66, No. 3, May 1972, pp. 47-55.

Zatek, J. E. "Instruments for Measuring Hazardous Atmospheres," *Fire Journal*, Vol. 64, Sept. 1970, pp. 76-80.

FIRE GUARD SERVICES AND FIRE PROTECTION SURVEILLANCE

Revised by Dean K. Wilson

Containment of fire to its area of origin is made possible by several factors, one of the most important of which is early fire detection.

During normal business hours, most areas of most facilities are occupied. By their presence, the occupants provide fire protection surveillance because they are able to detect a fire when it occurs. A carelessly discarded cigarette, for example, may ignite the contents of a wastebasket; that fire may spread until the entire building is involved. However, someone who discovers the fire while it is still in its incipient stage may be able to extinguish it with a portable fire extinguisher. In some cases, the occupants can also detect conditions such as a malfunctioning machine, that might lead to a fire.

Similarly, most companies have come to realize that increased security surveillance is vital in guarding against fires of incendiary origin, the numbers of which have already risen at an alarming rate.

Guard services generally serve three purposes to protect a property against fire loss: (1) to protect the property at times when the management is not present; (2) to facilitate and control the movement of persons into, out of, and within the property; and (3) to carry out procedures for the orderly conduct of some operations on the property. Guards may be facility employees or employees of outside firms established to provide these services on a contract basis. The duties of these individuals may be supplemented or replaced in part by various approved protective signaling systems.

GUARD SERVICE DIRECTION

In conjunction with fire and explosion protective systems and various other management programs for loss prevention and control, the fire protection and security surveillance provides a means of: (1) continually monitoring the facility for conditions which might lead to a fire or explosion, (2) promptly notifying the public fire department or private fire brigade that a fire or explosion has occurred, and (3) effectively inhibiting unauthorized access to the facility. Thus, management must determine how this surveillance can be best achieved.

Management should develop a written surveillance plan for both fire protection and security in the facility. In this regard, the property manager should determine which areas of the facility are unoccupied during both working and nonworking hours. The manager should designate a management representative who will be responsible for overseeing the surveillance program. It is very important that management retain supervisory control of surveillance to maintain program integrity. This representative should review surveillance reports daily and evaluate changes in the facility which might require modification of the surveillance plan.

The fire loss prevention and control manager should be consulted during the establishment of procedures to be followed by guard service personnel. Procedures and specific instructions to guards should be geared to the specific actions required. General instructions or superficial training are of little value. Meaningful, specific instructions cannot be prepared without an investment of time and thought by the management of the property.

Management should establish a clear line of succession in event of absences. Even when there are only two guards employed, one should be designated as the leader.

Supervision of guards from outside firms should be through the designated representatives of the company providing the guard service. In its contract or supplementary documents, that company should be given details as specific and complete as possible regarding the services expected.

When guards are assigned to patrol a property, management is responsible for providing them with adequate equipment and information with which to safeguard their own health and safety. Circumstances which must be considered include:

1. Sudden illness or injury while the guard is alone at the property.
2. The possibility of a guard being overpowered by an intruder.
3. The development of a situation requiring management decisions.

Mr. Wilson is Research Consultant at Industrial Risk Insurers (IRI), Hartford, CT.

The first eventuality can be provided for by equipping guards with portable two way radios, which would allow them to summon aid. In the second situation, an intruder may prevent the guard from using any equipment. It then becomes necessary to have a system or procedure whereby the guard's failure to transmit a signal or meet a predetermined schedule will be investigated promptly.

Situations requiring management decisions may arise at times when the property is attended by guards only. Such decisions may involve conditions ranging from the unscheduled arrival of merchandise or supplies to anonymous bomb threats. Guards must have instructions and the means for contacting management personnel.

COMMUNICATION EQUIPMENT

Guards should be provided with facilities for communication within and outside the property. A control center should provide a point with which guards may communicate, and the center should have communication facilities to points outside the property. Such a center is needed even when there is very limited guard service. For example, in a facility with only one or two guards, this center might simply be a room with a telephone. Even with central station service, a control center on the property is necessary.

Where the equipment for guard communications, including guards on watch patrol, requires that signals from guards be monitored, the control center should be provided with an operator. Additional operators and 24 hour operator service should be provided at the control center according to the character of guard service provided. For some services, runners or guards who can be dispatched to investigate signals should also be provided.

A directory of names and telephone numbers (including any other information to assist in making emergency calls to the outside) should be kept at the control center in a visible index or other form. This directory should give information about the public fire and police departments, key management personnel, and other outside agencies that may have to be contacted in an emergency.

GUARD PARTOL TOUR SUPERVISION

Guard supervisory services designed to continuously report the performance of a guard are found in connection with central station service, remote station service, and proprietary signaling systems. These services usually provide for supervised or compulsory tours.

Supervised Tours

In the first case, a series of patrol stations along the guard's intended route are successively operated by the guard with each station sounding a distinctive signal at a central headquarters. Customarily, the guard is expected to reach each of these stations at a definite time, and failure to do so within a reasonable grace period prompts the central station to investigate the guard's failure to signal. Frequently, manual fire alarm boxes that ordinarily transmit four or five rounds of signals for fire can also be actuated by a special watch key carried by the guard to transmit only a single round to the central station, thus signaling that the box has been visited.

By proper location of the stations, a fire or security guard can be compelled to take a definite route through the premises, and variations from that route would appear as misplaced signals on the recording tape. A further advantage is that the order of station operation can be varied from time to time in the interests of security or to meet special conditions within the building.

Compulsory Tours

In the second case, one or more stations are wired to the central station, and preliminary mechanical stations condition the guard's key to operate the wired station after, and only after, the preliminary stations have been operated in a prearranged order. This second arrangement is somewhat less flexible than the first, but has the advantages of the absence of interconnected wires between the preliminary stations and the reduction of signal traffic. The usual arrangement is to have the guard transmit only start and finish signals that must be received at the central point at programmed reception times.

Delinquency Indicators

Delinquency indicator systems contain a series of wired stations that will transmit a signal if the guard does not reach the particular station within the anticipated preset period.

Telephone System

Extension telephones that terminate in a central supervisory office are sometimes used. The guard reports successively from each of the extensions, and the route is timed and noted by the operating supervisor.

Guard Clock Systems

Where immediate supervision of guards is not imperative, portable watch clocks are widely used to record the progress of guard tours. A number of stations, each consisting of a key, are located throughout the premises, and a guard, upon reaching a station, operates the key in the portable clock. This records the station number on a paper dial or tape in the clock, indicating the time of visit. Each time a clock is opened or closed for any purpose, a tell-tale mark is punched on the dial or tape, so unauthorized tampering is readily detected. A representative portable watch clock is shown in Figure 16-6A.

A stationary guard clock system utilizes a clock installed at a central location, with electric wiring extending to stations throughout the property. The guard carries a small crank type key that is inserted at each station box. Turning the key operates a small magneto that generates sufficient current to actuate a recording mechanism in the central clock and indicate the time each station is visited.

EVALUATING GUARD SERVICE

There are several items to be considered when evaluating guard service.

Tour Supervisory System Recording Methods

Portable Watch Clock: The clock must be kept locked, and the key must be inaccessible to the guard. The clock records should be removed from the clock and checked daily by management's representative. Guards should never remove their own records from the clock. Even if contract guard service is employed, management's representative should remove the records from the clock and check them rather than relying upon a supervisor from that service.

Central Station Guard Patrol Tour Supervisory System: With this type of system, the written records of the guards' tours are maintained in the central station. NFPA 71, *Standard for Installation, Maintenance and Use of Central Station Signaling Systems*, requires the central station to notify both the facility and the authority having jurisdiction in writing if the guard is late in starting or finishing a tour. A specified grace period of 10 to 15 minutes is usually permitted. Upon the expiration of the grace period, the central station attempts to contact the guard by telephone. If unable to reach a guard, the central station immediately notifies the police and dispatches its own armed guard.

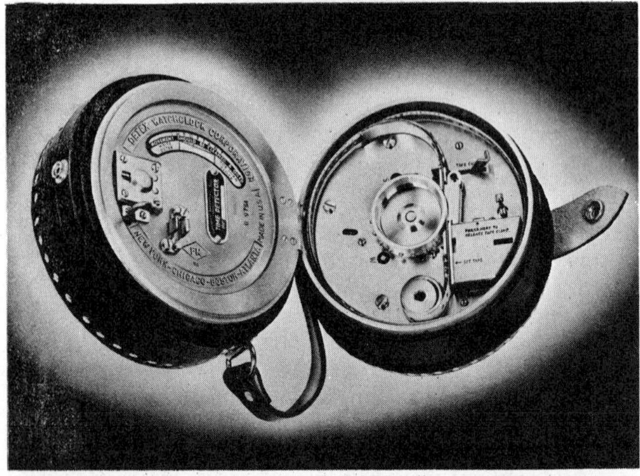

FIG. 16-6A. Front view (top), interior view (bottom) of a portable watch clock. This clock provides an embossed record tape, made directly from type on the recording key. The tape has 24 ruled segments and is synchronized with the clock mechanism. If a guard fails to punch in, the omission is indicated by a prominent white space on the tape. This clock indicates when it was opened. (Detex Corp.)

Proprietary Guard Patrol Tour Supervisory System and Stationary Watch Clocks: With these systems, the guards' tours are automatically recorded at a central location. Management's representative should review this record daily.

Tour Records: Management's representative should ensure that:

1. All unoccupied areas of the facility are included in each tour.
2. All key stations or tour supervisory transmitters in each tour have been recorded clearly in a regular hourly pattern at night and in a bihourly pattern during the day.
3. Tours last no longer than 45 minutes, allowing for a rest period of at least 15 minutes each hour.
4. Tours begin within one half-hour of the time the area becomes unoccupied and continue to within one half-hour of the resumption of occupancy.
5. The "tell tale" of a portable or stationary watch clock is recording each time the clock is opened. Look for indications that the clock has been opened more than once a day or at unusual times. This might indicate that unauthorized persons have access to a clock key.

Inspecting Tour Supervisory System Initiating Devices

Key stations, tour supervisory transmitters, or intermediate stations should be inspected once a month to ensure they are firmly attached and sealed with a "tamper" seal and have not been relocated or removed. The key should be checked for damage. If there is evidence of tampering, the key stations, tour supervisory transmitters, or intermediate stations should be checked more often and suitable action taken.

Guards

Management should expect, and is entitled to receive, guard service of the highest quality. Guards must be conscientious in the performance of their duties, note and report all infractions of company regulations, and closely follow the orders given to them.

1. Guards hold positions of trust which require individuals who are physically able, mentally alert, and morally responsible. Therefore, their physical and emotional stability should be evaluated by appropriate management personnel.
2. Guards must be sufficiently intelligent, calm in the face of an emergency, mature with sound judgment, and possess the physical stamina required for the job.
3. A sufficient number of guards should be provided to maintain proper surveillance in the facility. It is undesirable for guards to be assigned part time duties unrelated to surveillance. If they are so assigned, however, these duties must not interfere with surveillance responsibilities.
4. Guards should receive full support of management in the performance of their duties.
5. When the guards are facility employees, management should establish the scope of the service and provide necessary training and supervision for the guards.
6. If a contract guard service is used, management should not assume that it will be adequate. Rather, management should prepare detailed specifications and investigate the ability of prospective contractors to meet these specifications. When the contract has been met, management should make sure that its intent is being carried out.
7. Initial and continued training of guards should be by a formal, comprehensive, written program covering all applicable protection procedures. Each guard must be:

a. Acquainted with the general nature of the facility's operations and possess specific knowledge of any hazardous operations.
b. Familiar with all of the facility's manual and automatic fire protection equipment. They should be especially aware of the location of all sprinkler valves and know which area each controls. Guards should periodically accompany the person making fire protection equipment inspections in order to gain a working knowledge of facility protection features and hazards.
c. Familiar with the location and operation of manual fire alarm stations and other means of transmitting fire alarms. Such means should be provided throughout the facility so guards can easily report a fire.
d. Taught to notify the fire department before attempting to fight the fire.
e. Taught how to admit public fire apparatus to the property and how to direct fire department officers to the location of the fire.
f. Taught to properly notify company officials when an emergency occurs or when potential trouble is observed.
g. Taught to maintain a shift log and to prepare reports to management of observations made and action taken during tours.

8. Guard service should be integrated into the overall preemergency planning program.
9. General and special instructions and other data required by the guards should be written down and kept up-to-date. (See NFPA 601 and 601A, *Standard for Guard Service in Fire Loss Prevention*, and *Standard for Guard Operations in Fire Loss Prevention*, respectively; hereinafter referred to as NFPA 601 and NFPA 601A.)

While making regular tours throughout the facility, guards must be alert for all emergencies, paying special attention to known hazardous areas. Guards are in a position not only to detect and correct unsafe conditions which might develop into or contribute to a serious fire, but also to discover an incipient fire. Therefore, the guards should be familiar with the fundamentals of fire control and with the proper use of all available extinguishing equipment.

The importance of notifying the fire department before attempting to fight a fire should be stressed in guard training. Guards should report any situation which may endanger the facility, such as an exposing fire in adjacent properties. Any unusual condition which the guards cannot correct without assistance must be reported immediately to the proper official so the situation can be remedied without undue delay. Such situations include the interruption of sprinkler service or the failure of heating equipment.

In addition to carefully following written preemergency plans, guards must be resourceful and capable of applying common sense to any unusual conditions (such as "natural" hazard threats) they may encounter. If the facility is being subjected to freezing weather, for example, the guard should be alert to those areas of the facility where protective systems or process equipment might be vulnerable to freezing damage. Such areas should be checked frequently to assure the heating system is properly maintaining the temperature necessary to avoid damage. If a thermometer is unavailable, a resourceful guard might set out a small container of water to observe whether the area is approaching a dangerously low temperature. See NFPA 601 and NFPA 601A for further details.

GUARD SERVICE FUNCTIONS

A sufficient number of guards should be provided to accomplish the needed services. If guards are assigned to other part time duties in addition to their regular guard services, these duties should be chosen so they will not interfere with regular guard duties.

Guard service can facilitate and control the movement of persons within a property when the number of persons in the property requires such a service. Duties in this category include:

1. Preventing the entry of unauthorized persons who might set a fire or do damage to the facility.
2. Controlling the activities of people authorized to be on the property, but who may not be aware of procedures established for the prevention of fire.
3. Controlling pedestrian and vehicular traffic during exit drills, and evacuating the property or parts of it during emergencies.
4. Controlling gates and vehicular traffic to facilitate access to the property by the public fire department, members of a private fire brigade, and off duty management personnel in case of fire or emergency.

Guard service should be established to carry out certain procedures for the orderly conduct of the operations in the property, including procedures for fire loss prevention and control, both by personnel associated with the property and outside contractors. Duties in this category include:

1. Checking permits for "hot work" including cutting and welding, and standing by to operate fire extinguishing equipment where necessary.
2. Detecting conditions likely to cause a fire, such as leaks, spills, and faulty equipment.
3. Detecting conditions likely to reduce the effectiveness with which a fire may be controlled, such as portable fire extinguishers not in place, sprinkler valves not open, and water supplies impaired.
4. Performing operations to assure that fire equipment will function effectively. These may include testing automatic sprinkler and other fixed fire protection systems, including fire pumps, and other equipment related to these systems and assisting in the maintenance of this equipment; checking portable fire extinguishers and fire hose and assisting in pressure tests and maintenance service on these items; testing fire alarm equipment by actuating transmitting devices as required; and checking equipment provided on any motorized fire apparatus and carrying out the periodic tests and maintenance operations required for the apparatus.
5. Promptly discovering a fire and calling the public fire department (also the fire brigade of the property, where there is such a brigade).
6. Operating equipment provided for fire control and extinguishment after giving the alarm and before the response of other persons to the alarm.

7. Monitoring signals of protective signaling systems, such as alarms from manual fire alarm boxes, signals for water flow in sprinkler systems, and signals from systems for detecting fires and abnormal conditions, including trouble signals.
8. Patrolling routes chosen by management to assure surveillance of all the property at appropriate intervals.
9. Starting up and shutting down certain equipment when no other personnel are provided for this purpose.

PATROL ROUTES AND ROUNDS

Each route to be patrolled should be laid out by the responsible manager. The guard responsible for each route should be given instructions as to all details of the route, and what is expected in covering it. The route should be laid out to prevent shortcuts so the guard is required to pass through the entire patrol area. A rest period between rounds is reasonable.

Guards should make rounds at intervals determined for the particular situation by management. When operations in the property are normally suspended, rounds should be made hourly unless management is willing to accept rounds at less frequent intervals. When there are special conditions, such as the presence of exceptional hazards or when protection is impaired, management should institute as many additional rounds as necessary to meet the firesafety requirements of such conditions.

The first round of a patrol should begin as soon as possible after the end of activities of the preceding work shift. The guards should be instructed to make a thorough inspection of all buildings or spaces on the route during their first round. Their instructions should cover the following:

1. Outside doors and gates should be closed and locked. Windows, skylights, fire doors, and fire shutters should be closed.
2. All oily waste, rags, paint residue, rubbish, and similar items should be removed from buildings or placed in approved containers.
3. All fire apparatus should be in place and not obstructed.
4. Aisles should be clear.
5. Motors or machines carelessly left running should be shut off and reported.
6. All offices, conference rooms, and smoking areas should be checked for carelessly discarded smoking materials.
7. All gas and electric heaters, coal and oil stoves, and other heating devices on the premises should be checked.
8. All hazardous manufacturing processes should be left in a safe condition. The temperature of dryers, annealing furnaces, and similar equipment that continue to operate during the night, holidays, and weekends should be noted on all rounds.
9. Hazardous materials such as gasoline, rubber cement, and other flammable and highly volatile combustibles should be kept in proper containers or removed from buildings.
10. All sprinkler valves should be open with gages indicating proper pressures. If not open, the fact should be reported immediately.
11. During cold weather all rooms should be checked to determine if they are heated properly.
12. All water faucets and air valves found leaking should be closed. If leaks cannot be stopped, the condition should be reported.
13. Particular attention should be given to new construction or alterations which may be in progress.

SELECTION OF GUARDS

Management should require individuals considered for guard service to satisfactorily pass a character investigation. This investigation should attempt to evaluate the individual's reliability, self-control, and potential loyalty to the employer. Applicants for a guard position should be required to be finger printed and to give particulars of any police records. The local police should be furnished with this information, should corroborate it, and should ask for checks by other police agencies. The fingerprint data should be cleared with state, national, or appropriate international agencies that maintain clearing facilities for police records. All applicants for a position as a fire or security guard or patrolman should be required to state any military service record and to submit evidence of such service that may assist in an evaluation of the individual's suitabiilty for guard service.

Contracts for guard service should include a provision that the company furnishing guard service will replace any of its employees who, in the judgment of the company purchasing the service, are not qualified.

Management should be satisfied that individuals considered for guard service are mentally alert and have good powers of observation, intelligence, and judgment. Investigation should attempt to evaluate the individual's personality and temperament. Such an evaluation is more realistic than arbitrarily testing education or intelligence, or setting an age limit. Very young people may not qualify because they have not acquired a sense of responsibility or judgment. Very old people may have impaired alertness. Individuals should be sought who are known to be clear headed in an emergency.

Guards should be required to pass an annual written examination dealing with information about the property protected and procedures for fire loss prevention with which they are expected to be familiar.

Management should require that individuals considered for guard service pass an examination to determine whether they are physically able to perform the guard duties to which they will be assigned. Guards should also be required to pass an annual physical examination. The guards do not need to be athletes, but they should not have a heart condition or other physical ailment that might work to their disadvantage in moments of stress.

TRAINING

Management should establish a continuing training program for its guards. Its scope should be established by the manager or by a fire loss prevention and control manager acting for the manager. Courses for guards and fire fighters available through training programs of vocational training agencies, schools, universities, and other training and educational agencies should be considered in any training program for guards.

Management should require guards to have completed at least elementary courses of instruction in the use of portable fire equipment and emergency first aid. The time spent in such preliminary training should be a minimum of two working days in each subject. During service, guards should be given not less than the equivalent of two full working days per year of training to increase their knowledge and experience in the use of portable fire extinguishers, first aid and other useful training. Guards should be required as part of their training to participate in appropriate meetings devoted to prefire planning with operating personnel. Guards can help accomplish this by working together with the public fire department when it establishes its plan for the premises.

Management should require guards to know the location of portable fire extinguishers, hand hoses, standpipes and hydrants, valves controlling sprinkler systems, inside riser valves, post indicator valves, and sectional valves in the property's own water system. (They also should know how to start fire pumps.) Guards may also need to know the location and purpose of valves controlling water for purposes other than for fire protection, and valves controlling steam, gas, and other services. Management should require guards to know the locations of dangerous machinery or materials and inform them of hazardous manufacturing processes, especially those continuing during the night, holidays, or weekends.

Bibliography

NFPA Codes, Standards, Recommended Practices and Manuals. (See the latest *NFPA Codes and Standards Catalog* for availability of current editions of the following documents.)

NFPA 71, *Standard for the Installation, Maintenance and Use of Central Station Signaling Systems.*

NFPA 72A, *Standard for the Installation, Maintenance and Use of Local Protective Signaling Systems.*

NFPA 72C, *Standard for the Installation, Maintenance and Use of Remote Station Protective Signaling Systems.*

NFPA 72D, *Standard for the Installation, Maintenance and Use of Proprietary Protective Signaling Systems.*

NFPA 601, *Standard for Guard Service in Fire Loss Prevention.*

NFPA 601A, *Standard for Guard Operations in Fire Loss Prevention.*

Additional Readings

Accident Prevention Manual, National Safety Council, Chicago, IL, 1981.

Introduction to Fire Protection (training course for employees involved in facility safety programs), National Fire Protection Association, Quincy, MA, 1982.

Linville, J. L., ed., *Industrial Fire Hazards Handbook*, 2nd ed., National Fire Protection Association, Quincy, MA, 1984.

Planner, R. G., *Fire Loss Control*, Marcel Dekker, NY, 1979.

"Self Fire Inspection—A New Concept in Fire Safety," *Fire Chief*, Vol. 24, No. 5, May 1980, pp. 34-35.

Tuck, C. A. Jr., *NFPA Inspection Manual*, 4th ed. National Fire Protection Association, Quincy, MA, 1976.

SECTION 17

WATER AND WATER SUPPLIES FOR FIRE PROTECTION

WATER AND WATER ADDITIVES FOR FIRE FIGHTING

Revised by Robert M. Hodnett

Water is, and has long been, the most common extinguishing agent. This chapter discusses the properties of water as an extinguishing agent, both its advantages and limitations.

The basic principles of extinguishment of fire are contained in Section 4, Chapter 1, "Chemistry and Physics of Fire," and Section 4, Chapter 4, "Theory of Fire and Explosion Control." The systems and devices employed for the transportation and application of water as an extinguishing agent are contained in applicable chapters in Section 15, "Public Fire Protection;" Section 17, "Water and Water Supplies for Fire Protection;" Section 18, "Water-based Extinguishing Systems;" Section 19, "Special Fire Suppression Agents;" and Section 20, "Portable Fire Extinguishers." The use of water on specific types of materials, e.g., chemicals, flammable liquids, gases, metals, etc. is covered in more detail in the chapters of Section 4, "Fire Hazards of Materials." See Section 10, Chapter 17, "Nuclear Reactors, Radiation Machines, and Facilities Handling Radioactive Materials," for guidance on planning fire protection, including the use of water, for facilities containing radioactive materials.

PHYSICAL PROPERTIES

The physical properties that make water a good extinguishing agent are:

1. At ordinary temperature water is a heavy, relatively stable liquid.
2. The melting of 1 lb (0.45 kg) of ice into water at 32°F (0°C) absorbs 143.4 Btu (151.3 kJ), which is the heat fusion of ice.
3. One Btu is required to raise the temperature of 1 lb of water 1°F, which is the specific heat of water. Therefore, raising the temperature of 1 lb of water from 32°F to 212°F requires 180 Btu. (In S.I. units, the specific heat capacity of water is 4.186 kJ/kg K.)
4. The latent heat of vaporization of water, i.e., converting 1 lb (0.45 kg) of water to steam at a constant temperature is 970.3 Btu per lb (2254.8 kJ/kg) at atmospheric pressure.
5. When water is converted from liquid to vapor, its volume at atmospheric pressures increases about 1,600 times. This large volume of water (saturated steam) displaces an equal volume of air surrounding a fire, thus reducing the volume of air (oxygen) available to sustain combustion.

Other than water, there is no material easily available which has all these characteristics. Of course, water applied in the form of ice or snow would cool even better than plain water because it would take 143.4 Btu/lb (333.2 kJ/kg) to convert the ice or snow to water, but so far there is no practical way to do this.

EXTINGUISHING PROPERTIES

A fire can be extinguished only if an effective agent is applied at the point where combustion is occurring. For hundreds of years, the principal method of extinguishing fires has been to direct a solid stream of water (from a safe distance) into the base of the fire, and this method is still used widely today. A more efficient method, however, is to apply water in spray form.

Extinguishment by Cooling

In most cases, if the surface of the burning material is cooled below the temperature at which it will give off sufficient vapor to support combustion, the fire will be extinguished. Surface cooling is not usually effective on gaseous products and flammable liquids that have flash points below the temperature of the applied water, and water is generally not recommended for flammable liquids with a flash point below 100°F (37.8°C).

The amount of water required to extinguish a fire depends on how hot the fire is. How quickly a fire is extinguished depends on how quickly the water is applied, how much of it is applied, and what form of water is applied. It is best to apply water so that the maximum amount of heat will be absorbed. Water absorbs the most heat when it is converted into steam, and it will be

Mr. Hodnett is a consultant in fire protection and piping systems based in Providence, RI. He was NFPA staff liaison to the NFPA Water Extinguishing Systems Committees prior to his retirement in 1984.

converted into steam more easily from droplets than from a solid stream.

Much theoretical information is available on the factors that affect the rates of heat absorption and vaporization of water droplets. Because these factors cannot be closely controlled under most actual fire conditions, they cannot be used for accurate fireground calculations.

Water spray cools a fire according to the following principles:

1. The rate of heat transfer is proportional to the exposed surface of the liquid. For a given quantity of water, the surface is greatly increased by conversion to droplets.
2. The rate of heat transfer depends on the temperature difference between the water and the surrounding air or burning material.
3. The rate of heat transfer also depends on the vapor content of the air, particularly in regard to fire spread.
4. The heat absorbing capacity of water depends upon the distance it traveled and its velocity in the combustion zone. (This factor must take into account the necessity for projecting a suitable volume of water to the fire.)

Droplet Size: Calculations show that the optimum diameter of a water droplet is in the range of 0.01 to 0.04 in. (0.3 to 1.0 mm), and that the best results are obtained when the droplets are fairly uniform in size. At present there is no discharge device capable of producing completely uniform droplets, although many discharge devices spray droplets that are fairly uniform over a broad range of pressures. The droplet must be large enough to have sufficient energy to reach the point of combustion despite air resistance, the opposing force of gravity, and any air currents.

Wetting combustible materials is a method often employed to prevent ignition of unburned materials. If combustibles absorb water, it takes longer to ignite them because the water must be evaporated before they can get hot enough to burn.

Extinguishment by Smothering

If enough steam is generated, air can be displaced or excluded. Fires in certain materials can be extinguished by this smothering action, which is speedier if the steam generated can somehow be confined to the combustion zone. The process of heat absorption by steam ends when the steam starts to condense, a change which requires heat release from the steam. When this happens, visible clouds of water vapor form. When such condensation occurs above the fire, it has no cooling effect on the burning material. However, the steam may carry heat away from the fire if it can harmlessly dissipate itself into clouds of water vapor above the fire.

Fires in ordinary combustibles are normally extinguished by the cooling effect of water—not by the smothering effect created by the generation of steam. Although the latter might suppress flames, it usually cannot extinguish such fires.

Water may be used to smother a burning flammable liquid when the liquid has a flash point above 100°F (37.8°C), a specific gravity of 1.1 or heavier, and is not water soluble. To achieve this most effectively, a foaming agent is normally added to the water. The water must then be applied gently to the surface of the liquid.

In cases where oxygen is produced while a burning material decomposes, smothering by any agent is not possible.

Extinguishment by Emulsification

An emulsion is formed when immiscible liquids are agitated together and one of the liquids is dispersed throughout the other. Extinguishment by this process can be achieved by applying water to certain viscous flammable liquids, since the effect of cooling the surfaces of such liquids prevents the release of flammable vapors. With some viscous liquids (such as No. 6 fuel oil), the emulsification is a "froth" which retards the release of flammable vapors. Care must be used on liquids of appreciable depth, however, because frothing may spread the burning liquids over the sides of the container. A relatively strong, coarse water spray is normally used for emulsification. Avoid a solid stream of water as it will cause violent frothing.

Extinguishment by Dilution

Fires in water soluble flammable materials may, in some instances, be extinguished by dilution. The percentage of dilution necessary varies greatly, as will the volume of water and the time necessary for extinguishment. For example, dilution can be used successfully in a fire involving an ethyl or methyl alcohol spill if it is possible to get an adequate mixture of water and alcohol; however, dilution is not a common practice if tanks are involved. The danger of overflow because of the large amount of water required, and the danger of frothing should the mixture become heated to the boiling point of water, make this form of extinguishment seldom practical.

FREEZING TEMPERATURES AND ANTIFREEZE ADDITIVES

Because water freezes at 32°F (0°C), its use as an extinguishing agent is limited in climates or situations where freezing temperatures are encountered. There are several methods commonly used to prevent problems of freezing. These include using dry pipe sprinkler systems in lieu of wet pipe systems, circulating or heating tank water supplies held for fire protection purposes, adding freezing point depressants to the water, or a combination of these.

The water soluble freezing point depressant in fire equipment most widely used is calcium chloride with a corrosion inhibitor additive. Calcium chloride solutions are not used when fire protection systems are supplied by public water connections. Table 17-1A gives data on the amounts needed for various low temperatures.

Sodium chloride (common salt) is unsatisfactory because of its limited ability to depress the freezing point of water and because it is highly corrosive.

Sprinkler Systems: Chemically pure glycerine (U.S. Pharmacopoeia 96.5 percent grade) or pure propylene glycol can be used to depress the freezing point of water in portions of wet pipe systems connected to public water supplies if authorized by local health authorities. Diethylene glycol, ethylene glycol, or calcium chloride, as well as glycerine or propylene glycol, can be used for the same purposes where public water is not connected. Both ethylene and diethylene glycol are poisonous, and must never be permitted to contaminate drinking water. Tables 17-1A and 17-1B give data on the amounts needed for the various low temperatures.

Fire Pails or Casks: Calcium chloride-water solutions are used where continuous temperatures are below 40°F (4.4°C). Table 20-4A gives the amount of calcium chloride which should be mixed thoroughly with water to make 10 gal (38 L) of antifreeze solution.

Water type Extinguishers: Alkali-metal salt solutions provide protection against low temperatures. Only the solutions specified by the manufacturers should be used, because the extinguishers are examined by the testing laboratories using the particular solutions recommended for them, and they may not be reliable if used with other solutions. Glycol solutions should not be used in extinguishers because the amount required to reduce temperatures would be high, i.e., a 52.5 percent solution of ethylene glycol would be needed to achieve −40°F (−40°C) freezing point. Such amounts would alter the effectiveness of the extinguisher and could also lead to complications if the water should "boil off" and leave a strong concentration of the glycol which, under certain conditions, could be ignited.

Efforts have been made to develop other additives which can be mixed with water to lower the freezing point to -65°F (-54°C). This work has been largely stimulated by increased human activity in extremely low temperature areas, with most of the research being done by the U.S. armed forces. To date, formulas of lithium chloride, lithium chloride-calcium chloride, and lithium chloride-anhydrous sodium chromate have been used successfully. No commercial use of these solutions is known, although the U.S. Naval Research Laboratory has developed a lithium chloride solution for fire extinguishers exposed to low temperatures -65°F (-54°C).

SURFACE TENSION AND WETTING AGENT ADDITIVES

The relatively high surface tension of plain water slows its ability to penetrate burning combustibles and impedes its spread throughout any closely packed, baled, or stacked materials. Immersion of burning combustibles in water is rarely practical. If a fire originates or burrows in a mass of combustibles, it is necessary either to dismantle the mass or use a wetting agent additive to lower the surface tension of the water. Many chemicals can be used as wetting agents, but few are good extinguishing agents because they are toxic, corrosive, or unstable when mixed with water so it can penetrate the dense material. Minimum requirements for wetting agents for fire protection purposes are contained in NFPA 18, *Standard on Wetting Agents*. The purpose of the Wetting Agent Standard is to set forth requirements for wetting agent additives that will improve the fire fighting properties of plain water without making it harmful to personnel, property, or equipment.

Wetting agents are effective because they decrease the surface tension of the water, thus increasing the amount of free surface available for heat absorption. There is less water run-off, and the water is therefore more efficient. In the United States, UL lists wetting agents according to their qualities and their extinguishing abilities on Class A or Class B fires, or both. Wetting agents used in fire fighting

Table 17-1A. Solutions to Depress the Freezing Point of Water in Wet pipe Sprinkler Systems Not Connected to Public Water Supply Sources

Material	Solution (By Volume)		Sp Gr at 60°F (16°C)	Freezing Point °F	Freezing Point °C
Glycerine	If glycerine is used, see Table 17-1A				
Diethylene Glycol	50% Water		1.078	−13	−25
	45% Water		1.081	−27	−32
	40% Water		1.086	−42	−41
	Hydrometer Scale 1.000 to 1.120 (Subdivisions 0.002)				
Ethylene Glycol	61% Water		1.056	−10	−18
	56% Water		1.063	−20	−29
	51% Water		1.069	−30	−34
	47% Water		1.073	−40	−40
	Hydrometer Scale 1.000 to 1.120 (Subdivisions 0.002)				
Propylene Glycol	If propylene glycol is used, see Table 17-1B				
Calcium Chloride 80% Flake Fire Protection Grade*	Lbs $CaCl_2$ per gal. of water	(Kg $CaCl_2$ per liter of water)			
Add corrosion inhibitor of sodium dichromate ¼ oz per gal water	2.83	0.339	1.183	0	
	3.38	0.405	1.212	−10	−18
	3.89	0.466	1.237	−20	−29
	4.37	0.524	1.258	−30	−34
	4.73	0.560	1.274	−40	−40
	4.93	0.591	1.283	−50	−45

* Free from magnesium chloride and other impurities.

TABLE 17-1B. Solutions to Depress the Freezing Point of Water in Wet Pipe Sprinkler Systems Connected to Public Water Supply Sources if Authorized by Health Laws

Material	Solution (By Volume)	Sp Gr at 60°F (16°C)	Freezing Point °F	Freezing Point °C
Glycerine C.P. or U.S.P. Grade*	50% Water	1.133	−15	−26
	40% Water	1.151	−22	−30
	30% Water	1.165	−40	−40
Hydrometer Scale 1.000 to 1.200				
Propylene Glycol	70% Water	1.027	+ 9	−13
	60% Water	1.034	− 6	−21
	50% Water	1.041	−26	−32
	40% Water	1.045	−60	−51
Hydrometer Scale 1.000 to 1.120 (Subdivisions 0.002)				

* C.P.—Chemically Pure.
U.S.P.—United States Pharmacopoeia 96.5 percent.

and fire protection are no more corrosive than plain water to brass, bronze, and copper, although they tend to accelerate corrosion of some metals due to the cleaning and penetrating action of the concentrates. The use of containers made of materials such as cast iron, aluminum, zinc, galvanized steel, lead or lead-coated iron, die-cast alloys (such as white metal, zinc, etc.), or "air-dried" types of coatings (which may include plastics, oil paint, lacquers, or asphalt) should be avoided. Wetting agent solutions should only be used with appropriate equipment.

"Wet water" (plain water plus a wetting agent) is chiefly used to penetrate porous surfaces and allow the solution to reach hidden areas of burning combustibles (a "fire-packed" bale of cotton or stacked hay) and to penetrate the subsurface of ordinary combustibles to prevent rekindling. Wet water is sometimes used on grass, brush, or forest fires. Generally, these fires are more properly and easily handled by "thickened" water.

Wet water has the same limitations as plain water on fires in chemicals which react with water, such as sodium, calcium carbide, etc. The use of wetting agent solutions on flammable and combustible liquid fires is not common since Halon 1301, Halon 1211, carbon dioxide, dry chemicals, or foam are normally used on these types of fires. Wet water should not be used on flammable or combustible liquid fires if the liquids are water soluble (such as the alcohols, glycols, and some ketones).

In general, wet water solutions should not be used on live electrical equipment because of the conductivity of the solution. Spray or fog, however, might be used with caution. Wet water, due to its penetrating characteristics, may have more harmful effects on motors, transformers, and similar equipment than plain water. Any electrical equipment that has been penetrated by wet water should be thoroughly flushed and cleaned before it is returned to service.

Wetting agents may be either premixed with water or added to the water through suitable proportioning equipment at the time the water is being used. Mixing wetting agents from different manufacturers, or mixing a wetting agent with mechanical or chemical foam concentrates, is not recommended. Distinction must be made between wetting agents and wetting agent foams and other detergent type foams (high expansion foam) and aqueous film-forming agents. Methods of measuring the effect of the addition of a wetting agent to water to lower the surface tension are given in the appendix of the NFPA Wetting Agent Standard.

VISCOSITY AND ADDITIVES TO THICKEN WATER

The relatively low viscosity of water makes it tend to run off surfaces quickly and limits its ability to blanket a fire by forming a barrier on the surface of combustible materials. Additives to make water more viscous ("thickened" water) make the use of water more efficient on certain types of fires.

Viscous water has had one of several thickening agents added to it. In proper proportions, viscous water seems to have the following advantages over plain water in that it:

1. Sticks and clings more readily to the burning fuel.
2. Spreads itself out in a continuous coating over the fuel surface.
3. Develops in a layer several times the thickness of plain water.
4. Absorbs heat proportional to the amount of water present.
5. Projects somewhat further and higher from straight stream nozzles.
6. Forms a tough, dry film after drying which helps seal the fuel from oxygen.
7. Resists wind drift in some applications (as from aircraft in forest fire fighting).

Disadvantages of viscous water may be that it:

1. Does not penetrate the fuel as well as plain or wet water.
2. Increases friction loss in hose or pipe.
3. Increases water droplet size (where fine sprays are needed, they cannot be secured as readily).
4. Increases the slipperiness of surfaces, making it more difficult to walk in areas where it has been applied.
5. Increases handling problems and logistics of fire operations because it takes time to mix viscosity agents with water. (Under some conditions, stored solutions can lose viscosity, principally through water temperature changes and possible bacterial or chemical contamination.)

To date, most of the application research has been directed at fighting forest fires. A detailed report was prepared by the NFPA Forest Committee in cooperation with the U.S. Forest Service and other forest fire control specialists (NFPA 1977). Plain water is often capable of handling ordinary brush, grass, and forest fires; in some situations, such as burrowing fires in a mass of combustibles, "wet" water is preferred. Viscous water is most effective on very hot fires, whether in forests or structural fuels, and where radiation may spread the fire.

Two viscosity agents used in forest fire control are CMC (sodium carboxymethylcellulose) and Gelgard (a trade name product of the Dow Chemical Company). Other viscosity agents are available. The agents are desirable for the following reasons:

1. Only a small quantity of dry powder is necessary for a batch mix.
2. It takes only a few minutes to mix an adequate batch.
3. Near maximum viscosity is reached within the batch mix period, and the viscosity is retained indefinitely (assuming no contamination or spoilage).
4. Their cost is reasonable.
5. They are nontoxic and noncorrosive.
6. They pump easily.
7. They provide a good stream pattern.
8. They provide good coverage over the fuels.

Bentonite clay (montmorillonite) has been used as a short-term fire retardant and a slurry in water and produces a heavy coat of water. Longer term retardants are ammonium phosphates and ammonium sulfates. The phosphates are about 1.5 times as effective as ammonium sulfate since the former seem effective against both flaming and glowing ignition, while the sulfates seem to be effective only in flaming ignition. Two solid ammonium phosphate chemicals notable for their fire retardant and fire extinguishing capabilities are (DAP) diammonium phosphate and (MAP) monoammonium phosphate. Viscous retardant solutions employing DAP and CMC are available commercially, as are a liquid ammonium phosphate concentrate and an ammonium sulfate fire retardant.

The NFPA Forest Committee has prepared a special publication (NFPA 1975) covering the use of aircraft for reconnaissance, fire attack, and control of these types of fires.

ADDITIVES TO MODIFY FLOW CHARACTERISTICS

Friction loss in fire hose is always a problem for fire fighters. The longer the hose and the more water pumped through it, the greater the pressure loss. With good quality fire hose, most of the pressure loss is the result of friction between particles of water generated by the turbulence in the flowing stream. When flow is either smooth or laminar, the friction loss tends to be very low with a slow stream of water; however, the amount of water delivered under laminar flow is generally too low for fire fighting. Fire fighting requires high velocity streams that generate turbulence which, in turn, results in friction between water particles. This friction accounts for about 90 percent of the pressure loss in good fire hose. The friction between the flowing water and the interior hose wall accounts for only 5 to 10 percent of the loss.

Until 1948 it was generally believed that not much could be done to reduce friction loss. At that time trace quantities of certain polymers were found to reduce friction loss of turbulent streams. Most researchers report that linear polymers (polymers that form a single straight line chemical chain with no branches) are the most effective in reducing turbulent frictional losses and, of these, poly (polyethylene oxide) is the most effective. Friction reducing efficiency is a direct function of polymer linearity.

Poly is nontoxic, has no effects on plants or marine life, and will degrade in sunlight. It is a long linear chain, high molecular weight polymer, and is 2 to 3 times more effective as a friction reduction agent than other materials tested to date. It is an opaque, white slurry that has no odor and weighs 9.1 ppg (pounds per gallon), or 1.1 kg/L and must be kept within 0 to 120°F (−17 to 49°C). When it is injected into the hose stream, it dissolves completely and does not separate. It is compatible with all fire fighting equipment and is useful in both fresh and salt water. One gallon of additive treats 6,000 gal (22 710 L) of water and achieves at least a 40 percent greater water delivery.

Tests run by the New York City Fire Department and Union Carbide Corporation (Clough 1973) found that in present systems without poly, a 1 in. (25 mm) booster hose will deliver approximately 20 gpm (76 L/min). However, with poly and a change in nozzle design, 75 gpm (284 L/min) were delivered during the test. A 1½ in. (38 mm) hose delivered 250 gpm (946 L/min), or as much as a 2½ in. (64 mm) hose without the additive. With the additive, a 2½ in. (64 mm) hose was able to deliver more water than a 3 in. (76 mm) hose, and nearly as much water as a 3½ in. (89 mm) hose.

Fire fighters know from experience that a heavy, charged 2½ in. (64 mm) hose can be hard to move. A smaller, lighter, more mobile hose can be less difficult to manipulate when attacking fires, and it can also deliver 250 gpm (946 L/min) when poly is used. Therefore, it should be handled with the same respect as a 2½ in. (64 mm) hose.

Of equal significance, these tests showed that the additive nearly doubled the nozzle pressure. The stream's reach was increased by nearly 30 percent, and the stream was more coherent. Little effect was found on the fog nozzle sprays other than that the water spray seemed to be more dense.

OPACITY AND REFLECTIVITY

Tests conducted at UL using water spray to provide exposure protection to a sheet metal surface from a gasoline fire indicate that when the spray is applied as a thin film of water over the sheet metal, the temperature of the metal was contained within limits which protected the metal from significant damage. This was not true, however, when the water spray was adjusted so that it did not touch the sheet metal but did provide a water curtain between the metal and the fire. In this latter case, the temperatures of the metal were three to four times greater than when the water flowed over the metal. These tests may indicate that because of its lack of opacity, water does not prevent the passage of radiant heat well. The principal value of water used to protect exposures is from the cooling obtained by evaporation of a water film on the exposed surfaces.

NFPA 13, *Installation of Sprinkler Systems* (including outside sprinklers for protection against exposure fire), calls for sprinklers to be positioned so that the water will thoroughly wet exposed glass windows and run down over the window sash and the glass, wetting the entire window as much as possible. Similar recommendations require that as much of the cornice as possible be wetted when cornice sprinklers are installed. These rules reflect the experimental evidence.

In England, tests have been run to measure the transmission of radiant heat through water sprays from two types of nozzles (Heselden and Hinkley 1963). The tests proved that the transmission of this kind of heat depends mostly on nozzle design, and that with certain nozzles a water curtain of low transmission could be produced for water flows comparable to those of sprinkler installations. Fire fighters have, of course, used water curtains in situations where it was too hot or dangerous for them to remain exposed to flames and heat.

ELECTRICAL CONDUCTIVITY OF WATER

Water in its natural state contains impurities that make it conductive. If water is applied to fires involving live electrical equipment, a continuous circuit might be formed which would conduct electricity back to the user and cause a shock, especially if there are high voltages or potentials. Foam type extinguishing agents are very conductive. The amount of current rather than the voltage determines the extent of the shock. Principal variables, assuming contact with a live electrical charge, are:

1. The voltage and amount of current flowing.
2. The "break-up" of the stream as a result of the nozzle design, the pressures used, and the wind conditions. This break-up influences the conductivity of the stream because the air spaces formed between the droplets interrupt the electrical path to ground. Modern water spray nozzles and combination straight stream spray nozzles, the latter in the spray position, provide for effective dispersion of the water droplets. The hazards of these are less than those of solid streams of water.
3. The purity of the water and the relative resistivity of the water.
4. The length and cross-sectional area of the water stream.
5. The resistance to ground through a person's body as influenced by location (whether on wet ground or not), skin moisture, the amount of current the body can endure, the length of exposure to the current, and other factors, such as protective clothing.
6. The resistance to ground through the hose.

Conductivity and Hazard of Shock

There is usually little danger to fire fighters directing streams of water onto wires of less than 600 V to ground from any distance likely to be met under ordinary fire fighting conditions. But it is dangerous if fire fighters, standing either in puddles of water or on moist surfaces, come into contact with live electrical equipment. In such cases the fire fighters' bodies complete an electrical circuit, and the current from the electrical equipment relayed through their bodies is more readily grounded than if it were conveyed through dry, nonconductive surfaces. Rubber boots often contain enough carbon black to permit the passage of current through the body, and should not be relied upon for protection.

Research conducted by UL (Underwriters Laboratories Inc.) on electric fences indicates that there are differences in the electric current to which individuals may be safely subjected, and that the maximum continuous (uninterrupted) current to which an individual may be safely subjected is 5 ma (milliamperes) ac applied on the surface of the body (UL 1939).

Impurities in water (mostly the mineral content) also affect its conductivity. Tests of the resistivity of public water supplies in Indiana showed results ranging from 710 to 5,400 ohms per cc (cubic centimeter); the lowest values were found in supplies from deep wells. The resistivity of deep well supplies ranged between 1,000 and 2,000 ohms per cc, and the resistivity of river waters was about 4,000 ohms per cc. In tests conducted by the Commonwealth Edison Company in cooperation with the Chicago Fire Department, the resistivity of Chicago River water ranged from 1,671 to 2,393 ohms per cc (Commonwealth Edison Co. 1947). At the time the tests were made, the normal hydrant water in the Chicago area had a resistivity of about 3,800 ohms per cc.

Safe Distances from Live Equipment

From time to time many authorities have tried to determine safe distances between nozzles and live electrical equipment. The bibliography at the end of this chapter cites some of the more important papers on this subject.

The conductivity of water streams varies according to the type of equipment from which they are expelled, such as: (1) hand held or manually supported solid stream nozzles, (2) hand held water spray (water fog) nozzles, (3) fixed water spray systems for fire protection services, and (4) plain water and water solution portable fire extinguishers.

Data available on the minimum safe distances between manually supported solid stream hose lines and live electrical equipment carrying voltages higher than 600 V are not wholly consistent because the results of different tests vary. The reasons for these variances are the different testing methods used, the variances in the purposes of the tests, the limitations of the tests (necessitated by the physical circumstances and available equipment, and the fact that the same voltages were not used in all of the tests).

The American Insurance Association suggests the distances given in Table 17-1C between solid stream nozzles (dispensing fresh water) and electrical conductors or equipment carrying voltages higher than 600 V to be the minimum for safety (AIA 1963). The AIA data are based on tests run in the 1930s; thus the voltages given do not correspond to present day, commonly encountered ranges, nor does the data extend to the higher ranges now widely used. The Bulletin cautions that for streams from contaminated or salt water, or where additives have been introduced into the water, it is not possible to give simple rules due to the variances in the conductivity of the water.

TABLE 17-1C. Distances Between Solid Stream Nozzles (Fresh Water) and Electrical Conductors

		Diameter of Nozzle Orifice			
		1⅛ in. (29 mm)		1½ in. (38 m)	
		Minimum Safe Distance			
Voltage To Ground	Voltage Between Conductors	Feet	Meters	Feet	Meters
635	1,100	6	1.8	9	2.7
1,270	2,200	11	3.6	16	4.9
1,905	3,300	15	4.6	22	6.7
3,175	5,500	18	5.5	27	8.2
4,215	6,600	19	5.8	29	8.8
6,350	11,000	20	6.1	30	9.1
12,700	22,000	25	7.6	33	10.1
19,050	33,000	30	9.1	40	12.2

Some limited tests made in 1958 by the Hydroelectric Power Commission of Ontario in cooperation with the Office of the Fire Marshal of Ontario, Canada (Fitzgerald 1958), resulted in recommendations for minimum safe distances from live electrical equipment for a ⅝ in. (16 mm) solid stream nozzle. (See Table 17-1D.) The report recommends that solid streams greater than ⅝ in. (16 mm) should not be used near live electrical equipment, but the tests were limited to a maximum stream distance of 30 ft (9.1 m). Larger nozzles might produce sufficient stream dispersion over longer distances, which would permit their use.

Table 17-1D. Limit of Safe Approach to Live Electrical Equipment

Hydroelectric Power Commission of Ontario

Voltage to Ground	Voltage Between Conductors	⅝ in. (16 mm) Solid Stream Nozzle* Minimum Safe Distance Feet	Meters
2,400	4,160	15	4.6
4,800	8,320	20	6.1
7,200	12,500	20	6.1
8,000	13,800	20	6.1
14,400	24,900	25	7.6
16,000	27,600	25	7.6
25,000	44,000	30	9.1
66,000	115,000	30	9.1
130,000	230,000	30	9.1

* Nozzle pressure 100 psi (690 kPa) water resistance 600 ohms per cubic inch.

The results of tests made in 1934 for the Fire Brigade of Paris, France (Buffet 1934), present perhaps the most comprehensive guide (see Table 17-1E). The distances are based on preventing the transmission of a 1 ma current to a fire fighter in contact with a nozzle or hose. The tests covered only voltages to ground ranging from 115 to 150,000 V, and the groupings of the voltages do not correspond to current standard U.S. voltages. The maximum size of the nozzle used is also nonstandard in the United States.

The preceding information indicates that definite shock hazards exist unless adequate distances are kept, and that these distances can only be estimated from the available data. It is difficult for fire fighters in the field to know precisely what electrical potentials exist in any given situation. For this reason, and those given below, it is best, whenever possible, to use water spray streams rather than solid streams.

As noted before, water spray reduces the conductivity hazard. The design of the nozzle and the character of the spray determines the amount of leakage current that can actually flow in the stream, and each nozzle should be tested to determine precisely the characteristics it possesses. Tests which have been made on various commercial water spray nozzles indicate that a minimum distance of 4 ft (1.2 m) should be maintained for voltages to ground up to about 10 kV. This distance is actually no greater than is sensible to prevent an operator from getting dangerously close to live electrical equipment. Distances should be increased when attacking fires involving live electrical equipment operating above this voltage. Figure 17-1A shows the results of four researchers, as analyzed by the UK Fire Offices' Committee, Joint Fire Research Organization (O'Dogherty 1965).

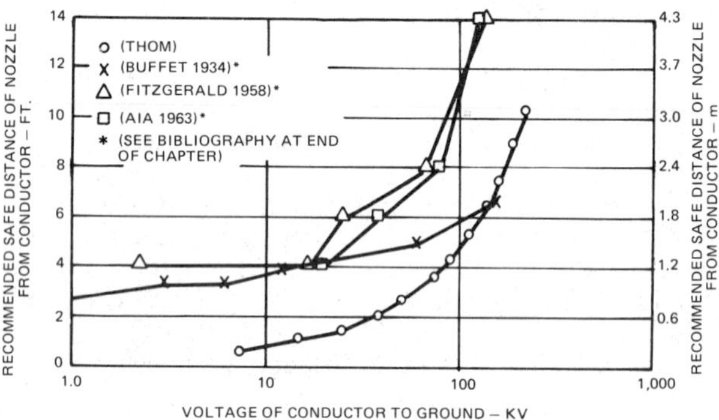

FIG. 17-1A. Variation of safe distance with conductor voltage for spray nozzles.

The Toledo Edison Company made tests in which water was discharged onto a screen at a potential to ground

TABLE 17-1E. Minimum Safe Distances Between Hose Nozzles and Live Electrical Equipment Recommended for the Paris, France, Fire Brigade

Voltage to Ground	Voltage Between Conductors	¼ in. (6 mm) Feet	¼ in. (6 mm) Meters	¾ in. (19 mm) Feet	¾ in. (19 mm) Meters	1¼ in. (32 mm) Feet	1¼ in. (32 mm) Meters
		Diameter of Nozzle Orifice — Safe Distance					
115	230	1.6	0.50	3.3	1.00	6.6	2.00
460	480	2.5	0.75	9.8	3.00	16.4	5.00
3,000	5,195	6.6	2.00	16.4	5.00	32.8	10.00
6,000	10,395	8.2	2.50	19.7	6.00	39.4	12.00
12,000	20,785	9.8	3.00	21.4	6.50	49.2	15.00
60,000	103,820	14.8	4.50	39.4	12.00	72.2	22.00
150,000	259,800	19.7	6.00	49.2	15.00	82.0	25.00

of 80,500 V (equivalent to a system or line voltage of 138 kV phase-to-phase). As a consequence the Edison Electric Institute in 1967 adopted the following safety rules (the distance in these rules will limit leakage currents to less than 1 ma):

1. Using all hand held water spray nozzles, the minimum approach distance is 10 ft (3 m).
2. Using hand held, 1½ in. (38 mm) straight (solid) stream nozzles, the minimum approach distance is 20 ft (6 m).
3. Using hand held, 2½ in. (64 mm) straight (solid) stream nozzles, the minimum approach distance is 30 ft (9 m).

It is better not to use combination nozzles to fight fires in live electrical equipment because a solid stream could accidentally be used instead of a spray. When such hand held nozzles must be used on live electrical equipment, fire fighters should be sure they have the desired spray pattern before applying the stream. The use of spray nozzles on "applicators" increases the possibility of accidental contact between the nozzle and live electrical equipment and most authorities recommend that they not be used.

Clearance from Fixed Water Spray Systems

Fixed water spray systems are used extensively to protect high value and/or critical electrical equipment such as transformers, oil switches, and motors. These systems are designed to provide effective fire control, extinguishment, prevention, or exposure protection. NFPA 15, *Standard for Water Spray Fixed Systems for Fire Protection*, hereinafter referred to as the NFPA Water Spray Standard, gives installation recommendations for such systems, and includes a table of recommended clearances between water spray equipment and unenclosed or uninsulated live electrical components at other than ground potential. (The NFPA Water Spray Standard should be consulted for more complete details.) Modern practice is to coordinate the required clearance with the electrical design. The BIL (Basic Insulation Level) values of the equipment are used as the basis, although the clearance between uninsulated live parts of the equipment and any portion of the water spray system should not be less than the minimum clearances provided elsewhere for electrical system insulation on any individual component (the minimum unshielded straight line distance from the exposed electrical parts to nearby grounded objects). The BIL (expressed in kilovolts) is the crest value of the full wave impulse test.

Portable Extinguishers and Hazard of Shock

Water based or water solution portable fire extinguishers are not recommended for use on fires involving live electrical equipment (Class C). NFPA 10, *Standard for Portable Fire Extinguishers*, recommends that extinguishers specifically tested for use on Class C fires be used for such fires. When electrical equipment is de-energized, extinguishers for Class A or B fires may be used safely. Tests on the conductivity of portable fire extinguishers containing water indicate that soda-acid, loaded-stream, foam, and antifreeze solution extinguishers, which produce solid streams and have a short range, are particularly hazardous. One test involving an antifreeze solution extinguisher showed a current of 157 ma for a potential of 1 kV with a 3/32 in. (2.4 mm) stream at a distance of 1 ft (0.3 m). To reduce the current to below 1 ma, it would be necessary to use the device from a distance at which the stream is dispersed; it is generally agreed that for such extinguishers the minimum distance should be 4 ft (1.2 m) for voltages up to 1 kV. Extinguishers containing plain water might in theory be used at shorter distances, but it is best to maintain a 4-ft (1.2 m) distance if it is necessary to use this kind of extinguisher.

Water on Electrical and Electronic Equipment

Automatic sprinkler protection and water spray fixed systems are valuable for fire control, even where electrical or electronic equipment may be exposed. There should be little concern about the possibility of shock, or of the water causing excessive damage to the equipment. Experience has proven that if a fire activates sprinklers, the sprinklers, if properly installed and maintained, provide effective protection with virtually no hazard to personnel and with no measurable increase in damage to the equipment as compared with the damage done by heat, flame, smoke, and the manual hose streams (Keigher 1968.)

USE OF WATER ON SPECIAL HAZARDS

While water is generally a universal extinguishing agent, there are certain prohibitions and precautions that must be observed when it is applied manually on some burning materials that either react chemically or explosively on contact with water. In other instances, the mechanical action of applying water must be carefully monitored so as not to create conditions that intensify the hazard rather than controlling it.

The following paragraphs give summary guidance on using water on different materials that can present problems if it is used indiscriminately as an extinguishing agent. A good reference to consult for recommendations on use of water on specific materials where problems could be encountered is the NFPA, *Fire Protection Guide on Hazardous Materials* (NFPA 1984).

Chemical Fires

As a general rule, water should not be used on materials, such as carbides, peroxides, etc., because the reaction may release flammable gases and heat. When wet, certain materials, such as unslaked lime, will heat sponstaneously over a period of time if heat cannot be dissapate due to storage conditions.

Combustible Metals

As a rule, water should not be used on fires involving combustible metals, such as magnesium, titanium, metallic sodium, hafnium, or metals that are combustible under certain conditions, such as calcium, zinc, and aluminum.

Radioactive Metals

Water should not be used continuously on radioactive metals. The requirements for fire protection for radioactive metals are generally consistent with their nonradioactive counterparts (for all practical purposes, radioactivity does not influence, nor is it influenced by, the fire properties of a metal). Control of contaminated runoff water is a complicating factor in using water on radioactive metals.

Gas Fires

Water used on gas fires emergencies is generally used for the control of heat from the fire while efforts are made to shut off, or stop, the flow of escaping gas. Water in the form of spray applied from hose lines or monitor nozzles or by fixed water spray systems is commonly used for disbursing or dilution of concentrations of flammable gases.

Combustible and Flammable Liquid Fires

Heavy fuel oil, lubricating oil, asphalt, and other high flash point liquids do not produce flammable vapors unless heated. Once ignited, the heat of the fire will cause enough vaporization for continued burning. If water in spray form is applied to the surface of such high flash point burning liquids, cooling will slow down the rate of vaporization—possibly enough to extinguish the fire. If water is applied to high flash point burning liquids by means of a coarse spray, extinguishment may be obtained by emulsification.

The ability of water without additives (foaming agents) to put out a fire is limited on low flash point flammable liquids, such as Class I flammable liquids (flash points below 100°F), as defined in NFPA 30, *Flammable and Combustible Liquids Code*. Any water that reaches the surface of a burning, low-flash-point flammable liquid in a tank will probably sink and may cause the tank to overflow. In the case of a spill fire, the water will probably cause the fire to spread. Professional handling of certain types of water spray nozzles can result in extinguishment of fires in these liquids or, at a minimum, effective fire control.

The uses of water on petroleum product fires can be summarized as follows (Johnson 1961):

1. As a cooling agent, water may be used to:
   (a) Cut off the release of vapor from the surface of a high flash point oil, thus extinguishing the fire.
   (b) Protect fire fighters from flame and radiant heat when closing a valve or doing other work requiring close approach to the fire.
   (c) Protect flame-exposed surfaces; most effective when the surface is above 212°F (100°C).
2. As a mechanical tool, a water stream can do work at a distance to:
   (a) Control leaks.
   (b) Direct the flow of petroleum product to prevent its ignition, or to move the fire to an area where it will do less damage.
3. As a displacing medium, water may be used to:
   (a) Float oil above a leak in a tank either before or during a fire.
   (b) Cut off fuel escape by pumping it into a leaking pipe ahead of a leak.

Bibliography

References Cited

AIA. 1963. "Fire Streams and Electrical Circuits." Special Interest Bulletin No. 91. American Insurance Association, NY.

Buffet. 1934. "Peut-on Employer les Lances d'Incendie sur des Conducteurs Electriques?" 1934.

Clough, T. C. 1973. "Research on Friction Reducing Agents." *Fire Technology*, Vol 9, No 1. Feb. 1975. pp 32-45.

Commonwealth Edision Co. 1947. "Conductivity of Electricity through Various Sizes and Types of Fire Streams." (An Engineering Report of Tests Conducted in Cooperation with the Chicago Fire Department.)

Fitzgerald, G. W. N. 1958. "Fire Fighting Near Live Electrical Apparatus." Research Division Report No. 58-160. Hydro-Electric Power Commission of Ontario.

Heselden, A. J. M., and Hinkley, P. L. 1963. "Measurement of the Transmission of Radiation Through Water Sprays." Fire Research Note No. 520. Dept. of Scientific and Industrial Research and Fire Offices' Committee. Joint Fire Research Organization. Boreham Wood, Herts, England.

Johnson, Oliver W. 1961. "Water on Oil Fires." *NFPA Quarterly*, Vol 55, No 2. pp 141-146.

Keigher, Donald J. 1968. "Water and Electronics Can Mix." *Fire Journal*. Vol 62. No 6. pp 68-72.

O'Dogherty, M. J. 1965. "The Shock Hazard Associated with the Extinction of Fires Involving Electrical Equipment." *Fire Research Technical Paper No. 13*. Ministry of Technology and Fire Offices' Committee, Joint Fire Research Organization. (Published by Her Majesty's Stationery Office).

NFPA. 1975. "Air Operations for Forest, Brush and Grass Fires." A Report of the NFPA Forest Committee. National Fire Protection Association, Quincy, MA.

NFPA. 1977. "Chemicals for Forest Fire Fighting." A Report of the NFPA Forest Committee, 3rd ed. National Fire Protection Association, Quincy, MA.

NFPA. 1984. *Fire Protection Guide to Hazardous Materials*. 7th ed. National Fire Protection Association, Quincy, MA.

UL. 1939. "Electric Shock as it Pertains to the Electric Fence." Bulletin of Research No. 14. Underwriters Laboratories, Inc., Chicago, IL.

NFPA Codes, Standards, Recommended Practices and Manuals. (See the latest *NFPA Publications and Visual Aids Catalog* for availability of current editions of the following documents.)

NFPA 10, *Standard for Portable Fire Extinguishers*.

NFPA 13, *Standard for the Installation of Sprinkler Systems*.

NFPA 15, *Standard for Water Spray Fixed Systems for Fire Protection*.

NFPA 18, *Standard on Wetting Agents*.

NFPA 30, *Flammable and Combustible Liquids Code*.

NFPA 75, *Standard for the Protection of Electronic Computer/Data Processing Equipment*.

Additional Readings

Aidun, A. R., and Grove, C. S., Jr., "Additives to Improve the Fire Fighting Characteristics of Water," Report No. Ch. E 504-6112F, Syracuse University Research Institute (sponsored by U.S. Dept. of the Navy), 1961.

Brown, H. F., "Report of Conductivity of Fire Streams Near 11,000 Volt Wires," NY, New Haven & Hartford Railroad Company, 1945.

Davis, J. B., Dibble, D. L., Richards, S. S., and Steck, L. V., "Gelgard—A New Fire Retardant for Air and Ground Attack", *Fire Technology*, Vol. 1, No. 3, Aug. 1965, pp. 216-224.

"Evaluation of Halogenated Hydrocarbon and Alkali-Earth-Metal Salt Fire Extinguishing Agents for Low Temperatures," Technical Memorandum M-108, 1 November 1955, U.S. Naval Civil Engineering Research and Evaluation Laboratory, Port Hueneme, CA.

Fryburg, George, "Review of Literature Pertinent to Fire-Extinguishing Agents and to Basic Mechanisms Involved in their Action," NACA TN 2102, May 1950, National Advisory Committee for Aeronautics, Washington, DC.

"High Voltage Electrical Conductivity Tests with Nu-Swift Dry Powder and Water Type Extinguishers," conducted at Institute of High Voltage and Measuring Technique, Technical High School, Darmstadt, Germany, 1962.

Hopping, R., and Knox, C., "Tests of Fire Fighting Streams," The

Electricity Commission of New South Wales, Fire Protection Organization, File No. 8610, 1960.

"Mechanism of Extinguishment of Fire by Finely Divided Water," NBFU Research Report No. 10, American Insurance Association, NY, 1955.

Reischl, Uwe, "Water Fog Stream Heat Radiation," *Fire Technology*, Vol. 15, Nov. 1979, pp. 262-270.

Sprague, C. S., and Harding, C. F., "Electrical Conductivity of Fire Streams," *Engineering Bulletin*, Purdue University, Vol. XX, No. 1, January 1936 (Research Series No. 53).

Stratta, J. J., "Ablative Fluids in the Fire Environment," *Fire Technology*, Vol. 1, No. 3, Aug. 1969, pp. 181-192.

"Study to Optimize Gellant Polymer-Water Systems for the Control of Hypergolic Spills and Fires," National Aeronautics and Space Administration, Report CR-2198, 1972.

"Use of Water on Electronic Equipment Fires," *NASA Safety Journal*, 68-9, Sept. 6, 1968.

Walker, H. S., "High Tension Wires and Hose Streams," *NFPA Quarterly*, October 1930. /et

HYDRAULICS

Revised by Robert Hodnett

This chapter describes the physical properties of water that are pertinent to hydraulic calculations, and explains the formulas used to calculate flow and the effects of flow through orifices, nozzles, and pipes.

HYDRAULIC PROPERTIES OF WATER

The density of water varies with temperature. At maximum density, occurring at 39.2°F (4°C), fresh water weighs 62.425 pcf (pounds per cubic foot) or 1000 kg/m^3 (kilograms per cubic meter) in vacuum or 62.35 pcf (998.7 kg/m^3) in air. It weighs 62.400 pcf (999.5 kg/m^3) at 52.72°F (11.5°C) in vacuum. For ordinary calculations, the approximate value of 62.4 pcf (1000 kg/m^3) is generally used. The term "water" as used in this text refers to fresh water unless otherwise specified. Average seawater weighs 64.1 pcf (1027 kg/m^3).

The following units are based on water at 62.4 pcf (1000 kg/m^3). One gallon (3.78 L) of water (U.S. Standard) is equivalent to 0.1337 cu ft or 231.03 cu in. (0.00378 m^3), and weighs 8.34 lbs (3.78 kg). One cu ft (0.28 m^3) equals 7.48 gal (28.2 L). All calculations in this text are in terms of U.S. gallons, unless otherwise indicated. An Imperial gallon equals 1.20 U.S. gallons (4.54 L).

Net or "Normal" Pressure: Normal pressure is the pressure exerted against the side of a pipe or container by a liquid in the pipe or container with or without flow. Without flow this would be known as "static pressure" or "static head." Pressure *(p)* is customarily measured in pounds per square inch (psi) or kilopascals (kPa) and head *(h)* in feet or meters (m). The pressure produced by a column of water 1 ft high is $\frac{62.4}{144} = 0.433$ psi, represented by w in the formula $p = wh = 0.433h$. The head *(h)* corresponding to a pressure p (psi) will be $h = \frac{p}{w} = \frac{p}{0.433} = 2.31p$.

For SI units, 1 m of water equals 9.81 k/Pa i.e. h(m) = 0.102p (kPa).

Pressure in a hydraulic system is measured in pounds per square inch (kPa) by a pressure gage.

A 1 in. (25.4 mm) head of mercury gives a pressure of 0.491 psi (3.39 kPa) or a 1.134 ft (0.3456 m) head of water.

Normal atmospheric pressure is taken as 14.70 psi (101.4 kPa), and is equivalent to a head of water of 33.96 ft (10.35 m) and a head of mercury of 29.94 in. (760.5 mm).

Velocity Head or Velocity Pressure

The velocity *(v)* produced in a mass of water by pressure acting upon it is the same as if the mass was to fall freely, starting from rest, through a distance equivalent to the pressure head in feet. This relation is represented by $v = \sqrt{2gh}$, v being the velocity produced in feet per second (m/sec), and h being the head in feet m producing the velocity. This is a reversible relation because not only can a pressure head produce velocity, but velocity can be converted to an equivalent pressure head. This relation is h_v (velocity head) $= \frac{v^2}{2g}$ (ft). Because $p_v = 0.433h_v$, the velocity pressure $p_v = 0.433\frac{v^2}{2g}$ (psi)

In SI units p_{vm} (kPa) $= 9.81\frac{v^2}{2g}$ for v in m/sec.

Values of velocity pressure for different rates of flow in various pipe sizes may be selected from Figure 17-2A. (For SI units see Figure 17-2B.) Velocity head or velocity pressure may be calculated by formulas involving velocity (rate of flow) and pipe diameter:

$$h_v = \frac{v^2}{64.4} \text{ or } p_v = \frac{.433v^2}{64.4} = \frac{v^2}{149}$$

In SI units: $h_{vm} = 0.0151\ v^2$ and $p_{vm} = 0.5\ v^2$.

A convenient equation for calculating velocity in feet per second (fps) from the rate of flow can be developed from the relationship $Q = av$ as follows:

$$v = \frac{Q}{a}$$

in which v is velocity in feet per second; Q is expressed as flow in cubic feet per second $\frac{\text{(gallons per minute)}}{60 \text{ sec/min} \times 7.48 \text{ gal/ft}^3}$;

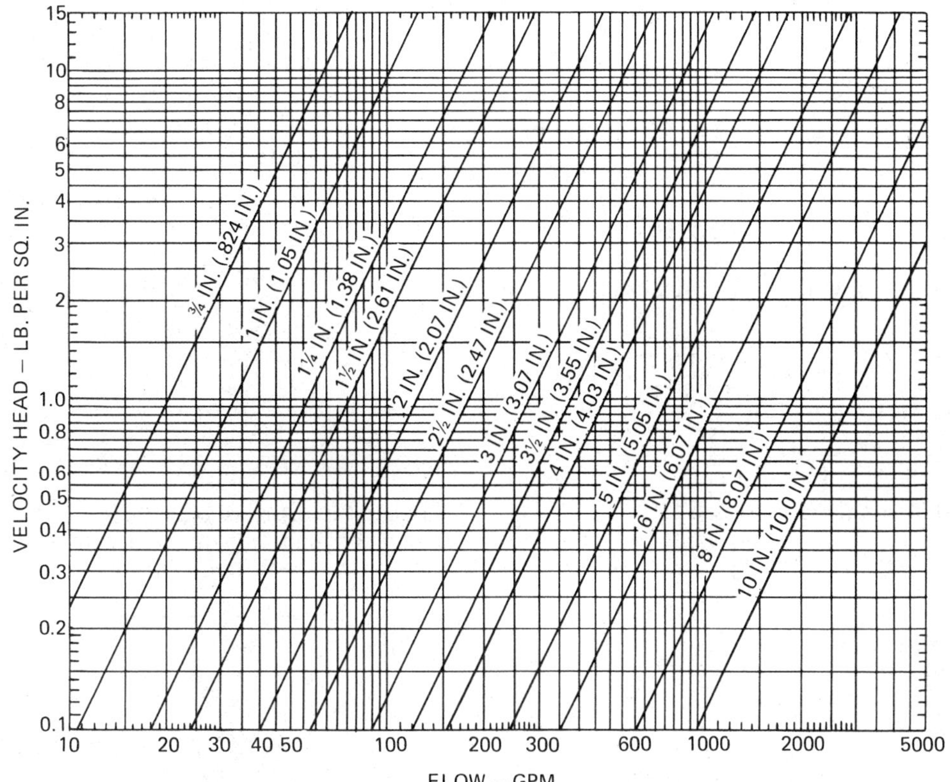

FIG. 17-2A. Graph for the determination of velocity pressure.

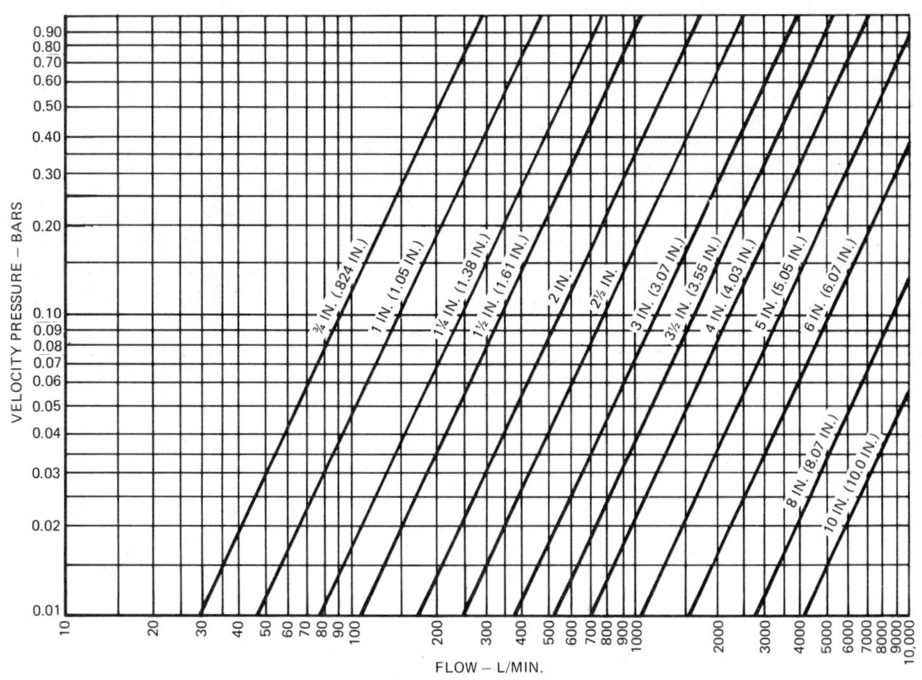

FIG. 17-2B. Graph for the determination of velocity pressure (SI units).

and area in square feet for a pipe with a diameter *(d)* in inches is given as $\frac{\pi d^2}{4} \times \frac{1}{144}$

Substituting as follows:

$$v = \frac{\text{gpm}}{60 \times 7.48} \div \frac{\pi d^2}{4 \times 144}$$

$$= \frac{\text{gpm} \times 4 \times 144}{60 \times 7.48 \times \pi d^2} = \frac{0.4085 \times \text{gpm}}{d^2}$$

In terms of Q (flow in gallons per minute) h_v and p_v are respectively:

$$h_v = \frac{v^2}{2g} = \frac{(0.4085Q)^2}{(d^2)^2} \div 64.4 = \frac{Q^2}{d^4}\frac{(.4085)^2}{64.4} = \frac{Q^2}{386d^4}$$

$$p_v = \frac{Q^2}{d^4} \times \frac{.433}{386} = \frac{Q^2}{891d^4}$$

In SI units the formula is expressed as:

$$p_{vm} = 225\,\frac{Q_m{}^2}{d_m{}^4}$$

where

p_{vm} = pressure, kPa
Q_m = flow, L/min
d_m = inside diameter, mm

As a liquid leaves a pipe, conduit, or container through an orifice, all pressure is converted to velocity pressure. This velocity pressure is sometimes referred to as a Pitot pressure when it is being measured by a Pitot tube inserted into the stream at the point of maximum contraction.

EXAMPLE 1: Find the velocity pressure in 1 in. pipe with 36 gpm flowing. The actual inside diameter of 1 in. pipe is 1.05 in.

SOLUTION:

$$p_v = \frac{Q^2}{891(d)^4} = \frac{(36)^2}{891(1.05)^4} = 1.20 \text{ psi}$$

EXAMPLE 2: Find the velocity pressure in a 25 mm pipe with 100 L/min flowing.

SOLUTION:

$$p_v = 225\,\frac{Q_m{}^2}{d_m{}^4} = \frac{225 \times (100)^2}{(25)^4} = 5.8 \text{ kPa}$$

Total Head. At any point within a piping system that contains water in motion, there is a pressure head h_p (normal pressure head) acting perpendicular to the pipe wall independent of velocity, and a velocity head h_v acting parallel to the pipe wall but exerting no pressure against the wall. Therefore, total head is $H = h_p + h_v$, and without flow (static condition) it is the static pressure head only. Total head expressed as pressure (psi) instead of feet is:

$$p_t = 0.433h_p + 0.433\,\frac{v^2}{2g}$$

In SI units, total head expressed in kPa is:

$$p_{tm} = 9.81\,h_{pn} + 9.81\,\frac{v^2}{2g}$$

When water discharges from an orifice or into a branch line from the side wall of a pipe through which water is flowing, the velocity pressure of the side discharge is equal to the *pressure* head in the pipe. When water is discharged through the open end of a pipe, the pressure of the jet is velocity pressure only, and is produced by the *total pressure* in the pipe at the *point of exit.*

Pressure Sources

The sources of pressure head at a specific location in a hydraulic system may be:

Gravity (elevated tanks, reservoirs, standpipes): Head is the elevation of the water supply surface above the point under consideration, measured directly in feet (m) or converted from a pressure gage reading.

Pumping: Head is the combination of pump discharge pressure and any difference in elevation between the pump discharge gage and the point under consideration.

Pneumatic (pressure tanks): Head is the tank air pressure combined with any difference in elevation of the tank water surface and the point under consideration.

Combinations: Any combination of above pressure sources.

BERNOULLI'S THEOREM

Bernoulli's Theorem expresses the physical law of the conservation of energy applicable to problems of incompressible fluid flow. The theorem may be defined as follows: "In steady flow without friction, the sum of velocity head, pressure head, and elevation head is constant for any incompressible fluid particle throughout its course." In other words, total head would be the same at all locations within the system.

In practice, pipe friction and other lost head is accounted for. Expressed mathematically, Bernoulli's Theorem when applied to locations "A" and "B," is:

$$\frac{v_A{}^2}{2g} + \frac{p_A}{w} + z_A = \frac{v_B{}^2}{2g} + \frac{p_B}{w} + z_B + h_{AB}$$

in which

v = velocity in feet per second (m/sec)
g = acceleration of gravity 32.2 feet per second per second (9.81 m/sec²)
p = pressure, pounds per square foot (kPa)
z = elevation head (distance above assumed datum), in feet (m)
w = specific weight of fluid in pounds per cubic foot (62.4 pcf or 9.81 kN/m³ for water)
$\frac{v^2}{2g}$ = velocity head, in feet (m)
$\frac{p}{w}$ = pressure head, in feet (m)
h_{AB} = lost head between location "A" and location "B" in feet (m)

(Note that in Bernoulli's Theorem all the individual head terms, i.e., velocity head, pressure head, elevation head, and lost head are expressed in feet (m). When using velocities in feet per second (m/sec) and gage pressure in psi (kPa), they must be converted to feet.)

Application of Bernoulli's Theorem

Consider a reservoir and a pipe line discharging water to atmosphere at "B" (Fig. 17-2C). Assume datum through

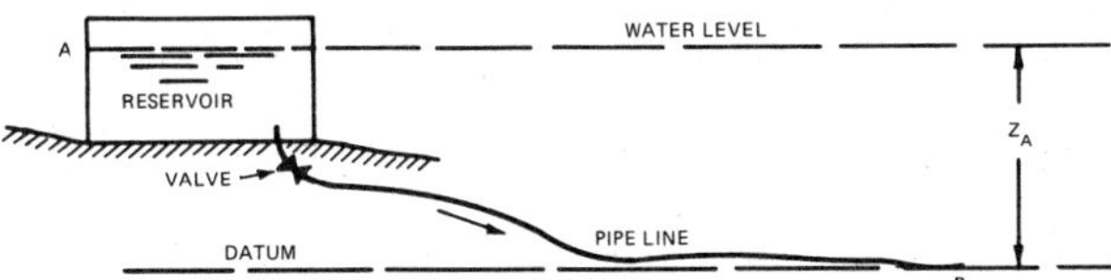

FIG. 17-2C. Graphic representation of the application of Bernoulli's Theorem to a reservoir and a pipe line.

"B," and write Bernoulli's Theorem between the water level at "A" and the outlet at "B."

$$z_A + \frac{v_A^2}{2g} + \frac{p_A}{w} = z_B + \frac{v_B^2}{2g} + \frac{p_B}{w} + h_{AB}$$

The velocity at "A" is practically zero because the diameter of the tank is very large, and the gage pressure at "A" is zero because only atmospheric pressure works on the water surface. At "A" the elevation is z, measured in feet (m) above the datum.

At "B," the elevation above the datum is zero; thus $z = 0$; likewise gage pressure p_B is zero and only velocity head is available as the water leaves the outlet. (A gage at right angles to the emerging stream would register zero pressure.)

Therefore,

$$z_A + 0 + 0 = 0 + \frac{v_B^2}{2g} + 0 + h_{AB}.$$

Therefore,

$$\frac{v_B^2}{2g} = z_A - h_{AB}.$$

Lost head is the sum of (1) hydraulic losses at the reservoir where water enters the pipeline, at the valve, and at the discharge outlet plus (2) the friction loss in the pipe line. The values of the components producing lost head can be estimated, as discussed later in this chapter. The theoretical total head at "B" can be obtained readily by taking pressure gage readings with the flow shut off, and converting these gage readings to feet (m). Another method is to obtain the difference in elevation between "A" and "B" from a topographic map. The actual total head at "B" is the pressure head (h_n). The normal head in the example is zero, however, so the total head is the same as the velocity head in this example.

As another example, determine what would be the head loss across 1,000 ft of 8 in. pipe with 750 gpm flowing from a $2^1/_2$ in. hydrant outlet at "B" and a residual pressure at hydrant "A" of 40 psi. With no flow, hydrant "A" has 60 psi static pressure and hydrant "B" has 80 psi static pressure. Assume datum through hydrant "B". Thus

$$\frac{v_A^2}{2g} + \frac{p_A}{w} + z_A = \frac{v_B^2}{2g} + \frac{p_B}{w} + z_B + h_{AB}.$$

Substituting

$$v_A = \frac{Q}{a} = \frac{Q}{\dfrac{\pi d^2}{4 \times 144}}$$

$$= \frac{\dfrac{750 \text{ gpm}}{7.48 \text{ gal/ft}^3 \times 60 \text{ sec/min}}}{\dfrac{3.1416 \times (8)^2}{4 \times 144 \text{ in.}^2/\text{ft}^2}} = 4.79 \text{ fps}$$

$$v_B = \frac{Q}{a} = \frac{\dfrac{750 \text{ gpm}}{7.48 \text{ gal/ft}^3 \times 60 \text{ sec/min}}}{\dfrac{3.1416 \times (2.5)^2}{4 \times 144 \text{ in.}^2/\text{ft}^2}} = 49.02 \text{ fps}$$

$$\frac{v_A^2}{2g} = \frac{(4.79 \text{ fps})^2}{64.4 \text{ fps/s}} = .36 \text{ ft} = 0.4 \text{ ft}$$

$$\frac{p_A}{w} = \frac{40 \text{ psi}}{62.4 \text{ pcf}} \times 144 \text{ in.}^2/\text{ft}^2 = 92.3 \text{ ft}$$

$$z_A = (80 \text{ psi} - 60 \text{ psi}) \times 2.31 \text{ ft/psi} \doteq 46.2 \text{ ft}$$

$$\frac{v_B^2}{2g} = \frac{(49.02)^2}{64.4} = 37.3 \text{ ft}$$

$\frac{p_B}{w}$ = 0 as there is no normal pressure because it is discharging to atmosphere.

z_B = 0 as datum is through Hydrant "B"

Thus $h_{AB} = 0.4 + 92.3 + 46.2 - 37.3 - 0 - 0 = 101.6$ ft.

One further problem expressed in metric practice: Water is pumped via a pipeline up a 5.0 m incline from "A" to "B." The pipeline at "A" has an inside diameter of 80 mm and a static pressure of 300 kPa. If the pipeline has changed in diameter to 70 mm at "B" and there is a frictional head loss over the length of the pipe (L_{AB}) of 21 m, determine the static pressure at "B" for a flow rate of 4200 L/min. The solution is expressed as:

$$\frac{p_B}{w} = \frac{v_A^2}{2g} + \frac{p_A}{w} + z_A - \frac{v_B^2}{2g} - z_B - h_{AB}$$

$$v_A = \frac{Q}{a_A} = \frac{\dfrac{4200 \text{ L/min}}{60 \text{ sec/min} \times 1000 \text{ L/m}^3}}{\dfrac{t\ (80)^2}{4 \times 10^6 \text{ mm}^2/\text{m}^2}} = 13.9 \text{ m/sec}$$

$$v_B = \frac{Q}{a_B} = \frac{\dfrac{4200 \text{ L/min}}{60 \text{ sec/min} \times 1000 \text{ L/m}^3}}{\dfrac{t\ (70)^2}{4 \times 10^6 \text{ mm}^2/\text{m}^2}} = 18.2 \text{ m/sec}$$

$$\frac{v_A^2}{2g} = \frac{(13.9)^2}{2 \times 9.81} = 9.9 \text{ m}$$

$$\frac{v_B^2}{2g} = \frac{(18.2)^2}{2 \times 9.81} = 16.9 \text{ m}$$

$$\frac{p_A}{w} = \frac{300 \text{ kPa}}{9.81 \text{ kN/m}^3} = 30.6 \text{ m.}$$

z_A = 0 (datum through "A")

z_B = 5 m.

h_{AB} = 12 m (frictional loss)

then $\frac{P_A}{w} = 9.9 + 30.6 + 0 - 16.9 - 5 - 12 = 6.6$ m.

In pressure terms: p_A (kPa) = 9.81 $h(w)$ = 9.81 × 6.6 = 65 kPa, i.e., static pressure at "B" is 65 kPa.

FLOW THROUGH ORIFICES

The rate of flow from an orifice can be expressed in terms of velocity and cross-sectional area of the stream, the basic relations being $Q = av$ where Q = rate of flow in cubic feet per second (m^3/sec); a = area of cross section in square feet (m^2); and v = velocity at the cross section in feet per second (m/sec). (See Table 17-2A, Theoretical Discharge Through Circular Orifices.) From the previous discussion in this chapter on velocity head, it is known that $v = \sqrt{2gh}$. Therefore, substituting in the basic formula for rate of flow through orifices, $Q = a\sqrt{2gh}$. It follows that with the orifice diameter in inches, Q in gallons per minute would equal:

$$60 \times 7.48 \times \frac{\pi d^2}{4 \times 144}\sqrt{64.4h}$$

Because $h = 2.31\ p$, the equation for Q in gallons per minutes becomes $Q = (448.8)(.00546d^2)(12.2)\sqrt{p_v}$.

Therefore, $Q = 29.83d^2\sqrt{p_v}$ (gallons per minute).

In SI units the flow formula is expressed as:

$$q_m = 0.0666{d_m}^2\sqrt{p_{vm}}$$

where

q_m = flow rate (L/min),
d_m = inside diameter (mm),
p_{vm} = pressure (kPa)

The above equation assumes that (1) the jet is a solid stream the full size of the discharge orifice, and (2) the available total head is converted to velocity head which is uniform over the cross section. This is a theoretical situation only, however, as these two conditions are not totally attainable, as the following discussion will show.

Coefficients of Flow

In actual flow from nozzles or orifices, there are two departures from the theoretical values. The actual velocity, considered to be the average velocity over the entire cross section of the stream, is somewhat less than the velocity calculated from the head. The reduction is due to friction and turbulence and is expressed as the coefficient of velocity, designated c_v. Values of c_v are determined experimentally by laboratory tests. With well designed nozzles, the coefficient of velocity is nearly constant and approximately equal to 0.98.

Some nozzles are so designed that the actual cross-sectional area of the stream is less than the calculated area of the orifice. This difference is usually considered a coefficient of contraction and designated as c_c. Coefficients of contraction vary greatly with the design and quality of the orifice or nozzle. For a sharp edge orifice, the value of c_c is about 0.62.

For practical use the coefficients of velocity and contraction can be combined as a single coefficient of discharge designated c_d; thus $c_d = c_v \times c_c$. If $Q = 29.83\,(c_v \times c_c)d^2\sqrt{p_v}$, the expression may be written:

$$Q = 29.83c_d d^2\sqrt{p_v}$$

For metric practice the formula is:

$$Q_m = 0.0666c_d d^2\sqrt{p_{vm}}$$

For any type of nozzle or orifice discharging in a solid stream to atmosphere, c_d may be defined as the ratio of the actual to the theoretical discharge. Approximate values of c_d are determined by standard test procedures, using carefully measured orifice or nozzle diameters. The rate of flow is measured by calibrated meters or "weigh tanks." The velocity pressure is measured by a Pitot gage. The theoretical flow assumes that $c_d = 1$, and is based on the measured diameter and the Pitot reading.

The drop in pressure produced by an orifice in a pipe or similar closed channel is used for the measurement of flow rate in some types of water meters, or as a measure for reducing objectionably high water pressure to fire hose lines from standpipes, or for other services. The coefficient applicable in this case differs from that of an orifice discharging into the open, and varies with several factors, including the ratio of diameters. Pressure reduction by orifices can be calculated only by making proper allowances for the specific conditions involved. Uncertainties in determining coefficients may make a solution by trial and error testing the most practical.

Such a method would involve the calculation of flow under theoretical conditions ($Q = 29.83d^2\sqrt{p_v}$) or ($q_m = 0.0666d_m^2\sqrt{p_{vm}}$) allowing unrestricted discharge from an open orifice directly into a measuring device. The actual flow divided by the theoretical flow is the coefficient of discharge. If this is not possible, then flow from a calibrated nozzle with accurate measurement of both velocity pressure and head pressure is used to establish conditions without any orifice restriction in the pipe. The same measurements are taken with the orifice in place. The flow with the orifice in place divided by the flow without the orifice is the coefficient for the orifice with the configuration as tested. Varying the discharge or flow rate will affect the coefficient for the orifice.

Standard Orifice

An orifice with a sharp entrance edge, shown as Form (1) in Figure 17-2D, is known as a standard orifice and is commonly used as a means of measuring water flow. In such orifices, the water as it leaves the orifice contracts to form a jet whose cross-sectional area is less than that of the orifice. The contraction is complete at the plane a' (see Figure 17-2E) which is located at a distance from the plane of the orifice equal to approximately half the diameter of the jet.

The quantity flowing is obviously the same at the orifice a as at the contracted section a', so the quantity flowing could be obtained by measurement of the velocity and area at either of these planes. Expressed in a formula, where Q is cubic feet per second (m^3/sec), v is velocity in

TABLE 17-2A. Theoretical Discharge Through Circular Orifices

This table is computed from the formula $Q = 29.83cd^2\sqrt{p}$ ($Q_m = 0.0666cd_m^2\sqrt{p_{vm}}$) with $c = 1.00$. The theoretical discharge of sea water, as from fireboat nozzles, may be found by subtracting 1 percent from the figures in the following table, or from the formula $Q = 29.47cd^2\sqrt{p}$ ($Q_m = 0.065cd_m^2\sqrt{p_{vm}}$).

When pressures are read with a Pitot gage at a nozzle, the nozzle discharge in most cases will correspond to the values in the tables within a range of 1 to 3 percent for nozzles up to 1⅜ in. in diameter. For larger diameter nozzles, the principles discussed in "The Nozzle Method of Measuring Flow" in this chapter of the HANDBOOK apply. Appropriate coefficients should be applied where it is read from a hydrant outlet. Where more accurate results are required, a coefficient appropriate to the particular nozzle must be selected and applied to the figures of the table.

The discharge from circular openings of sizes other than those in the table may readily be computed by applying the principle that quantity discharged under a given head varies as the square of the diameter of the opening.

Velocity Head, psi (kPa)*	Velocity Discharge, ft/sec (m/s)	Orifice Diameter, in (mm)																					
		⅜ (10)	½ (13)	⅝ (16)	¾ (19)	⅞ (22)	1 (25)	1⅛ (29)	1¼ (32)	1½ (38)	1¾ (44)	2 (51)	2¼ (57)	2⅜ (60)	2½ (64)	2⅝ (67)	2¾ (70)	3 (76)	3¼ (83)	3½ (89)	3¾ (95)	4 (101)	4½ (114)
1	12.20	4.20	7.46	11.7	16.8	22.9	29.9	37.8	46.7	67.2	91.4	119	151	168	187	206	226	269	315	366	420	478	604
(6.89)	(3.72)	(15.9)	(28.2)	(44.3)	(63.6)	(86.7)	(113)	(143)	(177)	(254)	(346)	(451)	(571)	(637)	(705)	(778)	(854)	(1020)	(1190)	(1390)	(1590)	(1810)	(2290)
2	17.25	5.94	10.5	16.5	23.7	32.3	42.2	53.4	66.0	95.0	129	169	214	238	264	291	319	380	446	517	594	676	854
(13.8)	(5.26)	(22.5)	(39.9)	(62.3)	(89.8)	(122)	(160)	(202)	(249)	(359)	(489)	(639)	(808)	(900)	(1000)	(1100)	(1210)	(1440)	(1690)	(1960)	(2250)	(2560)	(3230)
3	21.13	7.27	12.9	20.2	29.1	39.6	51.7	65.4	80.8	116	158	207	262	292	323	356	391	465	546	633	727	827	1045
(20.7)	(6.44)	(27.5)	(48.9)	(76.3)	(110)	(150)	(196)	(248)	(306)	(440)	(599)	(782)	(990)	(1100)	(1220)	(1350)	(1480)	(1760)	(2070)	(2400)	(2750)	(3130)	(3960)
4	24.39	8.40	14.9	23.3	33.6	45.7	59.7	75.6	93.3	134	183	239	302	337	373	411	452	537	631	731	840	955	1210
(27.6)	(7.43)	(31.8)	(56.4)	(88.2)	(127)	(173)	(226)	(286)	(353)	(508)	(692)	(930)	(1140)	(1280)	(1410)	(1560)	(1710)	(2030)	(2390)	(2770)	(3180)	(3610)	(4570)
5	27.26	9.39	16.7	26.1	37.6	51.1	66.7	84.5	104	150	204	267	338	376	417	460	505	601	705	817	938	1068	1350
(34.5)	(8.31)	(35.5)	(63.1)	(98.6)	(142)	(193)	(252)	(320)	(394)	(568)	(773)	(1010)	(1280)	(1420)	(1580)	(1740)	(1910)	(2270)	(2670)	(3090)	(3550)	(4040)	(5110)
6	29.87	10.3	18.3	28.6	41.1	56.0	73.1	92.5	114	164	224	292	370	412	457	504	553	658	772	896	1028	1170	1480
(41.4)	(9.10)	(38.9)	(69.1)	(108)	(156)	(212)	(277)	(350)	(432)	(622)	(847)	(1110)	(1400)	(1560)	(1730)	(1910)	(2090)	(2490)	(2920)	(3390)	(3890)	(4420)	(5600)
7	32.26	11.1	19.7	30.8	44.4	60.5	79.0	99.9	123	178	242	316	400	445	494	544	597	711	834	967	1111	1263	1600
(48.3)	(9.83)	(42.0)	(74.7)	(117)	(168)	(229)	(299)	(378)	(467)	(672)	(915)	(1190)	(1510)	(1680)	(1870)	(2060)	(2260)	(2690)	(3160)	(3660)	(4210)	(4780)	(6050)
8	34.49	11.9	21.1	33.0	47.5	64.6	84.4	107	132	190	259	338	427	476	528	582	638	760	892	1034	1187	1351	1710
(55.2)	(10.51)	(44.9)	(79.8)	(125)	(180)	(245)	(319)	(404)	(499)	(718)	(978)	(1280)	(1620)	(1800)	(2000)	(2200)	(2410)	(2880)	(3380)	(3910)	(4490)	(5110)	(6470)
9	36.58	12.6	22.4	35.0	50.4	68.6	89.5	113	140	201	274	358	453	505	560	617	677	806	946	1097	1259	1433	1815
(62.0)	(11.15)	(47.6)	(84.7)	(132)	(191)	(260)	(339)	(429)	(529)	(762)	(1040)	(1360)	(1710)	(1910)	(2120)	(2340)	(2560)	(3050)	(3580)	(4150)	(4770)	(5420)	(6860)
10	38.56	13.3	23.6	36.8	53.2	72.2	94.4	119	148	212	289	378	478	532	590	650	714	850	997	1156	1327	1510	1910
(68.9)	(11.75)	(50.2)	(89.3)	(139)	(201)	(273)	(357)	(452)	(558)	(803)	(1090)	(1430)	(1810)	(2010)	(2230)	(2460)	(2700)	(3220)	(3770)	(4380)	(5020)	(5710)	(7230)
11	40.45	13.9	24.7	38.6	55.8	75.7	99.0	125	155	223	303	396	501	553	619	682	759	891	1046	1213	1392	1584	2010
(75.8)	(12.33)	(52.6)	(93.6)	(146)	(211)	(287)	(375)	(474)	(585)	(842)	(1150)	(1500)	(1900)	(2110)	(2340)	(2580)	(2830)	(3370)	(3960)	(4590)	(5270)	(5990)	(7580)
12	42.24	14.5	25.7	40.3	58.3	79.1	103	131	162	233	317	414	524	583	646	712	782	931	1092	1267	1454	1655	2100
(82.7)	(12.87)	(55.0)	(97.8)	(153)	(221)	(299)	(391)	(495)	(611)	(880)	(1200)	(1560)	(1980)	(2210)	(2450)	(2690)	(2960)	(3520)	(4130)	(4800)	(5500)	(6260)	(7920)
13	43.97	15.1	26.7	41.9	60.7	82.3	108	136	168	242	330	431	545	607	673	741	814	969	1137	1318	1515	1722	2180
(89.6)	(13.40)	(57.2)	(102)	(159)	(229)	(312)	(407)	(515)	(636)	(916)	(1250)	(1630)	(2060)	(2300)	(2550)	(2800)	(3080)	(3670)	(4300)	(4990)	(5730)	(6520)	(8240)
14	45.63	15.7	27.7	43.5	63.0	85.4	112	141	175	251	342	447	566	630	698	769	845	1005	1180	1368	1572	1787	2260
(96.5)	(13.91)	(59.4)	(106)	(165)	(238)	(323)	(422)	(534)	(660)	(950)	(1290)	(1690)	(2140)	(2380)	(2640)	(2910)	(3200)	(3800)	(4470)	(5180)	(5950)	(6760)	(8550)
15	47.22	16.3	28.7	45.0	65.1	88.4	116	146	181	260	354	463	586	652	722	796	874	1040	1221	1416	1626	1849	2340
(103)	(14.39)	(61.4)	(109)	(171)	(246)	(335)	(437)	(553)	(683)	(984)	(1340)	(1750)	(2220)	(2470)	(2730)	(3010)	(3310)	(3940)	(4620)	(5360)	(6150)	(7000)	(8850)
16	48.78	16.8	29.7	46.5	67.2	91.3	120	151	187	269	366	478	605	673	746	822	903	1075	1261	1463	1679	1910	2420
(110)	(14.87)	(63.5)	(113)	(176)	(254)	(346)	(452)	(571)	(706)	(1020)	(1390)	(1810)	(2290)	(2550)	(2820)	(3110)	(3420)	(4070)	(4770)	(5540)	(6360)	(7230)	(9140)
17	50.28	17.3	30.7	47.9	69.3	94.1	123	156	192	277	377	493	623	694	769	848	931	1108	1300	1508	1731	1969	2500
(117)	(15.33)	(65.4)	(116)	(182)	(262)	(356)	(466)	(589)	(727)	(1050)	(1430)	(1870)	(2360)	(2630)	(2910)	(3210)	(3520)	(4190)	(4920)	(5710)	(6550)	(7540)	(9430)
18	51.73	17.8	31.6	49.3	71.3	96.8	127	160	198	285	388	507	642	714	791	872	958	1140	1338	1551	1781	2026	2570
(124)	(15.77)	(67.3)	(120)	(187)	(269)	(367)	(479)	(606)	(748)	(1080)	(1470)	(1920)	(2430)	(2700)	(2990)	(3300)	(3630)	(4310)	(5060)	(5870)	(6740)	(7670)	(9700)

Velocity Head, psi (kPa)*	Velocity Discharge, ft/sec (m/s)	Orifice Diameter, in (mm)																					
		3/8 (10)	1/2 (13)	5/8 (16)	3/4 (19)	7/8 (22)	1 (25)	1 1/8 (29)	1 1/4 (32)	1 1/2 (38)	1 3/4 (44)	2 (51)	2 1/4 (57)	2 3/8 (60)	2 1/2 (64)	2 5/8 (67)	2 3/4 (70)	3 (76)	3 1/4 (83)	3 1/2 (89)	3 3/4 (95)	4 (101)	4 1/2 (114)
19	53.15	18.3	32.5	50.7	73.3	99.5	130	165	203	293	399	521	659	733	813	896	984	1171	1374	1594	1830	2082	2640
(131)	(16.20)	(69.2)	(123)	(192)	(277)	(377)	(492)	(623)	(769)	(1110)	(1510)	(1970)	(2490)	(2770)	(3080)	(3390)	(3720)	(4430)	(5200)	(6030)	(6920)	(7870)	(9970)
20	54.54	18.8	33.4	52.0	75.3	102	134	169	209	300	409	534	676	753	834	920	1010	1201	1410	1635	1877	2136	2710
(138)	(16.62)	(70.9)	(126)	(197)	(284)	(386)	(505)	(639)	(789)	(1140)	(1550)	(2020)	(2560)	(2850)	(3160)	(3480)	(3820)	(4540)	(5330)	(6180)	(7100)	(8080)	(10200)
22	57.19	19.7	35.0	54.6	78.9	107	140	177	219	315	429	560	709	789	875	964	1059	1260	1479	1715	1969	2240	2840
(152)	(17.43)	(74.4)	(132)	(207)	(298)	(405)	(530)	(670)	(827)	(1190)	(1620)	(2120)	(2680)	(2290)	(3310)	(3650)	(4000)	(4770)	(5590)	(6490)	(7540)	(8470)	(10700)
24	59.74	20.6	36.4	57.0	82.4	112	146	185	229	329	448	585	741	824	914	1007	1106	1316	1545	1791	2056	2340	2970
(165)	(18.21)	(77.7)	(138)	(216)	(311)	(423)	(553)	(700)	(864)	(1240)	(1690)	(2210)	(2800)	(3120)	(3460)	(3810)	(4180)	(4980)	(5840)	(6770)	(7780)	(8850)	(11200)
26	62.18	21.4	38.0	59.4	85.8	116	152	193	238	343	466	609	771	858	951	1048	1151	1370	1608	1864	2140	2435	3090
(179)	(18.95)	(80.9)	(144)	(225)	(324)	(441)	(576)	(728)	(900)	(1300)	(1760)	(2300)	(2910)	(3250)	(3600)	(3970)	(4350)	(5180)	(6080)	(7050)	(8100)	(9210)	(11700)
28	64.52	22.2	39.5	61.6	89.0	121	158	200	247	356	484	632	800	890	987	1088	1194	1422	1668	1935	2221	2527	3210
(193)	(19.67)	(83.9)	(149)	(233)	(336)	(457)	(597)	(756)	(934)	(1340)	(1830)	(2390)	(3020)	(3370)	(3730)	(4120)	(4520)	(5380)	(6310)	(7320)	(8400)	(9560)	(12100)
30	66.79	23.0	40.9	63.7	92.2	125	164	207	256	368	501	654	828	922	1022	1126	1236	1472	1727	2003	2299	2616	3320
(207)	(20.36)	(86.9)	(155)	(241)	(348)	(473)	(618)	(782)	(966)	(1390)	(1890)	(2470)	(3130)	(3490)	(3860)	(4260)	(4680)	(5570)	(6530)	(7570)	(8700)	(9890)	(12500)
32	68.98	23.8	42.3	65.8	95.2	129	169	214	264	380	517	676	856	952	1055	1163	1277	1520	1784	2069	2375	2702	3430
(221)	(21.03)	(89.7)	(160)	(249)	(359)	(489)	(639)	(808)	(998)	(1440)	(1960)	(2550)	(3230)	(3600)	(3990)	(4400)	(4830)	(5750)	(6750)	(7820)	(8980)	(10200)	(12900)
34	71.10	24.5	43.6	67.8	98.2	133	174	220	272	392	533	697	882	981	1088	1199	1316	1566	1838	2132	2448	2785	3540
(234)	(21.67)	(92.5)	(165)	(257)	(372)	(504)	(658)	(833)	(1030)	(1480)	(2020)	(2640)	(3340)	(3710)	(4120)	(4540)	(4980)	(5930)	(6960)	(8070)	(9270)	(10540)	(13300)
36	73.16	25.2	44.8	69.8	101	137	179	226	280	403	548	717	908	1010	1119	1233	1354	1612	1892	2194	2519	2866	3640
(248)	(22.30)	(95.2)	(169)	(265)	(381)	(518)	(677)	(857)	(1060)	(1520)	(2070)	(2710)	(3440)	(3820)	(4240)	(4670)	(5120)	(6100)	(7160)	(8300)	(9530)	(10850)	(13800)
38	75.17	25.9	46.0	71.8	104	141	184	233	288	414	563	736	932	1037	1150	1267	1392	1656	1944	2254	2588	2944	3740
(262)	(22.91)	(97.8)	(174)	(272)	(392)	(533)	(696)	(881)	(1090)	(1570)	(2130)	(2790)	(3530)	(3930)	(4350)	(4800)	(5270)	(6270)	(7360)	(8530)	(9800)	(11140)	(14100)
40	77.11	26.5	47.2	73.6	106	144	189	239	295	425	578	755	956	1064	1180	1300	1428	1699	1994	2313	2655	3021	3840
(276)	(23.50)	(100)	(179)	(279)	(402)	(547)	(714)	(903)	(1120)	(1610)	(2190)	(2860)	(3620)	(4030)	(4470)	(4920)	(5400)	(6430)	(7550)	(8750)	(10050)	(11430)	(14500)
42	79.03	27.2	48.4	75.4	109	148	193	245	303	435	592	774	980	1091	1209	1332	1463	1741	2043	2370	2721	3095	3935
(290)	(24.09)	(103)	(183)	(285)	(412)	(560)	(732)	(926)	(1140)	(1650)	(2240)	(2930)	(3710)	(4130)	(4580)	(5040)	(5540)	(6590)	(7730)	(8970)	(10300)	(11710)	(14800)
44	80.88	27.8	49.6	77.2	112	151	198	251	310	445	606	792	1003	1116	1237	1364	1497	1782	2091	2426	2785	3168	4030
(303)	(24.65)	(105)	(188)	(292)	(421)	(573)	(749)	(947)	(1170)	(1690)	(2290)	(3000)	(3800)	(4220)	(4680)	(5160)	(5670)	(6740)	(7910)	(9180)	(10540)	(11990)	(15200)
46	82.70	28.5	50.8	79.0	114	155	203	256	317	455	620	810	1025	1141	1265	1394	1531	1822	2138	2480	2847	3239	4120
(317)	(25.21)	(108)	(192)	(299)	(431)	(586)	(766)	(969)	(1200)	(1720)	(2350)	(3070)	(3880)	(4320)	(4790)	(5280)	(5790)	(6900)	(8090)	(9390)	(10780)	(12260)	(15500)
48	84.48	29.1	51.8	80.7	117	158	207	262	324	465	633	828	1047	1166	1293	1424	1564	1861	2184	2533	2908	3309	4205
(331)	(25.75)	(110)	(196)	(305)	(440)	(598)	(782)	(989)	(1220)	(1760)	(2390)	(3130)	(3960)	(4410)	(4890)	(5390)	(5920)	(7040)	(8270)	(9587)	(11010)	(12520)	(15800)
50	86.22	29.7	52.8	82.3	119	161	211	267	330	475	646	845	1069	1190	1319	1454	1596	1900	2229	2586	2968	3377	4290
(345)	(26.28)	(112)	(200)	(312)	(449)	(611)	(798)	(1010)	(1250)	(1800)	(2450)	(3200)	(4050)	(4500)	(4990)	(5500)	(6040)	(7190)	(8440)	(9790)	(11230)	(12780)	(16200)
52	87.93	30.3	53.8	83.9	121	165	215	272	337	485	659	861	1091	1213	1345	1482	1628	1937	2274	2637	3027	3444	4375
(358)	(26.80)	(115)	(204)	(318)	(458)	(623)	(814)	(1030)	(1270)	(1830)	(2490)	(3260)	(4130)	(4590)	(5090)	(5610)	(6160)	(7330)	(8610)	(9980)	(11460)	(13040)	(16500)
54	89.61	30.9	54.9	85.5	124	168	219	277	343	494	672	878	1111	1237	1371	1511	1659	1974	2317	2687	3085	3510	4460
(372)	(27.31)	(117)	(208)	(324)	(467)	(635)	(830)	(1050)	(1300)	(1870)	(2540)	(3320)	(4200)	(4680)	(5190)	(5720)	(6280)	(7470)	(8770)	(10170)	(11680)	(13290)	(16800)
56	91.20	31.4	55.9	87.1	126	171	223	283	350	503	684	894	1132	1259	1396	1538	1689	2010	2359	2736	3141	3574	4540
(386)	(27.80)	(119)	(212)	(330)	(475)	(647)	(845)	(1070)	(1320)	(1900)	(2590)	(3380)	(4280)	(4770)	(5280)	(5820)	(6390)	(7610)	(8930)	(10360)	(11890)	(13530)	(17100)

58	92.87	32.0	56.9	88.6	128	174	227	288	356	512	696	909	1152	1282	1421	1566	1719	2046	2401	2785	3197	3637	4620
(400)	(28.31)	(121)	(215)	(336)	(484)	(658)	(860)	(1090)	(1340)	(1930)	(2630)	(3440)	(4350)	(4850)	(5370)	(5920)	(6500)	(7740)	(9080)	(10530)	(12090)	(13760)	(17400)
60	94.45	32.5	57.8	90.1	130	177	231	293	362	520	708	925	1171	1303	1445	1592	1749	2081	2442	2832	3252	3700	4700
(414)	(28.79)	(123)	(219)	(341)	(492)	(669)	(875)	(1110)	(1370)	(1970)	(2680)	(3500)	(4430)	(4930)	(5460)	(6030)	(6610)	(7870)	(9240)	(10710)	(12300)	(13990)	(17700)
62	96.01	33.0	58.8	91.6	133	180	235	297	368	529	720	941	1191	1325	1470	1619	1777	2115	2483	2879	3305	3761	4775
(427)	(29.26)	(125)	(222)	(347)	(500)	(680)	(889)	(1120)	(1390)	(2000)	(2720)	(3560)	(4500)	(5010)	(5560)	(6130)	(6720)	(8000)	(9390)	(10890)	(12500)	(14220)	(18000)
64	97.55	33.6	59.7	93.1	135	183	239	302	374	538	731	956	1210	1346	1493	1645	1806	2149	2522	2925	3358	3821	4850
(441)	(29.73)	(127)	(226)	(353)	(508)	(691)	(903)	(1140)	(1410)	(2030)	(2770)	(3610)	(4570)	(5090)	(5640)	(6220)	(6830)	(8130)	(9540)	(11060)	(12700)	(14450)	(18300)
66	99.07	34.1	60.6	94.6	137	186	243	307	379	546	742	971	1228	1367	1516	1670	1834	2183	2561	2971	3410	3880	4925
(455)	(30.20)	(129)	(229)	(358)	(516)	(702)	(917)	(1160)	(1430)	(2060)	(2810)	(3670)	(4640)	(5170)	(5730)	(6320)	(6940)	(8260)	(9690)	(11240)	(12900)	(14680)	(18600)
68	100.55	34.6	61.5	96.0	139	188	246	311	385	554	754	985	1247	1388	1539	1695	1862	2215	2600	3015	3462	3938	5000
(469)	(30.65)	(131)	(233)	(364)	(524)	(713)	(931)	(1180)	(1450)	(2090)	(2850)	(3720)	(4710)	(5250)	(5820)	(6420)	(7040)	(8380)	(9830)	(11400)	(13090)	(14900)	(18900)
70	102.03	35.1	62.5	97.4	141	191	250	316	391	562	765	999	1265	1408	1561	1720	1889	2248	2638	3059	3512	3996	5075
(483)	(31.10)	(133)	(236)	(369)	(531)	(723)	(945)	(1200)	(1480)	(2130)	(2890)	(3780)	(4780)	(5330)	(5900)	(6510)	(7140)	(8500)	(9980)	(11570)	(13280)	(15110)	(19100)
72	103.47	35.6	63.3	98.7	143	194	253	320	396	570	776	1014	1283	1428	1583	1745	1916	2280	2675	3103	3562	4053	5140
(496)	(31.54)	(135)	(239)	(374)	(539)	(733)	(958)	(1210)	(1500)	(2160)	(2930)	(3830)	(4850)	(5400)	(5990)	(6600)	(7250)	(8620)	(10120)	(11730)	(13470)	(15330)	(19400)
74	104.90	36.1	64.2	100	145	196	257	325	402	578	786	1028	1301	1448	1605	1769	1942	2311	2712	3146	3611	4109	5200
(510)	(31.97)	(136)	(243)	(379)	(546)	(743)	(971)	(1230)	(1520)	(2190)	(2970)	(3880)	(4920)	(5480)	(6070)	(6690)	(7350)	(8740)	(10260)	(11900)	(13660)	(15540)	(19700)
76	106.30	36.6	65.1	101	147	199	260	329	407	586	797	1041	1318	1467	1627	1792	1968	2342	2749	3188	3660	4164	5265
(524)	(32.71)	(138)	(246)	(384)	(554)	(753)	(984)	(1250)	(1540)	(2210)	(3010)	(3940)	(4980)	(5550)	(6150)	(6780)	(7440)	(8860)	(10400)	(12060)	(13840)	(15750)	(19900)
78	107.69	37.1	65.9	103	149	202	264	333	412	593	807	1055	1335	1486	1648	1816	1994	2373	2785	3230	3708	4218	5340
(538)	(32.82)	(140)	(249)	(389)	(561)	(763)	(997)	(1260)	(1560)	(2240)	(3050)	(3990)	(5050)	(5620)	(6230)	(6870)	(7540)	(8970)	(10530)	(12210)	(14020)	(15950)	(20200)
80	109.08	37.5	66.8	104	150	204	267	338	418	601	818	1068	1352	1505	1669	1839	2019	2403	2820	3271	3755	4272	5405
(552)	(33.25)	(142)	(252)	(394)	(568)	(773)	(1010)	(1280)	(1580)	(2270)	(3090)	(4040)	(5110)	(5690)	(6310)	(6960)	(7640)	(9090)	(10670)	(12370)	(14200)	(16160)	(20400)
82	110.42	38.0	67.6	105	152	207	270	342	423	608	820	1082	1369	1524	1689	1862	2044	2433	2855	3311	3801	4325	5470
(565)	(33.66)	(144)	(256)	(399)	(575)	(782)	(1020)	(1290)	(1600)	(2300)	(3130)	(4090)	(5180)	(5770)	(6390)	(7040)	(7730)	(9200)	(10800)	(12520)	(14380)	(16360)	(20700)
84	111.76	38.4	68.4	107	154	209	274	346	428	616	838	1095	1386	1542	1710	1884	2069	2462	2890	3351	3847	4377	5535
(579)	(34.06)	(145)	(259)	(404)	(582)	(792)	(1030)	(1310)	(1620)	(2330)	(3170)	(4140)	(5240)	(5840)	(6466)	(7130)	(7830)	(9310)	(10930)	(12670)	(14550)	(16560)	(21000)
86	113.08	38.9	69.2	108	156	212	277	350	433	624	848	1170	1402	1561	1730	1907	2094	2491	2924	3391	3893	4429	5600
(593)	(34.47)	(147)	(262)	(409)	(589)	(801)	(1050)	(1320)	(1640)	(2360)	(3210)	(4190)	(5300)	(5900)	(6540)	(7210)	(7920)	(9420)	(11070)	(12820)	(14720)	(16750)	(21200)
88	114.39	39.3	70.0	109	158	214	280	354	438	630	858	1120	1419	1579	1750	1929	2118	2520	2958	3430	3938	4480	5665
(607)	(34.87)	(149)	(265)	(414)	(596)	(811)	(1060)	(1340)	(1650)	(2380)	(3240)	(4240)	(5360)	(5970)	(6620)	(7300)	(8010)	(9530)	(11190)	(12970)	(14890)	(16950)	(21400)
90	115.68	39.8	70.8	110	160	217	283	358	443	637	867	1133	1434	1596	1770	1950	2142	2549	2991	3469	3983	4531	5730
(620)	(35.26)	(151)	(268)	(418)	(603)	(820)	(1070)	(1360)	(1670)	(2410)	(3280)	(4280)	(5420)	(6040)	(6690)	(7380)	(8100)	(9640)	(11310)	(13120)	(15060)	(17140)	(21700)
92	116.96	40.2	71.6	112	161	219	286	362	448	644	877	1146	1450	1614	1789	1972	2165	2577	3024	3507	4027	4581	5795
(634)	(35.65)	(152)	(271)	(423)	(609)	(829)	(1080)	(1370)	(1690)	(2440)	(3320)	(4330)	(5480)	(6110)	(6770)	(7460)	(8190)	(9750)	(11440)	(13260)	(15230)	(17330)	(21900)
94	118.23	40.7	72.4	113	163	221	289	366	453	651	886	1158	1466	1632	1809	1993	2189	2605	3057	3545	4070	4631	5865
(648)	(36.04)	(154)	(274)	(427)	(616)	(838)	(1090)	(1380)	(1710)	(2460)	(3350)	(4380)	(5540)	(6170)	(6840)	(7540)	(8280)	(9850)	(11560)	(13410)	(15390)	(17510)	(22000)
96	119.48	41.1	73.2	114	165	224	293	370	457	658	895	1170	1481	1649	1828	2014	2212	2632	3089	3583	4113	4680	5925
(662)	(36.42)	(155)	(277)	(432)	(622)	(847)	(1110)	(1400)	(1730)	(2490)	(3390)	(4420)	(5600)	(6240)	(6910)	(7620)	(8370)	(9960)	(11680)	(13550)	(15560)	(17700)	(22400)
98	120.71	41.6	73.9	115	167	226	296	374	462	665	904	1182	1497	1666	1847	2035	2235	2660	3121	3620	4156	4728	5985
(676)	(36.79)	(157)	(279)	(436)	(629)	(855)	(1120)	(1410)	(1750)	(2510)	(3420)	(4470)	(5660)	(6300)	(6980)	(7700)	(8450)	(10060)	(11810)	(13690)	(15720)	(17880)	(22600)
100	121.94	42.0	74.6	116	168	228	299	378	467	672	914	1194	1512	1683	1866	2056	2258	2687	3153	3657	4198	4776	6045
(689)	(37.17)	(159)	(282)	(441)	(635)	(864)	(1130)	(1430)	(1760)	(2540)	(3460)	(4520)	(5720)	(6370)	(7050)	(7780)	(8540)	(10160)	(11930)	(13830)	(15880)	(18060)	(22900)
102	123.15	42.4	75.4	118	170	231	302	381	472	679	923	1206	1527	1699	1884	2076	2280	2713	3184	3693	4240	4824	6100
(703)	(37.54)	(160)	(285)	(445)	(641)	(873)	(1140)	(1440)	(1780)	(2570)	(3490)	(4560)	(5770)	(6430)	(7130)	(7860)	(8620)	(10260)	(12040)	(13970)	(16040)	(18240)	(23100)
104	124.35	42.8	76.1	119	172	233	304	385	476	685	932	1218	1542	1716	1903	2097	2302	2740	3215	3729	4281	4871	6150
(717)	(37.90)	(162)	(288)	(450)	(648)	(881)	(1150)	(1460)	(1800)	(2590)	(3530)	(4610)	(5830)	(6490)	(7190)	(7930)	(8710)	(10360)	(12160)	(14100)	(16190)	(18420)	(23300)

Velocity Head, psi (kPa)*	Velocity Discharge, ft/sec (m/s)	Orifice Diameter, in (mm)																					
		3/8 (10)	1/2 (13)	5/8 (16)	3/4 (19)	7/8 (22)	1 (25)	1 1/8 (29)	1 1/4 (32)	1 1/2 (38)	1 3/4 (44)	2 (51)	2 1/4 (57)	2 3/8 (60)	2 1/2 (64)	2 5/8 (67)	2 3/4 (70)	3 (76)	3 1/4 (83)	3 1/2 (89)	3 3/4 (95)	4 (101)	4 1/2 (114)
106	125.55	43.4	76.9	120	173	235	307	389	481	692	941	1230	1556	1733	1921	2117	2324	2766	3246	3765	4322	4917	6200
(731)	(38.27)	(163)	(291)	(454)	(654)	(890)	(1160)	(1470)	(1820)	(2620)	(3560)	(4650)	(5890)	(6560)	(7260)	(8010)	(8790)	(10460)	(12280)	(14240)	(16350)	(18600)	(23500)
108	126.73	43.6	77.6	121	175	237	310	392	485	698	950	1241	1571	1749	1939	2137	2346	2792	3277	3800	4363	4963	6260
(745)	(38.63)	(165)	(293)	(458)	(660)	(898)	(1170)	(1480)	(1830)	(2640)	(3590)	(4690)	(5940)	(6620)	(7330)	(8080)	(8870)	(10560)	(12390)	(14370)	(16500)	(18770)	(23800)
110	127.89	44.0	78.3	122	177	239	313	396	490	705	959	1253	1586	1765	1957	2156	2368	2818	3307	3835	4403	5009	6320
(758)	(38.98)	(166)	(296)	(462)	(666)	(906)	(1180)	(1500)	(1850)	(2660)	(3630)	(4640)	(6000)	(6680)	(7400)	(8160)	(8960)	(10660)	(12510)	(14500)	(16650)	(18950)	(24000)
112	129.05	44.4	79.0	123	178	242	316	400	494	711	967	1264	1600	1781	1974	2176	2389	2843	3337	3870	4443	5054	6380
(772)	(39.33)	(168)	(299)	(467)	(672)	(914)	(1190)	(1510)	(1870)	(2690)	(3660)	(4780)	(6050)	(6740)	(7470)	(8230)	(9040)	(10750)	(12620)	(14640)	(16800)	(19120)	(24200)
114	130.20	44.8	79.7	124	180	244	319	403	499	717	976	1275	1614	1797	1992	2195	2410	2869	3367	3904	4482	5099	6440
(786)	(39.68)	(169)	(301)	(471)	(678)	(923)	(1210)	(1530)	(1880)	(2710)	(3690)	(4820)	(6100)	(6800)	(7530)	(8310)	(9120)	(10850)	(12730)	(14770)	(16950)	(19290)	(24400)
116	131.33	45.2	80.4	125	181	246	322	407	503	724	984	1286	1628	1812	2009	2214	2431	2894	3396	3938	4521	5144	6500
(800)	(40.03)	(171)	(304)	(475)	(684)	(931)	(1220)	(1540)	(1900)	(2740)	(3720)	(4860)	(6160)	(6860)	(7600)	(8380)	(9200)	(10940)	(12840)	(14890)	(17100)	(19460)	(24600)
118	132.46	45.6	81.1	126	183	248	324	410	507	730	993	1297	1642	1828	2027	2233	2452	2918	3425	3972	4560	5188	6560
(813)	(40.37)	(172)	(307)	(479)	(690)	(939)	(1230)	(1550)	(1920)	(2760)	(3760)	(4910)	(6210)	(6920)	(7660)	(8450)	(9280)	(11040)	(12950)	(15020)	(17250)	(19620)	(24800)
120	133.57	46.0	81.8	127	184	250	327	414	512	736	1001	1308	1656	1843	2044	2252	2473	2943	3454	4006	4599	5232	6620
(827)	(40.71)	(174)	(309)	(483)	(696)	(947)	(1240)	(1560)	(1930)	(2780)	(3790)	(4950)	(6260)	(6970)	(7730)	(8520)	(9350)	(11130)	(13060)	(15150)	(17390)	(19790)	(25000)
122	134.69	46.4	82.5	129	186	252	330	417	516	742	1010	1319	1670	1859	2061	2271	2494	2967	3483	4039	4637	5275	6680
(841)	(41.05)	(175)	(312)	(487)	(702)	(954)	(1250)	(1580)	(1950)	(2810)	(3920)	(4990)	(6310)	(7030)	(7790)	(8590)	(9430)	(11220)	(13170)	(15270)	(17540)	(19950)	(25300)
124	135.79	46.7	83.1	130	187	254	333	421	520	748	1018	1330	1684	1874	2077	2289	2514	2992	3511	4072	4675	5318	6740
(855)	(41.39)	(177)	(314)	(491)	(707)	(962)	(1260)	(1590)	(1960)	(2830)	(3850)	(5030)	(6370)	(7090)	(7860)	(8660)	(9510)	(11320)	(13280)	(15400)	(17680)	(20120)	(25500)
126	136.88	47.1	83.8	131	189	256	335	424	524	754	1026	1341	1697	1889	2094	2308	2534	3016	3539	4105	4712	5361	6800
(869)	(41.72)	(178)	(317)	(495)	(713)	(970)	(1270)	(1600)	(1980)	(2850)	(3880)	(5070)	(6420)	(7150)	(7920)	(8730)	(9580)	(11410)	(13390)	(15520)	(17820)	(20280)	(25700)
128	137.96	47.5	84.5	132	190	258	338	427	528	760	1034	1351	1711	1904	2111	2326	2554	3040	3567	4137	4749	5403	6850
(882)	(42.05)	(179)	(319)	(499)	(719)	(978)	(1280)	(1620)	(2000)	(2870)	(3910)	(5110)	(6470)	(7200)	(7980)	(8800)	(9660)	(11500)	(13490)	(15650)	(17960)	(20440)	(25900)
130	139.03	47.9	85.1	133	192	260	341	431	532	766	1042	1362	1724	1919	2127	2344	2574	3063	3595	4169	4786	5445	6900
(896)	(42.38)	(181)	(322)	(503)	(724)	(985)	(1290)	(1630)	(2010)	(2900)	(3940)	(5150)	(6520)	(7260)	(8040)	(8870)	(9740)	(11590)	(13600)	(15770)	(18100)	(20600)	(26100)
132	140.10	48.2	85.7	134	193	263	343	434	536	772	1050	1372	1736	1933	2144	2362	2594	3087	3623	4201	4823	5487	6950
(910)	(42.70)	(182)	(324)	(506)	(730)	(993)	(1300)	(1540)	(2030)	(2920)	(3970)	(5190)	(6570)	(7320)	(8110)	(8940)	(9810)	(11670)	(13700)	(15890)	(18240)	(20750)	(26300)
134	141.16	48.6	86.4	135	194	265	346	437	540	777	1058	1382	1749	1948	2160	2380	2613	3110	3650	4233	4860	5529	7000
(924)	(43.03)	(184)	(327)	(510)	(735)	(1000)	(1310)	(1650)	(2040)	(2940)	(4000)	(5230)	(6620)	(7370)	(8170)	(9010)	(9880)	(11760)	(13800)	(16010)	(18380)	(20910)	(26500)
136	142.21	49.0	87.0	136	196	267	348	441	544	783	1066	1392	1762	1962	2176	2398	2633	3133	3677	4625	4896	5570	7050
(938)	(43.35)	(185)	(329)	(514)	(741)	(1007)	(1320)	(1670)	(2060)	(2960)	(4030)	(5270)	(6670)	(7430)	(8230)	(9070)	(9960)	(11850)	(13910)	(16130)	(18520)	(21070)	(26700)

*For pressure in bars, multiply by 0.01.

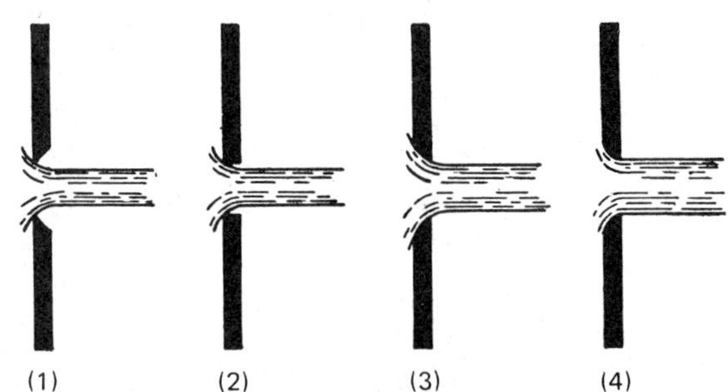

FIG. 17-2D. Orifices of various shapes. If the shape of the orifice is changed so as to decrease the contraction, its capacity will be increased. Form 1 in the illustration is a standard orifice having a sharp edge on the approach side. Form 2, when in a thin plate, gives the same stream characteristics as Form 1. Form 3 is the reverse of 1. In Form 4, the edge is rounded to conform to the shape of the stream. The coefficients of discharge of 3 and 4 are greater than those of the standard orifices, approaching a value of 1.0 in the case of 4.

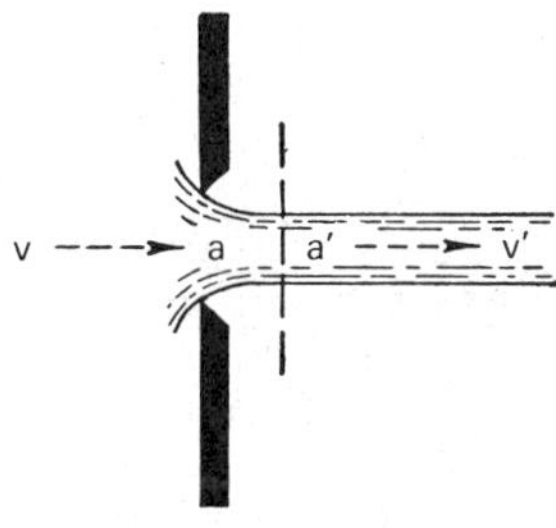

FIG. 17-2E. Flow through a standard orifice.

feet per second (m/sec), and a' is area in square feet (m^2):

$$Q = va = v'a'$$

The coefficient of discharge of a standard orifice would be the product of the coefficient of velocity and the coefficient of contraction, or $c = 0.98 \times 0.62 = 0.61$. Representative values for the coefficients of discharge are given in Table 17-2B.

TABLE 17-2B. Typical Discharge Coefficients of Solid Stream Nozzles

Standard sprinkler, average (nominal ½ in. dia.)	0.75
Standard orifice (sharp edge)	0.62
Smooth bore nozzles, general	0.96–0.98
Underwriter playpipes or equal	0.97
Deluge or monitor nozzles	0.997
Open pipe, smooth, well rounded	0.90
Open pipe, burred opening	0.80
* Hydrant butt, smooth and well-rounded outlet, flowing full	0.90
* Hydrant butt, square and sharp at hydrant barrel	0.80
* Hydrant butt, outlet square, projecting into barrel	0.70

* See Figure 17-2F.

FLOW IN SHORT TUBES

A tube attached to an orifice is known as a standard short tube when its length is 2½ or 3 times the diameter of the orifice and its diameter is the same as the orifice. A shorter tube will not flow full, and friction losses in a longer tube will affect results when used as a measuring device, hence the specified limit of length.

The characteristics of a standard short tube and a short conical converging tube are shown in Figures 17-2G and 17-2H. The principles of flow in orifices apply, but with different coefficients. With the conical tube the coefficients c_v and c_c vary with the angle β. When β is 0°, the converging tube becomes a cylindrical tube with $c_c = 1$ and $c_v = 0.82$; c_d is then 0.82. As the angle β increases, the coefficient of contraction (c_c) develops and the coefficient of velocity (c_v) increases, approaching the 0.98 value for a sharp edge orifice. Relations are such that the coefficient of discharge attains a maximum value of 0.94 with a β angle of about 13°.

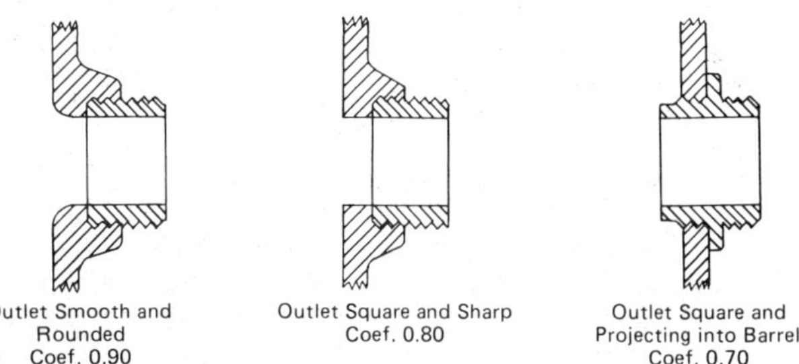

FIG. 17-2F. Three general types of hydrant outlets and their coefficients of discharge.

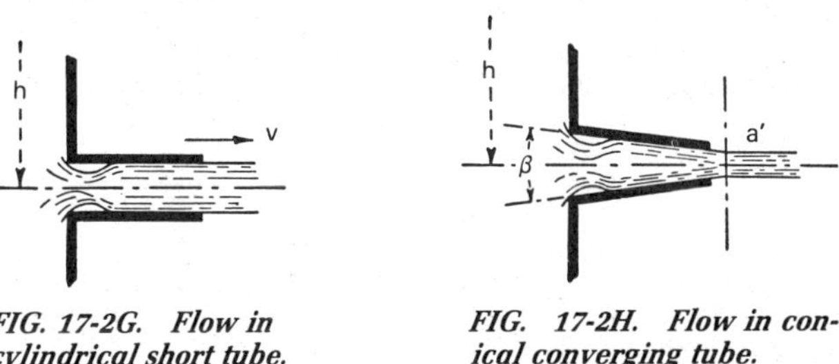

FIG. 17-2G. Flow in cylindrical short tube.

FIG. 17-2H. Flow in conical converging tube.

ENTRANCE LOSSES

Previous discussion in this chapter has shown that the actual velocity of the streams from orifices or short tubes is less than that which would be developed theoretically by the head producing the flow, the coefficient of velocity for a standard orifice being 0.98, and for a cylindrical short tube $c_v = 0.82$. The difference between the available head and the head equivalent to the actual velocity is the lost head. In terms of total head and coefficient of velocity, lost head equals:

$$h(1 - c_v^2) \text{ or } \frac{v^2}{2g}(1 - c_v^2)$$

This is evidence of the energy lost by turbulence where the character of the flow changes, as from a large tank or reservoir through an orifice or into a pipe, or at a sharp change in pipe diameters. In many hydraulic calculations, desired accuracy makes it necessary to determine the lost head and/or the velocity head at different locations. This type of lost head is commonly designated as an "entrance loss."

When calculating the lost head at a sharp reduction in pipe size, it is necessary to consider that the head producing the flow in the smaller pipe is the total head (pressure head and velocity head) of the approaching stream.

Tapered fittings, such as at the entrance or discharge of centrifugal pumps, or at a valve with reduced water passageway inserted in a pipe line, greatly reduce the head lost where marked velocity changes occur in piping systems.

VENTURI TUBE

The Venturi principle has a number of applications in fire protection. The Venturi tube is essentially a tapered constriction in a pipe. In the constricted part, the velocity must be greater than in the straight tube, and the pressure is correspondingly less in accordance with Bernoulli's Theorem. If the increase in velocity through the constricted portion is sufficient, the pressure at that point will be less than atmospheric, and a suction will be created at any opening into the side of the tube. The Venturi tube is illustrated by Figure 17-2I. The diverging portion of a Venturi meter serves only to restore the system pressure with a minimum of friction loss.

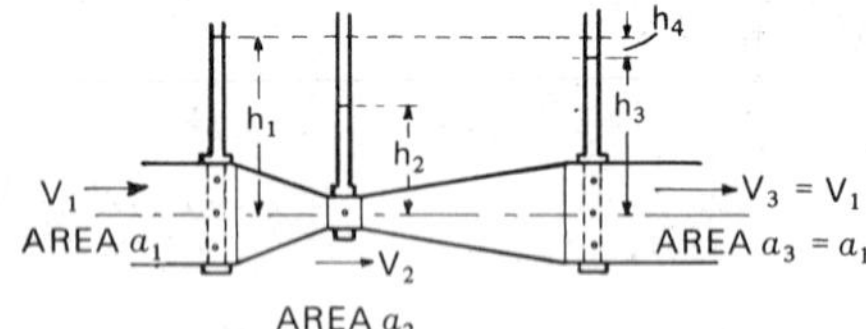

FIG. 17-2I. The Venturi tube.

Venturi Meter: The Venturi principle as applied in the Venturi meter for the measurement of flows in closed pipe lines under pressure is as follows:

With no elevation difference along the line of flow, Bernoulli's Theorem becomes:

$$\frac{v_1^2}{2g} + \frac{p_1}{w} + 0 = \frac{v_2^2}{2g} + \frac{p_2}{w} + 0$$

In Figure 17-2I, $\frac{p_1}{w}$ is represented by h_1, and $\frac{p_2}{w}$ by h_2, which are used in the following equations.

The quantity of liquid passing through all portions of the Venturi must be the same. Therefore,

$$Q = a_1v_1 = a_2v_2 \text{ or } v_1 = \frac{Q}{a_1} \text{ or } v_2 = \frac{Q}{a_2}$$

Substituting in Bernoulli's Theorem,

$$\frac{\left(\frac{Q}{a_1}\right)^2}{2g} + h_1 = \frac{\left(\frac{Q}{a_2}\right)^2}{2g} + h_2$$

$$\frac{\left(\frac{Q}{a_2}\right)^2}{2g} - \frac{\left(\frac{Q}{a_1}\right)^2}{2g} = h_1 - h_2$$

$$Q^2\left(\frac{1}{a_2^2} - \frac{1}{a_1^2}\right) = 2g(h_1 - h_2)$$

$$Q^2\left(\frac{a_1^2 - a_2^2}{a_1^2a_2^2}\right) = 2g(h_1 - h_2)$$

$$Q = \frac{a_1a_2}{\sqrt{a_1^2 - a_2^2}}\sqrt{2g(h_1 - h_2)}$$

For any specific Venturi, a_1 and a_2 are known constant values. There is also a friction loss coefficient which is usually determined by test and which does not remain constant with very low velocities. Combining the known constant values, the Venturi Meter Formula is generally expressed as:

$$Q = k\sqrt{h_1 - h_2}\text{, or } Q = k\sqrt{\frac{p_1}{w} - \frac{p_2}{w}}$$

By test, a value of k for any specific meter can be established with reasonable accuracy.

When used as a device for inducting gas or liquid into the stream, as is made possible by the reduced pressure in the throat section, the hydraulic performance will not be in full accordance with the above theoretical calculations because energy is expended on the induced substance.

PITOT TUBE METHOD OF MEASURING FLOW

The most used method of measuring the flow in an open stream from an orifice, nozzle, or open pipe is by a direct measurement of the total head which produces the flow in accordance with the equation $v = \sqrt{2gh}$. This measurement makes use of the well known Pitot tube and pressure gage combination of which representative forms are shown in Figure 17-8A. When the small opening, usually not over 1⁄16-in. (1.6 mm) diameter, is inserted in the center of a stream, with the opening directly in the line of flow, it will indicate, by water or mercury column, or by a pressure gage, the total head at that location. With the stream open to the atmosphere, there will be no pressure head so that the indicated reading will be velocity head alone, and the velocity of the stream can be calculated directly. Because the velocity at the surface of the stream is reduced slightly by friction against an orifice or nozzle, a coefficient of velocity of 0.97 is usually applied for nozzles of ordinary fire stream sizes. By knowing accurately the area of cross section of the stream at the location of the velocity measurement, the quantity flowing can be determined from the relation $Q = av = 29.83cd^2\sqrt{p}$ or $(q_m = 0.0666cd_m^2 (\sqrt{p_{vm}})$.

When measuring flow from a nozzle, the use of the Pitot tube method only holds with reasonable accuracy for tip sizes up to 1⅜ in. (35 mm) supplied by 2½ in. (64 mm) hose. Above that the error rate increases beyond acceptable limits and the more accurate method described next under "The Nozzle Method of Measuring Flow" should be used.

A typical Pitot tube as used in measuring the flow from a free stream nozzle is shown in Figure 17-2J.

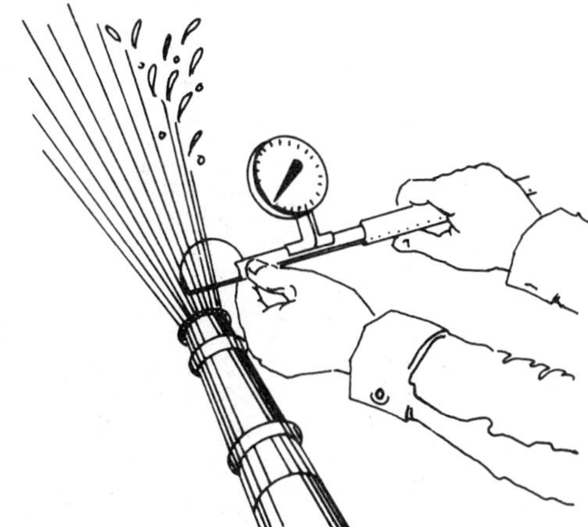

FIG. 17-2J. Taking nozzle pressure with a Pitot tube.

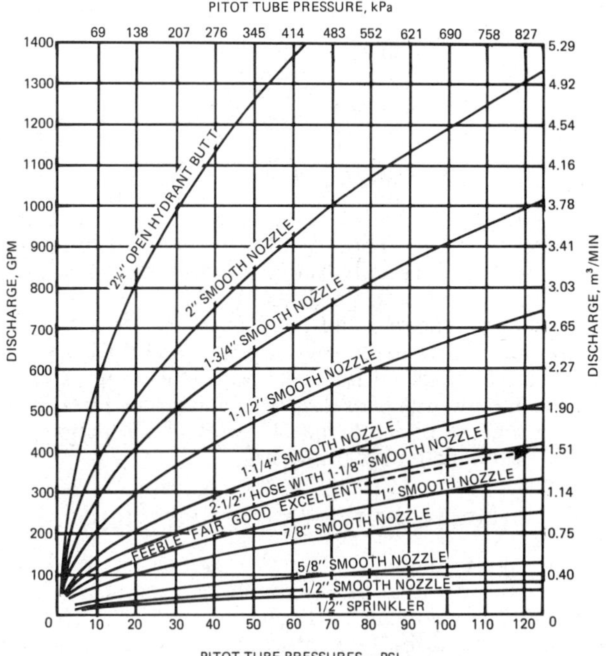

FIG. 17-2K. Relative discharge curves. (Chemical Engineers' Handbook, McGraw-Hill Book Co.)

The gage is normally calibrated in pounds per square inches (kPa). This is measuring the velocity pressure at the point of maximum stream contraction. As the normal pressure is zero at this point, this is also measuring total head.

For the usual forms of orifices and nozzles, the coefficient of contraction (c_c) is accurately known, so that $a = c_c \times$ actual discharge opening. With a sharp-edged orifice, the area of the stream may be determined from the actual diameter of the orifice opening and the use of the 0.62 coefficient of contraction. For a carefully made smooth nozzle the contraction is negligible, i.e., $c_c = 1.00$.

The rates of flow from various types and sizes of nozzles as given in Figure 17-2K are sufficiently accurate for most fire flow calculations.

THE NOZZLE METHOD OF MEASURING FLOW

The hydraulic characteristics of good solid-stream nozzles are consistent within a wide range of flow conditions. The rate of discharge can be calculated from gage pressure at the base of the nozzle, or from Pitot pressure at the discharge outlet.

The flow formula for using base pressure is:

$$Q = \frac{29.83cd^2\sqrt{p_1}}{\sqrt{1 - c^2\left(\frac{d}{D}\right)^4}}$$

in which

Q = flow in gallons per minute
c = coefficient of discharge
d = diameter of outlet, in inches
p_1 = gage pressure at base of nozzle, in psi
D = inside diameter of fitting to which gage is attached, in inches.

For SI units the formula is expressed as:

$$Q_m = \frac{0.0666cd_m{}^2\sqrt{P_{1m}}}{\sqrt{1 - c^2\left(\frac{d}{D}\right)^4}}$$

where

Q_m = flow (L/min)
d_m = diameter of outlet (mm)
D = inside diameter of filling (mm)
p_{1m} = gage pressure at base of nozzle (kPa).

This is the common formula for discharge from an orifice except that (1) gage pressure at the base of the nozzle is substituted for Pitot pressure, and (2) a factor is added which represents the ratio between gage pressure and total pressure at the nozzle base. (Total pressure is gage pressure plus velocity pressure.)

When base pressure is to be used, the gage is attached to a fitting close to the nozzle with a straight piece of approach pipe or hose to eliminate turbulence or unstable flow conditions. To obtain greater accuracy than provided by a simple fitting, a piezometer fitting may be used. With this device, the gage is connected to an annular tube or channel having a number of small holes drilled into the waterway around the circumference. The mean or resultant static pressure indicated by the gage is p_1 in the formula above.

EXAMPLE: Assume a rate of discharge from a 2 in. (51 mm) nozzle with a pressure measured by a 2½ in. (64 mm) piezometer ring and gage at 80 psi (552 kPa) at the base of the nozzle. Also assume a coefficient of discharge of .99.

Thus:

$$Q = \frac{29.83 \times .99 \times (2)^2\sqrt{80}}{\sqrt{1 - (.99)^2\left(\frac{2}{2.5}\right)^4}}$$

$$Q = 1364.9 \text{ gpm } (5166 \text{ L/min})$$

Discharge coefficients for ordinary nozzles and orifices are given in Table 17-2B, or can be determined satisfactorily by careful testing procedures, using calibrated meters, nozzles, weigh tanks, etc.

Although accurate and convenient for fixed test arrangements, the measurement of pressure at the base of a nozzle is not practical for usual hose stream operations. Because a Pitot gage is useless with spray type nozzles or other devices producing special types of discharge, the base pressure method is necessary.

DISCHARGE CALCULATIONS

The most common method of estimating nozzle or orifice discharge is to use the Table of Theoretical Flow (Table 17-2A). Knowing the diameter of the outlet; the flow corresponding to the Pitot pressure is multiplied by the applicable discharge coefficient (Table 17-2B).

EXAMPLE 1: Pitot reading was 20 psi at an open 2½ in. hydrant butt; assumed coefficient was 0.90. Tabular flow for 20 psi is 834 gpm; 90 percent of 834 = 751 gpm. (Or subtract 10 percent from 834 and get 751 gpm.)

EXAMPLE 2: Pitot reading was 800 kPa at a 38 mm pipe with a burred opening (discharge coefficient = 0.80). Table 17-2A for 800 kPa gives a theoretical flow of 2740 L/min. Actual flow would be 0.8 × 2740 = 2192 L/min.

Flow tables are available for certain nozzles, based on specific discharge coefficients.

Nozzle discharge also can be computed by the standard formulas previously discussed. (See "Flow Through Orifices," and "The Nozzle Method of Measuring Flow," in this chapter.)

To simplify calculations when it is desired to determine the flow in gallons per minute through an orifice or nozzle of a given diameter, and a known coefficient of discharge, the formula can be reduced to $Q = k\sqrt{p}$, where k combines the constants 29.83 (0.0666 in SI units), c, and d^2. (See Table 17-2C for k values of nozzles.) For SI, use k_m values.

Because $k = Q/\sqrt{p}$, the k values of spray nozzles can be calculated from data in testing laboratory listings of nozzles. Generally, the rate of flow at 50 and 100 psi base pressure, or both, is given in the listings for the various nozzles. Some nozzles are also listed at 25 and 125 psi base pressure.

EXAMPLE 1: A certain fire service spray nozzle is rated for 83 gpm at 50 psi. Therefore $k = 83/\sqrt{50} = 83/7.07 = 11.7$. At 25 psi base pressure the discharge would be $11.7\sqrt{25} = 11.7 \times 5 = 58.5$ gpm.

EXAMPLE 2: Determine the discharge from a 51 mm hydrant butt (c = 90) at a base pressure of 350 kPa.

$$k_m = 0.0666\ cd_m{}^2 = 0.0666 \times 0.90 \times (51)^2 = 156$$

Thus $Q \text{ (L/min)} = k_m\sqrt{p_m} = 156\ \sqrt{350}$, i.e.,

$$Q = 2918 \text{ L/min.}$$

TABLE 17-2C. Values of *k* for Various Discharge Orifices

Type of Orifice	Nominal Diameter (Inches)	(mm)	*k*	(S-I) km
Sprinkler	¼	7	1.3–1.5	1.9–2.2
Sprinkler	5⁄16	21	1.8–2.0	2.6–2.9
Sprinkler	3⁄8	10	2.6–2.9	3.7–4.2
Sprinkler	7⁄16	11	4.0–4.4	5.8–6.3
Sprinkler	½	13	5.3–5.8	7.6–8.4
Sprinkler	1 17⁄32	39	7.4–8.2	10.6–11.8
Nozzle	½	13	7.2	10.3
Nozzle	7⁄8	22	22.2	32.0
Nozzle	1	25	29.1	41.9
Nozzle	1 1⁄16	27	32.8	47.2
Nozzle	1⅛	29	36.8	53.0
Nozzle	1 3⁄16	30	41.0	59.0
Nozzle	1¼	32	45.4	65.4
Nozzle	1 5⁄16	33	50.1	72.1
Nozzle	1⅜	35	54.9	79.1
Nozzle (*c* = 0.97 for all nozzles)	1 7⁄16	37	60.0	86.4
Nozzle	1½	38	65.4	94.2
Nozzle	1 9⁄16	40	70.9	102.0
Nozzle	1⅝	41	76.8	110.6
Nozzle	1 11⁄16	43	82.8	119.2
Nozzle	1¾	44	89.0	128.2
Nozzle	1 13⁄16	46	95.5	137.5
Nozzle	1⅞	48	102.0	146.9
Nozzle	1 15⁄16	49	109.0	157.0
Nozzle	2	51	116.0	167.0
Hydrant butt (*c* = 0.90)	2	51	107.4	154.7
Hydrant butt (*c* = 0.90)	2¼	57	135.9	195.7
Hydrant butt (*c* = 0.90)	2½	64	167.8	241.6

FLOW OF WATER IN PIPES

The theory of liquid flow in pipes involves (1) Bernoulli's Theorem (taking into account friction losses), (2) the axiom that area multiplied by average velocity is the same at any cross section, assuming no significant discharge through branch lines or leaks, (3) lost head caused by friction and turbulence, (4) density of the liquid (fire protection hydraulics, with few exceptions, is based on a density of 62.4 pcf (1000 kg/m³) of ordinary fresh water), and (5) viscosity of the liquid. (Because ordinary water is used in fire protection systems, except for certain special applications, consideration of fluid viscosity is rarely needed.)

When water flows through a pipe there is almost always a drop in pressure. Theoretically, the lost head between two points is caused by (1) friction between the moving water and the pipe wall, and (2) friction between water particles, including that produced by turbulence when flow changes direction or when rapid increase or decrease of velocity takes place, such as at abrupt changes in pipe diameter. A change in velocity results in some conversion of velocity head to pressure head or vice versa.

At low velocity in a smooth pipe, very little turbulence is produced, and the flow is called "laminar." With this condition, all particles of water move along the pipe in essentially straight lines and in concentric layers; friction loss then occurs mainly in a thin layer at the pipe wall. The rate of loss is small compared to friction loss in turbulent flow.

In a pipe of specific diameter and roughness an increasing velocity changes laminar flow to unstable turbulent flow and then to complete turbulence. The range of flow condition from laminar to complete turbulence is called the transition zone.

Most fire protection systems and water distribution mains function under turbulent flow conditions, and friction losses within the pipe itself account for most of the lost head. Other losses are usually considered together and are called the "loss in fittings."

Experimental data have established that frictional resistance is:

1. Independent of pressure in the pipe.
2. Proportional to the amount and character of frictional surface.
3. Variable with the velocity of flow (nearly proportional to the second power of the velocity for velocities above the critical; if velocity is below critical, resistance varies as the first power).

FRICTION LOSS FLOW FORMULAS

Perhaps the best known and oldest expression relating velocity to friction loss in pipe is known as the Chezy formula expressed as:

$$v = c\sqrt{rs}$$

in which

c is a factor depending on kind and roughness of pipe

r (the hydraulic radius) = area/circumference = $d/4$ in which d = diameter of pipe

s is a hydraulic slope = h/l = slope of hydraulic gradient in which h is the friction head in length of pipe l (see Chap. 5 of this Section for a discussion of hydraulic gradients).

Therefore $v = c\sqrt{d/4 \times h/l}$ or $h = 4lv^2/c^2d$

The classical formula for friction loss in long, straight pipes of uniform diameter and roughness is ascribed to Darcy, Manning, Fanning and others. In modern textbooks it is derived by analysis of forces acting on a flowing particle of water in a pipe. Actually, it is a variation of Chezy's formula with a friction factor f replacing c, and expressed as:

$$h = \frac{flv^2}{d2g}$$

in which h = friction head; l = length of pipe; d = diameter of pipe; v = velocity; and g = acceleration of gravity.

The friction-flow formulas commonly used in the hydraulics of fire protection and water supply have been developed by experiment and experience. These formulas are exponential in the form $V = Cr^x s^y$, where V is velocity, C the coefficient of friction, r the hydraulic radius (area divided by circumference), and s the hydraulic slope (loss of head divided by length). The most popular exponential formula is the Hazen-Williams, its basic form being $V = 1.318Cr^{0.63}s^{0.54}$. The friction coefficients in formulas of this type are constant for a specific type or roughness of pipe, and are independent of velocity. Thus the accuracy of these formulas is variable. However, the fixed values generally assumed for viscosity and density are considered adequate.

Fluid motion in a pipe may be either laminar or turbulent. In laminar or "streamline" motion the fluid particles follow definite paths, and the resistance to flow is due only to the shear stress of adjacent layers. Laminar flow is usually associated with low velocity.

In turbulent motion the fluid moves in an eddying mass, and at any point may vary in an irregular manner from instant to instant.

With low velocity and laminar flow in either a smooth pipe or a rough pipe the flow would remain laminar until the velocity reaches what is called the critical velocity. Beyond that and until turbulent flow is definitely established, the fluid motion is in a transition zone.

Ordinary fire protection and utility flow rates are in the turbulent range. When fluids other than water are involved, the conventional exponential formulas are not suitable, and a universal flow formula, such as the Darcy-Weisbach, is needed:

$$\left(h = f\frac{l}{d}\frac{v^2}{2g}\right)$$

In the Darcy-Weisbach formula the friction factor f is dimensionless and variable, depending on (1) the roughness of the pipe and (2) on another dimensionless factor called the Reynolds number, which depends on the diameter, velocity, and kinetic viscosity.

The value of f can be computed by a formula known as the Colebrook-White equation which is neither completely empirical nor rigidly theoretical. This equation is usually written as follows:

$$\frac{1}{\sqrt{f}} = -2\log_{10}\left[\frac{\epsilon}{3.7D} + \frac{2.51}{R\sqrt{f}}\right]$$

where ϵ is a linear measure of roughness, f is the Darcy-Weisbach friction factor, D is the pipe diameter in feet, and R is the Reynolds number. Computation of f by the formula can be avoided by using tables and charts known as "Moody" diagrams (The Hydraulic Institute 1954).

Reynolds numbers can be calculated by the formula:

$$R = \frac{VD}{v}$$

where V = velocity, feet per second (m/sec); D = diameter, feet (m); and v = kinetic viscosity, square feet per second (m^2/sec). Figure 17-2L is a Moody diagram by which f can be determined for any kind or size of pipe. Knowing R and the relative roughness ϵ, the value of f can be read directly off the chart. The dimension of $\frac{\epsilon}{D}$ is difficult to obtain, and it may be necessary to assume a value for $\frac{\epsilon}{D}$ based on experience and judgment. The roughness factor of new pipe usually can be provided by the manufacturer.

TABLE 17-2D. Friction Loss in Pipe

Pounds per Square Inch per 100 Feet of Pipe
Hazen-Williams $C = 100$*

	Actual Diameter of Pipe ½ through 3½ in.†‡ Nominal Diameter of Pipe for 4 through 30 in.‡										
Gpm‡	½	¾	1	1¼	1½	2	2½	3	3½	4	Gpm
5	17.9	4.55	1.40	.369	.174	.052	—	—	—	—	5
10	64.5	16.4	5.06	1.33	.629	.186	.078	.030	—	—	10
15		34.7	10.7	2.82	1.33	.394	.166	.064	.028	—	15
20	***5***	59.1	18.2	4.89	2.27	.671	.282	.109	.048	.027	20
30	.019	***6***	38.6	10.2	4.80	1.42	.598	.231	.102	.057	30
40	.033	—	65.8	17.3	8.17	2.42	1.02	.393	.174	.097	40
50	.050	.020	***8***	26.2	12.3	3.66	1.54	.593	.263	.147	50
60	.069	.029	—	36.6	17.3	5.12	2.16	.831	.369	.206	60
70	.092	.038	—	48.7	23.0	6.81	2.87	1.11	.490	.274	70
80	.118	.049	—	62.4	29.4	8.72	3.67	1.41	.628	.350	80
90	.147	.060	—	77.6	36.6	10.8	4.56	1.76	.781	.435	90
100	.178	.074	—	***10***	44.5	13.1	5.55	2.14	.949	.529	100
120	.250	.103	—	—	62.3	18.5	7.77	3.00	1.33	.741	120
140	.333	.137	.034	—	82.9	24.6	10.3	3.98	1.77	.986	140
160	.426	.175	.043	—	106.0	31.4	13.2	5.10	2.26	1.26	160
180	.529	.218	.054	.018	***12***	39.1	16.5	6.34	2.81	1.57	180
200	.643	.265	.065	.022	—	47.5	20.0	7.71	3.42	1.91	200
220	.768	.316	.078	.026	—	56.7	23.9	9.19	4.08	2.28	220
240	.902	.371	.091	.031	.013	***14***	28.0	10.8	4.79	2.67	240
260	1.05	.430	.106	.036	.015	—	32.5	12.5	5.56	3.10	260
280	1.20	.493	.122	.041	.017	—	37.3	14.4	6.37	3.55	280
300	1.36	.562	.138	.047	.019	—	42.3	16.3	7.24	4.04	300
350	1.81	.746	.184	.062	.026	.012	***16***	21.7	9.63	5.37	350
400	2.32	.955	.235	.079	.033	.015	—	27.8	12.3	6.88	400
450	2.88	1.19	.292	.099	.041	.019	—	34.6	15.3	8.55	450
500	3.51	1.44	.353	.120	.049	.023	.012	42.0	18.6	10.4	500
550	4.18	1.72	.424	.143	.059	.028	.015	50.1	22.2	12.4	550
600	4.91	2.02	.498	.168	.069	.033	.017	58.8	26.1	14.6	600
650	5.70	2.34	.577	.195	.080	.038	.020	68.2	30.3	16.9	650
700	6.53	2.69	.662	.223	.092	.043	.023	***18***	34.7	19.4	700
750	7.42	3.05	.752	.254	.104	.049	.026	—	39.4	22.0	750
800	8.36	3.44	.848	.286	.118	.056	.029	—	44.5	24.8	800
850	9.35	3.85	.948	.320	.132	.062	.032	—	49.7	27.7	850
900	10.4	4.28	1.05	.356	.146	.069	.036	—	***20***	30.8	900
950	11.5	4.73	1.17	.393	.162	.076	.040	—	—	34.1	950
1,000	12.6	5.20	1.28	.432	.178	.084	.044	—	—	37.5	1,000
1,250	19.1	7.85	1.94	.653	.269	.127	.066	—	—	***24***	1,250
1,500	***30***	11.0	2.71	.914	.376	.178	.093	—	—	—	1,500
1,750	—	—	3.61	1.22	.501	.236	.123	—	—	—	1,750
2,000	.007	—	4.62	1.56	.641	.303	.158	.089	.053	.022	2,000
2,250	.009	—	—	1.94	.797	.376	.196	.111	.066	.027	2,250
2,500	.011	—	—	2.35	.969	.457	.239	.134	.081	.033	2,500
2,750	.013	—	—	2.81	1.16	.545	.285	.160	.096	.040	2,750
3,000	.016	—	—	3.30	1.36	.641	.334	.188	.113	.046	3,000
4,000	.027	—	—	—	2.31	1.09	.569	.321	.192	.079	4,000
5,000	.040	—	—	—	3.49	1.65	.860	.485	.290	.119	5,000

* To convert friction loss at $C = 100$ or other values of C, see Table 17-2E.
† Schedule 40 pipe sizes ½ through 3½ in. steel pipe.
‡ SI units: 1 psi = 6.895 kPa; 1 gpm = 0.378 L/min; 1 in = 25.4 mm.

Note: Actual inside diameter for sizes ½ in. through 3½ in. is given for greater accuracy as these sizes include sprinkler branch lines and the smaller sizes of cross mains. For sizes 4 in. and greater, the nominal diameters were used as a fairly safe average for the diameters of various types of underground pipes as follows: cast iron unlined and Enameline, greater than nominal; cast iron cement lines and Class 200 asbestos cement, less than nominal; Class 150 asbestos-cement sizes 6 and 8 in. less than nominal, and other sizes even nominal. (A 0.10 variation is true for Class 150 cement lined only—see ASHD FT-9 through 45 for actual IDs.)

This table will be usefu in approximating friction loss in flow through existing underground piping where the type, inside diameter, and condition are frequently unknown. However, in such cases, a flow test is recommended.

When the type, inside diameter, and condition are known, and in designing new systems for all sizes and types of pipes, the friction loss tables should be used. Friction tables based on Hazen-Williams formula are published in *Automatic Sprinkler Hydraulic Data* by "Automatic" Sprinkler Corporation of America, and tables based on Darcy-Weisbach formula are published in *Standards of the Hydraulic Institute.*

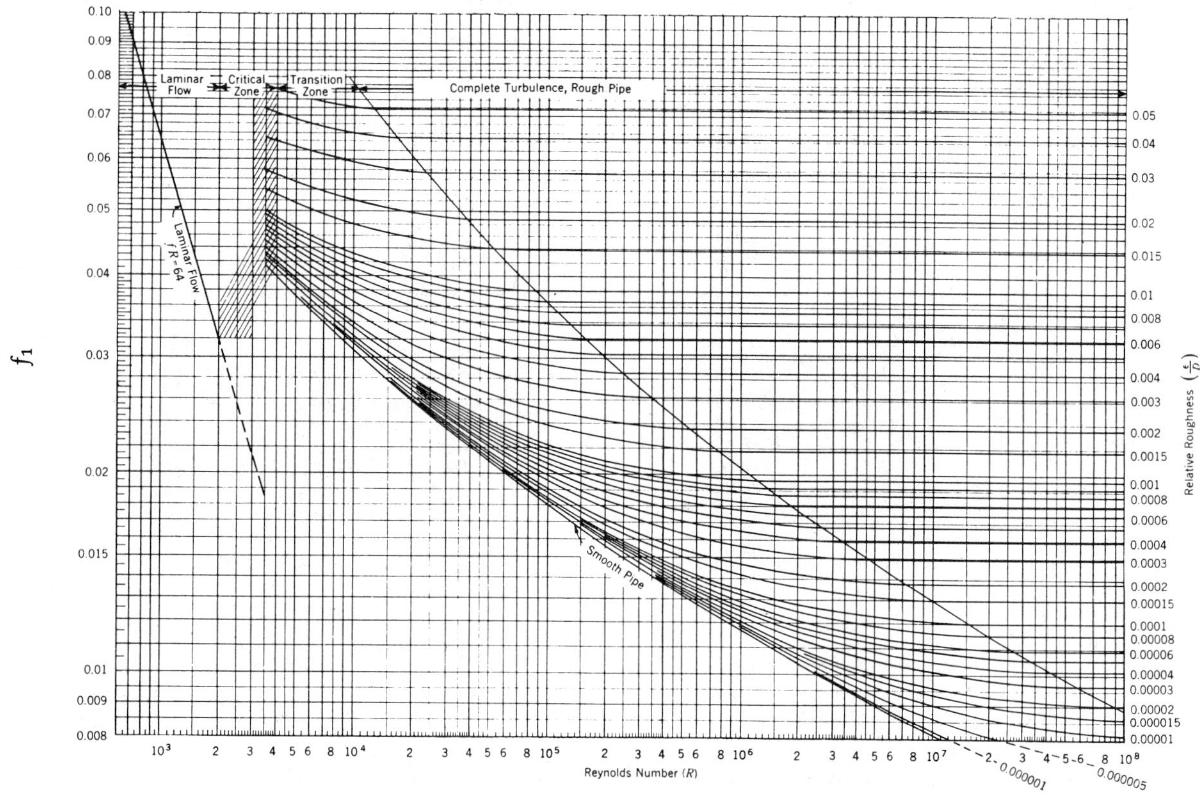

FIG. 17-2L. Moody Diagram for friction in pipe. The values for friction factor are on the vertical scale at left. (See Friction Manual © Hydraulic Institute.)

The Hazen-Williams Formula

The basic form of the Hazen-Williams formula ($V = 1.318Cr^{0.63}s^{0.54}$) is not practical for ordinary fire protection flow calculations. In terms of pressure loss and gallons per minute the formula becomes:

$$p = \frac{452Q^{1.85}}{C^{1.85}d^{4.87}}$$

in which p is the pressure loss in pounds per square inch per 100 ft of pipe; Q is the rate of flow in gallons per minute; and d is the inside diameter of the pipe in inches. (Numbers to the 1.85 power are given in Table 17-8C.)

TABLE 17-2E. Conversion Factors for Friction Loss in Pipe for Values of Coefficient Other Than 100

C	Factor	C	Factor	C	Factor
150	0.472	110	0.838	70	1.93
145	0.503	105	0.914	65	2.22
140	0.537	100	1.00	60	2.57
135	0.574	95	1.10	55	3.02
130	0.615	90	1.22	50	3.61
125	0.662	85	1.35	45	4.38
120	0.714	80	1.51	40	5.48
115	0.772	75	1.70	35	6.97

In metric practice the formula is:

$$p_m = 6.06 \times \frac{Q_m^{1.85}}{C^{1.85}d_m^{4.87}} \times 10^9$$

where

p_m = pressure loss (kPa) per 100 m of pipe
Q_m = rate of flow (L/min)
d = inside diameter (mm).

FRICTION LOSS CALCULATIONS

The solution of fire protection problems involving pipe flow and friction seldom requires direct calculation using formulas. However, in using the simplifying charts and tables, great care must be taken to identify the C value (coefficient of friction) on which the chart or table is based. Where the type or condition of a pipe necessitates the use of a different C value. The friction loss must be multiplied by an appropriate factor (See Table 17-2G).

By way of illustration, Table 17-2D gives values of p when C = 100 for Standard pipe sizes from ½ in. to 30 in. in diameter. For values of C other than 100, the tabular losses are multiplied by the corresponding factor in Table 17-2E.

EXAMPLE: Determine the friction loss with 700 gpm flowing in 700 ft of 8 in. cast iron pipe having a C value of 80.

SOLUTION: From Table 17-2D, the loss for 700 gpm per 100 ft of 8-in. pipe with C = 100 is .662 psi. From Table 17-2E, the factor for C = 80 is 1.51. Because friction loss is

TABLE 17-2F. Equivalent Pipe Length Chart

Fittings and Valves	Fittings and Valves Expressed in Equivalent Feet (m) of Pipe						
	¾ in.	1 in.	1¼ in.	1½ in.	2 in.	2½ in.	3 in.
45° Elbow	1 (0.3)	1 (0.3)	1 (0.3)	2 (0.6)	2 (0.6)	3 (0.9)	3 (0.9)
90° Standard Elbow	2 (0.6)	2 (0.6)	3 (0.9)	4 (1.2)	5 (1.5)	6 (1.8)	7 (2.1)
90° Long Turn Elbow	1 (0.3)	2 (0.6)	2 (0.6)	2 (0.6)	3 (0.9)	4 (1.2)	5 (1.5)
Tee or Cross (Flow Turned 90°)	4 (1.2)	5 (1.5)	6 (1.8)	8 (2.4)	10 (3.1)	12 (3.7)	15 (4.6)
Gate Valve	—	—	—	—	1 (0.3)	1 (0.3)	1 (0.3)
Butterfly Valve	—	—	—	—	6 (1.8)	7 (2.1)	10 (3.1)
Swing Check*	4 (1.2)	5 (1.5)	7 (2.1)	9 (2.7)	11 (3.4)	14 (4.3)	16 (4.9)

Fittings and Valves	Fittings and Valves Expressed in Equivalent Feet (m) of Pipe						
	3½ in.	4 in.	5 in.	6 in.	8 in.	10 in.	12 in.
45° Elbow	3 (0.9)	4 (1.2)	5 (1.5)	7 (2.1)	9 (2.7)	11 (3.4)	13 (4.0)
90° Standard Elbow	8 (2.4)	10 (3.1)	12 (3.7)	14 (4.3)	18 (5.5)	22 (6.7)	27 (8.2)
90° Long Turn Elbow	5 (1.5)	6 (1.8)	8 (2.4)	9 (2.7)	13 (4.0)	16 (4.9)	18 (5.5)
Tee or Cross (Flow Turned 90°)	17 (5.2)	20 (6.1)	25 (7.6)	30 (9.2)	35 (10.7)	50 (15.3)	60 (18.3)
Gate Valve	1 (0.3)	2 (0.6)	2 (0.6)	3 (0.9)	4 (1.2)	5 (1.5)	6 (1.8)
Butterfly Valve	—	12 (3.7)	9 (2.7)	10 (3.1)	12 (3.7)	19 (5.8)	21 (6.4)
Swing Check*	19 (5.8)	22 (6.7)	27 (8.2)	32 (9.8)	45 (13.7)	55 (16.8)	65 (19.8)

Use with Hazen and Williams C = 120 only. For other values of C, the figures in this table should be multiplied by the factors below:

Value of C	80	100	120	130	140	150
Multiplying factor	0.472	0.713	1.00	1.16	1.32	1.51

(This is based upon the friction loss through the fitting being independent of the C factor applicable to the piping.)
Specific friction loss values or equivalent pipe lengths for alarm valves, dry pipe valves, deluge valves, strainers, and other devices or fittings should be made available to the authority having jurisdiction.

* Due to the variations in design of swing check valves, the pipe equivalents shown in this table should be considered average.

NOTE: Use the equivalent feet (m) value for the "standard elbow" on any abrupt ninety-degree turn such as the screw-type pattern. Use the equivalent feet (m) value for the "long turn elbow" on any sweeping ninety-degree turn such as a flanged, welded or mechanical joint elbow type.

directly proportional to length of pipe, multiply .662 × 7 × 1.5 = 6.95 psi (answer).

TABLE 17-2G. Guide for Estimating Hazen-Williams *C*

Kind of Pipe	Value of *C*		
	1*	2†	3‡
Cast Iron, unlined:			
10 years old	110	90	75
15 years old	100	75	65
20 years old	90	65	55
30 years old	80	55	45
50 years old	70	50	40
Cast Iron, unlined, new		120	
Cast Iron, cement-lined		140	
Cast Iron, bitumastic enamel-lined		140	
Average steel, new		140	
Riveted steel, new		110	
Asbestos-cement		140	
Reinforced concrete		140	

* 1. Water mildly corrosive. Use same values for fire-protection mains having no mill-use or domestic draft.
† 2. Water moderately corrosive.
‡ 3. Water severely corrosive.
Note: *C* values chosen for design of piping systems should be based on applicable standards of NFPA or the authority having jurisdiction.

Equivalent Pipes

Problems involving piped water supplies and fire protection systems occasionally require substitution of one pipe for another. The term "equivalent pipe" usually means a pipe having the same friction loss characteristic as the pipe for which it is being substituted. The formula for using the friction loss table is applicable here.

EXAMPLE: What length of 8 in. pipe (C = 110) is equivalent to 700 ft of 6 in. pipe (C = 85)?

SOLUTION:

$$N_1 \times \frac{L_1}{100} \times T_1 = N_2 \times \frac{L_2}{100} \times T_2$$

Assume a rate of flow (say 1,000 gpm) and substitute known values:

$$1.35 \times \frac{700}{100} \times 5.20 = .838 \times \frac{L_2}{100} \times 1.28$$

Solving for L_2:

$$L_2 = \frac{1.35 \times 700 \times 5.20}{.838 \times 1.28} = 4{,}581 \text{ ft which can be rounded off to 4,600 ft.}$$

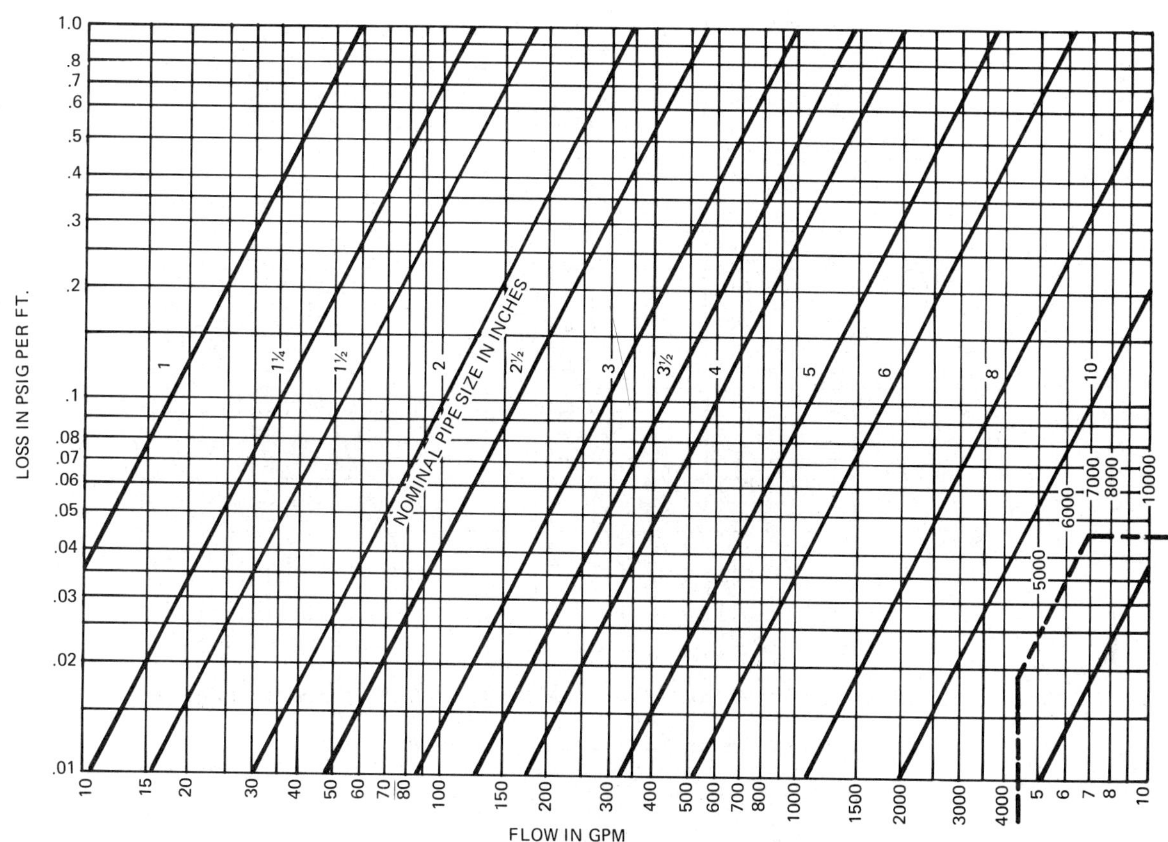

FIG. 17-2M. Friction loss in Schedule 40 steel pipe, Hazen & Williams C-120. For other c values use:

Value of c	*80*	*100*	*120*	*130*	*140*	*150*
Multiplying factor	*2.12*	*1.40*	*1.00*	*0.86*	*0.75*	*0.66*

Fittings and Valves

Friction losses resulting from flow of water through pipe valves and fittings are sometimes neglected if small compared with pipe friction. However, where significant, these losses may be determined from Table 17-2F which expresses the friction loss of the fitting as an "equivalent pipe length" having the same frictional loss as the fitting. This length is then added to the length of the pipe to which the fittings are connected to get the total functional loss of pipe and fittings.

EXAMPLE: Calculate the total friction loss for a 600 L/min flow through 200 m of nominal 2 in. (51 mm) pipe (c = 120) which incorporates three 90° elbows, one tee junction (90° turn), and two butterfly valves.

SOLUTION: Using Table 17-2F for equivalent lengths of fittings (C = 120).

Fitting or Pipe	Number	Equivalent length per fitting (m)	Total equivalent length (m)
Std. elbow	3	1.5	4.5
T-junction	1	3.1	3.1
Butterfly	2	1.8	3.6
Pipe	1	200.0	200.0
	Total equivalent pipe length		211.2 m

From Fig. 17-2N, the total friction loss for 600 L/min flow in 51 mm (2″ norm) pipe is 500 kPa per 100 m of pipe. Because frictional loss is directly proportional to length of pipe, total loss is given by:

$$(211.2/100) \times 500 = 1056 \text{ kPa}$$

The loss due to fittings is given by $(11.2/100) \times 500 = 56$ kPa which is 5.3% of the total friction loss.

The Hazen-Williams Diagram

Figure 17-2M is a graphical representation of Table 17-2D except that it is based upon a Hazen-Williams C = 120 instead of the C = 100 used in the table. It is limited in scope to pipes not over 10 in. in diameter and, because of the reduced scale, is less accurate than the table.

Figure 17-2N gives data in SI units for pipes up to 10 in. (254 mm) diameter, again based on Hazen-Williams C = 120.

EXAMPLE OF USE: What is the friction loss in 300 ft (90 m) of 8 in. cement-lined, cast iron pipe, at 1,500 gpm (5678 L/min)?

SOLUTION: From the intersection of the 1,500 gpm vertical line with the sloping 8 in. pipe diameter line in Figure 17-2M, read left horizontally for loss of head value, which is found to be 0.19 psi per ft. For 300 ft, the loss would be 300 × 0.19 psi or 5.7 psi. But the probable C value of cement-lined pipe is 140 (Table 17-2G), and the

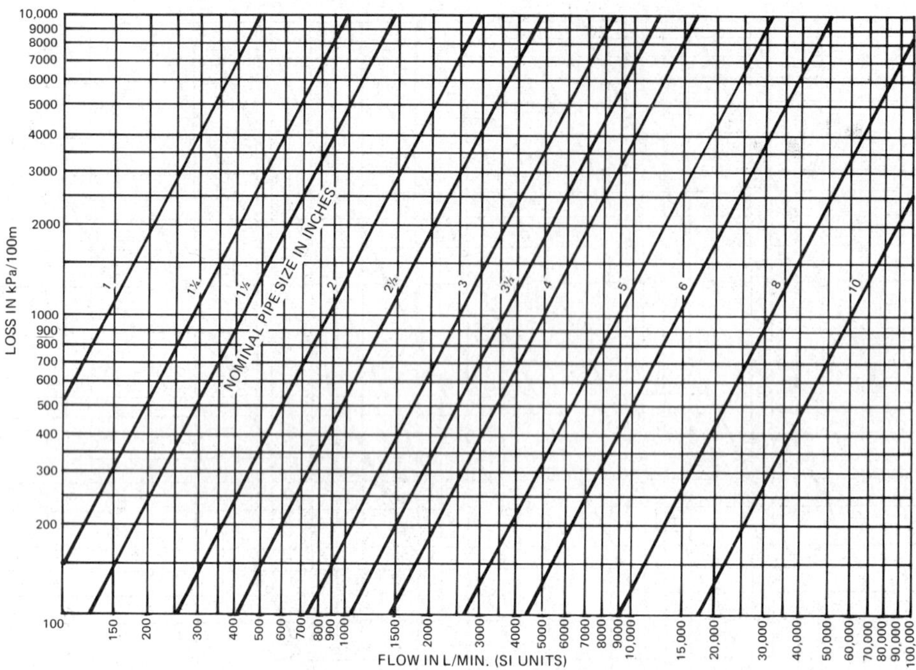

FIG. 17-2N. Friction Loss (SI units) in Schedule 40 Steel Pipe, Hazen & Williams C-120. For other c values use:

Value of c	***80***	***100***	***120***	***130***	***140***	***150***
Multiplying factor	***2.12***	***1.40***	***1.00***	***0.86***	***0.75***	***0.66***

conversion factor is 0.75 (Fig. 17-2M). Therefore, the friction loss is 5.7 × 0.75 = 4.3 psi.

METRIC EXAMPLE: What is the friction loss in 400 m of 6 in. (152 mm) 30 year old unlined cast iron pipe (C = 80) for a flow of 10,000 L/min

SOLUTION: From the intersection of the 10,000 L/min vertical line with the sloping 6 in. (152 mm) pipe diameter line in Figure 17-2N, read left horizontally for loss of head value, which is 480 kPa per 100 m. For 400 m, the loss would be 4 × 480 or 1920 kPa, based on C = 120. For the cast iron pipe C = 80 and the conversion factor is 2.12 (Figure 17-2N). Therefore, the friction loss is 1920 × 2.12 = 4070 kPa.

WATER HAMMER

Water hammer is the effect of pressure rise that may accompany a sudden change in the velocity of water flowing in a pipe. When deceleration of velocity is rapid or completely stopped, the kinetic energy of the moving water column is absorbed temporarily by elastic deformation of the pipe and by the compressibility of water. Consideration of water hammer and transient pressure surges is based on the elastic wave theories of Joukowski and Allievi. The force of water hammer is sometimes sufficient to rupture pipes, fittings, or hose lines. Theoretically, the resultant force could be infinite if the system were totally inelastic.

Pressure surges may be initiated by the closing of a valve, the stopping of a pump, or by the sudden development of an abnormal water demand when a water main breaks. Occasionally, the operation of automatic control valves in sprinkler systems may result in reversal of flow and a buildup of high pressure in the fire protection system.

The elasticity of hose tends to reduce the danger from water hammer, but the sudden closing of shutoff nozzles on long hose lines may cause a pressure rise sufficient to rupture the hose. Tests conducted by the New York City Fire Department indicated that pressure surges from closing of nozzles approximated twice the hydrant pressure. These tests point to the advantage of operating nozzle valves slowly.

Discharge lines from pumps are subject to water hammer caused by water column separation. This may occur when the pump suddenly stops (power failure, manual shutdown, etc.), or if the discharge valve is closed too quickly with the pump operating. Separation takes place somewhere downstream, especially at a summit, or where the downward slope of the pipe increases sharply. When forward movement becomes exhausted the flow reverses its direction and closes the gap.

When a pump is located at an elevation above the system outlet, a vacuum breaker in the line may provide effective control.

When there is static head on a pump at discharge, it is practically impossible to completely eliminate a watercolumn reversal. Surge suppressors may be effective, if used; likewise special type vacuum breakers designed to bypass a portion of the reverse flow water column around the check valve or control valve.

The restarting of a pump too quickly after a tripout may cause excessive surging, and installations subject to intermittent operation should be protected by time delay relays.

Simple relief valves are considered useless because

their operation is too slow to counteract the speed of the pressure rise.

The principal factors contributing to water column separation are: (1) rate of flow stoppage, either by rapid closing of a valve, or the fast deceleration of a pump; (2) length of pipe system (this determines the time that pressure continues to fall before positive pressure waves returning from the far end of the line counteracts the initial pressure drop); (3) the normal operating pressure at critical points, such as the crests of hills; and (4) the velocity of the water just before pump stoppage or valve closure occurs; (the greater the velocity, the larger the size of the void, reverse flow velocity, and the final pressure rise).

Elastic Wave Theory

The basic concepts of EWT are:

1. The magnitude of the pressure rise is proportional to the fluid velocity destroyed and to the velocity of the pressure wave.
2. The pressure rise is independent of the length and profile of the pipe.
3. The velocity of the pressure wave is the same as the velocity of sound through water.

The theoretical pressure rise when flow is stopped instantly may be calculated from the formula:

$$\Delta p = \frac{0.433av}{g}$$

in which Δp = pressure rise in pounds per square inch; a = velocity of pressure wave in feet per second; v = water flow velocity in feet per second; and g = acceleration of gravity in feet per second per second.

In metric practice the formula is:

$$\Delta p_m = \frac{9.80\ a_m v_m}{g_m}$$

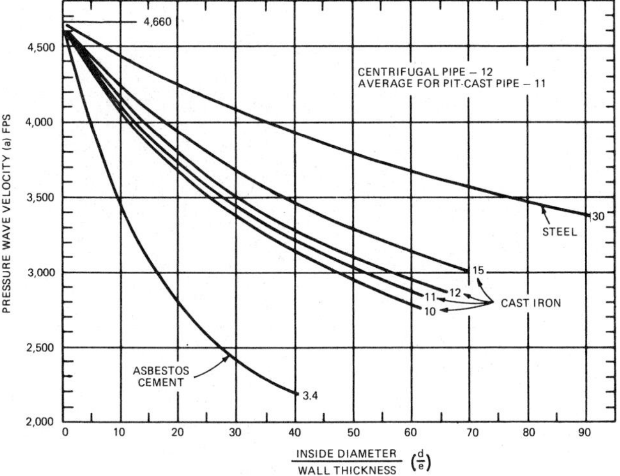

FIG. 17-2O. Surge wave velocity chart for water. The figures on the curves represent values of E in millions of pounds per square inch (bulk modules of elasticity).

in which Ap_m = pressure rise in kPa; v_m = velocity in m/sec, a_m = velocity of pressure wave mm/sec, and g_m = acceleration of gravity mm/sec^2.

In practice, the calculated Δp should be reduced to allow for valve closure characteristics, and friction loss in the pipe. Usually this is a matter of judgment and experience.

The pressure rise Δp is at maximum when the flow is stopped in a time equal to or less than the critical time of the pipe, which is the time required for the pressure wave to travel from the point of closure to the end of the pipe and return. The formula for critical time is:

$$t = \frac{2l}{a}$$

in which t = time in seconds; l = length of pipe in feet; and a = velocity of pressure wave in feet per second.

The value of a in the above formula can be calculated from the formula:

$$a = \frac{12}{\sqrt{\frac{w}{g}\left(\frac{1}{k} + \frac{d}{Ee}\right)}}$$

in which

- w = weight of water, pounds per cubic foot (L/m^3)
- g = acceleration of gravity, feet per second per second (m/sec^2)
- k = bulk modulus of compressibility of water, pounds per square inch (kPa)
- E = Young's modulus of elasticity of pipe wall material, pounds per square inch (kPa)
- e = thickness of pipe wall, inches (mm)
- d = inside diameter of pipe, inches (mm)

To avoid calculating a, use the chart in Figure 17-2O.

The calculated pressure rise in a 6 in. cast iron pipe is about 60 psi per foot (1357 kPa/m) of arrested velocity.

The water hammer potential of distribution systems, especially those with automatic pumps, should be examined, and practical steps taken to reduce the probability of destructive pressure surges. Valves and hydrants should be maintained in good condition, and operated carefully. Public utility networks with inadequate pipe capacity might be subject to flow reversals and surging. Remote-controlled power-operated valves should be carefully timed to prevent too-fast closing (never less than 5 sec).

Bibliography

References Cited

Hydraulic Institute. 1954. *Pipe Friction Manual*. Hydraulic Institute. New York.

Additional Readings

Albertson, Maurice L., Barton, James R., and Simons, Daryl B., *Fluid Mechanics for Engineers*, Prentice-Hall, Inc., Englewood Cliffs, N.J., 1960.

Benedict, Robert, P., *Fundamentals of Pipe Flow*, Wiley, New York, 1980.

Casey, James F., Ed., *Fire Service Hydraulics*, 2nd ed., Dun-Donnelley, New York, 1970.

Cozad, F. Dale, *Water Supply for Fire Protection*, Prentice-Hall, Englewood Cliffs, N.J., 1981.

"Data Sections," *Standards of Hydraulic Institute*, 11th ed., Hydraulic Institute, New York, 1965.

Davis, Calvin, V. and Sorensen, K. E., eds., *Handbook of Applied Hydraulics*, McGraw-Hill, New York, 1969.

"Flow of Fluids Through Valves, Fittings and Pipe," *Crane Technical Paper No. 410-C*, Crane Co., Chicago, 1969.

Freeman, John R., "Experiments Relating to Hydraulics of Fire Streams," *Transactions of American Society of Civil Engineers*, 1889, pp. 303–482.

Freeman, John R., "The Nozzle as an Accurate Water-meter," *Transactions of American Society of Civil Engineers*, Vol. 24, 1891, pp. 492–527.

Giles, Ranald V., *Fluid Mechanics and Hydraulics*, 2nd ed., Schaum's Outline Series, McGraw-Hill, New York, 1962.

Hickey, H. *Hydraulics for Fire Protection*, NFPA, Boston, 1980.

Howard, C. D. D., "Pipe Friction Coefficients from Measured Velocity Profiles," *American Water Works Association Journal*, Vol. 59, No. 5, May 1967, pp. 645–650.

Jedlicka, James R., "Pressure Surge Protection," *Fire Technology*, Vol. 2, No. 3, August, 1966, pp. 239–245.

King, Horace W., Wisler, Chester O., and Woodburn, James G., *Hydraulics*, 5th ed., John Wiley & Sons, New York, 1948.

"Hydraulics of Pipelines," Chapter 5 in *AWWA M11 Steel Pipe Manual*, American Water Works Association, New York, 1964.

King, H. W., and Brater, E. F., *Handbook of Hydraulics*, 5th ed., McGraw-Hill, New York, 1963.

Lescovich, J. E., "The Control of Water Hammer by Automatic Valves," *American Water Works Association Journal*, Vol. 59, No. 5, May 1967, pp. 632–644.

McJunkin, F. E., and Vesilind, P. A., "Practical Hydraulics for the Public Works Engineer," reprinted from *Public Works Magazine*, Sept., Oct., Nov., 1968.

Mahoney, Eugene F., *Fire Department Hydraulics*, Allyn and Bacon, Boston, 1980.

Purington, Robert G., *Fire-Fighting Hydraulics*, McGraw-Hill, New York, 1980.

Richard, R. T., "Water Hammer," *Transactions of American Society of Mechanical Engineers*, Vol. 78, No. 6, August 1956.

Rosenhan, A. K., "Water-Net-A Computerized Design Aid," *Fire Technology* Vol. 4, No. 3, August 1968, pp. 179–184.

Rouse, Hunter, ed., *Engineering Hydraulics*, John Wiley & Sons, New York, 1950.

Streeter, Victor L., ed., *Handbook of Fluid Dynamics*, McGraw-Hill, New York, 1961.

"Water Hammer and Surge," Chapter 7, *Ibid.*

Williams, G. S. and Hazen, A., *Hydraulic Tables*, John Wiley & Sons, New York, 1933.

Winn, W. P. "Techniques in Water Hammer Control and Surge Suppression," *American Water Works Association Journal*, Vol. 59, No. 5, May 1967, pp. 620–624.

Wood, Clyde M., *"Automatic" Sprinkler Hydraulic Data*, 1961 ed., "Automatic" Sprinkler Corporation of America (reprinted 1974).

Wood, Don J., "Calculation of Water Hammer Pressure Due to Valve Closure," *American Water Works Association Journal*, Vol. 60, No. 11, Nov. 1968, pp. 1301–1306.

WATER SUPPLY REQUIREMENTS FOR FIRE PROTECTION

Revised by John R. Anderson

The amount of water needed and the economics of supplying it are basic questions underlying good planning for fire suppression.

This chapter gives information on planning for good public water supply systems and how to evaluate them for their adequacy and reliability in supplying water in sufficient amounts overall to meet anticipated fire flows required for the severity of hazards encountered on a community-wide level. The chapter does not cover the specifics of water supply requirements for automatic extinguishing systems within individual properties. They are covered in the following chapters of Section 18: Chapter 2, "Water Supplies for Sprinkler Systems;" Chapter 4, "Residential Sprinkler Installations;" Chapter 6, "Standpipe and Hose Systems;" and Chapter 7, "Water Spray Protection."

The components that make up a water supply system are covered in Chapter 4, Water Distribution Systems while testing procedures to determine adequacy of flows will be found in Chapter 8, Test of Water Supplies, both in this section of the HANDBOOK.

Most public water supply systems serving a substantial number of customers are designed for a dual purpose: (1) to supply water for normal domestic demands, such as drinking and sanitary purposes, as well as for processing and industrial uses, and (2) to provide water to fire hydrants for emergency use by fire departments in supplying pumping engines and to supply fixed fire protection systems, such as automatic sprinklers, foam systems, and fire standpipe systems.

Water systems supplying both normal consumption demands and fire protection requirements must satisfy the design objective of providing a system with the capability of meeting the simultaneous demand rates for both purposes with reliability. To meet this goal it is necessary to focus upon the variations in normal consumption of water on the basis of the time of the year, the day of the week, and even during the hours of the day. Obviously, in a given system, as more water is used for normal consumption demands the less remains for fire protection. Normal consumption demands are usually expressed in the following terms:

1. Average daily consumption—the average of the total amount of water used each day during a 1 year period.
2. Maximum daily consumption—the maximum total amount of water used during any 24 hour period in a 3 year period. (Unusual situations which may have caused an excessive use of water, such as refilling a reservoir after cleaning, should not be considered when determining this figure.)
3. Peak hourly consumption—the maximum amount of water used in any given hour of a day.

The maximum daily consumption is normally about 1.5 times the average daily consumption. The peak hourly rate varies from two to four times a normal hourly rate. The effect these varying consumption rates have on the ability of the system to deliver required fire flows varies with the system design. But both maximum daily consumption and peak hourly consumption should be considered to ensure that water supplies and pressures do not reach dangerously low levels during these periods, and that adequate water will be available if there is a fire.

WATER FOR FIRE FIGHTING

Historically, water supply systems for cities and towns were developed primarily to provide for drinking water and water for sanitary purposes rather than for fire protection. However, it was found that in large cities requiring substantial amounts of water for domestic purposes there was usually sufficient water to provide a useful supply for fire fighting purposes.

All this led to inquiries into the cost of waterworks that could provide water for fire fighting as well as other uses. A number of distinguished engineers associated with individual waterworks examined the problem and presented their findings in technical papers at engineering society meetings. Papers by J. Herbert Shedd (Shedd 1889), J. T. Fanning (Fanning 1892), and Emil Kuichling (Kuichling 1897) give the details of discussions in which standards now followed in American and Canadian waterworks were first developed (Table 17-3A).

Mr. Anderson is a professional consulting engineer based in Marshfield Hills, MA.

TABLE 17-3A. Estimates of Fire Flow

	Number of Fire Streams Required Simultaneously				
Populations Thousands	Shedd 1889	Fanning 1892	Freeman 1892	Kuichling 1897	NBFU 1910
1			2–3	3	4
4		7		6	8
5	5		4–8	6	9
10	7	10	6–12	9	12
20	10		8–15	12	17
40	14		12–18	18	24
50		14		20	26
60	17		15–22	22	28
100	22	18	20–30	28	36
150		25		34	44
180	30			38	48
200			30–50	40	48

Sources (these authorities define streams slightly differently as described in accompanying text, but the streams were of the order of 200 gpm to 300 gpm):

Shedd, J. Herbert, discussion on a paper by Sherman, William, B., *Ratio of Pumping Capacity to Maximum Consumption* (Shedd 1889).

Fanning, J. T., *Distribution Mains and the Fire Service* (Fanning 1892).

Kuichling, E., *The Financial Management of Water Works* (Kuichling 1897).

Freeman, John R., *The Arrangement of Hydrants and Water Pipes for the Protection of a City Against Fire* (Freeman 1892).

Figures furnished by National Board of Fire Underwriters (Metcalf, Leonard, et al. 1911).

The Number of Hose Streams

The starting point for computing the cost of water for fire protection was to estimate the number of hose streams that a fire department might need for fire fighting. This was usually estimated on the basis of the central portion of the city where the largest buildings were located and where there was the greatest building congestion. The number of streams at that time (the late 19th Century) was found to be related, in a very rough way, to the population. Shedd's proposal, the first, used hose streams discharging 200 gpm (757 L/min). He suggested that a community of 5,000 would need about five such streams and that the needs of other cities could be graduated up to 30 streams in a city of 180,000. Fanning proposed streams requiring about 54 psi (372 kPa) pressure. His figures were similar to Shedd's, beginning at seven streams for a community of 4,000 and going up to 25 streams for a city of 150,000.

Kuichling suggested a formula in which the number of streams required would be the square root of the population in thousands multiplied by 2.8. There were arithmetical differences as to how these estimates worked out for individual cities, but they were of the same general order (Table 17-3A). Most important, they did provide a basis from which the waterworks designers could estimate costs.

The most important paper published during this time was John R. Freeman's "The Arrangement of Hydrants and Water Pipes for the Protection of a City Against Fire" (Freeman 1892). Freeman had done the fundamental work on water flow through hose and nozzles, so he was able to pin down the definition of a standard fire stream to one with a discharge of 250 gpm (946 L/min) at 40 to 50 psi (276 to 345 kPa) pressure. He said that the relationships suggested by Shedd and Fanning between population and the number of streams required were of the right order, but he did not think the needs of individual cities could be quite so definitely pinned down. He suggested two to three streams as a minimum at 1,000 population graduated up to thirty to fifty at 200,000 (Table 17-3A). Most significantly, he warned: "Ten streams, or as large a proportion thereof as the financial consideration will permit, may be recommended for a compact group of large, valuable buildings, irrespective of a small population."

Engineering: Distributing Network, Hydrant Spacing, Storage

Freeman noted a fundamental difference between systems designed to supply ordinary water needs and those for fire protection. Fire draft required concentration of the water, whereas domestic draft was a matter of distribution.

Freeman asserted that if a water system was to supply fire protection needs, the distribution system should be designed to concentrate the needed amounts of water. Small pipes were sufficient for distribution, but larger ones were needed for concentration of supply to fire streams. He suggested 6 in.* diameter pipe as the minimum for residential districts, and noted that 8 in. pipe was adequate only if it formed part of a network of distributing pipes whose intersections were not far apart.

Another important point Freeman made was that hydrants should be placed where they could concentrate streams at specific blocks or groups of buildings, rather than arbitrarily a certain number of feet apart on the street mains. His work on hose streams showed how long hose lines reduced the water that can be delivered promptly to a fire. He therefore suggested a working rule for hydrant spacing of 250 ft (76 m) between hydrants in compact mercantile and manufacturing districts, and 400 (122 m) to 500 ft (152 m) in residential districts. These working rules can still be used as guides for good design.

Freeman further insisted that fire supply should be in addition to maximum domestic consumption and laid the foundation for eventual recognition of this principle. He also calculated how much water should be stored in standpipes or elevated reservoirs. He figured that flow for all of the hose streams required should be supplied from a reliable source, such as an elevated storage reservoir, for at least 6 hr when the system was also furnishing maximum demands for domestic and other uses. He also calculated that to supply the combined fire and domestic needs in a system provided with reliable pump capacity, a 1 hr supply in a standpipe or elevated reservoir would be acceptable.

The Insurance Grading Schedule

As early as 1889, the NBFU (National Board of Fire Underwriters) began to make fire protection surveys of municipalities. This work was intensified in 1904 after a conflagration in Baltimore. Today communities in all but a few states are surveyed by the ISO (Insurance Services Office), successor to the NBFU. The ISO survey is called the "Fire Suppression Rating Schedule," and is discussed in more detail later in this chapter. It includes a procedure for establishing fire flows that is different from the original NBFU survey, which includes a cursory evaluation of a municipality's water system and its ability to supply water

*Nominal size; for convenience, one inch equals 25.4 mm.

for fire protection to specific locations. The ISO survey is solely for the purpose of establishing insurance rates.

FIRE PROTECTION REQUIREMENTS IN WATER SYSTEMS

The capacity of a water system is determined by the total amount of water it must furnish. This is the sum of: (1) water required for domestic or industrial uses, and (2) water required for fire service. In small towns, the requirements for fire protection exceeds other requirements.

In North American cities, a public water system is expected to furnish water for many uses. In some cities there may be a heavy industrial demand. At the same time, demands for air conditioning and lawn sprinkling, for example, can also affect the required capacity of the system. The adequacy of a public water system for fire protection cannot be taken for granted, and the other demands must be determined to estimate their effects on the capacity of the system.

A joint report of committees of the American Society of Civil Engineers, the American Water Works Association and others, suggested that the maximum general service demand on a waterworks system be taken as the peak hourly demand during a test year (ASCE 1951). This, the report noted, was the only figure which can fairly be compared with the maximum fire flow requirement.

Evaluating System Capacity

In most large cities, the peak hourly rate exceeds the maximum daily consumption rate plus fire flow, and therefore is the controlling factor in the supply system design. In smaller communities, however, the reverse is true with the maximum daily consumption rate plus fire flow being the controlling factor. For many years water consumption has been increasing in most municipalities, resulting in increased peak hourly rates. Consequently there has been an increase in the number of municipalities in which the peak hourly rate controls designs of the supply system.

In all areas served by the distribution system, fire flows are a very important consideration, and, in many instances, govern the size of pipe used in these locations. In all systems there should be a supply sufficient to provide for automatic sprinklers and other automatic fire protection systems in addition to the other demand rates imposed upon the system. For example, many smaller cities and large towns restrict lawn watering in summer months to specified periods, usually 2 to 4 hrs in the evening. In many water supply systems, the demand rates imposed by lawn watering are excessive, depleting storage facilities and reducing pressure throughout the system for many hours. In these situations, there would be little or no water available for fire protection systems, particularly at higher elevations.

Pressure Characteristics of Systems

The pressures for which systems are normally designed are the product of several practical considerations which attempt to provide pressures adequate for water supplies both for domestic consumption and for fire protection. If either demands special ranges of pressure, they too can be provided. Pipe and related fittings and methods of using them will allow almost any desired range.

San Francisco, for example, has a separate system, designated the "high pressure system," under the control of the fire department. All of the pipe is heavy cast iron, tar coated and lined, and tested on installation and repair to 450 psi (3103 kPa). Two steam operated pump stations can pump water from San Francisco Bay into the system, and 20,000 gpm (75 700 L/min) at 250 psi (1724 kPa) can be delivered to most of the principal mercantile district. San Francisco provided this system primarily because an earthquake might put the regular public water system out of service. A few other cities have provided similar high pressure systems.

Modern fire department pumpers make heavy streams and high pressures available from ordinary water systems where adequate volume is provided. Cities that formerly had separate systems of fire mains, operating at so-called high pressures, now generally have these operating at normal public water pressures. An advantage is retained because the second system, although not at high pressure, is still available.

Public water systems reflect a compromise on the question of pressures. Pressures in the range of 65 to 80 psi (448 to 552 kPa) are common. This range is adequate for ordinary consumption in buildings up to about ten stories. It will provide a good supply of water for automatic sprinkler systems in buildings of about four stories in which occupancies are classified as "ordinary." Where pressures of this order are provided, it is reasonably easy to compensate for local fluctuations in draft.

Because of the increased cost of energy, greater care is being given to the water pressure that systems should provide. Reduction in water pressure to lower values will substantially reduce pumping costs. Before a general reduction in water pressure is made, however, a study should be conducted as to the effects of a reduction upon fire supply to sprinklers and other fixed fire protection systems. Therefore, when a reduction in pressure is planned, it is imperative that the system be designed to meet anticipated excessive demand rates or steps be taken to lower the demand rates to a point where they will be within the capability of the system.

Under the conditions cited, it might be appropriate to design and construct the system to supply simultaneously peak hourly consumption in addition to fire flows for fixed fire extinguishing systems.

A minimum residual pressure of least 20 psi (138 kPa) should be maintained at hydrants delivering the required fire flow. Pumpers can be operated but with difficulty where hydrant pressures are less. Where hydrants are well distributed and of the proper size and type (so that friction losses in the hydrant and suction line will not be excessive), it may be possible to set 10 psi (69 kPa) as the minimum pressure. Sufficient hydrant pressure should be maintained to prevent developing a negative pressure in the street mains, which might cause back siphonage of polluted water from some interconnected source. Using residual pressures less than 20 psi (138 kPa) is prohibited by most state health departments.

Pressures in a public water system may be considered excessive as they approach 150 psi (1034 kPa). As pressures increase, they tend to cause leaks in domestic plumbing, and special attention is required to restrain pipelines in the ground. Pipe and fittings used in ordinary public water systems are designed for maximum working pressures of 150 psi (1034 kPa), but it is not good practice to operate with pressures that high. Pressure reducing valves can be used in some sections of a system where variations in topography result in excessive pressures. Individual water services to buildings may require pressure reducing valves to keep the pressure on domestic piping at safe levels.

Systems for Higher Elevations

When water must be supplied to high elevations, a separate water distribution system is usually provided for the elevated section so that normal pressures are maintained. In such cases, the elevated area should be provided with its own water storage facility, and pumps may be provided to boost the water from other parts of the system. Likewise, the upper stories of a high building should be provided with water supply systems in the building itself. These systems have the same requirements as areas on a hill. High rise structures are normally divided into a number of pressure zones. Zones of more than twelve stories tend to get outside the normal pressure ranges. In any case, each pressure zone must have storage of water in amounts needed for the sprinkler service or hose streams to be provided, and a system of pumps so that each zone is supplied from the zone below. Care should be taken to ensure that pumps will be able to operate even during power failures.

(For information and guidance concerning water supplies for high rise structures, see NFPA 13, *Standard for the Installation of Sprinkler Systems*; NFPA 14, *Standard for the Installation of Standpipe and Hose Systems*; and NFPA 20, *Standard for the Installation of Centrifugal Fire Pumps*.)

CALCULATING FIRE FLOWS

For many years the NBFU formula (see Table 17-3A) was commonly used as a guide to determine the fire flow required in the downtown business districts of municipalities. The formula

$$G = 1020 \sqrt{P}\,(1 - 0.01 \sqrt{P})$$

gave the fire flow, G, in gallons per minute as a function of the population, P, in thousands.

In making fire protection surveys, the fire flow requirements in the sections of the municipalities outside the downtown business district were estimated by the engineers of the NBFU and insurance bureaus.

As cities became more decentralized, the formula based on population became less reliable as a guide for the fire flow needed downtown. In addition, it became apparent that a guide to engineering judgment was needed for the other sections of the cities. In 1948, a paper by A. C. Hutson, assistant chief engineer of the NBFU, provided some specific suggestions for estimating fire flow requirements in these sections based on type of construction and area of building (Hutson 1948).

The Fire Suppression Rating Schedule

The latest developments in estimating fire flow requirements are found in the Insurance Services Office's "Fire Suppression Rating Schedule" (ISO 1980). It provides guidance for estimating fire flow requirements for specific structures for insurance rating purposes. The basic formula in the schedule is:

$$NFF_i = (C_i)(O_i)(X+P)_i$$

Where NFF_i is Needed Fire flow in gallons per minute (L/min); C_i is a construction factor which depends upon the type of construction of the structure under consideration; O_i is an occupancy factor which depends upon the determination of the type of occupancy in terms of rates of combustibility; and $(X+P)_i$ is an exposure factor which depends upon the extent of exposure from and to adjacent structures.

The needed flow should not exceed 12,000 gpm (45 425 L/min) nor be less than 500 gpm (1893 L/min). The practical reason for these figures is that manual fire fighting methods using hose streams and heavy stream appliances are not likely to need a larger supply considering the general arrangement of buildings and the availability of hydrants. However, the possibility of a second simultaneous fire in the largest cities is considered.

For groupings of one family and small two family dwellings not more than two stories high, the short method of determining required fire flow given in Table 17-3B may be used. The required fire flow should be available with consumption at the maximum daily rate. The number of hours during which the required fire flow should be available varies from 2 to 10 hours as indicated in Table 17-3C.

TABLE 17-3B. Fire Flows for Groups of Dwellings

Exposure Distances		Suggested Required Fire Flow	
Ft	m	gpm	L/min
Over 100	30	500	1893
31 to 100	9.5–30	750–1,000	2839–3785
11 to 30	3.4–9.1	1,000	3785
10 or less	3 or less	1,500	5678

There are fires where quantities of water in excess of the required fire flow are used. Water supplies of 50,000 gpm (190 000 L/min) or greater have been used in fire suppression, but to design systems capable of delivering flows of that magnitude in the average community for a possible unusual situation is too expensive.

ADEQUACY AND RELIABILITY OF SUPPLY

The adequacy of any given water supply system can be determined by engineering estimates. The source, including storage facilities in the distribution system, must be sufficient to furnish all the water that combined fire and domestic needs may call for at any one time. Arrangement of the supply works and details of the pumping facilities may limit the adequacy of the supply or affect its reliability.

In a pumping system, a common arrangement is to have one set of pumps that takes suction from wells or from a river, lake, or other body of water. If the water does not have to be filtered, the pumps may discharge directly into the distribution system. Where filtration or other treatment is necessary, pumps take suction from the primary or raw water source and discharge to sedimentation

TABLE 17-3C. Duration of Required Fire Flow

Required Fire Flow					Required Fire Flow				
gpm	L/min	Million gallons per day	Million liters per day	Duration hours	gpm	L/min	Million gallons per day	Million liters per day	Duration hours
1,000	3785	1.44	5.45	2	4,500	17 034	6.48	24.53	4
1,250	4732	1.80	6.81	2	5,000	18 927	7.20	27.25	5
1,500	5678	2.16	8.18	2	5,500	20 820	7.92	29.99	5
1,750	6624	2.52	9.54	2	6,000	22 712	8.64	32.71	6
2,000	7571	2.88	10.90	2	7,000	26 498	10.08	38.16	7
2,250	8517	3.24	12.26	2	8,000	30 283	11.52	43.61	8
2,500	9463	3.60	13.63	2	9,000	34 069	12.96	49.06	9
3,000	11 356	4.32	16.35	3	10,000	37 854	14.40	54.51	10
3,500	13 249	5.04	19.08	3	11,000	41 639	15.84	59.96	10
4,000	15 142	5.76	21.80	4	12,000	45 425	17.28	65.41	10

basins or other facilities and then to filter beds. After processing, the water flows to clear water reservoirs from where a second set of pumps takes suction and discharges the water directly into the supply system. Unfortunately, failure of any part of the equipment may affect the entire system. This is usually taken care of by duplication of units and by arrangement of the plant to facilitate repairs.

When assessing the reliability of the supply works, features which should be evaluated are: minimum yield; frequency and duration of droughts; condition of intakes; possibility of earthquakes, floods, and forest fires; ice formations, silting up or shifting of river channels; and absence of guards or watchmen where needed to protect the facility, from physical injury. Reservoirs out of service for cleaning, and the interdependence of parts of waterworks, also affect reliability. The condition, arrangement, and dependability of individual units of plant equipment, such as pumps, engines, generators, electric motors, fuel supply, electric transmission facilities and similar items, are also factors. Pumping stations of combustible construction are subject to destruction by fire unless protected by automatic sprinkler systems.

Duplication of pumping units and storage facilities, and arrangement of mains and distributors so that water may be supplied to them from more than one direction, are measures that can assure continuous operation. The importance of duplicate facilities is shown by the frequency of their use.

FUTURE REQUIREMENTS FOR DETERMINING FIRE FLOW

The amount of water needed to control and extinguish a fire in a given property cannot be established currently in precise terms. Better fire experience data bases should make it possible to tailor fire flows more specifically to conditions that might be expected at the time of a fire. Better analysis may indicate a need to increase fire flow beyond what is presently required, or it may result in a water system design based upon a balance between the risk involved and the economics of maintaining the water system.

The American Water Works Association (AWWA) at a meeting in St. Louis in 1981 formed a working committee on fire protection under the supervision of the AWWA's Distribution Division. The committee is in the process of developing a "Manual of Standard Practice for Considerations which Affect Fire Protection and Suppression." The document will be one of a comprehensive nature which will include all elements of water supply systems in terms of adequacy and reliability in addition to providing a procedure for determing required fire flows.

The Role of Codes and Ordinances

Fire prevention codes can effectively limit hazards and ignition sources within buildings which in turn not only limits the number of fires, but the size of fires through the control of combustibles in a fire area. A good building code further reduces the chance for a serious fire by requiring construction materials and building assemblies which will contain a developing fire to a given area. Codes alone can reduce considerably the amount of water needed for fire fighting. Zoning ordinances that establish distances between properties can be effective in controlling exposure situations.

The Role of Fire Detection and Extinguishing Systems

The increased use of automatic extinguishing systems, whether they use water or some other agent, will affect the quantities of water required. However, until more widespread use is made of early warning systems and automatic extinguishing systems, it will not be possible to equate the effect of these systems to required fire flow. Water supply requirements are just one factor in a system that determines what the potential for a fire is, how extensive the fire will be, and the measures needed to suppress it. Research will someday measure all these factors and permit establishing fire flows on the basis of thoroughly researched and documented principles.

Bibliography

References Cited

ASCE. 1951. "Fundamental Considerations in Rates and Rate Structures for Water and Sewage Works: A Joint Report of Committees of the American Society of Civil Engineers and the Section of Municipal Law of the American Bar Association and of Representatives of the American Water Works

Association, National Association of Railroad and Utilities Commissioners, Municipal Finance Officers Association, Federation of Sewage Works Association, American Public Works Association, and Investment Bankers Association of America." ASCE Bulletin No 2. American Society of Civil Engineers, NY.

Fanning, J. T. 1892. Distribution Mains and the Fire Service. *Proceedings of the American Water Works Association*. Vol 12. p 61.

Freeman, John R. 1892. "The Arrangement of Hydrants and Water Pipes for the Protection of a City Against Fire." *Journal of the New England Water Works Association*. Vol 7. p 49.

Hutson, A. C. 1948. "Water Works Requirements for Fire Protection." *Journal of the American Water Works Association*. Vol 40, No 9. p 936. Also reprinted in Special Interest Bulletin No. 266, National Board of Fire Underwriters (now American Insurance Association), NY.

ISO. 1980. *Fire Suppression Rating Schedule*. Insurance Services Office, New York.

Kuichling, E. 1897. "The Financial Management of Water Works." *Transactions of the American Society of Civil Engineers*. Vol 38. p. 16.

Metcalf, L, Kuichling, E., and Hawley, W. C. 1911. "Some Fundamental Considerations in the Determination of a Reasonable Return for Public Fire Hydrant Service." *Proceedings of the American Water Works Association*. Vol. 31. pp 55.

Shedd, J. Herbert. 1889. Discussion on a paper by William B. Sherman, "Ratio of Pumping Capacity to Maximum Consumption." *Journal of New England Water Works Association*. Vol 3. p 113.

NFPA Codes, Standards, Recommended Practices and Manuals. (See the latest *NFPA Codes and Standards Catalog* for availability of current editions of the following documents.)

NFPA 13, *Standard for the Installation of Sprinkler Systems*.

NFPA 14, *Standard for the Installation of Standpipe and Hose Systems*.

NFPA 20, *Standard for the Installation of Centrifugal Fire Pumps*.

Additional Readings

Babbit, Harold E., and Doland, James J., *Water Supply Engineering*, 6th ed., McGraw-Hill, NY, 1962.

Blake, Nelson M., *Water for the Cities*, Syracuse University Press, Syracuse, 1956.

Carl, Kenneth J., Young, Robert A., and Anderson, Gordon C., "Guide for Determining Fire Flow Requirements," *American Water Works Association Journal*, Vol. 65, 1973, pp. 335-344.

Carl, Kenneth J., and Anderson, Gordon C., "The 1973 Grading Schedule for Municipal Fire Protection," *American Water Works Association Journal*.

Engineering and Design, Water Supply for Fire Protection, Office of the Chief of Engineers, Department of the Army, Washington, DC, 1958.

Fair, G. M., Geyer, J. C., and Okum, D. A., *Water and Wastewater Engineering, Water Supply and Wastewater Removal*, Vol. 1, John Wiley & Sons, New York, 1966.

Laughlin, Jerry W., and Williams, Connie, eds., *Water Supplies for Fire Protection*, 3rd ed., IFSTA, Oklahoma State University, Stillwater, 1978.

Maatman, Gerald L., "Systems Approach to Evaluating Public Fire Fighting Operations," *Fire Technology*, Vol. 1, No. 3, Aug. 1965, pp. 187-193.

Patek, N. J., "Hydraulics for Fire Protection," *Fire Journal*, Vol. 67, No. 2, Mar. 1973, pp. 13ff.

Salzberg, F., Vodvarka, F. J., and Maatman, G. L., "Minimum Water Requirements for Suppression of Room Fires," *Fire Technology*, Vol. 6, No. 1, Feb. 1970, pp. 22-28.

Smith, Paul D., "Elements of Fire Protection Systems for Isolated Facilities," *Fire Journal*, Vol. 66, No. 4, July 1972, pp. 45-51.

Smith, Paul D., "What are the "Real" Fire Flow Requirements," *Fire Journal*, Vol. 69, No. 2, Mar. 1975, pp. 93-96.

WATER DISTRIBUTION SYSTEMS

Revised by John R. Anderson

Most public waterworks systems of any size are designed to provide water for fire protection and normal consumption demands. Many water distribution systems on private property providing water for fire protection also supply water for sanitary purposes. In some large private plant installations, extensive water distribution systems exist primarily to provide process water for elements of manufacturing or purification and additionally provide water for fire protection. In other private plants, large or small, water distribution systems exist for the sole purpose of providing fire protection.

This chapter provides information on components that make up a system for the distribution of water from sources of supply to specific areas of use for the purpose of providing for fire protection, and, in dual systems, simultaneously supplying water for normal consumption demands. Specifically, subjects covered are: the sources of supply, the distributions systems themselves including the rules for laying pipe, the types of pipe and equipment used in the systems for control and distribution purposes. The principles are the same whether the distribution system is owned by a municipality, a public utility, or whether it is a privately owned system providing water to a single property.

Water supply requirements—the amount of water necessary for adequate fire control—is covered in Chapter 3, "Water Supply Requirements for Fire Protection," of this Section, while test procedures for adequacy of supply for fire protection will be found in Chapter 8, "Test of Water Supplies."

A list of NFPA standards and recommended practices having application to water distribution systems will be found in the bibliography at the end of the chapter. The list also includes standards and manuals of the American Water Works Association, American National Standards Institute, Underwriters Laboratories Inc., and Underwriters Laboratories of Canada that refer to water distribution systems and their components.

Mr. Anderson is a professional consulting engineer based in Marshfield Hills, MA.

SEPARATE AND DUAL DISTRIBUTION SYSTEMS

There are many advantages to providing a completely separate distribution system for fire protection; there are also some disadvantages.

Among the advantages are:

1. Complete control over the system by those responsible for fire protection.
2. Proper design of the system to meet all fire demands.
3. The system is not subject to reduced supply when increases in population increase the use of water for normal demands or when processes using increased amounts of water are experienced.
4. There is little danger of introducing a nonpotable fire supply into a potable supply through faulting cross connections.

One of the disadvantages, however, is that pumps, prime movers, and other machinery and equipment are usually in a "non operating" mode. This equipment must be started, controlled and operated successfully.

In combined systems an advantage is that pumping machinery is in operation almost continuously and in many cases equipment is duplicated. Further, in combined systems, a malfunction surfaces quickly as customers will likely be without water. In addition, public water systems have trained personnel readily available and not just for emergency operations.

In systems providing for normal consumption demands and for fire protection, that portion of the system extending into private property should be isolated from the public portion of the system. This is done by the installation of listed or approved backflow prevention devices, thereby providing some measure of protection from contamination of the potable source.

Systems, on private property that provide water for process use should be made available to fire systems through the opening of normally closed valves or other cross ties. Process water systems usually have substantial capacities and could be of significant value in augumenting the adequacy and reliability of fire protection systems.

FUNCTION OF SUPPLY AND DISTRIBUTION SYSTEMS

It is generally accepted that water systems are subdivided into two divisions: supply systems and distribution systems. However, in small water systems there may be no way of differentiating between supply and distribution systems since the functions of both may be carried out in a single element of the system.

Supply System: The supply portion of a water system is usually that part of the system where the source or sources of supply are found. It also includes the storage and transmission of that supply through large conduits and aqueducts, and, in some cases, includes the arterial feeders extending to the distribution system.

Distribution Systems: The distribution system is that portion of the works that actually delivers water to the individual consumer connections and to which fire hydrants are attached.

SOURCES OF SUPPLY

Sources of supply have two major divisions—ground water supplies and surface water supplies.

Ground Water Supplies

Of the total quantity of fresh water available for use in the world, by far the large majority is available from ground water supplies. Ground water supply is that water that percolates into the ground from precipitation, and is stored in underground strata. The underground stratum holding water can be defined as an aquifer. The free surface of water in the underground stratum is referred to as the water table. The height of the water table varies throughout the year depending upon variations in precipitation, water movement in the aquifer, and water withdrawn from the aquifer through springs or wells.

There are locations, now greatly diminished in number because of usage, where water is stored in the underground strata under a positive head. When this aquifer is penetrated, water will rise to an elevation greater than that in the aquifer resulting, in free flow at the surface.

The amount of water in the underground strata and its availability depends upon the nature of the material of the strata. For example, the porosity of natural sand and gravel may exceed 40 percent while the porosity of igneous rocks is approximately 1 percent.

While porosity is important, perviousness, i.e., the ability of a substance to allow water to flow through it, is of greater importance insofar as the movement of water through the strata to drilled or dug wells is dependent upon the perviousness of the soil. For example, some clays have a porosity in excess of 40 percent but are impervious to waterflow.

The supply available from underground aquifers to wells is influenced, among other factors, by the precipitation falling on the area of recharge and the amount of water drawn from all of the wells penetrating the same aquifier. When the aquifier is of substantial size, static water levels usually lag the effects of drought and also lag the effects of increased preciptation. There are many different types of wells that may be constructed; however, high capacity wells that are used for municipal water supplies are usually drilled and are equipped with deepwell turbine or submersible pumps.

NFPA 20, *Standard for the Installation of Centrifugal Fire Pumps*, permits the installation of vertical turbine pumps in properly developed and tested wells under conditions described in the standard.

Surface Supplies

Surface supplies consist of rivers, lakes, streams, and impounded supplies. As with underground supplies, the availability and reliability of the supply is dependent upon precipitation falling within the drainage area or watershed of the supply. Usually, however, surface supplies respond more quickly to diminished precipitation or drought. Water levels may vary substantially between wet and dry periods. At certain times of the year, there may be excessive run off from the watershed resulting in highwater levels at intake structures and low lift pumping stations. At other times, under drought conditions, water levels in surface supplies could be so low as to require the installation of pumps in a reservoir discharging into normally gravity supplied intake structures. Large surface reservoirs supplied by substantial water sheds or runoff areas are reliable sources provided water consumption demands do not increase beyond the recharge capablities of the watershed.

There are several factors that may affect the operation of the intake structures of surface supplies, particularly in colder climates. For example, ice formation is a hazard. There are several kinds of ice which may form and affect the functioning or threaten the stability of intake structures or cribs. Anchor ice, which forms on submerged dark metallic pipe, fittings, or gratings can restrict flow into the intakes. Surface ice and surface ice forming ice dams may impose considerable thrust upon intake structures and cribs. Frazil ice, i.e., displaced anchor ice, or ice that has formed about small suspended particles, may clog intake parts.

When anchor or frazil ice begins to clog an intake, a relatively small change in temperature will free the obstruction. There are some intakes designed with a grid which is in essence a heating coil. In such cases, all that is required is the activation of the circuit which in turn will heat the intake element sufficiently to keep it free from ice. This method has been found economical in a number of locations, since it requires small amounts of electric power and is used only a relatively small number of times per year. When the intake is designed so that there is a continuous temperature recording, the imminence of the formation of anchor or frazil ice can be predicted.

A river can also be a reliable source of supply if the flow rates during drought periods are not seriously affected. A river can be susceptible to ice hazard, scouring of the bottom, changing of the channel, and silting. Before a river intake is constructed, a careful study must be made of the stream bottom, the degree of scour, the extent of formation of surface ice, and the likelihood of the formation of ice jams established. An intake can easily be destroyed by an ice jam, or the entire flow of a river may be stopped by ice. Provision must be made in the design of the intake to ensure that it can withstand the forces which will act upon it during times of flood, heavy silting, or ice conditions.

GRAVITY AND PUMPING SYSTEMS

There are two basic types of water distribution systems: gravity systems and direct pumping systems. Most water systems are a combination of the two types.

Gravity Systems

A true gravity system is one which delivers a supply from the source directly to the distribution system without the use of pumping equipment. This type of system is usually ideal for a fire supply providing pressures are adequate to supply fire demands and normal consumption rates. A gravity system is extremely reliable because the supply is not dependent upon the operation of mechanical equipment; however, the reliability of a well designed and safeguarded pumping system can be developed to the extent that no distinction is made between gravity and pumping systems.

Pumping Systems

When water cannot be obtained at an elevation sufficient to provide working pressures from the elevation head, it is necessary to provide pumps on the system. These pumps are normally located at the source of supply and are used to develop the pressure needed to overcome friction loss in the supply system and to provide satisfactory working pressures in the distribution system. Public systems sometimes have water treatment facilities associated with the pumping station. It is not the intent of this HANDBOOK to discuss water treatment; however, water treatment facilities will affect flow rates and quantities of water available to the distribution system. Many times limiting features of supply are due to some element of water treatment. Therefore it is imperative that the effects of water treatment on the availability of supply be thoroughly understood and considered.

Combination Systems

Often associated with pumping systems are water distribution storage facilities. These provide for storage of water during times of least demand and then supply water during times of peak demand.

Storage can be located so that pumps directly supply the storage facility, and water flows to the distribution system from the storage facility. Storage can also be provided at a remote location within the distribution system, and water can be pumped directly into the distribution system with any excess automatically dumping into the storage facility. The more water which can be maintained in elevated storage, the more reliable a system can be considered because water flowing from a storage facility is the same as a gravity system. Any failure of pumping equipment will not prevent this water from being available for fire protection purposes.

Distribution storage may also consist of large water tanks located at surface elevations equal to or even somewhat lower than the areas of the distribution system they serve. These tanks are filled during periods of relatively low consumption in the system. When demand rates are high, pumps deliver water from these storage facilities to the distribution system. Duplication of pumping equipment and proper design and operation can improve the reliability of these facilities to approach that provided by elevated storage.

SUPPLY CONDUITS, AQUEDUCTS, PIPELINES

Two terms sometimes used to describe the conveyor of water from the source of supply to the distribution system are "conduit" and "aqueduct." A conduit is an enclosed pipe capable of withstanding internal water pressures while an aqueduct is either a closed tube or an open trench, canal, or channel in which water flows but which has no pressure on the side or bottom except that caused by the weight of the water. Aqueducts are not usually designed to withstand internal pressure other than atmospheric.

Pipelines

Pipelines are designed to withstand pressure and to distribute water to the point of use. Three classes of pipelines, or distribution mains, in a large system are:

1. Primary feeders consisting of large pipes with relatively wide spacing. They convey large quantities of water to various points of the system for local distribution to the smaller mains.
2. Secondary feeders forming a network of pipes of intermediate size. They reinforce the distribution grid within the various panels of the primary feeder system and aid the concentration of the required fire flow at any point.
3. Distributors consisting of a gridiron arrangement of small mains. They serve the individual fire hydrants and blocks of consumers.

In order to provide for reliability, two or more primary feeders should extend by separate routes from the source of supply to the high value districts of the city. Similarly, secondary feeders should be arranged as much as possible in loops to give two directions of supply to any point. This practice increases the capacity of the supply at any given point and assures that a break in a feeder main will not completely cut off the supply.

Secondary feeders should generally be installed not over 3,000 ft (914 m) apart in builtup areas.

Where water systems are divided into pressure zones, water can be transferred from one zone to another by operating valves or by using fire department pumpers to pump from the hydrants in one zone to hydrants in the other. The same sort of thing can be done between the water systems of adjoining communities or between a private system and the public system. However, great care must be taken to prevent damage from occuring by subjecting parts of the system to excessive pressures and possible contamination. Usually this is not a good practice and should not be attempted without advance planning, adequate controls, and specific written approval from health authorities.

The Size of Pipe

Pipe less than 6 in.* in diameter is not recommended for fire service, and 6 in. (150 mm) pipes should only be

*Nominal pipe sizes are not directly convertible into metric sizes. The following metric equivalents for common pipe sizes (rounded off for convenience in use) may be useful as points of reference: 4 in. 100 mm; 5 in. 125 mm; 6 in. 150 mm; 8 in. 200 mm; 10 in. 250 mm; 12 in. 300 mm; 16 in. 400 mm; and 20 in. 500 mm.

used when looped in a gridiron where no leg is greater than 600 ft (183 m) in length. In congested districts, it is recommended that distributors should be not less than 8 in. (200 mm) in diameter and interconnected within every 600 ft (183 m). On principal streets and for all long lines, the distributors should be 12 in. (300 mm) or larger.

The cost of a line of pipe includes such factors as trenching (sometimes with piling), laying the pipe, backfilling, and testing. All of these factors are present regardless of the size of pipe used. To them is added the cost of the pipe delivered on the job. It is usually good practice, therefore, to install pipe for fire protection which is one or more sizes larger than the bare minimum might require. Increasing the pipe diameter only one size will often nearly double the possible flow. The figures in Table 17-4A show the relative capacity of pipe obtained by increasing sizes above 6 in. (150 mm).

TABLE 17-4A. Comparison of Pipe Capacity

Size of Pipe, Inches*	Relative Capacity
6	1.0
8	2.1
10	3.8
12	6.2
14	9.3
16	13.2

* Nominal pipe sizes are not directly convertible into metric sizes. For comparison, one inch equals 25.4 mm.

In designing a system, it is also important to consider the probable development of the area under consideration and to plan, in a general way at least, protection for its ultimate development; then install that part of the system for which there is immediate need.

The actual size of pipe needed is based on the rate of water flow required (domestic consumption plus fire flow) and the hydraulic gradient in the area.

Another way to show advantages of larger pipe is to indicate the better performance of four sizes by comparing their friction loss characteristics under a simple range of fire demands. (See Table 17-4B.)

Arrangements of Pipe Systems

Pipe systems should be arranged in loops wherever possible. This allows hydrants and other connections to be fed from at least two directions and greatly increases the possible delivery of water without excessive friction loss.

In private water systems where pipe lines supply hydrants only and where there are adequate pressures for good streams at the hydrants, it is general practice to use 6 in. pipe to supply two-outlet hydrants, and 8 in. or larger pipe under the following conditions: Dead end mains, if more than one hydrant is to be supplied or the distance is more than 500 ft (152 m); looped mains, if two hydrants are to be supplied on a loop with over 1,500 ft (457 m) of pipe; if three hydrants are to be supplied on a loop with over 1,000 ft (305 m) of pipe; or if four or more hydrants are to be supplied.

Where pressures are low, or where three-outlet or four-outlet hydrants only are installed, pipes should be larger. However, the design and layout of the pipe system must provide for the delivery of the maximum fire flow rates to all locations coincident with maximum daily consumption demands.

Internal Condition of Pipe Systems

In the course of time, the internal cross-sectional of unlined cast iron pipe may be reduced or its interior surface roughened because of tuberculation, incrustation, or sedimentation. Incrustations may be due to: (1) tubercular growth, (2) the depositing of chemical constituents normally in solution in the water, or (3) growth of biological or living organisms. Deposits in all kinds of pipe may be due to: (1) sediments, such as mud, clay, leaves, or vegetable decay, or (2) foreign matter other than sediment.

The existence of serious trouble can generally be detected by careful flushing tests. Flushing of the system will remove ordinary sediment. Operation of valves will sometimes show presence of sediment or corrosion. Local water conditions are taken into account in establishing a regular procedure of flushing and testing. Pipes can be cleaned by the use of a scraper or rotating auger. The cleaning device can be pulled through the pipe by a cable or forced through by water pressure.

When pipe has been cleaned in this manner, the rate of tuberculation following cleaning will usually be very rapid, quickly resulting in decreased carrying capacities. The addition of cement lining will retard or prevent further deterioration of capacity. Therefore a cost analysis should be made to determine if the pipe should be lined or the pipe cleaned periodically throughout its life.

TYPES OF PIPES

Underground pipe and fittings for fire protection should be suitable for the working pressures and the conditions under which the pipe is to be installed and in accordance with AWWA (American Water Works Associ-

Table 17-4B. Friction Loss in Cast-iron Pipe (Nominal Size)

Hazen Williams Coefficient C=100

		Loss per 1000 ft (305 m) of Pipe							
Flow		6 in*		8 in*		10 in*		12 in*	
GPM	L/min	psi	kPa	psi	kPa	psi	kPa	psi	kPa
500	1892	14.4	99	3.5	24	1.2	8.3	0.49	03.4
1000	3785	52.0	358	12.8	88	4.3	29.6	1.78	12.3
1500	5678	110.0	758	27.1	186	9.1	62.7	3.76	25.9

* Nominal pipe sizes.

ation) specifications. Pipe is mostly installed without blocks in flat bottomed trenches with selected tamped backfill. The required earth cover over pipe varies depending on pipe size, soil type, and geographical location. Generally because of frost penetration in the northern states, a minimum cover of 4 to 5 ft (1.2 to 1.5 m) is required for large diameter mains and as much as 7 or 8 ft (2.1 to 2.4 m) for small diameter mains. Classes of pipe for working pressures above 150 psi (1034 kPa) are often used if heavier and thicker pipe is desired, for example in unstable or corrosive soils or in locations that would be difficult to reach should leaks or breaks occur. Flexible construction is advantageous in difficult situations, such as under railroad tracks, in areas with heavy industrial machinery, in earthquake areas, or where steep slopes or unstable soil conditions are encountered.

Pipe and fittings used for underground fire service mains should be listed by a testing laboratory.

The titles and availability of AWWA and ANSI (American National Standards Institute) design and installation standards and manuals for the various types of pipe and fittings discussed in the following paragraphs will be found in the bibliography at the end of the chapter.

Asbestos Cement Pipe

Asbestos cement pipe is particularly well adapted for locations where ferrous types without special protective linings or coverings would be attacked by corrosive soil conditions, or by electrolysis. Where asbestos cement pipe must be buried in highly acid or alkaline soils, coatings can be provided that will protect the pipe from soil conditions.

The usual joint for asbestos cement pipe is an asbestos cement sleeve into which a specially shaped rubber gasket is inserted in a circumferential groove near each end of the sleeve. (See Fig. 17-4A.) Cast iron fittings may be used in asbestos cement pipelines.

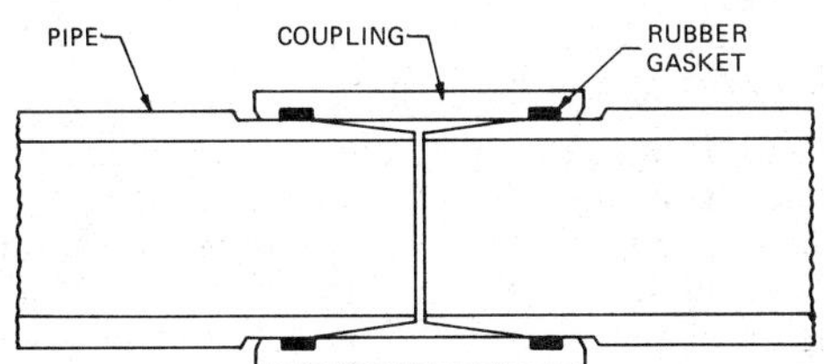

FIG. 17-4A. Coupling for asbestos cement pipes.

Polyvinyl Chloride (PVC) Pipe

PVC pressure pipe, manufactured in accordance with AWWA C900, is acceptable for fire service. PVC pipe is often specified in situations where severe corrosion problems are anticipated. PVC pipe is not subject to electrolytic corrosion, nor to tuberculation from corrosion byproducts.

Because of its flexibility and high impact strength, PVC pipe manufactured to AWWA C900 specifications is not susceptible to beam, sheer, or impact failure in unstable soil conditions. The product is considered particularly suitable where shifting or heaving soils, live loading, or earthquake shock is anticipated.

AWWA C900 specifies PVC pipe with three pressure classes—100, 150, and 200. The pressure class designations define the maximum working pressure ratings for the pipe. The product is available in two series of outside diameters—cast iron pipe (CI) and steel pipe (IPS). AWWA C900 PVC pipe is available in nominal diameters from 4 in. through 12 in. (approximately 100 to 300 mm).

PVC pipe used in potable water distribution and fire protection systems is commonly provided with gasketed joints. Either integral bell gasketed joints or gasketed couplers which satisfy the requirements of ASTM D 3139 may be specified. When installing PVC pipe with gasketed joints, thrust restraint or blocking should be provided as necessary to prevent movement of pipe or appurtenances in response to thrust.

Cast Iron Pipe

Table 17-4C gives dimensions and weights for cast iron pipes of Thickness Class 22, which is designed for standard laying conditions. The complete AWWA and ANSI specifications have additional data on pipe sizes, weights, pressure classes, and thickness classes for other laying conditions. Pipe should be selected on the basis of maximum working pressure with consideration for transient pressures and the laying condition.

There are several acceptable types of joints. The most common are single gasket push-on and standard mechanical. Poured bell-and-spigot joints are now seldom used. All these joints depend on friction between the parts and surrounding earth fill to prevent separation. (See Figs. 17-4B, 17-4C, and 17-4D.)

Push-on Joints: The push-on joint is made up by seating a circular rubber gasket of special cross section in the valve and then forcing the spigot end of the pipe past the gasket to the bottom of the valve socket. No packing or caulking is required.

Standard Mechanical Joints: These are joints in which a rubber ring gasket is held in place by a follower ring bolted to the bell. A mechanical joint provides for a limited amount of flexibility; a ball and socket joint of similar design provides a little more. Because of the flexibility, pipe with these joints is often selected for lines across bridges or in unstable soil.

Poured Bell-and-Spigot Joints: These joints have a ring packing of jute or other material and are caulked with lead. Special joint compounds are available which require no caulking.

When using bell-and-spigot pipe and fttings, changes in grade or direction should never be made by shifting the pipe in the joints. This would result in uneven packing and caulking, and such joints are likely to leak. Only a very slight variation from normal is tolerable.

Ductile Iron Pipe

Ductile iron has the corrosion resistance of cast iron and approaches the strength and ductility of steel. Ductile iron is now used in place of cast iron. Table 17-4D gives the minimum available thicknesses for ductile-iron pipe 4

TABLE 17-4C. Standard Dimensions and Weights of Cast Iron Pipe*

			Class 150 (150 Psi or 1034 kPa)				Class 200 (200 Psi or 1379 kPa)				Class 250 (250 Psi or 2069 kPa)			
	Outside Diameter		Thickness		Weight per 18 ft (5.5 m) Laying Length†		Thickness		Weight per 18 ft (5.5 m) Laying Length†		Thickness		Weight per 18 ft (5.5 m) Laying Length†	
Nominal Size In.	In.	mm	In.	mm	Lbs	kg	In.	mm	Lbs	kg	In.	mm	Lbs	kg
4	4.80	122	0.35	8.89	290	131	0.35	8.89	290	131	0.35	8.89	290	131
6	6.90	175	0.38	9.65	460	209	0.38	9.65	460	208	0.38	9.65	460	208
8	9.05	229	0.41	10.41	655	297	0.41	10.41	655	297	0.41	10.41	655	297
10	11.10	282	0.44	11.18	870	395	0.44	11.18	870	395	0.44	11.18	870	395
12	13.20	335	0.48	12.19	1,125	510	0.48	12.19	1.125	510	0.52	13.20	1,215	551
14	15.30	388	0.51	12.95	1,410	639	0.55	13.97	1,510	685	0.59	14.99	1,610	730
16	17.40	442	0.54	13.71	1,700	771	0.58	14.73	1,815	823	0.63	16.00	1,960	889
18	19.50	495	0.58	14.73	2,050	930	0.63	16.00	2,210	1002	0.68	17.27	2,370	1075
20	21.60	548	0.62	15.75	2,430	1102	0.67	17.02	2,610	1184	0.72	18.28	2,785	1263
24	25.80	655	0.73	18.54	3,415	1549	0.79	20.07	3,665	1662	0.79	20.07	3,665	1662

* ANSI A21.6-1970. Based on standard laying conditions (5 ft or 1.5 m cover, flat bottomed trench and tmaped backfill).
† Includes bell.

to 24 in. in diameter. Ductile iron pipe is made with push-on and mechanical joints.

Cast iron fittings are used with ductile iron pipe. Coating or cathodic protection is not needed, except where extremely corrosive conditions exist. Cement lined pipe is recommended for all new or replacement installations of cast iron or ductile iron pipe to offset the corrosive action of water. Portland cement is extensively used for lining. Coal-tar enamel linings are available but less common. A large percentage of all cast iron pipe is cement lined at the foundry.

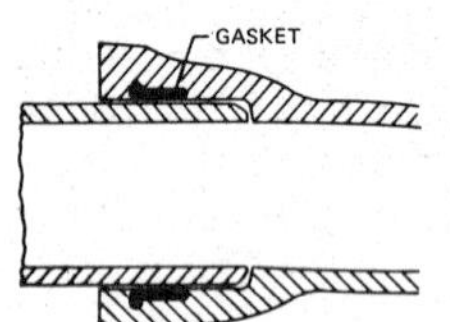

FIG. 17-4B. A push-on joint.

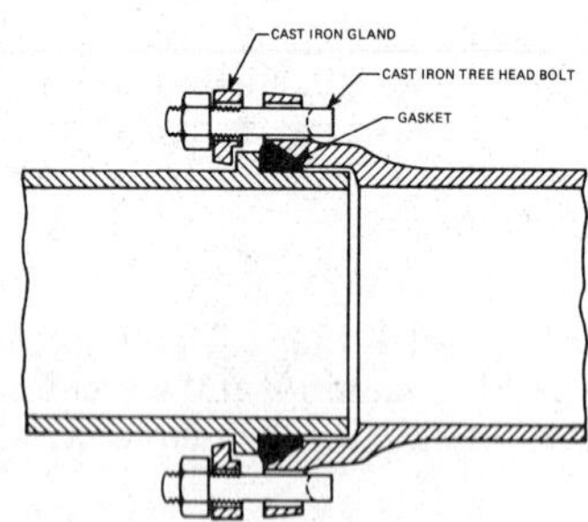

FIG. 17-4C. A standardized mechanical joint using anchoring fittings.

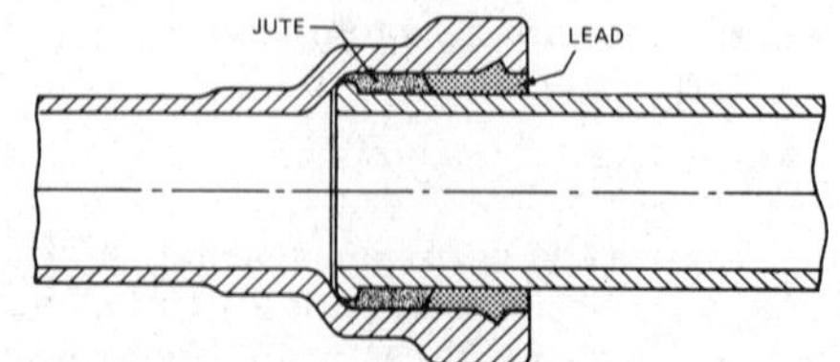

FIG. 17-4D. A poured bell-and-spigot lead joint.

Steel Pipe

Steel pipe of suitable wall thickness and manufacture, when lined and coated, can be used for fire service, both for underground mains and for supply lines in tunnels and buildings. Because of its high tensile strength, steel pipe is particularly suitable where it may be exposed to shock or to impact from railroad tracks, highways, drop-forge equipment, etc. The greater strength of steel is also advantageous in unstable soil or on steep slopes. Approximate dimensions and weights are given in Table 17-4E.

Steel pipe joints are welded, made with flanges or mechanical couplings. (See Fig. 17-4E.) Expansion joints may be needed in long runs of pipe in tunnels. Hangers and supports should conform to good engineering practice and applicable standards.

Reinforced Concrete Pipe

Several designs of pipe made of concrete and steel 24 in. and larger in diameter are available. Concrete pipe is often used for long supply and arterial feeders but it is not normally used in distribution systems. A "nonprestressed" design is a steel cylinder with one or two steel cage reinforcements encased in concrete. A "modified prestressed" design is a steel cylinder with a spirally wound steel rod reinforcement prestressed to provide a slight initial tension in the cylinder and concrete lining. "Prestressed" designs consist of a concrete lined cylinder or steel cylinder helically wrapped under tension with a high tensile strength wire. External coatings are cement mortar. Details of a reinforced concrete pipe joint are shown in Figure 17-4F.

Fittings

Fittings used should be appropriate for the same range of working pressures as the pipe with which they are used. A single class of cast iron fittings for sizes 3 to 12 in. is now used generally. These are for working pressures of 250 psi (1724 kPa). Cast iron fittings (bends and tees) are used with

TABLE 17-4D. Standard Dimensions and Weights of Ductile Iron Pipe

Nominal Size, In.	Outside Diameter		Working Pressure		Thickness		Weight per 18 ft (5.5 m) Laying Length‡	
	In.	mm	Psi	kPa	In.†	mm	Lbs	kg
4	4.80	122	350	2413	0.29	7.37	240	109
6	6.90	175	350	2413	0.31	7.87	380	172
8	9.05	229	350	2413	0.33	8.38	535	243
10	11.10	282	350	2413	0.35	8.89	700	317
12	13.20	335	350	2413	0.37	9.40	885	401
14	15.30	358	350	2413	0.36	9.14	1,005	456
16	17.40	442	350	2413	0.37	9.40	1,185	537
18	19.50	495	300	2068	0.38	9.65	1,370	621
20	21.60	548	250	1723	0.39	9.91	1,560	703
24	25.80	655	250	1723	0.41	10.41	1,975	896

* ANSI A21.51-1972. Based on standard laying conditions (5 ft or 1.5 m cover, flat-bottomed trench and tamped backfill).
† Minimum available.
‡ Includes bell.

asbestos cement pipe. They have bells designed to be used with the asbestos cement gasket type joint.

Cast iron and steel fittings for underground water lines for fire protection are approved by testing laboratories.

Pipe Corrosion

Water is corrosive to cast iron, ductile iron, and steel pipe, and to fittings. The initial rate of corrosion for steel pipe may be more rapid than for cast or ductile iron, but after several years exposure there is little difference.

External corrosion of buried iron and steel pipe is the direct result of complicated electrochemical reactions. Soil containing metallic salts, acids, or other substances in combination with moisture causes iron ions to separate from the pipe. The mass of the metal at the pipe surface is diminished, and the pipe becomes pitted or corroded. Iron or steel pipe should not be installed under coal piles, in cinderfill, or wherever acids, alkalies, pickling liquors, etc., can penetrate the soil.

Stray electric currents from external sources may reach and follow buried pipelines to locations where the resistance to ground is less than that of the pipeline. Ionization then occurs at points where the current leaves the pipe, producing an effect similar to that of soil corrosion.

When stray electric currents are suspected, the extent and origin should be determined by professional ground surveys. If the stray currents cannot be eliminated or diverted, the pipe, if not yet seriously corroded, can be protected by bonding all the joints and by providing direct low resistance metallic ground connections.

Cathodic methods are widely used for the external protection of iron and steel water mains. Cathodic protection is a technique of imposing direct electric current from a galvanic anode to the buried pipeline. In many instances, cathodic protection is more economical than coating and wrapping. Cast iron or ductile iron pipe used in water systems should be cement lined in accordance with *AWWA* Standards.

A smooth lining is necessary to minimize loss of carrying capacity. Buried piping needs a coating to protect against soil corrosion. An outside coating may be applied in the field if desired, but it is practical only on large jobs.

Table 17-4E. Recommended Minimum Dimensions and Weights of Steel Pipe for Fire Protection Mains

Nominal Diameter In.*	Outside Diameter		For Welded Joints: Minimum Wall Thickness		For Welded Joints: Weight per Ft†		For Flexible Couplings or Threaded Joints: Minimum Wall Thickness		For Flexible Couplings or Threaded Joints: Weight per Ft†	
	In.	mm	In.	mm	Lbs	kg	In.	mm	Lbs	kg
6	6.625	168	0.188	2.54	12.9	5.85	0.219	5.56	15.0	6.8
8	8.625	219	0.188	2.54	16.9	7.67	0.239	6.07	21.4	9.7
10	10.750	273	0.188	2.54	21.2	9.61	0.250	6.35	28.0	12. 7
12	12.750	324	0.188	2.54	25.1	11.38	0.281	7.14	37.0	16.78
14	14.000	355	0.239	5.07	35.1	15.92	0.281	7.14	41.2	18.69
16	16,000	406	0.250	6.35	42.0	19.05	0.312	7.92	52.4	23.77

† one foot equals 305 mm.

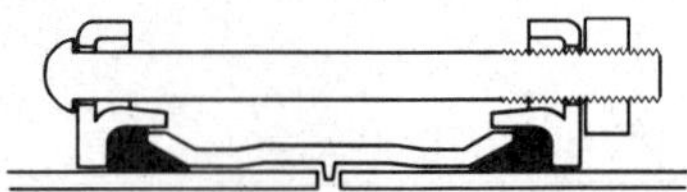

FIG. 17-4E. A steel mechanical coupling for plain-end steel pipe.

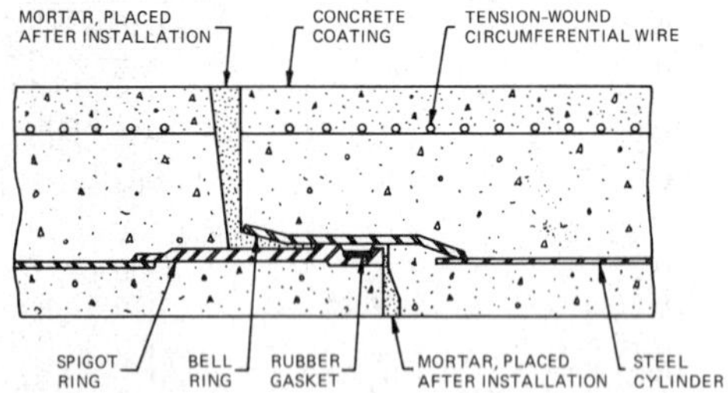

FIG. 17-4F. A reinforced concrete pipe joint.

Exposed piping should be painted or otherwise protected as required by atmospheric conditions. Nuts and bolts of buried joint assemblies should be heavily coated. Any damage resulting to lining or coating should be thoroughly repaired.

RULES FOR LAYING PIPE

The depth of cover to provide protection against freezing will vary from about 2½ ft (0.76 m) in the southern states to about 10 ft (3.05 m) in northern Canada. Because there is normally no circulation of water in fire protection mains, they require greater depth of covering than do public mains. The 2½ ft (0.76 m) minimum should always be maintained to prevent mechanical damage. Depth of covering should be measured, from top of pipe to ground level, and consideration should always be given to future or final grade and nature of soil. A greater depth is required in a loose, gravelly soil (or in rock) than in compact or clay soil. A safe rule to follow is to bury the top of the pipe not less than 1 ft (0.3 m) below the lowest frostline for the locality.

Placing pipes over raceways or near embankment walls should be avoided. Keep mains back a sufficient distance from the banks of streams or raceways to avoid any danger of freezing through the side of the bank. Where mains are laid in raceways or shallow streams, care should be taken that there will be running water over the pipe during all seasons of frost; a safer method is to bury the main 1 ft (0.3 m) or more under the bed of the waterway.

Protection Against Pipe Breakage

In general, it is advisable to avoid running pipe under buildings. Where mains necessarily pass under a building, the foundation walls can be arched over the pipe. Pipes passing under building walls with ground floors at grade are buried to the same depth as outdoors. Pipes under basement floors below grade may require less depth, but in no case should the cover be less than 2½ ft (0.76 m).

Any pipe that passes through a wall or foundation must be protected from fracture. This is done by keeping a clear 2 in. (51 mm) annular space around the pipe and sealing it with coal tar or asphalt. (See Fig. 17-4G.)

Special care is necessary in running pipes under railroad tracks, highways, large piles of iron, and under buildings housing heavy machinery that could subject the buried pipes to shock or vibration. Where subject to such breakage, pipes should be run in a covered pipe trench or be otherwise properly guarded. While flanged cast iron pipe with metallic gaskets is sometimes used under buildings, it is not recommended because of its lack of flexibility, higher cost, and almost impossible access.

Frost loading is also recognized as a substantial factor that must be considered in pipe installation. Frost loading may exceed earth and live loads for which the pipe was designed. Other loads must also be considered which result from seismic shifts and tremors, unequal settlement and expansion, and subsidence of specific clay soils with changes in moisture content.

Care in Laying Pipe

Pipes should be clean inside when put in trenches, and open ends should be plugged when work is stopped to prevent stones or dirt from entering.

Pipes should be supported throughout their length and not by the bell ends only. Superior support is obtained where the bottom of the trench is shaped to fit the pipe. If the ground is soft or of a quicksand nature, special provision must be made for supporting pipe. For ordinary conditions of soft ground, longitudinal wooden stringers with cross ties will give good results. A reinforced concrete mat 3 or 4 in. (75 to 102 mm) thick in the bottom of the trench can also be used. In extreme cases, the stringers and cross ties or concrete mat may have to be supported on piles. The most important aspect of laying pipe, though, is to follow the manufacturer's instructions for trench preparation, maximum deflection, and method of joining.

Pipe Anchorage

Most conventional pipe joints are not designed to resist unusual forces tending to pull them apart. Joints are

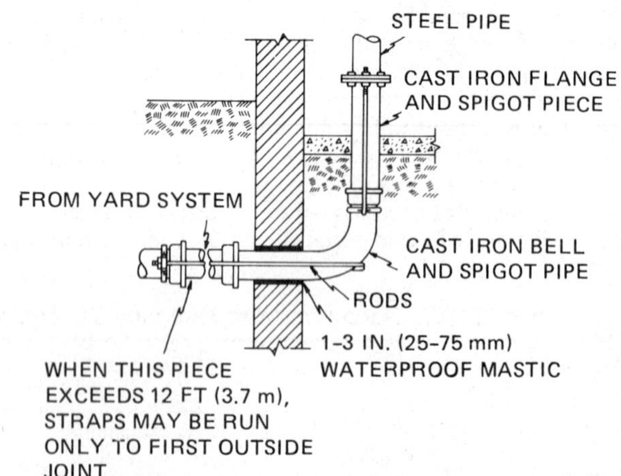

FIG. 17-4G. A common arrangement of pipe is through a foundation to feed a sprinkler or standpipe riser. The anchorage of the horizontal run is determined by soil conditions. In some cases, as in earthquake areas, considerable flexibility is desirable in this run and the first two joints outside the building. Often the run is brought into the building with the first joint of a flexible type, particularly if the pipe openings through foundation and floors are sealed with grouted concrete instead of mastic. The anchorages shown would be the usual ones.

expected to be kept in place by the soil in which the pipe is buried. It is necessary to determine, by tests in case of doubt, that a particular soil will actually do this. In unsatisfactory soils, trenches may have to be excavated below the final pipe level and filled to the pipe grade with soil suitable to give the pipe an even bearing throughout its length.

Forces acting on pipe laid in the ground that must be considered, among others, are: (1) internal static pressure of the water, (2) water hammer, (3) load from the backfill, and (4) load and impact from passing trucks and other vehicles. Static pressure and backfill loads are always present. Water hammer loads are considered separately from impact loads of passing vehicles on the theory that simultaneous action of the two would be a remote possibility. The thickness designated for the various pressure classes of pipe tends to reflect possible water hammer loads. The magnitude of water hammer depends upon the modulus of elasticity of the piping material, pipe wall thickness, pipe diameter, flow velocity, and the rate of change in velocity. Anchorages are needed to take care of additional loads due to water hammer and other forces. Trenches, which have to be unusually wide or deep, impose additional loads which may affect the choice of pipe thickness for the particular installation.

To prevent joints in cast iron pipes from coming apart, they should be securely braced or clamped unless anchoring fittings or locked joints are used. Typical methods of anchoring joints at elbows, tees and bends, and plugs at blanked openings are shown in Figures 17-4H to 17-4O. Clamps and rods and other steel fittings at anchors should be protected against corrosion by a thick covering of asphalt.

Thrust Blocks

It is also necessary to consider the loads imposed by water moving in the pipe. This is why at bends, tees, and pipe ends, and wherever the piping changes direction, the pipe assembly must actually bear on a surface that will resist the loads imposed. Thrust blocks are used to keep the pipe assemblies in place.

The usual thrust block is made of concrete and is placed between the fitting and the trench wall. Thrust blocks may be tied to foundations where these are of sufficient size to provide a bearing.

A thrust block under a hydrant or valve to prevent upward movement in newly laid soil would require rods bent over the bells to hold the valve or hydrant to the block. Blocks under hydrants should be located so as not to prevent the hydrant from draining properly when drains are installed. Thrust blocks are shown in Figures 17-4P and 17-4Q.

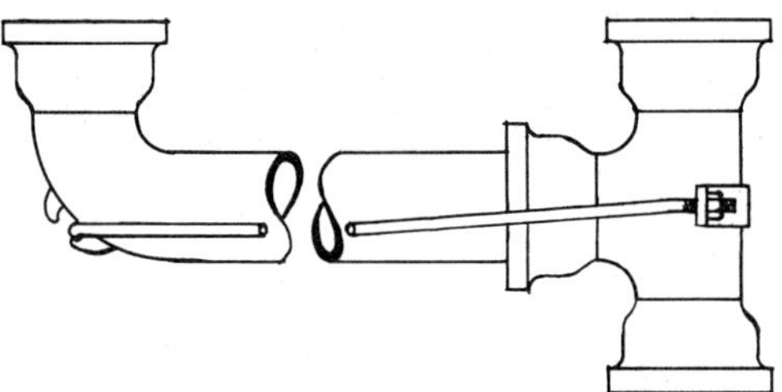

FIG. 17-4H. Strap for bend anchor to tee.

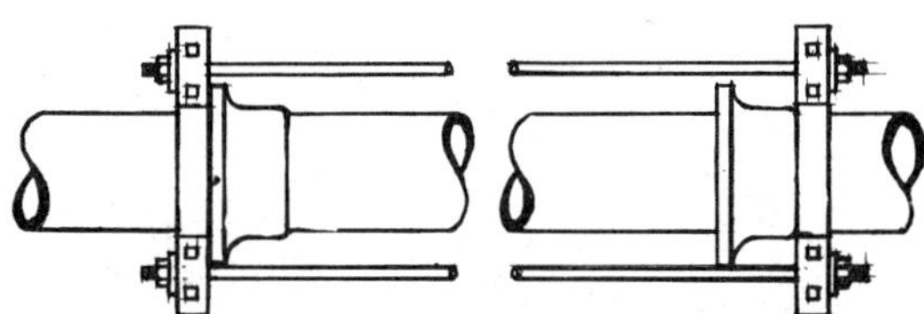

FIG. 17-4I. Anchor rods for bell-and-spigot pipe. When distance between bells is less than 12 ft (3.6 m), the next length of pipe is also anchored.

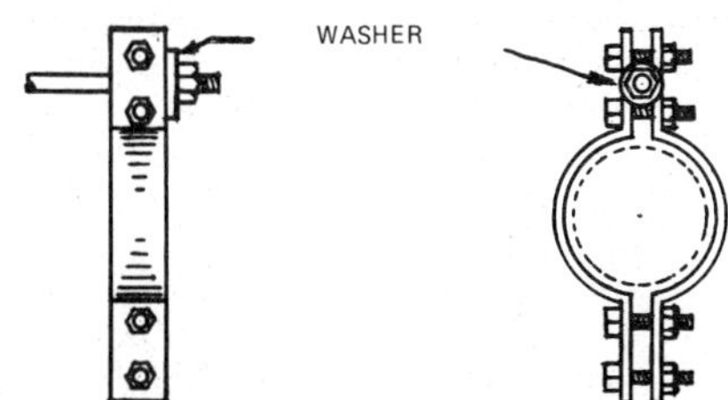

FIG. 17-4J. Clamps for bell-and-spigot and short-body fittings.

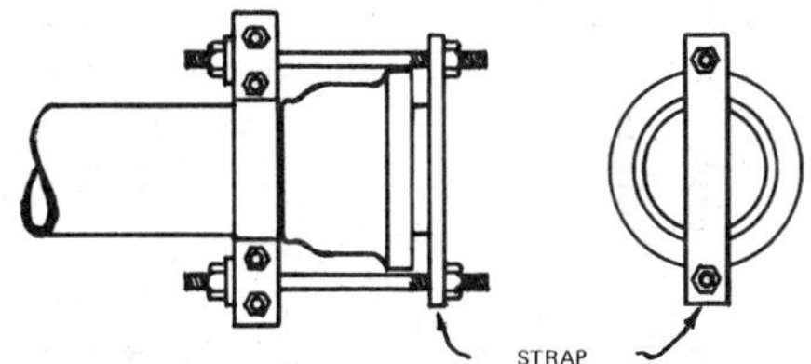

FIG. 17-4K. A plug for the bell end of a pipe.

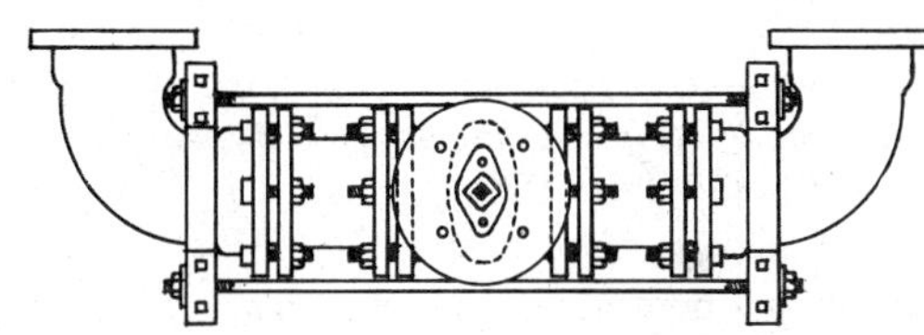

FIG. 17-4L. A general form of an anchor for indicator post valves. A mechanical joint pipe is illustrated.

Testing

All pipelines, of whatever material, should be subjected to hydrostatic test, either by sections as completed or as a whole after completion. Such testing is usually done after the trench has been partially backfilled. All tests should be conducted in accordance with *AWWA* standards, and NFPA 24, *Standard for the Installation of Private Fire Service Mains and Their Appurtenances.*

Backfilling

Earth should be well tamped under and around pipes (and puddled where possible) to prevent settlement or lateral movement, and should contain no ashes, cinders, or other corrosive materials. If the ground in which the pipe is laid is partly or wholly cinder fill, care should be taken that about a foot of cinder-free dirt is put in the trench below the pipe and no dirt containing cinders is used in backfilling around the pipe. Cinders stimulate galvanic actions which may cause the pipe to fail in a relatively short period. There are similar considerations in occupancies where salt or other industrial wastes, e.g., from cattle

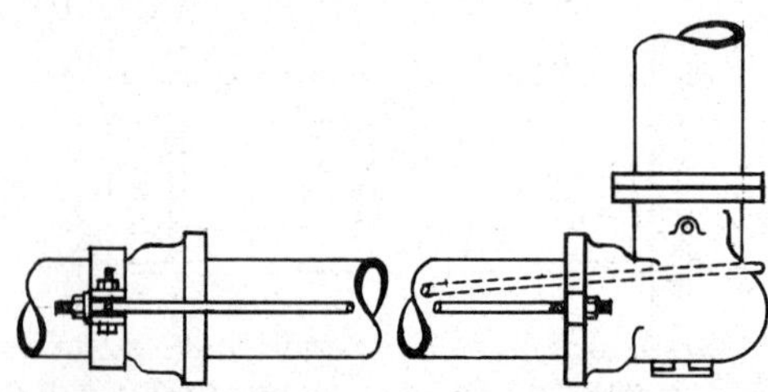

FIG. 17-4M. Hydrants are connected with a double spigot connection anchored as illustrated. Anchor fittings and locked joints are often more convenient alternatives for rodded anchors. When the hydrant does not have lugs, the anchor rods are arranged as indicated by the dotted lines.

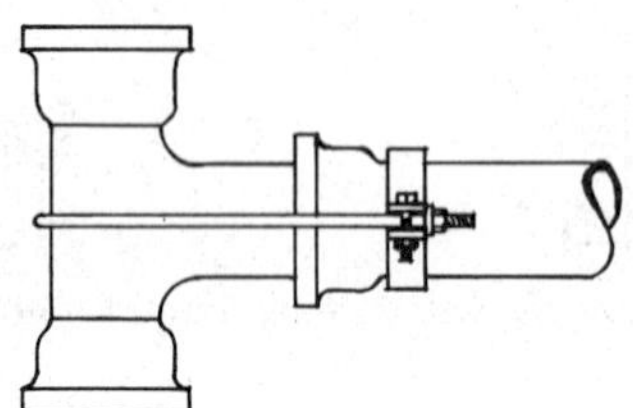

FIG. 17-4N. An anchor at a spigot end of a tee fitting.

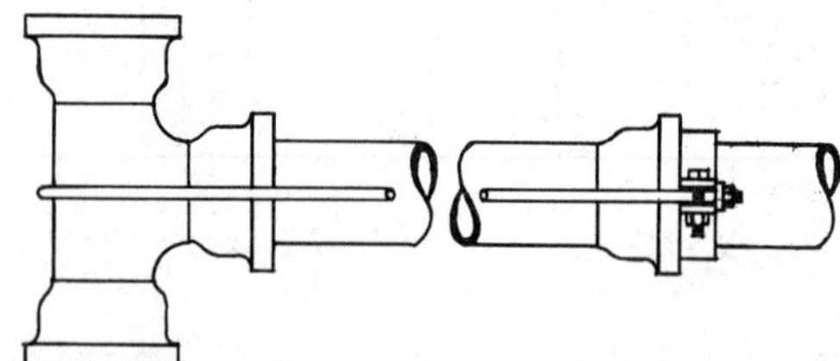

FIG. 17-4O. An anchor at a bell end of a tee fitting. If the disance between joints is less than 12 ft (3.6 m) rods should also be run to a clamp on the next bell.

sheds, packing operations, and chemical plants, might seep into the ground.

Rocks should not be rolled into trenches nor allowed to drop on pipes. In trenches cut through rock, backfilling is normally entirely of earth. In any case, earth should be used under and around pipe and at least 2 ft above it.

Flushing

After a system of underground pipe for fire protection has been completed and before it is permanently filled with water, the entire system should be thoroughly flushed out under pressure through hydrants or other outlets. Mains supplying sprinkler systems should be flushed at the following rates: 6 in. (150 mm) pipe, 750 gpm (2839 L/min); 8 in. (200 mm), 1,000 gpm (3785 L/min); 10 in. (250 mm), 1,500 gpm (5678 L/min); and 12 in. (300 mm), 2,000 gpm (7571 L/min).

NFPA 13, *Standard for the Installation of Sprinkler Systems,* contains information on the requirements for flushing underground connectors to insure their reliability.

During flushing, the valves to any inside sprinkler or standpipe equipment should be closed to avoid the washing of stones or other debris into the inside system. Branches from the outside system should be flushed before they are connected to sprinkler or standpipe risers.

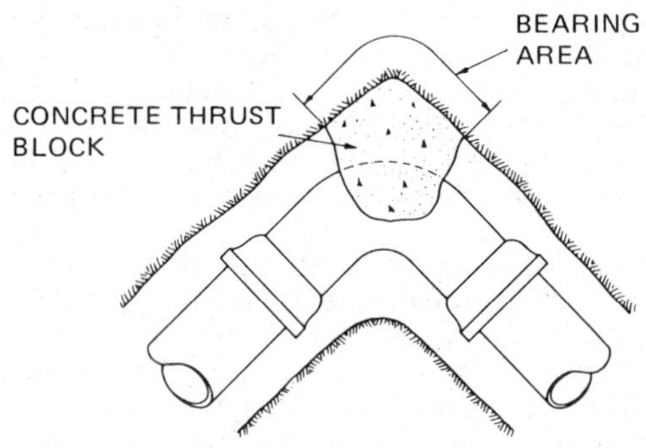

FIG. 17-4P. A thrust block at a one-quarter bend.

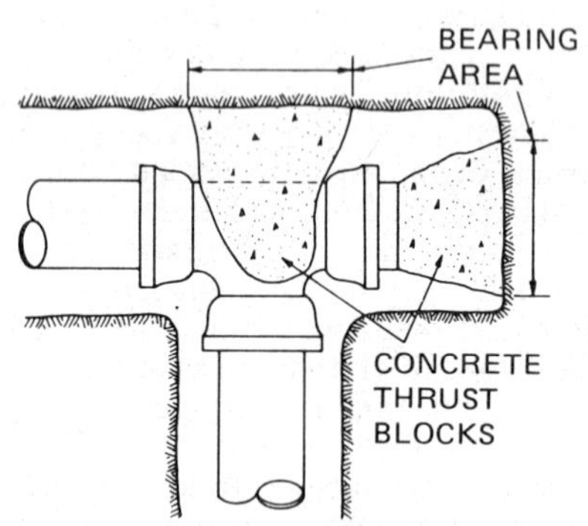

FIG. 17-4Q. A thrust block at a tee and plug.

HYDRANTS

The most important features of good hydrants for fire protection are:

1. Normal diameter of bottom valve opening at least 4 in. (100 mm) for two 2½ in. (64 mm) or larger outlets, 5 in. (125 mm) for three 2½ in. (64 mm) or larger outlets, and 6 in. (150 mm) for four 2½ in. (64 mm) or larger outlet hydrants. Hydrants having a bottom valve less than 4 in. (100 mm), or outlets less than two 2½ in. (64 mm), are usually not approved by testing laboratories. The connection between a water main and a hydrant should not be less than 6 in. (150 mm) in diameter.
2. The net area of the hydrant barrel and foot piece at the smallest part is not less than 120 percent that of the net opening of the main valve.
3. A liberal sized waterway and low friction loss. With the hydrant discharging 250 gpm (946 L/min) through each of two 2½ in. (64 mm) hose outlets, the total head loss at the hydrant should not exceed 2 psi (13.8 kPa). For a hydrant with a 4½ in. (114 mm) outlet, at a discharge rate of 1,000 gpm (3785 L/min), the maximum permisible head loss is 5 psi (34.5 kPa). For hydrants designed or intended to deliver more than 1,000 gpm (37 854 L/min), the maximum permissible head loss should not exceed 5 psi (34.4 kPa) when discharging at the intended flow rate.
4. A positive-operating, corrosion resistant drain or drip valve.
5. A uniform sized pentagonal operating nut measuring 1½ in. (38 mm) from point to flat at the base and 1⁷⁄₁₆ in. (36 mm) at the top. The faces should be tapered

uniformly, and the height of the nut should not be less than 1 in. (25 mm).

Hydrant bonnets, barrels, and foot pieces are generally made of cast iron with internal working parts of bronze. Valve facings should be of a suitable, yielding material such as rubber or a composition material. Hydrants are available with various configurations of outlets.

Types of Hydrants

There are two types of fire hydrants in general use today. The most common is the base valve (dry barrel) in which the valve controlling the water is located below the frost line between the foot piece and the barrel of the hydrant. (See Fig. 17-4R.) The barrel on this type hydrant is normally dry with water being admitted only when there is a need. A drain valve at the base of the barrel is open when the main valve is closed, allowing residual water in the barrel to drain out. This type of hydrant is used whenever there is a chance the temperature will go below freezing, because the valve and water supply are installed below the frost line.

The other type of hydrant is the wet barrel (California) type, which is sometimes used where the temperature remains above freezing. These hydrants usually have a compression valve at each outlet, but they may have another valve in the bonnet that controls the water flow to all outlets. (See Fig. 17-4S.)

Fire hydrants are covered by standards of the American Water Works Association (AWWA C502-80 and AWWA C503-82) and Underwriters Laboratories Inc. (UL 246), all listed in the Bibliography at the end of the chapter.

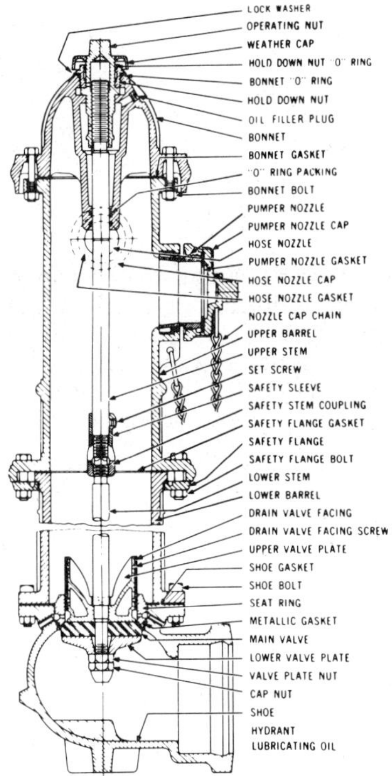

FIG. 17-4R. A base valve or "dry barrel" hydrant with nomenclature identified. When installed, the valve is below the frost line. This type of hydrant is also known as a "frost proof" hydrant. (Mueller Company)

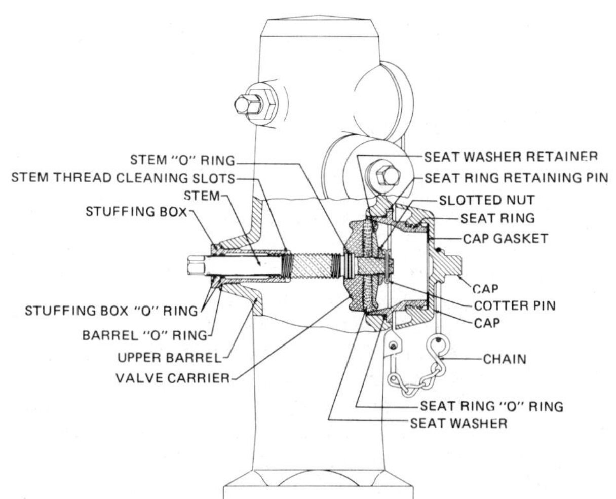

FIG. 17-4S. A wet barrel or "California" hydrant used where freezing is not encountered. There is a compression valve at each hydrant outlet. (Mueller Company)

Location

Hydrant spacing is usually determined by fire flow demand established on the basis of the type, size, occupancy, and exposure of structures. At present there is no universally accepted method for establishing fire flows for other than fixed extinguishing systems. ISO (Insurance Services Office) has a procedure that it has developed for insurance rate making purposes only. The procedure has been in affect for many years, and is of its value in arriving at hydrant spacing recommendations. Its use should be based upon sound engineering judgement that addresses the field situations that are encountered.

As a general rule, however, hydrant spacing should not exceed 800 ft (245 m) between hydrants. In closely built areas, 500 ft (150 m) or less between hydrants is more realistic. Hydrants should be located as close to a street intersection as possible with intermediate hydrants along the street to meet the area requirements.

Where hydrants are located on a private water system and hose lines are intended to be used directly from the hydrants, they should be so located as to keep hose lines short, preferably not over 250 ft (75 m). At a minimum, there should be enough hydrants to make two streams available at every part of the interior of each building not covered by standpipe protection. They should also provide hose stream protection for exterior parts of each building using only the lengths of hose normally attached to the hydrants. It is desirable to have a sufficient number of hydrants to concentrate the required fire flow about any important building with no hose line exceeding 500 ft (150 m) in length.

For average conditions, hydrants normally are placed about 50 ft (15.2 m) from the buildings to be protected. When that is impossible, they are set where the chance of injury by falling walls is small and where fire fighters are not likely to be driven away by smoke or heat. In crowded industrial yards, hydrants usually can be placed beside low buildings, near substantial stair towers, or at corners formed by masonry walls which are not likely to fall.

Hydrants that must be located in areas subject to heavy traffic need protection against damage from collision. The parking lots of shopping centers and mill yards are good examples.

Setting of Hydrants

Hydrants should be set plumb with outlets about 18 in. (0.46 m) above the ground and 18 in. (0.46 m) above the floor in hose houses. When hydrants are installed before grading is completed, the final grade line and accessibility should be considered. Most hydrants have a grade line indicated on the barrel of the hydrant.

Drainage is necessary for hydrants equipped with drain ports and can be provided by excavating a pit about 2 ft (0.61 m) in diameter and 2 ft (0.61 m) deep below the base of the hydrant and filling it compactly with coarse gravel or stones placed around the bowl of the hydrant to a level of 6 in. (152 mm) above the waste opening. If the drip valve of the hydrant is below ground water level, it may be plugged to exclude ground water. In that case, water remaining in the hydrant after use should be pumped out to prevent freezing.

The bowl of a hydrant should be secured to the next preceding bell with anchoring or locking joints or rods and clamps, or securely anchored by means of concrete backing. (See Fig. 17-4M.)

Maintenance and Testing

Well designed and properly installed hydrants present a minimum of maintenance difficulties. The dry barrel hydrant, for example, has a small drain near the base of the barrel arranged to permit water to drain out when the main valve is shut. When the main valve is opened several turns, this drain is closed. If this drain is working properly and the main valve is tight, the difficulty of water freezing in the barrel is avoided. Occasionally situations are found where ground drainage is unsatisfactory or where ground water may stand at dangerous levels. In those cases, drains may be closed entirely and hydrant barrels pumped out periodically.

The use of salt or salt solutions to prevent freezing is not recommended because of their corrosive effect and limited usefulness. If antifreeze is used in hydrant barrels, its use must be confined to hydrants that are not part of a system supplying water for domestic consumption.

Ethylene glycol is extremely toxic with as little as 0.1 mg/L ingested for a period of a week being fatal. This substance should not be used. Propylene glycol is not as toxic and may be used to prevent freezing but with proper precautions and in accordance with local health regulations.

Suggestions for detecting freezing in hydrants include:

1. Sounding by striking the hand over an open outlet. Water or ice shortens the length of the "organ tube" and raises the note.
2. Try turning hydrant stem. If solidly frozen, the stem will not turn. If only slightly bound by ice, placing a hydrant wrench on the nut and tapping smartly may release the stem. Blows should be moderate to prevent breaking the valve rod.
3. Lowering a weight on a stout string into the hydrant. It may strike ice or come up wet, showing water in the barrel.

Probably the most satisfactory method of thawing a hydrant is by means of a steam hose. A thawing device in which steam may be rapidly produced should be standard equipment for fire departments in cold weather climates. In either case, the steam hose is introduced into the hydrant through an outlet and pushed down, thawing as it goes.

A major item of periodic maintenance is a check for leaks in (1) the main valve when the hydrant is closed, (2) the drip valve when the main valve is open but outlets capped, and (3) the mains near the hydrant. Stethoscope like listening devices are available to make these checks. Maintenance routines provide for an operating test, repair of leaks, and pumping out of the hydrants where necessary. Threads of the outlet, caps, and valve stem should be lubricated with graphite. Hydrants should be kept painted but care should be taken to avoid accumulations of paint which might prevent easy removal of caps or operation of valve stems.

Uniform Marking of Fire Hydrants

Color coding of hydrants are based on water flow available from them and color coding is of substantial value to water and fire departments. A test of an individual hydrant does not give as complete and satisfactory results as group testing, but such a test is a start in the right direction. The colors signify the capacity of the individual hydrant as tested, not group hydrant effect.

NFPA 291, *Fire Flow Testing and Marking of Fire Hydrants*, recommends that hydrants be classified as follows:

Class	Flow	Color of bonnets and nozzle caps
A	1,000 gpm or greater	Green
B	500 to 1,000 gpm	Orange
C	Less than 500 gpm	Red

Capacities are rated by flow measurement and tests of individual hydrants during a period of ordinary water consumption. Rating is based upon the flow rate available at 20 psi (138 kPa) residual pressure.

The capacity indicating color scheme provides simplicity and consistency with colors used in signal work for

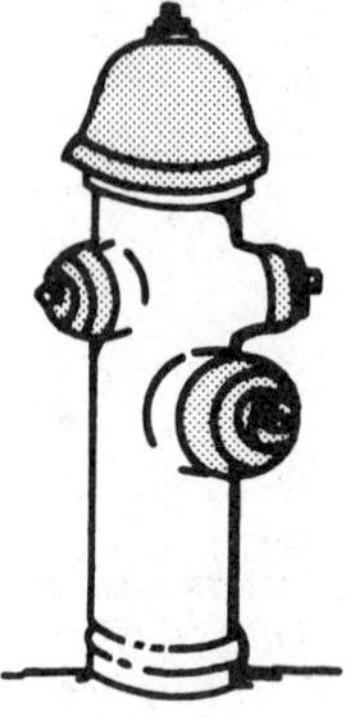

FIG. 17-4T. Shaded areas are color coded to show the hydrant's flow capacity.

safety, danger, and intermediate conditions. (See Fig. 17-4T.) Barrels of all public hydrants are normally chrome yellow except in cases where another color has already been adopted.

Within private enclosures, marking of private hydrants is at the discretion of the owner. Private hydrants on public streets are normally painted red to distinguish them from public hydrants.

Location markers for flush hydrants carry the same color background for class indication with such data stenciled or painted on them as may be necessary.

CONTROL VALVES FOR WATER DISTRIBUTION SYSTEMS

Water distribution systems require valves at strategic locations and intervals to control flow as circumstances dictate. Nonindicating gate valves, butterfly valves, check valves, and pressure reducing or altitude valves are the types of controlling valves used in supply systems.

Required features for valves and indicator posts listed by a testing laboratory include:

For Gate Valves:

1. Stems of bronze with a minimum tensile strength of 32,000 psi (220 640 kPa).
2. Stuffing boxes with good sized packing space.
3. Bronze lined gland and bonnet opening, and a valve capable of being repacked under pressure.
4. Yokes bolted on bonnets in valves larger than 4 in. (100 mm).

For Check Valves:

1. Large clearances between moving parts and valve body.
2. A clapper which moves entirely out of the waterway.
3. Bronze-to-bronze bearings having large wearing surfaces.
4. Valve of the straightaway type and iron body valve having a waterway equal to the area of the pipe.

For Indicator Posts:

1. A uniform flange to suit all sizes of gate valves.
2. Interchangeable operating stems.
3. Adjustable and interchangeable target plates.
4. Uniform sized square operating nut measuring 1 in. (25 mm) square by 1 in. (25 mm) high.

Gate Valves

Gate valves of the nonindicating type are provided in distribution systems to allow segments to be shut off for repairs or extensions without reducing protection over a wide area. Such valves are normally a nonrising stem type which requires a key wrench to operate. A valve box is located over the valve to keep dirt from the valve and to provide a convenient access point for the valve wrench to the valve nut. (See Fig. 17-4U.)

A complete record should be made for each valve in the system, including dated installed, make, size, direction of opening, number of turns to open, any maintenance performed, and location by triangulation from fixed bounds, preferably under the control of the water system management. Permanent valve records should be stored in a secure place and copies suitably indexed should be available in all water supply vehicles that respond to emergency calls. Copies should also be available to each fire station.

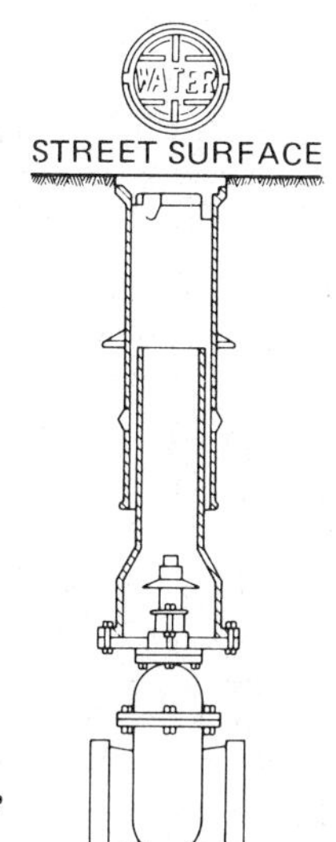

FIG. 17-4U. A nonindicating type gate valve for underground installation.

Good practice dictates that valves are provided so that no single accident, break, or repair will necessitate shutting down a length of pipe greater than 500 ft (150 m) in high value districts or greater than 800 ft (245 m) in other sections, and so that flow may be maintained through other arterial mains.

PRIVATE FIRE SYSTEMS

This part of the chapter covers the valving arrangement on a private fire protection water distribution system from the point where it connects with its supply.

Indicating Valves

The first valve in the fire line on private property should be a valve of the indicating type. It may be one of three kinds: (1) underground gate valve with indicator post attached, (2) underground butterfly indicating type with post and (3) O.S.&Y. (outside screw and yoke) gate valve in a pit.

The type of underground gate valve commonly used in domestic and industrial water lines (which requires a key wrench to operate) should be avoided in fire lines. The difficulties of finding such valves, getting up cover plates, and locating a key wrench that will fit are detriments to its use. If used, underground gate valves should open counter-clockwise and have the same size nut on all valve stems. The valve locations should be clearly marked on nearby buildings.

Indicator posts should have a metal plate showing what they control. Painting building names or numbers on them is advisable. The proper direction to turn for opening should be shown. Posts are often locked in the open position to prevent tampering. Lock shackles should, however, be brittle so that they can be removed readily if they are frozen and the keys will not work. Posts should have a handle, wheel, or wrench attached. Typical indicator post valves are shown in Figure 17-4V.

O.S.&Y. gate valves, if underground, should be in pits. Important junction points often warrant the use of pits to

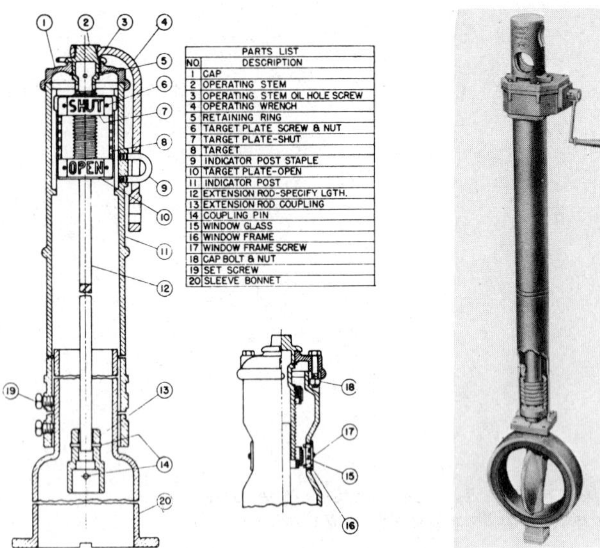

PARTS LIST	
NO	DESCRIPTION
1	CAP
2	OPERATING STEM
3	OPERATING STEM OIL HOLE SCREW
4	OPERATING WRENCH
5	RETAINING RING
6	TARGET PLATE SCREW & NUT
7	TARGET PLATE-SHUT
8	TARGET
9	INDICATOR POST STAPLE
10	TARGET PLATE-OPEN
11	INDICATOR POST
12	EXTENSION ROD-SPECIFY LGTH.
13	EXTENSION ROD COUPLING
14	COUPLING PIN
15	WINDOW GLASS
16	WINDOW FRAME
17	WINDOW FRAME SCREW
18	CAP BOLT & NUT
19	SET SCREW
20	SLEEVE BONNET

FIG. 17-4V. A standard post indicator valve. Nearly all manufacturers of valves furnish standard valves, and their catalogs should be consulted for details of each. At right is a butterfly post indicator valve. (Henry Pratt Company)

cover several valves. Indicating valves are used on all shut-offs. They are universally used on inside piping because they show the position of the gates at a glance. They are commonly strapped or otherwise sealed open. Where O.S.&Y. gate valves are located in pits, a metal sleeve with the upper end closed may be slipped over the valve stem to keep off dirt. An O.S.&Y. gate valve is shown in Figure 17-4W.

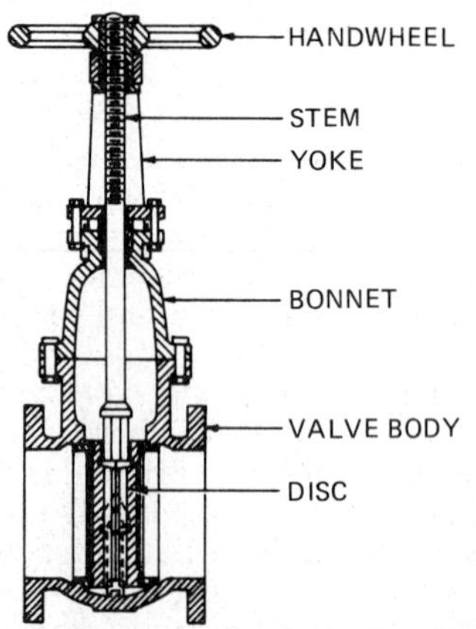

FIG. 17-4W. A typical outside screw and yoke (O.S.&Y) valve in the closed position. The spindle on the valve stem indicates whether the valve is open or shut.

Check Valves

A check valve is installed next to the main control valve inside the property. The check valve must be installed in a pit unless it is located inside a building, even when the main control valve is a conventional buried valve with an indicating post. The check valve also needs to be accessible, so the size of the pit, the arrangement of its manhole and ladder, and the size of working space around the valve or valves in the pit should be large enough for convenient access. (See Fig. 17-4X.)

A check valve allows water to flow from the public system to the private system but not from the private system into the public system. It also permits water at

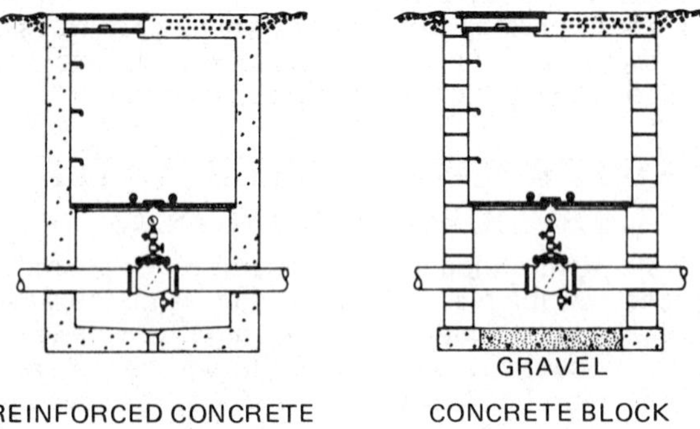

MOST SUBSTANTIAL DESIGN. APPROPRIATE WHERE THERE IS GROUND WATER.

A DESIGN WHICH CAN BE USED WHERE THERE IS GOOD GROUND DRAINAGE.

FIG. 17-4X. A valve spit is usually provided for a check valve (as shown) and for O.S.&Y, gate valves, meters, and other equipment on underground mains. Pit should be about 5 ft sq (0.46 m²) for one check valve. Clearance should be 18 in. (0.46 m) around all valves and equipment in the pit. Pits are as deep as necessary to conform to the location of the pipe in which the valve is located.

higher pressures in the private system than the normal pressures in the connecting public water mains in case of fire. Higher pressures can be provided in a private system through its own fire pumps, where these are provided, or by fire department pumpers when they pump through the fire department connections on automatic sprinkler systems and standpipe systems in buildings or on a private yard system.

Check valves are also installed at the base of gravity tanks and in the private or fire department pumper connections. The check valves at these points assure flow of water one way only.

Air Gap Connections

Connections between potable and nonpotable water systems should be avoided. Where one water system must be separated positively from another, an air gap is provided so that backflow cannot occur. An example of an air gap is where water from one system is supplied through a pipe which discharges into a "break" tank above the level of the water in the tank. Such an arrangement does not allow the receiving water system to have the advantage of the pressures available in the supplying system. With a connection through a check valve, practically all of the pressure in the supplying system may be available.

Double Check Valves

For situations where there is some question of leaks, double check valve equipment is available. (Double check valves are required for certain situations by the regulations of some health departments.) Check valves in double installations are made with bronze working parts and rubber facings so that they will have tight seats. For each connection, two of these valves are installed in series, thus increasing the efficiency because the probability that both will leak at the same time is extremely remote. The two check valves are installed between valves in a pit where they are readily accessible for examination. In addition, the valves are provided with pressure gages and test cocks so arranged that the tightness of each check valve may be verified in a few minutes. Figure 17-4Y shows a typical double check valve installation in a pit.

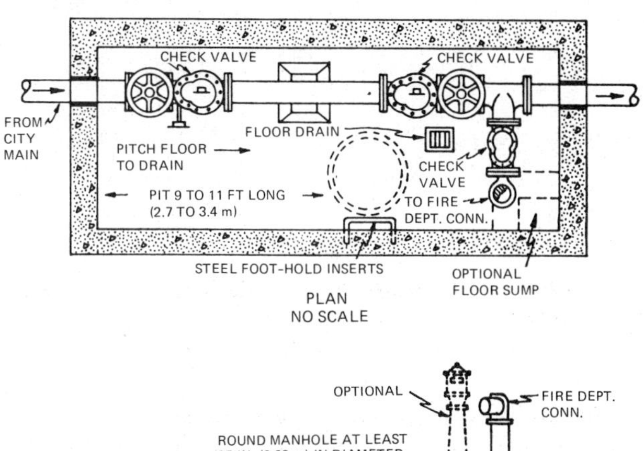

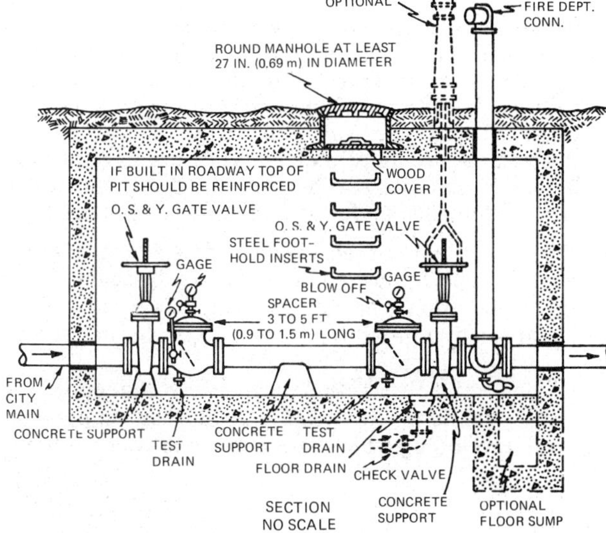

FIG. 17-4Y. Plan and section views showing the arrangment of double check valves in a connection for fire protection from a public water system. Indicating valves of the butterfly type designed for fire service may be used in place of the gate valves illustrated.

When the two check valves are bolted together, an 18 in. (0.46 m) space is provided between clappers, which is sufficient to prevent any ordinary material found in a pipe line from holding both clappers open at the same time. To improve further the efficiency of the equipment, a filling piece of pipe or spacer from 3 to 5 ft (0.9 to 1.5 m) long installed between the check valves is advised where space is available.

Note! Check valves do not always work properly. Whenever there is any possibility of backflow from a fire service main into a potable supply, a listed backflow prevention device should be installed or an air gap provided depending upon local health regulations.

Backflow Preventers: Figure 17-4Z shows a backflow preventer assembly which is intended to maintain pres-

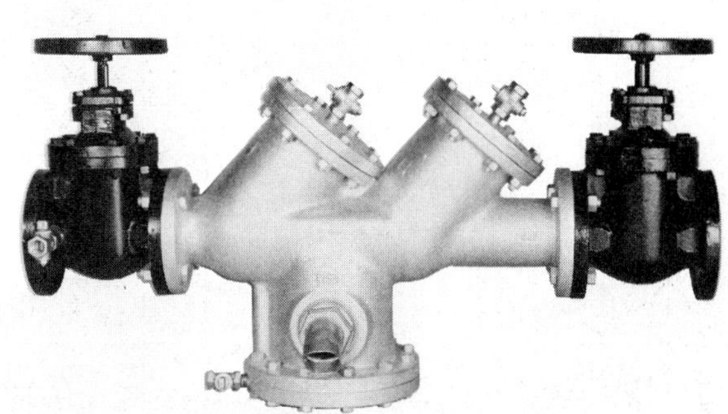

FIG. 17-4Z. A backflow preventer assembly for installation between two independently acting check valves. (Hersey Products, Inc.)

sure between two check valves at less than the supply pressure in the pipe line. The assembly consists of two independently acting approved check valves interspaced by an automatically operated pressure differential relief valve. In case of leakage at either check valve, the relief valve operates to maintain the pressure between the check valves at less than supply pressure. There is some loss of head in such backflow preventers, but there may be situations where sacrificing some loss of head is preferable to sacrificing all of it as would be necessary with an air gap separation.

Location of Valves in Private Fire Service Systems

Opinions vary on how many valves should be used in a system of underground mains. More plants have probably been destroyed because a sectional control valve was shut than because a large part of the system was out of commission by reason of too few sectional valves. Nevertheless, the modern tendency is to make fairly liberal use of valves. A few well established principles are shown in Figure 17-4AA. They include:

1. A city supply check valve (and meter, if required) located between indicating valves so it can be repaired without affecting city or plant systems.
2. A pump check valve located between pump and indicating valves so that the latter can be used to shut off the connection to the system when making check valve and pump repairs.
3. Three sectional valves (two, *G* and *H*, to take care of present loop and one, *J*, for a short branch supplying a small detached building) in addition to the main water supply valve (see Detail of Pit 1). The branch will ultimately be part of a second loop. There should be a loop valve on each side of every valvable water supply to permit cutting off a part of the loop without cutting the water supply off altogether. Best practice requires that post indicators be attached to valves in pits, with the posts cemented into the concrete tops.
4. Sectional control valves (indicator posts *C*, *E* and *F*) can cut the loop into four sections (in connection with

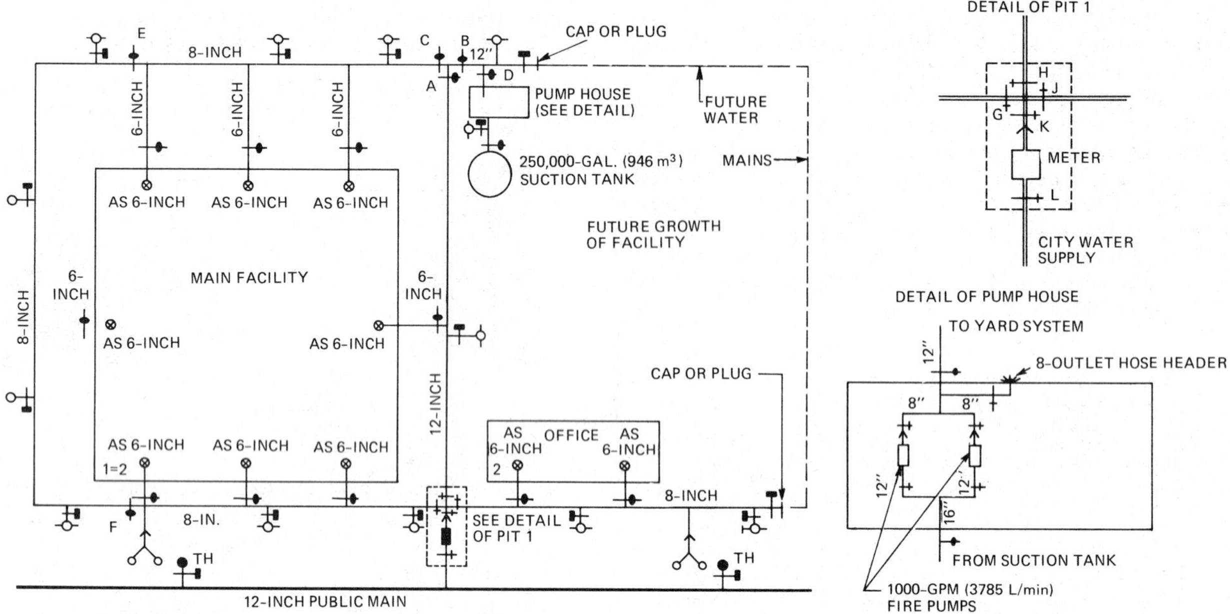

FIG. 17-4AA. Water piping for fire protection of an industrial site. Typical details shown are: connections to public mains and supplies for a private fire pump; looped water mains; sectional control valves (lettered); and hydrants.

valves H and G). In large or complicated underground systems, it is recommended that indicator posts controlling risers to sprinklers or standpipes be painted a different color from sectional control valves. Generally, no more than six hydrants or indicator posts should be located between section valves.

5. Gate valves must be provided on hydrant laterals to provide for isolation of the hydrant in the event it malfunctions or when repairs are necessary.

Records

Records for water systems should include the following:

1. A complete map and satisfactory sectional views of the distribution system made, on a suitable scale, showing all the features of the distribution system, including size and location of pipelines; location of gate, check valve, pressure reducing, and altitude valves; and hydrants. There should also be sectional maps showing the location of all gate valves by triangulation to fixed reference points. Copies of these maps should be distributed to the personnel who would be working on the system.

In event of a break in the underground system or the abandonment of a flowing hydrant during a fire, it is essential that water be promptly shut off to protect the supply to other hydrants or sprinkler connections and to avoid flooding. Operating valves correctly and rapidly at time of fire, especially at night and under adverse weather conditions, requires accurate knowledge of the system.

2. Complete records should be kept for each fire hydrant in the system. They should show the date of installation; location, number and size of outlets; size of the hydrant valve; size and location of the gate valve between the hydrant and street connection; the size of pipe to which it is connected, including the length of the branch connection; and, inspection dates. The records should note if the hydrant is equipped with drains, if the drains have been plugged or omitted, and if there has been any problem with drainage or with ground water entering the hydrant barrel. Records of all maintainance performed should also be kept.

Permanent hydrant records should be stored in a safe place with copies indexed for field use located in all vehicles responding to an emergency with copies available to each fire station providing protection in the area served by the water distribution system.

Copies of the distribution map and hydrant and valve records should be used by the fire department in preplanning and developing a familiarity with the water system.

Maintenance and Testing

Check valves should be checked and tested for tightness monthly, particularly where double check valves are used to prevent backflow between potable water and yard systems. The test drains and gages provide a means for determining where a leak may exist. Even though found tight when tested periodically, check valves should be thoroughly cleaned once a year, so they will remain tight.

Where there are several sets of check valves in fire connections from public mains, only one set of check valves should be overhauled and cleaned at a time, the others being left in service.

Gate valves should be operated periodically to ensure their proper working condition and that gate boxes are not silted up or tarred over.

METERS FOR FIRE CONNECTIONS

Fire flow meters are devices capable of measuring small and large flows with a minimum loss of head for heavy demands. They are offered in two types: (1) detector check valve meters that detect only small rates of flow and (2) full registration meters that measure the entire flow throughout the line in which they are installed. Meters of types other than the fire flow type have been found to be unsatisfactory for fire protection water supplies.

Detector Check Valves

These devices consist of a check valve with a weighted clapper in the main passage and a meter in a bypass around the check. In operation the smaller flows pass through the meter in the bypass and are accurately registered. Meters may be furnished up to 3 in. (76 mm) in size to serve specific needs. For heavy flows the check valve opens and a free unmetered waterway is provided. Beyond the point where the weighted check valve lifts, the bypass meter registers only a small part of the flow. In many situations, detector check valves give the water works assurance about the proper use of water. Figure 17-4BB shows a representative detector check valve.

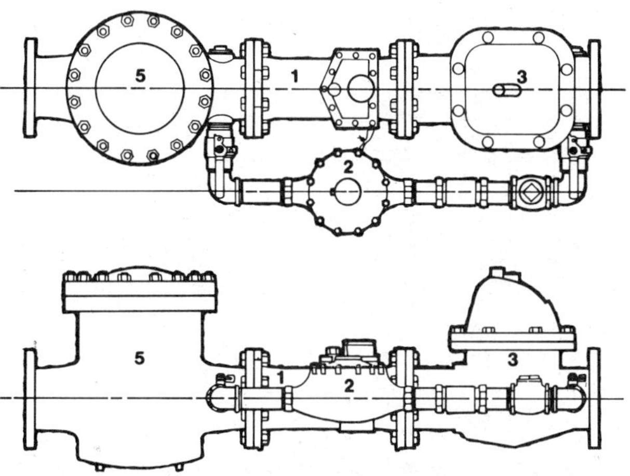

FIG. 17-4BB. A fireflow detector check valve assembly. Views show the location of the weighted check valve in relation to the bypass and its small capacity meter for smaller flows. Shown is the Rockwell International 6 in. Fire Line Meter Assembly. A full registration meter records major flows when the weighted check valve is open.

Full Registration Meters

These devices are of three general types, each produced by a different manufacturer, and they have been designed for small friction loss with large flows and for a main passageway practically unobstructed when open. The three types are: (1) proportional type meters, (2) meters of the displacement type in a bypass and (3) turbine type meters.

Proportional Type, Hersey Detector Meter, Model FM: This meter is a special meter of the compound type in which a "proportional meter" and an automatic valve in the main line of the meter are combined with a disc or compound meter in a bypass. (See Fig. 17-4CC.)

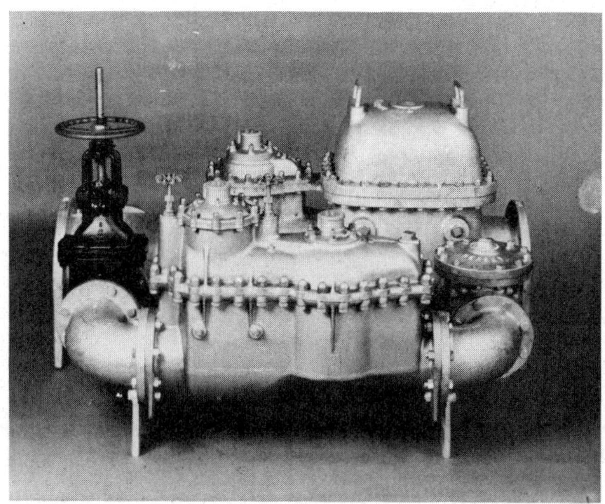

FIG. 17-4CC. A detector meter of the proportional type. (Hersey Products, Inc.)

In the Model FM, the smaller flows pass through and are measured by the bypass meter. When the demand for water reaches a rate of flow which causes a difference in pressure of 4 psi (28 kPa) in the bypass, the automatic check valve opens and provides a practically free waterway through the main line. When water begins to flow through the line in which the automatic valve has opened, it is slightly retarded by a restricting orifice placed a little upstream from the automatic valve, and a part of the water is diverted through a metering unit. This diverted flow is a fixed percentage of the total flow through the restricting orifice. The metering unit is calibrated to record the total quantity through the line—the sum of the readings of the bypass meter and the main line metering unit gives the total flow.

Displacement Type, Neptune Trident Protectus Meter: This meter has all of the working parts in one casing. A disc meter is installed in a bypass on one side of the main waterway, with a current meter on the other side. Small flows pass through the disc meter and are recorded when the check valve is closed. With larger flows the main check valve opens and gives a free waterway. The opening of the check valve stops the flow through the disc meter and opens the bypass to the current meter so that the flow through the open waterway is measured proportionately. The sum of the readings gives the total flow. (See Fig. 17-4DD.)

Turbo Type, Rockwell W-2000 Turbo Meter: This meter is based on the turbine principle of measurement. (See Fig. 17-4EE.) The meter is composed of two principal assemblies, the main case and the measuring chamber. The main case contains the flow straightening vane assembly. The measuring chamber includes rotor, adjusting vane, and register. The meter can be equipped with either a direct reading register or a pulse transmitting register (which also contains a direct reading odometer and test circle) for sending signals to remote reading registers and other instruments. The manufacturer recommends that, for fire service applications, the meter be used only with an approved fire service strainer placed immediately upstream of the meter.

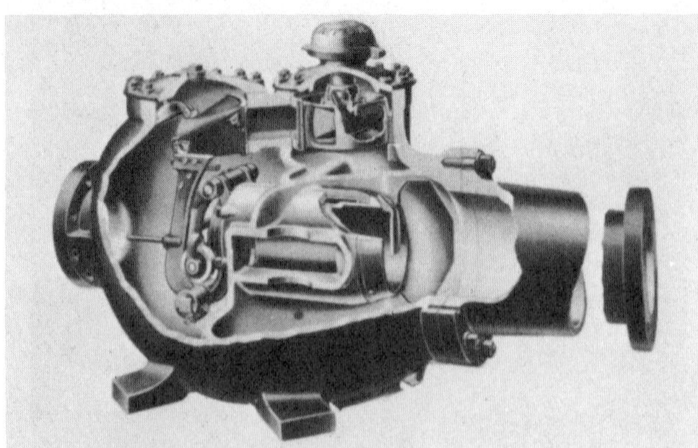

FIG. 17-4DD. A fire flow meter of the displacement type. (Neptune Water Meter Company)

Friction Loss in Fire Flow Meters and Detector Check Valves

The standard specifications for cold water meters adopted by both the American Water Works Association and the New England Water Works Association limit the friction loss for fire flow meters to 4 psi (28 kPa) at rated flow capacities. Table 17-4F gives friction loss values for the three currently available types of fire flow meters.

Friction loss values for meters of the disc, current, and compound type that are used commonly in waterworks systems for general purposes are relatively high and not suited for fire protection purposes. AWWA limits their friction loss values to 20 psi (138 kPa).

Detector Check Valves: The friction loss in detector check valves listed by testing laboratories is less than 3 psi (21 kPa) for the following flows:

Size (in.)	Flow (gpm)
4	750
6	1,500
8	3,000
10	4,500
12	6,500

The pressure required to open the clapper is less than 20 psi (138 kPa).

CONNECTION BETWEEN PUBLIC AND PRIVATE WATER

Connections from public water systems for fire protection are for the purpose of providing water supply to the following:

1. Automatic sprinkler systems.
2. Standpipes for hand hose or fire department use.
3. Open sprinklers
4. Yard systems with private hydrants.
5. Fire pumps.
6. Private storage reservoirs or tanks for fire protection.

A fundamental principle is that, within the property, piping systems from which water is used for fire shall be independent of systems supplying domestic and industrial service. Special conditions where this is not the case are rare and may be dealt with individually.

In public water systems, except in some unusual situations, economics favor a single water system for combined fire use and domestic service. In private properties, it is economically practicable to have separate systems; the advantages of a definite, dependable supply, free from interruptions of service inevitable in a domestic system, strongly favor the established practice. However, it should be remembered that a failure in any portion of the public system supplying the private and separate fire system may result in reduced flows or no flow into the private system.

Occasionally a property will have two sources of water supply for fire protection: one from the public supply; the other from a private source, many times nonpotable. When this happens, adequate measures must be followed in accordance with local health department regulations to prevent the public water supply from becoming contaminated. Means of achieving this have been discussed previously in this chapter. Some communities do not allow any direct connection, and if two supplies are required or desired, separate systems must be maintained with an air gap or backflow prevention device between the two systems.

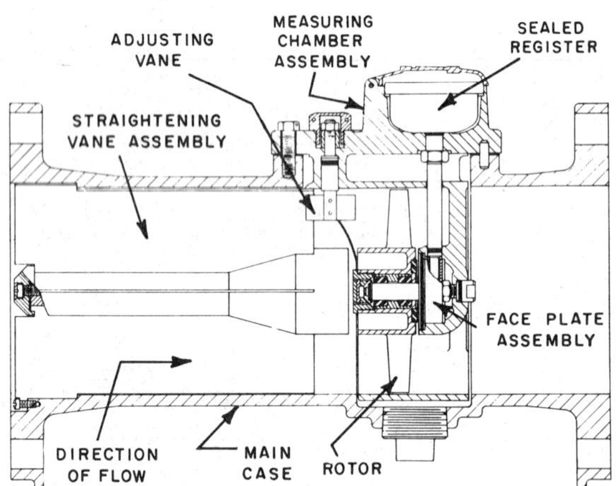

FIG. 17-4EE. A fire flow meter of the turbo type. (Rockwell International)

Charges for Connections to Private Fire Protection Systems

The question of annual standby charges for connections from a public water system to fire protection systems may be controversial largely because the nature of automatic sprinkler systems is not always understood by public officials. Whatever charges are made should be based on actual cost to the utility and not on the possible value of the installation to the customer. Annual charges for connections to fire protection systems are often established in water rate schedules for the sole purpose of obtaining additional revenue.

Although many waterworks, both public and private, make no standby charge for connections for fire protection, there may be no objections to charges that fairly reflect actual cost to the waterworks of the connection itself and the expense of servicing it including necessary pits, valves, and meters. There could be costs to the waterworks for amortization of costs of its part of the connection and for its maintenance and inspection. However, it is usual to charge for the installation when it is made, and if repairs, maintenance, and inspection are charged for when done, there is little other cost involved.

Since the amount of water used for fire fighting is usually very small when compared to normal consumption, it is almost universal practice to make no charge to property owners for the water actually used in the extinguishment of fire or for authorized hydrant flow testing.

Control of Connections During Fires

The safe location of control valves for connections for fire protection is important so that they may be shut off promptly after water extinguishing systems have put out the fire, or to conserve water and pressure if pipes have been broken.

Sprinkler systems have outside indicating valves where yard space is available. This is the preferable arrangement. In many city properties having no yard space, valves are installed in a stair tower or other sufficiently cut off area. Sometimes the control valves are located outside the area protected as, for example, on the other side of a fire wall. In a relatively few instances, dependence must be placed on the accessibility of the valve on the connection to the street main.

Experience has shown that in practically all fires, sprinkler control valves on private property have been adequate to serve their intended use and allow prompt and safe control of fire protection.

Control of Water Waste

There are a number of conditions under which flow of water from private fire protection equipment other than for fire or testing purposes may occur. These are:

1. Leakage
2. Wrongful use, such as wetting down roofs, watering lawns, or waste through ignorance
3. Emergency use
4. Theft

Fire protection equipment should be inspected several times a year to assure that the systems will be kept free from material leakage or unauthorized connections. Maintenance rules enforced by inspection authorities tend to discourage wrongful uses of these systems. These are further enforced in many places by fire department inspections, which are made to assure that valves controlling extinguishing systems and other fire supplies are kept open and that hose and other equipment are not used for any purpose other than that intended. Occasional inspections by waterworks inspectors are desirable.

In cases where inspections alone are not a sufficient safeguard against water waste, a plan involving the securing of valves and outlets would be effective. In such cases all sprinkler drains, hydrants and hose outlets are sealed by the water utility which may require:

1. A report from the owner when a seal is broken.
2. A notice from the user before testing through outlets on hydrants or sprinkler drains.
3. Payment of a nominal charge for resealing.

As a further safeguard, indemnity to the water utility may be secured by requiring the user to file a bond or deposit to be forfeited in the event of willful or persistent violation of the rules or abuse of the fire service furnished by the water utility.

The simplest cases of control of water waste to deal with are fire connections in single buildings such as those serving:

1. One or more standpipes for hand hose or fire department use.
2. Automatic extinguishing systems only.
3. Combined automatic sprinkler systems and standpipes.

Table 17-4F. Friction Loss in Fire Flow Meters

Compiled from data supplied by manufacturers

Meter Name and Type	Size of Meter Inches	(mm)*	Loss of Pressure Caused by Meter Pounds per Square Inceh (kPa) — Gallons per Minute Flowing (L/min) 250	(946)	500	(1892)	750	(2839)	1000	(3785)	1250	(4732)	1500	(5678)	2000	(7571)	2500	(9463)
Proportional	4	(100)	2.2	(15.1)	1.9	(13.1)	4.0	(27.6)										
Type	6	(150)	3.8	(26.2)	2.7	(18.6)	1.4	(09.7)	1.8	(12.4)	2.2	(15.1)	2.9	(20.0)	3.9	(26.9)		
(Hersey-FM)	8	(200)	3.0	(20.7)	3.7	(25.5)	1.9	(13.1)	1.0	(06.9)	0.8	(05.5)	0.8	(05.5)	1.3	(09.0)	2.2	(15.2)
	10		1.6	(11.0)	3.6	(24.8)	4.0	(27.6)	3.7	(25.5)	2.6	(17.9)	1.4	(09.7)	0.8	(05.5)	1.1	(07.9)
Differential	4	(100)	2.2	(15.2)	2.0	(13.8)	3.9	(26.9)										
Type	6	(150)	2.9	(20.0)	1.9	(13.1)	1.8	(12.4)	1.8	(12.4)	1.8	(12.4)	1.8	(12.4)	2.2	(15.2)	3.3	(22.8)
(Trident	8	(200)	3.2	(22.1)	1.8	(12.4)	1.5	(10.3)	1.5	(10.3)	1.5	(10.3)	1.6	(11.0)	1.8	(12.4)	2.2	(15.2)
Protectus)	10	(250)	3.0	(20.7)	2.3	(15.9)	1.8	(12.4)	1.3	(09.0)	1.2	(08.3)	1.1	(07.9)	1.1	(07.9)	1.2	(08.3)
Turbo Type (Rockwell W-2000 Turbo)	6	(150)	0.2	(01.4)	0.3	(02.1)	0.6	(04.1)	0.85	(05.9)	1.2	(08.3)	1.6	(11.0)	3.5	(24.1)		

* Rounded off for convenience of use.

In connections of these types, any regulation on the part of a water utility (for the ostensible purpose of preventing water waste) which involves an excessive annual charge or a costly meter installation, may impose an expense to the property owner so large as to discourage the installation of extinguishing systems or standpipes. This particularly affects a class of property—small industrials and mercantiles—in which the bulk of current fire losses occur, and which greatly needs the protection.

Ordinarily inspections, or at least a valve securing procedure, will give adequate protection to the utility. In extreme cases, special action may be taken by a utility against any individual violator. In adopting any procedure to apply to all users of private fire protection, the relative unimportance of a very small amount of water improperly used, as against the greater public good in the extensive installation of automatic sprinklers and other private fire equipment, should be considered.

Fire protection connections which serve several buildings with extinguishing systems and standpipes, or, in addition, a yard system with private hydrants, fire pump and tank supplies, may introduce a slightly more difficult problem of water waste control. In general, inspection or sealing procedures will suffice, and both may be appropriate in some cases. Where meters are employed on fire connections, the extent to which they reduce pressures and obstruct or reduce flows must be determined.

Bibliography

NFPA Codes, Standards, Recommended Practices and Manuals. (See the latest *NFPA Codes and Standards Catalog* for availability of current editions of the following documents.)

NFPA 13, *Standard for the Installation of Sprinkler Systems.*
NFPA 20, *Standard for the Installation of Centrifugal Fire Pumps.*
NFPA 24, *Standard for the Installation of Private Fire Service Mains and Their Appurtenances.*
NFPA 26, *Recommended Practice for the Supervision of Valves Controlling Water Supplies for Fire Protection.*
NFPA 291, *Recommended Practice for Fire Flow Testing and Marking of Fire Hydrants.*

Other Codes and Standards

Standards

AWWA Handbooks and Standards, American Water Works Association, 6666 W. Quincy Ave., Denver CO 80235 (see also ANSI A21 listings).
AWWA C200-80, Standard for Steel Water Pipe 6 in. and Larger.
AWWA C203-78, Standard for Coal-Tar Protective Coatings and Linings for Steel Water Pipe Lines—Enamel and Tape—Hot Applied.
AWWA C205-80, Standard for Cement-Mortar Protective Lining and Coating for Steel Water Pipe—4 in. and Larger—Shop Applied.
AWWA C300-82, Standard for Reinforced Concrete Pressure Pipe—Steel Cylinder Type, for Water and Other Liquids.
AWWA C301-84, Standard for Prestressed Concrete Pressure Pipe—Steel Cylinder Type, for Water and Other Liquids.
AWWA C302-74, Standard for Reinforced Concrete Pressure Pipe— Noncylinder Type, for Water and Other Liquids.
AWWA 303-78, Standard for Reinforced Concrete Pressure Pipe—Steel Cylinder Type, Pretensioned, for Water and Other Liquids.
AWWA C400-80, Standard for Asbestos Cement Distribution Pipe 4 in. through 16 in. for Water and Other Liquids.
AWWA C401-83, Standard Practice for the Selection of Asbestos Cement Distribution Pipe; 4 in. through 16 in. for Water and Other Liquids.
AWWA C500-80, Standard for Gate Valves 3 in. through 48 in. NPS for Water and Sewage Systems.
AWWA C502-80, Standard for Dry Barrel Fire Hydrants.
AWWA C503-82, Standard for Wet Barrel Fire Hydrants.
AWWA C506-78 (R83), Standard for Backflow Prevention Sevices—Reduced Pressure Principle and Double Check Valve Types.
AWWA C600-82, Standard for Installation of Ductile Iron Water Mains and Appurtenances.
AWWA C603-78, Standard for Installation of Asbestos Cement Pressure Pipe.
AWWA C703-79, Standard for Cold Water Meters—Fire Service Type.
AWWA C900-81, Standard for Polyvinyl Chloride (PVC) Pressure Pipe, 4 in. through 12 in. for Water.

Manuals

AWWA M6 (30006)-73, Water Meters—Selection Installation, Testing, and Maintenance.
AWWA M9 (30009)-79, Concrete Pressure Pipe.
AWWA M11 (30011)-64, Steel Pipe—Design and Installation.
AWWA M14 (30014)-70, Recommended Practice for Backflow Prevention and Cross-Connection Control.
AWWA M17 (30017)-70, Installation, Operation, and Maintenance of Fire Hydrants.
AWWA M23 (30023)-80, PVC Pipe.

ULC Publications, Underwriters' Laboratories of Canada, 7 Crouse Road, Scarborough, Ontario.
ULC No. C-246, Hydrants, 1974.
ULC No. C-312, Check Valves for Fire Protection, 1981.
ULC No. C-789, Indicator Posts for Fire Protection, 1975.
UL Standards, Underwriters Laboratories Inc., 333 Pfingsten Road, Northbrook, IL 60062.
UL No. 107, Asbestos Cement Pipe and Couplings, 1980.
UL No. 246, Hydrants for Fire Protection Service, 1979 (R82).
UL No. 262, Gate Valves for Fire Protection Service, 1980 (R-84).
UL No. 312, Check Valves for Fire Protection Service, 1980 (R-84).
UL No. 385, Play Pipes for Water Supply Testing in Fire Protection Service 1980 (R82).
UL No. 789, Indicator Posts for Fire Protection Service, 1982 (R-84).
UL No. 888, Steel Pipe for Underground Water Service, 1978.
UL No. 194, Gasketed Joints for Ductile-Iron and Gray-Iron Pressure Pipe and Fittings, for Fire Protection Service 1980 (R-84).
UL No. 753, Alarm Accessories for Automatic Water Supply Control Valves for Fire Protection Service, 1982 (R83).

American National Standards Institute, Sectional Committee A21. (This committee is sponsored jointly by American Gas Association, American Society for Testing and Materials, American Water Works Association, and New England Water Works Association.) The following manuals and standards are published by American Water Works Association, 6666 W. Quincy Ave., Denver, CO 80235:
AWWA C104/A21.4-80, American National Standard for Cement-Mortar Lining for Ductile-Iron and Gray-Iron Pipe and Fittings for Water.
AWWA C110/A21.10-82, American National Standard for Gray-Iron and Ductile-Iron Fittings, 3 in. through 48 in., for Water and Other Liquids.
AWWA C111/A21.11-80, American National Standard for Rubber Gasket Joints for Ductile-Iron and Gray-Iron Pressure Pipe and Fittings.

AWWA C150/A21.50-81, American National Standard for the Thickness Design of Ductile-Iron Pipe.

AWWA C151/A21.51-81, American National Standard for Ductile-Iron Pipe, Centrifugally Cast in Metal Molds or Sand-Lined Molds, for Water or Other Liquids.

Additional Readings

AWWA Committee on Financial Aspects of Fire Prevention and Protection, *A Business-like Approach to Fire Protection Changes*, AWWA, Denver, 1974.

Angele, G. J., "Backflow Prevention and Cross-Connection Control," *American Water Works Association Journal*, Vol. 62, No. 6, June 1970.

Babbit, Harold E., Cleasby, John L., and Doland, James, J., *Water Supply Engineering*, 6th ed., McGraw-Hill, NY, 1962.

Crocker, C., ed., *Piping Handbook*, 5th ed., McGraw-Hill, NY, 1967.

Fair, G. M., Geyer, J. C., and Okum, D. A., *Water and Wastewater Engineering, Vol. I, Water Supply and Wastewater Removal*, John Wiley & Sons, NY, 1966.

Handbook of Ductile Iron Pipe, Ductile Iron Pipe Research Association, Chicago, 1979.

Jensen, Rolf, ed., *Fire Protection for the Design Professional*, Cahners, Boston, 1975, pp. 41-57.

"Liability for Hydrant Maintenance," *Fire Journal*, Vol. 59, No. 5, Sept. 1965, p. 23.

WATER STORAGE FACILITIES AND SUCTION SUPPLIES

Revised by Robert M. Hodnett

In a broad sense, water storage facilities and suction supplies include all bodies of water available as sources of supply, whether they are contained by constructed or natural barriers. Elevated or ground level storage tanks of metal, wood, or rubberized fabric are examples of constructed storage facilities; rivers, ponds, and harbors are examples of natural storage facilities.

Open bodies of water, such as reservoirs created by the damming of streams, are sometimes used in private fire protection to supplement public water supplies or furnish the primary source of water for fire protection if public supplies are insufficient in volume or pressure, or both, or if they lack dependability. Such arrangements are, however, feasible only in special situations. A common method is to use elevated gravity tanks or ground level suction tanks with fire pumps. Pressure tanks, with their limited capacity, may be used where storage requirements are relatively small.

This chapter contains information on the design, installation and maintenance of man-made water storage facilities and the ways in which naturally occurring surface and ground water supplies can be used for fire protection. Of particular interest are the provisions described for preventing water in storage from freezing during inclimate months.

Water distribution systems and fire pumps (depending on the nature of the storage facilities), both integral parts of water supplies for fire protection systems, are covered in Chapters 4 and 6, respectively in this Section. Water based extinguishing systems are covered in the chapters of Section 18.

STORAGE TANKS (GRAVITY AND SUCTION)

At the outset, it is well to recognize that with the advent of hydraulically designed sprinkler systems in recent years, the use of elevated tanks for fire protection has declined; however, the use of ground level suction tanks, combined with fire pumps, has increased. Nevertheless, there are many elevated gravity tanks still in service used solely for fire protection service, and they require high standards of maintenance to continue their reliability as primary sources of water for extinguishing systems.

It is best if tanks for fire protection are not used for any other purpose. Tanks used for other purposes must usually be refilled frequently, and they become settling basins that collect large accumulations of sediment. When water is drawn from the tank, the sediment also is drawn into the yard or extinguishing system and may block the system.

If the tank is wooden and is frequently refilled, the alternate drying and wetting of the lumber may appreciably shorten the life of the tank; with a steel tank more frequent painting is required, which means not only greater expense, but more time out of service.

Another important consideration involving dual purpose tanks is the water available at the time of the fire. Such tanks will seldom be full, because domestic and industrial consumption constantly draws it down. In addition, the normal water level may drop lower and lower if the industry grows. If a fire occurs several years after the tank is installed, sufficient water at sufficient head might not be available.

Location

A gravity tank supported on an independent steel tower with foundations placed in the ground rather than on a building is the best arrangement. The tank should be built so that it will not be subject to fire exposure from adjacent buildings. If lack of yard room makes this impossible, the exposed steelwork should be suitably protected by fire resistant construction or coverings. The protection, when necessary, should include steelwork within 20 ft (6 m) of combustible buildings or openings from which fire might issue.

If the tank or supporting trestle is to be placed on the walls of a new building, the latter should be designed and built to carry the maximum loads.

Suction tanks should be located so as to minimize yard piping. The pump house is generally placed close to

---

Mr. Hodnett is a consultant in fire protection and piping systems based at Providence, RI. He was NFPA staff liaison to the NFPA Water Extinguishing Systems Committees prior to his retirement in 1984.

the tanks to minimize suction piping. The tanks should not be put where they will be exposed to fires in combustible construction or fires issuing from windows. (See Fig. 17-5A.)

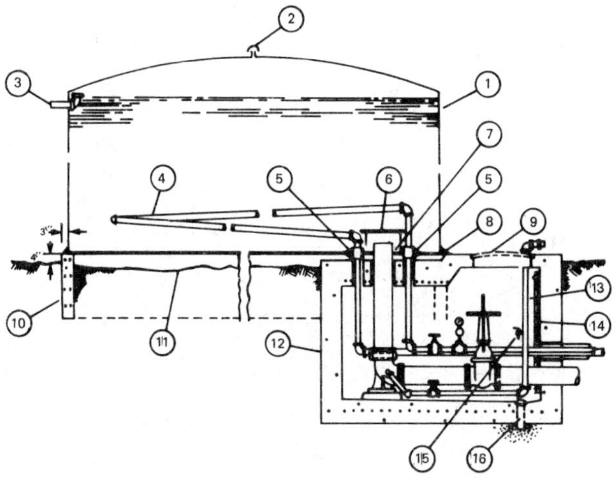

FIG. 17-5A. A ground level suction tank showing the discharge pipe connected to the bottom of the tank in a valve pit. Nomenclature: (1) pump suction tank, (2) screened vent, (3) stub overflow pipe, (4) steam coil for heating, (5) extra-heavy couplings welded to tank bottom, (6) vortex plate, (7) watertight lead slip joint (8) flashing around tank, (9) manhole with cover, (10) concrete ring wall, (11) sand or concrete pad (depending on soil condition), (12) valve pit, (13) drain pipe, (14) ladder, (15) drain cock, and (16) valve pit drain. (Factory Mutual System)

Design For Earthquake Resistance

Storage tanks can be designed to resist earthquakes. In an earthquake, the motion of the earth sets up a whip-like action in elevated tanks. These motions produce stresses beyond those provided for in a design which allows only for ordinary dead, live, and wind loads.

One method of earthquake resistant design allows an extra 10 percent horizontal load factor for elevated tanks. This theory was given a practical test in May 1940, when an earthquake struck the Imperial Valley in Southern California. There were eight steel gravity tanks in the area for municipal water systems. Two, designed with this extra 10 percent horizontal load factor, stood up without damage. After the earthquake, four towers not so designed were dangerous because the stresses set up by the earth's motion stretched or broke the brace rods, especially near the top of the tower. The other two of the tanks collapsed completely.

Suction tanks are affected less by earthquakes than elevated tanks; however, they should have good supports and foundations. Pipe connections between the supplies and the pumps should be designed to protect against earthquakes.

Tank Capacities

It is usually economical to install a gravity tank big enough and tall enough that it can be connected directly into the fire protection system, thus furnishing a supply for both hose streams from hydrants and automatic fire extinguishing systems. This means that a tank should hold at least 30,000 gal (110 m^3) and that its bottom should be at least 75 ft (23 m) above the ground if the tank is to supply adequately both hose lines from hydrants and the extinguishing systems. Tanks which hold less than 30,000 gal (110 m^3) for gravity tanks are acceptable only in special cases. For unsprinklered buildings, where hose streams are relied on, the bottom of the tank should be at least 40 ft (12 m) above the highest building, but not less than 75 ft (23 m) above the ground.

Gravity and suction tanks are generally erected in standard sizes. (See Tables 17-5A and 17-5B.) The capacity required is determined by the intended use of the tank and is specified in the number of gallons (cubic meters) available from the tank (1 gal = 0.00378 m^3). The capacities of cylindrical tanks of various diameters are given in Table 17-5C.

TABLE 17-5A. Standard Sizes of Gravity Tanks

Steel Tanks		Wooden Tanks		Standard Height	
gal	m^3*	gal	m^3*	ft	m*
5,000	20	5,000	20	75	22.9
10,000	40	10,000	40	100	30.5
15,000	60	15,000	60	125	38.1
20,000	80	20,000	80	150	45.7
25,000	100	25,000	100		
30,000	115	30,000	115		
40,000	150	40,000	150		
50,000	190	50,000	190		
60,000	230	60,000	230		
75,000	290	75,000	290		
100,000	380	100,000	380		
150,000	570				
200,000	760				
300,000	1100				
500,000	2000				

* Figures rounded off as approximations from nominal customary American tank sizes.

TABLE 17-5B. Common Sizes of Steel Pump Suction Tanks

gal	m^3	gal	m^3*
50,000	190	250,000	950
75,000	290	300,000	1100
100,000	380	400,000	1500
125,000	475	500,000	2000
150,000	575	750,000	3000
200,000	750	1,000,000	4000

* Figures rounded off as approximations from nominal customary American tank sizes.

Steel gravity tanks with suspended bottoms are usually erected on towers with four columns for capacities from 50,000 to 200,000 gal (190 to 750 m^3) inclusive, six for capacities from 200,000 to 300,000 gal (750 to 1100 m^3) inclusive, and eight for tanks over 300,000 gal (1135.50 m^3).

TABLE 17-5C. Capacities of Cylindrical Tanks

Diameter		Capacity per unit of depth		Capacity in Hemispherical Bottom		Capacity in Ellipsoidal Bottom	
ft	m	gal/ft	L/m	gal	L	gal	L
11	3.4	711	8 829	2610	9 879	1300	4 921
12	3.7	846	10 506	3390	12 831	1590	6 018
13	4.0	993	12 331	4300	16 276	2150	8 138
14	4.3	1152	14 306	5370	20 325	2690	10 182
15	4.6	1322	16 417	6610	25 019	3310	12 528
16	4.9	1504	18 677	8020	30 356	4010	15 178
17	5.2	1698	21 086	9620	36 412	4810	18 206
18	5.5	1904	23 644	11,400	43 149	5710	21 612
19	5.8	2121	26 339	13,400	50 719	6720	25 435
20	6.1	2350	29 182	15,700	59,425	7830	29 637
21	6.4	2591	32 175	18,100	65,509	9070	34 330
22	6.7	2844	35 317	20,900	79 107	10,400	39 364
23	7.0	3108	38 595	23,800	90,083	11,900	45 042
24	7.3	3384	42 022	27,100	102 574	13,500	51 098
25	7.6	3672	45 599	30,600	115 821	15,300	57 911
26	7.9	3972	49 324	34,400	130 204	17,200	65 102
27	8.2	4283	53 186	38,500	145 723	19,300	73 051
28	8.5	4606	57 197	43,000	162 755	21,500	81 378
29	8.8	4941	61 357	47,800	180 923	23,900	90 462
30	9.1	5288	65 666	52,900	200 227	26,400	99 924

Construction of Tanks

Gravity tanks are usually built of wood or steel and are supported by steel towers. Reinforced concrete towers are sometimes used, and tanks can also be placed directly on top of the structures they supply. In a few cases, concrete has also been used for the tank shells themselves. Typical gravity tanks are shown in Figures 17-5B and 17-5C.

Tanks should be designed and installed according to NFPA 22, *Water Tanks for Private Fire Protection*, hereinafter referred to as the NFPA Water Tank Standard. The Standard gives full requirements for construction materials, loads, unit stresses, details of design, foundations, accessories, and workmanship. Welding of towers should conform to code requirements for welding in building construction (AWS 1984).

Steel for tanks and towers should conform to the specifications in NFPA 22, *Standard for Water Tanks for Private Protection* hereafter referred to as the NFPA Water Tank Standard. Chief among these specifications is the American Waterworks Association standard for steel tanks (AWWA 1984), which gives the thickness of steel plates and the welding practices which should be followed. Other standards referenced in the NFPA Water Tank Standard cover steel shapes, plate materials, bolts, anchor bolts and rods, forgings, castings, reinforcing steel, and filler material for welding.

The NFPA Water Tank Standard also gives information on wood tanks, including types and dimensions of wood suitable for use, processes which should be used, and proper hoop materials and design.

Unit loads and unit stresses for steel tanks and towers, and working stresses for timber for wood tanks are given in the NFPA Water Tank Standard.

Steel tanks and towers should be riveted or welded. Unfinished bolts should only be used in field connections of nonadjustable tension members carrying wind stress and in-field connections of compression members and grillages in towers supporting tanks of 30,000 gal (110 m^3) or less capacity.

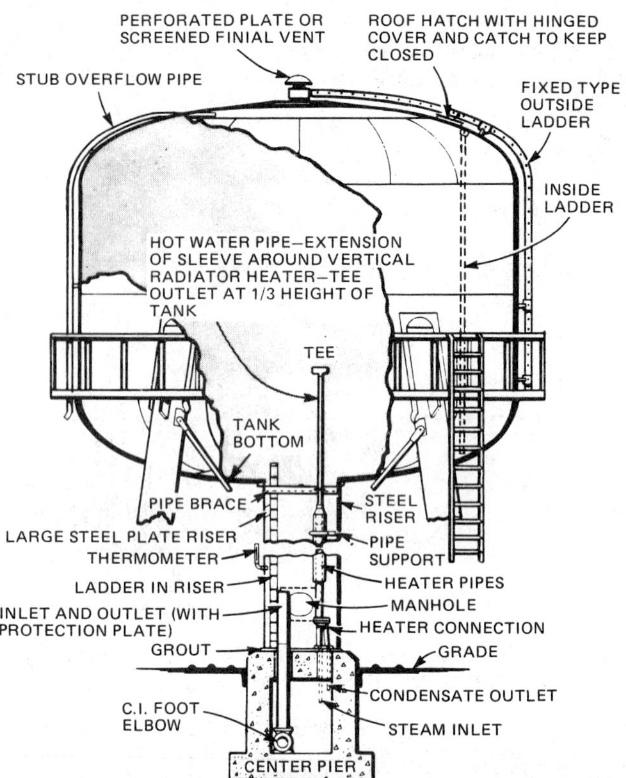

FIG. 17-5B. A typical tower-supported double ellipsoidal tank.

During assembly and erection, plates should be bolted firmly together before riveting. Drift pins should not be used to bring parts together or to enlarge unfair holes.

No waste material, such as boards, roofing, paint cans, etc., should be left in the tank or in the space at the top of the tank after its completion, because it may get into water and obstruct the piping.

Tanks should be promptly put in service after completion. Wooden tanks may be damaged by shrinkage if left empty.

TANK AND TOWER FOUNDATIONS

The following are principles of good foundations for suction tanks and gravity tank towers.

Foundations in the Ground

Material: Foundations should be built of concrete with compressive strength not less than 3,000 psi (20.69 mPa). The cement and aggregates should conform to the current American Concrete Institute requirements for reinforced concrete (ACI 1983). The maximum aggregate size for unreinforced concrete is 3 in. (75 mm) and 1½ in. (38 mm) for reinforced concrete.

Steel or wood pump suction tanks should be set upon crushed stone, sand, or concrete foundations. If the soil is good, at least 4 in. (100 mm) of crushed stone or sand is suggested. The material should then be saturated with oil and laid on moistened and compacted gravel after remov-

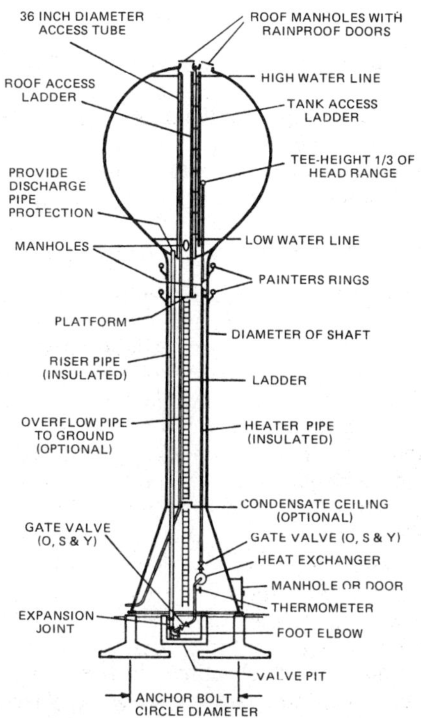

FIG. 17-5C. A typical pedestal tank. The higher the tank, the greater the gravity pressure.

ing soft surface. Special consideration (outlined below) must be given if soil is poor.

A concrete ring wall at least 2½ ft (760 mm) deep and 10 in. (250 mm) thick should surround the tank foundation. This ring normally projects 6 in. (150 mm) above grade and includes reinforcing steel equal to 0.25 percent of the cross sectional area. If the ring wall is outside of the tank shell, asphalt flashing should be installed between the tank and the ring wall at ground level.

For poor soil, an 8 in. (200 mm) reinforced concrete slab with a concrete ring wall directly beneath the tank sides and extended below the frost line is advised. For riveted tanks, a 1½ in. (38 mm) layer of a dry mixture of sand and cement should be placed on top of the concrete slab. For tanks of welded construction, no sand cushion is needed on the concrete slab. Piles can be used in addition to the reinforced concrete slab if the soil is very poor.

Form: The tops of foundations should be level and at least 6 in. (150 mm) above ground. The bottoms of foundation piers for towers should be located below the frost line, and in the case of piers, at least 4 ft (1.2 m) below grade, resting on thoroughly tamped soil or rock.

Piers: Pier foundations may be of any suitable shape and may be either plain or reinforced concrete. If they support a tower, their center of gravity should lie in the continued center of gravity line of the tower column or else be designed for eccentricity. The height of piers should not be less than the mean width. The top surface should extend at least 3 in. (75 mm) beyond the bearing plates on all sides and is generally chamfered at the edges.

Anchorage: The weight of piers should be sufficient to resist the maximum net uplift that occurs when the wind blows from any direction on the empty tank. The weight of earth directly above the base of the pier may be included in the calculations.

Anchor bolts should be arranged to engage a weight at least equal to the net uplift with the tank empty and the wind blowing from any direction. Their lower ends should be hooked or fitted with anchor plates.

Anchor bolts should be accurately located with sufficient free length of thread to fully engage their nuts. Expansion bolts are not acceptable. The minimum size of anchor bolts should be 1½ in. (38 mm).

Grouting and Flashing: Bearing or base plates should have complete bearing on the foundation or be laid on cement grout to secure a complete bearing. The stressed portion of anchor bolts should not be exposed except where necessary. If the stressed portions of anchor bolts must be exposed, they should be encased in cement mortar to protect against corrosion unless they are accessible for complete cleaning and painting. If structural shapes, plates, and bolts enter or are supported by masonry or concrete, the joint between the metal and masonry or concrete should be flashed with asphalt. (This does not refer to base plates under columns.)

Soil-bearing Pressure: In order to find the proper depth of the foundation, the soil-bearing pressure must be determined by analyzing the subsurface and checking the foundations of other structures in the area. Test borings should be made by or under the supervision of an experienced soils engineer or soils testing laboratory. The borings should be made deep enough to determine the adequacy of the support, which is usually a minimum of 20 to 30 ft (6 to 9 m).

Foundations: The foundations should be designed to carry the maximum loads without excessive settlement. If wooden piles are used above permanent low ground-water level, they should be protected as specified by the American Wood Preservers Association (AWPA 1983).

Foundations should not be constructed over buried pipes or immediately adjacent to existing or former deep excavations unless the foundation bases go below the excavation.

Center Pier: In addition to the weight of the water in a large plate riser, the weight of the column of water directly above the riser in the tank, and the weight of the steel plate, the center pier should also be considered as supporting a hollow cylinder of water in the tank. If the hemispherical or ellipsoidal bottom is rigidly attached to the top of the large riser by a flat horizontal diaphragm plate, the radius of the hollow cylinder of water should be determined as specified under large risers discussed elsewhere in this chapter.

TANK TOWERS

Steel is generally used for the construction of tank towers. Details on the specific types of steel used in tanks may be found in the NFPA Water Tank Standard.

Both live loads and dead loads should be considered when designing towers. The dead load is an estimate of the weight of the structure and all its fittings. The live load is considered as the weight of the water when the tank is filled and overflowing. Consideration should also be given to temporary loading such as ice and snow.

The weight of the water in the riser need not be considered when figuring loads unless the riser is suspended from the tank bottom. If the riser is used to support the tank bottom, the entire weight supported by the riser should be considered, including the weight of the water.

Additional factors to consider are wind, balcony, and earthquake loads. Wind loads are based on 30 psi (207 kPa) for vertical plain surfaces and 18 psi (124 kPa) over the vertical projection of cylindrical surfaces. These loads are applied to the center of gravity of the projected areas. The balcony load consists of the weight of the balcony railing and ladder material and, if exposed, the weight of snow. Earthquake loading must be designed specifically for local conditions.

Ladders and Balconies

All tanks should have ladders on both the outside and inside of tanks, with convenient passage from one ladder to the other. Ladders are for inspection and maintenance of the tank's interior and exterior surfaces; they should be constructed of materials compatible to those of the tank and tower, and they should be easily accessible. Ladders more than 20 ft (6 m) long should be equipped with a cage or other safety device designed to protect the climber.

Balconies and walkways are recommended for towers over 20 ft (6 m) high. Normally balconies should not be less than 24 in. (600 mm) wide and walkways not less than 18 in. (450 mm) wide. Railings should be at least 42 in. (1050 mm) high. Balconies and walkways should also be made of materials compatible with the tank and tower materials.

Tower Construction

During field erection the columns of towers should be built on thin metal wedges driven to equal resistance so that all columns will be loaded equally after the structure is completed. The spaces beneath the base plates and the anchor holes should be completely filled with cement mortar.

Sections of structural members for towers should be symmetrical and built of standard structural shapes or tubular sections. Structural shapes should be designed with open sections so that all corrosible surfaces which will be exposed to air or moisture can be painted. Tubular sections of columns and struts should be airtight.

TANK HEATING EQUIPMENT

Adequate heating of tank equipment ranks next to structural design in importance. An ice plug in a riser pipe may make the tank water unavailable in case of fire and may break the pipe. Ice in or on tank structures has been the direct cause of collapse in several cases. The heating system must therefore be reliable and allow convenient and economical operation to 50°F (10°C). However, overheating can seriously damage wooden tanks and the paint on steel tanks and is to be avoided.

Determination of Heater Capacity

To prevent freezing in any part of a tank during the coldest weather, the heating system must replace the heat lost from the tank and piping when the temperature of the coldest water is just safely above the freezing point and the mean atmospheric temperature for one day is at its lowest. (See Fig. 17-5D for lowest one-day mean temperatures.)

Tables 17-5D and 17-5E show the heat losses from tanks of common sizes exposed to various atmospheric temperatures. The following is an example of the method used to determine required heating capacities for various types and sizes of tanks:

Question: What heater capacity would be needed for a 75,000 gal (280 m^3) steel tank at Duluth, MN? Answer by interpolating from Figure 17-5D. The lowest mean temperature for one day at Duluth is −28°F (−33°C), and by interpolating from Table 17-5E it is found that the heat loss at −28°F (-33°C) for a 75,000 gal (284 m^3) steel tank is approximately 659,000 Btu per hr (193,000 W).

Answer: A heater capable of delivering 659,000 Btu per hr (193,000 W) under field conditions.

Selection of Heating Method

The selection of a heating method depends chiefly upon a tank's height, construction material, size, shape, and lowest temperature of exposure. The recommended methods of heating are covered in detail in the NFPA Water Tank Standard.

There are three basic methods of heating tank water: (1) gravity circulation of hot water, (2) steam coils inside tanks, and (3) direct discharge of steam into the water.

Gravity Circulation of Hot Water

Heating by gravity circulation is dependable and economical if correctly planned. Cold water received through a connection from the discharge pipe or from near the bottom of a suction tank or standpipe is heated and rises through a separate hot water pipe into the tank. Steam, coal burning, or oil heaters are ordinarily used; gas heaters or electric heaters are also satisfactory.

A steam heater ordinarily consists of a cast iron or steel shell through which water circulates by gravity

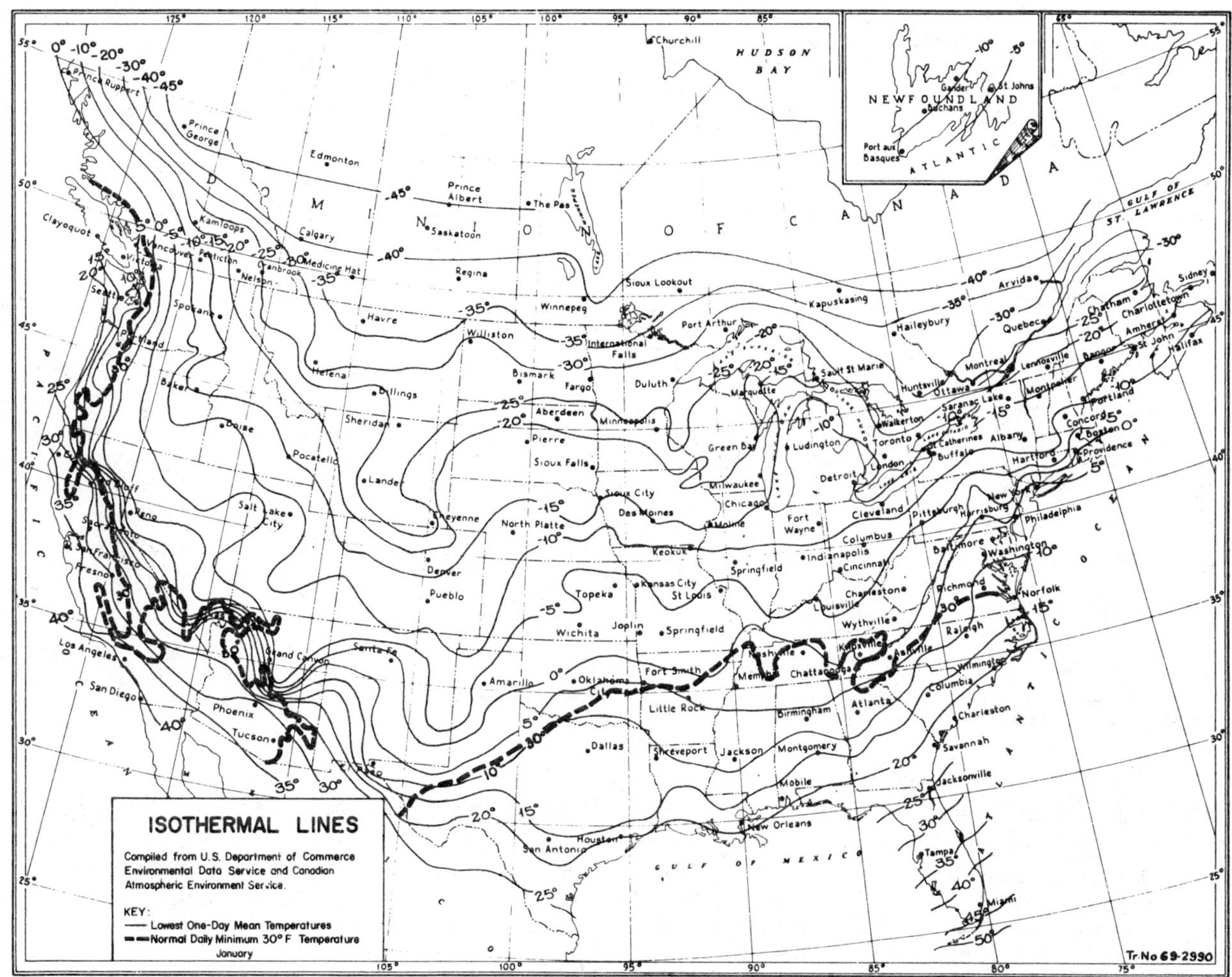

FIG. 17-5D. Isothermal lines—lowest one day temperature mean temperature.

around steam tubes or coils of brass or copper. It should be located in a valve pit, heater house, or in a nearby building at or near the base of the tank. When the tank is over a building, the steam heater should be located in the top story.

Heater Thermometer: The convenience of a gravity system is that it permits discovery of the temperature of the coldest water in the system. A recording thermometer should be placed in the cold water return pipe near the heater; it should be checked frequently to make sure that the temperature does not fall below 42°F (5.6°C). To fail to observe the temperature is to risk freezing the equipment. [Water has its maximum density at 39.2°F (4°C). When the temperature of the water falls below 39.2°F (4°C), there is an inversion and the warmer water settles to the bottom of the tank while the colder water rises. Therefore, if the circulation heater is to be effective, sufficient heat must be provided so that the temperature of the coldest water will be above 42°F (5.6°C).]

Water Circulating Pipes: The size of pipe used for the circulation of heating water is given in Table 17-5F. Pipe should be either copper water tubing or brass (85 percent copper) throughout. The hot water discharges into the tank through a tee fitting at the end of the hot water pipe at a height about one-third up from the bottom of the tank. The return pipe connects into the discharge pipe at a point that guarantees circulation throughout the portion of the discharge pipe subject to freezing. A typical arrangement of a circulating heater and piping in a valve pit is shown in Figure 17-5E.

Steam Coils Inside Tanks

Steam coils inside the tank do not permit convenient observation of water temperatures and have other faults which make them unsuited for heating elevated tanks except in areas of the South where only intermittent heating is necessary. This method may, however, be used for heating suction tanks and standpipes with flat bottoms supported near the ground level if coils are submerged continuously. The coil consists of at least 1¼ in. (32 mm) brass or copper pipe, pitched to drain, supplied with steam at not less than 10 psi (69 kPa) pressure through a pipe of sufficient size to furnish the required quantity of steam from a reliable source.

TABLE 17-5D. Heat Loss from Standpipes and Steel Suction Tanks

Thousands of British thermal units lost per hour when the temperature of the coldest water is 42°F (5.6°C). To determine capacity of heater needed, find the minimum mean atmospheric temperature for one day from the Isothermal Map, Figure 17-5D, and note the corresponding heat loss below.

Atmospheric Temperature Deg. F	TANK CAPACITIES—Thousands of Gallons							
	100	150	200	300	400	500	750	1,000
35	85	114	135	175	206	238	312	380
30	121	162	193	248	294	340	445	542
25	161	216	257	330	393	453	594	722
20	202	271	323	414	493	568	745	907
15	245	329	391	502	597	689	904	1,099
10	290	389	463	595	707	816	1,071	1,302
5	337	452	539	691	822	949	1,244	1,514
0	388	521	620	796	947	1,093	1,434	1,744
—5	441	592	705	905	1,076	1,241	1,628	1,981
—10	498	669	797	1,023	1,216	1,403	1,841	2,239
—15	557	748	891	1,143	1,360	1,569	2,058	2,503
—20	619	830	989	1,270	1,510	1,742	2,286	2,781
—25	685	920	1,096	1,406	1,673	1,930	2,532	3,080
—30	752	1,010	1,203	1,545	1,837	2,119	2,781	3,383
—35	825	1,108	1,320	1,694	2,015	2,325	3,050	3,710
—40	898	1,206	1,437	1,844	2,193	2,531	3,320	4,039
—50	1,059	1,422	1,694	2,175	2,586	2,984	3,915	4,762
—60	1,229	1,651	1,966	2,524	3,002	3,463	4,544	5,528

For S.I. Units: BTU/hr = 0.293 W; °C = 5/9 (F° −32); 1,000 gal = 3.785 m^3.

Direct Discharge of Steam

Steam from a reliable supply is blown directly into the tank water through a pipe that enters the tank through the bottom, extends to above the maximum water level, and then returns to a point 3 or 4 ft (0.9 or 1.2 m) below the normal fire service level. An air vent and check valve in the pipe above the surface of the water keeps the water from siphoning back down the steam line. This method is employed where the lowest mean temperature for one day is 5°F (−15°C) or above.

Solar Heating of Elevated Steel Tanks

With the advent of the energy crisis and the increase in fossil fuel prices, work was done in Canada on the feasibility of solar heating of elevated steel tanks (NRCC 1984). The basic technique is to insulate the entire tanks leaving a window on the south side through the insulation. The tank shell in the opening is painted flat black to absorb heat from the sun and the opening is double glazed.

Retrofit installations of solar systems on existing tanks involves special precautions in the attachment of components to the tank, such as the use of adhesives. The costs would be relatively high and the payback period long.

For new tanks, low cost welding techniques can be used for attachment of collector components to the tank. With substantially all energy being supplied by the sun, a backup heating system can be a low cost electrical system. Compared to a conventional tank with a fossil-fired boiler heating system, a solar heated tank with an electrical backup heating system is likely to have a lower capital cost.

The preliminary findings would indicate that solar heated water tanks would seem appropriate for new installations.

TANK EQUIPMENT AND ACCESSORIES

A complete installation of a gravity or suction tank includes the necessary pipe connections, valve enclosures, and where appropriate for the particular tank, such features as a frostproof casing for the riser.

Valve Pits

A pit 7 ft (2.1 m) deep and 6 ft by 9 ft (1.8 by 2.7 m) inside is usually large enough to house the necessary valves, tank heaters, and other fittings. Details of the construction of the pit and its arrangement, including clearance around equipment, water-proofing, manhole and ladder, and drainage, should conform to the recommendations given for valve pits.

Valve Enclosures

The tank heater and other fittings are sometimes installed in an enclosure above grade. In such a case, the indicating valve and the check valve are generally placed in the horizontal pipe below the frost line in a small pit heated sufficiently to maintain a temperature of at least 40°F (4.4°C) during the most severe weather.

The enclosure may be of concrete, brick, cement plaster on metal lath, or any other noncombustible material with suitable heat insulating properties. The roof should be strong enough to support the frostproof casing and other loads without excessive deflection.

Frostproof Casings

Except in the case of a large steel riser, a frostproof casing around all exposed tank piping is necessary in localities where the lowest mean atmospheric temperature for one day as shown by the isothermal map (Fig. 17-5D) is 20°F (−6.7°C) or lower. Tank piping subjected to temper-

TABLE 17-5E. Heat Loss from Elevated Tanks

Thousands of British thermal units lost per hour when the temperature of the coldest water is 42°F (5.6°C). To determine capacity of heater needed, find the minimum mean atmospheric temperature for one day from the Isothermal Map, Figure 17-5C, and note the corresponding heat loss below.

Atmospheric Temperature Deg. F	WOODEN TANKS—Capacities in Thousands of Gallons								
	10	15	20	25	30	40	50	75	100
35	8	10	11	13	14	19	21	28	33
30	11	14	16	19	21	27	31	40	49
25	15	20	21	25	28	36	42	54	65
20	19	25	27	32	35	46	54	69	83
15	24	31	34	39	44	57	66	85	102
10	28	36	40	46	51	68	78	100	121
5	33	43	47	54	60	78	92	117	142
0	38	49	53	62	69	90	106	135	164
—5	43	56	61	71	79	103	120	154	187
—10	49	63	69	80	89	116	136	174	211
—15	54	71	77	89	100	130	153	195	236
—20	61	79	86	99	111	145	169	217	262
—25	68	87	95	110	123	160	188	240	291
—30	74	96	104	121	135	176	206	264	319
—35	81	105	115	133	148	193	226	289	350
—40	88	114	125	144	162	210	246	317	382
—50	104	135	147	170	190	246	290	372	450
—60	122	157	171	197	222	266	307	407	490

Atmospheric Temperature Deg. F	STEEL TANKS—Capacities in Thousands of Gallons								See Note Below
	30	40	50	75	100	150	200	250	
35	43	51	59	77	92	120	145	168	69
30	62	72	83	110	132	171	207	242	192
25	82	96	111	146	175	228	275	323	340
20	103	120	139	183	220	287	346	405	506
15	125	146	169	222	267	347	419	491	692
10	147	172	200	263	316	411	496	582	893
5	171	200	233	306	367	478	577	676	1,092
0	197	231	268	352	423	551	664	779	1,309
—5	224	262	304	400	480	626	755	884	1,536
—10	253	296	344	452	543	707	853	1,000	1,771
—15	283	331	384	506	607	790	954	1,118	2,020
—20	314	368	427	562	674	878	1,059	1,241	2,291
—25	348	407	473	622	747	972	1,173	1,375	2,568
—30	382	447	519	683	820	1,068	1,288	1,510	2,860
—35	419	490	569	749	900	1,171	1,413	1,656	3,174
—40	456	534	620	816	979	1,275	1,538	1,803	3,494
—50	538	629	731	962	1,154	1,503	1,814	2,126	4,186
—60	624	730	848	1,116	1,340	1,745	2,105	2,467	4,936

Note: For each lineal foot of uninsulated riser 4 ft in diameter, add the number of Btu's in this column to the total Btu heat loss at the different temperatures for the various tank capacities.

For SI units: 1 Btu/hr = 0.293 W; °C = 5/9 (F° −32); 1,000 gal = 3.785 m^3.

atures below freezing within unheated buildings must also be adequately protected. Noncombustible frostproof casings should be used where there is danger of serious fire exposure.

Large Risers

Large steel plate riser pipes 3 ft (0.9 m) or more in diameter, without frostproof casings, are often desirable. The fire hazard and upkeep of the frost casing are thereby avoided, the expansion joint in the discharge pipe is eliminated, and it is not necessary to have a walkway to reach the valves. When the tank is on an independent tower, a concrete valve pit at the base of the discharge pipe and the pier are usually built as a single unit to support the riser. (See Fig. 17-5B.) On the other hand, the larger valve pit at the base of the riser makes the first cost more than for equipment with small risers.

Water Level Indicators

A water level gage of suitable design should be provided. A mercury gage is the most reliable water level indicator for tanks. A dependable, closed circuit, high and low water electrical alarm is a suitable substitute for the mercury gage in certain installations. The

TABLE 17-5F. Sizes of Circulating Pipes Required for Elevated Steel Tanks (inches)*

Minimum One-Day Mean Temp	TANK CAPACITY (Gallons)*									
Deg. F*	15,000	20,000	25,000	30,000	40,000	50,000	60,000	75,000	100,000	150,000
10	2	2	2	2	2	2	2	2	2	2½
5	2	2	2	2	2	2	2	2	2	2½
0	2	2	2	2	2	2	2	2	2½	2½
−5	2	2	2	2	2	2	2	2	2½	2½
−10	2	2	2	2	2	2	2	2½	2½	2½
−15	2	2	2	2	2	2	2½	2½	2½	3
−20	2	2	2	2	2	2½	2½	2½	2½	3
−25	2	2	2	2	2½	2½	2½	2½	3	3
−30	2	2	2	2	2½	2½	2½	2½	3	3
−35	2	2	2	2½	2½	2½	2½	3	3	3
−40	2	2	2	2½	2½	2½	2½	3	3	3

* For SI Units: °C = 5/9 (°F-32); 1000 gal = 3.785 m^3. Nominal pipe sizes are not directly convertible into metric sizes. For comparison, one inch equals 25.4 mm.

mercury gage is normally installed in a heated room, such as a boiler room, engine room, or office, where it will be readily accessible.

The gage should be accurately installed so that when the tank is filled to the level of the overflow, the mercury level is opposite the "FULL" mark on the gage board. The procedures for installing and testing mercury gages are given in detail in the NFPA Water Tank Standard.

Overflow Pipes

The overflow pipe located at the top capacity or high water line of the tank should not be less than 3 in. in diameter. If dripping water or a small accumulation of ice is not objectionable, the overflow may pass through the side of the tank near the top and extend not more than 4 ft (1.2 m), with a slight downward pitch to discharge beyond the balcony and away from the ladders.

When a stub pipe is undesirable, the overflow pipe can be extended down through the tank bottom and inside the frostproof casing or steel plate riser, discharging through the casing near the ground or roof level. The section of the pipe inside the tank should be brass except on tanks with steel plate risers, when overflow pipes 3½ in. (89 mm) or larger may be of extra-heavy wrought iron or of flanged cast-iron pipe.

Cathodic Protection

Cathodic protection can be used instead of painting to prevent interior corrosion on surfaces that are wetted by the tank's contents. Other surfaces inside the tank should be painted.

Internal corrosion is caused by galvanic current flowing from numerous anodic areas of the tank shell through the water to adjacent cathodic areas. Cathodic protection counteracts this process by passing sufficient direct current from an outside source (anodes suspended in the water) to the tank shell to keep the whole of the interior wetted surfaces of the tank at a negative potential. Low voltage direct current is supplied from a rectifier. Frequent and regular checking of the ammeter and voltmeter readings is advisable to determine whether the electrical system is functioning.

Aluminum anodes, which are common, require renewal annually. In unheated tanks, ice may damage the anodes in winter, and early servicing in the spring is necessary. Broken parts of the anodes should be removed from the tank.

For heated tanks, anodes of fine platinum wire (which is not deteriorated by the electric current) are sometimes used, but they cost much more than aluminum anodes.

EMBANKMENT SUPPORTED RUBBERIZED FABRIC SUCTION TANKS

ESRF (embankment supported rubberized fabric tanks) can be used as suction tanks for fire protection. The NFPA Water Tank Standard contains details on construction, installation, and maintenance of ESRF tanks.

ESRF tanks are available in 20,000 and 50,000 gal (75 and 190 m^3) sizes, and in 100,000 gal (378.5 m^3) increments up to 1 million gal (3785 m^3). The tank is usually composed of a reservoir liner with an integral flexible roof and is designed to be supported by earth underneath and on all four sides. The material of ESRF tanks is a nylon fabric coated with an elastomer compounded to provide resistance to abrasion and weather. Support is provided by a specifically prepared excavation and earthen berm, or both.

Preparation of the installation site is critical to the reliability of an ESRF tank. Usually, a shallow excavation the size of the bottom of the tank is made. The excavated earth is then graded to form the berm, or embankment, for the upper sides of the tank. The exterior of the berm should be graded to allow for rain and snow drainage and pipe connections, and the interior graded to match the contour of the tank including rounded corners. ESRF tanks may be placed underground with the top of the tank at grade level, or they may be placed aboveground where the earthen berm provides the entire support.

When the excavation meets the shape requirements of the tank and all sharp objects have been carefully removed from the floor of the tank excavation, a 6 in. (150 mm) layer of sand or clean soil is spread over a 3 in. (75 mm) underlayment of pea gravel which provides a firm base with good drainage. The tank is then placed in the foundation and all connections are made. (Tanks are shipped to

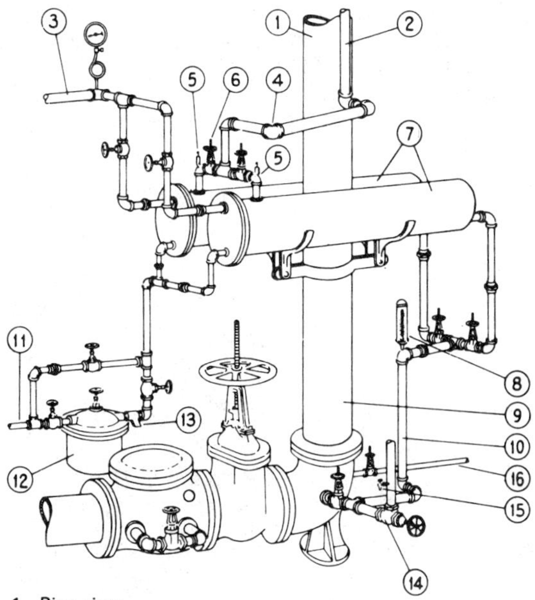

1. Pipe riser
2. Hot water circulating pipe
3. Steam supply pipe
4. Four-elbow swing joint
5. Relief valves
6. Approved OS&Y gate valves
7. Steam-heated water heaters
8. Thermometer
9. Pipe riser or discharge pipe from large steel-plate riser
10. Cold water circulating pipe
11. Condensate return
12. Steam trap
13. Strainer
14. Drain valve
15. Drain cock
16. From mercury gage

FIG. 17-5E. The piping arrangement for a multiunit steam heated water heater for a gravity circulation system. Nomenclature: (1) pipe riser, (2) hot water circulating system, (3) steam supply line, (4) four elbow joint, (5) relief valves, (6) O.S.&Y valve, (7) steam heated water heaters, (8) thermometer, (9) pipe riser, (10) cold water circulating pipe, (11) condensate return, (12) steam trap, (13) strainer, (14) drain valve, (15) drain cock, and (16) pipe from mercury gage. (Factory Mutual System)

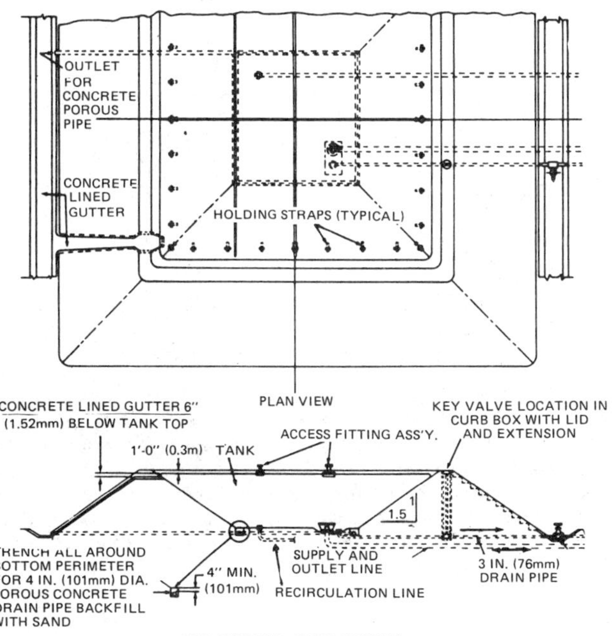

FIG. 17-5F. Installation details of a typical enbankment supported rubberized fabric tan, including fittings.

the site fully constructed.) After the tank is filled, a coating is applied to the exposed surface to protect it from the atmosphere. (See Fig. 17-5F for a typical ESRF installation.)

As with other types of storage tanks, water temperature in ESRF tanks should be maintained at not less than 42°F (5.6°C). An acceptable method of providing heat is a water recirculation system with a heat exchanger. (See Fig. 17-5G.) When the ambient temperature drops below 42°F (5.6°C), a thermostat activates a pump which draws water from the tank through an inlet-outlet fitting and pumps the heated water back into the tank through a recirculation fitting located in the bottom of the tank diagonally opposite from the inlet-outlet fitting. Heat loss in ESRF tanks are given in Table 17-5G.

PRESSURE TANKS

Pressure tanks are used for limited private fire protection services, such as sprinkler systems, standpipe and hose systems, and water spray systems. They are sometimes used with fire pumps and gravity tanks that are located at as high an elevation as possible for quicker discharge. A typical pressure tank installation is shown in Figure 17-5H.

Tank capacity is considered to be the total contents, both air and water, not including dished ends. For light hazard occupancies only, the tank may have a minimum capacity of 3,000 gal (11 m^3), of which 2,000 gal (7.5 m^3) is water, but the tank capacity ordinarily should be at least 4,500 gal (17 m^3) for ordinary hazard occupancies. Tanks for this service generally are not over 9,000 gal (34 m^3) capacity. For larger supplies more than one tank would be employed. (See Table 17-5H.)

The tank is normally kept two-thirds full of water, and an air pressure of at least 75 psig (517 kPa) maintained. As the last of the water leaves the pressure tank, the residual pressure shown on the gage should not be less than zero, and should give at least 15 psig (103 kPa) pressure at the highest automatic discharge device under the main roof of the building.

Air for pressure tanks is supplied by compressors capable of delivering not less than 16 cfm (0.045 m^3/min)

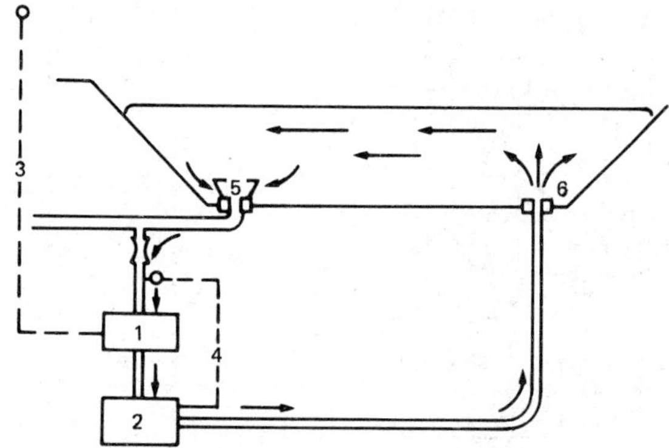

FIG. 17-5G. A schematic of a recirculation and heating system for an ESRF tank. Nomenclature and operation sequence are: (1) recirculation pump, (2) heat exchanger, (3) unit sensing atmosphere temperature starts pump and water recirculation enabling heat stored in the ground to transfer into water at a higher rate, (4) unit sensing water temperature starts heat exchanger when required, (5) inlet-outlet fitting, and (6) recirculation fitting.

of free air for tanks of 7,500 gal (28 m^3) total capacity, and not less than 20 cfm (0.057 m^3/min) for larger sizes. The compressors are located in the tank house.

Relationship of Air Pressure and Volume in Tanks

The volume of air in the tank at any time varies inversely with the pressure:

$$\frac{P_1}{P_2} = \frac{V_2}{V_1}.$$

Pressures are absolute pressures, not gage pressures, as used in the above general formula. Let us apply this formula to the special condition which exists in a tank used to supply water for sprinklers. Except where there is danger of air lock, it is desirable to have 15 psig (103 kPa) pressure in the tank when the last water leaves. Thus,

$$P_2 = \text{residual pressure required} + \text{atmospheric pressure},$$
$$= 15 \text{ psi} + 15 \text{ psi}$$
$$= 30 \text{ psi}.$$

Since at this condition the tank is full of air,

$$V_2 = 1.$$

If the tank is at or above the top line of sprinklers, the tank will be normally kept ⅔ full of water, so,

$$V_1 = 1/3.$$

Therefore,

$$P_1 = P_2 \times \frac{V_2}{V_1} = 30 \times \frac{1}{1/3} = 30 \times 3 = 90 \text{ psi}.$$

The corresponding gage pressure would be 90 psi (710 kPa) minus the atmospheric pressure, 15 psi (103 kPa), or 75 psi (517 kPa).

If the tank is below the top line of sprinklers, P_2, the required residual pressure would be increased 0.434 psi for every foot (0.912 kPa/m) of head represented in the distance between the base of the tank and the highest sprinkler. If we call this head H, we have

$$P_1 = P_2 \times \frac{V_2}{V_1} = (30 + 0.434H)\frac{V_2}{V_1}.$$

Gage pressure to be carried in tank would be P_1—15 or

$$(30 + 0.434H)\frac{V_2}{V_1} - 15.$$

For a tank ⅓ full of air $\frac{V_2}{V_1} - 3$, and gage pressure would be $3(30 + 0.434H) - 15 = 75 + 1.30H$.

For a tank ½ full of air $\frac{V_2}{V_1} - 2$, and gage pressure would be $2(30 + 0.434H) - 15 = 45 + 0.87H$.

Air Lock

A condition known as air lock can occur when a pressure tank and a gravity tank are connected into a sprinkler system through a common riser. Air lock occurs if the gravity water pressure at the gravity tank check valve is less than the air pressure trapped in the pressure tank and the common riser by a column of water in the sprinkler system, after water has been drained from the pressure tank. For instance, if the pressure tank is kept two-thirds full of water with an air pressure of 75 psi (517 kPa), and a sprinkler opens 35 ft or more above the point where connections from both tanks enter the common riser supplying the sprinkler system, the pressure tank drains, leaving an air pressure of 15 psi (103 kPa) balanced by a column of water of equal pressure (35 ft or 10.7 m head) in the sprinkler system. The gravity tank check valve is held closed unless the water pressure from the gravity tank is more than 15 psi (103 kPa) (35 ft or 10.7 m head).

Air lock can be prevented by increasing the volume of water and decreasing the air pressure in the pressure tank so that little or no air pressure remains after water has been exhausted. For example, if the pressure tank is kept four-fifths full of water, with an air pressure of 60 psi (414 kPa), the air pressure remaining in the tank after water has been drained is zero, and the gravity tank check valve opens as soon as the pressure at that point from the pressure tank drops below the static head from the gravity tank.

Air lock may be conveniently prevented in new equipment by connecting the gravity tank and pressure tank discharge pipes together 40 ft (12 m) or more below the bottom of the gravity tank. (See Fig. 17-5I.)

Construction of Pressure Tanks

Standard pressure tanks are constructed in accordance with the Boiler and Pressure Vessel Code (ASME 1983) with modifications given in the NFPA Water Tank

TABLE 17-5G. Heat Loss from Embankment Supported, Rubberized Fabric Suction Tanks

Thousands of British thermal units lost per hour when the temperature of the coldest water is 42 F. To determine capacity of heater needed, find the minimum mean atmospheric temperature for one day from the isothermal map, Figure 17-5D, and note the corresponding heat loss below.

Atmospheric Temperature Deg. F	Heat (BTU) Loss Per Sq Ft Tank Radiating Surface	TANK CAPACITIES—THOUSANDS U.S. GALLONS							
		100	200	300	400	500	600	800	1000
		Square Feet of Exposed Tank Surface							
		2746	4409	6037	7604	9139	10630	13572	16435
		BTU Lost Per Hour Thousands							
35	22.2	61	98	134	168	202	235	300	363
30	28.5	78	126	173	217	261	304	389	470
25	35.1	96	155	212	266	320	372	476	576
20	41.5	114	183	251	315	379	441	564	682
15	48.0	132	212	290	364	438	510	652	789
10	54.5	149	241	329	413	497	579	740	896
5	61.0	167	269	369	463	557	648	828	1003
0	67.5	185	298	408	512	616	717	916	1109
[1]5	73.9	203	326	447	561	675	786	1004	1216
[1]10	80.4	220	355	486	610	734	855	1092	1322
[1]15	86.8	238	384	525	659	793	924	1180	1429
[1]20	93.3	256	412	564	708	852	992	1268	1536
[1]25	99.9	273	441	604	758	912	1061	1356	1642
[1]30	106.2	291	469	643	807	971	1130	1444	1749
[1]40	119.3	327	526	721	905	1089	1268	1620	1962
[1]50	131.9	362	584	799	1003	1207	1406	1796	2175
[1]60	145.1	397	641	878	1102	1326	1544	1972	2389

Heat loss for a given capacity with a different radiating surface than shown above is obtained by multiplying the radiating surface by the tabulated heat loss per sq ft for the atmospheric temperature involved. The minimum radiation surface area is the wetted surface exposed to atmosphere. No heat loss is figured for tank bottoms resting on grade.

For SI Units: 1 ft = 0.3048 m; 1 ft^2 = 0.0929 m^2; 1 Btu/hr = 0.293 W; 1 Btu/ft^2 = 11.356 kJ/m^2; °C = $^5/_9$ (°F − 32); 1000 gal = 3.785 m^3.

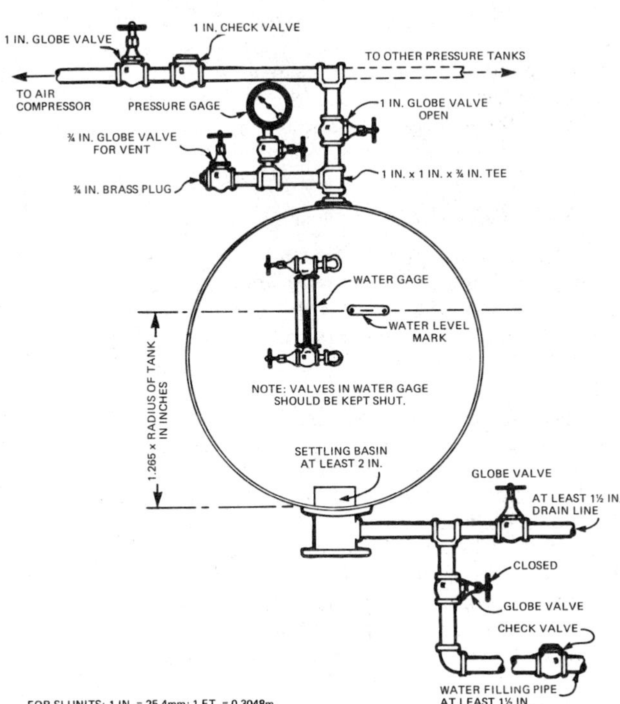

FIG. 17-5H. A typical pressure tank installation

Standard. Important modifications are a minimum hydrostatic test at 150 psi (1034 kPa) and a test for tightness.

Tanks should be located in substantial noncombustible housings unless they are within a heated room in a building. They should be large enough to provide free access to all connections, fittings, and manhole, with at least 3 ft (0.9 m) of clearance around the valves and gages and at least 18 in. (451 mm) around the rest of the tank. The distance between the floor and any part of a tank should be at least 3 ft (0.9 m).

The interiors of pressure tanks are inspected at 3 yr intervals to determine if corrosion is taking place and if repainting or repairing is needed. When necessary, tanks should be scraped, wire brushed, and repainted with an approved metal protective paint. Relief valves should be tested at least once each month.

Provisions should be made to drain each tank independently of all other tanks and have the sprinkler system drained by a pipe not less than 1½ in. in diameter.

The filling supply or pump should be reliable and capable of replenishing the water to be maintained in the tank against the normal tank pressure in not more than 4 hr.

NATURAL AND CONSTRUCTED SUCTION FACILITIES

The advantages of piped systems make them the first choice for a water supply for fire protection, but sometimes other water may be more convenient or less expensive.

Table 17-5H. Typical Dimensions of Horizontal Pressure Tanks of Standard Sizes

Approx Gross Capacity		Approx Net Cap ⅔ Full		Inside Diam		Inside Length		Approx Wt of Water ⅔ Full	
Gal	L	Gal	L	in.	m	Ft	m	Lbs	Kg
3,000	11355	2,000	7570	60	1.5	20.2	6.2	16,670	7568
3,000	11355	2,000	7570	66	1.7	17.0	5.2	16,670	7568
3,000	11355	2,000	7570	72	1.8	14.2	4.3	16,670	7568
4,500	17033	3,000	11355	66	1.7	25.4	7.7	25,000	11350
4,500	17033	3,000	11355	72	1.8	21.3	6.5	25,000	11350
4,500	17033	3,000	11355	78	2.0	18.2	5.5	25,000	11350
6,000	22710	4,000	15140	72	1.8	28.2	8.6	33,340	15136
6,000	22710	4,000	15140	78	2.0	24.2	7.4	33,340	15136
6,000	22710	4,000	15140	84	2.1	21.0	6.4	33,340	15136
7,500	28388	5,000	18925	78	2.0	30.3	9.2	41,670	18918
7,500	28388	5,000	18925	84	2.1	26.2	8.0	41,670	18918
7,500	28388	5,000	18925	90	2.3	22.7	6.9	41,670	18918
9,000	34065	6,000	22710	84	2.1	31.4	9.6	50,000	22700
9,000	34065	6,000	22710	90	2.3	27.3	8.3	50,000	22700
9,000	34065	6,000	22710	96	2.4	24.0	7.3	50,000	22700

Note: 4,500 gals (17 m^3) gross capacity is the minimum ordinarily accepted for pressure tanks for automatic sprinkler systems.
In the above table the length of the tank has been figured as though the ends were flat instead of dished.
The actual length of tanks of the above capacities and diameters will therefore be a trifle greater.

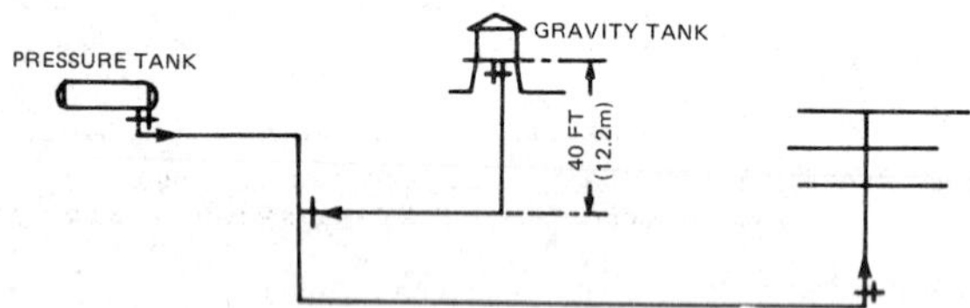

FIG. 17-5I. A gravity tank and a pressure tank showing how risers are connected to avoid air lock.

Examples are suburbs or farms and resorts which are not near a community with a piped water system for fire protection. In many of these, however, water is available for fire protection in the ocean, lakes and ponds, rivers, and streams. Cisterns or tanks can be used at locations distant from natural water sources. Water that is provided for other uses in fairly large quantities, such as the stock tanks on a farm, can be used for fire protection if the tanks are kept full. Conversely, small ponds in parks and swimming pools are not particularly good sources of emergency water because this competes with their regular purposes.

For water sources such as those mentioned, the general term "suction supplies" is used to distinguish these supplies from the conventional public water systems for fire protection. Static suction supplies must, in general, be used where they are found. While there have been instances where water in suburbs, farms, and forests have been used to provide effective fire streams at very long distances, laying hose lines more than 750 to 1,000 ft (225 to 300 m) seldom makes it possible to provide good protection because of the time required to lay the long lines and the greater difficulty in operating them.

Ground Tanks and Cisterns

If no hydrant on a water system is available, one way to assure the amount of water required for fire fighting is to store it in underground tanks or cisterns at locations where hydrants would normally be installed. The fire department pumper can then get water from a connection on the tank or by dropping a suction hose into the tank. Such tanks and cisterns are filled from domestic water sources too small for fire fighting. The water does not have to be changed so frequently that it presents difficulties.

Static water supplies are of obvious value in suburban, farm, and forest areas, but they are equally important for emergency conditions in cities. Before the present public water systems were developed, some cities provided underground reservoirs or cisterns for fire protection. When water was needed for fire fighting, a pumper suction hose was simply dropped through a manhole. Some of these cisterns are still used in San Francisco and Chicago. Even though San Francisco has a regular water system and a separate system of high pressure mains, it has been considered prudent, because of the threat of earthquake, to keep over 100 cisterns in service, most of which hold 75,000 gal (284 m^3). Some of the oldest cisterns have brick walls. Newer ones are reinforced concrete.

Rivers and Ponds

Rivers and ponds are important auxiliary water sources, but merely because the water can be seen does not mean that it can be used for fire fighting. These sources of water must have suitable approaches so that fire apparatus can get near the water without becoming mired. A cleared space and some fill in many cases will provide access. These sources must be studied to determine that the water is available all year round and to select the best possible approach. They should also be recorded and mapped and given periodic attention in much the same manner that hydrants are maintained.

At such locations it is usually necessary to provide a basin or cistern into which water can flow and the pumper suction strainer can be placed. There should be a screen or weir so that large objects will not interfere with the suction strainer. In some places a permanent suction pipe and strainer can be installed. The details of such an arrangement are the same as those for a fire pump operating under a lift. The suction pipe should be laid with enough cover so that it will not freeze. It can be connected at the top to a hydrant with one or more pumper suction connections.

Small Streams and Brooks

Where water supply comes from a brook or stream, it is usually best to drain the water into a sump rather than attempt to use it straight from the flowing stream. If there is a marshy bank or shore, it may be best to go out into the pond or stream to a point where water can be collected in a sump. From the sump in the pond or stream, a pipe should be laid for draining to another sump, the latter located near where the fire apparatus will stand.

In some sumps, and particularly in open water which freezes over in winter, covers may not be feasible at the point where the sump is provided for pumper suction. At such locations, an anchored plug of wood can be floated where the pumper suction should be taken. Such a plug should have a minimum diameter larger than the pumper suction hose and be tapered so that, when frozen in ice, it can be driven down and out of position with a sledge hammer.

In some parts of the country, streams do not flow continuously. By providing a covered cistern of generous capacity, it may be filled when water flows. Thus, the water may be stored for a longer period than it would keep in an open reservoir.

Water from streams and ponds is more important in areas without public water systems than in a city. There are small groups of farm buildings and small communities where even a brook, if dammed and with a proper pumper approach and a suction basin, can provide good protection at little cost.

For projects in which a stream will be dammed to form static water, experienced local engineers should be consulted. Details of the dam construction and decisions about the amount of water to be impounded depend upon many factors and may require a comprehensive engineering survey. Its effect on users of the stream, and flood conditions, also have to be considered.

Harbors and Rivers

The principal problem with using harbors and rivers as suction supplies is that their water levels change. In harbors, ocean tide levels vary throughout their entire range in 24 hr. River water levels are seasonal but the problem is generally the same.

Where there is little fluctuation in the level of a natural water source, a simple pier or platform may be all that is needed to make the water available. If a pumper can be driven onto a sturdy pier, it can go to work when its suction line is dropped over the side. But where the water level varies, special arrangements are needed. If the water is always deep enough for drafting, a hinged ramp leading down to a large float is feasible. But more common is a sloping beach down which the water recedes too far to be accessible at all times. Reservoirs which can be filled at high tide are one solution to this problem.

Where a bridge passes over tidewater, water can be obtained by permanently installing a deep-well pump with a discharge pipe to a point convenient for fire department pumper use. The most practical way to use tidewater supplies is to provide pumps on boats, supplementary to any fireboats in service. Relatively large quantities of water for fire fighting can be obtained from river and harbor sources because large capacity pumps are normally provided on fireboats, and where there are no fireboats, large pumps can be installed on a boat or barge and pressed into use to supply water for fire protection. To use the water on land, it is necessary to provide a place at which a fireboat, or a barge with pumps, can tie up and discharge into tanks, or a special pipe system with hydrants from which land based pumping equipment can take the water.

Wells

Wells have become increasingly popular for both domestic use and fire supply for industrial sites, shopping centers, etc., located beyond the reach of a piped water supply. Before a well is constructed, a thorough examination of the ground water must be made. This examination generally includes an aquifer performance analysis and a review of the history of nearby wells. The water in the ground must be of sufficient capacity and dependability, and of reasonably good quality.

A vertical shaft turbine pump is used to draw the water from wells into the fire system or into a storage tank.

Bibliography

References Cited

ACI. 1983. *Building Code Requirements for Reinforced Concrete.* ACI 318-1983. American Concrete Institute, Detroit, MI.

ASME. 1983. *Boiler and Pressure Vessel Code.* American Society of Mechanical Engineers, New York, NY.

AWPA. 1983. *Standard Specifications of the American Wood Preservers Association.* American Wood Preservers Association. Washington, DC.

AWS. 1984. *Structural Welding Code, Steel.* AWS D1.1-84. American Welding Society, Miami, FL.

AWWA. 1984. *Standard for Welded Steel Tanks for Water Storage.* AWWA D100-84 (AWS D5.2-84). American Waterworks Association, Denver, CO.

NRCC. 1984. *Testing and Monitoring Program for Solar Heated Water Storage Tanks.* Prepared for the National Research Council of Canada, Ottawa, Ontario.

NFPA Codes, Standards, Recommended Practices and Manual. (See the latest *NFPA Codes and Standards Catalog* for availability of current editions of the following document.)

NFPA 22, *Standard for Water Tanks for Private Fire Protection.*

Additional Readings

Factory Mutual Engineering Corp., *Embankment Supported Fabric Tanks,* Loss Prevention Data 3-4, April 1971, Factory Mutual Engineering Corp., Norwood, MA.

——*Isothermal Map for Europe,* Loss Prevention Data 3-2S, 1978.

——*Lined Earth Reservoirs for Fire Protection,* Loss Prevention Data 3-6, 1977.

——*Water Tanks for Fire Protection,* Loss Prevention Data 3-2, December 1977.

FIRE PUMPS

Revised by Robert M. Hodnett

Fire pumps are often used to supplement the supplies available from public mains, gravity tanks, reservoirs, pressure tanks, or other sources. The first modern fire pumps were the wheel-and-crank reciprocating type, belt-driven from mill machinery. If plant operations were stopped during a fire, the pump could not operate. At best, these pumps were inadequate.

Better water supplies became necessary as automatic sprinkler systems became more common, and the mill pumps were replaced by rotary displacement pumps driven by friction drive from the horizontal water wheels supplying power to the plant. As steam supplanted water power, the reciprocating steam pump was adopted for fire protection. For many years the Underwriter duplex, double acting, direct steam driven unit was universally accepted as the "standard" fire pump. Today the centrifugal fire pump is standard. (See Figs. 17-6A and 17-6B.) Its compactness, reliability, easy maintenance, hydraulic characteristics, and variety of available drivers (electric motors, steam turbines, and internal combustion engines) have made the Underwriter pump obsolete, although not entirely extinct.

This chapter covers the operating principles of stationary pumps used for fire protection, the methods of driving and controlling them, and the testing and maintenance procedures that should be followed to keep them in top operating condition.

The NFPA Standard on fire pumps is NFPA 20, *Installation of Centrifugal Fire Pumps*, referred to in this chapter as the NFPA Centrifugal Fire Pump Standard. Other NFPA Standards with information on fire pumps include NFPA 21, *Operation and Maintenance of Steam Fire Pumps*; NFPA 13, *Installation of Sprinkler Systems*; NFPA 14, *Installation of Standpipe and Hose Systems*; NFPA 15, *Water Spray Fixed Systems for Fire Protection*; NFPA 16, *Installation of Deluge Foam-Water Sprinkler and Foam-Water Spray Systems*; NFPA 22, *Water Tanks for Private Fire Protection*; and NFPA 24, *Installation of Private Fire Service Mains and Their Appurtenances*.

Mr. Hodnett is a consultant in fire protection and piping systems based in Providence, RI. He was NFPA staff liaison to the NFPA Water Extinguishing Systems Committees prior to his retirement in 1984.

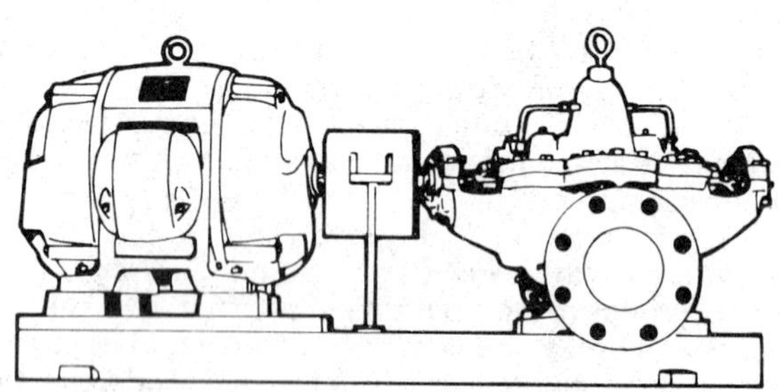

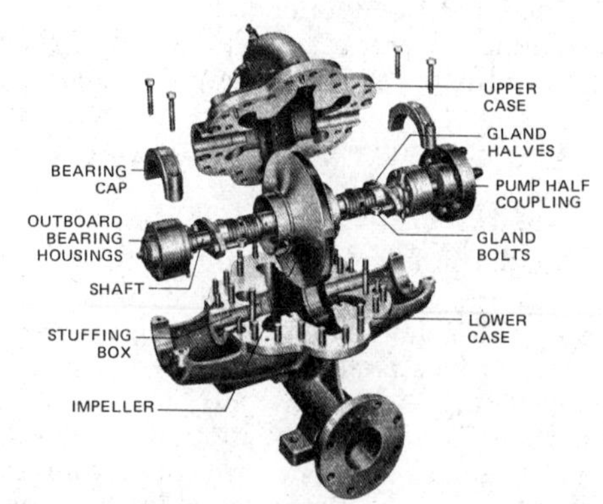

FIG. 17-6A. A horizontal shaft single-stage centrifugal pump, with cutaway view of pump. (Peerless Pump)

THE CENTRIFUGAL FIRE PUMP

An outstanding feature of a horizontal or vertical centrifugal pump is the relation of discharge to pressure at constant speed, insofar as when the pressure head is increased the discharge is reduced. With displacement pumps, however, the rated capacity can be maintained against any head if the power is adequate to operate the

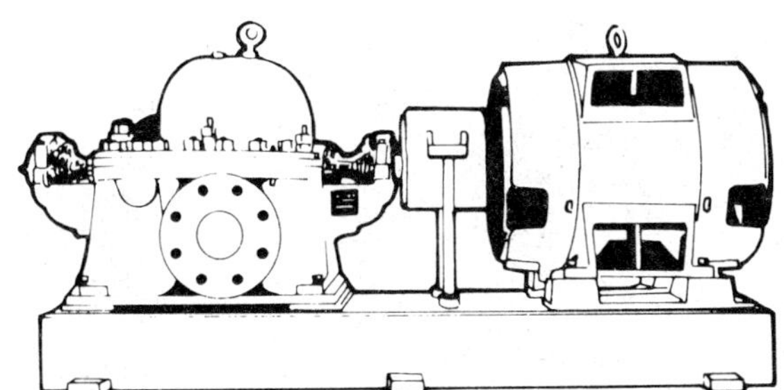

FIG. 17-6B. A horizontal multistage centrifugal pump. The housing would contain more than one impeller. (Peerless Pump)

pump at rated speed and if the pump, fittings, and piping can withstand the pressure.

Listed horizontal and vertical fire pumps are available with rated capacities up to 5,000 gpm (18 925 L/min). Pressure ratings range from 40 to 400 psi (276 to 2758 kPa) for horizontal pumps and 75 to 500 psi (517 to 3448 kPa) for vertical turbine pumps. (Vertical turbine pumps are centrifugal pumps with one or more impellers discharging into one or more bowls and a vertical educator or column pipe used to connect the bowls to the discharge head on which the pump driver is mounted.) It is anticipated that larger capacity fire pumps will be listed in the future.

The "size" of a horizontal centrifugal pump is generally the diameter of the discharge outlet; however, it is sometimes indicated by both suction and discharge pipe flange diameters. The "size" of a vertical turbine pump (Fig. 17-6C) is the diameter of the pump bowl.

PRINCIPLES OF OPERATION

The two major components of a centrifugal pump are a disc called the impeller, and the casing in which it rotates. (See Fig. 17-6D.) It operates by converting kinetic energy to velocity and pressure energy. Power from the driver (electric motor, internal combustion engine, or steam turbine) is transmitted directly to the pump through the shaft, rotating the impeller at high speed. The way that energy is converted varies with the class of pump. The major classes are known as radial flow and mixed flow. These pumps are identified by the direction of flow through the impeller with reference to the axis of rotation. (See Fig. 17-6E.)

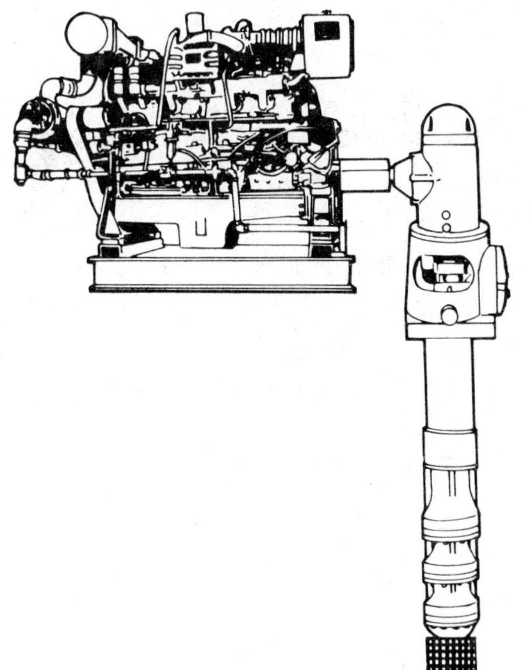

FIG. 17-6C. A vertical turbine fire pump. (Peerless Pump)

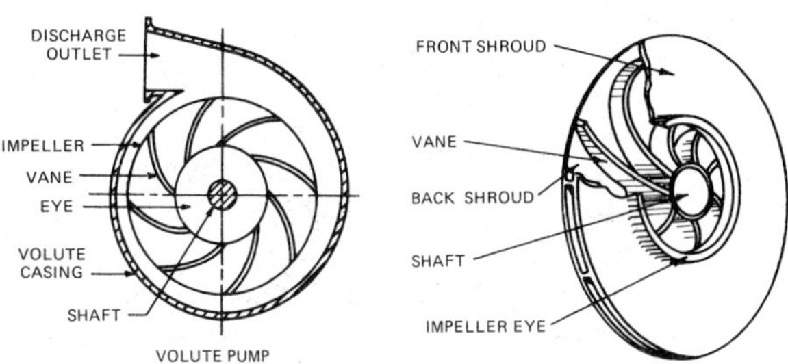

FIG. 17-6D. A volute casing and impeller.

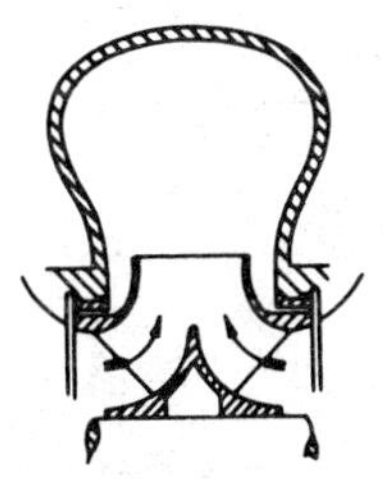

RADIAL FLOW
Pressure is developed principally by the action of centrifugal force. Liquid normally enters the impeller at the hub and flows radially to the periphery.

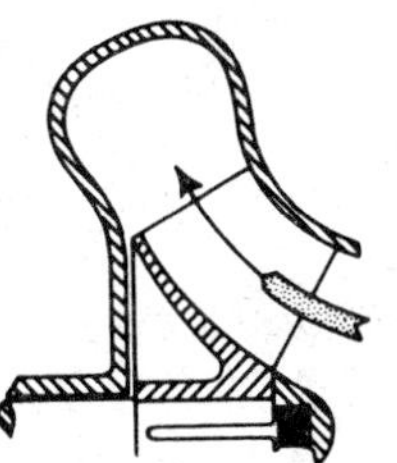

MIXED FLOW
Pressure is developed partly by centrifugal force and partly by the lift of the vanes on the liquid. The flow enters axially and discharges in an axial and radial direction.

FIG. 17-6E. The two major classes of fire pumps.

The horizontal shaft, one stage, double suction volute pump is the type most commonly applied to fire protection service or to commercial use. (See Fig. 17-6A.) In these pumps, waterflow from the suction inlet in the casing divides and enters the impeller from each side through an opening called the "eye." Rotation of the impeller drives the water by centrifugal force from the eye to the rim, and through the casing volute to the pump discharge outlet. The kinetic energy acquired by the water in its passage through the impeller is converted to pressure energy by gradual reduction of velocity in the volute.

Multistage Pumps

In order to give high pressure, two or more impellers and casings can be assembled on one shaft as a single unit, forming a multistage pump. (See Fig. 17-6B.) The discharge from the first stage enters the suction of the second stage; discharge from the second stage enters the suction of the third, and so on. The pump capacity is the rating in gallons per minute (L/min) of one stage; the pressure rating is the sum of the pressure ratings of the individual stages, minus a small head loss.

High Pressure Service Pumps

Single stage pumps can be designed for high pressure service by increasing the impeller diameter or the rated speed. Both of these methods offer certain undesirable features. The large diameter pumps may be less efficient, and high speed pumps may not be readily matched to the driver.

CHARACTERISTIC PUMP CURVES

The characteristic curves (Fig. 17-6F) of a horizontal centrifugal or a vertical turbine type pump are:

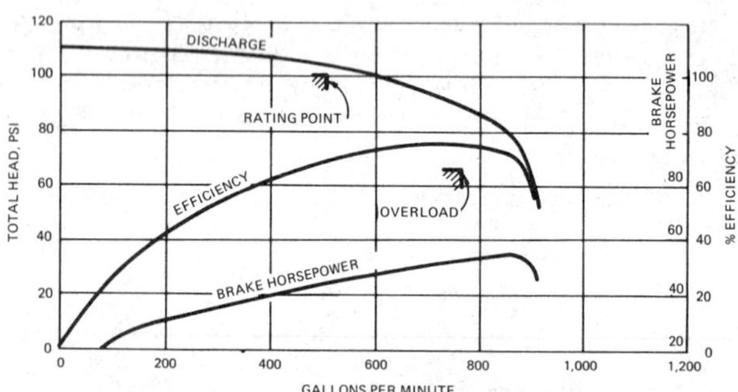

FIG. 17-6F. Typical fire pump characteristic curves. Illustrated curves are for a 500 gpm, 100 psi, 2,000 rpm, gasoline engine-drive pump with a 14 in. impeller and a maximum suction lift of 6.4 ft. Note that the shutoff point is 110 psi, maximum brake horsepower is 55, and the maximum efficiency is 75 percent at overload capacity, and head is 90 psi over the 65 percent minimum required. SI units: 1 psi = 6.89 kPa; 1 gpm = 0.38 L/min.

1. Total head vs discharge (feet of head or pounds per square inch of pressure vs gallons per minute).
2. Brake horsepower vs discharge.
3. Efficiency vs discharge $\left(\dfrac{\text{Water Hp}}{\text{Input Hp}} \text{ vs gpm}\right)$

These curves assume that the pump is operated at a constant speed equal to its rated rpm (revolutions per minute). In actual service, however, the speed of the driver may vary with changes in the load.

The flow and pressure ratings of commercial pumps are usually established on the basis of maximum efficiency and desired speed. Impellers can be designed for flat, medium, or steep head-discharge characteristics, as required for various uses. Figure 17-6G illustrates how the head-discharge curve is affected by the diameter of the eye, width of the impeller, number of vanes, and the shape or angle of the vanes.

TOTAL HEAD

The total head of a pump is the energy imparted to the liquid as it passes through the pump. It may be expressed in various units of pressure, but for fire protection it is

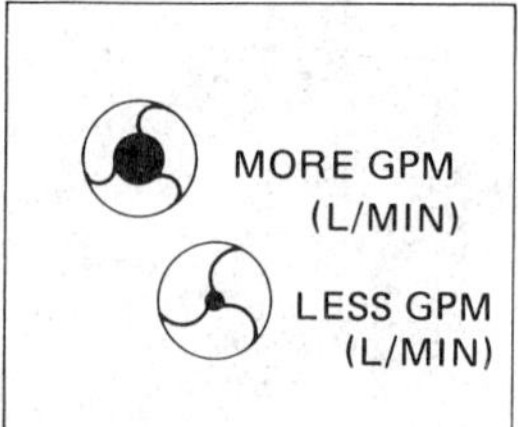

DIAMETER OF EYE

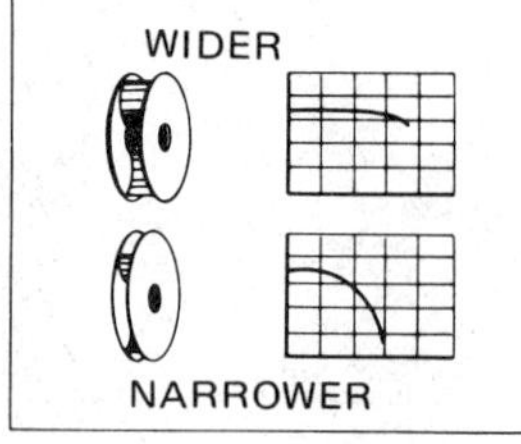

WIDTH OF IMPELLER

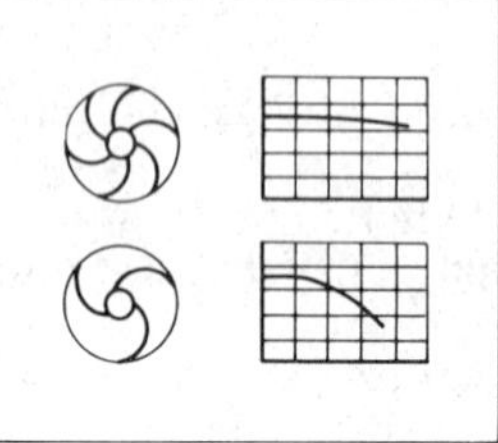

NUMBER OF VANES

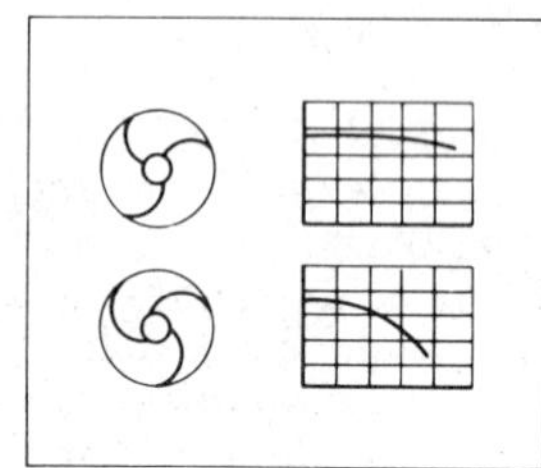

ANGLE OF VANES

FIG. 17-6G. Effect of impeller design on head-discharge curves for fire pumps.

generally given in psi (pounds per square inch) or kPa (kilopascals), or in ft (feet) or m (meters) of liquid measured vertically. The total head is calculated by subtracting the energy in the incoming liquid from the energy in the discharging liquid. Therefore, the total head (H) of a pump is calculated by the formula:

$$H = h_d + h_{vd} - h_s - h_{vs}$$

in which:

H = total head, feet (m)

h_d = discharge head, feet (m)

$h_{vd} = \dfrac{v_d^2}{2g}$ = discharge velocity head, feet (m)

h_s = suction head, feet (m)

$h_{vs} = \dfrac{v_s^2}{2g}$ = suction velocity head, feet (m)

v = average velocity, feet per second (m/sec)

g = acceleration due to gravity, 32.2 ft/sec^2 (9.81 m/sec^2).

For a horizontal split case pump, the individual heads (h_v) are measured at the pump discharge nozzle flange and at the suction flange. (See Fig. 17-6H.) The heads are read from pressure gages attached to the pump flanges. The velocity head must be calculated for the volume of liquid passing through the flanges:

$$\left(h_v = \frac{v^2}{2g}\right)$$

where:

v = average velocity

g = gravitational acceleration

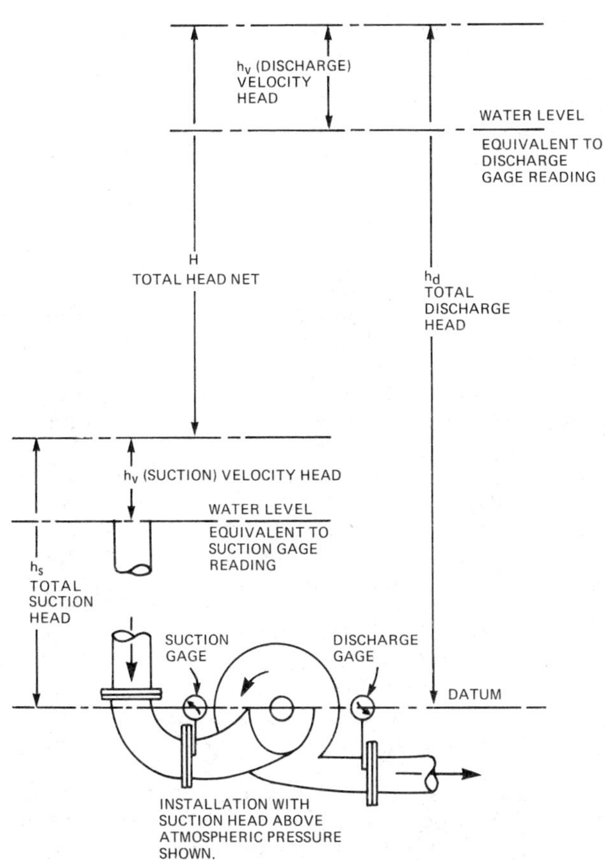

FIG. 17-6H. Typical head of horizontal shaft centrifugal fire pumps.

If the flanges have the same diameters, there will be no difference between the incoming and outgoing velocity and the calculation can be omitted.

For a vertical turbine pump, the discharge head is theoretically read at the discharge flange of the pump. Since this flange is usually inaccessible for gage readings, a gage is used at the discharge fitting at the top of the pump column pipe. (See Fig. 17-6I.) The discharge pressure at the pump discharge flange therefore equals the pressure at the gage above plus the pressure effect of the vertical distance between the two points plus the friction loss between the two points. In most cases the friction loss is so small that it may be disregarded.

The suction head is the vertical distance from water level to the pump discharge flange. The velocity head of the incoming liquid is assumed to be zero. Hence the formula would now take the following form:

$$H = h_d + h_{vd} - h_s$$
$$= (h_{gd} + L) + h_{vd} - h_s$$

where:

h_{gd} = discharge gage reading, feet (m)

L = gage to pump flange, feet (m)

H = gage to liquid level, feet (m).

However $L - h_s = h$ = the vertical distance between the discharge gage and water level.

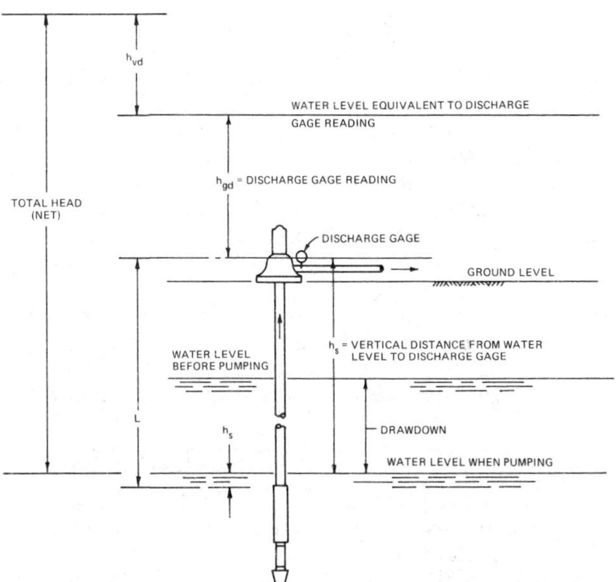

FIG. 17-6I. Total head of vertical turbine type fire pumps.

Therefore the formula becomes:

$$H = h_{gd} + h_{vd} + h$$

Hydraulic and power losses within the pump (turbulence, disc friction, shock, etc.) are represented by the efficiency rating.

Total head at rated capacity is used to establish the rated head of a pump. Actually, the rated head is the amount of energy given to the water.

The total head of a vertical turbine type pump also may be defined as the vertical water-to-water dimension of the system in which the pump operates. There is a difference, however, in the method of measuring total head. As shown in Figure 17-6I, it is the sum of the vertical distance between water level in the well or pit, the discharge head indicated by the gage on the pump outlet, and the velocity head at the gage connection.

SPECIFIC SPEED (N_s)

Specific speed is a number relating the head, capacity, and speed of a centrifugal pump for design purposes. Actually, specific speed is the revolutions per minute of a geometrically similar impeller that will discharge one gallon per minute (3.8 L/min) at one foot (0.3 m) total head. The formula for calculating specific speed of a centrifugal pump is:

$$N_s = \frac{\text{rpm} \times \text{gpm}^{1/2}}{H^{3/4}}$$

where:

N_s is the specific speed number, and

H is the head in feet (m).

When values of head, speed, and capacity in the formula correspond to pump performance at optimum efficiency, the specific speed is an index to the type of pump. Impellers for high heads usually have low specific speeds, and impellers for low heads have high specific speeds.

A pump of low specific speed will operate satisfactorily with greater suction lift than a pump of same head and capacity with a higher specific speed. Experience shows that specific speed is a useful guide for determining maximum suction lift or minimum suction head.

When suction lift exceeds 15 ft (4.5 m), it may be necessary to provide a larger pump at less speed; with low lift or positive head on the suction, a smaller pump operating at greater speed may be used. Abnormally high suction lifts may seriously reduce pump capacity and efficiency or cause excessive vibration and cavitation.

NET POSITIVE SUCTION HEAD

Net Positive Suction Head (NPSH) is the pressure head that causes liquid to flow through the suction pipe and fittings into the eye of a pump impeller. The pump itself has no ability to "lift," and the suction pressure depends on the nature of the supply.

If a pump is supplied from a pond, stream, open well, or uncovered reservoir where the water level is below the pump, the suction head is atmospheric pressure minus the lift. If the water level is above the pump, as from a water main, penstock, aboveground tank, etc., the suction head is atmospheric pressure plus static pressure.

Pressure readings at the inlet flange of a pump operating under lift are negative with respect to the gage, but positive when referred to absolute pressure—hence the expression, "net positive suction head." (Absolute pressure is gage pressure plus barometric pressure.)

There are two kinds of NPSH to consider. Pump NPSH is a function of the pump design. It varies with capacity and speed of any one pump, and with the designs of different pumps. Curves of NPSH vs gallons per minute usually can be obtained from pump manufacturers. (See Fig. 17-6J.) Available NPSH is a function of the system in

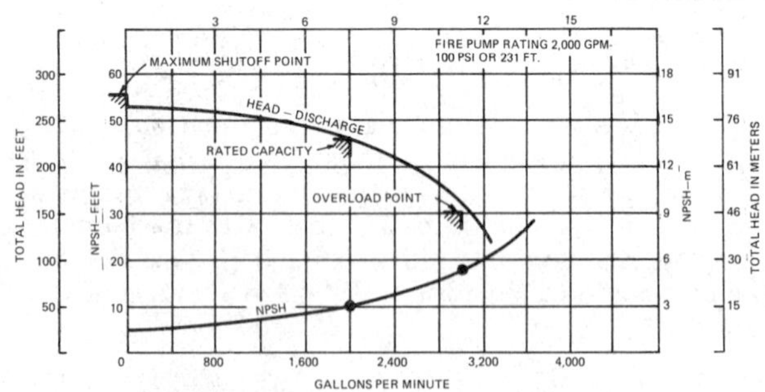

FIG. 17-6J. NPSH curve of a typical fire pump. Note that the required NPSH at 2,000 gpm is 10 ft and 18 ft at 3,000 gpm. SI units: 1 gpm = 0.38 L/min; 1 ft = 0.30 m.

which the pump operates, and can be calculated readily.

When the water source is above the pump, available NPSH = atmospheric pressure (ft or m) + static head on suction (ft or m) − friction and fitting losses in suction piping (ft or m) − vapor pressure of liquid (ft or m). (Note: The vapor pressure of water at 90°F (32°C) is 1.6 ft (0.48 m).

When the water source is below the pump, available NPSH = atmospheric pressure (ft or m) − static lift (ft or m) − friction loss in piping and head loss in the fittings (ft or m) − vapor pressure of liquid (ft or m).

For any pump installation, the available system NPSH must be equal to or greater than the pump NPSH at the desired operating conditions.

CAVITATION

Cavitation is a complex phenomenon that may take place in pumps or other hydraulic equipment. In a centrifugal pump, as liquid flows through the suction line and enters the eye of the impeller, the velocity increases and pressure decreases. If the pressure falls below the vapor pressure corresponding to the temperature of the liquid, pockets of vapor will form. When the vapor pockets in the flowing liquid reach a region of higher pressure, the pockets collapse with a hammer effect causing noise and vibration. Tests have shown that extremely high instantaneous pressures may be developed in this manner resulting in pitting various parts of the pump casing and impeller. Conditions may be mild or severe, and mild cavitation may occur without much noise. Severe cavitation can cause reduced efficiency and ultimate failure of the pump if it is not corrected.

AFFINITY LAWS

The mathematical relationships between head, capacity, brake horsepower, and impeller diameter are called affinity laws. Law 1 assumes constant impeller diameter with change of speed. Law 2 assumes constant speed with change in diameter of the impeller. These laws are expressed by proportion as follows:

Law 1.

$$\frac{Q_1}{Q_2} = \frac{N_1}{N_2} \qquad \frac{H_1}{H_2} = \frac{N_1^2}{N_2^2} \qquad \frac{bhp_1}{bhp_2} = \frac{N_1^3}{N_2^3}$$

Law 2.

$$\frac{Q_1}{Q_2} = \frac{D_1}{D_2} \qquad \frac{H_1}{H_2} = \frac{D_1^2}{D_2^2} \qquad \frac{bhp_1}{bhp_2} = \frac{D_1^3}{D_2^3}$$

The nomenclature for the relationships is: Q = capacity; H = head; N = speed; D = impeller diameter; and bhp = brake horsepower. Thus:

Q_1	= gpm (L/min) at N_1 or D_1	Q_2	= gpm (L/min) at N_2 or D_2
H_1	= head in ft (m) at N_1 or D_1	H_2	= head in ft (m) at N_2 or D_2
bhp_1	= brake horsepower (kW) at N_1 or D_1	bhp_2	= brake horsepower (kW) at N_2 or D_2.

Law 1 applies to common types of pumps, including horizontal centrifugal pumps and vertical turbine pumps. Law 2 applies to centrifugal pumps with reasonable close agreement between calculated and tested performance. Generally, pumps with low specific speeds show closer agreement than pumps with high specific speeds.

The affinity laws should be applied when proposed changes in a fire pump installation would increase the speed or significantly raise the pressure of the suction supply. Greater speed would increase the power demand, and high discharge pressure might be undesirable. In some instances, it is possible to trim the impeller or to install a speed reducing gear between the pump and driver. This should not be done, however, without the approval of the pump manufacturer. For pumps operating under lift, possible changes in performance should be studied carefully, because greater velocity in the suction line could cause cavitation and substantially alter the characteristic curve.

FIRE PUMP APPROVAL AND LISTING

NFPA standards for design and installation of various fire protection systems recommend the use of approved and listed equipment, or both, including fire pumps for installations requiring them.

Under the approval system, the manufacturer is responsible for providing a listed or approved shop tested pump that will perform satisfactorily when installed in conformance with the NFPA Centrifugal Fire Pump Standard. A contractor or others are responsible for installing the driver-pump combination in accordance with the provisions of the NFPA Centrifugal Fire Pump Standard, while it is the customer's obligation to provide adequate data about the pump driver, power supply, water supply, location, etc.

Fire pumps are designed to provide maximum reliability and specific net head-discharge characteristics. Except for periodic inspections and tests, fire pumps are idle most of the time. Pumps for commercial use, on the other hand, are chosen for maximum efficiency and economy.

To obtain official listing of a new pump, the manufacturer submits plans and specifications to a testing agency for review and comment. After any revisions or corrections have been agreed to, arrangements are made for representatives of the testing agencies involved to witness the required approval tests at the manufacturer's plant.

If the results are satisfactory, the new pumps are listed in the usual manner or with any restrictions considered desirable. It is the duty of the manufacturer to shop test every unit sold and to furnish certified curves of head, efficiency and brake horsepower vs discharge. Although not ordinarily shown, the NPSH curve of the pump, if available, should be provided upon request.

Many of the listed fire pumps used today are top quality commercial units. These pumps are upgraded when necessary, trimmed, and fitted to meet all the approval requirements for fire protection.

STANDARD HEAD DISCHARGE CURVES

The shape of the standard head discharge curve of a fire pump is determined by three limiting points as follows:

Shutoff

With the pump operating at rated speed and the discharge valve closed, the total head of a horizontal centrifugal pump at shutoff shall not exceed 120 percent or 140 percent, depending on the testing agency requirements, of the rated head at 100 percent capacity. For a vertical or end suction pump, the total head at shutoff shall not exceed 140 percent of the rated head at 100 percent capacity.

The shutoff point represents the maximum allowable total head pressure; otherwise the pump would have a rising or convex characteristic curve. Such pumps are not listed. With a convex curve there could be two flow points for one pressure.

Rating

The curve should pass through or above the point of rated capacity and head. (See Fig. 17-6K.)

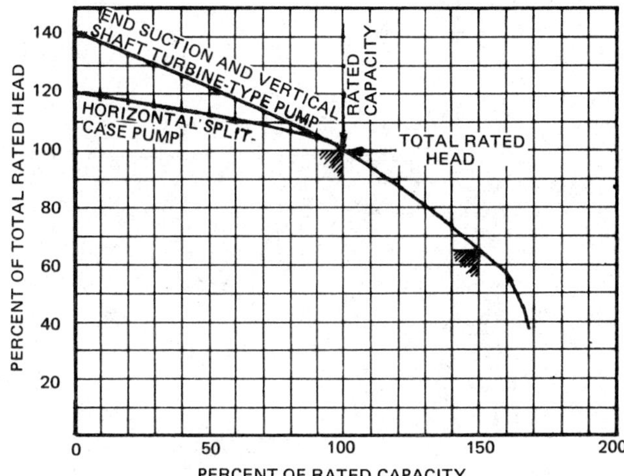

FIG. 17-6K. Standard head discharge curves for horizontal and vertical fire pumps.

Overload

At 150 percent of rated capacity, the total head should not be less than 65 percent of rated total head. Here, also, the curve should pass through or above the overload point. Most modern fire pumps have curves with a significant margin above the theoretical overload. Some models have a cavitation or "break" point in the curve just beyond overload.

HORIZONTAL SHAFT CENTRIFUGAL FIRE PUMPS

Horizontal shaft fire pumps should be installed to operate under positive suction head, especially with automatic or remote-manual starting. If the water supply is such that suction lift cannot be avoided, consideration should be given to installation of vertical turbine fire pumps.

Types of Pumps: Horizontal centrifugal fire pumps are of the split case (Fig. 17-6A) or end suction (Fig. 17-6L) types. The end section type is manufactured to ANSI specifications (ANSI 1973) for centrifugal pumps and are limited to capacities under 500 gpm (1893 L/min). There are no limits on the capacities of split case fire pumps, but at the present time the maximum capacity for listed fire pumps is 5,000 gpm (18 925 L/min).

Suction Facilities

When streams, ponds, and other open bodies of water are used, properly screened intakes should be provided to

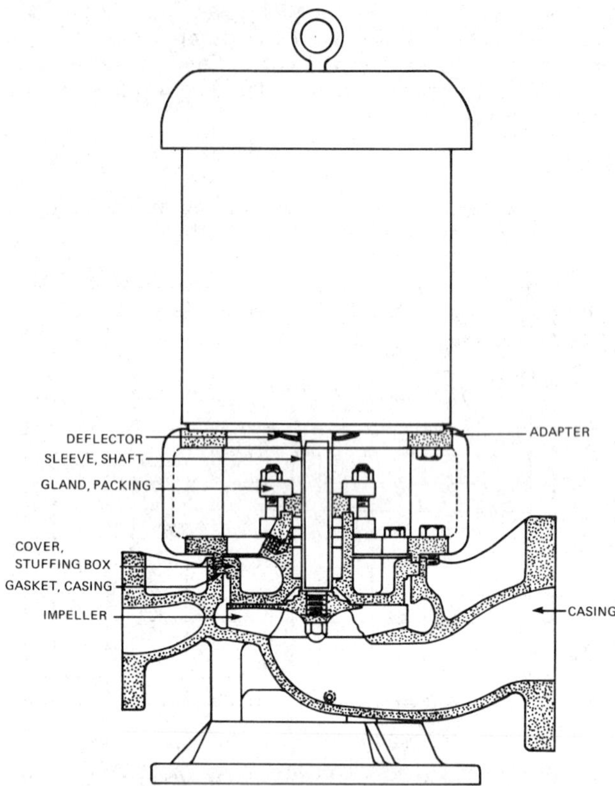

FIG. 17-6L. An overhung impeller, close coupled, single stage, inline, horizontal end suction fire pump.

prevent fish, eels, and foreign material from entering the pump and the fire protection system. A foot valve of proper design should be provided on each suction inlet. (See Fig. 17-6M.)

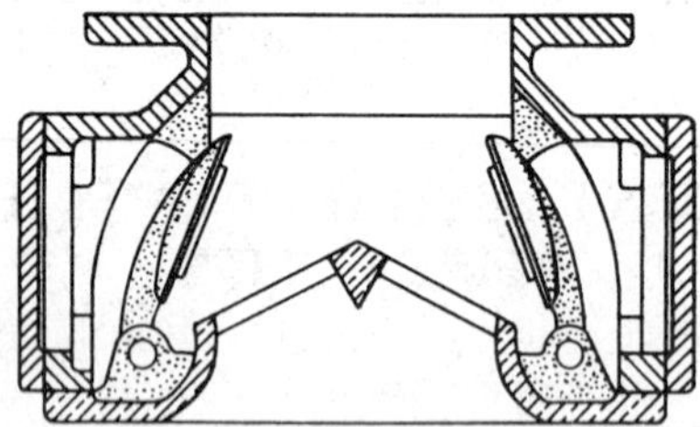

FIG. 17-6M. A typical foot valve of good design shown in open position.

Use of nonpotable water should be avoided when fire pumps discharge into a system that is also connected to public mains or other potable supplies. Otherwise, there will be cross-connections, which are prohibited, or in some locations restricted, by either health or water utilities authorities, or both, in most states and provinces. Aboveground covered tanks filled with potable water are recommended for supplying fire pumps.

The volume of suction storage should be sufficient to supply the pump at overload rate for the estimated duration of the water demand.

Break Tanks

At locations where a direct connection between a public water supply and a private fire protection system is prohibited, either for public health or hydraulic reasons, a break tank installation may be desirable. A break tank is an automatically filled tank which provides a suction supply for a fire pump without a direct connection to a public supply. (See Fig. 17-6N.) This is done by an actual physical break or gap, between the public supply and a private protection system. Water from the public supply enters the break tank from a height above the tank's overflow outlet and falls freely to the surface of the water in the tank. A fire pump is needed to take suction from the tank as the water is no longer under pressure from the public supply.

A separate break tank should be provided for each fire pump in an installation, and they should be large enough to supply pump operation at 150 percent of the pump rated capacity for 15 minutes. The flow into the tank from the public supply is controlled by an automatic fill mechanism; and since the water in the tank is considered potable, the top of the tank should be closed.

Reliability of break tanks is less than that for a good full sized suction tank as the automatic fill mechanisms could fail. There is no NFPA standard for the design and installation of break tanks, although recommendations have been made on them by an insurance engineering service (FMRC 1983).

Booster Pumps

These are fire pumps taking suction from public water mains, industrial systems, or power penstocks. (In a mechanical sense, all pumps are booster pumps.) As a prelude to installation, the available fire flow in the area is obtained by testing. Full overload capacity of the pump plus probable flow drawn from hydrants in the area by the fire department are calculated, and they should not drop the pressure in the water mains below 20 psi (138 kPa) or that allowed by the public health authorities. Head rating of the pump should be sufficient to meet all pipe friction in the connection plus pressure demand.

Pump Accessories

Auxiliary devices have an important bearing on the complete functioning of a pump as a fire protection water supply, and their provision or omission should never be decided solely on the basis of costs. The NFPA Centrifugal Fire Pump Standard gives detailed information concerning their installation; the following are worthy of special consideration:

Relief Valves: These are necessary on the pump discharge if excess pressure results from operation of the pump. Pumps having adjustable speed drivers need relief valves, as do those where suction pressure plus shutoff pressure would exceed the pressure rating of the fire protection equipment.

Hose Valves: Approved 2½ in. (63 mm) hose valves are used in testing pumps, and for hose stream fire protection. The valves should be attached to a header or manifold outside the pump room, or otherwise located to avoid water damage to the pump, driver, and controller. The number of valves needed depends on the pump capacity. For details see the NFPA Centrifugal Fire Pump Standard.

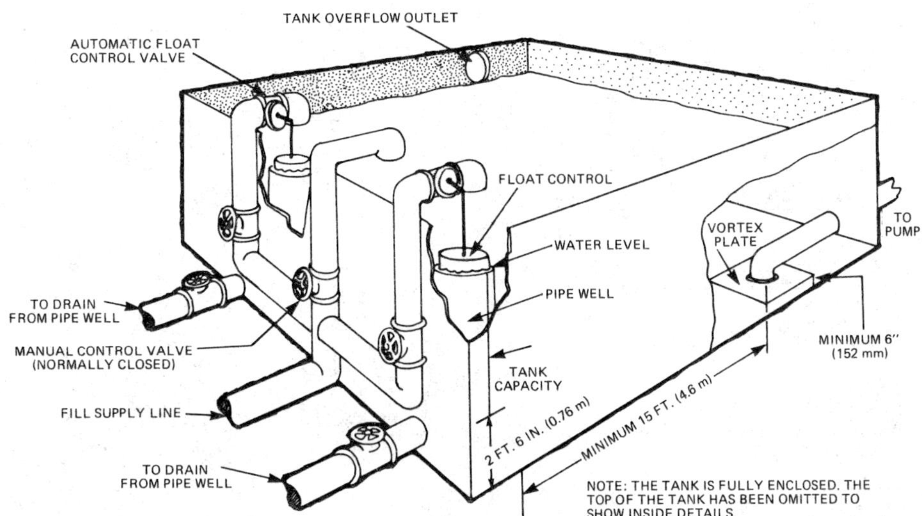

FIG. 17-6N. A break tank used in connection with a fire pump when a direct connection between a public water supply and a private fire protection system is prohibited. (Factory Mutual System)

Automatic Air Release Valves: These are necessary on the top of the casing of pumps arranged for automatic or remote control operation. An umbrella cock may be adequate for a pump which can be started only by manual means by an operator in the pump room, but an automatic air release is desirable on any pump with a casing that is normally full of water.

Circulation Relief Valves: These are necessary for pumps which may be started automatically or by remote control. Their function is to open at slightly above rated pressure, when there is little or no discharge, so that sufficient water is discharged to prevent overheating of the pump. These valves are not needed on engine driven pumps where cooling water is taken from the pump discharge.

VERTICAL TURBINE TYPE FIRE PUMPS

Vertical turbine pumps were originally designed to pump water from bored wells. As fire pumps, they are recommended in instances where horizontal pumps would operate with suction lift. An outstanding feature of vertical pumps is the ability to operate without priming. (See the NFPA Centrifugal Fire Pump Standard for required submergence.) Vertical pumps may be used to pump from streams, ponds, wet pits, etc., as well as in booster service.

Suction from wells is not recommended for fire service, although it is acceptable if the adequacy and reliability of the well is established, and the entire installation made in conformance with the NFPA Centrifugal Fire Pump Standard. In many instances, the cost of a deep well fire pump installation would be prohibitive, especially if the pumping level at maximum rate would be more than 50 ft (15 m) below ground level [200 ft (60 m) is the limit].

If the yield from a reliable well is too small to supply a standard fire pump, low capacity well pumps could be used to fill conventional ground level tanks or reservoirs for the fire pump supply.

A typical vertical fire pump consists essentially of a motor head or right-angle gear drive, a column pipe and discharge fitting, (Figs. 17-6O and 17-6P), an open or

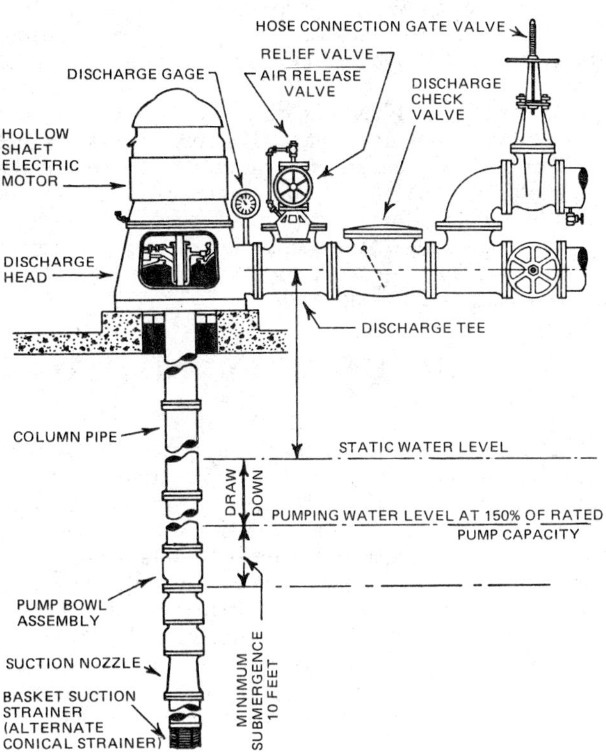

FIG. 17-6O. Vertical shaft turbine type pump installation.

enclosed drive shaft, a bowl assembly (containing the impellers), and a suction strainer. The principle of operation is comparable to that of a multistage horizontal centrifugal pump. Except for shutoff pressure, the charac-

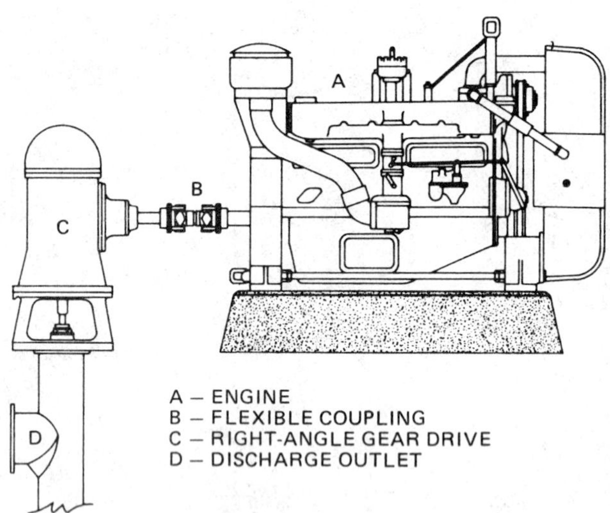

FIG. 17-6P. Engine-driven vertical fire pump.

teristic curve is the same as for horizontal pumps. (See Fig. 17-6K.)

Vertical Turbine Barrel or Can Pump: Sometimes vertical type multi stage pumps are installed in a casing called a "barrel" or "can" for high pressure installations. (See Fig. 17-6Q.) Where pumps with adequate pressure ratings are available, horizontal split case pumps are normally used.

Vertical pumps have the same standard capacity ratings as horizontal fire pumps. Pressure ratings are not standardized. By changing the number of stages and the impeller diameters or both, the pump manufacturer can provide a specific total head at rated speed.

For electrically driven pumps, hollow shaft motors are used. Diesel engines or steam turbines can be used by means of right angle gear heads.

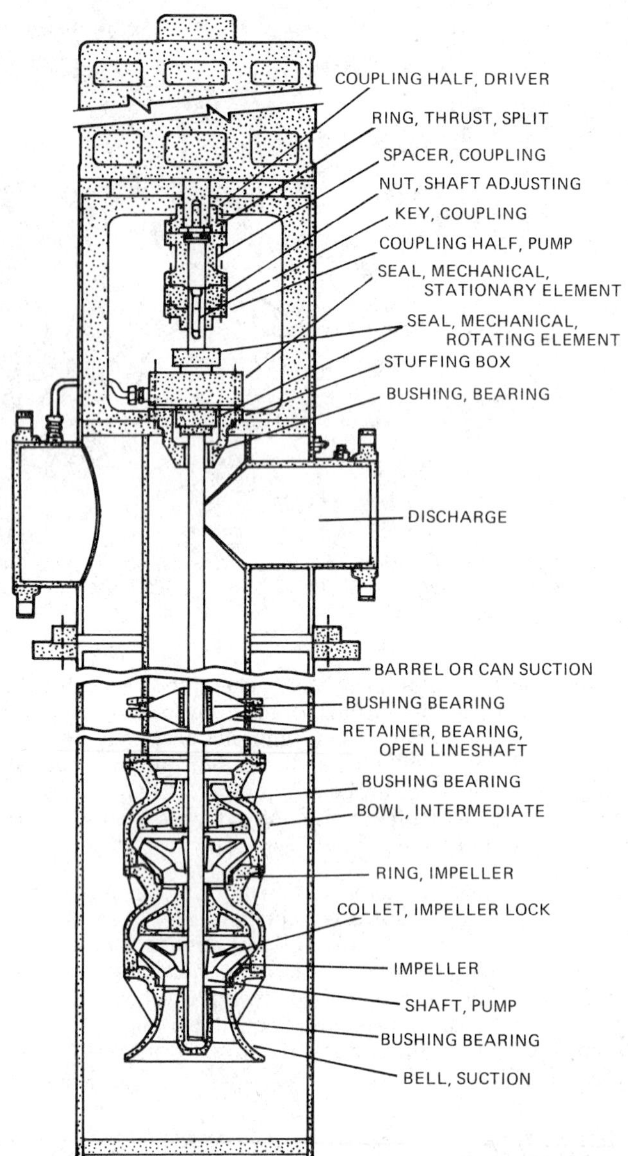

FIG. 17-6Q. A turbine type, vertical, multi stage, barrel or can pump.

FIRE PUMP CAPACITY AND HEAD RATING

The capacity and pressure ratings of fire pumps should be adequate to meet flow and pressure demands consistent with water supply requirements for the property in question. Fire pumps are designed to provide their rated capacity with a safety factor built in (150 percent of rated capacity at at least 65 percent of rated pressure) to provide some protection in case of greater than expected demand at time of fire. The following examples show one method of how rated capacity and pressure can be determined by using the standard fire pump curve for a typical manufacturer's pump characteristic curve. (See Fig. 17-6K.)

Example No. 1: Horizontal Centrifugal Pump

The estimated water demand for sprinklers and hose streams is 1,400 gpm at 90 psi pump discharge pressure. The suction supply is a pond, and the estimated lift is 5 psi at maximum flow.

PROBLEM: Determine the required rated capacity and pressure of the pump.

SOLUTION:

1. Meet the demand (1,400 gpm) with the overload capacity of the pump (150 percent of rated capacity);
2. Thus, 1,400 ÷ 150 percent = 933 gpm. (The nearest standard pump rating is 1,000 gpm);
3. Therefore, 1,400 gpm demand would be 140 percent of capacity;
4. From the manufacturer's pump characteristic curve, it is determined that at 140 percent capacity, the total pressure is 73 percent rated pressure;
5. Under lift condition of operation, the total pressure equals discharge pressure (90 psi) plus suction pressure (5 psi);
6. Therefore, net pressure at 1,400 gpm equals 90 + 5 = 95 psi, and rated pressure at 1,000 gpm = 95 ÷ 73 percent = 130 psi.

ANSWER: Pump rating should be not less than 1,000 gpm at 130 psi.

Example No. 2: Horizontal Centrifugal Pump (SI)
The estimated water demand for a sprinkler system is 2000 L/min at a pump discharge pressure of 400 kPa. The suction supply is a well with a lift of 5 m from water surface to the pump.

PROBLEM: Determine the required rated capacity and pressure of the pump.

SOLUTION:

1. Meet the demand (2000 L/min) with the overload capacity of the pump (150 percent of rated capacity);
2. Thus, 2000 ÷ 150 percent = 1330 L/min (The nearest standard pump rating is approximately 1500 L/min).
3. Therefore, 2000 L/min demand would be 2000/1500 × 100 percent = 133 per cent of rated capacity;
4. From the manufacturer's pump characteristic curve (in this case, Fig. 17-6K), it is determined that, at 133 percent capacity, the total pressure is 78 percent of rated pressure;
5. Under lift conditions of operations, the total pressure equals the discharge pressure (400 kPa) plus suction pressure (5 m head = 5 × 9.81 kPa = 50 kPa);
6. Therefore, net pressure at 2000 L/min equals 400 + 50 = 450 kPa and rated pressure at 1500 L/min = 450 ÷ 78 percent = 580 kPa.

ANSWER: Pump rating should be not less than 1500 L/min at 580 kPa.

Example No. 3: Vertical Turbine Pump (pump in a driven well):
The estimated water demand at the pump discharge gage (ground level) is 1,100 gpm at 100 psi. Tests and weather records show that the aquifer (underground water source) is reliable and adequate at all seasons. The static level is 45 ft below the surface. The draw down, or vertical distance between the static and pumping water levels, is 40 ft at 1,100 gpm pumping rate.

PROBLEM: Determine rated capacity and pressure of the pump.

SOLUTION:

1. Meet the demand (1,100 gpm) with the overload capacity of the pump (150 percent of rated capacity);
2. Thus, 1,100 ÷ 150 percent = 733 gpm. (The nearest standard pump rating is 750 gpm);
3. Therefore, the 1,100 gpm demand would be 147 percent capacity;
4. From the manufacturer's pump characteristic curve, it is determined that at 147 percent capacity, the total pressure is 70 percent rated pressure;
5. At 1,100 gpm, pressure demand at the surface = 100 psi;
6. Since the distance to water level at 1,100 gpm pumping rate is 45 + 40 = 85 ft × .434 = 37 psi; pumping pressure would be 100 + 37 = 137 psi which is 70 percent of rated pressure;
7. Therefore 137 ÷ 70 percent = 195 psi.

ANSWER: Pump rating should be not less than 750 gpm at 195 psi.

Example No. 4: Booster Pump on Public Water Connection:
A sprinklered building in a city has an estimated sprinkler demand of 750 gpm at 60 psi pump discharge. Based on fire flow tests from nearby street hydrants, 750 gpm at 27 psi would be available for sprinklers at the inlet flange of the pump (allowance already made for hose streams).

PROBLEM: Determine the rated capacity and pressure of the pump.

SOLUTION:

1. Meet the demand (750 gpm) with the overload capacity of the pump (150 percent of rated capacity);
2. Thus, 750 ÷ 150 percent = 500 gpm, a standard pump rating;
3. The total pressure at 150 percent capacity is 65 percent rated net pressure;
4. With positive head suction supply, the net pressure equals discharge pressure minus suction pressure; thus, at 750 gpm flow, net pressure equals 60 − 27 = 33 psi;
5. Therefore, 33 ÷ 65 percent = 51 psi.

ANSWER: Pump rating should be greater than 500 gpm at 50 psi.

(Note that all four examples utilize the pump curve out to 150 percent, which would be the absolute maximum. The NFPA Centifugal Fire Pump Standard recommends that the pump not be used at over 140 percent capacity. This leaves no reserve as pump curves drop off sharply after 150 percent. Generally a larger pump is recommended than the calculations indicate. Also, it must be considered that when the pump is used above its rated capacity the available pressure is reduced.)

HORSEPOWER OF FIRE PUMPS

Before matching a driver to a pump, it is necessary to know the maximum brake horsepower demand of the pump at rated speed. This can be determined directly from the horsepower curve provided by the pump manufacturer. Typical fire pumps reach maximum brake horsepower between 140 and 170 percent of rated capacity.

Horsepower can be calculated, if the curves are not available, by the formula

$$\text{bhp} = \frac{5.83QP}{10{,}000 \times E} \text{ or bhp} = \frac{QP}{1{,}710 \times E}$$

where:

bhp = brake horsepower

Q = gallons per minute

P = total head (psi) or net pressure

$$E = \text{efficiency} = \frac{\text{Water Horsepower}}{\text{Input Horsepower}}$$

The efficiency at maximum brake horsepower is usually 60 to 75 percent.

EXAMPLE: Find by formula the minimum horsepower needed to drive a 1,000 gpm, 100 psi, 1,760 rpm horizontal centrifugal fire pump.

SOLUTION:

1. Assume 65 percent efficiency at 160 percent capacity;
2. From standard pump curve (Fig. 17-6K) the pressure is 55 percent at 160 percent capacity, or 55 psi at 1,600 gpm;
3. By formula, bhp $= \dfrac{5.83 \times 1{,}600 \times 55}{10{,}000 \times .65} = 79$

ANSWER: Not less than 79 usable brake horsepower.

IN S.I. UNITS:

$$\text{Output power} = \frac{0.167 Q_m\, P_m}{10{,}000\, E}$$

where:

kW = power output (kilowatts)

Q_m = liters per minute

P_m = total head (kPa) or net pressure

$$E = \text{efficiency} = \frac{\textit{Water Power Output}}{\text{Input Power}}$$

EXAMPLE (SI): Find by formula the minimum power output needed to drive a 4000 L/min, 700 kPa, 1,760 rpm horizontal centrifugal fire pump.

SOLUTION:

1. Assume 65 percent efficiency at 160 percent capacity;
2. From standard pump curve (Fig. 17-6K) the pressure is 55 percent at 160 percent capacity, or 385 kPa at 6,400 L/min;
3. By formula, power output $= \dfrac{0.167 \times 6400 \times 385}{10{,}000 \times 0.65} = 63$ kW

ANSWER: Not less than 63 kW output.

FIRE PUMP DRIVERS

Power for driving fire pumps is selected on the basis of reliability, adequacy, safety, and economy. The reliability of utility electric power may be judged by the record of outages, and by review of the power sources and distribution layout of the system in question.

Gas utility systems may be subject to periods of restricted use because of high seasonal demand. To offset this, standby gas storage can be provided, or arrangements negotiated with utilities whereby the fire pump is supplied with gas even though the general use of gas is restricted.

Many public utilities operate steam distribution systems. When high pressure steam is available, it would be practical for large consumers to use turbine driven fire pumps.

Many industrial plants generate their own steam-electric and hydroelectric power, or both. Utility power also may be used possibly on a standby basis.

Diesel engines have the advantage of not being dependent upon outside sources of power.

Electric Motors

Electric motors for driving fire pumps are not specifically approved or listed, but they are required to be made by reliable manufacturers in accordance with specifications of (NEMA) National Electrical Manufacturers Association or (CEMA) Canadian Electrical Manufacturers Association. All electrical equipment and wiring in a fire pump installation should comply with the *National Electrical Code* except as modified by the NFPA Centrifugal Fire Pump Standard.

The pump manufacturer, or the installing contractor, is responsible for providing a motor of sufficient capacity to avoid overloading beyond the limit of the service factor at maximum brake horsepower and rated speed. The service factor is a numerical value, and it depends on the type of motor (open, splashproof, or totally enclosed) and resistance of the insulation on the windings to heat and breakdown.

When the service factor exceeds 1.0, the excessive amount is stamped on the nameplate along with the voltage and the full load ampere rating. For example, a 75 hp (56 kW) motor, with 1.15 service factor, could safely meet a demand of 75 × 1.15 = 86.25 bhp (64 kW).

Another use for the service factor is to estimate the maximum allowable ampere demand. For example, with a 40 amp full load rating and a 1.12 service factor, the maximum ammeter reading should not exceed 40 × 1.12 = 45 amps. Note also that for a given voltage, horsepower is proportional to amperes.

Only motors wound for 208 volts should be used on 208 volt services.

Direct current motors for pumps are either of the stabilized shunt type, or cumulative compound-wound type. The speed of the motor with no load at operating temperature shall not exceed the speed of the motor under full load at operating temperature by more than 10 percent.

The most commonly used alternating current motors are of the squirrel cage induction type. They usually are equipped with across-the-line starting equipment unless their starting characteristics would be objectionable to the company furnishing the power. In the latter case, primary resistance starting may be employed, or a wound rotor type of motor with appropriate starting equipment may be substituted. When squirrel cage motors are used, the voltage drop must not be so great as to prevent the motor from starting, i.e., not more than 10 percent below normal voltage at moment of start. While the motor is running at rated pump capacity, pressure, and speed, the line voltage should not drop more than five percent below motor nameplate voltage.

This type of motor should have normal starting and breakdown torque. The locked rotor currents for motors of various horsepower ratings are specified in the NFPA Centrifugal Fire Pump Standard.

Electric Motor Controllers

Motor controllers are available for alternating current fire pump motors operating at standard voltages up to 600 V. A controller is a complete, assembled unit, wired and tested, and ready for service by connecting to the power supply and the proper motor terminals. Detailed specifications are described in the NFPA Centrifugal Fire Pump Standard. Higher voltages are not recommended, but acceptable controllers conforming to special requirements of the NFPA Fire Pump Standard can be provided.

Controllers are available for combined manual and automatic operation, or for manual operation only. They also are available for either squirrel cage, wound rotor, or partwinding motors, and for two phase or three phase power. Across-the-line starting is recommended and preferred, but controllers for primary resistance, reduced voltage starting are also available.

The circuit breaker of a fire pump controller permits normal starting without tripping, and provides stalled rotor and instantaneous short circuit protection. The interrupting capacity of the breaker should be adequate for the circuit in which it is located, but not less than 15,000 amps in any case.

Automatic controllers known as limited service controllers are available for across-the-line squirrel cage motors of 30 hp (22 kW) or less driving special service booster fire pumps.

Steam Turbine

When adequate and reliable steam supplies are available, turbine driven fire pumps are acceptable. Only well built machines of good design with industrial records of proved reliability are used. Special arrangements are needed for automatic operation. Speed rating should not exceed 3,600 rpm, because this is the maximum speed of listed fire pumps. Detailed requirements for steam supply, speed governors, and controls are contained in the NFPA Centrifugal Fire Pump Standard.

Internal Combustion Engines

Engines that are powered by diesel fuel or natural gas are found in use for fire pump service; however, engines powered by gasoline, natural gas, or LP-Gas are not recognized by the NFPA Centrifugal Fire Pump Standard. Only the vapor withdrawal method of supplying an engine with LP-Gas should be allowed.

Natural gas and LP-Gas engines differ from gasoline engines only in design of the carburetors, intake manifolds, and cylinder heads. Large industrial consumers of natural or LP-Gas often prefer engine driven fire pumps using these fuels.

In addition to the Centrifugal Fire Pump Standard, reference should be made to NFPA 37, *Standard for the Installation and Use of Stationary Combustion Engines and Gas Turbines*; NFPA 54, *National Fuel Gas Code*; NFPA 58, *Standard for the Storage and Handling of Liquefied Petroleum Gases*; NFPA 31, *Standard for the Installation of Oil Burning Equipment*. Installations should be made in conformance to local codes.

The NFPA Standard for Stationary Combustion Engines and Gas Turbines specifically recognizes the need for special provisions covering engines driving fire pumps. In general, manual and automatic devices intended to limit or prevent the accidental discharge of flammable gases are needed, but the development of other unsatisfactory conditions, such as high cooling water temperature and low oil pressure, should be indicated by alarms and not by shutting down of the engine. The intent is to keep the pump operating just as long as possible. The importance of supervision, especially for automatic pumps, is obvious.

Cooling Systems: An adequate cooling system is vital to reliable operation of an internal combustion engine. A closed pipe system with a heat exchanger or a thermally insulated manifold are the usual cooling arrangements for a fire pump unit recognized in the NFPA Centrifugal Fire Pump Standard. Only clean or potable water should be circulated through the engine block. Raw water is piped from the fire pump through the heat exchanger tubes to free discharge in a visible location such as the cone of the fire pump relief valve. On some engines, the manifolds, oil coolers, and other parts are equipped with water jackets, as recommended by the engine manufacturer. (See Fig. 17-6R.) Most engines require a raw water flow of 15 to 30 gpm (57 to 114 L/min) or even more.

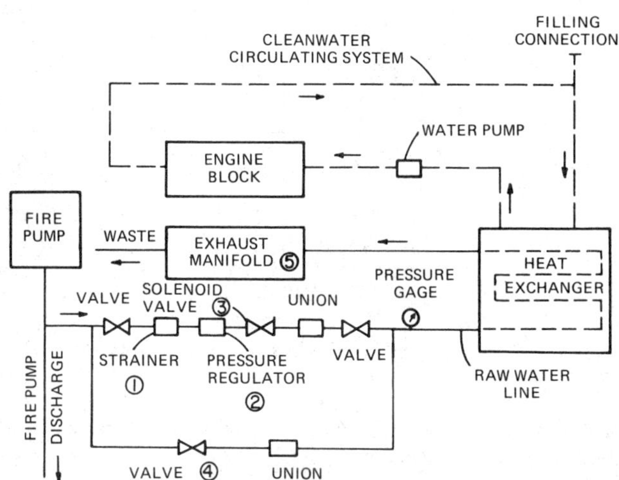

FIG. 17-6R. A typical heat-exchanger cooling system for an automatically controlled engine driven fire pump. Raw water from the fire pump enters the system through the strainer (1), which prevents sediment from entering the system, and the pressure regulator (2), which protects the heat exchanger from excessive pressure. The solenoid valve (3) is required with automatic control of the engine. Valve (4), normally closed, may be used to bypass the regulator and solenoid valve. The exhaust manifold (5) may be cooled by the clean water circulating system.

Fuel Tanks: The storage tank for liquid or gas fuel should contain at least an 8 hr supply; greater capacity should be provided if facilities for prompt refilling are not available. Tank capacity may be estimated by allowing 1 pt of diesel oil or LP-Gas per horsepower per hour. (See diagrams of typical fuel systems in the NFPA Centrifugal Fire Pump Standard.)

Engine Horsepower

Engines specifically designed for use with fire pumps are rated by measuring the horsepower developed with all accessories in operation and then making some allowance for wear and tear. Other engines used with fire pumps outside the power range and type of listed engines must have a horsepower capability, when equipped for fire pump driver service, of not less than 10 percent greater than the maximum brake horsepower required by the pump under any conditions of pump load. The engine must meet all the other requirements of listed engines. Typical bare engines and usable horsepower curves are shown in Figure 17-6S.

The engine manufacturer's test curves are based on barometric pressure of 29.61 in. (752 mm) Hg which approximates 300 ft (90 m) above sea level and 77°F (25°C). The usable horsepower of a fire pump engine should be reduced for each 1,000 ft (300 m) rise in altitude above 300

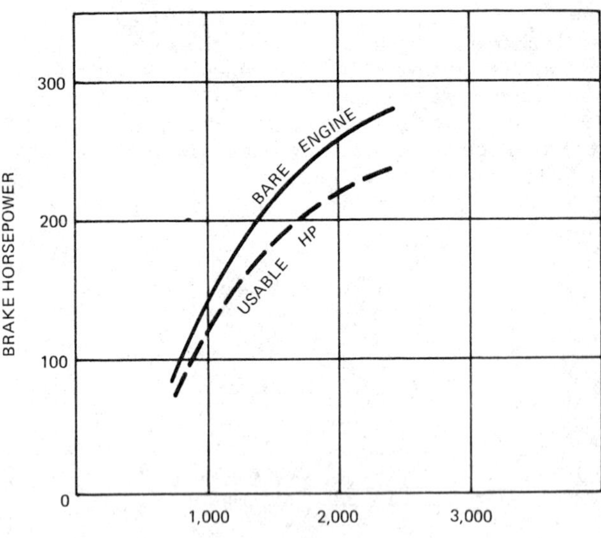

FIG. 17-6S. Typical engine horsepower curves. SI units: 1 BPH = 745.70 W.

ft (90 m) by 3 percent for a diesel engine and 1 percent for each 10°F (5.6°C) rise above 77°F (25°C).

Engine Controllers

Controllers are used for automatic operation of engine driven fire pumps. The specifications for construction, location, and methods of actuation of engine controllers are the same as for electric motor controllers. Automatic controllers are equipped with manual starting and stopping switches.

Alarm devices are provided to indicate low oil pressure in the lubrication systems, high engine jacket water temperature, failure of engine to start automatically, and shut down from overspeed (diesel only).

There are optional features for controllers as given in the NFPA Centrifugal Fire Pump Standard. A weekly program timer can be provided. This device may be arranged to start the unit automatically once a week, and run for a predetermined number of minutes. A recording pressure gage records this performance.

Controllers are operated on low voltage direct current power from the engine's batteries. The program timer, battery charger, or other auxiliary devices not essential to pump control are powered by the regular alternating current power supply at the property.

AUTOMATIC PUMP CONTROL

Most fire pump installations normally are arranged for automatic operation, preferably with automatic start and manual shutdown. The choice between manual and automatic shutdown depends on an evaluation of the specific conditions involved in a pump's installation and use. Horizontal centrifugal pumps under automatic control should always operate under a head to avoid the need of priming.

Each motor or engine controller is equipped with a pressure switch that actuates the pump unit when pressure in the water system piping drops to a preset level. Unless the normal water supply static pressure is higher than the pump starting pressure, an automatic jockey pump must be provided to maintain pressure in the system at the higher level.

Actuation of a pump by water flow instead of by pressure drop is desirable for certain installations, such as: (1) where the opening of a moderate number of sprinklers would not drop the system pressure enough to move the pressure switch, (2) where fire in a high hazard occupancy would demand fire pump service without delay, (3) in a combined fire protection and plant service system where a pressure maintenance pump would be impractical, and (4) where pressure fluctuates so much that a stable cut-in pressure could not be obtained.

The wiring system of a pump controller includes terminals for connection of a relay to an external alarm circuit from a sprinkler, deluge, or special fire protection system. To secure reliable pump actuation, external circuits should be installed in conformance with one of the following NFPA standards (short titles), depending upon the nature of the signaling system:

NFPA 71, *Central Station Signaling Systems*

NFPA 72A, *Local Protective Signaling Systems*

NFPA 72B, *Auxiliary Protective Signaling Systems*

NFPA 72C, *Remote Station Protective Signaling Systems*

NFPA 72D, *Proprietary Protective Signaling Systems*

Circuits for remote automatic starting of fire pumps should be powered from the controller power.

(See the bibliography at the end of the chapter for availability of these standards.)

FIELD ACCEPTANCE TESTS

After a new fire pump has been installed, it is general practice for a performance test to be made. Defects and faults can be discovered, and steps taken to remedy them. These tests enable the purchaser to determine that the contract has been properly concluded. They also demonstrate the need of future maintenance tests.

Details of the acceptance tests are given in the NFPA Centrifugal Fire Pump Standard. The test demonstrates the adequacy of the pump suction and the ability of the pump to deliver water in accordance with its head-capacity curve. The prime mover also is operated under various conditions and its performance noted. There are separate provisions for electric motors, steam turbines, and diesel engines. Repeated operations of the controlling equipment are required to ensure that full operation of the unit will result from either manual or automatic operation of the controller.

Flow tests are made to develop the pressure-discharge characteristic curve of the pump. The procedure followed is to run the pump at five or six different flows, including "shutoff" with no water flowing. The rate in gallons per minute is determined with a Pitot gage at the nozzles (preferably standard 30 in. (762 mm) long "Underwriter" playpipes) attached to hose lines from an outside hose-valve header. The discharge is varied by changing the number of lines and size of nozzle tips or both.

For every flow, pressure readings are taken at the suction gage and the discharge gage; the revolutions per minute are also measured, using a revolution counter or a tachometer, if available. (The tachometer on a fire pump engine is not sufficiently accurate.)

The net pressures are calculated from the pump gage readings, and the flows in gallons per minute corresponding to the Pitot readings are obtained from discharge tables.

The nozzles may be attached directly to outdoor headers, without hose lines, if water damage can be avoided. Disposal of test water is often a problem, and the length of hose lines would depend on the drainage facilities available, and exposure to property and people.

(With the growing problems of waste water disposal, many pump installations are equipped with water meters for the acceptance test and periodic service tests. The meters should be installed in conformance with the NFPA Centrifugal Fire Pump Standard in order to function properly and not interfere with the operation of the pump.)

Vertical turbine fire pumps are tested in the same manner as horizontal pumps, except that there is no suction gage. Also, the pumping water level should be recorded at the several test points unless it is more or less constant.

Figure 17-6T is a tabulation of data obtained by a typical field acceptance test of a horizontal 1,500 gpm, 100 psi, 1760 rpm (5678 L/min, 689 kPa, 1760 rpm) engine driven centrifugal pump. The net pressure and total flows

Acceptance Test 1,500 gpm, 100 psi, 1,760 rpm Fire Pump

				Streams					corrected to 1,760 RPM	
RPM	Discharge psi	Suction psi	*Net psi	No.	Size in.	Pitot Pressure psi	GPM*	Total*	GPM*	Net* psi
1,700	125	+16	109	0	—	—	—	0	0	118
1,695	120	+18	102	1	1¾	70	742	742	772	110
1,690	110	+16	94	2	1¾	60, 60	687, 687	1,374	1,420	101
1,686	95	+17	79	3	1¾	55, 55, 55	657, 657 657	1,971	2,060	85
1,675	85	+16	69	4	1¾	35, 37, 48, 48	525, 540 614, 614	2,293	2,410	76

* Calculated from observed data.

FIG. 17-6T. An example of a log for a fire pump acceptance test.

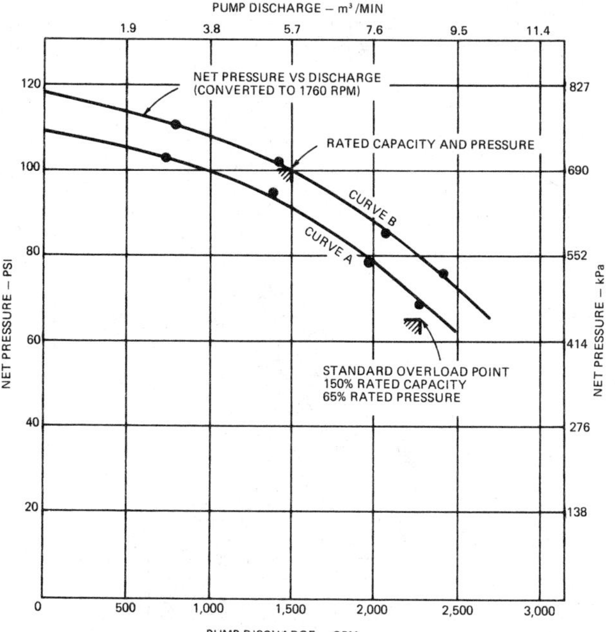

FIG. 17-6U. Head-capacity curves plotted from data compiled in a fire pump acceptance test. (See Fig. 17-6T for the data used to plot these curves.)

are calculated from the observed data and plotted. (See Fig. 17-6U.) The curve best fitting the plotted points is then drawn (Curve A). In this installation the engine governor appeared to be out of adjustment, restricting the average speed to 1,689 rpm, whereas the rated speed was 1,760 rpm.

Since the pump was tested at less than rated speed, the observed net pressures and flows were converted to what they would have been at the rated speed of 1,760 rpm. Curve B is the characteristic curve at rated conditions. Although the rating point was barely reached, the overload point exceeded the minimum by a good margin. With the engine adjusted to operate at full speed, the pump performance would be acceptable. The following is the conversion calculation procedure that was followed:

Flow is directly proportional to revolutions per minute
Net pressure is proportional to $(\text{rpm})^2$

EXAMPLE: Test flow 1,971 gpm at 1,686 rpm

Flow at 1,760 rpm $= 1.971 \left(\frac{1{,}760}{1{,}686}\right) = 2{,}060$ gpm

Net pressure for 1,971 gpm at 1,686 rpm is 78 psi

Net pressure for 2,060 gpm at 1,760 rpm is $78 \left(\frac{1{,}760}{1{,}686}\right)^2 = 85$ psi.

Similar calculations can be carried out in SI units using the affinity laws mentioned previously in this chapter.

In theory, the characteristic curve assumes operation at constant rated speed. Actually the speed of internal combustion engines and steam turbines is permitted to vary within a range of 8 to 10 percent between shutoff and maximum load. Electric motor speed is more nearly constant. Speed reduction may occur if the power supply is overloaded.

LOCATION AND HOUSING OF CENTRIFUGAL PUMPS

Fire pumps are housed in buildings of fire resistive or noncombustible construction. Even when the climate is so mild that there is no danger of freezing, sufficient enclosure is needed to protect against dirt, corrosion, and tampering. Structural separation of the pump room from other parts of a property is desirable.

Pump rooms and power facilities should be as free as possible from exposure to fire, explosion, flood, and windstorm damage.

Light, heat, ventilation, and floor drainage should be provided for pump rooms. An abovegrade, dry location is preferred. For an internal combustion engine driven unit, heat, ventilation, and abovegrade location are essential.

Fire pumps preferably are located as close as possible to those areas where protection is most important. In some large properties, it may be necessary to have water supplies at more than one point to obtain the most favorable distribution system. When this results in placing a pump in a somewhat isolated pump room, the requirements for housing and supervision are of special importance.

ANNUAL PUMP TESTS

A fire pump should be tested annually to make certain that the pump, driver, suction, and power supply function properly, and to correct faults that may be revealed. The hydraulic performance of the pump is measured by a flow test with hose and nozzles connected to the pump header or yard hydrants. Three points on the standard curve are checked: (1) shutoff, (2) overload (150 percent of rated capacity or more), and (3) a convenient rate of flow at or near capacity rating.

Automatic operation is tested by opening yard hydrants or sprinkler riser drains giving due consideration to the layout of the fire protection system (pressure drop or water flow actuation, jockey pump, etc.). It is not sufficient to initiate pressure drop by the test cock on the controller.

Water level of ponds and reservoirs, condition of suction screens and intakes, aboveground tanks, etc., should be carefully examined.

The history of power outages, low water, and failure of any kind involving pump, driver, or associated equipment, should also be investigated and gage records from engine controllers (when so equipped) examined.

PUMP OPERATION AND MAINTENANCE

A fire pump can be depended upon to work in an emergency only if it is properly operated and maintained. It is desirable to have someone at the property at all times who has been designated and instructed to operate the pump and its driver. A short test by the regular pump operators should be made each week by discharging water from some convenient outlet.

When a fire alarm is given or an alarm indicates an automatic fire pump is operating, the person responsible for the fire pump should proceed to its location immediately. The pump preferably should be put in manual operation and allowed to run until the emergency is over, when it may be shut down manually. During this and every other operating period, the equipment should be carefully checked to see whether it is performing properly.

To prevent too frequent starting and stopping, an electric motor controller has a timer to keep the motor running for at least 1 min for each 10 hp motor rating (not more than 7 min required). It is preferable with all types of pump drivers to permit the unit to run until it is shut down manually. When there is more than one automatic fire pump, the control is arranged for operation of the pumps in a predetermined sequence. Control of the pump from one or more remote push buttons, which will start but not stop the pump, may be provided if desired. Also, if there is deluge valve control of an open discharge device system, the pump may be started by a drop-out relay in a closed circuit.

The cooling and lubrication of a centrifugal fire pump is so dependent upon water that the pump must never be run without the pump casing full of water. Close attention should be given to the bearings and stuffing boxes during the first few minutes of running to see that there is no heating up and no need of adjustment. When water reaches the water seal, a small leak at the stuffing box glands is desirable. The suction inlet and discharge outlet pressure gages should be read occasionally to see that the inlet is not obstructed by a choked screen or foot valve.

With a vertical shaft turbine type fire pump, the water level can be observed if suction is from a visible supply. If the pump takes suction from a well, water level testing equipment must be used. The ground water level at the pump should be checked at intervals during the year and the draw down should be determined during the annual 150 percent capacity test. These tests should indicate any important change in the ground water supply.

The direction of rotation of the pump and the speed of operation should always be checked.

Power Supply Maintenance

The source of power for the pump should also be checked. With an electric motor drive this means current supply for the motor and its auxiliary equipment. For steam turbine drive it means the steam supply up to the control valve and the absence of condensate from supply, turbine, and exhaust. If the pump is driven by a diesel engine, there must be adequate fuel for 8 hr of operation. The batteries must be fully charged.

The starting equipment must be test operated and its functioning carefully checked. Any evidence of a drop in voltage to an electric motor or a drop in steam pressure to a turbine must be investigated.

With a diesel engine drive, the crankcase oil must be replenished or renewed as needed, the oil filter and air cleaner given necessary attention, the automatic battery charging equipment checked, and the specific gravity of battery electrolyte determined at least once a month.

RECIPROCATING STEAM FIRE PUMPS

Although few new reciprocating steam fire pumps have been installed in recent years, a number are still in service, and they should conform to NFPA 21, *Standard Operation and Maintenance of National Standard Steam Fire Pumps.* (These fire pumps are commonly referred to as Underwriter Steam Fire Pumps.)

The general features of a direct acting duplex steam pump are shown in Figure 17-6V. The size of the steam

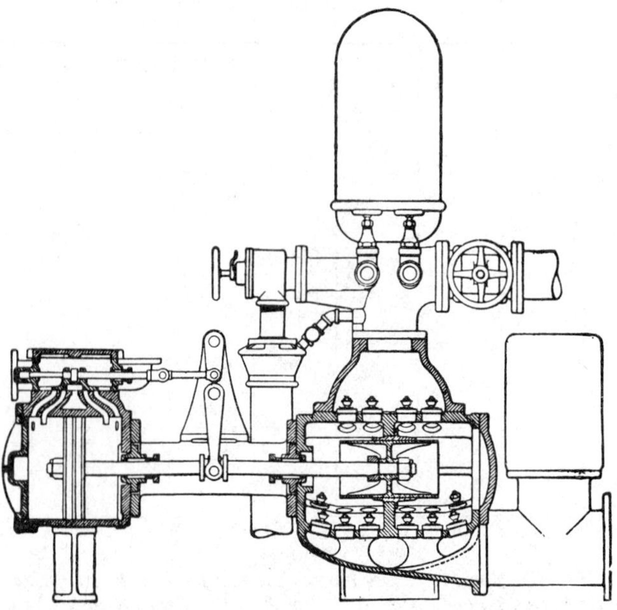

FIG. 17-6V. Sectional view of an approved duplex steam pump.

and exhaust ports is larger than in the general purpose steam pump, thus permitting higher speeds. The steam supply pipe should be an independent line run from boiler to pump in such a way that it will not be damaged by fire or other hazards.

Automatic Control

Automatic control can be provided by a pressure governor to regulate steam supply to the pump in accordance with the water pressure on the pump discharge. For successful operation, it is nearly always necessary to provide a small, automatically controlled jockey pump to maintain system pressure, control supply leakage, and avoid continuous operation of the large pump.

Bibliography

References Cited

ANSI. 1973. ANSI B73.1, Specifications for Horizontal End Suction Centrifugal Pumps. American National Standards Institute.

FMEC. 1983. Loss Prevention Data 3-251, May 1983, Break Tanks, Factory Mutual Engineering Corp. Norwood, MA.

NFPA Codes, Standards,Recommended Practices and Manuals. (See the latest *NFPA Codes and Standards Catalog* for availability of current editions of the following documents.)

NFPA 13, *Standard for the Installation of Sprinkler Systems.*

NFPA 14, *Standard for the Installation of Standpipe and Hose Systems.*

NFPA 15, *Standard for Water Spray Fixed Systems for Fire Protection.*

NFPA 16, *Standard for Foam-Water Sprinkler Systems and Foam-Water Spray Systems.*

NFPA 20, *Standard for the Installation of Centrifugal Fire Pumps.*

NFPA 21, *Standard for the Operation and Maintenance of National Standard Steam Fire Pumps.*

NFPA 22, *Standard for Water Tanks for Private Fire Protection.*

NFPA 24, *Standard for Outside Protection.*

NFPA 31, *Standard for the Installation of Oil Burning Equipment.*

NFPA 37, *Standard for the Installation and Use of Stationary Combustion Engines and Gas Turbines.*

NFPA 58, *Standard for the Storage and Handling of Liquefied Petroleum Gases.*

NFPA 71, *Standard for the Installation, Maintenance, and Use of Central Station Signaling Systems.*

NFPA 72A, *Standard for the Installation, Maintenance, and Use of Local Protective Signaling Systems for Watchmen, Fire Alarm, and Supervisory Service.*

NFPA 72B, *Standard for the Installation, Maintenance and Use of Auxiliary Protective Signaling Systems.*

NFPA 72C, *Standard for the Installation, Maintenance and Use of Remote Station Protective Signaling Systems.*

NFPA 72D, *Standard for the Installation, Maintenance and Use of Proprietary Protective Signaling Systems for Watchmen, Fire Alarm and Supervisory Service.*

Additional Readings

American Standard Specifications for Deep Well Vertical Turbine Pumps, AWWA, NY.

Cameron Hydraulic Data, 13th ed., Ingersoll Rand Co., NY.

FMEC, Loss Prevention Data Sheet 3-9, Dec. 1973, "Underwriter Steam Fire Pumps," Factory Mutual Engineering Corp., Norwood, MA.

Hydraulic Institute, "Centrifugal Pump Section," Standards of Hydraulic Institute, 14th ed., Hydraulic Institute, NY, 1969.

Jensen, Rolf, ed., *Fire Protection for the Design Professional*, Cahners, Boston, 1975, pp. 47-57.

Karrasik, Igor, and Carter, Roy, *Centrifugal Pumps, Selection, Operation and Maintenance*, F. W. Dodge Corporation, NY, 1960.

Rotary and Centrifugal Pump Theory and Design, Worthington Pump Corp., East Orange, NJ, 1971.

Smith, Paul D., "Talking Fire Protection Systems," *Fire Journal*, Vol. 73, No. 3, May 1979, pp. 11, 117.

FIRE STREAMS

Revised by Robert G. Purington

Water has always been and remains the primary fire extinguishing agent; therefore the movement of water from its source to the fire is a significant operation in most kinds of fire fighting. A typical fire department delivery system consists of suction hose, a pumper to increase the water pressure to overcome friction loss and provide nozzle pressure, a supply line to move the water from the pump to the fire scene, and "working lines" and nozzles for the final delivery. Water is used in a number of forms and with various additives, and there are numerous combinations in which the water can be delivered to the fire scene.

This chapter includes a description of the most important details of good fire streams, including pumps, nozzles, fire hose friction loss, pump pressure calculations, friction reducing additives, and reaction forces.

FIRE DEPARTMENT PUMPS

The primary purpose of fire department pumpers is to move water with adequate pressure from a source to the fire. Sources include hydrants, lakes, streams, rivers, or storage tanks in fixed locations or in vehicles. Pumpers serve other functions, including transporting fire fighters, water and other extinguishing agents and equipment. Pumps are also installed on specialized apparatus such as tank trucks, aerial ladders, elevating platforms, aircraft crash vehicles, forestry trucks, and fire boats.

Pumpers, or "engines," are generally equipped with a centrifugal pump and the necessary fittings for pressure regulation and priming operations. Piston, rotary gear, and rotary vane pumps are used for special purposes such as high pressure booster pumps and priming. Requirements for standard pumpers are spelled out in NFPA 1901, *Standard for Automotive Fire Apparatus*.

Two important characteristics of fire pumps are the discharge rate and the discharge pressure. They comprise the final output of the pump. The discharge rate, also known as the quantity, flow rate, or volume, is the amount of water pumped per unit time, such as gallons per minute (gpm) or liters per second (L/min). The discharge pressure, indicated by gages, is the pressure at which the pump is pumping water. The discharge pressure is in pounds per square inch (psi) or kilopascals (kPa). Pumps are rated according to their flow rate at 150 psi (1034 kPa) discharge and generally range from 500 gpm (1893 L/min) to 2,000 gpm (7571 L/min). These pumps are also required to provide 70 percent of their rated capacity at 200 psi (1379 kPa) and 50 percent of their rated capacity at 250 psi (1724 kPa). These ratings are summarized in Table 17-7A.

TABLE 17-7A. Standard Pumps

Rated Flow Rate		70 Percent of Rated Flow Rate		50 Percent of Rated Flow Rate	
gpm @ 150 psi	L/min @ 1034 kPa	gpm @ 200 psi	L/min @ 1379 kPa	gpm @ 250 psi	L/min @ 1724 kPa
500	1893	350	1325	250	946
750	2839	525	1987	375	1419
1000	3785	700	2650	500	1893
1250	4732	875	3312	625	2366
1500	5678	1050	3975	750	2839
1750	6624	1225	4637	875	3312
2000	7571	1400	5300	1000	3785

The centrifugal pump, a nonpositive displacement pump, consists essentially of a pair of rotating discs called shrouds, separated by curved partitions which are called vanes. This assembly makes up the impeller. Typically, fire apparatus pumps are multistage, usually two, thus permitting discharge at either higher flow rates with lower pressures or lower flow rates with higher pressures. The purpose of the multistage pump is to obtain the most effective operation at the lowest possible engine speed. However, single stage pumps are becoming more popular, particularly with the advent of the diesel engine for fire apparatus.

Pumping operations start with water from either a pressure source, such as a hydrant, or from a static water source, such as a pond. Drafting operations are required in

Mr. Purington is a retired fire chief and consulting engineer based in Livermore, CA.

the latter case. The factors which affect the height (lift) limitations to which water can be drafted and the flow rate are interrelated. These factors are: (1) atmospheric pressure, (2) water temperature, (3) friction loss in the suction hose and strainer, and (4) velocity energy.

The lift at a given flow rate can be determined by: Lift = atmospheric pressure − water vapor pressure − strainer and suction hose friction loss − velocity energy.

Atmospheric pressure varies from day to day according to the weather conditions and the elevation above sea level. Since drafting is essentially the creation of a vacuum or partial vacuum in the pump and suction hose, the total energy available for moving the water from the source to the pump is the atmospheric pressure. The nominal atmospheric pressure, 14.696 psia (101.333 kPa) at sea level, usually rounded to 14.7 psia (100 kPa), can theoretically provide a lift of almost 34 ft (10.5 m). Under actual conditions, a pump in excellent condition and pumping its rated capactiy at sea level should be able to draft water at least 25 ft (7.6 m).

Often atmospheric pressure is given in inches of mercury. If this is the case, it must be converted to feet of water using:

$$h_w = 1.13h'_{Hg}$$

Where:

h_w = feet of water
h'_{Hg} = inches of mercury.

The vapor pressure reduces the amount of vacuum that is possible and consequently reduces the total lift. Vapor pressure depends on the water temperature and is measured in terms of feet (meters) of water. Table 17-7B lists common values of vapor pressure as a function of water temperature.

Table 17-7B. Vapor Pressure

Water Temperature		Vapor Pressure	
°F	°C	Ft of water	kPa
60	16	0.59	1.7
80	27	1.20	3.5
100	38	2.20	6.6
120	49	3.90	11.6

Another factor affecting the total lift is the friction loss in the suction hose and the strainer. The friction loss in the suction hose depends on the hose diameter, hose length, and the flow rate of the water. Equations for the friction loss in various sizes of suction hose are shown in Table 17-7C. The friction loss through the strainer is a function of the flow rate and the strainer diameter. The friction equations for the various strainer sizes are shown in Table 17-7D.

The velocity head also affects the total lift of the pump. The velocity head is given by $V^2/2g$, where V is the water velocity in feet per second (meters per second) and g is the acceleration due to gravity and is equal to 32.2 ft/sec^2 (9.81 m/sec^2). The velocity may be found by using the continuity equation, $Q = VA$, where Q is the flow rate in cubic feet per second (cubic meters) and A is the cross-section of the suction hose in square feet (square meters). However, the flow rate is usually given in gallons (liters) per minute and the suction hose diameter is in inches (millimeters). Consequently, Table 17-7E can be used to determine the velocity head in the suction hose without tedious calculations.

TABLE 17-7C. Suction Hose Friction Losses per 10 ft (3 m) length

Suction Hose Diameter		Friction Loss (FL)			
in.	mm	psi	kPa	ft	m
3	76	$0.083q^2$*	$0.0400q^2$**	$0.191q^2$	$0.00408q^2$
3½	88	$0.0385q^2$	$0.0186q^2$	$0.0886q^2$	$0.00189q^2$
4	102	$0.0197q^2$	$0.0095q^2$	$0.0454q^2$	$0.00978q^2$
4½	114	$0.0109q^2$	$0.0053q^2$	$0.0251q^2$	$0.00054q^2$
5	127	$0.00646q^2$	$0.0031q^2$	$0.0149q^2$	$0.00033q^2$
6	152	$0.0026q^2$	$0.0014q^2$	$0.006q^2$	$0.00014q^2$

* q = 100s gpm (** 100s L/min)

In solving the lift equation, all units must be converted to the equivalent feet (or meters) of water. Atmospheric pressure is usually given in terms of inches of mercury (or kilo Pascals). Table 17-7F may be used to convert the atmospheric pressure to the equivalent feet (or meters) of water.

EXAMPLE: Convert atmospheric pressure to feet of water (lift).

Atmospheric pressure = 27.00 in. of mercury (91 kPa)
Water temperature = 60°F (15.5°C)
Flow rate = 1,100 gpm (4163 L/min)
Suction hose diameter (and strainer diameter) = 5 in. (127 mm)
Suction hose = two lengths of 10 ft (3 m) each.

SOLUTION: From Table 17-7F atmospheric pressure of 27.00 in. of mercury (91 kPa) equals 30.556 ft (9.31 m) of water. From Table 17-7B the vapor pressure (at 60°F or 15.5°C) equals 0.59 ft (0.18 m) of water. Using Table 17-7C, the friction loss for 10 ft (3 m) of 5 in. (127 mm) suction hose is $0.0149q^2$ in feet and $0.00033q^2$ in meters.

Q = 1,100 gpm (4163 L/min). However the equation is for Q in 100s of gpm or L/min, and $Q/100 = q$. Therefore use q = 11 (41.63). Thus:

$FL = (0.0149)(11)^2 = 1.8$ ft for 10 ft or $(2)(1.8) = 3.6$ ft for 20 ft.

In SI: $FL = (0.00033)\ (41.63)^2 = 0.57$ m for 3 m or $(2)(0.57) = 1.14$ m for 6 m.

From the suction strainer equation listed on Table 17-7D:

$FL = 0.0438q^2$ where q = 11 as in the suction equation.

$FL = (0.0438)(11)^2 = 5.3$ ft.
In SI: $FL = .00093q^2$ where $q = 41.63$
$FL = (00093)\ (41.63)^2 = 1.6$ m

The velocity head (from Table 17-7E) is 5 ft (1.53 m) of water. Therefore, the total lift is:

Lift = 30.56 − 0.59 − 3.6 − 5.3 − 5 = 16.1 ft.

Table 17-7D. Suction Strainer Friction Losses (Loss = Cq²)

Suction Hose Diameter		Friction Loss			
in.	mm	psi	kPa	ft	m
3	76	$0.147q^{2}$*	$0.0708q^{2}$**	$0.339q^2$	$0.00722q^2$
3½	88	$0.0793q^2$	$0.0353q^2$	$0.1827q^2$	$0.00390q^2$
4	102	$0.0465q^2$	$0.0225q^2$	$0.1071q^2$	$0.00229q^2$
4½	114	$0.0290q^2$	$0.0139q^2$	$0.0668q^2$	$0.00142q^2$
5	127	$0.0190q^2$	$0.0092q^2$	$0.0438q^2$	$0.00093q^2$
6	152	$0.00918q^2$	$0.0053q^2$	$0.0212q^2$	$0.00054q^2$

* q = 100s gpm (q** = 100s L/min)

IN SI: Lift = 9.31 − 0.18 − 1.14 − 1.6 − 1.53 = 4.9 m (16.1 ft).

NOZZLES

The purpose of a nozzle is to shape the stream and convert pressure energy to velocity (kinetic) energy. In this way the water can be applied to the fire in the appropriate quantity and from an adequate distance. There are many specialized nozzles. However they all can be classified in two ways: solid stream and fog (spray) nozzles. (See Fig. 17-7A.) They both have advantages and disadvantages. These characteristics will be discussed.

The governing hydraulic characteristics of nozzles are the flow rate, diameter, and the nozzle pressure. The flow rate is the amount of water flowing out of the nozzle per unit time and is measured in gallons per minute or liters per minute.

Nozzle pressure is defined as the Pitot pressure (the stagnation pressure of the water coming from the nozzle tip), or the pressure at the base of the nozzle (pressure energy only). The Pitot pressure is used only for solid stream nozzles while the pressure at the base of the nozzle can be used for both kinds of nozzles. Practically speaking there is usually little numerical difference in the two definitions. The unit of nozzle pressure is pounds per square inch or kilopascals (psi or kPa).

Solid Stream Nozzles

Solid stream nozzles are classified according to the nozzle diameter. The diameters range from ¼ in. to 2½ in. (6 mm to 64 mm) or larger. The nozzle sizes are generally available in ⅛ in. (3 mm) graduations. Nozzles up to 1⅛ in. (29 mm) or perhaps 1¼ in. (32 mm) are generally considered to be for "hand lines," i.e., those manually held. Nozzles larger than about 1¼ in. (32 mm) have such large reaction forces that they must be mechanically restrained. These are usually called master streams, monitors, deluge, or deck guns. The 1⅛ in. (29 mm) nozzle produces the so-called standard stream of 250 gpm (946 L/min) at about 45 psi (310 kPa) nozzle pressure.

Solid streams are useful where extreme range is desired, where it is necessary to penetrate soft material, such as peat, or when thermal degradation of spray streams prevents proper penetration.

Solid stream nozzles are also useful in measuring water flows since both their diameter and the Pitot pressure can be measured. From this information, the flow rate can be computed. It is related to the diameter and nozzle pressure by the equations:

$$Q = 29.7d^2\sqrt{p}$$

Where:

Q = flow rate, gpm
d = nozzle diameter, in.
P = nozzle pressure (Pitot), psi

For most field hydraulics, the equation can be rounded to:

$$Q = 30d^2\sqrt{p}$$

For SI units the flow equation is:

$$Q = 0.0666d_m{}^2\ \sqrt{P_m}$$

Where:

Q = flow rate, (L/min);
d_m = inside nozzle diameter, (mm);
P_m = nozzle pressure, (kPa).

Table 17-7E. Velocity Head (V) For Suction Hose, Ft (m) Of Water

FLOW, Q		Suction Hose Diameter, (D) in inches and mm											
		3.0 in.	76 mm	3.5 in.	88 mm	4.0 in.	102 mm	4.5 in.	114 mm	5.0 in.	127 mm	6.0 in.	152 mm
gpm	L/min	ft.	m	ft	m	ft	m	ft	m	ft	m	ft	m
100	378	0.32	0.09	0.17	0.05	0.10	0.03	0.06	0.02	0.04	0.01	0.02	0.006
200	757	1.28	0.39	0.69	0.21	0.40	0.12	0.25	0.05	0.17	0.05	0.08	0.02
300	1136	2.88	0.88	1.55	0.47	0.91	0.28	0.57	0.17	0.37	0.11	0.18	0.05
400	1514	5.11	1.56	2.76	0.84	1.62	0.49	1.01	0.31	0.66	0.20	0.32	0.09
500	1892	7.99	2.44	4.31	1.31	2.53	0.78	1.58	0.48	1.04	0.32	0.50	00.15
600	2271	11.51	3.51	6.21	1.90	3.64	1.11	2.57	0.78	1.49	0.45	0.72	0.22
700	2650			8.45	2.58	4.96	1.51	3.09	0.94	2.03	0.62	0.98	0.30
800	3028			11.04	3.36	6.47	1.97	4.04	1.23	2.65	0.81	1.28	0.39
900	3407					8.19	2.50	5.11	1.56	3.36	1.02	1.62	0.49
1000	3785					10.11	3.08	6.31	1.92	4.14	1.26	2.00	0.61
1100	4163					12.24	3.73	7.64	2.33	5.01	1.53	2.42	0.74
1200	4542							9.09	2.78	5.97	1.82	2.88	0.88
1300	4921							10.67	3.25	7.00	2.13	3.38	1.03
1400	5300							12.37	3.78	8.12	2.47	3.92	1.19
1500	1893									9.32	2.84	4.49	1.37

TABLE 17-7F. Inches of Mercury Vacuum Compared with Height of Water and Corresponding Pressure

Vacuum Inches of Mercury	Pressure psig	Pressure psia	Height of Water (ft)
0.00	0.000	14.696	.000
1.00	−0.491	14.205	1.132
2.00	−0.982	13.714	2.263
3.00	−1.473	13.223	3.395
4.00	−1.965	12.731	4.527
5.00	−2.456	12.240	5.658
6.00	−2.947	11.749	6.790
7.00	−3.438	11.258	7.922
8.00	−3.929	10.767	9.054
9.00	−4.420	10.276	10.185
10.00	−4.912	9.784	11.317
11.00	−5.403	9.293	12.449
12.00	−5.894	8.802	13.580
13.00	−6.385	8.311	14.712
14.00	−6.876	7.820	15.844
15.00	−7.367	7.329	16.975
16.00	−7.858	6.838	18.107
17.00	−8.350	6.346	19.239
18.00	−8.841	5.855	20.370
19.00	−9.332	5.364	21.502
20.00	−9.823	4.873	22.634
21.00	−10.314	4.382	23.766
22.00	−10.805	3.891	24.897
23.00	−11.297	3.399	26.029
24.00	−11.788	2.908	27.161
25.00	−12.279	2.417	28.292
26.00	−12.770	1.926	29.424
27.00	−13.261	1.435	30.556
28.00	−13.752	0.944	31,687
29.00	−14.244	0.453	32.819
29.92	−14.696	0.000	33.862

SI units: 1 in. of Mercury = 3.385 kPa; 1 psi = 6.895 kPa; 1 ft = 0.305 m.

Theoretical flows from various nozzles at different nozzle pressure are shown in the table, Theoretical Discharge Through Circular Orifices, in Chapter 2, "Hydraulics," of this Section.

Solid streams, with a lower surface-area-to-volume ratio, do not have as good heat-transfer characteristics as spray nozzles and consequently are not as effective in absorbing heat. Another disadvantage of a solid stream is that it is a better conductor of electricity than a spray.

Spray Nozzles

Spray nozzles, or fog nozzles as they are often called, produce varying degrees of water spray. They may have a fixed-spray angle, or may be adjustable from almost a straight stream to a very wide angle spray. The spray angle is customarily measured at the angle between the outer limits of the spray core. Many spray nozzles have predetermined pattern settings, usually at straight-stream, 30, 60, and 90 degree spray angles; however, most of them can also be set at intermediate angles.

Spray nozzles are generally classified according to the size of their hose coupling and their flow rate in gpm. Table 17-7G shows typical flow rates from spray nozzles.

The designated flow from a spray nozzle is usually rated at 100 psi (690 kPa) nozzle pressure. Some spray nozzles have different flows at various angles. Generally,

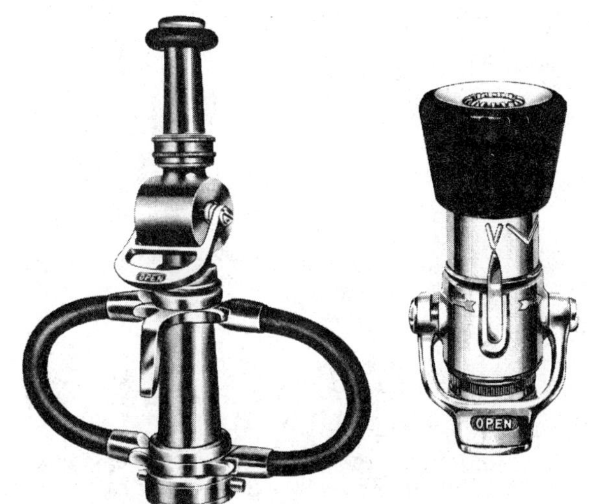

FIG. 17-7A. Straight stream nozzle (left) and fog nozzle (right). (W. S. Darley and Co.)

the flow increases as the angle widens. Other spray nozzles are "constant gallonage," that is, they have the same flow at all spray angles. In addition, some models have a flow rate setting; they can be set for several different flow rates.

Flow rates are determined experimentally, either by a flow meter or by flowing water into a tank and measuring the time for a given quantity. Flow rate information is obtained from the manufacturer, and is usually reported as the flow rate at 100 psi (690 kPa) nozzle pressure for the various spray angles.

Flows for other pressures may be obtained by using the flow equation:

$$Q = 29.7d^2 \sqrt{P}$$

and finding the equivalent nozzle diameter d_e, using

$$d_e = \sqrt{Q/29.7\sqrt{P}}$$

Where:

d_e = equivalent nozzle diameter, in inches.

Table 17-7G. Flows From Typical Fog Nozzles

Nozzle Type	Flow Rate gpm	Flow Rate L/mm
¾ to 1 in. (19 to 25 (mm) booster spray nozzle	10–40	38–151
1½ in. (38 mm) spray nozzle	70–150	265–568
2½ in. (64 mm) spray nozzle	200–300	757–1136
Master stream nozzles	500	1893
	750	2839
	1000	3785
	1250	4732
	1500	5678
	2000	7570

For SI units the flow formula is:

$$Q = 0.066\ d_m{}^2\ \sqrt{P_m}$$

Where:

Q = flow rate (L/min);
d_m = inside nozzle diameter, (mm);
P_m = nozzle pressure (kPa).

HOSE LINE FIRE STREAM CALCULATIONS

The pump pressure (commonly called "engine pressure") is usually the end result of hose stream calculations. The methods for obtaining pump pressure are described below. However, algebraic manipulation of the equations can yield any variable, providing the other quantities are known.

Special techniques known as "field hydraulics" or "rule-of-thumb" are used by pump operators for approximate results. The methods shown below give more precise results, but are generally too involved for field applications. Details on rule-of-thumb fire stream calculations may be found in the fire fighting hydraulic texts listed in the Bibliography of this chapter.

The calculation of the pump pressure depends on a number of variables: discharge pressure, hose friction loss, and back pressure due to elevation changes.

As shown previously, the discharge pressure is a function of the nozzle pressure and nozzle diameter. The hose friction loss is a function of the hose diameter and the hose length in addition to losses in various appliances such as monitors and ladder pipes.

Back pressure is the loss or gain of potential energy due to elevation changes between the pump and the nozzle. This is determined by the equations:

$$BP = 0.434h \text{ (SI Units: } BP_m = 9.81\ h_m)$$

Where:

BP = back pressure, psi (kPa)
h = elevation change or head, feet (m) of water

The basis of these calculation revolve around three fundamental hydraulic principles:

Bernouli's Equation: A special form of conservation of energy. For a pumper hose line system this becomes:

$$EP = NP + FL + BP$$

Where:

EP = pump pressure, psi (kPa)
NP = nozzle pressure, psi (kPa)
FL = friction loss, psi (kPa)
BP = back pressure, psi (kPa)

The Darcy-Weibach Equation:

$$FL = (fV^2L)/(2gD)$$

Where:
g = acceleration due to gravity, ft/sec^2 (m/s^2)
f = friction factor, a dimensionless number
L = hose length, ft (m)
D = hose diameter, in. (mm)

The Continuity Equation: This is a special form of conservation of mass.

$$Q = VA$$

Where:

Q = flow rate, ft^3/sec (m^3/sec)
V = velocity, ft/sec (m/sec)
A = area, ft^2 (m^2)

These equations can be manipulated to result in friction loss equations and a modern version of the Underwriters equation (Purington 1974).

Friction loss is the loss of energy resulting from the internal friction between the moving particles of water and the friction between the water and the lining of the hose. The friction loss depends on water viscosity, flow rate, hose diameter, hose length, and the roughness characteristics of the hose lining. While the viscosity is partly a function of the water temperature, the change is insignificant in the ranges of the temperature of water commonly used for fire fighting. The flow rate, hose diameter, and length are variables that are determined by the fire fighting requirements. The roughness of the hose lining cannot be easily measured and is generally determined experimentally and reported as the friction factor. Consequently, all variables except the flow rate, hose diameter, and hose length are all embodied in the friction factor f, of the Darcy-Weibach equation and in the continuity equation. Putting all constants into one constant factor results in $FL = cq^2L$ for a given hose diameter where L is traditionally given in hundreds of feet (meters) of hose and q is hundreds of gallons per minute or hundreds of liters per minute ($q = Q/100$).

For a standard 2½ in. (64 mm) double jacket, lined hose, the currently recommended constant is $c = 2$, or $FL = 2q^2L$. (For SI, $FL = 3.17q^2L$). Suggested constants for other hose sizes and friction loss equations are shown in Table 17-7H.

An approximate constant can be determined from $FL = (fQ^2L)/(2gD^5)$ by assuming the same friction characteristics for both hose sizes, so:

$$k_h = (D_1)^5/(D_2)^5$$

If D_1 is 2½ in., then:

$$k_h = (2.5)^5/(d_2)^5 = 97.9/(D_2)^5,$$

k_h can then be used to convert a given hose diameter (and length) to the 2½ in. equivalent length,

$$L' = k_hL$$

where L' is the equivalent 2½ in. length, k_h is the hose conversion factor and L is the given hose length.

[For SI units, $k_h = (64)^5/D_2)^5$ where D_2 is the given hose diameter required to be converted to the 64 mm equivalent length.]

The hose conversion factor, k_h, and the friction loss equation constant, c, are functions of both the hose diameter and the friction characteristics of the hose. These two

TABLE 17-7H. Recommended Friction Loss Equations

Hose Diameter and Type Inches (mm)	Friction Loss Equation psi	kPa
¾ (19) booster	$FL = 1100q^2L$*	$FL = 1741q^2L$*
1 (25) booster	$FL = 150q^2L$	$FL = 238q^2L$
1¼ (32) booster	$FL = 80q^2L$	$FL = 127q^2L$
1¼ (32) linen	$FL = 127q^2L$	$FL = 201q^2L$
1½ (38) rubber-lined	$FL = 24q^2L$	$FL = 38q^2L$
1½ (38) linen	$FL = 51.2q^2L$	$FL = 81q^2L$
1¾ (45) with 1½ (38) couplings	$FL = 15.5q^2L$	$FL = 24.6q^2L$
2 (51) rubber-lined	$FL = 8q^2L$	$FL = 12.7q^2L$
2 (51) linen	$FL = 12.5q^2L$	$FL = 19.8q^2L$
2½ (64) linen	$FL = 4.26q^2L$	$FL = 6.75q^2L$
2½ (64) rubber-lined	$FL = 2q^2L$	$FL = 3.17q^2L$
2¾ (70) with 3 (76) couplings	$FL = 1.5q^2L$	$FL = 2.36q^2L$
3 (76) with 2½ (64) couplings	$FL = 0.80q^2L$	$FL = 1.27q^2L$
3 (76) with 3 (76) couplings	$FL = 0.677q^2L$	$FL = 1.06q^2L$
3½ (88)	$FL = 0.34q^2L$	$FL = 0.53q^2L$
4 (102)	$FL = 0.2q^2L$	$FL = 0.305q^2L$
4½ (114)	$FL = 0.1q^2L$	$FL = 0.167q^2L$
5 (127)	$FL = 0.08q^2L$	$FL = 0.138q^2L$
6 (152)	$FL = 0.05q2L$	$FL = 0.083q^2L$

* q in 100s of gpm (100s of L/min)
L in 100s of ft (100s of meters)

constants are determined experimentally. The classical hose experimentation was done by John C. Freeman (Freeman 1889). Subsequent tests were made by the Underwriters' Laboratory Inc. (UL 1939), James Gaskill (Gaskill 1966), and many fire departments. The values in Table 17-7I are based on the best current information, and should be used until more current data becomes available.

The pump pressure can be found using the appropriate equation shown below:

Given *NP*, *L*, *D*, and *d*,

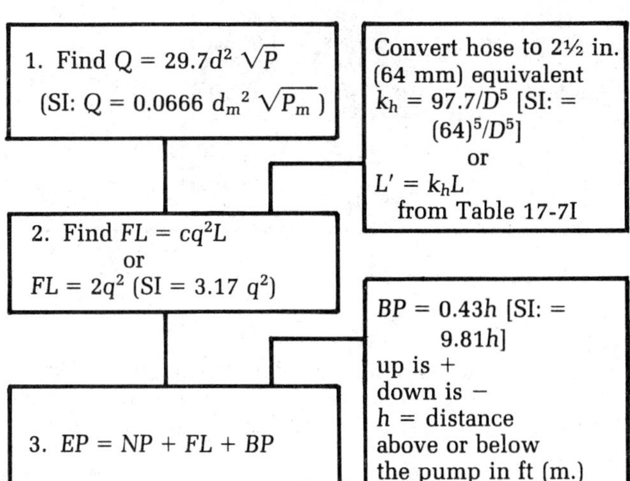
1. Find Q = 29.7d² √P
(SI: Q = 0.0666 dm² √Pm)
Convert hose to 2½ in. (64 mm) equivalent
kh = 97.7/D⁵ [SI: = (64)⁵/D⁵]
or
L′ = khL
from Table 17-7I
2. Find FL = cq²L
or
FL = 2q² (SI = 3.17 q²)
BP = 0.43h [SI: = 9.81h]
up is +
down is −
h = distance above or below the pump in ft (m.)
3. EP = NP + FL + BP

EXAMPLE: What pump pressure is required to establish a theoretical nozzle pressure of 10 psi (69 kPa) at the end of 400 ft (122 m.) of 4 in. (102 mm) hose if it rises up through an elevation of 10 ft (3.05 m) from engine to nozzle?

SOLUTION

US Customary Units:

$Q = 29.7\ (4.0)^2\ \sqrt{10}$

$Q = 1502$ gpm

i.e.,

$$q = \frac{1502}{100} = 15.02$$

$FL = 0.2\ q^2L$ (for 4 in. hose)
$FL = 0.2\ (1502)^2\ (400/100)$
$FL = 180$ psi

$BP = 0.434\ h$
$BP = 0.434 \times 10$
$BP = 4.34$ psi

$EP = NP + F + BP$
$EP = 10 + 180 + 4.3$
$EP = 194.3$ psi

SI units:

$Q = 0.0666\ (102)^2\ \sqrt{69}$
$Q = 5756$ (L/min)
$Q = 5756/100 = 57.6$

$FL = 0.305\ q^2L$ (for 102 mm hose)
$FL = 0.305\ (57.6)^2\ (122/100)$
$FL = 1234$ kPa

$BP = 9.81$ L
$BP = 9.81 \times 3.05$
$BP = 29.9$ kPa

$EP = NP + FL + BP$
$EP = 69 + 1234 + 29.9$
$EP = 1333$ kPa (this converts to 193 psi).

This scheme works well for simple straight lines, particularly if the flow rate is given. However, for more complex setups, the following is recommended (Purington 1974):

For 2½ in. (64 mm) rubber lined hose

$$EP = NP\ (1 + 1.76k_nL) + BP$$

Where:
EP = Engine pressure, psi
NP = Nozzle pressure, psi
k_n = nozzle constant (see Table 17-7K)
L = Hose length, 100s of feet
BP = back pressure due to elevation change, psi.

In SI units, for the equivalent 64 mm rubber lined hose

$$EP = NP\ (1 + 2.99\ k_{nm}\ L)$$

Where:
EP = Engine pressure, kPa
NP = Nozzle pressure, kPa

TABLE 17-7I. Factors for Converting Hose to 2½ in. (64 mm) Equivalent (k_n)

Hose Diameter in. (mm)	References*						Recommended Factors
	IFSTA 1980	Gaskill 1966	Erwin 1972	Holwege 1961	Other	$97.7/D^5$	
¾ (19)	344	550	590	345		412	500
1 (25)	91	75	78.5	90.9		97.7	75
1¼ (32)	40					32	40
1¼ (32) linen	63.6					32	63.6
1½ (38) rubber-lined	13.5	12	12.25	13.5	9.81	12.86	12
1½ (38)	25.6					12.86	25.6
1¾ (45) with 1½ (38)					7.76	5.95	7.76
2 (51) rubber-lined	2.94				4.0	3.05	4
2 (51) linen	6.25					3.05	6.25
2½ (64) linen	2.13					1.0	2.13
2¾ (70) with 3 (76)					0.75	0.62	0.75
3 (76) with 2½ (64)	0.4	0.4	0.4	0.4	0.4	0.4	
3 with 3 (76)	0.385	0.333		0.385		0.4	0.333
3½ (88)	0.17			0.172		0.186	0.17
4 (102)	0.09			0.09	0.1	0.095	0.1
4½ (114)	0.051			0.0513	0.05	0.053	0.05
5 (127)	0.031			0.031	0.03	0.033	0.04
					0.0446		
					0.333		
6 (152)						0.0125	0.0125

* See Bibliography.

TABLE 17-7J. Engine Pressure Formulas for Various Hose Sizes

Hose Size in. (mm)	Formulas	
	Customary Units	SI Units
¾ (19) booster	$EP = NP(1 + 968k_nL)$	$EP = NP(1 + 782k_{nm}L)$
1 (25) booster	$EP = NP(1 + 132k_nL)$	$EP = NP(1 + 106k_{nm}L)$
1¼ (32) booster	$EP = NP(1 + 70.4k_nL)$	$EP = NP(1 + 54.5k_{nm}L)$
1¼ (32) linen	$EP = NP(1 + 112k_nL)$	$EP = NP(1 + 86.3k_{nm}L)$
1½ (38) rubber-lined	$EP = NP(1 + 21.1k_nL)$	$EP = NP(1 + 16.8k_{nm}L)$
1½ (38) linen	$EP = NP(1 + 45k_nL)$	$EP = NP(1 + 35.9k_{nm}L)$
1¾ (45) with 1½ (38) couplings	$EP = NP(1 + 13.6k_nL)$	$EP = NP(1 + 11.0k_{nm}L)$
2 (51) rubber-lined	$EP = NP(1 + 7.04\ k_nL$	$EP = NP(1 + 5.63k_{nm})L$
2 (51) linen	$EP = NP(1 + 11k_nL)$	$EP = NP(1 + 8.78k_{nm}L)$
2½ (64) rubber-lined	$EP = NP(1 + 1.76k_nL)$	$EP = NP(1 + 2.99k_{nm}L)$
2½ (64) linen	$EP = NP(1 + 3.75k_nL)$	$EP = NP(1 + 1.40k_{nm}L)$
2¾ (70) w/3 (76) couplings	$EP = NP(1 + 1.32k_nL)$	$EP = NP(1 + 1.05k_{nm}L)$
3 (76) w/2½ (64) couplings	$EP = NP(1 + 0.704k_nL)$	$EP = NP(1 + 0.56k_{nm}L)$
3 (76) w/3 (76) couplings	$EP = NP(1 + 0.586k_nL)$	$EP = NP(1 + 0.47k_{nm}L)$
3½ (88)	$EP = NP(1 + 0.3k_nL)$	$EP = NP(1 + 0.235k_{nm}L)$
4 (102)	$EP = NP(1 + 0.17k_nL)$	$EP = NP(1 + 0.135k_{nm}L)$
4½ (114)	$EP = NP(1 + 0.88k_nL)$	$EP = NP(1 + 0.074k_{nm}L)$
5 (127)	$EP = NP(1 + 0.07k_nL)$	$EP = NP(1 + 0.061k_{nm}L)$
6 (152)	$EP = NP(1 + 0.022k_nL)$	$EP = NP(1 + 0.037k_{nm}L)$

k_{nm} = nozzle constant (see Table 17-7K)
L = Hose length in 100s of meters.
BP = back pressure due to elevation change, kPa.

For other sizes, use either 2½ in. (64 mm) equivalent length or a special equation for the particular hose size (shown in Table 17-7J).

The nozzle constant k_n is numerically equal to $k_n = d^4/10$ (In SI, $k_{nm} = d^4/10^6$). Table 17-7K shows values of k_n for solid-stream nozzles and the equivalent k_n for several spray nozzles.

For siamese or parallel lines, a single 2½ in. (64 mm) equivalent length must be determined:

$$L' = k_sL$$

Where:
L' = 2½ in. (64 mm) equivalent length
k_s = the siamese constant which is generally taken as the number of lines, assuming that all lines are the same diameter (Vennard 1961). (However some authorities use a slightly modified siamese constant to account for friction in fittings and appliances. Suggested values for k_s are given in Table 17-7L.)

Where siamesed lines are of unequal length and/or diameter, find the average length and the 2½ in. (64 mm)

TABLE 17-7K. Nozzle Factors

Nozzle Diameter in.	Customary Units k_n*	Nozzle Diameter mm	SI Units k_{nm}
1/16	1.53×10^{-6}	2	1.6×10^{-5}
1/8	2.44×10^{-6}	3	8.1×10^{-5}
3/16	1.24×10^{-4}	5	6.25×10^{-4}
1/4	3.91×10^{-4}	6	1.30×10^{-3}
5/16	9.54×10^{-4}	8	4.10×10^{-3}
3/8	1.98×10^{-3}	10	0.010
7/16	3.66×10^{-3}	11	0.015
1/2	6.3×10^{-3}	13	0.029
9/16	0.01	14	0.038
5/8	0.0153	16	0.066
11/16	0.0223	17	0.084
3/4	0.0316	19	0.130
13/16	0.0436	21	0.194
7/8	0.0586	22	0.234
1	0.1	25	0.416
1⅛	0.16	29	0.707
1¼	0.244	32	1.05
1⅜	0.357	35	1.50
1½	0.506	38	2.11
1⅝	0.697	41	2.83
1¾	0.938	44	3.90
1⅞	1.24	48	5.31
2	1.6	51	6.66
2½	3.9	64	16.3
3	8.1	76	33.7
3½	15.0	89	62.4
4	25.6	102	107.0
4½	41.0	114	171.0
5	62.5	127	260.0
6	130.0	152	539.0

* $k_n = \frac{d^4}{10}$ $(k_{nm} = \frac{d^4}{10^6})$

equivalent lengths respectively. For two lines, all of the above can be combined in one equation:

$$L'' = k_s\,(k_{h_1}L_1 + k_{h_2}L_2)/2.$$

L' ' is another way of expressing equivalent length. For more than two lines, the equation can be expanded.

If the lines are the same diameter, a precise equivalent length can be found from:

$$L'' = \frac{L_1}{\left[1 + \left(\sqrt{\frac{L_1}{L_2}}\right)\right]^2}$$

For more than two lines, the general equation is:

$$L'' = \frac{L_1}{\left[1 + \sqrt{\frac{L_1}{L_2}} + \sqrt{\frac{L_1}{L_3}} + \sqrt{\frac{L_1}{L_4}} + \cdots \sqrt{\frac{L_1}{L_n}}\right]^2}$$

Figure 17-7B is a schematic illustrating the solution of siamese line problems while Figure 17-7C includes hose sizes other than 2½ in. (64 mm).

A common hose lay is the wyed or branch lines. This set-up involves a 2½ in. (64 mm) or larger supply or lead line branched into two or more lines. The legs or branches are typically smaller diameter hoses than the supply line. A typical set-up involves a 2½ in. (64 mm) supply line and two 1½ in. (38 mm) branches, each 100 or 150 ft (30 or 45 m) in length.

TABLE 17-7L. Siamese Conversion Factors (K_s) To Single 2½ In. (63 mm) Equivalent Length

Number of Hose Lines	Siamese Conversion Factor To Single 2½ in. Equivalent Length
Two 2½ in.	0.25
Three 2½ in.	0.11
Four 2½ in.	0.0625
Five 2½ in.	0.04
Two 3 in. with 2½ in. couplings	0.08
Three 3 in. with 2½ in couplings	0.0307

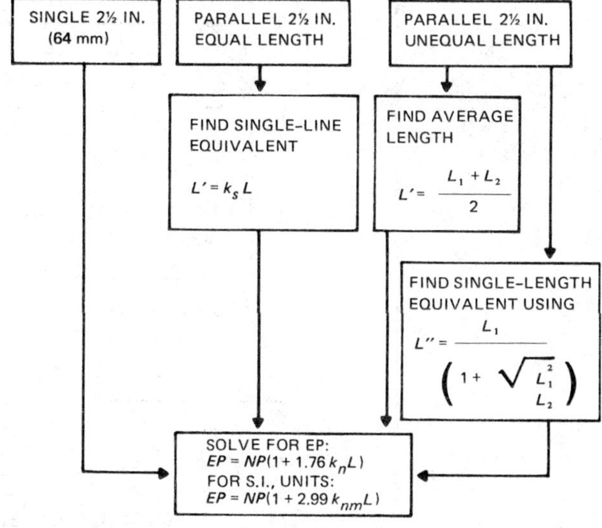

FIG. 17-7B. Solving for engine pressure, 2½ in. (64 mm) hose.

The solution to engine pressure problems depends on whether or not the branches or legs are symmetrical, i.e. hose length, diameters, and nozzles equal. (See Fig. 17-7D for outline of the solution.) If the legs are symmetrical, the solution can be easily determined by finding an equivalent nozzle diameter and equivalent nozzle constant from:

$$d_e = \sqrt{d_1^2 + d_2^2} \text{ and } k_n = d_e^4/100.$$

K_n can be found directly from k_n =

$$\left[\frac{d_1^2 + d_2^2 + \ldots .\, d_n^2}{10}\right]^2$$

or

$$k_n = \left[\frac{d_1^2 + d_2^2 + \ldots .\, d_n^2}{100}\right]^2$$

Note: For SI units, $k_{nm} = d^4/10^6$ for d in mm.

The hose line is reduced to the equivalent single 2½

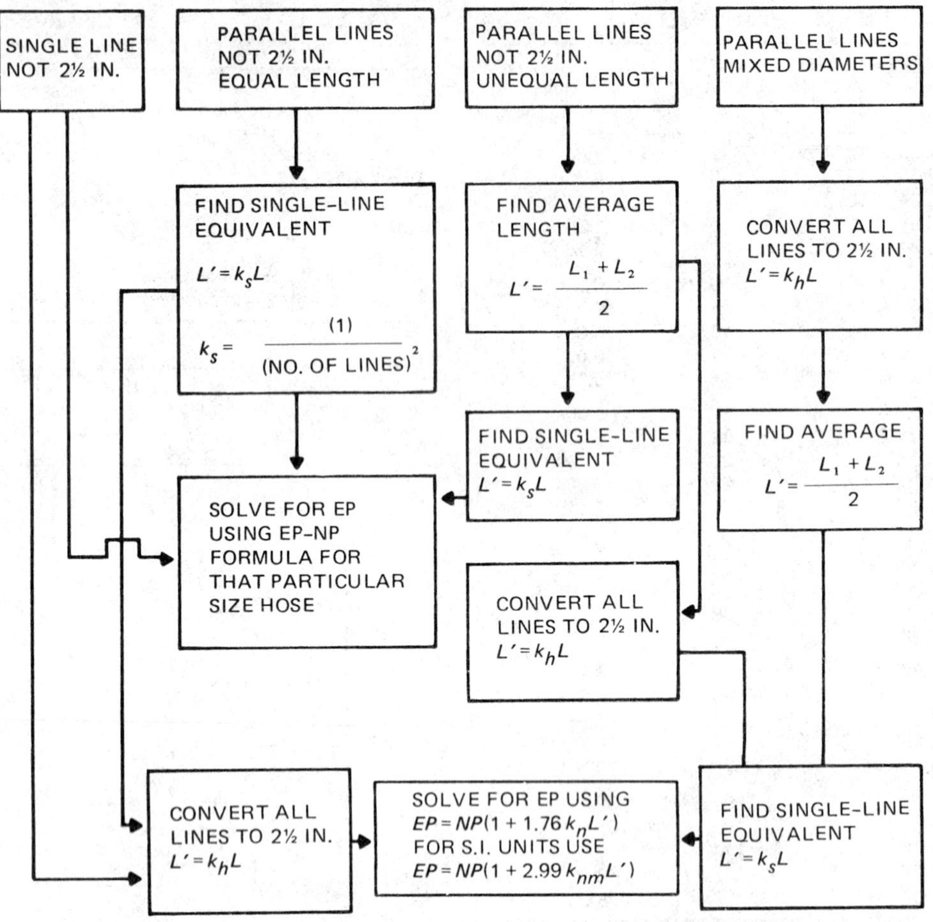

FIG. 17-7C. Solving for engine pressure, all hose sizes.

in. (64 mm) line; for example, for the branch line, the general formula (including the unsymmetrial case) is:

$$L' = L + k_s \left(\frac{k_h L_1 + k_h L_2 + \ldots\ldots k_h L_n}{n} \right)$$

Where L is the supply line, and n equals the number of legs.

If it is not 2½ in. (64 mm) hose, it must be converted to the 2½ in. (64 mm) equivalent by:

$$L' = k_h L.$$

If the lines are not symmetrical, a more involved calculation is required for an exact solution. (The method described above for symmetrical lines can be used for an approximate solution.)

The technique involves finding the flow through one of the legs and determining the friction loss in that leg and the pressure at the wye. Using the pressure at the wye as engine pressure, calculate the nozzle pressure on the other line and the flow from the second line and then sum up the flow from all the lines. With this flow known, the friction loss in the supply line can be found. The engine pressure is the supply line friction loss plus the pressure at the wye. This technique is also shown in Figure 17-7D.

LARGE DIAMETER HOSE

Many fire departments are now using large diameter hose. The most popular sizes in use include:

3 in. (76 mm) hose with 2½ in. (64 mm) couplings
3 in. (76 mm) hose with 3 in. (76 mm) couplings
4 in. (101 mm) hose with 4 in. (101 mm) couplings
5 in. (127 mm) hose with 5 in. (127 mm) couplings
6 in. (152 mm) hose with 6 in. (152 mm) couplings

The main advantage of large diameter hose is its lower friction loss. It is especially useful as a supply line with large flows. Because of the low friction loss, some fire departments use large diameter hose for supply lines into a portable hydrant, from which smaller working lines (2½ in. (64 mm), 1½ in. (38 mm), etc.) are then taken off.

Large diameter hose is also useful for relaying water, i.e., one pumper supplying a second.

There have been some complaints that the large diameter hose is difficult to handle, and in some cases wears out rapidly. Nevertheless, even if these disadvantages are not overcome, they appear to be overshadowed by the advantages, at least for certain applications.

FRICTION REDUCING ADDITIVES

Certain high molecular weight polymers injected into a pump or hose line have the capacity of reducing friction.

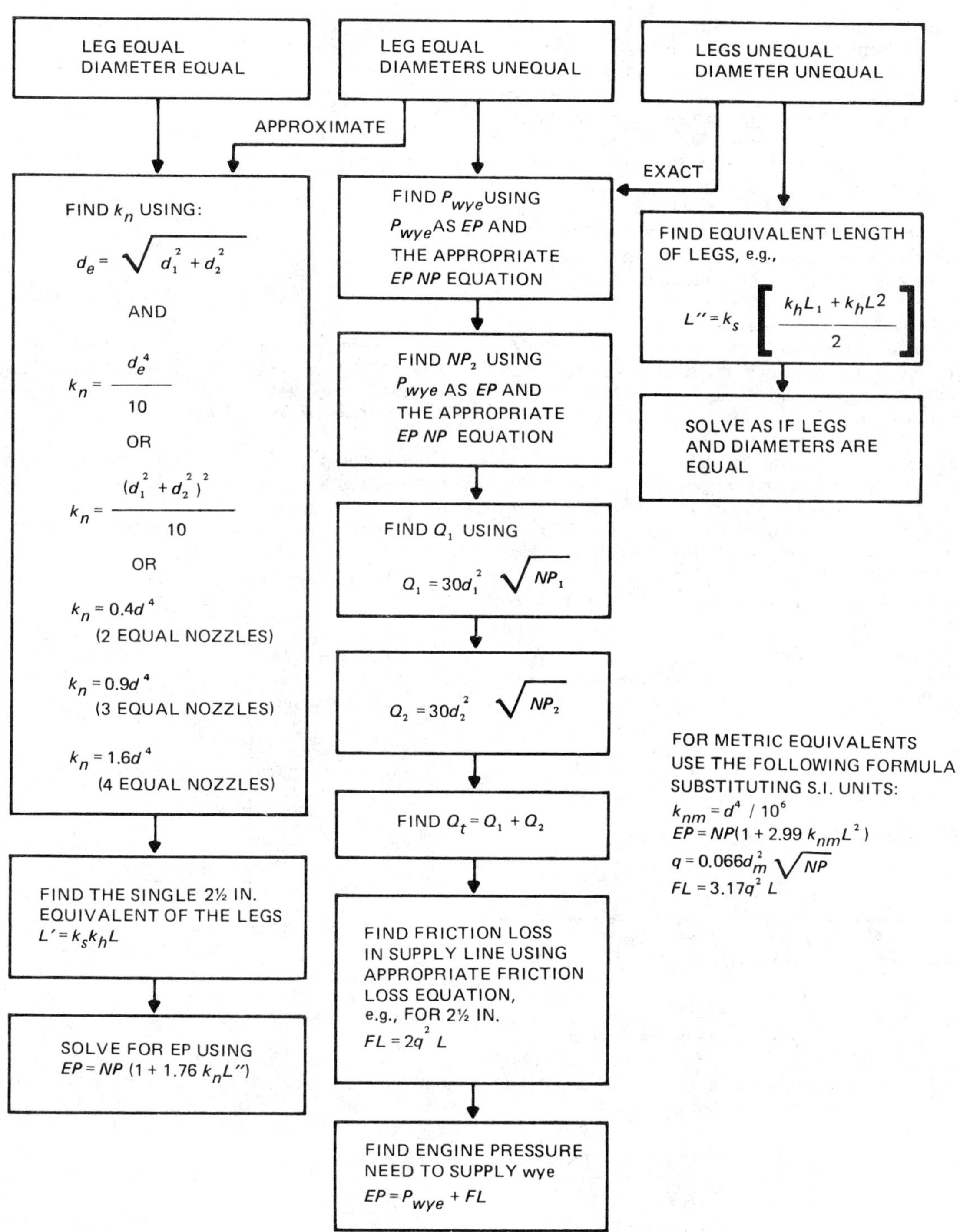

FIG. 17-7D. Solving for engine pressure for wyed lines.

These polymers include polyethylene oxide, polyacrylamide, polysulfonates and hydroxethyl cellulose, the former being the most prevalent in current use. Several trade names are used to describe friction reducing additives. Polymer additives reduce friction loss by as much as 50 percent. This permits flowing greater amounts of water in smaller diameter hose ("rapid water").

For example, 200 gpm (757 L/min) of plain water flowing in 1½ in. (38 mm) hose has friction loss of about 96 psi (662 kPa). With such high friction loss, and assuming a nozzle (spray type) pressure of 100 psi (690 kPa), the maximum distance 200 gpm (757 L/min) could be pumped through 1½ in. (38 mm) hose without exceeding the hose test pressure of 250 psi (1724 kPa) is about 150 ft (45.7 m). On the other hand, with polyethylene oxide additive the friction loss of 200 gpm (757 L/m) through 100 ft (30.5 m) of 1½ in. (38 mm) hose is about 25 psi (172 kPa). Adding the friction reducing polymer permits 1½ in. (38 mm) lines to provide 200 gpm (757 L/min) flows at 100 psi (690 kPa) nozzle pressures up to 600 ft (183 m) without exceeding pump pressure of 250 psi (1724 kPa). In other words, 1½ in. (38 mm) hose with the friction reducing additive can have almost the same "punch" as 2½ in. (64 mm) hose lines, for at least distances up to 600 ft (183 m).

The use of the smaller diameter hose can be a significant advantage since water in 2½ in. (64 mm) hose weighs about 2 lb/ft (3 kg/m) as compared to ¾ lb/ft (1.1 kg/m) for 1½ in. (38 mm) hose. For a 50 ft (15 m) length, this is 100 lb (45 kg) versus about 35 lb (16.8 kg) (neglecting the weight of the hose). This weight reduction is crucial to the fire fighter who is dragging a hose line into position.

The City of New York Fire Department has considerable experience using 1¾ in. (44 mm) hose [with 1½ in. (38 mm) couplings] in conjunction with a friction reducing agent. With this arrangement, they combined the fire extinguishing capability of a 2½ in. (64 mm) line with mobility of a 1½ in. (38 mm) line.

There are some technical difficulties with some of the additives that are apparently being overcome. For example, while the polymer is nontoxic, it is difficult to inject into the hose stream. Also, the polymer is subject to deterioration at temperatures exceeding 120°F (48.9°C) or less than 32°F (0°C).

Some equations are shown in Table 17-7M that can be used to establish the friction loss in hose with rapid water. These equations should be used with discretion, however, since the friction loss is a function of polyethylene oxide concentration, which is not reflected in these equations.

TABLE 17-7M. Friction Loss, Using "Rapid Water"

Hose Size in. (mm)	psi	kPa
1½ in. (38) rubber-lined hose*	$FL = 7.5q^2L$	$FL = 11.8q^2L$
1¾ in. (45) with 1½ in. (38) couplings	$FL = 6.2q^2L$	$FL = 9.8q^2L$
2½ in. (64) rubber-lined hose*	$FL = 0.7q^2L$	$FL = 1.1q^2L$
	q = 100s of gpm L = 100s of feet	q = 100s of L/min L = 100s of meters

* Derived from D. B. Brown, "Reducing Friction Loss in Fire Fighting Streams Using Poly (Ethylene Oxide) Slurry," presented at the 1974 NFPA Meeting, Miami Beach, Fla.

REACTION FORCES IN HOSE LINES AND NOZZLES

When water is discharged from a hose line, there are two reaction forces acting on the line: one is hose line reaction, and the other is nozzle reaction. Both contribute in varying degrees to the total reaction that is present in any hose nozzle flow configuration. For purposes of simplicity, each is treated separately in this chapter.

Hose Line Reaction

Reaction forces occur in hose lines when the direction of waterflow is changed by a bend in the hose. Between the hydrant and nozzle, the velocity is constrained in the hose and the acceleration is zero as shown by the equation $Q = AV$, where Q is the flow rate, A is the cross-section area, and V is the average velocity.

Figure 17-7E illustrates a typical hand line situation

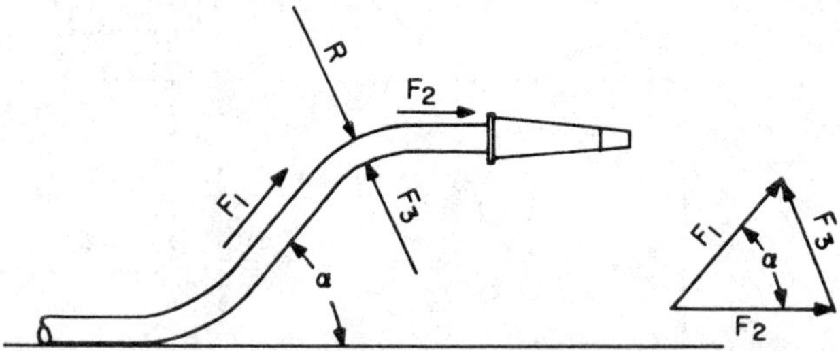

FIG. 17-7E. Reaction in a hose line. a Deflection angle at bend F_3 Resultant of forces F_1 and F_2 Reaction in pounds R Force needed to hold hose against F_3

where enough push must be applied at the bend to resist the reaction force F_3. There is no problem where the hose bends off the ground.

Figures 17-7F(a) and 17-7F(b) are charts by which

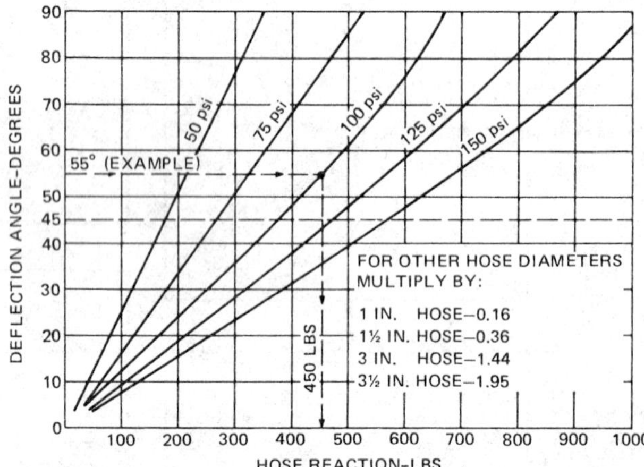

FIG. 17-7F(a). Reaction forces in 2½ in. hose. (See table within the graph for conversion factors to use in calculating reaction forces, expressed as pounds force, in hose of other diameters.

values of the reaction F_3 in pounds force, can be directly determined. To obtain reactions in other sizes of hose, Figures 17-7F(a) and 17-7F(b) include a table of conversion factors.

EXAMPLE: What is the reaction at 55 degree bend in 1½ in. (38 mm) hose if the pressure is 100 psi (690 kPa)?

SOLUTION: On the deflection angle side, move horizontally from 55 degree to intersection with the 100 psi (690 kPa) curve. From there go down vertically to the reaction scale and read 450 lb (204 kg). From Table 17-7F, note that the factor for 1½ in. (38 mm) hose is 0.36. Multiplying 450 (204 kg) by 0.36, the answer is 162 lb (73 kg).

Nozzle Reaction

Water discharging from a nozzle produces a reaction that is opposite the flow of water and known as nozzle reaction. As flow and pressure are increased, the nozzle

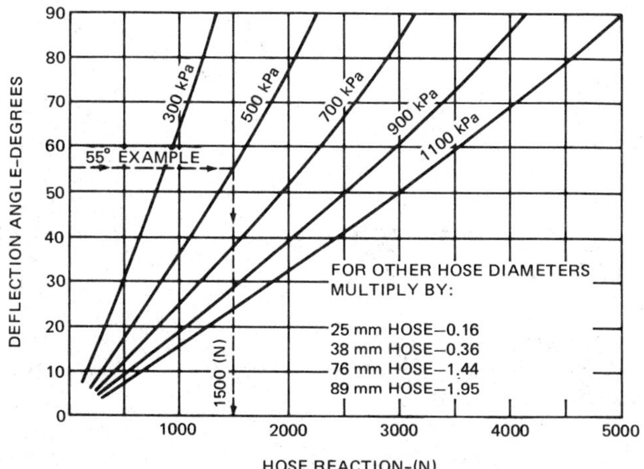

FIG. 17-7F(b). Reaction forces in 64 mm hose. (See table within the graph for conversion factors to use in calculating reaction forces, expressed in newtons, in hose of other diameters.

reaction increases. Nozzle reaction of solid stream nozzles may be calculated by means of the following formula:

$$NR = 1.5d^2NP$$

Where:

NR = Nozzle reaction, lb
d = Nozzle diameter, in.
NP = Nozzle pressure, psi

Another formula that may be used for both solid stream and spray nozzles is:

$$NR = 0.0505\ Q\sqrt{NP}$$

Where: Q = flow rate, gpm.

For SI Units:

$$NR = (1.6 \times 10\text{-}3)\ d^2NP$$

Where:

NR = Nozzle reaction, N (Newton)
d = Nozzle diameter, mm
NP = Nozzle pressure, kPa.

Reaction Forces In Ladder And Tower Streams

The reaction of heavy streams, either solid or spray, discharged from ladder pipes, platform booms, and towers can easily cause serious problems, such as leverage effects of large magnitude and displacement of centers of gravity. Thus the stability of the whole structure could be threatened. Likewise, a sudden loss of pressure when a hose line bursts or a pump stops may cause a whipping effect.

A ladder pipe or other large stream from a ladder (including 2½ in. [64 mm] hand lines) should always be operated in line with the main beams or trusses. Ladders have little resistance to torsional effects. The NFPA Committee on Fire Department Equipment recommends that ladder pipes be incapable of horizontal travel and, in any event, that maximum horizontal movement of the pipe must not exceed 15 degree. Rotation of the ladder turntable is the correct way to rotate the stream from ladder pipes. Also, care should be used elevating and lowering the stream because this changes the direction of the thrust on the ladder mechanism. The fly ladder should not be raised or lowered while a ladder pipe is discharging, and under no circumstances should the vehicle be moved with the ladder pipe discharging. All such streams should be gated to permit movement of the ladder and vehicle without shutting down the pump.

Because there are many variable forces acting on mobile ladders or towers in service, it is quite impractical to devise tables of specific reactions. The problem is one of design and careful operating procedures.

Bibliography

References Cited

Erwin, Lawrence W. 1972. *Fire Hydraulics*. Glencoe Press, Beverly Hills, CA.
Freeman, John R. 1889. "Experiments Relating to the Hydraulics of Fire Streams." *Transactions, American Society of Civil Engineers*. Vol 21. Nov, 1889.
Gaskill, James R. 1966. "Hydraulic Studies of Fire Hose." *Fire Technology*. Vol 2, No 1.
Holwege, J. 1961. "What Size Fire Hose." *Washington Comm. and School.*
UL. 1939. "Hydraulic Friction Losses in 1½ in. and 2½ in. Cotton Rubber-Lined Fire Hose." Underwriters' Laboratories, Inc., NY.
IFSTA. 1980. *Fire Stream Practices*. International Fire Service Training Association, Stillwater, OK.
Purington, Robert. 1974. *Fire-Fighting Hydraulics*. McGraw-Hill, NY.

Additional Readings

Bonadio, G. E., "Fire Hydraulics," Arco Publishing Co., NY, 1968.
Casey, J. F., ed., *Fire Service Hydraulics*, 2nd ed., Reuben H. Donnelley Corp., NY, 1970.
Cozad, Dale, *Water Supply for Fire Protection*, Prentice-Hall, Englewood Cliffs, NJ, 1981, pp. 79-160, 215-257.
"Developing Good Solid Fire Streams," American Insurance Association Bulletin 50, rev. ed., NY, 1970.
"Firefighters Hydraulics," Davis Publishing Co., Santa Cruz, CA, 1962.
"Fireground Hydraulics," California State Bureau of Industrial Education, Sacramento, 1970.
Gaskill, James R., "Hydraulics of Fire Hose Nozzles," Fire Technology, Vol. 3, No. 1, Feb. 1967, pp. 20-28.
Gaskill, James R., Henderson, R. L., and Purington, R. G., "Further Hydraulic Studies of Fire Hose," *Fire Technology*, Vol. 3, No. 2, May 1967, pp. 105-114.
"Heavy Stream Appliances and Their Effective Use," American Insurance Association Bulletin 147, rev. ed., NY, 1963.
Hickey, Harry E., *Hydraulics for Fire Protection*, National Fire Protection Association, Quincy, MA, 1980.
Kimball, Warren, Y., *Effective Streams for Fighting Fires*, National Fire Protection Association, Quincy, MA, 1961.
Mahoney, Eugene F., *Fire Department Hydraulics*, Allyn and Bacon, Boston, 1980, pp. 85-252.
"Operating Fire Department Aerial Ladders," National Fire Protection Association, Quincy, MA, 1967.
Purington, Robert G., *Fire Fighting Hydraulics*, McGraw-Hill, NY, 1974, pp. 127-286.
Purrington, Robert G., "Friction Loss Studies Using a 0.0127m Fire Hose," *Fire Technology*, Vol. 11, No. 3, Aug. 1975, pp. 184-190.
Salzberg, F., Vodvarka, F. J., and Maatman, G. L., "Minimum Water Requirements for Suppression of Room Fires," *Fire Technology*, Vol. 6, No. 1, Feb. 1970, pp. 22-28.
Vennard, John K., *Elementary Fluid Mechanics*, 4th Edition, John Wiley & Sons, NY, 1961.

TEST OF WATER SUPPLIES

Revised by John R. Anderson

An existing water supply system is one of the most important factors in public or private protection. Fire departments and fire protection engineers, as well as those responsible for the design, operation, and maintenance of water systems, are concerned with two aspects of the water supply: (1) its reliability, and (2) the rate at which water is available for fire suppression.

This chapter explains the objectives of water supply testing, the test procedures, and graphical solutions to test and flow problems.

The hydraulics principles underlying waterflow and pump testing will be found in Chapter 2, "Hydraulics," and Chapter 6, "Fire Pumps," in this Section. Information on water supply requirements and systems will be found in Chapter 3, "Water Supply Requirements for Fire Protection" and Chapter 4, "Water Supply Distribution Systems," also in this Section of the HANDBOOK.

TEST OBJECTIVES

Hydrant flow tests conducted on public water systems are usually made to determine the rate at which water is available at specific locations within the distribution system, and also to determine points of connection for pipe line extensions, booster pump applications, verification of the accuracy of distribution models, and for a variety of other purposes.

Of particular interest to fire departments and insurance companies is the rate at which water is available to concentrated high value areas, such as shopping centers, industrial parks, institutions, and built-up residential areas.

Private fire systems are usually tested for water flow annually; this includes testing the operating performance of fire pumps, standpipes or elevated tanks, municipal connections, other water sources, and the overall adequacy of the system.

Flow testing of public water systems should not be conducted without the consent and assistance of waterworks personnel, and possibly the local police and fire departments. Fire departments will usually aid in the test work; police may be needed to control vehicular and pedestrian traffic.

Hydrant and flow tests are made so frequently and the results obtained so important it is imperative for anyone charged with the responsibility for operation, management, or maintenance of a water supply system to be aware of how tests are conducted and to be able to interpret results correctly.

Hydrant flow tests should not be attempted until all the operational characteristics of a water system are known. Results may differ substantially depending upon the operation of pumping equipment, water levels in the system's storage facilities, rates of consumption, and points of demand in the system. Because of the many variables involved, even though it is possible to conduct accurate tests within acceptable tolerances, the results obtained will often vary from day to day and even at different periods during the same day.

TEST EQUIPMENT

In order to conduct hydrant flow tests, the following equipment should be provided calibrated in customary American units or appropriate SI units.

1. A 6 in. steel rule with 1/16 in. divisions.
2. A Pitot tube together with a test pressure gage, suitable for the pressures to be expected, with 1/2 lb graduations (usually a 60 psi gage is satisfactory). (See Fig. 17-8A.)
3. A 2½ in. hydrant cap with fittings together with a test pressure gage suitable for the pressures to be expected, with 1 lb. graduations (usually a 200-psi gage is satisfactory). (See Fig. 17-8B.)
4. A convenient form to record test data and to include a sketch of the location at which the test was conducted showing hydrants and any other salient features.

A Pitot tube and gage combination is indispensable in conducting flow tests from hydrants and nozzles. The small opening at the end of the tube, not over 1/16 in. (1.6 mm) in diameter is inserted in the center of the stream with the opening in direct line with the flow and a distance in front of the opening of one-half the diameter of

Mr. Anderson is a consulting professional engineer based in Marshfield Hills, MA.

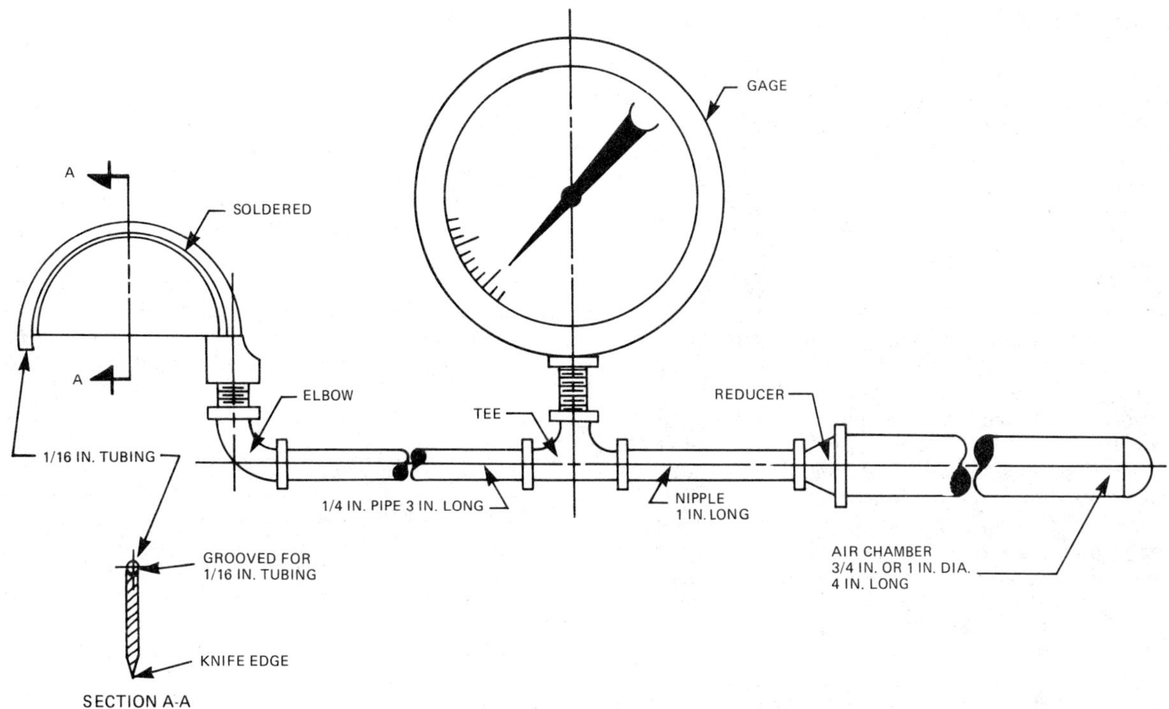

FIG. 17-8A. A typical Pitot tube assembly.

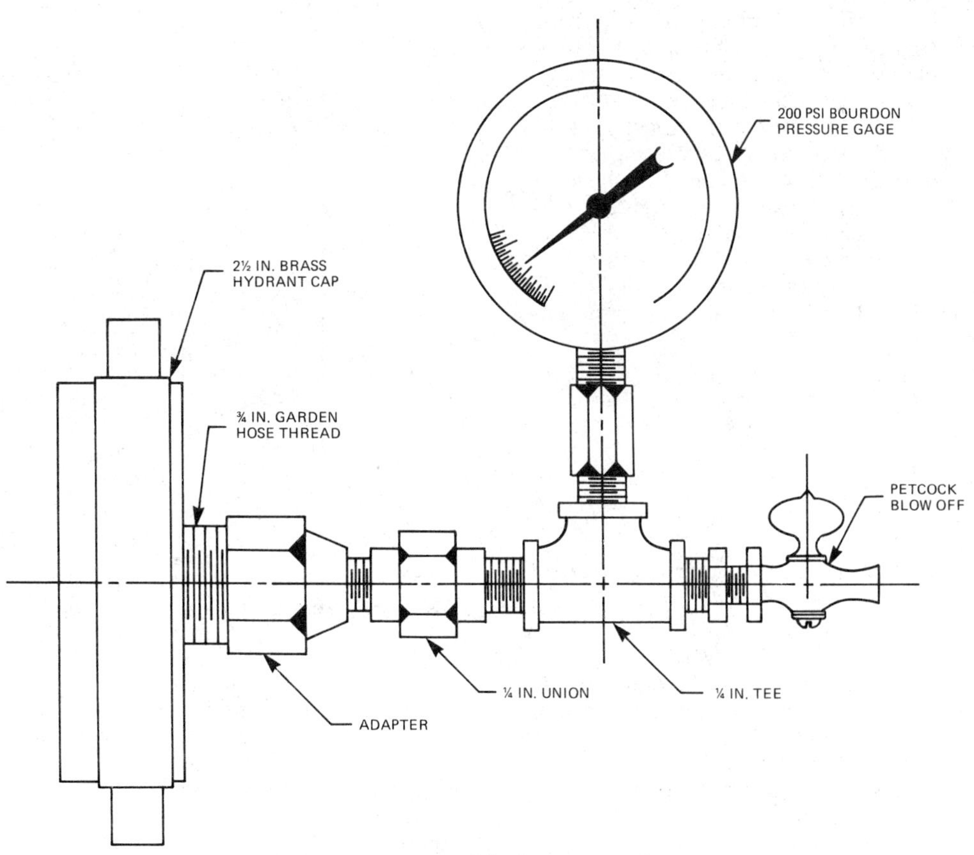

FIG. 17-8B. A typical hydrant cap and gage.

the opening. Velocity pressure is registered on the gage attached to the tube.

Note in Figure 17-8B the petcock blow off; its operation will permit air from the hydrant barrel to be vented when the hydrant is opened and will allow air to re-enter when the hydrant is closed. Failure to open the petcock during hydrant closing may subject the pressure gage to a partial vacuum which may introduce errors in future gage readings. Also note the 3/4 in. (19 mm) garden hose thread on the 2½ in. (64 mm) hydrant cap. Removal of the cap from the remainder of the fitting allows the gage assembly to be attached to a house silcock connection.

Test quality gages should be used in accordance with NAS Standard B40.1-1968, grade AA. The use of good quality test gages has produced results that are considered reasonably accurate within the scope of the test procedure. However, care should be taken to protect the gages from rough handling. Periodically they should be tested by means of a dead weight tester throughout the range of operation. Calibration sheets should be kept for each gage and correction factors affixed to the back of each gage prior to the start of a test series. All test data should be recorded in a neat, systematic fashion along with any system operating conditions that might affect the test results.

CONDUCTING FLOW TESTS

Hydrant flow tests are fairly simple to conduct and the results usually easily interpreted. Once the objective of the test has been determined, it is only necessary to discharge water at a known rate from one or more hydrants. Simultaneously the pressure drop the discharge produces is observed at a second hydrant or at other points of connection to the system that is supplying the flowing hydrants.

The usual procedure for conducting a flow test on a water system is to take Pitot readings on a sufficient number of hydrants to determine the capacity of the system in the area tested. Observed pressures without test hydrants flowing are called "static pressures"; those obtained with the test streams flowing are "residual pressures." One hydrant is chosen for observing the static and residual pressures, preferably located in the center of the group or where the best average pressure conditions might be expected. (See Fig. 17-8C.) Avoid using a flowing hydrant for this purpose.

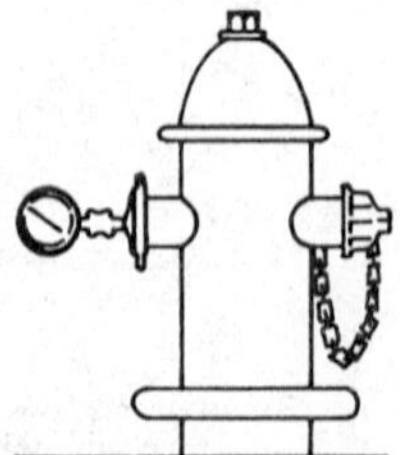

FIG. 17-8C. Gage on nonflowing hydrant to measure pressure.

Once flow and pressure has stabilized, Pitot readings are then taken simultaneously on the flowing hydrants. Residual pressure at the nonflow hydrant should also be recorded at the same time. After the flowing hydrants have been slowly shut off, the static pressure should be checked and averaged with the initial reading.

Pitot readings of less than 10 psi (69 kPa) or over 30 psi (207 kPa) at any open hydrant should be avoided. To keep within these pressure limits, the rate of flow can be controlled by throttling the hydrant or opening a second outlet, or both. However, the flow hydrants should be opened sufficiently so that hydrant drains are closed. If water is continuously discharged through the drains, it tends to erode the soil from the base of the hydrant. Avoid using the larger pumper connection on a hydrant for testing unless flow and pressure are sufficiently strong to produce a full stream. Also, when pumper outlets are used, a proper coefficient of discharge must be determined, based upon the extent to which the orifice is completely filled with water. On occasion, it may be desirable to obtain an average velocity pressure by moving the Pitot tube through the entire vertical dimension of the orifice.

Figure 17-8D is the record of a typical flow test.

LOCATION: Adams St. between Cox St. and Baker St.

HYDRANTS: 2 ½ In. Square Sharp Assumed C = 80

PRESSURE	PITOT PRESSURE				GPM	TOTAL GPM
HYD 1	HYD 2	HYD 3	HYD 4			
72	—	—	—		0	0
62	18	—	—		633	633
50	10 10		—		472 472	944

GAGE 103 AT HYDRANT No. 1 - GAGE 79 FOR PITOT READINGS

FIG. 17-8D. Log of a public waterflow test.

Hydrant No. 1 was the gaging point and Hydrant No. 2 was the point where flow tests were actually made. Figure 17-8E is a graphical plot of the results of the tests.

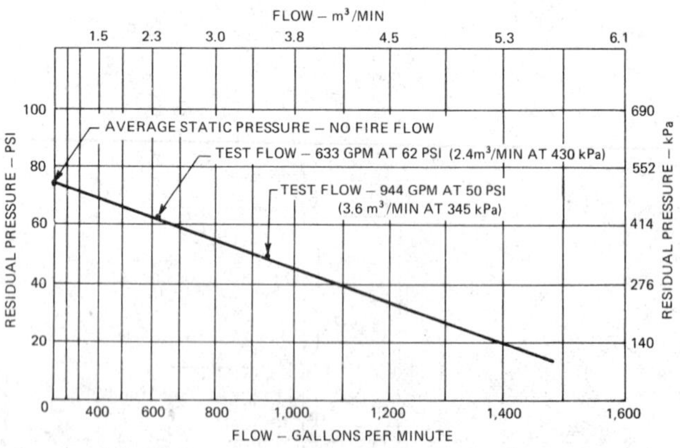

FIG. 17-8E. Curve plotted from flow test data given in Figure 17-8D.

Tests of this kind show the available flow over and above water consumption occuring during the test. To properly evaluate the adequacy and reliability of a system, consideration is given to the sources of supply, the levels of water in distribution storage, and the overall operating condition of the system. The hour of the day, the day of the week, the month of the year, the weather, an impairment made necessary by highway or other construction—all are factors that may affect the results of flow tests. Fireflow may be adequate one time and inadequate another because of variations in consumption or in the mode of operation of the system.

Sometimes unsuspected faults other than insufficient supplies are disclosed by hydrant flow tests. Valves may be partly or entirely shut; sometimes they are found broken or with bent valve stems. Valve boxes may be filled with muck or sand, or even paved over with concrete or asphalt cement. Silt, stones, fish, or other foreign material may be found in the hydrant streams. Usually many or all of these faults are due to lack of maintenance and operational control.

It is customary to report the results of hydrant flow tests conducted in public systems in gallons per minute (L/min) available at 20 psi (138 kPa) residual pressure, which is the minimum recommended residual pressure at hydrants for fire engine use. The observed flow and pressures can be converted to any desired residual pressure or flow rate by a simple proportion derived from the Hazen-Williams pipe flow formula. This formual shows the flow rate in gallons per minute is directly proportional to the 0.54 power of the head loss or the drop in pressure from static to residual observed during the test.

Head loss, which is mostly pipe friction, is the difference between the observed static pressure S and the residual pressure R. The gage value of S seldom indicates the true static (no flow) pressure; actually it is the residual pressure for the normal pipe flow occurring prior and during the test flow. Observed residual pressure is a function of total flow or $Q \pm \Delta Q$, where Q is the normal rate of flow, and ΔQ is the measured discharge during the test.

For common fire flow tests, the difference between the true and observed values of S is usually not significant. When meter readings or other consumption data are not available for the period of time during which flow tests were conducted, then rates of water consumption may be estimated considering records of consumption on preceeding days or by comparing similar water systems.

Examples of Calculating Flow and Pressure Conversion

Example A.

A 2,500-gpm test flow from a group of street hydrants dropped the pressure from 69 to 44 psi. (1) What flow would be available at 20 psi? (2) What would the residual pressure be if flow were increased to 3,000 gpm?

SOLUTION 1: Since the flow rate is directly proportional to the 0.54 power of the head (friction) loss, by proportion

$$\frac{Q_2}{Q_1} = \frac{(S - R_2)^{0.54}}{(S - R_1)^{0.54}} \text{ and } Q_2 = Q_1 \frac{(S - R_2)^{0.54}}{(S - R_1)^{0.54}}$$

Substituting the known values:

$S - R_2 = 69 - 20 = 49$ psi.

From Table 17-8A $49^{0.54} = 8.18$

$S - R_1 = 69 - 44 = 25$ psi.

From Table 17-8A $25^{0.54} = 5.69$

$$Q_2 = 2{,}500 \frac{8.18}{5.69} = 2{,}500 \times 1.445 = 3{,}620 \text{ gpm (answer).}$$

SOLUTION 2: Calculate R_2 when $W_2 = 3{,}000$ gpm.

$$3{,}000 = 2{,}500 \frac{(69 - R_2)^{0.54}}{69 - 44)^{0.54}}$$

or

$$(69 - R_2)^{0.54} = \frac{3{,}000}{2{,}500}(69 - 44)^{0.54}.$$

Since $(69 - 44)^{0.54} = 5.69$, then

$$(69 - R_2)^{0.54} = 1.20 \times 5.69 = 6.83.$$

From Table 17-8A it is found by interpolation that 6.83 is $35.1^{0.54}$. Therefore $69 - R_2 = 35.1$ and $R_2 = 69 - 35.1 = 33.9$ psi, residual pressure at 3,000 gpm (answer).

TABLE 17-8A. Numbers to 0.54 Power

h	$h^{0.54}$	h	$h^{0.54}$	h	$h^{0.54}$	h	$h^{0.54}$	h	$h^{0.54}$
1	1.00	36	6.93	71	9.99	106	12.41	141	14.47
2	1.45	37	7.03	72	10.07	107	12.47	142	14.53
3	1.81	38	7.13	73	10.14	108	12.53	143	14.58
4	2.11	39	7.23	74	10.22	109	12.60	144	14.64
5	2.39	40	7.33	75	10.29	119	12.66	145	14.69
6	2.63	41	7.43	76	10.37	111	12.72	146	14.75
7	2.86	42	7.53	77	10.44	112	12.78	147	14.80
8	3.07	43	7.62	78	10.51	113	12.84	148	14.86
9	3.28	44	7.72	79	10.59	114	12.90	149	14.91
10	3.47	45	7.81	80	10.66	115	12.96	150	14.97
11	3.65	46	7.91	81	10.73	116	13.03	151	15.02
12	3.83	47	8.00	82	10.80	117	13.09	152	15.07
13	4.00	48	8.09	83	10.87	118	13.15	153	15.13
14	4.16	49	8.18	84	10.94	119	13.21	154	15.18
15	4.32	50	8.27	85	11.01	120	13.27	155	15.23
16	4.48	51	8.36	86	11.08	121	13.33	156	15.29
17	4.62	52	8.44	87	11.15	122	13.39	157	15.34
18	4.76	53	8.53	88	11.22	123	13.44	158	15.39
19	4.90	54	8.62	89	11.29	124	13.50	159	15.44
20	5.04	55	8.71	90	11.36	125	13.56	160	15.50
21	5.18	56	8.79	91	11.43	126	13.62	161	15.55
22	5.31	57	8.88	92	11.49	127	13.68	162	15.60
23	5.44	58	8.96	93	11.56	128	13.74	163	15.65
24	5.56	59	9.04	94	11.63	129	13.80	164	15.70
25	5.69	60	9.12	95	11.69	130	13.85	165	15.76
26	5.81	61	9.21	96	11.76	131	13.91	166	15.81
27	5.93	62	9.29	97	11.83	132	13.97	167	15.86
28	6.05	63	9.37	98	11.89	133	14.02	168	15.91
29	6.16	64	9.45	99	11.96	134	14.08	169	15.96
30	6.28	65	9.53	100	12.02	135	14.14	170	16.01
31	6.39	66	9.61	101	12.09	136	14.19	171	16.06
32	6.50	67	9.69	102	12.15	137	14.25	172	16.11
33	6.61	68	9.76	103	12.22	138	14.31	173	16.16
34	6.71	69	9.84	104	12.28	139	14.36	174	16.21
35	6.82	70	9.92	105	12.34	140	14.42	175	16.26

For a graphical solution to the same two problems described above, see Figure 17-8H(a) which is a hydraulic flow curve plotted on semiexponential paper.

Example B. SI Units

A 5000 L/min test from a group of hydrants dropped the pressure from 500 kPa to 350 kPa. What would be the flows available at 140 kPa (approximately 20 psi)?

SOLUTION: Again, flow rate (Q) is directly proportional to 0.54 of the head (friction) on:

$$\frac{Q_2}{Q_1} = \frac{(S - R_2)^{0.54}}{(S - R_1)^{0.54}} \text{ and } Q_2 = Q_1 \frac{(S - R_2)^{0.54}}{(S - R_1)}.$$

Substituting known values:

$$S\text{-}R_2 = 500 - 140 = 360 \text{ kPa and } 360^{0.54} = 24.01$$
$$S\text{-}R_1 = 500 - 350 = 150 \text{ kPa and } 150^{0.54} = 14.97$$

$$Q_2 = 5000 \cdot \frac{24.01}{14.97} = 8019 \text{ L/min.}$$

See Figure 17-8H(b) for a graphical solution (on semi exponential paper) to the problem in SI units described above in Example B.

Flow Test of Public Main at Plant Site

A common procedure for using street hydrants to test the water supply for a connection to an industrial plant follows, using an example to describe the step-by-step testing procedures. The sprinkler system in the example is connected to the dead ended Adams Street main (Fig. 17-8F); therefore the entire supply comes from the 10 in. main in Baker Street. There are no yard hydrants. The test was conducted as follows:

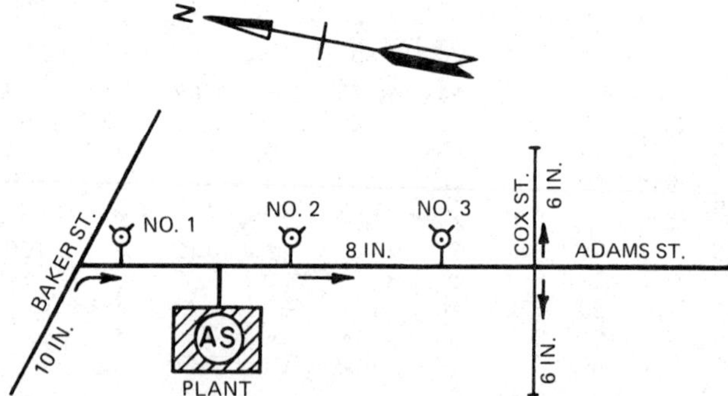

FIG. 17-8F. Data and sketch of a flow test of a public main at a plant site.

Step No. 1: A gage was attached to No. 1 Hydrant with a cap. The hydrant was opened and the static pressure, 72 psi (496 kPa), was recorded (Fig. 17-8D). No. 1 Hydrant was chosen for the gaging point. The gage could have been located on the sprinkler riser in the building, but the hydrant location probably was more convenient.

Step No. 2: The caps were removed from Hydrant No. 2. The diameter of the outlets was measured, and the outlets found to be square and sharp (Fig. 17-2F). After replacing one cap, the hydrant was opened, and a Pitot reading of 18 psi (124 kPa) was made and recorded. Before shutting off the flow, the residual pressure at No. 1 Hydrant was recorded as 62 psi (235 kPa).

Step No. 3: The second butt at Hydrant No. 2 was opened, and 10 psi (69 kPa) Pitot readings noted on both streams; at Hydrant No. 1 the residual pressure was now 50 psi (345 kPa). It is always desirable to obtain data for at least two rates of flow, one of which should be as heavy as facilities, conditions, and time will allow.

Step No. 4: Hydrant No. 2 was shut down slowly and carefully, and the caps replaced. The static pressure at No. 1 was read and found to be 72 psi (496 kPa), as before the test. The hydrant was then shut down, the gage cap removed, and the regular cap replaced.

Computing the flow: The rate of discharge is best determined by using the table of theoretical discharge (Table 17-2A) and by applying a suitable coefficient from Table 17-2B. In this instance the hydrant outlets justified use of c = 0.80 (square and sharp hydrant outlets in Figure 17-2F). Table 17-2A shows that when the velocity pressure (Pitot gage reading) from a 2½ in. (64 mm) orifice is 18 psi (124 kPa), the flow will be 791 gpm (2994 L/min). The actual flow then would be 791 × 0.8 or 634 gpm (2994 × 0.8 or 2395 L/min) at a residual pressure of 62 psi (235 kPa). With both outlets flowing, the Pitot reading on each was 10 psi (69 kPa), corresponding to 590 gpm (2233 L/min). The total actual flow was 2 × 590 × 0.8 = 944 gpm (2 × 2233 × 0.8 = 3573 L/min) at a residual pressure of 50 psi (345 kPa). It is now possible to calculate a flow at any residual pressure. There may be times when an open outlet of a hydrant cannot be flowed because of the damage it will do. In that case, additional or alternate hydrants should be flowed. When water supplies are very weak and the available water does not give a good Pitot reading with an open hydrant outlet flowing, a nozzle (with a diameter less than the hydrant outlet) may be attached directly to the hydrant outlet. Be sure to accurately measure the diameter of the nozzle when one is used.

Annual Tests

Insurance companies usually require annual tests of private water supplies. When there are multiple sources, each source should be tested separately and in combination to determine if the total supply is sufficient to meet the maximum required fire flow.

Many faults are disclosed by flow tests. Among them are:

1. Valves partly or wholly closed, or inoperative.
2. Stones, silt, and other foreign material in the mains.
3. Tuberculated mains causing high friction loss.
4. Empty or partly filled gravity tanks.
5. Check valves leaking, or installed backward.
6. Mains smaller than indicated on plans.
7. Broken meters or clogged strainers.
8. Existence of meters and valves not previously known.
9. Inoperative hydrants.

Hydrants and gaging points are carefully chosen. Tests should be conducted so that the available flow and pressure at high value or hazardous areas can be readily determined. Make sure that water from the test streams does not cause flooding or property damage. Do not use flowing hydrants for pressure readings; loss of head in hydrant and connection is not determined and difficult to estimate.

Care should be taken while testing a water supply system that serves automatic sprinklers to restrict the flow to a rate that will maintain not less than 10 or 15 psi (69 or 103 kPa) residual pressure on the top lines of sprinklers.

Because flow testing may involve valve operations, all control valves should be carefully checked after the tests to ensure that all are open and that the system is left in normal condition. Faults should be corrected as soon as possible, and recommendations made for desirable improvements.

Figure 17-8G illustrates how a private system supplied with public water would be tested.

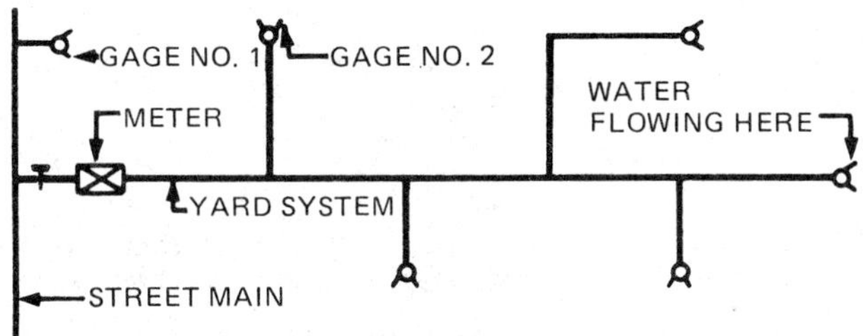

FIG. 17-8G. Method of testing city water supply to private system. Gage No. 1 shows street pressure and should, for gallons discharged, be consistent with data previously secured by street tests, if such tests have been made. Gage No. 2 on branch line will show flowing pressures in main at the point where the branch takes off. The difference in pressure between Gage No. 1 and No. 2 equals the loss in meter, pipe, and fittings. Friction losses to be expected in fittings and pipes can be estimated from Tables 17-2D and 17-2F. Fire flow meter friction loss can be checked to see if it is unusually high by consulting Table 17-4F.

TABLE 17-8B. Data for Making $N^{1.85}$ Hydraulic Graph Paper

1	2	3	1	2	3
Gpm N	$N^{1.85}$	Scale value, inches from 0	Gpm N	$N^{1.85}$	Scale value, inches from 0
1	1.00	0.05	10.5	77.48	3.87
1.5	2.12	0.11	11	84.44	4.22
2	3.60	0.18	11.5	91.68	4.59
2.5	5.45	0.27	12	99.19	4.96
3	7.63	0.38	12.5	107.0	5.35
3.5	10.15	0.50	13	115.0	6.75
4	13.00	0.65	13.5	123.3	6.16
4.5	16.16	0.81	14	131.9	6.59
5	19.64	0.98	14.5	140.8	7.04
5.5	23.42	1.17	15	149.9	7.49
6	27.52	1.37	15.5	159.2	7.96
6.5	31.90	1.60	16	168.9	8.44
7	36.60	1.83	16.5	178.8	8.95
7.5	41.58	2.08	17	189.0	9.45
8	46.85	2.34	17.5	199.5	9.96
8.5	52.40	2.62	18	210.0	10.50
9	58.26	2.91	18.5	221.0	11.05
9.5	64.39	3.22	19	232.1	11.60
10	70.80	3.54	19.5	243.5	12.18
			20	255.2	12.76

Column 1. Gallons per minute or rate of flow in other units.
Column 2. $N^{1.85}$ from Table 17-8C or by interpolation from that table.
Column 3. Column 2 multiplied by distance in inches from $N = 0$ to $N = 1$ (0.05 in.).
Examples:
Scale value for $N = 7$ is $36.60 \times 0.05 \times 1.83$ in.
Scale value for $N = 16$ is $168.9 \times 0.05 = 8.44$ in.
If scale value of $N = 1$ is 0.08, what is scale for $N = 9$?
Answer: $58.26 \times 0.08 = 4.66$ in.

HYDRAULIC FLOW CURVES

Many problems involving water tests and flow in pipes can be solved readily by graphs plotted on semiexponential, semilogarithmic, or linear cross-section paper. Semiexponential paper, commonly called $N^{1.85}$ or "hydraulic" paper, is recommended because a master sheet can be drawn easily and copies easily made by duplicating equipment.

The problems can be solved equally well in U.S. Customary units or SI units although only problem examples in U.S. Customary units will be presented in the remainder of this Chapter. The Factory Mutual Engineering Corporation has data sheets (FMEC 1977) with examples worked graphically in both sets of units.

The design of $N^{1.85}$ paper is based on the Hazen-Williams pressure-drop relation, i.e., head loss (static pressure less residual pressure) is proportional to the 1.85 power of flow. This relationship holds equally for head loss and flow calculations in SI units. Table 17-8B gives values of unit flows from 1 to 20 gpm in ½ gal increments raised to the 1.85 power, and the corresponding scale values when the distance from 0 to the 1 gpm point is 0.05 in. For plotting pressure, the vertical spacing is linear.

The unit values on either vertical or horizontal scales may be multiplied or divided by any constant that will best fit the problem and the paper is equally suited to U.S. Customary or SI units.

See Figure 17-8H(a) for an example of a flow curve in customary American units plotted on $N^{1.85}$ semiexponential scale. Figured 17-8H(b) is an example in SI units.

FLOW IN LOOP SYSTEMS

It is sometimes necessary to estimate the friction loss and flow characteristics of loops or parallel pipe systems. Such problems are readily solved graphically by procedures based on the principle that pressure drop through a simple loop is the same in every leg; this is true regardless of size, condition, and length of pipe. The method described is not applicable to grid or network systems, although it is sometimes possible to treat a grid as a loop system by making certain assumptions. Grids can be solved by network analyzers or computers, or calculated by the Hardy Cross Method and other relaxation formulas. Because these matters are outside the scope of this HANDBOOK, the words of John R. Freeman might well be recalled: "A day of (flow) testing is worth a week of calculating."

Figure 17-8I illustrates a graphical solution of a simple loop having two legs, one 800 ft of 8 in. pipe, and the other 1,200 ft of 6 in. pipe. The Hazen-Williams coefficient is

TABLE 17-8C. Numbers to 1.85 Power

N	$N^{1.85}$	N	$N^{1.85}$	N	$N^{1.85}$
5	19.64	39	877.9	230	23,400
6	27.52	40	920.1	240	25,320
7	36.60	41	963	250	27,300
8	46.85	42	1.007	260	29,360
9	58.26	43	1.052	270	31,480
10	70.80	44	1,097	280	32,910
11	84.44	45	1,144	290	34,310
12	99.19	46	1.192	300	38,250
13	115.0	47	1,240	350	50,880
14	131.9	48	1,289	400	65,150
15	149.9	49	1,339	450	80,990
16	168.9	50	1,390	500	98,440
17	189.0	55	1,658	550	117,400
18	210.0	60	1,948	600	137,900
19	232.1	65	2,259	650	159,900
20	255.2	70	2,591	700	183,400
21	279.3	75	2,944	750	208,400
22	304.4	80	3,317	800	234,800
23	330.5	85	3,710	850	262,700
24	357.6	90	4,124	900	291,900
25	385.7	95	4,558	950	322,700
26	414.7	100	5,012	1,000	354,800
27	444.7	110	5,979	1,200	497,200
28	475.6	120	7,022	1,400	661,100
29	507.5	130	8,144	1,600	846,400
30	540.4	140	9,339	1,800	1,053,000
31	574.1	150	10,610	2,000	1,279,000
32	608.9	160	11,960	2,200	1,526,000
33	644.5	170	13,370	2,400	1,792,000
34	681.2	180	14,870	2,600	2,079,000
35	718.7	190	16,440	2,800	2,384,000
36	757.1	200	18,070	3,000	2,708,000
37	794.6	210	19,770	4,000	4,611,000
38	836.7	220	21,550	5,000	6,968,000

Figures in this table are for use with the Hazen-Williams formula.

Determining the 1.85 power of numbers by proportional parts between the figures given (linear interpolation) will produce results with error of less than 2 percent.

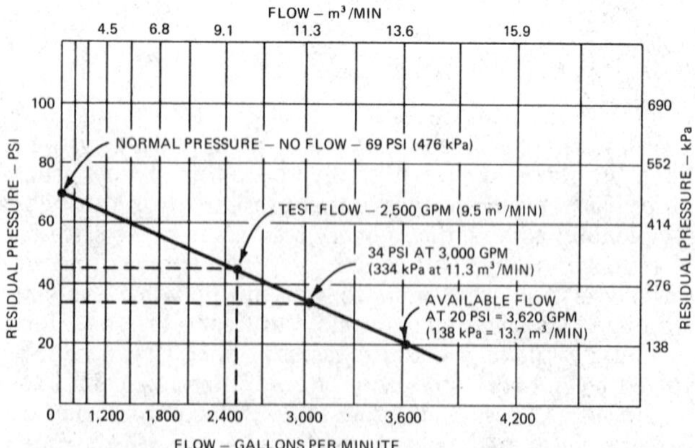

FIG. 17-8H(a). An example of a flow curve plotted on $N^{1.85}$ Semiexponential paper. The illustration is a graphical plot of data involved in Example A of a flow test of a public water system described under "Conducting Flow Tests" earlier in this chapter.

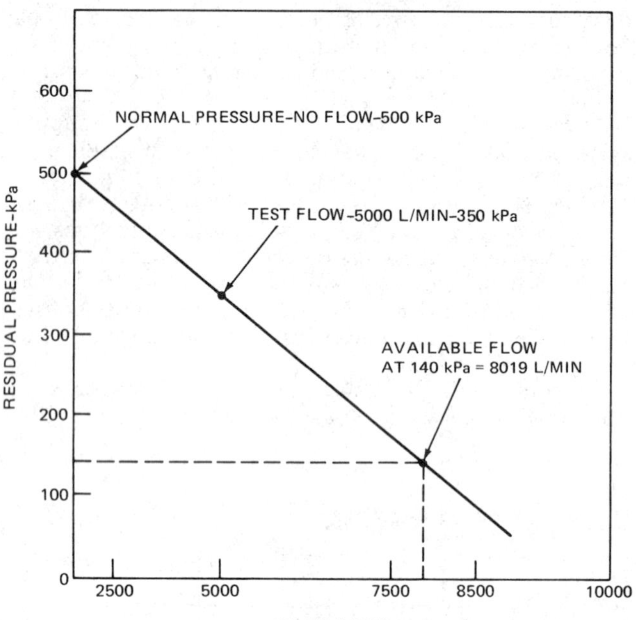

FIG. 17-8H(b). A graphical solution to the flow problem (Example B on page 17-106 expressed in SI units on $N^{1.85}$ semiexponential paper.

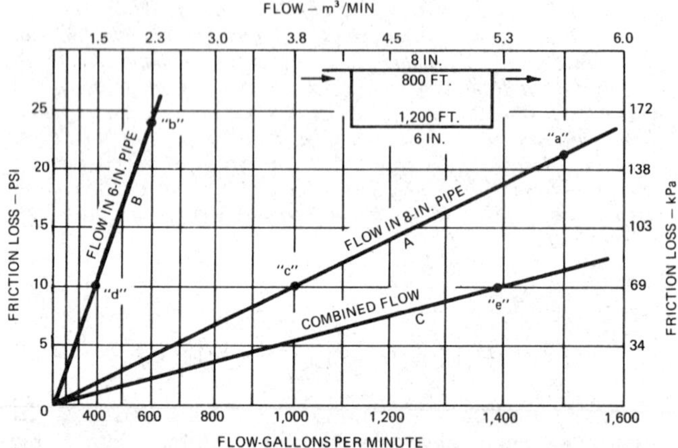

FIG. 17-8I. Graphic demonstration of flow in loop piping. The assumed friction coefficient is C = 100. Curve A represents flow in 800 ft of 8 in pipe; Curve B flow in 1,200 ft of 6 in. pipe.

C = 100 (unlined cast-iron pipe about 15 to 20 years old and in good condition). Using Table 17-2D the calculations would be as follows:

Step No. 1: Friction loss in the 8 in. pipe was calculated for an assumed flow of 1,500 gpm as follows: 1 × 8 × 2.71 = 21.6 psi in 800 ft of 8 in. pipe. Friction loss of 21.6 was plotted as Point "a" on the vertical 1,500 gpm line in Figure 17-8I. A straight line was drawn to connect Point "a" with 0, forming the Loss Curve "A" for the 8 in. leg.

Step No. 2: Assuming a flow of 600 gpm, follow the same procedure as followed in Step No. 1: 1 × 12 × 2.02 = 24.2 psi loss in 1,200 ft of 6 in. pipe. Friction loss of 24.2 was plotted as Point "b" on the vertical 600 gpm line and connected to 0, thus forming Loss Curve "B" for the 6 in. pipe.

Step No. 3: The flows corresponding to a convenient pressure of 10 psi on each curve (Points "c" and "d") were added together, and point "e" plotted at 1,400 gpm. This represents 400 gpm at Point "d" + 1,000 gpm at Point "c". The straight line connecting Point "e" with 0 is the Curve C for the whole loop.

GRAPHICAL DETERMINATION OF YIELD FROM COMBINED SUPPLIES

Determination of the yield from combined water supplies by actual testing is not always possible. A supply may be impaired, such as a gravity tank emptied for repairs or painting, or it may be necessary to estimate the yield when a new supply is to be added to a one-source system.

Figure 17-8J shows how to develop the combined yield curve from a gravity tank and a public water connection. Each source had been tested separately, but a combined test was impossible. The test flows had been taken at the yard hydrant, and the pressures on the nearby sprinkler riser.

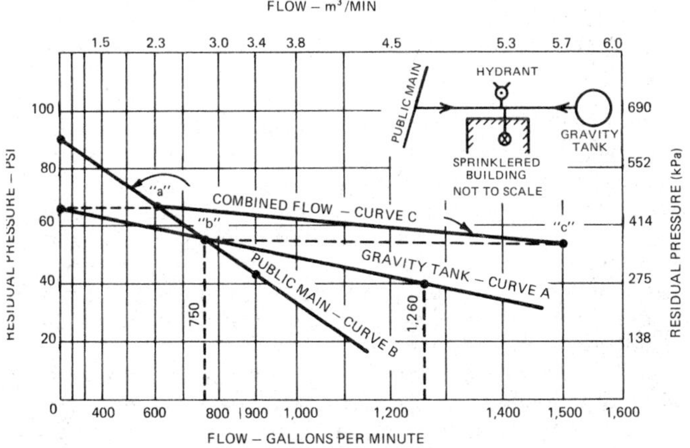

FIG. 17-8J. Graph of a combined water supply from a public system and a private gravity tank.

The curves were plotted from the test data as follows: 1,260 gpm flowing from the tank (Curve A) reduced pressure from 65 to 40 psi; 900 gpm flowing from the public supply (Curve B) reduced pressure from 90 to 43 psi.

All the water would come from the public main until the residual pressure drops to the static pressure of the tank, in this case to 65 psi at approximately 600 gpm plotted as Point "a." From then on, water would come from both sources.

The curves cross at 750 gpm at 55 psi plotted as Point "b"; with 750 gpm from each, the total flow at 55 psi would be 1,500 gpm plotted as Point "c." Connecting Points "a" and "c" with a straight line produces Curve C, which is the desired combined curve.

ANALYZING TEST DATA

A major purpose of flow testing is to determine whether or not the available water supply can meet the water demand required for acceptable protection. Plotting test data on $N^{1.85}$ hydraulic paper offers a simple and convenient method of analyzing water supply and demand (See "Hydraulic Flow Curves," in this chapter).

As an example of analyzing data, assume that flow tests have been made on a water supply to a sprinklered building and the following data recorded: no flow at 72 psi; 633 gpm at 62 psi; and 944 gpm at 50 psi. Curve A in Figure 17-8K was drawn by fitting a straight line to the three points. By extending the curve, residual pressures at larger flows can be read off directly. For example, the residual pressure at 1,200 gpm would have been 40 psi. It is not necessary, however, to extrapolate much beyond the maximum flow point, because tests from networks often produce supply curves that flatten out as the rate of flow increases.

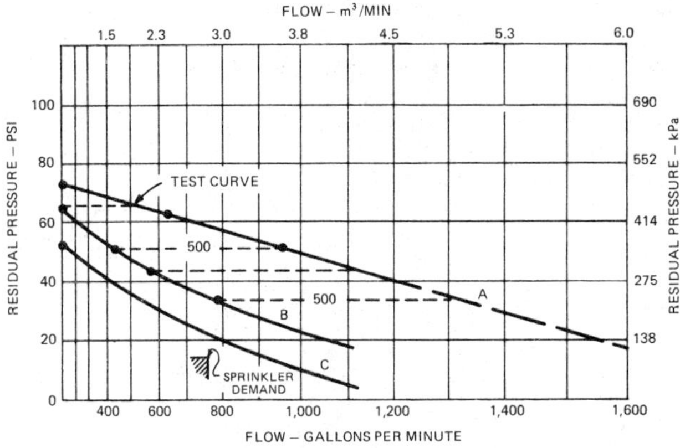

FIG. 17-8K. An example of a graphic water supply evaluation. Curve A is the test flow; Curve B the test flow less 500 gpm for hose streams; and Curve C is Curve B corrected for elevation (30-ft elevation in this example).

The fact that all three test points fix a practically straight line indicates that the observed static of 72 psi was close to a true figure and that the normal rate of flow was relatively small.

Curve B in Figure 17-8K represents the available supply at street level to the sprinkler system over and above a 500 gpm allowance for probable hose stream use by the fire department. Obviously, if more water were taken for hose streams, there would be less remaining for the sprinklers.

Curve B was developed by subtracting 500 gpm from Curve A at various pressures. The point at zero flow was obtained by moving horizontally from the intersection of Curve A and the vertical 500 gpm line. (See Fig. 17-8K.) The next point was found by plotting 444 gpm (944 − 500 gpm) at 50 psi, which was the residual pressure for the 944 gpm test. Other points were plotted in the same way.

Curves developed by this method are approximations at best, because the rate of flow is dependent on the 0.5 power of discharge pressure and the 0.54 power of head loss. However, this manner of evaluating the water supply to sprinkler systems is practical and acceptable.

Curve C in Figure 17-8K is actually Curve B corrected for a 30 ft (9.1 m) elevation difference between the residual pressure at street level and the top line of sprinklers. Therefore, every point of Curve C is 13 psi (89.6 kPa) below the corresponding point of Curve B. Friction loss in pipe and fittings between the city main and the top of the sprinkler riser in the building were not considered; however, the losses could have been estimated with certain details assumed as follows (no meter and no fire department connection to sprinklers):

System Component	Equivalent Length (ft)*
Eighty feet of 6 in. pipe, C = 120 (riser and connection)	80.0
One 8 × 8 × 6 in. tee (connection to main)	21.5
Two 6 in. standard elbows (top and bottom of riser)	20.0
Two 6 in. gate valves (waterworks valve and sprinkler valve)	04.0
One 6 in. check valve	23.0
	Total 148.5 ft.

$$\text{Friction loss at 750 gpm} = 0.714 \times \frac{148.5}{100} \times 3.05 = 3.24 \text{ psi}$$

(see "Friction Loss Calculations" in Chap. 2 of this Section).

It is also assumed in the foregoing example that the sprinkler system in the building requires a flow of 750 gpm at the point of supply to the sprinklers while at the same time residual pressure at the top line of sprinklers does not drop below 15 psi. Curve C of Figure 17-8K shows that 750 gpm would be available at approximately 20 psi residual pressure.

HYDRAULIC GRADIENT

A hydraulic gradient is a profile of residual pressure. Its function is to present graphically the flow characteristics of a pipe line. Hydraulic gradient is an important factor in the design of water supply conduits and trunk mains. The hydraulic gradient is a useful procedure for investigating the condition of a public or private main when tests produce less than the expected flow.

The relations between pressure and elevation in a pipe having uniform flow are shown in Figure 17-8L. The axioms accompanying the diagram should be noted carefully.

Hydraulic gradient tests of a private fire protection system usually involve shorter runs of pipe than tests of public mains. To reduce the number of tests, pipes should be chosen that are typical of the age and condition of the system. Relatively heavy flows should be induced through the test section to obtain a maximum pressure drop, thereby minimizing the effect of fluctuating pressure or inaccurate gage readings.

The data obtained from a hydraulic gradient are readily applied to calculating the C values (internal pipe roughness coefficient) of the pipes tested. The head loss in

*See Table 17-2F for head loss in values and fittings expressed as equivalent feet of straight pipe.

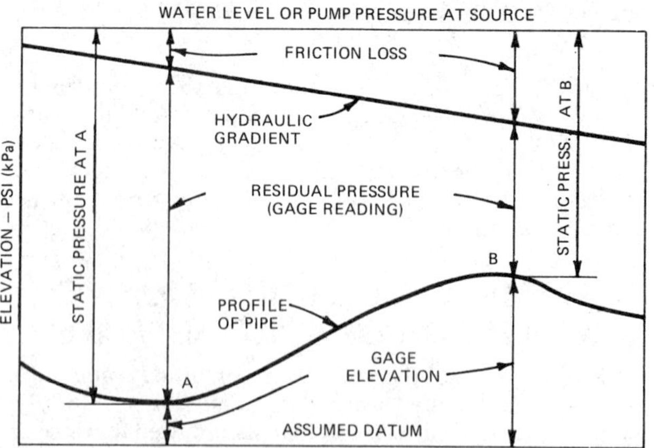

FIG. 17-8L. Principle of a hydraulic gradient. The following axioms apply: 1. Static-pressure readings measure distance below source. The higher the pipe elevation, the lower the static pressure. 2. Static pressure plus gage elevation (expressed in pounds per square inch) is constant for all points along pipe. 3. Static pressure minus residual pressure equals total friction loss from source to point of measurement. 4. Residual pressure plus gage elevation equals hydraulic-gradient elevation. 5. Friction loss is independent of elevation.

valves and fittings, if any, should be deducted from the observed pressure drop before calculating C, otherwise the value obtained would be too low.* However, in municipal water supply systems this deduction is not made as tested pipe lengths are relatively long and losses due to fittings are considered to be minor. Obviously, consideration of the C value is not part of gradient development; it is a convenient and widely used measure of determining the condition of the interior of the pipe.

If there are more than two gaging points, an attempt should be made to take simultaneous readings, but satisfactory results can usually be obtained (if rates of consumption are relatively low and steady) by moving the gage progressively from hydrant to hydrant while the test flow is maintained in the pipe. True static pressure obtained under no-flow conditions would plot as a horizontal line.

In a municipal system with normal flow, the observed static pressures are actually residual pressures, and would trace the normal gradient; therefore efforts should be made to find the true static pressure. If the system is supplied by gravity, it may be possible to determine the static by elevation differences, topographic maps, or other survey data. If it is necessary to depend on the static readings, it would be desirable to take readings between 1 a.m. and 3:30 a.m. when normal draft is at a minimum and observed static pressures are closer to the true levels.

Static pressure on fire protection mains usually can be obtained readily because there is little or no normal flow (except in properties with combined fire and industrial systems). With street mains it is often possible to close a valve below the downstream end of the section being tested, thereby reducing the normal flow temporarily.

It is usually desirable to plot the profile of the pipe being tested, along with the gradient. If a gradient falls be-

*See Table 17-2F for head loss in valves and fittings expressed as equivalent feet of straight pipe.

TABLE 17-8D. Data for Hydraulic Gradient

(1)	(2)	(3)	(4)	(5)	(6)	(7)	(8)	(9)
			Gage Pressure					
Gage Location	Actual Pipe Length (Ft)	Pipe Diameter (In.)	Static (Psi)	Residual (Psi)	Total Loss (Psi)	Loss Between Stations (Psi)	Gage Elevation Above Datum (Psi)	Gradient Elevation (Psi)
A	—	—	18	8	10	—	100	108
B	A − B = 2,000	8	110	78	32	A − B = 22	8	86
C	B − C = 800	8	95	54	41	B − C = 9	23	77
D	C − D = 280	6	100	35	65	C − D = 24	18	53
E	D − E = 750	8	118	43	75	D − E = 10	0	43

Explanation:
Columns 1–5: Data from actual gradient test.
Column 6: Static pressure − residual pressure = total loss.
Column 7: Difference in total loss, station to station.
Column 8: Gage elevation above datum = static pressure at datum (118 psi in this example) minus observed static at each location. Thus A = 118 − 18 = 100; B = 118 − 100 = 8; C = 118 − 95 = 23; D = 118 − 100 = 18; and E = 118 − 118 = 0.
Column 9: Gradient elevation = gage elevation + residual pressure. Thus A = 100 + 8 = 108; B = 8 + 78 = 86; C = 23 + 54 = 77; D = 18 + 53; E = 0 + 43 = 43.

TABLE 17-8E. Data for Calculated Hydraulic Gradient
(Rate of Flow − 750 Gpm − C = 100)

(1)	(2)	(3)	(4)	(5)	(6)	(7)	(8)	(9)
Gage Location	Actual Pipe Length (Ft)	Pipe Diameter (In.)	Gage Elevation Above Datum (Psi)	Static Pressure (Psi)	Calculated Loss Station to Station (Psi)	Total Loss (Psi)	Residual Pressure (Psi)	Gradient Elevation (Psi)
A	—	—	100	18	10	10	8	108
B	A − B = 2,000	8	8	110	15	25	85	93
C	B − C = 800	8	23	95	6	31	64	87
D	C − D = 280	6	18	100	9	40	60	78
E	D − E = 750	8	0	118	6	46	72	72

Explanation:
Columns 1–5: Data taken from Columns 1, 2, 3, 8, and 4 of Table 17-8D respectively.
Column 6: Calculated pressure loss from station to station (see Chap. 2 of this Sec. for friction loss calculations). The losses are:

$$\text{A to B} = 1 \times \frac{2{,}000}{100} \times .752 = 15$$

$$\text{B to C} = 1 \times \frac{800}{100} \times .752 = 6$$

$$\text{C to D} = 1 \times \frac{280}{100} \times 3.05 = 9$$

$$\text{D to E} = 1 \times \frac{750}{100} \times .752 = 6$$

Column 7: Total loss (accumulative loss at each location): Thus:
A = 10 psi
B = Loss A to B + Loss A = 15 + 10 = 25
C = Loss B to C + Loss B = 6 + 25 = 31
D = Loss C to D + Loss C = 9 + 31 = 40
F = Loss D to E + Loss D = 6 + 40 = 46
Column 8: Residual pressure: equals the static pressure (Column 5) minus the total loss (Column 7) at each location
Column 9: Gradient elevation is gage elevation (Column 4) plus the residual pressure (Column 8).

Table 17-8F. Calculations for C Values
(Rate of Flow–750 Gpm)

(1) Gage Location	(2) Actual Pipe Length (Ft)	(3) Pipe Diameter (In.)	(4) Loss Station to Station (Psi)	(5) Loss per 100 ft (Psi)	(6) C Factor	(7) C
A	—	—	—	—	—	—
B	A – B = 2,000	8	A – B = 22	1.10	1.46	82
C	B – C = 800	8	B – C = 9	1.125	1.49	81
D	C – D = 280	6	C – D = 24	8.6	2.82	57
E	D – E = 750	8	D – E = 10	1.35	1.76	74

Explanation:
Columns 1–4: Data taken from Columns 1, 2, 3, and 7 in Table 17-8D respectively.
Column 5: Pressure loss from station to station (Column 4) divided by the actual pipe length (Column 2) multiplied by 100.
Column 6: The actual pressure loss by test for each 100 ft of pipe (Column 5) divided by friction loss for 100 ft of pipe with $C = 100$ (Table 16-2E).
Column 7: Values for C interpolated from Table (Approximate linear interpolation is acceptable.) Example: For Station C the C factor is 1.49. Find C:

C	Factor	Factor
85	1.35	1.35
?		1.49
80	1.51	
5	14	16

$$\frac{14}{16} \times 5 = 4.4$$

$$85 - 4.4 = 80.6 = 81 \text{ (answer).}$$

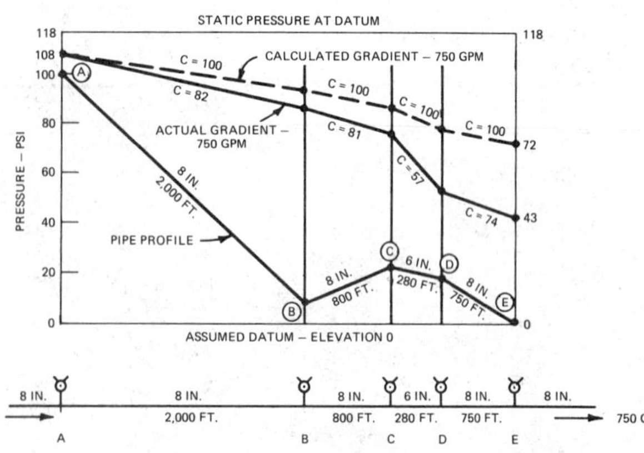

FIG. 17-8M. Hydraulic gradient and pipe profile.

low the pipe line, pressure in the pipe is less than atmospheric. This condition could impair the flow and cause dangerous pressure surges. When C values less than 80 are found by gradient tests, usually the pipe should be cleaned and lined by standard methods or replaced.

Regardless of C value, however, small-sized pipe causing a steep gradient slope should be replaced with pipe of adequate diameter.

The choice of method depends on relative costs of cleaning and lining as opposed to replacement, and on other practical considerations.

When improvements and changes are in order, it is desirable to plot a calculated gradient for comparison with the one tested. Both gradients must be based on the same rate of flow.

Figure 17-8M is a graph of a gradient test together with pipe profile and a calculated gradient. A uniform flow of 750 gpm is assumed.

Table 17-8D shows the data for Figure 17-8M, and explains the calculation.

Table 17-8E covers the calculated gradient.

Table 17-8F explains the calculation of C values.

Bibliography

References Cited

FMEC. 1977. "Hydraulics of Fire Protection Systems." *Loss Prevention Data 3-0.* Factory Mutual Engineering Corp. Norwood, MA.

Additional Readings

ANSI, B40.1, *Gages, Pressure and Vacuum, Indicating Dial Type-Elastic Element*, American National Standards Institute, Philadelphia, 1968.

Cozad, Dale, *Water Supply for Fire Protection*, Prentice-Hall, Englewood Cliffs, NJ, pp. 181-214.

"Fire Department Pumper Tests and Fire Stream Tables," National Board of Fire Underwriters, NY, 1959.

"Fire Flow Tests," National Board of Fire Underwriters, NY, 1963.

Mahoney, Eugene F., *Fire Department Hydraulics*, Allyn and Bacon, Boston, 1980, pp. 51-83.

Purington, R. G., "Fire Hydrant Efficiency Measurements," *Fire Technology*, Vol. 7, No. 1, Feb. 1971, pp. 57-60.

Shepperd, Fred, *The Fire Chiefs Handbook*, Case-Shepperd-Mann, NY, 1932.

Simplified Water Supply Testing, Alliance of American Insurers, Chicago, IL, 1982.

SECTION 18

WATER BASED EXTINGUISHING SYSTEMS

AUTOMATIC SPRINKLER SYSTEMS

Robert M. Hodnett

Automatic fixed extinguishing systems are the most effective means of controlling fires in buildings. In order to understand the capabilities of these systems, a thorough understanding of their components and uses is essential.

This chapter is an introduction to automatic sprinkler systems. It covers the fundamentals of good sprinkler protection by giving information on: (1) the value of sprinkler protection, (2) the standardizing of sprinkler protection, (3) building features that must be considered in planning sprinkler protection, (4) occupancy hazards and conditions that have a bearing, (5) the types of sprinkler systems that are available, and (6) an overview of installation practices that must be observed.

Chapter 2 of this section covers water supplies for sprinkler systems; Chapter 3, the sprinklers (heads) themselves, while Chapters 4 and 5 respectively give information on residential sprinkler installations (a relatively new development that deserves special attention), and sprinkler alarms and supervision devices. Maintenance of sprinkler systems is covered in Chapter 8 of this section.

Because of the wide application of sprinklers as suppression devices, it is difficult to single out other specific chapters in this HANDBOOK for additional guidance on the installation and use of sprinklers. Special mention must be made of Chapter 7, "Water Spray Protection," of this section where deluge and other special control valves associated with sprinklers are mentioned as part of the water spray systems for extinguishing flammable liquid and other intense fires, and of Section 16, Chapter 4, "Automatic Fire Detectors," for illustrations and descriptions of heat responsive devices that can be used to actuate preaction and deluge sprinkler systems.

DEVELOPMENT OF AUTOMATIC SPRINKLERS

Automatic sprinklers are devices for automatically distributing water upon a fire in sufficient quantity either to extinguish it entirely or to prevent its spread in the event that the initial fire is out of range of sprinklers, or is of a type that cannot be completely extinguished by water discharged from sprinklers.

The water is fed to the sprinklers through a system of piping, ordinarily suspended from the ceiling, with the sprinklers placed at intervals along the pipes. The orifice of the fusible link automatic sprinkler is normally closed by a disk or cap held in place by a temperature sensitive releasing element. Figure 18-1A shows in stop-action sequence the operation of a typical fusible link, upright automatic sprinkler. For more information on automatic sprinklers, see Chapter 3 of this Section.

The forerunners of the automatic sprinkler were the perforated pipe and the open sprinkler. These were installed in a number of mill properties from 1850 to 1880. The systems were not automatic, the discharge openings in the pipes often clogged with rust and foreign materials, and water distribution was poor.

Open sprinklers, an improvement over perforated pipes, consisted of metal bulbs with numerous perforations attached to piping and intended to give improved water distribution. This system was only slightly better than the perforated pipe.

The idea of automatic sprinkler protection, whereby heat from a fire opens one or more sprinklers and allows the water to flow, dates back to about 1860. Its practical application in the United States, however, began about 1878 when the Parmelee sprinkler was first installed. This sprinkler, while very crude when compared with modern devices, gave generally good results and proved conclusively that automatic sprinkler protection was both practical and valuable. (See Fig. 18-1B.)

VALUE OF AUTOMATIC SPRINKLER PROTECTION

Automatic sprinkler protection helped develop modern industrial, commercial, and mercantile practices. Large areas, high buildings, hazardous occupancies, large values, or many people in one fire area all tend to develop conditions which cannot be tolerated without automatic fixed fire protection.

Mr. Hodnett is a consultant in fire protection and piping systems based in Providence, RI. He was NFPA staff liaison to the NFPA Committee on Automatic Sprinkler Systems prior to his retirement in 1984.

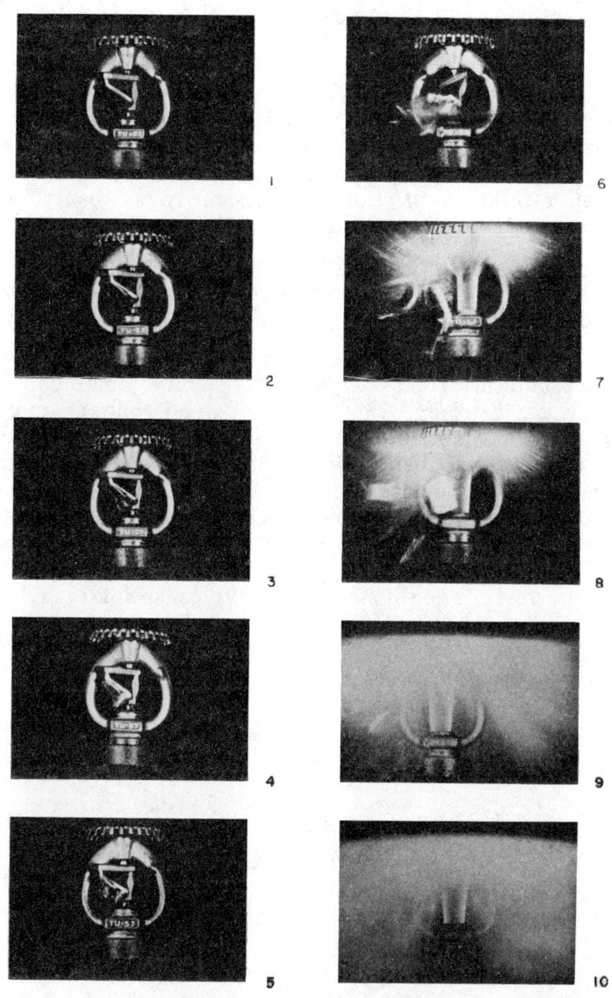

FIG. 18-1A. Operation of a typical fusible link automatic sprinkler is shown in this sequence of photos. As heat melts the solder, separation of members of the soldered link (the sloping side of the triangle in photos 1 to 5) is followed by complete separation of the link and lever arrangement (photo 6), which releases the cap over the sprinkler orifice, allowing water to escape and strike the deflector (photos 7 to 10).

Automatic sprinklers are particularly effective for life safety because they give warning of the existence of fire and at the same time apply water to the burning area. With sprinklers there are seldom problems of access to the seat of the fire or of interference with visibility for fire fighting due to smoke. While the downward force of the water discharged from sprinklers may lower the smoke level in a room where a fire is burning, the sprinklers also serve to cool the smoke and make it possible for persons to remain in the area much longer than they could if the room were without sprinklers.

Because of the research and development work going on, automatic sprinklers are now practicable for dwellings and other small properties. In country areas where water supplies are limited, a pressure tank can be provided with sufficient capacity to control the fire during evacuation. NFPA 13D, *Standard for the Installation of Sprinkler Systems in One- and Two-Family Dwellings and Mobile Homes*, gives guidance on sprinkler systems for dwellings.

NFPA 101, *Life Safety Code®*, recognizes sprinklers in numerous ways, particularly to offset deficiencies in existing buildings. For example, longer travel distances to exits and interior finish of a higher combustibility than would otherwise be permitted are allowed with sprinklers.

Automatic sprinklers, properly installed and maintained, provide a highly effective safeguard against the loss of life and property from fire. The NFPA has no record of a multiple death fire (a fire which kills three or more people) in a completely sprinklered building where the system was properly operating, except where an explosion occurred or flash fire killed victims prior to the system's operation. In most cases, victims of fatal fires in sprinklered properties were involved in the ignition of the fire and received their injuries prior to the operation of the sprinklers, or were unable to escape due to a physical or mental impairment.

For many years the NFPA published summaries of sprinkler performance. This practice was stopped after

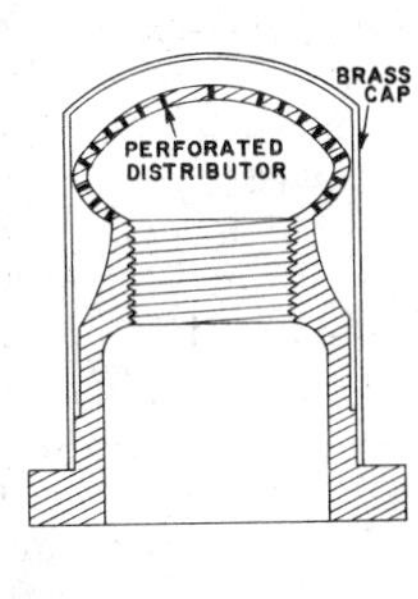

FIG. 18-1B. An early automatic sprinkler: water is shown discharging from a Parmelee No. 3 upright sprinkler, which was first used in 1875. It consisted of a brass cap soldered over a perforated distributor and was designed to screw onto a nipple. A cross-sectional view of the sprinkler is at right.

1970 when it became apparent that the data being used were biased by collection criteria which concentrated on fires causing large dollar loss. This bias led to an apparent and misleading decline in sprinkler effectiveness. Information of sprinkler effectiveness from insurers is also biased toward the larger property loss and the possible failure because of the widespread use of deductibles. Many fires which are extinguished by a small number of sprinklers and result in small property loss are never reported to the insurer.

While comprehensive statistics on sprinkler effectiveness are not currently available, documentation of their effectiveness in the protection of property can be found in the numerous examples of successful operation in National Fire Protection Association, United States Fire Administration, and insurer files.

In many situations, however, sprinkler protection is required by law for specific parts of the building only. Partial systems are generally not cost effective. Should the fire start remote to the system, sprinklers will have no effect on the growing fire. A fire burning into the protected area, will generally have developed sufficient intensity to overpower the sprinklers, thereby wasting water needed by the fire service to fight the fire.

Minimizing Business Interruption and Water Damage

In addition to the saving in direct fire losses due to sprinkler protection, there is a saving represented by the freedom from business interruption. There also is an undetermined but possibly even greater reduction in conflagration and exposure losses, which reasonably may be attributed to automatic sprinkler protection. The destruction of property and its adverse association and sometimes permanent effect upon business often is a great hardship not only to the owner, tenants, and employees but also to the community as a whole. Safeguarding a business from serious interruption by fire is often a determining factor in a decision to install sprinkler protection.

Standard sprinkler systems have devices which automatically give an alarm in case of sprinkler operation; thus, they not only apply water at the point most needed, but also give an audible signal on the premises and in many cases give an alarm at a remote location, such as the local fire department or a central station. This permits immediate check of fire conditions and minimizes water damage.

A properly installed sprinkler system operating in a timely manner will generate less water damage than the later application of hose streams by the fire service. Sprinklers are not hampered in their operation by smoke or heat as is the fire service. Sprinklers can apply water efficiently and promptly to the seat of the fire.

Another common misconception is that all sprinklers discharge water at the time of fire. This is not the case, most fires are controlled by a few automatic sprinklers in the immediate vicinity of the fire. Fear of water damage is sometimes offered as an objection to the installation of automatic sprinkler protection. This fear comes in part from the emphasis, often in ignorance, placed upon water damage in news reports of fires. Statements that a fire was of insignificant size, but that water damage was severe have been frequent. The probability of very severe destruction by fire in the absence of automatic sprinkler protection is seldom mentioned in these news accounts.

Accidental discharge of water from an automatic sprinkler system or other parts of a fire protection water system due to defects in sprinklers, water control devices, piping, or associated equipment, is very rare. Precautions to prevent unnecessary discharge of water as a result of mechanical injury, freezing or overheating, or corrosion are covered in Chapter 8 of this Section.

Economics of Sprinkler Protection

In addition to the protection against destruction of property values and interruption to business, the savings in insurance costs often make the expenditure for automatic sprinkler protection a sound business investment.

Many buildings do not have automatic sprinkler protection because the per dollar cost of the protection has appeared unjustifiably high to the building owners in relation to the value of the building. However, savings in insurance premiums alone could, in numerous cases, be adequate to finance, over a few years' time, the installation of automatic sprinkler protection. Of equal importance are the many building code "trade-offs" that are allowed when sprinklers are installed. These trade-offs permit an increase in undivided area and often less fire resistance for the building construction, and therefore less construction cost. No value can be placed on the life safety aspects of total sprinkler protection or the security occupants feel when such systems are installed.

STANDARDIZING SPRINKLER INSTALLATIONS

The terms sprinkler protection, sprinkler installations, and sprinkler systems usually signify a combination of water discharge devices (sprinklers), one or more sources of water under pressure, waterflow controlling devices (valves), distribution piping to supply the water to the discharge devices, and auxiliary equipment, such as alarms and supervisory devices. Outdoor hydrants, indoor hose standpipes, and hand hose connections are also frequently a part of the system that provides protection. Figure 18-1C is an illustration of a typical sprinkler installation with all common water supplies, outdoor hydrants, and underground piping.

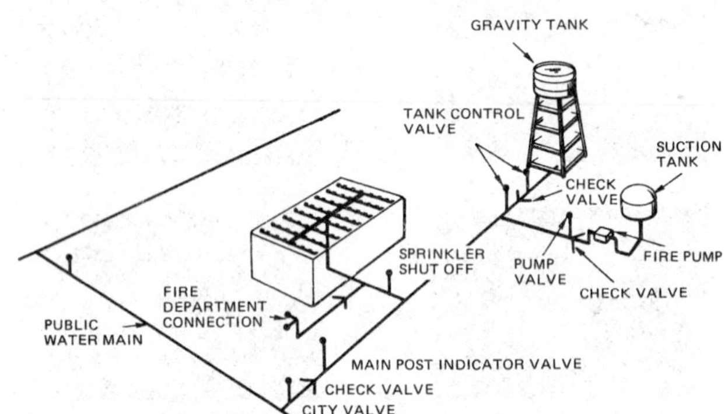

FIG. 18-1C. A typical sprinkler installation showing all common water supplies, outdoor hydrants, and underground piping.

When considering water supply problems, the performance of sprinklers, dry pipe or wet pipe systems, or special arrangements of sprinkler protection, the designation "sprinkler system" applies to the sprinklers controlled by a single water supply valve. Under this definition large buildings require several sprinkler systems, and a single water system may supply a number of sprinkler systems.

The fundamentals of sprinkler protection evolve around the principle of the automatic discharge of water in sufficient density to control or extinguish a fire in its incipiency. In planning for a system that fulfills this objective, many factors must be considered. They can, however, be broadly grouped into four categories: (1) the sprinkler system itself, (2) features of building construction, (3) hazards of occupancy, and (4) water supplies.

Automatic sprinkler systems of one type or another have been designed to extinguish or control practically every known type of fire in practically all materials in use today. It is essential, however, that for a given hazard the proper system be used. A sprinkler system designed to control and extinguish fire in an office occupancy with a relatively light amount of combustibles cannot be expected to have the same effectiveness in protecting a hazardous process involving considerable combustible materials, or a storage area where the fire loading is severe. On the other hand, it is not economical to overprotect by installing sprinkler equipment capable of controlling and extinguishing fire of a magnitude beyond any conceivable situation that could arise in the lifetime of a building.

The NFPA Sprinkler Systems Standard

NFPA 13, *Standard for the Installation of Sprinkler Systems*, covers the planning and design of sprinkler protection, the type of materials and components used in systems, and the operations carried on in making the installation. Compliance with the nationally recognized NFPA Sprinkler Systems Standard is often required by enforcement agencies, and it is used by insurance companies and insurance rating organizations. Property owners themselves often specify compliance with the Sprinkler Systems Standard in order that the protection provided will be in accordance with the best known practices.

While the NFPA Sprinkler Systems Standard is the primary standard for guidance on installation of sprinklers, other NFPA standards, recommended practices, and guides also have a direct bearing on certain phases of sprinkler protection. They should be referred to during design and construction of sprinkler systems. (See the Bibliography at the end of this chapter.)

Listing of sprinkler system devices by a testing laboratory is a separate procedure. The use of devices and equipment listed by such a laboratory after they have passed rigorous tests may be required by an authority having jurisdiction, or the authority itself may approve equipment.

CONSIDERATION OF BUILDING FEATURES

When sprinkler protection is being planned, it is necessary to evaluate construction, design, and certain features of location of a building in order to ensure effective operation of the sprinkler equipment. In older buildings, some typical modifications often needed are the following: (1) enclose vertical openings to divide multistoried structures into separate fire areas; (2) remove unnecessary partitions which could interfere with sprinkler discharge; (3) remove needless sheathing and shelving; (4) check concealed areas for the need for sprinkler protection or for the need for heat if such areas are to be sprinklered with wet pipe systems; and (5) pinpoint areas requiring high temperature sprinklers.

Older buildings often require some renovations to prepare them for sprinkler protection. On the other hand, designs for new buildings can incorporate features of construction and finish that are compatible to good sprinkler protection. Adding sprinklers as an afterthought is more expensive.

Further details concerning features of building design and construction as they relate to fire protection will be found in Section 7 of this HANDBOOK. The following paragraphs give information on some of the more prominent building features that deserve special mention.

Floor Cutoffs—Multistory Buildings

Sprinkler systems in multistory buildings are designed to extinguish fire in any one story, but not several stories at once. Therefore, the cutoffs for vertical openings must be complete to prevent the spread of heat to upper stories. Unprotected vertical openings can lead to upward spread of fire and to the opening of excessive numbers of sprinklers which, in turn, can overtax the water supply.

Substantial fire resistant enclosures around stairways, elevator shafts, utility shafts, etc., with fire doors at all interior openings into the enclosures are the preferred protection. Nevertheless, there may be instances where less substantial barriers to the passage of products of combustion may be acceptable. An example is the water curtain method of protecting escalator or conveyor openings. The NFPA Sprinkler Systems Standard suggests close spaced sprinklers for water curtains for this purpose.

High Ceilings

Sprinkler action may be delayed by excessive distance between sprinklers at the ceiling and combustible materials at the floor level. As the hot products of combustion from a flaming fire rise, air from the surrounding atmosphere mixes with the gases so that the temperature of the mixture decreases. When fire occurs in a room with an unusually high ceiling, the temperature at the ceiling directly over the fire is initially less than when under a low ceiling, and system activation is delayed.

Large clearances between sprinklers and combustibles below can also magnify the problem of obtaining the correct density and breakup of water discharge from sprinklers for maximum effect on fires, particularly in large masses of combustibles. The upward travel of combustion products creates temperature and draft conditions through which water droplets may have difficulty penetrating if the distance between a relatively intense flaming fire and the sprinklers is too great.

Fire tests have shown the relationship between varying clearances and varying water pressures, the latter governing the density and degree of atomization of the water droplets (Thompson 1964). The tests showed that high water pressures for sprinklers are of somewhat dubious advantage in compensating for extremes of height. Finely atomized water that must travel down through a strong fire draft is slowed by the upward velocity of the fire gases, and simultaneously the size of the droplets is continually reduced by evaporation.

Tests made in the sixties have also shown that a relatively coarse discharge density of 0.20 gpm per sq ft [8(L/min)/m^2] from sprinklers at ceiling heights of 30 ft and 50 ft (9 m and 15 m) was sufficient for fire control (Webb 1968). The test conditions simulated display booths in an exhibition hall representing a fire loading of 15 to 20 lb per sq ft (73 to 98 kg/m^2). The tests indicated that the ratio of sprinklers opening during a fire will increase proportionately with ceiling height, assuming a constant discharge density. Higher fire loadings would require correspondingly higher discharge densities. The concept of sprinklers with relatively coarse discharge was verified with the recent development of the large drop sprinkler which has been designed specificaly to produce large drops to penetrate the high velocity fire plume present in high challenge fires.

Concealed Spaces

Sprinklers should be installed in combustible spaces above ceilings. Fire may spread into these spaces which are shielded by construction from sprinklers in the main area. It is important to eliminate ignition sources in such spaces, and it is desirable to provide firestopping to prevent entrance and spread of fire. Before sprinklers are installed in an older building, it is important to closely examine the interior finish; some of the ceiling sheathing or hollow siding could be removed. Particularly objectionable are light flammable materials, such as paper, used for decorative effects.

Shielded Fires

Wide shelves or tables, partitions, conveyors, ducts, and other equipment that shield fire from sprinkler discharge may delay activation and cause an excessively large number of sprinklers to operate. In some cases, these conditions require additional sprinklers.

Building Location

A frequently unanticipated or neglected handicap to automatic sprinkler protection comes from exposure fires occurring outside the sprinklered building. Heat entering the building through improperly protected openings in the walls or through exterior walls that are combustible can easily open many sprinklers and tax the sprinkler water supply to the extent that fire can enter the building, particularly at upper stories. Interior sprinklers cannot be expected to be effective in protecting exterior combustible walls or roofs. When a nearby building offers a serious exposure to a sprinklered building, particularly if the latter has combustible walls or cornices and many unprotected exterior openings facing the exposure, the best protection is a complete installation of outside sprinklers.

Sprinkler systems in buildings subject to flood require special attention to the following: (1) the location and arrangement of piping so that it will not be washed out, nor its supports weakened; (2) location of valves so that they will be accessible during high water; (3) location of alarm devices so that they will remain operable during high water; and (4) location and arrangement of fire pumps and their power supply and controls, secondary water supplies, and other auxiliary equipment, to provide reasonable safeguards against interference with operation.

Sprinkler systems in buildings subject to earthquake require measures to prevent breakage of piping. Excessive movement is prevented by sway bracing. Unavoidable movement is met by providing flexible couplings between the major parts of the system, such as the top and bottom of risers, the joints between sections of buildings, and at pipe clearances at critical points, such as foundations, walls, and floors. For detailed requirements, consult the NFPA Sprinkler Systems Standard.

HAZARDS OF OCCUPANCY

The use made of a building is an essential consideration in designing a sprinkler system that is adequate to protect against the hazards inherent in the type of occupancy. For the purposes of evaluating hazards, three main classes of occupancy are recognized in the NFPA Sprinkler Systems Standard. Schedules of pipe sizes, spacing of sprinklers, sprinkler discharge densities, and water supply requirements differ for each in order to provide protection appropriate for the hazard—while also avoiding unnecessary expense.

Classification of Occupancies

The three main classifications are: light hazard, ordinary hazard, and extra hazard.

Light Hazard Class: Includes occupancies where the quantity and combustibility, or both, of materials is low, and fires with relatively low rates of heat release are expected. Examples are apartments, churches, dwellings, hotels, public buildings, office buildings, and schools.

Ordinary Hazard Class: This class is divided into three groups, mainly because each requires a somewhat different water supply for sprinklers. In general, this class includes ordinary mercantile, manufacturing, and industrial properties. Group 1 covers properties where combustibility is low, the quantity of combustibles is moderate, stockpiles of combustibles do not exceed 8 ft (2.4 m) in height, and fires with moderate rates of heat release are expected. Examples are canneries, laundries, and electronic plants. Group 2 includes properties where the quantity and combustibility of contents is moderate, stockpiles do not exceed 12 ft (3.6 m), and fires with moderate rates of heat release are expected. Examples are cereal mills, textile plants, printing and publishing plants, and shoe factories. Group 3 lists a small number of occupancies where the quantity and/or combustibility of the contents is high and fires of high rates of heat release are expected. Examples are flour mills, piers and wharves, paper manufacturing and processing plants, rubber tire maufacturing, and storage warehouses (paper, household furniture, paint, etc.).

Extra Hazard Class: Extra hazard occupancies involve a wide range of variables codified into two groups which may produce severe fires. Group 1 includes occupancies with little or no flammable or combustible liquids. Some examples are die casting, metal extruding, rubber production operations, sawmills, and upholstering operations using plastic foams. Group 2 includes occupancies with moderately substantial amounts of flammable or combustible liquids or where shielding of combustibles is extensive. Some examples are asphalt saturating, flammable liquid spraying, open oil quenching, solvent cleaning, varnish and paint dipping.

While classification of occupancies into three broad categories serves as a good basic guide, it does not rule out the necessity of evaluating separately certain portions of an occupancy that may contain hazards more severe than in the remainder of the building. For example, a hotel is listed in the NFPA Sprinkler Systems Standard as a light hazard occupancy. But certain areas in a hotel, such as kitchens and laundries, obviously are more hazardous than guest rooms. Consequently, the sprinkler protection for these more hazardous areas must be increased, and in this particular example, the design in the kitchen and laundry areas would conform to that required for ordinary hazard occupancies.

In each of the three broad groups, the system may either follow an appropriate piping schedule and spacing rules, or the system may be hydraulically designed. Hydraulically designed systems are preferable from a protection standpoint in extra hazard occupancies, and are desirable in all classifications.

SPECIAL OCCUPANCY CONDITIONS

Some conditions require more than ordinary sprinkler protection in order to provide dependable fire extinguishment and control. Sprinkler experience shows that occupancies which involve high piled combustible stocks, flammable and combustible liquids, combustible dusts and fibers, large quantities of light or loose combustible materials, and chemicals and explosives can permit rapid spread of fire and often cause the opening of excessive numbers of sprinklers with disastrous results. Complete automatic sprinkler protection with strong water supplies will usually control fires in occupancies containing these hazardous conditions, provided the severity of the hazards is plainly recognized and the sprinkler system is appropriately designed for the hazards.

High Piled Combustible Material

Practices in storage and warehousing lead to high piling for many combustible solid or packaged commodities (Thompson 1949). It is a matter of record that disastrous fires have occurred in sprinklered warehouses where combustible materials have been stored to heights over 50 ft (15 m). One of the principal factors contributing to the difficulty for sprinklers in controlling or extinguishing fires in high piled stock is that water discharged from the sprinklers cannot penetrate to the interior or lower portions of high piles, the most frequent areas where fires originate. High, closely packed piles, particularly over 15 ft (4.5 m) in height, have the inherent characteristics of shedding water; the bottom parts of the piles may not be adequately wetted to control fire spread.

Where racks are used for the storage of material, in-rack sprinklers are used in order that the water may get at the seat of the fire. For rack storage above 20 ft (6 m), in-rack sprinklers should be installed regardless of the material stored. For rack storage below 20 ft (6 m), in-rack sprinklers should be installed for the more hazardous materials such as wood and plastic products. For more details NFPA 231C, *Standard for Rack Storage of Materials,* should be consulted for rack storage and NFPA 231, *Standard for Indoor General Storage,* should be consulted for storage not on racks.

Flammable and Combustible Liquids

The effectiveness of automatic sprinklers on flammable and combustible liquid tank and spill fires depends on the flash point, physical and combustion characteristics, temperature, areas of burning surface, and quantity of liquid involved. Sprinklers are usually effective in extinguishing fires in combustible liquids with flash points of 200°F (93.3°C) and higher at normal room temperatures, in heavy flammable liquids (specific gravity greater than water), and in water soluble liquids. Control of fire, but not extinguishment, in low flash point [under 200°F (93.3°C)] flammable liquids, can be expected from automatic sprinklers.

Combustible Liquids: The basic action of water spray from sprinklers in extinguishing fire in liquids with a high flash point is considered to be cooling of the liquid surface to a point where an insufficient quantity of vapors are produced for combustion. Most of the cooling is obtained by the absorption of about 8,000 Btu (8440 kJ) of heat through conversion of each gallon (1 gal = 3.78 L) of water to steam. The vaporization of the spray should occur near the burning liquid surface; consequently, size of water droplets and the velocity of the droplets are critical factors in ensuring that the water does penetrate upward fire drafts and reaches the combustion zone at the right size for maximum cooling effect. Some of the many combustible liquids in this category are medium and heavy fuel oils, quench oils, asphalts, and lubricating oils.

"Heavy" Flammable Liquids: A few flammable liquids (i.e., carbon disulfide) are heavier than water and not soluble with water. As a consequence, spill and tank fires in such liquids are extinguished by vapor-air separation by a layer of water from sprinkler discharge floating on the liquid.

Water soluble Liquids: Certain flammable liquids, i.e., methyl alcohol and acetone, are soluble in water and may be extinguished by dilution with water from sprinklers. The resulting liquid-water solution is less volatile than the liquid alone, depending on the degree of dilution with water. This has the effect of eventually raising the flash and fire point temperatures higher than the actual solution temperatures. Combustion will cease because there is an insufficient quantity of flammable vapors. Extinguishment by dilution, however, is not commonly relied upon due to the amount of water required to make most liquids nonflammable, and there may be the danger of frothing if the burning liquid is heated to over 212°F (100°C). Alcohol, for example, may require as much as 4 gal (15 L) of water for each gallon (1 gal = 3.78 L) of alcohol.

Liquids with a Low Flash Point: Fires in flammable liquids with a low flash point that are not water soluble, and which have a specific gravity greater than water, cannot, as a rule, be extinguished with water, because the relationship between flash point and water temperature precludes adequate cooling. Water spray from sprinklers can, however, exert a degree of control short of extinguishment on fires involving flammable liquids with low flash points. Control is principally due to absorption of heat by water on conversion to steam. For example, a burning gasoline fire 100 sq ft (9.3 m^2) in area will liberate about one million Btu per minute (1055 MJ/min). (Theoretically, it would take only 120 gpm (454 L/min) of sprinkler water to absorb that amount of heat. Although 100 percent efficiency in converting water to steam in a combustion zone is rarely achieved, a substantial amount of water can be vaporized with absorption of considerable heat, provided the discharge from sprinklers is finely dispersed for maximum cooling effect.) Thus, in this example, a good portion of the million Btu per minute (1055 MJ/min) released from the fire is absorbed in the formation of steam.

Other factors contributing to the controlling effect of sprinkler discharge are the depositing of a protective film of water on exposed materials and the slowing down of the rate of combustion. The net result is a limiting of the zone of high temperature to a relatively small area in the immediate vicinity of the burning liquid.

Normally, all equipment containing flammable liquids or vapors is closed. Accidents or mistakes in operating procedures, however, may release vapors which, on ignition, may result in an explosion that can cripple a sprinkler system, or a flash fire that may open an excessively large number of sprinklers. Efficient and reliable ventilation and explosion protection equipment are important methods useful in reducing explosion hazards and consequent damage to sprinkler systems. Strong water supplies capable of supplying all the sprinklers in a single hazardous area, at higher than usual discharge densities, are essential. Densities ranging from 0.2 to 0.5 gpm per sq ft [8 to 20 (L/min)/m^2] are common. Excessively high discharge pressures, although yielding high discharge rates, may produce too fine a spray for effective extinguishment or control.

Combustible Dusts

Sprinkler systems may be crippled by severe dust explosions. A local dust explosion within equipment or a building can dislodge and ignite additional dust as the disturbance progresses to other areas. The explosion and flame may extend throughout a large area almost instantaneously.

Ordinary automatic sprinkler protection at properties where dust explosion hazards exist can be effective for fire control, provided it has been installed in a manner to reduce the probability that piping will be broken by an explosion. Also, to obtain full benefit from sprinkler protection, efficient dust removal systems and good housekeeping combined with explosion venting construction are essential. NFPA standards for the prevention of dust explosions cover essential precautions.

Chemicals and Explosives

The production and handling of hazardous chemicals and explosives frequently involve the use of materials that are unusually sensitive to shock or elevated temperatures. Ordinarily, to keep the hazards moderate, such materials are handled in small amounts and in closed or covered containers.

Fire in cellulose nitrate film, pyroxylin plastic, rocket propellants, and other chemicals subject to decomposition may generate much heat and also produce flammable or explosive vapors. Proper handling, process, and storage arrangements are thus primary operating principles to prevent excessive release of these vapors.

Conventional automatic sprinkler protection is not likely to arrest an explosion reaction once it starts. However, strong water supplies and closely spaced sprinklers [90 sq ft (8.4 m^2) or less] on piping designed and installed to resist damage, insofar as possible, are helpful in preventing extension of damage. Water from sprinklers can protect combustible construction and equipment in the vicinity of an explosion and can cool the atmosphere, reducing the spread of an initial fire or other heat producing disturbance. It is essential that good explosion venting practices are followed to minimize explosion damage not only to the building and process equipment that may be involved, but to the sprinkler system itself.

Rocket propellants used in the aerospace industry and in defense systems are examples of chemical hazards requiring more than ordinary sprinkler protection to guard against the explosive potential of high energy fuels. High speed water deluge systems employing special discharge devices and fast acting water control valves that respond to rate-of-pressure increase or visual detection (infrared or ultraviolet) are used in locations where rocket propellants are manufactured, on missile launching pads, and in missile silos.

LOCATION AND SPACING OF SPRINKLERS

The fundamental idea in locating and spacing sprinklers in a building is to make sure there is no unprotected place, however unexpected, where a fire can start. In other words, no matter where a fire starts, there must be one or more sprinklers located in relation to that particular point that will operate promptly and discharge water when heat from the fire reaches them. Furthermore, there should be no direction that fire can spread in which it will not encounter other sprinklers to stop its progress.

Complete Protection

It is obvious, in theory at least, that complete installation of sprinklers throughout a building is necessary for complete protection of life and property. No areas should be left unprotected. The NFPA Sprinkler Systems Standard treats specifically a number of locations where the need for sprinklers is sometimes questioned. These include locations such as stairways and vertical shafts; deep, blind, and concealed spaces; ducts; basements or subfloor spaces; attics and lofts; and under decks, tables, exhaust hoods, canopies, and outdoor platforms. It is risky to omit sprinklers from any single area because it is judged that the hazard is not sufficient to warrant them. Such omissions are the weak links in otherwise complete protection.

Frequently, building codes and ordinances require partial sprinkler protection for specific areas with the intent of providing limited protection for certain hazardous areas and as a life safety measure. But the limitations of partial protection often outweigh the supposed advantages they offer.

Area and Spacing Limitations

The location of sprinklers on a line of pipe, and the location of the lines in relation to each other determine the size of area protected by each sprinkler. The NFPA Sprinkler Standard gives a definite maximum area of coverage for each sprinkler, depending principally upon the severity of the occupancy hazard and, to a lesser degree, on the type of ceiling or roof construction above the sprinklers. [The three classes of hazard (light, ordinary, and extra) were discussed earlier in this chapter.] The four types of ceiling or roof construction and variations found in each type are: smooth ceiling, beam and girder, bar joist, and open wood joist.

The NFPA Sprinkler Systems Standard covers the maximum permissible distances between sprinklers on lines and between sprinkler lines and the maximum area of protection for each sprinkler. The maximum area of coverage permitted for each sprinkler must not be exceeded. However, the Sprinkler Standard allows for use of special sprinklers with greater areas of coverage when they have been tested and listed for this greater coverage.

On the other hand, sprinklers and lines should not be spaced too close together. If sprinklers are less than 6 ft (1.8 m) apart, baffles are required in order to prevent an operating sprinkler from wetting and thereby delaying operation of adjacent sprinklers.

Obstruction to Distribution

In addition to limits on the maximum distance between sprinklers on lines and between lines, certain limits of clearance have been established between sprinklers and structural members, such as beams, girders, and trusses, to avoid obstructing water being discharged from sprinklers. If a sprinkler is placed too closely to a beam that deflects the normal discharge pattern of the water, the area of protection for that sprinkler is considerably reduced and fire has a chance for additional growth. This causes more sprinklers to operate than should have been necessary. The NFPA Sprinkler Systems Standard is explicit in the limitations it places on distances between sprinklers and structural members to avoid obstruction to lateral distribution of water.

Clearance Between Sprinklers and Ceilings

The distance between sprinklers and the ceiling is important. The closer sprinklers are placed to the ceiling, the faster they will operate (see Fig. 18-1D). However,

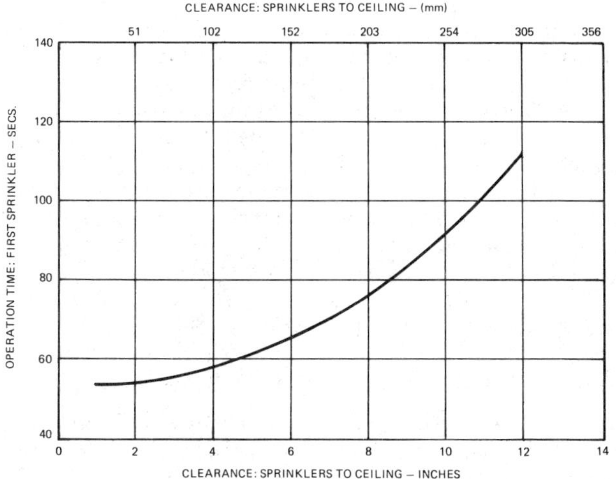

FIG. 18-1D. Effect of clearance between ceiling and sprinklers on operating time of sprinklers.

except for continuous smooth ceilings, locating them too close to the ceiling is more likely to result in serious interference to lateral distribution of water from sprinklers by structural members. Then too, when a combustible ceiling is broken up into bays by beams framed into girders or into narrow channels by beams, purlins, or joists, it is quite possible for a fire of moderate to severe intensity to ignite the ceiling and spread for considerable distances if the sprinklers are not located at the correct distance below the ceiling. The NFPA Sprinkler Systems Standard gives maximum distances below ceilings for a variety of construction types.

SPRINKLER PIPING

Sprinkler piping must be carefully planned and installed in accordance with the NFPA Sprinkler Systems Standard. Lines of pipe in which the sprinklers are directly placed are designated branch lines. The pipe directly supplying branch lines is designated as a cross

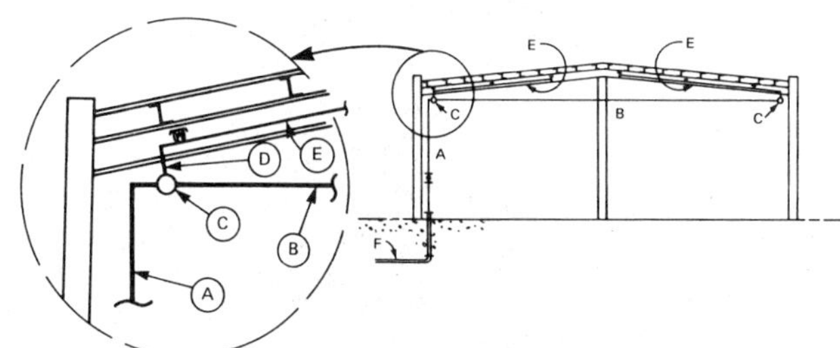

FIG. 18-1E. Building elevation showing parts of sprinkler piping system. (A) riser; (B) feed main; (C) cross main; (D) branch line; (E) risers; and (F) underground supply.

main. The pipe supplying a cross main is designated as a feed main. (See Fig. 18-1E.)

The size of piping supplying automatic sprinklers is determined either from the piping schedule discussed later in this chapter or on the basis of hydraulic calculations.

General Pipe Schedule Requirements

No practical sprinkler piping arrangement can produce a completely uniform protective water discharge from sprinklers in different locations or with various numbers of sprinklers simultaneously discharging water. The piping schedules listed in the NFPA Sprinkler Systems Standard are based upon extensive and carefully controlled tests and will provide consistently dependable sprinkler protection with practical economy in installation costs and water supply. Hydraulically designed systems will usually provide a more uniform water distribution with additional economy.

If conditions call for either unusually long runs of pipe or many bends, an increase in the size of risers or feed mains may be required in order to compensate for friction loss.

Wet pipe and dry pipe installations follow the same schedule of piping, except that the longer average time between the operation of sprinklers and the discharge of water required in dry pipe over wet pipe systems calls for specific restrictions on the air capacity of dry pipe system piping.

Arrangement of Sprinkler Supply Piping

Figure 18-1F illustrates various configurations for sprinkler system piping. Although it is permissible to supply water to sprinklers in one fire area by an overhead feed main which also supplies adjoining areas, this is not generally done except for small areas. Adjacent horizontal fire areas should usually have individual risers, each riser having its own control valve.

With the advent of hydraulically designed sprinkler systems many new installations employ gridded or looped piping arrangements in order to reduce friction loss in piping, which results in less costly piping systems to deliver a given quantity of water to the design area. Figure 18-1G illustrates both gridded and looped piping configurations.

Risers

The proper location, arrangement, and size of risers at any given property are problems that require skilled judg-

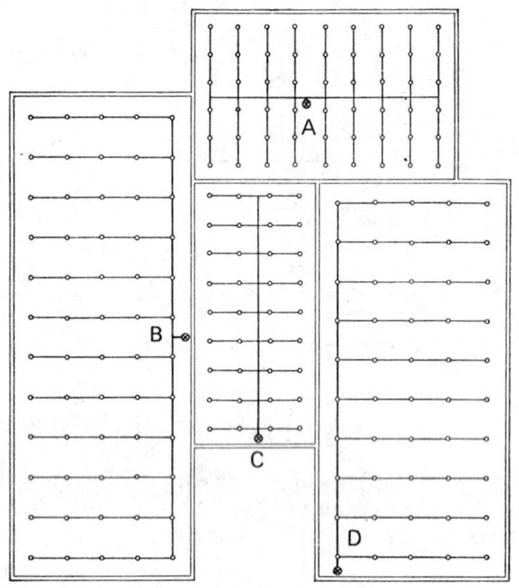

FIG. 18-1F. Location of risers. (A) central feed; (B) side central feed; (C) central end feed; and (D) side end feed.

ment. Construction, height, area, occupancy, and fire hazards must be carefully considered.

In a multistoried building having standard fire cutoffs between floors, the size of a riser supplying sprinklers on more than one floor is determined by the maximum number of sprinklers on any floor supplied by that riser or by hydraulic calculation of the maximum water demand on any one floor.

The NFPA Sprinkler Systems Standard covers methods of making riser connections to underground mains, the permissible location of connections to domestic supplies, and special piping arrangements needed when pressure tank supplies are used.

Water Supply Connections

Piping from the water supply to sprinkler risers should be at least as large as the riser. In private underground piping for buildings having other than light hazard occupancies, any dead-end pipe which supplies both sprinklers and hydrants should not be less than 8 in. in size. Underground pipe should conform to specifications and to rules for laying pipe as given in NFPA 24, *Standard for the Installation of Private Fire Service Mains and Their Appurtenances*. Steel pipes used underground may corrode and develop leaks in a short time unless special protection against corrosion is provided.

Each sprinkler system should have an accessible control valve to control flow of water to the system from all sources other than from fire department connections. If there is more than one source of water supply, a check valve is needed in the connection from each source.

A common arrangement in industrial properties is to locate water supply gate and check valves in a covered valve pit. An indicator post above ground is used with gate valves buried in the ground. An indicator post is also sometimes used even though the control valve is located in a pit. It is considered the best practice to have sprinkler shutoff valves equipped with supervisory devices that will give an alarm on closing of the valve.

In large plants, fire main systems have sectional control valves to improve flexibility in the use of water supplies. It is important that such valves be plainly marked to show the location of the system controlled.

Installation Standards for Sprinkler System Piping

Piping used in sprinkler systems should be of a type to withstand a working pressure of not less than 175 psi (1207 kPa) (ANSI 1975). Piping should conform to Table 18-1A. Other types of pipe or tube may be used, but only if they have been tested and listed by a testing laboratory as suitable for use as sprinkler piping (ASTM 1976).

At the present time, only metallic piping and tubing has been listed by testing laboratories for use in sprinkler systems installed according to the NFPA Sprinkler Standard. Plastic pipe has been listed by laboratories for use in systems installed in one and two family dwellings and mobile homes (NFPA 13D), and has been submitted to testing laboratories for use in systems installed according to the NFPA Sprinkler Systems Standard.

Pipe Fittings: If of cast iron, pipe fittings should be of extra heavy pattern for sizes larger than 2 in. if the normal pressure in the piping system exceeds 175 psi (1207 kPa). If fittings are of malleable iron, standard weight pattern is acceptable in sizes up to and including 6 in. if the normal pressure in the pipe system does not exceed 300 psi (2069 kPa).

Fittings should be of types designed for use in sprinkler systems. Reduction in pipe size should not be made by the use of bushings if it can be avoided.

Cast iron screwed fittings (Class 125 and 250) and malleable iron screwed fittings (Class 150 and 300) are

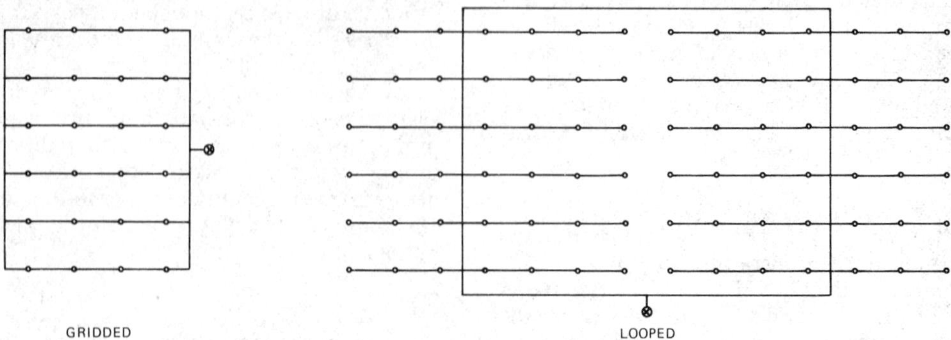

FIG. 18-1G. Low friction loss piping configurations showing gridded and looped systems.

TABLE 18-1A. Sprinkler System Pipe or Tube Materials

Material and Dimensions	Standard
Ferrous Piping (Welded and Seamless)	
Welded and Seamless Steel Pipe For Ordinary Uses, Spec. For Black and Hot-Dipped Zinc Coated (Galvanized)	ASTM A 120
Spec. for Welded Seamless Steel Pipe	ASTM A 53
Wrought-Steel and Wrought-Iron Pipe	ANSI B36.10
Copper Tube (Drawn, Seamless)	
Spec. For Seamless Copper Tube	ASTM B 75
Spec. For Seamless Copper Water Tube	ASTM B 88
Spec. For General Requirements for Wrought Seamless Copper and Copper-Alloy Tube	ASTM 5 251
Brazing Filler Metal (Classification BCuP-3 or BCuP-4)	AWS A 5.8
Solder Metal, 95–5 (Tin-Antimony-Grade 95TA)	ASTM B32

covered by American National Standard Institute Piping Standards (ANSI 1971a and 1971b).

All inside piping is installed by means of screwed, flanged, mechanical joint, or brazed fittings or, with specific approval, by welding or flexible couplings. Specifications for such welding are given in the American Welding Society Standard (AWS 1980).Where welding of joints is allowed, or where fittings are brazed, the fire hazard of indoor welding must be suitably safeguarded.

Approved flexible couplings are used for earthquake resistance. Couplings, bends, and tees of this type are sometimes employed in risers and feed mains if their use is of particular advantage.

Pipe Hangers and Clamps: These are used to attach sprinkler system piping to substantial structural elements of the building. The support offered by many forms of ceiling construction is inadequate to carry the load of sprinkler piping without additional connections.

The types of hangers necessary to meet various conditions of construction have been tested and listed by testing laboratories. Representative types are shown in Figure 18-1H. For the larger pipe, trapeze bars of steel angle or pipe supported by double hanger rods are frequently used. Expansion shields for attaching hangers to concrete are preferably installed horizontally, although vertical installation may be used in some instances. The adequate support of sprinkler piping is an important consideration. The Sprinkler Systems Standard provides detailed information.

Corrosive Conditions: Corrosive conditions call for the use of pipe, fittings, and hangers designed to resist the particular corrosive agent, or the application of protective coating over susceptible components. Threaded thin wall steel pipe should not be used where severe corrosive conditions exist. The choice depends on the kind and severity of the corrosive condition. Care must be taken that paint is not applied to the sprinklers when piping or other adjacent installations are being painted as protection against corrosion.

If it is necessary to use steel pipe underground as a connection from a system to sprinklers in a detached building, the pipe should be protected against corrosion before it is buried.

Test Equipment

It is essential that the condition of a sprinkler system and its water supply are known at all times. Actual tests of the supply and the system itself at stated intervals is the

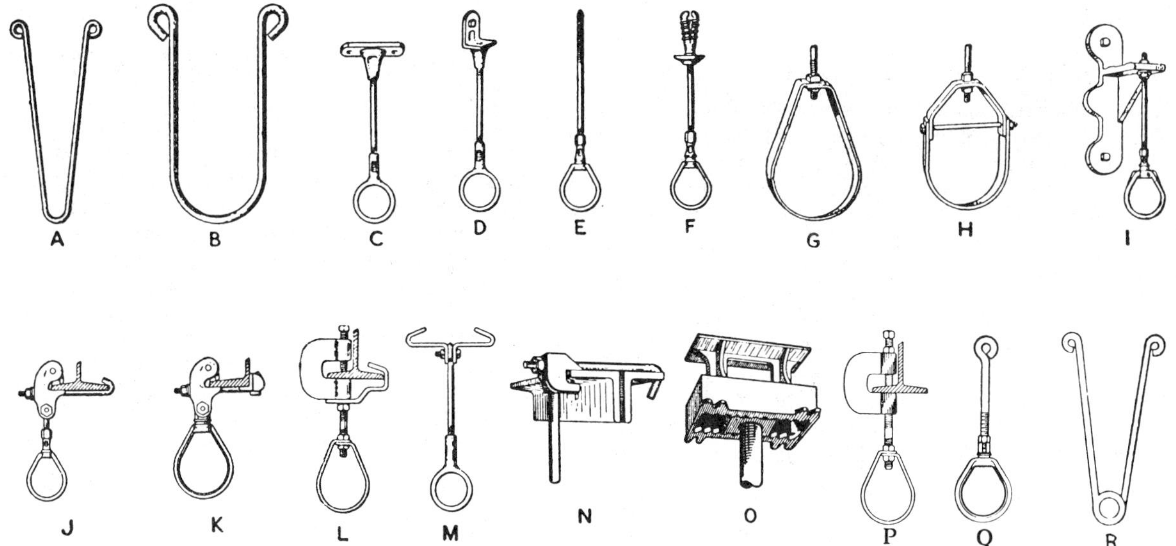

FIG. 18-1H. Common types of acceptable hangers. (A) U-type hanger for branch lines; (B) U-type hanger for cross mains and feed mains; (C) adjustable clip for branch lines; (D) side beam adjustable hanger; (E) adjustable coach screw clip for branch lines; (F) adjustable swivel ring hanger with expansion shield; (G) adjustable flat iron hanger; (H) adjustable clevis hanger; (I) cantilever bracket; (J) "universal" I-beam clamp; (K) "universal" channel clamp; (L) C-type clamp with retaining strap; (M) center I-beam clamp for branch line; (N) top beam clamp; (O) "CL Universal" concrete insert; (P) C-type clamp without retaining strap; (Q) eye rod and ring hanger; (R) wraparound U-hook.

best way of doing it. Special fittings and piping on the supply and system are required.

Water Supply Test Pipes and Pressure Gages: These are supplied for each sprinkler system installation. Test pipes, which may also serve as drain pipes, must be provided to permit flow tests. Test connections are not less than 2 in. in size, and are equipped with a shutoff valve. The arrangement permits a test with the system's main water control valve wide open without its discharge causing damage. A gage must be installed to show the pressure in the riser at or near the test connection. A typical arrangement of a test and drain connection with a pressure gage for a riser is shown in Figure 18-1I.

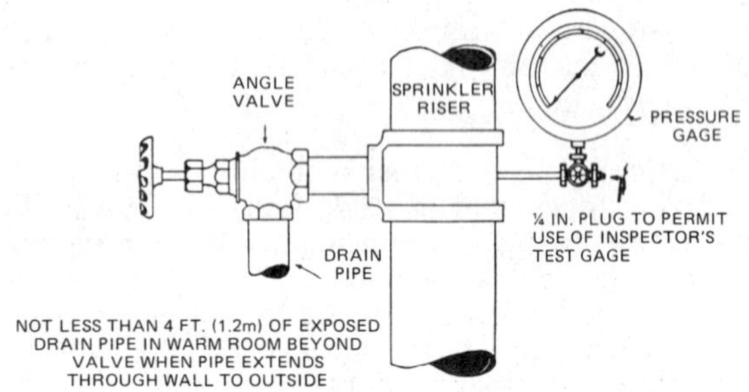

FIG. 18-1I. Test and drain connection for wet pipe sprinkler system riser.

Sprinkler System Test Pipe: A test pipe, not less than one inch in size and terminating in a corrosion resistant outlet which will give a flow equivalent to that from one sprinkler, should be installed in the top story. This provides a proper method of testing alarm devices and of tripping dry pipe valves and also shows that water can flow through the system. Typical arrangements are shown in Figure 18-1J.

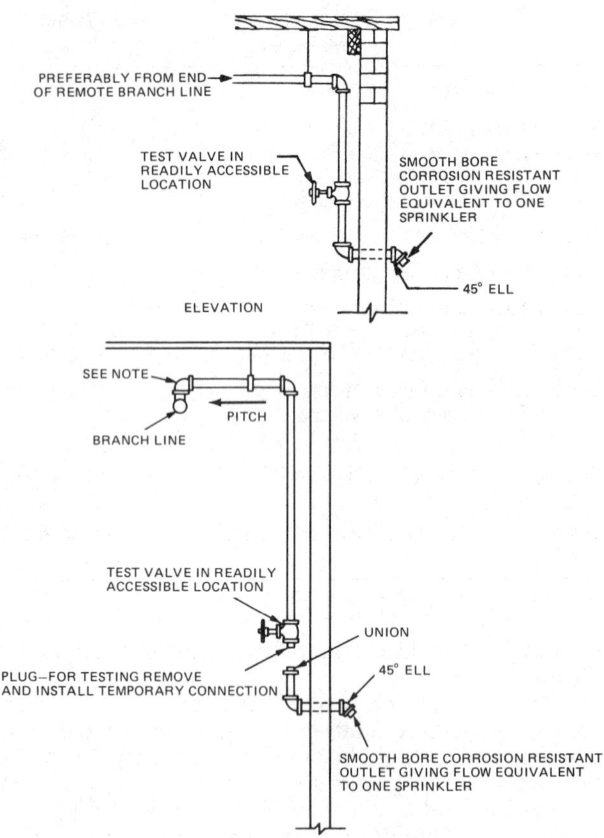

FIG. 18-1J. (Top) Wet pipe sprinkler system test pipes. (Bottom) One inch system test pipes on dry pipe system.

Other Connections and Features

Other connections to the sprinkler system piping should be limited to hand hose reserved exclusively for fire use. Circulation of water in sprinkler pipes is objectionable because of increased corrosion, which may impair the efficiency of the system. (See Circulating Closed Loop Systems later in this chapter for permitted use of sprinkler piping for circulation of water.) Sprinkler system piping must not be used in any way for domestic or utility water service because demands at peak periods could deplete the water supply required for fire service during a fire. Where hand hose is attached to sprinkler pipes within a room, it is done with the following restrictions: (1) hand hose should never be attached to a dry pipe sprinkler system; (2) piping and hose valve are at least 1 in., hose not larger than 1½ in.

Waterflow alarm valves, dry pipe valves, and other special features are covered separately in subsequent chapters of this Section.

Signs should be provided on all control, drain, test, and alarm valves to identify their purpose and function. Fire department connections should be properly identified to show whether they supply sprinkler systems, outside sprinklers, or hose standpipes. Manufacturers' instruction charts describing the operation and maintenance of equipment should be located near major sprinkler devices. Identification signs are shown in Figure 18-1K.

INSTALLING SPRINKLER SYSTEMS

Sprinkler system design and installation should be entrusted to only fully qualified and responsible parties. The installation of sprinkler systems is a trade by itself. Some large industrial properties, however, may have engineering and construction staffs to design and install automatic sprinkler systems and their water supplies.

Before a sprinkler system is installed or remodeled, a working plan is prepared. The plan identifies pertinent features of building construction and occupancy, water supplies, and system piping and equipment. The Sprinkler Systems Standard is quite precise on the data that must be shown on the plan because the plan is the basis for issuing approvals to proceed with installation where the approval of an authority having jurisdiction is required. Figure 18-1L illustrates a typical sprinkler system working plan.

Testing New Installations

Before connecting sprinkler risers for testing, underground connections to sprinkler installations must be flushed thoroughly. Obstructing materials in underground pipes can seriously impair sprinkler protection.

All piping and devices under pressure (including yard piping and fire department connections) should be tested

FIG. 18-1K. Identification signs.

hydrostatically for strength and leakage at not less than 200 psi (1379 kPa) pressure for 2 hr, or at 50 psi (345 kPa) pressure in excess of the maximum static pressure when that pressure is in excess of 150 psi (1034 kPa).

Any blank gasket used in testing should be of a special self-indicating type having red lugs protruding out beyond the flange in such a way as to clearly mark its presence. These should be numbered to assure their return after the work is completed.

Tests of drainage facilities are made by opening the main drain valve while the control valve is wide open. This provides assurance that main control valves are open and that the water will be disposed of safely.

When the weather is too cold for testing with water, tests of dry pipe systems are made by maintaining at least 50 psi (345 kPa) air pressure for two hr. During such a test, the clappers of differential type dry pipe valves should be held off their seats whenever a pressure in excess of 50 psi (345 kPa) is used, in order to prevent injury to the valves. When the weather warms up, it will be necessary to perform the hydrostatic test at a minimum of 200 psi (1379 kPa).

Dry pipe valves, quick opening devices, and water flow alarms should be given a complete working test.

TYPES OF SPRINKLER SYSTEMS

There are six major classifications of automatic sprinkler systems. Each type of system includes piping for carrying water from a source of supply to the sprinklers in the area under protection. The six major classifications of systems are:

Wet Pipe Systems: These systems employ automatic sprinklers attached to a piping system containing water under pressure at all times. When a fire occurs, individual sprinklers are actuated by the heat, and water flows through the sprinklers immediately.

Regular Dry Pipe Systems: These systems have automatic sprinklers attached to piping which contains air or nitrogen under pressure. When a sprinkler is opened by heat from a fire, the pressure is reduced to the point where water pressure on the supply side of the dry pipe valve can force open the valve. Then water flows into the system and out any opened sprinklers.

Preaction Systems: These systems are systems in which there is air in the piping that may or may not be under pressure. When a fire occurs, a supplementary fire detecting device in the protected area is actuated. This opens a water control valve which permits water to flow into the piping system before a sprinkler is activated. When sprinklers are subsequently opened by the heat of the fire, water flows through the sprinklers immediately—the same as in a wet pipe system.

Deluge Systems: These systems have all sprinklers open at all times. When heat from a fire actuates the fire detecting device, the deluge valve opens and water flows to, and is discharged from, all sprinklers on the piping system, thus deluging the protected areas.

Combined Dry Pipe and Preaction Systems: These include the essential features of both types of systems. The piping system contains air under pressure. A supplementary heat detecting device opens the water control valve and an air exhauster at the end of the unheated feed main. The system then fills with water and operates as a wet pipe system. If the supplementary heat detecting system should fail, the system will operate as a conventional dry pipe system.

Special Types: Special Types of systems depart from requirements of the NFPA Sprinkler Standard in such areas as special water supplies and reduced pipe sizes. They are installed according to the instructions that accompany their listing by a testing laboratory.

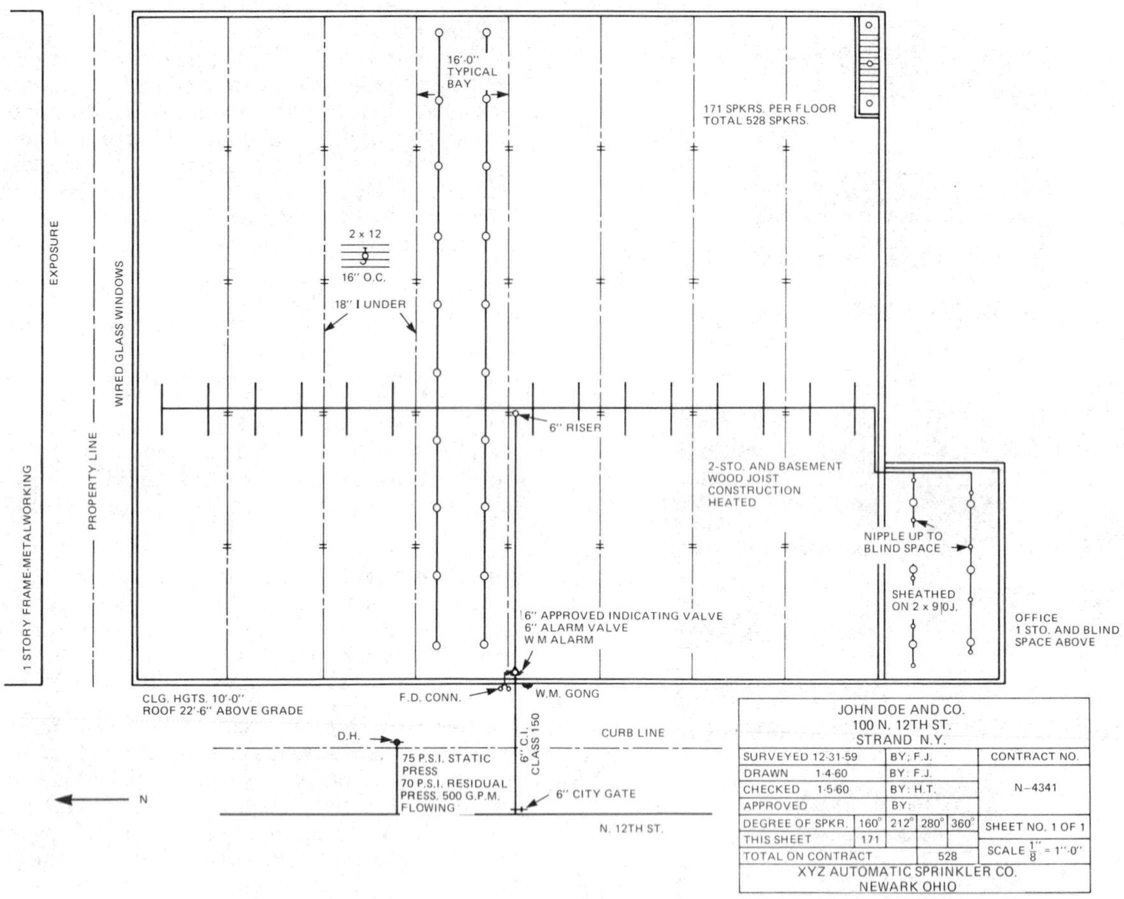

FIG. 18-1L. Typical sprinkler system working plan. SI units: 1 ft = 0.3048 m; 1 psi = 6.89 kPa.

Wet Pipe Sprinkler Systems

This type of system is generally used wherever there is no danger of the water in the pipes freezing, and wherever there are no special conditions requiring one of the other types of systems.

Where subject to temperatures below freezing, even for short periods, the ordinary wet pipe system cannot be used because the system contains water under pressure at all times. (See Fig. 18-1M.) There are two recognized methods of maintaining automatic sprinkler protection in such locations. One is the use of systems where water enters the sprinkler piping only after operation of a control valve (dry pipe, preaction, etc.), and the other by use of antifreeze solution in a portion of the wet pipe system.

Antifreeze Solutions: When a recommended antifreeze solution is maintained in the piping from the riser, the normal water supply does not flow except when the solution is discharged from an opened sprinkler. Because antifreeze solutions are costly and may be difficult to maintain, their use is usually limited to small, unheated areas served by a wet pipe system where the volume of the section involved is not more than 40 gal (150 L) and where the piping would otherwise have to be shut off and drained during cold weather. Where the section involved is more than 40 gal (150 L), the cost of refilling the system or even of replenishing losses from small leaks makes it advisable to use small dry pipe valves. In any event, antifreeze solutions should be used only in accordance with applicable local health regulations.

The antifreeze solution generally consists of water and a water-soluble liquid, such as glycerine or certain glycols. The proportions specified in the NFPA Sprinkler Systems Standard give the desired reduction in freezing temperature without producing a combustible mixture. Where the system is supplied from public water connections, the use of antifreeze solutions other than chemically pure glycerine (U.S. Pharmacopoeia 96.5 percent grade) or propylene glycol must not be permitted. In other systems, diethylene glycol, glycerine, propylene glycol, ethylene glycol, or a calcium chloride solution containing a corrosion inhibitor (such as sodium chromate) may be used.

Cold Weather Valves: Automatic sprinkler piping should not be shut off and drained as a regular practice to avoid freezing during cold weather. However, where the fire hazard is not severe, permission may be given to shut off not more than 10 sprinklers on a wet pipe system. Such shutoff valves are commonly referred to as cold weather valves.

Dry Pipe Sprinkler Systems

Dry pipe sprinkler systems, in which the piping contains air under pressure until the dry pipe valve operates, are used only in locations that cannot be properly heated.

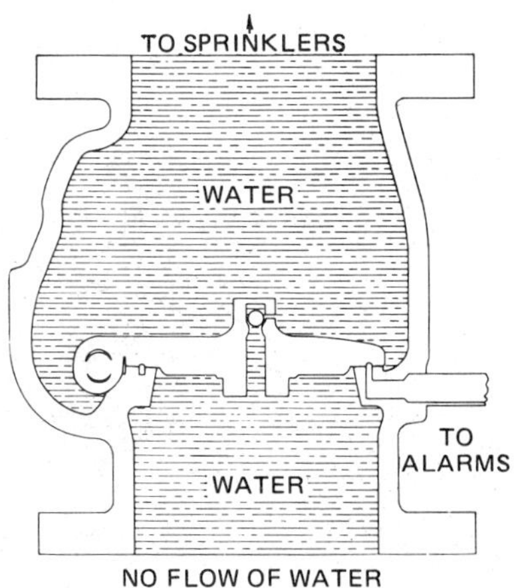

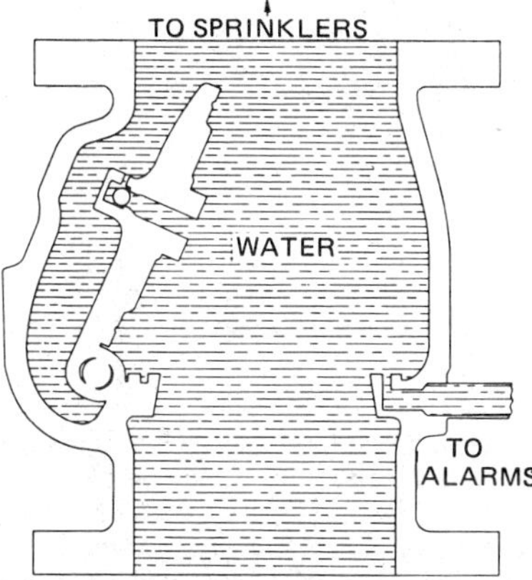

WATER FLOWING TO SPRINKLERS AND TO ALARMS

FIG. 18-1M. A wet pipe sprinkler system is under water pressure at all times so that water will be discharged immediately when an automatic sprinkler operates. The automatic alarm valve shown causes a warning signal to sound when water flows through the sprinkler piping.

The principle of a dry pipe system is illustrated in Figure 18-1N.

Dry pipe systems are often converted to wet pipe systems when they become unnecessary because adequate heat is provided.

Efficiency of Dry Pipe Systems: According to fire records, more sprinklers open on the average at fires with dry pipe than with wet pipe systems; this tends to show that the control of fire is not as prompt with dry pipe systems. However, in most classes of occupancy, and especially those of light and moderate hazard, dry pipe systems have shown generally good results and, when properly maintained, can be relied upon to satisfactorily extinguish or control fires.

Dry Pipe Valve Designs: Most dry pipe valves are designed so that a moderate air pressure in a dry pipe system will hold back a much greater water pressure. The difference between the air pressure and the water pressure, expressed as the ratio of these pressures when the air pressure is reduced to the value at which the valve opens, is called the differential.

If the differential is obtained by having a large diameter air clapper in a valve bear directly upon a smaller water clapper, the valve is often referred to as a differential type dry pipe valve. Figures 18-1O and 18-1P show the arrangement of parts and the relative sizes of air and water clappers in ordinary differential type dry pipe valves of recent and early manufacture. Large valve bodies are needed to accommodate the relatively large air clappers.

If the air pressure in the sprinkler system acts upon a small disc, diaphragm, or clapper which is arranged with levers, links, and latches to provide the necessary closing force upon the water clapper, the valve may be designated as a mechanical type or a latched clapper type. Mechanical dry pipe valves of early design are shown in Figure 18-1Q. Such valves were mechanically complicated and difficult to keep in good operating condition. A latched clapper type dry pipe valve is shown in Figure 18-1R.

Originally, a differential of 5 or 6 to 1 (water pressure to air pressure) was accepted practice; this has been altered with some designs so that higher air pressures may be used. In a low differential dry pipe valve, the differential is usually 1.0 and 1.2 to 1 rather than the 5 or 6 to 1 common with ordinary differential dry pipe valves. The low differential valve, which resembles a check valve or alarm valve, is kept closed by air pressure on the sprinkler system side of the valve clapper that exceeds the pressure on the water supply side of the clapper. The valve operates when the air pressure is reduced to only about 10 percent

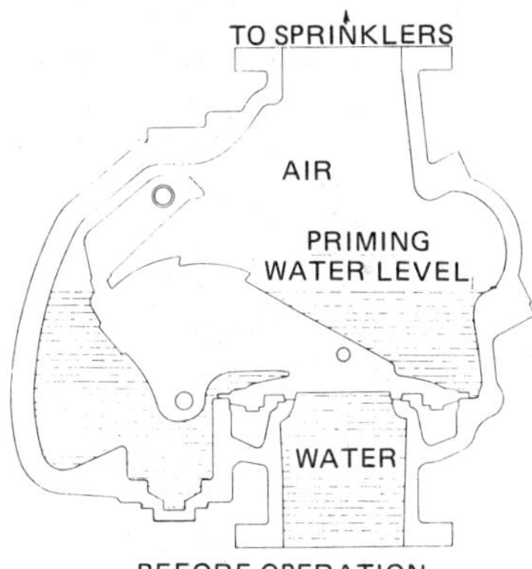

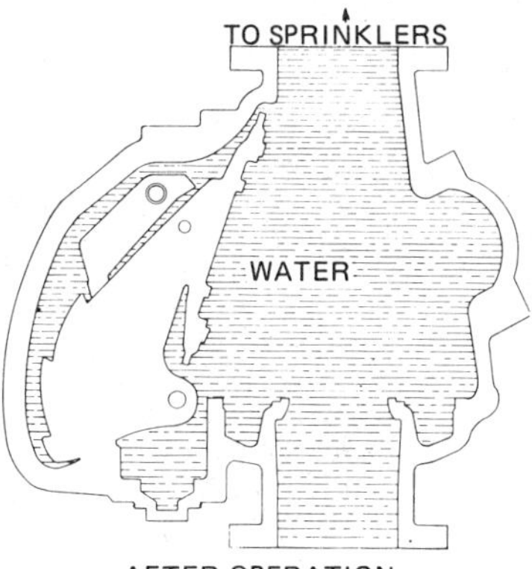

FIG. 18-1N. The principle of a dry pipe system is illustrated by these simplified drawings of a dry pipe valve. Compressed air in the sprinkler system holds the dry valve closed, preventing water from entering the sprinkler piping until the air pressure has dropped below a predetermined point.

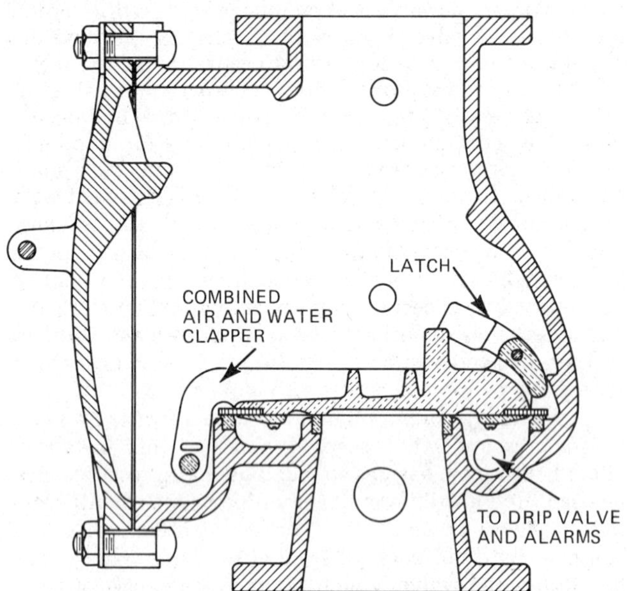

FIG. 18-1O. Differential type dry pipe valve. When the downward force exerted on the combined air and water clapper by the system air pressure is reduced to the value of the upward force exerted by the water supply upon the clapper, the valve will open to permit the flow of water into the sprinkler system. The latch prohibits the valve from returning to the closed position. Shown is the Hodgman Manufacturing Co., Inc., Model C dry pipe valve (no longer manufactured).

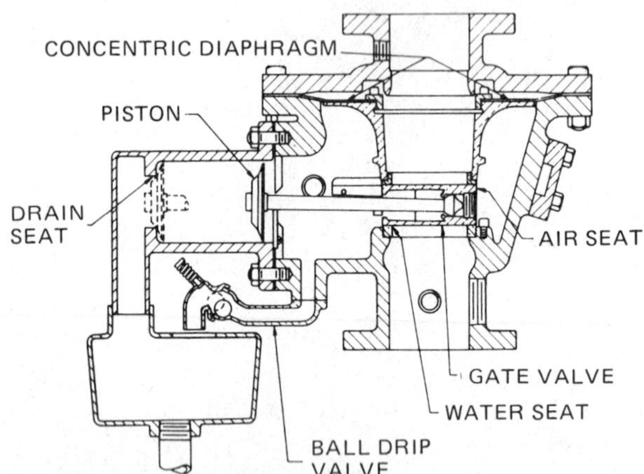

FIG. 18-1P. Early type differential dry pipe valve (manufactured 1905 to 1920). Air pressure on the concentric diaphragm and on the air seat of the gate valve holds the gate on the water seat. The ball drip valve prevents water leakage by the water seat from accumulating in the intermediate chamber. Release of air pressure allows the gate valve to lift slightly from the water seat. Pressure of water acts on the loosely fitting piston to withdraw the gate valve from the air seat and water seat. The piston against the drain seat prevents escape of water. (Grinnell Company, Inc., Models A and B)

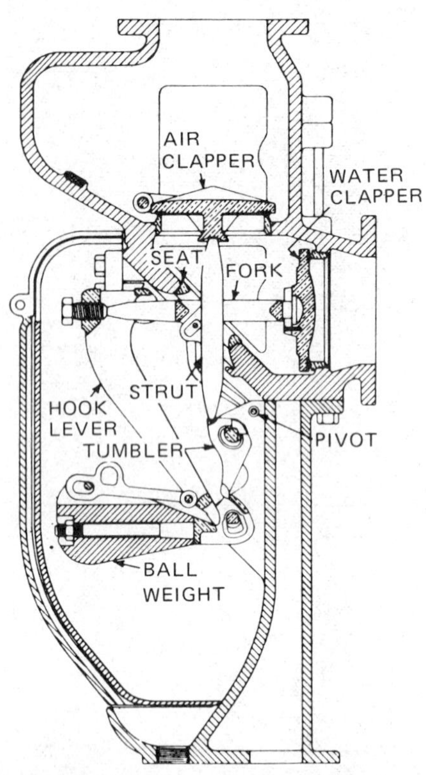

FIG. 18-1Q. Mechanical type dry pipe valve (manufactured 1909 to 1924). Water supply pressure on the water clapper is resisted by the fork hook lever, ball weight, tumbler, and strut, and by air pressure on the air clapper. When the valve trips, the strut and the fork hinge on the pivot, carrying the water clapper back to rest on the seat and thus preventing leakage into that portion of the valve containing the mechanical linkage assembly. ("Automatic" Sprinkler Corp. of America, International Models 4 and 5)

less than the water pressure at the point of operation. A low differential type dry pipe valve is shown in Figure 18-1S.

To reduce the danger of accidental tripping of ordinary dry pipe valves subjected to higher than normal water pressure or to water hammer, it is usual practice to maintain an air pressure well above the normal trip point pressure. For example, with a valve having a 6 to 1 differential and used with a normal maximum water supply pressure of 100 psi (689 kPa), the air pressure of the trip point would be approximately 100 ÷ 6 or 17 psi (117 kPa), but an air pressure of 30 to 35 psi (207 to 241 kPa) would be maintained. Unless a valve has been designed for especially low pressures (in which case the manufacturer's instructions regarding pressures should be followed), it is customary to maintain the dry system air pressure 15 to 20 psi (103 to 138 kPa) above the value at which the dry pipe valve would trip. About the same amount of excess pressure is carried in low differential dry pipe systems.

After a dry pipe valve has tripped and the piping has filled with water, the air and water clappers of ordinary differential valves must be prevented from returning to the set position should the water flow be interrupted, because the weight of the water column would restrict the valve. Reseating under temporary back flow can produce water hammer sufficient to break the valve or its fittings. An automatic mechanical latch is provided to prevent a clapper from returning to its seat after it has opened appreciably.

Quick Opening Devices: One characteristic of a dry pipe system is a delay in time between the opening of a

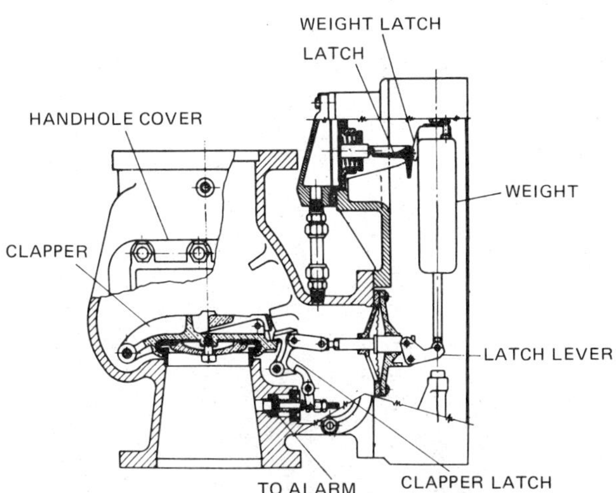

FIG. 18-1R. Latched clapper type mechanical dry pipe valve. The pressure of the water supply is resisted by a positive clapper latch which is released by the falling weight when a reduced system air pressure releases the diaphragm actuated weight latch. An automatic float drain may be provided to prevent water columning by possible leakage at the water valve seat or condensate drainage in the sprinkler piping. (See Fig. 18-1X.) Shown is the "Automatic" Sprinkler Corp. of American Model 141 dry pipe valve which is no longer manufactured.

sprinkler and the discharge of water, which may allow the fire to spread and more sprinklers to open. The delay is due to the time required to exhaust the air from the sprinkler piping. The difficulty may be partly overcome by installation of quick opening devices which either increase the rate of discharge of air from the piping or accelerate opening of the dry pipe valve when one or more sprinklers operate, depending upon the type of device used.

Dry pipe valves controlling systems having a capacity of more than 500 gal (1893 L) must have quick opening devices. They are generally designated as accelerators or exhausters and operate as a result of a prompt, but not large, drop in system air pressure produced by the opening of one or more sprinklers. The failure of an accelerator or exhauster to operate does not prevent normal tripping of a dry pipe valve.

In operating principle both accelerators and exhausters employ two air chambers. One chamber (often designated the inlet or lower chamber) has a connection always open to the dry pipe system, while the second chamber (usually called the upper or pressure chamber) is closed except for a small orifice which allows its internal pressure to equalize slowly with that in the inlet chamber, which is the normal dry pipe system air pressure. The two chambers are separated by a diaphragm which is deflected whenever the pressure in the system becomes less than that in the upper chamber due to the escape of system air pressure through an opened sprinkler. Movement of the diaphragm actuates valves and mechanisms that produce prompt tripping of the dry pipe valve.

Figure 18-1T shows an accelerator and gives an explanation of how it trips a dry pipe valve. See Figure 18-1U for a view of the same accelerator attached to a dry pipe valve as one of the latter's trimmings. An exhauster is shown in Figure 18-1V.

The reduction in operating time of a dry pipe valve due to the use of a quick opening device is recognized in the NFPA Sprinkler Systems Standard by allowing an increase in the allowable size and capacity of a dry pipe sprinkler system over that allowed if no accelerating or exhausting device is used.

Water Columning: Any leakage of water past the water valve of an ordinary dry pipe system, or water accumulating from slow drainage or condensation, must not be allowed in dry system piping above the valve. Due to the differential characteristics of most dry pipe valves, a relatively low head of water accumulated in the riser will exceed the normal trip point of the valve so that when the system air pressure is released by sprinkler operation the valve will not open due to the head of water on the air clapper. The valve is then "water columned." (Low differential dry pipe valves are not subject to failure due to water columning.)

Water columning can be avoided by making use of the

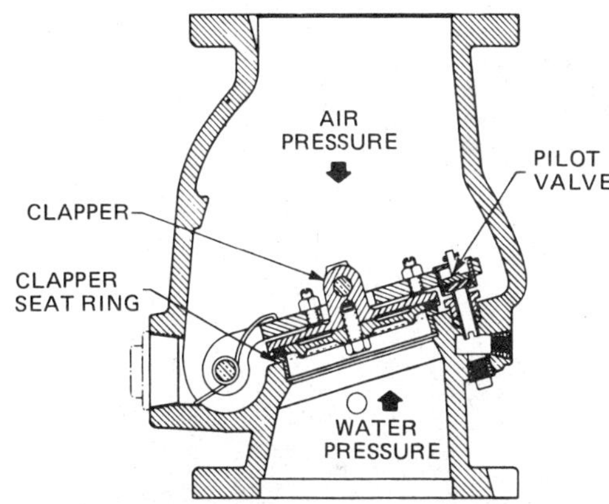

FIG. 18-1S. A low differential dry pipe valve. Air pressure 15 to 20 psi (103 to 140 kPa) higher than the water pressure supplying the sprinkler system holds the clapper shut on the clapper seat ring. When the air pressure is reduced to about 10 percent below the water pressure, the clapper is lifted off its seat and water enters the sprinkler system, and flows out through the pilot valve to operate alarm devices. Low differential valves are similar to alarm check valves used in wet pipe sprinkler systems. Shown is the "Automatic" Sprinkler Corp. of America Model 15 valve which is no longer manufactured.

intermediate chamber between the water and air clappers in the dry pipe valve. When the valve is set, the intermediate chamber is open to the atmosphere through an auxiliary drip valve; thus, water entering the chamber through a leaky water clapper can escape through the auxiliary valve. When the valve trips, the auxiliary valve closes automatically. Two designs of auxiliary valves are shown in Figures 18-1W and 18-1X. These valves reopen by gravity when the dry pipe valve is reset.

Location of Dry Pipe Valves: The dry pipe valve should be located in an accessible place as near as practicable to the sprinkler system it supplies. It should be protected from mechanical injury. When exposed to cold, it must be housed in a well constructed, lighted, and heated enclosure which will allow ready access to the valve. The water supply pipe below the dry pipe valve contains water at all times, and must be properly protected from freezing.

An example of construction for a dry pipe valve enclosure inside a sprinklered building is shown in Figure 18-1Y. If the water control valve is located at the dry pipe valve, the dry valve enclosure should be of fire resistant construction and located outdoors or accessible from outdoors.

In cold storage rooms where temperatures are maintained at 32°F (0°C) or lower, special arrangements of piping and devices are needed to prevent accumulation of

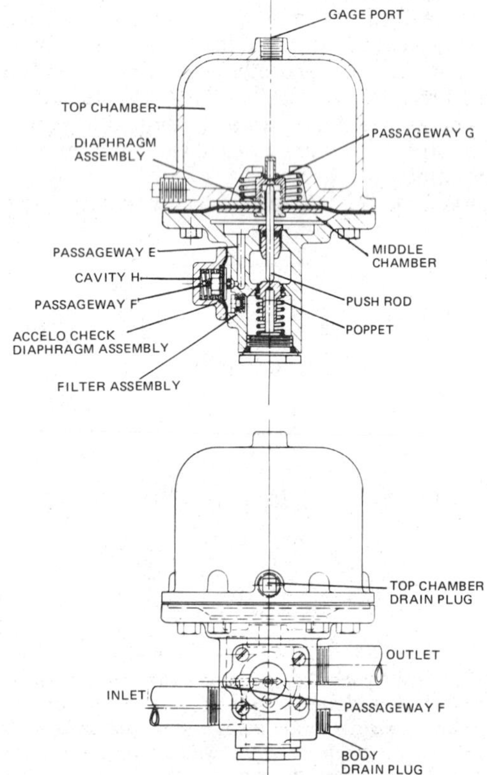

FIG. 18-1T. A dry pipe system accelerator with integral antiflooding device. System air pressure enters the top chamber from the inlet through passageways E and G. Once pressurized, the diaphragm assembly closes on the push rod restricting air flow from the top chamber. The opening of an automatic sprinkler reduces the air pressure in the sprinkler system and the middle chamber. The top chamber, being restricted from losing air, instead moves the diaphragm assembly and push rod. This movement opens the poppet which admits system air pressure to the outlet and dry pipe valve intermediate chamber. The buildup in pressure in the outlet backs up through passageway F to close the Accelocheck diaphragm assembly against passageway E, preventing any water or contamination from flowing upward to the middle and top chamber. Shown is the Reliable Automatic Sprinkler Co., Inc. Model B1 accelerator.

frost and ice inside the sprinkler piping and to permit ready inspection for these conditions. Replacing air with dry compressed nitrogen from cylinders will reduce the accumulation of moisture in the system, or propylene glycol or other suitable material may be substituted for the priming water. A small amount of mineral oil added to the surface of the priming water will prevent evaporation. Details are given in the NFPA Sprinkler Systems Standard.

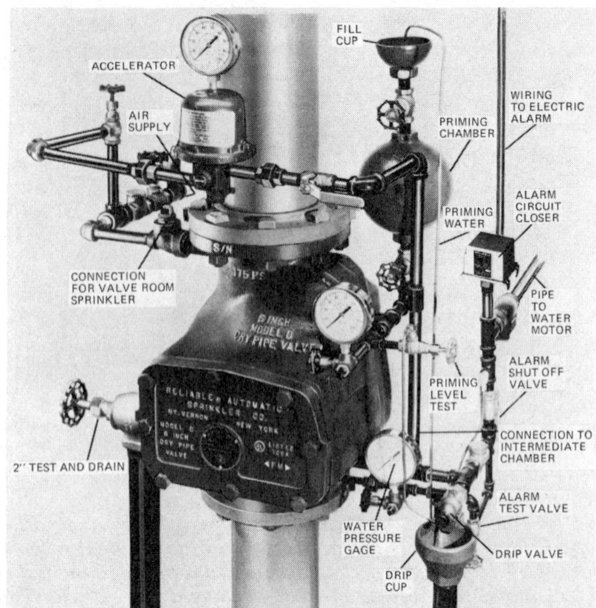

FIG. 18-1U. A representative dry pipe valve and trimmings. The arrangement shown includes some simplification in the use of copper tubing and fittings in place of standard steel pipe and fittings still commonly furnished by most manufacturers. The functions of the various components have not been changed, and the arrangement is recognized by testing laboratories. Shown is Reliable Automatic Sprinkler Co., Inc., Model D differential type dry pipe valve.

Preaction Sprinkler Systems

Preaction systems are designed primarily to protect properties where there is danger of serious water damage as a result of damaged automatic sprinklers or broken piping.

The principal difference between a preaction system and a standard dry pipe system is that in the preaction system, the water supply valve is actuated independently of the opening of sprinklers; that is, the water supply valve is opened by the operation of an automatic fire detection system and not by the fusing of a sprinkler. The valve can also be operated manually.

The preaction system has several advantages over a dry pipe system. The valve is opened sooner because the fire detectors have less thermal lag than sprinklers. The detection system also automatically rings an alarm. Fire and water damage is decreased because water is on the fire more quickly and the alarm is given when the valve is opened. Because the sprinkler piping is normally dry, preaction systems are nonfreezing and, therefore, applicable to dry pipe service.

The same heat responsive devices and release mechanisms used in preaction systems can also be used to operate water spray and foam extinguishing systems as well as to actuate alarm and supervisory systems (protective signaling systems).

The detection feature of a preaction system can also be added to a conventional dry pipe system in an arrangement that affords an acceptable means of supplying water through two dry pipe valves to a system of larger size than is permitted by the NFPA Sprinkler Systems Standard for

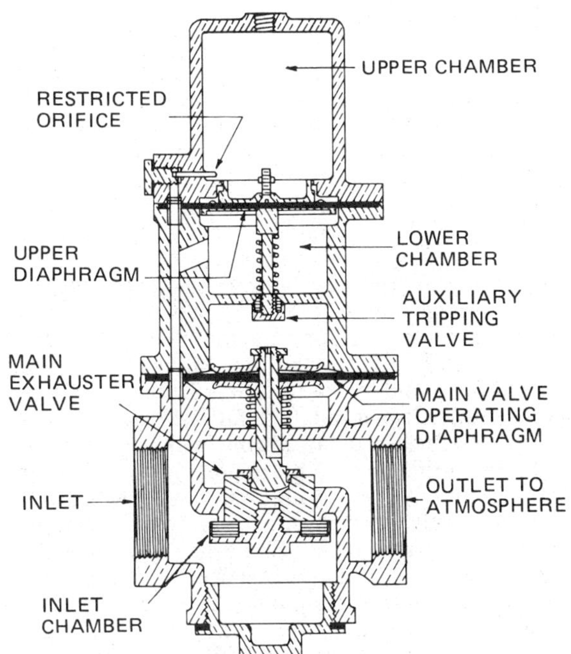

FIG. 18-1V. A dry pipe system exhauster. The inlet is connected to the sprinkler system side of the dry pipe valve. System air pressure enters the lower chamber by way of an open passage from the inlet chamber. Pressure can also build up in the upper chamber slowly through restricted orifice between the upper chamber and lower chamber, so that normally their pressures are equal. The opening of an automatic sprinkler starts a reduction in the air pressure in the lower chamber. The higher pressure remaining in the upper chamber moves the upper diaphragm in the direction to open the auxiliary tripping valve. This allows the pressure in the lower chamber and sprinkler system to enter the chamber above the main valve operating diaphragm where the pressure has been previoulsy kept at atmospheric pressure by a small passage to the outlet to atmosphere. The pressure below the main valve operating diaphragm is kept at atmospheric pressure through a piped connection to the intermediate chamber of the dry pipe valve. Because the area of the main valve operating diaphragm is much greater than that of the main exhauster valve, the latter opens and allows the system air pressure to escape rapidly down to the pressure at which the dry pipe valve trips. Water pressure through the connection to the intermediate chamber of the dry pipe valve then enters the chamber below the main valve operation diaghragm, balancing the pressure above it, so that the main exhauster valve is closed by the spring aided by the flow which has been passing through the valve. Shown is Central Automatic Sprinkler Company's Lewis Model A exhauster (no longer manufactured).

a single valve. The arrangements of devices used in combined dry pipe and preaction systems are described later in this chapter.

Supervision: Piping on early preaction systems contained air at atmospheric pressure and water was admitted to the system when the preaction valve was actuated by the fire detection system. Sprinklers or sections of piping could be removed without causing the system to operate. Subsequently supervision of the system was added by maintaining automatically a very low air pressure in the sprinkler piping. The rate of air supply is made low so that in case of air leakage, the supervisory air pressure will drop and cause a trouble signal without tripping the water control valve.

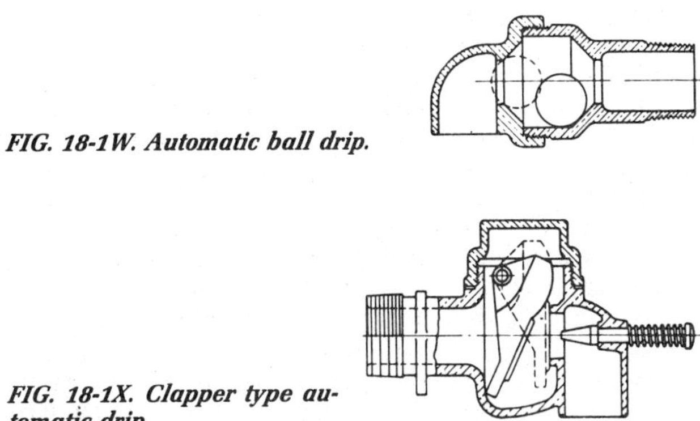

FIG. 18-1W. Automatic ball drip.

FIG. 18-1X. Clapper type automatic drip.

Devices and Equipment: The pipe schedules and the rules for sprinkler spacing, as given in the NFPA Sprinkler Systems Standard, are generally followed for preaction systems; however, not more than 1,000 automatic sprinklers can be controlled by one preaction valve.

Water control valves control the supply of water in the preaction system. Manufacturers use a wide variety of mechanical, pneumatic, hydraulic, and electrical devices for this purpose. In general, each manufacturer provides his particular complete combination of water control valve release, heat detection system, and supervisory equipment. A control valve for a preaction system is shown in Figure 18-1Z.

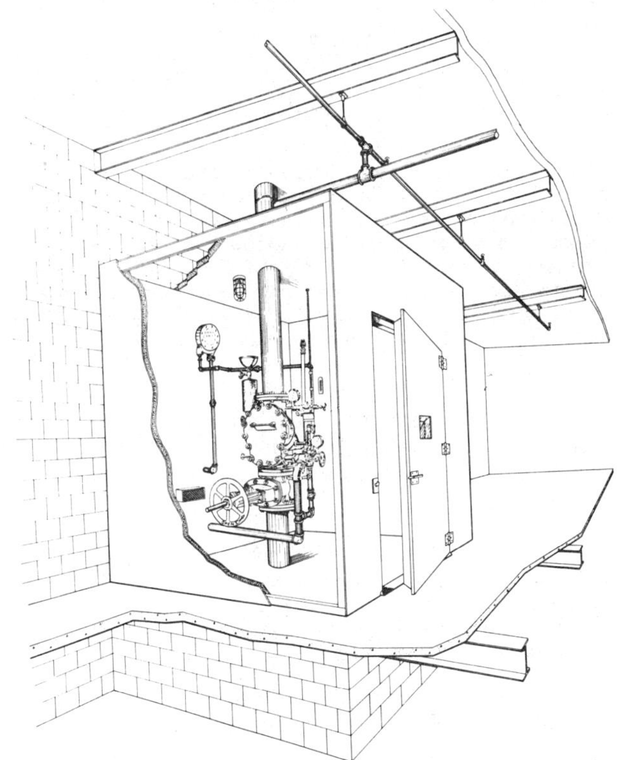

FIG. 18-1Y. A permissible type of indoor dry pipe valve enclosure. (Factory Mutual System)

Heat responsive devices are the most common means of actuating preaction valves, and the three prevalent methods of heat detection are: (1) devices actuated by a predetermined fixed temperature, (2) devices actuated by a predetermined rate of temperature increase (rate-of-rise), and (3) devices combining fixed temperature and rate-of-rise devices. The same devices also can be used to actuate deluge sprinkler systems.

Other means of actuating deluge and preaction valves are smoke detectors, combustible gas detecting systems, and automatic signals from process or other safety systems.

Alarms are standard accessory equipment on water control valves to provide an audible signal on the premises if the valve operates from any cause. Supervised preaction systems also give an audible alarm in case of loss of the supervisory means, or of an accident that would make the

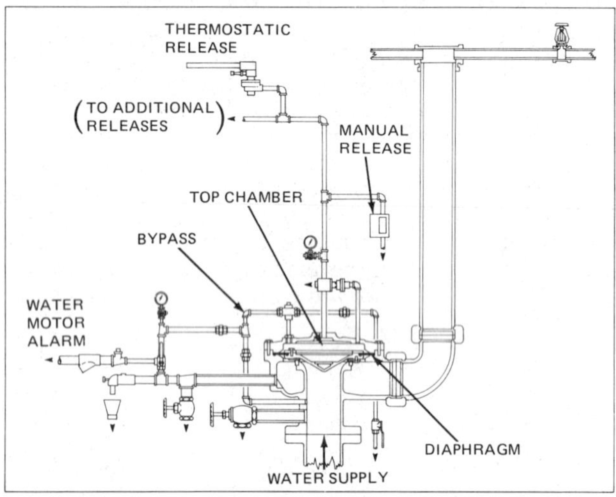

FIG. 18-1Z. A quick opending, differential diaphragm flood valve that can be used in deluge and preaction systems. The diaphragm is held closed by water pressure in the top chamber, which is admitted through a restricted orifice in the bypass from the water supply side of the flood valve. When either the thermostatic release or manual release operates the pressure in the top chamber is released, permitting the flood valve to open. Shown is The Viking Corporation Model E-1 valve.

detection system inoperative. Alarm systems giving signals at a central station office or which are connected to public fire alarm systems are advantageous and often required.

Tests: One detection device on each circuit must be accessible for test purposes and connected at a point that will assure a proper test of that circuit. Additional information is given in the NFPA Sprinkler Systems and Water Spray Fixed Systems Standards.

Preaction System with a Recycling Feature

A further refinement of the preaction principle is a recycling system for controlling sprinklers. It shuts off the water when the fire has been extinguished, reactivates itself if the fire rekindles, and continues cycling as long as fire persists.

Automatic sprinklers are used in the conventional manner. Supply water is held back by a flow control valve that is kept closed by water pressure. Operation of the flow control valve is controlled by an electrical panel which is activated by a system of heat detectors located in much the same way as sprinklers, but with a specific ratio of detectors to sprinklers, depending on the type and use of building and the degree of hazard.

Heat from a fire activates the detectors at 140°F (60°C) which, through closed circuitry, activates two solenoid valves. These valves, upon opening, exhaust water from the top chamber of the flow control valve causing that valve to open and water to flow into the system piping. Because sprinklers normally don't fuse until at least 160°F (71.1°C), there is a delay during which time water can reach the sprinklers in the fire area before the sprinklers fuse and begin discharging water on the fire. As the sprinklers bring the fire under control, the temperature decreases until at 140°F (60°C) the detectors again close the detector circuit. At this point a safety timer is activated, permitting the water to supply the system for a predetermined time. On completion of the timer cycle, the dual solenoid valves close, pressure builds up in the top chamber of the flow control valve and this valve closes. In case of a rekindle or of fire breaking out in another area, the detectors again turn on water when the temperature reaches 140°F (60°C) and continue to repeat the cycle as long as a temperature of 140°F (60°C) or higher persists. The system can be reset when the fire has been completely extinguished. Any damage to the closed detector circuit will automatically cause water to be supplied to the sprinklers which will then operate as a conventional wet pipe system. To insure uninterrupted detector circuit service, Type MI cable is used for the detector circuit. This cable consists of a copper wire surrounded by magnesium oxide insulating material and enclosed in a special copper sheath. A recycling preaction system is shown in Figure 18-1AA.

Combined Dry Pipe and Preaction Systems

The intended purpose of a combined dry pipe and preaction system is to provide an acceptable means of supplying water through two dry pipe valves connected in parallel to a sprinkler system of larger size than is permitted for a single dry pipe valve by the NFPA Sprinkler Systems Standard.

Although the NFPA Sprinkler Systems Standard does not restrict the use of combined systems to any particular classes of property, such systems were originally developed for protection of piers where long lines of supply piping could have been subject to freezing if a number of conventional dry pipe systems had been installed along the length of the pier. Due to the complications of combined dry pipe and preaction systems and the increased possibility of delayed water discharge, it is general practice to install them only in situations where it is difficult to protect a long supply main from freezing.

The main features of a combined system are as follows:

1. A dry pipe automatic sprinkler system usually with more than 600 sprinklers and supplied by a long feed main in an unheated area.

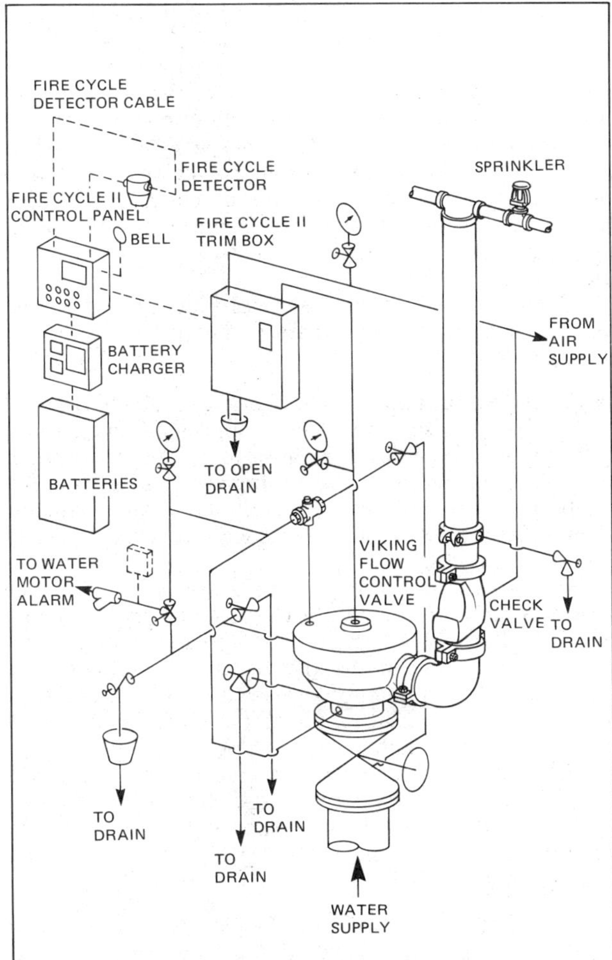

FIG. 18-1AA. The principal components of the Viking Corp. "Firecycle" system, a preaction system with a recycling feature for controlling water to sprinklers. Water is automatically shut off when heat is reduced below the detector operating temperature, and turns the water back on when the temperature is exceeded. A time delay reduces excessive cycling.

2. Two approved or listed dry pipe valves connected in parallel can be used to supply water to a single large sprinkler system. Two 6 in. dry pipe valves, interconnected with the tripping means for simultaneous operation, are required if a system has more than 600 sprinklers or more than 275 in one fire area. A combination system must have a quick opening device at the dry pipe valves.
3. A supplemental heat detection system of generally more sensitive characteristics than the automatic sprinklers themselves is installed in the same areas as the sprinklers. Operation of the heat detection system, as from fire, actuates tripping devices which open the dry pipe valves simultaneously without loss of air pressure in the system. The heat detection system is also used to give an automatic fire alarm.
4. Approved air exhaust valves installed at the end of the feed main are opened by the heat detection system to hasten the filling of the system with water, usually in advance of the opening of sprinklers.
5. Systems with more than 275 sprinklers in one fire area are divided into sections of 275 sprinklers or less by check valves at connections to the feed main. However, not more than 600 sprinklers can be supplied through a single check valve.
6. A means is provided for manual actuation of the system.

Advantages of a Combined System: Some advantages of a combined system include:

1. Elimination of the wrapping and heating required for long runs of exposed supply piping carrying water to dry pipe valves.
2. A substantial reduction in the number of dry pipe valves required for adequate protection for a given area, and a corresponding reduction in the number of dry pipe valve enclosures required.
3. Elimination of extensive air-line piping from the compressor to dry pipe valves.
4. Quick action of the rate-of-rise heat detection system enables water to enter the system piping by the time sprinklers operate.
5. Failure of the heat detection system does not prevent the system from operating properly as a conventional dry pipe system, while failure of the dry pipe system does not prevent the heat detection system from giving an automatic fire alarm.

Devices and Equipment: To assure the expected action of the system, the NFPA Sprinkler Systems Standard requires that after the action of the independent detection device, water must reach the farthest sprinklers within 1 min for each 400 ft (120 m) of common feed main, with the total time for the system not exceeding 3 min.

Combined dry pipe and preaction systems are not listed as single complete units by testing laboratories but are assembled from individually tested and approved components. Details of a typical installation supplying more than a total of 600 sprinklers or more than 275 in one fire area are shown in Figures 18-1BB, 18-1CC, and 18-1DD.

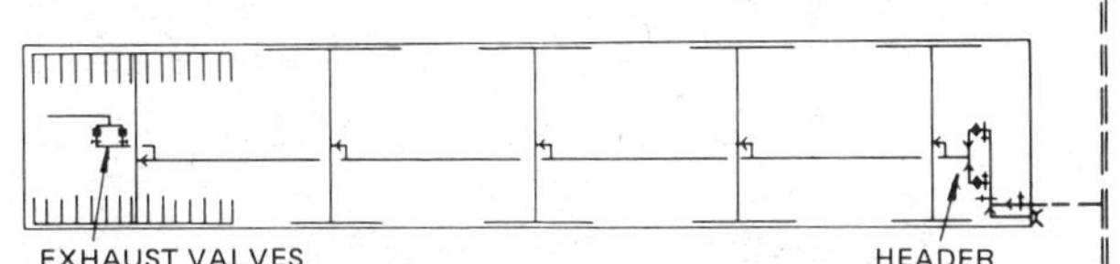

FIG. 18-1BB. Typical piping layout for combined dry pipe and preaction sprinkler system. See Figure 18-1CC for details of the exhaust valves and Figure 18-1DD for details of the header.

Figure 18-1EE shows an arrangement of devices which make up the basic combination of a single conventional dry pipe valve and an auxiliary heat responsive system for a sprinkler system of not more than 275 sprinklers.

Deluge Sprinkler Systems

The purpose of a deluge system is to wet down an entire fire area by admitting water to sprinklers that are open at all times. By using sensitive detectors operating on

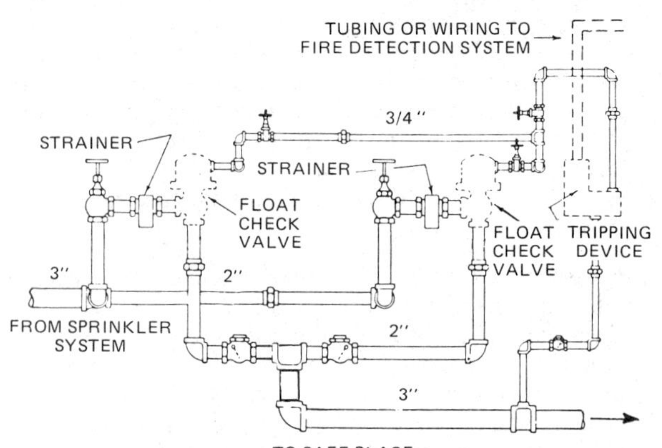

FIG. 18-1CC. Arrangement of air exhaust valves for combined dry pipe and preaction sprinkler system.

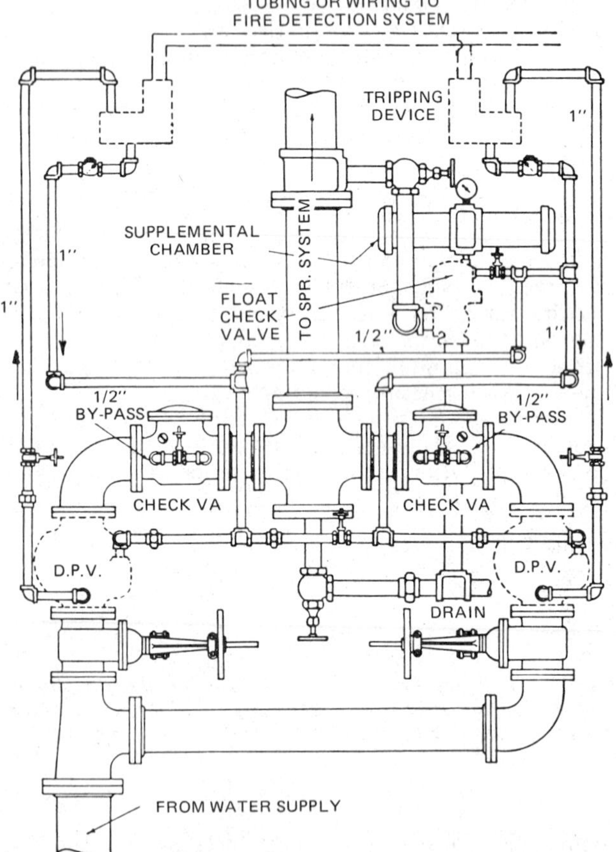

FIG. 18-1DD. Header for combined dry pipe and preaction sprinkler system. Standard trimmings for dry pipe valves are not shown.

the rate-of-rise or fixed temperature principle, or controls designed for individual hazards, it is possible to apply water to a fire more quickly and with wider distribution than with systems whose operation depends on opening of sprinklers only as the fire spreads.

Deluge systems are suitable for various extra hazard occupancies in which flammable liquids or other hazardous materials are handled or stored and where there is a possibility that fire may flash ahead of the operation of ordinary automatic sprinklers. They are also often used in aircraft hangars and assembly plants where ceilings are unusually high and where there is a likelihood that drafts, as from hangar doors, may deflect the direct rise of heat from an incipient fire so that ordinary sprinklers directly over the fire would not open promptly; however, others at some distance would open without effect on the fire. Deluge systems may also be used to automatically control the water supply to outside open sprinklers for protection against exposure fires.

Open sprinklers and closed sprinklers may be combined in a single system where deluge protection is not needed over the entire area.

Design and Installation: Where deluge systems are used to protect large areas, the water supply requirements may be heavy as compared with those for ordinary sprinkler systems. Also, the design of piping systems and the hydraulic problems involved, particularly where sprinklers are on different levels, as in an arched roof hangar, call for careful engineering.

Because all the sprinklers on a deluge system must be simultaneously supplied with water at effective pressure, the NFPA Sprinkler Systems Standard recommends that such systems be hydraulically designed. Systems with fewer than 20 sprinklers may use the extra hazard pipe schedule. Other special requirements given in the NFPA Sprinkler Systems Standard must also be followed.

Spacing of sprinklers closer than usual may be needed to give the desired density of discharge. The sprinkler piping and the water supply should be designed and coordinated to be appropriate for the specific density requirements.

Devices and Equipment: Valves controlling the water supply to deluge systems include a wide variety of mechanical, pneumatic, hydraulic, electrical, and explosive squib devices. In general, each manufacturer provides his particular combination of water control valve, releasing mechanism, heat detection system, and supervisory equipment.

Heat responsive devices should be located and spaced in accordance with their listing unless conditions call for closer spacing or special location. For unusual fire hazards special arrangements are often needed.

Alarms are required accessory equipment on control valves. Their purpose is to give an audible alarm on the premise if the valve operates for any reason. Alarm systems to give signals at a central station office are often advantageous.

Because of the use of heat responsive actuating devices, special testing facilities and procedures are necessary.

Circulating Closed Loop Systems

Sprinkler systems piping in some instances can be used to circulate water for heating and cooling purposes. The closed loop piping is used only to circulate the water; none is removed for manufacturing processes or other nonfire uses.

A prominent feature of a closed loop system is that water for sprinklers is not required to pass through any heating or cooling equipment when sprinklers operate. The system contains provisions for water to flow from the

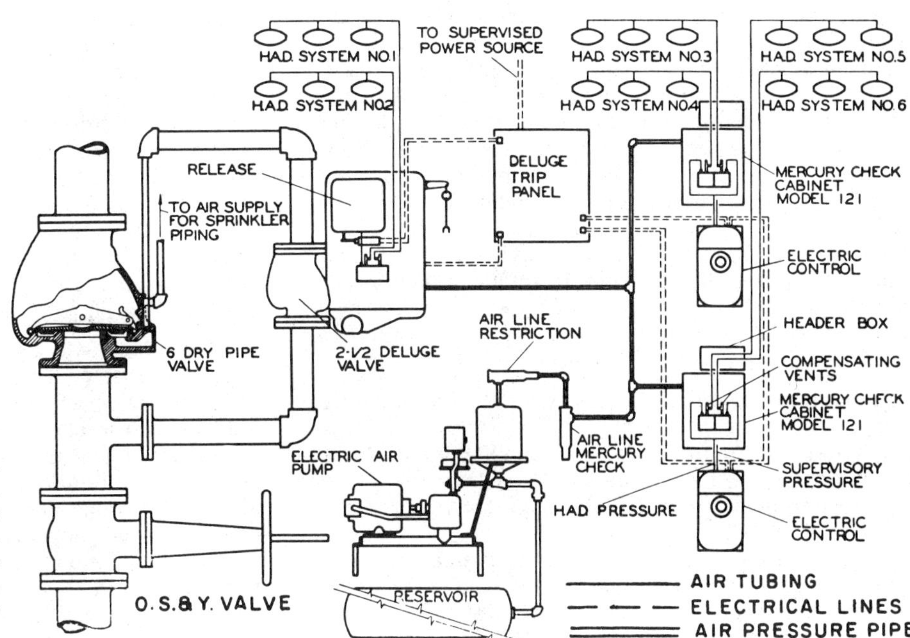

FIG. 18-1EE. Components of a combined dry pipe and preaction sprinkler system of less than 275 sprinklers. Manufacturers have various arrangements of proprietary equipment for operation of combined dry pipe and preaction systems.

sprinkler water supply to each sprinkler without compromising the sprinkler system design pressure or causing any loss of outflow of water from the system because of operation of heating or cooling equipment.

Another critical feature of a circulating system is that the temperature of the water must not have any affect at all on the operating temperature of sprinklers in the system. This protection is provided by tested or approved control devices that shut down ancillary heating equipment when the temperature of the water exceeds 120°F (49°C). Conversely, when cooling is involved, precautions must be taken to ensure that the temperature of the water does not fall below 40°F (4°C).

Sometimes additives are required for heating and cooling purposes. Care must be taken that the additives do not adversely affect the fire fighting properties of the water. Another hazard is that some additives may remove and suspend scale from older piping. This could lead to obstructions at some critical points in the systems.

Guidance for the installation of circulating closed loop systems will be found in the NFPA Sprinkler Systems Standard.

Systems Using Large Drop Sprinklers

Large drop sprinklers are special sprinklers that are designed to produce large drops to penetrate the strong updrafts generated by high challenge fires. They have a *k* factor betweens 11 and 15 (ordinary ½ in. orifice sprinklers have *k* factors ranging from 5.3 to 5.8, roughly half of that for large drop sprinklers). The use of these sprinklers is covered in the NFPA Sprinkler Systems Standard and they are described in Chapter 3 of this section of the HANDBOOK.

Special Types of Systems

There are many situations where the installation of sprinklers is advisable, especially for life safety, even though it is economically or otherwise impractical to meet all the requirements of the NFPA Sprinkler Systems Standard. These types of installations, commonly called nonstandard, involve features that depart from generally accepted practices. This does not, however, necessarily imply questionable reliability or capacity to handle the specific fire problems for which they are intended. Their use does, of course, require evaluation by qualified individuals to determine their suitability.

Nonstandard sprinkler installations may involve water supplies of limited capacity, reduced pipe sizes, partial protection, sprinklers with orifice sizes different from those generally used, and other features not typical of standard installations.

Small Capacity Pressure Tanks: A sprinkler system may have a single water supply from a pressurized tank that is of less capacity [2,000 gal (7570 L)] than is recognized by the NFPA Sprinkler Systems Standard for limited water supply systems, and may also depart from the standard by having reduced pipe sizes, small orifice sprinklers, or increased sprinkler spacing. These special systems may employ a water supply tank pressurized by air or by compressed inert gas, such as nitrogen or carbon dioxide, from cylinders. Manufacturers of fire protection equipment have supplied, and laboratories have tested, systems of the latter type which have the advantage of using the full water capacity of a tank, with the water being discharged at a preselected, nearly constant pressure.

Substandard Water Supplies: Automatic sprinklers having water supplies from public mains, domestic or industrial systems, or other sources not meeting NFPA Sprinkler Systems Standard requirements are sometimes installed to advantage. Such water supplies having pressure and capacity to effectively supply a few automatic sprinklers can provide valuable protection for light fire hazards in limited areas, provided the supply is continuously available. Public water supplies of limited capacity are likely not to be dependable due to varying supply-and-demand relations, small sizes of mains, and frequently long pipe lines. Occasionally, an automatic limited service fire pump can strengthen the supply pressure if sufficient volume is available.

Limited Water Supply Systems

Limited water supply systems are used where a public water supply or other conventional type of supply, such as a gravity tank or fire pump, is not available for sprinklers with sufficient volume or pressure to satisfy the water supply requirements in the NFPA Sprinkler Systems Standard.

A pressure tank of limited capacity is one source of supply in this type of system which, in other respects, is the same as a conventional system because standard sprinkler system piping and standard sprinklers are used. The minimum sizes of pressure tanks recognized by the NFPA Sprinkler Systems Standard for supplemental supply for limited supply systems contain 2,000 gal (7570 L) of water for light hazard occupancies, and 3,000 gal (11 355 L) for ordinary hazard occupancies. Approval of plans for all proposed limited supply systems, including the amount of water available for pressure tanks, should be obtained from the appropriate authority having jurisdiction.

Outside Sprinkler Systems

The use of a water curtain on the outside wall of a building probably antedates automatic sprinklers. In the early years of sprinkler protection, ordinary sprinklers with the caps removed were used at the peaks of combustible roofs and at the eaves of buildings, particularly wooden buildings. Special types of open sprinklers have since been designed to protect window openings in brick walls. Others have been designed to protect combustible cornices. These sprinklers are placed near the top of the window or under the cornice. The water is discharged against the glass and frame or cornice, thus providing the desired protection.

To be effective, outside sprinklers must wet the entire surface being protected. A water curtain not in contact with the window, the wall, or cornice is of little value because the water is broken up into drops through which radiated heat can pass. The water discharge patterns from open sprinklers are fan shaped or quarter-spherical, rather than hemispherical as with standard sprinklers.

Because of the large volume of water required to supply an open sprinkler system, this form of protection is ordinarily recommended only when sufficient water is available to supply both open and automatic sprinklers and any other needs for which there may be a demand at the same time. Where water supplies are limited, other methods of protection are preferable.

Outside sprinklers are usually needed at each floor level, except in the lower stories of multistory buildings. In the latter case, such a large volume of water is discharged from the open sprinklers on the upper stories that it is customary to assume that there will be ample water flowing down to protect the lower windows. Because window sills and other building structural details are commonly designed to deflect rain water away from the wall of the building, a larger quantity of water is required than might otherwise be necessary to wet the surface being protected. The NFPA Sprinkler Systems Standard's provisions on outside sprinkler installations are based on building construction of conventional type with recessed windows. That standard should be consulted for the details of installation.

Partial Installations: Installation of sprinklers throughout the premises is necessary for complete protection to life and property. However, in some cases partial sprinkler installations covering hazardous sections and other areas are specified in codes or standards for limited protection in the belief that they provide opportunity for safe exit from the building, help reduce fire spread, and improve access for manual fire control.

Just what portions of such buildings should be equipped depends on construction features as well as occupancy and fire hazards. For example, in multiple occupancy mercantile and apartment buildings, all portions occupied for stores or similar occupancies, as well as all basement areas, are frequently sprinklered to meet local codes. Many cities have passed special ordinances, some retroactive, calling for sprinklers in the basements of all mercantile buildings. These ordinances are predicated on the following: (1) the life hazard of fires originating in such areas, (2) the heavy concentration of storage frequently found, (3) the inaccessibility of many basement areas for manual fire fighting, and (4) the frequency of fires as shown by the fire records.

In apartment houses, tenements, dormitories, and similar properties of ordinary construction, partial sprinkler protection has been frequently called for to cover basements, kitchens, laundries, storerooms, halls, stairways, elevators, and other floor openings; however, residential "quick response" sprinklers are now being encouraged for use in the residential portions of these occupancies.

Attics or spaces between roofs and ceilings of top floors present a special problem because these portions may not be heated. A dry pipe system may be required if sprinklers are installed.

When installing partial sprinkler protection, reliance cannot be placed on such partial sprinklers to prevent the spread of fire when the fire originates in an unsprinklered area, except in unusual cases. NFPA fire records contain many case histories where such partial systems have been overtaxed by fires originating in such unsprinklered portions.

Self-contained Systems: So called package type systems, all supplied by pressure tanks, can in some instances meet NFPA Sprinkler Systems Standard requirements for a standard limited water supply sprinkler system. Other systems, because of the smaller capacity of pressure tanks supplying them and other features that depart from the NFPA Sprinkler Systems Standard requirements are limited to use only in situations where it is impractical to meet all provisions of the standard.

Special systems with pressure tank water capacities below the minimum requirements of the NFPA Sprinkler Systems Standard may be listed by testing laboratories and recognized by regulating agencies for the limited service for which they were designed. Some of the systems may employ a pressure source from separate compressed gas storage cylinders controlled by regulating valves for expelling water from the tank, while others have the gas supply approved and listed as a complete unit. In still other cases, the individual devices making up the system are listed.

Other Special Systems: Small orifice sprinklers with rates of discharge approximately one-half [3⁄8 in. (9.5 mm) orifice] and one-quarter [1⁄4 in. (6.4 mm) orifice] of that of the 1⁄2 in. (12.7 mm) sprinkler are listed for special service. They are used in small enclosures or for other special conditions for which a reduced density of discharge is effective.

The established discharge rate of approved large orifice [17⁄32 in. (13.5 mm)] sprinklers is 140 percent that of the 1⁄2 in. (12.7 mm) sprinkler. They are intended for use where high density of water discharge is needed. They require a special engineering study of spacing and pipe size.

Bibliography

References Cited

ANSI. 1971a. B16.4, "Cast Iron Screwed Fittings. (125 and 250 lb)." United States of America Standard Institute, NY.

ANSI. 1971b. B.16.3, "Malleable-Iron Screwed Fittings (150 and 300 lb)," United States of America Standards Institute, NY.

ANSI. 1975. B-36.10, *Wrought Steel and Wrought Iron Pipe.* United States of America Standards Institute, NY.

ASTM. 1976. A.120, *Black and Hot-Dipped Zinc-Coated (Galvanized) Welded and Seamless Steel Pipe for Ordinary Uses.* American Society for Testing and Materials, Philadelphia.

AWS. 1980. AWSD10.9, "Standard for Qualifications of Welding Procedures and Welders for Piping and Tubing." American Welding Society.

Thompson N. J. 1949. "Hazard of High Piled Combustible Stock." *NFPA Quarterly.* Vol 43, No 1. pp 38-46.

Thompson, N. J. 1964. *Fire Behavior and Sprinklers.* National Fire Protection Association, Quincy, MA. pp 104-110.

Webb, W. A. 1968. "Automatic Sprinklers in Exhibition Halls." *Fire Technology.* Vol 4, No 2. pp 115-125.

NFPA Codes, Standards, Recommended Practices and Manuals. (See the latest *NFPA Publications and Visual Aids Catalog* for availability of current editions of the following documents)

NFPA 13, *Standard for the Installation of Sprinkler Systems.*

NFPA 13A, *Recommended Practice for the Care and Maintenance of Sprinkler Systems.*

NFPA 13D, *Standard for the Installation of Sprinkler Systems in One- and Two-Family Dwellings and Mobile Homes.*

NFPA 14, *Standard for the Installation of Standpipe and Hose Systems.*

NFPA 15, *Standard for Water Spray Fixed Systems.*

NFPA 20, *Standard for the Installation of Centrifugal Fire Pumps.*

NFPA 22, *Standard for Water Tanks for Private Fire Protection.*

NFPA 24, *Standard for the Installation of Private Fire Service Mains and Their Appurtenances.*

NFPA 71, *Standard for the Installation, Maintenance, and Use of Central Station Signalling Systems.*

NFPA 72A, *Standard for the Installation, Maintenance, and Use of Local Protective Signaling Systems for Watchman, Fire Alarm, and Supervisory Service.*

NFPA 72B, *Standard for the Installation, Maintenance, and Use of Auxiliary Protective Signaling Systems for Fire Alarm Service.*

NFPA 72C, *Standard for the Installation, Maintenance, and Use of Remote Station Protective Signaling Systems.*

NFPA 72D, *Standard for the Installation, Maintenance, and Use of Proprietary Protective Signaling Systems for Guard, Fire Alarm, and Supervisory Service.*

NFPA 80, *Standard for Fire Doors and Windows.*

NFPA 204M, *Guide for Smoke and Heat Venting.*

NFPA 231, *Standard for Indoor General Storage.*

NFPA 231C, *Standard for Rack Storage of Materials.*

Additional Readings

Alvares, Norman J., "Fire Endurance of Soldered Copper Sprinkler Systems —Part I," *Fire Technology,* Vol. 13, No. 3, Aug. 1977, pp. 231-237.

Alvares, Norman J., "Fire Endurance of Soldered Copper Sprinkler Systems —Part II," *Fire Technology,* Vol. 13, No. 4, Nov. 1977, pp. 282-295.

Ault, Wayne, E., "The Use of Tight-Wall and Special Light-Weight Pipe in Automatic Sprinkler Systems," *Fire Journal,* Vol. 73, No. 6, Nov. 1979, pp. 62-64, 91.

"Automatic Sprinkler Performance Tables, 1970 Edition," *Fire Journal,* Vol. 64, No. 4, July 1970, pp. 35-39.

"Automatic Sprinkler System Performance and Reliability in United States Department of Energy Facilities," DOE-EP-0052, 1952-1980, U.S. Department of Energy, Washington, DC, June 1982.

Baldwin, R., and North, M. A., "The Number of Sprinklers Opening and Fire Growth," *Fire Technology,* Vol. 9, No. 4, Nov. 1973, pp. 245-253.

Bond, Horatio, "Sprinkler Protection for High-Rise Buildings," *Fire Journal,* Vol. 62, No. 6, Nov. 1968, p. 3.

Cote, A. E., "Fifty Editions of NFPA 13," *Fire Journal,* Vol. 77, No. 3, May 1983, pp. 36-38, 124-126.

Damon, Walter A., "Preventing Pitfalls in Fire Protection Systems," *Heating/Piping/Air Conditioning,* April 1985, pp. 61–64, 69–71.

Factory Mutual Approval Guide, Factory Mutual Engineering Corporation, Norwood, MA, published annually.

Fire Protection Equipment List, Underwriters Laboratories Inc., Northbrook, IL, published annually in January with one supplement in July.

Fleming, R., "Bracing for Earthquake," *Sprinkler Quarterly,* No. 49, Spring 1984, p. 43.

Foehl, J. M., "In Quest of an Economical Automatic Fire Suppression System for Single-Family Residences," *Fire Journal,* Vol. 68, No. 5, Sept. 1974, pp. 42-48.

———, "In Quest of an Economical Automatic Fire Suppression System for Multi-Family Residential Complexes," *Fire Journal,* Vol. 70, No. 15, Jan. 1976, pp. 48-57.

———, "A Life Safety Sprinkler System," *Fire Technology,* Vol. 9, No. 3, Aug. 1973, pp. 189-197.

Kravontka, Stanley J., "Fire Deluge Systems for School Theatrical Stages," *Fire Technology,* Vol. 13, No. 1, Feb. 1977, pp. 53-58.

Marryatt, H. W., *Fire-Automatic Sprinkler Performance in Australia and New Zealand, 1886-1968,* Australian Fire Protection Associations, Melborune, Aus, 1971.

———, "Suppression and Extinguishment of Fire-98 Years of Experience with Automatic Sprinkler Systems in Australia," *SFPE Bulletin,* Vol. 85, No. 1, Jan. 1985, pp. 19-23.

Nash, P., and Young, R. A., *Automatic Sprinklers for Fire Protection,* Victor Green, London, 1982.

O'Rourke, Gerald, "Polybutylene Piping for Automatic Sprinkler Systems," *Heating/Piping/Air Conditioning,* April 1985, pp. 54–58.

Powers, W. Robert, "Automatic Sprinkler Experience in High-Rise Buildings," *Fire Journal,* Vol. 66, No. 6, Nov. 1972, pp. 47-49.

Richardson, J. Kenneth, "An Assessment of the Performance of Automatic Sprinkler Systems" *Fire Technology,* Vol. 19, No. 4, Nov. 1983, pp. 275-279.

Stevens, Richard E., "For Architects and Builders: Misconceptions on Sprinklers and Life Safety," *Fire Journal*, Vol. 60, No. 4, July 1966, p. 28.

Troutman, J. E., "Fire Protection for High-Piled Combustible Stock in Warehouses," *NFPA Quarterly*, Vol. 57, No. 1, July 1963, pp. 15-24.

Watson, F., "Trip Times for Dry Pipe Valves in Grid Systems —A Theoretical Analysis," *Sprinkler Quarterly*, March 1982.

Yao, Cheng, "The ESFR Sprinkler System: A New Approach to High-Challenge Storage Protection, *Fire Journal*, Vol. 79, No. 2, March 1985, pp. 30–33, 70–72.

WATER SUPPLIES FOR SPRINKLER SYSTEMS

Revised by Robert M. Hodnett

It is vital that every automatic sprinkler system have at least one automatic water supply of adequate pressure, capacity, and reliability. An "automatic" supply is one that is not dependent on any manual operation, such as making connections, operating valves, or starting pumps, to supply water at the time of a fire. Both the rate of flow and the total volume that may be needed must be considered.

This chapter identifies the types of water supplies that are acceptable for sprinkler systems; the variables that must be considered, including the hazards of occupancies, in evaluating requirements for both sprinklers and supplementary hose line supplies; and the fundamentals of hydraulically designing sprinkler systems to match sprinkler piping sizes with the characteristics of available water supplies.

Other chapters in this HANDBOOK of interest in understanding the complexities of providing adequate water supplies for sprinkler systems are those chapters in Section 17 covering water supplies for fire protection. Of particular interest are Chapter 2, "Hydraulics" (the fundamentals of water flow through pipes and orifices); Chapter 5, "Water Storage Facilities;" and Chapter 6, "Fire Pumps." Chapter 7, "Water Spray Protection," in this section is also of value insofar as there is no sharp line between sprinkler systems and water spray systems, and many of the water supply requirements pertaining to sprinkler systems apply equally as well to spray systems.

TYPES OF SUPPLIES

Sprinkler systems may be supplied with water from one source or a combination of sources, such as street mains, gravity tanks, reservoirs, fire pumps, pressure tanks, rivers, lakes, wells, etc.

In theory, a single water supply would seem to be all that is necessary for satisfactory protection. However, a single supply may at times be temporarily out of service; it may be disabled at the time of a fire or before a fire is completely extinguished; or the pressure or the capacity may be below normal during an emergency. Therefore, a secondary supply may be necessary depending on the strength and reliability of the primary supply; the value and importance of the property; the area, height, and construction of the building; the occupancy; and the outside exposures. Occasionally, three supplies are needed, especially where neither the primary nor a single secondary supply is judged wholly satisfactory or reliable.

Connections to Public Water Works Systems

A connection from a reliable public water works system of adequate capacity and pressure is the preferred single or primary supply for automatic sprinkler systems. In determining its adequacy, consideration has to be given not only to the normal capacity and pressure of the system, but also to the probable minimum pressures and flows available at unfavorable times such as during summer months, during heavy demand on the system, or during impairment caused by flood or by winter conditions.

The size and arrangement of street mains and feeders from public water supplies are also important. Connections from large mains fed two ways or from two mains on a gridiron system may provide an excellent supply. Street mains less than 6 in. in diameter are usually inadequate and unreliable. Feeds from dead-end mains are also undesirable. Water meters, if required by the water supply authority, should be of types approved for fire service. Flow and pressure tests under varying conditions of demand are generally necessary to determine the amount of public water available for fire protection.

Cross-Connections Between Public and Private Supplies

Where a secondary supply is needed to supplement the public water supply, public and private supplies can be connected to feed into a single fire protection system. These systems are commonly referred to as being cross-connected.

In some localities, cross-connections may be prohibited by health authorities. Where they are not prohibited,

---

Mr. Hodnett is a consultant in fire protection and piping systems based in Providence, RI. He was NFPA staff liaison to the NFPA Committee on Automatic Sprinkler Systems prior to his retirement in 1984.

regulations and sound practices must be complied with in order to avoid the possibility of public health being endangered by water of questionable potability entering the public system.

In general, cross-connections are permitted if carefully supervised precautions, such as a special double check valve or other accepted devices for preventing backflow, are provided. In cases where one sprinkler supply is from public mains, health authorities usually permit, as a secondary source, either well-constructed and well-maintained covered steel tanks or concrete reservoirs that are filled with public water only.

Gravity Tanks

Gravity tanks of adequate capacity and elevation make a good primary supply and may be acceptable as a single supply. Details of the construction, heating, and maintenance of gravity tanks are given in NFPA 22, *Standard for Water Tanks for Private Fire Protection*, referred to in this chapter as the NFPA Water Tank Standard. In determining tank size and elevation, consideration should also be given to the number of sprinklers expected to operate, duration of operation, the arrangement of underground supply piping, and the provision of hose standpipes, hydrants, and fire department connections.

Suction Tanks

With the advent of hydraulically designed sprinkler systems, the use of gravity tanks for fire protection has declined. The use of fire pumps combined with suction tanks has increased. NFPA 22, *Standard for Water Tanks for Private Fire Protection*, should be consulted for details.

Fire Pumps

A fire pump having both a reliable source of power and a reliable suction water supply is a desirable piece of equipment. Fire pumps are used to a great extent because of the hydraulic advantages of having a water supply available at high pressure. With ample water a fire pump is capable of maintaining a high pressure over a long period of time and may be a necessary part of some installations requiring greater water pressure than would otherwise be available. For details of power sources, pump construction, installation, and methods of control and operation, NFPA 20, *Standard for Centrifugal Fire Pumps*, should be consulted.

Manually controlled pumps may be used if the primary water supply will last long enough to allow dependable starting of the fire pump and if there is an automatic waterflow signal to indicate the need for fire pump operation.

Automatic control of fire pumps is usually needed where a high water demand may occur immediately, as with a deluge system, or where a competent pump operator is not continuously present. Automatic fire pumps must have their suction under a positive head to avoid the delays and uncertainties of priming.

Most fire pumps are powered by electric motors or diesel engines. Where a reliable source of power is available at all times, an electrically driven installation might be the most desirable. Where the power supply might be questionable, a diesel driven fire pump would be used. In some critical installations, such as hospitals, a diesel driven emergency power generator might be used to supply power to the electric motor. The use of a diesel driven fire pump, in a critical installation, would eliminate the need for a good portion of the output of the emergency generator.

The automatic control of electrically driven centrifugal pumps must be arranged to prevent frequent repeated starting of the motor, either by initiating continuous running until stopped manually, or by a timing device that will stop the motor automatically only after a predetermined period of operation.

Pressure Tanks

Pressure tanks have several possible uses in automatic sprinkler protection. An important limitation is the small volume of water that can be stored in such tanks. Where a small pressure tank is accepted as the water supply, the system is classed as a Limited Supply System.

In situations where an adequate volume of water can be supplied by a public or private source but where the pressure is not sufficient to serve a sprinkler system directly, the pressure tank gives a good starting pressure for the first sprinklers that operate. The flow from it may be used while the fire pumps start automatically to increase the supply pressure.

In tall buildings where the public water pressure is too low for effective water distribution from the highest sprinklers, pressure tanks may be used to supply such sprinklers during the time required for a public fire department to begin supplying water through fire department connections.

Each proposed use of pressure tanks calls for special consideration and analysis of water capacity, location, and arrangement of the connection to the sprinkler system. Each installation is usually required to have specific approval. Details on the construction, installation, and maintenance of pressure tanks are given in the NFPA Water Tank Standard.

Fire Department Connections

Under fire conditions that result in a considerable number of sprinklers operating, public water or tank supplies may not provide water at sufficient pressure for effective sprinkler discharge and distribution. Also, the pressure in many public water supplies to sprinkler systems may be materially reduced by hose streams from hydrants. In such cases a connection through which the public fire department can pump water into the sprinkler system provides an important auxiliary supply. Fire department connections are therefore a standard part of sprinkler systems.

Fire department connections must be readily accessible, and properly marked. Each connection should be fitted with a check valve, but not with a gate valve, so that the connection will not be shut off inadvertently. There should be a proper drain, and a drip device between the check valve and the outside hose coupling. Figures 18-2A and 18-2B show the main features of a fire department connection. Other details of installation and pipe size are given in NFPA 13, *Standard for the Installation of Sprinkler Systems*, hereinafter referred to as the NFPA Sprinkler Systems Standard.

Where a sprinkler system has a single riser, the fire department connection should be attached to the system side of the controlling gate valve for a wet pipe system. For

FIG. 18-2A. Fire fighters attaching hose lines to a fire department (siamese) connection supplying a sprinkler system. The inset shows typical siamese connections for sprinkler systems and standpipes. A check valve allows the use of a single hose line.

a dry pipe system, the connection should be between the dry pipe valve and the gate valve. This makes it possible to pump water into the system even if the gate valve is closed. If there are two or more sprinkler system risers, each with its own separate connection to a public main, each system must have its own fire department connection. If more than one riser is connected to a yard system, the fire department connection should feed into the yard system on the supply side of all riser shutoff valves, and there must be a check valve in all other water supply connections into the yard system to prevent backflow and loss of water supplied through the fire department connection. If one riser is shut off, the fire department connection can still supply all other risers. If there are two or more sprinkler system risers fed by one sprinkler connection, one fire department connection can supply both risers.

In an emergency, a fire department can pump water from public hydrants or other sources of water into a sprinkler system through a hose connected to a yard hydrant or other hose connection using a double female hose coupling, if other supply connections have a check valve or a gate valve that can be closed.

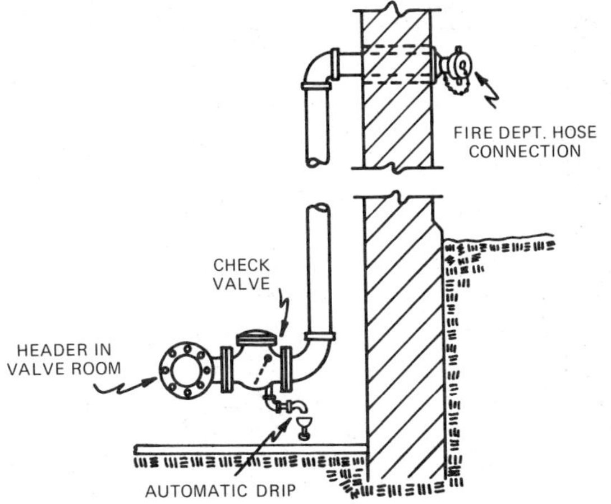

FIG. 18-2B. Typical fire department connection.

INFLUENCE OF VARIOUS FACTORS ON WATER SUPPLY NEED

Determining the water supply requirement for the majority of sprinkler systems is not always easy because of the many variables. If a water source is available that can supply all the sprinklers, there is no problem, but such a water supply is seldom practical except in the case of small systems. The water supply requirement for any sprinkler system is directly related to the number of sprinklers expected to operate, but this in turn depends on so many other variables and uncertain factors that no exact mathematical solution is possible.

Records that were kept by NFPA (prior to 1970) show that, in 93 percent of all fires in sprinklered buildings, 20 or fewer sprinklers opened. Experience shows that, with an adequate water supply, the percentage of unsatisfactory sprinkler performance is extremely small. Thus, water supply is a significant concern, particularly with large sprinkler systems and with systems protecting greater than ordinary hazards.

Establishing the water supply requirement for any particular sprinkler system requires good engineering judgment based on all the factors relating to sprinkler control. Where the cooling effect from the water discharged by sprinklers is greater than the heat liberated by the fire, the sprinklers can gain control. When the reverse situation occurs, as from an overtaxed water supply, the sprinklers cannot control the fire and the sprinkler system may fail. Where all conditions are favorable, the control of fire should be accomplished by the operation of only a small number of sprinklers. Because conditions vary with different classes of occupancy, areas, and types of buildings, the number of sprinklers expected to operate in order to control a fire may possibly range up to the total number in the area, and the water supply should be provided accordingly.

The primary factors affecting the number of sprinklers which might open in a fire, and therefore to be considered in a determination of the water supply requirement, include the following:

Hazard of Occupancy (Including Flash Fire Hazard and Potential Rate of Heat Liberation): This is the most important factor and one requiring experienced judgment to evaluate. Where the flash fire hazard is present, it is usually necessary to provide water sufficient for the operation of all the sprinklers in any individual fire area.

Initial Water Pressure: At a pressure of 15 psi (103 kPa), a standard sprinkler will discharge 22 gpm (83 L/min), or an average of approximately 0.17 gal per sq ft per min (L/min)/m^{2}on an area of 130 sq ft (12 m^2). At 30 psi (207 kPa), the discharge is 33 gpm (125 L/min); at 50 psi (345 kPa), 41 gpm (155 L/min); and at higher pressures the discharge is correspondingly greater. With a greater discharge, there is a better chance of fire control from a small number of sprinklers, and less need for large volumes of water to supply a large number of sprinklers.

Obstructions to Distribution of Water from Sprinklers: With obstructions, such as high piled stocks, bale tiering, pallets, racks, and shelving, there is less likelihood that fire will be controlled in its initial stages and a greater chance of opening a large number of sprinklers needing large water supplies.

High Ceilings and Draft Conditions: With ceilings of unusual height, there is greater chance that drafts will carry heat away from the sprinklers immediately over a fire, resulting not only in delay in the application of water, but also in the opening of sprinklers remote from the place of origin of the fire. More water is usually needed under such conditions. The same situation exists wherever there are drafts, such as in areas open on the sides to the weather where winds can divert heat from sprinklers over the fire.

Unprotected Vertical Openings: Sprinkler systems in multistory buildings are usually designed on the assumption that fire will be controlled on the floor of origin. Where there are unprotected openings up through which heat and fire may spread, it may be expected that more sprinklers will open, particularly in the case of a fire originating near the vertical opening. In the case of high combustibility, the interconnected floors may need to be considered as one fire area. This means more water and larger pipe sizes in risers and supply.

Wet Pipe vs Dry Pipe System: Because of the delay due to exhausting air from dry pipe systems, more sprinklers open on dry pipe systems than on wet pipe systems. This may call for greater water supplies.

Size of Undivided Areas: A large undivided area has a greater number of sprinklers, a possibility of a greater maximum number of sprinklers operating, and a consequently greater water demand than with a small area.

Configuration and Type of Ceiling Construction: These influence water demand, and include such factors as curtain boards or beams affording curtain board effects to retard fire spread. Conversely, there is the possibility that fire may spread in a concealed space above a combustible ceiling out of reach of sprinklers, and subsequently burn through.

Extent of Coverage and Exposures: Any fire in an unsprinklered space extending to an area with automatic sprinklers places an abnormal demand on the sprinkler system and requires increased water supplies for effective functioning of the system.

The preceding factors must be considered individually and collectively, and it is not feasible to derive any general formula or simple method of arriving at water supply requirements. There are, however, certain general statements on this subject that may be made. One is that any situation may be effectively protected with much less water where the water is applied automatically rather than manually. Another is that it is good practice to provide more water, at higher pressure, than will probably be needed to extinguish any fire. Hose streams may be used to supplement sprinklers, even when not necessary, and an ample supply of water provides a margin of safety.

With a very large fire area of low to moderate hazard, it is not reasonable to expect to supply all sprinklers simultaneously. Actually, the pipe sizes are not large enough to do so. An exception would be where very high supply pressures can produce a high discharge rate from sprinklers near the source of supply (as well as effective discharge from the most remote sprinkler).

WATER SUPPLY REQUIREMENTS FOR PIPE SCHEDULE SPRINKLER SYSTEMS

Notwithstanding the general problems involved in arriving at water supply requirements, the fire hazard represented by different building occupancies has made it possible to establish guides to water supply requirements for sprinkler systems. The occupancy factor is the primary consideration with latitude allowed for the contributing factors.

The established guide tables contained in NFPA Sprinkler Systems Standard divide hazards of occupancy, for the purpose of determining water supplies, into several groups with specified minimum water supplies for each group. (See Table 18-2A.)

The total water supply required is determined from Table 18-2A and the water for hose streams is added if required. In many installations, no water is allowed for hose streams. If stored water is used, the total water flow required would have to be multiplied by the duration of flow to determine if the capacity of the stored supply was adequate, a factor that would not be pertinent for a reliable municipal water source.

The guide should be used only with experienced judgment, but it can serve for all cases qualifying in the light hazard and ordinary hazard (Groups 1 and 2) occupancy classifications which constitute the larger percentage of sprinkler installations. The other occupancy classifications usually involve more complex factors and therefore require special consideration.

Where fire pumps contribute to the water supply, standard sizes of pumps used with adequate rates of discharge are required. A suction supply for a pump should preferably be large enough for continuous operation. Where pressure tanks furnish the water supply, the tank installation should comply with the appropriate provisions of the NFPA Water Tank Standard.

Where a combination of different water supplies is provided in the interest of reliability, it is good practice to have the rate of supply from each source at least equal to the minimum requirement for the system.

Light Hazard Occupancies

Examples of light hazard occupancies are apartment buildings, dormitories, office buildings, seating areas of restaurants, and hospitals. In these occupancies, the potential rate of heat liberation is low, areas are usually subdivided, and a small number of sprinklers should normally control any fire. Under these conditions, 500 gpm (1893 L/min) should generally be sufficient, with an upward range to 750 gpm (2839 L/min) where conditions are less favorable.

Ordinary Hazard Occupancies

Group 1: This ordinary hazard classification includes occupancies where the combustibility of contents is generally low, such as in garages, bakeries, laundries, and canneries, but is greater than for the light hazard classification. In this group the water supply requirement may be as low as 700 gpm (2650 L/min) where small areas, noncombustible construction, and very limited hazards

TABLE 18-2A. Guide to Water Supply Requirements for Pipe Schedule Sprinkler Systems

Occupancy Classification	Residual Pressure Required (See Note 1)	Acceptable Flow at Base of Riser (See Note 2)	Duration in Minutes (See Note 4)
Light Hazard	15 psi	500–750 gpm (See Note 3)	30–60
Ordinary Hazard (Group 1)	15 psi or higher	700–1000 gpm	60–90
Ordinary Hazard (Group 2)	15 psi or higher	850–1500 gpm	60–90
Ordinary Hazard (Group 3)	Pressure and flow requirements for sprinklers and hose streams to be determined by authority having jurisdiction.		60–120
Warehouses	Pressure and flow requirements for sprinklers and hose streams to be determined by authority having jurisdiction. Also see Chapter 7 of NFPA 13, NFPA 231, and NFPA 231 C.		
High-Rise Buildings	Pressure and flow requirements for sprinklers and hose streams to be determined by authority having jurisdiction. Also see Chapter 8 of NFPA 13.		
Extra Hazard	Pressure and flow requirements for sprinklers and hose streams to be determined by authority having jurisdiction.		

For SI Units: 1 psi = 6.89 kPa; 1 gpm = 3.785 L/min.

NOTES:

1. The pressure required at the base of the sprinkler riser(s) is defined as the residual pressure required at the elevation of the highest sprinkler plus the pressure required to reach this elevation.
2. The lower figure is the minimum flow including hose streams ordinarily acceptable for pipe schedule sprinkler systems. The higher flow should normally suffice for all cases under each group.
3. The requirement may be reduced to 250 gpm if building area is limited by size or compartmentation or if building (including roof) is noncombustible construction.
4. The lower duration figure is ordinarily acceptable where remote station water-flow alarm service or equivalent is provided. The higher duration figure should normally suffice for all cases under each group.

are encountered; it can range up to 1,000 gpm (3785 L/min) as these conditions become more adverse.

Group 2: This ordinary hazard classification includes occupancies such as clothing factories, mercantiles, pharmaceutical manufacturing, and shoe factories. With this group the features of combustibility of contents, ceiling heights, and obstruction are generally unfavorable, separately or jointly, and the water supply requirements may range as high as 1,500 gpm (5678 L/min); however, an 850 gpm (3217 L/min) minimum is retained for this group and would be applicable, of course, only under very favorable conditions.

Water supply requirements for light hazard and ordinary hazard (Groups 1 and 2) occupancies, as in all cases, call for a careful consideration of all factors concerned; but the figures given in Table 18-2A are of value in placing lower and upper limits for the classes concerned. While it is never advisable to provide less than the lower limit indicated, the upper limit will usually be sufficient for all situations within the group classification.

Group 3: This classification consists of occupancies where standard sprinkler spacing and pipe schedules are considered satisfactory, but where more than ordinary water supplies are advisable. This group includes certain woodworking and other occupancies such as flour and feed mills, paper mills, piers and wharves, and tire storage.

Extra Hazard Occupancies

Extra hazard occupancies consist of properties where flash fires opening all the sprinklers in a fire area are probable, calling for close sprinkler spacing and larger pipe sizes. They are divided into 2 groups. Extra hazard occupancies (Group 1) include occupancies which may produce severe fires and where little or no flammable or combustible liquids are present. These include such occupancies as die casting, plywood and particle board manufacturing, printing (using inks with below 100°F [37.8°C] flash points), rubber manufacturing operations, saw mills, textile operations, and plastic foam manufacturing.

Extra hazard occupancies (Group 2) include occupancies which may produce severe fires and have moderate to substantial amounts of flammable or combustible liquids present, or where shielding of combustibles is extensive. These include such occupancy hazards as asphalt saturating, flammable liquids spraying, flow coating, open oil quenching, solvent cleaning, and varnish and paint dipping.

In any treatment of hazards by general groups of occupancy, it must be noted that individual properties differ markedly and that buildings of the same nominal occupancy classification may show widely different individual hazards which should be considered in any determination of water supply.

WATER SUPPLY REQUIREMENTS FOR HOSE STREAM PROTECTION

The values given in Table 18-2A include hose stream requirements. In considering water requirements for hose streams, it should be realized that, if sprinklers perform effectively, little hose stream assistance is required. Although this is generally the case, a realistic viewpoint

TABLE 18-2B. Density, Area of Sprinkler Operation, and Water Supply Requirements for Hydraulically Designed Sprinkler Systems

Minimum Water Supplies

Hazard Classification	Sprinklers GPM†	Inside Hose GPM†	Combined Inside & Outside Hose—GPM†	Duration in Minutes
Light	See Density	50 or 100	100	30
Ord.—Gp. 1	Curves	50 or 100	250	*60–90
Ord.—Gp. 2	Below	50 or 100	250	*60–90
Ord.—Gp. 3		50 or 100	500	*60–120
Ex. Haz.—Gp. 1		50 or 100	500	*90–120
Ex. Haz.—Gp. 2		50 or 100	1,000	120

* The lower duration figure is ordinarily acceptable where remote station water flow alarm service or equivalent is provided.
† 1 GPM = 3.78 L/min.

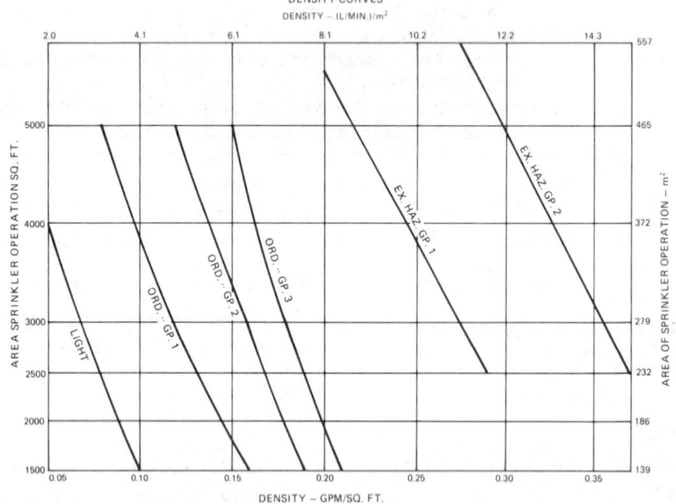

must be taken of possible contingencies and the amount of water that might be needed for hose stream protection under adverse conditions.

In evaluating hose stream requirements, several possibilities should be considered, such as the amount of water necessary for final extinguishment, clean-up operations, or in the event that sprinklers are retarding fire spread but are not fully effective in gaining control and extinguishment.

WATER SUPPLY REQUIREMENTS FOR HYDRAULICALLY DESIGNED SPRINKLER SYSTEMS

Planning new water supplies or evaluating existing supplies for sprinkler systems requires information regarding the hydraulic behavior of sprinkler piping systems.

Hydraulic Calculations

A hydraulically designed sprinkler system is one in which pipe sizes are selected on a pressure-loss basis to provide a prescribed density (gallons per minute per square foot [(L/min)/m^2]), distributed with a reasonable degree of uniformity over a specified area. This permits the selection of pipe sizes in accordance with the characteristics of the water supply available. The stipulated design density and area of application will vary with occupancy hazard. Table 18-2B is used to determine density, area of sprinkler operation, and water supply requirements for hydraulically designed sprinkler systems. Systems must be calculated to satisfy a single point on the appropriate design curve, and interior piping must be based on this design point. It is not necessary to meet all points on the selected curve. In some cases where a large water supply is available it might be advantageous, because of the elimination of a fire pump, to design the system using the top portion of the curve. In other cases it might be better to design with smaller areas of application and higher densities, because these tend to limit fire size, even though a fire pump is required. Total water supply available to the system at the base of the riser and at the residual pressure required by the design must not be less than shown in Table 18-2B; this total water supply need not be calculated through the overhead piping.

The same hazard occupancy classifications apply to hydraulically designed sprinkler systems as apply to pipe schedule sprinkler systems as mentioned in this chapter. The recommended water supply figures are, however, somewhat lower due to the greater efficiency of a calculated system.

The water allowances for inside hose and for outside hydrants may be combined and added to the system requirement at the system connection to the underground main. The total water requirement must be calculated through the underground main to the point of supply.

With deluge systems and water spray systems having open orifices, calculations are essential. Automatic sprinkler systems protecting high piled storage situations require a specific water density for fire control. Hydraulically calculated systems can, however, be used for all types of occupancies.

Flow Calculation Methods

Methods of making flow calculations for sprinkler systems are given in the following: (1) the NFPA Sprinkler Systems Standard, (2) NFPA 15, *Standard for Water Spray Fixed Systems*, (3) *Sprinkler Hydraulics* (Wass 1983), (4) *Hydraulic Design and Estimation of Fire Sprinkler Systems* (Crowley 1982), (5) *Automatic Sprinkler Hydraulic Data* (Wood 1961), (6) *Hydraulics of Fire Protection Systems* (FMEC 1977), and (7) *Water Flow Characteristics of Sprinkler Systems* (Nickerson 1954).

The design area for flow calculations is hydraulically the most demanding area for the sprinkler system. The area should have a dimension parallel to the branch line of 1.2 times the square root of the area of anticipated sprinkler operation (except gridded systems). The 1.2 times the square root of the area is added to ensure that the long side of the rectangle would be along the branch line. In gridded systems, the hydraulically most demanding area must be verified by using at least two additional sets of calculations.

Each sprinkler in the design area must discharge at a flow rate at least equal to the stipulated minimum water application rate (density). Begin calculations at the sprinkler hydraulically farthest from the supply connection. With common system configurations, this will be the end sprinkler on the end branch line. The minimum operating pressure for any sprinkler must not be less than 7 psi (48 kPa).

The Most Remote Sprinkler

Assuming a minimum pressure of 10 psi (69 kPa) at the most remote sprinkler and a discharge coefficient of 0.75 for a standard ½ in. (12.7 mm) orifice sprinkler (the coefficient varies; 0.78 is used elsewhere in this HANDBOOK), there is a discharge of 17.7 gpm (67 L/min) calculated from the formula:

$$Q = 29.8\ cd^2 \sqrt{P}$$

used in calculating flows through orifices and short tubes. The value for 29.83 cd^2 in this instance is 5.6, a figure commonly used as the sprinkler discharge constant K in the simplified formula:

$$Q = K \sqrt{P}.$$

Velocity pressure is not a factor at the most remote sprinkler, but it is considered at all the other sprinklers in the example that follows. Some organizations ignore velocity pressure in their calculations. The error introduced is on the safe side. NFPA 15, *Water Spray Fixed Systems*, recommends considering velocity only when it is more than 5 percent of the total pressure.

Assuming sprinklers 10 ft (3 m) apart on branch lines, with the end section of pipe 1 in. nominal diameter, the friction loss at 17.7 gpm (67 L/min) flow, with a Hazen and Williams formula coefficient of 120 (value for black steel pipe), will be 1.0 psi (7 kPa).

Second Sprinkler from the End

The total pressure at the second sprinkler will be 10.0 + 1.0 psi (69 +7 kPa) = 11.0 psi (76 kPa). Of this, velocity pressure based on a flow of 17.7 gpm (64 L/min) will be 0.3 psi (2 kPa). The normal pressure (pressure acting perpendicular to the pipe wall) acting on the second sprinkler is the total pressure of 11.0 psi (76 kPa) less the velocity pressure of 0.3 psi (2 kPa), which equals 10.7 psi (74 kPa). On all sprinklers except the end sprinkler, only normal pressure is considered as acting on the sprinklers. The discharge from the second sprinkler, at a pressure of 10.7 psi (74 kPa), will be 18.3 gpm (69 L/min).

The pipe between the second and third sprinkler, also 1 in. diameter, 10 ft (3 m) long, and with a flow of 17.7 gpm (67 L/min) + 18.3 gpm (69 L/min) = 36.0 gpm (136 L/min), will have a friction loss of 3.8 psi (26 kPa) and a velocity pressure of 1.2 psi (8 kPa). Total pressure at the third sprinkler is 10.7 psi (74 kPa) + 3.8 psi (26 kPa) − 1.2 psi (8 kPa) = 13.3 psi (92 kPa).

Other Sprinklers on a Branch Line

Up to this point, velocity pressure has been based on flow downstream from the sprinkler being considered; this has been confirmed by tests (Nickerson 1954). It has also been shown by those tests that, beyond the second sprinkler, velocity pressure should be figured from the flow on the upstream side of the sprinkler being considered. This is done by trial and error, assuming a flow from the sprinkler, calculating the velocity pressure from the total flow, determining a normal pressure, and calculating a flow from the normal pressure. If the calculated flow is not reasonably close to the assumed flow, assume a different flow and repeat the procedure until the two are close.

For example, assume a flow from the third sprinkler of 19.0 gpm (72 L/min) and also assume that the pipe between the third and fourth sprinkler is 1¼ in. Total flow is 36.0 gpm (136 L/min) + 19.0 gpm (72 L/min) = 55.0 gpm (208 L/min). Velocity pressure is 0.9 psi (6 kPa) and normal pressure at the third sprinkler is therefore 15.7 psi (108 kPa) − 0.9 psi (6 kPa), or 14.8 psi (102 kPa). Corrected flow then becomes 21.6 gpm (82 L/min), which is not close enough to the 19 gpm (72 L/min) assumed. Try an assumed flow of 21.4 gpm (81 L/min). Velocity pressure at 57.4 gpm (217 L/min) is 1.0 psi (7 kPa); normal pressure is 14.7 psi (101 kPa) and the new corrected flow is 21.5 gpm (81 L/min). Total flow at the third head then becomes 36.0 gpm (136 L/min) + 21.5 gpm (81.4 L/min) = 57.5 gpm (217.6 L/min). The calculating procedure for the other sprinklers on the branch line is the same as for the third sprinkler.

At this point it will be seen that the 15 psi (103 kPa) minimum riser pressure has been exceeded, unless, as is quite probable, the pressure with 57.5 gpm (217.6 L/min) flow is substantially higher than that with a 500 gpm (1893 L/min) flow. Whether or not the pressure with 57.5 gpm (217.6 L/min) flow is higher than 15 psi (103 kPa) depends on the characteristics of the water supply. However, in any case it appears that, with not many more sprinklers open, the pressure at the most remote sprinkler will be less than the 10 psi (69 kPa) selected in this example.

Branch Lines, Cross Mains, Risers, and Fittings

Cross Main Pressure at the Branch Line Connection: This is the normal pressure at the nearest open sprinkler increased by the friction loss and the velocity pressure in the intervening pipe. If the branch line is fed through a tee and nipple, additional friction loss allowances must be made except that the friction loss in nipples less than 6 in. (152 mm) long is customarily neglected.

Two Branches in One Line of Sprinklers: These may have the same or different numbers of sprinklers. The pressure at the entrance to the two branches will always be the same. The computations starting at the end sprinklers will be duplicated for the number of open sprinklers.After the discharge from any number of sprinklers on a branch line has been computed and the pressure to produce the flow has been determined, the entire branch line can be considered to have the discharge characteristics of a single orifice and the discharge constant k in the formule $Q = K\sqrt{P}$ can be determined, P being the net pressure where flows are taken from tees in the cross main.

Branches on Opposite Sides of a Cross Main: These branches may have different numbers of sprinklers open, in which case the cross main pressure must be the higher of the two computed values. This increases the discharge from the branch giving the lower computed pressure, and the actual discharge must be calculated for the higher pressure using the equation:

$$\frac{Q_1}{Q_2} = \sqrt{\frac{P_1}{P_2}}$$

in which P_2 is taken as the higher pressure, Q_2 the corresponding increased discharge to be determined, and P_1 and Q_1 the pressure and corresponding discharge from the branch requiring only the lower pressure.

After the appropriate increased discharge has been determined, the two rates of flow can be combined and K calculated for the combined branches.

When sprinklers on the second branch line are assumed to have opened, starting at the cross main sprinkler, the opened sprinkler most remote from the cross main is considered as the end sprinkler in the branch line computation, the next opened is the second, etc., regardless of nonoperating sprinklers on the outer end of the branch.

Cross Main Pressures: Cross main pressures are calculated by the same procedure as used for sprinklers on a single branch line, except that it is not necessary to use the trial and error procedure for the third and additional branch lines since the effect of change in velocity pressure with flows passing through tees in the cross main is usually negligible. The net head producing the flow in successive branch lines is taken as the normal pressure at the end branch line increased by the friction loss in the pipe between the branches.

Riser Pressure: Riser pressure is taken as the normal pressure at the nearest flowing branch increased by the total friction loss between this branch and the riser and by the velocity pressure in the cross main at the riser connection.

Friction Loss in Fittings: This is generally included in calculations only when the fitting involves a change in direction of flow. An exception to this is the fitting immediately preceding the sprinkler. Friction loss in control, gate and check valves, and in strainers, meters and similar devices is always included. The friction loss in piping between the source of supply and the opened sprinklers obviously must be included in all calculations.

Where there are differences in elevation, these must be allowed for on the basis that each foot (meter) of height represents 0.434 psi (9.82 kPa/m). In multistory buildings, this may be a substantial factor. Feed mains, cross mains, and branch lines within the same system may be looped or gridded to divide the total water flowing to the design area.

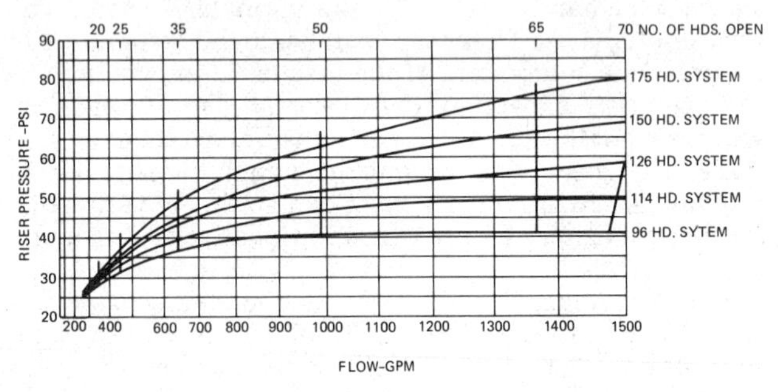

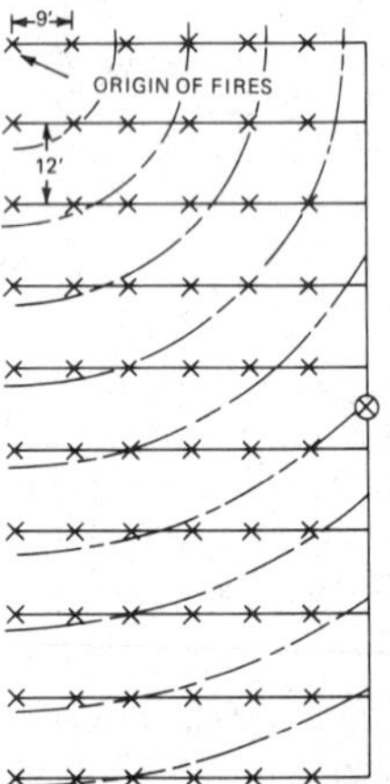

FIG. 18-2C. A flow curve for a side-central feed to sprinklers on a system having six sprinklers on each branch line is shown on the above graph. Below is the pattern of sprinklers opening on a side-central feed system. (Factory Mutual System)

Gridded Systems: Many gridded systems are currently being installed. The use of gridded systems means that every sprinkler is fed from several directions with a resulting reduction in pipe sizing. The hydraulic calculations required for a gridded system are complex and almost impossible without the use of computers.

Sprinkler System Water-Flow Curves

To avoid repetition of laborious computation of water flows and pressures when such information is needed in cases involving standard sprinkler, spray, or open head systems, it is possible to prepare diagrams or water-flow curves from which riser pressures and corresponding total sprinkler flows may be determined for different numbers of opened sprinklers. One such series of curves, as developed by the Factory Mutual Engineering Corporation, and the piping arrangement and assumed pattern of opened sprinklers is shown in Figure 18-2C. Programmable hand-held calculators are also useful, fast, and accurate.

Bibliography

References Cited

Crowley, John E. 1982. *Hydraulic Design and Estimation of Fire Sprinkler Systems.* Crowley Design Group, Inc. King of Prussia, PA.

FMEC. 1977. Loss Prevention Data 3-0. "Hydraulics of Fire Protection Systems." Factory Mutual Engineering Corp., Norwood, MA.

Nickerson, Malcolm H. 1954. "Water Flow Characteristics of Sprinkler Systems." *Proceedings of The Fifty-eighth Annual Meeting.* National Fire Protection Association. May 17-21. Washington, DC. pp 140-152.

Wood, Clyde M. 1980. *Automatic Sprinkler Hydraulic Data.* "Automatic" Sprinkler Corporation of America, Cleveland, OH.

Wass, Harold E. 1983. *Sprinkler Hydraulics.* IRM Insurance, White Plains, NY.

NFPA Codes, Standards, Recommended Practices and Manuals. (See the latest *NFPA Codes and Standards Catalog* for availability of current editions of the following documents.)

NFPA 13, *Standard for the Installation of Sprinkler Systems.*

NFPA 15, *Standard for Water Spray Fixed Systems for Fire Protection.*

NFPA 20, *Standard for the Installation of Centrifugal Fire Pumps.*

NFPA 22, *Standard for Water Tanks for Private Protection.*

NFPA 231, *Standard for Indoor General Storage.*

NFPA 231C, *Standard for Rack Storage of Materials.*

NFPA 231D, *Standard for Storage of Rubber Tires.*

NFPA 231E, *Recommended Practice for the Storage of Baled Cotton.*

NFPA 231F, *Standard for the Storage of Roll Paper.*

NFPA 291, *Recommended Practice for Fire Flow Testing and Marking of Hydrants.*

Additional Reading

Baldwin, R., and North, M. A., "The Number of Sprinklers Opening and Fire Growth," *Fire Technology,* Vol. 9, No. 4, Nov. 1973, pp. 245-253.

Beyler, Craig L., "An Evaluation of Sprinkler Discharge Calculation Methods," *Fire Technolgy,* Vol. 13, No. 2, May 1977, pp. 185-194.

Bond, Horatio, "Water for Fire Fighting in High-Rise Buildings," *Fire Technology,* Vol. 2, No. 2, May 1966, pp. 159-163.

Bratlie, Ernest H. Jr., "Automatic Sprinklers: How Much Water?," *Fire Journal,* Vol. 62, No. 1, Jan. 1968, pp. 27-31.

Brown, J. R., "Field Test to Compare Sprinkler Demands of Wet-Pipe and Dry-Pipe Systems," Factory Mutual Research Corp., Norwood, MA, May 1980.

Bryan, J. L., *Automatic Sprinkler and Standpipe Systems,* National Fire Protection Association, Quincy, MA, 1976, Chapter 4.

Chicarello, Peter Joseph Jr., "Analytical Methods for Calculating Sprinkler Discharge," *Fire Technology,* Vol. 8, No. 1, Feb. 1972, pp. 45-52.

Damon, Walter A., "Preventing Pitfalls in Fire Protection Systems," *Heating/Piping/Air Conditioning,* April 1985, pp. 61–64, 69–71.

Foehl, John, M. "Flow Characteristics —Copper Sprinkler Conductors," *Fire Technology,* Vol. 4, No. 3, Aug. 1968, pp. 169-178.

Gomberg, Alan, and Cote, A. E., "The Computer Approach to Fire Protection Problems," *Fire Technology,* Vol. 3, No. 1, Aug. 1967, pp. 202-212.

Heskestad, G., and Kung, H. C., "Relative Demands of Wet and Dry-Pipe Sprinkler Systems" Factory Mutual Research Corp., Norwood, MA, Nov. 1973.

Hoyle, H., and Bray, G., "Hydraulic Performance of Sprinkler Installations," *Fire Technology,* Vol. 3, No. 4, Nov. 1967, pp. 291-305.

Kalelkar, A. S., "Understanding Sprinkler Performance: Modeling of Combustion and Extinction," *Fire Technology,* Vol. 7, No. 4, Nov. 1971, pp. 293-305.

Kirkman, Hugh B., and Campbell, Layard E., "Velocity Pressure Effect on Sprinkler System Discharge," *Fire Technology,* Vol. 6, No. 1, Feb. 1970, pp. 68-72.

Merdinyan, P. H., "Friction Loss in Sprinkler Piping," *Fire Technology,* Vol. 4, No. 4, Nov. 1968, pp. 304-309.

Merry, J. T., and Schiffhauer, Earl J., "Hydraulic Sprinkler Systems Design—A Computer Approach," *Fire Technology,* Vol. 2, No. 2, May 1966, pp. 95-107.

Ng, Warren, *Fire Sprinkler System Hydraulic Calculations and Computer Aided Design Methods,* (privately published) available from author, 6341 Pinehaven Road, Oakland, CA, 1982.

O'Connor, T. Francis, "Pressure Losses in Deluge Sprinkler Fittings," *Fire Technology,* Vol. 2, No. 3, Aug. 1966, pp. 204-210.

Patek, N. J., "Hydraulics for Fire Protection," *Fire Journal,* Vol. 67, No. 2, March 1973, pp. 13-16, 40.

Patton, R. M., "Engineered Sprinkler Protection," *Fire Journal,* Vol. 60, No. 1, Jan. 1966, pp. 5-8, 18.

Reilly, Edward J., and Viniello, John A, "Sprinklers Cut Fresno's Fire Losses and Budget," *Fire Journal,* Vol. 73, No. 6, pp. 58-59, 91.

"Water vs Fire: The Essential Components of a Good Water Supply," *The Sentinel,* (Industrial Risk Insurers), Vol. 367, No. 4, July/Aug. 1980, pp. 3-9.

Webb, William A., "Effectiveness of Automatic Sprinkler Systems in Exhibition Halls," *Fire Technology,* Vol. 4, No. 2, May 1968, pp. 115-125.

Wood, Clyde M., *Study Guide for Automatic Sprinkler Hydraulic Data,* "Automatic" Sprinkler Corp. of America, Cleveland, OH, 1974, 64-pp.

Wright, Victor E., "Hydraulic Graphing on a Microcomputer for Fire Protection Analysis," Heating, Piping, *Air Conditioning,* April 1984, pp. 83-90.

AUTOMATIC SPRINKLERS

Revised by Robert Hodnett

Automatic sprinklers are thermosensitive devices designed to react at predetermined temperatures by automatically releasing a stream of water and distributing it in specified patterns and quantities over designated areas. The automatic distribution of water is intended to extinguish a fire or to prevent its spread in the event that the initial fire is out of range of the sprinklers or is of a type that cannot be extinguished by water discharged from sprinklers. The water is fed to the sprinklers through a system of piping, ordinarily overhead, with the sprinklers placed at intervals along the pipes.

This chapter covers the operating principles, including operating temperatures, of the various types of automatic sprinklers currently available and describes their construction in some detail. Sprinklers for special service conditions and sprinklers of early manufacture that may still be in use are also covered.

The National Fire Protection Association does not have standards that cover the manufacture of sprinklers. It does, however, insist through requirements in its many fire protection standards containing sprinkler protection requirements that only sprinklers listed by reputable product evaluation organizations be used in sprinkler systems meeting the requirements of NFPA 13, *Standard for the Installation of Sprinkler Systems* (referred to hereafter in this chapter as the NFPA Sprinkler Systems Standard). Listing of sprinklers in a national organization's listing service is an indication that the sprinklers have been scrutinized for their reliability and compliance with the organization's test criteria for sprinklers.

Since they were introduced in the latter part of the 19th century, the performance and the reliability of automatic sprinklers have been continually improved through experience and the efforts of manufacturers and testing organizations.

In 1952 and 1953, a radical change, which considerably improved its effectiveness, was made in the pattern of the sprinkler's water discharge. Originally, this improved sprinkler was called the spray sprinkler. In 1958, it became the standard sprinkler, and sprinklers of the older design became known as old style sprinklers. Redesign of the deflector was the principal feature of the new standard sprinklers.

Mr. Hodnett is a consultant in fire protection and piping systems based in Providence, RI. He was NFPA staff liaison to the NFPA Committee on Automatic Sprinkler Systems prior to his retirement in 1984.

OPERATING PRINCIPLES OF AUTOMATIC SPRINKLERS

In order to appreciate the ruggedness, mechanical simplicity, reliability in operation, and freedom from premature operation of an automatic sprinkler, a familiarity with the basic principles of its design, construction, and operation is necessary.

Operating Elements

Under normal conditions, the discharge of water from an automatic sprinkler is restrained by a cap or valve held tightly against the orifice by a system of levers and links or other releasing devices pressing down on the cap and anchored firmly by struts on the sprinkler.

Fusible Sprinklers: A common fusible style automatic sprinkler operates upon the fusing of a metal alloy of predetermined melting point. Various combinations of levers, struts, and links or other soldered members are used to reduce the force acting upon the solder so that the sprinkler will be held closed with the smallest practical amount of metal and solder. This minimizes the time of operation by reducing the mass of fusible metal to be heated. A fusible sprinkler of the link and lever design is shown in Figure 18-3A. A fusible sprinkler featuring solder under compression is shown in Figure 18-3B.

The solders used with automatic sprinklers are alloys of optimum fusibility composed principally of tin, lead, cadmium, and bismuth; all have sharply defined melting points. Alloys of two or more metals may have a melting point that is lower than that of the individual metal having the lowest melting point. The mixture of two or more metals that gives the lowest melting point possible is called a "eutectic" alloy.

FIG. 18-3A. Hodgman Model D upright link and lever sprinkler.

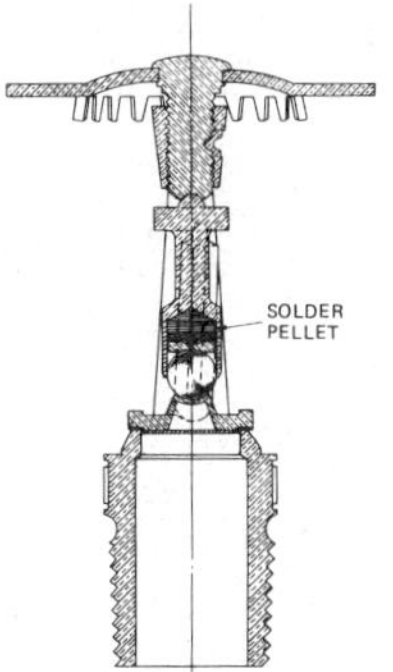

FIG. 18-3B. Chemetron Stargard, Model H.

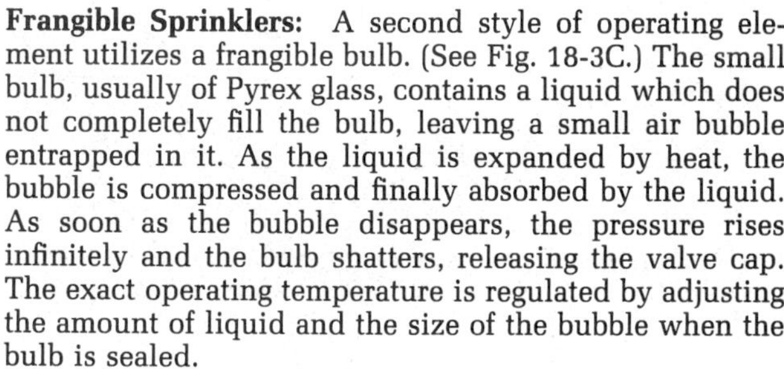

Frangible Sprinklers: A second style of operating element utilizes a frangible bulb. (See Fig. 18-3C.) The small bulb, usually of Pyrex glass, contains a liquid which does not completely fill the bulb, leaving a small air bubble entrapped in it. As the liquid is expanded by heat, the bubble is compressed and finally absorbed by the liquid. As soon as the bubble disappears, the pressure rises infinitely and the bulb shatters, releasing the valve cap. The exact operating temperature is regulated by adjusting the amount of liquid and the size of the bubble when the bulb is sealed.

Other Thermosensitive Elements: Other styles of thermosensitive operating elements may be, or have been, employed to provide automatic discharge, such as bimetallic discs (Fig. 18-3D), fusible alloy pellet (Fig. 18-3E), or chemical pellets (Fig. 18-3F).

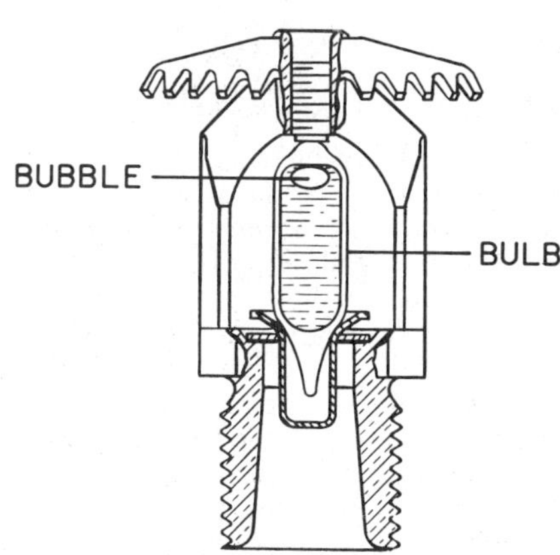

FIG. 18-3C. Grinnel Quartzoid, Issue D.

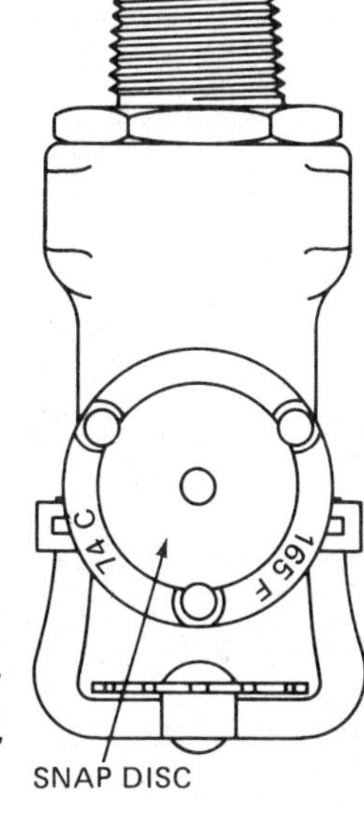

FIG. 18-3D. A snap disc that controls the flow of water on an on-off sprinkler. (See "Recycling Sprinklers" later in this chapter.)

FIG. 18-3E. Central Flow Control (on-off) sprinkler.

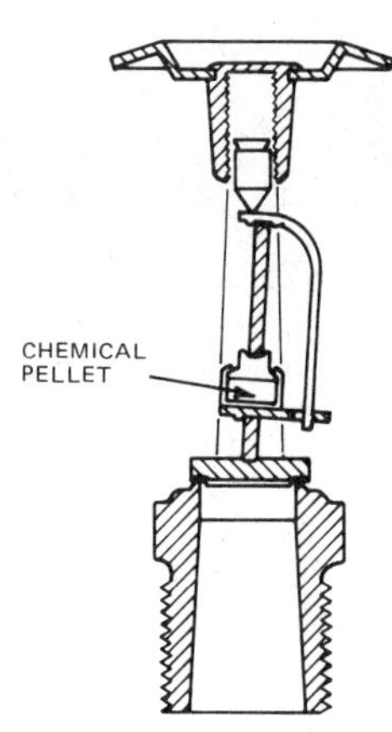

FIG. 18-3F. Globe Saveall, style D (no longer manufactured).

Sprinkler Dynamics

Figure 18-3G shows how the closing force exists in the link and lever style automatic sprinkler. The construction

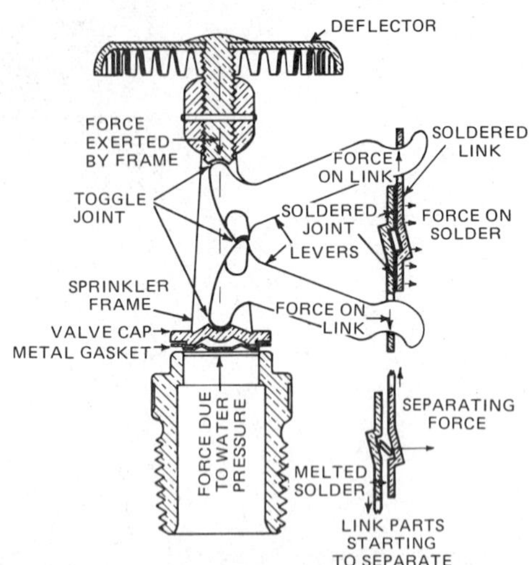

FIG. 18-3G. Representative arrangement of a soldered link and lever automatic sprinkler.

shown is diagrammatic and does not exactly represent any particular sprinkler.

The mechanical pressure normally exerted on the top of the cap or valve is many times that developed by the water pressure below, so that the possibility of leakage, even from water hammer or exceptionally high water pressure, is practically eliminated. The mechanical pressure in a link and lever sprinkler is produced in three stages: first by the toggle effect of the two levers, second by the mechanism of the link parts, and third by the load in the solder between the link parts. The last force resisted by the solder is made relatively low because solder of the composition needed to give the desired operating temperatures is subject to cold flow under high stress. The sprinkler frame or other parts possess a degree of elasticity

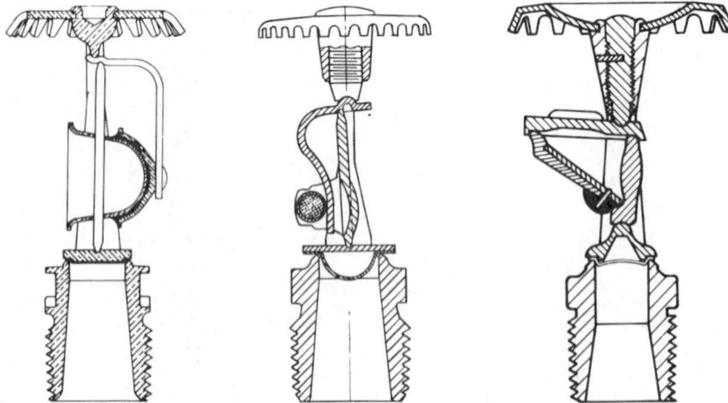

FIG. 18-3H. Some modifications of the common link and lever construction. At left is Grinnell F950 Duraspeed; at center is Reliable Model G; and at right is Firematic Type S.

to provide energy for producing a positive, sharp release of the operating parts.

Sprinklers illustrated in Figure 18-3H use modifications of the common link and lever construction, some of which employ solder under compression, tension, and tension and shear.

Deflector Design

Attached to the frame of the sprinkler is a deflector or distributor against which the stream of water is directed and converted into a spray designed to cover or protect a certain area. The amount of water discharged depends upon the flowing water pressure and the size of the sprinkler orifice. A flowing pressure of 7 psi (48 kPa) is generally considered a minimum for proper action. At this pressure a sprinkler having a nominal ½ in. (12.7 mm) orifice will discharge 15 gpm (58 L/min). At the same 7 psi (48 kPa) pressure, a nominal 17/32 in. (13.5 mm) orifice sprinkler will discharge 21 gpm (79 L/min). (See Figure 18-3I for the quantity of water discharged at various water pressures.)

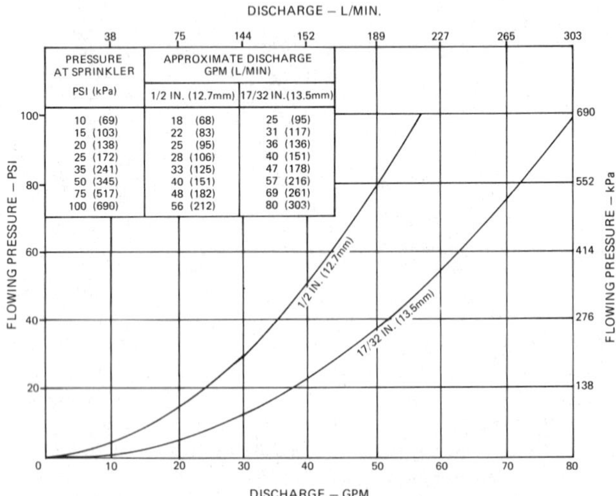

PRESSURE AT SPRINKLER PSI (kPa)	APPROXIMATE DISCHARGE GPM (L/MIN) 1/2 IN. (12.7mm)	17/32 IN. (13.5mm)
10 (69)	18 (68)	25 (95)
15 (103)	22 (83)	31 (117)
20 (138)	25 (95)	36 (136)
25 (172)	28 (106)	40 (151)
35 (241)	33 (125)	47 (178)
50 (345)	40 (151)	57 (216)
75 (517)	48 (182)	69 (261)
100 (690)	56 (212)	80 (303)

FIG. 18-3I. Water discharge rates of typical nominal ½ in. (12.7 mm) and 17/32 in. (13.5 mm) orifice automatic sprinklers.

In order to have even the minimum flowing pressure at sprinklers that are remote from the point of water supply, especially when a number of sprinklers are operating simultaneously, water supply pressures in the range of 50 to 100 psi (345 to 690 kPa) are customarily provided. Hydraulically calculated systems are designed around the normally available water supply volume and pressure.

TEMPERATURE RATINGS OF AUTOMATIC SPRINKLERS

Automatic sprinklers have various temperature ratings that are based on standardized tests in which a sprinkler is immersed in a liquid and the temperature of the liquid is raised very slowly until the sprinkler operates. (See Table 18-3A.)

The temperature rating of all solder style automatic sprinklers is stamped upon the soldered link. For other styles of thermosensitive elements, the temperature rating is stamped upon some one of the releasing parts.

The recommended maximum room temperature is generally closer to the operating temperature for frangible bulb than for soldered fusible-element sprinklers. This is because solder begins to lose its strength somewhat below

TABLE 18-3A. Temperature Ratings, Classification and Color Codings*

Max. Ceiling Temp. °F	Max. Ceiling Temp. °C	Temperature Rating °F	Temperature Rating °C	Temperature Classification	Color Code	Glass Bulb Colors
100	38	135 to 170	57 to 77	Ordinary	Uncolored	Orange or Red
150	66	175 to 225	79 to 107	Intermediate	White	Yellow or Green
225	107	250 to 300	121 to 149	High	Blue	Blue
300	149	325 to 375	163 to 191	Extra High	Red	Purple
375	191	400 to 475	204 to 246	Very Extra High	Green	Black
475	246	500 to 575	260 to 302	Ultra High	Orange	Black
625	329	650	343	Ultra High	Orange	Black

* Source: NFPA 13, Sprinkler Systems Standard.

its actual melting point. Premature operation of a solder sprinkler usually depends on the extent to which the normal room temperature is exceeded, the duration of the excessive temperature, and the load on the operating parts of the sprinkler.

The general rule of not using sprinklers of ordinary [135 to 170°F (57 to 77°C)] temperature rating [where temperatures exceed 100°F (38°C)] is necessary to provide a margin of safety. General practices regarding the use of automatic sprinklers of higher than the ordinary rating are given in Tables 18-3B and 18-3C. Sprinklers of ordinary temperature rating can be used inside buildings and other places where they are not subject to direct sun rays, except in monitors and in blind attics without ventilation, under metal or tile roofs, near or above heat sources, or in confined spaces where normal temperatures may be exceeded.

When there is doubt as to maximum temperatures at sprinkler locations, maximum reading thermometers should be used and the temperature determined under conditions which would show the highest readings to be expected.

Sprinklers with intermediate or high temperature ratings should be used in place of ordinary rated sprinklers in situations where a fast developing fire or a rapid rate of heat release can be anticipated. The sprinklers rated for

TABLE 18-3B. Distance of Sprinklers from Heat Sources*

Type of Heat Condition	Ordinary Degree Rating	Intermediate Degree Rating	High Degree Rating
1. Heating Ducts (a) Above	More than 2 ft. 6 in.	2 ft. 6 in. or less	—
(b) Side and Below	More than 1 ft. 0 in.	1 ft. 0 in. or less	—
(c) Diffuser (c) Downward Discharge (c) Horizontal Discharge	Any distance except as shown under Intermediate	*Downward:* Cylinder with 1 ft. 0 in. radius from edge, extending 1 ft. 0 in. below and 2 ft. 6 in. above *Horizontal:* Semicylinder with 2 ft. 6 in. radius in direction of flow, extending 1 ft. 0 in. below and 2 ft. 6 in. above.	—
2. Unit Heater (a) Horizontal Discharge	—	*Discharge Side:* 7 ft. 0 in. to 20 ft. 0 in. radius pie-shaped cylinder [see Figure 3-16.6.3 (a)] extending 7 ft. 0 in. above and 2 ft. 0 in. below unit heater; also 7 ft. 0 in. radius cylinder more than 7 ft. 0 in. above unit heater	7 ft. 0 in radius cylinder extending 7 ft. 0 in. above and 2 ft. 0 in. below unit heater
(b) Vertical Downward Discharge [Note: For sprinklers below unit heater, see Figure 3.16.6.3(a).]	—	7 ft. 0 in. radius cylinder extending upward from an elevation 7 ft. 0 in. above heater	7 ft. 0 in. radius cylinder extending from the top of the unit heater to an elevation 7 ft. 0 in. above unit heater
3. Steam Mains (Uncovered) (a) Above	More than 2 ft. 6 in.	2 ft. 6 in. or less	—
(b) Side and Below	More than 1 ft. 0 in.	1 ft. 0 in. or less	—
(c) Blow-off Valve	More than 7 ft. 0 in.	—	7 ft. 0 in. or less

For SI Units: 1 in. = 25.4 mm; 1 ft. = 0.3048 m.
* Source: NFPA 13, Sprinkler Systems Standard

TABLE 18-3C. Ratings of Sprinklers in Specified Locations*

Location	Ordinary Degree Rating	Intermediate Degree Rating	High Degree Rating
Skylights	—	Glass or plastic	—
Attics	Ventilated	Unventilated	—
Peaked Roof: Metal or thin boards; concealed or not concealed; insulated or uninsulated	Ventilated	Unventilated	—
Flat Roof: Metal; not concealed; insulated or uninsulated	Ventilated or unventilated	Note: For uninsulated roof, climate and occupancy may require Intermediate sprinklers. Check on job.	—
Flat Roof: Metal; concealed; insulated or uninsulated	Ventilated	Unventilated	—
Show Windows	Ventilated	Unventilated	—

NOTE: A check of temperatures by means of thermometers may be necessary.
*Source: NFPA 13, Sprinkler Systems Standard.

higher temperatures may have the advantage of reducing the number of sprinklers which would otherwise operate outside the fire area.

Automatic sprinklers may require a longer time period to operate when exposed to a slowly developing fire as opposed to the high heat release of a rapidly developing fire. They are designed to operate quickly enough to control fire and prevent its spread.

The speed of operation depends on the physical properties of the thermosensitive mechanism of the sprinkler. The time involved to operate depends, among other factors, upon the shape, size, and mass of the thermosensitive mechanism, and the temperature differential between the surrounding atmosphere and the operating temperature of the sprinkler.

In the testing of sprinklers, no distinction is made between makes of sprinklers as to their speed of operation as long as the speed comes within specified limits. In cases where extreme speed of operation is necessary owing to the likelihood of a rapidly developing and spreading fire as, for example, in explosives manufacturing, the practice is to use deluge systems. Such sprinklers are open without operating elements and consequently, there is no delay in their operation. The water is admitted to the system through a valve as a result of the operation of quick operating detection devices.

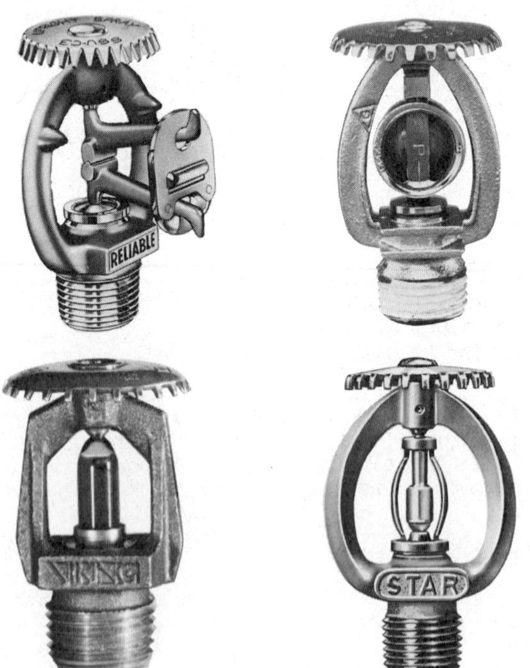

FIG. 18-3J. Standard sprinklers showing the various arrangements of releasing mechanisms. Clockwise from upper left is a fusible link and lever style (Reliable Issue C); a perforated heat collector style, which is a variation of the link and lever style (Grinnell F950 Duraspeed); a center strut style with solder under compression (Chemetron Stargard Model H); and a frangible bulb style (Viking Model M).

STANDARD AUTOMATIC SPRINKLERS

Standard sprinklers are generally similar in appearance to old style sprinklers and utilize the same style of frame and linkage or other release mechanism. The essential difference is in the deflector; seemingly minor differences in the deflector design make major differences in discharge characteristics. Several representative standard sprinklers are illustrated in Figure 18-3J.

On the assumption that discharge of water against the ceiling was essential to fire extinguishment, previous research on automatic sprinklers had been largely concerned with securing reasonably uniform distribution of water over the area protected by one sprinkler, and with wetting the ceiling. Later research showed that more effective extinguishment and a larger area of coverage could be secured by directing all the water downward and horizontally. Research further showed that, with this pattern, discharge is effective even in controlling fires on the ceiling above the sprinklers because of the improved cooling effect of the spray, better high-level water distribution, and decreased exposure to the ceiling because of more effective direct discharge of water on burning materials below.

Due to the design of the deflector, the solid stream of water issuing from the orifice of a standard sprinkler is broken up to form an umbrella shaped spray. The pattern is roughly that of a half sphere filled with spray. Relatively uniform distribution of the water at all levels below the sprinklers is characteristic of a standard sprinkler. At a distance of 4 ft (1.2 m) below the deflector, the spray covers a circular area having a diameter of approximately 16 ft (4.9 m) when the sprinkler is discharging 15 gpm (58 L/min).

Standard sprinklers are made for installation in an upright or pendent position and must be installed in the

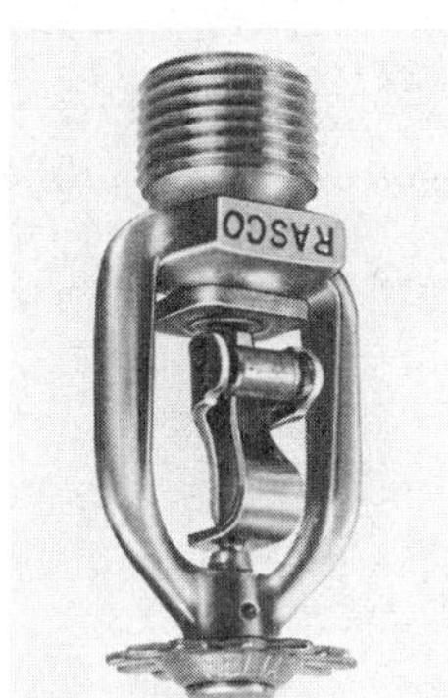

FIG. 18-3K. A listed sprinkler showing upright (right) and pendent (left) models of the same issue; note the difference in design of the deflectors. The sprinklers shown are Reliable Model G.

position for which they are designed. (See Fig. 18-3K.) It is customary to replace old style sprinklers with standard sprinklers in existing installations although the NFPA Sprinkler Systems Standard permits replacing old style sprinklers with similar devices. Most manufacturers, however, have discontinued producing old style sprinklers.

The general patterns of water discharge from the old style and the standard sprinklers are shown in Figures

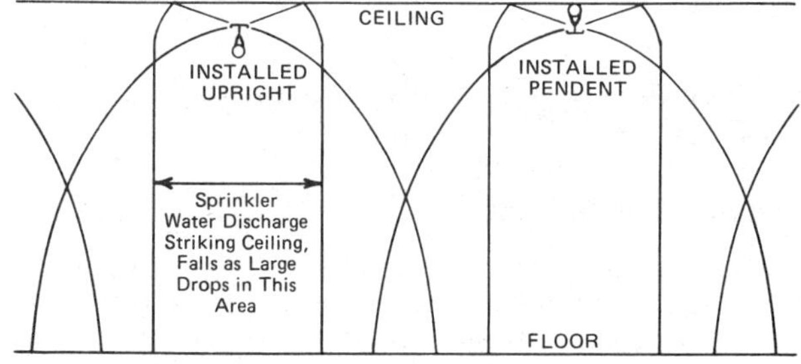

FIG. 18-3L. Principal distribution pattern of water from old style sprinklers (previous to 1953).

18-3L and 18-3M.

The water distribution characteristics of sidewall automatic sprinklers, picker trunk sprinklers, window sprinklers, and cornice sprinklers are described later in this chapter.

Experimentation, engineering judgment, and experience determined that for pipe schedule systems a favorable rate of water discharge from an automatic sprinkler would be that of a ½ in. (12.7 mm) diameter orifice. This is often not the case with hydraulically designed systems. Therefore, sprinklers of various orifice sizes are utilized.

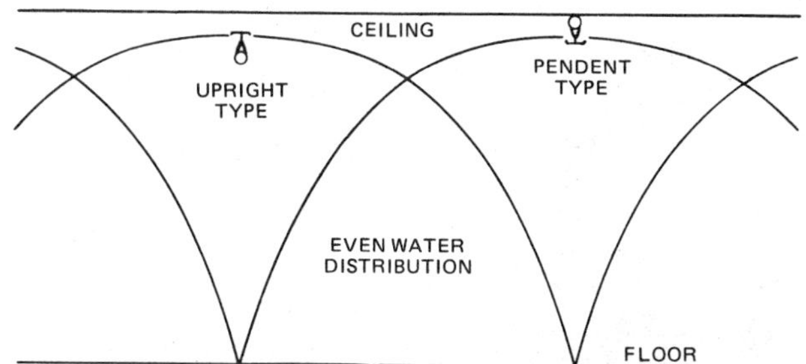

FIG. 18-3M. Principal distribution pattern of water from standard sprinklers (in use since 1953).

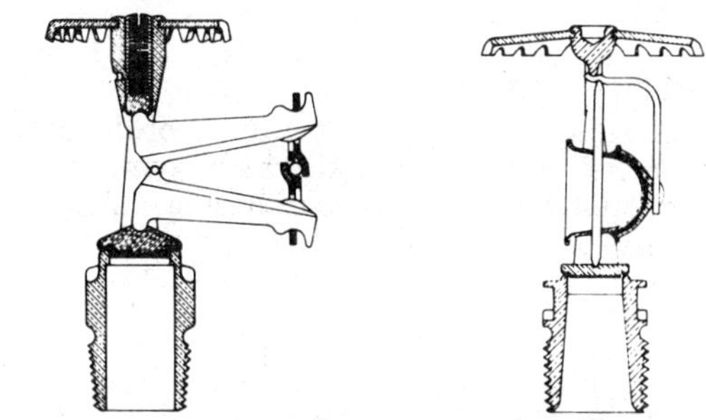

FIG. 18-3N. Typical sprinklers, showing a straight discharge orifice (ring style) at left and tapered nozzle orifice at right.

Standard automatic sprinklers have a nominal ½-in. (12.7 mm) orifice. The orifice may be of the ring nozzle or tapered nozzle style. (See Fig. 18-3N.)

The rate of water discharge from a sprinkler follows hydraulic laws and depends upon the size of the orifice or nozzle and the water pressure. Approximate rates of discharge at different pressures may be obtained from the plotted curve or the table given in Figure 18-3I.

Sprinklers discharging water through smaller or larger orifices are discussed later in this chapter. With similar forms of sprinkler nozzles and at the same water pressure, the discharge from these styles of sprinklers is approximately proportional to the nominal size of the orifice.

Listed Automatic Sprinklers

In order to obtain acceptance or approval of their sprinklers, manufacturers submit them to fire testing organizations. After extensive tests and verification of the manufacturer's ability to properly manufacture the product, sprinklers found satisfactory are listed. Acceptance of a sprinkler by inspection departments or other regulatory agencies is based on such a listing.

Standard sprinklers are designed to be installed and operated in their proper position, i.e., upright or pendent, as indicated by a stamping upon the deflector bearing the appropriate word or the letters "SSU" (Standard Sprinkler Upright) or "SSP" (Standard Sprinkler Pendent).

Double Deflectors: Several manufacturers developed a double deflector as their first design for use as standard upright sprinklers. A typical double deflector is shown in Figure 18-3O. Such sprinklers are no longer manufactured but, if previously "listed," are as acceptable as the presently approved upright standard sprinklers having a one-piece deflector.

FIG. 18-3O. Standard upright automatic sprinkler with typical double deflector.

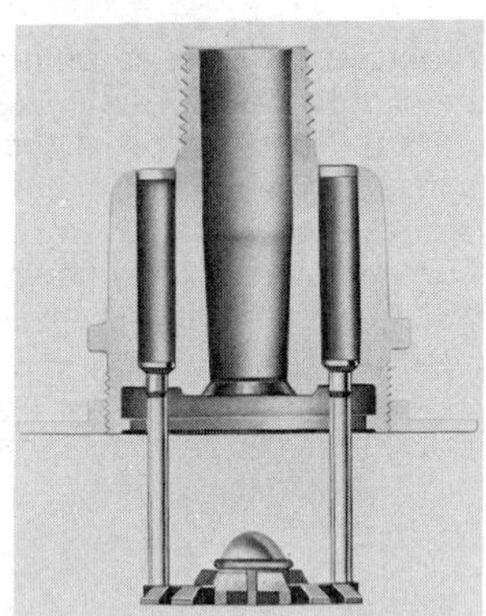

FIG. 18-3Q. A flush style ceiling sprinkler; the diagram at right shows the sprinkler after operation. This style of sprinkler is used where appearance is considered to be of prime importance. Shown is Reliable Model B.

Recessed Sprinklers

A recessed sprinkler has part or most of the body of the sprinkler, other than the part which connects to the piping, mounted within a recessed housing. Operation is similar to that of a standard pendent sprinkler. (See Fig. 18-3P.)

FIG. 18-3P. A Central Model H recessed sprinkler.

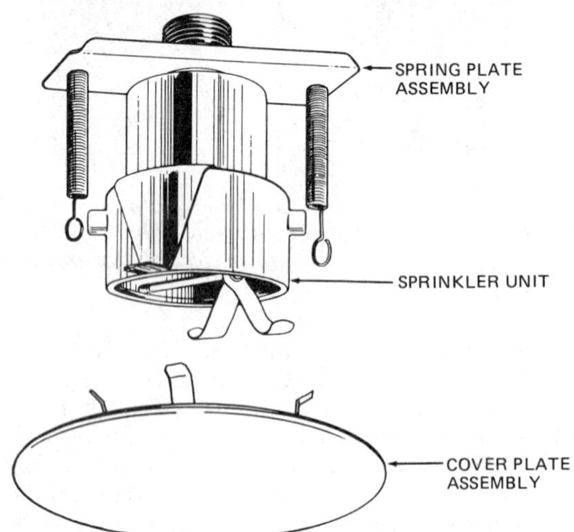

FIG. 18-3R. Concealed ceiling sprinkler. Cover plate drops away when heat is applied to bottom side of plate. (Stargard Model G)

Flush Type Sprinklers

Listed sprinklers of special designs but with the same water discharge pattern as standard pendent sprinklers are available for use with wet system piping concealed above ceilings in areas where appearance is important. A typical sprinkler of this style is shown in Figure 18-3Q. Testing organizations customarily list these special sprinklers as ceiling style or flush type sprinklers. The special design allows a minimum projection of the working parts of the sprinkler below the ceiling in which it is installed without adversely affecting the heat sensitivity or the pattern of water distribution.

In an effort to provide sprinkler protection in low risk occupancies where aesthetics are important, sprinklers have been designed that attractively blend with the ceiling. Only the ceiling plate and thermosensitive assembly are visible from the floor when these sprinklers are installed. When a fire occurs and the thermosensitive element operates, the deflector drops to a position below the ceiling and water discharge commences.

Concealed Sprinklers

A concealed sprinkler has its entire body, including the operating mechanism, above its concealing cover plate. When a fire occurs, the cover plate drops, exposing the thermosensitive assembly. (The subsequent operation of the thermosensitive assembly initiates discharge.) (See Fig. 18-3R.)

Ornamental Sprinklers

Ornamental sprinklers are automatic sprinklers that have been decorated by attachments or by plating or enameling to give desired surface finishes. Ornamentation or special decorative design must not unfavorably affect the operation or the water distribution. Listed styles of ornamental sprinklers are for pendent installation in accordance with the NFPA Sprinkler Systems Standard. (See Fig. 18-3S.)

Dry Pendent and Dry Upright Sprinklers

Dry pendent and dry upright sprinklers are used to provide sprinkler protection in unheated areas, such as freezers, where individual sprinklers are supplied from a drop or riser pipe from a wet pipe system outside the unheated area. To prevent damage to the system, the drop/riser must remain free of water. Dry pendent sprinklers may also be used in low areas of a dry pipe system

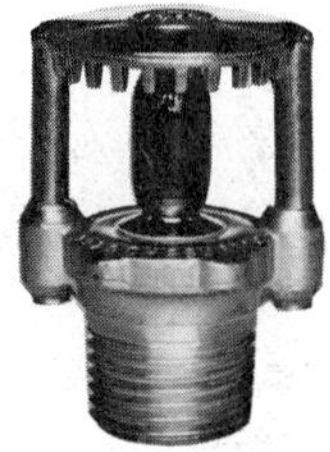

FIG. 18-3S. Prussag SFH glass bulb style decor sprinkler.

where water must not be allowed in the drop pipe supplying the sprinkler.

Sprinklers designated and listed by testing organizations as dry pendent or dry upright automatic sprinklers may be used. A seal is provided at the entrance of the dry sprinkler to prevent water from entering until the sprinkler fuses. The heat sensitive operating mechanisms are adaptations of those used with automatic standard sprinklers. See Figure 18-3T for examples of dry pendent automatic sprinklers.

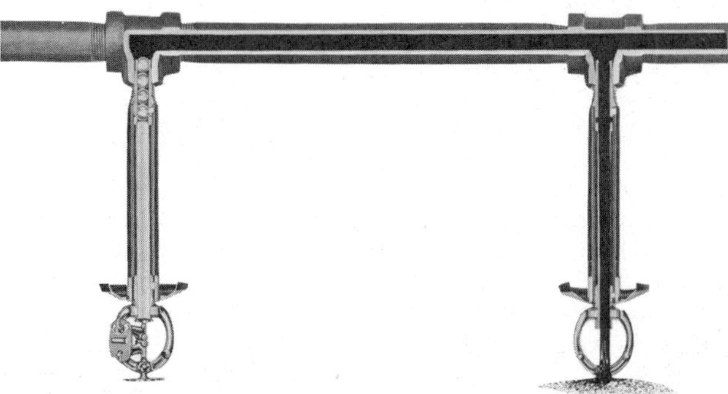

FIG. 18-3T. Representative dry pendent automatic sprinkler. When the ambient temperature rises beyond the operating temperature of the soldered link, solder melts and link plates separate on roller key. Levers held in place by deflector screw are released and the fixed tension of the frame, acting as a spring, ejects the levers and link parts clear of the sprinkler. Since the inner tube, which also serves as a discharge orifice, is no longer held in place by the levers, it moves to a predetermined position. With the support of the inner tube eliminated, the elements forming the watertight seal at the piping inlet pass through the inner tube and away from the sprinkler, allowing water to flow through the unobstructed waterway and strike the deflector which distributes it in a spray pattern comparable to that of a ½ in. (12.7 mm) standard sprinkler. Shown is Reliable's Model C Dry Pendent. Other manufacturers have similar arrangements of proprietary equipment for dry pendent automatic systems.

SPRINKLERS FOR SPECIAL SERVICE CONDITIONS

The standard sprinkler system using approved upright or pendent sprinklers is adaptable to a wide variety of conditions. However, there are situations for which special styles of sprinklers and special sprinkler arrangements are suited. Under all conditions of service, it is important to have the distribution of water equivalent in effectiveness to that of standard sprinklers. In some cases, such as with sidewall sprinklers, special patterns of water distribution are necessary; in others, unusual temperatures or corrosive atmospheres call for special design or construction features.

Residential Sprinklers

Residential sprinklers are sprinklers that have been specifically listed for use in residential occupancies, and installation requirements for them will be found in NFPA 13D, *Standard for the Installation of Sprinklers in One- and Two-family Dwellings and Mobile Homes*. These sprinklers have special low mass fusible links that result in the time of temperature actuation being much less than that for a sprinkler with a conventional style of fusible link. For some styles of residential sprinklers, the actuation time is one-fifth that of a sprinkler with a conventional link assembly. Because of their fast acting links and the life safety aspects of such fast action, these sprinklers are very often used in residential occupancies where a conventional system installed according to the provisions of the NFPA Sprinkler Systems Standard might not be as practical. A representative residential style of sprinkler is shown in Figure 18-3U. (Residential sprinkler installations are discussed in detail in Chapter 4 of this section.)

FIG. 18-3U. A representative residential style sprinkler.

Large Drop Sprinklers

Large drop sprinklers are special sprinklers with a K factor between 11.0 and 11.5. (K factors for standard ½ in. [12.7 mm] sprinklers range from 5.3 to 5.8.) The deflector of a large drop sprinkler is specially designed, and that, combined with the greater discharge, produces large drops of such size and velocity as to enable the spray to penetrate strong updrafts generated by high challenge fires. The installation requirements for large drop sprinklers are special and are covered in a separate chapter in the NFPA Sprinkler Systems Standard, because the requirements supersede the density/area requirements used for conventional sprinklers. A representative large drop sprinkler is shown in Figure 18-3V.

FIG. 18-3V. A representative large drop sprinkler.

Cycling Sprinklers

A cycling sprinkler cycles on and off as needed. The model shown in Figure 18-3W, manufactured by the Grinnell Company, operates on a pilot valve principle. Under normal conditions, the pilot valve is held closed by the snap disc. Water in the piston chamber holds the piston closed. When the snap disc is heated to the rated temperature, it opens the pilot valve and releases the water from the piston chamber. This allows the piston to open and water to flow from the sprinkler. When the snap disc cools to approximately 100°F (37.8°C), it closes the pilot valve. Water enters the piston chamber through a restricted orifice in the piston, and the difference in pressure closes the valve. The sprinkler is ready for repeated operation should the snap disc again be heated to the rated temperature. Both a pendent and a recessed model are available.

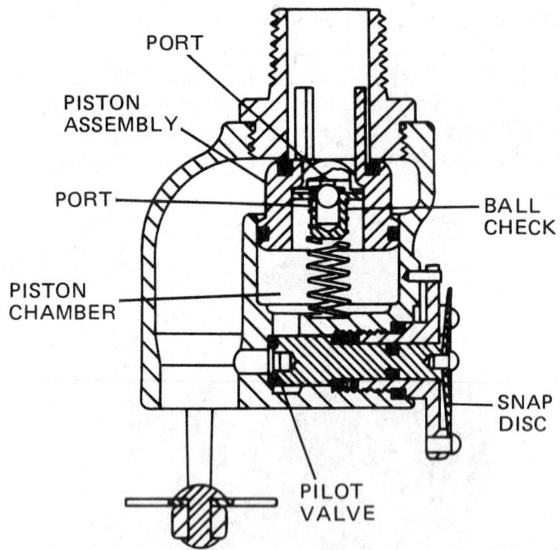

FIG. 18-3W. A cycling sprinkler showing its operating elements. In a fire the snap disc responds by opening the pilot valve which relieves pressure from the piston chamber faster than it can be replaced through the ports in the ball check. The system water pressure then forces the piston assembly down, thus permitting an unobstructed flow of water. (Grinnell Fire Protection Systems Company, Inc.)

Sprinklers for Corrosive Conditions

Measures have been developed to protect automatic sprinklers from corrosive conditions, and studies have been made by testing organizations of the value of each method. A complete covering of wax having a melting point slightly below the temperature at which the sprinkler operates is the most commonly used protective coating. A lead coating for the body of the sprinkler and the levers in combination with wax for protecting fusible elements is also common. Typical coatings include wax only (Fig 18-3X), asphalt only, lead only, wax over lead, and asphalt over lead for sprinklers with nominal temperature ratings from 135 to 212°F (57 to 100°C). Lead coatings are used on listed sprinklers for corrosive atmospheres for all rated sprinklers from 135 to 500°F (57 to 260°C). Other coatings are available and should be selected on the basis of the expected environment.

FIG. 18-3X. A wax coated upright sprinkler for corrosive atmospheres. ("Automatic" Sprinkler Corporation of America)

Whatever the protective measures taken, they must not delay the operation of the sprinkler, nor interfere with the release of operating parts, nor significantly alter the pattern of water distribution.

If a sprinkler system is to contain an additive to prevent freezing, the sprinklers on it must be made of carefully selected metals to avoid internal corrosion damage to the sprinklers.

Sidewall Sprinklers

Sidewall sprinklers have the components of standard sprinklers except for a special deflector which discharges most of the water toward one side in a pattern somewhat resembling one-quarter of a sphere. A small proportion of the discharge wets the wall behind the sprinkler. The forward horizontal range of about 15 ft (4.6 m) is greater than that of a standard sprinkler. Located and mounted either vertically or horizontally along the junction between a ceiling and sidewall, sidewall sprinklers provide protection adequate for light hazard occupancies, such as hotel lobbies, dining rooms, executive offices, and other areas where the usual sprinkler pipes would be objectionable in appearance. Some sidewall sprinklers have been tested and listed for use in ordinary hazard occupancies. Sidewall sprinklers are not used, however, in situations where a standard sprinkler system can be installed without detracting from decorative schemes. They have been used extensively in light hazard occupancies where special appearance and protection were desired.

The directional character of the discharge from sidewall sprinklers makes them applicable to occasional special protection problems. They may be installed to give discharge in any desired direction.

Figure 18-3Y shows typical sidewall sprinklers. Selection has been made to show the different shapes of a variety of deflectors. Many vertically mounted sidewall sprinklers may be installed in either the pendent or upright position.

Extended Coverage Sidewall Sprinklers: These are special sidewall sprinklers used in the horizontal position that have larger areas of coverage than allowed for conventional sidewall sprinklers. They are used in light hazard occupancies, particularly in hotels and similar occupancies where a sprinkler system can be installed in an existing building without having piping exposed in living areas, which could be objectionable aesthetically.

The water pressure required to obtain the greater coverage is specified in the listings for the sprinklers and is greater than that required for conventional sidewall sprin-

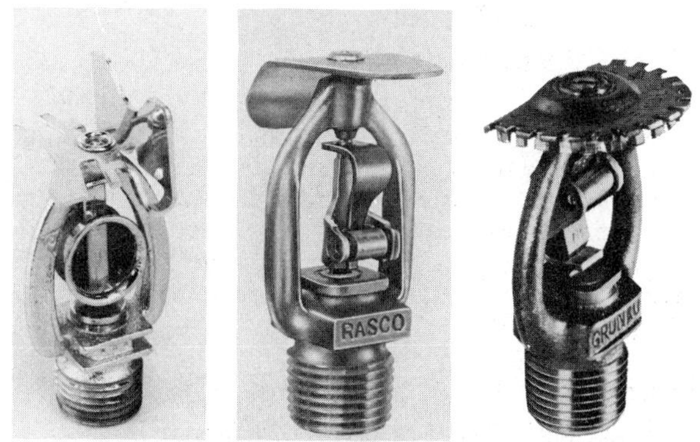

FIG. 18-3Y. A representative selection of listed sidewall sprinklers showing various shapes of deflectors.

klers. Installation requirements are also a part of the listing.

An extended coverage sidewall sprinkler with a conventional link release is shown in Figure 18-3Z.

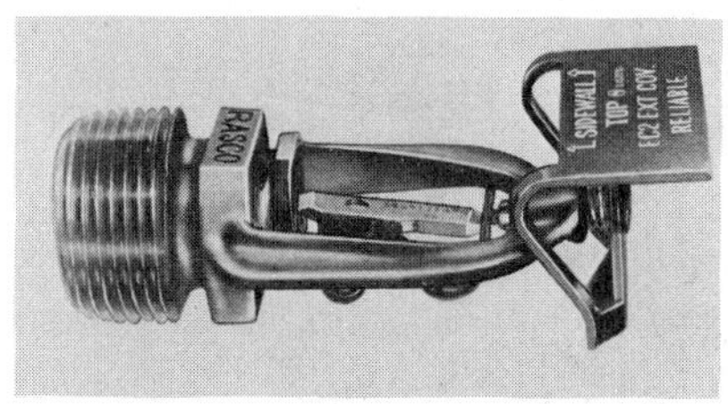

FIG. 18-3Z. A representative extended coverage sidewall sprinkler.

Sprinklers Without Operating Elements

Standard automatic sprinklers, or sidewall automatic sprinklers with the valve cap and heat responsive elements omitted, are used in deluge sprinkler systems where the water supply is controlled by an automatic water control valve actuated independently of automatic sprinklers. (See Fig. 18-3AA.) The water distribution pattern and the density of the discharge of the open head system are designed to be appropriate for the hazard to be protected.

Small and Large Orifice Sprinklers

Automatic sprinklers having a water discharge rate greater or less than that of a standard ½ in. (12.7 mm) orifice sprinkler operating at the same pressure, have characteristics desirable in protecting certain occupancies. High discharge, large orifice sprinklers operating at discharge densities of 0.45 gpm/ft^2 {18 (L/min)m^2} and greater provide a quantity discharge not available from standard nominal ½ in. (12.7 mm) orifice sprinklers. The pattern of the water discharge from small and large orifice sprinklers is similar to that of the standard ½-in. (12.7 mm) sprinkler. The NFPA Sprinkler Systems Standard, while recognizing large and small orifice automatic sprinklers, does not give

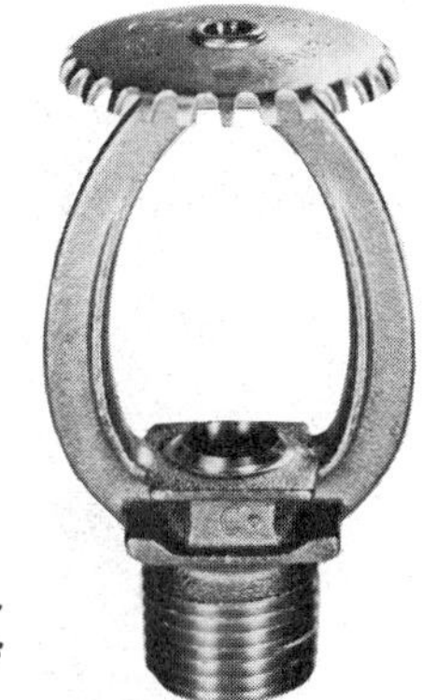

FIG. 18-3AA. A standard upright sprinkler with its operating elements removed.

pipe schedules or rules for spacing. Each installation calls for special study.

Whenever sprinklers are replaced, care must be taken to replace them with the proper styles. Replacement of correct orifice size is critical in hydraulically calculated systems.

Small Orifice Automatic Sprinklers: These sprinklers have been listed with orifices ranging from ¼ through 7/16 in. (6.4 through 11.1 mm). Small orifice sprinklers are identified by a pintle extending above the deflector and by the size stamped on the base of the sprinkler. (See Fig. 18-3BB.)

FIG. 18-3BB. A small orifice sprinkler in the pendent position (note pintle extending from deflector).

Large Orifice Automatic Sprinklers: These sprinklers have a discharge rate of 140 percent of that of the standard nominal ½ in. (12.7 mm) sprinkler operating at the same pressure. (See Fig. 18-3CC.) Large orifice sprinklers for use in new sprinkler installations are identified by a ¾ in. (19.1 mm) iron pipe thread connection. For use in existing installations, large orifice sprinklers are manufactured with a ½ in. (12.7 mm) iron pipe thread connection and an identifying pintle on the deflector and size marking on the sprinkler frame.

The discharge characteristics of various large and small orifice sprinklers are given in Table 18-3D.

Pull Type Sprinklers

A pull style sprinkler is a special purpose sprinkler combining in one unit a frangible bulb automatic sprinkler

FIG. 18-3CC. A representative large orifice sprinkler. (Viking Corporation)

with a supplementary pull style mechanical release that can be operated remotely by either manual or automatic means by connection to a pull chain or wire cord. (See Fig. 18-3DD.) This sprinkler may be mounted at any angle required for the protection of the interior of duct work, air filters, or other small enclosed spaces. It may be fitted with either standard or special deflectors which must be selected in advance, considering the pattern of water distribution that is desired.

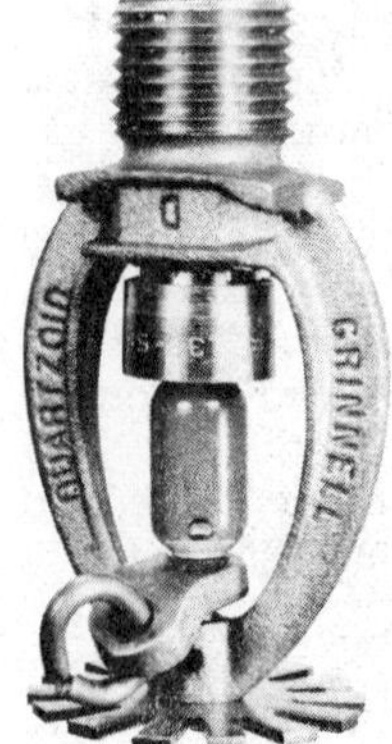

FIG. 18-3DD. A pull style sprinkler which was made with pendent or upright deflectors (note mechanical release between the deflector and frangible bulb). (No longer manufactured)

Picker Trunk Sprinklers

Picker trunk automatic sprinklers (Fig. 18-3EE) are sprinklers with a small, smooth deflector which aids in reducing collections of lint and fiber on the sprinklers when placed inside ducts or enclosures where moving air carries such foreign materials in suspension. Freedom from obstruction and a general breakup of the water stream are of more importance than any specific pattern of distribution.

Intermediate Level Sprinklers

Intermediate level or baffle sprinklers, sometimes referred to as rack storage sprinklers, have large discs (baffles) designed to shield the thermosensitive assembly from impingement from the spray of sprinklers suspended at higher levels. Without the protective discs, the impinging water could cool the thermosensitive element and retard sprinkler operation. Both upright and pendent rack storage sprinklers are listed by testing organizations. (See Fig. 18-3FF.)

TABLE 18-3D. Sprinkler Discharge Characteristics Identification

Nominal Orifice Size (in.)	Orifice Type	"K" Factor*	Percent of Nominal ½ In. Discharge	Thread Type	Pintle	Nominal Orifice Size Marked On Frame
¼	Small	1.3–1.5	25	½ in. NPT	Yes	Yes
5/16	Small	1.8–2.0	33.3	½ in. NPT	Yes	Yes
⅜	Small	2.6–2.9	50	½ in. NPT	Yes	Yes
7/16	Small	4.0–4.4	75	½ in. NPT	Yes	Yes
½	Standard	5.3–5.8	100	½ in. NPT	No	No
17/32	Large	7.4–8.2	140	¼ in. NPT	No	No
				or ½ in. NPT	Yes	Yes

For SI Units: $Q_m = K_m \sqrt{P_M}$, Where Q_m = Flow in L/min, P_m = Pressure in bars, and $K_m = 14\ K$

* K factor is the constant in the formula $Q = K\sqrt{P}$, where Q equals flow in gpm and P equals pressure in psi.

Universal Deflector Sprinklers

These are a style of sprinkler of current manufacture employing a "universal" deflector somewhat similar in appearance to an old style sprinkler deflector. (See Fig. 18-3GG.) The sprinkler produces a water distribution pattern with a portion of the discharge directed upward toward the ceiling and the majority of the discharge downward. Sprinklers of this style are found commonly in European countries and other foreign jurisdictions where sprinklers with either upright or pendent style deflectors are not specifically required as part of installation practices.

SPRINKLERS OF THE FUTURE

The development of the residential sprinkler with its fast acting link has led to test work being performed using links from residential sprinklers with conventional deflectors. The basic objective in the use of sprinkler systems for fire protection is to apply a minimum amount of water on the seat of the fire while the fire is still in its incipient stage. These new and faster acting sprinklers pursue that objective, and could very well change fire protection

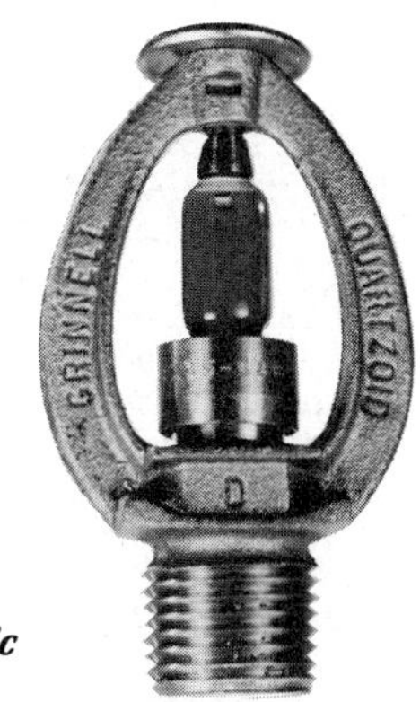

FIG. 18-3EE. A picker trunk automatic sprinkler. (Grinnell)

FIG. 18-3FF. Intermediate level (rack storage) sprinkers showing integral shields that protect operating elements from the discharge of sprinklers installed at higher levels. (Star Model LD)

FIG. 18-3GG. A sprinkler with a "universal" deflector design. This style of sprinkler finds service in foreign jurisdictions not requiring sprinklers of specific upright or pendent design. (Reliable Automatic Sprinkler Co.)

sprinkler systems as much as the advent of the spray sprinkler in the early fifties. An extended coverage sidewall sprinkler with a linkage arrangement representative of a residential style sprinkler is shown in Figure 18-3HH.

OLDER TYPE AUTOMATIC SPRINKLERS

Until the 1955 NFPA Sprinkler Systems Standard was adopted, the present old style automatic sprinklers were designated as standard or regular style sprinklers. These sprinklers were listed for use in either the upright or pendent position. Since 1955, under the designation of old style, their use has been restricted to the replacement of sprinklers of the same (old) style. Several sprinkler manufacturers can supply old style sprinklers of the last models they manufactured before the standard sprinkler was adopted. It is generally considered good practice to replace sprinklers over 50 years old. If not replaced, a representative sample must be tested.

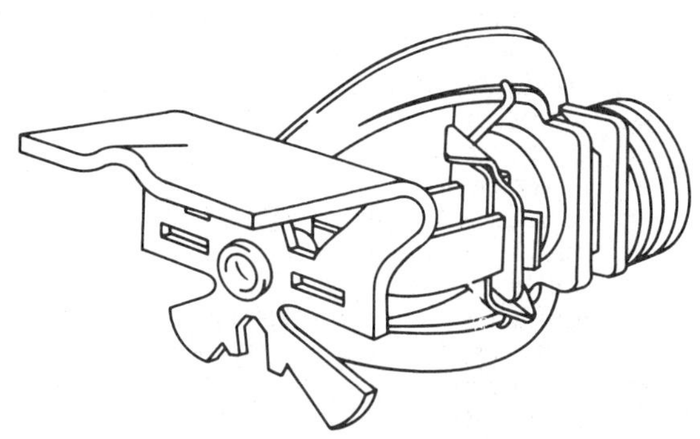

FIG. 18-3HH. An extended coverage sidewall sprinkler with a residential style fusible link. It is representative of the new generation of quick response sprinklers now being developed. (Grinnell)

With few exceptions, all parts of the old style sprinkler, other than the deflector, were retained for the then-new standard sprinkler. The model or issue designations also were usually continued. Some representative old style sprinklers are shown in Figure 18-3II.

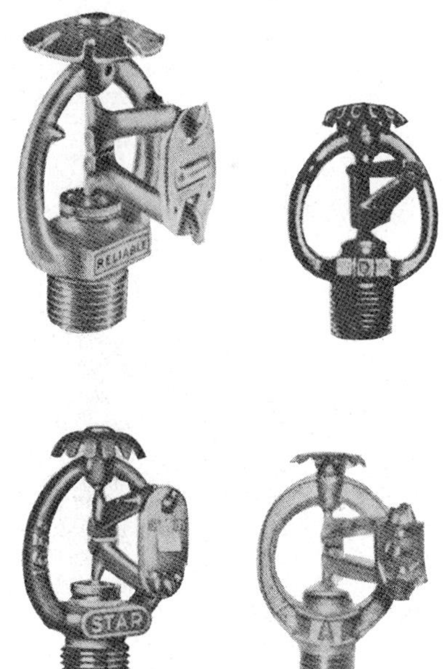

FIG. 18-3II. Four old style sprinklers. The principal difference between them and currently listed spray sprinklers is in the design of the deflector.

Sprinklers of earlier manufacture than the old style sprinklers may still be found in service, although as a general rule they have been replaced with standard sprinklers. In no case should they be allowed to remain in service unless representative samples have been removed and tested in a sprinkler testing laboratory. The remaining sprinklers should be considered dependable only if all of the tested samples operate within time limits considered satisfactory by the testing organization.

Many early styles of sprinklers have been produced and installed, and information concerning them is quite complete in the files of testing organizations. Arrangements can usually be made with these organizations for obtaining historical information and for testing representative sprinklers removed from systems still in service. A list of the more common styles of sprinklers of early manufacture that may be found in service today is given in Table 18-3E.

Table 18-3E. Automatic Sprinklers of Early Manufacture

Key: O = Obsolete: Q = Questionable; S = Serviceable

Acme B	1920 (S)	International I	1900 (O)
Associated A	1913 (S)	International A	1902 (O)
Associated B	1914 (S)	International B	1905 (S)
Automatic A-1920	1920 (S)	International C	1927 (S)
Automatic A	1921 (S)	J. Kane 3	1900 (O)
Cataract A	1906 (O)	J. Kane 4	1902 (O)
Cataract B	1907 (Q)	J. Kane 4½	1902 (O)
Central A	1920 (S)	Lapham 4-(1902)	1902 (O)
Clayton	1906 (O)	Lapham 4-(1903)	1903 (O)
Crowder A	1909 (S)	Lapham A	1910 (O)
C.S.B. A	1919 (S)	Lapham B	1911 (S)
Esty 5	1896 (S)	Manufacturers A	1895 (O)
Esty 6	1896 (S)	Manufacturers B	1903 (S)
Esty B	1910 (S)	Manufacturers C	1905 (S)
Evans 02	1902 (O)	Nacey A	1923 (S)
Evans A	1913 (S)	Nacey B	1944 (S)
Evans B	1914 (S)	Neracher 4	1895 (O)
FM A	1919 (S)	Neracher 5	1902 (O)
(FM) KME B	1919 (S)	Neracher 6	1902 (S)
F.P.C. B	1953 (S)	New York	1911 (O)
Garrett-Globe A	1911 (O)	Niagra B	1912 (S)
Garth A	1915 (S)	Niagra-Hibbard A	1902 (O)
Garth B	1918 (S)	Niagra-Hibbard B	1904 (S)
Globe A	1913 (O)	Phoenix A	1905 (O)
Globe B	1914 (S)	"Reliable" A	1919 (S)
Globe C	1916 (S)	"Reliable" B	1921 (S)
Globe E	1955 (S)	"Reliable" C	1944 (S)
Globe Saveall D	1930 (S)	Rockwood A	1906 (O)
Grimes A	1917 (O)	Rockwood B	1906 (S)
Grimes B	1922 (S)	Rockwood C	1910 (S)
Grimes C	1925 (S)	Rockwood D	1911 (S)
Grimes D	1935 (S)	Rockwood E	1934 (S)
Grinnell Glass Button	1890 (O)	Rundle Spence	1911 (O)
Grinnell Glass Button A	1897 (S)	Simplex	1902 (O)
Grinnell Quartz Bulb A	1923 (S)	Standard	1902 (O)
Grinnell Quartz Bulb B	1927 (S)	Star A	1924 (S)
Grinnell Duraspeed B	1932 (S)	Star B	1935 (S)
Grinnell Quartzoid Bulb C	1934 (S)	Star C	1955 (S)
Hibbard 2	1894 (O)	Star D	1954 (S)
Hibbard 3	1897 (O)	Superior Super-B	1954 (S)
Hibbard 3A	1897 (O)	United States A	1917 (O)
Hibbard 4	1901 (O)	United States B	1922 (S)
Hibbard 5	1909 (O)	Viking A	1919 (S)
Hibbard H	1911 (S)	Viking B	1935 (S)
Hibbard I	1912 (S)	Viking C	1949 (S)
Hodgman A	1918 (S)	Vogel A	1954 (S)
Hodgman B	1938 (S)	Vogel (Laconia)	1904 (Q)
Ideal A	1914 (O)	Witter E	1906 (Q)
Independent A	1916 (S)		

Bibliography

NFPA Codes, Standards, Recommended Practices and Manuals. (See the latest *NFPA Codes and Standards Catalog* for availability of current editions of the following documents.)

NFPA 13, *Standard for the Installation of Sprinkler Systems.*
NFPA 13A, *Recommended Practice for the Inspection, Testing, and Maintenance of Sprinkler Systems.*
NFPA 13D, *Standard for the Installation of Sprinkler Systems in One- and Two-Family Dwellings and Mobile Homes.*

Additional Readings

Beyler, Craig, L., "Effect of Selected Variables on the Distribution of Water from Automatic Sprinklers," NBS-GCR-77-105, National Bureau of Standards, Washington, DC, June 1977.
Bryan, John L., "The Automatic Sprinkler Head," *Automatic Sprinkler and Standpipe Systems*, NFPA, Quincy, MA, 1976, pp. 166-217.
Campbell, Layard E., "A Study of Sprinkler Sensitivity," *Fire Technology*, Vol. 5, No. 2, May 1969, pp. 93-99.
Evans, D. D., and Madrzykowski, D., "Characterizing the Thermal Response of Fusible-Link Sprinklers," NBSIR 81-2329, National Bureau of Standards, Washington, DC, Aug. 1981.
Evans, D. D., "Calculating Sprinkler Actuation Time in Compartments," National Bureau of Standards, Washington, DC, 1984.
Factory Mutual Approval Guide, Factory Mutual Engineering Corporation, (published annually) Norwood, MA.
"Final Report on Development of a Nitinol-Actuated Fire Sprinkler," Battelle Columbus Laboratories, Columbus, OH, Sept. 15, 1982.
"Fire Protection —Automatic Sprinkler Systems —Part I: Requirements and Test Methods for Sprinklers," DIS 6182/1, International Standards Organization, Geneva, Switzerland, 1984.
Fire Protection Equipment List, Underwriters Laboratories Inc., Northbrook, IL, published in January with a supplement in July.
Groos, Richard T., "Large-Drop Sprinklers: High Challenge," *Plumbing Engineer*, Vol. 13, No. 2, March/April 1985, pp. 32–36.
Hammerman, David M., "Project Home-A Pilot Program for a Low-Cost Residential Sprinkler System," *Fire Journal*, Vol. 75, No. 2, March 1981, pp. 66-69.
Heskestad, G., and Smith, H. F., "Investigation of a New Sprinkler Sensitivity Approval Test: The Plunge Test," FMRC Technical Report 22485, Factory Mutual Research Corporation, Norwood, MA, Dec. 1976.
Heskestad, G., "The Sprinkler Response Time Index (RTI)," Presented at the Technical Conference on Residential Sprinkler Systems, Factory Mutual Research Corp., Norwood, MA, April, 1981.
Hodnett, Robert A., ed., *Automatic Sprinkler Handbook*, 2nd ed., National Fire Protection Association, Quincy, MA, 1985.
Kung, H.C., et al., "Field Evaluation of Residential Prototype Sprinkler: Los Angeles Fire Test Program," *FMRC J.I. OEOR3.RA(1)*, Factory Mutual Research Corp., Norwood, MA, Feb. 1982.
Labes, Willis G., "Evaluation of Fire Protection Spray Devices: The State of the Art," NBS-GCR-76-72, National Bureau of Standards, Washington, DC, June, 1976.
Merdinyan, Philip H., "A Fully Approved On-Off Sprinkler," *Fire Journal*, Vol. 67, No. 1, Jan. 1973, p. 10.
"Standard for Automatic Sprinklers for Fire-Protection Service," Underwriters Laboratories Inc., Northbrook, IL, 1974.
Suchomel, M. R., "Factors Influencing the Use of High-Temperature Sprinklers," *Fire Technology*, Vol. 1, No. 1, Feb. 1965, pp. 15-22.

Suchomel, M. R., and Castino, G. T., "Sprinkler Performance Tests —The Extended Coverage Panacea," *Fire Technology*, Vol. 16, No. 2, May 1980, pp. 85-93.

Yao, C., and Kalelkar, A. S., "Effect of Drop Size on Sprinkler Performance," *Fire Technology*, Vol. 6. No. 4, Nov. 1970, pp. 254-268.

Yao, Cheng, "Early Fire Suppression with Fast-Response Sprinklers," *Sprinkler Quarterly* , Winter, 1983-84, pp. 19-23.

Yao, Cheng, "The ESFR Sprinkler System: A New Approach to High-Challenge Storage Protection," *Fire Journal*, Vol. 79, No. 2, March 1985, pp. 30–33, 70–72.

Yao, Cheng and Marsh, William S., "Early Suppression—Fast Response: A Revolution in Sprinkler Technology," *Fire Journal*, Vol. 78, No. 1, Jan. 1984, pp. 42–46.

RESIDENTIAL "QUICK-RESPONSE" SPRINKLERS

Arthur E. Cote, P.E.

Until the 1970s, automatic sprinkler systems had been installed almost exclusively in industrial and commercial properties for property protection and as a means of reducing fire insurance premiums. In 1973, however, a Presidential Commission on Fire Prevention and Control declared in its report, *America Burning*, that a more active, built-in automatic suppression system was needed in the home to detect and attack fire in its earliest stages. The Commission's concern for home firesafety was borne out in the ensuing years. Even though a decrease in home fire deaths has been experienced (Fig. 18-4A), fires in the home

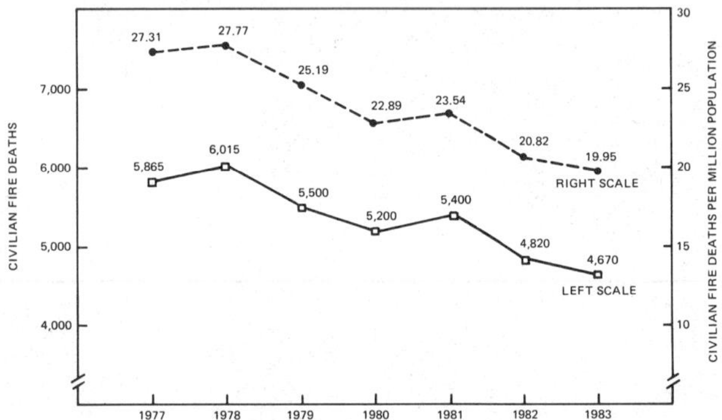

FIG. 18-4A. Civilian fire deaths and rates in the home in the United States (1977-1983).

still accounted for an average of approximately 80 percent of all fire fatalities from 1977 to 1983 in the United States.

This chapter covers the development of residential "quick-response" sprinklers as effective suppression devices in the residential environment. It also discusses some of the barriers encountered in the widespread use of residential sprinklers and some of the incentives that can be considered in encouraging their use.

Mr. Cote is Assistant Vice President, Standards, at NFPA and editor of the *Fire Protection Handbook*, 16th ed.

HISTORICAL BACKGROUND

In 1973, in response to recommendations in the Presidential Commission report, *America Burning*, the NFPA Committee on Automatic Sprinklers appointed a Subcommittee on Residential and Light Hazard Occupancies to prepare a residential sprinkler standard. In 1975, the first edition of NFPA 13D, *The Installation of Sprinkler Systems in One- and Two-Family Dwellings and Mobile Homes*, was published, based on expert judgment and the best available information at that time.

The purpose of the standard was "to provide a sprinkler system that will aid in the detection and control of dwelling fires and thus provide improved protection against injury, life loss, and property damage." The standard required the use of an NFPA 13, *Standard for the Installation of Sprinkler Systems*, light hazard water application density of 0.10 gpm/ft^2 [4 (L/min)/m^2], but permitted other concessions. The water supply could be based on the area of the largest room or 25 gpm (95 L/min), whichever was less, the total water supply required was to be only 250 gal (946 L), and sprinkler spacing of 256 sq ft (23.7 m^2) was permitted even though NFPA 13 allowed only 225 ft^2 (20.9 m^2). Further, 13D permitted sprinklers to be omitted from certain areas where the incidence of life loss from fires was shown statistically to be low. NFPA 13 had always required complete sprinkler protection in order to properly safeguard property. In departing from this ideal, the 1975 edition of NFPA 13D became the first attempt at a "life safety" sprinkler standard. In spite of these concessions, actual installations based on this standard were rare, primarily due to cost.

Residential Sprinkler Research

The standard sprinkler of the 1970s was really little changed from its 1870s ancestor. Except for the introduction of the spray sprinkler in the early 1950s to improve the sprinkler discharge pattern, sprinkler design, including the sensitivity of release mechanisms, remained relatively unchanged. Sprinklers were essentially devices for property protection, although their effectiveness in life saving was exemplary.

Beginning in 1976, the NFPCA (National Fire Prevention and Control Administration), which was later renamed the USFA (U.S. Fire Administration), acting on its mandate to reduce the nation's fire losses, funded research programs focusing on the residential fire problem in general and residential sprinkler protection in particular. The NFPCA/USFA programs included studies to assess the impact of using sprinklers to reduce deaths and injuries in residential fires (Halpin et al 1978). Other studies evaluated the design, installation, practical usage, and user acceptance factors that would have an impact on achieving reliable and acceptable systems (Yurkonis 1978); the minimum water discharge rates and automatic sprinkler flow required, and response sensitivity and design criteria (Kung et al 1978, Henderson et al 1978, Clark 1978); and full scale tests of prototype residential sprinkler systems (Kung et al 1980, Cote & Moore 1980, Moore 1980, Kung et al 1982 and Cote 1982.)

The research showed that a more sensitive sprinkler was needed to respond faster to both smoldering and fast developing residential fires if they were to be controlled with the water supplies typically available in residences, i.e., 20 to 30 gpm (76 to 114 L/min), and if low costs were to be achieved.

Full scale tests conducted by Factory Mutual Research Corporation resulted in the development of a prototype quick-response sprinkler which could control or suppress typical residential fires with the operation of not more than two sprinklers. It could also operate fast enough to maintain survivable conditions within the room of fire origin (Kung et al 1980). Survivable conditions were established as follows:

1. Maximum gas temperature at eye level —200°F (93°C).
2. Maximum ceiling surface temperature —500°F (260°).
3. Maximum carbon monoxide concentration —1,500 parts per million

Thus, the concept changed from the traditional one of property protection to one of life safety. Full scale field tests were then conducted in Los Angeles to establish system design parameters using the new prototype "quick-response" residential sprinkler developed by Grinnell Fire Protection Systems Company (Cote & Moore 1980, Moore 1980, Kung et al 1982, Cote 1982).

The data from these tests were studied by the National Fire Protection Association Technical Committee on Automatic Sprinklers, and were used to establish the criteria for the 1980 edition of NFPA 13D, *Standard for the Installation of Automatic Sprinklers in One- and Two-Family Dwellings and Mobile Homes*, referred to hereinafter in this chapter as the NFPA Residential Sprinkler Systems Standard.

NFPA 13D DESIGN REQUIREMENTS

The basic design requirements in the NFPA Residential Sprinkler Systems Standard are:

Performance Criteria: To prevent flashover in the room of fire origin, when sprinklered, and to improve the chance for occupants to escape or be evacuated.

Design Criteria:

1. Only listed residential sprinklers shall be used.
2. Deliver 18 gpm (68 L/min) to any single operating sprinkler, and 13 gpm (49 L/min) to all operating sprinklers in the design area up to a maximum of two sprinklers.
3. Maximum area protected by a single sprinkler is 144 sq ft (13.4 m^2).
4. Maximum distance between sprinklers is 12 ft (3.7 m).
5. Minimum distance between sprinklers is 8 ft (2.4 m).
6. Maximum distance from a sprinkler to a wall or partition is 6 ft (1.8 m).

Application rates, design areas, areas of coverage, and minimum design pressures other than those specified above may be used with special sprinklers that have been listed for such special residential installation conditions.

Sprinkler Coverage: Sprinklers shall be installed in all areas with the following exceptions:

Exception No. 1: Sprinklers may be omitted from bathrooms not exceeding 55 sq ft (5.1 m^2) with noncombustible plumbing fixtures.

Exception No. 2: Sprinklers may be omitted from small closets where the least dimension does not exceed 3 ft (0.9 m) and the area does not exceed 24 sq ft (2.2 m^2) and the walls and ceiling are surfaced with noncombustible materials.

Exception No. 3: Sprinklers may be omitted from open attached porches, garages, carports, and similar structures.

Exception No. 4: Sprinklers may be omitted from attics and crawl spaces that are not used or intended for living purposes or storage.

Exception No. 5: Sprinklers may be omitted from entrance foyers that are not the only means of egress.

Listed Residential Sprinklers

The design criteria, in the 1980 edition of the NFPA Residential Sprinkler Systems Standard, included for the first time the requirement that all sprinklers be listed "residential" sprinklers.

One of the major differences between standard sprinklers and "residential" sprinklers is their response or sensitivity. Residential sprinklers are designed for "quick response" and operate much faster than standard sprinklers because they are designed to have less of a "thermal lag."

Sprinkler response time as a function of the temperature rating of the fusible element or link is well understood, i.e., a 165°F (74°C) rated sprinkler will operate when its temperature becomes 165°F (74°C), i.e., plus or minus 5°F(15°C). However, because of thermal lag of the link mass, the air temperature may be as high as 1,000°F (538°C) before the element operates.

Residential sprinklers are listed by manufacturer in Table 18-4A.

Residential Sprinkler Sensitivity

The actual sensitivity requirements of residential sprinklers were arrived at somewhat by trial and error during the developmental test work on the NFPA Residential Sprinkler Systems Standard. To measure sensitivity, Factory Mutual researchers first developed the concept of the "tau" (τ) factor and later the Response Time Index (RTI).

Both the tau factor and RTI refer to the performance of a sprinkler or link in a standardized air oven tunnel test. The test is known as a "plunge" test because a sprinkler at room temperature is plunged into a heated air stream (Heskested and Smith 1976 and 1980). The tau factor is the time when the excess temperature of the sensing element of the sprinkler is approximately 63 percent of the excess gas temperature, i.e., when the temperature of the sprinkler link has risen 63 percent of the way to the higher temperature of the heated air. The smaller the tau factor, the faster the sprinkler sensing element heats up and operates.

The tau factor is independent of the air temperature used in the plunge test, but is inversely proportional to the square root of the air velocity. During the development of the 1980 edition of the NFPA Residential Sprinkler Systems Standard, a tau factor of 21 sec was considered to indicate the needed level of sensitivity, but this was associated with a specific velocity (5 ft/sec or 1.52 m/sec) used in the Factory Mutual plunge test. Since the tau factor changes with the velocity of heated air past the sprinkler, it is a fairly inconvenient measure of sprinkler sensitivity.

The RTI has replaced the tau factor as the measure of sensitivity, and is determined by simply multiplying the tau factor by the square root of the air velocity at which it is found. The RTI is therefore practically independent of both air temperature and air velocity. Comparisons of RTI give a good indication of relative sprinkler sensitivity.

The smaller the RTI, the faster the sprinkler operation. Standard sprinklers have RTIs in the range of 225 to 700 $sec^{1/2}/ft$ ($100^{1/2}$ to 400 $sec^{1/2}/m^{1/2}$),while the RTI for residential sprinklers is about 50 $sec^{1/2}/ft^{1/2}$ (28 $sec^{1/2}/m^{1/2}$).

The tau factor of 21 sec at 5 ft/sec converts to an RTI of 25.9 $sec^{1/2}/m^{1/2}$. Factory Mutual has set the maximum RTI for residential sprinklers at 55 $sec^{1/2}/ft^{1/2}$. (In English units an RTI is 1.81 times its metric value.)

Residential Sprinkler Distribution

In addition to the increased sensitivity, residential sprinklers differ from standard sprinklers in other ways. The most crucial is probably the distribution pattern (Kung et al 1980). Because the effective control of residential fires often depends on a single sprinkler in the room of fire origin, the distribution of residential sprinklers is required to be more uniform than that of standard sprinklers, which in large areas can rely upon the overlapping patterns of several sprinklers to make up for voids. Additionally, residential sprinklers are required to protect sofas, drapes and similar furnishings at the periphery of the room. The sprinklers' discharge spray patterns, therefore, must not only be capable of throwing water to the walls of their assigned areas, but high enough up on the walls to prevent the fire from getting "above" the sprinkler. The water delivered close to the ceiling not only protects the portion of the wall close to the ceiling but also enhances the capacity of the spray to cool gases at the ceiling level, thus reducing the likelihood of excessive sprinkler openings (Kung et al 1980).

Because of their differences, residential sprinklers are not listed by product evaluation organizations under the same product standards as standard sprinklers. Underwriters Laboratories Inc., for example, has developed UL 1626 for residential sprinklers, and Factory Mutual Research has published its Approval Standard 2030 for residential sprinklers. Both of these standards include a plunge test with specific sensitivity requirements and a distribution test that checks the spray pattern in the vertical plane as well as the horizontal plane. The product standards for standard sprinklers contain neither test.

Both UL 1626 and FM 2030 include fire test procedures that simulate a residential fire in the corner of a living room containing combustible materials that are representative of a living room environment. The fire test arrangements for UL 1626 are shown in Figure 18-4B.

The UL 1626 test procedure contains a time-temperature curve that is reproducible in actual tests. The curve parallels the time-temperature relationship developed during a series of tests of a prototype residential sprinkler in Los Angeles that was part of the on-going research effort (Cote & Moore 1980, Kung et al 1982, Cote 1982).

In summary, to meet test criteria, residential sprinklers, when installed in a fire test enclosure with an 8 ft (2.4 m) ceiling, are required to control a fire for 10 min with the following limits:

1. The maximum gas (air) temperature adjacent to the sprinkler—3 in. (76.2 mm) below ceiling and 8 in. (203 mm) horizontally away from sprinkler—shall not exceed 600°F (316°C).
2. The maximum temperature at 5 ft, 3 in. (1.6 m) above the floor and located at a distance of half the room length from each wall shall be less than 200°F (93°C) during the entire test. This temperature shall not exceed 130°F (54°C) for more than a 2 min period.
3. The maximum temperature at ¼ in. (6.3 mm) behind the finished surface of ceiling material directly above the test fire shall not exceed 500°F (260°C).
4. No more than two residential sprinklers in a test area shall operate.

In a typical test, the walls of one corner of the test enclosure are covered with combustible ⅛ in. (3.2 mm) thick cellulosic acoustical panels. Two combustible structures representing urethane-foam stuffed furniture are positioned at the same corner. (See Fig. 18-4B.) The enclosure is maintained at an initial ambient temperature of 80°F (27°C) ± 5°F or 3°C, and it is ventilated through two door openings on opposite walls. The fire source is to consist of a wood crib of 12 to 13 lb (5.4 to 5.9 kg) mass, measuring approximately 12 by 12 by 12 in. (0.3 by 0.3 by 0.3 m).

The wood crib is ignited by 8 oz (24 mL) of N-heptane in a pan directly below the crib. At 40 sec after ignition, the excelsior (¼ lb or 0.11 kg) located on the floor adjacent to the urethane foam furniture cushions (representing stuffed chairs) is ignited.

The fire test is conducted for 10 min after the ignition of the wood crib. The water flow to the first sprinkler that operates and the total water flow when the second sprinkler operates are specified as part of the listing limitations for the sprinklers in the test.

Water Distribution

The water distribution test requirements are based on the distribution pattern of the prototype residential sprinkler used in the Los Angeles test fires (Cote 1982). The distribution requirements involve both horizontal and vertical collection parameters.

All residential sprinklers in the test must discharge water at the flow rate specified by the manufacturer for a 10 min period simulating (1) one sprinkler operating, and

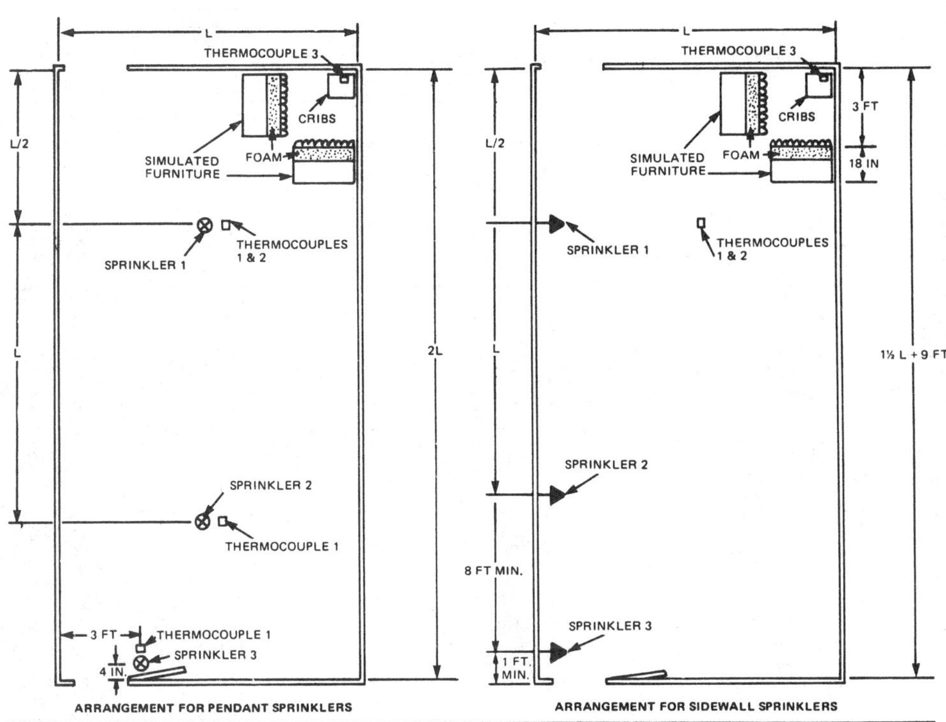

KEY: ⊗ RESIDENTIAL PENDANT SPRINKLER
▶ RESIDENTIAL SIDEWALL SPRINKLER
L SPRINKLER SPACING SPECIFIED BY MANUFACTURER

□ THERMOCOUPLE LOCATION
1. 3 IN. BELOW CEILING AND 8 IN. FROM SPRINKLER
2. 30 IN. BELOW CEILING, AND DIRECTLY BELOW SPRINKLER 1
3. DIRECTLY ABOVE WOOD CRIB AND EMBEDDED IN CEILING MATERIAL

FIG. 18-4B. Fire test arrangements for UL 1626 for pendent sprinklers (left) and sidewall sprinklers (right). The simulated furniture consists of 3 in. (76 mm) thick uncovered urethane foam cushions 30 in. (762 mm) high by 30 in. (762 mm) wide. The foam has a density of 1.25 lbs/ft³ (6.1 kg/m³) and is attached to a wood frame. The walls of the test room are covered with 4 by 8 ft by ⅛ in. (1.2 × 2.4 m × 3.2 mm) decorative plywood paneling (flame spread rating 200) attached to wood furring strips. The fire source consists of 12 by 12 by 12 in. (0.3 × 0.3 × 0.3 m) wood cribs weighing 12 to 13 lbs (5.4 to 5.9 kg). The ceiling of the test room is 8 ft (2.4 m) high and covered with 2 × 4 ft × ½ in. (6.1 × 1.2 m × 12.7 mm) thick acoustical panels attached to wood furring strips.

(2) two sprinklers operating. The quantity of water collected on both the horizontal and vertical surfaces is measured and recorded.

Sprinklers under test are required to discharge a minimum of 0.02 gpm per sq ft [0.8 (L/min/m²] at both flow rates over the entire design area, plus an area 2 ft (0.61 m) beyond the listed coverage area in each direction (only three directions for horizontal sidewall type sprinklers). They must also wet the walls of the test enclosure to a height not less than 28 in. (711 mm) below the ceiling with one sprinkler operating, and not less than 36 in. (914 mm) with two sprinklers operating. Each wall surrounding the coverage area is to be wetted with a minimum of 5 percent of the sprinkler application densities at both flow rates.

Figure 18-4C shows the first two listed residential "quick-response" sprinklers.

Table 18-4A lists residential sprinklers meeting UL 1626 requirements along with their pressure and flow requirements:

RESIDENTIAL SPRINKLERS IN OTHER OCCUPANCIES

Residential sprinklers may be installed in buildings other than one- and two-family dwellings and mobile homes under certain specified conditions. Those conditions are covered in NFPA 13, *Standard for the Installation of Sprinkler Systems*. Essentially, NFPA 13 allows residential sprinklers in dwelling units located in any occupancy provided they are installed in conformance with the requirements of their listing and the positioning requirements of the NFPA Residential Sprinkler Systems Standard. A dwelling unit is defined as one or more rooms arranged for the use of one or more individuals living together, as in a single housekeeping unit normally having cooking, living, sanitary, and sleeping facilities. Dwelling units include hotel rooms, dormitory rooms, sleeping rooms in nursing homes, and similar living units. Occupancies encompassing dwelling units include apartment buildings, board and care facilities, dormitories, condo-

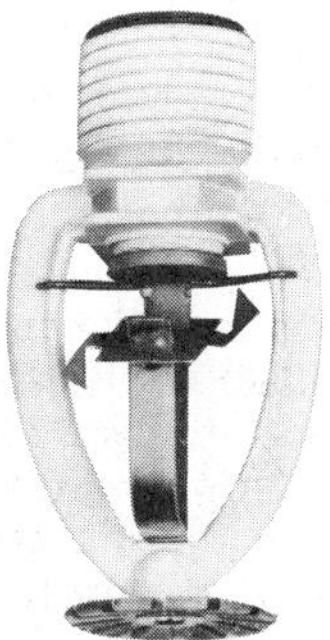

FIG. 18-4C. The first two listed residential "quick-response" sprinklers. Shown are the Grinnell Model F954 (left) and Central "Omega" Model R-1 (right) pendent sprinklers.

Table 18-4A. Residential Sprinklers Characteristics

(Residential sprinklers listed by Underwriters Laboratories Inc.)

				Temperature Rating		Maximum Spacing		Minimum 1 Sprinkler Flow		Minimum 1 Sprinkler Pressure		Minimum 2 Sprinkler Flow		Minimum 2 Sprinkler Pressure	
Manufacturer	Model	Style	K	°F	°C	ft	m	gpm	L/min	psi	kPa	gpm	L/min	psi	kPa
Grinnell Fire (GEM	F954	pendent	2.8	165	74	12×12	3.7×3.7	18	68	41.0	283	13	49	22.0	152
	F958	horizontal sidewall	4.2	165	74	14×14	4.3×4.3	24	91	33.0	228	17	64	16.5	114
	F991 Aquarius	flush pendent	4.2	160	71	14×14	4.3×4.3	18	68	18.4	127	13	49	9.6	66
						16×16	4.9×4.9	22	83	27.4	189	16	61	14.5	100
	F993 Aquarius	flush horizontal sidewall	5.4	160	71	14×14	4.3×4.3	30	114	30.9	213	21	79	25.0	172
Central Sprinkler	R-1 Omega	flush pendent	3.85	145	63	12×12	3.7×3.7	18	68	21.9	151	13	49	11.4	79
				160	71	14×14	4.3×4.3	24	91	38.9	268	17	64	19.5	134
	C-1 Omega	flush pendent	5.32	145	63	12×12	3.7×3.7	25	95	22.1	152	18	68	11.4	79
				160	71	14×14	4.3×4.3	30	114	31.8	219	21	79	15.6	108
				200	93										
	EC-20 Omega	flush pendent	5.32	145	63	14×14	4.3×4.3	23	87	18.7	129	16	61	9.1	63
				160	71	16×16	4.9×4.9	26	98	23.9	165	18	68	11.5	79
						18×18	5.5×5.5	32.5	123	37.3	257	22.5	85	17.9	123
	HEC-12 Omega	flush horizontal sidewall	5.56	145	63	12×12	3.7×3.7	27	95	23.6	163	19	72	11.7	81
				160	71	14×14	4.3×4.3	30	114	29.1	201	21	79	14.3	99
Reliable Automatic	A	pendent	4.15	165	74	14×14	4.3×4.3	18	68	18.8	130	13	49	9.8	68
						16×16	4.9×4.9	25	95	36.3	250	18	68	18.8	130
			5.5	165	74	12×12	3.7×3.7	21	79	14.6	101	15	57	7.4	51
						14×14	4.3×4.3	24	91	19.0	31	17	64	9.6	66

* Source: Flemming, R. P., The Special Listings, *Sprinkler Quarterly*, No. 52, Spring 1985.

miniums, lodging and rooming houses, and other multiple family dwellings. Figure 18-4D shows a typical residential sprinkler system for a single family dwelling.

Figure 18-4E shows acceptable valving arrangements for residential sprinklers connected to a public water supply.

Water Supply for Dwelling Units

The water supply requirements for dwelling units protected by residential sprinklers stipulate not less than 18 gpm (68 L/min) flow from any one operating sprinkler and not less than 13 gpm (44 L/min) per sprinkler from all operating sprinklers in the design area, i.e., the area of the system containing the four hydraulically most demanding sprinklers. Other discharge rates may be used with the flow rates indicated in the listings for individual residential sprinklers.

The Number of Design Sprinklers: All sprinklers within a compartment (room) to a maximum of four sprinklers are encompassed in the design area. (In determining the number of design sprinklers, compartment is defined as a space that is completely enclosed by walls and a ceiling. The compartment enclosure may have openings to an adjoining space if the openings have a minimum depth of 8 in. or 203 mm from the ceiling.)

When a compartment contains less than four sprinklers, the number of design sprinklers includes all sprinklers in that compartment plus sprinklers in adjoining compartments to a total of four sprinklers. Adjoining corridors may be considered as compartments for the purposes of the calculations. In all cases, the design area includes the four hydraulically most demanding sprinklers including those in dwelling units and their adjoining corridors. (See Fig. 18-4F.)

The water demand for a dwelling unit is determined by multiplying the design discharge by the number of design sprinklers.

Other areas, such as attics, basements, or other types of occupancies outside of dwelling units but within the same structure, are required to be protected in accordance with provisions of NFPA 13, including the appropriate water supply requirements. Thus, while residential sprinklers are not required in residential areas by NFPA 13, their use is encouraged through the four sprinkler design area.

Residential sprinklers installed in systems designed to NFPA 13 requirements are spaced and positioned in accordance with their residential listings and not with spacing requirements of NFPA 13. The water supply requirements for dwelling units are the same as in the NFPA 13D, except that the multiple sprinkler flow requirement is extended to four sprinklers rather than two as stipulated for one and two family dwellings and mobile homes in NFPA 13D. The more liberal piping, component, hanger, and water supply duration allowances of NFPA 13D, however, are not permitted in these systems.

The four sprinkler design area for dwelling units applies to wet systems only. One-half inch or larger residential sprinklers are allowed in dwelling units supplied by dry systems but a full water supply is required, as specified in NFPA 13, for the basic occupancy of the structure in question.

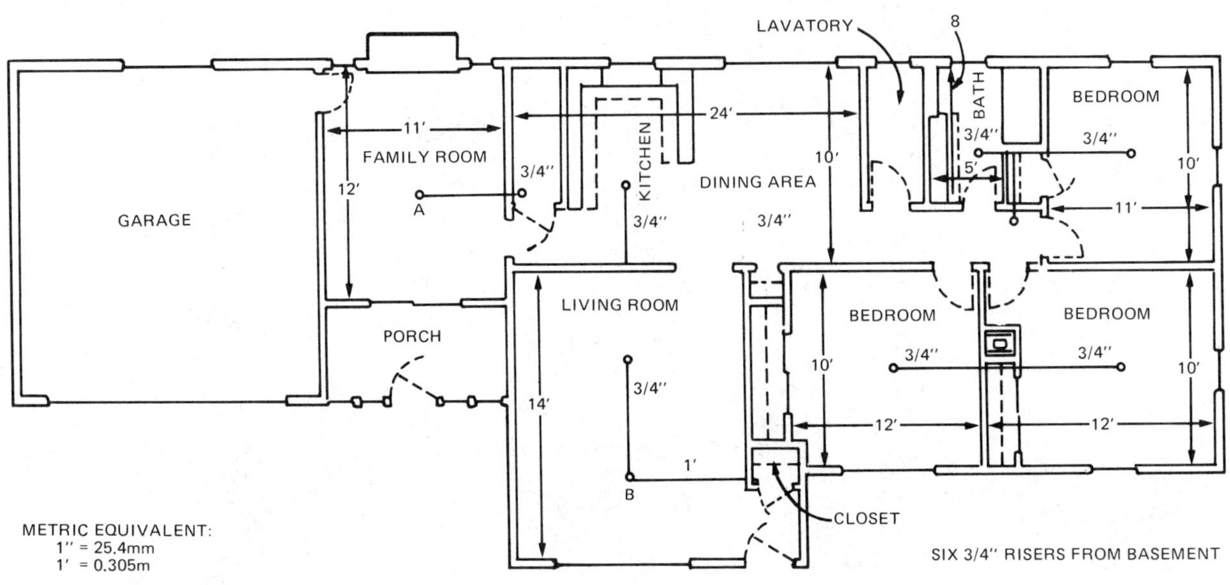

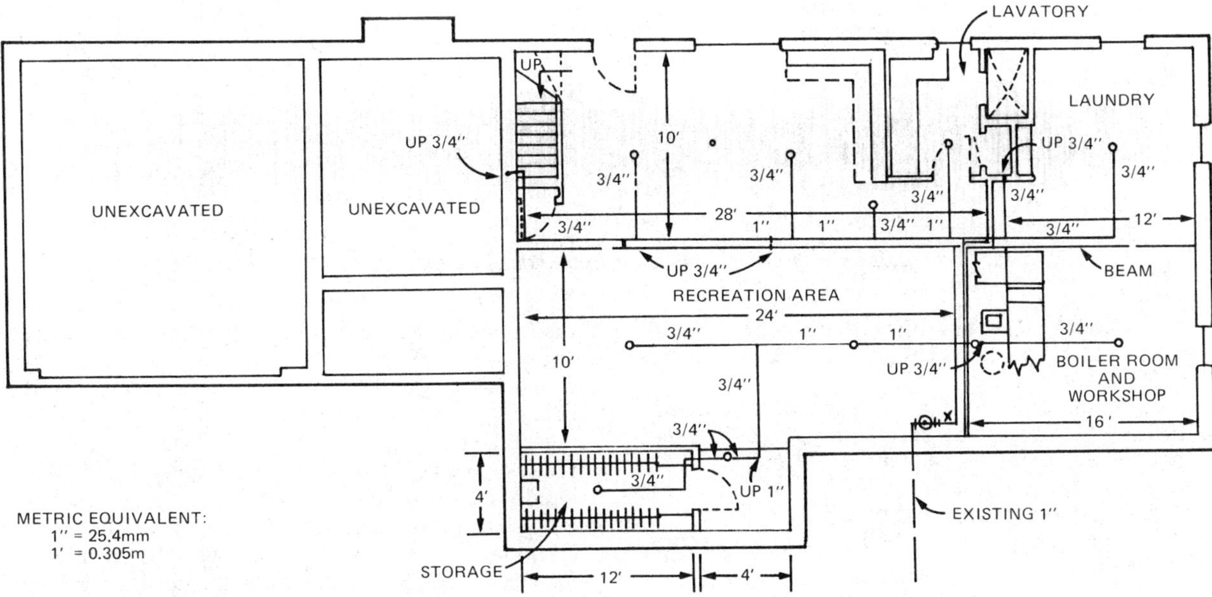

FIG. 18-4D. A typical residential sprinkler system for a single family dwelling. Top is first floor plan; bottom is basement floor plan.

QUICK-RESPONSE CONVENTIONAL SPRINKLERS

In addition to standard conventional sprinklers and residential quick-response sprinklers, a third type of sprinkler has been developed: the "quick-response standard sprinkler."

Quick-response standard sprinklers have been developed by manufacturers in two different ways. One way is to replace the actuating mechanism of standard sprinklers with the more sensitive heat responsive element used in residential sprinklers. This "quick-response" sprinkler is then resubmitted to a testing laboratory for testing and listing under a product standard for standard sprinklers such as UL 199. The other way is to simply submit a residential sprinkler to a testing laboratory for testing and listing under the provisions of a product standard for standard sprinklers.

These quick-response standard sprinklers are designed to be used in hotels, motels, office buildings, etc., where faster operation could enhance life safety. The design for sprinkler systems using these quick-response standard sprinklers, including water supply requirements, is required to be in accordance with the NFPA Sprinkler Systems Standard. The four sprinkler design option, however, is not allowed with the use of quick-response stan-

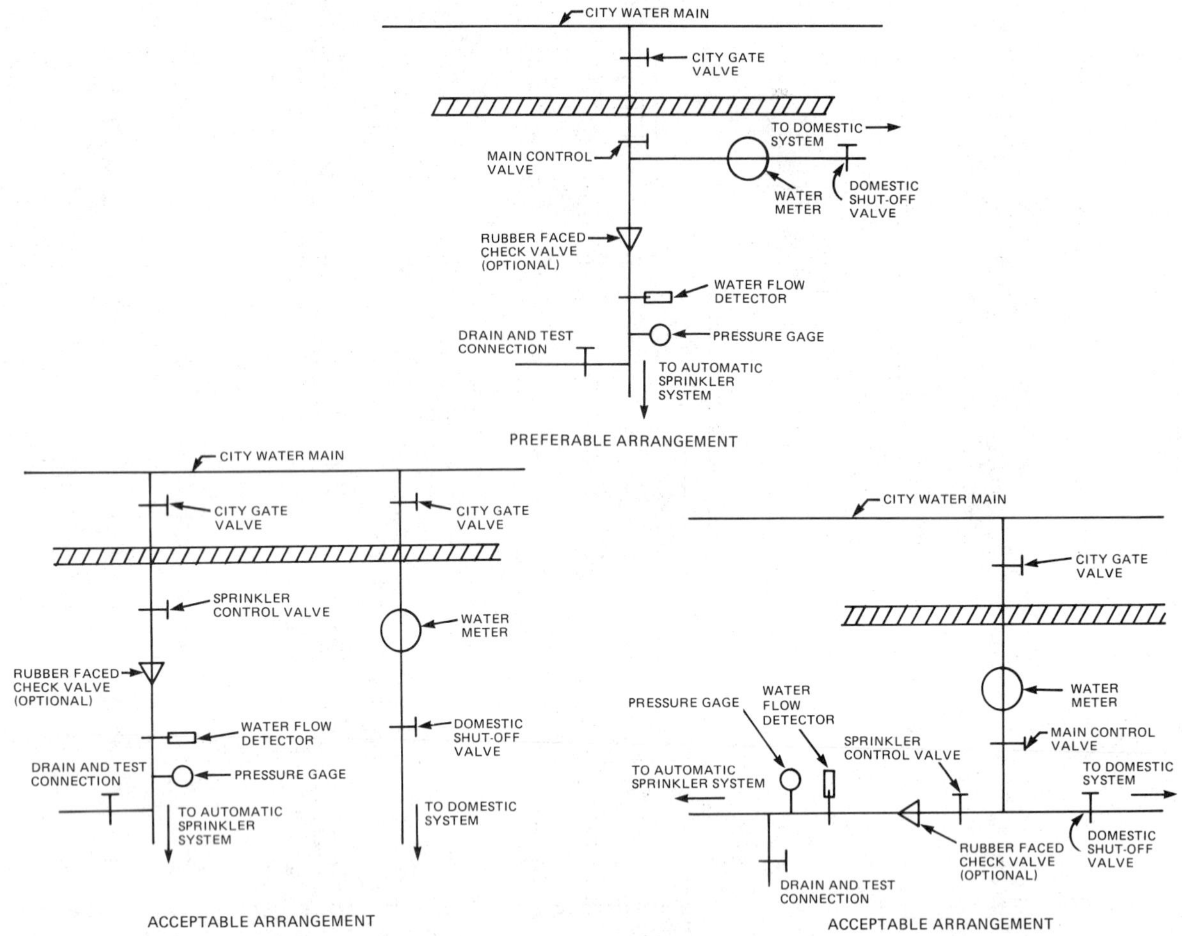

FIG. 18-4E. Acceptable valving arrangements for residential sprinkler systems (Source: NFPA 13D).

dard sprinklers. A series of full scale demonstration tests of these sprinklers in a hotel environment was conducted in 1982 (Cote 1983).

These quick-response standard sprinklers cannot be used in NFPA 13D residential sprinkler systems since they have not demonstrated satisfactory performance in the residential fire test environment of UL 1626 and/or FM 2030 at their listed spacing requirements.

Quick-response sprinklers are listed by manufacturer in Table 18-4B.

BARRIERS TO WIDESPREAD USE OF RESIDENTIAL SPRINKLERS

Unfortunately, residential sprinklers and residential sprinkler systems do not receive the wide public support they deserve for a variety of reasons. Some of those reasons are discussed in the following paragraphs.

Cost

The major obstacle to widespread use of residential sprinklers in one and two family dwellings and mobile homes throughout the United States is cost. In a study of user acceptance of the concept, 50 percent of the people interviewed said they would not have a sprinkler in their residence (Yurkonis 1978). Ten percent feared accidental discharge, ten percent rejected systems based on appearance, and 30 percent perceived no need for the system. Cost was a major factor in system acceptance, according to the study. Twenty percent said they would pay if the cost were \$.20 per sq ft or less (for a 2000-sq-ft (185 m^2) house, the system would have to cost \$400 or less), and 30 percent would not pay for a system but would accept it at no cost.

Costs for installation of an NFPA 13D residential sprinkler system in a new home vary between \$900 and \$2,000, depending on the size of the house and the person who is quoting the figure.

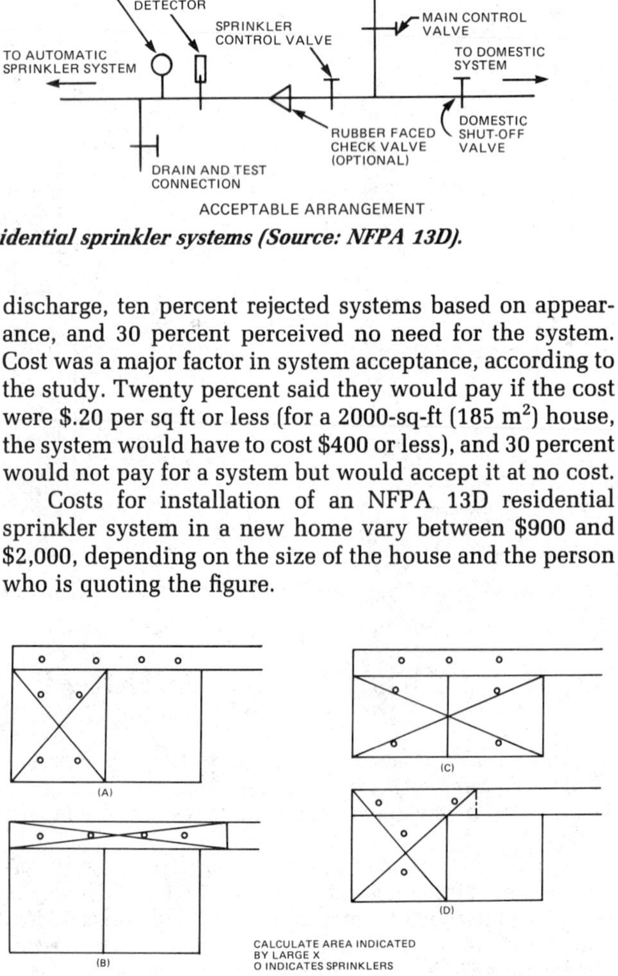

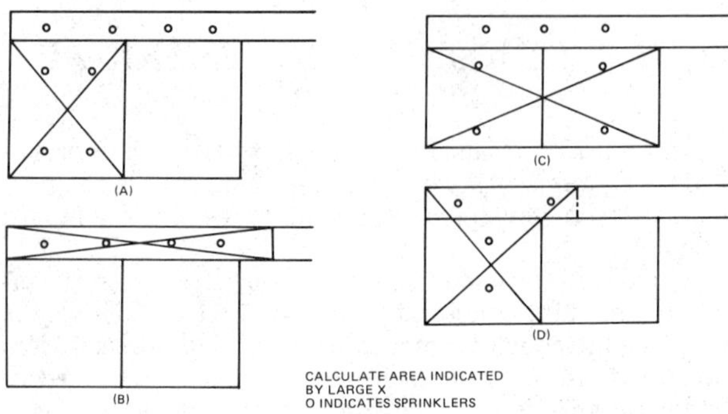

FIG. 18-4F. Examples of design areas for dwelling units.

Table 18-4B. Quick Response Sprinkler Characteristics
(Quick response* sprinklers listed by Underwriters Laboratories.)

Manufacturer	Model	Style	Temperature Rating °F	Temperature Rating °C	Orifice Size in.	Orifice Size mm	K
Grinnell Fire (GEM) Providence, RI	FR-1	Upright and pendent	140 165	60 74	⅜ and ½	9.5 and 12.7	5.5
	FR-1	Extended coverage Horizontal sidewall	140 165	60 74	½	12.7	5.5
Central Sprinkler Lansdale, PA	C-1	Flush pendent	135 145 160	57 63 71	½	12.7	5.3
Viking Corporation Hastings, MI	A-1	Upright	165	74	.66 in.	16.8	11.0 to 11.5

* Response Time of 0 to 13 sec.

Sprinklers vs Smoke Detectors

Since some smoldering fires produce lethal carbon monoxide levels before the activation of even a quick-response automatic sprinkler, smoke detectors are required in addition to residential sprinklers. NFPA 13D states: "This standard assumes that one or more smoke detectors will be installed in accordance with NFPA 74."

Installation/Supervision/Inspection

Sprinkler contractors in the United States are geared to installing sprinklers in large commercial, industrial, and institutional projects. Residential construction organizations and personnel have not been geared to the traditional and mandatory, quality controlled and supervised atmosphere that has been prevalent in the fire protection industry. If the residential sprinkler system is installed by a residential plumber rather than a sprinkler contractor, a mechanism for reviewing the plans and inspecting the installation will have to be established to insure that the sprinkler systems are properly designed and installed.

One of the reasons for the *minimum* failure rate of automatic sprinklers over the years has been the regulatory supervision and review process associated with sprinkler installations. The one and two family dwelling, on the other hand, is largely unregulated from a supervision/inspection standpoint. In addition, the typical homeowner may choose to act as a "do-it-yourselfer" and install a system without any outside guidance or assistance.

New Construction vs Retrofit

While the cost of installing a residential sprinkler system in a new home is higher than many homeowners will be willing to pay, it is still at an acceptable level for many others. In the case of retrofit, however, the cost to install a residential system has been at least three to five times greater—making it acceptable to only a few. This, coupled with the problem of aesthetics, has resulted in few installations of residential sprinklers in existing homes.

Plastic Pipe and Fittings

One way to reduce the cost of sprinkler system installation, especially in retrofitting, is to utilize a lightweight plastic piping material that is easier to install than the traditional piping materials. Both polybutylene and chlorinated polyvinal-chloride (CPVC) pipe and fittings have been listed for use in residential sprinkler systems for one and two family dwellings and mobile homes.

INCENTIVES TO MORE WIDESPREAD USE OF RESIDENTIAL SPRINKLERS

There are certain incentives that can stimulate interest in residential sprinklers despite the reluctance of the general public to accept them wholeheartedly. These incentives are discussed in the following paragraphs.

Reduction in Government Spending

Reduction in all forms of government spending, resulting from public pressure to reduce property taxes, is a prime factor in the future growth of the residential sprinkler concept. Many fire departments find themselves forced to protect larger areas and more subdivisions with the same number of (or even fewer) people since financial restrictions hamper a fire department's ability to grow with the community. As a result, alternate methods to traditional fire fighting techniques must be found. One of them is the use of residential sprinklers.

San Clemente, CA, was the first community in the United States to pass a Residential Sprinkler Ordinance in 1980 as part of the fire department's master plan. This ordinance requires automatic sprinkler systems to be installed in all new residential construction. The prime motivation for the passage of this ordinance was San Clemente's cutbacks in government spending brought about by Proposition 13, the state's tax capping measure. Many communities across the country face similar situations. Automatic sprinklers in residences may be the answer to fewer fire fighters responding and longer response times from the fire department.

Insurance Savings

Although the greatest benefit from widespread installation of residential sprinklers will be the lives saved and injuries prevented, lower property losses will be a secondary and substantial benefit. An ad-hoc committee from the insurance industry sponsored a number of the test fires in Los Angeles and concluded that residential sprinklers have the potential for reducing homeowners' claim payment expenses (Jackson 1980). As a result, the Insurance Services Office (ISO) Personal Lines Committee recommended that a 15 percent reduction in the homeowner's policy premium be given for installation of a NFPA 13D residential sprinkler system. While this would not pay for the system over a short period of time, as is the case in many commercial installations, the continuing increases in the cost of insuring a single family home make this nonetheless a significant incentive.

Real Estate Tax Reductions

In 1981, the State of Alaska enacted into law a significant piece of legislation that has a dramatic impact upon the installation of sprinkler systems throughout that state. The law provides that 2 percent of the assessed value of any structure would be exempt from taxation if the structure is protected with a fire protection system. The word "structure" is significant in the law, since it also applies to homes. In effect, if a home were assessed at $100,000 for purposes of taxation, the assessed value would be computed at $98,000, provided that it contained a fire protection system.

Buyers' Attitude

While one study indicated that 30 percent of the people interviewed perceived no need for a residential sprinkler system (Yurkonis 1980), a survey that was published by the National Association of Home Builders on luxury features that buyers want in a new home showed that 14.3 percent indicated "fire suppression systems" as a choice. For potential buyers with incomes over $50,000, the percentage rose to over 20 percent.

Zoning

Greater land use may be possible with zoning changes that would permit fully sprinklered residences to be built on smaller land parcels. The assumption is that the space between houses will not be as important, from a fire protection standpoint, if an entire street or neighborhood is fully sprinklered. One could argue, however, that if the sprinkler system fails, the resultant fire involving a number of residences could be much greater.

Sprinkler Legislation

In addition to the San Clemente residential sprinkler ordinance, a number of other California communities have passed residential sprinkler legislation including Orange County and Los Angeles County.

Greenburgh, NY, has a sprinkler ordinance which requires the installation of automatic sprinklers in virtually all new construction, including all new multiple and one and two family dwellings.

The State of Florida passed a law requiring that all public lodging and time shared units three stories or more in height in the state be sprinklered. It also requires that all existing units be sprinklered by 1988 unless an extension is approved by the State Fire Marshal.

In 1983, the City of Honolulu, Hawaii, adopted legislation which requires all high rise hotels (over 75 ft (23 m) above grade), both new and existing, to be sprinklered.

Trade-Offs

Many "authorities having jurisdiction" have used building code trade-offs as an incentive to the installation of sprinklers. Cobb County, Georgia, was one of the first communities to amend its Buildings and Construction Code to include trade-offs for multifamily structures equipped with residential sprinkler systems.

While these trade-offs can be a major incentive to the installation of residential sprinklers, the disaster potential must always be considered if the fire, for whatever reason, should overpower the sprinkler system. This is especially true if the system is designed with the minimal water supplies required by NFPA 13D.

The City of Dallas, TX, adopted a building code which requires all new buildings, or buildings under major renovation, having an area greater than 7,500 sq ft (697 m^2) to have automatic sprinklers. At the same time, this building code encourages trade-offs by allowing design options which trade "passive" fire protection features for "active" automatic sprinkler alternatives.

Mobile Homes

At least one major mobile home manufacturer is installing residential sprinkler systems in all its mobile homes at the factory.

Bibliography

References Cited

Clark, Graham. 1978. *Performance Specifications for Low-Cost Residential Sprinkler System.* Factory Mutual Research Corp., Norwood, MA. Jan.

Cote, A. E., and Moore, D. 1980. *Field Test and Evaluation of Residential Sprinkler Systems.* Los Angeles Test Series (A report for the NFPA 13D Subcommittee). National Fire Protection Association, Quincy, MA. April.

Cote, A. E. 1982. *Final Report on Field Test and Evaluation of Residential Sprinkler Systems.* National Fire Protection Association, Quincy, MA. July.

Cote, A. E. 1983. *Final Report on Field Test of a Retrofit Sprinkler System,* National Fire Protection Research Foundation, Washington, DC. Feb.

Halpin, B. M., Dinan, J. J., and Deters, O. J. 1978. *Assessment of the Potential Impact of Fire Protection Systems on Actual Fire Incidents.* Johns Hopkins University—Applied Physics Laboratory (JHU/APL), Laurel, MD. Oct.

Henderson, N. C., Riegel, P. S., Patton, R. M. and Larcomb, D. B. 1978. *Investigation of Low-Cost Residential Sprinkler Systems.* Battelle Columbus Laboratories, Columbus, OH. June.

Heskested, Gunner and Smith, Herbert. 1980. "Plunge Test for Determination of Sprinkler Sensitivity." *FMRC J.I.3AIE 2.RR.* Factory Mutual Research Corp. Norwood, MA.

Heskested, Gunner and Smith, Herbert, F. 1976. "Investigation of a New Sprinkler Sensitivity Approval Test: The Plunge Test." *FMRC 22485.* Factory Mutual Research Corp. Norwood, MA.

Jackson, R.J. 1980. *Report of 1980 Property Loss Comparison Fires.* FEMA/USFA. Washington, DC.

Kung, H. C., Haines, D, and Green, R. Jr. 1978. *Development of Low-Cost Residential Sprinkler Protection.* Factory Mutual Research Corp., Norwood, MA. Feb.

Kung, H. C., Spaulding, R. D., and Hill, E. E. Jr. 1980. *Sprinkler Performance in Residential Fire Tests.* Factory Mutual Research Corp., Norwood, MA. Dec.

Kung, H. C., Spaulding, R. D., Hill, E. E. Jr., and Symonds, A. P. 1982. *Technical Report, Field Evaluation of Residential*

Prototype Sprinkler Los Angeles Fire Test Program. Factory Mutual Research Corp., Norwood, MA. Feb.

Moore, D. 1980. *Data Summary of the North Carolina Test Series of USFA Grant 79027 Field Test and Evaluation of Residential Sprinkler Systems* (A report for the NFPA 13D Subcommittee). National Fire Protection Association, Quincy, MA. Sept.

Yurkonis, Peter. 1980. *Study to Establish the Existing Automatic Fire Suppression Technology for Use in Residential Occupancies.* Rolf Jensen & Associates, Inc., Deerfield, IL. Dec.

NFPA Codes, Standards, Recommended Practices and Manuals. (See the latest *NFPA Codes and Standard Catalog* for availability of current editions of the following documents.)

NFPA 13, *Standard for the Installation of Sprinkler Systems.*

NFPA 13D, *Standard for the Installation of Sprinkler Systems in One- and Two-Family Dwellings and Mobile Homes.*

Additional Readings

Benjamin/Clarke Associates, "Operation San Francisco Smoke/Sprinkler Test Technical Report," International Association of Fire Chiefs, Washington, April 1984.

Budnick, Edward K., "Estimating Effectiveness of State-of-the-Art Detectors and Automatic Sprinklers on Life Safety in Residential Occupancies," National Bureau of Standards/Center for Fire Research, NBSIR 84-2819, January 1984.

Chopra, I.S. "Subject 1626: Residential Automatic Sprinklers," Lab Data, Underwriters Laboratories, Northbrook IL Vol 15, No. 2 1984

Cote, A. E., *Field Test and Evaluation of Residential Sprinkler Systems, Part I, Fire Technology,* Vol. 19, No. 4, Nov. 1983, pp. 221-232.

———, Part II, Vol. 20, No. 1, Feb. 1984, pp. 48-58.

———, Part III, Vol. 120, No. 2, May 1984, pp. 41-46.

———, "Highlights of a Field Test of a Retrofit Sprinkler System," *Fire Journal,* Vol. 77, No. 3, May 1983.

———, "Update on Residential Sprinkler Protection," *Fire Journal,* Vol. 77, No. 6, Nov. 1983, p. 69.

———, "Use of Quick Response Sprinklers in Residential Occupancies—What You Should Know" *Sprinkler Quarterly* No 52, Spring 1985

"Development of an Experimental Prototype Low-Cost Electronic Sensor/Actuator for a Residential Automatic Sprinkler Head," Battelle Columbus Laboratories, Columbus, OH, 1982.

Evans, David D., "Calculating Sprinkler Activation Time in Compartments," National Bureau of Standards/Center for Fire Research, 1984.

Factory Mutual Research Corp., *Approval Standard—Residential Automatic Sprinklers,* Factory Mutual Research Corporation, Class No. 2030, September 1983.

Fleming, Russell P., "Quick Response Sprinklers—A State of the Art Technical Report," National Fire Protection Research Corporation, Quincy, MA, 1985.

———, "The Special Listings" *Sprinkler Quarterly* No. 52, Spring 1985.

———, "Understanding the New Sprinklers," *Sprinkler Quarterly,* National Fire Sprinkler Association, No. 47, Fall 1983.

Foehl, John M., "In Quest of an Economical Automatic Fire Suppression System for Multi-Family Residential Complexes," *Fire Journal,* Vol. 70, No. 1, Jan. 1976, pp. 48-57.

Kung, H. C., Spaulding, Robert D., and Hill, Edward E., "Sprinkler Performance in Living-Room Fire Tests: Effect of Link Sensitivity and Room Size," Factory Mutual Research Corporation, July 1983.

———, "Low-Cost Residential Sprinkler Protection," *Fire Technology,* Vol. 12, No. 2, May 1976, pp. 85.

Moore, David A., "Field Test and Evaluation of Residential Sprinkler Systems," *Fire Journal,* Vol. 74, No. 6, November 1980.

O'Neil, John G., "Fast Response Sprinklers in Patient Room Fires," *Fire Technology,* Vol. 17, No. 4, November 1981.

O'Rourke, Gerald W., "Automatic Sprinkler Systems: What's New?" *Heating/Piping/Air Condition,* Vol. 55, No. 4, April 1983, pp. 37-41.

Outline of Proposed Investigation for Residential Automatic Sprinklers for Fire Protection Service, Subject 1626, Underwriters Laboratories, October 27, 1980.

Pepi, Jerome S. "Concept and Development of the Residential and Fast Response Sprinklers" *Sprinkler Quarterly,* No. 52, Spring 1985

Proceedings of the Fourth Conference on Low-Cost Residential Sprinkler Systems, September 26–27, 1979, Federal Emergency Management Agency/U.S.; Fire Administration, August 1980.

Proceedings of the Third Conference on Low-Cost Residential Sprinklers, November 29–30, 1977, National Fire Prevention and Control Administration, June 1978.

Ruegg, R. T., and Fuller, S. K., "A Benefit-Cost Model of Residential Fire Sprinkler Systems," NBS Technical Note 1203, National Bureau of Standards, Washington, DC, Nov. 1984.

Shaw, Harry, "Fast Response Sprinklers—The Answer to the Toxicity Problem," *Sprinkler Quarterly,* National Fire Sprinkler Association, No. 47, Fall 1983.

Viriello, John A., "Residential & Quick Response Fire Sprinklers—You Need to Know the Difference!" *Sprinkler Quarterly,* No. 52, Spring 1985.

WATERFLOW ALARMS AND SPRINKLER SYSTEM SUPERVISION

Revised by Vincenzo Cirigliano

Sprinkler systems should have devices and equipment for giving an alarm notification when water flows through risers or mains supplying the systems. The flow may be due to fire, leakage, or accidental rupture of the piping.

This chapter describes the various types of alarm and supervisory devices that will be found on sprinkler systems, water tanks supplying them, and on valves controlling the water supplies. The established practices followed in providing sprinkler system waterflow alarms are given in NFPA 13, *Standard for the Installation of Sprinkler Systems*, referred to hereinafter in this chapter as the NFPA Sprinkler Systems Standard.

Insofar as alarms and supervisory signals may be the initiating signal for a protective signaling system, Chapter 2, "Protective Signaling Systems," of Section 16 in this HANDBOOK will be of interest in understanding the relationship between the actual sprinkler alarm and supervisory devices and the equipment and systems involved in responding to the signals.

FUNCTIONS OF ALARMS AND SUPERVISORY SIGNALS

A sprinkler system with a waterflow alarm serves two functions: that of an effective fire extinguishing system, and that of an automatic fire alarm. Immediate notification by an alarm of the operation of sprinklers is important to complete extinguishment of the fire and to ultimately returning the system to service. Under some conditions, the sprinklers do not immediately or completely extinguish the fire; therefore, it is vital to have backup manual fire fighting forces notified to complete extinguishment, either by portable extinguishing devices, private hose streams, or fire department equipment.

The amount of loss or damage by water after the fire has been extinguished may be held to a minimum by closing the control valve immediately after the need for sprinkler discharge has passed. One or two sprinklers may extinguish the fire, but water damage may be considerable unless the water is shut off as soon as it is safe to do so.

In addition to waterflow alarms, sprinkler systems frequently are equipped with devices to signal abnormal conditions which could make the protection inoperative or ineffective. In general, these devices, known as supervisory devices, give warning of troubles with equipment (shut valves, etc.), and require action by maintenance or security personnel. Waterflow alarms and fire alarms by themselves give warning of the actual occurrence of a fire or other conditions (broken pipes, etc.) causing water to flow through the system. Alarms alert occupants and summon the fire department. Any signal, whether waterflow or supervisory, may be used to give only an audible local sprinkler alarm, or it may be the initiating signal for a protective signaling system.

WATERFLOW SPRINKLER ALARMS

The various types of sprinkler alarms include: (1) those that operate with an actual flow of water, (2) those that activate a hydraulic or an electric alarm when a water control device, such as a dry pipe valve, trips to admit water to the alarm device or mechanically operates an electric switch whether or not water actually flows from sprinklers, and (3) those that not only signal the tripping of the control valve but may also give supplementary warning signals in case of damage that might impair the operation of the system, or if maintenance features need attention.

Sprinkler systems are usually required to have an approved water motor gong or an electric bell, horn, or siren on the outside of the building. An electric bell or other audible signal device may also be located inside the building. Water operated devices must be located near the alarm valve, dry pipe valve, or other water control valves in order to avoid long runs of connecting pipe.

All electric alarm devices, wiring, and power supplies must comply with NFPA 70, *National Electrical Code®*, and with NFPA Signaling Systems Standards, numbered 71, 72A, 72B, 72C, and 72D (see Bibliography at the end of this chapter).

Mr. Cirigliano is an engineer, Hydraulic Section of the Approval Division, of Factory Mutual Research Corporation in Norwood, MA.

Location of Alarm Signals

It is important that audible or visual signals be located for greatest effectiveness. Outdoor alarm gongs, either electrically or water motor operated, are needed especially if there is no dependable guard service or continuous supervision by central station or other protective signaling system services. However, outside alarms on buildings in remote areas or in urban areas which are sparsely populated at some times (as when residents are away or at work) may be of little value, making central station or other protective signaling system connections very desirable.

The types of waterflow alarm devices and equipment described in this chapter are mostly those ordinarily furnished and installed by sprinkler manufacturers or contractors as part of a sprinkler system conforming to the NFPA Sprinkler Systems Standard. They provide a complete local waterflow alarm system, and in many cases are basic units of more complete alarm and supervisory signaling systems.

Sprinkler System Supervisory Systems

The sprinkler system supervisory devices and equipment described in this chapter are considered to be a part of sprinkler installations, even though manufactured, installed, and maintained by a central station or other supervisory service organization.

Devices and Equipment Supervised

Sprinkler system supervision is commonly provided for: (1) water supply control valves, (2) low water level in water supply tanks, (3) low temperature in water supply tanks or ground level reservoirs, (4) high or low water level in pressure tanks, (5) high or low air pressure in pressure tanks, (6) high or low air pressure in dry pipe sprinkler systems, (7) failure of electric power supply to fire pumps, (8) automatic operation of electric fire pumps, and (9) fire detection devices used in conjunction with deluge and/or preaction and recycling systems.

Sprinkler system devices that give waterflow alarms or supervise the condition of the installation are shown schematically in Figure 18-5A.

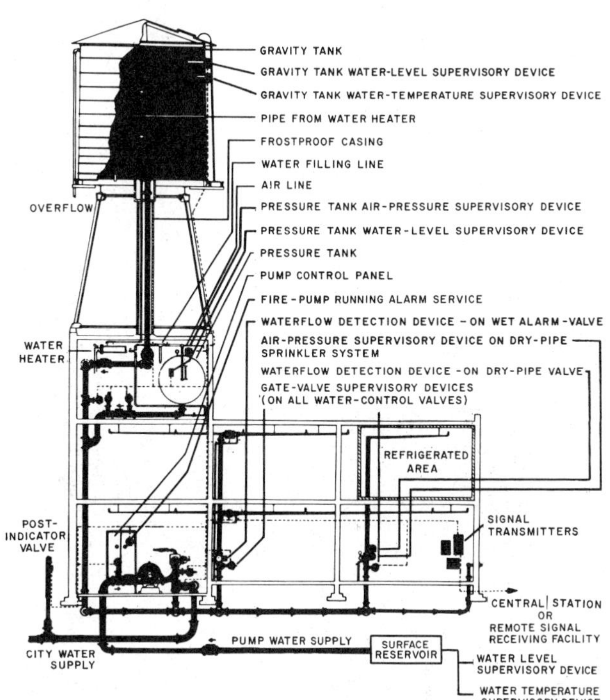

FIG. 18-5A. Sprinkler system waterflow alarm and supervisory devices.

Types of Systems

There are several recognized signaling systems for transmitting alarms when water actually flows in an automatic sprinkler system and for transmitting supervisory signals announcing abnormal conditions which could make the sprinkler protection inoperative or ineffective. The principal functions of each of these systems are described briefly in the following paragraphs.

Local Waterflow Alarm Systems: A local alarm unit is a device that will give an alarm on the premises on any flow of water from a sprinkler system equal to or greater than that from a single sprinkler.

Proprietary Signaling Systems: These systems are frequently provided at large properties to transmit both waterflow alarms and supervisory signals to a constantly attended central alarm headquarters at the property protected. The devices and equipment for proprietary systems are usually supplied and installed by a signaling equipment manufacturer, a central station service organization, or an electrical contractor. Some central station service companies furnish these systems with a maintenance or inspection contract, or both.

Central Station Signaling Systems: These systems transmit both waterflow alarm signals and supervisory signals to a constantly attended headquarters of a central station signaling service organization off the premises. The central station in turn notifies the fire department of all alarms and the property owner of supervisory signals.

Auxiliary Signaling Systems: Auxiliary signaling systems send signals directly to a municipal fire department over the municipal fire alarm system through master boxes which may be of the telegraph, telephone, or radio operated type.

Municipal alarm systems usually transmit only waterflow alarm signals to the fire department, and supervisory signals are transmitted over private circuits to some other location. Devices and equipment connected to the master box are generally installed by a supplier of signaling equipment or an electrical contractor.

Remote Station Signaling Service: This type of signaling service sends signals directly over private circuits to a fire department or other continually attended location where action will be taken immediately.

WET PIPE SPRINKLER SYSTEM ALARM DEVICES

Waterflow alarm devices have been used to some extent ever since automatic sprinklers were first installed. They are generally located at or near the base of sprinkler risers but may be used as floor or branch alarms. They are designed and adjusted to give an alarm if a water flow equal to the discharge of one or more automatic sprinklers occurs in the sprinkler system. The alarm signal may be

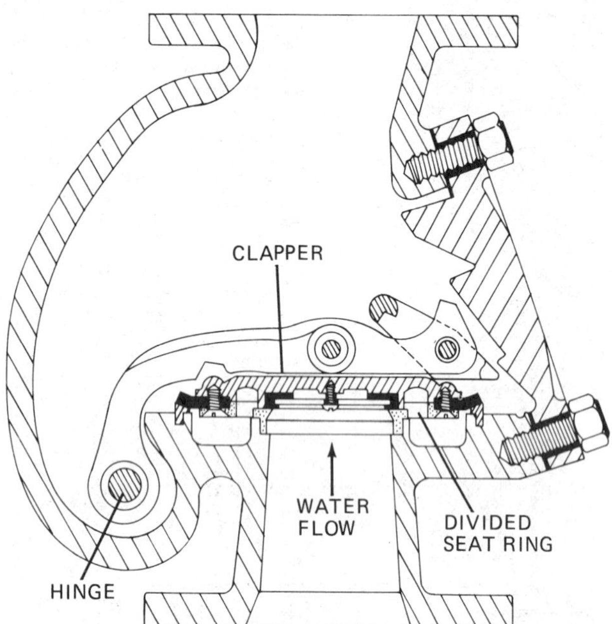

FIG. 18-5B. A waterflow alarm valve of the differential type. (Firematic Sprinklarm)

given electrically, or by water motor gongs, or both. By far the most common types are the waterflow alarm valve and the waterflow indicator.

Waterflow Alarm Valves

The basic design of most waterflow alarm valves is that of a check valve which lifts from its seat when water flows into a sprinkler system. The movement of the valve clapper is used in one of the following ways:

1. The valve seat ring can have a concentric groove with a pipe connection from the groove to the alarm devices. Such valves are commonly called differential type or divided-seat ring type valves. When the clapper of the alarm valve rises to allow water to flow to sprinklers, water also enters the groove in the divided-seat ring and flows through the pipe connection to an alarm giving device. An alarm check valve of this type is shown in Figure 18-5B.
2. The clapper of the alarm check may have an extension arm connected to a small auxiliary (pilot) valve having its own seat and a pipe connection to alarms. When the auxiliary valve is lifted by movement of the main valve clapper, water is admitted to alarm devices. An alarm check valve of this type is shown in Figure 18-5C.
3. Movement of the main clapper by a flow of water can operate mechanically an electric switch so located and arranged that it is not affected by the water pressure in the system. This method is limited to giving electric alarms and is unable to supply water under pressure

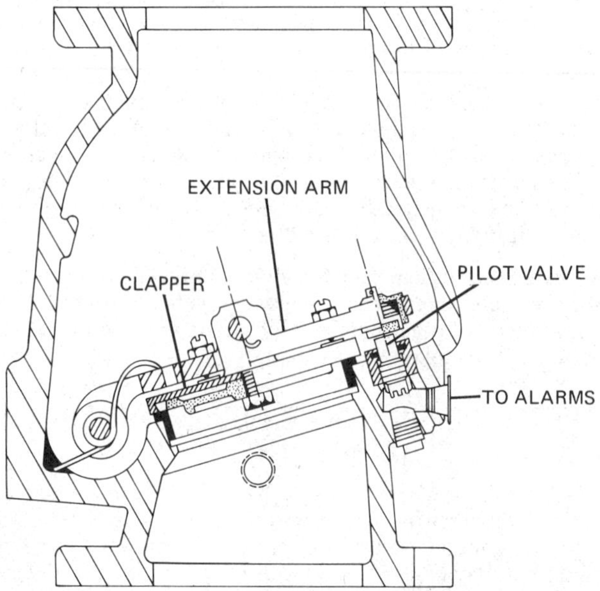

FIG. 18-5C. An alarm check valve with an auxiliary valve and pipe connection to alarms. Shown is the "Automatic" Sprinkler Co. Model 253 alarm check valve (no longer manufactured).

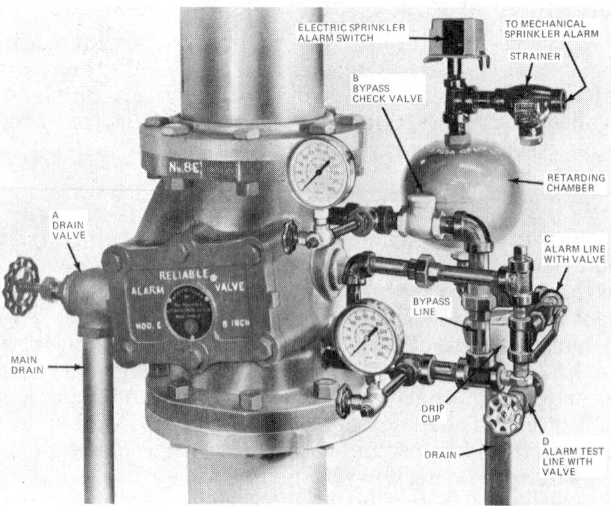

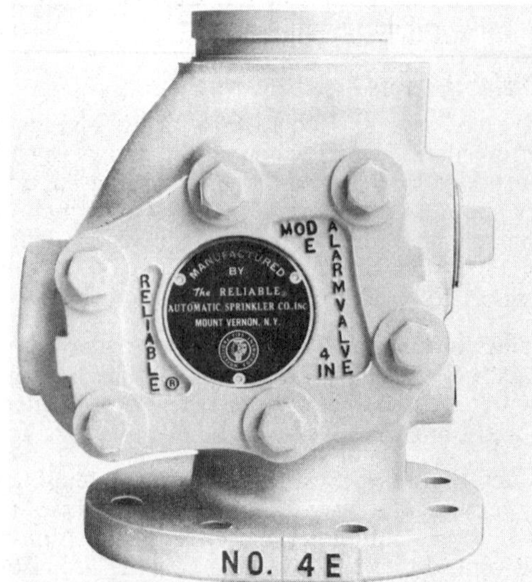

FIG. 18-5D. An alarm check valve installation with "trimmings." Shown at top is the Reliable Model E alarm valve. Below is the Reliable Model E valve with a grooved end connection.

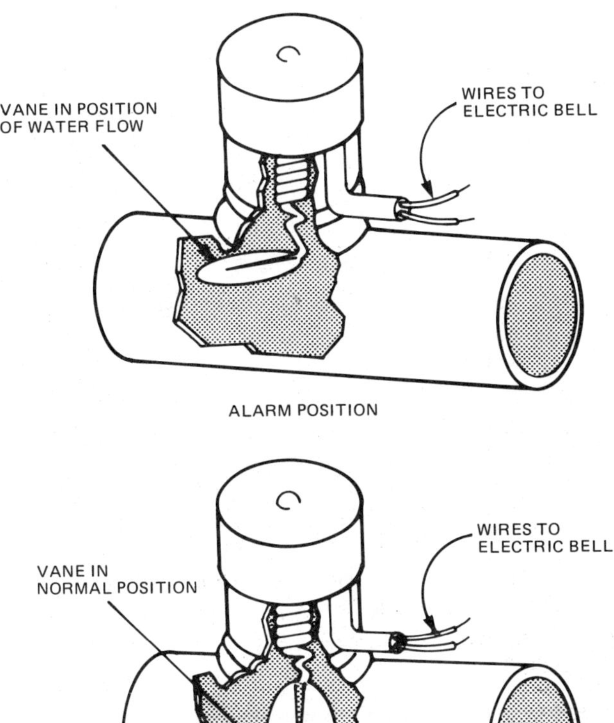

FIG. 18-5E. A vane type waterflow indicator showing the position of the normal and alarm positions of the vane.

for the operation of water motor gongs. It is being supplanted by other types of waterflow alarms, but a number of installations are still in service, mainly in central station signaling systems.

An alarm check valve installation with accessories is shown in Figure 18-5D.

Early types of alarm check valves with outside levers, weights, or movable pins operated by movements of a clapper are obsolete. Many features made these early check valves unreliable. For example, the free movement of the clapper was easily restricted. Also, such valves were not responsive to small flows of water.

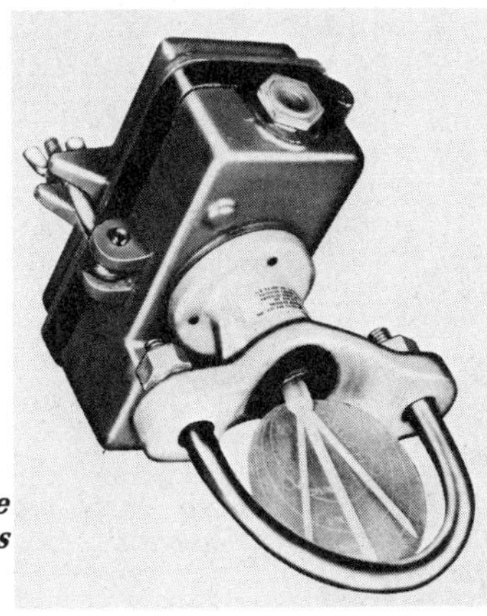

FIG. 18-5F. A vane type waterflow detector. Shown is ADT Model 6/27.

Improvement has been largely in mechanical and electrical design rather than basic operating principle. Some of the earliest devices, made previous to 1890, used the movement of a water clapper to actuate an electric switch. Grooved (divided) check valve seats and pilot valves, used to admit water to water motor gongs, were early developments.

Waterflow Indicators

A waterflow indicator of the paddle or vane type consists of a movable flexible vane of thin metal or plastic which is inserted through a circular opening cut in the wall of a sprinkler supply pipe. The vane extends into the waterway sufficiently to be deflected by any movement of water flowing to opened sprinklers. (See Fig. 18-5E.) Motion of the vane operates an alarm actuating electric switch or mechanically trips a signaling system transmitter. A mechanical, pneumatic, or electrical time delay feature in the detector or in the electric circuit or made a part of the signaling system transmitter prevents false alarms being given by fluctuating water pressures. The retard feature must be of the instantly recycling type or otherwise arranged so that the effect of a sequence of flows, each of less duration than the predetermined retard period, will not have a cumulative effect. Electrically heated thermal retards have not performed satisfactorily. Waterflow indicators have no provision for supplying water to water motor alarm gongs.

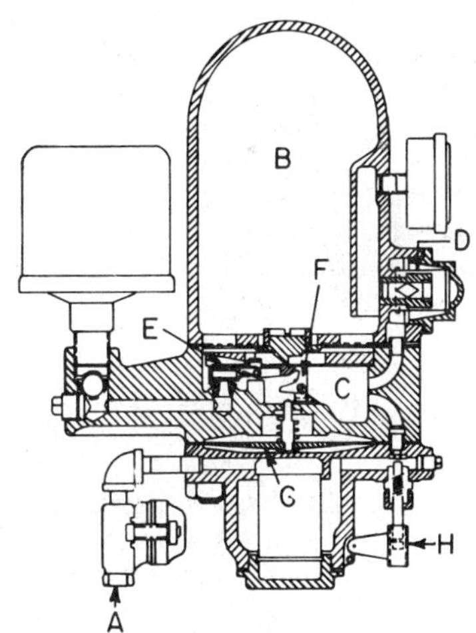

FIG. 18-5G. An alarm device actuated by a drop in sprinkler system water pressure. Inlet chamber (C) is connected to the sprinkler system above the alarm check valve through inlet (A). Pressure chamber (B) connects to chamber (C) through restricted passage (D) so that system pressure and pressure in chambers (B) and (C), separated by flexible metal diaphragm (G), are normally equal. A pressure difference above and below the alarm valve clapper, caused when a sprinkler is opened, is adequate to deflect the diaphragm and open water valve (E), leading to alarm devices, before water flows into the sprinkler system. Latch (F) prevents the flow valve from closing and is released by a plunger acted upon by diaphragm (G). The valve is reset by pushing up resetting valve (H), thus closing the passage into intermediate chamber (C). A target on the resetting valve shows "Normal Operating Position" or "Shut-Danger" as the case may be. Shown is the Firematic Sprinklarm, Model A.

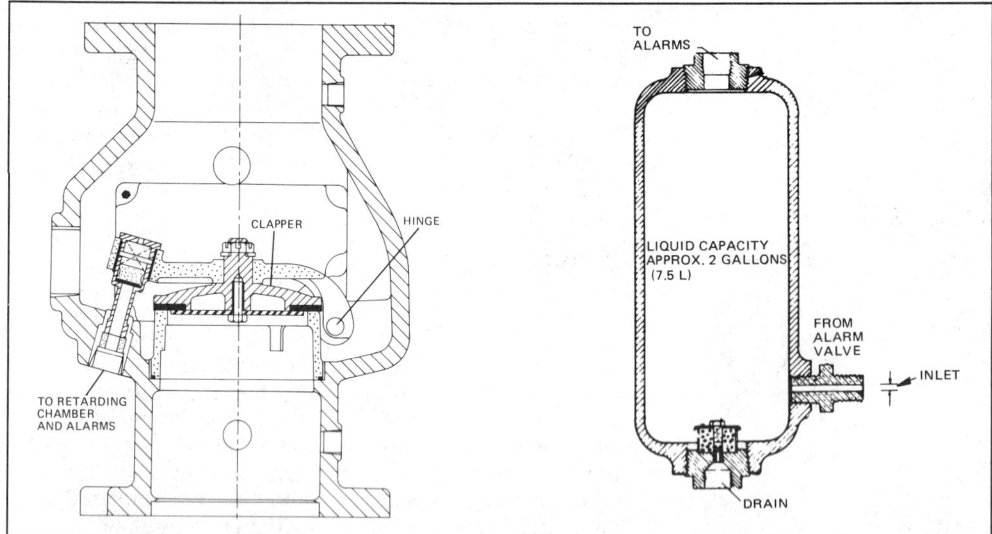

FIG. 18-5H. A waterflow alarm valve and retarding chamber. Shown are Viking Models F-1 and B-3.

It is important that the flexible vane of a waterflow indicator be of a design and material not subject to mechanical injury or corrosion so that it cannot become detached and possibly obstruct the sprinkler piping.

Waterflow indicators are commonly used on new systems. Situations where their use is most prevalent are: (1) where ease of installation and economy are important in installing a waterflow alarm, (2) where subdivision into several alarm areas is desired on a large sprinkler system, or (3) where a central station proprietary system or other remote signal station service is available which can receive only electric alarm signals from the protected property.

Waterflow indicators of the vane type cannot be used in dry pipe systems, deluge systems, or preaction systems because the vane and mechanism are likely to be damaged by the sudden rush of water when the control valve opens.

Waterflow indicators are supplied by manufacturers of both sprinkler equipment and signaling system equipment. A representative vane type indicator is shown in Figure 18-5F.

System Alarm Attachments (Water Pressure Type)

These alarm attachments are designed to initiate an alarm by a drop in water pressure in the sprinkler piping. They are used with conventional alarm check valves. Figure 18-5G shows an auxiliary attachment actuated by a drop in sprinkler system pressure.

Waterflow Detectors (Excess Pressure Type)

Another waterflow alarm principle depends upon the use of an ordinary type check valve to hold the pressure in the sprinkler system at the highest water supply pressure. A restricted bypass, also having a check valve, around the main clapper helps to produce and maintain the high pressure condition. The main clapper is weighted or has a small differential provided by a divided seat ring so that there must be a positive drop in sprinkler system pressure before water can flow through the main check. This drop in pressure operates a sensitive diaphragm valve which admits pressure to the alarm circuit closer or opener. Transient flow of water into the sprinkler piping does not give an alarm, so that this alarm arrangement is free from false alarms caused by variations in the supply pressure. This alarm arrangement is adapted only to pressure actuated electric switches and not the operation of water motor alarm gongs.

Alarm Retarding Devices

An alarm check valve that is subjected to fluctuating water supply pressures needs an alarm retarding device in order to prevent false alarms when the check valve clapper is lifted from its seat by a transient surge of increasing pressure.

Retarding Chambers: One type of device, usually designated as a retarding chamber, is essentially a chamber inserted in the water line from the alarm check valve to the water motor gong and electric circuit closer. Flows of short duration from the alarm check valve to the alarm device first accumulate in the retarding chamber which must become filled before water passes to the alarm device. The size of the chamber and of the water inlet are predetermined by the manufacturer to give a delay needed to make sure that the water flow is continuous before an alarm is given. The air displaced from the alarm piping in advance of water escapes at low pressure and does not operate the

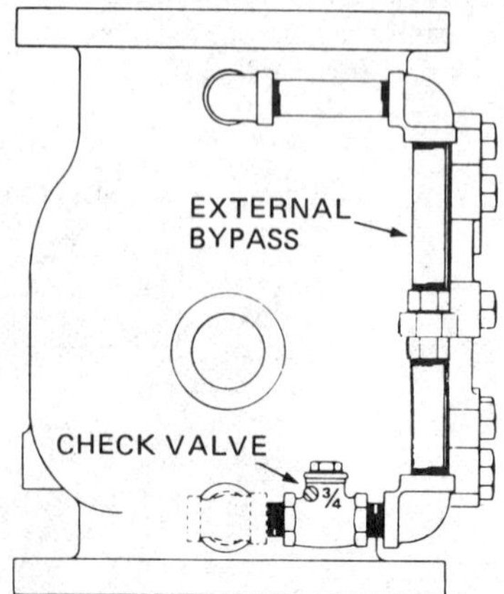

FIG. 18-5I. An alarm check valve with an external bypass. Shown is the Chemetron Model D alarm valve in vertical position.

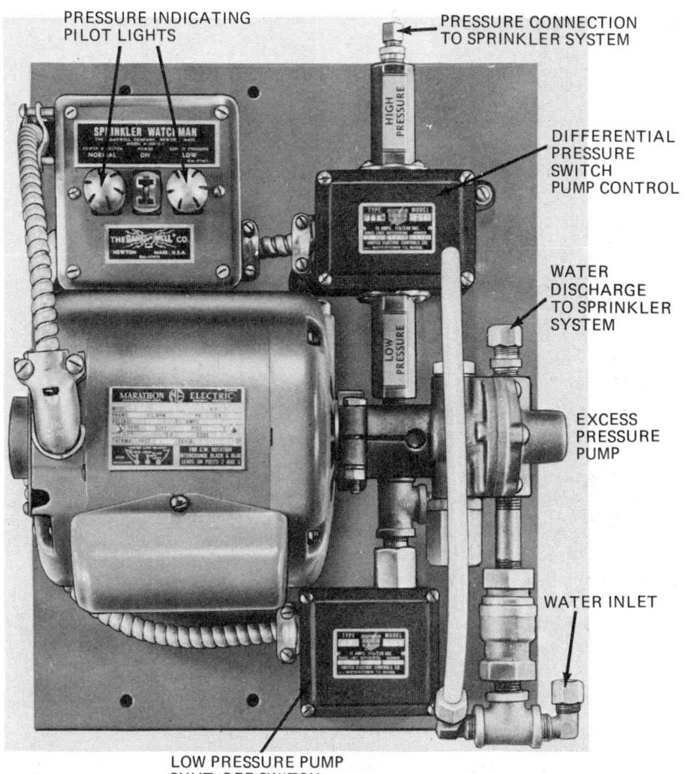

FIG. 18-5J. An excess pressure pump for installation on a wet pipe sprinkler system. The automatic excess pressure pump maintains a pressure 29 to 47 psi (200 to 325 kPa) in excess of the water supply pressure above an alarm check valve. A continuous white pilot light, controlled by a differential switch, indicates when the pressure in the sprinkler system is more than 29 psi (200 kPa) above the supply pressure. Should the water supply fail, a low pressure switch shuts off the current to the pump so that it will not run without water. This switch also controls a red pilot light to show when the supply pressure drops below 29 psi (200 kPa). A second pressure switch on the sprinkler system (not a part of the pump unit) is connected into the sprinkler system alarm circuit and gives an alarm if system pressure fails completely (e.g., a closed water control valve) when a sprinkler operates. A third pressure switch gives a warning by a local bell or other means in the event the pressure pump fails to maintain system pressure within the established excess pressure range. (Gamewell Corp.)

alarm devices. The retarding chamber is self-draining so that unless surges follow in close succession, the full retard interval is restored between surges. A retarding chamber is shown and described in Figure 18-5H.

Check Valve Bypass: To assist in avoiding false alarms in case of slow increases in water supply pressure, an alarm check valve is sometimes arranged with a small bypass opening through the main clapper, or piped through an exterior bypass. The bypass is restricted so that possible flow through it is small in comparison with the flow through a single automatic sprinkler, and it has a check valve intended to hold the pressure in the sprinkler piping at the highest value that occurs in the water supply system. An external bypass is shown in Figure 18-5I, and a bypass valve in the main clapper of an alarm valve is shown in Figure 18-5B.

Excess Pressure Pumps: Where great fluctuations in water pressures are encountered that exceed the normal adjustments of alarm valves, an excess pressure pump may be installed. An excess pressure pump is a small water pump of limited capacity which maintains pressure in the sprinkler system somewhat above the highest water supply pressure. Operation of an automatic sprinkler reduces the system pressure to the supply pressure. An electric pressure switch with its pressure setting slightly above the water supply pressure gives the waterflow alarm. A supervisory pressure switch with a setting slightly below the automatically maintained pressure, but above that of the flow alarm switch, is provided to give a trouble signal in case of failure of the automatic pressure pump. An excess pressure pump installation is shown in Figure 18-5J.

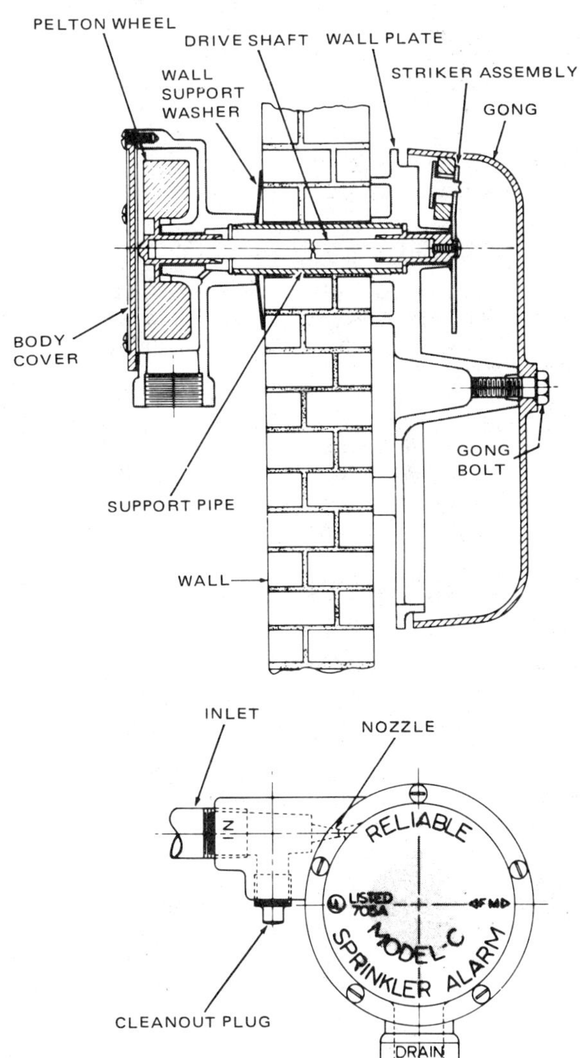

FIG. 18-5K. A representative water motor gong. Shown is the Reliable Automatic Sprinkler Co., Inc. Model C. Other manufacturers have similar arrangements of proprietary equipment for water motor gong installations.

DRY PIPE SPRINKLER SYSTEM ALARM DEVICES

It is relatively simple to arrange a connection from the intermediate chamber of a dry pipe valve to a pressure operated alarm device. The intermediate chamber of the dry pipe valve normally contains air only at atmospheric pressure. When the valve trips, the intermediate chamber immediately fills with water at supply pressure which is available to operate the alarm devices that are the same as those used with alarm check valves, except that a retarding device is not needed. Both an outdoor water motor gong and a pressure operated electric switch are usually provided.

ALARM DEVICES FOR DELUGE AND PREACTION SYSTEMS

The alarm devices used with deluge and preaction systems usually are of the same type as those used for dry pipe systems. They are connected to the sprinkler system side of the valve. It is also a common practice to have an electric alarm switch arranged so that an electric alarm will be given whether or not water flows through the valve after operation of the fire detecting system.

Whenever a preaction system has 20 or more sprinklers, the piping also is often supervised with a low air pressure switch along with the actuating device circuits.

WATER MOTOR GONGS

All manufacturers of waterflow alarm check valves and dry pipe valves can supply approved water driven alarm gongs for local outdoor alarms near dry pipe valves or alarm check valves.

Listed water motor gongs are similar in operating principle and mechanical construction, and are designed for mounting on the outside of building walls. Waste water may be discharged outdoors or drained off through a connection to the sprinkler system drain. A representative water motor gong is shown in Figure 18-5K.

PRESSURE ACTUATED ALARM SWITCHES

Electric switches, frequently called circuit closers, with contacts arranged to open or to close an electric circuit when subjected to increased or reduced pressure, are used in combination with dry pipe valves, alarm check valves, and some other types of water control valves to initiate an electric waterflow alarm signal when a flow of water to sprinklers occurs or to give a supervisory signal if pressures increase or decrease beyond established limits.

Manufacturers of dry pipe valves, alarm check valves, and special types of water control valves regularly furnish approved pressure operated switches to operate local electric waterflow alarms. In most cases, the motion to actuate a switch is obtained from a diaphragm exposed to the pressure on one side and opposed by a fixed adjustable spring on the other side. A typical waterflow alarm switch is shown in Figure 18-5L.

Mercury switches and other types of electric contacts may be used. Commercial pressure actuated switches, of which a representative design is shown in Figure 18-5M, may be adapted to waterflow alarm service.

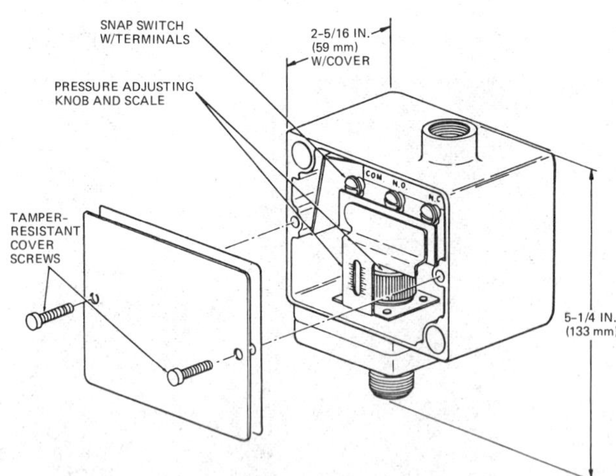

FIG. 18-5L. A waterflow alarm switch that can be used in combination with a conventional dry pipe valve or alarm check valve to actuate a fire alarm transmitter for waterflow signals in a central station, proprietary, or remote station signaling system. Shown is the Type 6159 switch of American District Telegraph Company (NJ) and Its Associated Companies.

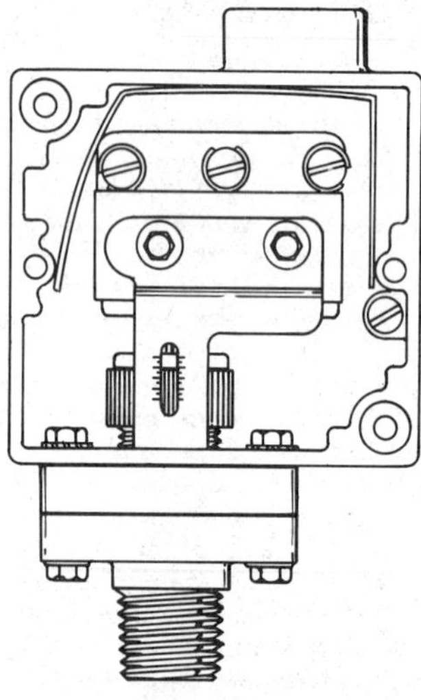

FIG. 18-5M. An internal view of a pressure switch for activating electric alarms on wet pipe, dry pipe, preaction, and deluge systems. Water pressure from the system acts upon a diaphragm in the housing at the base of the switch enclosure causing movable contacts to close with stationary contacts. The switch can be used for normally closed or normally opened circuits, and sensitivity to pressure differentials can be adjusted from 5 to 15 psi (34.0 to 103.0 kPa)/Shown is The Viking Corporation Model A-1.

OTHER SUPERVISORY DEVICES

Gate Valve Supervisory Switches

General requirements for supervisory signaling systems include electrical supervision of circuits to indicate conditions that could prevent the required operation of the

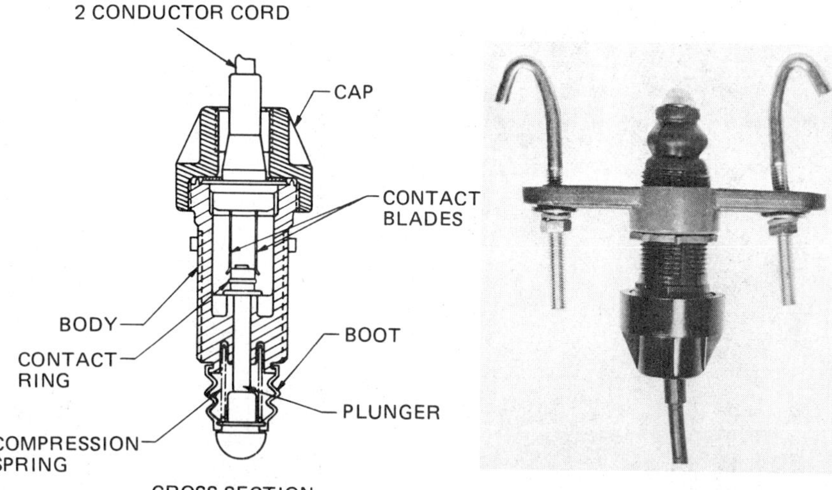

FIG. 18-5N. A gate valve supervisory switch. The switch is attached to the two sides of the valve yoke by the hook bolts shown in the photo at left. When the valve is wide open, the tip of the plunger enters a 1/8 in. (3 mm) deep depression drilled in the valve stem. The switch is adjusted so that in the open valve position the contact ring closes the electric circuit between the contact blades. When the valve is closed not over two turns, the tip of the plunger rides out of the depression in the valve stem and opens the electric circuit. Should the switch be tampered with or removed, the plunger takes the position shown in the sectional view and opens the electric circuit. Shown is the Model B-613 switch of American District Telegraph Company (NJ) and Its Associated Companies.

sprinkler system, the most common condition being shut control valves.

Electric switches for supervision of sprinkler system water supply control valves are of different mechanical designs for the different types of control valves.

Electric switches can be for open or closed circuit supervisory systems. Supervision may also be by means of signaling transmitters mechanically tripped by operation of the gate valve.

The signal to indicate valve operation is given within two turns of the valve wheel from the wide open position. The restoration signal is given when the valve is restored to its fully open position.

Figures 18-5N, 18-5O, and 18-5P are of representative approved devices for supervision of sprinkler valves.

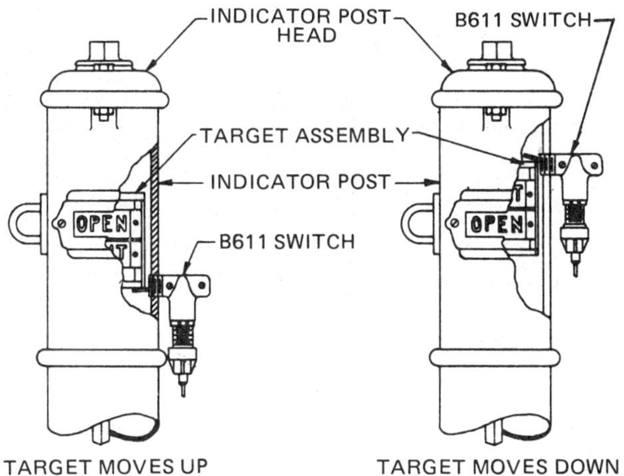

FIG. 18-5O. An indicator post valve supervisory switch. The operating stem of the switch is held against the movable target assembly by springs within the switch housing. The switch end of the operating stem carries an insulator which separates contact springs and opens the electric circuit when the target assembly is moved in the valve closing direction by about two turns of the valve stem. Shown is the Model B-611 switch of American District Telegraph Company (NJ) and Its Associated Companies.

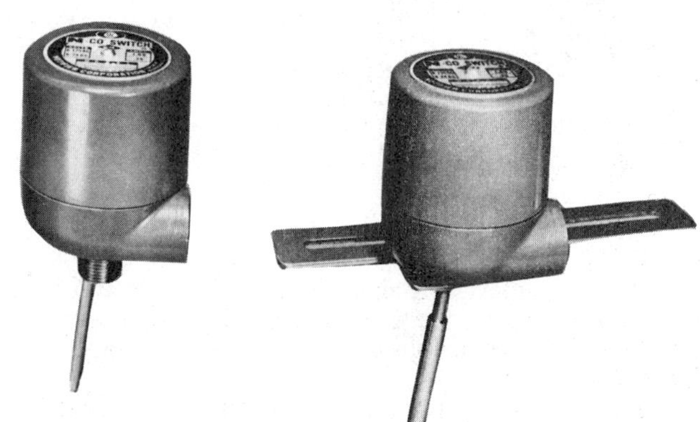

FIG. 18-5P. Water supply valve position indicator switch. The pivoted external control arm actuates the internal electric signal switch. When an outside-screw-and-yoke gate valve is in the wide open position, the control arm engages a shallow vee notch in the valve stem. If the stem is moved too close to the valve, the arm slides out of this notch and opens the electric signal circuit. If the switch is loosened or removed, the switch arm is moved by a spring and actuates the alarm switch. On indicator post valves, the arm extends into the indicator post and is adjusted to be operated by movement of the indicating target. Shown is the Notifier Company Model NGV/NIP switch.

Temperature Supervision for Water Tanks

The temperature of the water in fire service tanks exposed to cold weather is usually shown by a thermometer in the cold water return to a circulating type heater or near the bottom of a large riser.

Supervisory equipment is available for detecting dangerously low temperature near the surface of the water at the tank shell where freezing is most likely to start, as well as for checking manual supervision or automatic heat control.

Water Level Supervision

Devices for supervising the water level in pressure tanks differ from those used in water supply tanks in that level sensing elements are continuously subject to high

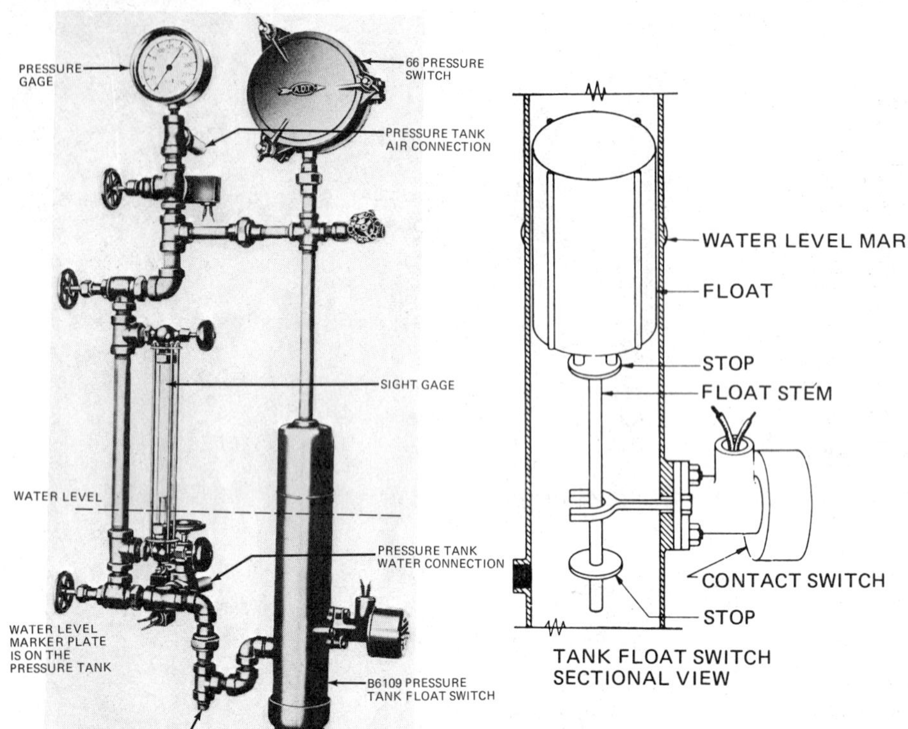

FIG. 18-5Q. A supervisory switch for pressure water tanks. If the water level rises or falls about 3 in. (76mm), one of the stops on the float stem in the tank float switch comes in contact with the forked lever which operates the contact switch. The switch is usually connected into the circuit of a supervisory signal transmitter. Shown is the Type B6109 switch of American District Telegraph Company (NJ) and Its Associated Companies.

pressure. The supervisory signal is given if the water level reaches 3 in. (76 mm) above or below the proper point, usually by means of an electrical connection to a separate signal transmitter.

Figure 18-5Q shows a water level supervisory switch for gravity tanks.

Bibliography

NFPA Codes, Standards, Recommended Practices and Manuals. (See the latest *NFPA Codes and Standards Catalog* for availability of current editions of the following documents.)

NFPA 13, *Standard for the Installation of Sprinkler Systems.*

NFPA 70, *National Electrical Code.*

NFPA 72A, *Standard for the Installation, Maintenance and Use of Local Protective Signaling Systems for Guard's Tour, Fire Alarm and Supervisory Service.*

NFPA 72B, *Standard for the Installation, Maintenance and Use of Auxiliary Protective Signaling Systems for Fire Alarm Service.*

NFPA 72C, *Standard for the Installation, Maintenance, and Use of Remote Station Protective Signaling Systems.*

NFPA 72D, *Standard for the Installation, Maintenance, and Use of Proprietary Protective Signaling Systems.*

Additional Readings

Bryan, John L., "Supervision of Automatic Sprinkler Systems," *Automatic Sprinkler and Standpipe Systems*, National Fire Protection Association, Inc., Quincy, MA, 1976, pp. 362-377.

Fire Protection Equipment List, published annually with bimonthly supplements, Underwriters Laboratories Inc., Northbrook, IL

Factory Mutual Approval Guide, published annually, Factory Mutual Engineering Corporation, Norwood, MA.

STANDPIPE AND HOSE SYSTEMS

Revised by Robert M. Hodnett

Standpipe and hose systems provide a means for manual application of water to fires in buildings. They do not take the place of automatic extinguishing systems which are generally the preferred form of protection. They are always needed where automatic protection is not provided and in areas of buildings not readily accessible to hose lines from outside hydrants.

Standpipe systems are designed for fire department use to provide quick and convenient means for obtaining effective fire streams on large low buildings or the upper stories of high buildings. Many jurisdictions have discontinued the requirement for occupant use hose systems in buildings that are completely protected by automatic sprinklers. The most effective use of standpipe systems is by fire departments or personnel who are trained in the use of 2½ in. (64 mm) hose streams at high pressures.

This chapter covers the various classifications for standpipe and hose systems, the different types of systems and their water supply requirements, the design of systems and the components required, the requirements for combined sprinkler and standpipe systems, and the requirements for outside hose systems.

NFPA 14, *Standard for the Installation of Standpipe and Hose Systems*, referred to in this chapter as the NFPA Standpipe and Hose Systems Standard, contains specific requirements for these systems and should be consulted for installation details. NFPA 13, *Standard for the Installation of Sprinkler Systems*, referred to in this chapter as the NFPA Sprinkler Systems Standard, is also an essential reference for installation of combined sprinkler and standpipe systems. Other NFPA standards and recommended practices that will be helpful can be found in the bibliography at the end of the chapter.

Other chapters in this HANDBOOK that contain additional information applicable to standpipe and hose systems are Section 17, Chapter 5, "Water Storage Facilities and Suction Supplies," and Chapter 6, "Fire Pumps"; Section 18, Chapter 8, "Care and Maintenance of Water-based Extinguishing Systems;" and Section 16, Chapter 2, "Protective Signaling Systems."

Mr. Hodnett is a consultant in fire protection and piping systems based in Providence, RI. He was NFPA staff liaison to the NFPA Committee on Automatic Sprinkler Systems prior to his retirement in 1984.

SYSTEM CLASSIFICATION

Class I Systems

Class I systems [2½ in. (64 mm) hose connections] are provided for use by fire departments and those trained in handling heavy fire streams. In nonsprinklered high rise buildings beyond the reach of fire department ladders, Class I systems can provide water supply for the primary means of fire fighting, i.e., manual attack on the fire.

Class II Systems

Class II systems [1½ in. (38 mm) hose lines] are provided for use by the building occupants until the fire department arrives. The hose is connected to ⅜ or ½ in. (9.5 to 12.7 mm) open nozzles or combination spray/straight stream nozzles with shutoff valves. Shutoff or spray nozzles are seldom provided unless the occupancy is one where hand hose would be used frequently. Normally the hose is kept attached to the shutoff valves at the outlets. Where the hose streams used by occupants can be properly supplied by connections to the risers of wet pipe automatic sprinkler systems, separate standpipes for these smaller streams are not required.

Class III Systems

Class III systems are provided for use by either fire departments and those trained in handling heavy hose streams or by the building occupants. Because of the multiple use, this type of system is provided with both 2½ in. (64 mm) hose connections (for use by fire departments or those trained in handling heavy hose streams) and 1½ in. (38 mm) hose connections (for use by the building occupants). One method for accommodating this multiple use is by means of a 2½ in. (64 mm) hose valve with an easily removable 2½ (64 mm) 1½ in. (38 mm) adapter, permanently attached to the standpipe.

The use of hose smaller than 1½ in. (38 mm) is permitted by the NFPA Standpipe and Hose Systems

Standard in Class II service in light hazard occupancies when listed for this service and approved by the authority having jurisdiction. The reasoning is that untrained building occupants may not be able to handle 100 ft (30 m) of 1½ in. (38 mm) hose with a residual pressure of 65 psi (448 kPa) at the outlet and a flow of 100 gpm (378 L/min). If smaller hose mounted on a reel with a flow smaller than 100 gpm (378 /minL) is provided, an untrained person might be less hesitant and more capable of using the equipment under fire conditions. Hard rubber ¾ in. (19 mm) and 1 in. (25 mm) fire hose is presently being used successfully in many foreign countries.

Water Supplies

The water supply for standpipe systems depends on the size and number of required streams and the length of time the systems may have to be operated, as well as the demand for automatic sprinklers using the same riser. The probable number of streams required should be ascertained before the water supply is decided upon. Water supplies for combined sprinkler and standpipe systems do not require that the water demand for sprinklers be added to the demand for standpipes if the occupancy is completely sprinklered. The greater amount that is required for the standpipe or the sprinkler system is sufficient. For a combined system in a partially sprinklered occupancy, the demands for each system must be added. Standpipe and hose systems should have water pressure maintained at all times. Where this is impractical, as in unheated buildings, the system should be arranged to admit water automatically by means of a dry pipe valve or other approved device.

Acceptable water supplies include the following:

1. City water works systems where pressure is adequate.
2. Automatic fire pumps.
3. Manually controlled fire pumps with pressure tanks.
4. Pressure tanks.
5. Gravity tanks.
6. Manually controlled fire pumps operated by remote control devices at each hose station.

Two independent sources of water supply are desirable. The primary water supply should be capable of supplying the streams first operated until the secondary sources can be brought into action. The secondary supply should be adequate for long periods.

Minimum city water and fire pump supplies for standpipes supplying 2½ in. (64 mm) hose that will be used by fire departments or specially trained personnel (Class I and III systems) are given as 500 gpm (1893 L/min) for a period of at least 30 min where only one standpipe is required. Where more than one standpipe is required, the minimum supply is 500 gpm (1893 L/min) for the first standpipe and 250 gpm (946 L/min) for each additional standpipe, with the total supply not exceeding 2,500 gpm (9463 L/min) for at least 30 min. For a completely sprinklered building, the water supply required need not exceed 1,500 gpm (5678 L/min) for a light hazard occupancy and 2,000 gpm (7571 L/min) for an ordinary hazard occupancy. In addition, the supply is required to be strong enough to maintain a residual pressure of 65 psi (448 kPa) at the topmost outlet of each standpipe (including outlets on the roof) with 500 gpm (1893 L/min) flowing from the topmost outlet of the most remote standpipe and 250 gpm from the topmost outlet of each standpipe up to the maximum flow required.

If a water source supplies more than one building or more than one fire area, the total supply may be reduced to 500 gpm (1893 L/min) for the first standpipe plus 250 gpm (946 L/min) for each additional standpipe in the building or fire area requiring the greatest number of standpipes.

The water supply requirement for Class II service [1½ in. (38 mm) hose] is 100 gpm (379 L/min) for a period of at least 30 min. The supply is also required to be strong enough to maintain a residual pressure of 65 psi (448 kPa) at the topmost outlet of each standpipe (including roof outlets) with 100 gpm (379 L/min) flowing.

Providing connections for a public fire department to pump water into a standpipe system makes a desirable auxiliary supply. One or more fire department connections is required for Class I and Class III systems. (See Fig. 18-6A.) However, it is essential that fire department connections be on the street side of buildings near fire hydrants and installed in a manner that permits hose lines to be attached to them easily and conveniently. Obstructions, such as adjacent buildings, fences, posts, shrubbery, etc., must be avoided.

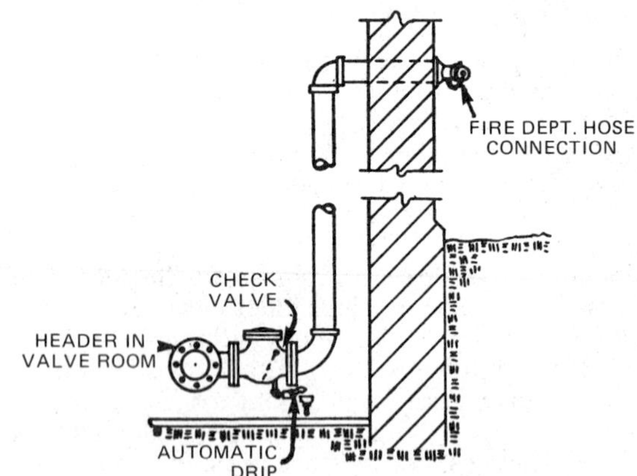

FIG. 18-6A. Typical fire department connection.

TYPES OF SYSTEMS

The four generally recognized types of standpipe systems are:

1. A wet standpipe system, having the supply valve open and water pressure maintained at all times. This is the most desirable type of system.
2. A dry standpipe system arranged to admit water to the system through manual operation of approved remote control devices located at each hose station. The water supply control mechanism introduces an inherent question of reliability which must be considered.
3. A dry standpipe system in an unheated building. The system should be arranged to admit water automatically by means of a dry pipe valve or other approved device. The depletion of system air at the time of use introduces a delay in the application of water to the fire and increases the level of competency required to control the pressurized hose and nozzle assembly during the charging period.

4. A dry standpipe system having no permanent water supply. This type would be used for reducing the time required for fire departments to put hose lines into action on upper floors of tall buildings. This type of system might also be used in buildings during construction when allowed in lieu of the wet standpipe in unheated areas.

SYSTEM COMPONENTS

The components of a typical standpipe and hose system are summarized in the following paragraphs.

Pipe and Tubing

Steel pipe assembled with welded joints, screwed fittings, flanged fittings, rubber gasketed fittings, or a combination of the above is the most common material used for standpipes. Ductile iron pipe and copper tubing, the latter assembled with brazed joints, are also used. For brazing or welding requirements refer to the NFPA Standpipe and Hose Systems Standard.

For Class I and III systems, standpipes not exceeding 100 ft (30 m) in height are a minimum of 4 in. nominal pipe size; standpipes over 100 ft (30 m) in height are a minimum of 6 in. nominal pipe size [The top 100 ft (30 m) may be 4 in. nominal pipe size.] In a completely sprinklered building, where 2½ in. (64 mm) hose outlets are provided on combined automatic sprinkler and standpipe risers, the risers may be calculated hydraulically in accordance with provisions of the NFPA Sprinkler Systems Standard. When pumps supplying two or more zones are located at the same level, each zone has a separate and direct supply—not smaller than the riser which it serves. Zones with two or more standpipes have at least two direct supply pipes—also not smaller than the risers served.

For Class II systems, standpipes less than 50 ft (15 m) in height are to be a minimum of 2 in. nominal pipe size, increased to 2½ in. nominal pipe size for those above 50 ft (15 m).

Fittings

The fittings must be rated for a minimum of 175 psi (1200 kPa). On parts of the standpipe system where the pressure can exceed 175 psi (1200 kPa), extra heavy pattern fittings must be used.

Nozzles, Hose, and Hose Cabinets

Each hose outlet for building occupant use may have up to 100 ft (30 m) of small hose attached to the outlet, preferably with a spray nozzle or a combination nozzle at the end for ready use. Excessive hose length may result in excessive kinking and other trouble during use. Lightweight woven jacket, rubber lined hose is preferred for standpipe or other inside service because of the lower friction loss, although unlined linen hose has been used extensively in the past as it takes up less space than the lined hose. Synthetic materials for jackets and linings have resulted in hose that takes up little more space than the unlined linen hose. When stored dry in heated atmospheres, unlined linen hose will last almost indefinitely, but under moist or wet conditions, it is subject to rapid deterioration. Hose should be kept on a standard rack

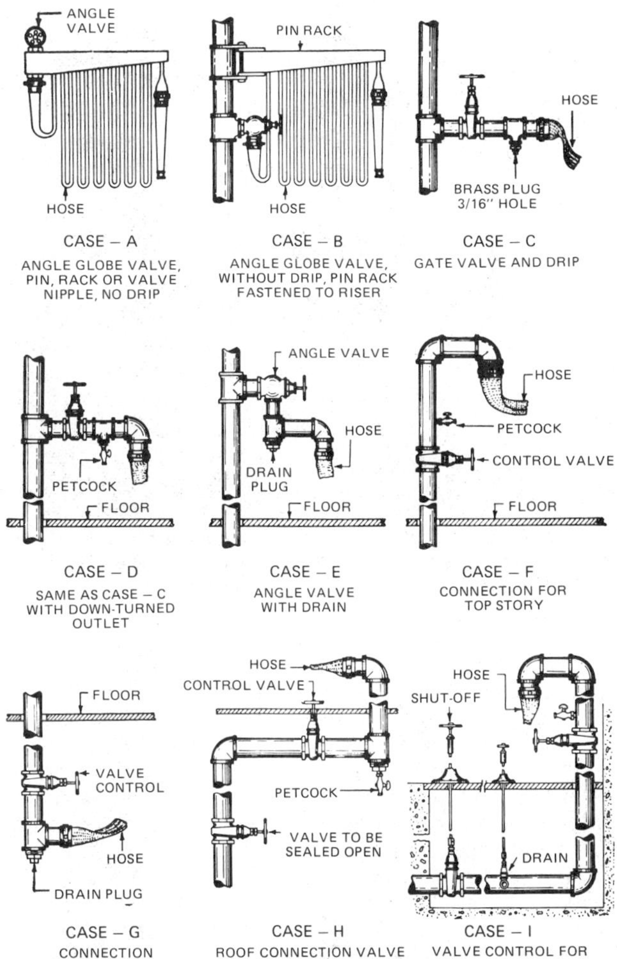

FIG. 18-6B. Hose valves and drip connections used to prevent leakage from entering hose. Racks of "semiautomatic" type, similar to that illustrated in Case A, are commonly used for 1½ in. (38 mm) hose. Hose, hung on pins, is released as nozzle is pulled away from rack. Valve may be opened wide before hose is pulled from rack; folds in hose will stop flow of water until hose is pulled from rack.

designed for use with the type of hose specified. For typical hose and valve connections, see Figure 18-6B.

Hose for occupant use should always be in a readily accessible location within convenient reach of a person standing on the floor. It should be clearly visible and in a place not likely to be obstructed. Where hose is in cabinets or closets, doors should open readily and have a glass panel or other form of easily recognized identification. Easily visible signs indicating the location of hose stations are valuable when the stations themselves are not directly visible from all directions. A typical cabinet for hose for occupant use is shown in Figure 18-6C.

The general practice is not to provide 2½ in. (64 mm) hose, but to depend upon hose brought in by the fire department at the time of fire. Nevertheless, some city codes require that 2½ in. (64 mm) hose be attached to standpipes. This presents the possibility of use of hose by untrained occupants of buildings, which introduces personal injury hazards and the probability of unnecessary water damage.

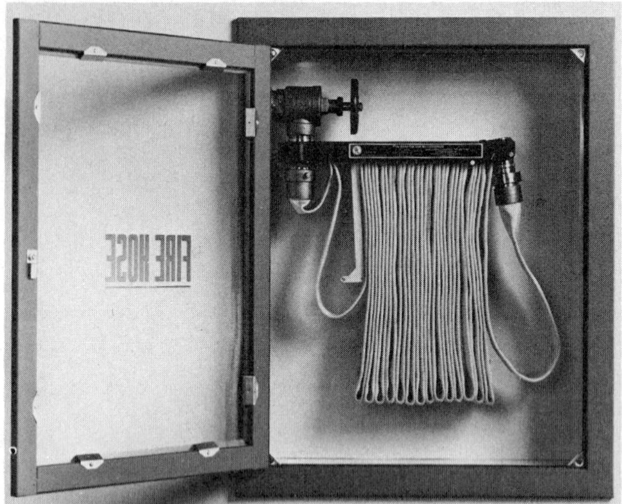

FIG. 18-6C. A cabinet containing hose for occupant use. The arrangement shown is of the "semiautomatic" type similar to that illustrated in Case A in Figure 18-6B.

Where hydrostatic pressure at any outlet for small hose exceeds 100 psi (690 kPa), a device should be installed at the outlet to reduce the pressure to such a value that the nozzle pressure will be limited to approximately 80 psi (552 kPa). One method of obtaining the pressure reduction under flow conditions is by use of an orifice disc. Another method is by the use of pressure reducing devices providing positive pressure control under both static and flow conditions. These devices are available for installation with 1½ (38 mm) and 2½ in. (64 mm) hose connections. They may be factory set for a specific inlet pressure to ensure a specific pressure from the discharge of the device, regardless of flow condition. These devices can be combination pressure reducers and shutoff valves.

Pressure reducing devices, hose valves, racks, and reels of suitable types for use on standpipes are listed by testing laboratories.

Fire hose must be maintained in proper position on the racks and in good condition. The hose, including gaskets, should be removed and inspected, then reracked, at intervals. Periodic hydrostatic testing of lined hose is required but unlined linen hose is not hydrostatically tested on a routine basis because it will rot or mildew if left even slightly damp. Thorough drying after it has been wet is extremely difficult. It should be carefully examined for cuts, loose couplings, and deterioration. Nozzles should be removed and examined for foreign objects.

Specific inspection and maintenance instructions for fire hose will be found in NFPA 1962, *Standard for the Care, Use, and Maintenance of Fire Hose Including Connections and Nozzles.*

Valves

A hose valve is provided at each outlet for attachment of hose. These valves are either straightway gate valves or globe valves with soft removable discs.

Indicating valves should be provided at the main riser for controlling branch lines to hose outlets so that, in the event a branch is broken during a fire, the fire department may shut off the branch and conserve water.

Drip connections, however, should be provided on either type of valve to protect hose against possible wetting and resultant deterioration.

Fire Department Connections

An approved fire department connection is provided for each Class I or Class III standpipe system. For high rise buildings having two or more zones, at least one fire department connection is needed for each zone. Fire department connections are on the street side of buildings near fire hydrants for easy connection to fire department pumpers, and are marked either "STANDPIPE" or "STANDPIPE AND AUTO. SPKR.," depending on the service.

Gages

A 3½ in. (89 mm) dial spring pressure gage is connected to each discharge pipe from fire pumps, to each supply connection from the public waterworks, at each pressure tank as well as at the air pump supplying the pressure tank, and at the top of each standpipe.

Alarm and Supervisory Equipment

Waterflow devices and tamper switches that give a signal either locally or at a central station are desirable additions to a standpipe system. The equipment used is the same as that used for a sprinkler system.

SYSTEM DESIGN

The Standard most widely used for standpipe system design is the NFPA Standpipe and Hose Systems Standard. Some large cities, notably New York and Chicago, have their own applicable codes. Federal building criteria usually supplements the NFPA Standard.

Number and Location of Outlets

The number of standpipes and the arrangement or distribution of equipment for proper protection are governed by local conditions such as occupancy, type and construction of building, exterior exposures, and accessibility. Standpipes supplying both 1½ in. (38 mm) and 2½ in. (64 mm) hose are located so that all portions of each story are within 30 ft (9 m) of a nozzle attached to 100 ft (30 m) of hose. Standpipes must be protected against mechanical and fire damage, with outlets for large hose in stairway enclosures and for small hose in corridors or adjacent to stairway enclosures. Location of occupant-use hose lines within an exit stair enclosure should be avoided because use of the hose could lead to infiltration of the stairway by smoke or heat, thus jeopardizing use of the exit and endangering those attempting to escape by these stairs.

Standpipes for large hose may furnish valuable protection against exposure fires. Care should be exercised in the choice of location to ensure that the window, door, or roof hatch—through which the stream would be operated—can be opened under conditions experienced during a fire. In addition to the use of hose streams from windows, roof connections are often employed. These may be equipped with hose houses or monitor nozzles.

Where a standpipe system is supplied by a fire pump, one 2½ in. (64 mm) hose outlet for each 250 gpm (946

L/min) pump capacity may be provided in the form of a wall outlet at ground level for fire department use on exposures.

Where buildings are within 60 ft (18 m) of exposing buildings, standpipes for large streams must be located to afford protection against exterior exposures as well as interior protection.

Zoning

The NFPA Standpipe and Hose Systems Standard limits zone height to 275 ft (84 m), except the height may be increased to 400 ft (120 m) when a pressure reducing device controlling nozzle pressure at both flow and no-flow conditions is installed at each outlet. When pressure regulating devices are used, they are set to give 100 psi (690 kPa) maximum at the hose valve outlet.

The best practice is to divide a tall building into pressure zones. Keeping a limit of about twelve stories in a zone prevents water pressures from becoming excessive. This simplifies fire protection measures because excessive pressure for hose lines does not have to be compensated for by pressure reducing devices and other complications. Water storage for fire protection should be calculated for each pressure zone, much as one calculates storage for areas of buildings on hills or other elevations supplied by municipal systems. Tanks providing storage in each pressure zone may be filled from the piping supplying water for other purposes in the building. Each pressure zone should have a gravity tank and a fire pump, the latter taking suction from the gravity tank in the next lower zone. Each zone should have its own fire department connection.

For a typical single zone system, see Figure 18-6D. In this system, a 1,000 gpm (3785 L/min) fire pump would be required. For a typical two zone system where the zones are independent (Fig. 18-6E), two 1,000 gpm (3785 L/min) fire pumps would be required. The first fire pump, connected to the city water system, supplies the low level zone including standpipe and storage, and furnishes high pressure suction to the second fire pump, which serves the high level zone.

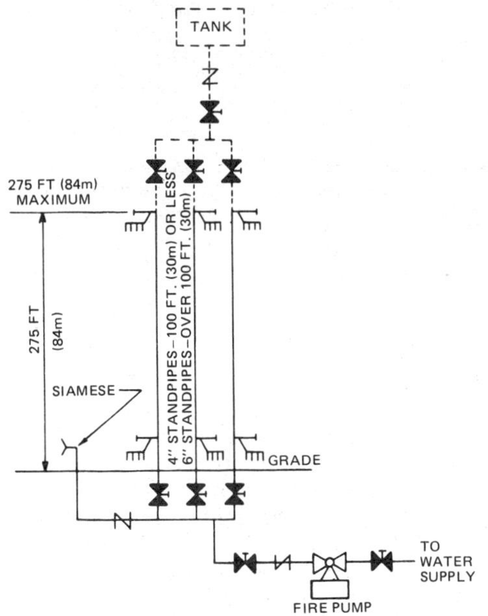

FIG. 18-6D. Typical single zone system.

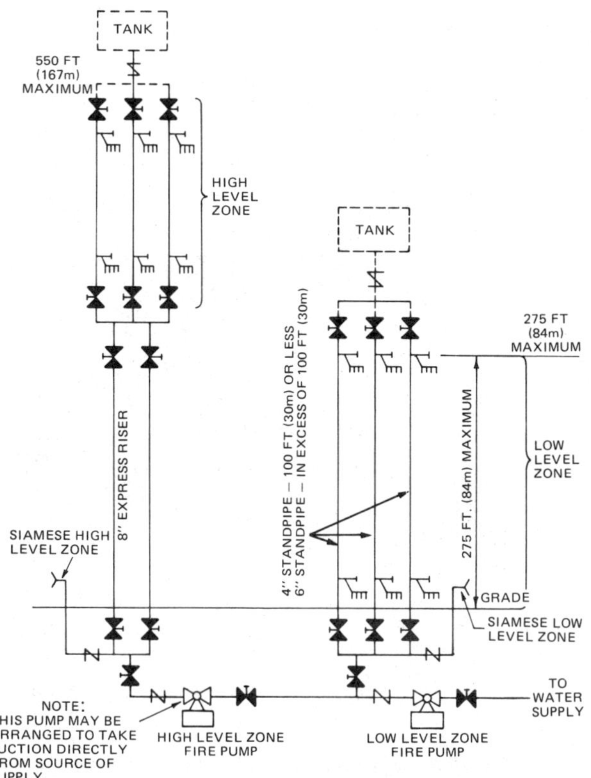

FIG. 18-6E. Typical two zone system.

For an alternate two zone system, see Figure 18-6F. The first fire pump serves the low level standpipe and the tank that serves as a source for the suction of the high level fire pump. Both fire pumps would have a 1,000 gpm (3785 L/min) rating so as to provide enough water for the three standpipes shown.

Standpipes in buildings with more than two zones would be designed as in Figure 18-6E, except there would be additional piping and pumps for the additional zones.

The choice of the location of the fire pumps would depend partially on the economics of the situation. Fire pumps at higher levels would require a lower pressure rating, but more protected wiring and less piping.

COMBINED SPRINKLER AND STANDPIPE SYSTEMS

In a combined sprinkler and standpipe system, the sprinkler risers can be used for feeding both the sprinkler system and the hose outlets. The outlets are 2½ in. (64 mm). If the building is completely sprinklered, 1½ in. (38 mm) hose for occupant use should be omitted.

The piping must comply with the requirements of the NFPA Sprinkler Systems Standard for the automatic sprinkler portions of the system, and with the NFPA Standpipe and Hose Systems Standard in regard to sizing of vertical risers and water supplies.

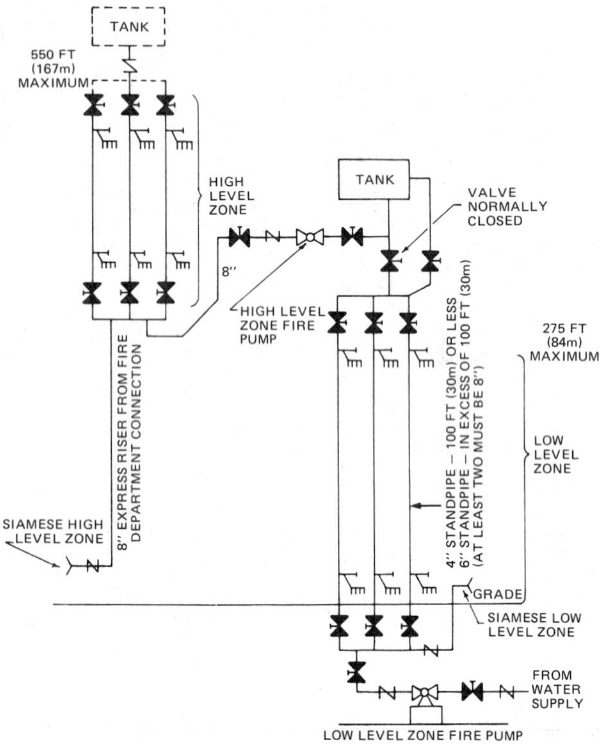

FIG. 18-6F. Alternate typical two zone system.

INSPECTIONS

Periodic inspection of all portions of standpipe systems is essential. The tanks must be kept properly filled, and where pressure tanks are employed at least 75 psi (517 kPa) pressure must be maintained.

Valves in the automatic sources of water supply must be open at all times. Where the system depends on such valves, they must be electromechanically supervised. Valves at the hose stations should be examined frequently for tightness. Leakage at the hose valves may be detected by inspection of the drips at the valves. Care should be taken to see that the connections and the drips are not clogged with dirt or sediment. Dry standpipes are not desirable and should be avoided to minimize maintenance difficulties.

OUTSIDE HOSE SYSTEMS

Where hose is kept connected to hydrants in hose houses, fire lines can be laid and water turned on in about half a minute, as compared with the two or three minutes required where a hose cart must be run up, hose coupled to hydrants, run out, and nozzle attached before water is turned on. In addition to the advantage of accessibility, hose kept in dry hose houses lasts longer than hose kept in heated buildings. In large plants having a fire department with trained fire fighters and hose carrying vehicles, hose houses may not be needed, but for most plants they provide the best means for storing hose.

Hose and Hydrant Houses and Equipment

NFPA 24, *Standard for the Installation of Private Fire Service Mains and Their Appurtenances*, gives requirements for construction and equipment for outside hose and hydrant houses. The principal features considered are ventilation and protection against weather. Ventilation obtained through slatted floor and shelves and through vent spaces under an overhanging roof is so arranged that rain cannot drive into the building.

Hose is generally attached to one of the hydrant outlets, leaving one or more outlets available for additional lines which can be laid from a hose cart or from reserve hose in the house. Sometimes equipment for two hose lines is connected to one hydrant. One shelf in a standard hose house will carry 150 ft to 200 ft (45 m to 60 m) of woven jacket hose folded forward and back. So placed, it will be ventilated and can be readily inspected.

With double outlet hydrants, it is desirable to have the hose attached to one outlet and a closed hose valve attached to the other. Hydrants having three and four outlets should have independent gates for each outlet.

Where the conditions are such that it is necessary to lock hose houses, this can be accomplished by using a special lock with a brittle shackle or by having a latch placed behind a glass plate to be broken at the time of fire.

Figures 18-6G and 18-6H are examples of hose houses

FIG. 18-6G. Hose house of the five sided design for installation over a yard hydrant. Such houses may be of wood or steel with a tight floor installed after construction. (W. D. Allen Mfg. Co.)

of suitable design which can be supplied by nearly all fire equipment dealers.

Amount of Hose

At least 100 ft (30 m) and preferably 150 ft (45 m) of 2½ in. (64 mm) woven jacket rubber lined hose should always be attached to the hydrant so that a fire stream can be put into action with minimum delay. The total amount of hose that should be maintained will depend on local conditions. If longer hose lays are anticipated, hose carts or other vehicles should be utilized.

In order to determine the length of hose required, a line should be laid in easy curves from the hydrant to the targeted building, through the door, up the stairs to the top story. Also, a line should be taken up an exterior fire ladder to the roof, allowing sufficient hose to have about one spare length on the roof. Where it is necessary to hoist hose to the roof, hoisting ropes and hose straps should be

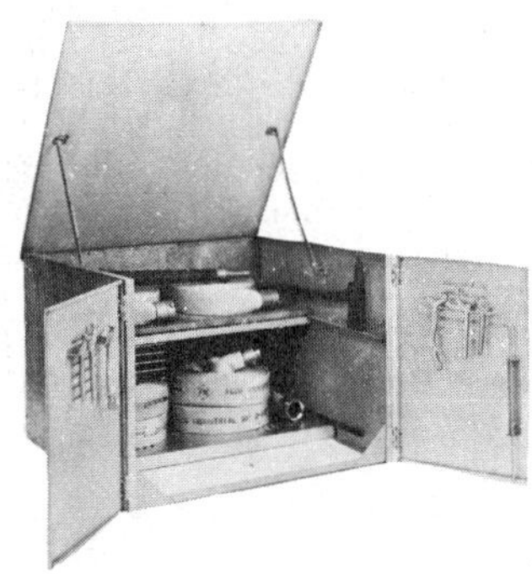

FIG. 18-6H. Steel hose house of compact dimensions. Top lifts up and doors on front side open for complete accessibility. Note ventilation louvers in the doors behind the tool racks.

available at the points where hoisting would likely take place.

Sufficient hose must be provided to direct a fire stream on all sides of the structure, allowing enough hose to hold the nozzle at a safe and effective distance from the building.

Accessories

The usual accessory equipment for a hydrant house includes nozzles, holders, spanners, a fire axe, and other tools.

Hose Carriers

Hose carriers or carts are useful where there are no hydrant houses with hose normally coupled up, where supply is needed to supplement hydrant houses, where long interior travel distances are anticipated (such as in large factories or warehouses), or where there are outlying buildings or yard storage areas to be reached. A carrier should have a capacity usually not exceeding 500 ft (150 m) of 2½ in. (64 mm) woven jacket hose, and it should be equipped with hydrant wrenches, spanners, extra gaskets, and other tools and equipment. One or two nozzles should be included. The carrier can be a cart pulled by manpower or a motor driven cart for in-plant service. It may be a compartment that is stored for pickup and transportation by a lift truck.

Carriers or carts should be stored in a separate building similar to a ventilated hose house and provided with a sloping approach to facilitate removal when needed. This incline can ordinarily be in position or it can be swung up into the building before the doors are closed. Electric lights outside or inside such a cart house should be available.

Power driven carriers have the advantage of better loading and greater carrying capacity for hose and appliances. Such carriers are occasionally provided with special appliances, water tanks, and some form of pumping equipment. Hose reels built on a roller platform to enable rapid transfer to pickup trucks or other automatic equipment can be used.

Monitor Nozzles

Where large amounts of combustible materials, such as log piles, lumber piles, or railway car or bus storage, are located in yards, it is necessary to provide a means of delivering large quantities of water at effective pressures. This can best be accomplished by installing permanent monitor nozzles around the piles, and occasionally where necessary on special trestles or roofs of buildings. (See Fig. 18-6I.) Portable deluge sets for use with siamesed hose lines are also valuable in many cases.

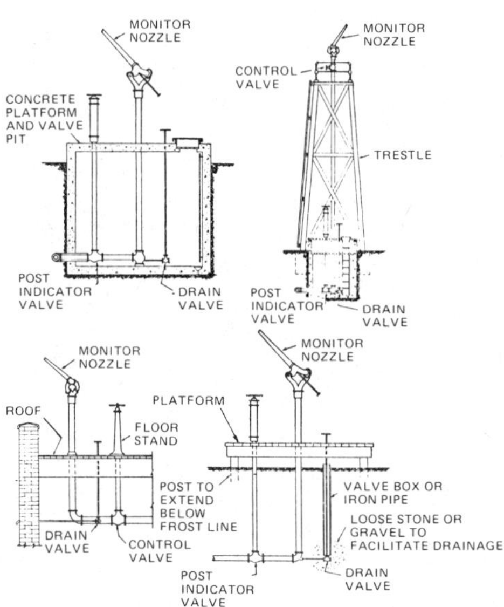

FIG. 18-6I. Four standard monitor nozzle installations. To keep the illustration simple, the monitors shown have levers for changing the position of nozzles and thus the direction of the stream. Geared wheels facilitate changing of the position of the monitor without shutting down the stream.

The location of this apparatus should be chosen so that the available water supplies are used efficiently. Hard-to-reach locations should be covered by the monitor nozzles. The piping and control valves usually require special consideration.

Bibliography

NFPA Codes, Standards, Recommended Practices and Manuals. (See the latest *NFPA Codes and Standards Catalog* for availability of current editions of the following documents.)

NFPA 13, *Standard for the Installation of Sprinkler Systems.*

NFPA 13E, *Recommendations for Fire Department Operations in Properties Protected by Sprinkler and Standpipe Systems.*

NFPA 14, *Standard for the Installation of Standpipe and Hose Systems.*

NFPA 20, *Standard for the Installation of Centrifugal Fire Pumps.*

NFPA 22, *Standard for Water Tanks for Private Fire Protection.*

NFPA 24, *Standard for the Installation of Private Fire Service Mains and Their Appurtenances.*

NFPA 72A, *Standard for the Installation, Maintenance and Use of Local Protective Signaling Systems for Guard's Tour, Fire Alarm and Supervisory Service.*

NFPA 1961, *Standard for Fire Hose.*

NFPA 194, *Standard for Screw Threads and Gaskets for Fire Hose Connections.*

NFPA 1962, *Standard for the Care, Use, and Maintenance of Fire Hose Including Connections and Nozzles.*

NFPA 1963, *Standard for Screw Threads and Gaskets for Fire Hose Connections.*

Additional Readings

Bond, Horatio, "Water for Fire Fighting in High-Rise Buildings," *Fire Technology*, Vol 2, No. 2, pp. 159-163.

Bryan, John L., *Automatic Sprinkler and Standpipe Systems*, NFPA, Quincy, MA, 1976, pp. 1-29.

Fire Protection Equipment List, Underwriters Laboratories Inc., Northbrook, IL, published annually with bimonthly supplements.

Factory Mutual Approval Guide, Factory Mutual Engineering Corporation, Norwood, MA., published annually.

Hammack, James M., "Combined Sprinkler System and Standpipes," *Fire Journal*, Vol. 63, No. 3, pp. 63-66.

Jensen, Rolf, ed., *Fire Protection for the Design Professional*, Cahners, Boston, 1975, pp. 66-71.

Kimball, Warren Y., "Can Pumpers Supply Standpipes in High-Rise Buildings," *Fire Journal*, Vol. 59, No. 5, Sept. 1965.

Lyons, Paul R., "Dry Standpipe Survey in Los Angeles," *Fire Journal*, Vol. 63, No. 3, pp. 63-66.

Nolan, J. W., "How to Approach Standpipe Design," *Actual Specifying Engineer*, Vol. 24, No. 7, July 1970, pp. 85-89. /et

WATER SPRAY PROTECTION

Revised by Robert M. Hodnett

The term water spray refers to the use of water that has a predetermined pattern, particle size, velocity, and density, and that is discharged from specially designed nozzles or devices. Water spray for fire protection has been called water fog, fog, or by trade name designations applied by equipment manufacturers. The use of such designations cannot be taken as indicative of any specific discharge pattern or spray characteristics of the nozzles so marketed, and has been discouraged.

There is no sharp line of demarcation between water spray protection and sprinkler protection. The discharge from nozzles or sprinklers producing a spray pattern differs only in the particular form of the spray and the other variables indicated in the above paragraph. In some cases, the same device may serve both purposes.

This chapter covers the uses and applications of water spray systems for fire prevention, control, and extinguishment; descriptions of components of spray systems; and specialized uses of the systems. Because of the similarities between sprinkler systems and water spray systems, their water supply requirements, some of the equipment used in the systems, and hydraulic calculations for determining water supplies, other chapters in this section of the HANDBOOK that should be consulted in connection with water spray systems are Chapter 1, "Automatic Sprinkler Systems," and Chapter 2, "Water Supplies for Sprinkler Systems."

Chapters in other sections of the HANDBOOK containing information applicable to water spray systems are Section 17, Chapter 1, "Water and Water Additives for Fire Extinguishment" (the extinguishing properties of water and safe clearances from electrical equipment), Section 16, Chapter 4, "Automatic Fire Detectors" (releasing devices that can be used with water spray systems), and Section 12, Chapter 7, "Materials Handling Equipment" (Protection of Conveyor Openings).

NFPA 15, *Standard for Water Spray Fixed Systems for Fire Protection*, hereinafter referred to in this chapter as the Water Spray Fixed Systems Standard, is the standard applicable to water spray systems and should be consulted on details of design and installation of the systems not covered in NFPA 13, *Standard for the Installation of Sprinkler Systems.*

USES FOR WATER SPRAY PROTECTION

Generally, water spray can be used effectively for any one or a combination of the following purposes: (1) extinguishment of fire, (2) control of fire, (3) exposure protection, and (4) prevention of fire.

Extinguishment: Extinguishment of fire by water spray is accomplished by cooling, smothering from the steam produced, emulsification of some liquids, dilution in some cases, or a combination of these factors.

Controlled Burning: With its consequent limitation of fire spread, controlled burning may be applied if the burning combustible materials are not susceptible to extinguishment by water spray, or if extinguishment is not desirable.

Exposure Protection: Exposure protection is accomplished by application of water spray directly to the exposed structures or equipment to remove or reduce the heat transferred to them from the exposing fire. Water spray curtains mounted at a distance from the exposed surface are less effective than direct application.

Prevention of Fire: It is sometimes possible to use water spray to dissolve, dilute, disperse, or cool flammable or combustible materials before they can ignite from an exposing ignition source.

APPLICATION OF WATER SPRAY SYSTEMS

Water spray protection is advantageous in meeting the previously listed purposes when it is applied to the following types of materials or equipment:

1. Ordinary combustible materials, such as paper, wood, and textiles, particularly to extinguish fires in such

Mr. Hodnett is a consultant in fire protection and piping systems based in Providence, RI. He was NFPA staff liaison to the NFPA Committee on Automatic Sprinkler Systems prior to his retirement in 1984.

materials rather than as a control measure.

2. Electrical equipment installations, such as transformers, oil switches, and rotating electrical machinery.
3. Flammable gases and liquids, particularly to control fires in these materials and to extinguish certain types of fires involving combustible liquids.
4. Flammable liquid and gas tanks, processing equipment, and structures, as protection for those installations against exposure fires.
5. Open cable trays and runs containing electrical cables or tubing.

Fixed water spray systems are specifically designed to provide optimum control, extinguishment, or exposure protection for special fire protection problems. They are not intended to replace automatic sprinkler systems, but they may be independent of, or supplementary to, other forms of protection. There are limitations to the use of water spray which should be recognized. Such limitations involve the nature of the equipment to be protected, the physical and chemical properties of the materials involved, and the environment of the hazard.

FIXED WATER SPRAY SYSTEMS

A water spray system is a special fixed pipe system connected to a reliable supply of fire protection water, and equipped with water spray nozzles for specific water discharge and distribution over the surface or area to be protected. The piping system is connected to the water supply through an automatically or manually actuated valve which initiates the flow of water.

Automatic water control valves for spray systems can be actuated electrically by operation of automatic detection equipment, such as heat detectors, relay circuits, gas detectors, etc., or mechnically by hydraulic or pneumatic systems, depending upon the operating mode of the individual valves. Generally, each manufacturer of valves, most of which can do dual service in deluge systems, provides its own particular combination of water control valve, releasing mechanism, heat detection system, and supervisory service. The operating principles of valves used in water spray systems shown in Figure 18-7A.

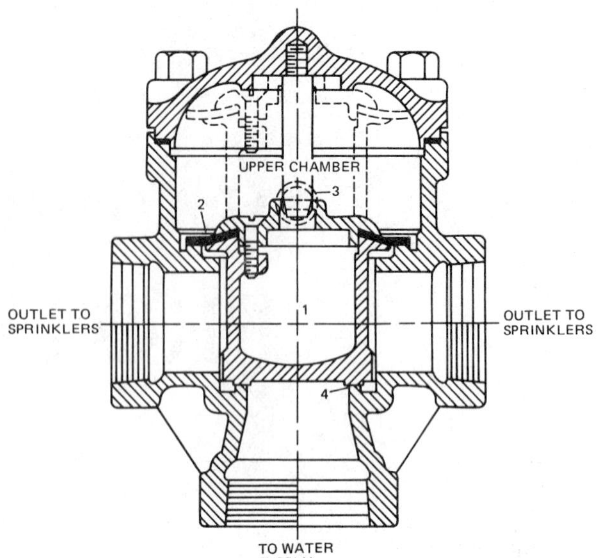

FIG. 18-7A. An automatic water control valve used primarily for control of small water spray protection systems. The center valve assembly (1) is held closed by water pressure acting on a rubber diaphragm (2) and retaining ring (3). Water pressure is admitted to the Upper Chamber by a restricting orifice in an external bypass line (not shown). Whenever water pressure is released to atmosphere (through operation of an activation device) faster than it can be supplied through the restricting orifice, the deluge valve will trip, lifting the valve from the lower valve seat (4). Shown is the Grinnell Flooding Valve, Model B.

Application of Systems

Fixed water spray systems are most commonly used to protect flammable liquid and gas tankage, piping and equipment; electrical equipment, such as transformers, oil switches, and rotating electrical machinery; and openings in firewalls and floors through which conveyors pass. The type of water spray required for any particular hazard will depend on the nature of the hazard and the purpose for which the protection is provided.

A water spray installation operating under test at a group of liquefied petroleum gas tanks is shown in Figure 18-7B. The spray system shown is designed to give complete surface wetting with a preselected water density, taking into consideration nozzle types, sizes and spacing, and the water supply. Ordinarily, it is neither expected nor desired that escaping liquefied petroleum gas be extinguished by the water spray. However, the rate of burning may be reduced and controlled by the cooling effect of the water on the tanks, and the severity of the exposure reduced until the gas supply to the fire is exhausted or can be shut off.

FIG. 18-7B. Water spray protection for LP-Gas tanks. The spray keeps the tanks cool in case of fire, prevents boiling away of the liquid contents, and protects the tank shells against rupture due to localized high temperature flame impingement.

Water spray protection for a group of oil-filled electric transformers is shown under test in Figure 18-7C. Due to the relatively high flash point and boiling point of transformer oil, transformer fires can be expected to be extinguished quickly by properly designed water spray systems.

FIG. 18-7C. A water spray system for oil-filled electric power transformers. A thick layer of crushed stone and subsurface drainage is provided around the base of the transformer installation to prevent the possibility that burning oil might flow beyond the ground area protected by the spray.

Design of Systems

The practical location of the piping and nozzles with respect to the surface to which the spray is to be applied, or to the zone in which the spray is to be effective, is determined largely by the physical arrangement and protection needs of the installation requiring protection. Once the criteria are established, the size (rate of discharge) of nozzles to be used, the angle of the nozzle discharge cone, and the water pressure needed can be determined.

The first factor to determine is the water density required to extinguish the fire or to absorb the expected heat from exposure or heat of combustion. When this is determined, a nozzle may be selected that will provide that density at a velocity adequate to overcome air currents and to carry the spray to the equipment to be protected. Each nozzle selected must also have the proper angle of discharge to cover the area to be protected by the nozzle.

The determination of the proper density to be used for extinguishment requires considerable engineering judgment and, in the case of flammable or combustible liquids, depends on such characteristics of the fuel as vapor pressure, flash point, viscosity, water solubility, and specific gravity. The density would vary between 0.2 gpm per sq ft to 0.5 gpm per sq ft [8.1 (L/min)m^2 to 20.4 (L/min)/m^2] of protected surface.

For exposure protection of vessels, a density of 0.25 gpm per sq ft [10.2 (L/min)/m^2] will provide sufficient cooling for a heat input of 6,000 Btu per hour per sq ft (18 930 W/m^2). Exposure protection for structures and miscellaneous equipment such as cable trays and runs, pipe racks, transformers, and belt conveyors varies from 0.10 gpm per sq ft [4.1 (L/min)/m^2] to 0.3 gpm per sq ft [12.2 (L/min)m^2]. The NFPA Water Spray Fixed Systems Standard should be consulted for more details on distribution densities.

Once the type of nozzle has been selected and the location and spacing to give the desired area coverage have been determined, hydraulic calculations are made to establish the appropriate pipe sizes and water supply requirements.

When water spray is to be used for the fire protection of oil-filled electrical equipment, such as transformers and large switch gear, special care must be taken to provide safe electrical clearances. To provide the high spray density needed, combined with good range, subject to minimum interference by wind, and with a simplified piping arrangement that does not need to be located close to live electrical parts, special fixed spray nozzles have been developed.

Size of Water Spray Systems

Many factors govern the size of a water spray system, including the nature of hazard or combustibles involved, amount and type of equipment to be protected, adequacy of other protection, and the size of the area which could be involved in a single fire. The size of the system needed may be minimized by taking advantage of possible subdivision by firewalls, by limiting the potential spread of flammable liquids by dikes, curbs, or special drainage; by water curtains or heat curtains; or by combinations of these features.

Because most water spray systems must perform as deluge type systems with all nozzles or devices open, and because a high density of water discharge is often needed, there is a heavy water demand. Each hazard should be protected by its own single system which should be adequate for dependable protection.

The NFPA Water Spray Fixed Systems Standard advises that the size of a single water spray system be limited only by the available water supply so that the designed discharge rate will be calculated at the minimum pressures for which the nozzles are effective. Experience has shown that in most installations a design discharge rate of 3,000 gpm (11 356 L/min) should not be exceeded for a single system. Separate fire areas should be protected by separate systems.

In cases where two or more systems are required for protection of a single fire area, the discharge capacity of each system is to be kept within the 3,000 gpm (11 356 L/min) limit.

Water Supplies

Fixed spray systems are usually supplied from one or more of the following:

1. Connections from a reliable waterworks system of adequate capacity and pressure.
2. Automatic fire pumps having reliable power and a water supply of adequate capacity and reliability.
3. An elevated (gravity) tank of adequate capacity and elevation.

The capacity of pressure tanks generally is inadequate to supply water spray systems. Pressure tanks, however, may be acceptable as water supplies to small systems whose water and pressure requirements do not exceed the capabilities of the tanks.

In some situations where the water supply is extremely limited, a cycle system, which collects and reuses water, may be acceptable. It is imperative, however, that foreign material and fuel be separated from the water before it is returned to the water spray system.

Water Demand Rate

The water supply must be adequate to supply at effective pressure all of the spray nozzles that may be expected to operate in a fire in the protected area. Additional water may be required for hose streams and should be considered when the system is designed. The duration of the discharge required will vary according to the nature of the hazard, the purpose for which the system is designed, and other factors which can be evaluated only for each installation.

Water demand is specified in terms of the density of a uniformly distributed spray measured in gallons per minute per square foot [(L/min)/m^2] of area protected. The discharge rate per unit of area depends on whether the spray system is installed for extinguishment of fire, control of fire, exposure protection, or prevention of fire, and upon the characteristics of the materials involved.

Pipe Sizes

Pipe sizes must be calculated for each system in order that the water at the spray nozzles will have adequate pressure. A procedure for making the hydraulic calculations for a fixed pipe water spray system is given in the Appendix to the NFPA Water Spray Fixed Systems Standard.

Selection and Use of Spray Nozzles

The selection of spray nozzles takes into consideration such factors as the character of the hazard to be protected, the purpose of the system, and possible severe wind or draft conditions.

High velocity spray nozzles, generally used in piped installations, discharge in the form of a spray filled cone, while low velocity spray nozzles usually deliver a much finer spray in the form of either a spray filled spheroid or cone. Due to differences in size of orifices in the various nozzles and the range of water particle sizes produced by each type, nozzles of one type cannot ordinarily be substituted for those of another type in an individual installation without the possibility of seriously affecting fire extinguishment. In general, the higher the velocity and the coarser the size of water droplets, the greater the effective "reach" or range of the spray.

Some open (nonautomatic) spray nozzles produce spray by giving the water high rotary motion in spiral passages inside the nozzle body. Sectional views of spray nozzles having internal spiral water passages are shown in Figure 18-7D.

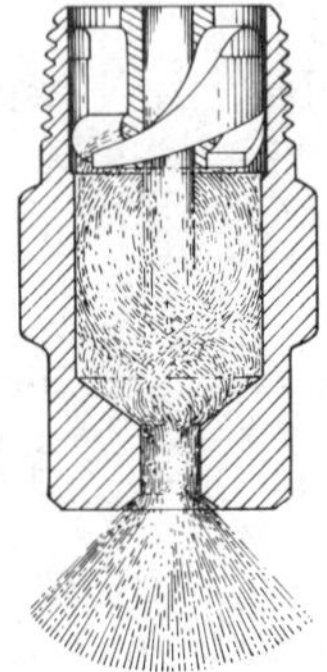

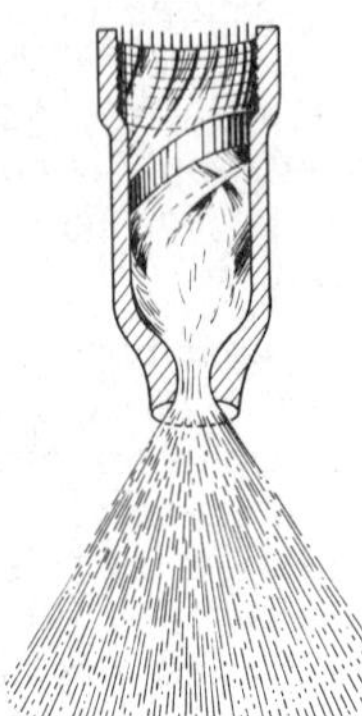

FIG. 18-7D. Water spray nozzles having internal spiral water passages.

Another type of water spray nozzle uses the deflector principle of the standard sprinkler (usually pendent type). The water discharge nozzle is unobstructed. The angle of the spray discharge cones is governed by the design of the deflector. Some manufacturers make spray nozzles of this type individually automatic by constructing them with heat responsive elements as used in standard automatic sprinklers. An open spray nozzle and an automatic nozzle of this type are shown in Figure 18-7E.

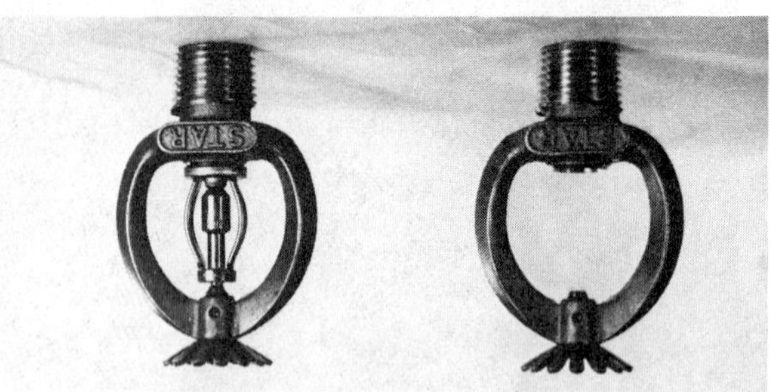

FIG. 18-7E. Water spray nozzles using the deflector principle of the standard automatic sprinkler. The nozzle at left has fusible elements for automatic operation.

One characteristically different type of water spray nozzle discharges water from its nozzle along the axis of a spiral of diminishing inside diameter. This spiral continuously peels off a thin layer of water from the surface of the cone. This thin layer of water breaks into spray as it leaves the spiral. (See Fig. 18-7F.)

FIG. 18-7F. A spiral type water spray nozzle.

Strainers

Strainers are ordinarily required in the supply lines of fixed piping spray systems to prevent clogging of the nozzles. (See Fig. 18-7G.) They should be selected with baskets having holes small enough to protect the smallest water passages in the nozzles used.

Water spray nozzles having very small water passages may have their own internal strainer as well as a supply line strainer to remove larger foreign material.

Drainage

Fixed pipe, open nozzle water spray systems discharge large quantities of water. To limit the spread of flammable liquids, special drainage and disposal facilities may be important. Pitched floors, curbs or dikes, and

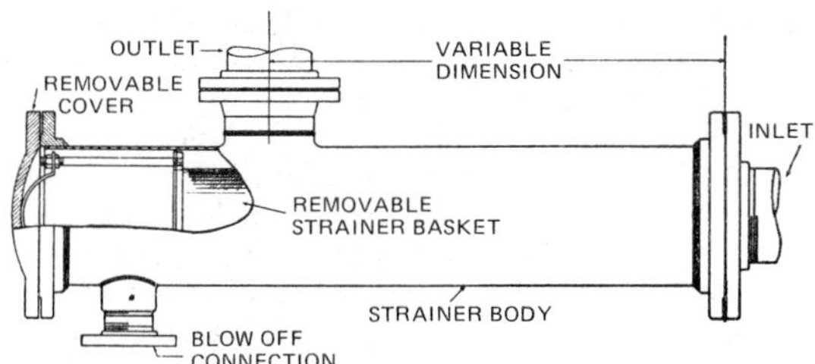

FIG. 18-7G. A Grinnell Model A strainer having a basket type screen of corrosion resistant metal. It is available in 3 in., 4 in., 6 in., 8 in., and 10 in. sizes. (1 in. = 25.4 mm.)

sumps or trenches designed for safe disposal may be used alone or in combination as best adapted to specific situations.

Maintenance

It is important that fixed water spray systems be inspected and maintained on a regularly scheduled program. Included in the program for checking are such items as strainers, piping, control valves, heat actuated devices, and the spray nozzles, particularly those equipped with strainers. Flow tests are frequently conducted to assure satisfactory operation. After tests, it is necessary to clean all strainers and to check all valves to be sure the system is in normal operating condition.

A problem in maintaining water spray protection systems using special spray nozzles or sprinklers is keeping small water passages clear. This may require special attention in the case of systems exposed to paint vapors and similar conditions. Nozzles with blowoff caps are available for protection against accumulation of foreign matter and attack from corrosive gases on nozzle orifices and on the interior of the piping system. (See Fig. 18-7H.)

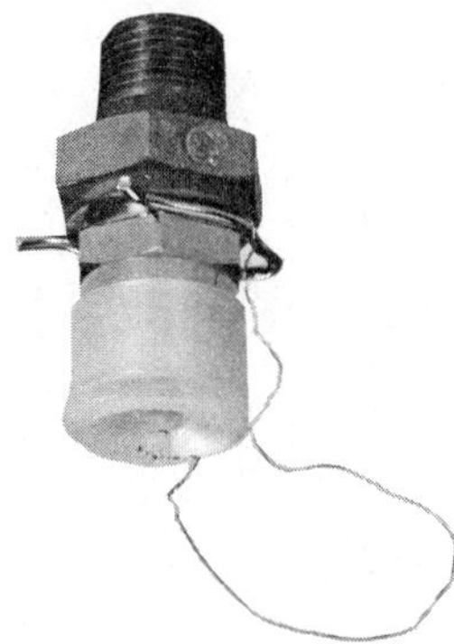

FIG. 18-7H. A Grinnell Mulsifyre nozzle with blowoff cap.

SPECIALIZED SYSTEMS UTILIZING WATER

Specialized systems utilizing water spray have been developed to fill particular fire control needs.

Ultra-High-Speed Water Spray Systems

These systems are designed to handle extremely rapid fires of the type that can occur in the handling of solid propellants, sensitive chemicals, and any industrial process or oxygen-enriched environment possessing this type of fire potential. The essential features of ultra-high-speed water spray systems that differ from conventional water spray systems are: the use of high speed detectors (normally photosensitive, infrared, or ultraviolet); the use of solid state devices (including an amplifier) to speed the signal from the detectors to the control panel; the employment of an explosively actuated valve; and the use of preprimed piping which speeds the water to the source of the fire upon the signal from the transistorized amplifier. Since the supply water and the priming water fill all cavities of the valve and all piping to the nozzle orifices (no air pockets anywhere in the system), the first movement of the valve results in water delivery within milliseconds from the time of initial detection. These systems use water spray nozzles which are fitted with caps to hold the priming water under a gravity head. These caps blow off as pressure is released into the system by the high speed valve. The number of milliseconds allowed for operation varies with the installation. In a group of hyperbaric test chambers where such ultra-high-speed water spray systems are installed, the time from fire ignition to delivery of water at design nozzle pressure was 96 milliseconds. Another installed system in a solid propellant installation operates in about 240 milliseconds. The time of water application for either system is very fast. Such installations must be carefully engineered for each special hazard to be protected and must be maintained by qualified experts. A number of fire protection equipment manufacturers can furnish and maintain this type of equipment.

Bibliography

NFPA Codes, Standards, Recommended Practices and Manuals. (See the latest *NFPA Codes and Standards Catalog* for availability of current editions of the following documents.)

NFPA 13, *Standard for the Installation of Sprinkler Systems.*
NFPA 14, *Standard for the Installation of Standpipe and Hose Systems.*
NFPA 15, *Standard for Water Spray Fixed Systems for Fire Protection.*
NFPA 18, *Standard on Wetting Agents.*
NFPA 20, *Standard for the Installation of Centrifugal Fire Pumps.*
NFPA 22, *Standard for Water Tanks for Private Fire Protection.*
NFPA 24, *Standard for the Installation of Private Fire Service Mains and Their Appurtenances.*
NFPA 30, *Flammable and Combustible Liquids Code.*
NFPA 69, *Standard on Explosion Prevention Systems.*
NFPA 70, *National Electrical Code®.*
NFPA 71, *Standard for the Installation, Maintenance and Use of Central Station Signaling Systems.*
NFPA 72A, *Standard for the Installation, Maintenance and Use of Local Protective Signaling Systems for Guard's Tour, Fire Alarm and Supervisory Service.*
NFPA 72B, *Standard for the Installation, Maintenance and Use of Auxiliary Protective Signaling Systems for Fire Alarm Service.*
NFPA 72C, *Standard for the Installation, Maintenance and Use of Remote Station Protective Signaling Systems for Fire Alarm and Supervisory Service.*
NFPA 72D, *Standard for the Installation, Maintenance and Use of Proprietary Protective Signaling Systems.*
NFPA 72E, *Standard on Automatic Fire Detectors.*

NFPA 80A, *Recommended Practice for Protection of Buildings from Exterior Fire Exposures.*
NFPA 251, *Standard Methods of Fire Tests of Building Construction and Materials.*
NFPA 321, *Standard on Basic Classification of Flammable and Combustible Liquids.*
NFPA 325M, *Fire Hazard Properties of Flammable Liquids, Gases and Volatile Solids.*

Additional Readings

Ault, Wayne E., and Carter, Donald I., "The Influence of Hyperbaric Chamber Pressure on Water-Spray Patterns," *Fire Journal*, Vol. 61, No. 6, Nov. 1967, pp. 48-49.
Braidech, M. M., and Neale, J. A., "The Mechanism of Extinguishment of Fire by Finely Divided Water," *NBFU Research Report No. 10*, Underwriters Laboratories Inc., Chicago, IL, 1955.
Bray, G., "Water Spray Protection of LPG Hazards," *Fire Protection*, Vol. 7, No. 4, Dec. 1980, pp. 3-10.
Cooper, L. Y., and O'Neill, J. G., "Fire Tests of Stairwell-Sprinkler Systems, " *NBSIR 81-2202*, National Bureau of Standards, Gaithersburg, MD, Feb. 1981.
Heselden, A. J. M., and Hinkley, P. L., "Measurements of the Transmission of Radiation Through Water Sprays," *Fire Technology*, Vol. 1, No. 2, May 1965, pp. 130-137.
Nakakuki, Atsushi, "Extinction of Fires in Hyperbaric Chambers: Part I, Water Spray Properties and Extinction with Single Spray Nozzles," *Fire Technology*, Vol. 8, No. 1, Feb. 1972, pp. 5-18.
O'Neill, J. G., "Sprinkler-Vent and Spray Nozzle Systems for Fire Protection of Openings in Fire Resistive Walls and Ceilings," *NBSIR 78-1571*, National Bureau of Standards, Gaithersburg, MD, 1978.
Turner, Harlan L., and Segal, Louis, "Fire Behavior and Protection in Hyperbaric Chambers," *Fire Technology*, Vol. 1, No. 4, Nov. 1965, pp. 269-277.
Waterman, T. E., "Use of Simplified Sprinkler Systems to Protect Wood Doors," *Fire Journal*, Vol. 67, No. 1, Jan. 1973, p. 42.

CARE AND MAINTENANCE OF WATER BASED EXTINGUISHING SYSTEMS

Revised by Robert M. Hodnett

The care and maintenance of water based extinguishing systems is vital to their performance in emergencies. Without well conceived and executed maintenance programs, the best of systems stand a chance of failure through neglect.

Water based extinguishing systems encompass all the different types of automatic sprinkler systems, water spray systems, and standpipe and hose systems discussed in earlier chapters in this section. While each of the systems has devices and equipment peculiar to its own function, much is common to all of them. Sprinkler systems have sprinklers, hose systems have hand held nozzles, and water spray systems have spray nozzles. But none of them will work if somewhere back in the water supply system a valve is closed that should be open.

Perhaps more has been written on good practices to follow in the care and maintenance of automatic sprinkler systems than on other water based systems for the simple reason that they are quite common and are relatively more sophisticated than, for example, a simple standpipe installation. There are, however, the common denominators of water supply, supply delivery system (piping and valves), and discharge devices that apply to all systems. The degree of dependency rests, of course, on the complexity and extent of the system.

This chapter discusses care and maintenance of water based systems with a principal focus on automatic sprinkler systems; however, the HANDBOOK user will find that many of the precepts discussed within the framework of sprinkler system care and maintenance will have application in planning and implementing maintenance for other types of water based systems.

The principles of good maintenance procedures are detailed in NFPA 13A, *Recommended Practice for the Inspection, Testing, and Maintenance of Sprinkler Systems*, referred to hereinafter in this chapter as NFPA 13A, Care and Maintenance of Sprinkler Systems. NFPA 13A is a companion to NFPA 13, *Standard for the Installation of Sprinkler Systems* (the NFPA Sprinkler Systems Standard), the basic document governing the installation of sprinkler systems.

Part of good inspection and maintenance procedures is the organization that administers them. Particularly important are on-site services not under the management of outside organizations, such as commercial inspection services. The chapters of Section 14, "Organization for Private Protection," in this HANDBOOK will be helpful in clarifying duties and responsibilities in the maintenance of fire protection equipment.

(For the purposes of this HANDBOOK, foam water sprinkler and spray systems are classified as special extinguishing systems and information on their care and maintenance will be found in Section 9, Chapter 4, "Foam Extinguishing Agents and Application Systems." But, again, much that applies to the care and maintenance of the water based systems covered in this chapter applies to foam systems as well.)

The maintenance of water supply equipment such as tanks, fire pumps, etc., is treated in the chapters relating to those items of equipment.

IMPORTANCE OF EXTINGUISHING SYSTEM MAINTENANCE

Care and maintenance includes more than just the inspection and testing of system devices and equipment. If protection is to be fully effective, the proper relation between hazard and protection must be maintained. This calls for careful consideration of the fire hazards to be protected and requires a decision regarding the adequacy of the protection for the varying conditions of the property. Further, inspection must include an assessment of whether the originally installed system is adequate for the current fire hazards.

Water based extinguishing systems employing standard devices and installed in accordance with established rules are sturdy and durable, and require minimum expenditure for maintenance. However, like other types of equipment, they may suffer deterioration or impairment through neglect or from certain conditions of service. Inspection is merely an organized, methodical procedure

---

Mr. Hodnett is a consultant in fire protection and piping systems based in Providence, RI. He was NFPA staff liaison to the NFPA Committee on Automatic Sprinkler Systems prior to his retirement in 1984.

for determining the operating condition of devices, equipment, and, in some cases, the qualifications of personnel for the detection of conditions calling for maintenance, repair, or remedy.

Maintenance of protection involves making the inspections as well as performing any special investigations or tests bearing on the performance of the devices and equipment, taking action to repair or keep devices and equipment in dependable operating condition, and assuring correct procedures by personnel responsible for the performance or use of the equipment.

Inspection and maintenance functions are closely related and may overlap in some features. Management is responsible for their correlation. Inspections frequently involve matters which may be classed as maintenance, and maintenance sometimes requires its own inspections and tests beyond those made routinely during so-called fire inspections.

RESPONSIBILITY FOR MAINTENANCE

Many of the troubles experienced with extinguishing systems have been due to lack of responsibility rather than lack of knowledge. The establishment of an appropriate organization is the responsibility of management. The types of organizations required to perform inspection, testing, and maintenance functions vary greatly. Size and value of the property are major considerations.

Inspection procedures may have some seasonal differences, most of which are indicated later in this chapter under the various items of equipment being considered. The following are examples of inspections governed by seasonal effects:

Spring Inspection: As soon as danger of freezing is past, the spring inspection will give attention to the opening of cold weather valves, testing, cleaning, and resetting dry pipe valves, testing water motor gongs, and conducting waterflow tests.

Fall Inspection: At the approach of freezing weather, the fall inspection will give special attention to such items as:

1. Closing cold weather valves and draining pipes exposed to freezing temperatures (drain valves on the exposed piping are left slightly open), and testing the specific gravity of the solution in antifreeze sprinkler systems.
2. Checking dry pipe valves to make sure that the systems are holding air properly and that the electric and water motor alarms are in order, checking drains at low points of the dry piping to make sure they are properly clear of water, and checking heating provisions for the dry valves.
3. Examining gravity tanks to determine if adequate protection against freezing is assured and that any heating system employed is in operative condition.
4. Checking the condition of fire pump reservoirs and the suction intakes from other water sources.
5. Looking over buildings to make sure that cold air will not enter or unduly expose piping to freezing.

TYPES OF SPRINKLER SYSTEM INSPECTIONS

In addition to the indispensable inspection procedures followed by the property owner, other inspection services are available.

Insurance Inspections

In insured properties, extinguishing systems are frequently given special attention by the insurance carrier. Routine testing of sprinkler systems and devices at regular intervals is a service extended by some insurance companies in the common interest of both the owner and the company. By these routine tests, the equipment can be shown to be in good operating condition, or any defects or impairments revealed. Since such tests are made at the owner's responsibility and risk, intelligent cooperation in conducting the tests serves the best interests of the owner.

Fire Department Inspections

Inspections are made of extinguishing system equipment by many fire departments at varying intervals. The inspection is principally to make sure that valves are open and to ensure familiarity with the use of the systems by the fire department. Inspections are customarily made by the fire company in whose district the installation is located.

Sprinkler Contractors' Services

Standardized sprinkler equipment inspection and maintenance services are offered by sprinkler manufacturers, sprinkler contractors, and inspection services companies. This service provides periodic examinations and reports, and is of value to the property owner not only for regular checkup of the condition of sprinkler equipment, but also because of valuable instruction that can be given to employees in the process. In addition to sprinkler devices and equipment, a contract can also cover other items (tanks, fire pumps, etc.) that are important in the fire protection of property.

Inspection and maintenance services offered by sprinkler contractors normally follow a form acceptable to most insurance interests. Table 18-8A lists the items that should be inspected and tested and the time intervals that should be observed. Table 18-8B contains a listing of pertinent information that should be included in reports on inspections and tests that have been conducted.

Central Station Supervisory Service

Central station supervision of sprinkler alarm and control devices provided under contract is an especially valuable aid to maintenance. The reporting of each incident involving waterflow or gate closure or other supervised action keeps a constant check on the condition of the equipment and stimulates care on the part of the plant fire organization.

GENERAL MAINTENANCE OF SPRINKLERS AND SPRINKLER PIPING

Automatic sprinklers installed where they are not subjected to abnormal conditions such as corrosion, abnormally high temperatures, or mechanical abuse, will give continued satisfaction over a great many years. Some sprinklers accepted by insurance organizations when first installed proved unreliable after years of service. Those falling into this category are so identified in Table 18-3E. Sprinklers listed by leading testing laboratories since 1900 have generally given satisfactory service even though superseded by new or improved sprinklers. NFPA 13A, Care and Maintenance of Sprinkler Systems, however, recommends testing representative samples of all sprinklers that are 50 or more years old at 10 year intervals. Sprinklers made previous to 1920 should be replaced.

Where sprinklers are subject to loading (accumulations of foreign material) or corrosion, even to only a moderate or slight extent, they should be carefully and frequently examined. If the condition of the sprinklers appears to be doubtful, a representative sample (six or more) should be removed, carefully packed to avoid injury in transit, and sent for testing to testing laboratories or to the sprinkler manufacturer. The results of their testing will indicate the condition of similar sprinklers remaining in the property.

To prevent mechanical injury and distortion when installing or removing sprinklers to be cleaned and reinstalled, the special wrenches provided by manufacturers (for their size and shape sprinklers) should be used.

Accumulation of Foreign Material on Sprinklers

In many classes of properties, conditions exist which cause an accumulation of foreign material on automatic sprinklers so that operation of the sprinkler may be retarded or prevented. This condition is commonly called loading. (See Fig. 18-8A.)

Any accumulation of foreign material on sprinklers tends to retard their operation as a result of the heat insulating effect of the loading material. If the deposit is hard, it may physically retard or prevent the sprinkler from operating. The best practice is to replace loaded sprinklers with new sprinklers rather than to attempt to clean them. Attempts at cleaning, particularly in instances where deposits are hard, are likely to damage the sprinkler, rendering it inoperative or possibly causing leakage.

Deposits of light dust, such as may be found on sprinklers in woodworking plants and grain elevators, are less serious than hard deposits. Dust may be expected to delay the operation of sprinklers but ordinarily will not prevent the eventual discharge of water. Dust deposits can be blown or brushed off, but blowing by compressed air should not be undertaken where it can create a dust explosion or ignition hazard. If a brush is used, it should be soft to avoid possible injury to sprinkler parts.

Water solution cleaning liquids of caustic or acid type are likely to be destructive to sprinklers and should not be used for cleaning. No hot solution of any kind should be used.

Sprinklers are sometimes protected when ceilings or sprinkler piping are being painted by temporarily placing small, lightweight paper or plastic bags over them, and securing the bags with a rubber band. Bags, however, are likely to delay the operation of the sprinklers, and should be removed immediately after the painting is completed.

There is no known method whereby paint under the water cap or on the fusible link can be removed adequately. Sprinklers that have been painted other than by the manufacturer must be replaced with new units.

Table 18-8A. Summary: Minimum Inspection - Testing - Maintenance

Records - Inspection = Visual Observation
Testing = Handling Equipment, etc.
Maintenance = Periodic Servicing and Repair

For Guidance on Specific Valves, Pumps, Hydrants, etc., Refer to the Manufacturer's Instructions.

Parts	Activity	Frequency
Flushing Piping	Test	5 Years 10 Years
Fire Department Connections	Inspection	Monthly
Indicator Post Valve or Tamper Switch	Inspection Test	Weekly-Sealed Monthly-Locked Quarterly
Main Drain	Flow Test	Quarterly
Open Sprinklers	Test	Annual
Pressure Gage	Calibration Test	5 Years
Sprinklers	Test	50 Years
Sprinklers-High Temp.	Test	5 Years
Valves in Roadway Boxes	Inspection Test	Weekly-Sealed Monthly-Locked Quarterly
Valves on Systems	Inspection Inspection Maintenance	Weekly-Sealed Monthly-Locked Yearly
Water Flow Alarms	Inspection	Monthly
Hydrants	Inspection Test (Open and Close) Maintenance	Monthly Semi-Annually
Antifreeze Solution	Test	Annually
Cold Weather Valves	Open and Close Valves	Fall, Close; Spring, Open
Dry Pipe Valve		
Air Pressure, Water Pressure	Inspection	Weekly
Enclosure	Inspection	Daily-Cold Weather
Priming Water Level	Inspection	Quarterly
Dry Pipe Valves	Trip Test	Annual-Spring
Dry Pipe Valves	Full Flow Trip Test	3 Years-Spring
Low Point Drains	Test	Fall
Quick-Opening Devices	Test	Semi-Annually
Gravity Tank		
Water Level	Inspection	Monthly
Heat	Inspectioin	Daily-Cold Weather
Condition	Inspection	Bi-Annual
Pressure Tank		
Water Level & Pressure	Inspection	Monthly
Heat Enclosures	Inspection	Daily-Cold Weather
Condition	Inspection	3 Years
Pump Centrifugal	Test Operate Test-Reduced Water Pressure Start Test-Capacity	Weekly Weekly Weekly Yearly
Steam Pump	Test Operate Test Capacity	 Yearly

Table 18-8B. A Sample Report of Inspection

(Based on Exhibit 1, Report of Inspection, as recommended in NFPA 13A. Recommended Practice for the Care and Maintenance of Sprinkler Systems.)

Owner's Section (To be answered by Owner or Occupant)

1. Explain any occupany hazard changes since the previous inspection.
2. Describe fire protection modifications made since last inspection.
3. Describe any fires since last inspection.
4. When was the system piping last checked for stoppage, corrosion or foreign material?
5. When was the dry piping system last checked for proper pitch?
6. Are dry pipe valves adequately protected from freezing?

Inspector's Section (All responses reference current inspection)

1. General
   a. Is the building occupied?
   b. Are all systems in service?
   c. Is there a minimum of 18 in. (457 mm) clearance between the top of the storage and the sprinkler deflector?
   d. In areas protected by wet system, does the building appear to be properly heated in all areas, including blind attics and perimeter areas, where accessible? Do all exterior openings appear to be protected against freezing?
   e. Does the hand hose on the sprinkler system appear to be satisfactory?
2. Control Valves (*See Item 14.*)
   a. Are all sprinkler system control valves and all other valves in the appropriate open or closed position?
   b. Are all control valves in the open position and locked, sealed or equipped with a tamper switch?
3. Water Supplies (*See Item 15.*)
   a. Was a water flow test of main drain made at the sprinkler riser?
4. Tanks, Pumps, Fire Department Connections
   a. Are fire pumps, gravity tanks, reservoirs and pressure tanks in good condition and properly maintained?
   b. Are fire department connections in satisfactory condition, couplings free, caps in place, and check valves right? Are they accessible and visible?
5. Wet Systems (*See Item 13.*)
   a. Are cold weather valves (O.S.&Y.) in the appropriate open or closed position?
   b. Have antifreeze system solutions been tested?
   c. Were the antifreeze test results satisfactory?
6. Dry Systems (*See Items 10 to 14.*)
   a. Is the dry valve in service?
   b. Are the air pressure and priming water level in accordance with the manufacturer's instructions?
   c. Has the operation of the air or nitrogen supply been tested? Is it in service?
   d. Were low points drained during this inspection?
   e. Did quick opening devices operate satisfactorily?
   f. Did the dry valve trip properly during the trip pressure test?
   g. Did the heating equipment in the dry pipe valve room operate at the time of inspection?
7. Special Systems (*See Item 16.*)
   a. Did the deluge or pre-action valves operate properly during testing?
   b. Did the heat responsive devices operate properly during testing?
   c. Did the supervisory devices operate during testing?
8. Alarms
   a. Did water motor and gong test satisfactorily?
   b. Did electric alarm test satisfactorily?
   c. Did supervisory alarm service test satisfactorily?
9. Sprinklers
   a. Are all sprinklers free from corrosion, loading or obstruction to spray discharge?
   b. Are sprinkers over 50 years old, thus requiring sample testing?
   c. Is stock of spare sprinklers available?
   d. Does the exterior condition of sprinkler system appear to be satisfactory?
   e. Temperature. Are sprinklers of proper temperature ratings for their locations?
10. Date dry pipe valve trip tested (control valve partially open) __________ (*See Trip Test Table which follows.*)
11. Date dry pipe valve trip tested (control valve fully open) __________ (*See Trip Test Table which follows.*)
12. Date quick opening device tested __________ (*See Trip Test Table which follows.*)
13. Date deluge or preaction valve tested __________ (*See Trip Test Table which follows.*)

Trip Test Table

DRY VALVE MAKE	DRY VALVE MODEL	DRY VALVE SERIAL NO.	Q.O.D. MAKE	Q.O.D. MODEL	Q.O.D. SERIAL NO.

DRY PIPE OPERATING TEST	TIME TO TRIP THRU TEST PIPE		WATER PRESSURE	AIR PRESSURE	TRIP POINT AIR PRESSURE	TIME WATER REACHED TEST OUTLET		ALARM OPERATED PROPERLY	
	MIN.	SEC.	PSI	PSI	PSI	MIN.	SEC.	YES	NO
Without Q.O.D.									
With Q.O.D.									

IF NO, EXPLAIN

DELUGE & PREACTION VALVES

OPERATION ☐ PNEUMATIC ☐ ELECTRIC ☐ HYDRAULIC

PIPING SUPERVISED ☐ YES ☐ NO | DETECTING MEDIA SUPERVISED ☐ YES ☐ NO

DOES VALVE OPERATE FROM THE MANUAL TRIP AND/OR REMOTE CONTROL STATIONS ☐ YES ☐ NO

IS THERE AN ACCESSIBLE FACILITY IN EACH CIRCUIT FOR TESTING ☐ YES ☐ NO | IF NO, EXPLAIN

MAKE	MODEL	DOES EACH CIRCUIT OPERATE SUPERVISION LOSS ALARM		DOES EACH CIRCUIT OPERATE VALVE RELEASE		MAXIMUM TIME TO OPERATE RELEASE	
		YES	NO	YES	NO	MIN.	SEC.

14. See Control Valve Maintenance Table.

Control Valve Maintenance Table

Control Valves	Number	Type	Open	Secured	Closed	Signs	Explain Abnormal Condition
City Connection Control Valve							
Tank Control Valves							
Pump Control Valves							
Sectional Control Valves							
System Control Valves							
Other Control Valves							

15. Water Flow Test at Sprinkler Riser

Water Supply Source		City		Tank	Pump
Date	Test Pipe Location	Size Test Pipe	Static Pressure	Residual Pressure	(Flow)
Last Water Flow Test					
This Water Flow Test					

16. Heat Responsive Devices

Test Method ______________________________

Type of Equipment ______________________________

Manufacturer ______________________________

Test Results:

Valve No. A B C D E F Valve No. A B C D E F

Valve No. A B C D E F Valve No. A B C D E F

Valve No. A B C D E F Valve No. A B C D E F

Valve No. A B C D E F Valve No. A B C D E F

Auxiliary Equipment: No.? ___ Type? ___ Location? ___ Test Result? ___

17. Explain any "No" answers and comments. ______________

____

18. Adjustments or corrections made during this inspection:

____

19. Although these comments are not the result of an engineering review, the following desirable improvements are recommended:

____

Signature: ______________________________

Date: ______________________________

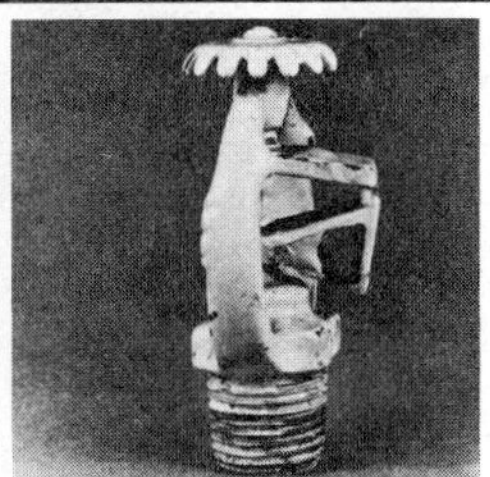

FIG. 18-8A. Examples of loaded automatic sprinklers. (Factory Mutual Engineering Corp.)

Sprinklers in spray booths present a special problem for which there is no wholly satisfactory solution except to conduct the spraying process in such a manner that no spray will reach the sprinklers. If so located as to minimize deposits, and if cleaned very frequently, conveniently accessible sprinklers may be cleaned without removal. Using a coating of grease, motor oil, or soft neutral soap facilitates washing or wiping off deposits. If grease is used, it should be a grease with a low melting point (vaseline, etc.). Unless cleaning is done very carefully, deposits are likely to accumulate to such an extent as to interfere seriously with sprinkler operation. The use of paper, polyethylene, or cellophane bags to protect sprinklers in spray booths is fairly common.

Corrosion of Automatic Sprinklers

Corrosive conditions are likely to make automatic sprinklers inoperative or retard the speed of their operation. They can also seriously impede the waterways of spray system nozzles. Corrosive vapors may seriously affect not only the heat actuated element and the valve retaining members of an automatic sprinkler, but also may be severe enough to weaken or destroy other portions of the sprinkler. In most instances such corrosive action is slow but sure, and thus must be vigilantly watched. Illustrations of some typical corroded sprinklers are shown in Figure 18-8B.

Some types of sprinklers are less susceptible than others to corrosive conditions. Nonferrous metal is used for sprinkler parts, but special protective coatings are necessary for all types when exposed to extreme corrosive conditions. Approved corrosion resistant or special coated

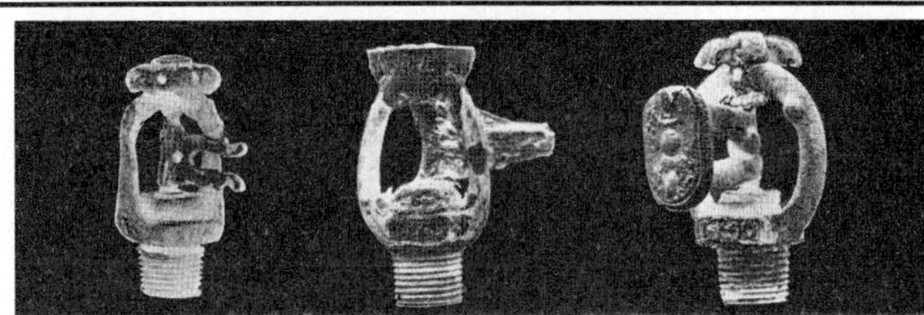

FIG. 18-8B. Examples of corroded automatic sprinklers.

sprinklers are needed in locations where chemicals, excessive moisture, or corrosive vapors exist.

Protection of Pipe Against External Corrosion

Under some conditions, corrosive vapors may cause rapid deterioration of steel pipe and hangers as well, necessitating frequent replacement unless the proper protection is provided. However, under most conditions cast iron fittings will not be seriously affected.

Under severe corrosive conditions, protective methods are not wholly satisfactory. Copper or special alloy noncorrosive pipe will give the best results.

Galvanized steel, under some conditions, may be the best and most economical method of obtaining reasonably long life for the piping system. This might apply to chemical plants, salt works, or similar properties where corrosion may be severe. Stainless steel and copper piping have also been used in some cases.

When corrosion of existing equipment becomes a maintenance problem, replacement or the application of a recognized type of protective coating are remedial measures.

Internal corrosion conditions leading to pipe obstructions are discussed later in this chapter.

Emergency Measures for Maintaining Protection During Repairs or Alterations

Fire records show the seriousness of having extinguishing systems shut off when a fire starts. There is danger if the water supply is shut off for (1) extensions or alterations to piping, (2) repairs due to accidental damage to piping or sprinklers and spray nozzles, (3) replacement of sprinklers after a fire, or (4) maintenance or replacement of sprinklers and other system devices. When, of necessity, protection is interrupted, every effort must be made to limit the extent and duration of the interruption. A cardinal rule is to notify the fire department whenever an impairment exists so they will not place false reliance on the systems. Most insurance companies also request that owners advise them when there is interruption of protection so that alternate means of protection can be arranged if judged necessary or desirable.

Advance Preparations Before Shutoff

When work requires systems to be shut off, the work should be planned for a time when the least hazard exists. In industrial plants, about three times as many fires occur during operations as during idle periods; therefore, work should be done on a weekend or other idle period. Special guard service may be required, however, to help in detection of any fire which might develop while the systems are shut off.

Sectional valves, rather than main valves, should be used to reduce to a minimum the number of shutoff systems and to take advantage of multiple water supplies. All personnel, materials, and tools should be made ready before the protection is impaired.

If underground mains are involved, tapping machine should be used when possible to avoid shutting off the water. If mains are to be opened, wooden or other plugs or caps and clamps to close the end of pipes quickly should be prepared. Emergency measures should be taken to maintain the maximum possible water supply. One possibility is to make a temporary hose connection from a hydrant still in service or from a domestic or industrial supply to the riser or risers. These connections are normally made to the nominal 2 in. drain with the drain valve and hydrant valve left open. Such arrangements are shown in Figure 18-8C. Adapters for connecting 2½ in. (64 mm) hose to sprinkler systems should be kept on hand.

Definite procedures should be followed for supervising closed valves, for notifying the fire department and insurance companies, and for making waterflow tests after the work is completed.

FIG. 18-8C. If it is necessary to shut off the water supply to a sprinkler system when repairs or extensions are to be made, a temporary supply can be provided by a hose line, keeping all or part of the system in service.

CARE AND MAINTENANCE OF SPECIFIC COMPONENTS

The following specific items of care and maintenance are treated in relation to routine inspection, test, and maintenance procedures. They are independent of the organization performing the inspection and maintenance functions, whether guard service, fire inspector, maintenance personnel, supervisor, or others making up the fire organization at any particular property, or outside contractors.

Public and Private Water Supply Equipment

The inspection and test of public water supplies and of private water supply equipment, such as tanks and fire pumps, is covered in chapters of Section 17.

Control Valves and Meters

The inspection and supervision of valves in fire protection systems is covered in NFPA 26, *Recommended Practice for the Supervision of Valves Controlling Water Supplies for Fire Protection.* The following paragraphs summarize some of the salient points of valve inspection and supervision.

Service Valves: Service valves at private fire system connections to public systems are usually under the control of the water department and are seldom operated. Their condition is usually indicated adequately by waterflow tests made from the private protection system as covered later in this chapter.

Meters: Meters in public water system connections are also generally under the control of the water utility but are sometimes located in pits on the protected property.

Check Valves: Check valves in public water connections, where needed to prevent backflow from private systems into public systems, are usually a part of the protection system for which the property owner is responsible. Tightness of check valves should be determined periodically by proper tests, depending on water supplies.

Check valves and their control valves should be properly arranged and located in accordance with the appropriate NFPA Standards (Sprinkler Systems, Water Spray Protection, and Standpipe and Hose Systems).

Gate Valves: All control valves should be readily accessible and unobstructed so they can be closed promptly and, in addition, can be examined to see that they are open and in good operative condition, turn easily, and do not leak.

Pits for gate valves (and check valves) should be kept reasonably dry and clean so that valves can be tested, examined, and maintained in good condition. Manhole covers should be kept clear of snow and ice. An example of a well arranged check valve pit with access for inspection is shown in Figure 17-4Y.

Each control valve should be numbered at the valve and listed (giving location, use, or portions of the system controlled) on a plant fire inspection report form. A plan showing valve locations should be posted at a central point known to plant and public fire officials.

Each valve should have a sign showing what it controls, with the legend "Must Be Open at All Times" or other proper wording. Underground valve locations should be shown by distance markings on nearby buildings, and on accurate plans of the property.

All control valves should be sealed or locked open unless there is central station supervisory service. Wrenches should be kept at indicator post valves or at locations where they are readily accessible.

There should be someone on the premises at all times who knows the use and location of all of the control and drain valves. This includes the person on watch or any others who may be on duty at night.

Sprinklers and Sprinkler Piping

In the inspection of sprinkler systems, the following are important considerations.

Observe whether there is any building or room from which sprinklers have been omitted, including such places as basements, lofts, show windows, concealed spaces, towers, under stairs, under skylights, and inside elevator wells, vertical shafts, small enclosures (such as drying and heating enclosures), and closets (unless open at the top). Observe also whether there are sprinklers under large air ducts, shelves, benches and tables, overhead storage racks, platforms, and similar surfaces which might obstruct distribution of water from sprinklers above.

Additional branches and sprinklers installed after the original installation was made should be examined to make sure that the system, particularly the smaller distributing pipes, has not become overloaded by supplying too many sprinklers for the size of the supply piping and water pressure available. Where the system is hydraulically designed, calculations should be kept available on additions to the system to help in checking on the effect of these additions.

Sprinklers must not be obstructed by high piled stock or other material, or by partitions or walls which might prevent free and proper water distribution. A clear space of not less than 18 in. (457 mm) below the deflectors of sprinklers is required. Observe whether the proper schedule of spacing is followed by referring to the NFPA Sprinkler Systems Standard for spacing recommendations given for the various occupancy classes.

Note whether all pipes in dry pipe systems have the proper pitch. This feature is of special importance because water remaining in pockets or low places is likely to freeze and cripple the system.

Are any hangers loose or pipes not properly supported? Observe whether sprinkler piping is used for the support of stock, clothing, etc. Sprinkler piping should not be used for such purposes.

Note whether the sprinklers are placed in an upright or pendent position. Sprinklers must be installed in the position for which they are designed and marked.

The distance of the deflectors from the ceiling or bottom of beams or joists should conform to the NFPA Sprinkler Systems Standard.

Note the type and design of the sprinklers, the year of their manufacture, and the date of installation (sometimes old sprinklers are used in a new installation).

Are all sprinklers of the proper temperature rating? Ordinary degree sprinklers should be substituted for high degree sprinklers when the latter are unnecessary. Wherever the temperature around sprinklers exceeds 100°F (37.8°C), intermediate degree sprinklers should be used.

Are any sprinklers corroded or loaded? Are sprinklers of a proper type, or properly protected against corrosion? If any sprinklers are in doubtful condition, samples should be removed and tested.

Sprinklers having coatings of paint, excessive deposits or incrustations, whitewash, bronzing, or other coating should be replaced by new sprinklers.

A supply of extra sprinklers should be kept in a sprinkler cabinet so that any sprinklers that have operated or have been damaged in any way may be replaced promptly. These sprinklers should correspond in type and in temperature ratings to the sprinklers in the property. The cabinet should be situated in a cool location. The number of sprinklers stocked for replacement purposes is governed by the size and number of systems, the location of the protected property relative to the source of supply for replacement sprinklers, and the number of sprinklers likely to be opened by extraordinary conditions such as a flash fire.

Ordinarily, under average conditions, the stock of emergency sprinklers should be as follows:

—for systems having not over 300 sprinklers—6 sprinklers

—for systems having 300 to 1,000 sprinklers—12 sprinklers

—for systems having above 1,000 sprinklers—24 sprinklers

—for systems aboard vessels or in isolated locations, a greater number of sprinklers should be carried to permit restoring equipment to service promptly after a fire.

A special sprinkler wrench should be kept in the cabinet to be used in the removal and installation of sprinklers. This wrench should always be used for installing new sprinklers.

System Waterflow Tests

The NFPA Sprinkler Systems Standard calls for a water supply test pipe and pressure gages to be provided at locations that will permit flowing tests to be made to determine whether water supplies and connections are in order. A 2 in. (51 mm) drain at the sprinkler riser may suffice as a water supply test pipe if installed so that the valve may be opened wide for a sufficient time to assure a proper test without causing water damage. A similar main drain valve in a water spray system can serve the same purpose.

There should be provision for checking the system's gage and its pressure recordings with an inspector's test gage.

The pressure on the sprinkler system side of alarm or check valves may be higher than that of the water supply because any momentary high pressure on the supply will be transmitted to the system and retained by the check valves. This excess pressure is relieved when a waterflow test is made.

The drop in pressure below the normal static pressure with the 2 in. (51 mm) water supply test pipe (drain) wide open can be noted, and general flow conditions evaluated. For instance, if the normal static pressure is 50 psi (345 kPa), and if previous tests have shown a pressure drop to 45 psi (310 kPa) when the drain valve is opened, it will be apparent that if at inspection the pressure drops to, say, 35 psi (241 kPa) or under, there is some obstruction to the flow which should be removed, or some other defect which should be located and remedied.

At each inspection, a flow test should be made separately for each water supply and each connection from a supply. This can be accomplished by closing the water supplies temporarily, except the one under test. Too much emphasis cannot be placed on the value of flow tests to determine whether there is any obstruction to full flow. It is mandatory that all water supplies be returned to service immediately after testing.

If there is waterflow supervisory service, tests should not be made without first notifying the central station or other alarm headquarters.

There should be a 1 in. nominal test pipe having a standard brass ½ in. (51 mm) outlet at the remote point of each sprinkler system. This should be operated at each inspection of a wet system to make sure that there is free flow at good pressure and to test the waterflow alarm. In dry pipe systems, the test pipe is used to trip the dry pipe valve (as discussed later).

Dry Pipe Sprinkler Systems

The best practice is to keep dry pipe systems in the dry condition throughout the year. When water flows into sprinkler piping, rust and foreign materials tend to be carried into the smaller pipes where accumulation and obstruction to flow can occur. Annual tripping of the dry pipe valve to make the system wet accelerates this behavior.

When systems are wet, with the dry pipe valve tripped, the waterflow alarm feature of the dry valve is out of service.

Old, obsolete, and unapproved dry pipe valves, and any in poor mechanical condition, are likely to be unreliable and should be replaced with currently approved or acceptable types.

Instruction charts are provided for the maintenance of dry pipe valves by the sprinkler installing company and should be posted at or near these valves.

All dry pipe valves should be numbered and listed on inspection report forms.

Check the air pressure on each dry pipe system at least once a week and pump up the systems when necessary. A daily check is recommended for the first week after a dry pipe valve is reset. If pressure is lost rapidly, requiring frequent pumping, the piping system should be gone over and made tight. Avoid air pressure higher than that called for by the valve manufacturer's instructions.

Make sure that the priming water is maintained at the proper level above the dry pipe valve.

Slight freezing of a dry pipe valve may cause it to be inoperative. It is extremely important to make sure that both adequate provision is made for the heating of the valve enclosure or the room in which it is located, and that the heating equipment is safe and in order.

Dry pipe system piping should be thoroughly drained before freezing weather and kept clear of water during the winter. The freezing of a small amount of water in the piping may cause rupture of the piping or sprinklers and the operation of the valve. Make sure that all low point drains of the system are kept free of water and that the automatic drip or drain is clear and free to operate.

Dry pipe valves should be examined externally at frequent intervals to detect evidences of deterioration.

Thoroughly clean and reset each dry pipe valve once yearly during the warm weather. On such occasions, the valve body should be thoroughly washed out, preferably with warm water, with care being taken to make sure that the small ports and piping leading to the alarm connections are free from obstructions.

Operating tests of dry pipe valves, including quick opening devices, if any, should be made from time to time. Such tests may be combined with the annual cleaning and resetting, and with any necessary service with respect to renewal of rubber parts or the adjustment of gages, alarm devices and connecting piping, and quick opening devices.

When dry pipe valves are tripped for testing purposes, the procedure should be such that only the minimum flow of water needed to trip the valve is admitted to the sprinkler riser. This can usually be done by opening the control valve only one or two turns and by closing it immediately after the dry pipe valve trips.

Operating tests and servicing of dry pipe valves, including quick opening devices, should be conducted either by a sprinkler installing company or other fully qualified personnel.

No grease or other sealing material should be used on seats of dry pipe valves in an effort to stop leaks. Force should not be used in an attempt to make dry pipe valves tight.

Dry pipe valves should carry a tag or card showing the date on which the valve was last tripped and the name of the organization making the test. Such tags are usually available from the installing company or the insurance authorities.

Quick Opening Devices

The operation of quick opening devices usually can be tested either with or without operating the dry pipe valve itself. The manufacturer's instructions for testing and resetting the device should be carefully followed.

If the device does not operate properly on test, it can, if necessary, be removed and the sprinkler system kept in operative condition. Repair parts can be ordered from the manufacturer, or the device sent to the manufacturer for repair or adjustment.

Pressure Gages

Water pressure and air pressure gages should be tested for accuracy whenever the dry pipe valve is cleaned and reset and at other times if the pressure indication is questionable.

Waterflow Alarm Services

Waterflow alarm devices should be tested at regularly scheduled inspections of fire protection equipment. Electric alarms should be tested by the bypass test valve at dry pipe valves and alarm valves, or by test switches. Actual waterflow alarm tests should be made when cold weather does not make the outdoor discharge of water from test or drain valves objectionable. Watermotor gongs should not be tested in freezing weather.

Keep the small valve or cocks controlling the water supply to alarm devices sealed open.

Deluge and Preaction Systems

Complete charts are furnished by installing companies showing in detail the proper method of operating and testing thermostatically controlled systems. Only competent persons fully instructed with respect to the details and operation of such systems should be employed in their repair and adjustment. It is highly advisable for the owner to arrange with the installing company for regular periodic inspection and testing of the equipment.

The automatic valves controlling the flow of water into these systems operate through the effect of fire temperatures on actuating devices. Ordinarily, when it is necessary to repair the actuating system, as distinguished from the piping system itself, the piping system is run "wet" and automatic sprinkler protection is thus maintained on a preaction system, provided there is no danger of freezing.

Sprinkler System Supervisory Services

Central station, proprietary, remote station, fire alarm, and supervisory services for sprinkler systems interpret sprinkler system waterflow signals as fire alarms and immediately initiate fire department response. Signals of other types, such as valve supervision, pressure, or temperature supervision, etc., do not serve as fire alarms, but initiate appropriate courses of action as prearranged between the supervisory service and the property owner. It is extremely important that the headquarters of any of these services be notified and any special arrangements be made in advance of any inspection or tests of equipment that will cause a signal to be transmitted.

Fire Department Connections

Inspect each fire department connection regularly to make sure that caps are in place, threads are in good condition, the ball drip or drain is in order, and that the check valve is not leaking. A hydrostatic test should be conducted periodically on old fire department connection piping to assure that it will withstand the required pressure.

Open Sprinkler Equipment

Outside or open sprinkler equipment should be tested once each year during warm weather. These tests should preferably be made in conjunction with the inspection department having jurisdiction and, if desired, with representatives of the fire department.

Before making operating tests, care should be exercised to make sure that all windows and doors through which water might enter are tightly closed. Proper precautions must be taken to prevent damage from discharge or accumulation of water to sidewalks, streets, areaways, or adjoining buildings.

Determine by test whether the sprinklers and the system piping are in good condition and free from plugging. Any piping or sprinklers found clogged should be removed, cleaned, and replaced at once.

OBSTRUCTIONS IN SYSTEM PIPING

Obstructions in system piping or yard mains will reduce or cut off the flow of water from a part or all of the system and will not give the protection intended. (Obstructed piping is one of the well recognized causes of unsatisfactory sprinkler performance.) One source of obstruction is foreign material in the water supply system, and a second is foreign material originating within the system piping itself. Figure 18-8D is a graphic example of obstruction in system piping.

FIG. 18-8D. A stone lodged in the tee of a dry pipe system's cross main. Good maintenance programs enable discovery of such obstructions before they can become a problem. (Factory Mutual Engineering Corp.)

Foreign material, such as sand, gravel, stones, pieces of wood, etc., may enter underground mains as a result of carelessness of workers when laying or repairing the mains. Sand, silt, or wood chips may also enter the mains through inadequately protected fire pump suction inlets. Sometimes when gravity tanks are cleaned, foreign material enters the drop pipes unless care is taken to prevent it.

Keeping foreign material out of piping for water spray systems is particularly critical. If the waterways in the discharge nozzles are less than ⅜ in. (9.5 mm) in size or if it is suspected that the water is likely to contain obstructive material, then a strainer is required in the pipeline supplying the system. The small nozzles are prone to clogging if water borne debris reaches them; thus, it is imperative that the strainers are maintained in good condition.

Scale, corrosion, and incrustations can also form in the piping. In dry pipe sprinkler systems, for example, the condensation of moisture in the air supply may result in hard scale forming along the bottom of piping. Deep well water and water containing natural salts tend to corrode pipe interiors, and the piping of wet systems may be found obstructed by rust and corrosion. Obstructing material is carried into piping when systems operate or are refilled after draining. It may plug one or more fittings or obstruct a number of sprinklers at the end of the system. Pipe capacities may be reduced by incrustation in localities where water contains lime or magnesia.

If the water is badly discolored during flow tests, or if gravel or stones are discharged with the water, foreign material is probably present in the piping. Finding foreign material in a fire pump suggests that there may also be similar material in the system piping. Finding foreign material in resetting dry pipe valves or examining check valves indicates a possible serious impairment of the entire fire protection system.

Obstructions to the flow of water from sprinklers usually comprise concentrations of the lighter materials such as silt, sand, small pebbles, etc., piling up in the ends of the cross mains and in the nearby branch lines, with the heaviest solids collecting nearer the system risers. (See Fig. 18-8E.) There are some cases where the foreign material deposits extend so far back into the system from the ends of the cross mains that complete cleaning of the entire system is necessary. More often, however, a preliminary investigation will disclose that, while foreign material may cause complete stoppage of water flow from sprinklers on branch lines connected to the ends of the cross mains, there may be no stoppage at any point on branch lines farther back in the system.

FIG. 18-8E. Silt buildup in this branch sprinkler line has totally obstructed the waterway. (Factory Mutual Engineering Corp.)

Conditions Indicating Possible Obstruction in Piping

Evidence of possible obstruction in piping is given by any one of the following:

1. Plugging of test connections, or discharge of dirty or colored water.
2. Discharge of foreign material during routine water tests.
3. Foreign material in dry pipe valves, check valves, or fire pumps.
4. Obstructing material found in piping dismantled for building or piping alterations.
5. Defects found in fire pump suction screens when source is from open bodies of water.
6. Repairs made in public water mains in vicinity.
7. Underground piping not flushed before connecting systems.
8. Plugged sprinklers or spray nozzles at time of fire.
9. Old equipment, especially dry pipe systems.

Investigation of Severity and Extent of Obstruction

An investigation of conditions must be made when evidence of foreign materials in systems is found or obstruction is suspected. Where needed, cleaning or flushing procedures must subsequently be carried out to remove obstructing materials. Inspection and flushing methods outlined in this chapter are treated more completely in NFPA 13A, Care and Maintenance of Sprinkler Systems.

Flushing Feed Main

Existing yard piping suspected of containing obstructing material should be thoroughly flushed through hydrants at dead ends of the system, blowoff valves, or groups of riser drains, by opening simultaneously as many outlets as possible and allowing the water to run until clean. If the water is supplied from more than one direction or through a looped system, divisional valves should be closed to produce a high velocity flow through each single line.

It is usually desirable to learn the nature and extent of foreign material that may be in the piping. This can best be done by fastening burlap bags securely to the flowing outlets.

The connections to system risers should next be flushed through drain valves. If the foreign material is too large to pass through a riser drain valve, some ingenuity must be used to procure a larger opening. Fire department connections can be used by removing the clappers. Flanged check valves or check valve covers, or flanged fittings can be removed and a siamese fitting substituted.

Flushing System Piping

In choosing the critical points for examination, consideration should be given to the arrangement of the piping and the probable pattern of the flow of water when a system is rapidly filled (as in a dry pipe system), or when a large flow of water from the system occurs due to the operation of many sprinklers on a wet pipe system.

After selection of the area in which ordinarily the heaviest deposit of foreign material would be expected, the investigation should take into consideration how much of the system's equipment in that area might be made ineffective by the obstructing material at a time of fire. It is important that the preliminary investigation should indicate the extent of cleaning that may be necessary.

Flow Tests

Flow tests at critical points in a sprinkler system provide a positive means of determining whether there is foreign material in the system that may cause stoppage of flow from any of the sprinklers, and also of indicating how much of the equipment is affected. Valved hose connections are made to sprinkler fittings at the ends of two, three, or four branch lines on one side of and at the end of the cross main in the selected area. Flows are then taken simultaneously from these points.

If stoppage occurs at all of the branch lines tested simultaneously, similar tests may be necessary at points farther back on the system toward the system riser. Tests may next be extended to other floors to help make a judgment on the extent of cleaning operations that may be necessary for removing all foreign material that might cause stoppage.

Foreign deposits in sprinkler systems tend to be crowded toward the ends of the cross mains and in the end branch lines. Of course, there will be exceptions to this general statement. For example, where highly corrosive material has been introduced into the system, severe corrosion and scaling of the interior surfaces of the piping will be evident uniformly throughout the system.

Visual Examination

Visual inspection of the interior of piping is not a safe substitute for flow tests because it is difficult to judge the behavior of a deposit within the piping from its appearance in place. However, after obstructing material has been found by flow tests, inspection of the interior may assist in determining the extent and type of flushing needed. To examine the interior of a cross main, back out several upright branch line nipples, remove the end elbow of the cross main, and inspect the interior of the pipe by means of a flashlight at the openings. A further check is to take down some of the suspected piping and clean the interior, carefully noting the character of the material removed.

EXTENT OF CLEANING REQUIRED

When preliminary tests show evidence of obstruction in sprinkler system piping, the next step is to undertake a cleaning program to remove the obstructing material. There are three degrees of cleaning: (1) complete cleaning, (2) limited cleaning, and (3) flushing extremities only. The selection of the degree required is dictated by the character, extent, and location of the obstruction.

Generally, when stoppage occurs only in the branch lines at extremities of the system, the flushing extremities only procedure can be used to clean the system. If stoppage occurs at several test points, but indications are that free flows are available on the other lines back toward the riser, a limited cleaning procedure may be followed. If the stoppage is more extensive, then a complete cleaning procedure should be undertaken.

Before any cleaning is undertaken, however, written specifications or procedure plans covering the entire proposed operations should be understood by all concerned. After completion of the work, the sprinkler contractor or other party performing the cleaning operation should furnish a written statement to those concerned indicating that all operations have been completed in accordance with the approved procedure, that the specified examinations and tests have been made, and that the system has been left in normal condition.

Complete Cleaning

There are some cases where the foreign material is of such character and extends so far back into the system from the ends of the cross mains that complete cleaning of the entire system will be necessary. In such cases, it is of prime importance that a plan of the piping system be prepared and be marked numerically to show the proper sequence of the cleaning operations, the size and length of hose to be used, the points where the hose is to be attached, and the number of flows to be made at each point.

Limited Cleaning

Where points of impairment of protection can be identified as occurring within restricted areas rather than throughout the system, and if these points exist in any appreciable amount only in the ends of cross mains and in the branch lines connected at or near those points, not all branch lines need to be cleaned. The cleaning procedure may be limited to the branch lines at or near these points and to the cross mains on each floor.

Limited cleaning operations serve to minimize the time during which sprinkler protection is impaired; they lessen the disturbance to normal activities within the premises where the cleaning operations are under way; and they materially lower the cost of the work as compared with the cost of more extensive cleaning. Since only a fraction of the total number of branch lines are being cleaned directly under the limited procedure, periodic flow tests should be made as the cleaning progresses in order to determine the effectiveness of the work. The limited procedure takes into account the purging effect in branch lines not actually flushed, caused by high velocity water and air flow along the cross main.

Cleaning System Extremities Only

This cleaning method should be used only in cases in which impairment of dry pipe sprinkler protection has been found to occur solely at the system extremities, and it relates to every floor protected, regardless of the number of floors. Extreme care should be exercised before this cleaning method is employed. Adequate examinations and flow tests are made to make certain that only the sprinklers located at the extremities of the system are subject to plugging. It should not be used where the deposits extend back into the cross main beyond the second branch line from the end of the cross main.

CLEANING METHODS

There are two methods in general use for cleaning sprinkler systems: water flushing and hydropneumatic. Convenience is a factor that may make the hydropneumatic method preferable to the water flushing method. With the former, there is no open discharge of water outside the building; there is less disturbance of plant operations; frequent fillings and drainings of the sprinkler system are unnecessary; there is considerably less pipe work; and a complete operation is performed at a given flush point.

Both the water flushing and hydropneumatic methods are appropriate for complete cleaning and limited cleaning procedures. Good practice is first to flush the underground piping and then to clean all private sources of water supply that might cause silting in the underground main.

Both of these methods are thoroughly covered in NFPA 13A, *Care and Maintenance of Sprinkler Systems.*

Branch Line Testers

If it is anticipated that a sprinkler system may need periodic testing for obstruction or repeated flushing, a branch line tester may be installed permanently as a nippled fitting at the end of branch lines or between the last sprinkler and the end fitting on branch lines. The tester has a normally closed and capped outlet connection for hose. Turning a wrench head opens a passage to the outlet which has a 1 in. (25 mm) hose coupling thread. It is not necessary to remove the sprinkler.

SPRINKLER LEAKAGE

The term "sprinkler leakage" is ordinarily intended to cover all leakage or discharge of water from the sprinkler system except in case of fire. The system includes sprinklers, piping, tanks, and any other water supplies used for fire protection.

The danger of accidental water discharge is often exaggerated, because a sprinkler system is very rugged. The pipe, sprinklers, and fittings are made to stand far greater water pressures than are met with in practice. The supports are strong also, and the system is so installed that it should remain intact even under severe mechanical strain, including any anticipated earthquake shock or movement.

There is no reasonable ground for omission of sprinklers from areas in hospitals and other institutions occupied by patients. While it is true that accidental discharge of cold water from sprinklers might, for example, have an adverse effect upon certain patients in a hospital, the probability of loss of life in event of fire is infinitely more serious than any remote chance of an accidental cold shower from the sprinklers. Conditions which might result in accidental discharge of water from sprinklers are highly unlikely where ordinary hospital care would preclude the existence of either freezing temperatures or temperatures so high as to cause the opening of sprinklers except in the case of fire.

It has been estimated, based on sprinkler leakage loss reports, that the chance that a sprinkler may open accidentally because it is defective is less than one in a million each year.

Automatic sprinkler protection for property particularly vulnerable to water damage, such as computers, finely made complicated mechanical equipment, works of art, valuable books, etc., should not be omitted due to fear of premature operation of sprinklers if reasonable precautions are taken against freezing, overheating, and mechanical injury.

Some of the more important conditions responsible for sprinkler leakage and water loss are explained in the following paragraphs.

Mechanical Injury

Automatic sprinklers are designed to withstand at least 500 psi (3448 kPa) pressure without injury or leakage. If properly installed, there is little danger of the sprinkler breaking apart unless it is injured in some way. All of the listed sprinklers are rugged and will stand considerable abuse. The frames can be bent somewhat without opening the sprinkler.

Where sprinklers are located in areas where they may suffer mechanical injury, they can be protected by commercially available metal guards. (See Fig. 18-8F.)

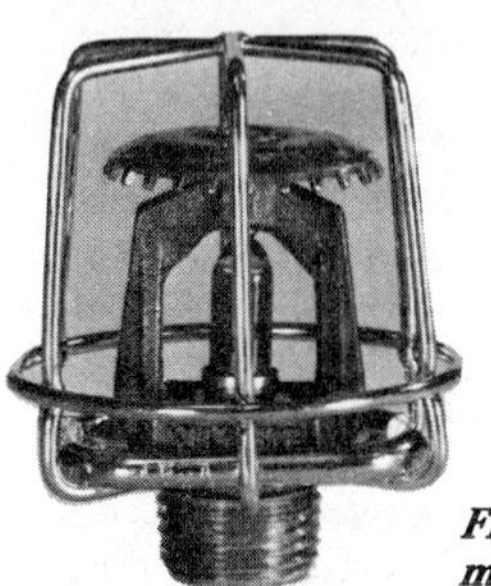

FIG. 18-8F. Metal guard protection against mechanical injury.

The use of lift trucks or tractors in handling materials indoors must be carefully planned and supervised to avoid hitting sprinklers and sprinkler piping.

Improper Installation and Maintenance

Improper installation includes lack of proper supports for risers and feed mains, lack of proper tie rods where inside piping is connected to cast iron fittings, and defective pipe hangers or supports.

Breaks or leaks sometimes occur in underground work due to improper laying of underground mains or poor workmanship in making joints. Sometimes breaks occur in properly installed systems due to undue stresses on the piping system, such as those caused by settling of a building or foundations or lack of clearance around pipes where they pass through foundations.

Every piping system should be frequently inspected and properly maintained. The condition of hangers and supports should be watched to determine whether they are in good order, or loose, or with screws or fastenings pulled out.

Sprinkler piping should never be used as a support for ladders, stock, or other material. Any system of piping under constant internal pressure should receive proper care and not be subject to mechanical abuse.

Freezing

The freezing of water in pipes is a common cause of trouble and the remedy in most cases is self-evident. Where this is a local condition and not unduly severe, ice in the sprinkler or fittings may rupture the sprinkler without damage to the pipe or fittings. When the ice melts, water is discharged. In other cases, the pipe or fittings may burst, causing serious damage and expensive repairs.

In some types of sprinklers a moderate amount of freezing may not break the sprinkler open, but may cause it to leak at the seat after ice melts.

Some of the conditions which cause trouble from freezing are:

1. Insufficient heat for severe weather conditions, in certain portions such as concealed spaces under roofs or blind attics, large open doorways for trucks or railway cars, entryways, and space under buildings.
2. Windows left open during severely cold weather.
3. Insufficient heat because of shortage of fuel supply.
4. Branch lines of dry pipe systems not correctly pitched to drain.
5. Feed mains or piping in unheated areas, or not properly protected to prevent freezing.
6. Freezing of tank risers due to lack of heat or lack of adequate protection; obstructed circulating pipes of heating systems.
7. Freezing of water in underground mains not buried deep enough.

Overheating

Overheating is the most common cause of accidental water damage from sprinklers and is the term used when an automatic sprinkler operates as a result of abnormally high temperatures but without the presence of fire. This condition is usually caused by hot manufacturing processes, artificial heating, or lack of ventilation. It may result from a sudden increase in temperature which operates the heat responsive element, or from longer exposure to temperature sufficiently high to cause gradual weakening. In the solder type of sprinkler the latter condition, sometimes termed "cold flow," is indicated by partial separation of the soldered members.

In practice there are a number of conditions which cause unsafe temperatures, such as:

1. Climatic condition causing temperatures well above normal (above 110°F or 43.3°C).
2. Ordinary low temperature sprinklers located too near steam mains, unit heaters, heating ducts, etc. Temperatures higher than anticipated may occur after installation of sprinklers. Changes in heating methods, installation of unit heaters, changes in manufacturing processes or equipment, etc., after sprinklers are installed, frequently occur.
3. Changes in temperatures due to changes in occupancy or processes which cause higher temperatures than contemplated. These changes may be general or local in portions of a building.
4. In some cases a fire that opens a number of sprinklers may weaken other sprinklers in that vicinity. This is not a common occurrence.

The NFPA Sprinkler Systems Standard gives some information on sprinkler temperature ratings near unit heaters. Unit heaters should never be used without a careful study of conditions and, in most cases, sprinklers near unit heaters must be replaced by sprinklers with a higher temperature rating in order to avoid accidental operation.

When conditions occur which require sprinklers higher than "ordinary" degree rating, the maximum temperature at the sprinkler level must be determined and sprinklers of the correct temperature rating installed if trouble is to be avoided.

REDUCING WATER DAMAGE

It is evident that if leakage occurs when there is no one in the building and no proper means of alarm notification to outsiders, the leakage may continue for a considerable length of time and cause serious loss, even if the rate of water discharge is relatively small, as with one sprinkler operating.

While good or "standard" guard service is of value, it may not be adequate unless there is a sprinkler alarm gong which will notify the guard if water is flowing.

Every sprinkler system should have a waterflow alarm device and properly located alarm gong, not only to give alarm in case sprinklers operate because of fire, but also to minimize loss from sprinkler leakage.

Most large losses from sprinkler leakage are due to lack or failure of local sprinkler alarm devices or gongs, or failure to notify promptly some authorized person who will investigate immediately and shut off the water if there is no fire. In some cases water runs for many hours without anyone knowing about it, and the operation of even one sprinkler on an upper floor of a building containing expensive merchandise may cause a very heavy loss.

Sprinkler supervisory service is of great value in the prompt discovery of any leakage from the system, and in securing appropriate action to minimize the water loss.

Various provisions or safeguards may be taken to lessen the damage to stock from water used to extinguish fire. Similar provisions are of value in case of sprinkler leakage.

Bibliography

NFPA Codes, Standards, Recommended Practices and Manuals. (See the latest *NFPA Codes and Standards Catalog* for availability of current editions of the following documents.)

NFPA 13, *Standard for the Installation of Sprinkler Systems.*

NFPA 13A, *Standard for the Care and Maintenance of Sprinkler Systems.*

NFPA 14, *Standard for the Installation of Standpipe and Hose Systems.*

NFPA 15, *Standard for Water Spray Fixed Systems for Fire Protection.*

NFPA 24, *Standard for the Installation of Private Fire Service Mains and Their Appurtenances.*

NFPA 26, *Recommended Practices for the Supervision of Valves Controlling Water Supplies for Fire Protection.*

Additional Readings

Bryan, John L., *Automatic Sprinkler and Standpipe Systems,* National Fire Protection Association, Quincy, MA, 1976, p. 76, 81.

Brannigan, Franics L, "Automatic Sprinklers-Prelude to Disaster," *Fire Engineering,* Vol. 138, No. 5, May 1985, pp. 56-62-64.

Damon, Walter A., "Preventing Pitfalls in Fire Protection Systems," *Heating/Piping/Air Conditioning,* April 1985, pp. 61-64, 69-71.

Factory Mutual Engineering Corporation, "Notes on Dry Pipe Valves," *Loss Prevention Data* 2-10, Nov. 1977, Factory Mutual System, Norwood, MA.

Factory Mutual Engineering Corporation Loss Prevention Data Sheets Nos. 2-11 to 2-95 (various titles), information on the care and maintenance of sprinkler system components (dry pipe valves, quick-opening devices, deluge and preaction system valves, etc.) by manufacturers' names and model numbers (various dates).

SECTION 19

SPECIAL FIRE SUPPRESSION AGENTS AND SYSTEMS

CARBON DIOXIDE AND APPLICATION SYSTEMS

Revised by H. V. Williamson

Carbon dioxide (CO_2) has been used for many years in the extinguishment of flammable liquid fires, gas fires, fires involving electrically energized equipment, and—to a lesser extent—fires in ordinary combustibles such as paper, cloth, and other cellulosic materials. CO_2 will effectively suppress fire in most combustible materials; exceptions are a few active metals and metal hydrides, and materials such as cellulose nitrate that contain available oxygen. Further practical limitations of CO_2 are related to the method of application and to restrictions imposed by the hazard itself.

This chapter will discuss the extinguishing properties of CO_2 and its proper use in fixed fire extinguishing systems. NFPA 12, *Standard on Carbon Dioxide Extinguishing Systems* (hereinafter referred to as NFPA 12), provides guidance to those responsible for the purchase, design, installation, testing, inspection, operation and maintenance of carbon dioxide systems.

PROPERTIES OF CARBON DIOXIDE

Carbon dioxide has a number of properties that make it a desirable fire extinguishing agent: it is noncombustible, it does not react with most substances, and it provides its own pressure for discharge from the storage container. Also, since carbon dioxide is a gas, it can penetrate and spread to all parts of the fire area. As a gas, or as a finely divided solid called "snow" or "dry ice," it will not conduct electricity and therefore can be used on energized electrical equipment. It leaves no residue, thus eliminating cleanup due to the agent itself.

Thermodynamic Properties

Under normal conditions, carbon dioxide is a gas. It is easily liquefied by compressing and cooling, and with further compressing and cooling it can be converted to a solid. The effect of changes of temperature on compressed carbon dioxide in a closed container is shown in Figure 19-1A.

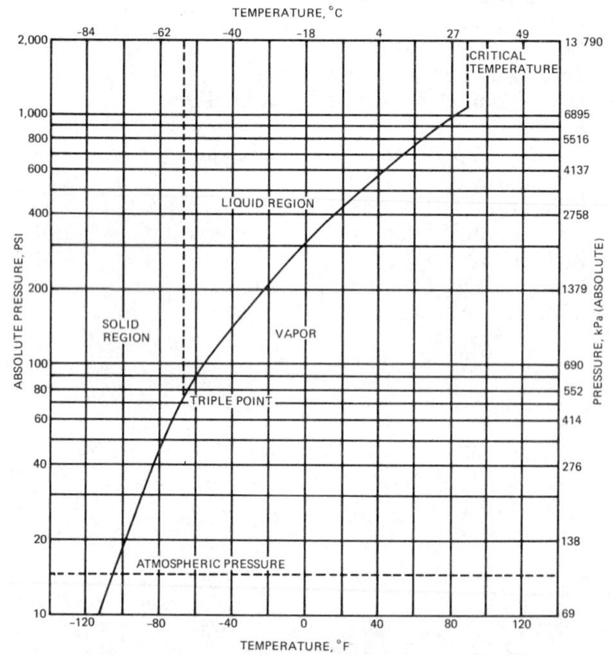

FIG. 19-1A. Effect of pressure and temperature change on the physical state of carbon dioxide. Above the critical temperature of 87.8°F (31°C), irrespective of pressure, it is entirely gas. Between 87.8°F (31°C) and the temperature of the triple point, which is −69.9°F (−56.6°C), in a closed container, it is part liquid and part gas. Below the triple point it is either a solid or gas, depending upon the pressure and temperature.

On that part of the curve between −69.9°F (−57°C) and the critical temperature of 87.8°F (31°C), carbon dioxide in a closed container may be a gas or a liquid. The pressure is related to the temperature as long as both vapor (gaseous) and liquid states are present. As the temperature and pressure increase, the density of the vapor phase increases while the density of the liquid phase decreases. At 87.8°F (31°C), the density of the vapor becomes equal to the density of the liquid, and the clear demarcation between the two phases disappears. Above the critical tem-

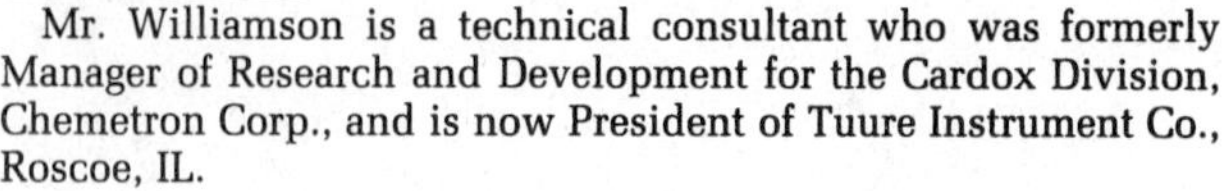
Mr. Williamson is a technical consultant who was formerly Manager of Research and Development for the Cardox Division, Chemetron Corp., and is now President of Tuure Instrument Co., Roscoe, IL.

perature, high pressure carbon dioxide exists only in a gaseous form with properties somewhat between normal vapor and liquid states.

When the temperature is reduced to −69.9°F (−57°C) at 75 psia (517 kPa), carbon dioxide may be present in vapor, liquid, and solid forms in equilibrium with each other; hence the term "triple point" to describe this condition. Below the triple point, only vapor and solid phases can exist. Thus, when liquid carbon dioxide is discharged to atmospheric pressure, a portion instantly flashes to vapor while the remainder is cooled by evaporation and converted to finely divided snow (dry ice) at a temperature near −110°F (−79°C). The proportion of CO_2 converted to dry ice depends upon the temperature of the stored liquid. Approximately 46 percent of the liquid stored at 0°F (−18°C) will be converted to dry ice, compared to approximately 25 percent for liquid stored at 70°F (21°C).

Storage

Liquid carbon dioxide may be stored in high pressure cylinders at normal ambient temperature or in low pressure refrigerated containers designed to maintain a storage temperature near 0°F (−18°C). Freezing (solidifying) in storage is not a problem; however, any substantial reduction in storage temperature (and corresponding storage pressure) could reduce the discharge flow rate below acceptable design limits. High pressure systems normally are designed to operate properly with storage temperatures ranging from 32 to 120°F (0 to 49°C). Low pressure systems, which operate normally at 0°F (−18°C), would not be affected unless the ambient temperature surrounding the storage container dropped well below this level for a prolonged time. Extreme climatic environments may require special design considerations to assure proper operation.

Discharge Properties

A typical discharge of liquid carbon dioxide has a white cloudy appearance due to finely divided dry ice particles carried along with the flash vapor. Because of the low temperature, some water vapor will condense from the atmosphere, creating additional fog which will persist for a time after the dry ice particles have settled out or sublimed. The cooling effect of the dry ice normally is beneficial in reducing temperatures after a fire; however, when protecting extremely temperature sensitive equipment, direct impact of heavy discharge streams should be avoided.

Vapor Density

Carbon dioxide gas has a density of one and one-half times the density of air at the same temperature. The cold discharge has a much greater density, which accounts for its ability to replace air above burning surfaces and maintain a smothering atmosphere. Because any carbon dioxide and air mixture will be heavier than air at the same temperature, an atmosphere containing the highest concentration of carbon dioxide will settle to the lowest level, while the lowest concentration remains on top.

Toxicity

Although carbon dioxide is only mildly toxic, it can produce unconsciousness and death when present in fire extinguishing concentrations. The reaction in such cases is more closely related to suffocation than to any toxic effect of the carbon dioxide itself. A concentration of nine percent is about all most people can withstand without losing consciousness within a few minutes. Breathing a higher concentration of carbon dioxide could render a person helpless almost immediately.

EXTINGUISHING PROPERTIES OF CO_2

Carbon dioxide is effective as an extinguishing agent primarily because it reduces the oxygen content of the atmosphere by dilution to a point where the atmosphere no longer will support combustion. Under suitable conditions of control and application, the available cooling effect also is helpful, especially where carbon dioxide is applied directly on the burning material.

Extinguishment by Smothering

In any fire, heat is generated by rapid oxidation of a combustible material. Some of this heat raises the unburned fuel to its ignition temperature, while a large part of the heat is lost by radiation and convection—especially in the case of surface burning materials. If the atmosphere that supplies oxygen to the fire is diluted with carbon dioxide vapor, the rate of heat generation (oxidation) is reduced until it is below the rate of heat loss. When the fuel is cooled below its ignition temperature, the fire dies out and is completely extinguished.

The minimum concentration of carbon dioxide needed to extinguish surface burning materials, such as liquid fuels, can be accurately determined, since the rate of heat loss by radiation and convection is reasonably constant. Table 19-1A lists the minimum concentrations of CO_2 for some common liquid and gaseous fuels as determined by the U. S. Bureau of Mines. It is difficult to obtain similar data for solid materials because the rate of heat loss by radiation and convection can vary widely, depending upon shielding effects caused by the physical arrangement of the burning material.

Extinguishment by Cooling

The value of rapid cooling is most apparent where the agent is discharged directly on the burning material, such as a liquid filled dip tank. A massive application quickly covering the entire surface area prevents reignition when the discharge ends and normal air again contacts the fuel area. The presence of dry ice particles in the discharge stream helps to promote fast surface cooling.

LIMITATIONS OF CO_2 AS AN EXTINGUISHING AGENT

The use of carbon dioxide on general Class A fires is limited mostly by (1) its low cooling capacity (particles of dry ice do not "wet" or penetrate), and (2) enclosures incapable of retaining an extinguishing atmosphere. True surface burning fires are extinguished easily because natural cooling takes place quickly. On the other hand, if the fire penetrates below the surface, or under materials that provide thermal insulation that slows down the rate of heat loss (generally referred to as "deep seated burning"), a higher concentration of carbon dioxide and a much longer holding time are needed for complete extinguishment.

TABLE 19-1A. Minimum Carbon Dioxide Concentrations for Extinguishment

Material	Theoretical Min. CO_2 Concentration (%)
Acetylene	55
Acetone	26*
Benzol, Benzene	31
Butadiene	34
Butane	28
Carbon Disulfide	55
Carbon Monoxide	53
Coal Gas or Natural Gas	31*
Cyclopropane	31
Dowtherm	38*
Ethane	33
Ethyl Ether	38*
Ethyl Alcohol	36
Ethylene	41
Ethylene Dichloride	21
Ethylene Oxide	44
Gasoline	28
Hexane	29
Hydrogen	62
Isobutane	30*
Kerosene	28
Methane	25
Methyl Alcohol	26
Pentane	29
Propane	30
Propylene	30
Quench, Lubricating Oils	28

Note: The theoretical minimum extinguishing concentrations in air for the above materials were obtained from Bureau of Mines, Bulletin 503 (Coward and Jones 1952). Those marked * were calculated from accepted residual oxygen values.

Liquid fuel fires frequently are extinguished by discharging CO_2 directly on the burning material. No enclosure is needed and a 30 second discharge usually is adequate to cool everything below the reignition temperature of the fuel. This use of carbon dioxide is limited mainly to when there is serious overheating of massive metal objects or a substantial quantity of glowing embers from carbonaceous materials. A much longer discharge time may be needed to prevent reignition.

Oxygen Containing Materials and Reactive Chemicals

Carbon dioxide is not an effective extinguishing agent for fires involving chemicals such as cellulose nitrate that contain their own oxygen supply. Fires involving reactive metals such as sodium, potassium, magnesium, titanium, zirconium, and the metal hydrides cannot be extinguished by carbon dioxide, because the metals and hydrides decompose CO_2.

Life Safety Considerations

Carbon dioxide should not be used in normally occupied spaces unless arragements can be made to assure evacuation before discharge. The same restriction applies to spaces that are not normally occupied but where personnel may be present for maintenance or other purposes. It may be difficult to assure evacuation if the space is large or if egress is in any way impeded by obstacles or complicated passageways. Escape is even more difficult after the discharge starts, because of possible confusion due to noise and greatly reduced visibility.

Consideration also should be given to any possibility of large volumes of carbon dioxide vapor leaking or flowing into unprotected lower levels such as cellars, tunnels, or pits. In this event, the suffocating atmosphere would not be visible and might not be detected until too late.

METHODS OF APPLICATION

Two basic methods are used to apply carbon dioxide in extinguishing fires. One method is to discharge a sufficient amount of the agent into an enclosure to create an extinguishing atmosphere throughout the enclosed area. This is called "total flooding." The second method is to discharge the agent directly on the burning material without relying on an enclosure to retain the carbon dioxide. This is called "local application."

Total Flooding

In total flooding systems, carbon dioxide is applied through nozzles designed and located to develop a uniform concentration of CO_2 in all parts of an enclosure. (See Figs. 19-1B and 19-1C.) Calculation of the quantity of

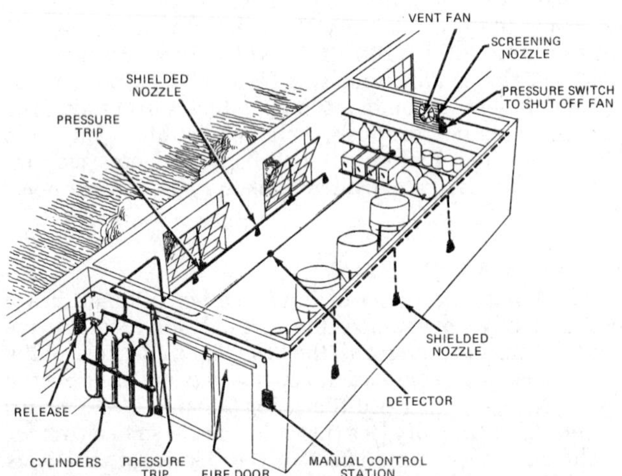

FIG. 19-1B. Diagram of high pressure carbon dioxide system for total flooding.

carbon dioxide required to achieve an extinguishing atmosphere is based upon the volume of the room and the concentration of CO_2 required for the combustible materials therein.

The integrity of the enclosure is a very important part of total flooding. If the room is tight, especially on the sides and bottom, the atmosphere can be retained for a long time to assure complete control of the fire. However, if there are openings on the sides and bottom, the heavier mixture of carbon dioxide and air may leak out of the room rapidly and be replaced with air entering at higher openings. If the extinguishing atmosphere is lost too rapidly, glowing embers may remain and cause reignition when air reaches the fire zone. Therefore, it is important to close all openings to minimize leakage, or to compensate for the open-

FIG. 19-1C. A carbon dioxide extinguishing system that has activated. (The Ansul Company)

ings by discharging additional carbon dioxide. Because of the relative weight of carbon dioxide, an opening in the ceiling helps to relieve internal air pressure during the discharge, with very little effect on leakage rate after the discharge.

The minimum concentration used in total flooding systems is 34 percent carbon dioxide by volume for surface burning materials such as some liquid fuels. Electrical wiring hazards, including small electrical machines, require 50 percent concentration; bulk paper requires 65 percent, and fur storage vaults and dust collectors 75 percent. These are specific hazards for which there is a background of test experience. Other materials should be tested to determine minimum CO_2 concentrations and holding time.

Local Application

In local application systems, carbon dioxide is discharged directly on the burning surfaces through nozzles designed for this purpose. The intent is to cover all combustible areas with nozzles located so they will extinguish all flames as quickly as possible. Any adjacent area to which fuel may spread must also be covered, because any residual fire could cause reignition after the CO_2 discharge is terminated. The discharge should be continued for a minimum of 30 seconds, or longer if required for cooling.

Discharge nozzles usually are designed for relatively low velocity to avoid splashing and air entrainment. Automatic detection is a necessity to provide fast response and minimize heat buildup. Although not essential, an enclosure would help retain carbon dioxide in the fire area. Local application of CO_2 can be used for fast extinguishment, even in an enclosure where final total flooding can provide absolute assurance that extinguishment will be complete.

Extended Discharge

An extended discharge of CO_2 is used when an enclosure is not tight enough to retain an extinguishing concentration as long as is needed. The extended discharge normally is at a reduced rate, following a high initial rate used to develop the extinguishing concentration in a reasonably short time. The reduced rate of discharge should be a function of the leakage rate (which can be calculated on the basis of leakage area), or the flow rate, through ventilating ducts that cannot be shut down.

Extended discharge is particularly applicable to enclosed rotating electrical equipment such as generators, where it is difficult to prevent leakage until rotation stops. Extended discharge can be applied to ordinary total flooding systems as well as to local application systems where a small hot spot may require prolonged cooling.

Hand Hose Lines

Carbon dioxide systems can consist of hand hose lines permanently connected by means of fixed piping to a fixed supply of CO_2. Such systems frequently are provided for manual protection of small localized hazards. Although not a substitute for a fixed system, a hose line may be used to supplement a fixed system where the hazard is accessible for manual fire fighting, as well as to supplement portable equipment.

Hand hose lines are required to have a minimum bursting pressure of 5,000 psi (34.5 MPa) if connected to a high pressure supply, or 1,800 psi (12.4 MPa) if the supply is low pressure. The lines are not under pressure until the valve actuating the system is opened. There must be a sufficient carbon dioxide supply to permit use of the hand line nozzle for at least 1 minute. Figure 19-1D shows a hand hose line system.

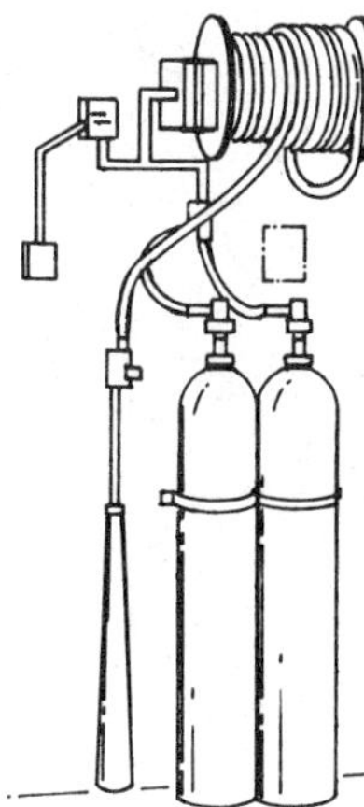

FIG. 19-1D. Carbon dioxide hand hose extinguishing system with hose mounted on a reel. This system offers flexibility for attacking fires. (Walter Kidde and Company, Inc.)

Standpipe Systems and Mobile Supply

Total flooding, local application, and hand hose line systems without a permanently connected carbon dioxide supply are known as gas standpipe systems. They are supplied by containers of carbon dioxide mounted on mobile units that can be moved and quickly coupled to a standpipe in case of fire.

Gas standpipe systems can be used (1) as a supplement to complete fixed fire protection systems, or (2) as the only protection of hazards under certain circumstances, i.e., when extinguishment would not be adversely delayed while the mobile supply is moved to the scene and attached to the standpipe. Mobile supplies may be equipped with hand hose lines for the protection of scattered hazards.

COMPONENTS OF CARBON DIOXIDE SYSTEMS

The main components of a carbon dioxide system are the carbon dioxide supply, the discharge nozzles, and the piping system. These components, along with control valves and other operating devices, dispense the carbon dioxide and provide effective fire extinguishment.

Carbon Dioxide Storage

The CO_2 supply may be stored in high or low pressure containers. Because of the differences in pressure, system design is influenced by the storage method.

High Pressure Storage: High pressure containers (usually cylinders) are designed to store liquid carbon dioxide at atmospheric temperature. Since the maximum pressure in the cylinder or other container is affected by the ambient temperature, it is important that the container be designed to withstand the maximum expected pressure. Figure 19-1E shows the change in gage pressure with temperature.

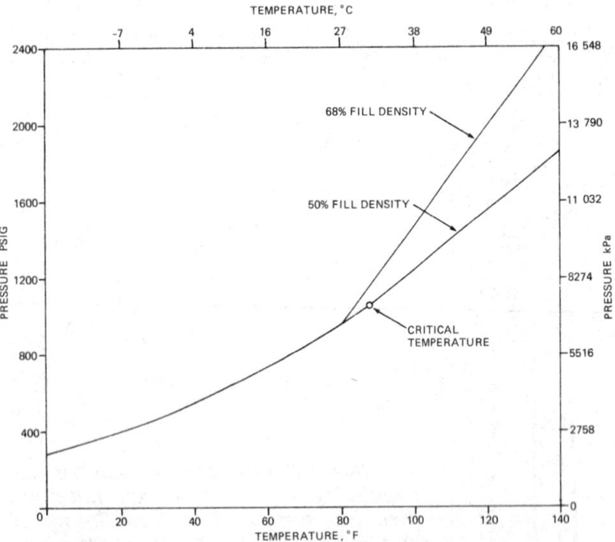

FIG. 19-1E. Variation in pressure of carbon dioxide with change in temperature, showing the effect of filling density in high pressure storage.

Storage cylinders are designed, tested, and filled to U.S. Department of Transportation (DOT) specifications. The maximum permitted filling density is equal to 68 percent of the weight of water that the container can hold at 60°F (16°C). Fire extinguishing cylinders are fitted with an internal dip tube so that liquid will be discharged from the bottom when the cylinder is upright and the valve is opened.

Abnormally low storage temperatures would adversely affect the rate of discharge. For this reason, NFPA 12 does not permit storage temperatures below 0°F (−18°C) for total flooding systems or below 32°F (0°C) for local application systems. Lower storage temperatures may be permitted only if the design includes special features to compensate for the reduced pressure.

Low Pressure Storage: Low pressure storage containers are pressure vessels with a design working pressure of at least 325 psi (2240 kPa). These containers are maintained at a temperature of approximately 0°F (−18°C) by use of insulation and mechanical refrigeration. At this temperature, the pressure is approximately 300 psi (2069 kPa). A compressor, controlled by a pressure switch in the tank, circulates refrigerant through coils near the tank top. Tank pressure is controlled by condensation of carbon dioxide vapor by the coils. In the event of refrigeration failure, pressure relief valves bleed off some of the vapor to keep the pressure within safe limits. This permits some of the liquid to evaporate, creating a self-refrigerating effect that keeps the emergency loss rate at a low level. Figure 19-1F shows a typical low pressure container.

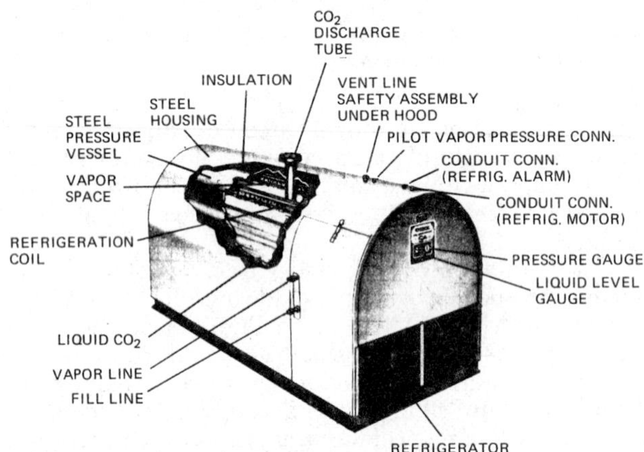

FIG. 19-1F. A low pressure storage unit. The carbon dioxide is maintained at 0°F (−18°C) by refrigeration and the pressure is 300 psi (2069 kPa). (Cardox Division of Chemetron Corporation)

With low pressure storage, it is common practice to protect multiple hazards from one central storage unit. The quantity of carbon dioxide discharged on a particular hazard is controlled by opening and closing the discharge valves in a preset timed sequence. Central storage units may have capacities ranging from less than one ton to a hundred tons or more. For large hazards, the distance between hazard and storage may be several hundred feet or more (100 ft equals approximately 30.5 m).

Discharge Nozzles

Many nozzle types are available for fire extinguishment applications. Nozzles used in total flooding may be simply orifices giving high velocity jet streams, or may be partially shielded to achieve reduced velocity or a specific discharge pattern. High velocity types provide substantial convection mixing to assure uniform concentration of carbon dioxide throughout an enclosure. Low velocity types have a tendency to create high concentrations in the lower levels, which may be desirable under certain conditions.

Nozzle types used in local application systems normally are designed for relatively low discharge velocity. This helps to avoid splashing of liquid fuels and to minimize turbulence and air entrainment. All such nozzles must be tested for fire extinguishing characteristics and listed or approved by a testing laboratory on the basis of test performance.

Figure 19-1G shows a typical performance curve for a

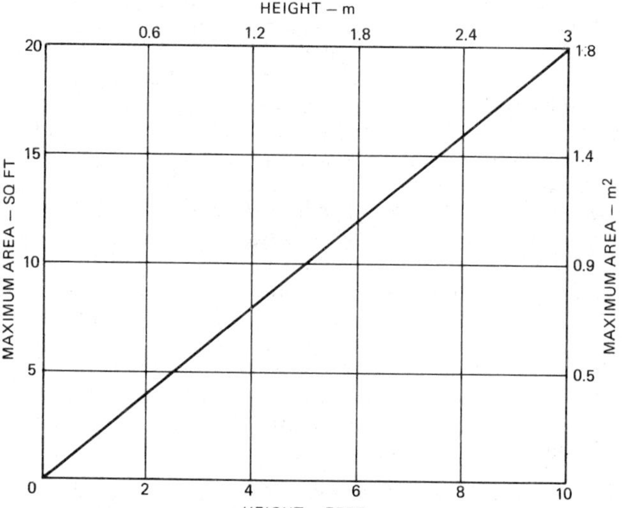

FIG. 19-1G. A listing or approval curve of a typical carbon dioxide nozzle showing maximum area versus height or distance from liquid surface.

nozzle designed for overhead mounting. The rate of discharge used is based on the height of the nozzle above the hazard or fuel surface and is given in listing cards or approvals. (Testing is based on the maximum rate that will not splash liquid fuels.) The maximum fire area that the nozzle will extinguish also is based on the height above the hazard surface (using the design flow rate).

Performance information enables the system designer to select and properly locate the number of nozzles necessary to cover the entire fire area of the hazard, and establishes the design flow rate that must be used on the hazard. The effective design flow rate must continue for a minimum of 30 seconds, which is part of the criteria used in performance testing.

Figures 19-1H and 19-1I show typical performance curves for nozzles designed for mounting on the sides of open tanks containing flammable liquids. In such cases, the design flow rates are based on the area of coverage for the tank side nozzles, or the width of the hazard for linear nozzles mounted to extend the full length of one side. In each case, the design flow rate must be in the shaded area of the curves.

Piping

Piping systems, normally empty, convey carbon dioxide from the storage container to open nozzles where there is a fire. Since the proper rate of flow is a critical require-

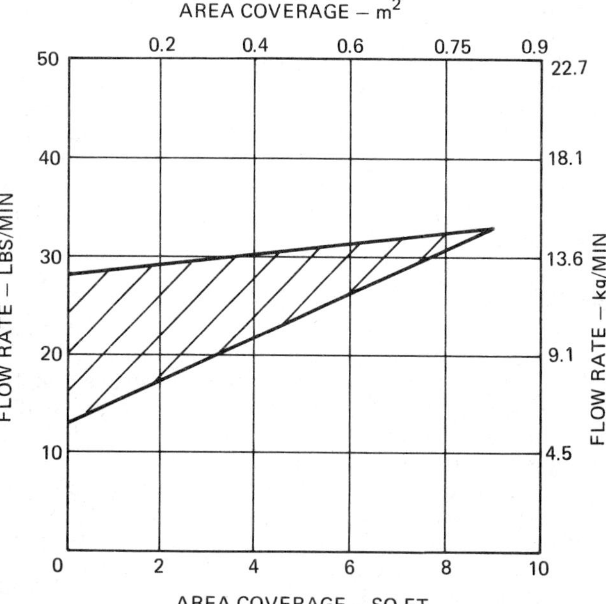

FIG. 19-1H. A typical listing or approval curve of a tankside carbon dioxide nozzle showing the flow rate versus area coverage.

ment for fire extinguishment, it is important that the piping be designed accurately.

Carbon dioxide drawn from the bottom of the storage container enters the piping as a liquid. Friction causes loss in pressure. As pressure drops, the liquid boils, resulting in a mixture of liquid and vapor in the piping. The vapor increases in volume as the mixture passes through the piping, with a further drop in pressure. Thus, the flow is two phase (a mixture of liquid and gas), a fact that pressure drop calculations must take into account.

NFPA 12 covers the calculation of flow conditions in some detail and provides pertinent equations and data tables. The calculating procedure is too complex to be carried out readily by manual methods; computer programs are available to simplify the process of designing

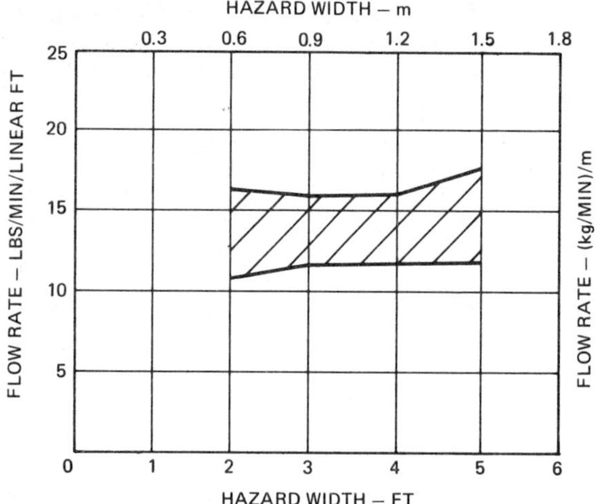

FIG. 19-1I. A typical listing or approval curve of a linear carbon dioxide nozzle showing flow rate versus hazard width.

complete piping systems for optimum flow rates at each nozzle.

The piping must be adequately supported to prevent movement during the discharge, and provision must be made for its contraction and expansion. Because liquid carbon dioxide is a refrigerant, it will substantially reduce the pipe temperature during discharge. Low pressure liquid, in particular, starts at 0°F (−18°C) and may reach temperatures as low as −50°F (−46°C) in the piping before the discharge ends.

Valves and Operating Devices

Valves for controlling the discharge of carbon dioxide must withstand the maximum operating pressure, be absolutely bubble tight when closed, and be capable of both manual and automatic operation. Valves and allied devices such as timers and pressure switches must be listed or approved for use in CO_2 systems.

To resist maximum anticipated pressure, valves in high pressure systems are required to have a minimum bursting pressure of 5,000 psi (34.5 MPa); valves constantly under pressure must have a minimum bursting pressure of 6,000 psi (41.4 MPa). Valves for low pressure systems must be capable of withstanding hydrostatic testing to 1,800 psi (12.4 MPa) without permanent distortion. Corrosion resistance, noncombustibility, and ability to withstand the expected temperature extremes are other characteristics required of the valves.

CO_2 SYSTEM DESIGN CONSIDERATIONS

To organize basic components into a CO_2 fire protection system, a number of factors must be considered. These include the quantity of stored carbon dioxide, the method of actuation, the use of predischarge alarms, ventilation shutdown, pressure venting, and anything else involved in assuring safe, prompt, and effective fire extinguishment.

Quantity Requirements

The quantity of carbon dioxide required for extinguishment depends upon the type of fire, the type of extinguishing system, and conditions in the fire area. Where continuous protection is required, the quantity of carbon dioxide on hand should be at least twice that needed for extinguishment. Detailed methods of determining the quantity required for extinguishment are covered in NFPA 12.

Total Flooding Systems: The quantity of carbon dioxide must be sufficient to achieve a minimum design concentration and hold it until the fire is extinguished. In an ideal enclosure, the volume of carbon dioxide vapor per unit volume of enclosure can be expressed by the equation:

$$X = l_n\ (100/(100 - C))$$

where:
X = Volume of vapor injected
C = Concentration of carbon dioxide
l_n = Natural log.

This equation assumes that carbon dioxide is injected with high velocity nozzles so that complete mixing of CO_2 with the room air is maintained throughout the discharge period. This means that some of the carbon dioxide will be lost along with air displaced from the room. The weight of carbon dioxide required is determined by dividing the volume of vapor injected, by the specific volume of carbon dioxide (near 9 cu ft per lb or 0.56 m^3/kg) and multiplying the result by the volume of the room.

Alternatively, a series of specific flooding factors has been established to simplify the calculation for quantity by including an allowance for distributed leakage based on room size. The factors are greater for smaller rooms because the anticipated leakage would be greater relative to volume. The flooding factors are further multiplied by conversion factors to achieve design concentrations for the type of combustible involved. This procedure does not eliminate the need to compensate for leakage through obvious openings that cannot be closed, or through ventilating systems that cannot be shut off in the event of a fire.

Surface burning fires, such as flammable liquid fires, are normally extinguished during carbon dioxide discharge, which should not be longer than one minute. Deep seated fires require higher concentrations and much longer holding times. The rate of discharge must be high enough to develop a concentration of 30 percent in not more than two minutes, and the final design concentration must be achieved in not more than seven minutes. Leakage compensation must be in addition to the basic quantity. Enclosures for deep seated fires must be relatively tight or it quickly becomes uneconomical to maintain the CO_2 discharge design concentration.

Local Application Systems: Preferably, these are designed using the rate by area method. The rate of discharge from each nozzle is determined from listing information, and the total discharge rate for the system is the sum of the design rates required for each nozzle. A minimum 30 second effective discharge time is required. A discharge of vapor is not considered effective, a fact which calculation of the total quantity must take into account.

In high pressure systems, only approximately 70 percent of the stored quantity will be discharged as liquid. Therefore, the quantity calculated by rate and time must be increased by a factor of 1.4 to assure a 30 second liquid discharge. In low pressure systems, the time of the initial vapor discharge while the pipe is cooling down must be added to the 30 second liquid discharge.

Where the hazard consists of three dimensional irregular objects that cannot easily be reduced to equivalent surface areas, the rate-by-volume method may be used. In this case, the discharge rate of the system is based on the volume of an assumed enclosure entirely surrounding the hazard. Because this method is not as precise as the rate-by-area procedure, it usually results in more carbon dioxide discharge unless the hazard is nearly enclosed with walls.

Extended Discharge Systems: The most efficient way to compensate for leakage where the concentration must be maintained for a substantial time is to use an extended discharge at a reduced rate equal to the rate of leakage. The quantity of carbon dioxide required is based on the rate of discharge multiplied by the time that the concentration must be maintained. This is in addition to the quantity required to develop the design concentration.

Hand Hose Line Systems: In theory, the carbon dioxide supply depends upon the type and size of the hazard to be protected; but as a practical matter, carbon dioxide should be available to counteract possible waste by inexperienced users. In any case, there should be sufficient carbon dioxide to permit operation of the system for at least one minute.

Venting Requirements

When liquid carbon dioxide is discharged into a closed room, the atmosphere will shrink initially due to the sudden refrigerating effect. At a later flooding stage, the combined volume of the carbon dioxide and air will become greater than the initial room volume. Thus, while the first result will be to create a vacuum or draw in air, the final result must be to increase the pressure or to exhaust the excess volume through vent openings. Actual air temperatures are greatly reduced during the discharge, but soon return to normal as heat is absorbed from solid surfaces in the room.

Experience has shown that most ordinary rooms have sufficient leakage (through cracks around doors and windows and general porosity) to prevent noticeable vacuum or pressure buildup. Even so, from the viewpoint of efficient total flooding, it is better to have major vent openings in or near the ceiling rather than near the floor. In rooms that may be tightly sealed, a safe vent area for light structures can be estimated on the basis of calculated discharge flow rate:

$$X = Q/6.5$$

where:
X = Vent area in sq in
Q = Flow rate in lbs/min

For SI units:

$$X_m = 219\ Q_m$$

where:
X_m = Vent area in sq mm
Q_m = Flow rate in kg/min

More information on room venting can be found in NFPA 12.

Methods of Actuation

Total flooding and local application carbon dioxide systems normally are designed to operate automatically. The detection device may be any of the listed or approved devices that are actuated by heat, smoke, flame, flammable vapors, or other abnormal process conditions that could lead to a fire or explosion. Automatically operated systems are required to have an independent means of manual actuation.

Actuation of the system may be by an automatic electric switch or other means to open a cylinder valve. For full control of all functions, the control system may use a programming means to operate switches in the proper sequence and at the proper time. The switches serve to give the alarm, shut down equipment, close fire doors, start and stop the carbon dioxide flow, and perform other functions necessary to extinguish the fire safely and quickly or to correct dangerous conditions.

Whether supplemental to automatic actuation or the sole means of placing a system in operation, manual controls must be easy to operate, accessible in case of fire, and located close to the valves they control. Remote manual controls usually are located near an exit for a total flooding installation, or near a hazardous area for local application installations.

Supervision of Systems

Depending upon their complexity, carbon dioxide systems require different degrees of supervision. As a minimum requirement, every system should be supervised to the extent that it has an audible or visual signal to indicate that the system has operated and should be placed back in service. Where the detection and control systems are so extensive and complex that they cannot be visually checked, the supervisory system should be capable of giving prompt and positive indication of any failure.

Life Safety Considerations

Every precaution must be taken to assure that personnel in a protected area are evacuated before the area is flooded with carbon dioxide. This is the primary purpose of a predischarge alarm, which should provide a suitable time delay before the CO_2 discharge. Additional life safety steps include posting warning signs of the protected status, and informing personnel about operating procedures. In large systems, watch procedures should be established if the protected area must be entered for maintenance or other work.

Consideration should be given to any possibility that substantial volumes of carbon dioxide vapor could leak or flow into adjacent low level areas such as cellars, pits, or narrow hallways—spaces not included in the protected area. If this potential exists, alarms and warning signs should be extended to include these areas. Finally, especially for very large rooms, safely ventilating the flooded space following the emergency should be considered.

TESTING AND MAINTAINING CO_2 SYSTEMS

An elaborate fire protection system could be worthless unless it is properly tested to demonstrate adequate performance and maintained in good working condition.

Acceptance Testing

No newly installed system should be accepted until it has been properly inspected and tested to prove performance in accordance with design specifications. The inspection should ascertain whether the system is installed in accordance with approved plans. The piping is particularly important, because any changes might affect the flow rates which were calculated on the basis of the pipe sizes shown on the plans.

The final proof of performance is to discharge the system and, for total flooding hazards, measure the concentrations of CO_2 and the holding time in the fire zone. This is sometimes the only positive way of uncovering leakage conditions that might jeopardize the effectiveness of the system. Although local application systems are not subject to concentration testing, their effectiveness can be judged reasonably well by observation during the test discharge.

Inspections

To maintain a system in proper operating condition, it is essential to conduct periodic inspections. The status of the carbon dioxide supply should be checked at six month intervals, or more frequently, and immediately after the system has operated. The entire system should be inspected on an annual basis. In particular, inspectors should look for changes in the scope of the hazard, changes in the condition of any enclosure, and other changes that could affect the adequacy of the fire protection system.

Maintenance Testing

All operating devices should be tested on an annual basis. Although much of this testing can be accomplished without discharging the system, in some cases a partial discharge may be necessary. Any required maintenance found by testing should be performed without delay.

Bibliography

References Cited

Coward, H. W. and Jones, G. W. 1952. "Limits of Flammability of Gases and Vapors." *Bulletin 503*. USDI Bureau of Mines. Pittsburgh, PA.

NFPA Codes, Standards, Recommended Practices and Manuals. (See the latest *NFPA Codes and Standards Catalog* for availability of current edition of the following documents.)

NFPA 12, *Standard on Carbon Dioxide Extinguishing Systems.*

Additional Readings

"Carbon Dioxide Fire Extinguishing Systems," Walter Kidde and Co., Wake Forest, NC, 1981, rev. Aug. 1984.

Fire Extinguishing Systems Design Manual: Carbon Dioxide, Walter Kidde and Co., Wake Forest, NC, 1982.

Noronha, J. A., and Schiffhauer, E. J., "Explicit Equations for Two-Phase Carbon Dioxide Flow," *Fire Technology*, Vol. 10, No. 2, May 1974, pp. 101-109.

Strasser, A., Liebman, I., and Kuchta, J. M., "Methane Flame Extinguishment with Layered Halon or Carbon Dioxide," *Fire Technology*, Vol. 10, No. 1, Feb. 1974, pp. 25-34.

HALOGENATED AGENTS AND SYSTEMS

Revised by Daniel W. Moore

Halogenated extinguishing agents are hydrocarbons in which one or more hydrogen atoms have been replaced by atoms from the halogen series: fluorine, chlorine, bromine, or iodine. This substitution confers not only nonflammability, but flame extinguishment properties to many of the resulting compounds. Halogenated agents are used both in portable fire extinguishers and in extinguishing systems.

This chapter discusses the properties of halogenated fire extinguishing agents and their use in extinguishing systems. Their use in portable fire extinguishers is discussed in Section 20.

Prior to 1945, three halogenated fire extinguishing agents were widely used: carbon tetrachloride (Halon 104), methyl bromide (Halon 1001) and chlorobromomethane (Halon 1011). The earliest, carbon tetrachloride, became available in the early 1900s and found immediate wide use in portable hand pump extinguishers. Its main advantages were electrical nonconductivity and lack of residue following application (Wharry and Hirst 1974).

In the late 1920s, methyl bromide (Halon 1001) was found to have greater extinguishing potential than carbon tetrachloride. It was used extensively in German aircraft and ships and in British aircraft during World War II, but because of its high vapor toxicity it was never widely used in portable extinguishers. Chlorobromomethane (Halon 1011) was developed in Germany in 1939-1940 as a replacement for methyl bromide, but its use did not become widespread until after World War II (Strasiak 1954).

For toxicological reasons, however, concern about using these three early halogenated agents gained significant momentum during the early 1960s. Except for a few Halon 1001 and Halon 1011 systems that may still be in service on several older models of European and United States military aircraft, and except for certain explosion suppression applications, systems containing these three halogenated agents have been removed from service.

In 1947, the Purdue Research Foundation performed a systematic evaluation of more than 60 new extinguishing agents. Simultaneously, the U.S. Army Chemical Center undertook toxicological investigations of these same compounds. From these tests, four halogenated agents—bromotrifluoromethane (Halon 1301), bromochlorodifluoromethane (Halon 1211), dibromodifluoromethane (Halon 1202), and dibromotetrafluoromethane (Halon 2402)—were selected for further evaluations in specific applications (NFPA 1954).

From these further tests, Halon 1301 was determined to be the second most effective and least toxic of the group, while Halon 1202 was the most effective but also the most toxic. As a result, Halon 1301 was selected by the U.S. Army for use in portable extinguishers and by the Federal Aviation Administration (FAA) for use in commercial aircraft engine nacelles. Halon 1202 was selected by the U.S. Air Force to protect military aircraft engines. In a similar evaluation program in England, Halon 1211 was selected for military and civilian aircraft systems and for portable fire extinguishers.

The concept of using halogenated agents in commercial total flooding systems seems to have originated between 1962 and 1964. Between 1964 and 1968, a number of Halon 1301 total flooding systems using carbon dioxide equipment and technology were installed in the U.S. Such a system installed in Winterthur Museum was described in the November 1969 issue of *Fire Journal* (Dowling and Ford 1969).

In 1966, NFPA organized a Technical Committee on Halogenated Fire Extinguishing Agent Systems to develop standards covering installation, maintenance, and use of such systems. NFPA 12A, *Standard on Halon 1301 Fire Extinguishing Systems*, and NFPA 12B, *Standard on Halon 1211 Fire Extinguishing Systems* (hereinafter referred to as NFPA 12A and NFPA 12B), were approved in the early 1970s, NFPA 12C-T is a tentative standard for Halon 2402. Because of its rather recent recognition, there is limited experience with Halon 2402 in the U.S. and,

Mr. Moore is Program Manager, Halon 1301 Fire Extinguishants, Chemicals and Pigments Department, E. I. duPont de Nemours & Company. He is a member of the NFPA Technical Committee on Halogenated Fire Extinguishing Agent Systems; Chairman of the U.S. Technical Advisory Group (TAG) to the International Organization for Standardization (ISO) for Halon Extinguishing Systems, and a member of the ISO Subcommittee on Extinguishing Media for Fire Fighting.

therefore, it cannot be treated as extensively as the other agents in this HANDBOOK.

During 1966, attention began to focus on the use of Halon 1301 to protect computer rooms and electronic data processing (EDP) equipment. In 1972, following extensive testing by several major companies on the effects of Halon 1301 decomposition products on electronic equipment (Ford 1972a), the NFPA Committee on Electronic Computer/Data Processing Equipment recognized Halon 1301 total flooding systems as suitable for protection of electronic computer/data processing equipment. With suitable precautions such as time delays, Halon 1211 total flooding systems are now used in Europe.

CHEMICAL COMPOSITION AND CLASSIFICATION

The halogenated extinguishing agents are currently known simply as halons, and the halon system for naming the halogenated hydrocarbons was devised by the U.S. Army Corps of Engineers (NFPA 1954). This simplified system of nomenclature describes the chemical composition of the materials without the use of chemical names or possibly confusing abbreviations (i.e., "BT" for bromotrifluoromethane and "DB" for dibromodifluoromethane). Examples of this system are shown in Table 19-2A. The first digit of the number represents the number of carbon atoms in the compound molecule; the second digit, the number of fluorine atoms; the third digit, the number of chlorine atoms; the fourth digit, the number of bromine atoms; and the fifth digit, the number of iodine atoms (if any). If the fifth digit is a zero, it is not expressed; bromotrifluoromethane ($BrCF_3$), for example, is referred to as Halon 1301 (not 13010), although its chemical formula shows one carbon atom, three fluorine atoms, no chlorine atoms, one bromine atom, and no iodine atoms.

TABLE 19-2A. Sample Halon Numbers for Various Halogenated Fire Extinguishing Agents

Chemical Name	Formula	Halon No.
Methyl bromide	CH_3Br	1001
Methyl iodide	CH_3I	10001
Bromochloromethane	CH_2BrCl	1011
Dibromodifluoromethane	CF_2Br_2	1202
Bromochlorodifluoromethane	CF_2BrCl	1211
Bromotrifluoromethane	CF_3Br	1301
Carbon tetrachloride	CCl_4	104
Dibromotetrafluoroethane	$C_2F_4Br_2$	2402

The three halogen elements commonly found in extinguishing agents are fluorine (F), chlorine (Cl), and bromine (Br). Substitution of a hydrogen atom in a hydrocarbon with these three halogens influences the relevant properties in the following manner:

Fluorine: Imparts stability to the compound, reduces toxicity, reduces boiling point, increases thermal stability.

Chlorine: Imparts fire extinguishing effectiveness, increases boiling point, increases toxicity, reduces thermal stability.

Bromine: Same effects as chlorine, but to a greater degree.

Thus, compounds containing combinations of fluorine, chlorine, and bromine can possess varying degrees of extinguishing effectiveness, chemical and thermal stability, volatility, and toxicity.

The halogen is linked to the carbon atom by a "covalent" chemical bond, which means that unlike inorganic halogen compounds such as common table salt (sodium chloride), there is no tendency to ionize or become electrically conductive in the presence of water. The presence of a fluorine atom in the molecule increases the C-Cl and C-Br strength and improves the overall chemical stability of the compound.

Because they are either gases or liquids that rapidly vaporize in fire, halons leave no corrosive or abrasive residue after use. They are nonconductors of electricity and have high liquid densities which permit use of compact storage containers. The areas of major halon use are for the protection of electrical and electronic equipment, petroleum production facilities, engine compartments (ships, military vehicles, and aircraft), and other areas where rapid extinguishment is important, or where damage to equipment or materials or cleanup after use must be minimized.

PROPERTIES OF HALOGENATED AGENTS

Some of the physical properties of the common halogenated fire extinguishing agents referred to in this chapter are shown in Table 19-2B. More details on bromotrifluoromethane and bromochlorodifluoromethane are given in the following paragraphs.

Physical Properties

Halon 1301 is a gas at 70°F (21°C) with a vapor pressure of 199 psig (1372 kPa). Although this pressure would adequately expel the material, it decreases rapidly to 56 psig (386 kPa) at 0°F (−17°C) and to 17.2 psig (119 kPa) at −40°F (−40°C). It is, therefore, usual to increase the pressure on portable extinguishers or systems with nitrogen either to 360 psig (2482 kPa) or to 600 psig (4137 kPa), which ensures adequate performance at all temperatures. The high volatility of Halon 1301 has led to its use in total flooding systems, where rapid vaporization is an asset.

Halon 1211 is also a gas at 70°F (21°C), with a vapor pressure of 22 psig (152 kPa) and a boiling point of 25°F (−4°C). Its relatively high boiling point allows it to be projected as a liquid stream, thus enabling portable extinguishers and local application systems to have a greater range than possible with other gaseous materials. For portable extinguishers, the normal superpressurization with nitrogen is approximately 100 psig at 70°F (689 kPa at 21°C) which permits the use of inexpensive hardware. Halon 2402, with a boiling point of 117°F (47°C), is a liquid at room temperature.

Corrosion and Other Effects on Materials

The early nonfluorinated halogenated agents had significant corrosion problems. The fluorinated agents currently in use are chemically more stable, and neither Halon 1301, Halon 1211 nor Halon 2402 has any significant corrosive action on the commonly used construction metals unless free water is present. (Free water is defined as the presence of a separate water phase in the liquid

TABLE 19-2B. Some Physical Properties of the Common Halogenated Fire Extinguishing Agents

Agent	Chemical Formula	Halon No.	Type of Agent	Approx. Boiling Point °F†	Approx. Freezing Point °F†	Specific Gravity of Liquid at 68°F† (Water = 1)	Approx. Critical Temp. °F†	Estimated Pressure psig‡ At 130°F†	Estimated Pressure psig‡ At Critical Temp.	Latent Heat of Vaporization cal/g Water = 540 cal/g CO_2 = 138 cal/g
Carbon tetrachloride	CCl_4	104	Liquid	170	−8	1.595	—	—	—	46
Methyl bromide	CH_3Br	1001	Liquid	40	−135	1.73	—	—	—	62
Bromochloromethane	CH_2BrCl	1011	Liquid	151	−124	1.93	—	—	—	—
Dibromodifluoromethane	CF_2Br_2	1202	Liquid	76	−223	2.28	389	23	585	29
Bromochlorodifluoromethane	CF_2BrCl	1211	Liquefied Gas*	25	−257	1.83	309	75	580	32
Bromotrifluoromethane	CF_3Br	1301	Liquefied Gas	−72	−270	1.57	153	435	560	28
Dibromotetrafluoroethane	$C_2F_4Br_2$	2402	Liquid	117	−167	2.17	—	3.8	—	25

* May be kept as a liquid at reduced temperatures.
† 5/9 (°F −32) = °C
‡ 1 psig = 6.895 kPa

halon. When present in a small quantity, free water can provide a site for concentrating acid impurities into a corrosive liquid. It has been determined that dissolved water is not a problem.)

The effect of these agents on plastics and elastomers depends upon the precise grade or composition of the plastics and elastomers. In general, normally satisfactory effects are obtained with rigid plastics such as polytetrafluoroethylene, nylon, and acetal copolymers. Elastomers of particular suitablity are neoprene, Buna N, and "Viton" fluoroelastomer rubbers. More precise guidelines should be obtained from the manufacturers of the materials. Halon 1301 generally has considerably less effect on plastic and elastomeric materials than does Halon 1211 or Halon 2402.

EXTINGUISHING CHARACTERISTICS

The extinguishing mechanism of the halogenated agents is not clearly understood. However, a chemical reaction undoubtedly occurs which interferes with the combustion processes. The agents act by removing the active chemical species involved in the flame chain reactions (a process known as "chain breaking"). While all the halogens are active in this way, bromine is much more effective than chlorine or fluorine.

A number of possible mechanisms by which bromine inhibits combustion have been proposed. The most active species in hydrocarbon combustion are oxygen and hydrogen atoms, and hydroxyl radicals (O·, H·, and OH·). Under fire conditions, the halogenated agent—for example, bromotrifluoromethane (Halon 1301)—will release a bromine atom:

$$CBrF_3 \rightarrow CF_3^{\cdot} + Br^{\cdot}$$

The Br atom can react with a hydrocarbon molecule to form hydrogen bromide, HBr:

$$R\text{-}H + Br^{\cdot} \rightarrow R^{\cdot} + HBr$$

The HBr then reacts with an active hydrogen or hydroxyl radical, releasing the bromine atom for further inhibition reactions.

$$OH^{\cdot} + HBr \rightarrow H_2O + Br^{\cdot}$$

In this way, chain carriers are removed from the system while the inhibiting HBr is continuously regenerated.

Another "ionic" theory supposes that the uninhibited combustion process includes a step in which oxygen ions are formed by the capture of electrons which come from ionization of hydrocarbon molecules. Since bromine atoms have a much higher cross section for the capture of slow electrons than has oxygen, the bromine inhibits the reaction by removing the electrons that are needed for activation of the oxygen (Hough 1960).

Fire Extinguishing Effectiveness

The effectiveness of the different halons as extinguishing agents has been the subject of many studies. The initial results of some of the early studies are contained in the October 1954 issue of the *NFPA Quarterly* (NFPA 1954). It is now recognized that comparison of effectiveness of agents depends upon whether portable extinguishers or fixed systems are being considered and, particularly in tests on portable units, whether the agents were evaluated with hardware of optimum nozzle and value design. Therefore, the following information should be used only as a general guide because other considerations such as range of the unit (which varies on the gas or liquid characteristic of the individual halon and the total weight and volume of the equipment), may be of equal importance, depending upon use.

Table 19-2C gives the approximate pounds of agent required per unit of Class B rating obtained with small portable extinguishers currently listed with Underwriters' Laboratories, Inc. (UL) or approved by Factory Mutual Research.

In total flooding systems, the effectiveness of the halogenated agents on flammable liquid and vapor fires is

TABLE 19-2C. Relative Effectiveness of Halogenated Agents on Small UL Class B Fires

Agent	Halon No.	Weight of Agent* (pounds)	UL B:C	Pounds Per Unit B Rating*
Bromotrifluoromethane	1301	4.0	5	0.8
Bromochlorodifluoromethane	1211	2.5	5	0.5
Halon Mixture	1301/1211	0.9	1	0.9
Carbon Dioxide	—	5.0	5	1.0

* one qt = 4 lb; one lb = 0.45 kg

quite dramatic. Rapid and complete extinguishment is obtained with low concentrations of agent. On a worldwide basis the development of systems using Halon 1301 and Halon 1211, together with the development of realistic test methods which have been applied to both these agents and recognized by the NFPA, has shown that in total flooding application for flame extinguishment or inerting, Halon 1301 requires an average of ten percent less material on a gas volume basis than does Halon 1211 for any given fuel. Table 19-2D shows a comparison of flame extinguishment values included in NFPA 12A and NFPA 12B. It is generally recognized that on a weight of agent basis, both agents are approximately two and one-half times more effective than carbon dioxide.

TABLE 19-2D. Comparison of Flame Extinguishment Values for Halon 1301 and Halon 1211

Fuel	Average Percent by Volume of Agent in Air Required for Flame Extinguishment: Halon 1301	Halon 1211
Methane	3.1	3.5
Propane	4.3	4.8
n-Heptane	4.1	4.1
Ethylene	6.8	7.2
Benzene	3.3	2.9
Ethanol	3.8	4.2
Acetone	3.3	3.6

Note: Design flame extinguishment concentrations for Halon 1301 and 1211 total flooding systems are calculated from the tested value by adding a 20 percent safety factor. However, they are never less than 5 percent.

The flame extinguishing ability of Halon 2402 vapors is very good and quite similar to that for Halon 1301 and Halon 1211. However, because it is a liquid at room temperature and is intended mainly for local application situations, its performance is not directly comparable to the data obtained for Halon 1301 and Halon 1211 with total flooding in mind.

The effectiveness of halogenated agents on Class A fires is less predictable. It depends to a large extent upon the specific burning material, its configuration, and how early in the combustion cycle the agent is applied. Most plastics behave as flammable liquids—they can be extinguished rapidly and completely with 4 to 6 percent concentrations of Halon 1211 or Halon 1301. Other materials, particularly cellulosic products, can in certain forms develop deep seated fires in addition to flaming combustion. The flaming portion of such fires can be extinguished with low 4 to 6 percent concentrations of agent, but the deep seated portion may continue glowing under some circumstances (Ford 1962). Even so, the deep seated fire will be controlled in that its rate of burning and consequent heat release will be reduced. Considerably higher agent concentrations (18 percent to 30 percent) are required to achieve complete extinguishment, and these levels are seldom economical to apply. However, the concept of controlling deep seated fires with halogenated agents has been accepted in the respective NFPA standards.

TOXIC AND IRRITANT EFFECTS

The toxicology of Halon 1301, Halon 1211, and Halon 2402 has been studied extensively in both animals and humans. As a result, safety guidelines for these agents can be written.

Early animal studies by the U.S. Army Chemical Center, summarized in Table 19-2E, determined the Approximate Lethal Concentration (ALC) for 15 minute exposures to Halon 1301, Halon 1211 and Halon 2402 to be 83 percent, 32 percent, and 13 percent by volume, respectively. The ALC for decomposed vapors of these agents is also shown (NFPA 1954).

Animals exposed to Halon concentrations below lethal levels exhibit two distinct types of toxic effects: (1) central nervous system changes, such as tremors, convulsions, lethargy, and unconsciousness at high airborne concentrations (above 30 percent by volume for Halon 1301 and 10 percent for Halon 1211); and (2) cardiovascular effects including hypotension, decreased heart rate, and occasional cardiac arrhythmia (lack of rhythm in the heartbeat). Effects are transitory and disappear rapidly after exposure.

The inhalation of many halocarbons and hydrocarbons can make the heart abnormally sensitive to elevated adrenaline levels, resulting in cardiac arrhythmia and possibly death. This phenomenon has been referred to as cardiac sensitization. The halon compounds can also sensitize the heart. In the case of Halon 1301, this occurs only at high exposure levels. For example, in standard cardiac sensitization screening studies in dogs using 5 min exposures and large doses of injected adrenaline, the threshold for sensitization is in the 7.5 to 10 percent range (duPont 1969). The sensitization level for Halon 1211 is 1 to 2 percent and for Halon 2402 is 1,000 to 2,500 ppm (duPont 1970; Beck et al 1973).

Human exposures to both Halon 1301 and Halon 1211 have shown that Halon 1301 concentrations up to 7 percent by volume, and Halon 1211 concentrations of 2 to 3 percent by volume have little noticeable effect on the subject. At Halon 1301 concentrations between 7 and 10 percent and Halon 1211 concentrations between 3 and 4 percent, subjects experienced dizziness and tingling of the extremities, indicating mild anesthesia. At Halon 1301 concentrations above 10 percent and Halon 1211 concentrations above 4 to 5 percent, the dizziness becomes pronounced, the subjects feel as if they will lose consciousness (although none have), and physical and mental dexterity is reduced (Hine 1968; duPont 1966a; duPont 1966b). In human subjects exposed to 7 percent Halon 1301 for periods up to 30 minutes, the effects appeared within the first 5 to 10 minutes of exposure, remained constant throughout the remainder of the exposure, then

TABLE 19-2E. Approximate Lethal Concentrations for 15 min Exposure to Vapors of Various Fire Extinguishing Agents*

Agent	Formula	Halon No.	Approximate Lethal Concentration in Parts Per Million	
			Natural Vapor	Decomposed Vapor
Bromotrifluoromethane	CF_3Br	1301	832,000	14,000†
Bromochlorodifluoromethane	CF_2ClBr	1211	324,000	7,650
Carbon Dioxide	CO_2	—	658,000	658,000
Dibromodifluoromethane	CF_2Br_2	1202	54,000	1,850
Bromochloromethane	CH_2ClBr	1011	65,000	4,000
Dibromotetrafluoroethane	$C_2F_4Br_2$	2402	126,000	1,600‡
Carbon tetrachloride	CCl_4	104	28,000	300
Methyl bromide	CH_3Br	1001	5,900	9,600

* Based on tests with white rats by the Medical Laboratories, U.S. Army Chemical Center.

† Subsequent tests by Kettering Laboratory of the University of Cincinnati (unpublished data) with a commercial Halon 1301 of improved quality indicated that the lethal concentration of decomposed vapor is at least 20,000 ppm. Other tests have given the ALC (Approximate Lethal Concentration) value of Halon 1301 decomposition products as low as 2,500 ppm. The variance is based on differing analytical procedures.

‡ This figure does not agree with manufacturer's data on their product.

disappeared quickly after exposures were stopped (Stewart et al 1978). When used as intended, no significant adverse health effects have been reported from the use of Halon 1301 or Halon 1211 as a fire extinguishant since their introduction into the marketplace 30 years ago.

Halon 1301 has been tested for effects on pregnant rats during the critical stages of gestation. Results showed this agent was not embryotoxic or teratogenic. In addition, Halon 1301 was not mutagenic when tested in the salmonella/microsome assay (Ames test).

A mutagen is a chemical that causes changes in the DNA structure of a cell. A standardized test, called the Ames test or assay, performed on a special strain of salmonella bacteria, provides a preliminary indication that a chemical may be a mutagen and possibly a carcinogen in humans. A high proportion of known carcinogens do exhibit mutagenic activity in the Ames assay. Halon 1301 concentrations up to 40 percent by volume did not exhibit mutagenic activity in this test (duPont 1976).

A teratogen is a chemical that causes permanent structural or functional alteration of the normal processes of fetal development. Pregnant female rats were exposed (by inhalation) to five percent Halon 1301 by volume. No evidence of teratogenic or embryotoxic (growth retardation or death of the fetus) effects was seen in their offspring (duPont 1978).

From the extensive medical data available, the following exposure guidelines have been produced for use of Halon 1301, Halon 1211 and Halon 2402. (See Table 19-2F.)

NFPA 12A permits Halon 1301 design concentration up to 10 percent in normally occupied areas, and up to 15 percent in areas not normally occupied. Because the required extinguishing concentration of Halon 1211 is near or above its limit for safe exposures, Halon 1211 systems are not recognized in NFPA 12B for use in normally occupied areas.

NFPA 12C-T sets acceptable concentrations for human exposure to Halon 2402 vapors at 0.05 to 0.10 percent by volume. Additionally, the standard recognizes only local application systems which may be outdoors with appropriate personnel safeguards, or indoors in unoccupied areas.

DECOMPOSITION PRODUCTS OF HALONS

Consideration of life safety during use of halogenated agents also must include the effects of breakdown products, which have a relatively higher toxicity to humans. Decomposition of halogenated agents takes place on exposure to flame, or to surface temperatures above approximately 900°F (482°C). In the presence of available hydrogen (from water vapor or the combustion process itself), the main decomposition products of Halon 1301 are hydrogen fluoride (HF), hydrogen bromide (HBr), and free bromine (Br_2). Although small amounts of carbonyl halides (COF_2, $COBr_2$) were reported in early tests, more recent studies have failed to confirm the presence of these compounds. The decomposition products of Halon 1211 and Halon 2402 are similar, but in the case of Halon 1211 include hydrogen chloride (HCl) and free chlorine (Cl_2) as well.

The approximate lethal concentrations for 15 minute exposures to some of these compounds are given in Column 1 of Table 19-2G. Column 2 gives the concentrations of these materials that have been quoted by Sax as "dangerous" for short exposures (Sax 1984).

Even in minute concentrations of only a few parts per million, the decomposition products of the halogenated

TABLE 19-2F. Permitted Exposure Times to Halon 1301, Halon 1211 and Halon 2402

	Concentration Percent by Volume	Permitted Time of Exposure
Halon 1301	Up to 7	15 min
	7–10	1 min
	10–15	30 sec
	Above 15	Prevent exposure
Halon 1211	Up to 4	5 min
	4–5	1 min
	Above 5	Prevent exposure
Halon 2402	0.05	10 min
	0.10	1 min

agents have characteristically sharp, acrid odors. This characteristic provides a built in warning system for the agent, and at the same time creates a noxious, irritating atmosphere for those who must enter the hazard area following a fire. It also serves as a warning that other potentially toxic products of combustion (such as carbon monoxide) will be present.

It is necessary to establish the concentrations likely to be encountered when extinguishing fires with halogenated agents and to relate them not only to the absolute toxicity of the agent, but also to the toxic effects of the normal products of combustion.

The expected concentration of decomposition products depends upon many factors (i.e., size of room, size of fire, presence of large quantities of hot surfaces, and elapsed time before extinguishment). Test results, which may help to put the situation into perspective, have been made for both Halon 1301 and Halon 1211 in total flooding and portable applications. As an example of total flooding, Halon 1301 was used to extinguish 0.1 ft^2, 1.0 ft^2 and 10 ft^2 (0.009 m^2, 0.09 m^2 and 0.9 m^2) n-heptane fires in a 1,695 cu ft (48 m^3) test enclosure, at agent discharge times of 15, 10, and 5 seconds. The results are shown in Table 19-2H, relating the resulting decomposition products to the ratio of fuel area to enclosure volume and flame extinguishment time (Ford 1972b). It is obviously advantageous to attack the fire at an early stage while it is still small, and to release the extinguishing agent as rapidly as practical.

For a similar example in portable extinguishers, bromochlorodifluoromethane (Halon 1211) was used in 3 lb (1.36 kg) portable units to extinguish a 2.5 sq ft (0.23 m^2) n-heptane fire in a 2,500 cu ft (70.8 m^3) room. The results of two tests, one normal extinguishment in 1 second, and the second deliberately extended extinguishment in 10 seconds, are given in Table 19-2I. In the case of normal extinguishment, the levels of breakdown products are below the limits allowed by Occupational Safety and Health Administration (OSHA) regulations for 8 hour continuous exposure (NFPA 1972).

The extensive studies on the toxicological effects of halogenated agents indicate that little, if any, risk is attached to the use of Halon 1301 or Halon 1211 when used in accordance with provisions of NFPA standards governing use of these agents.

TABLE 19-2G. Approximate Lethal Concentrations for Predominant Halon 1301 and Halon 1211 Decomposition Products

Compound	ALC for 15-min Exposure ppm by Volume in Air	Dangerous Concentrations ppm by Volume in Air*
Hydrogen Fluoride, HF	2500	50–250
Hydrogen Bromide, HBr	4750	—
Hydrogen Chloride, HCl	—	—
Bromine, Br_2	550	—
Chlorine, Cl_2	—	50
Carbonyl Fluoride, COF_2	1500	—
Carbonyl Chloride, $COCl_2$	100–150	—
Carbonyl Bromide, $COBr_2$	—	—

* (Sax 1984)

TABLE 19-2H. Halon 1301 Decomposition Produced by n-Heptane Fires†

Enclosure volume: 1695 cu ft (48 m^3)
Halon 1301 concentration: 4% by volume

Fuel Surface Area Area Per Unit Enclosure Volume ft^2/ft^3 or m^2/m^3	Flame Extinction Time, Seconds	Total Decomposition Products* ppm
0.06	11.5–15.4	4.5–5.6
0.06	7.1–7.6	2.8–4.2
0.06	4.0–4.8	3.3–4.5
0.6	20–37	94–289
0.6	11.5–13.5	64–284
0.6	4.7–6.7	11.5–169
6.0	20–22	2252–2304
6.0	13.0–16.3	1292–1590
6.0	5.2–10.0	358–778

* Sum of HF, HBr and Br_2.
† (Edmonds 1972)

TABLE 19-2I. Concentrations of Breakdown Products Obtained from Extinguishing 2.5 sq ft (0.23 m^2) n-Heptane Fires with Halon 1211 Portable Extinguishers

Concentration of Breakdown Products—ppm by Volume

Compounds	Test I Normal Extinction In One Second	Test II Prolonged Extinction In Ten Seconds
HCl + HBr	2	50
HF	0.5	10
Cl_2 + Br_2	Not detected*	2.5
$CoCl_2$	Not detected†	Not detected†

* Limit of detection 0.1 ppm.
† Limit of detection 0.25 ppm.

APPLICATION SYSTEMS

A system differs from a portable or mobile appliance primarily in that the agent discharge stream is not directed by a person. The discharge stream or pattern is usually determined in advance, as is either the quantity of agent or discharge rate, or both, and the number and types of nozzles provided. A system consists of a supply of agent, a means for releasing or propelling the agent from its container, and one or more discharge nozzles to apply the agent into the hazard or directly onto the burning object. A system may also contain other elements, such as one or more detectors, remote and local alarms, a piping network, mechanical and electrical interlocks to close fire doors, and shut down ventilation, directional control valves, installed reserve agent supplies, etc. The extent of the auxiliary functions of a system is usually dependent upon the nature of the hazard, in keeping with the desires and resources of the user.

A system is usually a fixed or stationary apparatus, but some portable or mobile systems have been designed. A portable or mobile system may be moved from one hazard to a similar one, but it is placed in a stationary position while it is in service. In this regard, all systems may be considered to be "fixed," but not necessarily permanently placed.

Halogenated agent systems are broadly classified by their method of applying agent to the hazard. The two main types recognized in NFPA 12A and NFPA 12B are total flooding and local application systems. Another category, termed "specialized systems" here, includes those systems which are designed to protect special or unique hazards, and which have been tested and approved under these specific conditions.

Total Flooding Systems

These systems protect enclosed, or at least partially enclosed, hazards. A sufficient quantity of extinguishing agent is discharged into the enclosure to provide a uniform fire extinguishing concentration of agent throughout the entire enclosure. Examples of total flooding systems using halogenated agents are found in computer rooms, electrical switchgear rooms, magnetic tape storage vaults, and electronic control rooms; storage areas for art work, books, and stamps; aerosol filling rooms; machinery spaces in ships; cargo areas in large transport aircraft; processing and storage areas for paints, solvents, and other flammable liquids, etc. Halon 1301, by virtue of its lower toxicity, higher volatility, and lower molecular weight, offers particular advantages for use in total flooding systems. Halon 1211 total flooding systems, with suitable safeguards, are used outside the U.S.

Total flooding systems may be further distinguished by their method of design or installation. An engineered system is custom designed for a particular hazard, using components that are approved or listed only for their broad performance characteristics. Components may be arranged into an almost unlimited variety of configurations. (See Fig. 19-2A.)

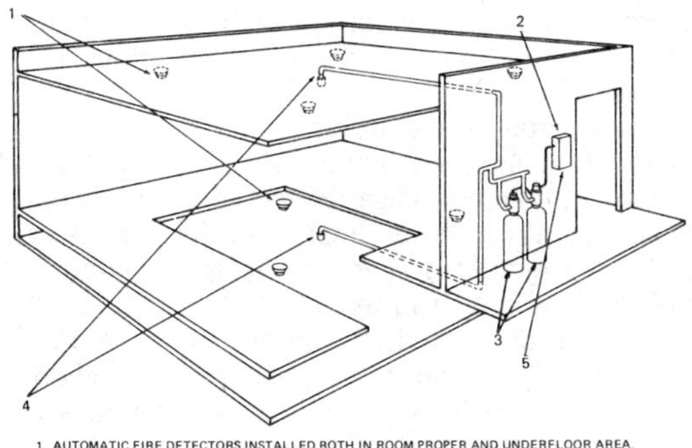
1
2
3
4
5
1. AUTOMATIC FIRE DETECTORS INSTALLED BOTH IN ROOM PROPER AND UNDERFLOOR AREA.
2. CONTROL PANEL CONNECTED BETWEEN FIRE DETECTORS AND CYLINDER RELEASE VALVES.
3. STORAGE CONTAINERS FOR ROOM PROPER AND UNDERFLOOR AREA.
4. DISCHARGE NOZZLES INSTALLED BOTH IN ROOM PROPER AND UNDERFLOOR AREA.
5. CONTROL PANEL MAY ALSO SOUND ALARMS, CLOSE DOORS, AND SHUT OFF POWER TO THE AREA.

FIG. 19-2A. A total flooding system installed in a room with a raised floor. (The Ansul Company)

In pre-engineered systems the number of components and configurations are determined in advance and included in the description of the system's approval or listing. While the degree of pre-engineering can differ from one system to another, the following limits of components and configurations must be considered:

1. Maximum number of cylinders per manifold.
2. Maximum and minimum size and length of piping.
3. Maximum and minimum size and number of elbows, tees, and discharge nozzles.
4. Container volume, fill density, and level of nitrogen superpressurization.

Modular systems consist of single containers connected to discharge nozzles, with minimal piping. A group of container-nozzle assemblies may be distributed throughout the protected area and interconnected electrically. Modular systems permit a more attractive and less expensive initial installation, but often result in higher maintenance costs. Figure 19-2B illustrates the installation of a typical modular system.

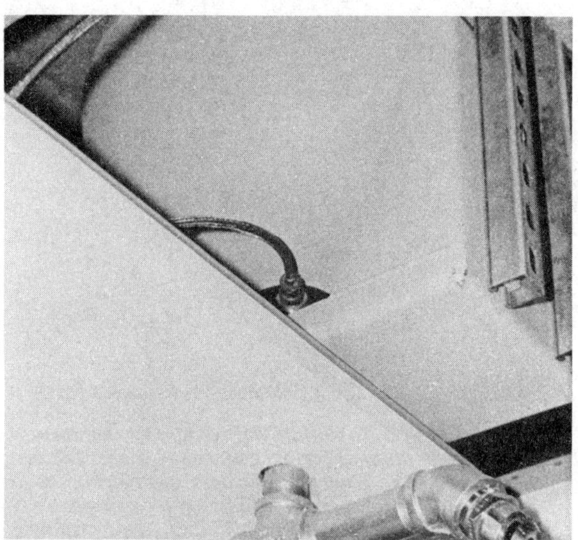

FIG. 19-2B. A modular Halon 1301 installation in a computer room. The spherical container holds 105 lb (47.6 kg) of Halon 1301 and is mounted above a false ceiling. Only a short length of pipe protrudes into the room, and it connects to two discharge nozzles at lower left and lower right. The mounting bracket of container is visible at upper right, and braided cable in the center is the electrical connector to release valve on container outlet. Six identical container assemblies are installed in this room; they are interconnected to discharge simultaneously.

Central storage systems locate all agent containers in one centralized location. The agent can be stored in multiple tanks connected by a manifold arrangement (Fig. 19-2C) or in a single bulk storage tank. A piping network distributes the agent to the various discharge nozzles. Ease of maintenance is the primary advantage of central storage systems.

Local Application Systems

As the name implies, these systems discharge extinguishing agent in such a manner that the burning object is surrounded locally by a high concentration of agent to extinguish the fire. In local application systems, neither the quantity of agent nor the type or arrangement of discharge nozzles is sufficient to achieve total flooding of the enclosure containing the object. Often, too, a local application system is required because the enclosure itself may not be suitable to provide total flooding. Examples of areas protected by local application are printing presses,

FIG. 19-2C. A Halon 1301 centralized storage system installed in the control room of an electrical generating plant. The large manifold has two 90 lb (40.8 kg) cylinders and discharges into the room proper. The smaller manifolds, each composed of two 56 lb (25.4 kg) cylinders, protect two individual underfloor areas. Control panels are visible at upper center and upper right. Cylinder valves are activated pneumatically through small hose connected to top center of valve. The large hose connected to the side of the valve is a Halon 1301 outlet.

dip and quench tanks, spray booths, oil filled electric transformers, vapor vents, etc. (See Fig. 19-2D.)

Because of its lower volatility, Halon 1211 is well suited for local application systems. The lower volatility, plus a high liquid density, permit the agent to be sprayed as a liquid and thus propelled into the fire zone to a greater extent than is possible with other gaseous agents. Halon 2402 also is well suited for local application systems, but its major usage has been outside the U.S.

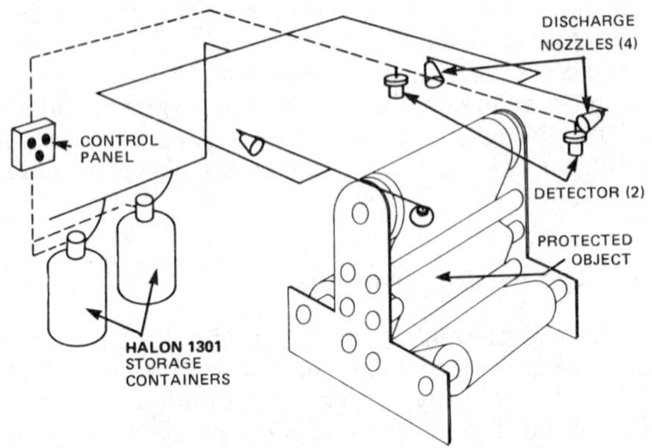

FIG. 19-2D. A local application system. The protected object is not enclosed, therefore, discharge nozzles and rate of application must be capable of enveloping the object. The agent supply must be sufficient to maintain flow for required time of protection, usually several minutes. Nozzle design is critical and must be determined by extensive testing.

The material relating to local application systems in NFPA 12A and NFPA 12B is largely theoretical, and intended for use by equipment manufacturers and testing laboratories in designing and evaluating components for local application systems. The component of overriding importance is the discharge nozzle, and its performance characteristics must be known with a high degree of reliability. As of 1984, no local application nozzles for Halon 1301 or Halon 1211 had been listed or approved by the testing laboratories.

Specialized Systems

These systems, using both Halon 1301 and Halon 1211, are widely used throughout the world. Systems to protect aircraft engine nacelles, racing cars, military vehicles, emergency generator motors, etc., all fall into this category. The distinguishing characteristic of a specialized system is that it can be applied only to the specific hazard for which it was designed and tested. For example, a system designed to protect the jet engines of a Boeing 757 aircraft cannot be applied, per se, to a M-1 battle tank. For each such application, the system is developed by a comprehensive test program.

Systems protecting racing cars are somewhat less specialized, although they properly fit into this category. Sanctioning bodies such as the *Federation International de l'Automobile*, the United States Automobile Club, and the National Hot Rod Association have formulated general rules governing these systems. One study showed a wide variety of sizes and performance characteristics of commercially available Halon 1301 systems available in the U.S. (Curry 1973). Virtually none of these systems is listed or approved by testing laboratories. In spite of this, they have been credited with saving a large number of lives in the racing field.

Another type of specialized system is a "thermatic" unit. This unit consists of a container connected directly to a thermally actuated release device which is similar to an automatic sprinkler. When heat from the fire activates the valve, the extinguishing agent is released to attack the fire. Sizes of thermatic units range from a 1½ lb (0.68 kg) Halon 1301 unit to protect engine compartments of small pleasure boats to 150 lb (68 kg) units containing Halon 1301 and Halon 1211 designed for total flooding of small rooms.

DESIGN CONSIDERATIONS

This section highlights many of the aspects and requirements contained in NFPA 12A and NFPA 12B.

Uses and Limitations

Halogenated agent systems are generally considered useful for the following types of hazards:

1. Where a clean agent is required.
2. Where live electrical or electronic circuits exist.
3. For flammable liquids or flammable gases.
4. For surface burning flammable solids such as thermoplastics.
5. Where the hazard contains a process or objects of high

value, and where use of other extinguishing agents could cause extensive damage or downtime.

6. Where the area is normally occupied by personnel (use of Halon 1301 only).
7. Where availability of water, or space for systems using other agents, is limited.

There are several types of flammable materials on which halogenated agents are ineffective. These are:

1. Fuels that contain their own oxidizing agent such as gun powder, rocket propellants, cellulose nitrate, organic peroxides, etc.
2. Reactive metals, such as sodium, potassium, NaK eutectic alloy, magnesium, titanium, and zirconium.
3. Metal hydrides such as lithium hydride.
4. Chemicals capable of autothermal decomposition, such as organic peroxides and hydrazine.

In the first category—in which the compound contains its own oxygen supply, often built into the fuel molecule—the halogenated agent is unable to penetrate into the reaction zone quickly enough to put out the fire. The oxidizer is in too close physical proximity to the fuel to permit interaction with the extinguishing agent. Often the only effective agent for such fuels is a water deluge to dilute the fuel and attempt to remove heat ahead of the combustion front. In the second category, the reactive metals and metal hydrides are too reactive at flame temperatures for the halogenated agent to operate effectively. Also, the flame chemistry of metal fires is quite different from that of hydrocarbon fires (Wharry and Hirst 1974).

A more commonly encountered limitation of the capabilities of an agent is its limited effectiveness on deep seated Class A fires at concentrations below ten percent by volume. When the concept of control is applied to using halogenated agent systems on Class A fires, rapid response by outside help is necessary. Otherwise, a reflash potential will exist once the extinguishing concentration is dissipated.

Safety

While both Halon 1301 and Halon 1211 have low vapor toxicity, there are hazards in exposing personnel to high concentrations of either agent (above ten percent Halon 1301 or above four percent Halon 1211). Further, the inhalation hazard produced by the fire itself, such as heat, smoke, oxygen depletion, and toxic combustion and decomposition products, may be substantial. Several general safety precautions are listed in the appendices of NFPA 12A, NFPA 12B, and NFPA 12C-T.

Detection and Actuation

Halogenated agent systems generally have operation requirements similar to systems using other types of agents. However, there are strong recommendations in NFPA 12A and NFPA 12B for the use of automatically actuated systems. The primary reason behind these recommendations is to limit the size and severity of fire with which the system must deal, thus minimizing decomposition of the agent during extinguishment. In a Class A fire, automatic actuation, coupled with sensitive detectors, often can prevent the fire from becoming deep seated.

The detector portion of the system is therefore of paramount importance in system design. The detector must be sensitive enough to the fuel in question to rapidly respond to the fire at an early stage. But too great a sensitivity may produce false actuations, thus imposing an economic burden on the owner, causing unnecessary outage of fire protection, and creating a lack of confidence in the system. Designers have successfully combined sensitivity and reliability by utilizing multiple detectors connected in a double circuit or "cross-zoned" mode of operation. (See Fig. 19-2E.) Actuation of either zone alone

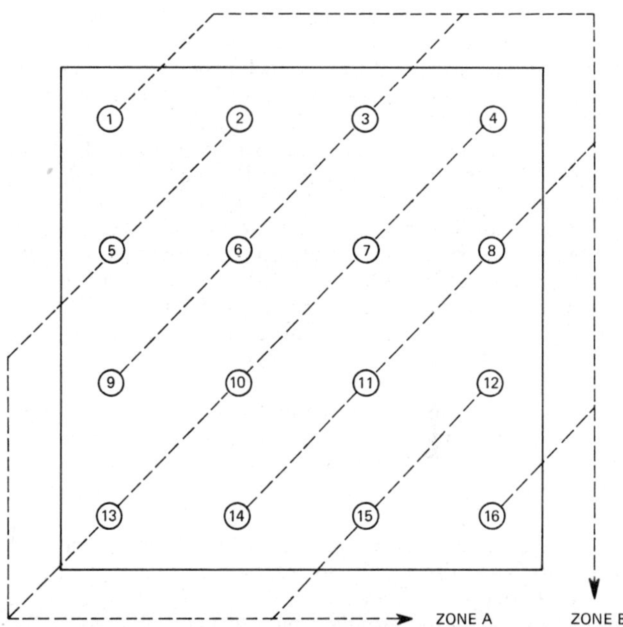

FIG. 19-2E. A cross-zoned detector circuit. Sixteen detectors in a 4 by 4 array are connected so that adjacent detectors are on different circuits, or "zones." For example, consider detector No. 7, connected to Zone A. Adjacent detectors, Nos. 3, 6, 8, and 11 are connected to Zone B.

activates local and remote alarms, but does not discharge the extinguishing agent. Actuation of both zones simultaneously causes the agent to discharge. This arrangement prevents a false signal by an individual detector from discharging the entire system. The two zones may contain the same or different types of detectors. Ionization and photoelectric detectors are popularly used in computer and EDP areas.

In another arrangement, appropriate for generally attended locations, one zone of ionization detection provides early warning, and one zone of rate compensated thermal detection releases the extinguishing agent (Grabowski 1972).

Particular attention must be given to locating detectors within a hazard. The amount and location of combustibles, the listed sensitivity spacing relationship of the detectors considered, and ventilation characteristics of the hazard are all important factors.

Agent Supply

The relatively high cost of agents and the specialized nature of systems using them dictate that a specific supply of agent be provided to protect against a given hazard or set of hazards. Conventionally, the agent is contained in one or more pressurized vessels which are installed near or within the protected area. Steel pressure cylinders or spheres containing from a few pounds to several hundred pounds of agent (1 lb equals 0.45 kg) are commonly used. Larger "bulk" tanks containing several tons (1 ton equals 907 kg) of agent are used in Europe and Canada and have been proposed in the U.S. as well.

Extra pressurization of storage containers with nitrogen is required by NFPA 12A and NFPA 12B. For Halon 1211, this extra pressurization is necessary to expel the contents of the storage container at temperatures below the atmospheric boiling point of 25°F (−4°C) and at a practical rate at higher temperatures. Two levels of extra pressurization for systems are recognized in NFPA 12B: 150 psig (1034 kPa) and 360 psig (2482 kPa) at total pressure at 70°F (21°C). Although the vapor pressure of Halon 1301 at all reasonable temperatures above −30°F (−34°C) is sufficient to completely expel the contents of the container, extra pressurization is used to flatten the pressure versus temperature characteristics during storage, provide a higher average pressure in the storage container during discharge, and permit the use of smaller piping. Two levels of extra pressurization of Halon 1301 are permitted by NFPA 12A: 360 psig (2482 kPa), and 600 psig (4137 kPa) total pressure at 70°F (21°C).

The fill density of a container is defined as the weight of agent divided by the internal volume of the container. NFPA 12A and NFPA 12B permit maximum fill densities of 70 pcf (1121 kg/m^3) for Halon 1301, and 102 pcf (1634 kg/m^3) for Halon 1211. These limits are set by Department of Transportation (DOT) regulations for shipping containers, which require that the container not become liquid-full at temperatures below 130°F (54°C). The maximum fill densities permitted in the standards coincide with the liquid densities of the two agents at 130°F (54°C) in the presence of nitrogen at the highest level of extra pressurization, again allowing for a 5 percent maximum error in filling the container. Most manufacturers utilize a variable fill density in their design to obtain maximum economy in the system. There is no lower limit on permissible fill densities in NFPA 12A and NFPA 12B.

Flow Characteristics

The flow of Halon 1301 and Halon 1211 through pipe and tubing has been determined experimentally. For Halon 1301, the flow is two phase liquid-vapor. Although the agent exists as a single phase liquid in the storage container, it begins to vaporize as soon as its pressure drops during flow. By the time the agent reaches the discharge nozzle, this has substantially reduced agent density. Therefore, classical single phase flow equations of the Moody or Fanning type cannot be used to predict the frictional pressure losses during flow of Halon 1301.

There are two methods in NFPA 12A to estimate flow characteristics. One is a simplified chart method to be used for balanced piping systems only. (A balanced system is one in which the actual pipeline lengths, the equivalent pipeline lengths, and the flow rates to each nozzle are equal.) The other is a rigorous two phase equation which may be used either for balanced or unbalanced piping systems. The chart method can be applied to complex balanced systems using hand calculations, but even relatively simple unbalanced systems calculated by the two phase equation require a computer solution. Most U.S. manufacturers of Halon 1301 systems have developed computer programs to help perform these flow calculations.

Because the solubility of nitrogen in the liquid phase of Halon 1211 is only one-half as great as for Halon 1301, the flow properties of Halon 1211 are treated as single phase liquid flow in NFPA 12B.

Total Flooding Systems

The great majority of Halon 1301 and Halon 1211 systems throughout the world are of the total flooding type. As described earlier, a total flooding system is one which develops a uniform extinguishing concentration of agent throughout an enclosure, such as a room. This system is capable of extinguishing a fire within the enclosure, regardless of the location of the fire. Wickham outlines the main design elements of a total flooding system as follows (Wickham 1972):

1. Define the hazard, which includes the dimensions and configuration of the enclosure, the maximum and minimum net volumes, the fuels involved, the expected temperature range in the hazard area, ventilation and unclosable openings, and occupancy status.
2. Establish a minimum design concentration based upon the fuels involved.
3. Calculate the minimum agent quantity based upon the minimum design concentration, the maximum net volume, the minimum expected temperature of the enclosure, and compensation for losses from ventilation and unclosable openings.
4. Calculate the maximum possible concentration which could occur if the hazard is at the conditions of minimum net volume and maximum temperature. This concentration must not be greater than that permitted by NFPA 12A and NFPA 12B for the occupancy status of the hazard.
5. Select the agent storage container, based upon the design quantity of the agent and the sizes of standard containers available from the equipment manufacturers.
6. Determine the minimum agent flow rate by dividing the design agent quantity by the maximum permissible discharge time. This time is currently set by NFPA 12A and NFPA 12B at 10 seconds, but is subject to some discretion.
7. Determine the size of piping, considering the location of the agent storage containers and locations of the discharge nozzles. This step must be performed concurrently with the following Step 8.
8. Determine the number, size, and locations of discharge nozzles. Discharge rate and nozzle area coverage data must be obtained from laboratory listings or from design manuals of the equipment manufacturers.

Jensen regards the first of the preceding elements as the single most important (Jensen 1972). Certainly it is the most important one in which the end user can usually participate. The remainder of the system is generally designed by the equipment manufacturer.

In establishing the minimum design concentration (Step 2 above), one must consider the conditions under which the fuels are used. For flammable liquids or vapors where an explosion is unlikely, a flame extinguishment concentration may be used. If an explosion potential is considered likely, a higher inerting concentration must be used. An inerting concentration is sufficient to prevent combustion of any fuel-air ratio which might occur in the enclosure. It therefore must be applied upon development of a dangerous condition, and not after ignition has occurred. Minimum design concentrations for both flame extinguishment and inertion for some flammable liquids and gases are included in NFPA 12A and NFPA 12B. Guidelines for determining which level should be used are also given in these standards.

For combustible solids or Class A fires, the possibility and consequences of developing a deep seated fire must be considered. Rapid detection and prompt application of agent also can help prevent a Class A fire from becoming deep seated. Remote alarms to summon outside assistance should be an integral part of a fire protection system.

Enclosures with fresh air make-up ducts and exhaust ducts to the outside must have a means of self closure at the time of agent discharge. It is not necessary to shut down closed loop ventilation systems in which all the exhaust air is returned to the room. In fact, continued operation of such systems will improve agent distribution and reduce the rate at which agent is lost through unclosable openings. Losses through unclosable openings are more insidious in that they often are undetected in a casual inspection of a plan of the enclosure. Because of the high density of the vapors of halogenated agents, there is a definite tendency of agent-air mixtures to find their way out of openings, particularly ones in the lower portion of the enclosure. As this mixture leaves the enclosure, it is replaced by fresh air which enters and collects near the top of the enclosure. This often gives the impression that the agent is separating itself from the air, which it is not. Also, the tendency frequently is to compensate for such losses by providing a higher concentration initially. This, of course, simply aggravates the situation by increasing the rate at which the loss occurs. An extended agent discharge or provision for mechanical mixing during the soaking period are the only effective ways to overcome a serious loss of agent through unclosable openings. Fortunately, since the fire hazard in most applications is located in the lower half of the room, some agent losses in this manner can be tolerated.

Local Application Systems

Design methods for local application systems for both Halon 1301 and Halon 1211 have not yet become well established. The information currently appearing in the NFPA halon standards on local application systems is useful only as a guide to the equipment manufacturer or testing laboratory.

The principle of local application differs greatly from total flooding. Local application is similar to a portable fire extinguisher in that the discharge nozzle is directed at the surface or object on which the fire is anticipated. For this reason, the single most important element in a local application system is the discharge nozzle. The discharge velocity and rate must be sufficient to penetrate the flames and produce extinguishment, but not so great as to cause splashing of fuel and thus increase the fire hazard. The performance of the discharge nozzle must be determined in advance by laboratory testing; the location, position, and orientation of nozzles in the installation must be in strict accordance with these listings. To date, there is no listed or approved equipment for local application systems for either Halon 1301 or Halon 1211 in the U.S.

TESTING AND MAINTENANCE SYSTEMS

Following installation, a system must first be tested to ensure that it will perform in accordance with its design, and thereafter it must be maintained to ensure its continued performance at some later time. Although minimum requirements for testing and maintenance of halogenated agent systems are given in the NFPA halon standards, some elaboration on both topics seems in order.

Testing of Halogenated Agent Systems

While the performance of individual components comprising the system has been determined by laboratory testing and listings, certain tests must be made on the complete installation to ensure that it has been installed properly and that as a whole the system performs according to design. The extent to which tests are conducted on a completed system depends in part upon the complexity of the system, the novelty of the application, the number of unknown variables which might be present in the hazard, and the experience of individual personnel involved. The testing required may range from a checkout of the detection and actuation circuit to a full scale discharge test.

The first stage of a system test is a thorough visual inspection of the system and the hazard. Nameplate data on the storage containers should be compared to those specified by the design. The piping, operational equipment, and discharge nozzles should be inspected for proper size and location. The nozzles should be free of oily contamination. The locations of alarms and manual emergency releases should be confirmed. The configuration of the hazard should be compared to the original hazard specification. The hazard should also be inspected closely for unclosable openings and sources of agent loss which may have been overlooked in the original specification. Any serious deviation of these factors from design must be corrected before conducting any actual test.

The second stage of testing a system is to check the operation of the detection and actuation circuits with the agent release mechanism disconnected. The performance of detectors, local and remote alarms, and interlocks to shut down processes or ventilating equipment is determined to see if they will be suitable. Manual emergency releases are also checked in this stage. Interlock switches, which are operated from agent pressure in the piping system, may be checked at this stage by tripping them manually. Compressed air may be connected to the piping system and the piping blown out to be sure it is free from obstructions. In fact, everything short of the discharging agent is tested. Often, these two stages of testing are sufficient.

The third stage of testing a system, if required, is to conduct a full scale discharge test. Because it is a somewhat special test, it should be conducted only after careful planning. In the first place, a full scale test is expensive, even if a substitute agent is used. Secondly, the time and personnel necessary to perform a full scale discharge test justify sufficient planning to make sure that all objectives of the test are accomplished.

A full scale discharge test is usually required when conditions in the hazard are such that the agent discharge time, design concentration, distribution of concentration, or maintenance of concentration during a required soaking time are in doubt. Further, these are the only items that should be tested in a discharge test. Other items as described above should be tested separately and in advance of the discharge test. Placing a tape recorder near a discharge nozzle is helpful in accurately determining the discharge time of liquid from the nozzle. Monitoring agent concentrations at several locations within the hazard (three sampling points are sufficient for most enclosures) with continuous recording analyzers provide a satisfactory means of measuring initial agent concentrations, distribution, and maintenance of concentration.

Maintenance of Halogenated Agent Systems

NFPA 12A and NFPA 12B specify certain items that must be checked at semiannual and annual intervals. Semiannually, the system should be visually inspected for evidence of corrosion or other damage, and the storage containers checked for loss of agent. This latter item involves a two fold check. First, the pressure corrected for temperature should be measured to ensure no loss of pressurizing gas, and second, each container must be weighed to determine loss of agent. Neither check alone is sufficient; both are required. Liquid level indicators are sometimes used to indicate the quantity of agent, rather than weighing, which is often cumbersome and time consuming.

At least annually, the operational characteristics of the system should be retested. This generally involves repeating Stages 1 and 2 as previously outlined. A full scale discharge test is rarely required as a part of an annual inspection. Both the semiannual and annual inspections should be performed by knowledgeable and qualified personnel. Inspection reports should be filed with the owner of the system. Needless to say, all impediments found in these inspections must be corrected promptly.

Bibliography

References Cited

Beck, Phillippa S., Clark, D. G., Tinston, D. J. 1973. "The Pharmacologic Actions of Bromochlorodifluoromethane (BCF)." *Toxicology and Applied Pharmacology*. pp 24, 20-29.

Curry, Thomas H. 1973. "Halon 1301 Protects Racing Cars." *Fire Journal*. Vol 67, No 2, Mar 1973.

Dowling, John, and Ford, C. 1969. "Halon 1301 Total Flooding System for Winterthur Museum." *Fire Journal*. Vol 63, No 6.

duPont. 1966a. duPont Haskell Laboratory. Unpublished data.

duPont. 1966b. duPont Haskell Laboratory. Unpublished data.

duPont. 1969. duPont Haskell Laboratory. Unpublished data.

duPont. 1970. duPont Haskell Laboratory. Unpublished data.

duPont. 1976. duPont Haskell Laboratory. Unpublished data.

duPont. 1978. duPont Haskell Laboratory. Unpublished data.

Ford, Charles. 1962. "Overview of Halon 1301 Systems." *Symposium on the Mechanism of Halogenated Extinguishing Agents*. ACS Symposia Series.

Ford, Charles. 1972a. *Halon 1301 Computer Fire Test Program—Interim Report*. E. I. duPont de Nemours & Co. Inc., Wilmington, DE.

Ford, Charles. 1972b. *Extinguishment of Surface and Deep-Seated Fires with Halon 1301*. An Appraisal of Halogenated Fire Extinguishing Agents. National Academy of Sciences, Washington, DC.

Grabowski, George J. 1972. *Fire Detection and Actuation Devices for Halon Extinguishing Systems*. An Appraisal of Halogenated Fire Extinguishing Agents. National Academy of Sciences, Washington, DC.

Hine. 1968. *Clinical Toxicologic Studies on Freon® FE 1301*. Report No. 1. The Hine Laboratories, Inc., San Francisco, CA. Unpublished.

Hough, Ralph. 1960. *Determination of a Standard Extinguishing Agent for Airborne Fixed Systems*. WADD Technical Report 60-552. Wright Air Development Division. Wright-Patterson, Air Force Base, OH.

Jensen, Rolf. 1972. "Halogenated Extinguishing Agent Systems." *Fire Journal*. Vol 66, No 3. May 1972. pp 37-39.

NFPA 1954. "The Halogenated Extinguishing Agents." *NFPA Quarterly*. Vol 48, No 8. Part 3. Quincy, MA.

NFPA. 1972. *Guide to OSHA Fire Protection Regulations*. Vol 1. 2nd ed. National Fire Protection Association, Quincy, MA, pp 22140-22142.

Sax, N. 1984. *Dangerous Properties of Industrial Materials*. 6th ed. Van Nostrand, Reinhold, NY.

Stewart, Richard D., Newton, Paul E., Wu, Anthony, Hake, Carl L., and Krivanek, Neil D., 1978. "Human Exposure to Halon 1301." Medical College of Wisconsin, Milwaukee, WI. Unpublished, 1978.

Strasiak, Raymond. 1954. *The Development of Bromochloromethane (CB)*. WADC Technical Report 53-279. Wright Air Development Center, OH.

Wharry, David, and Hirst, Ronald. 1974. *Fire Technology: Chemistry and Combustion*. Institution of Fire Engineers, Leicester, UK.

Wickham, Robert T. 1972. *Engineering and Economic Aspects of Halon Extinguishing Equipment*. An Appraisal of Halogenated Fire Extinguishing Agents. National Academy of Sciences, Washington, DC.

NFPA Codes, Standards, Recommended Practices and Manuals. (See the latest *NFPA Codes and Standards Catalog* for availability of current editions of the following documents.)

NFPA 10, *Standard for Portable Fire Extinguishers*.

NFPA 12A, *Standard on Halon 1301 Fire Extinguishing Systems*.

NFPA 12B, *Standard on Halon 1211 Fire Extinguishing Systems*.

NFPA 12C-T, *Standard on Halon 2402 Extinguishing Systems*.

NFPA 75, *Standard for the Protection of Electronic Computer/Data Processing Equipment*.

Additional Readings

Atallah, S., and Buccigross, H. L., "Extinction of Fire by Halogenated Compounds—A Suggested Mechanism," *Fire Technology*, Vol. 8, No. 2. May 1972. pp. 131-141.

Atallah, S., and Buccigross, H. L., "Testing the Performance of Halon 1301 on Real Computer Installations," *Fire Journal*, Vol. 66, No. 5. Sept. 1972. pp 105-108.

Bischoff, B. G., "Gaseous Extinguishing Agents," *Heating, Ventilation and Air Conditioning*, Oct. 1978.

Breen, David E., "Interactions in Binary Halon Mixtures Used as Fire Suppressants," *Fire Technology*, Vol. 13, No. 4, Nov. 1977, pp. 261-265, 281.

Brenneman, James J., and Charney, Marvin, "Testing a Total Flooding Halon 1301 System in a Computer Installation," *Fire Journal*, Vol. 68, No. 6 Nov. 1974.

Chemetron Corp., "Halon System Testing," *Hot Spots*, Cardox Products, Chicago, IL, 1974.

Clarke, D. G., *The Toxicity of Bromotrifluoromethane (FE 1301) in Animals and Man*, Industrial Hygiene Research Laboratory, Imperial Chemical Industries, Alderley Park, Chesire, UK, 1970.

Coll, John P., "Inerting Characteristics of Halon 1301 and 1211 with Various Combustibles," *Report PSR 661*, Fenwall, Inc., Ashland, MA, July 16, 1976.

Creitz, E. C., "Inhibition of Diffusion Flames by Methyl Bromide and Trifluoromethyl Bromide Applied to the Fuel and Oxygen Sides of the Reaction Zones," *Journal of Research of National Business Standards*, Vol. 65, No. 4, 1961.

"Designing Halon Systems," *Plumbing Engineer*, Vol. 13, No. 2, Mar./Apr. 1985, pp. 27-31.

Dyer, J. H. et al., "The Extinction of Fires in Aircraft Jet Engines—(Parts I through IV)," *Fire Technology*, Vol. 12, No. 4, Nov. 1976, pp. 266-275, 289; Vol. 13, No. 1, Feb. 1977, pp. 59-68; Vol. 13, No. 2, May 1977, pp. 126-138; Vol. 13, No. 3, Aug. 1977, pp. 223-230.

Echternacht, John E., "Halon Extinguishing Systems Design Criteria," *Fire Journal*, Vol. 65, No. 6, Nov. 1971, pp. 51-55, 66.

Edmonds, Albert, "Use of Halon 1211 in Hand Extinguishers and Local Application Systems," *An Appraisal of Halogenated Fire Extinguishing Agents*, National Academy of Sciences, Washington, DC, 1972.

"Flow of Nitrogen-Pressurized Halon 1301 in Fire Extinguishing Systems," *JPL Pub. 84-62*, Jet Propulsion Laboratory, California Institute of Technology, Pasadena, CA, 1984.

Ford, Charles L., "Halon 1301 Fire Extinguishing Agent: Properties and Applications," *Fire Journal*, Vol. 64, No. 6, Nov. 1970.

Ford, Charles L., "Halon 1301 Update: Research, Application, New Standard," *Specifying Engineer*, May, 1977.

Ford, Charles L., "Where and Why to Use Halon 1301 Systems," *Specifying Engineer*, Jan. 1972.

Franck, Thomas E., "Clean Room Protection Using Halon 1301," *Fire Journal*, Vol. 65, No. 2, Mar. 1971, pp. 77-79.

Gaskill, J. R., Leonhart, E. C., and Sanborn, E. N., *A Halon 1301 Fire Extinguishing System for Trailers*, Lawrence Radiation Laboratory, Livermore, CA, Aug. 19, 1969.

Gassman, Julius J., and Hill, Richard G., *Fire Extinguishing Methods for New Passenger-Cargo Aircraft*, National Aviation Facilities Experimental Center, Atlantic City, NJ, 1971.

Gassman, Julius J., and March, John F., "Application of Halon 1301 to Aircraft Cabin and Cargo Fires," *An Appraisal of Halogenated Fire Extinguishing Agents*, National Academy of Sciences, Washington, DC, 1972.

Hammack, James M., "More About Halon 1301," *Fire Journal*, Vol. 66, No. 4 July 1972, pp. 43-44.

McDaniel, Dale E., "Evaluation of Halon 1301 for Shipboard Use," *An Appraisal of Halogenated Fire Extinguishing Agents*, National Academy of Sciences, Washington, DC, 1972.

National Academy of Sciences, "An Appraisal of Halogenated Fire Extinguishing Agents," *Proceedings of a Symposium*, 1972.

Paulet, G., "Etude toxicologique et physiopathologique du monobromotrifluoromethane (CF_3Br)." *Arch. Mal. Prof. Med. Trav. Secur. Soc.* 23:341-348. (Chem. Abstr. 60:7358e), 1962.

Peterson, Parker E., "A Systems Approach to Optimum Damage Control," *Fire Journal*, Vol. 67, No. 2, Mar. 1973, pp. 70-73.

Poeschl, Paul M., "Large-Scale Halon 1301 Fire Test Program," *Fire Journal*, Vol. 67, No. 6, Nov. 1973, pp. 35-38.

Robinson, Victor B., "Partial Flooding of Volumes with Halon 1301," *Fire Technology*, Vol. 14, No. 2, May 1978, pp. 97-109.

Serb, Thomas J., "The Magic of Halon Protects Computers," *Security World*, Oct. 1983.

Sheehan, Daniel F., *An Investigation into the Effectiveness of Halon 1301 (Bromotrifluoromethane, $CBrF_3$) as an Extinguishing Agent for Shipboard, Machinery Space Fires*, U.S. Coast Guard, Washington, DC, 1972.

Steinberg, Marshall, "Toxic Hazards from Extinguishing Gasoline Fires Using Halon 1301 Extinguishers in Armored Personnel Carriers," *An Appraisal of Halogenated Fire Extinguishing Agents*, National Academy of Sciences, Washington, DC, 1972.

Strasser, A., Liebman, I., and Kuchta, J. M. "Methane Flame Extinguishment with Layered Halon or Carbon Dioxide," *Fire Technology*, Vol. 10, No. 1, Feb. 1974, pp. 25-34.

Van Stee, E. W., and Back, K. C., "Short-term Inhalation Exposure to Bromotrifluoromethane," *Tox. & Appl. Pharm.* 15:164-174, 1969.

Wiersman, Steve J., "Flow Characteristics of Halon 1301 in Pipelines," *Fire Technology*, Vol. 14, No. 1, Feb. 1978, pp. 5-14.

Williamson, H. V., "Halon 1301—Minimum Concentrations for Extinguishing Deep-Seated Fires," *Fire Technology*, Vol. 8, No. 4, Nov. 1972.

Williamson, H. V., "Halon 1301 Flow in Pipelines," *Fire Technology*, Vol. 11, No. 4, Nov. 1975, pp. 18-32.

Wilson, Rexford, "That Extinguishing Thing (Firepac)," *Fire Journal*, Vol. 64, No. 1, Jan. 1970.

DRY CHEMICAL AGENTS AND APPLICATION SYSTEMS

Revised by Walter Haessler

Dry chemical is a powder mixture which is used as a fire extinguishing agent. It is intended for application by means of portable extinguishers, hand hose line systems, or fixed systems. Borax and sodium bicarbonate based dry chemical were the first such agents developed. Sodium bicarbonate became the standard because of its greater effectiveness as a fire extinguishing agent. About 1960, sodium bicarbonate base dry chemical was modified to render it compatible with protein based low expansion foams to permit a dual agent attack. Multipurpose (monoammonium phosphate base) and "Purple-K" (potassium bicarbonate base) dry chemicals then were developed for fire extinguishing use. Shortly thereafter, "Super-K" (potassium chloride base) was developed to equal "Purple-K" in effectiveness. In the late 1960s, the British developed urea-potassium bicarbonate base dry chemical. Presently, there are five basic varieties of dry chemical extinguishing agents.

This chapter contains information on the properties of dry chemicals affecting their use as extinguishing agents, their uses and limitations, methods of storage and handling, and quality control. Portable dry chemical extinguishers are discussed in Section 20 of this HANDBOOK. Dry powder extinguishing agents for combustible metal fires are discussed in Chapter 5 of this section, "Combustible Metal Agents and Application Techniques," and fire and explosion control are discussed in Section 4, Chapter 4, "Theory of Fire and Explosion Control."

The terms "regular dry chemical" and "ordinary dry chemical" generally refer to powders that are listed for use on Class B and Class C fires (UL annually). "Multipurpose dry chemical" refers to powders that are listed for use on Class A, B, and C fires. The terms "regular dry chemical," "ordinary dry chemical," and "multipurpose dry chemical" should not be confused with "dry powder" or "dry compound," which are used to identify powdered extinguishing agents developed primarily for use on combustible metal fires (FMEC annually).

Mr. Haessler is a consultant based in Santa Rosa, CA. He is a member of the NFPA Technical Committee on Dry and Wet Chemical Extinguishing Systems. His activities also include teaching and writing in the field of fire protection.

Dry chemical is unusually efficient in extinguishing fires in flammable liquids. It can also be used on fires involving some types of electrical equipment. Regular dry chemical has certain limited applications in extinguishment of flash surface fires with ordinary combustibles, but the chemical requires water to put out deep seated smoldering fires. Multipurpose dry chemical can be used on fires in flammable liquids, fires involving energized electrical equipment, and fires in ordinary combustible materials. Multipurpose dry chemical seldom needs the help of water to completely extinguish fires in Class A materials.

PHYSICAL PROPERTIES OF DRY CHEMICAL

The principal base chemicals used in the production of currently available dry chemical extinguishing agents are sodium bicarbonate, potassium bicarbonate, potassium chloride, urea-potassium bicarbonate, and monoammonium phosphate. Various additives are mixed with these base materials to improve their storage, flow, and water repellency characteristics. The most commonly used additives are metallic stearates, tricalcium phosphate, or silicones, which coat the particles of dry chemical to make them free flowing and resistant to the caking effects of moisture and vibration.

Stability

Dry chemical is stable at both low and normal temperatures. However, since some of the additives may melt and cause sticking at higher temperatures, an upper storage temperature limit of 120°F (49°C) is recommended for dry chemical. Up to 150°F (66°C) may be acceptable for very short durations. At fire temperatures, the active ingredients either disassociate or decompose while performing their function in fire extinguishment. Of extreme importance is the danger caused by indiscriminate mixing of the various dry chemicals. For example, mixing multipurpose (monoammonium base) dry chemical, which is acidic, with an alkaline dry chemical (most of the other dry chemicals), will result in an undesirable reaction that releases free carbon dioxide gas and causes caking. Extinguisher shells have been known to explode because of this phenomenon.

Toxicity

The ingredients presently used in dry chemical are nontoxic. However, the discharge of large quantities may cause temporary breathing difficulty during and immediately after discharge and may seriously interfere with visibility.

Particle Size

Particles of dry chemical range in size from less than 10 microns up to 75 microns (1 micron = 0.000039 in.). Particle size has a definite effect on extinguishing efficiency, and careful control is necessary to prevent particles from exceeding the upper and lower limits of this performance range. The best results are obtained by a heterogeneous mixture with a median particle in the order of 20 to 25 microns. Underwriters' Laboratories (UL) and Factory Mutual (FM) listings and ratings are based upon the specified use of the particular type of dry chemical set forth by the equipment manufacturer. In no case are different dry chemicals to be mixed in recharging the equipment.

EXTINGUISHING PROPERTIES

Fire tests on flammable liquids have shown potassium bicarbonate base dry chemical to be more effective than sodium bicarbonate base dry chemical in extinguishment. Similarly, monoammonium phosphate has been found equal to or better than sodium bicarbonate in extinguishment effectiveness (Guise 1962). The effectiveness of potassium chloride is about equivalent to potassium bicarbonate, and urea-potassium bicarbonate exhibits the greatest effectiveness of all the dry chemicals tested.

When introduced directly to the fire area, dry chemical causes the flame to go out almost at once. Smothering, cooling, and radiation shielding contribute to the extinguishing efficiency of dry chemical, but studies suggest that a chain-breaking reaction in the flame is the principal cause of extinguishment (Haessler 1974).

Smothering Action

For many years it was widely held that regular dry chemical extinguishing properties relied primarily on the smothering action of the carbon dioxide released when sodium bicarbonate was heated by fire. The carbon dioxide does undoubtedly contribute to the effectiveness of dry chemical, as does the like volume of water vapor released when dry chemical is heated. However, tests have generally disproved the belief that these gases are a major factor in extinguishment.

When multipurpose dry chemical is discharged into burning ordinary combustibles, the decomposed monoammonium phosphate leaves a sticky residue (metaphosphoric acid) on the burning material. This residue seals glowing material from oxygen, thus helping to extinguish the fire and prevent reignition.

Cooling Action

It cannot be substantiated that the cooling action of dry chemical is an important reason for its ability to promptly extinguish fires. More information on this subject is contained in a paper based on studies of the heat capacities of various powders tested for extinguishing effectiveness. The heat energy required to decompose dry chemicals plays an undeniable role in contributing toward their individual extinguishing abilities, but the effect, per se, is minor. To be effective, any dry chemical must be heat sensitive and, as such, absorbs heat in order to become chemically active (McCamy et al 1956).

Radiation Shielding

Discharge of dry chemical produces a cloud of powder between the flame and the fuel; this cloud shields the fuel from some of the heat radiated by the flame. Tests to evaluate this factor concluded that the shielding factor is of some significance (McCamy et al 1956).

Chain-Breaking Reaction

The preceding extinguishing actions, each to a certain degree, contribute to the extinguishing action of dry chemical. However, studies reveal that still another factor is present, which makes an even greater contribution than that of the other factors combined.

The chain reaction theory of combustion has been advanced by some investigators to provide the clue to the identity of this unknown extinguishing factor. This theory assumes that free radicals are present in the combustion zone and that the reactions of these particles with each other are necessary for continued burning. The discharge of dry chemical into the flames prevents reactive particles from coming together and continuing the combustion chain reaction. The explanation is referred to as the chain-breaking mechanism of extinguishment (Guise 1960; Haessler 1962).

USES AND LIMITATIONS

Dry chemical is primarily used to extinguish flammable liquid fires. Because it is electrically nonconductive, it can also be used on flammable liquid fires involving live electrical equipment. Regular dry chemical extinguishers have been tested and found suitable for use on flammable liquid and electrical fires (Class B and C fires) by fire equipment testing laboratories.

Due to the rapidity with which dry chemical extinguishes flame, dry chemical is used on surface fires involving ordinary combustible materials (Class A fires). There are several areas in the textile industry, notably opener-picker rooms and carding rooms in cotton mills, where regular dry chemical has been used effectively. However, wherever regular dry chemical is provided for use on surface type Class A fires, it should be supplemented by water spray for extinguishing smoldering embers or in case the fire gets beneath the surface. In some baled cotton storage areas, the tops of bales can be covered with regular dry chemical to prevent surface spread should fire break out. This preventive measure does not eliminate the need for automatic sprinkler protection in such areas. Since multipurpose dry chemical becomes sticky when heated, it is not recommended for textile card rooms or other locations where removal of the residue from fine machine parts may be difficult.

Dry chemical does not produce a lasting inert atmosphere above the surface of a flammable liquid; consequently, its use will not result in permanent extinguishment if reignition sources, such as hot metal surfaces or persistent electrical arcing are present.

Dry chemical should not be used in installations where relays and delicate electrical contacts are located (e.g., in telephone exchanges and computer equipment rooms), since the insulating properties of dry chemical might render such equipment inoperative. Because some dry chemicals are slightly corrosive, they should be removed from all undamaged surfaces as soon as possible after fire extinguishment.

Regular dry chemical will not extinguish fires that penetrate beneath the surface, or fires in materials that supply their own oxygen for combustion. Dry chemical may be incompatible with mechanical (air) foam unless the dry chemical has been specially prepared to be reasonably foam compatible.

Specifications have been established by fire equipment testing laboratories to assure the positive and consistent performance of dry chemical as an extinguishing agent. These specifications control moisture content, water repellency, electrical resistivity, storage at elevated temperatures, flow capability, caking resistivity, and abrasive action. The discharge characteristics of the device in which the dry chemical is to be used are also evaluated. Extinguishing effectiveness is determined by performance tests of the application to standard fires under conditions recommended by the manufacturer.

Though dry chemical had been used for many years in fire extinguishers, it was not until 1954 that the first dry chemical extinguishing system was tested and listed by a testing laboratory. In 1952 the NFPA Committee on Dry Chemical Extinguishing Systems was established, and in 1957 the first edition of NFPA 17, *Standard for Dry Chemical Extinguishing Systems*, was adopted.

DRY CHEMICAL EXTINGUISHING SYSTEMS

Dry chemical extinguishing systems can be used in those situations where quick extinguishment is desired and where reignition sources are not present. Dry chemical systems are used primarily for flammable liquid fire hazards such as dip tanks, flammable liquid storage rooms, and areas where flammable liquid spills may occur. Systems have been designed for kitchen range hoods, ducts, and associated rangetop hazards such as deep fat fryers. Where it is necessary to extinguish a flammable liquid or gas fire being fed by fuel under pressure, dry chemical hand hose line systems can be used, followed by closure of fuel shutoff valves.

Since dry chemical is electrically nonconductive, extinguishing systems using this agent can be used on electrical equipment that is subject to flammable liquid fires such as oil filled transformers and oil filled circuit breakers. Dry chemical system protection is not recommended, however, for delicate electrical equipment, such as telephone switchboards and electronic computers. Such equipment is subject to damage by dry chemical deposit and, because of the insulating properties of the dry chemical, may require excessive cleaning to restore operation.

Hand hose line systems containing regular or ordinary dry chemical have been used to a limited extent for quick spreading surface fires on ordinary combustible material. In such applications, the dry chemical system only stops or prevents a rapid surface spread, and must be supplemented by a water type extinguishing device to put out deep seated smoldering fires. Fixed systems containing multipurpose dry chemical are available and are suitable for the protection of ordinary combustibles, provided the dry chemical can reach all burning surfaces.

METHODS OF APPLICATION

The two basic types of dry chemical systems are referred to as fixed systems and hand hose line systems. Other methods of applying dry chemical are by portable and wheeled type extinguishers.

Fixed Systems

Fixed dry chemical systems consist of a supply of dry chemical, an expellant gas, an actuating method, fixed piping, and nozzles through which the dry chemical can be discharged into the hazard area. Fixed dry chemical systems are of two types: total flooding and local application.

In total flooding, a predetermined amount of dry chemical is discharged through fixed piping and nozzles into an enclosed space or enclosure around the hazard. (See Fig. 19-3A.) Total flooding is applicable only when the hazard is totally enclosed or when all openings surrounding a hazard can be closed automatically when the system is discharged. Total flooding can be used only where no reignition is anticipated because the extinguishing action is transient.

FIG. 19-3A. This 15 by 24 ft (4.6 by 7.3 m) flammable liquids storage building is protected by a total flooding dry chemical system. Upon actuation of the system, by a heat detector, nitrogen is discharged into the 150 lb (68 kg) storage container and dry chemical is expelled through eight nozzles in the roof.

Local application differs from total flooding in that the nozzles are arranged to discharge directly into the fire. Local application is practical in those situations where the hazard can be isolated from other hazards so that fire will not spread beyond the area protected, and where the entire hazard can be protected. The principal use of local application systems is to protect open tanks of flammable liquids. As with total flooding systems, local application is ineffective unless extinguishment can be immediate and there are no reignition sources.

Hand Hose Line Systems

Hand hose line systems consist of a supply of dry chemical and expellant gas with one or more hand hose lines to deliver the dry chemical to the fire. (See Fig. 19-3B.) The hose stations are connected to the dry chemical container either directly or indirectly by means of intermediate piping. They can provide a large quantity of extinguishing agent for quick knockdown and extinguishment of relatively large fires such as might be experienced at gasoline loading racks, flammable liquid storage areas, diesel and gas turbine locomotives, and aircraft hangars.

FIG. 19-3B. A dry chemical hand hose line system consisting of an expellant gas assembly, dry chemical storage tank assembly, and a discharge assembly. Dry chemical systems of this type range from 125 to 2,000 lb (57 to 907 kg); the 2,000 lb (907 kg) system shown has four nitrogen cylinders and two ¾ in. (19 mm) hoses with shutoff nozzles. (The Ansul Company)

DESIGN OF DRY CHEMICAL SYSTEMS

Usually, dry chemical systems consist of dry chemical and expellant gas storage tanks, piping and/or hose to carry the agent to the fire area, nozzles to assure proper distribution of the agent into the fire or area to be protected, and automatic and/or manual actuating mechanisms. (See Fig. 19-3C.)

Dry chemical systems are called either engineered or pre-engineered depending upon how the quantity of dry chemical, rate of flow, size and length of piping, and number and size of fittings are determined. An engineered system is one in which individual calculation and design is needed to determine the flow rate, nozzle pressures, pipe sizes, quantity of dry chemical, and the number, types, and placement of nozzles for the hazard being protected. A pre-engineered system, sometimes called a package system, is one in which the size of the system (i.e., the quantity of dry chemical, pipe sizes, maximum and minimum pipe lengths, number of fittings, and number and types of nozzles) are all predetermined by fire tests for specific sizes and types of hazards. Installation within these limits of hazard and system design assures adequate flow rate, nozzle pressure, and pattern coverage without individual calculation. (See Fig. 19-3D.)

Pre-engineered systems are very frequently used for kitchen range and hood fire protection, including deep fat fryers. Only alkaline dry chemicals can be used in these cases (sodium bicarbonate, potassium bicarbonate, etc.) in order to saponify fats and oils. Multipurpose dry chemical, monoammonium phosphate, must never be used. (See Fig. 19-3E.)

Storage of Dry Chemical and Expellant

Dry chemical is stored in a pressure container, usually of welded steel construction, either under atmospheric pressure until the system is actuated or under the pressure of the internally stored expellant gas.

Containers in which dry chemical is stored separately under atmospheric pressure are equipped with an expellant gas inlet, a moisture sealed fill opening, and a dry chemical outlet. The gas inlet leads to an internal gas tube arrangement constructed so that when it flows into the tank it agitates and permeates the powder, making it fluidlike. The dry chemical outlet is provided with a rupture disc or valve to permit buildup of proper operating pressure in the tank before the dry chemical can start to flow. The expellant gas assembly consists of a pressure storage vessel together with necessary valves, pressure regulators, and piping to deliver the expellant gas to the dry chemical storage tank at the correct pressure and rate of flow. (See Fig. 19-3A.) The expellant gas is usually nitrogen, but carbon dioxide is used in some of the smaller systems. The volume and storage pressure of the expellant are dictated by the gas used and the requirements of the system. Containers in which dry chemical and the expellant gas are stored together are equipped with a moisture sealed fill opening, a valve with integral discharge outlet and expellant gas charging inlet, and a pressure gage. (See Fig. 19-3E.) The expellant gas is usually nitrogen, but dry air can be used.

It is desirable to locate the dry chemical expellant gas assemblies as near as practicable to the hazard to be protected. An area in which temperatures stay between −40 and +120°F (−40 and +49°C) is desirable to maintain the quality of the dry chemical.

Dry chemical is commercially available in various sized packages of 10 lb (4.5 kg) or more in weight, or in metal drums, plastic lined paperboard containers, or plastic bags. Whatever the container, it should be kept tightly closed and stored in a dry location to prevent absorption of moisture. Storage in a dry location is also essential for extinguisher recharging. Once the powder has lost its free flowing characteristic, it should be discarded.

System Actuation

Actuation of fixed systems is initiated by automatic mechanisms that incorporate sensing devices located in the hazard area and automatic, mechanical, or electrical releases which initiate the flow of dry chemical, actuate alarms, and shut down process equipment. In systems with separate dry chemical and expellant gas containers, the flow of dry chemical is started by releasing the expellant gas, which pressurizes the dry chemical chamber to the point where the rupture disc in the dry chemical outlet

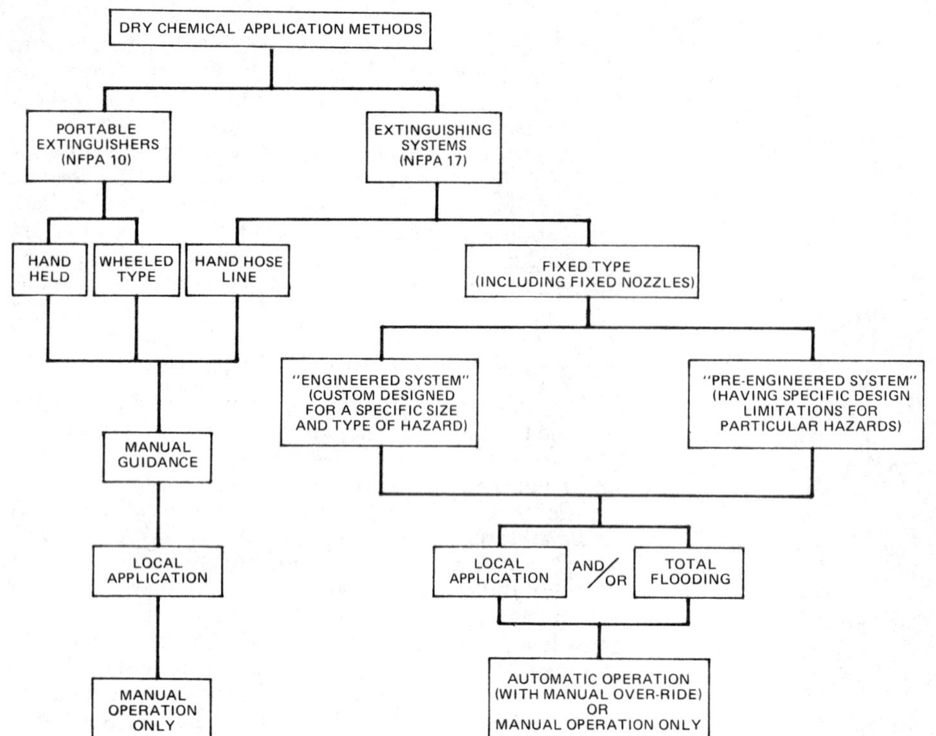

FIG. 19-3C. Methods of dry chemical application.

FIG. 19-3D. A double 30 lb (13.6 kg) stored-pressure dry chemical system with pneumatic release for automatic actuation. Each cylinder contains 30 lb (13.6 kg) of dry chemical pressurized to 350 psi (2413 kPa) with nitrogen.

operates. The dry chemical is then carried by the expellant gas through the distribution system to the hazard. In stored-pressure systems, however, the flow of dry chemical is started by merely opening the valve on the dry chemical chamber. An easily accessible device for manual operation is required for all automatically operated systems.

When automatically actuated systems are used, consideration should be given to the reduced visibility and temporary breathing difficulty sometimes caused by quick discharge of large amounts of dry chemical in a restricted space. In all cases where there is a possibility that personnel may be stationed in such locations, suitable alarms and safeguards should be incorporated in the system to assure adequate warning and prompt evacuation.

Operation of a hand hose line system requires two or, at the most, three steps. Step one: pressurizing the dry chemical chamber by opening the expellant gas valve (if the dry chemical and expellant gas containers are separate), or opening the main discharge valve if dry chemical is under stored pressure. Step two: operating the nozzle at the end of the hose line. Step three: if multiple hose stations are supplied by the same dry chemical supply, a distribution valve must be opened to direct the flow to the particular hose station to be used. It is extremely important to play out the hose fully (without kinks) to ensure the proper chemical discharge rate.

Fixed piping systems normally blow themselves clear. However, in hand hose line systems where the operator exercises control over the amount of dry chemical discharged, it is vital that all pipes and hose be independently blown clear before recharging to prevent blockage.

Distribution System

For fixed systems, the provisions for conveying the dry chemical to the hazard area and discharging it properly consist of piping, or for hand hose line systems, piping and hose lines or hose lines alone. Nozzles designed to emit jets or wide, flat, or round streams in desired patterns are available to meet specific hazard requirements. Adjustable nozzles permitting an operator

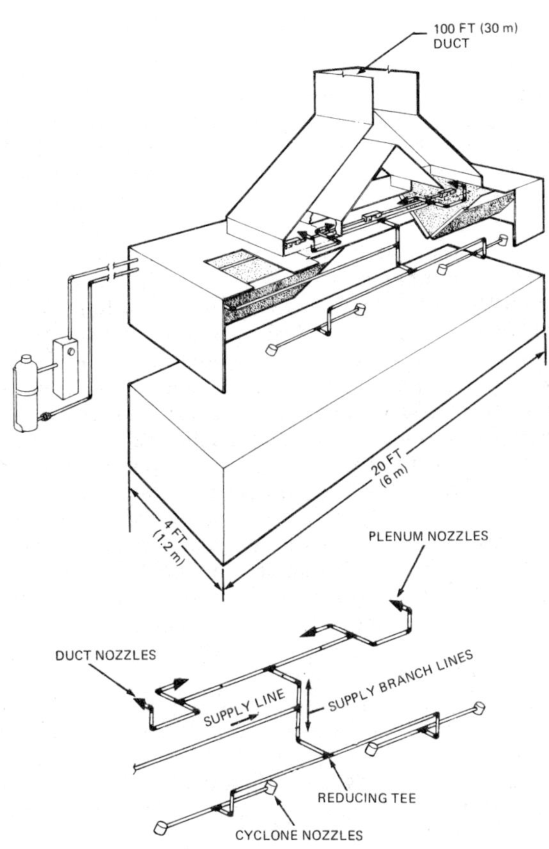

FIG. 19-3E. A typical single 30 lb (13.6 kg) cartridge operated dry chemical system with fusible links for automatic operation, it is pre-engineered for kitchen range, hood, duct, and fryer fire protection. (The Ansul Company)

to vary the range and shape of the discharge are also available for use with hand hose line systems.

The piping and valving for dry chemical systems are of special design because dry chemical, while tending to behave as a fluid, has distinct flow characteristics.

Control of dry chemical flow starts at the dry chemical storage tank. In systems of separate dry chemical and expellant gas containers, expellant gas must be admitted to the tank in such a way as to properly fluidize the dry chemical while the pressure builds up equally throughout the entire volume of the tank before the dry chemical is released from the tank. Should the pressure increase too rapidly above the dry chemical, the powder will not be properly fluidized and, upon release, the pressure drop in the piping or hose line will be excessive and result in a rate of dry chemical flow too low for proper extinguishing effectiveness.

If the expellant gas is permitted to channel to the outlet before the top of the dry chemical storage tank is properly pressurized, insufficient dry chemical will be carried by the flowing stream of gas, and the fire extinguishing effectiveness of the system again will be greatly reduced.

After release from the storage tank, dry chemical is carried through the piping at high velocities by the expellant gas and is thrown to the wall of the piping by centrifugal force whenever the direction of flow is sharply altered, as by an elbow. Should an elbow be directly connected to a tee and lie in the same plane as the tee, it is obvious that the two branches of the tee will carry appreciably different proportions of gas and dry chemical. This condition is overcome by installing all tees in planes perpendicular to the planes of adjacent elbows, by allowing sufficient length of straight pipe between tees and elbows, or by inserting special venturi devices between tees and elbows to insure proper redistribution of dry chemical in the stream of gas before the mixture enters the tee.

Another critical factor in dry chemical distribution system design is the pressure drop through various lengths of pipe, hose, or fittings. Consequently, the length and size of piping and hose, and the number and size of fittings must be selected to provide the required rate of discharge and nozzle distribution. In pre-engineered systems this is already assured by virtue of the limitations established for piping and fittings. In engineered systems, this selection must be made on the basis of actual calculation of pressure drop and subsequent flow rate.

For engineered systems, pressure drop data have been obtained at various rates of flow for various sizes of pipe and fittings. In one manufacturer's system, the pressure drop through a standard elbow is equivalent to the pressure drop through 20 ft (6 m) of straight pipe of the same size (Guise and Lindlof 1955). Pressure drop through a tee is equivalent to that of 45 ft (14 m) of straight pipe of the same size.

Nozzles

The selection of the proper types and sizes of nozzles for fixed systems is necessary to obtain proper coverage of the hazard.

Nozzles for hand hose line systems may be of either a one or two position type. One position nozzles provide a modified straight stream, while two position nozzles provide a straight stream or a fan. Since the straight stream has a longer reach than the fan discharge, it is considered better for initial attack. The shorter, wider fan shaped stream is usually preferred for a close range attack to complete the extinguishment. However, the present state of the art seems to prefer the use of the modified straight stream only. It is more effective and less confusing in the hands of an experienced operator.

Quantity and Rate of Application of Dry Chemical

The quantity of dry chemical and rate of flow must be sufficient to create a fire extinguishing concentration throughout all parts of the enclosure protected by a total flooding system, or over the specific fire area protected by a local application system. It must be realized that the fire extinguishing concentration accomplished throughout an enclosure (total flooding) can occur only during discharge. Following discharge, dry chemical rapidly settles out, and diminishes. This is in contrast to gaseous diffusion, where concentrations persist following discharge such as in carbon dioxide and halon systems. Minimum rate of flow is critical because dry chemical will not extinguish a fire if applied too slowly. In local application systems, application at too high a rate may result in uneven discharge, or complete discharge of dry chemical before extinguishment is accomplished.

Optimum quantitites of dry chemical and application rates have been determined by experiment for engineered total flooding and local application systems. For pre-engineered systems, the maximum sizes of hazards that can be protected by the various sizes of systems within specific piping limitations have been determined.

From tests with one manufacturer's engineered total flooding system, it has been determined that the weight in pounds (kg) of dry chemical required is obtained by multiplying the number of cubic feet (m^3) in the net volume of the open space to be flooded by 0.0385 (in SI 0.616). Net volume is the gross volume less the volume of machinery or other permanently located bulky objects.

The minimum flow rate in pounds per second (kg/sec) is determined by multiplying the net volume in cubic feet (m^3) by 0.00125 [in SI 0.020). These volume factors apply only to this manufacturer's equipment and are cited merely to illustrate a procedure used (Guise 1955). Quantities and flow rates determined in this manner are contingent upon a nozzle arrangement that provides even distribution throughout the volume. The rates also assume that devices will be installed to close automatically all doors, windows, ventilators, or other openings through which dry chemical could escape from the enclosure.

In one manufacturer's pre-engineered total flooding system, a 1,000 cu ft (28 m^3) space not longer than 20 ft (6 m) can be protected by a 30 lb (14 kg) system with 4 nozzles and no more than 90 ft of ¾ in. pipe (27 m of 19 mm pipe), 13 elbows, 3 tees, and 3 venturi devices (Safety First undated). As with engineered systems, successful extinguishment is contingent upon proper nozzle location, shutdown of ventilation, and the closing of all openings such as doors and windows in addition to closure of fuel valves.

Although simultaneous shutdown of ventilation with the actuation of a total flooding system is the normal procedure, one exception exists. This exception is the special case of kitchen range hood and duct protection wherein pre-engineered systems are specifically designed to provide extinguishment, regardless of whether the ventilation is operating or not. Actually, this latter type of system is a combination of both total flooding and local application.

Quantities of dry chemical and flow rates for engineered local application systems are obtained from graphs plotted from tests using different flammable liquid surface areas, weights of dry chemical, and rates of application. These graphs are available from the manufacturer. For pre-engineered local application systems, the limitations of hazard size, system size, and piping arrangement are established by test and used in the same manner as for total flooding systems.

The minimum recommended quantity of dry chemical for hand hose line systems is sufficient to permit use of the system for 30 seconds. Capacities of hand hose line systems range from 125 to 2,000 lb (57 to 900 kg), and as many as 4 hose lines can be operated from a single system. As with fixed systems, a minimum flow rate must be maintained to prevent surging or interruption of flow. Minimum rates are determined by the equipment manufacturer and depend upon the equipment used.

MAINTENANCE, INSPECTION, AND TEST PROCEDURES

In general, all dry chemical systems, including alarms and shutdown devices, should be thoroughly inspected and checked at least annually for proper operation. The frequency of inspection depends upon the type of hazard being protected. Systems subjected to process deposits (such as paint and dust) and corrosive conditions will require more frequent inspection of components.

The amount of expellant gas should be checked semi-annually to ensure that there is sufficient gas to provide an effective discharge, and to automatically clean out the piping after the dry chemical has dissipated. In systems with separate expellant gas containers, this is accomplished by checking the pressure (if nitrogen) or weight (if carbon dioxide) against the manufacturer's recommended minimums. In stored-pressure systems, the pressure gage is checked to see that it is in the operable range.

At least semiannually, the quantity of dry chemical should also be checked either by visual inspection of the powder level in the separate dry chemical chamber, or by weighing the stored-pressure chamber. Except in stored-pressure systems, the dry chemical should be checked annually for any evidence of caking which could subsequently prevent proper flow.

Most dry chemicals are alkaline (sodium and potassium bicarbonate and potassium carbonate). One is neutral (potassium chloride) and the multipurpose type (monoammonium phosphate) is acidic. Inadvertent mixing of different base dry chemicals can initiate undesirable reactions, generating carbon dioxide gas and double decomposition resulting in equipment failure or loss of discharge capability.

During the periodic inspections, the nozzles should be examined to see that they are properly aimed, free of obstruction, and in good operating condition. Nozzles in kitchen range hood and duct protection systems must be fitted with grease seals that are tight, yet easily blown clear. The system actuating devices such as fusible links, pneumatic heat detectors, or electric thermostats should also be checked to ensure that they are not loaded with residues or otherwise impaired.

The inspection and maintenance of hand hose line systems will vary with the location and climate conditions. Equipment located in extremely hot or humid areas will require more frequent checking because the heat can cause cylinder pressure to increase and possibly cause leakage. Inspection of hand hose line systems consists primarily of checking the pressure of the expellant gas container, or the pressure of the unit itself (if it is of the pressurized type). It is also advisable to inspect all hose lines and nozzles to be sure that they are unobstructed and in good operating condition.

Systems containing dry chemical chambers of less than 150 lb (69 kg) nominal capacity must be hydrostatically tested every 12 years. The hydrostatic test should be applied to the auxiliary pressure containers, valve assemblies, hoses, fittings (not including field piping), check valves, directional valves, manifolds, and hose nozzles, in addition to the dry chemical chambers. The dry chemical removed from the chambers should be discarded. Care must be exercised to ensure that all equipment tested is thoroughly dried prior to recharging. If there is no automatic, connected reserve supply, alternate fire protection should be provided.

Bibliography

References Cited

FMEC. annually. *Factory Mutual Approval Guide*, Norwood, MA.

Guise, A. B. 1962. "Potassium Bicarbonate-Base Dry Chemical." *NFPA Quarterly*. Vol 56, No 1. pp 21-27.

Guise, A. B. 1960. "The Chemical Aspects of Fire Extinguishment." *NFPA Quarterly*. Vol 53, No 4. pp 330-336.

Guise, A. B., and Lindlof, J. A. 1955. "A Dry Chemical Extinguishing System." *NFPA Quarterly*. Vol 49, No 1. pp 52-60.

Haessler, W. M. 1974. *The Extinguishment of Fire*. Revised ed. National Fire Protection Association, Quincy, MA.

Haessler, W. M. 1962. "Fire and Its Extinguishment." *NFPA Quarterly*. Vol 56, No 1. pp 89-96.

McCamy, C. S., Shoub, H., and Lee, T. G. 1956. "Fire Extinguishment by Means of Dry Powder." *Sixth Symposium on Combustion*. The Combustion Institute. Van Nostrand, Reinhold, NY. pp 795-801.

Safety First. undated. *Safe-T-Meter Automatic Dry Chemical Fixed System*. Bulletin UP-30M. Safety First Products Corp., Elmsford, NY.

Underwriters Laboratories Inc. annually. *Fire Protection Equipment List*. Northbrook, IL. Published annually with bimonthly supplement.

NFPA Codes, Standards, Recommended Practices and Manuals. (See the latest *NFPA Codes and Standards Catalog* for availability of current editions of the following documents.)

NFPA 10, *Standard for Portable Fire Extinguishers*.

NFPA 17, *Standard for Dry Chemical Extinguishing Systems*.

Additional Readings

Fire Control Engineering Co., *Dry Chemical Fire Extinguishing Characteristics*, expanded 2nd ed., Ft. Worth, TX.

Guise, A. B., "Extinguishment of Natural Gas Pressure Fires," *Fire Technology*, Vol. 3, No. 3, Aug. 1967, pp. 175-193.

Hird, D., and Fippes, D. W., *The Effect of Various Powdered Materials on the Stability of Protein Foams*, British Joint Fire Research Organization, 1960.

Jensen, R. H., "Compatibility of Mechanical Foam and Dry Chemical," *NFPA Quarterly*, Vol. 57, No. 3, Jan. 1964, pp. 296-303.

Lee, T. G., and Robertson, A. F., "Extinguishing Effectiveness of Some Powdered Materials on Hydrocarbon Fires," *Fire Research Abstracts and Reviews*, Vol. 2, No. 1, Jan. 1960.

McGarry, J. E., "Improved Fire Safety for the Indianapolis 500," *Fire Journal*, Vol. 60, No. 2, Mar. 1966, pp. 32-36.

Meldrum, D. N., "Combined Use of Foam and Dry Chemical," *NFPA Quarterly*, Vol. 56, No. 1, July 1962, pp. 28-34.

Stevens, R. E., "Dealing with the Grease Duct Fire Problem," *NFPA Quarterly*, Vol. 56, No. 3. Jan. 1963. pp. 216-220.

Tuve, R. L., "Light Water and Potassium Bicarbonate Dry Chemical—A New Two Agent Extinguishing System," *NFPA Quarterly*, Vol. 58, No. 1, July 1964, pp. 64-69.

Underwriters Laboratories Inc., "The Compatibility Relationship Between Mechanical Foam and Dry Chemical Fire Extinguishing Agents," *UL Bulletin of Research No. 54*, July 1963, Chicago, IL.

Wesson, H. R., "Studies of the Effects of Particle Size on the Flow Characteristics of Dry Chemical," *Fire Technology*, Vol. 8, No. 3, Aug. 1972, pp. 173-180.

Woolhouse, R. A., and Sayers, D. R., "Monnex Compared with Other Potassium-Based Dry Chemicals," *Fire Journal*, Vol. 671, No. 1, Jan. 1973, pp. 85-88.

FOAM EXTINGUISHING AGENTS AND SYSTEMS

Revised by Norman R. Lockwood, P. E.

Fire fighting foam is an aggregate of gas filled bubbles formed from aqueous solutions of specially formulated concentrated liquid foaming agents. The gas used is normally air, but in certain applications may be an inert gas. Since foam is lighter than the aqueous solutions from which it is formed, and lighter than flammable liquids, it floats on all flammable or combustible liquids, producing an air excluding, cooling, continuous layer of vapor sealing, water bearing material that halts or prevents combustion.

Foam is produced by mixing a foam concentrate with water at the appropriate concentration, and then aerating and agitating the solution to form the bubble structure. Some foams are thick and viscous and form tough, heat resistant blankets over burning liquid surfaces and vertical areas; other foams are thinner and spread more rapidly. Some foams are capable of producing a vapor sealing film of surface active water solution on a liquid surface. Some, such as medium or high expansion foam, are meant to be used as large volumes of wet gas cells for inundating surfaces and filling cavities.

Foams are defined by their expansion ratio, which is the ratio of final foam volume to original foam solution volume before adding air. They are arbitrarily subdivided into three ranges: (1) Low Expansion Foam—expansion up to 20:1; (2) Medium Expansion Foam—expansion 20 to 200:1; and (3) High Expansion Foam—expansion 200 to 1,000:1.

There are various methods of generating and applying foams. This chapter covers the basic characteristics of various foaming agents and the methods for producing fire fighting foams, as well as specific applications of equipment and systems useful for different types of hazards.

USES AND LIMITATIONS OF FIRE FIGHTING FOAMS

Low expansion foam is principally used to extinguish burning flammable or combustible liquid spill or tank fires by application to develop a cooling, coherent blanket. Foam is the only permanent extinguishing agent used for fires of this type. Its application allows fire fighters to extinguish fires progressively. A foam blanket covering a tank's liquid surface can prevent vapor transmission for some time, depending upon the stability and depth of the foam. Fuel spills are quickly rendered safe by foam blanketing. The blanket may be removed after a suitable period of time; often it has no detrimental effect on the product with which it comes into contact.

Foams may be used to diminish or halt the generation of flammable vapors from nonburning liquids or solids and may be used to fill cavities or enclosures where toxic or flammable gases may collect.

Foam is of great importance where aircraft are fueled and operated. Sudden large fuel spills resulting from aircraft accidents or malfunction require rapid foam application. Hangar fire protection is best accomplished by foam-water sprinkler systems and portable foam equipment.

Foams of the medium or high expansion type (20 to 1,000 times) may be used to fill enclosures such as basement room areas or holds of ships where fires are difficult or impossible to reach. Here foams act to halt convection and access to air for combustion. Their water content also cools and diminishes oxygen by steam displacement. Foams of this type (with expansion ratios of 400 to 500) may be used to control liquefied natural gas (LNG) spill fires and help disperse the resulting vapor cloud.

Many foams are generated from solutions with very low surface tension and penetration characteristics. Foams of this type are useful where Class A combustible materials are present. In such instances, the water solution draining from the foam cools and wets the solid combustible.

Foam breaks down and vaporizes its water content under attack by heat and flame. It therefore must be applied to a burning liquid surface in sufficient volume and rate to compensate for this loss, with an additional amount applied to guarantee a residual foam layer over the extinguished liquid. Foam is unstable, and may be easily broken down by a physical or mechanical force such as a water hose stream. Certain chemical vapors or fluids may also quickly destroy foam. When certain other extinguishing agents are used in conjunction with foam, severe

Mr. Lockwood is a consultant in fire protection engineering based in Washington Crossing, PA.

breakdown of the foam may occur. Turbulent air or violently uprising combustion gases from fires may divert foam from the burning area.

Foam solutions are conductive and therefore not recommended for use on electrical fires. If foam is used, a spray is less conductive than a straight stream. However, because foam is cohesive and contains materials that allow water to conduct electricity, foam spray is more conductive than water spray.

Engineering design requirements and recommended application methods must be followed for successful use of foams. These requirements can be found in NFPA 11, *Standard for Low Expansion Foam and Combined Agent Systems*; NFPA 11A, *Standard for Medium and High Expansion Foam Systems*; NFPA 11C, *Standard for Mobile Foam Apparatus*; NFPA 16, *Standard for the Installation of Deluge Foam-Water Sprinkler Systems and Foam-Water Spray Systems*, and NFPA 403, *Recommended Practice for Aircraft Rescue and Fire Fighting Services at Airports and Heliports*. (Hereinafter, these standards will be referred to as NFPA 11, NFPA 11A, NFPA 11C, NFPA 16, and NFPA 403 respectively.)

In general, the following criteria for the hazardous liquid must be met for a foam to be fully effective:

1. The liquid must be below its boiling point at the ambient conditions of temperature and pressure.
2. Care must be taken in application of foam to liquids with a bulk temperature higher than 212°F (100°C). At these fuel temperatures and above, foam forms an emulsion of steam, air, and fuel. This may produce a four fold increase in volume when applied to a tank fire, with dangerous frothing or slopover of the burning liquid.
3. The liquid must not be unduly destructive to the foam used, or the foam must not be highly soluble in the liquid to be protected.
4. The liquid must not be water reactive.
5. The fire must be a horizontal surface fire. Three dimensional (falling fuel) or pressure fires cannot be extinguished by foam unless the hazard has a relatively high flashpoint and can be cooled to extinguishment by the water in the foam.

TYPES OF FOAM

There are a number of types of foaming agents available, known as foam concentrates, some of which are designed for specific applications. Some are suitable for extinguishing all types of flammable liquids, including water soluble and foam destructive liquids. Descriptions of the common types of foam follow.

Aqueous Film-forming Foaming Agents (AFFF)

Aqueous film-forming foam agents are composed of synthetically produced materials that form air foams similar to those produced by the protein based materials. In addition, these foaming agents are capable of forming water solution films on the surface of flammable hydrocarbon liquids, hence the term "aqueous film-forming foam" (AFFF). AFFF concentrates are available for proportioning to a final concentration of either one percent, three percent, or six percent by volume with either fresh water or seawater.

The air foams generated from AFFF solutions possess low viscosity, have fast spreading and leveling characteristics, and, like other foams, act as surface barriers to exclude air and halt fuel vaporization. These foams also develop a continuous aqueous layer of solution under the foam maintaining a floating film on hydrocarbon fuel surfaces to help suppress combustible vapors and cool the fuel substrate. This film, which can also spread over fuel surfaces not fully covered with foam, is self-healing following mechanical disruption and continues to spread as long as there remains a reservoir of nearby foam. Film effectiveness may be reduced on hot surface and aromatic hydrocarbons. To ensure fire extinction, an AFFF blanket should entirely cover the fuel surface, as with other types of foam.

The double action of aqueous film-forming foams yields an efficient foam extinguishing agent, in terms of water and concentrate needed and the rapidity with which it acts on hydrocarbon spill fires.

AFFF concentrates contain fluorinated, long chain synthetic hydrocarbons with particular surface active properties. They are nontoxic and biodegradable in their diluted form of use. AFFF concentrate may be stored for long periods of time without degradation of its characteristics.

Because of the extremely low surface tension of the solutions draining from AFFF, these foams may be useful under mixed class fire situations (Class A and B) where deep penetration of water is needed in addition to the surface spreading action of foam itself.

Foam generating devices yielding stable, homogeneous foams are not necessarily needed in the employment of AFFF. Less sophisticated foaming devices such as water spray nozzles and sprinklers (unlike with most other foaming agents) may be used because of the inherent rapid and easy foaming capability of AFFF solutions. However, such foams are relatively rapid draining. AFFF also may be used, without compatibility problems, in conjunction with dry chemical agents . Although AFFF concentrates must not be mixed with other types of foam concentrates, foams made from them do not break down other types of foams in fire fighting operations.

Fluoroprotein Foaming Agents (FP)

The concentrates utilized for generating fluoroprotein foams are similar in composition to protein foam concentrates, but in addition to protein polymers they contain fluorinated surface active agents that confer a "fuel shedding" property to the foam generated. This makes them particularly effective for fire fighting conditions where the foam becomes coated with fuel, such as in the method of subsurface injection of foam for tank fire fighting, and nozzle or monitor foam applications where the foam may often be plunged into the fuel. Fluoroprotein foams are very effective for in-depth crude petroleum or other hydrocarbon fuel fires because of this fuel shedding property. In addition, these foams demonstrate better compatibility with dry chemical agents than do the regular protein type foams. They also possess superior vapor securing and burnback resistance characteristics. Fluoroprotein type concentrates are available for proportioning to a final concentration of either three percent or six percent by volume using either fresh water or seawater. They are nontoxic and biodegradable after dilution. The normal use temperature range for these agents is 20 to 120°F (−7 to 49°C).

Film-Forming Fluoroprotein Agents (FFFP)

Film-forming fluoroprotein agents are composed of protein together with film-forming fluorinated surface active agents, which make them capable of forming water solution films on the surface of flammable liquids, and of conferring a fuel shedding property to the foam generated.

Air foams generated from FFFP solutions have fast spreading and leveling characteristics and, just as other foams, act as surface barriers to exclude air and prevent vaporization. Like AFFF, they also generate a self-healing continuous floating film on hydrocarbon fuel surfaces which helps suppress combustible vapors. However, to ensure fire extinction, an FFFP blanket, as with other types of foam, should cover the entire fuel surface.

Because of the rapid and easy foaming capability of FFFP solutions, water spray devices may be used in many situations. However, the foams produced drain rapidly and should not be relied on for postfire security.

Film forming fluoroprotein type concentrates are available for proportioning to a final concentration of either three or six percent by volume, using either fresh or seawater. They may be used in conjunction with dry chemical agents without compatibility problems, and are nontoxic and biodegradable after dilution.

Protein Foaming Agents (P)

Protein type air foams utilize aqueous liquid concentrates proportioned with water for their generation. These concentrates contain high molecular weight natural proteinaceous polymers derived from a chemical digestion and hydrolysis of natural protein solids. The polymers give elasticity, mechanical strength, and water retention capability to foams generated from them. The concentrates also contain dissolved polyvalent metallic salts which aid the protein polymers in their bubble strengthening capability when the foam is exposed to heat and flame. Protein type concentrates are available for proportioning to a final concentration of either three percent or six percent by volume using either fresh water or seawater. In general, these concentrates produce dense, viscous foams of high stability, high heat resistance, and good resistance to burnback. They are nontoxic and biodegradable after dilution. The normal use ambient temperature range for these concentrates is 20 to 120°F (−7 to 49°C).

Low Temperature Foaming Agents

This type of foam concentrate is protected for storage and use at low temperature by the inclusion of freezing point depressants. Low temperature foaming agents may be used at ambient temperatures as low as −20°F (−29°C). They are available for use at either three percent or six percent by volume concentration in either fresh or seawater, and may be of the AFFF or protein base type.

Alcohol Type Foaming Agents (AR)

Air foams generated from ordinary agents are subject to rapid breakdown and loss of effectiveness when they are used on fires that involve fuels which are water soluble, water miscible, or of a "polar solvent" type. Examples of this type of fuel are alcohols, enamel and lacquer thinners, methyl ethyl ketone, acetone, isopropyl ether, acrylonirile, ethyl and butyl acetate, and the amines and anhydrides. Even small amounts of these substances mixed with common hydrocarbon fuels such as gasohol may cause rapid breakdown of ordinary fire fighting foams.

Certain special foaming agents, called alcohol type concentrates, have therefore been developed. These alcohol resistant concentrates are proprietary compositions of several types, some containing a protein, fluoroprotein, or an aqueous film-forming foam concentrate base. The most common are usually described as polymeric alcohol resistant AFFF concentrates which produce foams suitable for application to spill or in-depth fires of either hydrocarbon or water miscible flammable liquids by any foam generating device. They exhibit AFFF characteristics on hydrocarbons and produce a floating gel like mass for foam buildup on water miscible fuels. Agents of this type have no transit time limitations. Normal use temperatures for any of the alcohol type agents are 35 to 120°F (1.7 to 49°C).

Medium and High Expansion Foaming Agents (SYNDET)

Medium and high expansion foams are agents for control and extinguishment of Class A and some Class B fires and are particularly suited as a flooding agent for use in confined spaces. The foam is an aggregation of bubbles mechanically generated by aspiration or a blower-fan which forces air or other gas through a net or screen that is wetted by an aqueous solution of surface active foaming agents. Under proper conditions, fire fighting foams of expansions from 20 to 1 up to 1,000 to 1 can be generated.

Medium or high expansion foam is a unique vehicle for transporting wet foam masses to inaccessible places, for total flooding of confined spaces, and for volumetric displacement of vapor, heat, and smoke. Tests have shown that when used under certain circumstances in conjunction with water from automatic sprinklers, the foam will provide more positive control and extinguishment than either extinguishing agent by itself. Optimum efficiency in any one type of hazard is dependent upon the rate of application and the foam expansion and stability.

Liquid concentrates for producing medium and high expansion foams consist of synthetic hydrocarbon surfactants of a type that will foam copiously with a small input of turbulent action. They are generally used in approximately two percent proportion in water solution. Medium expansion foam may also be generated from solutions of fluoroprotein, protein, or AFFF concentrates at 3 or 6 percent proportion.

Medium or high expansion foams are particularly suited for indoor fires in confined spaces. Their use outdoors may be limited because of the effects of weather. These foams have several effects on fires:

1. When generated in sufficient volume, they can prevent the air necessary for continued combustion from reaching the fire.
2. When forced into the heat of a fire, the water in the foam is converted to steam, which reduces the oxygen concentration by diluting the air.
3. The conversion of the water to steam absorbs heat from the burning fuel. Any hot object exposed to the foam will continue breaking down the foam, converting the

water to steam, and being further cooled.

4. Because of its relatively low surface tension, foam solution not converted to steam will tend to penetrate Class A materials. However, deep seated fires may require overhaul.
5. When accumulated in depth, high expansion foam can provide an insulating barrier for protection of exposed materials or structures not involved in a fire, thereby preventing fire spread.

Research has shown that using air from inside a burning building to generate high expansion foam has an adverse effect on the volume and stability of the foam produced. Combustion and pyrolysis products, when they react chemically with the foaming agent, can reduce the volume of foam produced and increase the drainage rate. The high temperature of the air breaks down the foam as it is generated. Physical disruption also takes place, apparently caused by vapor and solid particles from the combustion process. These factors that cause foam breakdown may be compensated for by higher rates of foam generation.

Entry to a foam filled passage must not be attempted without use of self-contained breathing apparatus. The foam mass also reduces vision and hearing, and life lines must be used for personnel entering into the foam.

Tests have shown that foam of approximately 500 to 1 expansion can be successfully used for control of fires and reduction of vaporization from LNG spills. As water slowly drains from the foam in small amounts, it forms a thin ice layer which floats on the LNG and supports the high expansion foam blanket. Fixed and portable high expansion foam facilities of this type have been provided for LNG storage and manufacturing plants.

Other Synthetic Hydrocarbon Surfactant Foaming Agents

There are many synthetically produced surface active compounds which foam copiously in water solution. When these are properly formulated, they may be used as wetting agents or as fire fighting foams and employed in much the same manner as other types of foam. (See NFPA 18, *Standard on Wetting Agents.*)

Hydrocarbon surfactant foam liquid concentrates are employed in 1 to 6 percent proportions in water. When these solutions are used in conventional foam making devices, the resulting air foam possesses low viscosity and spreads quickly over liquid surfaces. Its fire fighting characteristics depend upon the volume of the foam layer on the burning surface, which halts access to air and controls combustible vapor production; and the minor cooling effect of the water in the foam, which becomes available due to a relatively rapid breakdown of the foam mass. This water solution does not possess film forming characteristics on the flammable liquid surface, although under some conditions it may produce a temporary water emulsion due to its wetting agent or "detergent type" properties. Because of the low surface tension and wetting properties of the water solutions of these foams, they may also be used as extinguishing agents for Class A fires, although they were primarily developed with medium or high expansion foam in mind.

Synthetic hydrocarbon surfactant foams are generally less stable than other fire fighting foams. Their water solution content drains away rapidly, leaving a bubble mass which is highly vulnerable to heat or mechanical disruption. Usually these foams must be applied at higher rates than other fire fighting foams to achieve extinction. Many formulations of this type of foam concentrate break down other foams if used simultaneously or sequentially.

Chemical Foam Agents and Powders

These foam producing materials have become obsolete because of the superior economics and ease of handling of the liquid foam forming concentrates previously discussed. Chemical foam is formed from the chemical reaction in aqueous solution between aluminum sulfate ("A," acidic) and sodium bicarbonate ("B," basic) which also contains proteinaceous foam stabilizers. Foam is formed by the generation of carbon dioxide gas trapped in the bubbles of the foaming solution.

GUIDELINES FOR FIRE PROTECTION WITH FOAMS

The following general rules apply to the application and use of ordinary air foams:

1. The more gently the foam is applied, the more rapid the extinguishment and the lower the total amount of agent required.
2. Successful use of foam is also dependent upon the rate at which it is applied. Application rates are described in terms of the amount (by volume) of foam solution reaching the fuel surface (in terms of total area) every minute. If the foam has an expansion rate of 1 to 8, an application rate of 0.1 gpm per sq ft [4.1 (L/min)m^2] will provide 0.8 gal per sq ft (32.8 L/m^2) of finished foam every minute. Increasing the foam application rate over the recommended minimum will generally reduce the time required for extinguishment. However, little time advantage is gained if application rates are increased more than three times the minimum recommended. If application rates are less than the minimum, extinguishment time will be prolonged or may not be accomplished at all. If application rates are so low that the rate of foam loss by heat or fuel attack equals or exceeds the rate of foam application, the fire will not be controlled or extinguished.
3. The minimum recommended application rate is the rate found by test to be the most practical in terms of speed of control and amount of agent required. The general curve in Figure 19-4A illustrates the rate-time relationship for foam application to a hazard. The curve may be displaced right or left depending upon fuel and method of application; hence the need for carefully engineered systems based on the appropriate standard and actual test information.
4. In general, air foams will be more stable when they are generated with water at ambient temperature. Preferred water temperatures range from 35 to 80°F (2 to 27°C). Either fresh or seawater may be used. Water containing known foam contaminants such as detergents, oil residues, or certain corrosion inhibitors may adversely affect foam quality.
5. Foams are adversely affected by air containing certain combustion products. While the effect is minor with ordinary air foam and ordinary hydrocarbon fuels, it is desirable to locate fixed foam makers on the sides of,

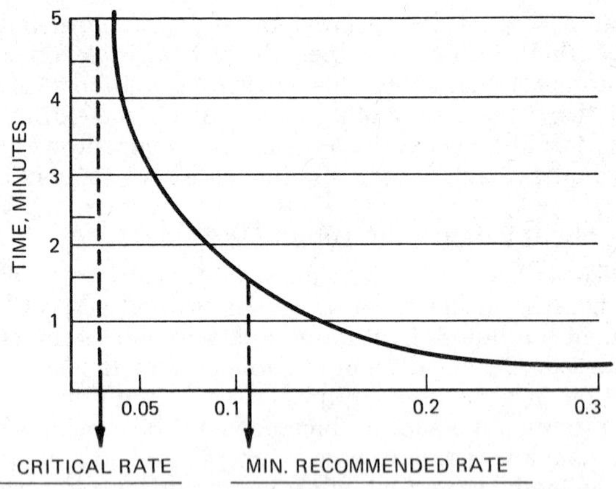

FIG. 19-4A. General relationship of foam application rate to time of application necessary for extinction. (1 gpm per sq ft equals 40.746 [(L/min)/m²].

rather than directly over the hazard.

6. Recommended pressure ranges should be observed for all foam making devices. Foam quality will deteriorate if these limits (high and low), are exceeded.
7. Many air foams are adversely affected by contact with vaporizing liquid extinguishing agents, their vapors, and by some dry chemical agents. Gases from combustion have a similar breakdown effect on foams.

FOAM GENERATING METHODS

The process of producing and applying fire fighting air foams to hazards requires three separate operations, each of which consumes energy. They are: (1) the proportioning process, (2) the foam generation phase, and (3) the distribution method. A flow diagram illustrating the relationship of the three operations is given in Figure 19-4B.

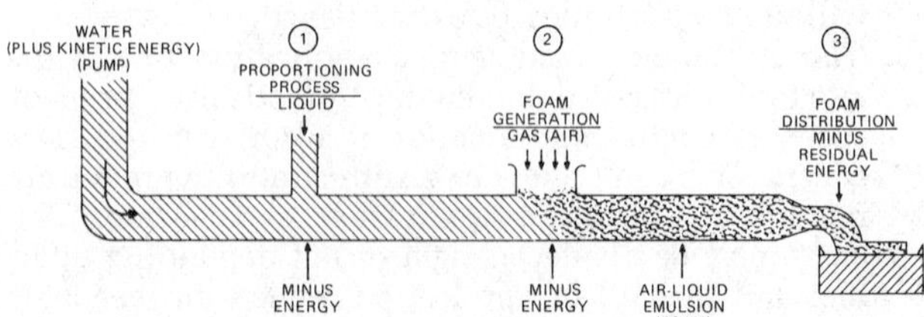

FIG. 19-4B. A diagram illustrating the steps in air foam generation.

In general practice, air foam generation and distribution occur nearly simultaneously within the same device. There are also many types of proportioning. In certain portable devices, all three functions are combined into a single device. The design and performance requirements of foam systems dictate the choice of types of proportioning, generating, and distributing equipment for the protection of specific hazards.

Foam Concentrate Proportioners

So that a predetermined volume of liquid foam concentrate may be mixed with a water stream to form a foam solution of fixed concentration, the following two general methods are used:

1. Methods that utilize the pressure energy of the water stream by venturi action and orifices to induct concentrate.
2. Methods that utilize external pumps or pressure heads to inject concentrate into the water stream at a fixed ratio to flow.

Figures 19-4C and 19-4D illustrate the general principles of the two different proportioning methods. Specific

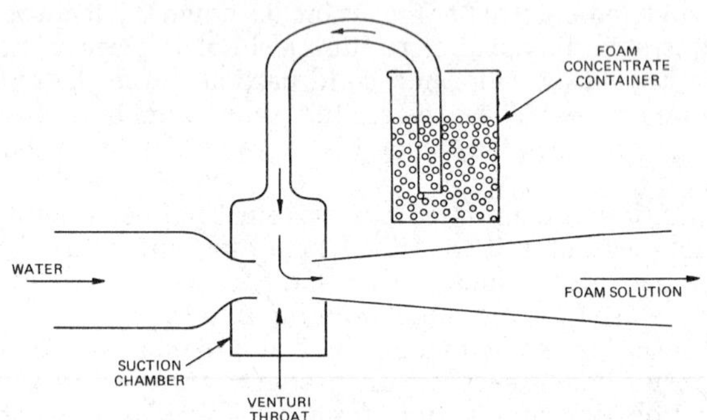

FIG. 19-4C. Venturi induction (in line) proportioner.

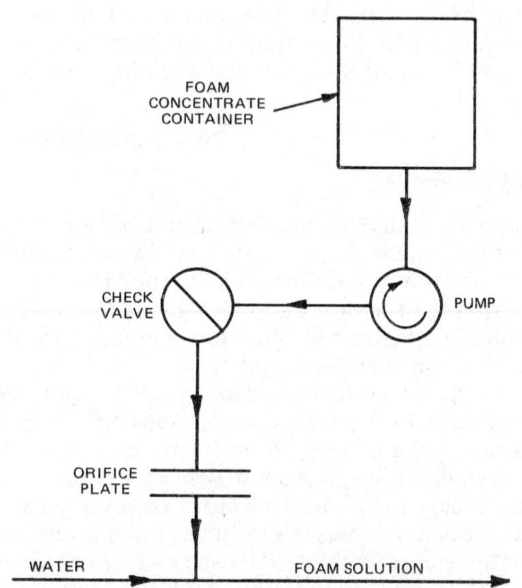

FIG. 19-4D. Foam concentrate pump proportioner.

system designs of proportioning equipment are given below.

The Nozzle Eductor: This type of foam concentrate proportioner is of simple design and is widely used in portable foam making nozzles where foam concentrate is available in 5 gal (19 L) pails or drums as shown in Figure 19-4E. Incorporating a modified venturi within the foam making nozzle section, the nozzle eductor drafts concentrate from a portable container through a pickup tube. Using a properly sized orifice or pipe section at the low

FIG. 19-4E. Air foam nozzle with built in eductor.

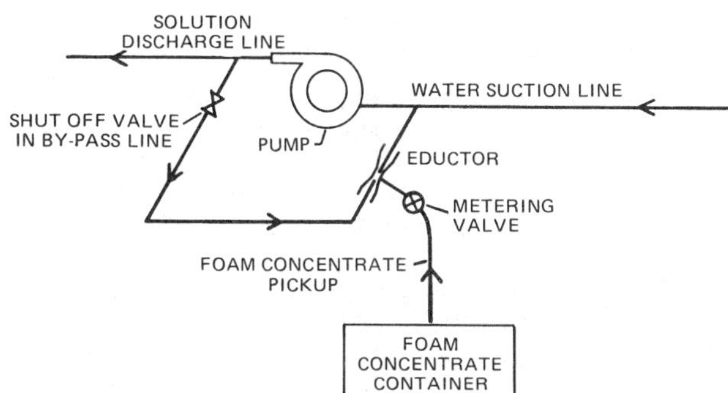

FIG. 19-4F. An around the pump proportioner.

pressure cavity, the concentrate is mixed in proper proportion to the fixed flow and operating pressure of the nozzle, and foam generation proceeds.

In Line Eductor: This type of proportioner educts or drafts foam concentrate from a container or tank by venturi action, utilizing the operating pressure of the hose water stream on which it is installed, and injecting concentrate into that flow of water. Its correct operation is very sensitive to throughput water rates and pressures. Changes in either of these factors from those for which the eductor was designed will result in incorrect proportioning. Distances of more than 6 ft (1.8 m) elevation from the eductor to the lowest liquid level of the foam concentrate container also may result in incorrect proportioning. Foam generation devices and maximum downstream lengths of hose recommended for each eductor must be carefully adhered to. This proportioning device may be used in the hose line leading to the foam generation device. An eductor of this type may also be installed at the foam concentrate tank in a fixed system, or at the pump discharge of a mobile pumper.

Some designs of this proportioning device incorporate metering valves in the foam concentrate intake line so various volume percentages of concentrate in the water stream may be obtained. A check valve is usually placed in this intake so that water cannot flow back into the foam concentrate container if a blockage occurs or a valve is closed in the downstream hose. The use of this eductor requires an allowance for a pressure loss of approximately 30 percent in the system or layout.

Around the Pump Proportioner: This type of proportioner also operates on the venturi principle, except that this proportioner must be situated at the pump and connected to both suction and pressure sides. Its advantage is that pressure recovery of the venturi action is attained, and pump delivery pressures to the foam making device or devices downstream require no compensation for pressure loss except for that in the layout hose length. The net delivery volume of the pump will be decreased by 10 to 40 gpm (38 to 150 L/min) in this method of proportioning. (See Fig. 19-4F.) The small portion of the pump discharge flows through a bypass line to the suction side of the pump. A venturi eductor in this line produces a negative pressure on the foam concentrate pickup line from the foam concentrate container. Foam concentrate is led to the eductor, where it mixes with water and is delivered to the suction side of the pump.

Multiple foam makers may be supplied with foam solution by this type of proportioner when it is supplied with a multiported metering valve designed for the required flows.

Pressure Proportioning Tank: This type of proportioner may consist of one tank, or two tanks separately connected to the water and foam solution lines. The tank or tanks in the system may each be fitted with a flexible diaphragm or bladder to separate the "driving" water from foam concentrate. Tanks may rely simply on differences in density of the two liquids to retard mixing during operation (without diaphragm or bladder), when proportioning protein or fluoroprotein concentrate.

The principle of this device is simple. A small amount of the flowing water volumetrically displaces foam concentrate into the main water stream. The design pressure of the vessel must be above the maximum static water pressure encountered in the system.

Water is allowed to enter the foam tank from the main stream with as little friction loss as possible. An orifice meters liquid in the tank into the low pressure area. Advantages of this system are its low pressure drop, automatic proportioning over range of flows and pressures, and its freedom from external power. Its disadvantages are a long refill time (since it is a "batch" method, the tank must be drained of its driving water and the tank or the bladder refilled with concentrate), and an economic limit on size.

Coupled Water Motor-Pump Proportioner: This proportioner consists of two positive displacement rotary pumps mounted on a common shaft. Water delivered to the larger pump (motor) causes it to drive the smaller pump, which is used to draft concentrate from a container and deliver it (at line pressure) to the water discharge line from the larger pump. By proportioning the sizes of the two pumps, the correct volume of concentrate is delivered to the water stream.

This proportioner is manufactured in only two sizes for water throughputs of 60 to 180 gpm (227 to 680 L/min) and 200 to 1,000 gpm (757 to 3800 L/min). Both sizes are designed for proportioning foam concentrates recommended for use at 6 percent concentration. A pressure drop of 25 to 30 percent in the water stream supplied to the device is required for its operation.

Balanced Pressure Proportioner: The use of a separate pump for pressurizing foam concentrate to be delivered in the correct proportions to a flowing water stream offers the greatest advantage for reliable and accurate operation of a foam concentrate proportioning system which must function at varying rates of volume or pressure during its use. (See Fig. 19-4G for a fixed design and Fig. 19-4H for a foam truck mounted design.) Many systems require manual attention to flow meters, duplex gages, or other measuring devices during system operation. These designs have been replaced by automatically operated proportioning systems which do not require continuous manual attention.

Balanced pressure proportioning systems are of two types: the bypass system and the variable flow demand system. These systems are simple in principle and reliable.

Bypass System: This type of balanced pressure system relies on a hydraulically monitored pressure control valve in a bypass foam concentrate line from a pump back to the concentrate supply tank; a correctly sized venturi type proportioning controller in the water line for correct and automatic proportioning over wide ranges of flow; and a fixed or variable metering orifice between the pump and the proportioning controller.

Variable Flow Demand System: This type of balanced pressure system utilizes a variable speed mechanism to drive a pump delivering foam concentrate to the system. Only a correctly sized venturi type proportioning controller in the water line (for correct and automatic proportioning over wide ranges of flow) and a fixed or variable metering orifice between the pump and proportioning controller are used. The variable drive mechanism is controlled by a feedback system which can be designed to operate hydraulically or electronically. Once activated, the pump output is monitored automatically so foam concentrate flow and pressure is always proportional to system demand for foam solution. (See Fig. 19-4I.) The variable flow demand system may be fixed or truck mounted.

Variable Orifice, Variable Flow Demand Proportioner: This type of balanced pressure system utilizes a specially designed, flow sensitive, movable piston section in the water supply that controls a variable orifice in the foam concentrate line. A pump supplies concentrate at monitored pressures to the metering orifice, which changes in size proportionally to the system demand for foam solution. The device is especially designed for large capacity systems. Its design provides accurate proportioning over ranges of flow of water of approximately 2.3 to 1.

Premixed Foam Solution: This method of proportioning is a batch type of mixing of concentrate with water, usually in a container that can be pressurized. The measured volume of concentrate is poured into a measured volume of water to yield a foam solution of the recommended strength; i.e., for a 3 percent solution in a container which holds 100 gal (378.5 L) of liquid, 97 gal (367 L) of water are poured into it and 3 gal (11.5 L) of foam concentrate are mixed with it to give a solution of 3 percent by volume. The final solution mixture is then educted from a tank to a pump or placed in a pressurizing vessel. In many cases, a time degradation of effectiveness of stored premix solutions of foam concentrates will be experienced. Many foam concentrates that are specifically designed for use on alcohol or polar solvent flammable liquids cannot be used as a premix.

Portable Aspirating Equipment and Systems

Because of the difficulties associated with pumping or transporting generated foam in pipes or hoses and the

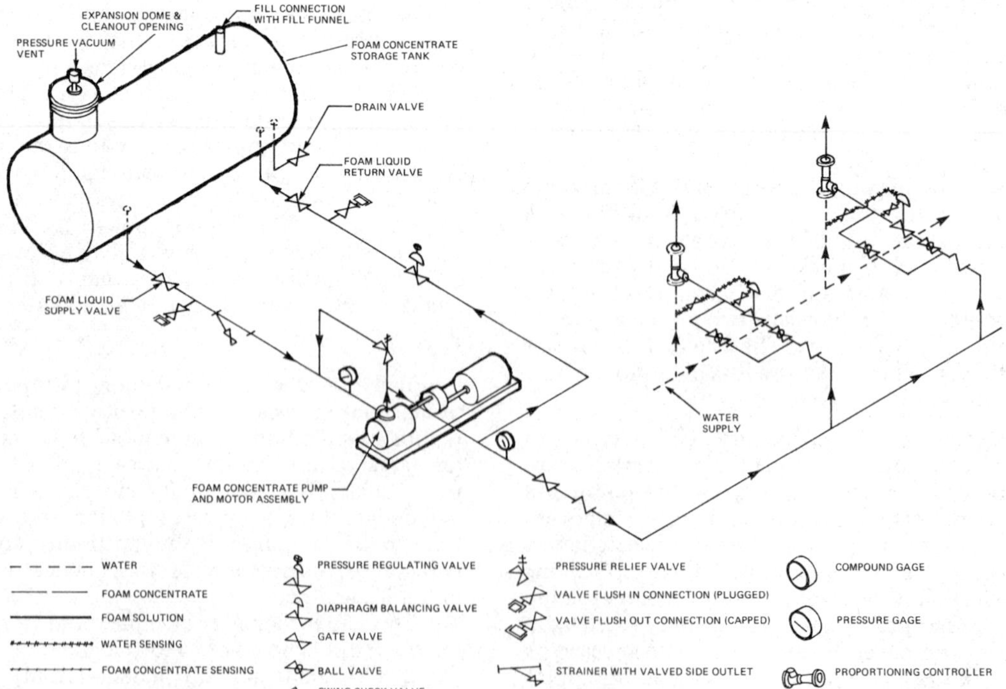

FIG. 19-4G. Balanced pressure proportioning with multiple injection points (metered proportioning).

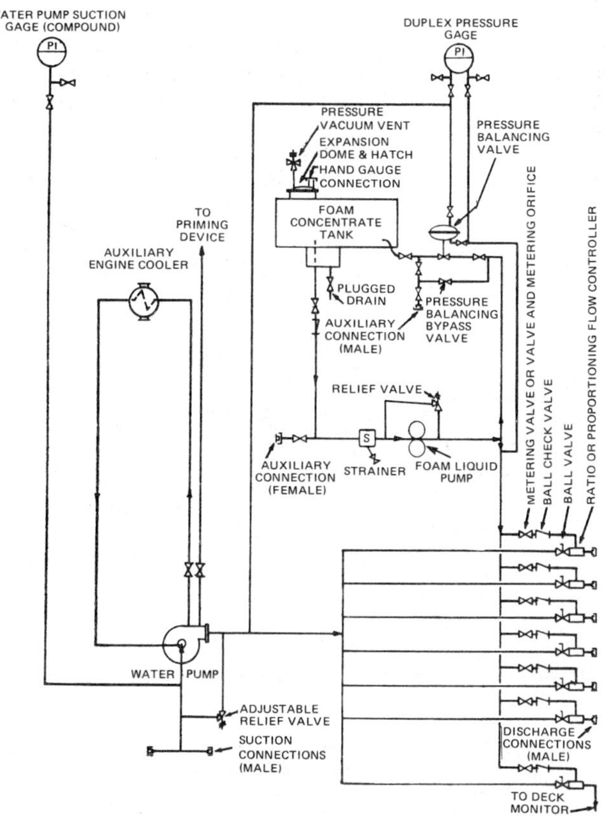

FIG. 19-4H. Balanced pressure proportioning.

familiarity of fire fighters with nozzles, the earliest designs of air foam generators were devised to be used in much the same manner as water nozzles. They incorporated a crude venturi design whereby a jet or jets of foam solution enter an open contracted portion of a large diameter foam tube. This action lowers the atmospheric pressure surrounding the jets, and air is drawn or aspirated into the throat of the tube. Downstream of the contracted portion of the tube, a high turbulence and mixing of air and foam solution occurs. This turbulence may be increased by internal turbulence accelerating devices such as screens or baffles. The kinetic energy of the fluid contributes to this mixing action so that a usefully stable foam exits the tube at a relatively low pressure. (See Fig. 19-4J.)

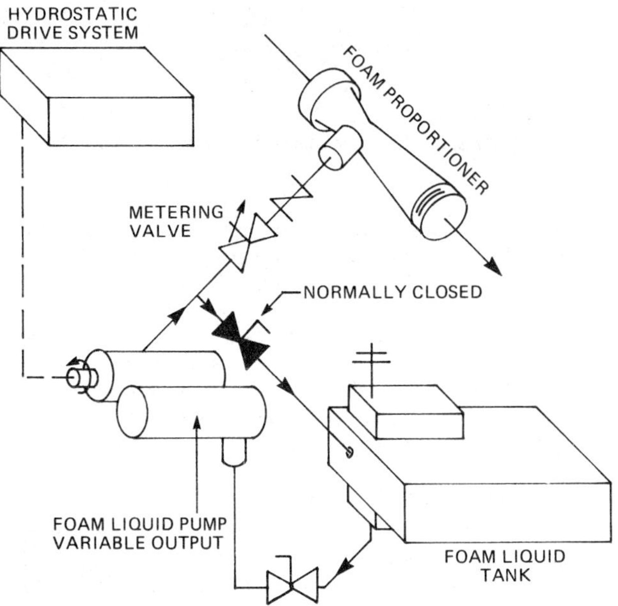

FIG. 19-4I. Balanced pressure proportioning demand system. (National Foam Systems, Inc.)

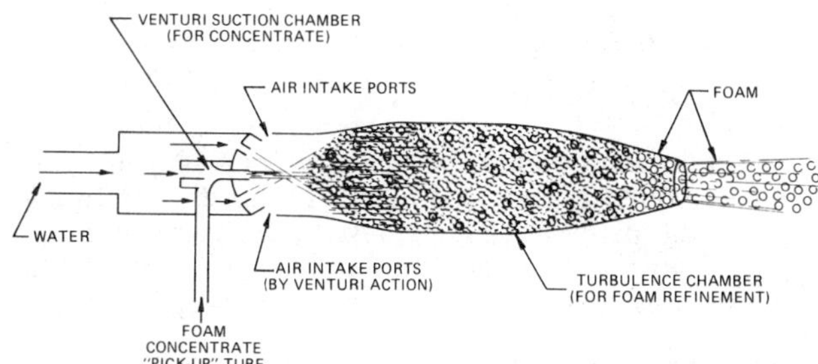

FIG. 19-4J. Cross section of an aspirating foam maker with a concentrate pickup tube.

The basic design principles of this method of producing air foam have been changed in many ways to yield, for many purposes, foams with greatly differing characteristics. However, all types of nozzles incorporating foam solution jets leading into free air mixing cavities, followed by discharge apertures of one kind or another, use the aspirating action for making foam.

Hose Line Foam Nozzle: This is the most universally used portable, air aspirating foam device for flammable liquid fire fighting. It is manufactured in a variety of capacities up to approximately 350 gpm (1325 L/min). (See Fig. 19-4K.) Supplied with foam solution from a

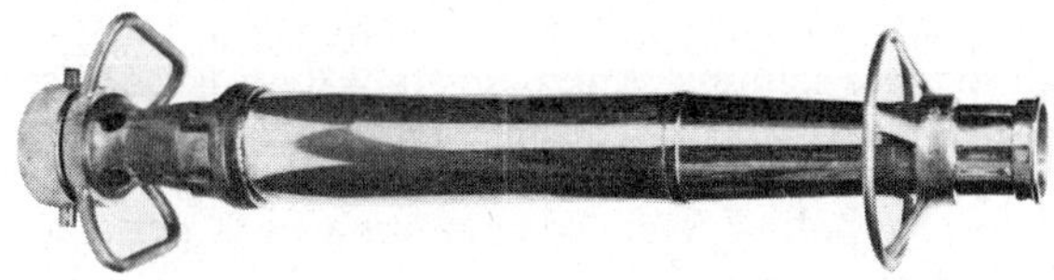

FIG. 19-4K. Hose line foam nozzle. (National Foam Systems)

proportioner or by means of a pickup tube, it successfully combats fires resulting from spills of flammable liquid, or fires in tanks or fuel pits. To provide a variety of foam stream patterns which may be needed for the extinguishment operation, these nozzles often contain built in devices that allow continuous foam pattern variation from a solid straight stream to an inverted filled umbrella shape. (See Fig. 19-4L.) As with water streams, foam is employed in a solid straight stream for range or reach; a flat or wide bushy shape is used for gentle "snowstorm" application on the burning fuel surface; a very wide, circular bushy shape is used for radiance shielding of the operator during fire extinguishment or penetration into the fire area.

Another type of hose line foam nozzle is especially designed for quick, portable, one person use during emer-

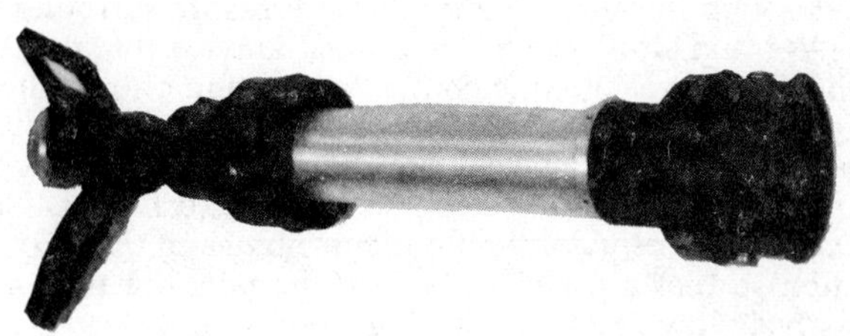

FIG. 19-4L. Adjustable hose line foam nozzle. (National Foam Systems)

gency operations from airport crash-rescue vehicles. Customarily called a handline foam nozzle, it is equipped with a foam pattern changing device, and some types are supplied with a valve control for diverting the water or solution stream into a water spray for cooling purposes.

Foam and Foam Water Monitors: In large scale fuel-fire firefighting operations it may become necessary to position a foam making nozzle with a high discharge rate at an advantageous position for continuous application of foam to one point or over an area. Devices for this purpose are available in a variety of types, as part of mobile truck apparatus, on a trailer, wheeled, or permanently mounted. The foam pattern change is accomplished by moving a deflector into the foam stream or by opening or closing the "jaws" at the exit end of the large tube.

High Backpressure Foam Maker: Certain circumstances necessitate that foam be generated and supplied under pressure for transmission in pipes under a definite pressure head. The subsurface foam injection method for fuel tank fire extinguishment requires this type of foam maker. The high backpressure foam maker or "forcing foam maker" is a venturi device which is carefully designed to make foam by air aspiration, and to supply it under pressure at a carefully selected ratio (from 2 to 1 to 4 to 1) of air to foam solution. Approximately 20 to 40 percent of the inlet pressure is recoverable. In use, this foam device is usually brought to the fixed piping foam inlet, installed, and then supplied with foam solution by a portable or mobile pumper. (See Fig. 19-4M.)

Medium and High Expansion Foam Generating Devices: There are two principal methods used for the generation of these types of fire fighting foams. One method utilizes a modified venturi action with air aspiration flow, while the other requires use of a blower and screen to form the finished foam. The latter system produces high expansion foam containing sufficient residual kinetic energy to enable it to be forced through large tubes and passageways.

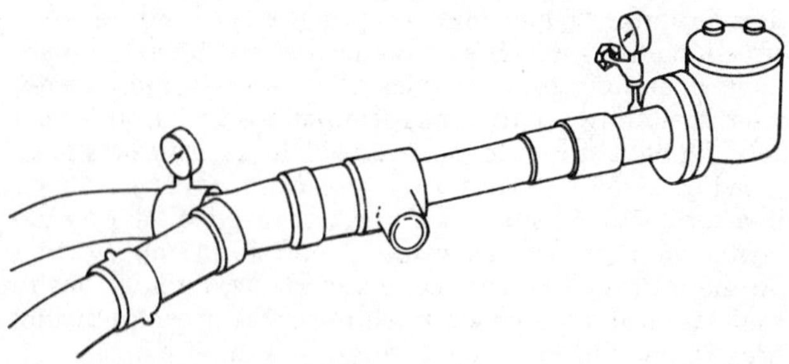

FIG. 19-4M. A "high backpressure" (or "forcing") type foam maker.

Figure 19-4N illustrates the operating principles of high expansion foam generating devices.

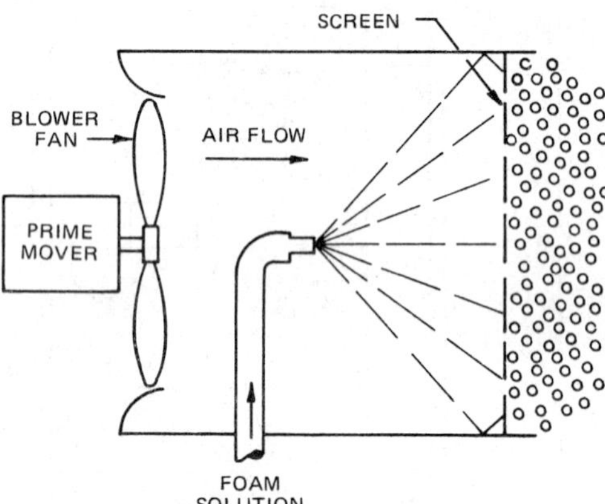

FIG. 19-4N. A fan-blower type high expansion foam generator.

Water Fog or Spray Nozzles: Several types of adjustable water fog or spray nozzles for portable use provide an acceptable fire fighting foam of adequate characteristics when supplied with certain foam concentrates, the most universal design of which is shown in Figure 19-4O. An adjustable water spray monitor nozzle designed for this purpose is shown in Figure 19-4P.

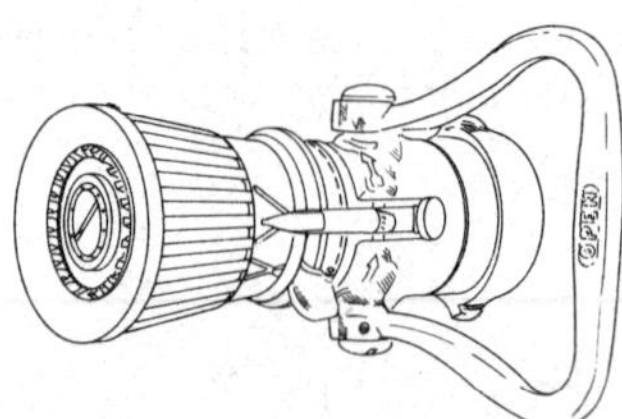

FIG. 19-4O. Variable pattern water fog nozzle foam maker for use with AFFF.

These portable water spray nozzles are used with aqueous film forming foam solutions (AFFF) for combatting flammable liquid tank and spill fires and are used in this manner on crash-rescue vehicles. The foam resulting from the discharge of AFFF solution devices that do not aspirate air is generally fast draining and does not impart the same degree of burnback resistance as the foam produced from AFFF agents when foam generating devices of an air aspirating type are used.

Foam Generating Equipment

Where flammable liquid fire protection is required for permanently installed hazards, such as fuel storage tanks or dip tanks containing flammable or combustible liquids, air foam generating and distributing devices are installed integrally with the hazard. These fixed devices which are piped to a source of foam solution, may be arranged for

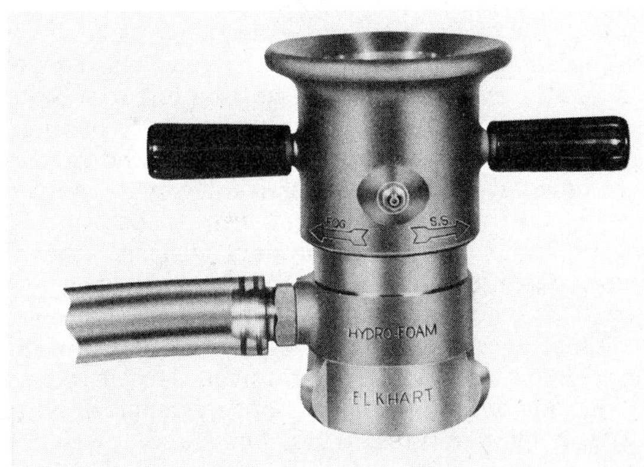

FIG. 19-4P. Adjustable 350 to 500 gpm (1325 to 1893 L/min) water spray foam monitor nozzle. (Elkhart Brass Mfg. Co.)

manual control or automatic activation by fire detectors in the event of fire.

Open Dip Tank, Quench Tank System: This system consists of a small aspirating foam maker supplied by a water line and foam concentrate educted to the foam maker. Foam discharges into a mixing box which also acts as a surface distributor for gentle foam application.

Foam Chambers for Large Fuel Storage Tanks: Fire protection of large outdoor fuel storage tanks requires that several foam chambers with foam makers be installed at equally spaced positions slightly below the curb angle on the top periphery of the tanks. These chambers are connected to lines on the ground that supply foam solution to each foam maker simultaneously in case of ignition of the flammable contents of the tank. Frangible seals at the discharge outlet of the foam chamber prevent vapor from entering the foam piping. These seals are designed to burst when foam pressure is applied. A screen for the air inlet to the aspirating foam maker prevents clogging from foreign matter such as bird nesting material. A universal or swing pipe joint is installed at ground level (Fig. 19-4Q) in the foam solution inlet pipe to prevent fracturing of the supply piping if an explosion precedes a tank fire.

Internal Tank Foam Distributing Devices: A prime requirement for efficient fuel tank extinguishment by topside foam devices has always been that the foam must be applied to the burning surface without undue plunging into the fuel, or allowing the foam to become coated with burning fuel. This gentle application of foam must be accomplished at any level of the contents of the tank. Many devices have been developed to gently apply foam from one point, regardless of burning fuel level. These devices are listed as "Type I" foam discharge outlets for tanks and are required for some alcohol type foams. When foam discharge into a tank is deflected to run down the inside tank shell to the burning fuel surface, it is called a "Type II" outlet for foam application.

Central "Foam House" Distributing Systems: These systems, as shown in Figure 19-4R, consist of an enclosure housing a foam concentrate supply tank and a proportioning device of an automatic or balanced pressure type. Foam solution is supplied under adequate pressure from this foam house to the piping system and controlled by

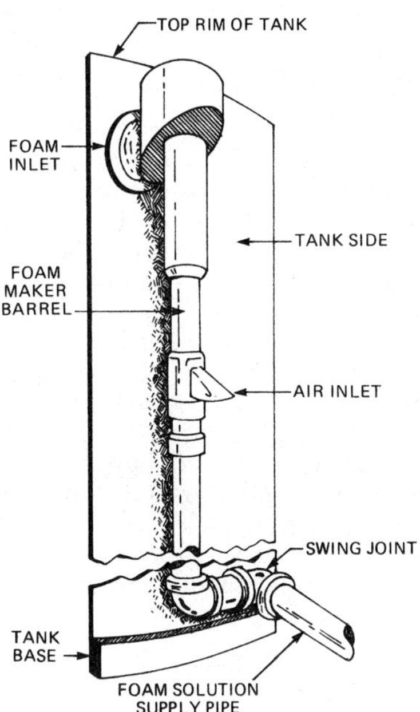

FIG. 19-4Q. Air foam chamber at top of storage tank.

appropriate valves so that the foam chambers with foam makers on the burning tank receive foam solution.

Semifixed systems of similar design are more frequently used with mobile foam concentrate supply from foam trucks. The truck proportions and pumps foam solution to the pipe laterals feeding the foam makers from a safe location outside the dike.

Intermediate Backpressure System: Although this system of tank protection is similar to the preceding central foam house distributing systems, it utilizes strong and well braced foam delivery pipes on the side of the tank which act as supports to prevent buckling of the tank from heat. This system also uses a foam truck to proportion and pump foam solution to the piping outside the dike or firewall of the burning tank.

Subsurface Foam Injection Systems: The problems inherent with the application of foam from above the burning surface (topside) are sometimes difficult to combat. Problems may consist of explosion or fire damage to the foam makers or tankside piping; forceful upward fire-drafted air currents that prevent the falling foam from reaching the burning surface; hazard to workers attempting to erect portable foam distributing devices near the

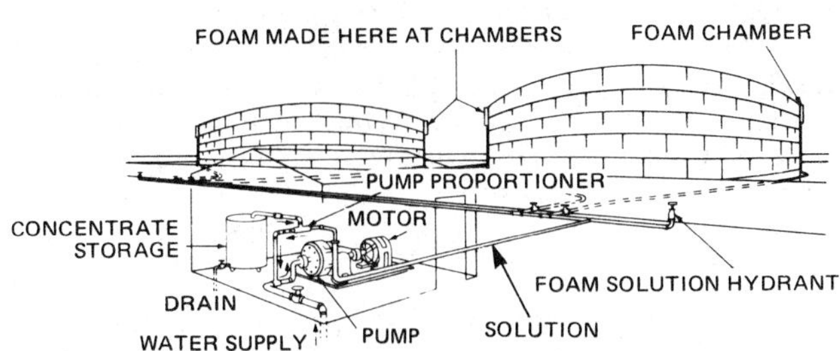

FIG. 19-4R. Schematic arrangement of air foam protection for storage tanks.

burning tank; or inability of foam applied from the periphery of a large tank [greater than about 200 ft (61 m) in diameter] to flow and form a complete center seal during fire attack.

The obvious solution to these problems is to apply foam from the underside of the fire, causing it to come up through the contents of the tank. The subsurface foam injection system accomplishes this by injecting foam under the pressure of the head of fuel in the tank, using the high backpressure foam maker referred to previously in this chapter.

Entry of foam may be provided at several points at the base of the tank (base injection) or it may be accomplished by means of the product line. Where large tanks are involved, a branched pipe foam distributor may be installed on or slightly above the floor of the tank above any expected water level and connected to a central foam injection point outside the tank.

Mobile foam concentrate proportioning equipment is used with these systems, pumping from a protected position outside the tank dike. (See Fig. 19-4S.)

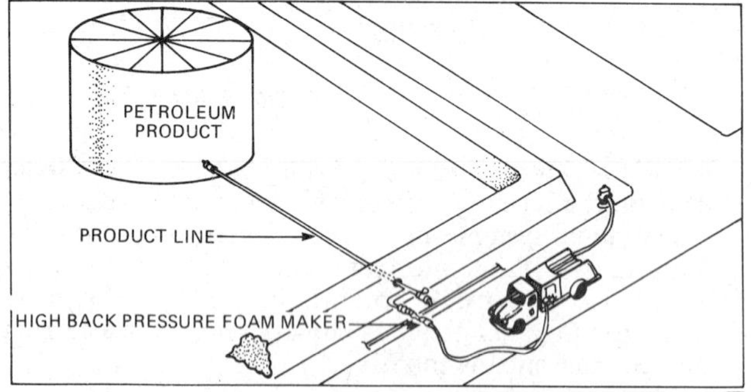

FIG. 19-4S. Semifixed subsurface foam installation.

These topside and subsurface injection systems require careful design. Their minimum installation requirements are found in NFPA 11.

Portable Foam Devices for Tank Protection: Mobile foam monitors of high capacity discharge (foam cannon) may be used to direct a stream of foam over the open top rim of a burning tank so foam will fall into the burning area. These devices waste foam because of cross winds, fire updrafts, and inability to place the equipment in an advantageous position. A 60 percent higher foam application rate therefore must be included in the design; details are contained in NFPA 11. Nozzles of this type are also used to extinguish fires in the overflow space inside the dike surrounding the tank. Care is needed in directing the foam stream from such devices, as shown in Figure 19-4T.

The employment of special foam making and distributing devices for use during fires in fuel storage tanks is extremely hazardous because of the need for personnel to approach the tank. The extendable pipe hydraulic foam tower is hydraulically raised into position at the top edge of the burning tank using water pressure. It is usually supplied with foam solution from a foam truck.

Other Considerations for Fuel Storage Tanks: The floating-roof storage tank has an excellent record of freedom from fire; consequently, fixed foam systems are usually not required for their protection. Under certain circumstances, however, there may be a need for a foam flooding system to flood the rim area.

As indicated by its name, the "covered" floating roof tank is totally enclosed above the floating roof with a properly vented steel roof. Usually Class IB flammable liquids such as gasolines and crudes are stored in "open top" or "covered" floating-roof tanks. During storage operation there is no vapor space between the bottom of the floating roof and the stored product surface. However, during periods of initial fill, a flammable vapor space will exist until the floating roof is buoyant. In the case of covered floating-roof tanks, the space between the fixed and the floating roof will be within the flammable range during periods of initial fill and longer, depending upon atmospheric temperatures, wind, and vapor pressure of the stored product. Class II and III combustible liquids are usually stored in fixed-roof tanks.

In an open top floating-roof tank, a rim of fire is all that can be expected. Rim fires may be caused by atmospheric disturbances, such as lightning, but usually do not occur if the floating roof is properly bonded to the shell as specified in NFPA 78, *Lightning Protection Code*. Rim fires due to atmospheric disturbances do not occur in covered floating-roof tanks because of the "Faraday effect" of the fixed roof. Rim fires may occur in either type floating-roof tank due to a serious exposure fire.

FIG. 19-4T. Portable (trailer) foam monitor (cannon). (Angus Fire Armour Corporation)

Fixed foam fire protection for the open top floating-roof consists of aspirating type foam makers installed so that when supplied with foam solution, their foam discharge floods the annular area covered by the seal around the tank periphery. A metal foam dam may be needed to restrict the foam to this area. Further details of such construction are contained in the Appendix to NFPA 11.

When foam protection is desired for covered floating-roof tanks, a foam system may be provided similar to those described for fixed-roof tanks. Subsurface injection systems are not generally recommended for open top or covered floating-roof tanks because of the possibility of tilted or sunken roofs resulting in improper foam distribution.

Fixed foam systems employing foam spray nozzles or monitors are used to protect large oil-water separators, pump areas, and oil piping manifolds. Fixed foam systems are not generally used to protect tank diked areas. Large wheeled monitors or trailer mounted foam cannons, as shown in Figure 19-4T, are used for this purpose.

Certain hazardous areas, such as aircraft hangars may require additional foam making and distributing devices such as foam monitors mounted at near floor level. Often, these devices are provided with an automatic oscillator so that foam is continually distributed over the floor for extinguishing burning spilled fuel under obstacles, such as aircraft wings.

Fixed foam systems are used to protect petroleum piers and wharves where products and petroleum crude are handled. Foam monitors of 1,000 gpm (3800 L) and greater capacity are mounted on towers and remotely operated to cover both the pier and the tanker deck in the pipe manifold area. Below deck, foam spray systems or oscillating monitors are installed to keep the pier tenable in the event of a spill fire floating on the water surface under the pier.

Fixed systems consisting of automatically operated combinations of foam spray systems and foam monitors are often installed to protect chemical processing plants. Special foams, some of which require gentle application, are frequently used. In these designs, where there may be a high risk, process vessels, pumps, and piping are often all included within the foam distribution pattern for overall protection. The system can be automatically activated by heat or fire detectors. See NFPA 11 for design of foam spray systems.

Foam Water Sprinkler Systems: In areas where flammable and combustible liquids are processed, stored, or handled, a water discharge may be ineffective for controlling or extinguishing fires. The foam making sprinklers (aspirating type) and deluge or spray nozzles using AFFF foams have successfully replaced water sprinkler nozzles for such systems so that fires in these occupancies may be controlled and property safeguarded.

When supplied with foam solution, sprinkler system piping grids provided with foam water nozzles generate air foam in essentially the same water sprinkler pattern as when water is discharged from the same nozzle. This dual capability affords the system Class A and B extinguishment ability. Design requirements can be found in NFPA 16.

Obviously, fixed sprinkler systems using these nozzles require that foam concentrate tanks, proportioners, and suitable pumps be provided to supply the system with foam solution or water. Detection devices may also be used to activate the system, or the system may be activated manually.

Ordinary water sprinkler nozzles may be employed effectively in such fixed systems if they are provided with AFFF concentrates.

DELIVERING FOAM FROM VEHICLES

The majority of the mobile fire protection vehicles using foam consist of airport crash-rescue trucks and industrial foam trucks used at oil refineries and petrochemical plants. Many of these vehicles are also equipped to discharge dry chemical in combination with foam. These have combined or "twinned" agent equipment, and are described later in this chapter.

Crash-Rescue Trucks

Mobile foam trucks, developed by military and municipal authorities in the United States and other countries, are large, custom designed vehicles with oversize running gear which allows them to travel over all types of terrain. The trucks carry their own water supply as well as foam concentrate, usually for six percent proportioning. Automatic balanced pressure proportioning is usually provided, although in some designs an around the pump proportioning system is used. The trucks are equipped with separate engine driven water pumps and foam concentrate pumps so that the turret monitor nozzles and roadway foam nozzles can be discharged while the truck is in motion. Recent designs accomplish this using one or two engines for all necessary power requirements. Crash-rescue trucks are equipped with large capacity, adjustable foam monitors of 500 to 1,500 gpm (1900 to 5700 L/min) discharge, depending upon truck size. The foam monitors are usually installed on top of the cab with remotely operated controls. The trucks are also equipped with portable handlines. Although limited, their supply of foam concentrate and water is sufficient for fire fighters to form a passageway for access to a burning aircraft for rescue. Their primary purpose is rescue of people, and not necessarily a total extinguishment of fire. Aircraft crash-rescue foam vehicles are specially designed and require special attention to detail in their fabrication and performance requirements. NFPA 414, *Standard for Aircraft Rescue and Fire Fighting Vehicles* (hereinafter referred to as NFPA 414), provides standards for such vehicles.

Industrial Foam Truck Design

Foam trucks are manufactured by vendors who make a specialty of this design, and by some vendors of fire department pumpers. Generally, using NFPA 11C, *Standard for Mobile Foam Apparatus*, and NFPA 1901, *Standard for Automotive Fire Apparatus*, as a design basis, these trucks are fabricated on a suitable commercial or custom truck chassis. In the U.S. they are available with water pumps in sizes from 750 to 2,000 gpm (2839 to 7570 L/min) using gasoline or diesel engine drives.

Balanced pressure proportioning is provided for maximum flexibility and simplification. A positive displacement type foam concentrate pump takes suction from a 500 to 1,000 gal (1900 to 3800 L) foam concentrate storage tank. The foam concentrate pump is driven by a power takeoff or hydraulic motor. The water pump is driven by a transfer gear from the main drive shaft behind the transmission. A metering valve at each truck outlet can vary foam proportioning from 3 to 6 percent.

In addition to the foam concentrate, each foam truck usually is equipped with a 500 to 1,000 gpm (1900 to 3800 L/min) monitor and carries fire hose, foam nozzles and adjustable water spray nozzles, hose adapters, and other accessories. The maximum truck height is limited to permit passage beneath overhead obstructions. When the truck is equipped with a large capacity nozzle on a telescopic or articulated boom, a second power takeoff is used to power the boom hydraulic system.

Foam trucks for special hazard application, instead of fixed foam systems, are popular for refinery and petrochemical plant use. Their advantages are:

1. Capability of discharging their maximum capacity at any hazard in the plant, rather than only to the limited areas covered by a fixed system.
2. Improved reliability, because their equipment is easily maintained; thus, operating procedures can be simplified and fire fighters more easily trained to use the equipment.

Industrial foam trucks are used to extinguish spill fires in process areas, piping runs, and tank diked areas, as well as in fighting tank fires. Their 500 to 1,000 gpm (1900 to 3800 L/min) foam water monitors and 1½ in. (38 mm) and 2½ in. (65 mm) handlines are used for fighting major spill fires. Designs that are equipped with monitor nozzles on articulated or telescopic booms enable the operator to discharge foam at various elevations in and around process equipment. In some cases, a tank fire can be extinguished with this equipment. Figure 19-4U illustrates a typical industrial foam truck with monitor. Figure 19-4V illustrates a typical foam truck with a telescopic boom, ladder, and basket. The foam trucks with booms may or may not be equipped with a ladder and/or basket.

FIG. 19-4U. A typical industrial foam truck. (National Foam Systems, Inc.)

FIG. 19-4V. Foam truck with telescopic boom and nozzle with ladder and basket. (National Foam Systems, Inc.)

COMBINED AGENT OR "TWINNED" EQUIPMENT

The superior capability of dry chemical agent (especially potassium salt types—Purple K) for very fast flame control and three dimensional flowing fuel extinguishment is well documented. Its fire fighting deficiency of reflash protection over fuel surfaces that have been extinguished is also well known. With dry chemical compatible foam concentrates, it is possible to apply a coating of vapor securing foam to a burning fuel surface which has been freshly extinguished by the chemical action of dry chemical discharges.

The logical extension of these developments has been incorporated into portable and mobile devices with dual trigger valve nozzles and monitor nozzles discharging AFFF and dry chemical. Figure 19-4W illustrates a type of twinned pistol grip trigger valve dual nozzle device primarily used by the U.S. Navy for dual hose line use. Figure 19-4X shows an improved industrial design used on offshore oil rigs. The use of this combined agent attack allows three dimensional as well as spill fire extinguishment of flammable liquid with both speed and freedom from reignition. The fire fighting vehicle shown in Figure 19-4Y

FIG. 19-4W. A twinned nozzle applicator for combined agent use—the AFFF nozzle is in operator's left hand and the dry chemical nozzle in right hand.

FIG. 19-4X. Industrial twinned nozzle applicator for combined agent use. The AFFF nozzle is below and the dry chemical nozzle is above. (The Ansul Company)

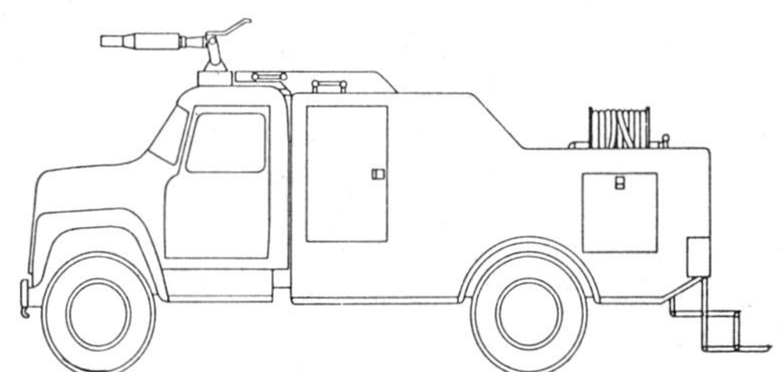

FIG. 19-4Y. A combined agent vehicle. (The Ansul Company)

is for aircraft crash-rescue purposes and for extinguishment on freeways or in plant, using combined agents.

In use, the twinned hose from the hose reel supplies dry chemical and AFFF to a dual nozzle for one person operation. Similarly, the turret nozzles above the cab discharge dry chemical on one side and AFFF on the other. With a sweeping motion, the flames are controlled with the dry chemical, quickly followed by a vapor securing, cooling layer of foam to prevent reflash. This rapid action halts flame radiation and makes advancement over the fire area safe and cool.

A design especially for petroleum refinery plant protection involves the combined agent concept in a triple use arrangement consisting of a 2,000 lb (908 kg) dry chemical tank, a 200 gal (757 L) AFFF premixed solution tank, nitrogen bottles for pressurization of both tanks, a 500 gal (1900 L) foam concentrate tank, a 1,500 gpm (5700 L/min) water pump, and balanced pressure proportioning up to 2,000 gpm (7600 L/min). Two monitors mounted behind the truck cab are twinned for operation as a unit and consist of a 50 lb per sec (23 kg/sec) dry chemical turret nozzle and a 500 gpm (1900 L/min) AFFF or foam nozzle as shown in Figure 19-4Z. A different configuration uses

FIG. 19-4Z. Triple Agent Truck. (National Foam Systems, Inc.)

three monitors mounted behind the truck cab—twinned 50 lb per sec (23 kg/sec) dry chemical and 180 gpm (680 L/min) AFFF nozzles, and one adjacent 1,000 gpm (3800 L/min) adjustable foam monitor. In addition, the truck carries two hose reels, each with 100 ft (30 m) of twinned hose, and hose nozzles for dry chemical and AFFF arranged for twinned operation. The hose reels are equipped with an electric motor rewind. Foam nozzles, adjustable water fog nozzles, fire hose, and other accessories are carried on the truck.

The dry chemical turret nozzle will extinguish three dimensional or pressure fires at a distance of 100 ft (30.5 m) and spill fires at a distance of approximately 150 ft (46 m). Both pressure and spill fires can be knocked down, controlled, and extinguished in less than 20 seconds and immediately secured using AFFF. In addition, for long range operation with lengthy duration, the 1,000 gpm (3800 L/min) foam-water monitor has a range of more than 200 ft (61 m). The truck is capable of fighting fires in fixed-roof oil storage tanks with a maximum diameter of 160 ft (49 m). The proportioning system is designed to permit an external tank, such as a tank truck or trailer to be used for foam concentrate supply without interrupting foam solution proportioning. (See Fig. 19-4W.)

Twinned agent hose reels are often used on offshore oil platforms for equipment protection and in naval shipboard engine rooms. These systems have suitably sized potassium bicarbonate (Purple K) dry chemical containers that are pressurized by nitrogen when put into operation, and AFFF central pumped systems using balanced pressure proportioning. Similar twinned agent skid mounted systems are used in refineries and petrochemical plants.

MEDIUM AND HIGH EXPANSION FOAM GENERATING EQUIPMENT AND SYSTEMS

Medium or high expansion foam is an aggregation of bubbles resulting from the mechanical expansion of a foam solution by air (or by other gases), with expansion ratios in the range of from 20 to 1 to approximately 1,000 to 1. There are three types of systems: total flooding, local application, and portable.

The foam generators for these systems are of two types: aspirator and blower. The (portable or fixed) aspirator type device utilizes jet streams of water foam solution, entraining suitable amounts of air to produce foam. The aspirator type produces foam of expansion up to 200:1. With the blower type, foam solution is discharged onto a screen through which an air stream, developed by a fan or blower, is passing. As the air passes through the screens (which are wetted with the foam solution), large masses of bubbles or foam are formed. The blower may be powered by a hydraulic or water motor, compressed air or gas, an electric motor, or an internal combustion engine. This type of generator produces foam of expansion up to 1,000:1.

System Design and Use

Detailed design information on medium and high expansion foam systems can be found in NFPA 11A.

Basically, medium and high expansion foam systems are used to control or extinguish fires involving surface fires in flammable and combustible liquids and solids, and deep seated fires involving solid materials subject to smoldering. Three dimensional fires in flammable liquids (falling or flowing under pressure) with flash points below 100°F (38°C) generally cannot be extinguished with this technique, although they may be kept under control. Key factors to consider in determining the design adequacy of medium and high expansion foam systems are:

1. The quality and adequacy of water supply, the adequacy of supply of foam liquid concentrate, and the source of the air supply.
2. Suitability of the generator and foam delivery system (piping, fittings, valves, ducts).
3. Needed submergence volume and time as influenced

by the space being protected, the nature of the hazards involved, leakage of the foam, and similar factors.

4. If outdoors, the effect of wind, since some of the foam may be dissipated by air currents.

Tests have shown that high expansion foam may be useful in controlling LNG fires and unignited spills, by forming an ice layer on the liquid and by helping disperse the vapor cloud.

Total Flooding Systems: A total flooding system may be used where there is an enclosure surrounding the hazard being protected which will permit the required amount of medium or high expansion foam to be built up to extinguish or control the fire. Examples of such enclosures are rooms, vaults, pits, and basement areas. Even an entire building may be so protected where the foam generators are of sufficient capacity and steps are taken to assure effective distribution of the foam and its retention. Since the efficiency of the system depends upon the development and maintenance of a suitable quantity of foam within the particular enclosure, leakage of the foam from the enclosure must be avoided. Thus, it is important that doorways or windows be designed to close automatically with consideration being given to the evacuation of personnel. High level venting is required for the air which is displaced by the foam.

For adequate protection, sufficient high expansion foam must be discharged at a rate to fill the space to an effective depth above the hazard before an unacceptable degree of damage occurs. The depth of the foam above the hazard will vary, depending upon the type of materials creating the hazard. Generally, the minimum depth above the hazard should be 2 ft (0.6 m). The time allowed to cover the hazard will likewise vary depending upon the type of material involved, construction features, and whether the enclosure also has an automatic sprinkler system or similar protection. Consideration has to be given to the disintegration of the foam by the heat of the fire; by normal foam shrinkage; by leakage around doors and windows, and through unclosable openings; and by the effects of sprinkler discharge where sprinkler protection is provided.

Local Application Systems: Local application systems can be used where total flooding systems may be impractical or unnecessary. Such hazards may be indoors or outdoors (where air currents are not likely to be severe). These systems are best adapted to the protection of flammable or combustible liquids in dip tanks and associated drainboards, and for pits and trenches. Medium or high expansion foam may be used for extinguishing spill fires where it is feasible to apply the foam from fixed or portable nozzles at adequate rates of discharge to develop a foam blanket for purposes of achieving complete extinguishment.

Precautions to be Observed with Medium or High Expansion Foam

A space filled with medium or high expansion foam is normally nontoxic to persons who may be trapped in the space, since the air entrained in the foam is generally not contaminated. However, because of the foam bubbles, some difficulty may be experienced in breathing, and breathing discomfort will increase with reduction of foam expansion. Air is usually taken from a clean source, but where products of combustion are introduced, the foam quality may be substandard because of the contaminates present and the temperature of the heated air. Entering a foam filled space should be avoided unless adequate precautions are taken because loss of vision and disorientation introduce life and injury hazards. A coarse water spray may be used to "cut" a path in the foam. Personnel should wear self-contained breathing apparatus and employ a lifeline when entering an area filled with high expansion foam.

FOAM EQUIPMENT TESTING AND SURVEILLANCE

The continued effective emergency performance of foam equipment depends entirely upon fully adequate maintenance procedures, with periodic testing where possible. The many variations in system design and equipment applications for hazards requiring foam make it impossible to establish anything other than general procedures for periodic inspection. Because air foam concentrates, foam equipment, and fire protection systems are subject to change, variation, and even malfunction over long unattended spans of time, they must be kept adequate for the purpose for which they are designed.

Foam Concentrate Surveillance

Because all air foam concentrates are water solutions of organic and inorganic chemicals of one sort or another, they must be carefully observed for changes in constitution and characteristics. They must be stored in shipping containers and in storage tanks according to the manufacturer's recommendations. Exposure to extreme heat, cold, contamination, or mixing with other materials must be avoided. Sedimentation or precipitate formation in containers or tanks of concentrate should be carefully checked periodically. The manufacturer or the manufacturer's representative is best qualified to test and determine the extent of reliability of foam concentrates under questionable conditions of deterioration of these liquids.

Equipment Testing

The performance of all foam equipment under emergency conditions is best guaranteed by the initial acceptance test and periodic inspection of the equipment. Original design specifications of fixed installations should include provision for periodic testing of proportioners, pumps, and other ancillary equipment, without the trial distribution of foam to the hazard being protected.

Problems of corrosion, clogged orifices, sticky valves, and electric circuitry malfunction may be detected by suitable means without full system activation.

In the absence of actual fire, and without full system activation, the complete testing of foam equipment performance for fire protection adequacy may be accomplished by various means. In the absence of actual fire, complete testing may be accomplished by several physical test methods. However, because these tests are similar in nature to "fingerprint" tests, their results can only be interpreted by comparison with similar tests conducted at the installation acceptance of the equipment, or as provided for under the performance guarantee of the manufacturer.

Foam equipment performance can be tested for the following physical characteristics of foam:

1. The dimensions of the discharge pattern or patterns of the device or system.
2. The percent concentration of foam concentrate in the finished foam solution.
3. The degree of expansion of the finished foam.
4. The rate at which water drains from the foam, i.e., 25 percent drainage time. This correlates with its viscosity or rate of spreading over fuel surfaces.
5. Film-forming capacity of the foam concentrate.

The preceding tests require specially designed test equipment and standardized techniques. They require qualified operators and cannot be easily carried out in the field. Detailed information concerning the equipment required, methods used, and some interpretations of the test results will be found in the Appendices of NFPA 11, Tests for the Physical Properties of Foam and NFPA 412, *Standard for Evaluating Foam Fire Fighting Equipment on Aircraft Rescue and Fire Fighting Vehicles* (Tests for foam discharge patterns, AFFF type foams, and protein base foams).

The test for the concentration of the foam concentrate is very important because it indicates the efficiency of operation of the concentrate proportioning device in the system. It is easily performed in the field with a hand refractometer and volume measuring device.

Bibliography

NFPA Codes, Standards, Recommended Practices and Manuals. (See the latest *NFPA Codes and Standards Catalog* for availability of current editions of the following documents.)

NFPA 11, *Standard for Low Expansion Foam and Combined Agent Systems.*

NFPA 11A, *Standard for Medium and High Expansion Foam Systems.*

NFPA 11C, *Standard for Mobile Foam Apparatus.*

NFPA 18, *Standard on Wetting Agents.*

NFPA 78, *Lightning Protection Code.*

NFPA 403, *Recommended Practice for Aircraft Rescue and Fire Fighting Services at Airports and Heliports.*

NFPA 412, *Standard for Evaluating Foam Fire Fighting Equipment on Aircraft Rescue and Fire Fighting Vehicles.*

NFPA 414, *Standard for Aircraft Rescue and Fire Fighting Vehicles.*

NFPA 1901, *Standard for Automotive Fire Apparatus.*

Additional Readings

Aircraft Fire Protection and Rescue Procedures, IFSTA 206, 2nd ed., International Fire Service Training Association, Oklahoma State University, Stillwater, OK, 1978.

"Air Foam Equipment and Liquid Concentrates," *UL-162*, Underwriters Laboratories, Northbrook, IL, 1969.

Alvares, N. J., and Lipska, A. E., "The Effect of Smoke on the Production and Stability of High-Expansion Foam," *Journal of Fire and Flammability*, Vol. 3, Apr. 1972, pp. 88-114.

"Aqueous Foams—Their Characteristics and Applications," *Industrial and Engineering Chemistry*, American Chemical Society, Vol. 48, Nov. 1956.

Avant, C. E., "Subsurface Foam KOs Fire," *Fire Engineering*, Apr. 1973.

Ballas, Thomas, "Electric Shock Hazard Studies of High Expansion Foam," *Fire Technology*, Vol. 5, No. 1, Feb. 1969, pp. 38-42.

Beers, R. J., "High Expansion Foam Fire Control for Records Storage," *Fire Technology*, Vol. 2, No. 2, May 1966, pp. 108-117.

Beiggs, A. A., "Use of Nitrogen-Filled High Expansion Foam to Protect a 500-Ton Fuel Tank," *Fire Research Note 1074*, Fire Research Station, Boreham Wood, Herts, England, Aug. 1977.

Breen, D. E., "Hangar Fire Protection with Automatic AFFF Systems," *Fire Technology*, Vol. 9, No. 2, May 1973, pp. 119-131.

Burford, R. R., "The Use of AFFF in Sprinkler Systems," *Fire Technology*, Vol. 12, No. 1., Feb. 1976.

Butlin, R. N., "High-Expansion Air Foam, A Survey of Its Properties and Uses," *Fire Research Note No. 669*, Fire Research Station, Boreham Wood, Hertsfordshire, England, May 1967.

Casey, James F., ed., "Foam," *Fire Service Hydraulics*, 2nd ed., Part 4, R. H. Donnelley Corporation, NY, 1970.

"Comparative Nozzle Study for Applying AFFF on Large Scale Fires," *CEEDO-TR-78-22*, Civil and Engineering Development Office, Tyndall Air Force Base, FL.

Cray, E. W., "High-Expansion Foam," *NFPA Quarterly*, Vol. 58, No. 1, July 1964, pp. 57-63.

Erriksson, Lars, "Semi-Subsurface Foam System," *NFPA Quarterly*, Vol. 58, No. 1, July 1964, pp. 54-56.

"Fighting Fuel Storage Fires with Subsurface Foam," *Fire International*, Vol. 3, No. 26, Oct. 1969.

"Fire Extinguishing Agent, Aqueous Film Forming Foam (AFFF) Liquid Concentrate, Six Percent for Fresh and Sea Water," *Military Specification MIL-F-24385 (NAVY)*, Amendment 8, Naval Ship Engineering Center, Department of the Navy, Hyattsville, MD.

Fittes, D. W., Griffiths, D. J., and Nash, P. "Extinction of Experimental Aircraft Fires with Light Water," *Fire*, Vol. 62, No. 773, Nov. 1969, pp. 315-317.

Fittes, D. W., Griffiths, D. J., and Nash, P. "The Use of Light Water for Major Aircraft Fires," *Fire Technology*, Vol. 5, No. 4, Nov. 1969, pp. 284-298.

"Flammable Liquid Storage Tank Protection," National Foam System, Inc., Lionville, PA.

"Foam Blanket Rapidly Quells Fire in Gas Storage Tank," *Volunteer Firefighter*, Mar. 1966.

"Foam Liquid Concentrates," National Foam System, Inc., Lionville, PA, 1982.

"Foam Liquid Proportioning," National Foam System, Inc., Lionville, PA, 1983.

Geyer, G. B. "Extinguishing Agents for Hydricarbon Fuel Fires," *Fire Technology*, Vol. 5, No. 2, May 1969, pp. 151-159.

———, "Foam and Dry Chemical Application Experiments," National Aviation Facilities Experimental Center, Federal Aviation Administration, Atlantic City, NJ, Dec. 1968.

Hammack, James M., "Talking Extinguishing Equipment: Effects of Low Temperatures and Combustion Products on High Expansion Foam," *Fire Journal*, Vol. 64, No. 4, Sept. 1970, pp. 93-94.

Herzog, G. R., "Recent Major Floating Roof Tank Fires and Their Extinguishment," *Fire Journal*, Vol. 68, No. 4, July 1974, p. 93.

Hird, D., "The Use of Foaming Agents for Aircraft Crash Fires," *Fire*, Sept. 1974, pp. 179-180.

Hird, D., Rodrigues, A., and Smith, D., "Foam—Its Efficiency in Tank Fires," *Fire Technology*, Vol. 6, No. 1, Feb. 1970, pp. 5-11.

Hoshino, M., and Hayashi, K., "Evaluation of Extinguishing Abilities of Fire-fighting Foam Agents Applied Alone or Together," *Report of the Fire Research Institute of Japan*, No. 46, 1978.

King, Carl J., "Maintenance of Low- and High-Expansion Foam Systems," *Fire Journal*, Vol. 67, No. 2, Mar. 1973, pp. 41-43, 48.

Lev. Y., "The Use of Foam for LNG Fire Fighting," *Fire Technol-*

ogy, Vol. 17, No. 1, Feb. 1981, pp. 17-24.

"Light Water, AFFF/ATC," Fire Protection Systems, Commercial Chemicals Division, 3M Company, St. Paul, MN, 1981.

Mahley, H. S., "Fire Tank Fires Subsurface," *Hydrocarbon Processing*, Aug. 1975.

Martin, G. T. O., "Fire Fighting Foam," *The Institution of Fire Engineers Quarterly*, Vol. 32, No. 86, June 1972, pp. 165-176.

Meldrum, D. N., "Aqueous Film-Forming Foam—Facts and Fallacies," *Fire Journal*, Vol. 66, No. 1, Jan. 1972, pp. 57-64.

Meldrum, D. N., "Fire Fighting with Foam: Basics of Effective Systems," *SFPE TR 79-2*, Society of Fire Protection Engineers, Boston, MA, 1979.

Meldrum, D. N., and Williams, J. R., "Dry Chemical-Compatible Foam," *Fire Journal*, Vol. 59, No. 6, Nov. 1965, pp. 18-22.

Meldrum, D. N., Williams, J. R., and Gilroy, D., "Foam Fire Protection of Liquid Propellants," *Fire Technology*, Vol. 2, No. 3, Aug. 1966, pp. 234-238.

Nash, P., and Whittle, J., "Fighting Fires in Oil Storage Tanks Using Base Injection of Foam," *Fire Technology*, Part I, Feb. 1978, Part II, May 1978.

Nash, P., and Yount, R. A., *Automatic Sprinkler Systems for Fire Protection*, 1st ed., Ch. 15, Victor Green Publications Ltd., London, England.

Peterson, H. F., "Suppression Capability of Foam on Runways," *NFPA Aviation Bulletin No. 250*, National Fire Protection Association, Quincy, MA.

Peterson, H. B., et al., "Full-Scale Fire Modeling Test Studies of Light Water and Protein Type Foams," *NRL Report 6573*, 1967, U. S. Naval Research Laboratory, Washington, DC.

Pignato, J. A., "Enhancement of Fire Protection with Directional Cooling Spray Nozzles by Use of AFFF," *Fire Technology*, Vol. 19, No. 1, Feb. 1983, pp. 5-13.

Pignato, J. A., "Storage Tank Protection Using AFFF," *Fire Journal*, Vol. 71, No. 6, Nov. 1978, p. 32.

Rasbash, D. J., "Notes for Specification of High Expansion Foam Liquid," *Fire Research Note No. 706*, Fire Research Station, Boreham Wood, Hertfordshire, England, Apr. 1968.

"Results of Tests Using Low Expansion Foams to Control and Extinguish Simulated Industrial Fires," prepared for 3M Company by Wesson and Associates, Inc., Norman, OK, Aug. 1983.

Sarkinen, C. F., and Nellis, C. L., "Electrical Tests on High Expansion Foams," *Technical Report No. ERJ-78-67*, U. S. Dept. of Energy, Bonneville Power Administration, Vancouver, WA, May 26, 1978.

Sherad, Shirley E., ed., *Fire Fighting Foams and Foam Systems*, National Fire Protection Association, Quincy, MA, 1977.

"Suffocation in High Expansion Foam," *Fire International*, Vol. 18, Oct. 1967, pp. 27-30.

"Tank Fire Extraction Tests," *Fire Protection Review*, Jan. 1977.

Tuve, R. L., "Light Water and Potassium Bicarbonate—A New Two-Agent Extinguishing System," *NFPA Quarterly*, Vol. 58, No. 1, July 1964, pp. 64-69.

Tuve, R. L., Peterson, H. B., Jablonski, E. J., and Neill, R. R., "A New Vapor-Securing Agent for Flammable-Liquid Fire Extinguishment," U. S. Naval Research Laboratory, Washington, DC, March 13, 1964.

———. "Twinned Agent Extinguishing System Uses," *Fire Engineering*, Vol. 117, No. 5, May 1964, pp. 358-360.

Wesson, H. R., Walker, J. R., and Brown, L. E., "Control LNG Spill Fires," *Hydrocarbon Processing*, Dec. 1972.

Woodman, Alan L., et al., "AFFF Spreading Properties at Elevated Temperatures," *Fire Technology*, Vol. 14, No. 4, Nov. 1978, pp. 265-272.

Young, R. A., and Corrie, J. G., "The Performance of a Foam-Sprinkler Installation on Simulated Oil Rig Fires," *Report CP 98/75*, Borehamwood, Hertfordshire, England, 1975.

Young, J. R., and Fitzgerald, P. M., "The Feasibility of Using Light Water Brand AFFF in a Closed-head Sprinkler System for Protection Against Flammable Liquid Spill Fires," *Factory Mutual Research Report No. 22352*, Factory Mutual Research Corp., Norwood, MA, Jan. 1975.

Zuber, K., "LNG Facilities-Engineered Fire Protection Systems," *Fire Technology*, Vol. 12, No. 1, Feb. 1976.

COMBUSTIBLE METAL AGENTS AND APPLICATION TECHNIQUES

Revised by Andrew S. Prokopovitsh

A variety of metals burn, particularly those in finely divided form. Some metals burn when heated to high temperatures by friction or exposure to external heat; others burn from contact with moisture or in reaction with other materials. Because accidental fires may occur during the transportation of these materials, it is important to understand the nature of the various fires and the hazards involved.

The hazards involved in the control or complete extinguishment of metal fires include extremely high temperatures, steam explosions, hydrogen explosions, toxic products of combustion, explosive reaction with some common extinguishing agents, breakdown of some extinguishing agents with the liberation of combustible gases or toxic products of combustion, and, in the case of certain nuclear materials, dangerous radiation. Some agents displace oxygen, especially in confined spaces. Therefore, extinguishing agents and methods for their specific application must be selected with care. Some metal fires should not be approached without suitable self-contained breathing apparatus and protective clothing, even if the fire is small. Other metal fires may be readily approached with minimum protection; still others may have to be fought with unmanned, fixed equipment.

Numerous agents have been developed to extinguish combustible metal (Class D) fires, but a given agent does not necessarily control or extinguish all metal fires. Although some agents are valuable in working with several metals, other agents are useful in combating only one type of metal fire. Despite their use in industry, some of these agents provide only partial control and cannot be considered actual extinguishing agents. Certain agents that are suitable for other classes of fires should be avoided for metal fires, because violent reactions may result (e.g., water on sodium; vaporizing liquids on magnesium fires).

Certain combustible metal extinguishing agents have been used for years, and their success in handling metal fires has led to the designation "approved extinguishing powder" and "dry powder." These designations have appeared in codes and other publications where it was not possible to employ the proprietary names of the powders. These terms have been accepted in describing extinguishing agents for metal fires and should not be confused with the name "dry chemical," which normally applies to an agent suitable for use on flammable liquid (Class B) and live electrical equipment (Class C) fires. Other extinguishing agents discussed herein have been used only experimentally in limited areas or at specific installations, and require much judgment in application.

The successful control or extinguishment of metal fires depends to a considerable extent upon the method of application and the training and experience of the fire fighter. Practice drills should be held on the particular combustible metals on which the agent is expected to be used. Prior knowledge of the capabilities and limitations of agents and associated equipment is always useful in emergency situations. Fire control or extinguishment will be difficult if the burning metal is in a place or position where the extinguishing agent cannot be applied in the most effective manner. In industrial plant locations where work is performed with combustible metals, public fire departments and industrial fire brigades have the advantage of fire control drills conducted under the guidance of knowledgeable individuals.

The transportation of combustible metals creates unique problems, in that a fire could occur in a location where necessary fire fighting knowledge and suitable extinguishing agents are not readily available. The U.S. Department of Transportation (DOT) has anticipated such situations and specifies cargo limitations, labeling, and placarding for the various means of transportation. Storage and handling of metals are discussed in Section 5, Chapter 10, "Metals," which also describes methods of fighting metal fires based on the types of metals involved.

APPROVED COMBUSTIBLE METAL EXTINGUISHING AGENTS

A number of proprietary combustible metal extinguishing agents have been submitted to testing agencies for approval or listing. Others have not, particularly those agents developed for special metals in rather limited commercial use. Those extinguishing agents described as

Mr. Prokopovitsh is a metallurgist and staff engineer at the U.S. Department of Interior (DOI), Bureau of Mines, Washington, DC.

follows have been approved or listed for use on fires in magnesium, aluminum, sodium, potassium, and sodium-potassium alloy.

G-1 Powder

"Pyrene" G-1 powder is composed of screened graphitized foundry coke to which an organic phosphate has been added. A combination of particle sizes is used to provide good packing characteristics when applied to a metal fire. The graphite acts as a heat conductor and absorbs heat from the fire to lower the metal temperature below the ignition point, which results in extinguishment. The closely packed graphite also smothers the fire, and the organic material in the agent breaks down with heat to yield a slightly smoky gas that penetrates the spaces between the graphite particles, excluding air. The powder is nontoxic and noncombustible.

G-1 powder is stored in cardboard tubes or metal pails, and can be stored for long periods of time without deterioration or caking. It is applied to the metal fire with a hand scoop or a shovel. The packing characteristics of the powder prevent its discharge from a fire extinguisher.

The powder is applied by spreading it evenly over the surface of the fire to a depth sufficient to smother the fire. A layer at least ½ in. (12.5 mm) deep is recommended for fires involving fines (finely divided particles) of magnesium and magnesium alloys. Larger chunks of metal require additional powder to cover the burning areas.

Where burning metal is on a combustible surface, the fire should be extinguished by: (1) first covering it with powder, (2) shoveling the burning metal onto another 1 or 2 in. (25 or 50 mm) layer of powder that has been spread out on a nearby noncombustible surface, and (3) adding more powder as needed.

G-1 powder is effective for fires in magnesium, sodium, potassium, titanium, lithium, calcium, zirconium, hafnium, thorium, uranium, and plutonium, and has been recommended for special applications on powder fires in aluminum, zinc, and iron. It is listed by Underwriters Laboratories Inc. (UL) for use only on magnesium and magnesium alloys (dry fines and moist fines that are not moistened or wetted with water or water soluble cutting oils) and is approved by the Factory Mutual System (FM) for use on fires in magnesium, aluminum, sodium, potassium, and sodium-potassium alloy. When plans call for use of G-1 powder on those metals mentioned in this paragraph, practice fire drills should be held in advance. The products of combustion of thorium, uranium, beryllium, and plutonium can be a health hazard, and precautions should be observed consistent with the usual procedures in combating fires in radioactive material.

MetalGuard Powder

MetalGuard powder is identical to G-1 powder in composition, and is simply a trade name variation.

Met-L-X Powder

This dry powder, with its particle size controlled for optimum extinguishing effectiveness, is composed of a sodium chloride base with additives. The additives include tricalcium phosphate to improve flow characteristics and metal stearates for water repellency. A thermoplastic material is added to bind the sodium chloride particles into a solid mass under fire conditions.

Met-L-X powder is noncombustible, and secondary fires do not result from its application to burning metal. No known health hazard results from the use of this agent. It is nonabrasive and nonconductive.

Stored in sealed containers or extinguishers, Met-L-X powder is not subject to decomposition or a change in properties. Periodic replacement of extinguisher charges is unnecessary. Extinguishers range from 30 lb (14 kg) portable hand units (carbon dioxide cartridge propellant), through 150 and 350 lb (16 and 160 kg) wheeled units, to 2,000 lb (900 kg) for stationary or piped systems. The wheeled units and piped systems employ nitrogen as the propellant.

The powder is suitable for fires in solid chunks (such as castings) because of its ability to cling to hot vertical surfaces. To control and then extinguish a metal fire, the nozzle of the extinguisher is fully opened and, from a safe distance (in order to prevent blowing the burning metal into other areas), a thin layer of agent is cautiously applied over the burning mass. Once control is established, the nozzle valve is used to throttle the stream to produce a soft, heavy flow. The metal can then be completely and safely covered from close range with a heavy layer. The heat of the fire causes the powder to cake, forming a crust which excludes air and results in extinguishment.

Met-L-X extinguishers are available for fires involving magnesium, sodium (spills or in depth), potassium, and sodium-potassium alloy (NaK). In addition, Met-L-X has been successfully used where zirconium, uranium, titanium, and powdered aluminum present serious hazards.

Comparison of G-1 and Met-L-X Powders

Based upon their past usage and known value as extinguishing agents for metal fires, the two agents previously discussed (G-1 and Met-L-X powders) are the most notable. Continuous experience with these agents has provided sufficient information to list, in Table 19-5A, the capabilities and limitations of each when applied to certain metal fires (Zeratsky 1960).

Na-X Powder

This powder was developed to satisfy the need for a low chloride content agent that could be used on sodium metal fires. Na-X has a sodium carbonate base with various additives incorporated to render the agent nonhygroscopic and easily fluidized for use in pressurized extinguishers. It also incorporates an additive which softens and crusts over an exposed surface of burning sodium metal.

Na-X is noncombustible, and does not cause secondary fires when applied to burning sodium metal above temperatures ranging from 1,200 to 1,500°F (649 to 849°C). No known health hazard results from the use of this agent on sodium fires, and it is nonabrasive and nonconductive.

Stored in 50 lb (23 kg) pails, 30 lb (14 kg) hand portables, and 150 and 350 lb (68 and 160 kg) wheeled and stationary extinguishers, Na-X is listed by UL for fires involving sodium metal (spills and in depth) at a fuel temperature of 1,200°F (649°C). Na-X has been tested on sodium metal (spills and in depth) at fuel temperatures as high as 1,500°F (816°C). Stored in the supplier's metal pails and extinguishers, Na-X is not subject to decomposition, so periodic replacement of the agent is unnecessary.

TABLE 19-5A. Comparison of G-1 and Met-L-X Powders

	G-1 Powder			Met-L-X		
Type of Fire	Capable of Complete Extinguishment	Capable of Control Only	Unsatisfactory	Capable of Complete Extinguishment	Capable of Control Only	Unsatisfactory
Dry or oily magnesium chips or turnings	X			X		
Magnesium castings and wrought forms	X*			X†		
Dry or oily titanium turnings	X‡			X		
Uranium turnings and solids	X			X		
Zirconium chips and turnings coated with water soluble oil	X			X		
Moist zirconium chips and turnings		X			X	
Sodium spills or in depth	X			X		
Sodium sprayed or spilled on vertical surfaces			X	X§		
Potassium or sodium-potassium alloy spill	X			X		
Potassium or sodium-potassium alloy fire in depth			X#			X#
Lithium spill	X			X		
Lithium fire in depth	X					X‖
Aluminum powder	X			X		

Notes:

* Requires sufficient powder to cover the burning pieces. More agent required than with Met-L-X.

† Powder clings to vertical surfaces. Unnecessary to bury burning parts.

‡ More effective pound for pound than Met-L-X.

§ Adheres to molten sodium on vertical surfaces.

\# Extinguished with difficulty.

‖ Powder sinks into molten metal, the sodium chloride reacting with lithium to form lithium chloride and sodium. If continued until sodium is in excess, the fire can then be extinguished.

OTHER COMBUSTIBLE METAL EXTINGUISHING AGENTS

Foundry Flux

In magnesium foundry operations, molten magnesium is protected from contact with air by layers of either molten or crust type fluxes. These fluxes, which are also used as molten metal cleaning agents, consist of various amounts of potassium chloride, barium chloride, magnesium chloride, sodium chloride, and calcium fluoride. The fluxes are stored in covered steel drums. When applied to burning magnesium, these fluxes melt on the surface of the solid or molten metal, excluding air. The thin layer of protection can be provided by properly applying relatively small amounts of flux.

Fluxes are valuable in extinguishing magnesium spill fires from broken molds or leaking pots, and in controlling and extinguishing fires in heat treating furnaces. In open fires, the flux is applied with a hand scoop or a shovel. Areas of furnaces that are difficult to reach can be coated by means of a flux throwing device similar to those used to throw concrete onto building forms.

While fluxes would rapidly extinguish chip fires in machine shops, such use is not recommended. The fluxes are hygroscopic and the water picked up from the air, combined with the salts, causes severe rusting of equipment.

Lith-X Powder

This dry powder is composed of a special graphite base with additives. The additives render it free flowing so it can be discharged from an extinguisher. The technique used to extinguish a metal fire with this agent is the same as that used with Met-L-X. Lith-X does not cake or crust over when applied to the burning metal. It excludes air and conducts heat away from the burning mass to effect extinguishment. It does not cling to hot metal surfaces, so it is necessary to completely cover the burning metal.

Lith-X will successfully extinguish lithium fires and is suitable for the control and extinguishment of magnesium and zirconium chip fires. It will extinguish sodium spill and sodium fires in depth. Sodium-potassium alloy spill fires are extinguished, and fires in depth are controlled.

TMB Liquid

TMB is the chemical abbreviation for trimethoxyboroxine (Tuve et al 1957). The agent contains an excess of methanol to render it free flowing. It is classed as a flammable liquid for shipping purposes. The liquid is colorless and hydrolyzes readily to form boric acid and methanol. Contact with moist air or other sources of water must be avoided to prevent hydrolysis.

This agent is applied with a specially adapted 2½ gal (9.5 L) stored-pressure extinguisher which delivers either spray or straight stream. Typical application of TMB to a metal fire yields a heat flash due to the breakdown of the chemical compound and ignition of the methanol. The white metal fire is rapidly extinguished, and the secondary greenish flame is of short duration. A molten boric oxide coating on the hot metal prevents contact with air. A stream of water may be used to cool the mass as soon as metal flames are no longer visible; this should be done cautiously to avoid rupture of the coating. Indoor application (such as in machine shops) is not recommended due to the large volume of boric oxide smoke produced.

While TMB has been used primarily on magnesium fires, it has shown value in application to fires in zirconium and titanium. Although TMB applied as a spray has been used to control small sodium and sodium-potassium alloy fires, it is not recommended for fires in sodium, sodium-potassium alloy, and lithium. TMB reacts violently with lithium and sodium-potassium alloy. It will extinguish sodium in depth, but the protective coating formed by the TMB absorbs moisture very rapidly and in time may penetrate through to the sodium, resulting in a violent reaction. Field experience has been limited to aircraft fires.

Pyromet Powder

Pyromet powder is composed of specially processed sodium chloride, diammonium phosphate, protein, and a waterproofing and flow promoting agent. The powder is discharged under pressure provided by a carbon dioxide gas cartridge. The unit contains 25 lb (11 kg) of powder. The applicator consists of a tubular extension from the control valve, terminating in a cone shaped nozzle. A mechanism in the nozzle absorbs the discharge pressure by swirling the powder as it is expelled. This enables the operator to let the powder fall gently on the burning metal rather than to scatter burning material under the blast of a jet of powder.

Pyromet has proven effective in handling fires involving sodium, calcium, zirconium, and titanium, as well as magnesium and aluminum in the form of powder or chips.

T.E.C. Powder

T.E.C. (Ternary Eutectic Chloride) powder is a mixture of potassium chloride, sodium chloride, and barium chloride that is effective in extinguishing fires in certain combustible metals. The powder tends to seal the metal, excluding air. On a hot magnesium chip fire its action is similar to that of foundry flux. In tests reported in *Fire Technology*, T.E.C. powder was the most effective salt for control of sodium, potassium, and sodium-potassium alloy fires (Rodgers and Everson 1965).

Small uranium and plutonium fires within scientific glove boxes have been extinguished with T.E.C. A small plastic bag filled with the powder is simply placed directly on the metal fire. The barium chloride in the mixture is toxic, but this is of no practical concern if the glove box remains intact. In other locations, the operator should avoid inhalation of the airborne powder.

NONPROPRIETARY COMBUSTIBLE METAL EXTINGUISHING AGENTS

Talc (Powder)

Talc, which has been used industrially on magnesium fires, acts to control rather than extinguish fire. Talc acts as an insulator to retain the heat of the fire, rather than as a coolant. It does, however, react with burning magnesium to provide a source of oxygen. The addition of organic matter (such as protein) to talc assists in the controlling action, but does not prevent the reaction which releases oxygen to the fire.

Graphite Powder

Graphite powder (plumbago) has been used as an extinguishing agent for metal fires. Its action is similar to that of G-1 powder in that the graphite acts as a coolant. Unless the powder is finely divided and closely packed over the burning metal, some air does get through to the metal and extinguishment is not as rapid as with G-1 powder.

Sand

Dry sand has often been recommended as an agent for controlling and extinguishing metal fires. At times it seems to be satisfactory, but usually hot metal (such as magnesium) obtains oxygen from the silicone dioxide in the sand and continues to burn under the pile. Sand is seldom completely dry. Burning metal reacting with the moisture in the sand produces steam and, under certain conditions, may produce an explosive metal-water reaction. By laying the sand around the perimeter of the fire, fine, dry sand can be used to isolate incipient fires of aluminum dust.

Cast Iron Borings

Cast iron borings or turnings are frequently available in the same machine shops as the various combustible metals. Clean iron borings applied over a magnesium chip fire cool the hot metal and help extinguish the fire. This agent is used by some shops for handling small fires where, with normal good housekeeping, only a few combustible metal chips are involved. Contamination of the metal chips with iron may be an economic problem. Oxidized iron chips must be avoided to prevent possible thermite reaction with the hot metal, and the iron chips must be free from moisture.

Sodium Chloride

Alkali metal fires can be extinguished by sodium chloride, which forms a protective blanket that excludes air over the metal so that the metal cools below its burning temperature. Sodium chloride is an agent that is used for extinguishing sodium and potassium fires. It can also be used to extinguish magnesium fires.

Soda Ash

Sodium carbonate or soda ash (not dry chemical) is recommended for extinguishing sodium and potassium fires. Its action is similar to that of sodium chloride.

Lithium Chloride

Lithium chloride is an effective extinguishing agent for lithium metal fires. However, its use should be limited to specialized applications because the chemical is hygroscopic to a degree and may present problems because of the reaction between the moisture and the lithium.

Zirconium Silicate

This agent has been used successfully to extinguish lithium fires.

Dolomite

If zirconium or titanium in the form of dry powder becomes ignited, neither can be extinguished easily. Control can be effected by spreading dolomite (a carbonate of calcium and magnesium) around the burning area and then adding more powder until the burning pile is completely covered.

Boron Trifluoride and Boron Trichloride

Boron trifluoride and boron trichloride have both been used to control fires in heat treating furnaces containing magnesium. The fluoride is considerably more effective. In the case of small fires, the gases provide complete extinguishment. In the case of large fires, the gases effect control over the flames and rapid burning, but reignition of the hot metal takes place on exposure to air. A combined attack of boron trifluoride gas followed by application of foundry flux completely extinguishes the fire. For details of gas application, see NFPA 48, *Standard for the Storage, Handling and Processing of Magnesium* (hereinafter referred to as NFPA 48).

Inert Gases

In some cases, inert gases (such as argon and helium) will control zirconium fires if they can be used under conditions that will exclude air. Gas blanketing with argon has been effective in controlling lithium, sodium, and potassium fires (Rodgers and Everson 1965). Caution should be exercised when using the agent in confined spaces because of the danger of suffocation to personnel.

Water

In a shop where magnesium or other combustible metals (except the alkali metals and fissionable materials) are being machined or fabricated, if a fire gets out of control to the point where automatic sprinklers open, the large volume of water from the sprinklers normally will extinguish both the Class A and magnesium fires. Where automatic sprinkler protection is provided, a deflecting shield or hood should be provided over furnaces, reactors, or other places where hot or molten metal may be present. Additional information on the use of sprinkler protection in shops handling magnesium or titanium can be found in NFPA 48 and NFPA 481, *Standard for the Production, Processing, Handling and Storage of Titanium.*

When burning metals are spattered with limited amounts of water, the hot metal extracts oxygen from the water and promotes combustion. At the same time, hydrogen is released in a free state and ignites readily. Since small amounts of water do accelerate combustible metal fires (particularly where chips or other fines are involved), use of common portable extinguishers containing water is not recommended except to control fires in adjacent Class A materials.

Water, however, is a good coolant and can be used on some combustible metals under proper conditions and applications to reduce the temperature of the burning metals below the ignition point. The following paragraphs discuss the advantages and limitations of using water on fires involving various combustible metals.

Water on Sodium, Potassium, Lithium, NaK, Barium, Calcium, and Strontium Fires: Water applied to sodium, potassium, lithium, sodium-potassium alloys (NaK), barium, and probably calcium and strontium will induce chemical reactions that will lead to fire or explosion even at room temperature. Therefore, water must not be used on fires involving these metals.

Water on Zirconium Fires: Powdered zirconium wet with water is more difficult to ignite than the dry powder. However, once ignition takes place, wet powder burns more violently than dry powder. Powder containing about five to ten percent water is considered to be the most dangerous. Small volumes of water should not be applied to burning zirconium, but large volumes of water can be successfully used to completely cover solid chunks or large chips of burning zirconium (e.g., by drowning the metal in a tank or barrel of water). Hose streams applied directly to burning zirconium chips may yield violent reactions.

Water on Plutonium, Uranium, and Thorium Fires: Limited amounts of water add to the intensity of a fire in natural uranium or thorium, and greatly increase the contamination cleanup required after the fire. A natural uranium scrap fire can be fought with water by personnel (wearing face shields and gloves and using long handled shovels) shoveling the burning scrap into a drum of water in the open. The hydrogen formed may ignite and burn off above the top of the drum. The radioactivity hazard of natural uranium is extremely low (in reality, uranium is a metal poison, although considerably less toxic than lead). The use of water on enriched uranium or plutonium (fissionable materials) is generally prohibited. If ingested, plutonium is considerably more hazardous to humans than uranium.

Water on Magnesium Fires: Although water in small quantities accelerates magnesium fires, rapid application of large amounts of water is effective in extinguishing magnesium fires because of the cooling effect of water. Automatic sprinklers will extinguish a typical shop fire where the quantity of magnesium is limited. However, water should not be used on any fire involving a large number of magnesium chips when it is doubtful that there is sufficient water to handle the large area. (A few burning chips can be extinguished by dropping them into a bucket of water.) Small streams from portable extinguishers will violently accelerate a magnesium chip fire.

Burning magnesium parts such as castings and fabricated structures can be cooled and extinguished with coarse streams of water applied with standard fire hoses. A straight stream scatters the fire, but coarse drops (produced by a fixed nozzle operating at a distance or by use of an adjustable nozzle) flow over and cool the unburned metal. Some temporary acceleration normally takes place with this procedure, but rapid extinguishment follows if the technique is pursued. Well advanced fires in several hundred pounds (100 lb equals 45 kg) of magnesium scrap have been extinguished in less than 1 minute with two 1½ in. (37.5 mm) fire hoses. Water fog, on the other hand, tends to accelerate rather than cool such a fire. Application of water to magnesium fires must be avoided where quantities of molten metal are likely to be present; the steam formation and possible metal-water reactions may be explosive.

Water on Titanium Fires: Water must not be used on fires in titanium fines and should be used with caution on other titanium fires. Small amounts of burning titanium (other than fines) can be extinguished and considerable salvage realized by quickly dumping the burning material into a large volume of water to completely submerge it. Hose streams have been used effectively on fires in outside piles of scrap, but violent reactions have been reported in other cases where water was applied to hot or burning titanium, resulting in serious injury to personnel.

Miscellaneous Agents

Agents that have been tested for cooling capacity and ability to extinguish magnesium fires include the following: (1) boric acid dissolved in triethylene glycol, followed by foam to extinguish the secondary fire (McCutchan 1954); (2) diisodecyl phthalate combined with chlorobromomethane (Greenstein and Richman 1955); and (3) tricresyl phosphate followed by foam to handle the secondary fire. The Joint Fire Research Organization has developed a suitable powder for extinguishing burning magnesium, which consists essentially of powdered polyvinyl chloride and sodium borate (Nash 1963). The powder forms a coating on the molten metal and enables a fine water spray to be used to cool the metal to extinction. Research has been conducted at two U.S. laboratories on optimum methods of controlling zirconium and uranium fires with mixed halogen organic derivatives (Lawrence 1978).

Among the more recent literature on combustible metals is a report on computer modeling of titanium combustion. The report describes an extensive program of analytical and experimental investigation to establish the parameters that govern the ignition and self-sustained combustion of metals such as titanium. The analytical model allows prediction of frictional heating and exposure to high ambient temperatures (Glickstein 1980). Other pertinent publications include suggestions for safely machining magnesium metal (Morales 1980), and for determining the pyrophorosity of the liquid mixing of combustible metal powders. The flammability of metal charges dried after liquid and dry mixing is stated to be different and is connected with a substitution of oxygen adsorbed on the metal surface. From the viewpoint of firesafety, maximum diluents are suggested to minimize the opportunity for fire or explosion (Afanasieva et al 1981).

Bibliography

References Cited

Afanasieva, L. F., Chernenko, E. V., and Rozenband, V. I. 1981. "Pyrophorosity of Liquid Mixing of the Charge for Titanium Carbide Production." *Poroshk, Metall.* Vol 5. May 1981. pp 90-94.

Glickstein, M. R. 1980. "Computer Modeling of Titanium Combustion," *Tri-Service Conference on Corrosion, Vol II.* U.S. Air Force Wright Aeronautical Laboratories, Wright-Patterson AFB, OH, 5-7 Nov 1980. pp 93-122.

Greenstein, L. M., and Richman, S. I. 1955. "A Study of Magnesium Fire Extinguishing Agents." *WADC Technical Report No. 55-170.* U.S. Air Force Wright Air Development Center (by Francis Earle Laboratories, Inc.). Wright-Patterson AFB, OH.

Lawrence, Kenneth D. 1978. "New Agents for the Extinguishment of Magnesium Fires," *CEEDO Technical Report No. 78-19.* Civil and Environmental Engineering Development Office, Tyndall Air Force Base, FL.

McCutchan, R. T. 1954. "Investigation of Magnesium Fire Extinguishing Agents." *WADC Technical Report No. 55-5.* U.S. Air Force Wright Air Development Center (by Southwest Research Institute). Wright-Patterson AFB, OH.

Morales, H. J. 1980. "Machining Magnesium Safely." *American Machinist.* Vol 124, No 10. Oct 1980. pp 154-155.

Nash, P. 1963. "A Dry Powder Extinguishing Agent for Magnesium Fires." *Institution of Fire Engineers Quarterly.* Vol 23, No 51, Sept 1963. pp 275-276.

Rodgers, S. J., and Everson, W. A. 1965. "Extinguishments of Alkali Metal Fires." *Fire Technology.* Vol 1, No 2. May 1965. pp 103-111.

Tuve, R. I., Gipe, R. L., Peterson, H. B., and Neil, R. R. 1957. "The Use of Trimethoxyboroxine for the Extinguishment of Metal Fires." *NRL Report No. 4933.* Naval Research Laboratory, Washington, DC.

Zeratsky, E. D. 1960. "Extinguishing Agents for Combustible Metals and Special Chemicals," *Safety Maintenance.* Vol 120, No 2. Aug 1960. pp 28-32.

NFPA Codes, Standards, Recommended Practices and Manuals. (See the latest *NFPA Codes and Standards Catalog* for availability of current editions of the following documents.)

NFPA 48, *Standard for the Storage, Handling, and Processing of Magnesium.*

NFPA 481, *Standard for the Production, Processing, Handling, and Storage of Titanium.*

NFPA 482, *Standard for the Production, Processing, Handling, and Storage of Zirconium.*

Additional Readings

"Fire Protection for Combustible Metals," *Data Sheet 567,* National Safety Council, Chicago, IL, 1965.

"Lithium," *Data Sheet No. D-566,* National Safety Council, Chicago, IL, 1978.

"Magnesium Data Sheet," *No. 426A,* National Safety Council, Chicago, IL, 1975.

"Methods Used by an AEC Contractor for Handling Sodium, NaK, and Lithium," *Accident and Prevention Information,* Feb. 1960, U.S. Atomic Energy Commission, Washington, DC.

Petersen, M. E., "Dry Powder for Combustible Metals," *National Safety News,* Vol. 92, No. 1, Jan. 1963.

Riley, John F., "Na-X, a New Fire Extinguishing Agent for Metal Fires," *Fire Technology,* Vol. 10, No. 4, Nov. 1974.

Stout, E. L., "Safety Considerations for Handling Plutonium, Uranium, Thorium, the Alkali Metals, Zirconium, Titanium, Magnesium, and Calcium," *Report No. LA-2147,* Los Alamos Scientific Laboratory, Los Alamos, NM.

"Zirconium Powder Data Sheet," *No. D-382A,* National Safety Council, Chicago, IL, 1974.

SPECIAL SYSTEMS AND EXTINGUISHING TECHNIQUES

Revised by Peter F. Johnson

A range of fire extinguishing and control systems, agents, devices, and techniques have been developed and used with varying degrees of success, but do not fit into those commonly recognized categories of systems described in previous chapters of this section. Many of the systems and techniques are novel or still in the experimental stage. Often they are directed toward special applications such as aircraft or coal mines for which traditional extinguishing methods are not entirely suited.

This chapter describes some of these special systems and techniques by grouping them as follows:

1. Techniques based on solids and powders.
2. Systems using water or water solutions for fire control.
3. Inerting gases used for fire or explosion suppression.
4. Combined agent systems.

For a complete discussion of the more conventional extinguishing systems refer to the other chapters in this section: Chapter 1, "Carbon Dioxide and Application Systems;" Chapter 2, "Halogenated Agents and Systems;" Chapter 3, "Dry Chemical Agents and Application Systems;" Chapter 4, "Foam Extinguishing Agents and Systems;" and Chapter 5, "Combustible Metal Agents and Application Techniques." In addition, reference should be made to Section 17, "Water Based Extinguishing Systems," and for removal of emulsifying agents, to Section 6, Chapter 5, "Hazardous Waste Control."

SOLIDS AND POWDERS

Solid materials in powder or granular form have been used for many years to extinguish fires. Use of solids to cover and suppress fires by excluding oxygen is a common technique; rows of sand filled buckets for fire fighting were once a familiar sight. Solids in the form of dry powder systems are still a widely used extinguishing technique. Special proprietary powders and a range of common materials such as talc, graphite, and soda ash are capable of suppressing various metal fires, as well.

Mr. Johnson is supervising materials scientist at Central Investigation and Research Laboratory (CIRL), Department of Housing and Construction, Melbourne, Australia.

A number of more recent developments have taken place in the use of powders and other solid materials to suppress special fire hazards.

Granules for Flammable Liquid Pool Fires

Problems associated with fighting large fires of flammable liquids, such as major spills from storage tanks, have led to considerable research in this area. One of the more dangerous fire situations occurs when a major spill of liquefied natural gas (LNG) flows into a dike surrounding a storage tank. Conventional fire fighting methods for large LNG fires have serious limitations which include the following:

1. Vast quantities of dry chemical are required to guarantee total LNG fire extinguishment.
2. Because of their high water content, low expansion foams are unsuitable for LNG fires.
3. High and medium expansion foams can control (but not extinguish) LNG fires, and large quantities of foam are required to continually replace foam destroyed by fire.

A novel concept, therefore, has been to use a "solid foam" consisting of ¾ to 1½ in. (20 to 40 mm) cubes of cellular glass in a layer about 8 in. (200 mm) deep to cover the full extent of an LNG spill where it is contained in a storage dike (Lev 1981). The granules are of nonflammable, low density solid material which is stored within the dike in bags. Loose filled material (with a suitable binder on the top) is also used. When a LNG spill occurs, the granules float to the top of the liquid and control the rate of burning.

Both small tray and large scale tests have shown the glass granules to be most effective. The material shields the LNG surface from back radiation from the flames and reduces the evaporation of the fuel from the liquid surface. This limits the fuel burning rate and overall size of the fire. In addition, radiation at a distance from the spill fire, which is the major contribution to fire spread, is reduced by approximately 90 to 95 percent, compared with LNG fires without the cellular glass granules.

Carbon Microspheres for Metal Fires

Specially formulated extinguishing powders are generally used to suppress fires involving metals. Because of the reaction of many metals with water, sprinklers (and the use of other water based agents) are not appropriate, and in some cases quite dangerous.

However, many of the special agents for metal fires are at times unsatisfactory because they are (1) corrosive, (2) applied manually rather than by an extinguishing system, (3) capable of clogging extinguishing nozzles, and (4) expensive.

Studies have been undertaken to examine the effectiveness of carbon microspheres or microspheroids to extinguish fires involving alkali metals, such as sodium, sodium-potassium, and lithium (McCormick and Schmitt 1974). These microspheres are petroleum-coke based particles with a diameter of approximately 100 to 500 microns. The particles possess high thermal conductivity, chemical inertness, and excellent flow characteristics and are capable of being directed onto fires from dry chemical type extinguishers and conventional nozzles.

Tests have shown that carbon microspheres compare favorably in performance to other metal extinguishing agents (McCormick and Schmitt 1974). In particular, experiments with carbon microspheroids incorporating neutron absorbers have been effective in extinguishing fires involving nuclear fissionable materials such as uranium metal powder (Schmitt 1975). The excellent flow characteristics and noncaking properties of these microspheres suggest an effective way to extinguish radioactive metal fires within the inert atmosphere glove box enclosures which are used in the nuclear industry.

Ground Cover Materials—Aircraft Fuel Fires

Aircraft fuel spill fires are another significant hazard in which solid materials have been used for extinguishment. Research in the U.S. suggests that crushed rock graded to approximately ⅜ in. (10 mm) can reduce fire growth and spread of flames in hydrocarbon spill fires (Geyer 1973). The suppression mechanism seems to be similar to that for cellular glass granules—the solid material largely blocks back-radiation of the flames to the liquid surface and prevents heat from being distributed to the fuel in front of the flame by convection currents. Ground rock materials do not float and must cover the fuel surface by 1 to 2 in. (25 to 50 mm) to effectively retard flame growth.

Dry Chemical Projectile

Since the 1940s, a number of interesting suppression techniques have been developed using projectiles of dry chemical that can be fired from a distance at a target fire.

One such system developed in the U.S. for coal mine fires used a cannon like dispersal system. Upon an alarm from an ultraviolet detector, mono-ammonium phosphate powder was propelled from the cannon by high pressure nitrogen to quench potential coal explosions and methane-air ignitions (NAS 1980). In reality the devices proved somewhat inpractical because of space limitations.

Dr. John L. Bryan of the University of Maryland reports on a related extinguishing technique developed some years ago that utilized a device known as a "gren gun." This gun consisted of cylinders that rotated in turn into the firing position. The projectiles contained 1¾ lb (0.8 kg) of dry chemical that, when fired off with compressed air, could reach distances of 200 ft (60 m) (Bryan 1982). The "gren gun" was designed to operate with a minimum air pressure of 150 psi (1030 kPa) from a compressed air cylinder carried on the back of the operator. The gun itself was constructed of lightweight aluminum, weighed approximately 10 lb (4½ kg), and was fired from the shoulder. Being portable, it was ideal as a first attack suppression device that could place dry chemical on a target from a safe, remote distance.

SYSTEMS UTILIZING WATER OR WATER SOLUTIONS

A number of specialized systems using water or water based agents have been developed to fill particular fire control needs. These extinguishing agents remove heat from the combustion process or form a noncombustible emulsion with the fuel.

Ultra Highspeed Water Spray Systems

These systems are designed to handle extremely rapid fires such as those which can occur during the handling of solid propellants or sensitive chemicals, or in any industrial process or oxygen enriched environment.

Wetting Agent Systems

Most wetting agent systems (as opposed to wetting agents used through manually operated hose lines from mobile tank supplies through portable proportioners) are designed to dispense wetting agent foam. The design, volume, and purpose of each installation depends upon the hazard being protected. Normally, wetting agent systems meeting the requirements of NFPA 18, *Standard on Wetting Agents*, can use standard water spray, sprinkler, or foam system equipment; however, this should be carefully checked to assure maximum system efficiency. The detergent action of a wetting agent solution may introduce some special problems. Wet water foam is a mixture of wet water with air that forms a cellular structure foam. This foam breaks down into its original liquid state at temperatures below the boiling point of water. Breakdown is proportional to the heat exposure and is intended to sufficiently cool the surface on which the foam is applied. These properties improve the ability of water to protect exposed properties against heat transfer, and water requirements are thus appreciably reduced. Wet water systems can introduce the potential of increased water damage to exposed stocks (due to the high absorption ability of wet water), and increase the load on any flooring affected because of the retention of large volumes of wet water. These factors must be considered in the design of such installations.

Viscous ("Thickened") Water Systems

Systems for applying viscous water vary in detailed components, depending upon whether the equipment is designed for application from the ground or from aircraft. Most ground applications are from conventional fire tankers with some modifications. The NFPA manual *Chemicals for Forest Fire Fighting* lists the following three basic systems for ground applications (NFPA Forest Committee 1977):

Injector-recirculating Ground Tanker System: In this system, water is drawn from the tank bottom through a centrifugal pump and bypassed through an added return line to the injector-disperser, where the liquid and dry powder combine and flow into the tank. The centrifugal pump serves as a mixing device, continuing until the correct amount of powder is added.

Demand Viscous Water Tanker-Mixer: This system consists of an auxiliary mixing tank, an auger feed mechanism, and a positive displacement rotary meter which acts as a power unit for the feed system. This system uses the tank, centrifugal pump, and discharge lines of a conventional tanker.

Slip on Chemical Tanker-Mixer: This system is a skid mounted, all metal unit for mounting onto a heavy duty tractor-trailer for off road assignments. The major assemblies consist of a mixing tank, chemical hopper, hose reel, engine, pump, and pump priming tank. This system has only been used experimentally, utilizing a modified ordinary agricultural sprayer.

Conventional fire nozzles are used for viscous water or light gels. Those nozzles giving coarse droplets normally give satisfactory results, but fine spray nozzles should be avoided. Slurry type fire retardant chemicals need special nozzles because these chemicals quickly erode most conventional brass and aluminum nozzles. Cast alumina inserts (a ceramic aluminum oxide) seem to be satisfactory, however, and can be manufactured inexpensively.

Another NFPA manual, *Chemicals for Forest Fire Fighting*, gives additional helpful information on equipment for mixing the chemicals, pumps for handling the chemicals, distribution systems, storage of mixed and unmixed chemicals, and means for judging the quality of the solutions in the field (NFPA Forest Committee 1977).

The NFPA publication *Air Operations for Forest, Brush, and Grass Fires* contains information on the use of aircraft for fire control operations. This manual covers air operations plans, airports, heliports, helistops, suitable aircraft, operating principles and procedures, and amphibious operations. Special equipment is needed to utilize aircraft for these fire control purposes (NFPA Forest Committee 1975).

Emulsifying Agents—Detergents

Various types of emulsifying agents or detergents have been used with varying degrees of success in extinguishing or controlling flammable liquid spill fires either on land or on water. These agents also have been used on unignited spills of such liquids to minimize the possible ignition hazard. In general, these agents mix the oil and water to form a non-combustible emulsion.

A major problem with the use of emulsifying agents and detergents in lakes, rivers, harbors, etc., is that the resultant emulsions often have a more severe effect on aquatic forms of life than does oil alone; however, relatively nontoxic emulsifying agents may be developed in the future.

One interesting example of the use of local application emulsifying agent in cooking range hood protection was a system that used an aqueous potassium carbonate solution with minor amounts of sodium dichromate added to inhibit corrosion, and ethylene glycol to prevent freezing. It was dispersed in the form of a dense liquid spray from conventional nozzles located in the range hood. The alkaline solution not only removed heat by steam generation, but underwent a saponification reaction with burning fat or oil to produce a soap foam or emulsified layer that in turn produced inerting carbon dioxide. Similar aqueous potassium carbonate extinguishing systems for range hoods on naval vessels have also been used (Neill and Peterson 1969).

INERTING GAS SUPPRESSION

For many years a range of gases, including carbon dioxide, nitrogen and helium, has been used successfully to prevent ignition of potentially flammable mixtures or to extinguish fires, particularly those involving flammable liquids or gases. These methods have been based on the knowledge that fires can be extinguished if a sufficient volume of an inert gas is introduced into an enclosed space where a fire is burning. The concentration must be maintained in the space for an adequate period of time to prevent rekindling of the burned material.

Traditionally, this method of extinguishment has been explained as a reduction of the fuel or oxygen concentration to the point where combustion is prevented. Modern fire dynamics show that the mechanism is related to the fact that the added inerting gas acts as "thermal ballast" to reduce the temperature of the flame/vapor mixture below the limiting adiabatic flame temperature necessary to sustain combustion (Drysdale 1985). The gaseous extinguishing agent actually absorbs the combustion energy and its thermal capacity is the important factor in inerting explosive or burning materials.

The thermal capacities of the more commonly used inerting agents are as follows.

Table 19-6A shows why carbon dioxide is an important gaseous extinguishing agent and why helium, with its lower heat absorbing capacity, is not widely used. Halon extinguishing agents have not been included in the table. While they have even higher thermal capacity than carbon dioxide, their extinguishment mechanism is more related to chemical interaction in the combustion process.

TABLE 19-6A. Thermal Capacities of the More Commonly Used Inerting Agents.

		Thermal Capacity at 340°F (727°C)	
Agent	Symbol	BTU/MOL °F	J/MOL-K
Carbon Dioxide	CO_2	12.9	54.3
Steam	$H_2O(g)$	9.8	41.2
Nitrogen	N_2	7.8	32.7
Helium	He	5.0	20.8

Nitrogen

Nitrogen has been used in a number of inerting applications, mostly those related to aircraft fire suppression systems.

Nitrogen has been considered for aircraft engine nacelle fires, but halons are generally far superior for this application (Dyer et al 1977). The use of nitrogen has been directed more toward fuel tank inerting where cryogenic nitrogen in Dewar vessels is directed into the fuel tank ullage. Nitrogen up to an additional 80 percent of the gross volume to be protected is added so that a maximum oxygen concentration of 9 percent by volume is not exceeded in the atmosphere above the fuel (NAS 1977).

Nitrogen fire fighting systems have also been considered for cargo compartment underfloor areas, wheel wells, and wing areas in both military and civil aircraft (Dyer et al 1977). This technique has been used on U. S. Air Force C-5A transport aircraft; consideration is being given to using nitrogen in parked aircraft to reduce oxygen concentrations to below ten percent by volume and eliminate cabin fires in unattended aircraft.

One of the problems with an inerting gas is to ensure that it is well mixed and that all parts of the volume being protected have the minimum concentration required for fire suppression. This is particularly true in enclosures where the amount of extinguishing agent required for suppression must be balanced by the potential damage to equipment and injuries to personnel that could arise from too much inerting gas. Scale modeling studies of nitrogen distribution within pressurizable enclosures have proven that poor distribution patterns of nitrogen can be identified and corrected (Corlett et al 1980).

Steam Inerting Systems

Steam may be used to smother fires in the same manner as other inert gases. It effectively reduces the concentrations of fuel vapor and oxygen and absorbs sufficient heat to stop the combustion process. The use of steam systems for fire extinguishment precedes the use of other modern smothering systems such as carbon dioxide and foam extinguishing systems, and is rarely used today. It is clearly an impractical method except where a large steam supply is continuously available, and this supply can be tapped effectively and efficiently when a fire emergency arises. The possible burn hazard to personnel must be considered in any steam extinguishing installation.

Steam extinguishing systems are not recommended for fire protection purposes in any current NFPA standards. However, the Appendix to NFPA 86, *Standard for Ovens and Furnaces—Design, Location, and Equipment,* does offer suggestions for fire protection of steam systems which may be followed "where steam flooding is the only alternative" after automatic sprinklers or water spray systems and approved types of supplementary fire protection (carbon dioxide, foam, or dry chemical systems) have already been considered. To protect ovens, this standard calls for steam outlets to supply at least 8 lb (3.6 kg) of steam per minute for each 100 cu ft (2.8 m^2) of oven volume. One lb (0.45 kg) of saturated steam at 212°F (100°C) and normal atmospheric pressure has a volume of 26.75 cu ft (or 0.76 m^3). Steam outlets should be located near the bottom of the oven, but may be located at the top, pointing downward if the oven is not more than 20 ft (6.0 m) high. Pipe sizes and the steam supply should be sufficient to deliver the required amount of steam and maintain a minimum pressure of 15 psi (100 kPa) at the outlet. The standard recommends manual release devices for such systems, and that controls be arranged to close down oven outlets as far as practicable. Additional guidance offered includes arrangements to reduce hazards to personnel.

Steam smothering systems used to be employed for the protection of cargo spaces and the holds of steamships. This method is no longer recommended. Tests indicating the relative inefficiency of such systems to control cotton cargo fires were conducted by the U.S. Coast Guard during "Operation Phobos" (NFPA 1947).

One more interesting example of the use of steam inerting is the system used to prevent explosions in certain coal pulverizers in the U.S. (DeGabriele et al 1980). In essence, the inerting system introduces steam into the pulverizing system to reduce the oxygen concentration below the flammability limit for ignition of a coal dust atmosphere. Steam was chosen for this application rather than carbon dioxide or nitrogen because it: (1) was readily available, (2) needed no storage system, (3) was abundant from a reliable source, (4) did not require tight enclosure of the inerted space, (5) had no toxic effects, and (6) was inexpensive. On site tests and plant experience have shown the system to be quite effective in preventing coal dust explosions.

Use of Combustion Gases For Fire Extinguishment

Experiments have been conducted on the use of the gases of combustion to achieve extinguishment. Inert gas was generated by an experimental turbojet engine specially designed for the test. The engine burned most of the oxygen in the air it used, and the hot combustion gases were cooled by vaporizing water introduced as a fine spray. The most efficient inert gas produced by the experimental appliance contained 46 percent nitrogen, 44 percent water vapor, 3 percent carbon dioxide, and 7 percent oxygen. The highest rate of delivery used was 45,000 cfm (21 m^3/sec) when the appliance consumed 70 imperial gal per minute (320 L/min) of water and 6.5 imperial gal per minute (29.5 L/min) of kerosene. The emergent gas was cooled to 194 to 248°F (90 to 120°C). In this experimental work, either the inert gas alone was used or a high expansion foam was made with the inert gas. By spraying a suitable detergent solution on a screen and blowing the inert gas through the wetted screen at a velocity not exceeding 5 fps (1.5 m/sec), an inert gas high expansion foam was generated with an expansion ratio of 1,000 to 1. Between 50 and 80 percent of the gas was made into foam. To convey the gas or foam to the fire site, a flexible ducting was used (made of Terylene with a neoprene coating to make it impermeable).

In similar investigations (Bryan 1982; McGuire 1964), a smaller inert gas generator was constructed which produced approximately 3,300 cfm (1.6 m^3/sec), of the mixture, consisting of 68 percent water vapor, 28 percent nitrogen, and 4 percent carbon dioxide at a temperature of 194°F (90°C). Depending upon the fuel involved, it was reported that the inert gas concentration must reduce the oxygen level to between 9 and 12 percent for effective extinguishment. In addition, a quantity of inert gas equal to twice the building volume involved must be produced within 15 minutes for successful extinguishment.

Air Agitation for Oil Tank Fire Control

Fuel tanks are often protected by topside foam protection systems or by subsurface foam injection into the fuel itself. Fire in an oil storage tank may, under some conditions, be controlled or extinguished by introducing air under pressure near the bottom of the tank. The principle is founded upon the fact that flammable and combustible liquids require a temperature greater than their flash point temperature to sustain burning. If a tank of oil with a flash point above the actual temperature of the liquid becomes ignited when the surface temperature is raised above the flash point, or the oil has become contaminated with a low flash point liquid, the resulting fire may be controlled or extinguished by agitation of the oil mass. By agitating the mass, cool oil circulated at a proper rate decreases the temperature of burning oil at the surface by displacement and mixing to a point below its flash point, thus slowing or inhibiting further combustion. The time element during which such control operations must be actively pursued is critical. If the oil mass heats too quickly, the cooling effect might be inadequate, and if the burning liquid is crude oil, the time before boiling occurs can be shortened considerably. This technique is not considered a standard method of combating oil tank fires. Under most circumstances where fire protection is provided for such storage tanks, foam fire fighting equipment meeting the recommendations contained in NFPA 11, *Standard for Low Expansion Foam and Combined Agent Systems*, should be employed.

COMBINED AGENT SYSTEMS

It is common fire fighting practice to use two or more agents simultaneously or in rapid sequence, taking advantage of the best qualities of any two agents in achieving fast fire control and permanent extinguishment. Some common combinations used in manual fire control include: (1) water (usually water spray) and foam, (2) carbon dioxide and foam, and (3) certain dry chemicals and foam. Aircraft rescue and fire fighting operations have long utilized such combined agent methods using water spray, carbon dioxide, or dry chemical to secure rapid flame knockdown, plus simultaneous or sequential use of foam to permanently smother the fire. During World War II, the combined use of bulk applications of carbon dioxide from mobile low pressure carbon dioxide trucks, with simultaneous applications of foam from specialized crash trucks achieved many significant personnel rescues in aircraft fire accidents, as well as achieving quick fire extinguishment or control. This technique, and the combined use of foam compatible dry chemicals and foam, is still being used. While these techniques do involve simultaneous and/or sequential use of two agents, each agent system is independent of the other, most often located on separate mobile vehicles.

Recent developments involving dual agent systems combine light water with potassium salt based dry chemical, or mechanical (air) foam with potassium salt based dry chemical, providing the foam is compatible with dry chemical. These are systems in the sense that they are designed to provide a balanced discharge of the agents, although the agents themselves are not premixed. The two agents are discharged through independent but twinned nozzles (both nozzles independent but under the control of one fire fighter). Some experiments, with encouraging results, have entrained certain dry chemicals in the water-foam concentrate turret stream.

Apart from the combined agent systems using foam, other combinations of halons and water, and halons with dry chemical (potassium bicarbonate) have been tried and found effective (NAS 1980).

Bibliography

References Cited

Breen, D. E. 1977. "Interactions in Binary Halon Mixtures Used as Fire Suppressants." *Fire Technology*. Vol 13, No 4. Nov 1977. pp 261-265.

Bryan, J. L. 1982. *Fire Suppression and Detection Systems*. Glencoe Press, Encino, CA.

Corlett, R. C., Stone, J. P., and Williams, F. W. 1980. "Scale Modeling of Inert Pressurant Distribution." *Fire Technology*. Vol 16, No 4. Nov 1980. pp 259-273.

DeGabriele, R. M., Causilla, H., and Herschel, J. P. 1980. "Dynamic Steam Inerting System for a Ball Tube Mill Pulverizing Subbituminous Coal." *Fire Technology*, Vol 16, No 3. Aug 1980. pp 212-226.

Drysdale, D. D. 1985. *Fire Dynamics*. Wiley & Sons., Chichester, UK.

Dyer, J. H., Marjoram, M. J., and Simmons, R. F. 1977. "The Extinction of Fires in Aircraft Jet Engines —Part III, Extinction of Fires at Low Airflows." *Fire Technology*. Vol 13, No 2. May 1977. pp 126-130.

Geyer, G. B. 1973. "The Use of Ground Cover Materials to Suppress Fuel Fires at Airports." *Fire Technology*. Vol 9, No 4. Nov 1973. pp 254-262.

Lev. Y. 1981. "A Novel Method for Controlling LNG Pool Fires." *Fire Technology*. Vol 17, No 4. Nov 1981. pp 275-284.

McCormick, J. W., and Schmitt, C. R. 1974. "Extinguishment of Selected Metal Fires Using Carbon Microspheroids." *Fire Technology*. Vol 10, No 3. Aug 1974. pp 197-199.

McGuire, J. H. 1964. "Large Scale Use of Inert Gas to Extinguish Building Fires." *Research Paper 246*. National Research Council of Canada. Division of Building Research, Ottawa, Canada.

NAS. 1977. "Fire Safety Aspects of Polymeric Materials—Volume 6—Aircraft: Civil and Military." *Publication NMAB 318-6 Appendix C, D*. National Academy of Sciences, Washington, DC.

NAS. 1980. "Fire Safety Aspects of Polymeric Materials—Volume 10—Mines and Bunkers." *Publication NMAB 318-10*. National Academy of Sciences, Washington, DC.

Neill, R. R., and Peterson, H. B. 1969. "An Investigation of the Heating and Fire Extinguishment of Deep Frying Fat." *N. R. L. Memorandum Report 2080*. Naval Research Laboratory, Washington, DC.

NFPA Forest Committee. 1975. "Air Operations for Forest, Brush and Grass Fires." *Report*. rev. ed. National Fire Protection Association, Quincy, MA.

NFPA Forest Committee. 1977. "Chemicals for Forest Fire Fighting." *Report*. 3rd ed. National Fire Protection Association, Quincy, MA.

NFPA. 1947. *Proceedings of the Fifty-first Annual Meeting of the National Fire Protection Association*. May 26-29, 1947. National Fire Protection Association, Quincy, MA. pp 119-123.

Schmitt, C. R. 1975. "Carbon Microspheres as Extinguishing Agents for Fissionable Material Fires." *Fire Technology*. Vol 11, No 2. May 1975. pp 95-98.

NFPA Codes, Standards, Recommended Practices and Manuals. (See the latest *NFPA Codes and Standards Catalog* for availability of current editions of the following documents.)

NFPA 11, *Standard for Low Expansion Foam and Combined Agent Systems*.

NFPA 18, *Standard on Wetting Agents*.

NFPA 86, *Standard for Ovens and Furnaces—Design, Location, and Equipment*.

SECTION 20
PORTABLE FIRE EXTINGUISHERS

THE ROLE OF EXTINGUISHERS IN FIRE PROTECTION

Marshall E. Petersen, P. E.

Virtually all fires are small at first and might be easily extinguished if the proper type and amount of extinguishing agent were promptly applied. Portable fire extinguishers are designed for this purpose, but their successful use depends on the following conditions:

1. The extinguisher must be properly located and in good working order.
2. The extinguisher must be the proper type for the fire which occurs.
3. The fire must be discovered while still small enough for the extinguisher to be effective.
4. The fire must be discovered by a person ready, willing, and able to use the extinguisher.

Fire extinguishers are the first line of defense against unfriendly fires, and should be installed regardless of other fire control measures. However, the fire department should be notified as soon as a fire is discovered; notification should never be delayed in hope that the extinguisher will be sufficient.

This chapter extensively discusses the role of portable fire extinguishers. It begins by giving historical background concerning development of extinguishers, and then discusses their reliability and design safety. Finally, current extinguisher identification and fire extinguishment in the home are detailed.

For a complete discussion of the allied topics of portable fire extinguishers, refer to the remaining chapters in this section: "Selection, Operation, and Distribution" (Chapter 2); "Inspection and Maintenance" (Chapter 3); and "Auxiliary Devices" (Chapter 4). Also, consult Section 19 for an explanation of the extinguishing properties of the agents used in portable fire extinguishers.

HISTORICAL BACKGROUND

The first real portable fire extinguishers were developed in the late 1800s; they contained glass bottles of acid which, when broken, dumped acid into a soda solution and produced a mixture with sufficient gas pressure to expel the solution. Cartridge operated water extinguishers (inverting type) were introduced in the late 1920s. In 1928, a nonfreeze alkali metal-salt solution called "loaded stream" was developed for use in cartridge operated extinguishers. Stored pressure water extinguishers were developed in 1959, and over the next ten years, they gradually replaced cartridge operated models. (In 1969, the manufacture of all inverting extinguishers was discontinued in the United States, and these extinguishers are no longer listed or approved by testing laboratories.)

The first foam extinguisher was developed in 1917, and it looked and worked much like the soda-acid extinguisher. The use of foam extinguishers steadily increased over the years until, during the 1950s, dry chemical extinguishers gained widespread acceptance.

Vaporizing Liquids

Carbon tetrachloride (CCl_4) was one of the first chemicals used in portable fire extinguishers (1908). However, it was subsequently found that its vapors were toxic, and when used on a fire, this substance could produce highly toxic hydrogen chloride and phosgene. Slightly less toxic chlorobromomethane (CH_2ClBr) was produced after World War II, when the term "vaporizing liquid" was first used to designate these extinguishers. However, some federal agencies banned vaporizing-liquid extinguishers in the 1950s because they were poisonous. By the mid 1960s, many states, cities, and industrial firms had also banned them. In the late 1960s, listings of vaporizing liquids by testing laboratories were discontinued.

Halogenated Agents

Although vaporizing liquids proved unacceptable, less toxic halogenated hydrocarbon chemicals found use in the form of compressed or liquefied gases. Bromotrifluoromethane (Halon 1301) was first introduced in 1954 as a high vapor pressure compressed gas extinguisher for use on fires in flammable liquids and live electrical equipment. A medium vapor pressure extinguisher using bromochlorodifluoromethane (Halon 1211) became available in 1973. In 1974 extensive testing was begun on extinguishers containing low vapor pressure,

Mr. Petersen is an Assistant Vice President—Engineering at Rolf Jensen & Associates, Inc., Deerfield, IL.

dibromotetrafluoromethane (Halon 2402), which is a liquid at room temperature. The tests indicated that this agent could be used on all types of fires, but its use in portable extinguishers in the United States has been limited thus far due to high costs and potential toxicity problems.

Carbon Dioxide

The first carbon dioxide extinguishers were produced during World War I, and during World War II they were the leading extinguisher for flammable liquids fires. By 1950, however, dry chemical agents had frequently replaced them as the preferable agent for flammable liquids fires. A further decline in the use of carbon dioxide extinguishers occurred as the popularity of halogenated agent extinguishers increased.

Dry Chemicals

Although the extinguishing ability of sodium bicarbonate was recognized as early as the late 1800s, it wasn't until 1928 that an effective cartridge operated dry chemical extinguisher was developed. Research and development produced an improved, finely granulated agent in 1943 and a further improved model in 1947.

As the use of flammable liquids increased, so did the development of effective dry chemical agents. In 1959, a potassium bicarbonate base agent about twice as effective as the sodium bicarbonate (ordinary) base agent was introduced.

In 1961, manufacturers introduced a new kind of agent called "multipurpose dry chemical." It had the added advantage of being 50 percent more effective than ordinary dry chemical on flammable liquid and electrical fires and also capable of extinguishing fires in ordinary combustibles. Originally diammonium phosphate was used because it was less expensive, but monoammonium phosphate soon became preferred because it is considerably less hygroscopic.

In 1968, an agent with a potassium chloride base was introduced. It was 80 percent more effective than ordinary dry chemical, but more corrosive and hygroscopic than potassium bicarbonate. A urea potassium bicarbonate base agent (potassium carbamate) was developed in Europe in 1967 and brought to America in 1970. It has been judged as at least 2½ times as effective as ordinary dry chemical.

Dry Powder

The increased use of combustible metals (magnesium, sodium, lithium, etc.) established the need for a special agent to extinguish fires involving them. The designation "dry powder" was specifically chosen to indicate an agent's suitability for use on Class D (combustible metal) fires; the term "dry chemical" was reserved for agents effective on Class A:B:C or B:C fires. In 1950, a dry powder extinguisher using sodium chloride as a base agent was first marketed.

RELIABILITY AND DESIGN SAFETY OF FIRE EXTINGUISHERS

Portable fire extinguishers may remain idle for many years, but they must always be capable of functioning at any time with maximum efficiency and without hazard to the user. Because most extinguishers are pressure vessels, they may rupture if not properly designed, constructed, and maintained. The initial responsibility for safe extinguishers belongs to the manufacturer, who is subject to the design standards, and testing, inspection, and labeling procedures of responsible fire testing laboratories. But the owner is responsible for maintaining an extinguisher once it is in service, and may retain the services of a qualified service company.

NFPA Extinguisher Standard

NFPA 10, *The Standard for Portable Fire Extinguishers*, hereinafter called NFPA Extinguisher Standard or NFPA 10, contains recommendations for the selection, installation, inspection, maintenance, and testing of extinguishers which have been widely adopted by property owners, sales/servicing agencies, and enforcing officials.

The first edition of this standard was published in 1921 from data assembled by the NFPA Committee on Field Practices (predecessor of the present committee). This standard is continuously reviewed, and revised editions are published every few years. Starting with the 1978 edition, more detailed information, with specific examples, was added to give guidance on selection and distribution.

Extinguisher Testing, Labeling, and Inspection

In North America, manufacturers have traditionally submitted their extinguishers to UL (Underwriters Laboratories Inc.), ULC (Underwriters Laboratories of Canada) and FM (Factory Mutual) for evaluation. Both UL and ULC had their own standards for construction, testing, and rating of extinguishers. Although FM had "Approval Standards" for evaluating extinguishers, they did not use the rating system advocated by the NFPA Extinguisher Standard.

To assist purchasers of extinguishers in the United States and Canada, UL, ULC, and FM authorize manufacturers to affix a label to extinguishers which have been constructed and evaluated according to their standards. (See Fig. 20-1A.) UL and ULC also require periodic examinations and tests of samples of extinguishers taken from the manufacturer's current production and/or stock. FM requires periodic follow up inspections of the manufacturer's facilities and quality control procedures to ensure that their standards are maintained. NFPA Standards call for the use of listed or approved extinguishers, and most responsible authorities and property owners rely on this requirement.

Typically, extinguishers manufactured in the United States have been both listed by UL and approved by FM. The combined effect of the standards of these two testing agencies resulted in maintaining a quality and reliability consistent with the objectives of the NFPA Extinguisher Committee.

The NFPA Extinguisher Standard only required that "organizations concerned with product evaluation" be acceptable to the authority having jurisdiction.

This requirement establishes the potential that other organizations could develop their own standards and evaluate/approve fire extinguishers. In an effort to establish uniformity and maintain levels of high quality and reliability that have been achieved, the NFPA Extinguisher Committee incorporated specific standards that must be used by any organization conducting extinguisher product

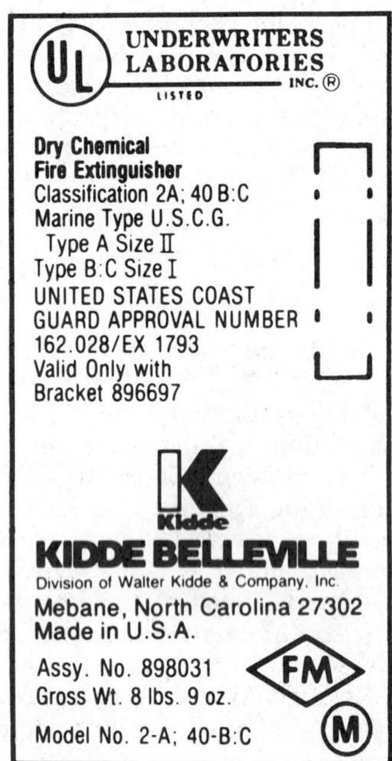

FIG. 20-1A. A close-up of an extinguisher label, showing approvals from both Underwriters Laboratories (UL) (top) and Factory Mutual (FM)(bottom right.)

evaluation. The following United States and Canadian standards are included in the requirements of the 1984 edition of NFPA 10.

Fire Test Standards

ANSI/UL 711, Rating and Testing of Fire Extinguishers CAN 4-S508-M83.

Performance Standards*

Carbon Dioxide (CO_2) Types: ANSI/UL 154, CAN 4-S503-M83

Dry Chemical Types: ANSI/UL 299, ULC-S504

Water Types: ANSI/UL 626, CAN 4-S507- M83

Halogenated Agent Types: ANSI/UL 1093, ULC-S512

Foam Types: ANSI/UL 8

According to NFPA 10, extinguishers manufactured after January 1, 1986 shall have the identification of the listing and labeling organization, the fire test, and the performance standard which the extinguisher meets or exceeds clearly marked on each extinguisher.

Substandard Extinguishers

Extinguishers are designed for use in emergencies, and it is vital that they operate effectively. Testing laboratories have maintained standards for the construction and performance of extinguishers; they have also tested extinguishers submitted to them to be listed or labeled. Only extinguishers which are now tested in accordance with ANSI Standards should be purchased, because otherwise it is difficult to know whether an extinguisher is reliable and effective. An inadequate extinguisher gives its owner a false sense of security.

Some common types of untested extinguishers are aerosol cans, "glass bulb grenades," and dry chemical "shaker units." Their instructions may claim that they can be used on all types of fires, both indoors and outdoors. They are unlikely, however, to list the exact composition and amount of extinguishing agent. Even if unlisted extinguishers do contain recognized extinguishing agents, it cannot be assumed that they are as safe and effective as those tested in accordance with ANSI Standards.

RELATION OF EXTINGUISHERS TO CLASSES OF FIRES

Because the effectiveness of various extinguishing agents are not uniform on different fires, the NFPA Extinguisher Standard classifies fires into the following four types:

Class A: Fires in ordinary combustible materials (wood, cloth, paper, rubber, and many plastics) which require the heat absorbing (cooling) effects of water or water solutions, the coating effects of certain dry chemicals which retard combustion, or the interrupting of the combustion chain reaction by halogenated agents (medium vapor pressure agents also have some cooling capability).

Class B: Fires in flammable or combustible liquids, flammable gases, greases, and similar materials, which must be put out by excluding air (oxygen), inhibiting the release of combustible vapors, or interrupting the combustion chain reaction.

Class C: Fires in live electrical equipment; safety to the operator requires the use of electrically nonconductive extinguishing agents. (Note: when electrical equipment is deenergized, extinguishers for Class A or B fires may be used.)

Class D: Fires in certain combustible metals (magnesium, titanium, zirconium, sodium, potassium, etc.) which require a heat absorbing extinguishing medium that does not react with the burning metals.

Some portable extinguishers will put out only one class of fire, and some are suitable for two or three, but none is suitable for all four. Most extinguishers are labeled so that users may quickly identify the class of fire for which they may be used. This classification is contained in the NFPA Extinguisher Standard, and it gives the applicable picture symbol or symbols. Color coding is also used. (See Figs. 20-1B and 20-1C) The NFPA classification is required on the extinguisher label in accordance with ANSI Standards.

Rating numerals are also used on the labels of extinguishers for Class A and Class B fires; the rating numeral gives the relative extinguishing effectiveness of the extinguisher. For example, an extinguisher rated 4-A; 20-B:C indicates: (1) It should extinguish approximately twice as much Class A fire as a 2-A rated extinguisher. (2) It should extinguish approximately twenty times as much Class B fire as a 1-B rated extinguisher. (3) It is suitable for use on energized electrical equipment. Class C and D extinguishers have no numeral ratings.

Extinguishers rated for Class B fires also have numeral ratings, in this case based on the relative quantity of

*ANSI-American National Standards Institute Inc. UL-Underwriters Laboratories Inc. ULC-Underwriters Laboratories of Canada CAN-Standards Council of Canada

burning flammable liquid in a flat pan that can be extinguished during a laboratory test. Again, the point is that an extinguisher rated 20-B can put out much more fire than one rated 5-B.

No rating numerals are used for extinguishers labeled for Class C fires. Since electrical equipment has either ordinary combustibles or flammable liquids, or both, as part of its construction, an extinguisher for Class C fires should be chosen according to the nature of the combustibles in the immediate area.

Extinguishers for Class D fires contain different dry powders that are effective on fires in different kinds of combustible metals. One extinguisher for magnesium fires may not work on a sodium fire, or at least not with the same effectiveness. For that reason, general numerical ratings are not used; instead, each extinguisher for Class D fires has a nameplate detailing the type of metal the particular agent will extinguish a fire in.

Extinguishers that are effective on more than one class of fire have multiple "letter" and "numeral-letter" classifications and ratings. Fractional ratings are not recognized in the NFPA Extinguisher Standard and are not included in the requirements of the ANSI Standards.

The most recently recommended marking system is one that combines pictographs of both uses and non uses on a single label. (See Fig. 20-1B.) Letter shaped symbol markings, as shown in Figure 20-1C, are recommended for use until full conversion to the newer pictographs is completed.

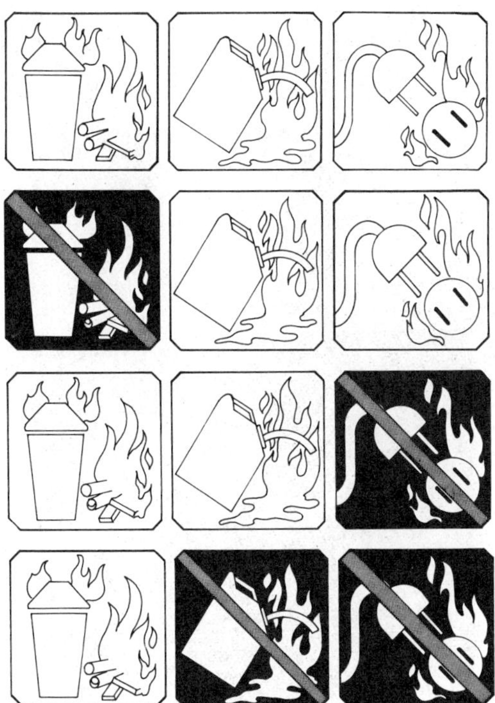

FIG. 20-1B. These pictographs are designed so that their proper use may be determined at a glance. When an application is prohibited, the background is black and the slash is bright red. Otherwise the background is light blue. Top row of picture symbols indicates an extinguisher for Class A:B:C: fires; second row indicates an extinguisher for Class B:C fires; third row indicates an extinguisher for Class A:B: fires; and bottom row indicates an extinguisher for Class A fires.

FIG. 20-1C. Extinguisher markings that can be used until conversion to pictographs is complete. Color coding is part of the identification system, and the triangle (Class A) is colored green, the square (Class B) red, the circle (Class C) blue, and the five-pointed star (Class D) yellow.

Sometimes extinguishers produced by different manufacturers (or even by the same manufacturer) have the same quantity of the same extinguishing agent, but have different ratings. In such cases, the difference in performance is usually due to differences of design (including rates of discharge, nozzle design, discharge patterns, etc.)

IDENTIFICATION OF EXTINGUISHERS

Clearly marked extinguishers and extinguisher locations are of utmost importance. In an emergency it is essential that extinguishers be located quickly and used while the fire is still small.

The rating class and numeral of an extinguisher should be visible so that the proper unit can be selected for the particular fire at hand. Manufacturers are required to provide permanently attached markings that describe not only the type and rating but also how the unit is operated. ANSI Standards specify general requirements for the form of the identification label to be placed on the front of the extinguisher. (See Figs. 20-1D and 20-1E.) Maintenance and other information can only be included on a rear label. A stenciled inventory number on extinguishers and mountings may also aid in inventory and maintenance record keeping.

FIG. 20-1D. Example of the type of instruction/identification label required by ANSI Standards for cartridge operated dry chemical extinguishers (ANSUL).

FIG. 20-1E. Example of the type of instruction/identification label required by ANSI standards for stored-pressure water extinguishers (AMEREX).

If the pictograph marking system is used, the decal should be visible from the front as the extinguisher hangs. If class and rating markings are applied to wall panels in the vicinity of extinguishers, they should permit easy legibility at a distance of 15 ft (4.5 m).

The locations of wall or column mounted extinguishers can easily be marked by painting a red rectangle or band about 8 to 10 ft (2.4 to 3.0 m) above them. (See Fig. 20-1F.) It is also good practice to paint the background red on which it is mounted.

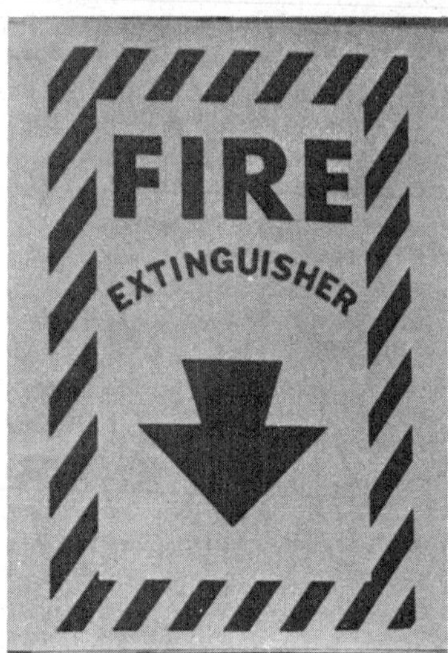

FIG. 20-1F. An extinguisher marker. It should be placed so that it is clearly visible from a distance. (The Badger Company)

If separate extinguishers for different classes of fire are mounted at the same location, extra attention should be given to distinguishing between them. Where extinguishers can easily be blocked by storage or equipment, barriers or floor markings can be employed. If the problem persists, it may be more practical to find another location.

Extinguishers installed in recessed cabinets or wall recesses are generally more difficult to locate unless clearly marked. Where possible, the cabinet frame or wall recess should be painted red. However, when such installations are in long halls or corridors, they may still be very difficult to find unless a sign is mounted perpendicular to the cabinet wall at a height of about 8 ft (2.4 m).

In many public and commercial buildings, extinguisher locations are camouflaged for aesthetic reasons or installed in hidden areas such as closets or stairwells. Unless one knows where to look or happens to see a small sign on a door, these extinguishers are of little value in an emergency.

FIRE EXTINGUISHMENT IN THE HOME

The cardinal rules when fire occurs in the home are: (1) make sure that everyone gets out of the house before exits become blocked by heat or smoke, (2) be certain that the fire department in called immediately, and (3) if the fire is small, try to control or extinguish it. To achieve the latter, every home should have readily available, easy to use fire fighting equipment. Listed or approved portable fire extinguishers are the most reliable first defense against fire. In addition, a garden hose may be useful. (See Fig. 20-4B.)

Selection of Extinguishers

Since most dwelling fires are Class A fires, the most practical and economical extinguishing agent is water. Although stored-pressure water extinguishers or pump tanks could be used, they are infrequently selected because they are large, unattractive, heavy, expensive, and are limited to Class A fires. Small, efficient, dry chemical extinguishers are ideally suited for use on any small fire that may occur in the home. Of these, the most readily available are stored-pressure models in capacities ranging from 1 to 5 lb (0.5 to 2.3 kg). The most popular size contains about 2½ lb (1.1 kg) of agent (Fig. 20-1G), and both refillable and disposable models are sold. Many home owners prefer the disposable type because they can keep a

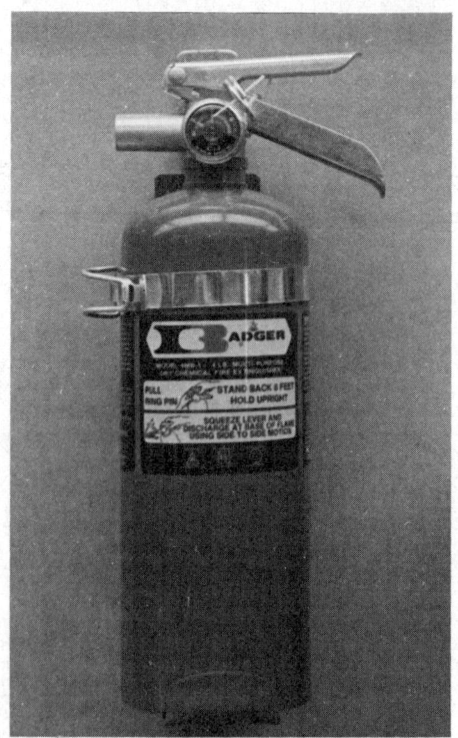

FIG. 20-1G. A small dry chemical home fire extinguisher suitable for home protection. (The Badger Company)

replacement on hand. The disposable type also requires less maintenance. Currently these small extinguishers can be purchased with ordinary (bicarbonate base) agent or multipurpose agent. It is important to note that the multipurpose agent can cause corrosion and may seriously damage stoves, heaters, appliances, and vehicles unless *all* agent deposits are promptly and thoroughly removed after extinguishment. The 5 lb (2.3 kg) and larger sizes of ordinary and multipurpose dry chemical extinguishers can also be obtained in cartridge operated models.

In general, dry chemical extinguishers will provide adequate protection for small Class B fires involving greases or cooking oils in the kitchen, paints and solvents in the basement or utility area, and oils or gasoline in the garage or outdoors. When multipurpose types are used on fires of ordinary combustibles, the small sizes will be of limited value, and the larger sizes (given higher ratings by the testing laboratories) are more suitable. Multipurpose dry chemicals are surface coating agents. Even though extinguishers of this type may rapidly extinguish the flames in combustible materials, it is important that the deep seated burning embers (especially in furniture cushions, pillows, or bedding) be thoroughly wetted with water.

It is also essential that once the extinguisher is used, it then be refilled or replaced as soon as possible. Even if only a short burst of agent is released, the extinguisher will probably lose the rest of its pressure very shortly.

Possibly the best approach to home fire extinguishment is to provide fire extinguishers such as ordinary or dry chemical for Class B:C protection and the means for rapid flame "knock-down" of Class A fires. A water extinguisher or permanently connected hose should then be used for final extinguishment of the Class A fire. If the homeowner decides to install a hose, he or she may also decide to forego the Class A capability of the multipurpose agent and purchase an extinguisher with the more powerful potassium base Class B:C agents.

Home owners who have extensive home workshops, gasoline powered equipment, etc., should consider installing additional conveniently located fire extinguishers in order to control any fire which might occur in such work areas. A logical choice would seem to be a dry chemical extinguisher with a capacity of 5 lb (2.3 kg) or greater.

Although more expensive, extinguishers containing halogenated agents (Halon 1211 and/or Halon 1301) are also suitable for home fire protection. Small sizes containing 2 to 2½ lb (0.9 to 1.1. kg) are well suited for small Class B:C fires. For Class A capability, it is recommended that sizes of at least 9 lb (4.1 kg) be selected. One advantage of the halogenated agents is that they do not leave an agent residue and are not corrosive.

For private automobiles, a multipurpose dry chemical extinguisher is effective on engine fires and any electrical fires that also include Class A insulating materials. For small trucks, NFPA committees have recommended extinguishers in the 5- to 20-B:C range, depending on the type of vehicle. Due to the corrosive characteristics of multipurpose dry chemical, it is important to promptly and thoroughly clean all agent deposits off metal surfaces.

Location of Extinguishers

Experience clearly indicates that most dwelling fires start in the living room and kitchen. Thus, at least one extinguisher should be located where it can be quickly reached from both of these rooms. It is best to locate each extinguisher near the path of exit travel so that if the fire cannot be readily controlled with the device, a quick escape can be made. In bedrooms, the extinguisher should be located in a handy closet or cabinet. In basements, the head of the basement stairs is preferred except when a basement workshop may dictate otherwise. It is important that the location of each extinguisher be known to each member of the family who is able to use the device. A permanently mounted bracket on which to hang the extinguisher will help to ensure that it remains in place and is not tucked away in a closet or other place difficult to reach.

Use of Extinguishers

The mere presence of an extinguisher in the home is not worthwhile unless the home owner is willing to: (1) learn how to use the device properly, (2) instruct family members who may have to use it, and (3) maintain and recharge it according to the manufacturer's instructions.

It is important for home owners to understand that extinguishers of the sizes discussed have a discharge time of only 8 to 15 sec; in actual use, no time can be wasted determining the best way to use the device. Operating instructions are given on each listed or approved extinguisher, and most of them are easily understood. When possible, families should practice using extinguishers in a safe outdoor location. Instructional help on fire extinguisher use may also be obtained from local fire department personnel.

Maintenance of Extinguishers

Maintenance requirements vary for each extinguisher, and the information is given on each device. Most small modern extinguishers have a pressure gage which should be checked on a regular basis. Also, to be sure that the device can be readily removed from its bracket and is in good physical condition, the extinguisher should occasionally be lifted from its bracket and handled. The family should periodically review how each extinguisher should be operated and used.

Bibliography

NFPA Codes, Standards, Recommended Practices and Manuals. (See the latest *NFPA Codes and Standards Catalog* for availability of current editions of the following documents.)

NFPA 10, *Standard for Portable Fire Extinguishers.*

NFPA 10L, *Model Enabling Act for the Sale or Leasing and Servicing of Portable Fire Extinguishers.*

Additional Readings

"A Computerized Study of Fire Extinguisher Effectiveness," National Association of Fire Equipment Distributors, Chicago, IL, 1979, p. 11.

"Carbon Tetrachloride," *Fire Journal*, Vol. 65, No. 3, May 1971, p. 19.

"Fire Extinguishers," *Consumer Reports*, Vol. 44, No. 10, Oct. 1979, pp. 607-609.

"Portable Fire Extinguishers," *Handbook of Industrial Loss Prevention*, Factory Mutual Engineering Corp., McGraw-Hill, NY, 1967.

Underwriters Laboratories Inc.,
Safety Standards

UL 154, 3rd ed., 1971, Carbon Dioxide Fire Extinguishers.

UL 299, 5th ed., April 1971, Dry Chemical Fire Extinguishers.

UL 626, 3rd ed., April 1971, 2½-Gallon Stored Pressure, Water Type Fire Extinguishers.

UL 711, 2nd ed., April 1973, Rating and Fire Testing of Fire Extinguishers.

UL 715, 2nd ed., May 1973, 2½-Gallon Cartridge Operated, Water Type Fire Extinguishers.

Bulletin of Research

No. 54, July 1963, The Compatibility Relationship Between Mechanical Foam and Dry Chemical Fire Extinguishing Agents.

SELECTION, OPERATION, AND DISTRIBUTION OF FIRE EXTINGUISHERS

Marshall E. Petersen, P. E.

Before choosing an extinguisher, it is important to know: (1) the nature of the fuels present, (2) who will use the extinguisher, (3) the physical environment in which the extinguisher will be placed, and (4) whether any chemicals present in the area will react adversely with an extinguishing agent. When choosing from among various extinguishers, one should consider: (1) whether it is effective on the specific hazards present, (2) whether it is easy to operate, and (3) what maintenance and upkeep it requires.

This chapter specifically discusses these important fire extinguisher concerns within the following topics: Matching Extinguishers to Hazards; Available Personnel; Physical Environment; Health and Operational Safety; Operation and Use; and Distribution. For a complete discussion of the allied topics of portable fire extinguishers, see the remaining chapters in this section: "The Role of Extinguishers in Fire Protection" (Chapter 1); "Inspection and Maintenance" (Chapter 3); and "Auxiliary Devices" (Chapter 4).

More specific information about Special Fire Suppression Agents and Systems is contained in the six chapters of Section 19. Also, see Section 18, Chapter 6, "Standpipe and Hose Systems."

MATCHING EXTINGUISHERS TO HAZARDS

By far the most important consideration when selecting extinguishers is the nature of the area to be protected. NFPA 10, *Standard for Portable Fire Extinguishers*, hereinafter called NFPA Extinguisher Standard or NFPA 10, classifies fires as either Class A, Class B, Class C, or Class D, according to the fuel involved. Extinguishers are designated for use on one or more of these types of fires.

In addition, the relative hazard in a building varies according to the amount of combustibles (fire load) it contains. The NFPA Extinguisher Standard establishes three types of hazards. A light (low) hazard exists when there are few combustibles and only small fires need be expected. Light hazards include offices, churches, schoolrooms, assembly halls, etc. An ordinary (moderate) hazard exists when the amount of combustibles is such that a medium sized fire may be expected. Examples of ordinary hazards are mercantile storage and display areas, auto showrooms, and parking garages. Some offices, schools, etc. may contain a sufficient amount of combustible materials to be classified as ordinary hazard. An extra (high) hazard exists in areas where a severe fire may be expected, such as woodworking areas, aircraft servicing areas, and warehouses with high piled combustibles.

Hazard type is an important factor when selecting extinguishers. For example, a 2½ gal (9.5 L) water extinguisher may only be suitable in low and moderate hazard areas. In a high hazard area, multipurpose dry chemical extinguishers with ratings of 3-A to 40-A may be required.

Class A rated extinguishers are most often used for ordinary building protection. Among the agents classified for Class A use are water, loaded stream, AFFF (aqueous film forming foam), multipurpose (ammonium phosphate base) dry chemical, and halogenated types. Class A rated extinguishers are not, however, the only type needed for building protection. For example, in most areas of a restaurant, the principal combustibles are wood, paper, and fabrics. In the kitchen, however, the essential hazard is cooking grease, which requires a Class B rated extinguisher. In hospitals, there is a general need for Class A rated extinguishers in rooms, corridors, and offices, but Class B: C rated extinguishers should be placed in laboratories, kitchens, generator rooms, and areas where anesthetics are stored or used. In short, the extinguishers in any one area should correspond to the hazards of that area.

Class B fires are fires in flammable liquids. Class B rated extinguishing agents include carbon dioxide, dry chemicals, AFFF, and halogenated types. There are three general types of flammable liquids fires: (1) fires in liquids of appreciable depth (deeper than ¼ in. or 6.4 mm), such as those in industrial dip tanks and quench tanks; (2) spill fires or running fires in liquids of no real depth (¼ in. or 6.4 mm or less); and (3) pressurized flammable liquid or gas fires from damaged vessels or product lines. Each of these fires is very different from the others, and the differences may be further complicated by atmospheric

Mr. Petersen is an Assistant Vice President—Engineering, Rolf Jensen & Associates, Inc., Deerfield, IL.

conditions. Portable fire extinguishers should not be relied upon when the surface area of an indoor open tank is in excess of 10 sq ft (1 m^2); fires in such areas give off so much heat and smoke that it is dangerous for anyone to remain in the area.

Pressurized flammable liquids and pressurized gas fires present special hazards. Only dry chemical types have proven to be effective. In addition, special nozzles and rates of agent application are required. Extinguishers should therefore be selected on the basis of manufacturer's recommendations. *Caution:* No attempt should be made to extinguish these fires unless there is reasonable certainty that the source of fuel can be promptly shut off.

Class C rated extinguishers are for use on electrical fires. They should be chosen according to: (1) how the electrical equipment is constructed; (2) the degree of agent contamination that can be tolerated; and (3) the nature of other combustibles in the area. Of these, the last is particularly important; a power panel will, for example, contain more Class A insulating materials than an oil filled transformer, which will contain mostly Class B material. The agents classified for Class C use include carbon dioxide, dry chemicals, and the halogenated types.

Class D fires are fires in combustible metals. These require special agents, generally known as dry powder.

Once an area has been analyzed for hazards, appropriate extinguishers can be chosen. For the Class A fire, there are three basic types of extinguishers which can be used: water base extinguishers, multipurpose (ammonium phosphate base) dry chemical extinguishers, or halogenated agent extinguishers. For Class B fires, there are carbon dioxide, dry chemical, halogenated agent, and AFFF extinguishers. (Some small extinguishers are rated less than 5-B and, as indicated by their omission from Table 20-2A, may not be permitted as a required extinguisher.) For Class C fires there are carbon dioxide, dry chemical, or halogenated agent extinguishers. The various types of extinguishing agents for Class D combustible metal fires are discussed separately.

AVAILABLE PERSONNEL—EASE OF USE

Before extinguishers are selected, consideration should be given to who will use them. Evaluation should include the potential user's physical abilities, reaction under stress, and previous training. The more choices the user must make in an emergency, the greater the chance for error. Many firms have standardized their extinguishers so that employees need only learn one set of instructions. (In companies employing trained fire brigades, there may be more variation.)

An individual's emotional reaction to a fire will be largely influenced by familiarity with the extinguisher, experience in using it or observing its use, and training and self confidence. Training is, therefore, very important. Many companies have employees practice with extinguishers when scheduled for recharging.

Sometimes the size or weight of an extinguisher is important. This might be the case if it must be carried up or down stairs or through an obstruction. Some extinguisher models have light weight shells or use agents with more extinguishing capacity per unit of weight.

The most common fire extinguisher holds 2½ gal (9.5 L) of water and weighs about 30 lb (13.6 kg). The weight of carbon dioxide and dry chemical extinguishers varies; for example, one manufacturer offers two dry chemical extinguishers with a rating of 40-B:C. The sodium bicarbonate base model has a 20 lb (9.1 kg) capacity and weighs 27 lb (12.3 kg), but the potassium bicarbonate base model has a 9 lb (41 kg) capacity and weighs only 14 lb (6.4 kg).

PHYSICAL ENVIRONMENT

Still another factor which influences extinguisher selection is the area in which the extinguishers will be placed, for example, if the extinguisher is affected by extreme temperatures. ANSI Standards require evaluation of water base extinguishers at temperatures between 40° (4°C) and 120°F (49°C) and all other types between −40 and 120°F (−40 and 49°C). If extinguishers are installed in areas subject to higher or lower temperatures, they should be listed or approved for those areas or put in an enclosure where the proper temperature is maintained. Some stored pressure extinguishers for Class B fires use nitrogen rather than carbon dioxide as the pressurizing force. For units which must perform in temperature as low as −65°F (−54°C), nitrogen pressurization is used.

Other conditions that may affect extinguisher performance are direct sunlight, snow, rain, airborne debris, and corrosive fumes. If extinguishers must be placed outdoors, installing them in cabinets and sheltered areas or placing a protective cover over them will help prevent damage or premature deterioration. (See Figs. 20-2A and 20-2B.)

Corrosives can cause extinguishers to fail, and UL maintains separate lists of extinguishers capable of withstanding corrosive conditions. (For example: Extinguishers which may be used in salt air atmospheres appear under the heading: "Marine Type, USCG.") Where corrosive fumes exist at industrial sites, special analysis should be made before extinguishers are chosen.

Extinguishers may also be affected by the vibration found in places like a drop forge or hammer mill, trains, vehicles, and power boats. In such cases, they should be of rugged design, mounted securely, and inspected at frequent intervals.

HEALTH AND OPERATIONAL SAFETY CONSIDERATIONS

Potential health hazards should always be considered when selecting extinguishers; manufacturers normally provide prominent labels of caution on extinguishers which could produce toxic vapors or decomposition vapors. Sometimes, however, the danger resides not in the extinguisher but in the area in which it will be used. Useful safeguards include: posting warning signs at the entries to confined areas, providing extinguishers with long range nozzles, installing special ventilation, or supplying employees in the area with self-contained breathing apparatus.

All water base extinguishers are rated only for Class A fires, except AFFF models which are also used on Class B fires. If the nonfoam water types are used on Class B fires, the fire may flare up, spread, or injure the operator in some way. If water base extinguishers are used on fires in or near live electrical equipment, the water stream may transmit a fatal shock to the operator.

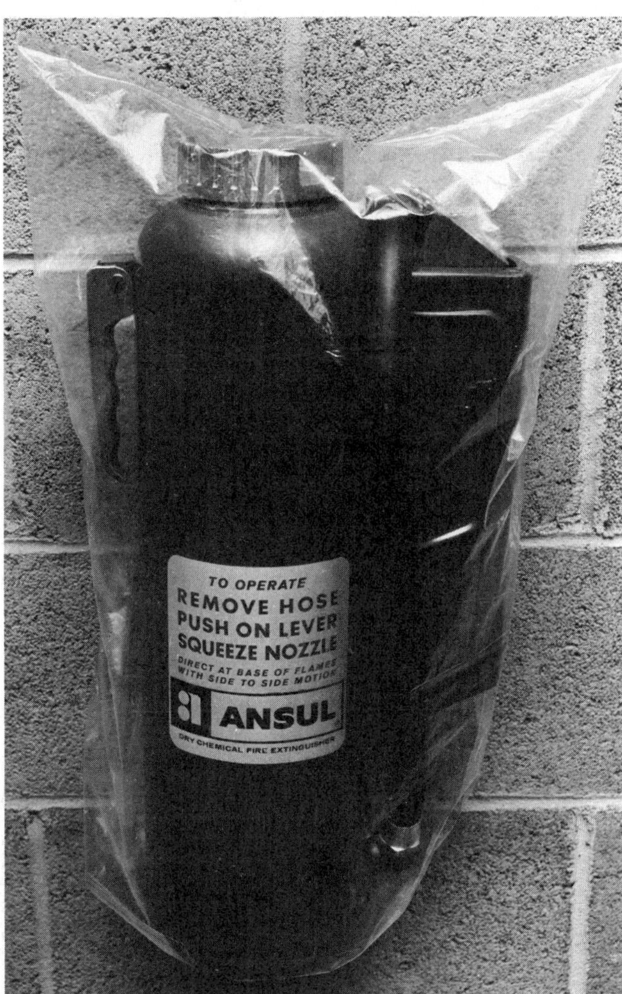

FIG. 20-2A. An extinguisher mounted on an outside wall, covered with plastic to shield it from substances which would interfere with operation. (The Ansul Company)

Although carbon dioxide is not itself toxic, it will not support life when used in concentrations high enough to extinguish a fire. If a carbon dioxide extinguisher is used in an unventilated area, it dilutes the oxygen supply, and anyone remaining in the area may become unconscious or even die if they do not receive oxygen. In addition, the thick cloud of carbon dioxide agent that forms upon discharge may cause persons to become disoriented.

Older carbon dioxide extinguishers may have metal horns which transmit a shock to the user when the horns touch live electrical equipment. These metal horns should be replaced with nonmetallic horns. Occasionally the operator may receive a shock even when no contact has been made; these "shocks" result from built up static electricity and are generally more annoying than hazardous. They are usually caused by the turbulence of the agent discharging through a damaged or defective pickup tube which can easily be replaced.

Dry chemical extinguishers are not considered toxic, but their discharge can be irritating if breathed for a long time. Monoammonium phosphate is the most irritating, followed by potassium base agents. Sodium bicarbonate is the least irritating. If dry chemicals are discharged in a confined area, they may reduce visibility and cause disorientation. Because dry chemical agents are nonconductors, deposits left on electrical contacts can reduce or prevent the ability of the contacts to conduct. They may also clog air conditioning and air cleaning filters if discharged nearby. Multipurpose dry chemical (monoammonium phosphate base) is acidic and, if mixed with even a small amount of water, will corrode some metals unless all agent residues are promptly and thoroughly removed. Finally, the initial discharge of agent from an extinguisher has considerable force; if it is aimed at close range on a small flammable liquid or grease fire, it may cause extensive spreading before the fire can be brought under control.

Extinguishers containing bromochlorodifluoromethane (Halon 1211) present only slight toxicity under normal operating conditions. However, the decomposition products of this agent can be hazardous. When using these

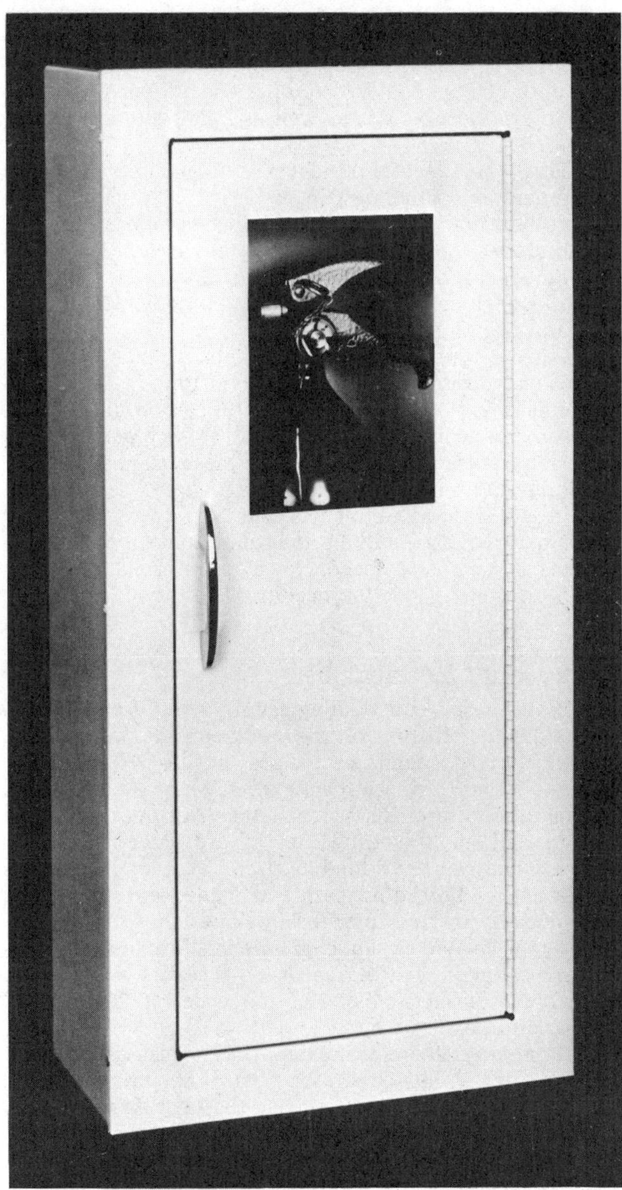

FIG. 20-2B. A surface mounted extinguisher cabinet with a hinged door. (Larsen's Manufacturing Company)

extinguishers in unventilated places (small rooms, closets, motor vehicles, or other confined spaces), it is best to avoid breathing the vapors or the gases produced by thermal decomposition.

Class D fires involving burning metal chips can be scattered if the full force of a dry powder extinguisher is used at close range. To avoid spreading the fire, the nozzle should be opened slowly at a safe distance.

Virtually every fire produces toxic decomposition products and some burning materials create highly toxic gases. Until the fire has been extinguished and the area well ventilated, it is important to either stay out of the area or wear protective breathing apparatus.

OPERATION AND USE

Whether or not an extinguisher is effective often depends on who is using it. One person may be able to extinguish a fire that someone else, using the same equipment, could not. Many extinguishers discharge their entire contents in 8 to 15 seconds, leaving little time for experimentation. Occasionally, improper use of an extinguisher may injure the operator as well as delay putting out the fire.

There are several kinds of extinguishers. Because differences exist among extinguishers it is imperative that people be trained to use extinguishers properly. Ideally, this includes the general public. In any case, fire fighters and others responsible for fire protection (e.g., industrial fire brigades) should be thoroughly trained in the operation and use of extinguishers.

Currently listed fire extinguishers are classified into six major groups based on the extinguishing medium each contains. They are (1) water type, (2) carbon dioxide, (3) halogenated agent, (4) dry chemical, (5) dry powder, and (6) foam. Information about each of these may be found in the following sections and in Table 20-2A.

In addition, extinguishers which are now no longer manufactured may still be found in some areas. These models should be replaced. However, information is provided on them under the heading "Obsolete Extinguishers."

Water Base Extinguishers

Water base extinguishing agents include water, antifreeze, loaded stream, wetting agent, soda-acid, and foam. All except AFFF foam are for use on Class A fires only. Soda-acid and ordinary foam extinguishers are no longer available and are discussed later. Antifreeze, loaded stream, AFFF, and wetting agent all use water as a base to which chemicals are added to improve the extinguisher's performance. Both antifreeze and loaded stream models are specially treated to withstand low temperatures; the additive in loaded stream extinguishers is an alkali metal-salt solution. In wetting agents, a material is added to reduce the surface tension of the water so that it will spread and penetrate better.

Originally, there were three basic designs of water base extinguishers: 1) stored-pressure, 2) pump tank, 3) and inverting. In 1969, however, the manufacture of all inverting extinguishers was discontinued. Consequently, the agents used exclusively in inverting extinguishers (soda-acid and foam) also became obsolete. The two designs that remain, stored-pressure and pump tank, are discussed later.

If ordinary (nonfoam) water base extinguishers are used on flammable liquids or electrical fires, they may spread the fire or injure the operator or both. After activating an extinguisher, point the stream at the base of the flames and work from side to side or around the fire When flames are high use the range of the stream (about 30 ft or 9 m) to best advantage. As the flames diminish and it is possible to move closer and change the solid stream to a spray by holding a fingertip over the end of the nozzle. A spray stream is more effective on burning embers. Make certain to wet deep seated smoldering or glowing areas thoroughly. If necessary, kick or poke apart burning materials in order to do this.

Stored-Pressure: For the capacities, ratings, discharge times, stream range, and temperature requirements of the various stored-pressure water base extinguishers available, see Table 20-2A. The most common stored-pressure water base extinguisher contains 2½ gal (9.5 L) and weighs about 30 lb (13.6 kg). It can be operated intermittently, is rechargeable, and has a comparatively long stream range and discharge time. It consists of a single chamber which contains both the agent and expellant gas. The cap (head) assembly includes a siphon tube, combination carrying handle/operating lever, discharge valve, air pressure valve and gage, discharge hose, and nozzle. (See Fig. 20-2C.) The

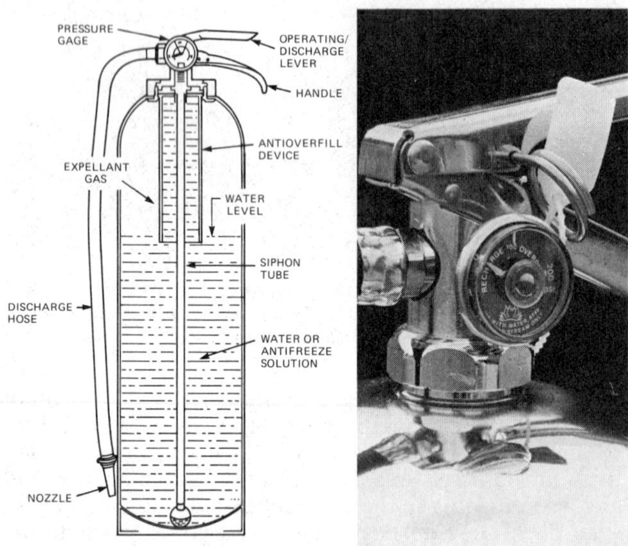

FIG. 20-2C. A stored-pressure water extinguisher with close-up of cap assembly. (Badger-Powhatan)

extinguisher is pressurized with air or inert gas by means of an automobile tire type valve. Charging pressures range from 90 to 125 psi (620 to 862 kPa). There is a "fill mark" stamped on many older stored-pressure water base extinguishers about 6 in. (1.5 mm) from the top, and when refilling, this level should never be exceeded. Some models use a special tube device to prevent overfilling.

In most models, the operating lever is locked by a ring pin which prevents accidental discharge. To activate the extinguisher, set it on the ground. Hold the combination handle loosely in one hand and pull out the ring pin (or release a small latch) with the other hand. Then grab the hose in one hand and squeeze the discharge lever with the other.

Pump Tanks: Two different types of pump tanks are available: one is cylindrical and the other is made to be used as a back pack. The sizes, ratings, discharge times, stream ranges, and weather requirements of the various pump tanks are given in Table 20-2A. The cylindrical model has carrying handles either attached to the container or built into the pump handle, and the water is discharged by a built-in, hand operated vertical piston pump with an attached short rubber hose. The pump is double acting and discharges water on both the up and the down strokes. (See Fig. 20-2D.) The cylindrical models are available with copper, steel, or plastic shells.

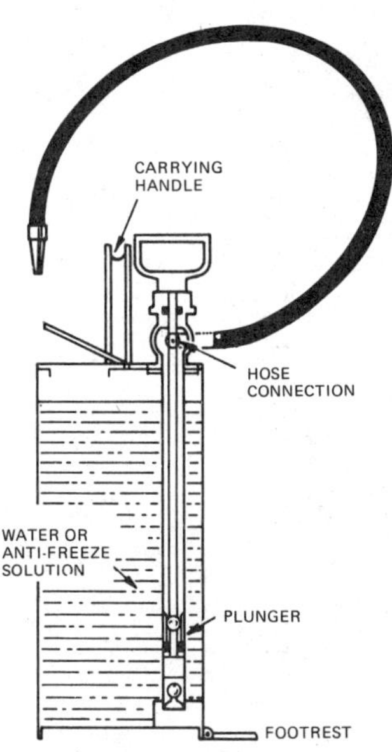

FIG. 20-2D. A pump tank fire extinguisher.

To operate a cylindrical pump tank, place it on the ground and put one foot on the extension bracket at the base to steady the unit. To force water through the hose, pump the handle up and down. Cylindrical pump tanks have two characteristics which may be slight disadvantages: (1) to move the extinguisher, the operator must stop pumping, and (2) the force, range and duration of the stream are dependent, in part, on the operator.

Cylindrical pump tanks may be filled with either plain water or antifreeze charges recommended by the manufacturer. Common salt or other freezing depressants may corrode the extinguisher or damage the pump assembly. Copper or plastic shell models do not corrode as easily as steel and are recommended for use with antifreeze.

Back pack pump tanks are chiefly used for fighting outdoor brush and wildland fires. As the name implies, they are carried on the operator's back. The most common back pack pump tank holds 5 gal (19 L) and weighs about 50 lb (23 kg) when full. (See Figs. 20-2E and 20-2F.) Although listed, back pack pump tanks do not have a designated rating. Generally, the tank is filled with plain water, though antifreeze agents, wetting agents, and other

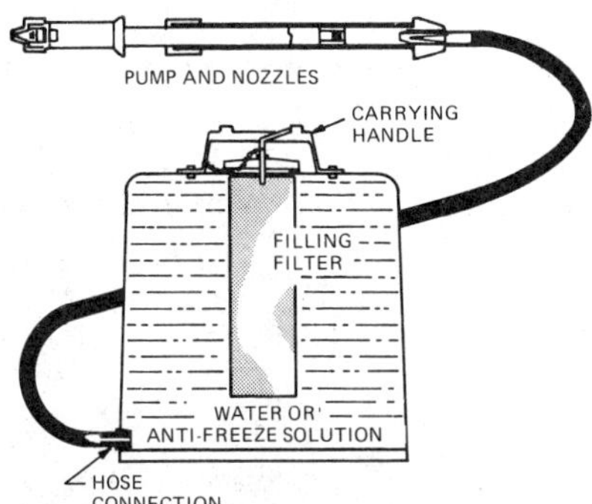

FIG. 20-2E. A back pack pump tank fire extinguisher.

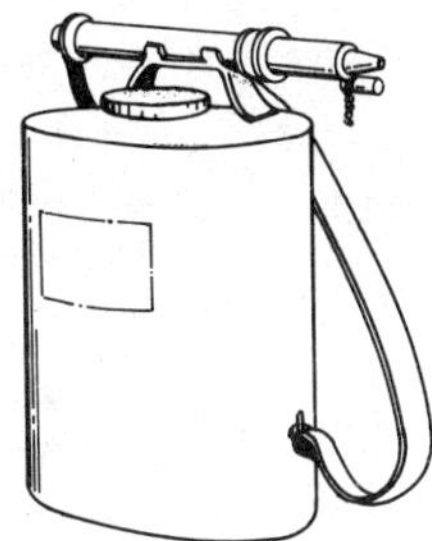

FIG. 20-2F. Back pack pump tank extinguisher with 5 gal (19 L) tank made of fiberglass. Trombone type pump action is at nozzle.

special water base agents may be used. The tank may be made of fiberglass, stainless steel, galvanized steel, brass, or it may be a flexible water bag. Some models have a large opening with a tight fitting filter which allows speedy refilling and prevents foreign matter from entering and clogging the pump. (This design also permits refilling from nearby sources such as ponds, lakes, and streams.)

The most common back pack pump tank has a trombone type, double acting piston pump connected to the tank by a short length of rubber hose. To discharge the device, the operator holds the pump in both hands and moves the piston back and forth. Other models have compression pumps mounted on the right side of the tank. In these, the expellant pressure is built up with about 10 strokes of the handle, and then maintained by slow, easy pumping. The left hand controls discharge by means of a lever operated shutoff nozzle at the end of the hose.

Carbon Dioxide Extinguishers

Carbon dioxide (CO_2) is a compressed gas agent. Though intended for use on Class B and C fires, it may be used on Class A fires until water or some other Class A rated agent can be obtained.

Carbon dioxide prevents combustion by displacing the oxygen in the air surrounding a fire. Its principal advantage is that it does not leave a residue, a consideration which may be important in laboratories, areas where food is prepared, or where there is electronic equipment. It is also very cold and has a refrigerating effect on the area around a fire. However, carbon dioxide extinguishers have a relatively short range because the agent is expelled in the form of a gas/snow cloud; they are also affected by wind or drafts. If a carbon dioxide extinguisher is used in a confined or unventilated area, precautions should be taken so that people are not overcome from lack of oxygen. For

TABLE 20-2A. Characteristics of Extinguishers

Extinguishing Agent	Method of Operation	Capacity*	Horizontal Range of Stream†	Approximate Time of Discharge	Protection Required Below 40°F (4°C)	UL or ULC Classifications‡
Water/Antifreeze	Stored-Pressure or Cartridge	2½ gal	30–40 ft	1 min	Yes	2-A
	Pump	2½ gal	30–40 ft	1 min	Yes	2-A
	Pump	4 gal	30–40 ft	2 min	Yes	3-A
	Pump	5 gal	30–40 ft	2–3 min	Yes	4-A
Water (Wetting Agent)	Stored-Pressure	1½ gal	20 ft	30 sec	Yes	2-A
	Carbon Dioxide Cylinder	25 gal (wheeled)	35 ft	1½ min	Yes	10-A
	Carbon Dioxide Cylinder	45 gal (wheeled)	35 ft	2 min	Yes	30-A
	Carbon Dioxide Cylinder	60 gal (wheeled)	35 ft	2½ min	Yes	40-A
Water (Soda Acid)	Chemically generated expellant	2½ gal	30–40 ft	1 min	Yes	2-A
	Chemically generated expellant	17 gal (wheeled)	50 ft	3 min	Yes	10-A
	Chemically generated expellant	33 gal (wheeled)	50 ft	3 min	Yes	20-A
Loaded Stream	Stored-Pressure or Cartridge	2½ gal	30–40 ft	1 min	No	2 to 3-A:1-B
	Carbon Dioxide Cylinder	33 gal (wheeled)	50 ft	3 min	No	20-A
Foam	Chemically generated expellant	2½ gal	30–40 ft	1½ min	Yes	2-A: 4 to 6-B
	Chemically generated expellant	17 gal (wheeled)	50 ft	3 min	Yes	10-A: 10 to 12-B
	Chemically generated expellant	33 gal (wheeled)	50 ft	3 min	Yes	20-A: 20 to 40-B
AFFF	Stored-Pressure	2½ gal	20–25 ft	50 sec	Yes	3-A: 20 to 40-B
	Nitrogen Cylinder	33 gal	30 ft	1 min	Yes	20-A: 160-B
Carbon Dioxide	Self Expelling	2½ to 5 lb	3–8 ft	8 to 30 sec	No	1 to 5-B:C
	**	10 to 15 lb	3–8 ft	8 to 30 sec	No	2 to 10-B:C
		20 lb	3–8 ft	10 to 30 sec	No	10-B:C
		50 to 100 lb (wheeled)	3–10 ft	10 to 30 sec	No	10 to 20-B:C
Dry Chemical (Sodium Bicarbonate)	Stored-Pressure	1 to 2½ lb	5–8 ft	8 to 12 sec	No	2 to 10-B:C
	Cartridge or Stored-Pressure	2¾ to 5 lb	5–20 ft	8 to 20 sec	No	5 to 20-B:C
	Cartridge or Stored-Pressure	6 to 30 lb	5–20 ft	10 to 25 sec	No	10 to 160-B:C
	Nitrogen Cylinder or Stored-Pressure	75 to 350 lb (wheeled)	15–45 ft	20 to 105 sec	No	40 to 320-B:C
Dry Chemical (Potassium Bicarbonate)	Cartridge or Stored-Pressure	2 to 5 lb	5–12 ft	8 to 10 sec	No	5 to 20-B:C
	Cartridge or Stored-Pressure	5½ to 10 lb	5–20 ft	8 to 20 sec	No	10 to 80-B:C
	Cartridge or Stored-Pressure	16 to 30 lb	10–20 ft	8 to 25 sec	No	40 to 120-B:C
	Cartridge	48 lb	20 ft	30 sec	No	120-B:C
	Nitrogen Cylinder or Stored-Pressure	125 to 315 lb (wheeled)	15–45 ft	30 to 80 sec	No	80 to 640-B:C
Dry Chemical (Potassium Chloride)	Cartridge or Stored-Pressure	2 to 5 lb	5–8 ft	8 to 10 sec	No	5 to 10-B:C
	Cartridge or Stored-Pressure	5 to 9 lb	8–12 ft	10 to 15 sec	No	20 to 40-B:C
	Cartridge or Stored-Pressure	9½ to 20 lb	10–15 ft	15 to 20 sec	No	40 to 60-B:C
	Cartridge or Stored-Pressure	19½ to 30 lb	5–20 ft	10 to 25 sec	No	60 to 80-B:C
	Cartridge or Stored-Pressure	125 to 200 lb	15–45 ft	30 to 40 sec	No	160-B:C

TABLE 20-2A. Characteristics of Extinguishers *(cont)*

Extinguishing Agent	Method of Operation	Capacity*	Horizontal Range of Stream†	Approximate Time of Discharge	Protection Required Below 40°F (4°C)	UL or ULC Classifications‡
Dry Chemical (Ammonium Phosphate)	Stored-Pressure	1 to 5 lb	5–12 ft	8 to 10 sec	No	1 to 2-A§ and 2 to 10-B:C
	Stored-Pressure or Cartridge	2½ to 8½ lb	5–12 ft	8 to 15 sec	No	1 to 4-A and 10 to 40-B:C
	Stored-Pressure or Cartridge	9 to 17 lb	5–20 ft	10 to 25 sec	No	2 to 20-A and 10 to 80-B:C
	Stored-Pressure or Cartridge	17 to 30 lb	5–20 ft	10 to 25 sec	No	3 to 20-A and 30 to 120-B:C
	Cartridge	45 lb	20 ft	25 sec	No	20-A:80-B:C
	Nitrogen Cylinder or Stored-Pressure	110 to 315 lb (wheeled)	15–45 ft	30 to 60 sec	No	20 to 40-A and 60 to 320-B:C
Dry Chemical (Foam Compatible)	Cartridge or Stored-Pressure	4¾ to 9 lb	5–20 ft	8 to 10 sec	No	10 to 20-B:C
	Cartridge or Stored-Pressure	9 to 27 lb	5–20 ft	10 to 25 sec	No	20 to 30-B:C
	Cartridge or Stored-Pressure	18 to 30 lb	5–20 ft	10 to 25 sec	No	40 to 60-B:C
	Nitrogen Cylinder or Stored-Pressure	150 to 350 lb (wheeled)	15–45 ft	20 to 150 sec	No	80 to 240-B:C
Dry Chemical (Potassium Chloride)	Cartridge or Stored-Pressure	2½ to 5 lb	5–12 ft	8 to 10 sec	No	10 to 20-B:C
	Cartridge or Stored-Pressure	9½ to 20 lb	5–20 ft	8 to 25 sec	No	40 to 60-B:C
	Cartridge or Stored-Pressure	19½ to 30 lb	5–20 ft	10 to 25 sec	No	60 to 80-B:C
	Stored-Pressure	125 to 200 lb (wheeled)	15–45 ft	30 to 40 sec	No	160-B:C
Dry Chemical (Potassium Bicarbonate Urea Based)	Stored-Pressure	5 to 11 lb	11–22 ft	13 to 18 sec	No	40 to 80-B:C
	Stored-Pressure	9 to 23 lb	15–30 ft	17 to 33 sec	No	60 to 160-B:C
		175 lb (wheeled)	70 ft	62 sec	No	480-B:C
Halon 1301 (Bromotrifluoromethane)	Stored-Pressure	2½ lb	4–6 ft	8 to 10 sec	No	2-B:C
Halon 1211 (Bromochlorodifluoromethane)	Stored-Pressure	1 lb	6–10 ft	8 to 10 sec	No	1-B:C
		2 lb	6–10 ft	8 to 10 sec	No	2-B:C
		2½ lb	6–10 ft	8 to 10 sec	No	5-B:C
		5½ to 9 lb	9–15 ft	8 to 15 sec	No	1-A:10-B:C
		13 to 22 lb	14–16 ft	10 to 18 sec	No	1 to 4-A and 20 to 80-B:C
		150 lb	20–30 ft	30 to 35 sec	No	30-A:160-B:C
Halon 1211/1301 (Bromochlorodifluoromethane/Bromotrifluoromethane) Mixtures	Self Expelling	1 to 5 lb	3–12 ft	8 to 10 sec	No	1 to 10-B:C
	Stored-Pressure	9 to 20 lb	10–18 ft	10 to 22 sec	No	1-A and 10-B:C to 4-A and 80-B:C

Notes: * 1 gal = 3.78 L; 1 lb = 0.45 kg. † 1 ft = 0.30 m.

‡ UL and ULC ratings checked as of December 9, 1983. Readers concerned with subsequent ratings should review the pertinent "lists" and "supplements" issued by these laboratories: Underwriters Laboratories Inc., 333 Pfingsten Road, Northbrook, IL 60062, or Underwriters' Laboratories of Canada, 7 Crouse Road, Scarborough, Ont., Canada M1R 3A9.

**Carbon dioxide extinguishers with metallic horns do not carry a "C" classification.

§ Some small extinguishers containing ammonium phosphate base dry chemical do not carry an "A" classification.

N.B.: Vaporizing liquid extinguishers (carbon tetrachloride or chlorobromomethane base) are not recognized in NFPA10.

the various sizes, stream ranges, discharge times, temperature requirements, and ratings of carbon dioxide extinguishers available, see Table 20-2A.

In all carbon dioxide extinguishers, the agent is retained as a liquid at 800 to 900 psi (5516 to 6205 kPa) at temperatures below 88°F (31°C), and it is self-expelling. The extinguisher design consists of a pressure cylinder (shell), a siphon tube and valve for releasing the agent, and a discharge horn or horn-hose combination. The siphon tube extends from the valve almost to the bottom of the shell, so that normally only liquid CO_2 reaches the discharge horn until about 80 percent of the content has been released. The remaining 20 percent enters the siphon tube as a gas. The rapid expansion from liquid to gas when most of the CO_2 is discharged converts about 30 percent of the liquid into a very cold "snow" or "dry ice" which then sublimes into a gas.

To operate a carbon dioxide extinguisher, hold it upright by its carrying handle, remove the locking pin, and squeeze the operating lever. (See Fig. 20-2G.) Smaller

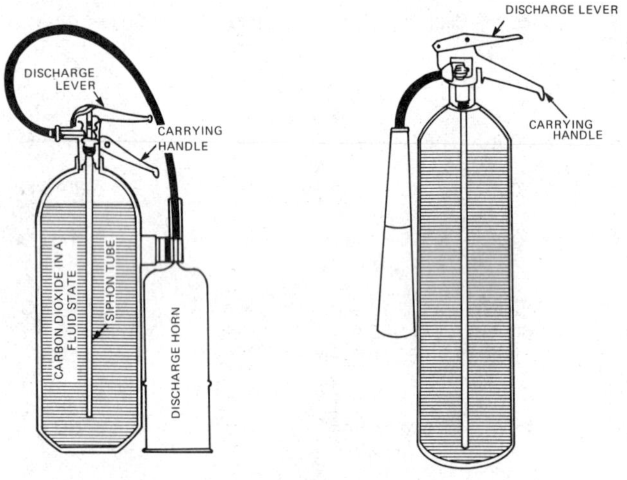

FIG. 20-2G. Carbon dioxide fire extinguisher.

portable models often have the discharge horn connected to the valve assembly by a metal tube/swing joint connector, and may be operated with one hand. Larger portable models require two hands and have discharge horns attached to several feet of hose. Wheeled extinguishers have a cylinder valve with a locking ring pin, a long hose (15 to 40 ft or 4 m to 12 m), and a projector which consists of a horn, long handle, and a control valve. Once the cylinder valve has been opened, the operator controls the discharge with the valve on the projector handle.

Care should be taken not to touch the discharge horn during operation, because it can become extremely cold. If carbon dioxide extinguishers are used in subzero temperatures, the valve must remain open at all times or the discharge may become blocked (unless a special low temperature charge has been added).

Because carbon dioxide extinguishers have a limited range and are affected by draft or wind, they should be applied as close to the base of a fire as possible. Agent should be applied even after the flames are extinguished in order to allow time for cooling and to prevent reflash. Due to the relatively short discharge range, carbon dioxide extinguishers are of limited effectiveness on large Class B fires. For flammable liquid fires, the usual method is to begin at the near edge and sweep from side to side towards the back of the fire. There is, however, another method, called overhead application. In this method, the discharge horn is pointed downward (at an angle of about 45 degrees) toward the center of the burning area. Usually the horn is not moved and the agent spreads out in all directions. The side to side sweeping method may give better results on spill fires, while the overhead method may be best for confined fires.

For fires involving electrical equipment, the discharge should be directed at the source of the flames. It is important to deenergize the equipment as soon as possible in order to prevent possible reignition.

Halogenated Agent Extinguishers

In general, bromochloroidifluoromethane (Halon 1211) is similar to carbon dioxide by nature of being a "clean agent." (See Fig. 20-2H.) Though intended for use mostly

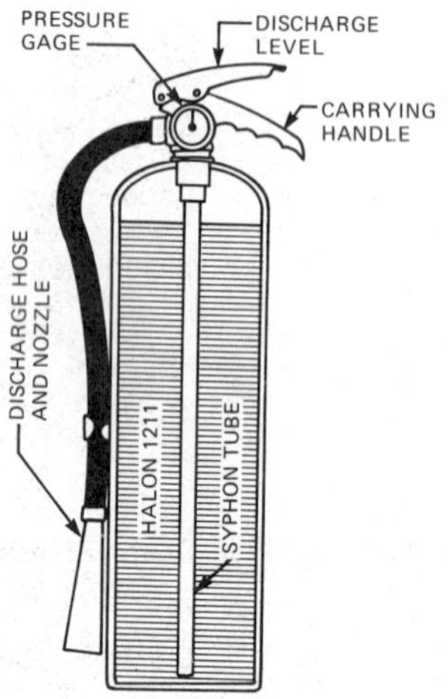

FIG. 20-2H. Halon 1211 Stored-pressure Fire Extinguisher.

on Class B:C fires, Halon 1211 is effective on Class A fires. Sizes of 9 lb (4.1 kg) capacity and greater have Class A ratings. For listed ratings, capacities, discharge times, stream range, and weather requirements for Halon 1211, see Table 20-2A.

Halon 1211 leaves no residue, is virtually noncorrosive and nonabrasive, and is at least twice as effective on Class B fires as carbon dioxide, compared on a weight of agent basis. It also has about twice the range of carbon dioxide. Like carbon dioxide, it needs no cold weather protection. Though Halon 1211 does not have the cooling effect that carbon dioxide has, it is not affected by wind as much as carbon dioxide. (Strong air currents, however, may disperse the agent too rapidly.) The chief disadvantage of Halon 1211 is its relative toxicity; inhalation of 4 to 5 percent for 1 minute is the maximum that can safely be inhaled. When Halon 1211 is used on a fire, the decomposition products include hydrogen chloride, hydrogen fluoride, hydrogen bromide, and traces of free halogens. Normally, only small quantities of these chemicals are formed, and as a warning of their presence they give off acrid odors. Although Halon 1211 is retained under pressure in a liquid state (40 psi at 70° or 276 kPa at 21°C), a booster charge of nitrogen is added to ensure proper operation.

Several manufacturers have produced Halon 1301 extinguishers during the past 30 years in a 2½ lb. (1.1 kg) size. (See Fig. 20-2I.) These extinguishers are only rated for

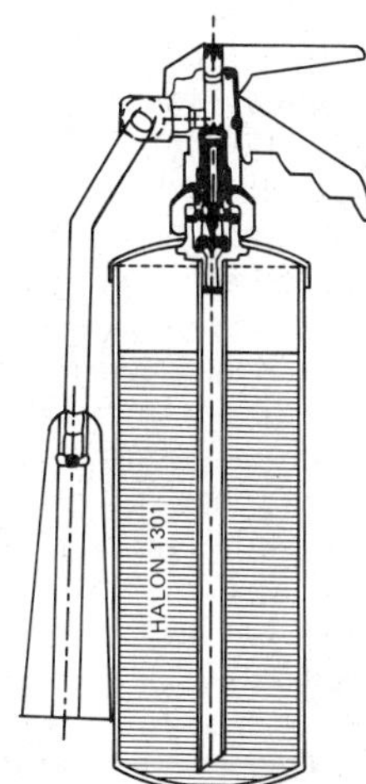

FIG. 20-2I. Halon 1301 Fire Extinguisher.

small Class B:C fires and have a significantly less toxicity hazard. Because of its low vapor pressure, the discharge stream of Halon 1301 is principally an invisible gas which may complicate effective application.

Halogenated agent extinguishers are also available containing a blend of Halon 1211 and Halon 1301. Halogenated agent extinguishers should be operated and applied to fires in the same general way that carbon dioxide extinguishers are.

Dry Chemical Extinguishers

There are two basic kinds of dry chemical agents. Ordinary dry chemicals may be used on Class B:C fires, and include sodium bicarbonate, potassium bicarbonate, urea potassium bicarbonate, and potassium chloride base agents. Multipurpose dry chemicals may be used on Class A:B:C fires. An ammonium phosphate base agent is the only multipurpose dry chemical currently manufactured. There are also two basic designs of dry chemical extinguishers: one uses a separate pressurized cartridge to expel the agent, and the other pressurizes the agent chamber for the same purpose. The stored-pressure (rechargeable) type is the most widely used. It is best suited where infrequent use is anticipated and where skilled personnel with professional recharge equipment are available. By contrast, the cartridge operated type can be refilled quickly in remote locations without special equipment.

The size, rating, and method of operation of any particular extinguisher depends both on its design and the agent. Table 20-2A gives specific information on the capacities, ratings, discharge times, and stream range of the dry chemical extinguishers now available.

Each of the five dry chemical agents is different from the others. Using sodium bicarbonate for a base comparison (circa 1960), the relative ability of each to extinguish a fire is as follows: ammonium phosphate base, 1.5 times as effective; potassium chloride base, 1.8 times as effective; potassium bicarbonate base, 2.0 times as effective; and urea potassium bicarbonate base, 2.5 times as effective. The cost of the agents varies considerably.

In areas where corrosion caused by agent residue is not a problem, potassium chloride may be included in this group, though it is not in any specific way superior to the potassium bicarbonate agents, and is not widely used.

In areas where Class A:B:C protection is needed, water base extinguishers may be omitted if multipurpose dry chemical extinguishers are installed. However, they are not replacements for water base extinguishers under all conditions.

When choosing a dry chemical extinguisher for Class C protection, it is important to remember that potassium chloride is more corrosive than the other dry chemicals, and that a multipurpose agent (ammonium phosphate) is also corrosive and will be more difficult to remove because it hardens when it cools.

The various dry chemical extinguishers possess certain common characteristics. Dry chemical extinguishers can be discharged while being carried, and all can be momentarily stopped from discharging in order to conserve the agent. Dry chemicals can also be used simultaneously with water (either straight stream or spray). Special long range nozzles may be necessary for some very hazardous areas, such as those with pressurized gases and flammable liquids. Dry chemicals are dispersed less by wind than either carbon dioxide or halogenated agents, though during windy conditions dry chemicals are most effective if the wind is at the operator's back.

If dry chemicals are used on wet energized electrical equipment (such as rain soaked utility poles, high voltage switch gear, and transformers), electrical leakage problems may be aggravated. The combination of dry chemical with moisture provides an electrical path which can keep insulation from being effective. In such cases, all traces of dry chemical should be removed after the fire has been put out.

Although all dry chemicals are treated for water repellancy, they may harden if exposed to water. It is therefore important to avoid exposing them to any moisture during storage, handling, and recharging.

Ordinary dry chemicals are rated for use on Class B:C fires, but they may be used on Class A fires to rapidly knock down flames until something more suitable can be obtained. When used on flammable liquid fires, the stream should be directed at the base of the flame. Attack the near edge of the fire and move discharge towards the back of the fire while sweeping the nozzle rapidly from side to side. Do not direct the initial discharge directly at the burning surface at close range (less than 5 to 8 ft or 1.5 to 2.4 m) because the high velocity of the stream may splash or scatter the burning material.

Multipurpose dry chemical extinguishers should be used in exactly the same manner as ordinary dry chemical agents on Class B fires. On Class A fires, the multipurpose agent softens and sticks to hot surfaces. In this way, it forms a coating which smothers and isolates the fuel from the fire. When applying the agent to Class A fires, try to coat all burning areas in order to eliminate small embers which could reignite. The agent itself has little cooling effect and cannot penetrate below the burning surface. For this reason, it may be hard to extinguish deep seated fires unless the agent is discharged below the surface or the material is broken apart and spread out.

Cartridge Operated: Cartridge type dry chemical extinguishers have a chamber in which the agent is kept at atmospheric pressure; the chamber has a large opening at the top through which the extinguisher may be filled. Attached to the side of the extinguisher, on most models, is a small cartridge of propellant gas (carbon dioxide or nitrogen) threaded into a puncture valve and gas tube assembly. The agent is discharged through a hose attached to the bottom edge of the shell, and the discharge is controlled by a squeeze grip nozzle at the end of the hose. (See Fig. 20-2J.) Cartridge operated extinguishers should

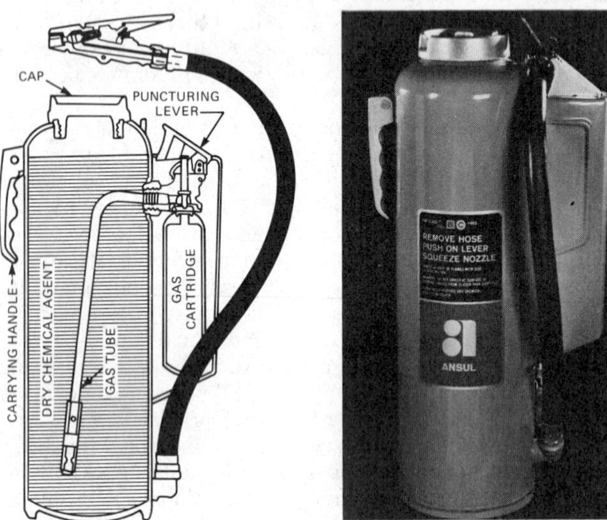

FIG. 20-2J. A cartridge operated dry chemical fire extinguisher. (The Ansul Company)

be recharged once they have been pressurized. Even if no agent has been released, the propellant gas can leak away in several hours, leaving a "dead" extinguisher.

To activate a cartridge operated extinguisher, hold it upright or place it on the ground. Remove the nozzle from its holder and hold it in one hand while first pushing down the puncture lever before squeezing the nozzle. (Pushing the puncture lever releases the propellant from the cartridge and pressurizes the agent in the cylinder. On some models, it may be necessary to remove a locking pin before pushing the puncture lever.) The operation requires two hands: one holds the device, and the other releases and directs the discharge.

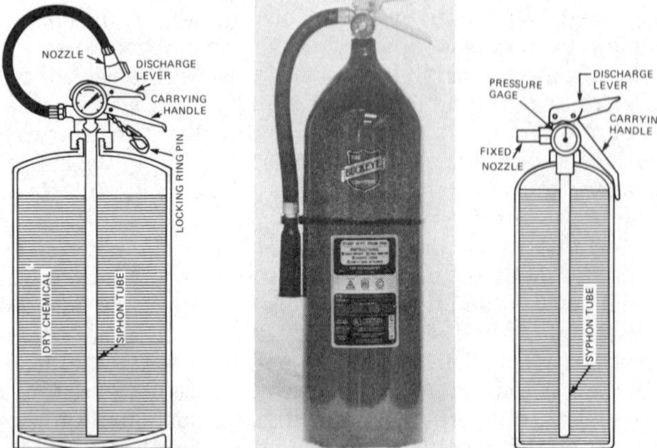

FIG. 20-2K. Stored-pressure dry chemical fire extinguishers. (Photo: Buckeye Fire Equipment)

Stored-Pressure: Dry chemical extinguishers with this expellant method are available in both rechargeable and disposable shell models. Most disposable models have a disposable factory sealed cylinder containing agent and propellant gas which is threaded into the valve and nozzle assembly. Some small models are designed so that the entire device is discarded after use. Once this type of device has been used, it should be replaced, even if only a small amount of agent has been discharged, because the propellant gas will leak leaving a "dead" extinguisher.

Rechargeable stored-pressure extinguishers come in two types. In both, the agent and propellant gas are mixed in the extinguisher shell. When the extinguisher is activated, the agent is forced up the siphon tube, where its release is controlled by the operator. One type has a threaded valve assembly and combination carrying handle/operating lever which screws into an opening on the top of the shell. To activate, release the locking device and squeeze the operating lever. (See Fig. 20-2K.) On smaller sizes of this extinguisher, only one hand is needed to operate the device because the nozzle is part of the valve assembly. Larger sizes require two hands for operation: one squeezes the lever and carries the device while the other directs the agent discharge from the hose.

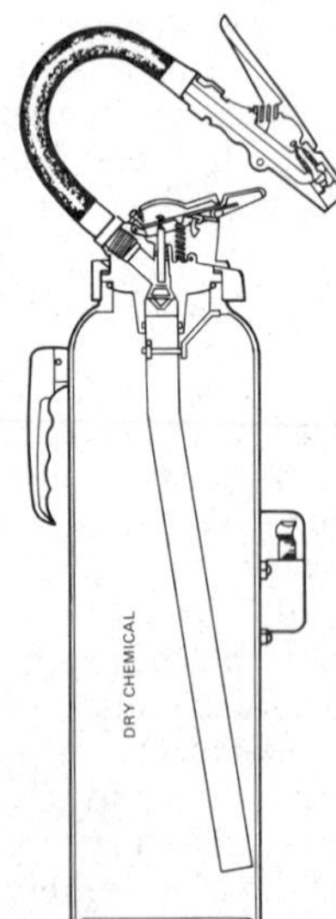

FIG. 20-2L. A stored-pressure dry chemical fire extinguisher with a rechargeable shell; this type of equipment has not been made since 1975.

The other rechargeable stored-pressure dry chemical extinguisher (pre-1975) has a release lever and a hose attached to the cap assembly that covers the fill opening. The hose has a squeeze grip nozzle for controlling the discharge. (See Fig. 20-2L.) To activate, pull the ring pin and push down the release lever; the agent will travel down the hose to the nozzle, where its discharge can be controlled by squeezing the nozzle. Two hands are needed to operate this extinguisher: one carries the extinguisher while the other releases and aims the discharge stream.

Dry Powder Extinguishers

Dry powder extinguishers are intended for use on Class D fires, which involve combustible metals. The agent, extinguisher, and method of application should be chosen according to the manufacturer's recommendations. The agent may be applied to the fire from an extinguisher or with a scoop and shovel, depending on the type of both agent and metal. In any case, the agent should be applied so that it covers the fire and provides a smothering blanket. More agent may be necessary on hot spots. Care should be taken not to scatter the burning material, and it should be left undisturbed until it has cooled.

If there is a fire in finely divided combustible metal or combustible metal alloy scrap that is wet, or if such a fire is wet with water or water soluble machine lubricants, the metal is likely to burn rapidly and violently. Such fires may become so hot that it is impossible to get close enough to apply extinguishing agent. If burning metal is on a combustible surface, it should be covered with dry powder. Then a 1 or 2 in. (25 or 50 mm) layer of powder should be spread out nearby and the burning metal shoveled into it.

One cartridge operated portable (30 lb or 13.6 kg) dry powder extinguisher is currently available. (See Fig. 20-2M.) Wheeled models are available in 150 and 350 lb (68

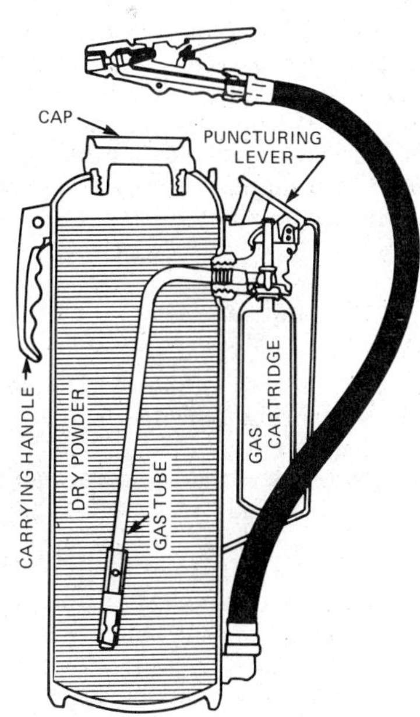

FIG. 20-2M. Dry powder extinguisher.

and 159 kg) sizes. The extinguishing agent is sodium chloride with additives which render it free flowing; thermoplastic material is also added to bind the sodium chloride particles into a solid crust when applied to a fire. With the nozzle fully open, the portable model has a range of 6 to 8 ft (1.8 to 2.4 m).

The method of agent application depends on the type, quantity, and physical form of the burning metal. If the fire is very hot, discharge should begin at the maximum range with the nozzle fully open. Once control is established, the nozzle should be partly closed to produce a soft heavy flow so that complete coverage can be accomplished at close range. More complete details should be obtained from the manufacturer.

In bulk form, dry powder agents are available in 40 or 50 lb (18 or 23 kg) pails and 350 lb (159 kg) drums. The two most common agents are sodium chloride and G-1 powder. The latter consists of graded, granular graphite to which compounds containing phosphorous have been added. The sodium chloride can be used in an extinguisher or applied by hand. The G-1 agent must be applied by hand. When G-1 is applied to a metal fire, the heat of the fire causes the phosphorous compounds to generate vapors which blanket the fire and prevent air from reaching the burning metal. The graphite, being a good conductor of heat, cools the metal.

Aqueous Film-Forming Foam (AFFF) Extinguishers

This extinguishing agent is an aqueous film forming surfactant. When added to water AFFF forms a solution which when discharged through an aspirating nozzle creates a foam. On Class A fires, the agent both cools and penetrates to reduce temperatures below the ignition level. On Class B fires, it acts as a barrier to exclude oxygen from the surface of the fuel.

AFFF is currently available in a 2½ gal (9.5 L) stored-pressure model rated at 3-A:20-B. It discharges its contents in approximately 55 sec., and has a range of 20 to 25 ft (6 to 8 m). It should only be installed in areas not subject to temperatures below 40°F (4°C).

This type of extinguisher closely resembles the stored-pressure water extinguisher except for its special aspirating nozzle. (See Fig. 20-2N.) AFFF extinguishers are also

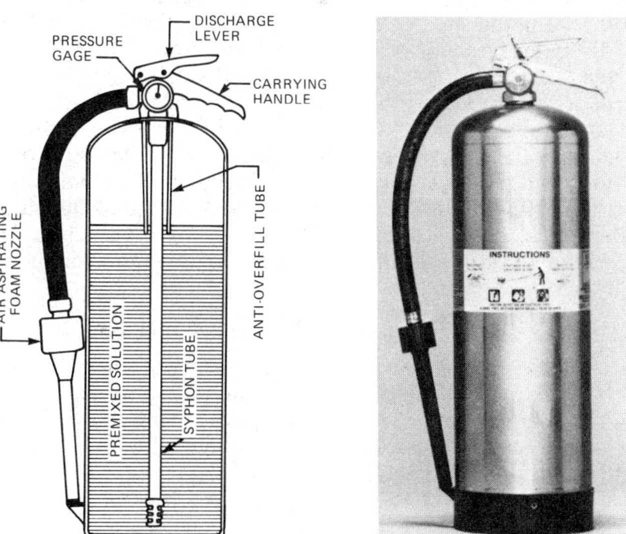

FIG. 20-2N. Aqueous film forming foam (AFFF) extinguishers. (Photo: Amerex)

available in another model which contains the agent in a solid form. When the water from the extinguisher shell is discharged, it passes through the solidified agent that is contained in a special nozzle. (See Fig. 20-2O.) On flammable liquid fires of appreciable depth, best results are obtained when the discharge is played against the inside of

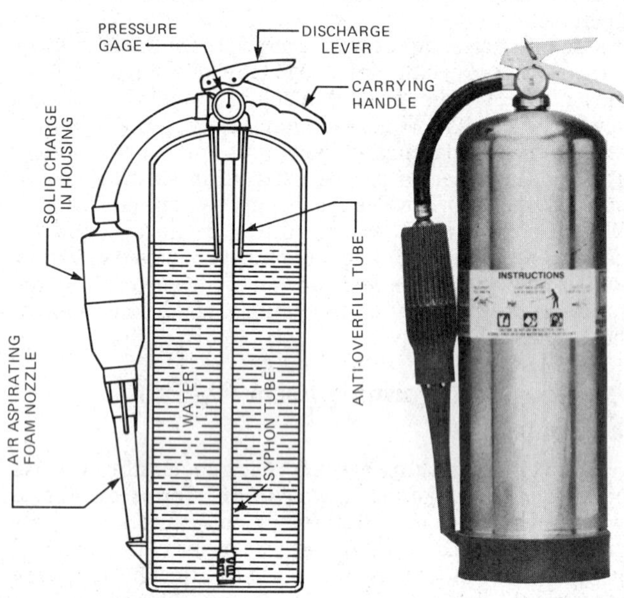

FIG. 20-2O. Stored-pressure AFFF solid charge extinguisher.

the back wall of the vat or tank just above the burning surface, which should permit the natural spread of the foam back over the burning liquid. If this is not possible, the operator should stand far enough away from the fire to allow the foam to fall lightly on the burning surface instead of splashing it into the burning liquid. Where possible, the operator should walk around the fire while directing the stream, to get maximum coverage during the discharge period. For flammable liquid spill fires, the foam may be flowed over a burning surface by bouncing it off the floor just in front of the burning area. For fires in ordinary combustibles, the foam may be used to coat the burning surface directly. Foam is not effective on flammable liquids and gases escaping under pressure, nor is it suitable for fires involving ethers, alcohols, esters, acetone, lacquer thinners, carbon disulphide, and other flammable liquids which either break down or penetrate the foam blanket.

OBSOLETE EXTINGUISHERS

In 1969 the manufacture and testing of all inverting type extinguishers (soda-acid, foam, and cartridge operated water and loaded stream) was discontinued in the United States. It was not only the difficult and unorthodox "upside down" method of actuation that brought about the discontinuance of these extinguishers. Of greater importance was the fact that after 10 to 15 years, many inverting types tended to fail the minimum test pressure requirements of the NFPA Extinguisher Standard (original factory test pressure). When pressurized, the failure record, including hydrostatic test failures, was alarmingly high. Since the inverting types are not normally pressurized, potential failures are generally not evident until the time of operation or hydrostatic testing.

When inverted, normal operating pressures are about 100 psi (690 kPa). Should the discharge elbow or hose become blocked, however, pressures in excess of 300 psi (2069 kPa) may occur. Container failures have at times seriously injured the operator. Other disadvantages of these extinguishers are:

1. The agent is an extremely good conductor of electricity.
2. They cannot be turned off once actuated.
3. The agent is more corrosive than plain water.
4. They are more costly to inspect, maintain, and recharge.
5. They are potentially dangerous to the operator during use.

For all of these reasons, the NFPA Extinguisher Standard was revised in 1978. It stated that extinguishers with copper or brass shells joined by soft solder or rivets could no longer be hydrostatically tested, which would result in their removal from use within five years from the latest hydrostatic test date. Tests showed that the reliability and safety of this type of construction could not be determined by following standard hydrostatic test procedures. It is estimated that there may be as many as a million inverting extinguishers still in use, with soda-acid models accounting for about 85 percent. As it becomes harder to find replacement parts and recharge materials, it will become more and more difficult to maintain these extinguishers in good operating condition. For these reasons, they should be replaced as quickly as possible.

Vaporizing liquid extinguishers with CCl_4 (carbon tetrachloride) and CBM (chlorobromomethane) became obsolete in the late 1960s. The toxic properties of these agents expose the operator to unwarranted health hazards, and their use was supplanted by safer, more effective extinguishers. If any extinguishers of this type are found in use, they should be promptly destroyed and replaced with other currently available extinguishers.

Soda-acid Extinguishers

This extinguisher was commonly manufactured in the 2½ gal (9.5 L) size, weighing about 30 lb (13 kg) fully charged, with a listed rating of 2-A. A few 1¼ and 1½ gal (4.7 and 5.6 L) portable models were manufactured, as were a few 17 and 33 gal (64.4 and 124.9 L) wheeled models.

As its name implies, this extinguisher contains two chemicals: sodium bicarbonate and sulfuric acid. Near the top inside the extinguisher shell, there is an acid bottle or container supported in an upright position by a wire cage. This container has a loose lead or ceramic stopple and contains a quantity of sulfuric acid; 4 oz (0.1 L) in the 2½ gal (9.5 L) units. The extinguisher shell is filled to a prescribed level with water in which a specific quantity of sodium bicarbonate has been thoroughly dissolved. (See Fig. 20-2P.) When the extinguisher is inverted, the loose stopple moves to allow the acid to mix with the soda solution to form carbon dioxide. At this point the extinguisher becomes a pressure vessel, and it is important that the extinguisher shell is in no way weakened. When these extinguishers are located in areas subject to temperatures below 40°F (4°C), they must be placed in heated enclosures to prevent freezing. Antifreeze additives should never be put into the bicarbonate solution, because this interferes with the chemical reaction and renders the extinguisher unsafe and inoperative.

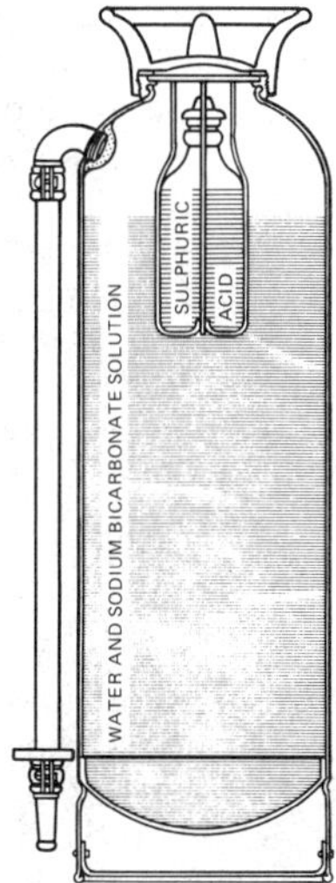

FIG. 20-2P. Soda-acid fire extinguisher. (Photo: Badger-Powhatan)

Cartridge-Operated Water Extinguishers

This type of extinguisher closely resembles the soda-acid extinguisher. The most common model manufactured was the 2½ gal (9.5 L) size, which had a rating of 2-A. Some 1¼ gal (4.7 L) models rated 1-A were also manufactured.

The extinguisher shell contains water and a small cylinder of carbon dioxide expellant. (See Fig. 20-2Q.)

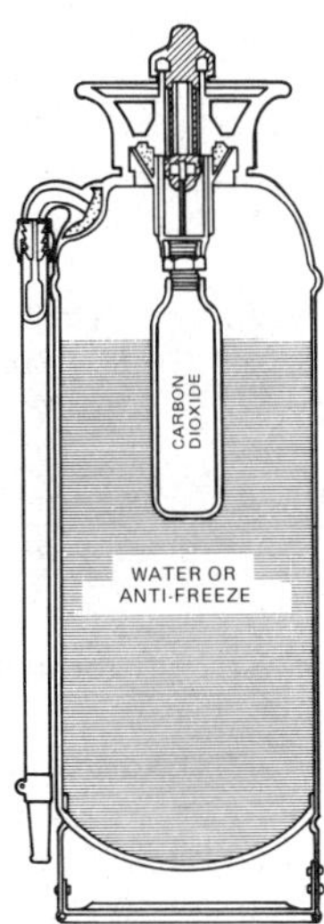

FIG. 20-2Q. Cartridge operated water fire extinguisher.

There are two basic designs of this extinguisher. One expels the agent by means of a sliding pin plunger which is part of the cap. To operate, the extinguisher must be inverted and bumped on the ground, which causes the pin to break the seal of the CO_2 cylinder. The other design has the CO_2 cylinder in a cage inside the shell just below the cap. When the extinguisher is inverted, the CO_2 cylinder strikes a rupture pin. Sometimes, however, this model must also be bumped on the ground.

Some models of this type of extinguisher may be installed in areas with freezing temperatures if an anti-freeze agent is added. Potassium carbonate is usually added if the shell is stainless steel, and calcium chloride if the shell is brass or copper.

Foam Extinguishers

Foam extinguishers look much like soda-acid extinguishers. They were more popular in the 2½ gal (9.5 L) size. A typical rating for a 2½ gal (9.5 L) size was 2-A:4-B. Other sizes manufactured included 1 ¼ and 1½ gal (4.7 and 5.9 L) portables, and 17 and 33 gal (65 and 125 L) wheeled models.

This extinguisher has an inner chamber or cylinder fitted with a loose stopple, which contains an aluminum sulphate "A" solution. The main extinguisher shell is filled with the "B" solution, which contains sodium bicarbonate and a foam stabilizing agent. (See Fig. 20-2R.) When the extinguisher is inverted, the stopple moves just enough to allow the "A" solution to intermix with the "B" solution and produce carbon dioxide. Thus the expellant gas evolves from the mixture, and produces a liquid foam that expands in a ratio of approximately 1 to 8; i.e., 2½ gal (9.5 L) of solution produces a minimum of 18 gal (68 L) of foam.

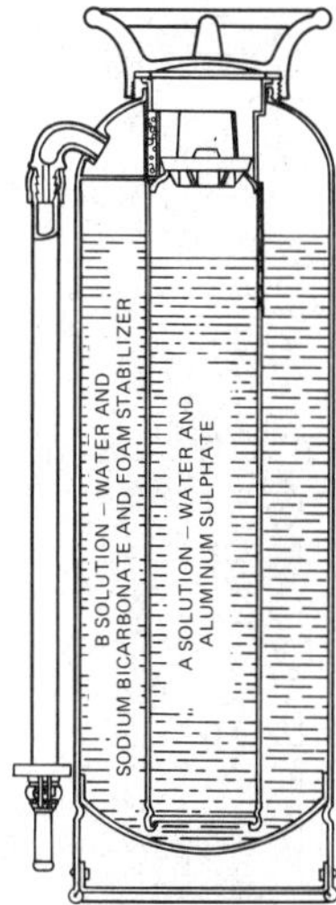

FIG. 20-2R. Foam type fire extinguisher.

When these extinguishers are located in areas subject to temperatures below 40°F (4°C), they must be placed in heated enclosures. Antifreeze additives should never be used because they interfere with the needed chemical reaction and render the extinguisher inoperative and even unsafe.

Although foam extinguishers were primarily designed for use on Class B fires, they could be used on Class A fires with about the same effectiveness as soda-acid. For fires in ordinary combustible materials, the foam should be used to coat the burning surface. For liquid spill fires, the foam can be spread over the burning surface by bouncing it off the floor just ahead of the flame front, or by standing back and aiming the stream upward so that it will fall lightly on the burning area. On flammable liquid fires of appreciable depth, the discharge should be played against the inside of the back wall of the vat or tank just above the burning surface to permit the natural spread of the foam over the burning liquid. If this cannot be done, the operator should stand far enough away from the fire to allow the foam to fall lightly upon the burning surface, because discharging the stream directly into the burning liquid may cause the fire to spread to other areas. Preferably, this type should be replaced with a newer AFFF type.

Vaporizing-Liquid Extinguishers

Two basic designs of this extinguisher were produced; in one, the agent is expelled by means of a hand pump. In the other, the expellant force is a gas (either carbon dioxide or nitrogen) or air, stored in either the agent chamber or an attached chamber. When the agent stream is directed at the base of a fire, it rapidly evaporates to form a smothering blanket of vapor that is heavier than air and will not support combustion. The extinguishing agent was rated for Class B:C fires and consists of either specially treated CCl_4 or CBM. Although both are liquids, they vaporize too quickly to be effective on deep seated Class A fires.

The models with hand pumps came in sizes ranging from 1 qt to 2 gal (0.5 to 9.5 L), the most common being 1 qt (0.5 L). (See Fig. 20-2S.) Pressure operated models came in sizes ranging from 1 qt (0.5L) to 3½ gal (13 L).

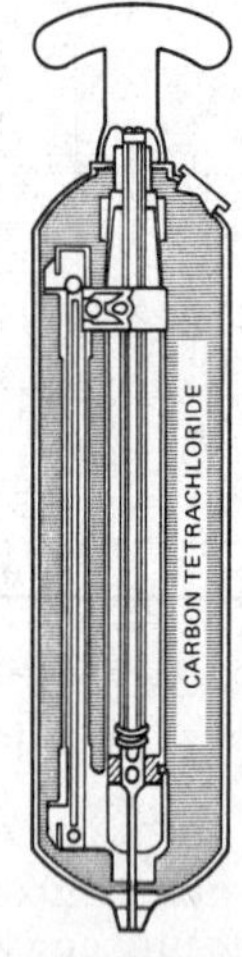

FIG. 20-2S. Vaporizing-liquid fire extinguisher.

Due to the toxicity of the agent and the agent decomposition products when used on a fire, vaporizing extinguishers are banned from use.

DISTRIBUTION OF FIRE EXTINGUISHERS

No matter how carefully extinguishers are chosen to match the hazards of an area and the capability of the people who will use them, they will not be effective unless they are readily available. Sometimes extinguishers are kept at hand, as in welding operations, but more often one has to travel from the fire to the extinguisher and back before beginning to put the fire out. In such cases travel distance to the nearest extinguisher is very important. Travel distance is the actual distance (i.e., around partitions, through doorways and aisles, etc.) that someone must walk to reach the extinguisher.

When placing extinguishers, select locations that will: (1) provide uniform distribution, (2) provide easy access and be relatively free from temporary blockage, (3) be near normal paths of travel, (4) be near exits and entrances, (5) be free from the potential of physical damage, and (6) be readily available.

Mounting Extinguishers

If an extinguisher falls, it may injure someone or be so damaged that it must be replaced. Most extinguishers are mounted to walls or columns by securely fastened hangers so that they are adequately supported. Some extinguishers are mounted in cabinets or wall recesses. When this is the case the operating instructions should face outward and the extinguisher should be placed so that it can be easily removed. Cabinets should be kept clean, dry, and ventilated if installed in outdoor locations subject to the direct rays of the sun.

Where extinguishers may become dislodged, brackets specifically designed to cope with this problem are available. In areas where they are subject to physical damage (such as warehouse aisles), protection from impact is important. In large open areas (such as aircraft hangars), extinguishers may be mounted on moveable pedestals or wheeled carts. In order to maintain some pattern of distribution and specify intended placement, locations should be marked on the floor.

The NFPA Extinguisher Standard specifies floor clearance and mounting heights, based on extinguisher weight, as follows:

1. Extinguishers with a gross weight not exceeding 40 lb (18 kg) should be installed so that the top of the extinguisher is not more than 5 ft (1.5 m) above the floor.
2. Extinguishers with a gross weight greater than 40 lb (18 kg) (except wheeled types) should be installed so that the top of the extinguisher is not more than 3½ ft (1 m) above the floor.
3. In no case shall the clearance between the bottom of the extinguisher and the floor be less than 4 in (100 mm).

When extinguishers are mounted on industrial trucks, vehicles, boats, aircraft, trains, etc., special mounting brackets (available from the manufacturer) should be used. Although cabinets or lockers are sometimes used, these locations frequently become cluttered with items that hamper quick retrieval. It is also important that the extinguisher be located at a safe distance from the hazard so that it will not become involved in the fire.

Extinguisher Distribution for Class A Combustibles

Table 20-2B is a guide to determining the minimum number and rating of extinguishers for Class A fires needed in any particular area. Sometimes extinguishers with ratings higher than what the table indicates may be necessary because of process hazards, building configura-

TABLE 20-2B. Fire Extinguisher Size and Placement for Class A Hazards

Basic Minimum Extinguisher Rating for Area Specified	Maximum Travel Distances to Extinguishers ft (m)	Areas to be Protected per Extinguisher		
		Light Hazard Occupancy sq ft (m^2)	Ordinary Hazard Occupancy sq ft (m^2)	Extra Hazard Occupancy sq ft (m^2)
1-A	75 (23)	3,000 (279)	—	—
2-A	75 (23)	6,000 (557)	3,000 (279)	2,000 (186)
3-A	75 (23)	9,000 (836)	4,500 (418)	3,000 (279)
4-A	75 (23)	11,250 (1045)	6,000 (557)	4,000 (372)
6-A	75 (23)	11,250 (1045)	9,000 (836)	6,000 (557)
10-A	75 (23)	11,250 (1045)	11,250 (1045)	9,000 (836)
20-A	75 (23)	11,250 (1045)	11,250 (1045)	11,250 (1045)
40-A	75 (23)	11,250 (1045)	11,250 (1045)	11,250 (1045)

Note: 11,250 sq ft (1045 m^2) is considered a practical limit.

tion, etc., but in no case should the recommended maximum travel distance be exceeded.

The first step when calculating how many Class A extinguishers are needed is to determine whether an occupancy is light, ordinary, or extra hazard according to the NFPA Extinguisher Standard. Next, extinguisher rating should be matched with occupancy hazard to determine the maximum area that an extinguisher can protect. Table 20-2B also specifies the maximum travel distance (actual walking distance) allowed; for Class A extinguishers, it is 75 ft (23 m). For example, each 2½ gal (9.5 L) stored pressure water extinguisher rated 2-A will protect an area of 3,000 ft^2 (279 m^2) in an ordinary hazard occupancy, but only 2000 ft^2 (186 m^2) in an extra hazard occupancy.

The figure of 11,250 ft^2 (1044 m^2) in Table 20-2B is used instead of 12,000 ft^2 (1115 m^2) (which would appear to be a normal progressive increment). This is because if a circle with a 75 ft (2.3 m) radius is drawn, the largest square inside that circle would be 106 ft x 106 ft (32.1 m × 32 m), or 11,250 ft^2 (1044 m^2). Because buildings are usually rectangular, this is the largest open area one can have and still comply with the 75 ft (23 m) travel distance rule. (See Fig. 20-2T for an illustration of this concept.)

The NFPA Extinguisher Standard also provides that up to one half of the complement of extinguishers for Class A fires, as specified in Table 20-2B, may be replaced by uniformly spaced small hose (1½ in. or 3.8 cm) stations. The hose stations and extinguishers, however, should be located so that the hose stations do not replace more than every other extinguisher.

The following examples illustrate the number and placement of extinguishers according to occupancy hazard and extinguisher rating. The sample building is 150 ft × 450 ft (46 × 137 m), which gives a floor area of 67,500 ft^2 (6270 m^2). Although several different ways of placing extinguishers are given, a number of other locations could have been used with similar results.

The first example demonstrates placement at the maximum protection area limits (11,250 ft^2 or 1044 m^2) allowed per extinguisher in the NFPA Extinguisher Standard for each class of occupancy. Installing extinguishers with higher ratings will not affect distribution or placement.

Example 1:

$$\frac{67,500}{11,250} = 6 \begin{cases} \text{4-A Extinguishers for Light Hazard Occupancy} \\ \text{10-A Extinguishers for Ordinary Hazard Occupancy} \\ \text{20-A Extinguishers for Extra Hazard Occupancy} \end{cases}$$

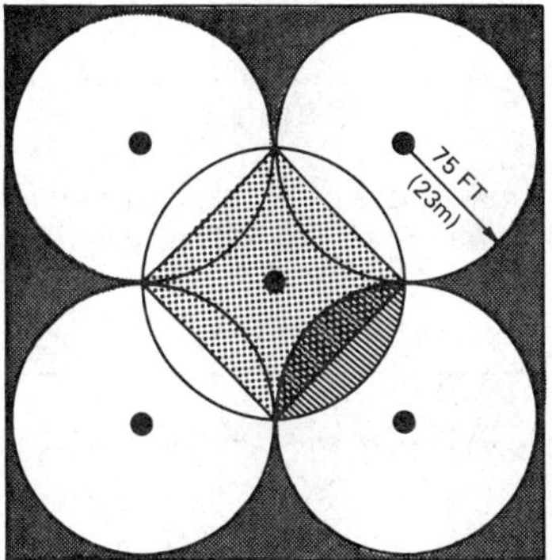

FIG. 20-2T. The dotted square shows the maximum area (11,250 sq ft or 1045 m²) that an extinguisher can protect within the limits of a 75 ft (23 m) radius.

This placement along outside walls would not be acceptable because the travel distance rule is clearly violated. (See Fig. 20-2U.) Instead, relocation and/or additional extinguishers are needed.

Examples 2 and 3 are for extinguishers having ratings which correspond to protection areas of 6,000 ft^2 (557 m^2) and 3,000 ft^2 (279 m^2) respectively. The examples show only one of many ways these extinguishers could be placed. As the number of lower rated extinguishers increases, meeting the travel distance requirement generally becomes less of a problem. Similar examples could be worked out for protection areas of 4,000 (371 m^2) and 4,500 ft^2 (418 m^2) as required by Table 20-2B.

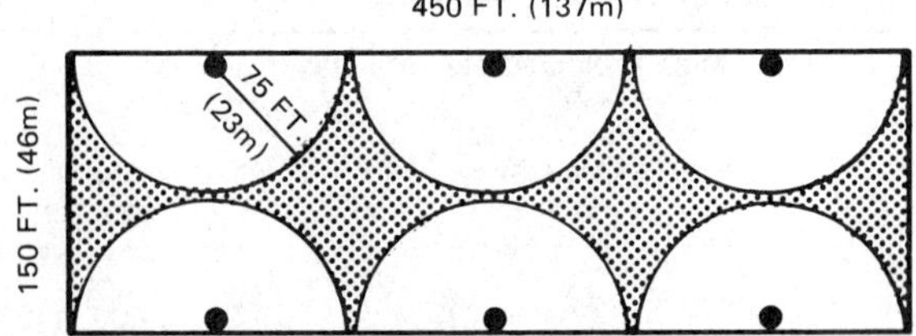

FIG. 20-2U. A diagrammatic representation of extinguishers located along the outside walls of a 150 ft by 450 ft (46 m by 137 m) building. (The large dots represent extinguishers. The shaded areas indicate "voids" which are farther than 75 ft [23 m] from the nearest extinguisher.)

Example 2:

$\frac{67,500}{6,000} = 12$

- 2-A Extinguishers for Light Hazard Occupancy
- 4-A Extinguishers for Ordinary Hazard Occupancy
- 6-A Extinguishers for Extra Hazard Occupancy

Example 3:

$\frac{67,500}{3,000} = 24$

- 1-A Extinguishers for Light Hazard Occupancy
- 2-A Extinguishers for Ordinary Hazard Occupancy
- 3-A Extinguishers for Extra Hazard Occupancy

Extinguishers could be mounted on exterior walls or, as shown in Figure 20-2V, on building columns or interior walls, and conform to both distribution and travel distance rules.

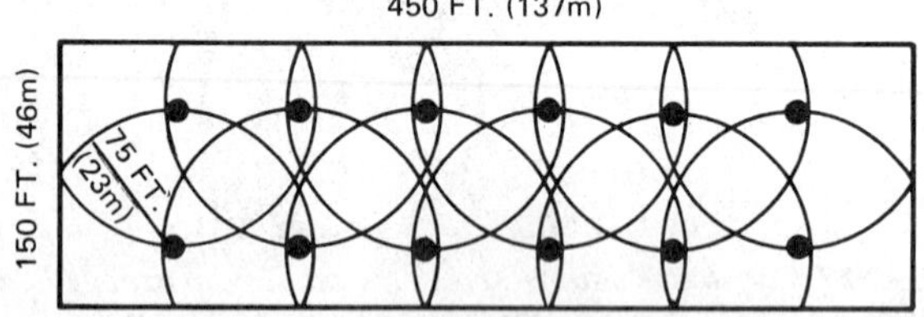

FIG. 20-2V. Requirements for both travel distance and extinguisher distribution are met in this configuration of twelve extinguishers mounted on building columns or interior walls.

The arrangement, illustrated in Figure 20-2W, shows extinguishers grouped together on building columns or interior walls in a manner that still conforms to distribution and travel distance rules.

Extinguisher Distribution for Class B Combustibles

As noted earlier, Class B fire hazards fall into two distinct categories; the first includes liquids ¼ in. (6.4 mm) deep or less, and the other includes liquids deeper than ¼ in (6.4 mm).

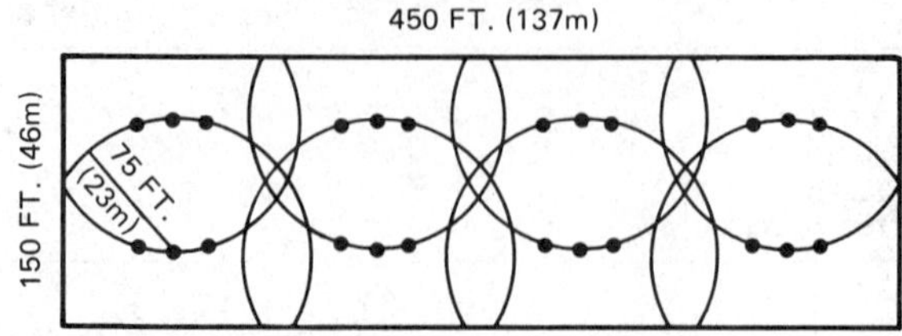

FIG. 20-2W. Extinguishers grouped together.

In areas where liquids do not reach an appreciable depth, extinguishers should be provided according to Table 20-2C. The reason why the basic maximum travel distance to Class B extinguishers is 50 ft (15.2 m), as opposed to 75 ft (23 m) for Class A extinguishers, is that flammable liquids fires reach their maximum intensity almost immediately, and thus the extinguisher must be nearer to hand. With lower rated extinguishers, the travel distance is reduced to 30 ft (9 m).

TABLE 20-2C. Fire Extinguisher Size and Placement for Class B Hazard Excluding Protection of Deep Layer Flammable Liquid Tanks

Type of Hazard	Basic Minimum Extinguisher Rating	Maximum Travel Distance to Extinguishers ft (m)
Low	5-B	30 (9)
	10-B	50 (15)
Moderate	10-B	30 (9)
	20-B	50 (15)
High	40-B	30 (9)
	80-B	50 (15)

Even though Table 20-2C specifies maximum travel distances to extinguishers for Class B fires, judgment should be used when actually placing the devices. The closer the extinguisher to the hazard, the better—up to the point at which a fire might damage or obstruct access to the device. When an entire room is a Class B hazard (such as an auto repair garage), extinguishers should be placed at regular intervals so that the travel distances do not exceed those in Table 20-2C.

When flammable liquids reach an appreciable depth, the extinguisher's rating number should be (except for foam types) at least twice the number of square feet of surface area of the largest tank in the area (assuming that other requirements are met). The travel distances specified by Table 20-2C should also be used to locate extinguishers for protection of spot hazards. Sometimes one extinguisher can be installed to provide protection against several different hazards, provided that travel distances are not exceeded. Where hazards are widely separated and travel distances are exceeded, individual protection should be installed according to the square foot requirements.

Where there are open process tanks of flammable liquids with surface areas greater than 10 ft^2 (1 m^2), complete dependence should not be placed on fire extinguishers. Fires in tanks as large as these rapidly become so large and intense that it is impossible to approach them, and some appropriate fixed protection should be installed. If a fixed protection system suitable for Class B fires is installed, then the extinguisher requirement may be waived for the hazard, or hazards, it protects, though not for the complete structure or any other special hazards in the area. But portable extinguishers are always desirable. They may be useful even when fixed protection systems have been installed, in case a burning tank spills liquid outside the range of fixed equipment or a fire begins adjacent to a tank rather than in it.

Pressurized flammable liquids and gases are not stored in open containers, and it is not possible to select extinguishers for them according to the square foot requirements. Instead, extinguishers with special nozzles and rate of agent application should be chosen. The manufacturers of these specialized extinguishers usually recommend what equipment they are suited for. In general, however, no attempt to extinguish pressurized fuel fires should be made unless the source of fuel can be promptly shut off, because otherwise an explosion may occur. As before, the travel distances for portable extinguishers should not exceed those specified by Table 20-2C.

Because Class B fires become so intense so quickly, the flow rate and duration of discharge of an agent are very important. For these reasons, the NFPA Extinguisher Standard does not permit substituting two or more extinguishers of lower rating for the minimum ratings give in Table 20-2C, except for certain foam extinguishers. The exception permits the substitution because foam progressively "secures" a fire, and the foam from one extinguisher is effective during the time it takes to bring other units into operation. With carbon dioxide and dry chemical extinguishers, however, four extinguishers rated 5-B are *not* equal to one extinguisher rated 20-B; the larger extinguisher has a higher flow rate in pounds per second and a longer continuous discharge time than smaller models. Smaller models are appropriate only if access to an area is limited or a trained fire brigade is available to attack a fire simultaneously from several positions.

Wheeled extinguishers with ratings from 20-B to 480-B are available. They are designed chiefly for outdoor fire fighting and should be used only by trained employees. They should be distributed according to the 50 ft (15 m) travel distance rule, though longer travel distances may be authorized by the proper authorities.

Extinguisher Distribution for Class C Fires

Extinguishers for Class C fires are required wherever there is live electrical equipment. This sort of extinguisher contains a nonconducting agent, usually carbon dioxide, dry chemical, or a halogenated agent.

Once the power to live electrical equipment is cut off, the fire becomes a Class A, Class B, or Class A:B fire, depending on the nature of the burning electrical equipment and the burning material in the vicinity. Extinguishers for Class C fires should be selected according to (1) the size of the electrical equipment, (2) the configuration of the electrical equipment (particularly the enclosures of units, which influence agent distribution), and (3) the range of the extinguisher's stream. At large installations of electrical equipment where continuity of power is critical, fixed fire protection is desirable. But even when fixed fire protection is present, it is recommended that some Class C extinguishers be provided to handle incipient fires.

Class D Extinguisher Distribution

It is particularly important that the proper extinguishers be available for Class D fires. Because the properties of combustible metals differ, even an agent for Class D fires may be hazardous if used on the wrong metal. Agents should be carefully chosen according to the manufacturer's recommendations; the amount of agent needed is normally figured according to the surface area of the metal plus the shape and form of the metal, which could contribute to the severity of the fire and cause "bake-off" of the agent. For example, fires in magnesium filings are more difficult to put out than fires in magnesium scrap and more agent is needed for magnesium filings. The maximum

SAMPLE PROBLEM

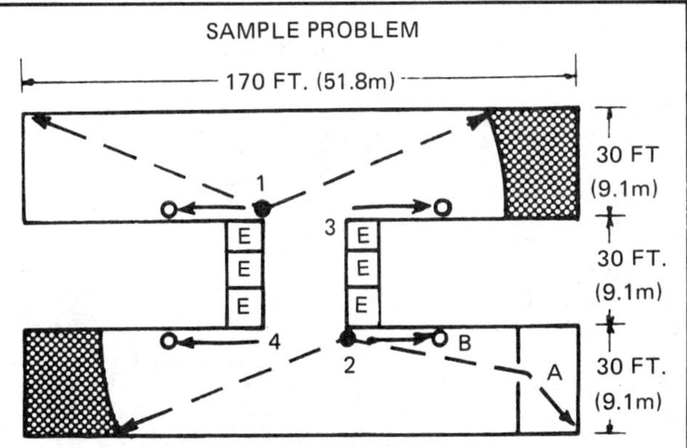

A light-occupancy office building needs to be protected by portable fire extinguishers. The floor area is 11,100 ft² (1031m²) and of unusual design.

The most common extinguisher selection would be 2½ gal (9.5L) stored-water pressure models rated 2-A. According to the requirements, two extinguishers are needed (11,100 divided by 6,000 = 2). Travel distance requirements are 75 ft. (23m) maximum.

The two units are placed at Points 1 and 2, and a check is made on the travel distance requirement. Because of the area's unusual shape, it is found that the shaded areas exceed the 75 ft. (23m) distance. Two additional extinguishers (at Points 3 and 4) are needed. The additional extinguishers afford more flexibility in placement, and alternate locations are indicated. It is important to consider any partitions, walls or other obstructions in determining the travel distance.

As an additional item, consider that Area A contains a small printing and duplicating department that uses flammable liquids. This area is judged to be an ordinary Class B hazard. A 10-B:C or 20-B:C extinguisher should be specified to protect this area.

There are now two alternatives to be considered. First, a fifth extinguisher, either carbon dioxide or ordinary dry-chemical, with a rating of 10-B:C or 20-B:C could be specified. Second, the water extinguisher at Point 2 could be replaced with a multi-purpose dry-chemical extinguisher that has a rating of at least 2-A:10-B:C. It should be located near Point B, keeping in mind the 75 ft. (23m) travel distance for the 2-A protection and the 30 or 50 ft. (9 or 15m) travel distance required for the Class B protection that this extinguisher provides.

FIG. 20-2X. Sample problem for protection by portable extinguishers.

travel distance to all extinguishers for Class D fires is 75 ft (23 m).

Figure 20-2X provides a sample problem for protection by portable extinguishers.

Bibliography

NFPA Codes, Standards, Recommended Practices and Manuals. (See the latest *NFPA Codes and Standards Catalog* for availability of current editions of the following documents.)

NFPA 10, *Standard for Portable Fire Extinguishers.*

Additional Readings

"Checklist for Fire Extinguisher Applications," *Cat. No. CK 84*, National Association of Fire Equipment Distributors, Chicago, IL, 1984, p. 10.

Katzel, Jeanine A., "Selecting, Maintaining and Using Portable Fire Extinguishers," *National Safety News*, Vol. 119, No. 6, June 1979, pp. 52-58.

Peterson, H. B. and Gipe, R. L., "Discharge Characteristics of Potassium Bicarbonate Dry Chemical Fire Fighting Agents from Cartridge and Stored Pressure Extinguishers," *NRL Report 5853*, U.S. Naval Research Laboratory, Washington, DC, Dec. 13, 1962.

"Portable Extinguishers: Selection, Placement, Use, (for use with NFPA *Slide Set No. SL-47*)," National Fire Protection Association, Quincy, MA, 1979.

Sayers, David R., "Halon 1211, Areas of Particular Effectiveness," *Fire Journal*, Vol. 67, No. 6, Nov. 1973, pp. 14-15.

"Training Your Fire Brigade to use First Aid Extinguishers," Walter Kidde & Co., Belleville, NJ.

Tuve, Richard L., "Dry Chemical Fire Extinguishers," *Security World*, July-August 1966.

"When Fire Extinguishers are Dangerous," *Cat. No. DNG-84*, National Association of Fire Equipment Distributors, Chicago, IL, 1984, p. 14.

INSPECTION AND MAINTENANCE OF FIRE EXTINGUISHERS

Marshall E. Petersen, P. E.

Once a fire extinguisher has been purchased, it becomes the responsibility of the purchaser or an assigned agent to maintain the device. Adequate maintenance consists of: (1) periodically inspecting each extinguisher, (2) recharging each extinguisher following discharge, and (3) performing hydrostatic tests as needed.

A fire equipment servicing agency is usually the most reliable way for the general public to maintain extinguishers, but large industries often train employees to handle this maintenance themselves.

This chapter specifically details the inspection and maintenance of fire extinguishers. Also included within this chapter are discussions of hydrostatic testing and maintenance services of extinguishers.

For a complete discussion of the allied topics of extinguishers, refer to the remaining parts of this section: Chapter 1, "Role of Portable Extinguishers;" Chapter 2, "Selection Operation and Distribution;" and Chapter 4, "Auxiliary Devices." Also, consult Section 19 for an explanation of the extinguishing agents used in portable fire extinguishers.

INSPECTION

An "inspection" is a quick check that visually determines that the fire extinguisher is properly placed and will operate. Its purpose is to give reasonable assurance that the extinguisher is fully charged and will function effectively if needed. An inspection should determine that the extinguisher: (1) is in its designated place, (2) is conspicuous, (3) is not blocked in any way, (4) has not been activated and partially or completely emptied, (5) has not been tampered with, (6) has not sustained any obvious physical damage or been subjected to an environment which could interfere with its operation (such as corrosive fumes), and (7) if the extinguisher is equipped with a pressure gage and/or tamper indicators, that each shows conditions to be satisfactory. In addition, the maintenance tag can be checked to determine the date of the last thorough maintenance check.

In order to be effective, inspections must be frequent, regular, and thorough. In a small building with few extinguishers, the manager, property owner, or some designated person could check extinguishers at the beginning of each workday. If a building is large enough to employ security guards or watchmen, extinguishers should be inspected at least once during each 8 hr shift. In industrial plants, the plant fire brigade or fire inspector often inspects extinguishers either daily, weekly, or monthly.

An individual evaluation must be made of each property to determine how frequent inspections should be made. If a particular operation is fire prone or is crucial to the use of the property, more frequent inspections should be made. For example, if all the products manufactured in a particular industrial plant had to be painted, a fire in the painting room could be disastrous, and, consequently, inspections should be very frequent.

Some important considerations are: (1) the nature of the hazards present (which influence the potential use of the equipment), (2) the exposure of the extinguisher to tampering, vandalism, and malicious mischief, (3) extraordinary weather conditions, (4) the likelihood of accidental damage to the equipment , and (5) the possibility of visual or physical obstructions to the accessibility of extinguishers.

MAINTENANCE

"Maintenance," as distinguished from inspection, means a complete and thorough examination of each extinguisher. A maintenance check involves disassembling the extinguisher, examining all its parts, cleaning and replacing any defective parts, and reassembling, recharging, and, where appropriate, repressuring the extinguisher.

Maintenance checks sometimes reveal the need for special testing of extinguisher shells or other components. They may, for example, reveal that the extinguisher container should be hydrostatically tested or even replaced.

Maintenance should be performed periodically, but at least once every 12 months, after each use, or when an inspection shows that the need is obvious. For example, if during an inspection there is evidence of serious damage

Mr. Petersen is an Assistant Vice President—Engineering at Rolf Jensen & Associates, Inc., Deerfield, IL.

by corrosion, the extinguisher should be subjected to a thorough maintenance check even though it may have recently undergone one. Similarly, if an inspection shows evidence of tampering, agent leakage, or physical damage, a complete maintenance check should be initiated. NFPA 10, *Standard for Portable Fire Extinguishers* contains specific details related to maintenance.

Tags, Seals, and Tamper Indicators

For many years extinguisher tags have been used as a convenient means for recording maintenance checks. For routine maintenance, a tied-on tag or pressure sensitive label is used to record the date and the inspector's initials. Seals and tamper detectors should also be used; the seal or tamper indicator often consists of a wire, band, plastic insert, or other device that conforms to ANSI standards. Lead and wire seals were commonly used until plastic seals were introduced in 1972. As long as the seal remains intact, one can be reasonably sure that the extinguisher has not been used. (It is important, however, to note that a stored-pressure extinguisher can develop a leak and lose its pressure even though the tamper indicator remains intact or the pressure gage reads normal.)

Maintenance Operations

In any maintenance test of a portable fire extinguisher, there are three basic items that need to be checked: (1) the mechanical parts of the device (that is, the extinguisher shell and other component parts); (2) the amount and condition of the extinguishing agent; and (3) the condition of the means for expelling the agent.

A record of the date of purchase as well as maintenance dates should be kept for each extinguisher. A separate record is also desirable and should include: (1) the maintenance date and the name of the person or agency performing the maintenance; (2) the date when last recharged and the name of the person or agency performing the recharge; (3) the hydrostatic retest data and the name of the person or agency performing the hydrostatic test; (4) description of dents remaining after passing a hydrostatic test, and (5) the date of the 6 year maintenance for certain stored-pressure dry chemical and halogenated agent extinguishers.

Individuals who own extinguishers often neglect them because there is no planned periodic follow up program. It is recommended that owners become familiar with their extinguishers so that they can detect telltale signs which may suggest the need for maintenance. An alternative is to have the dealer from whom the extinguisher was purchased establish an annual follow up maintenance program.

On properties where extinguishers are maintained by the occupant, a supply of recharging materials should be kept on hand. When recharging extinguishers other than those with plain water, use only those recharging materials specified on the extinguisher's nameplate. Other recharging materials may impair the efficiency of the extinguisher or cause it to malfunction and injure the operator. Special precautions are necessary for certain types of extinguishers, and they are discussed separately in this chapter.

The NFPA Extinguisher Standard has an appendix with a checklist of items that require maintenance examination. One section of the checklist is arranged to pinpoint the mechanical parts (containers and their compounds) common to most extinguishers.

Water Type Extinguishers

Inspection and maintenance for stored-pressure models is too often thought of as quite simple. Unless the extinguisher has been used or a routine inspection reveals a defect, maintenance is sometimes left until the 5 year hydrostatic test is performed. This practice has resulted in many failures and malfunctions. Principal items that need to be checked during an inspection are: loose, worn, or damaged hose; plugged nozzle; dented shell; damaged indicator gage; and a damaged or jammed ring pin. Normal maintenance service, as specified in the NFPA Extinguisher Standard, should be performed annually.

Mechanical pump extinguishers, such as pump tanks and back packs, are easy to inspect and maintain. The tanks containing the water or antifreeze solution must be checked to ensure they are in good condition, have not been weakened by corrosive action, and do not contain sediment which might block the stream during the pumping operation. If the device has not been used in the last 12 months, remove the liquid from the tank, flush the tank, and fill it with new liquid. Pumps may require lubrication from time to time. They should be checked annually to ensure that the plungers are in proper condition, that seals and washers have not deteriorated, and that hoses are not clogged. Antifreeze solutions may need to be checked to ensure that they can withstand the temperatures in the area and that they are not contaminated.

Dry Chemical Extinguishers

Dry chemical extinguishers should be inspected monthly, and they should also undergo normal annual maintenance. The quantity of agent for a cartridge operated model can be checked by weighing or by removing the fill cap and checking it visually. The gas cartridge may also be checked by weighing. On stored-pressure models, the pressure gage will indicate if adequate pressure is maintained, and the agent quantity can be checked by weighing. During annual maintenance of cartridge operated types, all agent should be dumped or vacuumed from the shell.

Dry chemical extinguishers should be promptly refilled after use, even if they have been only partially discharged. When refilling, extreme caution must be taken to ensure that no water, moisture, or foreign material enters the cylinder. Even though dry chemical agents are treated for moisture repellency, they can eventually harden if moisture is present. When these extinguishers are hydrostatically tested, they must be thoroughly dried so that no trace of water or moisture remains. Before new agent is added, all of the unused agent should be removed by dumping or vacuuming, and any residue should be removed from the hose.

The type of dry chemical used in recharging must be the one recommended by the manufacturer. For example, if an extinguisher contains a Class B:C dry chemical (bicarbonate base or potassium chloride), it should not be replaced with a Class A:B:C agent (monoammonium phosphate base). Mixing different dry chemical agents can result in malfunction or damage to the extinguisher or both. The bicarbonate base agent is chemically alkaline and will react with the acidic ammonium phosphate. This reaction is aggravated by exposure to heat or the presence of moisture in amounts as small as 0.1 percent. One result is caking (hardening) of the agent; another possible result is the internal corrosion of the extinguisher. Under certain conditions, the reaction will cause excess pressure to build up within the shell, which may damage or rupture the extinguisher.

Substituting another manufacturer's dry chemical of the same type is not recommended unless it has the same chemical and physical characteristics, or has been tested and found to give equivalent performance. The problems involved in substitution include:

1. An altered flow rate of the dry chemical, which could influence extinguishing efficiency;
2. The amount of agent that could be placed in the extinguisher, which could influence discharge time; and
3. Incompatibility of one dry chemical to another, including chemical reactions, caking, and jamming of hose or nozzles. The flow rate can change with the particle size and density of the agent. A discharge nozzle designed to dispense one dry chemical efficiently might well be less efficient for another. An extinguisher designed to hold a given weight of one formula may not hold the same amount of another. Dip tubes, hoses, or valves could be plugged.

Recharge operations for the cartridge operated models are relatively easy. The rubber sleeves on the gas tube should be checked for cracking or overstretching. The gasket and the gasket seats on the shell and cap should be wiped clean. The cap should be screwed on hand tight. When replacing the cartridge, care should be taken that the threads are not dirty, cross threaded, or otherwise damaged. For stored pressure models, the entire valve body, stem, and "O" ring should be thoroughly cleaned (carefully wiped clean and blown out with dry nitrogen) to rid all traces of agent residue which might cause a pressure leak.

Nitrogen gas used for pressurizing should be of the standard industrial grade with a dew point of −70°F (−56.6°C) or lower so that it is free of moisture. Once the extinguisher is pressurized, it should be given a leak test or allowed to stand for about 12 hours and then checked for leaks before it is returned to service.

Some stored-pressure extinguishers are factory sealed and are not rechargeable. No effort should be made to recharge these units in the field. Normally the shell of the extinguisher is discarded following use; only the valve and nozzle assembly are retained. Refill units should be obtained from the manufacturer, and detailed instructions for replacement are included on the extinguisher nameplate.

The general inspection and maintenance procedures for dry powder (Class D) extinguishers are the same as for cartridge operated dry chemical extinguishers.

Carbon Dioxide Extinguishers

Weighing is the only way to determine whether carbon dioxide extinguishers are fully charged. They should be weighed at least semiannually for loss of weight, and inspected for deterioration, and/or physical damage. Any carbon dioxide extinguisher that has a weight loss of 10 percent or more should be recharged and tested for leaks. Recharging is generally done by an extinguisher servicing company. However, should on site recharging be desirable, the manufacturer should be contacted for assistance in setting up a recharging station.

Recharging extinguishers with carbon dioxide from dry ice converters should not be attempted unless stringent restrictions are enforced to prevent the introduction of moisture into the cylinder. The vapor phase of carbon dioxide must not be less than 99.5 percent carbon dioxide. The water content of the liquid phase must not be more than 0.01 percent by weight [−30°F dew point (−34.4°C)]. Oil content of the carbon dioxide must not exceed 10 ppm by weight.

Specific recommendations are given for such recharging in the NFPA Extinguisher Standard. The preferable source of carbon dioxide is a low pressure carbon dioxide supply (300 psi at 0°F or 206.8 kPa at −17.8°C), either directly from the source or through dry cylinders used as intermediaries. It cannot be emphasized too strongly that internal moisture within an extinguisher can create serious corrosion, and that the most likely time when moisture is most easily introduced into a cylinder is either during recharging or following a hydrostatic test.

Halogenated Agent Extinguishers

The general inspection and maintenance procedures for halogenated agent extinguishers are similar to the requirements for other extinguishers. Bromochlorodifluoromethane (Halon 1211) extinguishers are physically similar to stored pressure dry chemical extinguishers. The specific requirements of each manufacturer should be followed.

HYDROSTATIC TESTING OF EXTINGUISHERS

The purpose of hydrostatic testing of portable fire extinguishers that are subject to internal pressures is to protect against unexpected, in-service failure due to: (1) undetected internal corrosion caused by moisture in the extinguisher, (2) external corrosion caused by atmospheric humidity or corrosive vapors, (3) damage caused by rough handling (which may or may not be obvious by external inspection), (4) repeated pressurizations, (5) manufacturing flaws in the construction of the extinguisher, (6) improper assembly of valves or safety relief discs, or (7) exposure of the extinguisher to abnormal heat, as after exposure in a fire.

The NFPA Extinguisher Standard requires that hydrostatic pressure tests be performed on extinguishers according to Table 20-3A. The first hydrostatic retest may be conducted between the fifth and sixth year for those with a designated test interval of 5 years. Carbon dioxide types, however, require a strict 5 year interval according to federal law. These test intervals, based on experience, have been established in an effort to present general guidance. Hydrostatic tests should also be conducted immediately upon discovery of any mechanical injury or corrosion to the extinguisher shell.

In 1978, the NFPA Extinguisher Standard called for the removal from service, by or before 1983, of all riveted and soft solder copper and brass shell extinguishers of inverting type. The one exception is stainless steel shells and brass shells that are brazed (as opposed to soft solder construction). Also in 1978 stored-pressure water and/or

TABLE 20-3A. Hydrostatic Test Interval for Extinguishers

Extinguisher Type	Test Interval (Years)
Soda-acid	5*
Cartridge operated water and/or antifreeze	5*
Stored-pressure water and/or antifreeze	5
Wetting agent	5
Foam	5*
AFFF (Aqueous Film Forming Foam)	5
Loaded stream	5
Dry chemical with stainless steel shells	5
Carbon dioxide	5
Dry chemical, stored-pressure, with mild steel shells, brazed brass shells, or aluminum shells	12
Dry chemical, cartridge or cylinder operated, with mild steel shells	12
Halogenated agents	12
Dry powder, cartridge or cylinder operated, with mild steel shells	12

* No longer recognized in the NFPA Extinguishing Systems Standard.

antifreeze fiberglass shell types were required to be immediately removed from service because they had been recalled by the manufacturer.

In the United States, various rules of the U.S. Department of Transportation (DOT) apply to specific types of cylinders used for fire extinguishers. Title 49 of the Code of Federal Regulations calls, with some exceptions, for the hydrostatic retesting every 5 years of compressed gas cylinders offered for interstate transportation in a charged condition. Paragraph 173.34(e) calls for periodic retesting of DOT compressed gas cylinders whether or not the cylinders are shipped interstate.

A great many different DOT specification cylinders are used for dry chemical extinguishers, including Types DOT-4B and DOT-4BA. The requirements for these types are peculiar to them.

The U.S. Coast Guard Regulations Title 46, Code of Federal Regulations, Part 147, Paragraph 147.04-1(a)(2) and (3) specify that compressed gas cylinders brought aboard a vessel shall have been tested hydrostatically within 5 years. After that, a cylinder continuously installed in place on board a vessel as part of the vessel's equipment for a period of time exceeding 5 years, shall, after 12 years have elapsed from the date of previous test and marking, be removed from the vessel, its contents discharged, and the cylinder retested and remarked. The Board of Transport Commissioners of Canada has published similar rules covering these same subjects.

The preferred method of hydrostatic testing of compressed gas cylinders is the water jacketed, volumetric expansion method. (It is recommended in the NFPA Extinguisher Standard.) The guide to follow in conducting this type of test is Pamphlet C-1, "Methods for Hydrostatic Testing of Compressed Gas Cylinders," published by the Compressed Gas Association, Inc. (CGA). For visually evaluating the condition of cylinders made to DOT specifications, the CGA's Pamphlet C-6, "Standard Visual Inspection of Compressed Gas Cylinders," is helpful.

Procedures for testing extinguishers other than compressed gas extinguishers (or affixed compressed gas cylinders) are detailed in the NFPA Extinguisher Standard. Table 20-3B gives the basic data on the hydrostatic test pressure requirements. Figures 20-3A and 20-3B illustrate equipment used in hydrostatic testing work.

It is not necessary to hydrostatically test certain extinguishers such as pump tanks, back packs, and similar devices. Factory sealed, nonrefillable, disposable fire extinguishers cannot be hydrostatically tested. When such extinguishers are damaged they should be replaced.

Proper hydrostatic testing requires competent personnel and suitable testing equipment and facilities. The preparation for testing and the precautions to remove all traces of water and moisture by using special drying equipment after testing are very important, and reinforce the recommendation that only those experienced in this type of work should undertake such servicing. For hydrostatic testing of carbon dioxide extinguishers, it is a DOT requirement that the testing agency be certified by the DOT.

TABLE 20-3B. Hydrostatic Test Pressure Requirements

Self-generating (soda-acid and foam)	350 psi (2413 kPa)
Carbon dioxide extinguishers Carbon dioxide and nitrogen cylinders (used with wheeled extinguishers)	5/3 service pressure stamped on cylinder
Carbon dioxide extinguishers with cylinder specification ICC3	3,000 psi (20 685 kPa)
All stored-pressure and Bromochlorodifluoromethane (1211)	Factory test pressure not to exceed 2 times the service pressure
Carbon dioxide hose assemblies	1,250 psi (8619 kPa)
Dry chemical and dry powder hose assemblies	300 psi (2068 kPa)

Note: The factory test pressure is the pressure at which the shell was tested at time of manufacture. This pressure is shown on the nameplate.

The service pressure is the normal operating pressure as indicated on the gage and nameplate.

Extinguisher hoses of certain types also need to be hydrostatically tested, and the details of these tests may be found in the NFPA Extinguisher Standard.

Because hydrostatic test records are of major importance, they must be recorded on the extinguisher. For compressed gas cylinders and cartridges passing a hydrostatic test, the month and year is stamped into the cylinder in accordance with Pamphlet No. C-1, CGA, or the Canadian Transport Commission regulation. It is important that the recording (stamping) be placed only on the shoulder, top head, neck, or footing (when so provided) of the cylinder. For noncompressed gas extinguisher shells, the test information should be recorded on metal or some equally durable material. The label (Fig. 20-3C) should be affixed by a heatless process to the shell, and should be self-destructive if removal is attempted. The label must include the following information:

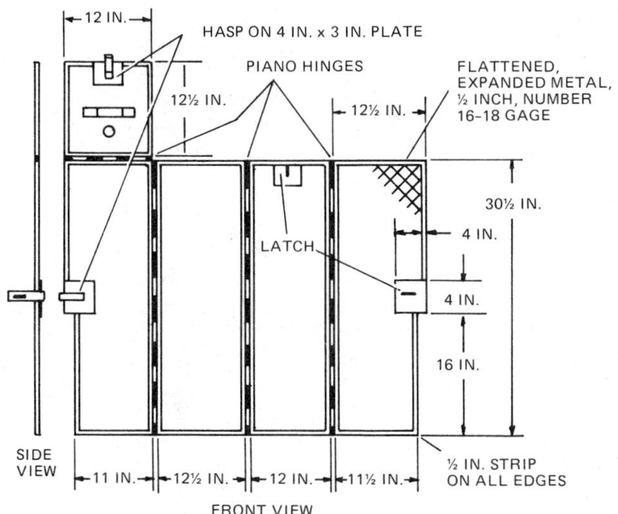

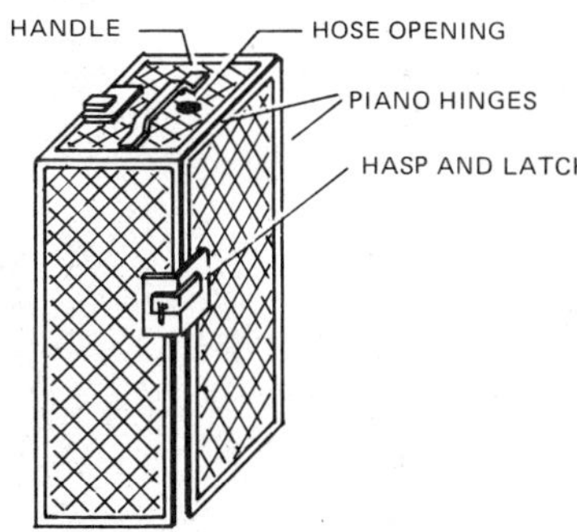

FIG. 20-3A. A low pressure portable hydrostatic test cage used for hydrostatic tests of noncompressed gas extinguishers. (Cage is not used for hydrostatic testing of compressed gas extinguishers.) Cage should not be anchored to floor during testing. Such cages can be made by any metal fabricator.

1. Month and year the test was performed, indicated by a perforation, such as by a hand punch;
2. Test pressure used; and
3. Name or initials of person performing the test, or name of agency performing the test.

Above all, remember the following: allow only competent personnel with proper equipment and facilities to

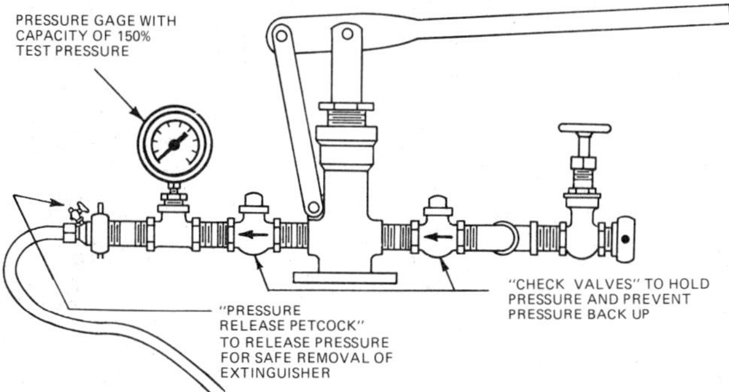

FIG. 20-3B. Hydrostatic testing equipment (test pump) for use with other than compressed gas cylinders.

Gorham Fire Appliance Co.
288 Willard St. Quincy, Mass
472-5785

6 YR. MAINT.						HYDRO TEST							1983
MO.	1	2	3	4	5	6	7	8	9	10	11	12	1984
TEST PSI	100	125	150	175	200	225	250	275	300	325	350	375	1985
	400	425	450	475	500	525	550	575	600	625	650	675	1986

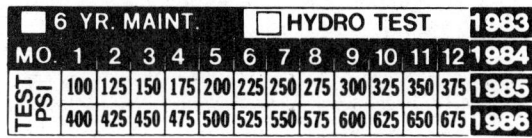

FIG. 20-3C. A typical hydrostatic test label.

do the testing. Do not use air or gas for pressure testing because of the hazard of a violent rupture occurring in cases of failure. Place any extinguisher undergoing a test in a protective cage before applying the test pressures. Remove all traces of moisture from dry chemical, carbon dioxide, halogenated agent, and dry powder extinguishers after each test before refilling; a heated air stream (temperature not exceeding 150°F or 65°C) is recommended for such drying. Destroy any extinguisher shell which fails a hydrostatic test—do not attempt to repair it.

Destroy any cylinder or shell if it has one or more of the following conditions—do not hydrostatically test it:

1. Repairs by soldering, welding, brazing, or use of patching compounds. (For welding or brazing on mild steel shells, consult the manufacturer of the extinguisher);
2. Damaged cylinder or shell threads;
3. Pitting caused by corrosion, even under the removable nameplate band assemblies;
4. Burns from a fire; and
5. Calcium chloride agent in a stainless steel extinguisher.

EXTINGUISHER MAINTENANCE SERVICES

Fire extinguisher maintenance (particularly hydrostatic testing) is a specialized activity and should be performed by competent, dependable people. Fire extinguishers are provided to protect life and property; this means that there should be no doubt as to their reliability or safe use in an emergency. Extinguisher owners are thus urged to seek out the services of reliable fire extinguisher maintenance firms. Such firms should be able to show proof that they are competent, that their facilities are adequate, and that they are licensed or registered in accordance with any local or federal law.

Bibliography

NFPA Codes, Standards, Recommended Practices, and Manuals. (See the latest *NFPA Codes and Standards Catalog* for availability of current editions of the following documents.)

NFPA 10, *Standard for Portable Fire Extinguishers.*
NFPA 10L, *Model Enabling Act for the Sale or Leasing and Servicing of Portable Fire Extinguishers.*

Additional Readings

"Beware of the Fire Extinguisher Servicing Racket," *Fire Journal*, Vol. 60, No. 3, May 1966, pp. 22-23.

Katzel, Jeanine A., "Selecting, Maintaining and Using Portable Fire Extinguishers," *National Safety News*, Vol. 119, No. 6, June 1979, pp. 52-58.

Thorne, P. S., "Why Extinguishers Burst," *Fire Journal*, 1968.

Tryon, G. H., "Hydrostatic Testing of Fire Extinguishers," *Fire Journal*, Vol. 61, No. 1, Jan. 1967, pp. 32-35.

"When Fire Extinguishers are Dangerous," *Cat. No. DNG-84*, National Association of Fire Equipment Distributors, Chicago, IL, 1984, p. 14.

AUXILIARY PORTABLE FIRE EXTINGUISHING DEVICES

Marshall E. Petersen, P. E.

Occasionally quantities of water or other agents are desirable in places where commercial extinguishers are not practical or else need to be replaced. The most common reasons for supplying auxiliary portable fire extinguishing devices are that property owners have had trouble with theft or vandalism of standard extinguishers, property value does not justify providing standard extinguishers, or standard extinguishers cannot be effectively handled and used at a particular location. As far as fire blankets are concerned, blankets are easier and safer to use should a person's clothes catch fire. Information about the following auxiliary devices is presented in this chapter: (1) covered buckets of water, (2) barrels and pails—bucket tanks, (3) sand buckets, (4) fire blankets, and (5) garden hose.

NFPA 10, *Standard for Portable Fire Extinguishers* does not recognize any of these auxiiary devices as a substitute for the normally required complement of fire extinguishers.

COVERED BUCKETS OF WATER

A standard fire bucket holds 10 or 12 quarts (9 or 11 L). It differs from ordinary metal buckets in that it is made of heavy galvanized metal and has a rounded or pointed bottom that makes it unsuitable for general use because it cannot remain upright when set down. Because fire buckets are intended to be carried to a fire and the contents thrown over the burning material, their effective range is limited to about 10 ft (3 m). Fire buckets should be placed on racks no more than 5 ft (1.5 m) and no less than 2 ft (0.6 m) above the floor. They are intended primarily for use inside buildings.

Fire buckets should have lids or covers to exclude foreign matter such as used cigarette packages, cigarette and cigar butts, dust accumulations, trash, etc. They should be painted bright red and the word "Fire" should be stenciled on them in contrasting paint. Frequent inspection and refilling is required in order to take care of normal evaporation.

Mr. Petersen is an Assistant Vice President—Engineering at Rolf Jensen & Associates, Inc., Deerfield, IL.

BARRELS AND PAILS—BUCKET TANKS

Outdoors, some property owners provide 55 gal (208 L) barrels or drums filled with water. These should have lids or covers to exclude foreign matter. Pails should be hung on hangers around the top edges of the barrels, either nested on racks above the barrels or immersed (but easily accessible) inside the barrels or bucket tanks.

Barrels and drums should be painted bright red, and the buckets, if not immersed, should have the word "Fire" stenciled on them in contrasting paint. The following combinations are considered the equivalent of an extinguisher rated 2-A:

1. Five 12 qt or six 10 qt (11 or 9 L) water filled standard fire pails.
2. Five 12 or six 10 qt (11 or 9L) standard fire pails immersed in a 25 to 55 gal (95 to 208 L) capacity water filled bucket tank.
3. Three standard fire pails hung on, or racked above, one 55 gal (208 L) water filled drum, cask, or barrel.

Because the extinguishing agent used is plain water, it should be converted to an antifreeze solution (where necessary) by adding calcium chloride (free from magnesium chloride). Table 20-4A gives the amounts of calcium

Table 20-4A. Calcium Chloride Required to Make 10 Gal (37.85L) of Antifreeze Solutions for Fire Pails, Drums, and Bucket Tanks

Probable Lowest Temp. °F	(°C)	Water Quantity gal	(L)	Calcium Chloride lb	(kg)	Specific Gravity of Solution	Degrees Baumé of Solution
+10	(12)	9	(34)	20	(9)	1.139	17.7
0	(−18)	8½	(32)	25	(11)	1.175	21.6
−10	(−23)	8	(30)	29½	(13)	1.205	24.7
−20	(−29)	8	(30)	33½	(15)	1.228	26.9
−30	(−34)	8	(30)	36½	(16)	1.246	28.6
−40	(−40)	8	(30)	40	(18)	1.263	30.2

Note: Use only the manufacturer's recommended antifreeze solution for a water type extinguisher that is to be located where freezing may occur.

chloride required to prepare a proper antifreeze solution, based upon probable lowest temperatures.

Considerable heat develops when calcium chloride is dissolved in water. Therefore, when planned for use in galvanized containers, such solutions should be mixed separately and allowed to cool before being placed in galvanized barrels or drums.

SAND BUCKETS

Although their value as extinguishing agents is almost nil, buckets of sand are still used on a limited basis. They are not considered the equivalent of any fire extinguisher. However, they may be used to control the flow of spilled flammable liquids and are sometimes useful on certain combustible metal fires and small low voltage electrical fires. Because it is abrasive, sand should not be used where there are moving parts.

A bucket or box of sand can be used to extinguish incipient fires caused by small pieces of burning metal or sparks from cutting or welding operations. (See Fig. 20-4A.) Frequently, large efflux particles fall on or become lodged in combustible or flammable materials. Spreading sand on the hot spot and wetting it down with water provides a convenient means of extinguishing the fire.

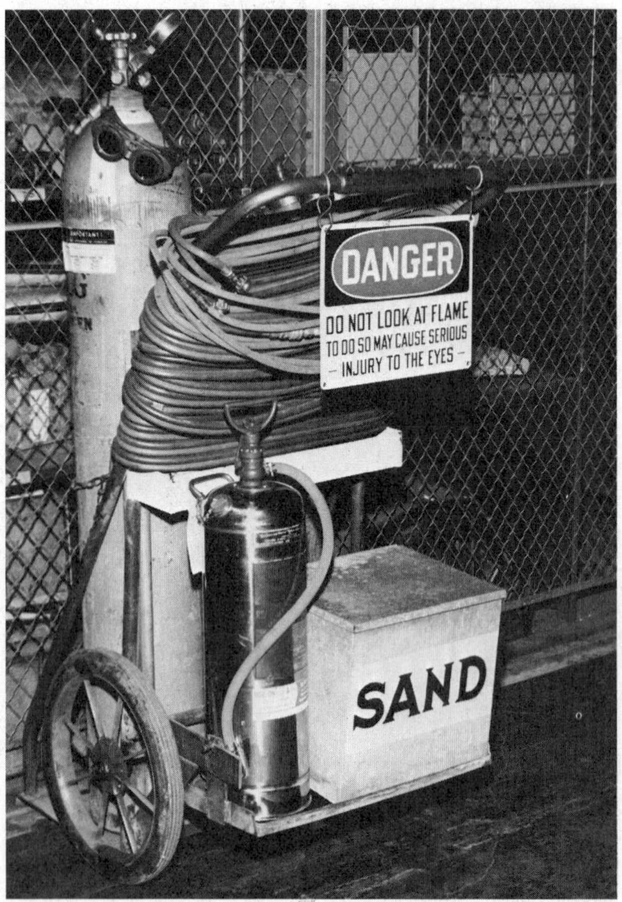

FIG. 20-4A. Water pump tank extinguisher and container of sand (inert material) mounted on oxyacetylene cart, intended for control of "hot spot" fires. Larger Class B extinguishers are kept at full charge as standby protection for use when needed.

FIRE BLANKETS

Fire blankets are made of high grade woven flame retardant treated fabric, aluminized fabrics, or flameproof wool. They are available in various sizes (62 x 82 in. or 157 208 cm is the most common size) and are usually folded in metal wall cases or in portable canvas bags. While fire blankets are principally used for smothering fires in clothing, they can also be used on other small Class A fires if no fire extinguisher is available. Treated fabric curtains are frequently used to prevent sparks from reaching combustibles around cutting and welding operations, and aluminized fabrics are sometimes provided as heat shields.

SMALL HOSE

Many industrial plants and construction firms have effectively used small diameter hose as a supplemental extinguishing means. Where convenient outlets are available in permanent or temporary water piping systems, small diameter (¾ to 1 in. or 19 to 25 mm) commercial hose or ½- to ¾-in. (13 to 19 mm) garden hose can be installed. The water system supply and pressure should be sufficient to give a straight stream discharge, of at least 30 ft (9 m) through a nozzle.

Even though small, such a hand hose has the advantage of virtually unlimited discharge time. Because of its length, however, the hose is limited in range, whereas an extinguisher can be carried directly to the fire.

Many hose nozzles are available. Some operate by twisting the nozzle from shutoff to spray to straight stream; others have a squeeze-grip device. Several have been approved by Factory Mutual System as combination water-spray nozzles for garden hose, and may deliver a flat spray, a cone spray and solid stream, or a solid cone spray. Specifications for garden hose threads and couplings are included in ANSI Standard B2.4-1966, "Hose Coupling Screw Threads."

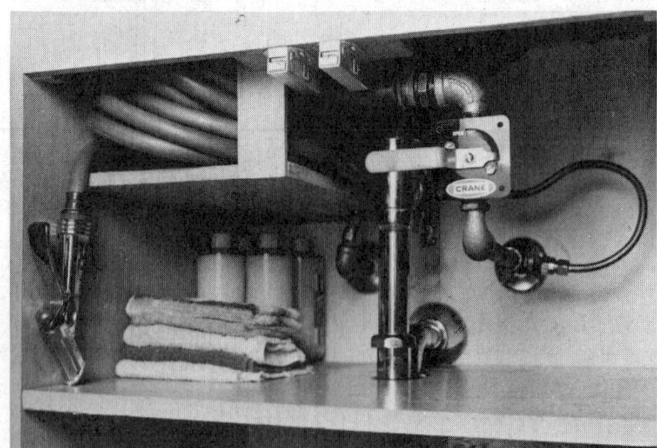

FIG. 20-4B. A garden hose connected to an ordinary domestic water supply and stored beneath a bathroom sink. To operate, the lever must be pushed to "open" and the nozzle of the hose squeezed.

Garden Hose in the Home

An ordinary garden hose connected to a domestic water system is an effective and inexpensive means for combatting fires in ordinary combustible materials. (See Fig. 20-4B.) With a special adaptor, it can be connected either to the cold water supply pipe to kitchen, bathroom, or utility sinks, or near the entry point to the hot water heater. Because of its central location, the bathroom is generally an ideal place. In large multilevel homes it may be advantageous to have several hose installations. The hose can be coiled up and placed on a shelf under the sink, or the pipe connection can be extended through a wall into an adjoining closet or stairway where the hose can be hung on a bracket. The special hose reels available for garden use are also ideal for indoor installation. Any lightweight plastic hose, at least ½ in. (13 mm) inside diameter, would be a good choice; however, it should be equipped with an adjustable squeeze-grip nozzle, and should be long enough to reach all intended areas. Such hand hose equipment, even though small, can be very effective provided it remains connected, is periodically checked for deterioration, and is reserved for use in fires only.

Bibliography

NFPA Codes, Standards, Recommended Practices and Manuals. (See the latest NFPA Codes and Standards Catalog for availability of current editions of the following documents.)

NFPA 10, *Standard for Portable Fire Extinguishers.*

SECTION 21

FIRE MODELING AND ANALYSIS

FIRE HAZARD ASSESSMENT

Frederic B. Clarke, Ph.D.

Risk and hazard are different, they are not interchangeable concepts. The concept of fire risk is better established: it is the probability that a given severity of fire will occur. For example, the risk of fire in a given environment may be very high, but owing to the presence of a sprinkler system, the risk of a *serious* fire may be quite low. Describing a fire risk, or any class of risk, therefore, carries with it the need to describe both the likelihood and the severity of the event.

"Hazard" has different meanings to different groups. The American Society of Testing and Materials, for example, defines a fire hazard as an unacceptably high level of fire risk (ASTM 1982). Such a definition does not permit discussions of comparative or relative degrees of danger from fire. Fire hazard as used in this chapter means the expected level of harm associated with the exposure to a fire or its products. The relative fire hazard posed by two materials, therefore, is the expected relative harm they offer if they both were to occur under the same circumstances. This definition means that it is impossible to discuss fire hazard (or fire risk) without knowing the circumstances in which a product is used and the fire conditions to which it will be exposed. These circumstances together comprise the fire "scenario" (Clarke and Ottoson 1976) for which hazard is to be assessed.

Quantifying Hazard

Fire scenarios may be categorized by increasing complexity. The simplest category is a fire involving a limited quantity of combustible, and occurring in a closed room. There is, for example, a soundly sleeping occupant in the compartment. The toxic smoke* accumulates and mixes more or less uniformly with the room air. A maximum level of toxic smoke is achieved as the combustible is consumed. The occupant, who is presumed to be unable to escape, may or may not be able to survive the exposure. To a first approximation, such a model involves simple computations using the mass of combustible involved, the volume of the room, and the lethal concentration level of the combustion products, assuming a substantial exposure time. Given valid data for the toxicity of the smoke, such calculations can be performed on the back of an envelope; no computer is needed. The result will show simply that the maximum acceptable quantity of combustible is inversely proportional to the toxicity of its smoke. From another viewpoint, one sees that there will be cases where a small increase in smoke toxicity will make the difference between survival and death.

Because the critical acceptable quantity of combustible for most materials and residential size rooms comes out to be of the order of a pound or two (1 lb = 0.45 kg), this simple fire scenario is probably not applicable to most lethal fires. Most dangerous fires involve more combustible than this, since typical residential fire loads are in excess of 2 or 3 lb per sq ft (10 or 15 kg/m^2). In other words, to survive most potentially lethal fires, it is necessary either to stop the fire or get away from it. The remaining categories of fire scenarios will deal with escape possibility. (The possibility of suppression is not considered in this chapter.)

Figure 21-1A shows a generalized fire growth curve, where the ordinate is a measure of the intensity, or size, of the fire. For example, the figure could represent the upper

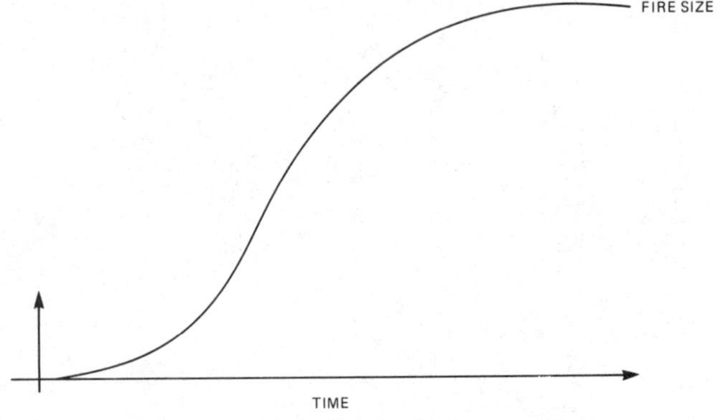

FIG. 21-1A. Typical growth of a room fire.

Dr. Clarke is President, Benjamin/Clarke Associates, Inc. in Kensington, MD.

*"Smoke" is used throughout this chapter to mean the *total* airborne effluent from the fire. It includes the gaseous, aerosol and sootlike products of the fire.

room temperature in a room in which the fuel is ignited with a match. Little energy is generated at the outset, but the fire eventually becomes large enough to begin heating the room. The room temperature, which reflects the size of the fire, begins to increase rapidly. This corresponds to the steep middle portion of the curve in Figure 21-1A. Finally, the fire will reach a limit in size either because the entire surface of the item is involved, or because air cannot enter the doorway any faster. In either case, the temperature will approach a limiting value, governed by the relative size of the fire, and the rate at which hot gases escape from the doorway. If Figure 21-1A were drawn for a longer period of time, the temperature would eventually decline again as the fuel burns itself out.

The burning itself is a time dependent process and, even after a fire has reached steady state conditions, the concentration of smoke will still be changing in most of the building. Hence, it is natural to use time as a basis for evaluating the relative hazard of different fires, and of the same fire in different locations. This is generally done by identifying some level of temperature and smoke concentration which is unacceptable for life safety, and determining how long the subject fire takes to reach that level.

Figure 21-1B shows the growth of temperature in the upper part of a room in which two different fuel packages are burned. As the temperature increases, it reaches a point at which escape is no longer possible, shown here somewhat arbitrarily as 100°C (212°F). The sofa represented by Curve 2 reaches this temperature at some time, *T2*, and the slower growing fire, represented by Curve 1, would reach this temperature after a somewhat greater time, *T1*. The difference between these two times is then a measure of the relative hazard posed by these two sofas in this particular room environment and fire growth scenario. It is worth noting that if the critical temperature were chosen to be much higher (or much lower) than 100°C (212°F), the difference in the two times would be different. This illustrates that the relative performance of materials can depend heavily upon how the hazard in a given scenario is characterized.

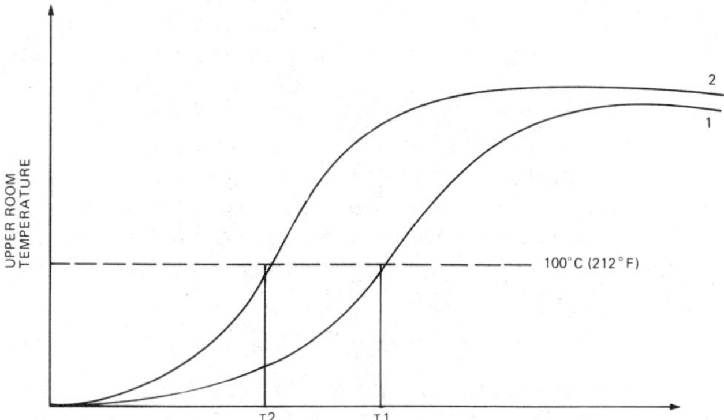

FIG. 21-1B. Comparative growth of two fires burning in a room, showing the time taken in each case to reach 100°C (212°F).

When hazards associated with smoke are evaluated, the situation is somewhat different. Figure 21-1C shows the same fire growth curve as Figure 21-1A, but a dotted curve showing smoke production has been added. Note that the smoke produced continues to increase steeply even after the temperature has reached a constant value. In effect, the smoke produced is proportional to the area under the fire size curve so it continues to increase until the fire has gone out.

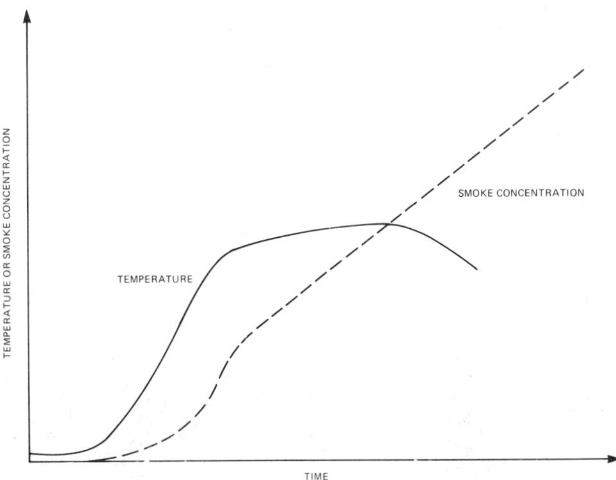

FIG. 21-1C. Growth of temperature and of smoke concentration as a function of time.

Unlike thermal hazard, the level of smoke which represents unacceptable toxicity is different for each material. The toxicity of the smoke must be measured by some appropriate method and the "toxic" dose for a given material determined. The dose can generally be related to concentration and time (Alexeeff and Packham 1984), so it is possible in principle to identify the point on the smoke concentration-time curve which corresponds to the arrival of unacceptably toxic conditions. Then the two materials can be compared. (See Fig. 21-1D.)

One way of viewing the development of a dangerous fire scenario is as a race between the fire and its potential victims. As the fire grows, conditions approach those where human survival is no longer possible. At some point, those exposed are alerted to the threat and begin to take the steps which will result in escape. The time at which the fire is detected is the starting point of the race.

The essentials of mechanical smoke and heat detection are reviewed elsewhere in this HANDBOOK. A few general observations, however, are helpful:

1. The human senses are, if unimpaired, good detectors of the signatures of fire, especially smoke. The fact that most fire fatalities occur when the victims are asleep, incapacitated, or physically unable to escape underscores this point.
2. Of the mechanical detection devices, smoke detectors usually give warning at an earlier stage of the fire than do heat detectors. Obviously, however, the location of the detector and the nature of the fire will be the dominant factors in detector performance in any given scenario.
3. Except in special circumstances, the fire characteristics which trigger detectors are the same ones which are threats to life, i.e., heat and/or smoke. Therefore, the kinds of calculation needed to predict detector response times are also relevant to predicting the development of life threatening conditions. With reference to detector response, the following data are needed:

   a. Ceiling height above fire.

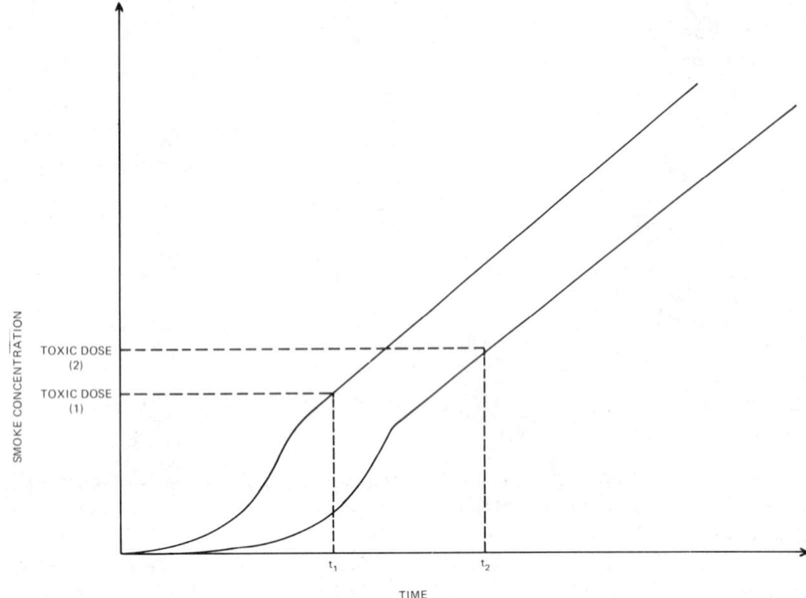

FIG. 21-1D. Comparison of smoke production and development of a toxic smoke dose for two different materials.

b. Distance from fire axis to detector.
c. Rate of fire growth.
d. Specific characteristics of the detector.
e. Specific characteristics of the smoke (if a smoke detector).
f. Ventilation characteristics.
g. Vertical temperature gradients in the compartment.
h. Layout of compartments between the fire and the detector.

This means that if temperature and smoke build up toward untenability can be estimated, detector response can also be estimated if the appropriate data are available. In the discussions which follow, the zero point in the fire growth will be assumed to be the time at which detection occurs, rather than the ignition of the fire. It bears repeating, however, that early detection of a fire is the key to survival. In the race between the fire and its victims, the odds increasingly favor the fire as the starting point is advanced.

Two kinds of information must be available before the time difference can be determined:

1. The unacceptable level of temperature or smoke concentration.
2. The fire or smoke growth curve.

The former is directly or indirectly derived from the response of an animal on exposure to hostile heat or smoke conditions. The latter can be obtained either from a full scale burn experiment or, in many cases, analytically from small scale laboratory data and knowledge of the fire scenario. Calculations of fire growth (fire modeling) will be discussed in more detail later.

Computation of the "time available for escape" (TAE) permits a comparison of the relative hazard posed by products in the same application. A ranking of TAEs is a ranking of hazard for the scenario under consideration. This parameter alone, however, does not determine whether a difference in hazard is significant, or to what degree a given product is safe. For example, product A may offer two more minutes of available escape time than product B, but if product B already provides 50 minutes of escape time, then the difference would be relatively unimportant. In another case, product A may offer five minutes and B three minutes of escape time; but if ten minutes is the time needed, then neither product is acceptable.

Determining if a situation is safe requires knowing how much escape time is needed as well as how much is available. The difference between the time available for escape (TAE) and the time needed for escape (TNE) is a measure of safety (Cooper 1984).

$$\text{Escape Margin} = \text{TAE} - \text{TNE} \quad (1)$$

A negative escape margin is bad; it means that the fire is fatal. The larger the positive value of the margin the better, although, as discussed previously, it is possible to reach a point of diminished significance.

The escape time needed (TNE) depends almost entirely upon the nature of the structure, the capabilities of those exposed to the fire, and the vagaries of human behavior. Usually, TNE is not an important function of the fire or smoke properties of a material. Hence, even though it is critical to the proper use of fire hazard assessment results, it is of indirect interest in this discussion.

Figure 21-1E illustrates the conceptual framework of fire hazard assessment modeling (NBS undated). The five components at the top of the diagram, labeled from *N1* to *N5*, describe the time needed for escape; the seven components labeled from *A1* to *A7* permit computation of the time available. Two inputs (the building layout and protection system models), are used in computing both values.

N1, the occupant location and condition, includes data such as the distance of occupants from exits or refuge areas, whether they can escape unaided, and their estimated age. *N2*, the behavior/decision model, describes how the occupants will behave once alerted to the fire; how and when they are alerted is dependent upon *N3* the protection system. The evacuation model, *N4*, predicts how long the building population will take to reach safety

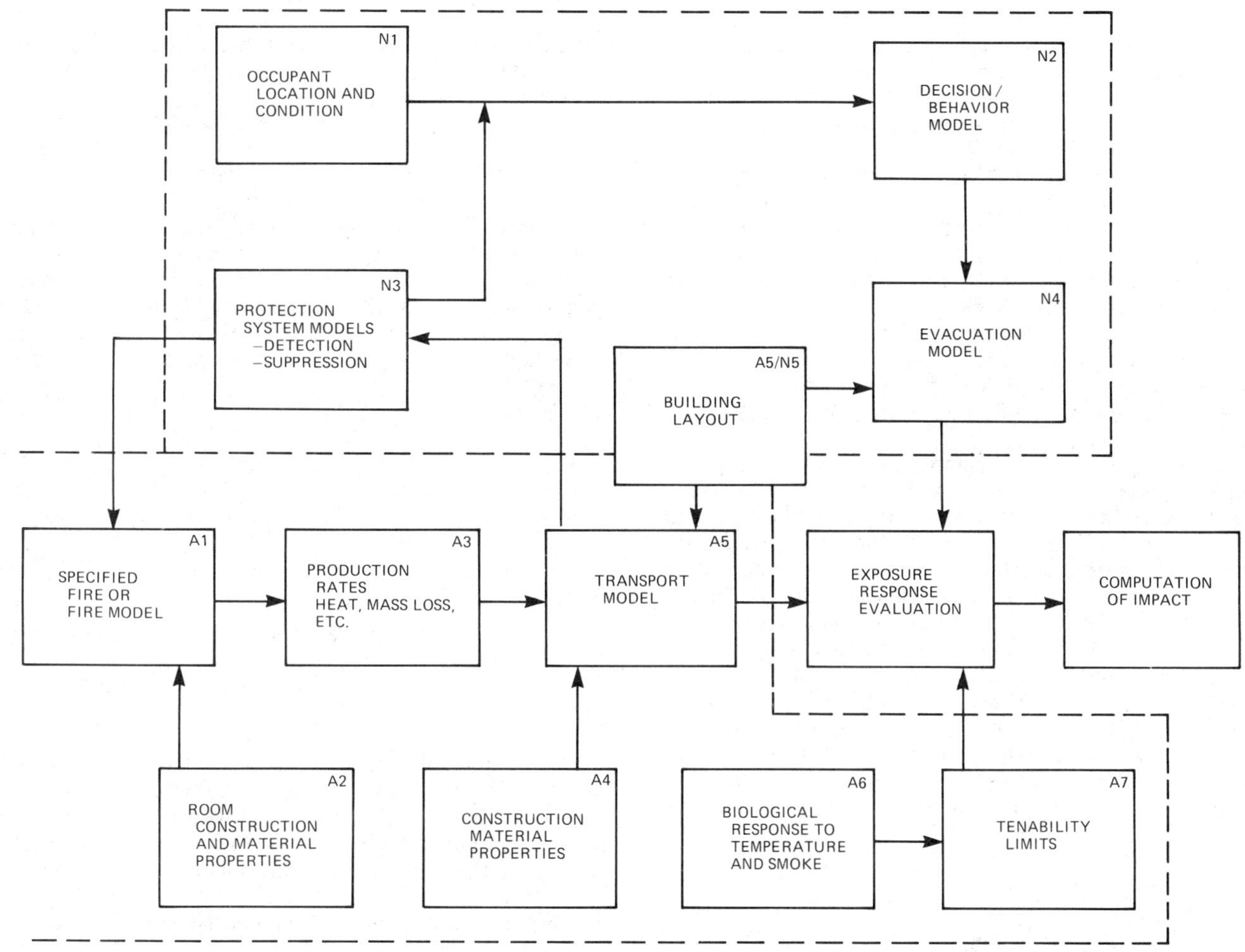

FIG. 21-1E. Major components of fire hazard model.

in a given layout, *N*5. The output of *N*4 is thus the time needed for escape, TNE. The "exposure response evaluation" compares TNE with the time available, TAE, which is computed from the "*A*" components. The escape margin differs for occupants in different locations, and the overall effect of a given fire is determined by computing its impact on the entire building population.

Time Available for Escape

The details of fire hazard differ, depending upon the kind of fire growth scenario. A rapidly developing flaming fire whose products accumulate in a relatively confined space such as a small apartment, produces a well defined upper layer which descends rapidly from the ceiling. There is little opportunity for the smoke to lose energy to the upper room surfaces, so the layer is quite hot. Once the hot layer has descended to a few feet from the floor, it is difficult for occupants to escape without coming in direct contact with it. Regardless of its chemical composition, this layer poses an immediate threat to life safety because of its temperature.

In a larger space, or if the fire is burning more slowly or perhaps smoldering, the layer is cooler. A distinct upper layer may not even be apparent, since it is the elevated temperature which gives the layer its buoyancy and results in stratification. In such cases, the toxic properties of the smoke can become important, since then these properties, not the temperature, determine the tenability of the compartment.

In larger structures, it is standard practice to provide barriers to the free passage of smoke or fire between floors, into exitways and, often, between groups of rooms. (Modern apartment buildings, for example, have fire walls between apartments but not as interior partitions.) The smoke from a fire may reach areas well removed from the fire, but seldom is it the hot buoyant steam which exists near the fire. This means that at distances well removed from the fire, the primary threat is usually smoke toxicity, not heat.

In a scenario where only one item is burning, it is possible to relate TAE to the fire and smoke toxicity properties of that item. In cases where multiple items are burning, TAE is influenced by the properties of each contributor. Scenarios involving multiple items are applicable when considering the hazard attributable to an item which can burn only after exposure to a fairly large ignition source. An example of each kind of scenario is given below.

Burning of a Single Item

The first case to consider is a fire ignited in a compartment of 250 m^3 (8800 cu ft) total volume, which corresponds roughly to 1,000 sq ft (93 m^2) of floor area with a normal 8 ft (2.5 m) ceiling. The fire is restricted to one relatively large item, such as a heavily upholstered chair or loveseat. Because the fire is so restricted, the fire properties of the rest of the room are relatively unimportant, although the thickness and thermal conductivity of the walls and ceiling should be known in order to determine how much of the heat energy of the fire is lost to these surfaces. As for the furniture itself, its burning rate (heat release and mass loss rate) must be measured or calculated from small scale test results. These data constitute the input to *A2* in Figure 21-1E. Because the scenario is concerned only with conditions in the room fire, the layout of the building, *A5*, is not very important to this calculation; neither are construction material properties, *A4*, other than those already mentioned for *A2*.

Assume that air temperature of 100°C (212°F) or higher represents the upper limit under which humans can escape from the compartment. Suppose that there are available smoke toxicity data, *A6*, on the furniture material which fairly reflects its smoke toxicity under actual burning conditions. The most useful measurement is of the dose of smoke, i.e., the amount inhaled that is necessary to cause incapacitation or death. One such formulation is the $L(Ct)_{50}$, the concentration-time product required for death to occur in 50 percent of the animals exposed to the smoke. In a laboratory smoke toxicity test, this quantity is obtained by continuously monitoring the smoke concentration to which the animals are exposed and plotting that concentration against time. The $L(Ct)_{50}$ is the area under

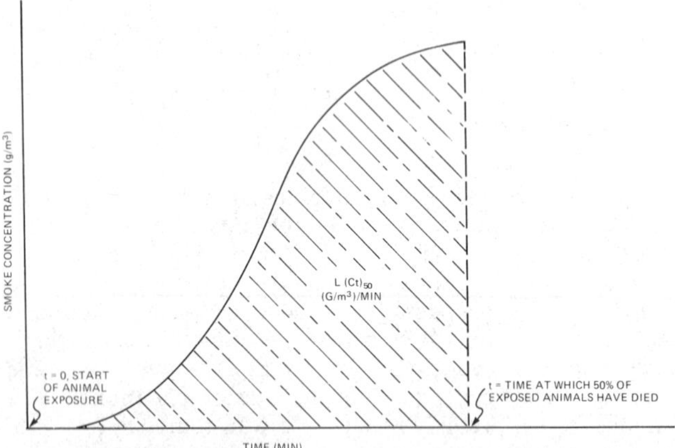

FIG. 21-1F. Plot of smoke concentration against time where test animals are exposed to the smoke. The $L(Ct)_{50}$ is the area under the curve from t=0 to the point in time where 50 percent of the animals have expired.

the curve at the time the animals expire. (See Fig. 21-1F.)

The simplest burning scenario is one in which fire is started on the furniture and does not spread appreciably. If the fire size is 100 kW, i.e., about 2 ft (0.6 m) in diameter, the hot smoke will have filled the room to a depth of 1 m (3.28 ft) from the floor in about 6 minutes; the temperature of the hot layer will have reached 100°C (212°F) after 11 minutes (Cooper 1982). Hence, by the temperature criterion mentioned above, the environment will become lethal 11 minutes after ignition.

Whether smoke toxicity becomes a problem before this time depends upon whether the occupants have been exposed to the smoke throughout the course of the fire and upon toxicity of the smoke. For smoke toxicity to be an important threat in this scenario, the atmosphere must become lethally toxic before it becomes lethally hot:

$$TAE \text{ (toxicity)} \leq TAE \text{ (temperature)}$$

Knowing the burning rate of the fire and the associated mass loss rate, it is a simple matter to compute the smoke concentration in the hot layer. Assuming that the occupants have been exposed to the smoke beginning when the hot layer was at the 5 ft (1.5 m) level, the time to take on a lethal dose of smoke, (*TAE*), is given by the integral over time, *dt*:

$$L(Ct)_{50} = \int_{t_5}^{TAE} C_s dt \qquad (2)$$

where

$L(Ct)_{50}$ is the lethal dose determined from a laboratory toxicity measurement;

t_5 is the time at which the smoke reaches the 5 ft (1.52 m) level;

C_s is the smoke concentration.

This is simply the mathematical equivalent of determining when the $L(Ct)_{50}$ of the smoke, as measured in the laboratory, is reached in the room. The concentration room smoke is followed with time, and the area under the curve (the integral) is set equal to the $L(Ct)_{50}$ observed in the laboratory.

For this scenario, TAE is plotted as a function of

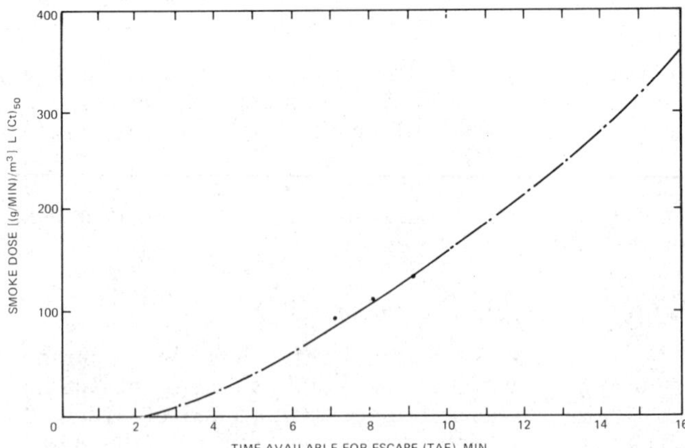

FIG. 21-1G. Time available for escape (TAE), in minutes.

$L(Ct)_{50}$ in Figure 21-1G. It can be seen if the furnishings material has a smoke $L(Ct)_{50}$ below about 200 (g/min)/m^3, then death from smoke toxicity could be expected to occur before conditions were thermally untenable; otherwise the thermal hazard is more immediate in this scenario.

If instead of a flaming fire, the fire on the furniture is a smoldering one, there usually is not enough heat generated to maintain a stable upper layer, and the smoke disperses more or less uniformly through the compartment volume.

One would intuitively expect smoke which is twice as toxic, or is produced twice as fast, to produce untenable conditions in half the time. In fact, owing to the mathematics of the scenario, the escape time is not actually that sensitive to changes in smoke toxicity or burn rate; doubling either toxicity or burn rate results in substantially less than a two fold reduction in escape time.

For a smoldering fire which does not change in size ($\dot{m}$ is constant)

$$TAE = [2\ L(Ct)_{50}\ V/\dot{m}]^{1/2} - t_d \qquad (3)$$

where t_d = time at which fire is detected.

For a fire which grows linearly with time ($\dot{m} = kt$),

$$TAE = [6\ L(Ct)_{50}\ V/k]^{1/3} - t_d \qquad (4)$$

This trend continues, so that the higher the order of time dependence upon mass loss rate, i.e., if it increases as the square or cube of time, the less sensitive TAE is to changes in the value of $L(Ct)_{50}$. This is the case for flaming as well as smoldering fires, although the former are complicated by the hot layer descent.

Burning of Multiple Items

Most real fire scenarios involve the thermal exposure of a number of items. In some cases, the sequence of ignitions is almost arbitrary. It is as easy to envision drapes igniting from a burning chair as the reverse. In other cases, however, it is far more likely that the sequence will be fixed: combustible materials such as plastic pipe or wiring, behind a wall are much more likely to be exposed to heat from a fire in the room than to be ignited directly by a small ignition source.

Figure 21-1H shows the buildup of temperature in a room as the result of a known fire, the standard time temperature curve for the *ASTM E-119 Fire Endurance Test* (ASTM 1983). It also shows the buildup of temperature behind a wall of ⅝ in. (15 mm) gypsum wallboard. These two curves are inputs A1 and A2, respectively, to determining the smoke production rate of the room fire and the concealed combustibles.

It should be clear that the temperature in the room becomes untenable long before the region behind the wall warms appreciably. If, instead, one focuses on conditions outside the room, it can be shown that smoke issues only from the room fire for the first fifteen minutes or so. When the temperature behind the wall is high enough, the combustibles will begin to decompose and eventually to ignite. The data necessary to predictwhen this will happen include mass loss versus temperature and the total amount of fuel they comprise as part of A2.

The curves in Figure 21-1I show the contribution of the in-room and behind-wall fuel packages to the total smoke produced. It is necessary to know the toxicity of the smoke from both fuel sources to predict the effect of the behind-wall material on TAE. It is obvious, however, that its toxicity must be substantially greater than that of the smoke produced in the room before it will have much effect. Formalisms have been proposed to compute the TAE from fires involving multiple components. The TAE is given by the expression:

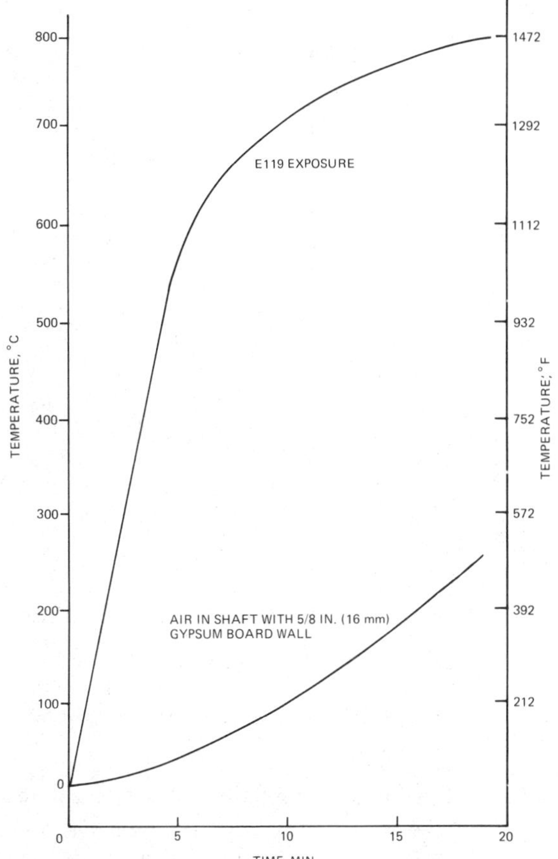

FIG. 21-1H. Time-temperature profile of a fire simulating ASTM E-119 Fire Endurance Test and of cavity behind a gypsum board wall.

$$\frac{1}{L(Ct)_{50(1)}} \int_0^{TAE} C_{(1)}\, dt + \frac{1}{L(Ct)_{50(2)}} \int_0^{TAE} C_{(2)}\, dt = 1$$

where the subscripts (1) and (2) denote the in-room and behind-wall fuels, respectively. In practice, the smoke production curves are followed, integrated, and normalized with respect to the toxic dose (determined in advance by small scale test). At the time that the normalized contributors sum to unity, TAE is deemed to have been reached. This "additive" approach to smoke toxicity is probably sound when the materials produce the same kind of major toxicants such as carbon monoxide. It also appears to hold for a combination of toxicants such as carbon monoxide and hydrogen cyanide. It has not yet been shown, however, to be generally applicable to all toxicants.

The smoke concentration depends both upon the mass loss rate of the fuel and on the point in the building at which fire hazard is being assessed. Hence, knowledge of the building layout, A5, and the construction material properties, A4, may also be needed to make this computation.

In comparing two alternative materials for the same use behind the wall, it is possible to compute a difference in the TAE associated with the change from one material to another. The significance of this difference, however, will be affected by the contribution of the in-room fire, as well

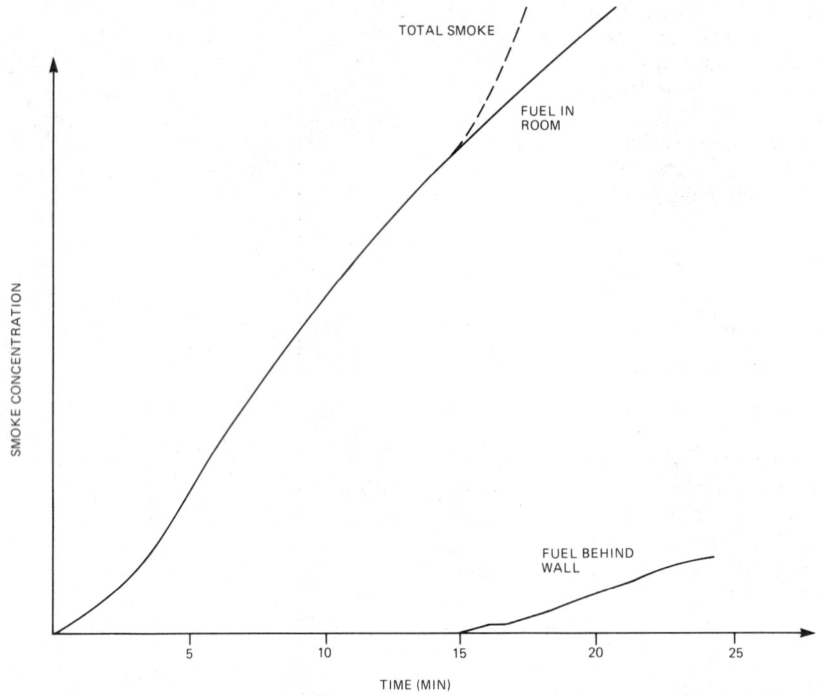

FIG. 21-11. Smoke production for a fire burning in a room and igniting combustibles behind the wall.

as by the TNE, which is essentially independent of material composition.

Fire Modeling

Most fire scenarios are too complex for the calculations of escape time to be expressed in analytical form. Computer based numerical solutions are usually required. "Burning" a fire on a computer, or fire modeling, has made great advances in the past decade. A number of different computational systems have been developed, each for a slightly different purpose. Each system is called a fire model, but provides the phenomenological basis for assessing hazard. The heart of a fire model is that part which describes the burning process itself. It is here that that flammability properties of the materials and products under study (measured in a laboratory or taken from the literature) are coupled with the characteristics of the compartment such as its geometry, the ventilation, and the thermal conductivity of the floors, walls, and ceiling.

The output of the burn calculation can include the fire size, temperature, smoke production rate, and, if sufficient data are available, the concentration of specific toxicants in the smoke. All this information is usually time dependent, and many models recalculate the data at regular intervals as the fire grows. The rest of the computation is concerned with tracking the spread of fire and the smoke components through whatever kind of compartment, or network of compartments, is under study.

Fire models have been extensively reviewed (Jones 1983) and are gaining increasing acceptance as practical everyday tools. For most situations, a minicomputer is required, although some simple models can be accommodated on home microcomputers. The more a model is readily adaptable and easy to use the larger the number of simplifying assumptions and approximations it is likely to contain. All models are thus imperfect—they can simulate real fire experience quite well under the right circumstances, but they fail to do so in other situations. For this reason they will probably remain a tool for the professional, in whose hands they can make hazard assessment an everyday part of fire protection.

Summary

Whether hazard assessment is performed with the aid of a computer or by hand, the essentials of the method are the same. The fire properties and smoke toxicity characteristics of a material or product are related to the development of thermally and toxically hazardous conditions in a given scenario. This allows one to see how changes in these input variables affect the outcome.

The relative influence of heat release rate, flame spread, smoke toxicity, etc., on the buildup of hazard depends upon the scenario under study. In general, however, the flammability properties of a material are at least as important as, and often more important than, its smoke toxicity. This is because the flammability properties alone govern thermal hazard, which is often more acute, and they, in concert with smoke toxicity, govern the hazard from smoke.

Bibliography

References Cited

ASTM. 1983. "Fire Tests of Building Constructions and Materials." *ASTM E-119-83*. American Society of Testing and Materials, Philadelphia, PA.

ASTM. 1982. "Standard Terminology Relating to Fire Standards."

ASTM-176-82. American Society of Testing and Materials, Philadelphia, PA.

Alexeeff, G., and Packham, S. 1984. "Evaluation of Smoke Toxicity Using Concentration-Time Products." *Journal of Fire Science*. Vol 2. p 362.

Clarke, F., and Ottoson, J. 1976. "Fire Scenarios and Fire Safety Planning." *Fire Journal*. May 1976.

Cooper, L. 1982. "Calculating Available Safe Egress Time (ASET) —A Computer Program and User's Guide." *NBSIR 82-2578*. National Bureau of Standards, Washington, DC.

Cooper, L. 1981. "Estimating Safe Available Egress Time From Fires." *NBSIR 80-2172*. National Bureau of Standards, Washington, DC.

Jones, W. 1983. "A Review of Compartment Fire Models." *NBSIR 83-2684*. National Bureau of Standards, Washington, DC.

NBS. Undated. Adapted from material supplied by H.E. Nelson. Center for Fire Research. National Bureau of Standards, Gaithersburg, MD.

Additional Readings

Alpert, R. L., "Calculation of Response Time of Ceiling-Mounted Fire Detectors," *Fire Technology*, Vol. 18, 1982, pp. 181-195.

Bukowski, R., "The Application of Models to the Assessment of Fire Hazard From Consumer Products," *NBSIR 85-3219*, National Bureau of Standards, Washington, DC, Aug. 1985.

Emmons, H. W., "The Prediction of Fire in Buildings," *17th Symposium (Intl) on Combustion*, The Combustion Institute, Pittsburgh, PA, 1978, pp. 1101-1111.

Fowell, A. J., "Assessing Toxic Hazard as It Relates to Overall Fire Hazard," *Fire Technology*, Vol. 21, pp. 199-212.

Friedman, R., "Quantification of Threat from a Rapidly Growing Fire in Teams of Relative Material Properties," *Fire and Materials*, Vol. 2, 1978.

Heskestad, G., "Physical Modeling of Fire," *Journal of Fire and Flammability*, Vol. 6, 1975, p. 253.

FIRE RISK ANALYSIS

John R. Hall, Jr., Ph.D.

In the rapidly developing field of scientific research on fire, one of the most useful and unusual branches is fire risk analysis. It is useful because it provides a flexible framework for estimating the impact of any type of firesafety program or strategy in terms of actual reductions in losses—deaths, injuries, property damage—and in terms that can be compared to the costs of those programs and strategies. While fire risk analysis may lack the depth of detail required for complete evaluation of building design alternatives, there is no other method as well suited for the analysis of strategic options affecting large numbers of properties and their occupants. Product development, related research and marketing, and regulatory decisions can benefit from fire risk analysis, as can any program to manage and oversee firesafety from a financial point of view.*

Fire risk analysis is unusual because the framework it uses is taken not from the hard sciences of physics, chemistry, biochemistry and engineering, which underlie the rest of firesafety research, but from statistical decision theory, which is based upon the fields of economics and operations research. Because fire risk analysis comes from a setting with which much of the fire community may be unfamiliar, this chapter will focus on a discussion of basic concepts, definitions, and approaches.

What is Fire Risk Analysis?

A measure of fire risk always has two parts: (1) a measure of the expected severity (e.g., how many deaths, injuries, dollars of damage per fire) for all fires or for a particular type of fire, and (2) a measure of the probability of occurence of all fires or of that particular type of fire. In general, a fire risk measure will be a product of an expected severity term and a probability term or a sum of such products. In the latter case, each product addresses the risk of a particular type of fire and the sum of products gives the overall risk associated with fires of all types. A risk measure, then, gives expected losses due to fire under a certain set of assumptions (e.g., no detector present, functional detector present).

Risk analysis refers to a systematic examination of how these risk measures change as the assumptions change, in conjunction with a parallel examination of changes in costs. For example, if no detector is present, a household can expect to experience fires with a certain probability, and these fires will have a certain expected severity in terms of deaths, injuries and property damage. On the other hand, if a detector is present, the household will pay purchase and maintenance costs not paid by the household without detectors; but the detector household can expect on average to achieve lower severities of losses—and in particular, fewer deaths per fire. Risk analysis examines whether the reductions in deaths, injuries, and damage are great enough to justify the cost; it is an analysis of cost versus risk-reduction benefits in which the risk portion is measured explicitly.

The term "risk analysis," and similar terms such as "hazard analysis," may be used to describe this kind of analysis, but other kinds of analysis are also labeled with some of these terms. The three key elements of the fire risk analysis approach are: (1) the explicit estimation of fire ignition probabilities, (2) the use of a single scale of fire severity (possibly, but not necessarily, monetary) so that reductions in risk can be compared to the cost of achieving those reductions, and (3) the explicit consideration of the uncertainty attached to all estimates developed in the analysis. Some modeling approaches that are called risk assessment omit one or more of these key elements.

More specifically, the term "hazard analysis" usually refers to engineering analyses using fault tree models and other systems concepts approaches developed by fire protection engineers. Risk analysis differs from this type of hazard analysis in several ways. For instance, risk analysis looks at all unwanted fires, not just the sometimes larger fires defined by "established burning," and measures all degrees and types of damage, not just deaths or spread beyond a fire rated compartment; hazard analysis often limits its attention to major failures of fire protection and rarely addresses prevention strategies. Risk analysis mea-

*Much of the material in this chapter was first developed for an in-house concept paper at the National Bureau of Standards, Center for Fire Research (NBS/CFR). The author expresses his appreciation for the valuable support accorded this work by NBS/CFR.

Dr. Hall is Director of the Fire Analysis Division, NFPA.

sures expected deaths, injuries, and property damage; it does not stop with measures of how many rooms or square feet reached a particular temperature or had smoke or gas levels over a particular threshold. Risk analysis must address whether people and property are actually damaged, which means implicit or explicit attention to the locations, decisions, movements, and vulnerabilities of occupants—actual, not just potential—and the damageability of property. Risk analysis also requires estimation of the factors that contribute to reliability of systems, such as the probability of features and systems being rendered inoperative or ineffective due to human error or negligence; hazard analysis does not always address field reliability in all its aspects.

Hazard analysis addresses the core concerns of fire development, smoke spread, and sometimes toxicity; it often does so in far greater detail than is possible with risk analysis. Risk analysis includes victim decisions, locations, and characteristics; ignition factors; reliability of systems and building features; and costs. In order to be this broad, fire risk analysis cannot always make use of the full power and sensitivity to detail of the hard science models of hazard analysis; the only models that can be used are those that can be fitted into the complete risk analysis framework. Thus, hazard analysis has the advantage of being able to examine details of design that fire risk analysis cannot now handle. In the long term, however, work is proceeding on model integration techniques capable of combining the breadth of fire risk analysis with the depth of hazard analysis.

Risk analysis can be separated into two parts: (1) *risk estimation*, the estimation of probabilities, severities, and attendant uncertainties for all relevant consequences of various alternatives; and (2) *risk evaluation*, the translation of those probabilities and severities into a single scale suitable for comparison with the costs associated with those same alternatives. Risk estimation—the process of estimating parameters—is the part most similar to hazard analysis, but it may draw on sources of information different from those commonly used in hazard analysis.

Because real fires reflect all the factors that affect ignition probability and fire severity, fire risk analysis usually begins with calculations from data bases on actual fires. (Two to three million fires are reported to the fire service each year.) For most analyses of classes of properties, then, one can identify initial data bases of historical fires that will give statistically meaningful probabilities and severities of fires. For some analyses, it may not be necessary to go beyond these data bases.

Many situations, however, may dictate going beyond incident data bases in performing a fire risk analysis. First, in analyzing new strategies, products, or systems, there will be no data base of fires reflecting the influence of these innovations. Second, there may be interest in so many product or building characteristics that the data base does not have enough fires to cover all possible combinations. Third, more detail may be needed than is available in the typical data base.

If available fire experience data is insufficient, a fire risk analysis will begin to look more like a hazard analysis, with tree diagrams to capture alternative sequences of events, probabilities for the various possible starting conditions, and probabilities for transitions from one stage to the next in a fire. What is measured and in how much detail may differ from a hazard analysis, in the ways noted earlier.

Risk evaluation, the other part of risk analysis, rarely has a counterpart in hazard analysis. It should be familiar, however, to anyone who has had to make business decisions, because it essentially involves using analysis to determine whether you will get what you pay for. The most common approach to risk evaluation is *cost-benefit analysis*, a technique in which all benefits of risk reduction are translated into monetary equivalents. This technique permits a proposed new product or fire suppression system to be assessed in terms of its net profit or loss, total cost plus loss, or ratio of profit to loss. In such a context, "profit" means saved lives, avoided injuries, and reduced property damage, all combined in one monetary scale, and "loss" means the cost—both initial and ongoing—of the new product or system. A variation is *cost-effectiveness analysis*, in which the benefits of risk reduction are translated into a single non-monetary scale. For example, it is possible to derive the cost per life saved for a new system or product.

Cost-benefit and cost-effectiveness analyses require the explicit estimation or derivation of some controversial parameters. Examples include the value of reduced risk of death or injury and the discount rate by which future consequences are compared to present consequences (both discussed in more detail later). Because these parameters are controversial, risk evaluation sometimes uses methods that downplay the role of these parameters.

Acceptable risk is a term used when the method of risk evaluation involves treating risk as a constraint. This method may seem attractive because it refuses to consider costs until or unless a sufficient degree of firesafety has been provided. In an acceptable risk approach, a certain level of risk is defined as acceptable; then all alternatives meeting that level are evaluated strictly on the basis of cost.

This approach can produce unsatisfactory results. If risk is greater than the acceptable level by even a small fraction, no cost is too great to reach acceptable risk. If risk already is acceptably low, not a nickel more should be spent, no matter how much more firesafety could be purchased for very little money. This means that the selected level of acceptable risk is often set with an eye toward affordability and may be reset if technology changes. In effect, this makes the acceptable risk approach a kind of back door cost-benefit analysis and runs counter to most approaches to decision making in business.

When acceptable risk is not defined in terms of affordable risk, it is often defined in terms of (a) historically acceptable risk (i.e., anything in use for a long time is all right), which may be overturned if public understanding of the magnitude of the risk changes dramatically, or (b) unavoidable risk, such as the use of background radiation levels as a guide for acceptable exposure to medical x-rays. In fire protection, acceptable risk has sometimes been inferred from provisions of NFPA codes and standards. The most extreme version of an acceptable approach is a *minimum risk* approach, in which cost is not considered unless *all* feasible safety improvements have been made.

A logical complement to the acceptable risk approach would be an *acceptable cost* approach, in which the greatest risk reduction available within the fixed cost budget, (but no more), would be sought. Although this approach is rarely mentioned in the literature, it almost certainly describes the way some decisions are made.

Note that the acceptable risk and acceptable cost approaches both can be very sensitive to the initial selection of a single key parameter. If these approaches are used, it is necessary to conduct a *sensitivity analysis* to see how conclusions would be changed if the reference level of acceptable risk or acceptable cost were different. Sensitivity analysis of key parameters is essential to all other forms of risk analysis as well, and is one way in which the uncertainty surrounding estimates can be addressed systematically.

Overview of a Risk Analysis Conceptual Framework

Figure 21-2A provides an overview of the kind of conceptual framework used to identify necessary models and necessary data in risk analysis. The major aspects to be modeled can be grouped into six models, as shown in the ovals: (1) Decision Model, (2) Ignition Initiation Model, (3) Post Ignition Model, (4) Loss Evaluation Model, (5) Cost Model, and (6) Cost-Benefit Comparison Model. The rectangles indicate numbers, either inputs or outputs, derived from the models. These numbers may in turn be supplied as inputs to other models later in the modeling sequence. The model refers to a "proposed change," which is a general term for anything that might be modeled—a new sprinkler system, increased compartmentation requirements, no smoking areas, mandatory self-extinguishing cigarettes, a training program for staff, or any other change that could make fires more or less likely or more or less severe.

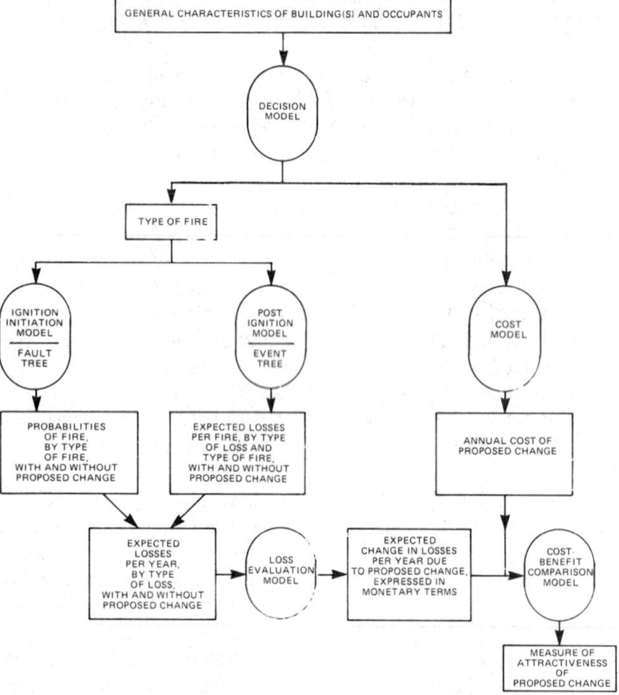

FIG. 21-2A. A risk analysis conceptual framework.

A Decision Model is used to describe what the building and its occupants would look like if a proposed change were or were not made. An Ignition-Initiation Model is used to estimate the probabilities of occurrence per year for each type of fire, while a Post-Ignition Model is used in parallel to estimate the expected losses per fire, for each type of fire and each type of loss (i.e., deaths, injuries, property damage). The types of fires are defined by the requirements of the two models. For example, if the Post-Ignition Model uses different parameters for smoldering and flaming fires, then that distinction must be reflected in the separation of fires into types for the Ignition-Initiation Model. When the outputs of these two models are combined, they produce estimates of expected losses per year by type of loss and for all types of fires. Then a Loss Evaluation Model is used to convert all types of loss to a common scale and produce year by year projections of overall expected loss, with and without the proposed change.

Meanwhile, on the right side of the framework, the purchase, installation, maintenance, inspection, operating, replacement, and other costs of the new system are combined by a Cost Model into year by year figures of the cost impact of the proposed change.

Finally, a Cost-Benefit Comparison Model produces a *measure of attractiveness* for the proposed change. This is simply a comparison of annualized costs and benefits, with the benefits consisting of changes in risk.

The discussion that follows will address the six component models in more detail. With respect to two terms introduced earlier, "risk estimation" involves the Decision Model, the Ignition Initiation Model, and the Post-Ignition Model, while "risk evaluation" covers the other three models.

General Characteristics and Fire Types

Before the models shown in Figure 21-2A come into play, there must be an initial structure to the problem that describes the type of building, the characteristics of its occupants, and the types of fires to be studied. Probability distributions may be needed for any of these factors. Even if only a single building is being analyzed, that building's behavior with respect to fire may vary as a function of randomly varying conditions such as the positions of doors and windows (open or closed) or of forced air heating and cooling systems (on or off). Occupant locations and conditions also may vary in random or patterned ways. At present, fire risk analysis is able to capture only a fraction of these factors, but useful results are still possible. It is also useful in setting up a problem to be aware of any factors that can not be explicitly addressed. This helps in interpreting the results.

The most important point to remember in defining types of fires is that *all* possible fire ignitions must be covered. Fires may have to be grouped into classes that are not entirely homogeneous, but it is not sound practice to exclude certain categories of fires. Suppose that the Post-Ignition Model will require each fire type to be specified in great detail (e.g., location and heat release rate of first material ignited). Then there may be either an unmanageable number of fire types to be modeled or a small number of specific fire types each used to represent a much larger, less well defined class.

The classification scheme should distinguish among fire scenarios that are affected differently by the systems or strategies being studied. Suppose fast response sprinklers are being evaluated for possible home use. Then the key properties of fires affecting their response to sprinklers might be identified by answering these questions: Which areas of origin would not be accessible to sprinklers (e.g., the outside of the house, concealed spaces, unsprinklered living areas such as bathrooms)? Could the fire ignition energy best be characterized as smoldering, flaming, or fast flaming, high energy? The distinction among smoldering, flaming, and fast flaming fires is made to recognize the fact that fire growth (which activates the sprinkler) and smoke spread (which is the principal cause of death) do not proceed according to a relationship that is the same for all fires.

Decision Model

In addition to defining types of fires and specifying general building and occupant characteristics, it is necessary to specify all the differences that would occur if the proposed change were made. If the change involves new built-in smoke detection systems, for example, specifications for the systems would be needed. If the systems were available in two or more versions (e.g., ionization and photoelectric smoke detectors), the probability of each version being used should be estimated (e.g., from data on usage patterns or projected shares of market). Probabilities also would be needed for the variations in status of a system (e.g., fully operational, operational but blocked in one room, turned off, or removed). These specifications dictate precisely what fire protection and fire related features would be in place to influence the course of fire if one occurs.

If the proposed change involves training staff or educating occupants, similar questions would be asked: What are the specifics of the program(s)? What is the likelihood that each version of the program would be used? If fire occurs, what would be the status of the program (e.g., everyone was trained and acted accordingly, some missed the training, some forgot, some panicked, etc.)?

It is sometimes useful to separate the parts of the Decision Model into those characteristics that were selected by a building owner or manager (e.g., a sprinkler was installed) and those that came into play after the selection (e.g., the sprinkler was installed incorrectly). The latter parts are sometimes collectively referred to as the *Implementation Model*. The Implementation Model isolates those factors that are hardest to control, either because they require continued alertness (e.g., without scheduled testing and maintenance, a sprinkler system's reliability will decrease) or because they are not directly under the control of the building owner or manager (e.g., a manager can order a sprinkler system, but he has to work through others to make sure it is installed properly).

Neither the Implementation Model nor the larger Decision Model needs every bit of information on the proposed changes. All the models need are those details that will affect the likelihood of having a fire, the development of that fire, or the reactions of people and property to the fire. For example, some factors will incapacitate the system so that the fire will develop as if no system were present, and that is all the information needed. Some factors will degrade system performance but still permit some impact on fire development; it may be possible to model such reduced impact as simple modifications to the impact expected if the system were fully functional.

Ignition-Initiation Model

The Ignition-Initiation Model is needed to produce estimates of the probability of ignition per year, by type of fire, given a structure of fire scenarios and fire relevant characteristics of the building and its occupants.

The Ignition-Initiation Model is a device for combining known probabilities in order to infer unknown probabilities. A simple example will illustrate the kinds of calculations that are possible and desirable. Suppose there are n brands of cigarettes. Define p_i as the share of total product usage contributed by product i; in this case p_i is the share of all cigarettes smoked that are brand i. Let r_i be the proportion of brand i cigarettes discarded each year in such a way as to make a fire possible. Let q_i be the probability of ignition given such an exposure for brand i; q_i is a measure of the relative self-extinguishing tendencies of brand i, but it can be affected by changes in the ignitability of materials that are typically ignited. Let F be the number of cigarette-related fires per year and let N be the number of cigarettes smoked per year.

$$\text{Then } F = \sum_{i=1}^{n} N p_i r_i q_i$$

Note that there are three ways to cut down on fires from brand i cigarettes: (1) see that almost no one buys them (reduce p_i); (2) see that the people who buy them almost always dispose of used cigarettes properly (reduce r_i) and (3) see that discarded cigarettes usually self-extinguish (reduce q_i).

This model can be used to examine several different types of strategies. A ban on all cigarettes that do not meet a self-extinguishment standard would change the p_i values, because the prohibited brands would have their p_i values cut to zero, and the p_i values of all the other brands would rise to fill the gap in the market. A campaign to educate the public not to discard cigarettes would lower the r_i values; or, a requirement to reduce the flammability of couches would lower all the q_i values, because more cigarettes would self-extinguish if couches were harder to ignite.

To perform the analysis, one must set initial values for the parameters. The purpose of this example is to show how to derive values for the parameters through a combination of direct measurement, reasonable assumptions, and mathematical inference. The values of p_i might be measured through sales records or a survey of product users. The values of q_i might be measured in laboratory tests. The most elusive will be the r_i values, the measures of how often cigarettes are being discarded, whether or not fires result.

If it is assumed or can be proven that all types of cigarettes are equally likely to be discarded, then there is only one common r_i value, and it can be solved for. Or data can be obtained on numbers of cigarette fires per year by brand of cigarette, then individual r_i values can be solved for.

In a more complex case, suppose the analyst had at least as many years of fire data as there were brands of cigarettes. Also suppose that it can be assumed or shown that year to year variations in fire experience were due solely to changes in market share. (This is equivalent to assuming that neither behavior in discarding cigarettes nor the propensity of each brand to start fires if discarded has changed over the years.) Then if p_i values (market shares) could be obtained for each year, it would be possible to solve for the r_i values using the years of fire data as a set of n equations in n unknowns.

These diverse approaches illustrate a general point. To estimate unknown probabilities, it is necessary to (a) find direct data sources for the unknowns (e.g., survey of couch usage); (b) develop reasonable assumptions as to the values of the unknowns (e.g., why the r_i might all be equal); (c) find valid formulas relating the unknowns to other variables on which there are sufficient data to make inferences back to the unknown probabilities; or (d) fill the gaps with expert estimates.

A full Ignition-Initiation Model needs to combine far more factors than were used in the simple illustration and to accommodate a multitude of logical and probabilistic interdependencies among the variables. There may be many different factors, none necessary for fire to occur and none sufficient to cause fire except in combination with certain other factors. The standard format for constituting such models is the *fault tree* or *success tree*, which is discussed in more detail in Section 7, Chapter 2, "Systems Concepts for Building Design."

Post-Ignition Model

The Post-Ignition Model is used for projecting the expected losses (deaths, injuries, property damage) of specified types of fires, given specified building and occupant characteristics and the status of all systems, features, products, and other changes being analyzed. The methodology for tracking the development and final consequences of a specified fire under specified building and occupant conditions may be deterministic, probabilistic, or, more likely, a mix of the two.

The use of existing deterministic models tends to be constrained by two requirements of the Post-Ignition Model: (1) it cannot accommodate a more finely divided typology of fires than can be handled by the Ignition-Initiation Model (i.e., for every distinct type of fire whose development and final impact is modeled, it must also be possible to estimate the likelihood of that type of fire); and (2) it must produce not only physical descriptions of flame, gas and smoke extent but also estimates of impact in terms of deaths, injuries and dollar value property damage.

The Post-Ignition Model—a model of fire development and outcomes—consists of a set of linked time functions of fire characteristics and their effects. For example, the model might need to specify, at each time, the extent of flame, smoke, and gases by type, the status of fire protection systems and other building characteristics, the locations of occupants, the tenability of various locations, etc. Stated in this way, no such complete model exists.

To develop a practical version of the model, one must ask: (1) how many of these characteristics need be known to develop reasonably good estimates of deaths, injuries, and damage from a given fire? and (2) at which points in time will the development of a particular type of fire change as a function of the different input values to be modeled (e.g., different initial statuses of fire protection systems)? In other words, one must identify a limited number of fire descriptors to track and a limited number of points in time at which to check them.

One type of model with which to do this is an *event tree*, the standard modeling format of statistical decision theory. In an event tree, each point in time is characterized by a set of events. These events indicate all the states in which the fire might be at given points in time (with the states defined by the limited number of fire descriptors previously cited). Then, for each event, one must specify the probabilities that—starting from that event—the fire will develop into each of the possible states (or events) that could characterize the fire at the next point in time. Event trees work well if all the factors being analyzed operate at relatively well defined points in time or stages of the fire. Experience in using these models indicates that the performance of most fire protection systems and features can be tied to events, which are themselves defined by characteristics of the fire.

Eventually, the event tree approach may be replaced by an integrated system of fire and smoke growth models. Such a system would be based on physics and would thereby model fire development deterministically rather then probabilistically. However, the integrated system would have to be capable of tracking the inherently probabilistic aspects of people movement and of providing outputs measured in deaths, injuries, and property damage. For now, the event tree approach is workable and can produce useful, albeit much less detailed results, than might be desirable.

An event tree, like any tree model, consists of nodes and arcs. Each node represents an event or stage in the development of the fire—typically a point where a particular system or feature will activate if it is operational or when the speed of the fire changes dramatically (e.g., flashover). The condition of the fire at extinguishment is captured by the *outcome events*, which are the various terminal events of the tree. Each outcome event is assigned values of expected losses per fire (i.e., average severity—deaths, injuries, property damage) that are the estimated losses if the fire ends in the way described by the outcome event. However, the only losses covered by these values are those that occur during the final stage of the fire or are otherwise correlated with the ultimate size of the fire. For computational reasons, deaths and some other losses that take place early in the fire must be associated with the stage of the fire where they occur. For example, suppose the first event in a fire is ignition and the second is the point where a fast response sprinkler would activate. Fatal injuries occurring before that second event (e.g., due to clothing ignitions) would be captured in a loss-per-fire figure assigned to that second critical event and would be understood to occur no matter how the fire does or does not grow after that event. These losses are sometimes referred to as *tolls* because any fire that passes through the event acquires those losses, just as a car pays a toll on an express highway.

At each event (except outcome events), there will be two or more directions in which the fire may subsequently develop, and there will be *transitional probabilities* associated with each of these directions. All transitional probabilities from an event add up to one. They are conditional probabilities that a fire will reach a particular event, given that it has already reached the immediately preceding event. A *path* is a possible sequence of events for the fire from the first event (ignition) to one of the outcome events. Each path describes the growth and extinguishment of the fire with as much detail as the model can provide. A *path probability* is the probability of a complete path, calculated as the product of the transitional probabilities for all the events along the path. Because several outcome events often are identical except for the paths used to reach them, it is sometimes useful to compute an *outcome probability*. This is the probability of a set of similar outcome events and is equal to the sum of the path probabilities for the paths ending in those outcome events. The expected loss per fire is calculated by multiplying transitional probabilities for branches leading to outcomes by the losses associated with those outcomes, then assigning those losses to the events from which the branches originated and adding any tolls associated with those events. This process is repeated (and is called *rolling back the tree*) until an expected loss figure has been computed for the first event (ignition). That value then becomes the expected loss for the event tree.

Loss Evaluation Model

The explicit or implicit assignment of monetary values to lives saved and injuries averted is the key element of this model. It is a difficult step which many people find distasteful or even immoral. The first and most important point to make is that individuals are not being asked to name a price for which they would be willing to die or suffer crippling injury. Instead, they are being asked to name a price they would be willing to pay to allow their current low risk of incurring death or injury in fire to increase or what they would pay to make that risk still smaller. With a resident population of about 230 million and an annual fire death toll in the range of 5,000 to 6,000, an average U.S. citizen has less than one chance in 40,000 each year of dying in a fire. Even for the highest risk groups, the risk is probably less than one chance in 5,000 each year or less than one chance in 65 over an entire lifetime. A person could rationally attach a price to a 10 or 50 percent change in such a risk and still be consistent in believing life (i.e., the certainty of losing it) is beyond price. A rational person would pay much more to reduce the risk of dying from 1.0 to 0.8 than he or she would pay to cut that risk from 0.3 to 0.1.

If that point is made, the next task is identifying what particular figures should be used for the value of life and the value of injury when considering alternatives that change risks in the range characteristic of fire risk. In the 1960s and earlier, the value of life was generally calculated on the basis of discounted foregone future earnings. This approach implicitly assigned no value to the lives of retired people and full time housewives, and negligible value to the lives of older workers and young children. Such distinctions were philosophically objectionable. Even for prime wage earners, the methodology did not afford any guarantee that the value obtained would match the price people wanted to pay for risk reduction. In recent years, this approach has been largely abandoned in favor of calculations of willingness to pay to reduce risk of death. Practically speaking, the shift in approach roughly tripled the standard values of life (Graham and Vaupel 1981).

For all the philosophical disagreements, the actual values attached to lives saved, however calculated, tend to be concentrated within two orders of magnitude (Graham and Vaupel 1981). Most studies estimate the value of life in hundreds of thousands of dollars. A few studies go as high as millions of dollars; some of these studies infer values of life from such actions as jury awards that compensate deaths. No estimates go as high as tens of millions of dollars or as low as the low end of tens of thousands of dollars.

It is difficult to set up fully persuasive methodologies to assess a popular consensus on value of life because people do not like to think about death. If asked about the value of a whole life, they refer to the sanctity of life and say the value is infinite. If asked about the value of a shift in the risk of dying, they find it difficult to relate to such a choice. If presented with forced choice situations that contain implicit values of life, they give answers that can reflect the way the questions were posed. Nevertheless, the use of a range of $250,000 to $1,000,000 is sufficient to capture most estimates of the value of a life.

Another alternative is to use a value per year of life saved. A large number of regulatory cases were examined and the ranking of alternatives found was not drastically affected by the use of life-year value versus life value (Graham and Vaupel 1981). However, use of life-year value tends to give more credit to saving children (by up to double, since their expected life spans are about double those of the population at large) and less credit to saving the elderly (by a factor of four or more). Firesafety in schools would be boosted and firesafety in nursing homes might be abandoned if life-year value calculations were used.

Even after deciding to use willingness to pay as the standard for value of life, some difficult technical problems remain. One is the question of whether to calculate separately the willingness to pay for each individual (or each major group) affected by a proposed change. In an analysis aimed at the individual property owner or manager, such differentiation is unavoidable and should be an explicit, or at least implicit, part of any analysis of the market for a new product, system or approach.

There also have been several studies of factors that affect willingness to pay. Willingness to pay is lower for the poor, the elderly, the seriously ill or handicapped, and the risk takers. For the poor, of course, ability to pay is lower, too. For the elderly and the seriously ill, the lower value given to life seems to reflect the fact that the quantity (for the elderly) or the quality (for the sick) of life remaining is well below the national norm. However, all these groups with lower willingness to pay also tend to have relatively high risks of becoming fire fatalities. They are precisely the groups to target if total lives saved were the criterion of choice. Conversely, the people most willing to pay—affluent, healthy, risk-averse, young heads of families—are the ones least likely to benefit because their current risks of dying in fire are already below average.

Another reason for variations in the willingness to pay involves the nature of the risks rather than the characteristics of those who experience these risks. Risks that are *voluntary, non essential, occupational*, or *results of product misuse* are deemed less serious than risks that are *involuntary, essential, public*, or *results of normal product use*. A risk of death to someone who lives near a nuclear reactor is valued more highly than an equal risk of death to someone who works in a coal mine. The difference is based on the assumption that occupational risks are more likely to be voluntary and more likely to be financially compensated. (Both of these assumptions are questionable. Workers in hazardous occupations such as mining may have few realistic occupational alternatives, while residents of hazardous areas, like flood plains, may have many alternative places to live and may have received financial compensation in the form of lower housing costs that at least equal any financial benefits received by the workers.) Similarly, risks of death associated with voluntary nonessential activities, such as smoking and hang gliding, are valued less than equal risks associated with voluntary but essential activities, such as driving a car. In fire risk, this argument appears in the debate over the fairness of imposing flame resistance standards (and accompanying costs) on all mattresses to protect people who choose to smoke in bed.

Deaths occurring in major multi fatality incidents are valued differently—and generally more highly—than deaths occurring in smaller incidents. Major incidents are termed *dread hazards* in the risk analysis literature; it is the factor of dread—the greater fear of death occurring in a major incident—that inflates the value of risk in such cases. The effect of major incidents on families and communities has been used to argue for both higher and lower weighting of such deaths—higher because familial blood lines may be extinguished, lower because multiple deaths in one family mean fewer survivors to mourn per fatality (Starr and Whipple 1980).

Dread incidents constitute an especially dramatic example of the phenomenon of *risk aversion*. For example, most people feel that if loss A is ten times as great as loss B but only one-tenth as likely, losses A and B still are not equally onerous. The general public tends to be more concerned about fire scenarios that may kill, say, 100 people once every three years than they are of fire scenarios that kill one person at a time every week, year after year.

Technical adjustments can be made to incorporate some risk aversion into a benefit calculation. Such adjustments will have less effect on dwelling fire risk calculations, where really large incidents are impossible, than on risk calculations for large residential (hotel), institutional, or public assembly properties.

Values for injuries avoided can be estimated more directly than values for fatalities avoided because direct costs such as medical expense and lost wages seem more appropriate as indicators of value. A survey was used to estimate direct injury-related costs for residential fires (Munson and Ohls 1980). Based on their figures, after adjusting for inflation since their study and for the fact that their cost-per-injury figures are dominated by very small injuries from unreported fires, an estimate of $5,000 is obtained for actual costs per injury received in a reported fire. Willingness to pay to avoid an injury is greater because of pain and suffering considerations, so a range of $5,000 to $20,000 per injury seems reasonable.

This average is based on a highly skewed distribution. The vast majority of injuries can be valued in the low hundreds of dollars or less, but a small number of serious burn injuries each year—considerably fewer than the corresponding number of fire fatalities—can cost hundreds of thousands or even millions of dollars in medical expenses. These few injuries account for most of the overall cost average. This suggests that analyses of expected impacts of new systems or programs on injuries should, if possible, separate serious and non serious injuries.

For property damage, the losses probably will already have emerged from the Post-Ignition Model expressed in monetary terms, but that is not necessarily so, and even if it is, some conversions may still be necessary. If fire growth and smoke spread models have been used, property damage may have been calculated in such terms as rooms exposed to fire or smoke for various lengths of time, or areas where structural integrity has been lost. Converting such descriptions to monetary equivalents would require data and models that do not now exist, despite repeated efforts to develop them (Hall 1982).

Even if damage is expressed in dollars, one may wish to take account of the fact that the loss faced by the property owner may be mediated by such things as insurance. In that case, it would be necessary to estimate the likely reduction in out of pocket, uninsured damage plus insurance premiums, rather than the likely reduction in total direct damage achievable.

Another consideration in the Loss Evaluation Model is whether to include an adjustment for *indirect loss*—such items as lost wages, costs of a temporary location, and lost revenue for days that a business is closed. These losses can be very large in individual cases, such as the 1980 MGM Grand Hotel fire, but in the aggregate they tend to be an order of magnitude smaller than the direct losses.

Cost Model

Costs may be divided into (a) initial costs of the proposed changes being studied; (b) the ongoing costs of these changes once they have been made; and (c) the ripple effects on other costs, such as the need to increase the water supply to support a sprinkler system. The latter could involve cost increases or cost reductions, including calculation of costs for many years into the future. To make this task manageable, the analysis can be set up in terms of the normal periods of maintaining, repairing and replacing the items being analyzed. This is called *life-cycle costing*. An overview of major components of each of these three types of costs is shown below:

A. Initial Costs of Changes Being Studied

1. Equipment Costs. For new products, it may be necessary to estimate what costs will be when mass production is under way. In many cases, the mass production cost continues to drop as further development occurs. (Smoke detectors have shown this pattern, for example.)
2. Installation Costs. Estimation of costs of installation may require an analysis of the steps required for installation, because the person-hours and skills required for those steps may be higher or lower than for comparable products already in use. (For example, plastic pipe may be faster to install than iron pipe, and it may require less time consuming effort to protect carpets and furniture from soiling during installation.)

Labor costs per hour may vary considerably from one place to another, as may overhead rates; these variations argue in favor of a serious effort to collect representative data.

3. Financing Costs. These will be relevant if the systems

are financed through time-payment plans, e.g., as part of what is covered by the building mortgage.

4. Permit/License Costs. There may be some one time fees required to install the systems.
5. If the new systems and features add to the resale value of the property, this will partially offset the initial costs.

B. Ongoing Costs of Changes Being Studied

1. Operating Costs. A new system or product may need labor, power, or some other continuing input to operate. These costs need to be included.
2. Inspection and Testing Costs. Many systems require periodic inspection and testing after installation. These costs should be included. Labor usually will be the main cost element, but some tests (such as sprinkler tests) may involve materials costs, and other tests may require destruction of a sample of system components that would need replacement.
3. Repair, Maintenance and Replacement Costs. Most systems will require repair and maintenance, and—if the study period is long enough—periodic replacement will need to be considered.
4. Costs of Non Fire Damage Caused by the Systems. An example would be water damage due to accidental discharge of a sprinkler.
5. Permit/License Costs. An example would be the standby water charge levied in some jurisdictions on buildings equipped with sprinklers.
6. Equipment that is replaced may be resellable. If so, salvage revenues help reduce net system costs.

C. Ripple Effects on Other Building Costs

1. Costs of Supporting Systems. Many new products may require replacement, modification, or addition of critical supporting systems (e.g., extra water supply for home sprinkler system in a rural area). The equipment and installation costs of these changes in supporting systems need to be identified and included. So do any changes in operating costs, repair and maintenance costs, inspection and testing costs, etc., for the modified supporting systems, and any changes in these ongoing costs for unmodified supporting systems.
2. Special Incentives or Credits. Insurance premium reductions that reflect the expected reduction in direct loss should be counted in the Loss Evaluation Model. Extra reductions offered as inducements to buy systems, as well as incentives or credits in property or income taxes, should be counted here.
3. Changes in property taxes reflecting changed property value assessments. There may be tax consequences if the features add value to the property.
4. Changes in land costs or required building features. Added safety features may permit trade offs in the form of increased density or reduced requirements for other building features. These need to be accommodated as costs, and any trade offs in other safety features need to be addressed in the Loss Evaluation Models as well.
5. Changes in costs of public fire protection. If buildings in a group receive similar modifications, it may be possible to accept longer response times or reduced sizes of fire suppression teams, resulting in reduced costs of public fire protection.

These lists are not exhaustive, but they indicate the need to estimate the effects of different decisions and assess their cost impacts.

Cost-Benefit Comparison Model

The Cost Model and Loss Evaluation Model produce time streams of costs and risk-reduction benefits—that is, year by year estimates of costs and of reductions in fire deaths, injuries, and property damage, with the latter expressed as total monetized losses. To compare the costs to the benefits, the two time streams need to be combined into a single, manageable, indicator of net benefits.

To compare future and present costs and benefits, it is necessary to decide what the future costs and benefits are worth in the present. This involves the concept of *opportunity cost.* Suppose $20 was spent now on a firesafety system and $20 received back 10 years later in the form of reduced property damage in a fire. This would not be a break-even proposition because alternative investments could pay interest over that period.

Assumptions about the attractiveness of such investments are reflected in an assumed *discount rate*—a proportion between 0 and 1 used to reduce the value of future costs and benefits. Most fire risk reducing strategies involve greater costs than benefits in the near years and greater benefits than costs in the later years; this makes the discount rate a critical factor in overall assessment of whether the benefits justify the costs. Also, even if opportunity costs were not involved, there would be a cost associated with delayed consumption. All other things being equal, people usually prefer to have goods and services now rather than later, and a discount rate reflects that fact.

If a cost is incurred 10 years from now, for instance, the discount rate must be applied 10 times to translate that cost into a figure comparable with today's costs. This figure is called the *present value* of a future cost or benefit. It is calculated as the discount rate raised to a power equal to the number of years in the future when the cost or benefit will occur, then multiplied by the value of that cost or benefit.

A reasonable discount rate can be assumed for the purpose of analysis or can be calculated as the discount rate required to just balance benefits and costs. If the latter is done, the derived discount rate is called the *internal rate of return.* It can be used to compare alternatives in the same way that a benefit-cost ratio can be used.

The two principal objections to discounting of future safety benefits are (1) the possibility of very large, perhaps even irreversible, effects at a remote point in the future, and (2) the cumulative effects of the short term biases induced by rigorous application of discounted assessments. The first objection is not a great concern for fire-risk problems because fire does not produce irreversible effects on the scale contemplated by this argument. At most, several small towns could be wiped out by a wild fire (ignoring, for the moment, the possibility of wartime firestorms). Nevertheless, as a technical matter, it is worth considering the possibility that discount rates undervalue the real value people assign to events beyond the next decade or so. This approach has been suggested with an example noted that most people would regard benefits in 105 years as equal to benefits in 100 years; but under constant discounting of say, ten percent, the former would be only 59 percent of the latter (Whipple 1981).

As for the cumulative problems of short term bias, this has been discussed more in the context of business research, development, and innovation in general, than in regard to safety innovations in particular. In business, investments are expected to balance benefits and costs within three to seven years, but many analysts believe such requirements are too demanding and tend, over time, to choke off truly dramatic breakthroughs. The result, in business, can be eventual loss of competitive edge to a competitor willing to take a longer view. One pertinent article (Hayes and Garvin 1982) was particularly forceful on this point, arguing that the implied opportunity cost model underlying a short payback period requirement assumes a standard reference alternative investment which, contrary to the model's assumptions, is not itself immune to the cumulative effects of a stream of choices driven by short term considerations. The fallacy, then, is in assuming that there always is an alternative investment that pays back in three to seven years; the short term driven decisions may have the cumulative effect of eroding all such alternatives.

The technical approach to addressing this concern is to check the sensitivity of any conclusions to the use of a lower discount rate. Any innovation that year by year, after the initial cost period, produces more benefits than costs, can be made to look attractive through the selection of a sufficiently low discount rate. It is risky, however, to use too low a discount rate, because that will give a misleading picture of what people will be willing to pay.

Other approaches such as using a higher discount rate for costs than for benefits, can produce perverse results. For example, such an approach could mean that an attractive safety program would seem even more attractive if its implementation were delayed. In this way, a program can be made to seem attractive but may never be implemented because further delay will always make it seem even more attractive (Keeler and Cretin 1983).

Summary of Conclusions

Fire risk analysis is a technique still in its infancy but already capable of pulling together, reducing, and providing perspectives on large quantities of data from many sources. Recent rapid advances in the size and quality of fire related data bases and in the power and affordability of computers have helped turn fire risk analysis into a practical tool for individual decision making and especially for large scale decision making by governments and companies. The accelerating pace of development of computerized models of fire development and occupant behavior will eventually produce integrated deterministic/probabilistic models, providing even more powerful tools for fire risk analysis.

At present there are only a few abbreviated published examples of fire risk analysis (e.g., Ruegg and Fuller 1984; Helzer et al 1979), but other more elaborate examples, providing more detailed illustrations and more broadly applicable models, are in progress. The largest single source of developmental studies in general purpose fire risk analysis in the U.S. has been the Center for Fire Research (CFR) of the National Bureau of Standards NBS). The Nuclear Regulatory Commission (NRC) has sponsored some of the best work in the field, but it is highly specialized to the problems of nuclear power plants and may need more work to translate to other settings. Individual fire risk analyses have appeared in materials prepared as background for regulatory decisions (e.g., Hall and Stiefel 1984), but these also tend to be highly specialized.

Until better materials are available to demonstrate the specific analytic techniques of fire risk analysis, it is useful to recognize the essence of fire risk analysis: creation of the simplest possible framework for identifying the available choices and estimating their consequences. Whether the estimates consist of nothing more than a series of guesses or are the result of an elaborate network of sophisticated models, laboratory tests, and fire experience, the framework in principle remains the same. Its purpose is to assemble all the information available, however slight or extensive, and focus that information on how and how much things would change as a result of the choices made. Identifying choices, predicting consequences, evaluating those consequences, and finally making choices—those are the steps of fire risk analysis. However complicated the methods used, the purpose is always to perform these familiar tasks that are the basics of rational decision making.

Bibliography

References Cited

Graham, John K., and Vaupel, James W. 1981. "Value of a Life: What Difference Does It Make?" *Risk Analysis*. pp 89-95.

Hall, John R. "Reduce Fire Loss Guesstimates." 1982. *Fire Service Today*. Nov 1982. pp 11-13.

Hall, John R. Jr., and Stiefel, S. Wayne. 1984. "Decision Analysis Model for Passenger-Aircraft Fire Safety With Application to Fire-Blocking of Seats." *NBSIR 84-2817*. National Bureau of Standards, Washington, DC.

Hayes, Robert H., and Garvin, David A. 1982. "Managing As If Tomorrow Mattered." *Harvard Business Review*. May-June 1982. pp 70ff.

Helzer, Susan Godby, et al. 1979. "Decision Analysis of Strategies for Reducing Upholstered Furniture Fire Losses." *NBS Technical Note 1101*. National Bureau of Standards, Washington, DC.

Keller, Emmet B., and Cretin, Shan. 1983. "Discounting of Life-Saving and Other Non-Monetary Effects." *Management Science*. pp 300-306.

Munson, Michael J., and Ohls, James C. 1980. *Indirect Costs of Residential Fires*. FA-61. Federal Emergency Management Agency, Washington, DC.

Ruegg, Rosalie T., and Fuller, Sieglinde K. 1984. "A Benefit-Cost Model of Residential Fire Sprinkler Systems." *NBS Technical Note 1203*. National Bureau of Standards, Gaithersburg, MD.

Starr, Chauncey, and Whipple, Chris. 1980. "Risks of Risk Decisions." *Science*. pp 1, 114ff.

Whipple, Chris. 1981. "Energy Production Risks: What Perspective Should We Take?" *Risk Analysis*. pp 29-35.

Additional Readings

Lawrence, William W., *Of Acceptable Risk*, Kaufman Publishing, Los Altos, CA, 1976.

Rowe, William D., *Ananatomy of Risk*, John Wiley and Sons, New York, NY, 1977.

HAND CALCULATIONS FOR ENCLOSURE FIRES

Edward K. Budnick, P.E., and David D. Evans, Ph.D., P.E.

Frequently it is desirable to estimate specific enclosure fire effects. Typical examples are temperature rise, radiant flux to target fuels, the quantity of smoke produced, the occurrence and time to flashover, and the minimum burning rate and fuel load to produce flashover. Also important for many analyses is the estimation of when fire protection devices such as heat detectors or automatic sprinklers will activate for specific fire conditions. Equations are available, based principally on experimental correlations, which permit the user to estimate these effects for enclosure fires.

The heat of combustion is a material property and is discussed in Chapter 6 of this section, "Pool Fires: Burning Rates and Heat Fluxes" and tabulated in Section 5, Chapter 11, "Tables and Charts." Values of the stoichimetric air/fuel mass ratio (X_r) also will be found in Section 5, Chapter 11, "Tables and Charts."

In this chapter, a brief discussion of enclosure fire effects is presented, along with equations that can be evaluated using hand calculators to provide estimates of particular effects. Generally, the equations presented are well documented, and are widely used for such estimates. However, the user is cautioned that most of the equations were developed based on data from experiments that were conducted for very specific, and sometimes idealized, conditions. Therefore, some judgment must be exercised when applying these equations to complex conditions occurring in enclosure fires of general interest.

The equations in this chapter are intended to be used in evaluating fire conditions in enclosures during the pre-flashover fire growth period. Most of the methods presented do not apply to fully developed room fires, e.g., post flashover conditions. In addition, these shorthand calculations apply only to the room of fire origin, and for a single burning fuel package. More complicated methods are available for multiroom analysis, but they are beyond the scope of this chapter. Methods to address multiple fuel package involvement are under development but are not yet available.

For some of the effects, more than one equation is presented. In these cases, one equation may be preferred over the other based on the best match of the experimental basis for the equation to the specific case of interest.

ENERGY RELEASE RATE

Calculation procedures for fire effects in enclosures require knowledge of the energy release rate of the burning fuel. The term energy release rate is frequently used interchangeably with "heat release rate," and is usually expressed in units of kilowatts (kW) and symbolized by $\dot{q}$

Energy release rates from burning fuels cannot be predicted from basic measurements of material properties. It depends on the fire environment, the manner in which the fuel is volatilized, and the efficiency of the vapor combustion. Therefore, one must rely on available laboratory test data for the specific or similar fuels. In addition, a knowledge of the complete energy release rate history may be required for many situations. This is particularly desirable where the fuel package exhibits unsteady burning. (See Fig. 21-3A.) For those cases where only limiting conditions or worst case analysis is required, it may be reasonable to assume that the fuel is burning at a constant rate, which simplifies the calculation considerably.

For the equations presented here, the more simplified condition of constant energy release rate is generally assumed. However, techniques are available which represent a growing fire by a series of constant energy release rate fires. This approach can require a great deal of calculation time, depending upon the desired accuracy. Such analysis is generally more suited to computer analysis.

For the complete combustion of a fuel, energy release rate and mass loss rate are related by the equation:

$$\dot{q} = \Delta h_c \cdot \dot{m} \quad (1)$$

where $\dot{q}$ = energy release rate (kJ/sec) or (kW)
Δh_c = heat of combustion (kJ/kg)
$\dot{m}$ = mass loss rate (kg/sec)

Mr. Budnick is a fire protection engineer, and Dr. Evans is a mechanical engineer in the Fire Growth and Extinction Group, U.S. National Bureau of Standards, Center for Fire Research, Gaithersburg, MD.

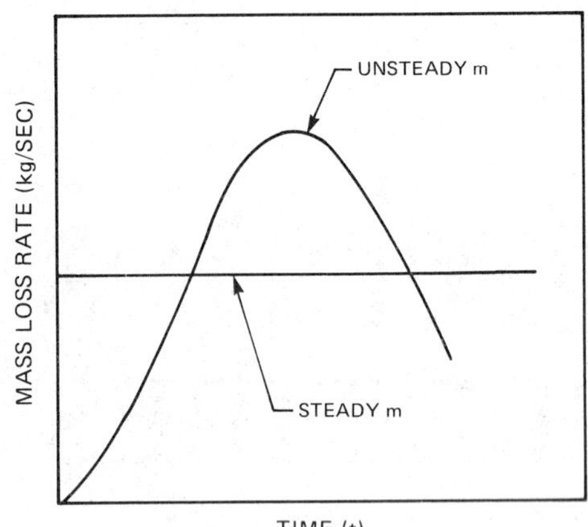

FIG. 21-3A. An illustration of steady and unsteady burning rates.

(The heat of combustion is a material property, and is tabulated for selected materials in Section 5, Chapter 11, "Tables." The mass loss rate is typically found experimentally.)

It should be recognized that most enclosure fires of interest do not exhibit constant energy release rates. Rather, as illustrated in Figure 21-3B for selected furniture items, the mass loss rate, and therefore the energy release rate, varies over time. Depending upon the detail required, one might select a constant mass loss rate, e.g., a peak value or an average value, as the basis for analysis. Data on mass loss rates for selected fuel packages are available in publications of the National Bureau of Standards (Babrauskas and Krasny 1985; Lawson et al 1984; Babrauskas et al 1982).

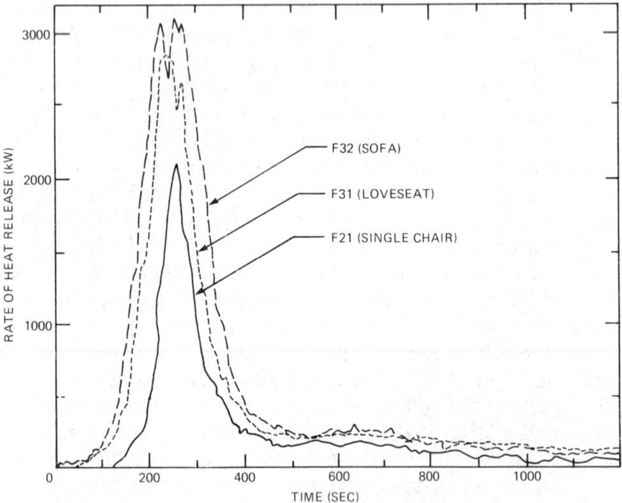

FIG. 21-3B. Free burn heat release rates for selected furniture items (Babrauskas et al 1982).

At this point, it should be noted that most information available on fuel package burning rates is reported for "free burn" conditions, i.e., the data is collected for items burning in the open rather than in an enclosure. While enclosure effects are less important in evaluating early fire growth, they may be very important in fully developed room fires that may be dominated by radiation feedback to the fuel from the hot smoke and enclosure linings, and ventilation conditions.

The following equation for the stoichiometric fuel pyrolysis can be used to estimate the mass loss rate at which these effects begin to dominate:

$$\dot{m}_{st} = \frac{1}{r} \cdot 0.5\, A_v \sqrt{h_v} \qquad (2)$$

where $\dot{m}_{st}$ = stoichiometric mass loss rate (kg/sec)
r = stoichiometric air/fuel mass ratio
A_v = area of ventilation opening (m^2)
h_v = height of ventilation opening (m)

For wood fuel, r = 5.7. Values of r for other materials can be found in Section 5, Chapter 11, "Tables."

An estimate of the maximum burning rate possible for an enclosure with a particular opening can be determined from Equation (2). If the mass loss rate for a particular fuel package is less than this value, the condition is referred to as fuel controlled, and results from Equation (1) provide a reasonable estimate of the energy release rate. If the free burn mass loss rate is higher than the stoichiometric rate, e.q., from equation (2), then the rate determined for stoichiometric conditions should be used.

A more rigorous treatment of energy release rates is available for selected material types, e.g., wood cribs, wood and plastic slabs, and liquid pool fires, where experimental correlations have been established. Chapter 6 of this Section, "Pool Fires," provides a detailed discussion of the prediction of burning rates for liquid pool fires. Detailed discussions of energy release rates for specific fuels are available elsewhere (Babrauskas 1985; Lawson and Quintiere 1985).

PRE-FLASHOVER TEMPERATURE ESTIMATES

Temperature rise is an important condition in evaluating the effects of a fire in an enclosure. Extensive experiments have been conducted to study temperature rise in compartments, leading to useful correlations. Equation (3) is a method for predicting fire temperature rise and may be used for single or multiple vent openings, with ventilation from natural convection only. The general form of the equation, based on experimental correlations, is:

$$\frac{T - T_o}{T_o} = 1.6 \left(\frac{\dot{q}}{\sqrt{g}\, C_p \rho_o T_o A_v \sqrt{h_v}} \right)^{2/3} \left(\frac{h_k A}{\sqrt{g}\, C_p \rho_o A_v \sqrt{h_v}} \right)^{-1/3} \qquad (3)$$

where T = upper gas temperature (K)
T_o = ambient gas temperature (K)
$\dot{q}$ = heat release rate (kJ/sec or kW)
g = gravitational acceleration (m/sec^2)
C_p = specific heat of air at constant pressure [(kJ/kg)/K]
ρ_o = density of ambient air (kg/m^3)
A = total surface area of enclosure interior (m^2)
A_v = vent area (m^2)
h_o = vent height (m)
h_k = effective enclosure conductance [(kW/m)/K]

$$h_k = \sqrt{k\rho c/t} \qquad t \leq t_p$$

$$h_k = k/\delta \qquad t > t_p$$

k = thermal conductivity of the enclosure material [(k/W/m)/K]
ρ = density of the enclosure material (kg/m^3)
c = specific heat of enclosure material [(kJ/kg)/K]
δ = enclosure material thickness (m)
t = time (sec)

$$t_p = \frac{\rho c}{k}\left(\frac{\delta}{2}\right)^2, \text{ thermal penetration time}$$

The terms $(h_k A)$ and $(A_v \sqrt{h_v})$ should be summed in Equation (3) for multiple structural materials and openings, respectively. In addition, while it is recognized that the enclosure gas temperature varies within the compartment, this equation is based on the assumption that an average upper layer temperature and an average lower layer temperature (ambient) reasonably approximate temperature conditions in the enclosure. By substituting values for key variables in Equation (3):

C_p = 1.0 (kJ/kg)/K, at one atmosphere
ρ_o = 1.18 kg/m^3, density of ambient air
T_o = 290 K, absolute ambient temperature
g = 9.81 m/sec^2, gravitational constant

$$h_k = \sqrt{\frac{0.18}{\delta}} \text{ for gypsum board [(kW/m)K]}$$

A simplified expression can be provided for estimating temperature rise in an enclosure lined with gypsum board, or a material with similar heat transfer properties. The form of this equation is:

$$T = T_o\left(1 + \frac{0.0236\,\dot{q}^{2/3}}{(h_k A A_v \sqrt{h_v})^{1/3}}\right) \qquad (4)$$

RADIANT HEAT FLUX TO A TARGET

For many enclosure fires, it is of interest to estimate the radiation transmitted from a burning fuel array to a target fuel positioned at some distance from the fire to determine if secondary ignitions are likely. Figure 21-3C depicts the configuration utilized in developing the expression:

$$\dot{q}''_o \cong \frac{P}{4\pi R_o^2} \cong \frac{x_r\,\dot{q}}{4\pi R_o^2} \qquad (5)$$

where: $\dot{q}''_o$ = incident radiation on the target (kW/m^2)
R_o = distance to target fuel (m)
P = total radiative power of the flame (kW)
x_r = radiative fraction
$\dot{q}$ = total heat release rate (kW)

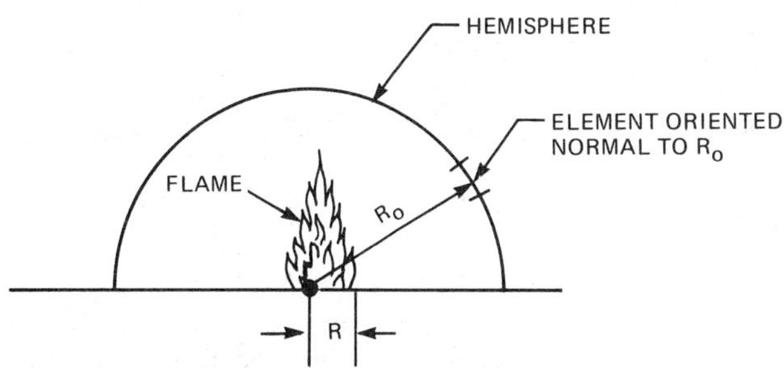

FIG. 21-3C. Illustration of radiant heat transfer to a target fuel (Lawson and Quintiere 1985).

Usually, x_r ranges from 20 to 45 percent, depending upon the fuel type. See Table 21-6A for values for x_r for selected liquid pools. Experimental measurements indicate that Equation (5) has good accuracy for

$$\frac{R_o}{R} > 4$$

where R (in meters) is the radius of the base of the fire. For radiation at

$$\frac{1}{2} < \frac{R_o}{R} < 4,$$

refer to Chapter 6 of this section for a more exact analysis.

PREDICTION OF FLASHOVER

A critical point in room fire growth is an event often referred to as flashover. While a universal definition does not exist, this event is generally associated with rapid transition in fire behavior from localized burning of fuel to involvement of all combustibles in the enclosure. Experimental work indicates that this transition can occur when upper room temperatures are between 400 and 600°C (750 and 1112°F) (Thomas 1981). Consistent with this, a general equation has been proposed for predicting the necessary burning rate to achieve flashover (Lawson and Quintiere 1985). This expression, based on equating flashover to 500°C (932°F) upper room temperature, is:

$$\dot{q}_{fo} = 610\,(h_k A A_v \sqrt{h_v})^{1/2} \qquad (6)$$

where $\dot{q}_{fo}$ = heat release rate at flashover (kW)
h_k = enclosure conductance [(kW/m^2)/K]
A = total enclosure area (m^2), excluding vent opening
A_v = area of vent opening (m^2)
h_v = height of vent opening (m)

An alternative expression, based on a flashover temperature of 600°C (1112°F) and enclosures with materials having densities similar to gypsum board, is:

$$\dot{q}_{fo} = 7.8\,A + 378\,A_v \sqrt{h_v} \qquad (7)$$

Time to Flashover

For a given steady energy release rate, Equation (3) for estimating upper room temperature can be rearranged and solved for the time to flashover, e.g., time to 500°C (932°F). For constant burning rate:

$$t = .059\ (\sqrt{g}\ C_p\ \rho_o\ AA_v\ \sqrt{h_v}\ )^2 \left(\frac{\Delta T^6}{T\ (\dot{q})^4}\right) \cdot k\rho c \qquad (8)$$

where t is in seconds, $\Delta T = T_0$ and $\dot{q} \geq \dot{q}_f$.

Required Fire Load for Flashover

For some problems, it is desirable to estimate the fire load necessary to reach flashover in an enclosure. Based on experimental correlations from compartment fires with wood fuel, the following formula can be used for such an estimate (Thomas 1974; Ove Arup 1977). Under natural ventilation conditions, the upper limit enclosure air temperature is approximated by:

$$\theta_f = 6000\ \frac{(1 - e^{-0.10\eta})}{\eta^{1/2}}\ (1 - e^{-0.05\tau}) + T_o \qquad (9)$$

where θ_f = upper limit air temperature (°C)

T_o = ambient temperature (°C)

$$\eta = \frac{A_T}{A_w \sqrt{h}}\ (m^{-1/2})$$

$$\tau = \frac{L}{(A_w A_T)^{1/2}},\ (kg/m^2)$$

A_T = total enclosure area (m^2), excluding vent opening

A_w = area of vent opening (m^2)

h = height of opening (m)

L = wood fuel mass (kg)

By setting θ_f equal to a temperature criterion for flashover, e.g., 500°C (932°F), one can solve Equation (9) for L, the wood fuel mass necessary to reach flashover conditions in the compartment under steady burning conditions.

SMOKE PRODUCTION RATE

The volume of smoke-filled gas produced in a fire depends on the material and the rate of burning. The rate of smoke-filled gas produced by a fire is nearly equal to the amount of air entrained into the rising fire plume, so the mass production rate can be estimated by (Butcher and Parnell 1979):

$$\dot{m}_s = 0.096\ P\rho_o y^{3/2} \left(g\ \frac{T_o}{T_{fl}}\right)^{1/2} \qquad (10)$$

where: $\dot{m}$ = rate of smoke production (kg/sec)
P = perimeter of fire (m)
y = distance from floor to bottom of smoke layer (m)
T_o = ambient temperature (K)
T_{fl} = flame temperature (K)
ρ_o = density of ambient air (kg/m^3)
g = gravitational acceleration (9.81 m/sec^2)

Since the expression assumes a constant or steady burning rate, its application has limitations. Yet it will provide a reasonable estimate of smoke generation rate for many enclosure configurations of practical interest.

This expression can be further simplified, based on value assignments for selected parameters. That is, for:

ρ_o = 1.22 kg/m^3 at 17°C
T_o = 290°K
T_{fl} = 1100 K
g = 9.81 m/sec^2

Equation (10) is reduced to:

$$\dot{m} = 0.188\ Py^{3/2} \qquad (11)$$

Figure 21-3D provides graphical results based on the calculation of the smoke-filled gas mass production rate in Equation (11) for selected values of P and y. The mass rate of smoke-filled gas production can be changed to a volume rate by dividing by the density of air at the appropriate gas temperature.

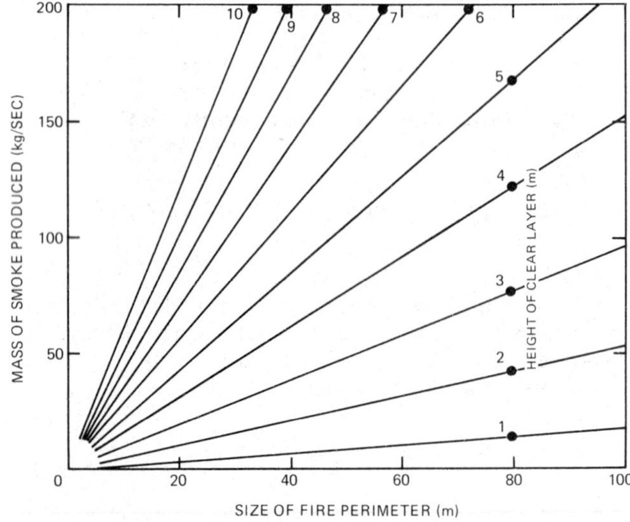

FIG. 21-3D. Smoke production rate for steady fires (Butcher and Parnell 1979).

THERMAL FIRE DETECTOR RESPONSE

Computer programs have been developed to calculate the response time of heat detectors and sprinklers installed below ceilings in large rooms (Evans and Stroup 1985). These programs can determine the time to operation for a user specified fire energy release rate history. They are convenient to use because the tedious repetitive calculations needed to analyze a growing fire can be avoided. However, the same calculations can be performed easily (with a hand calculator) for steady fires that have a constant energy release rate. In cases where a more detailed analysis of a fire that has important changes in energy release rate over time is required, the fire may be represented as a series of steady fires occurring one immediately after the other.

A useful calculation directly related to thermal detection is to find the temperature at positions directly above the flame produced by burning materials. This is done by using the following correlation (Evans 1985):

(12) $$T_m = \left(0.0577 \frac{\dot{q}^{2/3}}{H^{5/3}} + 1\right) T_o$$

where:
H = distance above fuel surface (m)
$\dot{q}$ = fire energy release rate (kW)
T_m = gas temperature above fire (K)
T_o = ambient room temperature (K)

This equation was developed from analysis of experiments with large scale fires having energy release rates from 670 kW to 100 MW (Alpert 1972).

As an example, using Equation (12), the gas temperature, (T_m), 5 m, (H), above the fuel surface producing a 500 kW, ($\dot{q}$), fire is found to be 366°K (93°C) for an ambient room temperature (T_o), of 293 K (20°C).

For the case of fixed temperature detectors, the minimum fire energy release rate $\dot{q}$ needed to operate a fixed temperature detection or suppression device located directly above the fire can be calculated using Equation (12), solving for $\dot{q}$ with T_m set equal to the activation temperature of the thermal device. In this form Equation (12) becomes:

(13) $$\dot{q} = 72.2 \left(\frac{T_m - T_o}{T_o}\right)^{3/2} \times H^{5/2}$$

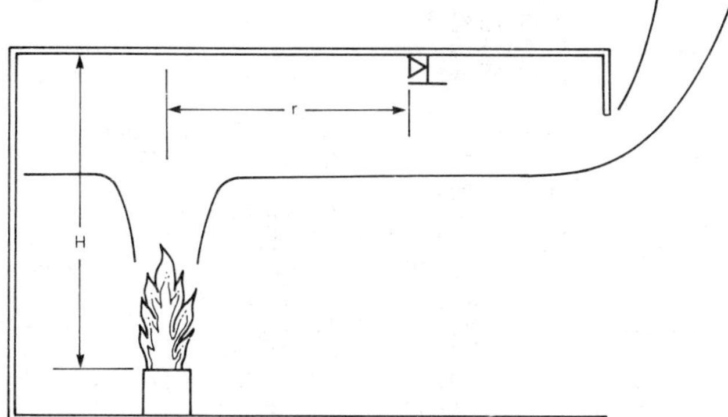

FIG. 21-3E. Parameters H and r both related to calculation of sprinkler or heat detector actuation time.

Based on cases where the hot gases have begun to spread under a ceiling located above the fire, Equation (12) also applies for a small radial distance, r, from the impingement point. (See Fig. 21-3E.) Over this distance (up to $r/H = 0.18$) where the gas is turning to flow out under the ceiling, the highest temperature in the flow remains equal to the value at the impingement point directly over the fire, calculated using Equation (12).

At radial distances greater than $r/H = 0.18$, the maximum temperature in the ceiling jet flow depends upon the distance from the impingement point according to:

$$T_m = \left(0.0184 \frac{(\dot{q}/r)^{2/3}}{H} + 1\right) T_o \qquad (15)$$

Correlations are also available for maximum velocities in the ceiling jet flow (U_m) under a ceiling. As with the temperature correlations there are two regions, one close to the impingement point where velocities are nearly constant, and the other farther away where velocities vary with radial position. The two correlations are:

$$U_m = 0.052 \left(\frac{\dot{q}}{H}\right)^{1/3} \text{ for } r/H < 0.15, \text{ and} \qquad (15)$$

$$U_m = 0.196 \left(\frac{\dot{q}^{1/3} H^{1/2}}{r^{5/6}}\right) \text{ for } r/H > 0.15 \qquad (16)$$

The previous equations can be used to determine if the fire driven gas flow past a detection device is at a high enough temperature to operate the device. However, more information is needed to calculate the amount of time needed to heat the detector or sprinkler sensing element to the operating temperature. Often these elements are made from metal, such as the ordinary solder type fusible link used in the link and level sprinkler. These metal elements require some time to absorb heat that is transferred from the hot gas flowing around the device.

The ease of heating of detector elements is measured using a plunge test that rates devices according to the values of the response time index (RTI). Small RTI values are associated with devices that heat quickly; devices with large RTI values increase in temperature slowly. RTI values for sprinklers range from 15 $m^{1/2}$ $sec^{1/2}$ to 400 $m^{1/2}$ $sec^{1/2}$ (Heskestad and Smith 1976).

For steady fires, the time required to heat the sensing element of a thermal detection or suppression device from room temperature to operation temperature is given by

$$t_{operation} = \frac{RTI}{\sqrt{U_m}} \log_e \left(\frac{T_m - T_o}{T_m - T_{operation}}\right) \qquad (17)$$

In the previous example, it was found using Equation (12) that a 500 kW ($\dot{q}$) fire would produce a gas temperature of 93°C (T_m) at 5 m (H) above the fuel surface in a room with 20°C (T_o) ambient temperature. From Equation (15), the gas velocity at this position would be 4.4 m/sec (U_m). For a sprinkler with an RTI of 200 m½ sec½ and operation temperature of 74°C, ($T_{operation}$), the time to operation in response to the steady fire can be calculated from Equation (17) as:

$$t_{operation} = \frac{200}{\sqrt{4.4}} \log_e \left(\frac{93 - 20}{93 - 74}\right) = 128 \text{ seconds}$$

All calculations in use today for calculating times to operation only consider the convective heating of sensing elements by the hot fire gases, and do not account for any direct heating by radiation from the flames. Research is continuing to improve and generalize these types of calculations.

Bibliography

References Cited

Alpert, R. L. 1972. "Calculation of Response Time of Ceiling-Mounted Fire Detectors." *Fire Technology*. Vol 8. pp 181-195.

Babrauskas, V. 1985. "Free Burning Fires." *Proceedings, SFPE Symposium: Quantitative Methods for Fire Hazard Analysis*. University of Maryland, College Park, MD.

Babrauskas, V. 1981a. "A Closed-Form Approximation for Post-Flashover Compartment Fire Temperatures." *Fire Safety Journal*. Vol 4. pp 63-73.

Babrauskas, V. 1981b. "Applications of Predictive Smoke Measurements." *Journal of Fire and Flammability*. Vol 12. p 51.

Babrauskas, V., and Krasny, J. F. 1985. "Fire Behavior of Upholstered Furniture." *NBS Monograph*. National Bureau of Standards, Washington, DC.

Babrauskas, V., et al. 1982. "Upholstered Furniture Heat Release Rates Measured with a Furniture Calorimeter." *NBSIR 82-2604*. National Bureau of Standards, Washington, DC.

Butcher, E. G., and Parnell, A. C. 1979. *Smoke Control in Fire Safety Design*. E. and F. N. Spon, Ltd., London, England.

Evans, D. D. 1985. "Calculating Sprinkler Actuation Time in Compartments." *Fire Safety Journal*. Vol 9. pp 147-155.

Evans, D. D., and Stroup, D. W. 1985a. "Methods to Calculate the Response Time of Heat and Smoke Detectors Installed Below Large Unobstructed Ceilings." *NBSIR 85-3167*. National Bureau of Standards, Washington, DC.

Heskestad, G., and Smith, H. 1976. "Investigation of a New Sprinkler Sensitivity Approval Test: The Plunge Test." *FMRC Serial No. 22485*. Factory Mutual Research, Norwood, MA.

Jin, T. 1971. "Visibility Through Fire Smoke (Part 2)." *Report of the Fire Research Institute of Japan*, Tokyo, Japan. Nos 31, 33.

Lawson, J. R., and Quintiere, J. G. 1985. "Slide-Rule Estimates of Fire Growth." *NBSIR 85-3196*. National Bureau of Standards, Washington, DC.

Lawson, J. R., et al. "Fire Performance of Furnishings as Measured in the NBS Furniture Calorimeter, Part I." *NBSIR 83-2787*. National Bureau of Standards, Washington, DC.

Lopez, E. L. 1975. "Smoke Emission from Burning Cabin Materials and the Effect on Visibility in Wide-Bodied Jet Transports." *Journal of Fire and Flammability*. Vol 6.

McCaffrey, B. J., Quintiere, J. G., and Harkleroad, M. F. 1981. "Estimating Room Temperatures and the Likelihood of Flashover Using Fire Test Data Correlations." *Fire Technology*. Vol 17, No 2, p 98.

Modak, A. T. 1977. "Thermal Radiation From Pool Fires." *Combustion and Flame*. Vol 29, pp 177-192.

Ove Arup and Partners. 1977. *Design Guide for Fire Safety of Bare Exterior Structural Steel*, London, England. Sponsored by American Iron and Steel Institute, Washington, DC.

Quintiere, J. G. 1983. "A Simple Correlation for Predicting Temperature in a Room Fire." *NBSIR 83-2712*. National Bureau of Standards, Washington, DC.

Quintiere, J. G. 1982. "Smoke Measurements: An Assessment of Correlations Between Laboratory and Full-Scale Experiments." *Fire and Materials*. Vol 6, Nos 3 and 4.

Rasbash, D. J. 1967. "Smoke and Toxic Products Produced at Fires." *Plastics Institute Transaction and Journal*. pp 55-61. Jan 1967.

Tewarson, A. 1980. "Physico-Chemical and Combustion/Pyrolysis Properties of Polymeric Materials." *NBS-GCR-80-295*. National Bureau of Standards, Washington, DC.

Thomas, P. H. 1981. "Testing Products and Materials for Their Contribution to Flashover in Rooms." *Fire and Materials*. Vol 5, No 3. pp 103-111.

Thomas, P. H. 1974. *Fire in Model Rooms*. CIB Research Program. Building Research Establishment. Borehamwood, Hertfordshire, England.

Additional Readings

American Society for Testing and Materials (ASTM), *Standard Test Method for Heat and Visible Smoke Release Rates for Materials and Products (E-906-83)*, Philadelphia, PA, 1983.

Babrauskas, V., "Estimating Room Flashover Potential," *Fire Technology*, Vol. 16, May 1980, pp. 94-103, 112.

Babrauskas, V., "Upholstered Furniture Room Fires—Measurements, Comparison with Furniture Calorimeter Data, and Flashover Predictions," *Journal of Fire Sciences*, Vol. 2, Jan./Feb. 1984, pp 5-19.

Babrauskas, V., "Development of the Cone Calorimeter—A Bench-Scale Heat Release Rate Apparatus Based on Oxygen Consumption," *NBSIR 82-2611*, National Bureau of Standards, Washington, DC, 1982.

Delichatsios, M. A., "Fire Growth Rates in Wood Cribs," *Combustion and Flame*, Vol. 27, 1976, pp. 267-278.

Heskestad, G., "Modeling of Enclosure Fires," *14th International Symposium on Combustion*, The Combustion Institute, Pittsburgh, PA, p. 1021, 1973.

Heskestad, G., *A Fire Products Collector for Calorimetry into the MW Range*, Factory Mutual Research Corp., Norwood, MA, 1981.

Huggett, C., "Estimation of Rate of Heat Release by Means of Oxygen Consumption Measurements," *Fire and Materials*, Vol. 4, 1980, pp. 61-65.

Kawagoe, K., and Sekine, T., *Estimation of Fire Temperature-Time Curve for Rooms*, Japanese Ministry of Construction Building Research Institute, June 1963.

Krasner, L. M., *Burning Characteristics of Wooden Pallets as a Test Fuel*, Factory Mutual Engineering, Norwood, MA, 1968.

Law, M., "A Relationship Between Fire Grading and Building Design and Contents," *Fire Research Note 877*, British Fire Research Station, 1971.

Law, M., "Radiation from Fires in a Compartment," *Fire Research Tech. Paper No. 20*, British Fire Research Station, 1968.

Law, M., "Notes on the External Fire Exposure Measured at Lehrte," *Fire Safety Journal*, Vol. 4, 1981/82, pp. 243-246.

Modak, A. T., and Croce, P. A., "Plastic Pool Fires," *Combustion and Flame*, Vol. 30, 1977, pp. 251-265.

Quintiere, J. G., et al., "An Analysis of Smoldering Fires in Closed Compartments and Their Hazard Due to Carbon Monoxide," *Fire and Materials*, Vol. 16, Nos. 3 and 4, 1982.

Williams, F. A., "Mechanisms of Fire Spread," *16th International Symposium on Combustion*, The Combustion Institute, 1976, pp. 1281-1294.

Yamashika, S., and Kurimoto, H., "Burning Rate of Wood Cribs," *Report of the Fire Research Institute of Japan*, Tokyo, No. 41, Mar. 1976, pp. 8-15.

Zukoski, E. E., "Development of a Stratified Ceiling Layer in the Early Stages of a Closed-Room Fire," *Fire and Materials*, Vol. 2, No. 2, 1978.

COMPUTER FIRE MODELS

Edward K. Budnick, P.E. and William D. Walton, P.E.

Computer models are simply computer programs which model or simulate a process or phenomenon. Computer models have been used for some time in the design and analysis of fire protection hardware. The use of computer models, commonly known as design programs, has become the industry's standard method for designing water supply and automatic sprinkler systems. These programs perform large numbers of tedious and lengthy calculations and provide the user with accurate, cost optimized designs in a fraction of the time required for manual procedures.

In addition to the design of fire protection hardware, computer models may also be used to evaluate the effects of fire on people and property. These computer fire models can provide a faster and more accurate estimate of the impact of a fire, and the measures used to prevent or control the fire, than many of the methods previously used. While manual calculation methods provide good estimates of specific fire effects, e.g., prediction of time to flashover, they are not well suited for comprehensive analyses involving the time dependent interactions of multiple physical and chemical processes present in developing fires.

In recent years, increasing attention has been given to the development and use of computer fire models. They have been used by engineers and architects for building design, by building officials for plan review, by the fire service for prefire planning, by investigators for post fire analysis, by groups writing fire codes, and by materials manufacturers, fire researchers, and educators. While these models are not a replacement for the building and fire codes they can be a valuable tool for fire professionals.

The state of the art in computer fire modeling is changing rapidly. The understanding of the processes involved in fire growth is improving and thus the technical basis for the models is improving. The capabilities, documentation, and support for a given model can change dramatically over a short period of time. In addition, computer technology (both software and hardware) is advancing rapidly. A few years ago a large mainframe computer was required to run most of the computer fire models. Today, almost all of the models can be run on minicomputers, and a substantial number on microcomputers. Therefore, rather than provide an exhaustive discussion of rapidly changing state of the art computer models, this chapter focuses on a representative selection of these models. Refer to the references at the end of this chapter for an in depth review of a particular model.

In general, computer models for fire hazard prediction can be grouped into two categories: enclosure fire models and special purpose fire models.

ENCLOSURE FIRE MODELS

Major advances have occurred in developing computational models structured to predict the interaction of multiple fire processes involving heat transfer, fluid mechanics, and combustion chemistry occurring simultaneously in an enclosure. These models provide estimates of particular elements of hazard development such as fire growth, temperature rise, and smoke generation and transport. Some models are able to address multiple rooms while others are confined to the room of fire origin. Generally, the large number of mathematical expressions to be solved simultaneously in any of these models necessitates the use of a computer.

There are two general classes of computer models, probabilistic and deterministic, for analyzing enclosure fire development.

Probabilistic Models

Probabilistic models treat fire growth as a series of sequential events or states. These models are sometimes referred to as state transition models. Mathematical rules are established to govern the transfer from one event to another, e.g., from ignition to established burning. Probabilities as functions of time are assigned to each transfer point, based on analysis of relevant experimental data and historical fire incident data. These models do not normally make direct use of the physical and chemical equations describing the fire processes.

Mr. Budnick is a fire protection engineer in the Fire Growth and Extinction Group, and Mr. Walton is a fire protection engineer in the Fire Simulation Group of the U.S. National Bureau of Standards, Center for Fire Research, Gaithersburg, MD.

Deterministic Models

In constrast, deterministic models represent the processes encountered in a compartment fire by interrelated mathematical expressions based on physics and chemistry. These models may also be referred to as room fire models, computer fire models, or mathematical fire models. Ideally, such models represent the ultimate capability: discrete changes in any physical parameter could be evaluated in terms of the effect on fire hazard. While the state of the art in understanding fire processes will not yet support the "ultimate" model, a number of computer models are available that provide reasonable estimates of selected fire effects. These models are described below:

Zone Models: The most common type of physically based fire model is the "zone" or control volume model, which solves the conservation equations for distinct regions (control volumes). A number of zone models exist, varying to some degree in the detailed treatment of fire phenomena. The dominant characteristic of this class of model is that it divides the room(s) into a hot upper layer and a lower cooler layer. (See Fig. 21-4A.). The model calculations provide estimates of key conditions for each of the layers as a function of time. Zone models have proven to be a practical method at providing first order estimates of fire processes in enclosures.

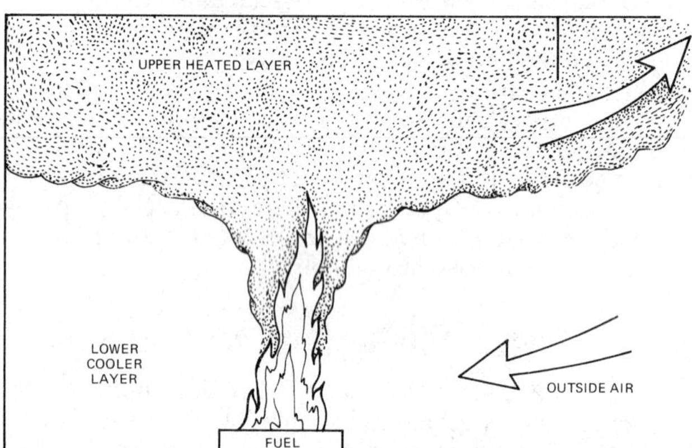

FIG. 21-4A. Conceptualization of Two Layer Zone Model.

Field Models: The other general type of deterministic model is the "field" model. This type of model solves the fundamental equations of mass, momentum, and energy at each element in a compartment space which has been divided into a grid of small elements. For example, in an enclosure filled with a three dimensional grid of tiny cubes, a field model will calculate the physical conditions in each cube as a function of time. The calculation will account for physical changes generated within the cube and changes on the cube from surrounding cubes. This will permit the user to determine the conditions at any point in the compartment.

Currently, field models are not sufficiently developed for most applied uses. The computational demands are beyond most computer capabilities, and the description of the fire process is incomplete. Considerable research and development is required before this type of model can be used for routine analysis.

Table 21-4A provides a listing of a number of zone type enclosure fire models which are in the public domain. Several of the models from the list are described below. In general, these are the models which have been most widely used and are of the greatest interest to the fire community as a whole.

Overview of Leading Zone Models

ASET: ASET (Available Safe Egress Time) is a program for calculating the temperature and position of the hot smoke layer in a single room with closed doors and windows. ASET can be used to determine the time to the onset of hazardous conditions for both people and property. The required program inputs are the heat loss fractions, the height of the fuel above the floor, criteria for hazard and detection, the room ceiling height, the room floor area, a heat release rate, and (optional) species generation rate of the fire. The program outputs are the temperature, thickness, and (optional) species concentration of the hot smoke layer as a function of time, and the time to hazard and detection. ASET can examine multiple cases in a single run. ASET was written in FORTRAN (Cooper and Stroup 1985).

ASET-B: ASET-B is a program for calculating the temperature and position of the hot smoke layer in a single room with closed doors and windows. ASET-B is a compact version of ASET which is designed to run on personal computers. The required program inputs are a heat loss fraction, the height of the fire, the room ceiling height, the room floor area, the maximum time for the simulation, and the rate of heat release of the fire. The program outputs are the temperature and thickness of the hot smoke layer as a function of time. Species concentrations and time to hazard and detection calculated by ASET are not calculated in the compact ASET-B version. ASET-B was written in BASIC (Walton 1985).

BFSM: BFSM (Building Fire Simulation Model) is a stochastic finite state transition model that allows the examination of the interrelationships among fire development, spread of combustion products, and people movement in residential occupancies. Rather than an explicit set of mathematical expressions, which describe the physical processes, this model consists of: (1) a set of process states, and (2) a set of rules governing transition from one state to another. The required program inputs include building geometry, burn characteristics of the contents, and occupant locations. Also necessary are assignments of probabilities for fire transitions among the discrete states established for the model. These probabilities are based on test data and experience. The output from the BFSM program includes time dependent temperatures and smoke concentrations, fire realm (state) history, and egress route availability. The current version of this model is appropriate as a research tool, but is not adequately developed for use in engineering applications. The model was written in FORTRAN (Swartz et al 1983).

COMPF2: COMPF2 is a computer program for calculating the characteristics of a post flashover fire in a single building compartment, based on fire induced ventilation through a single door or window. It is intended both to perform design calculations and to analyze experimental burn data. Wood, thermoplastic, and liquid fuels can be

TABLE 21-4A. Listing of Enclosure Computer Fire Models

Model Name	Author	Maintaining Organization	Appropriate Resources: Computer Size	Appropriate Resources: Computer Language	Applications and Special Features
*ASET	L. V. Cooper D. W. Stroup	NBS	Micro	Fortran	Single room enclosure fire model
*ASET-B	W. D. Walton	NBS	Micro	Basic	Single room enclosure fire model
*BFSM	J. A. Swartz G. Berlin R. F. Fahy E. M. Connelly D. P. Demers	Nat. Fire Protection Assn.	Mini	Fortran	State transition model based on statistical likelihood of events
BRI	T. Tanaka	Building Research Inst., Japan NBS	Mini	Fortran	Multi-room, multi-floor enclosure fire model
CALTECH	E. E. Zukoski T. Kubota	Calif. Inst. of Technology	Mini	Fortran	Two room enclosure fire model
COMPBURN	N. Siu	UCLA	Mini	Fortran	Single room enclosure fire model, developed for nuclear power facilities
*COMPEZ	V. Babrauskas	NBS	Mini	Fortran	Single room, post-flashover enclosure fire model
DACFIR-3	C. D. MacArthur	FAA	Mainframe	Fortran	Enclosure fire model for aircraft cabin geometries
DSLAYI	B. Hagglund	Nat'l. Defense Research Inst., Sweden	Mini	Fortran	Single room enclosure fire model, smoke filling
*FAST	W. W. Jones	NBS	Mini	Fortran	Multi-room enclosure fire model
FLASHOVER	B. Hagglund	Nat'l. Defense Research Inst., Sweden	Mini	Fortran	Single room enclosure fire model
*HARVARD	H. W. Emmons H. E. Mitler	NBS	Mini	Fortran	Enclosure fire model, version 5 single room, version 6 multi-room
*OSU	E. E. Smith S. Satija	Ohio State Univ.	Mini	Fortran	Single enclosure fire model, input from ASTM E 906 calorimeter
RFIRES	R. Pape T. Waterman	Ill. Inst. of Tech. Research Inst.	Mini	Fortran	Single enclosure fire growth model

* A more detailed narrative can be found in the text for these models.

treated. A comprehensive output format is provided which gives gas temperatures, heat flow terms, and flow variables. The documentation includes input instructions, sample problems, and a listing of the program. The program was written in FORTRAN (Babrauskas 1979).

FAST: FAST (Fire and Smoke Transport) is a multiroom fire model which predicts the conditions within a structure resulting from a user specified fire. FAST (version 17) can accommodate up to ten rooms with multiple openings between the rooms and to the outside. The required program inputs are the geometrical data describing the rooms and connections, the thermophysical properties of the ceiling, walls and floors, the fire as a rate of mass loss, and the generation rates of the products of combustion. The program outputs are the temperature and thickness of, and species concentrations in, the hot upper layer and the cooler lower layer in each compartment. Also given are surface temperatures and heat transfer and mass flow rates. FAST was written in FORTRAN (Jones 1985).

Harvard: The Harvard fire model predicts the development of a fire and the resulting conditions within a room (version 5) or multiple rooms (version 6), resulting from a user specified fire or user specified ignition. Version 5 predicts the heating and possible ignition of up to four targets due to the original fire. The room must have at least one opening to the outside. Version 6 does the same for up to five rooms connected to each other with openings; at least one of the rooms must have an opening to the outside. The required program inputs are the geometrical data describing the rooms and openings, and the thermophysical properties of the ceiling, walls, burning fuel, and of the targets. The generation rate of soot must be specified and the generation rates of other species may be specified. The fire may be entered either as a mass loss rate or in terms of fundamental properties of the fuel. Among the program outputs are the temperature and thickness of, and species concentrations in, the hot upper layer and the cooler lower layer in each compartment. This program gives surface temperatures and heat transfer and mass flow rates. The Harvard program was written in FORTRAN by H.W. Emmons and H.E. Mitler (Mitler 1985; Rockett 1983; Handa et al 1983).

OSU: OSU (Ohio State University Compartment Fire Model) is a program for calculating heat release rate, smoke generation rate, and smoke and heat venting from single compartments. The OSU model takes into account an initiating fire, and both horizontal and vertical fire spread, providing a prediction of how combustible materials in the compartment influence the course of a developing fire. The required program inputs are material thermal properties, ignition point, rate of heat and smoke release, and flame propagation parameters. In addition, certain plume properties including dimensions, temperature, and emissivity are required. Typical model outputs include upper layer air temperature, smoke generation rate, and heat release rate. OSU was written in FORTRAN (Smith and Satija 1981).

SPECIAL PURPOSE MODELS

Special purpose computer models include computer models designed for special purpose analyses, such as structural fire resistance, prediction of response time of heat detectors and automatic sprinklers, wild fire development, egress design for building occupants, the design of sprinkler systems, and the performance of smoke control or ventilation systems. These models may require and/or permit coupling with other special purpose models, or with a more general enclosure fire development model.

Table 21-4B is a listing of a number of special purpose fire models which are in the public domain. Several of the models from the list are described below. Again, as for enclosure fire models, these are the models that have been most widely used and are of the greatest interest to the fire community as a whole.

ASCOS: ASCOS (Analysis of Smoke Control Systems) is a program for steady air flow analysis of smoke control systems. This program can analyze any smoke control system that produces pressure differences with the intent of limiting smoke movement in building fire situations. The input consists of the outside and inside building temperatures, a description of the building flow network, and the flows produced by the ventilation or smoke control system. The output consists of the steady state pressures and flows throughout the building. ASCOS was written in FORTRAN (Klote 1982).

DETACT-QS: DETACT-QS is a program for calculating the actuation time of thermal devices below unconfined ceilings. It can be used to predict the actuation time of fixed temperature heat detectors and sprinklers subject to a user specified fire. DETACT-QS assumes that the thermal device is located in a relatively large area; that is, there is no accumulated hot gas layer in the room so only the fire ceiling flow heats the detection device. The required program inputs are the height of the ceiling above the fuel, the distance of the thermal device from the axis of the fire, the actuation temperature of the thermal device, the response time index (RTI) for the device, and the rate of heat release of the fire. The program outputs are the ceiling gas temperature at the device location and the device temperature, both as a function of time and the time required for device actuation. DETACT-QS was written in BASIC (Evans and Stroup 1985).

EVACNET+: EVACNET+ (Evacuation Network Computer Model) is an interactive computer program that models emergency building evacuation. It consists of a network of nodes and arcs to represent building locations and connecting paths. Input for this model includes a detailed geometry of the building and contents (multiple rooms and floors can be addressed) and information regarding the initial location of the occupants. Output can be selected from a menu, and generally takes the form of optimum evacuation times and identification of impediments to smooth, orderly evacuation. Two versions of the program are available, one in FORTRAN and one in BASIC (Kisko and Francis 1985).

FASBUS-II: FASBUS-II (Fire Analysis of Steel Building Systems) is a finite element model designed to analyze the structural response of steel framed floors exposed to fire conditions. Temperature effects on the material properties of the structural components are considered in predicting the deflections of the assembly. Input requirements include the geometry and arrangement of components of the structural assembly, the material properties as a function of temperature, applied loads and end conditions (restraints), and the temperature profile within structural components as a function of time. The outputs include deflections and rotations of floor systems, and stresses and strains induced in structural members. FASBUS-II is written in FORTRAN (Wiss 1982).

FIRES-T3: (Fire Response of Structures—Thermal, Three-Dimensional Version) is a finite element computer model designed to analyze heat transfer through structural assemblies. A wide variety of structural assemblies can be examined with FIRES-T3, including columns, walls, beams, floor/ceiling assemblies, and roof/ceiling assemblies. The structural assembly may be solid or include air cavities. The input requirements include a description of the structural assembly and the fire exposure. The information necessary to describe the column assembly includes geometric factors (dimensions, shape of assembly) and material property values (thermal conductivity, specific heat, and density) as a function of temperature. The fire exposure is characterized in terms of the temperature of the surrounding media and appropriate heat transfer coefficients. The output is a tabulation of the temperatures within the structural assembly as a function of time. FIRES-T3 was written in FORTRAN (Iding et al 1977).

RELATED ACTIVITIES

The development of computer fire models is very active today. The foregoing review is intended to offer a perspective on what is currently available. One should also recognize that the information in this chapter is not exhaustive. For example, little attention has been given to fire modeling efforts and programs available outside the U.S. Selected papers which provide some review of such activities are listed in the Bibliography at the end of this chapter.

Most of the computer fire models discussed in this chapter are only available from the authors. There are, however, efforts underway by organizations such as the Society of Fire Protection Engineers (SFPE), to distribute public domain software on a cost recovery basis. The user should be aware that there may be significant differences between computers and between versions of so called standard computer languages. As a result, programs may not always be transferred directly from one computer to another.

There are a number of organizations in the U.S. which are very active in developing and coordinating computer fire models. The Ad Hoc Working Group on Mathematical

TABLE 21-4B. Special Purpose Computer Fire Models

Model Name	Author	Maintaining Organization	Model Type	Appropriate Resources: Computer Size	Appropriate Resources: Computer Language	Applications and Special Features
*ASCOS	J. H. Klote	NBS	Smoke control	Micro	Fortran	Steady state network flow model for smoke control evaluation, no fire condition
*DETACT-QS	D. D. Evans	NBS	Thermal device actuation	Micro	Basic	Calculates actuation time for heat detectors and sprinklers, unconfined ceilings
*EVACNET+	T. M. Kisko R. L. Francis	Univ. of Florida	Building egress	Mini/Micro	Fortran/Basic	Network model for calculating evacuation time; multi-rooms, multi-floors
*FIRES-T3	R. H. Iding B. Bresler Z. Nizamuddin	Amer. Iron & Steel Inst.	Structural heat transfer	Mini	Fortran	3-dimensional heat transfer analysis through structural assemblies
*FASBUS-II	Chiapetta B. Bresler R. H. Iding D. Jeanes N. Breese	Amer. Iron & Steel Inst.	Structural response	Mini	Fortran	Finite element analysis of structural response of steel framed floors
FEES/MB	D. M. Alvord	NBS	Emergency escape	Mini	Fortran	Simulation of emergency escape and rescue times, developed for board and care facilities
FOREST FIRES	R. C. Rothermel	US Dept. of Agriculture	Fire spread	TI-59		Prediction of spread and intensity of forest and range fires
HPO 10	I. A. Hansever	Technical Univ. of Braunschweig, Germany	Structural response	Mainframe	Fortran	Finite-element analysis of reinforced concrete beam structural and thermal response
HSLAB	L. Abrahamsson B. Hagglund K. Janzon	Nat'l. Defense Research Inst., Sweden	Heat transfer	Mini	Fortran	One-dimensional calculation of transient temperature in concrete slabs
MINE VENT	R. E. Greuer	Michigan Technical Univ.	Shaft vent	Mini	Fortran	Simulation of underground mine ventilation, mine fire interaction

* A more detailed narrative can be found in the text for these models.

Fire Modeling (see: Levine, NBS) has been active for over ten years and coordinates the development of the technical basis for fire models, and serves as a forum for active discussion and exchange of ideas. The American Society of Testing and Materials (ATSM) has recently formed a new subcommittee (ASTM E 5.39, Subcommittee on Fire Modeling) to coordinate and develop ASTM's role in the evolution of computer fire modeling. The SFPE has an active computer committee which maintains a current list of available computer models. The serious user of computer models will need to be acquainted with the activities of all of these groups.

Bibliography

References Cited

Babrauskas, V. 1979. "COMPF2 —A Program for Calculating Post-Flashover Fire Temperatures." *NBS TN 991*. National Bureau of Standards, Washington, DC.

Cooper, L. Y., and Stroup, D. W. 1985. "ASET —A Computer Program for Calculating Available Safe Egress Time." *Fire Safety Journal*. Vol 9. pp 29-45.

Evans, D. D., and Stroup, D. W. 1985. "Methods to Calculate the Response Time of Heat and Smoke Detectors Installed Below Large Unobstructed Ceilings." *NBSIR 85-3167*. National Bureau of Standards, Washington, DC.

Handa, T. et al. 1983. "Some Examples of Application of Harvard V Fire Computer Code to Fire Investigation." *Fire Science and Technology*. Vol 3, No 1. pp 63-72.

Iding, R. H. et al. 1977. "FIRES T3 —A Computer Program for the Fire Response of Structures —Thermal Three Dimensional Versions." *UCB FRG 77-15*. University of California, Berkeley, CA.

Jones, W. W. 1985. "A Multicompartment Model for the Spread of Fire, Smoke and Toxic Gases." *Fire Safety Journal*. Vol 9. pp 55-79.

Kisko, T. M., and Francis, R. L. 1985. "EVACNET+: A Computer Program to Determine Optimal Building Evacuation Plans." *Fire Safety Journal*. Vol 9. pp 211-220.

Klote, J. H. 1982. "Computer Program for Analysis of Smoke Control Systems." *NBSIR 82-2512*. National Bureau of Standards, Washington, DC.

Mitler, H. E. 1985. "The Harvard Fire Model." *Fire Safety Journal*. Vol 9. pp 7-16.

Rockett, J. A., et al. 1983. "Using the Harvard Fire Simulation." *Fire Science and Technology*. Vol 3, No 1. pp 57-62.

Smith, E. E., and Satija, S. 1981. "Release Rate Model for Developing Fires." *20th Joint ASME/AIChE National Heat Transfer Conference*, Milwaukee, WI.

Swartz, J. A., et al. 1983. *Final Technical Report on Building Fire Simulation Model*, Vol. 1 and 2. National Fire Protection

Association, Quincy, MA.
Walton, W. D. 1985. "ASET-B A Room Fire Program for Personal Computers." *NBSIR 85-3144*. National Bureau of Standards, Washington, DC.
Wiss, Janney, Elstner and Associates, Inc. 1982. "Effect of Fire Exposure on Steel Frame Buildings." *Final Report WJE No. 78124*. American Iron and Steel Institute, Washington, DC.

Additional Readings

Abrahamsson, L., et al, *HSLAB –An Interactive Program for One Dimensional Heat Flow Problems*, FOA C 20288-D6(A3), National Defense Research Institute, Sweden, Jan. 1979.
Alvord, D. M., "Status Report on the Escape and Rescue Model and the Fire Emergency Evacuation Simulation for Multifamily Buildings," NBS-GCR-85-496, National Bureau of Standards, Washington, DC, 1985.
Emmons, H. D., "The Prediction of Fires in Buildings," *Seventeenth Symposium on Combustion*, The Combustion Institute, Pittsburgh, PA, 1979, p. 1101.
Evans, D. D., "Calculating Sprinkler Actuation Time in Compartments." *Fire Safety Journal*, Vol. 9, 1985, pp 147-155.
Friedman, R., "Quantification of Threat From a Rapidly Growing Fire in Terms of Relative Material Properties," *Fire and Materials*, Vol. 2, No. 1, 1978.
Greuer, R. E., "Modeling the Movement of Smoke and the Effect of Ventilation Systems in Mine Shaft Fires," *Fire Safety Journal*, Vol. 9, Nos. 1 and 2, May/June 1985, p 81.
Hagglund, B., "A Room Fire Simulation Model," *FOA C 20501-D6*, National Defense Research Institute, Sweden, June 1983.
Hagglund, B., "Simulation the Smoke Filling in Single Enclosures," *FOA C 20513-D6*, National Defense Research Research Institute, Sweden, Oct. 1983.
Haksever, I. A., *A Computer Program HPO 10*, Technical University of Braunschweig, Germany.
Jones, W. W., "A Review of Compartment Fire Models," *NBSIR 83-2684*, National Bureau of Standards, Washington, DC, 1983.
Jones, W. W., and Quintiere, J. G., "Prediction of Corridor Smoke Filling by Zone Models." *Combustion Science and Technology*, Vol. 11, p. 111, 1983.
Klote, J. K., "Computer Modeling for Smoke Control Design." *Fire Safety Journal*, Vol. 9, pp. 181-188, 1985.
Klote, J. H., "A Computer Program for Analysis of Pressurized Stairwells and Pressurized Elevator Shafts, *NBSIR 80-2157*, National Bureau of Standards," Washington, DC, 1981.
Kisko, T. M., and Francis, R. L., "Network Models of Building Evacuation: Development of Software System," *NBS-GCR-85-489*, National Bureau of Standards, Washington, DC, 1985.
Kisko, T. M., and Francis, R. L., "Network Models of Building Evacuation: Development of Software System," *NBS-GCR-84-457*, National Bureau of Standards, Washington, DC, 1984.
Levine, R. S., "Mathematical Modeling of Fires," *NBSIR 80-2107*, National Bureau of Standards, Washington, DC, 1980.
MacArthur, C. D., and Reeves, J. S., "Dayton Aircraft Cabin Fire Model," *FAA-RD-76-120*, University of Dayton, June 1976.
Parikh, J. S., and Beyreis, J. R., "Survey of the State of the Art of Mathematical Fire Modeling," *SFPE Bulletin*, Mar. 1985.
Quintiere, J. G., "A Perspective on Compartment Fire Growth," *Combustion Science and Technology*, Vol. 39, 1984, pp. 11-54.
Rasbash, D. J., *Fire Safety Journal*, Vol. 9, Nos. 1 and 2, May, June 1985.
Rothermel, R. C., "How to Predict the Spread and Intensity of Forest and Range Fires," *INT 143*, U.S. Department of Agriculture, Forest Service, June 1983.
Sander, D. M., "FORTRAN IV Program to Calculate Air Infiltration in Buildings," *DBR Computer Program #37*, National Research Council, Canada, May 1974.
Sander, D. M., and Tamura, G. T., "FORTRAN IV Program to Simulate Air Movement in Multistory Buildings," *DBR Computer Program #35*, National Research Council, Canada, Mar. 1973.
Siu, N. O., "COMPBRN—A Computer Code for Modeling Compartment Fires," *UCLA-ENG-8257*, University of California, Los Angeles, CA, Aug. 1982.
Tanaka, T., "A Model of Multiroom Fire Spread," *NBSIR 83-2718*, National Bureau of Standards, Washington, DC, 1983.
Underwriters Laboratories, Inc., *Survey of the State of the Art of Mathematical Fire Modeling*, File NC554, Project 82NK1618, 1983.
Yoshida, H., et al, "A FORTRAN IV Program to Calculate Smoke Concentrations in a Multistory Building," *DBR Computer Program #45*, National Research Council, Canada, June 1979.
Zukoski, E. E., and Kubota, T., "Two-Layer Modeling of Smoke Movement in Building Fires," *Fire and Materials*, Vol. 4, No. 1, Mar. 1980, p. 17.

ROOM FIRE TEMPERATURE COMPUTATIONS

Dr. Vytenis Babrauskas

For many fire engineering and design purposes it is desirable to be able to compute the temperature of a fire in a room. A major application, for instance, is to predict whether flashover can occur in a given room with a given fuel configuration. This can be viewed as a temperature prediction problem, since flashover is normally considered to occur when the temperature reaches 600°C. (1,112°F) In any real fire, the temperatures not only vary with time, but differ at various locations within the room. Nonetheless, significant advances in the mathematical modeling of room fires became possible only when it was realized that it is a good approximation to consider that there are only two average temperatures: (1) an upper gas temperature, T_u, and (2) a lower gas temperature, T_l. At the very start of a fire, the entire room is assumed to be at a uniform ambient temperature. As the fire begins, a hot gas layer, of temperature T_u, extends down from the ceiling. When and if flashover finally occurs, the hot gas layer is down to near the floor level. For most applications, it is sufficient to assume that the lower gas temperature remains at its initial ambient value. The fire temperature history can then be characterized solely by determining T_u as a function of time.

A simpler analysis is possible when it is realized that for many practical applications only the temperature history after flashover is of consequence. Such applications as the determination of fire endurance for building structural members are invariably concerned solely with post flashover performance. The preflashover fire may be threatening to humans because of combustion product concentrations, and may threaten fire spread to other areas. A beam which would collapse prior to flashover or a wall which would burn through, however, would represent conditions of such minimal fire endurance that they would be considered to have no endurance. To emphasize this point, the standard fire endurance test's starting point at zero time on the standard time-temperature curve does not correspond to the start of the fire, but to the time of flashover occurrence.

Dr. Babrauskas is Head, Flammability and Toxicity Measurement, U.S. National Bureau of Standards, Center for Fire Research, Gaithersburg, MD.

CALCULATION OF FIRE TEMPERATURES

For applications where a detailed temperature history prediction is required over the entire course of the fire and to the highest accuracy possible, computer programs are available (Jones 1984 and 1983). These are not necessarily simple to use, nor is the amount of their detail necessary for most practical design work. For some applications, it may be required to accurately calculate a fire temperature. But in the post flashover regime simpler programs can be used (Babrauskas 1979). Even such programs may represent an unwarranted degree of calculational complexity. One main reason for this can be the fact that heat release rates (burning rates) for the combustibles being used are poorly known. Each calculational procedure, no matter how simple, requires knowledge of the burning rates of the combustibles in the room. These rates cannot, as of yet, be theoretically determined for practical items, and must be based on laboratory measurements. Good laboratory measurements are lacking for many categories of combustibles. Thus the ability to perform detailed calculations cannot be usefully exploited in such circumstances.

It bears emphasis to point out that the standard time-temperature curve is merely a testing strategem and bears no relationship to actual fire temperature. Since fire endurance testing cannot be conducted for an infinite variety of potential fire histories, a standard furnace test curve was necessary and was developed in 1918.

RATES OF HEAT RELEASE

Since the rate of heat release is a crucial variable for all fire temperature determinations, it is appropriate to summarize the status of available data.

Liquid (or Thermoplastic) Pools

The burning rates for pool fires given in Section 21, Chapter 6, "Pool Fires," are computed for free-burn conditions, and are applicable only to the very initial stages of a room fire. In later stages, for a pool (unlike most other combustibles), the effect of the ongoing room fire is substantial in modifying the pool burning rate. Because of this strong coupling, isolated rate data cannot be quoted. Instead, an integrated room fire pool burning procedure is needed.

Cribs and Slabs of Wood or Plastics

For *thin slabs* (thickness $\leq$ 0.05 m),

$$\dot{q} = (\Delta h_c)\, v_p \rho_f A_f \text{ (kW)}$$

$v_p = 2.2 \times 10^{-6} D^{-0.6}$ (m/sec)	wood
$= 1.4 \times 10^{-6} D^{-0.6}$	PMMA
$= 3.1 \times 10^{-6} D^{-0.6}$	thermosetting polyester
$= 3.8 \times 10^{-6} D^{-0.6}$	rigid polyurethane foam

where

Δh_c = heat of combustion (kJ/kg), (See Table 5-11A.)
v_p = regression rate (m/sec)
ρ_f = density (kg/m^3)
A_f = surface area (m^2)
D = thickness (m)

For *thick slabs*, (thickness > 0.005 m),

$$v_p = 8.5 \text{ to } 10.0 \times 10^{-6} \text{ (m/sec) wood,}$$

with the calculation for the heat release rate, as given above.

For wood cribs, which are regular arrays of square sticks, three regimes of burning are possible. For sticks far apart, fuel surface burning results:

$$\dot{q} = (\Delta h_c)\, v_p \rho_f A_f\, 4 \left(1. - \frac{2 v_p t}{D}\right)$$

with v_p as given for thin slabs above. Note that this expression is now time-dependent, with t = time (sec). A_f refers to the total surface area for one side only, of one stick, multiplied by the total number of sticks.

With sticks close together to give wood crib porosity control

$$\dot{q} = (\Delta h_c)\, \rho_f A_f\, 4.4 \times 10^{-4}\, (S/H)$$

where S = clear spacing between sticks (m) and H = crib height (m).

With room ventilation limit for wood cribs

$$\dot{q} = (\Delta h_c)\, 0.12\, A \sqrt{h}$$

where A = area (m^2) of window or open door providing ventilation during fire, and h = height (m) of that opening.

The actual crib burning rate is the *least* of the three previous computations.* Sufficient data for thick plastic slabs and plastic cribs are not available for design purposes.

Furniture

A discussion of the burning rates of upholstered furniture is given in Section 12, Chapter 15, and a small amount of additional data on other types of furniture is also available (Lawson 1984).

APPROXIMATE TEMPERATURE CALCULATION

Pre-flashover

For estimating the upper gas temperature in the early stages of the fire, prior to flashover, the following form, based on experimental correlations, has been suggested (Quintiere 1983).

$$T_u = T_o \left[1 + \frac{1.6\, \dot{q}^{2/3}}{(\sqrt{g}\, C_p \rho_o)^{1/3}\, T_o^{2/3}\, (A\sqrt{h})^{1/3} (h_k A_w)^{1/3}}\right]$$

where:

g = the gravitational constant, 9.8 m/sec^2,
C_p = specific heat of gas, 1.0 (kJ/kg)/K
ρ_o = ambient air density, 1.18 kg/m^3
T_o = ambient temperature, 290 K
A_w = wall area, m^2

Inserting the values of the constants given,

$$T_u = T_o \left[1 + \frac{0.0236\, \dot{q}^{2/3}}{(h_k A_w A \sqrt{h})^{1/3}}\right]$$

and the effective heat transfer coefficient, h_k [(kW/m)/K], is taken as

$$h_k = \sqrt{\frac{0.18}{t}}$$

valid for gypsum wallboard walls during the early stages of the fire (Quintiere 1983).

Post Flashover

The following calculation method is based on previous research (Babrauskas 1981). The upper gas temperature, T_u, is expressed according to a series of factors, each one accounting for a different physical phenomenon.

$$T_u = T_o + (T^* - T_o) \cdot \theta_1 \cdot \theta_2 \cdot \theta_3 \cdot \theta_4 \cdot \theta_5$$

T^*, an empirical constant, = 1725 K, and the factors θ are listed below.

Burning Rate Stoichiometry θ_1 : The dimensionless stoichiometric coefficient ϕ is defined as

$$\phi = \frac{\dot{m}_f}{\dot{m}_{f,st}}$$

where $\dot{m}_f$ is the fuel mass pyrolysis rate (kg/sec) and at stoichiometric (i.e., no excess fuel and no excess oxygen) conditions

$$\dot{m}_{f,st} = \frac{0.5\, A \sqrt{h}}{r}$$

*For more details, see Babrauskas, V. 1981. A Closed-Form Approximation for Post-Flashover Compartment Fire Temperatures. *Fire Safety Journal*. Vol 4. pp 63-73.

where the ratio r is such that 1 kg fuel + r kg air → (1 + r) kg products. The value of r is readily computable for fuels containing carbon, hydrogen and oxygen from the chemical formula of the fuel, taking the products to be CO_2, H_2O, and N_2.

At stoichiometry $\phi = 1$, and it is greater than 1 for fuel rich burning and less than 1 for fuel lean conditions.

The effect of ϕ on gas temperatures was evaluated by numerical computations using the COMPF2 program (Babrauskas 1979). The efficiency factor, θ_1, accounts for deviation from stoichiometry and is shown in Figure 21-5A. It is seen that the fuel lean and the fuel rich regimes

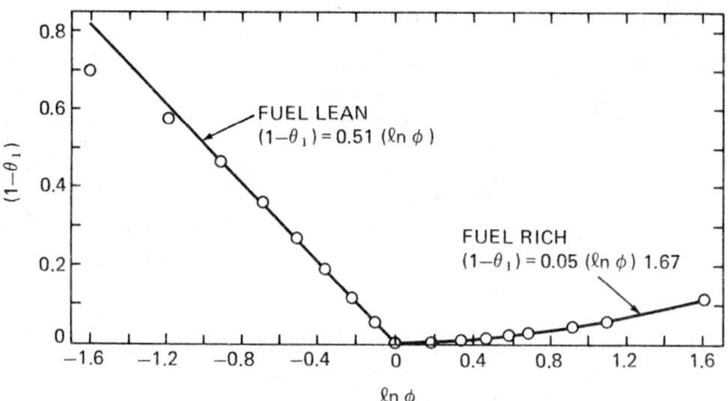

FIG. 21-5A. Effect of equivalence ratio θ.

exhibit a very different dependence. For the fuel lean regime the results can be approximated by

$$\theta_1 = 1.0 + 0.51 \ln \phi \qquad [\phi < 1]$$

Similarly, in the fuel rich regime a suitable approximation is

$$\theta_1 = 1.0 - 0.05 (\ln \phi)^{5/3} \qquad [\phi > 1]$$

If heat release rates, $\dot{q}$, rather than mass loss rates, $\dot{m}$, are used, then

$$\phi = \frac{\dot{q}}{\dot{q}_{st}}$$

and since the stoichiometric heat release rate is

$$\dot{q}_{st} = 1500\, A\sqrt{h} \qquad \text{(kW)}$$

$$\phi = \frac{\dot{q}}{1500\, A\sqrt{h}}$$

The value of $\dot{q}$ can be determined from the section above.

A separate procedure is necessary for *pool* fires, due to the strong radiative coupling. Here,

$$\theta_1 = 1.0 - 0.092\,(-\ln \eta)^{1.25}$$

where

$$\eta = \frac{A\sqrt{h}}{A_f}\,\frac{0.5\,\Delta h_p}{r\sigma\,(T_u^4 - T_b^4)}$$

and:

A_f = pool area (m^2)
Δh_p = heat of vaporization of liquid (kJ/kg), (Table 21-6A)
σ = Stefan-Boltzmann constant (5.67×10^{-11} [kW/m²]/K⁴)
T_b = liquid boiling point (K)

This expression unfortunately requires an estimate for T_u to be made, so for the pool case a certain amount of iteration is necessary. The relationship above is plotted in

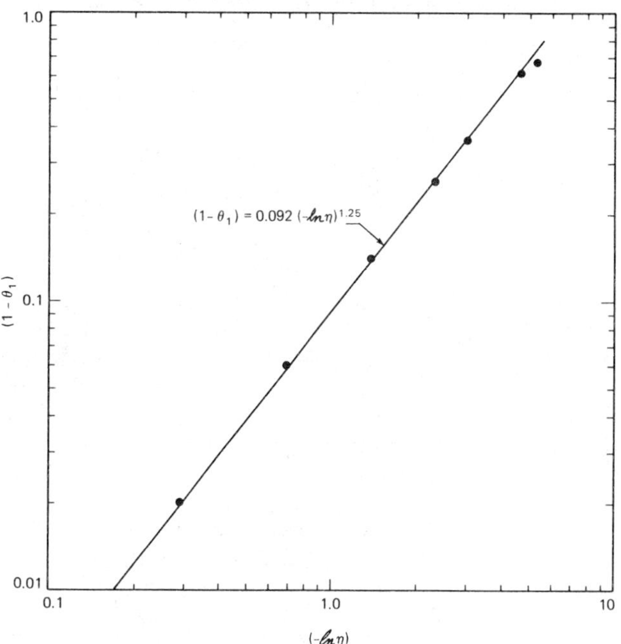

FIG. 21-5B. Effect of pool parameter θ.

Figure 21-5B.

Wall Steady State Losses θ_2: The next efficiency factor, θ_2, accounts for variable groups of importance involving the wall surface (which is defined to include the ceiling) properties: area A_w (m^2), thickness L (m), density ρ (kg/m³), thermal conductivity k (kW/m)/K, and heat capacity C_p (kJ/kg)/K. This factor is given as

$$\theta_2 = 1.0 - 0.94 \exp\left(-54\left[\frac{A\sqrt{h}}{A_w}\right]^{2/3}\left[\frac{L}{k}\right]^{1/3}\right)$$

and as shown in Figure 21-5C.

Wall Transient Losses θ_3: For the transient case, the above relationship predicts the asymptotic temperature value. An additional time-dependent factor, however, is needed. (See Fig. 21-5D).

$$\theta_3 = 1.0 - 0.92 \exp\left(-150\left[\frac{A\sqrt{h}}{A_w}\right]^{0.6}\left[\frac{t}{k\rho C_p}\right]^{0.4}\right)$$

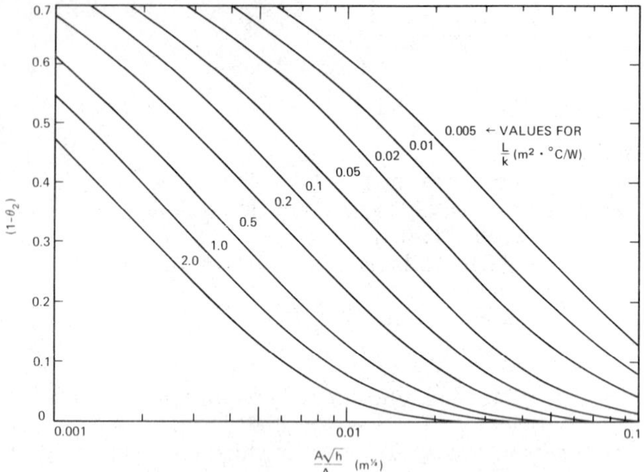

FIG. 21-5C. Effect of wall steady state losses.

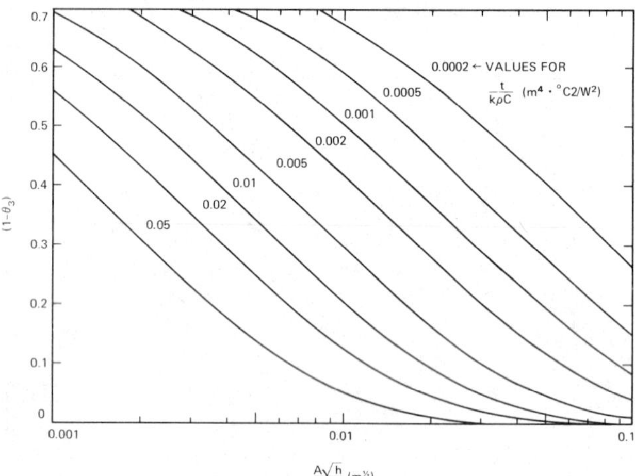

FIG. 21-5D. Effect of wall transient losses.

If only steady state temperatures need to be evaluated, then $\theta_3 \equiv 1.0$.

Wall effects for t just slightly greater than zero are not well modeled with the above relationships for $\theta_2 \times \theta_3$; however, this is not a serious limitation since the method is only designed for post flashover fires.

For transient fires, the possibility of two separate effects must be considered. First, the wall loss effect, represented by the equation for θ_3 in all fires, exhibits a nonsteady character. Second, the fuel release rate may not be constant. Since in the calculational procedure we do not store the previous results it is appropriate to restrict consideration to fires where $\dot{m}_f$ does not change drastically over the time scale established by θ_3. This "natural" time scale can be determined as the time when the response has risen to 63 percent of its ultimate value, i.e., at $\theta_3 = 0.63$, and is

$$t = 2.92 \times 10^{-6}\,(k\rho C_p)\left[\frac{A_w}{A\sqrt{h}}\right]^{1.5}$$

Opening Height Effect θ_4: The normalization of burning rate and wall loss quantities with the ventilation factor $A\sqrt{h}$ does not completely determine the total heat balance. An opening of a given $A\sqrt{h}$ can be made tall and narrow or short and squat. For the shorter opening, the area will have to be larger. Radiation losses are proportional to opening area and will, therefore, be higher for the shorter opening. By slight simplification a representation for θ_4 can be made as

$$\theta_4 = 1.0 - 0.205\,h^{-0.3}$$

as shown in Figure 21-5E.

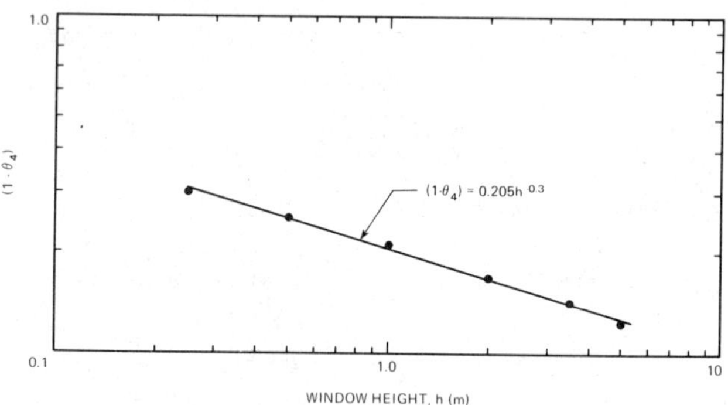

FIG. 21-5E. Effect of window height.

Combustion Efficiency θ_5: The fire compartment is viewed as a well, but not perfectly, stirred reactor. Thus, a certain "unmixedness" is present. A maximum combustion efficiency, b_p, can be used to characterize the unmixedness. Since the model assumes infinitely fast kinetics, any limitations can also be included here. Data have not been available to characterize b_p in real fires, but agreement with measured fires can generally be obtained with b_p values in the range 0.5-0.9. The effect of b_p variation can be described by

$$\theta_5 = 1.0 + 0.5 \ln b_p$$

as shown in Fig. 21-5F.

Examples of Use

Case 1: The steady state temperature for a thermoplastic pool fire burning under stoichiometric conditions will be determined . Since burning is stoichiometric, $\eta \equiv 1$, making $\theta_1 \equiv \theta$ 1. Wall area $A_w = 200$ m^2, window area $A = 2.0$ m^2, window height $h = 1.0$ m, wall thermal conductivity $k = 0.17$ (W/m)/K (similar to that of gypsum wallboard), and thickness $L = 0.04$ m. The steady state wall loss factor is then $\theta_2 = 0.800$. Since a steady state solution is sought, $\theta_3 \equiv 1.0$. For the given h, $\theta_4 = 0.795$. Taking $b_p = 0.9$, gives $\theta_5 = 0.947$. Assuming $T_{\infty} = 25$°C, the estimate becomes $T_f = 25 + 1700\,(1.0) \times (0.800)(1.0)(0.795)(0.947) = 1049$°C.

Considering a polyethylene fuel, $r = 14.7$ and $\Delta h_p = 2.4 \times 10^6$ J/kg, gives $A_f \simeq 1.04$ m^2.

Case 2: Transient pool fire, $\eta = 0.1$, $t = 300$ seconds, wall density $\rho = 790$ kg/m^3, $C_p = 840$ (J/kg); otherwise same as Case 1. Approximate solution: 507°C.

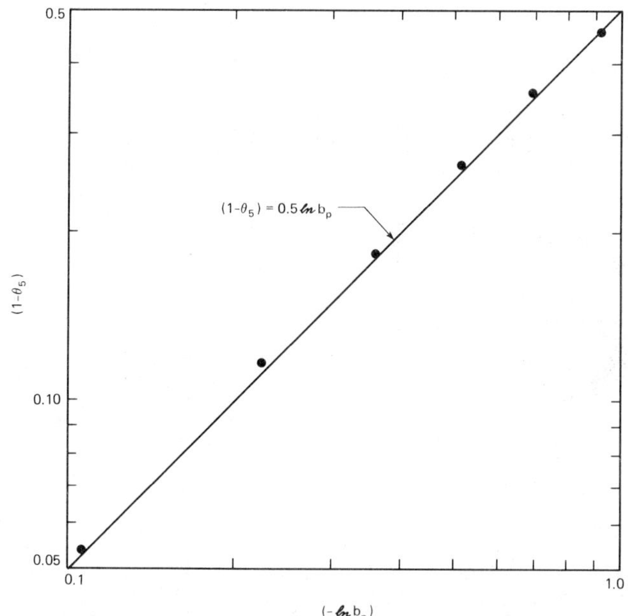

FIG. 21-5F. Effect of b_p, the maximum combustion efficiency.

Case 3: Steady fire, specified mixed $\dot{m}_f = 0.12$ kg/sec; otherwise as in Case 1. The approximate solution is 1028°C.

Flashover Determination

For many applications the determination whether flashover is or is not expected to occur is the single most important fire computation. On the simplest level, it has been shown that at the time of flashover various heat losses from the fire amount to approximately 50 percent of fire output (Babrauskas 1980), thus the estimate is that the minimum fire heat release rate required to cause flashover,

$$\dot{q}_{fo} = 0.5\ \dot{q}_{stoich}$$

where:

$$\dot{q}_{stoich} = 1500\ A\ \sqrt{h}$$

giving:

$$\dot{q}_{fo} = 750\ A\ \sqrt{h}\ \text{(kW)}$$

For an improved estimate, several relationships have been proposed which account for the character of the heat losses in more detail. These different methods have been evaluated and compared (Babrauskas 1984). A method seen to be of best utility has been developed (Thomas 1981). It is given as:

$$\dot{q}_{fo} = 378\ A\ \sqrt{h} + 7.8\ A_w$$

This is useful for all cases where walls are made of materials generally similar in density to gypsum wallboard.

Methods are not currently available for estimating flashover conditions when numerous fuel packages exist, although this is the subject of ongoing research.

Bibliography

References Cited

Babrauskas, V. 1984. "Upholstered Furniture Room Fires—Measurements, Comparison with Furniture Calorimeter Data, and Flashover Predictions." *Journal of Fire Sciences*. Vol 2. pp 5-19.

Babrauskas, V. 1981. "A Closed-Form Approximation for Post-Flashover Compartment Fire Temperatures." *Fire Safety Journal*. Vol 4. pp 63-73.

Babrauskas, V. 1980. "Estimating Room Flashover Potential." *Fire Technology*. Vol 16. pp 94-103, 112.

Babrauskas, V. 1979. "COMPF2—A Program for Calculating Post-Flashover Fire Temperatures." *TN 991*. National Bureau of Standards, Washington, DC.

Jones, W. W. 1984. "A Model for the Transport of Fire, Smoke, and Toxic Gases (FAST)." *NBSIR 84-2934*. National Bureau of Standards, Washington, DC.

Jones, W. W. 1983. "A Review of Compartment Fire Models." *NBSIR 83-2684*. National Bureau of Standards, Washington, DC.

Lawson, J.R., et al. 1984. "Fire Performance of Furnishings as Measured in the NBS Furniture Colorimeter, Part I." NBSIR 84-2787. National Bureau of Standards. Gaithersburg, MD.

Quintiere, J. G. 1983. "A Simple Correlation for Predicting Temperature in a Room Fire." *NBSIR 83-2712*. National Bureau of Standards, Washington, DC.

Thomas, P. H. 1981. "Testing Products and Materials for Post-Flashover Compartment Fire Temperatures." *Fire and Materials*. Vol 5. pp 103-111.

POOL FIRES: BURNING RATES AND HEAT FLUXES

Dr. Vytenis Babrauskas

INTRODUCTION

Pool burning is probably the simplest form of combustion applicable to a wide range of industrial fire protection concerns. Typically, this is conceived of as a fire in an open topped, circular flammable liquid tank or as a bounded spill of combustible liquid. More generally, both liquefied gases and melting plastics materials, horizontally placed, conform to the same pattern. Somewhat related, but computationally different, are problems of pools burning in enclosed spaces. These solutions (Babrauskas 1979) must consider the limit where the enclosure effects dominate the fire. In this chapter, only "free" pools, will be considered, not inside an enclosure nor in the vicinity of another fire. The burning of pool fires presents a rich field for inquiry into flame chemistry, radiation, fluid mechanics and other aspects. To a fire protection engineer, however, two questions are primary: How fast is the fire burning? and what is its temperature (or heat flux) distribution?

The larger fires are of greatest practical concern. A fire of 100 kW can be typically produced by a fuel pool ~ 0.2 m in diameter. As will be shown, such a restriction to "large" pools simplifies the data analysis considerably. Thus, except where otherwise specified, all the discussion will pertain to pools with $D \gtrsim 0.2$ m.

BURNING RATES

Theory

How to systematically analyze pool burning data according to basic heat transfer principles was probably first suggested in 1958 (Hottel 1958). By conservation of energy for the liquid we have:

$$\dot{m}''\Delta h_g = \dot{q}_r'' + \dot{q}_c'' - \dot{q}_{rr}'' - \dot{q}_{misc}'' \qquad (1)$$

Dr. Babrauskas is Head, Flammability and Toxicity Measurement, U.S. National National Bureau of Standards, Center for Fire Research, Gaithersburg, MD.

where $\dot{m}$ is the mass loss rate per unit area (assumed identical to the burning rate); Δh_g is the total heat of gasification, i.e., the heat to bring a liquid fuel at 298 K to its boiling temperature and then to change it to vapor; $\dot{q}_r''$ is the radiant flux absorbed by the pool; $\dot{q}_c''$ is the heat received convectively; $\dot{q}_{rr}''$ is the re-radiant heat loss, due to the surface of the pool being at an elevated temperature; and into $\dot{q}_{misc}''$ are lumped wall conduction losses and nonsteady terms. (Units for all are given in the "Nomenclature" section at the end of this chapter.) Quantitative expressions for $\dot{q}_{misc}''$ are usually not available, while $\dot{q}_{rr}''$ is usually small. For simple analysis both are customarily dropped. Hottel's analysis of the data showed that two basic regimes are possible: (1) radiatively dominated burning for large D, and (2) convectively dominated burning for small d. Furthermore, in the convective regime the flow can be either laminar or turbulent (always being turbulent for radiatively driven pools), while the flames can be optically thin or thick in the radiative regime. These distinctions can, in the simplest analysis, be made solely on the basis of pool diameter. Thus,

D(m)	Burning Mode
<0.05	convective, laminar
0.05 to 0.2	convective, turbulent
0.2 to 1.0	radiative, optically thin
>1.0	radiative, optically thick

In the convective limit—small pools—then we would expect that

$$\dot{m}'' \simeq \dot{q}_c''/\Delta h_g \qquad (2)$$

Behavior in the convective laminar mode has not been fully correlated although there are functional relations of the form

$$\dot{m}'' = aD^{-n} + b \qquad (3)$$

where $\frac{1}{2} \leq n \leq \frac{3}{2}$. For the convective turbulent mode, the $\dot{m}''$ values are independent of D and at their lowest.

In the radiative mode both the optically thick and thin regimes might be modeled if we let

$$\dot{m}'' = \frac{\sigma T_f^4 (1 - e^{-k\beta D})}{\Delta h_g} \qquad (4)$$

Here σ is the Stefan-Boltzmann constant and $\Delta\ h_g$ is easily determinable, at least for pure liquids. T_f is an effective equivalent gray gas flame temperature. It should be related to the measured temperatures in the hottest zone, but a predictive relationship is not available. The effective flame volume emissivity is represented by $(l - e^{-k\beta\ D})$, where k is the absorption-extinction coefficient of the flame, D is pool diameter and β is a "mean-beam-length corrector." This emissivity expression is for radiation to the base of the fire, which is what is needed to determine fuel vaporization.

For most fuels, reliable measurements exist only for $\dot{m}''$ as a function of D, and not for T_f, k, or β separately. The data can be presented in predictive form as

$$\dot{m}'' = \dot{m}''_\infty (1 - e^{-k\beta D}) \qquad (5)$$

This form requires determining two empirical factors: $\dot{m}''\infty$ and $(k\beta)$—not separated into k and β. For a few fuels, independent measurements of T_f, k, and β are reported. In those cases one could examine the quantity

$$\frac{\dot{m}''_\infty \Delta h_g}{\sigma T_f^4}$$

which should go to 1.0. Instead, values of 0.05 to 0.25 can be computed, based on data in Table 21-6A. This computation illustrates several pitfalls: (1) the assumption of grey gas radiation, while fruitful as a functional form for correlation, is too simplified when computed from fundamental constants; (2) the flame volume should not, in fact, be represented as being at a mixed mean temperature. The volume right above the pool surface contains mostly low temperature pyrolysate gases.

If $\dot{m}''\infty$ has to be obtained experimentally, instead of from theory, how about β? Can values of k measured through flames in a laboratory fire be used with a fixed β to produce $k\beta$? Polymethylmethacrylate is a material for which adequate values of $k\ \beta$ and of k exist. These imply $\beta = 2.6$. For other fuels, the tabulated data range over $0.7 \le \beta \le 3.8$. This probably can be attributed to widely differing measurement techniques for k. Nonetheless, a common β does not emerge.

Analysis of Data: From experimental data points, values of $k\beta$ and $\dot{m}''$ were determined by using a numerical algorithm for nonlinear curve fitting. In each case, Equation 5 was used for fitting, with the exception of the alcohols. For alcohols, a functional form of

TABLE 21-6A. Data for Large Pool Burning Rate Estimates

Material	Density (kg/m^3)	Δh_g (kJ/kg)	Δh_c (MJ/kg)	$\dot{m}''_\infty$ ($kg/m^2 - s$)	$k\beta$ (m^{-1})	k (m^{-1})	T_f (K)	$\chi_r^{(c)}$ (—)
Cryogenics								
Liquid H_2	70	442	120.0	0.017 (±0.001)	6.1 (±0.4)	—	1600	0.25
LNG (mostly CH_4)	415	619	50.0	0.078 (±0.018)	1.1 (±0.8)	0.5	1500	0.16–0.23
LPG (mostly C_3H_8)	585	426	46.0	0.099 (±0.009)	1.4 (±0.5)	0.4	—	0.26
Alcohols								
Methanol (CH_3OH)	796	1195	20.0	0.017 (±0.001)	[a]	—	1500	0.17–0.20
Ethanol (C_2H_5OH)	794	891	26.8	0.015 (±0.001)	[a]	0.4	1490	0.20
Simple Organic Fuels								
Butane (C_4H_{10})	573	362	45.7	0.078 (±0.003)	2.7 (±0.3)	—	—	0.27–0.30
Benzene (C_6H_6)	874	484	40.1	0.085 (±0.002)	2.7 (±0.3)	4.0	1460	0.14–0.38
Hexane (C_6H_{14})	650	433	44.7	0.074 (±0.005)	1.9 (±0.4)	—	1300	0.20–0.40
Heptane (C_7H_{16})	675	448	44.6	0.101 (±0.009)	1.1 (±0.3)	—	—	
Xylenes (C_8H_{10})	870	543	40.8	0.090 (±0.007)	1.4 (±0.3)	—	—	
Acetone (C_3H_6O)	791	668	25.8	0.041 (±0.003)	1.9 (±0.3)	0.8	—	
Dioxane ($C_4H_8O_2$)	1035	552	26.2	0.018[b]	5.4[b]	—	—	
Diethyl Ether ($C_4H_{10}O$)	714	382	34.2	0.085 (±0.018)	0.7 (±0.3)	—	—	
Petroleum Products								
Benzine	740	—	44.7	0.048 (±0.002)	3.6 (±0.4)	—	—	
Gasoline	740	330	43.7	0.055 (±0.002)	2.1 (±0.3)	2.0	1450	0.18
Kerosene	820	670	43.2	0.039 (±0.003)	3.5 (±0.8)	2.6	1480	
JP-4	760	—	43.5	0.051 (±0.002)	3.6 (±0.1)	—	1250	0.35
JP-5	810	700	43.0	0.054 (±0.002)	1.6 (±0.3)	0.5	1250	
Transformer oil, hydrocarbon	760	—	46.4	0.039[b]	0.7[b]	—	1500	
Fuel oil, heavy	940–1000	—	39.7	0.035 (±0.003)	1.7 (±0.6)	—	—	
Crude oil	830–880	—	42.5–42.7	0.022–0.045	2.8 (±0.4)	—	—	0.18
Solids								
Polymethylmethacrylate $(C_5H_8O_2)_n$	1184	1611	24.9	0.020 (±0.002)	3.3 (±0.8)	1.3	1260	0.40
Polyoxymethylene $(CH_2O)_n$	1425	2430	15.7			—	1200	0.15
Polypropylene $(C_3H_6)_n$	905	2030	43.2			1.8	1200	0.40
Polystyrene $(C_8H_8)_n$	1050	1720	39.7			5.3	1200	0.44

(a)—Value independent of diameter in turbulent regime.
(b)—Only two data points available.
(c)—For diameters ca. 1 m. Decreases for small and for very large diameters.

$$\dot{m}'' = \dot{m}''_{\infty}, \qquad D \geq 0.2 \text{ m}$$

is appropriate. The results are listed in Table 21-6A, along with values of some relevant thermochemical properties, taken largely from Section 5, Chapter 11, "Tables." To illustrate, the experimental data points and the curve fit of three fuels are shown in Figures 21-6A, 21-6B, and 21-6C.

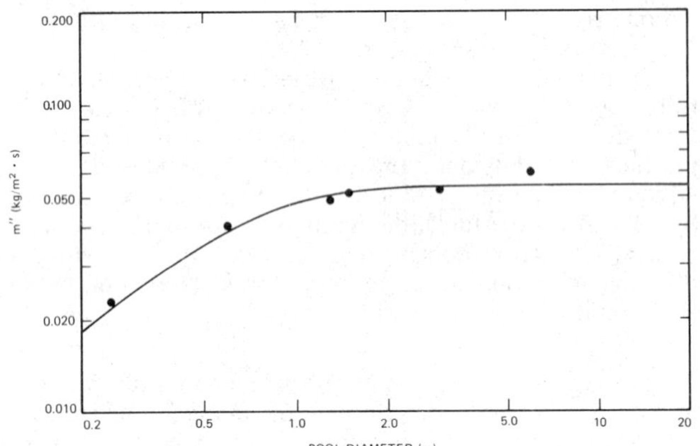

FIG. 21-6A. Pool burning rates for gasoline.

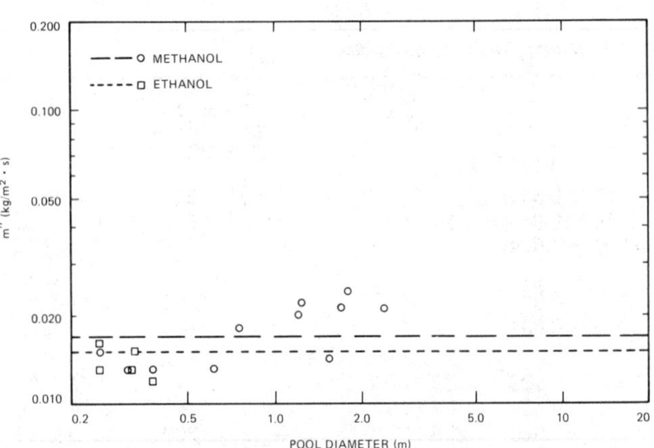

FIG. 21-6B. Pool burning rates for LNG, shown as an example of a fuel where experimental data show large variations.

Figure 21-6A shows the results for gasoline, a typical fuel. Figure 21-6C gives results for LNG, chosen to illustrate the larger degree of scatter associated with cryogenic fuel measurements. Finally, Figure 21-6C illustrates the behavior of alcohol fuels. The fit in these curves, along with the standard deviations indicated in Table 21-6A, can be used to gauge the expected uncertainty of predictions.

Use of Tabulated Data: The data in Table 21-6A can be used to directly estimate the burning rates of pools with $D \gtrsim 0.2$ m. The values pertain to steady state burning in a wind free environment, and in a vessel without excessive lip height (freeboard). When these assumptions are not met, weaker predictions are to be expected. Some of these complications are discussed below.

Boilover: A few fuels do not show steady burning properties. Instead, they start to boil rapidly at a certain temperature. This results in a significant expansion and may cause overflow of the vessel, with attendant flame

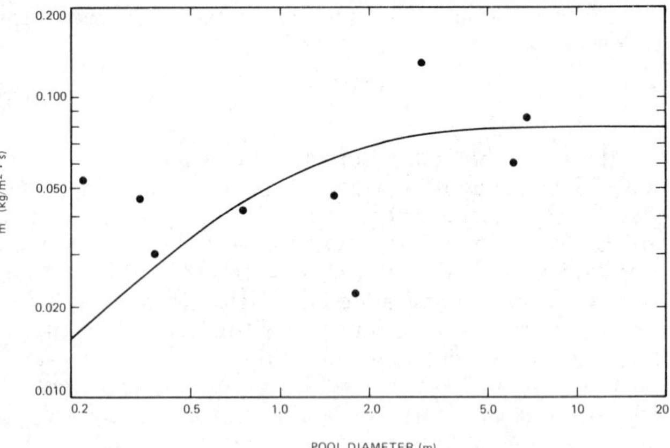

FIG. 21-6C. Pool burning rates for alcohols.

spread hazards. This problem has been associated primarily with certain crude oils and with petroleum products with a significant amount of moisture.

Transient Effects: Systematic burning rate data are available only for steady state burning. A pool fire, however, does not reach a steady state immediately after ignition for several reasons: (1) the heat conduction losses into the liquid are still changing, (2) edge heating effects may still be present, (3) the bottom of the vessel may be progressively heated if the fuel layer is thin, and (4) lip effects on convective and radiative fluxes may progressively change if the liquid level is allowed to run down significantly in the vessel. In some short burning fires there may indeed not be a steady state, but other fires, however, do show eventually a fairly steady burning rate. Some experimenters have reported transient periods as short as one minute. On the whole, however, ten or more minutes may be expected before fully steady state burning results. A general model of the various losses which could predict these transient effects in large pool fires has not been developed.

Bonding Materials and Layer Thickness Effects: Pool fires in vessels made of different materials can be expected to show different burning rates, primarily due to conduction losses. Quite substantial effects for small pools have been shown (Blinov and Khudiakov 1961). Systematic studies are not available for large pools.

The layer thickness for a pool has an effect if it is less than that required to reach steady state burning conditions. A fuel spilled in relatively small volume on a surface is a problem often encountered (Modak 1981).

For thin, diathermanous layers, the boundary materials present under the pool can show radiative differences, depending upon their reflectivity. The practical effects of this have not been explored.

Lip Height Effects: The effects of having a nonzero freeboard height, d, are significant for all pool fires (especially tank fires) and involve numerous phenomena (Hall 1972). For large pools, there are convective, conductive, and radiative effects: (1) a lip of significant height can initiate turbulence closer to the pool edge, and thereby raise convective heat transfer; (2) it can change the temperature distribution of the vessel walls, and thereby change conduction heat terms; and (3) it tends to promote a stubbier, more emissive flame volume. A study of the combined effects of this heat redistribution on one fuel, PMMA, has shown that between $d/D = 0$ and $d/D = 0.07$, the burning rate is roughly raised by 60 percent, then slowly rises to twice the zero-lip rate at $d/D = 0.20$, and then slowly falls (Orloff 1980). Earlier data are available for liquid fuels burned at large freeboard heights (Blinov and Khudiakov 1961; Magnus 1961). These do not show burning rate increasing with lip height but, rather, monotonically decreasing, down to as low as 12 percent of the zero-lip condition. It is unfortunate that further data do not exist on the lip height effects. Orloff's findings, however, could explain up to a factor of two, inconsistencies among reported experimental data.

Effects of Wind: The effects of wind on a pool fire are complex. In the small diameter limit, the main effect could be taken simply as convective heat transfer enhancement. Here, however, we are concerned with the effects at 'large' diameters, where, there still is an effect of convective enhancement, but two other phenomena also appear. The flame temperature is raised due to improved mixing and combustion, and the radiant heat fluxes are redistributed. It would seem that the flux change would make the flame volume smaller and less well centered, and thereby lower the radiative heating (Capener and Alger 1972). Indeed such an effect for 1 and 3 m JP-5 pools has been found to show that the burning rate of a 1 m pool in a 6 m/sec wind drops to about half of its still air value.

Unfortunately, there is somewhat more documentation showing an *increased* burning rate for large diameter pools in wind. A doubling of the burning rate of a hexane pool in a 4 m/sec wind, with no further increase for greater velocities has been observed (Lois and Swithenbank 1978). Data leading to an estimation evaluation have also been given (Blinov and Khudiakov 1961).

$$\frac{\dot{m}''_{\text{windy}}}{\dot{m}''_{\text{still}}} = 1 + 0.15\,\frac{u}{D}$$

The equation appears to be the best formula available for use, with the restriction that it is inappropriate for alcohol fuels, nor for wind velocities sufficient to lead to blow off. Beyond approximately 5 m/sec some fuels can be blown out, but the exact value depends upon whether a flameholder action can be obtained.

Effects at very large diameters: Some experimental data show a slight decrease in the burning rate of very large pools ($D \gtrsim 5$ or 10 m). Not enough systematic, precise data exist to provide a numeric model here beyond assuming independence of $\dot{m}''$ on D in this regime. Qualitatively, this is presumed to be due to poorer mixing, leading to a larger cool vapor zone, lower flame temperatures, and cooler smoke (which can act to shield a fire base from its flames). In any case, this effect is not likely to be larger than approximately 20 percent.

Recommendations: Burning rates for pools with $D > 0.2$ m can be estimated on the basis of the equations

$$\dot{m}'' = \dot{m}''_{\infty}\,(1 - e^{-k\beta D})$$

and

$$\dot{q} = \Delta h_c \cdot \dot{m}'' \cdot A$$

where A is the pool area, with appropriate values for $\dot{m}''_{\infty}$, $k\beta$, Δhc taken from Table 21-6A is the pool area.

The largest causes of uncertainty are believed to stem from effects of wind and lip height. In the worst case this can introduce an uncertainty on the order of a factor of two. Additional investigations are needed to provide better estimates in these areas.

HEAT FLUXES

It is often important to compute the radiation to a target object from a pool fire. The most common desired configuration is radiation to a vertically oriented target, outside the fire plume itself, and at the same height as the base of the pool. Figure 21-6D shows the geometry consid

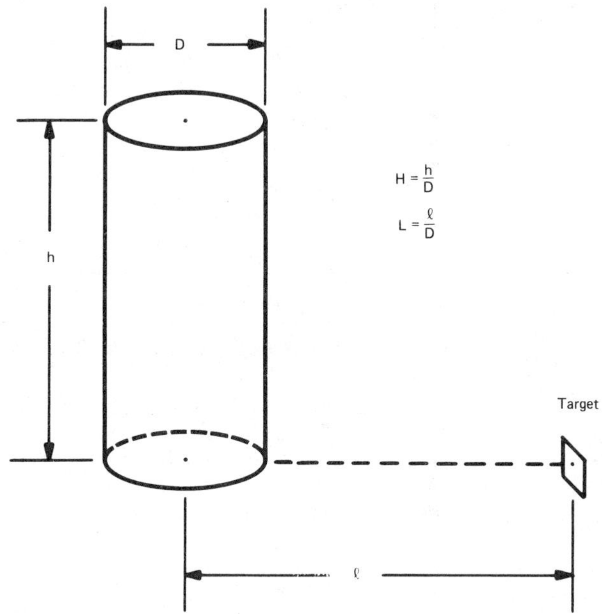

FIG. 21-6D. Cylinder representation of flame geometry for calculating heat fluxes to a target object.

ered. For $\ell/D \equiv L$ greater than about 4, it is a suitable approximation to assume a point fire radiating uniformly to a spherical environment. Then the radiative power, P (kW), is equal to

$$P = 4\pi\, l^2 \dot{q}''_i,$$

where $\dot{q}''_i$ (kW/m^2) is the radiation incident on the target. P is also equal to

$$P = X_r\,\Delta h_c\,\dot{m}''\,\pi D^2/4,$$

where χ_r is the fraction of the power appearing as radiation, giving

$$\dot{q}''_i = \frac{X_r\,\Delta h_c \dot{m}''}{16\,L^2}$$

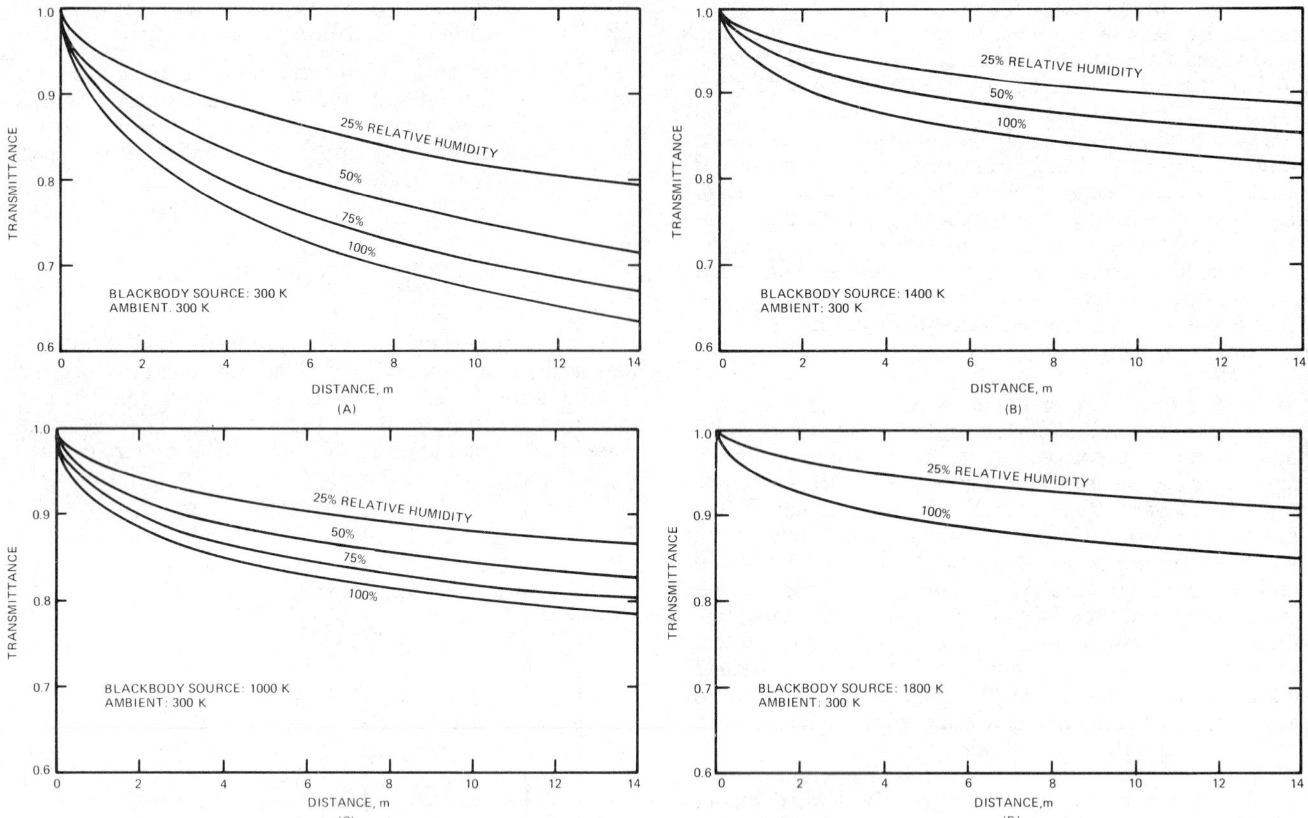

FIG. 21-6E. Curves for determining atmospheric attenuation for various source temperatures and relative humidity values.

The radiant fraction value, χ_r, is reasonably constant for a given fuel type and is listed in Table 21-6A. For very long distances, ℓ , the value of $\dot{q}''$ calculated should be multiplied by τ, the atmospheric transmittance, Figure 21-6E. It can, however, be ignored in arriving at a conservative estimate.

For radiation at $\frac{1}{2} < L < 4$, a more exact analysis is needed. A suitable expression is

$$\dot{q}_i'' = F\tau\, \epsilon_f\, \sigma T_f^4$$

where F is the view factor and ϵ_f is the flame emissivity for this geometry and is not identical to $(l - e^{-k\,\beta\,D})$ for radiation to the base. Figure 21-6F gives the values of F.

To use the view factor, the flame height, h, must be known. Data show $H \simeq 5$ for $D = 0.01$, dropping to $H \simeq 2$ for $D = 0.10$ and constant at about 1.5 for larger D (Blinov and Khudiakov 1961). While refined theories for computing flame heights are available, such calculations are not needed due to the approximations inherent in the assumption that the flame can be represented at a cylinder of uniform radiance. For practical purposes setting $H \simeq 1.5$ for $D > 0.02$ should be sufficient. For JP-4 pools

$$H \simeq 2.6\ D^{-1/3}$$

is a more refined estimate (Hagglund and Persson 1976).

The view factor expression must also hold for $L > 4$. A suitable approximation is

$$F \simeq \frac{1}{\pi}\frac{H}{L^2}$$

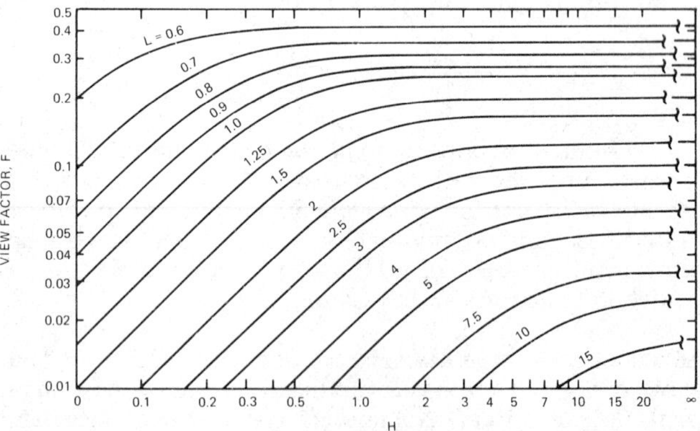

FIG. 21-6F. View factor for the geometry indicated in Figure 21-6D.

Equating the two expressions for $_i$

$$(\epsilon_f\ \sigma\ T_f^4) = \frac{\pi X_r \Delta h_c \dot{m}''}{16\ H}$$

For $D \rightarrow \infty$ this can be expressed as

$$(\epsilon_f \sigma T_f^4)_\infty = 0.131\ X\ \Delta h_c\ \dot{m}''_\infty$$

which can readily be obtained from the data in Table 21-6A. For smaller pools ϵ_f drops off but likewise does $\dot{m}''$. Even though the form of $E_f = E_f$ (D) may not be identical to that of $\dot{m}'' = \dot{m}''$ (D), high accuracy data do not exist to be able to make that distinction. Thus, it is appropriate to assume that χ_r is exactly a constant and that, for a given L,

$$\dot{q}_i'' (D) = \frac{\chi_r \Delta h_c}{16L^2} \dot{m}_\infty'' (1 - e^{-k\beta D}) \qquad \text{for } L \geq 4$$

$$= F (0.131 \chi_r \Delta h_c) \dot{m}_\infty'' (1 - e^{-k\beta D}) \qquad \text{for } 1/2 \leq L \leq 4$$

The two expressions above are the best ones available based on existing data, for estimating the radiant heat flux.

With very large pools, smoke blockage effects on target irradiance can be much greater than the slight decrease in Data for JP-4 (Hagglund and Persson 1976) show:

D(m)	χ_r
1.0	0.35
1.5	0.39
2.0	0.34
3.0	0.31
5.0	0.16
10.0	0.10

However, a similar study is not available for other fuels, and there is no general expression for describing irradiance dropoff due to smoke blockage. For a conservative estimate, this effect along with atmospheric transmission losses can be neglected. Wind can have a significant effect on local irradiance values; but little data are available.

Sometimes estimates of radiation to targets in a different orientation than the one shown in Figure 21-6D are desired, e.g., targets in a horizontal orientation, or ones located at some height above the base. For such cases, the view factor F in Equation 17b should not be taken from Figure 21-6F, but from the relevant geometry.

Nomenclature

A = pool area (m^2)
d = lip height (m)
D = pool diameter (m), = $\sqrt{4A/\pi}$ for noncircular pools
F = view factor (—)
h = flame height (m)
H = normalized flame height, h/D (—)
Δh_c = lower heat of combustion (kJ/kg)
Δh_g = total heat of vaporization or gasification (kJ/kg)
k = extinction coefficient (m^{-1})
l = distance to target from pool center (m)
L = normalized distance, l/D (—)
$\dot{m}''$ = pool mass loss rate [(kg/m^2)/sec]
$\dot{m}_\infty''$ = infinite-diameter pool mass loss rate [(kg/m^2)/sec]
P = power (kW)
$\dot{q}_c''$ = convective heat flux (kW/m^2)
$\dot{q}_i''$ = incident heat flux at target (kW/m^2)
$\dot{q}_{misc}''$ = miscellaneous heat loss flux (kW/m^2)
$\dot{q}_{rr}''$ = reradiant heat flux (kW/m^2)
T_f = flame temperature (K)
β = mean beam length corrector (—)
ϵ_f = flame emissivity, to external target (—)

Bibliography

References Cited

Babrauskas, V., and Wickstrom, U. G. 1979. "Thermoplastic Pool Compartment Fires." *Combustion and Flame*. Vol 34. pp 195-201.

Blinov, V. I., and Khudiakov, G. N. 1961. "Diffusion Burning of Liquids." *NTIS No. AD296762*. U.S. Army Translation. National Technical Information Service, Washington, DC.

Capener, E. L., and Alger, R. S. 1977. "Characterization and Suppression of Aircraft and Fuel Fires." (WSCI 72-26). Paper presented at Western States Section Meeting of the Combustion Institute, Monterey, CA.

Hagglund, B., and Persson, L. E. 1976. "The Heat Radiation from Petroleum Fires." *Rapport C 20126-06-A3*. Forsvarets Forkningsanstalt, Stockholm, Sweden.

Hall, A. R. 1972. "Pool Burning: A Review." *NTIS No. AD781347*. Rocket Propulsion Establishment, Westcott, England.

Hottel, H. C. 1958. "Review—Certain Laws Governing Diffusive Burning of Liquids, by V. I. Blinov and G. N. Khudiakov." *Fire Research Abstracts and Review*. Vol 1. pp 41-44.

Lois, E., and Swithenbank, J. 1978. "Fire Hazards in Oil Tank Arrays in a Wind." *Seventh Symposium (International) on Combustion*. The Combustion Institute, Pittsburgh, PA. pp. 1087-1098.

Magnus, G. 1961. "Tests on Combustion Velocity of Liquid Fuels and Temperature Distribution in Flames and Beneath Surface of Burning Liquid." *International Symposium on the Use of Models in Fire Research 1959*. W. G. Berl, ed. National Academy of Sciences, Washington, DC.

Modak, A. T. 1981. "Ignitability of High-Fire-Point Liquid Spills." *EPRI Report NP-1731*. Prepared by Factory Mutual Research Corp. for Electric Power Research Institute, Palo Alto, CA.

Orloff, L. 1980. "Simplified Radiation Modeling of Pool Fires." *Eighteenth Symposium (International) on Combustion*. The Combustion Institute, Pittsburgh, PA.

Additional Readings and Sources of Experimental Data

Atallah S., and Raj, P. P. K., "Radiation from LNG Fires," in AGA Project IS-3-1, *LNG Safety Program, Interim Report on Phase II Work*, American Gas Association, Alexandra, VA, 1974.

Babrauskas, V., "Estimating Large Pool Fire Burning Rates," *Fire Technology*, Vol. 19, 1983, pp. 251-261.

Becker, H. A., and Liang, D., "Visible Length of Vertical Free Turbulent Diffusion Flames," *Combustion and Flame*, Vol. 32, 1978, pp. 115-137.

Burgess, D. S., and Zabetakis, M. G., "Fire and Explosion Hazards Associated with Liquified Natural Gas," *R.I. 6099*, Bureau of Mines, Pittsburgh, PA, 1962.

Burgess, D., and Hertzberg, M., "Radiation from Pool Flames," Ch. 27 in *Heat Transfer in Flames*, N. H. Afgan and J. M. Beer, eds., John Wiley and Sons, NY, 1974.

Burgoyne, J. H., and Katan, L. L., "Fires in Open Tanks of Petroleum Products: Some Fundamental Aspects," *Journal of the Institute of Petroleum*, Vol. 33, London, England, 1947, pp. 158-191.

Byram, G. M., et al., *Project Fire Model, an Experimental Study of Model Fires, Final Report*. U.S. Forest Service, Southeastern Forest Experiment Station, Macon, GA, 1966.

Corlett, R. C., and Fu, T. M., "Some Recent Experiments with Pool Fires," *Pyrodynamics*, Vol. 1, 1966, pp. 253-269.

Cragoe, C. S., "Thermal Properties of Petroleum Products," *Misc. Publ. 97*, National Bureau of Standards, Washington, DC, 1929.

Dayan, A., and Tien, C. L., "Radiant Heating From a Cylindrical Fire Column," *Combustion Science and Technology*, Vol. 9, 1974, pp. 41-7.

deRis, J., "Fire Radiation-A Review," *Seventeenth Symposium (International) on Combustion*, The Combustion Institute, Pittsburgh, PA, 1978, pp. 1003-1016.

deRis, J., and Orloff, L., "A Dimensionless Correlation of Pool Burning Data," *Combustion and Flame*, Vol. 18, 1972, pp. 381-8.

Fons, W. L., "Rate of Combustion from Free Surfaces of Liquid Hydrocarbons," *Combustion and Flame*, Vol. 5, 1961, pp. 283-7.

Fu, T. T., "Aviation Fuel Fire Behavior Study," *AD-A014-224*, U.S. Naval Civil Engineering Laboratory, 1972.

Graves, K. W., "Fire Fighter's Exposure Study (Tech. Report AGFSRS 71-2)," *NTIS No. AD722774*, Cornell Aeronautical Lab., Ithaca, NY, 1970.

Hamilton, D. C., and Morgan, W. R., "Radiant-Intercharge Configuration Factors," *Tech. Note 2836*, National Advisory Committee for Aeronautics, Washington, DC, 1952.

Hertzberg, M., "The Theory of Free Ambient Fires. The Convectively Mixed Combustion of Fuel Reservoirs," *Combustion and Flame*. Vol. 21, 1973, pp. 195-209.

Johnson, D. W., et al., "Control and Extinguishment of LPG Fires." *NTIS No. DOE/EV/06020-T3*. Dept. of Energy, Washington, DC, 1980.

Kuchta, D. W., et al., "Crash Fire Hazard Rating System for Controlled Flammability Fuels Report NA-69-17," *NTIS No. AD684089*, Bureau of Mines, Pittsburgh, PA, 1969.

Kung, H. C., and Stavrianidis, P., "Buoyant Plumes of Large-Scale Pool Fires," *Nineteenth Symposium (International) on Combustion*, 1982, pp. 905—912.

May, W. G., and McQueen, W., "Radiation From Large Liquefied Natural Gas Fires," *Combustion Science and Technology*, Vol. 7, 1973, pp. 55-56.

Modak, A. T., "Atmospheric Absorption," *Unpublished Report*, Factory Mutual Research Corp., Norwood, MA, 1978.

Modak, A. T., and Croce, P. A., "Plastic Pool Fires," *Combustion and Flame*, Vol. 30, 1977, pp. 251-265.

Ndubizu, C. C., et al., "A Model of Freely Burning Pool Fires," *Combustion Science and Technology*, Vol. 31, 1983, pp. 233-247.

Orloff, L., and deRis, J., *Froude Modeling of Pool Fires* (FMRC J.I. OHON3.BU), Factory Mutual Research Corp., Norwood, MA, 1983.

Pagni, P. J., and Bard, S., "Particulate Volume Fractions in Diffusion Flames," *Seventeenth Symposium (International) on Combustion*, The Combustion Institute, Pittsburgh, PA, 1978, pp. 1017-1028.

Raj, P. K., et al., "LPG Spill Tests on Water, An Overview of the Results," *Paper presented at American Gas Association Transmission Conference*, New Orleans, LA, May 1979.

Rasbash, D. J., et al, "Properties of Fires of Liquids," *Fuel*, Vol. 35, 1956, pp. 94-107,.

Russell, L. H., and Canfield, J. A., "Experimental Measurement of Heat Transfer to a Cylinder Immersed in a Large Aviation-Fuel Fire," *Journal of Heat Transfer*. Vol. 95, 1973, pp. 397-404.

Siegel, R., and Howell, J. R., *Thermal Radiation Heat Transfer*, McGraw-Hill Inc., NY, 1980.

Tarifa, C. S., *Open Fires*, Instituto Nacional de Tecnica Aeroespacial Esteban Terradas, Madrid, Spain 1967.

Welker, J. R., "Radiation from LPG Fires (WSCI 82-38)," *Paper presented at Western States Section Meeting of the Combustion Institute*, Salt Lake City, UT, 1982.

Wood, B. D., and Blackshear, P. L. Jr., "An Experimental Study of the Heat Transfer to the Surface of a Burning Array of Fuel Elements (WSCI 69-38)," *Paper presented at Western States Section Meeting of the Combustion Institute*, La Jolla, CA, 1969.

Wood, B. D., and Blackshear, P. L., Jr., "Some Observations on the Mode of Burning of a One and One-Half Meter Diameter Pan of Fuel," *NTIS No. AD704713*, University of Minnesota, 1969.

Yumoto, T., "Heat Transfer from Flame to Fuel Surface in Large Pool Fires," *Combustion and Flame*, Vol. 17, 1971, pp. 108-110.

Zabetakis, M. G., and Burgess, D. S., "Research on the Hazards Associated with the Production and Handling of Liquid Hydrogen," *R.I. 5707*, Bureau of Mines, Pittsburgh, PA, 1961.

SECTION 22

MISCELLANEOUS DATA

INSPECTING, SURVEYING, AND MAPPING

Revised by Carl E. Peterson

The reasons for inspecting a property are: (1) to evaluae the danger to life from fire, and (2) to evaluate and determine ways for minimizing fire danger to buildings and contents. Inspections are made by fire protection engineers, building and fire officials, insurance representatives, and others similarly qualified.

In order to properly evaluate the protection for a property or specific hazard, a tour or inspection of the premises must be made. Three essential results of the inspection should be:

1. A permanent, precise, and complete narrative report describing the fire protection features and fire hazards of the property.
2. A plan indicating the physical characteristics and layout of the premises.
3. Recommendations for improvement (if necessary).

This chapter provides a list of the key items contained in the written narrative report, a list of the important features to observe when surveying a property, and some guidelines for drawing (mapping) a site plan. Also included are a chart of standard plan symbols and a legend of standard abbreviations.

INSPECTION

Functions performed during an inspection, and the compilation of the narrative report resulting from the inspection, are discussed in detail in the *NFPA Inspection Manual* (NFPA 1982). The following list forms an outline of the key items that should appear in the written report:

1. Property identification.
   a. Name of company and address.
   b. Date of report.
   c. Name of inspector.
   d. Building identification.
2. Property use.
   a. General property use; e.g., educational, mercantile, industrial, warehouse.
   b. Specific uses. State each principal use and its location. For storage areas, identify the product(s) stored.
   c. Names of tenants in a building of multiple occupancy, including their location and amount of space occupied.
3. Construction.
   a. Types of construction. For mixed construction, show a table of buildings and report percentage of each type construction.
   b. Exterior exposure.
      1. Other buildings.
      2. Outside storage, processes, parking.
      3. Grass, brush.
   c. Interior exposure.
   d. Fire areas, walls, and partitions.
   e. Protection or lack of protection of horizontal openings.
   f. Protection or lack of protection of vertical openings.
4. Life safety.
   a. Exit facilities.
      1. Adequacy.
      2. Deficiencies.
   b. Exit illumination and emergency lighting.
   c. Interior finish.
   d. Evacuation plan and drills.
5. Common hazards.
   a. Heat, light, power, air conditioning.
   b. Housekeeping, brush and grass.
   c. Ordinary combustibles.
   d. Electrical appliances.
   e. Smoking.
6. Special hazards.
   a. Finishing processes.
   b. Flammable liquids.
   c. Flammable gases.
   d. Welding and cutting.
   e. Cooking.
   f. Chemicals and plastics, etc.
   g. Electronic equipment.
   h. Others.
7. Water supply.
   a. General description including adequacy and reliability.

Mr. Peterson is Manager, Fire Service Management Systems on the staff of NFPA.

b. Fire flow requirements, availability determined by tests.
c. Storage requirements.
d. Sources of supply.
e. Storage facilities.
f. Pumps.
g. Distribution system and appurtenances.
h. Maintenance, inspections, and tests.

8. Extinguishing systems and devices.
   a. Automatic sprinkler systems.
      1. Types.
      2. Coverage, temperature rate of sprinklers, pipe schedules, spacing of sprinklers.
      3. Adequacy.
      4. Tests and maintenance.
   b. Special hazard systems—carbon dioxide, foam, dry chemical, Halon.
      1. Types.
      2. Hazard protected.
      3. Adequacy.
      4. Tests and maintenance.
   c. Portable extinguishers.
      1. Type and coverage.
      2. Tests and maintenance.
   d. Standpipes and fire hose systems.
      1. Type.
      2. Tests and maintenance.
9. Alarm and detection systems.
   a. Automatic fire detection systems.
      1. Type, coverage, connected to fire alarm.
      2. Power supply.
      3. Maintenance and test.
   b. Local evacuation alarm systems.
      1. Type, coverage, connected to fire alarm.
      2. Power supply.
      3. Maintenance and test.
10. Fire prevention.
   a. Building inspection program.
      1. Frequency, scope, recording forms.
   b. Fire protection system inspection program.
      1. Type, frequency, scope, recording forms.
   c. Employee firesafety training.
      1. Adequacy and frequency of training.
      2. Reference material.
   d. Fire brigade program.
      1. Type of brigade.
      2. Training.
      3. Special equipment provided.
11. Recommendations.

SURVEYING

There are many different routes to choose when touring a property; a particular route is selected according to personal preference and convenience. The inspector should choose a standard touring procedure with which he/she is comfortable, and should adopt that pattern for normal use. It is important for the inspector to pass through the building and the plant yard systematically without leaving any space uncovered.

A thorough understanding of the features depicted on a complete plan is absolutely necessary. The features of any property can be placed in one of four categories: Construction -C-; Occupancy -O-; Protection -P-; and Exposure -E- (COPE). These major classifications of important features are comprised of many components, as shown here.

1. Construction.
   a. Building(s) identification.
   b. Dimensions.
      1. Buildings.
      2. Distances to exposures, other properties.
      3. Width of streets, natural barriers.
   c. Date of construction.
   d. Height of building(s); i.e., number of floors, ceiling heights, concealed spaces, etc.
   e. Construction material of:
      1. Walls.
      2. Floors.
      3. Roofs.
   f. Location of fire division walls and type of material.
   g. Location of fire doors and/or other horizontal protected/unprotected opening(s).
   h. Location of stairways, elevators, and other vertical protected/unprotected openings.
   i. Type of windows; i.e., plain glass, wire glass, etc.
   j. Location and type of suspended ceilings.
   k. Location of roof parapets.
   l. Chimney location, construction, and height.
   m. Location of fire escapes.
2. Occupancy.
   a. Floor by floor listing of major occupancy processes.
   b. Location, size, content, and construction of tanks and/or cylinders.
   c. Ratings for boilers, furnace.
   d. Cooling towers, dust collector, silos, cranes, conveyors, and other special equipment.
   e. Electrical transformers.
      1. Capacity.
      2. Ownership.
      3. Protection.
      4. Cutoffs.
3. Protection.
   a. Automatic sprinklers.
      1. Type of System—wet pipe, dry pipe with/without accelerator or exhauster, deluge, preaction.
      2. Diameter.
      3. Approximate location.
   b. Control valves for water supplying sprinkler system fire pump.
      1. Location.
      2. Size.
      3. Type of valve.
      4. Label status of valve, only if it is normally "shut."
   c. Fire pumps—label symbol with following items.
      1. Capacity rated—gpm, psi (L/min, kPa).
      2. Type—centrifugal, vertical turbine, rotary, or steam.
      3. Name of manufacturer.
      4. Type of motor drive—diesel, electric, gas, gasoline, or team.
      5. Suction pressure—head or lift (in feet, or meters).
      6. Starting mechanism—automatic or manual.
   d. Water tank, gravity tanks, suction tank for fire pumps.
      1. Capacity—gpm, psi (L/min, kPa).
      2. Height.
      3. Pipe arrangement to yard main or fire pump.
      4. Percent of capacity for fire protection use.

e. Public water supplies.
   1. Type of system—gravity, direct pumping.
   2. Size of all water mains and locations in relation to plant.
   3. Location, size, and type of control valves.
   4. Type of water service—high and low pressure or fire service only.

f. Yard water mains by symbols only.
   1. Location.
   2. Size.
   3. Connection arrangement to all supply sources available.
   4. Type of pipe—cement line, cast iron, etc.

g. Fire alarm and detection systems.
   1. Water flow detection on sprinkler risers.
   2. Location of main control valve/panel type detection units.
   3. Products of combustion—rate of heat rise, fixed temperature.
   4. Fire department alarm box.

h. Location and type of check valves and water meters.

i. Hydrants.
   1. Public.
   2. Private.
   3. Frost proof.

j. Fire department pumper connections.

4. Exposure.
   a. Indicate proximity, construction, and occupancy of buildings or structures facing the four sides of the building being surveyed.
   b. Location of rivers or other natural exposures (i.e., forests).
   c. Railroad tracks.
   d. Yard storage.

By formulating a short checklist similar to the preceding for each COPE feature, the inspector will be assured of having all the necessary information for preparing a good sketch after the tour. Utilizing such checklists to make rough, hand drawn sketches of every individual section, floor, or building will reduce the difficulty of preparing the finished plan.

The symbols in Figure 22-1A are used to represent the general location and type of the COPE features of a facility. Occasionally, the use of a symbol alone will not sufficiently describe the situation. In that case, it is also necessary to also label that point on the plan. Table 22-1A gives a list of standard abbreviations that can be used. Although brief, the list nevertheless provides all of the abbreviations commonly needed. *Abbreviations for Use on Drawings and in Text*, prepared by the American National Standards Institute (ANSI 1972), lists additional, less commonly used abbreviations.

MAPPING

Once the necessary information has been collected, a site plan can be readily drawn. A good, practical approach is to split the preparation into three parts:

1. Building sketch.
2. Addition of fire protection equipment.
3. Detailing occupancy and specific processes or hazards.

The initial step is to outline the shape of the building, using an appropriate architect/engineer scale on an adequately sized sheet of paper. Once the shell is complete, the other important construction features referenced in the preceding section on surveying should be added, and various construction material labelled.

The inspector should draw sufficient sectional views to show all vertical areas of the plant. Figure 22-1B illustrates what a typical, simple, "site sketch" looks like when completed.

The second step is to plot the fire protection facilities and equipment on the finished building sketch. Two important points should be noted. First, water supplies—public or private (yard) mains—are continuous from a point of origin to the individual automatic sprinkler risers. Conversely, the automatic sprinkler systems within the buildings are not entirely shown. (NOTE: Sprinkler scheduling and spacing is referred to in the body of the narrative inspection report.) Secondly, when space limitations prohibit drawing all the equipment on the plan at approximate actual locations (due to size, quantity of symbols, or labeling requirements), a small side sketch (not drawn to scale) can be used effectively.

The last step in completing the plan is to label the internal occupancy or processes of each area of the plant, and any yard storage or processes of interest.

The symbols and plans in this HANDBOOK will enable any inspector to develop a building plan in accordance with widespread convention. These symbols are based on NFPA 174, *Standard for Fire Protection Symbols for Risk Analysis Diagrams*. However, many organizations use their own, slightly different set of symbols and layouts. When trying to take information from a plan developed from another source, it is essential to refer to the legend of symbols used by that organization to accurately interpret the information.

At one time, colors were used to describe the type of construction materials used for walls. Today, color-coded plans are being replaced by black and white drawings to facilitate preparation and for economic reasons, so construction type is being noted narratively.

Table 22-1B identifies colors that should be used if the plan is to be colored. As can be seen from the table, only four basic construction types are coded. To avoid confusion on complicated plans, construction materials can be labeled in accordance with the abbreviations in Table 22-1A.

TABLE 22-1B. Color Code for Denoting Construction Materials for Walls

Color	Interpretation
Brown	Fire resistive protected steel
Red	Brick, hollow tile
Yellow	Frame-wood, stucco
Blue	Concrete, stone, or hollow concrete block
Gray	Noncombustible (sheet metal or metal lath and plaster) unprotected steel

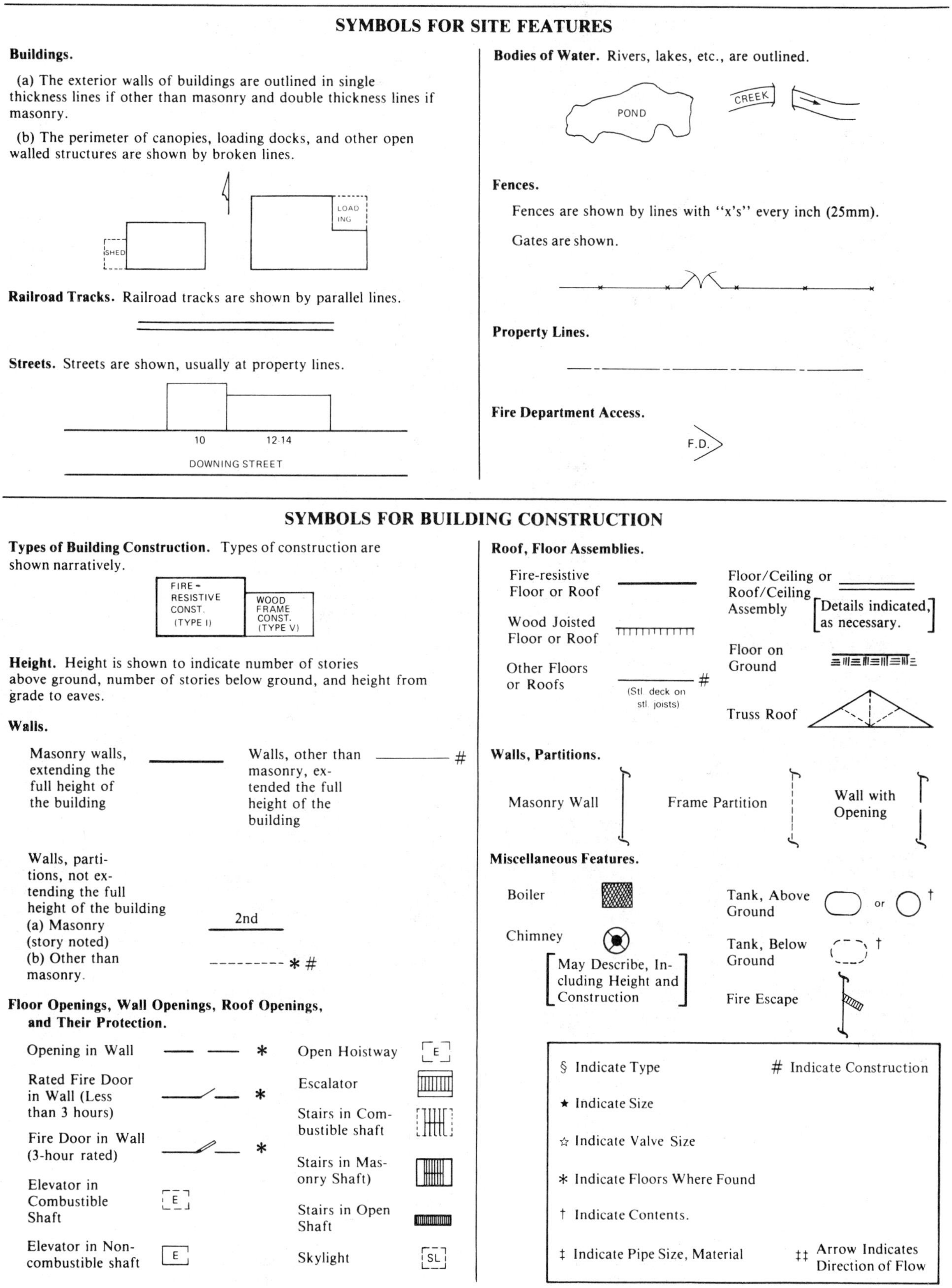
SYMBOLS FOR SITE FEATURES
Buildings.
(a) The exterior walls of buildings are outlined in single thickness lines if other than masonry and double thickness lines if masonry.
(b) The perimeter of canopies, loading docks, and other open walled structures are shown by broken lines.
LOAD ING
SHED
Railroad Tracks. Railroad tracks are shown by parallel lines.
Streets. Streets are shown, usually at property lines.
10
12-14
DOWNING STREET
Bodies of Water. Rivers, lakes, etc., are outlined.
POND
CREEK
Fences.
Fences are shown by lines with "x's" every inch (25mm).
Gates are shown.
Property Lines.
Fire Department Access.
F.D.
SYMBOLS FOR BUILDING CONSTRUCTION
Types of Building Construction. Types of construction are shown narratively.
FIRE-RESISTIVE CONST. (TYPE I)
WOOD FRAME CONST. (TYPE V)
Height. Height is shown to indicate number of stories above ground, number of stories below ground, and height from grade to eaves.
Walls.
Masonry walls, extending the full height of the building
Walls, other than masonry, extended the full height of the building #
Walls, partitions, not extending the full height of the building
(a) Masonry (story noted)
2nd
(b) Other than masonry.
* #
Floor Openings, Wall Openings, Roof Openings, and Their Protection.
Opening in Wall *
Rated Fire Door in Wall (Less than 3 hours) *
Fire Door in Wall (3-hour rated) *
Elevator in Combustible Shaft
E
Elevator in Non-combustible shaft
E
Open Hoistway
E
Escalator
Stairs in Combustible shaft
Stairs in Masonry Shaft)
Stairs in Open Shaft
Skylight
SL
Roof, Floor Assemblies.
Fire-resistive Floor or Roof
Wood Joisted Floor or Roof
Other Floors or Roofs #
(Stl. deck on stl. joists)
Floor/Ceiling or Roof/Ceiling Assembly
[Details indicated, as necessary.]
Floor on Ground
Truss Roof
Walls, Partitions.
Masonry Wall
Frame Partition
Wall with Opening
Miscellaneous Features.
Boiler
Chimney
[May Describe, Including Height and Construction]
Tank, Above Ground or †
Tank, Below Ground †
Fire Escape
§ Indicate Type
# Indicate Construction
★ Indicate Size
☆ Indicate Valve Size
* Indicate Floors Where Found
† Indicate Contents.
‡ Indicate Pipe Size, Material
‡‡ Arrow Indicates Direction of Flow

FIG. 22-1A. Standard plan symbols used to represent the general location and type of COPE (Construction -C-; Occupancy -O-; Protection -P-; and Exposure -E-) features of a facility.

SYMBOLS FOR WATER SUPPLY AND DISTRIBUTION

Mains, Pipe.

Public Water Main ‡

Water Main Under Building ‡

Private Water Main ‡

Suction Main ‡

Hydrants.

Private Hydrant, One Hose Outlet ★

Wall Hydrant, Two Hose Outlets ★

Public Hydrant, Two Hose Outlets ★

Private Housed Hydrant, Two Hose Outlets ★

Public Hydrant Two Hose Outlets and Pumper Connection. ★

Stored Water.

Water Tower or Tank-Above Ground — or — or

[Indicate Type, Construction, Size and Height via Notations]

Pressure Tank — or

[Indicate Type, Construction and Size via Notations. Symbol Orientation Must Not Be Changed.]

Meter §

Valves.

Post Indicator and Valve ☆

Non-indicating Valve (Non-rising-Stem Valve) ☆

Key-operated Valve ☆

OS&Y Valve (Outside Screw and Yoke, Rising Stem) ☆

Valve In Pit (OS&Y Shown) ☆

Check Valve — or — ‡ ‡‡

Fire Department Connections.

Two-way Fire Department Connection (Siamese Connection) (Specify Size and Angle)

Freestanding Fire Department Connection (Siamese Connection) (Sidewalk or Pit Type, Specify Size)

Fire Pump.

Fire Pump with Drives

SYMBOLS FOR SPRINKLER SYSTEMS[1]

Piping, Valves, Control Devices.

Sprinkler Riser

Deluge Valve ★ §

Check Valve, General — or — ‡‡

Alarm Check Valve — or — ★ ★

Dry Pipe Valve — or — ★

Dry Pipe Valve with Quick Opening Device (Accelerator or Exhauster) — or — ★ §

Alarm/Supervisory Devices.

Flow Detector/Switch (Flow Alarm) — or — ★

Pressure Detector/Switch § (Specify Type – Water, Low Air, Hi Air, etc.)

Water Motor Alarm (Water Motor Gong) (Shield Optional)

Electric Alarm Bell (Electric Alarm Gong)

SYMBOLS FOR EXTINGUISHING SYSTEMS

Wet (Charged) System.

(a) Automatically Actuated — or AS

(b) Manually Actuated

Dry System.

(a) Automatically Actuated — or AS

(b) Manually Actuated

Foam System.

(a) Automatically Actuated

(b) Manually Actuated

For Liquid—, Gas—, and Electrical-type Fires.

(b) Automatically Actuated

(b) Manually Actuated

For Fires of All Types, Except Metals.

(a) Automatically Actuated

(b) Manually Actuated

Carbon Dioxide System.

(a) Automatically Actuated

(b) Manually Actuated

Halon System

(a) Automatically Actuated

(b) Manually Actuated

Supplementary Symbols

Nonsprinklered Space NS

Partially Sprinklered Space AS

[1] These symbols are intended for use in identifying the type of installed system protecting an area within a building.

TABLE 22-1A. Legend of Common Abbreviations*

Above	ABV	Liquid	LIQ
Accelerator	ACC	Liquid Oxygen	LOX
Acetylene	ACET	Manufacture	MFR
Aluminum	AL	Manufacturing	MFG
Asbestos	ASB	Maximum Capacity	MAX CAP
Asphalt Protected Metal	APM	Mean Sea Leves	MSL
Attic	A	Metal	MT
Automatic	AUTO	Mezzanine	MEZZ
Automatic Fire Alarm	AFA	Mill Use	MU
Automatic, Sprinklers	AS	Normally Closed	NC
Avenue	AVE	Normally Open	NO
Basement	B	North	N
Beam	BM	Number	No
Board on Joist	BDOJ	Open Sprinklers	OS
Brick	BR	Outside Screw & Yoke Valve	OS & Y
Building	BLDG	Partition (Label Composition)	PTN (ie WD PTN)
Cast Iron	CI	Plaster	PLAS
Cement	CEM	Plaster Board	PLAS BD
Centrifugal Fire Pump	CFP	Platform	PLATF
Cinder Block	CB	Pound (Unit of Force)	LB
Composition Roof	COMPR	Pressure	PRESS
Concrete	CONC	Unit of Pressure (Pounds Per Square Inch)	PSI
Construction	CONST	Protected Steel	PROT ST
Corrugated Iron	COR IR	Private	PRIVATE
Corrugated Steel	COR ST	Public	PUB
Diameter	DIA	Railroad	RR
Diesel-Engine	D ENG	Reinforced Concrete	RC
Domestic	DOM	Reinforcing Steel	RST
Double Hydrant	DH	Reservoir	RES
Dry Pipe Valve	DPV	Revolutions per Minute	RPM
East	E	Roof	RF
Electric Motor Driven	EMD	Room	RM
Elevator	ELEV	Slate Shingle Roof	SSR
Engine	ENG	Space	SP
Exhauster	EXH	South	S
Feet	FT	Stainless Steel	SST
Fibre Board	FBR BD	Steam Fire Pump	SFP
Fire Escape	FE	Steel	ST
Fire Department Pumper Connection	FDPC	Steel Deck	ST DK
Fire Detection Units	FDU	Stone	STONE
Products of Combustion	POC	Story	STO
Rate of Heat Rise	RHR	Street	STREET
Fixed Temperature	FTEP	Stucco	STUC
Fire Pump	FP	Suspended Acoustical Plaster Ceiling	SAPL
Floor	FL	Suspended Acoustical Tile Ceiling	SATL
Frame	FR	Suspended Plaster Ceiling	SPC
Fuel Oil (Label with Grade Number)	FO #____	Suspended Sprayed Acoustical Ceiling	SSAL
Gallon	GAL	Tank (Label Capacity in Gallons)	TK
Gallons Per Day	GPD	Tenant	TEN
Gallons Per Minute	GPM	Tile Block	TB
Galvanized Iron	GALVI	Timber	TMBR
Galvanized Steel	GALVS	Tin Clad	TIN CL
Gas, Natural	GAS	Triple Hydrant	TH
Gasoline	GASOLINE	Truss	TR
Gasoline Engine Driven	GED	Under	UND
Generator	GEN	Vault	VLT
Glass	GL	Veneer	VEN
Glass Block	GLB	Volts (Indicate Number Of)	450v
Gypsum	GYM	Wall Board	WLBD
Gypsum Board	GYM BD	Wall Hydrant	WLH
High Voltage	HV	Water Pipe	WP
Hollow tile	HT	West	W
Hose Connection	HC	Wire Glass	WGL
Hydrant	HYD	Wire Net	WN
Inch, Inches	IN	Wood	WD
Iron	IR	Wood Frame	WD FR
Iron Clad	IR CL	Yard	YD
Iron Pipe	IP		
Joist, Joisted	J		

* Some words that have a common abbreviation, e.g. "ST" for "street," are spelled out fully to avoid confusion with similar abbreviations used herein for other terms.

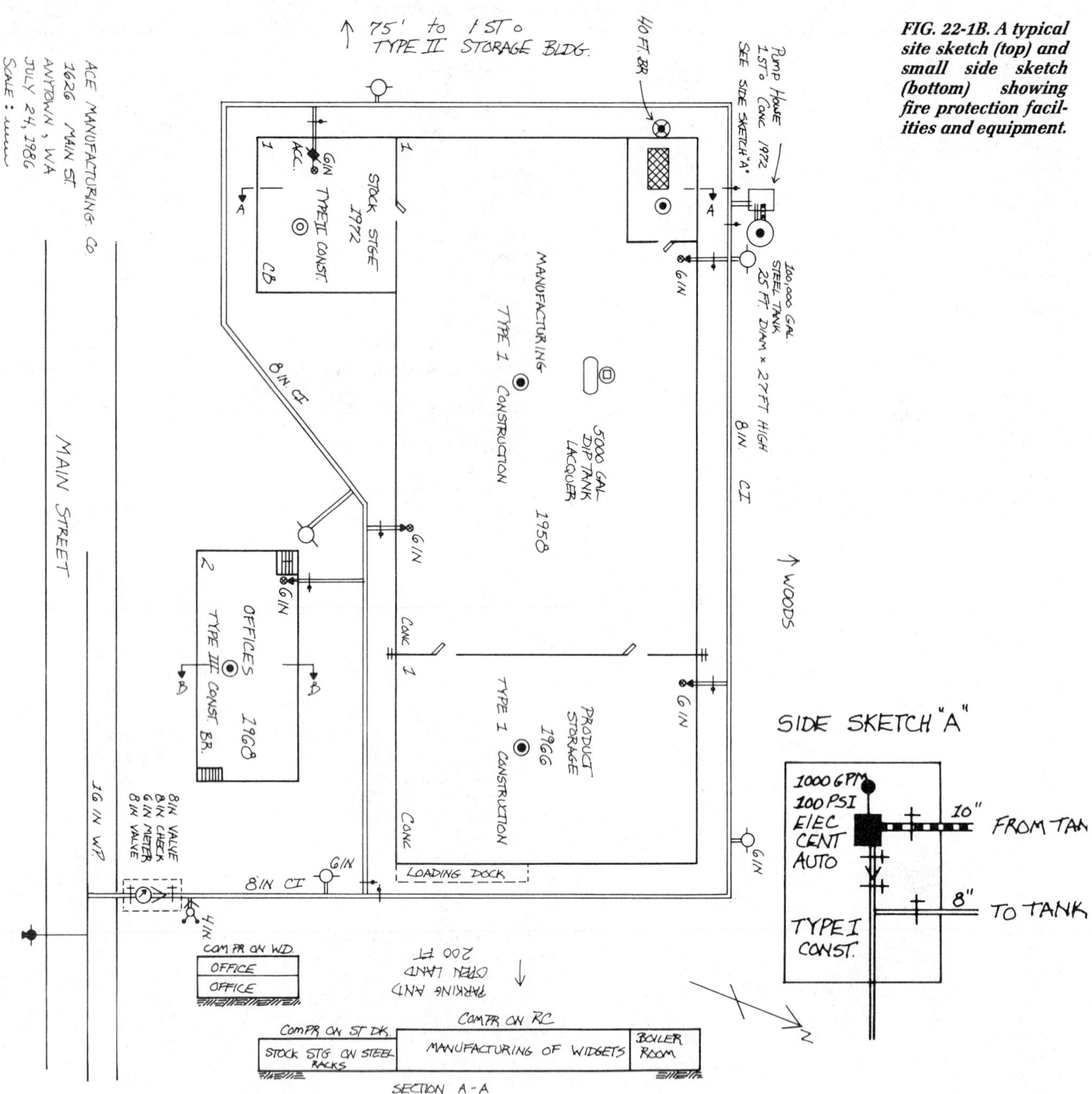

FIG. 22-1B. A typical site sketch (top) and small side sketch (bottom) showing fire protection facilities and equipment.

Bibliography

References Cited

ANSI. 1972. *ANSI Y1.1, Abbreviations for Use on Drawings and in Text.* American National Standards Institute, NY.

NFPA. 1982. *NFPA Inspection Manual.* 5th ed. National Fire Protection Association, Quincy, MA.

NFPA Codes, Standards, Recommended Practices and Manuals. (See the latest *NFPA Codes and Standards Catalog* for availability of current editions of the following documents.)

NFPA 174, *Standard for Fire Protection Symbols for Risk Analysis Diagrams.*

NFPA 220, *Standard on Types of Building Construction.*

Additional Readings

Conducting Fire Inspections. National Fire Protection Association, Quincy, MA, 1982.

SI UNITS AND CONVERSION TABLES

Peter F. Johnson

Introduction to SI

This HANDBOOK serves as a major fireprotection reference not only in the United States but throughout many other parts of the world. Almost all other countries use metric units for measurement and calculation, and the U. S. is expected to make the transition to metric units in the future. It is therefore appropriate and timely that this HANDBOOK be metricated as far as possible, while still retaining the customary units with which most people in the U.S. are familiar.

There are several versions of the metric system, but the one used by NFPA for this HANDBOOK is *Le Systeme International d'Unites* commonly known as the International System of Units or simply "SI." The SI system constitutes a logical set of units based on powers of 10 that permits ease of manipulation and avoids the use of confusing numerical factors such as 1,760 yards (or 5,280 feet) in a mile.

The extent to which SI units have been incorporated varies throughout the HANDBOOK. Generally, a quantity expressed in customary units has been followed by the appropriate SI value and unit in parentheses, e.g., 14.7 psi (101 kPa). In many chapters, however, there are formulas that are unit dependent; they only give the correct answer when customary units are used. These formulas have been supplemented (where possible) with new formulas for use with SI units and the inclusion of relevant SI examples illustrating the use of these formulas. In many cases, supplemental columns and curves representing metric units have been added to existing tables and graphs. In some chapters, however, supplemental SI graphs or tables are either not warranted or have not yet been generated. In these cases, appropriate conversion factors have been included in the captions (or as footnotes) to existing graphs and tables.

Mr. Johnson is supervising materials scientist at Central Investigation and Research Laboratory (CIRL), Department of Housing and Construction, Melbourne, Australia. Mr. Johnson served as metrication consultant for this edition of the HANDBOOK.

Base units for SI

The use of SI units has generally followed the approach outlined in ASTM E-380—82, *Standard for Metric Practice* (ASTM 1982). The seven base units defined in E-380—82 are:

Quantity	Unit	Symbol
length	meter	m
mass	kilogram	kg
time	second	s
electric current	ampere	A
thermodynamic temperature	kelvin	K
amount of substance	mole	mol
luminous intensity	candela	cd

Temperature Scales

One base unit worthy of further mention is the degree kelvin (K). This is the SI unit for thermodynamic (or absolute) temperature and it has been properly used throughout the HANDBOOK for expressing thermodynamic and heat transfer quantities. For example, heat capacity is expressed as J/(kg · K), and radiant power is proportional to the fourth power of the absolute temperature T_k.

The degree Celsius (°C) also has been used extensively. Degree Celsius is the SI unit for expressing measured temperature or temperature intervals. Thus, the flashpoint of a flammable liquid is expressed in degrees Celsius. Similarly, a sprinkler head set to operate at 60°C (rather than 343 K) is appropriate.

The relationship between the measured temperature T (°C) and the thermodynamic temperature T_k is:

$$T_K = T\ (°C) + 273.15$$

A similar relationship exists between the measured temperature (°F) and the thermodynamic degree Rankine (R):

$$T_R = T\ (°F) + 459.7$$

Prefixes

To get multiples or submultiples of any SI unit, a range of standard prefixes is used, and the usual practice of choosing a prefix to limit the numerical value of a quantity to four or less digits has been followed, e.g., 28 000 m is better written as 2.8 km; 0.0017 g should be 1.7 mg.

Name	Symbol	Multiplication Factor
tera	T	10^{12}
giga	G	10^{9}
mega	M	10^{6}
kilo	K	10^{3}
milli	m	10^{-3}
micro	μ	10^{-6}
nano	n	10^{-9}
pico	p	10^{-12}
femto	f	10^{-15}
atto	a	10^{-18}

Special Comments

Meter versus Metre: One departure from the SI system has been to use the spelling "meter" and "liter" throughout the HANDBOOK instead of the more international "metre" and "litre." This was done because the former terms are more commonly used in the U.S. and to maintain consistency with other NFPA publications.

Minute versus Second: A second difference is the choice of liter per minute and cubic meter per minute (L/min and m^3/min) for fluid flow instead of the more international L/s or m^3/s, particularly for hydraulics and water supplies. The minute was chosen in preference to the second because m^3/min or L/min more nearly equate to gallons per minute (gpm), and fire protection practitioners in many countries find gallons per minute an easier quantity to conceptualize and use in calculations. In addition, L/min and m^3/min are used in other NFPA publications, so consistency is again maintained.

Force and Mass: One area of confusion for those new to SI units is the conversion of force and mass terms. In the customary system a body may have a mass (amount of material) of 20 lb and a gravitational force of 20 lb wt on it (sometimes 20 lbf). The unit of pound (lb) appears in both terms. In SI totally different units are used for mass and force. The SI unit for mass is the kilogram (kg) and the unit for force is the newton (N). The appropriate conversion factors are:

Mass	1 lb = 0.453 25 kg
Force	1 lb wt (1bf) = 4.448 N

The newton (N) and not the kilogram (kg) consequently appear in force related terms like pressure (N/m^2 = Pa), energy (N · m = J), and power (N · m/sec = W).

ROUNDING OFF

The approach to conversion from customary units to SI has been to retain both the number and its precision. For example, "It is *about* 8 miles" translates properly to "It is *about* 13 kilometers." Conversion to "It is *about* 12.88245 kilometers" would be totally inappropriate.

Conversions have been chosen to reflect the precision with which the quantity can be measured within the context of the chapter where the unit is used. As an illustration, a 125 ft length of sprinkler pipe measured to the nearest foot would convert to 38.1 m. However, if measured to the nearest 5 ft, 38 m would be the appropriate conversion. This consistent approach to rounding off has been taken throughout the HANDBOOK.

The SI units into which quantities have been converted have, for the most part, followed the SI practice of expressing a quantity so that its numerical value falls between 0.1 and 1000. An example found widely is the conversion of atmospheric pressure of 14.7 psi into 101 kPa, not 101 000 Pa. Under certain circumstances, however, permitted deviations from this practice are used. In particular, this applies where a range of values of the same quantity is being compared or presented in tabular form, e.g., if four samples had masses of 500 kg, 800 kg, 900 kg, and 1200 kg, it would not be appropriate to express the last mass as 1.2 Mg.

Conversion Factors

Conversion factors used throughout this HANDBOOK are detailed in Tables 22-2A and 22-2B. They include conversions within customary units as well as into SI units.

Bibliography

References Cited

ASTM. 1982. *Standard for Metric Practice.* ASTM E-380 −82. American Society for Testing and Materials, Philadelphia, PA.

Additional Readings

Metric Practice Guide −SI Units and Conversion Factors for the Steel Industry, 3rd ed., American Iron & Steel Institute, Washington, DC, 1978.

SI Units in Fire Protection Engineering, SFPE-MFP Report—1980, Society Of Fire Protection Engineers, Boston, MA, 1980.

TABLE 22-2A. Conversion Factors

Length

1 inch = 0.08333 foot, 1,000 mils, 25.40 millimeters.
1 foot = 0.3333 yard, 12 inches, 0.3048 meter, 304.8 millimeters.
1 yard = 3 feet, 36 inches, 0.9144 meter.
1 rod = 16.5 feet, 5.5 yards, 5.029 meters.
1 mile (U.S. and British) = 5,280 feet, 1.609 kilometers, 0.8684 nautical mile.
1 millimeter = 0.03937 inch, 39.37 mils, 0.001 meter, 0.1 centimeter, 100 microns.
1 meter = 1.094 yards, 3.281 feet, 39.37 inches, 1,000 millimeters.
1 kilometer = 0.6214 mile, 1.094 yards, 3,281 feet, 1,000 meters.
1 nautical mile = 1.152 miles (statute), 1.853 kilometers.
1 micron = 0.03937 mil, 0.00003937 inch.
1 mil = 0.001 inch, 0.0254 millimeters, 25.40 microns.
1 degree = 1/360 circumference of a circle, 60 minutes, 3,600 seconds
1 minute = 1/60 degree, 60 seconds.
1 second = 1/60 minute, 1/3600 degree.

Area

1 square inch = 0.006944 square foot, 1,273,000 circular mils, 645.2 square millimeters.
1 square foot = 0.1111 square yard, 144 square inches, 0.09290 square meter, 92,900 square millimeters.
1 square yard = 9 square feet, 1,296 square inches, 0.8361 square meter.
1 acre = 43,560 square feet, 4,840 square yards, 0.001563 square mile, 4,047 square meters, 160 square rods.
1 square mile = 640 acres, 102,400 square rods, 3,097,600 square yards, 2.590 square kilometers.
1 square millimeter = 0.001550 square inch, 1,974 circular mils.
1 square meter = 1.196 square yards, 10.76 square feet, 1,550 square inches. 1,000,000 square millimeters.
1 square kilometer = 0.3861 square mile, 247.1 acres, 1,196,000 square yards, 1,000,000 square meters.
1 circular mil = 0.7854 square mil, 0.0005067 square millimeter, 0.0000007854 square inch.

Volume (Capacity)

1 fluid ounce = 1.805 cubic inches, 29.57 milliliters, 0.03125 quarts (U.S.) liquid measure.
1 cubic inch = 0.5541 fluid ounce, 16.39 milliliters.
1 cubic foot = 7.481 gallons (U.S.) 6.229 gallons (British), 1,728 cubic inches, 0.02832 cubic meter, 28.32 liters.
1 cubic yard = 27 cubic feet, 46,656 cubic inches, 0.7646 cubic meter, 746.6 liters, 202.2 gallons (U.S.), 168.4 gallons (British).
1 gill = 0.03125 gallon, 0.125 quart, 4 ounces, 7.219 cubic inches, 118.3 milliliters.
1 pint = 0.01671 cubic foot, 28.88 cubic inches, 0.125 gallon, 4 gills, 16 fluid ounces, 473.2 milliliters.
1 quart = 2 pints, 32 fluid ounces, 0.9464 liter, 946.4 milliliters, 8 gills, 57.75 cubic inches.
1 U.S. gallon = 4 quarts, 128 fluid ounces, 231.0 cubic inches, 0.1337 cubic foot, 3.785 liters (cubic decimeters), 3,785 milliliters, 0.8327 Imperial gallon.
1 Imperial (British and Canadian) gallon = 1.201 U.S. gallons, 0.1605 cubic foot, 277.3 cubic inches, 4.546 liters (cubic decimeters), 4,546 milliliters.
1 U.S. bushel = 2,150 cubic inches, 0.9694 British bushel, 35.24 liters.
1 barrel (U.S. liquid) = 31.5 gallons (various industries have special definitions of a barrel)
1 barrel (petroleum) = 42.0 gallons.
1 millimeter = 0.03381 fluid ounce, 0.06102 cubic inch, 0.001 liter.
1 liter (cubic decimeter) = 0.2642 gallon, 0.03532 cubic foot, 1.057 quarts, 33.81 fluid ounces, 61.03 cubic inches, 1,000 milliters.
1 cubic meter (kiloliter) = 1.308 cubic yards, 35.32 cubic feet, 264.2 gallons, 1,000 liters.
1 cord = 128 cubic feet, 8 feet × 4 feet × 4 feet, 3.625 cubic meters.

Weight

1 grain = 0.0001428 pound.
1 ounce (avoirdupois) = 0.06250 pound (avoirdupois) = 0.06250 pound (avoirdupois), 28.35 grams, 437.5 grains.
1 pound (avoirdupois) = the mass of 27.69 cubic inches of water weighed in air at 4°C (39.2°F) and 760 millimeters of mercury (atmospheric pressure), 16 ounces (avoirdupois), 0.4536 kilogram, 453.6 grams, 7,000 grains.
1 long ton (U.S. and British) = 1.120 short tons, 2,240 pounds, 1.016 metric tons, 1016 kilograms.
1 short ton (U.S. and British) = 0.8929 long ton, 2,000 pounds, 0.9072 metric ton, 907.2 kilograms
1 milligram = 0.001 gram, 0.000002205 pound (avoirdupois).
1 gram = 0.002205 pound (avoirdupois), 0.03527 ounce, 0.001 kilogram, 15.43 grains.
1 kilogram = the mass of 1 liter of water in air at 4°C and 760 millimeters of mercury (atmospheric pressure), 2.205 pounds (avoirdupois), 35.27 ounces (avoirdupois), 1,000 grams.
1 metric ton = 0.9842 long ton, 1.1023 short tons, 2,205 pounds, 1,000 kilograms.

Velocity

1 foot per second = 0.6818 mile per hour, 18.29 meters per minute, 0.3048 meters per second.
1 mile per hour = 1.467 feet per second, 1.609 kilometers per hour, 26.82 meters per minute, 0.4470 meters per second.
1 kilometer per hour = 0.2778 meter per second, 0.5396 knot per hour, 0.6214 mile per hour, 54.68 feet per minute.
1 meter per minute = 0.03728 mile per hour, 0.05468 foot per second, 0.06 kilometer per hour, 16.67 millimeters per second, 3.281 feet per minute.
1 knot per hour = 1.152 miles per hour, 1.689 feet per second, 1.853 kilometers per hour.
1 revolution per minute = 0.01667 revolution per second, 6 degrees per second.
1 revolution per second = 60 revolutions per minute, 360 degrees per second.

Acceleration

Standard gravity = 32.17 feet per second per second, 9.807 meters per second per second.

Density

1 gram per millimeter = 0.03613 pound per cubic inch, 8.345 pounds per gallon, 62.43 pounds per cubic foot, 998.9 ounces per cubic foot.
Mercury at 0°C = 0.1360 grams per millimeter, basic value used in expressing pressures in terms of columns of mercury.
1 pound per cubic foot = 16.02 kilograms per cubic meter.
1 pound per gallon = 0.1198 grams per milliter.

Flow

1 cubic foot per minute = 0.1247 gallon per second, 0.4720 liter per second, 472.0 milliliters per second.
1 gallon per minute = 0.06308 liter per second, 1,440 gallons per day, 0.002228 cubic foot per second.
1 liter per second = 2.119 cubic feet per minute, 15.85 gallons (U.S.) per minute.
1 liter per minute = 0.0005885 cubic foot per second, 0.004403 gallon per second.
1 gallon per minute per square foot = 40.746 liter per minute per square meter.

Pressure

Absolute pressure = the sum of the gage pressure and the barometric pressure.
1 atmosphere = pressure exerted by 760 millimeters of mercury of standard density at 0°C, 14.70 pounds per square inch, 29.92 inches of mercury at 32°F, 33.90 feet of water at 39.2°F, 101.3 kilopascal.
1 millimeter of mercury (at 0°C) = 0.001316 atmosphere, 0.01934 pound per square inch, 0.04460 foot of water (4°C or 39.2°F), 0.0193 pound per square inch, 0.1333 kilopascal.
1 inch of water (at 39.2°F) = 0.00246 atmosphere, 0.0361 pound per square inch, 0.0736 inch of mercury (at 32°F). 0.2491 kilopascal.
1 foot of water (at 39.2°F) = 0.02950 atmosphere, 0.4335 pound per square inch, 0.8827 inch of mercury (at 32°F), 22.42 millimeters of mercury, 2.989 kilopascal.
1 inch of mercury (at 32°F) = 0.03342 atmosphere, 0.4912 pound per square inch, 1.133 feet of water, 13.60 inches of water (at 39.2°F), 3.386 kilopascal.
1 millibar (1/1000 bar) = 0.02953 inch of mercury. A bar is the pressure exerted by a force of one million dynes on a square centimeter of surface.
1 pound per square inch = 0.06805 atmosphere, 2.036 inches of mercury, 2.307 feet of water, 51.72 millimeters of mercury, 27.67 inches of water (at 39.2°F), 144 pounds per square foot, 2,304 ounces per square foot, 6.895 kilopascal.
1 pound per square foot = 0.00047 atmosphere, 0.00694 pound per square inch, 0.0160 foot of water, 0.391 millimeter of mercury, 0.04788 kilopascal.
1 ton (short) per square foot = 0.9451 atmosphere, 13.89 pounds per square inch, 9.765 kilograms per square meter.
1 Torr = 0.013 Pascal.

Power

1 British thermal unit per hour = 0.293 watts.
1 British chemical unit per minute = 1.054 kW.
1 British chemical unit per second per square foot = 11.33 kilowatts per square meter.
1 horsepower = 746 watts, 1.014 metric horsepower, 10.69 kilograms-calories per minute, 42.42 British thermal units per minute, 550 pound-feet per second, 33,000 pound-feet per minute.
1 kilowatt = 1.341 horsepower, 1.360 metric horsepower, 14.33 kilogram-calories per minute, 56.90 British thermal units per minute, 1,000 watts.

Heat (Mean Values)

1 British thermal unit = 0.2520 kilogram-calorie, 1,055 joules (absolute).
1 kilogram-calorie = 3.969 British thermal units, 4,187 joules.
1 British thermal unit per pound = 0.5556 kilogram-calorie per kilogram, 2.325 joules per gram.
1 gram-calorie per gram = 1.8 British thermal units per pound, 4.187 joules per gram.

Electrical

1 volt = potential required to produce current flow of 1 ampere through a resistance or impedance of 1 ohm, or current flow of 2 amperes through resistance of ½ ohm, etc.
1 ampere = current flow through a resistance or impedance of 1 ohm produced by a potential of 1 volt, or current flow through a resistance of 100 ohms produced by a potential of 100 volts, etc.
1 milliampere = 0.001 ampere.
1 ohm = resistance or impedance through which current of 1 ampere will flow under a potential of 1 volt.
1 microhm = 0.000001 ohm.
mho = Unit of conductance. In a direct current, circuit conductance in mhos is the reciprocal of number of ohms resistance.
1 watt = power developed by current flow of 1 ampere under potential of 1 volt. (DC, or AC with power factor unity.) See also Power.
1 joule = 1 watt second. A flow of 1 ampere through a resistance of 1 ohm for 1 second (see also Heat).
1 millijoule = 0.001 joule.

Radiation

1 curie = the emission of 3.70×10^{10} beta particles per second (the particles emitted per ;second from 1 gram of radium).
1 roentgen = the quantity of X-rays which will produce 2.08×10^{9} ion pairs in 1 cubic centimeter of dry air at 0°C and standard atmospheric pressure.

TABLE 22-2B. Temperature Conversion, Celsius—Fahrenheit

Temp. Celsius = 5/9 (Temp. F − 32 deg.)
Temp. Fahrenheit = 9/5 × Temp. C + 32 deg.
Rankine (Fahrenheit Absolute) = Temp. F + 459.67 deg.
kelvin (Celsius Absolute) = Temp. C + 273.15 deg.
Freezing point of water: Celsius = 0 deg.; Fahr. = 32 deg.
Boiling point of water: Celsius = 100 deg.; Fahr. = 212 deg.
Absolute zero: Celsius = −273.15 deg.; Fahr. = −459.67 deg.

Celsius	Fahrenheit	Celsius	Fahrenheit	Celsius	Fahrenheit	Celsius	Fahrenheit
−273.15	−459.67	21	69.8	48.9	120	98	208.4
−200	−328	21.1	70	49	120.2	98.9	210
−100	−148	21.7	71	50	122	99	210.2
0	32	22	71.6	51	123.8	100	212
0.56	33	22.2	72	52	125.6	120	248
1	33.8	22.8	73	53	127.4	121.1	250
1.11	34	23	73.4	54	129.2	140	284
1.67	35	23.3	74	54.4	130	148.9	300
2	35.6	23.9	75	55	131	160	320
2.22	36	24	75.2	56	132.8	176.7	350
2.78	37	24.4	76	57	134.6	180	356
3	37.4	25	77	58	136.4	200	392
3.33	38	25.6	78	59	138.2	204.4	400
3.89	39	26	78.8	60	140	250	482
4	39.2	26.1	79	61	141.8	260	500
4.44	40	26.7	80	62	143.6	300	572
5	41	27	80.6	63	145.4	315.8	600
5.56	42	27.2	81	64	147.2	350	662
6	42.8	27.8	82	65	149	371.1	700
6.11	43	28	82.4	65.6	150	400	752
6.67	44	28.3	83	66	150.8	426.7	800
7	44.6	28.9	84	67	152.6	450	842
7.22	45	29	84.2	68	154.4	482.2	900
7.78	46	29.4	85	69	156.2	500	932
8	46.4	30	86	70	158	537.8	1000
8.33	47	30.6	87	71	159.8	600	1112
8.89	48	31	87.8	71.1	160	648.9	1200
9	48.2	31.1	88	72	161.6	700	1292
9.44	49	31.7	89	73	163.4	760	1400
10	50	32	89.6	74	165.2	800	1472
10.6	51	32.2	90	75	167	871.1	1600
11	51.8	32.8	91	76	168.8	900	1652
11.1	52	33	91.4	76.7	170	982.2	1800
11.7	53	33.3	92	77	171.6	1000	1832
12	53.6	33.9	93	78	172.4	1093.3	2000
12.2	54	34	93.2	79	174.2	1100	2012
12.8	55	34.4	94	80	176	1200	2192
13	55.4	35	95	81	177.8	1204.4	2200
13.3	56	35.6	96	82	179.6	1300	2372
13.9	57	36	96.8	82.2	180	1315.6	2400
14	57.2	36.1	97	83	181.4	1400	2552
14.4	58	36.7	98	84	183.2	1428	2600
15	59	37	98.6	85	185	1500	2732
15.6	60	37.2	99	86	186.8	1537.8	2800
16	60.8	37.8	100	87	188.6	1600	2912
16.1	61	38	100.4	87.8	190	1648.9	3000
16.7	62	39	102.2	88	190.4	1700	3092
17	62.6	40	104	89	192.2	1760	3200
17.2	63	41	105.8	90	194	1800	3272
17.8	64	42	107.6	91	195.8	1871.1	3400
18	64.4	43	109.4	92	197.6	1900	3452
18.3	65	43.3	110	93	199.4	1982	3600
18.9	66	44	111.2	93.3	200	2000	3632
19	66.2	45	113	94	201.2	2204.4	4000
19.4	67	46	114.8	95	203	2500	4532
20	68	47	116.6	96	204.8	2760	5000
20.6	69	48	118.4	97	206.6	3000	5432

MICROCOMPUTER APPLICATIONS IN FIRE PROTECTION

John M. Watts, Jr., Ph.D.

In a manner similar to the way the industrial age produced the automatic sprinkler system as a powerful tool to deal with industrial fire problems, the information age has brought the computer to enhance firesafety decision making. This chapter will present some of the fundamental concepts associated with applications of microcomputers in fire protection. Because this represents an interface with a different technology, a brief glossary is included at the end of the chapter. Terminology not fully explained in the text will be found in the glossary.

INTRODUCTION

Before the advent of large-scale computers, man had to rely on manual methods and slow mechanical calculators to perform arithmetic operations. No doubt the first devices used as an aid to computation were fingers, sticks, stones, and similar objects. These items served only as reminders or indicators of a particular quantity and could not themselves do any work; the actual computation was carried out in the mind of the individual.

The word "computer" comes from the Latin verb "computare," which means to reckon or think. Thus a computer is a machine which reckons or thinks. Although it is often asserted that a machine cannot think, this is more a philosophical matter than a technical one.

The first significant electronic computer was the Electronic Numerical Integrator and Calculator, or ENIAC, built in 1946 at the University of Pennsylvania. The ENIAC used 18,000 vacuum tubes as storage elements. It occupied 4,000 cu ft (113 m^3) of space and cost $480,000. Today's desktop microcomputers are faster, more powerful, and easier to use. (See Fig. 22-3A.)

Vacuum tubes, used in calculating machines, became the "first generation" of electronic computers. When transistors were developed in the 1950s, they became the "second generation." This was followed by the "chip" or IC (Integrated Circuit), whereby an entire electronic circuit of transistors, resistors, capacitors, and other devices could be etched on a tiny bit of silicon. These educated grains of sand represent the third generation. (See Fig. 22-3B.) Now, as a fourth generation, are Very Large Scale Integrated (VLSI) circuits. These chips contain logic circuits which have given rise to sophisticated hand calculators and, most recently, microcomputers. All the computing circuitry of ENIAC is now available an a single chip. This is the technology that made the "personal computer" a household word and, for many, a household appliance. In commerce and industry, the microcomputer has become a standard business tool.

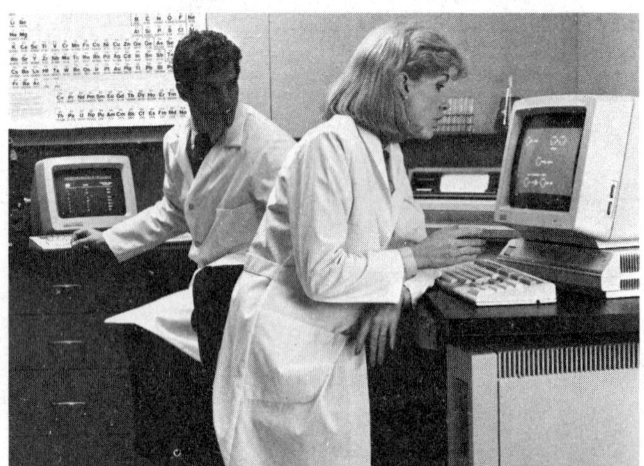

FIG. 22-3A. Microcomputers. (Digital Equipment Corp.)

Characteristic of these developments is a geometric increase in speed of operation. Today, computer functions take place in a time span of a few nanoseconds. A nanosecond is one billionth of a second. To understand the significance of a nanosecond, consider that electrical impulses in a computer travel at the speed of light (186,000 mph or 300,000 kph). A nanosecond is about how long it takes electricity to travel one foot. Thus, computer components which are several feet apart can double the time of a computer operation. These speeds may increase even more as new materials like gallium arsenide or Josephson

Dr. Watts is Director of the Fire Safety Institute, Middlebury, VT, a not-for-profit information, research, and educational corporation.

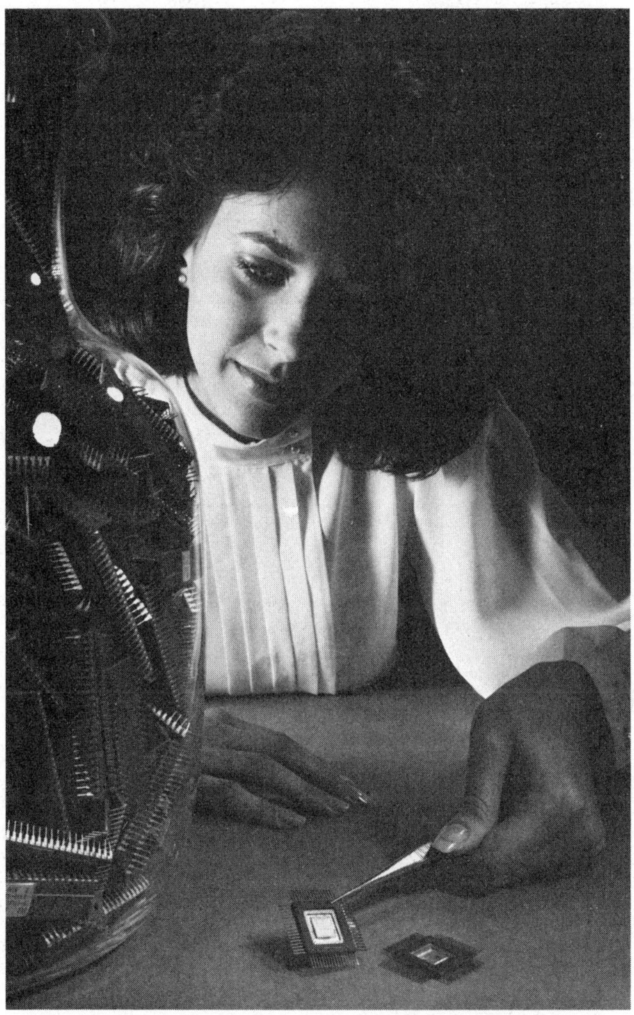

FIG. 22-3B. An integrated circuit. (Digital Equipment Corp.)

junctions replace silicon in the next generation of computer chips.

COMPUTER PRINCIPLES

To understand how a computer performs its many tasks, it is necessary to be aware of some fundamental operating principles. Computers function by reducing all information, no matter how complex, to simple electrical signals grouped together in "bits," "bytes," and "words."

Bit: Computer operations are performed with "bits" (contraction of BInary digiT). A bit holds one of two pieces of information—either a one or a zero. This information is stored as an electric charge on a piece of semiconductor material. If the charge is positive, the bit has a value of one; if the charge is negative, then the value of the bit is zero. A bit may also be thought of as an electronic switch that is either on or off. On indicates a value of one and off indicates zero.

Computers manipulate numbers in the binary system. In a sequence of four bits, each bit represents a value of 2^n where n corresponds to the position of the bit. That is, the first or right most bit indicates a value of $2^0 = 1$ if it is "on," or a value of 0 if it is "off." The next bit corresponds to $2^1 = 2$ if "on," and 0 if "off." Similarly, the value $2^2 = 4$ is associated with the third bit and $2^3 = 8$ for the left most of the four bits. Adding these values allows the computer to represent the numbers from zero (0000) to fifteen (1111) using just four bits. For example, the number four is 0100 and the number nine is 1001; adding these together gives 1101 or thirteen. (Fig. 22-3C.) All of the

BINARY	DECIMAL
0100	4
1001	+9
1101	13

FIG. 21-3C. Binary arithmetic.

operations in a computer are performed in this manner, by combining values of electronic switches which are either on or off.

Byte: While four bits are adequate to represent decimal digits, more are required to represent all the alphanumeric characters that are convenient to use with a computer. Most often, a string of eight bits is used for this purpose and is referred to as a "byte." Thus each character in the memory of a computer uses eight bits or one byte. Storage capacity of computers is measured in bytes. Since most computers can hold thousands of bytes, the common term for expressing memory size is the "kilobyte," indicated by the symbol K. A kilobyte is defined as $2^{10} = 1{,}024$ bytes. For example, a 64K memory has the capacity to hold $64 \times 1{,}024 = 65{,}536$ characters or bytes (524,288 electronic switches).

Word: A computer "word" is the number of bits that it can handle at one time. Most of the first popular microcomputers used eight bits or one byte at a time. Later models used a 16-bit word which means they can handle 16 bits or two bytes at a time. Minicomputers use 32- or 64-bit words. Inasmuch as the size of the word is an important factor affecting the speed of operation of a computer, it is likely that 32-bit word or larger microcomputers will become prevalent.

Elements of the Computer

In order for a computer to operate, it requires certain elements for the proper handling of information. A block diagram illustrating the various elements of a computer is shown in Figure 22-3D.

The input section is capable of accepting information in a variety of forms and converting it to the standard format used by the computing portion of the equipment. The most common types of input units are keyboards and magnetic tapes or disks. The input section obtains information from the various input devices in the form of data required to solve a problem or programmed instructions which tell the computer what to do.

The control section serves as the focal point. It directs input to an appropriate location in the computer's mem-

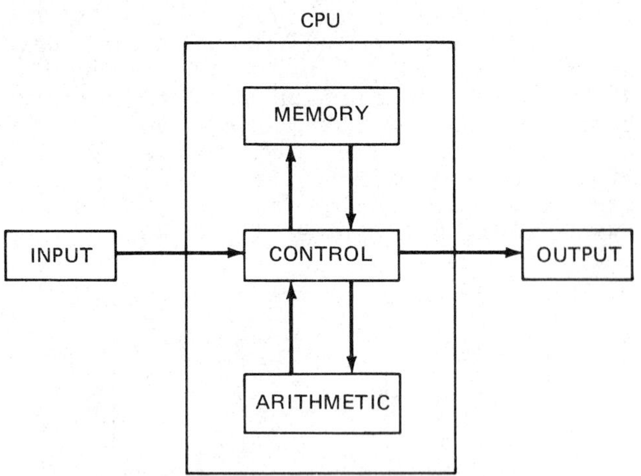

FIG. 22-3D. Components of a computer.

ory. The control section interprets or decodes instructions stored in the memory section and then sends signals to other parts of the computer to tell it what to do.

The memory section is made up of a large number of storage locations in which information can be kept until it is needed by one of the other sections. The information may be numerical data or program instructions to be executed. The computer's memory is partitioned into specific areas for data and instructions.

Actual computations specified by input or in a stored program are carried out by the arithmetic section. This section of the computer basically performs addition only. Other arithmetic functions are variations of the addition function; for example, multiplication is repetitive addition, and subtraction is the addition of a positive and a negative number. Some microcomputers utilize an arithmetic co-processor, which is a separate chip used to speed up the arithmetic operations.

The output section has the function of recording, or writing, the results of computer processing. This section is capable of writing anything located in memory. The results are usually presented on a Cathode Ray Tube (CRT) display (television screen), printed pages, or magnetic tapes or disks.

Hardware

The actual physical components of a computer that can be touched and felt are the hardware. In some small desktop microcomputers, the hardware may consist of a single piece of equipment. More commonly, a computer has several hardware components. There is usually a Central Processing Unit (CPU) and several input, output, storage, communications, etc., devices which are collectively referred to as peripherals. Today there are three general sizes of CPUs: mainframes, minicomputers (minis), and microcomputers (micros).

The mainframe computer is the largest and most powerful. It has the greatest memory capacity and the fastest speed of computation. Advances in computer technology have enabled smaller machines—minicomputers—to perform a great many functions formerly limited to mainframes. While less powerful, minicomputers are more cost-effective for certain applications and now have a very significant position in the computer world.

The latest newcomer, however, is the micro or personal computer. Now a device with many times the computing capacity of the giant ENIAC sits on a desktop or can even be carried around. The personal computer has revolutionized the business world because it provides ready access of significant computing capability. Many advances in fire protection during the next decade will evolve from the availability of microcomputers.

Table 22-3A shows a number of computer attributes

TABLE 22-3A. Compute Attributes*

	Mainframe	Minicomputer	Microcomputer
Cost	$1,000,000	$100,000	$10,000
Floor Area	1000 ft^2 (90 m^2)	100 ft^2 (9 m^2)	10 ft^2 (0.9 m^2)
Users	100	10	1
Memory	10 Mbyte	1 Mbyte	0.1 Mbyte
Mass Storage	1,000 Mbyte	100 Mbyte	10 Mbyte

Note: All values are approximate.
(* Walton 1985b)

and typical values for the three categories of computers. As a very general approximation, there is an order of magnitude difference in each attribute among the types of computers. These distinctions are becoming blurred as more powerful microcomputers are rapidly being developed. At the larger end of the scale, new "super computers" are being developed which employ "parallel processing" to better emulate human thought processes. One form of "super computer" is being constructed by interconnecting 128 microcomputers. At the opposite end, hand-held microcomputers are being used for fire protection engineering calculations and to assist fire prevention inspections.

Memory (Storage Media)

Internal memory, also called main memory, is the information storage capacity of a computer which can be accessed without any mechanical interface. It is made up of memory chips directly connected to the microprocessor. Internal memory is of two types, ROM and RAM. Read only memory (ROM), permanently installed in the computer, contains information necessary to the computer's operation and whatever other functions the manufacturer desires. Random access memory (RAM) is the portion of internal memory available to the user for temporary storage of information being used by the computer for a particular operation. RAM has the characteristic of being "volatile," which means that when the computer is shut off (for whatever reason) the electronic switches are also shut off and all information contained in RAM is lost. Typical microcomputers have RAM capacities of from 16K to 640K. The larger the RAM, the easier it is for the computer to work with large amounts of information. However, the volatility of RAM means that other storage media are also necessary.

External or mass memory is information storage which is not electronically directly connected to the microprocessor. It is accessed by the CPU by an electromechanical device. Unlike internal memory, mass memory does not use electronic switches but relies instead on magnetic charges. Since magnetic charges are stable without additional power, mass storage is considered nonvolatile; i.e., the information is retained on the media when the computer is shut off. This makes the storage of information more permanent and portable. The primary devices for external storage are the tape cassette, floppy disk, and hard disk. These have different characteristics of capacity, speed, cost, reliability, and portability. Mass storage devices have capacities of from 100 kilobytes to more than 100 megabytes (100 million bytes). Because mechanical devices are typically 10 times less reliable than electrical devices and take much longer to access by the computer, mass storage has disadvantages. However, because mass storage is less expensive and nonvolatile, most computer installations use a combination of internal and external storage.

Peripherals

A peripheral is a device that is external to the CPU and main memory but electrically connected to it, and to some degree is controlled by the computer. In addition to external storage units, the most common peripherals are input/output devices.

The most common form of input device is the keyboard. Each key or combination of keys sends the computer a coded signal which is interpreted as a particular symbol, command, or alphanumeric character. Other forms of input devices include optical scanners, light pens, "mouse," track balls, joy sticks, optical readers, touch screens, voice recognition, and digitizers.

Almost all microcomputers use a Cathode Ray Tube (CRT) display as the primary output device. Referred to as monitors, CRTs vary in their resolution and graphics presentation, as well as in their color capability.

Most microcomputer installations also include a printer. Output among the wide variety of printers varies in appearance and speed. The faster "dot matrix" printers generally yield the squarish characters recognized as computer output. Slower, "letter quality" printers form characters from individual impact hammers. Newer laser printers produce letter quality hard copy at very high speeds. Plotters use one or more ink filled pens to produce graphic output.

A "modem" is a peripheral device which functions as both an input and output unit. Modems enable microcomputers to exchange data and information as described under "telecommunications" in the following section.

Figure 22-3E illustrates some of the numerous hardware configurations available for one type of microcomputer.

Software

Computer hardware, no matter how big, fast, powerful, or expensive, is useless without proper instructions in the form of software. Software is a loosely used term for programs or other forms of instructions which enable a computer to perform its desired functions. There are generally three types of computer software: operating systems, programming languages, and applications software.

An operating system is the most basic set of instructions necessary for a computer to function. It controls the transfer of information among the computer's components and translates various input and output signals. A variety of operating systems apply to different types of microcomputers, varying according to the microprocessor chip used and the manufacturer. Typical microcomputer operating systems are CP/M, DOS, and UNIX.

Like the elemental hardware units, computer programming languages are classified by "generation." First generation languages, more commonly called "machine language," are the only ones that don't need to be translated for the computer to understand them. In machine language, instructions to the computer are given as a series of 0s and 1s. The sequence is in two parts: the first part is an "address" which identifies a location in the computer's memory; the second part tells how the electronic switches at that location are to be set, i.e., on or off.

"Assembly language" is the name given to the second generation languages which use English words such as "add, move, store, branch," etc., or abbreviations, acronyms or other symbols in place of the string of 0s and 1s. These more readily recognizable expressions make assembly language easier for people to read and write than machine language. However, the computer must use an internal program called an "assembler" to translate the expressions into machine language. Machine languages and assembly languages are different for each type of computer.

Third generation languages, referred to as "high-level," are the more commonly known languages such as BASIC, COBOL, and FORTRAN. These languages use English words and phrases in a manner that is almost readable by anyone. They are the languages most often used by programmers for specific applications. High-level languages require internal "compilers" or "interpreters" to translate them into machine language.

The relationship of these three levels or generations of computer languages and their accompanying internal software is shown in Figure 22-3F.

Fourth generation languages can be more easily used and understood by a much larger population. They are quickly learned and allow an average microcomputer user to write programs in about one-tenth the time required by third generation languages. The fourth generation incorporates concepts of artificial intelligence to make programming a computer simple and fast.

GENERAL PURPOSE APPLICATIONS SOFTWARE

One of the principal causes of the microcomputer revolution is the widespread availability of general purpose software. These packages serve many applications within their specific functional areas. Compared to software available on mainframes or minicomputers, they are easy to learn, hence have the descriptive phrase "user friendly." The primary areas of general purpose applications are word processing, electronic spreadsheets, data base management, telecommunications, and graphics.

Word Processing

Word processor packages create, revise, edit, and print just about any type of document—correspondence, reports, form letters, mailing labels, invoices, and even large handbooks. They speed up processing of "paperwork," saving time and money for clerical effort. These savings result from the ability of word processors to store and edit electronically rather than on paper. Word processing is particularly applicable to updating firesafety inspection

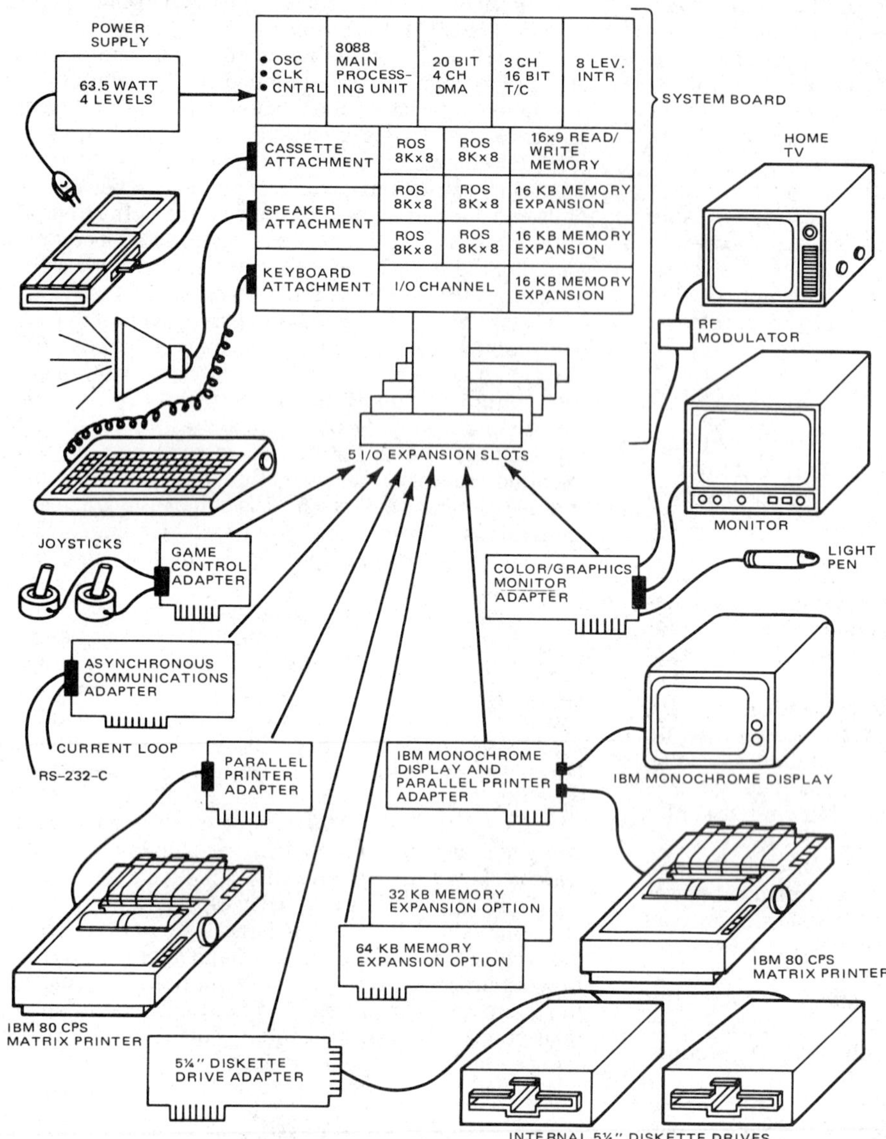

FIG. 22-3E. System Block Diagram from the Personal Computer's Technical Reference Manual. (IBM)

surveys. Standard recommendations can be added or deleted with a few key strokes. Proposals, specifications, and procedures can be similarly updated with ease. For instance, word processing has greatly facilitated the National Fire Protection Association's standards making procedures.

Electronic Spreadsheets

No factor is more responsible for the proliferation of microcomputers among small businesses than the "electronic spreadsheet." These programs emulate the columnar page of many rows and columns referred to by accountants as a spreadsheet. The intersection of each row and column, called a cell, can contain written text, numbers, or a formula. The appeal is the almost immediate recalculation of all cell values when any one cell value is altered. While there are many applications for these spreadsheet programs, one of the greatest uses is budget preparation. It is a simple matter to make changes, additions, or deletions of budget items and have all the totals and subtotals recomputed automatically. With an electronic spreadsheet, experiments with reallocations can be done quickly and easily. In project management, spreadsheet analysis is commonly used to estimate costs. Many types of engineering applications, from hydraulic calculations to code requirements, have been adapted to a spreadsheet program. The full range of applications is limited only by the user's imagination.

Data Base Management

Data base management programs provide a means of storing, retrieving, and rearranging large amounts of similar data. A "data base" is a collection of information that has been assembled, organized, and presented to serve some specific purpose. Data base management software helps create more orderly, manageable record storage and

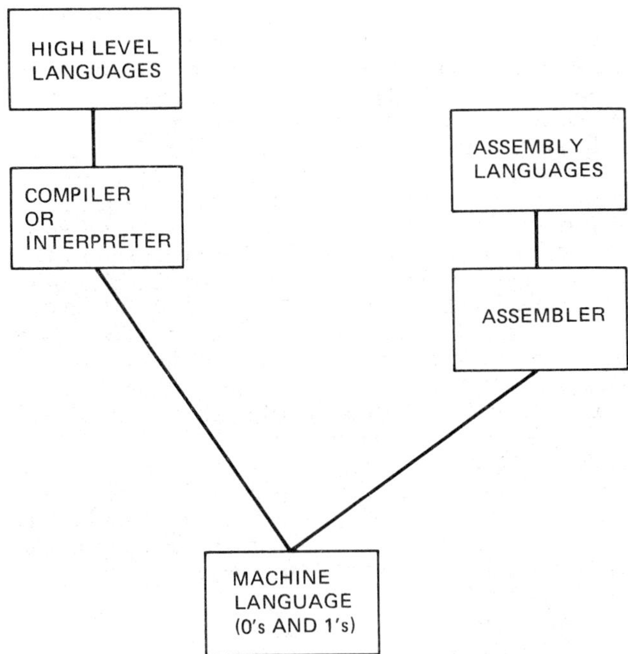

FIG. 22-3F. Hierarchy of computer languages.

retrieval systems. Files become instantly accessible and easy to update. Records of inspections, inventory, personnel, customers, training, maintenance, etc., become useful management tools rather than just requirements. Computer data base management provides capability to sort data in any sequence, select specific categories of items, and generate reports with totals, subtotals, and various statistical measures.

Telecommunications

Telecommunications is a process by which computers transmit data from one to another using telephone lines, satellites, or other transmission devices. (See Fig. 22-3G.)

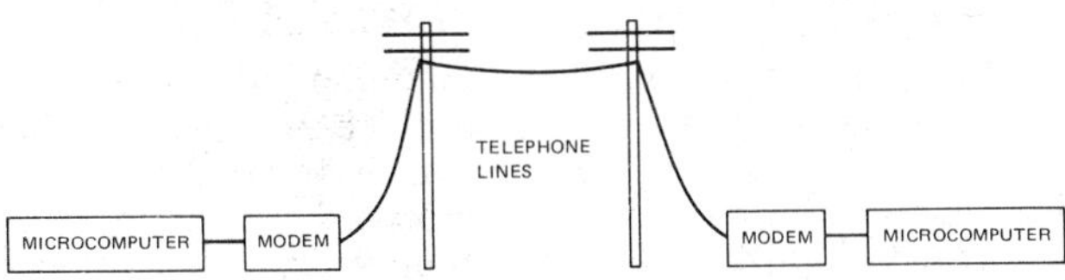

FIG. 22-3G. Ground telecommunications.

Telecommunications requires both special hardware and special software. The hardware required is a "modem," a contraction of MOdulator-DEModulator. This device converts (modulates) a computer's digital signal into analog signals which can be transmitted like a telephone call to another modem, which then demodulates the signal for the receiving computer. Communicating computers may be located in the same room or at opposite sides of the world. Special communications software controls the modem and facilitates the use of telecommunications for many purposes.

Information Retrieval: Many people consider the computer's ability to provide access to information as its most beneficial aspect. Information retrieval implies user access to a wide variety of information sources. The computer facilitates information retrieval through telecommunications access to publicly available data bases.

Several types of data bases are of interest to the firesafety professional. One is a data base of bibliographic information. Sources such as the National Technical Information Service (NTIS) can be accessed remotely by computer to search for published information on a particular topic. Another type of data base contains records of fires. NFIRS, the National Fire Information Reporting System, has a record of every fire reported by participating states. While NFIRS information is not presently available by remote access, it is hoped that it will be in the near future. Another fire data base is FIDO, Fire Incident Data Organization, operated by the National Fire Protection Association. FIDO contains in-depth data on the more severe fires which occur throughout the United States and, in particular, fatal fires on which detailed technical information has been collected. As with NFIRS, FIDO data is not now available by remote access.

Computer Networks: Networking is the interconnection of a number of computers so they can communicate among themselves. When the computers are relatively close together, they may be hardwired to form a Local Area Network (LAN). In large networks, telephone lines are used to interconnect computers over great distances. Several commercial companies operate large national multipurpose computer networks. In addition, many smaller public computer "bulletin boards" link people with special interests through their computers. Fire protection networks, sometimes operated by local fire departments, exist on both large commercial systems and smaller bulletin boards.

Other applications of telecommunications include electronic mail, teleconferencing, on-line publishing (electronic newsletters), videotex, and downloading software. The field of telecommunications is one of the fastest growing, combining computer science with emerging technologies such as fiber optics and direct broadcast satellites.

Graphics

Computer graphics consist of visual displays of data composed of a series of basic entities such as line segments. The graphic is a representation of digital information and is made up of many small parts. (See Fig. 22-3H.)

These features are associated with the two most im-

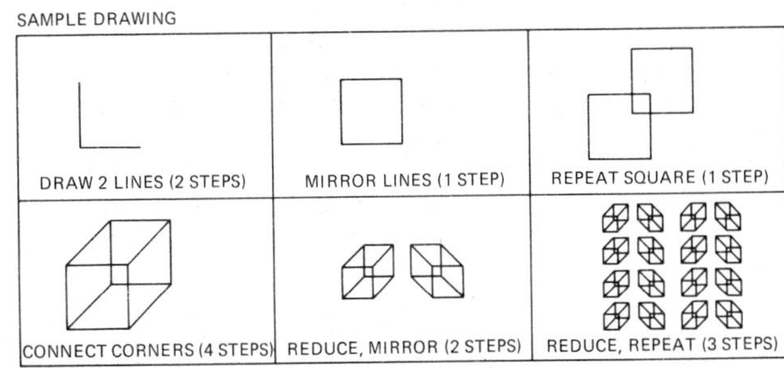

FIG. 22-3H. Repetitive elements in a drawing can be replicated and changed easily with a CAD system.

portant applications of computer graphics: business graphics and computer aided design.

Business graphics refers to presentation-quality charts, graphs, and diagrams on video displays, paper, transparencies, or slides. Because images communicate faster and more dramatically than reams of computer printout, using graphics effectively can mean a dynamic increase in productivity and impact. Until recently, it was costly and time consuming to produce business graphics; now, with microcomputers, they can be produced in the office quickly, easily, and cost-effectively.

Computer Aided Design, or CAD, is the generic term for any system using computer graphics as a replacement for the traditional pen of a designer/drafter. CAD produces plans, detailed layouts, drawings, and visualizations. In the design of structures and mechanical systems, CAD may include interfaces with engineering routines for various load, stress, and sizing calculations. Visualizations may include static views or dynamic simulations in two and three dimensions. (See Fig. 22-3I.) A number of limited

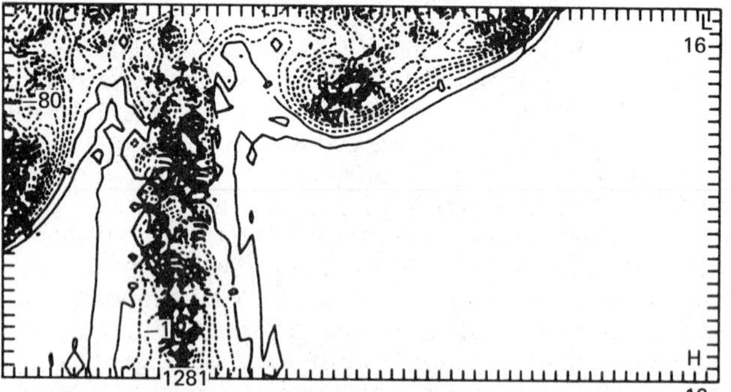

FIG. 22-3I. Computer graphics illustration in which a field model developed by NBS researchers shows that for a live fire to one side of a room, the hot gases will hit the ceiling and the near wall and then plunge towards the floor while the rest of the hot gas progresses along the ceiling toward the far wall.

CAD programs are now available for microcomputers. As memory and speed increase, CAD will become a more common application. CAD has significant potential use in the drawing of fire risk analysis diagrams and the layout of fire protection systems. (Note: In fire service terminology CAD is also used to mean Computer Aided Dispatch.)

SPECIAL PURPOSE FIRE PROTECTION SOFTWARE

As use of microcomputers become more widespread, an increased number of specific fire protection applications are developing. These programs may be revised from mainframe and minicomputer versions or may be new applications stimulated by the availability of microcomputers. Specialized fire protection software can be categorized by three areas of application: research, management, and engineering.

Research

Computers have long played an essential part in fire research, on both mainframes and minicomputers. Now, significant computing power is more readily available to researchers, and the computational capability to conduct research is extended to any microcomputer user. For example, a microcomputer's capacity to produce sophisticated statistical analyses increases the importance of the collection of meaningful data in areas such as human behavior in fires. Fire dynamics and deterministic fire modeling is an area of research that has been heavily dependent on computers, perhaps best characterized by the Harvard Fire Code (Mitler 1985). Another similar type of fire model is used in the computer program ASET (Available Safe Egress Time) (Cooper 1985). Originally a research project on a mainframe computer, ASET is now available for microcomputers and will develop into an engineering design tool (Walton 1985a). A computer program has been written to analyze fire fighting turnout coats and other items of protective clothing (Veghte and Smedley 1982). The program considers environmental climate, underclothing, physical workload, and thermal workload. The power of a computer is illustrated by the capability of this program to consider more than 700 variable combinations of underclothing.

Management

Management applications of microcomputers in fire protection are, at present, mostly adaptations of business software. However, the available computational power will likely be utilized to develop better means of strategic planning for firesafety. One example is in the health care area, where fire protection has been combined with economic analysis in a computer program to provide a management decision making tool (NFPA 101; Chapman 1985). A more rational generalized approach to firesafety decision support is also under development (Watts 1985). In the area of code enforcement computers have not yet made a very great impact (Klein). The potential, however, is very great. One of the most active areas of computer science is artificial intelligence (AI). This area includes such topics as natural languages, robotics, and expert systems. The practice of fire protection is an expert system. Using a microcomputer to assist in this practice can reduce much of the routine effort expended in the application of firesafety codes and standards.

FIRE SERVICE APPLICATIONS

Nowhere has the impact of computers on fire protection been more widely felt than in the fire service. The computer has been compared with the internal combustion engine, the centrifugal pump, and the radio as a technological development to improve fire department operations. A study by the Consortium of Western Fire Agencies on Integrated Information Systems (CSULA 1982) indicates fire service involvement with computers is aready extensive in large fire departments. With time, the low cost availability of microcomputers will enable even the smallest of volunteer fire departments to become "computerized."

In addition to typical management functions such as payroll, inventory, and budget preparation, data base management systems have been developed to provide tactical information on high risk buildings (Sojka 1983), track the number and types of fires that occur (Schaenman unpublished manuscript), evaluate Emergency Medical Service (EMS) advanced life support systems (Hicks 1982), and myriad other fire service functions. Among the more pronounced applications in fire departments are computer aided dispatching, fire prevention activities, and training.

Computer Aided Dispatch (CAD)

One of the early applications of computers to fire department operations was Computer Aided (or Assisted) Dispatch (CAD). Basically, a CAD system uses the speed and storage capabilities of the computer to monitor fire apparatus location and to dispatch the closest units to the scene of an emergency. Experience with CAD has generally been very good, resulting in reduced dispatcher workload, savings in personnel costs, and fast response to major fires and simultaneous accidents (Phillips and Russell 1979). CAD is not limited to large main frame computers or even minicomputers, but has been successfully implemented on microcomputers (Miller 1982). (Note: CAD also is used in computer/engineering/architectural terminology to signify Computer Aided Design. See "Graphics" topic of this chapter.)

Fire Prevention Activities

Two characteristic capabilities of computers are handling large amounts of data and scheduling. Merging these abilities enables a fire department to greatly enhance its fire prevention activities. The computer can keep track of deficiencies and violations, and schedule properties for follow-up or periodic inspections. Interconnected with a CAD system, the dispatcher can be informed of uncorrected conditions which may influence decisions by the fire ground officer in an emergency. In a less technical application, there are educational computer games which emphasize firesafety.

Training

Fire service training provides one of the most interesting applications of microcomputers. Computer Assisted Instruction (CAI)—also known as Computer Band Training (CBT)—has significant advantages to fire service personnel. Advantages include freeing the instuctor to concentrate on field exercises, letting students progress at their own pace, and providing reinforcement. Some of the variety of applications that have been implemented include (1) Emergency Medical Technician (EMT) case studies of victim care, (2) familiarization with the building code, (3) fire hydraulics drills, and (4) fire ground simulation (Parrott 1982). Interactive training programs may use computers alone or they may be interconnected with video tape players or with laser disk players.

The Federal Emergency Management Agency (FEMA) is actively supporting the evolution of microcomputer technology in the fire service. This support is in the form of courses at the National Fire Academy (Weaver 1983), and studies sponsored by the United States Fire Administration. These studies include a hazardous materials incident simulation and a development demonstration program in eight different fire departments to measure usefulness of computer generated management information.

Some fire department pumpers are already equipped with microprocessors for data communications, computer graphic displays, and pumping information. Future applications will include interaction of microcomputers with satellites for dispatch and vehicle location.

Engineering

One of the most commercial areas of microcomputer applications in fire protection is the development of system design and evaluation software. This is particularly the case with automatic sprinkler systems. More than two dozen commercial programs are available for sprinkler system design and/or evaluation and an even larger number of privately developed and used programs. These packages come in many forms and capabilities for various hardware configurations; some even come bundled with their own hardware system. There are minimal programs written in relatively slow running interpretive BASIC, and more complex programs written in standard FORTRAN and newer microcomputer languages. Certain programs will interface with plotters to produce working drawings of sprinkler installations. Some are used to merely check plans, while others have the capacity to optimize pipe sizes to design a minimum cost system.

Computerized design of other types of extinguishing systems is not as widely available for microcomputers. Halon systems, for example, rely on fluid dynamics that vary with the manufacturer's equipment, and design programs are generally proprietary.

In the area of static or passive protection, there are computer programs for the design/evaluation of fire endurance of protected and unprotected steel and reinforced concrete structural assemblies. Like the special hazard extinguishing systems, they are not widely available for microcomputers.

EVACNET+ is a computer program to plan and evaluate the evacuation of large buildings (Kisko and Francis 1985). Originally written for a mainframe, it has been adapted for microcomputer use. New microcomputer software to aid in providing firesafety in buildings, other structures, transportation facilities, mines, forests, and outer space is in various stages of development and application.

THE FUTURE

Microcomputer technology and applications are growing so fast that there will be many new features by the time this is read that did not exist when the chapter was written.

The future of microcomputers holds many exciting innovations in fire protection. New semiconductor materials combined with super-fast parallel processing may soon provide each fire science practitioner with computing power greater than any in existence today. Computational problems such as finite element analysis of complex structures may be reduced to a matter of a few key strokes. Laser technology may even supersede all that. In the area of software, the field of artificial intelligence, with its knowledge-based and expert systems, may replace the inscrutable building code with a computer program. This chapter was composed on a microcomputer in Middlebury, VT and transmitted via telephone lines to the word processing system at NFPA headquarters in Quincy, MA. Some day it may be transmitted directly to you, the reader.

One needs to make a time commitment to keep up with new developments in the field of microcomputers and their applications. Many courses are offered by adult education programs, colleges and universities, computer dealers, and the National Fire Academy. There are excellent periodicals available at all levels of technical detail. Professional societies such as the Society of Fire Protection Engineers (SFPE) Committee on Computers in Fire Protection Engineering are also sources of information. Perhaps the most effective way to keep up is through local computer clubs. The Boston Computer Society is one such club that has become so popular it now serves members from throughout the United States.

Bibliography

References Cited

Chapman, Robert E. 1985. "The FSESCM Model: A Planning Tool for Health Care Facilities." *Computer Applications in Fire Protection: Analysis, Modeling, & Design*. Society of Fire Protection Engineers, Boston, MA.

CSULA. 1982. Consortium of Western Fire Agencies on Integrated Information Systems. "Fire Department Information Technology in Nine Western Communities." The Center for Information Resource Management. California State University, Los Angeles, CA.

Cooper, L. Y., and Stroup, D. W. 1985. "ASET—A Computer Program for Calculating Available Safe Egress Time." *Fire Safety Journal*. Vol 9, No 1. May 1985. pp 29-45.

Francis, Richard L., 1985. "EVACNET+: A Computer Model of Emergency Building Evacuation." Proceedings, SFPE Symposium *Computer Applications in Fire Protection: Analysis, Modeling, & Design*. Society of Fire Protection Engineers, Boston, MA.

Hicks, James E. 1982. "Fire Service Use of Microcomputers." *Fire Chief Magazine*. Aug 1982. pp 59-60.

Kisko, T. M., and Francis, R. L. 1985. *Network Models of Building Evacuation: Development of Software System*. National Bureau of Standards, Washington, DC. 1985.

Klein, Marshall A. "Computer Software for the Codes and Standards Practitioner." Unpublished manuscript.

Miller, Jonathan. 1982. "PC Brings Alarm to Madison." *Softalk for the IBM-PC*. Vol 1, No 5. Oct 1982. pp 30-33.

Mitler, H. E. 1985. "The Harvard Fire Model." *Fire Safety Journal*. Vol 9, No 1. May 1985. pp 7-16..

Parrott, Joe. 1982. "Computer-Aided Instruction: It Has a Place in the Fire Service." *Fire Engineering*. Aug 1982. pp 43-44, and 46.

Phillips, Stan, and Russell, Dave. 1979. "The NFPA CADS Project." *Fire Command*. June 1979. pp 21-23.

Schaenman, Phillip S. *Data Collection, Processing and Analysis*. Unpublished manuscript. In preparation.

Sojka, Deborah. 1983. "Soft-Selling Software." *Datamation*. Vol 29, No 6. June 1983. pp 68,70,72-73.

Veghte, James H., and Smedley, David C. 1982. "S.A.F.E. Analyses Protective Gear." *Fire Service Today*. Sept. 1982. pp 26-28.

Walton, W. D. 1985a. "ASET-B: A Room Fire Program for Personal Computers." *Fire Technology*. Vol 21, No 4. Nov. 1985. pp 293-309.

Walton, W. D. 1985b. "The Users' View of Computer Hardware and Software." *Fire Safety Journal*. Vol 9, No 1. May 1985. pp 3-6.

1985. "ASET-B—A Room Fire Program for Personal Computers." *Fire Technology*. Vol 21, No 4. Nov. 1985. pp 293-309.

Watts, John M., Jr. 1985. "Fire Safety Decision Support: A Microcomputer Spreadsheet Model of a Complex System." Paper presented at The Institute of Management Sciences and the Operations Research Society of America, Boston, MA. Apr 29, 1985.

Weaver, Betsy. 1983. "Computer Assisted Management: Learning the Basics at the National Fire Academy." *The International Fire Chief*. Vol 49, No 2. Feb. 1983. pp 14-17.

Additional Readings

Boston Computer Society. Miscellaneous publications. 1 Center Plaza, Boston, MA 02108.

Fire Service Resource Directory for Microcomputers. National Fire Protection Association, Quincy, MA. 1983.

On-Line Resources. P.O. Box 140, Emmitsburg, MD 27727. (Monthly newsletter EMS and fire service applications of computer.)

Osborne, Adam. "An Introduction to Microcomputers." *The Beginner's Book*, Vol. 0, Adam Osborne and Associates, Berkeley, CA, 1977.

Roemmelt, Bruce E., "Your Friend, the Microcomputer," *JEMS*, July 1982, pp. 26-33).

Watts, John M., Jr. "Report of Committee on Computers in Fire Protection Engineering," Society of Fire Protection Engineers, Boston, MA. Unpublished report.

GLOSSARY

ALPHANUMERIC—Any letter of the alphabet, numeral, punctuation mark, or other symbolic character.

ASCII—American Standard Code for Information Interchange. Most common convention for representing alphanumeric data for transmission or storage.

ASSEMBLER—Program which translates assembly language instructions into machine language.

ASSEMBLY LANGUAGE—Machine specific programming language which uses symbols for fundamental computer instructions.

BACK-UP—Duplicate files on a separate piece of storage media.

BASIC—Beginners' All-purpose Symbolic Instruction Code. Programming language developed at Dartmouth College. Widely available on microcomputers.

BAUD—Unit of speed of data transmission about equal to bits per second.

BINARY—Number system using only two symbols, one and zero. Inside a computer, the one is represented by an electric charge being on, and the zero by the absence of such a charge (off).

BIT—Short for Binary digit. Smallest unit of information in a computer. Has a value of one or zero.

BOOT—To start up a microcomputer or a program.

BUFFER—Temporary storage area used during data transmission.

BUG—Any error or malfunction in a program or equipment that causes an undesired outcome.

BYTE—A group of eight bits. Each byte can represent any of 256 different alphanumeric characters. Unit of memory capacity.

CAD—Computer Aided Design.

CAD—Computer Aided Dispatch.

CAI—Computer Aid Intelligence.

CAM—Computer Aided Manufacturing.

CARD—Removable internal circuit board in a computer containing electronics to perform a specific function.

CBT—Computer Based Training.

CELL—Location on a spreadsheet, usually identified by a column letter and a row number.

CHIP—Integrated circuit.

COBOL—Common Business Oriented Language. High-level computer programming language widely used for business applications.

CODE—Symbols which represent instructions to a computer. A program.

COMPILER—Program that translates a high-level language into machine language.

CO-PROCESSOR—Additional microprocessor which handles specific tasks in conjunction with the CPU.

CP/M—Control Program/Microprocessors. Operating system of many microcomputers.

CPS—Characters Per Second. Measure of printing speed.

CPU—Central Processing Unit. The heart of the computer which contains the control, memory, and arithmetic units. A microprocesser.

CRASH—Abnormal termination of a program or loss of data from storage media.

CRT—Cathode Ray Tube. TV-like screen or monitor providing visual dsplay of computer output.

CURSOR—Flashing mark on CRT indicating the position on the screen where the next entry or deletion will be made.

DAISY WHEEL—Removable print head in many letter quality printers.

DATA BASE—File of records, each containing comparable information on different items.

DATA SECURITY—Protection of computerized information.

DEBUG—To find and remove errors or malfunctions from a program or device.

DENSITY—Compactness of information stored on magnetic media.

DISK—Circular magnetic storage device.

DISKETTE—see Floppy Disk.

DOS—Disk Operation System. Widely used microcomputer operating system.

DOT MATRIX—Printer that forms symbols as a pattern of closely positioned dots.

DOWNLOAD—Transfer a file to a computer system from a remote computer via telecommunications.

DRIVE (or Disk Drive)—Component to read and write data on disks.

FILE—Named collection of data or instructions.

FLOPPY DISK—Flexible plastic disk. Most common form of external storage on microcomputers. Commercial software is typically provided on one or more "floppies."

FORTRAN—Formula Translator. A high-level computer language originally designed for scientific and mathematical use.

FULL DUPLEX—Mode of telecommunications in which data is transmitted in two directions simultaneously.

HALF DUPLEX—Mode of telecommunications in which data is transmitted in one direction at a time.

HARD COPY—Computer output printed on paper.

HARD DISK—Inflexible disk storage component with much greater capacity than a floppy disk.

HARDWARE—Physical equipment forming a microcomputer system.

HARDWIRED—Direct connection of components by wires or cables.

HIGH-LEVEL LANGUAGE—Problem oriented, human readable, programming language, generally usable on many different types of computers.

I/O—Input/Output. Communication of information to and from a computer or peripheral device.

INSTRUCTION—Programmed action to be performed by a computer.

INTEGRATED CIRCUIT (or IC)—Small "chip" of silicon on which miniaturized electronic components have been etched.

INTEGRATED SOFTWARE—Programs for different applications which have common features and can exchange data with one another.

INTERFACE—Hardware or software required to connect peripherals to a computer.

INTERPRETER—Program which translates a high-level language one statement at a time.

KEYBOARD—Typewriter-like component used to enter information into a microcomputer.

K—Kilo byte. 1024 bytes. (2 to the 10th power).

LAN—Local Area Network. Interconnection of geographically close microcomputers.

LANGUAGE—System for writing computer programs that has a highly specific, inflexible vocabulary of commands and syntax.

LISTING—Printed copy of a program.

LOAD—Transfer a file into the computer's memory.

MACHINE LANGUAGE—Directly executable binary codes.

MACRO—Sequence of commands executed with a small number of keystrokes.

MAINFRAME—Large scale, fast operating computer.

MB—MegaByte. 1,048,576 (2&20) bytes.

MEMORY—Electronic storage area for data and programs.

MENU—List of programs or options available for selection by a microcomputer user.

MICROCOMPUTER—Smallest size full function computer.

MICROPROCESSOR—A CPU on a single chip.

MINICOMPUTER—Midsize computer.

MODEM—Modulator/Demodulator. Device to transform signals for transmission of data via telephone lines.

MONITOR—CRT and its housing when not an integral part of the microcomputer.

MOUSE—One type of microcomputer input device.

ON-LINE—Connected to a computer system.

OPERATING SYSTEM (also OS)—Program that manages a microcomputer and its peripherals, and controls the operation of a program.

PARITY—Parameter used to validate transmitted data.

PERIPHERAL—Device connected to a microcomputer.

PERSONAL COMPUTER—Microcomupter.

PIXEL—Picture element. One of the dots which form the picture on a CRT sreeen.

PLOTTER—Mechanical device for drawing lines under computer control.

PROGRAM—Series of instructions to the computer in the form of a step-by-step procedure.

PROTOCOL—Set of parameters defining compatability for telecommunications.

RAM—Random Access Memory. Microcomputer internal memory available to the user for storage of data or instructions.

READ—Accept data for processing or transfer.

RESOLUTION—Quality of image on a CRT.

ROM—Read Only Memory. Permanent information stored in the microcomputer.

SCROLL—Move the contents of a screen display up or down.

SOFTWARE—Operating systems and other programs used by computers.

STRING—Group of alphanumeric characters.

TELECOMMUNICATIONS—Use of telephone communications facilities to transmit data and other information between computers.

UPLOAD—Transfer a file from a computer system to some remote computer via telecommunications.

USER-FRIENDLY—A computer system or component that is easy and non-threatening to use and understand.

UTILITY—Program used to assist in operation of the computer.

VOLATILE—Characteristic of memory that loses its contents when the power supply is removed or disrupted.

VSLI—Very Large Scale Integration. Technology for putting many (i.e., thousands) electronic circuits on a single chip.

WORD—Series of bits procesesed simultaneously by a computer.

WRITE—Transfer a file from internal to external memory. /etET

Appendix A

STRENGTH AND SIZE OF MATERIALS

Ultimate Strength of Common Metals

American Institute of Steel Construction, Inc.

Material	Stress in Kips per Square Inch‡					Modulus of Elasticity (psi§)	Elongation (percent)
	Tension Ultimate	Elastic Limit	Compression Ultimate	Bending Ultimate	Shearing Ultimate		
Aluminum, Alloy 2014	62–70	42–60			38–42	10,600,000	†20–13
Aluminum, Alloy 6061	35–41	21–40			24–30	10,000,000	†22–12
Brass, 50% Zn	31	17.9	117	33.5	....	....	5.0
Brass, cast, common	18–24	6	30	20	36	9,000,000	
Brass, wire, hard	80						
Brass, wire, annealed	50	16	....	....	....	14,000,000	
Bronze, aluminum 5 to 7½%	75	40	120				
Bronze, Tobin, cast (38% Zn, 1½% Sn, ⅓% Pb)	66						
Bronze, Tobin, rolled (38% Zn, 1½% Sn, ⅓% Pb)	80	40	....	....	....	14,500,000	
Bronze, Tobin, c. rolled (38% Zn, 1½% Sn, ⅓% Pb)	100						
Copper, plates, rods, bolts	32–35	10	32				
Iron, cast, gray	18–24	....	....	25–33			
Iron, cast, malleable	27–35	15–20	46	30	40		
Iron, wrought, shapes	48	26	Tensile	Tensile	⅚ Tens.	28,000,000	*1,500,000 / Tensile Strength
Steel, plates for cold pressing	48–58	½ Tens.	Tensile	Tensile	¾ Tens.	29,000,000	
Steel, cars	50–65	½ Tens.	Tensile	Tensile	¾ Tens.	29,000,000	
Steel, locos., stat. boilers	55–65	½ Tens.	Tensile	Tensile	¾ Tens.	29,000,000	
Steel, bridges and bldgs., ships	60–72	33	Tensile	Tensile	¾ Tens.	29,000,000	
Steel, structural silicon	80–95	45	Tensile	Tensile	¾ Tens.	29,000,000	
Steel, struc. nickel (3.25% Ni)	85–100	50	Tensile	Tensile	¾ Tens.	29,000,000	
Steel, rivet, boiler	45–55	½ Tens.	Tensile	Tensile	¾ Tens.	29,000,000	
Steel, rivet, br., bldg., loco., cars	52–62	28	Tensile	Tensile	¾ Tens.	29,000,000	
Steel, rivet, ships	55–65	30	Tensile	Tensile	¾ Tens.	29,000,000	
Steel, rivet, high-tensile	70–85	38	Tensile	Tensile	¾ Tens.	29,000,000	
Steel, cast, soft	60	27	Tensile	Tensile	¾ Tens.	29,000,000	
Steel, cast, medium	70	31.5	Tensile	Tensile	¾ Tens.	29,000,000	
Steel, cast, hard	80	36	Tensile	Tensile	¾ Tens.	29,000,000	†24
Steel wire, unannealed	120	60					†20
Steel wire, annealed	80	40					†17
Steel wire, bridge cable	215	95					

* 8 in. gage length.
† 2 in. gage length.
‡ 1 Kip per sq in. = 6895 MPa
§ 1 psi = 6.895 kPa

Capacity of Cylinders or Cylindrical Tanks

U.S. Gallons and Barrels (Petroleum) per Foot‡ of Length or Height

$$\text{Capacity per foot, U.S. gallons*} = \frac{\pi D^2}{4} \times 7.481$$

D = Inside diameter in feet‡
For capacity in barrels (petroleum), divide by 42

Diameter, ft‡	Capacity per ft‡		Diameter, ft‡	Capacity per ft‡		Diameter, ft‡	Capacity per ft*	
	Gals*	Bbls†		Gals*	Bbls†		Gals*	Bbls†
1	5.8751	0.13989	51	15277.3	363.826	101	59932.6	1426.91
2	23.5007	0.55981	52	15883.9	378.235	102	61125.3	1455.30
3	52.8766	1.258	53	16503.3	392.921	103	62329.7	1483.98
4	94.0029	2.2389	54	17131.9	407.888	104	63549.5	1512.94
5	146.879	3.49699	55	17792.4	423.135	105	64773.7	1542.17
6	211.506	5.03565	56	18424.5	438.662	106	66013.4	1571.68
7	287.883	6.85409	57	19088.4	454.468	107	67264.8	1601.48
8	376.011	8.95228	58	19764.1	470.554	108	68528.0	1631.55
9	475.889	11.3302	59	20451.5	486.920	109	69802.9	1661.90
10	587.517	13.9879	60	21150.6	503.565	110	71089.5	1692.54
11	710.892	16.9254	61	21861.5	520.491	111	72387.9	1723.45
12	846.022	20.1425	62	22584.2	537.696	112	73698.1	1754.64
13	992.902	23.6396	63	23318.6	555.182	113	75019.7	1786.12
14	1151.53	27.4164	64	24064.7	572.946	114	76353.5	1817.87
15	1321.92	31.4729	65	24822.6	590.991	115	77699.2	1849.90
16	1504.04	35.8091	66	25592.2	609.314	116	79056.2	1882.21
17	1697.92	40.425	67	26373.6	627.918	117	80425.1	1914.80
18	1903.56	45.3209	68	27166.8	646.802	118	81805.9	1947.68
19	2120.94	50.4965	69	27971.7	665.966	119	83198.1	1980.83
20	2350.07	55.9517	70	28788.3	685.409	120	84602.2	2014.26
21	2590.95	61.6868	71	29616.7	705.132	121	86018.3	2047.97
22	2843.58	67.7016	72	30456.9	725.134	122	87438.8	2081.97
23	3107.89	73.9963	73	31308.8	745.418	123	88885.5	2116.24
24	3384.02	80.5704	74	32172.4	765.980	124	90261.9	2150.79
25	3671.98	87.4247	75	33047.8	786.821	125	91799.2	2185.61
26	3971.69	94.5584	76	33935.0	807.944	126	93274.4	2220.73
27	4283.00	101.972	77	34833.9	829.346	127	94760.7	2256.12
28	4606.13	109.665	78	35744.5	851.026	128	96259.1	2291.79
29	4941.02	117.638	79	36666.9	872.987	129	97768.6	2327.73
30	5286.36	125.891	80	37601.1	895.228	130	99290.2	2363.96
31	5646.04	134.424	81	38547.0	917.749	131	100824	2400.45
32	6016.18	143.237	82	39504.6	940.549	132	102369	2437.26
33	6398.06	152.328	83	40474.1	963.630	133	103927	2474.33
34	6791.70	161.701	84	41455.2	986.989	134	105495	2511.67
35	7197.09	171.352	85	42448.1	1010.63	135	107075	2549.31
36	7614.25	181.284	86	43452.7	1034.55	136	108667	2587.21
37	8043.11	191.494	87	44469.2	1058.75	137	110271	2625.39
38	8483.71	201.984	88	45497.3	1083.23	138	111887	2663.86
39	8936.13	212.756	89	46537.2	1107.99	139	113514	2702.61
40	9400.30	223.807	90	47588.9	1133.02	140	115153	2741.64
41	9876.13	235.136	91	48652.3	1158.34	141	116804	2780.94
42	10363.8	246.747	92	49727.4	1183.94	142	118467	2820.53
43	10863.2	258.636	93	50814.4	1209.82	143	120141	2860.39
44	11374.3	270.806	94	51913.1	1235.79	144	121827	2900.53
45	11897.2	283.256	95	53023.4	1262.41	145	123525	2940.97
46	12431.8	295.984	96	54145.6	1289.13	146	125235	2981.66
47	12978.2	308.992	97	55280.1	1316.14	147	126957	3022.66
48	13536.4	322.283	98	56425.1	1343.40	148	128690	3063.93
49	14106.3	335.850	99	57582.6	1370.96	149	130434	3105.46
50	14688.0	349.699	100	58751.7	1398.79	150	132192	3147.29

* 1 U.S. Gallon = 3.78 L
† 1 Barrel = 0.159 m^3
‡ 1 ft = 0.305 m

Ultimate Strength PSI (Average) of Common Materials Other Than Metals

American Institute of Steel Construction, Inc.

Material	Average Ultimate Stress (psi*)			Safe Working Stress (psi*)			Modulus of Elasticity (psi*)
	Compression	Tension	Bending	Compression	Bearing	Shearing	
Masonry, granite	....	....	....	420	600		
Masonry, limestone, bluestone	....	....	....	350	500		
Masonry, sandstone	....	....	....	280	400		
Masonry, rubble	....	....	....	140	250		
Masonry, brick, common	10,000	200	600				
Ropes, cast steel hoisting	....	80,000					
Rope, standing, derrick	....	70,000					
Rope, manila	....	8000					
Stone, bluestone	12,000	1200	2500	1200	1200	200	7,000,000
Stone, granite, gneiss	12,000	1200	1600	1200	1200	200	7,000,000
Stone, limestone, marble	8000	800	1500	800	800	150	7,000,000
Stone, sandstone	5000	150	1200	500	500	150	3,000,000
Stone, slate	10,000	3000	5000	1000	1000	175	14,000,000

* 1 psi = 6.985 kPa

Sizes and Capacity of Steel Pipe

Nominal Pipe Size (inches)*	Outside Diameter (inches)*	Inside Diameter (inches)*	Capacity of One Foot Length of Pipe§: Cubic Feet†	Capacity of One Foot Length of Pipe§: Gallons‡	Nominal Pipe Size (inches)*	Outside Diameter (inches)*	Inside Diameter (inches)*	Capacity of One Foot Length of Pipe: Cubic Feet†	Capacity of One Foot Length of Pipe: Gallons‡
Schedule 40 Pipe					Schedule 80 Pipe				
½	0.840	0.622	0.0021	0.0158	½	0.840	0.546	0.0016	0.012
¾	1.050	0.824	0.0037	0.0276	¾	1.050	0.742	0.0030	0.022
1	1.315	1.049	0.0060	0.0449	1	1.315	0.957	0.0050	0.037
1¼	1.660	1.380	0.0104	0.0774	1¼	1.660	1.278	0.0089	0.066
1½	1.900	1.610	0.0142	0.106	1½	1.900	1.500	0.0123	0.092
2	2.375	2.067	0.0233	0.174	2	2.375	1.939	0.0205	0.153
2½	2.875	2.469	0.0332	0.248	2½	2.875	2.323	0.0294	0.220
3	3.500	3.068	0.0513	0.383	3	3.500	2.900	0.0548	0.344
3½	4.000	3.548	0.0686	0.513	3½	4.000	3.364	0.0617	0.458
4	4.500	4.026	0.0883	0.660	4	4.500	3.826	0.0798	0.597
5	5.563	5.047	0.139	1.04	5	5.563	4.813	0.126	0.947
6	6.625	6.065	0.200	1.50	6	6.625	5.761	0.181	1.35
8	8.625	7.981	0.3474	2.60	8	8.625	7.625	0.3171	2.38
10	10.75	10.020	0.5475	4.10	10	10.75	9.564	0.4989	3.74
12	12.75	11.938	0.7773	5.82	12	12.75	11.376	0.7058	5.28
14	14.0	13.126	0.9397	7.03	14	14.0	12.500	0.8522	6.38
16	16.0	15.000	1.2272	9.16	16	16.0	14.314	1.1175	8.36
18	18.0	16.876	1.5533	11.61	18	18.0	16.126	1.4183	10.61
20	20.0	18.814	1.9305	14.44	20	20.0	17.938	1.7550	13.13
24	24.0	22.626	2.7920	20.87	24	24.0	21.564	2.536	19.0

* 1 in. = 25.4 mm
† 1 cu ft = 0.0283 m^2
‡ 1 gallon = 3.78 L
§ 1 ft = 0.305 m

Coefficients of Expansion

American Institute of Steel Construction, Inc.

The coefficient of linear expansion (ε) is the change in length, per unit of length, for a change of one degree of temperature. The coefficient of surface expansion is approximately two times the linear coefficient, and the coefficient of volume expansion, for solids, is approximately three times the linear coefficient.

A bar, free to move, will increase in length with an increase in temperature and will decrease in length with a decrease in temperature. The change in length will be εtl, where ε is the coefficient of lienar expansion, t the change in temperature, and l the length. If the ends of a bar are fixed, a change in temperature (t) will cause a change in the unit stress of $E\varepsilon t$, and in the total stress of $AE\varepsilon t$, where A is the cross-sectional area of the bar and E the modulus of elasticity.

The following table gives the coefficient of linear expansion for 100°, or 100 times the value indicated above.

Example: A piece of medium steel is exactly 40 ft long at 60°F. Find the length at 90°F assuming the ends free to move.

$$\text{Change of length} = \varepsilon tl = \frac{0.00065 \times 30 \times 40}{100} = 0.0078 \text{ ft.}$$

The length at 90°F is 40.0078 ft.

Example: A piece of medium steel is exactly 40 ft long and the ends are fixed. If the temperature increases 30°F, what is the resulting change in the unit stress?

$$\text{Change in unit stress} = E\varepsilon t = \frac{29{,}000{,}000 \times 0.00065 \times 30}{100} = 5655 \text{ psi.}$$

Coefficients of Expansion for 100 Degrees = 100ε

Materials	Linear Expansion		Materials	Linear Expansion	
	Celsius	Fahrenheit		Celsius	Fahrenheit
Metals and Alloys			Stone and Masonry		
Aluminum, wrought	0.00231	0.00128	Ashlar masonry	0.00063	0.00035
Brass	0.00188	0.00104	Brick masonry	0.00061	0.00034
Bronze	0.00181	0.00101	Cement, portland	0.00126	0.00070
Copper	0.00168	0.00093	Concrete	0.00099	0.00055
Iron, cast, gray	0.00106	0.00059	Granite	0.00080	0.00044
Iron, wrought	0.00120	0.00067	Limestone	0.00076	0.00042
Iron, wire	0.00124	0.00069	Marble	0.00081	0.00045
Lead	0.00286	0.00159	Plaster	0.00166	0.00092
Magnesium, various alloys	0.00290	0.00160	Rubble masonry	0.00063	0.00035
Nickel	0.00126	0.00070	Sandstone	0.00097	0.00054
Steel, mild	0.00117	0.00065	Slate	0.00080	0.00044
Steel, stainless, 18-8	0.00178	0.00099			
Zinc, rolled	0.00311	0.00173			
Timber			Timber		
Fir (parallel to fiber)	0.00037	0.00021	Fir (perpendicular to fiber)	0.0058	0.0032
Maple (parallel to fiber)	0.00064	0.00036	Maple (perpendicular to fiber)	0.0048	0.0027
Oak (parallel to fiber)	0.00049	0.00027	Oak (perpendicular to fiber)	0.0054	0.0030
Pine (parallel to fiber)	0.00054	0.00030	Pine (perpendicular to fiber)	0.0034	0.0019

Expansion of Water
Maximum Density = 1

°C	Volume	°C	Volume	°C	Volume	°C	Volume	°C	Volume	°C	Volume
0	1.000126	10	1.000257	30	1.004234	50	1.011877	70	1.022384	90	1.035829
4	1.000000	20	1.001732	40	1.007627	60	1.016954	80	1.029003	100	1.043116

1 ft = 304.8 mm
°C = 5/9 (°F − 32)
°F = 9/5 (°C + 32)
1 psi = 6.895 kPa

Approximate Minimum Thickness (Inch†) for Carbon Sheet Steel Corresponding to Manufacturers Standard Gage and Galvanized Sheet Gage Numbers

Carbon Sheet Steel			Galvanized Sheet		
Manufacturers Standard Gage No.	Decimal & Nominal Thickness Equivalent (Inch†)	Recommended Minimum Thickness Equivalent* (Inch†)	Galvanized Sheet Gage No.	Decimal & Nominal Thickness Equivalent (Inch†)	Recommended Minimum Thickness Equivalent (Inch†)
8	.1644	.156	8	.1681	.159
9	.1495	.142	9	.1532	.144
10	.1345	.127	10	.1382	.129
11	.1196	.112	11	.1233	.114
12	.1046	.097	12	.1084	.099
13	.0897	.083	13	.0934	.084
14	.0747	.068	14	.0785	.070
15	.0673	.062	15	.0710	.065
16	.0598	.055	16	.0635	.058
17	.0538	.050	17	.0575	.053
18	.0478	.044	18	.0516	.047
19	.0418	.038	19	.0456	.041
20	.0359	.033	20	.0396	.036
21	.0329	.030	21	.0366	.033
22	.0299	.027	22	.0336	.030
23	.0269	.024	23	.0306	.027
24	.0239	.021	24	.0276	.024
25	.0209	.018	25	.0247	.021
26	.0179	.016	26	.0217	.019
27	.0164	.014	27	.0202	.017
28	.0149	.013	28	.0187	.016
			29	.0172	.014
			30	.0157	.013

* Minimum Thickness is the difference between the Thickness equivalent of each gage and the maximum negative tolerance for the widest rolled width.
† 1 in. = 25.4 mm

COMPLETE TITLES OF ALL OFFICIAL NFPA DOCUMENTS

NFPA 1, *Fire Prevention Code*

NFPA 10, *Standard for Portable Fire Extinguishers*

NFPA 10L, *Model Enabling Act for the Sale or Leasing and Servicing of Portable Fire Extinguishers (Including Recommended Rules and Regulations for the Administration of the Act)*

NFPA 11, *Standard for Low Expansion Foam and Combined Agent Systems*

NFPA 11A, *Standard for Medium and High Expansion Foam Systems*

NFPA 11C, *Standard for Mobile Foam Apparatus*

NFPA 12, *Standard on Carbon Dioxide Extinguishing Systems*

NFPA 12A, *Standard on Halon 1301 Fire Extinguishing Systems*

NFPA 12B, *Standard on Halon 1211 Fire Extinguishing Systems*

NFPA 12C, *Standard on Halon 2402 Fire Extinguishing Systems*

NFPA 13, *Standard for Installation of Sprinkler Systems*

NFPA 13A, *Recommended Practice for the Inspection, Testing and Maintenance of Sprinkler Systems*

NFPA 13D, *Standard for the Installation of Sprinkler Systems in One- and Two-Family Dwellings & Mobile Homes*

NFPA 13E, *Recommendations for Fire Department Operations in Properties Protected by Sprinkler & Standpipe Systems*

NFPA 14, *Standard for the Installation of Standpipe and Hose Systems*

NFPA 15, *Standard for Water Spray Fixed Systems for Fire Protection*

NFPA 16, *Standard for the Installation of Deluge Foam-Water Sprinkler Systems and Foam-Water Spray Systems*

NFPA 16A, *Recommended Practice for the Installation of Closed-Head Foam-Water Sprinkler Systems*

NFPA 17, *Standard for Dry Chemical Extinguishing Systems*

NFPA 17A, *Standard for Wet Chemical Extinguishing Systems*

NFPA 18, *Standard on Wetting Agents*

NFPA 20, *Standard for the Installation of Centrifugal Fire Pumps*

NFPA 21, *Standard Operation and Maintenance of National Standard Steam Fire Pumps*

NFPA 22, *Standard for Water Tanks for Private Fire Protection*

NFPA 24, *Standard for the Installation of Private Fire Service Mains and Their Appurtenances*

NFPA 26, *Recommended Practice for the Supervision of Valves Controlling Water Supplies for Fire Protection*

NFPA 27, *Recommendations for Organization, Training and Equipment of Private Fire Brigades*

NFPA 30, *Flammable and Combustible Liquids Code*

NFPA 30A, *Automotive and Marine Service Station Code*

NFPA 31, *Standard for the Installation of Oil Burning Equipment*

NFPA 32, *Standard for Drycleaning Plants*

NFPA 33, *Standard for Spray Application Using Flammable and Combustible Materials*

NFPA 34, *Standard for Dipping and Coating Process Using Flammable or Combustible Liquids*

NFPA 35, *Standard for the Manufacture of Organic Coatings*

NFPA 36, *Standard for Solvent Extraction Plants*

NFPA 37, *Standard for the Installation and Use of Stationary Combustion Engines and Gas Turbines*

NFPA 40, *Standard for the Storage and Handling of Cellulose Nitrate Motion Picture Film*

NFPA 40E, *Code for the Storage of Pyroxylin Plastic*

NFPA 43A, *Code for the Storage of Liquid and Solid Oxidizing Materials*

NFPA 43C, *Code for the Storage of Gaseous Oxidizing Materials*

NFPA 43D, *Code for Storage of Pesticides in Portable Containers*

NFPA 45, *Standard on Fire Protection for Laboratories Using Chemicals*

NFPA 46, *Recommended Safe Practice for Storage of Forest Products*

NFPA 48, *Standard for the Storage, Handling and Processing of Magnesium*

NFPA 49, *Hazardous Chemicals Data*

NFPA 50, *Standard for Bulk Oxygen Systems at Consumer Sites*
NFPA 50A, *Standard for Gaseous Hydrogen Systems at Consumer Sites*
NFPA 50B, *Standard for Liquefied Hydrogen Systems at Consumer Sites*
NFPA 51, *Standard for the Design and Installation of Oxygen-Fuel Gas Systems for Welding, Cutting and Allied Processes*
NFPA 51A, *Standard for Acetylene Cylinder Charging Plants*
NFPA 51B, *Standard for Fire Prevention in Use of Cutting and Welding Processes*
NFPA 52, *Standard for Compressed Natural Gas (CNG) Vehicular Fuel Systems*
NFPA 53M, *Manual on Fire Hazards in Oxygen-Enriched Atmospheres*
NFPA 54, *National Fuel Gas Code*
NFPA 56F, *Standard for Nonflammable Medical Gas Systems*
NFPA 58, *Standard for the Storage and Handling of Liquefied Petroleum Gases*
NFPA 59, *Standard for the Storage and Handling of Liquefied Petroleum Gases at Utility Gas Plants*
NFPA 59A, *Standard for the Production, Storage and Handling of Liquefied Natural Gas (LNG)*
NFPA 61A, *Standard for Prevention of Fire and Dust Explosions in Facilities Manufacturing and Handling Starch*
NFPA 61B, *Standard for the Prevention of Fires and Explosions in Grain Elevators and Facilities Handling Bulk Raw Agricultural Commodities*
NFPA 61C, *Standard for the Prevention of Fire and Dust Explosions in Feed Mills*
NFPA 61D, *Standard for the Prevention of Fire and Dust Explosions in the Milling of Agricultural Commodities for Human Consumption*
NFPA 65, *Standard for the Processing and Finishing of Aluminum*
NFPA 68, *Guide for Explosion Venting*
NFPA 69, *Standard on Explosion Prevention Systems*
NFPA 70, *National Electrical Code*
NFPA 70A, *Electrical Code for One- and Two-Family Dwellings*
NFPA 70B, *Recommended Practice for Electrical Equipment Maintenance*
NFPA 70E, *Standard for Electrical Safety Requirements for Employee Workplaces*
NFPA 70L, *Model State Law Providing for Inspection of Electrical Installations*
NFPA 71, *Standard for the Installation, Maintenance, and Use of Central Station Signaling Systems*
NFPA 72A, *Standard for the Installation, Maintenance and Use of Local Protective Signaling Systems for Guard's Tour, Fire Alarm and Supervisory Service*
NFPA 72B, *Standard for the Installation, Maintenance and Use of Auxiliary Protective Signaling Systems for Fire Alarm Service*
NFPA 72C, *Standard for the Installation, Maintenance and Use of Remote Station Protective Signaling Systems*
NFPA 72D, *Standard for the Installation, Maintenance and Use of Proprietary Protective Signaling Systems*
NFPA 72E, *Standard on Automatic Fire Detectors*
NFPA 72F, *Standard for the Installation, Maintenance and Use of Emergency Voice/Alarm of Communication Systems*
NFPA 72G, *Standard for the Installation, Maintenance and Use of Notification Appliances for Protective Signaling Systems*
NFPA 72H, *Guide for Testing Procedures for Local, Auxiliary, Remote Station and Proprietary Protective Signaling Systems*
NFPA 74, *Standard for the Installation, Maintenance, and Use of Household Fire Warning Equipment*
NFPA 75, *Standard for the Protection of Electronic Computer/Data Processing Equipment*
NFPA 77, *Recommended Practice on Static Electricity*
NFPA 78, *Lightning Protection Code*
NFPA 79, *Electrical Standard for Industrial Machinery*
NFPA 80, *Standard for Fire Doors and Windows*
NFPA 80A, *Recommended Practice for Protection of Buildings from Exterior Fire Exposures*
NFPA 81, *Standard for Fur Storage, Fumigation and Cleaning*
NFPA 82, *Standard on Incinerators, Waste and Linen Handling Systems and Equipment*
NFPA 85A, *Standard for Prevention of Furnace Explosions in Fuel Oil- and Natural Gas-Fired Single Burner Boiler-Furnaces*
NFPA 85B, *Standard for Prevention of Furnace Explosions in Natural Gas-Fired Multiple Burner Boiler-Furnances*
NFPA 85D, *Standard for Prevention of Furnace Explosions in Fuel Oil-Fired Multiple Burner Boiler-Furances*
NFPA 85E, *Standard for Prevention of Furnace Explosions in Pulverized Coal-Fired Multiple Burner Boiler-Furnances*
NFPA 85F, *Standard for the Installation and Operation of Pulverized Fuel Systems*
NFPA 85G, *Standard for the Prevention of Furnace Implosions in Multiple Burner Boiler-Furnaces*
NFPA 86, *Standard for Ovens and Furances—Design, Location and Equipment*
NFPA 86C, *Standard for Industrial Furances Using a Special Processing Atmosphere*
NFPA 86D, *Standard for Industrial Furnances Using Vacuum as An Atmosphere*
NFPA 88A, *Standard for Parking Structures*
NFPA 88B, *Standard for Repair Garages*
NFPA 90A, *Standard for the Installation of Air Conditioning and Ventilating Systems*
NFPA 90B, *Standard for the Installation of Warm Air Heating and Air Conditioning System*
NFPA 91, *Standard for the Installation of Blower and Exhaust Systems for Dust, Stock and Vapor Removal or Conveying*
NFPA 96, *Standard for the Installation of Equipment for the Removal of Smoke and Grease-Laden Vapors from Commercial Cooking Equipment*
NFPA 97M, *Standard Glossary of Terms Relating to Chimneys, Vents and Heat Producing Appliances*
NFPA 99, *Standard for Health Care Facilities*
NFPA 99A, *Manual on Home Use of Respiratory Therapy*
NFPA 101, *Code for Safety to Life from Fire in Buildings and Structures*
NFPA 102, *Standard for Assembly Seating, Tents, and Air-Supported Structures*

NFPA 105, *Recommended Practice for the Installation of Smoke and Draft Control Door Assemblies*
NFPA 110, *Standard on Emergency Power Supplies*
NFPA 120, *Standard for Coal Preparation Plants*
NFPA 121, *Standard on Fire Protection for Mobile Surface Mining Equipment*
NFPA 122, *Standard on the Storage and Handling of Flammable and Combustible Liquids Within Underground Mines Other Than Coal*
NFPA 130, *Standard for Fixed Guideway Transit Systems*
NFPA 150, *Standard on Fire Safety in Racetrack Stables*
NFPA 172, *Standard Fire Protection Symbols for Architectural and Engineering Drawings*
NFPA 174, *Standard for Fire Protection Symnbols for Risk Analysis Diagrams*
NFPA 178, *Standard Symbols for Fire Fighting Operations*
NFPA 203M, *Manual on Roof Coverings and Roof Deck Constructions*
NFPA 204M, *Guide for Smoke and Heat Venting*
NFPA 211, *Standard for Chimneys, Fireplaces, Vents and Solid Fuel Burning Appliances*
NFPA 214, *Standard on Water-Cooling Towers*
NFPA 220, *Standard on Types of Building Construction*
NFPA 224, *Standard for Homes and Camps in Forest Areas*
NFPA 231, *Standard for General Storage*
NFPA 231C, *Standard for Rack Storage of Materials*
NFPA 231D, *Standard for Storage of Rubber Tires*
NFPA 231E, *Recommended Practice for the Storage of Baled Cotton*
NFPA 231F, *Standard for the Storage of Roll Paper*
NFPA 232, *Standard for the Protection of Records*
NFPA 232AM, *Manual for Fire Protection for Archives and Record Centers*
NFPA 241, *Standard for Safeguarding Building Construction and Demolition Operations*
NFPA 251, *Standard Methods of Fire Tests of Building Construction and Materials*
NFPA 252, *Standard Methods of Fire Tests of Door Assemblies*
NFPA 253, *Standard Method of Test for Critical Radiant Flux of Floor Covering Systems Using a Radiant Heat Energy Source*
NFPA 255, *Standard Method of Test of Surface Burning Characteristics of Building Materials*
NFPA 256, *Standard Methods of Fire Tests of Roof Coverings*
NFPA 257, *Standard for Fire Tests of Window Assemblies*
NFPA 258, *Standard Research Test Method of Determining Smoke Generation of Solid Materials*
NFPA 259, *Standard Test Method for Potential Heat of Building Materials*
NFPA 260A, *Standard Methods of Tests and Classification System for Cigarette Ignition Resistance of Components of Upholstered Furniture*
NFPA 260B, *Standard Method of Test for Determining Resistance of Mock-Up Upholstered Furniture Material Assemblies to Ignition by Smoldering Cigarettes*
NFPA 262, *Standard Method of Test for Fire and Smoke Characteristics of Electrical Wires and Cables*
NFPA 263, *Standard Test Method for Heat Release Rates of Materials*
NFPA 291, *Recommended Practice for Fire Flow Testing and Marking of Hydrants*
NFPA 295, *Standard for Wildfire Control*
NFPA 296, *Guide on Air Operations for Forest, Brush and Grass Fires*
NFPA 297, *Guide on Telecommunications—Systems Principles and Practices for Rural and Forestry Services*
NFPA 302, *Fire Protection Standard for Pleasure and Commercial Motor Craft*
NFPA 303, *Fire Protection Standard for Marinas and Boatyards*
NFPA 306, *Standard for the Control of Gas Hazards on Vessels*
NFPA 307, *Standard for the Construction and Fire Protection of Marine Terminals, Piers, and Wharves*
NFPA 312, *Standard for Fire Protection of Vessels During Construction, Repair and Lay-Up*
NFPA 321, *Standard on Basic Classification of Flammable and Combustible Liquids*
NFPA 325M, *Fire Hazard Properties of Flammable Liquids, Gases, and Volatile Solids*
NFPA 327, *Standard Procedures for Cleaning or Safeguarding Small Tanks and Containers*
NFPA 328, *Recommended Practices on the Control of Flammable and Combustible Liquids and Gases in Manholes, Sewers, and Similar Underground Structures*
NFPA 329, *Recommended Practice for Handling Underground Leakage of Flammable and Combustible Liquids*
NFPA 385, *Standard for Tank Vehicles for Flammable and Combustible Liquids*
NFPA 386, *Standard for Portable Shipping Tanks for Flammable and Combustible Liquids*
NFPA 395, *Standard for the Storage of Flammable and Combustible Liquids on Farms and Isolated Construction Projects*
NFPA 402M, *Manual for Aircraft Rescue and Fire Fighting Operational Procedures*
NFPA 403, *Recommended Practice for Aircraft Rescue and Fire Fighting Services at Airports and Heliports*
NFPA 407, *Standard for Aircraft Fuel Servicing*
NFPA 408, *Standard for Aircraft Hand Fire Extinguishers*
NFPA 409, *Standard on Aircraft Hangars*
NFPA 410, *Standard on Aircraft Maintenance*
NFPA 412, *Standard for Evaluating Foam Fire Fighting Equipment on Aircraft Rescue and Fire Fighting Vehicles*
NFPA 414, *Standard for Aircraft Rescue and Fire Fighting Vehicles*
NFPA 415, *Standard on Aircraft Fueling Ramp Drainage*
NFPA 416, *Standard on Construction and Protection of Airport Terminal Buildings*
NFPA 417, *Standard on Construction and Protection of Aircraft Loading Walkways*
NFPA 418, *Standard on Roof-top Heliport Construction and Protection*
NFPA 419, *Guide for Master Planning Airport Water Supply Systems for Fire Protection*
NFPA 421, *Recommended Practice on Aircraft Interior Fire Protection Systems*
NFPA 422M, *Manual for Aircraft Fire and Explosion Investigators*
NFPA 423, *Standard for Construction and Protection of Aircraft Engine Test Facilities*

NFPA 424, *Recommended Practice for Airport/Community Emergency Planning*
NFPA 481, *Standard for the Production, Processing, Handling and Storage of Titanium*
NFPA 482, *Standard for the Production, Processing, Handling, and Storage of Zirconium*
NFPA 490, *Code for the Storage of Ammonium Nitrate*
NFPA 491M, *Manual of Hazardous Chemical Reactions*
NFPA 493, *Standard for Intrinsically Safe Apparatus and Associated Apparatus for Use in Class I, II, and III, Division 1 Hazardous Locations*
NFPA 495, *Code for the Manufacture, Transportation, Storage, and Use of Explosive Materials*
NFPA 496, *Standard for Purged and Pressurized Enclosures for Electrical Equipment in Hazardous (Classified) Locations*
NFPA 497, *Recommended Practice for Classification of Class I Hazardous Locations for Electrical Installations in Chemical Plants*
NFPA 497M, *Manual for Classification of Gases, Vapors and Dusts for Electrical Equipment in Hazardous (Classified) Locations*
NFPA 498, *Standard for Explosives Motor Vehicle Terminals*
NFPA 501A, *Standard for Firesafety Criteria for Mobile Home Installations, Sites and Communities*
NFPA 501C, *Standard on Firesafety Criteria for Recreational Vehicles*
NFPA 501D, *Standard for Firesafety Criteria for Recreational Vehicle Parks and Campgrounds*
NFPA 502, *Recommended Practice on Fire Protection for Limited Access Highways, Tunnels, Bridges, Elevated Roadways, and Air Right Structures*
NFPA 505, *Fire Safety Standard for Powered Industrial Trucks Including Type Designations, Areas of Use, Maintenance and Operations*
NFPA 512, *Standard for Truck Fire Protection*
NFPA 513, *Standard for Motor Freight Terminals*
NFPA 550, *Guide to the Fire Safety Concepts Tree*
NFPA 601, *Standard for Guard Service in Fire Loss Prevention*
NFPA 601A, *Standard for Guard Operations in Fire Loss Prevention*
NFPA 650, *Standard for Pneumatic Conveying Systems for Handling Combustible Materials*
NFPA 651, *Standard for the Manufacture of Aluminum and Magnesium Powder*
NFPA 654, *Standard for the Prevention of Fire and Dust Explosions in the Chemical, Dye, Pharmaceutical, and Plastics Industries*
NFPA 655, *Standard for Prevention of Sulfur Fires and Explosions*
NFPA 664, *Standard for the Prevention of Fires and Explosions in Wood Processing and Woodworking Facilities*
NFPA 701, *Standard Methods of Fire Tests for Flame-Resistant Textiles and Films*
NFPA 702, *Standard for Classification of the Flammability of Wearing Apparel*
NFPA 703, *Standard for Fire Retardant Impregnated Wood and Fire Retardant Coatings for Building Materials*
NFPA 704, *Standard System for the Identification of the Fire Hazards of Materials*
NFPA 801, *Recommended Fire Protection Practice for Facilities Handling Radioactive Materials*
NFPA 802, *Recommended Fire Protection Practice for Nuclear Research Reactors*
NFPA 803, *Standard for Fire Protection for Light Water Nuclear Power Plants*
NFPA 850, *Recommended Practice for Fire Protection for Fossil Fuel Steam Electric Generating Plants*
NFPA 901, *Uniform Coding for Fire Protection*
NFPA 902M, *Fire Reporting Field Incident Manual*
NFPA 903M, *Fire Reporting Property Survey Manual*
NFPA 904M, *Incident Follow-up Report Manual*
NFPA 907M, *Manual on the Investigation of Fires of Electrical Origin*
NFPA 910, *Recommended Practice for the Protection of Libraries and Library Collections*
NFPA 911, *Recommended Practice for the Protection of Museums and Museum Collections*
NFPA 1001, *Standard for Fire Fighter Professional Qualifications*
NFPA 1002, *Standard for Fire Apparatus Driver/Operator Professional Qualifications*
NFPA 1003, *Standard for Airport Fire Fighter Professional Qualifications*
NFPA 1004, *Standard on Fire Fighter Medical Technicians Professional Qualifications*
NFPA 1021, *Standard for Fire Officer Professional Qualifications*
NFPA 1031, *Standard for Professional Qualifications for Fire Inspector, Fire Investigator and Fire Prevention Education Officer*
NFPA 1041, *Standard for Fire Service Instructor Professional Qualifications*
NFPA 1121L, *Model State Firewords Law*
NFPA 1122, *Code for Unmanned Rockets*
NFPA 1123, *Standard for Public Display of Fireworks*
NFPA 1124, *Code for the Manufacture, Transportation, and Storage of Fireworks*
NFPA 1141, *Standard on Fire Protection in Planned Building Groups*
NFPA 1201, *Recommendations for the Organization for Fire Services*
NFPA 1202, *Recommendations for Organization of a Fire Department*
NFPA 1221, *Standard for the Installation, Maintenance and Use of Public Fire Service Communication Systems*
NFPA 1231, *Standard on Water Supplies for Suburban and Rural Fire Fighting*
NFPA 1301, *Guide to Public Fire Prevention Criteria*
NFPA 1401, *Recommended Practice for Fire Protection Training Reports and Records*
NFPA 1402, *Guide to Building Training Centers*
NFPA 1410, *A Training Standard on Initial Fire Attack*
NFPA 1452, *Guide for Training Fire Department Personnel to Make Dwelling Fire Safety Surveys*
NFPA 1461, *Standard for Criteria for Accreditation of Fire Protection Education Programs*
NFPA 1501, *Standard for Fire Department Safety Officer*
NFPA 1901, *Standard for Automotive Fire Apparatus*
NFPA 1904, *Standard for Testing Fire Department Aerial Ladders and Elevating Platforms*
NFPA 1921, *Standard for Fire Department Portable Pumping Units*

NFPA 1931, *Standard on Design, and Design Verification Tests for Fire Department Ground Ladders*
NFPA 1932, *Standard on Use, Maintenance, and Service Testing of Fire Department Ground Ladders*
NFPA 1961, *Standard for Fire Hose*
NFPA 1962, *Standard for the Care, Use, and Maintenance of Fire Hose Including Connections and Nozzles*
NFPA 1963, *Standard for Screw Threads and Gaskets for Fire Hose Connections*
NFPA 1971, *Standard on Protective Clothing for Structural Fire Fighting*
NFPA 1972, *Standard on Structural Fire Fighters' Helmets*
NFPA 1973, *Standard on Gloves for Structural Fire Fighters*
NFPA 1975, *Standard on Station/Work Uniforms*
NFPA 1981, *Standard on Self-Contained Breathing Apparatus for Fire Fighters*
NFPA 1982, *Standard on Personal Alert Safety Systems (PASS) for Fire Fighters*
NFPA 1983, *Standard on Fire Service Life Safety Rope, Harnesses, and Hardware*

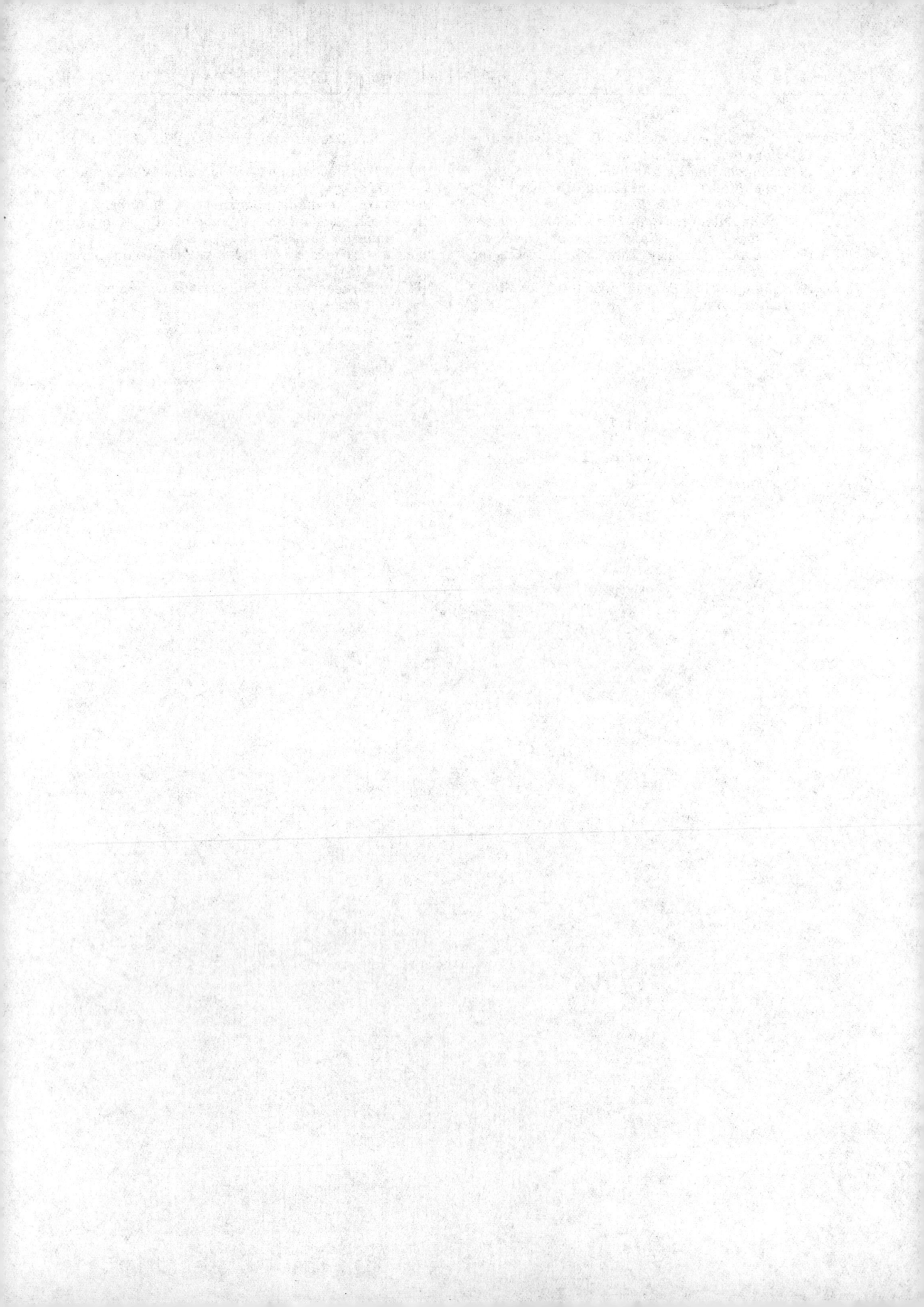

INDEX

C

D

G

H

I

M

O

P

Q

R

S

T

U

X

Z